Lora Hooper
Separating gut from microbe
Page 930, Ch 23

James Versalovic
Microbiome affects pain
Page 524, Ch 13

Throughout this book, examples of the gut microbiome are indicated with this icon to make them easier to find.

FIFTH EDITION

Microbiology

An Evolving Science

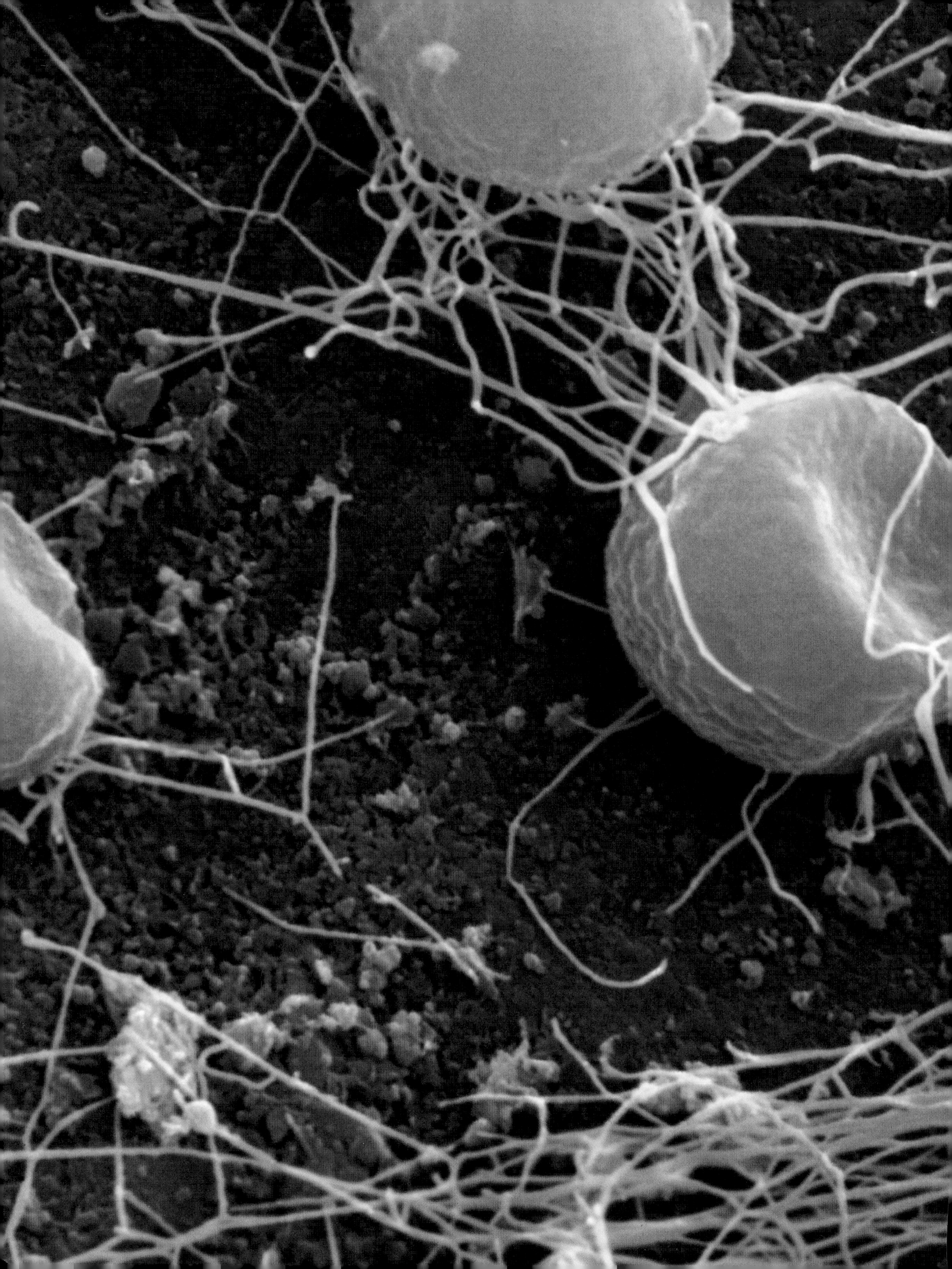

FIFTH EDITION

Microbiology

An Evolving Science

Joan L. Slonczewski

Kenyon College

John W. Foster

University of South Alabama

Erik R. Zinser

University of Tennessee, Knoxville

W. W. NORTON & COMPANY

Independent Publishers Since 1923

W. W. Norton & Company has been independent since its founding in 1923, when William Warder Norton and Mary D. Herter Norton first published lectures delivered at the People's Institute, the adult education division of New York City's Cooper Union. The firm soon expanded its program beyond the Institute, publishing books by celebrated academics from America and abroad. By midcentury, the two major pillars of Norton's publishing program—trade books and college texts—were firmly established. In the 1950s, the Norton family transferred control of the company to its employees, and today—with a staff of five hundred and hundreds of trade, college, and professional titles published each year—W. W. Norton & Company stands as the largest and oldest publishing house owned wholly by its employees.

Printed in Canada
Fifth Edition

Editor: Betsy Twitchell
Associate Editor: Katie Callahan
Developmental Editor: Michael Zierler
Senior Project Editor: Thomas Foley
Copyeditor: Stephanie Hiebert
Assistant Editor: Danny Vargo
Managing Editor, College: Marian Johnson
Associate Director of Production, College: Benjamin Reynolds
Media Editor: Kate Brayton
Content Development Specialist: Todd Pearson
Associate Media Editor: Jasmine Ribeaux
Media Project Editor: Jesse Newkirk
Assistant Media Editor: Katie Daloia
Managing Editor, College Digital Media: Kim Yi
Ebook Production Manager: Michael Hicks
Marketing Manager: Katie Sweeney
Design Director: Rubina Yeh
Designer: Jillian Burr
Director of College Permissions: Megan Schindel
College Permissions Assistant: Patricia Wong
Photo Editor: Catherine Abelman
Composition: MPS North America LLC
MPS Project Manager: Jackie Strohl
Illustrations: Dragonfly Media Group
Manufacturing: Transcontinental—Beauceville, QC

Library of Congress Cataloging-in-Publication Data

Names: Slonczewski, Joan, author. | Foster, John Watkins, author. | Zinser, Erik R., author.
Title: Microbiology : an evolving science / Joan L. Slonczewski, John W. Foster, Erik R. Zinser.
Description: 5th edition. | New York : W.W. Norton & Company, [2020] |
Includes bibliographical references and index.
Identifiers: LCCN 2019046961 | ISBN 9780393664584 (hardcover) | ISBN 9780393420104 (epub)
Subjects: MESH: Microbiological Phenomena | Genetics, Microbial | Microbiota
Classification: LCC QH434 | NLM QW 4 | DDC 579/.135—dc23
LC record available at https://lccn.loc.gov/2019046961

W. W. Norton & Company, Inc., 500 Fifth Avenue, New York, NY 10110
wwnorton.com

W. W. Norton & Company Ltd., 15 Carlisle Street, London W1D 3BS

1 2 3 4 5 6 7 8 9 0

Dedication

We dedicate this Fifth Edition to the memory of two extraordinary microbiologists: Stanley Falkow (1934–2018), founder of the field of molecular microbial pathogenesis; and Jennifer Moyle (1921–2016), who demonstrated the proton motive force in microbial metabolism. We, the authors, are grateful for their contributions and deeply moved by their passing.

BRIEF CONTENTS

eTopic and eAppendix Contents xvi
Preface xviii
About the Authors xxxvi

PART 1
The Microbial Cell

1 Microbial Life: Origin and Discovery 1
2 Observing the Microbial Cell 37
3 Cell Structure and Function 75
4 Bacterial Culture, Growth, and Development 119
5 Environmental Influences and Control of Microbial Growth 159
6 Viruses 195

PART 2
Genes and Genomes

7 Genomes and Chromosomes 239
8 Transcription, Translation, and Protein Processing 277
9 Genetic Change and Genome Evolution 315
10 Molecular Regulation 357
11 Viral Molecular Biology 401
12 Biotechniques and Synthetic Biology 443

PART 3
Metabolism and Biochemistry

13 Energetics and Catabolism 485
14 Electron Flow in Organotrophy, Lithotrophy, and Phototrophy 533
15 Biosynthesis 581
16 Food and Industrial Microbiology 617

PART 4
Microbial Diversity and Ecology

17 Origins and Evolution 659
18 Bacterial Diversity 701
19 Archaeal Diversity 747
20 Eukaryotic Diversity 785
21 Microbial Ecology 827
22 Element Cycles and Environmental Microbiology 883

PART 5
Medicine and Immunology

23 The Human Microbiome and Innate Immunity 921
24 The Adaptive Immune Response 963
25 Pathogenesis 1011
26 Microbial Diseases 1057
27 Antimicrobial Therapy and Discovery 1115
28 Clinical Microbiology and Epidemiology 1159

APPENDIX: Reference and Review A-1

Answers to Thought Questions AQ-1
Glossary G-1
Credits C-1
Index I-1

CONTENTS

eTopic and eAppendix Contents xvi
Preface xviii
About the Authors xxxvi

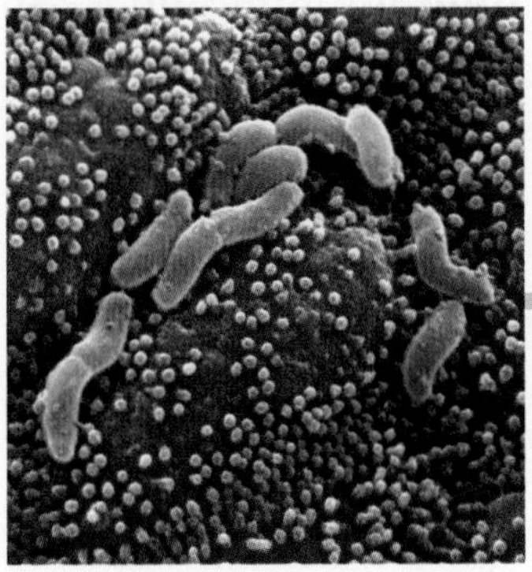

PART 1
The Microbial Cell

CHAPTER 1
Microbial Life: Origin and Discovery 1

1.1 From Germ to Genome: What Is a Microbe? 4
1.2 Microbes Shape Human History 7
1.3 Medical Microbiology 14
1.4 Environment and Ecology 20
Special Topic 1.1: Gut Bacteria Fight Cancer 24
1.5 The Microbial Family Tree 26
1.6 Cell Biology and the DNA Revolution 29

CHAPTER 2
Observing the Microbial Cell 37

2.1 Observing Microbes 38
2.2 Optics and Properties of Light 42
2.3 Bright-Field Microscopy 46
2.4 Fluorescence Microscopy, Super-Resolution Imaging, and Chemical Imaging 53
Special Topic 2.1: Biogeography of a Gut Pathogen 59
2.5 Dark-Field and Phase-Contrast Microscopy 61
2.6 Electron Microscopy, Scanning Probe Microscopy, and X-Ray Crystallography 63

CHAPTER 3
Cell Structure and Function 75
3.1 The Bacterial Cell: An Overview 77
3.2 The Cell Membrane and Transport 82
3.3 The Envelope and Cytoskeleton 87
3.4 Bacterial Cell Division 97
3.5 Cell Polarity, Membrane Vesicles, and Nanotubes 103
Special Topic 3.1: Turrets and Horseshoes: What Are They For? 108
3.6 Specialized Structures 108

CHAPTER 4
Bacterial Culture, Growth, and Development 119
4.1 Microbial Nutrition 120
4.2 Nutrient Uptake 124
4.3 Culturing and Counting Bacteria 130
Special Topic 4.1: Antibiotic Hunters Culture the "Unculturable" 136
4.4 The Growth Cycle 140
4.5 Biofilms 147
4.6 Cell Differentiation 151

CHAPTER 5
Environmental Influences and Control of Microbial Growth 159
5.1 Environmental Limits on Growth: Temperature and Pressure 160
5.2 Osmolarity 167
5.3 Hydronium (pH) and Hydroxide Ion Concentrations 168
5.4 Oxygen 173
5.5 Nutrient Deprivation and Starvation 176
5.6 Physical, Chemical, and Biological Control of Microbes 180
Special Topic 5.1: Phage "Smart Bombs" Target Biofilms 190

CHAPTER 6
Viruses 195
6.1 Viruses in Ecosystems 196
6.2 Virus Structure 202
6.3 Viral Genomes and Classification 208
6.4 Bacteriophages: The Gut Virome 215
Special Topic 6.1: Phages Go Everywhere 222
6.5 Animal and Plant Viruses 222
6.6 Culturing Viruses 233

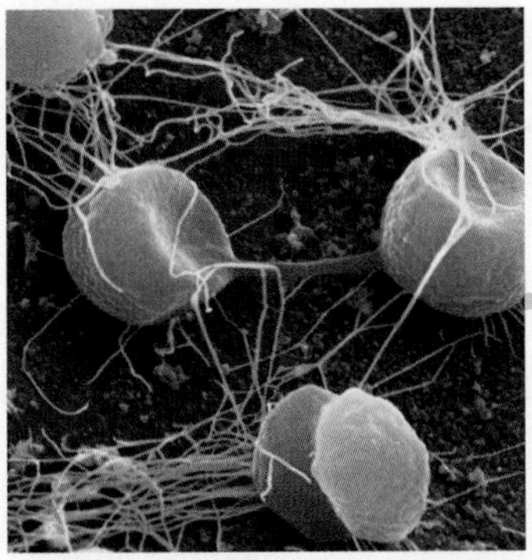

PART 2
Genes and Genomes

CHAPTER 7
Genomes and Chromosomes 239

7.1 DNA: The Genetic Material 240
7.2 Genome Organization 241
Special Topic 7.1: DNA as Digital Storage 246
7.3 DNA Replication 251
7.4 Plasmids and Secondary Chromosomes 261
7.5 Eukaryotic and Archaeal Chromosomes 266
7.6 Microbiomes and Metagenomes 268

CHAPTER 8
Transcription, Translation, and Protein Processing 277

8.1 RNA Polymerases and Sigma Factors 278
8.2 Transcription of DNA to RNA 281
8.3 Translation of RNA to Protein 288
Special Topic 8.1: Translocation: EF-G Gets Physical 298
8.4 Protein Modification, Folding, and Degradation 302
8.5 Secretion: Protein Traffic Control 306

CHAPTER 9
Genetic Change and Genome Evolution 315

9.1 Mutations 316
9.2 DNA Repair 322
Special Topic 9.1: DNA as a Live Wire: Using Electrons to Find DNA Damage 324
9.3 Gene Transfer: Mechanisms and Barriers 329
9.4 Mobile Genetic Elements 343
9.5 Genome Evolution 348

CHAPTER 10
Molecular Regulation 357

10.1 Transcription Repressors and Activators 358
10.2 Alternative Sigma Factors and Anti-Sigma Factors 370
10.3 Regulation by RNA 372
10.4 Second Messengers 380
10.5 Clocks, Thermometers, and Switches 388
Special Topic 10.1: Inteins, Exteins, and "Spliced-Up" Regulation 390
10.6 Chemotaxis: Posttranslational Regulation of Cell Behavior 395

CHAPTER 11
Viral Molecular Biology 401

11.1 Phage Lambda: Enteric Bacteriophage 402
11.2 Influenza Virus: (−) Strand RNA Virus 410
Special Topic 11.1: Designing a Pandemic Flu 418
11.3 Human Immunodeficiency Virus (HIV): Retrovirus 420
11.4 Endogenous Retroviruses and Gene Therapy 431
11.5 Herpes Simplex Virus: DNA Virus 435

CHAPTER 12
Biotechniques and Synthetic Biology 443

12.1 DNA Amplification and Sequence Analysis 444
12.2 Genetic Manipulation of Microbes 457
Special Topic 12.1: Constructing the Smallest Genome for Cellular Life 462
12.3 Gene Expression Analysis 466
12.4 Applied Biotechnology 472
12.5 Synthetic Biology: Biology by Design 475

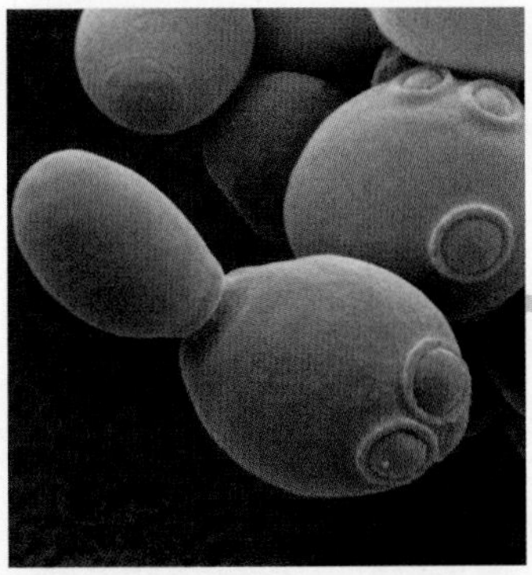

PART 3
Metabolism and Biochemistry

CHAPTER 13
Energetics and Catabolism 485

13.1 Energy for Life 487
13.2 Energy Carriers and Electron Transfer 493
13.3 Catabolism: The Microbial Buffet 501
13.4 Glucose Fermentation and Respiration 505
13.5 The Gut Microbiome: Friends with Benefits 520
13.6 Aromatic Catabolism and Syntrophy 523
Special Topic 13.1: Gut Bacteria Rule Host Behavior 524

CHAPTER 14
Electron Flow in Organotrophy, Lithotrophy, and Phototrophy 533

14.1 Electron Transport Systems and the Proton Motive Force 535
14.2 The Respiratory ETS and ATP Synthase 544
14.3 Anaerobic Respiration 551
14.4 Nanowires, Electron Shuttles, and Fuel Cells 553
14.5 Lithotrophy and Methanogenesis 557
Special Topic 14.1: The Ocean Floor Is a Battery 558
14.6 Phototrophy 566

CHAPTER 15
Biosynthesis 581

15.1 Overview of Biosynthesis 582
15.2 CO_2 Fixation: The Calvin Cycle and Other Pathways 586
15.3 Fatty Acids and Antibiotics 594
15.4 Nitrogen Fixation and Regulation 600
Special Topic 15.1: Mining Bacterial Genomes for Antibiotics 602
15.5 Amino Acids and Nitrogenous Bases 608

CHAPTER 16
Food and Industrial Microbiology 617

16.1 Microbial Foods 618
16.2 Acid- and Alkali-Fermented Foods 623
16.3 Ethanolic Fermentation: Bread and Wine 629
16.4 Food Spoilage and Preservation 633
16.5 Industrial Microbiology 643
Special Topic 16.1: Microbial Vitamins for Sale 644
16.6 Microbial Gene Vectors for Plants and Human Gene Therapy 652

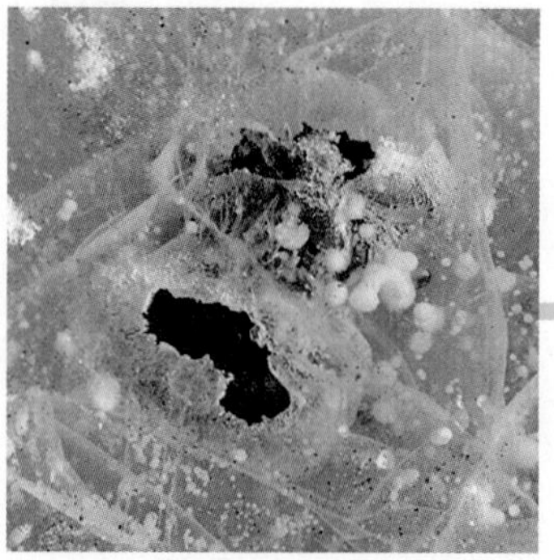

PART 4
Microbial Diversity and Ecology

CHAPTER 17
Origins and Evolution 659

17.1 Origins of Life 660
17.2 Forming the First Cells 669
17.3 Evolution: Phylogeny and Gene Transfer 673
17.4 Natural Selection and Adaptation 681
Special Topic 17.1: A Giant Petri Dish and the Race to Resistance 682
17.5 Microbial Species and Taxonomy 690
17.6 Symbiosis and the Origin of Mitochondria and Chloroplasts 694

CHAPTER 18
Bacterial Diversity 701

18.1 Bacterial Diversity at a Glance 702
18.2 Cyanobacteria: Oxygenic Phototrophs 708
18.3 Firmicutes, Tenericutes, and Actinobacteria (Gram-Positive) 714
Special Topic 18.1: Gut Bacterial Hair Balls 720
18.4 Proteobacteria (Gram-Negative) 726
18.5 Spirochetes, Acidobacteria, Bacteroidetes, and Chlorobi (Deep-Branching Gram-Negative) 738
18.6 Planctomycetes, Verrucomicrobia, and Chlamydiae (PVC Superphylum) 741

CHAPTER 19
Archaeal Diversity 747

19.1 Archaeal Diversity at a Glance 748
19.2 TACK Hyperthermophiles Eat Sulfur 756
19.3 Thaumarchaeota: Ammonia Oxidizers and Animal Symbionts 763
19.4 Euryarchaeota: Methanogens from Gut to Globe 766
Special Topic 19.1: Methanogens for Dinner 772
19.5 Haloarchaea and Other Euryarchaeotes: Underground and Under Ocean 774
19.6 DPANN Symbionts, Altiarchaeales, and Asgard: Branch to Eukaryotes? 780

CHAPTER 20
Eukaryotic Diversity 785

20.1 Phylogeny of Eukaryotes 786
20.2 Fungi 793
Special Topic 20.1: Yeast: A Single-Celled Human Brain 796
20.3 Amebas and Slime Molds 805
20.4 Algae 808
20.5 Alveolates: Ciliates, Dinoflagellates, and Apicomplexans 815
20.6 Parasitic Protozoa 821

CHAPTER 21
Microbial Ecology 827

21.1 Microbial Communities: Metagenomes and Single-Cell Sequencing 829
21.2 Functional Ecology 838
21.3 Symbiosis 845
Special Topic 21.1: Antarctic Lake Mats: Have Ecosystem, Will Travel 846
21.4 Animal Digestive Microbiomes 851
21.5 Marine and Freshwater Microbes 857
21.6 Soil and Plant Microbial Communities 867

CHAPTER 22
Element Cycles and Environmental Microbiology 883

22.1 The Carbon Cycle and Climate Change 885
22.2 The Hydrologic Cycle and Wastewater Treatment 893
22.3 The Nitrogen Cycle 899
22.4 Sulfur, Phosphorus, and Metals 905
22.5 Our Built Environment 911
Special Topic 22.1: A Microbial Jungle: The Kitchen Sponge 914
22.6 Astrobiology 914

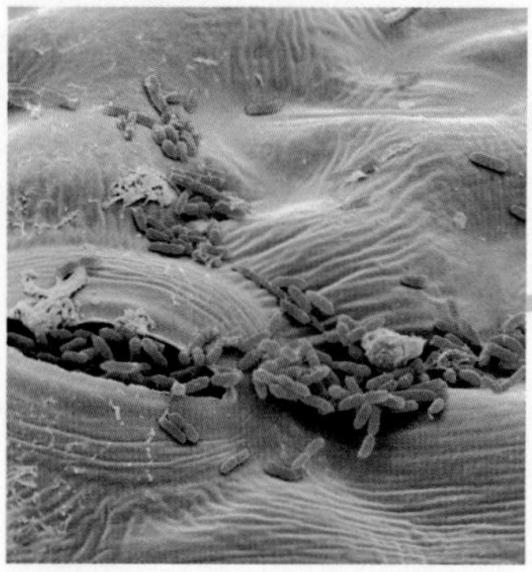

PART 5
Medicine and Immunology

CHAPTER 23
The Human Microbiome and Innate Immunity 921

23.1 The Human Microbiome 922
23.2 Benefits and Risks of Microbiota 931
23.3 Overview of the Immune System 936
Special Topic 23.1: Why Do Tattoos Last Forever? 940
23.4 Physical and Chemical Defenses against Infection 945
23.5 Innate Immunity: Surveillance, Cytokines, and Inflammation 949
23.6 Complement and Fever 957

CHAPTER 24
The Adaptive Immune Response 963

24.1 Overview of Adaptive Immunity 964
24.2 Antibody Structure, Diversity, and Synthesis 970
24.3 T Cells Link Antibody and Cellular Immune Systems 982
24.4 Complement as Part of Adaptive Immunity 993
24.5 Gut Mucosal Immunity and the Microbiome 994
24.6 Immunization 997
24.7 Hypersensitivity and Autoimmunity 1001
Special Topic 24.1: A Monoclonal Magic Bullet for Ebola? 1002

CHAPTER 25
Pathogenesis 1011

25.1 Host-Pathogen Interactions 1012
25.2 Microbial Attachment: First Contact 1020
25.3 Toxins Subvert Host Functions 1025
Special Topic 25.1: Chronic Staph Infections Work with a NET 1028
25.4 Deploying Toxins and Effectors 1036
25.5 Surviving within the Host 1040
25.6 Tools Used to Probe Pathogenesis 1051

CHAPTER 26
Microbial Diseases 1057

26.1 Skin, Soft-Tissue, and Bone Infections 1059
26.2 Respiratory Tract Infections 1064
26.3 Gastrointestinal Tract Infections 1072
26.4 Genitourinary Tract Infections 1083
26.5 Cardiovascular and Systemic Infections 1092
Special Topic 26.1: How Neutrophils Ambush *Staphylococcus aureus* in a Lymph Node 1094
26.6 Central Nervous System Infections 1104

CHAPTER 27
Antimicrobial Therapy and Discovery 1115

27.1 Fundamentals of Antimicrobial Therapy 1116
27.2 Antibiotic Mechanisms of Action 1123
27.3 Challenges of Drug Resistance and Discovery 1135
Special Topic 27.1: Are Designer Antibodies the Next Antibiotics? 1146
27.4 Antiviral Agents 1147
27.5 Antifungal Agents 1153

CHAPTER 28
Clinical Microbiology and Epidemiology 1159

28.1 Clinical Specimen Collection and Handling 1160
28.2 Pathogen Identification by Culture and Phenotype 1166
28.3 Molecular and Serological Identification of Pathogens 1174
Special Topic 28.1: Next-Generation Diagnostics: CRISPR Launches a "Flare" 1180
28.4 Epidemiology 1187
28.5 Detecting Emerging Microbial Diseases 1193

APPENDIX
Reference and Review A-1

A.1 A Periodic Table of the Elements A-2
A.2 Chemical Functional Groups A-2
A.3 Amino Acids A-4
A.4 The Genetic Code A-5
A.5 Calculating the Standard Free Energy Change, ΔG°, of Chemical Reactions A-5
A.6 Generalized Cells A-6
A.7 Semipermeable Membranes A-6
A.8 The Eukaryotic Cell Cycle and Cell Division A-8

Answers to Thought Questions AQ-1
Glossary G-1
Credits C-1
Index I-1

eTOPIC AND eAPPENDIX CONTENTS

Access to the eTopics and eAppendices is available through both the ebook and for Instructor download via digital .wwnorton.com/microbio5

eTOPICS

1.1 Rita Colwell: The Global Impact of Microbiology—An Interview
1.2 Clifford W. Houston: From Aquatic Pathogens to Outer Space—An Interview
2.1 Confocal Microscopy
2.2 Differential Interference Contrast Microscopy
3.1 Isolation and Analysis of the Ribosome
3.2 Christine Jacobs-Wagner: The Thrill of Discovery in Molecular Microbiology—An Interview
3.3 Senior Cells Make Drug-Resistant Tuberculosis
4.1 Transport by Group Translocation: The Phosphotransferase System
4.2 Eukaryotes Transport Nutrients by Endocytosis
4.3 Sharks and Biofilms Don't Mix
4.4 Biofilm Quorum Sensing Triggers Antibiotic Resistance
5.1 The Arrhenius Equation
5.2 It's Raining Bacteria
5.3 Membrane-Permeant Organic Acids Alter Cell pH
5.4 Signaling Virulence
5.5 Evolution in Aging Colonies
5.6 Oligotrophs
5.7 The Phenol Coefficient
6.1 How Did Viruses Originate?
6.2 Virus to the Rescue
6.3 West Nile Virus, an Emerging Pathogen
7.1 Trapping a Sliding Clamp
7.2 Nucleoid Occlusion Factors and the Septal "Guillotine"
8.1 Building the Ribosome Machine
8.2 Discovering the mRNA Ribosome-Binding Site
8.3 Stalking the Lone Ribosome
8.4 Unsticking Stuck Ribosomes: tmRNA and Protein Tagging
8.5 The Shifty Chaperone: GroEL-GroES
8.6 Ubiquitination: A Ticket to the Proteasome
9.1 Repair of UV Damage by Nucleotide Excision Repair and Homologous Recombination
9.2 Homologous Recombination and RecA
9.3 Mapping Bacterial Chromosome Gene Position by Conjugation
9.4 F-Prime
9.5 How Nonreplicative and Replicative Transposons Hop into New Locations
9.6 Integrons and Gene Capture
9.7 There's a Bacterial Genome Hidden in My Fruit Fly
10.1 Glucose Transport Alters cAMP Levels
10.2 In Sporulation, Different Sigma Factors Are Activated in the Mother Cell and Forespore
10.3 Virulence Gene Control by sRNA in *Staphylococcus aureus*
10.4 Slipped-Strand Mispairing
11.1 Phage T4: The Classic Molecular Model
11.2 The Filamentous Phage M13: Vaccines and Nanowires
11.3 Poliovirus: (+) Strand RNA Virus: A Research Model for Non-Polio Enterovirus That May Cause Acute Flaccid Myelitis
11.4 Hepatitis C: (+) Strand RNA Virus
12.1 Acid Survival: A Research Case Study
12.2 Equilibrium Density Gradient Centrifugation
12.3 Error Correction for Single-Molecule, Real-Time Sequencing
12.4 Site-Directed Mutagenesis Helps Us Probe Protein Function
12.5 Mapping Transcriptional Start Sites with RNAseq
12.6 Green Fluorescent Proteins Track Cell Movements in Biofilms
12.7 Mapping the *E. coli* Interactome
12.8 DNA Vaccines
12.9 Gene Therapy and Gene Delivery Systems
12.10 Directed Evolution through Phage Display Technology
12.11 DNA Shuffling Enables In Vitro Evolution
12.12 Bacteria "Learn" to Keep Time and Signal Danger
13.1 Genomic Analysis of Metabolism
13.2 Pyruvate Dehydrogenase Connects Sugar Catabolism to the TCA Cycle
14.1 Environmental Regulation of the ETS
14.2 Caroline Harwood: A Career in Bacterial Photosynthesis and Biodegradation—An Interview
15.1 Dan Wozniak: Polymer Biosynthesis Makes a Pathogenic Biofilm—An Interview
15.2 The Discovery of ^{14}C
15.3 Calvin Cycle Intermediates
15.4 The 3-Hydroxypropionate Cycle

16.1 From Barley and Hops to Beer
16.2 Microbial Enzymes Make Money
16.3 Caterpillar Viruses Produce Commercial Products
17.1 The RNA World: Clues for Modern Medicine
17.2 Phylogeny of a Shower Curtain Biofilm
17.3 Jump-Starting Evolution of a Hyperthermophilic Enzyme
17.4 Richard Lenski: Evolution in the Lab—An Interview
17.5 The Dichotomous Key
17.6 Leaf-cutter Ants with Partner Fungi and Bacteria
18.1 Karl Stetter: Adventures in Microbial Diversity Lead to Products in Industry—An Interview
18.2 Carbon Monoxide: Food for Bacteria?
19.1 Haloarchaea in the Classroom
19.2 *Archaeoglobus* Partly Reverses Methanogenesis
20.1 Oomycetes: Lethal Parasites That Resemble Fungi
20.2 A Ciliate Model for Human Aging
20.3 The Trypanosome: A Shape-Shifting Killer
21.1 Mapping Bermuda Phytoplankton
21.2 Cold-Seep Ecosystems
22.1 Bioremediation of Weapons Waste
22.2 Metal Contamination and Bioremediation
22.3 Subterranean Arsenic Bioremoval
23.1 Are NETs a Cause of Lupus?
23.2 Do Defensins Help Determine Species Specificity for Infection?
23.3 Cathelicidins
24.1 The "Bubble Boy"
24.2 Factors That Influence Immunogenicity
24.3 ABO Blood Groups: Antigens, Antibodies, and Karl Landsteiner
24.4 How T Cells Meet B Cells in Lymph Nodes
24.5 MHC Restriction, Organ Donation, and Transplant Rejection
24.6 Microbiota Minimize Inflammation
24.7 Do Vaccines Stimulate Cell-Mediated Immunity?
24.8 Case Studies in Hypersensitivity
25.1 Finding Virulence Genes: Signature-Tagged Mutagenesis
25.2 Finding Virulence Genes: In Vivo Expression Technologies
25.3 Caught in the Act: *Streptococcus agalactiae* Evolved through Conjugation
25.4 Pilus Tip Proteins Tighten Their Grip
25.5 Normal G-Factor Control of Adenylate Cyclase
25.6 Diphtheria Toxin
25.7 Identifying New Microbial Toxins
25.8 Type VI Secretion: Poison Darts
26.1 Human Papillomavirus
26.2 The Respiratory Tract Pathogen *Bordetella* Binds to Lung Cilia
26.3 Case History: Blastomycosis
26.4 The Common Cold versus Influenza
26.5 Sprouts and Emerging *Escherichia coli*
26.6 Intracellular Biofilm Pods Are Reservoirs of Infection
26.7 Human Immunodeficiency Virus: Pathogenesis
26.8 Atherosclerosis and Coronary Artery Disease
26.9 Spongiform Encephalopathies
27.1 Antibiotic Spectrum of Activity
27.2 Antibiotic Biosynthesis Pathways
27.3 Anti-Quorum Sensing Drug Blocks Pathogen "Control and Command"
27.4 Resurrection, Analysis and Treatment of the 1918 Pandemic Flu Virus
28.1 API Reactions and Generating a Microbe ID Code
28.2 DNA-Based Detection Tests
28.3 Microbial Pathogen Detection Gets Wired Up
28.4 Whipple's Disease
28.5 What's Blowing in the Wind?
28.6 SARS: An Epidemiological Success Story

eAPPENDICES

eAppendix 1 Biological Molecules
eAppendix 2 Introductory Cell Biology: Eukaryotic Cells
eAppendix 3 Laboratory Methods for Microbiology

PREFACE

Microbiology: An Evolving Science is the defining core text of our generation—the book that inspires undergraduate science majors to embrace the microbial world. This Fifth Edition expands our focus on the intestinal microbiome—including remarkable new discoveries of the gut-brain axis, how our microbial communities may modulate brain function. We also present a new breathtaking view of marine microbiology. We highlight marine discoveries, from the molecular basis of the tiniest phototrophs to the global expanse of our ocean ecosystem, including the crucial roles of marine microbes in climate change.

Our new coauthor, Erik Zinser, presents a fresh take on microbial genetics (Chapters 7–10 and 12). For example, we show how certain bacteria expanded their genetic code to include new amino acids. Our discussion of posttranslational protein modifications has been expanded to include protein glycosylation in gut microbes and the use of mass spectrometry to identify posttranslational protein modifications. A new section (Section 7.6) presents advances in analyzing genomes, transcriptomes, and proteomes through current research on human gut microbiomes and marine microbiomes.

This Fifth Edition maintains our signature balance between cutting-edge ecology and medicine, including the use of case histories in the medical section (Part 5). Our balanced depiction of women and minority scientists, including young researchers, continues to draw enthusiastic responses from our adopters. Our focus on evolution, along with our modern organization of topics, has set the standard that other textbooks follow. Unlike other texts, we maintain our consistent chapter organization to facilitate year-to-year course transitions for instructors.

In many chapters we relate topics to current events, to keep students interested in and informed on the role of microbiology in the world today. One example is the invention of microbial fuel cells that use environmental microbes to generate electricity (Chapter 14, Electron Flow in Organotrophy, Lithotrophy, and Phototrophy). Another example is the development of lentiviral treatments for cancer and inherited disorders, including the landmark FDA approval of CAR-T therapy for pediatric leukemia (presented in Chapter 11, Viral Molecular Biology; and in Chapter 16, Food and Industrial Microbiology).

The Fifth Edition continues our vision of this text as a community project, drawing not only on the authors' experience as researchers and educators, but also on the input of hundreds of colleagues from around the world to create a comprehensive microbiology book for the twenty-first century. We present the full story of molecular microbiology and microbial ecology from its classical history of Koch, Pasteur, and Winogradsky right up to the research of twenty-first-century researchers Stanley Falkow and Ariane Briegel. We have included countless contributions recommended by colleagues from around the globe, at institutions such as Washington University; University of California, Davis; University of Wisconsin–Madison; Cornell University; Florida State University; University of Toronto; University of Edinburgh; University of Antwerp; Seoul National University; Chinese University of Hong Kong; and many more. We are grateful to you all.

In order to contain length while adding new material, we continue to transfer certain topics online as "eTopics." The eTopics are called out in the text and hyperlinked in the ebook, and they are indexed in the printed book. Therefore, returning adopters can be confident of retaining access to all of the material they taught from the previous editions, but now they also have new topics on *Mycobacterium tuberculosis* cell aging and drug resistance (Chapter 3) and on bacteria that convert phage genes into toxin secretion systems (Chapter 25), and much more.

Major Features

Our book targets the science major in biology, microbiology, or biochemistry. Several important features make it the best text available for undergraduates today:

Updated and restructured coverage of microbial genetics and genomics. New author Erik Zinser brings his experience as an active teacher and researcher to capture how these central topics are being studied by microbiologists today. Chapters 7–10 and 12 have been comprehensively updated to reflect the current state of the field and thoughtfully restructured to improve pedagogy and the flow of topics.

Themes of discovery: marine microbiology and our intestinal microbiome. The Fifth Edition features new content on two compelling and relevant themes. One is the new theme of marine microbiology, an exciting area of research that features cutting-edge approaches such as metagenomics and single-cell whole-genome sequencing to investigate how microbes contribute to the vast ocean ecosystem. And for the popular theme of the gut microbiome, new examples have been added throughout the Fifth Edition, connecting engaging examples to important concepts in medical and ecological microbiology. Icons in the margin make it easy to locate examples of each theme, such as:

- In Chapter 21, research by Olivia Mason (Pacific Northwest National Laboratory, U.S. Department of Energy) that assessed marine microbes following the 2010 *Deepwater Horizon* oil well blowout in the Gulf of Mexico. Mason's research analyzed metagenomes to reveal which marine microbes had the ability to degrade petroleum. These findings suggest the potential for bioremediation of oil spills.
- In Chapter 13, research by James Versalovic (Baylor University) and colleagues that explored whether gut bacteria might regulate function in the brain. Their research showed that certain gut bacteria had a probiotic effect to suppress the activity of pain sensory neurons in rats.

A new mini-interview at the start of each chapter, offering a total of twenty-eight new perspectives from cutting-edge researchers. Examples include:

- A Chapter 12 interview with Roberto Di Leonardo (University of Rome), who genetically engineered *E. coli* to produce a replica of Leonardo da Vinci's *Mona Lisa*.
- A Chapter 13 interview with Julie Biteen (University of Michigan) explaining how gut bacteria enable humans to digest sugar chains that human cells cannot break down on their own.

Research on contemporary themes such as evolution, genomics, metagenomics, molecular genetics, and biotechnology enriches students' understanding of foundational topics and provides comprehensive coverage of the rapidly evolving field of microbiology. Every chapter presents numerous current research examples within the up-to-date framework of molecular biology. Examples of current research include exploring the native functions and genetic engineering applications of

CRISPR-Cas systems, tools to explore evolution in aging bacterial colonies, and the simultaneous profiling of gene expression patterns in host and microbe during an infection.

Student-friendly presentation of core concepts that motivates learning. Ample Thought Questions throughout every chapter challenge students to think critically about core concepts, the way a scientist would, and to draw connections between concepts across sections and chapters. "To Summarize" bulleted lists at the end of every chapter section distill important concepts into brief, memorable statements. Abundant eye-catching illustrations with accessible bubble captions support the text.

An innovative media package, including an updated Smartwork5 online homework course and new active learning resources, promotes learning and retention. Smartwork5, the new Interactive Instructor's Guide active learning resources, and Squarecap classroom response content work with the textbook to address the challenges that instructors and students face. Smartwork5 includes new Tutorial Lessons in addition to review, critical thinking, visual, and animation questions, as well as questions based on the gut microbiome and marine microbiology themes for every chapter. Smartwork5 questions and activities help students master complicated microbiology topics and prepare for class and exams. The new Interactive Instructor's Guide is an all-in-one teaching resource for instructors that want to integrate active learning in their course. Book-specific questions are available for use with Squarecap classroom response technology.

Additional Features of the Fifth Edition

- **Genetics and genomics are presented as the foundation of microbiology.** Molecular genetics and genomics are thoroughly integrated with core topics throughout the book. This approach gives students an understanding of how genomes reveal potential metabolic pathways in diverse organisms, and how genomics and metagenomics reveal the character of microbial communities.
- **Microbial ecology and medical microbiology receive equal emphasis,** with particular attention paid to the merging of these fields. Throughout the book, phenomena are presented with examples from both ecology and medicine; for example, when introducing metagenomics we present a study that directly compares metagenomes of the ocean with that of the human gut (Chapter 7).
- **Size scale information appears for nearly every micrograph**—a feature often absent or inconsistent in other microbiology textbooks.
- **Viruses are presented in molecular detail and in ecological perspective.** In marine ecosystems, for example, viruses play key roles in limiting algal populations while selecting for species diversity (Chapter 6). Similarly, a constellation of bacteriophages influences enteric flora.
- **Microbial diversity is made clear and easy to grasp.** We present microbial diversity in a manageable framework that enables students to grasp the essentials of the most commonly presented taxa, in the context of the continual discovery of organisms ranging from anammox bacteria to emerging pathogenic *Escherichia* strains.
- **Appendices are provided for review and further study.** Our book assumes a sophomore-level understanding of introductory biology and chemistry, with online eAppendices for those in need of review.

Organization

The topics in this book are arranged so that students can progressively develop an understanding of microbiology from key concepts and research tools. The chapters of Part 1 present key foundational topics: history, visualization, the bacterial cell, microbial growth and control, and virology.

The six chapters in Part 1 present many topics that are then developed in further detail throughout Parts 2 through 5. Part 2 presents modern genetics and genomics. Part 3 presents cell metabolism and biochemistry, although the chapters in Part 3 are written in such a way that they can be presented before the genetics material if so desired. Part 4 explores microbial ecology and diversity and discusses the roles of microbial communities in local ecosystems and global cycling. And then the chapters of Part 5 (Chapters 23–28) present medical and disease microbiology from an investigative perspective, founded on the principles of genetics, metabolism, and microbial ecology.

What's New in the Fifth Edition?

The Fifth Edition of *Microbiology: An Evolving Science* presents significant updates throughout the book, especially in the areas of molecular and genomic microbiology. The art has been revised, and numerous Thought Questions and Special Topic boxes have been added or updated. Every chapter opens with a new research topic that features the work of established scientists, postdocs, and graduate students from around the world.

Many of these changes are featured in the following list.

CHAPTER 1: Microbial Life: Origin and Discovery. The chapter opener describes research on the exotic chemistry carried out by *Hormoscilla spongeliae*, a cyanobacterium that lives within marine sponges. This is one example of the vast potential of microorganisms that inhabit the oceans. A new Special Topic box considers how our gut microbiota help fight cancer.

CHAPTER 2: Observing the Microbial Cell. In the chapter opener, cryo-electron tomography and genetic techniques reveal the essential role of gut bacteria in the development of the mammalian immune system. A new Special Topic box describes techniques used to map the biogeography of intestinal pathogens. The section on cryo-electron microscopy and tomography has been updated with a new example on modeling the molecular evolution of bacterial flagellar motors.

CHAPTER 3: Cell Structure and Function. In the new chapter opener, we learn that some bacteria have a chemoreceptor "nose" that they use to find food and avoid trouble. Again, cryo-electron tomography helps us visualize the components of the chemoreceptors. Coverage of cell division by septation has been updated with new figures. Section 3.5 now presents cellular nanotubes and secreted membrane vesicles. A new Special Topic box focuses on cryo-electron microscopy, with the amazing discovery of entirely new cell components. The presentation of chemotaxis signaling includes a new figure and text.

CHAPTER 4: Bacterial Culture, Growth, and Development. The chapter opener reveals how bacteria in biofilms control nutrient distribution and availability through an electrochemical mechanism akin to neural transmission. A new discussion in Section 4.4 looks at research probing the genetics of cellular dormancy. The chapter's final section includes some of the latest research on the exploratory growth phase of the antibiotic-producing *Streptomyces*.

CHAPTER 5: Environmental Influences and Control of Microbial Growth. Microbial assault and thievery are the topics of the chapter opener. Nanotubes that run between competing *Bacillus* species are used to introduce toxins and steal nutrients. The discussion of barophiles in Section 5.2 includes new examples of microbes found in the ocean abyss. Coverage of oxygen and microorganisms (in Section 5.4) has been extensively revised and updated. As in Chapter 4, the topic of cellular dormancy is presented, this time as a response to starvation. The final section has been updated with current examples of antimicrobial agents such as the proprietary antimicrobial coating AGXX. The Special Topic box on phage therapy has been updated with research from 2018.

CHAPTER 6: Viruses. In the chapter opener, cryo-electron tomography again plays a starring role, helping scientists visualize the mechanism of infection of *Prochlorococcus* by bacteriophage P-SSP7. This and other examples highlight the role of marine viruses in controlling microbial populations (especially blooms) and in cycling food molecules throughout the ocean ecosystem. Cryo-electron tomography reveals the structure of an Ebola virion (Figure 6.9). The new Special Topic box reports the surprising evidence of bacteriophages found throughout the human body.

PART 2: Genes and Genomes. With the addition of Erik Zinser to the author team, this part of the book has been extensively updated. Chapters 7–10 and 12 now contain the latest advances in our understanding of microbial processes and the experimental techniques used to study them. Examples of these techniques and processes highlight the microbiome and marine microbiology themes. These chapters have been trimmed, revised, and updated to improve pedagogy and avoid being overly detailed. Material has been moved within and between chapters to improve the flow of topics. New thought questions and new discussions have been added that challenge the reader to understand microbial function from an evolutionary perspective.

CHAPTER 7: Genomes and Chromosomes. The chapter opener looks at *Epulopiscium*, a giant bacterium that can grow to 300 mm long. A single *Epulopiscium* cell contains tens of thousands of copies of its chromosome. The chapter opener describes how these chromosomes are packed in the cell and how that organization changes, depending on the developmental stage of the organism. Secondary chromosomes, which expand the genetic repertoire of microbes, share a section of this chapter with plasmids. Their method of replication and segregation, and their evolutionary origins, are discussed. The introduction to archaeal genomes has been completely revised and updated. The new Special Topic box looks at some provocative research on storing information in DNA code. The final section of the chapter provides several current examples of research on human gut microbiomes and marine microbiomes.

CHAPTER 8: Transcription, Translation, and Protein Processing. The chapter opener looks at a new technique to follow *E. coli* cells as they move about in the mammalian gastrointestinal tract. Genes from bacteria that produce gas vesicles were transferred to *E. coli*, which transcribed the foreign genes and ultimately produced gas vesicles in its cytoplasm. The vesicles and hence the bacteria could be followed through the use of ultrasound. New details of translation are presented; for example, we describe the spatial separation of mRNA in bacteria after transcription has been completed. We present bacteria, such as *Acetohalobium arabaticum*, that have expanded their genetic code to include new amino acids. The discussion of post-translational protein modifications has been greatly expanded to include protein glycosylation in gut microbes and the use of mass spectrometry to comprehensively identify such modifications.

CHAPTER 9: Genetic Change and Genome Evolution. The chapter has been extensively revised, as indicated by the new chapter title. The main topics are mutations in the genome, DNA repair and recombination, horizontal gene transfer, and mobile genetic elements. The final section examines genome evolution. The revised introduction of Section 9.3, on gene transfer, will help students better understand gene transfer mechanisms as survival tools and in an evolutionary context. The discussion on transformation notes that it has now been observed in archaea. *Vibrio cholerae* and its type IV pilus is used as a new detailed example of transformation. The intestinal microbiome is used as a detailed example of multiple gene transfer mechanisms operating in a complex environment. The discussion of CRISPR function has been revised to improve pedagogy and includes a new figure (Figure 9.29). New material is included on conjugative transposons and mobilizable genetic islands in archaea and bacteria, including an example in *V. cholerae*. There is new material on the use of transposable elements in genetic analysis, including examples of mutagenesis to investigate purple antimicrobial molecules made by the marine bacterium *Phaeobacter* and TnSeq high-throughput screening of the gut bacterium *Bacteroides*. There are new figures and text on gene duplication. Also included is an expanded explanation of genome reduction with new figures illustrating gene loss and a new example in *Pelagibacter*. The discussion of horizontal gene transfer has been revised so that the process is described first and then the evidence is explained. A new figure (Figure 9.43) helps clarify how scientists identify genome islands within chromosomes.

CHAPTER 10: Molecular Regulation. The chapter opener reveals another fascinating observation about our gut microbiome: Some enteric bacteria have melatonin receptors and exhibit circadian behavior in response to host melatonin levels. The chapter organization is different from that in the previous edition. Traditional repressors and activators of transcription are covered first, followed by a section on sigma factors, anti-sigma factors, and anti-anti-sigma factors. Regulation of transcription and translation by RNA is next. Section 10.5 presents new information and examples of control by circadian cycles, RNA thermometers, and DNA switches. The chapter closes with a look at bacterial chemotaxis as an example of posttranslational control of proteins. The Special Topic box introduces the mobile genetic elements known as inteins. An intein is a fragment within a protein that inactivates protein function but, when given the appropriate signal, will excise itself and splice the remaining protein back together, thus activating the protein's function.

CHAPTER 11: Viral Molecular Biology. The chapter opener captures the measles virus as it assembles its genome within a host cell. New imaging reveals the influenza virus replicating in its host cells. Transmission electron microscopy and fluorescence microscopy were used to show the replication process. Updated information is presented on the inhibitory mechanisms of host cell protein APOBEC3G against HIV. The description of how HIV integrates into the host genome has been updated to reflect recent discoveries. Section 11.4 expands the presentation of human gene therapy using viral vectors. The new Special Topic box discusses a controversial set of experiments that tested which mutations could transform an avian influenza strain into a "pandemic" strain.

CHAPTER 12: Biotechniques and Synthetic Biology. The chapter opener presents an image of the *Mona Lisa* "painted" in *E. coli*. The image illustrates the ability of scientists to control the swimming speed, and hence density, of a population of *E. coli* with varying amounts of light. This chapter has been greatly revised. Material previously in Chapters 7 and 8 has been moved into this chapter, including DNA isolation and purification, restriction endonucleases, PCR, DNA sequencing, and annotation of DNA sequences. There is a new example of quantitative PCR used to study the distributions of the photosynthetic bacterium *Prochlorococcus* in the Pacific and Atlantic oceans. The latest sequencing methods are described, including single-molecule sequencing and sequencing that detects methylation of DNA. In addition, the chapter includes new material on cloning by PCR (Gibson Assembly), new sections on random and site-directed mutagenesis, updates on the use of CRISPR in bacteria and archaea, and a new Special Topic box, which describes the effort to create a minimum functional genome.

CHAPTER 13: Energetics and Catabolism. The chapter opener presents evidence using advanced fluorescence imaging techniques that show how members of the gut microbiome digest large starch molecules that we cannot digest ourselves. In Section 13.1, an example is given of entropy-driven metabolism, in this case by the soil archaeon *Methanosarcina barkeri*. Section 13.5 (The Gut Microbiome: Friends with Benefits) is new to this chapter. This section looks at the roles our gut microbes play in the host's ability to move into new nutritional niches, their essential role in aiding human infant digestion, and their role in producing select neurotransmitters. The Special Topic box presents evidence that gut microbes might regulate development and function of the mammalian brain.

CHAPTER 14: Electron Flow in Organotrophy, Lithotrophy, and Phototrophy. The chapter opener shows how cable bacteria conduct electricity across long distances, creating underwater batteries. Section 14.4 is a new section about nanowires and bacterial electron shuttles. Recent research on pili that conduct electricity (known as nanowires) is presented that supports their existence and reveals details of their electron-conducting mechanism. A diverse array of bacteria has been shown to use various mechanisms and structures to conduct electrons out of the cell and shuttle them to extracellular electron acceptors. The Special Topic box continues the chapter's look at bioelectricity by showing how scientists are tapping into the electricity that microbes produce in the ocean sediments. This electricity is being used to power measuring devices that monitor ocean conditions. In Section 14.6, the structure of the light-harvesting complex from *Blastochloris viridis*, based on new data using cryo-electron microscopy, is shown.

CHAPTER 15: Biosynthesis. The stars of the chapter opener are marine actinomycetes, whose genomes are being mined in the search for new antibiotics. Section 15.3, on fatty acids and antibiotics, has been revised with a new discussion of nonribosomal peptide antibiotics. The Special Topic box on antibiotic discovery has been updated with recent work from Pieter Dorrestein's lab.

CHAPTER 16: Food and Industrial Microbiology. The chapter opener looks at the complex and poorly understood process of cocoa bean fermentation. The new Special Topic box examines how scientists are attempting to engineer bacteria to improve the industrial-scale production of vitamin B_{12}. Updates are included about the TB antibiotic sutezolid, now in clinical trials. The innovative new Section 16.6 covers the commercial production of microbial gene vectors for agriculture and human gene therapy, with a focus on lentiviral vectors for chimeric antigen receptor T-cell therapy (CAR-T therapy).

CHAPTER 17: Origins and Evolution. The chapter opener presents an exciting hypothesis and supporting data positing that bacteria can induce multicellularity in single-celled eukaryotes. Raman spectroscopy is discussed as a method for confirming the presence of organic biosignatures in rock samples. Raman spectroscopy, isotope ratio determination, and NanoSIMS mass spectrometry are discussed as important methods for confirming the presence of microfossils. Section 17.2 (now called Forming the First Cells) has been revised to provide a more linear progression of ideas about precellular conditions and early life. The discussion of early oxidation-reduction reactions has been expanded. The new Special Topic box describes a vivid demonstration of natural selection in a giant (2 feet × 4 feet) Petri dish. Section 17.4 has been updated with new data on antibiotic resistance formation and new results from Richard Lenski's Long-Term Evolution Experiment.

CHAPTER 18: Bacterial Diversity. The chapter opener introduces ectosymbiotic bacteria that live attached to marine nematodes. These very long bacteria include the longest bacteria known to divide by classic Z-ring constriction. The phylogenetic tree of bacteria (now Figure 18.1) has been updated to include genomic data from over 1,000 uncultivated and little-known organisms sequenced by the latest methods. Section 18.1 and Table 18.1, which summarize bacterial diversity, have been revised to reflect our latest understanding of bacterial phylogenetic groupings. The Special Topic box on the human gut microbiome has been updated with recent data from 2018.

CHAPTER 19: Archaeal Diversity. The chapter opener presents structural evidence that the hyperthermophilic *Ignicoccus hospitalis* has an archaeal endomembrane system. Archaeal phylogeny has been updated to incorporate the latest discoveries from metagenomic and single-cell analysis. The chapter is organized to reflect the latest groupings of archaeal organisms, such as the recent TACK and DPANN superphyla. New discoveries are included, such as the aerobic Aigarchaeota phylum and the strictly anaerobic *Halanaeroarchaeum*. The Special Topic box has been updated with recent work showing that administering probiotics reduces the abundance of *Methanobrevibacter* in the gut microbiome and consequently reduces flatulence. Section 19.6 introduces the *Altiarchaeum*, with its pilus-like grappling hooks, and presents the Asgard superphylum, which some scientists speculate may be closely related to the early Eukarya.

CHAPTER 20: Eukaryotic Diversity. The eukaryotic diversity chapter opens with the disturbing phenomenon known as coral bleaching. Warming temperatures melt the photosynthetic membranes of *Symbiodinium*, the dinoflagellate symbiont found in coral. Without the ability to photosynthesize, the microbes are expelled and the coral subsequently dies. This topic is taken up again in Section 20.5 (see Figure 20.36). In Table 20.1, Figure 20.4, and throughout the chapter, names and groupings of eukaryotic clades reflect our current knowledge based on genome sequencing, protein phylogeny, and genetic analysis. The Special Topic box on *Saccharomyces cerevisiae* has been updated to include some of the final research performed by the late Susan Lindquist and her colleagues on "humanized" yeast cells as a model of beta-amyloid metabolism.

CHAPTER 21: Microbial Ecology. This chapter expands on what we are learning by sequencing metagenomes from a wide variety of microbial communities. The chapter opener describes a recently discovered microbial endosymbiont, *Pantoea carbekii*, found in the gut of a stinkbug. The explanation of performing a metagenome sequencing project has been extensively updated. Additional methods for analyzing microbial communities, such as metatranscriptomics, single-cell sequencing, flow cytometry, and fluorescence-activated cell sorting are discussed. Numerous new soil, ocean, and gut examples of microbial communities analyzed by these methods are presented (see examples throughout the chapter). Section 21.2, on functional ecology, presents a bacterium that can hydrolyze polyester, *Ideonella sakaiensis*, isolated from soil at a bottle-recycling factory. The Special Topic box on cyanobacterial mats in Antarctic lakes has been updated with recent data.

CHAPTER 22: Element Cycles and Environmental Microbiology. The new chapter title reveals this chapter's greater emphasis on the roles of environmental microbiologists. The chapter includes new information on climate change and on the human built environment. The chapter opener presents evidence that Arctic microbes are accelerating global temperature rise. The chapter introduction and Section 22.1 have been extensively revised with a focus on global climate change, as reported in 2019 by the United Nations Intergovernmental Panel on Climate Change. The explanation of wastewater treatment has been expanded. Section 22.3, on the nitrogen cycle, includes a new table and color-coded highlights to aid students in understanding the numerous oxidation states of nitrogen. A similar approach is applied to the sulfur cycle. The new section on microbiology in the human built environment (Section 22.5) introduces the microbiomes found in our building and transportation interiors, and looks at how our human microbiomes interact with those in the built environment. The new Special Topic box looks at life in a kitchen sponge. Recent evidence of subsurface liquid water on Mars is presented, raising new questions about whether life might exist on Mars.

CHAPTER 23: The Human Microbiome and Innate Immunity. While we know that the human microbiome is essential for human health, the chapter opener presents evidence that at least one member of our microbiome, *Bacteroides fragilis*, is a factor in forming colonic tumors. Numerous examples of current research on human microbiome–host interactions are presented throughout the chapter. For example, Section 23.1 talks about the presence of microbial metabolites in our blood and their possible role in human health and disease. This section also mentions how the method by which a baby is delivered affects initial colonization of its microbiome and how smoking negatively impacts the lung microbiome. Recently developed culture enrichment methods are presented that reveal greater diversity in our gut microbiome than was previously known. Links between our microbiome and obesity continue to be explored (Section 23.2). The Special Topic box presents new evidence about the tantalizing question: Why do tattoos last forever? The answer involves aspects of innate immunity. New research on natural killer cells shows how one NK-produced granzyme kills intracellular bacteria. The discussion of interferons has been updated to include type III interferons and the importance of interferon-stimulated genes.

CHAPTER 24: The Adaptive Immune Response. The chapter opener explains that live vaccines are better than dead vaccines at generating an immune response at least in part because of bacterial RNA's effect on host cytokines. This knowledge is being used to develop more effective vaccines. In Section 24.2, mechanisms are described for eliminating immature B cells that target self antigens. In Section 24.3, explanations of various aspects of T-cell biology have been revised, including the genetics of T-cell receptor formation and T-cell education. The myriad effects of interferon-gamma on the immune system are discussed and illustrated in a new figure, also in Section 24.3. Section 24.6 has been extensively revised to expand the concept of building better vaccines. The topic of whether vaccines are dangerous has been updated to include consideration of the impact of the anti-vaxxer movement. The Special Topic box looks at the development of anti-Ebola monoclonal antibodies. Section 24.7 presents new research on using T cells with reengineered T-cell receptors to restore self-tolerance—a novel strategy for curing autoimmune diseases.

CHAPTER 25: Pathogenesis. The bag of tricks that pathogens possess continues to amaze us. In the chapter opener, we learn that *Streptococcus pyogenes* uses a toxin to inflict pain far greater than what you would expect from the initial tissue damage. The pain induces host responses that assist bacterial pathogenesis. In the section on microbial deployment of toxins and effector molecules, the illustrations of bacterial secretion systems have been revised to include our latest understanding of these molecular machines. An innovative therapy using bead-bound copies of the bacterial adhesion molecule MAM7 is discussed. Applied to the skin of burn victims, this form of MAM7 inhibits the binding and spread of *Pseudomonas*. The Special Topic box summarizes current research on the ability of *Staphylococcus aureus* biofilms to block innate immunity, using neutrophil extracellular traps, or NETs. A new figure (Figure 25.32) illustrates the differing ways in which intracellular and extracellular pathogens impact host cell cytokines. Current research is presented on how *Neisseria gonorrhoeae* triggers apoptosis of mucosal macrophages.

CHAPTER 26: Microbial Diseases. The chapter opener examines the transformation of *Cryptococcus neoformans* cells, following inhalation by a host, from small yeast cells to Titans. This transformation might be due to the yeast cell interactions with the lung microbiome. The chapter introduction has been thoroughly revised to emphasize the importance of a patient history and variable susceptibility to disease based on immunocompetence and pangenomic differences among pathogen strains. A new case history presents a challenging example of a fungal lung infection, a jumping-off point to discuss respiratory fungal pathogens. The section on tuberculosis has been revised and updated. New information about the formation and prevention of UTIs is presented, and the two cases of people "cured" of HIV are discussed. A new Special Topic looks at how neutrophils ambush bacterial pathogens in regional lymph nodes. The use of CRISPR as an antimalaria strategy is presented.

CHAPTER 27: Antimicrobial Therapy and Discovery. This chapter emphasizes the war that humans wage against infectious diseases and, in particular, the threat of antibiotic resistance. A novel form of antibiotic resistance is described in the chapter opener. The new mechanism dislodges certain antibiotics from the ribosome. There is a new discussion of antibiotics that target aminoacyl-tRNA synthetases in Section 27.2, and in Section 27.3, a new classification scheme of antibiotic resistance strategies is presented and mechanistic details are expanded. The section titled "How did we get into this mess?" has been rewritten to include some of the latest research on antibiotic resistance in vulnerable populations, the effects of antibiotics on our microbiome, and the need for antibiotic stewardship. Also in Section 27.3, the search for new antibiotics and efforts to counter drug resistance are two topics that have been updated. The latest FDA-approved antiviral agents, like ibalizumab and Xofluza, are discussed. Novel antiviral agents that target host pattern recognition receptors are also presented. Orotomides, a new class of antifungal drugs, are described.

CHAPTER 28: Clinical Microbiology and Epidemiology. A surprising and perhaps ominous discovery is presented in the chapter opener. A strain of *Enterococcus faecium* has been found that contains a botulism neurotoxin–like gene cluster. Discussion of pathogen identification methods has been expanded and is now divided into two sections: Section 28.2 focuses on classical detection methods, while Section 28.3 presents expanded and updated coverage of molecular and serological methods like mass spectrometry, nucleic acid amplification, next-generation sequencing, and programmable RNA sensors. A new Special Topic box describes a CRISPR-Cas12a pathogen identification system that utilizes a reporter molecule that fluoresces when the pathogen DNA is detected. The section on emerging microbial diseases has been updated to include a more extensive introduction, examples of outbreaks in 2018, and a new case study on acute flaccid myelitis, a rare disease of unknown origin.

Resources

SMARTWORK5 (digital.wwnorton.com/microbio5). Norton's powerful and accessible online homework platform features answer-specific feedback, a variety of engaging question types, and integration of the stunning art from the book and animations to help students master microbiology concepts and come to class prepared. New Tutorial Lessons break down complex topics, promoting deeper understanding that goes beyond simple memorization. Smartwork5 integrates with campus LMSs such as Blackboard and Canvas and features a simple, intuitive interface, making it the easiest-to-use online homework system for instructors and students.

EBOOK. The Norton ebook for *Microbiology: An Evolving Science* includes dynamic features that engage students, including animations and dropdown answers to in-text Thought Questions. Instructors can focus student reading by sharing notes with their classes, including embedded images and video. Art expands, pop-up key terms are linked to definitions, and students can search, highlight, and take notes with ease. The ebook can be viewed on any device and will sync across devices.

INTERACTIVE INSTRUCTOR'S GUIDE. Searchable by chapter, phrase, and topic, the Interactive Instructor's Guide compiles the many valuable teaching and active learning resources available with *Microbiology: An Evolving Science.* This sortable database includes resources to implement active learning that include animations with discussion questions, learning objectives, primary literature suggestions, in-class activities, sample lecture outlines, clicker questions available for use with Squarecap, and more.

SQUARECAP CLASSROOM RESPONSE QUESTIONS. Squarecap is a next-generation classroom response system that promotes a high level of student engagement in the classroom through quality assessment in a variety of question types. Students can answer questions directly from their smartphones; the Squarecap system offers instructors in-depth analytics on student performance in real time. *Microbiology: An Evolving Science* can be packaged with review questions specific to each chapter.

PRESENTATION TOOLS. Every figure and photograph in the textbook is available for instructor use and presentation, in both labeled and unlabeled versions, as JPEGs and PowerPoint images. Chapter-specific Lecture PowerPoint decks include key figures from the text, links to animations, and clicker questions, and are available for use in the classroom or for student self-study.

TEST BANK. Each chapter of the Test Bank consists of five question types classified according to the first five levels of Bloom's taxonomy of knowledge types: Remembering, Understanding, Applying, Analyzing, and Evaluating. Questions are further classified by section, difficulty, and learning objective, making it easy to construct tests and quizzes that are meaningful and diagnostic according to instructors' needs. Questions are multiple-choice and short-answer. The Test Bank is available in ExamView and downloadable Word and PDF formats from wwnorton.com/instructors.

COURSEPACKS. Coursepacks make it easy to add high-quality Norton digital resources to your online, hybrid, or lecture course. You can access all activities in the coursepacks within your existing LMS, and many components are customizable. Resources include learning objectives, animations, flashcards, and links to Smartwork5, the Interactive Instructor's Guide, eTopics and eAppendices, the Test Bank, and the ebook.

eAPPENDICES. eAppendices on three topics—biological molecules, introductory cell biology, and laboratory methods—written by the book's authors, supplement the text. The eAppendices are linked in the ebook and are available as downloadable PDFs.

ANIMATIONS. More than forty-five animations depicting key processes of microbiology were developed under the careful supervision of the book's authors. Instructors can assign the animations with assessment in Smartwork5, and students can access the animations right from the ebook.

Animation topics include:

Microscopy
Replisome Movement in a Dividing Cell
Chemotaxis
Phosphotransferase System (PTS) Transport
Dilution Streaking Technique
Biofilm Formation
Endospore Formation
Lysis and Lysogeny
Supercoiling and Topoisomerases
DNA Replication
Rolling-Circle Mechanism of Plasmid Replication
PCR
Protein Synthesis
Protein Export
SecA-Dependent General Secretion Pathway
ABC Transporters
Bacterial Conjugation
Recombination
DNA Repair Mechanisms: Methyl Mismatch Repair
DNA Repair Mechanisms: Nucleotide Excision Repair
DNA Repair Mechanisms: Base Excision Repair
Transposition
The *lac* Operon
Transcriptional Attenuation
Chemotaxis: Molecular Events
Quorum Sensing
Influenza Virus Entry into a Cell
Influenza Virus Replication
HIV Replication
Herpesvirus Replication
Construction of a Gene Therapy Vector
Tagging Proteins for Easy Purification
Real-Time PCR
A Bacterial Electron Transport System
ATP Synthase Mechanism
Oxygenic Photosynthesis
Agrobacterium: A Plant Gene Transfer Vector
Phylogenetic Trees
DNA Shuffling
Listeria Infection
Light-Driven Pumps and Sensors
Malaria: A Cycle of Transmission between Mosquito and Human
The Basic Inflammatory Response
Phagocytosis
The Activation of the Humoral and Cell-Mediated Pathways
Cholera Toxin Mode of Action
Process of Type III Secretion
Retrograde Movement of Tetanus Toxin to an Inhibitory Neuron
DNA Sequencing

Acknowledgments

We are very grateful for the help of many people in developing and completing this book, including Norton editors John Byram, Vanessa Drake-Johnson, Mike Wright, and especially Betsy Twitchell and associate editor Katie Callahan, whose efforts assured completion of the Fifth Edition. Our developmental editor, Michael Zierler, contributed greatly to the clarity of presentation. Catherine Abelman did a heroic job of tracking down all kinds of images from sources all over the world with assistance from Michele Riley, Stacey Stambaugh, Tommy Persano, Dena Betz, Jane Miller, and Julie Tesser. Kate Brayton's coordination of electronic media development has resulted in a superb suite of resources for students and instructors alike. The Fifth Edition's resources have also benefited greatly from content development specialist Todd Pearson's expertise. We thank associate media editor Jasmine Ribeaux and media assistant editor Katie Daloia for producing the new Interactive Instructor's Guide and the Test Bank, as well as contributing in many other ways to development of the digital resources. Without senior project editor Thom Foley's incredible attention to detail, the innumerable moving parts of this project would never have become a finished book. We also thank Stephanie Hiebert, our superb copyeditor, whose value cannot be underestimated. Both she and Thom have worked on every edition of this title, going all the way back to the First Edition; bringing their experience and expertise to the pages of this book over many, many years. Marian Johnson, Norton's managing editor in the college department, helped coordinate the complex process involved in shaping the manuscript over the years. Ben Reynolds ably and calmly managed the manufacturing of this book. Assistant editor Danny Vargo coordinated the transfer of many drafts among many people. We thank marketing manager Katie Sweeney for ensuring that microbiology instructors know about our exciting Fifth Edition. Finally, we thank Roby Harrington, Drake McFeely, Julia Reidhead, and Ann Shin for their support of this book over its nearly decade in print.

For the quality of our new illustrations in the Fifth Edition, we thank the many artists at Dragonfly Media Group, who developed attractive and accurate representations and showed immense patience in getting the details right.

We thank the numerous colleagues over the years who encouraged us in our project, especially the many attendees at the Microbial Stress Response Gordon conferences. We greatly appreciate the insightful reviews and discussions of the manuscript provided by our colleagues, and the many researchers who contributed their micrographs and personal photos. We thank the American Society for Microbiology journals for providing many valuable resources. Reviewers Lynn Thomason and Robert Barrington offered particularly insightful comments on the metabolism and genetics sections, and Richard Lenski and Zachary Blount provided particularly insightful comments on experimental evolution. We would also like to thank the following reviewers:

Fifth Edition Reviewers

Camilla Ambivero, University of Central Florida
Shivanthi Anandan, Drexel University
Gregory Anderson, Indiana University–Purdue University Indianapolis
Michael Angell, Eastern Michigan University
Daniel Aruscavage, Kutztown University
Dennis Arvidson, Michigan State University
Perrin Beatty, University of Alberta
Douglas Bernstein, Ball State University
Michael Bidochka, Brock University
Christopher Blackwood, Kent State University
Dwayne Boucaud, Quinnipiac University
Laurie Bradley, Hudson Valley Community College
Suzanna Brauer, Appalachian State University
Andrea Castillo, Eastern Washington University
Joseph Corsini, Eastern Oregon University
Francisco Cruz, Georgia State University
Ashley Day, Isothermal Community College
Peter Dunfield, University of Calgary
Elizabeth Emmert, Salisbury University
William Ettinger, Gonzaga University
Babu Z. Fathepure, Oklahoma State University
Melinda Faulkner, Bradley University
Kathleen Feldman, University of Connecticut
Pat Fidopiastis, California Polytechnic State University
Erin Field, East Carolina University
Clifton Franklund, Ferris State University
Chiara Gamberi, Concordia University
Marcos García-Ojeda, University of California, Merced
Jason Gee, East Carolina University
Enid T. Gonzalez, California State University, Sacramento
Diane Hartman, Baylor University
Bethany Henderson-Dean, University of Findlay
Andrew Herbig, Washburn University
Marina Kalyuzhnaya, San Diego State University
Robert Kearns, University of Dayton
Christine Kirvan, California State University, Sacramento
Doug Lake, Arizona State University
Beth Lazazzera, University of California, Los Angeles
Emily Ledgerwood, Le Moyne College
Lee Lee, Montclair State University
Manuel Llano, University of Texas, El Paso
Richard Long, Florida A&M University
Betsy Martinez-Vaz, Hamline University
Jake McKinlay, Indiana University
Javier Ochoa-Reparaz, Eastern Washington University
Florence Okafor, Alabama A&M University
Jessica Parilla, Georgia State University
Hyun Woo Park, California Baptist University
Samantha Parks, Georgia State University
Sladjana Prisic, University of Hawaii
Ines Rauschenbach, Rutgers University
James Riordan, University of South Florida
Chandan Robbins, Georgia State University
Ben Rowley, University of Central Arkansas
Dennis Sampson, Cleveland State University
Kristen Savage-Ashlock, University of Oklahoma
Pratibha Saxena, University of Texas, Austin
Sebastian Schmidl, Texas A&M International University

Jared Schrader, Wayne State University
Sheila Schreiner, Salem State University
Paul Schweiger, University of Wisconsin–La Crosse
Mark Silby, University of Massachusetts Dartmouth
Adam Silver, University of Hartford
Marek Sliwinski, University of Northern Iowa
Emily Stowe, Bucknell University
Matthew Swearingen, Florida Gulf Coast University
Wesley Swingley, Northern Illinois University
Nikhil Thomas, Dalhousie University
Juliette Tinker, Boise State University
Beth Traxler, University of Washington
Brian Trewyn, Colorado School of Mines
Tricia Van Laar, California State University, Fresno
Helene Ver Eecke, Metropolitan State University of Denver
Kristi Whitehead, Clemson University
Stephen Winans, Cornell University
Alejandra Yep, California Polytechnic State University
Noha Youseff, Oklahoma State University

Fourth Edition Reviewers

Emma Allen-Vercoe, University of Guelph
Alexandra Armstrong, University of Arizona, Pima Community College
Daniel Aruscavage, Kutztown University
Dennis Arvidson, Michigan State University
Nazir A. Barekzi, Old Dominion University
Miriam Barlow, University of California, Merced
Suzanne S. Barth, University of Texas, Austin
Hazel Barton, University of Akron
Yan Boucher, University of Alberta
Linda Bruslind, Oregon State University
Kathleen L. Campbell, Emory University
John Carmen, Northern Kentucky University
Carlton Rodney Cooper, University of Delaware
Vaughn Cooper, University of Pittsburgh, School of Medicine
John Dennehy, Queens College
Kathleen A. Feldman, University of Connecticut
Kelly A. Flanagan, Mount Holyoke College
Clifton Franklund, Ferris State University
Heather Fullerton, Western Washington University
Bethany Henderson-Dean, University of Findlay
Karen Huffman, Genesee Community College
Edward Ishiguro, University of Victoria
Mack Ivey, University of Arkansas
Ece Karatan, Appalachian State University
Robert J. Kearns, University of Dayton
Alexandra M. Kurtz, Georgia Gwinnett College
Manuel Llano, University of Texas, El Paso
Shawn Massoni, Mount Holyoke College
Ann G. Matthysse, University of North Carolina at Chapel Hill
William R. McCleary, Brigham Young University
James A. Nienow, Valdosta State University
C. O. Patterson, Texas A&M University
Ronald D. Porter, Pennsylvania State University
Veronica Riha, Madonna University
Benjamin G. Rohe, University of Delaware
Joseph Romeo, San Fransciso State University
Pratibha Saxena, University of Texas, Austin
Richard Seyler, Virginia Tech
Alastair Simpson, Dalhousie University
Marek Sliwinski, University of Northern Iowa
Amy Springer, University of Massachusetts Amherst
Vincent J. Starai, University of Georgia
Nikhil Thomas, Dalhousie University
Mitch Walkowicz, University of Massachusetts Amherst
Susan Wang, Washington State University
Cheryl Whistler, University of New Hampshire
Adam C. Wilson, Georgia State University
Erik Zinser, University of Tennessee

Pre-revision Survey Reviewers

Eric Allen, University of California, San Diego
Emma Allen-Vercoe, University of Guelph
Jason Andrus, Meredith College
Catalina Arango Pinedo, St. Joseph's University
Alexandra Armstrong, University of Arizona, Pima Community College
Nazir Barekzi, Old Dominion University
Miriam Barlow, University of California, Merced
Prakash H. Bhuta, Eastern Washington University
Cheryl Boice, Florida Gateway College
Blaise Boles, University of Iowa
Suzanna Bräuer, Appalachian State University
Alison Buchan, University of Tennessee
Robert Carey, Lebanon Valley College
Christian Chauret, Indiana University Kokomo
Cindy Cisar, Northeastern State University
Jeff Copeland, Eastern Mennonite University
Bela Dadhich, Delaware County Community College
Jaiyanth Daniel, Indiana University–Purdue University Fort Wayne
Diane Davis, Rutgers University
Sandra G. Devenny, Delaware County Community College
Eugene Dunkley, Greenville College
Kathleen A. Feldman, University of Connecticut
Pat M. Fidopiastis, California Polytechnic State University
David Fulford, Edinboro University of Pennsylvania
Heather Fullerton, Western Washington University
Michelle Furlong, Clayton State University
Eileen Gregory, Rollins College
Julianne Grose, Brigham Young University
Haidong Gu, Wayne State University
Julie Harless, Lone Star College–Montgomery
Geoffrey Holm, Colgate University
Edward Ishiguro, University of Victoria
Mark Kainz, Ripon College
Dubear Kroening, University of Wisconsin–Fox Valley
Douglas F. Lake, Arizona State University
Maia Larios-Sanz, University of St. Thomas

Craig Laufer, Hood College
Maureen Leonard, Mount Mary University
Alex Lowrey, University of North Georgia–Gainesville
Aaron Lynne, Sam Houston State University
Ann Matthysse, University of North Carolina at Chapel Hill
Brendan Mattingly, University of Kansas Edwards Campus
William R. McCleary, Brigham Young University
Robert McLean, Texas State University
Aaron Mitchell, Carnegie Mellon University
Naomi Morrissette, University of California, Irvine
Annika Mosier, University of Colorado, Denver
Jacalyn Newman, formerly of University of Pittsburgh
Tanya Noel, University of Windsor
Florence Okafor, Alabama A&M University
Lorraine Olendzenski, St. Lawrence University
Samantha Oliphant, Nevada State College, Marian University
Hyun-Woo Park, California Baptist University
Todd Primm, Sam Houston State University
Veronica Riha, Madonna University
Joseph Romeo, San Francisco State University
Silvia Rossbach, Western Michigan University
Natividad Ruiz, Ohio State University
Robert Rychert, Boise State University
Pratibha Saxena, University of Texas, Austin
Matthew M. Schmidt, Stony Brook University (SUNY)
Adam Silver, University of Hartford
David Singleton, York College of Pennsylvania
Marek Sliwinski, University of Northern Iowa
Amy Springer, University of Massachusetts Amherst
Vincent Starai, University of Georgia
Sang-Jin Suh, Auburn University
James R. Walker, University of Texas, Austin
Dara L. Wegman-Geedey, Augustana College
Elizabeth Wenske-Mullinax, University of Kansas
Gordon Wolfe, California State University, Chico
Marie Yeung, California Polytechnic State University, San Luis Obispo
Virginia Young, Mercer University
Noha Youssef, Oklahoma State University
Fanxiu Zhu, Florida State University

Third Edition Reviewers

Emma Allen-Vercoe, University of Guelph
Gregory Anderson, Indiana University–Purdue University Indianapolis
Lisa Antoniacci, Marywood University
Bruce M. Applegate, Purdue University
Dennis Arvidson, Michigan State University
Vicki Auerbuch Stone, University of California, Santa Cruz
Tom Beatty, University of British Columbia
Melody Bell, Vernon College
Prakash Bhuta, Eastern Washington University
Blaise Boles, University of Michigan
Suzanna Bräuer, Appalachian State University
Ginger Brininstool, Louisiana State University Baton Rouge
Kathleen L. Campbell, Emory University
Jeff Cardon, Cornell College
Rob Carey, Lebanon Valley College
Maria Castillo, New Mexico State University
Todd Ciche, Michigan State University
Sharron Crane, Rutgers University
Nicola Davies, University of Texas, Austin
Angus Dawe, New Mexico State University
Janet Donaldson, Mississippi State University
Erastus Dudley, Huntingdon College
Kathleen Dunn, Boston College
Valerie Edwards-Jones, Manchester Metropolitan University
Lehman Ellis, Our Lady of Holy Cross College
David Esteban, Vassar College
Xin Fan, West Chester University
Babu Fathepure, Oklahoma State University
Michael Gadsden, York University
Veronica Godoy-Carter, Northeastern University
Stjepko Golubic, Boston University
Vladislav Gulis, Coastal Carolina University
Ernest Hannig, University of Texas, Dallas
Julian Hurdle, University of Texas, Arlington
Edward Ishiguro, University of Victoria
Choong-Min Kang, Wayne State University
Bessie Kebaara, Baylor University
John Lee, City College of the City University of New York
Manuel Llano, University of Texas, El Paso
Aaron Lynne, Sam Houston State University
Sladjana Malic, Manchester Metropolitan University
Gregory Marczynski, McGill University
Ghislaine Mayer, Virginia Commonwealth University
Bob McLean, Texas State University
Naomi Morrissette, University of California, Irvine
Kenneth Murray, Florida International University
Kari Naylor, University of Central Arkansas
Tracy O'Connor, Mount Royal University
Rebecca Parales, University of California, Davis
Roger Pickup, University of Lancaster
Robert Poole, University of Sheffield
Geert Potters, Antwerp Maritime Academy
Ines Rauschenbach, Rutgers University
Veronica Riha, Madonna University
Marie-Claire Rioux, John Abbott College
Jason A. Rosenzweig, Texas Southern University
Ronald Russell, University of Dublin
Matt Schrenk, East Carolina University
Gary Schultz, Marshall University
Chola Shamputa, Mount Saint Vincent University
Nilesh Sharma, Western Kentucky University
Donald Sheppard, McGill University
Garriet Smith, University of South Carolina Aiken
Vincent J. Starai, University of Georgia
Lisa Stein, University of Alberta

Karen Sullivan, Louisiana State University
Kapil Tahlan, Memorial University of Newfoundland
Liang Tang, University of Kansas
Tzuen-Rong Jeremy Tzeng, Clemson University
Claire Vieille, Michigan State University
James R. Walker, University of Texas, Austin
Susan C. Wang, Washington State University
Chris Weingart, Dennison College
John Zamora, Middle Tennessee State University
Stephanie Zamule, Nazareth College
Fanxiu Zhu, Florida State University

Second Edition Reviewers

Michael Allen, University of North Texas
Gladys Alexandre, University of Tennessee, Knoxville
Suzanne S. Barth, University of Texas, Austin
Hazel Barton, Northern Kentucky University
Barry Beutler, College of Eastern Utah
Michael J. Bidochka, Brock University
Dwayne Boucaud, Quinnipiac University
Derrick Brazill, Hunter College
Graciela Brelles-Mariño, California Polytechnic State University, Pomona
Jay Brewster, Pepperdine University
Marion Brodhagen, Western Washington University
Linda Bruslind, Oregon State University
Alison Buchan, University of Tennessee, Knoxville
Jeffrey Byrd, St. Mary's College of Maryland
Silvia T. Cardona, University of Manitoba Andrea Castillo, Eastern Washington University
Miguel Cervantes-Cervantes, Rutgers University
Tin-Chun Chu, Seton Hall University
Paul Cobine, Auburn University
Tyrrell Conway, University of Oklahoma
Scott Dawson, University of California, Davis
Thomas W. DeLany, Kilgore College
Jose de Ondarza, SUNY Plattsburgh
Donald W. Deters, Bowling Green State University
Clarissa Dirks, Evergreen State College
William T. Doerrler, Louisiana State University
Janet R. Donaldson, Mississippi State University
Xin Fan, West Chester University
Babu Z. Fathepure, Oklahoma State University
Katrina Forest, University of Wisconsin–Madison
Clifton Franklund, Ferris State University
Gregory D. Frederick, University of Mary Hardin-Baylor
Christopher French, University of Edinburgh
Jason M. Fritzler, Stephen F. Austin State University
Kimberley Gilbride, Ryerson University
Stjepko Golubic, Boston University
Enid T. Gonzalez, California State University, Sacramento
John E. Gustafson, New Mexico State University
Lynn E. Hancock, Kansas State University
Martina Hausner, Ryerson University
J. D. Hendrix, Kennesaw State University
Michael C. Hudson, University of North Carolina at Charlotte
Jane E. Huffman, East Stroudsburg University
Michael Ibba, Ohio State University
Gilbert H. John, Oklahoma State University
John A. Johnson, University of New Brunswick St. John
Mark C. Johnson, Georgetown College
Carol Ann Jones, University of California, Riverside
Ece Karatan, Appalachian State University
Daniel B. Kearns, Indiana University Bloomington
Robert J. Kearns, University of Dayton
Peter Kennedy, Lewis & Clark College
Greg Kleinheinz, University of Wisconsin–Oshkosh
Susan Koval, University of Western Ontario
Deborah Kuzmanovic, University of Michigan
Jesse J. Kwiek, Ohio State University
Andrew Lang, Memorial University of Newfoundland
Margaret Liu, University of Michigan
Maia Larios-Sanz, University of St. Thomas
Beth Lazazzera, University of California, Los Angeles
Dr. Lee H. Lee, Montclair State University
Mark Liles, Auburn University
Jun Liu, University of Toronto
Manuel Llano, University of Texas, El Paso
Zhongjing Lu, Kennesaw State University
Aaron Lynne, Sam Houston State University
John C. Makemson, Florida International University
Donna L. Marykwas, California State University, Long Beach
Ann G. Matthysse, University of North Carolina at Chapel Hill
Robert Maxwell, Georgia State University
Ghislaine Mayer, Virginia Commonwealth University
William R. McCleary, Brigham Young University
Nancy L. McQueen, California State University, Los Angeles
Scott A. Minnich, University of Idaho
Philip F. Mixter, Washington State University
Christian D. Mohr, University of Minnesota
Craig Moyer, Western Washington University
Scott Mulrooney, Michigan State University
Kari Murad, College of Saint Rose
William Wiley Navarre, University of Toronto
Ivan J. Oresnik, University of Manitoba
Cleber Costa Ouverney, San Jose State University
Deborah Polayes, George Mason University
Pablo J. Pomposiello, University of Massachusetts Amherst
Joan Press, Brandeis University
Todd P. Primm, Sam Houston State University
Sharon R. Roberts, Auburn University
Michelle Rondon, University of Wisconsin–Madison
Silvia Rossbach, Western Michigan University
Ben Rowley, University of Central Arkansas
Matthew O. Schrenk, East Carolina University
Chad R. Sethman, Waynesburg University
Anthony Siame, Trinity Western University
Lyle Simmons, University of Michigan

Daniel R. Smith, Seattle University
Garriet W. Smith, University of South Carolina Aiken
Geoffrey B. Smith, New Mexico State University
Ruth Sporer, Rutgers University–Camden
Virginia Stroeher, Bishop's University
Anand Sukhan, Northeastern State University
Karen Sullivan, Louisiana State University
Dorothea K. Thompson, Purdue University
Wendy C. Trzyna, Marshall University
Bernard Turcotte, McGill University
Dave Westenberg, Missouri University of Science and Technology
Ann Williams, University of Tampa
Charles F. Wimpee, University of Wisconsin–Milwaukee
Jianping Xu, McMaster University

First Edition Reviewers

Laurie A. Achenbach, Southern Illinois University, Carbondale
Stephen B. Aley, University of Texas, El Paso
Mary E. Allen, Hartwick College
Shivanthi Anandan, Drexel University
Brandi Baros, Allegheny College
Gail Begley, Northeastern University
Robert A. Bender, University of Michigan
Michael J. Benedik, Texas A&M University
George Bennett, Rice University
Kathleen Bobbitt, Wagner College
James Botsford, New Mexico State University
Nancy Boury, Iowa State University of Science and Technology
Jay Brewster, Pepperdine University
James W. Brown, North Carolina State University
Whitney Brown, Kenyon College undergraduate
Alyssa Bumbaugh, Pennsylvania State University, Altoona
Kathleen Campbell, Emory University
Alana Synhoff Canupp, Paxon School for Advanced Studies
Jeffrey Cardon, Cornell College
Tyrrell Conway, University of Oklahoma
Vaughn Cooper, University of New Hampshire
Marcia L. Cordts, University of Iowa
James B. Courtright, Marquette University
James F. Curran, Wake Forest University
Paul Dunlap, University of Michigan
David Faguy, University of New Mexico
Bentley A. Fane, University of Arizona
Bruce B. Farnham, Metropolitan State College of Denver
Noah Fierer, University of Colorado, Boulder
Linda E. Fisher, late of the University of Michigan, Dearborn
Robert Gennis, University of Illinois, Urbana-Champaign
Charles Hagedorn, Virginia Polytechnic Institute and State University
Caroline Harwood, University of Washington
Chris Heffelfinger, Yale University graduate student
Joan M. Henson, Montana State University
Michael Ibba, Ohio State University
Nicholas J. Jacobs, Dartmouth College
Douglas I. Johnson, University of Vermont
Robert J. Kadner, late of the University of Virginia
Judith Kandel, California State University, Fullerton
Robert J. Kearns, University of Dayton
Madhukar Khetmalas, University of Central Oklahoma
Dennis J. Kitz, Southern Illinois University, Edwardsville
Janice E. Knepper, Villanova University
Jill Kreiling, Brown University
Robert Lausch, University of South Alabama
Donald LeBlanc, Pfizer Global Research and Development (retired)
Petra Levin, Washington University in St. Louis
Elizabeth A. Machunis-Masuoka, University of Virginia
John Makemson, Florida International University
Stanley Maloy, San Diego State University
Scott B. Mulrooney, Michigan State University
Spencer Nyholm, Harvard University
John E. Oakes, University of South Alabama
Oladele Ogunseitan, University of California, Irvine
Anna R. Oller, University of Central Missouri
Rob U. Onyenwoke, Kenyon College
Michael A. Pfaller, University of Iowa
Joseph Pogliano, University of California, San Diego
Martin Polz, Massachusetts Institute of Technology
Robert K. Poole, University of Sheffield
Edith Porter, California State University, Los Angeles
S. N. Rajagopal, University of Wisconsin–La Crosse
James W. Rohrer, University of South Alabama
Michelle Rondon, University of Wisconsin–Madison
Donna Russo, Drexel University
Pratibha Saxena, University of Texas, Austin
Herb E. Schellhorn, McMaster University
Kurt Schesser, University of Miami
Dennis Schneider, University of Texas, Austin
Margaret Ann Scuderi, Kenyon College
Ann C. Smith Stein, University of Maryland, College Park
John F. Stolz, Duquesne University
Marc E. Tischler, University of Arizona
Monica Tischler, Benedictine University
Beth Traxler, University of Washington
Luc Van Kaer, Vanderbilt University
Lorraine Grace Van Waasbergen, University of Texas, Arlington
Costantino Vetriani, Rutgers University
Amy Cheng Vollmer, Swarthmore College
Andre Walther, Cedar Crest College
Robert Weldon, University of Nebraska, Lincoln
Christine White-Ziegler, Smith College
Jianping Xu, McMaster University

Finally, we offer special thanks to our families for their support. Joan's husband, Michael Barich, offered unfailing support. John's wife, Zarrintaj ("Zari") Aliabadi, contributed to the text development, especially the sections on medical microbiology and public health.

To the Reader: Thanks!

We greatly appreciate your selection of this book as your introduction to the science of microbiology. As our textbook continues to evolve, it benefits greatly from the input of its many readers, students as well as professors. We truly welcome your comments, especially if you find text or figures that are in error or unclear. Feel free to contact us at the addresses listed below.

Joan L. Slonczewski
slonczewski@kenyon.edu
John W. Foster
jwfoster@southalabama.edu
Erik R. Zinser
ezinser@utk.edu

ABOUT THE AUTHORS

JOAN L. SLONCZEWSKI received her BA from Bryn Mawr College, and her PhD in molecular biophysics and biochemistry from Yale University, where she studied bacterial motility with Robert M. Macnab. Since completing postdoctoral work at the University of Pennsylvania, she has taught undergraduate microbiology in the Department of Biology at Kenyon College, where she earned a Silver Medal in the National Professor of the Year program of the Council for the Advancement and Support of Education. She has published numerous research articles with undergraduate coauthors on bacterial pH regulation, as well as five science fiction novels, including *The Highest Frontier* and *A Door into Ocean*, both of which earned the John W. Campbell Memorial Award. She conducted fieldwork on microbial ecosystems in Antarctica, sponsored by the National Science Foundation. She has served as At-Large Member representing Divisions on the Council Policy Committee of the American Society for Microbiology, and as a member of the editorial board of the journal *Applied and Environmental Microbiology.*

JOHN W. FOSTER received his BS from the Philadelphia College of Pharmacy and Science (now the University of the Sciences in Philadelphia), and his PhD from Hahnemann University (now Drexel University School of Medicine), also in Philadelphia, where he worked with Albert G. Moat. After postdoctoral work at Georgetown University, he joined the Marshall University School of Medicine in West Virginia; he is currently teaching in the Department of Microbiology and Immunology at the University of South Alabama College of Medicine in Mobile, Alabama. Dr. Foster has coauthored three editions of the textbook *Microbial Physiology* and has published over 100 journal articles describing the physiology and genetics of microbial stress responses. He has served as chair of the Microbial Physiology and Metabolism division of the American Society for Microbiology and as a member of the editorial advisory board of the journal *Molecular Microbiology.*

ERIK R. ZINSER received his AB from Kenyon College, where he worked in the lab of coauthor Joan Slonczewski. He received his PhD in microbiology from Harvard Medical School for his research with Roberto Kolter on the evolution in *Escherichia coli* during prolonged starvation. He performed his postdoctoral research at MIT with Penny Chisholm on the ecology of the marine cyanobacterium *Prochlorococcus.* He currently teaches in the Department of Microbiology at the University of Tennessee, Knoxville. He and his students study phytoplankton and associated bacteria from marine and freshwater environments. Laboratory and field studies by his group led to development of the Black Queen Hypothesis, an evolutionary theory that explains the adaptive nature of genome streamlining in free-living microbes. He has been a member of the American Society for Microbiology since college, and has served on the editorial boards of the journals *Applied and Environmental Microbiology* and *Environmental Microbiology.*

CHAPTER 1

Microbial Life: Origin and Discovery

1.1 From Germ to Genome: What Is a Microbe?

1.2 Microbes Shape Human History

1.3 Medical Microbiology

1.4 Environment and Ecology

1.5 The Microbial Family Tree

1.6 Cell Biology and the DNA Revolution

Microbes grow everywhere, from the deepest ocean to the human body. Our own bodies contain more microbes than human cells, including 100 trillion bacteria in the digestive tract. Throughout history, humans have had a hidden partnership with microbes ranging from food production to mineral mining. Microscopes revealed the tiny organisms at work in our bodies and in our environment. In the twentieth century, microbial genetics led to recombinant DNA and sequenced genomes. Today, microbes yield lead discoveries in medicine and global ecology.

CURRENT RESEARCH highlight

Microbes are nature's chemical factories. Marine sponges harbor microbes that make exotic molecules—such as polybrominated toxins. A microbe that could make such toxins might also break them down. At the Scripps Institution of Oceanography, Michelle Schorn mines the genomes of microbial partners of sponges such as *Lamellodysidea*. Remarkably, the sponge-associated bacteria produce polybrominated toxins similar to human-made fire retardants that now pollute our soil and water. Schorn analyzed the genomes of the cyanobacterium *Hormoscilla spongeliae* (inset), a symbiont that conducts photosynthesis for the sponge. The cyanobacterial genomes show genes to produce toxins as well as potentially new antibiotics.

Source: Michelle Schorn, et al. 2019. *mBio* **10**:1.

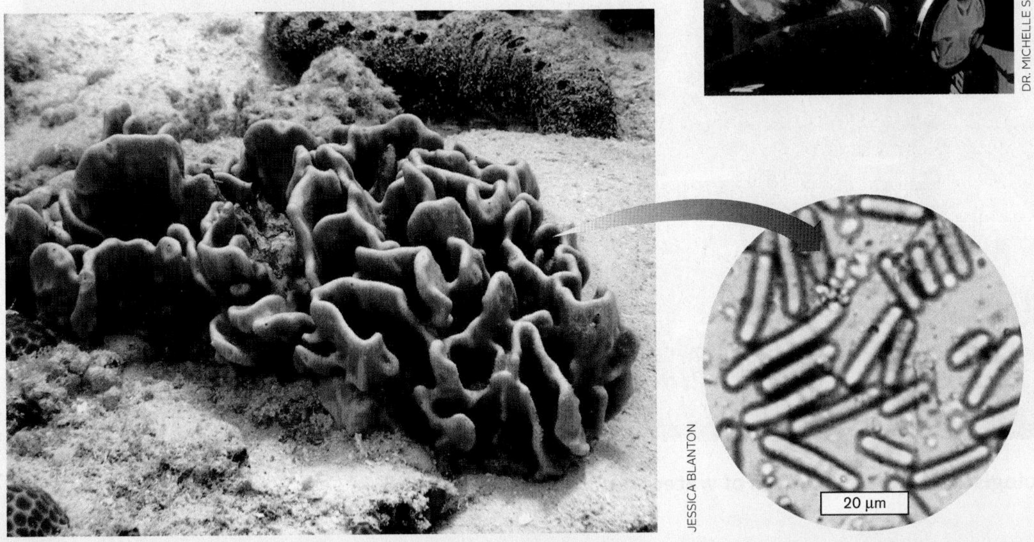

PROFESSOR JASON BIGGS

JESSICA BLANTON

AN INTERVIEW WITH

MICHELLE SCHORN, MARINE MICROBIOLOGIST, SCRIPPS INSTITUTION OF OCEANOGRAPHY, UC SAN DIEGO

DR. MICHELLE SCHORN

What is most surprising about finding cyanobacteria that make polybrominated toxins in sponges?

Polybrominated toxins may defend sponges against predation. But these chemically rich sponges are populated with diverse bacteria and archaea. The toxins may shape the holobiont (host sponge with its microbes) by excluding organisms that find polybrominated products toxic but allowing those that are unaffected to prosper. Confirming that cyanobacterial symbionts produce toxins in these sponges leads to new questions about how these compounds may be involved in the sponge biology.

Life began early in the history of planet Earth, with microscopic organisms, or "microbes." Over the eons, those microbes evolved to shape our atmosphere, our geology, and the energy cycles of all ecosystems. For the first 2 billion years, all life was microbial. Then, as now, a vast realm of that microbial diversity resided in Earth's oceans. Oceans constitute the largest biome on our Earth's surface, a community where the biomass is dominated by microbes. Marine microbes contribute 50% of Earth's oxygen gas (O_2), they are key players in the global carbon cycle, and their vast diversity suggests great potential for biotechnology. Microbiologists sail the seas to find organisms that make novel anticancer agents, or whose chemistry impacts climate change. In **Figure 1.1**, marine microbiologists retrieve collection bottles designed to trap water at designated ocean depths, revealing the secrets of marine microbes.

Much of Earth's microbial inventory remains a mystery. Barely 0.1% of the microbes in our biosphere can be cultured in the laboratory; even the digestive tract of a newborn infant contains species of bacteria unknown to science. All around us, microbes exhibit diverse forms and lifestyles. Rod-shaped bacteria such as *Escherichia coli* are among the 100 trillion inhabitants of our intestines, where they help digest our food. Alternatively, *E. coli* may colonize the plants we eat (**Fig. 1.2A**). Microbial eukaryotes (cells with nuclei) such as the voracious *Stentor* engulf aquatic prey (**Fig. 1.2B**). Archaea are a life form distinct from both bacteria and eukaryotes. Some archaea grow in extreme environments, such as concentrated salt (**Fig. 1.2C**). And all kinds of life host viruses. For example, herpes simplex virus infects human cells (**Fig. 1.2D**).

Yet, before we devised microscopes in the seventeenth century, we humans were unaware of the unseen living organisms that surround us, that float in our air and water, and that inhabit our own bodies. Microbes generate the very air we breathe, including nitrogen gas and much of the oxygen and carbon dioxide. They fix nitrogen for plants, and they make vitamins, such as vitamin B_{12}. In the ocean, microbes produce biomass for the food web that feeds the fish we eat, and microbes consume toxic wastes such as the oil from the *Deepwater Horizon* spill in the Gulf of Mexico in 2010. At the same time, virulent pathogens take our lives—and researchers risk their lives to study them. Working with pathogens such as Ebola virus requires sealed suits and respiratory equipment (**Fig. 1.3**). Despite all our advances in medicine and public health, humans continue

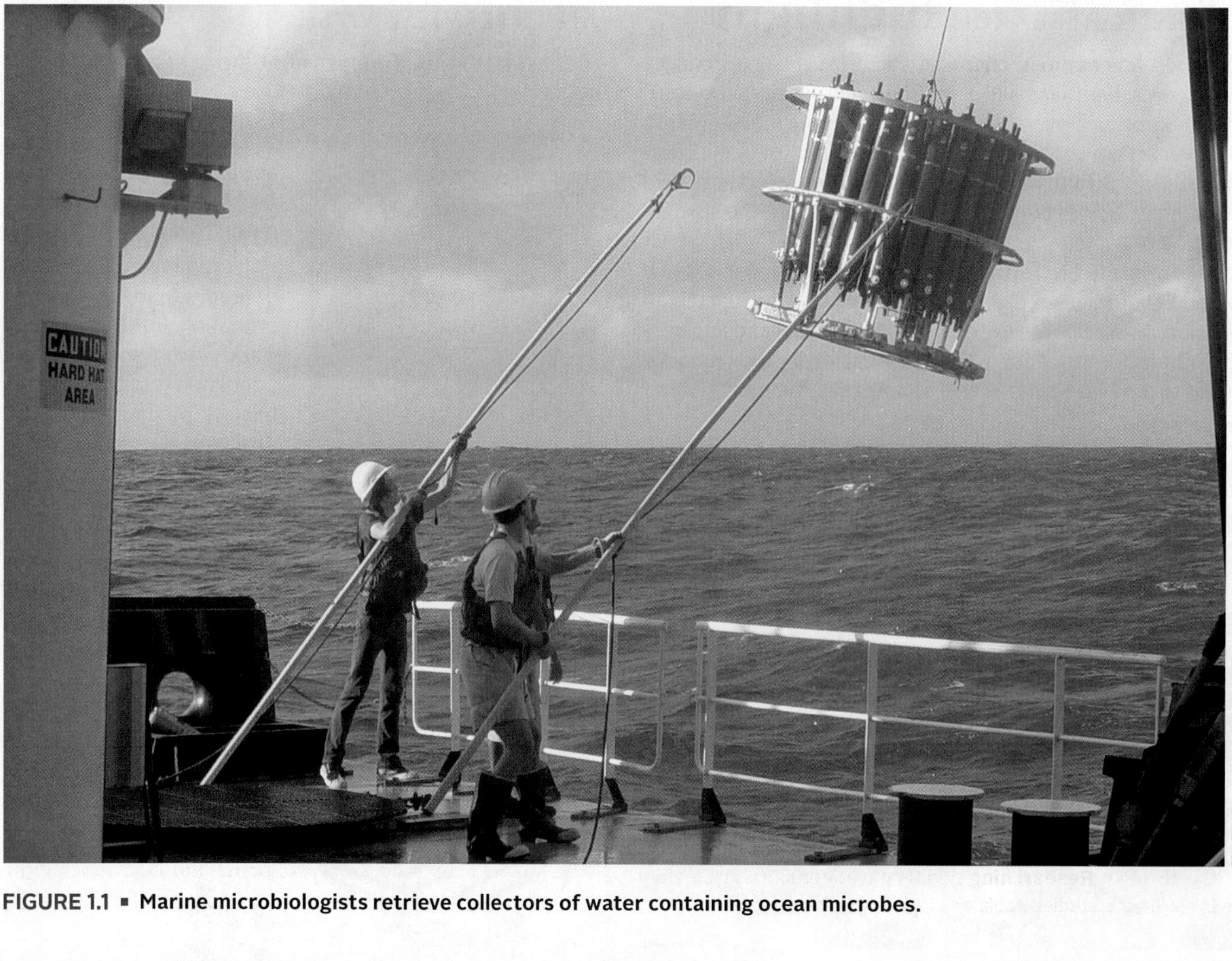

FIGURE 1.1 ▪ Marine microbiologists retrieve collectors of water containing ocean microbes.

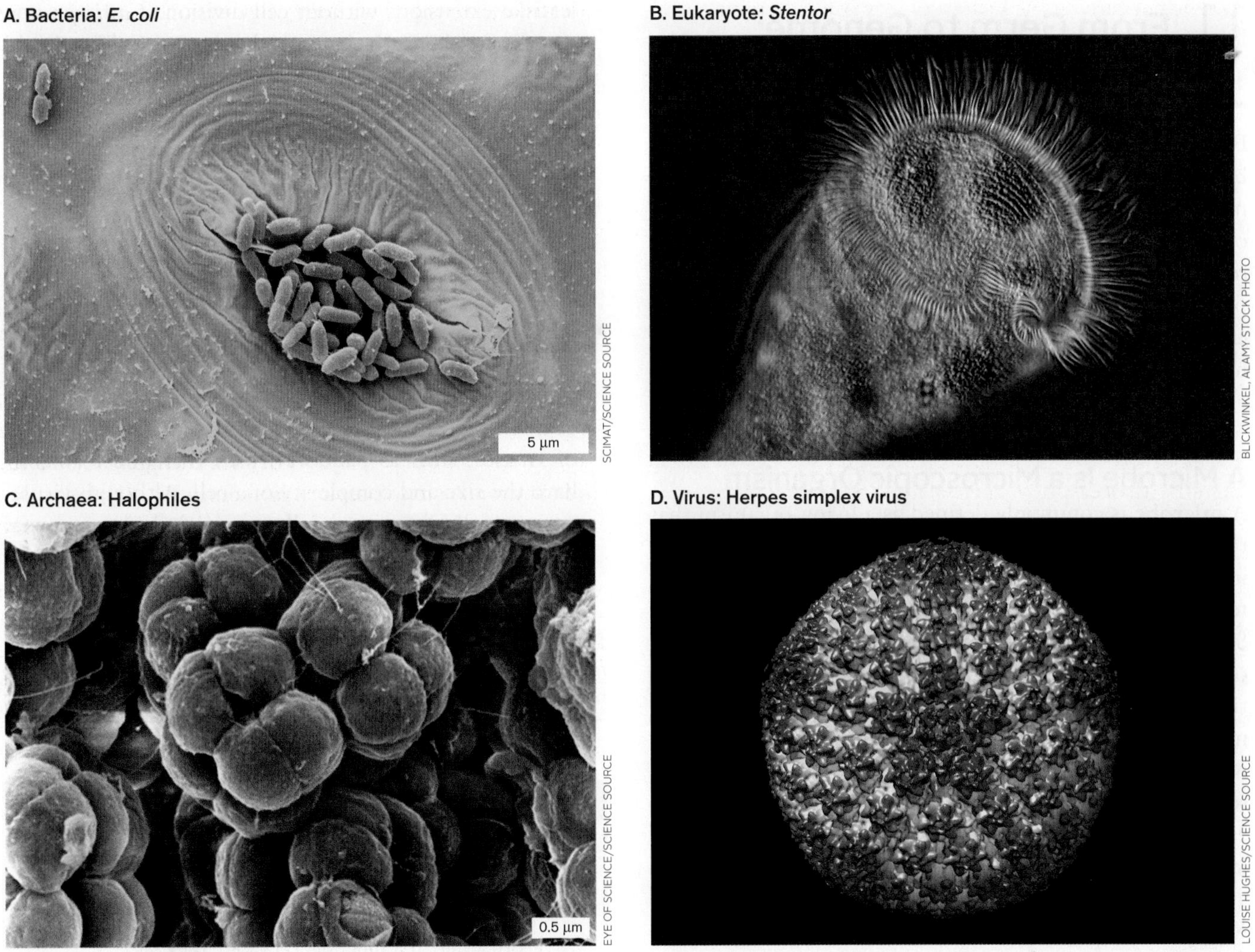

FIGURE 1.2 ■ Representative kinds of microbes. A. Bacterium: *E. coli* on leaf stomate. B. Eukaryote: *Stentor*, a stalked protist. C. Archaeon: *Halococcus*. D. Virus: Herpes simplex virus.

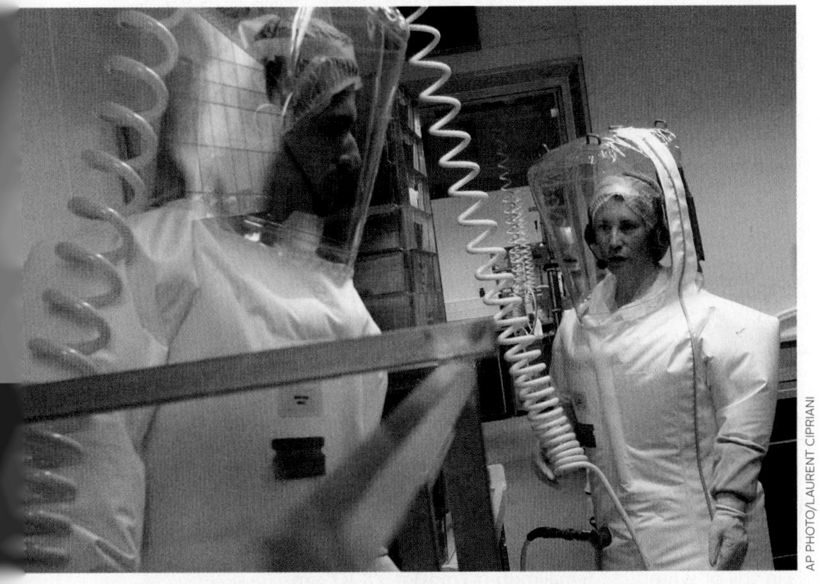

FIGURE 1.3 ■ Researching deadly pathogens. This laboratory in Lyon, France, studies Ebola virus and influenza virus.

to die of microbial diseases. Each year, millions of children succumb to waterborne pathogens and respiratory infections.

Today, microbes provide new tools that impact human society. For example, the use of heat-stable bacterial DNA polymerase (a DNA-replicating enzyme) in a technique called the **polymerase chain reaction (PCR)** enables us to detect minute amounts of DNA in traces of blood or fossilized bone. Microbial technologies led us from the discovery of the double helix to the sequence of the human genome, the total genetic information that defines our species.

In Chapter 1 we introduce the concept of a microbe, and we survey the history of human discovery. We explain how to show which pathogen causes a disease. Finally, we address the exciting century of molecular microbiology, in which microbial genetics, genomics, and evolution have transformed the practice of medicine and our understanding of the natural world.

1.1 From Germ to Genome: What Is a Microbe?

From early childhood, we hear that we are surrounded by microscopic organisms, or "germs," which we cannot see. What are these microbes? Our modern concept of a microbe has deepened through the use of two major research tools: advanced microscopy and the sequencing of genomic DNA. Microscopy reveals the cell, which we examine in Chapter 2, and we explore microbial genomes in Chapter 7.

A Microbe Is a Microscopic Organism

A **microbe** is commonly defined as a living organism that requires a microscope to be seen. Microbial cells range in size from millimeters (mm) down to 0.2 micrometer (μm), and viruses may be tenfold smaller (**Table 1.1**). Some microbes consist of a single cell, the smallest unit of life, a membrane-enclosed compartment of water solution containing molecules that carry out metabolism. Each microbe contains a genome used to reproduce its own kind. Microbial cells acquire food, gain energy to build themselves, and respond to environmental change. Microbes evolve at rapid rates—often fast enough to observe in the laboratory (discussed in Chapter 17).

Our simple definition of a microbe, however, leaves us with contradictions:

- **Super-size microbial cells.** Most single-celled organisms require a microscope to render them visible, and thus they fit the definition of a microbe. Nevertheless, some species of protists, such as giant amebas, grow to sizes large enough to see with the unaided eye. The marine sulfur bacterium *Thiomargarita namibiensis*, called the "sulfur pearl of Namibia," grows to 0.7 mm, larger than the eye of a fruit fly (**Fig. 1.4**). Even more surprising, a single cell of the "killer alga" *Caulerpa taxifolia* grows up to several meters long. The alga expands and forms leaflike extensions without cell division. *Caulerpa* now covers acres beneath the coastal waters of California.
- **Microbial communities.** Many microbes form complex multicellular assemblages, such as mushrooms, kelps, and biofilms. In these structures, cells are differentiated into distinct types that complement each other's functions, as in multicellular organisms. Yet, some multicellular worms and arthropods require a microscope for us to see but are not considered microbes.
- **Viruses.** A **virus** is a noncellular particle containing genetic material that takes over the metabolism of a cell to generate more virus particles. Some viruses consist of only a short chromosome packed in protein. Other kinds of viruses, such as pandoraviruses that infect amebas, have the size and complexity of a cell. Although viruses are not fully functional cells, some viral genomes may have evolved from cells.

Note: Each section of text contains Thought Questions that may have various answers. Possible responses are provided at the back of the book.

Thought Questions

1.1 The minimum size of known microbial cells is about 0.2 μm. Could even smaller cells be discovered? What factors may determine the minimum size of a cell?

1.2 If viruses are not functional cells, are they "alive"?

In practice, our definition of a microbe derives from tradition as well as genetic considerations. In this book we consider microbes to include **prokaryotes** (cells lacking a nucleus, including bacteria and archaea), as well as certain classes of **eukaryotes** (cells with a nucleus), such as algae, fungi, and protozoa (discussed in Chapter 20). The bacteria, archaea, and eukaryotes—known as the three "domains"—evolved from a common ancestral cell (see Section 1.5 and

TABLE 1.1 Sizes of Some Microbes

Microbe	Description	Approximate size
Varicella-zoster virus 1	Virus that causes chickenpox and shingles	100 nanometers (nm) = 10^{-7} meter (m)
Prochlorococcus	Photosynthetic marine bacterium	500 nm = 5×10^{-7} m
Escherichia coli	Bacterium growing within human intestine	2 micrometers (μm) = 2×10^{-6} m
Spirogyra	Aquatic alga that forms long filaments of cells	100 μm = 10^{-5} m (cell length)
Pelomyxa	Ameba (a protist) that consumes bacteria in soil or water	5 millimeters (mm) = 5×10^{-3} m

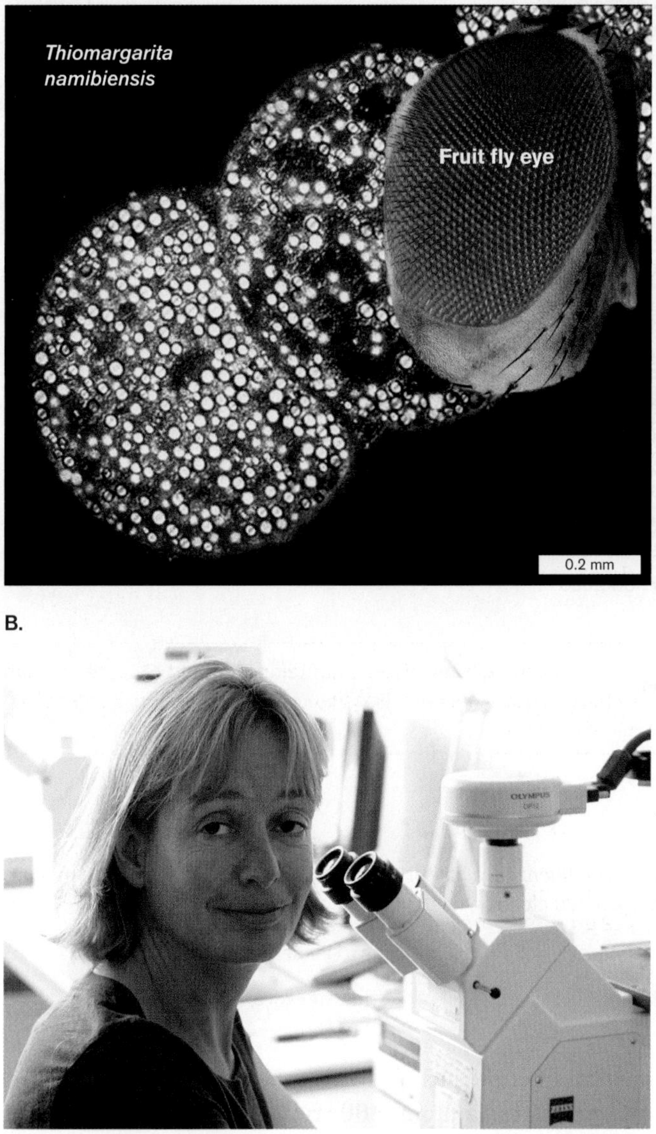

FIGURE 1.4 ▪ A giant microbial cell: *Thiomargarita namibiensis*. **A.** *T. namibiensis*, a marine sulfur bacterium (light microscopy). **B.** Heide Schulz-Vogt of the Leibniz Institute for Baltic Sea Research, Warnemünde.

Chapter 21). Viruses and even smaller infectious particles are discussed in Chapters 6 and 11.

Note: The formal names of the three domains are **Bacteria**, **Archaea**, and **Eukarya**. Members of these domains are called **bacteria** (singular, **bacterium**), **archaea** (singular, **archaeon**), and **eukaryotes** (singular, **eukaryote**), respectively. The microbiology literature includes alternative spellings for some of these terms, such as "archaean" and "eucaryote."

Microbial Genomes Are Sequenced

How have we learned how microbes work? A key tool is the study of microbial genomes. A **genome** is the total genetic information contained in an organism's chromosomal DNA. The genes in a microbe's genome and the sequence of DNA tell us a lot about how that microbe grows and associates with other species. For example, if a microbe's genome includes genes for nitrogenase, a nitrogen-fixing enzyme, that microbe probably can fix nitrogen from the atmosphere into proteins—its own proteins and those of associated plants. And by comparing DNA sequences of different microbes, we can figure out how closely related they are and how they evolved.

The first method of DNA sequencing that was fast enough to sequence large genomes was developed by Fred Sanger (1918–2013) at the University of Cambridge (**Fig. 1.5**). This achievement—which jump-started the study of molecular biology—earned Sanger the 1980 Nobel Prize in Chemistry, together with Walter Gilbert and Paul Berg. Sanger and his colleagues used the new method to sequence DNA containing tens of thousands of base pairs, such as the DNA of the human mitochondrion (plural, mitochondria).

But most genomes of cells contain millions, or even billions, of base pairs. In 1995, scientists completed the first genome sequence of a cellular microbe, the bacterium *Haemophilus influenzae* (**Fig. 1.6**). *H. influenzae* causes several diseases including meningitis in children, a disease now prevented by the Hib vaccine. The *H. influenzae* genome has nearly 2 million base pairs, which specify about 1,700 genes. The sequence was determined by a large team of scientists led by Craig Venter, Hamilton Smith, and Claire Fraser (**Fig. 1.7**) at The Institute for Genomic Research (TIGR). The TIGR team devised a special computational strategy for

FIGURE 1.5 ▪ Fred Sanger devised the method of DNA sequence analysis used to sequence the first genomes.

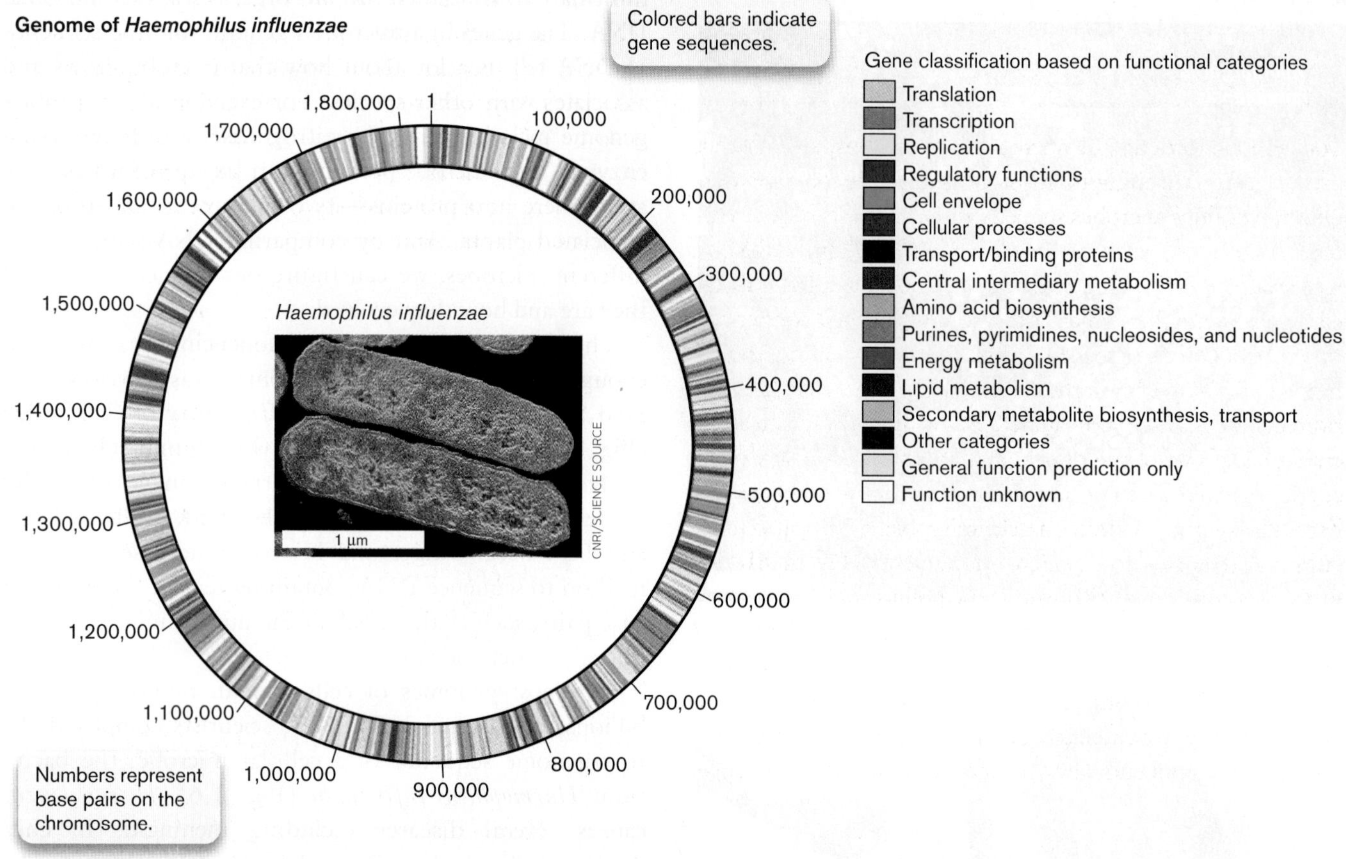

FIGURE 1.6 ■ The first sequenced genome. The genome of *Haemophilus influenzae,* a bacterium that causes ear infections and meningitis, was the first DNA sequence completed for a cellular organism. **Inset:** Colorized electron micrograph.

FIGURE 1.7 ■ Claire Fraser, past president of The Institute for Genomic Research (TIGR), sequenced numerous microbial genomes.

assembling large amounts of sequence data—a strategy later used to sequence the human genome.

Today we sequence new bacterial genomes daily. In addition to sequencing individual genomes, computational strategies are used to sequence thousands of genomes of microbes sampled from a natural environment, such as the acid drainage from an iron mine. The collection of sequences taken directly from the environment is called a **metagenome**. The first metagenome of an acid mine was sequenced by Jillian Banfield and co-workers at UC Berkeley in 2004. Now, metagenomes are sequenced for microbial communities of medical interest, such as that of the human colon. Human gut microbes contain 100 times more genes in their metagenomes than the human genome contains—and many of these microbial genes contribute to our health!

Comparing genomes has revealed a set of core genes shared by all organisms. These core genes add further evidence that all living beings on Earth, including humans, share a common ancestry. Genomes are discussed further in Chapter 7, and the evolution of genomes and metagenomes is discussed in Chapters 17 and 21.

To Summarize

- **A microbe is a living organism that requires a microscope to be seen.** Some organisms exist in both microbial and macroscopic forms.
- **Microbes grow in communities.** A community may include both microbial and macroscopic species.
- **Microbial capabilities are defined by their genome sequences.**

1.2 Microbes Shape Human History

Today our knowledge of microbes is enormous, and it keeps growing. Yet throughout most of human history we were unaware of how microbes shaped our culture. Yeasts and bacteria made foods such as bread and cheese (**Fig. 1.8A**), as well as alcoholic beverages (discussed in Chapter 16). "Rock-eating" bacteria, known as "lithotrophs," leached copper and other metals from ores exposed by mining, enabling ancient human miners to obtain these metals. The lithotrophic oxidation of minerals for energy generates strong acids, which accelerate breakdown of the ore. Today, about 20% of the world's copper, as well as some uranium and zinc, is produced by bacterial leaching. Unfortunately, microbial acidification also consumes the stone of ancient monuments (**Fig. 1.8B**)—a process intensified by airborne acidic pollution.

How did people find out about microbes? **Table 1.2** (pages 8–9) lists the discoveries throughout history that have brought us to our current level of knowledge. Microscopists in the seventeenth and eighteenth centuries formulated key concepts about microbes and their existence, including their means of reproduction and death. In the nineteenth century, the "golden age" of microbiology, scientists established the fundamental principles of disease pathology and microbial ecology that are still in use today. This period laid the foundation for modern biology, in which genetics and molecular biology provide powerful tools for scientists to manipulate microorganisms for medicine, research, and industry.

A.

BETTMANN/GETTY IMAGES

B.

JOHNER IMAGES/GETTY IMAGES

FIGURE 1.8 ▪ Production and destruction by microbes. **A.** Roquefort cheeses ripening in France. **B.** Statue at the cathedral of Cologne, Germany, decaying from the action of lithotrophic microbes. The process is accelerated by acid rain.

Microbial Disease Devastates Human Populations

Microbial diseases such as the bubonic plague and AIDS have had a profound affect on human history (**Fig. 1.9**).

FIGURE 1.9 ▪ Microbial disease in history and culture. **A.** Medieval church procession to ward off the Black Death (bubonic plague). **B.** The AIDS Memorial Quilt spread before the Washington Monument for display in 1992. Each panel of the quilt memorializes an individual who died of AIDS.

A.

CHRONICLE OF AEGIDIUS LI MUISIS/PRIVATE COLLECTION/BRIDGEMAN IMAGES

B.

VANESSA VICK/GETTY IMAGES

TABLE 1.2 Microbes and Human History

Date	Microbial discovery	Discoverer(s)
Microbes impact human culture without detection		
10,000 BCE	Food and drink are produced by microbial fermentation.	Egyptians, Chinese, and others
1500 BCE	Tuberculosis, polio, leprosy, and smallpox are evident in mummies and tomb art.	Egyptians
50 BCE	Copper is recovered from mine water acidified by sulfur-oxidizing bacteria.	Roman metal workers under Julius Caesar
1546 CE	Syphilis and other diseases are observed to be contagious.	Girolamo Fracastoro (Padua)
Early microscopy and the origin of microbes		
1676	Microbes are observed under a microscope.	Antonie van Leeuwenhoek (Netherlands)
1688	Spontaneous generation is disproved for maggots.	Francesco Redi (Italy)
1717	Smallpox is prevented by inoculation of pox material, a rudimentary form of immunization.	Turkish women taught Lady Mary Montagu, who brought the practice to England
1765	Microbe growth in organic material is prevented by boiling in a sealed flask.	Lazzaro Spallanzani (Padua)
1798	Cowpox vaccination prevents smallpox.	Edward Jenner (England)
1835	Fungus causes disease in silkworms (first pathogen to be demonstrated in animals).	Agostino Bassi de Lodi (Italy)
1847	Chlorine as antiseptic wash for doctors' hands decreases pathogens.	Ignaz Semmelweis (Hungary)
1881	Bacterial spores survive boiling but are killed by cyclic boiling and cooling.	John Tyndall (Ireland)
"Golden age" of microbiology: principles and methods established		
1855	Sanitation shows statistical correlation with mortality (Crimean War).	Florence Nightingale (England)
1857	Microbial fermentation produces lactic acid or alcohol.	Louis Pasteur (France)
1864	Microbes fail to appear spontaneously, even in the presence of oxygen.	Louis Pasteur (France)
1866	Microbes are defined as a class distinct from animals and plants.	Ernst Haeckel (Germany)
1867	Antisepsis during surgery prevents patient death.	Joseph Lister (England)
1877	Bacteria are a causative agent of anthrax.	Robert Koch (Germany)
1881	First artificial vaccine is developed (against anthrax).	Louis Pasteur (France)
1882	First pure culture of colonies, *Mycobacterium tuberculosis*, is grown on solid medium.	Robert Koch (Germany)
1884	Koch's postulates are published, based on anthrax and tuberculosis.	Robert Koch (Germany)
1884	Gram stain is devised to distinguish bacteria from human cells.	Hans Christian Gram (Denmark)
1886	Intestinal bacteria include *Escherichia coli*, the future model organism.	Theodor Escherich (Austria)
1889	Bacteria oxidize iron and sulfur and fix CO_2 (lithotrophy).	Sergei Winogradsky (Russia)
1889	Bacteria isolated from root nodules are proposed to fix nitrogen.	Martinus Beijerinck (Netherlands)
1892, 1899	The concept of a virus is proposed to explain tobacco mosaic disease.	Dmitri Ivanovsky (Russia) and Martinus Beijerinck (Netherlands)
Cell biology, biochemistry, and genetics		
1908	Antibiotic chemicals are synthesized and identified (chemotherapy).	Paul Ehrlich (USA)
1911	Viruses are found to be a cause of cancer in chickens.	Peyton Rous (USA)
1917	Bacteriophages are recognized as viruses that infect bacteria.	Frederick Twort (England) and Félix d'Herelle (France)
1924	The ultracentrifuge is invented and used to measure the size of proteins.	Theodor Svedberg (Sweden)
1928	*Streptococcus pneumoniae* bacteria are transformed by material from dead cells.	Frederick Griffith (England)
1929	Penicillin, the first widely successful antibiotic, is made by a fungus. The molecule is isolated in 1941.	Alexander Fleming (Scotland), Howard Florey (Australia), and Ernst Chain (England)
1933–1945	The transmission electron microscope is invented and used to observe cells.	Ernst Ruska and Max Knoll (Germany)
1937	The tricarboxylic acid cycle is discovered.	Hans Krebs (Germany)
1938	The microbial "kingdom" is subdivided into prokaryotes (Monera) and eukaryotes.	Herbert Copeland (USA)
1938	*Bacillus thuringiensis* spray is produced as the first bacterial insecticide.	Insecticide manufacturers (France)
1941	One gene encodes one enzyme in *Neurospora*.	George Beadle and Edward Tatum (USA)
1941	Poliovirus is produced in human tissue culture.	John Enders, Thomas Weller, and Frederick Robbins (USA)
1944	DNA is the genetic material that transforms *S. pneumoniae*.	Oswald Avery, Colin MacLeod, and Maclyn McCarty (USA)
1945	The bacteriophage replication mechanism is elucidated.	Salvador Luria (Italy) and Max Delbrück (Germany), working in the USA
1946	Bacteria transfer DNA by conjugation.	Edward Tatum and Joshua Lederberg (USA)
1946–1956	X-ray diffraction crystal structures are obtained for the first complex biological molecules: penicillin and vitamin B_{12}.	Dorothy Hodgkin, John Bernal, and co-workers (England)
1950	Anaerobic culture technique is devised to study anaerobes of the bovine rumen.	Robert Hungate (USA)
1950	The *E. coli* K-12 genome carries a latent bacteriophage lambda.	Esther Lederberg (USA) and André Lwoff (France)
1951	Transposable elements in DNA are discovered in maize and later shown in bacteria.	Barbara McClintock (USA)
1952	DNA is injected into a cell by a bacteriophage.	Martha Chase and Alfred Hershey (USA)

TABLE 1.2 Microbes and Human History (*continued*)

Date	Microbial discovery	Discoverer(s)
Molecular biology and recombinant DNA		
1953	Overall structure of DNA is identified by X-ray diffraction analysis as a double helix.	Rosalind Franklin and Maurice Wilkins (England)
1953	Double-helical DNA consists of antiparallel chains connected by the hydrogen bonding of AT and GC base pairs.	James Watson (USA) and Francis Crick (England)
1959	Expression of the messenger RNA for the *E. coli lac* operon is regulated by a repressor protein.	Arthur Pardee (England); François Jacob and Jacques Monod (France)
1960	Radioimmunoassay for detection of biomolecules is developed.	Rosalyn Yalow and Solomon Bernson (USA)
1961	The chemiosmotic theory, which states that biochemical energy is stored in a transmembrane proton gradient, is proposed and tested.	Peter Mitchell and Jennifer Moyle (England)
1966	The genetic code by which DNA information specifies protein sequences is deciphered.	Marshall Nirenberg, H. Gobind Khorana, and others (USA)
1967	Bacteria can grow at temperatures above 80°C in hot springs at Yellowstone National Park.	Thomas Brock (USA)
1968	Serial endosymbiosis is proposed to explain the evolution of mitochondria and chloroplasts.	Lynn Margulis (USA)
1969	Retroviruses contain reverse transcriptase, which copies RNA to make DNA.	Howard Temin, David Baltimore, and Renato Dulbecco (USA)
1972	Inner and outer membranes of Gram-negative bacteria (*Salmonella*) are separated by ultracentrifugation.	Mary Osborn (USA)
1973	A recombinant DNA molecule is made in vitro (in a test tube).	Stanley Cohen, Annie Chang, Robert Helling, and Herbert Boyer (USA)
1974	A rotary motor drives the bacterial flagellum.	Howard Berg, Michael Silverman, and Melvin Simon (USA)
1975	mRNA-rRNA base pairing initiates protein synthesis in *E. coli*.	Joan Steitz and Karen Jakes (USA); Lynn Dalgarno and John Shine (Australia)
1975	The dangers of recombinant DNA are assessed at the Asilomar Conference.	Paul Berg, Maxine Singer, and others (USA)
1975	Monoclonal antibodies are produced indefinitely in tissue culture by hybridomas, antibody-producing cells fused to cancer cells.	George Köhler (Germany) and Cesar Milstein (UK)
1977, 1980	A DNA sequencing method is invented and used to sequence the first genome of a virus.	Fred Sanger, Walter Gilbert, and Allan Maxam (England and USA)
1977	Archaea are identified as a third domain of life, the others being eukaryotes and bacteria.	Carl Woese (USA)
1978	The first protein catalog, based on 2D gels, is compiled for *E. coli*.	Fred Neidhardt, Peter O'Farrell, and colleagues (USA)
1978	Biofilms are a major form of existence of microbes.	William Costerton and others (Canada)
1979	Smallpox is declared eliminated—a global triumph of immunology and public health.	World Health Organization
Genomics, structural biology, and molecular ecology		
1981	Invention of the polymerase chain reaction (PCR) makes available large quantities of DNA.	Kary Mullis (USA)
1981–1986	Self-splicing and self-replicating RNA is discovered in the protist *Tetrahymena*.	Thomas Cech, Sidney Altman, Jennifer Doudna, and Jack Szostak (USA)
1982	Archaea are discovered with optimal growth above 100°C.	Karl Stetter (Germany)
1982	Viable but noncultured bacteria contribute to ecology and pathology.	Rita Colwell and Norman Pace (USA)
1982	Prions, infectious agents consisting solely of protein, are characterized.	Stanley Prusiner (USA)
1983	Human immunodeficiency virus (HIV) is discovered as the cause of AIDS.	Françoise Barré-Sinoussi and Luc Montagnier (France); Robert Gallo (USA)
1983	Genes are introduced into plants by use of *Agrobacterium tumefaciens* plasmid vectors.	Eugene Nester, Mary-Dell Chilton, and colleagues (USA)
1984	Acid-resistant *Helicobacter pylori* grow in the stomach, where they cause gastritis.	Barry Marshall and J. Robin Warren (Australia)
1987	*Geobacter* bacteria that can generate electricity are discovered.	Derek Lovley and colleagues (USA)
1988	*Prochlorococcus* is identified as Earth's most abundant marine phototroph.	Sallie Chisholm and colleagues (USA)
1993	A giant bacterium (*Epulopiscium*)—large enough to see—is identified.	Esther Angert and Norman Pace (USA)
1995	First genome is sequenced for a cellular organism, *Haemophilus influenzae*.	Craig Venter, Hamilton Smith, Claire Fraser, and others (USA)
2004	Mimivirus genome shows that large DNA viruses evolved from cells.	Didier Raoult and colleagues (France)
2006	First metagenomes are sequenced, from Iron Mountain acid mine drainage and from the Sargasso Sea.	Jillian Banfield, Craig Venter, and others (USA)
2006	Vaccine prevents genital human papillomavirus (HPV), the most common sexually transmitted infection.	Patented by Georgetown University and other institutions (USA and Australia)
2012	CRISPR-Cas9 bacterial self-defense mechanism is used for programmable gene editing.	Jennifer Doudna (USA) and Emmanuelle Charpentier (France)
2013	A genetically modified form of HIV cures a person of cancer.	Michael Kalos, Stephan Grupp, Carl June, and colleagues (USA)
1988–2020	*Escherichia coli* long-term evolution experiment reaches 50,000 generations and continues.	Richard Lenski, Zachary Blount, and colleagues (USA)

The plague, which wiped out a third of Europe's population in the fourteenth century, was caused by *Yersinia pestis*, a bacterium spread by fleas of rats and humans. Ironically, the plague-induced population decline enabled the social transformation that led to the Renaissance, a period of unprecedented cultural advancement. In the nineteenth century, the bacterium *Mycobacterium tuberculosis* stalked overcrowded cities, and tuberculosis was so common that the pallid appearance of tubercular patients became a symbol of tragic youth in European arts, such as Puccini's opera *La Bohème*. Today, strains of tuberculosis that resist all known antibiotics stalk human communities throughout the world. As the leading infectious cause of death, *M. tuberculosis* infects one-third of the world's population.

Historians traditionally emphasize the role of warfare in shaping human destiny, and the brilliance of leaders, or the advantage of new technology, in determining which civilizations rise or fall. Yet the fate of human societies is often determined by microbes. For example, much of the native population of North America was exterminated by smallpox introduced by European invaders. Throughout history, more soldiers have died of microbial infections than of wounds in battle.

The significance of disease in warfare was first recognized by the British nurse and statistician Florence Nightingale (1820–1910; **Fig. 1.10A**). Better known as the founder of professional nursing, Nightingale also founded the science of medical statistics. She used methods invented by French statisticians to demonstrate the high mortality rate due to disease among British soldiers during the Crimean War. To show the deaths of soldiers due to various causes, she devised the "polar area chart" (**Fig. 1.10B**). In this chart, blue wedges represent deaths due to infectious disease, red wedges represent deaths due to wounds, and black wedges represent all other causes of death. Infectious disease accounts for more than half of all mortality.

Before Nightingale, no one understood the impact of disease on armies, or on other crowded populations, such as in cities. Nightingale's statistics convinced the British government to improve army living conditions and to upgrade the standards of army hospitals. In modern epidemiology, statistical analysis continues to be a crucial tool in determining the causes of disease.

Microscopes Reveal the Microbial World

The seventeenth century was a time of growing inquiry and excitement about the "natural magic" of science and patterns of our world, such as the laws of gravitation and motion formulated by Isaac Newton (1642–1727). Robert Boyle (1627–1691) performed the first controlled experiments on the chemical conversion of matter. Physicians attempted new treatments for disease involving the application of "stone and minerals" (that is, chemicals)—what today we would call "chemotherapy." Minds were open to consider the astounding possibility that our surroundings, indeed our very bodies, were inhabited by tiny living beings.

Robert Hooke observes the microscopic world. The first microscopist to publish a systematic study of the world

A.

B.

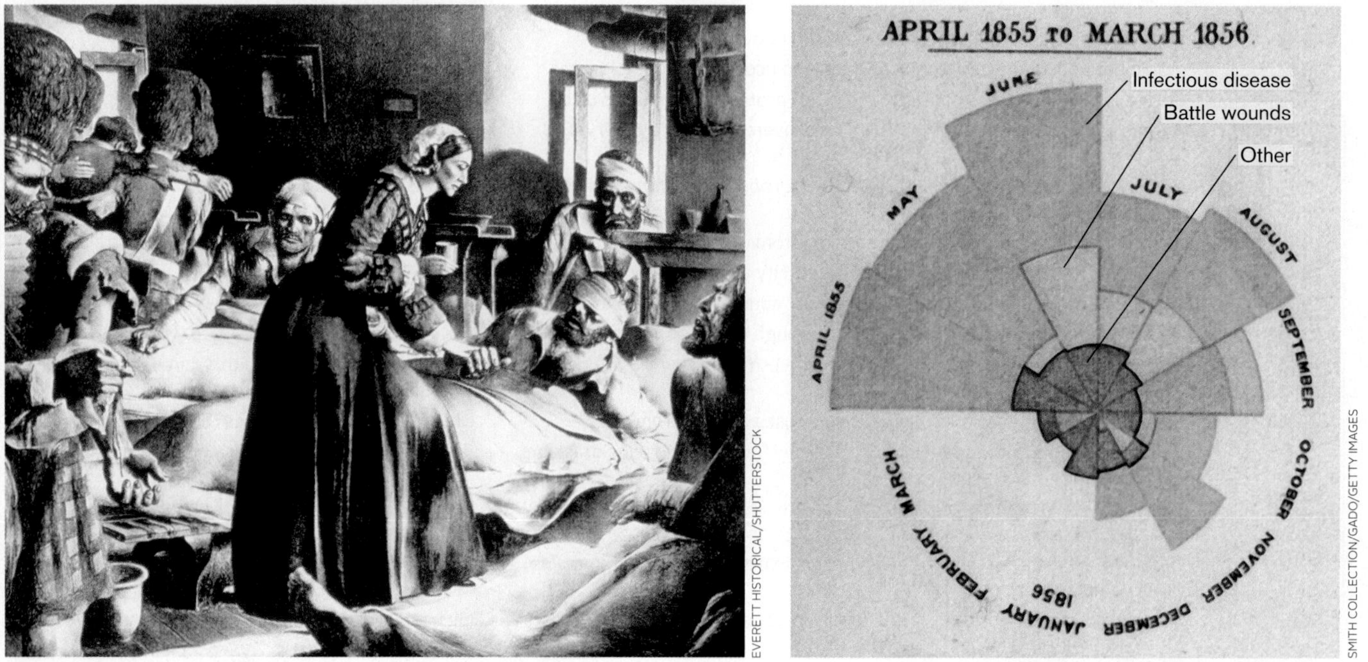

FIGURE 1.10 ▪ Florence Nightingale, founder of medical statistics. **A.** Florence Nightingale was the first to use medical statistics to demonstrate the significance of mortality due to disease. **B.** Nightingale's polar area chart of mortality data during the Crimean War.

as seen under a microscope was Robert Hooke (1635–1703). As curator of experiments for the Royal Society of London, Hooke built a compound microscope—a magnifying instrument containing two or more lenses that multiply their magnification in series. With his microscope, Hooke observed biological materials such as nematode "vinegar eels," mites, and mold filaments. Hooke published drawings of these microbes in *Micrographia* (1665), the first publication of objects observed under a microscope (**Fig. 1.11**).

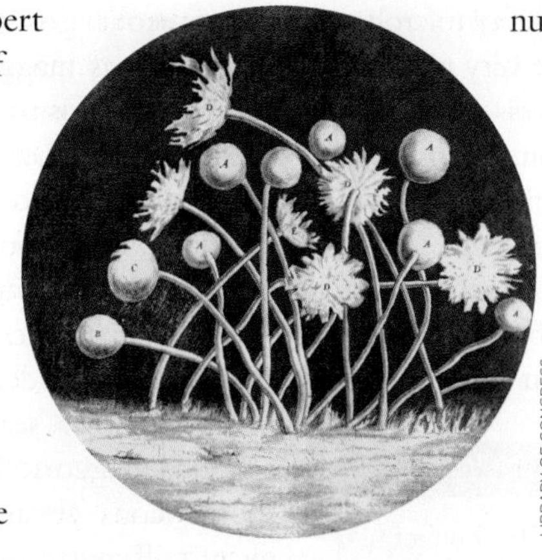

FIGURE 1.11 ■ Robert Hooke's *Micrographia*. Mold sporangia, drawn by Hooke in 1665 from his observations of objects using a compound microscope.

LIBRARY OF CONGRESS

Hooke was the first to observe distinct units of living material, which he called "cells." Hooke first named the units cells because the shape of hollow cell walls in a slice of cork reminded him of the shape of monks' cells in a monastery. But his crude lenses achieved at best 30-fold power (30×), and he never observed single-celled bacteria.

Antonie van Leeuwenhoek observes bacteria with a single lens. Hooke's *Micrographia* inspired other microscopists, including Antonie van Leeuwenhoek (1632–1723), who became the first individual to observe single-celled microbes (**Fig. 1.12A**). As a young man, Leeuwenhoek lived in the Dutch city of Delft, where he worked as a cloth draper, a profession that introduced him to magnifying glasses. The magnifying glasses were used to inspect the quality of the cloth, enabling the worker to count the number of threads. Later in life, Leeuwenhoek took up the hobby of grinding ever-stronger lenses to see into the world of the unseen.

Leeuwenhoek ground lenses stronger than Hooke's, which he used to build single-lens magnifiers, complete with sample holder and focus adjustment (**Fig. 1.12B**). First he observed insects, including lice and fleas; then the relatively large single cells of protists and algae; then, ultimately, bacteria. One day he applied his microscope to observe matter extracted from between his teeth. He wrote, "To my great surprise [I] perceived that the aforesaid matter contained very many small living Animals, which moved themselves very extravagantly."

Over the rest of his life, Leeuwenhoek recorded page after page on the movement of microbes, reporting their size and shape so accurately that in many cases we can determine the species he observed (**Fig. 1.12C**). He performed experiments, comparing, for example, the appearance of "small animals" from his teeth before and after drinking hot coffee. The disappearance of microbes from his teeth after drinking a hot beverage suggested that heat killed microbes—a profoundly important principle for the study and control of microbes ever since.

Leeuwenhoek is believed to have died, ironically, of a disease contracted from sheep whose bacteria he had observed. Historians have often wondered why it took so many centuries for Leeuwenhoek and his successors to determine the link between microbes and disease. Although observers such as

A.

BETTMANN/GETTY IMAGES

B.

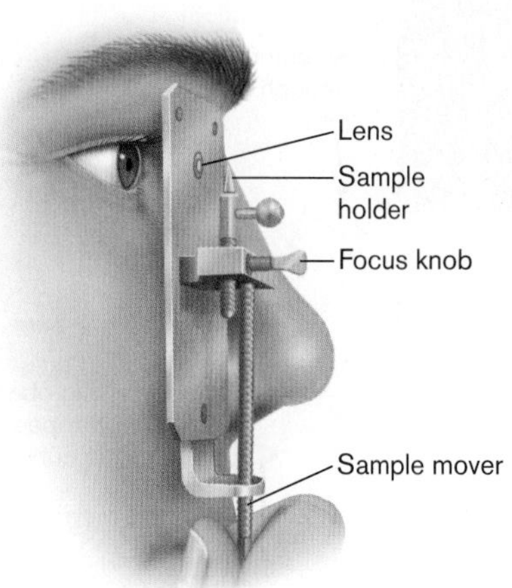

C.

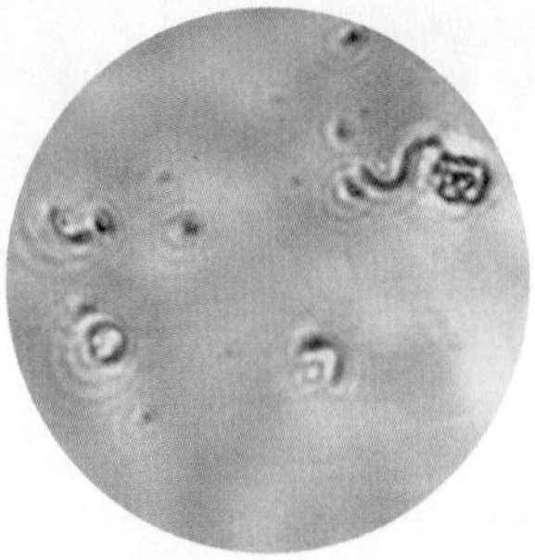

BRIAN J. FORD

FIGURE 1.12 ■ Antonie van Leeuwenhoek. **A.** A portrait of Leeuwenhoek, the first person to observe individual microbes. **B.** "Microscope" (magnifying glass) used by Leeuwenhoek. **C.** Spiral bacteria viewed through a replica of Leeuwenhoek's instrument.

Agostino Bassi de Lodi (1773–1856) noted cases of microbes associated with pathology (see **Table 1.2**), the very ubiquity of microbes—most of them actually harmless—may have obscured their more deadly roles. In addition, it was hard to distinguish between microbes and the single-celled components of the human body, such as blood cells and sperm. It was not until the nineteenth century that human tissues could be distinguished from microbial cells by the application of differential chemical stains (discussed in Chapter 2).

Thought Question

1.3 Why do you think it took so long for humans to connect microbes with infectious disease? What innovations helped make the connection?

Spontaneous Generation: Do Microbes Have Parents?

The observation of microscopic organisms led priests and philosophers to wonder where these tiny beings came from. In the eighteenth century, scientists and church leaders intensely debated the question of **spontaneous generation**. Spontaneous generation is the concept that living creatures such as maggots could arise spontaneously, without parental organisms. Chemists of the day tended to support spontaneous generation, as it appeared similar to the way chemicals changed during reaction. Christian church leaders, however, supported the biblical view that all organisms have "parents" going back to the first week of creation.

The Italian priest Francesco Redi (1626–1697) showed that maggots in decaying meat were the offspring of flies. Meat kept in a sealed container, excluding flies, did not produce maggots. Thus, Redi's experiment argued against spontaneous generation for macroscopic organisms. The meat still putrefied, however, producing microbes that seemed to arise "without parents."

To disprove spontaneous generation of microbes, another Italian priest, Lazzaro Spallanzani (1729–1799), showed that a sealed flask of meat broth sterilized by boiling failed to grow microbes. Spallanzani also noticed that microbes often appeared in pairs. Were these two parental microbes coupling to produce offspring, or did one microbe become two? Through long and tenacious observation, Spallanzani watched a single microbe grow in size until it split in two. Thus he demonstrated cell fission, the process by which cells arise by the splitting of preexisting cells.

Even Spallanzani's experiments, however, did not put the matter to rest. Proponents of spontaneous generation argued that the microbes in the priest's flask lacked access to oxygen and therefore could not grow. The pursuit of this question was left to future microbiologists, including the famous French microbiologist Louis Pasteur (1822–1895; **Fig. 1.13A**). In addressing spontaneous generation and related questions, Pasteur and his contemporaries laid the foundations for modern microbiology.

Louis Pasteur reveals the biochemical basis of microbial growth. Pasteur began his scientific career as a chemist and

A.

B.

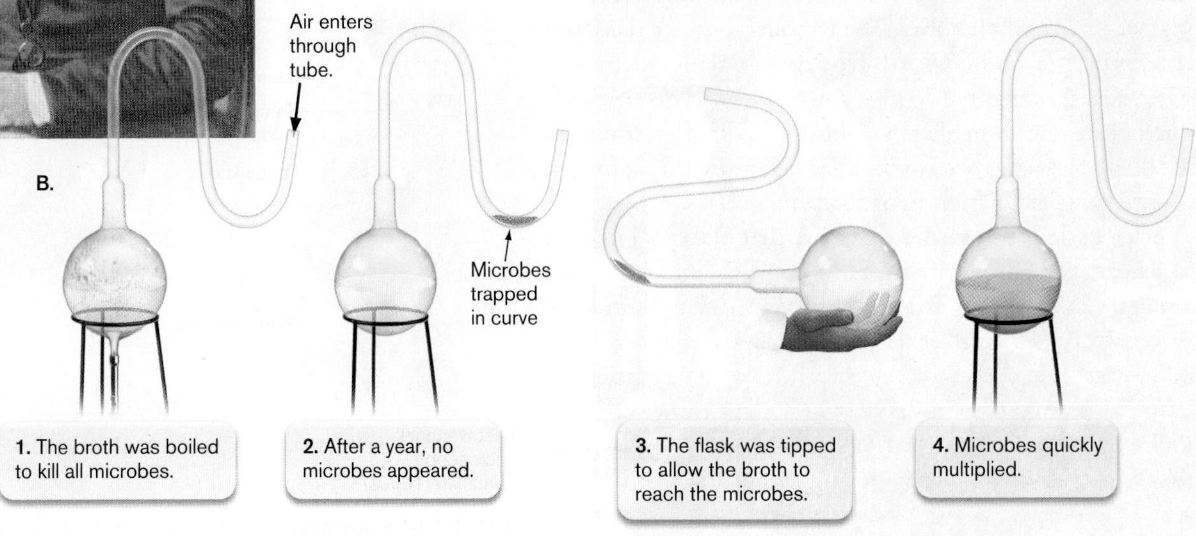

FIGURE 1.13 ■ Louis Pasteur, founder of medical microbiology and immunology. A. Pasteur's contributions to the science of microbiology and immunology earned him lasting fame. B. Swan-necked flask. Pasteur showed that, after boiling, the contents in such a flask remain free of microbial growth, despite access to air.

wrote his doctoral thesis on the structure of organic crystals. He discovered the fundamental chemical property of chirality, the fact that some organic molecules exist in two forms that differ only by mirror symmetry; in other words, the two structures are mirror images of one another, like the right and left hands. Pasteur found that when microbes were cultured on a nutrient substance containing both mirror forms, only one mirror form was consumed. He concluded that the metabolic preference for one mirror form was a fundamental property of life. Subsequent research has confirmed that most molecules of organisms, such as DNA and proteins, are found in only one of their mirror forms.

As a chemist, Pasteur was asked to help with a widespread problem encountered by French manufacturers of wine and beer. The alcohol in beverages comes from **fermentation**, a process by which microbes gain energy by converting sugars into alcohol. In the time of Pasteur, however, the conversion of grapes or grain to alcohol was believed to be a spontaneous chemical process. No one could explain why some fermentation mixtures produced vinegar (acetic acid) instead of alcohol. Pasteur discovered that fermentation is actually caused by living yeast, a single-celled fungus. In the absence of oxygen, yeast produces alcohol as a terminal waste product. But when the yeast culture is contaminated with bacteria, the bacteria outgrow the yeast and produce acetic acid instead of alcohol. (Fermentative metabolism is discussed in Chapter 13.)

Pasteur's work on fermentation led him to test a key claim made by proponents of spontaneous generation. The proponents claimed that Spallanzani's failure to find spontaneous appearance of microbes was due to lack of oxygen. From his studies of yeast fermentation, Pasteur knew that some microbial species do not require oxygen for growth. So he devised an unsealed flask with a long, bent "swan neck" that admitted air but prevented the passage of dust that carried microbes (**Fig. 1.13B**). When beef broth in the flask was boiled, the sterile broth remained clear, showing no growth of microbes. The famous swan-necked flasks remained free of microbial growth for many years. But when a flask was tilted so that the broth reached the dust, microbes grew immediately. Thus, Pasteur disproved the idea that lack of oxygen was the reason for the failure of spontaneous generation in Spallanzani's flasks.

Even Pasteur's work did not prove that microbial growth requires preexisting microbes. The Irish scientist John Tyndall (1820–1893) attempted an experiment similar to Pasteur's but sometimes found the opposite result. Tyndall found that some kinds of broth, such as hay infusion, gave rise to microbes no matter how long they were sterilized by boiling. The microbes appear because hay infusion is contaminated with a heat-resistant form of bacteria called "endospores" (or "spores"). The spore form can be eliminated only by repeated cycles of boiling and resting, in which the spores germinate to the growing, vegetative form that is killed at 100°C.

It was later discovered that endospores could be killed by boiling under pressure, as in a pressure cooker, which generates higher temperatures than can be obtained at atmospheric pressure. The steam pressure device called the **autoclave** became the standard way to sterilize materials for the controlled study of microbes. (Microbial control and antisepsis are discussed further in Chapter 5.)

How Did Life Originate?

Spontaneous generation was discredited as a continual source of microbes. Yet at some point in the past, the first living organisms must have originated from nonliving materials. How did the first microbes arise?

The earliest fossil evidence of cells in the geological record appears in rock that formed as long ago as 4 billion years (discussed in Chapter 17). Although the nature of the earliest reported fossils remains controversial, it is generally accepted that "microfossils" from over 2 billion years ago were formed by living cells. Moreover, the living cells that formed microfossils looked remarkably similar to bacterial cells of today, forming chains of simple rods or spheres (**Fig. 1.14**).

The exact composition of the first environment for life is controversial. The components of the first living cells may have formed from spontaneous reactions sparked by ultraviolet absorption or electrical discharge. American chemists Stanley Miller (1930–2007) and Harold C. Urey (1893–1981) argued that the environment of early Earth contained mainly reduced compounds—compounds that

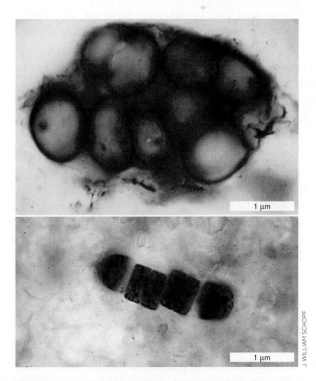

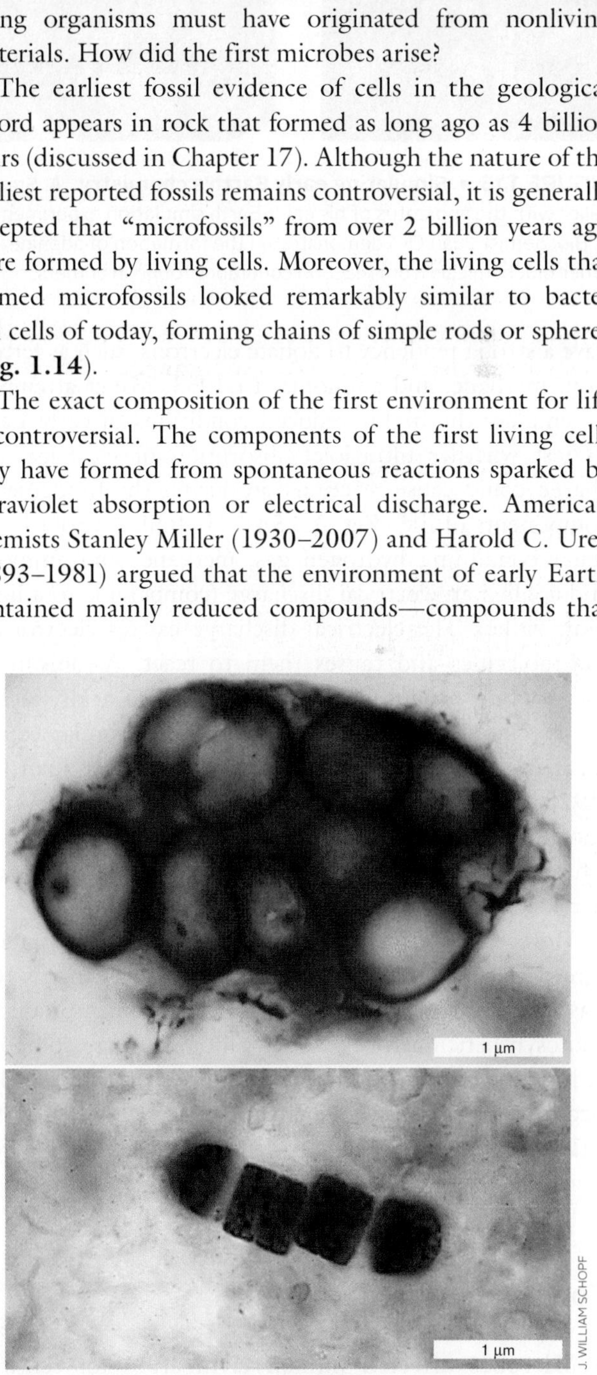

FIGURE 1.14 ■ Microfossils of ancient cyanobacteria. These fossils, from the Bitter Springs Formation, Australia, are about 850 million years old.

A. B.

JIM SUGAR/GETTY IMAGES

ALBUM/ALAMY STOCK PHOTO

FIGURE 1.15 ■ Simulating early Earth's chemistry. A. Stanley Miller with the apparatus of his early-Earth simulation experiment. B. Biochemist Juan Oró demonstrated the formation of adenine and other biochemicals from reaction conditions found in comets.

have a strong tendency to donate electrons, such as ferrous iron, methane, and ammonia. In 1953, Miller attempted to simulate the highly reduced conditions of early Earth to test whether ultraviolet absorption or electrical discharge could cause reactions producing the fundamental components of life (**Fig. 1.15A**). He boiled a solution of water containing hydrogen gas, methane, and ammonia and applied an electrical discharge (comparable to a lightning strike). The electrical discharge excites electrons in the molecules and causes them to react. Astonishingly, the reaction produced a number of amino acids, including glycine, alanine, and aspartic acid. A similar experiment in 1961 by Spanish-American researcher Juan Oró (1923–2004; **Fig. 1.15B**) combined hydrogen cyanide and ammonia under electrical discharge to obtain adenine, a fundamental component of DNA and of the energy carrier adenosine triphosphate (ATP).

More recent evidence has modified this view, but it is agreed that the strong electron acceptor oxygen gas (O_2) was absent until the evolution of the first oxygen-producing photosynthetic microbes. Today, all our cells are composed of highly reduced molecules that are readily oxidized (lose electrons to O_2). This seemingly hazardous composition may reflect our cellular origin in the chemically reduced environment of early Earth. **The experimental basis for the origin of life and evolution is discussed in detail in Chapter 17.**

To Summarize

- **Microbes affected human civilization** for centuries before humans guessed at their existence through their contributions to our environment, food and drink production, and infectious diseases.
- **Florence Nightingale** statistically quantified the impact of infectious disease on human populations.
- **Robert Hooke** and **Antonie van Leeuwenhoek** were the first to record observations of microbes through simple microscopes.
- **Spontaneous generation** is the theory that microbes arise spontaneously, without parental organisms. **Lazzaro Spallanzani** showed that microbes arise from preexisting microbes and demonstrated that heat sterilization can prevent microbial growth.
- **Louis Pasteur** discovered the microbial basis of fermentation. He also showed that providing oxygen does not enable spontaneous generation.
- **John Tyndall** showed that repeated cycles of heat were necessary to eliminate spores formed by certain kinds of bacteria.
- **Earth's first living organisms arose from nonliving materials.** Evidence from microfossils and chemical simulations supports the origin of microbial life within the first 100 million years of Earth's existence.

1.3 Medical Microbiology

Over the centuries, thoughtful observers such as Girolamo Fracastoro and Agostino Bassi de Lodi (see **Table 1.2**) noted a connection between microbes and disease. Ultimately, researchers developed the **germ theory of disease**, the theory that many diseases are caused by microbes. Research today pursues the secrets of many microbial diseases, such as cholera, the focus of Rita Colwell, who was the first microbiologist to direct the National Science Foundation (**eTopic 1.1**).

The first scientific basis for determining that a specific microbe causes a disease was devised by the German physician Robert Koch (1843–1910; **Fig. 1.16**). As a college student, Koch conducted biochemical experiments on his own digestive system. Koch's curiosity about the natural world led him to develop principles and methods crucial to microbial investigation, including the pure-culture technique and the famous Koch's postulates for identifying the causative agent of a disease. He applied his methods to numerous lethal diseases around the world, including anthrax and tuberculosis in Europe, bubonic plague in India, and malaria in New Guinea (**Fig. 1.17**).

Growth of Microbes in Pure Culture

Unlike Pasteur, who was a university professor, Koch took up a medical practice in a small Polish-German town. To make space in his home for a laboratory to study anthrax

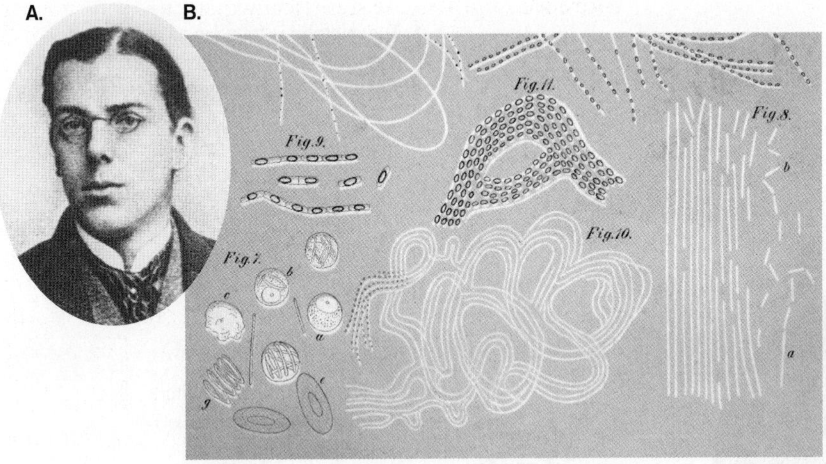

MUSEUM IN THE ROBERT KOCH INSTITUTE, BERLIN

SCIENCE HISTORY IMAGES/ALAMY STOCK PHOTO

FIGURE 1.16 ▪ Robert Koch, founder of the scientific method of microbiology. **A.** Koch as a university student. **B.** Koch's sketch of anthrax bacilli in mouse blood.

SUEDDEUTSCHE ZEITUNG PHOTO/ALAMY STOCK PHOTO

FIGURE 1.17 ▪ Robert Koch visits New Guinea. Koch (center, standing) investigated malaria in New Guinea in 1899.

and other deadly diseases, his wife curtained off part of his patient examination room.

Anthrax interested Koch because its epidemics in sheep and cattle caused economic hardship among local farmers. Today, anthrax is no longer a major problem for agriculture, because its transmission is prevented by effective environmental controls and vaccination. It has, however, gained notoriety as a bioterror agent because anthrax bacteria can survive for long periods in the dormant desiccated form of an endospore. In 2001, anthrax spores sent through the mail contaminated post offices, as well as an office building of the U.S. Senate, causing several deaths.

To investigate whether anthrax was a transmissible disease, Koch used blood from an anthrax-infected cow carcass to inoculate a rabbit. When the rabbit died, he used its blood to inoculate a second rabbit, which then died in turn. The blood of the unfortunate animal had turned black with long, rod-shaped bacilli. Upon introduction of these bacilli into healthy animals, the animals became ill with anthrax. Thus, Koch demonstrated an important principle of epidemiology: the **chain of infection**, or transmission of a disease. In retrospect, his choice of anthrax was fortunate, because anthrax microbes generate disease very quickly, multiply in the blood to high numbers, and remain infective outside the body for long periods.

Koch and his colleagues then applied their experimental logic and culture methods to a more challenging disease: tuberculosis. In Koch's day, tuberculosis caused one-seventh of all reported deaths in Europe; today, tuberculosis bacteria continue to infect millions of people worldwide. Koch's approach to anthrax, however, was less applicable to tuberculosis, a disease that develops slowly after many years of dormancy. Furthermore, the causative bacterium, *Mycobacterium tuberculosis*, is small and difficult to distinguish from human tissue or from different bacteria of similar appearance associated with the human body. How could Koch prove that a particular bacterium caused a particular disease?

What was needed was to isolate a **pure culture** of microorganisms, a culture grown from a single "parental" cell. Previous researchers had achieved pure cultures by a laborious process of serially diluting suspended bacteria until a culture tube contained only a single cell. Alternatively, inoculating a solid surface such as a sliced potato could produce isolated **colonies**—distinct populations of bacteria, each grown from a single cell. For *M. tuberculosis*, Koch inoculated serum, which then formed a solid gel after heating. Later he refined the solid-substrate technique by adding gelatin to a defined liquid medium, which could then be chilled to form a solid medium in a glass dish. A covered version called the **Petri dish** (or "Petri plate") was invented by a colleague, Julius Richard Petri (1852–1921). The Petri dish is a round dish with vertical walls covered by an inverted dish of slightly larger diameter. Today the Petri dish, generally made of disposable plastic, remains an indispensable part of the microbiological laboratory.

Another improvement in solid-substrate culture was the replacement of gelatin with materials that remain solid at higher temperatures, such as the gelling agent **agar** (a polymer of the sugar galactose). The use of agar was recommended by Angelina Hesse (1850–1934), a microscopist and illustrator, to her husband, Walther Hesse (1846–1911), a young medical colleague of Koch (**Fig. 1.18**). Agar comes from red algae (seaweed), which is used by East Indian birds to build nests; it is the main ingredient in the delicacy "bird's nest soup." Dutch colonists used agar to make jellies and preserves, and a Dutch colonist from Java introduced it to Angelina Hesse. The Hesses used agar to develop the first effective growth medium for tuberculosis

FIGURE 1.18 ■ Angelina and Walther Hesse. **A.** Portrait of the Hesses, who first used agar to make solid-substrate media for bacterial growth. **B.** Colonies from a streaked agar plate.

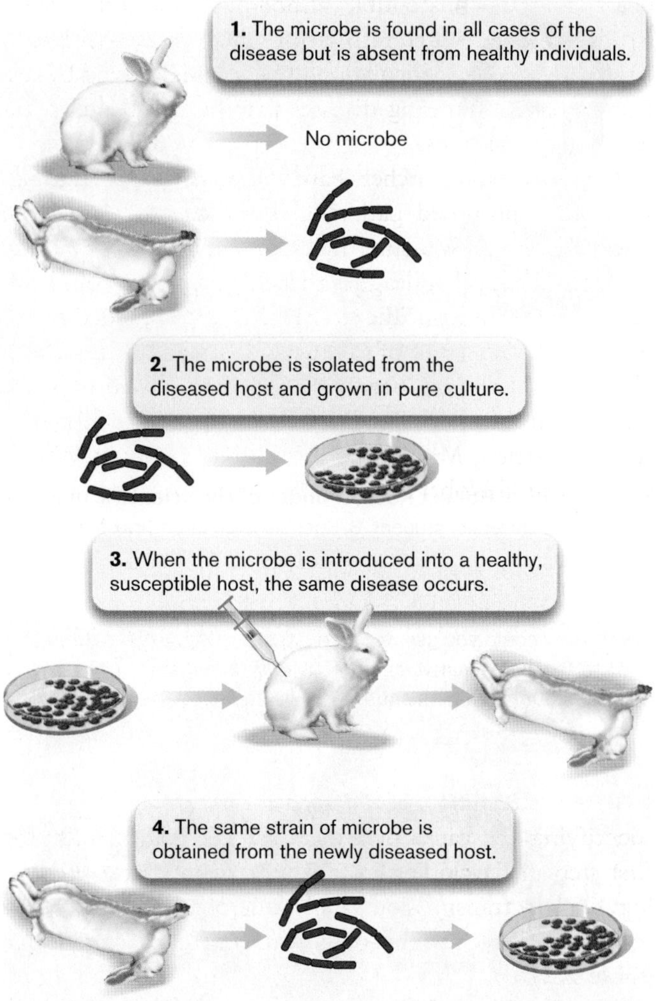

FIGURE 1.19 ■ Koch's postulates defining the causative agent of a disease.

bacteria. Pure culture and growth conditions are discussed further in Chapters 4 and 5.

Note that some kinds of microbes cannot be grown in pure culture—that is, without other organisms. For example, intracellular pathogens such as *Chlamydia trachomatis*, the cause of trachoma and of infections of the reproductive tract, must be cultivated in human host tissue culture. And all viruses can be cultured only within their host cells (see Chapter 6). The discovery of viruses is explored at the end of this section.

Koch's Postulates

For his successful determination of the bacterium that causes tuberculosis, *Mycobacterium tuberculosis*, Koch was awarded the 1905 Nobel Prize in Physiology or Medicine. Koch formulated his famous set of criteria for establishing a causative link between an infectious agent and a disease. These four criteria are known as **Koch's postulates** (**Fig. 1.19**):

1. The microbe is found in all cases of the disease but is absent from healthy individuals.
2. The microbe is isolated from the diseased host and grown in pure culture.
3. When the microbe is introduced into a healthy, susceptible host (or animal model), the host shows the same disease.
4. The same strain of microbe is obtained from the newly diseased host. When cultured, the strain shows the same characteristics as before.

Koch's postulates continue to be used to determine whether a given strain of microbe causes a disease. An example is Lyme disease (Lyme borreliosis), a tick-borne infection first described in New England, and shown in 1981 to be caused by the bacterium *Borrelia burgdorferi.* Nevertheless, the postulates remain only a guide; individual diseases and pathogens may confound one or more of the criteria. For example, tuberculosis bacteria are now known to cause symptoms in only 10% of the people they infect. If Koch had been able to detect these silent bacilli, they would not have fulfilled his first criterion. In the case of AIDS, the concentration of HIV is so low that initially no virus could be detected in patients with fully active symptoms. It took the invention of the polymerase chain reaction (PCR), a method of producing any number of copies of DNA or RNA sequences, to detect the presence of HIV. Modern research on microbial disease is presented in Chapter 26.

Another difficulty with AIDS and many other human diseases is the absence of an animal host that exhibits the same disease. For AIDS, even chimpanzees (our closest relatives) are not susceptible, although they acquire a similar disease from related retroviruses. For diseases without a cure, experimental inoculation of humans is banned by law. Curable or self-limiting diseases may be tested on volunteers in clinical trials.

In rare cases, researchers have voluntarily exposed themselves to a proposed pathogen. For example, Australian researcher Barry Marshall ingested *Helicobacter pylori* to convince skeptical colleagues that this organism could colonize the extremely acidic stomach. *H. pylori* turned out to be the causative agent of gastritis and stomach ulcers, conditions that had long been thought to be caused by stress rather than infection. For the discovery of *H. pylori* and its role in gastritis, Marshall and colleague J. Robin Warren won the 2005 Nobel Prize in Physiology or Medicine.

Thought Question

1.4 How could you use Koch's postulates **(Fig. 1.19)** to demonstrate the causative agent of influenza? What problems not encountered with anthrax would you need to overcome?

Immunization Prevents Disease

Identifying the cause of a disease is, of course, only the first step in developing an effective therapy and preventing further transmission. Early microbiologists achieved some remarkable insights on how to control pathogens (see **Table 1.2**).

The first clue of how to protect an individual from a deadly disease came from the dreaded smallpox. In the eighteenth century, smallpox virus (also called variola virus) infected a large fraction of the European population, killing or disfiguring many people. The existence of the virus was then unknown, but in countries of Asia and Africa the incidence of smallpox was decreased by the deliberate inoculation of children with material from smallpox pustules. Inoculated children usually developed a mild case of the disease and were protected from smallpox thereafter.

The practice of smallpox inoculation (or variolation) was introduced from Turkey to Europe in 1717 by Lady Mary Montagu, a smallpox survivor (**Fig. 1.20A**). While traveling in Turkey, Lady Montagu learned that many elderly women there had perfected the art of variolation: "The old woman comes with a nut-shell full of the matter of the best sort of small-pox, and asks what vein you please to have opened." During a period outside the host, the virus becomes "attenuated"—that is, loses some of its molecular structure required for infection. The attenuated virus stimulates the immune system with much lower mortality than does the fully virulent virus. Lady Montagu arranged for the procedure on her own son and then brought the practice back to England. A similar practice of smallpox inoculation was introduced to the American colonies by a slave, Onesimus, from the Coromantee people of Africa. Onesimus convinced his master, Reverend Dr. Cotton Mather, to promote smallpox inoculation as a defense against an epidemic that was devastating Boston.

Preventive inoculation with smallpox was dangerous, however, because some infected individuals still contracted serious disease and were contagious. Thus, doctors continued to seek a better method of prevention. In England, milkmaids claimed that they were protected from smallpox after they contracted cowpox (caused by vaccinia virus), a related but much milder disease. English physician Edward Jenner (1749–1823) confirmed this claim by deliberately infecting patients with matter from cowpox lesions (**Fig. 1.20B**). The practice of cowpox inoculation

A. B. C.

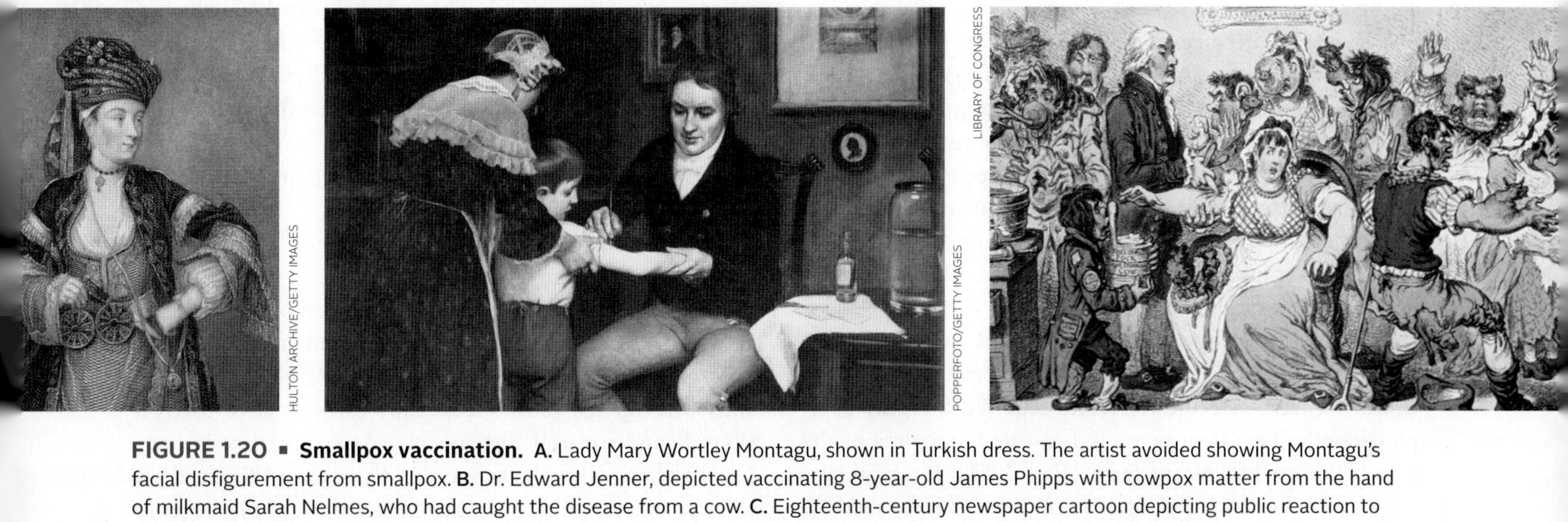

FIGURE 1.20 ■ Smallpox vaccination. **A.** Lady Mary Wortley Montagu, shown in Turkish dress. The artist avoided showing Montagu's facial disfigurement from smallpox. **B.** Dr. Edward Jenner, depicted vaccinating 8-year-old James Phipps with cowpox matter from the hand of milkmaid Sarah Nelmes, who had caught the disease from a cow. **C.** Eighteenth-century newspaper cartoon depicting public reaction to cowpox vaccination.

was called **vaccination**, after the vaccinia virus, which was derived from the Latin word *vacca*, meaning "cow." At the time, the practice was highly controversial, as people feared they would somehow turn into cows (**Fig. 1.20C**). Today, unfortunately, modern immunizations still raise irrational concerns. Failure to accept immunization leads to outbreaks of preventable disease, such as the measles outbreak in 2019, with more than 1,200 cases across the U.S.

Pasteur was aware of vaccination as he studied the course of various diseases in experimental animals. In the spring of 1879, he was studying fowl cholera, a transmissible disease of chickens with a high death rate. He had isolated and cultured the bacteria that had killed the chickens, but he left his work during the summer for a long vacation. No refrigeration was available to preserve cultures, and when he returned to work, the aged bacteria failed to cause disease in his chickens. Pasteur then obtained fresh bacteria from an outbreak of disease elsewhere, as well as some new chickens. But the fresh bacteria failed to make the original chickens sick (those that had been exposed to the aged bacteria). All of the new chickens, exposed only to the fresh bacteria, contracted the disease. Grasping the clue from his mistake, Pasteur had the insight to recognize that an attenuated strain of microbe, altered somehow to eliminate its potency to cause disease, could still confer immunity to the virulent disease-causing form.

Pasteur was the first to recognize the significance of attenuation and extend the principle to other pathogens. We now know that the molecular components of pathogens generate **immunity**, the resistance to a specific disease, by stimulating the **immune system**, an organism's exceedingly complex cellular mechanisms of defense (see Chapters 23 and 24). Understanding the immune system awaited the techniques of molecular biology a century later, but nineteenth-century physicians developed several effective examples of **immunization**, the stimulation of an immune response by deliberate inoculation with an attenuated pathogen.

The way to attenuate a strain depends on the pathogen. Heat treatment or aging for various periods often turns out to be the most effective approach. The original success of prophylactic smallpox inoculation was due to natural attenuation of the virus during the time between acquisition of smallpox matter from a diseased individual and inoculation of the healthy patient. A far more elaborate treatment was required to combat the most famous disease for which Pasteur devised a vaccine: rabies.

The rabid dog loomed large in folklore, and rabies was dreaded for its particularly horrible and inevitable course of death. Pasteur's vaccine for rabies required a highly complex series of heat treatments and repeated inoculations. Its success led to his instant fame (**Fig. 1.21**). Grateful survivors of rabies founded the Pasteur Institute for medical research, one of the world's greatest medical research institutions, whose scientists in the twentieth century discovered HIV.

FIGURE 1.21 ■ Pasteur cures rabies. This cartoon in a French newspaper depicts Louis Pasteur protecting children from rabid dogs.

Antiseptics and Antibiotics

Before the work of Koch and Pasteur, many patients died of infections transmitted unwittingly by their own doctors. In 1847, Hungarian physician Ignaz Semmelweis (1818–1865) noticed that the death rate of women in childbirth due to puerperal fever was much higher in his own hospital than in a birthing center run by midwives. He guessed that the doctors in his hospital were transmitting pathogens from cadavers that they had dissected. So he ordered the doctors to wash their hands in chlorine, an **antiseptic** agent (a chemical that kills microbes). The mortality rate fell, but this revelation displeased other doctors, who refused to accept Semmelweis's findings.

In 1865, the British surgeon Joseph Lister (1827–1912) noted that half of his amputee patients died of sepsis. Lister knew from Pasteur that microbial contamination might be the cause. So he began experiments to develop the use of antiseptic agents, most successfully carbolic acid, to treat wounds and surgical instruments. After initial resistance, Lister's work, with the support of Pasteur and Koch, drew widespread recognition. In the twentieth century, surgeons developed fully **aseptic** environments for surgery—that is, environments completely free of microbes.

The problem with most antiseptic chemicals that killed microbes was that if taken internally, they would also kill the patients. Researchers sought a "magic bullet," an **antibiotic** molecule that would kill only microbes, leaving their host unharmed.

An important step in the search for antibiotics was the realization that microbes themselves produce antibiotic compounds. This conclusion followed from the famous accidental discovery of penicillin by the Scottish medical researcher Alexander Fleming (1881–1955; **Fig. 1.22A**). In 1929, Fleming was culturing *Staphylococcus*, which infects wounds. He found that one of his plates of *Staphylococcus* was contaminated with a mold, *Penicillium notatum*, which he noticed was surrounded by a clear region free of *Staphylococcus* colonies (**Fig. 1.22B**). Following up on this observation, Fleming showed that the mold produced a substance that killed bacteria. Today we know this substance as penicillin.

In 1941, biochemists Howard Florey (1898–1968) and Ernst Chain (1906–1979) purified the penicillin molecule, which we now know inhibits formation of the bacterial cell wall. Penicillin saved the lives of many Allied troops during World War II, the first war in which an antibiotic became available to soldiers.

A.

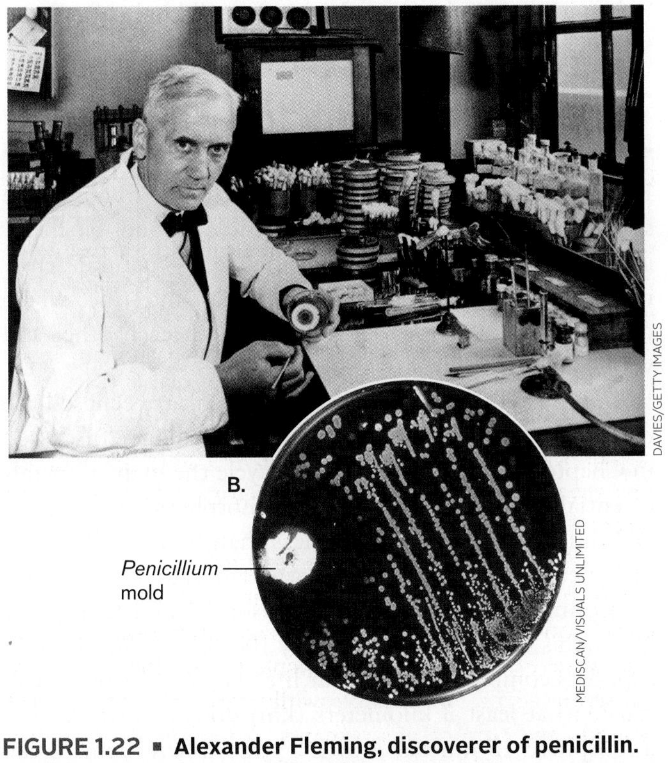

FIGURE 1.22 ■ Alexander Fleming, discoverer of penicillin. **A.** Fleming in his laboratory. **B.** Fleming's original plate of bacteria with *Penicillium* mold inhibiting the growth of bacterial colonies.

The second half of the twentieth century saw the discovery of many new and powerful antibiotics. Most of the new antibiotics, however, were made by little-known bacteria and fungi from endangered ecosystems—a circumstance that focused attention on wilderness preservation. Furthermore, the widespread and often indiscriminate use of antibiotics selects for pathogens to evolve resistance to antibiotics. As a result, antibiotics have lost their effectiveness against certain strains of major pathogens. For example, multidrug-resistant *Mycobacterium tuberculosis* and methicillin-resistant *Staphylococcus aureus* (MRSA) are now serious threats to public health. To combat evolving drug resistance, we continually need to research and develop new antibiotics. Microbial biosynthesis of antibiotics is discussed in Chapter 15, and the medical use of antibiotics is discussed in Chapter 27.

Thought Questions

1.5 Why do you think some pathogens generate immunity readily, whereas others evade the immune system?

1.6 How do you think microbes protect themselves from the antibiotics they produce?

The Discovery of Viruses

Viruses are much smaller than the host cells they infect; most are too small to be seen by a light microscope. So how were they discovered? In 1892, the Russian botanist Dmitri Ivanovsky (1864–1920) studied tobacco mosaic disease, a condition in which the leaves become mottled and the crop yield is decreased or destroyed altogether. Ivanovsky knew that some kind of microbe from the affected plants transmitted the disease, and he wondered how small it was. He was surprised to find that the agent of transmission could pass through a porcelain filter having a pore size (0.1 μm) that blocked known microbes. Later, the Dutch plant microbiologist Martinus Beijerinck (1851–1931) conducted similar filtration experiments. Beijerinck concluded that because the agent of disease passed through a filter that retained bacteria, it could not be a bacterial cell.

The "filterable agent" of disease was ultimately purified by the American scientist Wendell Stanley (1904–1971), who processed 4,000 kilograms (kg) of infected tobacco leaves. Stanley obtained a sample of infective virus particles pure enough to crystallize, in a 3D array comparable to crystals composed of inert chemicals. The crystal was analyzed by X-ray crystallography (discussed in Chapter 2) to reveal the molecular structure of tobacco mosaic virus (**Fig. 1.23A**)—a feat that earned Stanley the 1946 Nobel Prize in Chemistry. The fact that an object capable

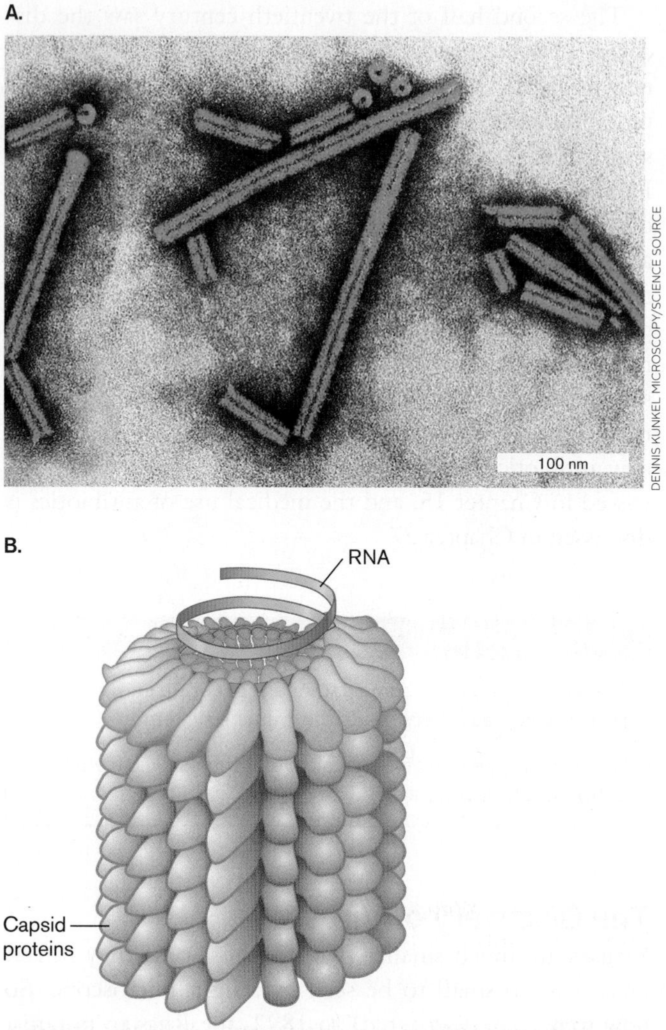

FIGURE 1.23 ▪ Tobacco mosaic virus (TMV). **A.** Particles of tobacco mosaic virus (colorized transmission electron micrograph). **B.** In TMV, a capsid of proteins surrounds an RNA chromosome.

of biological reproduction could be stable enough to be crystallized amazed scientists, ultimately leading to a new, more mechanical view of living organisms. Today, we consider viruses "subcellular organisms."

The individual particle of tobacco mosaic virus consists of a helical tube of protein subunits containing its genetic material coiled within (**Fig. 1.23B**). Stanley thought the virus was a catalytic protein, but colleagues later determined that it contained RNA as its genetic material. The structure of the coiled RNA was solved through X-ray crystallography by the British scientist Rosalind Franklin (1920–1958). Other viruses that have RNA genomes include influenza virus and HIV (the virus that causes AIDS); viruses with DNA genomes include human papilloma virus (HPV) and herpesviruses. We now know that all kinds of animals, plants, and microbial cells can be infected by viruses—and carry endogenous viruses that may benefit their hosts. Viral function and disease are discussed in Chapters 6, 11, and 26.

To Summarize

- **Robert Koch** devised techniques of pure culture to study a single species of microbe in isolation. A key technique is culture on solid medium using agar, as developed by Angelina and Walther Hesse, in a double-dish container devised by Julius Petri.
- **Koch's postulates** provide a set of criteria to establish a causative link between an infectious agent and a disease.
- **Edward Jenner** established the practice of vaccination, inoculation with cowpox to prevent smallpox. Jenner's discovery was based on earlier observations by Lady Mary Montagu and others that a mild case of smallpox could prevent future cases.
- **Louis Pasteur** developed the first vaccines based on attenuated strains, such as the rabies vaccine.
- **Ignaz Semmelweis** and **Joseph Lister** showed that antiseptics could prevent the transmission of pathogens from doctor to patient.
- **Alexander Fleming** discovered that the *Penicillium* mold generates a substance that kills bacteria.
- **Howard Florey** and **Ernst Chain** purified the substance penicillin, the first commercial antibiotic to save human lives.
- **Dmitri Ivanovsky** and **Martinus Beijerinck** discovered viruses as filterable infective particles. Viruses, which may be harmful or beneficial, infect all kinds of cells.

1.4 Environment and Ecology

Koch's growth of microbes in pure culture founded the systematic study of microbial physiology and biochemistry. But how does pure culture relate to "natural" environments, such as a forest or a human intestine, where countless kinds of microbes interact with each other and with multicellular hosts?

In hindsight, the invention of pure culture eclipsed the equally important study of microbial ecology (discussed in Chapters 21 and 22). Microbes cycle the many minerals essential for all life, including all atmospheric nitrogen gas and much of the oxygen. Yet, less than 0.1% of all microbial species can be cultured in the laboratory. In natural environments, uncultured microbes make up the majority of Earth's entire biosphere. Only the outer skin of Earth supports complex multicellular life. The depths of Earth's crust, to at least 3 kilometers (km) down, as well as the atmosphere 15 km out into the stratosphere, remain the domain of microbes. So, to a first approximation, Earth's ecology is microbial ecology.

Environmental Microbes Support Ecosystems

The first microbiologists to culture microbes in the laboratory selected the kinds of nutrients that feed humans, such as beef broth or potatoes. Some of Koch's contemporaries, however, suspected that other kinds of microbes living in soil or wetlands consume more exotic fare. Soil samples were known to oxidize hydrogen gas, and this activity was eliminated by treatment with heat or acid, suggesting microbial origin. Ammonia in sewage was oxidized by donating electrons to oxygen, forming nitrate. Nitrate formation was eliminated by antibacterial treatment. These findings suggested the existence of microbes that "eat" hydrogen gas or ammonia instead of beef or potatoes, but no one could isolate these microbes in culture.

Among the first to study microbes in natural habitats was the Russian scientist Sergei Winogradsky (1856–1953). Winogradsky waded through marshes to discover microbes with metabolisms quite alien from human digestion. For example, he discovered that species of the bacterium *Beggiatoa* oxidize hydrogen sulfide (H_2S) to sulfuric acid (H_2SO_4). *Beggiatoa* fixes carbon dioxide into biomass without consuming any organic food. Organisms that feed solely on inorganic minerals are known as chemolithotrophs, or lithotrophs (discussed further in Chapters 4 and 14).

The lithotrophs studied by Winogradsky could not be grown on Koch's plate media containing agar or gelatin. The bacteria that Winogradsky isolated can grow only on inorganic minerals; in fact, some species are actually poisoned by organic food. For example, nitrifiers convert ammonia to nitrate, forming a crucial part of the nitrogen cycle in natural ecosystems. Winogradsky cultured nitrifiers on a totally inorganic solution containing ammonia and silica gel, which supported no other kind of organism. This experiment was an early example of **enrichment culture**, the use of selective growth media that support certain classes of microbial metabolism while excluding others.

Instead of isolating pure colonies, Winogradsky built a model wetland ecosystem containing regions of enrichment for microbes of diverse metabolism. This model is called the **Winogradsky column** (**Fig. 1.24**). The model consists of a glass tube containing mud (a source of wetland bacteria) mixed with shredded newsprint (an organic carbon source) and calcium salts of sulfate and carbonate (an inorganic carbon source for autotrophs). After exposure to light for several weeks, several zones of color develop, full of mineral-metabolizing bacteria. At the top, cyanobacteria conduct **photosynthesis**, using light energy to split water and produce molecular oxygen. Below, purple sulfur bacteria use photosynthesis to

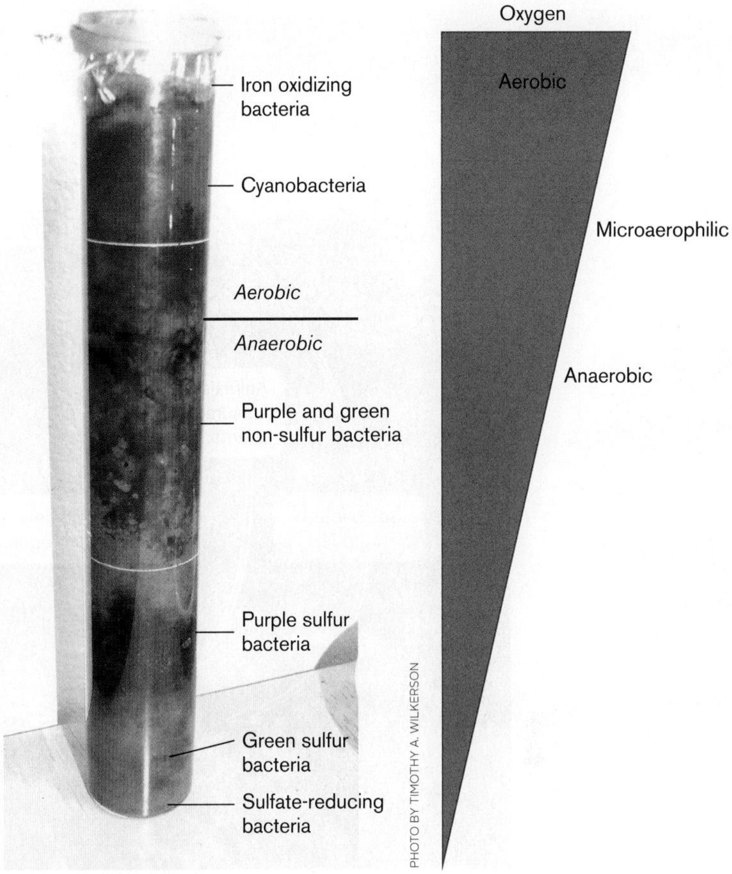

FIGURE 1.24 ▪ Winogradsky column. A wetland model ecosystem designed by Sergei Winogradsky.

split hydrogen sulfide, producing sulfur. At the bottom, with O_2 exhausted, bacteria reduce (donate electrons to) alternative electron acceptors such as sulfate. Sulfate-reducing bacteria produce hydrogen sulfide and precipitate iron.

Like a battery cell, the gradient from oxygen-rich conditions at the surface to highly reduced conditions below generates a voltage potential. We now know that the entire Earth's surface acts as a battery—for humans, a potential source of renewable energy. A fuel cell to generate electricity is described in Chapter 14.

Winogradsky and later microbial ecologists showed that bacteria perform unique roles in **geochemical cycling**, the global interconversion of inorganic and organic forms of nitrogen, sulfur, phosphorus, and other minerals. Without these essential conversions (nutrient cycles), no plants or animals could live. Bacteria and archaea fix nitrogen (N_2) by reducing it to ammonia (NH_3), the form of nitrogen assimilated by plants. This process is called **nitrogen fixation** (**Fig. 1.25**). Other bacterial species oxidize ammonium ions (NH_4^+) in several stages back to nitrogen gas.

Nitrogen fixation and geochemical cycling are discussed further in Chapters 21 and 22.

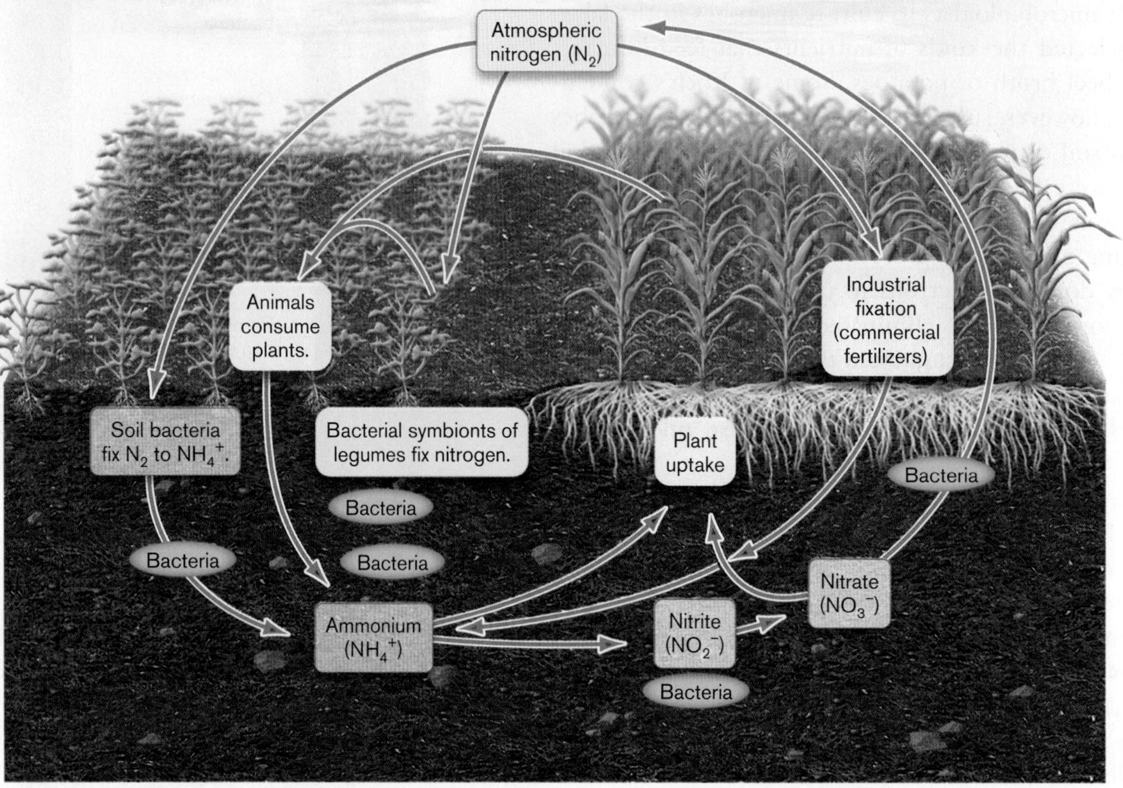

FIGURE 1.25 ▪ The global nitrogen cycle. All life depends on these oxidative and reductive conversions of nitrogen—most of which are performed only by microbes.

Thought Question

1.7 Why don't all living organisms fix their own nitrogen? Consider the structure of a dinitrogen molecule, N≡N.

Today, microbes with unusual properties, such as the ability to digest toxic wastes or withstand extreme temperatures, have valuable applications in industry and bioremediation. For this reason, microbial ecology is a priority for funding by the National Science Foundation (NSF). Microbial ecologist Rita Colwell (**Fig. 1.26A**) directed the NSF from 1998 to 2004 and founded the Biocomplexity Initiative to study complex interactions between microbes and other life in the environment. Such research includes the discovery of **extremophiles**, microbes from environments with extreme heat, salinity, acidity, or other factors. One such extremophile microbe is the archaeon *Geogemma* (**Fig. 1.26B**), which reduces rust (iron oxide, Fe_2O_3) to the magnetic mineral magnetite (Fe_3O_4) while growing in an autoclave at 121°C (250°F), a temperature high enough to kill all other known organisms. Microbiologist Kazem Kashefi uses a magnet to show that *Geogemma* converts nonmagnetic iron oxide (rust, Fe_2O_3) to the magnetic mineral magnetite (Fe_3O_4; **Fig. 1.26C**).

Microbial Endosymbiosis with Plants and Animals

The pure-culture model of microbiology, a powerful tool of discovery, nonetheless disregarded the fact that outside the laboratory, all microbes live in the presence of other kinds of life. Many live in **endosymbiosis** with multicellular organisms. Endosymbiosis is the partnership of a host organism with its associated **endosymbionts**, microbes that grow within a host body, or within a host cell. The first person to describe endosymbiosis was Martinus Beijerinck, who observed nitrogen-fixing bacteria called rhizobia (singular, rhizobium) within the cells of plants. Rhizobia induce the roots of legumes such as soybean plants to form special nodules to fix nitrogen into biomass, which is shared with the plant cells.

Microbial endosymbiosis occurs everywhere. Endosymbiotic microbes make essential nutritional contributions to host animals. Invertebrates such as hydras and corals harbor endosymbiotic phototrophs that provide products of photosynthesis in return for protection and nutrients. Other endosymbionts produce antibiotics or toxins that thwart predators, such as the toxin-producing cyanobacteria of the sponge *Lamellodysidea* (presented in the Current Research Highlight). Among vertebrates, ruminant animals such as

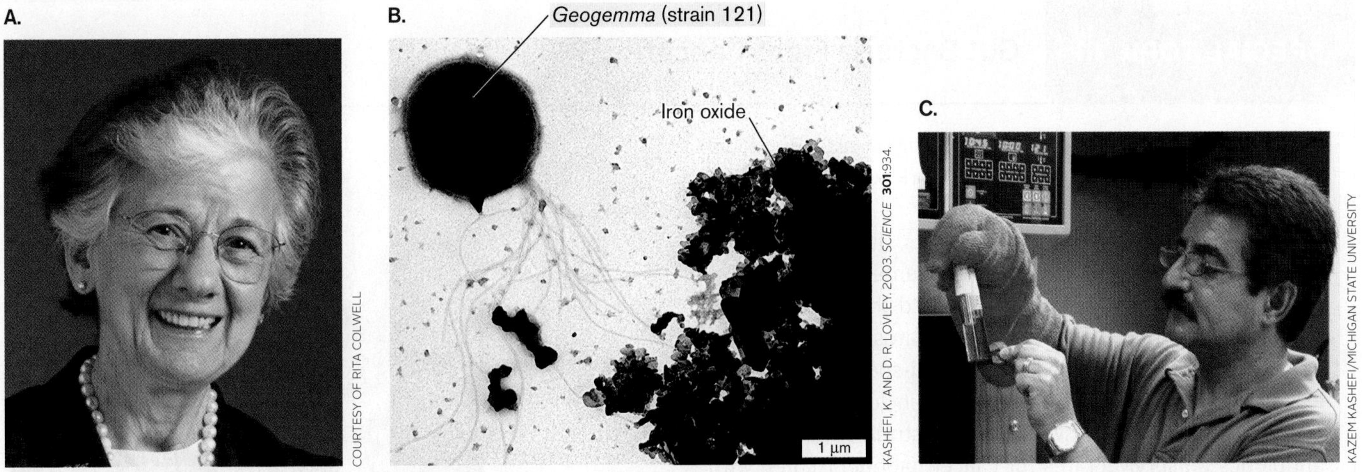

FIGURE 1.26 ■ An extreme thermophile reduces iron oxide to magnetite. **A.** While serving as NSF director, Rita Colwell promoted the study of environmental microbiology. See **eTopic 1.1** for an interview with her. **B.** *Geogemma* is a round archaeon with a tuft of flagella (transmission electron micrograph). **C.** Kazem Kashefi, now at Michigan State University, pulls a live culture of "strain 121" (*Geogemma*) out of an autoclave generally used to kill all living organisms at 121°C (250°F).

cattle, as well as insects such as termites, require digestive bacteria such as *Bacteroides* to break down cellulose and other plant polymers. Even humans obtain about 15% of their nutrition from bacteria growing within the colon (**Fig. 1.27**). Some intestinal bacteria, such as *E. coli*, grow as **biofilms**, organized multispecies communities adhering to a surface—in this case the surface of colonic epithelial cells. Biofilms play major roles in all ecosystems and within parts of the human body (discussed in Chapters 4, 13, and 21). The biofilm shown magnified in Figure 1.27 is attached to the surface of a digested food particle.

Today we know that all multicellular organisms possess a **microbiota**, or **microbiome**, the collection of all microbes associated with an organism or habitat. Remarkably, our concept of "multicellular organism" has changed to include its microbiome as a functional part of the organism. Physicians consider the human microbiome to be a part of the body, as essential as a limb or an organ. Bacteria that normally inhabit the human intestine and skin protect our bodies from infection by pathogens. Gut bacteria regulate the development of our immune system, and even send signals to the brain (discussed in Chapter 13). **Special Topic 1.1** describes research showing how our gut microbiome modulates immune system protection from cancer. In 2016, the U.S. government announced the National Microbiome Initiative to advance understanding of how microbiomes contribute to our health and the environment. Microbiomes are discussed further in Chapter 21.

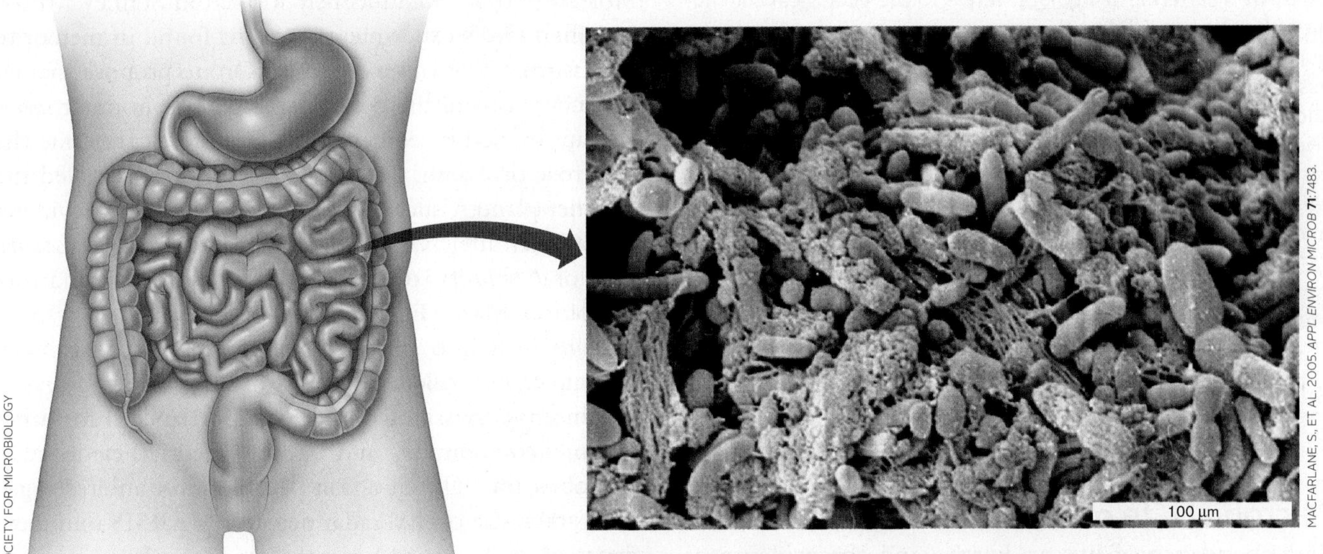

FIGURE 1.27 ■ Intestinal microbiota form a biofilm.
Blowup: Bacterial biofilm on the surface of a residual food particle.

SPECIAL TOPIC 1.1 Gut Bacteria Fight Cancer

We commonly think of bacteria as causing disease. But could our own intestinal bacteria actually work with our immune system to protect us from cancer?

An international group of scientists, led by Laurence Zitvogel at the Gustave Roussy cancer center, addresses this question (**Fig. 1**). The scientists reasoned that if human gut bacteria help fight cancer, then survival rates should be lower for cancer patients treated with broad-spectrum antibiotics, which kill a wide range of bacteria, including many beneficial gut species. So Zitvogel's group studied the effect of antibiotics on survival rates for cancer patients treated with immunotherapy, which uses special forms of antibodies that target cancerous cells. Some of the patients happened to have taken antibiotics for unrelated conditions, such as dental or urinary tract infections. In fact, statistics showed that the patients taking antibiotics survived for shorter times than those who did not.

The researchers then sought to establish a cause-effect relationship between gut microbiome and cancer progression. They asked: "Could the gut bacteria from 'responder' patients (those whose survival was improved by immunotherapy) increase survival of a new host treated for cancer?" This question was tested in a mouse model (**Fig. 2**). Mice were treated with broad-spectrum antibiotics that kill gut bacteria, then injected with tissue culture of cancer cells that form tumors. The cancer cells had been modified to express luciferase, an enzyme that catalyzes a reaction emitting light; thus, light intensity reveals the extent of the tumor within the mouse.

EUROPEAN SOCIETY FOR MEDICAL ONCOLOGY VIA AP IMAGES

FIGURE 1 ■ Laurence Zitvogel, scientific director, Gustave Roussy cancer center, France.

The mice were then divided into two groups that received fecal bacteria from human cancer patients: patients who were

Note: The term "microbiota" refers to the ecological community of microbes living within or upon an organism, such as the human body. "Microbiome" refers to the community of microbes associated with an organism, or with a different defined habitat, such as soil or plants; the term emphasizes the microbes' collective DNA sequences. Despite these subtle distinctions, we use the terms interchangeably in this book. Another term, **metagenome**, refers specifically to the collective DNA sequences found in all the microbes of a microbiome.

Thought Question

1.8 Could endosymbiosis occur today; that is, could a small microbe be engulfed by a larger one and evolve into an endosymbiont, and then into an organelle? Explain.

Microbial Life on Other Planets

The abundance of life on Earth—and the evidence of life's appearance early in Earth's history—leads us to ask whether microbial life has emerged on other planets. Molecules that spontaneously formed in Stanley Miller's and Juan Oró's experiments are also found in meteorites and comets. This observation led Oró to propose that the first chemicals of life could have come from outer space, perhaps carried by comets. Furthermore, at the time that life arose on Earth, Earth's geochemistry resembled that of other planets, such as Mars. In 2012, to seek evidence for Martian life, NASA landed the *Mars Science Laboratory*, or *Curiosity* rover, near the base of a mountain on the planet Mars (**Fig. 1.28**). The car-sized rover has a laser to drill into rock, X-ray and fluorescence analyzers, and camera microscopes. As of this writing, *Curiosity* continues its mission, testing the Martian soil for water, organic compounds, and other potential evidence of microbial life. The question became more interesting in 2018 when the orbital radar detector MARSIS found evidence of water buried beneath Mars's southern ice cap—similar to ice-buried lakes that support microbial life on Earth.

responders (showing improved survival with immunotherapy), or patients who were nonresponders (showing no improvement). These two groups were then subdivided into groups that received immunotherapy and groups that did not (control). What happened to their tumors? The mice that were fed fecal bacteria from responders and treated with immunotherapy showed statistically longer survival and lower tumor growth than did the other groups represented in **Figure 2**. Subsequent experiments associated improved cancer treatment with particular species of bacteria, such as *Akkermansia muciniphila*, a gut inhabitant that feeds on intestinal mucus and may help prevent obesity and diabetes. Future studies may help physicians recruit microbial partners for many kinds of medical therapy.

RESEARCH QUESTION

What experiments might identify types of gut bacteria that inhibit tumors? Assume that you have identified two candidate bacteria; propose experiments that could show their effect on tumors.

Bertrand Routy et al. 2018. Gut microbiome influences efficacy of PD-1-based immunotherapy against epithelial tumors. *Science* **359:**91–97.

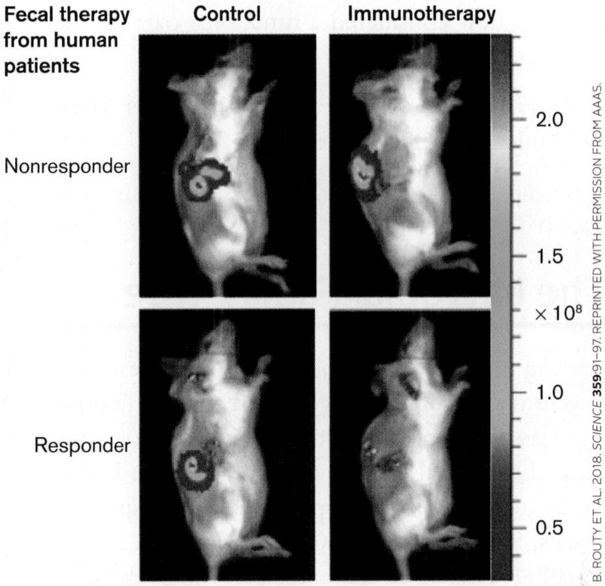

FIGURE 2 ■ Mice with tumors receive fecal microbial therapy from human patients. Fecal microbes from human patients who responded to immune system cancer inhibitors were transplanted into mice, which were then inoculated with cancer cells. The mice treated with fecal microbes from human nonresponders developed tumors (colored regions). But fecal therapy from responders (human patients who responded to inhibitors) prevented tumor formation in the mouse. Color scale shows intensity of light emission from the tumor.

FIGURE 1.28 ■ Mars *Curiosity* rover. The rover explores Mars for evidence of microbial life.

To Summarize

- **Sergei Winogradsky** developed the first system of enrichment culture, called the Winogradsky column, to grow microbes from natural environments.
- **Chemolithotrophs (or lithotrophs)** metabolize inorganic minerals, such as ammonia, instead of the organic nutrients used by the microbes isolated by Koch.
- **Geochemical cycling** depends on bacteria and archaea that cycle nitrogen, phosphorus, and other minerals throughout the biosphere.
- **Endosymbionts** are microbes that live within host organisms and may provide essential functions for their hosts, such as nitrogen fixation, digestion of food molecules, or protection from predation.

- **Martinus Beijerinck** was the first to demonstrate that nitrogen-fixing rhizobia grow as endosymbionts within leguminous plants.
- **The microbiome**, or microbial community, of a multicellular host is now considered a functional part of the host organism.
- **NASA's *Curiosity* rover** now explores the planet Mars for signs of microbial life outside Earth.

1.5 The Microbial Family Tree

The bewildering diversity of microbial life forms presented nineteenth-century microbiologists with a seemingly impossible task of classification. So little was known about life under the lens that natural scientists despaired of ever learning how to distinguish microbial species. The famous classifier of species, Swedish botanist Carl von Linné (Carolus Linnaeus, 1707–1778), called the microbial world "chaos."

Microbes Are a Challenge to Classify

Early taxonomists faced two challenges as they attempted to classify microbes. First, the resolution of the light microscope visualized little more than the outward shape of microbial cells, and vastly different kinds of microbes looked more or less alike (discussed in Chapter 2). This challenge was overcome as advances in biochemistry and microscopy made it possible to distinguish microbes by metabolism and cell structure, and ultimately by DNA sequence.

Second, microbes do not readily fit the classic definition of a species—that is, a group of organisms that interbreed. Unlike multicellular eukaryotes, microbes generally reproduce asexually. When they do exchange genes, they may do so with related strains or with distantly related species (discussed in Chapter 9). Nevertheless, microbiologists have devised working definitions of microbial species that enable us to usefully describe populations while being flexible enough to accommodate continual revision and change (discussed in Chapter 17). The most useful definitions are based on genetic similarity. For example, two distinct species generally share no more than 95% similarity between their DNA sequences.

Note: The names of microbial species are occasionally changed to reflect new understanding of genetic relationships. For example, the causative agent of bubonic plague was formerly called *Bacterium pestis* (1896), *Bacillus pestis* (1900), and *Pasteurella pestis* (1923), but it is now called *Yersinia pestis* (1944). The older names, however, still appear in the literature—a point to remember when carrying out research. Names of bacteria and archaea are compiled in the List of Prokaryotic Names with Standing in Nomenclature (LPSN).

Microbes Include Eukaryotes and Prokaryotes

In the nineteenth century, taxonomists had no DNA information. As they tried to incorporate microbes into the tree of life, they faced a conceptual dilemma. Microbes could not be categorized as either animals or plants, which since ancient times had been considered the two "kingdoms" or major categories of life. Taxonomists attempted to apply these categories to microbes—for example, by including algae and fungi with plants. But German naturalist Ernst Haeckel (1834–1919) recognized that microbes differed from both plants and animals in fundamental aspects of their lifestyle, cell structure, and biochemistry. Haeckel proposed that microscopic organisms constitute a third kind of life—neither animal nor plant—which he called Monera.

In the twentieth century, biochemical studies revealed profound distinctions even within the Monera. In particular, microbes such as protists and algae contain a nucleus enclosed by a nuclear membrane, whereas bacteria do not. Herbert Copeland (1902–1968) proposed a system of classification that divided Monera into two groups: the eukaryotic protists (protozoa and algae) and the prokaryotic bacteria. Copeland's four-kingdom classification (plants, animals, eukaryotic protists, and prokaryotic bacteria) was later modified by Robert Whittaker (1920–1980) to include fungi as another kingdom of eukaryotic microbes. Whittaker's system thus generated five kingdoms: bacteria, protists, fungi, and the plants and animals.

Eukaryotes Evolved through Endosymbiosis

The five-kingdom system was modified dramatically by Lynn Margulis (1938–2011) at the University of Massachusetts (**Fig. 1.29**). Margulis tried to explain how it is that eukaryotic cells contain mitochondria and chloroplasts, membranous organelles that possess their own chromosomes. She proposed that eukaryotes evolved by merging with bacteria to form composite cells by intracellular endosymbiosis, in which one cell internalizes another that grows within it. The endosymbiosis may ultimately generate a single organism whose formerly independent members are now incapable of independent existence.

Margulis proposed that early in the history of life, respiring bacteria similar to *E. coli* were engulfed by pre-eukaryotic cells, where they evolved into mitochondria, the eukaryote's respiratory organelles. Similarly, she proposed that a phototroph related to cyanobacteria was taken up by a eukaryote, giving rise to the chloroplasts of phototrophic algae and plants.

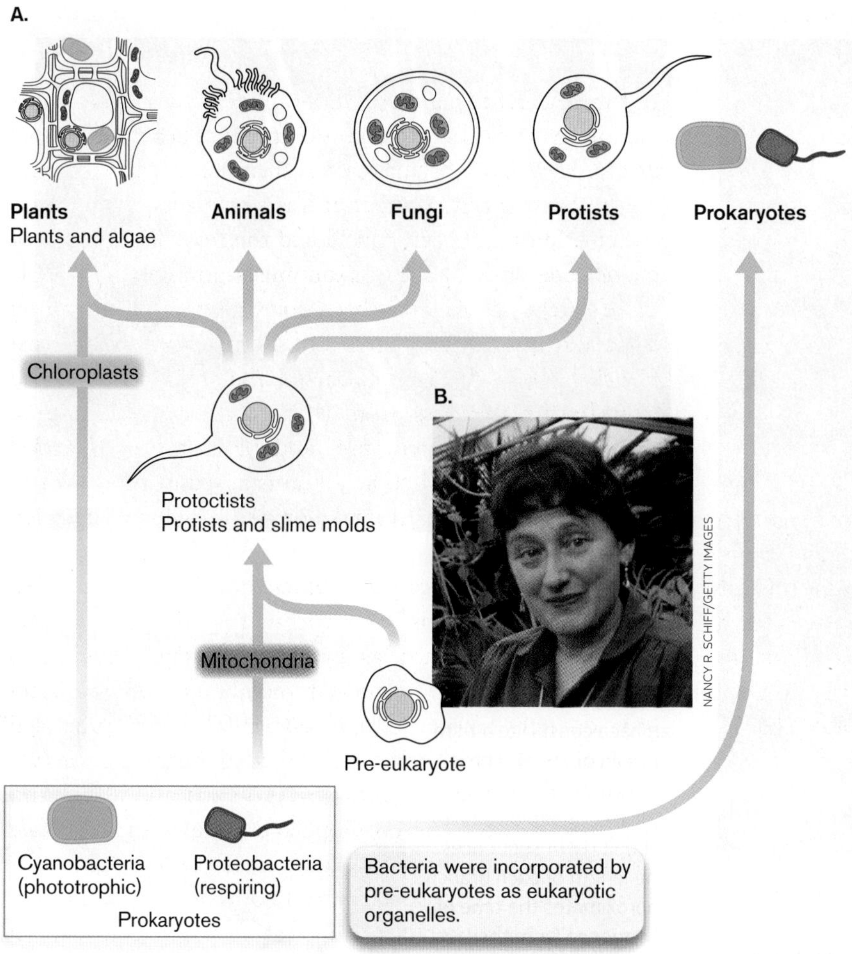

FIGURE 1.29 ■ Lynn Margulis and the serial endosymbiosis theory. **A.** Five-kingdom scheme, modified by the endosymbiosis theory. **B.** Margulis proposed that organelles evolve through endosymbiosis.

The endosymbiosis theory was highly controversial because it implied a **polyphyletic**, or multiple, ancestry of living species, inconsistent with the long-held assumption that species evolve only by divergence from a common ancestor (**monophyletic** ancestry). Ultimately, DNA sequence analysis produced compelling evidence of the bacterial origin of mitochondria and chloroplasts. Both of these classes of organelles contain circular molecules of DNA, whose sequences show unmistakable homology (similarity) to those of bacteria. DNA sequences and other evidence established the common ancestry between mitochondria and respiring bacteria, and between chloroplasts and cyanobacteria. The symbiotic origins and evolution of mitochondria and chloroplasts are discussed further in Chapter 17.

Archaea Differ from Bacteria and Eukaryotes

Are there cellular microbes that differ from both bacteria and eukaryotes? Gene sequence analysis led to another startling advance in our understanding of how cells evolved. In 1977, Carl Woese (1928–2012), at the University of Illinois, was studying a group of recently discovered prokaryotes that live in seemingly hostile environments, such as the boiling sulfur springs of Yellowstone, or that conduct unusual kinds of metabolism, such as production of methane (methanogenesis). Woese used the sequence of the gene for 16S ribosomal RNA (16S rRNA) as a "molecular clock," a gene whose sequence differences can be used to measure the time since the divergence of two species (discussed in Chapter 17). The divergence of rRNA genes showed that the newly discovered prokaryotes were a distinct form of life: **archaea** (**Fig. 1.30**).

The archaea resemble bacteria in their relatively simple cell structure, and in their lack of a nucleus; thus, both archaea and bacteria are called "prokaryotes." But the genetic sequences of archaea differ as much from those of bacteria as from those of eukaryotes; in fact, their gene expression machinery is more similar to that of eukaryotes. Archaea are found in a wide range of environments, and certain species, such as the autoclave-cultured archaeon *Geogemma*, grow in environments more extreme than any that support bacteria. Other kinds of archaea grow alongside bacteria in common soil or water—or even within the human gut or skin (discussed in Chapter 19).

Woese's discovery replaced the classification scheme of five kingdoms with three equally distinct groups, now called the three "domains": Bacteria, Archaea, and Eukarya (**Fig. 1.31**). In the three-domain model, the bacterial ancestor of mitochondria derives from ancient proteobacteria (shaded pink in the figure), whereas chloroplasts derive from ancient cyanobacteria (shaded green). The three-domain classification is largely supported by the sequences of microbial genomes, although some genes are transferred both within and between the domains (discussed in Chapter 17).

Thought Question

1.9 What arguments support the classification of Archaea as a third domain of life? What arguments support the classification of archaea and bacteria together, as prokaryotes, distinct from eukaryotes?

A.

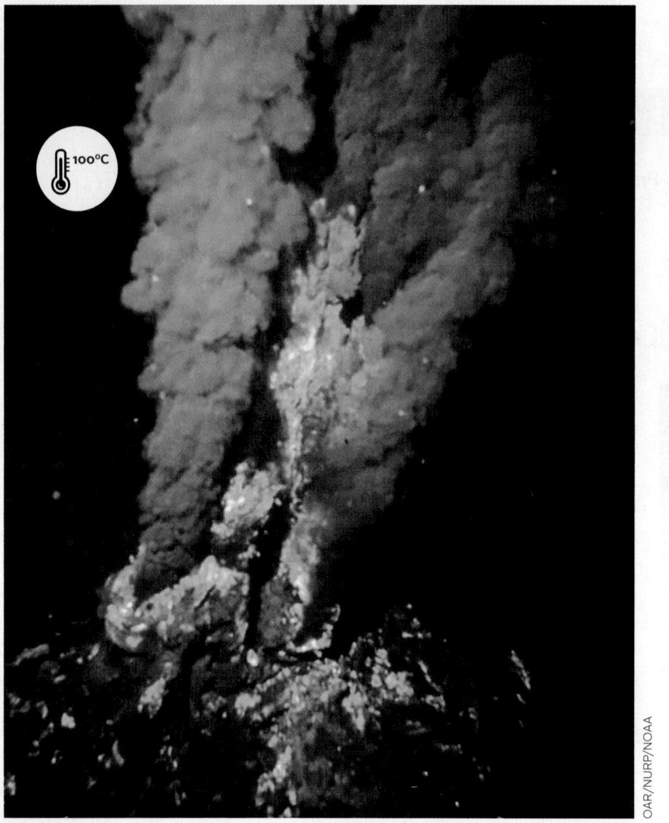

B.

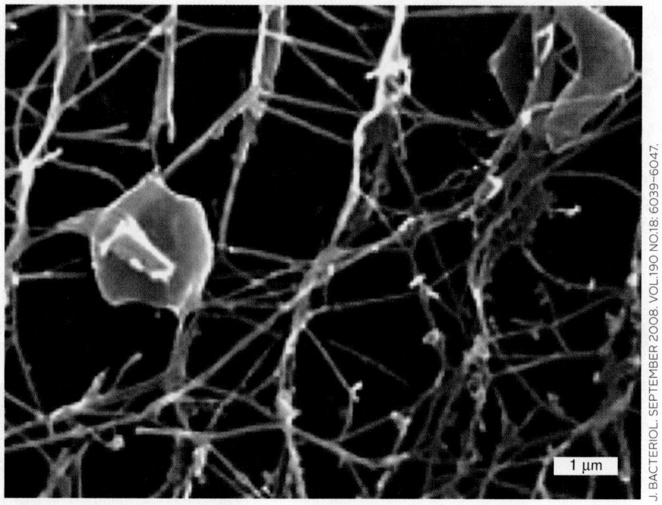

FIGURE 1.30 ▪ Archaea include extremophiles such as *Pyrodictium abyssi*. **A.** Thermal vents of superheated water and sulfides rise from the ocean floor. **B.** The vent sulfides feed networks of interconnected *P. abyssi* cells, at temperatures above 100°C (scanning electron micrograph).

FIGURE 1.31 ▪ Carl Woese and the three domains of life. **A.** Woese proposed that archaea constitute a third domain of life. **B.** Three domains form a monophyletic tree based on small-subunit rRNA sequences. The length of each branch approximates the time of divergence from the last common ancestor. For a more detailed tree, see Chapter 17.

A.

B.

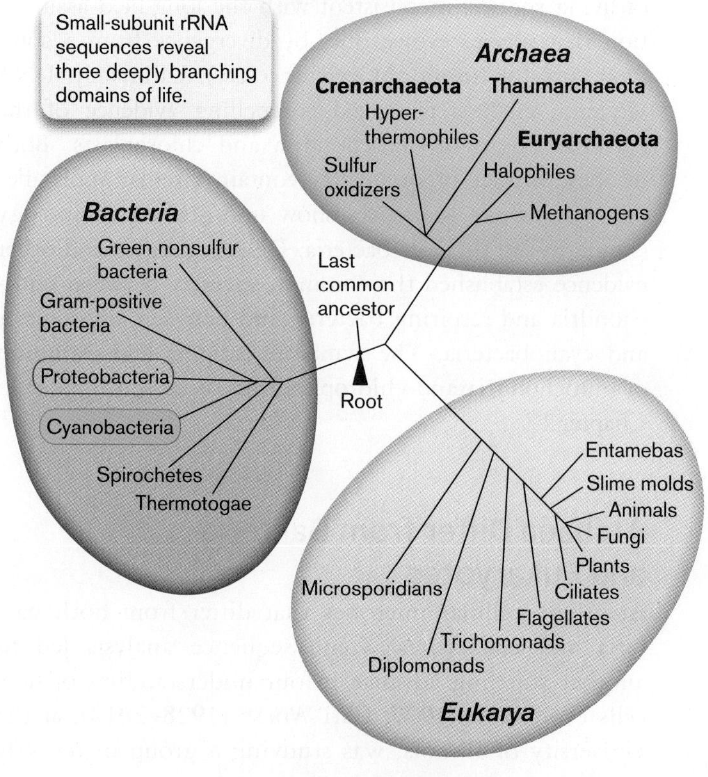

To Summarize

- **Classifying microbes** was a challenge historically because of the difficulties in observing distinguishing characteristics of different categories.
- **Ernst Haeckel** recognized that microbes constitute a form of life distinct from animals and plants.
- **Herbert Copeland** and **Robert Whittaker** classified prokaryotes as a form of microbial life distinct from eukaryotic microbes such as protists.
- **Lynn Margulis** proposed that eukaryotic organelles such as mitochondria and chloroplasts evolved by endosymbiosis from prokaryotic cells engulfed by pre-eukaryotes.
- **Carl Woese** discovered a domain of prokaryotes, Archaea, whose genetic sequences diverge equally from those of bacteria and those of eukaryotes. Archaea grow in a wide range of environments; some species grow under conditions that exclude bacteria and eukaryotes.

1.6 Cell Biology and the DNA Revolution

During the twentieth century, amid world wars and societal transformations, the field of microbiology exploded with new knowledge (see **Table 1.2**). More than 99% of what we know about microbes today was discovered after 1900 by scientists too numerous to cite in this book. Advances in biochemistry and microscopy revealed the fundamental structure and function of cell membranes and proteins. The revelation of the structures of DNA and RNA led to the discovery of the genetic programs of model bacteria, such as *E. coli*, and the lambda bacteriophage. Beyond microbiology, these advances produced the technology of "recombinant DNA," or genetic engineering, the construction of molecules that combine DNA sequences from unrelated species. These microbial tools offered unprecedented applications for human medicine and industry (discussed in Chapters 7–12).

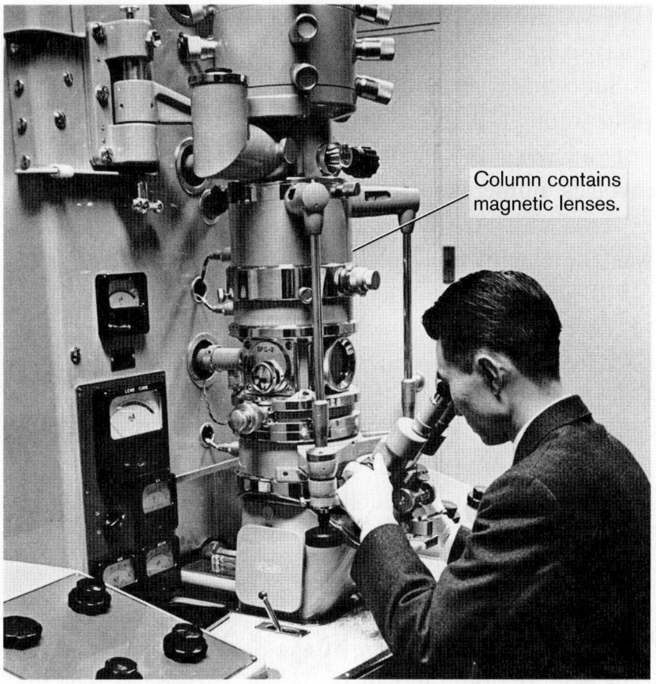

FIGURE 1.32 ■ An early transmission electron microscope.

Cell Membranes and Macromolecules

In 1900, the study of cell structure was still limited by the resolution of the light microscope and by the absence of tools that could take apart cells to isolate their components. Both of these limitations were overcome by the invention of powerful instruments. Just as society was being transformed by machines ranging from jet airplanes to vacuum cleaners, the study of microbiology was also being transformed by machines. Two instruments had exceptional impact: The **electron microscope** revealed the internal structure of cells (Chapter 2), and the **ultracentrifuge** enabled isolation of subcellular parts (Chapter 3).

The electron microscope. In the 1920s, at the Technical University in Berlin, student Ernst Ruska (1906–1988) was invited to develop an instrument for focusing rays of electrons. Ruska recalled, from his childhood, that his father's microscope could magnify fascinating specimens of plants and animals, but that its resolution was limited by the wavelength of light. He was eager to devise lenses that could focus beams of electrons, with wavelengths far smaller than that of light, to reveal living details never seen before. Ultimately, Ruska built lenses to focus electrons using specially designed electromagnets. Magnetic lenses were used to complete the first electron microscope in 1933 (**Fig. 1.32**). Early transmission electron microscopes achieved about tenfold greater magnification than the light microscope, revealing details such as the ridged shell of a diatom. Further development steadily increased magnification, to as high as a millionfold.

For the first time, cells were seen to be composed of a cytoplasm containing macromolecules and bounded by a phospholipid membrane. For example, the electron micrograph in **Figure 1.33** shows a "thin section" of the photosynthetic bacterium *Chlorobium*, including its nucleoid

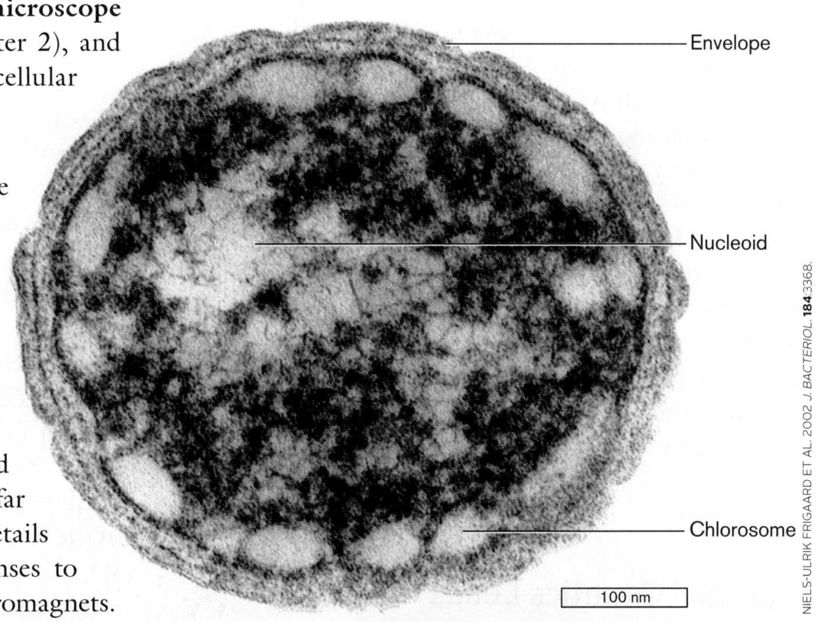

FIGURE 1.33 ■ Electron micrograph of *Chlorobium* species, a photosynthetic bacterium. The thin section reveals the nucleoid (containing DNA), the light-harvesting chlorosomes, and envelope membranes.

(DNA) and its light-harvesting chlorosomes. Electron microscopy is discussed further in Chapter 2.

Subcellular structures, however, raised many questions about cell function that visualization alone could not answer. Biochemists showed that cell function involves numerous chemical transformations mediated by enzymes. A milestone in the study of metabolism was the elucidation by German biochemist Hans Krebs (1900–1981) of the tricarboxylic acid cycle (TCA cycle, or Krebs cycle), by which the products of sugar digestion are converted to carbon dioxide. The TCA cycle provides energy for many bacteria and for the mitochondria of eukaryotes. But even Krebs understood little of how metabolism is organized within a cell; he and his contemporaries considered the cell a "bag of enzymes." The full understanding of cell structure required experiments on isolated parts of cells.

The ultracentrifuge. Centrifugation can separate whole cells from the fluid in which they are suspended. The first centrifuges spun samples in a rotor with centrifugal force of a few thousand times that of gravity. In the nineteenth century, biochemists proposed that even greater centrifugal forces could separate components of lysed cells, even macromolecules such as proteins. The Swedish chemist Theodor Svedberg (1884–1971), at the University of Uppsala, built such a machine: the ultracentrifuge. By the twentieth century, ultracentrifuges had achieved rates so high that they required a vacuum to avoid burning up like a space reentry vehicle. Ultracentrifuges isolated protein complexes such as ribosomes, and DNA molecules such as plasmids (small circular pieces of DNA).

Experiments combining electron microscopy and ultracentrifugation revealed how membranes govern energy transduction within bacteria and within organelles such as mitochondria and chloroplasts. In the 1960s, English biochemists Peter Mitchell (1920–1992) and Jennifer Moyle (1921–2016) proposed and tested a revolutionary idea called the **chemiosmotic theory**. The chemiosmotic theory states that the reduction-oxidation (redox) reactions of the electron transport system store energy in the form of a gradient of protons (hydrogen ions) across a membrane, such as the bacterial cell membrane or the inner membrane of the mitochondrion. The energy stored in the proton gradient, in turn, drives the synthesis of ATP (discussed in Chapter 14).

Microbial Genetics Leads the DNA Revolution

As the form and function of living cells emerged in the early twentieth century, a largely separate line of research revealed patterns of heredity of cell traits. In eukaryotes, the Mendelian rules of inheritance were rediscovered and connected to the behavior of subcellular structures called chromosomes. Frederick Griffith (1879–1941) showed in 1928 that an unknown substance from dead bacteria could carry genetic information into living cells, transforming harmless bacteria into a strain capable of killing mice—a process called **transformation**. Some kind of "genetic material" must be inherited to direct the expression of inherited traits. Biochemists thought the inherited material might be protein, because of the tremendous variety of amino acid sequences.

Then, in 1944, Oswald Avery (1877–1955) and colleagues showed that the genetic material for transformation is deoxyribonucleic acid, or DNA. An obscure acidic polymer, DNA had been previously thought too uniform in structure to carry information; its precise structure was unknown. As World War II raged among nations, scientists embarked on an epic struggle: the quest for the structure of DNA.

The double helix. The tool of choice to discover the structure of molecules was X-ray crystallography, a method developed by British physicists in the early 1900s. The field of X-ray analysis included an unusual number of women, including Dorothy Hodgkin (1910–1994), who later won a Nobel Prize for the structures of penicillin and vitamin B_{12} (discussed in Chapter 2). In 1953, crystallographer Rosalind Franklin joined a laboratory at King's College London to study the structure of DNA (**Fig. 1.34A**). As a woman and as a Jew who supported relief work in Palestine, Franklin felt socially isolated at the male-dominated Protestant university; her work was disparagingly called "witchcraft." Nevertheless, her exceptional X-ray micrographs (**Fig. 1.34B**) revealed for the first time that the common form of DNA was a double helix.

Without Franklin's knowledge, her colleague Maurice Wilkins (1916–2004) showed her data to a competitor, James Watson at the University of Cambridge. The pattern led Watson and Francis Crick (1916–2004) to propose that the four bases of the DNA "alphabet" were paired in the interior of Franklin's double helix (**Fig. 1.34C**). They published their model in the journal *Nature*, while denying that they had used Franklin's data. The discovery of the double helix earned Watson, Crick, and Wilkins the 1962 Nobel Prize in Physiology or Medicine. Franklin died of ovarian cancer before the prize was awarded. Before her death, however, she had turned her efforts to the structure of ribonucleic acid (RNA). She determined the helical form of the RNA chromosome within tobacco mosaic virus, the first viral RNA to be characterized.

Modern X-ray crystallography (discussed in Chapter 2) reveals with atomic precision the structure of DNA, including its complementary base pairs (**Fig. 1.35A**). The complementary pairing of DNA bases led to the development of techniques for **DNA sequencing**, the reading of a

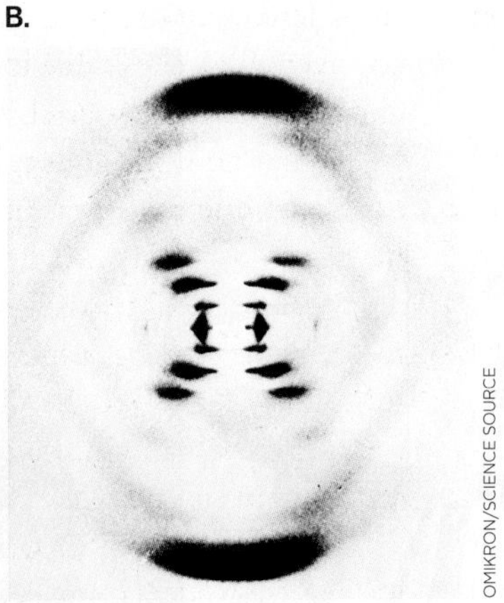

FIGURE 1.34 ■ The DNA double helix. **A.** Rosalind Franklin discovered that DNA forms a double helix. **B.** X-ray diffraction pattern of DNA, obtained by Franklin. **C.** James Watson (left) and Francis Crick discovered the complementary pairing between bases of DNA and the antiparallel form of the double helix.

sequence of DNA base pairs. **Figure 1.35B** shows a portion of the DNA sequence from bacterial DNA isolated by an undergraduate student. (The sequencing process is described in Chapter 7.) In the data, each color represents a fluorescent signal from one of the four bases: adenine (A), guanine (G), cytosine (C), or thymine (T). Each peak represents a DNA fragment terminating in that particular base. The order of fragment lengths yields the sequence of bases in one strand. Reading the DNA sequence enabled microbiologists to determine the beginning and endpoint of microbial genes, and ultimately entire genomes, as discussed in Section 1.1.

Reading the genomes enabled microbiologists to see the history of microbial evolution, reaching back to a time even before the advent of DNA—to a pre-DNA world when the cell's chromosomes were actually composed of ribonucleic acid, RNA. This hypothetical world without DNA is called the **RNA world**. How did life function in the RNA world? We hypothesize that cells used RNA for all the functions of DNA and protein, including

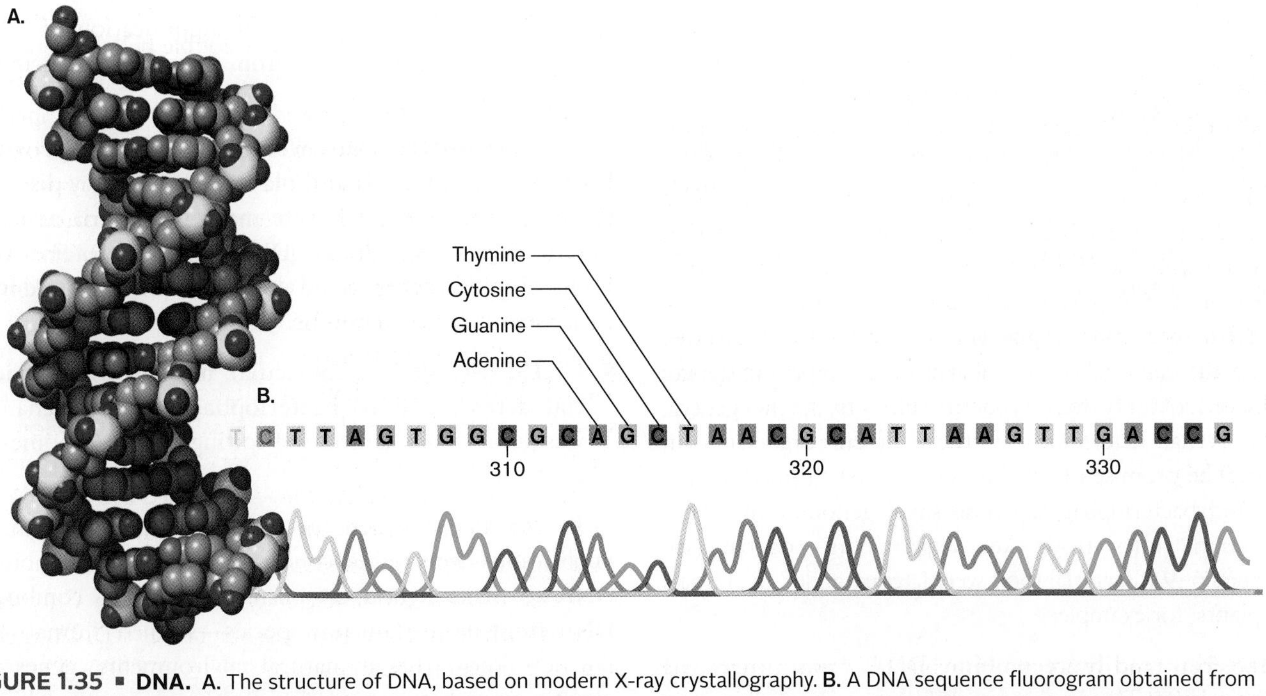

FIGURE 1.35 ■ DNA. **A.** The structure of DNA, based on modern X-ray crystallography. **B.** A DNA sequence fluorogram obtained from bacterial genomic DNA. Each colored trace represents the fluorescence of one of the four bases terminating a chain of DNA. Units represent number of DNA bases.

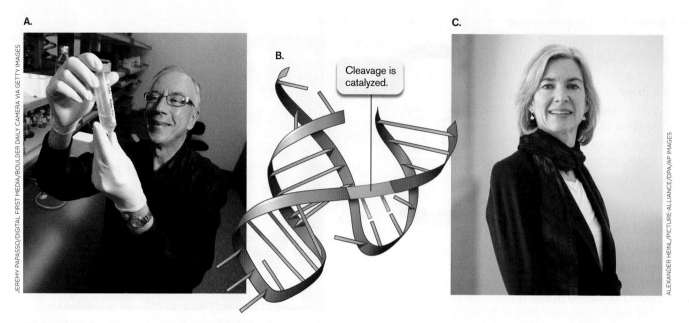

FIGURE 1.36 ▪ **Discovery of catalytic RNA.** **A.** Tom Cech holds a flask containing protists that make catalytic RNA, the kind of molecule that in early cells may have served both genetic and catalytic functions. **B.** Diagram of a catalytic RNA, where horizontal bars represent bases. The RNA catalyzes cleavage of itself. **C.** Jennifer Doudna, now at UC Berkeley, discovered a self-replicating catalytic RNA.

information storage and replication, and for biochemical catalysis. RNA molecules capable of catalysis, called ribozymes, were discovered in 1982 by Thomas Cech at the University of Colorado, and Sidney Altman at Harvard University, who jointly earned the Nobel Prize in Chemistry in 1989. That same year, Jennifer Doudna and Jack Szostak at Harvard showed how an RNA molecule from the protozoan *Tetrahymena* could catalyze its own replication (**Fig. 1.36**). In 2009, Gerald Joyce and Tracey Lincoln at the Scripps Research Institute constructed the first self-replicating ribozyme, an RNA that can catalyze reactions and copy itself indefinitely. These achievements support the theory that early organisms were composed primarily of RNA.

How do DNA and RNA sequences convey information in the cell? To read the DNA language required deciphering the genetic code—how triplets of DNA "letters" specify the amino acid units of proteins. This story is discussed in Chapter 8.

The DNA revolution began with bacteria. What amazed the world about DNA was that such a simple substance, composed of only four types of subunits, is the genetic material that determines all the different organisms on Earth. The promise of this insight was first fulfilled in bacteria and bacteriophages, whose small genomes and short generation times made key experiments possible (see Chapters 6–9). Bacterial tools were later extended to animals and plants; for example:

▪ **Bacteria readily recombine DNA from unrelated organisms.** The mechanisms of bacterial recombination led to construction of artificially recombinant DNA, or "gene cloning." Recombinant DNA ultimately enabled us to transfer genes between the genomes of virtually all types of organisms.

▪ **Bacterial DNA polymerases are used for polymerase chain reaction (PCR) amplification of DNA.** A hot spring in Yellowstone National Park yielded the bacterium *Thermus aquaticus*, whose DNA polymerase could survive many rounds of cycling to near-boiling temperature. The Taq polymerase formed the basis of a multibillion-dollar industry of PCR amplification of DNA, with applications ranging from genome sequencing to forensic identification.

▪ **Gene regulation discovered in bacteria provided models for animals and plants.** The first key discoveries of gene expression were made in bacteria and bacteriophages. Regulatory DNA-binding proteins were discovered in bacteria and then subsequently found in all classes of living organisms.

▪ **CRISPR-Cas9 is a molecular mechanism of bacterial defense against bacteriophages.** This mechanism was developed as a means of editing human genomes for gene therapy.

In the 1970s, when the DNA revolution began, its implications drew public concern. The use of recombinant DNA to make hybrid organisms—organisms combining DNA from more than one species—seemed "unnatural." We now know that in natural environments, genes frequently move between species. Furthermore, recombinant

DNA technology raised the specter of placing deadly genes that produce toxins such as botulin into innocuous human-associated bacteria such as *E. coli*.

The unknown consequences of recombinant DNA so concerned molecular biologists that in 1975 they held a conference to assess the dangers and restrict recombinant DNA experimentation. The conference, led by Paul Berg and Maxine Singer at Asilomar (Pacific Grove, California) was possibly the first time in history that a group of scientists organized and agreed to regulate and restrict their own field.

On the positive side, the emerging world of molecular biology excited the imagination of young scientists and entrepreneurs. The pioneering biotechnology company Genentech was founded in 1976 by Robert Swanson and bacterial geneticist Herbert Boyer, from UC San Francisco. Growing numbers of students entered the field of molecular biology, seeking to invent medical cures—or even to clone dinosaurs, as in Michael Crichton's novel and film *Jurassic Park* (1993). While the idea of cloning a dinosaur remains science fiction, the tools of microbial genetics opened a window into the past by letting us read the DNA of long-dead organisms preserved in museums. In 1980, the first patent was upheld for a living organism—a bacterium genetically modified to catabolize oil components from petroleum spills. Since then, it has become accepted to patent modified mice for cancer studies.

Thought Question

1.10 State an argument in favor of patenting a microbial isolate or a gene sequence. What argument can be made against patenting microbes or genes?

Microbial Discoveries Transform Medicine and Industry

Twentieth-century microbiology transformed the practice of medicine and generated entire new industries of biotechnology and bioremediation. Following the discovery of penicillin, Americans poured millions of dollars of private and public funds into medical research. In 1938, the March of Dimes was founded to discover a polio vaccine; today, polio has been nearly eliminated. Research on microbes and other aspects of biology has grown with support from U.S. government agencies, such as the National Institutes of Health and the National Science Foundation, as well as from governments of other countries, particularly the European nations and Japan. Further support comes from private foundations, such as the Pasteur Institute, the Wellcome Trust, and the Howard Hughes Medical Institute. Medical research generates astonishing advances, such as the use of human immunodeficiency virus (HIV, the cause of AIDS) to devise gene therapy agents that cure cancer (discussed in Chapter 11).

Research in microbiology includes fields as diverse as medicine and space science (**Table 1.3**). (A scientist whose career combines medicine and space science is profiled in **eTopic 1.2**.) These fields all recruit microbiologists (**Fig. 1.37**). Industrial and applied biology (see Chapter 16) use bacteria to clone and produce therapeutic proteins, such as insulin for diabetics. Recombinant viruses make safer vaccines. At the frontiers of science, we use microbes for "synthetic biology," the construction of novel organisms with useful functions (see Chapter 12). For example, synthetic biology may design bacteria with an on/off switch to report the presence of arsenic in environmental samples. On a global level, the management of our planet's biosphere, with the challenges of pollution and global warming, increasingly depends on our understanding of microbial populations (see Chapter 22).

TABLE 1.3 Fields of Research in Microbiology

Field	Subject of study
Experimental microbiology	Fundamental questions about microbial form and function, genetics, and ecology
Medical microbiology	The mechanism, diagnosis, and treatment of microbial disease
Epidemiology	Distribution and causes of disease in humans, animals, and plants
Immunology	The immune system and other host defenses against infectious disease
Food microbiology	Fermented foods and food preservation
Industrial microbiology	Production of drugs, cloned gene products, and biofuels
Environmental microbiology	Microbial diversity and microbial processes in natural and artificial environments
Bioremediation	The use of microbial metabolism to remediate human wastes and industrial pollutants
Forensic microbiology	Analysis of microbial strains as evidence in criminal investigations
Astrobiology	The origin of life in the universe and the possibility of life outside Earth

- **Molecular microbiology generated key advances**, such as the cloning of the first recombinant molecules and the invention of DNA sequencing technology.
- **Genome sequence determination and bioinformatic analysis** became the tools that shape the study of biology in the twenty-first century.
- **Microbial discoveries transformed medicine and industry.** Biotechnology produces new kinds of pharmaceuticals and industrial products. Synthetic biology engineers new kinds of organisms with useful functions.

Concluding Thoughts

Advances in microbial science raise important questions for society. How can medical research control emerging diseases? How do microbes contribute to global cycles of carbon and nitrogen? How can microbial metabolism clean up polluted environments, such as the vast oil spill from the explosion of the *Deepwater Horizon* wellhead in 2010? How can we engineer viruses to treat human diseases, such as the leukemia cured by viral gene therapy in 2013?

This book explores the current explosion of knowledge about microbial cells, genetics, and ecology. We introduce the applications of microbial science to human affairs, from medical microbiology to environmental science. Most important, we discuss research methods—how scientists make the discoveries that will shape tomorrow's view of microbiology. Chapter 2 presents the imaging tools that make possible our increasingly detailed view of the structures of cells (Chapter 3) and viruses (Chapter 6). Chapters 4 and 5 introduce microbial nutrition and growth in diverse habitats, including what we discover from microbial genomes. Throughout, we invite readers to share with us the excitement of discovery in microbiology.

FIGURE 1.37 ▪ **Microbiologists at work.** Students at Kenyon College conduct research on bacterial gene expression.

To Summarize

- **Genetics of bacteria, bacteriophages, and fungi** in the early twentieth century revealed fundamental insights about gene transmission that apply to all organisms.
- **Structure and function of the genetic material, DNA**, emerged from a series of experiments in the twentieth century.

CHAPTER REVIEW

Review Questions

1. Explain the apparent contradictions in defining microbiology as the study of microscopic organisms or as the study of single-celled organisms.
2. What is the genome of an organism? How do genomes of viruses differ from those of cellular microbes?
3. Under what conditions might microbial life have originated? What evidence supports current views of microbial origin?
4. List the ways in which microbes have affected human life throughout history.

5. Summarize the key experiments and insights that shaped the controversy over spontaneous generation. What questions were raised, and how were they answered?
6. Explain how microbes are cultured in liquid and on solid media. Compare and contrast the culture methods of Koch and Winogradsky. How did their different approaches to microbial culture address different questions in microbiology?
7. Explain how a series of observations of disease transmission led to the development of immunization to prevent disease.
8. Summarize key historical developments in our view of microbial taxonomy. What attributes of microbes have made them challenging to classify?
9. Explain how various discoveries in "natural" bacterial genetics were used to develop recombinant DNA technology.

Thought Questions

1. How do Earth's microbes contribute to human health? Include examples of environmental microbes outside the human body, as well as microbes associated with the human body.
2. When space scientists seek evidence for life on Mars, why do you think they expect to find microbes rather than creatures like the "alien monsters" often depicted in science fiction?
3. Why do you think so many environmental microbes cannot be cultured in laboratory broth or agar media?
4. Outline the different contributions to medical microbiology and immunology of Louis Pasteur, Robert Koch, and Florence Nightingale. What methods and assumptions did they have in common, and how did they differ?
5. Outline the different contributions to environmental microbiology of Sergei Winogradsky and Martinus Beijerinck. Why did it take longer for the significance of environmental microbiology to be recognized, as compared with pure-culture microbiology?
6. What kinds of evidence support the common ancestry of life from cells with RNA chromosomes? Could cells with RNA chromosomes exist today? Why or why not?

Key Terms

agar (15)
antibiotic (19)
antiseptic (18)
Archaea (5)
archaea (5, 27)
aseptic (19)
autoclave (13)
Bacteria (5)
bacteria (5)
biofilm (23)
chain of infection (15)
chemiosmotic theory (30)
colony (15)
DNA sequencing (30)
electron microscope (29)
endosymbiont (22)
endosymbiosis (22)
enrichment culture (21)
Eukarya (5)
eukaryote (4, 5)
extremophile (22)
fermentation (13)
genome (5)
geochemical cycling (21)
germ theory of disease (14)
immune system (18)
immunity (18)
immunization (18)
Koch's postulates (16)
metagenome (6, 24)
microbe (4)
microbiota (microbiome) (23)
monophyletic (27)
nitrogen fixation (21)
Petri dish (15)
photosynthesis (21)
polymerase chain reaction (PCR) (3)
polyphyletic (27)
prokaryote (4)
pure culture (15)
RNA world (31)
spontaneous generation (12)
transformation (30)
ultracentrifuge (29)
vaccination (18)
virus (4)
Winogradsky column (21)

Recommended Reading

Albers, Sonja-Verena, Patrick Forterre, David Prangishvili, and Christa Schleper. 2013. The legacy of Carl Woese and Wolfram Zillig: From phylogeny to landmark discoveries. *Nature Reviews. Microbiology* **11**:713–719.

Anguela, Xavier M. and Katherine A. High. 2019. Entering the modern era of gene therapy. *Annual Review of Medicine* **70**:273–288.

Blaser, Martin, Peer Bork, Claire Fraser, Rob Knight, and Jun Wang. 2013. The microbiome explored: Recent insights and future challenges. *Nature Reviews. Microbiology* **11**:213–217.

Blount, Zachary D., Christina Z. Borland, and Richard E. Lenski. 2008. Historical contingency and the evolution of a key innovation in an experimental population of *Escherichia coli*. *Proceedings of the National Academy of Sciences USA* **105**:7899–7906.

Brock, Thomas D. 1999. *Robert Koch: A Life in Medicine and Bacteriology.* ASM Press, Washington, DC.

Brokowski, Carolyn and Mazhar Adli. 2019. CRISPR ethics: moral considerations for applications of a powerful tool. *Journal of Molecular Biology* **431**:88–101.

Dinc, Gulten, and Yesim I. Ulman. 2007. The introduction of variolation "A La Turca" to the West by Lady Mary Montagu and Turkey's contribution to this. *Vaccine* **25**:4261–4265.

Dubos, Rene. 1998. *Pasteur and Modern Science.* Translated by Thomas Brock. ASM Press, Washington, DC.

Fleishmann, Robert D., Mark D. Adams, Owen White, Rebecca A. Clayton, Ewen F. Kirkness, et al. 1995. Whole-genome random sequencing and assembly of *Haemophilus influenzae* Rd. *Science* **269**:496–512.

Gann, Alexander, and Jan Witkowski. 2010. The lost correspondence of Francis Crick. *Nature* **467**:519–524.

Hesse, Wolfgang. 1992. Walther and Angelina Hesse—early contributors to bacteriology. *ASM News* **58**:425–428.

Luef, Birgit, Kyle R. Frischkorn, Kelly C. Wrighton, Hoi-Ying N. Holman, Giovanni Birarda, et al. 2015. Diverse uncultivated ultra-small bacterial cells in groundwater. *Nature Communications* **6**:6372.

Maddox, Brenda. 2002. *The Dark Lady of DNA*. HarperCollins, New York.

Margulis, Lynn. 1968. Evolutionary criteria in Thallophytes: A radical alternative. *Science* **161**:1020–1022.

Raoult, Didier. 2005. The journey from *Rickettsia* to Mimivirus. *ASM News* **71**:278–284.

Schinke, Claudia, Thamires Martins, Sonia C. N. Queiroz, Itamar S. Melo, and Felix G. R. Reyes. 2017. Antibacterial compounds from marine bacteria, 2010–2015. *Journal of Natural Products* **80**:1215–1228.

Sherman, Irwin W. 2006. *The Power of Plagues.* ASM Press, Washington, DC.

Thomas, Gavin. 2005. Microbes in the air: John Tyndall and the spontaneous generation debate. *Microbiology Today* (November 5): 164–167.

Ward, Naomi, and Claire Fraser. 2005. How genomics has affected the concept of microbiology. *Current Opinion in Microbiology* **8**:564–571.

CHAPTER 2

Observing the Microbial Cell

2.1 Observing Microbes

2.2 Optics and Properties of Light

2.3 Bright-Field Microscopy

2.4 Fluorescence Microscopy, Super-Resolution Imaging, and Chemical Imaging

2.5 Dark-Field and Phase-Contrast Microscopy

2.6 Electron Microscopy, Scanning Probe Microscopy, and X-Ray Crystallography

Microscopy reveals the vast realm of microorganisms invisible to the unaided eye. The **microscope** enables us to count the number of microbes in the human bloodstream or in dilute natural environments such as the open ocean. It shows us how microbes swim and respond to signals such as a new food source. Fluorescence microscopy captures single molecules within a living cell. Electron microscopy explores the cell's interior and models viruses, even catching a virus in the act of infection.

CURRENT RESEARCH highlight

Cryo-electron tomography of a human gut bacterium. Our immune system needs developmental signals from gut bacteria. The bacteria deliver their signaling molecules packaged in membrane vesicles. Sarkis Mazmanian used cryo-electron tomography to build a 3D model of *Bacteroides fragilis* forming vesicles. The outer membrane (green) pinches off the vesicles (yellow); also shown are the inner membrane (cyan) and ribosomes (pink). Mazmanian then showed how mice require certain genes to detect the bacterial signals. The mice need to "hear" their bacteria calling to develop T lymphocytes that regulate immune responses.

Source: Hiutung Chu et al. 2016. *Science* **352**(6289):1116–1120.

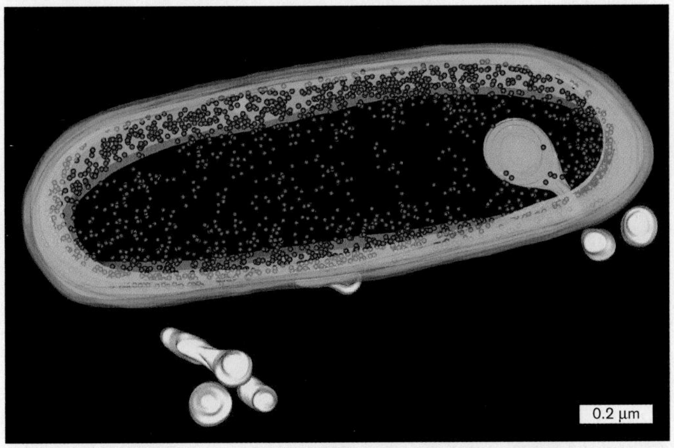

MARK S. LADINSKY, CALIFORNIA INSTITUTE OF TECHNOLOGY

AN INTERVIEW WITH

SARKIS MAZMANIAN, MICROBIOLOGIST, CALIFORNIA INSTITUTE OF TECHNOLOGY

DEBORAH WILLIAMS-HEDGES; CALTECH

How do *Bacteroides* forming membrane vesicles influence inflammatory bowel disease?

Outer membrane vesicles of the human commensal *Bacteroides fragilis* deliver immune-modulating signals during colonization of the gut. These signals dampen the uncontrolled inflammation in disorders such as Crohn's disease. Unlike current FDA-approved therapies, *Bacteroides* signals have no side effects such as systemic immunosuppression. Further, the bacterial signals delivered by vesicles from *B. fragilis* function in specific subsets of patients based on their genetic makeup. Thus, for the first time, gut bacteria offer personalized medicine.

How did people first see microbes? As we saw in Chapter 1, the microscope of Antonie van Leeuwenhoek first revealed the tiny life forms on his teeth; his superior lenses were key to his success. Since the time of Leeuwenhoek, microscopists have devised ever-more-powerful instruments to search for microbes in familiar and unexpected habitats.

An example of such a habitat is the colonic crypts, intestinal glands that form deep pits in the gut epithelium. These pits harbor bacteria, such as *Bacteroides fragilis*, that produce inflammatory signals essential for development of the immune system (see the Current Research Highlight). The crypts and their hidden bacteria can be visualized by fluorescence microscopy (**Fig. 2.1**). In fluorescence microscopy, key parts of cells act as light sources, labeled with fluorophores that absorb light and emit photons of a different wavelength. In **Figure 2.1**, the crypt tissue is visualized by DAPI stain of DNA in the cell nuclei (blue), and by phalloidin stain of actin filaments that line the crypt (green). The bacteria (red) are marked by a fluorophore attached to an antibody that is specific for *B. fragilis*. Fluorescence microscopy and cryo-electron tomography are two amazing tools that enable us to explore the world of the unseen.

Chapter 2 begins with light microscopy, which is an essential tool for every student and professional in the field or clinic. We then explore exciting advanced tools for research. Fluorescence and super-resolution imaging, electron microscopy, and scanning probe microscopy push ever farther the frontiers of our ability to observe the microbial world. And we continue to invent new kinds of microscopy, such as chemical imaging microscopy, to reveal microbial metabolism at work.

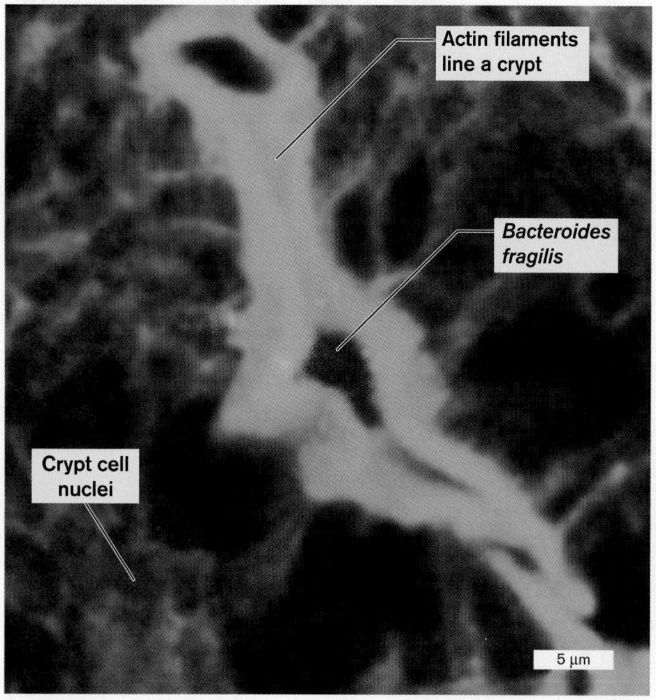

MELANIE LEE, ET AL. 2013. *NATURE* **501**:426.

FIGURE 2.1 ■ Confocal fluorescence microscopy of a mouse intestinal crypt. The crypt pocket of mouse tissue shelters symbiotic *Bacteroides fragilis* bacteria, whose molecular signals direct the development of the host immune system. Different fluorophores mark the bacteria (red), host actin filaments (green), and host nuclei (blue).

2.1 Observing Microbes

Most microbial cells are too small to be seen; by definition, they are microscopic, requiring the use of a microscope to be seen. But why can't we see microbes without magnification? The answer is surprisingly complex. In fact, our definition of "microscopic" is based on the properties of our eyes. We define what is visible and what is microscopic in terms of the human eye.

Resolution of Objects by Our Eyes

What determines the smallest object we can see? The size at which objects become visible depends on the eye's ability to resolve detail. **Resolution** is the smallest distance between two objects that allows us to see them as separate objects. The eyes of humans and other animals observe an object by focusing its image on a retina packed with light-absorbing photoreceptor cells (**Fig. 2.2** ▶). The image appears sharp, in **focus**, if the eye's lens and cornea bend all the light rays from each point of the object to converge at one point on the retina. Nearby points are then resolved as separate.

In the human eye, the finest resolution of two separate points is perceived by the fovea, the portion of the retina where the photoreceptors are packed at the highest density. The foveal photoreceptors are cone cells, which detect primary colors (red, green, or blue) and finely resolved detail. A group of cones with their linked neurons forms one unit of detection, comparable to a pixel on a computer screen. The distance between two foveal "pixels" (groups of cones with neurons) limits our resolution to 100–200 micrometers (µm)—that is, one- or two-tenths of a millimeter. So, a tenth of a millimeter is about the smallest object that most of us can see (resolve distinctly) without a magnifier.

What if our eyes were formed differently? The retinas of eagles have cones packed more closely than ours, so an eagle can resolve objects eight times as small (or eight times as far away) as a human can; hence, the phrase "eagle-eyed" means "sharp-sighted." On the other hand, insect compound eyes have photoreceptors farther apart than ours, so insect eyes have poorer resolution. The best they can do is resolve objects 100-fold larger than those we can resolve. If a science-fictional giant ameba had eyes with photoreceptors 2 meters apart, it would perceive humans as "microscopic."

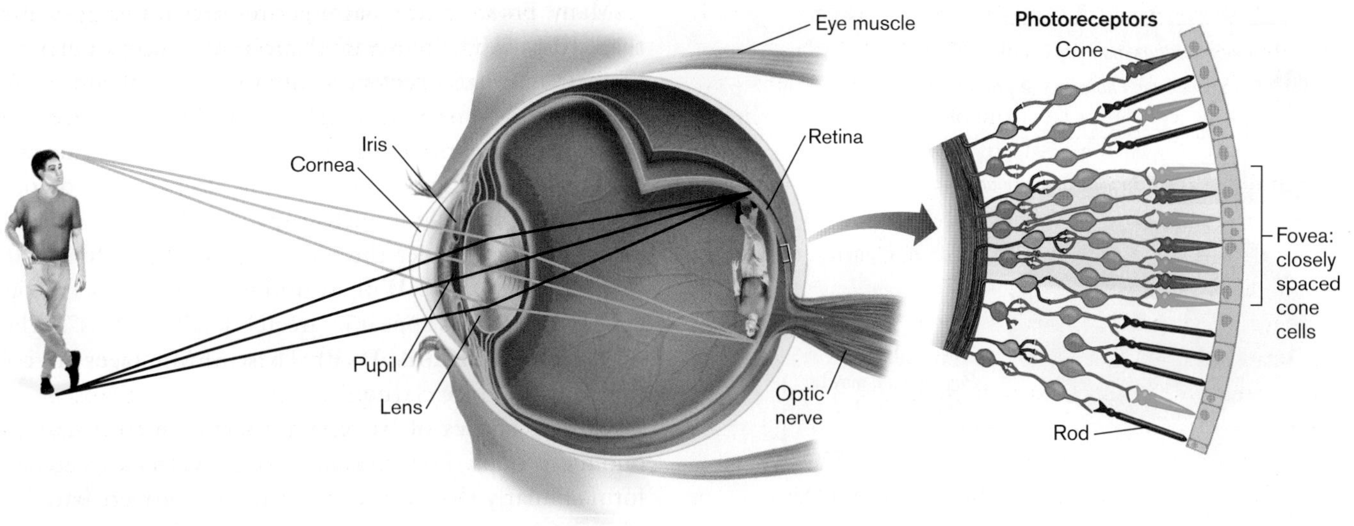

FIGURE 2.2 ▪ Defining the microscopic. Within the human eye, the lens focuses an image on the retina.

> **Thought Question**
>
> **2.1** As shown in **Figure 2.2**, the image passing through your cornea and lens is inverted on your retina. Why, then, does the world appear right side up?

A. B.

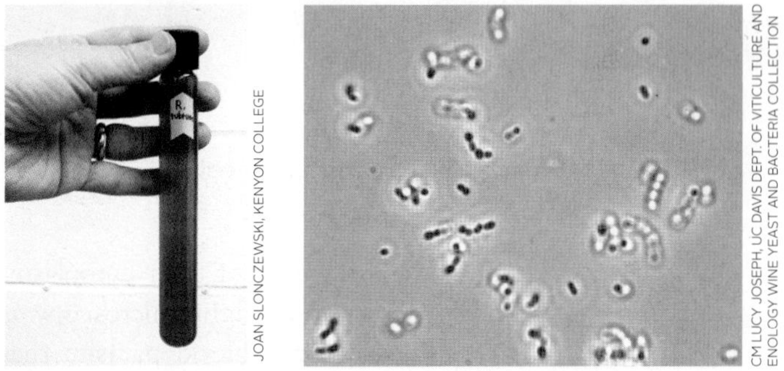

FIGURE 2.3 ▪ Detecting and resolving bacteria. **A.** A tube of bacterial culture, *Rhodospirillum rubrum*. The presence of bacteria is detected, though individual cells are not resolved. **B.** Individual cells of *Oenococcus oeni* are resolved by light microscopy.

Note: In this book, we use standard metric units for size:

1 millimeter (mm)	= one-thousandth of a meter (m) = $\mathbf{10^{-3}}$ **m**
1 micrometer (μm)	= one-thousandth of a millimeter = $\mathbf{10^{-6}}$ **m**
1 nanometer (nm)	= one-thousandth of a micrometer = $\mathbf{10^{-9}}$ **m**
1 picometer (pm)	= one-thousandth of a nanometer = $\mathbf{10^{-12}}$ **m**

Some authors still use the traditional unit angstrom (Å), which equals a tenth of a nanometer, or 10^{-10} meter.

Resolution Differs from Detection

Can we detect the presence of objects whose size we cannot resolve? Yes, we can detect their presence as a group. For example, our eyes can detect a large population of microbes, such as a spot of mold on a piece of bread (about a million cells) or a cloudy tube of bacteria in liquid culture (a hundred million cells per milliliter; **Fig. 2.3A**). **Detection**, the ability to determine the presence of an object, differs from resolution. When the unaided eye detects the presence of mold or bacteria, it cannot resolve distinct cells.

To resolve most kinds of microbial cells, our eyes need assistance—that is, **magnification**. Magnification reveals the shapes of individual bacteria such as the grape-fermenting bacterium *Oenococcus oeni* (**Fig. 2.3B**). Magnifying an object means increasing the object's apparent dimensions. As the distance increases between points of detail, our eyes can now resolve the object's shape as a magnified image.

Microbial Size and Shape

Different kinds of microbes differ in size, over a range of several orders of magnitude, or powers of ten (**Fig. 2.4**). Eukaryotic microbes are found across the full range of cell size, from photosynthetic picoeukaryotes abundant in the oceans (0.2–2.0 μm) to giant amebas that reach nearly a centimeter and marine xenophyophores that may reach 20 cm (discussed in Chapter 20).

Within a eukaryotic cell, a student's light microscope may resolve intracellular compartments such as the nucleus and vacuoles containing digested food (**Fig. 2.5**). Protists show complex shapes and appendages. For example, an ameba from a freshwater ecosystem shows a large nucleus and pseudopods to engulf prey (**Fig. 2.5A**). Pseudopods

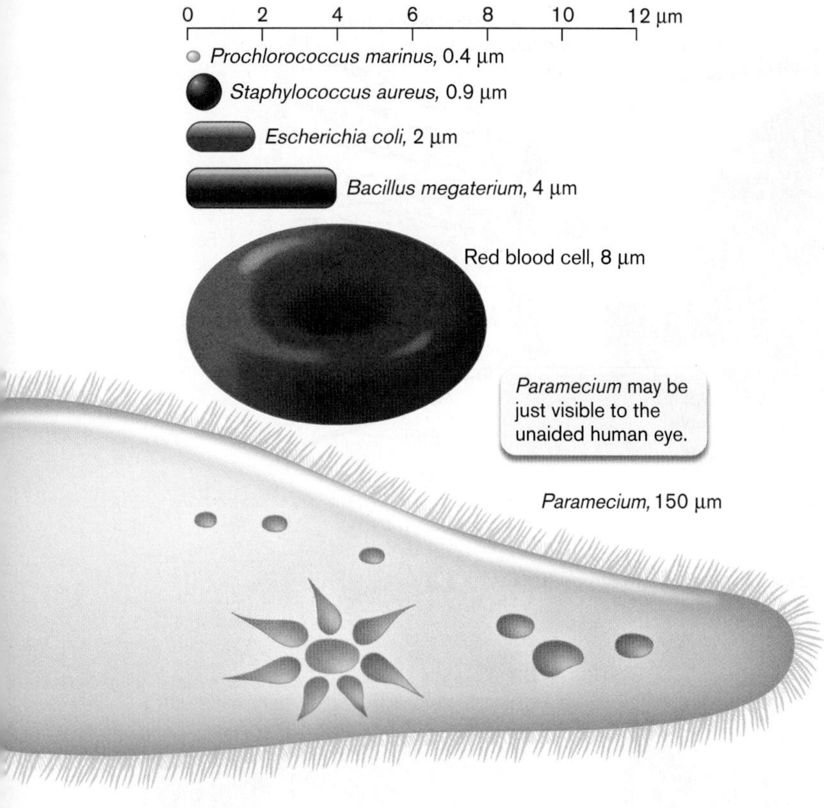

FIGURE 2.4 ■ Relative sizes of different cells.

can be seen moving by the streaming of their cytoplasm. Another protist readily observed by light microscopy is *Trypanosoma brucei*, an insect-borne blood parasite that causes African sleeping sickness (**Fig. 2.5B**). In the trypanosome, we observe a nucleus and a flagellum. Eukaryotic flagella propel the cell by a whiplike action. For more on microbial eukaryotes, see Chapter 20.

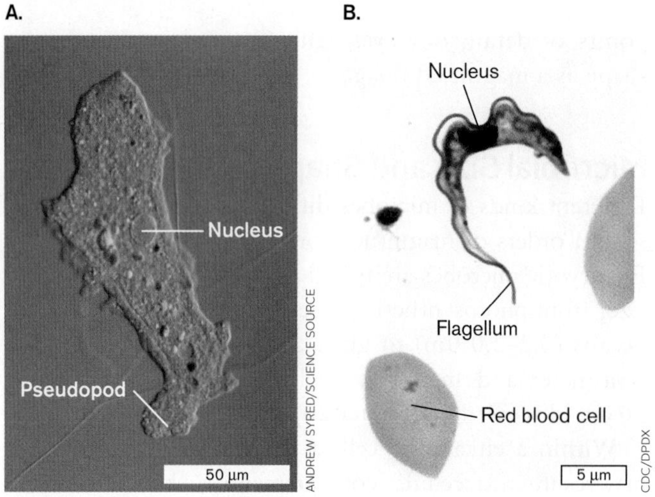

FIGURE 2.5 ■ Eukaryotic microbial cells. Eukaryotic microbes are large enough that details of internal and external organelles can be seen under a light microscope. **A.** *Amoeba proteus.* **B.** *Trypanosoma brucei* (cause of sleeping sickness) among blood cells.

Many prokaryotes (bacteria and archaea) are smaller than 10 µm. Their overall shape can be seen, but most of their internal structures (discussed in Chapter 3) are too small to resolve by light microscopy. **Figure 2.6** shows some common cell shapes of bacteria, as visualized by light microscopy or by scanning electron microscopy. With bright-field light microscopy (LM), the cell shape is just discernible under the highest power, usually 1,000× (**Fig. 2.6A, C, E**). With scanning electron microscopy (SEM), cell shapes appear in greater detail—that is, higher resolution (**Fig. 2.6B, D, F**). These SEM images are colorized to enhance clarity.

Certain shapes of bacteria are common to many taxonomic groups. For example, both bacteria and archaea form similarly shaped rods, or **bacilli** (singular, **bacillus**; **Fig. 2.6A** and **B**), and **cocci** (spheres; singular, **coccus**; **Fig. 2.6E** and **F**). Thus, rods and spherical shapes evolved independently within different taxa. In contrast, an example of a unique bacterial shape that evolved in only one taxon is the **spirochete**, a tightly coiled spiral (**Fig. 2.6C** and **D**). Species of spirochetes cause diseases such as syphilis and Lyme borreliosis. The spiral form of the spirochete cell is maintained by internal axial filaments and flagella, as well as an outer sheath. (For more on spirochetes, see Section 18.5.) A different, unrelated spiral form is the "spirillum" (plural, spirilla), a wide, rigid spiral cell that is similar to a rod-shaped bacillus.

Note: The genus name *Bacillus* refers to a specific taxonomic group of bacteria, but the term "bacillus" (plural, bacilli) refers to any rod-shaped bacterium or archaeon.

Microscopy at Different Size Scales

To resolve microbes and microbial structures of different sizes requires different kinds of microscopes. **Figure 2.7** shows the different techniques used to resolve microbes and structures of various sizes. For example, a paramecium can be resolved under a light microscope, but an individual ribosome (20 nm in diameter) requires electron microscopy.

- **Light microscopy (LM)** resolves images of individual bacteria by their absorption of light. The specimen is commonly viewed as a dark object against a light-filled field, or background; this is called **bright-field microscopy** (seen in **Fig. 2.7A** and **B**). Advanced techniques, based on special properties of light, include fluorescence, dark-field, and phase-contrast microscopy.
- **Electron microscopy (EM)** uses beams of electrons to resolve details several orders of magnitude smaller than those seen under light microscopy. In **scanning electron microscopy (SEM)**, the electron beam is

A. Filamentous rods (bacilli).
Lactobacillus lactis, Gram-positive bacteria (LM).

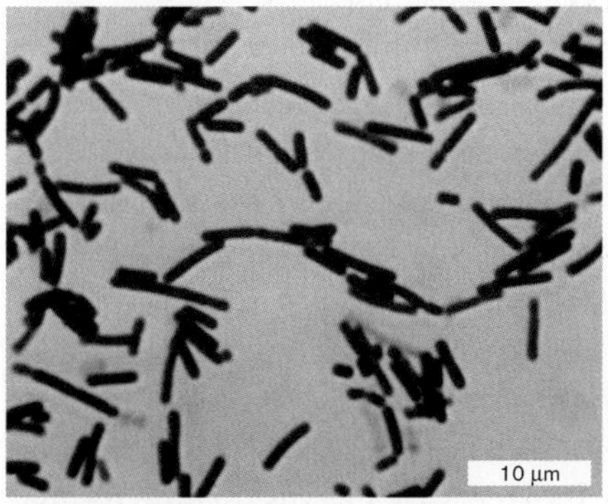

DR. A. M. SIEGELMAN/VISUALS UNLIMITED, INC.

B. Rods (bacilli).
Lactobacillus acidophilus, Gram-positive bacteria (SEM).

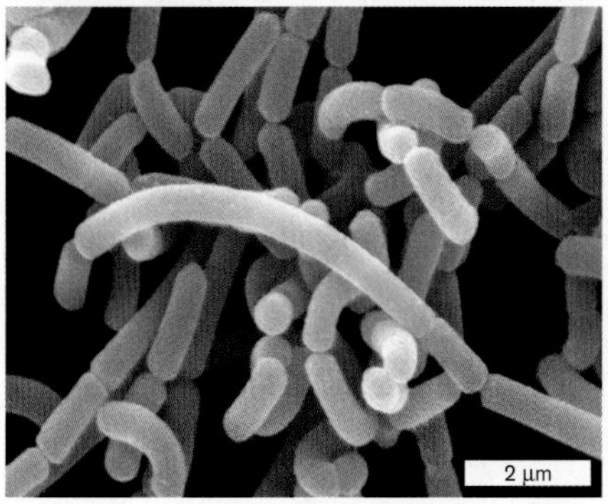

DENNIS KUNKEL MICROSCOPY/SCIENCE SOURCE

C. Spirochetes.
Borrelia burgdorferi, cause of Lyme disease, among human blood cells (LM).

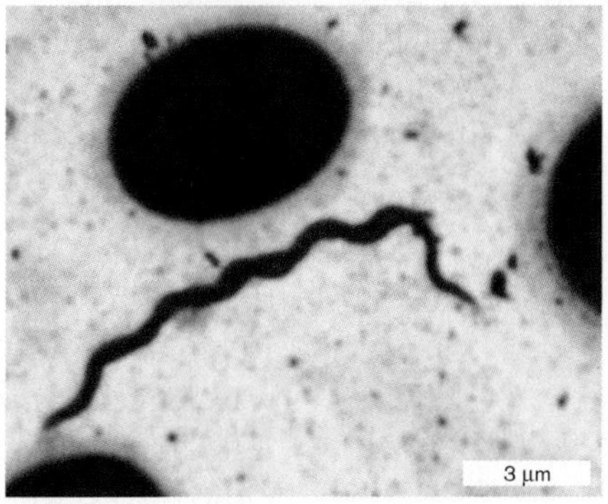

MICHAEL ABBEY/VISUALS UNLIMITED

D. Spirochetes.
Leptospira interrogans, cause of leptospirosis in animals and humans (SEM).

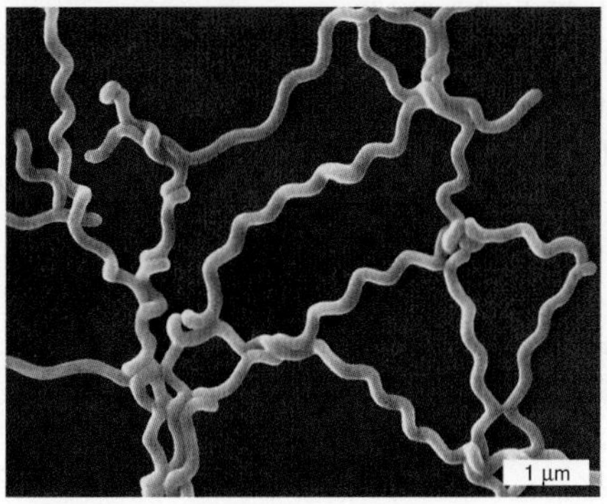

DENNIS KUNKEL MICROSCOPY/SCIENCE SOURCE

E. Cocci in chains.
Streptococcus pneumoniae, a cause of pneumonia. Methylene blue stain (LM).

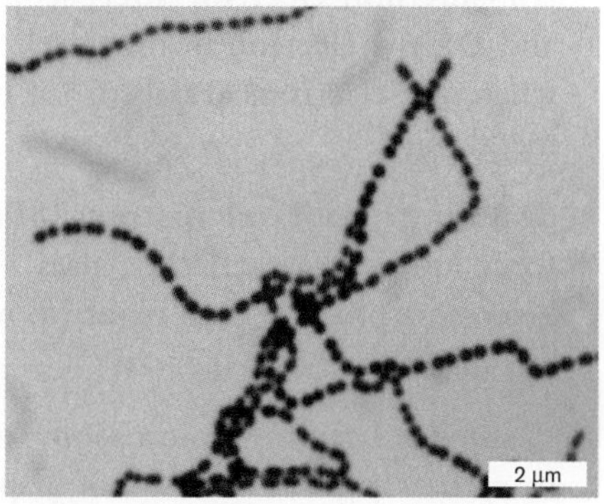

DR. EDWARD CHAN/VISUALS UNLIMITED, INC.

F. Cocci in chains.
Streptococcus salivarius, found in the human mouth. Gram stain (positive). (LM).

DENNIS KUNKEL MICROSCOPY/SCIENCE SOURCE

FIGURE 2.6 ■ Common shapes of bacteria. A, C, E. The shapes of most bacterial cells can be discerned with light microscopy (LM), but their subcellular structures and surface details cannot be seen. **B, D, F.** Surface detail is revealed by scanning electron microscopy (SEM). These SEM images are colorized to enhance clarity.

A. **B.** **C.** **D.** **E.**

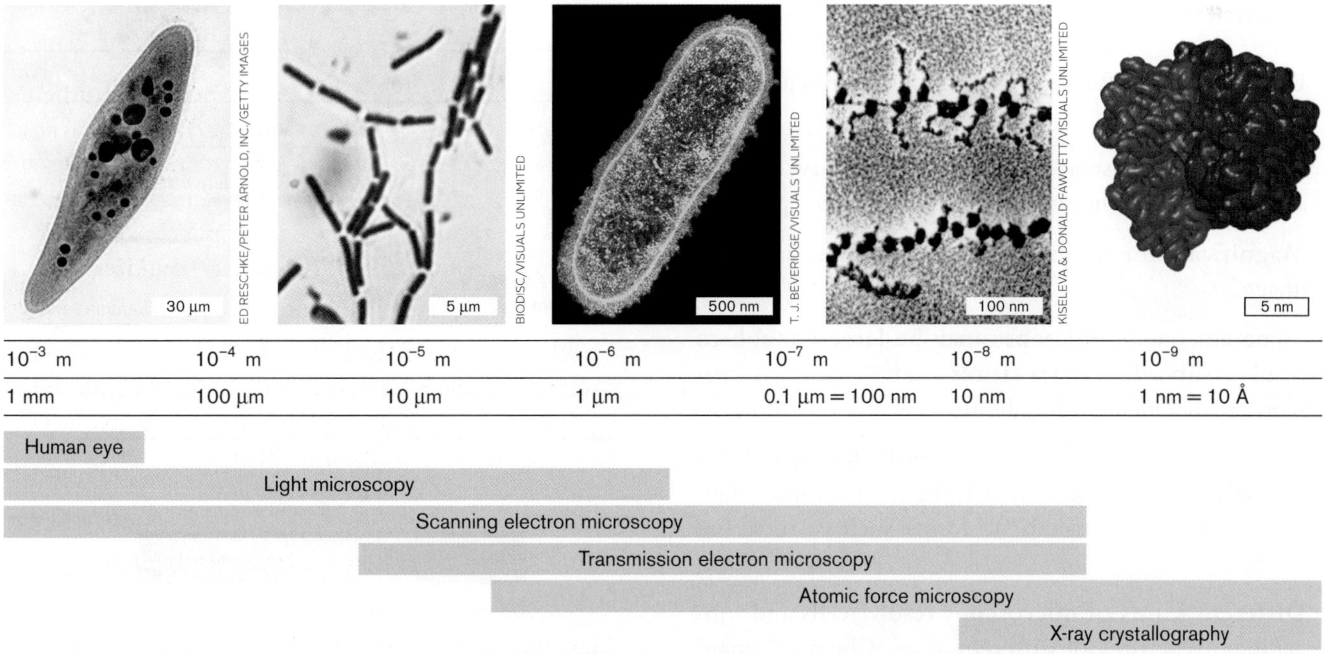

FIGURE 2.7 ■ Microscopy and X-ray crystallography, range of resolution. **A.** *Paramecium* (stained light microscopy, LM). **B.** *Bacillus* sp. (stained LM). **C.** *Escherichia coli* (transmission electron microscopy, TEM). **D.** Ribosomes on messenger RNA (TEM). **E.** Ribosome model (X-ray crystallography).

scattered from the metal-coated surface of an object, generating an appearance of 3D depth. In **transmission electron microscopy** (**TEM**; **Fig. 2.7C** and **D**), the electron beam travels through the object, where the electrons are absorbed by an electron-dense metal stain.

- **Scanning probe microscopy** (**SPM**), such as **atomic force microscopy** (**AFM**), uses intermolecular forces between a probe and an object to map the 3D topography of a cell, or of cell parts.
- **Chemical imaging microscopy** uses spectroscopy to map the chemical contents of a specimen, such as the distribution of nitrogen and carbon compounds.
- **X-ray crystallography** (also called **X-ray diffraction analysis**) detects the interference pattern of X-rays entering the crystal lattice of a molecule. From the interference pattern, researchers build a computational model of the structure of the individual molecule, such as a protein or a nucleic acid, or even a molecular complex such as a ribosome (**Fig. 2.7E**).

Thought Question

2.2 (refer to **Fig. 2.7**) You have discovered a new kind of microbe, never observed before. What kinds of questions about this microbe might be answered by light microscopy? What questions would be better addressed by electron microscopy?

To Summarize

- **Detection** is the ability to determine the presence of an object.
- **Resolution** is the smallest distance by which two objects can be separated and still be distinguished as separate.
- **Magnification** is an increase in the apparent size of an image.
- **Some eukaryotic microbes may be large enough to resolve subcellular structures** under a light microscope. Other eukaryotic cells are as small as bacteria.
- **Bacteria and archaea are generally too small for subcellular resolution by a light microscope.** Their shapes include characteristic forms such as rods and cocci.
- **Different kinds of microscopy** resolve cells and subcellular structures of different sizes. Chemical imaging microscopy reveals the chemical composition of a cell.

2.2 Optics and Properties of Light

How do light rays magnify an image? Light microscopy directly extends the lens system of our own eyes. Light is part of the spectrum of **electromagnetic radiation** (**Fig. 2.8**), a form of energy propagated as waves that are associated with electrical and magnetic fields. Regions of the electromagnetic spectrum are defined by wavelength, which for visible light is about 400–750 nm. Radiation of longer wavelengths includes infrared and radio waves, whereas shorter wavelengths include ultraviolet rays and X-rays.

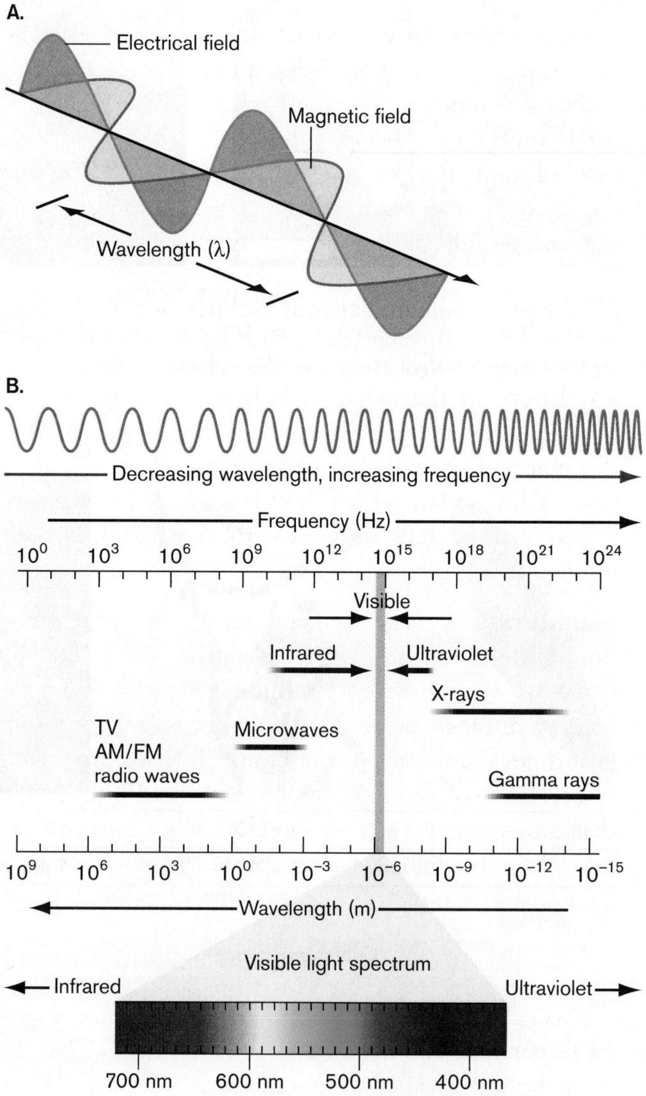

FIGURE 2.8 ■ Electromagnetic energy. **A.** Electromagnetic radiation is composed of electrical and magnetic waves perpendicular to each other. **B.** The electromagnetic spectrum includes the visible range of light.

Light Carries Information

All forms of electromagnetic radiation carry information from the objects they interact with. The information carried by radiation can be used to detect objects; for example, radar (using radio waves) detects a speeding car. All electromagnetic radiation travels through a vacuum at the same speed: about 3×10^8 meters per second (m/s), the speed of light. The speed of light (c) is equal to the wavelength (λ) of the radiation multiplied by its frequency (ν), the number of wave cycles per unit time:

$$c = \lambda\nu$$

Because c is constant, the longer the wavelength λ is, the lower the frequency ν is. Frequency is usually measured in hertz (Hz), reciprocal seconds (1/s).

The wavelength λ limits the size of objects that can be resolved as separate from neighboring objects. Resolution requires:

- **Contrast between the object and its surroundings. Contrast** is the difference in light and dark. If an object and its surroundings absorb or reflect radiation equally, then the object will be undetectable. It is hard to observe a cell of transparent cytoplasm floating in water, because the aqueous cytoplasm and the extracellular water tend to transmit light similarly, producing little contrast.
- **Wavelength smaller than the object.** For an object to be resolved, the wavelength of the radiation must be equal to or smaller than the size of the object. If the wavelength of the radiation is larger than the object, then most of the wave's energy will simply pass through the object, like an ocean wave passing around a dock post. Thus, radar, with a wavelength of 1–100 centimeters (cm), cannot resolve microbes, though it easily resolves cars and people.
- **Magnification.** The human retina absorbs radiation within a range of wavelengths, 400–750 nm (0.40–0.75 μm), which we define as visible light. But the smallest distance our retina can resolve is 150 μm, about 300 times the wavelength of light. Thus, we are unable to access all of the information contained in the light that enters our eyes. To use more of the information carried within the light, we must spread the light rays apart far enough for our retina to perceive the resolved image.

Light Interacts with an Object

The physical behavior of light resembles in some ways a beam of particles and in other ways a waveform. The particles of light are called "photons." Each photon has an associated wavelength that determines how the photon will interact with a given object. The combined properties of particle and wave enable light to interact with an object in several different ways:

- **Absorption** means that the absorbing object gains the photon's energy (**Fig. 2.9A** ▶). The energy is converted to a different form, usually heat. (That is why a live specimen eventually "cooks" on the slide if observed for too long.) When a microbial specimen absorbs light, it can be observed as a dark spot against a bright field, as in bright-field microscopy. Some molecules that absorb light of a specific wavelength reemit energy as light with a longer wavelength; this is called **fluorescence**. Fluorescence microscopy is discussed in Section 2.4.
- **Reflection** means that the wavefront redirects from the surface of an object at an angle equal to its incident angle (**Fig. 2.9B**). The reflection of light waves is analogous to the reflection of water waves. Reflection from a silvered mirror or a glass surface is used in the optics of microscopy.
- **Refraction** means that light bends as it enters a substance that slows its speed (**Fig. 2.9C**). Such a substance is said to be refractive and, by definition, has a higher **refractive index** than air has. Refraction is the key property that enables a lens to magnify an image.
- **Scattering** means that a portion of the wavefront is converted to a spherical wave originating from the object (**Fig. 2.9D**). If a large number of particles simultaneously scatter light, we see a haze—for example, the haze of bacteria suspended in a culture tube. Special optical arrangements (such as dark-field microscopy, discussed

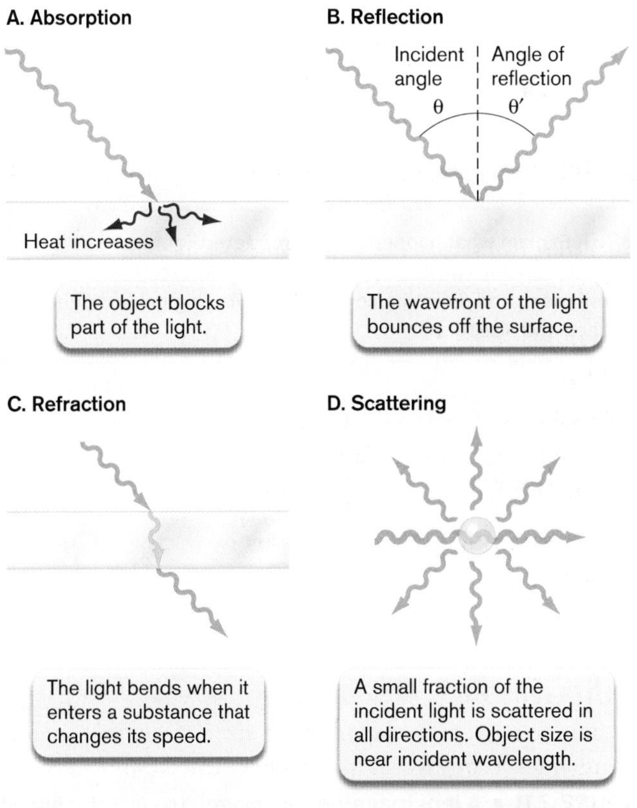

FIGURE 2.9 ■ Interaction of light with matter. ▶

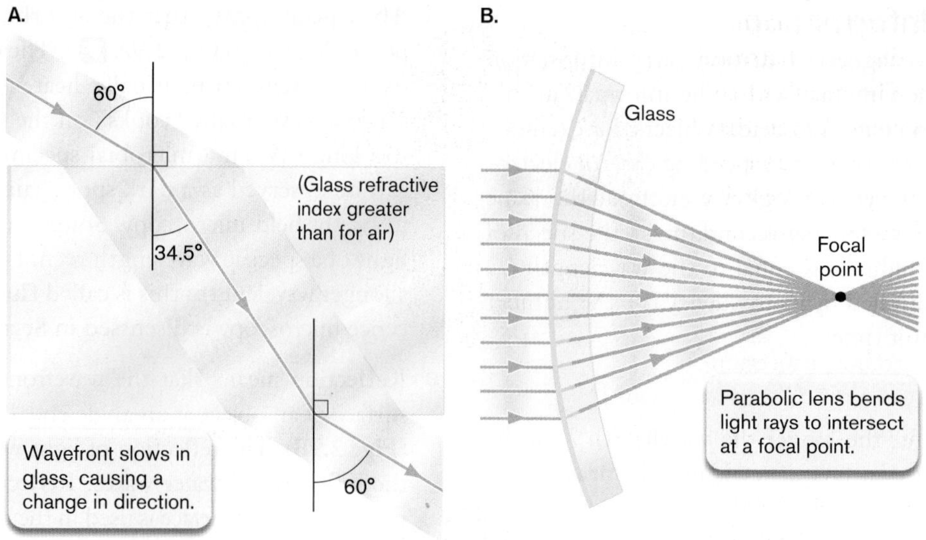

FIGURE 2.10 ▪ Refraction of light waves. **A.** Wavefronts of light shift direction as they enter a substance of higher refractive index, such as glass. **B.** Glass with parabolic curvature (a lens) bends light rays to intersect at a focal point.

in Section 2.5) can use scattered light to detect (but not resolve) microbial shapes smaller than the wavelength of light.

Magnification by a Lens

Magnification requires the bending of light rays, as in refraction. As a wavefront of light enters a refractive material, such as glass, the region of the wave that first reaches the material is slowed, while the rest of the wave continues at its original speed until it also passes into the refractive material (**Fig. 2.10A**). As the entire wavefront passes through the refractive material, its path continues, bent at an angle from its original direction.

Thought Question

2.3 Explain what happens to the refracted light wave as it emerges from a piece of glass of even thickness. How do its new speed and direction compare with its original (incident) speed and direction?

How does refraction accomplish magnification? Refraction magnifies an image when light passes through a refractive material shaped so as to spread its rays. One shape that spreads light rays is a parabolic curve. When light rays enter a **lens** of refractive material with a parabolic surface (**Fig. 2.10B**), parallel rays each bend at an angle such that all of the rays meet at a certain point, called the **focal point**. From the focal point behind the lens, the light rays continue, spreading out with an expanding wavefront. This expansion magnifies the image carried by the wave. The distance from the lens to the focal point (called the focal distance) is determined by the degree of curvature of the lens, and by the refractive index of its material.

In **Figure 2.11**, the object under observation is placed near the focal point (*F*) in front of a lens. The light rays trace a path opposite to that of **Figure 2.10B**. The rays expanding from point *F* are bent by the lens into a nearly parallel path entering the eye. The eye perceives the expanded light rays as a virtual image—that is, an image that appears to represent a much larger object farther away.

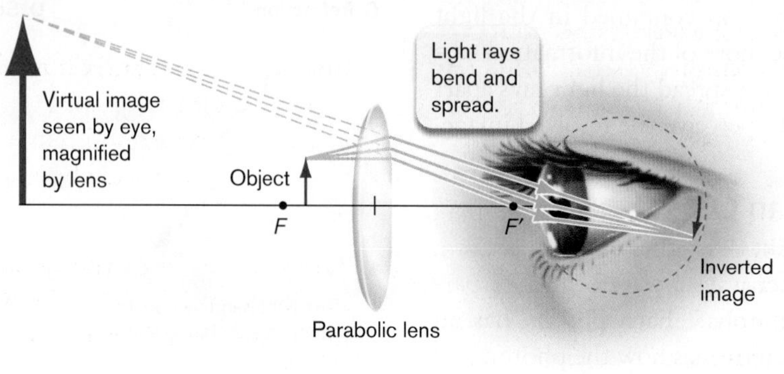

FIGURE 2.11 ▪ A lens magnifies an image. The object is placed near the focal point (*F*) in front of the lens. The lens bends and spreads the light rays, inverting the magnified image.

The expansion of light rays, or magnification, increases the distances between points of the image. The details of the magnified image become larger than the spacing of photoreceptor units in the retina. Thus our eye can perceive details that the unaided eye cannot see.

Resolution of Detail

What limits the effect of magnification? The spreading of light rays does not in itself increase resolution. For example, an image composed of dots does not gain detail when enlarged on a photocopier, nor does an image composed of pixels gain detail when enlarged on a computer screen. In these cases, magnification fails to show details because the individual details of the image expand in proportion to the expansion of the overall image. Magnification without increasing detail is called **empty magnification**.

The resolution of detail in microscopy is limited by the wave nature of light. In theory, a perfect lens that focuses all of the light from an object should form a perfect image as its rays converge through the focal point. But light rays actually form wavefronts of infinite extent. Since the width of the lens is finite, only part of the wavefront enters, causing **interference**. The converging edges of the wave interfere with each other to form alternating regions of light and dark (**Fig. 2.12** ▶). Thus, a point source of light (such as a point of detail in a specimen) forms an image of a bright central peak surrounded by interference rings of light and dark. Even a well-focused bright object appears as a bright disk surrounded by faint rings.

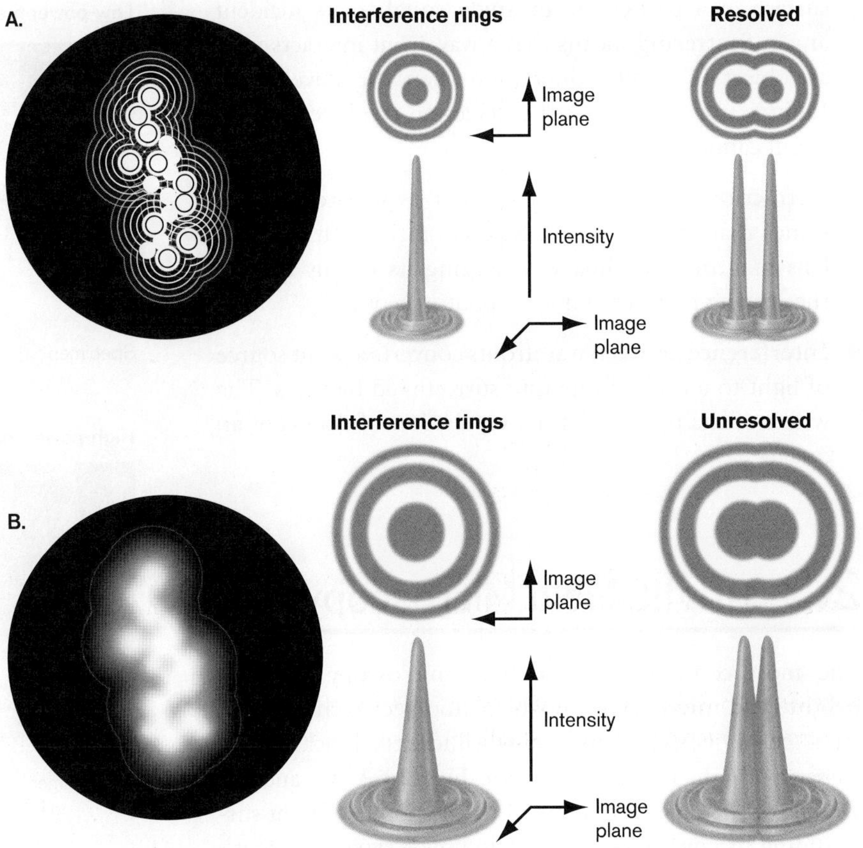

FIGURE 2.12 ▪ Interference of light waves at the focal point generates concentric rings surrounding the peak intensity. **A.** Broad wavefronts generate narrow interference rings with peaks well resolved. **B.** Narrow wavefronts generate wide interference rings that are unresolved. ▶

Suppose an object consists of a collection of point sources of light. Each point source generates a central peak of intensity. The width of this central peak will define the resolution, or separation distance, between any two points of the object (**Fig. 2.12**). This resolution determines the degree of detail that can be observed. In practice, any object, such as a stained microbe against a bright field, can be considered a large collection of points of light that act as partly resolved peaks of intensity.

What factors limit resolution of an image? The wavelength of light limits the sharpness of the peak intensity of a point of detail. The finite width of the wavefront captured by the lens leads to interference and widens the peak intensity. Thus, bright-field light microscopy resolves only details that are greater than half the wavelength of light, about 200 nm (0.2 μm). Nevertheless, in advanced optical methods such as fluorescence microscopy (see Section 2.4), computation can extract positional detail from light rays. Such methods, called **super-resolution imaging**, enable us to track cellular molecules at a precision of 20–40 nm (discussed in Section 2.4).

To Summarize

- **Electromagnetic radiation** interacts with an object and acquires information we can use to detect the object. **Contrast** between object and background makes it possible to detect the object and resolve its component parts.
- **The wavelength of the radiation** must be equal to or smaller than the size of the object for a microscope to resolve the object's shape.
- **Absorption** means that the energy from light (or other electromagnetic radiation) is acquired by the object. **Reflection** means that the wavefront bounces off the

surface of a particle at an angle equal to its incident angle. **Scattering** means that a wavefront interacts with an object of smaller dimension than the wavelength. Light scattering enables the detection of objects whose detail cannot be resolved.

- **Refraction** is the bending of light as it enters a substance that slows its speed. Refraction through a curved lens **magnifies** an image, enlarging its details beyond the spacing between our eye's photoreceptors.
- **Interference** between wavefronts converts a point source of light to a peak of intensity surrounded by rings. The width of the peak limits the **resolution** of details of an image.

2.3 Bright-Field Microscopy

The most common kind of light microscopy is called **bright-field microscopy**, in which an object such as a bacterial cell is perceived as a dark silhouette blocking the passage of light (for examples, see **Fig. 2.6A**, **C**, and **E**). Details of the object are defined by the points of light surrounding its edge. Here we explain how a typical student's microscope works, and how to use it to image microbes.

Magnification

How do the optics of a bright-field microscope maximize the observation of detail? We consider the following factors:

- **Wavelength and resolution.** Our eyes can resolve a distance as small as 100–200 μm, while the resolution limit from the wavelength of light is 200 nm (0.2 μm)—that is, 500- to 1,000-fold smaller. Thus, the greatest magnification that can improve our perception of detail is about 1,000×. Any greater magnification expands the image size, but the peaks expand without resolution between them (see **Fig. 2.12**). As we noted in the previous section, this expansion without increasing resolution is called empty magnification.
- **Light and contrast.** For any given lens system, a balanced amount of light yields the highest contrast between the dark specimen and the light background. High contrast is needed to perceive the full resolution at a given magnification.
- **Lens quality.** All lenses contain inherent **aberrations** that detract from perfect curvature. Optical properties limit the perfection of a single lens, but manufacturers construct microscopes with a series of lenses that multiply each other's magnification and correct for aberrations.

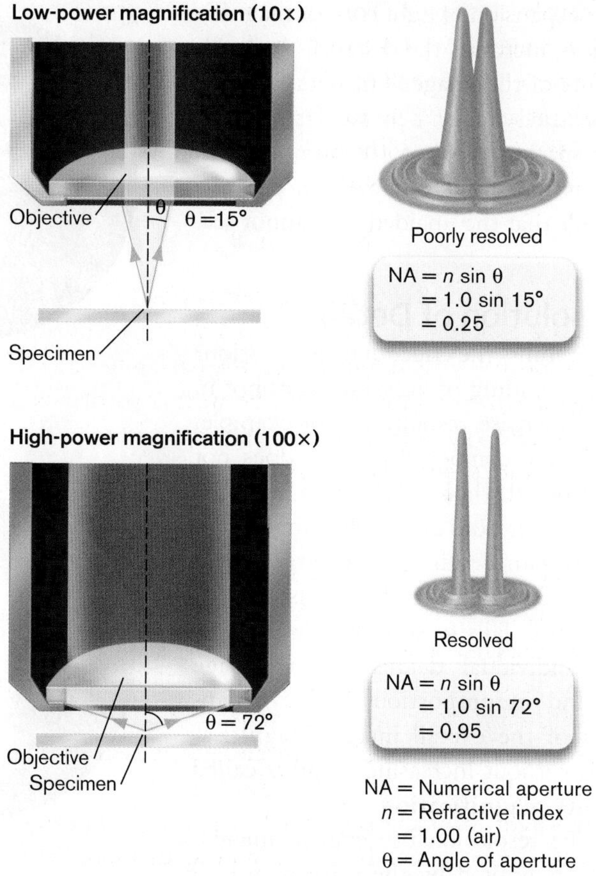

FIGURE 2.13 ▪ Numerical aperture and resolution. The numerical aperture (NA) equals the refractive index (n) of the medium (air) containing the light cone, multiplied by the sine of the angle (θ) of the light cone. Higher NA allows greater resolution.

Let's first consider magnification of an image by a single lens. **Figure 2.13** shows an **objective lens**, a lens situated directly above an object or specimen that we wish to observe at high resolution. How can we maximize the resolution of details?

An object at the focal point of a lens sits at the tip of a cone of light formed by rays from the lens converging at the object. The angle of the light cone is determined by the curvature and refractive index of the lens. The lens fills an aperture, or hole, for the passage of light; and for a given lens the light cone is defined by an angle theta (θ) projecting from the midline, known as the **angle of aperture**. As θ increases and the horizontal width of the light cone ($\sin \theta$) increases, a wider cone of light passes through the specimen. The wider the cone of light rays, the less the interference between wavefronts—and the narrower the peak intensities in the image. Thus, a wider light cone enables us to resolve smaller details. The greater the angle of aperture of the lens ($\sin \theta$), the better the resolution.

Resolution also depends on the refractive index of the medium that contains the light cone, which is usually air. The refractive index (n) is the ratio of the speed of light in a vacuum to its speed in another medium. For air, n is extremely close to 1. For water, n = 1.33; for lens material, n ranges from 1.4 to 1.6. As light passing through air or water enters a lens of higher refractive index, the light bends, at angles (θ) up to a maximum. The product of the refractive index (n) of the medium multiplied by sin θ is the **numerical aperture** (NA):

$$\mathrm{NA} = n \sin \theta$$

In **Figure 2.13** we see the calculation of NA for an objective lens of magnification 10× and for a lens of magnification 100×. As NA increases, the peak intensities of an image narrow and the distance between two objects that can be resolved decreases. The minimum resolution distance R varies inversely with NA:

$$R = \frac{\lambda}{2(\mathrm{NA})}$$

where λ represents the wavelength of incident light. Notice that this equation limits resolution to approximately half the wavelength of light ($\lambda/2$).

As the lens strength increases and the light cone widens, the lens must come nearer the object. Defects in lens curvature become more of a problem, and focusing becomes more challenging. As θ becomes very wide, too much of the light from the object is lost from refraction at the glass-to-air interface. To collect and focus more light, we need to increase the refractive index of the medium (air) between the object and the **objective lens** (the lens nearest the object or specimen). For the highest-power objective lens, generally 100×, we can replace air with **immersion oil** between the object and the lens. Immersion oil has a refractive index (n = 1.5) comparable to that of the lens (**Fig. 2.14**). Immersion oil minimizes the loss of light rays by refraction and makes it possible to reach 100× magnification with minimal distortion. The 100× objective with immersion oil is generally the most powerful lens available on a student's microscope.

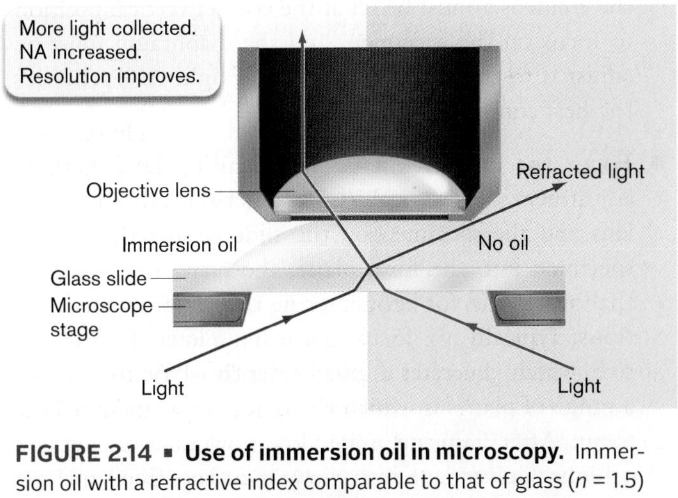

FIGURE 2.14 ■ Use of immersion oil in microscopy. Immersion oil with a refractive index comparable to that of glass (n = 1.5) prevents light rays from bending away from the objective lens.

Thought Question

2.4 (refer to **Fig. 2.13**) For a single lens, what angle θ might offer magnification even greater than 100×? What practical problem would you have in designing such a lens to generate this light cone?

The Compound Microscope

A **compound microscope** is a system of multiple lenses designed to correct or compensate for lens **aberrations** (deviations from perfect curvature). Why do we use a compound microscope instead of a single perfect lens? The manufacture of high-power lenses is difficult because as the glass curvature increases, the effects of aberration increase faster than the magnification. Instead of one thick lens, a series of lower-power lenses can multiply their magnification with minimal aberration. **Figure 2.15** shows a typical arrangement of a compound microscope: the light source is placed at the bottom, shining upward through a series of lenses, including the condenser, objective, and ocular lenses.

Between the light source and the condenser sits a diaphragm, a device to cut the diameter of the light column. Lower-power lenses require lower light levels because the excess light makes it impossible to observe the darkening effect of specimen absorbance. This difference between the dark (absorbing) specimen and the bright (transparent) field is called **contrast**. Higher-power lenses spread the light rays farther and thus require an open diaphragm to collect sufficient light for contrast.

Above the diaphragm, the **condenser** consists of one or more lenses that collect a beam of rays from the light source onto a small area of the slide, where light may be absorbed by the object or specimen. Condenser lenses increase light available for contrast but do not participate in magnification.

The **objective lens** is the first to form a magnified image (I) of the object (**Fig. 2.15A**). As the image forms, each light ray traces a path toward a position opposite its point of origin; thus, the image is mirror-reversed. Keep this mirror reversal in mind when exploring a field of cells.

The first image of the object (I) is then amplified by a secondary magnification step through the **ocular lens** within the eyepiece. The final image (I') is comparable to the virtual image of **Figure 2.11**, but I' includes the total magnification of the object by both objective and ocular lenses. The magnification factor of the ocular lens is multiplied by the magnification factor of the objective lens to generate the **total magnification** (power). Thus, a 10× ocular multiplied by a 40× objective generates 400× total magnification. A 100× objective with immersion oil is multiplied by 10× ocular magnification to yield 1,000× total magnification.

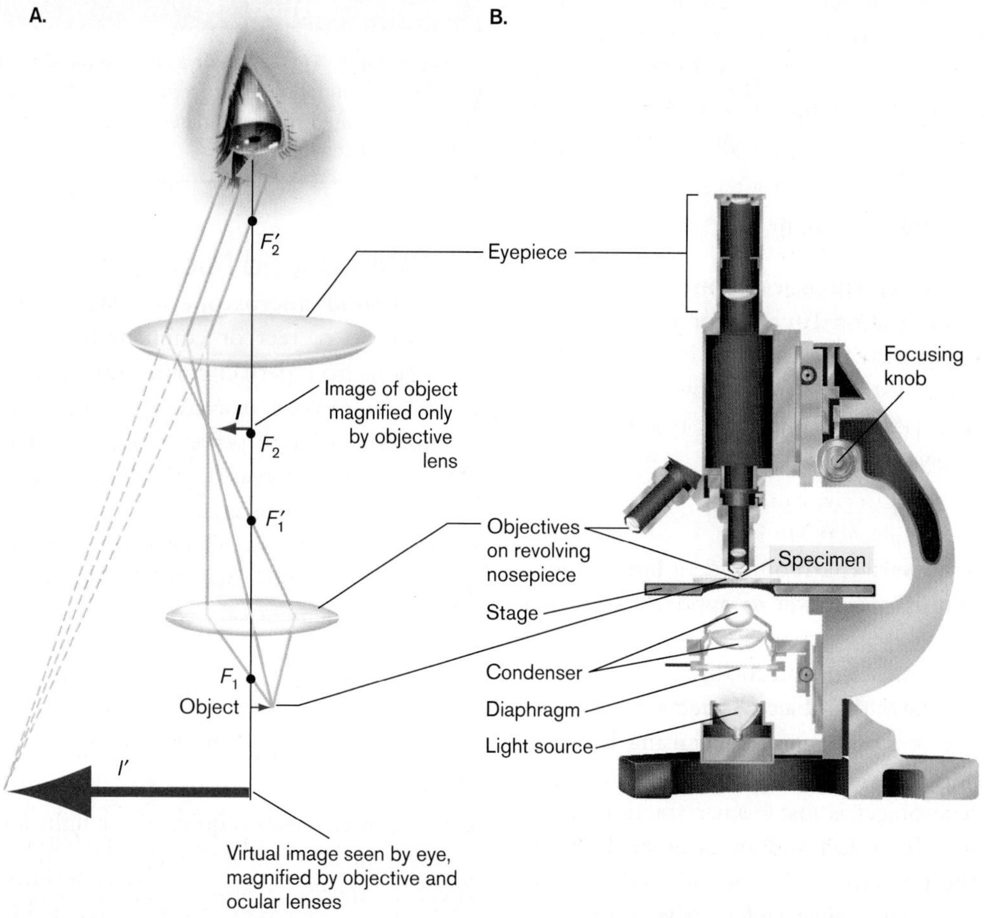

FIGURE 2.15 ▪ Anatomy of a compound microscope. **A.** Light path through the microscope. **B.** Cutaway view.

The nosepiece of a compound microscope typically holds three or four objective lenses of different magnifying power, such as 4×, 10×, 40×, and 100× (requiring immersion oil). These lenses are arranged so that they rotate in turn into the optical column. In a high-quality instrument, the lenses are set at different heights from the slide so as to be **parfocal**. In a parfocal system, when an object is focused with one lens, it remains in focus, or nearly so, when another lens is rotated to replace the first.

Note: Objective lenses can be obtained in several different grades of quality, manufactured with different kinds of correction for aberrations. Lenses should feature at minimum the following corrections: "plan" correction for field curvature, to generate a field that appears flat; and "apochromat" correction for spherical and chromatic aberrations.

Observing a specimen under a compound microscope requires several steps:

- **Position the specimen centrally in the optical column.** Only a small area of a slide can be visualized within the field of view of a given lens. The higher the magnification, the smaller the field of view that will be seen.
- **Optimize the amount of light.** At lower power, too much light will wash out the light absorption of the specimen. At higher power, more light needs to be collected, lest everything appear dark. To optimize light, the condenser must be set at the correct vertical position to focus on the specimen, and the diaphragm must be adjusted to transmit the amount of light that produces the best contrast.
- **Focus the objective lens.** The focusing knob permits adjustment of the focal distance between the objective lens and the specimen on the slide so as to bring the specimen into the focal plane, the plane that contains the focal points for light entering the lens from all directions. Typically, we focus first using a low-power objective, which generates a greater **depth of field**—that is, a range of planes in which the object appears in or near focus. After focusing under low power, we can rotate a higher-power lens into view and then fine-tune the adjustment.

Preparing a Specimen for Microscopy

A simple way to observe microbes is to place them in a drop of water on a slide with a coverslip. This is called a **wet mount** preparation. The advantage of the wet mount is that the organism is viewed in as natural a state as possible, without artifacts resulting from chemical treatment; and we can observe live behavior such as swimming (see **Figs. 2.3B** and **2.17**). The disadvantage of the wet mount is that most living cells are transparent and therefore show little contrast with the external medium. With limited contrast, the cells can barely be distinguished from background, and both detection and resolution are minimal.

Another disadvantage of the wet mount is that the sample rapidly converts absorbed light to heat, thus tending to overheat and dry out. To avoid overheating, researchers use a temperature-controlled flow cell, in which fresh medium passes through the specimen (**Fig. 2.16**). The microbe to be observed must adhere to a specially coated slide within the flow cell. The adherent cells may grow and multiply as a biofilm, nourished continually by fresh medium.

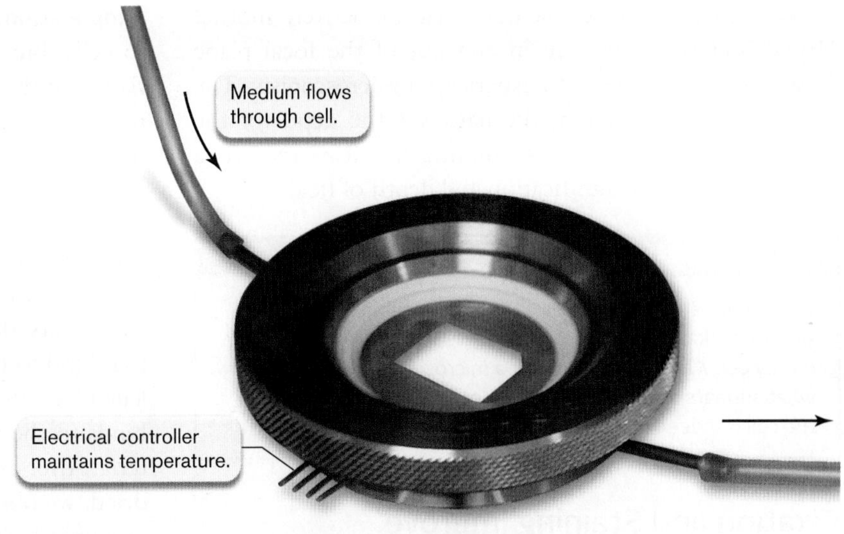

FIGURE 2.16 ■ A flow cell enables extended observation of living microbes. In a flow cell, culture medium flows through an inlet tube into the slide chamber and then exits through the outlet.

Focusing the Object

An object appears in focus (that is, it is situated within the focal plane of the lens) when its edge appears sharp and distinct from the background. The shape of the dark object is actually defined by the points of light surrounding its edge. At higher power, as we reach the resolution limit, these points of light are only partly resolved. The partial resolution of these points of light generates interference effects, such as extra rings of light surrounding an object.

In **Figure 2.17** we observe *Oenococcus oeni*, bacteria that ferment the malic acid of grapes during wine production. As chains of *O. oeni* drift in and out of the focal plane, their appearance changes through optical effects. When a bacterium drifts out of the focal plane too close to the lens, resolution declines and the image blurs (**Fig. 2.17A**). When the bacterium lies within the focal plane, its image appears sharp, with a bright line along its edge (**Fig. 2.17B**). When the chain of cells lies too far past the focal plane, the bright interference lines collapse into the object's silhouette, which now appears bright or "hollow," or surrounded by rings (**Fig. 2.17C**). In fact, the bacterium is not hollow at all; only its image has changed.

When the chain extends across several focal planes, different portions appear out of focus (either too near or too far from the lens) (**Fig. 2.17D**). In addition, when the end of a cell points toward the observer, light travels through the length of the cell before reaching the observer, so the cell absorbs more light and appears dark.

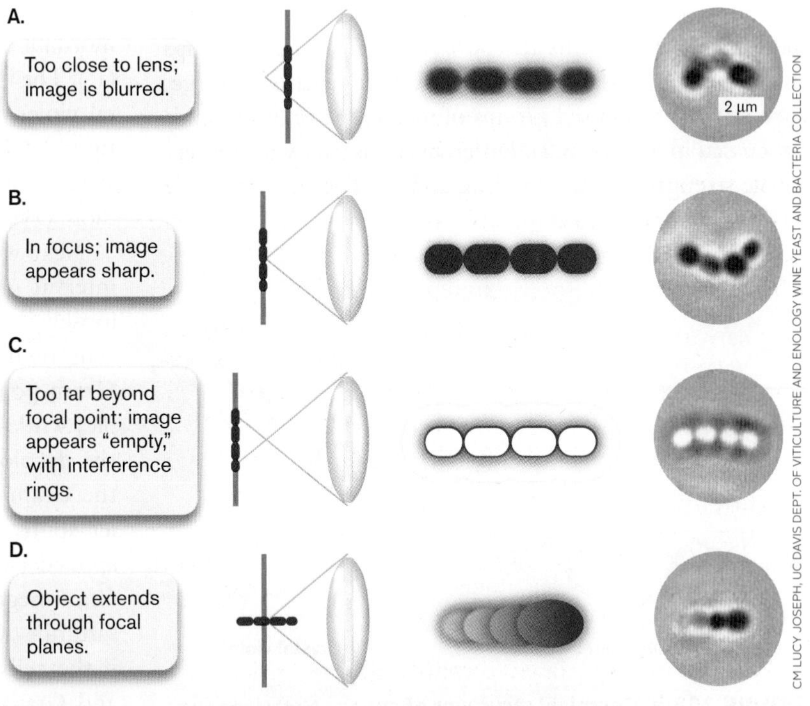

FIGURE 2.17 ■ Bacteria observed at different levels of focus.

How do we observe microbes that are actively motile? Motile bacteria swimming in and out of the focal plane present a challenge even to experienced microscopists. The higher the magnification, the narrower the depth of the focal plane; thus, observing swimming organisms requires a trade-off between magnification and depth of field.

Thought Question

2.5 Under starvation, a soil bacterium such as *Bacillus subtilis* packages its cytoplasm into a spore, leaving behind an empty cell wall. Suppose, under a microscope, you observe what appears to be a hollow cell. How can you tell whether the cell is indeed hollow or is simply out of focus?

Fixation and Staining Improve Resolution and Contrast

Detection and resolution of cells under a microscope are enhanced by **fixation** and **staining**, procedures that usually kill the cell. Fixation is a process by which cells are made to adhere to a slide in a fixed position. We can fix cells with methanol or by heat treatment to denature the cell's proteins, whose exposed side chains then adhere to the glass. A stain absorbs much of the incident light, usually over a wavelength range that results in a distinctive color. The use of chemical stains was developed in the nineteenth century, when German chemists used organic synthesis to invent new coloring agents for clothing. Clothing was made of natural fibers such as cotton or wool, so a substance that dyed clothing would be likely to react with biological specimens.

How do stains work? Most stain molecules contain conjugated double bonds or aromatic rings that absorb visible light (**Fig. 2.18**). Stains also have positively charged groups that bind cell-surface components with negative charge, such as the phosphoryl groups of membrane phospholipids (discussed in Section 3.2). Different stains vary with respect to the strength of their binding and the degree of binding to different parts of a cell.

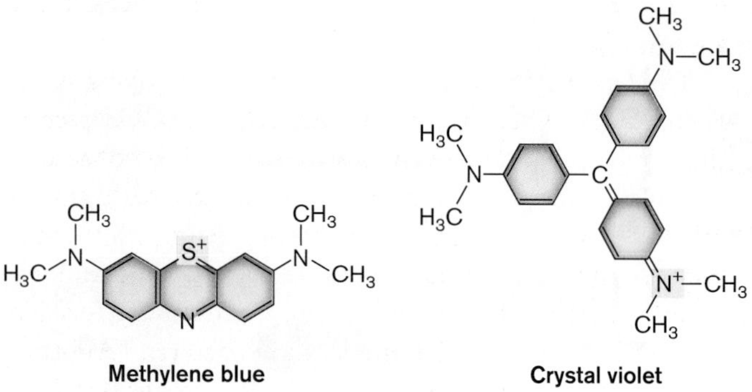

FIGURE 2.18 ■ Chemical structure of stains. Methylene blue and crystal violet are cationic (positively charged) dyes.

Simple stains. A **simple stain** adds dark color specifically to cells, but not to the external medium or surrounding tissue (in the case of pathological samples). The most commonly used simple stain is methylene blue (see **Fig. 2.6E**), originally used by Robert Koch in the nineteenth century to stain bacteria. A typical procedure for fixation and staining is shown in **Figure 2.19**. First we fix a drop of culture on a slide by treating it with methanol or by heating it on a slide warmer (steps 1–4 in the figure). Either of these treatments denatures cell proteins, exposing side chains that bind to the glass. We then flood the slide with methylene blue solution (step 5). The positively charged molecule binds to the negatively charged cell envelope of fixed bacteria. After excess stain is washed off and the slide has been dried, we observe it under high-power magnification using immersion oil (steps 6–8).

Differential stains. A **differential stain** colors one kind of cell but not another. The most famous differential stain is the **Gram stain**, devised in 1884 by the Danish physician Hans Christian Gram (1853–1938). Gram first used this stain to distinguish pneumococcal bacteria (*Streptococcus pneumoniae*) from human lung tissue. In **Figure 2.20A**, Gram-stained *S. pneumoniae* bacteria appear dark purple among unstained white blood cells. Other species of bacteria, such as *Proteus mirabilis* (a cause of urinary infections), fail to retain the purple stain (**Fig. 2.20B**). Different bacterial species are classified as Gram-positive or Gram-negative, depending on whether they retain the stain.

In the Gram stain procedure (**Fig. 2.21A**), a dye such as crystal violet binds to the bacteria; it also binds to the surface of human cells, but less strongly. After the excess stain is washed off, we apply a **mordant**, or binding agent. The mordant used is iodine solution, which contains iodide ions (I^-). The iodide complexes with the positively charged crystal violet molecules trapped inside the cells (step 3 in the figure). The crystal violet–iodide complex is now held more strongly within the cell wall. The thicker the cell wall, the more crystal violet–iodide molecules are held.

Next we add a decolorizer, ethanol, for a precise time interval (typically 20 seconds). The decolorizer removes loosely bound crystal violet–iodide, but Gram-positive cells retain the stain tightly (**Fig. 2.21A**, step 4). The **Gram-positive** cells that retain the stain appear dark purple, while the **Gram-negative** cells are colorless. Timing the decolorizer step is critical because if it lasts too long, the Gram-positive cells, too, will release their crystal violet stain. In the final step, a **counterstain**, safranin, is applied (step 5). This process allows the visualization of Gram-negative material, which is stained pale pink by the safranin. Gram-positive cells also retain safranin; thus, if the cells are decolorized too long, both Gram-positive and Gram-negative cells will appear pink because of the safranin.

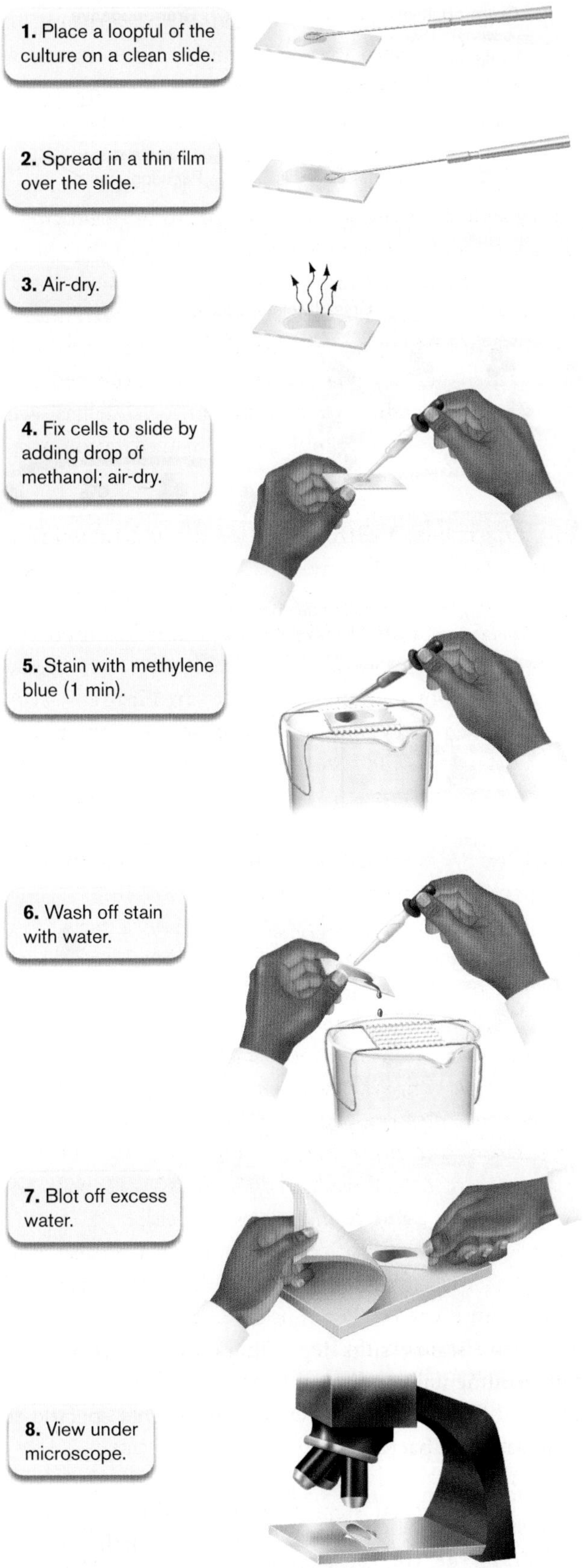

FIGURE 2.19 ■ Procedure for simple staining with methylene blue.

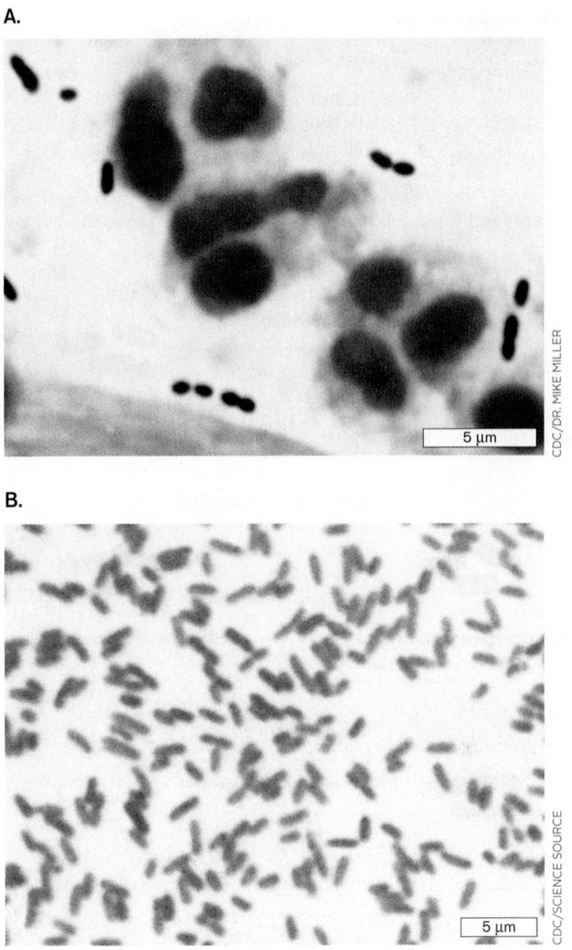

FIGURE 2.20 ■ Gram staining of bacteria. A. Gram stain of a sputum specimen from a patient with pneumonia, containing Gram-positive *Streptococcus pneumoniae* (purple diplococci) among white blood cells in pus. The white blood cell nuclei stain pink (counterstain). **B.** Gram-negative *Proteus mirabilis* (pink rods).

How does the Gram stain distinguish different cell types? Most Gram-negative species of bacteria possess a cell wall that is thinner and more porous than that of Gram-positive species (discussed in Chapter 3). A Gram-negative cell wall has only one to three layers of peptidoglycan (sugar chains cross-linked by peptides), whereas a Gram-positive cell has five or more layers (**Fig. 2.21B**). The multiple layers of peptidoglycan retain enough stain complex that the cell appears purple.

The Gram stain became a key tool for identifying pathogens in the clinical laboratory. As we'll see in Chapter 3, the Gram stain effectively distinguishes Proteobacteria (a diverse group of Gram-negative bacteria with a thin cell wall and an outer membrane) from Firmicutes (Gram-positive bacteria with a thick cell wall and no outer membrane). Proteobacteria include *Escherichia coli* and many related intestinal bacteria. Another phylum that stains Gram-negative is Bacteroidetes, which work with Proteobacteria to digest our food (discussed in Chapters 13 and 21).

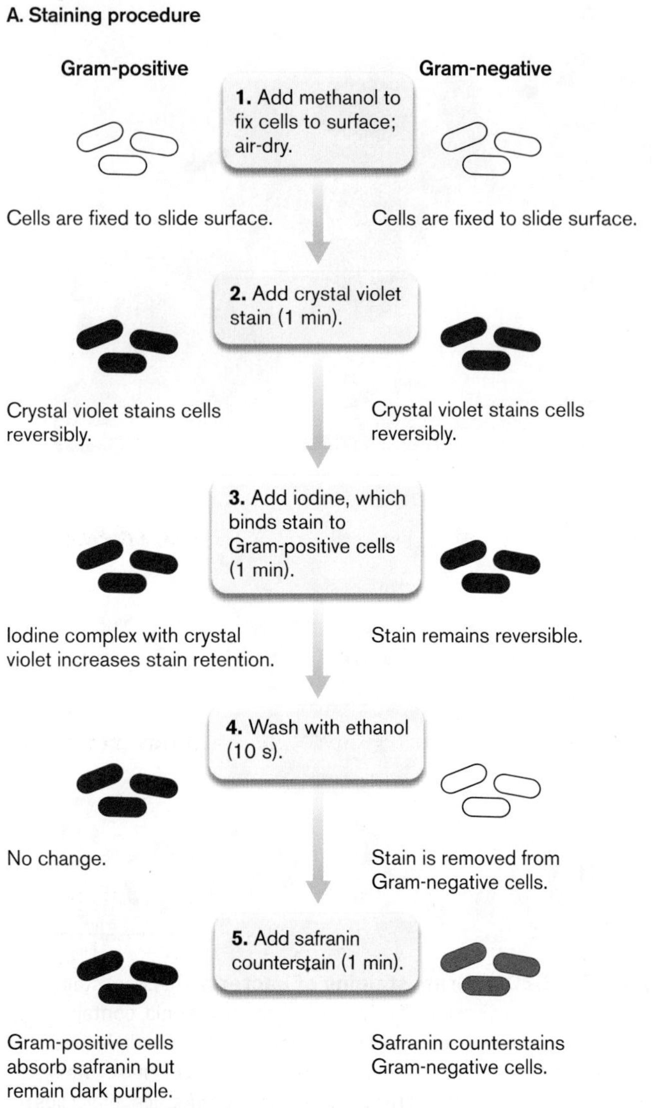

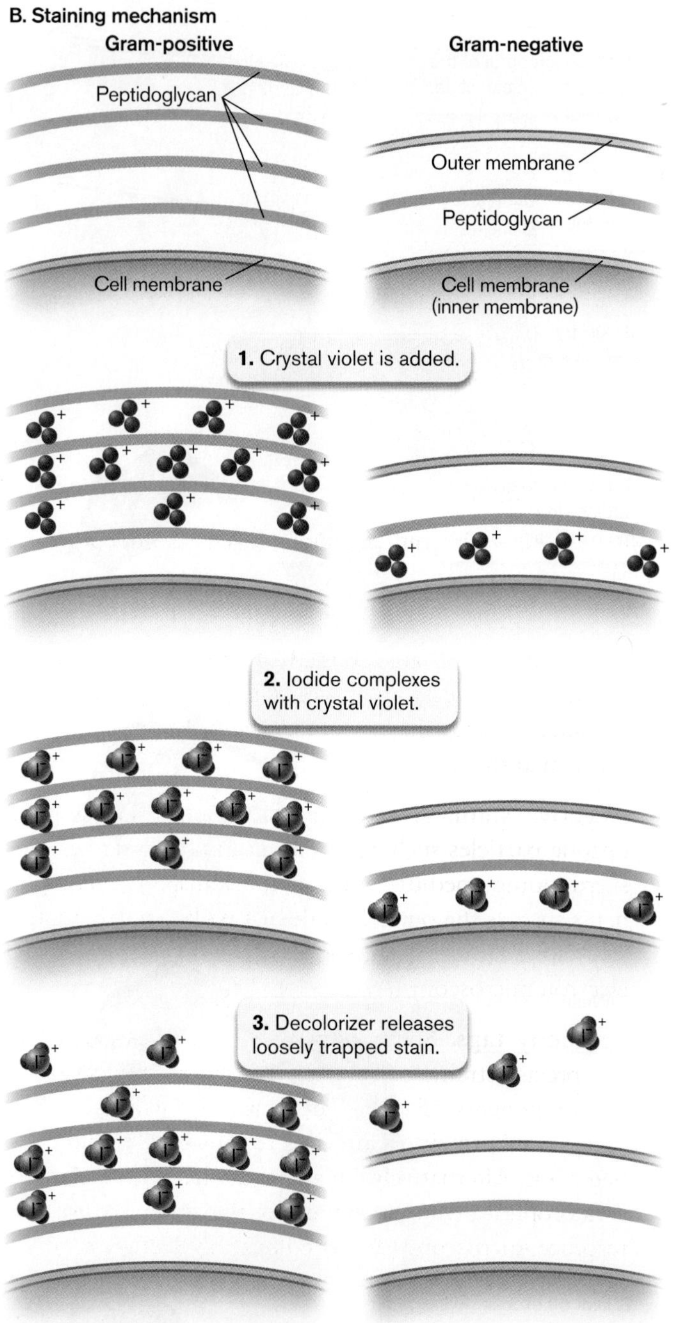

FIGURE 2.21 ■ The Gram stain. **A.** Gram-positive cells have thick cell walls that retain the crystal violet stain. Gram-negative cells have thinner cell walls, which lose the crystal violet stain but are counterstained by safranin. **B.** In a Gram-positive cell, multiple layers of cell-wall peptidoglycan retain the crystal violet–iodide complex. In a Gram-negative cell, the stain leaks out.

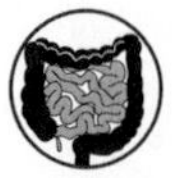

Our colon also contains Gram-positive Firmicutes such as species of *Clostridium* and *Enterococcus.* Most intestinal bacteria are mutualists; that is, they share positive contributions with their host (the human body). However, the gut community may be invaded by deadly pathogens, such as the pathogenic *Escherichia coli* strain O157:H7, or the Gram-positive *Enterococcus faecalis* and *Clostridioides difficile.*

Still other groups of bacteria and archaea have different kinds of cell walls that may stain Gram-positive, Gram-negative, or variable (discussed in Chapters 18 and 19). Moreover, even Firmicutes such as *Bacillus* species show variable stain results depending on their growth state and environmental conditions.

Other differential stains reveal components specific to certain classes of bacteria:

- **Acid-fast stain** (Ziehl-Neelsen). Carbolfuchsin specifically stains mycolic acids of *Mycobacterium tuberculosis* and *M. leprae*, the causative agents of tuberculosis and leprosy, respectively (**Fig. 2.22A**).
- **Spore stain.** When samples are boiled with malachite green, the stain binds specifically to the endospore coat. **Figure 2.22B** shows cells of *Bacillus* species.

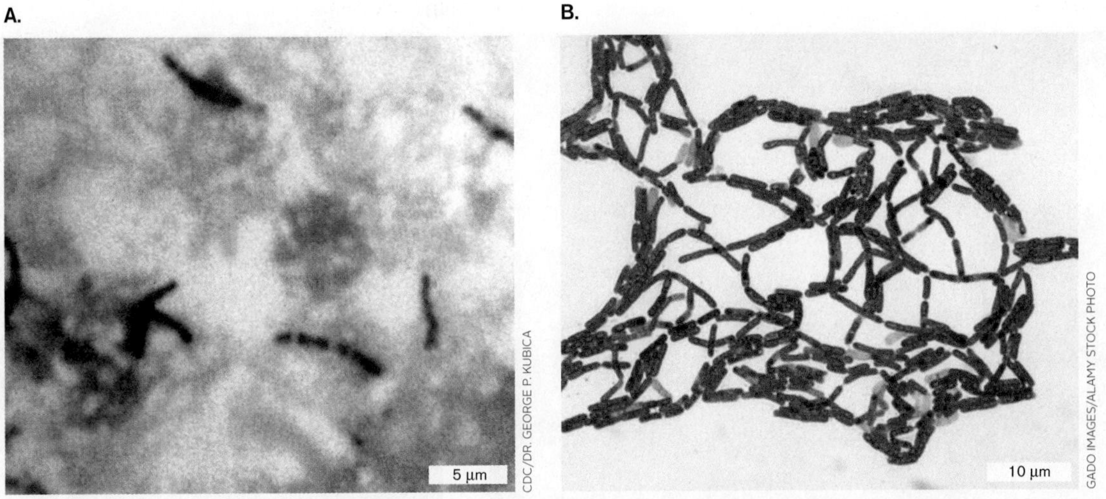

FIGURE 2.22 ■ Differential stains. **A.** Acid-fast stain of *Mycobacterium tuberculosis* (red) in sputum. **B.** Endospore stain of *Bacillus* species.

The endospores are stained green. The malachite green stain detects spores of *Bacillus* species such as *B. thuringiensis* (the insecticide) and *B. anthracis* (the cause of anthrax).

- **Negative stain.** A negative stain is a suspension of opaque particles such as India ink added to darken the surrounding medium and reveal transparent components such as the outer capsule of a pathogen (presented in Chapter 3). Other kinds of negative stains are used for electron microscopy (see Section 2.6).
- **Antibody tags.** Stains linked to antibodies can identify precise strains of bacteria or even specific molecular components of cells. The antibody (which binds a specific cell protein) is attached to a reactive enzyme for detection. Alternatively, an antibody may be attached to a fluorophore (fluorescent molecule) for immunofluorescence microscopy (discussed next, in Section 2.4).

To Summarize

- **In bright-field microscopy**, image quality depends on the wavelength of light; on the magnifying power of a lens; and on the position of the focal plane, the region where the specimen is in focus (that is, where the sharpest image is obtained).
- **A compound microscope** achieves magnification and resolution through the objective and ocular lenses.
- **A wet mount** specimen contains living microbes.
- **Fixing and staining** a specimen kills it but improves contrast and resolution.
- **The Gram stain** differentiates between two major bacterial taxa, which stain either Gram-positive or Gram-negative. Eukaryotes stain Gram-negative, and Archaea stain Gram-variable.
- **Acid-fast, spore, negative, and antibody stains** are other kinds of differential stains.

2.4 Fluorescence Microscopy, Super-Resolution Imaging, and Chemical Imaging

Fluorescence microscopy (also called epifluorescence microscopy) is a powerful tool for identifying specific kinds of microbes, such as pathogens or members of environmental communities. Fluorescence also reveals specific cell parts at work, such as division proteins in the act of accomplishing cell fission. The profound importance of fluorescence for microbial discovery was acknowledged by the awarding of the 2008 Nobel Prize in Chemistry for the discovery and development of green fluorescent protein, GFP (to Osamu Shimomura, Martin Chalfie, and Roger Tsien). Fluorescence microscopy can now be coupled to new tools of chemical imaging, which reveals the actual chemistry of cell parts under microscopy.

What Is Fluorescence?

In fluorescence microscopy, the specimen absorbs light of a defined wavelength and then emits light of lower energy, hence longer wavelength; thus, the specimen is said to "fluoresce." Some microbes, such as cyanobacteria and algae, fluoresce on their own (autofluorescence), owing to endogenous fluorescent molecules such as chlorophyll. For other aims, specific parts of the

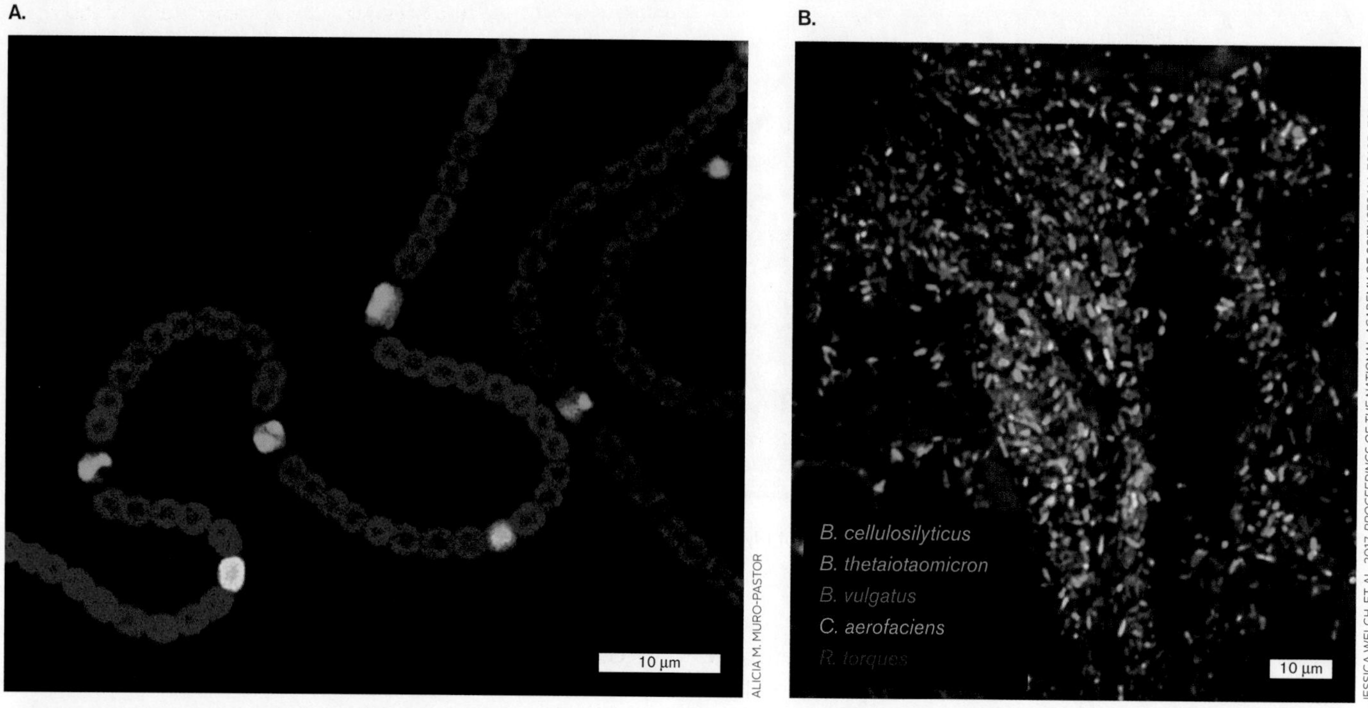

ALICIA M. MURO-PASTOR

JESSICA WELCH, ET AL. 2017. *PROCEEDINGS OF THE NATIONAL ACADEMY OF SCIENCES* 114: E9105–E9114; DOI 10.1073/PNAS.1711596114.

FIGURE 2.23 ■ Fluorescence microscopy. A. Cyanobacteria show chlorophyll-based autofluorescence (red) and fluorescence from heterocysts (green) expressing a nitrogen-stress gene fused to GFP. **B.** Mouse gut bacteria labeled by FISH. The species detected include *Bacteroides cellulosilyticus, B. thetaiotaomicron, B. vulgatus, Collinsella aerofaciens*, and *Ruminococcus torques*.

cell are labeled with a **fluorophore**, a fluorescent dye or protein. In **Figure 2.23A**, cells of the cyanobacterium *Nostoc* sp. PCC7120 (formerly *Anabaena* sp. PCC7120) show autofluorescence (red) arising from their chlorophyll. Every tenth cell or so, however, develops as a nitrogen-fixing heterocyst that lacks chlorophyll. In the sample shown, the bacteria are engineered such that heterocysts express a nitrogen-stress gene fused to a gene that encodes green fluorescent protein. Thus, the two different colors of fluorescence distinguish between the cyanobacterial cells conducting photosynthesis and the heterocysts conducting nitrogen fixation.

Fluorescence microscopy is also used by marine ecologists to reveal tiny bacteria and plankton growing in seawater, a highly dilute natural environment (discussed in Chapter 21). Such observations support the study of microbial responses to climate change. Microbes, including viruses, bacteria, and protists, are detected by fluorescence of DNA-specific stains such as DAPI (4′,6-diamidino-2-phenylindole). The advantage of DAPI fluorescent stain is that it detects only live cells whose DNA is intact, distinguishing them from environmental debris.

For medical studies, fluorescence reveals microbes that colonize the relatively large cells and organ systems of the vertebrate body. **Figure 2.23B** displays the bacterial community of the mouse gut epithelium. Bacteria such as *Bacteroides cellulosilyticus* and *B. thetaiotaomicron* help digest our food (discussed in Chapter 13). In the micrograph, five different species of bacteria are labeled distinctly by **fluorescence in situ hybridization** (**FISH**). The FISH colors show the dense packing of the gut population and reveal which species preferentially sort with each other.

In the FISH method, short, species-specific DNA probes hybridize with the ribosomal RNA (16S rRNA) of different bacteria. Each DNA probe is attached to a fluorophore that emits light over a distinct range of wavelength. The use of FISH for microbial ecology is discussed further in Chapter 21.

Excitation and Emission

How and when does a molecule fluoresce? Fluorescence occurs when a molecule absorbs light of a specific wavelength (the **excitation wavelength**) that has just the right energy needed to raise an electron to a higher-energy orbital (**Fig. 2.24**). Because this higher-energy electron state is unstable, the electron decays to an orbital of slightly lower energy, while losing some energy as heat. The electron then falls to its original level by emitting a photon of less energy and longer wavelength (the **emission wavelength**). The emitted photon has a longer wavelength (less energy) because part of the electron's energy of absorption was lost as heat.

The optical system for fluorescence microscopy uses filters to limit the source light to the wavelength range of excitation, and the specimen's emitted light to the wavelength range of emission (**Fig. 2.25**). The wavelengths of excitation and emission are determined by the choice of fluorophore; in **Figure 2.25B** we show excitation light as

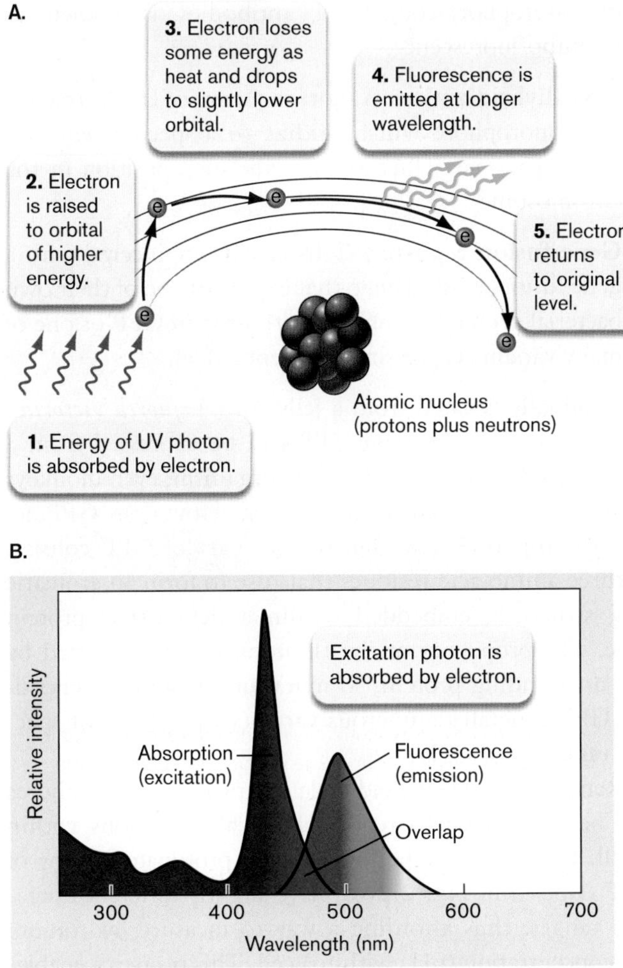

FIGURE 2.24 ■ Fluorescence. Energy gained from UV absorption is released as heat and as a photon of longer wavelength in the visible region. **A.** Fluorescence on the molecular level. **B.** Comparison of absorption and emission spectra for a fluorophore.

green, and emission as red. Because only a small portion of the spectrum is used, fluorescence requires a high-intensity light source such as a tungsten arc lamp. The light passes through a filter that screens out all but the peak wavelengths of excitation. The excitation light (green) is then reflected by a dichroic mirror (dichroic filter), a material that reflects light below a certain wavelength but transmits light above that wavelength.

The reflected green light enters the objective lens, which focuses it onto the specimen, where it excites fluorophores to fluoresce red. The fluorescence emanates in all directions from the specimen, like a point source. Since the light rays point in all directions, a small portion of the emitted light (red) returns through the objective lens to reach the dichroic mirror. The red light now has a longer wavelength, above the penetration limit of the mirror, so it continues through to the ocular lens. The ocular lens focuses the emitted light onto the photodetectors of a digital camera.

Fluorescence can be observed in live organisms. The fluorescent organisms are commonly held in place on the slide—for example, by a pad of agarose gel.

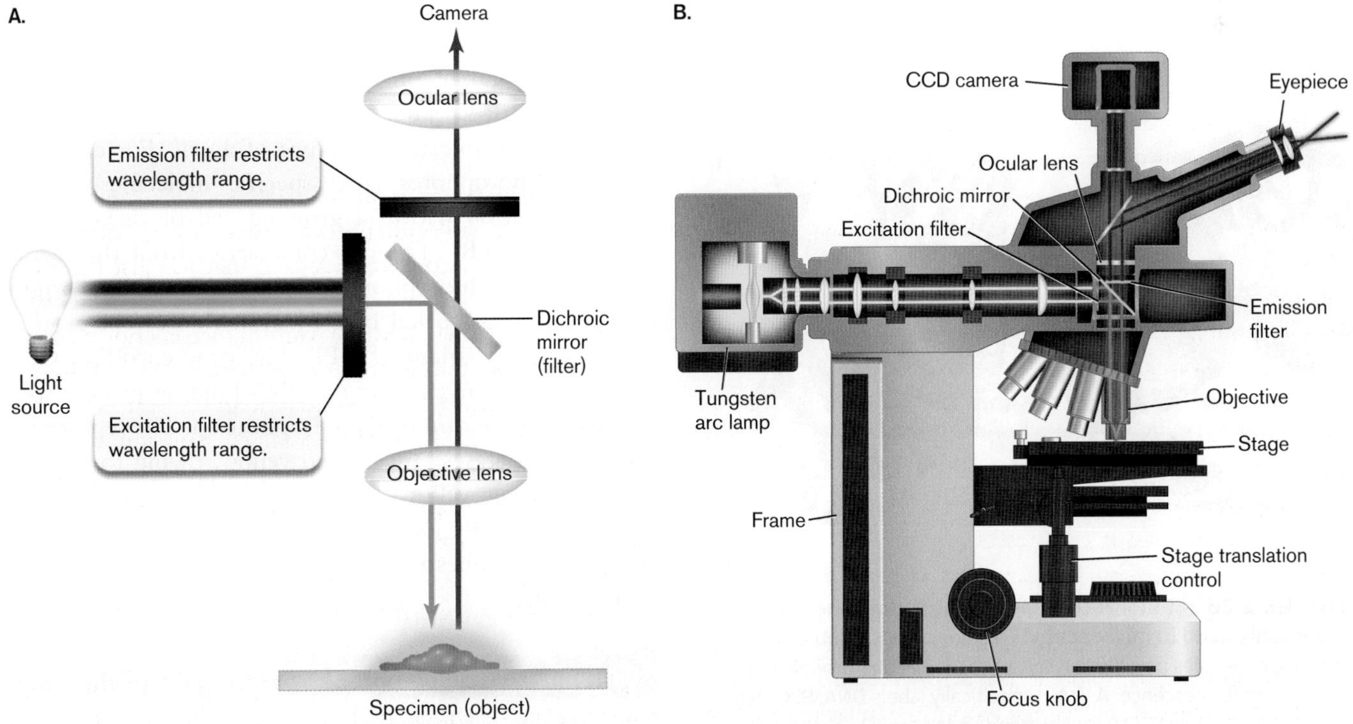

FIGURE 2.25 ■ Fluorescence microscopy. A. The light path of a fluorescence microscope separates the excitation beam from light emitted by the specimen. **B.** Diagram of a fluorescence microscope.

Fluorophores for Labeling

What determines the properties of a fluorophore? The molecular structure of each fluorophore determines its peak wavelengths of excitation and emission, as well as its binding properties. For example, the aromatic rings of DAPI mimic a base pair (**Fig. 2.26A**), enabling intercalation between base pairs of DNA. DAPI absorbs in the UV spectrum and emits in the blue range. Note, however, that the computer driving the optical system can convert the fluorescence signal to any color chosen by the microscopist.

The cell specificity of the fluorophore can be determined by:

- **Chemical affinity.** Certain fluorophores have chemical affinity for certain classes of biological molecules. For example, the fluorophore FM4-64 (green excitation, red emission) specifically binds membranes.
- **Labeled antibodies.** Antibodies that specifically bind a cell component are covalently attached to a fluorophore, forming a "conjugated antibody." The use of fluorophore-conjugated antibodies is known as immunofluorescence.
- **DNA hybridization.** A short sequence of DNA attached to a fluorophore will hybridize to a specific sequence in the genome. Thus we can label one position in the chromosome.
- **Gene fusion reporter.** Cells can be engineered with a gene fusion, a fused gene that expresses one of their own bacterial proteins combined with protein GFP or one of many variants expressing different colors.

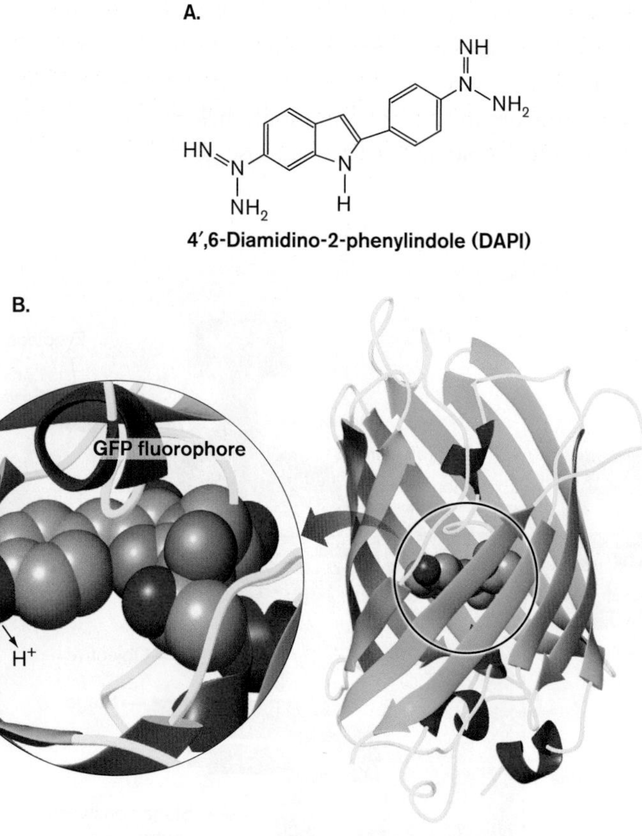

FIGURE 2.26 ■ Fluorescent molecules (fluorophores) commonly used in microscopy. The conjugated double bonds in these molecules provide closely spaced molecular orbitals that give rise to fluorescence. **A.** DAPI specifically labels DNA. **B.** Green fluorescent protein (GFP) is expressed endogenously by the cell. **B blowup:** Three GFP amino acid residues (serine, tyrosine, and glycine) condense to form the fluorophore.

Originally isolated from a jellyfish (*Aequorea victoria*), the Nobel prize–winning GFP can be expressed from a gene spliced into the DNA of any organism; even monkeys have been engineered to glow green. How does GFP act as a fluorophore? The fluorophore part of GFP consists of three amino acid residues that fuse to form an aromatic ring structure, embedded within a beta barrel protein tube. The properties of the fluorophore are modified by the surrounding protein, so mutation of the gene encoding GFP generates numerous variants with different spectral ranges.

Remarkably, GFP and related protein fluorophores can be engineered to report chemical conditions within a cell. For example, the ionized and protonated forms of GFP (labeled in **Fig. 2.26B**) have slightly different excitation ranges, thus affording a way to measure hydronium ion concentration (pH) within a cell. This property enables GFP to report on a cell's response to pH stress. Other GFP sensors are designed to report concentrations of chloride or calcium, second messengers, redox level, or even protease activity.

The wide range of colors and environmental sensitivities of GFP variants provides an extraordinary set of probes for the internal structure of a cell. For example, different fluorophores label specific molecules during DNA replication within a growing cell of *Bacillus subtilis* (**Fig. 2.27B**). The red color arises from the membrane-specific fluorophore FM4-64. The DNA origin of replication is labeled blue by cyan fluorescent protein (CFP), a color variant of GFP. The gene encoding CFP is fused to a gene encoding a DNA-binding protein specific to the *B. subtilis* origin of replication. The replisomes (DNA polymerases) are labeled yellow, owing to a fused gene encoding yellow fluorescent protein (YFP). The replisomes usually locate together near the center of the cell, but sometimes they separate, and are visible as two yellow spots.

Thought Question

2.6 What experiment could you devise to determine the order of events in *Bacillus subtilis* DNA replication?

A.

B.

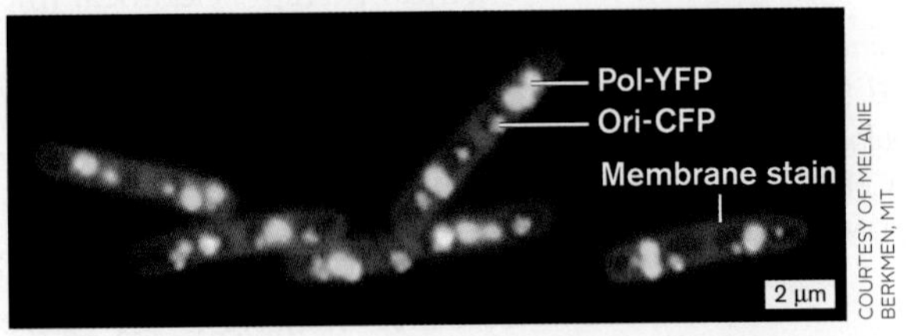

FIGURE 2.27 ■ The replisome and the DNA origin of replication. **A.** Melanie Berkmen, working in the laboratory of Alan Grossman, obtains the fluorescence micrograph shown in panel B. **B.** Fluorescence microscopy reveals the DNA origin, labeled blue by a protein fused to cyan fluorescent protein, binding at a sequence near the origin (Ori-CFP). Replisomes are labeled yellow by fusion of a DNA polymerase subunit to yellow fluorescent protein (Pol-YFP) in dividing cells of *Bacillus subtilis*.

A concern with the use of GFP fluorescence is that proteins fused to GFP may behave differently from the original nonfused protein. In some cases, the GFP portion causes fusion proteins to form complexes at the cell poles that are absent in non-GFP cells. Thus, in cellular biology it is always important to confirm data with the results of a different kind of technique—for example, localization of the target protein with a labeled antibody.

Note also in **Figure 2.27B** that the labeled membranes and DNA origin appear diffuse—that is, unresolved. The emitted light travels in all directions from the point source of the object, and its resolution is limited by the wavelength. Thus, fluorescence cannot resolve the detailed shape of a protein, or distinguish two proteins that are close together. But we can detect the location of a DNA-binding protein within a cell and resolve it as distinct from another fluorescent object located elsewhere. Furthermore, computational techniques called **super-resolution imaging** enable us to pinpoint the protein's location with a precision tenfold greater than the resolution of ordinary optical microscopy.

Super-Resolution Imaging

Cell function requires the interaction of key single molecules, such as the chromosomal DNA with a protein binding its origin. William Moerner at Stanford University was the first to demonstrate the possibility of single-molecule tracking of fluorescent proteins in bacteria (**Fig. 2.28**). The 2014 Nobel Prize in Chemistry honored Eric Betzig, Stefan Hell,

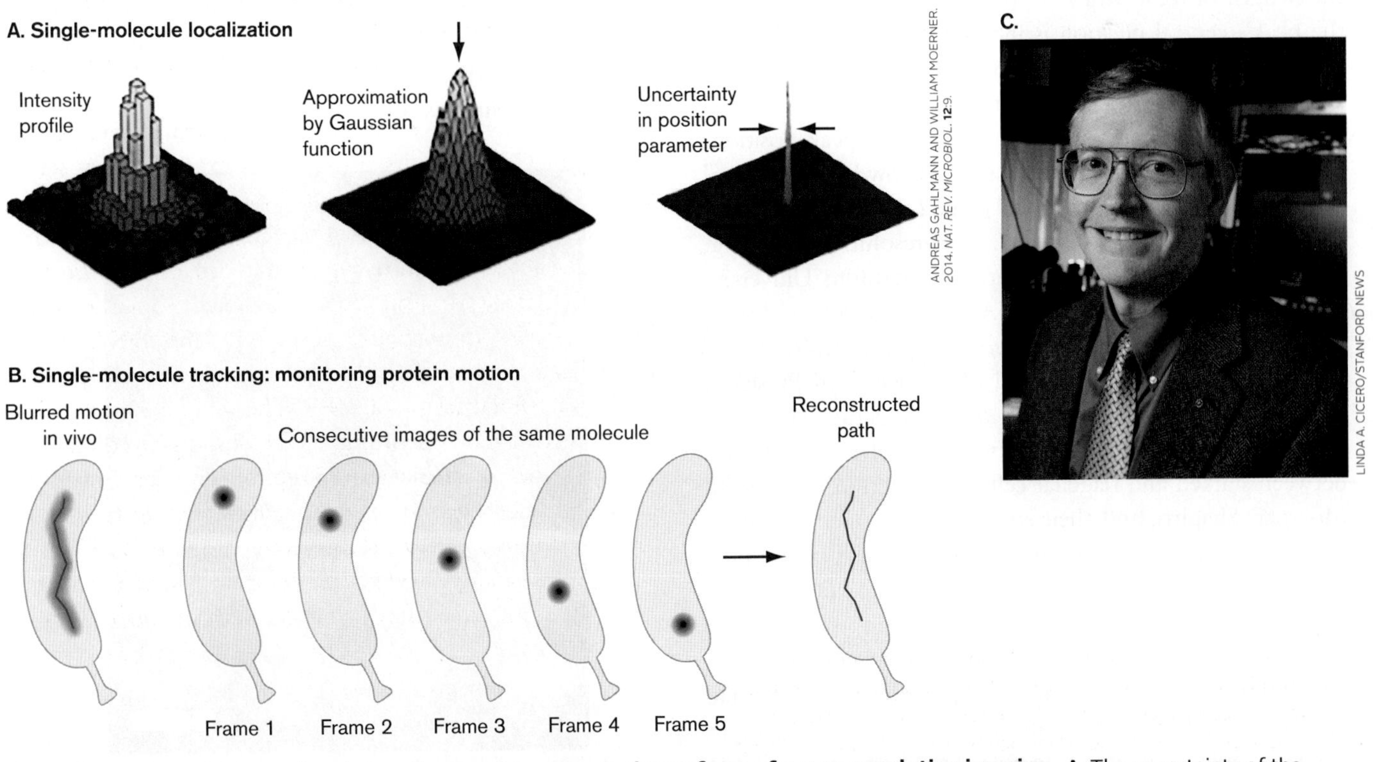

FIGURE 2.28 ■ Single-molecule localization by computation: a form of super-resolution imaging. **A.** The uncertainty of the central peak position. **B.** Tracking a single molecule in a cell. **C.** William Moerner was the first to demonstrate the possibility of single-molecule tracking of fluorescent proteins in bacteria.

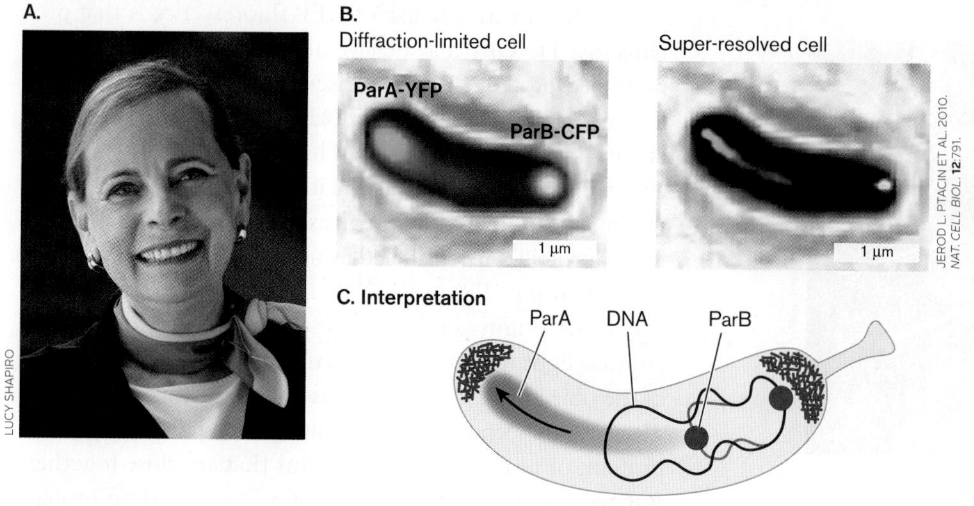

FIGURE 2.29 ■ Super-resolution imaging reveals movement of origin-binding proteins across a cell. A. Lucy Shapiro, winner of the National Medal of Science in 2013. **B.** Super-resolution imaging reveals the migration of the ParA cell fission protein across the cell of *Caulobacter crescentus*. The gene encoding ParA is fused to YFP, and the gene encoding ParB is fused to CFP. **C.** ParA migrates clear across the cell, generating a track to guide ParB to pull the newly replicated DNA origin across from one pole to the opposite pole. The arrow indicates movement.

and William Moerner for the development of super-resolved fluorescence microscopy, or super-resolution imaging.

How can we track a single molecule, which is far smaller than the resolution limit of light ($\lambda/2$ = 200 nm)? Recall the shape of the magnified image of a point source of light (**Fig. 2.12**). Upon magnification, each image of a point source appears as a peak intensity surrounded by rings of much lower intensity. The sharpness of the main peak is limited by the wavelength of light. But the precision with which we know the peak's central position is much narrower (**Fig. 2.28A**). In other words, the uncertainty of the central peak position is about a tenth the width of the intensity profile. Computation based on the intensity profile can reveal the peak positions with high precision. The peak positions show how individual proteins move within a living cell (**Fig. 2.28B**).

In an early application of super-resolution imaging, Moerner worked with Lucy Shapiro at Stanford University to track the movement of DNA-binding proteins during cell fission of *Caulobacter crescentus* (**Fig. 2.29**). Shapiro received the National Medal of Science from President Obama in 2013 for her studies of the intriguing developmental cycle of this bacterium, which involves a transition between stalked and flagellar cells (presented in Chapter 3). Moerner, Shapiro, and their students used super-resolution imaging to track the migration of the ParA cell fission protein (**Fig. 2.29B** and **C**). In this experiment, a gene encoding ParA is fused to a gene for yellow fluorescent protein (YFP), whereas ParB is fused to cyan fluorescent protein (CFP). ParA migrates clear across the cell, generating a spindle-like track to guide ParB pulling the newly replicated DNA origin from one pole to the opposite pole. This dramatic feat is one of many intracellular mechanisms we explore in Chapter 3.

Some advanced forms of fluorescence microscopy use laser beams to resolve subcellular details, and even to visualize cells in 3D. One such method is **confocal laser scanning microscopy** (or confocal microscopy). In confocal microscopy, a microscopic laser light source scans across the specimen. **Figure 2.30** shows a biofilm composed of the pathogenic bacterium *Pseudomonas aeruginosa* treated with the antibiotic tobramycin. The biofilm is treated with fluorophores that reveal live cells (green) beneath the dead cells, killed by tobramycin (red). The hidden live cells cause problems for medical therapy. Confocal microscopy enables us to visualize the 3D structure of the biofilm. The method of confocal microscopy is explained in **eTopic 2.1**.

An advanced form of 3D fluorescence imaging is the microbial identification after passive CLARITY technique (MiPACT). This remarkable method generates optical clarity in background tissue, revealing the spatial distribution of bacteria that colonize a host tissue (see **Special Topic 2.1**).

Chemical Imaging Microscopy

Chemical imaging uses mass spectrometry (analysis of molecular fragments by mass) to visualize the distribution of chemicals within living cells. The combination of fluorescence and chemical imaging offers extraordinary opportunities to map the structure and function of cells in natural communities, such as soil or the intestinal microbiome.

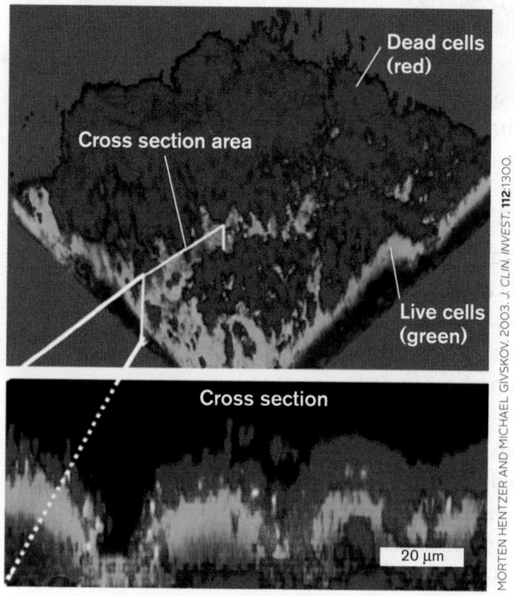

FIGURE 2.30 ■ Biofilm with live/dead fluorophore, observed by confocal laser scanning microscopy. *Pseudomonas aeruginosa* cells growing in a biofilm treated with the antibiotic tobramycin. Dead cells fluoresce red; live cells fluoresce green.

SPECIAL TOPIC 2.1 Biogeography of a Gut Pathogen

What do microbial pathogens really look like within the host organ they infect? The host tissue constitutes a vast geography, where the microbe's location may determine its ability to cause disease. William DePas (**Fig. 1**), in the laboratory of Dianne Newman at the California Institute of Technology, combined several state-of-the-art technologies to image the biogeography of pathogens. For the example shown here, DePas worked with Fitnat Yildiz, at UC Santa Cruz, to image the pathogen *Vibrio cholerae* within the lumen of the mouse gut (**Fig. 2**).

COURTESY OF WILLIAM DEPAS

FIGURE 1 ■ William DePas, California Institute of Technology. Depas developed the MiPACT method in the laboratory of Dianne Newman.

The first technology used is the CLARITY technique, which can render tissues transparent for whole organs. The aim of CLARITY is to selectively remove the tissue's lipid components, which are the main cause of opacity. DePas devised MiPACT, a microscale version of CLARITY that allows microscopic visualization.

To render the tissue transparent, the sample is soaked in acrylamide monomers that cross-link with each other, as well as with the protein components of the tissue. The lipid components are then dissolved by sodium dodecyl sulfate (SDS) detergent, the same detergent used for gel electrophoresis. Once the tissue clears, a final solution treatment removes the SDS and adds a preservative chemical.

The clarified tissue retains structural proteins as well as DNA. A lysozyme treatment then breaks down bacterial cell walls, enabling the entry of DNA-hybridizing probes. To detect specific types of bacteria, DePas used an advanced version of FISH called the hybridization chain reaction (HCR). In this technique, specific bacteria are detected by hybridization of their DNA with a species-specific DNA probe. The DNA oligomer has no detectable label—but its formation triggers the unfolding of a DNA hairpin loop that triggers other hairpins to unfold and hybridize with each other. The hybridized hairpins then shift their attached fluorophores into a form that fluoresces.

MiPACT with HCR reveals the presence of *V. cholerae* bacteria within the landscape of mouse intestinal villi. In **Figure 2**, the nuclei of intestinal villi are stained blue with DAPI fluorophore; their surrounding cytoplasm is transparent. The villi secrete mucin, which is stained red with a lectin (carbohydrate-binding protein). The mucin fills the gut lumen, where it harbors *V. cholerae* (green) identified by HCR. We can now see the *V. cholerae* cells as individuals, and as they form a dense microbial community tightly associated with mucin. Images such as this reveal 3D spatial patterns that help us simulate the infection environment in laboratory experiments.

RESEARCH QUESTION

How can we use MiPACT imaging to study the mechanism and effectiveness of antibiotic therapies on host infections?

William H. DePas, Ruth Starwalt-Lee, Lindsey Van Sambeek, Sripriya Ravindra Kumar, Viviana Gradinaru, et al. 2016. Exposing the three-dimensional biogeography and metabolic states of pathogens in cystic fibrosis sputum via hydrogel embedding, clearing, and rRNA labeling. *mBio* **7**:e00796-16.

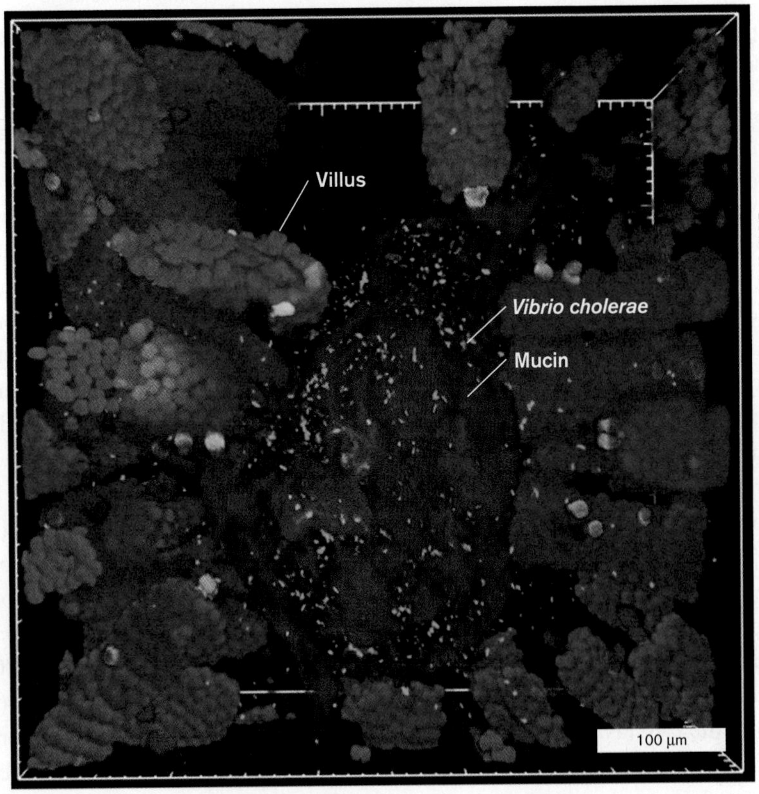

COURTESY OF WILLIAM DEPAS, ANA GALLEGO-HERNANDEZ DIANNE NEWMAN, AND FITNAD YILDIZ

FIGURE 2 ■ MiPACT imaging reveals *Vibrio cholerae* within the mouse gut epithelium. Mouse intestinal tissue was clarified by the MiPACT procedure. DAPI (blue) stains nuclei of intestinal villi. Villi secrete mucin (red, stained with wheat germ agglutinin lectin) into the gut lumen. Embedded in mucin, *Vibrio cholerae* (green) are identified by the hybridization chain reaction (HCR).

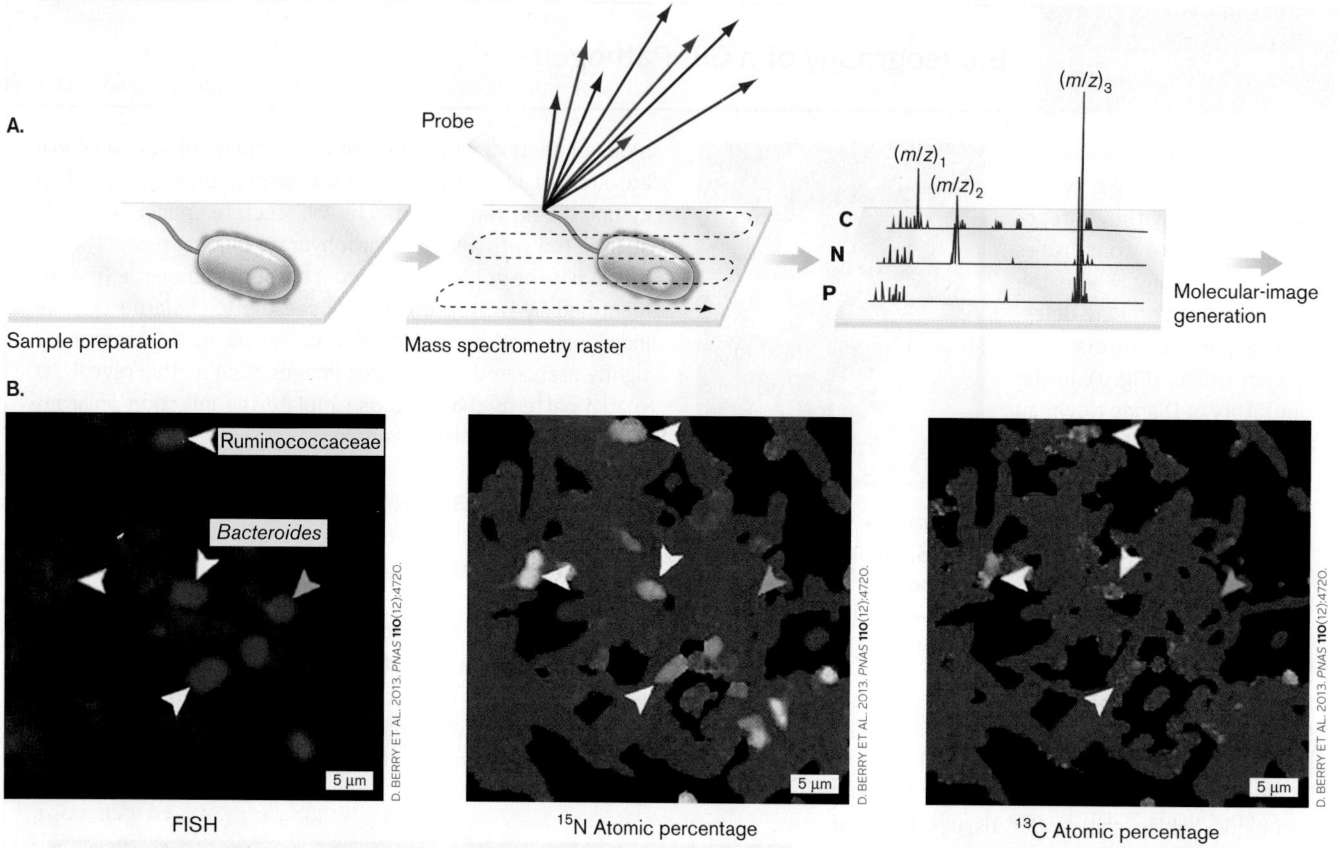

FIGURE 2.31 ■ Imaging mass spectrometry. Mass spectra are obtained from thousands of locations throughout the sample surface. **A.** Molecular fragments are selected for analysis of mass-to-charge ratio (m/z). Selected isotopes may label specific atoms—for example, C, N, or P. The relative intensities of individual compounds are visualized using false-color gradients. **B.** FISH (left) and NanoSIMS (middle and right) of bacteria feeding on secretions from the intestinal epithelium, showing atomic percentages (at%) of intracellular nitrogen (^{15}N) and carbon (^{13}C).

A high-resolution method for chemical imaging is called **nanoscale secondary ion mass spectrometry (NanoSIMS)**. The NanoSIMS process starts with an ionizing probe, a source of energy that breaks up the large organic molecules of a sample (**Fig. 2.31A**). The molecular fragments, called "secondary ions," fly off from the source and are captured by a mass spectrometer. This instrument measures fragment masses of the secondary ions, generating a mass spectrum. Mass spectra are taken from thousands of locations, scanned across a bacterial cell.

For NanoSIMS, the microbial sample is prepared by growth on nutrients labeled with a heavy isotope. For example, the cell's uptake of nitrogen-rich protein can be detected by incorporation of the heavy isotope ^{15}N. The increased weight of ^{15}N compared to the normally predominant ^{14}N (as indicated by the mass ratio $^{15}N/^{14}N$) is detected quantitatively in the molecular-fragment masses. Alternatively, we can detect isotopes of carbon ($^{13}C/^{12}C$).

At the University of Vienna, Michael Wagner and colleagues used NanoSIMS to address the question: Which of our intestinal bacteria feed on our own secretions, instead of the food we consume? Our intestinal epithelium secretes a mucus layer, which helps prevent colonization by pathogens. But the mucus also "farms" bacteria, whose feeding helps moderate the buildup of secretions.

Wagner's group used different types of bacteria to colonize germ-free mice. The mice were injected intravenously with a nutrient (the amino acid threonine) containing the heavy-atom isotopes ^{13}C and ^{15}N. The threonine is readily incorporated into mouse proteins and secretions; however, the bacteria inoculated into the mouse's intestinal lumen have no direct access to the threonine, but instead only to mouse secretions containing it. Thus, NanoSIMS of the intestinal bacteria reveals which ones are snacking on mucus.

In **Figure 2.31B**, various species of bacteria on the mouse epithelium are identified by FISH. In the FISH panel shown, the fluorescent probes identify single cells of *Bacteroides acidifaciens* (blue) and Ruminococcaceae OTU_5807 (red). The parallel panels show the atomic percentages (at%) of the given isotopes ^{15}N and ^{13}C, imaged on a color scale quantifying the amount of isotope enrichment. The panels show that most of the bacterial cells corresponding to *B. acidifaciens* and Ruminococcaceae acquire nitrogen from the host (gold or yellow color, white arrowhead). The third panel shows that some but not all of the bacteria show carbon uptake. Overall, out of the mixed gut microbes, the *B. acidifaciens* and Ruminococcaceae strains were shown to be the main snackers on host mucus. These observations will help medical researchers characterize the composition and dynamics of a healthy gut microbiome.

To Summarize

- **Fluorescence microscopy** uses fluorescence by a fluorophore to reveal specific cells or cell parts.

- **The specimen absorbs light at one wavelength and then emits light at a longer wavelength.** Color filters allow only light in the excitation range to reach the specimen, and only emitted light to reach the photodetector.

- **A fluorophore can label a cell part by** chemical affinity for a component such as a membrane, attachment to an antibody stain, or attachment to a short nucleic acid that hybridizes to a DNA sequence.

- **Fluorescent proteins such as GFP can be fused to a specific protein expressed by the cell.** Endogenous GFP-type proteins can track intracellular movement of cell parts and can report environmental stress responses.

- **Super-resolution imaging** can define the position of a fluorescent protein with a precision of 20–40 nm, tenfold better than the resolution limit of light magnification.

- **Chemical imaging microscopy** maps the distribution of compounds within a cell or a microbial community. NanoSIMS combines fluorescence microscopy (FISH) with mass spectral analysis of bacterial components labeled by heavy isotopes.

2.5 Dark-Field and Phase-Contrast Microscopy

Advanced optical techniques enable us to visualize structures that are difficult or impossible to detect under a bright-field microscope, either because their size is below the limit of resolution of light or because their cytoplasm is transparent. These techniques take advantage of special properties of light waves, such as light scattering (dark field) and phase contrast.

Dark-Field Microscopy Detects Unresolved Objects

Dark-field microscopy enables microbes to be visualized as halos of bright light against the darkness, just as stars are observed against the night sky. A tiny object whose size is well below the wavelength of light, such as a virus particle, can be detected by light scattering. Dark-field microscopy is used in research to detect pathogenic spirochetes whose cells are so narrow (0.1 μm) that their shape cannot be resolved by bright-field microscopy. Examples are *Treponema pallidum*, the cause of syphilis; and *Borrelia burgdorferi*, the cause of Lyme borreliosis. These spirochetes can be detected by dark-field microscopy because the cells scatter light, although the cell's narrow width is not resolved.

The optics of dark-field microscopy involve light scattering. The wavefront of scattered light is spherical, like a wave emitted by a point source (see **Fig. 2.9D**). The scattered wave has a much smaller amplitude than that of the incident (incoming) wave. Therefore, with ordinary bright-field optics, scattered light is washed out. But the optical arrangement of a dark-field microscope excludes all light that is transmitted directly. A hollow cone of light focuses on the object. The incident light converges at the object and then generates an inverted hollow cone radiating outward. The objective lens is positioned in the central region, where it completely misses the directly transmitted light. For this reason, the field appears dark. However, light scattered by the object radiates outward in a spherical wave. A sector of this spherical wave enters the objective lens and is detected as a halo of light.

Dark-field optics also detects unresolved cell components, such as **flagella** (singular, **flagellum**), the bacterial swimming apparatus. Each flagellum consists of a helical filament that is rotated by a motor device embedded in the bacterial cell wall (for flagellar structure, see Section 3.6). The "swimming strokes" of bacteria were first elucidated by Howard Berg and Robert Macnab (1940–2003), using dark-field optics to view the helical flagella. The flagella are too narrow to resolve, but dark-field optics can detect them, as seen in the dark-field micrograph of the gut pathogen *Salmonella enterica* (**Fig. 2.32**). Detecting flagella requires such high light intensity that the bacterial cell itself appears "overexposed"; its shape is unresolved.

A. Low light intensity

B. High light intensity

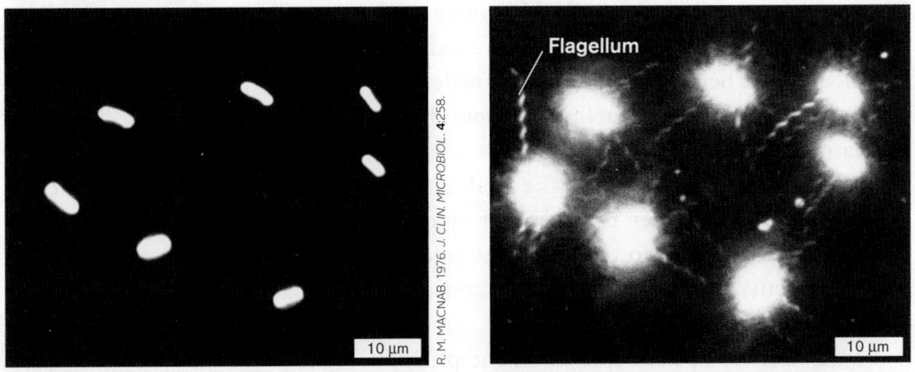

FIGURE 2.32 ▪ **Motile bacteria observed under dark-field microscopy.** **A.** Flagellated *Salmonella enterica* observed with low light, which limits scattering. Only cell bodies are detected; no flagella. **B.** At higher light intensity the cell bodies are overexposed. Flagella are detected, but their actual width is not resolved, because it is much less than the wavelength of light.

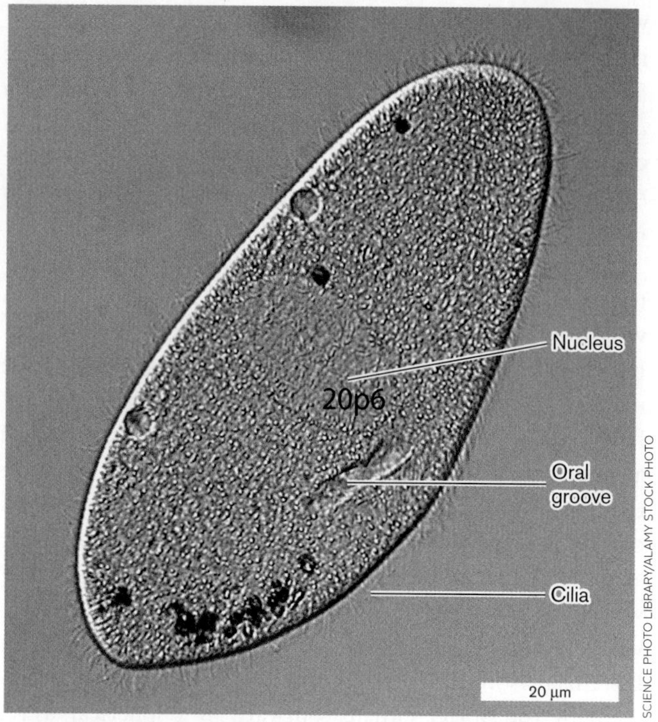

FIGURE 2.33 ▪ Phase-contrast microscopy (PCM) of a paramecium. Differences in refractive index reveal the nucleus, oral groove, and cilia.

Thought Questions

2.7 Some early observers claimed that the rotary motions observed in bacterial flagella could not be distinguished from whiplike patterns, comparable to the motion of eukaryotic flagella. Can you design an experiment to distinguish the two and prove that the flagella rotate? *Hint:* Bacterial flagella can get "stuck" to the microscope slide or coverslip.

2.8 Compare and contrast fluorescence microscopy with dark-field microscopy. What similar advantage do they provide, and how do they differ?

Phase-Contrast Microscopy

Phase-contrast microscopy (PCM) exploits differences in refractive index between the cytoplasm and the surrounding medium or between different organelles. **Figure 2.33** shows a paramecium, a ciliated protist found in pond water. Under phase contrast, the cell outlines appear dark because light passes entirely through the cell envelope, whose refractive index is higher than that of cytoplasm. Differences in refractive index reveal the shape of organelles such as the nucleus, oral groove (mouth), and cilia (whip-like structures for motility).

Phase-contrast optics depends on the principle of interference (introduced in Section 2.2). In interference, two wavefronts (or two portions of a wavefront) interact with each other by addition (amplitudes in phase) or subtraction (amplitudes out of phase; **Fig. 2.34**). The result of interference between two waves is a pattern of alternating zones of constructive and destructive interference (brightness and darkness).

The optical system for phase contrast was invented in the 1930s by the Dutch microscopist Frits Zernike (1888–1966), for which he earned the 1953 Nobel Prize in Physics. In this system, slight differences in the refractive index of the various cell components are transformed into differences in the intensity of transmitted light. Zernike's scheme makes use of the fact that living cells have relatively high contrast because of their high concentration of solutes. Given the size and refractive index of commonly observed cells, light is retarded by approximately one-quarter of a wavelength when it passes through the cell. In other words, after passing through a cell, light exits the cell about one-quarter of a wavelength behind the phase of light transmitted directly through the medium.

The Zernike optical system is designed to retard the refracted light by an additional quarter of a wavelength, so that the light refracted through the cell is slowed by a total of half a wavelength compared with the light transmitted through the medium. When two waves are out of phase by half a wavelength, they produce destructive interference, canceling each other's amplitude (**Fig. 2.34B**). The result is a region of darkness in the image of the specimen.

In phase-contrast microscopy, the light transmitted through the medium needs to be separated from the light interacting with the object, where it is slowed by refraction. The transmitted light is separated by a ring-shaped slit (annular ring). The annular ring generates a hollow cone of light, which is focused through the specimen and

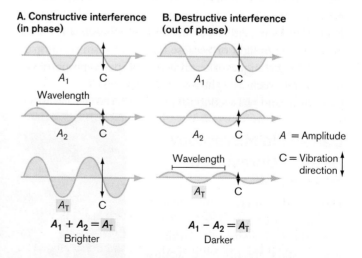

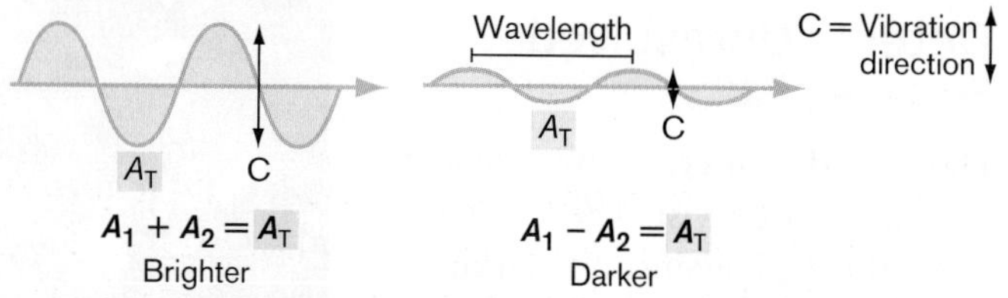

FIGURE 2.34 ▪ Phase interference. A. In constructive interference, the peaks of the two wave trains rise together; their amplitudes are additive ($A_1 + A_2$), forming a wave of greater total amplitude (A_T). **B.** In destructive interference, the peaks of the waves are opposite one another, so their amplitudes cancel ($A_1 - A_2$), forming a wave of lesser amplitude.

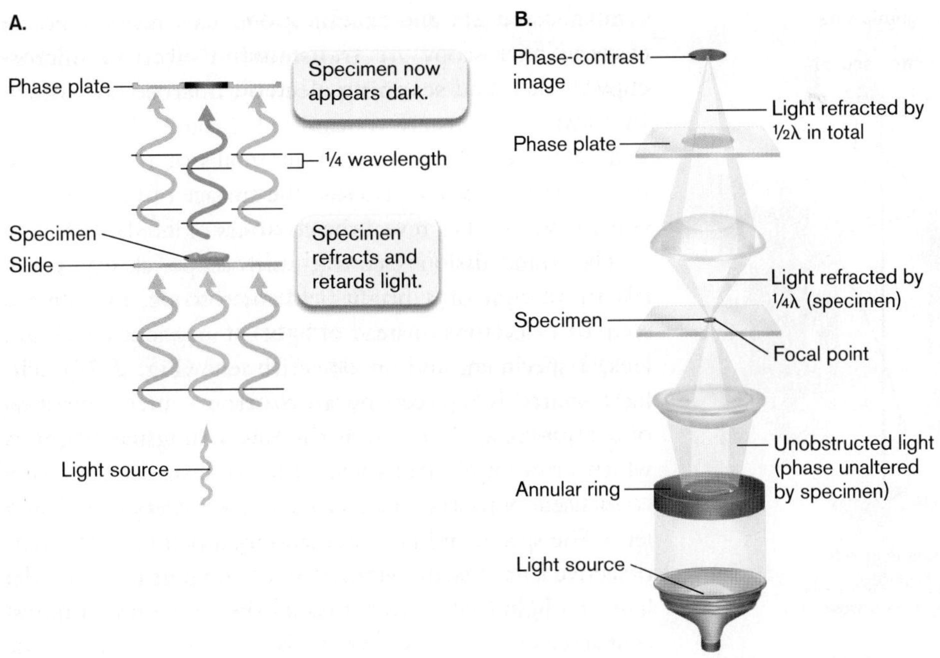

FIGURE 2.35 ■ Phase-contrast optics. A. The specimen retards light by approximately one-quarter of a wavelength. The phase plate contains a central disk of refractive material that retards light from the specimen by another quarter wavelength, increasing the phase difference to half a wavelength. The light from the specimen and the transmitted light are now fully out of phase; they cancel, making the specimen appear dark. **B.** In the phase-contrast microscope, the annular ring forms a hollow cone of light that passes through the refractive material of the specimen. When the transmitted and refracted light cones re-join at the focal point, they are out of phase; their amplitudes cancel each other, and that region of the image appears dark against a bright background.

generates an inverted cone above it (**Fig. 2.35**). Light passing through the specimen, however, is refracted and thus bent into the central region within the inverted cone.

Both the refracted light from the specimen and the outer cone of transmitted light enter the phase plate. The phase plate consists of refractive material that is thinner in the region met by the outer (transmitted) light cone. The refracted light passing through the center of the phase plate is retarded by an additional one-quarter wavelength compared with the transmitted light passing through the thinner region on the outside; the overall difference approximates half a wavelength. When the light from the inner and outer regions focuses at the ocular lens, the amplitudes of the waves cancel and produce a region of darkness. In this system, small differences in refractive index can produce dramatic differences in contrast between the offset phases of light.

Other optical systems use phase interference in different ways. Differential interference contrast microscopy (DIC) with Nomarski optics superimposes interference bands on an image, accentuating small differences in refractive index (discussed further in **eTopic 2.2**).

To Summarize

- **Dark-field microscopy** uses scattered light to detect objects too small to be resolved by light rays. Extremely small microbes and thin structures can be detected. The shapes of objects are not resolved.
- **Phase-contrast microscopy** with Zernike optics superimposes refracted light and transmitted light shifted out of phase so as to reveal differences in refractive index as patterns of light and dark. Live cells with transparent cytoplasm, and the organelles of eukaryotes, can be observed with high contrast.
- **High refractive index of the specimen** causes light to bend and pass through the highly refractive phase plate. Retarded phase partly cancels the phase of transmitted light, generating a dark edge with high contrast.
- **Differential interference contrast microscopy (DIC)** with Nomarski optics superimposes interference bands on an image, accentuating small differences in refractive index.

2.6 Electron Microscopy, Scanning Probe Microscopy, and X-Ray Crystallography

All cells are built of macromolecular structures. The foremost tool for observing the shapes of these structures is **electron microscopy (EM)**. In electron microscopy, magnetic lenses focus beams of electrons to image cell membranes, chromosomes, and ribosomes with a resolution a thousand times that of light microscopy. Other kinds of microscopy are emerging, such as scanning probe microscopy, which images the contours of live bacteria. For atom-level detail of a macromolecule, the tool of choice is X-ray crystallography.

Electron Microscopy

How does an electron microscope work? Electrons are ejected from a metal subjected to a voltage potential. Like photons, the electrons travel in a straight line, interact with matter,

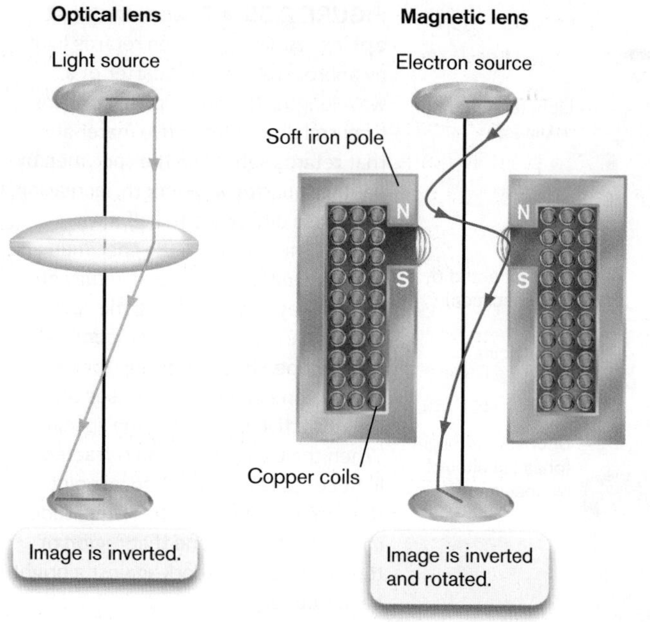

FIGURE 2.36 ▪ A magnetic lens. The beam of electrons spirals around the magnetic field lines. The U-shaped magnet acts as a lens, focusing the spiraling electrons much as a refractive lens focuses light rays.

and carry information about their interaction. And also like photons, electrons can exhibit the properties of waves. The wavelength associated with an electron is 100,000 times smaller than that of a photon; for example, an electron accelerated over a voltage of 100 kilovolts (kV) has a wavelength of 0.0037 nm, compared with 400–750 nm for visible light. However, the actual resolution of electrons in microscopy is limited not by the wavelength, but by the aberrations of the lensing systems used to focus electrons. The magnetic lenses that focus the electrons never achieve the precision required to utilize the full potential resolution of the electron beam.

Electrons are focused by means of a magnetic field directed along the line of travel of the beam (**Fig. 2.36**). As a beam of electrons enters the field, it spirals around the magnetic field lines. The shape of the magnet can be designed to generate field lines that will focus the beam of electrons in a manner analogous to the focusing of photons by a refractive lens. The electron beam, however, forms a spiral because electrons travel around magnetic field lines. Magnetic lenses generate large aberrations; thus, we need a series of corrective magnetic lenses to obtain a resolution of about 0.2 nm. This resolution is a thousand times greater than the 200-nm resolution of light microscopy.

Thought Question

2.9 Like a light microscope, an electron microscope can be focused at successive powers of magnification. At each level, the image rotates at an angle of several degrees. Given the geometry of the electron beam (see **Fig. 2.36**), why do you think the image rotates?

Transmission EM and scanning EM. Two major types of electron microscopy are **transmission electron microscopy (TEM)** and **scanning electron microscopy (SEM)**. In TEM, electrons are transmitted through the specimen as in light microscopy to reveal internal structure. In SEM, the electron beams scan across the surface of the specimen and are reflected to reveal the contours of its 3D surface.

The transmission electron microscope closely parallels the design of a bright-field microscope, including a source of electrons (instead of light), a magnetic condenser lens, a specimen, and an objective lens (**Fig. 2.37**). The light source is replaced by an electron source consisting of a high-voltage current applied to a tungsten filament, which gives off electrons when heated. The electron beam is focused onto the specimen by a magnetic condenser lens. The specimen image is then magnified by a magnetic objective lens. The projection lens, analogous to the ocular lens of a light microscope, focuses the image on a fluorescent screen.

The scanning electron microscope is arranged somewhat differently from the TEM, in that a series of condenser lenses focuses the electron beam onto the surface of the specimen. Reflected electrons are then picked up by a detector (**Fig. 2.38**).

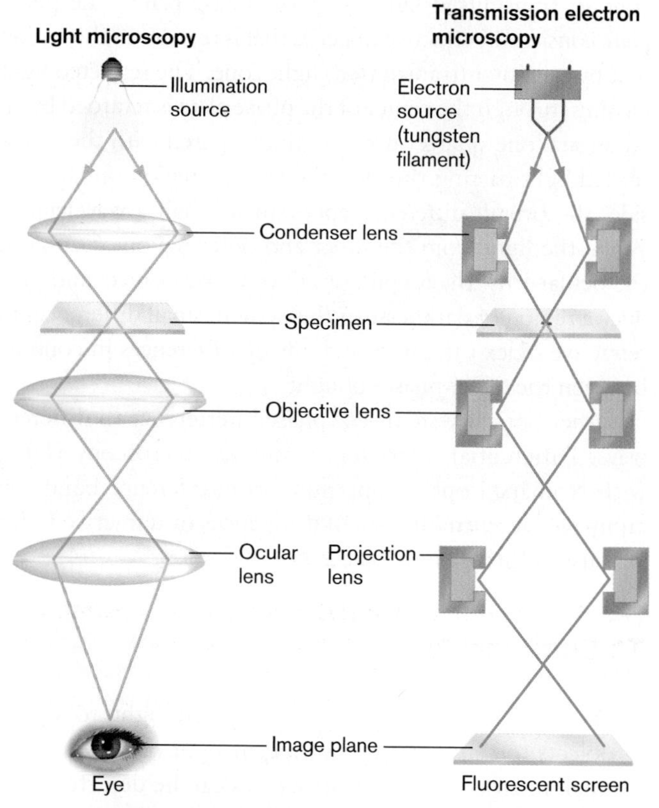

FIGURE 2.37 ▪ Transmission electron microscopy (TEM). In the transmission electron microscope (right), the light source is replaced by an electron source consisting of a high-voltage current applied to a tungsten filament, which gives off electrons when heated. Each magnetic lens shown (condenser, objective, projection) actually represents a series of lenses.

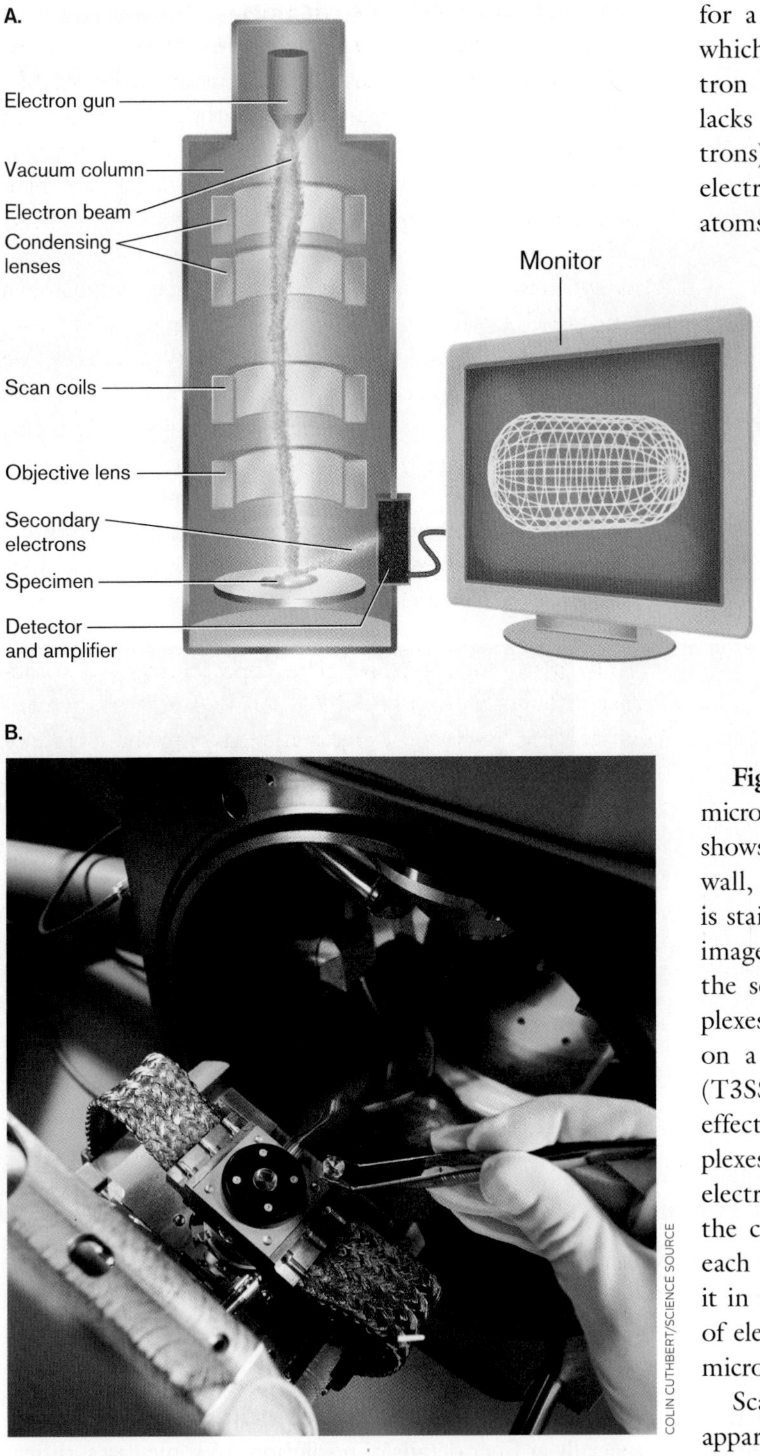

FIGURE 2.38 ■ Scanning electron microscopy (SEM). **A.** In the scanning electron microscope, the electron beam is scanned across a specimen coated in gold, which acts as a source of secondary electrons. The incident electron beam ejects secondary electrons toward a detector, generating an image of the surface of the specimen. **B.** Loading a specimen into the vacuum column.

Sample preparation for EM. Standard electron microscopy of biological specimens at room temperature poses special problems. The entire optical column must be maintained under vacuum to prevent the electrons from colliding with the gas molecules in air. The requirement for a vacuum precludes the viewing of live specimens, which in any case would be quickly destroyed by the electron beam. Moreover, the structure of most specimens lacks sufficient electron density (ability to scatter electrons) to provide contrast. Thus, the specimen requires an electron-dense **negative stain** using salts of heavy-metal atoms such as tungsten or uranium. The heavy atoms collect outside the surfaces of cell structures such as membranes, where their electron scatter reveals the outline of the structure. Staining, however, can be avoided for cryo-electron microscopy (see the next section).

We can prepare a specimen by embedding it in a polymer for thin sections. A special knife called a microtome cuts slices through the specimen, each slice a fraction of a micrometer thick. Alternatively, a specimen consisting of, for example, virus particles or isolated organelles can be sprayed onto a copper grid. In either case, the electron beam penetrates the object as if it were transparent. The electrons are actually absorbed by the heavy-atom stain, which collects at the edges of biological structures.

Figure 2.39 shows examples from transmission electron microscopy. The TEM of *Bacillus anthracis* in **Figure 2.39A** shows a thin section through a bacillus, including a cell wall, membranes, and glycoprotein filaments. The section is stained with uranyl acetate (a salt of uranium ion). The image includes electron density throughout the depth of the section. In **Figure 2.39B**, *Salmonella* protein complexes called "injectisomes" have been isolated and spread on a grid. Injectisomes, or type III secretion systems (T3SS), are used by *Salmonella enterica* to inject virulence effector proteins into a host cell. The motor protein complexes are negatively stained with phosphotungstate, an electron-dense material that deposits in the crevices around the complexes on the grid. The TEM reveals details of each complex, including the individual rings that anchor it in the cell envelope. The details resolved by the beam of electrons are far smaller than those resolved by a light microscope.

Scanning electron microscopy can show whole cells in apparent 3D view, with much greater resolution than light microscopy can accomplish. SEM is particularly effective for visualizing cells within complex communities such as a biofilm. **Figure 2.40A** shows a biofilm of archaea with "hami," grappling hooks that enable the cells to encase filaments of a bacterial partner. The shapes of the archaea and their hami appear distinct in SEM.

In a clinical example, **Figure 2.40B** shows the pathogen *Helicobacter pylori* colonizing the gastric epithelium (stomach lining). *H. pylori* bacteria are helical rods (colorized green). Note that the colorizing consists of interpretation by a photo artist; no actual colors are observed by electron microscopy, since colors are defined not by electrons but

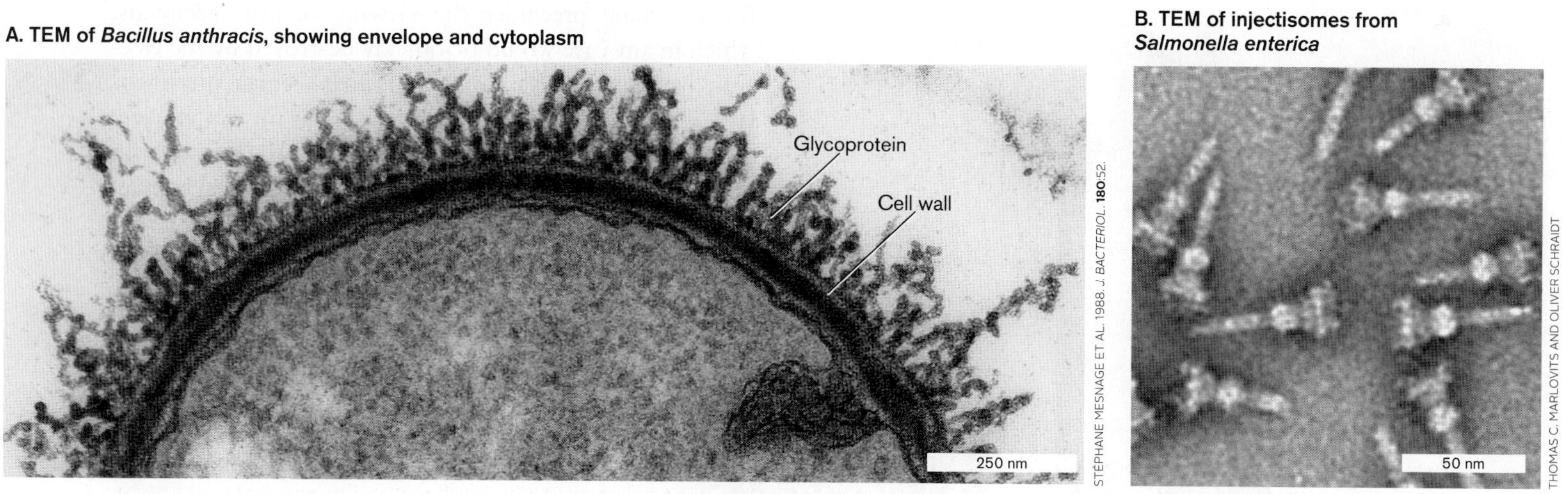

FIGURE 2.39 ■ Transmission electron micrographs. A. *Bacillus anthracis* thin section, showing envelope and cytoplasm. Stained with uranyl acetate. B. "Injectisome" toxin injection devices from *Salmonella enterica*, with phosphotungstate negative stain.

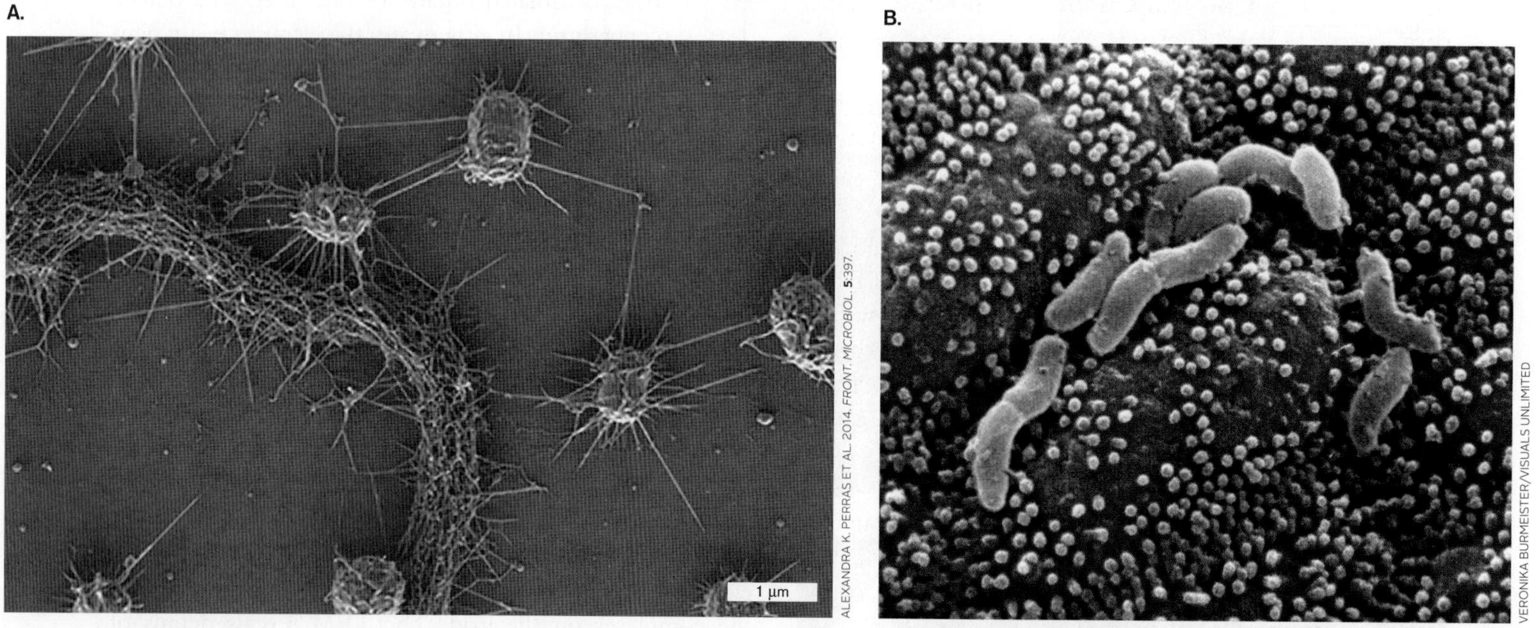

FIGURE 2.40 ■ Scanning electron micrographs. A. Archaea of a wetland biofilm, with numerous hami. The archaea encase bacterial filaments. From Sippenauer Moor, Germany. B. *Helicobacter pylori* adheres to the villi (small bulges) of the gastric epithelium. Bacteria are colorized green.

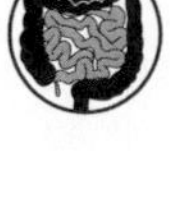

by visible light. The bacterium *Helicobacter pylori* was first reported by Australian scientist Barry Marshall, but it proved difficult to isolate and culture. Ultimately, electron microscopy confirmed the existence of *H. pylori* in the stomach and helped to document its role in gastritis and stomach ulcers.

Thought Question

2.10 What kinds of research questions could you investigate using SEM? What questions could you answer using TEM?

An important limitation of traditional electron microscopy, whether TEM or SEM, is that the fixatives and heavy-atom stains can introduce artifacts into the image, especially at finer details of resolution. In some cases, different preparation procedures have led to substantially different interpretations of subcellular structure. For example, an oval that appears hollow might be interpreted as a cell when, in fact, it represents a deposit of staining material. A microscopic structure that is interpreted incorrectly is termed an **artifact**. Avoiding artifacts is an important concern in microscopy.

Cryo-Electron Microscopy and Tomography

How can electron microscopy achieve finer resolution and avoid artifacts due to staining? High-strength electron

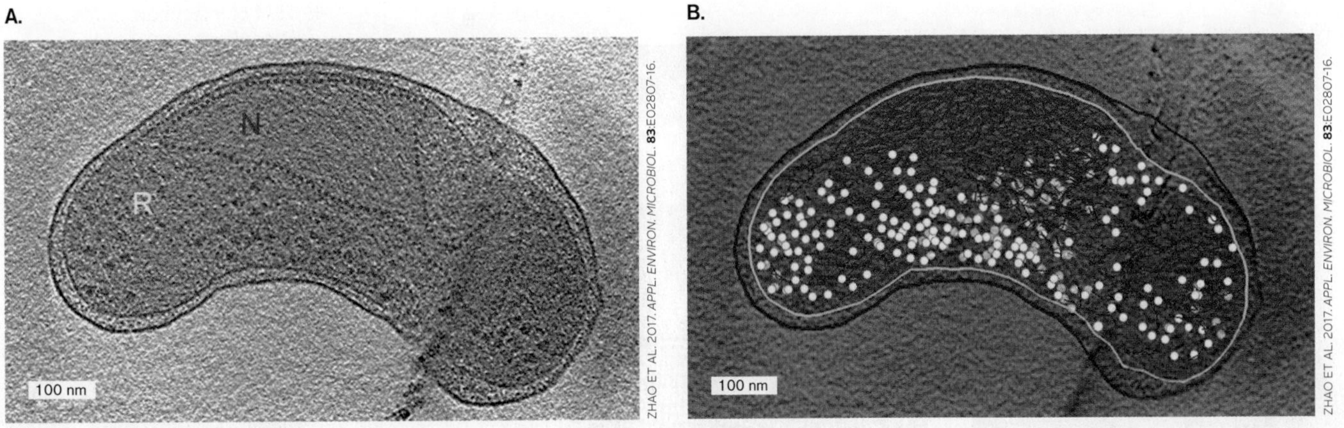

FIGURE 2.41 ▪ Cryo-electron tomography of the marine bacterium *Pelagibacter*. **A.** A single cryo-EM scan lengthwise through *Pelagibacter*. **B.** 3D model of *Pelagibacter* based on multiple scans, showing nucleoid DNA (red), ribosomes (yellow), inner membrane (cyan), and outer membrane (blue).

beams now permit low-temperature **cryo-electron microscopy (cryo-EM)**, also known as **electron cryomicroscopy**.

Cryo-EM avoids staining. In cryo-EM, the specimen does not require staining, because the high-intensity electron beams can detect smaller signals (contrast in the specimen) than earlier instruments could. The specimen must, however, be flash-frozen—that is, suspended in water and frozen rapidly in a refrigerant of high heat capacity (ability to absorb heat). The rapid freezing avoids ice crystallization, leaving the water solvent in a glass-like amorphous phase. The specimen retains water content and thus closely resembles its living form, although it is still ultimately destroyed by electron bombardment.

The cryo-EM micrograph in **Figure 2.41A** shows *Pelagibacter*, an ultrasmall bacterium found throughout the open oceans. *Pelagibacter* is a heterotroph adapted to the lowest concentrations of nutrients. It is one of the world's most abundant life forms, and one of the smallest free-living cells (about 0.01 cubic micrometer). Little is known about its means of survival, with its streamlined genome and cell structure. The cell section reveals in detail the form of its Gram-negative outer membrane (blue), inner membrane (cyan), and periplasmic space between (**Fig. 2.41B**). Within the cytoplasm, we can see the DNA strands of nucleoid (red) and individual ribosomes (yellow). As small as the cell is, we can see its distinctive asymmetry of shape, a curved cell with one end pinched. The function of the asymmetry is unknown.

Note: *Pelagibacter ubique* is the first of several cultured isolates of the "SAR11" cluster of bacteria, originally identified by analysis of DNA sequences of bacteria in the Sargasso Sea near Bermuda. The SAR11 cluster has since been expanded beyond genus to the rank of order: Pelagibacterales, phylum Alphaproteobacteria.

Tomography. Another innovation made possible by cryo-EM is **tomography**, the acquisition of projected images from different angles of a transparent specimen (see the Current Research Highlight). **Cryo-electron tomography**, or **electron cryotomography**, avoids the need to physically slice the sample for thin-section TEM. The images from tomography are combined digitally to visualize objects in 3D, such as the *Pelagibacter* nucleoid and ribosomes (**Fig. 2.41B**). Repeated scans can be summed computationally to obtain an image at high resolution (**Fig. 2.42**). The scans are taken either at different angles or within different focal planes. Each different scan images a slightly different part of the cell. The summed scans then generate a 3D model.

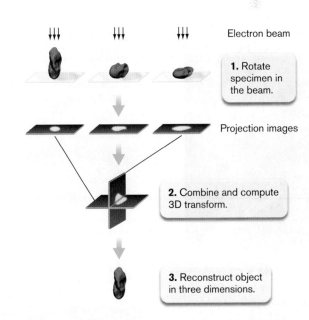

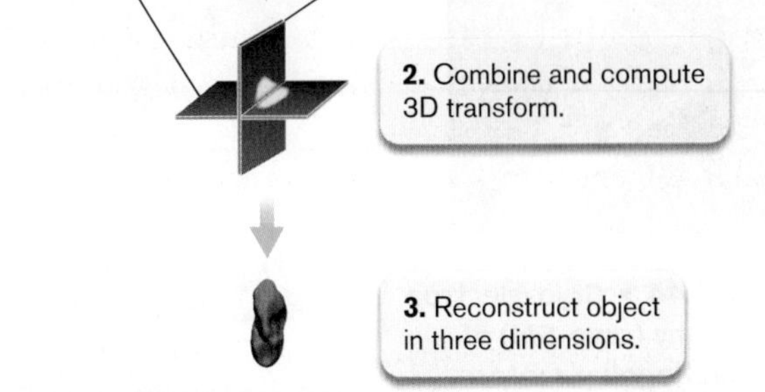

FIGURE 2.42 ▪ 3D image construction in cryo-electron tomography. Cryo-EM images are obtained in multiple focal planes throughout an object. The images are combined through a mathematical transformation to model the entire object in 3D.

A.

B.

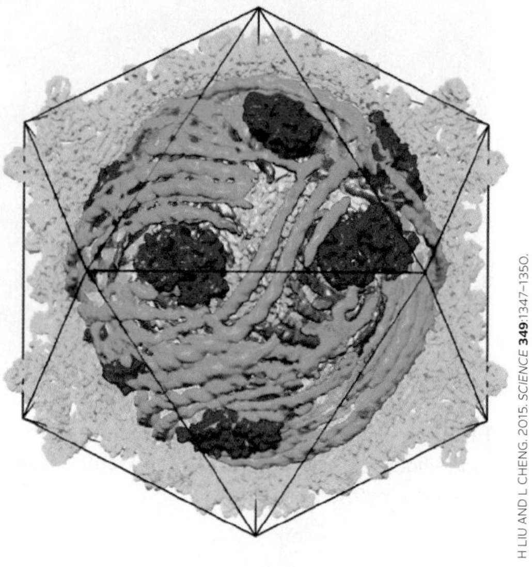

FIGURE 2.43 ■ Cryo-electron tomography reveals virus structure. **A.** Hongrong Liu and Lingpeng Cheng used cryo-EM and symmetry-based computation to model the structure of a cypovirus. **B.** Cypovirus model shows the double-stranded RNA genome (blue) packed inside the capsid, along with viral RNA-dependent RNA polymerases (red).

Modeling cell parts. One use of cryo-electron tomography is to generate high-resolution models of complex particles, such as viruses. For a symmetrical virus, particle images can be rotated for averaging. In addition, images of multiple particles can be averaged together. The digitally combined images can achieve high resolution, nearly comparable to that of X-ray crystallography. Chinese microscopists Hongrong Liu, at Hunan Normal University, and Lingpeng Cheng, at Tsinghua University, used cryo-EM to model cypovirus, a virus that infects silkworms and butterfly larvae (**Fig. 2.43**). Other viruses modeled recently include herpesvirus and human immunodeficiency virus (HIV) (presented in Chapter 11). Cryo-EM is especially useful for particles that cannot be crystallized for X-ray diffraction analysis, the most common means of molecular visualization.

Another example is the modeling of rotary flagellar motors. Bonnie Chaban (**Fig. 2.44A**) and co-workers at Imperial College London used cryo-EM digital combination to obtain detailed models of the motors of Campylobacterales, a clade of bacteria including enteric pathogens. The models allowed reconstruction of each motor interior, including each axle

A.

B.

FIGURE 2.44 ■ Cryo-electron tomography (cryo-EM) of bacterial flagellar motors. **A.** Bonnie Chaban, now at University of Saskatchewan. **B.** The flagellar motor structures of the bacteria *Bdellovibrio bacteriovorus* and *Wolinella succinogenes*.

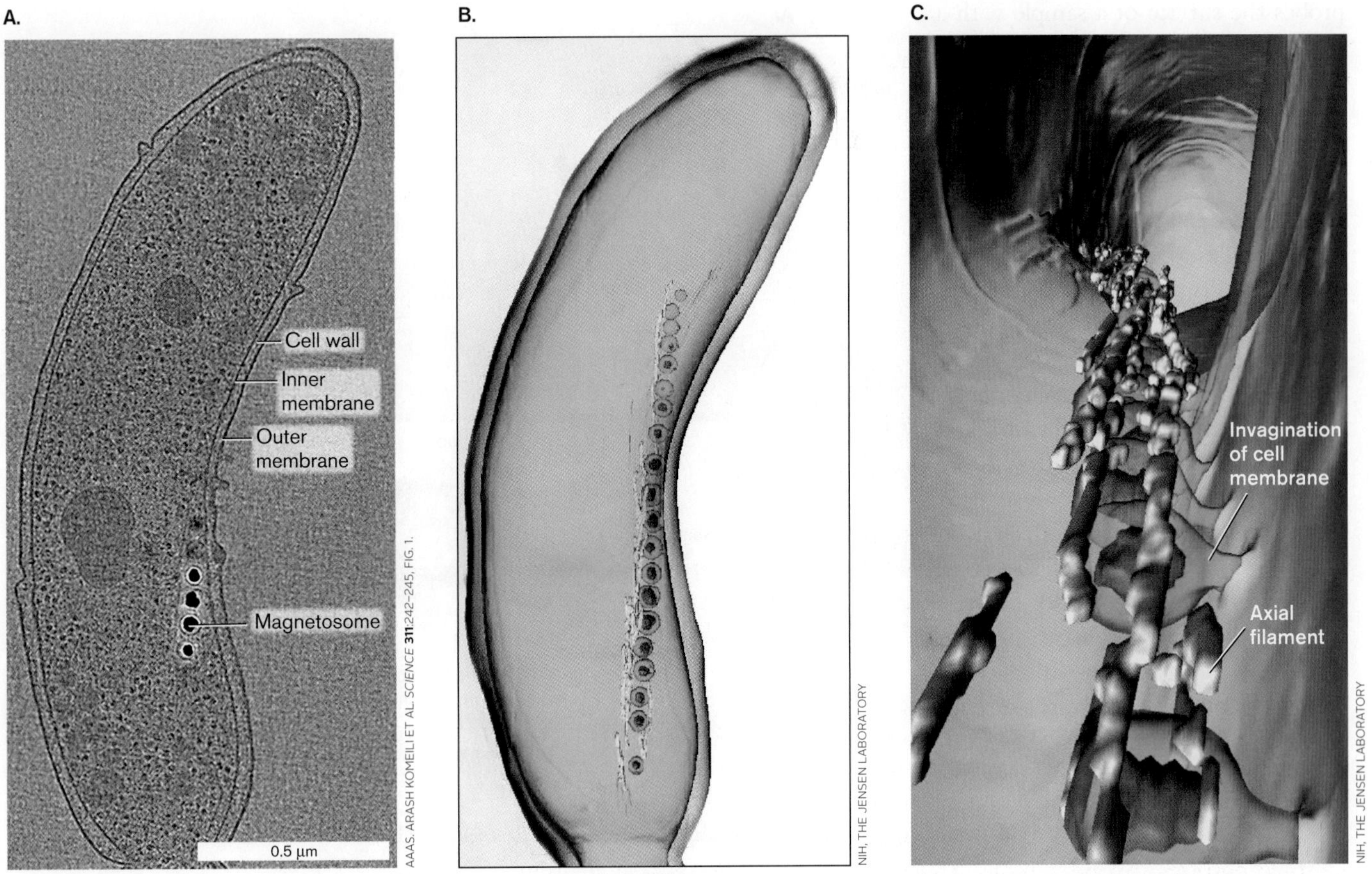

FIGURE 2.45 ■ Magnetotactic cell visualized by cryo-electron tomography. **A.** A single cryo-EM scan lengthwise through *Magnetospirillum magneticum*. **B.** 3D model of *M. magneticum* based on multiple scans. **C.** Expanded view of the cell interior.

(shown as violet). It was possible to compare the motors from related bacteria, such as the *E. coli* predator *Bdellovibrio* and the rumen symbiont *Wolinella* (**Fig. 2.44B**). The comparison enabled computation of a phylogenetic tree of motor evolution—something once thought impossible because of the precision requirements of a rotary device.

Model of a cell. Cryo-EM models of a motor are impressive, but can we build a 3D model of an entire cell? Grant Jensen and colleagues at the California Institute of Technology use cryo-electron tomography to visualize an entire flash-frozen bacterium. Such a model includes all the cell's parts and their cytoplasmic connections—and reveals new structures never seen before.

The cell modeled in **Figure 2.45** is *Magnetospirillum magneticum*, a bacterium that can swim along magnetic field lines because the cell contains a string of magnetic particles composed of the mineral magnetite (iron oxide, Fe_3O_4). A cryo-EM section through the bacterium (**Fig. 2.45A**) shows fine details, including the inner membrane (equivalent to the cell membrane), peptidoglycan cell wall, and outer membrane, an outer covering found in Gram-negative bacteria. Four dark magnetosomes (particles of magnetite) appear in a chain, each surrounded by a vesicle of membrane.

Figure 2.45B models the magnetosomes, reconstructed using multiple cryo-EM scans across the volume of the cell. The magnetosomes are colorized red, each surrounded by a membrane vesicle (green). The vesicles are organized within the cell by a series of protein axial filaments, colorized yellow. **Figure 2.45C** shows an expanded image of the magnetosomes viewed from the cell interior. This expansion reveals that the magnetosome vesicles consist of invaginations from the cell membrane. Thus, the 3D model shows how the magnetite particles are fixed in position by invaginated membranes and held in a line by axial filaments.

Scanning Probe Microscopy

Scanning probe microscopy (SPM) enables nanoscale observation of cell surfaces. Unlike electron microscopy, some forms of SPM can be used to observe live bacteria in water or exposed to air.

SPM techniques measure a physical interaction, such as the "atomic force" between the sample and a sharp tip. **Atomic force microscopy (AFM)** measures the van der Waals forces between the electron shells of adjacent atoms of the cell surface and the sharp tip. In AFM, an instrument probes the surface of a sample with a sharp tip a couple of micrometers long and often less than 10 nm in diameter (**Fig. 2.46A**). The tip is located at the free end of a lever that is 100–200 μm long. The lever is deflected by the force between the tip and the sample surface. Deflection of the lever is measured by a laser beam reflected off a cantilever attached to the tip as the sample scans across. The measured deflections allow a computer to map the topography of cells in liquid medium with a resolution below 1 nm.

In **Figure 2.46B**, AFM was used to observe live bacteria collected on a filter, from seawater off the coast of California. Two round bacteria and a helical bacterium can be seen (raised regions, green-white). The cells were observed in water suspension, without stain. Thus, AFM can help assess the ecological contributions of marine bacteria that cannot be cultured.

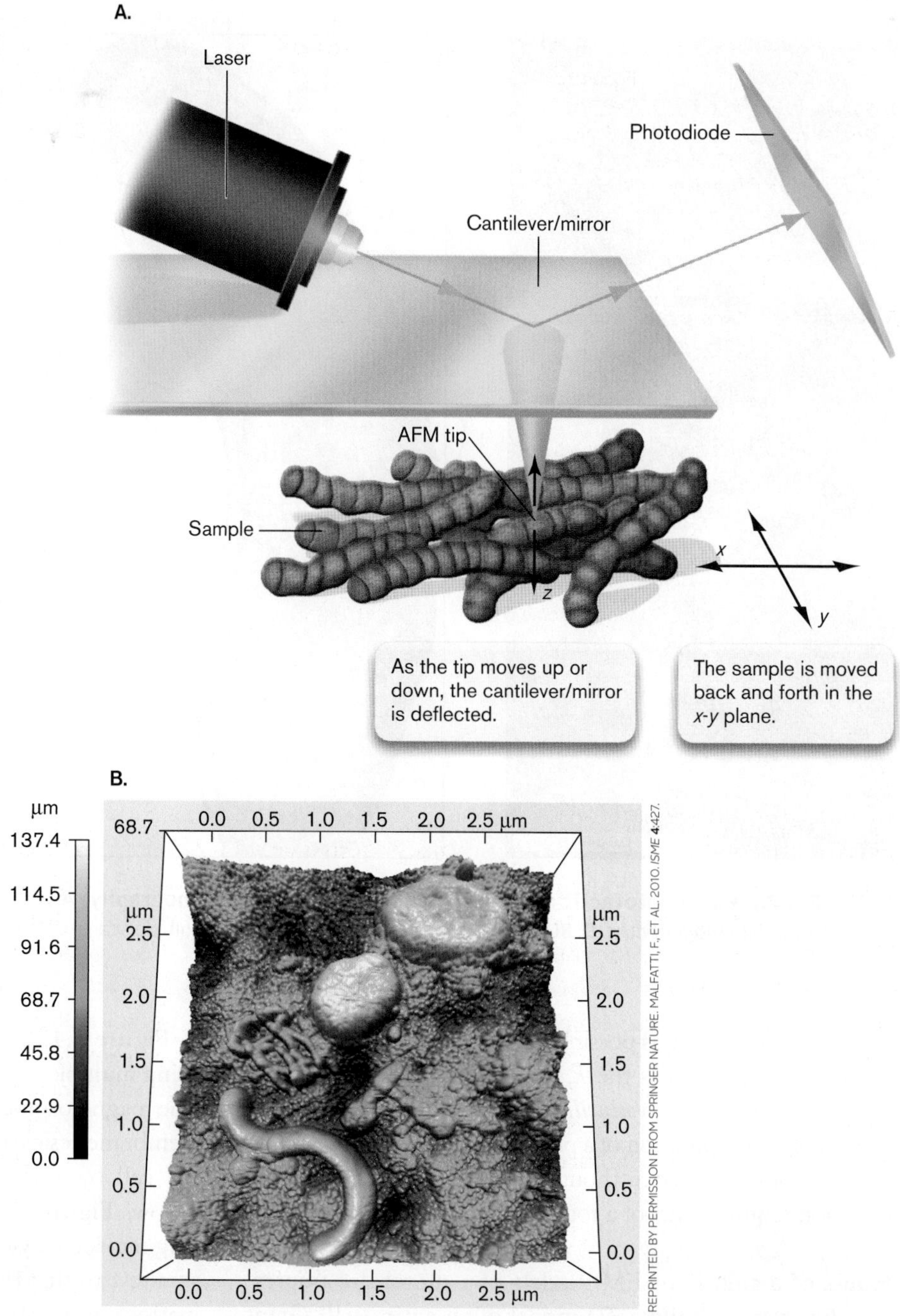

FIGURE 2.46 ▪ Atomic force microscopy enables visualization of untreated cells. **A.** The atomic force microscope (AFM) has a fine-pointed tip attached to a cantilever that moves over a sample. The tip interacts with the sample surface through atomic force. As the tip is pushed away, or pulled into a depression, the cantilever is deflected. The deflection is measured by a laser light beam focused onto the cantilever and reflected into a photodiode detector. **B.** This AFM image shows live bacteria collected on a filter, from seawater off the coast of California. Two round bacteria and a helical bacterium can be seen (raised regions, green-white).

Thought Question

2.11 How could you use atomic force microscopy to study the effect of an antibiotic on *Pseudomonas aeruginosa* contamination of medical catheters?

X-Ray Diffraction Analysis

To know a cell, we need to isolate the cell's individual molecules. The major tool used at present to visualize a molecule is **X-ray diffraction analysis**, or **X-ray crystallography**. Much of our knowledge of microbial genetics (Chapters 7–12) and metabolism (Chapters 13–16) comes from crystal structures of key macromolecules.

Unlike microscopy, X-ray diffraction does not present a direct view of a sample, but instead generates computational models. Dramatic as the models are, they can only represent particular aspects of electron clouds and electron density that are fundamentally "unseeable." That is why we represent molecular structures in different ways that depend on the context—by electron density maps, as models defined by van der Waals radii, or as stick models. Proteins are frequently presented in a cartoon form that shows alpha helix and beta sheet secondary structures.

For a molecule that can be crystallized, X-ray diffraction makes it possible to fix the position of each individual atom in the molecule. Atomic resolution is possible because the wavelengths of X-rays are much shorter than the wavelengths of visible light and are comparable to the size of

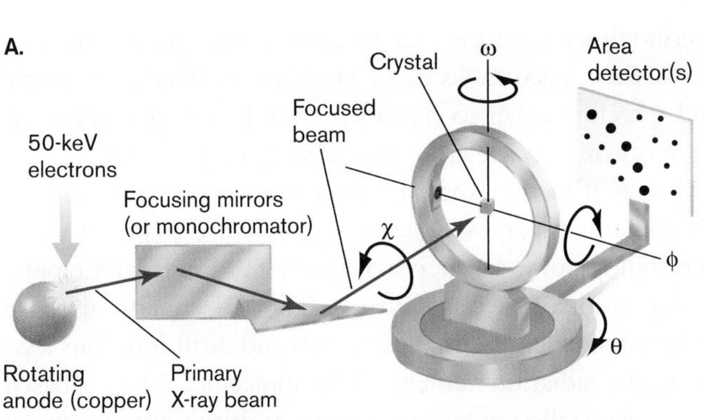

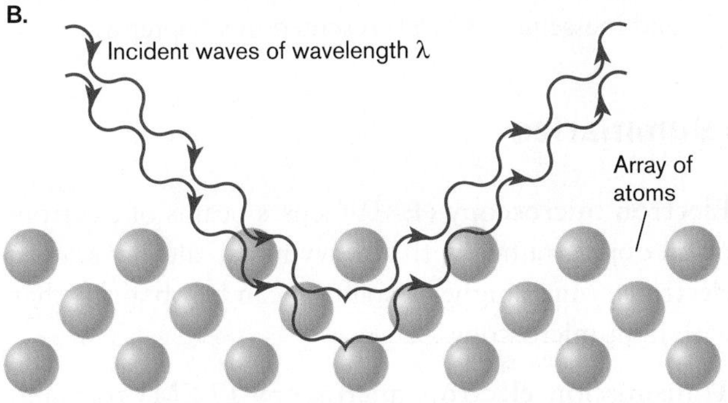

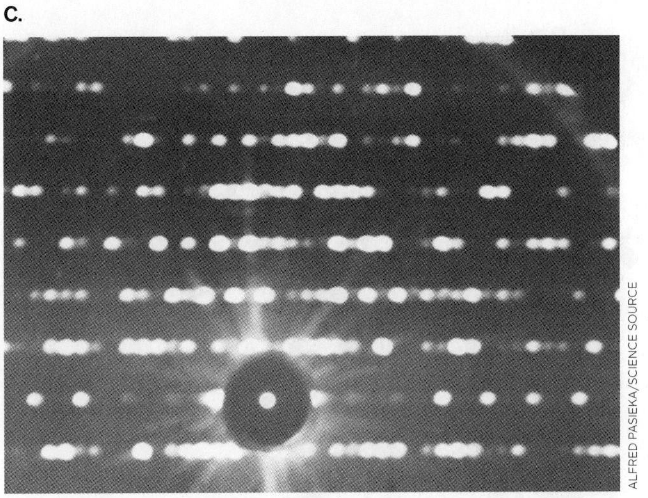

FIGURE 2.47 ▪ Visualizing molecules by X-ray crystallography. **A.** Apparatus for X-ray crystallography. The X-ray beam is focused onto a crystal, which is rotated over all angles to obtain diffraction patterns. The intensity of the diffracted X-rays is recorded on film or with an electronic detector. **B.** X-rays are diffracted by rows of identical molecules in a crystal. The diffraction pattern is analyzed to generate a model of the individual molecules. **C.** Diffraction pattern from a crystal.

atoms. X-ray diffraction is based on the principle of wave interference (see **Fig. 2.34**). The interference pattern is generated when a crystal containing many copies of an isolated molecule is bombarded by a beam of X-rays (**Fig. 2.47A**). The wavefronts associated with the X-rays are diffracted as they pass through the crystal, causing interference patterns. In the crystal, the diffraction pattern is generated by a symmetrical array of many sample molecules (**Fig. 2.47B**). The more copies of the molecule in the array, the narrower the interference pattern and the greater the resolution of atoms within the molecule. Diffraction patterns obtained from the passage of X-rays through a crystal (**Fig. 2.47C**) can be analyzed by computation to develop a precise structural model for the molecule, detailing the position of every atom in the structure.

The application of X-ray crystallography to complex biological molecules was pioneered by the Irish crystallographer John Bernal (1901–1971; **Fig. 2.48A**). Bernal was particularly supportive of women students and colleagues, including Rosalind Franklin (1920–1958), who made important discoveries about DNA and RNA, and Nobel laureate Dorothy Crowfoot Hodgkin (1910–1994).

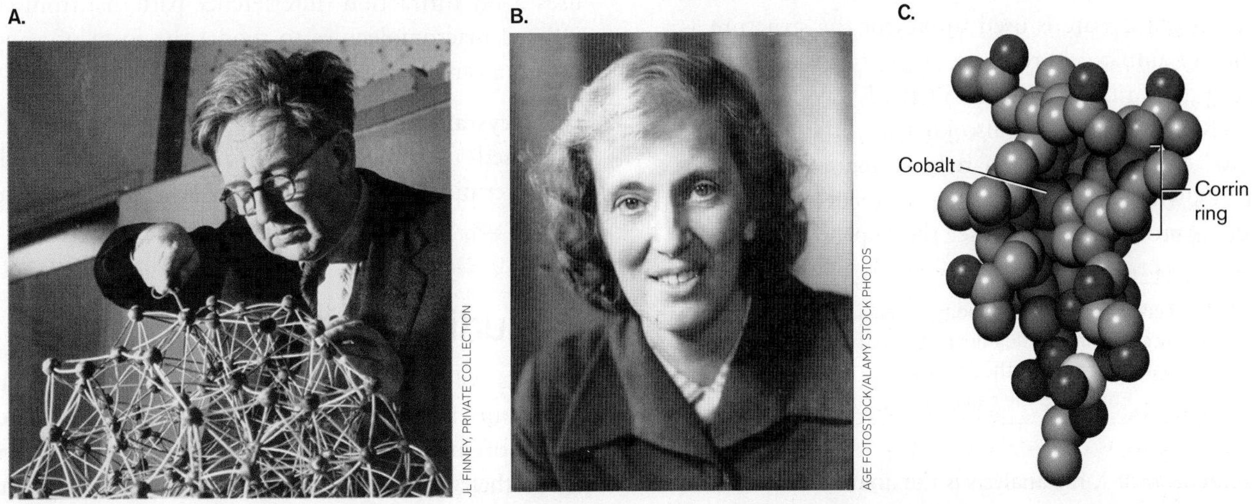

FIGURE 2.48 ▪ Pioneering X-ray crystallography. **A.** John Bernal at Cambridge University developed X-ray crystallography to solve the structure of complex biological molecules. **B.** Dorothy Hodgkin at Oxford University was awarded the 1964 Nobel Prize in Chemistry for her work in X-ray crystallography. **C.** Vitamin B_{12}, whose structure was originally solved by Hodgkin. The corrin ring structure is built around an atom of cobalt (pink). Carbon atoms here are gray; oxygen, red; nitrogen, blue; phosphorus, yellow. Hydrogen atoms are omitted for clarity.

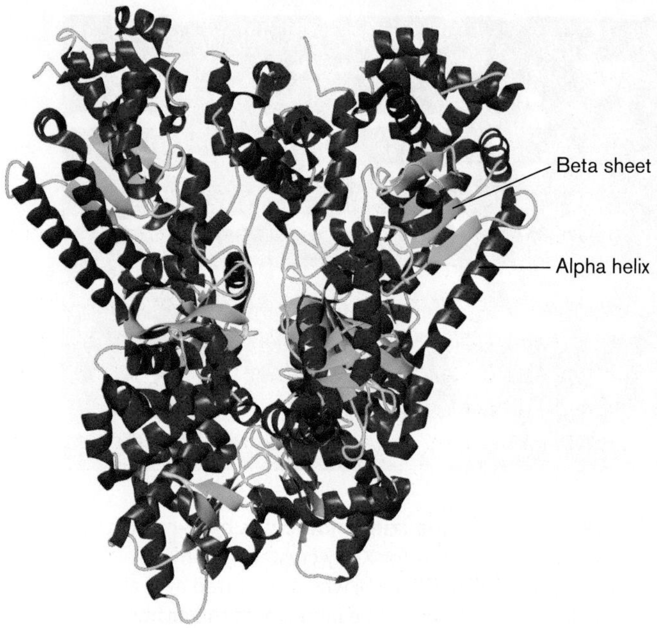

FIGURE 2.49 ■ X-ray crystallography of a protein complex, anthrax lethal factor. The toxin consists of a butterfly-shaped dimer of two peptide chains. This cartoon model is based on X-ray-crystallographic data, showing alpha helix (red coils) and beta sheet (blue arrows). (PDB code: 1J7N)

Hodgkin (**Fig. 2.48B**) solved the crystal structures of penicillin and vitamin B_{12} (**Fig. 2.48C**), as well as one of the first protein structures—that of the hormone insulin.

Today, X-ray data undergo digital analysis to generate sophisticated molecular models, such as the one seen in **Figure 2.49** of anthrax lethal factor, a toxin produced by *Bacillus anthracis* that kills the infected host cells. The model for anthrax lethal factor was encoded in a Protein Data Bank (PDB) text file that specifies coordinates for all atoms of the structure. The Protein Data Bank is a worldwide database of solved X-ray structures, freely available on the Internet. Visualization software is used to present the structure as a "ribbon" of amino acid residues, color-coded for secondary structure. The ribbon diagram method was developed by Jane S. Richardson, and is also known as the "Richardson diagram." In **Figure 2.49**, the red coils represent alpha helix structures, whereas the blue arrows represent beta sheets (for a review of protein structures, see the Appendix).

Note: Molecular and cellular biology increasingly rely on visualization in 3D. Many of the molecules illustrated in this book are based on structural models deposited in the Protein Data Bank, as indicated by the PDB file code. Each PDB file can be viewed in 3D in the browser.

A limitation of X-ray analysis is the unavoidable deterioration of the specimen under bombardment by X-rays. The earliest X-ray diffraction models of molecular complexes such as the ribosome relied heavily on components from thermophilic bacteria and archaea that grow at high temperatures. Because thermophiles have evolved to grow under higher thermal stress, their macromolecular complexes are often more stable and therefore easier to crystallize than homologous proteins from organisms that grow at moderate temperatures.

X-ray diffraction analysis of crystals from a wide range of sources was made possible by cryocrystallography. In cryocrystallography, as in cryo-EM, crystals are frozen rapidly to liquid-nitrogen temperature. The frozen crystals have greatly decreased thermal vibrations and diffusion, thus lessening the radiation damage to the molecules. Models based on cryocrystallography can present multisubunit structures such as the bacterial ribosome complexed with transfer RNAs and messenger RNA (presented in Chapter 8).

To Summarize

- **Electron microscopy (EM)** focuses beams of electrons on an object stained with a heavy-metal salt that scatters electrons. Much higher resolution can be obtained than with light microscopy.
- **Transmission electron microscopy (TEM)** transmits electron beams through a thin section.
- **Scanning electron microscopy (SEM)** involves scanning of a 3D surface with an electron beam.
- **Cryo-electron microscopy (cryo-EM)** involves the observation of samples flash-frozen in water solution. **Tomography** combines multiple images by computation to achieve high resolution.
- **Atomic force microscopy (AFM)**, a form of scanning probe microscopy (SPM), uses intermolecular force measurement to observe cells in water solution.
- **X-ray diffraction analysis**, or **X-ray crystallography**, uses X-ray diffraction (interference patterns) from crystallized macromolecules to model the form of a molecule at atomic resolution.
- **Cryocrystallography** uses frozen crystals with greatly decreased thermal vibrations and diffusion, enabling the determination of structures of large macromolecular complexes, such as the ribosome.

Concluding Thoughts

The tools of microscopy and molecular visualization described in this chapter have shaped our current understanding of microbial cells—how they grow and divide, organize their DNA and cytoplasm, and interact with other cells. Our current models of cell structure and function are explored in Chapter 3. In Chapter 4 we discuss how cells use their structures to obtain energy, reproduce, and develop dormant forms that can remain viable for thousands of years.

CHAPTER REVIEW

Review Questions

1. What principle defines an object as microscopic?
2. Explain the difference between detection and resolution.
3. How do eukaryotic and prokaryotic cells differ in appearance under the light microscope?
4. Explain how electromagnetic radiation carries information and why different kinds of radiation can resolve different kinds of objects.
5. Describe how light interacts with an object through absorption, reflection, refraction, and scattering.
6. Explain how refraction enables magnification of an image.
7. Explain how magnification increases resolution and why empty magnification fails to increase resolution.
8. Explain how angle of aperture and resolution change with increasing lens magnification.
9. Summarize the optical arrangement of a compound microscope.
10. Explain how to focus an object and how to tell when the object is in or out of focus.
11. Explain the relative advantages and limitations of wet mount and stained preparations for observing microbes.
12. Explain the significance (and limitations) of the Gram stain for bacterial taxonomy.
13. Explain the basis of dark-field, phase-contrast, and fluorescence microscopy. Give examples of applications of these advanced techniques.
14. Explain how super-resolution imaging enables tracking of intracellular molecules.
15. Explain the difference between transmission and scanning electron microscopy, and the different applications of each.
16. Explain how cryo-EM reveals the structure of cells and viruses.

Thought Questions

1. Explain which features of bacteria you can study by (a) light microscopy; (b) fluorescence microscopy; (c) scanning EM; (d) transmission EM.
2. Explain how resolution is increased by magnification. Why can't the details be resolved by your unaided eye? Explain why magnification reaches a limit. Why can it not go on resolving greater detail?
3. Explain why artifacts appear in microscopic images, even with the best lenses. Explain how you can tell the difference between an optical artifact and an actual feature of an image.
4. How can "detection without resolution" be useful in microscopy? Explain with specific examples.
5. In D. Berry et al., 2013, *PNAS* **110**:4720, the number of *Bacteroides* foraging on host secretions changes when other species are introduced. Why might this happen? Can you propose an experiment using NanoSIMS to test your hypothesis?

Key Terms

aberration (46, 47)
absorption (43)
acid-fast stain (52)
angle of aperture (46)
antibody tag (53)
artifact (66)
atomic force microscopy (AFM) (42, 69)
bacillus (40)
bright-field microscopy (40, 46)
chemical imaging microscopy (42)
coccus (40)
compound microscope (47)
condenser (47)
confocal laser scanning microscopy (58)
contrast (43, 47)
counterstain (50)
cryo-electron microscopy (cryo-EM) (electron cryomicroscopy) (67)
cryo-electron tomography (electron cryotomography) (67)
dark-field microscopy (61)
depth of field (48)
detection (39)
differential stain (50)
electromagnetic radiation (42)
electron microscopy (EM) (40, 63)
emission wavelength (54)
empty magnification (45)
excitation wavelength (54)
fixation (50)
flagellum (61)
fluorescence (43)
fluorescence in situ hybridization (FISH) (54)
fluorophore (54)
focal point (44)
focus (38)
Gram-negative (50)
Gram-positive (50)
Gram stain (50)

immersion oil (47)
interference (45)
lens (44)
light microscopy (LM) (40)
magnification (39, 43)
microscope (37)
mordant (50)
nanoscale secondary ion mass spectrometry (NanoSIMS) (60)
negative stain (53, 65)
numerical aperture (47)
objective lens (46, 47)
ocular lens (47)
parfocal (48)
phase-contrast microscopy (PCM) (62)
reflection (43)
refraction (43)
refractive index (43)
resolution (38)
scanning electron microscopy (SEM) (40, 64)
scanning probe microscopy (SPM) (42, 69)
scattering (43)
simple stain (50)
spirochete (40)
spore stain (52)
staining (50)
super-resolution imaging (45, 57)
tomography (67)
total magnification (47)
transmission electron microscopy (TEM) (42, 64)
wet mount (49)
X-ray crystallography (X-ray diffraction analysis) (42, 70)

Recommended Reading

Berry, D., B. Stecher, A. Schintlmeister, J. Reichert, S. Brugiroux, et al. 2013. Host-compound foraging by intestinal microbiota revealed by single-cell stable isotope probing. *Proceedings of the National Academy of Sciences USA* **110**:4720.

Gahlmann, Andreas, and William E. Moerner. 2014. Exploring bacterial cell biology with single-molecule tracking and super-resolution imaging. *Nature Reviews. Microbiology* **12**:9–22.

Jiang, W., J. Chang, J. Jakana, P. Weigele, J. King, et al. 2006. Structure of epsilon15 bacteriophage reveals genome organization and DNA packaging-injection apparatus. *Nature* **439**:612–616.

Komeili, A., Z. Li, D. K. Newman, and G. J. Jensen. 2006. Magnetosomes are cell membrane invaginations organized by the actin-like protein MamK. *Science* **311**:242–245.

Matias, Valério R. F., Ashruf Al-Amoudi, Jacques Dubochet, and Terry J. Beveridge. 2003. Cryo-transmission electron microscopy of frozen-hydrated sections of *Escherichia coli* and *Pseudomonas aeruginosa*. *Journal of Bacteriology* **185**:6112–6118.

Murphy, Douglas B. 2001. *Fundamentals of Light Microscopy and Electronic Imaging.* Wiley-Liss, Hoboken, NJ.

Oikonomou, Catherine M., Yi-Wei Chang, and Grant J. Jensen. 2016. A new view into prokaryotic cell biology from electron cryotomography. *Nature Reviews. Microbiology* **14**:205–220.

Popescu, Aurel, and R. J. Doyle. 1996. The Gram stain after more than a century. *Biotechniques in Histochemistry* **71**:145–151.

Ptacin, Jerod L., Steven F. Lee, Ethan C. Garner, Esteban Toro, Michael Eckart, et al. 2010. A spindle-like apparatus guides bacterial chromosome segregation. *Nature Cell Biology* **12**:791–798.

Rodriguez, Erik A., Robert E. Campbell, John Y. Lin, Michael Z. Lin, Atsushi Miyawaki, et al. 2017. The growing and glowing toolbox of fluorescent and photoactive proteins. *Trends in Biochemical Sciences* **42**:111–129.

Söderström, Bill, Helena Chan, and Daniel O. Daley. 2019. Super-resolution images of peptidoglycan remodeling enzymes at the division site of *Escherichia coli*. *Current Genetics* **65**:99–101.

Su, Zhaoming, Chao Wu, Liuqing Shi, Priya Luthra, Grigore D. Pintilie, et al. 2018. Electron cryo-microscopy structure of Ebola virus nucleoprotein reveals a mechanism for nucleocapsid-like assembly. *Cell* **172**:966–978.

CHAPTER 3

Cell Structure and Function

3.1 The Bacterial Cell: An Overview

3.2 The Cell Membrane and Transport

3.3 The Envelope and Cytoskeleton

3.4 Bacterial Cell Division

3.5 Cell Polarity, Membrane Vesicles, and Nanotubes

3.6 Specialized Structures

A bacterial cell contains surprising devices, such as a chemoreceptor "nose" to find food, and motorized flagella to swim toward it. Unlike eukaryotes, a bacterium continually builds new parts while copying its DNA in time for cell division. Cytoplasmic vesicles and interconnecting nanotubes share nutrients with the microbial community. The bacterial cell's machines give us tools for technology, as well as targets for antibiotics and vaccines.

CURRENT RESEARCH highlight

Bacteria have a chemoreceptor "nose." Bacteria live in a world of "feast or famine," where they need to hunt for food. Ariane Briegel, microbiologist, University of Leiden, studies chemoreceptors of *Bdellovibrio bacteriovorus*. At one pole of the bacterium is a crystalline array of chemoreceptors that each extend through the cell membrane into the periplasm. When the periplasmic receptor binds a higher concentration of nutrients, its cytoplasmic end sends a phosphorylated signal protein through the cell to keep the cell swimming toward the source of nutrients.

Sources: Catherine M. Oikonomou and Grant Jensen. 2017. *Annu. Rev. Biochem.* **86:**873; Ariane Briegel et al. 2014. *Biochemistry* **53:**1575.

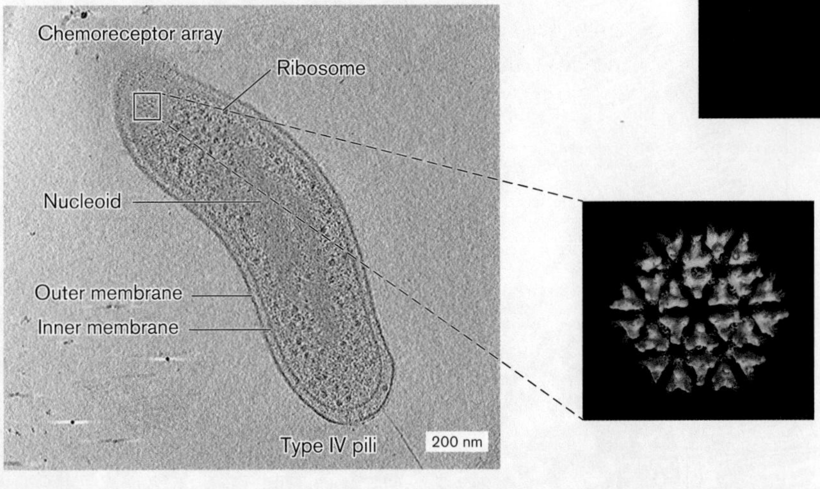

OIKONOMOU AND JENSEN. 2017. *ANNU REV BIOCHEM* **86**:873–876.

AN INTERVIEW WITH

ARIANE BRIEGEL, MICROBIOLOGIST, UNIVERSITY OF LEIDEN

INSTITUTE OF BIOLOGY LEIDEN

What is your main research interest?

What fascinates me most is how bacteria detect signals from their environment. How can they find sugars, avoid toxins, and adapt to changing environments? We already know that bacteria use certain signals, like the presence of particular chemical substances, to determine where they need to go. But how do they detect those signals? What does their "nose" look like? The images we make with cryo-electron tomography show us that many types of bacteria have large chemoreceptors.

To grow and multiply, a microbial cell must obtain nutrients faster than its competitors do—or else share nutrients for mutual advantage. At the same time, each cell's envelope must exclude toxins such as your intestinal bile salts. To manage all these survival tasks, microbes have evolved an amazing set of molecular parts, from chemoreceptors that process signals (see the Current Research Highlight) to toxin secretion machines. Chapter 3 explores the key components of bacteria and archaea; the parts of microbial eukaryotes are presented in Chapter 20.

Most bacteria share these traits:

- **Thick, complex outer envelope.** The envelope protects the cell from environmental stress and mediates exchange with the environment.
- **Compact genome.** Prokaryotic genomes are compact, with relatively little noncoding DNA. Small genomes maximize the production of cells from limited resources.
- **Tightly coordinated functions.** The cell's parts work together in a highly coordinated mechanism, which may enable a high rate of reproduction.

Beyond the fundamentals, bacteria evolve amazing diversity, which we will explore in Chapter 18. We also discuss the diversity of archaea in Chapter 19, and of microbial eukaryotes such as fungi and protists in Chapter 20. Overall, the three domains show these traits:

- **Archaea, like bacteria, are prokaryotes (cells that lack a nucleus).** Archaea have unique membrane and envelope structures that enable survival in extreme environments. Many archaea also live in moderate environments, such as soil, water, or human skin.
- **Eukaryotic cells have extensive membranous organelles.** Organelles such as the endoplasmic reticulum and Golgi complex are reviewed in eAppendix 2. The mitochondria and chloroplasts of eukaryotic cells evolved by endosymbiosis with engulfed bacteria (see Chapter 17).

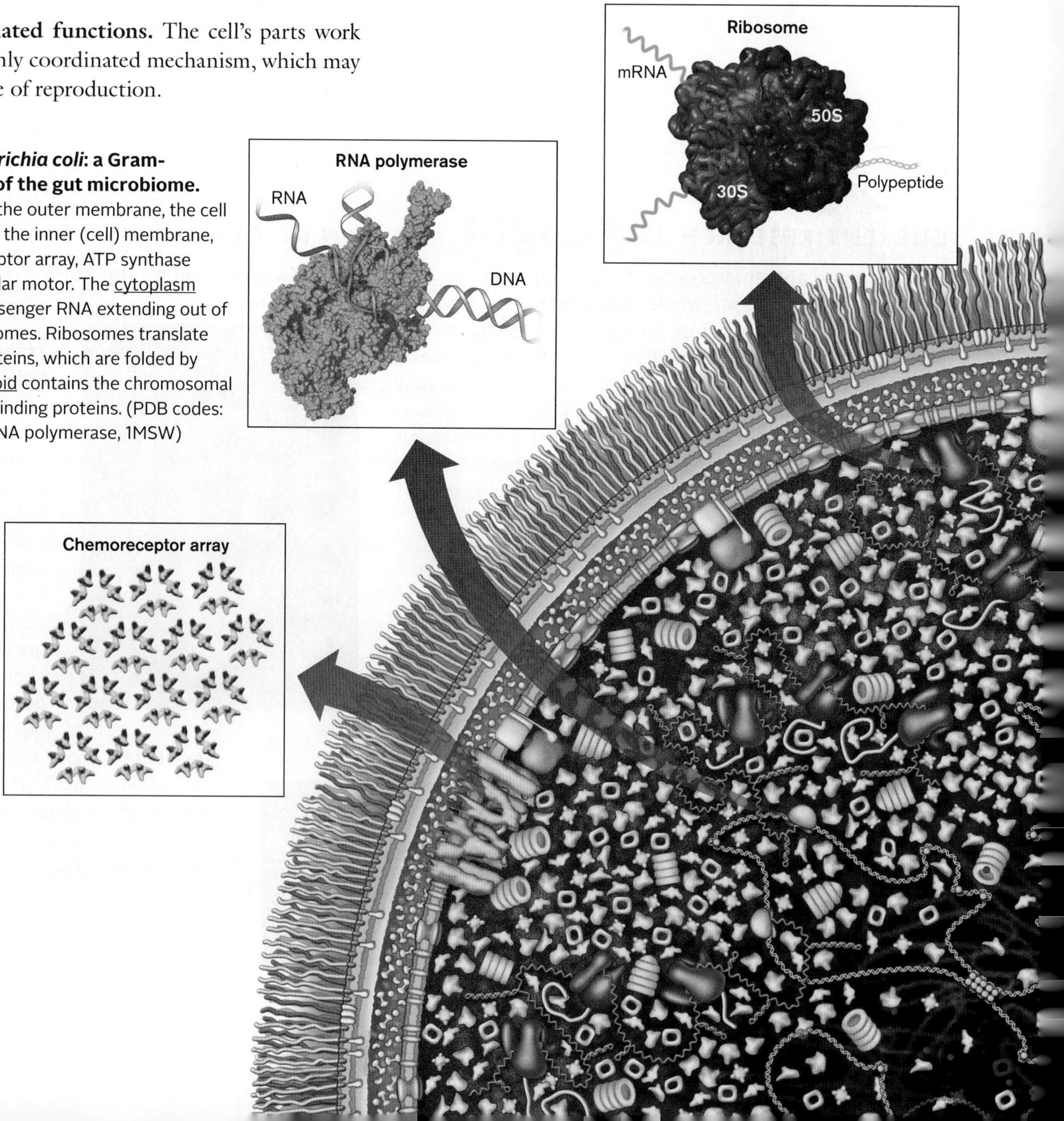

FIGURE 3.1 ▪ *Escherichia coli*: a Gram-negative bacterium of the gut microbiome. The envelope includes the outer membrane, the cell wall and periplasm, and the inner (cell) membrane, as well as a chemoreceptor array, ATP synthase complexes, and a flagellar motor. The cytoplasm contains enzymes, messenger RNA extending out of the nucleoid, and ribosomes. Ribosomes translate the mRNA to make proteins, which are folded by chaperones. The nucleoid contains the chromosomal DNA wrapped around binding proteins. (PDB codes: ribosome, 1GIX, 1GIY; RNA polymerase, 1MSW)

Diverse microbial eukaryotes, such as fungi and protists, are explored in Chapter 20.

Now we embark on a tour of a typical Gram-negative bacterial cell—a common resident of your gut, *Escherichia coli* (**Fig. 3.1**). Along the way, we learn how our understanding of this model cell has emerged from microscopy, cell fractionation, and genetic analysis.

Note: This chapter assumes an understanding of introductory biology and chemistry. For review, see eAppendix 1 ("Biological Molecules") and eAppendix 2 ("Introductory Cell Biology").

3.1 The Bacterial Cell: An Overview

A cell is more than a "soup" full of ribosomes and enzymes. In fact, the cell's parts fit together in a structure that is ordered, though flexible. Our model of the bacterial cell (**Fig. 3.1**) offers an interpretation of how the major components of one cell fit together. The model represents *Escherichia coli.* Its general features apply to many kinds of bacteria, particularly the Gram-negative inhabitants of your colon, such as Proteobacteria and Bacteroidetes. Remember that we cannot literally "see" all molecules within a cell, but microscopy (Chapter 2) and subcellular analysis (this section) provide the evidence for our model.

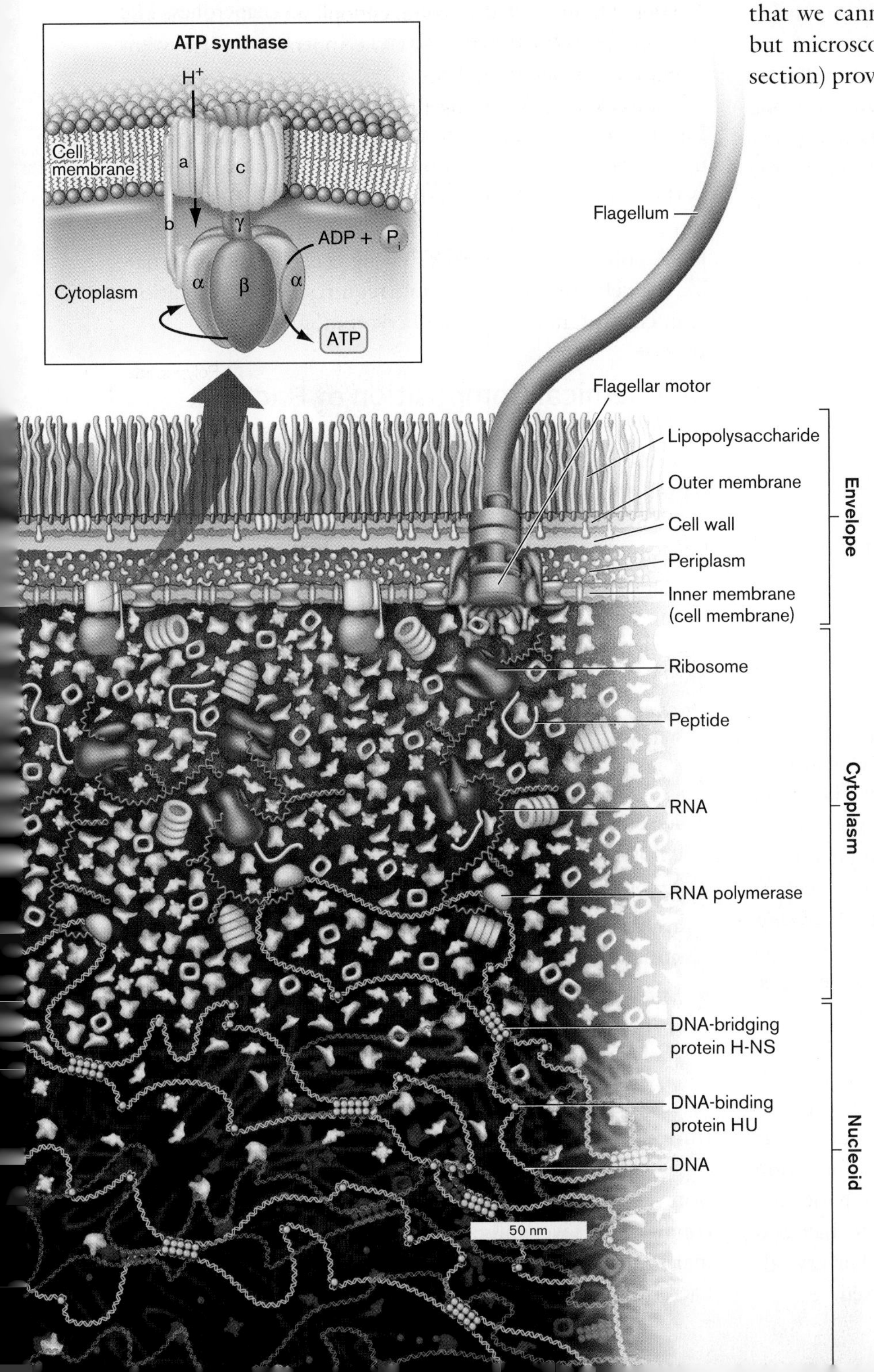

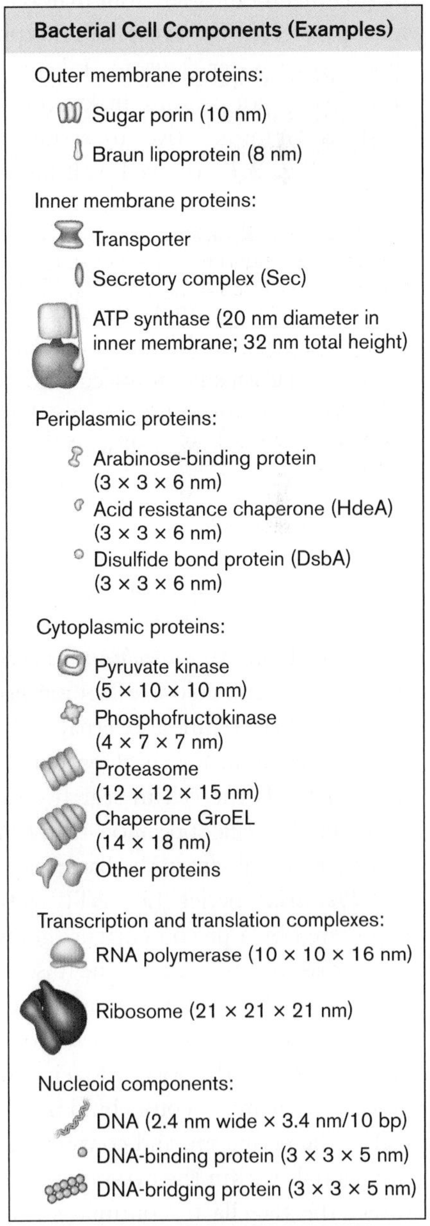

Model of a Bacterial Cell

Within a cell, the cytoplasm consists of a gel-like network composed of proteins and other macromolecules. The cytoplasm is contained by a **cell membrane**, or **plasma membrane**. For *E. coli*, a Gram-negative bacterium (discussed in Section 3.3), the plasma membrane is called the **inner membrane**, in order to distinguish it from the additional **outer membrane**. The inner membrane is composed of phospholipids, transporter proteins, and other molecules. This membrane prevents cytoplasmic proteins from leaking out and maintains gradients of ions and nutrients.

Between the inner and outer membranes lies the **cell wall**, a fortress-like structure composed of sugar chains linked covalently by peptides (peptidoglycan). The cell wall forms a single molecule that surrounds the cell. A Gram-positive species would have the cell wall outside its one plasma membrane. The wall material is flexible, but it limits expansion of the cytoplasm, keeping the cell membrane intact when water flows in. Like air inside a balloon, the resulting turgor pressure makes the cell rigid.

In a Gram-negative bacterium, such as our model *E. coli* (**Fig. 3.1**), the cell wall lies within the **periplasm**, a water-filled space containing nutrient-binding proteins and secretion machines. Outside the cell wall lies the outer membrane of phospholipids and **lipopolysaccharides (LPS)**, a class of lipids attached to long polysaccharides (sugar chains). The LPS layer may be surrounded by a thick capsule. The capsule polysaccharides form a slippery layer that inhibits phagocytosis by amebas or white blood cells. The inner membrane, cell wall, and outer membrane constitute the Gram-negative cell **envelope**.

The bacterial envelope includes cell-surface proteins that enable the bacterium to interact with specific host organisms. For example, *E. coli* cell-surface proteins help the bacterium colonize the human intestinal epithelium. The nitrogen-fixing symbiont *Sinorhizobium* has cell-surface proteins that help the bacteria colonize legume plants for nitrogen fixation. Another common external structure is the **flagellum** (plural, **flagella**), a helical protein filament whose rotary motor propels the cell in search of a more favorable environment. Flagella appear on many Proteobacteria, such as *E. coli*, but are absent in the Bacteroidetes, such as *Bacteroides* and *Prevotella* species. Like ATP synthase, flagellar rotation is powered by a proton current (gradient of hydrogen ions, H^+) across the inner membrane (discussed in Chapter 14).

To sense the environment and to direct flagellar motion, motile bacteria possess an array of chemoreceptors, usually situated at the forward-facing pole opposite the flagella. The chemoreceptors bind molecules from outside the cell or the periplasm, and convert this binding information into signals within the cytoplasm. The signaling molecules direct the flagella to continue swimming, or to halt and turn the cell in a new direction (Section 3.6). The crystalline array of receptors amplifies the signal received.

Within the cell, the cell membrane and envelope provide an attachment point for one or more chromosomes. The chromosome is organized within the cytoplasm as a system of looped coils called the **nucleoid**. Unlike the round, compact nucleus of eukaryotic cells, the bacterial nucleoid is <u>not</u> enclosed by a membrane. Instead, loops of DNA extend throughout the cytoplasm. The DNA is transcribed by RNA polymerase to form messenger RNA (mRNA), as well as transfer RNA (tRNA) and ribosomal RNA (rRNA). As the mRNA transcripts extend, they bind ribosomes to start synthesizing polypeptide chains. As the polypeptides grow, they fold into their three-dimensional shape. This process may require help from protein complexes called chaperones. The concept of information flow from DNA to RNA to protein is presented in Chapters 7–10.

How can we know how all the cell parts shown in **Figure 3.1** interact and work together? Fluorescence microscopy can show us how the nucleoid DNA expands as the cell grows. As we learned in Chapter 2, fluorescence can pinpoint a molecule or complex within a cell, such as DNA polymerase fused to YFP (Fig. 2.27B). But to visualize the cell's interior as a whole requires higher resolution, as in electron microscopy and cryotomography.

Biochemical Composition of Bacteria

The bacterial cell model in **Figure 3.1** represents the shape and size of cell parts but tells us little about the chemistry of the cell, or of the cell's environment. Chemistry explains, for example, why wiping a surface with ethanol kills microbes, whereas water has little effect. Water is a universal constituent of cytoplasm but is excluded by cell membranes. Ethanol, however, dissolves both polar and nonpolar substances; thus, ethanol disintegrates membranes and destroys the folded structure of proteins. For a review of elementary chemistry, see eAppendix 1.

All cells share common chemical components:

- **Water**, the fundamental solvent of life
- **Essential ions**, such as potassium, magnesium, and chloride ions
- **Small organic molecules**, such as lipids and sugars, that are incorporated into cell structures and that provide nutrition by catabolism
- **Macromolecules**, such as nucleic acids and proteins, that contain information, catalyze reactions, and mediate transport, among many other functions

The details of bacterial chemistry emerged in the 1950s through pioneering studies by Fred Neidhardt at the University of Michigan, and many colleagues. Cell composition varies with species, growth phase, and environmental conditions (as we will see in later chapters). **Table 3.1** summarizes the chemical components of a cell for the model bacterium *Escherichia coli* during exponential growth.

TABLE 3.1 Molecules of a Bacterial Cell, *Escherichia coli*, during Balanced Exponential Growth[a]

Component	Percentage of total weight[b]	Approximate number of molecules/cell	Number of different kinds
Water	70	20,000,000,000	1
Proteins	16	2,400,000	4,000[c]
RNA:			
rRNA, tRNA, and other small RNA (sRNA) molecules	6	250,000	200
mRNA	0.7	4,000	2,000[c]
Lipids:			
Phospholipids (membrane)	3	25,000,000	50
Lipopolysaccharides (outer membrane)	1	1,400,000	1
DNA	1	2[d]	1
Metabolites and biosynthetic precursors	1.3	50,000,000	1,000
Peptidoglycan (murein sacculus)	0.8	1	1
Inorganic ions	0.1	250,000,000	20
Polyamines (mainly putrescine and spermidine)	0.1	6,700,000	2

[a]Values shown are for a hypothetical "average" cell cultured with aeration in glucose medium with minimal salts at 37°C.

[b]The total weight of the cell (including water) is about 10^{-12} gram (g), or 1 picogram (pg).

[c]The number of different kinds is difficult to estimate for proteins and mRNA because some genes are transcribed at extremely low levels and because proteins and RNA include kinds that are rapidly degraded.

[d]In rapidly growing cells, cell fission typically lags approximately one generation behind DNA replication—hence, two identical DNA copies per cell.

Source: Modified from F. Neidhardt and H. E. Umbarger. 1996. Chemical composition of *Escherichia coli*, p. 14. In F. C. Neidhardt (ed.), *Escherichia coli and Salmonella: Cellular and Molecular Biology*, 2nd ed. ASM Press, Washington, DC.

Small molecules and ions. The *E. coli* cell consists of about 70% water, the essential solvent required to carry out fundamental metabolic reactions and to stabilize proteins. The water solution contains inorganic ions, predominantly potassium, magnesium, and phosphate. Inorganic ions store energy in the form of transmembrane gradients, and they serve essential roles in enzymes. For example, a magnesium ion is required at the active site of RNA polymerase to help catalyze the linking of ribonucleotides into RNA.

The cell also contains many kinds of small, charged organic molecules, such as phospholipids and enzyme cofactors. A major class of organic cations is the polyamines, molecules with multiple amine groups that are positively charged when the pH is near neutral. Polyamines balance the negative charges of the cell's DNA and stabilize ribosomes during translation.

Macromolecules. Many cells have similar content of water and small molecules, but their specific character is defined by their macromolecules, especially their nucleic acids (DNA and RNA) and their proteins. DNA and RNA molecules can be isolated by size using agarose-gel **electrophoresis**, in which the negatively charged molecules migrate in an electrical field (see eAppendix 3). The nucleic acid content of bacteria is relatively high, nearly 8% for *E. coli*—much higher than in multicellular eukaryotes. For microbes, the high nucleic acid content allows the cell to maximize reproduction of its chromosome while minimizing cell resources.

A high proportion of nucleic acids in food is a problem for us, because humans lack the enzymes to digest the uric acid waste product of nucleotides. That's why we cannot eat most kinds of bacteria as a major part of our diet. Nevertheless, we do consume bacteria within vegetables, because all plants have bacteria growing within their transport tissues.

The cell's genomic DNA directs expression of its proteins (discussed in Chapters 7–10). A given cell uses different genes to make different proteins, depending on environmental conditions such as temperature, nutrient levels, and entry into a host organism. Individual proteins are made in very different amounts, from 10 per cell to 10,000 per cell. The proteins expressed by a cell under given conditions are known collectively as a proteome. (We discuss the proteome in Chapter 8, as a function of the DNA genome and RNA transcriptome.) Other kinds of macromolecules are found in the cell wall and outer membrane. The bacterial cell wall consists of **peptidoglycan**, an organic polymer of peptide-linked sugars that constitutes nearly 1% of the cell mass, approximately the same mass as that of DNA. Peptidoglycan limits the volume of the enclosed cell, so water rushing in generates turgor pressure. This investment of biomass in the cell wall shows the importance (for most species) of maintaining turgor pressure in dilute environments, where water would otherwise enter by osmosis, causing osmotic shock (see the Appendix and Section 3.2).

Thought Question

3.1 Which chemicals do we find in the greatest number in a bacterial cell? The smallest number? Why does a cell contain 100 times as many lipid molecules as strands of RNA?

Cell Fractionation

Cell fractionation (**Fig. 3.2**) is how we separate cell components such as membranes, ribosomes, and flagella. We can study these isolated parts in detail, though we lose information about interactions with other parts of the cell. Cell fractionation also provides purified proteins that act as antigens for candidate vaccines. For example, a vaccine against *Neisseria meningitidis* type B (meningococcus) contains a highly immunogenic outer membrane protein.

How can we disassemble a cell to isolate its parts? Early-twentieth-century microbiologists wondered how to separate cell parts, without all the molecules mixing together. The answer was ultracentrifugation, the separation of molecules subject to high *g* forces (explained in eAppendix 3). Mary Jane Osborn at the University of Connecticut Health Center discovered that the inner and outer membranes of Gram-negative bacteria have different densities, and thus can be separated by density gradient ultracentrifugation.

Cell fractionation requires techniques that **lyse** (break open) the cell. The lysis method must generate enough force to separate the membrane lipids (held together by hydrophobic force) but not enough to disintegrate complexes of protein and RNA. For a Gram-negative cell, the method requires further specificity to separate "compartments" containing different sets of proteins: the inner and outer membranes, and the aqueous cytoplasm and periplasm.

Cell wall lysis and spheroplast formation. First we suspend the cells in a sucrose solution containing ethylenediaminetetraacetic acid (EDTA), which disrupts the outer membrane by removing Mg^{2+} and Ca^{2+} ions (**Fig. 3.2**, step 1). The disrupted membrane allows sucrose to cross. Sucrose fills the periplasm, maintaining an osmotically stable solution. Next, lysozyme cleaves peptidoglycan and thus breaks down the cell wall (step 2). Lacking the turgid cell wall, the cell swells into a sphere called a **spheroplast**.

To isolate the periplasmic contents, the spheroplasts are transferred to distilled water (**Fig. 3.2**, step 3). Water rushes in through the EDTA-weakened outer membrane, causing osmotic shock of the periplasmic compartment, while the inner membrane remains intact. The periplasm leaks out into the extracellular medium, where its proteins can be collected. The periplasm is a valuable source of interesting proteins such as sugar transporters and chaperones (proteins that help other proteins fold under stress).

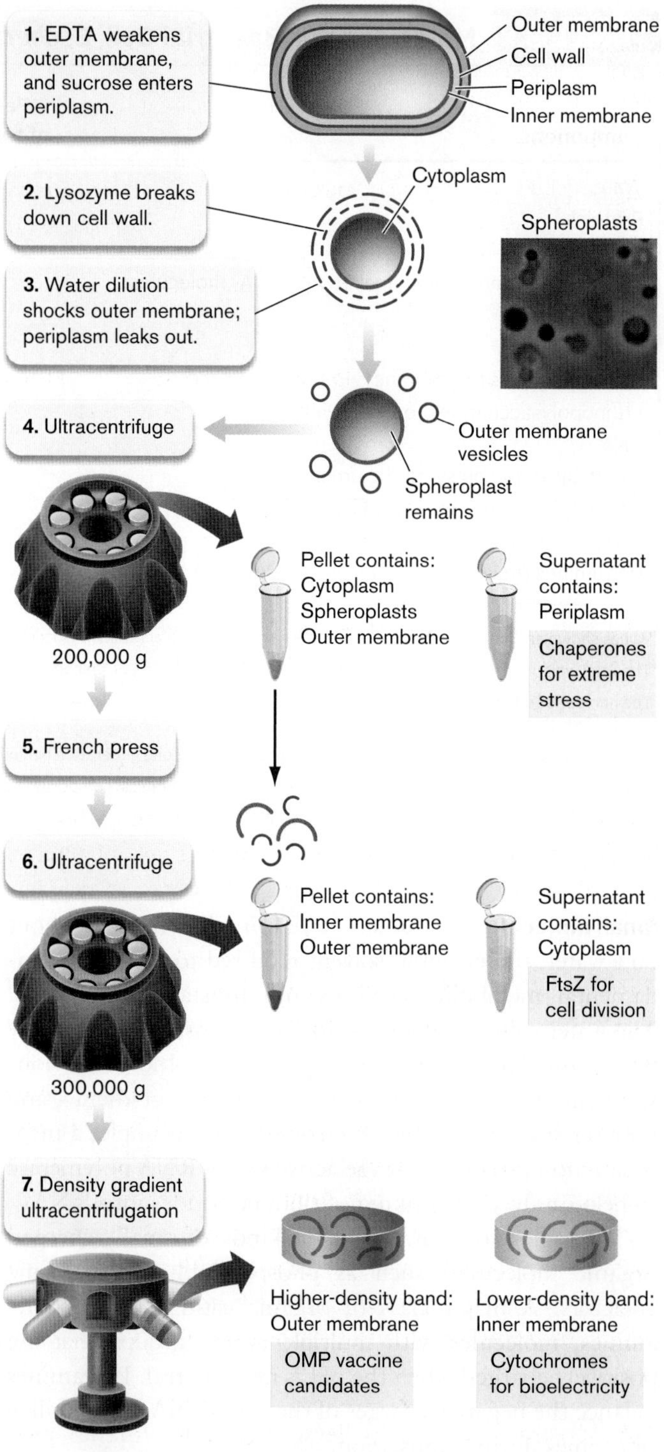

FIGURE 3.2 ▪ **Fractionation of Gram-negative cells.** Cell periplasm fills with sucrose, and lysozyme breaks down the cell wall. Dilution in water causes osmotic shock to the outer membrane, and periplasmic proteins leak out. Subsequent centrifugation steps separate the proteins of the periplasm, cytoplasm, and inner and outer membranes. EDTA = ethylenediaminetetraacetic acid; OMP = outer membrane protein.

Note that other means of cell disruption may work better for other kinds of cells. Mild-detergent lysis can dissolve membranes without denaturing proteins. Alternatively, sonication is a way to lyse a cell by intense

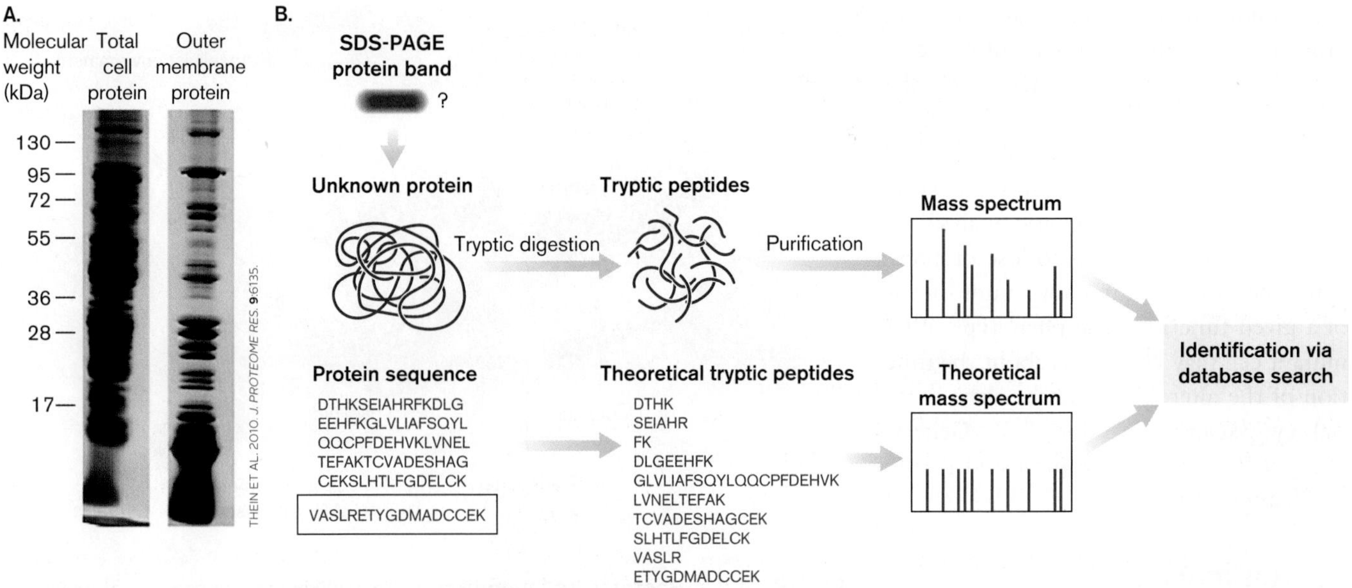

FIGURE 3.3 ■ Protein analysis. **A.** Gel electrophoresis of total cell proteins compared to outer membrane proteins from cell fractionation. **B.** Outer membrane proteins are identified by tryptic digestion and mass spectrum analysis. The resulting peptide sequence is compared with those predicted from genomic data.

ultrasonic vibrations that are above the range of human hearing. For especially tough cells, such as cyanobacteria, a "bead beater" with microscopic glass beads can tear the cells open.

Ultracentrifugation. Following osmotic shock, the spheroplasts undergo ultracentrifugation (**Fig. 3.2**, step 4) to separate the periplasmic contents from the other three types of cell compartments (cytoplasm, inner membrane, and outer membrane fragments, which form vesicles). The **ultracentrifuge** is a device in which tubes containing solutions of cell components are spun at very high speed. The high rotation rate generates centrifugal forces strong enough to separate subcellular particles (see eAppendix 3). The particles are collected in fractions of sample from the tube. The fractions are observed by electron microscopy.

Spheroplast lysis. The pellet can now be further processed by a French press (**Fig. 3.2**, step 5), a device that squeezes cell contents through a narrow tube to break open the membranes. The broken membranes coalesce into tiny vesicles. A second step of ultracentrifugation now pellets the inner and outer membrane vesicles, while removing the cytoplasm in the supernatant (liquid above the pellet; step 6). The cytoplasm provides many types of proteins for study, such as FtsZ, as well as complexes (multiprotein structures) such as the ribosomes and DNA polymerases.

Finally, the membrane fraction contains a mixture of very different components of the inner membrane (such as electron transport proteins) and the outer membrane (such as cell-surface proteins that we could use to make vaccines). We can separate the inner and outer membranes by density gradient ultracentrifugation (**Fig. 3.2**, step 7). The gradient of solution density is generated by forming a gradient of sucrose concentration. In the gradient, the inner and outer membrane vesicles separate because of differences in their density, not their particle size. The lower-density fractions contain inner membrane vesicles, whereas the higher-density fractions contain outer membrane vesicles.

The membrane vesicle proteins are analyzed on electrophoretic gels (**Fig. 3.3**). We can identify the protein bands on the gel by enzyme digestion and mass spectroscopy. The enzyme trypsin cleaves proteins only at aminoacyl residues lysine or arginine, thus generating peptides of defined composition. The peptide mass sizes are used to identify the cleaved protein, by comparison with the protein sequence that the organism's genome predicts.

Different kinds of experimental methods may complement each other, to confirm or extend the conclusions from each. Another example is the analysis of the ribosome, presented in **eTopic 3.1**.

Thought Question

3.2 Suppose we wish to isolate flagellar motors, which are protein complexes that span the envelope from inner membrane to outer membrane (**Fig. 3.1**). How might we modify the cell fractionation procedure to achieve such isolation?

A limitation of cell fractionation is that it provides little information about processes that require an intact cell, such as cell division. How can we remove or alter a part of a cell without breaking it open? An approach that is complementary to cell fractionation is genetic analysis. In genetic analysis, we can mutate a strain so as to lose or alter a gene; then we select mutant strains for loss of a given function. The phenotype of the mutant cell may yield clues about the function of the altered part, such as the bacterial cytoskeleton (Section 3.3). Genetics and genetic analysis are discussed further in Chapters 7–12.

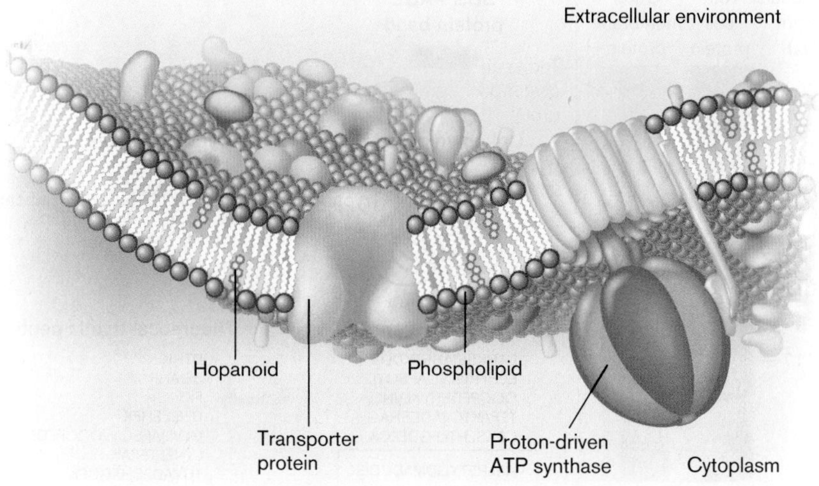

FIGURE 3.4 ■ Bacterial cell membrane. The cell membrane consists of a phospholipid bilayer, with hydrophobic fatty acid chains directed inward, away from water. The bilayer contains stiffening agents such as hopanoids. Half the membrane volume consists of proteins.

To Summarize

- **Bacterial cells are protected by a thick cell envelope.** The envelope includes a cell membrane and a peptidoglycan cell wall. A Gram-negative cell includes an outer membrane, and the cell membrane is called the inner membrane.
- **Bacteria are composed of nucleic acids, proteins, phospholipids, and other organic and inorganic chemicals.** Proteins in the cell vary, depending on the species and environmental conditions.
- **The bacterial cytoplasm is highly structured.** DNA replication, RNA transcription, and protein synthesis occur coordinately within the cytoplasm.
- **Microscopy reveals cell structure.** Transmission electron microscopy shows how cell parts fit within the cell as a whole. Fluorescence microscopy reveals the location and dynamics of individual components.
- **Cell fractionation isolates cell parts for structural and biochemical analysis.** The compartments of a Gram-negative cell can be separated by cell lysis and ultracentrifugation.

3.2 The Cell Membrane and Transport

How does a cell distinguish itself from what is outside? The structure that defines the existence of a cell is the cell membrane (**Fig. 3.4**). Overall, the membrane contains the cytoplasm within the external medium, mediating exchange between the two. The cell membrane consists of a phospholipid bilayer containing lipid-soluble proteins. It behaves as a two-dimensional fluid, within which proteins and lipids can diffuse. The proteins form about half the mass of the membrane and provide specific functions, such as nutrient transport. For a review of elementary cell structure, see eAppendix 2.

Membrane Lipids

Most membrane lipids are **phospholipids**. A phospholipid possesses a charged phosphoryl "head" that contacts the water interface, as well as a hydrophobic "tail" of fatty acids packed within the bilayer. Lipid biosynthesis is a key process that is vulnerable to some antibiotics. For example, the bacterial enzyme enoyl reductase, which synthesizes fatty acids (discussed in Chapter 15), is the target of triclosan, a common antibacterial additive in detergents and cosmetics.

A typical phospholipid consists of glycerol with ester links

$$\text{(C—}\overset{\displaystyle \text{O}}{\overset{\|}{\text{C}}}\text{—O—C)}$$

to each of two fatty acids, and a phosphoryl polar head group, which at neutral pH is deprotonated (negatively charged; **Fig. 3.5**). This kind of phospholipid is called a phosphatidate. The negatively charged head group of the phosphatidate can contain various organic groups, such as glycerol to form phosphatidylglycerol (**Fig. 3.5A**). In other lipids, the polar head group has a side chain with positive charge. The positive charge commonly resides on an amine group, such as ethanolamine in phosphatidylethanolamine

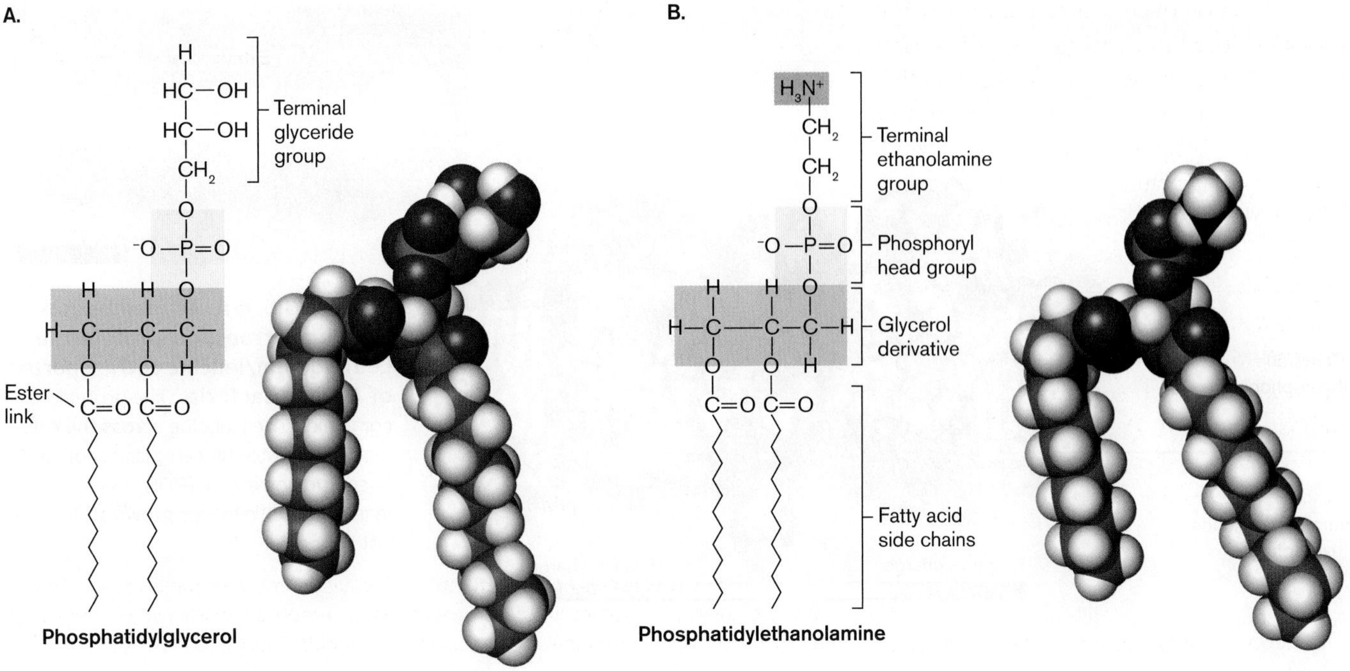

FIGURE 3.5 ■ Phospholipids. **A.** Phosphatidylglycerol consists of glycerol with ester links to two fatty acids, and a phosphoryl group linked to a terminal glyceride. **B.** Phosphatidylethanolamine contains a glycerol linked to two fatty acids, and a phosphoryl group with a terminal ethanolamine. The ethanolamine carries a positive charge.

(**Fig. 3.5B**). Phospholipids with positive charge or with mixed charges are concentrated in portions of the membrane that interact with DNA, which has negative charge.

In the bilayer, all phospholipids face each other tail to tail, keeping their hydrophobic side chains away from the water inside and outside the cell. The two layers of phospholipids in the bilayer are called **leaflets**. One leaflet of phospholipids faces the cell interior; the other faces the exterior. As a whole, the phospholipid bilayer imparts fluidity and gives the membrane a consistent thickness (about 8 nm).

Membrane Proteins

Membrane proteins serve functions such as transport, communication with the environment, and structural support.

- **Structural support.** Some membrane proteins anchor together different layers of the cell envelope (discussed in Section 3.3). Other proteins attach the membrane to the cytoskeleton, or form the base of structures extending out from the cell, such as flagella.
- **Detection of environmental signals.** In *Vibrio cholerae*, the causative agent of cholera, the membrane protein ToxR detects acidity and elevated temperature—signs that the bacterium is in the host's digestive tract. The ToxR domain facing the cytoplasm then binds to a DNA sequence, activating expression of cholera toxin.
- **Secretion of virulence factors and communication signals.** Membrane protein complexes export toxins and cell signals across the envelope. For example, symbiotic nitrogen-fixing rhizobia require membrane proteins NodI and NodJ to transport nodulation signals out to the host plant roots, inducing the plant to form root nodules containing the bacteria.
- **Ion transport and energy storage.** Transport proteins manage ion flux between the cell and the exterior. Ion transport generates gradients that store energy.

An example of a bacterial membrane protein is the leucine transporter LeuT (**Fig. 3.6**). LeuT drives uptake of leucine, coupled to a gradient of sodium ions. The protein complex was purified for X-ray diffraction from *Aquifex aeolicus*, a thermophile whose heat-stable proteins form durable crystals. Remarkably, LeuT is homologous (shares common ancestry) with a human neuron protein that transports neurotransmitters. Thus, this bacterial protein serves as a model for the study of neuron function.

Proteins embedded in a membrane require a hydrophobic portion that is soluble within the membrane. Typically, several hydrophobic alpha helices thread back and forth through the membrane. Other peptide regions extend outside the membrane, containing charged and polar amino acids that interact favorably with water. **Figure 3.6** shows the LeuT charge distribution. Hydrophobic amino acid residues (white) make the

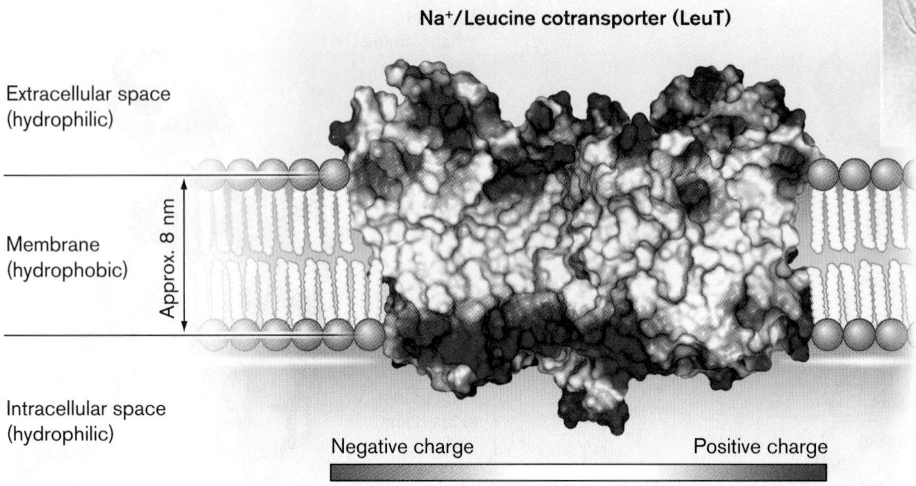

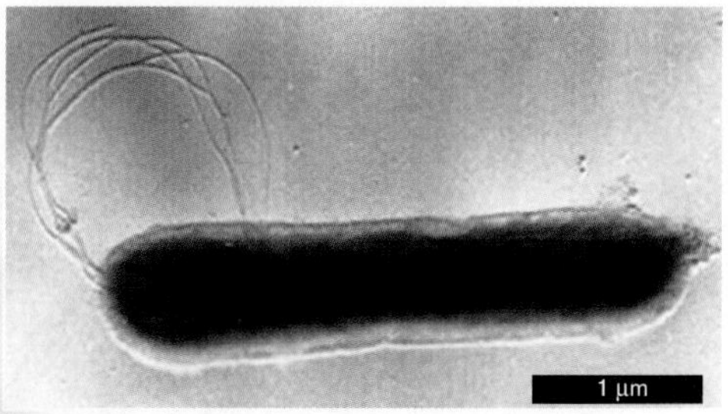

KARL STETTER AND REINHARDT RACHEL, U. REGENSBURG, GERMANY

FIGURE 3.6 ■ A cell membrane–embedded transport protein: the LeuT sodium/leucine cotransporter of *Aquifex* bacteria. The protein complex carries leucine across the cell membrane into the cytoplasm, coupled to sodium ion influx. (PDB code: 3F3E) **Inset:** *Aquifex aeolicus* grows at 96°C in hot springs.

protein soluble in the membrane, while portions with negative charge (red) and positive charge (blue) lock the protein in.

Transport across the Cell Membrane

The cell membrane acts as a barrier to keep water-soluble proteins and other cell components within the cytoplasm. But how do nutrients from outside get into the cell—and how do secreted products such as toxins get out? Specific membrane proteins transport molecules across the membrane between the cytoplasm and the outside. Selective transport is essential for cell survival; it means the ability to acquire scarce nutrients, exclude waste, and transmit signals to neighbor cells.

Passive diffusion. Small, uncharged molecules, such as O_2, CO_2, and water, easily permeate the membrane. Some molecules, such as ethanol, also disrupt the membrane—an action that can make such molecules toxic to cells. By contrast, large, strongly polar molecules such as sugars, and charged molecules such as amino acids, generally cannot penetrate the hydrophobic interior of the membrane, and thus require transport by specific proteins. Water molecules permeate the membrane, but their rate of passage is increased by protein channels called aquaporins (discussed in Chapter 4).

Osmosis. Most cells maintain a concentration of total solutes (molecules in solution) that is higher inside the cell than outside. As a result, the internal concentration of water is lower than the concentration outside the cell. Because water can cross the membrane but charged solutes cannot, water tends to diffuse across the membrane into the cell, causing the expansion of cell volume, in a process called osmosis. The resulting pressure on the cell membrane is called **osmotic pressure** or **turgor pressure** (see Figure A.6 in the Appendix). Osmotic pressure will cause a cell to burst, or lyse, in the absence of a countering pressure such as that provided by the cell wall. That is how penicillin kills bacteria—by disrupting cell wall synthesis.

Membrane-permeant weak acids and bases. A special case of movement across cell membranes is that of **membrane-permeant weak acids** and **weak bases** (**Fig. 3.7**), which exist in equilibrium between charged and uncharged forms:

$$\text{Weak acid: } HA \rightleftharpoons H^+ + A^-$$
$$\text{Weak base: } B + H_2O \rightleftharpoons BH^+ + OH^-$$

Membrane-permeant weak acids and weak bases cross the membrane in their uncharged form: HA (weak acid) or B (weak base). On the other side, entering the aqueous cytoplasm, the acid dissociates (HA to A^- and H^+) or the base reassociates with H^+ (B to BH^+). In effect, membrane-permeant acids conduct acid (H^+) across the membrane, causing acid stress; similarly, membrane-permeant bases conduct OH^- across the membrane, causing alkali stress. If the H^+ concentration (acidity) outside the cell is greater than inside, it will drive weak acids into the cell.

Many key substances in cell metabolism are membrane-permeant weak acids and bases, such as acetic acid. Most pharmaceutical drugs—therapeutic agents delivered

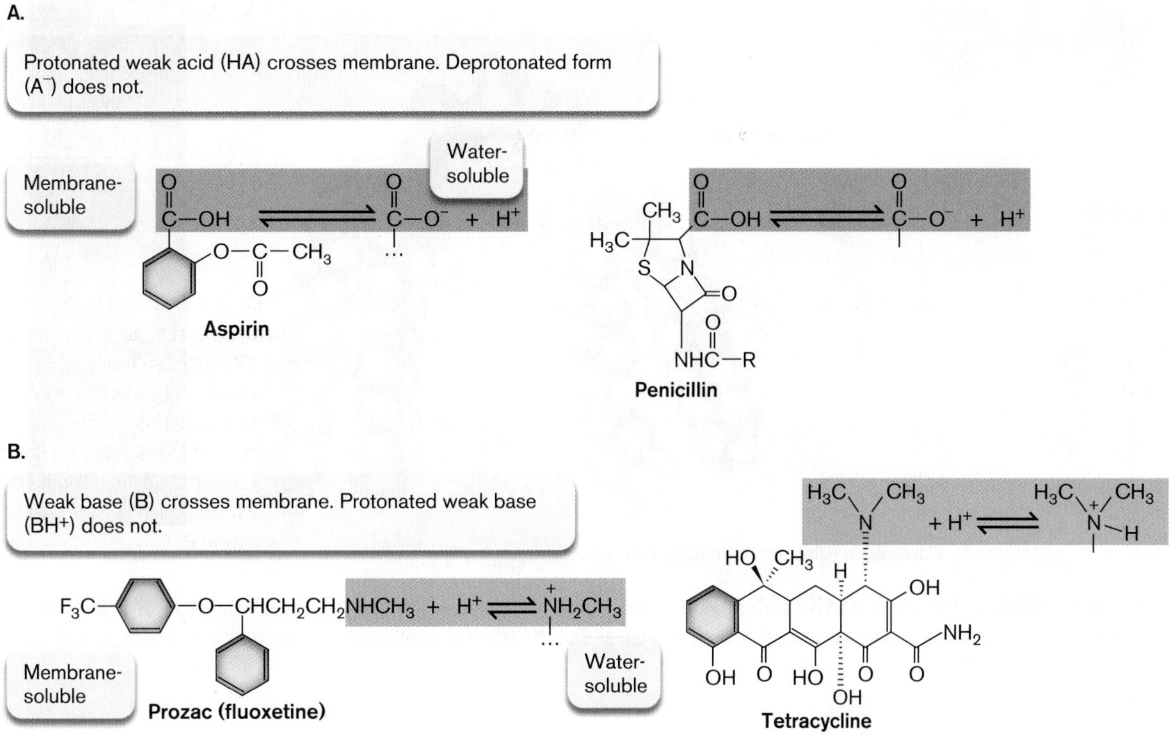

FIGURE 3.7 ■ Common drugs are membrane-permeant weak acids and bases. **A.** In its charged form (A^- or BH^+), each drug is soluble in the bloodstream. **B.** The uncharged form (HA or B) is hydrophobic and penetrates the cell membrane.

to our tissues via the bloodstream—are weak acids or bases whose uncharged forms exist at sufficiently low concentration to cross the membrane without disrupting it. Examples of weak acids that deprotonate (acquiring negative charge) at neutral pH include aspirin (acetylsalicylic acid) and penicillin (**Fig. 3.7A**). Examples of weak bases that protonate (acquiring positive charge) at neutral pH include Prozac (fluoxetine) and tetracycline (**Fig. 3.7B**).

Thought Question

3.3 Amino acids have acidic and basic groups that can dissociate. Why are they not membrane-permeant weak acids or weak bases? Why do they fail to cross the phospholipid bilayer?

Transmembrane ion gradients. Molecules that carry a fixed charge, such as hydrogen and sodium ions (H^+ and Na^+), cannot cross the phospholipid bilayer. Such ions usually exist in very different concentrations inside and outside the cell. An **ion gradient** (ratio of ion concentrations) across the cell membrane can store energy for nutrition or to drive the transport of other molecules. Inorganic ions require transport through specific **transport proteins**, or **transporters**. So, too, do organic molecules that carry a charge at cytoplasmic pH, such as amino acids and vitamins. Transport may be passive or active. In **passive transport**, molecules accumulate or dissipate along their concentration gradient. **Active transport**—that is, transport from lower to higher concentration—requires cells to spend energy. A transporter protein obtains energy for active transport by cotransport of another substance down its gradient from higher to lower concentration, or by coupling transport to a chemical reaction (discussed in Chapter 4).

Membrane Lipid Diversity

Membranes require a uniform thickness and stability to maintain structural integrity and function. So why do individual membrane lipids differ in structure? Different environments favor different forms of membrane lipids. For example, lipid structure helps determine whether an organism can grow in a hot spring, or whether it can colonize human lungs.

Environmental stress. Starvation stress increases bacterial production of lipids with an unusual type of phosphoryl head group. **Cardiolipin**, or diphosphatidylglycerol, is actually a double phospholipid linked by a glycerol (**Fig. 3.8A**). Cardiolipin concentration increases in bacteria grown to starvation or stationary phase (discussed in Chapter 4). Within a cell, cardiolipin does not diffuse at random; it concentrates in patches called "domains" near the cell poles. The polar localization of cardiolipin was demonstrated by fluorescence microscopy, in which a

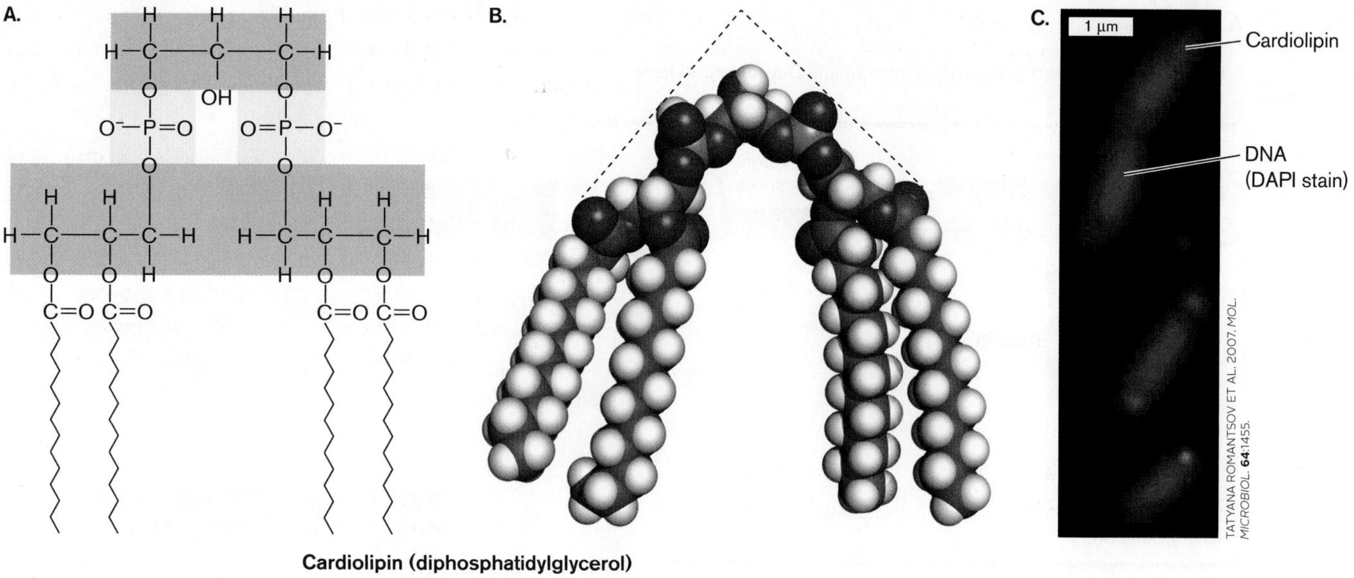

FIGURE 3.8 ▪ Cardiolipin localizes to the poles. A. Cardiolipin is a double phospholipid joined by a third glycerol. B. A space-filling model of cardiolipin shows its triangular shape. C. Cardiolipin localizes to the bacterial cell poles, as shown by microscopy with a cardiolipin-specific fluorophore (orange).

cardiolipin-specific fluorophore localized to the poles of *E. coli*. The "wedge" shape of cardiolipin (**Fig. 3.8B**), with its narrow head group and wide fatty acid group, is thought to form concave domains of lipid that stabilize the curve of the polar membrane. Cardiolipin may enhance the formation of smaller cells during starvation. At the cell pole (**Fig. 3.8C**), cardiolipin binds certain environmental stress proteins, such as a protein that transports osmoprotectants when the cell is under osmotic stress. Thus, a phospholipid can have specific functions associated with specific membrane proteins.

The fatty acid component of phospholipids also varies. The most common bacterial fatty acids are hydrogenated chains of varying length, typically between 6 and 22 carbons. But some fatty acid chains are partly unsaturated (possess one or more carbon-carbon double bonds). Most unsaturated bonds in membranes are *cis*, meaning that both alkyl chains are on the same side of the bond, so the unsaturated chain has a "kink," as in the *cis* form of oleic acid (**Fig. 3.9**). Because the kinked chains do not pack as closely as the straight hydrocarbon chains do, the membrane is more "fluid." This is why, at room temperature, unsaturated vegetable oils are fluid, whereas highly saturated butterfat is solid. The enhanced fluidity of a kinked phospholipid improves the function of the membrane at low temperature; hence, bacteria can respond to cold and heat by increasing or decreasing their synthesis of unsaturated phospholipids.

Another interesting structural variation is cyclization of part of the chain to form a stiff planar ring with decreased fluidity. The double bond of unsaturated fatty acids can incorporate a carbon from *S*-adenosyl-L-methionine to form a three-membered ring, generating a cyclopropane fatty acid (**Fig. 3.9**). Bacteria convert unsaturated fatty acids to cyclopropane during starvation and acid stress, conditions under which membranes require stiffening. Cyclopropane conversion is an important factor in the pathogenesis of *Mycobacterium tuberculosis* and in the acid resistance of food-borne toxigenic *E. coli*.

In addition to phospholipids, membranes include planar molecules that fill gaps between hydrocarbon chains. These stiff, planar molecules reinforce the membrane, much as steel rods reinforce concrete. In eukaryotic membranes, the reinforcing agents are sterols, such as **cholesterol**. In some bacteria, the same function is filled by pentacyclic (five-ring) hydrocarbon derivatives called **hopanoids**, or hopanes (**Fig. 3.10**). Like cholesterol, hopanoids fit between the fatty acid side chains of membranes and limit

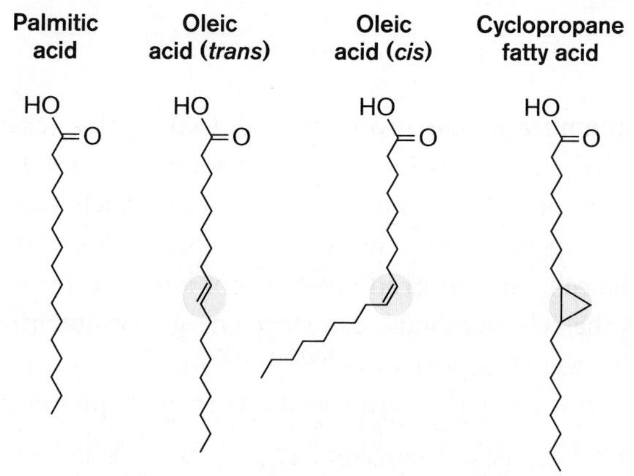

FIGURE 3.9 ▪ Phospholipid side chains.

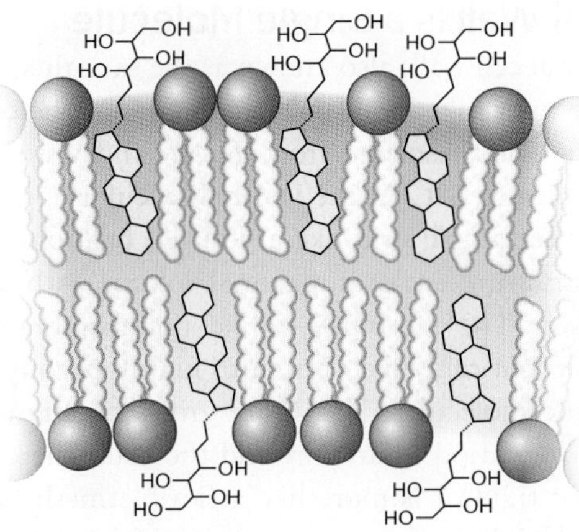

FIGURE 3.10 ■ Hopanoids add strength to membranes. Hopanoids limit the motion of phospholipid tails, thus stiffening the membrane.

their motion, thus stiffening the membrane. Hopanoids provide biomarkers for petroleum exploration, as signs of potential petroleum formation in ancient rock.

Archaea have unique membrane lipids. The membrane lipids of archaea differ fundamentally from those of bacteria and of eukaryotes. All archaeal phospholipids replace the ester link between glycerol and fatty acid with an ether link, C–O–C (**Fig. 3.11**). Ethers are much more stable than esters, which hydrolyze easily in water. This is one reason why some archaea can grow at higher temperatures than all other forms of life. Another modification is that archaeal hydrocarbon chains are branched **terpenoids**, polymeric structures derived from isoprene, in which every fourth carbon extends a methyl branch. The branches strengthen the membrane by limiting movement of the hydrocarbon chains.

The most extreme hyperthermophiles, which live beneath the ocean at 110°C, have terpenoid chains linked at the tails, forming a tetraether monolayer. In some species, the terpenoids cyclize to form cyclopentane rings. These planar rings stiffen the membrane under stress to an even greater extent than do the cyclopropane chains of bacteria. For more on archaeal cells, see Chapter 19.

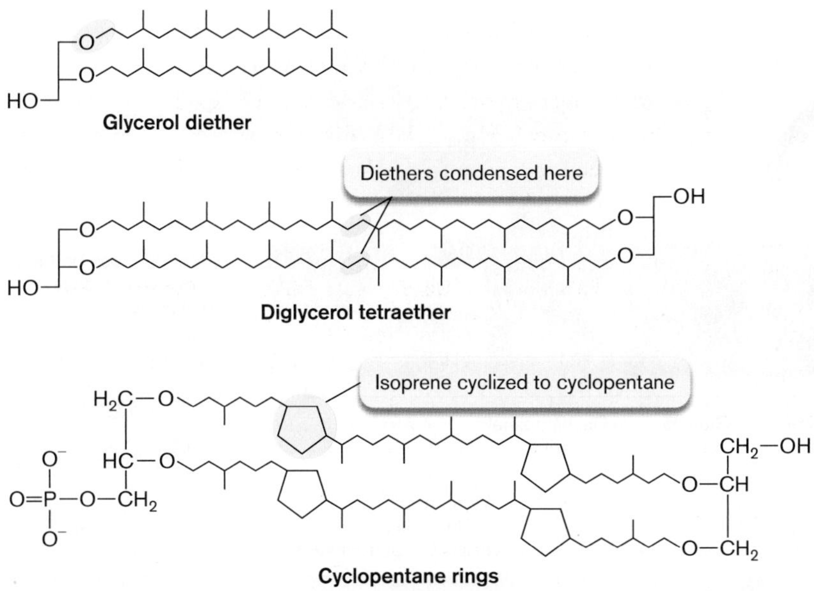

FIGURE 3.11 ■ Terpene-derived lipids of archaea. In archaea, the hydrocarbon chains are ether-linked to glycerol, and every fourth carbon has a methyl branch. In some archaea, the tails of the two facing lipids of the bilayer are fused, forming tetraethers; thus, the entire membrane consists of a monolayer.

To Summarize

- **The cell membrane** consists of a phospholipid bilayer containing hydrophobic membrane proteins. Membrane proteins serve diverse functions, including transport, cell defense, and cell communication.
- **Small uncharged molecules**, such as oxygen, can penetrate the cell membrane by diffusion.
- **Membrane-permeant weak acids and weak bases** exist partly in an uncharged form that can diffuse across the membrane and increase or decrease, respectively, the H^+ concentration within the cell.
- **Large polar molecules and charged molecules** require membrane **proteins to mediate** transport. Such facilitated transport can be active or passive.
- **Ion gradients** generated by membrane pumps store energy for cell functions. **Active transport** (up a concentration gradient) requires input of energy, whereas **passive transport** (down the gradient) does not.
- **Diverse fatty acids** are found in different microbial species and in microbes adapted to different environments.
- **Archaeal membranes have ether-linked terpenoids**, which confer increased stability at high temperature and extreme acidity. Some archaea have diglycerol tetraethers, which form a monolayer.

3.3 The Envelope and Cytoskeleton

How do bacteria and archaea protect their cell membrane? For most species, the cell envelope includes at least one structural supporting layer, like an external skeleton, located outside the cell membrane. The most common structural support is the cell wall (see **Fig. 3.1**). Many species possess additional coverings, such as an outer membrane or an S-layer. Nevertheless, a few prokaryotes, such as the mycoplasmas, have a cell membrane with no outer layers, depending on host fluids for osmotic balance. Some archaea with only a cell membrane grow in extreme acid (pH zero); for example, *Ferroplasma*, found in iron mines. How they survive is unknown.

The Cell Wall Is a Single Molecule

The bacterial cell wall, also known as the **sacculus**, consists of a single interlinked molecule that envelops the cell. The sacculus has been isolated from *E. coli* and visualized by TEM; in **Figure 3.12A**, the isolated sacculus appears flattened on the sample grid like a deflated balloon. Its geometrical structure encloses maximal volume with minimal surface area. The sacculus—unlike the membrane—is a single-molecule cage-like structure, highly porous to ions and organic molecules. The mesh grows by strand insertion and elongation in arcs around the cell. The cage-like form is not rigid; it is more like a flexible mesh bag with unbreakable joints. Turgor pressure within the enclosed cytoplasm fills out the cell's shape, whether elongated rod, spherical coccus, or other.

Peptidoglycan structure. Most bacterial cell walls are composed of peptidoglycan, a polymer of peptide-linked chains of amino sugars. Peptidoglycan is synonymous with **murein** ("wall molecule"). The molecule consists

A.

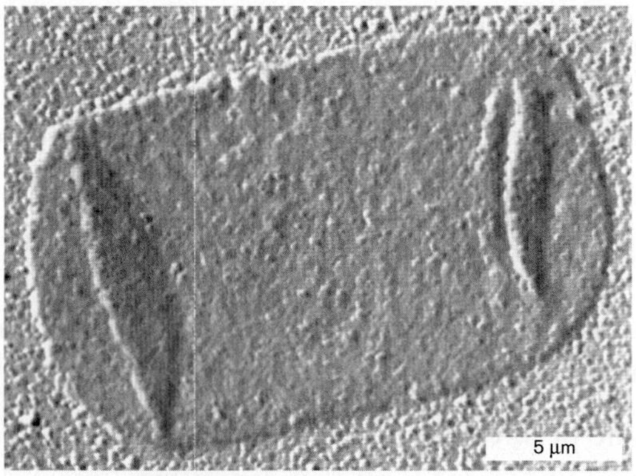

W. VOLLMER ET AL. 2008. *FEMS MICROBIOL. REV.* **32**:149, FIG. 3A.

B.

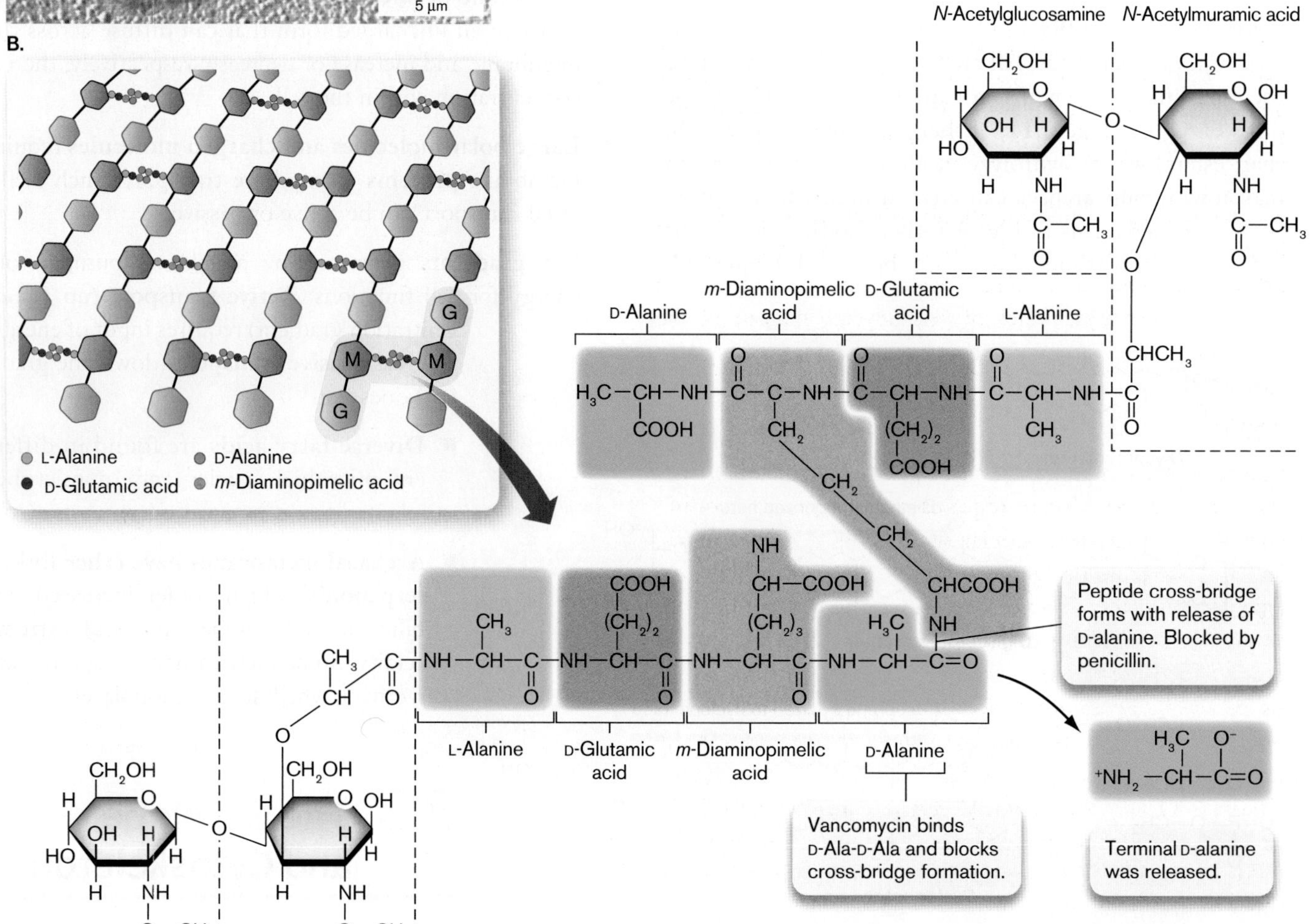

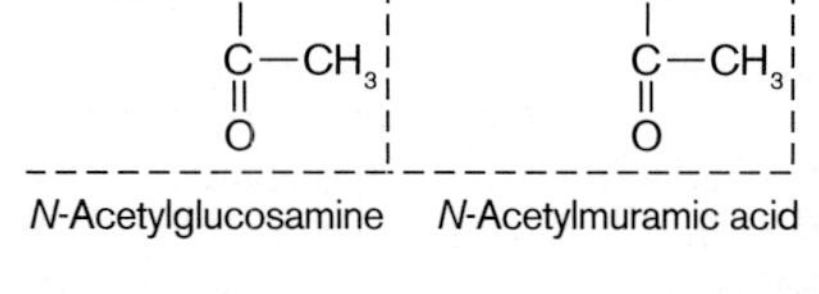

FIGURE 3.12 ■ The peptidoglycan sacculus and peptidoglycan cross-bridge formation. A. Isolated sacculus from *Escherichia coli* (TEM). **B.** A disaccharide unit of glycan has an attached peptide of four to six amino acids.

of parallel polymers of disaccharides called **glycan** chains cross-linked with peptides of four amino acids (**Fig. 3.12B**). Peptidoglycan is unique to bacteria, although some archaea build analogous structures whose overall physical nature is similar. (Archaeal cell walls are presented in Chapter 19.)

The long chains of peptidoglycan consist of repeating units of a disaccharide composed of *N*-acetylglucosamine (an amino sugar derivative) and *N*-acetylmuramic acid (glucosamine plus a lactic acid group; **Fig. 3.12B**). The lactate group of muramic acid forms an amide link with the amino terminus of a short peptide containing four to six amino acid residues. The peptide extension can form **cross-bridges** connecting parallel strands of glycan.

The peptide contains two amino acids in the unusual D mirror form: D-glutamic acid and D-alanine. The third amino acid, *m*-diaminopimelic acid, has an extra amine group, which forms an amide link to a cross-bridged peptide. The amide link forms with the fourth amino acid of the adjacent peptide, D-alanine (**Fig. 3.12B**). Removal of a second D-alanine at the end of the chain forms the cross-bridge. The cross-linked peptides of neighboring glycan strands form the cage of the sacculus.

Note: Amino acids have two forms that are mirror opposites, D and L, of which only the L form is incorporated by ribosomes into protein. The D-form amino acids, however, are used by microbes for many nonprotein structural molecules.

The details of peptidoglycan structure vary among bacterial species. Some Gram-positive species, such as *Staphylococcus aureus* (a cause of toxic shock syndrome), have peptides linked by bridges of pentaglycine instead of the D-alanine link to *m*-diaminopimelic acid. In Gram-negative species, the *m*-diaminopimelic acid is linked to the outer membrane, as will be discussed shortly.

Peptidoglycan synthesis as a target for antibiotics. Synthesis of peptidoglycan requires many genes encoding enzymes to make the special sugars, build the peptides, and seal the cross-bridges (see **Fig. 3.12B**). Many of these enzymes bind the antibiotic penicillin and are thus known as **penicillin-binding proteins**. Because peptidoglycan is unique to bacteria, enzymes of peptidoglycan biosynthesis make excellent targets for new antibiotics. For example, the transpeptidase that cross-links the peptides is the target of penicillin. Vancomycin, a major defense against *Clostridioides difficile* and drug-resistant staphylococci, prevents cross-bridge formation by binding the terminal D-Ala-D-Ala dipeptide. Vancomycin binding prevents release of the terminal D-alanine.

Unfortunately, the widespread use of such antibiotics selects for evolution of resistant strains. One of the most common agents of resistance is the enzyme beta-lactamase, which cleaves the lactam ring of penicillin, rendering it ineffective as an inhibitor of transpeptidase (also called penicillin-binding protein 2). In a different mechanism, strains resistant to vancomycin contain an altered enzyme that adds lactic acid to the end of the branch peptides in place of the terminal D-alanine. The altered peptide is no longer blocked by vancomycin. As new forms of drug resistance emerge, researchers continue to seek new antibiotics that target cell wall formation (discussed in Chapter 27).

How does peptidoglycan grow, overall, throughout the elongating cell? A peptidoglycan synthesis complex extends the chains of amino-sugars. So-called penicillin-binding proteins catalyze the formation of peptide cross-bridges (**Fig. 3.13**). These proteins were named for their property of binding the antibiotic penicillin, which disables cross-bridge formation (details in Chapter 27). The overall direction of cell wall extension is organized by a protein complex that includes MreB. MreB polymerizes in a helical direction along an arc beneath the plasma membrane (inner membrane, for a Gram-negative bacterium).

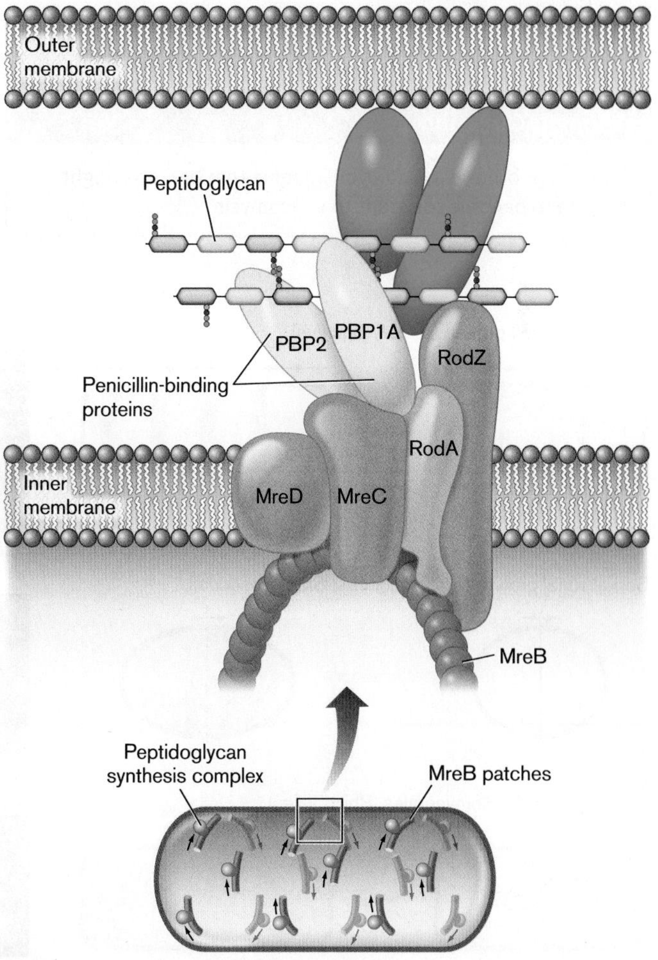

FIGURE 3.13 ■ Peptidoglycan synthesis is organized by penicillin-binding proteins (PBP2, PBP1A) and by cytoskeletal proteins. Protein MreB guides the direction of synthesis in helical arcs around the cell.

The RodA and RodZ proteins are needed to shape the cell as a rod (bacillus).

Thought Question

3.4 What genetic experiments could you propose to test the model of envelope expansion shown in **Figure 3.13**?

Interestingly, different kinds of bacteria have evolved to organize their cell wall growth differently. Yves Brun and his student Erkin Kuru, at Indiana University, devised an ingenious way to reveal the growth pattern (**Fig. 3.14**). Kuru designed fluorophore-tagged D-amino acids that the growing cell wall incorporates (such as D-alanine; see **Fig. 3.12**). Because these are D-amino acids and not L-amino acids, ribosome-directed translation does not use them. Thus, the fluorophore-linked D-amino acids label cell wall only, not proteins. Kuru found that bacteria such as *E. coli* and *Bacillus subtilis* synthesize cell wall in zones dispersed throughout the cell. Gram-positive cocci, however, such as *Streptococcus* and *Staphylococcus*, synthesize cell wall only at the midpoint (or septum), where the cell is about to fission (discussed in the next section). Still other bacteria, such as the actinomycete *Streptomyces*, form new cell wall only at the poles.

Thought Question

3.5 What other ways can you imagine that bacteria might mutate to become resistant to vancomycin?

Cell Envelope of Bacteria

Most bacteria have additional envelope layers that provide structural support and protection from predators and host defenses (**Fig. 3.15**). Additional molecules are attached to the cell wall and cell membrane, and some thread through the layers. Here we present the envelope composition of three major kinds of bacteria, two of which (Firmicutes and Proteobacteria) are distinguished by the Gram stain. The third, Mycobacteria, is distinguished by the acid-fast stain.

- Firmicutes (Gram-positive) have a thick cell wall with 3–20 layers of peptidoglycan, interpenetrated by teichoic acids (**Fig. 3.15A**). The phylum Firmicutes consists of Gram-positive species such as *Bacillus thuringiensis* and *Streptococcus pyogenes*, the cause of strep throat.
- Proteobacteria (Gram-negative) have a thin cell wall with one or two layers of peptidoglycan, enclosed by an outer membrane (**Fig. 3.15B**). Included among the Proteobacteria are Gram-negative species such as *Escherichia coli* and nitrogen-fixing *Sinorhizobium meliloti.*
- Mycobacteria of the phylum Actinomycetes have a complex, multilayered envelope that includes defensive structures such as mycolic acids. Examples include *Mycobacterium tuberculosis* (the cause of tuberculosis) and *M. leprae* (the cause of leprosy).

Note that other important kinds of bacteria, such as cyanobacteria and spirochetes, have very different

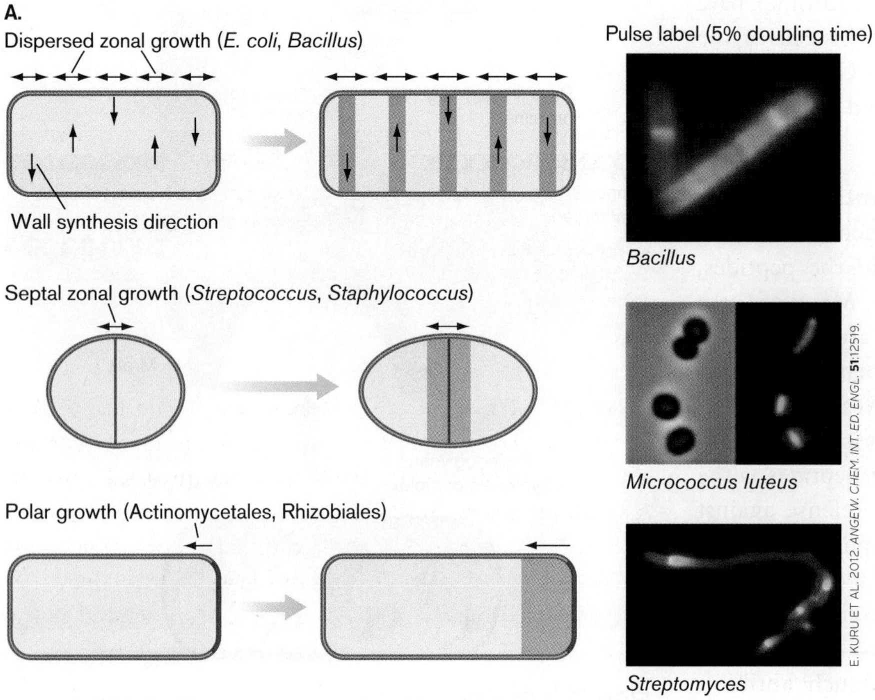

FIGURE 3.14 ▪ Peptidoglycan growth in different species. **A.** Different species of bacteria synthesize new peptidoglycan in dispersed zones (top), at the septum only (middle), or at the poles (bottom). Fluorescent D-amino acids are added for short periods (pulse labeling) to reveal the growth zones. **B.** Erkin Kuru, in the laboratory of Yves Brun at Indiana University, devised the fluorescent D-amino acids to probe cell wall growth.

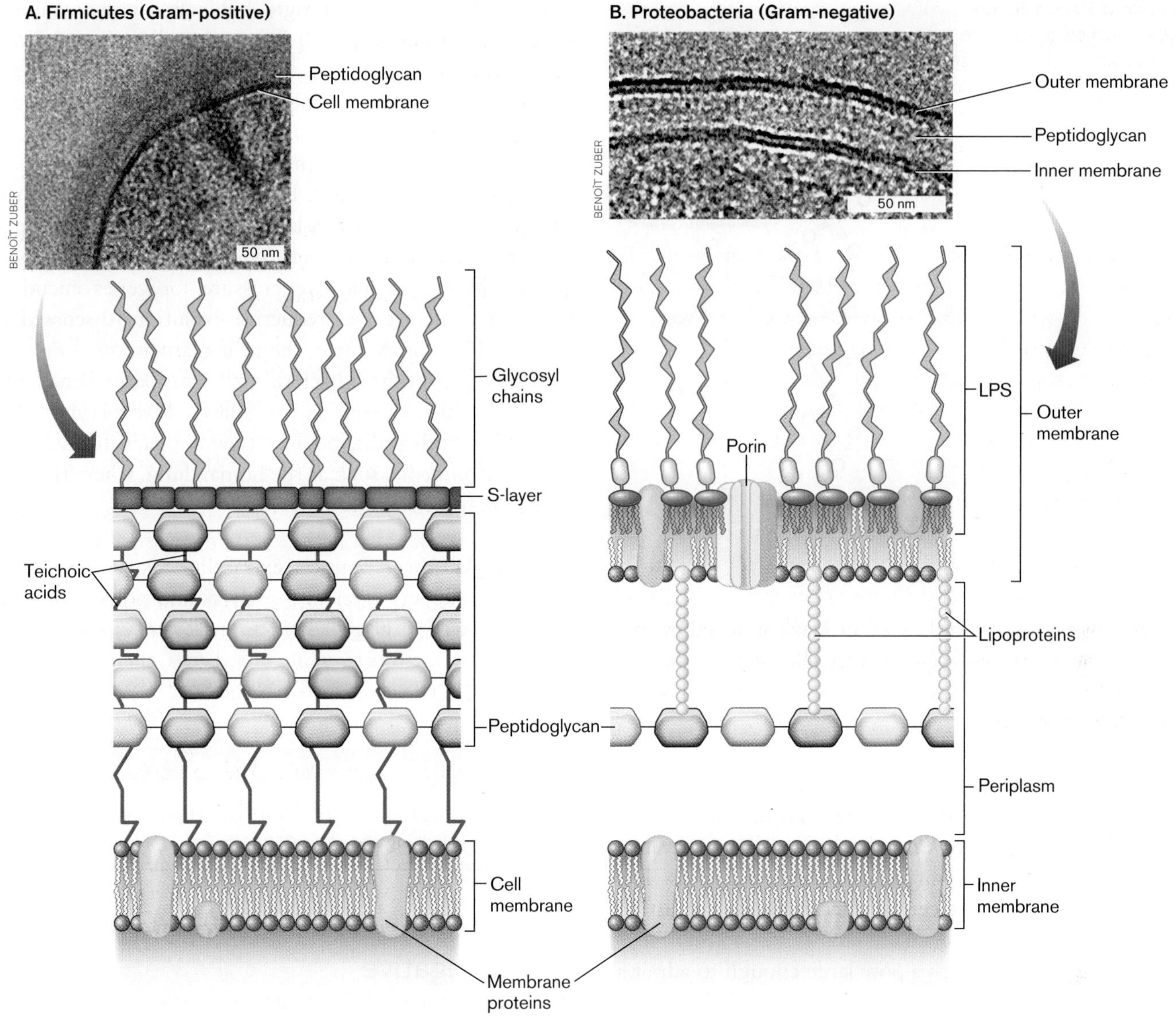

FIGURE 3.15 ■ Cell envelope: Gram-positive (Firmicutes) and Gram-negative (Proteobacteria). **A.** Firmicutes (Gram-positive) cells have a thick cell wall with multiple layers of peptidoglycan, threaded by teichoic acids. The micrograph shows the Gram-positive envelope of *Bacillus subtilis* (TEM). **B.** Proteobacteria (Gram-negative) cells have a single layer of peptidoglycan covered by an outer membrane; the cell membrane is called the inner membrane. The micrograph shows the Gram-negative envelope of *Pseudomonas aeruginosa* (TEM).

envelopes. These different envelope structures may stain Gram-positive, Gram-negative, or variable (discussed in Chapter 18). Archaea have yet other diverse kinds of envelopes that cannot be distinguished by Gram stain (see Chapter 19).

Thought Question

3.6 Figure 3.15 highlights the similarities and differences between the cell envelopes of Gram-negative and Gram-positive bacteria. What do you think are the advantages and limitations of a cell having one layer of peptidoglycan (Gram-negative) versus several layers (Gram-positive)?

Firmicute Cell Envelope—Gram-Positive

A section of a firmicute cell envelope (Gram-positive) is shown in **Figure 3.15A**. The multiple layers of peptidoglycan are reinforced by **teichoic acids** threaded through its multiple layers. Teichoic acids are chains of phosphodiester-linked glycerol or ribitol, with sugars or amino acids linked to the middle –OH groups (**Fig. 3.16**). The negatively charged cross-threads of teichoic acids, as well as the overall thickness of the Gram-positive cell wall, help retain the Gram stain.

How does the cell wall attach extracellular structures? Gram-positive bacteria have a type of enzyme called sortase that forms a peptide bond from a cell wall cross-bridge to

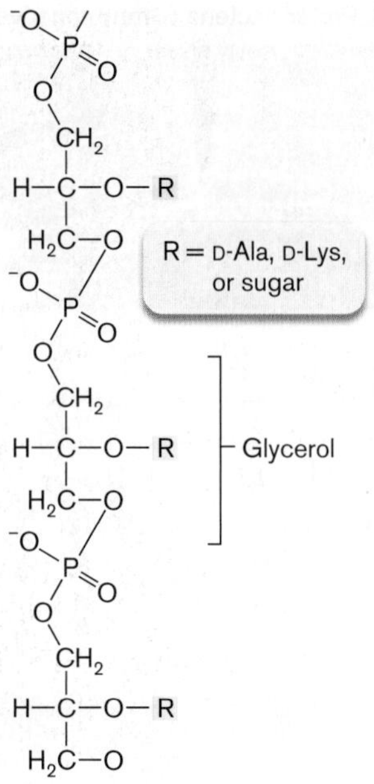

FIGURE 3.16 ▪ Teichoic acids. Teichoic acids in the Gram-positive cell wall consist of glycerol or ribitol phosphodiester chains.

a protein extending from the cell. Proteins attached by sortases can help the cell acquire nutrients, or help the cell adhere to a substrate. Sortases are now used in the protein engineering industry.

S-layer. Free-living bacteria and archaea often possess a tough surface layer called the **S-layer**. **Figure 3.17** shows the S-layer of *Lysinibacillus sphaericus*, a Gram-positive bacterium found on the surface of beets and carrots. The S-layer is a crystalline sheet of thick subunits consisting of protein or glycoprotein (proteins with attached sugars). Each subunit contains a pore large enough to admit a wide range of molecules (**Fig. 3.17**). As modeled by cryo-EM tomography (see Chapter 2), the S-layer proteins are arranged in a highly ordered array, either hexagonally or tetragonally. The S-layer is rigid, but it also flexes and allows substances to pass through it in either direction. S-layers help pathogens such as *Bacillus anthracis* bind and attack host cells. An S-layer contributes to biofilm formation (the periodontal bacterium *Tannerella forsythia*) and swimming (the aquatic cyanobacterium *Synechococcus* species).

The functions of the S-layer are hard to study in the laboratory because the S-layer is often lost by bacteria after repeated subculturing. Traits commonly disappear in the absence of selective pressure for genes encoding them—a process called reductive evolution (discussed in Chapter 17). For example, the mycoplasmas are close relatives of Gram-positive bacteria that have permanently lost their cell walls, as well as the S-layer. Mycoplasmas have no need for cell walls, because they are parasites living in host environments, such as the human lung, where they are protected from osmotic shock.

Capsule. Another common extracellular structure is the **capsule**, a slippery layer of loosely bound polysaccharides. The capsule of pathogens such as *Staphylococcus aureus* can prevent phagocytosis by white blood cells, thereby enabling the pathogen to persist in the blood.

Thought Question

3.7 Why would laboratory culture conditions select for evolution of cells lacking an S-layer?

Proteobacterial Cell Envelope—Gram-Negative

A cell envelope of Proteobacteria (Gram-negative) is shown in **Figure 3.15B**. The envelope includes one or two layers of peptidoglycan covered by an outer membrane

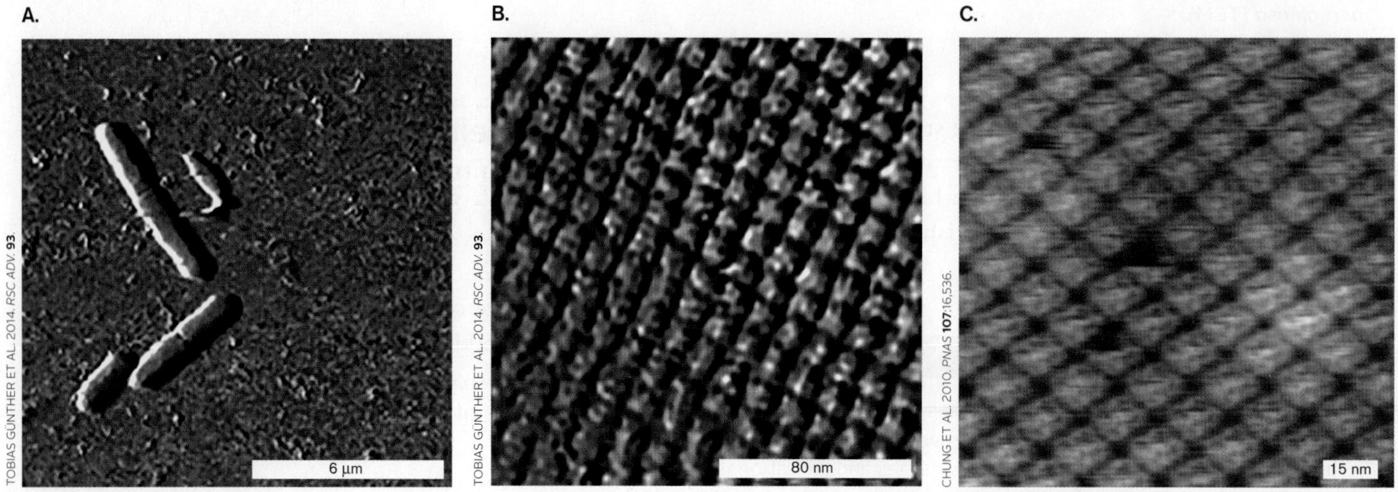

FIGURE 3.17 ▪ S-layer of *Lysinibacillus* bacteria. **A.** *Lysinibacillus* imaged by atomic force microscopy (AFM). **B.** S-layer surface imaged by AFM. **C.** Computational model of S-layer based on cryo-EM.

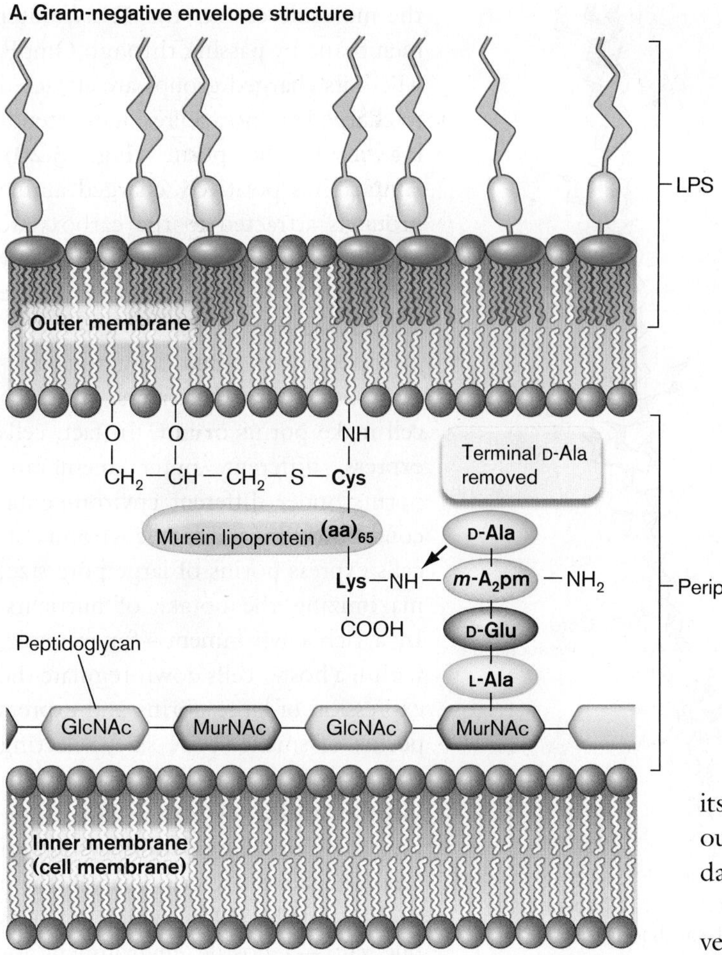

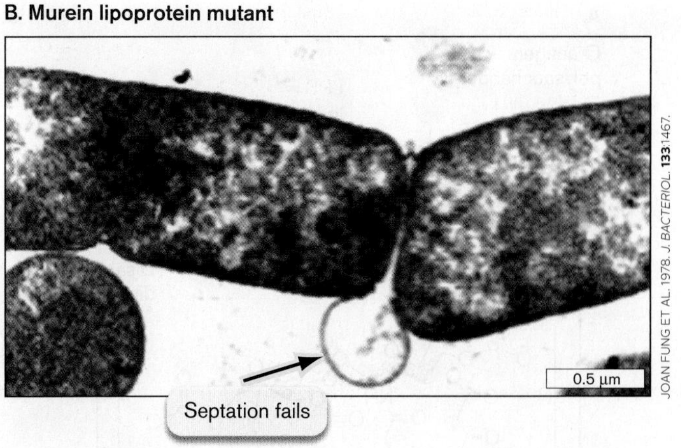

FIGURE 3.18 ▪ Gram-negative cell envelope. **A.** Murein lipoprotein has an N-terminal cysteine triglyceride inserted in the inward-facing leaflet of the outer membrane. The C-terminal lysine forms a peptide bond with the *m*-diaminopimelic acid of the peptidoglycan (murein) cell wall. **B.** Lack of murein lipoprotein in mutant *Salmonella* causes the outer membrane to balloon out (arrow) when the cell tries to divide (TEM).

(**Fig. 3.18**). The Gram-negative outer membrane confers defensive abilities and toxigenic properties on many pathogens, such as *Salmonella* species and enterohemorrhagic *E. coli* (strains that cause hemorrhaging of the colon). Between the outer and inner (cell) membranes, the aqueous compartment (containing the cell wall) is called the periplasm.

Lipoprotein and lipopolysaccharide (LPS). The inward-facing leaflet of the outer membrane has a phospholipid composition similar to that of the cell membrane (which in Gram-negative species is called the **inner membrane** or inner cell membrane). The outer membrane's inward-facing leaflet includes lipoproteins that connect the outer membrane to the peptide bridges of the cell wall. The major lipoprotein is called **murein lipoprotein**, also known as Braun lipoprotein (**Fig. 3.18A**). Murein lipoprotein consists of a protein with an N-terminal cysteine attached to three fatty acid side chains. The side chains are inserted in the inward-facing leaflet of the outer membrane. The C-terminal lysine forms a peptide bond with the *m*-diaminopimelic acid of peptidoglycan (murein). What happens to a mutant cell that fails to make murein lipoprotein? As the cell grows and divides, it fails to attach its outer membrane to the growing cell wall, causing the outer membrane to balloon out in the region where the daughter cells separate (**Fig. 3.18B**).

The outward-facing leaflet of the outer membrane has very different lipids from the inner leaflet. The main outward-facing phospholipids are called **lipopolysaccharides** (**LPS**; **Fig. 3.19**). LPS are of crucial medical importance because they act as **endotoxins**. Endotoxins are cell components that are harmless as long as the pathogen remains intact; but when released by a lysed cell, endotoxins overstimulate host defenses, inducing potentially lethal shock (discussed in Chapter 25). Thus, antibiotic treatment of an LPS-containing pathogen can kill the cells but can also lead to death of the patient.

The membrane-embedded anchor of LPS is **lipid A**, a molecule shaped like a six-legged giraffe (**Fig. 3.19A**). The lipid A moiety is the endotoxic part of LPS. The molecule's six fatty acid "legs" have shorter chains than those of the inner cell membrane, and two pairs are branched. The fatty acids have ester or amide links to the "body," a dimer of glucosamine (an amino sugar also found in peptidoglycan). Analogous to the glycerol of glyceride phospholipids, each glucosamine has a phosphoryl group whose negative charge interacts with water. One glucosamine extends the long "neck" of the core polysaccharide, a sugar chain that reaches far outside the cell (**Fig. 3.19B**).

The core polysaccharide consists of five to ten sugars with side chains such as phosphoethanolamine. It extends to an **O antigen** or O polysaccharide, a chain of as many as 200 sugars. The O polysaccharide may be longer than

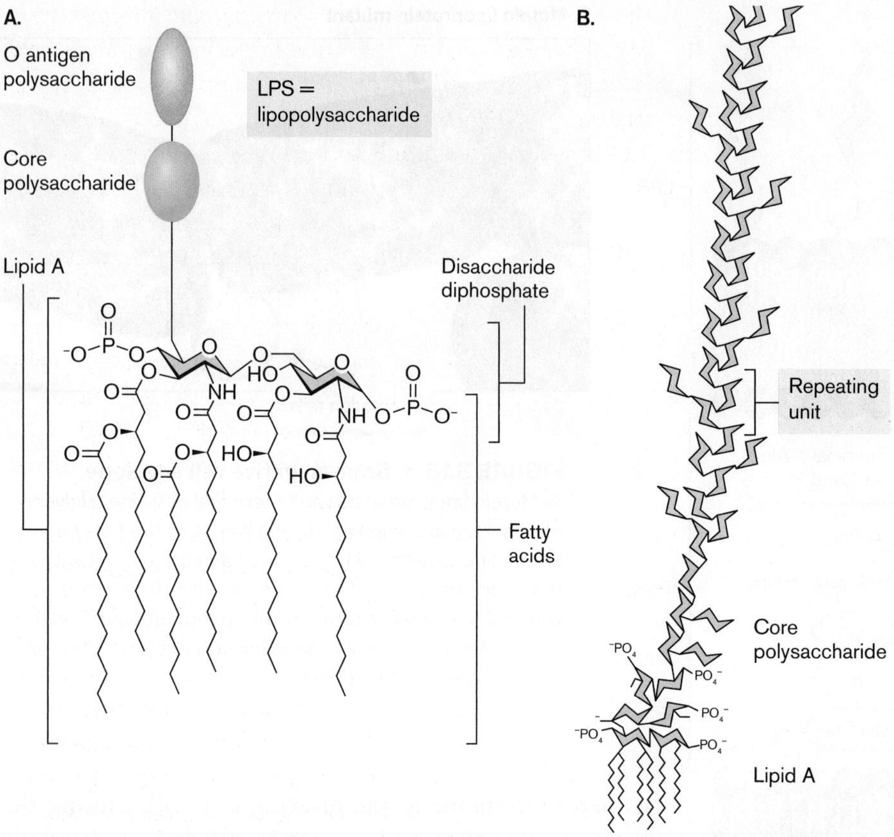

FIGURE 3.19 ■ Lipopolysaccharide (LPS). A. Lipopolysaccharide (LPS) consists of core polysaccharide and O antigen linked to a lipid A. Lipid A consists of a dimer of phosphoglucosamine esterified or amidated to six fatty acids. B. Repeating polysaccharide units of O antigen extend from lipid A.

the cell itself. These chains of sugars form a layer that helps bacteria resist phagocytosis by white blood cells. The combination of sugar units in the O antigen varies greatly; *E. coli* clinical isolates show hundreds of different O-antigen sugar chains, which are recognized by distinct antibodies. Thus, O-antigen diversity offers one means by which pathogens evade the host immune system.

Outer membrane proteins. The outer membrane contains unique proteins not found in the inner membrane. Outer membranes contain a class of transporters called **porins** that permit the entry of nutrients such as sugars and peptides (**Fig. 3.20**). Outer membrane porins such as OmpF have a distinctive cylinder of beta sheet conformation (reviewed in eAppendix 1), also known as a beta barrel. A typical outer membrane porin exists as a trimer of beta barrels, each of which acts as a pore for nutrients.

Outer membrane porins have limited specificity, allowing passive uptake of various molecules—including antibiotics such as ampicillin. Ampicillin is a form of penicillin, which must get through the outer membrane to access the cell wall in order to block the formation of peptide cross-bridges. Ampicillin contains two charged groups and is thus unlikely to diffuse through a lipid bilayer. But the molecule crosses the *E. coli* outer membrane by passing through OmpF, where its charged groups are attracted to charged amino acid residues extending inside the porin (**Fig. 3.20**). Ampicillin's positively charged amine group is attracted to the carboxylate of glutamate-117, and its negatively charged carboxylate is attracted to the arginine amines.

If porins can admit dangerous molecules as well as nutrients, should a cell make porins or not? In fact, cells express different outer membrane porins under different environmental conditions. In a dilute environment, cells express porins of large pore size, maximizing the uptake of nutrients. In a rich environment—for example, within a host—cells down-regulate the expression of large porins and express porins of smaller pore size, selecting only smaller nutrients and avoiding the uptake of toxins. For example, the porin regulation system of Gram-negative bacteria enables them to grow in the colon, which contains bile salts—a hostile environment for Gram-positive bacteria, which lack an outer membrane.

Periplasm. The outer membrane is porous to most ions and many small organic molecules, but it prevents the passage of proteins and other macromolecules. Thus, the region between the inner and outer membranes of Gram-negative cells, including the cell wall, defines a separate membrane-enclosed compartment of the cell known as the periplasm (see **Fig. 3.18**). The periplasm contains specific enzymes and nutrient transporters not found within the cytoplasm, such as periplasmic transporters for sugars, amino acids, or other nutrients. Periplasmic proteins are subjected to fluctuations in pH and salt concentration because the outer membrane is porous to ions. Some periplasmic proteins help refold proteins unfolded by oxidizing agents or by acidification.

Overall, the outer membrane, periplasm, inner membrane, and cytoplasm define four different cell compartments within a Gram-negative cell: two membrane-soluble compartments (outer and inner membranes), and two aqueous compartments (periplasm and cytoplasm). Each type of protein is typically found in only one of these locations. For example, the proton-translocating ATP synthase is found only in the inner membrane fractions, whereas sugar-accepting porins are only in the outer membrane.

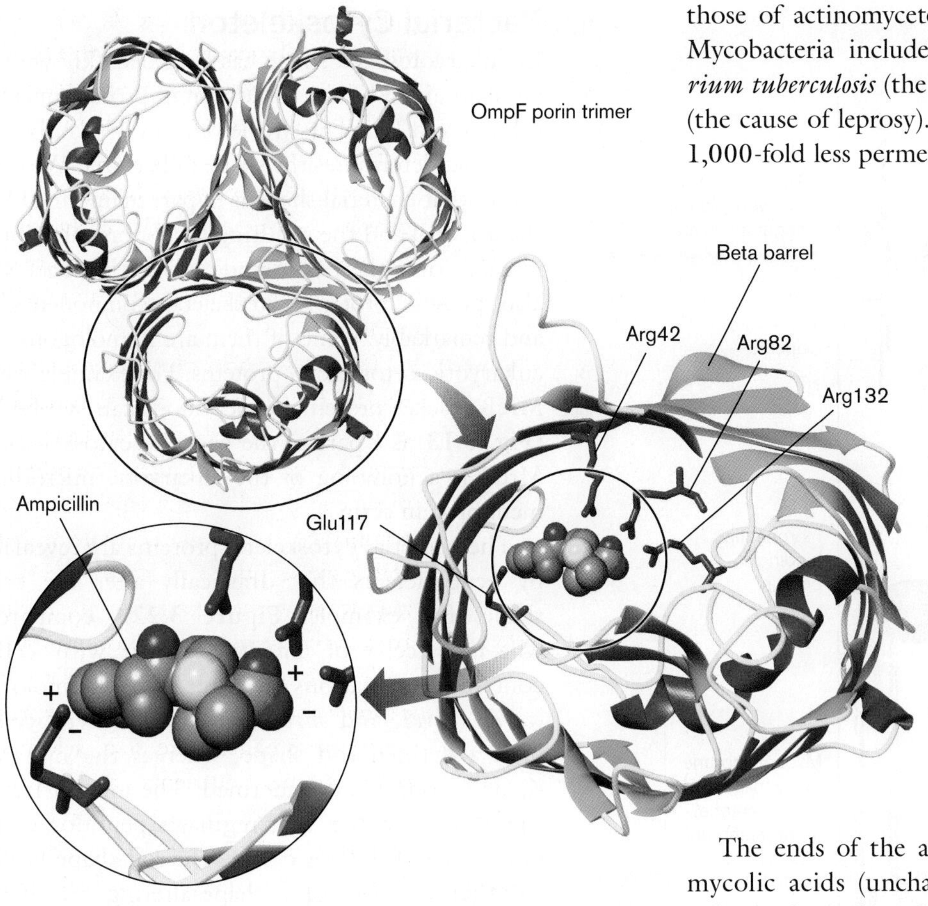

FIGURE 3.20 ■ OmpF porin transports ampicillin. This model of the OmpF trimer is based on X-ray crystallography. (PDB code: 2OMF) Within each tubular porin monomer, charged amino acid residues contact ampicillin. *Source:* Modified from Ekaterina M. Nestorovich et al. 2002. *PNAS* **99**:9789–9794.

S-layer and capsule. Some Gram-negative bacteria also form an S-layer and/or a capsule exterior to the outer membrane. For example, an S-layer is found in *Caulobacter*, an aquatic proteobacterium; and in species of the genus *Campylobacter*, which cause diarrheal disease and systemic infections of immunocompromised patients. A capsule is found in virulent strains of *Haemophilus influenzae*, which was the leading cause of childhood meningitis before development of the Hib vaccine (see Chapter 24).

Thought Question

3.8 Why would proteins be confined to specific cell locations? Why would a protein not be able to function everywhere in the cell?

Mycobacterial Cell Envelope

Exceptionally complex cell envelopes are found in actinomycetes, a large and diverse family of soil bacteria that produce antibiotics and other industrially useful products (discussed in Chapter 18). The most complex envelopes known are those of actinomycete-related bacteria, the mycobacteria. Mycobacteria include the famous pathogens *Mycobacterium tuberculosis* (the cause of tuberculosis) and *M. leprae* (the cause of leprosy). The mycobacterial envelope may be 1,000-fold less permeable to nutrients and toxins than the envelope of *E. coli* is. Thus, mycobacteria must grow slowly—but they effectively resist host defenses.

The mycobacterial envelope includes features of both Gram-positive and Gram-negative cells, as well as structures unique to mycobacteria (**Fig. 3.21**). In mycobacteria, the peptidoglycan is linked to chains of galactose, called galactans. The galactans are attached to arabinans, polymers of the five-carbon sugar arabinose. The arabinan-galactan polymers are known as arabinogalactans. Arabinogalactan biosynthesis is inhibited by two major classes of anti-tuberculosis drugs: ethambutol and the benzothiazinones.

The ends of the arabinan chains form ester links to mycolic acids (uncharged mycolates). Mycolic acids provide the basis for acid-fast staining, in which cells retain the dye carbolfuchsin, an important diagnostic test for mycobacteria and actinomycetes (described in Chapter 28). Mycolic acids contain a hydroxy acid backbone with two hydrocarbon chains—one comparable in length to typical membrane lipids (about 20 carbons), the other about threefold longer. The long chain includes ketones, methoxyl groups, and cyclopropane rings. Hundreds of different forms are known.

The mycolic acids form a bilayer interleaved with sugar mycolates—a kind of outer membrane, or "mycomembrane," analogous to the Gram-negative outer membrane. This mycomembrane even contains porins homologous to Gram-negative beta barrel porins such as OmpA. Other mycolate-embedded proteins include virulence factors such as fibronectin-binding protein (Fbp). Fbp enhances the ability of *M. tuberculosis* to invade macrophages.

The outer ends of the sugar mycolates are interleaved with phenolic glycolipids, which include phenol groups linked to sugar chains. The extreme hydrophobicity of the phenol derivatives generates a waxy surface that prevents phagocytosis by macrophages. Overall, the thick, waxy envelope excludes many antibiotics and offers exceptional protection from host defenses, enabling the pathogens of tuberculosis and leprosy to colonize their hosts over long periods. However, the thick envelope also retards uptake of nutrients. As a result, *M. tuberculosis* and *M. leprae* grow extremely slowly and are a challenge to culture in the laboratory.

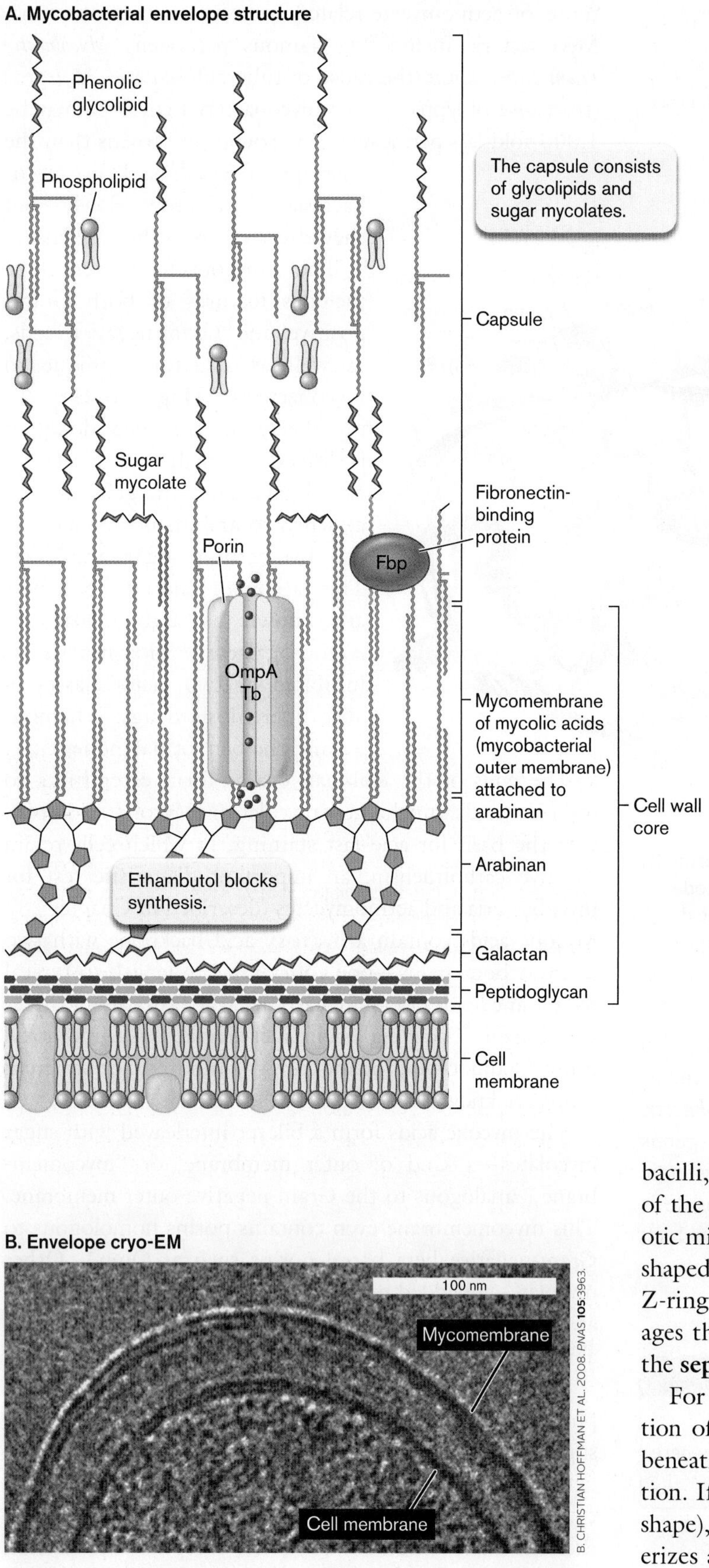

FIGURE 3.21 ▪ Mycobacterial envelope structure. **A.** The peptidoglycan layer is linked to a chain of galactose polymer (galactan) and arabinose polymer (arabinan). Arabinan forms ester links to mycolic acids, which form an outer bilayer with phenolic glycolipids. **B.** Mycobacterial envelope (cryo-TEM).

Bacterial Cytoskeleton

In eukaryotes, cell shape has long been known to be maintained by a cytoskeleton of protein microtubules and filaments (reviewed in eAppendix 2). But what determines the shape of bacteria? We saw earlier that bacterial shape is in part maintained by the cell wall and the resulting turgor pressure. But research over the past decade shows that bacteria also possess protein cytoskeletal components—and remarkably, some of them are homologous to eukaryotic cytoskeletal proteins. For example, the MreB tracker protein for peptidoglycan synthesis (**Fig. 3.13**) is a part of the bacterial cytoskeleton. MreB is a homolog of the eukaryotic microfilament protein actin.

The bacterial cytoskeletal proteins are revealed by gene defects that drastically alter the cell shape. For example, **Figure 3.22A** compares wild-type cells of *Bacillus subtilis* with cells containing mutations in three *mreB* homologs (*mreB*, *mreI*, and *mreBH*). The wild-type cells have a defined rod shape, whereas the mutant shape is round and undefined. The mutant lacks the MreB complex that regulates peptidoglycan synthesis and thereby defines the rod-shaped cell. Another example of a shape-altering mutation affects the comma-shaped cell of *Caulobacter crescentus* (**Fig. 3.22B**). A mutation in the gene *creS* results in cells that are straight instead of curved. The *creS* gene expresses the cytoskeletal protein CreS (crescentin).

How do the various cytoskeletal proteins work together to generate the overall shape of a bacterial cell? The functions of cytoskeletal proteins are probed by fluorescent protein fusions (**Fig. 3.23**). In both spherical bacteria (cocci) and rod-shaped bacilli, cell division requires the protein FtsZ, a homolog of the eukaryotic protein tubulin (the subunit of eukaryotic microtubules). The bacterial protein FtsZ forms a ring-shaped complex around the middle of the cell, called the Z-ring. The Z-ring determines the cell diameter and manages the growth of the dividing partition, which is called the **septum** (plural, **septa**).

For a rod-shaped cell, elongation requires polymerization of MreB (**Fig. 3.23B**). MreB travels in a helical arc beneath the cell membrane, guiding peptidoglycan elongation. If the rod shape is curved (called a vibrio, or crescent shape), the third cytoskeletal protein, crescentin, polymerizes along the inner curve of the crescent (**Fig. 3.23C**). The cell's outer curve is visualized by a membrane-specific fluorophore. These cytoskeletal proteins, and their variants that have evolved in other species, work together within cells to generate the shapes of bacteria.

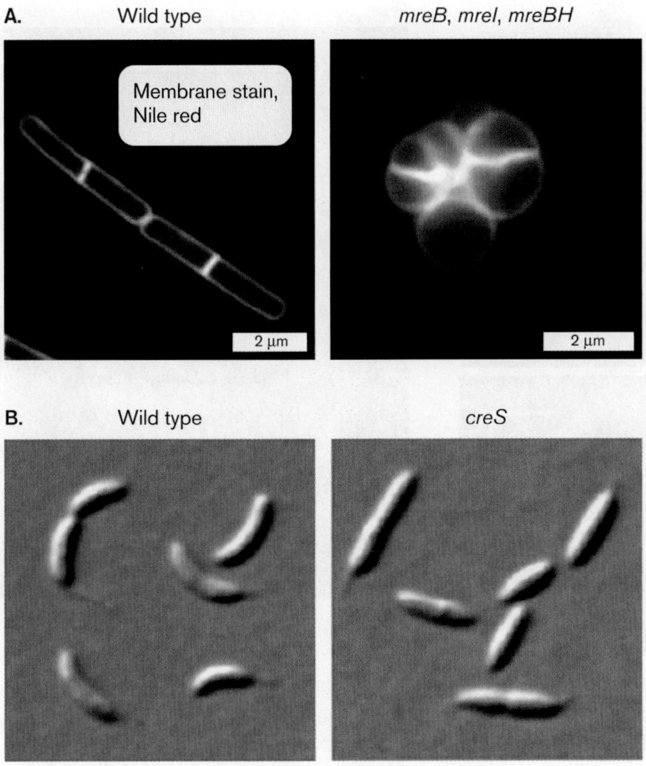

FIGURE 3.22 ■ Cytoskeletal mutants. Wild-type cells compared with mutants in **(A)** *Bacillus subtilis* (fluorescence microscopy) and **(B)** *Caulobacter crescentus* (DIC). *Source* for A: https://www.nature.com/articles/ncomms4442; https://creative commons.org/license/by/3.0/us/legalcode.

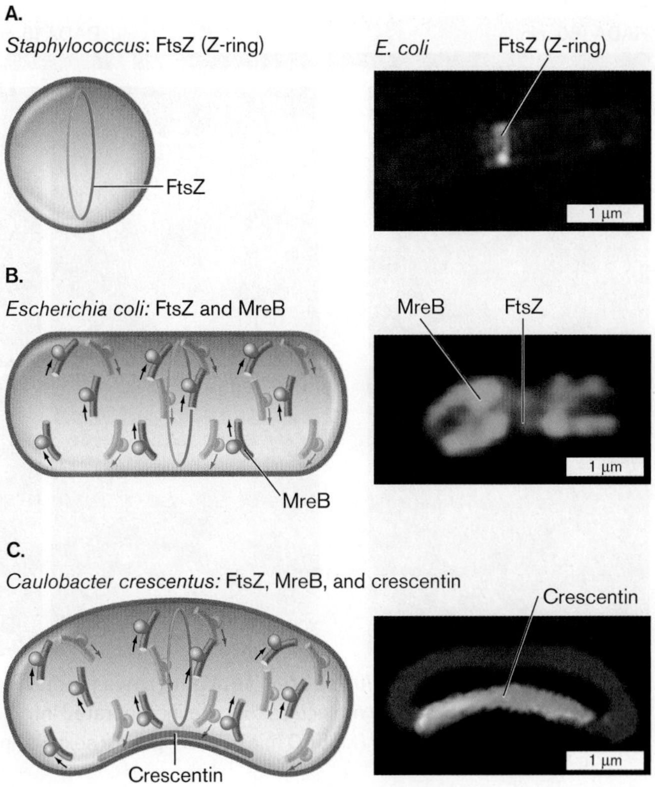

FIGURE 3.23 ■ Shape-determining proteins. **A.** Cell diameter is maintained by FtsZ polymerization to form the Z-ring. **B.** Elongation of a rod-shaped cell requires MreB proteins. MreB polymerizes around an *E. coli* cell (MreB-YFP fluorescence) along with a Z-ring of FtsZ (fluorescent anti-FtsZ antibody). **C.** Crescent-shaped cells possess a third shape-determining protein, CreS (crescentin), which polymerizes along the inner curve of the crescent. Crescentin protein fused to green fluorescent protein (CreS-GFP) localizes to the inner curve of *Caulobacter crescentus*. Membrane-specific stain FM4-64 (red fluorescence) localizes to the membrane around the cell.

To Summarize

- **The cell wall maintains turgor pressure.** The cell wall is porous, but its network of covalent bonds generates turgor pressure that protects the cell from osmotic shock.
- **The Gram-positive cell envelope** has multiple layers of peptidoglycan, threaded by teichoic acids.
- **The S-layer of proteins** is highly porous but can prevent phagocytosis and protect cells in extreme environments. In archaea, the S-layer serves the structural function of a cell wall.
- **The capsule,** composed of polysaccharide and glycoprotein filaments, protects cells from phagocytosis. Both Gram-positive and Gram-negative cells may possess a capsule.
- **The Gram-negative outer membrane** regulates nutrient uptake and excludes toxins. The outer membrane contains LPS and protein porins of varying selectivity.
- **The mycobacterial cell wall includes features of both Gram-positive and Gram-negative cells.** The arabinogalactan layer adds thickness to the cell wall. The mycolate outer membrane and phenolic glycolipids limit uptake of nutrients and antibiotics.
- **The bacterial cytoskeleton** includes proteins that regulate cell size, play a role in determining the rod shape of bacilli, and generate curvature in vibrios.

3.4 Bacterial Cell Division

How does a growing bacterial cell divide, or fission, into daughter cells? Bacterial cell fission requires highly coordinated growth and formation of all the cell's parts. Unlike eukaryotes, prokaryotes synthesize RNA and proteins continually while the cell's DNA undergoes replication. Bacterial DNA replication is coordinated with the expansion of the cell wall and ultimately with the separation of the cell into two daughter cells. **Bacteria do <u>not</u> undergo mitosis or meiosis** (these eukaryotic processes are reviewed in eAppendix 2). Bacterial DNA replication is outlined here as it relates to cell division; the genetic aspects of DNA replication are discussed in Chapter 7.

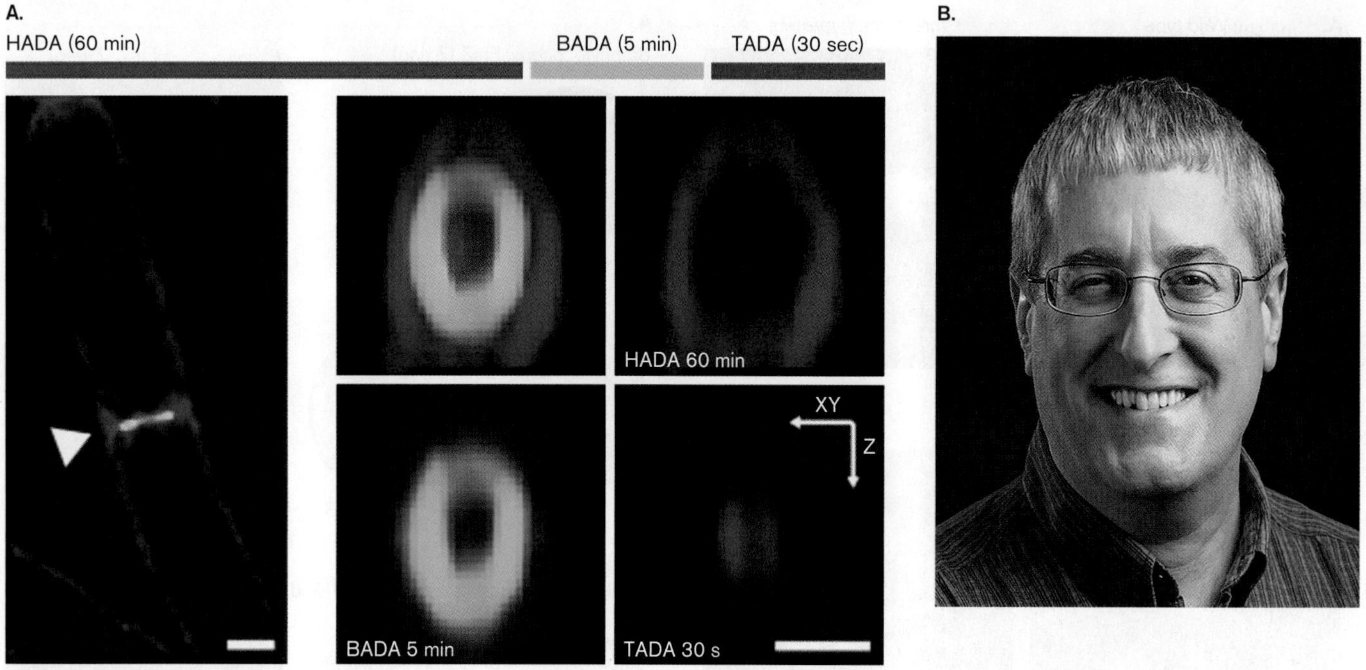

ALEXANDRE BISSON-FILHO ET AL. 2017 *SCIENCE* **355**:739, FIG. 1A.

PHOTO BY SANDEE MILHOUSE

FIGURE 3.24 ■ The *Bacillus subtilis* septum grows from the outer ring inward. **A.** Cells were pulse-labeled with different-colored fluorescent D-alanine molecules that are incorporated into peptidoglycan, catalyzed by penicillin-binding proteins. The D-alanine fluorophores are: HADA, blue (first 60 min); BADA, green (next 5 min); TADA, red (final 30 sec). **B.** Yves Brun.

Cell Division by Septation

In rod-shaped cells, the envelope elongates to a consistent length by extension of peptidoglycan chains in spiral tracks around the cylindrical cell, as we saw in **Figure 3.13**. But as DNA synthesis terminates, the cell divides by a process called **septation**, the formation of the septum, the partition that divides the envelope. How does the septum actually form, completing two entire cell envelope layers back-to-back? The laboratories of Ethan Garner at Harvard University and of Yves Brun at Indiana University revealed the progression and timing of septum growth (**Fig. 3.24**).

The septum grows inward from the sides of the cell, at last constricting and sealing off the two daughter cells. The inward growth of peptidoglycan can be seen in **Figure 3.24A**, where a cell of *Bacillus subtilis* incorporates D-alanine fluorophores. D-Alanine is a key amino acid that gets incorporated into peptidoglycan cross-bridges (**Fig. 3.12**) but not into proteins, which use only L-form amino acids. The D-alanine fluorophores emit one of three different colors, in successive periods of pulse labeling. The initial, longest period (60 minutes) is required to grow the outermost ring of cell wall pinching in; then shorter pulses (5 minutes, 30 seconds) rapidly complete the septum. Garner's team showed that FtsZ subunit assembly circles around the septum in a "treadmilling" pattern, stepwise around the cell, that directs septal growth.

Septation requires rapid biosynthesis of all envelope components, including membranes and cell wall (**Fig. 3.25**). Envelope expansion must coordinate the extension of all layers—and regulate the placement and timing of the septum. As we saw earlier (**Fig. 3.13**), the enzymes of cell wall biosynthesis must coordinate the formation of new links as closely as possible with subunit insertion. The site of septation is uniquely vulnerable because two new large enclosures must form simultaneously. Thus, the enzymes of septum biosynthesis are of great interest as antibiotic targets.

The overall process of septation is managed by a protein complex called the **divisome** (**Fig. 3.25C**). The divisome manages assembly of the septum with its two envelopes back-to-back. One component of the divisome is FtsZ, which polymerizes to form the Z-ring, as seen previously, in **Figure 3.23**. Mutations in a gene such as *ftsZ* cause *E. coli* to form long filaments instead of dividing normally. Another divisome component, FtsN, helps regulate the timing of constriction of the septum. A cell was constructed that requires an inducer molecule to express FtsN (**Fig. 3.25C**). Cells that lack FtsN fail to constrict their septum, and their membranes balloon out at cell division. Such divisome components could be targets for new antibiotics, just as penicillin-binding proteins are targets for penicillin.

The bacterial process of cell fission must solve a key problem: how to coordinate septation with the replication of DNA. Indeed, mutations exist that lead to septum formation across DNA with replication incomplete. The result is to "guillotine" the cell. To avoid this disastrous situation, septation is coordinated with DNA replication.

DNA Is Organized in the Nucleoid

The genetic functions of microbial DNA are discussed in detail in Chapters 7–12. Here we focus on the physical organization of DNA within the nucleoid of bacterial and archaeal cells.

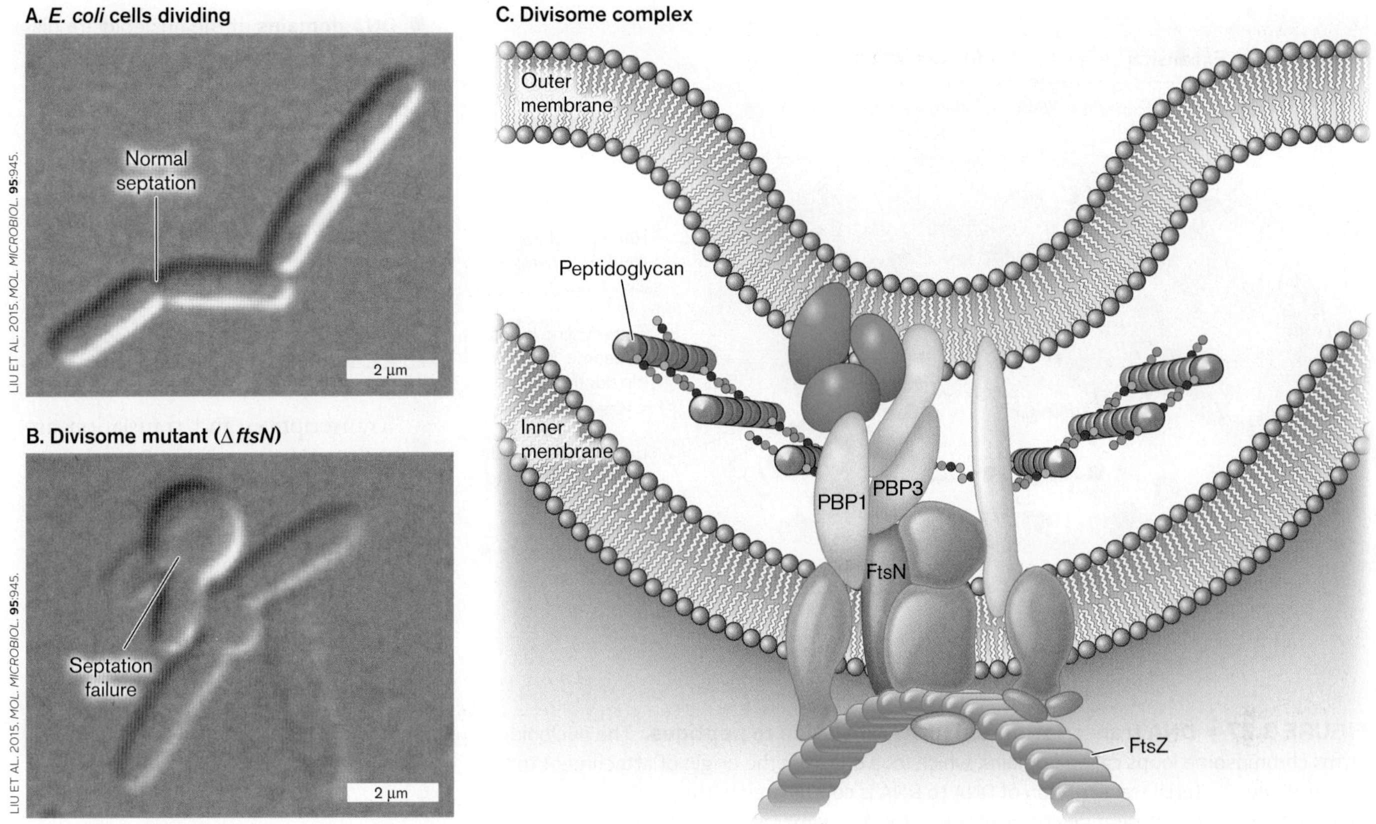

FIGURE 3.25 ▪ Divisome enables cell division. A. *E. coli* cells divide normally. **B.** Cells that lack FtsN (divisome component) balloon out at the septum. **C.** The divisome complex coordinates the extension of all envelope layers while the septum constricts. *Source:* Part C modified from C. Typas et al. 2012. *Nat. Rev. Microbiol.* 10:123, fig. 2.

Bacteria organize their DNA very differently from eukaryotes. For example, **Figure 3.26** shows enteropathogenic *E. coli* cells growing on a cultured human cell. Enteropathogenic *E. coli* (EPEC) are diarrheal pathogens that attach to the host cell membrane and inject toxins (discussed in Chapter 25). In this thin-section TEM, each bacterium contains a filamentous nucleoid region that extends through the cytoplasm. In contrast, the nucleus of the eukaryotic cell (not shown), is many times larger than the entire bacterial cell, and the chromosomes it contains are separated from the cytoplasm by the nuclear membrane.

Note: In bacteria and archaea, the genome typically consists of a single circular chromosome, but some species have a linear chromosome or multiple chromosomes. In this chapter we focus on the simple case of a single circular chromosome.

In a bacterial cell, the DNA is organized in loops called **domains**, which extend throughout the cytoplasm. The midpoint on the DNA is the origin of replication, which is attached to the cell envelope at a point on the cell's equator, halfway between the two poles (**Fig. 3.27**). To initiate DNA replication, the DNA double helix at the

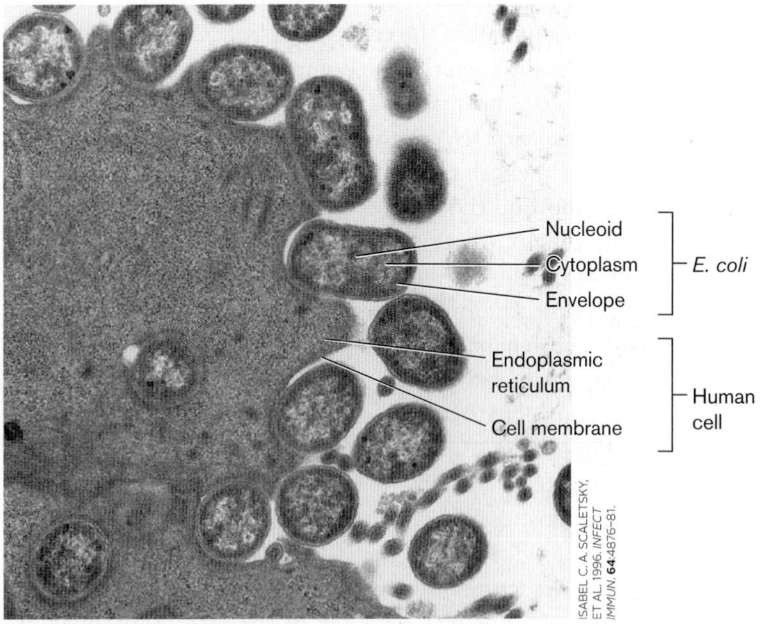

FIGURE 3.26 ▪ Bacteria invading a human cell. Enteropathogenic *Escherichia coli* bacteria attached to the surface of a tissue-cultured human cell (TEM). The bacterial nucleoid appears as a lighter region with few ribosomes.

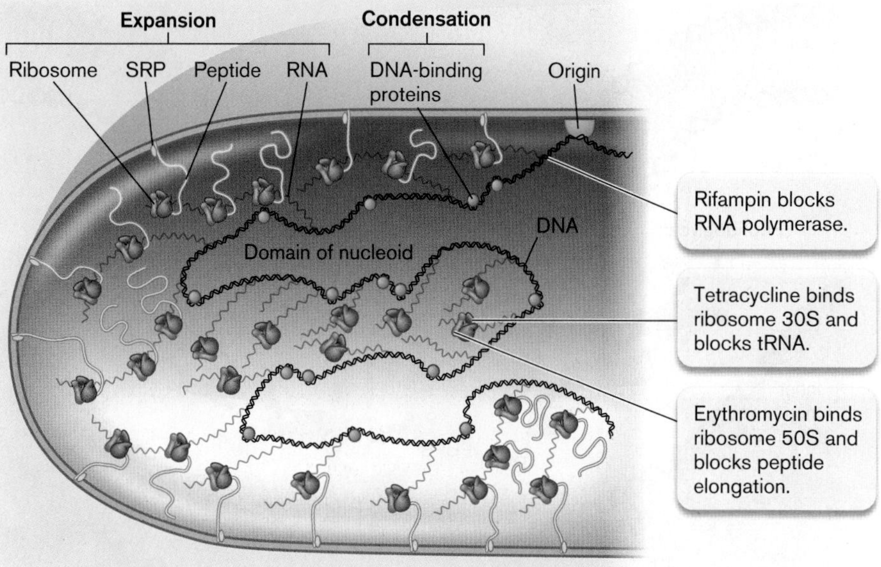

FIGURE 3.27 ■ DNA transcription and RNA translation to peptides. The nucleoid forms chromosome loops called domains, which loop out from the origin of attachment to the cell envelope. Bacterial transcription of DNA to RNA is coordinated with translation of RNA to make proteins. Growing peptide chains destined for the membrane bind the signal recognition particle (SRP) for membrane insertion.

origin is melted open by binding proteins, and then DNA polymerase synthesizes new strands in both directions (bidirectionally). The origin and other aspects of DNA replication are covered in detail in Chapter 7.

How does all of the cell's DNA fit neatly into the nucleoid? In some bacteria, the domains loop back to the center of the cell, near the origin of replication. Within the domains, the DNA is compacted by supercoils. Supercoils (or superhelical turns) are extra twists in the chromosome, beyond those inherent in the structure of the DNA double helix (discussed in Chapter 7). The supercoiling causes portions of DNA to double back and twist upon itself, resulting in compaction of the chromosome. Supercoils are generated by enzymes such as gyrase, which are a major target for antibiotics such as quinolones. DNA is also compacted by **DNA-binding proteins** (green spheres in **Fig. 3.27**). Binding proteins can respond to the state of the cell; for example, under starvation conditions, when most RNA synthesis ceases, the binding protein Dps is used to organize the DNA into a protected crystalline structure. Such "biocrystallization" by Dps and related proteins may be a key to the extraordinary ability of microbes to remain viable for long periods in stationary phase or as endospores.

Note: In biology, the word "domain" is used in several different ways, each referring to a defined portion of a larger entity.

- **DNA domains** of the nucleoid are distinct loops of DNA that extend from the origin.
- **Protein domains** are distinct functional or structural regions of a protein.
- **Lipid domains** are clusters of one type of lipid within a membrane.
- **Taxonomic domains** are genetically distinct classes of organisms, such as Bacteria, Archaea, and Eukarya.

Transcription and translation are coupled (discussed in Chapter 8). The information encoded in DNA is "read" by the processes of transcription and translation to yield gene products. In bacteria and archaea, some translation is tightly coupled to transcription; the ribosomes bind to mRNA and begin translation even before the mRNA strand is complete. Thus, a growing bacterial cell is full of mRNA strands dotted with ribosomes (**Fig. 3.27**).

In rapidly growing bacteria, the DNA is transcribed and the messenger RNA is translated to proteins while the DNA itself is being replicated. This remarkable coordination of replication, transcription, and translation explains why some bacterial cells can divide in as little as 10 minutes. An example is the hot-spring bacterium *Geobacillus stearothermophilus* cultured at 60°C. DNA synthesis, transcription to RNA, and translation to proteins are discussed further in Chapters 7 and 8.

Some of the newly translated proteins are destined for the cell membrane or for secretion outside. Proteins destined for the membrane are synthesized in association with the membrane, directed there by signal recognition particles (**Fig. 3.27**). This coupling of transcription and translation to membrane insertion has the effect of expanding the nucleoid into distal parts of the cell, partly counteracting the condensation of DNA by DNA-binding proteins. Membrane protein maturation and secretion are discussed in Chapter 8.

DNA Replication Regulates Cell Division

The process of synthesizing daughter cells begins at the origin of replication, a unique DNA sequence in the chromosome. In bacteria, the origin is attached to a site on the envelope—most commonly, at a point on the cell's equator (**Fig. 3.28** ▶). At the origin sequence, the DNA double helix begins to unzip, forming two replication forks. At each replication fork, DNA is synthesized by DNA polymerase. The complex of DNA polymerase with

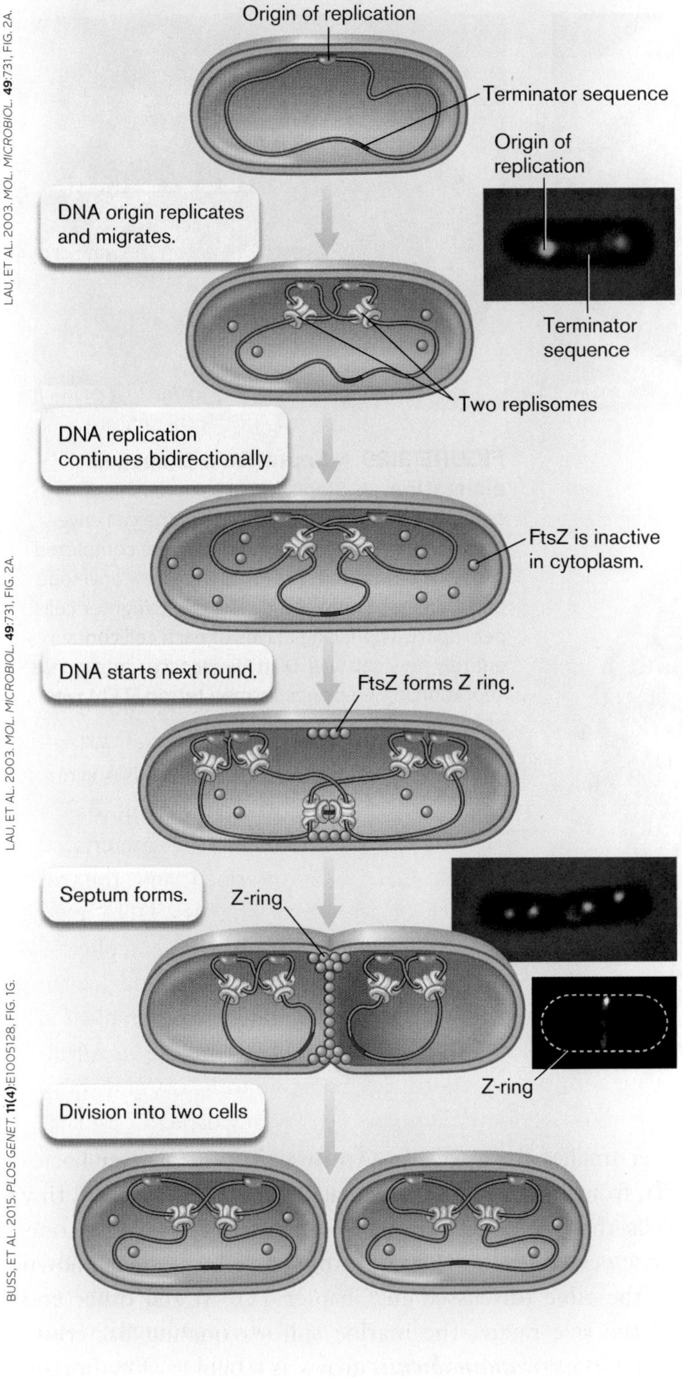

FIGURE 3.28 ■ Replisome movement within a dividing cell. The DNA origin-of-replication sites (green) move apart in the expanding cell as the two replisomes (yellow) stay near the middle, where they replicate around the entire chromosome, completing the terminator sequence last (red). As the terminator sequence nears completion, FtsZ proteins assemble the Z-ring organizing septum formation. *Source:* (Top 2 insets): Ivy Lau, et al. 2003. Mol. Microbiol. **49**:731. (Bottom inset): Jackson Bust, et al. 2015. *PLoS Genet.* **11**(4).

its accessory components is called a **replisome**. The replisome actually includes two DNA polymerase enzymes: one to replicate the "leading strand," the other for the "lagging strand." The lag time is short compared with the overall time of replication; thus, as the replisome travels along the DNA, it converts one helix into two progeny helices almost simultaneously.

Note: Each replisome contains two DNA polymerases (for leading and lagging strands). Each dividing nucleoid requires two replisomes (for bidirectional replication), and thus four DNA polymerases overall.

Within the cell, replication proceeds outward in both directions around the genome. Thus, bidirectional replication requires two replisomes, one for each replicating fork. Fluorescent probes show that two replisomes are located near the middle of the growing cell (**Fig. 3.28**). Each replisome forms a replicating fork that directs two daughter strands of DNA toward opposite poles. The two copies of the DNA origin of replication (green in the figure), attached to the cell envelope, move apart as the cell expands. The termination site (red) remains in the middle of the cell, where the two replisomes continue replication at both forks. Finally, as the termination site replicates, the two replisomes separate from the DNA. At each new origin site, however, two pairs of new replisomes have formed. In a fast-growing cell, the new origin sites begin a second round of replication, even before termination of the previous round.

Completion of replication triggers Z-ring formation. For the cell to divide, DNA replication must be complete. During the process of replication, the cytoplasm contains several kinds of proteins that will determine septation, among them FtsZ (**Fig. 3.28**). Other septation-related proteins are bound to the cell membrane or to DNA. Replication of the DNA termination site triggers several proteins to form the divisome. For simplicity, only FtsZ is shown, as the subunits assemble to form the Z-ring.

Ultimately, septation completes cell division, and the two envelope ends come apart. In some species, such as filamentous cyanobacteria, the actual separation of cells may occur long after septation, forming extended filaments of individual cells. Others, such as *Bacillus megaterium*, commonly separate two or three generations after septation. Nutrient conditions may affect cell separation and cell size.

Thought Question

3.9 Suppose a cell has a defect in its *ftsZ* gene. What might happen to the cell during growth? How could such a mutant strain be maintained in the laboratory?

Septation of spherical cells. In spherical cells (cocci), such as *Staphylococcus aureus*, the process of septation generates most of the new cell envelope to enclose the expanding cytoplasm (**Fig. 3.29A–C**). Furrows form in the cell envelope, in a ring all around the cell equator, as the completed envelope layers

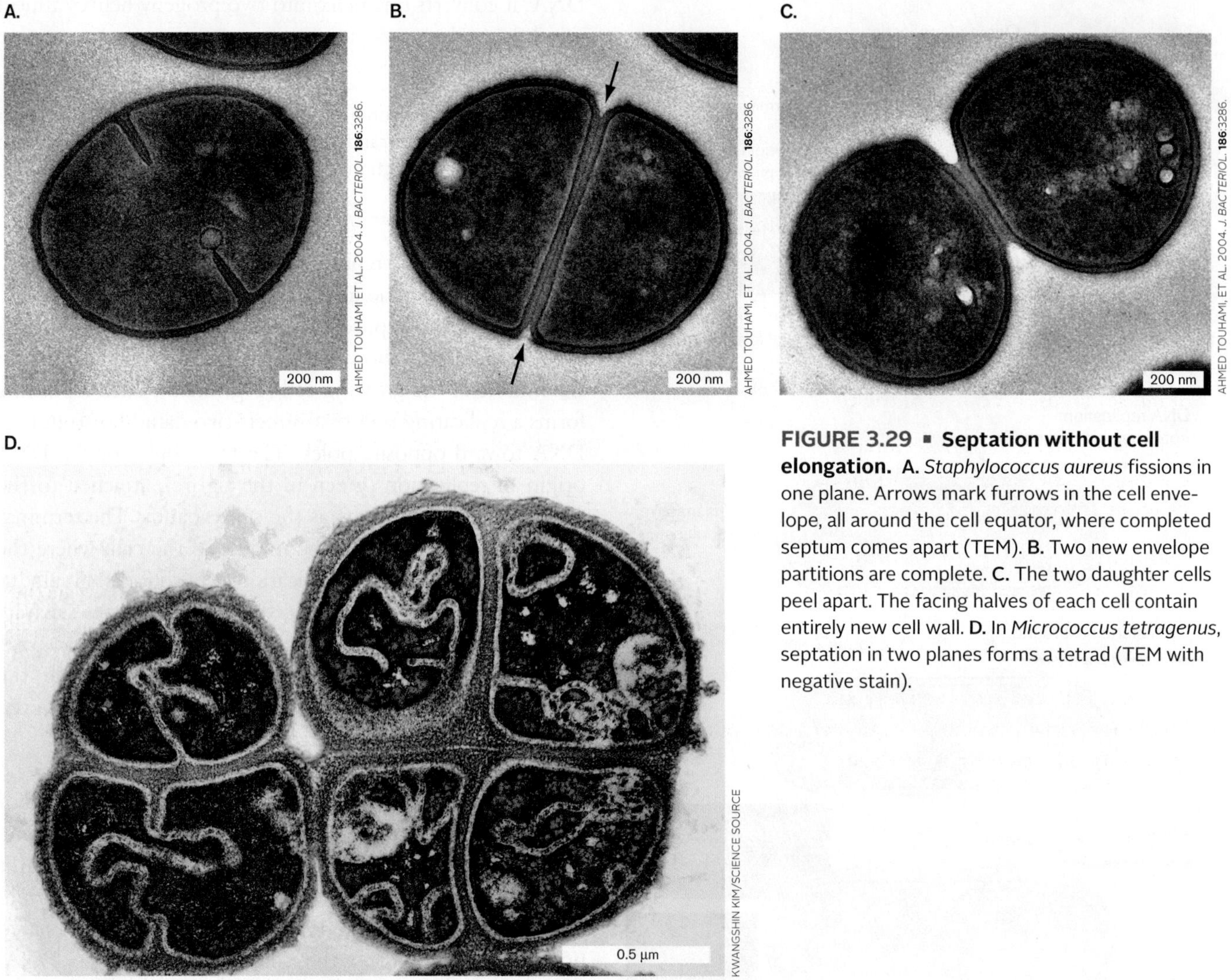

FIGURE 3.29 ■ Septation without cell elongation. **A.** *Staphylococcus aureus* fissions in one plane. Arrows mark furrows in the cell envelope, all around the cell equator, where completed septum comes apart (TEM). **B.** Two new envelope partitions are complete. **C.** The two daughter cells peel apart. The facing halves of each cell contain entirely new cell wall. **D.** In *Micrococcus tetragenus*, septation in two planes forms a tetrad (TEM with negative stain).

of the septum peel apart. Unlike rod-shaped cells (which elongate between divisions), the facing halves of each spherical cell form out of septum envelope as the two halves peel apart.

The spatial orientation of septation has a key role in determining the shape and arrangement of cocci. If the cell always septates in parallel planes, as in *Streptococcus* species, cells form chains. If, however, the cell septates in random orientations, or if cells reassociate loosely after septation, they form compact hexagonal arrays similar to the grape clusters portrayed in classical paintings—hence the Greek-derived term **staphylococci** (*staphyle* means "bunch of grapes"). Such clusters are found in colonies of *Staphylococcus aureus.* If subsequent septation occurs at right angles to the previous division, the cells may form tetrads and even cubical octads called "sarcinae" (singular, sarcina). Tetrads are formed by *Micrococcus tetragenus*, a cause of pulmonary infections (**Fig. 3.29D**).

Bacterial cell size. How do cells "know" how large to grow? This question is hard to answer, as we are still discovering ever-smaller forms of life. An investigation of river bacteria, from Jillian Banfield's lab at UC Berkeley, revealed tiny cells that pass through a 0.2-µm filter. These cells comprise new kinds of life accounting for 15% of all taxa known at the time (discussed in Chapter 18). At the other end of the size range, the marine sulfur-oxidizing bacterium *Thiomargarita namibiensis* grows as a bubble of cytoplasm 200 µm across (presented in Chapter 1).

For a given species, cell size depends on genetic regulators and environmental constraints. When a bacterial population is first diluted into fresh medium, with abundant nutrients, cells elongate faster and reach larger sizes before septation and division. As nutrients become scarce, cell growth slows, and early division produces smaller cells. Thus, cell size is one factor in the phases of growth of bacterial populations (discussed in Chapter 4). Yet, repeated cycles of growth and starvation in minimal glucose medium lead *E. coli* populations to undergo selection for cells that are larger. Experimental evolution (discussed in Chapter 17) thus enables us to test models of cell size development.

To Summarize

- **Bacterial cell division includes elongation and septation.**
- **DNA is organized in the nucleoid.** In most bacterial species, the DNA is attached to the envelope at the origin of replication, on the cell's equator. Loops of DNA called domains are supercoiled and bound to DNA-binding proteins.
- **During transcription, the ribosome translates RNA to make proteins.** Proteins are folded by chaperones and in some cases secreted at the cell membrane.
- **DNA is replicated bidirectionally by the replisome.** During bacterial DNA replication, genes continue transcription and translation.
- **Completion of DNA replication triggers Z-ring formation and septation.** Septation may occur in one plane (forming a chain of cells) or at right angles to the previous septation (forming a tetrad).
- **Bacterial cell size varies widely among taxa.** Within a population, environmental parameters such as nutrient availability may determine cell size.

3.5 Cell Polarity, Membrane Vesicles, and Nanotubes

Do dividing bacteria produce symmetrical offspring? Even superficially symmetrical bacilli such as *E. coli* show underlying chemical and physical asymmetry, such as possession of a chemoreceptor array at the "forward" pole. Other species, such as *Caulobacter crescentus*, develop different structures at either pole, and their cell division generates two different cell types. And many kinds of bacteria extend their cytoplasm in surprising ways, by forming extracellular membrane vesicles and nanotubes. Such cell extensions complicate the very definition of an individual cell.

Bacterial Cell Differentiation

Bacteria whose poles have different structures generate two different forms of progeny. The wetland bacterium *Caulobacter crescentus* has one plain pole and one pole with either a flagellum or a cytoplasmic extension called a stalk (**Fig. 3.30**). A flagellated cell swims about freely in an aqueous habitat, such as a pond or a sewage bed. After swimming for about half an hour, if the bacterium finds a place with enough nutrients, the cell sheds its flagellum and replaces it with a stalk. The stalked cell attaches to sediment and then immediately starts to replicate its DNA and divides, producing a flagellated daughter cell, as well as a daughter cell containing the original stalk.

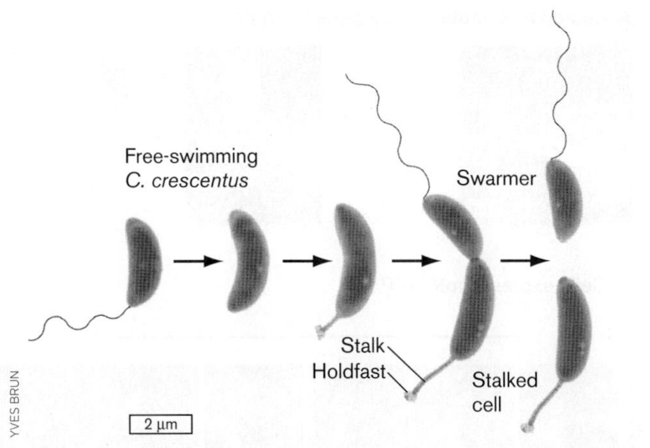

FIGURE 3.30 ▪ Asymmetrical cell division: a model for development. A swarmer cell of *Caulobacter crescentus* loses its flagellum and grows a stalk. The stalked cell divides to produce a swarmer cell (TEM).

How does *C. crescentus* organize itself to produce two different cell types, each with a different organelle at one pole? The process is a rudimentary form of cell differentiation, comparable to the differentiation processes that animal cells undergo in the embryo. The *C. crescentus* life cycle is governed by regulator proteins such as TipN, studied by students of Christine Jacobs-Wagner at Yale University (see **eTopic 3.2**). Mutants lacking TipN make serious mistakes in development. Instead of making a single flagellum at the correct cell pole, the cell makes multiple flagella at various locations, even on the stalk (**Fig. 3.31A**). Jacobs-Wagner proposed that TipN is a landmark protein that correctly marks the site of a new cell pole and directs the polar placement of flagella. **Figure 3.31B** shows cells expressing TipN fused to the fluorescent protein GFP, which is then detected by fluorescence microscopy. The fluorescent fusion protein TipN-GFP localizes to the cell pole opposite the stalk. As the cell prepares to divide, TipN leaves the pole, delocalizing around the cell. Eventually, the TipN protein relocalizes at the septum, where the new poles appear.

Cell development involves many such proteins working together. **Figure 3.32** shows how TipN interacts with two other polar proteins: the flagellar marker PodJ, and the stalk marker DivJ. Each young cell (swarmer cell at top of cycle) has a new pole containing TipN. PodJ migrates to the pole with TipN. The stalk marker DivJ is produced at the pole where PodJ was previously. These redistributions of proteins cause the flagellated swarmer cell to lose its flagellum, replacing it with a stalk as the cell settles in a favorable environment. As the stalked cell grows, TipN delocalizes around the cell; then it localizes again at the middle, where the cell septates and divides. Once division is complete,

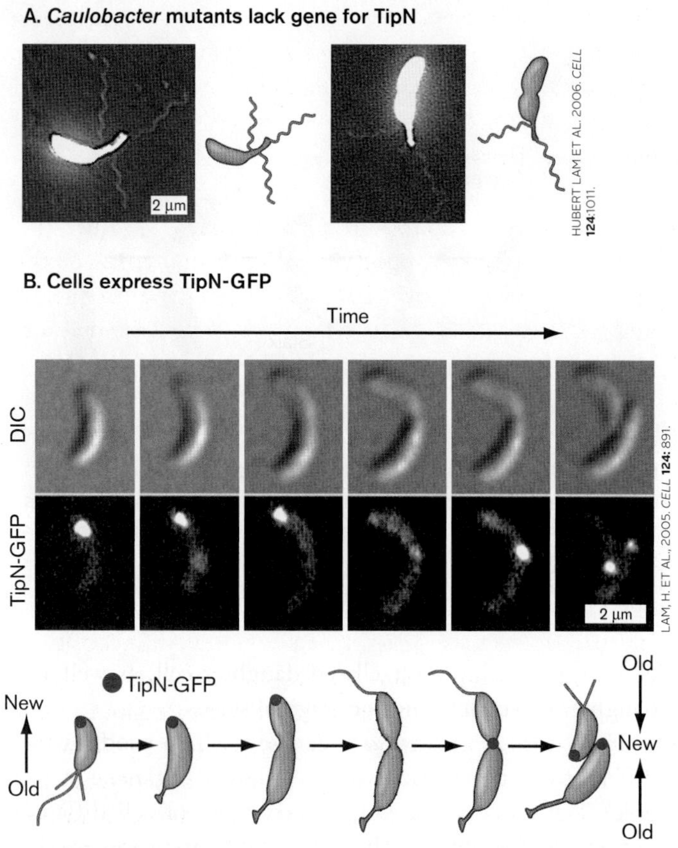

FIGURE 3.31 ■ A landmark protein for the cell pole. **A.** *Caulobacter* mutants lacking TipN protein make mistakes: flagella grow out of stalks (left), or at the stalked pole (right; fluorescence microscopy). **B.** The protein TipN appears at the pole of a *Caulobacter* stalk cell, visualized as TipN-GFP [differential interference contrast microscopy, or DIC (top row); and fluorescence microscopy (bottom row)]. As the cell grows, TipN delocalizes and then localizes again at the septum. Septation yields two daughter cells with TipN at the pole of each. **C.** Christine Jacobs-Wagner, 2011 winner of the American Society for Microbiology's Eli Lilly Award for her studies of *Caulobacter* development.

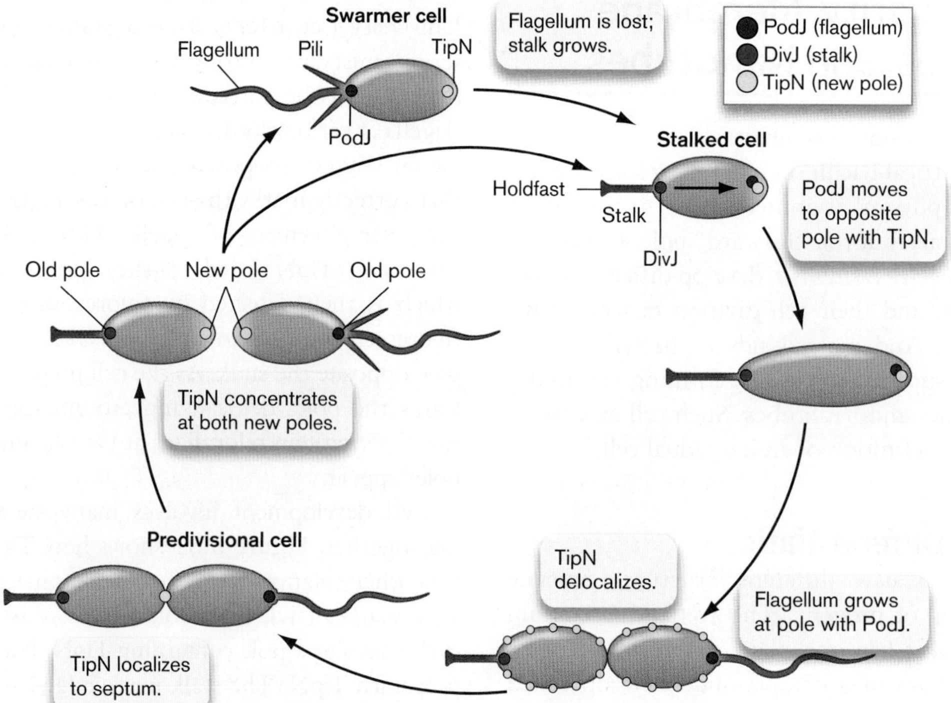

FIGURE 3.32 ■ Cell cycle of *Caulobacter*. A swarmer cell loses its flagellum and grows a stalk. PodJ protein (purple) is at the flagellar pole, while DivJ protein (red) is at the stalk. TipN (yellow) is found at "new" poles (newly septated). TipN delocalizes and then localizes at the cell equator, midway between poles. The pole with PodJ grows a flagellum. The cell septates, forming two new poles, each containing TipN. The stalked cell still has DivJ at the stalk, and the new swarmer cell has PodJ at the flagellum. *Source:* Modified from Melanie Lawler and Yves Brun. 2006. *Cell* **124**:891.

TipN concentrates at both new poles. As PodJ moves to the new pole, that pole grows a flagellum. Overall throughout the cycle, a series of polar proteins localize and delocalize to define the polar functions.

Thought Question

3.10 **Figure 3.31** presents data from an experiment that allows the function of the TipN protein of *Caulobacter* to be visualized by microscopy. Can you propose an experiment with mutant strains of *Caulobacter* to test the hypothesis that one of the proteins shown in **Figure 3.32** is required for one of the cell changes shown?

Growth Asymmetry and Polar Aging

Does an apparently symmetrical cell such as *E. coli* actually show two different polar forms? In fact, cell division generates daughter cells with chemically different poles (**Fig. 3.33**). Each cell starts out with one "old" pole (red in the figure) and one "new" pole (blue) where the parental cell septated. As the next cell divides, two daughter cells form, each with another "new" pole. But meanwhile, the "old" poles continue to age. With each generation, the polar cell wall material degrades slightly, increasing the chance of cell lysis. In a population of *E. coli* under environmental stress, at each cell division some members of the population die—of polar old age.

The cause of polar aging in stressed *E. coli* is the preferential accumulation of protein aggregates, which are nonfunctional and cannot be unfolded or degraded. For unknown reasons, proteins aggregate more frequently under a stressful condition, such as low pH or the presence of an antibiotic. Proteins damaged by a stressful condition are packed away in the cell's older pole, allowing the new-pole cells to remain intact and grow faster. This asymmetrical cell provisioning may represent a form of "altruism" in which the older half cell promotes faster growth of the younger half cell.

Yet other kinds of cells grow by extending one pole only. The actinomycete *Corynebacterium glutamicum*, a soil bacterium useful for industrial production, positions its replisome at one cell pole. As DNA replication begins, a second replisome moves to the opposite pole, while new cell wall forms at the poles. In the next generation, the opposite pole possesses the replisome and undergoes extension. Unequal or unipolar cell extension is common among actinomycetes and mycobacteria, such as *Mycobacterium tuberculosis*.

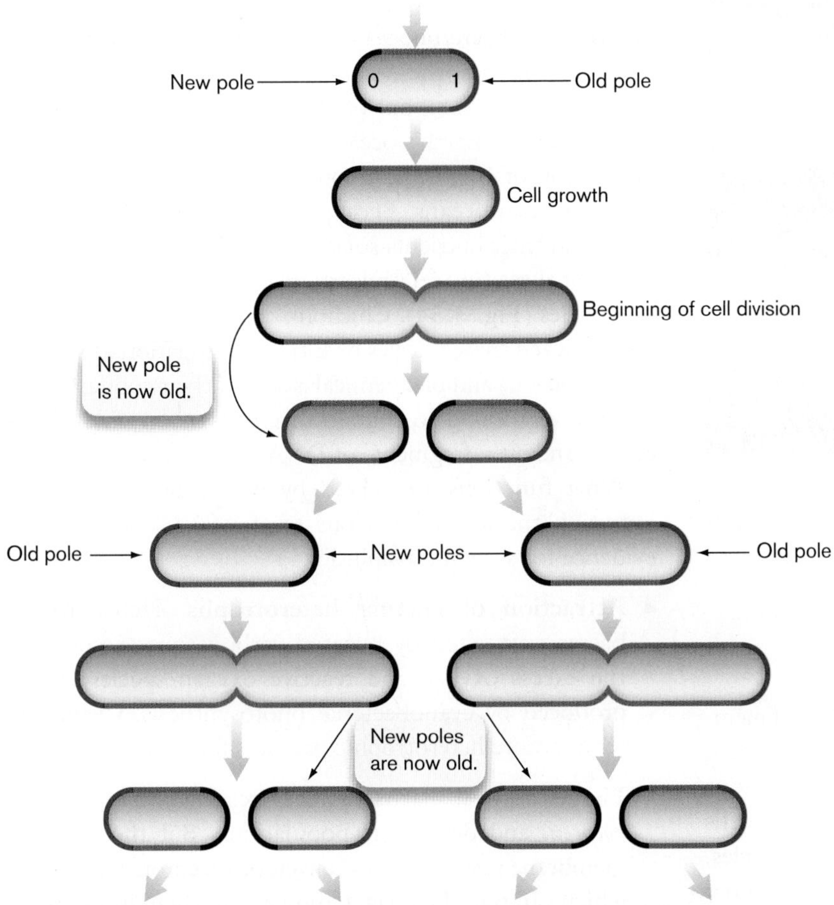

FIGURE 3.33 ▪ **Bacterial cell division generates cells with an old pole and a new pole.** Succeeding generations have cells with diverse combinations of new poles (blue), old poles (red), and very old poles (two or more generations, also red).
Source: Modified from Eric Stewart et al. 2005. *PLoS Biol.* **3**:e45.

Why does polar aging matter? One consequence of polar aging is that cells of different polar ages may differ in their resistance to antibiotics. This phenomenon could cause problems for antibiotic therapy (discussed in **eTopic 3.3**). In *Mycobacterium tuberculosis*, alternate polar aging generates variable resistance to antibiotics. The result may give tuberculosis bacteria the opportunity to "try out" resistance to various antibiotics applied in chemotherapy.

A major form of asymmetrical growth is endospore formation by Firmicutes such as *Bacillus* and *Clostridium* species. An endospore is an inert but viable cell form, having no active metabolism but capable of germination under the right conditions. Under starvation, desiccation, or other stress conditions, a bacterium can undergo an asymmetrical cell division to develop an endospore at one end. Endospore formation requires an extreme form of cellular altruism, in which the mother cell sacrifices itself for the spore-forming cell. The process generates an endospore

FIGURE 3.34 ■ **Marine bacteria release membrane vesicles.** **A.** *Prochlorococcus* cyanobacteria release vesicles (arrows) of cell membrane into the open ocean (SEM). **B.** Vesicles collected from *Prochlorococcus* (TEM). **C.** Sallie Chisholm, professor of environmental studies, pioneered the study of *Prochlorococcus* and its significance for marine ecology.

capable of remaining dormant but viable for thousands of years. **Endospore formation is covered in detail in Chapter 4.**

Membrane Vesicles and Nanotubes

Our concept of the cell assumes a defined boundary of membrane that encloses the cell's contents and separates them from the external space. The cell's cytoplasm is a precious limited resource. Yet surprisingly, isolated microbial cells continually export bits of cytoplasm in membrane vesicles. Some kinds of microbes share their materials with other cells—even cells of other species—via intercellular nanotubes. How does this cytoplasmic sharing serve the cell?

Membrane vesicles carry proteins and nucleic acids. An example of cytoplasmic export via membrane vesicles in the marine cyanobacterium *Prochlorococcus* was documented by Sallie Chisholm at the Massachusetts Institute of Technology. *Prochlorococcus* is one of the smallest yet most abundant phototrophs in Earth's oceans, believed to include a global population of 10^{27} cells, accounting for half the photosynthesis in open oceans. Because *Prochlorococcus* cells are so small and their nutrients so scarce, it is particularly remarkable that these tiny cells release their cytoplasm by pinching off vesicles (**Fig. 3.34**). Chisholm and her postdoctoral fellow analyzed these vesicles by ultracentrifugation (discussed in Section 3.1) and biochemical assays of their contents. The vesicles were found to contain diverse proteins, RNA molecules, and even fragments of DNA.

What functions are served by vesicle production that outweigh the loss of precious resources? Chisholm finds evidence for several possibilities:

- **Attraction of partner heterotrophs.** Heterotrophic bacteria attracted by released carbon sources consume the excess oxygen and reactive oxygen species (ROS) produced by cyanobacterial photosynthesis. Cyanobacteria require heterotrophic partners for optimal growth.

- **Phage decoys.** Bacteriophages readily infect *Prochlorococcus* and deplete its populations. But the bacterial membrane vesicles possess envelope receptors for phages, which can trap the phages and prevent them from infecting cells.

- **DNA transfer.** The DNA released by *Prochlorococcus* may provide useful genetic traits for other members of the population.

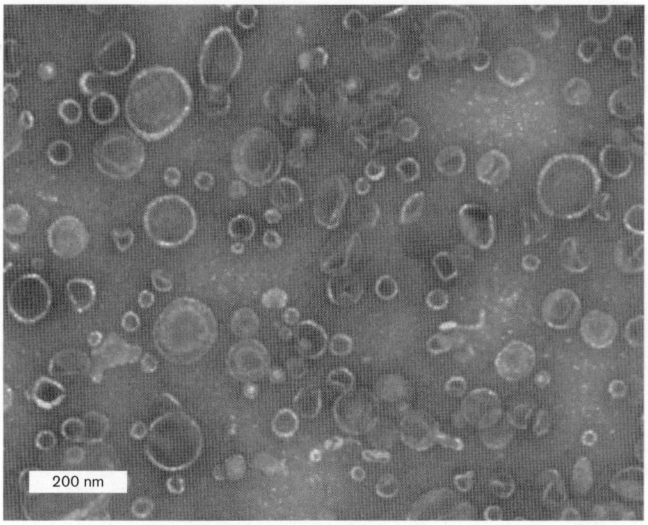

Another system in which membrane vesicles are shared is that of human gut bacteria such as *Bacteroides* and related anaerobes. One way the gut environment differs from the open ocean is in the abundance of nutrients available. In the gut, many anaerobes release vesicles of partly digested complex polysaccharides for further catabolism by other species. Often, both community members benefit as a result. Gut microbial interactions are discussed further in Chapter 21.

Intercellular nanotubes mediate bacterial cooperation. Some bacteria and archaea can form membrane extensions that merge directly with the membranes of neighboring cells. Sigal Ben-Yehuda and students at the Hebrew University of Jerusalem revealed such extensions, called intercellular **nanotubes**, between cells of *Bacillus subtilis* (**Fig. 3.35**). *B. subtilis* is a Gram-positive bacterium common in soil, a highly complex environment full of diverse nutrients and antimicrobial toxins (discussed in Chapter 21). The nanotubes enable bacteria to directly share proteins and messenger RNA encoding products useful under hostile conditions, such as exposure to antibiotics (**Fig. 3.35B**). Ben-Yehuda showed that two *Bacillus* cells encoding resistance to two different antibiotics—chloramphenicol (Cat protein) and lincomycin (Erm protein)—could share their resistance proteins and messenger RNA via nanotubes. The connected bacteria resist both antibiotics.

A similar experiment showed that even bacteria of different species can share beneficial components of cytoplasm. Christian Kost and students at Max Planck Institute for Chemical Ecology in Jena, Germany, used fluorescence microscopy to show that *Escherichia coli* bacteria can form nanotubes with the Gram-negative bacterium *Acinetobacter baylyi*. The nanotubes facilitate exchange of different amino acids between these two species. Remarkably, the nanotubes form only when the two types of cells each produce an amino acid lacking in the other. Thus, nanotubes facilitate metabolic cross-feeding.

A third kind of sharing via nanotubes is that of electron transfer, or "bacterial electricity." Bacterial electron transfer is presented in Chapter 14.

Thought Question

3.11 Could two bacteria share protein complexes via nanotubes? What about hydrogen molecules (H_2) as electron donors?

To Summarize

- **The poles of a bacterial cell may differ in form and function.** *Caulobacter crescentus* has one plain pole and one pole that has either a flagellum or a stalk. A stalked cell fissions to produce one stalked cell and one flagellar cell.
- **The two bacterial poles differ in age.** One pole arises from the septum of the parental cell, whereas the other pole arises from a parental pole. In *E. coli*, successive cell divisions yield progeny with a mixture of polar ages. Cells with a very old pole may cease replication and die.
- **Polar aging is increased by stress.** Environmental stress, such as an antibiotic or low pH, causes protein aggregates to collect at the cell's older pole. Actinobacterial cells extend at alternating poles.
- **Membrane vesicles transmit cytoplasmic contents.** Vesicles share proteins, nucleic acids, and other cytoplasmic contents with the exterior environment and other cells.
- **Intercellular nanotubes directly share cytoplasmic contents.** Nanotubes between individual bacteria share

A.

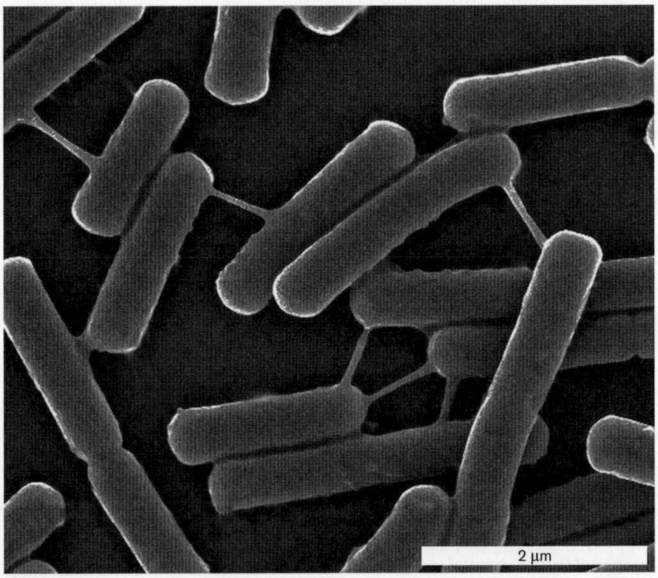

B.

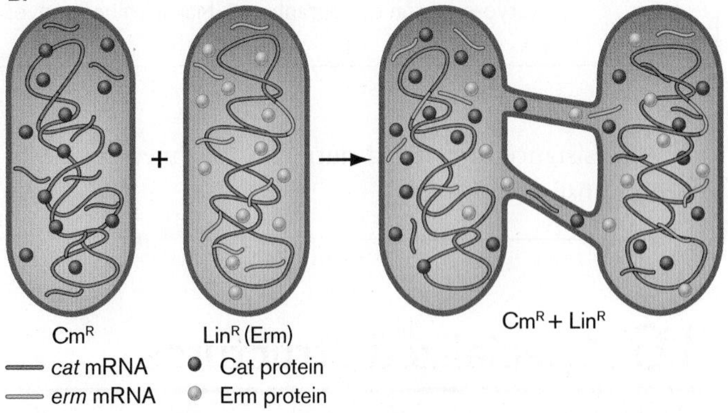

FIGURE 3.35 ▪ Intercellular nanotubes. A. *Bacillus subtilis* bacteria connected by intercellular nanotubes, which pass material from one cell to the next. **B.** Nanotubes connect bacteria with two different genes encoding resistance to an antibiotic: chloramphenicol (Cat) or lincomycin (Lin). The connected bacteria share mRNA and resistance proteins for both Cat and Lin resistance.

SPECIAL TOPIC 3.1 Turrets and Horseshoes: What Are They For?

What if you found an intricate subcellular structure four times the size of a ribosome and had no idea what it does? Martin Pilhofer, at the ETH Zürich, Switzerland, and his colleagues found something surprising in their cryotomograms of *Prosthecobacter debontii*, a Gram-negative bacterium found in water and soil (**Fig. 1**). *P. debontii* possesses a long extension

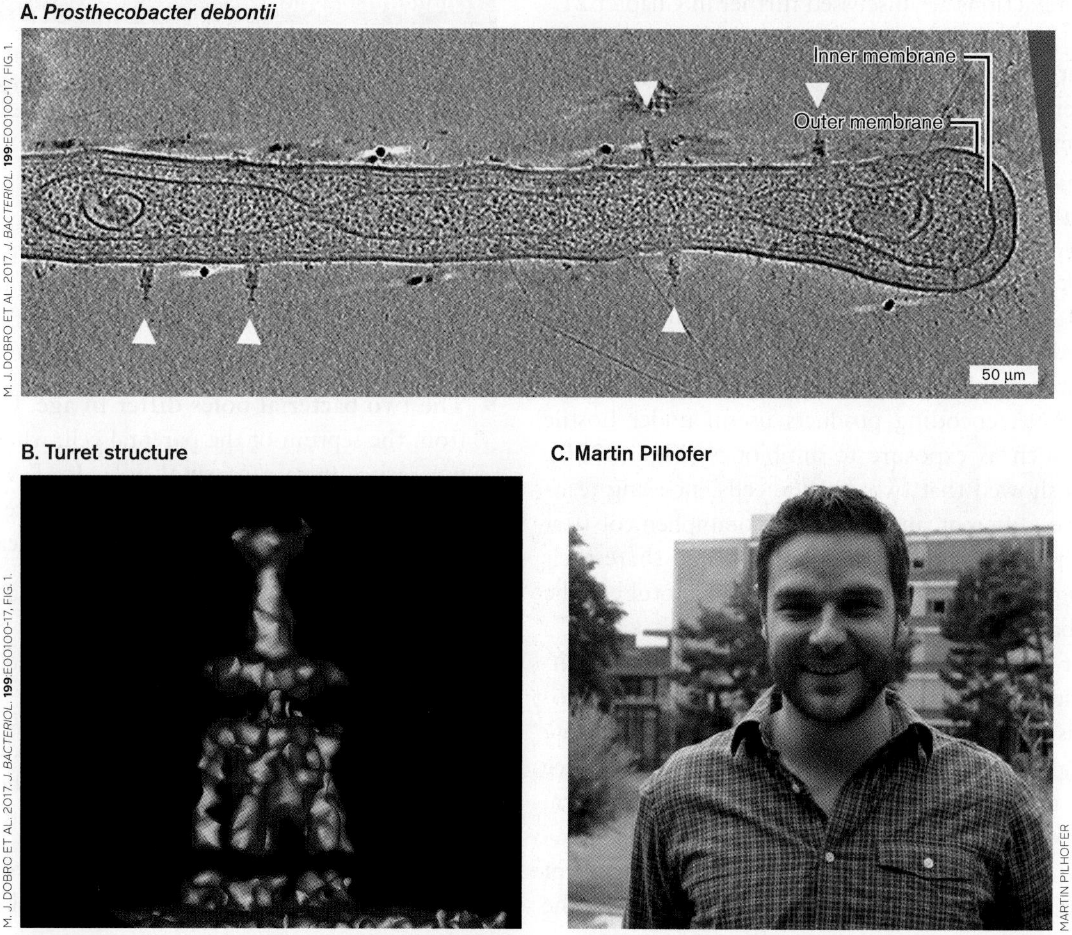

FIGURE 1 ■ Mysterious turrets on a bacterium. A. *Prosthecobacter debontii* cell studded with turret-shaped structures (arrowheads) of unknown function (cryo-TEM). **B.** 3D model of turret structure (cryo-electron tomography). **C.** Martin Pilhofer investigates mystery structures using cryo-EM.

drug resistance, cross-feed nutrients, and mediate electron transfer.

3.6 Specialized Structures

We have introduced the major structures that cells need to contain and organize their contents, maintain their DNA, and synthesize new parts. Besides these fundamental structures, different species have evolved specialized devices adapted to diverse metabolic strategies and environments. And microscopy continually reveals new microbial structures whose functions remain a mystery (**Special Topic 3.1**).

Thylakoids, Carboxysomes, and Storage Granules

Cyanobacteria are phototrophs that produce food and oxygen for marine and freshwater ecosystems. In the water, cyanobacteria must absorb sufficient amounts of light to drive photosynthesis (see Section 14.6). To maximize the collecting area of their photosynthetic membranes, cyanobacteria have evolved specialized systems of extensively folded intracellular membrane called **thylakoids** (**Fig. 3.36A**).

of cytoplasm called a prostheca, similar to the stalk of *Caulobacter*. Electron microscopy of the prostheca revealed numerous turret-shaped structures. Each turret had five legs supporting a platform, upon which stood a long neck with a knob and two antennae (not seen in 3D). What is the function of this structure, and why would it specifically decorate the prostheca? Pilhofer and his team propose that the structure might acquire nutrients from a nutrient-scarce habitat, or it could secrete toxins into a host, or it could even be the capsid of an unknown bacteriophage. For now, all such ideas are speculative, lacking good evidence.

Pilhofer's team used cryo-EM to survey more than 80 kinds of bacteria for novel subcellular structures. Besides the "turrets" of *P. debontii*, they found a unique kind of vesicle in cells of *Ralstonia eutropha*, an aquatic Gram-negative bacterium known for bioremediating aromatic pollutants. Surprisingly, the EM images revealed vesicles in the shape of a horseshoe (**Fig. 2**). We do not know what substance these vesicles contain, or what "lucky" advantage the horseshoe shape confers. We only know that the more we investigate microbial cells, the more amazing and unexpected components we find.

RESEARCH QUESTION

What experiments might you perform to reveal the function of a mysterious bacterial structure?

Megan J. Dobro et al. 2017. Uncharacterized bacterial structures revealed by electron cryotomography. *Journal of Bacteriology* **199**:e00100-17.

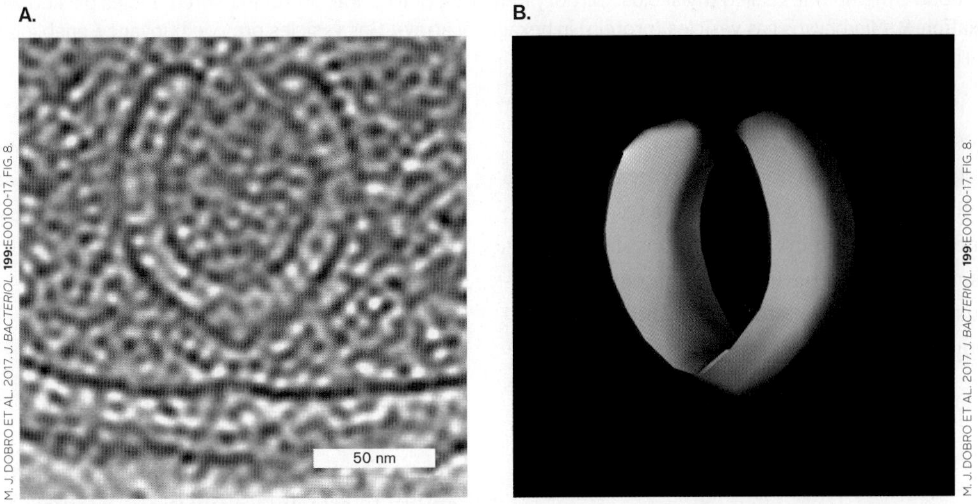

FIGURE 2 ▪ Horseshoe-shaped vesicles. *Ralstonia eutropha* vesicle shaped like a horseshoe. **A.** Cryo-TEM. **B.** 3D model by cryo-electron tomography.

Thylakoids consist of layers of folded sheets (lamellae) or tubes of membranes packed with chlorophylls and electron carriers. Cyanobacteria containing thylakoids structurally resemble eukaryotic chloroplasts, which evolved from a common ancestor of modern cyanobacteria.

The thylakoids conduct only the "light reactions" of photon absorption and energy storage. The energy obtained is rapidly spent to fix carbon dioxide—a process that occurs within **carboxysomes**. Carboxysomes are polyhedral, protein-covered bodies packed with the enzyme Rubisco for CO_2 fixation.

How do phototrophs keep themselves at the top of the water column? Some bacteria and archaea form **gas vesicles** to increase buoyancy and keep the cell afloat. **Figure 3.36B** shows a cross section of *Microcystis*, a cyanobacterium that forms toxic algal blooms in lakes polluted by agricultural runoff. *Microcystis* shows typical gas vesicles, which are hollow protein structures that collect gases. The gases are hydrogen or carbon dioxide produced by the cell's metabolism. Each vesicle consists of a tube of protein with two conical ends. The tubes pack in hexagonal arrays. Gas vesicles turn out to have surprising medical use for ultrasound trackers in our gut (see the Current Research Highlight in Chapter 8).

When light is scarce, cyanobacteria may digest their thylakoids for energy and as a source of nitrogen. Alternatively,

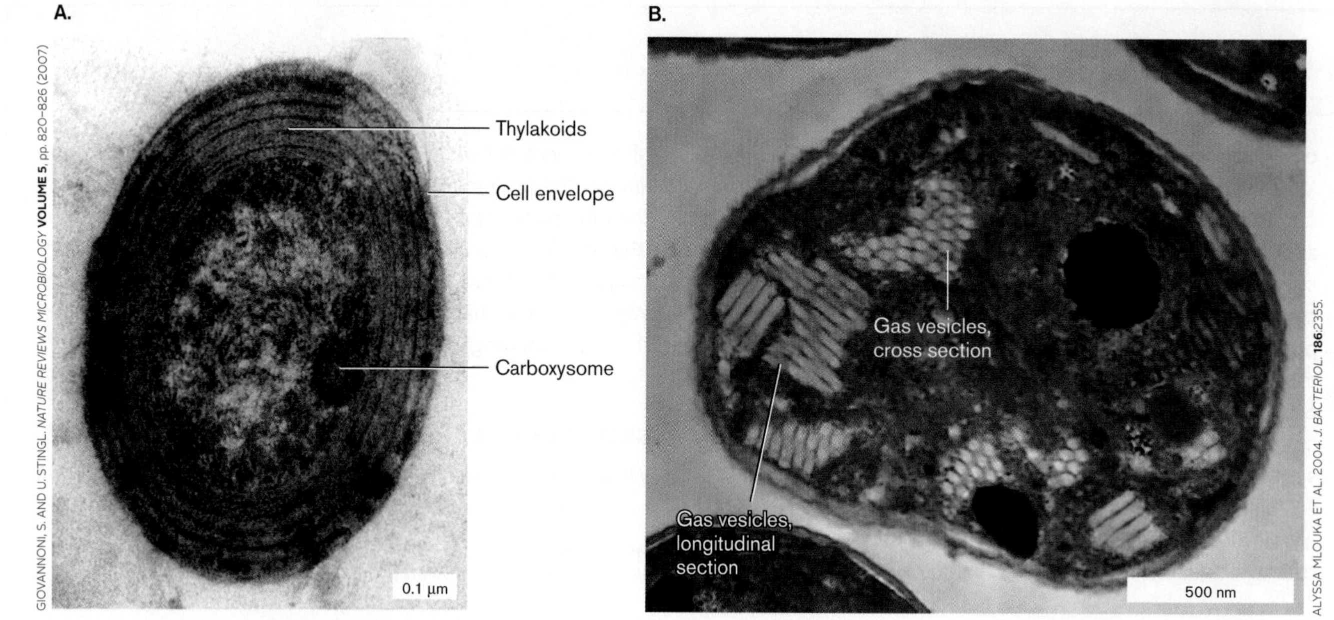

FIGURE 3.36 ■ Organelles of phototrophs. **A.** The marine phototroph *Prochlorococcus marinus* (TEM). Beneath the envelope lie the photosynthetic double membranes called thylakoids. Carboxysomes are polyhedral, protein-covered bodies packed with the Rubisco enzyme for CO_2 fixation. **B.** *Microcystis* gas vesicles (protein) in hexagonal arrays. Gas vesicles provide buoyancy, enabling the phototroph to remain at the surface of the water, exposed to light.

the cell may digest energy-rich materials from storage granules composed of glycogen or other polymers, such as polyhydroxybutyrate (PHB) and poly-3-hydroxyalkanoate (PHA). PHB and PHA polymers are of interest as biodegradable plastics, and bacteria have been engineered to produce them industrially. Similar storage granules are also produced by nonphototrophic soil bacteria.

Another type of storage device is sulfur—granules of elemental sulfur produced by purple and green phototrophs through photolysis of hydrogen sulfide (H_2S). Instead of disposing of the sulfur, the bacteria store it in granules, either within the cytoplasm (purple phototrophs) or as "globules" attached to the outside of the cell. Sulfur-reducing bacteria also make extracellular sulfur globules (**Fig. 3.37**). The sulfur may be usable as an oxidant when reduced substrates are available, or as a source of electrons for photosynthesis (Chapter 14). And the presence of potentially toxic sulfur granules may help cells avoid predation.

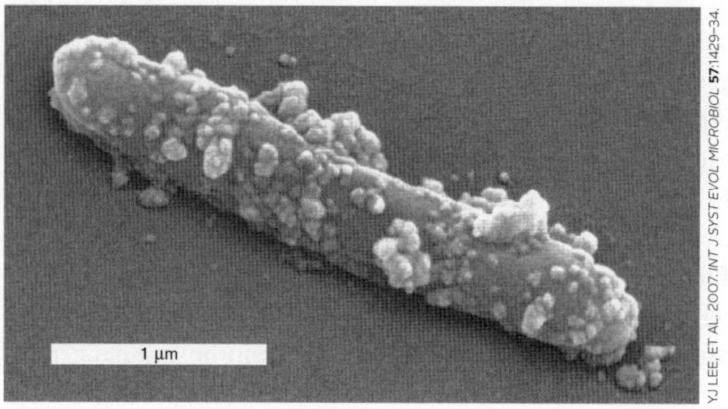

FIGURE 3.37 ■ External sulfur particles. Sulfur globules dot the surface of *Thermoanaerobacter sulfurigignens*, an anaerobic, thermophilic bacterium that gains energy by reducing thiosulfate ($S_2O_3^{2-}$) to elemental sulfur (S^0).

Pili and Stalks

In a favorable habitat, such as a running stream full of fresh nutrients or the epithelial surface of a host, it is advantageous for a cell to adhere to a substrate. Adherence, the ability to attach to a substrate, requires specific structures. A common adherence structure is the **pilus** (plural, **pili**), also called fimbria (plural, fimbriae), which is constructed of straight filaments of protein monomers called pilin. For example, the sexually transmitted pathogen *Neisseria gonorrhoeae* uses pili to attach to the mucous membranes of the reproductive tract (**Fig. 3.38**). Pili can also provide a form of motility called "twitching," in which the pili act as limbs to "walk" the bacterium across a substrate (presented in Chapter 4). Bacteria such as *Pseudomonas aeruginosa* use twitching motility to begin biofilm formation (discussed in Section 4.5).

In Gram-negative enteric bacteria, pili of a different kind, also called the sex pili, attach a donor cell to a recipient cell for transfer of DNA. This process of DNA transfer is called conjugation. The genetic consequences of conjugation are discussed in Chapter 9.

A different kind of attachment organelle is a membrane-embedded extension of the cytoplasm called a **stalk**, seen earlier in the stalked cell of *Caulobacter* (**Fig. 3.30**).

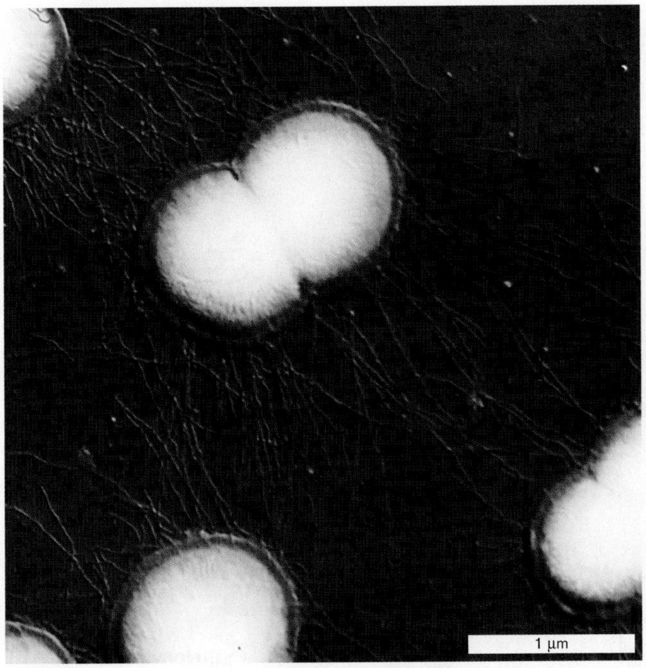

FIGURE 3.38 ▪ Pili are protein filaments for attachment. *Neisseria gonorrhoeae*, cause of the sexually transmitted infection gonorrhea, use pili to attach to the host mucous membrane (SEM).

A.

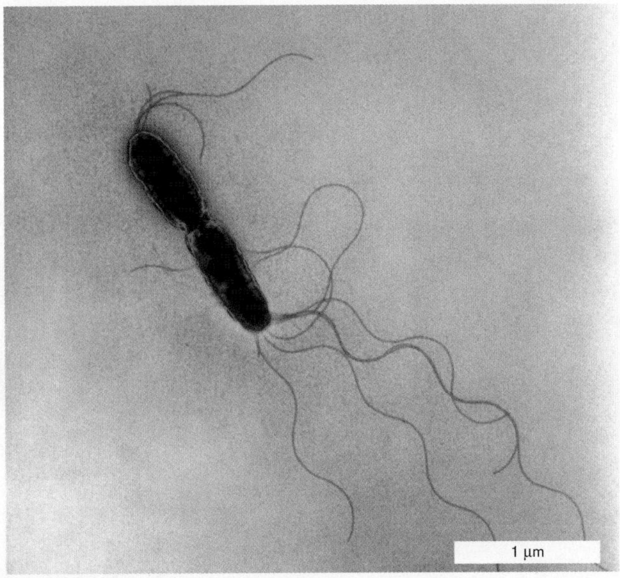

B.

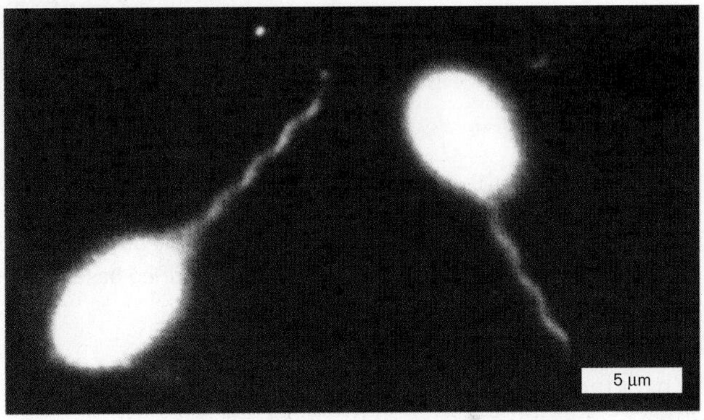

FIGURE 3.39 ▪ Flagellated *Salmonella* bacteria. **A.** *Salmonella enterica* has multiple flagella (colorized TEM). **B.** The flagella collect in a bundle behind a swimming cell. Under dark-field microscopy, the cell body appears overexposed, about five times as large as the actual cell.

The tip of the stalk secretes adhesion factors that form a "holdfast," which firmly attaches the bacterium in an environment that has proved favorable. A stalk and holdfast enable iron-oxidizing bacteria to form large biofilms in streams contaminated by iron drainage. The biofilms become coated by orange iron hydroxides, tinting the stream orange.

Rotary Flagella

What happens when the cell's environment runs out of nutrients, or becomes filled with waste? In rapidly changing environments, cell survival requires **motility**, the ability to move and relocate. Many bacteria and archaea can swim by means of rotary **flagella** (singular, **flagellum**). Flagellar motility benefits the cell by causing the population to disperse, decreasing competition. Motility also enables cells to swim toward a favorable habitat (by chemotaxis, discussed shortly).

Flagellar motility. Flagella are helical propellers that drive the cell forward like the motor of a boat. Howard Berg at the California Institute of Technology originally described the bacterial flagellar motor, which was the first rotary device to be discovered in a living organism. Different bacterial species have different numbers and arrangements of flagella. Peritrichous cells, such as *E. coli* and *Salmonella* species, have flagella randomly distributed around the cell (**Fig. 3.39A**). The flagella rotate together in a bundle behind the swimming cell (**Fig. 3.39B**). Lophotrichous cells, such as *Rhodospirillum rubrum*, have flagella attached at one or both ends. In monotrichous (polar) species, such as the *Caulobacter* swarmer cell (see **Fig. 3.30**), the cell has a single flagellum at one end.

How does a rotary flagellum work? Each flagellum has a spiral filament of protein monomers called flagellin (protein FliC). The filament actually rotates by means of a motor driven by the cell's transmembrane proton current—the same proton potential that drives the membrane-embedded ATP synthase (presented in Chapter 14). The flagellar motor is embedded in the layers of the cell envelope (**Fig. 3.40**). The motor possesses an axle and rotary parts, all composed of specific proteins. For example, protein MotB forms part of the ion channel whose flux of hydrogen ions powers rotation, similarly to the proton flux that drives ATP synthase. Another protein, FliG, forms part of the device that generates torque (rotary force). Much of the motor's structure and function was elucidated by Scottish microbiologist Robert Macnab (1940–2003) at Yale University.

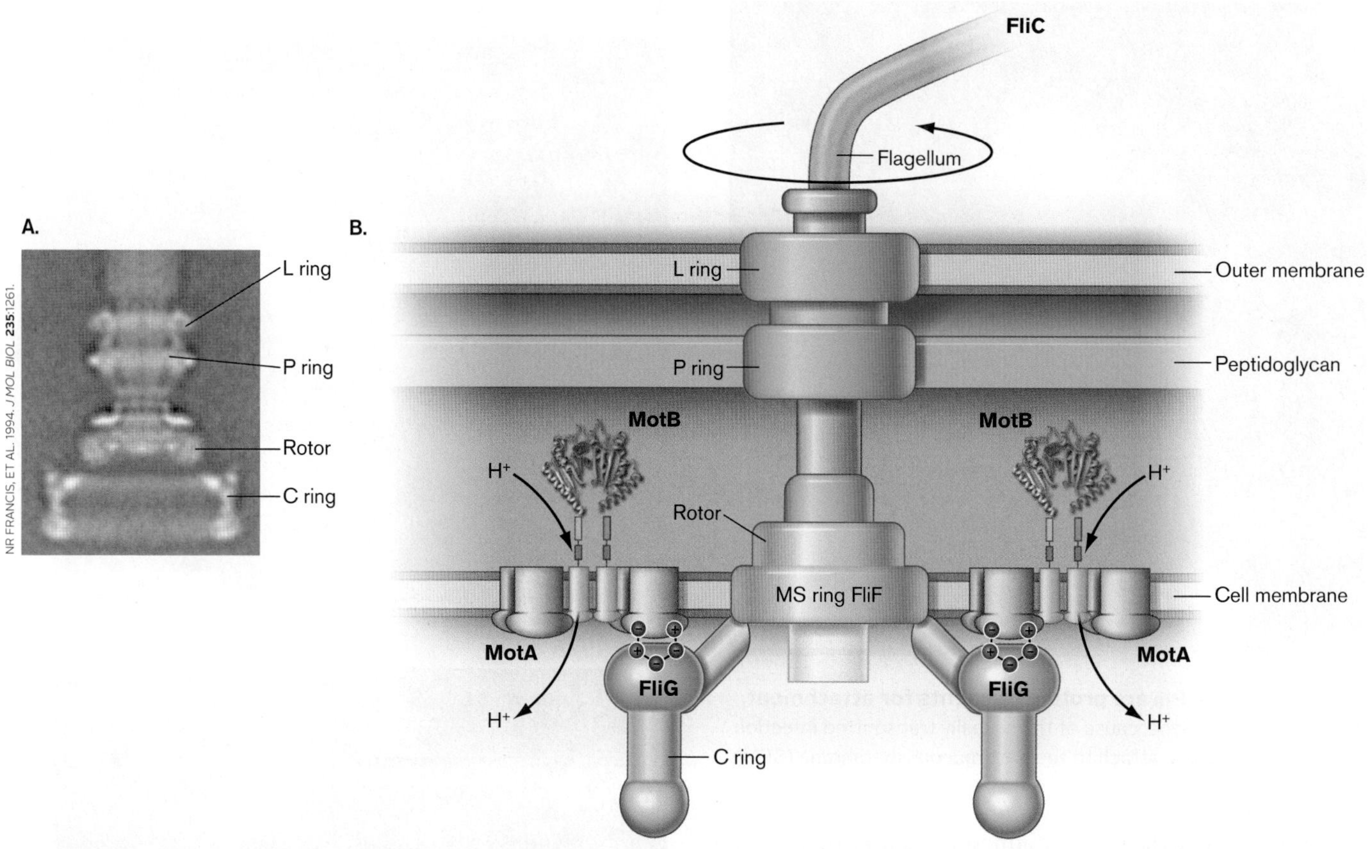

FIGURE 3.40 ■ The flagellar motor. **A.** The basal body, or motor, of the bacterial flagellum (TEM). This image is based on digital reconstruction, in which electron micrographs of purified basal bodies were rotationally averaged. **B.** H^+ flow through the MotA-MotB complex drives rotation of the flagellar motor.

What kinds of experiments reveal the motor components? Results from an experiment dissecting the flagellar motor are shown in **Figure 3.41**. Japanese microbiologist Tohru Minamino (Osaka University) and colleagues constructed strains of *Salmonella enterica* in which the gene that encodes fluorescent GFP is fused to a gene encoding a flagellar protein, MotB or FliG, each of which is proposed to be a part of the motor. For each flagellar construct strain,

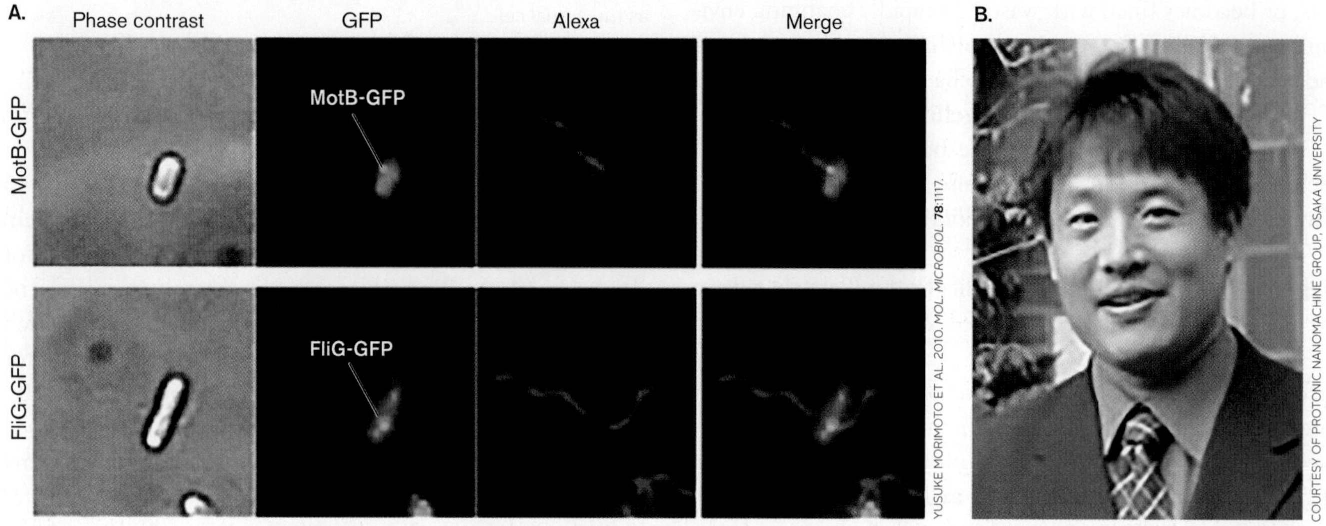

FIGURE 3.41 ■ Flagellar motor proteins localized by fluorescence microscopy. **A.** Cells of *Salmonella enterica* express a GFP fused to flagellar motor protein MotB or to FliG. Bright green dots (white arrows) indicate MotB-GFP or FliG-GFP complexed with a flagellar motor. The flagellar filament is visualized via the red fluorophore Alexa, conjugated to an anti-flagellin antibody. The merged image shows how each flagellar filament (red) extends from a motor containing either MotB or FliG (green, protein fused to GFP). **B.** Tohru Minamino investigates the structure and function of the flagellar motor.

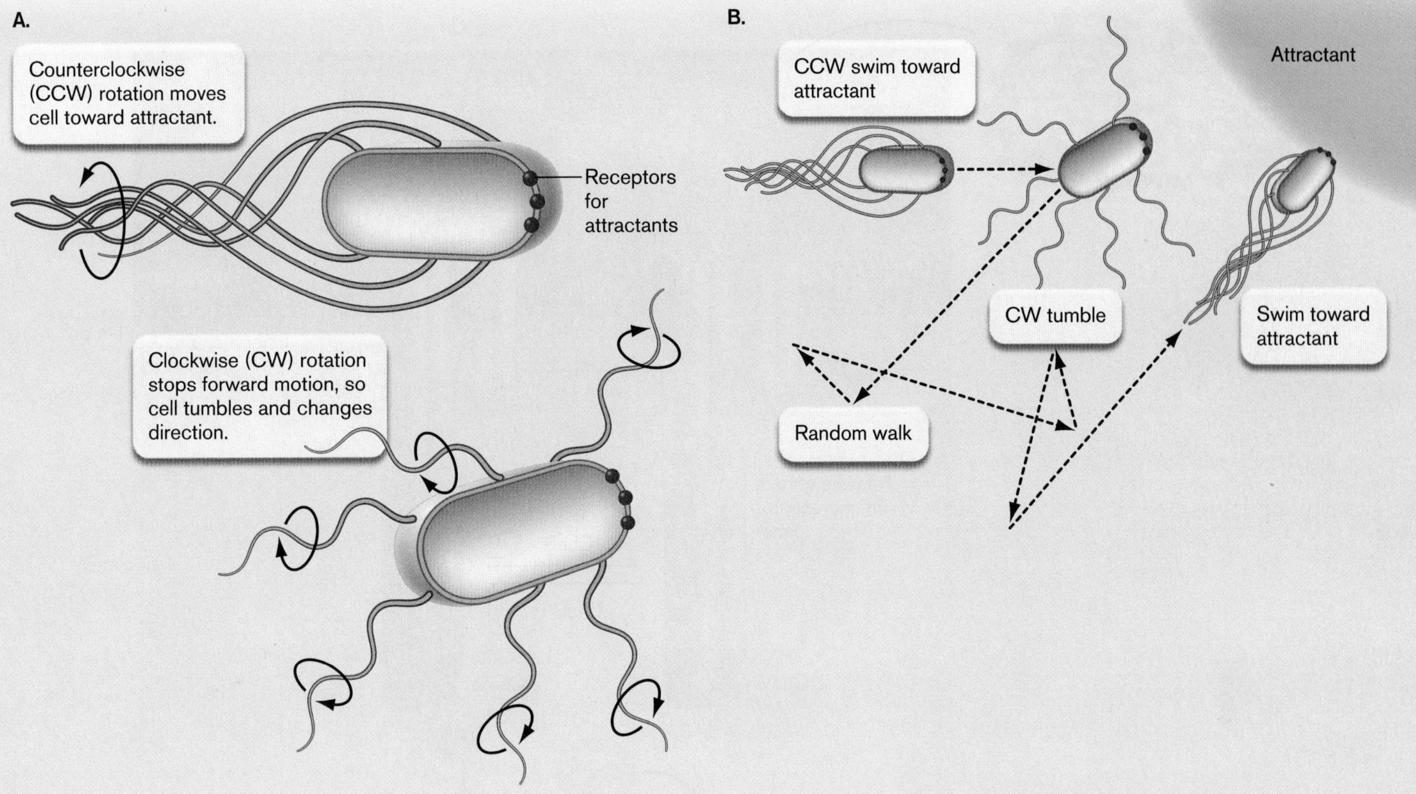

FIGURE 3.42 ■ Chemotaxis. **A.** Flagella are oriented in a bundle extending behind one pole. When the cell veers away from the attractant, the receptors send a signal that allows one or more flagella to switch rotation from counterclockwise (CCW) to clockwise (CW). This switched rotation disrupts the bundle of flagella, causing the cell to tumble briefly before it swims off in a new direction.
B. The resulting pattern of movement is a "biased random walk" in which the cell sometimes moves randomly but overall tends to migrate toward the attractant. ▶

fluorescence microscopy reveals the GFP fluorescence at one or two positions within the cell (green dots). A second fluorophore, Alexa, is conjugated to an anti-flagellin antibody. The Alexa fluorescence (red) reveals the flagellar filament. When the green and red fluorescence images are merged, the red flagellar filaments appear to extend from the motor positions that contain either MotB or FliG. Further experiments dissect the roles of key amino acid residues in the function of these proteins.

Note: Bacterial flagella differ completely from the whiplike flagella and cilia of eukaryotes. Eukaryotic flagella are much larger structures containing multiple microtubules enclosed by a membrane (shown in Chapter 20). They move with a whiplike motion, powered by ATP hydrolysis all along the flagellum.

Chemotaxis. How do cells decide where to swim? Most flagellated cells have an elaborate sensory system for taxis, the ability to swim toward favorable environments (attractant signals, such as nutrients) and away from inferior environments (repellent signals, such as waste products). Taxis to specific chemicals is called **chemotaxis**. For chemotaxis, the attractants and repellents are detected by a polar array of chemoreceptors, as seen in the Current Research Highlight. Chemoreceptors act like a "nose," telling the bacterium when it is swimming toward a source of attractant such as a sugar or an amino acid. Other chemoreceptor systems (not shown) respond to oxygen (aerotaxis) or to light (phototaxis, discussed in Chapter 19).

How does the chemoreceptor "tell" the bacterium to start swimming, or to switch direction? Chemotaxis requires a mechanism for the rotary flagella to propel the cell toward attractants or away from repellents. This movement is accomplished by flagellar rotation either clockwise or counterclockwise relative to the cell (**Fig. 3.42A** ▶). When a cell is swimming toward an

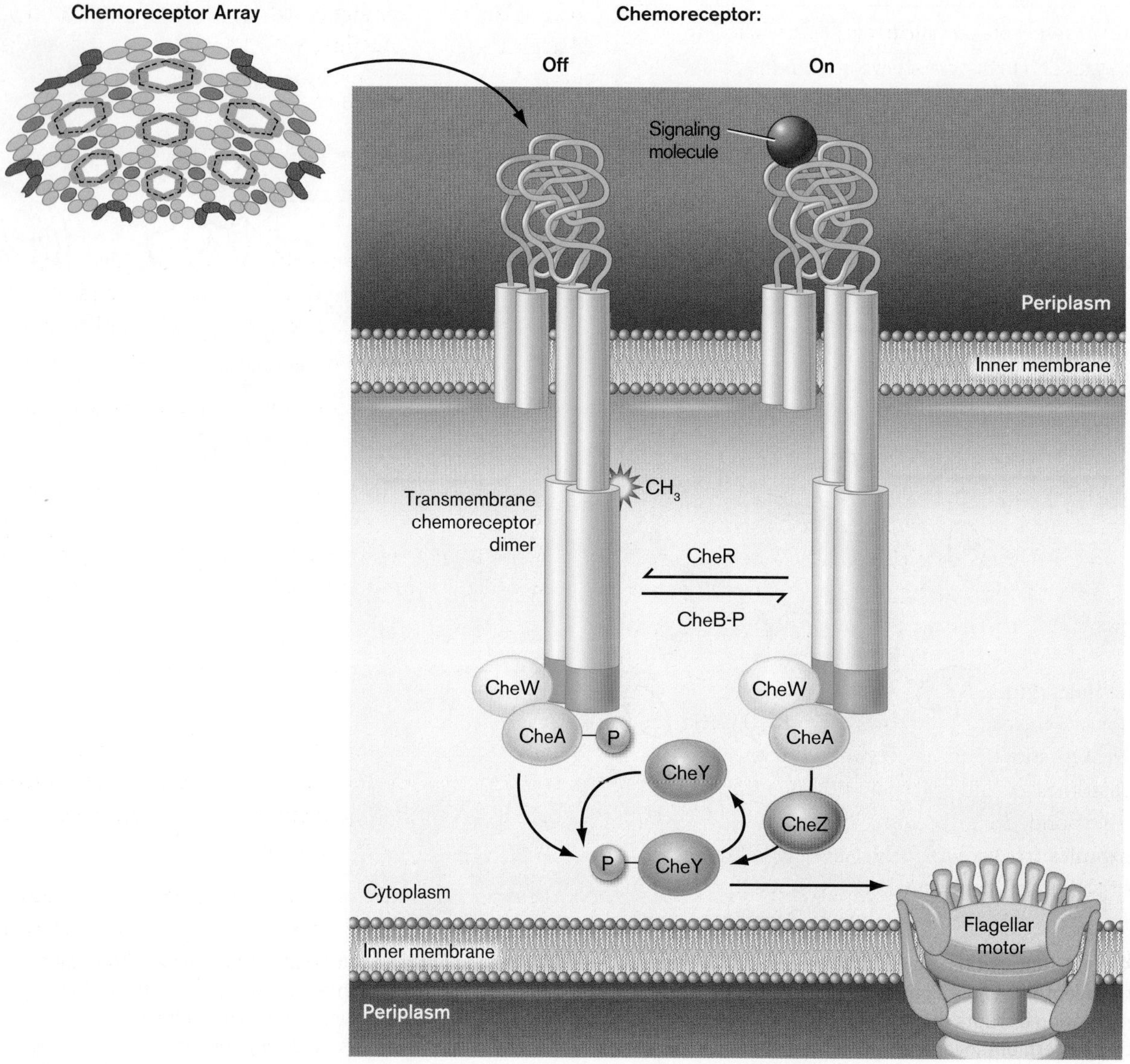

FIGURE 3.43 ▪ Chemoreceptors receive signals and transmit them to the flagellar motor. *Source:* Modified from Kieran D. Collins et al. 2014. *Microbiol. Mol. Biol. Rev.* **78:**672–674, fig. 2.

attractant chemical, the flagella rotate counterclockwise (CCW), enabling the cell to swim smoothly for a long stretch. When the cell veers away from the attractant, receptors send a signal that allows one or more flagella to switch rotation clockwise (CW), against the twist of the helix. This switch in the direction of rotation disrupts the bundle of flagella, causing the cell to tumble briefly and end up pointed in a random direction. The cell then swims off in the new direction. The resulting pattern of movement generates a "biased random walk" in which the cell tends to migrate toward the attractant (**Fig. 3.42B**).

How do bacteria connect their chemoreceptor signal reception to the flagellar switching mechanism? Bacteria don't have a brain, but they do have a complex network of signal proteins within the cell. A small portion of the chemotaxis signaling network is shown in **Figure 3.43**. The chemotaxis proteins CheW and CheA, located at the base of the chemoreceptor, interact with an intracellular messenger CheY. As an attractant signal is received (such as an amino acid nutrient), the phosphorylation of CheA is blocked, and CheA does not phosphorylate CheY. The flagellar motor keeps turning, and the cell swims smoothly. But when the attractant signal decreases in concentration, then CheA becomes phosphorylated and transfers a phosphoryl group to CheY. The phosphorylated CheY then binds the flagellar motor to switch its direction, causing reverse rotation, and the cell tumbles. Finally, CheY-phosphate is dephosphorylated by CheZ, returning it to the state ready to receive another attractant signal and allow smooth swimming.

To adapt the sensor to a new level of attachment, CheR methylates active receptors. The phosphorylated CheR demethylates receptors to allow signal response again. Additional chemotaxis proteins combine to coordinate the cell's "switch" between swimming up a gradient of attractant or tumbling to avoid swimming in an unfavorable direction.

Note: Later chapters present further aspects of chemotaxis.

- Chemotaxis control circuits are shown in detail in Section 10.4.
- Internalized flagella of spirochetes are shown in Section 18.5.
- Phototaxis (taxis toward light) for haloarchaea is presented in Section 19.5.

Thought Question

3.12 Most laboratory strains of *E. coli* and *Salmonella* commonly used for genetic research lack flagella. Why and how do bacterial strains evolve to lose flagella? How can a researcher maintain a motile strain?

Another form of taxis, related to chemotaxis, is **magnetotaxis**, the ability to sense and respond to magnetism. Found in pond water, magnetotactic bacteria orient themselves along the Earth's lines of magnetic field. Magnetotactic bacteria can be collected from the environment by placing a magnet in a jar of pond water; bacteria orienting by the field lines collect nearby. The bacteria align their cell axis along the magnetic field using linear arrays of **magnetosomes**, microscopic membrane-enclosed crystals of the magnetic mineral magnetite, Fe_3O_4 (shown in the previous chapter, Fig. 2.45). The magnetosomes orient bacterial swimming toward the bottom of the pond. Magnetotactic bacteria are anaerobes or microaerophiles (requiring low oxygen), which prefer the lower part of the water column, where oxygen concentration is lowest. In the northern latitudes, where Earth's magnetic field lines point downward, bacteria that are magnetotactic swim "downward" toward magnetic north.

Thought Question

3.13 How would a magnetotactic species have to behave if it were in the Southern Hemisphere instead of the Northern Hemisphere?

Besides motility and chemotaxis, surprisingly, flagella have also evolved an alternate function: adherence of cells to a substrate to begin forming a biofilm (discussed in Chapter 4). Thus, an organism can evolve a structure that serves one function but later evolves to serve another function.

In addition to flagellar rotation, other forms of bacterial motility are just beginning to be understood, such as pili-dependent twitching motility (discussed in Section 4.5). Another kind of motility, called "gliding," is observed in cyanobacteria and in myxobacteria.

To Summarize

- **Cyanobacteria** possess thylakoid membrane organelles packed with photosynthetic apparatus and carboxysomes for carbon dioxide fixation. Gas vesicles provide buoyancy in the water column.
- **Storage granules** store sulfur or carbon polymers for energy.
- **Adherence structures** enable prokaryotes to remain in an environment with favorable environmental factors. Major adherence structures include pili or fimbriae (protein filaments), and the holdfast (a cell extension).
- **Flagellar motility** involves rotary motion of helical flagella. Flagellar rotation is driven by the transmembrane proton motive force.
- **Chemoreceptors** provide information that directs flagellar motility. Chemotaxis is a mechanism by which the cell migrates up a gradient of an attractant substance or down a gradient of repellents.
- **Magnetotaxis** is a process of flagellar motility directed along magnetic field lines. Taxis involves magnetosomes, crystals of magnetite encased by intracellular membrane vesicles.

Concluding Thoughts

The form and function of cells continue to amaze us with each new discovery, such as the polar aging of bacteria and the intricate mechanism of their flagellar motors. The elaborate cell forms evolved by microbes challenge the inventors of antibiotics, as well as designers of molecular machines. As a journalist observed in *Science*, "When it comes to nanotechnology, physicists, chemists, and materials scientists can't hold a candle to the simplest bacteria."

CHAPTER REVIEW

Review Questions

1. What are the major features of a bacterial cell, and how do they fit together for cell function as a whole?
2. What fundamental traits do most prokaryotes have in common with eukaryotic microbes? What traits are different?
3. Explain how cell fractionation enables us to separate proteins of the Gram-negative outer membrane, periplasm, inner membrane, and cytoplasm.
4. Outline the structure of the peptidoglycan sacculus, and explain how it expands during growth. Cite two different kinds of experimental data that support our current views of the sacculus.
5. Compare and contrast the structure of Gram-positive and Gram-negative cell envelopes. Explain the strengths and weaknesses of each kind of envelope.
6. Explain how the process of DNA replication is coordinated with cell wall septation.
7. Explain how the asymmetry of a bacterial cell generates daughter cells with different structure and function. Explain the significance for environmental adaptation and for antibiotic therapy.
8. What kinds of subcellular structures are found in certain cells with different functions, such as photosynthesis or magnetotaxis?
9. Compare and contrast bacterial structures for attachment and motility.
10. Explain how a chemical signal is detected and converted to bacterial motility, in the process of chemotaxis.

Thought Questions

1. The aquatic bacterium *Caulobacter crescentus* alternates between two cell forms: a cell with a flagellum that swims, and a stalked cell that adheres to particulate matter. The flagellar cell can discard its flagellum to grow a stalk and adhere, and then the stalked cell divides to give one stalked cell and one flagellated cell. What would be the adaptive advantage of this alternating morphology?
2. Suppose that one cell out of a million has a mutant gene blocking S-layer synthesis, and suppose that the mutant strain can grow twice as fast as the S-layered parent. How many generations would it take for the mutant strain to constitute 90% of the population?
3. Explain two different ways that an aquatic phototroph might use its subcellular structures to maximize its access to light. Explain how an aerobe (an organism requiring molecular oxygen for growth) might remain close to the surface, with access to air.
4. How do pathogenic bacteria avoid being engulfed by phagocytes of the human bloodstream? How do you think various aspects of the cell structure can prevent phagocytosis?

Key Terms

active transport (85)
capsule (92)
carboxysome (109)
cardiolipin (85)
cell fractionation (80)
cell membrane (78)
cell wall (78)
chemotaxis (113)
cholesterol (86)
cross-bridge (89)
divisome (98)
DNA-binding protein (100)
domain (of proteins) (99)
electrophoresis (79)
endotoxin (93)
envelope (78)
flagellum (78, 111)
gas vesicle (109)
glycan (89)
hopanoid (hopane) (86)
inner membrane (78, 93)
ion gradient (85)
leaflet (83)
lipid A (93)
lipopolysaccharide (LPS) (78, 93)
lysis (lyse) (80)
magnetosome (115)
magnetotaxis (115)
membrane-permeant weak acid (84)
membrane-permeant weak base (84)
motility (111)
murein (88)
murein lipoprotein (93)
nanotube (107)
nucleoid (78)
O antigen (93)
osmotic pressure (84)
outer membrane (78)
passive transport (85)

penicillin-binding protein (89)
peptidoglycan (79)
periplasm (78)
phospholipid (82)
pilus (110)
plasma membrane (78)
porin (94)
replisome (101)
S-layer (92)
sacculus (88)
septation (98)
septum (96)
spheroplast (80)
stalk (110)
staphylococcus (102)
teichoic acid (91)
terpenoid (87)
thylakoid (108)
transport protein (transporter) (85)
turgor pressure (84)
ultracentrifuge (81)

Recommended Reading

Aldridge, Bree B., Marta Fernandez-Suarez, Danielle Heller, Vijay Ambravaneswaran, Daniel Irimia, et al. 2012. Asymmetry and aging of mycobacterial cells lead to variable growth and antibiotic susceptibility. *Science* **335**:100–103.

Belin, Brittany J., Nicolas Busset, Eric Giraud, Antonio Molinaro, Alba Silipo, et al. 2018. Hopanoid lipids: From membranes to plant–bacteria interactions. *Nature Reviews. Microbiology* **16**:304–315.

Bisson-Filho, Alexandre W., Yen-Pang Hsu, Georgia R. Squyres, Erkin Kuru, Fabai Wu, et al. 2017. Treadmilling by FtsZ filaments drives peptidoglycan synthesis and bacterial cell division. *Science* **355**:739–743.

Celler, Katherine, Roman I. Koning, Abraham J. Koster, and Gilles P. van Wezel. 2013. Multidimensional view of the bacterial cytoskeleton. *Journal of Bacteriology* **195**:1627–1636.

Cesar, Spencer, and Kerwyn C. Huang. 2017. Thinking big: The tunability of bacterial cell size. *FEMS Microbiology Reviews* **41**:672–678.

Clark, Michelle W., Anna M. Yie, Elizabeth K. Eder, Richard G. Dennis, Preston J. Basting, et al. 2015. Periplasmic acid stress increases cell division asymmetry (polar aging) of *Escherichia coli. PLoS One* **10**:e0144650.

Dubey, Gyanendra P., and Sigal Ben-Yehuda. 2011. Intercellular nanotubes mediate bacterial communication. *Cell* **144**:590–600.

Lam, Hubert, Whitman B. Schofield, and Christine Jacobs-Wagner. 2006. A landmark protein essential for establishing and perpetuating the polarity of a bacterial cell. *Cell* **124**:1011–1023.

Libby, Elizabeth A., Manuela Roggiani, and Mark Goulian. 2012. Membrane protein expression triggers chromosomal locus repositioning in bacteria. *Proceedings of the National Academy of Sciences USA* **109**(19):7445–7450.

Moore, Jeremy P., Haofan Li, Morgan L. Engmann, Katrina M. Bischof, Karina S. Kunka, Mary E. Harris, et al. 2019. Inverted regulation of multidrug efflux pumps, acid resistance and porins in benzoate-evolved *Escherichia coli* K–12. *Applied and Environmental Microbiology* **85**: e00966.

Oikonomou, Catherine M., Yi-Wei Chang, and Grant J. Jensen. 2016. A new view into prokaryotic cell biology from electron cryotomography. *Nature Reviews. Microbiology* **14**:205–221.

Pande, Samay, Shraddha Shitut, Lisa Freund, Martin Westermann, Felix Bertels, et al. 2015. Metabolic cross-feeding via intercellular nanotubes among bacteria. *Nature Communications* **6**:6238.

Partridge, Jonathan D., Vincent Nieto, and Rasika M. Harshey. 2015. A new player at the flagellar motor: FliL controls both motor output and bias. *mBio* **6**:e02367.

Renner, Lars D., and Douglas B. Weibel. 2011. Cardiolipin microdomains localize to negatively curved regions of *Escherichia coli* membranes. *Proceedings of the National Academy of Sciences USA* **108**:6264–6269.

Schwechheimer, Carmen, and Meta J. Kuehn. 2015. Outer-membrane vesicles from Gram-negative bacteria: Biogenesis and functions. *Nature Reviews. Microbiology* **13**:605–619.

Wagstaff, James, and Jan Löwe. 2018. Prokaryotic cytoskeletons: Protein filaments organizing small cells. *Nature Reviews. Microbiology* **16**:187–201.

CHAPTER 4

Bacterial Culture, Growth, and Development

4.1 Microbial Nutrition

4.2 Nutrient Uptake

4.3 Culturing and Counting Bacteria

4.4 The Growth Cycle

4.5 Biofilms

4.6 Cell Differentiation

Like all living things, bacteria need nutrients to grow. But where do they find them? In natural habitats, bacteria struggle to survive because they constantly compete for food. Over eons, however, bacteria gradually evolved to find new nutritional niches that required consuming strange foods (mothballs, for instance), and some developed mechanisms to harness energy from light or enable survival in extreme environments such as boiling temperatures around deep-sea thermal vents. In the process, single-celled organisms developed ways to chemically "talk" to each other and form intricate multicellular communities called biofilms.

CURRENT RESEARCH highlight

Nutrient time-sharing in biofilms. Biofilms are complex aggregates of bacteria growing on surfaces. A biofilm's edge easily accesses passing nutrients (e.g., glutamate); its center does not. The solution, discovered by Gürol Süel and colleagues, is nerve-like signaling within the biofilm. Interior cells deprived of glutamate release potassium (K^+), causing neighboring cells to depolarize, release more K^+, and import less glutamate. This electrical wave propagates to the biofilm's edge. Less glutamate transport at the edge improves glutamate availability to interior cells. The signal also travels between biofilms, making them take turns consuming a limited nutrient. The figure shows alternating waves of polarization (blue) to depolarization (black) in adjacent biofilms grown in limiting glutamate.

Source: J. Liu et al. 2017. *Science* **356**:638–642.

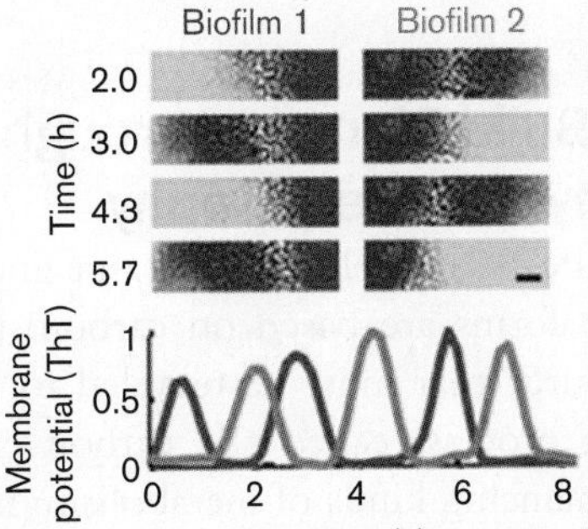

J. LIU ET AL. 2017. *SCIENCE* **356**:638–642, FIG. 2D.

AN INTERVIEW WITH

GÜROL SÜEL, MICROBIOLOGIST, UC SAN DIEGO

COURTESY OF GÜROL SÜEL

What do you find the most exciting about your discovery?

The most exciting aspect of our discovery is the unexpected connection between micro- and neurobiology. Our work shows that, instead of coping with stress independently, as unicellular individuals do, a community of densely packed bacteria acts as a multicellular organism that orchestrates millions of individual stress responses to function as a single unit. Using ion channels to execute long-range electrical signaling, bacteria in a community behave similarly to neurons in the brain.

Like all living things, microorganisms need sources of carbon and energy to grow. As those resources are depleted, organisms can do one of the following: they can die (most do), evolve to better use what little resource remains, or cannibalize other, less fortunate members of the community. The Current Research Highlight demonstrates that two competing communities of bacteria fixed to solid surfaces (biofilms) can also cooperate with each other to time-share access to limiting nutrients. And they do it using nerve-like communications. Nutrient sources can also influence the disease-causing potential of some pathogens. A recent report shows that *Clostridioides difficile* (formerly *Clostridium difficile*), which produces a potentially life-threatening bloody diarrhea, has somehow coupled the use of the sugar trehalose to increased toxin production and virulence.

Understanding how bacteria use food to increase cell mass and, ultimately, cell number enables us to control their growth and even manipulate them to make useful products. Yeasts, for example, consume glucose and break it down to ethanol and carbon dioxide gas. These end products are mere waste to the yeast, but extremely important to humans who enjoy beer.

Chapter 4 provides a broad perspective of microbial growth, introducing topics that will be discussed further in later chapters. We start by discussing the nutrients that bacteria need to grow and the ways nutrients are used. For instance, how do different bacteria obtain carbon and nitrogen, where do they get their energy, and what mechanisms do they use to gather the nutrients they need? (The details of metabolism emerge in Chapters 13, 14, and 15.) Next we explain how scientists use our knowledge of microbial nutrition to culture bacteria in the laboratory and measure their growth. And we consider why most bacterial species growing in natural environments fail to grow in the lab. We end by describing how bacteria form interactive communities and how some species differentiate to survive starvation or explore for new sources of food—all so they can grow again.

4.1 Microbial Nutrition

Bacterial cells look simple but are remarkably complex and efficient replication machines. One cell of *Escherichia coli*, for example, can divide to form two cells every 20–30 minutes. At a rate of 30 minutes per division, one cell could potentially multiply to over 1×10^{14} cells in 24 hours. That's 100 trillion organisms! Although these 100 trillion cells would weigh only about 1 gram altogether, the mass of cells would explode to 10^{14} grams (that is, 10^7 tons) after another 24 hours (two days total) of replicating every 30 minutes. Why, then, are we not buried under mountains of *E. coli*?

Nutrient Supplies Limit Microbial Growth

One factor limiting growth is the finite supply of nutrients. Microbes, in fact, often live where nutrients are scarce because of limited availability or competition. **Essential nutrients** are compounds that a microbe must have but cannot make. The organism needs to find and import these nutrients from the immediate environment. If an essential nutrient becomes depleted, microbes stop growing. How organisms cope with these periods of starvation will be discussed later.

All microorganisms require a minimum set of **macronutrients**, nutrients needed in large quantities (as discussed in Chapter 3). Six macronutrients—carbon, nitrogen, phosphorus, hydrogen, oxygen, and sulfur—make up the carbohydrates, lipids, nucleic acids, and proteins of the cell. Four other macronutrients are cations that serve as **cofactors** for specific enzymes (Mg^{2+}, Fe^{2+}, and K^+) or act as regulatory signaling molecules (Ca^{2+}). All cells also require very small amounts of certain trace elements, called **micronutrients**. These include cobalt, copper, manganese, molybdenum, nickel, and zinc, which are ubiquitous trace contaminants on glassware and in water. As a result, these six elements are not added to laboratory media unless heroic measures have been taken to first remove the elements from the medium. Cells require micronutrients as essential components of enzymes or cofactors. Cobalt, for instance, is part of the cofactor vitamin B_{12}.

All cells require nutrients to increase biomass and generate energy. Some organisms, such as the common laboratory bacterium *E. coli*, make all their cell wall and membrane components, proteins, nucleic acids, and lipids using a very simple recipe: a carbohydrate such as glucose (the source of carbon, hydrogen, and oxygen), plus ammonia (nitrogen), sodium phosphate, and potassium phosphate. For other microbes, this basic set of nutrients is insufficient. For example, *Borrelia burgdorferi*, the cause of Lyme disease, requires an extensive mixture of complex organic supplements to grow. We now provide a brief overview of the many ways in which different microbes use nutrients to increase biomass and produce energy.

Microbes Build Biomass through Autotrophy or Heterotrophy

Maintaining life on this planet is an amazing process. All of Earth's life forms are based on carbon, but carbon is a limited resource that must be recycled to maintain life. The recycling process, called the carbon cycle, involves two counterbalancing kinds of metabolism: **heterotrophy**, which breaks down multicarbon nutrients (organic compounds) to carbon dioxide; and **autotrophy**, which reassembles CO_2 molecules into multicarbon nutrients (**Fig. 4.1**).

The metabolic details of these pathways are discussed later, in Chapters 13 and 14.

Heterotrophs (such as *E. coli*) rely on other organisms (autotrophs) to form organic compounds that heterotrophs use as carbon sources (glucose, for instance). Most heterotrophs are organotrophs. **Organotrophy** is a form of metabolism in which organic carbon sources are broken down in ways that generate energy (oxidation) (**Fig. 4.1A**). Organotrophy converts a large amount of the organic carbon source to CO_2, which is then released to the atmosphere. Thus, left on their own, organotrophs would deplete the world of organic carbon sources (converting them to unusable CO_2) and then starve to death. For life to continue, CO_2 must be recycled into organic compounds by autotrophs.

Autotrophs (such as cyanobacteria) assimilate CO_2 gas as a carbon source via CO_2 fixation. The process reduces CO_2 (adding hydrogen atoms) to generate complex, organic cell constituents made up of C, H, and O (for example, carbohydrates, which have the general formula CH_2O; **Fig. 4.1B**). When autotrophs later die or are eaten, these organic compounds can be used as carbon sources by heterotrophs. Autotrophs are classified as photoautotrophs or chemolithoautotrophs by how they obtain energy. **Photoautotrophs** use light energy to fix CO_2 into biomass, whereas **chemolithoautotrophs** fix CO_2 using chemical reactions without light. Most chemolithoautotrophs gain energy by oxidizing inorganic substances such as iron or ammonia (described next). In addition, many microorganisms (for example, phototrophic soil bacteria) can use both organotrophy and autotrophy to gain carbon.

Thought Question

4.1 In a mixed ecosystem of autotrophs and organotrophs, what happens if the autotroph begins to outgrow the organotroph, producing more and more organic food?

Microbes Obtain Energy through Phototrophy or Chemotrophy

Although the macronutrients mentioned earlier (C, N, P, H, O, and S) provide the essential building blocks to make proteins and other cell structures, all synthetic processes require an energy source. Depending on the organism, energy can be obtained from chemical reactions triggered by the absorption of light (**phototrophy**—for example, photosynthesis) or from oxidation-reduction reactions that remove electrons from preformed high-energy compounds to make products of lower energy (**chemotrophy**), capturing the energy difference to do work. Chemotrophic organisms fall into two classes that use different sources of electron donors: **lithotrophs** (also called chemolithotrophs) and **organotrophs** (chemoorganotrophs). Lithotrophs oxidize (remove electrons from) inorganic chemicals

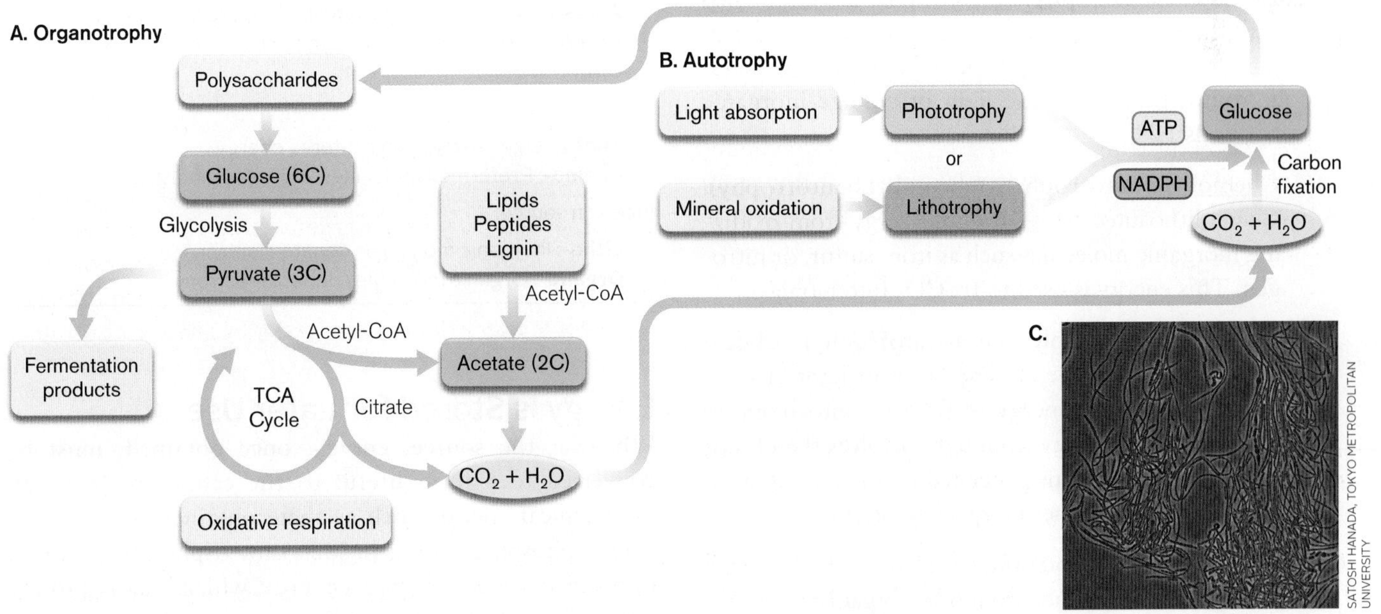

FIGURE 4.1 ■ The carbon cycle. The carbon cycle requires both autotrophs and heterotrophs. **A.** Heterotrophs gain energy from degrading complex organic compounds (such as polysaccharides) to smaller compounds (such as glucose and pyruvate). The carbon from pyruvate moves through the tricarboxylic acid (TCA) cycle and is released as CO_2. In the absence of a TCA cycle, the carbon can end up as fermentation products, such as ethanol or acetic acid. **B.** Autotrophs use light energy or energy derived from the oxidation of minerals to capture CO_2 and convert it to complex organic molecules. **C.** *Chloroflexus aggregans*, originally isolated from hot springs in Japan, possesses extraordinary metabolic versatility. It grows anaerobically (without oxygen) as a photoheterotroph and aerobically (with oxygen) as a chemoheterotroph.

(for example, H_2, H_2S, NH_4^+, NO_2^-, and Fe^{2+}) for energy, whereas organotrophs oxidize organic compounds (for example, sugars). Oxidation and reduction are reviewed in eAppendix 1. The metabolic reactions introduced here are presented in detail in Chapters 13 and 14.

What if a microbe conducts more than one type of metabolism? Many free-living soil and aquatic bacteria can obtain energy from lithotrophy and phototrophy, all in one cell. Such microbes are called mixotrophs. A **photoheterotroph** is a type of mixotroph that has multiple gene systems expressed under different conditions to yield products that carry out different functions. For example, *Rhodospirillum rubrum* grows by photoheterotrophy when light is available and oxygen is absent, but by respiration, without absorbing light, when O_2 is available.

In chemotrophy, the amount of energy harvested from oxidizing a compound depends on the compound's reduction state. The more reduced the compound is, the more electrons it has to give up and the higher its potential energy yield. A reduced compound, such as glucose, can donate electrons to a less reduced (more oxidized) compound, such as nicotinamide adenine dinucleotide (NAD), releasing energy (in the form of donated electrons). Glucose, itself becomes oxidized in the process. NAD is a cell molecule critical to energy metabolism and is discussed, along with oxidation-reduction reactions, in Chapter 13.

In short, microbes are classified on the basis of their carbon and energy acquisition as follows:

- **Autotroph (autotrophy)**. Autotrophs build biomass by fixing CO_2 into complex organic molecules. Autotrophs gain energy through one of two general metabolic routes that either use light or oxidize inorganic compounds.

  Chemolithoautotroph (chemolithoautotrophy). Chemolithoautotrophs produce energy from oxidizing inorganic molecules such as iron, sulfur, or nitrogen. This energy is used to fix CO_2 into biomass.

  Photolithoautotroph (photoautotrophy). Photolithoautotrophs generate energy from light absorption and use that energy to fix CO_2 into biomass. Capturing the energy from light involves the photoexcitation of electrons generated by the photolysis of H_2O, H_2S, or another inorganic molecule.

- **Heterotroph (heterotrophy)**. Heterotrophs break down organic compounds from other organisms to gain energy and to harvest carbon for building their own biomass. Heterotrophic metabolism can be divided into two classes, also based on whether light is involved.

  Chemoorganoheterotrophy. Chemoorganoheterotrophs obtain energy and carbon for biomass solely from organic compounds. Chemoheterotrophy is also called just heterotrophy.

  Photoorganotroph (photoorganotrophy). Photoorganotrophs obtain energy from the catabolism (breakdown) of organic compounds and through light absorption. Organic compounds are broken down and used to build biomass.

The survival and metabolism of any one group of organisms depends on the survival and metabolism of other groups of organisms. For example, the cyanobacteria, a type of photosynthetic microorganism that originated 2.5–3.5 billion years ago, produce about half of the oxygen we breathe. Cyanobacteria also depend on heterotrophic bacteria to consume the molecular oxygen that the cyanobacteria produce, since oxygen by-products can be toxic to cyanobacteria.

Today, cyanobacteria (and other phytoplankton) form the base of Earth's marine food chain. The autotrophic cyanobacteria fix carbon in the ocean and are eaten by heterotrophic protists. The protists are then devoured by fish, and the fish produce the CO_2 fixed by the cyanobacteria. And eventually, we eat the fish.

Note: In biology, the suffix "-trophy" refers to the acquisition of nutrients. The following prefixes for "-trophy" terms help distinguish different forms of biomass-building (carbon source) and energy-yielding (energy source) metabolism.

Carbon source for building biomass:

Auto-: CO_2 is fixed and assembled into organic molecules.
Hetero-: Preformed organic molecules (having two or more carbon atoms).

Energy source:

Photo-: Light absorption captures energy.
Chemo-: Preformed molecules; organic or inorganic.

Electron source:

Litho-: Inorganic molecules donate electrons.
Organo-: Organic molecules donate electrons.

Energy Is Stored for Later Use

Whatever the source, energy, once obtained, must be converted to a form useful to the cell. This form can be chemical energy, such as that contained in the high-energy phosphate bonds in adenosine triphosphate (ATP), or it can be electrochemical energy, which is stored in the form of an electrical potential generated between compartments separated by a membrane (see Chapter 14). Energy stored by an electrical potential across the membrane is known as the **membrane potential** (for most cells the membrane potential is more negative inside than outside).

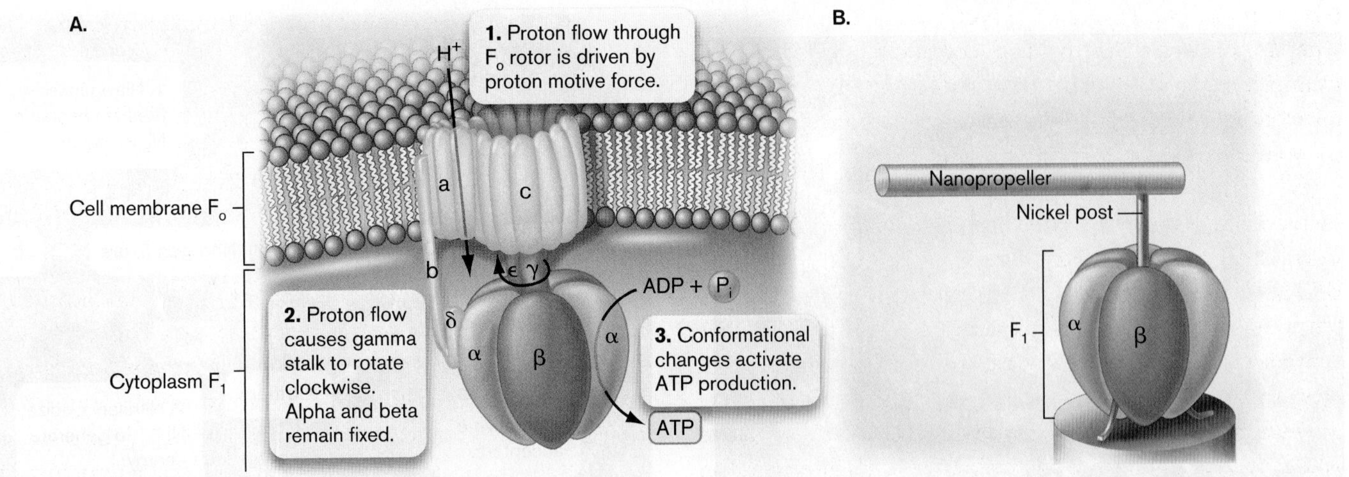

FIGURE 4.2 ■ Bacterial membrane ATP synthase. A. The F_o portion of the F_1F_o complex of ATP synthase is embedded in the cell membrane. **B.** An artificial "biomolecular motor" was built from an ATP synthase F_1 unit attached to a nickel post and a nanopropeller.

A membrane potential is generated when chemical (or light) energy is used to pump protons (H^+), Na^+, or K^+ to the outside of the cell, making the cation concentration (positive charges) greater outside the cell than inside. For example, membrane proteins such as cytochrome oxidases use energy from respiration to pump protons across the cell membrane, and out of the cell, generating a proton gradient. The proton gradient (ΔpH) plus the charge difference (voltage potential) across the membrane form an **electrochemical potential**. When this electrochemical potential includes a proton gradient, it is also called the **proton potential**, or **proton motive force** (detailed in Chapter 14). The energy stored in the proton motive force can be used by specific transport proteins to move nutrients into the cell (see Section 3.3), to directly drive motors that rotate flagella, and to drive the synthesis of ATP by a membrane-embedded **ATP synthase** (**Fig. 4.2A**).

The membrane-embedded F-ATP synthase, also called F_1F_o ATP synthase, provides most of the ATP for aerobic respiring cells such as *E. coli*. Essentially the same complex mediates ATP generation in our own mitochondria. ATP synthase is a complex of many different proteins. The enzyme includes a channel (F_o) that allows H^+ to move across the membrane and drive rotation of the ATP c-ring. Rotation of the c-ring causes changes in the F_1 complex that mediate formation of ATP. The role of the proton potential in metabolism is discussed in detail in Chapters 13 and 14.

The idea that a living organism could contain rotating parts was controversial when such parts were first discovered in bacterial flagella (discussed in Section 3.6). The discovery of rotary biomolecules has inspired advances in nanotechnology, the engineering of microscopic devices. For example, a "biomolecular motor" was devised using an ATP synthase F_1 complex to drive a metal submicroscopic propeller (**Fig. 4.2B**). In the future, such biomolecular motors may be used to build microscopic robots that enter the bloodstream to deliver small drug molecules to specific locations.

The Nitrogen Cycle

Nitrogen is an essential component of proteins, nucleic acids, and other cell constituents, and as such, living cells require it in large amounts. So, how do bacteria get nitrogen? Nitrogen gas (N_2) makes up nearly 79% of Earth's atmosphere, but most organisms are unable to use nitrogen gas, because the triple bond between the two nitrogen atoms is highly stable and requires considerable energy to be broken. For nitrogen to be used for growth, it must first be "fixed," or converted to ammonium ions (NH_4^+). Nitrogen gas is converted to NH_4^+ by **nitrogen-fixing bacteria**. The ammonium is then used by all microbes to make amino acids and other nitrogenous compounds needed for growth (presented in Chapter 15).

Nitrogen-fixing bacteria may be free-living in soil or water, or they may form symbiotic associations with plants or other organisms (see Chapter 17). A **symbiont** is an organism that lives in intimate association with a second organism. For example, *Rhizobium*, *Sinorhizobium*, and *Bradyrhizobium* species are nitrogen-fixing symbionts of leguminous plants such as soybeans, chickpeas, and clover (**Fig. 4.3**). Although symbionts are the most widely known nitrogen-fixing bacteria, the majority of nitrogen in soil and aquatic environments is fixed by free-living bacteria and archaea.

Once fixed into organic compounds, how does nitrogen get back into the atmosphere? As with the carbon cycle, various groups of organisms collaborate to recycle ammonium ions and nitrate ions (NO_3^-) into nitrogen gas in what

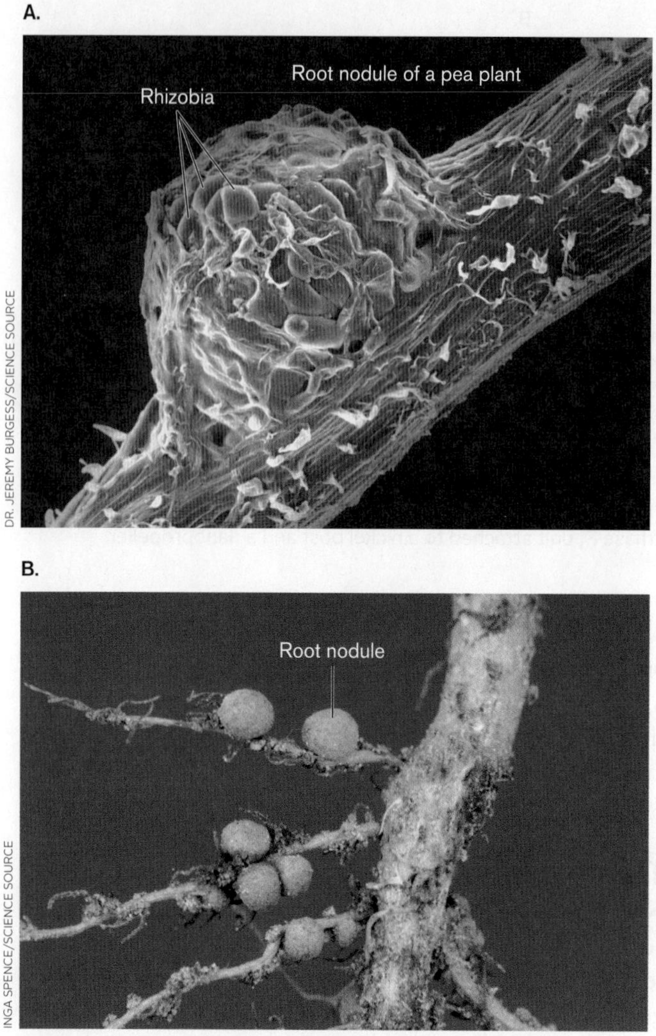

FIGURE 4.3 ▪ *Rhizobium* and a legume. **A.** Symbiotic *Rhizobium* cells forming a nodule on a pea plant root (SEM). The rhizobia (about 0.9 µm × 3 µm) invade the root and begin a symbiotic partnership that will benefit both organisms. **B.** Root nodules. After the rhizobia invade the plant root, symbiosis between plant and microbe produces nodules.

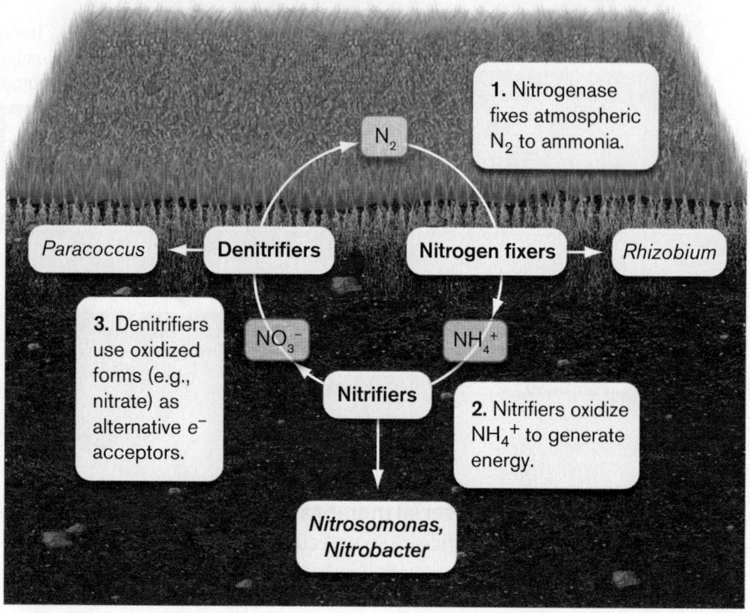

FIGURE 4.4 ▪ The nitrogen cycle. Different sets of bacteria fix nitrogen from atmospheric nitrogen gas (N_2) to form ammonia (NH_3) or ammonium (NH_4^+), convert NH_4^+ to nitrate (NO_3^-), and return N_2 to the atmosphere.

is called the nitrogen cycle, and they collect energy in the process (**Fig. 4.4**). One group of bacteria (the nitrifiers) gains energy by converting, or oxidizing, ammonia in two steps to form nitrate in a process called **nitrification** (*Nitrosomonas*, *Alcaligenes*, and *Nitrobacter*). Nitrification is a form of lithotrophy. Other heterotrophic microbes (for example, *Paracoccus*) can reduce nitrate to N_2 via **denitrification**, a process that uses nitrate and related inorganic forms of nitrogen as terminal electron acceptors for certain electron transport chains. Denitrifying bacteria send an amount of nitrogen into the atmosphere that roughly balances the amount removed by nitrogen fixation. Like the carbon cycle, the nitrogen cycle illustrates how nature manages to replenish planet Earth. For the environmental significance of nitrogen metabolism, see Chapter 22.

To Summarize

- **Microorganisms** require certain essential macro- and micronutrients to grow.
- **Autotrophs** use CO_2 as a carbon source, either through photosynthesis or through lithotrophy, and make organic compounds as biomass.
- **Heterotrophs** consume the organic compounds made by autotrophs to gain carbon.
- **Energy gained by phototrophy or chemotrophy** is stored either as proton motive force or as chemical energy (ATP).
- **Nitrogen fixers (only bacteria and archaea)** incorporate nitrogen into biomass and contribute organic nitrogen to the rest of the ecosystem.
- **Chemotrophic nitrifying bacteria gain energy** by converting NH_4^+ (made by nitrogen-fixing bacteria) into nitrate and nitrite.
- **Heterotrophic denitrifying organisms** use nitrate and nitrite as electron acceptors to make nitrogen gas.

4.2 Nutrient Uptake

How do bacteria gather nutrients? Whether a microbe swims by flagella toward a favorable habitat or, lacking motility, drifts through its environment, the organism must be able to find nutrients and move them across the membrane into the cytoplasm. The membrane, however, presents a daunting obstacle. Membranes separate what is outside the cell from what is inside. So, for a cell to gain sustenance from the environment, the membrane must be selectively permeable to nutrients the cell can use. A few compounds, such as oxygen and carbon dioxide, can passively diffuse across the membrane, but most cannot.

Selective permeability is achieved in three ways:

- Substrate-specific carrier proteins (**permeases**) in the cytoplasmic membrane
- Nutrient-binding proteins that stick to the cell wall of Gram-positive bacteria and sample the environment, or that freely patrol the periplasmic space of Gram-negative bacteria
- Membrane-spanning protein channels, or pores, that discriminate between substrates

Microbes must also overcome the problem of low nutrient concentrations in the natural environment. If the intracellular concentration of a nutrient never rose higher than the extracellular concentration, the cell would starve in low-nutrient environments. To solve this dilemma, most organisms have evolved efficient transport systems that concentrate nutrients inside the cell relative to outside. However, moving molecules against a concentration gradient requires some form of energy.

In contrast to environments where nutrients are available but exist at low concentrations (for example, most aquatic environments), certain habitats have plenty of nutrients but those nutrients are locked in a form that cannot be transported into the cell. Starch, a large, complex carbohydrate, is but one example. Many microbes unlock these nutrient "vaults" by secreting digestive enzymes that break down complex carbohydrates or other molecules into smaller compounds that are easier to transport. The amazing mechanisms that cells use to extrude these large digestive proteins through the membrane and into their surrounding environment are discussed in Chapter 8.

Facilitated Diffusion

Although most transport systems use cellular energy to bring compounds into the cell, a few do not. **Facilitated diffusion** transports nutrients across a membrane only from a compartment of higher concentration to a compartment of lower concentration. The best these passive systems can do is to equalize the internal and external concentrations of a solute; facilitated transport cannot move a molecule against its gradient. Facilitated diffusion systems are used for compounds that are either too large or too polar to diffuse on their own. The most important facilitated diffusion transporters are those of the aquaporin family that transport water and small polar molecules such as glycerol (which is used for energy and for building phospholipids). Glycerol transport is performed by an integral membrane protein in *E. coli* called GlpF. The structure of a glycerol channel is shown in **Figure 4.5A**, where the complex is viewed from the outer face of the membrane. The complex is a tetramer of four channels, each of which transports a glycerol molecule.

In the membrane, GlpF reversibly (and randomly) assumes two conformations. One form exposes the glycerol-binding site to the external environment; the second form exposes this site to the cytoplasm. When the concentration of glycerol is greater outside than inside the cell, the form with the binding site exposed to the exterior is more likely to find and bind glycerol. After binding glycerol, GlpF changes

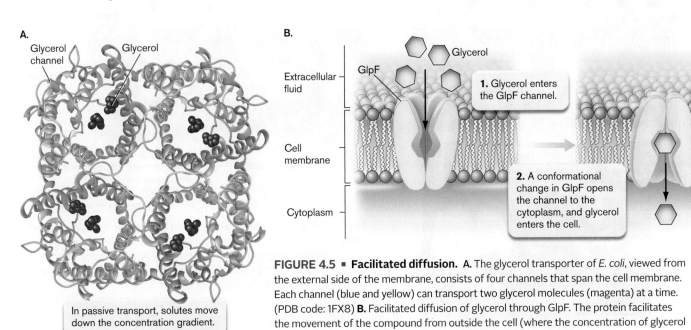

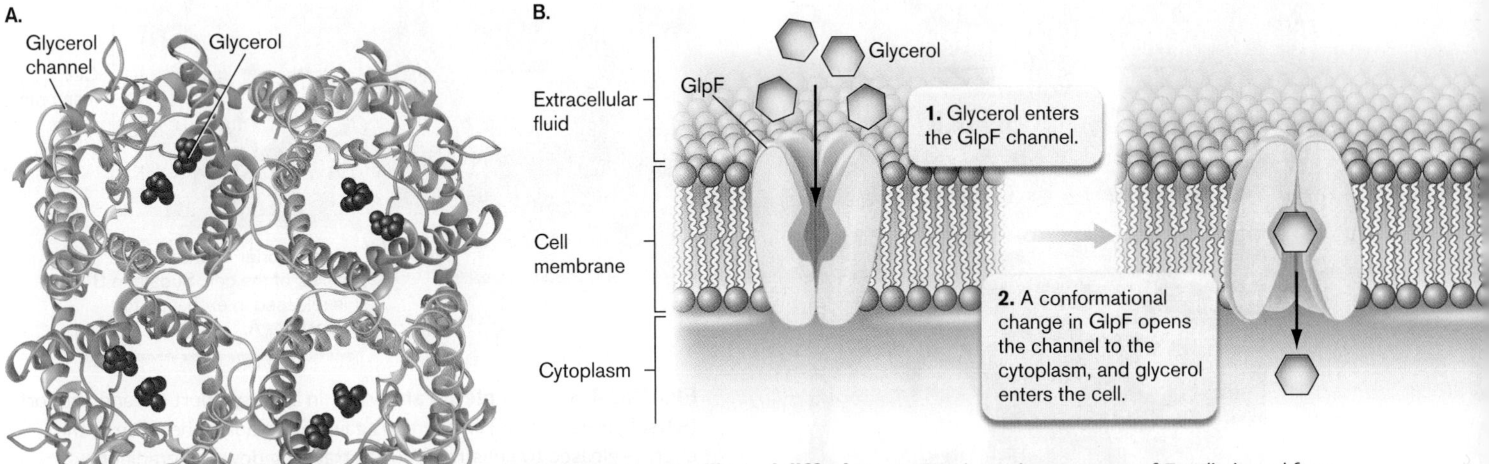

FIGURE 4.5 ■ Facilitated diffusion. **A.** The glycerol transporter of *E. coli*, viewed from the external side of the membrane, consists of four channels that span the cell membrane. Each channel (blue and yellow) can transport two glycerol molecules (magenta) at a time. (PDB code: 1FX8) **B.** Facilitated diffusion of glycerol through GlpF. The protein facilitates the movement of the compound from outside the cell (where the concentration of glycerol is high) to inside the cell (where the concentration of glycerol is low).

shape, closing itself to the exterior and opening to the interior (**Fig. 4.5B**). Bound glycerol is then released (diffuses) into the cytoplasm (influx). Of course, this form of GlpF could also bind a glycerol molecule in the cytoplasm and release it outside the cell. Thus, once the cytoplasmic concentration of glycerol equals the concentration outside the cell, bound glycerol can be released to either compartment. However, facilitated diffusion normally promotes glycerol influx, because the cell consumes the compound as it enters the cytoplasm, keeping cytoplasmic concentrations of glycerol low.

Active Transport Requires Energy

Most forms of transport expend energy to take up molecules from outside the cell and concentrate them inside. The ability to import nutrients against their natural concentration gradients is critical in aquatic habitats, where nutrient concentrations are low, and in soil habitats, where competition for more plentiful nutrients is fierce because of high cell numbers.

The simplest way to move molecules against their gradient across a membrane is to exchange the energy of one chemical gradient for that of another. The most common chemical gradients used are those of ions, particularly the positively charged ions Na^+ and K^+. These ions are kept at different concentrations on either side of the cell membrane. When an ion moves down its concentration gradient (from high to low), energy is released. Some transport proteins harness that free energy and use it to drive transport of a second molecule up, or against, its concentration gradient in the process called **coupled transport**.

The two types of coupled transport systems are **symport**, in which the two molecules travel in the same direction (**Fig. 4.6A**), and **antiport**, in which the actively transported molecule moves in the direction opposite that of the driving ion (**Fig. 4.6B**). An example of a symporter is the lactose permease LacY of *E. coli*, one of the first transport proteins to have its function elucidated. This work was carried out by the pioneering membrane biochemist H. Ronald Kaback of UCLA (**Fig. 4.6C**). LacY moves lactose inward, powered by a proton that is also moving inward (symport). LacY proton-driven transport is said to be electrogenic because an unequal distribution of charge results (for example, symport of an uncharged lactose molecule with a H^+ results in net movement of positive charge).

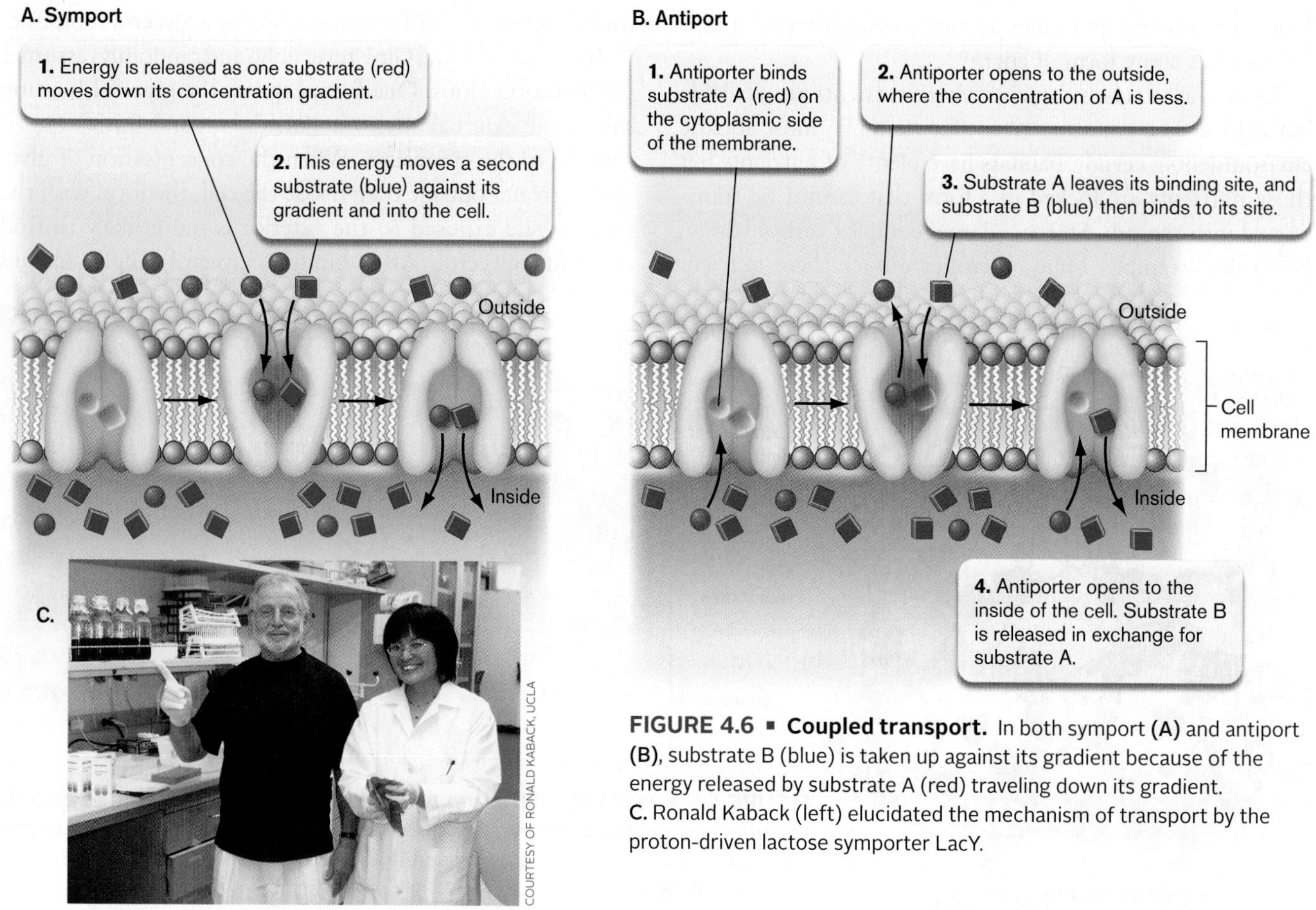

FIGURE 4.6 ■ Coupled transport. In both symport **(A)** and antiport **(B)**, substrate B (blue) is taken up against its gradient because of the energy released by substrate A (red) traveling down its gradient. **C.** Ronald Kaback (left) elucidated the mechanism of transport by the proton-driven lactose symporter LacY.

An example of electroneutral coupled transport, in which there is no net transfer of charge, is that of the Na^+/H^+ antiporter, which couples the export of Na^+ with the import of a proton (antiport). Because molecules of like charges are merely exchanged, there is no net movement of charge. Sodium exchange is important for all organisms and is particularly critical for organisms living in high-salt habitats (see Section 5.2).

Thought Questions

4.2 How could a symport transporter produce electroneutral coupled transport?

4.3 How might mutations in transporter gene sequences influence bacterial survival under different conditions—for example, normal versus very low glucose concentrations?

Symport and antiport transporter proteins function by alternately opening one end or the other of a channel that spans the cell membrane. The channel contains solute-binding sites (**Fig. 4.6A** and **B**). When the channel is open to the high-concentration side of the membrane, the driving ion (solute) attaches to the binding sites. The transport protein then changes shape to open that site to the low-concentration side of the membrane, and the ion leaves. When and where the second (cotransported) solute binds depends on whether the transport protein is an antiporter or a symporter.

With all this ion traffic across the membrane going on, a careful accounting must be kept of how many ions are inside the cell relative to outside. The cell must recirculate ions back and forth across the membrane to maintain certain gradients if the organism is to survive. Because key ATP-producing systems require an electrochemical gradient across the membrane, it is especially important to keep the interior of the cell negatively charged relative to the exterior. However, because the movement of many compounds is coupled to the import of positive ions, the electrochemical gradient will eventually dissipate, or depolarize, unless positive ions are also exported. Complete depolarization must be avoided, because a depolarized cell loses membrane integrity and cannot carry out the transport functions needed to sustain growth. A healthy cell maintains a proper charge balance by using the electron transport chain to move protons out of the cell and by using antiporters to exchange negatively and positively charged ions as needed.

While each symport and antiport protein links the transport of two different ions, the movement of two different ions can be linked indirectly by two different transport systems. For example, proton concentrations are typically greater outside the cell than inside. The inwardly directed proton gradient is called proton motive force. Proton motive force can impel the exit of Na^+ through the Na^+/H^+ antiporter. The resulting Na^+ gradient can then drive the symport of amino acids into the cell. In this case, Na^+ moves back into the cell down its gradient, and the energy released is tied to the import of an amino acid against its gradient. In the Current Research Highlight, *Bacillus subtilis* uses a glutamate/Na^+ symporter to import glutamate (which is negatively charged) as a carbon source.

ABC Transporters Are Powered by ATP

As we pointed out in Section 3.2, a major function of proton transport is to form the proton motive force that powers ATP synthesis (for details on the proton motive force, see Chapter 14). The energy stored in ATP can then drive membrane transport of nutrients.

The largest family of energy-driven transport systems is the ATP-binding cassette superfamily, also known as **ABC transporters** (**Fig. 4.7**). These transporters are found in bacteria, archaea, and eukaryotes. All of them appear to have arisen from a common ancestral porter, so they share a considerable amount of amino acid sequence homology. Many different ABC transporters mediate the import or export of a wide variety of substrates.

It is impressive that nearly 5% of the *E. coli* genome is dedicated to producing the components of 70 different varieties of uptake and efflux ABC transporters. The uptake ABC transporters are critical for transporting simple carbohydrates such as maltose, arabinose, and galactose, as well as the amino acid histidine, as a carbon source. Other uptake ABC transporters can import long-chain carbohydrates such as the acidic polysaccharide alginate. For instance, the marine hyperthermophilic bacterium *Thermotoga maritima* utilizes the ABC transporter AguEFG to import alginate as a carbon source. Alginate, a major component of brown seaweed, is being examined as an alternative source of bioethanol. Bacteria that import alginate and ferment the carbohydrate to ethanol will be critical to this effort.

Efflux ABC transporters are generally used as multidrug efflux pumps that enable microbes to survive exposures to hazardous chemicals. *Pseudomonas aeruginosa*, a pathogenic bacterium found in soil and water habitats, uses numerous multidrug efflux pumps (including ABC transporters) to export a broad range of antibiotics, including tetracyclines, streptogramins, quinolones, macrolides, and aminoglycosides, thus conferring resistance to those drugs (see Section 27.3).

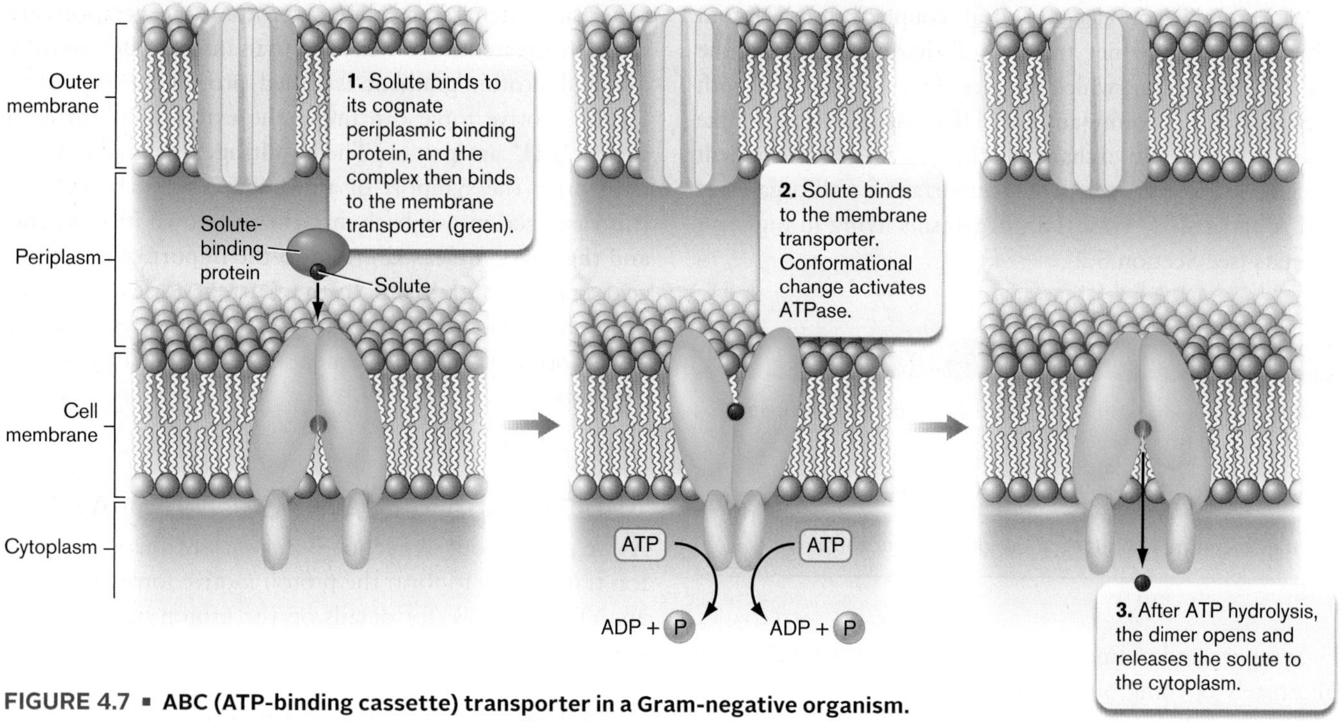

FIGURE 4.7 ■ ABC (ATP-binding cassette) transporter in a Gram-negative organism.

An ABC transporter typically consists of two hydrophobic membrane proteins and two cytoplasmic proteins that contain a highly conserved amino acid motif, called the ATP-binding cassette, that binds ATP (**Fig. 4.7**). The uptake systems (but not the efflux systems) possess an additional, extracytoplasmic protein, called a substrate-binding protein, that initially binds the substrate. In Gram-negative bacteria, these substrate-binding proteins float in the periplasmic space between the inner and outer membranes. In Gram-positive bacteria, which lack an outer membrane, the proteins must be tethered to the cell surface.

ABC transport (**Fig. 4.7**) starts with the substrate-binding protein snagging the appropriate solute, either as it floats by a Gram-positive microbe or as the molecule enters the periplasm of a Gram-negative cell. Most substrates nonspecifically enter the periplasm of Gram-negative organisms through the outer membrane pores, although some high-molecular-weight substrates, like vitamin B_{12}, require the assistance of a specific outer membrane protein to move the substrate into the periplasm. Because substrate-binding proteins have a high affinity for their cognate (matched) solutes, their use increases the efficiency of transport when concentrations of solute are low.

Once united with its solute, the binding protein binds to the periplasmic face of the channel protein and releases the solute, which now moves to a site on the channel protein. This interaction triggers a structural (or conformational) change in the channel protein (the green "membrane transporter" in **Fig. 4.7**) that is telegraphed to the nucleotide-binding proteins on the cytoplasmic side (yellow). On receiving this signal, the nucleotide-binding proteins start hydrolyzing ATP and send a return conformational change through the channel, signaling the channel to open its cytoplasmic side and allow the solute to enter the cell.

Siderophores Are Secreted to Scavenge Iron

Iron, an essential nutrient of most cells, is largely locked up in nature as insoluble $Fe(OH)_3$, which is unavailable for transport. Many bacteria and fungi have solved this transport dilemma by synthesizing and secreting specialized molecules called **siderophores** (Greek for "iron bearer"), which have a very high affinity for the small amounts of soluble ferric iron available in the environment. These iron scavenger molecules are produced and secreted by cells when the intracellular iron concentration is low (**Fig. 4.8A**).

In most Gram-negative organisms, the siderophore (for example, enterochelin in *E. coli*) binds iron in the environment, and the siderophore-iron complex then attaches to specific receptors in the outer membrane. At this point, either the iron is released directly and is passed to other transport proteins, or the complex is transported across the cytoplasmic membrane by a dedicated ABC transporter. The iron is released intracellularly and reduced from Fe^{3+} to Fe^{2+} for biosynthetic use.

Other Gram-negative microorganisms, such as *Neisseria gonorrhoeae* (the causative agent of gonorrhea), do not use siderophores at all, but employ receptors on their surface that bind human iron complexes (for example, transferrin or lactoferrin) and wrest the iron from them. Because microbial iron-transporting proteins are critical to the pathogenesis of

A.

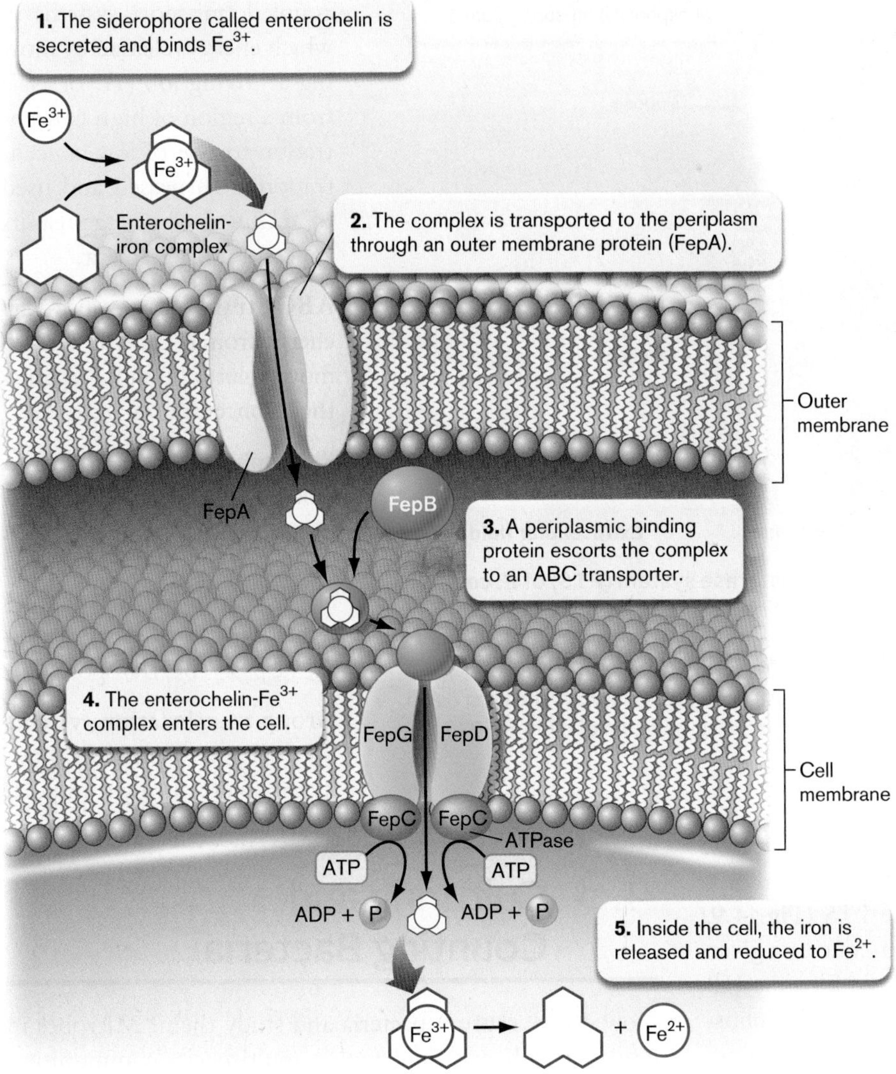

B.

FIGURE 4.8 ▪ Siderophores and iron transport. A. This model shows the *E. coli* siderophore called enterochelin. B. Tim Bugni discovered the siderophore thalassosamide from a marine bacterium.

many microbes, solving their molecular structures of these proteins is of great interest. Such structural studies can expose vulnerabilities in the transport proteins that we can use in designing new antibiotics that stop iron transport.

Scientists are also interested in using siderophores directly as antimicrobial agents. Tim Bugni from the University of Wisconsin–Madison (**Fig. 4.8B**) has purified a novel siderophore called thalassosamide from the marine bacterium *Thalassospira profundimaris*. The organism uses thalassosamide to scavenge for soluble iron in the western Pacific Ocean and employs a siderophore-specific transport system to capture the siderophore-iron complexes. How could this siderophore be used as an antimicrobial agent? Pathogens such as *Pseudomonas aeruginosa* lack uptake systems for thalassosamide-iron complexes. Consequently, adding purified thalassosamide to a culture of *P. aeruginosa* will trap extracellular iron in a form unusable by *Pseudomonas*. Starved of iron, *Pseudomonas* growth stalls. Preliminary tests using thalassosamide to treat *P. aeruginosa*–infected animals suggest that this siderophore, or a future derivative of it, could become a useful therapeutic agent.

Group Translocation Avoids "Uphill" Battles

The uptake transporters that we have just considered increase a solute's concentration inside the cell relative to the outside. They move nutrients "uphill" against a concentration gradient. An entirely different system, known as **group translocation**, cleverly accomplishes the same result but without really moving a substance uphill. Group translocation alters the substrate during transport by attaching a new group (for example, phosphate) to it. Because the modified nutrient inside the cell is chemically different from the related compound outside, the parent solute entering the cell is always moving down its concentration gradient, regardless of how much solute has already been transported. Note that this process uses energy to chemically alter the solute. ABC transporters and group translocation systems both involve active transport, but group translocation systems are not ABC transporters.

The **phosphotransferase system** (**PTS**) is a well-characterized group translocation system present in many bacteria. It uses energy from phosphoenolpyruvate (PEP), an intermediate in glycolysis, to attach a phosphate to specific sugars during their transport into the cell (**Fig. 4.9** ▶). Glucose, for example, is converted during transport to glucose 6-phosphate. The system has a

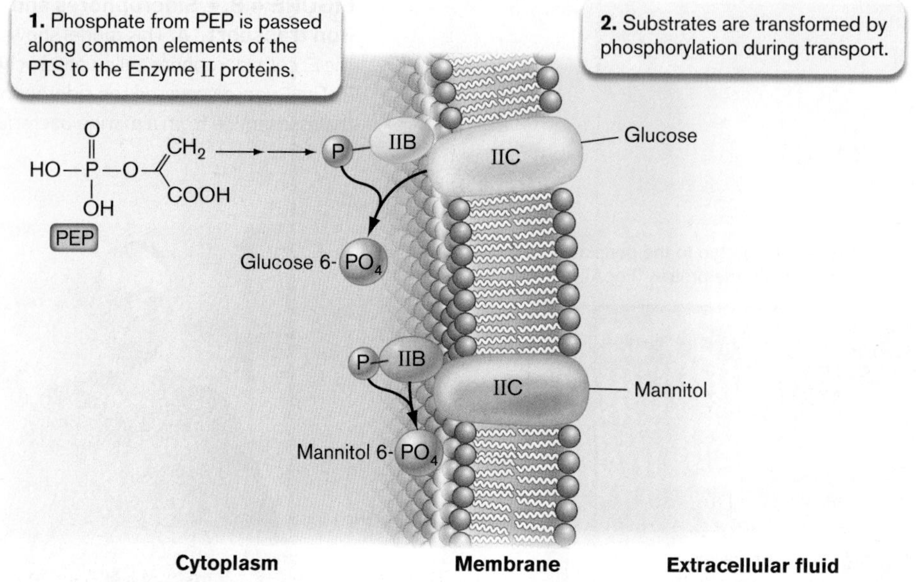

FIGURE 4.9 ■ Group translocation: the phosphotransferase system (PTS) of *E. coli*. The phosphate group from phosphoenolpyruvate (PEP) is ultimately passed along proteins common to all PTS sugar transport systems and to Enzyme II components of the PTS that are specific to individual substrates (such as glucose or mannitol). The substrate is then phosphorylated during transport, making it different from the external sugar. As a result, the sugar is always traveling down its concentration gradient into the cell. A more detailed look at the cell location of different PTS proteins is found in **eTopic 4.1.**

modular design that accommodates different substrates. Protein elements that initiate phosphotransfer from PEP are used for all sugars transported by the PTS (**Fig. 4.9**, step 1). Later PTS proteins, however, are unique to a given carbohydrate (step 2). For example, different Enzyme IIB and membrane-embedded IIC proteins transfer phosphate to glucose and mannitol during transport. Because the sugar (glucose, for instance) is phosphorylated during transport, glucose itself never accumulates in the cytoplasm. As a result, the sugar transported via the PTS always travels down its concentration gradient into the cell. More details about the PTS are given in **eTopic 4.1**. In Chapter 10 we will see how this physiological system impacts the genetic control of many other systems.

Like prokaryotic cells, eukaryotic cells possess antiporters and symporters and use ABC transport systems as multidrug efflux pumps. But eukaryotes also employ another process, called endocytosis, which often precedes nutrient transport across membranes. Endocytosis is discussed in **eTopic 4.2** and eAppendix 2.

To Summarize

- **Transport systems** move nutrients across phospholipid bilayer membranes.
- **Facilitated diffusion** helps solutes move across a membrane from a region of high concentration to one of lower concentration.
- **Antiport** and **symport** are coupled transport systems in which energy released by moving a driving ion (H^+ or Na^+) from a region of high concentration to one of low concentration is harnessed and used to move a solute against its concentration gradient.
- **ABC transporters** use the energy from ATP hydrolysis to move solutes "uphill," against their concentration gradients.
- **Siderophores** are secreted to bind ferric iron (Fe^{3+}) and transport it into the cell, where it is reduced to the more useful ferrous form (Fe^{2+}). Siderophore-iron complexes enter cells with the help of ABC transporters.
- **Group translocation systems** use energy to chemically modify the solute during transport.

4.3 Culturing and Counting Bacteria

How do we capture bacteria and study them? Microbes in nature usually exist in complex, multispecies communities. For detailed studies of a single species, however, cells of the species are usually grown in pure culture. In this section we will describe how that is done. But after almost 140 years of trying to grow microbes in the laboratory, the vast majority of the microbial world (>99%) has yet to be cultured. At the end of this section we will discuss some new, innovative methods that enable us to culture at least some of these uncultured microbes.

Bacteria Are Grown in Culture Media

For those organisms that can be cultured, we have access to a variety of culturing techniques that can be used for different purposes. Bacterial culture media may be either liquid or solid. A liquid, or broth, medium, in which organisms can move about freely, is useful for studying the growth characteristics of a single strain of a single species (that is, a **pure culture**). Liquid media are also convenient for examining growth kinetics and microbial biochemistry at different phases of growth. Solid media, usually gelled with agar, are useful for trying to separate mixtures of different organisms as they are found in the natural environment or in clinical specimens.

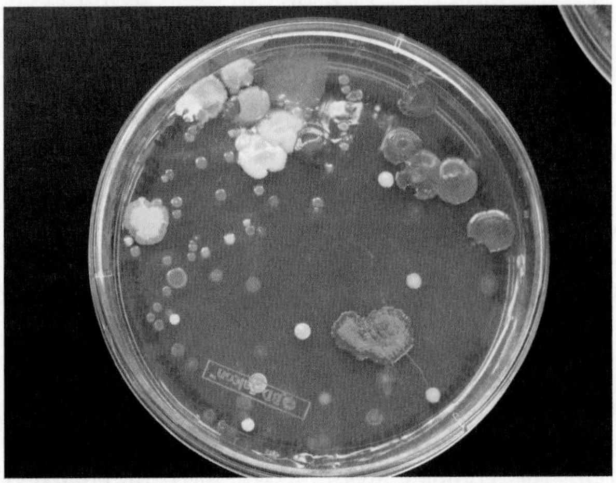

FIGURE 4.10 ■ Separation and growth of microbes on an agar surface. Agar plate culture made from swabbing patient elevator buttons and escalator rails at a university hospital. Note the various colony shapes and colors made by different species of microbes. https://commons.wikimedia.org/wiki/File:VCU_agar_plate_colonies.jpg https://creativecommons.org/licenses/by-sa/3.0/legalcode

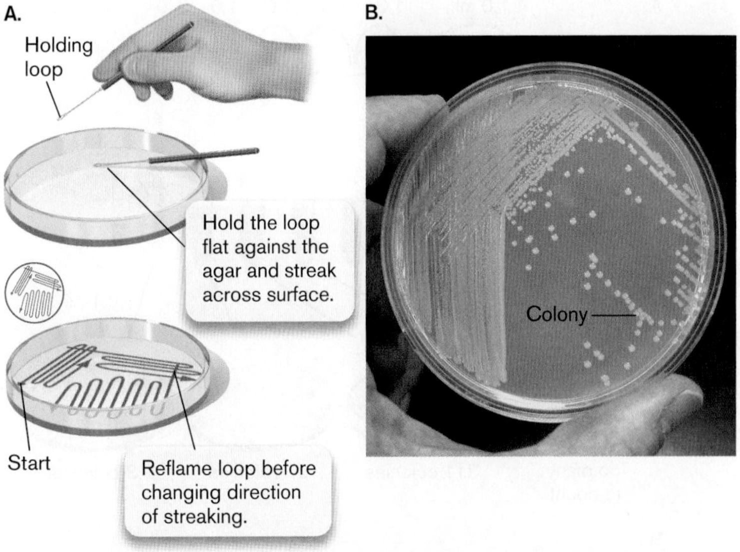

FIGURE 4.11 ■ Dilution streaking technique. **A.** A liquid culture is sampled with a sterile inoculating loop and streaked across the plate in three or four areas, with the loop flamed between areas to kill bacteria still clinging to it. Dragging the loop across the agar diminishes the number of organisms clinging to the loop, until only single cells are deposited at a given location. **B.** *Salmonella enterica* culture obtained by dilution streaking. ▶

Dilution Streaking and Spread Plates

Solid media are basically liquid media to which a solidifying agent has been added. The most versatile and widely used solidifying agent is agar (discussed in Section 1.3). Derived from seaweed, agar forms an unusual gel that liquefies at 100°C but does not solidify again until cooled to about 40°C. Liquefied agar medium poured into shallow, covered Petri dishes cools and hardens to provide a large, flat surface on which a mixture of microorganisms can be streaked to separate individual cells. Each cell will divide and grow to form a distinct, visible colony of cells (**Fig. 4.10**).

As shown in **Figure 4.11A** ▶, a drop of liquid culture is collected with an inoculating loop and streaked across the agar plate surface in a pattern called **dilution streaking** or streak plating. Organisms fall off the loop as the loop moves along the agar surface. Toward the end of the streak, few bacteria remain on the loop, so individual cells will land and stick to different places on the agar surface. If the medium contains the proper nutrients and growth factors, a single cell will multiply into many millions of offspring, forming a microcolony. At first visible only under a microscope, the microcolony (**Fig. 4.12**) grows into a visible droplet called a **colony** (**Fig. 4.11B**). A pure culture of the species or of one strain of a species can be obtained by touching a single colony with a sterile inoculating loop and inserting that loop into fresh liquid medium.

Another way to isolate pure colonies is the **spread plate** technique. Starting from a liquid culture of bacteria, a series of tenfold dilutions is made, and a small amount of each dilution is placed directly on the surface of separate agar plates (**Fig. 4.13**). The sample is spread over the surface of the plate with an alcohol flame-sterilized, bent glass rod. The early dilutions, those containing the most bacteria, will

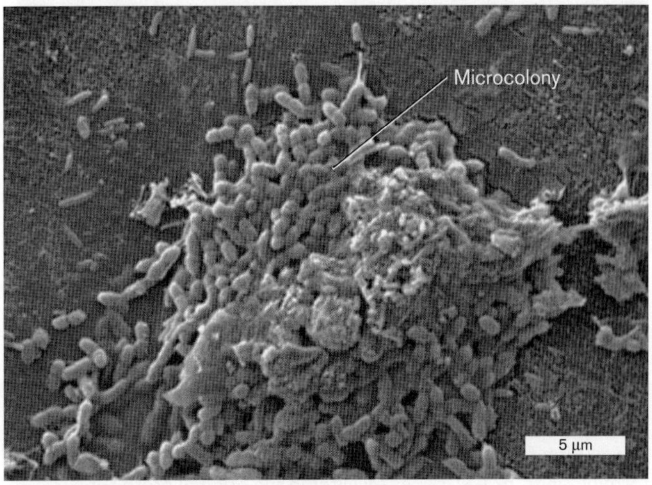

FIGURE 4.12 ■ Bacterial microcolonies. Typical microcolony formed on the surface of a membrane after 16 days. (SEM)

produce **confluent** growth that covers the entire agar surface. Subsequent dilutions, containing fewer and fewer organisms, yield individual colonies. As we will see later, spread plates are often used to enumerate the number of **viable** bacteria in the original growth tube. A viable organism, also called culturable, is one that successfully replicates to form a colony. **Thus, each colony on an agar plate represents one viable organism (or colony-forming unit, CFU) present in the original liquid culture.** Figure 4.13 illustrates how to enumerate viable cells using dilutions and spread plates.

It is important to realize that the "one cell equals one colony" paradigm does not hold for all bacteria. Organisms

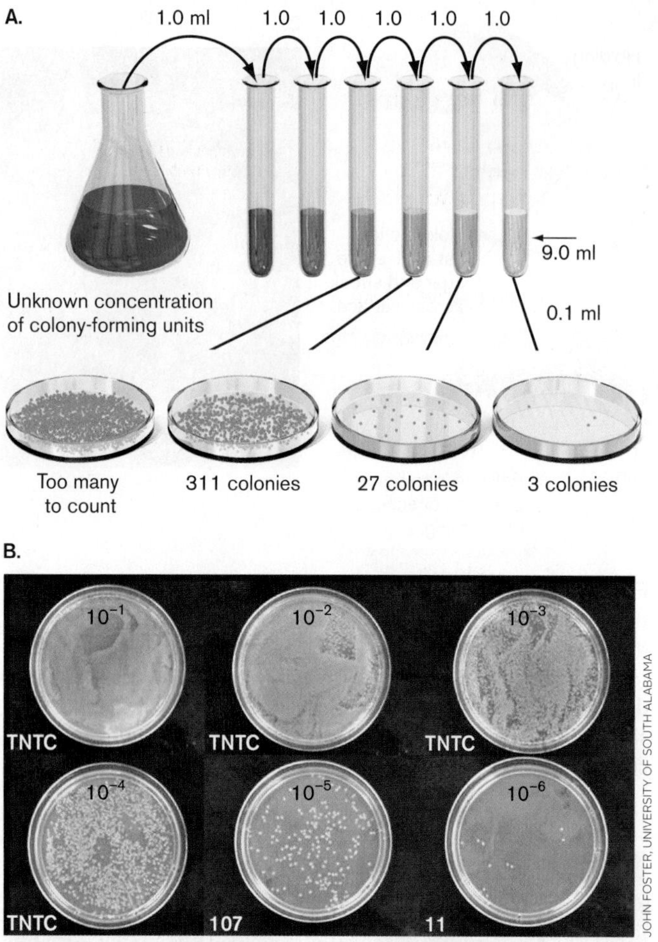

FIGURE 4.13 ▪ Tenfold dilutions, plating, and viable counts. A. A culture containing an unknown concentration of cells is serially diluted. First 1 milliliter (ml) of culture is added to 9 ml of diluent broth and mixed, and then 1 ml of this 1/10 dilution is added to another 9 ml of diluent (10^{-2} dilution). These steps are repeated for further dilution, each of which lowers the cell number tenfold. After dilution, 0.1 ml of each dilution is spread onto an agar plate. B. Plates prepared as in (A) are incubated at 37°C to yield colonies. By multiplying the number of countable colonies (107 colonies on the 10^{-5} plate) by 10, you get the number of cells in 1.0 ml of the 10^{-5} dilution. By multiplying that number by the reciprocal of the dilution factor, you can calculate the number of cells (colony-forming units, or CFUs) per milliliter in the original broth tube ($107 \times 10^1 \times 10^5 = 1.1 \times 10^8$ CFUs/ml). TNTC = too numerous to count.

such as *Streptococcus* and *Staphylococcus* usually do not exist as single cells, but grow as chains or clusters of several cells. Thus, a cluster of ten *Staphylococcus* cells will form only one colony on an agar medium and so is also called a colony-forming unit.

Complex versus Synthetic Media

We often culture bacteria in a type of medium called **complex medium** (or "rich medium") that is nutrient-rich but has poorly defined components, such as yeast extract (**Table 4.1**). Alternatively, we can select a defined or synthetic medium in which all of the chemical components and their concentrations are known. Bacteria grow more slowly on defined media because the organisms must synthesize more of their own components. A **minimal defined medium** contains only those nutrients that are essential for growth of a given microbe. However, not every microbe that can be cultured on a complex medium can be grown on a defined medium. Recipes for complex media usually contain several ingredients, such as yeast extract or beef extract, that provide a rich variety of amino acids, peptides, nucleosides, vitamins, and some sugars. Some organisms are particularly fastidious, requiring that components of blood be added to a basic complex medium. With such additions, the complex medium is called an **enriched medium**.

Complex media provide many of the chemical building blocks that a cell would otherwise have to synthesize on its own. For example, instead of making proteins that synthesize tryptophan, all the cell needs is a membrane transport system to harvest tryptophan already present in complex culture medium. Likewise, fastidious organisms that require blood in their media may reclaim the heme released from red blood cells as their own, using it as an "enzyme prosthetic group," a group critical to enzyme function (for example, the heme group in cytochromes). All of this saves the scavenging cell a tremendous amount of energy, and as a result, bacteria tend to grow fastest in complex media.

Complex medium is not useful, however, to scientists trying to determine the metabolism of a microbe. How would you know whether *E. coli* possesses the ability to make tryptophan if the bacterium grew only in complex media? Questions like this can be asked of organisms able to grow in fully defined synthetic media. In preparing a synthetic medium, we start with water and then add various salts, carbon, nitrogen, and energy sources in precise amounts. For self-reliant organisms like *E. coli* or *Bacillus subtilis*, that is all that's needed. Other organisms, such as *Shigella* species or mutant strains of *E. coli* or *B. subtilis*, require additional ingredients to satisfy requirements imposed by the absence of specific metabolic pathways (growth factors are discussed shortly). However, newer molecular tools, such as whole-genome sequencing, can reveal missing biosynthetic pathways and help predict which nutrients are needed for a particular species to grow in a synthetic medium.

Thought Question

4.4 Describe the phenotype (growth characteristic) of a cell that lacks the *trp* genes (genes required for the synthesis of tryptophan). What would be the phenotype of a cell missing the *lac* genes (genes whose products catabolize the carbohydrate lactose)?

TABLE 4.1 Composition of Commonly Used Media

Medium	Ingredients per liter		Organisms cultured
Luria Bertani (complex)	Bacto tryptone[a]	10 g	Many Gram-negative and Gram-positive organisms (such as *Escherichia coli* and *Staphylococcus aureus*, respectively)
	Bacto yeast extract	5 g	
	NaCl	10 g	
	Adjust to pH 7		
M9 medium (defined)	Glucose	2.0 g	Gram-negative organisms such as *E. coli*
	Na_2HPO_4	6.0 g (42 mM)	
	KH_2PO_4	3.0 g (22 mM)	
	NH_4Cl	1.0 g (19 mM)	
	NaCl	0.5 g (9 mM)	
	$MgSO_4$	2.0 mM	
	$CaCl_2$	0.1 mM	
	Adjust to pH 7		
Sulfur oxidizers (defined)	NH_4Cl	0.52 g	*Acidithiobacillus thiooxidans*
	KH_2PO_4	0.28 g	
	$MgSO_4 \cdot 7H_2O$	0.25 g	
	$CaCl_2$	0.07 g	
	Elemental sulfur	1.56 g	
	CO_2	5%	
	Adjust to pH 3		

[a]Bacto tryptone is a pancreatic digest of casein (bovine milk protein).

Selective and Differential Media

Microorganisms are remarkably diverse with respect to their metabolic capabilities and resistance to certain toxic agents. These differences are exploited in **selective media**, which favor the growth of one organism over another, and in **differential media**, which expose biochemical differences between two species that grow equally well. For example, Gram-negative bacteria (such as Proteobacteria), with their outer membrane, are much more resistant than Gram-positive bacteria to detergents like bile salts and certain dyes, such as crystal violet (Gram-positive and Gram-negative cell types were described in Chapter 2). A solid medium containing bile salts and crystal violet is considered selective because it favors the growth of Gram-negative over Gram-positive bacteria.

On the other hand, a differential medium is needed to distinguish between organisms that differ not in their ability to grow, but in a particular biochemical function that they possess. For example, *E. coli* and *Salmonella enterica*, a major cause of diarrhea, are both Gram-negative, but only *E. coli* can ferment lactose. Both organisms will grow on a solid medium (MacConkey agar; **Fig 4.14**) that contains a nonfermentable carbon source like peptone, a dye called "neutral red," and lactose. *E. coli* will ferment the lactose and produce acidic end products. These products lower the pH surrounding the colony, so the neutral red dye stays red. The red dye enters cells, so the growing colony of *E. coli* becomes red. However, *S. enterica* cannot ferment lactose but grows well on the nonfermentable peptides in MacConkey agar to make alkaline products. Neutral red is colorless at alkaline pH, so the colonies of *S. enterica* remain white, their natural color. In this example, growth in differential media easily distinguishes colonies of lactose fermenters (red) from nonfermenters (white).

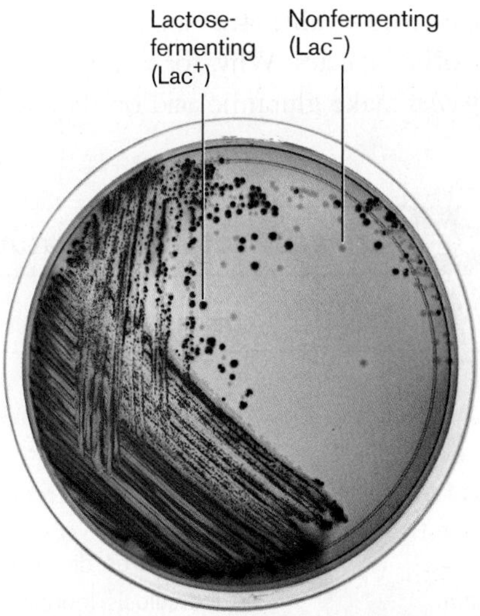

JOHN FOSTER, UNIVERSITY OF SOUTH ALABAMA

FIGURE 4.14 ■ MacConkey medium, a culture medium both selective and differential. Only Gram-negative bacteria grow on lactose MacConkey (selective). Only a species capable of fermenting lactose produces pink colonies (differential), because only fermenters can take up the neutral red and peptones that are also in the medium. Gram-negative nonfermenters appear as uncolored or white colonies.

Several media used in clinical microbiology are both selective and differential. The MacConkey medium described here also contains bile salts and crystal violet to prevent the growth of Gram-positive bacteria and allow the growth of Gram-negative bacteria (**Fig. 4.14**). This medium is often used to identify bacteria that cause diarrheal disease because most normal microbiota that grow on this medium are lactose fermenters, whereas two important Gram-negative pathogens, *Salmonella* and *Shigella*, are lactose nonfermenters.

Thought Question

4.5 If lactose were left out of MacConkey medium (**Fig. 4.14**), would lactose-fermenting *E. coli* bacteria grow, and if so, what color would their colonies be?

Growth Factors and Uncultured Microbes

Why do some bacterial species fail to grow in minimal medium, or fail to grow at all in the laboratory? Such failures are a consequence of evolution and of the organism's natural growth environment. If an ecological niche continually provides a compound that a microbe would otherwise have to make, the relevant biosynthetic pathway for the compound becomes unnecessary. Random mutations can slowly degrade the pathway as long as the organism remains in that environment. However, once the defective species is removed from its natural niche and cultured in a laboratory, the organism will require that compound, or growth factor, to grow.

Growth factors (**Table 4.2**) are specific nutrients not required by other species. Why, for example, should *Streptococcus pyogenes* make glutamic acid or alanine if both are readily available in its normal environment, the human oral cavity? Because *S. pyogenes* never needs to make glutamic acid or alanine in its natural habitat, it lost the genes needed to synthesize these amino acids. For *S. pyogenes* to grow in the laboratory, the culture medium must contain alanine and glutamic acid along with the essential macro- and micronutrients mentioned earlier.

Some species have adapted so specifically to their natural habitats that we do not yet know how to grow them in the laboratory. As mentioned earlier, over 99% of bacterial species that are present in water or soil, or that grow in or on animals, will not form colonies on an agar plate. These organisms are referred to as **uncultured** or **unculturable**. "Uncultured" at present is preferable because future methods may be found to grow such organisms. Some of these uncultured organisms depend on growth factors, such as siderophores, provided by other species that cohabit their natural environment. Some of these factors even appear to act like hormones, somehow stimulating replication of the uncultured organism. For example, **Figure 4.15** shows a marine organism (MSC33) that will not grow in the laboratory unless you include a peptide made by a companion microbe from the same natural environment. The peptide is not a nutrient, but instead has a signaling function that induces cell division. Today, new methods of culturing these uncultured species are being used to discover new antibiotic-producing microbes (**Special Topic 4.1**).

If a microbe cannot be cultured, how do we know it exists? All known microorganisms have a set of genes that encode the RNA molecules present in ribosomes. The ribosomal RNA molecules are highly conserved across the phylogenetic tree. A DNA-amplifying procedure called the polymerase chain reaction (PCR, described in Section 7.6 and 12.1) can screen for the presence of these genes in soil and water samples. Comparing the DNA sequences of the PCR products from environmental samples with the

TABLE 4.2 Growth Factors and Natural Habitats of Organisms Associated with Disease

Organism	Diseases	Natural habitats	Growth factors
Abiotrophia	Osteomyelitis	Humans and other animal species	Vitamin K, cysteine
Bordetella	Whooping cough	Humans and other animal species	Glutamate, proline, cysteine
Francisella	Tularemia	Wild deer, rabbits	Complex, cysteine
Haemophilus	Meningitis, chancroid	Humans and other animal species, upper respiratory tract	Hemin, NAD
Legionella	Legionnaires' disease	Soil, refrigeration cooling towers	Cysteine
Mycobacterium	Tuberculosis, leprosy	Humans	Nicotinic acid (NAD),[a] alanine (*M. leprae* is unculturable)
Shigella	Bloody diarrhea	Humans	Nicotinamide (NAD)[a]
Staphylococcus	Boils, osteomyelitis	Widespread	Complex requirement
Streptococcus pyogenes	Pharyngitis, rheumatic fever	Humans	Glutamate, alanine

[a]Both nicotinamide and nicotinic acid are derived from NAD, nicotinamide adenine dinucleotide.

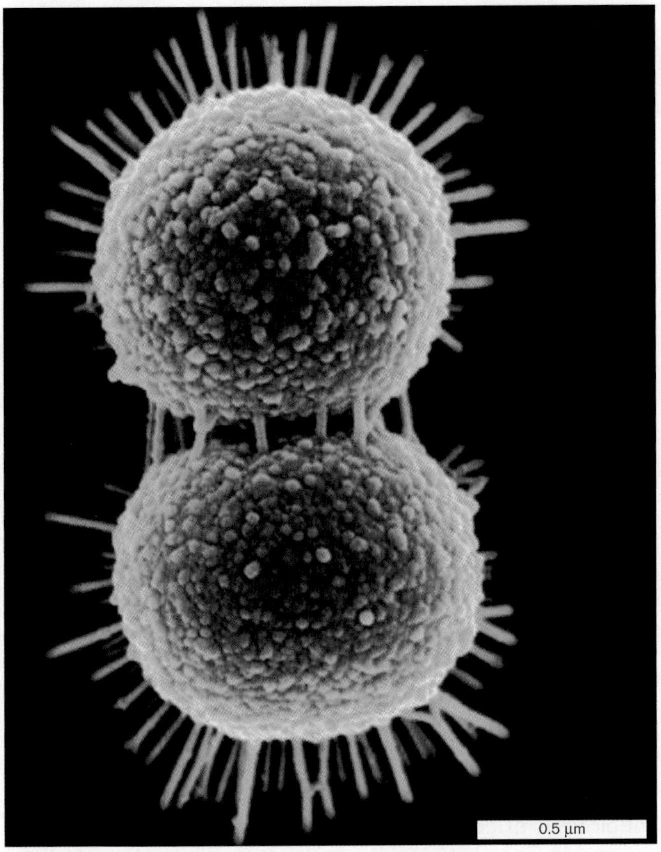

D. NICHOLS ET AL. 2008. *APPL. ENVIRON. MICROBIOL.* **74**:4889–4897. 2008, AMERICAN SOCIETY FOR MICROBIOLOGY.

FIGURE 4.15 ■ "Unculturable" marine organism MSC33. To grow in natural environments, many bacterial species rely on factors produced by other species within their niche. The microbe shown, MSC33, will not grow in laboratory media unless a peptide growth factor from another species is included.

DNA sequences of similar genes from known, culturable organisms reveals that nature harbors many undiscovered microbes. Even though we don't know the growth and nutritional requirements of these phantom microbes, modern genomic techniques can expose their existence and can even provide remarkable insight into their physiologies (see Chapter 8).

Another class of bacteria, known as obligate intracellular bacteria, could also be called unculturable because they, too, will not grow on laboratory media. These species first evolved to penetrate and then grow only within a eukaryotic cell. For example, an unknown ancestor of *Rickettsia prowazekii*, the cause of epidemic typhus fever, adapted to grow within the cytoplasm of eukaryotic cells (**Fig. 4.16**). As it evolved, this obligate intracellular bacterium lost key biochemical pathways needed for independent growth because the host cell supplied them. We still do not know what those factors are, but we can grow *Rickettsia* in animal cell tissue culture or in chicken eggs (the bacteria grow inside the endothelial cells of blood vessels formed in fertilized eggs). Despite extensive efforts to grow this bacterium outside of a host cell (called axenic growth), *R. prowazekii* has proved uncooperative.

A.

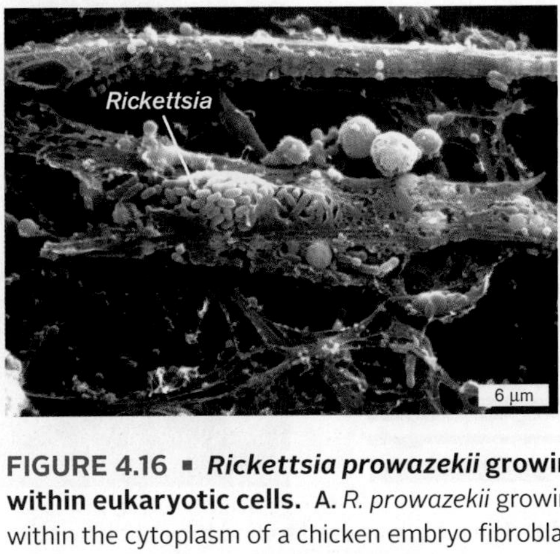

D. J. SILVERMAN, ET AL. 1975. *INFECTION AND IMMUNITY* **11**:1391.

B.

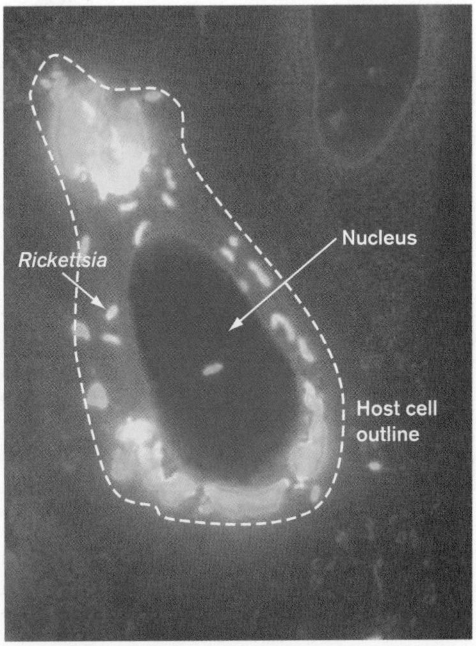

JON AUDIA, U. OF SOUTH ALABAMA COLLEGE OF MEDICINE

FIGURE 4.16 ■ *Rickettsia prowazekii* growing within eukaryotic cells. **A.** *R. prowazekii* growing within the cytoplasm of a chicken embryo fibroblast (SEM). **B.** Fluorescent stain of *Rickettsia prowasekii* (approx. 0.5 µm long) growing within a cultured human cell (outline marked by dotted line). The rickettsias are green (FITC labeled antibody, arrow), the host cell nucleus is blue (Hoechst stain) and the mitochondria are red (Texas Red MitoTracker). The bacterium grows only in the cytoplasm.

Techniques for Counting Bacteria

Growing bacteria in laboratory media is important for studying their physiology, but knowing how many bacteria are present in the medium is critical for interpreting results. Counting bacteria is how we determine whether a lake is contaminated with fecal bacteria and whether our peanut butter is contaminated with *Salmonella*. Counting or quantifying microorganisms is surprisingly difficult because each of the available techniques measures a different physical or biochemical aspect of growth. Thus, a cell density value (given as cells per milliliter) derived from one technique will not necessarily agree with the value obtained by a different method. Here are some commonly used methods for counting microbes.

Direct counting of living and dead cells. Microorganisms can be counted directly using a microscope. A dilution of a bacterial culture is placed on a special microscope slide called a hemocytometer (or, more specifically for bacteria, a Petroff-Hausser counting chamber; **Fig. 4.17**).

SPECIAL TOPIC 4.1 Antibiotic Hunters Culture the "Unculturable"

There is a growing panic among infectious disease doctors, who worry that we will soon run out of effective antibiotics to treat deadly infections. Since the introduction of antibiotics over 70 years ago, bacterial pathogens have continually evolved to resist their effects. Scientists, for their part, have struggled to find new antibiotics to replace the old, ineffective ones. Chapter 27 will discuss this "arms race" in more detail. The classic approach for finding new antibiotics is to screen bacteria and fungi snatched from exotic environments such as the Amazon for secreted antimicrobial compounds. This method worked for decades, but over the past 25 years it has yielded very few new antibiotics. A new way of thinking was needed if we are to maintain an advantage over microbes.

Fewer than 1% of the microbes in the world whose DNA we can detect can be cultured in the laboratory. This means we have missed capturing the antibiotic potential of over 99% of living microbes. These uncultured organisms are present in soil, water, and even the human body. Why do the microbes grow there but not in the lab? They grow in their natural environments because of interspecies cooperation. In this environmental "buddy system," one species provides a growth factor to another species that cannot make its own. Can we find a way to tame such uncooperative organisms and screen them for new antibiotics?

Kim Lewis at Northeastern University in Boston (see **Fig. 2C**) and colleagues devised an elegant method to do just that. To harvest previously uncultured soil microbes, samples prepared from soil were diluted such that one bacterial cell was delivered into each channel of a multichannel iChip (isolation chip; **Fig. 1A**). The channels contained agar plugs in which a microbe could grow. Both sides of the device were covered with semipermeable membranes and placed back into the soil (**Fig. 1B**). Nutrients and growth factors made by organisms in the soil diffused into the chamber and allowed many of the uncultured microbes to form colonies. Once a colony formed, many previously uncultured organisms became "domesticated," able to grow on agar medium without further assistance. The mechanism of domestication is not yet known.

The scientists made extracts from 10,000 iChip isolates and tested each one for antibiotic activity against *Staphylococcus aureus*, an important pathogen that can cause infections ranging from boils to sepsis. One extract was highly effective. The Gram-negative organism that produced this antibiotic was sequenced and named *Eleftheria terrae* (**Fig. 2**). The structure of the antibiotic, teixobactin, is shown in **Figure 2B**. It is a nonribosomal peptide synthesized by enzymes instead of a ribosome (nonribosomal peptide synthesis is described in Section 15.3). The new antibiotic is effective against Gram-positive but not Gram-negative bacteria, because the peptide cannot penetrate outer membranes.

Teixobactin is unusual among antibiotics because attempts to generate antibiotic resistance to it have failed. Most antibiotics bind to bacterial proteins whose sequence can easily change by mutation to become antibiotic resistant. Teixobactin, however, binds to two highly conserved,

A.

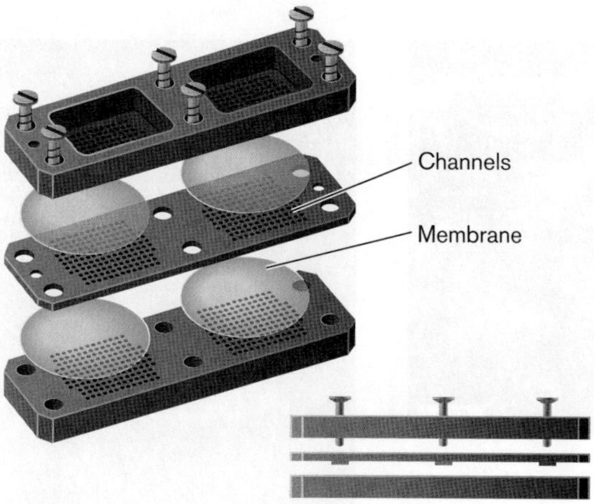

B.

FIGURE 1 ■ The multichannel iChip. The iChip was used to find the antibiotic teixobactin.

nonprotein (lipid carrier) components of the bacterial cell. One lipid carrier (bactoprenol) ferries peptidoglycan precursors (*N*-acetylglucosamine and *N*-acetylmuramic acid) to sites of peptidoglycan synthesis (discussed in Chapters 3 and 27). The other is a cell wall teichoic acid precursor. Altering these components to teixobactin resistance would require the bacterium to make variants of enzymes in two different pathways. Thus, the likelihood that pathogens can develop resistance is low. This feature makes this drug highly attractive as a potential therapeutic agent. Whether or not teixobactin proves useful in treating human infections, the new in situ culturing strategies used to discover this drug have reenergized the field of antibiotic discovery and should help scientists solve the mysteries of the uncultured.

RESEARCH QUESTION

Propose a mechanism by which uncultured microbes can become domesticated. Why would "domestication" not happen in the natural environment?

Ling, Losee L., Tanja Schneider, Aaron J. Peoples, Amy L. Spoering, Ina Engels, et al. 2015. A new antibiotic kills pathogens without detectable resistance. *Nature* **517**:455–459.

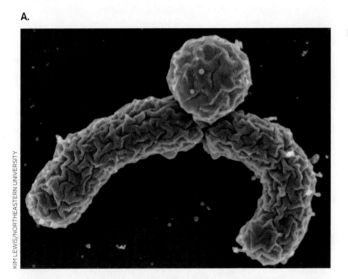

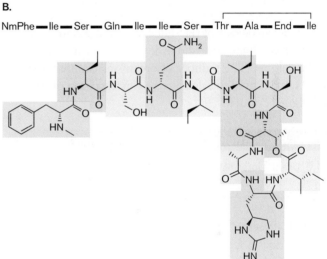

FIGURE 2 ■ ***Eleftheria terrae* and teixobactin.** A previously uncultured Gram-negative bacterium, *Eleftheria terrae* (**A**), makes teixobactin (**B**), a new antibiotic. Color highlights mark each residue. End = enduracididine, a nonprotein amino acid; modified from arginine. **C.** Kim Lewis (right), with postdoctoral researcher Brian Conlon.

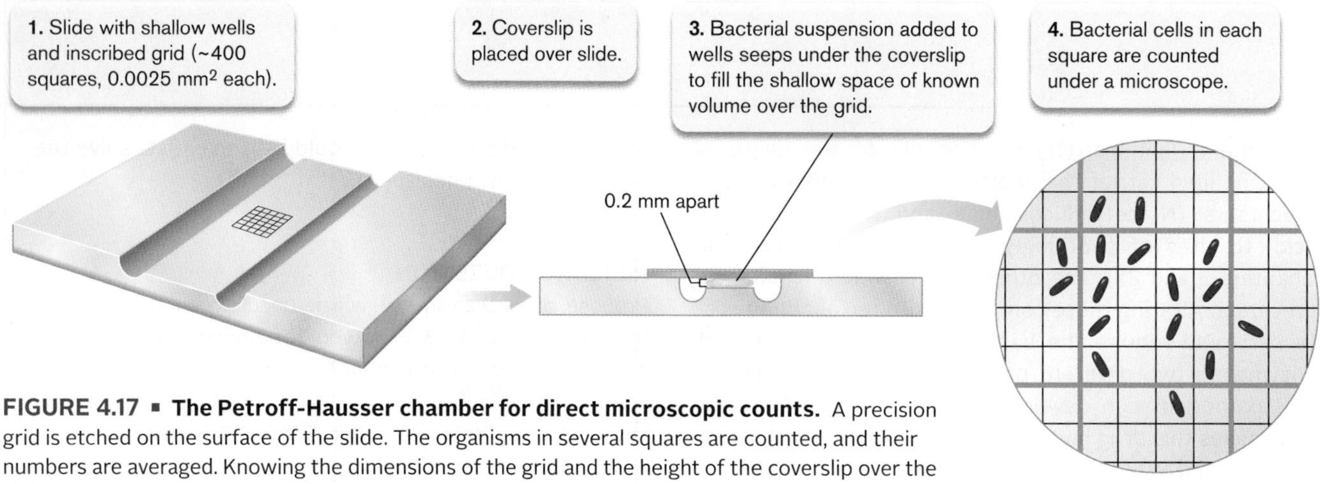

FIGURE 4.17 ■ The Petroff-Hausser chamber for direct microscopic counts. A precision grid is etched on the surface of the slide. The organisms in several squares are counted, and their numbers are averaged. Knowing the dimensions of the grid and the height of the coverslip over the slide makes it possible to calculate the number of organisms in a milliliter.

Etched on the surface of the slide is a grid of precise dimensions, and placing a coverslip over the grid forms a space of precise volume. The number of organisms counted within that volume is used to calculate the concentration of cells in the original culture.

Thought Question

4.6 Use the information in **Figure 4.17** to determine the concentration (in cells per milliliter) of bacteria shown.

Simply "seeing" an organism under the microscope, however, does not mean that the organism is alive, because living and dead cells are indistinguishable by this basic approach. Living cells may be distinguished from dead cells by fluorescence microscopy using fluorescent chemical dyes, as discussed in Chapter 2. For example, propidium iodide, a red dye, intercalates between DNA bases but cannot freely penetrate the energized membranes of living cells. Thus, only dead cells stain red under a fluorescence scope. Another dye, Syto-9, enters both living and dead cells, staining them both green. By combining Syto-9 with propidium iodide, living and dead cells can be distinguished: Living cells stain green, whereas dead cells appear orange (**Fig. 4.18**).

Direct counting without microscopy can be achieved using an electronic technique, called **flow cytometry**, that counts individuals within populations of bacterial cells and distinguishes classes of cells with different properties. The instrument is called a flow cytometer. In flow cytometry, bacterial cells that synthesize a fluorescent protein (such as cyan fluorescent protein; see Section 2.4), or that have been labeled with a fluorescent antibody or chemical, are passed single file through a small orifice and then through a laser beam (**Fig. 4.19A**). Detectors measure light scatter in the forward direction (an indicator of particle size) and to the side (which indicates shape or granularity). In addition, the laser activates the fluorophore in the fluorescent antibody, and a detector measures fluorescence intensity.

Within a single culture, one subpopulation of cells may fluoresce more than or less than another because of the presence or absence of a targeted protein. Thus, flow cytometry enables us to use cell size and the level of fluorescence to identify and count different populations of cells. For example, flow cytometry can determine the

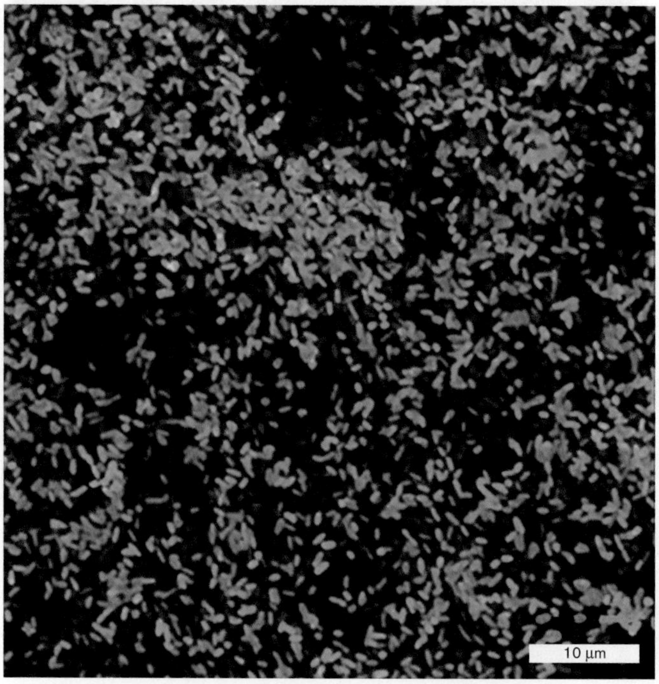

SANS-SERRA MITJANA, ET AL. 2017. 12:4409.

FIGURE 4.18 ■ Live/dead stain. LIVE/DEAD *Bac*Light Bacterial Viability Kit. Dead bacterial cells fluoresce orange because propidium iodide (red) can enter the cells and intercalate the base pairs of DNA. Live cells fluoresce green because Syto-9 (green) enters the cell.

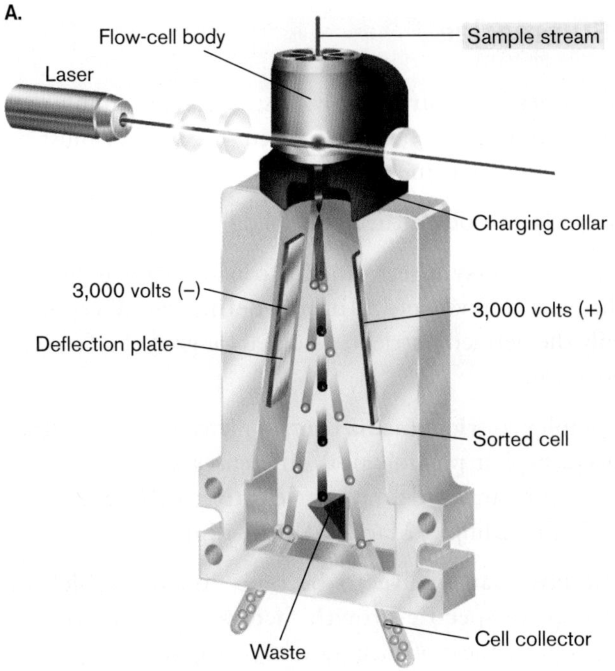

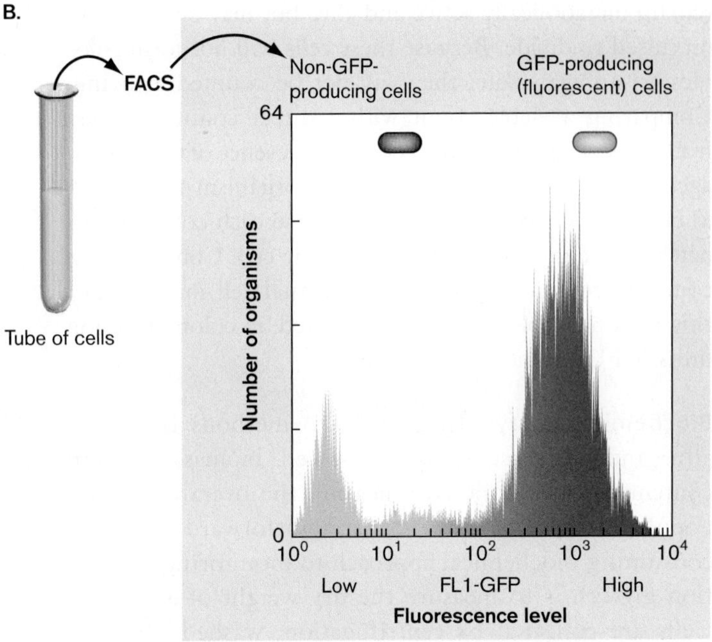

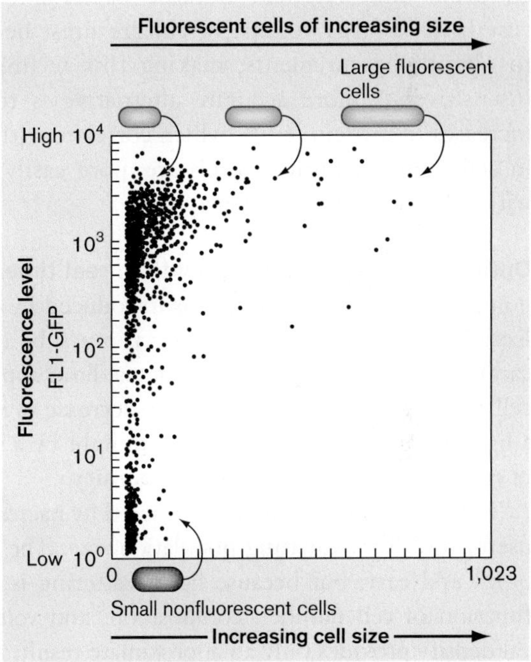

FIGURE 4.19 ■ Fluorescence-activated cell sorting. **A.** Schematic of a fluorescence-activated cell sorter (FACS) conducting bidirectional sorting. **B.** Counting and separation of GFP-producing *E. coli* and non-GFP-producing *E. coli*. The low-level fluorescence in the cells on the left is baseline fluorescence (autofluorescence). The scatterplot displays the same FACS data, showing the size distribution of cells (*x*-axis) with respect to the level of fluorescence (*y*-axis). The larger cells may be cells that are about to divide.

expression of a gene in different subpopulations of cells in culture. This can be done by placing the green fluorescent protein gene (*gfp*) under the control of a specific bacterial gene (making a gene fusion). Researchers can then use flow cytometry to count cells expressing the gene, and even sort high-expressing cells from low expressing cells (**fluorescence-activated cell sorting**, or **FACS**). The sorted cells can then be used for additional applications, such as selection of recombinant clones for molecular biology. Flow cytometry and FACS analysis make it possible to determine which environmental conditions trigger expression of the gene and whether all cells in the population express that gene at the same time and to the same extent (**Fig. 4.19B**).

Viable counts. Viable cells, as noted previously, are those that can replicate and form colonies on a plate. To obtain a viable cell count, dilutions of a liquid culture can be plated directly on an agar surface (as in **Fig. 4.13**) or added to liquid agar cooled to 42°C–45°C. The agar is subsequently poured into an empty Petri plate (this is called the **pour plate technique**), where the agar cools further and solidifies. Because many bacteria resist short exposures to that temperature, individual cells retain the ability to form colonies on, and in, the pour plate. After colonies form, they are counted, and the original cell number is calculated. For example, if 100 colonies are observed in a pour plate made with 100 microliters (μl) of a 10^{-3} dilution of a culture, then there were 10^6 organisms per milliliter in the original culture. **Figure 4.13** illustrates these types of calculations.

Although viable counts are widely used in research, this method is problematic for measuring cell number. Colony counts usually underestimate the number of <u>living</u> cells in a culture. Cells damaged for one reason or another can remain metabolically active and alive but may be too compromised to divide. Because these cells will not form colonies on an agar plate, they will not be counted as living. Comparing a viable count with a direct count obtained from a live/dead stain can expose the presence of these damaged cells. Another challenge is any organism that grows in chains, such as *Streptococcus*, because each colony originates from a group of cells, not a single cell. Consequently, counting colonies underestimates actual cell number. For this reason, viable counts are reported as colony-forming units (CFUs) rather than as cells.

Biochemical assays. In contrast to methods that visualize individual cells, assays of cell biomass, protein content, or metabolic rate measure the overall size of a population of cells. The most straightforward but time-consuming biochemical approach to monitoring population growth is to measure the dry weight of a culture. Cells are collected by centrifugation, washed, dried in an oven, and weighed. Because bacterial cells weigh very little, a large volume of culture must be harvested to obtain measurements, making this technique quite insensitive. A more accurate alternative is to measure increases in protein levels, which correlate with increases in cell number. Protein levels are more easily measured with relatively sensitive assays.

Optical density measures growth in real time. The phenomenon of light scattering was introduced in Chapter 2. Recall that the presence of bacteria in a tube of medium can be detected by how cloudy the medium appears as the cells scatter light (see Fig. 2.3). The decrease in intensity of a light beam due to the scattering of light by a suspension of particles is measured as **optical density**.

The optical density of light scattered by bacteria is a very useful tool for estimating population size. The method is quick and easy, but because light scattering is a complex function of cell number, composition, and volume, optical density provides only an approximate result. A standard curve that compares viable counts to optical density is typically made to improve accuracy.

Optical density as a measure of cell numbers has limitations. These include changes in the light-scattering properties of a cell as it grows, and the fact that dead cells also scatter light.

Clearly, using optical density to estimate viable count can be misleading, especially when measuring populations in stationary phase. A molecular method that can quantify cell number based on DNA content—the polymerase chain reaction (described in Section 12.1)—also suffers from limitations. Dead cells contain DNA, and fast-growing cells can contain more than one copy of their chromosome—both of which can lead to overestimations in cell number.

To Summarize

- **Microbes in nature** usually exist in complex, multispecies communities, but for detailed studies they must be grown separately in pure culture.
- **Bacteria can be cultured** on solid or in liquid media.
- **Defined or synthetic media** contain known lists of chemical components. **Minimal defined media** contain only the defined nutrients essential for growth of a given organism.
- **Complex**, or **rich**, **media** contain many nutrients. Other media exploit physiological differences between organisms and can be defined as **selective**, **differential**, or both (for example, **MacConkey medium**).
- **Microbes can evolve to become "unculturable" or to require specific growth factors** depending on the nutrient richness of their natural ecological niche.
- **Obligate intracellular bacteria** lose metabolic pathways provided by their hosts and develop requirements for growth factors supplied by their hosts.
- **Microorganisms in culture may be counted directly** under a microscope (with or without staining) or by use of fluorescence-activated cell sorting (FACS).
- **Microorganisms can be counted indirectly** by viable counts or by the measurement of dry weight, protein levels, or optical density.
- **A viable bacterial organism** is defined as being capable of replicating.

4.4 The Growth Cycle

How do microbes grow? What determines their rate of growth? And when can growth resume in a nongrowing population? It is difficult to answer these questions in natural environments where most microbes exist in complex, mixed communities affixed to solid surfaces. Yet these same multicellular communities can also send off **planktonic cells**, free-living individual organisms that grow and multiply on their own. Growth of these planktonic cells is easier to measure.

All species at one time or another exhibit rapid growth, nongrowth, and many phases in between. For clarity, we present here the principles of rapid growth, while bearing in mind the diversity of growth situations in nature. We care about growth for many reasons. For instance, how rapidly a microbe grows will influence how fast a pathogen causes disease, how quickly contaminated food spoils, and how fast an oil-consuming bacterial species will remediate an oil spill.

Survival of any species ultimately depends on its ability to generate offspring. A typical bacterium grows by increasing in length and mass, which facilitates expansion of its nucleoid as its DNA replicates (see Section 3.4). As DNA replication nears completion, the cell, in response to complex genetic signals, begins to synthesize a midcell septum that separates the two daughter cells. In this overall process, called **binary fission**, one parental cell splits into two equal daughter cells (**Fig. 4.20A**).

Although a majority of culturable bacteria divide symmetrically into two equal halves, some species divide asymmetrically. For example, the bacterium *Caulobacter* forms a stalked cell that remains fixed to a solid surface but reproduces by budding from one end to produce small, unstalked motile cells (see Fig. 3.30). The marine organism *Hyphomicrobium* also replicates asymmetrically by budding, releasing a smaller cell from a stalked parent (**Fig. 4.20B**).

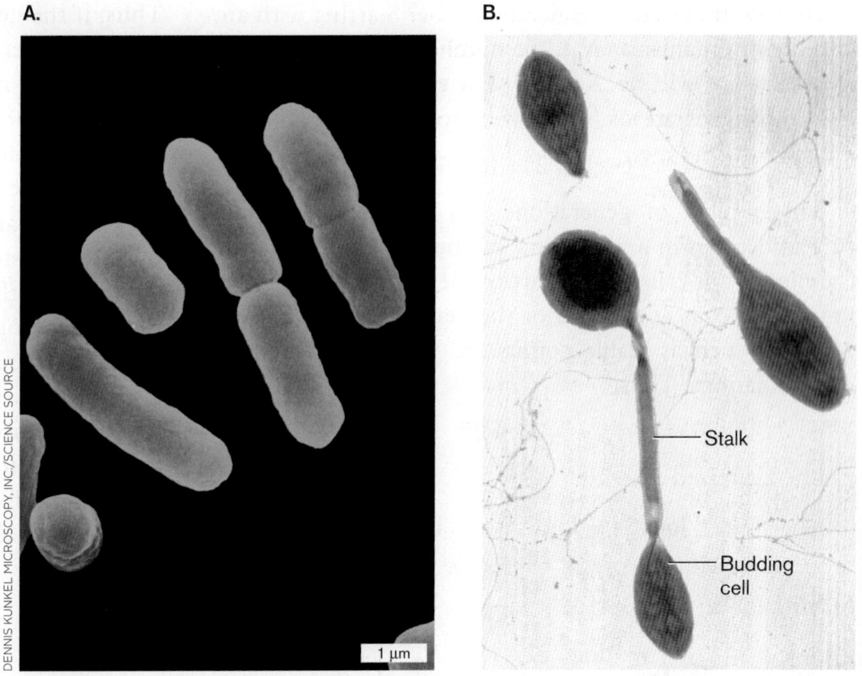

FIGURE 4.20 ■ Symmetrical and asymmetrical cell division. **A.** Symmetrical cell division, or binary fission, in *Lactobacillus* sp. (SEM). **B.** Asymmetrical cell division via budding in the marine bacterium *Hyphomicrobium* (approx. 4 µm long).

Eukaryotic microbes divide by a special form of cell fission involving mitosis, the segregation of pairs of chromosomes within the nucleus (see Section 20.2 and eAppendix 2). Some eukaryotes also undergo more complex life cycles involving budding and diverse morphological forms.

Thought Question

4.7 A virus such as influenza virus might produce 800 progeny virus particles from one host cell infected by one virus. How would you mathematically represent the exponential growth of the virus? What practical factors might limit such growth?

Exponential Growth

If we assume that microbial growth is unbounded, what happens to the population? The unlimited growth of any population obeys a simple law: The **growth rate**, or rate of increase in cell numbers or biomass, is proportional to the population size at a given time. Such a growth rate is called "exponential" because it generates an exponential curve, a curve whose slope increases continually.

How does binary fission of cells generate an exponential curve? If each cell produces two cells per generation, then the population size at any given time is proportional to 2^n, where the exponent n represents the number of generations (that is, cell divisions in which offspring replace parents) that have taken place between two time points. Thus, cell number rises exponentially. Many microbes, however, have replication cycles based on numbers other than 2. For example, some cyanobacteria form cell aggregates that divide by multiple fissions, releasing dozens of daughter cells. The cyanobacterium enlarges without dividing, and then suddenly divides many times without separating. The cell mass breaks open to release hundreds of progeny cells.

Generation Time

In an environment with unlimited resources, bacteria divide at a constant interval called the **generation time** (but not all cells in the population divide at the same instant). The generation time varies with respect to many parameters, including the bacterial species, type of medium, temperature, and pH. The generation time for cells in culture is also known as the **doubling time**, because the population of cells doubles over one generation. For example, one cell of *E. coli* placed into a complex medium will divide every 20 minutes. After 1 hour of growth (three generations), that one cell will have become eight (1 to 2, 2 to 4, 4 to 8). Because cell number (N) doubles with each division, the increase in cell number over time is exponential, not linear. A linear increase would occur if cell number rose by a fixed amount after every generation (for example, 1 to 2, 2 to 3, 3 to 4).

Why do we care about generation time? As noted already, generation time is important when we're trying to understand how rapidly a pathogen can cause disease symptoms. In biotechnology, how fast a producing organism grows will affect how quickly a commercially useful by-product can be made.

How do we calculate generation time? Starting with any number of organisms (N_0), the number of organisms after n generations will be $N_0 \times 2^n$. For example, a single cell after three generations ($n = 3$) will produce

$$1 \text{ cell} \times 2^3 = 8 \text{ cells}$$

The number of generations that an exponential culture undergoes in a given time period can be calculated if the number of cells at the start of the period (N_0) and the number of cells at the end of the period (N_t) are known. Methods such as viable counts are used to make those determinations. Thus,

$$N_t = N_0 \times 2^n$$

can be expressed as

$$\log_2 N_t = \log_2 N_0 + n \log_2 2$$
$$= n + \log_2 N_0$$

Solving for n:

$$n = \log_2 N_t - \log_2 N_0$$
$$= \log_2 (N_t/N_0)$$

Once the number of generations (n) over a given time is known, generation time (g) is calculated as

$$g = t/n$$

For example, if 2 hours (120 minutes) yielded 6 generations, then g = 20 minutes.

The rate of exponential growth is expressed as the mean **growth rate constant** (k), which is the number of generations (n) per unit time (usually generations per hour). This is written as

$$k = n/t, \text{ where } t \text{ is 1 hour; or } k = 1/g$$

Thus, if the generation time is 20 minutes (0.33 hour) for a given bacterial species in a given medium, the growth rate constant k = 1/0.33 hour = 3 generations per hour. If the generation time is 2 hours, the growth rate constant is 0.5 generation per hour.

In practice, exponential growth lasts for only a short period when all nutrients are in full supply and the concentration of waste products has not become a limiting factor.

The growth rate constant can also be calculated from the slope of $\log_2 N$ over time, where N is a relative measure of culture density, such as the optical density measured in a spectrophotometer (**Fig. 4.21A**). The units of N do not matter, because we are always looking at ratios of cell numbers relative to an earlier level (N_1/N_0). For example, we can use the following series of optical density (OD) measurements to determine N:

Time (min)	OD_{600}	$\log_2 OD_{600}$
0	0.05	−4.32
15	0.08	−3.65
30	0.13	−2.94
45	0.20	−2.32
60	0.33	−1.59

If we plot $\log_2 OD_{600}$ versus time, we obtain a line with a slope of 0.0452 per minute. The growth rate constant, in doublings per hour, becomes

$$k = (0.0452/\text{min})(60 \text{ min/h})$$
$$= 2.7 \text{ generations per hour}$$

The steeper the slope, the faster the organisms are dividing.

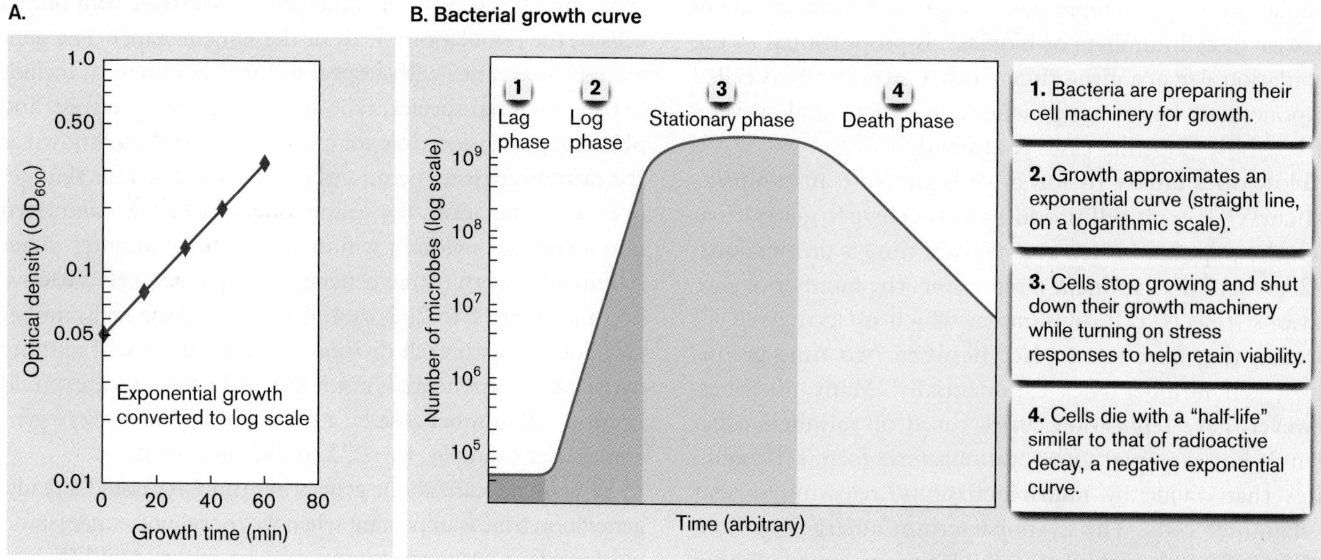

FIGURE 4.21 ▪ Bacterial growth curves. **A.** Theoretical growth curve of a bacterial suspension measured by optical density (OD) at a wavelength of 600 nm. **B.** Phases of bacterial growth in a typical batch culture.

Note: An alternative formula uses logarithms to base 10, requiring conversion to base 2:

$$\log_{10} N_t = \log_{10} N_0 + n \log_{10} 2$$
$$= \log_{10} N_0 + n0.301$$
$$n = 3.3 \log_{10} (N_t/N_0)$$

Thought Question

4.8 Suppose one cell of the nitrogen fixer *Sinorhizobium meliloti* colonizes a plant root. After 5 days (120 hours), there are 10,000 bacteria fixing N_2 within the plant cells. What is the bacterial doubling time?

The mathematics of exponential growth is relatively straightforward, but remember that microbes grow differently in pure culture (very rare in nature) than they do in mixed communities, where neighboring cells produce all kinds of substances that may feed or poison other microbes. In mixed communities, the microbes may grow planktonically (floating in liquid), as in the open ocean, or as a biofilm on solid matter suspended in that ocean. In each instance, the mathematics of exponential growth applies, at least until the community reaches a density at which different species begin to compete or nutrients become scarce.

Thought Question

4.9 It takes 40 minutes for a typical *E. coli* cell to completely replicate its chromosome and about 20 minutes to prepare for another round of replication. Yet the organism enjoys a 20-minute generation time growing at 37°C in complex medium. How is this possible? *Hint:* How might the cell overlap the two processes?

Stages of Growth in Batch Culture

Exponential growth never lasts indefinitely when cells are grown in a closed system such as a flask. Nutrient consumption and toxic by-products eventually slow the growth rate until it halts altogether. This is called **batch culture**. In batch culture, no fresh medium is added during incubation; thus, nutrient concentrations decline and waste products accumulate during growth.

The progressively deteriorating conditions of a batch culture profoundly affect bacterial physiology and growth. These changes illustrate the remarkable ability of bacteria to adapt to their environment. As medium conditions worsen, alterations occur in membrane composition, cell size, and metabolic pathways, all of which impact generation time. Microbes possess intricate, self-preserving genetic and metabolic mechanisms that slow growth before their cells lose viability. Plotting culture growth (as represented by the logarithm of the cell number) versus incubation time makes it possible to see the effect of changing conditions on generation time and reveals the stages of growth shown in **Figure 4.21B**—namely, lag phase, log (or exponential) phase, stationary phase, and death phase.

Lag phase. Cells transferred from an old culture to fresh growth media typically experience a lag period, or **lag phase**, during which they do not divide. Several factors influence lag phases. Cells taken from an aged culture may be damaged and require time for repair. Carbon, nitrogen, or energy sources different from those originally used by the seed culture must be sensed, and the appropriate enzyme systems must be synthesized. The length of the lag phase also varies with changes in temperature, pH, and salt concentration, as well as nutrient richness. For example, transferring cells from one complex medium to a fresh complex medium results in a very short lag phase, whereas cells grown in a complex medium and then plunged into a minimal defined medium experience a protracted lag phase. During lag phase in the latter case, cells must synthesize all the amino acids, nucleotides, and other metabolites previously supplied by the complex medium.

Early log, or exponential, phase. Once cells have retooled their physiology to accommodate the new environment, they begin to grow exponentially and enter what is called **exponential**, or **logarithmic (log), phase**. Exponential growth is balanced growth, in which all cell components are synthesized at constant rates relative to each other—the assumption behind the generation time calculation given in the previous section. At this stage, represented by the linear part of the growth curve, cells are growing and dividing at the maximum rate possible in the medium and growth conditions provided (such as temperature, pH, and osmolarity). Cells are largest at this stage of growth. If cell division were synchronized and all cells divided at the same time, the growth curve during this period would appear as a series of steps with cell numbers doubling instantly after every generation time. But batch cultures are not synchronous. Every cell has an equal generation time, but each cell divides at a slightly different moment, making the cell number rise smoothly.

Cells enjoying balanced, exponential growth are temporarily thrown into metabolic chaos (unbalanced growth) when their medium is abruptly changed. Nutritional downshift (moving cells from a good carbon source such as glucose to a poorer carbon source such as succinate) or nutritional upshift (moving cells to a better carbon source) casts cells into unbalanced growth. Downshifting to a carbon source with a lower energy yield means not only that a different set of enzymes must be made and employed to use the carbon source, but also that the previous high rate of macromolecular synthesis (such as ribosome synthesis) used to support a fast generation time is now too rapid relative to the lower energy yield. Failure to adjust will lead to increased mistakes in RNA, protein, and DNA synthesis, depletion of key energy stores, and ultimately death.

Note that the nice, smooth exponential growth phase shown in **Figure 4.21B** does not always hold for microbes growing in natural environments or in complex laboratory media containing multiple carbon and energy sources. Some bacterial species produce odd-looking growth curves in complex media as the population depletes one carbon source and must switch physiology to use another. Even *E. coli* doesn't really experience a uniform log-phase metabolism growing in complex medium; instead, it smoothly transitions through a series of metabolic states.

Late log phase. As cell density (number of cells per milliliter) rises during log phase, the rate of doubling eventually slows, and a new set of growth phase–dependent genes is expressed. At this point, some species can also begin to detect the presence of others by sending and receiving chemical signals in a process known as quorum sensing (discussed in Chapter 10).

Stationary phase. Eventually, cell numbers stop rising, owing to the lack of a key nutrient or the buildup of waste products. At this point the growth curve levels off and the culture is said to be in **stationary phase** (**Fig. 4.21B**). It was thought that in stationary phase, the rate of cell division equaled the rate of cell death. That is, cells were either dead or they were alive. That model has recently been challenged by Nathalie Balaban and colleagues from the Hebrew University of Jerusalem. They elegantly show that the vast majority of cells thought to have died in stationary phase are actually alive but growth-arrested—and can still make protein. To prove growth arrest, *E. coli* cells that expressed a red fluorescent protein were grown to stationary phase for 15 hours and then trapped in a microfluidic channel (**Fig. 4.22A**). A chemical inducer molecule was added to observe the percentage of cells that could still synthesize a green fluorescent protein encoded by a second gene. Nine hours later, none of the cells had divided but at least 90% of cells had produced the new protein, meaning that these cells are alive and growth-arrested. What controls a cell's entry into growth arrest is unclear but may involve the molecule guanosine tetraphosphate, which is associated with the stringent response (discussed later and in Chapter 10).

If they did not change their physiology, microbes entering stationary phase would be very susceptible to damage from oxygen radicals and the toxic by-products of metabolism. As an avoidance strategy, some bacteria differentiate into very resistant spores in response to nutrient depletion (see Section 4.6), while other bacteria undergo less dramatic but very effective molecular reprogramming. The microbial model organism *E. coli*, for example, adjusts to stationary phase by decreasing its size, thus minimizing the volume of its cytoplasm compared to the volume of its nucleoid. Fewer nutrients are then required to sustain the smaller cell. New stress resistance enzymes are also synthesized to handle oxygen radicals, protect DNA and proteins, and increase cell wall strength through increased peptidoglycan cross-linking. As a result, *E. coli* cells in stationary phase become more resistant to heat, osmotic pressure, pH changes, and other stresses that they might encounter while waiting for a new supply of nutrients.

A.

B.

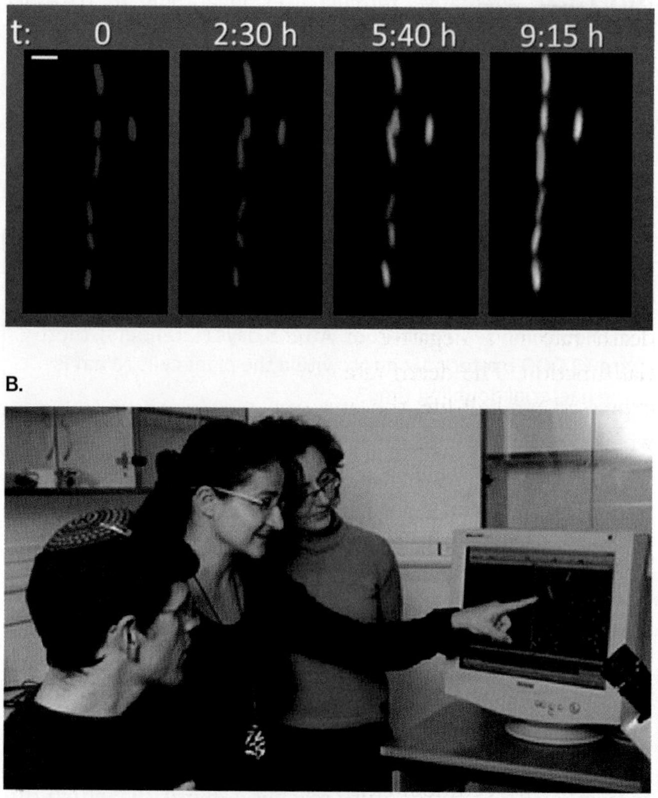

O. GEFEN ET AL. 2014. *PNAS* **111**:556–561.

COURTESY OF HEBREW UNIVERSITY

FIGURE 4.22 ■ Observing dormant *E. coli* making protein in stationary phase. **A.** Stationary-phase *E. coli* cells expressing a red fluorescent protein were trapped in a microfluidic channel (t = 0) and induced to make a new, green fluorescent protein. **B.** Nathalie Balaban (center) with student Eitam Rotem (left) and lab manager Irene Ronin (right), viewing fluorescent bacteria growing and dying.

Thought Question

4.10 The bacterium *Acidithiobacillus thiooxidans* is an extremophile that grows using sulfur as an energy source.
(a) Draw the approximate growth curves that you would expect to see, extending from log phase to stationary phase, if four cultures with different starting numbers of bacteria were grown in the same concentration of sulfur. Use **Figure 4.21B** as the model, and 4×10^5, 4×10^6, 4×10^7, and 4×10^8 as the starting cell densities. Maximum growth yield is 10^9 cells per milliliter. (b) Draw a second graph showing how the curves would change if the initial population density were constant but the concentration of sulfur varied.

Death phase. Without reprieve in the form of new nutrients, cells in stationary phase will eventually succumb to toxic chemicals present in the environment. Cells begin to die, and the culture enters the death phase. Like the growth rate, the **death rate**—the rate at which cells die—is logarithmic. In **death phase**, the number of cells that die in a given time period is proportional to the number that existed at the beginning of the time period. Thus, the death rate is a negative exponential function. The death rate can be expressed as a half-life, the time over which a population declines by half.

Determining microbial death rates is critical to the study of food preservation and to the development of antibiotics (further discussed in Chapters 5, 16, and 27). Although death curves are basically logarithmic, exact death rates are difficult to define because mutations arise that promote survival, and some cells grow by cannibalizing others. Consequently, the death phase is extremely prolonged. In fact, a portion of the cells will often survive for months in dormant or growth-arrested states, discussed next.

A

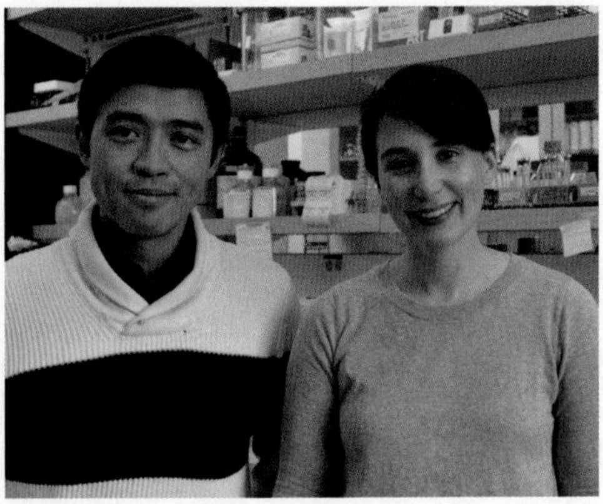

TROY MORRIS

B.

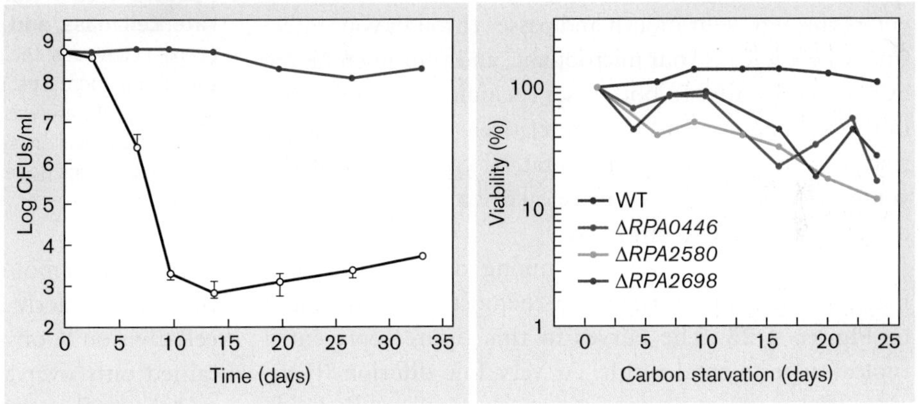

FIGURE 4.23 ■ Identification of longevity genes in growth-arrested *Rhodopseudomonas palustris*. A. Kieran Pechter (right) and Liang Yin studied growth arrest in the phototroph *Rhodopseudomonas*. **B.** They determined longevity by initiating growth of *R. palustris* at low acetate concentrations without oxygen (day 0) and then plating for viable count at weekly intervals on rich agar medium. **Left:** Cultures were incubated with light (red line) or in the dark (black line). **Right:** Wild-type (WT) and selected deletion mutants (Δ*RPA*) with decreased longevity were tested as in the left panel (in light only). *Source:* Part B modified from Pechter et al. 2017. *mBio* **8**:e01726-17.

Longevity genes, dormancy, and persistence. As described earlier, many non-spore-forming bacterial species can enter into dormant or growth-arrested states during nutrient or energy limitation. Persistence, a special form of dormancy that develops in the presence of antibiotics and other stresses, will be discussed in Chapter 27. Growth-arrested bacteria can survive for months or even years. How is such longevity possible? Two students, Kieran Pechter and Liang Yin (**Fig. 4.23A**), working in Caroline Harwood's laboratory at the University of Washington asked this question about *Rhodopseudomonas palustris*, a phototrophic Gram-negative bacterium found in a variety of aquatic environments, including marine coastal sediments. *R. palustris* can enter a growth-arrested state for months during carbon or nitrogen restriction, but only in the presence of light (**Fig. 4.23B**).

To remain viable, growth-arrested cells of any species must somehow maintain an electrochemical membrane potential to provide energy. *R. palustris* uses photosynthesis to maintain membrane potential during nutrient restriction. The ability to separate energy production from carbon use in this organism enabled the investigators to ask which genes, other than those needed for photosynthesis, were required for long-term growth arrest and recovery. Two approaches were used. Random insertions into the *R. palustris* genome were screened to determine which genes were needed to survive dormancy. In addition, RNA sequencing was used to identify which RNA molecules were made during growth arrest (**Fig. 4.23B**). These approaches revealed 117 longevity genes. The identities of the genes suggest that growth-arrested cells must remain metabolically active, carry out protein synthesis, and be proficient at DNA repair. Capturing light energy to maintain a proton motive force coupled with longevity gene products provides *R. palustris* with a distinct survival advantage in nutrient-depleted bodies of water such as the ocean.

Thought Questions

4.11 What can happen to the growth curve when a culture medium contains two carbon sources, if one is a preferred carbon source of growth-limiting concentration and the second is a nonpreferred source?

4.12 How would you modify the equations describing microbial growth rate to describe the rate of death?

4.13 Why are cells in log phase larger than cells in stationary phase?

Continuous Culture

In the classic growth curve that develops in closed systems, the exponential phase spans only a few generations. In open systems, however, where fresh medium is continuously added to a culture and an equal amount of culture is constantly siphoned off, bacterial populations can be maintained in exponential phase at a constant cell mass for extended periods of time. In this type of growth pattern, known as **continuous culture**, all cells in a population achieve a steady state, which permits a detailed analysis of microbial physiology at different growth rates.

The **chemostat** is a continuous culture system in which the diluting medium contains a limiting amount of one essential nutrient (**Fig. 4.24**). You could think of your gastrointestinal tract as a kind of crude chemostat. Nutrient enters through your mouth and passes through your intestine, where it feeds your microbiome, and your microbiome exits in fecal waste. In both cases the numbers of microbes in the chamber (or gut) remain relatively constant. The GI tract is different from a chemostat, of course, in that the volumes of nutrient intake and fecal exit are not continuous or equal, and water is absorbed.

The complex relationships among dilution rate, cell mass, and generation time in a chemostat are illustrated in **Figure 4.25**. The curves in this figure represent a typical experimental result. At very low dilution (flow) rates, the nutrient is so limiting that cells will divide very slowly and cell mass will remain low. Any increase in flow rate, however, will increase the availability of the limiting nutrient such that cells grow faster and cell mass increases. The rate of cell division and the cell mass are kept constant when the flow rate is kept constant,

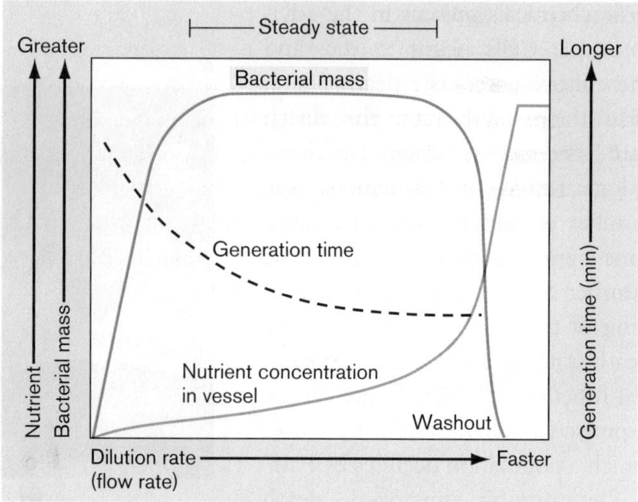

FIGURE 4.25 ■ **Relationships among chemostat dilution rate, cell mass, and generation time.** As the dilution rate (*x*-axis) increases, the generation time decreases and the mass of the culture increases. When the rate of dilution exceeds the division rate, cells are washed from the vessel faster than they can be replaced by division, and the cell mass (bacterial mass) decreases. The *y*-axis varies depending on the curve, as labeled.

because the amount of culture (and cells) removed from the vessel exactly compensates for the increased rate of cell division. Constant cell mass, or density, can be maintained only over a certain range of flow rates. At faster and faster flow rates, cells are eventually removed more quickly than they can be replenished by division, so cell density (cell mass) decreases in the vessel—a phenomenon called "washout."

Continuous culture is used to study large numbers of cells at a constant growth rate and cell mass for both research

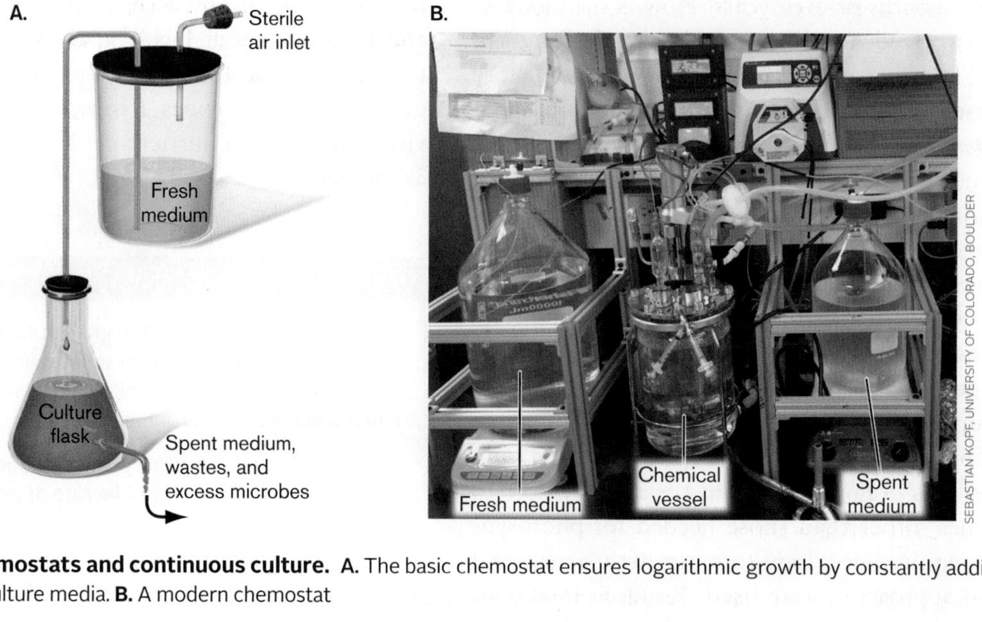

FIGURE 4.24 ■ **Chemostats and continuous culture.** **A.** The basic chemostat ensures logarithmic growth by constantly adding and removing equal amounts of culture media. **B.** A modern chemostat

and industrial applications. Its advantage over batch culture is that the physiology of cells in continuous culture is homogeneous. Consequently, continuous cultures are used in industry to optimize the production of antibiotics, beer, and other microbial products. In research, continuous culture is used to examine what happens to metabolic flux (essentially the rate at which molecules move through metabolic pathways) before and after a biochemical step is altered, and to conduct long-term studies of bacterial evolution.

To Summarize

- **Generation time** is the length of time it takes for a population of cells to double in number.
- **The generation time for a single species will vary** as culture conditions change. **During exponential growth**, generation time remains constant.
- **The growth cycle** of organisms grown in liquid batch culture consists of lag phase, log (exponential) phase, stationary phase, and death phase.
- **The physiology of a bacterial population** changes with growth phase.
- **Continuous culture** can be used to sustain a population of bacteria at a specified growth rate and cell density.

4.5 Biofilms

Can bacteria collaborate? Bacteria are typically thought of as unicellular, but many, if not most, bacteria in nature form specialized, surface-attached, collaborative communities called **biofilms**. Indeed, within aquatic environments bacteria are found mainly associated with surfaces—a fact that underscores the importance of biofilms in nature. Soil biofilms help filter groundwater through the soil, where the bacteria break down organic waste. This process is especially important in wetlands, which provide an important "ecosystem service" for human communities (discussed in Chapter 22).

Other biofilms play critical roles in microbial pathogenesis and environmental degradation, which costs billions of dollars each year in equipment damage, product contamination, and medical infections. For example, pseudomonad or staphylococcal biofilms can damage ventilators used to assist respiration. Biofilms can also form inside indwelling catheter tubes that deliver fluids and medications to patients. In both of these examples, the biofilms serve as direct sources of infection, so developing ingenious ways to prevent biofilm formation on medical instrumentation is a major goal of biomedical research. Making catheter surfaces mimic the architecture of sharkskin is one intriguing approach (**eTopic 4.3**).

The Biofilm Life Cycle

Biofilms can be constructed by a single species or by multiple, collaborating species, as in the mixture of facultative and anaerobic bacteria that form dental plaque. Biofilms can form on a range of organic and inorganic surfaces (**Fig. 4.26** ▶). The Gram-negative bacterium *Pseudomonas aeruginosa*, for example, can form a single-species biofilm on the lungs of cystic fibrosis patients or on medical implants. Distinct stages in biofilm development (the biofilm "life cycle") include initiation, maturation, maintenance, and dissolution (or dispersal). Bacterial biofilms form when nutrients are plentiful. The goal of biofilms in nature is to stay where food is abundant. Why should a microbe travel off to hunt for food when it is already available? Once nutrients become scarce, however, individuals detach from the community to forage for new sources of nutrients.

Biofilms in nature can take many different forms and serve different functions for different species. Confocal images of biofilms are shown in **Figure 4.27** (inset) and Figure 2.30. Biofilm formation can be cued by different environmental signals in different species. These signals include pH, iron concentration, temperature, oxygen availability, and the presence of certain amino acids. Nevertheless, a common pattern emerges in the formation of many kinds of biofilms: adhesion of single cells to a solid surface, microcolony formation, biofilm maturation, and dispersal (**Fig. 4.27** ▶).

First, the specific environmental signal induces a genetic program in planktonic cells. The planktonic cells then start to attach to nearby inanimate surfaces by means of flagella, pili, lipopolysaccharides, or other cell-surface appendages or charge interactions, and they begin to coat that surface

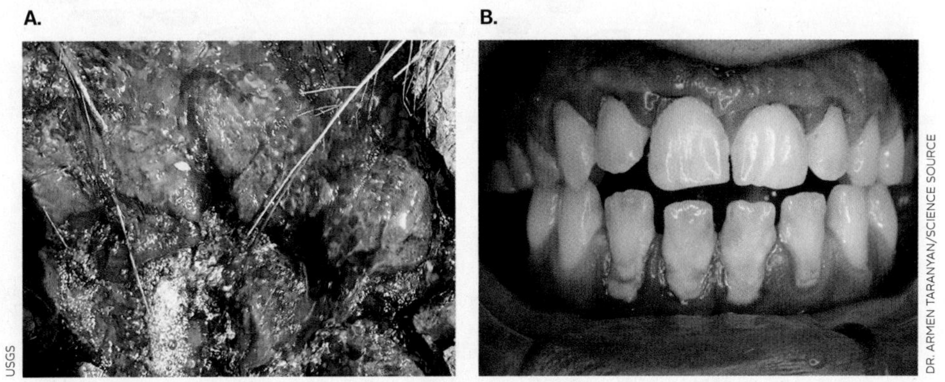

FIGURE 4.26 ■ **Biofilms.** **A.** A greenish brown slime biofilm found on cobbles of the streambed in High Ore Creek, Montana. **B.** The biofilm that forms on teeth is called plaque. ▶

with an organic monolayer of polysaccharides or glycoproteins to which more planktonic cells can attach. At this point, cells of some species may move along surfaces using a **twitching motility** that involves the extension and retraction of a specific type of pilus. Ultimately, they stop moving and firmly attach to the surface.

As more and more cells bind to the surface, they can begin to communicate with each other by sending and receiving chemical signals in a process called **quorum sensing**. These chemical signaling molecules are continually made and secreted by individual cells. Once the population reaches a certain number (analogous to an organizational "quorum"), the chemical signal achieves a specific concentration that the cells can sense (further discussed in Chapter 10). This concentration triggers genetically regulated changes that cause cells to bind tenaciously to the substrate and to each other. Quorum sensing serves the biofilm in many ways. Among other functions, quorum sensing triggers the increased resistance of biofilms to antibiotics (**eTopic 4.4**) and to phagocytosis by white blood cells.

Once established, the cells in a microcolony form a thick extracellular matrix of polysaccharide polymers and entrapped organic (DNA and proteins) and inorganic materials (**Fig. 4.27**). These **exopolysaccharides**, or **extracellular polymeric substances** (both abbreviated as **EPSs**), such as alginate produced by *P. aeruginosa* and colanic acid produced by *E. coli*, increase the antibiotic resistance of residents within the biofilm. As the biofilm matures, the amalgam of adherent bacteria and matrix takes on complex 3D shapes such as columns and streamers, forming channels through which nutrients flow. For many bacteria, sessile (nonmoving) cells in a biofilm chemically "talk" to each other in order to build microcolonies and keep water channels open. *Bacillus subtilis* also spins out a fibril-like amyloid protein called TasA, which tethers cells and strengthens the biofilm. The effect of TasA on a floating biofilm formed on liquid medium is evident in **Figure 4.28**.

Biofilm Differentiation and Communication

Bacteria growing in biofilms also exhibit a type of cell differentiation initiated by physiological conditions that develop in different layers of the biofilm. For example, oxygen does not penetrate deep into biofilms. Imagine a colony growing on agar. Cells at the colony surface will be exposed to oxygen, whereas cells near the agar will not. Colonies of microbes like *E. coli* that do not need oxygen to grow will have an actively growing zone at the agar-colony interface. Cells there can, without oxygen, ferment the nutrients diffusing up from the agar. Cells at the colony-air interface will be in stationary phase because

FIGURE 4.27 ▪ Biofilm development. The stages of biofilm development in *Pseudomonas*, which generally apply to the formation of many kinds of biofilms. **Inset:** A mucoid environmental strain of *P. aeruginosa* produces uneven, lumpy biofilms in a continuous-flow cell. Cells in the biofilm were stained green with Live/Dead viability stain (3D confocal laser scanning microscopy).
Source: O. E. Petrova and K. Sauer. 2011. *J. Bacteriol.* **193**:6614-6628.

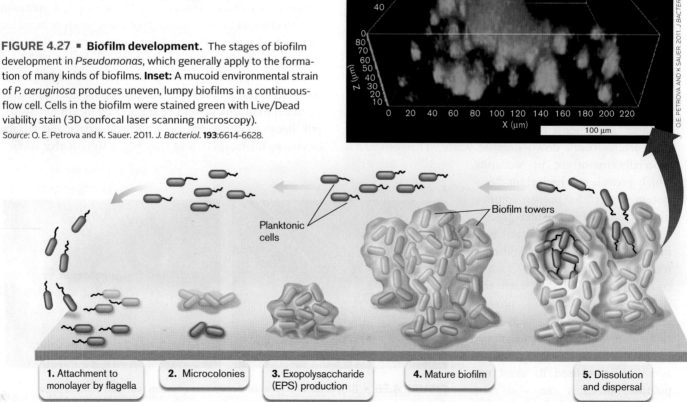

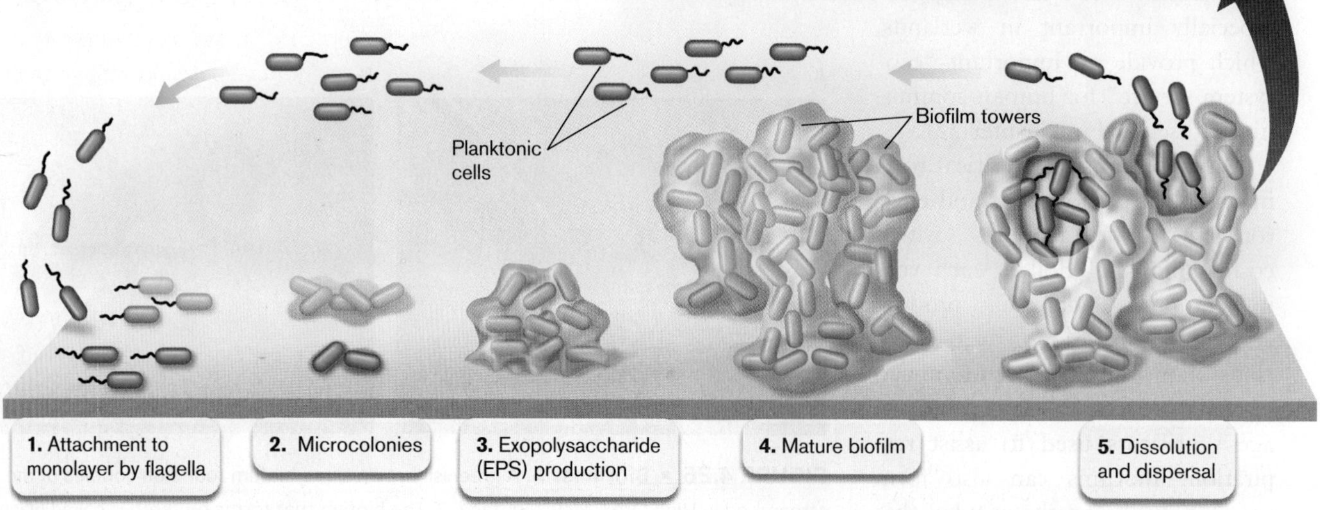

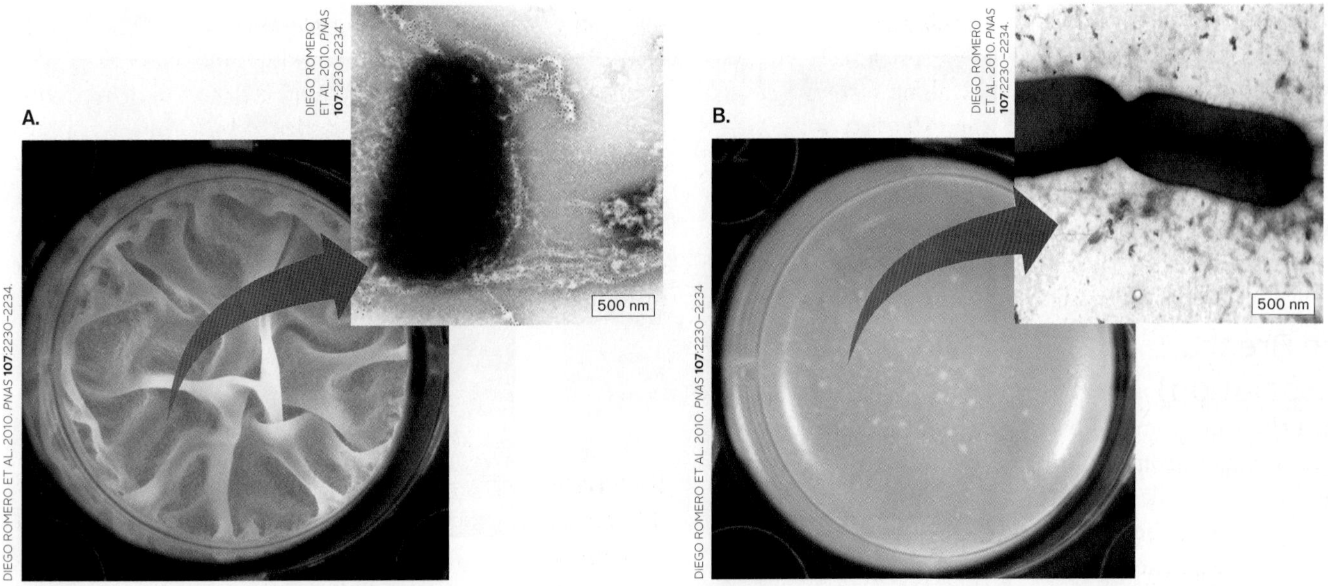

FIGURE 4.28 ■ Floating biofilm (pellicle) formation of *Bacillus subtilis*. Cells were grown in a broth for 48 hours without agitation at 30°C. The pellicles formed by wild-type and *tasA* mutant *B. subtilis* are strikingly different. Wild-type pellicles are extremely wrinkly **(A)**, whereas *tasA* mutant pellicles are flat and fragile **(B)**. **Blowups:** Electron micrographs of wild-type (A) and *tasA* mutant (B) cells.

the nutrients seeping up from below are consumed before reaching them (**Fig. 4.29**). The reverse pattern happens for oxygen-requiring bacteria such as *Pseudomonas*. Cells deep in the colony at the agar surface are in stationary phase in part because they can't get oxygen. Nutrients from the agar are not consumed, so they diffuse toward the oxygen-rich colony surface to feed a growth zone.

We previously described how cells within a biofilm can communicate chemically via quorum signaling. However, when conditions turn dire (as during carbon limitation) cells within different zones of a biofilm can generate nerve-like signals that coordinate behaviors among different sections of the biofilm. The Current Research Highlight describes this novel coping mechanism used by carbon-starved cells of *Bacillus subtilis* buried deep within a biofilm. The biofilms were grown on the walls of a flow cell under a constant flow of low-carbon liquid medium (not agar). The carbon-starved cells at the base of the biofilm, far from the nutrient source at the surface, send intermittent waves of K^+-dependent electrical signals that tell cells feeding at the biofilm's surface to temporarily stop importing the limited carbon source (glutamate). As a result, more glutamate periodically diffuses into the biofilm to temporarily feed cells at the center. The electrical signal can also pass back and forth between separate, nearby biofilms to initiate a community-wide time-sharing of a limiting nutrient. Meanwhile,

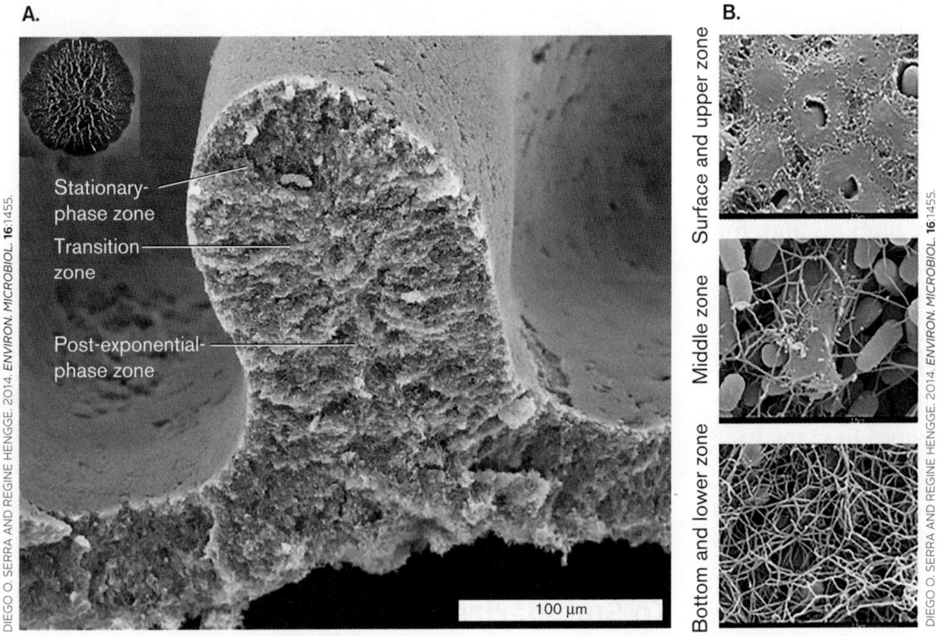

FIGURE 4.29 ■ Two-layer differentiation in *Escherichia coli* biofilms. **A.** Side view of a cross section through a ridge of a macrocolony (inset) grown on salt-free LB agar plates (a complex medium) for 5 days (SEM). Areas false-colored in red and blue represent zones of cells exhibiting stationary-phase and post-exponential-phase physiologies, respectively. The narrow purple area represents the physiological transition zone between the lower and upper layers. **B.** SEM images showing stationary-phase cells on the macrocolony surface covered with secreted cellulose (top), transition zone cells covered with pili and cellulose next to "naked" cells (middle), and mesh-entangled flagella of post-exponential-growth cells (bottom).

a failsafe mechanism meant to ensure survival is triggered. The stressed biofilm disperses single scout cells whose mission is to establish new biofilms in more bountiful environs.

The Breakup (Dissolution)

When the time comes, how do cells escape from a biofilm? When a sessile biofilm (or a part of it) begins to starve or experiences oxygen depletion, some cells start making enzymes that dissolve the matrix. *Pseudomonas aeruginosa*, for instance, produces an alginate lyase that can strip away the EPS. Cells of *Bacillus subtilis* sever their links to TasA fibers in some as yet unknown way. The sessile biofilm then sends out "scouts" called dispersal cells to initiate new biofilms. For scouts to form, genes whose products make EPSs must be turned off, and for bacteria capable of motility, genes needed for flagella must be activated. Dispersed cells have properties different from typical planktonic cells, such as increased expression of adhesion factors needed to establish new biofilms and, for pathogens, virulence factors that release host nutrients or inhibit host immune systems.

Recall that biofilms are important for chronic infections, so preventing or reversing their formation could prove therapeutic. Two intracellular signaling molecules critical to biofilm development and dispersal in many bacteria are the unusual nucleotides cyclic dimeric guanosine monophosphate (c-di-GMP) and guanosine tetra- and pentaphosphate [(p)ppGpp]. High concentrations of either molecule promote biofilm formation, whereas low levels promote dispersal. Robert Hancock and colleagues from the University of British Columbia may have found a way to undo a biofilm's architecture by targeting one of these signal nucleotides. They discovered two synthetic antibiofilm peptides that bind the signaling molecule (p)ppGpp and promote its degradation in various pathogens. When used in vitro, these peptides caused the death and dispersal of biofilm cells (**Fig. 4.30**). The antibiofilm peptides were also effective in treating *Pseudomonas* infections in mice, suggesting that these peptides could play a role in treating human disease.

A. No peptide

B. + 1018 peptide (0.8 μg/ml)

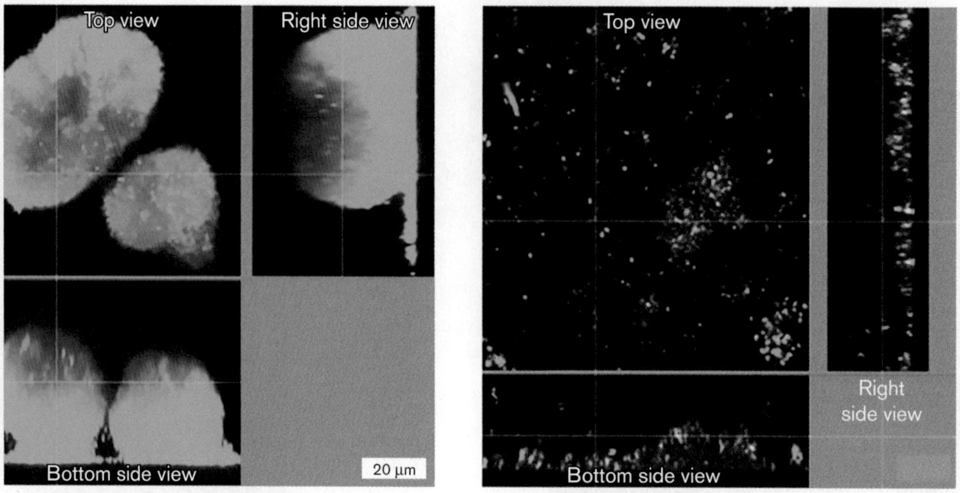

CESAR DE LA FUENTE-NUNEZ ET AL. 2014. *PLOS PATHOG.* **10**(5):E1004152.

CESAR DE LA FUENTE-NUNEZ ET AL. 2014. *PLOS PATHOG.* **10**(5):E1004152.

C.

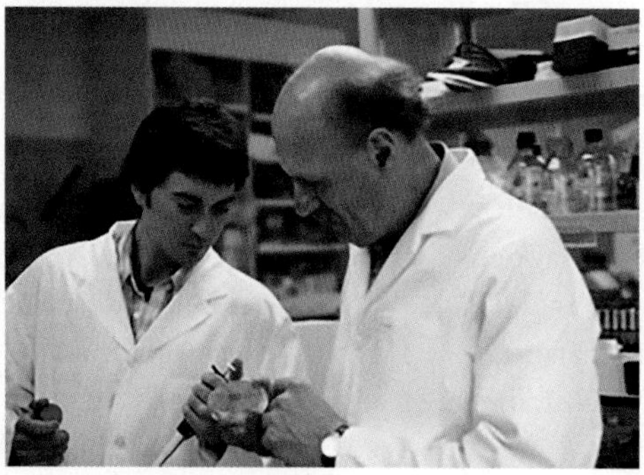

TAMEA BURD PHOTOGRAPHY

FIGURE 4.30 ■ Effect of an antibiofilm peptide on a *Pseudomonas* biofilm. Confocal images of a 2-day-old *Pseudomonas* biofilm (top view and two side views) after 23 hours of no peptide **(A)** or with antibiofilm peptide treatment **(B)**. The peptide promotes degradation of the intracellular signaling molecule (p)ppGpp. Cells were stained with live/dead stain in which live cells fluoresce green and dead cells stain red. Most cells from the treated biofilm were dead or had dispersed. **C.** Robert Hancock (right) and master's student Pat Taylor.

Organisms adapted to life in extreme environments also form biofilms. Archaea form biofilms in acid mine drainage (pH 0), where they contribute to the recycling of sulfur, and cyanobacterial biofilms are common in thermal springs. In the ocean there are suspended particles of biofilm called "marine snow" comprising many unidentified organisms (discussed in Chapter 21). The particles seem to be capable of methanogenesis, nitrogen fixation, and sulfide production, indicating that the architecture of biofilms enables anaerobic metabolism to occur in an otherwise aerobic environment.

To Summarize

- **Biofilms** are complex, multicellular, surface-attached microbial communities.
- **Chemical signals** enable bacteria to communicate (**quorum sensing**) and in some cases to form biofilms.
- **Biofilm development** involves the adherence of cells to a substrate, the formation of microcolonies, and,

ultimately, the development of complex channeled communities that generate new planktonic cells.

4.6 Cell Differentiation

Can bacteria change shape? Many bacteria faced with environmental stress undergo complex molecular reprogramming that includes changes in cell structure. Some species, like *E. coli*, experience relatively simple changes in cell structure, such as the formation of smaller cells or thicker cell surfaces. However, select species, such as *Caulobacter crescentus*, undergo elaborate cell differentiation processes. *Caulobacter* cells convert from the swimming form to the holdfast form before cell division (see Section 3.5). Each cell cycle then produces one sessile cell attached to its substrate by a holdfast, while its sister cell swims off in search of another habitat.

Eukaryotic microbes also have highly complex life cycles and can, in some instances, communicate, move, and assemble to become multicellular collectives. For example, *Dictyostelium discoideum* is a seemingly unremarkable ameba that grows as separate, independent cells. When challenged by adverse conditions such as starvation, however, *D. discoideum* secretes chemical signaling molecules that choreograph a massive interaction of individuals to form elaborate multicellular structures. The developmental cycle of this organism may be compared to that of other eukaryotic microbes that are human parasites (discussed in Chapter 20).

Endospores Are Bacteria in Suspended Animation

Certain Gram-positive genera, including important pathogens such as *Clostridium tetani* (tetanus), *C. botulinum* (botulism), and *Bacillus anthracis* (anthrax), have the remarkable ability to develop dormant spores that are heat and desiccation resistant. Spores are particularly hearty because they do not grow and do not need nutrients until they germinate. Resistance to heat and desiccation (and its lethal toxin) makes *B. anthracis* spores a potential bioweapon.

Most of what we know about bacterial sporulation comes from the Gram-positive soil bacterium *Bacillus subtilis*. When growing in rich media, this microbe undergoes normal vegetative growth and can replicate every 30–60 minutes. However, starvation initiates an elaborate 8-hour genetic program that directs an asymmetrical cell division process and ultimately yields a spore (eTopic 10.2).

As shown in **Figure 4.31C**, sporulation is divided into seven discrete stages based primarily on cell morphology. Stage 0 (not shown) represents the point at which the vegetative cell "decides" to use one of two potential polar division sites to begin septum formation instead of the central division site used for vegetative growth. In stage I, the genome is replicated and the duplicate chromosomes are stretched into a long axial filament that spans the length of the cell. Ultimately, one of the polar division sites wins out and forms a septum. In stage II, the septum divides the cell into two unequal compartments: the smaller **forespore**, which will ultimately become the spore, and the larger **mother cell**, from which the forespore is derived. Each compartment eventually contains one of the replicated chromosomes. (Most of the chromosome in the forespore has to be pumped in after septum formation.)

In stage III of sporulation, the mother cell membrane engulfs the forespore. Next, the mother cell chromosome is destroyed (stage IV), and a thick peptidoglycan layer (cortex) is placed between the two membranes surrounding the forespore protoplast (stage V). Layers of coat proteins are then deposited on the outer membrane, also in stage V. Stage VI completes the development of spore resistance to heat and chemical insults. This last process includes the synthesis of dipicolinic acid (which stabilizes and protects spore DNA) and the uptake of calcium into the coat of the spore. Finally, the mother cell, now called a sporangium, releases the mature spore (stage VII).

Spores resist many environmental stresses that would kill vegetative cells. Spores owe this resistance, in part, to their desiccation (they have only 10%–30% of a vegetative cell's water content). But, as discovered by Peter Setlow and colleagues, spores are also packed with small acid-soluble proteins (SASPs) that bind to and protect DNA. The SASP coat protects the spore's DNA from damage by ultraviolet light and various toxic chemicals.

A fully mature spore can exist in soil for at least 50–100 years, and spores have been known to last thousands of years. Once proper nutrient conditions arise, another genetic program, called **germination**, is triggered to wake the dormant cell, dissolve the spore coat, and release a viable vegetative cell.

While sporulation is an effective survival strategy, bacteria that sporulate actually go out of their way not to sporulate—even to the point of cannibalism. During nutrient limitation, these bacteria will secrete proteins that can kill their siblings and then use the released nutrients to prevent starvation and, thus, sporulation. When (and only when) that strategy fails, the ill-fated cells sporulate.

Cyanobacteria Differentiate to Fix Nitrogen

Some autotrophic cyanobacteria, such as *Anabaena*, not only carry out photosynthesis but also "fix" atmospheric nitrogen to make ammonia. This is surprising because nitrogenase, the enzyme required to fix nitrogen, is very sensitive to oxygen, which is produced as a by-product of

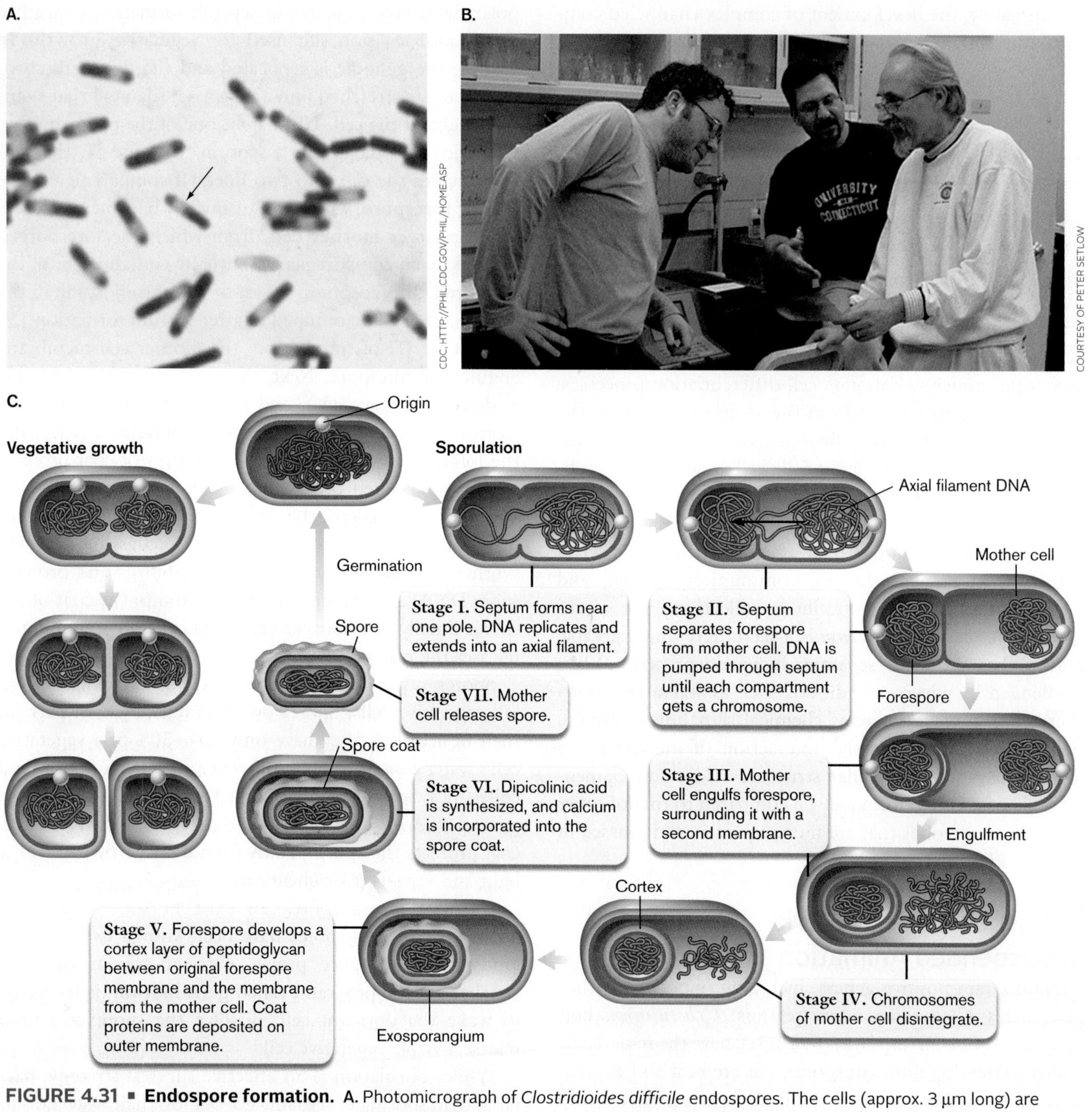

FIGURE 4.31 ▪ Endospore formation. **A.** Photomicrograph of *Clostridioides difficile* endospores. The cells (approx. 3 μm long) are stained with crystal violet. The cells are blue, whereas the spores are colorless and located inside the cells (arrow). A spore stain in which spores appear blue-green is shown in Figure 2.22B. **B.** Peter Setlow (right) of the University of Connecticut figured out how proteins regulate the process of endospore differentiation. **C.** The seven stages of endospore formation.

photosynthesis. So, one might expect that photosynthesis and nitrogen fixation would be two mutually exclusive physiological activities. *Anabaena* solved this dilemma by developing specialized cells, called heterocysts, that can fix nitrogen (**Fig. 4.32**). A tightly regulated genetic program converts every tenth photosynthetic, vegetative cell to a heterocyst. As part of the differentiation process, heterocysts make nitrogenase, produce three additional cell walls, and form a specialized envelope that provides a barrier to atmospheric O_2. In addition, these cells degrade the photosynthesis machinery that produces O_2. The heterocyst then supplies nitrogen compounds to the adjacent vegetative cells, which, in turn, send carbon sugars to the heterocyst. The precise spacing of heterocysts relies on the ratio of inhibitors relative to activators produced by different cells in the chain. Inhibitors predominate in vegetative cells, whereas cells fated to become heterocysts contain higher levels of an activator.

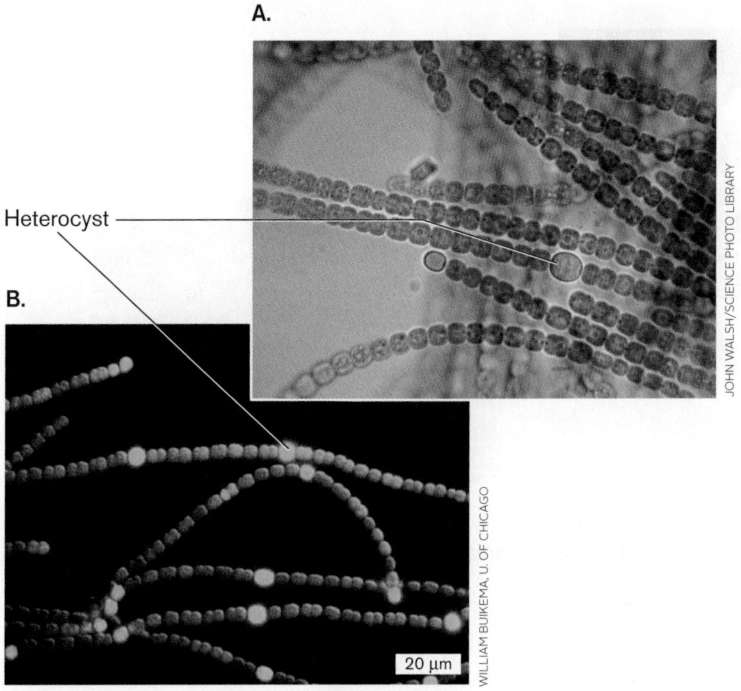

FIGURE 4.32 ■ Cyanobacteria and heterocyst formation. A. Light-microscope image of the cyanobacterium *Phanizomenon*. **B.** The cyanobacterium *Anabaena*. The expression of genes in heterocysts is different from their expression in other cells. All cells in the figure contain a cyanobacterial gene to which the gene for green fluorescent protein (GFP) has been spliced. Only cells that have formed heterocysts are expressing the fused gene, which makes the cell fluoresce bright green.

Myxococcus Differentiation Is a "Family" Gathering

Certain species of bacteria, in the microbial equivalent of a barn raising, produce architectural marvels called fruiting bodies. The Gram-negative, soil-dwelling species *Myxococcus xanthus* uses **gliding motility** (involving a type of pilus, not a flagellum) to travel on surfaces as individuals or to move together as a mob (**Fig. 4.33**). Starvation triggers a developmental cycle in which 100,000 or more individuals attract each other, aggregate, and rise into a mound called a fruiting body. To be successful, only *M. xanthus* cells from the same clonal line (siblings) are permitted to join the swarm. Specific cell-surface receptors ensure kinship and make possible the exchange of outer membranes between individuals. The outer membrane exchanges aid the swarm to transition from unconnected individuals into an interactive multicellular organism. Myxococci within the interior of the fruiting body differentiate into thick-walled, spherical spores that are released into the surroundings. The random dispersal of spores is an attempt to find new sources of nutrients. This differentiation process requires many cell-cell interactions and a complex genetic network that is slowly being unraveled.

Actinomycetes: The Fungus-like Bacteria

The actinomycetes, such as *Streptomyces* (within the order Actinomycetales; Chapter 18), are bacteria that form mycelia and sporangia analogous to the filamentous structures of eukaryotic fungi (**Fig. 4.34**). Several developmental programs tied to nutrient availability are at work in this process (**Fig. 4.35**). Under favorable nutrient conditions, a germ tube emerges from a germinating spore, grows from its tip (tip extension), and forms branches that grow along, and within, the surface of its food source (**Fig. 4.35**, step 1). This type of growth produces an intertwined network of long multinucleate filaments (**hyphae**; singular, **hypha**) collectively called substrate **mycelia** (singular, **mycelium**). After a few days, a signaling molecule made by the organisms accumulates to a level that activates a new set of genes, including one encoding a surfactant, that allow hyphae to grow into the atmosphere, rising above the surface to form aerial mycelia (**Fig. 4.35**, steps 2a and 2b). Compartments at the tips of these aerial hyphae contain 20–30 copies of the genome. Aerial hyphae stop growing as nutrients decline, triggering a developmental program that synthesizes antibiotics. Meanwhile, the older ends of the filaments senesce, and their decomposing cytoplasm attracts scavenger microbes—which are killed by the antibiotics. The younger streptomycete cells then feast on the dead scavengers.

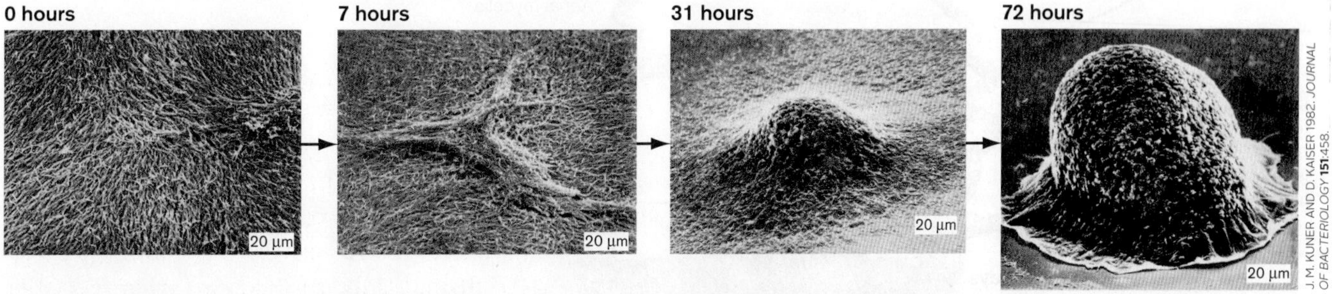

FIGURE 4.33 ■ *Myxococcus* swarm, erecting a fruiting body. Approximately 100,000 cells begin to aggregate, and over the course of 72 hours they erect a fruiting body.

FIGURE 4.34 ■ **Mycelia. A.** *Streptomyces* substrate mycelia. **B.** Filamentous colonies of *Streptomyces coelicolor*, an actinomycete known for producing antibiotics (blue pigment in water droplets). **C.** *Streptomyces* aerial hyphae. The arrow points to a hyphal spore (approx. 1 μm each).

The aerial hyphae ultimately produce spores (arthrospores) that are fundamentally different from bacterial endospores. This program lays down multiple septa that subdivide the compartment into single-genome prespores (**Fig. 4.35**, step 3). The shape of the prespore then changes, its cell wall thickens, and deposits are made in the

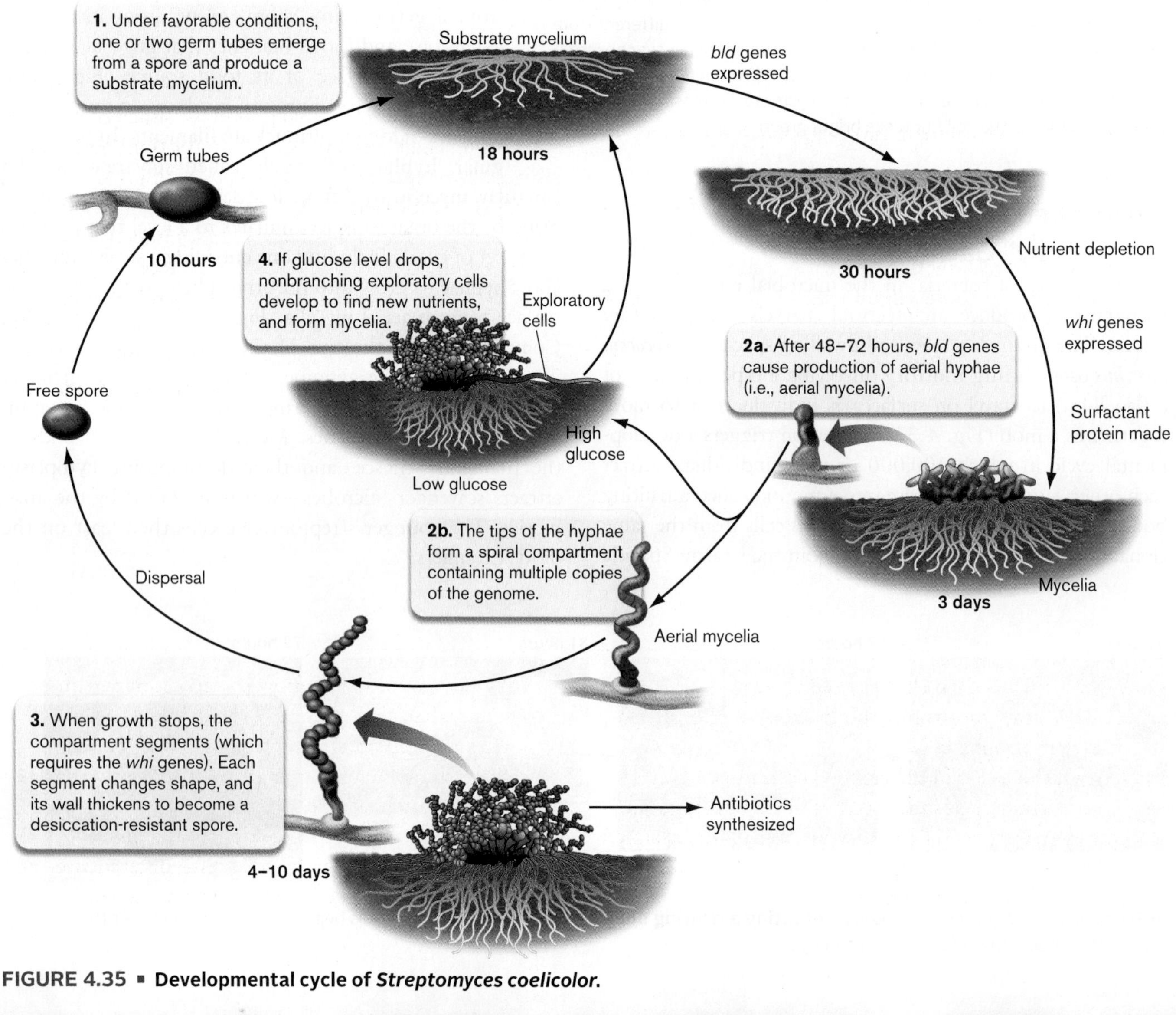

FIGURE 4.35 ■ **Developmental cycle of *Streptomyces coelicolor*.**

spore that all increase resistance to desiccation. The result is a quiescent form of the species that can withstand a variety of environmental assaults.

A new, fourth stage of *Streptomyces* development was recently described by Marie Elliot at McMaster University (**Fig. 4.35**, step 4). Called exploratory growth, this fourth stage develops when nutrient profiles in the soil change (low glucose level and alkaline pH). Such changes can happen following the growth of competing soil organisms such as yeast. The nutrient and pH changes initiate the growth of nonbranching filaments (explorer cells) that emerge from the mass of sporulating cells to find new, but distant, sources of food. The explorer cells also emit an airborne signal compound, trimethylamine, that stimulates explorer growth in distant *Streptomyces* colonies. This newly described developmental stage is a bet-hedging strategy to scavenge nutrients for the group while the sporulating cells provide a highly resistant genetic repository that ensures colony survival in case exploration fails.

Streptomyces species remain of tremendous scientific interest, both for their fascinating developmental programs and for their ability to make antibiotics.

To Summarize

- **Microbial development** involves complex changes in cell forms.
- **Endospore development** by *Bacillus*, *Clostridium*, and *Clostridioides* species is a multistage process that includes asymmetrical cell division to make a forespore and a mother cell, forespore engulfment by the mother cell, deposition of coat proteins around the forespore, and steps that increase chemical and heat resistance of the endospore.
- **Heterocyst development** enables cyanobacteria to fix nitrogen anaerobically while maintaining oxygenic photosynthesis.
- **Multicellular fruiting bodies** in *Myxococcus* and **mycelia** in actinomycetes develop in response to starvation, dispersing dormant cells to new environments.

Thought Question

4.14 How might members of the Actinomycetales such as *Streptomyces* species avoid "committing suicide" when they make their antibiotics?

Concluding Thoughts

Despite their unassuming appearance, bacteria perform incredibly complex and highly orchestrated processes to achieve growth. The gathering of nutrients is coordinated with the synthesis of biomass, and then adjusted as nutrients change in the local environment, or when the organism is challenged by stress. Most, if not all, bacteria also undergo elegant developmental processes ranging from biofilms and swarming, to sporulation or heterocyst formation. These developmental programs launch in response to environmental pressures such as starvation, desiccation, changes in temperature, and even crowding. All of these processes help perpetuate the species. The environmental influence over these microbial processes impacts the composition and interspecies collaboration among aquatic and soil ecosystems, as well as within the microbial communities that inhabit the human body. In Chapter 5 we build on these concepts and explore the remarkable ways that microbes respond to environmental stress and change.

CHAPTER REVIEW

Review Questions

1. What nutrients do microbes need to grow?
2. Explain how autotrophy, heterotrophy, phototrophy, and chemotrophy differ.
3. Explain the basics of the carbon and nitrogen cycles.
4. Describe the various mechanisms of transporting nutrients in prokaryotes and eukaryotes. What are facilitated diffusion, coupled transport, ABC transporters, group translocation, and endocytosis?
5. Why is it important to grow bacteria in pure culture?
6. Under what circumstances would you use a selective medium? A differential medium?
7. What factors define the growth phases of bacteria grown in batch culture?
8. Describe the important features of biofilms.
9. Name three kinds of bacteria that differentiate, and give highlights of the differentiation processes.

Thought Questions

1. Bile salts are used in certain selective media. What are bile salts, and why might they be more harmful to Gram-positive organisms than to Gram-negatives?
2. Why is *Rickettsia prowazekii*, which can grow only in the cytoplasm of a eukaryotic cell, considered a living organism but viruses are not?
3. Suppose 1,000 bacteria are inoculated in a tube containing a minimal salts medium, where they double once an hour, and 10 bacteria are inoculated into rich medium, where they double in 20 minutes. Which tube will have more bacteria after 2 hours? After 4 hours?
4. An exponentially growing culture has an optical density at 600 nm (OD_{600}) of 0.2 after 30 minutes and an OD_{600} of 0.8 after 80 minutes. What is the doubling time?
5. Microbes that grow at high NaCl concentrations are called halophiles. Describe how you would isolate a halophilic organism from the natural environment.
6. Mercuric ions (Hg^{2+}) and methylmercury [$(CH_3Hg)^+$] are major, human-generated, toxic contaminants of water and soil. Some species of bacteria can bioremediate these compounds by transporting them into the cell and reducing them to elemental mercury (Hg^0). One of the transport proteins is called MerC. Use an Internet search engine to determine how many organisms have a MerC homolog.
7. Environmental bacteria were isolated from the water reservoirs of insect-digesting pitcher plants. The isolated strains were cultured in tryptone–yeast extract broth, which contains many different peptide and carbohydrate nutrients. The different growth curves obtained are shown in the graph. Explain how these curves differ from the "standard" growth curve and propose hypotheses as to why they differ.

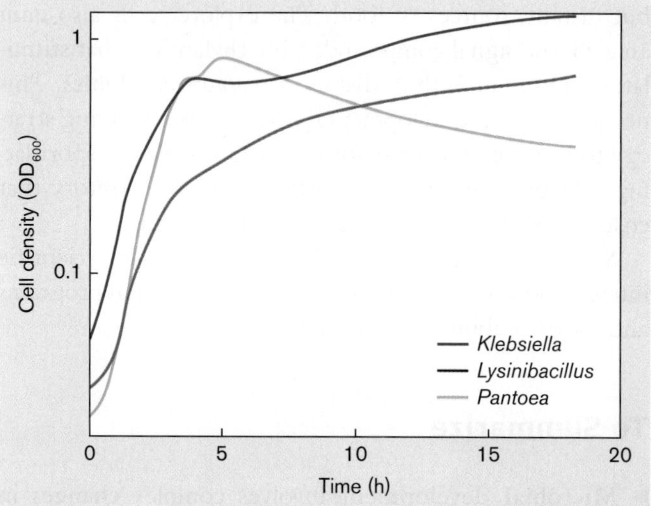

Key Terms

ABC transporter (127)
antiport (126)
ATP synthase (123)
autotroph (121)
autotrophy (120, 122)
batch culture (143)
binary fission (141)
biofilm (147)
chemoorganoheterotrophy (122)
chemolithoautotroph (121, 122)
chemolithoautotrophy (122)
chemostat (146)
chemotrophy (121)
cofactor (120)
colony (131)
complex medium (132)
confluent (131)
continuous culture (146)
coupled transport (126)
death phase (145)
death rate (145)
denitrification (124)
differential medium (133)
dilution streaking (131)
doubling time (141)
electrochemical potential (123)
enriched medium (132)
essential nutrient (120)
exopolysaccharide (EPS) (148)
exponential phase (143)
extracellular polymeric substance (EPS) (148)
facilitated diffusion (125)
flow cytometry (138)
fluorescence-activated cell sorting (FACS) (139)
forespore (151)
generation time (141)
germination (151)
gliding motility (152)
group translocation (129)
growth factor (134)
growth rate (141)
growth rate constant (142)
heterotroph (122)
heterotrophy (120, 122)
hypha (153)
lag phase (143)
lithotroph (121)
logarithmic (log) phase (143)
macronutrient (120)
membrane potential (122)
micronutrient (120)
minimal defined medium (132)
mother cell (151)
mycelium (153)
nitrification (124)
nitrogen-fixing bacterium (123)
optical density (140)
organotroph (121)
organotrophy (121)
permease (125)
phosphotransferase system (PTS) (129)
photoautotroph (121)
photoautotrophy (122)
photoorganotroph (122)

photoorganotrophy (122)
photolithoautotroph (122)
phototrophy (121)
planktonic cell (140)
pour plate technique (139)
proton potential (proton motive force) (123)
pure culture (130)
quorum sensing (147)
selective medium (133)
siderophore (128)
spread plate (131)
stationary phase (144)
symbiont (123)
symport (126)
twitching motility (147)
uncultured (unculturable) (134)
viable (131)

Recommended Reading

Asfahl, Kyle L., and Martin Schuster. 2017. Social interactions in bacterial cell-cell signaling. *FEMS Microbiology Reviews* **41**:92–107.

Bergauera, Kristin, Antonio Fernandez-Guerrab, Juan A. L. Garcia, Richard R. Sprengerd, Ramunas Stepanauskase, et al. 2018. Organic matter processing by microbial communities throughout the Atlantic water column as revealed by metaproteomics. *Proceedings of the National Academy of Sciences USA* **115**:E400–E408.

Collins, J., C. Robinson, H. Danhof, C. W. Knetsch, H. C. van Leeuwen, et al. 2018. Dietary trehalose enhances virulence of epidemic *Clostridium difficile. Nature* **553**:291–294.

Errington, John. 2010. From spores to antibiotics. *Microbiology* **156**:1–13.

Eswara, Prahathees J., and Kumaran S. Ramamurthi. 2017. Bacterial cell division: Nonmodels poised to take the spotlight. *Annual Review of Microbiology* **71**:393–411.

Gefena, Orit, Ofer Fridmana, Irine Ronina, and Nathalie Q. Balabana. 2014. Direct observation of single stationary-phase bacteria reveals a surprisingly long period of constant protein production activity. *Proceedings of the National Academy of Sciences USA* **111**:556–561.

Gonzalez-Pastor, Jose E. 2011. Cannibalism: A social behavior in sporulating *Bacillus subtilis. FEMS Microbiology Reviews* **35**:415–424.

Guilhen, Cyril, Christiane Forestier, and Damien Balestrino. 2017. Biofilm dispersal: Multiple elaborate strategies for dissemination of bacteria with unique properties. *Molecular Microbiology* **105**:188–210.

Kaur, Amandeep, Neena Capalash, and Prince Sharma. 2019. Communication mechanisms in extremophiles: Exploring their existence and industrial applications. *Microbiological Research* **221**:15–27.

Laverty, Garry, Sean P. Gorman, and Brendan F. Gilmore. 2014. Biomolecular mechanisms of *Pseudomonas aeruginosa* and *Escherichia coli* biofilm formation. *Pathogens* **3**:596–632.

Lopian, Livnat, Yair Elisha, Anat Nussbaum-Shochat, and Orna Amster-Choder. 2010. Spatial and temporal organization of the *E. coli* PTS components. *EMBO Journal* **29**:3630–3645.

Pechter, Kieran B., Liang Yin, Yasuhiro Oda, Larry Gallagher, Jianming Yang, et al. 2017. Molecular basis of bacterial longevity. *mBio* **8**:e01726-17.

Pengbo, Cao, and Daniel Wall. 2017. Self-identity reprogrammed by a single residue switch in a cell surface receptor of a social bacterium. *Proceedings of the National Academy of Sciences USA* **114**:3732–3737.

Pletzer, Daniel, Heidi Wolfmeier, Manjeet Bains, and Robert E. W. Hancock. 2017. Synthetic peptides to target stringent response-controlled virulence in a *Pseudomonas aeruginosa* murine cutaneous infection model. *Frontiers in Microbiology* **8**:1867. https://doi.org/10.3389/fmicb.2017.01867.

Romero, Diego, Hera Vlamakis, Richard Losick, and Roberto Kolter. 2011. An accessory protein required for anchoring and assembly of amyloid fibres in *B. subtilis* biofilms. *Molecular Microbiology* **80**:1155–1168.

Saha, Maumita, Subhasis Sarkar, Biplab Sarkar, Bipin Kumar Sharma, Surajit Bhattacharjee, et al. 2016. Microbial siderophores and their potential applications: A review. *Environmental Science and Pollution Research International* **23**:3984–3999.

Serra, Diego O., and Regina Hengge. 2014. Stress responses go three dimensional—The spatial order of physiological differentiation in bacterial macrocolony biofilms. *Environmental Microbiology* **16**:1455–1471.

Setlow, Peter. 2014. Germination of spores of *Bacillus* species: What we know and do not know. *Journal of Bacteriology* **196**:1297–1305.

Skaar, Eric P. 2010. The battle for iron between bacterial pathogens and their vertebrate hosts. *PLoS Pathogens* **6**:e1000949.

Teschler, Jennifer K., David Zamorano-Sánchez, Andrew S. Utada, Christopher J. A. Warner, Gerard C. L. Wong, et al. 2015. Living in the matrix: Assembly and control of *Vibrio cholerae* biofilms. *Nature Reviews. Microbiology* **13**:255–268.

photoheterotroph (122)
photolithotroph (122)
[illegible] (141)
planktonic cell (140)
pour plate technique (139)
[illegible] (12?)
pure culture (120)
quorum sensing (15?)
selective medium (1??)
siderophore (128)
spread plate (139)
stationary phase (14?)
[illegible] (125)
[illegible] (120)
swarming motility (145)
[illegible] (134)
[illegible] (131)

Recommended Reading

Abisado, Rhea G., [illegible]. 201?. Social interactions in bacterial cell-cell signaling. *FEMS Microbiology Reviews* 41:92–107.

[illegible], Rubén, Ana de [illegible], Joan A. [illegible], [illegible]. 2018. [illegible] [illegible] *Nature* [illegible]

[illegible] 2019 [illegible] *Microbiology* [illegible]

[illegible], [illegible], and [illegible] 2016 [illegible] *Annual Review of Microbiology* 7?:[illegible]

[illegible] [illegible] *Applied and Environmental Microbiology* [illegible]

[illegible] 201? [illegible] *Microbiology Reviews* [illegible]

[illegible] 2017. Biofilms [illegible] properties. *[illegible] Microbiology* 10?:188–210.

Kain, Amanda, [illegible], and Bruce Shapiro. 2019. [illegible] *Microbiology* 2?:[illegible]

[illegible] 2015. [illegible] *Science* [illegible]

[illegible] and [illegible]

organization [illegible] components [illegible] 29 [illegible]

[illegible] 2015 [illegible] *Cell* [illegible]

[illegible] 2017 [illegible] *Nature Reviews Microbiology* [illegible]

[illegible] 2012 [illegible] *mBio* [illegible]

Romero, Diego, [illegible] Richard [illegible], and Roberto Kolter. 2010. An accessory protein required for anchoring and assembly of amyloid fibres in *B. subtilis* biofilms. *Molecular Microbiology* 80:1155–1168.

[illegible] *Environmental Science and Technology* [illegible]

[illegible], Diego O., and Regina [illegible]. 2011. Stress responses [illegible] *Microbiology and Molecular Biology Reviews* [illegible]

Setlow, Peter. 2014. Germination of spores of *Bacillus* species: What we know and do not know. *Journal of Bacteriology* 196:1297–1305.

[illegible] 2010. [illegible] *Cell* [illegible]

[illegible], Jennifer R., [illegible], [illegible] 201? [illegible] *Nature* [illegible]

CHAPTER 5

Environmental Influences and Control of Microbial Growth

5.1 Environmental Limits on Growth: Temperature and Pressure

5.2 Osmolarity

5.3 Hydronium (pH) and Hydroxide Ion Concentrations

5.4 Oxygen

5.5 Nutrient Deprivation and Starvation

5.6 Physical, Chemical, and Biological Control of Microbes

Microbes have both the fastest and the slowest growth rates of any known organisms. Some hot-springs bacteria can double their population in as little as 10 minutes, whereas deep-sea-sediment microbes may take as long as 100 years. What determines these differences in growth rate? Nutrition is one factor, but niche-specific physical parameters such as temperature, pH, and osmolarity are equally important. In Chapter 5 we explore the limits of microbial growth and show how this knowledge helps us control the microbial world.

CURRENT RESEARCH highlight

Microbial assault and thievery. Bacterial species cooperate if necessary but assert dominance if possible. One ploy involves nanotubes, tubelike membranous structures formed between adjacent cells. Sigal Ben-Yehuda and colleagues discovered that *Bacillus subtilis* (green) uses nanotubes to deliver a tRNase toxin (WapA) to its competitor, *B. megaterium*. When mixed in equal numbers, *B. megaterium* (red) cannot grow (panel A); but will grow alongside a toxin-negative mutant (panel B). SEMs localized WapA to nanotubes (not shown). *B. subtilis* also used nanotubes to steal nutrients from its victim.

Source: O. Stempler et al. 2017. *Nat. Commun.* **8**:315.

A.

B.

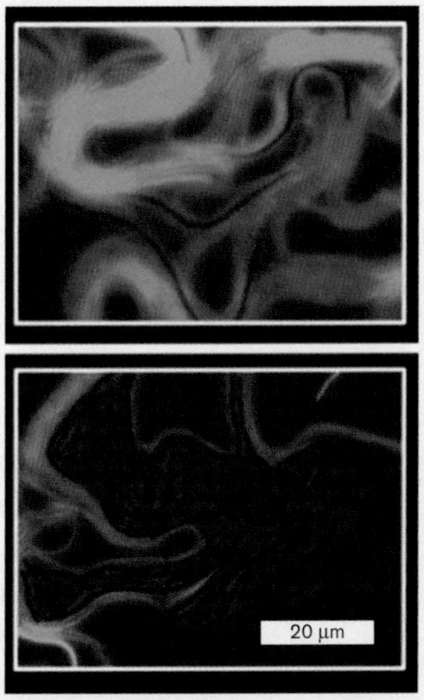

COURTESY OF SIGAL BEN-YEHUDA

AN INTERVIEW WITH

SIGAL BEN-YEHUDA, MICROBIOLOGIST, HEBREW UNIVERSITY HADASSAH MEDICAL SCHOOL

COURTESY OF SIGAL BEN-YEHUDA, HEBREW UNIVERSITY HADASSAH MEDICAL SCHOOL

What led to your discovery of nanotubes? Serendipity, phenotype, or intuition?

Serendipity! While studying sporulation, we mixed GFP-expressing cells (*gfp*$^+$) with wild-type cells lacking GFP (*gfp*$^-$) and microscopically followed fluorescence over time. Remarkably, *gfp*$^-$ cells lying close to *gfp*$^+$ cells became weakly fluorescent, suggesting that adjacent bacterial cells could somehow exchange cytoplasmic molecules in an ordered manner. The observation triggered a search for intercellular structures that might mediate the transfer. The original thinking of my postdoctoral student, Gyanendra Dubey, led to electron microscopy that revealed the nanotube connections.

A microbe's physiology typically operates within a very narrow range of physical parameters. Yet in nature, an environment surrounding a microbe can change quickly and dramatically. The dangers faced include extremes in temperature, pH, pressure, and osmolarity. Many marine microbes, for instance, move within seconds from deep-sea cold to the searing heat of a thermal vent. How do organisms survive these stresses? Most species have stopgap measures, called stress responses, that temporarily protect the organism from brief forays into potentially lethal environments. But some organisms evolve to thrive, not just survive, in extreme environments. How do these so-called extremophiles grow under conditions that would kill most living things? In addition to environmental threats, microbes in nature wage wars among themselves by secreting antimicrobial compounds, as illustrated in the Current Research Highlight. Why doesn't the toxin-producing microbe kill itself? Is it somehow immune to its own toxin?

We begin Chapter 5 by describing how physical and chemical changes in the environment impact the growth of different groups of microbes. We also explore how microorganisms adapt to different environments in ways both transient (involving temporary expression of genes) and permanent (modifications of the gene pool). The permanent genetic changes have led to biological diversity. Finally, we examine the different ways humans use physical, chemical, and biological means to limit microbial growth and protect plants, animals, and ourselves from disease.

As you proceed through this chapter, you will encounter two recurring themes: that different groups of microbes live in vastly different environments, and that microbes can respond in diverse ways when confronted by conditions outside their **niche**, or suitable environment for growth. Chapter 21 builds on this information to discuss how communities of microbes interact with their environment, and presents current research methods that are used to study microbial ecology.

5.1 Environmental Limits on Growth: Temperature and Pressure

What is normal? With our human frame of reference, we tend to think that "normal" growth conditions are those found at sea level with a temperature between 20°C and 40°C, a near-neutral pH, a salt concentration of 0.9%, and ample nutrients. Any habitat outside this window is labeled "extreme," and the organisms (bacteria, archaea, and some eukaryotes) populating them are called **extremophiles**. One group of extremophiles can grow at temperatures above the boiling point of water (100°C), while another group requires a strongly acidic (pH 2) environment to grow. Actually, a single environment can simultaneously encompass multiple extremes. In Yellowstone National Park an acid pool can be found next to an alkali pool, both at extremely high temperatures. Thus, extremophiles typically evolve to survive multiple extreme environments. These organisms are called polyextremophiles.

When did extremophiles first appear on Earth? According to our perspective of what is "normal," the conditions on Earth when life began were certainly extreme, which means that the earliest microbes likely grew in these extreme environments. As Earth's environment slowly changed to present-day conditions, some of the ancient "extremophiles" adapted to the change and became what could be called today's "normalophiles." Species that did not adapt to the change were constrained to live in environments on Earth that remained extreme.

Some scientists study extremophiles to gain insight into the physiology of extraterrestrial microbes that we may one day encounter. NASA biochemist Robert MacElroy first coined the term "extremophile" in his quest for life forms that might inhabit other planets. Astrophysicists predict that some extreme environments on Earth could be similar to those of known planets in our galaxy. For instance, the hypersaline Gypsum Hill spring system on Axel Heiburg Island in Canada is an analog for the putatively habitable subsurface briny aquifers on Mars. The Gypsum Hill spring system harbors a rich yet unique microbial ecosystem, suggesting that Mars's cold aquifers may do the same.

Our experiences with extremophiles should also caution us when handling extraterrestrial samples brought back to Earth. For instance, we should not assume that irradiation will sterilize samples from future planetary or interstellar missions. Such treatments do not even kill *Deinococcus radiodurans*, an extremophile found on Earth. Scientists have also taken extreme steps to avoid contaminating other planets with Earth microbes. For example, select components of the Mars rovers were heated to over 110°C (230°F) for up to 144 hours.

How do we even begin to study organisms that grow in boiling water or sulfuric acid solutions, or organisms that we cannot culture in the laboratory? New DNA sequencing technologies, such as metagenomics, have been instrumental in this task. Metagenomic analysis includes the random screening for, and sequencing of, genes that encode ribosomal RNA from a mixed environmental sample. Ribosomal RNA genes from all known life forms have conserved sequences, which permit us to find and selectively amplify these genes from a complex mixture of DNA. These genes also contain unique DNA sequences (you can think of them as "fingerprints") that are specific to different genera. Metagenomics, then, allows scientists to inventory all the culturable and uncultured organisms in any given niche.

Bioinformatic analysis uses the DNA sequence of a gene to identify the probable function of its protein product. This ability enables scientists to make predictions about the biology of extremophiles and even uncultured species. The techniques of bioinformatics are discussed further in Chapters 7, 8, and 12. In addition to bioinformatics, transcriptomic and proteomic techniques can reveal all of the gene and protein expression that takes place in bacteria as they adjust to environmental changes such as shifts in temperature or pH.

We mentioned the fundamental physical conditions (temperature, pH, osmolarity) that define an environment and favor (select for) the growth of specific groups of organisms. But what enables different microbes to grow in such physically diverse environments? Species find their niche, in part, because every protein and macromolecular structure within a cell is affected by changes in environmental conditions. For example, every enzyme works best under a unique set of temperature, pH, and salt conditions because those conditions favor folding of an enzyme into its optimal shape, or conformation. Deviations from these optimal conditions cause the protein to fold a little differently and become less active. While not all enzymes within a given cell boast the same physical optima, these optima must at least be similar and matched to the organism's environment for the organism to function effectively.

As the preceding discussion suggests, microbes are commonly classified by their environmental niche. **Table 5.1** summarizes these environmental classes.

Temperature Effects on Physiology

How do microbes react to hot and cold? Unlike humans (and mammals in general), microbes cannot control their temperature; thus, bacterial cell temperature matches that of the immediate environment. Because temperature affects the average rate of molecular motion, changes in temperature impact every aspect of microbial physiology, including membrane fluidity, nutrient transport, DNA stability, RNA stability, and enzyme structure and function. Every organism has a temperature optimum at which it grows most quickly, as well as minimum and maximum temperatures that define the limits of growth.

Growth temperature limits are imposed, in part, by the thousands of proteins in a cell, all of which must function within the same temperature range. A species grows most quickly at temperatures where all of the cell's proteins work most efficiently as a group to produce energy and synthesize cell components. Growth stops when rising temperatures cause critical enzymes or cell structures (such as the cell membrane) to fail. At cold temperatures, growth ceases because enzymatic processes become too sluggish and the cell membrane becomes too rigid. The membrane needs to remain fluid so that it can expand as cells grow larger and so that proteins needed for solute transport can be inserted into the membrane.

The different branches of life reflect narrowing tolerance to heat. Different archaeal species, for example, can grow in extremely hot or extremely cold temperatures, and some can grow in the middle range. Bacteria, for the most part, tolerate temperatures between the archaeal extremes. Eukaryotes are even less temperature tolerant than bacteria, with individual species capable of growth between 10°C and 65°C.

Growth Rate and Temperature

In general, microbes that grow at higher temperatures can achieve higher rates of growth (**Fig. 5.1A**). Remarkably, for any one species the relationship between growth temperature and the growth rate constant k (the number of generations per hour; see Section 4.4) obeys the Arrhenius equation for simple chemical reactions (**eTopic 5.1**). The general result of the Arrhenius equation is that growth

TABLE 5.1 Basic Environmental Classification of Microorganisms

Environmental parameter	Classification (optimal growth condition)			
Temperature	Hyperthermophile* (above 80°C)	Thermophile* (between 50°C and 80°C)	Mesophile (between 15°C and 45°C)	Psychrophile* (below 15°C)
pH	Alkaliphile* (above pH 9)	Neutralophile (between pH 5 and pH 8)	Acidophile* (below pH 3)	
Osmolarity	Halophile* (high salt, >2-M NaCl)	Halotolerant (high salt not required, but able to grow at up to 2-M NaCl)		
Oxygen	Strict aerobe (only in O_2)	Facultative microbe (with or without O_2)	Microaerophile (only in small amounts of O_2)	Strict anaerobe (only without O_2)
Pressure	Barophile* (high pressure, greater than 380 atm)		Barotolerant (between 10 and 500 atm)	

*Considered extremophiles.

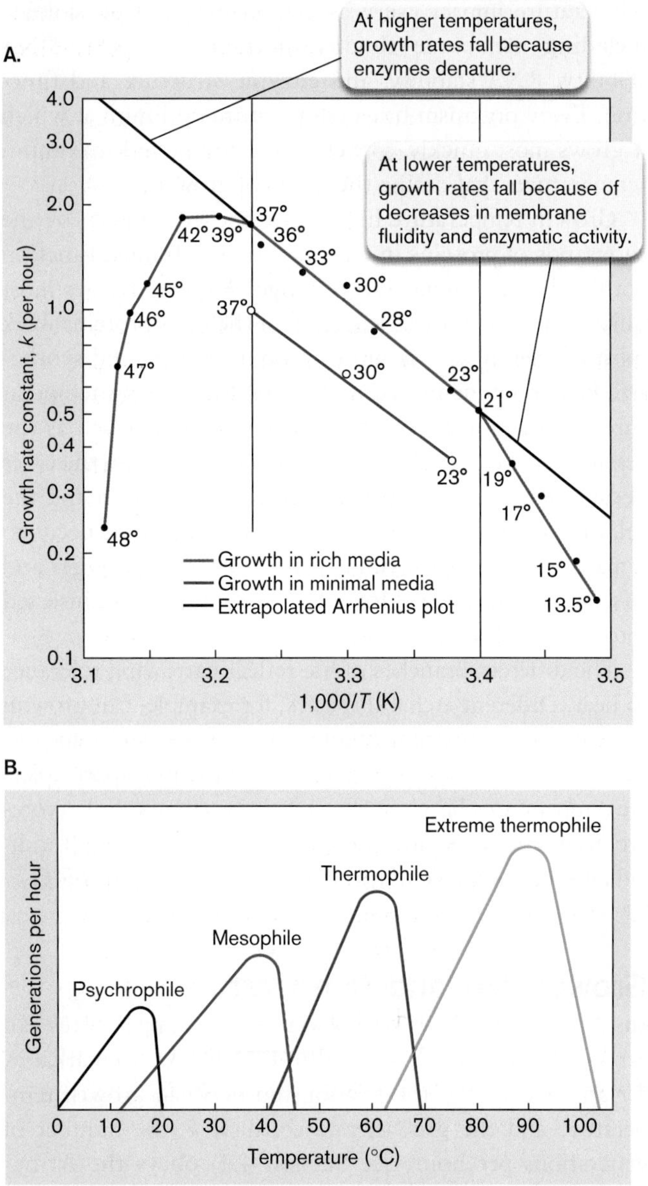

FIGURE 5.1 ▪ Relationship between temperature and growth rate. A. The growth rate constant (k) of the enteric organism *Escherichia coli* is plotted here against the inverse of the growth temperature on the Kelvin scale (1,000/T is used to give a convenient scale on the x-axis). As temperature in celsius rises above or falls below the optimal range, growth rate decreases faster than is predicted by the Arrhenius equation. **B.** The relationship between temperature and growth rate for different groups of microbes. Note that the peak growth rate increases linearly with temperature and obeys the Arrhenius equation. *Source:* Part A from Sherrie L. Herendeen et al. 1979. *J. Bacteriol.* **139**:185.

rate roughly doubles for every 10°C rise in temperature (**Fig. 5.1A**). This same relationship is observed for most chemical reactions with activation energies of about 50 kilojoules per mole (kJ/mol).

At the upper and lower limits of the growth range, however, the Arrhenius effect breaks down. Critical proteins denature at high temperatures, whereas lower temperatures decrease membrane fluidity and limit the conformational mobility of enzymes, thereby lowering their activity. As a result, growth stops at temperature extremes. The typical temperature growth range for most bacteria spans the organism's optimal growth temperature by 30°C–40°C, but some organisms have a much narrower tolerance. Even within a species, we can find mutants that are more sensitive to one extreme or the other (heat sensitive or cold sensitive). These mutations often define key molecular components of stress responses, such as the heat-shock proteins (discussed in Chapter 10).

Thermodynamic principles limit a cell's growth to a narrow temperature range. For example, heat increases molecular movement within proteins. Too much or too little movement will interfere with enzymatic reactions. A great diversity exists among microbes because different groups have evolved to grow within very different thermal ranges. A species grows within a specific thermal range because its proteins have evolved to tolerate that range. Outside that range, proteins will denature or function too slowly for growth. The upper limit for protists (eukaryotic microbes) is about 50°C, while some fungi can grow at temperatures as high as 60°C. Prokaryotes (Bacteria and Archaea), however, have been found to grow at temperatures ranging from below 0°C to above 100°C. Temperatures over 100°C are usually found near thermal vents deep in the ocean. Vent water temperature can rise to 400°C (750°F), but the pressure is sufficiently high to keep the water in a liquid state.

Thought Question

5.1 Why haven't cells evolved so that all their enzymes have the same temperature optimum? If they did, wouldn't they grow even more rapidly?

Microbial Classification by Growth Temperature

According to their temperature ranges for optimal growth, microorganisms can be classified as mesophiles, psychrophiles, or thermophiles (**Fig. 5.1B**).

Mesophiles include the typical "lab rat" microbes, such as *Escherichia coli* and *Bacillus subtilis.* Their growth optima range between 20°C and 40°C, with a minimum of 15°C and a maximum of 45°C. Because they are easy to grow and because most human pathogens are mesophiles, much of what we know about protein, membrane, and DNA structure came from studying this group of organisms. However, detailed 3D views of protein structures are frequently based on studies of two other classes of organisms whose optimal growth temperature ranges flank that of the mesophiles—namely, psychrophiles (on the low-temperature side) and thermophiles (on the high-temperature side). For instance, proteins from thermophiles are more stable than proteins

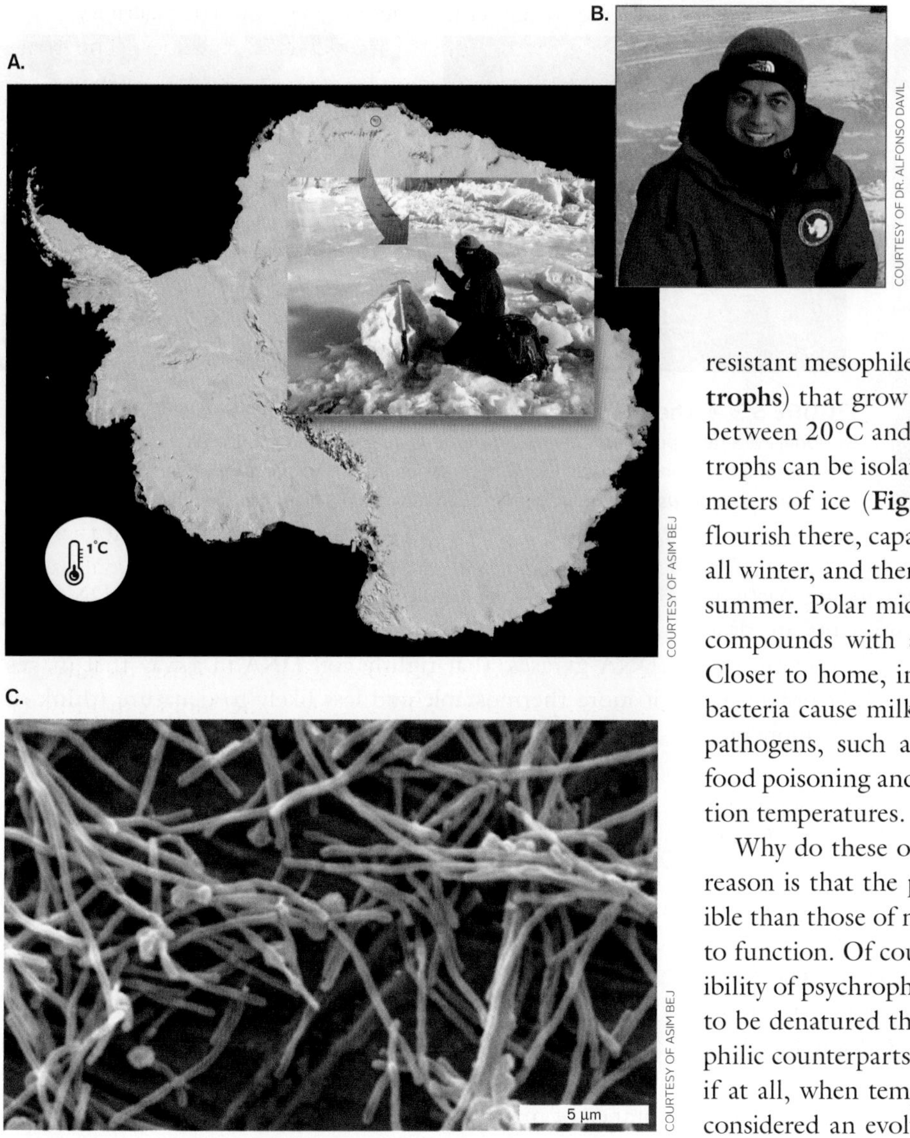

FIGURE 5.2 ▪ Psychrophilic environments and microbes. **A.** The continent of Antarctica is an extreme environment populated by many species of psychrophilic microorganisms, most of them unknown. In addition to being brutally cold, this extreme environment is nutrient-poor and subject to high levels of solar UV irradiation. (inset) Asim Bej, University of Alabama at Birmingham, collects samples from the ecosystem at Schirmacher Oasis (location as marked). Genome sequences from the captured microbes, like those shown in (C), reveal the composition and metabolic capabilities of the South Pole microbiome. **B.** Asim Bej. **C.** Psychrotolerant *Flavobacterium* (grows between 0°C and 22°C) from a South Pole lake made from glacial meltwater in summer (high temperature = 0.9°C; SEM). Novel compounds made by members of the polar microbiome are screened for anticancer and antimicrobial potential.

from mesophiles and, so, are easier to crystallize. These protein crystals are essential for X-ray crystallography studies of protein structure (see Section 2.6).

An underappreciated fact is that Earth's biosphere is predominantly cold and permanently exposed to temperatures below 5°C. **Psychrophiles** are microbes that can grow in extreme cold temperatures, as low as –10°C (or even –20°C in the brine between polar sea ice crystals). Consequently, psychrophiles are prominent members of microbial communities beneath icebergs in the Arctic and in Antarctic soil and lakes (**Fig. 5.2**). Despite their abilities to grow in extreme cold, the optimal growth temperature of psychrophiles is usually about 15°C. In addition to true psychrophiles, there are cold-resistant mesophiles (psychrotolerant bacteria, or **psychrotrophs**) that grow between 0°C and 7°C but have optima between 20°C and 35°C. Both psychrophiles and psychrotrophs can be isolated from Antarctic lakes, beneath several meters of ice (**Fig. 5.2B**). Entire microscopic ecosystems flourish there, capable of surviving (not growing) at –40°C all winter, and then growing at near zero during the sunlit summer. Polar microorganisms may be screened for novel compounds with anticancer and antimicrobial potential. Closer to home, in human environments, psychrotolerant bacteria cause milk to spoil in the refrigerator. Even some pathogens, such as *Listeria monocytogenes* (one cause of food poisoning and septic abortions), can grow at refrigeration temperatures.

Why do these organisms grow so well in the cold? One reason is that the proteins of psychrophiles are more flexible than those of mesophiles and require less energy (heat) to function. Of course, the downside to the increased flexibility of psychrophilic proteins is that they require less heat to be denatured than do their more thermotolerant mesophilic counterparts. As a result, psychrophiles grow poorly, if at all, when temperatures rise above 20°C. This can be considered an evolutionary trade-off. Another reason psychrophiles favor cold is that their membranes are more fluid at low temperatures (because they contain a high proportion of unsaturated fatty acids such as oleic acid; Fig. 3.9); at higher temperatures their membranes are too flexible and fail to maintain cell integrity. Finally, bacteria and archaea that grow at 0°C in glaciers also contain antifreeze proteins and other cryoprotectants (such as trehalose) that can lower the freezing point by 2°C. So, these organisms can grow in ice but will not freeze. Interestingly, some psychrophilic and psychrotolerant bacteria actually stimulate ice formation in their surrounding environment (**eTopic 5.2**).

Psychrophilic enzymes are of commercial interest because their ability to carry out reactions at low temperature is useful in food processing and bioremediation. Enzymes help brew beer more quickly, break down lactose in milk, and can remove cholesterol from various foods. The production of foods at lower temperatures is beneficial too, because the lower processing temperatures minimize the growth of typical mesophiles that degrade and spoil food. Genetically engineered psychrophilic organisms can safely degrade toxic organic contaminants (for example,

A. B. C.

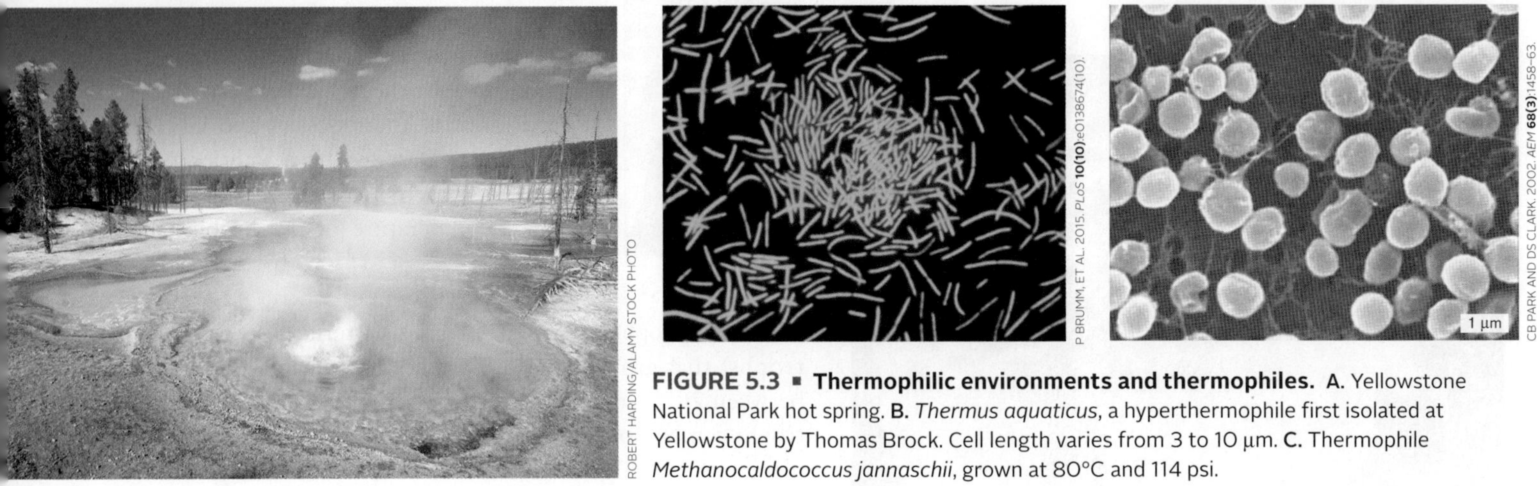

FIGURE 5.3 ▪ Thermophilic environments and thermophiles. **A.** Yellowstone National Park hot spring. **B.** *Thermus aquaticus*, a hyperthermophile first isolated at Yellowstone by Thomas Brock. Cell length varies from 3 to 10 µm. **C.** Thermophile *Methanocaldococcus jannaschii*, grown at 80°C and 114 psi.

petroleum) in the cold. Arctic environments are particularly sensitive to pollution because contaminants are slow to degrade in the freezing temperatures. Consequently, the ability to seed Arctic oil spills with psychrophilic organisms armed with petroleum-degrading enzymes could more rapidly restore contaminated environments.

Thermophiles (**Fig. 5.3**) are species adapted to growth at high temperatures (typically 50°C and higher). **Hyperthermophiles** (also called extreme thermophiles), which grow at temperatures as high as 121°C, are found near thermal vents that penetrate Earth's crust on the ocean floor and on land (for example, hot springs). The thermophile *Thermus aquaticus* was the first source of a high-temperature DNA polymerase used for PCR amplification of DNA. *T. aquaticus* was discovered in a hot spring at Yellowstone National Park by microbiologist Thomas Brock, a pioneer in the study of thermophilic organisms. Its application to the polymerase chain reaction, which requires periods of high temperature, revolutionized molecular biology (discussed in Chapter 12).

Extreme thermophiles often have specially adapted membranes and protein sequences. The thermal limits of these structures determine the specific high-temperature ranges in which various species can grow. Because enzymes in thermophiles (thermozymes) do not unfold as easily as mesophilic enzymes do, they more easily hold their shape at higher temperatures. Thermophilic enzymes are stable, in part, because they contain relatively low amounts of glycine, a small amino acid that contributes to an enzyme's flexibility. Glycines do not contain side chains, so they cannot form stabilizing intramolecular bonds. In addition, the amino termini of proteins in these organisms often are "tied down" by hydrogen bonding to other parts of the protein, making them harder to denature.

Like all microbes, thermophiles have chaperone proteins that help refold other proteins that may undergo thermal denaturation. Thermophile genomes are also packed with numerous DNA-binding proteins that stabilize DNA. In addition, these organisms possess special enzymes, like DNA gyrases, that tightly coil DNA in a way that makes it more thermostable and less likely to denature (think of a helically coiled phone cord that has twisted and bunched up on itself).

Special membranes also help give cells additional stability at high temperatures. Unlike the typical lipid bilayers of mesophiles, the membranes of thermophiles manage to "glue" together parts of the two hydrocarbon layers that point toward each other, making them more stable. They do this by incorporating more saturated linear lipids into their membranes. Saturated lipids form straight hydrocarbon tails that align well with neighboring lipids and form a highly organized structure that is stable in heat. The membranes of mesophiles are composed mostly of unsaturated lipids that bend against each other and align poorly. Consequently, the membranes of mesophiles are more fluid at lower temperatures.

The membranes of hyperthermophilic archaea impart an amazing level of heat resilience by being lipid monolayers, not bilayers (see Fig. 3.11 and eAppendix 2). Lipid bilayers peel apart under withering heat. Monolayers, built for extremophile living, do not. Monolayer membranes are heat stable because long hydrocarbon chains directly tether glycerophosphates on opposite sides of the membrane. The chains (40 carbons long) do not contain fatty acids, but are made of isoprene units bonded by ether linkages to glycerol phosphate. More on thermophiles can be found in Chapter 19.

How might the study of thermophiles enhance human existence? Earlier we explained how the heat-stable DNA polymerase from thermophiles helped commercialize the polymerase chain reaction. PCR can be used to amplify DNA directly from a natural environment—a feat that helps identify the unique 16S ribosomal RNA sequences of uncultured extremophiles, including those of new

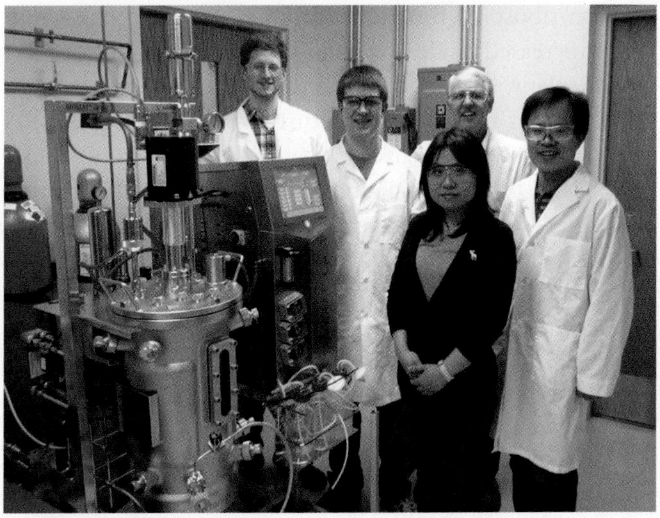

FIGURE 5.4 ▪ **Bioreactor used to grow thermophilic microorganisms.** Robert Kelly and students stand next to a 20-liter bioreactor that they use for the engineering analysis of biofuel-producing microbes. Left to right: Aaron Hawkins, Andrew Loder, Hong Lian, Kelly, and Yejun Han.

> **Thought Question**
>
> **5.2** If microbes lack a nervous system, how can they sense a temperature change?

Adaptation to Pressure

Living creatures at Earth's surface (sea level) are subjected to a pressure of 1 atmosphere (atm), which is equal to 0.101 megapascal (MPa) or 14 pounds per square inch (psi). At the bottom of the ocean, however—thousands of meters deep—hydrostatic pressure averages a crushing 400 atm and can reach as high as 1,000 atm (101 MPa, or 14,600 psi) or more in ocean trenches (**Fig. 5.5**). Organisms adapted to grow at these overwhelmingly high pressures are called **barophiles** or **piezophiles**. From the curves in **Figure 5.6**, notice that barophiles actually require elevated pressure to grow, while barotolerant organisms grow well in the range of 1–50 MPa, but their growth falls off thereafter.

thermophiles and hyperthermophiles. But beyond simply identifying the uncultured microbes, random PCR amplification and sequencing of DNA fragments from an environment enables scientists to see the entire repertoire of genes, regardless of species, present in that environment. The ability to peek at the genomes of hyperthermophiles and predict what the various genes do led Robert Kelly and colleagues at North Carolina State University to actually mix genes, and thus enzymes, from different species to invent new metabolic pathways (**Fig. 5.4**). Their quest is to engineer a new organism that can convert abundantly available CO_2 and H_2 directly into high-energy liquid fuels that could further decrease dependence on dwindling fossil fuels.

The Heat-Shock Response

What happens to a bacterium that accidentally encounters a temperature above its comfort zone? Most microorganisms possess quick-response genetic programs that remodel their physiology into one that can temporarily survive this inhospitable condition, which is called heat shock. Rapid temperature changes experienced during growth activate batches of stress response genes, resulting in the **heat-shock response** (discussed in Chapter 10). The protein products of these heat-activated genes include chaperones that maintain protein shape and enzymes that change membrane lipid composition. The heat-shock response, first identified in *E. coli* by Tetsuo Yamamori and Takashi Yura in 1982, has since been documented in almost all living organisms examined thus far. Antarctic marine organisms that live under extremely stable temperatures (−1.9°C) are an exception.

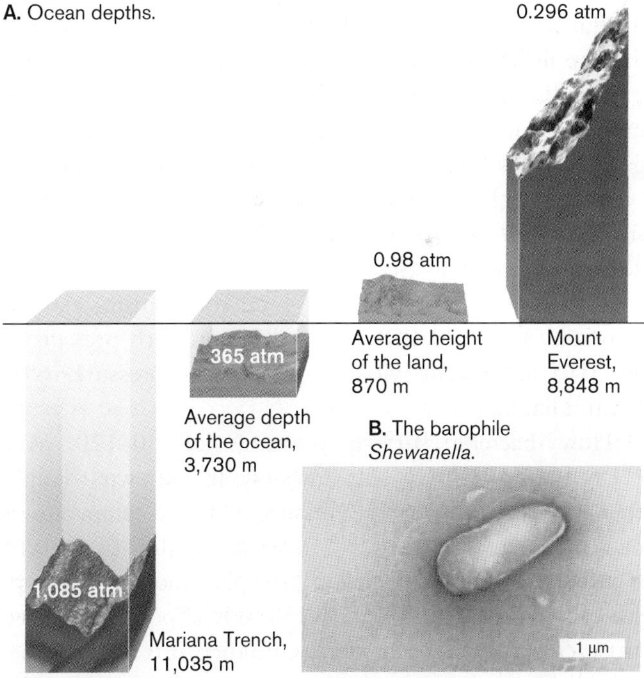

FIGURE 5.5 ▪ **Barophilic environments and piezophiles.**
A. The deepest part of the ocean is at the bottom of the Mariana Trench, a depression in the floor of the western Pacific Ocean, just east of the Mariana Islands. The Mariana Trench is 2,500 km (1,554 miles) long and 70 km (44 miles) wide. Near its southwestern extremity, about 340 km (210 miles) southwest of Guam, lies the deepest point on Earth. This point, referred to as the Challenger Deep, plunges to a depth of 11,035 meters (nearly 7 miles). The pressure there (110 MPa) is over 1,000 times higher than what we experience on land (0.1 MPa). **B.** A Gram-negative barophile in the genus *Shewanella*, isolated from sea sediment located 6.8 miles below sea level.

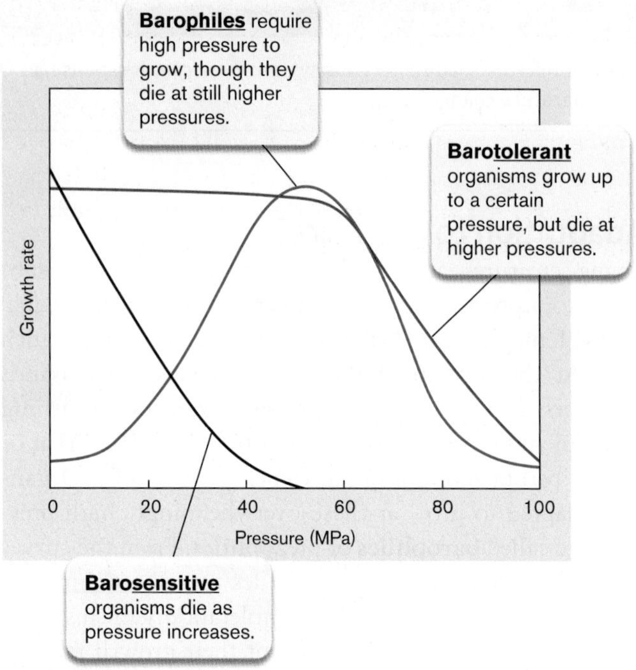

FIGURE 5.6 ■ Relationship between growth rate and pressure.

Many barophiles are also psychrophilic, because the average temperature at the ocean's floor is 2°C. However, barophilic hyperthermophiles form the basis of thermal vent communities that support symbiotic worms and giant clams (see Chapter 21). A recently identified thermophilic piezophile, *Thermococcus piezophilus*, was isolated from a hydrothermal vent 4,969 meters below sea level. This archaeon has the broadest known pressure range for growth ever described for a microorganism, extending from 1 to 120 MPa. Its optimal growth pressure is 50 MPa, consistent with the hydrostatic pressure in its natural habitat.

How bacteria survive pressures of 80–120 MPa (11,600–17,404 psi) is still a mystery. It is known, though, that increased hydrostatic pressure and cold temperatures decrease membrane fluidity. Because fluidity of the cell membrane is critical to survival, the phospholipids of deep-sea bacteria commonly have high levels of polyunsaturated fatty acids to increase membrane fluidity. It is thought that in addition to these membrane changes, internal structures must be pressure adapted. For example, ribosomes in the barosensitive organism *E. coli* (maximum growth pressure 50 MPa) dissociate at pressures above 60 MPa, so barophiles must contain uniquely designed ribosome structures that can withstand even higher pressures.

A study using the obligate piezophile *Pyrococcus yayanosii* CH1 revealed that either above or below optimal growth pressure (50 MPa), this archaeon will alter the expression of many ribosomal genes, some of which increase while others decrease. What effect these changes have on ribosome function is unclear. One protein whose levels increased under high and low pressures has a predicted role in ribosome recycling—a finding that supports the idea that ribosomal structure and function are especially sensitive to pressure.

Clues to other physiological changes needed for growth at high pressure may come from *E. coli* adaptive-evolution studies. Over a period of 500 generations (126 days), Douglass Bartlett and colleagues at the Scripps Institution of Oceanography gradually increased the pressure under which *E. coli* was grown to 62 MPa. One *E. coli* mutant evolved to successfully grow at this pressure. The evolved strain developed a mutation in a key fatty acid synthesis protein, which changed the ratio of its membrane fatty acids. Further studies could uncover mutations in additional genes and improve our understanding of the evolutionary steps necessary for microbial growth under extreme pressure.

To Summarize

- **Different species** exhibit different optimal growth values of temperature, pH, and osmolarity.
- **Extremophiles** inhabit fringe environments with conditions that do not support human life.
- **The environmental habitat** (for example, the temperature or pressure condition) of a particular species is defined by the tolerance of that organism's proteins and other macromolecular structures to the physical conditions within that niche.
- **Global approaches** used to study gene expression help us view how organisms respond to changes in their environment.
- **The Arrhenius equation** applies to the growth of microorganisms: within a specific growth temperature range, the growth rate doubles for every 10°C rise in temperature.
- **Mesophiles, psychrophiles, and thermophiles** are groups of organisms that grow at moderate, low, and high temperatures, respectively.
- **Membrane fluidity** varies with the composition of lipids in a membrane, which in turn dictates the temperature and pressure at which an organism can grow.
- **The heat-shock response** produces a series of protective proteins in organisms exposed to temperatures near the upper edge of their growth range.
- **Barophiles (piezophiles)** can grow at pressures up to 1,000 atm or more but fail to grow at low pressures. Growth at high pressure requires specially designed membranes and protein structures.

Thought Question

5.3 Predict how hyperthermophilic microorganisms colonize a newly formed hydrothermal vent (black smoker), which is sterile at birth. How do the microbes get there through ice-cold water (0°C–3°C)?

5.2 Osmolarity

Water is critical to life, but environments differ in the amount of water actually available to growing organisms. Water availability in any solution is measured as **water activity** (a_w), a quantity affected by solute concentration. Interactions between water and solutes in solution will lower water activity. So, the more solutes there are in a solution, the less water there is available for microbes to use for growth. Water activity is typically measured as the ratio of the solution's vapor pressure in a sealed chamber relative to that of pure water. If the air above the sample is 97% saturated relative to the moisture present over pure water, the relative humidity is 97% and the water activity is 0.97. Most bacteria growing on land or in freshwater habitats require water activity to be greater than 0.91 (the water activity of seawater). Fungi can tolerate water activity levels as low as 0.86. Thus, fungi predominate in spoilage of dry foods or on damp walls.

Osmotic Stress

Osmolarity is a measure of the number of solute molecules in a solution and is inversely related to a_w. The more particles there are in a solution, the greater the osmolarity and the lower the water activity. Osmolarity is also important for a cell because of the cell's semipermeable cytoplasmic membrane. This membrane allows the osmolarity inside the cell to differ from the osmolarity outside. The principles of physical chemistry dictate that a solute present at different concentrations in two chambers separated by a semipermeable membrane will tend to equilibrate. But if the semipermeable membrane does not allow solutes to move through the membrane, water will move through the membrane. Water leaves the chamber with the lower solute concentration and moves into the chamber with the higher solute concentration (see eAppendix 2).

Water does not move across cell membranes primarily by simple diffusion. Instead, special membrane water channels formed by proteins called aquaporins enable water to traverse the membrane much faster than by unmediated diffusion (**Fig. 5.7**). Rapid movement of water helps protect cells against osmotic stress. However, too much water moving in or out of a cell is detrimental. Cells may ultimately explode or implode, depending on the direction the water moves. Even bacteria with a rigid cell wall suffer. They may not explode like a human cell, but the forces placed on the cell membrane are great. Membrane transport systems, for example, can be inactivated.

Protection against Osmotic Stress

In addition to moving water, microbes have at least two other mechanisms to minimize osmotic stress across membranes. When stranded in a hypertonic medium (higher osmolarity than the cell), bacteria try to protect their internal water from leaving the cell by synthesizing or importing compatible solutes that increase intracellular osmolarity. **Compatible solutes** are small molecules that do not disrupt normal cell metabolism even at high intracellular concentrations. Increasing the intracellular levels of these compounds (such as proline, glutamic acid, potassium, or betaine) elevates cytoplasmic osmolarity without any detrimental effects, making it unnecessary for water to leave the cell. In contrast, ions such as Na^+ are not compatible solutes and will disturb metabolism at high intracellular concentrations.

Cells also contain pressure-sensitive (mechanosensitive) channels that can be used to leak solutes out of the cell. Internal pressure rises in cells immersed in a hypotonic medium (having lower osmolarity than the cell has). The mechanosensitive channels open in response to increased internal pressure. The opened channels allow small solutes to escape, thereby lowering internal osmolarity and preventing too much water from entering the cell.

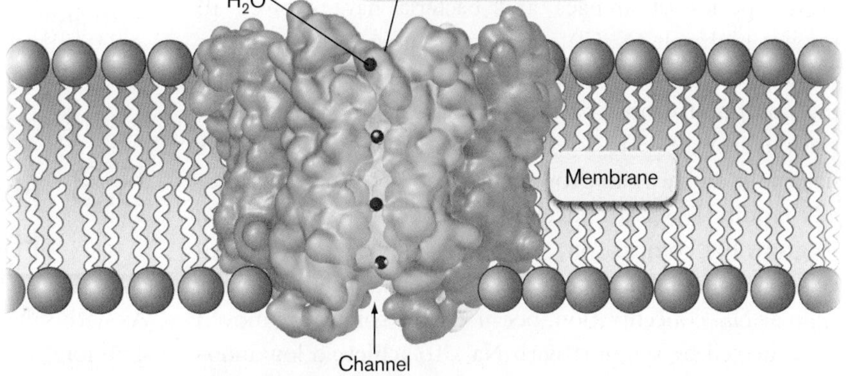

FIGURE 5.7 ■ Aquaporin. Transverse view of the channel through which water molecules move. Complementary halves of the channel are formed by adjacent protein monomers. (PDB code: 1J4N)

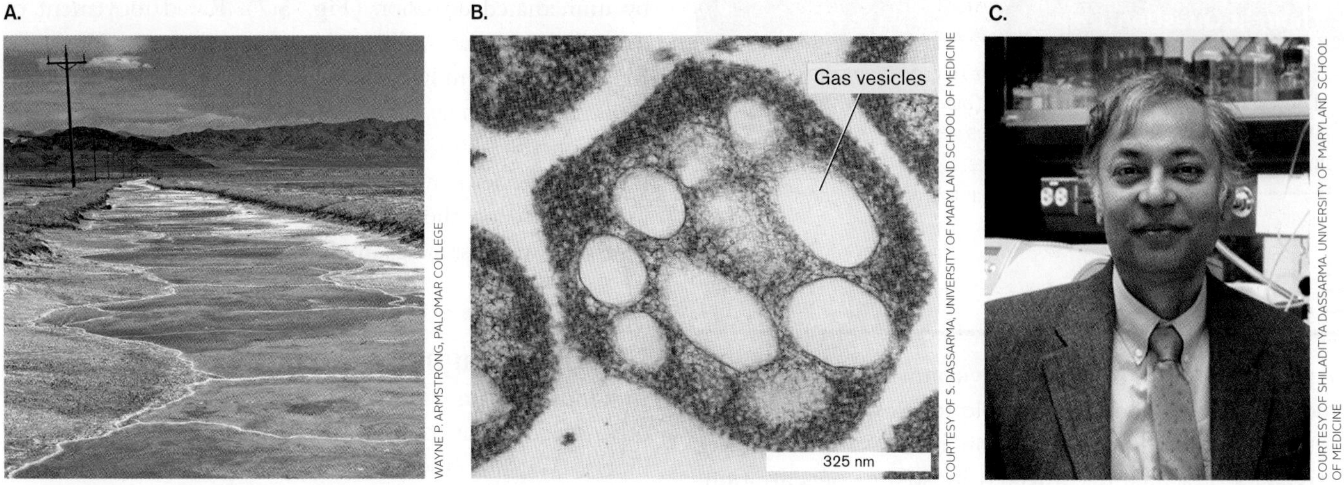

FIGURE 5.8 ▪ Halophilic salt flats and halophilic bacteria. **A.** The halophilic salt flats along Highway 50 east of Fallon, Nevada, are colored pinkish red by astronomical numbers of halophilic bacteria. **B.** Cross section of the archaeon *Halobacterium* species (TEM). Gas vesicles allow the organism to float in liquid and acquire more oxygen. **C.** Shiladitya DasSarma and colleagues at the University of Maryland sequenced the genome of *Halobacterium* species NRC-1 and are developing gas-vesicle nanoparticles as a novel vaccine delivery system.

Outside a certain range of external osmolarity, the aquaporin and compatible solute strategies become ineffective at controlling internal osmolarity. To adapt, microbes launch a larger global response in which cell physiology is transformed to tolerate brief encounters with potentially lethal salt (or other solute) concentrations. Some changes are similar to those provoked by heat shock, such as the increased synthesis of chaperones that protect critical cell proteins from denaturation. Other changes include alterations in outer membrane pore composition (for Gram-negative organisms).

Thought Question

5.4 How might the concept of water availability be used by the food industry to control spoilage?

Halophiles Require High Salt

Some species of archaea and bacteria have evolved to require high salt (NaCl) concentrations in order to grow. These microbes are called **halophiles** (**Fig. 5.8**). In striking contrast to most bacteria, which require salt concentrations from 0.05 to 1 M (0.2%–5% NaCl), the extremely halophilic archaea can grow at an a_w of 0.75, and they actually require NaCl at concentrations of 2–4 M (10%–20%) to grow. For comparison, seawater is about 3.5% NaCl. All cells, even halophiles, prefer to keep a relatively low intracellular Na^+ concentration, because some solutes are moved into the cell by symport with Na^+. To achieve a low internal Na^+ concentration, halophilic microbes use special ion pumps to excrete sodium and replace it with other cations, such as potassium, which is a compatible solute. In fact, the proteins and cell components (for example, ribosomes) of halophiles require remarkably high intracellular potassium levels to maintain their structure. Halophilic archaea are presented in Chapter 19.

To Summarize

- **Water activity** (a_w) is a measure of how much water in a solution is available for a microbe to use.
- **Osmolarity** is a measure of the number of solute molecules in a solution and is inversely related to a_w.
- **Aquaporins** are membrane channel proteins that allow water to move quickly across membranes to equalize internal and external pressures.
- **Compatible solutes** are used to minimize pressure differences across the cell membrane.
- **Mechanosensitive channels** can leak solutes out of the cell when internal pressure rises.
- **Halophilic organisms** require high salt concentrations to grow.

5.3 Hydronium (pH) and Hydroxide Ion Concentrations

As with salt and temperature, the concentration of hydrogen ions (H^+)—actually, hydronium ions (H_3O^+)—also has a direct effect on the cell's macromolecular structures. Extreme concentrations of either hydronium or hydroxide ions (OH^-) in a solution will limit growth and kill cells. Despite this sensitivity to pH extremes, living cells tolerate a greater range in environmental concentration of H^+

than of virtually any other chemical substance. *E. coli*, for example, tolerates a pH range from 2 to 10, a 100-million-fold difference (but grows only between pH 4.5 and 9). For a brief review of pH, refer to eAppendix 1.

Note: Recall that pH = $-\log_{10}$ of H^+ concentration in moles per liter. Thus, a solution containing 1×10^{-6} mole/liter of H^+ ions has a pH of 6. Likewise, pOH = $-\log_{10}$ of OH^- concentration. The thermodynamic properties of water dictate that pH + pOH always equals 14.

It is also important to note that pH can have a dramatic effect on the availability of some nutrients. The classic example is iron. Iron hydroxide [$Fe(OH)_3$], the predominant form of iron in nature, is very insoluble above pH 7. Organisms that thrive under moderate to severe alkaline conditions have evolved highly efficient iron transport systems that can scavenge what little soluble $Fe(OH)_3$ remains in those environments. Iron transport is discussed in Section 4.2.

pH Optima, Minima, and Maxima

The charges on various amino or carboxyl groups within a protein help forge the intramolecular bonds that dictate protein shape and thus protein activity. Because shifting H^+ concentration, [H^+], will affect the protonation of these ionizable groups, changing the pH will alter protein structure and activity. The result is that all enzyme activities exhibit pH optima, minima, and maxima. As we saw with temperature, groups of microbes have evolved to inhabit diverse niches, for which pH values can range from 0 to 11.5 (**Fig. 5.9**). However, species differences in optimal growth pH are not dictated by the pH limits at which critical cell proteins function.

Generally speaking, the majority of enzymes, regardless of the pH at which their source organism thrives, tend to operate best between pH 5 and 8.5 (which, if you think about it, is still a 3,000-fold range in hydrogen ion concentration). Yet many microbes grow in even more acidic or more alkaline environments.

Unlike its temperature, the intracellular pH of a microbe, as well as its osmolarity, is not necessarily the same as that of its environment. Biological membranes are relatively impermeable to protons—a fact that allows the cell to maintain an internal pH compatible with protein function when growing in extremely acidic or alkaline environments. When the difference between the intracellular and extracellular pH (ΔpH) is very high, protons can leak through proteins that thread the membrane. Excessive influx or efflux of protons can cause problems by altering internal pH.

Membrane-permeant organic acids, also called weak acids (discussed in Chapter 3), can accelerate the leakage of protons into a cell. Unlike H^+, the uncharged form of an organic acid (HA) can freely permeate cell membranes and dissociate intracellularly, releasing a proton that then acidifies the internal pH (**eTopic 5.3** describes how). Lactic acid produced and secreted by lactobacilli during the formation of yogurt is an example of self-imposed organic acid stress. The buildup of lactic acid (via fermentation) limits bacterial growth, leaving yogurt with plenty of food value. The food industry has taken advantage of this phenomenon by preemptively adding citric acid or sorbic acid to certain foods. This practice controls microbial growth under pH conditions that do not destroy the flavor or quality of the food. Food microbiology is discussed further in Chapter 16.

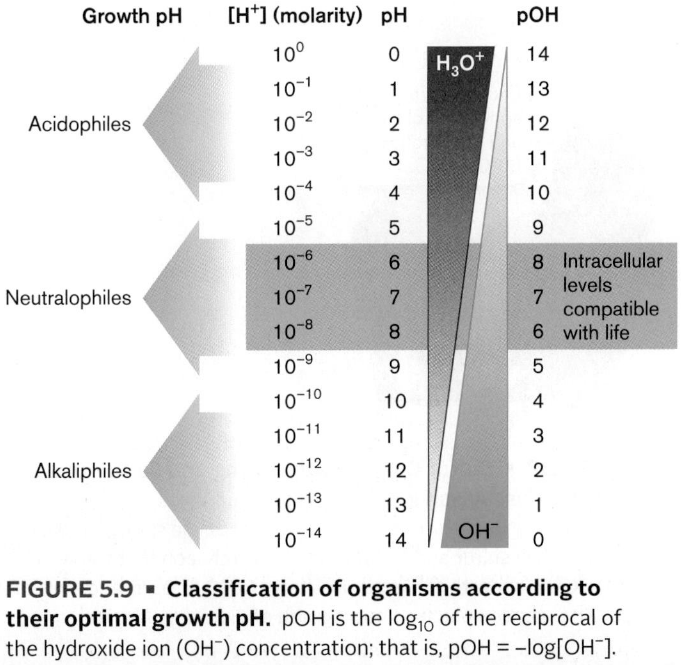

FIGURE 5.9 ■ Classification of organisms according to their optimal growth pH. pOH is the $\log_{10}$ of the reciprocal of the hydroxide ion (OH^-) concentration; that is, pOH = $-\log[OH^-]$.

Neutralophiles, Acidophiles, and Alkaliphiles Grow in Different pH Ranges

Defined groups of cells have evolved to live under different pH conditions. They do not do this by drastically changing the pH optima of their enzymes. Instead they use novel pH homeostasis strategies that maintain intracellular pH between pH 5 and pH 8, even when the cell is immersed in pH environments well above or below that range.

Three classes of organisms are differentiated by the pH range at which they grow: neutralophiles, acidophiles, and alkaliphiles (see **Fig. 5.9**).

Neutralophiles, which generally grow between pH 5 and pH 8, include most human pathogens. Many neutralophiles, including *E. coli* and *Salmonella enterica*, adjust their metabolism to maintain an internal pH slightly above neutrality, which is where their enzymes work best. They maintain this pH even in the presence of moderately acidic

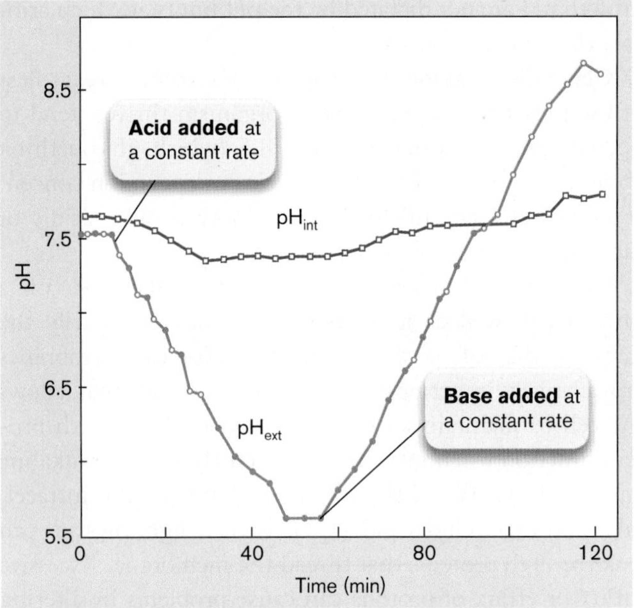

FIGURE 5.10 ■ Maintaining internal pH (pH homeostasis) over a wide range of external pH. Internal pH (pH_{int}) of the neutralophile *Escherichia coli* measured after the addition of acid to change external pH (pH_{ext}) and the subsequent addition of base. Internal pH was determined using nuclear magnetic resonance to measure changes in methyl phosphate. The two phosphate species titrate over different pH ranges. *Source:* Joan L. Slonczewski et al. 1981. *PNAS* **78**:6271.

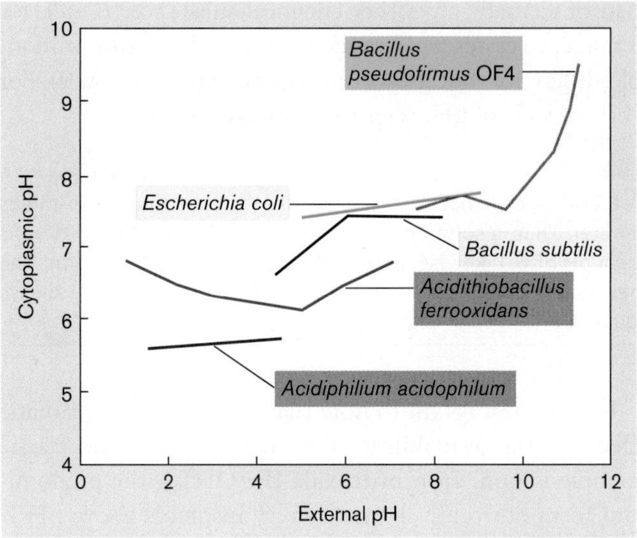

FIGURE 5.11 ■ Cytoplasmic pH as a function of the external pH among acidophiles (purple), neutralophiles (yellow), and alkaliphiles (blue). *Source:* Modified from Joan L. Slonczewski et al. 2009. *Adv. Microb. Physiol.* **55**:1–79.

or basic external environments (**Fig. 5.10**). Other neutralophiles allow their internal pH to fluctuate with external pH but usually maintain a pH difference (ΔpH) of about 0.5 pH unit across the membrane at the upper and lower limits of growth pH. The ΔpH value is an important component of the transmembrane proton potential, a source of energy for the cell (see Chapter 14).

Acidophiles are bacteria and archaea that live in extreme acidic environments. They are often lithotrophs (chemolithoautotrophs) that oxidize reduced metals and generate strong acids, such as sulfuric acid. Consequently, they grow between pH 0 and pH 5. Acidophiles generally maintain an internal pH that is considerably more acidic than that of neutralophiles but still less acidic than their growth environment (**Fig. 5.11**). The ability to grow at this pH is due partly to altered membrane lipid profiles (high levels of tetraether lipids) that decrease proton permeability, and partly to ill-defined proton extrusion mechanisms. Often, an organism that is an extremophile with respect to one environmental factor is an extremophile with respect to others as well. *Sulfolobus acidocaldarius*, for example, is a thermophile and an acidophile (**Fig. 5.12**). It grows in acidic hot springs rich in sulfur, and oxidizes reduced sulfur as an energy source.

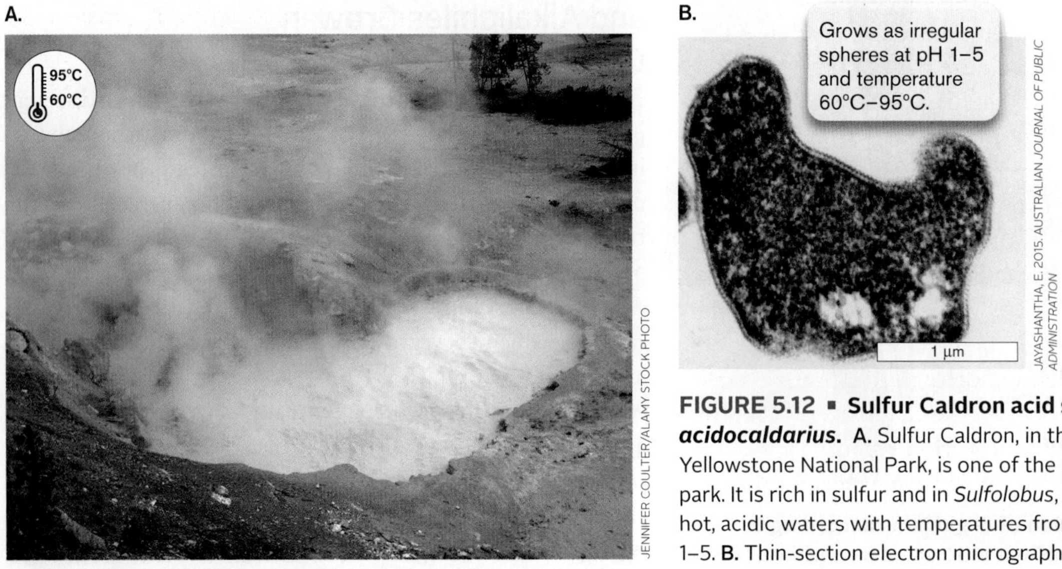

FIGURE 5.12 ■ Sulfur Caldron acid spring and *Sulfolobus acidocaldarius*. **A.** Sulfur Caldron, in the Mud Volcano area of Yellowstone National Park, is one of the most acidic springs in the park. It is rich in sulfur and in *Sulfolobus*, an archaeon that thrives in hot, acidic waters with temperatures from 60°C to 95°C and a pH of 1–5. **B.** Thin-section electron micrograph of *S. acidocaldarius*.

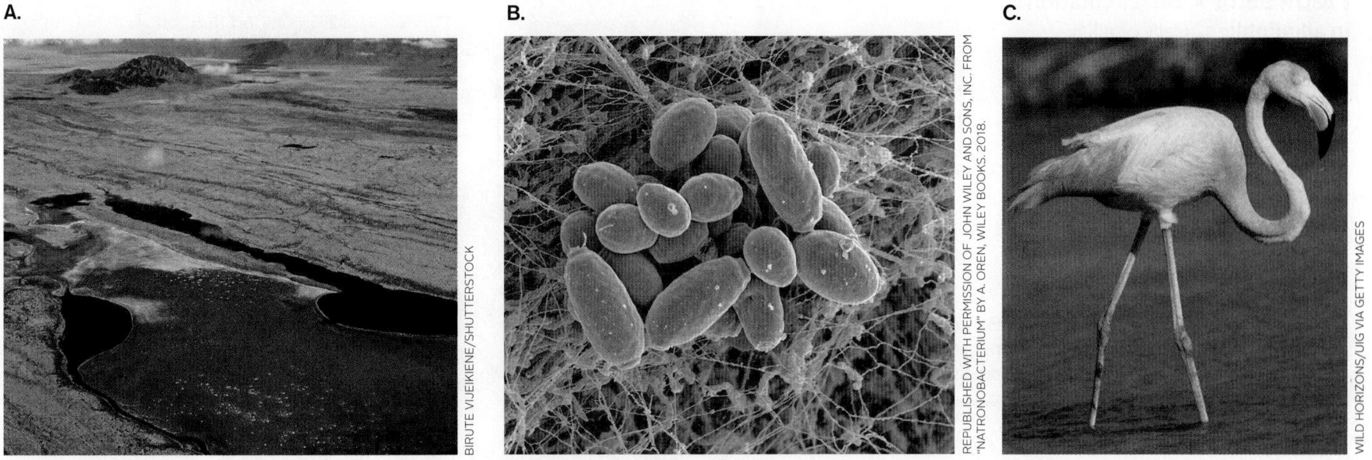

FIGURE 5.13 ■ A soda lake ecosystem. **A.** Lake Magadi in Kenya. Its pink color is due to the cyanobacterium *Spirulina*. **B.** Alkaliphile *Natronobacterium gregoryi*. Cell size, approx. 1 µm × 3 µm. **C.** Pink flamingos turn pink because they ingest large quantities of *Spirulina*.

Alkaliphiles occupy the opposite end of the pH spectrum, growing best at values ranging from pH 9 to pH 11. They are commonly found in saline soda lakes, which have high salt concentrations and pH values as high as pH 11. Soda lakes, like Lake Magadi in Kenya's Great Rift Valley (**Fig. 5.13A**), are steeped in carbonates, which explains their extraordinarily alkaline pH. An alkaliphilic organism first identified in Lake Magadi is the halophilic archaeon *Natronobacterium gregoryi* (**Fig. 5.13B**).

The cyanobacterium *Spirulina* is another alkaliphile that grows in soda lakes. Its high concentration of carotene gives the organism a distinctive pink color (note the color of the lake in **Fig. 5.13A**). *Spirulina* is also a major food for the famous pink flamingos indigenous to these African lakes and is, in fact, the reason pink flamingos are pink. After the birds ingest these organisms, digestive processes release the carotene pigment to the circulation, which then deposits it in the birds' feathers, turning them pink (**Fig. 5.13C**). Humans also consume *Spirulina*, as a health food supplement, but they do not turn pink, because the cyanobacteria are only a small component of their diet.

The internal enzymes of alkaliphiles, like those of acidophiles, exhibit rather ordinary pH optima (around pH 8). The key to the survival of alkaliphiles is the cell-surface barrier that sequesters fragile cytoplasmic enzymes away from harsh extracellular pH. Key structural features of the cell wall, such as the presence of acidic polymers and an excess of hexosamines in the peptidoglycan, appear to be essential. The reason is unclear. At the membrane, some alkaliphiles also possess a high level of diether lipids (more stable than ester-linked phospholipids), which prevent protons from leaking out of the cell (see Section 3.2).

Because external protons are in such short supply at alkaline pH, most alkaliphiles use a sodium motive force in addition to a proton motive force to do much of the work of the cell (see Section 14.1). They also rely heavily on Na^+/H^+ antiporters (see Section 4.1) to bring protons into the cell. This H^+ influx keeps the internal pH well below the extremely alkaline external pH. The Na^+/H^+ antiporters partly explain why many alkaliphiles are resistant to high salt (NaCl) concentrations: Sodium ions are expelled while protons are sucked in. Some important aspects of sodium circulation in alkaliphiles are depicted in **Figure 5.14**.

In contrast to proteins within the cytoplasm, enzymes secreted from alkaliphiles are able to work in very alkaline environments. The inclusion of base-resistant enzymes like proteases, lipases, and cellulases in laundry detergents helps get our "whites whiter and our brights brighter." See Chapter 16 for more on industrial microbiology.

Thought Question

5.5 Recall from Section 4.2 that an antiporter couples movement of one ion down its concentration gradient with movement of another molecule uphill, against its gradient. For Na^+/H^+ symporters, that means more sodium inside than out and more protons out than in. If this is true, how could a Na^+/H^+ antiporter work to bring protons into a haloalkaliphile growing in high salt at pH 10? In this situation (high salt and high pH media) there will be more sodium outside than there is in, and more H^+ inside than out. The opposite of what you'd think the cell would need. Both ions would have to move AGAINST their concentration gradients. Sodium moves out, protons move in.

pH Homeostasis and Acid Resistance

When cells are placed in pH conditions below their optimum, protons can enter the cell and lower internal pH to lethal levels. Microbes can prevent the unwanted influx of protons in a variety of ways (**Fig. 5.15**). *E. coli*, for example, can counter proton influx by transporting a

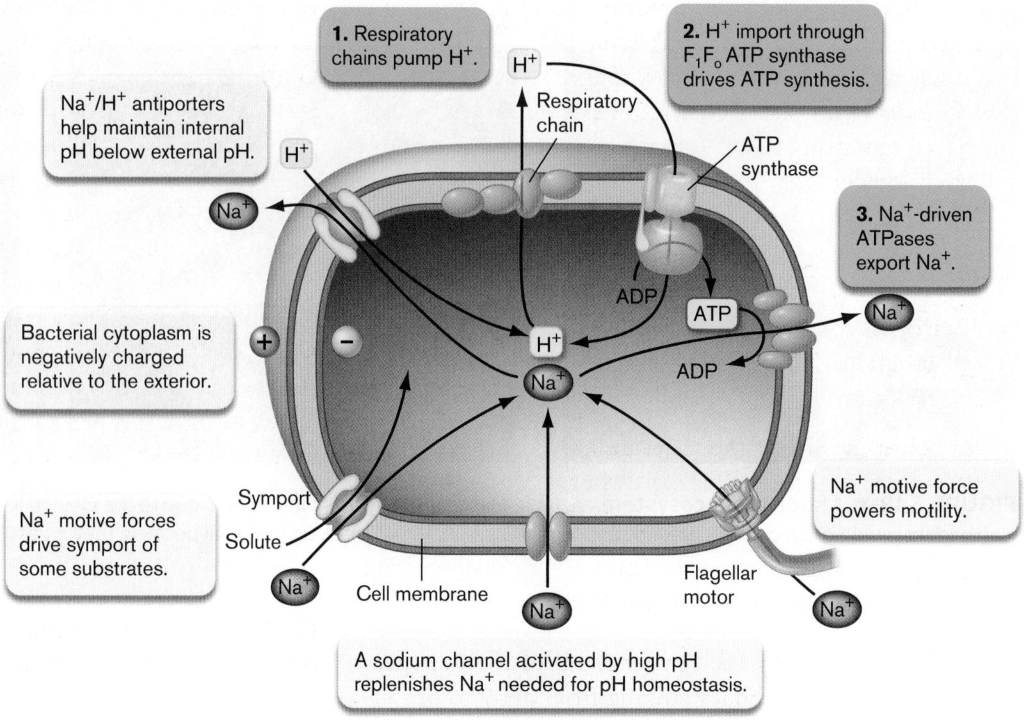

FIGURE 5.14 ■ Na^+ circulation in alkaliphiles. Cells of alkaliphiles use Na^+ in place of H^+ to do some of the work of the cell. They require an inwardly directed sodium gradient to rotate flagella and transport nutrient solutes. The Na^+/H^+ antiporter is also used to keep internal pH lower than external pH.

variety of cations, such as K^+ or Na^+. How cation transport accomplishes H^+ efflux is unclear. Some evidence suggests a link to the role of K^+ in osmoprotection. At the other extreme, under extremely alkaline conditions, the cells can use the Na^+/H^+ antiporters mentioned previously (and in Section 4.1) to recruit protons into the cell in exchange for expelling Na^+.

Some organisms can also change the pH of the medium by using various amino acid decarboxylases and deaminases. For instance, *E. coli* consumes organic acids when growing at low pH, but produces these acids while trying to grow under alkaline conditions. *Helicobacter pylori*, the causative agent of gastric ulcers, employs an exquisitely potent urease to generate massive amounts of ammonia, which neutralizes the acid pH environment. These acid stress and alkali stress protection systems are usually not made or at least do not become active until the cell encounters an extreme pH.

Many, if not all, microbes also possess an emergency global response system referred to as acid tolerance or acid resistance. In a process analogous to the heat-shock response, bacterial physiology undergoes a major molecular reprogramming in response to hydrogen ion stress. The levels of a large number of proteins increase, while the levels of others decrease. Many of the genes and proteins involved in the acid stress response overlap with other stress response systems, including the heat-shock response. These

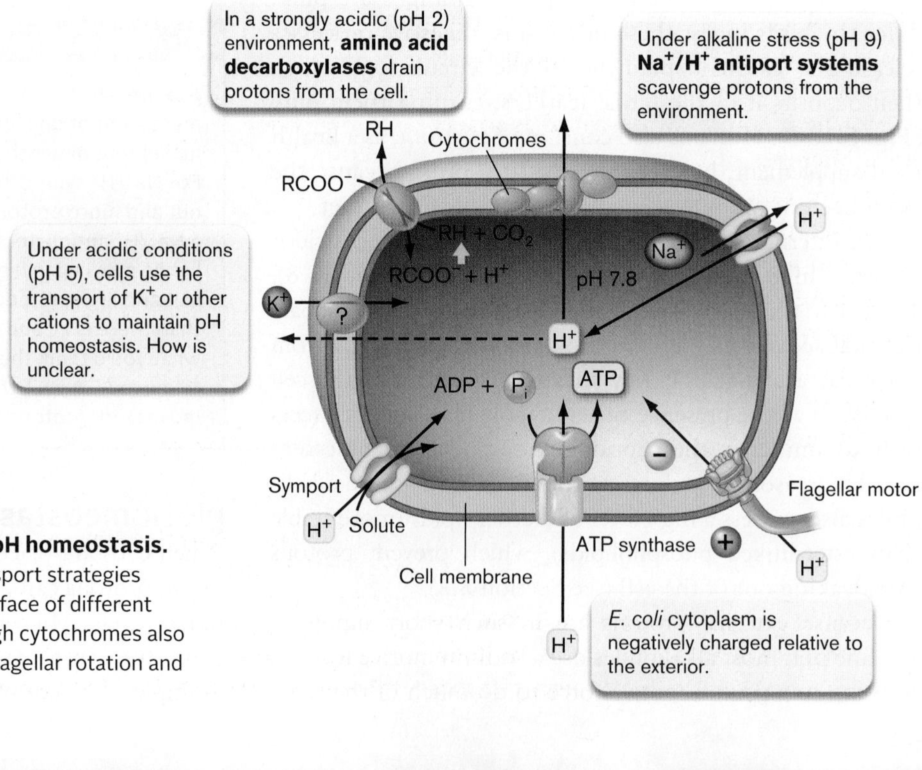

FIGURE 5.15 ■ Proton circulation and pH homeostasis. A typical *E. coli* cell uses various proton transport strategies to maintain an internal pH near pH 7.8 in the face of different external pH stresses. Proton pumping through cytochromes also establishes a proton gradient, which drives flagellar rotation and solute transport.

physiological responses include modifications in membrane lipid composition, enhanced pH homeostasis, and numerous other changes with unclear purpose. Some pathogens, such as *Salmonella*, sense a change in external pH as part of the signal indicating that the bacterium has entered a host cell environment (see **eTopic 5.4**).

To Summarize

- **Hydrogen ion concentration** affects protein structure and function. Thus, enzymes have pH optima, minima, and maxima.
- **Microbes use pH homeostasis mechanisms** to keep their internal pH near neutral when in acidic or alkaline media.
- **Adding weak acids** to certain foods undermines bacterial pH homeostasis mechanisms, thereby preventing food spoilage and killing potential pathogens.
- **Neutralophiles, acidophiles, and alkaliphiles** prefer growth under neutral, low, and high pH conditions, respectively.
- **Acid and alkali stress responses** result when a given species is placed under pH conditions that slow its growth. The cell increases the levels of proteins designed to mediate pH homeostasis and protect cell constituents.

5.4 Oxygen

Imagine having the ability to swim underwater without breathing oxygen. This sounds like science fiction for humans, but many microorganisms can grow in the absence of molecular oxygen (O_2). Environments that lack oxygen, such as sediments at the bottom of the ocean, hot springs, and vertebrate intestines, are described as **anaerobic**. Microbes that grow in those environments are called **anaerobes**. One class of anaerobes, designated as **facultative anaerobes**, can grow with or without molecular oxygen. **Aerobes**, however, grow only in the presence of oxygen. In the next section we describe the aerobes' need for oxygen, why anaerobes cannot grow in oxygen, and how facultative anaerobes grow with or without oxygen.

Oxygen Has Benefits and Risks

The key to understanding the relationships between microbes and oxygen is to know how organisms gain energy. Some microbes gain energy (ATP) only through fermenting carbohydrates, an oxygen-free process. Others require a series of membrane proteins and lipids known as an **electron transport system (ETS)** to extract intrinsic energy from electrons pulled from an energy source. Note that the term "cytochrome system" is synonymous with "ETS." The energy released from electrons moving down an ETS is then used to pump protons (H^+) out of the cell. The resulting unequal distribution of H^+ across the membrane produces a transmembrane electrochemical gradient, a kind of "biobattery" called the proton motive force (details are discussed in Section 14.1). The movement of electrons along an ETS, coupled with the generation of proton motive force, is called respiration if the electron donor is from an organic source, and lithotrophy if the source is inorganic (discussed in Chapters 13 and 14).

Once the ETS has drained as much energy as possible from an electron, that electron must be passed to a final (terminal) electron acceptor molecule, such as oxygen gas (O_2), which, in its reduced form (H_2O), diffuses away in the medium. This clears the way for another electron to be passed down the chain (**Fig. 5.16**). Aerobes use O_2 as the terminal electron acceptor for their ETS in a process called **aerobic respiration** (see Chapter 14). **Strict aerobes** use only aerobic respiration to gain energy. Anaerobes, as you will see later, have options.

We just described how O_2 benefits some organisms as a terminal electron acceptor. But why is oxygen also a risk to living systems? Regardless of whether a microbe utilizes oxygen as a terminal electron acceptor, oxygen's breakdown

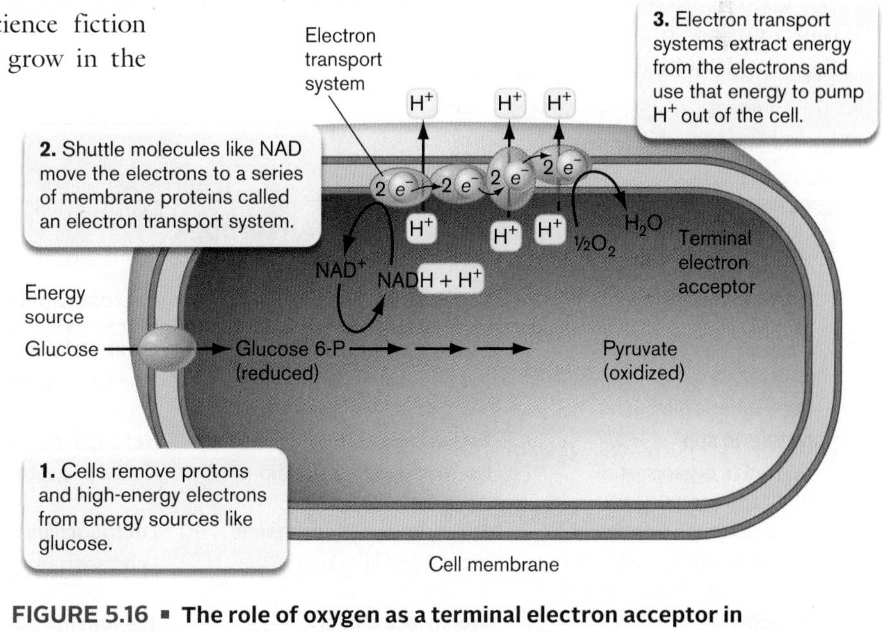

FIGURE 5.16 ▪ The role of oxygen as a terminal electron acceptor in respiration. The pumping of protons out of the cell by electron transport systems produces more positive charges outside the cell than inside, resulting in an electrochemical gradient (also called proton motive force). At the end of the ETS, the electron must be passed to a final (terminal) electron acceptor (for example, O_2), thus clearing the path for the next electron. This net process is called respiration.

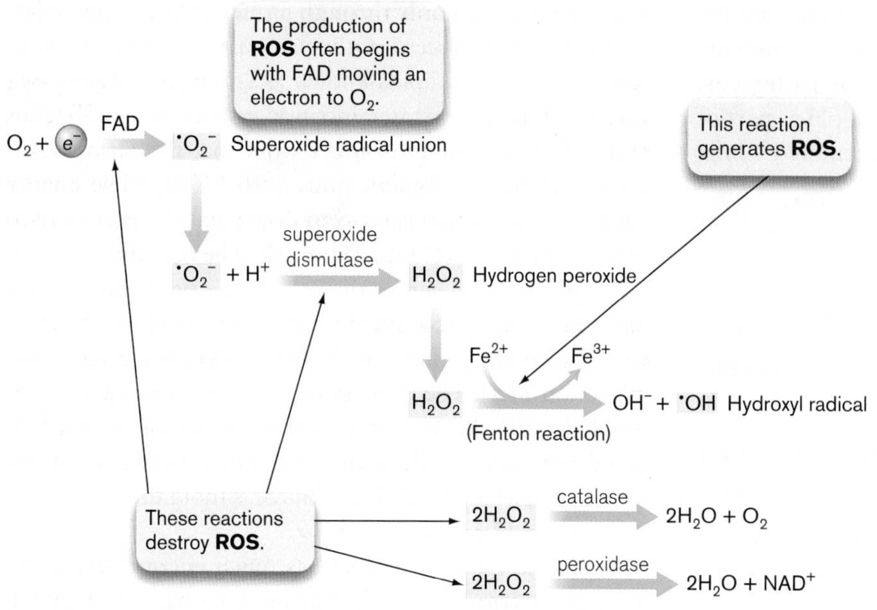

FIGURE 5.17 ■ Generation and destruction of reactive oxygen species (ROS). ROS are marked yellow. The autooxidation of flavin adenine dinucleotide (FAD) and the Fenton reaction occur spontaneously to produce superoxide and hydroxyl radicals, respectively. The other reactions require enzymes. FAD is a cofactor for a number of enzymes (for example, NADH dehydrogenase 2). Catalase and peroxidase detoxify hydrogen peroxide.

products, called reactive oxygen species (ROS), can severely damage cells. As a result, different bacteria have evolved to either tolerate or avoid oxygen altogether.

How are reactive oxygen species made? Any organism, including bacteria, that possesses NADH dehydrogenase 2—aerobe or anaerobe—will, in the presence of oxygen, inadvertently autooxidize the flavin adenine dinucleotide (FAD) cofactor within the enzyme and produce dangerous amounts of superoxide radicals ($^{\bullet}O_2^-$; **Fig. 5.17**). Superoxide will degrade to hydrogen peroxide (H_2O_2), another reactive molecule. Iron, present as a cofactor in several enzymes, can then catalyze a reaction with hydrogen peroxide to produce the highly toxic hydroxyl radical ($^{\bullet}OH$). All of these reactive oxygen species seriously damage DNA, RNA, proteins, and lipids by stripping them of electrons.

Consequently, oxygenated environments are toxic, and organisms that live in them require special talents to survive. Aerobes, for instance, destroy reactive oxygen species with an ample supply of enzymes such as superoxide dismutase (to remove superoxide) and peroxidase and catalase (to remove hydrogen peroxide). Aerobes also have resourceful enzyme systems that detect and repair macromolecules damaged by oxidation.

Aerobes versus Anaerobes

Table 5.2 gives examples of microbes that grow at different levels of oxygen. Consider where these different classes of microbes would grow in a standing test tube containing growth medium (**Fig. 5.18**). The top of the tube, closest to air, is oxygenated; the bottom of the tube has almost no oxygen (5 μM or less dissolved O_2). In between is a gradient of decreasing dissolved oxygen. Some microbes grow only at the top of the tube (aerobes), while others grow only at the bottom (anaerobes) and die (or stop growing) if

TABLE 5.2 Examples of Aerobes and Anaerobes

Aerobic microbes	Facultative microbes	Microaerophilic microbes	Anaerobic microbes
***Azotobacter* spp.** Soil microorganisms; fix atmospheric nitrogen	***Bacillus anthracis*** Cause of anthrax	***Campylobacter* spp.** One cause of gastroenteritis	***Actinomyces* spp.** Soil microorganisms; synthesize antibiotics
***Neisseria* spp.** Causative organisms of meningitis, gonorrhea	***Escherichia coli*** Normal gut biota; additional pathogenic strains	***Helicobacter pylori*** Cause of gastric ulcers	***Azoarcus tolulyticus*** Degrades toluene
Pseudomonas fluorescens Found in soil; degrades TNT and aromatic hydrocarbons	***Saccharomyces cerevisiae*** Yeast; used in baking	***Lactobacillus* spp.** Ferment milk to form yogurt	***Bacteroides* spp.** Normal gut biota
***Rhizobium* spp.** Soil microorganisms; plant symbionts	***Staphylococcus* spp.** Found on skin; cause boils	***Treponema pallidum*** Cause of syphilis	***Clostridium* spp.** Soil microorganisms; causative agents of tetanus and botulism
	Vibrio cholerae Cause of cholera		***Desulfovibrio* spp.** Reduce sulfate

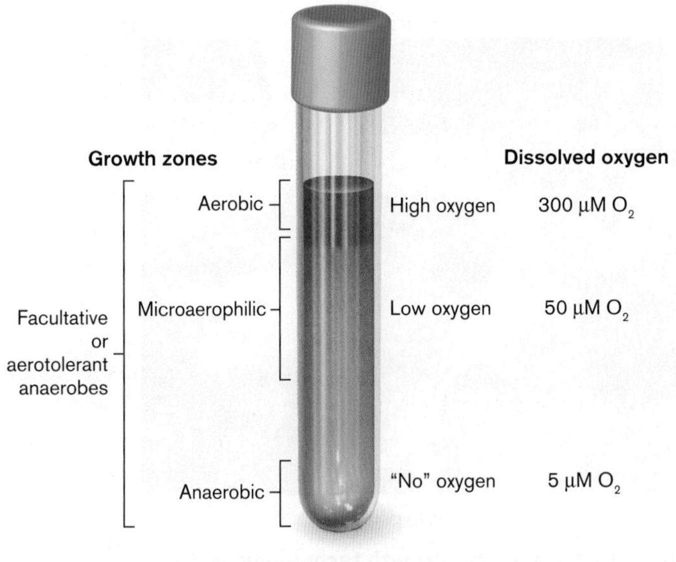

FIGURE 5.18 ■ Oxygen-related growth zones in a standing test tube containing culture medium.

they enter higher levels in the tube. Facultative anaerobes, however, can grow from top to bottom.

Why do anaerobes perish in oxygen? Some anaerobes die in oxygen because they lack the enzymes needed to destroy ROS molecules produced by their own metabolism. Other anaerobes have enzymes that can protect them from ROS, but the dissolved oxygen raises the redox potential to a point that interferes with the use of alternative (non-oxygen) electron acceptors that the organism needs to make energy. Consequently, **strict anaerobes** grow only at the bottom of the tube shown in **Figure 5.18**. As we will discuss later, some bacteria previously considered strict anaerobes are not so strict.

Classes of anaerobes. Anaerobic microbes fall into several categories. Some anaerobes actually do respire by means of electron transport systems, but instead of using oxygen, they rely on alternative terminal electron acceptors like nitrate (NO_3^-) to conduct **anaerobic respiration** and produce energy. Anaerobes of another ilk do not possess cytochromes (members of various ETSs), thus cannot respire, and so must rely on carbohydrate **fermentation** for energy; that is, they conduct **fermentative metabolism**. In fermentation, ATP energy is produced through substrate-level phosphorylation. In either case, tolerance for ROS is low.

Facultative anaerobes (such as *E. coli*) possess enzymes that destroy toxic oxygen by-products, but they have both fermentative and respiratory potential. Whether a member of this group uses aerobic respiration, anaerobic respiration, or fermentation depends on the availability of oxygen or other terminal electron acceptor and the amount of carbohydrate present. **Aerotolerant anaerobes** use only fermentation to provide energy but contain superoxide dismutase and catalase (or peroxidase) to protect them from ROS. These enzymes enable aerotolerant anaerobes to grow in air (containing oxygen) while retaining a fermentation-based (anaerobic) metabolism. Aerotolerant anaerobes will grow throughout the tube in **Figure 5.18**. Microorganisms that possess decreased levels of superoxide dismutase and/or catalase will be **microaerophilic**, meaning they will grow only at low oxygen concentrations.

Some organisms previously considered anaerobes (for example, *Bacteroides fragilis*) are really transiently aerotolerant. These "anaerobes" tolerate oxygen because they possess low levels of ROS-protective enzymes and can even use very low levels of oxygen as terminal electron acceptors. *B. fragilis*, part of the normal gastrointestinal microbiota, may even help lower O_2 levels in the intestine.

The fundamental composition of all cells reflects their evolutionary origin as anaerobes. Lipids, nucleic acids, and amino acids are all highly reduced—which is why our bodies are combustible (not spontaneously, however). We never would have evolved that way if molecular oxygen had been present from the beginning. Even today, the majority of all microbes are anaerobic, growing buried in the soil, within our anaerobic digestive tract, or within biofilms on our teeth.

Thought Questions

5.6 If anaerobes cannot live in oxygen, how do they incorporate oxygen into their cell components?

5.7 How can anaerobes grow in the human mouth, where there is so much oxygen?

Culturing Anaerobes in the Laboratory

Many anaerobic bacteria cause horrific human diseases, such as tetanus, botulism, and gangrene. Some of these organisms or their secreted toxins are even potential weapons of terror (for example, *Clostridium botulinum*). Because of their ability to wreak havoc on humans, culturing these microorganisms was an early goal of microbiologists. Despite the difficulties involved, conditions were eventually contrived in which all, or at least most, of the oxygen could be removed from a culture environment.

Three oxygen-removing techniques are used today. Special reducing agents (for example, thioglycolate) or enzyme systems (such as Oxyrase) that eliminate dissolved oxygen can be added to ordinary liquid media. Anaerobes can then grow beneath the culture surface. A second, very popular way to culture anaerobes, especially on agar plates, is to use an anaerobe jar (**Fig. 5.19A**). Agar plates streaked with the organism are placed into a sealed jar with a foil packet that releases H_2 and CO_2 gases.

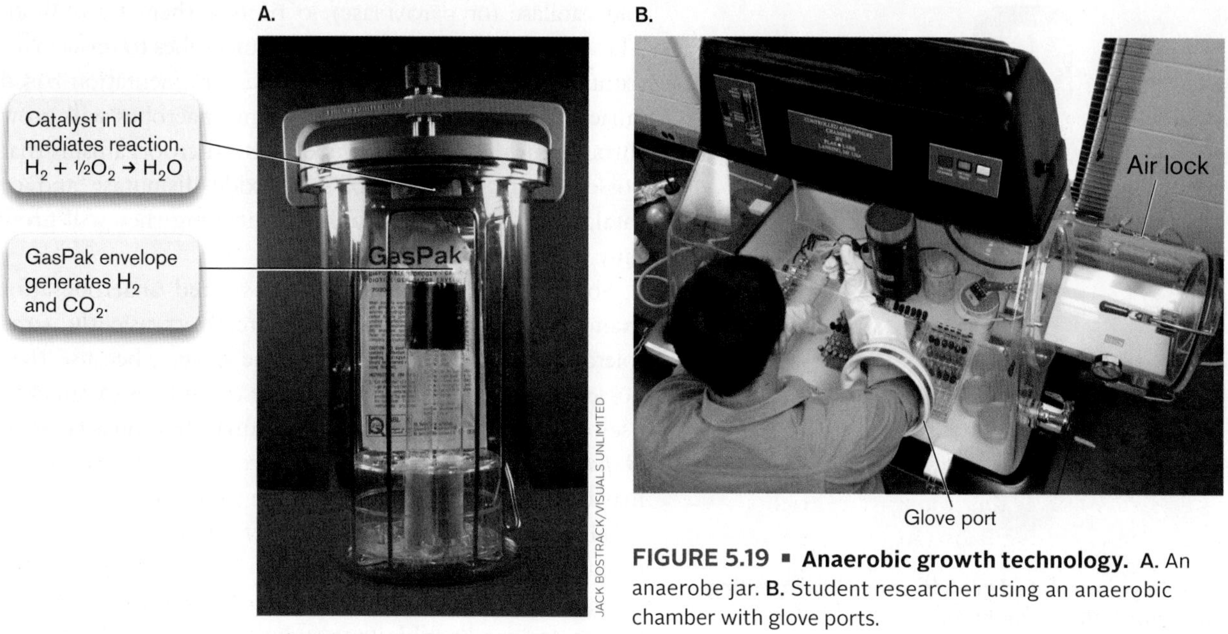

FIGURE 5.19 ■ Anaerobic growth technology. **A.** An anaerobe jar. **B.** Student researcher using an anaerobic chamber with glove ports.

A palladium packet hanging from the jar lid catalyzes a reaction between the H_2 and O_2 in the jar to form H_2O, and effectively removes O_2 from the chamber. The CO_2 released is required by some reactions to produce key metabolic intermediates. Some microaerophilic microbes, like the pathogens *Helicobacter pylori* (the major cause of stomach ulcers) and *Campylobacter jejuni* (a major cause of diarrhea), require low levels of O_2 but elevated amounts of CO_2. Another type of gas-generating packet can produce these conditions.

Because strict anaerobes are exquisitely sensitive to oxygen, even more heroic efforts are required to establish an oxygen-free environment. A special anaerobic glove box must be used in which the atmosphere is removed by vacuum and replaced with a precise mixture of N_2 and CO_2 gases (**Fig. 5.19B**).

Thought Question

5.8 What evidence led people to think about looking for anaerobes? *Hint:* Look up "Spallanzani," "Pasteur," and "spontaneous generation" on the Internet.

To Summarize

- **Oxygen is a benefit to aerobes**, organisms that can use it as a terminal electron acceptor to extract energy from nutrients.
- **Oxygen is toxic** to all cells—for example, anaerobes—that do not have enzymes capable of efficiently destroying reactive oxygen species (ROS).
- **Anaerobic metabolism** can be either **fermentative** or **respiratory**. Anaerobic respiration requires the organism to possess cytochromes that can transfer electrons to terminal electron acceptors other than oxygen. Fermentative metabolism uses substrate-level phosphorylation to generate ATP in a process that does not involve cytochromes.
- **Aerotolerant anaerobes** grow in either the presence or the absence of oxygen, but they use fermentation as their primary, if not only, means of gathering energy. These microbes also have enzymes that destroy ROS, allowing them to grow in oxygen.
- **Facultative anaerobes** grow with or without oxygen and have enzymes that destroy ROS. Some utilize only fermentative metabolism, while others can ferment and respire via anaerobic and/or aerobic means. Those that aerobically respire use oxygen as a terminal electron acceptor.

5.5 Nutrient Deprivation and Starvation

Obviously, limiting the availability of a carbon source or other essential nutrient will limit growth. Less obvious are the dramatic molecular events that cascade through a starving cell. Optimizing growth rate when nutrient levels are suboptimal is an important aim of free-living bacteria, given that intestinal, soil, and marine environments rarely offer excess nutrients.

Starvation Activates Survival Genes

Numerous gene systems are affected when nutrients decline (see Sections 10.1 and 10.3). Growth rate slows, and daughter cells become smaller and begin to experience what is called a "starvation" response, in which the microbe senses a dire situation developing but still strives to find new nourishment. The resulting metabolic slowdown generates increased concentrations of critically important small signaling molecules, such as cyclic adenosine monophosphate (cyclic AMP, or cAMP) and guanosine tetraphosphate (ppGpp), which globally transform gene expression. The highly soluble nature of these small molecules means they can quickly diffuse throughout the cell, promoting a fast response. During this metabolic retooling, transport systems for potential nutrients are produced even if the matching substrates are unavailable. Cells begin to make emergency internal energy stores such as glycogen, in case no other nutrient is found. Some organisms growing on nutrient-limited agar plates can even form colonies with intricate geometrical shapes that help the population cope, in some unknown way, with nutrient stress (**Fig. 5.20**).

As nutrient conditions worsen, the organism prepares for famine by activating many different stress survival genes. The products of these genes afford protection against stressors such as reactive oxygen radicals or temperature and pH extremes. No cell can predict the precise stresses it might encounter while incapacitated, so it is advantageous to be prepared for as many as possible. As described in Section 4.6, some species undergo elaborate developmental processes that ultimately produce dormant spores.

When severely stressed by starvation, some members of a bacterial population appear to sacrifice themselves to save others by undergoing what is termed **programmed cell death**. The dying cells release nutrients that neighboring cells use to survive. One of the mechanisms for programmed cell death involves so-called toxin-antitoxin (TA) systems. For each TA pair, the toxin protein will stop growth or kill the cell, but the antitoxin (sometimes a protein, sometimes a small RNA molecule) can inactivate the toxin.

An important toxin-antitoxin system in *E. coli* is the MazE (antitoxin)–MazF (toxin) module (**Fig. 5.21A**). Because toxin and antitoxin are simultaneously made, healthy cells live. However, MazE is unstable (degraded by the ClpAP protease) and must continually be replenished by synthesis to inactivate MazF. If cells are starved, they stop making MazE and F. MazE antitoxin is degraded, leaving the more stable MazF free to cleave many cellular mRNA molecules. As a result, the cell first enters stasis, from which it can recover if more MazE is made. But if MazE is not forthcoming, the cell dies and releases nutrients. The dying cell will also signal nearby cells to undergo programmed cell death. A peptide cleaved from glucose 6-phosphate dehydrogenase is released from the dying cell and enters nearby cells. The peptide binds to MazE and prevents it from neutralizing MazF. The MazF toxin, now active, will eventually kill the bystander cell. Combined, enough nutrients are released to rescue a subset of the population. Nancy Woychik at Rutgers University studies how toxin-antitoxin systems contribute to latency of the pathogen *Mycobacterium tuberculosis* (**Fig. 5.21B**).

When all else fails, a small subset of cells in a starving bacterial culture or biofilm will enter a dormant state (different from sporulation) in which cells remain viable but do not grow. As growth slows during starvation, ribosome numbers initially decrease as protein synthesis demand slows. To become truly dormant, cells must stabilize the few remaining ribosomes in a state of hibernation that can be reversed. One protein critical to this process in many species is called hibernation promoting factor (HPF). The binding of HPF to ribosomes halts

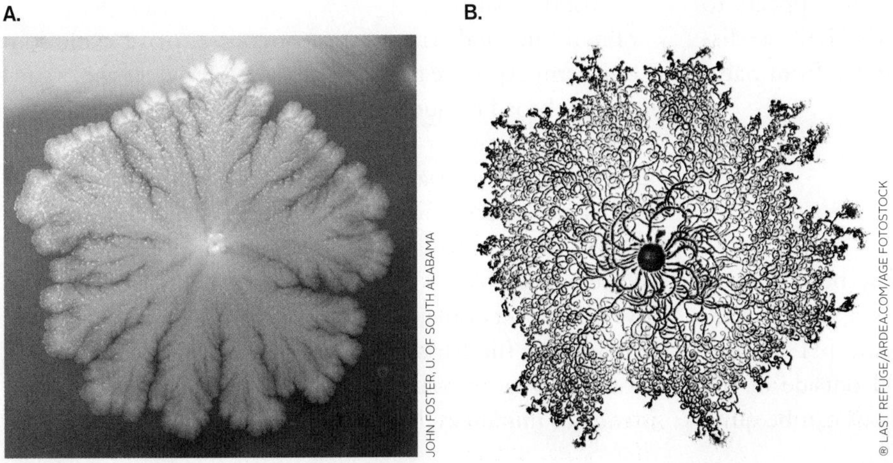

JOHN FOSTER, U. OF SOUTH ALABAMA

© LAST REFUGE/ARDEA.COM/AGE FOTOSTOCK

FIGURE 5.20 ■ Effects of starvation on colony morphology. **A.** Starving *E. coli* colony (diameter, 6 cm). **B.** *Paenibacillus dendritiformis* grown under starvation conditions. The colony consists of branches with chiral twists, all with the same handedness.

A.

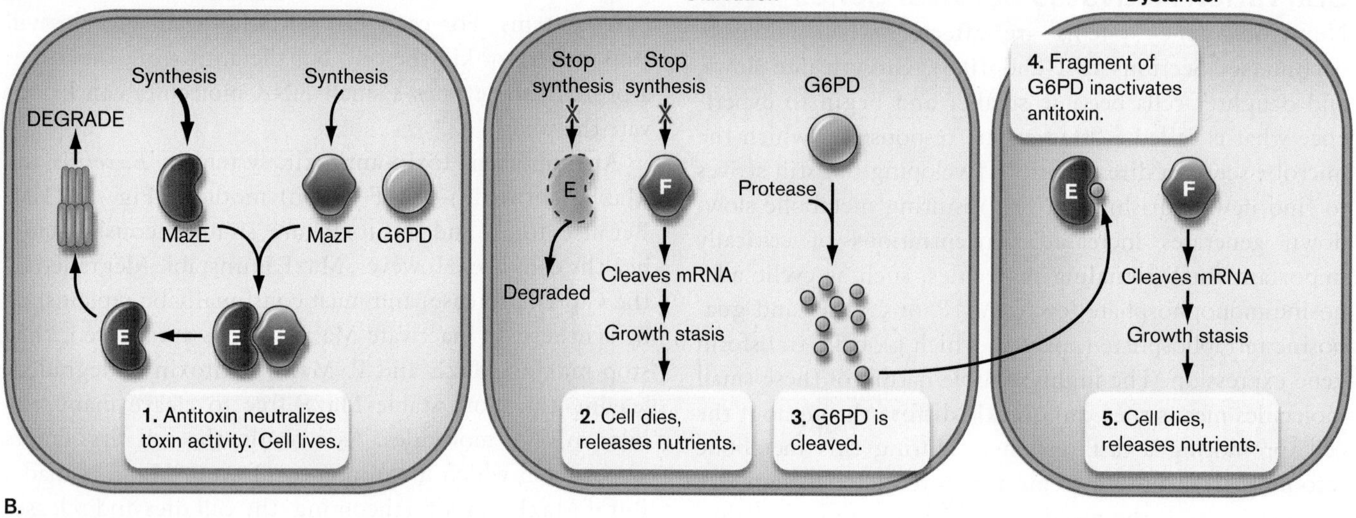

B.

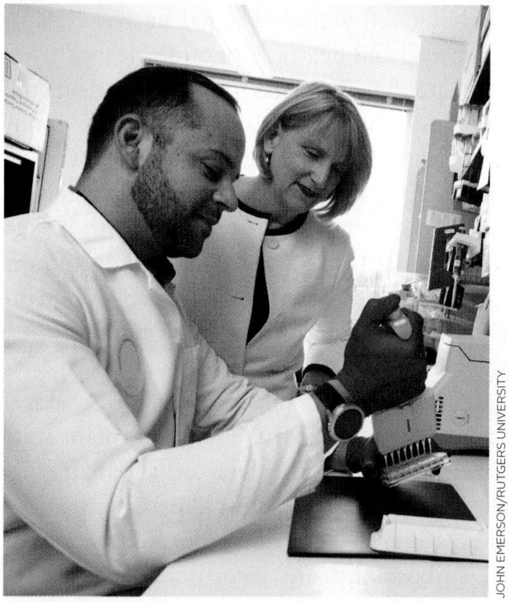

FIGURE 5.21 ■ Programmed cell death in response to starvation. A. The *E. coli* MazEF toxin-antitoxin system is thought to play an important role in bacterial survival during stress. MazE antitoxin is continually degraded by ClpAP and must be replenished to neutralize MazF toxin. Other toxin-antitoxin systems are found in other bacterial species. Stresses such as starvation, oxidative stress, or antibiotics can activate this system. G6PD = glucose 6-phosphate dehydrogenase. B. Nancy Woychik and graduate student Valdir Barth study homologs of MazEF toxin-antitoxin systems in *Mycobacterium tuberculosis*, the cause of tuberculosis. Her lab explores how these toxin-antitoxin systems contribute to the latency of this pathogen.

protein synthesis and stabilizes ribosome pairs as inactive 100S dimers (**Fig. 5.22A**). This state of ribosome hibernation prevents ribosome turnover during dormancy and provides a reservoir of ribosomes that can reactivate when conditions improve. How ribosome dimers dissociate when conditions improve is unclear but appears to require a "wake-up" GTPase protein called HflX, as discovered by Mee-Ngan F. Yap and colleagues from Saint Louis University (**Fig. 5.22B**).

Microbes Encounter Multiple Stresses in Real Life

Bacterial stress responses have traditionally been studied by exposing organisms to <u>individual</u> stresses such as starvation, high or low osmolarity, or extreme pH in controlled laboratory situations. In the world outside of the laboratory, however, environmental situations can be quite complex, involving multiple, not just single, stressors. An organism could simultaneously undergo carbon starvation in a high-salt, low-pH environment. Thus, caution is advised when trying to use controlled laboratory studies that alter a single parameter to predict cellular responses to real-world situations.

Environmental stresses can, of course, drive evolution. Long-term exposure to multiple stresses provides the organism with broad evolutionary opportunities. As described in Chapter 4, a variety of stresses can develop within a single colony of bacteria as that colony ages. Starvation, reactive oxygen species, toxic end products of metabolism, and changes in pH can all build to a point that threatens survival of the entire colony. Stress response systems alone cannot save them. Yet some cells survive. Ivan Matic and Claude Saint-Ruf at the Paris Descartes University discovered that 10-day-old aging colonies of *E. coli* contain numerous tiny islands of mutant cells that evolved to survive (**eTopic 5.5**).

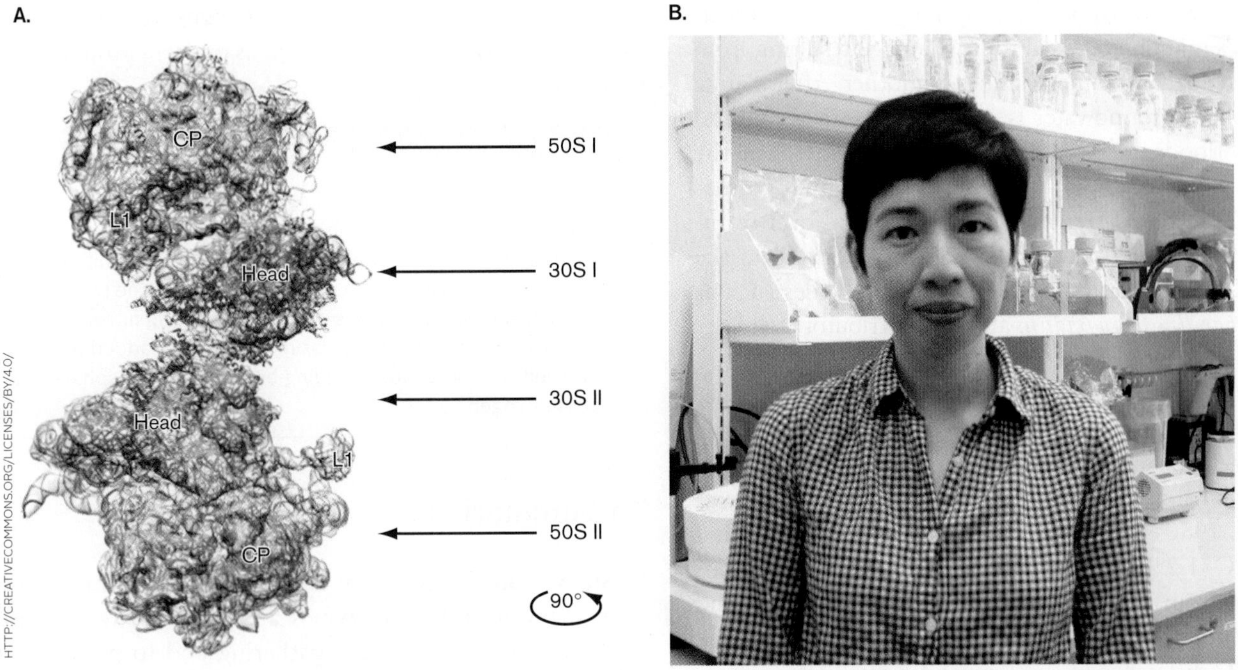

FIGURE 5.22 ▪ Ribosome dimerization protects ribosomes and stops protein synthesis in dormant cells. **A.** Structure of 100S ribosome dimer from *Staphylococcus aureus*. Two 70S ribosomes (each represented by a box) are linked at their 30S subunits (pink), in a process promoted by HPF. Dimerization prevents protein synthesis in dormant bacteria but stores functional ribosomes that reemerge when stress is relieved. **B.** Mee-Ngan F. Yap discovered that dissociation of ribosome dimers requires the GTPase HflX.

The altered physiology of these mutants provides them with an ability to outgrow their parental strain when both are mixed on fresh agar medium. Investigations into which aspects of bacterial physiology are affected by these mutations are under way.

Humans Influence Microbial Ecosystems

Human activities have striking impacts on microbial ecosystems. A major example of this impact is eutrophication. Natural ecosystems are typically low in nutrients (oligotrophic; **eTopic 5.6**) but teem with diversity, so numerous species compete for the same limiting nutrients. Maximum diversity in a given ecosystem is maintained, in part, by the different nutrient-gathering profiles of competing microbes. However, the sudden infusion of large quantities of a formerly limiting nutrient, a process called **eutrophication**, can lead to a "bloom" of microbes, typically autotrophic cyanobacteria (formerly called blue-green algae; see Section 18.2). One species initially held in check by the limiting nutrient now exhibits unrestricted growth, consuming other nutrients as it grows to a degree that threatens the existence of competing species.

Humans cause eutrophication in several ways. Fertilizer runoff from agricultural fields, urban lawns, and golf courses is one source. Untreated or partially treated domestic sewage is another. Spilling large amounts of phosphates or nitrogen into lakes—Lake Erie, for example (**Fig. 5.23**)—powerfully stimulates cyanobacterial growth. The resulting bacterial blooms (wrongly called "algal blooms") can deplete the oxygen in the water and lead to fish kills. Native fish species can disappear, to be replaced by species more tolerant of the new conditions. Some cyanobacteria, such as *Microcystis aeruginosa*, produce liver toxins such as microcystin. A *Microcystis* bloom in 2014 shut down the entire water supply of Toledo, a city on the shores of Lake Erie. The concept of limiting nutrients in ecosystems is covered further in Chapter 21.

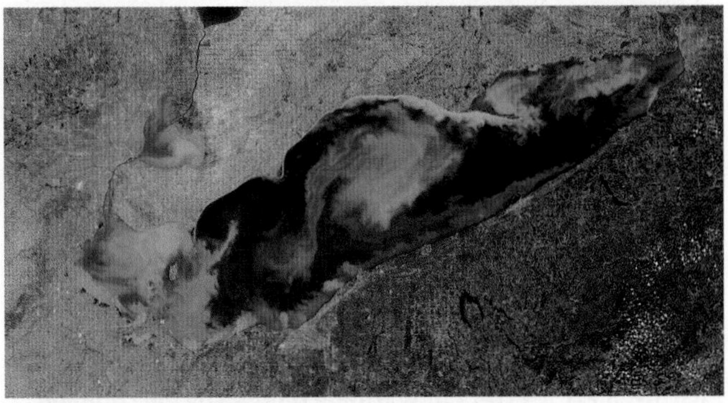

FIGURE 5.23 ▪ Eutrophication in Lake Erie. Cyanobacterial bloom (bright blue-green color) in Lake Erie caused by excessive phosphorous eutrophication.

Eutrophication is not the only human activity affecting Earth's microbiota. Acid mine drainage resulting from abandoned coal and mineral mines is another. After a mine is abandoned, groundwater is no longer pumped out and the mine floods. Acid mine drainage develops from the oxidation of pyrite (FeS_2) unearthed by the mining operations. The exposed pyrite oxidizes in air to form sulfuric acid that, along with soluble Fe^{2+}, can drain from the mine and destroy natural ecosystems. Acidophiles such as *Acidithiobacillus ferrooxidans* are key contributors to pyrite oxidation.

Climate change caused by human activity is one other process that will gradually alter microbial ecosystems. Put simply, the spewing of heat-trapping CO_2 into the atmosphere by burning hydrocarbons is speeding Earth's warming. Since 1980, the average atmospheric temperature has risen 0.6°C (1.2°F). A 10-year study by Konstantinos Konstantinidis from the Georgia Institute of Technology, and Jizhong Zhou from the University of Oklahoma found that a mere 2°C difference in soil temperature will alter a community of soil microorganisms. Using an infrared light, the team warmed a patch of Oklahoma prairie soil 2°C above that of an adjacent control patch of soil. They then used DNA-based techniques to catalog the microbial genera present (**Fig. 5.24**). Some taxonomic groups of organisms became more dominant (Actinobacteria), while others became less abundant (Proteobacteria and Acidobacteria).

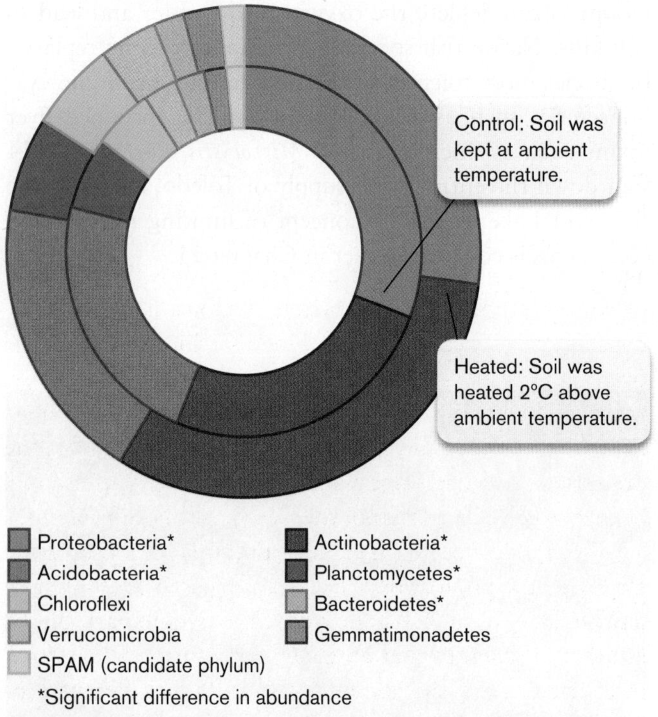

FIGURE 5.24 ■ Effect of warming on the soil microbiome. Two patches of Oklahoma soil were monitored for microbial taxa over 10 years. The two rings illustrate the average abundances of various taxa. SPAM = candidate division, microbes discovered in spring in alpine and meadow soils.

What future effects these kinds of changes will have on the carbon and nitrogen cycles, as well as on farm productivity, remain to be seen.

Thought Question

5.9 Given a mixture of two microbes, A and B—where organism A can utilize limiting phosphate more efficiently than organism B, but organism B can utilize limiting nitrogen better than organism A—what would happen to the relative growth of the two organisms placed in a limiting nitrogen and phosphate medium if excess nitrogen were added to the mixed culture? What about adding both excess phosphate and excess nitrogen?

To Summarize

- **Starvation** is a stress that can elicit a molecular response in many microbes. Enzymes are produced to increase the efficiency of nutrient gathering and to protect cell macromolecules from damage.
- **The starvation response** is usually triggered by the accumulation of small signaling molecules, such as cyclic AMP or guanosine tetraphosphate.
- **Human activities can cause eutrophication**, which damages delicately balanced ecosystems by introducing nutrients that can allow one member of the ecosystem to flourish at the expense of other species.

5.6 Physical, Chemical, and Biological Control of Microbes

We have seen how adept microbes are at surviving environmental stresses; now, how can we kill them? A primary goal of our health care system is to control or kill microbes that can potentially harm us. Within the recent past, infectious disease was an imminent and constant threat to most of the human population. The average family in the United States prior to 1900 had four or five children, but parents could expect half of them to succumb to deadly infectious diseases. Improvements in sanitation procedures and antiseptics and the advent of antibiotics have, to a large degree, curtailed the incidence and lethal effects of many infectious diseases. These advancements have played a major role in extending life expectancy.

A variety of terms are used to describe antimicrobial control measures. The terms convey subtle, yet vitally important, differences in control strategies and outcomes.

- **Sterilization** is the process by which all living cells, spores, and viruses are destroyed on an object.

- **Disinfection** is the killing, or removal, of disease-producing organisms from inanimate surfaces; it does not necessarily result in sterilization. Pathogens are killed, but other microbes may survive.
- **Antisepsis** is similar to disinfection, but it applies to removing pathogens from the surface of living tissues, like the skin. Antiseptic chemicals are usually not as toxic as disinfectants, which frequently damage living tissues.
- **Sanitation** is closely related to disinfection. It consists of reducing the microbial population to safe levels and usually involves both cleaning and disinfecting an object.

Antimicrobials can also be classified on the basis of the specific groups of microbes destroyed, leading to the terms "microbicide," "bactericide," "algicide," "fungicide," and "virucide." These agents can be classified further as either "-static" (inhibiting growth) or "-cidal" (killing cells). For example, antibacterial agents may be **bacteriostatic** or **bactericidal**. Chemical substances are **germicidal** if they kill pathogens (and many nonpathogens), but germicidal agents do not necessarily kill spores.

Although these descriptions emphasize the killing of pathogens, it is important to note that antimicrobial agents can also kill or prevent the growth of nonpathogens. Many public health standards are based on total numbers of microorganisms on an object, regardless of pathogenic potential. For example, to gain public health certification, the restaurants we frequent must demonstrate low numbers of bacteria (pathogenic or not) wherever food is prepared.

Thought Question

5.10 Bacteriostatic antibiotics do not kill bacteria; they only inhibit their growth. Why are they nevertheless effective at treating bacterial infections? *Hint:* Is the human body a quiet bystander during an infection?

Cells Treated with Antimicrobials Die at a Logarithmic Rate

Exposing microbes in a solution or on a surface to lethal chemicals or conditions will not instantly kill all the microorganisms. Microbes die according to a negative exponential curve, where cell numbers decrease in equal fractions at constant intervals. The efficacy of a given lethal agent or condition is measured as **decimal reduction time** (**D-value**), which is the length of time it takes that agent (or condition) to kill 90% of the population (a drop of one log unit, or a drop to 10% of the original value). **Figure 5.25** illustrates the exponential death profile of a bacterial culture heated to 100°C. The D-value (called D_{100} in this instance) is a little over 1 minute. The food industry uses D-value and several other parameters to evaluate the efficiency of killing (see Section 16.4).

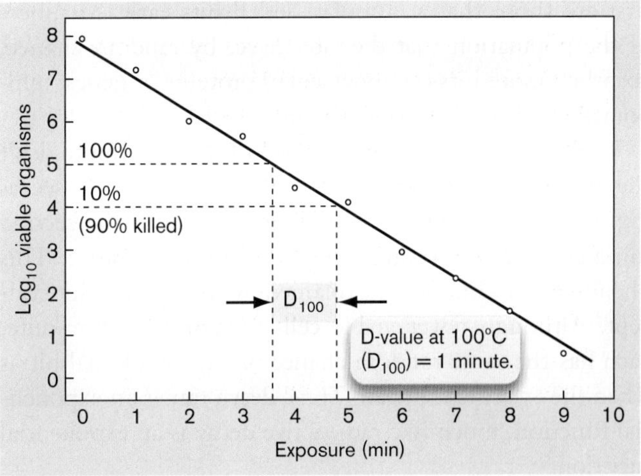

FIGURE 5.25 ■ The death curve and the determination of D-values. Bacteria were exposed to a temperature of 100°C, and survivors were measured by viable count. The D-value (D_{100}) is the time required for 100°C to kill 90% of cells (that is, the time it takes for the viable cell count to drop by one $\log_{10}$ unit).

Thought Question

5.11 If a disinfectant is added to a culture containing 1×10^6 CFUs per milliliter and the D-value of the disinfectant is 2 minutes, how many viable cells will be left after 4 minutes of exposure?

Several factors influence the ability of an antimicrobial agent to kill microbes. These include the initial population size (the larger the population, the longer it takes to reduce it to a specific number), the population composition (are spores involved?), the concentration of the antimicrobial agent, and the duration of exposure. Although the effect of concentration seems intuitively obvious, an increase in concentration is matched by an increase in death rate only over a narrow concentration range. Increases above a certain level might not accelerate killing at all. For example, 70% ethanol is actually better than pure ethanol at killing organisms, because some water is needed to help ethanol penetrate cells. The ethanol then dehydrates cell proteins.

Why, then, is death an exponential function? Why don't all cells in a population die instantly when treated with lethal heat or chemicals? The reason is based, in part, on the random probability that an agent will cause a lethal "hit" in a given cell. Cells contain thousands of different proteins and thousands of molecules of each one. Not all proteins and not all genes in a chromosome are damaged by an agent at the same time. Damage accumulates. Only when enough molecules of an essential protein or a gene encoding that protein are damaged will the cell die. Cells that die first are those that accumulate lethal hits early. Members of the population that die later have, by random chance, absorbed more hits on nonessential proteins or genes, temporarily sparing the essential ones.

Why, if 90% of a population is killed in 1 minute, isn't the remaining 10% killed in the next minute? It seems logical that all should have perished. Yet after the second minute, 1% of the original population remains alive. This phenomenon can also be explained by the random-hit concept. Although fewer viable cells remain after 1 minute, each has the same random chance of having a lethal hit as when the treatment began. Thus, death rate is an exponential function, much like radioactive decay is an exponential function.

A final consideration is the overall fitness of individual cells. It is a mistake to assume that all cells in a population are identical. At any given time, for instance, one cell may express a protein that another cell has just stopped expressing (for example, superoxide dismutase). In that instant, the first cell might contain a bit more of that protein. If the protein is essential or confers a level of stress protection (such as against superoxide), the cell with more of that protein can absorb more punishment before it dies. The presence of lucky individuals expressing the right repertoire of proteins might also explain why death curves commonly level off after a certain point.

Physical Agents That Kill Microbes

Physical agents are often used to kill microbes or control their growth. Commonly used physical control measures include temperature extremes, pressure (usually combined with temperature), filtration, and irradiation.

High temperature and pressure. Even though microbes were discovered nearly 350 years ago, thermal treatment of foods to render them safe has been practiced for over 5,000 years. Moist heat is a much more effective killer than dry heat, thanks to the ability of water to penetrate cells. Many bacteria, for instance, easily withstand 100°C dry heat but not 100°C boiling water. We humans are not so different, finding it easier to endure a temperature of 32°C (90°F) in dry Arizona than in humid Louisiana.

While boiling water (100°C) can kill most vegetative (actively growing) organisms, spores are built to withstand this abuse, and thermophiles prefer it. Killing spores and thermophiles usually requires combining high pressure and temperature. At high pressure, the boiling point of water rises to a temperature rarely experienced by microbes living at sea level. Even endospores quickly die under these conditions. This combination of pressure and temperature is the principle behind sterilization using the steam autoclave (**Fig. 5.26**). Standard conditions for steam sterilization are 121°C (250°F) at 15 psi for 20 minutes—a set of conditions that experience

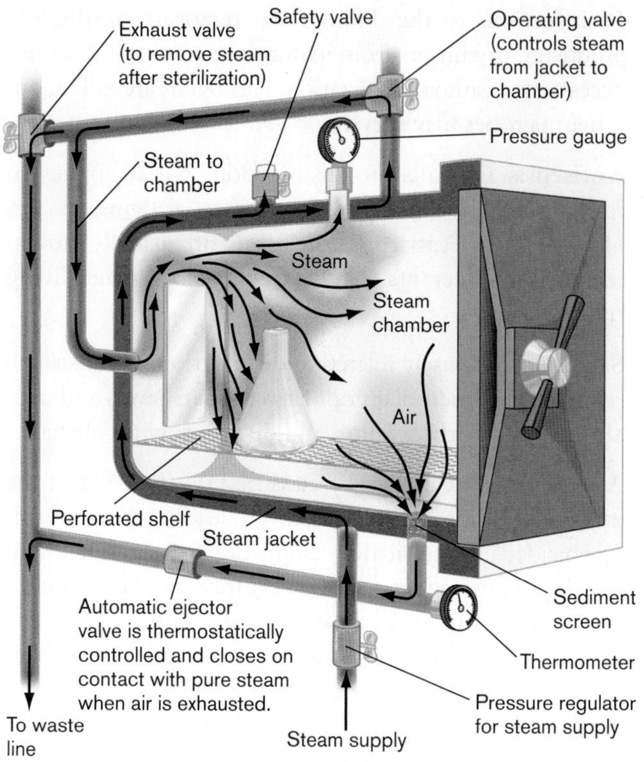

FIGURE 5.26 ■ Steam autoclave.

has taught us will kill all spores except those of some thermophiles. (Thermophiles, however, do not affect food or human health.) Standard sterilization conditions are also produced in pressure cookers used for home canning of vegetables and fruits.

Failure to adhere to sterilization heat and pressure parameters can have deadly consequences, even in your own home. For instance, *Clostridium botulinum* is a spore-forming soil microbe that commonly contaminates fruits and vegetables used in home canning. The improper use of a pressure cooker while canning these goods will allow spores of this pathogen to survive. Once the can or jar is cool, the spores will germinate and begin producing their deadly toxin. All of this happens while the canned goods sit on a shelf waiting to be opened and consumed. Once ingested, the toxin makes its way to the nervous system and paralyzes the victim. Several incidents of this disease, called botulism, occur each year in the United States. (For more on food poisoning, see Chapter 16.)

Thought Question

5.12 How would you test the killing efficacy of an autoclave?

Pasteurization. Originally devised by Louis Pasteur to save products of the French wine industry from devastating bacterial spoilage, **pasteurization** today involves heating a particular food (such as milk) to a specific temperature long enough to kill *Coxiella burnetii*, the causative agent of

Q fever, the most heat-resistant non-spore-forming pathogen known. In the process, pasteurization also kills other disease-causing microbes.

Three U.S. government–approved time and temperature combinations can be used for pasteurization of milk:

- **LTLT (low temperature, long time)**. In this method, the milk is brought to a temperature of 63°C (145°F) for 30 minutes.
- **HTST (high temperature, short time)**. This method (also called "flash pasteurization") brings the milk to a temperature of 72°C (161°F) for only 15 seconds.
- **UHT (ultra-high temperature)**. In this method, the milk is brought to a temperature of 134°C (273°F) for 1–2 seconds.

Both the LTLT and HTST methods accomplish the same thing—the destruction of *C. burnetii* and other bacteria—but they do not sterilize milk. UHT pasteurization decreases bacterial content even more than the LTLT and HTST methods, producing nearly sterile milk with an unrefrigerated shelf life of up to 6 months. This is important, especially in developing countries, where refrigeration is not always available.

Cold. Low temperatures have two basic purposes in microbiology: to slow growth and to preserve strains. Bacteria not only grow more slowly in cold, but also die more slowly. Refrigeration temperatures (4°C–8°C, or 39°F–43°F) are used for food preservation because most pathogens are mesophilic and grow poorly, if at all, at those temperatures. One exception is the Gram-positive bacillus *Listeria monocytogenes*, which can grow reasonably well in the cold and causes disease when ingested.

Long-term storage of bacteria usually requires placing solutions in glycerol at very low temperatures (–70°C). Glycerol prevents the production of razor-sharp ice crystals that can pierce cells from without or within. This deep-freezing suspends growth altogether and keeps cells from dying. Another technique, called **lyophilization**, freeze-dries microbial cultures for long-term storage. In this technique, cultures are quickly frozen at very low temperatures to limit ice crystal formation and placed under vacuum, where the resulting sublimation removes all water from the media and cells, leaving just the cells in the form of a powder. These freeze-dried organisms remain viable for years. Finally, viruses and mammalian cells must be kept at extremely low temperatures (–196°C), submerged in liquid nitrogen. Liquid nitrogen freezes cells so quickly that ice crystals do not have time to form.

Filtration. Filtration through micropore filters with pore sizes of 0.2 μm can remove microbial cells, but not viruses, from solutions. Samples from 1 milliliter to several liters can be drawn through a membrane filter by vacuum or can be forced through it with a syringe (**Fig. 5.27**). Filter sterilization avoids the use of heat, which can damage certain materials. Strictly speaking, though, the solutions are not really sterile, because these filters do not trap viruses.

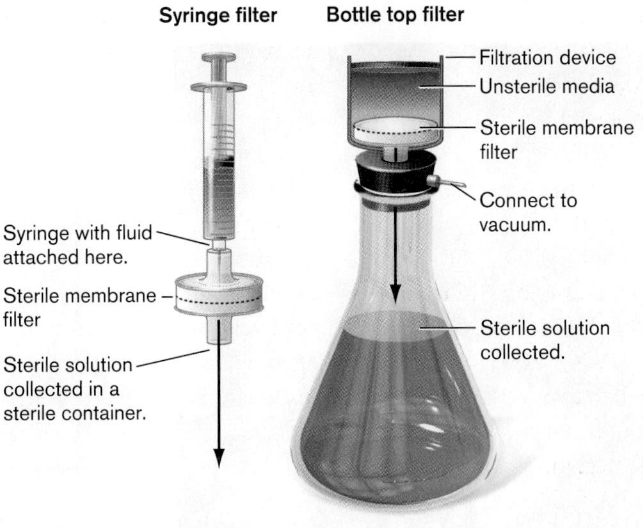

FIGURE 5.27 ■ Membrane filtration devices.

Air can also be sterilized by filtration. This process forms the basis of several personal protective devices. A surgical mask is a crude example, while **laminar flow biological safety cabinets** are more elaborate (and more effective). These cabinets force air through high-efficiency particulate air (HEPA) filters and remove over 99.9% of airborne particulate material 0.3 μm in size or larger. Biosafety cabinets are critical to protect individuals working with highly pathogenic material (**Fig. 5.28A** and **B**). Newer technologies have been developed that embed antimicrobial agents or enzymes directly into the fibers of the filter (**Fig. 5.28C**). Organisms entangled in these fibers are not just trapped; they are attacked by the antimicrobials and lysed.

HEPA filters have also been tested in automobiles, whose small, shared spaces place occupants at higher risk of exposure to airborne infectious agents. A recent, controlled study found that placing HEPA filter devices in cars can decrease airborne CFUs 1,000-fold.

Irradiation. Public health authorities worldwide are constantly worried about food being contaminated with pathogenic microorganisms such as *Salmonella* species, *E. coli* O157:H7, *Listeria monocytogenes*, and *Yersinia enterocolitica*. Irradiation, the bombardment of foods with high-energy electromagnetic radiation, has long been a potent strategy for sterilizing food after harvesting. The food consumed by NASA astronauts, for example, has for some time been sterilized by irradiation as a safeguard against food-borne illness in space. Early public concerns that irradiation might produce dangerous toxic by-products have largely

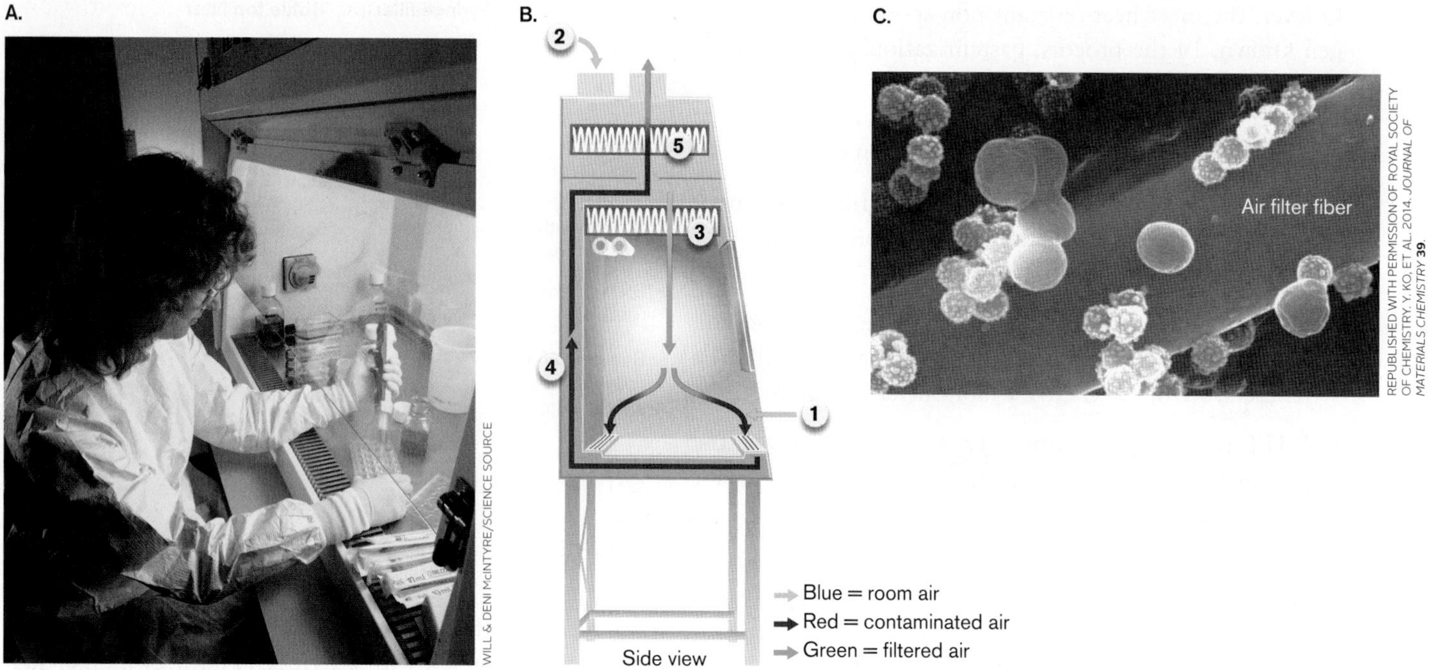

FIGURE 5.28 ▪ Biological safety cabinet. **A.** A scientist examines a sample under the hood. **B.** Schematic of the safety cabinet. Air from the room enters the cabinet through the cabinet opening (1), or is pumped in (2) through a HEPA filter (3). It then passes behind the negative-pressure exhaust plenum (4) and is passed from the cabinet through another HEPA filter (5). **C.** Antibacterial activity of silver-silica coated particles on an air filtration unit. The primary function of this filter is to kill airborne microorganisms caught on the surface of the filter, thus protecting against secondary contamination by microorganisms in air filtration systems. The photo shows that *Staphylococcus epidermidis* cells attached to the smaller silver-silica beads on the filter fiber have an altered morphology.

disappeared. Numerous studies have proved that foods do not become radioactive when irradiated, nor are long-lived reactive molecules produced that are dangerous to humans.

Aside from ultraviolet light, which, owing to its poor penetrating ability, is useful only for surface sterilization, there are three other sources of irradiation: gamma rays, electron beams, and X-rays. Radiation dosage is usually measured in a unit called the gray (Gy), which is the amount of energy transferred to the food, microbe, or other substance being irradiated. A single chest X-ray delivers roughly half a milligray (1 mGy = 0.001 Gy). To kill *Salmonella*, freshly slaughtered chicken can be irradiated at up to 4.5 kilograys (kGy)—about 7 million times the energy of a single chest X-ray. The Food and Drug Administration has also approved the use of irradiation (4 kGy) on beef, pork, fruits, vegetables, oysters, seeds, shell eggs, and spices.

How does radiation kill bacteria? When microbes present in food are irradiated, water and other intracellular molecules absorb the energy and form very short-lived reactive chemicals, typically reactive oxygen species, that damage DNA and proteins. Unless this damage is prevented or repaired, the organism will die. Microbes differ greatly in their sensitivity to irradiation, depending on the size of their genome, the rate at which they can repair damaged DNA, and other factors. Whether the food to be irradiated is frozen or fresh also matters, as it takes a higher dose of radiation to kill microbes in frozen foods.

The size of the DNA "target" is a major factor in radiation efficacy. Parasites and insect pests, which have large amounts of DNA, are rapidly killed by extremely low doses of radiation, typically with D-values of less than 0.1 kGy (in this instance, the D-value is the dose of radiation needed to kill 90% of the organisms). It takes more radiation to kill bacteria (D-values in the range of 0.3–0.7 kGy) because they have less DNA per cell unit (less target per cell). It takes even more radiation to kill bacterial spores (D-values on the order of 2.8 kGy) because they contain little water, the source of most ionizing damage to DNA.

Viral pathogens have the smallest amount of nucleic acid, making them resistant to irradiation doses approved for foods (viruses have D-values of 10 kGy or higher). Infectious agents that do not contain nucleic acids are an even bigger problem. Prions, for example, are misfolded brain proteins that "self-replicate" and cause neurodegenerative diseases (see Section 26.6). Because prions do not contain nucleic acids, the agent can be inactivated by irradiation only at extremely high doses. Thus, irradiation of food is effective in eliminating parasites and bacteria but is woefully inadequate for eliminating viruses or prions.

Note: Electromagnetic radiation emitted by microwave ovens does not directly kill bacteria. However, the heat generated when electromagnetic radiation excites water molecules in an organism will kill the organism if the temperature attained is high enough.

A. B.

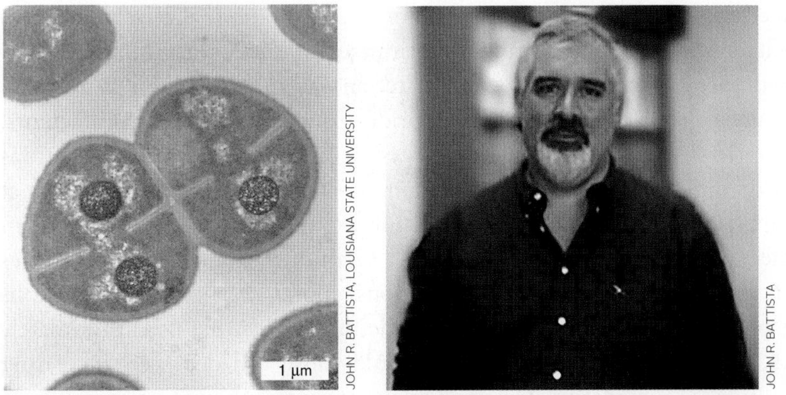

FIGURE 5.29 ■ *Deinococcus radiodurans*. **A.** The amount of radiation that *D. radiodurans* can survive is equivalent to that of an atomic blast. The nature of the dark inclusion bodies in three of the four cells in this quartet is currently not known. **B.** John Battista of Louisiana State University showed that *D. radiodurans* has exceptional capabilities for repairing radiation-damaged DNA.

Resistance to ionizing radiation. *Deinococcus radiodurans* could be branded the "Incredible Hulk of Microbes" and the mascot of extremophile (**Fig. 5.29**). *D. radiodurans* was discovered in 1956 in a can of meat that had spoiled despite having been sterilized by radiation. The microbe, which is also a moderate thermophile, has the greatest ability of any known organism to survive radiation. It could probably even survive an atomic blast. The bacterium's ability to withstand radiation may have evolved as a side effect of developing resistance to extreme drought, since desiccation and radiation produce similar types of DNA damage.

What properties account for this microbe's amazing resistance to radiation? One is that *D. radiodurans* possesses an unusual capacity for repairing damaged DNA. Many mechanisms are involved, including a highly efficient double-strand-break DNA repair system requiring homologous recombination (see Section 9.2 and **eTopic 9.2**). Each cell of this organism contains four to six copies of its two chromosomes and two plasmids. So even when its DNA is irradiated and broken into thousands of fragments, overlapping intact fragments can be found and spliced together in proper order.

Many of the DNA repair genes needed to survive radiation are part of a gene set, collectively called the radiation and desiccation response regulon, whose transcriptional expression increases during radiation or desiccation conditions. In addition to this transcriptional regulation, Lydia Contreras working out of the University of Texas at Austin recently discovered an RNA sequence that can sense radiation in the mRNA transcript encoding DNA gyrase. DNA gyrase is a DNA-winding enzyme needed for DNA replication and repair (Chapters 7 and 9). The regulatory RNA sequence in DNA gyrase mRNA somehow senses ionizing radiation and acts to increase the translation of mRNA into protein.

A second property that contributes to radiation resistance is the ability of *D. radiodurans* to aggressively protect its proteins, which are even more susceptible to radiation than is DNA. This mechanism, discovered by Michael Daly at the Uniformed Services University of the Health Sciences in Maryland, involves the intracellular accumulation of large amounts of manganese-metabolite complexes that can nonenzymatically remove highly damaging free radicals generated by radiation. This mechanism limits damage to macromolecules (proteins, DNA, RNA, lipids), including DNA repair proteins needed to fix macromolecules that are damaged.

Because of its incredible level of radiation resistance, *D. radiodurans* was genetically engineered to treat radioactive mercury–contaminated waste from nuclear reactors in a process called **bioremediation** (discussed in Chapter 22). The genes for mercury conversion were spliced from a strain of *E. coli* that is resistant to particularly toxic forms of mercury and inserted into *D. radiodurans*. The genetically altered superbug was able to withstand the ionizing radiation and convert toxic waste into forms that could be removed safely. Fortunately, there is little need to worry about *D. radiodurans* becoming a superpathogen, because the organism does not cause disease and is susceptible to antibiotics.

Chemical Agents

Disinfection by physical agents is very effective, but in numerous situations their use is impractical (kitchen countertops) or plainly impossible (skin). In these instances, chemical agents are the best approach. A number of factors influence the efficacy of a given chemical agent. These include:

- **The presence of organic matter**. A chemical placed on a dirty surface will bind to any inert organic material present, lowering the agent's effectiveness against microbes. It is not always possible to clean a surface prior to disinfection (as in a blood spill), but the presence of organic material must be factored into estimates of how long to disinfect a surface or object.
- **The kinds of organisms present**. Ideally, the agent should be effective against a broad range of pathogens.
- **Corrosiveness**. The disinfectant should not corrode the surface or, in the case of an antiseptic, damage skin.
- **Stability, odor, and surface tension**. The chemical should be stable during storage, possess a neutral or pleasant odor, and have a low surface tension so that it can penetrate cracks and crevices.

The phenol coefficient. Phenol, first introduced by Joseph Lister in 1867 to reduce the incidence of surgical infections, is no longer used as a disinfectant, because of its toxicity, but its derivatives, such as cresols and orthophenylphenol, are still in use. The household product Lysol is a mixture of phenolics. Phenolics are useful disinfectants because they denature proteins, are effective in the presence of organic material, and remain active on surfaces long after application. Although phenol is no longer used as a disinfectant, its potency makes it the benchmark against which other disinfectants are measured. Details of calculating phenol coefficients for disinfectants are provided in **eTopic 5.7**.

Commercial Disinfectants

Ethanol, iodine, chlorine, and surfactants (for example, detergents) are all used to decrease or eliminate microbial content from commercial products (**Fig. 5.30**). The first three are compounds that damage proteins, lipids, and DNA. Highly reactive iodine complexed with an organic carrier forms an iodophor, a compound that is water-soluble, stable, nonstaining, and capable of releasing iodine slowly to avoid skin irritation. Wescodyne and Betadine (trade names) are iodophors used, respectively, for the surgical preparation of skin and for wounds. Chlorine is another highly reactive disinfectant with universal application. It is recommended for general laboratory and hospital disinfection and kills HIV, the AIDS virus. Sodium hypochlorite (liquid bleach) is a common disinfectant in bathroom cleaners such as Tilex.

Detergents can also be antimicrobial agents. The hydrophobic and hydrophilic ends of detergent molecules (the coexistence of which makes the molecules amphipathic) will emulsify fat into water. Cationic (positively charged) detergents are useful as disinfectants because the positive charges can gain access to the negatively charged bacterial cell and disrupt membranes. Anionic (negatively charged) detergents are not antimicrobial but do help remove bacteria from surfaces. Quaternary ammonium compounds such as benzalconium chloride are useful cationic disinfectants. These compounds target membranes at high concentrations but can also inflict damage to internal cell components at lower concentrations. For example, a 2018 study with *Acinetobacter baumannii* (a Gram-negative bacillus that causes hospital-derived infections) found that low concentrations of benzalconium chloride, similar to what might ultimately penetrate a biofilm, could cause cell death by increasing oxidative stress and by interfering with protein homeostasis mechanisms. The result is the intracellular accumulation of protein aggregates that are dangerous to the microbe.

Low-molecular-weight aldehydes such as formaldehyde (H-CHO) are very potent disinfectants. The aldehydes are highly reactive, combining with and inactivating proteins and nucleic acids. However, the noxious odor of aldehydes and the labeling of formaldehyde as a suspected carcinogen limit their use in sterilizing medical equipment.

Disposable plasticware (including Petri dishes, syringes, sutures, and catheters) cannot undergo heat sterilization or liquid disinfection because the items will melt or dissolve. These materials are best sterilized using antimicrobial gases. Ethylene oxide gas (EtO) is a very effective sterilizing agent; it destroys cell proteins, is microbicidal and sporicidal, and rapidly penetrates packing materials, including plastic wraps. Via an instrument resembling an autoclave, EtO at 700 milligrams per liter (mg/l) will sterilize an object after 8 hours at 38°C or 4 hours at 54°C if the relative humidity is kept at 50%. Unfortunately, EtO is explosive. A less hazardous gas sterilant is betapropiolactone. It does not penetrate as well as EtO, but it decomposes after a few hours, which makes it easier to dispose of than EtO.

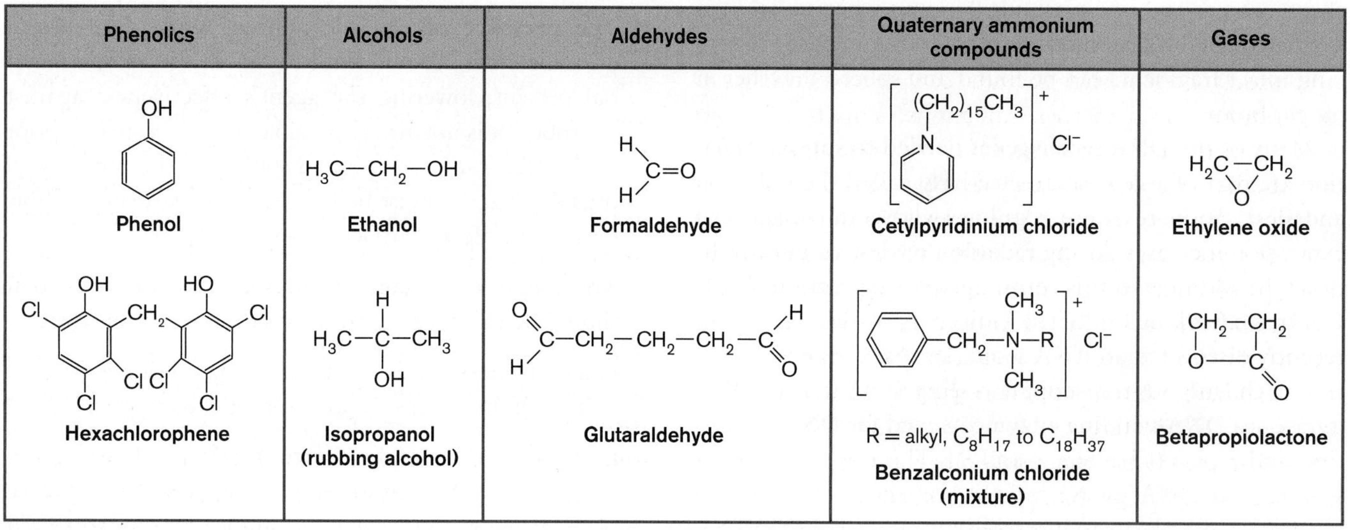

FIGURE 5.30 ■ Structures of some common disinfectants and antiseptics.

A relatively new procedure, known as gas discharge plasma sterilization, may replace EtO because it is less harmful to operators. Gas discharge plasma is made by passing certain gases through an electrical field to produce highly reactive chemical species that can damage membranes, DNA, and protein. It is not yet widely used.

Antimicrobial touch surfaces. Despite widespread use of disinfectants and antibiotics by medical personnel, hospital-acquired infections remain a major concern. A promising antimicrobial technology that can help reduce infections in hospitals and elsewhere involves embedding antimicrobial compounds such as copper (Cu) in the surfaces that people touch. For example, a bed rail made with copper can kill pathogens deposited by one person before the organism can be transmitted to a second person touching the same surface. Upon contact with bacteria, metallic copper releases toxic Cu^+ ions that trigger the lysis of bacterial membranes within minutes, although the mechanism involved is unclear (neither DNA damage nor reactive oxygen species are involved). Companies are incorporating metallic copper into objects such as handrails, door releases, and hospital bed rails.

In a newer development, an antimicrobial coating composed of silver plus rubidium (trade name AGXX) was found to prevent pathogenic methicillin-resistant *Staphylococcus aureus* (MRSA) from forming biofilms on metal surfaces. The coating limited the expression of *S. aureus* attachment proteins (adhesins), toxins, and other virulence factors. AGXX may be useful in hospital environments, but it is also being developed for air-conditioning cooling towers, which can harbor pathogens such as *Legionella pneumophila*, washing machines, dishwashers, and even drinking-water purifiers.

Bacteria Can Develop Resistance to Disinfectants

It is widely known that bacteria can develop resistance to antibiotics used to treat infections. This is a serious concern in the medical community. So, one might wonder whether bacteria can also develop resistance against chemical disinfectants used to prevent infections. The answer is yes—and no. It is difficult for a bacterium to develop resistance to chemical agents that have multiple targets and can easily diffuse into a cell. Iodine, for example, has both of these characteristics. However, disinfectants that have multiple targets at high concentrations may have only a single target at lower concentrations—a situation that can foster the development of resistance. For instance, triclosan (a halogenated bisphenol compound used in many soaps, deodorants, and toothpastes) targets several cell constituents, making it nicely bactericidal at high concentrations. However, at low concentrations triclosan only inhibits fatty acid synthesis and is merely bacteriostatic. Organisms have developed resistance to triclosan at low concentrations by altering the fatty acid synthesis protein normally targeted by triclosan. Consequently, in 2016 the FDA banned triclosan in consumer antiseptic washes.

Bacteria can achieve low-level resistance to disinfectants through membrane-spanning, multidrug efflux pumps (described in Section 4.2). For instance, the MexCD-OprJ efflux system of *Pseudomonas aeruginosa*, a Gram-negative bacterium that causes infections in burn and cystic fibrosis patients, can pump several different biocides, detergents, and organic solvents out of the cell, thereby reducing their efficacy. This finding and other reports of *Pseudomonas* gaining resistance to disinfectants have led many clinicians to advocate caution in the widespread use of certain chemical disinfectants.

Biofilm formation is another ingenious way that bacteria survive exposures to disinfectants. A biofilm is a 3D community of bacterial cells attached to a solid surface (see Section 4.5). A biofilm can protect cells in several ways. For example, the extracellular matrix proteins and polysaccharides that hold biofilms together also bind disinfectants, slowing their penetration into the deeper recesses of the structure. Slower penetration means that cells deep in the biofilm have time to activate protective stress response systems before destructive levels of disinfectant reach them. Biofilms also exhibit stratified growth patterns based in part on nutrient access (see Fig. 4.29). Cells at the periphery have ample access to oxygen and nutrients, while cells within do not. Nutrient starvation worsens, the farther a cell is from the surface. Because nutrient starvation activates stress response systems, each biofilm has stratified layers of increasingly stress-resistant cells that can better tolerate chemical insults.

Finally, biofilms, which contain multiple species in nature, are opportunities for protective, interspecies collaborations. Protective enzymes from one species could protect a nonproducing species from a chemical insult, much as a big brother protects a little brother from a bully. For example, **Figure 5.31** shows a biofilm of a *Bacillus subtilis* strain (green) isolated from an endoscope washer-disinfector, mixed with pathogenic *Staphylococcus aureus* (red). Romain Briandet and colleagues from the French National Institute of Agricultural Research observed that *B. subtilis* in this biofilm protected *S. aureus* from the disinfectant peracetic acid. Multispecies biofilms can also be more massive than monospecies biofilms. The food pathogen *Escherichia coli* O157:H7 forms a biofilm with 400 times more volume when grown with *Acinetobacter calcoaceticus*, an organism found in meatpacking plants. Increased volume alone will slow the penetration of the disinfectant and protect the collective.

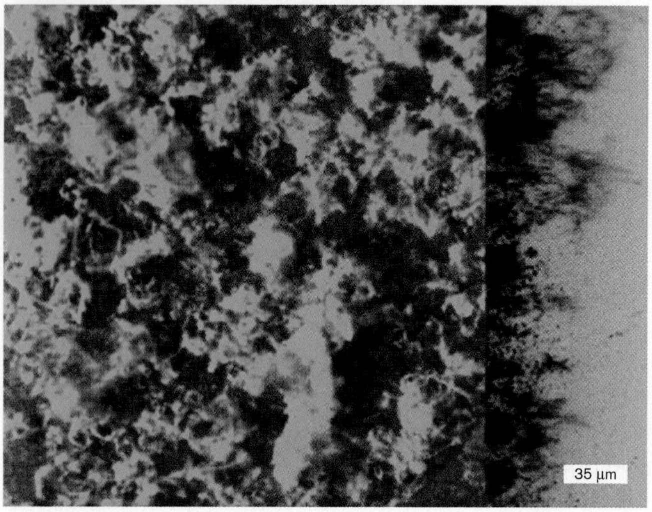

ROMAIN BRIANDET. FROM KHALIFA, L., ET AL. 2015. *APPLIED AND ENVIRONMENTAL MICROBIOLOGY* **81:** 2696.

FIGURE 5.31 ■ Mixed biofilm. Three-dimensional projection of a mixed 24-hour biofilm of *Bacillus subtilis* expressing green fluorescent protein (GFP; green) and *Staphylococcus aureus* expressing mCherry fluorescent protein (red). The presence of *B. subtilis* protects *S. aureus* from peracetic acid treatment. Right side of the image is a black and white photo of the biofilm's edge.

Antibiotics Selectively Control Bacterial Growth

Antibiotics as made in nature are chemical compounds synthesized by one microbe to either kill (bactericidal) or stop the growth of (bacteriostatic) other competing microbial species. Chapter 27 describes the various antibiotic modes of action. When purified and administered to patients suffering from an infectious disease, antibiotics can produce seemingly miraculous recoveries.

As we saw in Chapter 1, penicillin, produced by *Penicillium notatum*, was discovered serendipitously in 1929 by Alexander Fleming. Because the penicillin structure mimics a part of the cell wall, this drug binds to biosynthetic proteins involved in peptidoglycan synthesis and prevents cell wall formation. The drug is bactericidal, in part, because actively growing cells lyse without the support of the cell wall (**Fig. 5.32**). Other antibiotics target protein synthesis, DNA replication, cell membranes, and various enzyme reactions. Antimicrobials that affect macromolecular synthesis are described throughout Parts 1–4 of this text.

So how do antibiotic-producing microbes avoid suicide? In some instances, the producing organism lacks the target molecule. *Penicillium* mold, for instance, lacks peptidoglycan and is immune to penicillin by default. Some bacteria produce antimicrobial compounds that target other members of the same species. In this case, the producing organism can modify its own receptors to no longer recognize the compound (as with some bacterial colicins).

Another strategy is to modify the antibiotic if it reenters the cell. This is the case with streptomycin produced by *Streptomyces griseus*. Streptomycin inhibits bacterial protein synthesis and does not discriminate between the protein synthesis machinery of *Streptomyces* and that of others.

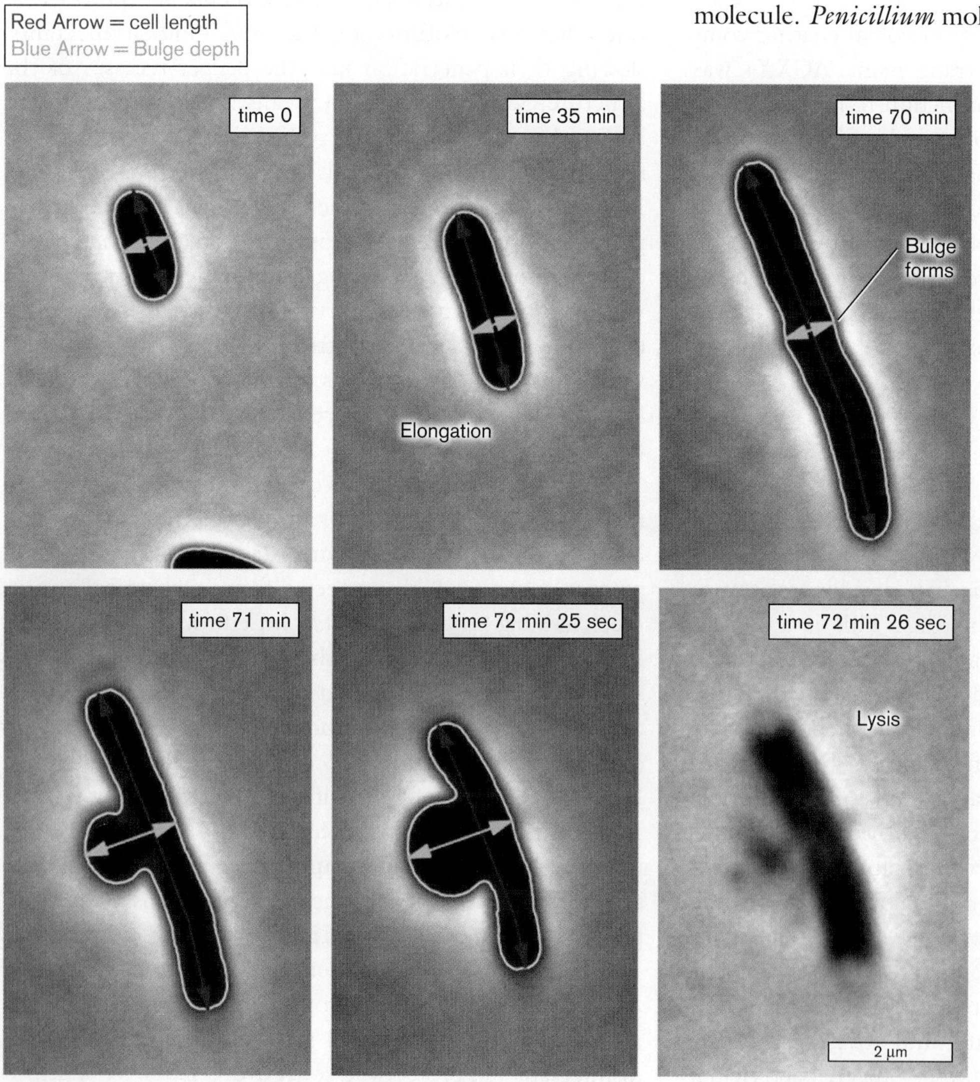

Z. YAO, ET AL. 2012. *MOLCELL* **48:** 705-712.

FIGURE 5.32 ■ Effect of cephalexin (a penicillin-like antibiotic) on *E. coli*. Time lapse light microscopy follows the antibiotic's effect over 73 minutes of growth. Arrows indicate directions of cell and bulge expansions. Cephalexin weakens the cell wall of growing cells and causes a bulge to form. The bulge enlarges until the cell lyses.

However, while making enzymes that synthesize and secrete streptomycin, *S. griseus* simultaneously makes the enzyme streptomycin 6-kinase, which remains locked in the cell. If any secreted streptomycin reenters the cell, this enzyme renders the drug inactive by attaching a phosphate group to it.

Because many microorganisms have become resistant to commonly used antibiotics, pharmaceutical companies continually search for new antibiotics, using a variety of approaches (see Special Topic 4.1). Traditional procedures include scouring soil and ocean samples collected from all over the world for new antibiotic-producing organisms, and chemically redesigning existing antibiotics so that they can bypass microbial resistance strategies. Newer techniques allow scientists to "mine the genomes" of microbes for potential drug targets and use computer-based methods to predict the structure and function of potential new antibiotics.

Biological Control of Microbes

Pitting microbe against microbe is an effective way to prevent disease in humans and animals. One of the hallmarks of a healthy ecosystem is the presence of a diversity of organisms. This is true not only for tropical rain forests and coral reefs, but also for the complex ecosystems of human skin and the intestinal tract. In these environments, the presence of harmless microbial communities can retard the growth of undesired pathogens. The pathogenic fungus *Phytophthora cinnamomi*, for example, causes root rot in plants but is biologically controlled by *Myrothecium* fungi. *Staphylococcus* species that are normally present on human skin produce short-chain fatty acids that retard the growth of pathogenic strains. Another illustration is the human intestine, which is populated by as many as 500 microbial species. Most of these species are nonpathogenic organisms that exist in symbiosis with their human host. Vigorous competition with members of the normal intestinal microbiota and the production of permeant weak acids by fermentation helps control the growth of numerous pathogens. A prominent example of this phenomenon is the pathogen *Clostridioides difficile*, whose growth is normally kept in check by gut microbiota.

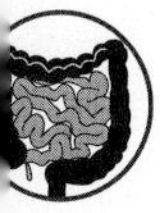

Microbial competition has been widely exploited for agricultural purposes and to improve human health through the intake of **probiotics**. In general, a probiotic is a food or supplement that contains live microorganisms and improves intestinal microbial balance. Newborn baby chicks, for instance, are fed a microbial cocktail of normal gut microbes designed to quickly colonize the intestinal tract and prevent colonization by *Salmonella*, a frequent contaminant of factory-farmed chicken. In another example, *Lactobacillus* and *Bifidobacterium* have been used to prevent and treat diarrhea in children.

Russian biologist Ilya Mechnikov, winner of the 1908 Nobel Prize in Physiology or Medicine, was the first to suggest that a high concentration of lactobacilli in the gut microbiome is important for health and longevity in humans. Yogurt is a probiotic that contains *Lactobacillus acidophilus* and a number of other lactobacilli. It is often recommended as a way to restore a normal balance to gut microbiota (for example, after the microbiota has been disturbed by antibiotic treatment), and it appears useful in the treatment of some forms of inflammatory bowel disease. Other probiotic microbes are described in Chapter 23.

Phage therapy, another **biocontrol** method, was first described in 1907 by Félix d'Herelle at France's Pasteur Institute, long before antibiotics were discovered. Bacteriophages are viruses that prey on bacteria (discussed in Section 6.1). Every bacterial species is susceptible to a limited number of specific phages. Because a phage infection often causes bacterial lysis, it was considered feasible to treat infectious diseases with a phage targeted to the pathogen. At one time, doctors used phages as medical treatment for illnesses ranging from cholera to typhoid fever. In some cases, a liquid containing the phage was poured into an open wound. In other cases, phages were given orally, introduced via aerosol, or injected. Sometimes the treatments worked; sometimes they did not.

When antibiotics came into the mainstream, phage therapy largely faded. Now that strains of antibiotic resistant bacteria are on the rise, the idea of phage therapy has enjoyed renewed interest from the worldwide medical community. The treatment is being considered for many infections, including respiratory, gastrointestinal, and pregnancy-related infections, and it is even being tested for potential dental applications (**Special Topic 5.1**).

A dramatic example of the power of phage therapy played out in 2016. Tom Patterson, a professor at UC San Diego, had been in a coma for months; the result of an infection by the drug resistant Gram-negative bacterium *Acinetobacter baumannii*. Granted emergency approval by the FDA, an experimental phage cocktail targeting *A. baumannii* was injected intravenously. Almost immediately Dr. Patterson began to improve and soon emerged from his coma. Based in part on this success, the FDA, in 2019, approved a clinical trial to study the safety and efficacy of a different experimental phage therapy for patients with ventricular assist devices who have developed *Staphylococcus aureus* infections.

Commercial phage products are currently available to target the food-borne pathogens *E. coli* O157:H7, *Salmonella enterica*, and *Listeria monocytogenes*. *E. coli* O157:H7, for instance, is a pathogen that can contaminate hamburger and cause bloody diarrhea. The phage product contains several different phages and is sprayed onto the hides of cattle 1–4 hours before slaughter. Cattle carcasses are steam-pasteurized and acid-washed to diminish bacterial contamination. *E. coli* cells that manage to survive these treatments will be infected and killed by the phage, reducing the consumer's risk of disease. Using multiple phages in each preparation nearly eliminates the risk that phage resistance will develop in the pathogen.

SPECIAL TOPIC 5.1 Phage "Smart Bombs" Target Biofilms

Biofilms play an important role in many infectious disease processes. Examples include *Pseudomonas aeruginosa* lung infections that plague cystic fibrosis patients and life-threatening heart valve infections caused by *Enterococcus faecalis*. It is difficult to cure these types of infections, because many of the pathogens responsible are resistant to antibiotics and because biofilm architecture may hinder antibiotic and immune cell access to bacteria nestled in the deep recesses of the structure. What's more, biofilms provide a reservoir of bacteria for chronic infections throughout the body. With antibiotics failing, is there an alternative? Recent studies by Ronen Hazan and colleagues from Hebrew University (**Fig. 1**) suggests that phage therapy can be used to destroy biofilm infections without antibiotics.

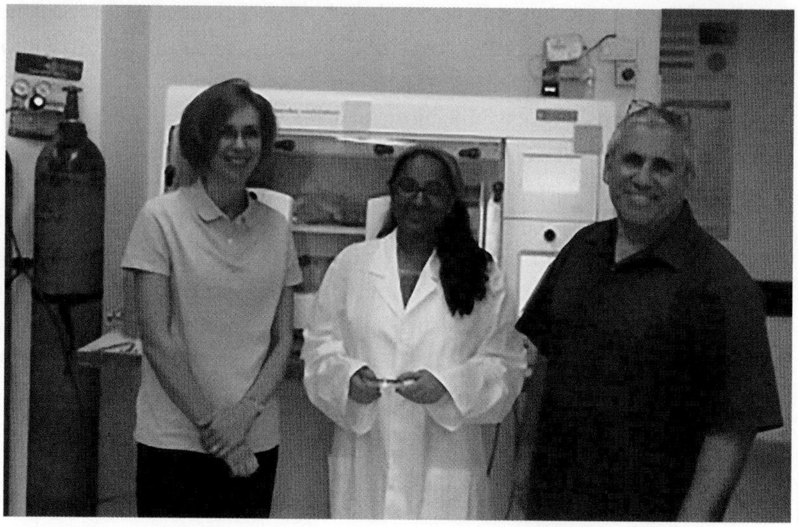

RONEN HAZAN

FIGURE 1 ■ Nurit Beyth, Leron Khalifa, and Ronen Hazan.

The first model they used was tooth infection following endodontic dental procedures. Endodontic procedures remove the nerve of a tooth because the pulp around it is infected. Drilling to remove infected pulp and the inflamed nerve will inadvertently allow entry of oral microbiota, such as *Enterococcus faecalis*, into dentinal tubules. Dentinal tubules are thin, branching tubes that extend radially from the pulp (the center of the tooth) through the dentine, and then stop at the enamel. Once it is within the tubules and the canal is sealed, *E. faecalis* can form a biofilm infection of the dentine.

The scientists used an *Enterococcus*-specific bacteriophage called EFDG1 (**Fig. 2**) to first determine whether the phage could destroy a 2-week-old biofilm of *E. faecalis* grown in a plastic dish. The confocal fluorescent images in **Figure 3** show a biofilm without phage treatment and a similar biofilm 7 days after phage treatment. The results suggest that phages are promising candidates for killing well-established *E. faecalis* biofilms.

The scientists next tested whether phage treatment would prevent root canal infections. Extracted teeth with exposed nerve canals were autoclaved and contaminated with a suspension of *E. faecalis*. Root canal procedures were performed on the teeth, and the canals were sealed. After 2 days, cross sections were made and stained with live/dead stain to show bacteria in a tooth prepared without any contamination (**Fig. 4A**), in a tooth prepared with *E. faecalis* contamination (**Fig. 4B**), and in a contaminated tooth simultaneously treated with phage (**Fig. 4C**). Live bacteria (green) are seen within the dentinal tubules in **Figure 4B** but not in the tubules treated with phage (**Fig. 4C**). Note that the root canal itself stained nonspecifically, even in the absence of bacteria (**Fig. 4A**). The results show that EFDG1 is an efficient killer of *E. faecalis* biofilms in this ex vivo model of root canal infection and may be a complementary strategy to antibiotic treatment, especially for antibiotic-resistant microbes.

In a subsequent study, published in 2018, Hazan's lab used a cocktail of two different phages to treat *E. faecalis* in a

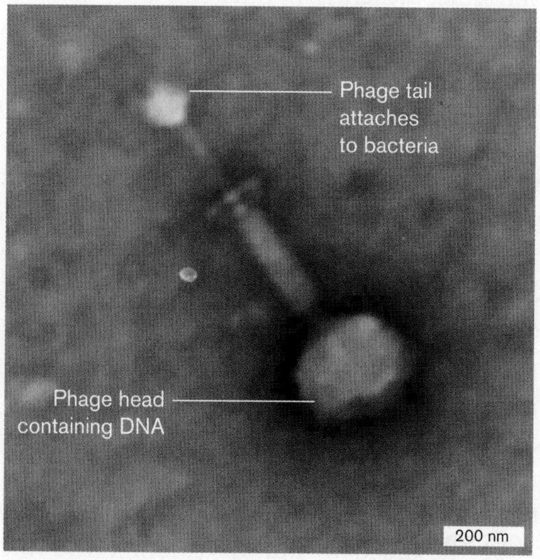

L. KHALIFA ET AL. 2015. *APPL. ENVIRON. MICROBIOL.* **81**:2696.

FIGURE 2 ■ *Enterococcus* phage EFDG1.

fibrin-clot model of infection. This cocktail achieved a sixfold reduction of *E. faecalis* CFUs in these clots, providing further evidence that these or other phages could be crucial for treating dangerous antibiotic-resistant *Enterococcus* infections.

RESEARCH QUESTION

How would you test whether *Enterococcus faecalis* can develop resistance to the EFDG1 phage? List other potential drawbacks to using phage as a treatment for human infections.

Khalifa, Leron, Yair Brosh, Daniel Gelman, Shunit Coppenhagen-Glazer, Shaul Beyth, et al. 2015. Targeting *Enterococcus faecalis* biofilms with phage therapy. *Applied and Environmental Microbiology* **81**:2696–2705.

Khalifa, Leron, Daniel Gelman, Mor Shlezinger, Axel Lionel Dessal, Shunit Coppenhagen-Glazer, et al. 2018. Defeating antibiotic- and phage-resistant *Enterococcus faecalis* using a phage cocktail *in vitro* and in a clot model. *Frontiers in Microbiology* **9**:326. https://doi.org/10.3389/fmicb.2018.00326.

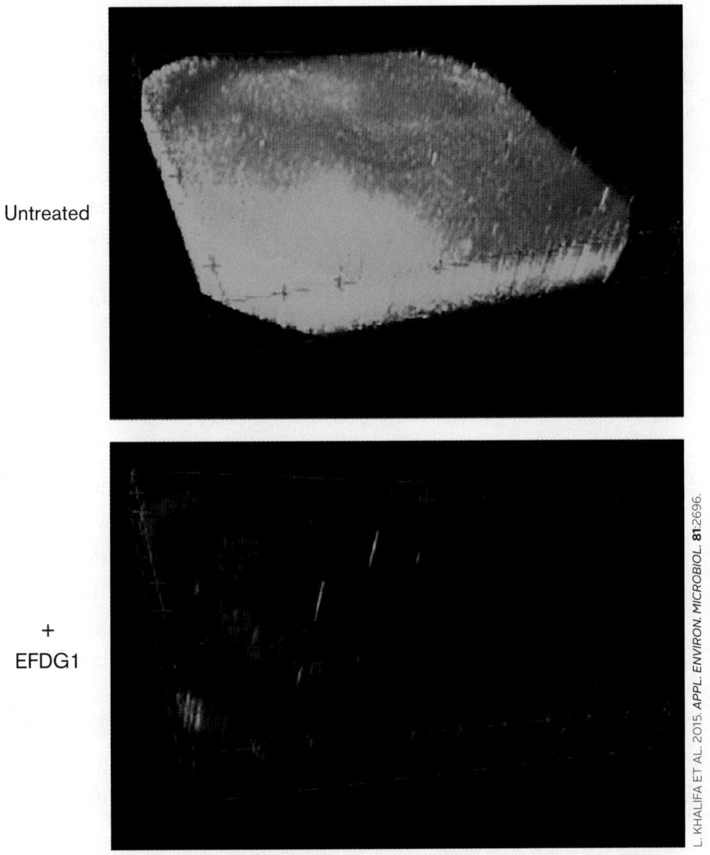

FIGURE 3 ■ Two-week-old biofilm of *Enterococcus faecalis* was either untreated (top) or treated with phage EFDG1 for 7 days (bottom). Confocal fluorescence microscopy.

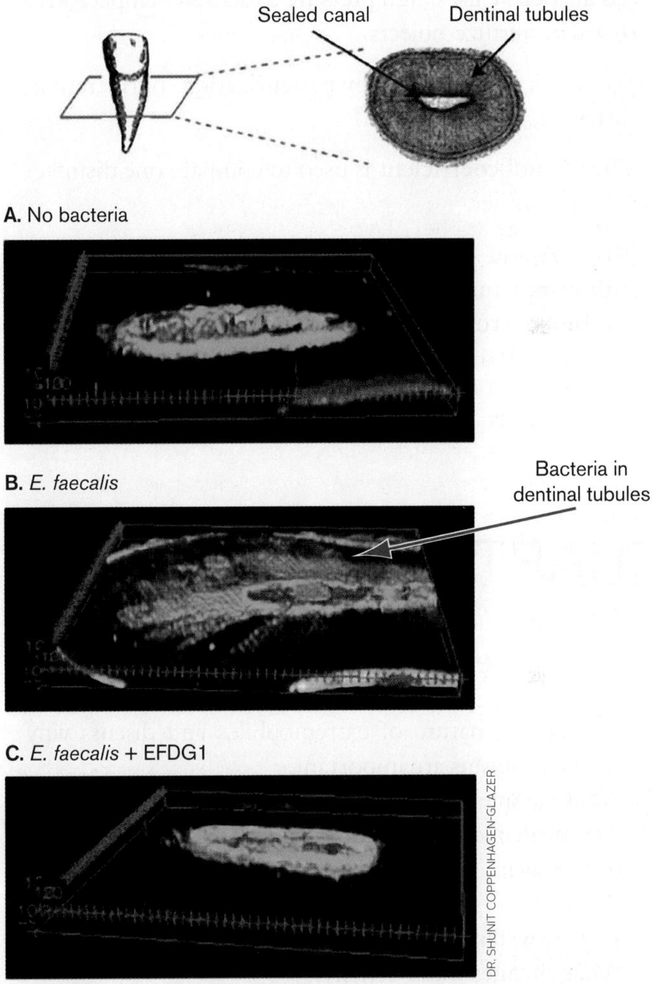

FIGURE 4 ■ Horizontal root sections of teeth subjected to endodontic treatment. Confocal fluorescence microscopy. **A.** Tooth without bacteria or phage. **B.** Tooth irrigated with bacteria only. **C.** Tooth irrigated with bacteria and phage EFDG1. Stained bacteria (green from live/dead viability stain) are depicted in the dentinal tubules surrounding the root canal (red arrow). Note that the root canal is stained nonspecifically, even in the absence of bacteria.

To Summarize

- **Antisepsis** is the removal of potential pathogens from the surfaces of living tissues, while **disinfection** kills pathogens on inanimate objects. **Sterilization** kills all living organisms.
- **Antimicrobial** compounds can be **bacteriostatic** or **bactericidal**.
- **The D-value** is the time (or dose, for irradiation) needed to decrease the number of viable cells to 10% of the original number.
- **An autoclave** uses high pressure to achieve temperatures that will sterilize objects.
- **Food can be preserved** by pasteurization, refrigeration, filtration, and irradiation.
- **The phenol coefficient** is used to compare one disinfectant to another.
- **Physical and chemical agents** kill microbes by denaturing proteins or DNA, or by disrupting lipid bilayers. **Biocontrol** is the use of one microbe to control the growth of another.
- **Antibiotics** are compounds produced by one living microorganism that kill other microorganisms.
- **Probiotics** contain certain microbes that, when ingested, aim to restore balance to the intestinal microbiome.
- **Phage therapy** offers a possible alternative to antibiotics in the face of rising antibiotic resistance.

Concluding Thoughts

Microbiology as a science was founded on the need to understand and control microbial growth. The imperative was to control human diseases, as well as the diseases of plants and animals. As it turned out, the field impacts so much more. Early microbiologists never realized the ubiquitous distribution of microbes throughout Earth's many harsh environments, and they couldn't imagine how vital these creatures are in maintaining all life on our planet. We will see in upcoming chapters that microbiology was, and still is, a crucial tool for probing the very processes that define life. Later chapters will also revisit and refine the core concepts of biological diversity, the physiology of stress, food microbiology, disease prevention, and antibiotics.

CHAPTER REVIEW

Review Questions

1. Explain the nature of extremophiles and discuss why these organisms are important.
2. What parameters define any growth environment?
3. List and define the classifications used to describe microbes that grow in different physical growth conditions.
4. What do thermophiles have to do with PCR technology?
5. Why is water activity important to microbial growth? What changes water activity?
6. How do cells protect themselves from osmotic stress?
7. Why do changes in H^+ concentration affect cell growth?
8. How do acidophiles and alkaliphiles manage to grow at the extremes of pH?
9. If an organism can live in an oxygenated environment, does that mean the organism uses oxygen to grow? If an organism can live in an anaerobic environment, does that mean it cannot use oxygen as an electron acceptor? Why or why not?
10. What happens when a cell exhausts its available nutrients?
11. List and briefly explain the various means by which humans control microbial growth. What is a D-value?
12. How do microbes prevent the growth of other microbes?

Thought Questions

1. Given a natural lake environment with 100 species of bacteria, why does the species with the fastest generation time not overwhelm the others? Or does it?
2. *Escherichia coli* is a facultative species, able to grow with or without oxygen. What would it take to make this organism a strict anaerobe?

3. Two spore formers—*Geobacillus thermophilus* and *Bacillus coagulans*—have D-values at 121°C of 5 minutes and 0.07 minute, respectively. How could the spores from these organisms have such different D-values? *Hint:* Find the organisms' optimal growth temperatures.
4. Phage therapy is touted by some as a solution to antibiotic resistance. Discuss three possible problems we could encounter with phage therapy and how we could overcome them.
5. With respect to bacteria, is the physiological effect of hydrochloric acid (HCl) at pH 4 the same as that of an organic acid at pH 4?

Key Terms

acidophile (170)
aerobe (173)
aerobic respiration (173)
aerotolerant anaerobe (175)
alkaliphile (171)
anaerobe (173)
anaerobic (173)
anaerobic respiration (175)
antibiotic (188)
antisepsis (181)
bactericidal (181)
bacteriostatic (181)
barophile (165)
biocontrol (189)
bioremediation (185)
compatible solute (167)
decimal reduction time (D-value) (181)
disinfection (181)
electron transport system (ETS) (173)
eutrophication (179)
extremophile (160)
facultative anaerobe (173, 175)
fermentation (fermentative metabolism) (175)
germicidal (181)
halophile (168)
heat-shock response (165)
hyperthermophile (164)
laminar flow biological safety cabinet (183)
lyophilization (183)
mesophile (162)
microaerophilic (175)
neutralophile (169)
niche (160)
osmolarity (167)
pasteurization (182)
piezophile (165)
probiotic (189)
programmed cell death (177)
psychrophile (163)
psychrotroph (163)
sanitation (181)
sterilization (180)
strict aerobe (173)
strict anaerobe (175)
thermophile (164)
water activity (167)

Recommended Reading

Dhakar, Kusum, and Anita Pandey. 2016. Wide pH range tolerance in extremophiles: Towards understanding an important phenomenon for future biotechnology. *Applied Microbiology and Biotechnology* **100**:2499–2510.

Furfaro, Lucy L., Barbara J. Chang, and Matthew S. Payne. 2018. Applications for bacteriophage therapy during pregnancy and the perinatal period. *Frontiers in Microbiology* **8**:2660. https://doi.org/10.3389/fmicb.2017.02660.

Gohara, David W., and Mee-Ngan F. Yap. 2017. Survival of the drowsiest: The hibernating 100S ribosome in bacterial stress management. *Current Genetics* **64**:753–760. https://doi.org/10.1007/s00294-017-0796-2.

Hoehler, Tori M., and Bo Barker Jørgensen. 2013. Microbial life under extreme energy limitation. *Nature Reviews. Microbiology* **11**:83–94.

Hori, Hiroyuki. 2019. Regulatory factors for tRNA modifications in extreme-thermophilic bacterium *Thermus thermophiles*. *Frontiers in Genetics* **10**: article 204 doi: 10.3389/fgene.2019.00204.

Luo, Chengwei, Luis M. Rodriguez-R, Eric R. Johnston, Liyou Wu, Lei Cheng, et al. 2014. Soil microbial community responses to a decade of warming as revealed by comparative metagenomics. *Applied and Environmental Microbiology* **80**:1777–1786.

Poli, Annarita, Ilaria Finore, Ida Romano, Alessia Gioiello, Licia Lama, et al. 2017. Microbial diversity in extreme marine habitats and their biomolecules. *Microorganisms* **5**:25. https://doi.org/10.3390/microorganisms5020025.

Ranawat, Preeti, and Seema Rawat. 2017. Radiation resistance in thermophiles: Mechanisms and applications. *World Journal of Microbiology and Biotechnology* **33**:112. https://doi.org/10.1007/s11274-017-2279-5.

Sapers, Haley M., Jennifer Ronholm, Isabelle Raymond-Bouchard, Raven Comrey, Gordon R. Osinski, et al. 2017. Biological characterization of microenvironments in a hypersaline cold spring Mars analog. *Frontiers in Microbiology* **8**:2527. https://doi.org/10.3389/fmicb.2017.02527.

Sattar, Syed A., Bahram Zargar, Kathryn E. Wright, Joseph R. Rubino, and M. Khalid Ijaz. 2017. Airborne pathogens inside automobiles for domestic use: Assessing in-car air decontamination devices using *Staphylococcus aureus* as the challenge bacterium. *Applied and Environmental Microbiology* **83**:e00258-17

Shuryak, Igor, Vera Y. Matrosova, Elena K. Gaidamakov, Rok Tkav, Olga Grichenko, et al. 2017. Microbial cells can cooperate to resist high level chronic ionizing radiation. *PLoS One* **12**:e0189261.

Slonczewski, Joan, James A. Coker, and Shiladitya DasSarma. 2010. Microbial growth with multiple stressors. *Microbe* **5**:110–116.

Stubbendieck, Reed M., and Paul D. Straight. 2016. Multifaceted interfaces of bacterial competition. *Journal of Bacteriology* **198**:2145–2155.

Vaishampayan, Ankita, Anne de Jong, Darren J. Wight, Jan Kok, and Elisabeth Grohmann. 2017. A novel antibacterial coating represses biofilm and virulence-related genes in methicillin-resistant *Staphylococcus aureus*. *Frontiers in Microbiology* **9**:221. https://doi.org/10.3389/fmicb.2018.00221.

CHAPTER 6
Viruses

6.1 Viruses in Ecosystems

6.2 Virus Structure

6.3 Viral Genomes and Classification

6.4 Bacteriophages: The Gut Virome

6.5 Animal and Plant Viruses

6.6 Culturing Viruses

Viruses infect a host cell and use the cell's machinery to form progeny virions. In ecosystems, viruses cycle nutrients, control host populations, and promote host diversity. Viruses may kill their host cells, or they may copy themselves into their host genome. In humans, endogenous viral DNA evolved into many portions of our genome. Deadly viruses can be engineered as vectors for lifesaving gene therapy.

CURRENT RESEARCH highlight

Bacteriophages infect *Prochlorococcus*. The ocean plays an epic drama on a tiny scale, in which the water's smallest and most abundant photosynthetic cells, *Prochlorococcus* bacteria, are attacked by something smaller: bacteriophage P-SSP7. Matthew Sullivan, Sally Chisholm, and Wah Chiu used cryo-electron tomography to visualize how phage P-SSP7 infects *Prochlorococcus*. Each phage particle binds specific proteins on the host cell. Then the phage inserts its DNA through the envelope, reprogramming the cell to form hundreds of progeny phages. Trillions of marine viruses cycle nutrients among organisms throughout the ocean.

Sources: Modified from Murata et al. 2017. *Sci. Rep.* **7**:44176. See also Hurwitz et al. 2017. *Genome Biol.* **14**:123.

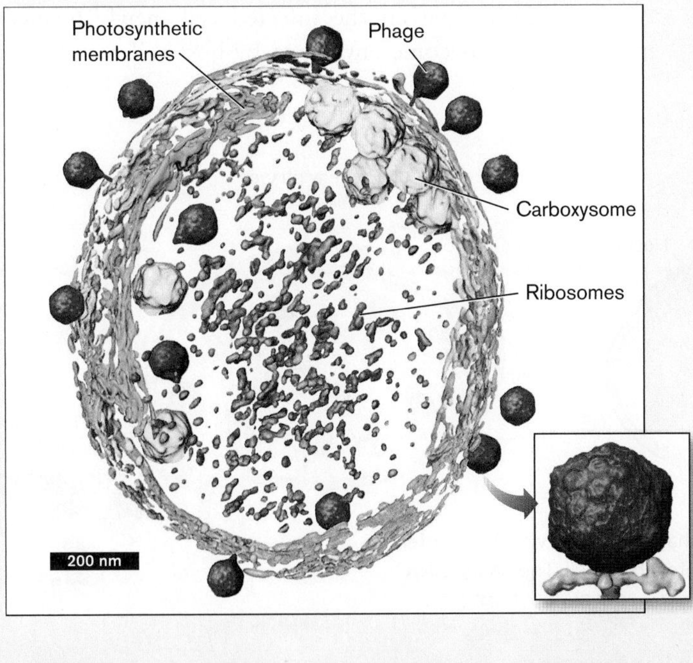

K. MURATA ET AL. 2017. *SCI. REP.* **7**:44176, FIG. 1B.

K. MURATA ET AL. 2017. *SCI. REP.* **7**:44176, FIG. 5A.

AN INTERVIEW WITH

MATTHEW SULLIVAN, PROFESSOR, THE OHIO STATE UNIVERSITY

PHOTO COURTESY OF THE OHIO STATE UNIVERSITY

What is the global significance of marine viruses?

Viruses were once thought to be inconsequential in marine systems because researchers used host strains to try to isolate viruses and did not get any. But the host strains used are not common in marine waters. In the late 1980s, researchers started concentrating seawater and looking at it—microscopy revealed some 50 million virus-like particles per teaspoon of water!

Our work has tried to paint a clearer picture of "who" these viruses are. Instead of just "counting dots," we now have genomes for thousands of these viruses to go with the abundance data. This helps us link viruses to which hosts they infect, scan their genomes to identify ways they might manipulate their host cells during infection, and find new ways to classify viruses. Many viruses do not share even a single gene. Yet we now have developed automated approaches to establish taxonomy. I never thought we'd reach the point where we might have a relatively complete catalog of surface ocean viruses—globally—but here we are!

A **virus** is defined as a noncellular particle that infects a host cell and directs it to produce progeny particles (**Fig. 6.1**). The virus particle, also called a **virion**, generally consists of a viral genome (DNA or RNA) contained within a protein **capsid**. The capsid may or may not contain appendages such as a spike and tail fibers.

Viruses are everywhere, in all ecosystems. In the oceans, the number of viruses infecting bacteria and algae can reach 10^7 (10 million) per milliliter. Viruses are the dominant consumer of marine microbes. These include viruses of bacteria, called **bacteriophages** or **phages**. For example, phage P-SSP7 infects the tiny cyanobacterium *Prochlorococcus* (see the Current Research Highlight). As *Prochlorococcus* fixes CO_2 by photosynthesis and its population grows, phage infections spread and break down the cells, releasing their organic molecules, which other forms of life can use. Matthew Sullivan's research shows how marine viruses play a global role in cycling carbon, nitrogen, and sulfur (discussed in Chapter 22).

All cellular organisms are infected by viruses. In humans, the viruses we hear about are those that cause epidemics, such as seasonal influenza outbreaks and the AIDS pandemic. But most human-associated viruses go unnoticed, and some actually enhance our health. For example, our gut contains numerous bacteriophages that control our populations of gut bacteria (discussed in Section 6.4).

In research, viruses provide tools and model systems for our discovery of the fundamental principles of molecular biology. Vectors for gene cloning and gene therapy are derived from viruses (see Chapter 12) and the CRISPR antiviral defense in bacteria gives us tools to edit human DNA. Our understanding of bacteriophages provides a background for the molecular biology we will encounter in Part 2 of this book (Chapters 7–12). And remarkably, we now engineer human viruses to deliver gene therapy and kill cancers (discussed in Chapter 11).

In Chapter 6 we introduce the major themes of virus structure and function, and the ways that viruses manipulate host cells for their own reproduction. The molecular biology of viral infection and replication is explored further in Chapter 11. Viral disease pathology and epidemiology are discussed in Chapters 25–27.

6.1 Viruses in Ecosystems

Different viruses infect different hosts, and their impact profoundly shapes their host's evolution and ecology. Host organisms must continually evolve novel variants that resist viruses. Thus, viruses generally increase population diversity.

Viruses Infect Specific Hosts

Different types of viruses infect different kinds of host cells. For example, bacteriophage T2 infects a specific strain of the bacterium *E. coli*. The T2 and T4 phages are "tailed" phages, whose capsid has a tail that inserts the viral genome into the host cell. Within the host, the phage genes direct production and assembly of progeny virions (**Fig. 6.2A**). Virions are released when the host cell lyses. As cells lyse, their disappearance can be observed as a **plaque**, a clear spot within a lawn of bacterial cells (**Fig. 6.2B**). Each plaque arises from a single **virion**, or phage particle, that lyses a host cell and spreads progeny to infect adjacent cells. The plaque count represents the number of individual infective virions from the phage suspension that was spread on the plate.

A virus that infects humans is the measles virus (**Fig. 6.2C**). The measles virus has an envelope that is derived from the host cell plasma membrane as the virus exits the host cell. When the virus infects a new host cell, the envelope fuses with the host cell plasma membrane, releasing the viral contents into the cytoplasm of the cell. After replicating within the infected cell, newly formed measles virions become enveloped by host cell membrane

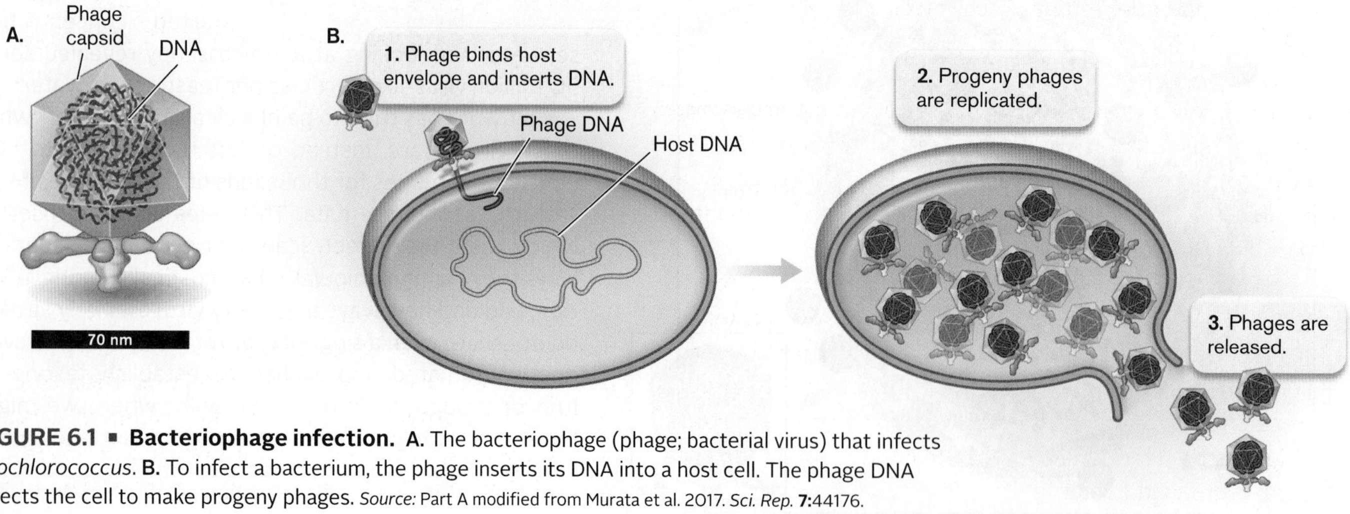

FIGURE 6.1 ■ Bacteriophage infection. A. The bacteriophage (phage; bacterial virus) that infects *Prochlorococcus*. B. To infect a bacterium, the phage inserts its DNA into a host cell. The phage DNA directs the cell to make progeny phages. *Source:* Part A modified from Murata et al. 2017. *Sci. Rep.* **7**:44176.

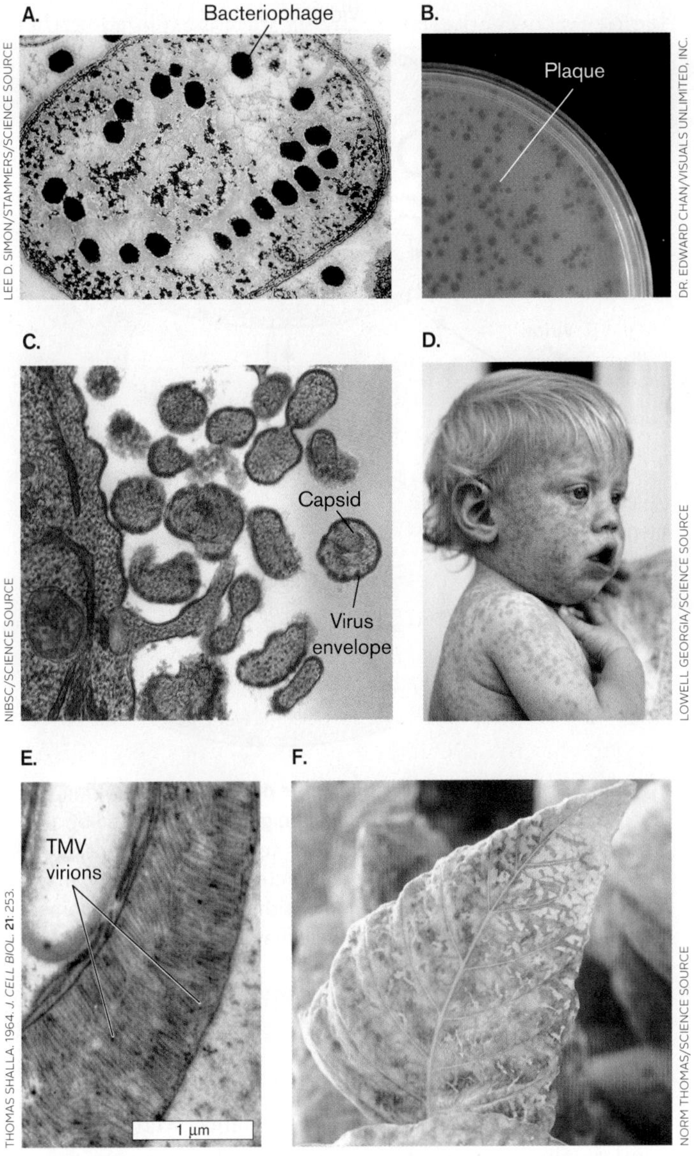

FIGURE 6.2 ▪ Virus infections and disease. **A.** Bacteriophage T2 particles pack in an array within an *E. coli* cell (TEM). **B.** Bacteriophage infection forms plaques of lysed cells on a lawn of bacteria. **C.** Measles virions bud out of human cells in tissue culture (TEM). **D.** Child infected with measles shows a rash of red spots. **E.** Tobacco leaf section is packed with tobacco mosaic virus (TMV) particles. **F.** Tomato leaf infected by TMV shows mottled appearance.

as they bud out of the host cell. The spreading virus causes an immune reaction with a rash of red spots on the skin of infected patients (**Fig. 6.2D**). Measles illness generally resolves without treatment, but infections can be fatal (one in 500 cases).

Plants are infected by viruses such as tobacco mosaic virus (TMV). Within the plant cell, virions accumulate to high numbers (**Fig. 6.2E**) and travel through interconnections to neighboring cells. Infection by tobacco mosaic virus results in mottled leaves and stunted growth (**Fig. 6.2F**). Plant viruses cause major economic losses in agriculture worldwide.

Are Viruses Alive?

Historically, viruses were defined as nonliving particles, because at first the virion was the only form to be visualized and understood. The Russian botanist Dmitri Ivanovsky (1864–1920) and the Dutch microbiologist Martinus Beijerinck (1851–1931) first proposed the existence of viruses as infectious agents that passed through a filter too small for cells to pass. Certain kinds of these infectious agents were actually crystallized from solution. Wendell Stanley (1904–1971) earned the 1946 Nobel Prize in Chemistry for the first crystallization of a virus, tobacco mosaic virus (TMV). He later crystallized poliovirus. The crystallization of viruses reinforced their definition as nonliving.

But what happens after a virus infects a cell? The viral components interact with the cell's parts in a dynamic way, fully part of the cell's living processes. The replication cycles of bacteriophages were first studied by English bacteriologist Frederick William Twort (1877–1950) and by French microbiologist Félix d'Herelle (1873–1949). To actually reveal what went on in the host cell, Ernst Ruska (1906–1988) used the newly invented electron microscope (discussed in Chapter 2).

Today, our concept of the virus has been upended by the discovery of "giant viruses" with genomes of up to 2.5 megabases—larger than the genomes of some microbial cells. Some of these giant viruses, such as Mimivirus and Klosneuvirus, possess genes for ribosomal components and transfer RNA (discussed in Section 6.3). As of this writing, no known virus encodes ribosomes with a full translation system.

Integrated Viral Genomes

Some viruses do more than replicate within a cell: they can integrate their own genomes into the host genome. In effect, such viruses become a part of the host organism. A virus that integrates its genome into the DNA of a bacterial genome is called a **prophage**. Viruses transfer enormous quantities of genes among microbial populations, such as the genes for toxin production acquired by human-associated bacteria.

Within a human cell, an integrated viral genome is called a **provirus**. A permanently integrated provirus transmitted via the germ line is called an **endogenous virus**. Remarkably, our human genome includes endogenous viral genomes that express essential human genes.

How were integrated viral genomes discovered? In the 1950s, genetic analysis of bacteriophage lambda in *E. coli* showed evidence of phage gene expression from within host genomes. **Bacteriophage lambda is discussed at length in Chapter 11.** Early in the twentieth century, American cancer researcher Peyton Rous (1879–1970) discovered that some avian viruses cause tumors. The tumor generation results from integration of a viral genome. For this discovery, Rous won the 1966 Nobel Prize in Physiology or Medicine.

But the biggest surprise emerged from decades of study of the genomes of host organisms. Many host cell traits turn out to be expressed by integrated viral genomes that actually confer benefits on their host. For example, *Nostoc* cyanobacteria possess a prophage that encodes nitrogenase, an enzyme for nitrogen fixation. Many human pathogens, such as *Staphylococcus aureus*, contain prophages that encode toxins. Human genomes show evidence of much genetic material originating from viral genomes that integrated into host DNA during the course of our primate evolution. Over time, these viral sequences mutated and got "stuck" in the host. As much as half of the human genome may have originated from viruses or from virus-like elements of DNA called transposons (discussed in Chapter 9). For example, some integrated viral genomes express placental proteins that are essential for early development of human embryos.

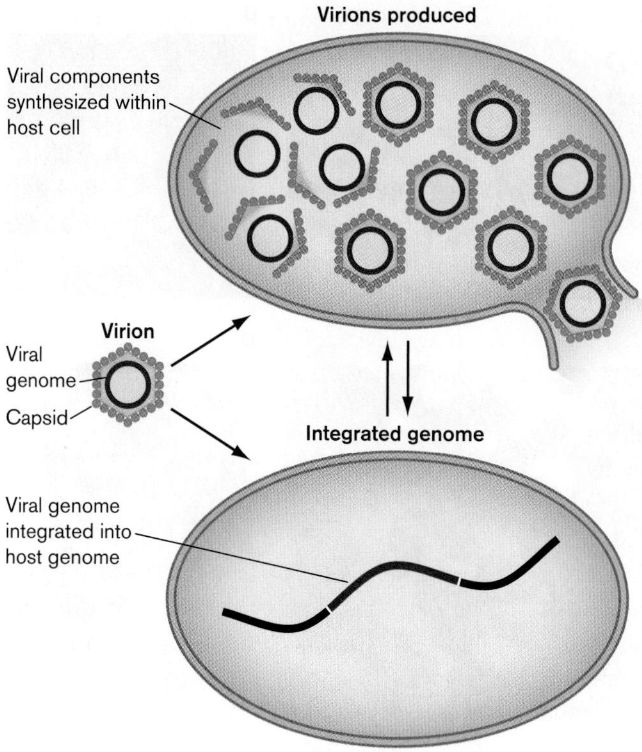

FIGURE 6.3 ■ Virus as a subcellular organism. The virion, or virus particle, consists of a nucleic acid genome contained by a protein capsid. A virion may infect a host cell and cause the cell's enzymes to synthesize progeny virions. Alternatively, infection may lead to integration of the viral genome within the host cell genome. Integration may last indefinitely, or else may lead to production of virions.

Dynamic Nature of Viruses

Emerging evidence supports a dynamic view of viral existence and leads us to revisit the question: What is a virus? Some viruses may interconvert among three very different forms (**Fig. 6.3**):

- **Virion, or virus particle.** The **virion** is an inert particle consisting of nucleic acid enclosed by a protein **capsid**. Some viruses package enzymes and possess a lipid envelope. A virion does not carry out any metabolism or energy conversion.
- **Intracellular replication complex.** Within a host cell, the viral gene products direct the cell's enzymes to assemble progeny virions at "virus factories" called replication complexes (discussed in Section 6.5). Virus assembly requires recruitment of host ribosomes, as well as intricate collaboration between host and viral proteins.
- **Viral genome integrated within host DNA.** Some types of viral genomes may integrate within a host chromosome as a provirus and replicate as part of the host. Integration may be permanent; alternatively, the viral genome may be reactivated to start assembling virions.

The inert nature of the virion (particle), which lacks metabolism, argues that viruses are nonliving. However, we know that cellular life forms such as bacteria can convert into inert forms, such as endospores, that remain viable for thousands of years. The virion assembly process argues that viruses are living organisms. Assembly includes metabolism and production of progeny, processes we see for obligate intracellular bacteria such as chlamydias. Furthermore, the genomes of large viruses such as chloroviruses show evidence of reductive evolution (evolutionary loss of genes) from a cell.

On the other hand, genome integration argues for a view of viruses as host cell components. Many bacteria regulate phage-encoded genes along with their own genes (discussed in Chapter 10). For example, *Vibrio cholerae*, the cause of cholera, uses bacterial genes to regulate expression of the cholera toxin, which is encoded by an integrated prophage. For the *V. cholerae* bacterium, the viral prophage is a functional part of the cell. Viruses that enter and leave microbial genomes can transfer valuable genes among different species, conferring traits such as the ability to metabolize new food sources. For example, bacteriophages can transfer genes encoding catabolic enzymes for complex carbon sources with soil or gut microbiomes. In the ocean, viruses transfer genes among host microbes on a daily basis, including genes for photosynthesis and catabolism that enhance the production of viruses.

How and where did viruses originate? Did they form within host cells, or did they arise as independent entities? Giant viral genomes encode numerous metabolic enzymes, including aminoacyl-tRNA synthetases (enzymes that attach an amino acid to the transfer RNA for translation). The size and complexity of giant viruses argue that these viruses evolved by reductive evolution of an intracellular parasitic bacterium. Similar arguments are made for large viruses such as the smallpox viruses and herpesviruses. On the other hand, small viruses with small RNA genomes, such as influenza virus and HIV, look more like something that arose from parts of a cell. For example, the key retroviral enzyme reverse transcriptase (which copies RNA to DNA) shows homology to telomerase, a host cell enzyme that maintains chromosome ends. Viral origins are discussed further in **eTopic 6.1**.

Note: Some scientists consider viruses "not alive." Other scientists consider viruses a noncellular life form that requires a host cell for replication.

Ecological Roles of Viruses

Viruses are the tiniest of biological entities, but they play starring roles in ecosystems. Acute viruses (which rapidly kill their hosts) act as predators or parasites to limit host population density. They also recycle nutrients from their host bodies. An increase in host population density increases the rate of transmission of viral pathogens. As the host population declines, viruses are less likely to find a new host before they lose infectivity, while many of the remaining hosts have undergone selection for resistance. Thus, viruses can limit host density without causing extinction of the host.

The sum of viral populations in an ecosystem is called the **virome**. In a host community, the overall effect of the virome is to increase host diversity. Each virus species has a limited host range and requires a critical population density to sustain the chain of infection. In marine phytoplankton, a virus limits its particular host species to a population density far lower than what would be sustainable by the available resources, without competition. The resources then support other species resistant to the given virus (but susceptible to other ones). Thus, overall, marine viruses prevent the dominance of any one species and foster the evolution of many distinct host species.

The continual threat of marine viral epidemics explains the great diversity of phytoplankton ranging from silica-shelled diatoms to toxin-producing dinoflagellates. Viruses can dissipate marine algal blooms large enough to see from outer space. **Figure 6.4** shows an example of a marine virus infecting cells of the alga *Emiliana huxleyi*. These algal cells, called coccolithophores, are known for their calcite plates, or "coccoliths," which coat each cell. The virus particle (**Fig. 6.4A**) consists of a polyhedral protein capsid with outer and inner membranes, containing a coiled DNA genome. The virus attaches one of its corner proteins to the surface of a coccolith and inserts its DNA.

When coccolithophores overgrow, they can generate a bloom that covers the ocean for thousands of square kilometers (**Fig. 6.4C**). The pale clouds in the water are the reflected light from billions of coccoliths. Yet within a few days, this gigantic bloom is wiped out by viruses.

Throughout the ocean, viruses play a decisive role in controlling algal blooms. By lysing the algae as they grow,

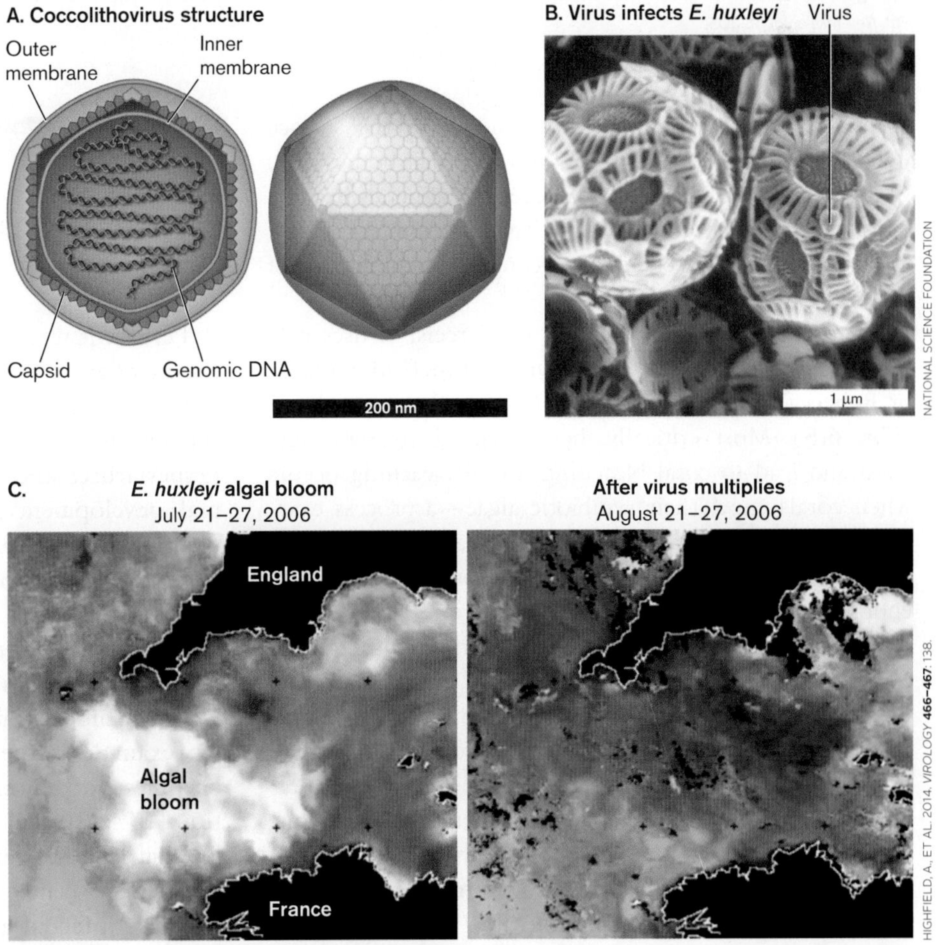

FIGURE 6.4 ■ Virus infection controls marine algal bloom. **A.** Coccolithovirus structure. This virus infects the coccolithophore, a eukaryotic alga, *Emiliana huxleyi*. **B.** *E. huxleyi* infected by Coccolithovirus. **C.** Bloom of *E. huxleyi* off Plymouth, England, detected by the MODIS imaging sensor aboard a NASA satellite while it was remote-sensing light reflectance from chlorophyll. Shown before and after virus infection. *Source:* Part A modified from https://viralzone.expasy.org/589.

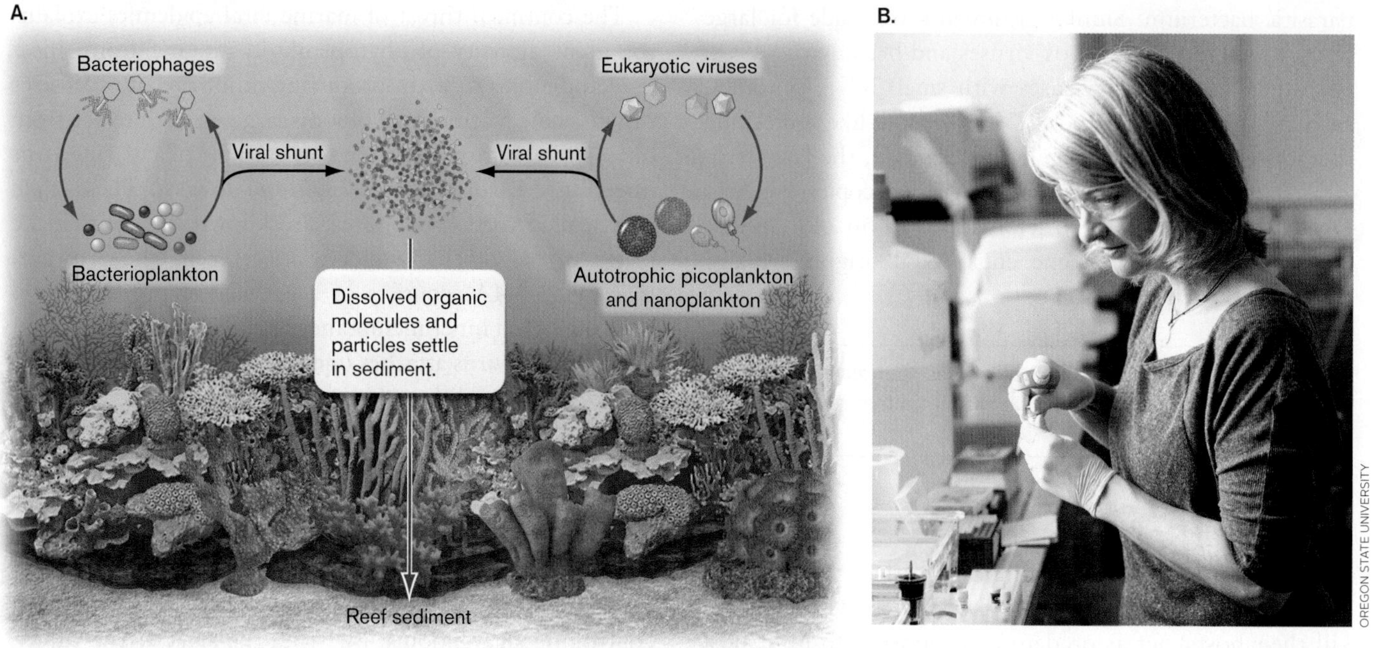

FIGURE 6.5 ■ Viruses cycle nutrients in coral reef ecosystems. **A.** Bacteriophages and eukaryotic viruses convert marine organisms into dissolved organic molecules and particles. The organic particles settle to the benthos (marine sediment), where they provide carbon and nitrogen for corals and other marine life. **B.** Rebecca Vega Thurber studies the virome of coral reefs. *Source:* Part A modified from Thurber et al. 2017. *Nat. Rev. Microbiol.* **15**:205.

viruses return algal carbon and minerals to the surface water before the algae starve to death and their bodies sink. When biomass sinks to the ocean deep, its minerals become unavailable for phototrophs and their consumers. Thus, viruses are major players in the cycling and sequestering of atmospheric CO_2.

In marine ecosystems such as coral reefs, viruses play several important roles, as shown by the field research of Rebecca Vega Thurber from Oregon State University (**Fig. 6.5**). Most critically, herpes-related viruses infect coral and lead to coral bleaching. Coral bleaching occurs when corals expel their symbiotic algae—a process exacerbated by global warming. The coral-associated virome includes different viruses infecting coral tissues, coral endosymbionts, and coral-associated bacteria.

The most widespread role of marine viruses is the cycling of food molecules (**Fig. 6.5A**). Viral infection and lysis convert the bodies of phytoplankton (photosynthetic microbes) and consumers into detritus consisting of small organic particles and soluble molecules. The virus-generated detritus gets taken up by consumers (heterotrophs), and minerals are absorbed by phototrophs. This viral generation of detritus is called the **viral shunt**. The viral shunt returns some organic matter to microbial consumers in the upper region of the ocean (discussed in Chapter 21). Some carbon is thus diverted from the ocean sink and returns to the CO_2 cycle. The viral shunt also plays a major part in marine cycling of phosphorus, nitrogen, and iron.

Thought Question

6.1 Suppose a certain virus depletes the population of an algal bloom. What will happen if some of the algae are genetically resistant?

Persistent viruses (viruses that don't immediately kill the host) may benefit the host population overall. If a virus can evade the host defenses, it may replicate, and thus persist, in the hosts. Humans harbor many obscure herpesviruses and cold-type viruses that stimulate normal development of the immune system. In natural mammal populations, viruses that have been shed from persistently infected mammals can infect and kill uninfected competitors of the same species. The loss of competitors means more food and territory for the infected host. Thus, a persistent virus can act as a bioweapon that kills off competitor populations susceptible to the virus. An example is the koala retrovirus, KoRV, which has become endogenous (persistent) in many koala populations of Australia. When KoRV is transmitted to koalas that lack the retrovirus, it kills most of those newly infected—except for the few in which the virus persists. In these infected animals, the immune system controls the virus but cannot eradicate it.

Integrated viral genomes can evolve complex symbiotic interactions with multiple hosts, combining mutualism and parasitism. An example is the relationship of polydnaviruses,

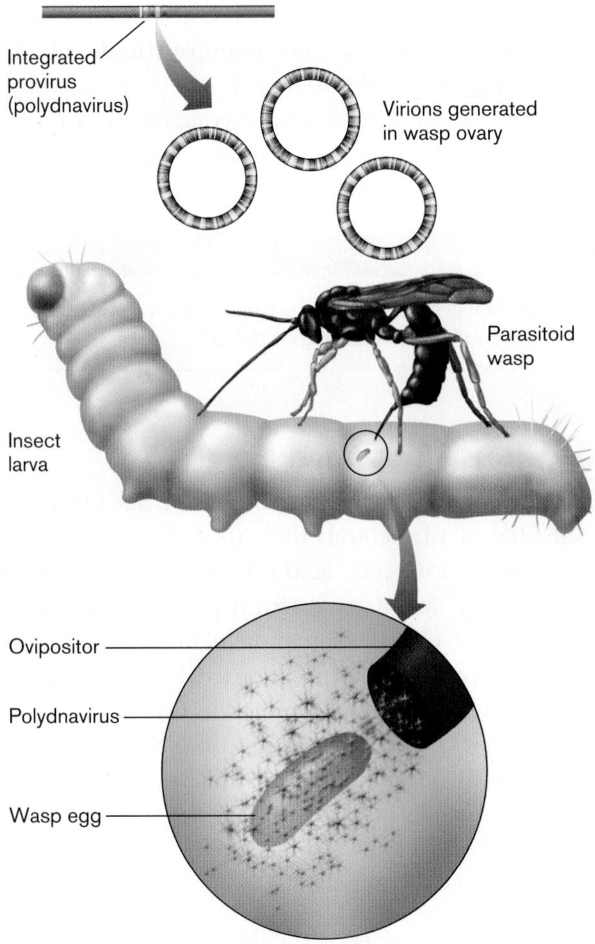

FIGURE 6.6 ■ The relationship of polydnaviruses, wasps, and caterpillars. Parasitoid wasps lay their eggs inside a living insect caterpillar. When a female wasp deposits her eggs inside the caterpillar, she also deposits her symbiogenic polydnavirus virions. The virions express wasp genes in the caterpillar, where they prevent the encapsulation process that would otherwise wall off the wasp egg and kill it.

wasps, and caterpillars (**Fig. 6.6**). Parasitoid wasps lay their eggs inside a living insect caterpillar, where the larvae must hatch and feed without stimulating the caterpillar's defense. The wasp genome contains an integrated genome (or **provirus**) of a polydnavirus. The polydnavirus forms its own genomes and virions only within the wasp ovary. When the wasp deposits her eggs inside the host caterpillar, she also deposits her mutualistic virions. The virions express proteins that prevent the caterpillar cells from undergoing encapsulation, a process that would otherwise wall off the wasp eggs and kill them.

Viral Disease

When viral infection causes harm to an animal or plant, the host has a disease. The range of host species infected by a given virus is known as its **host range**. Some viruses can infect only a single host species; for example, HIV infects only humans. Close relatives of humans, such as the chimpanzee, are not infected by HIV, although they are susceptible to a highly related virus, simian immunodeficiency virus (SIV). By contrast, West Nile virus, transmitted by mosquitoes, infects many species of birds and mammals. West Nile virus has a much broader host range than HIV and SIV have.

For humans, viruses cause many forms of illness. The impact of viral diseases on human history and culture would be hard to overstate. More people died of influenza in the global epidemic of 1918 than in the battles of World War I. Poliovirus is famous for causing the outbreaks of poliomyelitis that swept the United States during the first half of the twentieth century. President Franklin Roosevelt, himself a victim of the disease, established a national foundation called the March of Dimes to develop a vaccine. The spectacular success of the March of Dimes set the pattern for future public support of research on cancer and AIDS.

Chronic viral infections, however, are more common than acute disease, and are an everyday part of our lives. The most frequent infections of college students are due to respiratory pathogens such as rhinovirus (the common cold) and Epstein-Barr virus (infectious mononucleosis), as well as sexually transmitted viruses such as herpes simplex virus (HSV) and papillomavirus (genital warts). Some viruses, such as rhinoviruses, are eliminated rapidly by our immune system, whereas others, such as herpesviruses, establish a lifelong latent infection. Viruses also impact human industry; for example, bacteriophages (literally, "bacteria-eaters") infect cultures of *Lactococcus* during the production of yogurt and cheese. Plant pathogens such as cauliflower mosaic virus and rice dwarf virus cause substantial losses in agriculture.

In contrast to our vast arsenal of antibiotics (effective against bacteria), the number of antiviral drugs remains small. Because the machinery of viral growth is largely that of the host cell, viruses present relatively few targets that can be attacked by antiviral drugs without harming the host. An exception is human immunodeficiency virus (HIV), the cause of AIDS. Molecular studies of HIV have yielded several major classes of antiviral drugs, such as AZT and protease inhibitors. The molecular biology behind anti-HIV drugs is discussed in Chapters 11 and 27.

Thought Question

6.2 Search for some specific viruses on the Internet. Which viruses do you think have a narrow host range, and which have a broad host range?

To Summarize

- **A virus particle or virion has a nucleic acid genome enclosed by a protein capsid.** Within a host cell, the viral genome can generate an intracellular replication complex, or the genome sequence can become integrated within host cell DNA.
- **All classes of organisms are infected by viruses.** Usually the hosts are limited to a particular host range of closely related strains or species.
- **Are viruses living or nonliving?** A virion consists of an inert particle lacking metabolism. But some viral genomes resemble those of cells, and viral replication directs a metabolic program within the host cell.
- **Acute virus infection limits host population density.** Virus-associated mortality may increase the genetic diversity of host species.
- **Persistent viruses remain in hosts**, where they may evolve traits that confer positive benefits in a virus-host mutualism.
- **Marine viruses** infect most phytoplankton, releasing their minerals in the upper water, where they are available for other phototrophs. Viral activity substantially impacts the global carbon balance.
- **Viruses transfer genes between host genomes.** Viral gene transfer is a major source of cellular genome evolution.

6.2 Virus Structure

The structure of a virion keeps the viral genome intact, and it enables infection of the appropriate host cell. First, the stable capsid protects the viral genome from degradation and enables the genome to be transmitted outside the host. Second, in order for the viral genome to reproduce, the virion must either insert its genome into the host cell or disassemble within the host. In the process, the original particle loses its stable structure and its own identity as such. But if viral reproduction succeeds, then it yields numerous progeny virions.

A virion possesses a genome of either DNA or RNA, contained by proteins that compose the capsid. The shape of the virion depends on the species of virus. The virion may be symmetrical or asymmetrical, or combine aspects of both. The papillomavirus virion, for example, consists solely of a capsid containing a genome. Some viruses, such as the tailed bacteriophages, have an elaborate delivery device to transfer the viral genome into the host cell. Other viruses possess a membrane envelope, derived from membrane of the previous host cell. The protein capsid of an enveloped virus is called a core particle.

Understanding virus structure is crucial for devising vaccines and drug therapies. For example, the Gardasil vaccine for human papillomavirus, HPV (recommended for all children before adolescence), is composed of capsid proteins from nine different HPV strains. The molecular basis of viral structure and drugs is presented in Chapter 11.

Thought Question

6.3 What will happen if a virus particle remains intact within a host cell and fails to release its genome?

Symmetrical Virions

The capsid of a symmetrical virus may be one of two types: icosahedral or filamentous (helical). Some capsids have an intermediate form, such as the HIV core. The advantage of geometrical symmetry is that it provides a way to form a package out of repeating protein units generated by a small number of genes and encoded by a short chromosomal sequence. The smaller the viral genome, the more genome copies can be synthesized from the host cell's limited supply of nucleotides. Nevertheless, some symmetrical viruses, such as herpesvirus and Mimivirus, have much larger genomes. Large genomes offer a greater range of functions for viral components.

Icosahedral viruses. Many viruses package their genome in an **icosahedral** (20-sided) **capsid**. An icosahedral capsid takes the form of a polyhedron with 20 identical triangular faces. In the capsid, each triangle can be composed of three identical but asymmetrical protein units. An icosahedral capsid is found in the herpes simplex virus (**Fig. 6.7A**). Each triangular face of the capsid is determined by the same genes encoding the same protein subunits. The actual form of viral subunits can vary greatly, generating very different complex shapes for different viral species. But no matter what the pattern of subunits in the triangular unit, the structure overall shows the rotational symmetry of an icosahedron (**Fig. 6.7B**): threefold symmetry around the axis through two opposed triangular faces, fivefold symmetry around an axis through opposite points, and twofold symmetry around an axis through opposite edges.

Thought Question

6.4 For a viral capsid, what is the advantage of an icosahedron (20-sided solid), as shown in **Figure 6.7**, instead of some other polyhedron, such as a cube or a tetrahedron?

Virus particles can be observed by standard transmission electron microscopy (TEM), but the details of capsid structure (as in **Fig. 6.7A**) require digital reconstruction

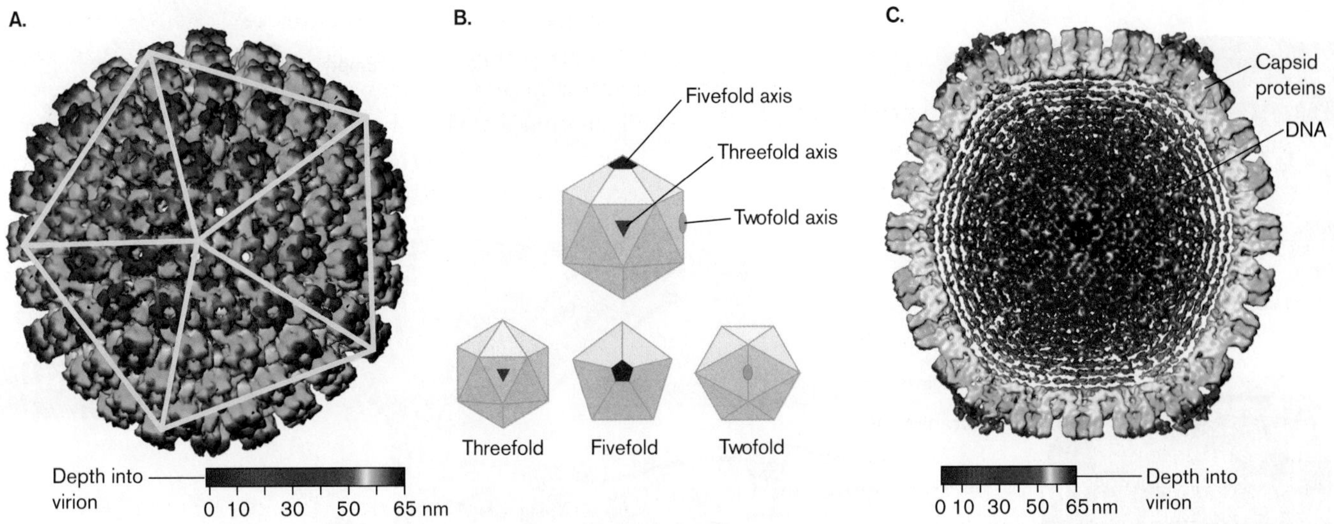

FIGURE 6.7 ▪ Herpesvirus: icosahedral capsid symmetry. **A.** Icosahedral capsid of herpes simplex 1 (HSV-1), with envelope removed. Imaging of the capsid structure is based on computational analysis of cryo-TEM. Images of 146 virus particles were combined digitally to obtain this model of the capsid at 2-nm resolution. **B.** Icosahedral symmetry includes fivefold, threefold, and twofold axes of rotation. **C.** The icosahedral capsid contains spooled DNA. *Source:* Parts A and C modified from C. Z. Hong Zhou et al. 1999. *J. Virol.* **73**:3210.

of cryo-EM images (discussed in Section 2.6). Recall from Chapter 2 that in cryo-EM, the viral samples for TEM are flash-frozen, preventing the formation of ice crystals. Flash-freezing enables observation without stain. The electron beams penetrate the object; thus, images of individual capsids actually provide a glimpse of the virus's internal contents. By digitally combining and processing cryo-TEM images from a number of capsids, we build a 3D reconstruction for the entire virus particle. Within the icosahedral capsid, the herpesvirus genome is spooled tightly (**Fig. 6.7C**). During virion synthesis, the viral DNA is packaged under high pressure. A molecular motor powered by ATP drives viral DNA into the viral capsid (discussed in Chapter 11).

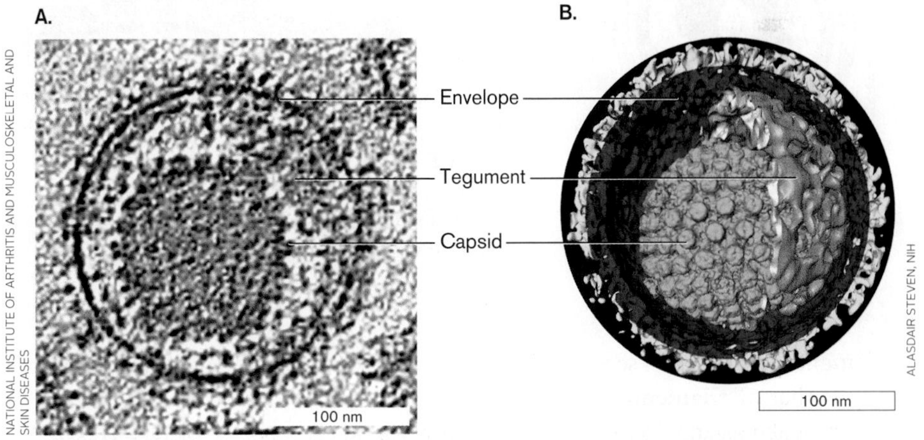

FIGURE 6.8 ▪ Envelope and tegument surround the herpesvirus capsid. **A.** Section showing envelope and tegument proteins surrounding the capsid (cryo-EM). **B.** Cutaway reconstruction of the herpes virion (cryo-EM tomography).

In some icosahedral viruses, such as HIV, the capsid (called a core particle) is enclosed in an **envelope** membrane. The envelope is composed of virion membrane from the host cell in which the virion formed; it also contains proteins specified by the viral genome. HIV derives its envelope from host plasma membrane, whereas other viruses, such as herpesvirus, derive their envelope from intracellular membranes such as the nuclear membrane or endoplasmic reticulum. The envelope and capsid contents of herpesvirus are shown in **Figure 6.8**.

The space between the capsid and the envelope contains "tegument" proteins, such as enzymes that interact with the host cell to depress its defense responses. Tegument proteins are expressed during infection of a host cell, and then get packaged in the virion. Both viral and host proteins may be packaged as tegument. The mature envelope bristles with glycoprotein **spike proteins** that attach it to the capsid. The spike proteins enable the virus to attach to and infect the next host cell. Further details of herpesvirus structure and function are presented in Section 11.5.

Note: Distinguish the viral envelope (phospholipid membrane derived from a host cell membrane) from the bacterial cell envelope (protective layers outside the bacterial cell membrane). The bacterial envelope is discussed in Chapter 3.

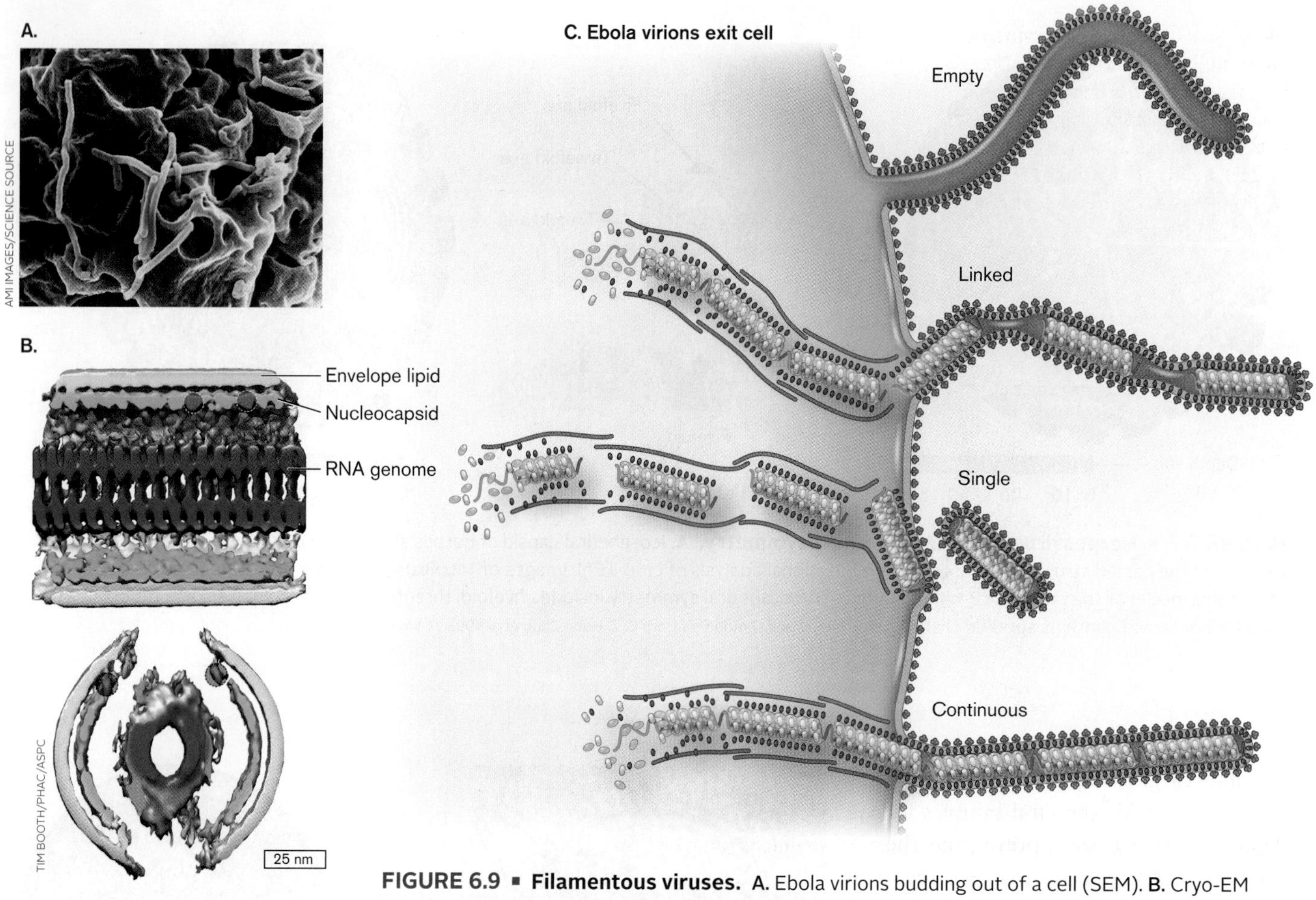

FIGURE 6.9 ▪ Filamentous viruses. **A.** Ebola virions budding out of a cell (SEM). **B.** Cryo-EM tomogram showing 3D structure of the Ebola virion (top: side view; bottom: end view). **C.** Ebola virions budding out as tandem arrays of multiple RNA genomes packaged in the nucleocapsid, held together within a flexible lipid envelope.

Filamentous viruses. A second major category of virus structure is that of **filamentous viruses**. Unlike an icosahedral capsid, which usually has a fixed size, a helical capsid can vary in length to accommodate different lengths of nucleic acid.

A well-known filamentous virus is Ebola virus (**Fig. 6.9**), which causes a fatal disease of humans and related primates. Ebola virus caused a major epidemic in western Africa in 2014, and it sparked another outbreak in 2018 (discussed in Chapter 28). The virus has become known for the iconic micrographs of its long, twisted filaments. Cryo-EM tomography reveals the helical form of the RNA genome, packaged in nucleocapsid proteins (**Fig. 6.9B**). Tandem arrays of multiple encapsidated Ebola genomes are contained loosely within the viral envelope, which can therefore twist into the flexible forms seen in **Figure 6.9A**. Timothy Booth's lab at the National Microbiology Laboratory, Winnipeg, Canada, showed how the process of enveloping nascent Ebola virions generates filaments that may contain a single genome, multiple linked genomes, or even an empty filament (see **Fig. 6.9C**).

At the molecular level, filamentous viruses show helical symmetry. The pattern of capsid monomers forms a helical tube around the genome, which usually winds helically within the tube. In a helical capsid, the genome is a single-stranded RNA (as in Ebola virus or tobacco mosaic virus) or DNA (as in bacteriophage M13). **Figure 6.10** shows how the RNA strand of tobacco mosaic virus (TMV) winds in a spiral within a tube of capsid monomers laid down in a spiral array. Unlike Ebola, TMV has no lipid envelope—just the RNA genome wound within a tube of protein. Such a tube can be imagined as a planar array of subunits that coils around such that each row connects to the row above, generating a spiral. The length of the helical capsid may extend up to 50 times its width, generating a flexible filament.

Filamentous phages cause problems for human medicine and industry. One infects the Gram-positive species *Propionibacterium freudenreichii*, a key fermenting agent for Swiss cheese. Another filamentous phage, CTXphi (CTXφ), integrates its sequence into the genome of *Vibrio cholerae*, where it carries the deadly toxin genes required for cholera. On the other hand, filamentous phages such as M13 have been used for biotechnology, to clone products as proteins fused to the subunits of the helical capsid. Another application, in nanotechnology, is the use of a filamentous phage to nucleate the growth of crystalline "nanowires" for electronic devices.

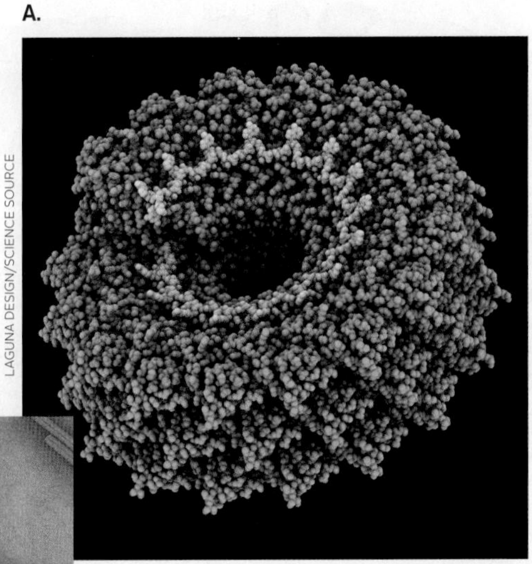

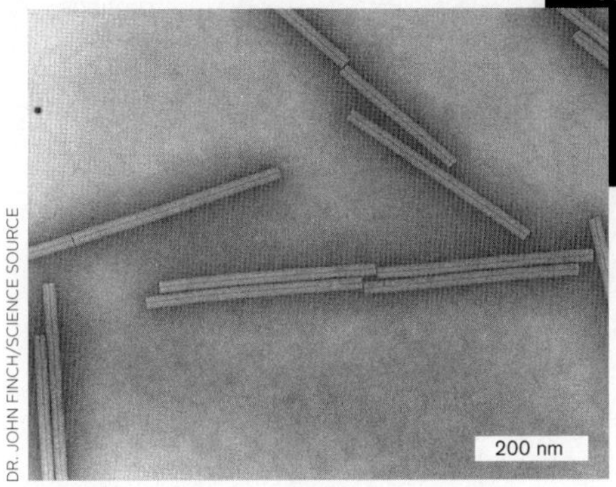

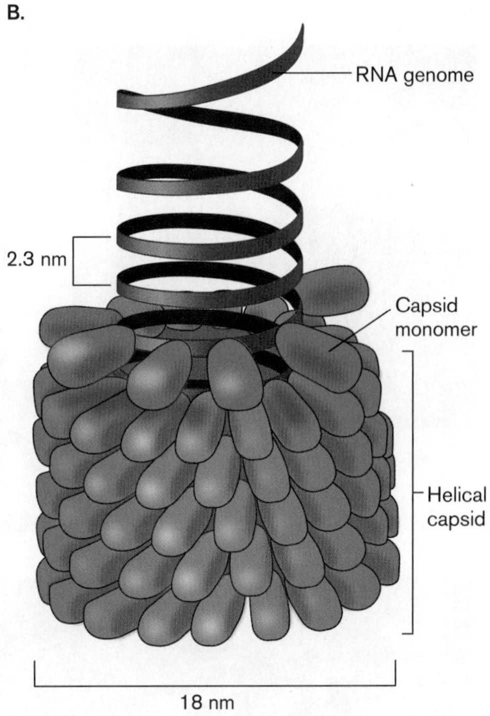

FIGURE 6.10 ▪ Tobacco mosaic virus: helical symmetry. A. The helical filament of tobacco mosaic virus (TMV) contains a single-stranded RNA genome coiled inside (X-ray model). **Inset:** TMV virions (TEM). B. Components of the TMV virion.

Asymmetrical Virions

Many viruses have a capsid, or core particle, that is asymmetrical. An important group of viruses with asymmetrical capsids is the poxviruses, such as vaccinia (the source of smallpox vaccine, derived from cowpox virus). The virion of a poxvirus consists of an oval-shaped particle nearly as large as a cell (200–300 nm). Vaccinia has a relatively large double-stranded DNA genome of 190 kilobases (kb), encoding 250 genes. The DNA is stabilized by covalent connection of its two strands in a loop at each end (**Fig. 6.11A** and **B**). The core envelope encloses the nucleocapsid-coated DNA, as well as a large number of accessory proteins. The accessory proteins are needed early in viral infection, such as initiation proteins for the transcription of viral genes and RNA-processing enzymes that modify viral mRNA molecules. The proteins may be found either inside the capsid or in the tegument between the core envelope and the outer membrane, which is coated with surface tubules of protein. Large asymmetrical viruses contain so many enzymes that they appear to have evolved from degenerate cells.

Vaccinia virus offers opportunities for new applications, such as conversion to vaccines for other viruses, and engineering to combat tumors. For example, students in Paulo Verardi's laboratory at the University of Connecticut (**Fig. 6.11C**) engineered an on/off switch for vaccinia that will make the virus safer for immunocompromised patients. The switch involves a recombinant tetracycline repressor gene, whose product confers resistance to tetracycline. Repressor controls and engineering are discussed in Chapters 10 and 12.

Another class of asymmetrical viruses is influenza (molecular details are presented in Chapter 11). Influenza viruses are RNA viruses in which a set of unique RNA segments is coated with nucleocapsid proteins. The influenza virus packages the different RNA segments into separate helical packages of different sizes, contained together within a membrane envelope. Separate chromosome packaging enables influenza virus to make "mistakes" and pack different numbers of RNA segments into different virions. The process enables rapid evolution of new strains (Chapter 11).

Tailed Viruses

Many bacteriophages supplement the icosahedral capsid or head coat with an elaborate delivery device. For example, bacteriophage T4 (**Fig. 6.12**) has an icosahedral "head" containing the pressure-packed DNA, attached to a helical "neck" that channels the nucleic acid into the host cell. The tail baseplate has six jointed tail fibers that stabilize the structure on the host cell surface. After phage infection, the production of virions within a cell requires a factorylike assembly line of phage tail parts (described for phage T4 in **eTopic 11.1**).

Viroids

Are there infectious agents even simpler than small viruses? For some infectious agents, the nucleic acid genome is itself the entire infectious particle; there is no protective capsid.

A. B.

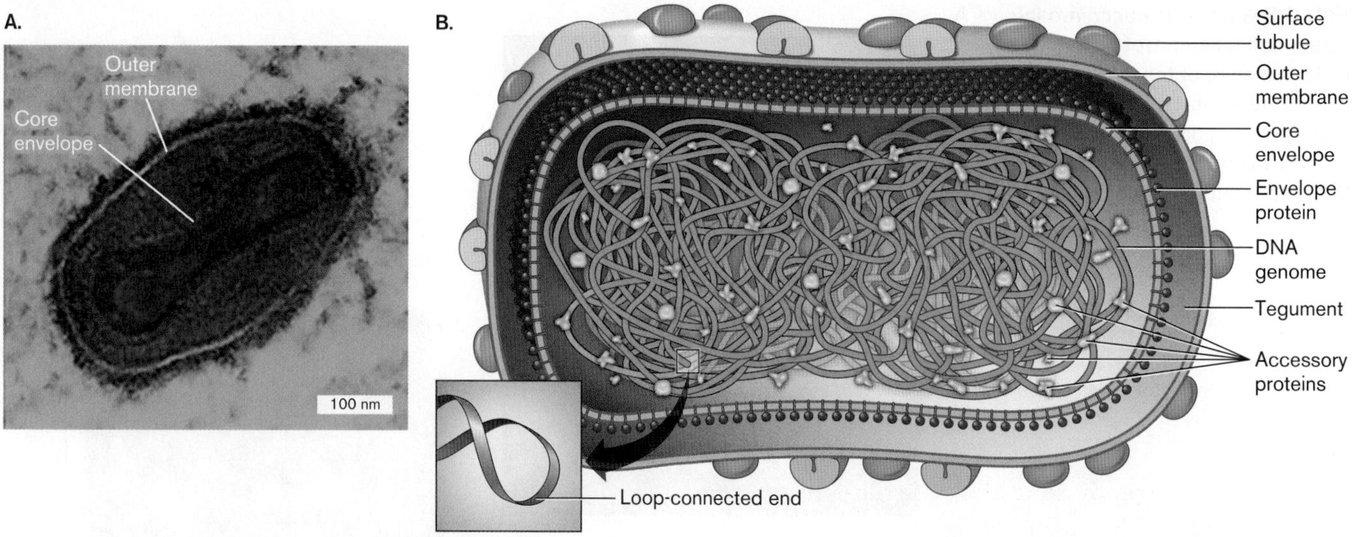

C.

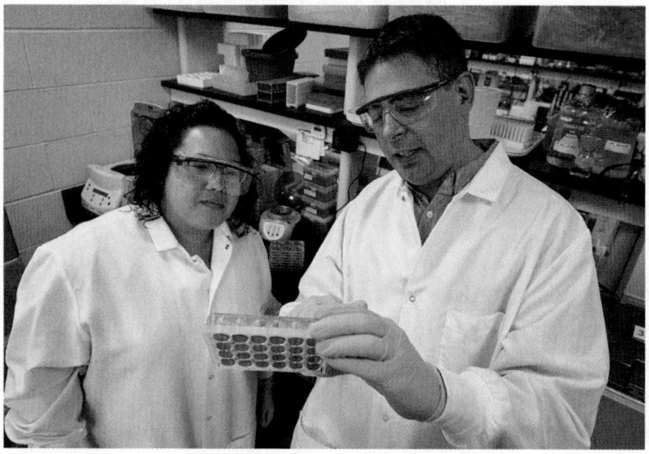

FIGURE 6.11 ■ Vaccinia poxvirus. **A.** Vaccinia virion observed in aqueous medium by atomic force microscopy (AFM). **B.** A pox virion includes an outer membrane and a core envelope membrane containing envelope proteins enclosing the double-stranded DNA genome and accessory proteins. The DNA is stabilized by a hairpin loop at each end. **C.** Allison Titong and Paulo Verardi examine vaccinia virus in tissue culture, to develop advanced vaccines and cancer treatments.

Such agents are called **viroids**, which are infectious agents of plants. A viroid encodes no genes, but it hijacks the plant cell's RNA polymerase to replicate itself. Viroids infect many kinds of fruits and vegetables, entering the plant cell through a damaged cell wall. For example, citrus viroids cause economic losses in the citrus industry.

The potato spindle tuber viroid (**Fig. 6.13**) consists of a circular, single-stranded molecule of RNA that doubles back on itself to form base pairs interrupted by short, unpaired loops. This unusual circularized form avoids breakdown by host RNase enzymes. The RNA folds up into a globular structure that interacts with host cell proteins. Host RNA polymerase replicates the RNA. The host RNA polymerase normally requires a DNA template, but the replication process is modified during viroid infection. During infection, the RNA polymerase thus replicates progeny copies of the viroid, which encodes no products other than itself. Viroids can cause as much host destruction as "true viruses," and some authors, particularly in plant pathology, classify them as "viruses without capsids."

A. B.

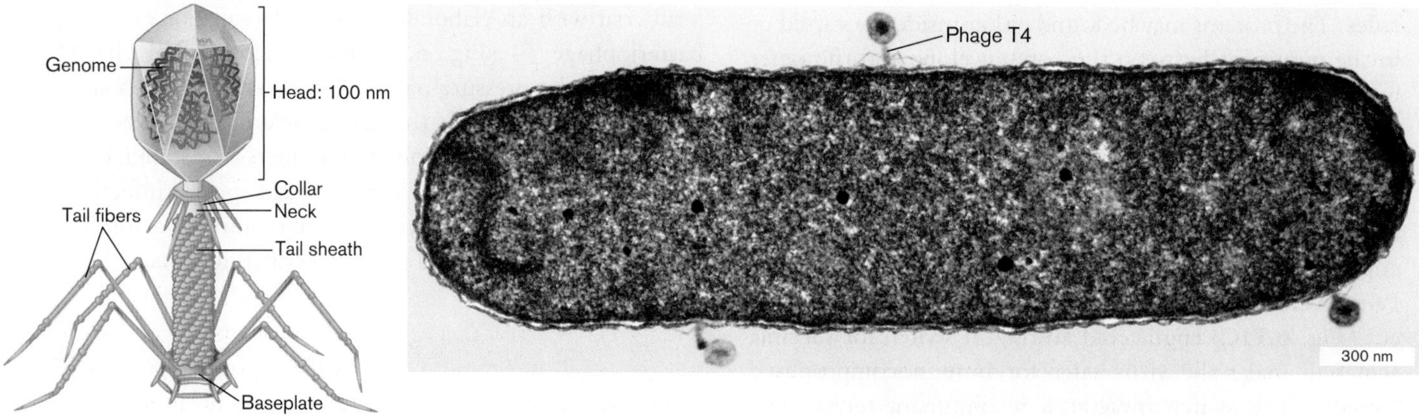

FIGURE 6.12 ■ Bacteriophage T4 capsid. **A.** Phage T4 particle with protein capsid containing packaged double-stranded DNA genome. The capsid has a sheath with tail fibers that facilitate attachment to the surface of the host cell. After attachment, the sheath contracts and the core penetrates the cell surface, injecting the phage genome. **B.** *E. coli* infected by phage T4 (colorized blue; TEM).

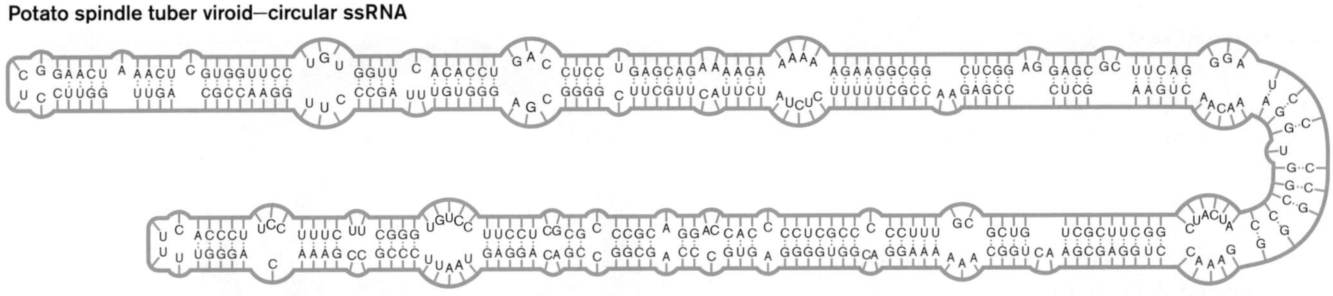

FIGURE 6.13 ▪ Viroids: infective RNA. The potato spindle tuber viroid consists of a circular single-stranded RNA (ssRNA) that hybridizes internally.

Some viroids have catalytic ability, comparable to that of enzymes made of protein. An RNA molecule capable of catalyzing a reaction is called a ribozyme (discussed in Chapter 9). Viral ribozymes may be able to cleave themselves or other specific RNA molecules. Their ability to cleave very specific RNA sequences has applications in medical research, such as cleaving human mRNA involved in cancer.

Prions

Can an infectious agent propagate without a genome of its own? A remarkable class of infectious agents is believed to consist solely of protein. These agents, known as **prions**, are thought to be aberrant proteins arising from the host cell. Prions gained notoriety when they were implicated in brain infections such as Creutzfeldt-Jakob disease, a variant form of which is known as "mad cow" disease because it may be transmitted through defective proteins in beef from diseased cattle. Other diseases believed to be caused by prion transmission include scrapie, a disease of sheep; and kuru, a degenerative brain disease that was found in a tribe of people who customarily consumed the brains of deceased relatives.

In prion-associated diseases, the infective agent is unaffected by treatments that destroy RNA or DNA, such as nucleases or UV irradiation. A prion is an aberrant form of a normal cell protein that assumes an abnormal conformation or tertiary structure (**Fig. 6.14A**). The prion form of the protein acts by binding to normally folded proteins of the same class and altering their conformation to that of the prion. The multiplying prion then alters the conformation of other normal subunits, forming harmful aggregates in the cell and ultimately leading to cell death. In the brain, prion-induced cell death leads to tissue deterioration and dementia (**Fig. 6.14B**). Prion diseases are unique because they can be transmitted by an infective protein instead of by DNA or RNA, and they propagate the conformational change of existing molecules without synthesizing entirely new infective molecules.

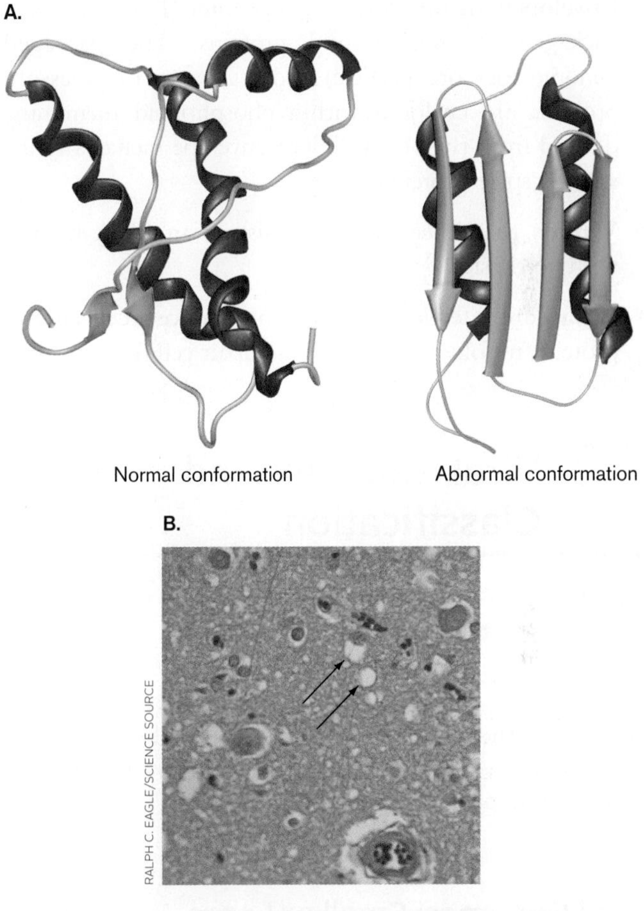

FIGURE 6.14 ▪ Prion disease. A. The normal conformation of a protein, compared to the abnormal (prion) conformation. The abnormal form "recruits" normally folded proteins and changes their conformation into the abnormal form. (PDB code: 1AG2) **B.** Section of a human brain showing "spongiform" holes typical of Creutzfeldt-Jakob disease.

A prion disease can be initiated by infection with an aberrant protein. More rarely, the cascade of protein misfolding can start with the spontaneous misfolding of an endogenous host protein. The chance of spontaneous misfolding is greatly increased in individuals who inherit certain alleles encoding the protein; thus, Creutzfeldt-Jakob disease can be inherited genetically from one person with one mutant allele.

To Summarize

- A **viral capsid** is composed of repeated protein subunits—a structure that maximizes the structural capacity while minimizing the number of genes needed for construction.
- **The capsid, or core particle, packages the viral genome** and delivers it into the host cell.
- **Icosahedral capsids** have regular, 20-sided symmetry.
- **Filamentous (helical) capsids** have uniform width, generating a flexible filamentous virion.
- **Enveloped virions** consist of a genome (RNA or DNA) packaged by nucleocapsid proteins. The packaged genome (or core particle) and tegument or accessory proteins are enclosed within phospholipid membrane derived from the host cell. The envelope includes virus-specific spike proteins.
- **Viroids** that infect plants consist of an RNA hairpin with no capsid.
- **Prions** are infectious proteins that induce a cell's native proteins to fold incorrectly and impair cell function.

6.3 Viral Genomes and Classification

Viral genomes are structurally more diverse than those of cells. A given type of virus may have a genome that consists of either RNA or DNA, single- or double-stranded, linear or circular. The form of the genome has key consequences for the mode of infection, and for the course of a viral disease. Viral genomes are used as the basis of virus classification.

Viral Genomes: Small or Large

Small viruses commonly have a small genome, encoding fewer than ten genes. For example, cauliflower mosaic virus (diameter 50 nm) has a genome encoding only seven genes (**Fig. 6.15A**), which actually overlap each other in sequence. This overlap in sequence is made possible by the use of different reading frames—start positions to define the first base of the codon for translation to an amino acid sequence (discussed in Chapter 8).

The smallest of viral genomes generally consist of RNA. RNA viruses include some of today's most important human pathogens, such as influenza virus, hepatitis C virus, and human immunodeficiency virus (HIV). The RNA genome of avian leukosis virus (**Fig. 6.15B**) has

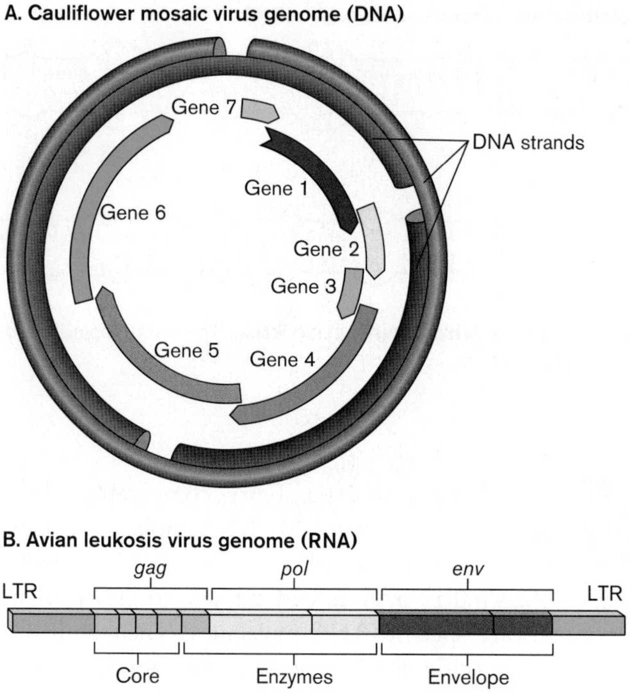

FIGURE 6.15 ▪ Simple viral genomes. **A.** Cauliflower mosaic virus has a circular genome of double-stranded DNA, whose strands are interrupted by nicks. The genome encodes seven overlapping genes. **B.** Avian leukosis virus is a single-stranded RNA retrovirus resembling eukaryotic mRNA. Three genes (*gag*, *pol*, and *env*) encode polypeptides that are eventually cleaved to form a total of nine functional products. LTR = long terminal repeat.

protein-encoding genes grouped by functional categories of core capsid, replication enzymes, and envelope proteins (proteins embedded in the envelope phospholipid bilayer).

At the opposite end of the scale, the "giant viruses" have genomes of double-stranded DNA comprising 300–2,500 genes. A giant virus called Mimivirus (diameter 300 nm), which infects amebas and may cause human pneumonia, is itself as large as some bacteria (**Fig. 6.16A**) and has a bacterium-sized genome of 1.2 million base pairs, encoding more than 1,000 genes. Discovered in 2003 by French virologist Didier Raoult and colleagues, mimivirus has a genome that encodes numerous cell functions, including DNA repair and protein folding by chaperones. The virus gains entry to its ameba host by phagocytosis, because its large particle size makes the virus particle resemble a bacterium that the ameba could engulf for food. Once taken up, the mimivirus capsid opens to release viral enzymes and DNA. Within the ameba's cytoplasm, the viral enzymes generate a replication complex that produces progeny mimiviruses. Mimiviruses are so large that they can actually become infected themselves by smaller viruses, such as the Sputnik virophage, or "virus-eater" (**Fig. 6.16B**).

A surprising source of giant viruses is the frozen polar environments of the Arctic and Antarctic regions.

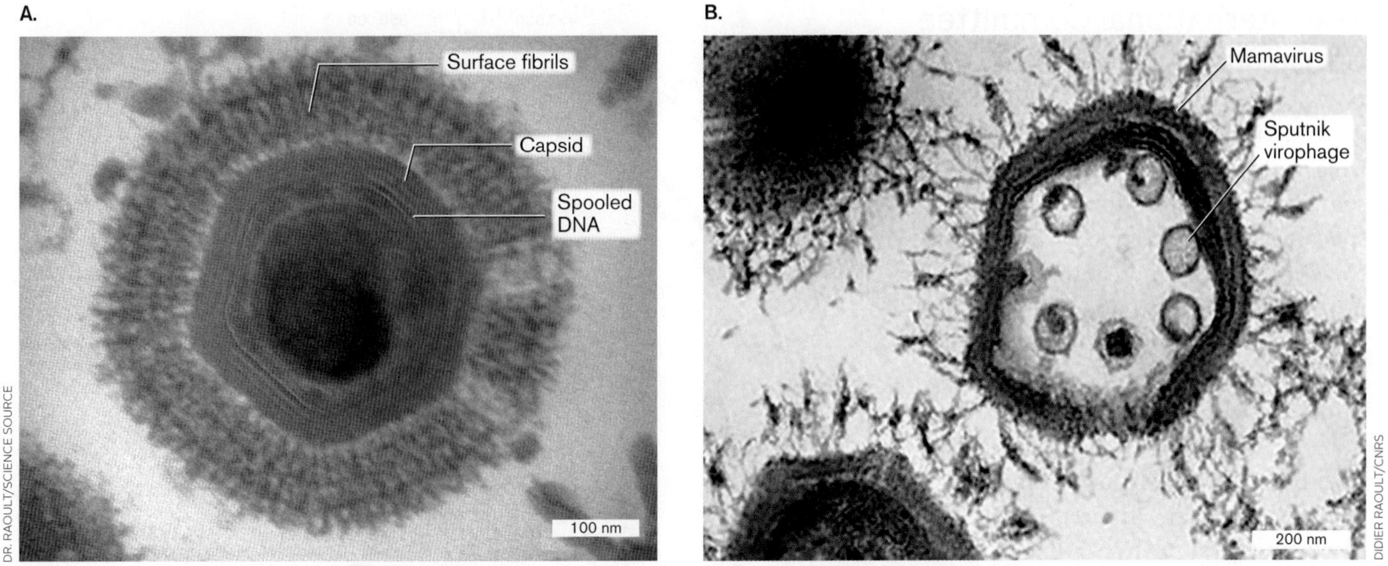

FIGURE 6.16 ■ Giant virus infecting an ameba. **A.** Mimivirus is larger than some bacteria (TEM). **B.** Sputnik virophage infects Mamavirus, a relative of Mimivirus (TEM).

Antarctic lakes reveal giant viruses related to Mimivirus and the Sputnik virophage. The Arctic tundra of Siberia reveals even more remarkable viruses, such as Pithovirus (**Fig. 6.17**). Pithovirus is an unusual lozenge-shaped virus, nearly the size of bacteria such as *E. coli*. Jean-Marie Claverie and colleagues from Aix-Marseille University isolated Pithovirus from Siberian tundra that had been frozen for 30,000 years. Remarkably, this ancient virus was successfully inoculated into an ameba host, where it generated progeny virions. The revival of ancient viruses leads to concern that as tundra melts during inevitable climate change, the melting permafrost will release human pathogens from long-dead hosts such as buried victims of smallpox.

Another Siberian giant virus, Mollivirus sibericum, was revived and found to package ribosomal proteins from its host ameba—the first virus ever shown to contain parts of a ribosome. An even larger virus, Klosneuvirus, was isolated from an Austrian wastewater treatment plant. Klosneuvirus expresses transfer RNA molecules for all 20 amino acids of the genetic code. Giant viruses are intriguing because their genomes specify so many enzymes with housekeeping cell functions, such as nutrient transport, mRNA translation, and even cell motility (**Fig. 6.18**). Such large, cell-like genomes suggest the likelihood that a virus evolved from a parasitic cell.

Thought Question

6.5 Klosneuvirus shows evidence of integrating genes from cellular hosts. What kind of fitness advantage might favor acquisition of host genes?

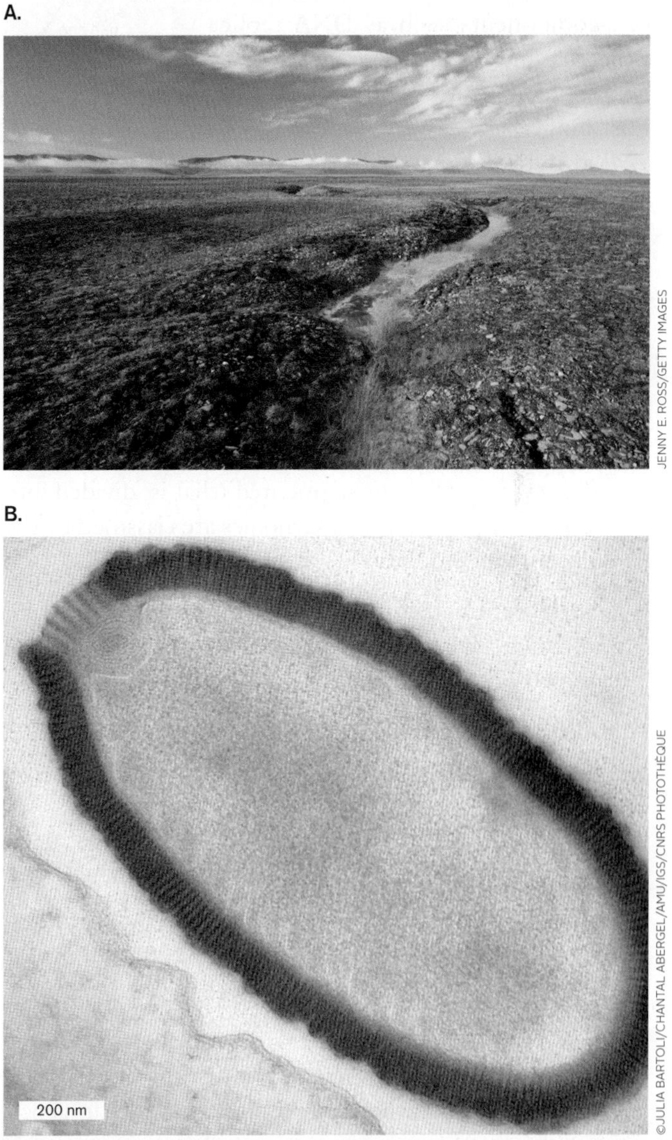

FIGURE 6.17 ■ Giant virus from Siberia. **A.** The Siberian tundra is melting. **B.** Pithovirus, a virus as large as *E. coli*, was isolated from tundra frozen for 30,000 years (TEM).

The International Committee on Taxonomy of Viruses

How do we classify the bewildering diversity of viruses? Today we classify organisms by the relatedness of their gene sequences. The definition of a virus species, however, is problematic, given the small size and high mutability of viral genomes and the ability of different viruses to recombine their genome segments within an infected host cell. Furthermore, not all viruses are monophyletic—that is, descended from a common ancestor. In fact, different classes of viruses appear to have evolved from different sources—for example, from parasitic cells or from host cell components such as DNA replication enzymes. Viruses are classified by genome composition, virion structure, and host range.

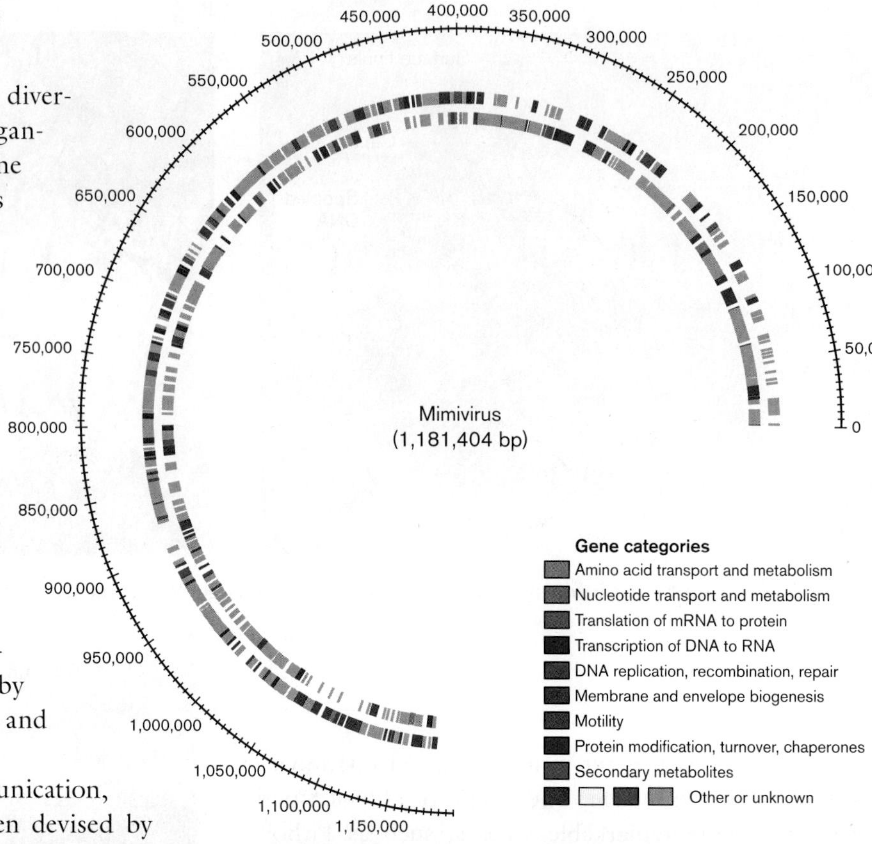

FIGURE 6.18 ■ Genome of Mimivirus. The genome of this giant virus specifies numerous enzymes with cell functions.

For purposes of study and communication, a working classification system has been devised by the International Committee on Taxonomy of Viruses (ICTV). The ICTV classification system is based on several criteria:

- **Genome composition.** The nucleic acid of the viral genome can vary remarkably with respect to physical structure: It may consist of DNA or RNA, it may be single- or double-stranded, it may be linear or circular, and it may be whole or **segmented** (that is, divided into separate "chromosomes"). Genomes are classified by the Baltimore method (discussed next).
- **Capsid symmetry.** The protein capsid, or core particle, may be helical or icosahedral, with various levels of symmetry.
- **Envelope.** The presence of a host-derived envelope, and the envelope structure, if present, are characteristic of related viruses.
- **Size of the virus particle.** Related viruses generally share the same size range; for example, enteroviruses such as poliovirus are only 30 nm across (about the size of a ribosome), whereas poxviruses are 200–400 nm, as large as a small bacterium.
- **Host range.** Closely related viruses usually infect the same or related hosts. However, viruses with extremely different hosts can show surprising similarities in genetics and structure. For example, both rabies virus and potato yellow dwarf virus are enveloped, bullet-shaped viruses of the rhabdovirus family.

Note: In nomenclature, families of viruses are designated by Latin names with the suffix "-viridae": for example, *Papillomaviridae*. Nevertheless, the common forms of such family names are also used—for example, "papillomaviruses." Within a family, a virus species is simply capitalized, as in "Papillomavirus."

The Baltimore Virus Classification

Given the wide variety of viral structures, how do we determine their relatedness? In general, viruses of the same genome class (such as double-stranded DNA) show greater evidence of shared ancestry with each other than with viruses of a different class of genome (such as RNA). In 1971, David Baltimore proposed that the primary distinctions among classes of viruses should be the genome composition (RNA or DNA) and the route used to express messenger RNA (mRNA). The form of the genome is critical for mechanisms such as virus involvement in carcinogenesis. Baltimore, together with Renato Dulbecco (1914–2012) and Howard Temin (1943–1994), was awarded the 1975 Nobel Prize in Physiology or Medicine for discovering how tumor viruses cause cancer.

All cells and viruses need to make messenger RNA to produce their fundamental protein components. The

A.

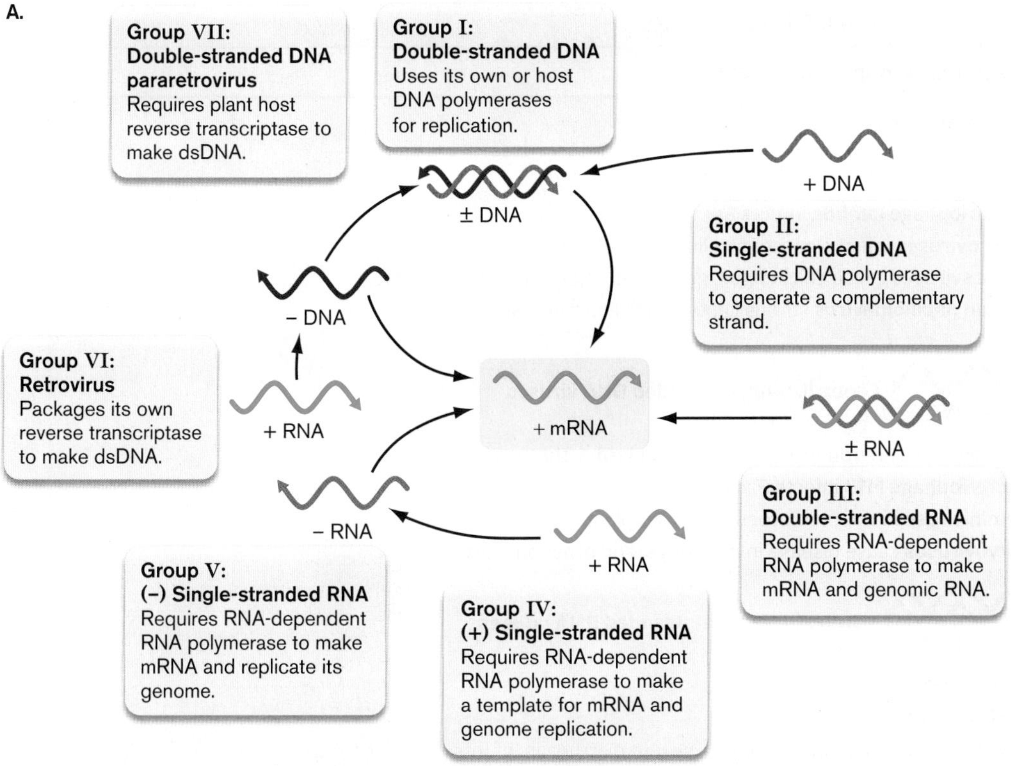

B.

SLADE PAUL/CONTRIBUTOR/GETTY IMAGES

FIGURE 6.19 ■ **Baltimore classification of viral genomes.** **A.** Seven categories of viral genome composition and replication mechanism. **B.** David Baltimore studies viral sequence DNA at the California Institute of Technology.

production of mRNA from the viral genome is central to a virus's ability to propagate its kind. Cellular genomes always make mRNA by copying double-stranded DNA. For viruses, however, different kinds of genomes require fundamentally different mechanisms to produce mRNA. The different means of mRNA production generate distinct groups of viruses with shared ancestry.

So far, the genome composition and mechanisms of replication and mRNA expression define seven fundamental groups of viral species (**Fig. 6.19**). Examples of these seven fundamental groups are given in **Table 6.1**.

Group I. Double-stranded DNA viruses such as the herpesviruses and smallpox viruses make their own DNA polymerase or use that of the host for genome replication. Their genes can be transcribed directly by a standard RNA polymerase, in the same way that a cellular chromosome would be transcribed. The RNA polymerase used can be that of the host cell, or it can be encoded by the viral genome.

Group II. Single-stranded DNA viruses such as canine parvovirus require the host DNA polymerase to generate the complementary DNA strand. The double-stranded DNA can then be transcribed by host RNA polymerase.

Group III. Double-stranded RNA viruses require a viral **RNA-dependent RNA polymerase** to generate messenger RNA by transcribing directly from the RNA genome. Since the RNA polymerase is required immediately upon infection, such viruses package a viral RNA polymerase with their genome before exiting the host cell. A major class of double-stranded RNA viruses is the reoviruses, including rotavirus, a cause of diarrhea in children. Another reovirus has been engineered to destroy human tumors without infecting normal cells. This "oncolytic" reovirus, called Reolysin, is now in clinical trials for cancer therapy.

Group IV. (+) sense single-stranded RNA viruses consist of a positive-sense (+) strand (the coding strand) that can serve directly as mRNA to be translated to viral proteins. Replication of the RNA genome, however, requires synthesis of the complementary (–) strand by a viral RNA-dependent RNA polymerase. The (–) strand then serves as a template for (+) strand synthesis. Positive-sense (+) RNA viruses include enteroviruses such as poliovirus, the coronaviruses, and the flaviviruses that cause West Nile encephalitis and hepatitis C.

Group V. (–) sense single-stranded RNA viruses such as influenza virus have genomes that consist of template, or "negative-sense," RNA. Thus, they need to package a viral RNA-dependent RNA polymerase for transcribing (–) RNA to (+) mRNA. The (–) strand

TABLE 6.1 Groups of Viruses—Baltimore Classification

Virus example	Taxonomic group with examples
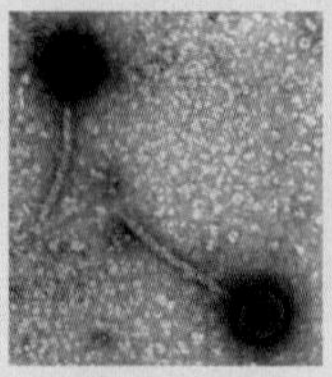 BIOPHOTO ASSOCIATES/SCIENCE SOURCE Phage lambda	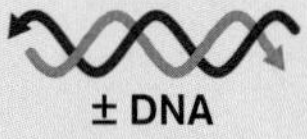 **Group I. Double-stranded DNA viruses** **Bacteriophage lambda** infects *Escherichia coli*. **Chloroviruses** infect algae, controlling algal blooms. **Herpesviruses** cause chickenpox, genital infections, and birth defects. **Human papillomavirus** strains cause warts and tumors.
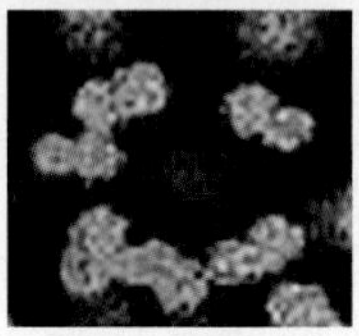 CINDY L. MCKENZIE/USDA Geminivirus	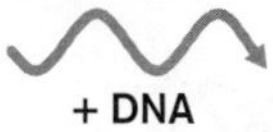 **Group II. Single-stranded DNA viruses** **Anelloviruses** are found in human blood plasma; they cause no known harm. **Bacteriophage M13** infects *E. coli*. **Geminiviruses** infect tomatoes and other plants. **Parvoviruses** cause disease in cats, dogs, and other animals.
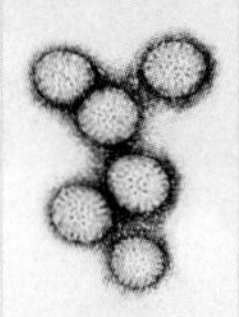 JAMES GATHANY/CDC Rotavirus	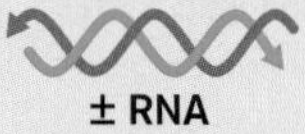 **Group III. Double-stranded RNA viruses** **Birnaviruses** infect fish. **Cystoviruses** infect bacteria. **Reoviruses** such as rotavirus cause severe diarrhea in infants. Other reoviruses are in clinical trials to fight tumors (oncolysis).
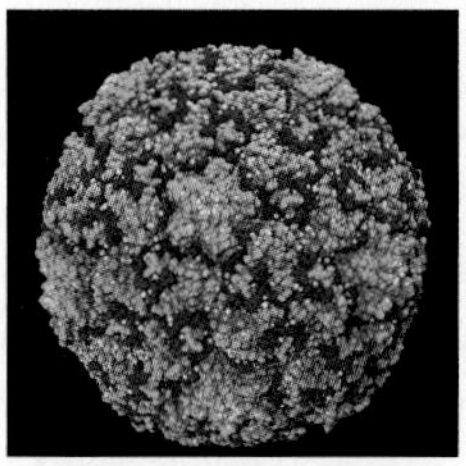 SCOTT CAMAZINE/VISUALS UNLIMITED, INC. Rhinovirus	 **Group IV. (+) sense single-stranded RNA viruses** **Coronaviruses** such as SARS virus cause severe respiratory disease. **Flaviviruses** cause hepatitis C, Zika fever (Zika virus), West Nile disease, yellow fever, and dengue fever. **Poliovirus** infects human intestinal epithelium and nerves. **Tobacco mosaic virus** infects plants.
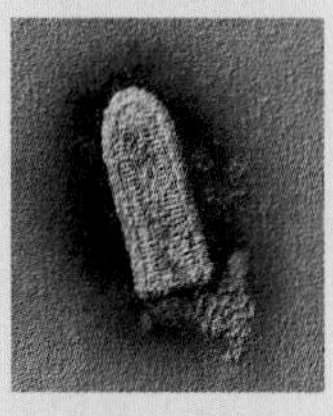 EYE OF SCIENCE/SCIENCE SOURCE Rabies virus	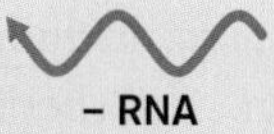 **Group V. (–) sense single-stranded RNA viruses** **Filoviruses** such as Ebola virus cause severe hemorrhagic disease. **Orthomyxoviruses** cause influenza. **Paramyxoviruses** cause measles and mumps. **Rhabdovirus** causes rabies.
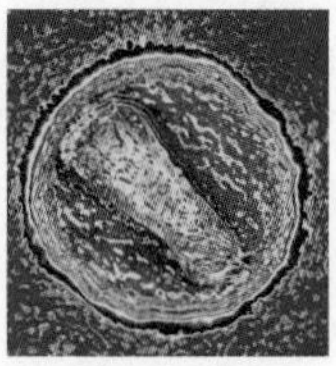 JAMES CAVALLINI/SCIENCE SOURCE Human immuno-deficiency virus	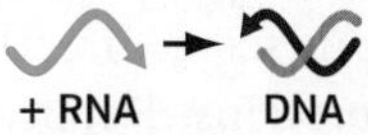 **Group VI. Retroviruses (RNA reverse-transcribing viruses)** **Feline leukemia virus (FeLV)**, **Rous sarcoma virus (RSV)**, and **avian leukosis virus (ALV)** cause cancer. **Lentiviruses** include **human immunodeficiency virus (HIV)**, the cause of AIDS. Engineered "lentivectors" are used for gene therapy.
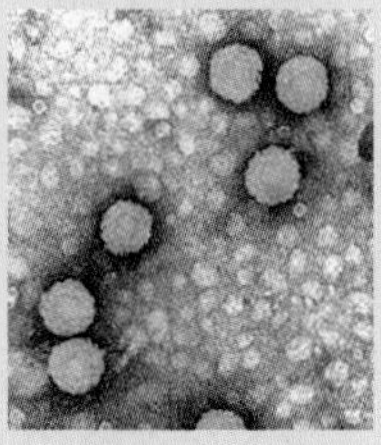 SCIENCE PHOTO LIBRARY/SCIENCE SOURCE Caulimovirus	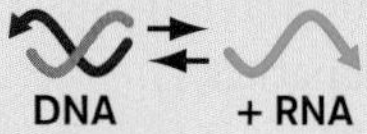 **Group VII. Pararetroviruses (DNA reverse-transcribing viruses)** **Caulimoviruses** (such as cauliflower mosaic virus, or CaMV) infect many kinds of vegetables. CaMV provides the best vector tools for plant biotechnology. **Hepadnaviruses** such as hepatitis B virus infect the human liver.

RNA viral genomes are often segmented; that is, they consist of multiple separate linear chromosomes—a key factor in the evolution of killer strains of influenza (see Section 11.2).

Group VI. Retroviruses, or **RNA reverse-transcribing viruses**, such as HIV and feline leukemia virus, have genomes that consist of (+) strand RNA. Instead of RNA polymerase, they package a **reverse transcriptase**, which transcribes the RNA into a double-stranded DNA (for details, see Section 11.3). The double-stranded DNA is then integrated into the host genome, where it directs the expression of the viral genes.

Group VII. Pararetroviruses, or **DNA reverse-transcribing viruses**, have a replication cycle that requires reverse transcriptase. For example, hepatitis B virus (a hepadnavirus) first copies its double-stranded DNA genomes into RNA, and then reverse-transcribes the RNA to progeny DNA using a reverse transcriptase packaged in the original virion. In contrast, plant pararetroviruses, such as cauliflower mosaic virus (CaMV, a caulimovirus), generate an RNA intermediate that replicates using a reverse transcriptase made by the host cell. Many plant genomes include a gene for reverse transcriptase. Cauliflower mosaic virus is of enormous agricultural significance for its use as a vector to construct pesticide-resistant food crops.

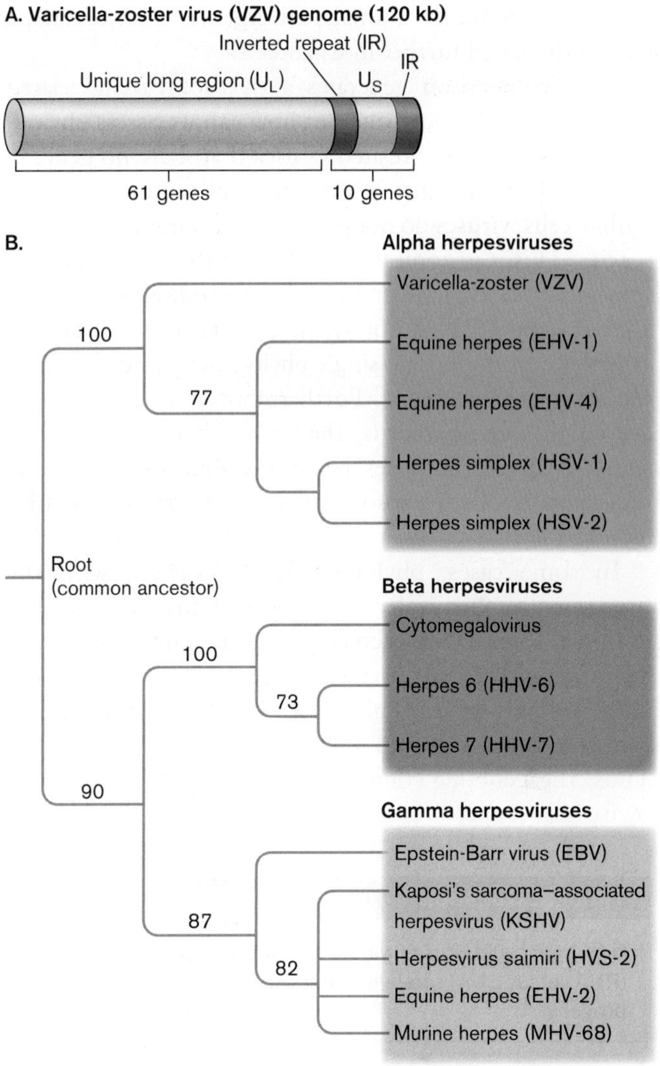

FIGURE 6.20 ■ Phylogeny of herpesvirus genomes. **A.** Genome structure of human varicella-zoster virus (VZV), the causative agent of chickenpox. **B.** Phylogeny of human and animal herpesviruses, based on whole-genome sequence analysis comparing clusters of orthologous groups of genes. Numbers measure percentage of DNA sequence shared by two divergent genomes.

Molecular Evolution of Viruses

The phylogeny, or genetic relatedness, of viruses can be determined within families. For example, the herpesvirus family includes double-stranded DNA viruses that cause several human and animal diseases, such as chickenpox, oral and genital herpes infection, and respiratory and genital infections in horses. Herpesvirus genomes consist of double-stranded DNA, 120–220 kilobases (kb) encoding about 70–200 genes; an example is that of varicella-zoster virus, the causative agent of chickenpox (**Fig. 6.20A**). The genome includes two "unique" segments of genes, one long and one short (U_L and U_S), joined by two inverted repeats (IRs). Other herpesvirus genomes share similar structure, though they differ in gene order and IR position.

The large size of herpesvirus DNA genomes allows effective measurement of genetic divergence based on sequence comparison. The relatedness of different herpesviruses that evolved from a common ancestor can be measured using orthologous genes, or **orthologs**. Orthologs are genes of common ancestry in two genomes that share the same function (a topic discussed in Chapter 8). For cellular organisms, we often use the ribosomal RNA gene sequences to measure relatedness. Viruses have no ribosomal RNA, but closely related viruses share other orthologous genes. In pairs of orthologs, the amount of difference in sequence correlates approximately with the time following divergence from a common ancestor (a topic discussed in Chapter 17).

A tree of genomic divergence (or phylogeny) was devised for herpesviruses. The comparison of all gene pairs places herpes strains into three related classes designated alpha, beta, and gamma (**Fig. 6.20B**). The alpha class includes human varicella-zoster virus and the oral and genital herpesviruses (HSV-1 and HSV-2), as well as two equine herpesviruses. The beta class includes cytomegalovirus, a common cause of congenital infections (present at birth), as well as two lesser-known viruses. The gamma class includes Epstein-Barr virus (the cause of infectious mononucleosis), Kaposi's sarcoma–associated herpesvirus, and several viruses of nonhuman animals. Divergence of phylogenetic trees is discussed further in Chapter 17.

Gene comparison generates a tree for closely related viruses such as herpesvirus. But how can we assess phylogeny of more distantly related viruses that share no genes—and even have genomes of different kinds of nucleic acids? Unlike cells, viruses do not possess genes universal for all species. This is in contrast to cellular organisms, which universally carry at least one gene encoding 16S or 18S ribosomal RNA. For viruses, no gene is universally shared, and thus no single phylogenetic tree of life can be made for viruses. Furthermore, viral genomes are highly mosaic; that is, they evolved from multiple sources. Mosaic genomes result from recombination or reassortment of chromosomes from different viruses coinfecting a host.

In some cases, phylogeny is inconsistent with the fundamental chemical composition of the genome. For example, some DNA bacteriophages actually share closer ancestry with RNA bacteriophages than they do with DNA animal viruses. This is because two or more viruses can coinfect a cell and exchange genetic components. Thus, the genomic content of a virus can be influenced by its host range.

Thought Question

6.6 How could viruses with different kinds of genomes (RNA versus DNA) combine and share genetic content in their progeny?

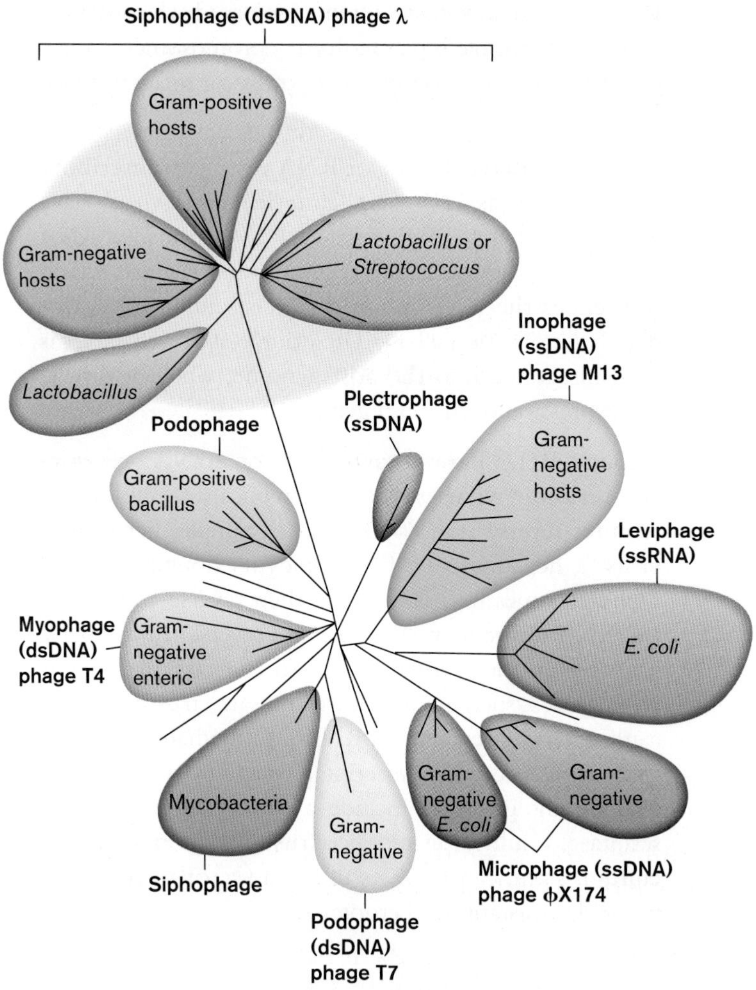

FIGURE 6.21 ■ A bacteriophage proteomic tree. Comparison of all proteins encoded by each genome predicts distinct groups of bacteriophages. Within each group there are subgroups of phages with shared hosts, since sharing of hosts facilitates genetic recombination and horizontal transfer of genes between different phages. *Source:* Based on Forest Rohwer and Rob A. Edwards. 2002. *J. Bacteriol.* **184**:4529.

As more viral genomes are sequenced, classification methods are devised to take advantage of sequence information without requiring gene products common to all species. One promising approach is that of proteomics, analysis of the **proteome**, the proteins encoded by genomes (discussed in Chapters 9 and 12). Proteins are identified through biochemical analysis of virus particles and through bioinformatic analysis of protein sequences encoded in the genomes. Proteomic analysis is useful for distantly related viruses because protein sequences often show relatedness that is obscured in the nucleic acid sequence by silent mutations (discussed in Chapter 9). Even if only a few proteins are shared by any two viruses, statistical comparison of all proteins in a set of viral species reveals underlying degrees of relatedness.

An example of virus classification based on proteomic analysis is that of the "proteomic tree" of bacteriophages proposed by Forest Rohwer and Rob Edwards (**Fig. 6.21**). Unlike earlier trees based on a single common gene sequence, the proteomic tree is based on the statistical comparison of phage protein sequences predicted by the genomic DNA of many different species of phages. The proteomic analysis predicts major evolutionary categories of phage species that share common host bacteria. For example, phages that infect Gram-negative hosts will show more genetic commonality with each other than with phages that infect Gram-positive hosts. Shared hosts have a significant impact on phage evolution because coinfecting a host enables phages to exchange genes.

To Summarize

- **Viruses contain infective genomes of RNA or DNA.** A viral genome may be single- or double-stranded, linear or circular.
- **Giant viruses** have 300–2,500 genes and may have evolved from intracellular parasitic cells.

- **Smaller viral genomes** comprise fewer than ten genes. Such viruses might have evolved from cell parts.
- **Classification of viruses** is based on a variety of criteria, including genome composition, virion structure, and host range.
- **The Baltimore virus classification** emphasizes the form of the genome (DNA or RNA, single- or double-stranded) and the route to generate messenger RNA.
- **Phylogeny of closely related viruses** can be calculated by comparing all related pairs of genes from the viral genomes.
- **Proteomic classification of distantly related viruses** includes information from all viral proteins. Statistical analysis reveals common descent of viruses infecting a common host.

6.4 Bacteriophages: The Gut Virome

Viruses display a remarkable diversity of ways to replicate within a host cell. Here, we discuss the key modes of bacteriophage replication, and their consequences for host cells. We focus on the best-known bacteriophages, those of the mammalian intestinal community. Gut bacteriophages, or "coliphages," are part of a microbial community that modulates human digestion, the immune system, and mental health.

Historically, bacteriophages have provided some of the most fundamental insights in molecular biology. In 1952, Alfred Hershey and Martha Chase showed that the transmission of DNA by a bacteriophage to a host cell led to the production of progeny bacteriophages, thus confirming that DNA is the hereditary material. In 1950, André Lwoff and Antoinette Gutman showed that a phage genome could integrate itself within a bacterial genome—the first recognition that genes could enter and leave a cell's genome. Other fundamental concepts of the genetic unit and the basis of gene transcription came from experiments on bacteriophages, as discussed in Chapters 7–9.

Bacteriophages Infect a Host Cell

To commence an infection cycle, bacteriophages need to contact and attach to the surface of an appropriate host cell. Contact and attachment are mediated by **cell-surface receptors**, proteins on the host cell surface that are specific to the host species and that bind to a specific viral component. A cell-surface receptor for a virus is actually a protein with an important function for the host cell, but the virus has evolved to take advantage of the protein. For example, phages that infect *Salmonella enterica* can use outer membrane proteins such as OmpF or TolC, which is part of a drug efflux complex (**Fig. 6.22**). Alternatively, a phage might bind to LPS (lipopolysaccharide; see Chapter 3). The phage-receptor binding is usually highly specific; a bacterium can evolve resistance through single-amino-acid mutations in its protein. The lambda phage receptor protein (maltose porin) of *E. coli* is described in Chapter 11.

Most bacteriophages (phages) deliver only their genome into a cell through the cell envelope, thus avoiding the need for the capsid to penetrate the molecular barrier of the cell wall. For example, the phage T4 virion has a sheath that contracts, bringing the head near the cell surface to inject its DNA (**Fig. 6.23A** ▶). The pressure of the spooled DNA—as high as 50 atmospheres (atm)—is released, expelling the DNA into the cell.

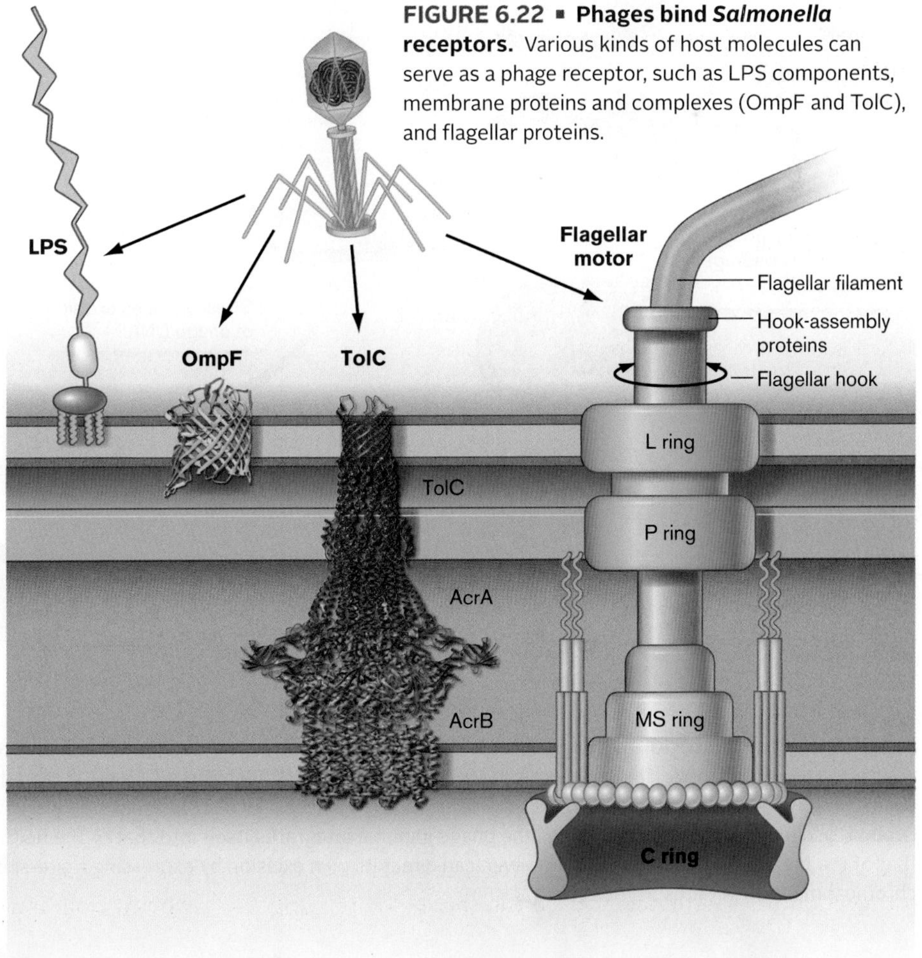

FIGURE 6.22 ■ Phages bind *Salmonella* receptors. Various kinds of host molecules can serve as a phage receptor, such as LPS components, membrane proteins and complexes (OmpF and TolC), and flagellar proteins.

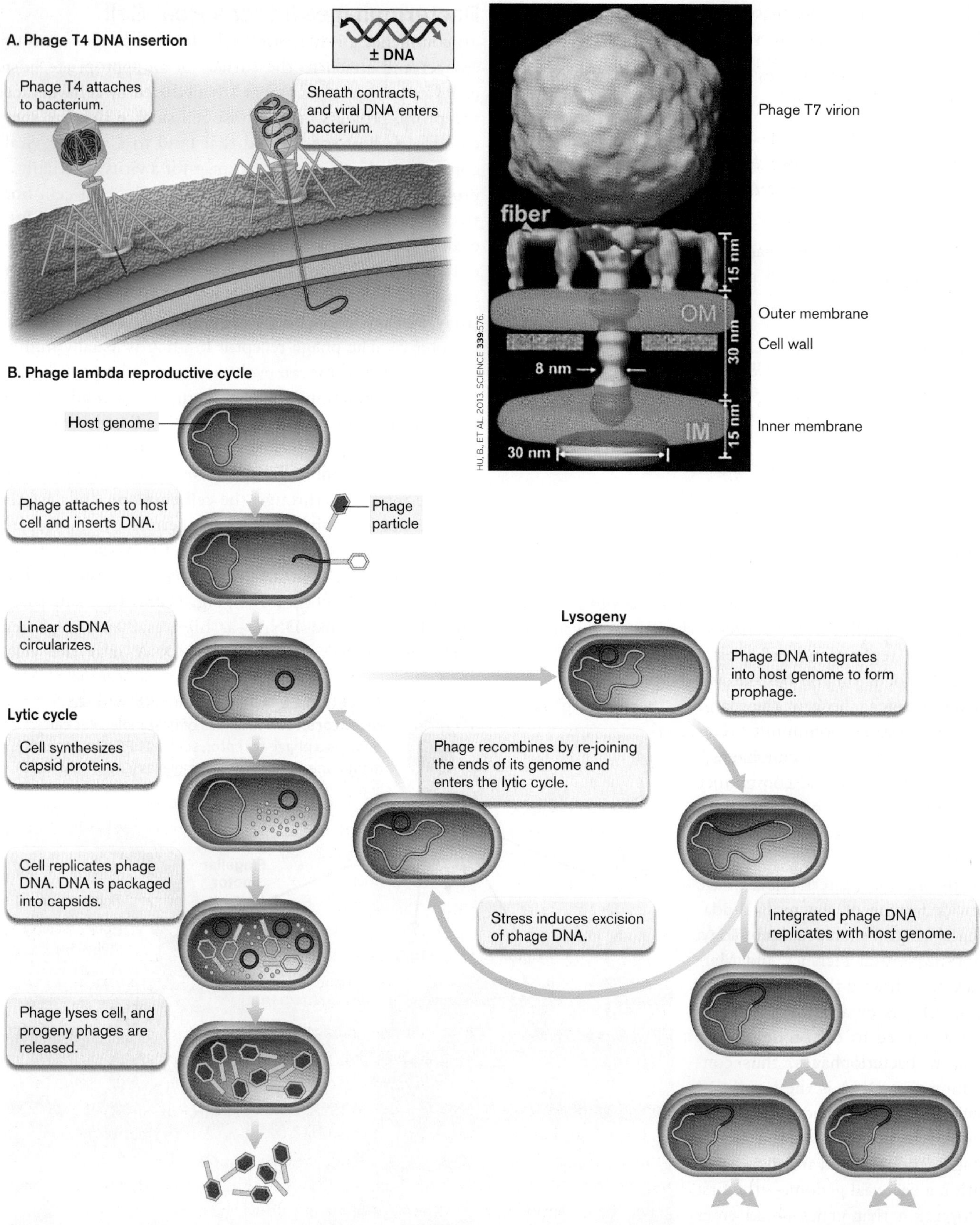

FIGURE 6.23 ■ Bacteriophage reproduction: lysis and lysogeny. **A.** Phage T4 attaches to the cell surface by its tail fibers and then contracts to inject its DNA. The empty capsid remains outside as a "ghost." **Inset:** Cryo-EM model of phage T7 injecting DNA. **B.** Lysis (left) occurs when the phage genome reproduces progeny phage particles, as many as possible, and then lyses the cell to release them. In phage lambda, lysogeny (right) can occur when the phage genome integrates itself into that of the host. The phage genome is replicated along with that of the host cell. The phage DNA, however, can direct its own excision by expressing a site-specific DNA recombinase. This excised phage chromosome then initiates a lytic cycle.

After the genome has been inserted, the phage capsid remains outside, attached to the cell surface. The empty capsid is termed a "ghost" because of its pale appearance in an electron micrograph.

The lytic cycle. A lytic cycle of replication generates a large number of progeny phages and then lyses the defunct cell. The lytic replication cycle requires these steps:

- **Host recognition and attachment.** A phage particle must contact a receptor molecule and adhere to a host cell.
- **Genome entry.** The phage genome must enter the host cell and gain access to the cell's machinery for gene expression.
- **Assembly of phages.** Phage components must be expressed and assembled. Components usually "self-assemble"; that is, the joining of their parts is favored thermodynamically.
- **Exit and transmission.** Progeny phages must exit the host cell, and then reach new host cells to infect.

In a lytic cycle, when a phage particle delivers its genome into a cell, it immediately reproduces as many progeny phage particles as possible (**Fig. 6.23B**). The process of reproduction involves replicating the phage genome, as well as expressing phage mRNA to make enzymes and capsid proteins. Some phages, such as T4, digest the host DNA to increase the efficiency of phage production. Phage particles assemble, and the host cell lyses, releasing progeny phages.

The phage adsorbs to its host receptor and inserts its double-stranded DNA into the host cytoplasm. The phage genes are then expressed by the host cell RNA polymerase and ribosomes. "Early genes" are expressed first in the lytic cycle. Other phage-expressed proteins then work together with the cellular enzymes and ribosomes to replicate the phage genome and produce phage capsid proteins. The capsid proteins self-assemble into capsids and package the phage genomes—a process that takes place in defined stages, like a factory assembly line. At last, a "late gene" from the phage genome expresses an enzyme that lyses the host cell wall, releasing the mature virions. **Lysis** is also referred to as a burst, and the number of virus particles released is called the **burst size**.

Lysogeny. A **temperate phage**, such as phage lambda, can infect and lyse cells like a virulent phage, but it also has an alternative pathway: to integrate its genome within the host chromosome (see **Fig. 6.23B**). The phage is said to "lysogenize" the host, in a cycle called **lysogeny**. Phage lambda has a linear genome of double-stranded DNA, which circularizes upon entry into the cell. The circularized genome then recombines into that of the host by **site-specific recombination** of DNA. In site-specific recombination, a recombinase enzyme aligns the phage genome with the host DNA and exchanges the DNA backbone linkages with those of the host genome. (This process of DNA backbone recombination is explained in Chapter 9.) The DNA recombination event thus integrates the phage genome into that of the host. Now integrated, the phage genome is called a **prophage**. The presence of the prophage prevents further infection (superinfection) by other virions of the same type.

In lysogeny, the prophage DNA is replicated along with that of the host cell as the host reproduces (**Fig. 6.23B**). The host gains the benefit of resistance to superinfection. Implicit in the term "lysogeny," however, is the ability of such a strain to generate a lytic burst of phage. For a lysogen to enter lysis, the prophage directs its own excision from the host genome by an intramolecular process of site-specific recombination. The two ends of the phage genome exchange their DNA linkages so as to come apart from the host DNA, which now closes its circle with the prophage removed. As the phage DNA exits the host genome, it circularizes and initiates a lytic cycle, destroying the host cell and releasing phage particles.

How does a lysogen "decide" to reactivate and begin a lytic cycle? The decision between lysogeny and lysis is determined by proteins that bind DNA and repress the transcription of genes for virus replication (discussed in Chapter 11). Exit from lysogeny into lysis can occur at random times, or it can be triggered by environmental stress such as UV light, which damages the cell's DNA. The regulatory switch of lysogeny responds to environmental cues indicating the likelihood that the host cell will survive and continue to propagate the phage genome. If a cell's growth is strong, it is more likely that the phage DNA will remain inactive, whereas events that threaten host survival will trigger a lytic burst. Similarly, in animal viral infections such as herpes, an environmental stress triggers reactivation of a virus that was dormant within cells (a latent infection). Reactivation of a latent herpes infection results in painful outbreaks of skin lesions.

During the exit from lysogeny, the virus can acquire host genes and pass them on to other host cells. The process of transferring host genes is known as **transduction**. Sometimes a transducing bacteriophage picks up a bit of host genome and transfers it to a new host cell. In another kind of transduction, the entire phage genome is replaced by host DNA packaged in the phage capsid, resulting in a virus particle that transfers only host DNA. Host DNA transferred by viruses can become permanently incorporated into the infected host genome (**Table 6.2**). These former viral genes evolve into host genes that express products with new functions useful to the host.

The mechanisms of phage-mediated transduction are discussed in Chapter 9. In natural environments, phage transduction mediates much of the recombination of bacterial genomes. In the laboratory, the ability of phages to transfer genes provided some of the first vectors for recombinant DNA technology (see Chapter 12).

TABLE 6.2 Integrated Viral Genomes That Provide Host Traits

Prophage (host infected)	Bacterial product from prophage gene	Human endogenous retrovirus (HERV)	Human protein or regulator provided by HERV
Phage C1 (*Clostridium botulinum*)	Botulinum toxin (*c1*)	HERV-W	Syncytin-1 (retroviral Env protein; placental fusion)
Beta phage (*Corynebacterium diphtheriae*)	Diphtheria toxin (*tox*)	HERV-FRD	Syncytin-2 (retroviral Env protein; placental fusion)
Lambda (*E. coli* O157:H7)	Cell envelope protein (*bor*)	HERV	INSL4 (insulin-like protein 4; placental development)
Phage 933 (*E. coli* O157:H7)	Shiga toxin (*stx*)	HERV-E	MID1 (midline development; prevents Opitz syndrome)
Epsilon 34 (*Salmonella enterica*)	LPS synthesis enzyme (*rfb*)	HERV-E	Apolipoprotein C1 (liver function)
Epsilon 34 (*Salmonella enterica*)	Type III secreted toxin (*sopE*)	HERV-E	Endothelin type B receptor (placenta function)
TSST-1 (*Staphylococcus aureus*)	Toxic shock syndrome toxin (*speA*)	HERV-L	Beta-1,3-galactosyltransferase (colon and mammary gland function)
CTSφ (*Vibrio cholerae*)	Cholera toxin (*ctxAB*)	LINE-1	ATRN (soluble attractin; modulates inflammation)

The slow-release cycle. A slow-release replication cycle differs from lysis and lysogeny in that phage particles reproduce without destroying the host cell (**Fig. 6.24**). Slow release is performed by filamentous phages such as phage M13. In slow-release replication, the single-stranded circular DNA of M13 serves as a template to synthesize a double-stranded intermediate. The double-stranded intermediate slowly generates single-stranded progeny genomes, which are packaged by supercoiling and coated with capsid proteins. The phage particles then extrude through the cell envelope without lysing the cell. The host cell continues to reproduce, though more slowly than uninfected cells do, because many of its resources are diverted to virus production.

Thought Question

6.7 What are the relative advantages and disadvantages (to a virus) of the slow-release strategy, compared with the strategy of a temperate phage, which alternates between lysis and lysogeny?

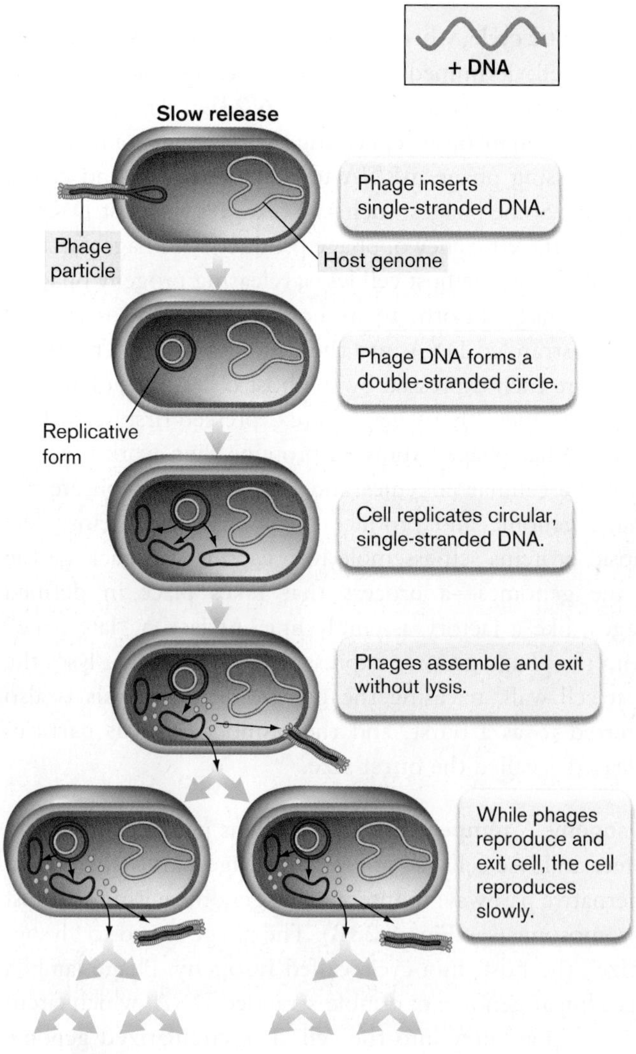

FIGURE 6.24 ■ Bacteriophage replication cycle: slow release. In the slow-release replication cycle, a filamentous phage produces phage particles without lysing the cell. The host continues to reproduce itself, but more slowly than uninfected cells do because many of its resources are being used to make phages.

Integrated Viruses as Host Cell Parts

Many prophages and endogenous viruses function as part of their host cell. **Table 6.2** gives examples of prophages that express virulence genes for pathogenic bacteria—a benefit to the pathogen, enabling it to better colonize the host animal. For example, phage C1 in *Clostridium botulinum* expresses the peptide botulinum toxin. This toxin causes the paralysis associated with botulism (and is also the basis of the Botox cosmetic treatment). In *Corynebacterium diphtheriae*, the cause of diphtheria, the diphtheria toxin is expressed by beta phage. Other virulence factors (proteins that enhance disease) expressed by prophages include envelope proteins that help defend the bacterium from the immune system.

Analogous to bacterial prophages, many human genes and gene control elements have evolved from human viruses, particularly human endogenous retroviruses (HERVs). Retroviruses are a category of virus that includes HIV, the cause of AIDS (discussed in Section 6.5, and in greater detail in Chapter 11). HERVs are reverse-transcribing RNA viruses whose DNA copies became permanently fixed in our chromosomes. For example, placental proteins called "syncytin" mediate cell fusion during an early stage of placental development, allowing the fused syncytium to implant in the uterine lining and access the maternal blood supply. Two genes for syncytin evolved from HERV genes encoding retroviral envelope proteins. The details of the remarkable story of endogenous retroviruses—and the related story of retroviral gene therapy—are discussed in Chapter 11.

The virus-host relationship often evolves in response to environmental stress. For example, a virus-infected fungus can increase the growth of a plant at higher temperature. The virus is required for the fungus to increase thermal tolerance of the plant (see **eTopic 6.2**).

Bacterial Defenses

In natural environments, viruses commonly outnumber cellular microbes by tenfold or more. So, how have host cells evolved to defend themselves? Several remarkable resistance mechanisms have evolved. Their molecular basis is explained in greater detail in Chapter 9.

Genetic resistance. All bacteria acquire random mutations in their genomes as they reproduce (see Chapters 7 and 9). When attacked by bacteriophages, bacterial populations undergo natural selection; mutants that happen to be harder to infect will survive. Bacteria resist phage infection by expressing a gene that encodes an altered host receptor protein, which fails to bind the viral coat protein. Alternatively, a different cell protein evolves to block phage binding to the receptor. An evolutionary "arms race" ensues, in which phages may evolve enzymes that cleave the host defense molecules.

Restriction endonucleases. Bacteria modify their DNA by adding methyl groups to bases within certain sequences. The bacteria then express restriction endonucleases, enzymes that cleave DNA lacking the methylated patterns—which includes potential viral DNA (see Chapter 9). However, phage genomes composed of RNA or of modified DNA (such as phage T4, discussed in Chapter 11) escape cleavage by these enzymes.

CRISPR: a bacterial immune system. Amazingly, bacteria possess an adaptive defense against viruses that is analogous to an immune system. (Adaptive immunity of humans is presented in Chapter 24.) The bacterial adaptive defense involves short DNA sequences homologous to DNA of phages that could infect the cell. The sequences are called clustered regularly interspaced short palindromic repeats (**CRISPR**). These series of short sequences were first characterized in the 1990s by Francisco Mojica at the University of Alicante, Spain, who proposed that they might represent an adaptive defense against phages. Many other research groups figured out components of the mechanism.

Where do the CRISPR short repeats come from? When a phage attacks a bacterium, if bacterial enzymes succeed in destroying the phage DNA, they may copy a tiny piece of it as a CRISPR segment, inserted as a spacer at the head of a long line of about 30 CRISPR sequences (**Fig. 6.25**). Now the adapted host cell "remembers" infection by the specific phage—along with many other previous phages from previous infections that had inserted other spacers. The next time the adapted host cell is attacked by the same phage, all of its genomic CRISPR sequences are expressed as RNA. The CRISPR RNA is cleaved into small sequences (crRNA) containing one spacer from the original bacterial CRISPR DNA. The crRNA binds to the Cascade protein complex (or Cas complex), which now detects phage DNA homologous to its virus-derived crRNA. The Cas-crRNA complex proceeds to cleave the phage DNA, preventing phage replication.

Diverse bacteria show many variations on the CRISPR theme. For example, in 2019, Jonathan Strecker and coworkers discovered a CRISPR-associated transposase (DNA transfer enzyme) within the cyanobacterium *Scytonema hofmanii*. This CRISPR transposase could be used for editing large pieces of DNA in human genomes. For more on CRISPR see Chapter 9.

In response to CRISPR, how do phages adapt? Not surprisingly, phages have evolved several means of defense. One mechanism involves the "anti-CRISPR" phage protein family Acr. Acr proteins inhibit the CRISPR-Cas system of specific host bacteria. Different Acr proteins act by various mechanisms, one of which is to bind the bacterial host's Cas complex (**Fig. 6.25B**). An Acr protein binding to a Cas subunit can prevent the binding of Cas to phage DNA, in some cases by acting as a DNA mimic for the Cas DNA-binding site. Thus, phage infection may proceed despite the presence of homologous CRISPR sequences in the host genome.

Thought Question

6.8 How else, besides Acr proteins, might a phage evolve resistance to the CRISPR host defense (outlined in **Fig. 6.25**)?

The Gut Bacteriophage Community

How do bacteriophages impact microbial communities? The best-understood phage community is that of the gut virome. A **virome** is a community of viruses within a host ecosystem (for example, the human or other animal intestinal tract). We present some of the major

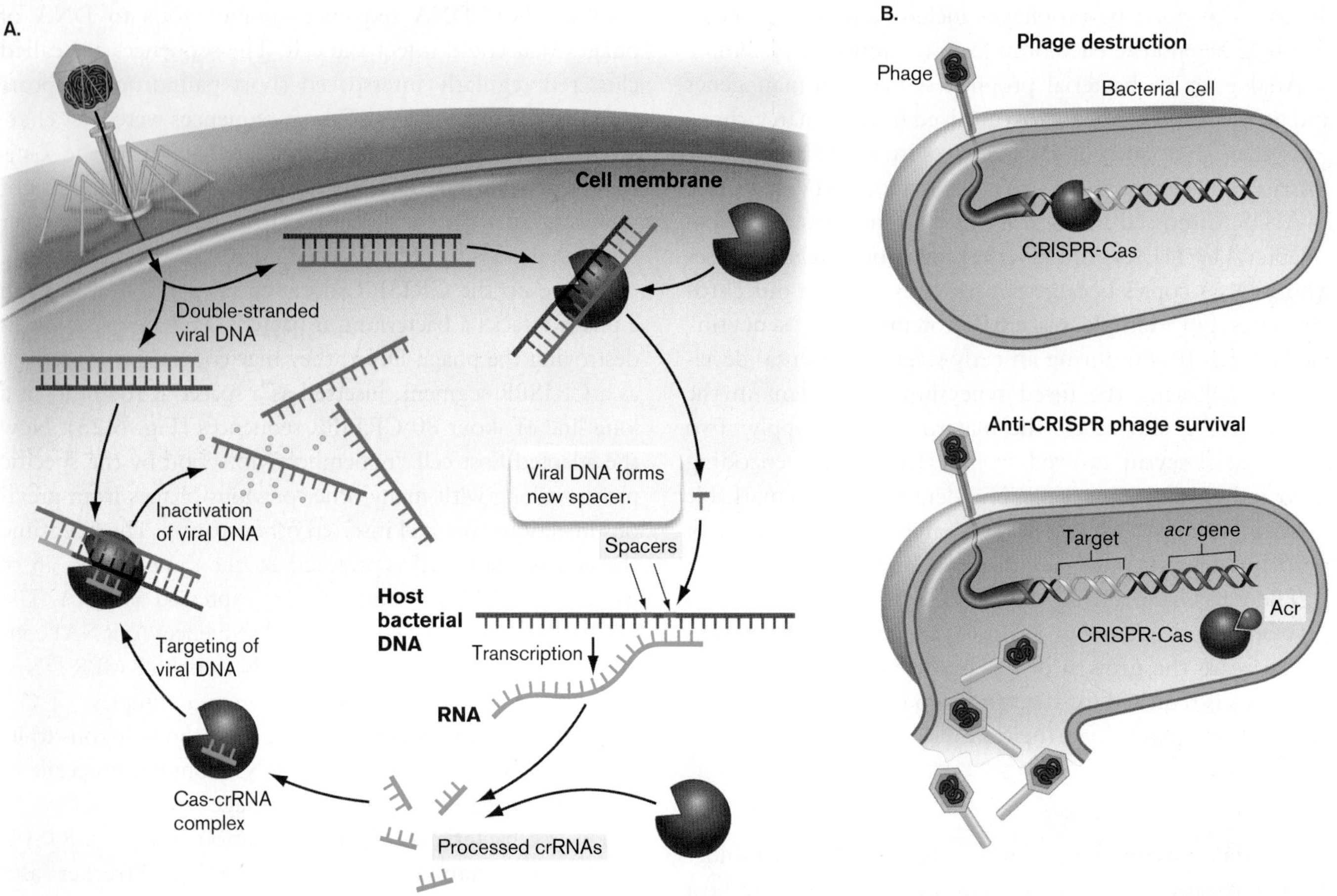

FIGURE 6.25 ■ CRISPR defense of a bacterial cell and anti-CRISPR counterdefense. **A.** A piece of phage DNA gets copied as a "spacer" into the host genome. If the bacterium survives infection, later reinfection by the same kind of phage causes transcription of the spacers into CRISPR RNA. A processed spacer (crRNA) joins the Cas complex to recognize and cleave the phage DNA. **B.** Phage may carry an anti-CRISPR gene (*acr*) encoding a protein Acr that blocks the host bacterial CRISPR-Cas from binding phage DNA.

phage-host interactions within the human intestinal tract in **Figure 6.26**. Note that this simplified diagram omits the viruses that infect human body cells; we focus here on the phages that infect gut bacteria. Surprisingly, phages are so important to the human host that human gut tissues take up phages by a process called **transcytosis** (**Special Topic 6.1**). The effects of transcytosis remain a mystery.

The lumen of the human intestines contains a remarkably dense microbial community. Bacterial and archaeal cells are estimated at 10^{11}–10^{13} cells per gram, whereas phage particles may be 10^9 per gram. The phage count is likely underestimated, as it is based on detection by electron microscopy and plate counts of viability (discussed in Section 6.4). In addition, more than half the genomes sequenced from gut bacteria show lysogeny, commonly with several different types of prophages in a given genome. Prophages may protect the gut bacteria from superinfection—that is, infection by other phages in the gut that cause lysis.

The gut community receives continual influx of new bacteria and phages, while shedding present members (**Fig. 6.26**). When a new type of phage enters, what happens? If the phage encounters a susceptible host bacterium, it may undergo a lytic cycle, lysogeny, or slow release, depending on the phage's genetic program. Cell lysis may have the effect of depopulating a dominant population of the bacterial community. Bacterial depopulation by phages might lead to intestinal "dysbiosis," deterioration of health due to loss of health-enhancing bacteria. Other bacterial species, less healthy for the host, might increase in population. But now, a different phage species (shown red in **Fig. 6.26**) may depopulate the newly risen species, restoring equilibrium in the community. Still other phages may transfer useful genes from newcomer bacteria, by a process called transduction (see Chapter 9). Such genes may encode enzymes to metabolize new kinds of food molecules.

Some prophages of pathogenic bacteria express virulence factors, such as the Shiga toxin of *Shigella* and of *E. coli* O157:H7 (**Table 6.2**). At the same time, other effects of phages may be positive:

- **Phages may limit the bacterial numbers to levels that the human immune system can tolerate.** Lysogenized bacteria may use quorum sensing to detect host cell populations and "decide" whether to start a lytic cycle.

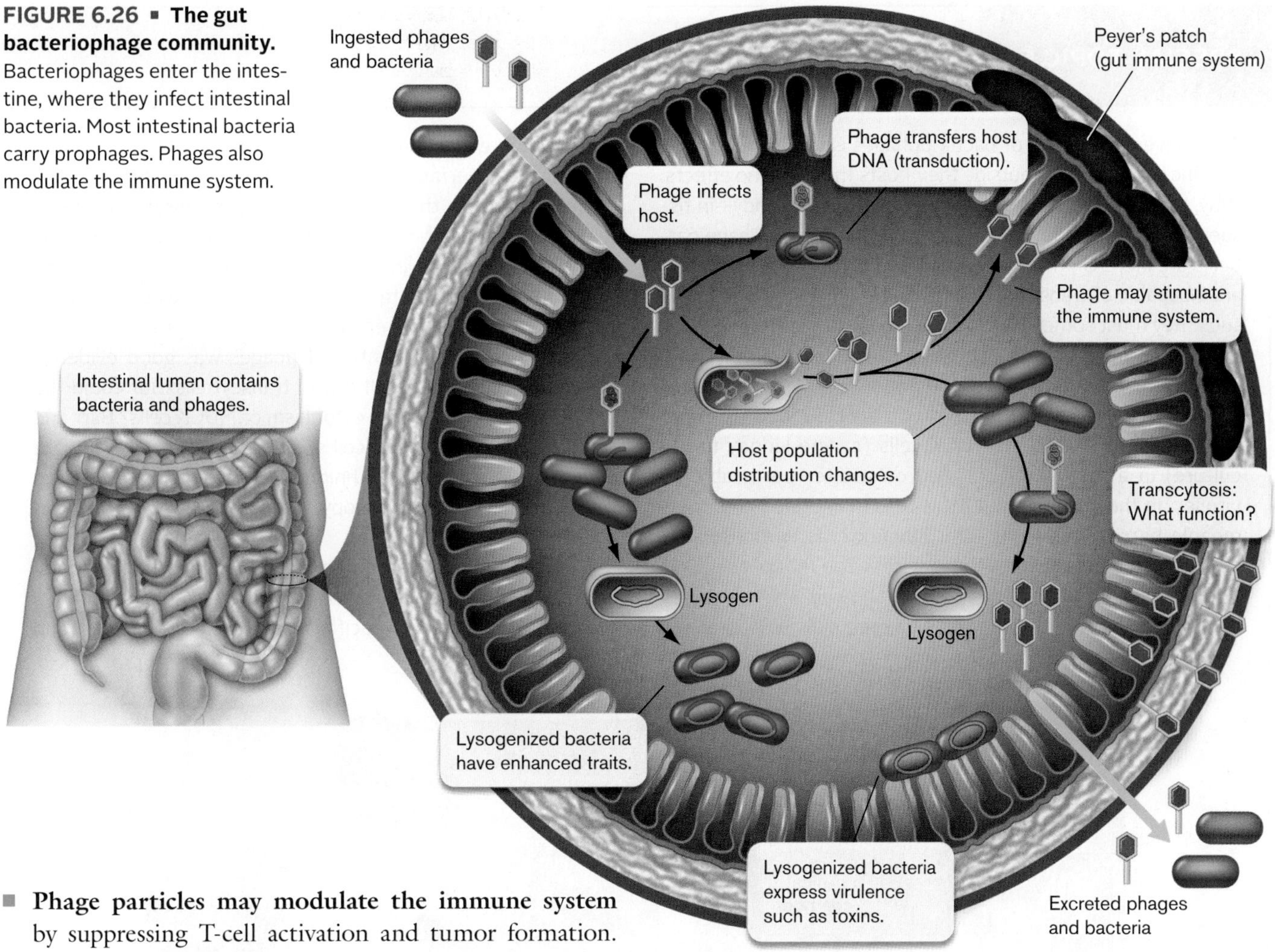

FIGURE 6.26 ■ The gut bacteriophage community. Bacteriophages enter the intestine, where they infect intestinal bacteria. Most intestinal bacteria carry prophages. Phages also modulate the immune system.

- **Phage particles may modulate the immune system** by suppressing T-cell activation and tumor formation. (T cells are discussed in Chapter 24.)
- **Phages may attack biofilms.** Biofilms of pathogens such as *Pseudomonas aeruginosa* may be eroded by phage infection.

Finally, the fact that our bodies tolerate phage transcytosis—the uptake of phages throughout our tissues—implies that phages have more direct positive benefits for our tissues (**Special Topic 6.1**).

The positive potential of bacteriophages has led researchers to investigate the engineering of phages for phage therapy. An idea dating back to the early twentieth century, phage therapy was eclipsed by the rise of antibiotics. Today, as we face growing antibiotic resistance in pathogens, we are taking another look at therapeutic uses of bacteriophages (Chapters 5 and 11).

To Summarize

- **Host cell-surface receptors** mediate the attachment of bacteriophages to a cell and confer host specificity.
- **Lytic cycle.** A bacteriophage injects its DNA into a host cell, where it utilizes host gene expression machinery to produce progeny virions.
- **Lysogeny.** Some bacteriophages can insert their genome into that of the host cell, which then replicates the phage genome along with its own. A lysogenic bacterium can initiate a lytic cycle.
- **Gene transfer.** Genes are transferred by phage processes of transduction and lysogeny.
- **Slow release.** Some bacteriophages use the host machinery to make progeny that bud from the cell slowly, slowing growth of the host without lysis.
- **Bacterial host defense.** Bacteria have evolved several forms of defense against bacteriophage infection, such as altered receptor proteins, restriction endonucleases, and CRISPR integration of phage DNA sequences.
- **The gut bacteriophage community** includes phages that infect, lyse, or lysogenize bacterial hosts. Gut phages impact bacterial community structure, transfer genes among bacteria, and modulate the human gut immune system. For unknown reasons, phages are taken up by host tissues—a process called transcytosis.

SPECIAL TOPIC 6.1 Phages Go Everywhere

The conventional view of bacteriophages is that they infect a specific host, and that outside their hosts they have no effects. Why, then, are they found throughout the human body—in tissues such as the blood, liver, lungs, even the brain? Phage particles are inert: On their own, they conduct no metabolism and are incapable of transport. Thus, to enter and cross human tissues they must somehow get taken up by cellular trafficking.

At the University of California-San Diego, Jeremy Barr conducts experiments to observe the process of phage transcytosis, how phage particles enter and cross the cytoplasm (**Fig. 1**). Barr established MDCK cells (canine kidney tissue culture) in a monolayer cultured on a filter permeable to viruses. In the culture, the MDCK cells maintained an impermeable barrier, sealed by tight junctions, as demonstrated by the high electrical resistance across the monolayer. This finding was important because it showed that particles could not be crossing the layer between cells. When phage particles were placed in the culture fluid on one side (above the apical cell surface), they subsequently were found in the chamber beneath the filter (the cell layer's basal side). After 2 hours of incubation, approximately 10,000 phages per milliliter were found in the basal chamber. The phage particles must have been taken up actively by the cultured cells and released on the other side.

Finding this high quantity of phages was good evidence of phage transport by cell-driven transcytosis. But could the phages be detected in the cytoplasm of MDCK cells? Barr and colleagues analyzed their cell samples using correlative light electron microscopy, a technique in which cells stained and visualized by light microscopy are correlated with sections observed by EM.

First, the phage-treated cells were stained with fluorescent dyes to reveal various compartments: Hoechst stain, a blue fluorescent stain specific for DNA of the nucleus; and

A. Apical-to-basal transcytosis

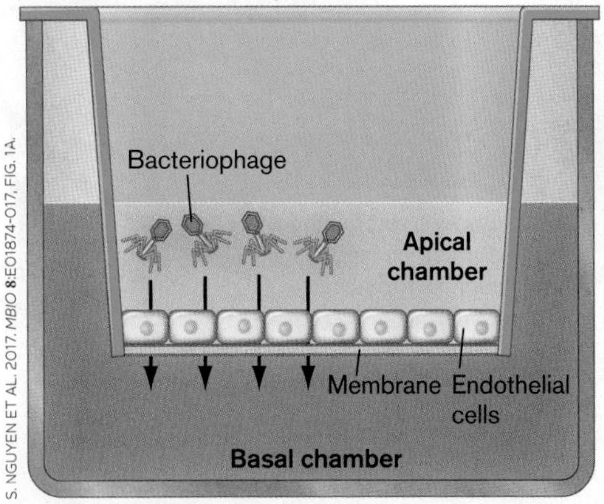

B.

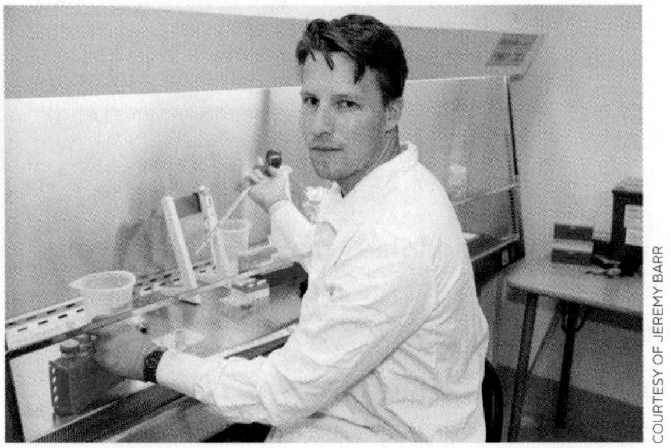

FIGURE 1 ■ Observing transcytosis in tissue culture. **A.** Cell monolayer is cultured on a membrane in the apical chamber submerged in the basal chamber. Phage particles inoculated in the apical chamber are detected later in the basal chamber. **B.** Jeremy Barr investigates transcytosis at University of California-San Diego.

6.5 Animal and Plant Viruses

Animal and plant viruses solve problems similar to those faced by bacteriophages: host attachment, genome entry and gene expression, virion assembly, and virion release. The more complex structure of eukaryotic cells, however, requires viral replication cycles that are more complex. Viral reproduction may involve intracellular compartments such as the nucleus or secretory system and may depend on tissue and organ development in multicellular organisms. Studies of animal virus replication reveal potential targets for antiviral drugs, such as protease inhibitors for HIV (discussed in Chapter 11).

Note: The virus replication cycles in this chapter are simplified. For greater molecular detail of selected viruses, see Chapter 11.

Animal Viruses Show Tissue Tropism

Like bacteriophages, animal viruses evolve by the fitness advantage of binding specific receptor proteins on their host cell. An example of a human virus-receptor interaction is that of rhinovirus (see **Table 6.1**, Group IV), which causes the common cold. Rhinovirus attaches to ICAM-1, a human glycoprotein (protein with sugar chains) needed for intercellular adhesion (**Fig. 6.27A**). The rhinovirus binds to a domain of ICAM-1 essential for ICAM-1 to bind a lymphocyte protein called integrin.

CellMask, a red fluorophore that binds membranes (**Fig. 2A**). The cells had been incubated with phage particles labeled with SYBR Gold fluorophore (green in the image). Cells that showed numerous puncta (spots) of green fluorescence in their cytoplasm were then correlated with sections for TEM. At higher resolution, TEM revealed the hexagonal electron-dense particles that appear to be phages (**Fig. 2B**).

Why would human cells actively take up phage particles? The reasons are not yet known, although Barr and colleagues suggest several hypotheses. An intriguing idea is the possibility that phages throughout the body provide an ever-present defense against invading bacteria, which might be susceptible to phage infection. Alternatively, the phages might modulate the immune system, enhancing long-term tolerance of the gut microbiome. Phages also offer an extraordinary range of genetic diversity, and phage genes may be accessible to the human genome. At this point we do not know the answers, but it's clear that phages get all around our bodies.

RESEARCH QUESTION

Can you propose an experiment to test the hypothesis that genes of phages taken up by transcytosis might be incorporated in the genome of host cells?

Sophie Nguyen et al. 2017. Bacteriophage transcytosis provides a mechanism to cross epithelial cell layers. *mBio* **8**:e01874-17.

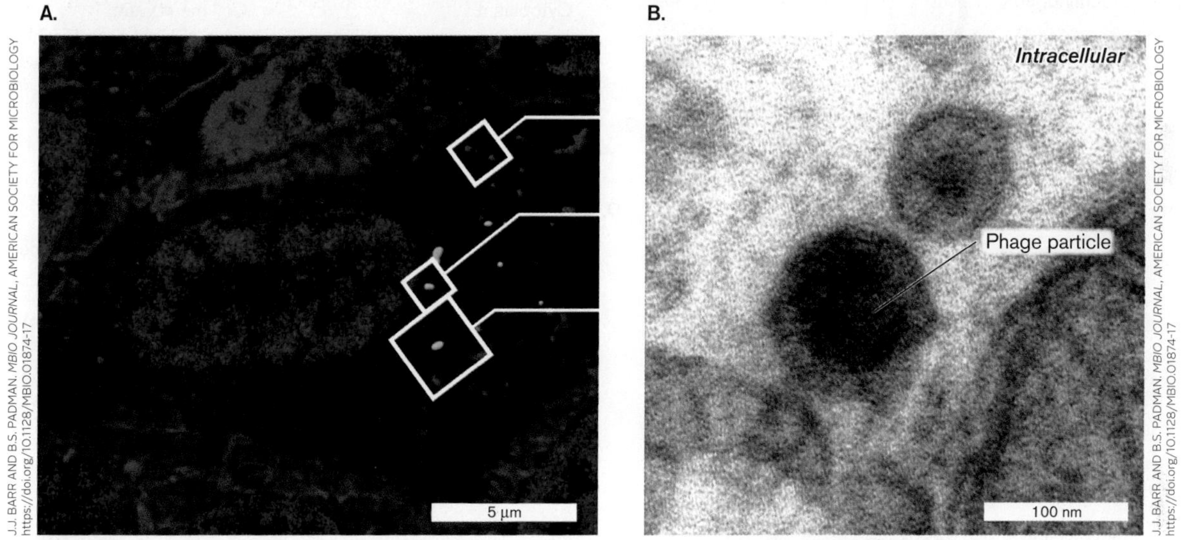

FIGURE 2 ■ Phages within the cytoplasm. **A.** MDCK cells (canine kidney tissue culture) stained with Hoechst (blue, nuclei) and CellMask (red, membranes). Phage T4 was labeled with SYBR Gold fluorophore (green) before incubation with cells. **B.** Correlated images were visualized by TEM at higher resolution, revealing individual phages (six-sided electron-dense particles) within the cytoplasm.

The host receptors play a key role in determining the host range, the group of host species permitting infection. Within a host, receptor molecules can also determine the viral **tropism**, or ability to infect a particular tissue type within a host. Some viruses, such as Ebola virus, exhibit broad tropism, infecting many kinds of host tissues, whereas others, such as papillomavirus, show tropism for only one type (in the case of papillomavirus, the epithelial tissues). Tropism may depend on the virus's ability to interact with the cytoplasm, or it may require the presence of an appropriate host cell receptor protein that can bind the viral surface attachment protein. For example, poliovirus infects only a specific class of human cells that display the immunoglobulin-like receptor protein PVR. Mice lack the PVR protein on their cell surfaces, and they are not normally infected by polio, but when transgenic mice were engineered to express PVR on their cells, the mice could then be infected.

Tissue-specific receptors determine the host tropism of avian influenza strain H5N1. The H5N1 strain infects birds by binding to a glycoprotein receptor on cell surfaces of the avian respiratory tract. The H5N1 strain requires a receptor protein with a sialic acid sugar chain terminating in galactose linked at the C-3 position (alpha-2,3), common in the avian respiratory tract. In humans, however, most nasal upper respiratory cells have receptors with galactose linked at the C-6 position (alpha-2,6). Human cells with the C-3 linkage are more common in the lower

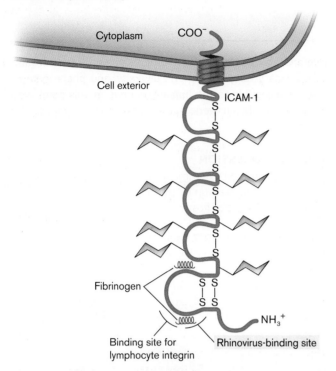
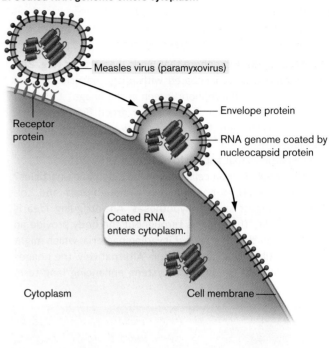
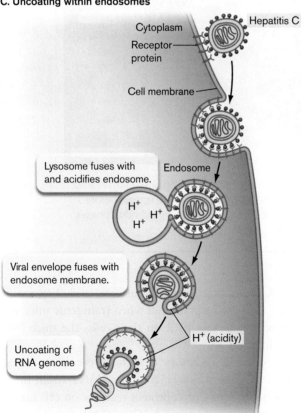
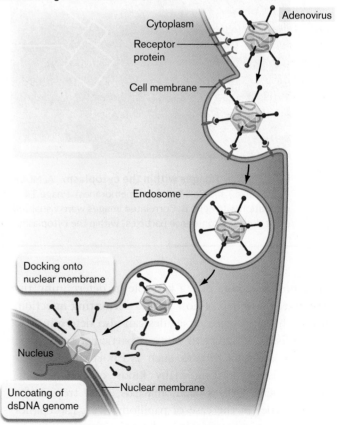

FIGURE 6.27 ■ Receptor binding and genome uncoating. A. Rhinovirus attaches to the intercellular adhesion molecule (ICAM-1), a glycoprotein required by the host cell to bind a lymphocyte integrin, a cell-surface matrix protein required for cell-cell adhesion. After binding a specific receptor on the host cell membrane, an animal virus enters the cell, where its genome is uncoated. **B.** Measles virus: The coated RNA genome enters the cytoplasm. **C.** Hepatitis C: The genome is uncoated within an endosome. **D.** Adenovirus: The genome is uncoated at the nuclear membrane. *Source:* Part A based on J. Bella et al. 1998. *PNAS* **95**:4140.

respiratory tract. That is why avian influenza H5N1 infection of humans has been relatively rare. However, a small mutation in the H5N1 envelope protein could enable it to bind to the alpha-2,6 receptor more effectively, allowing rapid transmission between humans. The molecular basis of influenza infection is discussed further in Chapter 11.

> **Thought Question**
>
> **6.9** How might humans undergo natural selection for resistance to rhinovirus infection? Is such evolution likely? Why or why not?

Genome entry and uncoating. Most animal viruses, unlike bacteriophages, enter the host cell as intact virions. The contents of the virion (genome and matrix proteins) may then interact with the cell in several different ways. For example, measles virus, a paramyxovirus, enters the cell by binding host receptor proteins, thereby causing the viral envelope to fuse with the host cell membrane (**Fig. 6.27B**). The measles RNA genome coated by nucleocapsid proteins is released directly into the cytoplasm.

For other types of viruses, the entire virion is internalized within an endosome. The internalized capsid undergoes **uncoating**, a process in which the capsid comes apart, releasing the viral genome into the cytoplasm. Uncoating is shown for a flavivirus, hepatitis C virus, in **Figure 6.27C**. The hepatitis C virion is first taken up by **endocytosis**. In endocytosis, the cell membrane forms a vesicle around the virion and engulfs it, forming an endocytic vesicle. The endocytic vesicle fuses with a lysosome, whose acidity activates entry of the capsid into the cytoplasm. The capsid then comes apart, and the viral genome is uncoated.

Yet other kinds of viruses, such as adenovirus, enter the cell by endocytosis but require transport to the nucleus (**Fig. 6.27D**). Following endocytosis and lysosome fusion, the adenoviral genome loses some of its capsid proteins (partial uncoating). A capsid protein then disrupts the endocytic membrane, allowing the remaining capsid to exit. The capsid then docks at a nuclear pore and injects its DNA genome into the nucleus. The adenoviral DNA is replicated by its own adenoviral DNA polymerase (carried in the virion), but it uses host nuclear histones for DNA packing, as well as host transcription factors available in the nucleus.

Animal Virus Replication Cycles

How does a virus replicate within an animal cell? Many animal viruses induce the cell's endoplasmic reticulum (ER) to form membranous organelles called **virus factories** or replication complexes. An example of virus factories is shown for coronavirus replication in **Figure 6.28A**. These virus factories support the synthesis of viral proteins and genomes.

The details of a replication cycle vary considerably, depending on the given virus. An important factor is the form of the viral genome. A DNA genome can use some or all of the host replication enzymes. An RNA genome, however, must encode either an RNA-dependent RNA polymerase to generate an RNA template or, in the case

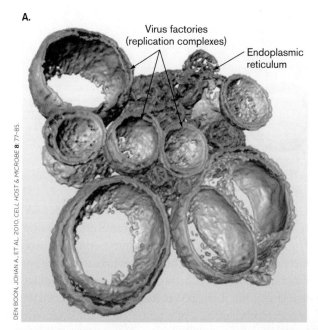

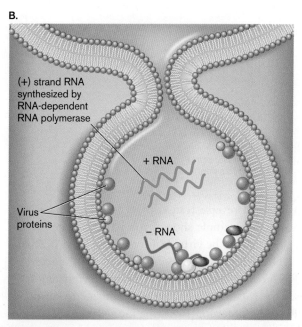

FIGURE 6.28 ■ **Virus factories (replication complexes) of a (+) strand RNA virus.** **A.** Membrane contours of virus factories derived from coronavirus endoplasmic reticulum (cryo-EM tomography). **B.** Virus proteins assemble upon ER-derived membranes. The viral RNA-dependent RNA polymerase synthesizes (+) strand RNA from a (−) strand template.

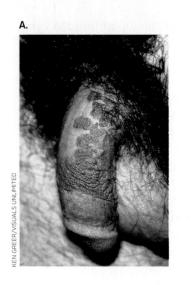
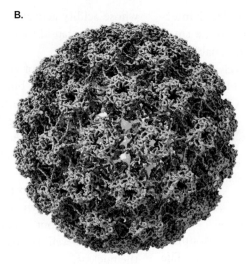
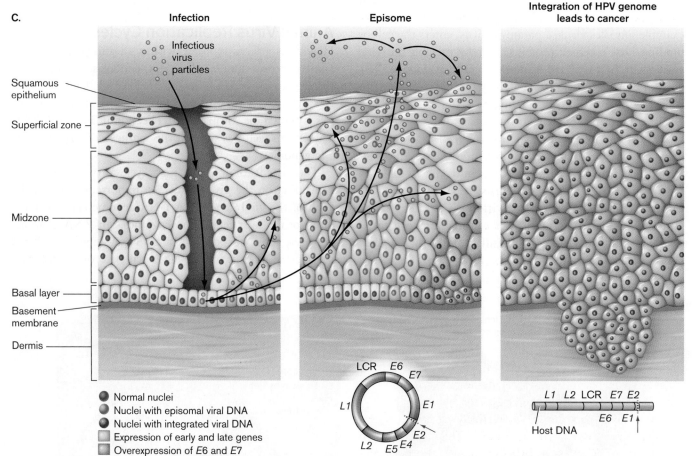

FIGURE 6.29 ▪ **Human papillomavirus.**
A. Certain strains of human papillomavirus (HPV) cause warts on the genitals or anus. **B.** HPV virion model (cryo-EM). **C.** HPV enters the cervical epithelium, where it infects basal epithelial cells. The DNA uncoats and forms episome (circularized). As cells differentiate, new virions are synthesized and released by shedding cells. Some HPV genomes may integrate into host DNA. Integrated HPV genomes express proteins that transform host cells into cancer cells.

of retroviruses, a reverse transcriptase to generate a DNA template. In the example shown, coronavirus proteins are concentrated within the virus-factory vesicles (**Fig. 6.28B**). A viral RNA-dependent RNA polymerase synthesizes (−) strand RNA complementary to the (+) strand viral genome, then uses the (−) strand genome as a template to form (+) strand RNA genomes for packaging into virions.

DNA virus replication. An example of a double-stranded DNA virus is human papillomavirus (HPV; see **Table 6.1**, Group I), the cause of genital warts (**Fig. 6.29A**). HPV is the most common sexually transmitted disease in the United States and one of the most common worldwide. Certain strains infect the skin, whereas others infect the mucous membranes through genital or anal contact (sexual transmission). Like phage lambda, papillomavirus has an active reproduction cycle and a dormant cycle within the host cell.

HPV initially infects basal epithelial cells, requiring access through a cut or abrasion in the skin or mucous membrane (**Fig. 6.29C**). In the basal cells, papilloma virions enter the cytoplasm by receptor binding and membrane fusion. The virion then gets uncoated, losing its protein capsid, and releases its genome of circular double-stranded DNA (**Fig. 6.30**). The uncoated viral DNA enters the host cell nucleus, where it is called an episome. The episome gets replicated by the host DNA polymerase and transcribed by the host RNA polymerase. The viral mRNA molecules are exported to the cytoplasm for translation of capsid proteins, which return to the nucleus for assembly of virions.

The process of HPV replication is complicated by the developmental progression of basal cells into keratinocytes (mature epithelial cells), and ultimately cells to be shed or sloughed off from the surface (see **Fig. 6.29B**). Viral replication is largely inhibited until the basal cells start to differentiate into keratinocytes. Host cell differentiation induces the viral DNA to replicate and undergo transcription by host polymerases. The mRNA transcripts then exit the nuclear pores, as do host mRNAs, for translation in the cytoplasm. The translated capsid proteins, however, return to the nucleus for assembly of the virion. Nuclear virion assembly is typical of DNA viruses (with the exception of poxviruses, which replicate entirely in the cytoplasm).

How do HPV virions disseminate? As the keratinocytes containing HPV complete differentiation, the cells start to come apart and are shed from the epithelial surface (**Fig. 6.29B**). Cells release HPV virions during this shedding process. Virion-releasing cells show partly accelerated growth, leading to formation of warts.

Basal cells containing viral episomes generally do not cause cancer. However, the integrated HPV genome up-regulates expression of oncogenes (cancer-causing genes) to form oncoproteins E6 and E7. These oncoproteins can transform host cells into cancer cells, which invade the dermis and grow out of control (see **eTopic 26.1**). The oncoproteins also inhibit the expression of host tumor suppressor genes. In some cases, the HPV genome integrates into the genome of cancer-transformed host cells.

Certain HPV strains are more likely to cause cancer, whereas others more likely cause warts. The most common strains that cause warts or that cause cancer are both preventable by the Gardasil vaccine.

Note: Different kinds of viruses may cause cancer by different mechanisms. These mechanisms include:

- Expression of unique viral oncogenes to form proteins that deregulate host cell growth (such as HPV oncoproteins E6 and E7).
- Expression of viral oncogenes that are mutated forms of a host proto-oncogene; the latter is an essential cellular gene involved in growth regulation.
- Integration of the viral genome near a cellular proto-oncogene, leading to inappropriate expression of the cellular gene.

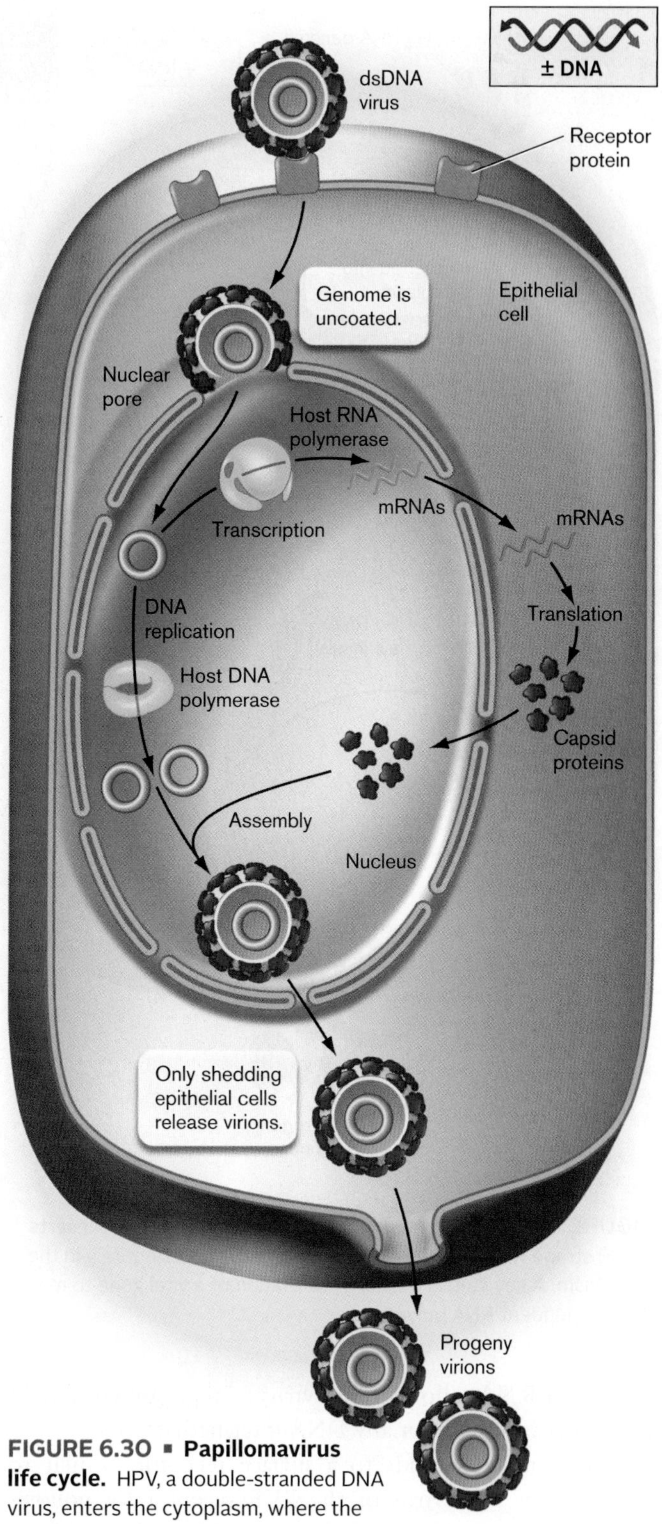

FIGURE 6.30 ■ Papillomavirus life cycle. HPV, a double-stranded DNA virus, enters the cytoplasm, where the protein coat disintegrates. The viral DNA enters the nucleus for replication and transcription by host polymerases.

RNA virus replication. The picornaviruses include poliovirus, the cause of paralytic poliomyelitis; and rhinoviruses, which cause the common cold (see **Table 6.1**, Group IV). In 2018–2019, an enterovirus similar to polio caused a record number of cases of polio-like paralysis in children (acute flaccid myelitis). Picornavirus genomes contain

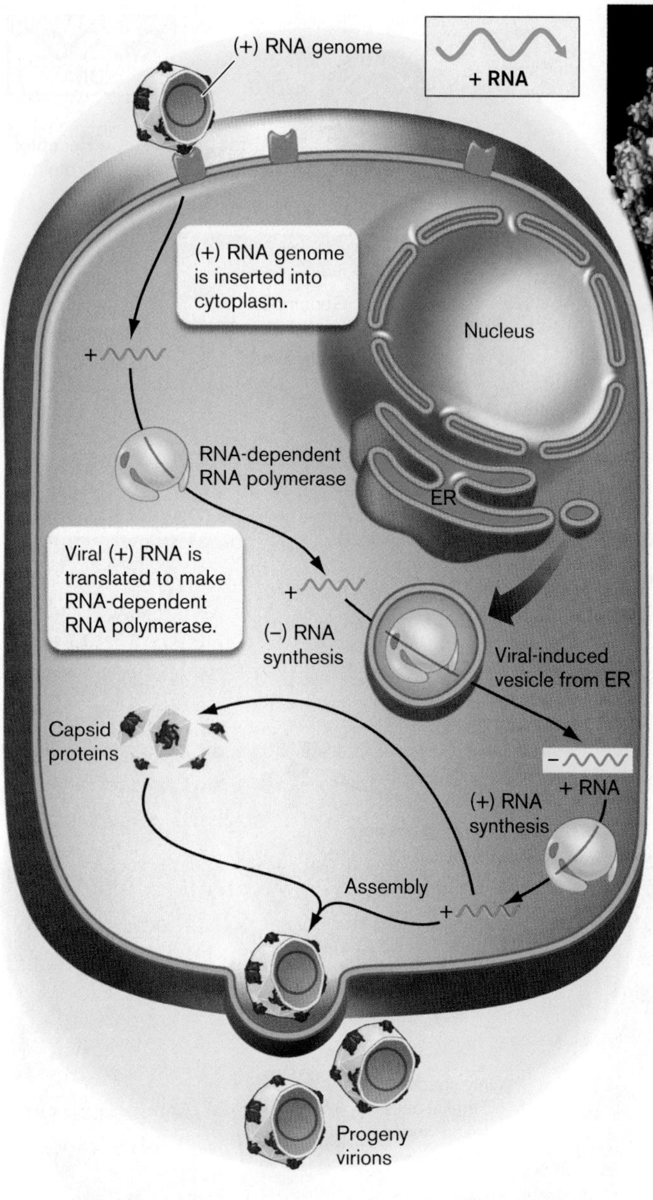

FIGURE 6.31 ■ **Picornavirus life cycle.** A picornavirus inserts its (+) strand RNA into the cell. Reproduction occurs entirely in the cytoplasm. A key step is the early translation of a viral gene to make RNA-dependent RNA polymerase.

(+) strand RNA, allowing the virus to replicate entirely in the cytoplasm, without any DNA intermediates.

A picornavirus binds to a surface receptor, similar to ICAM-1 for rhinovirus or the PVR receptor for poliovirus. The (+) strand RNA is uncoated by insertion through the cell membrane into the cytoplasm (**Fig. 6.31**)—much as a bacteriophage inserts its genome into a cell. The role of endocytosis is debated; poliovirus requires no endocytosis, but the rhinovirus genome may require endocytosis and low-pH induction.

After uncoating, the (+) strand RNA genome of picornavirus can be translated directly by host ribosomes to form viral proteins. One picornavirus gene is translated to make RNA-dependent RNA polymerase. The polymerase uses

the viral RNA template to make (–) strand RNA. The (–) strand RNA then serves as a template for other viral mRNAs, as well as for progeny genomic (+) RNA. These RNA molecules are synthesized within virus factories (replication complexes) formed from the endoplasmic reticulum, similar to those shown for coronavirus (**Fig. 6.28**). Capsid proteins are synthesized by host ribosomes, and the capsids self-assemble in the cytoplasm. Virions then assemble at the cell membrane and are released by subverting lysosomes, which attempt to digest them. For greater detail of the poliovirus replication cycle, see **eTopic 11.3**.

Note that other kinds of RNA viruses, such as influenza virus, encapsidate a (–) strand genome. In this case, the (–) strand must serve as the template to generate mRNA, as well as a (+) secondary template for (–) strand progeny genomes. Influenza virus replication includes other interesting molecular complications, discussed in Chapter 11.

RNA retroviruses. Retroviruses include human immunodeficiency virus (HIV), the causative agent of AIDS; and feline leukemia virus (FeLV), the cause of feline leukemia, a disease that commonly afflicts domestic cats (see **Table 6.1**, Group VI). A retrovirus such as HIV uses a reverse transcriptase to make a DNA copy of its RNA genome (**Fig. 6.32**). Instead of being translated from an early gene, the reverse transcriptase protein is actually carried within the virion, bound to the RNA genome with a primer in place. The virion contains two (+) strand RNA copies of the HIV genome, each carrying its own reverse transcriptase. Virions infect helper T cells and other lymphocytes of the active immune system (discussed in Chapter 24). After uncoating in the cytoplasm, the viral RNA is copied into double-stranded DNA.

The DNA copy then enters the nucleus, where a viral integrase enzyme integrates the DNA into the host genome. To generate virions, a host RNA polymerase transcribes the viral genome into viral mRNA and viral genomic RNA. The reverse transcriptase has a high error rate, generating slightly different versions of the virus—some of which can evade host defenses and antiviral drugs. This is the reason why HIV treatment requires multiple antiviral agents that target different parts of viral replication.

The HIV viral mRNA reenters the cytoplasm for translation to produce coat proteins, reverse transcriptase, and envelope proteins. The coat proteins are transported by the endoplasmic reticulum (ER) to the cell membrane, where virions self-assemble and bud out. Alternatively, a retrovirus may transmit from cell to cell through cytoplasmic nanotubes (introduced in Chapter 3), which avoid viral destruction by the immune system.

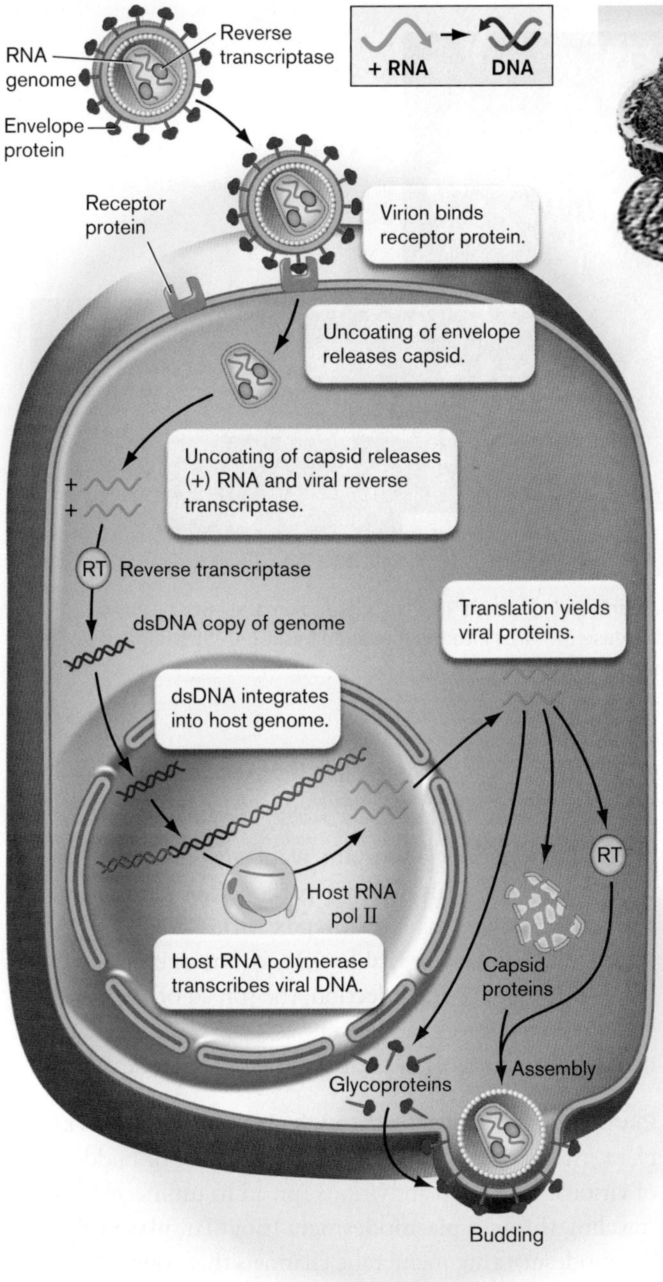

FIGURE 6.32 ■ Retrovirus life cycle. A retrovirus such as human immunodeficiency virus (HIV) uses reverse transcriptase to copy its RNA into double-stranded DNA. The DNA then enters the nucleus to recombine in the host genome, where a host RNA polymerase generates viral mRNA and viral genomic RNA.

At a certain point, infected cells can suddenly begin to generate large numbers of virions, destroying the immune system. The cause of accelerated reproduction is poorly understood, although it involves cytokines released under stress conditions such as poor health or pregnancy. HIV replication is activated via protein regulators encoded by the virus. The full process of HIV reproduction is described in Chapter 11.

MATTEI, S., ET AL. 2016. *SCIENCE*

There is no vaccine for HIV, but transmission can now be prevented nearly 100% by taking a daily dose of antiviral drugs. This dose is called pre-exposure prophylaxis (PrEP).

Thought Question

6.10 From the standpoint of a virus, what are the advantages and disadvantages of replication by the host polymerase, compared with using a polymerase encoded by its own genome?

Oncogenic viruses. As many as 20% of human cancers are caused by **oncogenic viruses**, such as Epstein-Barr virus (which causes lymphomas) and hepatitis C virus (which causes liver cancer). (Hepatitis C replication is discussed in **eTopic 11.4**.) When these viruses infect a cell, instead of destroying it the virus may **transform** the cell to divide and grow out of control. For a virus, the advantage of cancer transformation is that it expands the population of infected cells that produce virus particles, or that replicate a viral genome hidden within a host chromosome.

How do oncogenic viruses cause cancer? Several different mechanisms enable different types of viruses to transform normal host cells so that they proliferate abnormally and form tumors.

- **Oncogenes.** A retrovirus such as feline leukemia virus (FeLV) may carry an oncogene, which can transform the host cells. Usually an oncogene encodes an abnormal form of a host protein that controls cell proliferation.
- **Genome integration.** Certain viruses can integrate their genome into a host chromosome. The integrated viral genome expresses proteins that stimulate host cell division and may ultimately lead to growth of tumors.
- **Cell cycle control.** Oncogenic viruses such as papillomaviruses express viral proteins that interact with host cell cycle controls and can stimulate uncontrolled growth.

Viral capacities for gene transfer and host genome control may be manipulated artificially and used for gene therapy. In fact, some of the most dangerous viruses, such as HIV, are being engineered to make nonvirulent gene delivery devices (discussed in Section 11.4). HIV-derived vectors form the basis of CAR-T therapy for leukemia and other cancers, first approved by the US Food and Drug Administration in 2017.

Chronic viral infections. Some human viruses appear in our bodies indefinitely, causing either mild disease or none. For example, anelloviruses commonly appear in our

FIGURE 6.33 ▪ Plum pox is caused by potyvirus. **A.** Potyvirus, a filamentous (+) strand RNA virus, approximately 800 nm in length (TEM). **B.** Potyvirus is transmitted by aphids, which suck the plant sap and release the virus into the damaged tissues. **C.** Streaking of flowers caused by potyvirus infection. **D.** Ring-shaped pockmarks appear on the infected fruit.

blood plasma, without known disease. Children frequently shed enteroviruses (viruses of the intestinal tract, related to poliovirus) with no known effects. Some evidence supports the idea that low-level chronic viral infections might actually enhance the function of our immune system.

Plant Virus Replication Cycles

All kinds of plants are subject to viral infection. Plant viruses pose enormous challenges to agriculture, especially where the concentrated growth of a single strain of food crop (monoculture) provides ideal conditions for a virus to spread.

Plant virus entry to host cells. In contrast to animal viruses and bacteriophages, plant viruses infect cells by mechanisms that do not involve specific membrane receptors. The reason may be that plant cell membranes are covered by thick cell walls impenetrable to virion uptake or genome insertion. Thus, the entry of plant viruses usually requires **mechanical transmission**—nonspecific access through physical damage to tissues, such as abrasions of the leaf surface caused by a feeding insect. Mechanical transmission of plant viruses is limited by the cell wall. Most plant viruses gain entry to cells by one of three routes:

- **Contact with damaged tissues.** Viruses such as tobacco mosaic virus appear to require nonspecific entry into broken cells.
- **Transmission by an animal vector.** Insects and nematodes transmit many kinds of plant viruses. For example, the geminiviruses are inoculated into cells by plant-eating insects such as aphids, beetles, and grasshoppers.
- **Transmission through seed.** Some plant viruses enter the seed and infect the next generation.

An economically important plant virus is the potyvirus called plum pox virus, a major pathogen of plums, peaches, and other stone fruits. Plum pox virus, a Group IV (+) strand RNA virus (see **Table 6.1**), is transmitted by aphids (**Fig. 6.33**). Following infection, the spread of the virus generates streaked leaves and flowers, as well as ring-shaped pockmarks on the surfaces of the fruit and of the stone within.

Plant virus transmission through plasmodesmata. Within a plant, the thick cell walls prevent a lytic burst or budding out of virions. Instead, plant virions spread to uninfected cells by traveling through **plasmodesmata** (singular, **plasmodesma**). Plasmodesmata are membrane channels that connect adjacent plant cells (**Fig. 6.34**). The outer part of the channel connects the cell membranes of the two cells; the inner part connects the endoplasmic reticulum of one cell to that of the other. Passage through the plasmodesmata requires action by movement proteins whose expression is directed by the viral genome. In some cases, the movement proteins transmit the entire plant virion; in other cases, only the nucleic acid itself is small enough to pass through. The infection strategies of plant viral genomes may have features in common with those of viroids, which lack capsids altogether.

DNA pararetroviruses. Pararetroviruses possess a DNA genome that requires transcription to RNA in the cytoplasm, followed by reverse transcription to form DNA genomes for progeny virions. Some pararetroviruses, such as hepadnavirus, infect humans, but the best-known pararetrovirus is cauliflower mosaic virus (CaMV), a caulimovirus

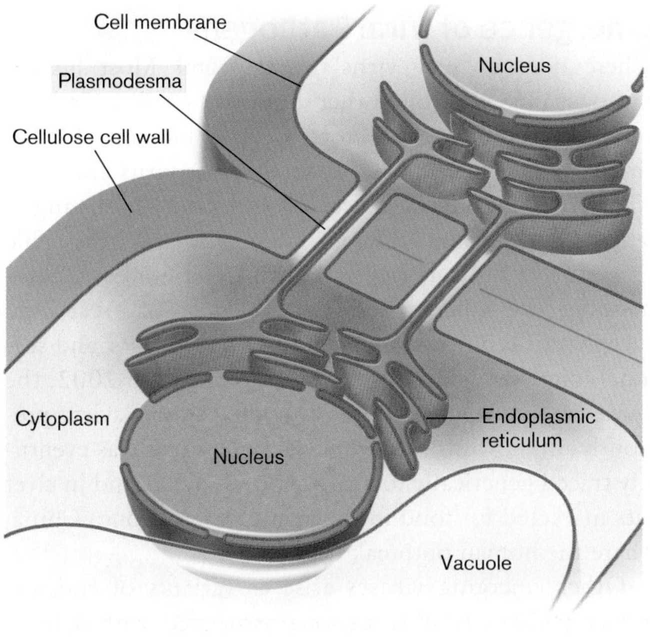

FIGURE 6.34 ■ Plant cells connected by plasmodesmata. Plasmodesmata offer a route for plant viruses to reach uninfected cells.

(see **Table 6.1**, Group VII). CaMV is an important tool for biotechnology because it has a highly efficient promoter for gene transcription, enabling high-level expression of cloned genes. Vectors derived from caulimoviruses are used to construct transgenic plants.

CaMV is transmitted by secretions from an insect whose bite damages plant tissues, providing access to the cytoplasm (**Fig. 6.35**). The CaMV genome moves from the cytoplasm to the cell nucleus through a nuclear pore. Within the nucleus, two promoters on its DNA genome direct transcription to RNA. The two RNA transcripts exit the nucleus for translation by host ribosomes to make viral proteins. A host reverse transcriptase, present in plant cells, copies the RNA into DNA viral genomes. After virions are assembled in the cytoplasm, virus-encoded proteins, called movement proteins, help transfer the virions through plasmodesmata into an adjacent cell.

A CaMV promoter sequence is commonly used in gene transfer vectors for plant biotechnology because transcription of the gene (such as one that confers pesticide resistance on the host) linked to the viral promoter is very efficient. In the field, 10% of cruciferous vegetables are typically infected with cauliflower mosaic virus. Some critics of gene technology fear that the prevalence of the

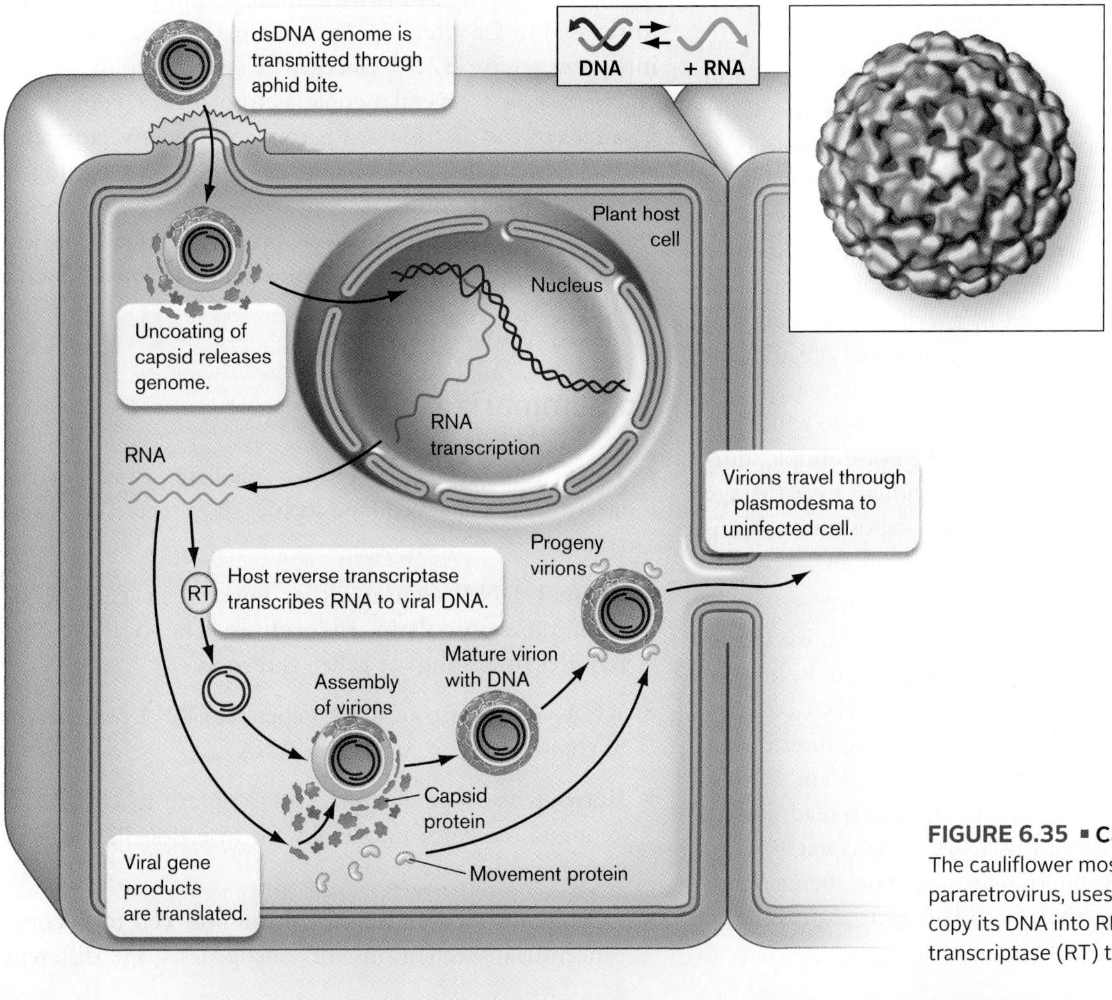

FIGURE 6.35 ■ Caulimovirus life cycle. The cauliflower mosaic virus (CaMV), a DNA pararetrovirus, uses host RNA polymerase to copy its DNA into RNA and uses host reverse transcriptase (RT) to make DNA copies.

CaMV promoter in transgenic crops may lead to the evolution of new pararetroviruses.

Animal and Plant Host Defenses

How do animals and plants defend themselves from virus infection? Since viruses are ubiquitous, a wide range of defense mechanisms have evolved. Defenses important for humans are part of our immune system, presented in Chapters 23 and 24.

Genetic resistance. As we saw for bacteria, animal and plant hosts continually experience mutations, some of which lead to strains that resist viral infection by halting adsorption or some other key step of the virus's replication cycle. When a virus becomes widespread, natural selection favors resistant strains. But commercial livestock and crops are typically a monoculture in which no resistant variants are available. Thus, when an outbreak arises, an entire crop may be destroyed. To save a crop, it may be interbred with a wild strain that possesses genes conferring resistance.

An example of genetic resistance in humans is resistance to HIV/AIDS. The basis of this resistance is a defective allele encoding a T-lymphocyte cell-surface protein that acts as a coreceptor, which is required for binding of virus HIV-1. The role of coreceptors in HIV infection and resistance is discussed further in Chapter 11.

Immune system. The immune systems of humans and other animals possess extensive cell machinery to thwart viral infection. A component of our innate immunity is the class of proteins called interferons, which recognize general signs of viral infections, such as the presence of double-stranded RNA (see Chapter 23). For adaptive immunity, viral proteins expressed in the cell membrane of an infected cell are recognized by specific antibodies that stimulate immune cells to destroy the infected cell and halt its viral production. The antibodies recognize a specific virus strain, such as a strain of influenza virus to which the individual has been exposed previously. For example, during the 2009 flu epidemic, many individuals over the age of 50 had some protection arising from exposure to a similar strain in an earlier epidemic (see Chapter 24).

RNA interference. RNA interference, or RNAi, is a mechanism by which mRNA molecules expressed by a viral genome are recognized by a host protein-RNA complex that shuts down further expression. RNA interference was first discovered in plants, where the system is most extensive, but it is now known to be widespread among all eukaryotes and archaea (discussed in Chapter 9). The mechanisms of RNA interference are now being engineered for use in gene therapy to halt gene expression in cancer and in inherited diseases.

Emergence of Viral Pathogens

Where does a "new virus" come from? Most human pathogens come from other humans, or from related animals referred to as vectors. Certain human-infecting viruses persist in the wild, such as rabies virus and West Nile virus. Their persistence requires broad host ranges: rabies infects many different mammals, and West Nile virus infects birds as well as humans and horses. Understanding the epidemiology of rabies or of West Nile encephalitis requires understanding the behavior and seasonal migration patterns of wild organisms. In 2002, the SARS respiratory disease was caused by a coronavirus previously unknown to science. The coronavirus was eventually traced genetically to viruses found in bats and in civet cats marketed for food in the area of Guangdong, China, where the human outbreak began.

Other emerging viruses arise as variants of endemic milder pathogens. Viruses long associated with a host, such as the common-cold viruses (rhinoviruses), tend to have evolved a moderate disease state that provides ample opportunities for host transmission. A virus that "jumps" from an animal host, however, may cause a more acute syndrome with higher mortality. The best-known cases are the exceptionally virulent emerging strains of influenza, which generally result from intracellular recombination of human strains with strains from pigs or ducks (discussed in Chapter 11). For example, in 2013 the avian influenza strain H7N9 emerged from poultry in China, where it killed several people before it was contained. Changes in the distribution patterns of insect vectors and animal hosts can generate new epidemics of a pathogen in regions where the virus could not spread before. Such changes in distribution can be brought about by many factors, including global climate change (**eTopic 6.3** and Chapter 28).

To Summarize

- **Host cell-surface receptors** mediate animal virus attachment to a cell and confer host specificity and tropism.
- **Animal DNA viruses** either inject their genome or enter the host cell by endocytosis. The viral genome requires uncoating for gene expression.
- **RNA viruses** use an RNA-dependent RNA polymerase to transcribe their messenger RNA.
- **Retroviruses** use a reverse transcriptase to copy their genomic sequence into DNA for insertion in the host chromosome.
- **Oncogenic viruses** transform the host cell to become cancerous. Mechanisms of oncogenesis by different

types of viruses include insertion of an oncogene into the host genome, integration of the entire viral genome, and expression of viral proteins that interfere with host cell cycle regulation.

- **Plant viruses** enter host cells by transmission through a wounded cell surface or an animal vector, and they travel to adjacent cells through plasmodesmata.
- **Pararetroviruses** contain DNA genomes but generate an RNA intermediate that requires reverse transcription to DNA for progeny virions.
- **Emerging viral pathogens** increase during environmental change.

6.6 Culturing Viruses

To learn how microbes grow, we culture them in the laboratory. So how do we culture a virus? A complication of virus culture is the need to grow the virus within a host cell. Therefore, any virus culture system must be a double culture of host cells plus viruses. Culturing viruses of multicellular animals and plants involves additional complications, because viruses show tropism for particular tissues or organs. Viruses may replicate in tissue culture, but the tissue culture does not show all the properties of an organ within a living organism. Therefore, a virus propagated in tissue culture will evolve to lose some of the virulence factors needed to infect an animal.

Batch Culture

Batch culture, or culture in an enclosed vessel of liquid medium, enables growth of a large population of viruses for study. Bacteriophages can be inoculated into a growing culture of bacteria, usually in a culture tube or a flask. The culture fluid is then sampled over time and assayed for phage particles. The growth pattern usually takes the form of a step curve (**Fig. 6.36**).

To observe one cycle of phage reproduction, phages are added to host cells at a high **multiplicity of infection** (**MOI**, ratio of phage to cells) such that every host cell is infected. The phage particles immediately adsorb to surface receptors of host cells and deliver their DNA. As a result, intact virions are virtually undetectable in the growth medium. This short period after infection is called the **eclipse period**. For some species, it is possible to distinguish between the eclipse period and a **latent period**, the time during which the inserted phage genome directs production of progeny virions. The virions accumulate within the cell but have not yet emerged in the medium. In animal viruses, the latent period is less distinct because large numbers of virions usually generate progeny through budding out of the host cell (**Fig. 6.37**).

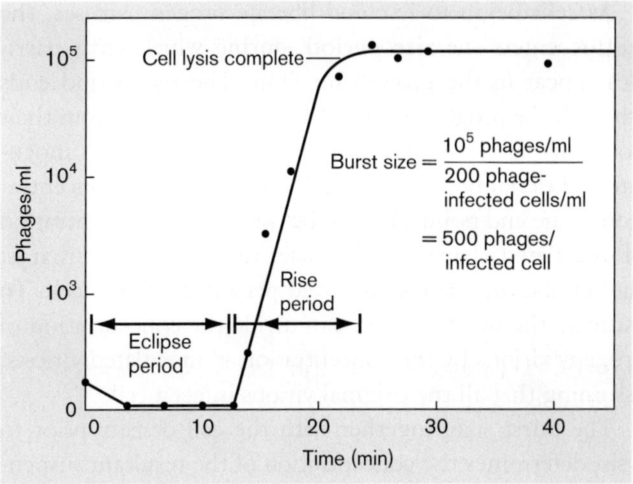

FIGURE 6.36 ■ One-step growth curve for a bacteriophage. After initial infection of a liquid culture of host cells, the titer of virus drops to near zero as all virions attach to the host. During the eclipse period, progeny phages are being assembled within the cell. As cells lyse (the rise period), virions are released until they reach the final plateau.

Note: The "latent period" of a lytic virus is the period between initial phage-host contact and the first appearance of progeny phage. This period must be distinguished from the "latent infection" of a virus that maintains its genome within a host cell without reproducing virions.

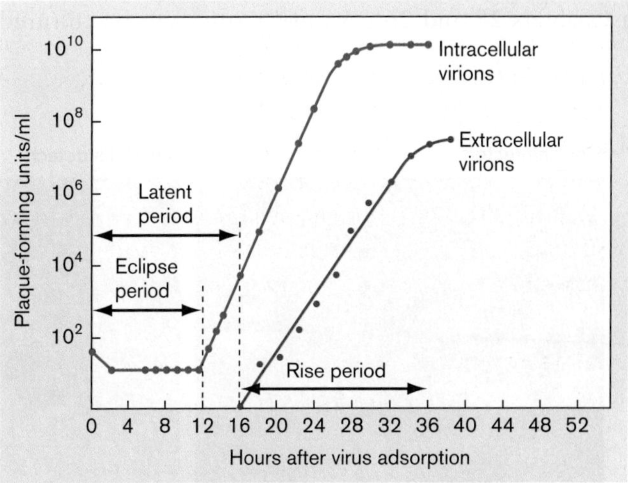

FIGURE 6.37 ■ One-step growth curve for a virus. The titer of extracellular virus drops to near zero during the latent period, as all virions adsorb to the host. Then progeny virions begin to emerge by budding out from the infected cell. For an animal virus, the growth curve may take hours to level off; the "burst" event is not defined as clearly as for phages.

As cells begin to lyse and liberate progeny viruses, the culture enters the **rise period**, during which virus particles appear in the growth medium. The rise period ends when all the progeny viruses have been liberated from their host cells. If the number of viruses that go on to inoculate additional host cells is small, then the virus concentration at the end point, divided by the original concentration of inoculated phage, approximates the **burst size**—that is, the number of viruses produced per infected host cell. To estimate the burst size, we can divide the concentration of progeny virions by the concentration of inoculated virions, assuming that all the original virions infect a cell.

The burst size, together with the cell density prior to lysis, determines the concentration of the resultant suspension of virus particles, called a **lysate**. In the case of bacteriophages, a lysate of phage particles can be extremely stable, remaining infective at room temperature for many years. Eukaryotic viruses, however, tend to be less stable and need to be maintained in culture or deep freeze.

Thought Question

6.11 Why does bacteriophage reproduction give a step curve, whereas cellular reproduction generates an exponential growth curve? (Compare **Fig. 6.36** with Fig. 4.21.) Could you design an experiment in which viruses generate an exponential curve? Under what conditions does the growth of cellular microbes give rise to a step curve?

Tissue Culture of Animal Viruses

In the case of animal and plant viruses, the multicellular nature of the host is an important factor in the pathology and transmission of the pathogen (discussed in Chapters 25 and 26). Animal viruses can be cultured within whole animals by serial inoculation, where virus is transferred from an infected animal to an uninfected one. Culture within animals ensures that the virus strain maintains its original **virulence** (ability to cause disease). But the process is expensive and laborious, involving the large-scale use of animals.

A historic event in 1949 was the first successful growth of a virus in tissue culture. Poliovirus, the causative agent of the devastating childhood disease poliomyelitis, was grown in human cell tissue culture (**Fig. 6.38**) by John F. Enders, Thomas J. Weller, and Frederick Robbins at Children's Hospital in Boston. As heralded that year in *Scientific American*: "It means the end of the 'monkey era' in poliomyelitis research. . . . Tissue-culture methods have provided virologists with a simple in vitro method for testing a multitude of chemical and antibiotic agents." Since then, tissue culture has remained the most effective way to study the molecular biology of animal and plant viruses and to develop vaccines and antiviral agents.

Some viruses can be propagated in a tissue culture of cells growing confluently on a surface. The cells must be immortalized—that is, genetically altered—to continue cell division indefinitely. The fluid bathing the tissue layer is sampled for virus concentration. As in the case of bacteriophage batch culture, we can define an eclipse period, a latent period before appearance of the first progeny virions in the culture fluid, and a rise period. In tissue culture, the time course of animal virus replication is usually much longer (hours or days) than that of bacteriophages (typically less than an hour under optimal conditions). The burst size of animal viruses, however, is typically several orders of magnitude larger than that of phages. The reason for the larger burst size is that the volume of a host cell is much larger than that of a bacterial host, thus providing a larger supply of materials to build virions.

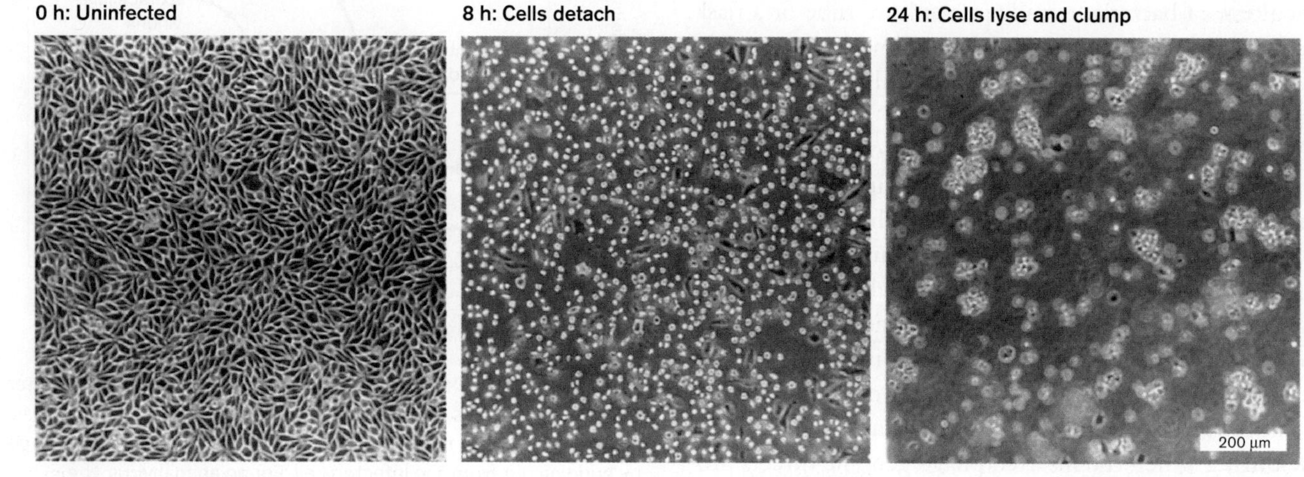

FIGURE 6.38 ■ Poliovirus replication in human tissue culture. Before infection (0 hours), the cultured cells grow in a smooth layer. At 8 hours, infected cells have detached from the culture dish. By 24 hours, cells have lysed or in some cases clumped with other cells.

Thought Question

6.12 What kinds of questions about viruses can be addressed in tissue culture, and what questions require infection of an animal model?

Plaque Isolation and Assay of Bacteriophages

For the investigation of cellular microbes, an important tool is the culturing of individual colonies on a solid substrate that prevents dispersal throughout the medium, as described in Chapters 1 and 4. Plate culture of colonies enables us to isolate a population of microbes descended from a common progenitor. But viruses cannot be isolated as "colonies." The reason is that although viruses can be obtained at incredibly high concentrations, they disperse in suspension. Even on a solid medium, viruses never form a solid visible mass comparable to the mass of cells that constitutes a cellular colony.

In viral plate culture, viruses from a single progenitor lyse their surrounding host cells, forming a clear area called a **plaque**. Each plaque arises from a single infected bacterium that bursts, its phage particles diffusing to infect neighboring cells. To perform a plaque assay of bacteriophages, a diluted suspension of phages is mixed with bacterial cells in soft agar, and the mixture is then poured over a nutrient agar plate (**Fig. 6.39**). Where no bacteriophages are present, the bacteria grow homogeneously as a "lawn," an opaque sheet over the surface (confluent growth). Where bacteriophages are included, each one infects a cell, replicates, and spreads progeny phages to adjacent cells, killing them as well (**Fig. 6.40**). The loss of cells results in a round, clear area seemingly cut out of the bacterial lawn. Plaques can be counted and used to calculate the concentration of phage particles, or **plaque-forming units** (**PFUs**), in a given suspension of liquid culture. The liquid culture can be analyzed by serial dilution in the same way one would analyze a suspension of bacteria.

Plaques offer a convenient way to isolate a recombinant DNA molecule contained in a bacteriophage vector (discussed in Chapter 12). In **Figure 6.40B**, the blue plaques result from a phage vector carrying the gene that encodes the enzyme beta-galactosidase.

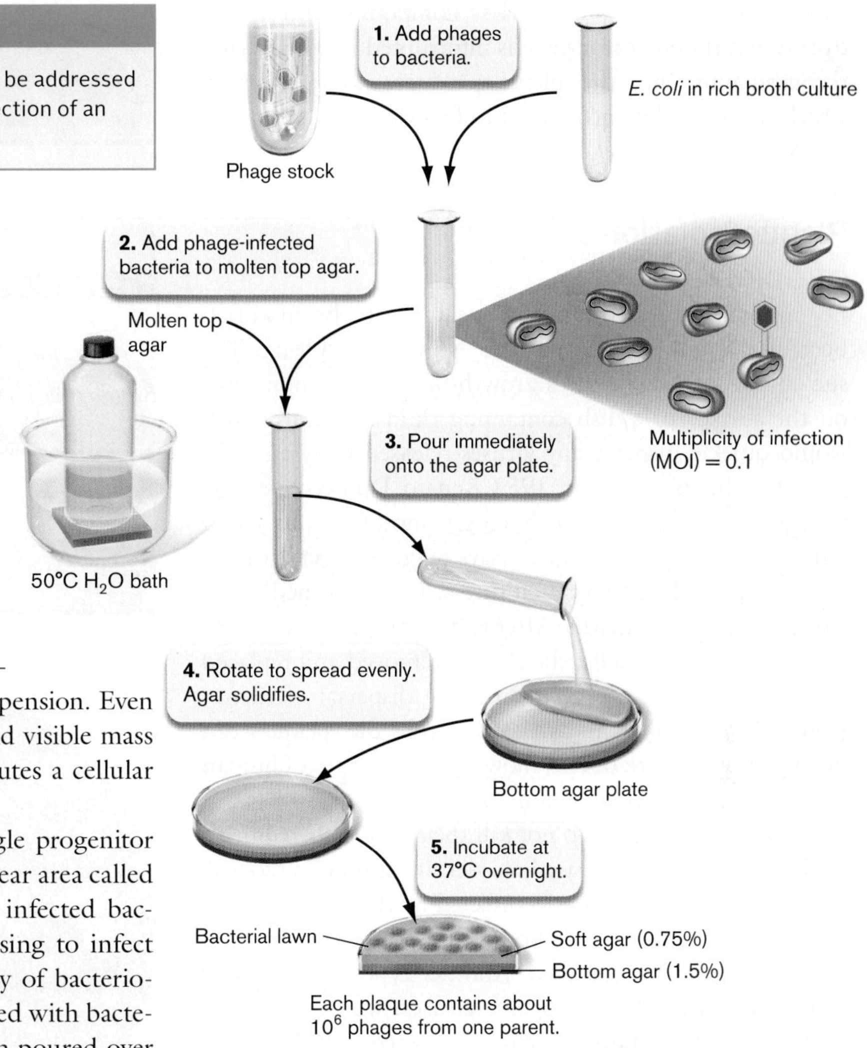

FIGURE 6.39 ▪ **Plating a phage suspension to count isolated plaques.** A suspension of bacteria in rich broth culture is inoculated with a low proportion of phage particles (multiplicity of infection is approximately 0.1). Each plaque arises from a single infected bacterium that bursts, its phage particles diffusing to infect neighboring cells.

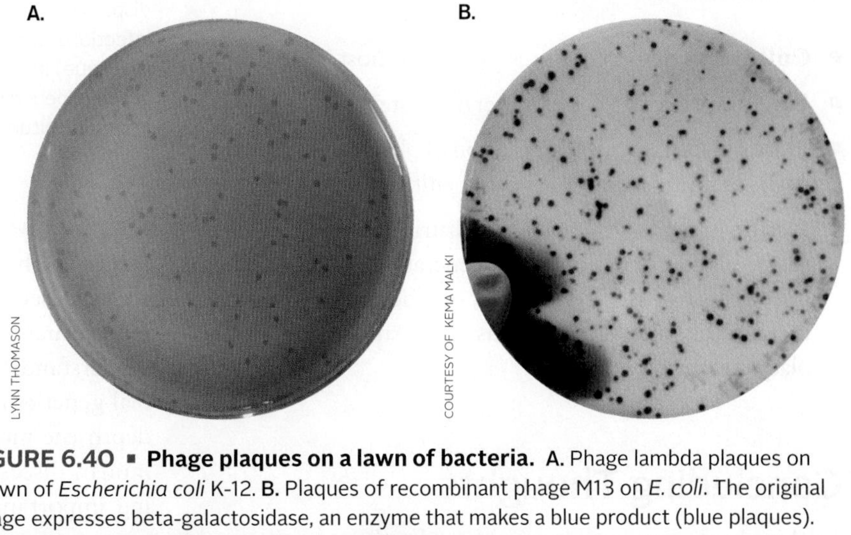

FIGURE 6.40 ▪ **Phage plaques on a lawn of bacteria.** **A.** Phage lambda plaques on a lawn of *Escherichia coli* K-12. **B.** Plaques of recombinant phage M13 on *E. coli*. The original phage expresses beta-galactosidase, an enzyme that makes a blue product (blue plaques). White plaques are produced by phage particles whose genome is recombinant (contains a cloned gene that interrupts the gene for beta-galactosidase).

This enzyme converts a colorless compound into a blue dye. When the indicator gene is interrupted by an inserted recombinant gene, the phage produces white plaques, which indicate the successful production of recombinant DNA phages.

Plaque Isolation and Assay of Animal Viruses

For animal viruses, the plaque assay has to be modified because it requires infection of cells in tissue culture. Tissue culture usually involves growth of cells in a monolayer on the surface of a dish containing fluid medium, which would quickly disperse any viruses released by lysed cells. To solve this problem, in 1952 Renato Dulbecco, at the California Institute of Technology, modified the tissue culture procedure for plaque assays (**Fig. 6.41A**). In Dulbecco's method, the tissue culture with liquid medium is first inoculated with virus. After sufficient time to allow for viral attachment to cells, the fluid is removed and replaced by a gel medium. The gel retards the dispersal of viruses from infected cells, and as the host cells die, plaques can be observed. **Figure 6.41B** shows a plate culture of human coronavirus infection of colon carcinoma cells.

Animal viruses that do not kill their host cells require a different kind of assay, based on identification of a "focus" (plural, foci), a group of cells infected by the virus. A focus may be identified using a fluorescent antibody—a method called fluorescent-focus assay. Another type of focus assay can be used to isolate oncogenic viruses, which transform their host cells into cancer cells. The cancer cells lose contact inhibition; they grow up in a pile instead of remaining in the normal monolayer. These piles of transformed cells, or transformed foci, can easily be visualized and counted. This procedure is known as the transformed-focus assay.

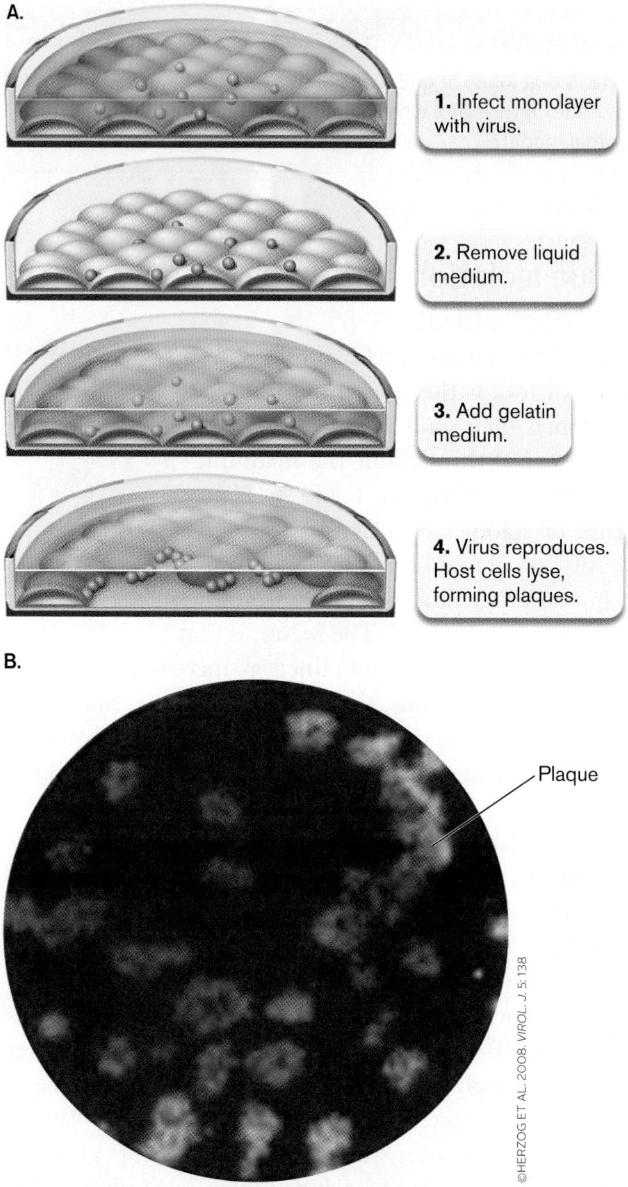

FIGURE 6.41 ■ Plate culture of animal viruses. **A.** Modified plaque assay for animal viruses. The gelled medium retards the dispersal of progeny virions from infected cells, restricting new infections to neighboring cells. The result is a visible clearing of cells (a plaque) in the monolayer. **B.** Plaque assay in which human coronavirus suspension was plated on a monolayer of colon carcinoma cells in tissue culture.

To Summarize

- **Culturing viruses** requires growth in host cells.
- **Batch culture** of viruses generates a step curve.
- **Plate culture** involves replication of phages on a bacterial lawn, or animal viruses in tissue culture.
- **Plaque assay** is a method of culturing viruses in which single phages or virions each generate an isolated clearing of host cells. The plaques can be counted to enumerate the infectious virions in a suspension, called plaque-forming units.

Concluding Thoughts

In this chapter we have covered broadly the structure and function of all kinds of viruses. As devastating as viruses can be to their hosts, they also provide engines of genomic change (see Chapter 9) and ecological balance, including mutualistic relationships with a host (see Chapter 21). Understanding virus function prepares us to discuss microbial genetics in Chapters 7–10. Chapter 11 then explores in depth the molecular biology of selected viruses that cause human disease. Viral molecular biology is a field of growing importance for medical and agricultural research, for genetic engineering and gene therapy, and for understanding natural ecosystems.

CHAPTER REVIEW

Review Questions

1. Compare and contrast the forms of icosahedral and filamentous (helical) viruses, citing specific examples.
2. How do viral genomes gain entry into cells in bacteria, plants, and animals?
3. Explain the key structural features that define the seven Baltimore groups of viral genomes. Explain the consequences of each structure for viral replication.
4. How do viral genomes interact with host genomes, and what are the consequences for host evolution?
5. Compare and contrast the lytic, lysogenic, and slow-release replication cycles of bacteriophages. What are the strengths and limitations of each?
6. Compare and contrast the replication cycles of RNA viruses and DNA viruses in animal hosts. What are the strengths and limitations of each?
7. Explain the plate count procedure for enumerating viable bacteriophages. How must this procedure be modified to measure the concentration of animal viruses? Oncogenic viruses?
8. Explain how a pure isolate of a virus can be obtained. How do the procedures differ from those used for isolating bacteria?
9. Explain the generation of the step curve of virus proliferation. Why is virus proliferation generally observed as a single step, or generation, in contrast to the life cycles of cellular microbes, outlined in Chapter 4?
10. Explain the key contributions of viruses to natural ecosystems. What may happen in an ecosystem where viruses are absent or fail to cause significant infection?

Thought Questions

1. Discuss the functions of different structural proteins of a virion, such as capsid, nucleocapsid, tegument, and envelope proteins. How do these functions compare and contrast with functions of cell proteins?
2. What are the relative advantages of the virulent phage replication cycle of phage T4, the lysis/lysogeny options of phage lambda, and the slow-release replication of phage M13? Under what conditions might each strategy be favored over the others?
3. Given the basis of viral tropism, how might an animal evolve traits that confer resistance to a virus infection?
4. If viruses return a substantial fraction of marine CO_2 to the atmosphere, can you imagine any ways to modulate virus proliferation so as to divert the carbon into sedimenting biomass?

Key Terms

bacteriophage (196)
batch culture (233)
burst size (217, 234)
capsid (196, 198)
cell-surface receptor (215)
CRISPR (219)
DNA reverse-transcribing virus (213)
eclipse period (233)
endocytosis (225)
endogenous virus (197)
envelope (203)
filamentous virus (204)
host range (201)
icosahedral capsid (202)
latent period (233)
lysate (234)
lysis (217)
lysogeny (217)
mechanical transmission (230)
multiplicity of infection (MOI) (233)
oncogenic virus (229)
ortholog (orthologous gene) (213)
pararetrovirus (213)
phage (196)
plaque (196, 235)
plaque-forming unit (PFU) (235)
plasmodesma (230)
prion (207)
prophage (197, 217)
proteome (214)
provirus (197, 201)
retrovirus (213)
reverse transcriptase (213)
rise period (234)
RNA-dependent RNA polymerase (211)
RNA reverse-transcribing virus (213)
segmented genome (210)
site-specific recombination (217)
spike protein (203)
temperate phage (217)
transcytosis (220)
transduction (217)
transform (229)
tropism (223)
uncoating (225)
viral shunt (200)
virion (196, 198)
viroid (206)
virome (199, 219)
virulence (234)
virus (196)
virus factory (225)

Recommended Reading

Abergel, Chantal, Matthieu Legendre, and Jean-Michel Claverie. 2015. The rapidly expanding universe of giant viruses: Mimivirus, Pandoravirus, Pithovirus and Mollivirus. *FEMS Microbiology Reviews* **39**:779–796.

Chaturongakul, Soraya, and Puey Ounjai. 2014. Phage–host interplay: Examples from tailed phages and Gram-negative bacterial pathogens. *Frontiers in Microbiology* **5**:442.

Chaudhry, Rabia M., J. E. Drewes, and Kara Nelson. 2015. Mechanisms of pathogenic virus removal in a full-scale membrane bioreactor. *Environmental Science & Technology* **49**:2815–2822.

Denner, Joachim, and Paul R. Young. 2013. Koala retroviruses: Characterization and impact on the life of koalas. *Retrovirology* **10**:108.

Grigg, Patricia, Allison Titong, Leslie A. Jones, Tilahun D. Yilma, and Paulo H. Verardi. 2013. Safety mechanism assisted by the repressor of tetracycline (SMART) vaccinia virus vectors for vaccines and therapeutics. *Proceedings of the National Academy of Sciences USA* **110**:15407–15412.

Harris, Audray, Giovanni Cardone, Dennis C. Winkler, J. Bernard Heymann, Matthew Brecher, et al. 2006. Influenza virus pleiomorphy characterized by cryoelectron tomography. *Proceedings of the National Academy of Sciences USA* **103**:19123–19127.

Horvath, Phillippe, and Randolphe Barrangou. 2010. CRISPR/Cas, the immune system of bacteria and archaea. *Science* **327**:167–170.

Knoops, Kèvin, Marjolein Kikkert, Sjoerd H. E. van den Worm, Jessika C. Zevenhoven-Dobbe, Yvonne van der Meer, et al. 2008. SARS-coronavirus replication is supported by a reticulovesicular network of modified endoplasmic reticulum. *PLoS Biology* **6**:e226.

Legendre, Matthieu, Julia Bartoli, Lyubov Shmakova, Sandra Jeudy, Karine Labadie, et al. 2013. Thirty-thousand-year-old distant relative of giant icosahedral DNA viruses with a pandoravirus morphology. *Proceedings of the National Academy of Sciences USA* **111**:4274–4279.

Maggie, Fabricio, and Mauro Bendinelli. 2009. Immunobiology of the Torque teno viruses and other anelloviruses. *Current Topics in Microbiology and Immunology* **331**:65–90.

Manrique, Pilar, Michael Dills, and Mark J. Young. 2017. The human gut phage community and its implications for health and disease. *Viruses* **9**:141.

McBride, Alison A., and Alix Warburton. 2017. The role of integration in oncogenic progression of HPV-associated cancers. *PloS Pathogens* **13**:e1006211.

Mills, Susan, Fergus Shanahan, Catherine Stanton, Colin Hill, Aidan Coffey, et al. 2013. Movers and shakers. *Gut Microbes* **4**:4–16.

Raoult, Didier, Stéphane Audic, Catherine Robert, Chantel Abergel, and Patricia Renesto. 2004. The 1.2-megabase genome sequence of Mimivirus. *Science* **306**:1344–1350.

Roossinck, Marilyn J. 2011. The good viruses: Viral mutualistic symbioses. *Nature Reviews. Microbiology* **9**:99.

Srinivasiah, Sharath, Jaysheel Bhavsar, Kanika Thapar, Mark Liles, Tom Schoenfeld, et al. 2008. Phages across the biosphere: Contrasts of viruses in soil and aquatic environments. *Research in Microbiology* **159**:349–357.

Strecker, Jonathan, Alim Ladha, Zachary Gardner, Jonathan L. Schmid-Burgk, Kira S. Makarova, Eugene V. Koonin, and Feng Zhang. 2019. RNA-guided DNA insertion with CRISPR-associated transposases. *Science* **365**: 48–63.

Thurber, Rebecca Vega, Jérôme P. Payet, Andrew R. Thurber, and Adrienne M. S. Correa. 2017. Virus-host interactions and their roles in coral reef health and disease. *Nature Reviews. Microbiology* **15**:205–216.

Trask, Shane D., Sarah M. McDonald, and John T. Patton. 2012. Structural insights into the coupling of virion assembly and rotavirus replication. *Nature Reviews. Microbiology* **10**:165–177.

CHAPTER 7

Genomes and Chromosomes

7.1 DNA: The Genetic Material

7.2 Genome Organization

7.3 DNA Replication

7.4 Plasmids and Secondary Chromosomes

7.5 Eukaryotic and Archaeal Chromosomes

7.6 Microbiomes and Metagenomes

A genome is all of the genetic information that defines an organism. For a bacterial species, this can include one or more chromosomes, plasmids, and viruses. In Chapter 7 we discuss how bacterial genomes replicate, organize, and migrate through the dividing bacterial cell. We compare these bacterial genome properties with those of archaea and eukaryotic microbes. We also describe how innovations in molecular biology help researchers explore the genomes in natural microbial communities.

CURRENT RESEARCH highlight

The many, many chromosomes of giant bacteria. Most bacteria have few chromosome copies, loosely organized as nucleoids at midcell. But what about *Epulopiscium*, a giant bacterium about 1,000,000 times larger than *Escherichia coli*? Elizabeth Hutchison and colleagues at SUNY Geneseo and Cornell University discovered that these cells contain thousands of chromosome copies. Hutchison used a fluorescent DNA stain to locate the chromosomes in a mother cell containing two offspring. The chromosomes were found primarily at the poles and just below the cell membranes along the lengths of the two daughter cells.

Source: E. Hutchison. 2018. *Mol. Microbiol.* **107**:68–80.

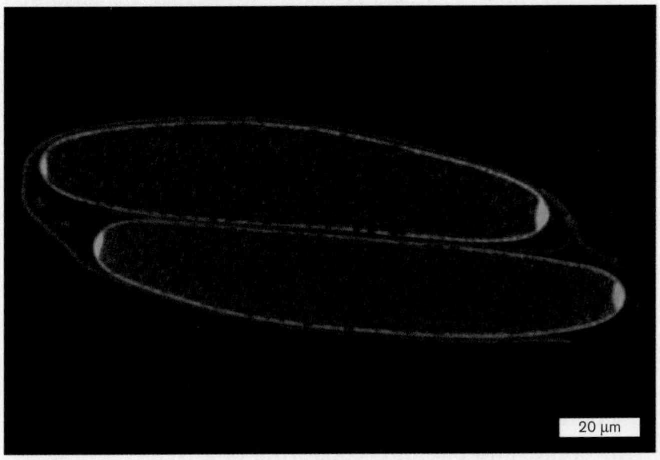

AN INTERVIEW WITH

ELIZABETH HUTCHISON, MICROBIOLOGIST, SUNY GENESEO

What are the potential benefits for having thousands of copies of the chromosome positioned along these massive cells?

Because of its size, this *Epulopiscium* species likely requires thousands of chromosome copies to support its large cell volume. Spacing between chromosomes and chromosome packing density in *Epulopiscium* were not greater than the values observed in smaller bacterial model systems, suggesting that large cell size does not affect these features. Having thousands of chromosome copies may also have interesting and important implications for genome evolution.

The **genome** is the total of all genes in an organism, whether they are found on a single chromosome or distributed among multiple replicons. With the exception of RNA viruses, microbial genomes are encoded by DNA. DNA is an ideal molecule to encode the genome because (1) it is stable and not readily degraded within the lifetime of the cell; (2) it is mutable, and thus can allow the organism to evolve through heritable genetic change; and (3) as a double-stranded polymer it is readily replicated, using the original strands as the template for making copies during cell replication.

In 1977, there was but one fully sequenced genome: that of the *Escherichia coli*–infecting phage phiX174 (φX174). Forty years later, we have over 35,000 well-curated genomes, and many more on the way. We now know that genome organization varies widely in the microbial world. While most of the prokaryotic genomes sequenced have a single chromosome, about 10% have more than one. Many genomes also include extrachromosomal elements such as plasmids and viruses, some of which can insert into the host chromosome. There are many questions that researchers continue to ask about microbial genomes. To what end does an organism maintain genes on a single chromosome or on multiple chromosomes? How does a microbe acquire a second chromosome? What is the advantage of gene placement on plasmids instead of chromosomes? How does a bacterium replicate its chromosomes and plasmids, and how is this replication coordinated with cell division? And finally, do the answers to these questions differ between bacteria and archaea?

In Chapter 7 we explore the physical structure of DNA, how it is replicated, and how the genome is studied in isolated strains and in mixed communities. In subsequent chapters we describe how genes are coded and "read" to make products (for example, RNA, protein) for the cell (Chapter 8), how genes and genomes can change through mutation (Chapter 9), and how gene expression is regulated by internal and external stimuli (Chapter 10). The remaining chapters of Part 2 discuss the specialized genetic mechanisms of viruses (Chapter 11) and the various molecular techniques and technologies that spawned the field of synthetic biology (Chapter 12).

7.1 DNA: The Genetic Material

Genes are the units of heredity, and most genes are located on chromosomes. Early in the twentieth century, chromosomes were known to contain protein as well as nucleic acid, but which molecule served as the genetic material was uncertain. For many years scientists believed erroneously that the protein component is what served as the genetic material, while the DNA might serve simply as a scaffold for the proteins. In support of the protein hypothesis was the fact that there are far more amino acid variants (20) than nucleic acid variants (5, if RNA's uracil and DNA's thymine are considered). The larger array of protein-building components suggested the possibility of a more diverse set of genes because the genetic alphabet would have more letters.

Studies of bacteria and their viruses in the 1920s–1950s finally overturned the protein hypothesis and revealed DNA as the genetic material. In 1928, Frederick Griffith (1879–1941) discovered that he could kill mice with live but seemingly harmless (avirulent) *Streptococcus pneumoniae*, but only if he coinjected the mice with dead cells from a virulent strain of the bacteria. Something from the dead bacteria transformed the innocuous live bacteria into killers. Through careful biochemical analysis in 1944, Oswald Avery, Colin Macleod, and Maclyn McCarty demonstrated that this transforming agent was DNA, and they coincidentally discovered the first of several mechanisms of horizontal gene transfer between microbes: **transformation**.

Overturning a long-standing model often requires multiple lines of evidence, and in the case of DNA, the evidence came from the 1952 study by Alfred Hershey and Martha Chase on the *E. coli*–infecting virus T2. Using radioisotopes of sulfur and phosphorus to track the protein and DNA components of the virus, respectively, Hershey and Chase showed that during viral adsorption and replication, the viral protein coat remains on the cell exterior, while the DNA penetrates into the interior and serves as the code to produce new viral DNA and protein.

Thought Question

7.1 Before the studies by Avery, Hershey, and others, some scientists believed that, unlike plants and animals, bacteria lacked genes. Considering what little was known about the modes of reproduction and the recombination of alleles in bacteria, why was this a reasonable, albeit incorrect, assumption?

To Summarize

- **A genome** is all of the genetic information that defines an organism.
- **Genomes** of bacteria and archaea are made up of chromosomes and plasmids consisting of DNA.
- **Studies in bacteria** provided the critical evidence that the genetic material is composed of DNA, not protein.

7.2 Genome Organization

Microbes are tremendously diverse (see Chapters 18–20 for examples). They are found in every imaginable habitat (for example, ocean sediments, the human colon), carrying out all kinds of nutrition (photosynthesis and anaerobic respiration, for instance), and performing an extraordinary variety of activities, such as swimming, fruiting-body formation, or obligate intracellular living. These lifestyle differences are often reflected in the size and content of microbial genomes. In addition, the physical arrangement of genes within a genome can vary from microbe to microbe. As we will see, certain genome arrangements provide clear advantages for the cell, whereas in other cases the advantages are not yet known.

Genomes Vary in Size

Bacterial and archaeal chromosomes range in size from approximately 106 to 16,000 kilobase pairs (kb). For comparison, eukaryotic chromosomes range from 2,900 kb (Microsporidia) to over 100,000,000 kb (flowering plants). The human genome is over 3,000,000 kb.

Note: The designation "kb" can refer to the length of either a double-stranded or a single-stranded DNA molecule. A bacterial genome is, by definition, double-stranded. Some viral genomes can be single-stranded.

One of the smallest cellular genomes sequenced thus far is that of the candidate bacterial species *Tremblaya princeps* (**Table 7.1**). The complete genome of *T. princeps* consists of

TABLE 7.1 Genomes of Representative Bacteria and Archaea

Species (strain)	Chromosome(s)* (kilobase pairs, kb)	Plasmid(s)* (kb)	Total (kb)
	Circular and linear	Circular	
Bacteria			
Tremblaya princeps Endosymbiont of mealybugs	139		139
Mycoplasma genitalium Normal flora, human skin	580		580
***Vibrio cholerae* (El Tor N16961)** Cholera	2,961 + 1,072		4,033
Mycobacterium tuberculosis Tuberculosis	4,400		4,400
***Escherichia coli* K-12 (W3110)** Model strain for *E. coli* research	4,600		4,600
Agrobacterium tumefaciens Tumors in plants; genetic engineering vector	2,840 + 2,070	214 + 542	5,666
***Anabaena* species (PCC 7120)** Cyanobacteria: major photosynthetic producer of carbon source for freshwater ecosystems	6,370	110 + 190 + 410	7,080
Borrelia burgdorferi Lyme disease	911	21; sizes 9 to 58	>1,250
Archaea			
Methanocaldococcus jannaschii Methanogen from thermal vent	1,660	16 + 58	1,734
Haloarcula marismortui Halophile from volcanic vent	3,130 + 288	33 + 33 + 39 + 50 + 155 + 132 + 410	4,270
	1,000 kb	500 kb	

*Purple circles and lines indicate relative sizes of genomic elements and whether these are circular or linear. Size bars are provided under each column.

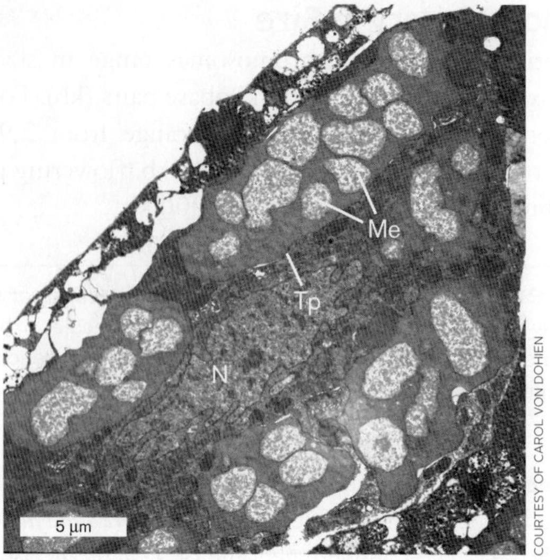

FIGURE 7.1 ■ The mealybug endosymbiont *Tremblaya princeps*. *T. princeps* cells (Tp; arrow points to envelope of the dark, irregular cells), living inside mealybug *Planococcus citri* cells (N = nucleus), are themselves host to other endosymbiotic bacteria, the candidate species *Moranella endobia* (Me, light cells).

only 139 kb and encodes a mere 120 proteins. It lacks the genes required for many biosynthetic functions, including several genes for the translation of mRNA into protein (Chapter 8). With the absence of so many genes, how does this organism survive? Survival depends on the peculiar symbiotic lifestyle of *T. princeps*. It lives as an endosymbiont inside mealybug insects, and in turn, *T. princeps* plays host to smaller microbes living inside its own cytoplasm (**Fig. 7.1**). Evolution likely permitted the dramatic reduction in the *T. princeps* genome because its symbiotic partners could provide many of those lost gene functions. For additional examples of genomic reduction of host-associated microbes, see the discussions of the evolution of mitochondria and chloroplasts (Section 17.6) and of obligate intracellular pathogens (Chapter 23).

In contrast, many bacteria that grow as free-living cells have larger genomes and dedicate many genes to the synthesis or acquisition of amino acids or TCA cycle intermediates. One example is *Vibrio cholerae*, the causative agent of cholera. Its 4,033-kb genome is divided into two chromosomes, of 2,961 and 1,072 kb (**Fig. 7.2**). Are genomes like *V. cholerae* divided into two chromosomes because there is a size restriction per chromosome, perhaps because of difficulties associated with replication or packaging? This does not appear to be the case, since genomes of much larger size have been found on single chromosomes; for example, the single chromosome of the bacterium *Myxococcus xanthus* is over 9,000 kb. We will consider other possible explanations later in the chapter.

Because eukaryotic chromosomes are linear, scientists initially expected that bacterial chromosomes would be linear too. However, the early genetic maps for bacteria such as *Escherichia coli* just would not fit together in a manner consistent with a linear model. We now know that most bacteria (including *E. coli*) and archaea have circular chromosomes. Some

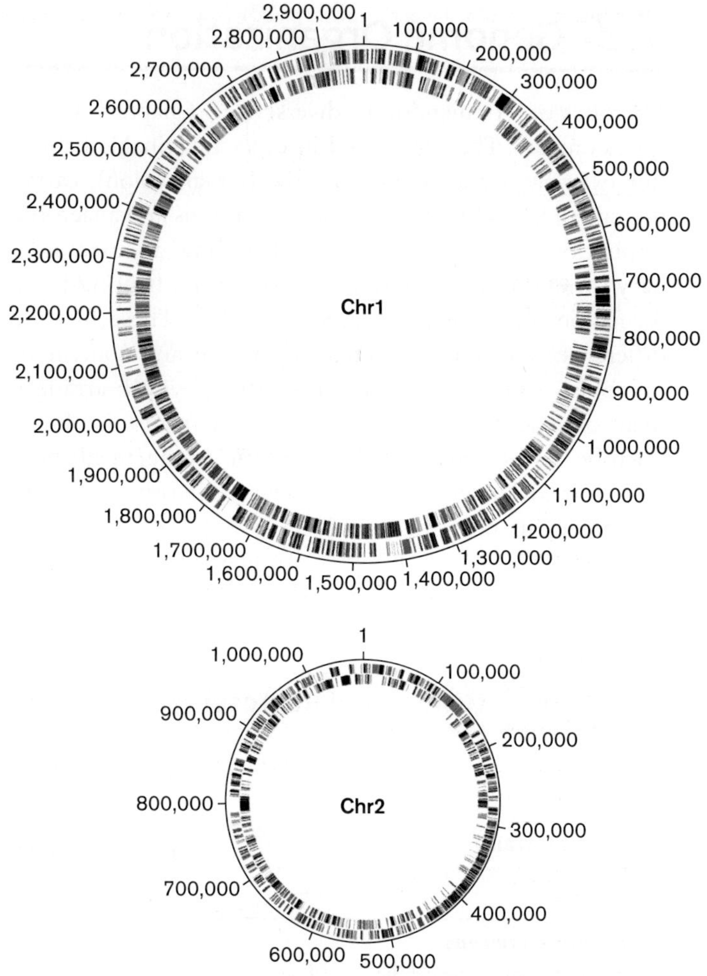

FIGURE 7.2 ■ The two chromosomes of the *Vibrio cholerae* genome. In each chromosome, the numbers refer to the position in base pairs, starting at the origin of replication and extending clockwise. Genes encoded on the plus and minus strands of the chromosome are depicted on the outer and inner rings, respectively. Chr1 = primary chromosome; Chr2 = secondary chromosome.
Source: Modified from J. F. Heidelberg et al. 2000. *Nature* **406**:477–483, fig. 2.

species, however, do have linear chromosomes, or even a mix of linear and circular chromosomes. Examples include the Lyme disease agent *Borrelia burgdorferi*, and the plant tumor agent *Agrobacterium tumefaciens* (**Table 7.1**). We will revisit the structure of chromosomes later in this chapter, and consider the advantages of linear versus circular forms.

In addition to size, a feature that distinguishes bacterial and archaeal genomes from those of eukaryotes is the amount of so-called noncoding DNA (DNA that does not encode proteins). Many eukaryotes contain huge amounts of noncoding DNA scattered between genes (**Fig. 7.3**). In some species (such as humans), over 90% of the total DNA is noncoding. Some noncoding regions include **enhancer** sequences needed to drive transcription of eukaryotic promoters and DNA expanses that separate enhancers. Enhancer sequences can function at large distances from the genes they regulate. A **promoter** is the DNA sequence immediately in front of, and sometimes within, a gene which is needed to activate the gene's expression. Most of

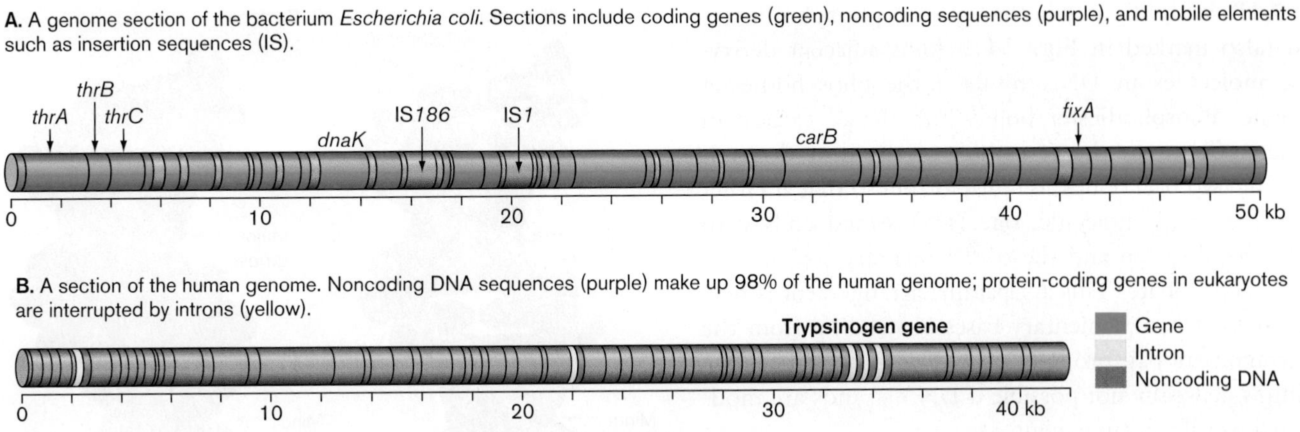

FIGURE 7.3 ■ Genome structure in a bacterium (A) and a eukaryote (B).

the noncoding regions appear to be either remnants of genes lost over the course of evolution or pieces of defunct viral genomes. Noncoding DNA, however, may provide raw material for future evolution.

In contrast to many eukaryotes, bacteria and archaea tend to have very little noncoding DNA (typically less than 15% of the genome). Archaeal genomes do, however, contain a few genes with internal noncoding DNA sequences that resemble the introns of eukaryotes.

DNA Function Depends on Its Chemical Structure

DNA is composed of four different nucleotides linked by a phosphodiester backbone (**Fig. 7.4A**). Each nucleotide consists of a **nucleobase** (also called a **nitrogenous base**) attached through a ring nitrogen to carbon 1 of 2-deoxyribose in the phosphodiester backbone. The 2-deoxy position that distinguishes DNA from RNA

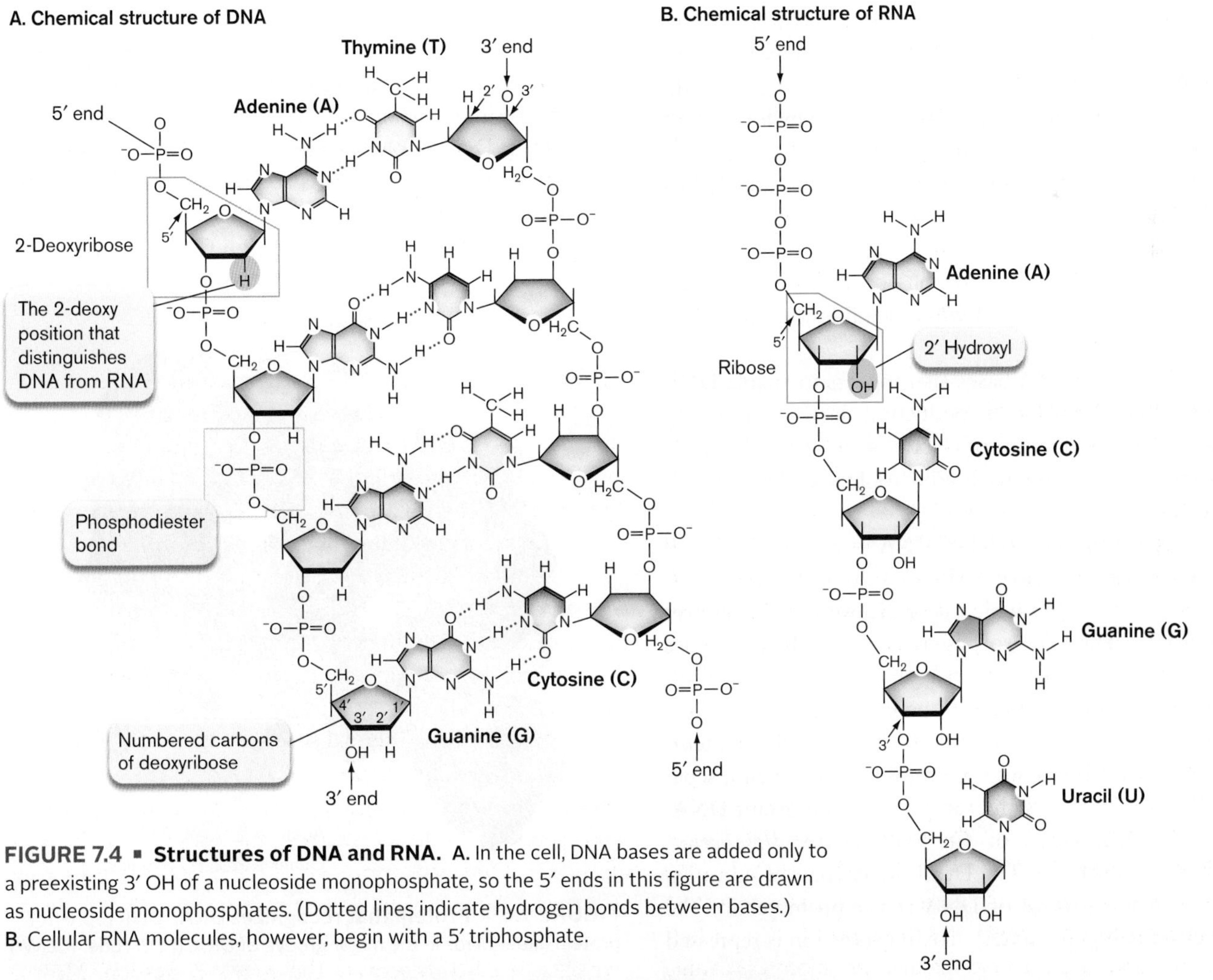

FIGURE 7.4 ■ Structures of DNA and RNA. A. In the cell, DNA bases are added only to a preexisting 3′ OH of a nucleoside monophosphate, so the 5′ ends in this figure are drawn as nucleoside monophosphates. (Dotted lines indicate hydrogen bonds between bases.) **B.** Cellular RNA molecules, however, begin with a 5′ triphosphate.

(**Fig. 7.4B**) is highlighted in the figure. A **phosphodiester bond** (also marked in **Fig. 7.4A**) joins adjacent deoxyribose molecules in DNA to form the phosphodiester backbone. Phosphodiester bonds link the 3′ carbon of one deoxyribose to the 5′ carbon of the next deoxyribose. The two backbones are **antiparallel** so that at either end of the DNA molecule, one DNA strand ends with a 3′ hydroxyl group and the complementary strand ends with a 5′ phosphate. This antiparallel arrangement is necessary so that complementary bases protruding from the two strands can pair properly via hydrogen bonding. Base pairing is generally not possible if DNA strands are modeled in a parallel arrangement.

The nucleobases in DNA are planar heteroaromatic structures arranged perpendicular to the phosphodiester backbone and parallel to each other like a stack of coins. **Purines** (adenine or guanine) pair with **pyrimidines** (thymine or cytosine). Under physiological conditions of salt (about 0.85% NaCl) and pH (pH 7.8), the hydrogen bonding of the bases permits adenine to pair only with thymine (via two hydrogen bonds) and likewise guanine with cytosine (via three hydrogen bonds). These complementary base interactions enable the two phosphodiester backbones to wrap around each other to form the classic double helix, or duplex.

The thousands of H-bonds that form between purines and pyrimidines along the interior of a DNA duplex (**Fig. 7.4**) make the bonding of the two complementary strands of DNA highly specific, so that a duplex forms only between complementary strands. Although H-bonds govern the specificity of strand pairing, the thermal stability of the helix is due predominantly to the stacking of the hydrophobic base pairs. Stacking of base pairs allows water and ions to interact with the hydrophilic, negatively charged phosphate backbone of DNA while avoiding the hydrophobic interior of the helix.

The stacking of DNA bases also appears to enable DNA to act as a "wire" capable of conducting electrons over long distances. The conduction of electrons along DNA can be sensed by certain proteins bound to DNA (discussed in Chapter 9).

In the space-filling model of DNA in **Figure 7.5A** and the contour map in **Figure 7.5B**, notice that the DNA double helix has grooves: a wide major groove and a narrower minor groove. The two grooves are generated by the angles at which the paired bases meet each other. These grooves provide DNA-binding proteins access to base sequences buried in the center of the molecule, so that the proteins can interact with the bases without the strands being separated. **Figure 7.6** shows an example of an important DNA-binding protein, DtxR of *Corynebacterium diphtheriae*, the cause of diphtheria. The DtxR dimer (pair of subunits) binds the major groove of DNA at the promoter for the diphtheria toxin gene (*dtxA*). Toxin expression is repressed in the presence of a high iron concentration, which signals to the bacteria that they are in an environment outside a human host. Gene regulation is discussed further in Chapter 10.

At high temperatures (50°C–90°C), the hydrogen bonds in DNA break and the duplex falls apart, or **denatures**, into two single strands. The temperature required to denature a DNA molecule depends on the GC/AT ratio of a sequence. More energy is required to break the three H-bonds of a GC base pair than the two H-bonds of an AT base pair. Thus, DNA with a high GC content requires

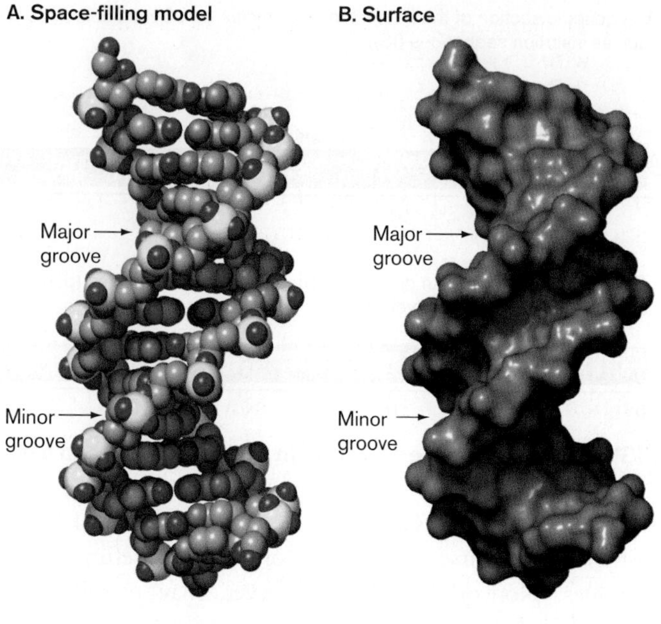

FIGURE 7.5 ■ Models of DNA. **A.** Space-filling model of DNA. **B.** DNA surface, modeled using nuclear magnetic resonance. (PDB code: 1K8J)

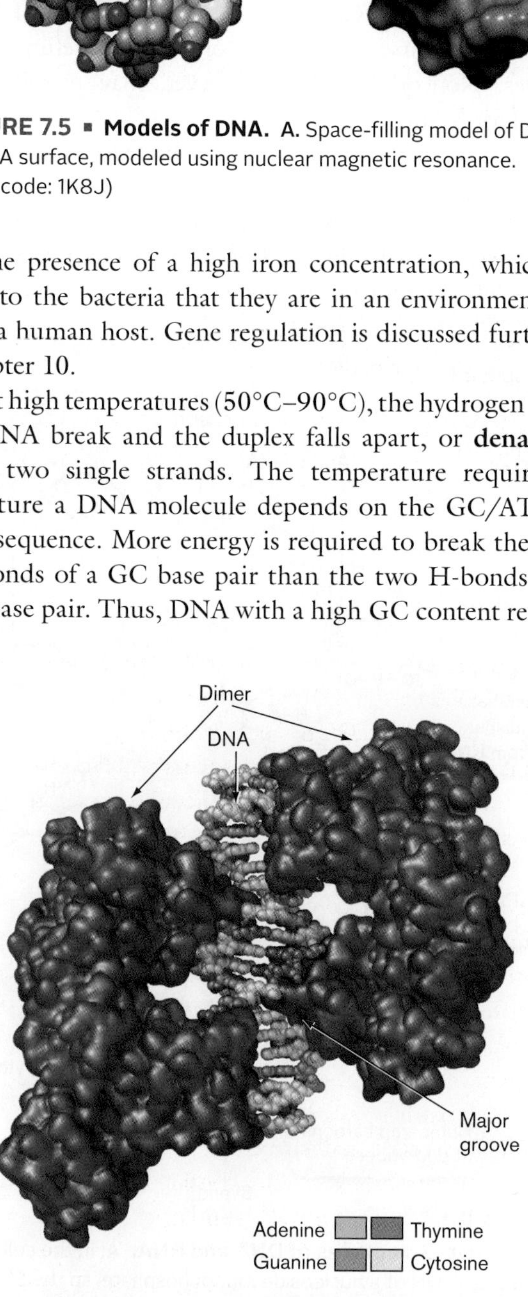

FIGURE 7.6 ■ Protein that recognizes DNA. DtxR repressor protein dimer binds in the major groove of DNA. (PDB code: 1COW)

a higher denaturing temperature than does similar-sized DNA with a lower GC content.

When DNA has been heated to the point of strand separation, lowering the temperature permits the two single strands to find each other and reanneal into a stable double helix. The kinetics of DNA renaturation is much slower than that of denaturation, because renaturation is a random, hit-or-miss process of complementary sequences finding each other. This melting/reannealing property of DNA is exploited in a number of molecular techniques (see the discussion of the polymerase chain reaction in Chapter 12). Note, however, that bacteria and archaea growing at extreme pH or temperature protect their DNA from denaturation through the use of remarkable DNA-binding proteins, such as the archaeal histone proteins Hmf and Htz. The role of DNA-binding proteins in microbial survival is the subject of considerable research.

Thought Questions

7.2 What do you think happens to two single-stranded DNA molecules isolated from different genes when they are mixed together at very high concentrations of salt? *Hint:* High salt concentrations favor bonding between hydrophobic groups.

7.3 How do the kinetics of denaturation and renaturation depend on DNA concentration?

RNA Differs from DNA

Why do cells make both RNA and DNA? As we learned in Chapter 3, DNA and RNA have different missions. The cell uses DNA to stably archive the information needed to make a functional cell. In fact, the high stability of DNA makes it possible for human engineers to record and store digital information on this biomolecule, much like a silicon-based hard drive. This novel use of DNA to archive data is discussed in **Special Topic 7.1**.

The growing cell continually accesses the information stored in DNA by making temporary copies of its genes in the form of RNA (ribonucleic acid) molecules that direct the synthesis of proteins. The cell also makes small RNA molecules that adjust information flow by modulating DNA gene expression. To keep their roles separate, DNA and RNA must have slightly different structures.

DNA in a cell usually consists of two complementary strands, whereas RNA usually consists of a single strand. DNA and RNA are chemically similar, except that in RNA the sugar ribose replaces deoxyribose and the pyrimidine base uracil replaces thymine (see **Fig. 7.4B**). Functionally, these two differences prevent enzymes meant to work on DNA, such as DNA polymerases, from acting on RNA. They also prevent RNA nucleases (RNases) from degrading DNA. However, uracil can base-pair with adenine, which means that hybrid RNA-DNA double-stranded molecules can form (hybridize) when base sequences are complementary. In fact, this **hybridization** is a necessary step in the decoding of genes to make proteins.

Although RNA molecules are commonly thought of as single-stranded, each single-stranded RNA molecule has regions that loop back to form "hairpins." Hairpin structures form when complementary nucleotide sequences within the primary RNA sequence bend back and hybridize. These double-stranded hairpins have a variety of biological functions.

Bacterial Chromosomes Are Compacted into a Nucleoid

The chromosome of *E. coli* has over 4.6 million bases in one strand, or over 9 million, counting both strands. This is a huge molecule. At the normal pH of the cell (7.8), all the phosphates in the backbone (all 9 million of them) are unprotonated and negatively charged, so this one molecule contributes greatly to the overall negative charge of the cytoplasm.

Figure 7.7 shows DNA spewing out of a damaged bacterial cell. Laid out, the chromosome is 1,500 times longer than the cell. It is obvious from this photomicrograph

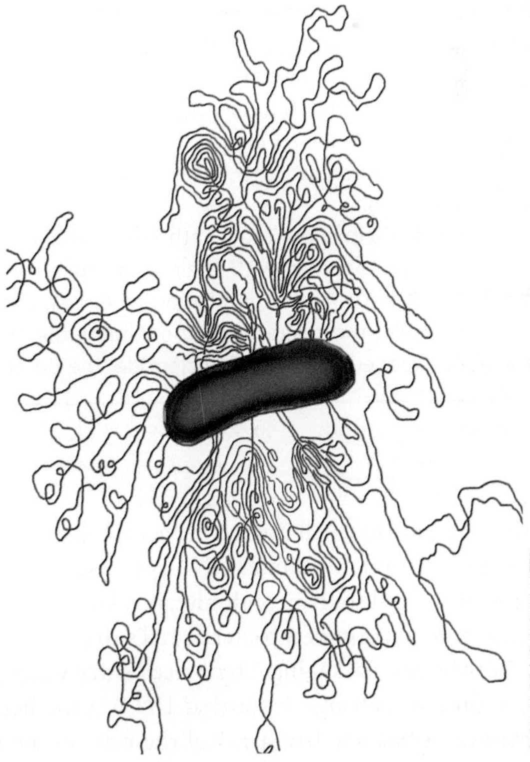

FIGURE 7.7 ■ Osmotically disrupted bacterial cell with its DNA released. Length of the bacterium is approximately 2 μm.

SPECIAL TOPIC 7.1 DNA as Digital Storage

Currently, this textbook is available in paper and silicon-based digital formats; in the near future, you may be able to access a version that is stored as DNA. This DNA copy would occupy a space not much larger than the head of a pin, and it could survive in legible condition for thousands of years or longer.

Just as the sequence of the four bases is used to code genes in living organisms, those bases can be exploited as digits for information storage. Each base can theoretically serve as 2 bits of digital information; for example, using the binary code of 0's and 1's: A = 00, C = 01, G = 10, and T = 11. To store data as DNA, the digital code is translated into a code involving the four nucleotides. The nucleotide sequence is then synthesized by machine to become the storage molecule. It can be stored as an isolated molecule or introduced into the genome of a microbial host cell. The DNA code is then read, usually with the polymerase chain reaction (PCR) to amplify the DNA, followed by sequencing of the DNA amplicons (see Chapter 12 for descriptions of these techniques). Finally, the sequences are decoded back to the digital code.

DNA has been hailed as a possible improvement to silicon-based storage for several key reasons. DNA is a very stable molecule and, under ideal conditions, can be preserved for centuries or longer, as evidenced by retrieval of DNA from ancient prokaryotes and Neanderthals in permafrost. This is a significant improvement over the durability of silicon-based storage, which is limited to decades or less. DNA has greater potential information density as well. A maximum of 455 exabytes (455×10^{18}) per gram has been estimated, which is orders of magnitude higher than silicon, tape, or optical storage capacities, in the range of terabytes to petabytes (10^{12}–10^{15} bytes). Finally, the decrease in costs for DNA synthesis and sequencing is outpacing the cost reduction for silicon-based digital storage, suggesting that it may soon be more cost-effective to store information on DNA than on silicon.

As a proof-of-concept study, Seth Shipman and George Church (**Fig. 1**), together with their colleagues, recorded a movie into the DNA of a population of *E. coli*, and then played it back. For the image pixels, the tones within the gray scale were encoded by a triplet nucleotide: the 64 possible triplets generated a redundant set to encode 21 different tones (**Fig. 2A**). The pixel locations within the image were determined by the position of the triplet along an oligonucleotide (**Fig. 2B**). Multiple oligonucleotides were needed to encode the entire image, and these different DNA molecules were tagged with unique bar codes (pixets; **Fig. 2B**). In total, five sets of oligonucleotides were synthesized, each encoding one frame of a five-frame movie of a galloping mare from Eadweard Muybridge's *Human and Animal Locomotion*.

FIGURE 1 ■ Seth Shipman (left) and George Church.

The CRISPR-Cas system (see Chapters 6, 9, and 12) was used to incorporate the oligonucleotides into the chromosome of *Escherichia coli*. Because new oligonucleotides are almost always added adjacent to the leader sequence of the CRISPR array, the researchers were able to load the frames sequentially into the *E. coli* genome. Thus, just as a camera records and stores images over time, they were able to "record" and store the images of a movie in DNA. To play back the video, the researchers used PCR to amplify the oligonucleotides within the population of *E. coli* cells, and high-throughput sequencing to obtain the nucleotide sequence, which was then back-translated into pixel information (**Fig. 2A** and **B**) to obtain the images (**Fig. 2C**).

that an intact, healthy cell must compact a huge bundle of DNA into a very small volume. DNA is the second-largest molecule in the cell (only peptidoglycan is larger) and constitutes a large portion of a bacterial cell's dry mass, about 3%–4%. Although packaging 3% of a cell's dry weight may not seem like a challenge, note that DNA is further confined only to ribosome-free areas of the cell, so the chromosome-packing density reaches about 15 mg/ml. In a test tube, DNA at 15 mg/ml is almost a gel, so how can anything move inside a cell? And how does all of this DNA keep from getting hopelessly tangled?

As introduced in Chapter 3, cells pack their DNA into a manageable form that still allows ready access to DNA-binding proteins. Although bacteria lack a nuclear membrane, they pack their DNA into a series of protein-attached domains collectively called the **nucleoid** (see Section 3.4). Unlike the compact nucleus of eukaryotes, the bacterial nucleoid is distributed throughout the cytoplasm.

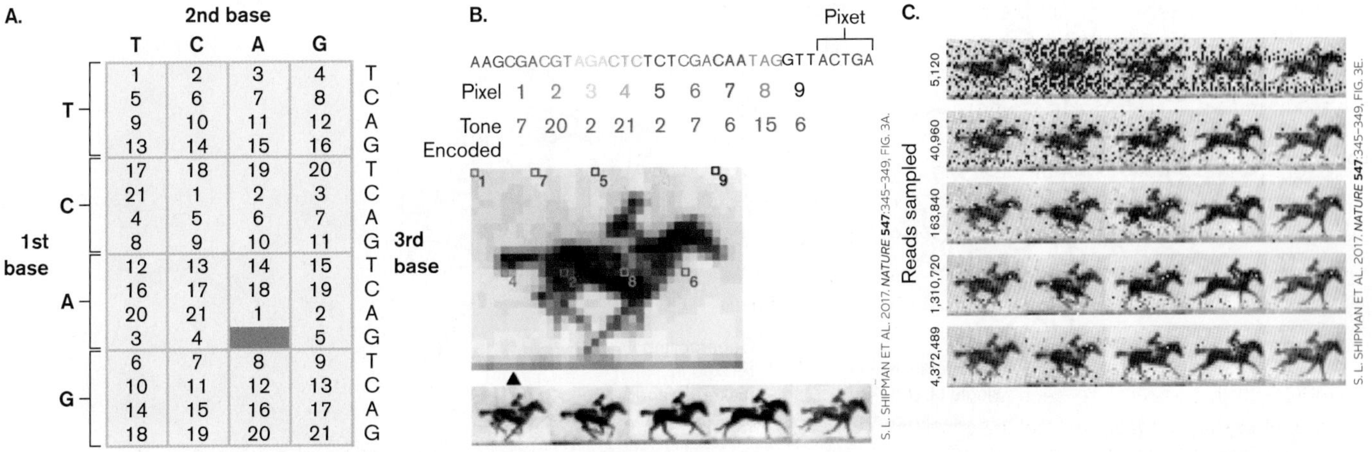

FIGURE 2 ■ Recording a movie as DNA in the *E. coli* genome. A. The triplet codes used to convert the 21 pixel tones (numbers 1–21) into a DNA sequence. **B.** A representative oligonucleotide, with its own unique identification tag (pixet), and the locations of the nine coded pixels on the encoded image. **C.** Playback of the DNA-encoded movie by increasing amounts of sequencing leads to greater recovery of the five movie frames. *Source:* Part A modified from S. L. Shipman et al. 2017. *Nature* **547**:345–349, fig. 1b.

Increasing the number of sequence reads led to more accurate recall of the images, though perfect recall was not achieved in this initial study.

For DNA to become a viable option for data storage, several practical limitations need to be dealt with. As indicated in the study just described, perfect data recall is a major challenge, because of technical aspects of DNA sequencing, and the potential for DNA mutation when encoded in a live organism. Recently, other groups have shown that adaptation of the error-correcting codes, such as the one used to prevent video dropouts during streaming, could provide error-free data recovery from DNA. The long recording time (weeks) and retrieval time (days) of current technologies limits the application of DNA as a rapid-retrieval storage system for daily use.

The high capacity and durability, but relatively long access time, currently make DNA storage most applicable to long-term archiving of information. This capability is highly relevant to the medical field, where secure, long-term storage of patient records is critical. There is even consideration of DNA as a means of safeguarding human knowledge via deep, ultracold storage in outer space, away from any future catastrophes on Earth.

RESEARCH QUESTION

The first genome to be fully sequenced was that of the *E. coli*–infecting virus phiX174. Shortly after the genome sequence was published in 1977, some scientists searched for possible messages from extraterrestrial beings embedded within the genetic code. This effort, while ultimately futile, inspired the concept that DNA could be used to communicate to alien life, if placed on our space probes the way the Golden Records were on *Voyager 1* and *2*. What are the possible limitations of such forms of biochemical communication, given the potentially different chemistries and biological origins on different planets?

Shipman, Seth L., J. Nivala, J. D. Macklis, and George M. Church. 2017. CRISPR-Cas encoding of a digital movie into the genomes of a population of living bacteria. *Nature* **547**:345–349.

DNA Supercoiling Compacts the Chromosome

A nucleoid gently released from *E. coli* appears as 30–100 tightly wound loops (**Fig. 7.8** ▶). The boundaries of each loop are defined by anchoring proteins called histone-like proteins for their similarity to histones, the DNA-binding proteins of eukaryotes. The double helix within each domain is itself helical, or supercoiled. The easiest way to envision supercoiling is to picture a coiled telephone cord. After much use, a phone cord twists, or supercoils, upon itself. Supercoiled phone cords are quite compact, taking up less space than a relaxed cord. Circular DNA works the same way, using supercoiling to tightly pack its chromosome.

Note that DNA cannot form supercoils unless its ends are tethered. In a circular chromosome, the DNA ends are tethered to each other. Introducing an extra twist by breaking one or both strands, twisting one end, and then resealing the strands means that the increased, or decreased,

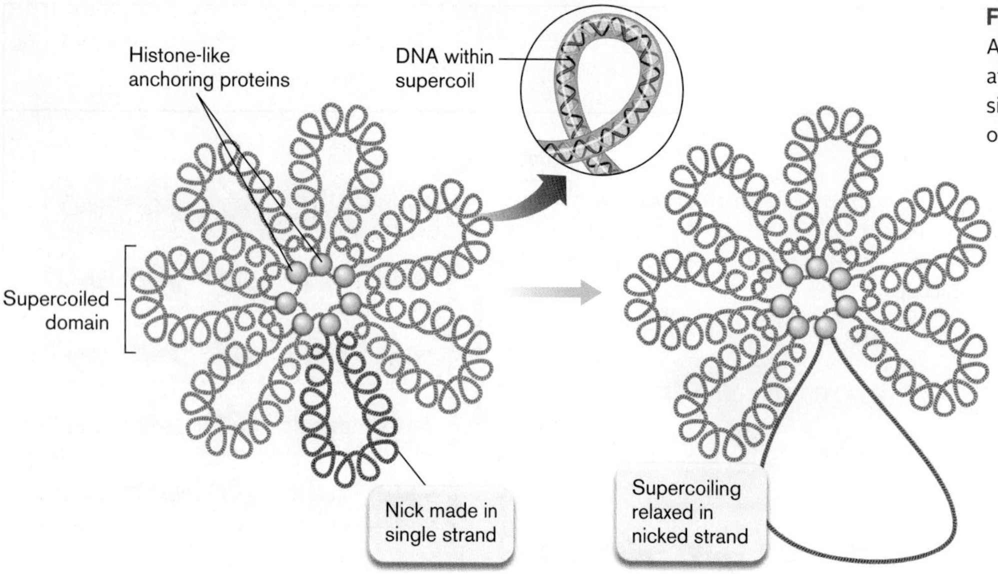

FIGURE 7.8 ▪ Bacterial nucleoid. A nucleoid, showing domain loops after gentle release from cells. The single-strand nick unwinds (relaxes) only one loop.

torsional (twisting) stress is trapped in the final circular molecule. It cannot then spontaneously unwind.

Remarkably, the nucleoid with its 30–100 loops (or "domains") can maintain different loops at different superhelical densities. The independence of supercoiled domains was demonstrated by introduction of one single-strand nick in the phosphodiester backbone of one domain (see **Fig. 7.8**). You can do this by adding very small amounts of a nuclease (an enzyme that cleaves a nucleic acid). The ends of the broken strand, driven by the energy inherent in the supercoil, rotate about the unbroken complementary strand of the duplex and relax the supercoil. A single nick in a genome, however, removes supercoils from only one domain. How is this possible if the chromosome is one circular molecule? The unaffected chromosomal domains remain supercoiled because they are constrained at their bases by anchoring proteins, such as HU and H-NS (histone-like proteins), that prevent rotation (for nucleoid organization, see Fig. 3.27).

How does DNA achieve the supercoiled state? The bacterial cell produces enzymes that can twist DNA into supercoils, and other enzymes that relieve supercoils. A single twist introduced into a small (300-bp) circular DNA molecule forms a single supercoil as shown in **Figure 7.9.** To introduce the DNA twist, a supercoiling enzyme makes a double-strand break at one point in the circle, passes another part of the DNA through the break, and reseals it. The result is the same as if one end of the broken circle were twisted one full turn.

To put this in context, most DNA in nature is right-handed. Right-handed, helical DNA turns clockwise when you look down the length of the double strand. Now think of a right-handed helical phone cord. Twist the cord so as to decrease the number of helix twists (underwinding the cord). As you keep underwinding, the torsional stress of

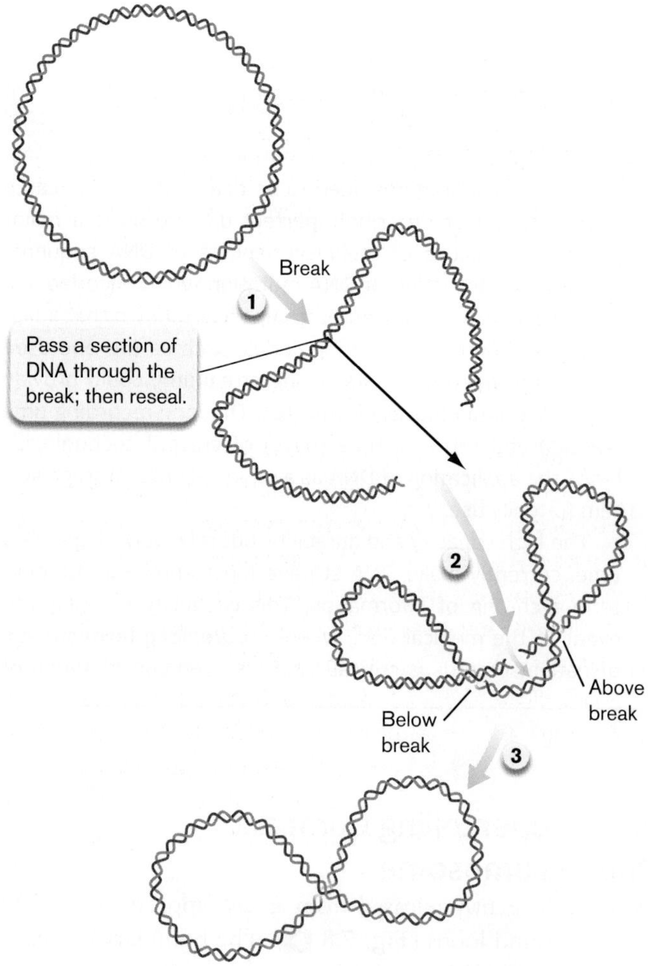

FIGURE 7.9 ▪ Supercoiling of 300-bp circular DNA. To introduce a supercoil into a double-stranded, circular DNA molecule, both strands are cleaved at one site in the molecule (step 1), an intact part of the molecule is passed between ends of the cut site (step 2), and the free ends are reconnected (step 3).

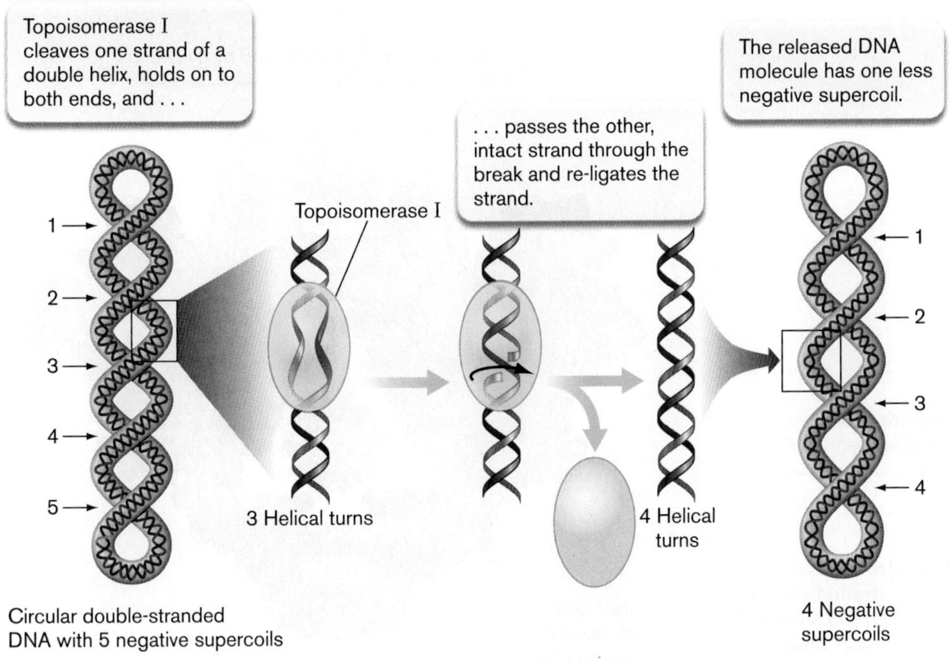

FIGURE 7.10 ■ Mechanism of action for type I topoisomerases (Topo I of *E. coli*). Topoisomerase I relaxes a negatively supercoiled DNA molecule by introducing a single-strand nick.

removing clockwise helical twists is relieved when the phone cord (or DNA) flips around itself (supercoils) in a clockwise rotation. Because the DNA is underwound, the compensatory supercoil is called a negative supercoil. Overwinding a clockwise helix (like DNA) introduces additional clockwise helical twists, the stress of which is relieved by the DNA supercoiling in a counterclockwise direction. Because the DNA helix is overwound, the compensatory supercoil is called a positive supercoil. Either positive or negative supercoils require energy to form, because the molecule is constrained and entropy is decreased.

The nucleoids of bacteria and most archaea, as well as the nuclear DNA of eukaryotes, are kept negatively supercoiled. Because the DNA is underwound, the two strands of negatively supercoiled DNA are easier to separate than those of positively supercoiled DNA. This is important for transcription enzymes, such as RNA polymerase, that must separate strands of DNA to make RNA.

Note, however, that some archaeal species living in acid at high temperature have nucleoids that are positively supercoiled to keep DNA double-stranded in these inhospitable environments (discussed next). Positively supercoiled DNA is harder to denature, because it takes excess energy to separate overwound DNA.

Topoisomerases Supercoil DNA

Supercoiling changes the topology of DNA. Topology is a description of how spatial features of an object are connected to each other. Thus, enzymes that change DNA supercoiling are called **topoisomerases**. To maintain proper DNA supercoiling levels, a cell must delicately balance the activities of two types of topoisomerases. Type I topoisomerases are typically single proteins that cleave only one strand of a double helix, while type II topoisomerases have multiple subunits that cleave both strands of a DNA molecule. Type I enzymes relieve or unwind supercoils. As shown in **Figure 7.10**, topoisomerase I cleaves one strand of a negatively supercoiled double helix and holds on to both ends of the break. The release of the energy stored in the negative supercoil allows the enzyme to pass the intact strand through the break and religate the strand, thereby reintroducing a helical turn. The molecule is released with one less negative supercoil.

The action of DNA gyrase is more involved (**Fig. 7.11**). DNA gyrase is an example of a type II topoisomerase whose function is to introduce negative supercoils in DNA (see **Fig. 7.11A**). Adding a supercoil requires spending energy, by hydrolysis of ATP. The active gyrase complex is a tetramer composed of two GyrA and two GyrB proteins. The gyrase B subunits first grab a section of the double helix. Then GyrA, in an ATP-dependent process catalyzed by GyrB, introduces a double-strand break, passes a different part of the double helix through the break, and reseals the break. The other end of GyrA then opens to release the DNA, now with one more negative supercoil. The 3D representation in **Figure 7.11B** shows DNA gyrase in the midst of generating a supercoil.

Enzymes that make or manage bacterial DNA, RNA, and proteins are common targets for antibiotics. For instance, the **quinolone** antibiotics specifically target bacterial type II topoisomerases. These antibiotics do not affect eukaryotic topoisomerases. A modern quinolone, ciprofloxacin, was the treatment of choice for anthrax pneumonia during the 2001 domestic terrorism attacks that used letters containing anthrax spores to kill 5 people and infect 17 others in the United States. The quinolones nalidixic and oxolinic acid inhibit DNA gyrase and were used to map the location of the gene that encodes the GyrA subunit in *E. coli*, *gyrA*. The modern successors of these drugs, the fluoroquinolones, are among the most widely used antimicrobials in the world. These drugs do not block topoisomerase action but stabilize the complex in which DNA gyrase is covalently attached to DNA (see **Fig. 7.11**). The stuck complex forms a physical barrier in front of the DNA replication complex, and the bacterial cell dies.

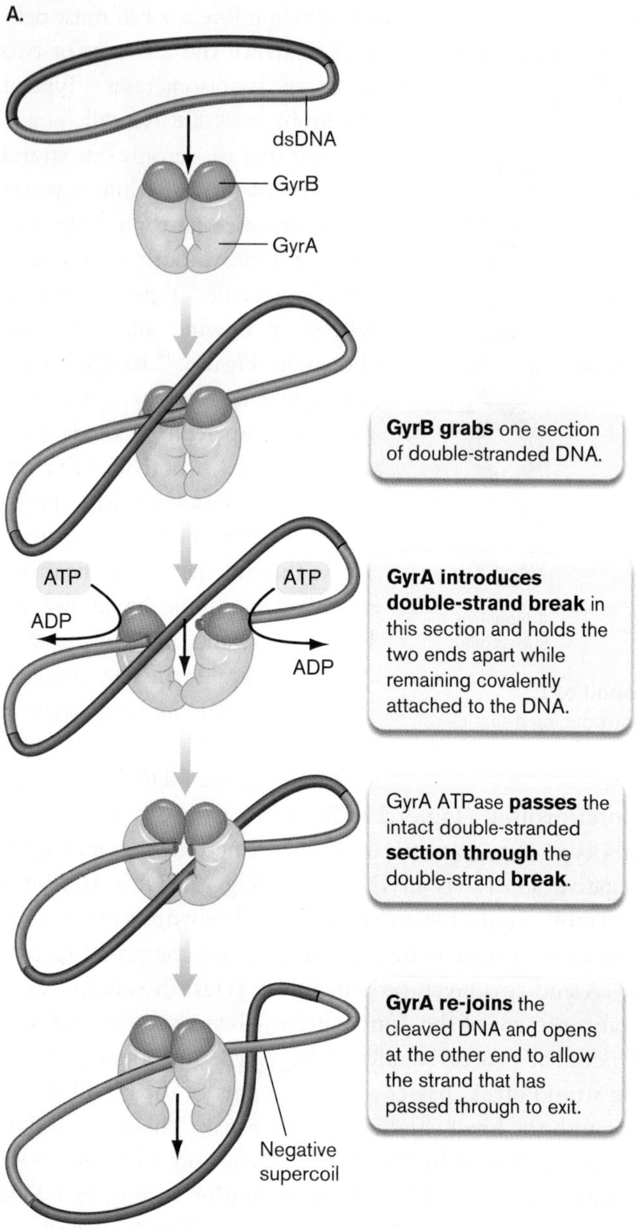

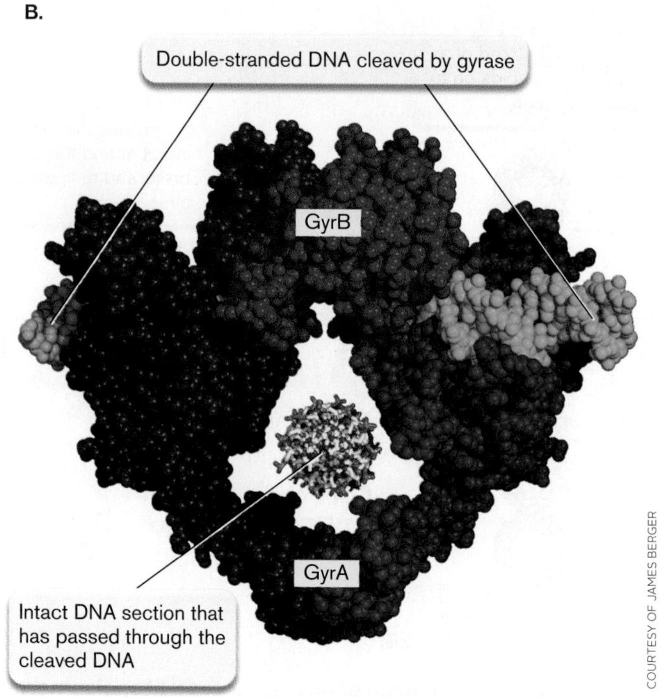

FIGURE 7.11 ■ Mechanism of action for type II topoisomerases. **A.** Mode of action of DNA gyrase from *E. coli*. **B.** Three-dimensional representation of DNA gyrase. The gyrase complex grips a broken DNA duplex (shown in green) and has transported a second duplex (the multicolored rosette) through the break.

Extreme thermophiles (hyperthermophilic archaea) possess an unusual gyrase called reverse DNA gyrase. In contrast to the DNA gyrase from mesophiles, reverse gyrase introduces positive supercoils into the chromosome. It is proposed that tightening the coil helps protect the chromosome against thermal denaturation. Because the DNA has extra turns, it takes more energy (heat) to separate the strands.

Thought Questions

7.4 DNA gyrase is essential to cell viability. Why, then, are nalidixic acid–resistant cells that contain mutations in *gyrA* still viable?

7.5 Bacterial cells contain many enzymes that can degrade linear DNA. How, then, do linear chromosomes in organisms like *Borrelia burgdorferi* (the causative agent in Lyme disease) avoid degradation?

To Summarize

- **Noncoding DNA** can constitute a large amount of a eukaryotic genome, while prokaryotes have very little noncoding DNA.
- **DNA is composed of two antiparallel chains** of purine and pyrimidine nucleotides in which phosphate links the 5′ carbon of one nucleotide with the 3′ carbon of its neighbor. The result is a double helix containing a deep major groove and a more shallow minor groove.
- **Hydrogen bonding and interactions between the stacked bases** hold together complementary strands of DNA.
- **Supercoiling** by topoisomerases compacts DNA into an organized nucleoid.
- Bacteria, eukaryotes, and most archaea possess **negatively supercoiled DNA**. Archaea living in extreme environments have **positively supercoiled genomes**.
- **Type I topoisomerases** cleave one strand of a DNA molecule and relieve supercoiling; **type II topoisomerases** cleave both strands of DNA and use ATP to introduce supercoils.

7.3 DNA Replication

How do bacteria replicate their DNA quickly and minimize errors? Quick and accurate replication of DNA can help a microorganism grow and compete with other species. Replication efficiency is one reason why bacterial pathogens such as *Salmonella* can cause disease so quickly after ingestion. In this respect, bacteria differ from multicellular organisms, which need to regulate cell division carefully within their tissues because unregulated growth within tissues leads to cancer. The process of bacterial replication involves over 20 proteins coming together in a complex machine. Operation of the replication complex is all the more remarkable, considering that some bacteria, such as thermophilic *Bacillus* species that live in hot springs, can double their population in less than 15 minutes.

The molecular details of bacterial DNA replication are important to understand because they provide targets for new antibiotics and tools for biotechnology, such as the polymerase chain reaction (PCR; see Chapter 12). In addition, the bacterial proteins of DNA repair have homologs in the human genome, defects in which produce inherited human diseases, such as xeroderma pigmentosum, that predispose the carrier to certain cancers.

Overview of Bacterial DNA Replication

To replicate a molecule containing millions of base pairs poses formidable challenges. How does replication begin and end? How is accuracy checked and maintained?

Semiconservative replication. Replication of cellular DNA is **semiconservative**, meaning that each daughter cell receives one parental strand and one newly synthesized strand (**Fig. 7.12**). At the **replication fork**, the advancing DNA synthesis machine separates the parental strands while extending the new, growing strands. The semiconservative mechanism provides a means for each daughter duplex to be checked for accuracy against its parental strand.

Enzymes that synthesize DNA or RNA can connect nucleotides only in a 5′-to-3′ direction (see **Fig. 7.4**). That is, the nucleic acid polymer elongates only at the 3′ end. A polymerase (a chain-lengthening enzyme complex) forms a phosphodiester bond between the 3′ end of the growing chain and the alpha-phosphate located at the 5′ end of an incoming nucleoside triphosphate. (The alpha-phosphate is the phosphate closest to the sugar.) This reaction releases the beta- and gamma-phosphates of the incoming nucleoside as a diphosphate molecule called pyrophosphate. Pyrophosphate is subsequently cleaved by pyrophosphatase; removal of this product of the polymerization reaction prevents that reaction from working in the reverse direction. Polymerization thus results in the eventual breaking of both phosphoryl bonds within the triphosphate moiety of the incoming nucleoside. Given the high cost for the formation of these phosphoryl bonds, polymerization is a very expensive, albeit essential, process for the cell.

The 5′-to-3′ enzymatic constraint of polymerization produces an interesting mechanistic puzzle: If polymerases can synthesize DNA only in a 5′-to-3′ direction and the two phosphodiester backbones of the double helix are antiparallel, then how are both strands of a moving replication fork synthesized simultaneously? One strand presents no problem, because it is synthesized in a 5′-to-3′ direction toward the fork, but synthesizing the other, new strand in a 5′-to-3′ direction would seem to dictate that it move away from the fork (**Fig. 7.12**). How is the cell able to copy both parental strands during semiconservative replication?

Note: The term "nucleoside triphosphate" (nucleobase-ribose condensed with three phosphoryl groups) is also commonly written "nucleotide triphosphate."

DNA replication proceeds in three phases: (1) initiation, which is the melting (unwinding) of the helix and the loading of the DNA polymerase enzyme complex; (2) elongation, which is the sequential addition of deoxyribonucleotides to a growing DNA chain, followed by proofreading; and finally (3) termination, in which the DNA duplex is completely duplicated, the negative supercoils are restored, and key sequences of new DNA are methylated. The basic process of chromosome replication is outlined in **Figure 7.13**.

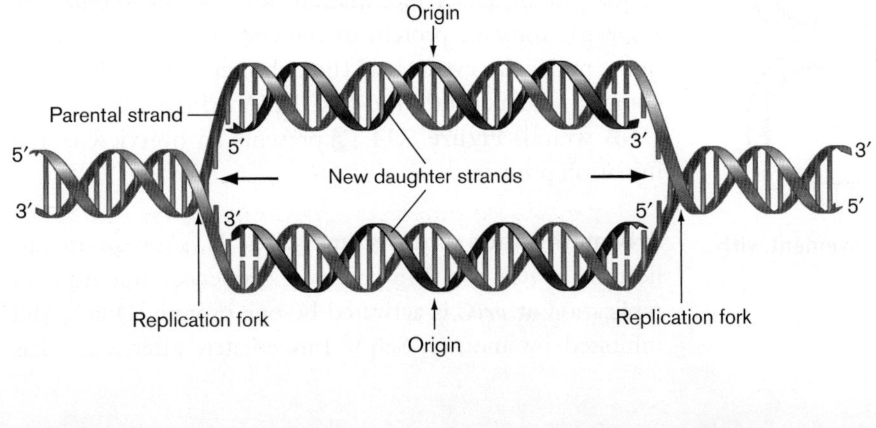

FIGURE 7.12 ■ Semiconservative replication. A replication bubble with two replication forks. DNA replication is termed "semiconservative" because each of the resulting double-stranded chromosomes contains one of the original parental strands and one newly synthesized strand. It is called "bidirectional" because it begins at a fixed origin and progresses in opposite directions.

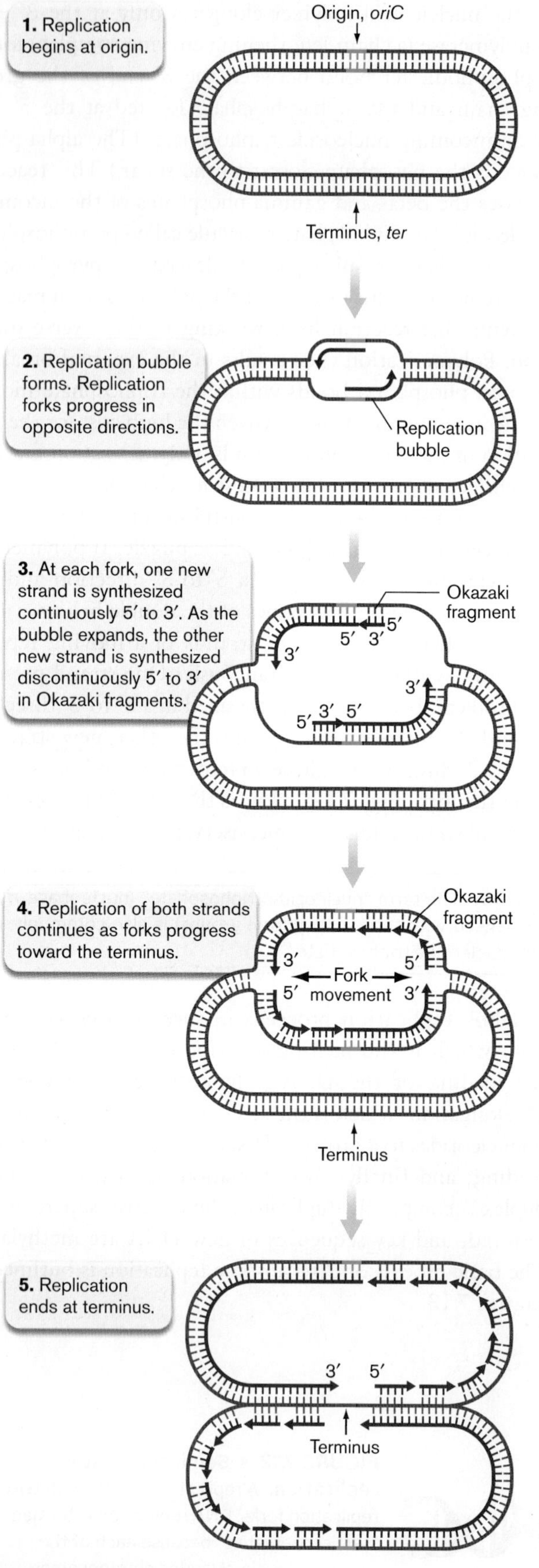

FIGURE 7.13 ■ Comparing direction of fork movement with direction of DNA synthesis.

Replication from a single origin. Replication in bacteria begins at a single defined DNA sequence called the **origin** (***oriC***; **Fig. 7.13**, step 1). Following initiation, a circular bacterial chromosome replicates bidirectionally (in both directions away from the origin) until it terminates at defined **termination** (***ter***) **sites** located on the opposite side of the molecule. Once the process has begun, the cell is committed to completing a full round of DNA synthesis. As a result, the decision of when to start copying the genome is critical. If that process starts too soon, the cell accumulates unneeded chromosome copies; if it starts too late, the dividing cell's septum severs the chromosome, killing both daughter cells. Consequently, elaborate fail-safe mechanisms link the initiation of DNA replication with cell mass, generation time, and cellular health, making the timing of initiation remarkably precise.

Fundamentals of DNA replication. After initiation of replication, a replication bubble forms at the origin. The bubble contains two replication forks that move in opposite directions around the chromosome (**Fig. 7.13**, step 2). DNA polymerases synthesize DNA in a 5′-to-3′ direction. At each fork, therefore, one new DNA strand can extend continuously until the terminus region (step 3). However, because the two DNA strands are antiparallel and the DNA polymerases synthesize only in the 5′-to-3′ direction, the other daughter strand has to be synthesized discontinuously, in stages—seemingly backward relative to the moving fork (step 4). The fragments of DNA formed on this discontinuously synthesized strand are called **Okazaki fragments**, after the Japanese scientists Reiji and Tsuneko Okazaki, a married couple, who discovered them. As we will discuss later, the Okazaki fragments are progressively stitched together to make a continuous, unbroken strand. Ultimately, the two replicating forks meet at the terminal sequence (step 5), and the two daughter chromosomes separate.

Now let's examine each step in molecular detail, to answer some important questions about this mechanism.

Initiating Replication

What determines when replication begins? Initiation is controlled by DNA methylation, and by the binding of a specific initiator protein to the origin sequence. Subsequent molecular events load the elaborate DNA polymerase complex and generate the first RNA primer for the new DNA strand. **Figure 7.14** ▶ presents an overview of the initiation process.

DNA methylation controls timing. The *E. coli* origin of replication (*oriC*) is a 245-base-pair sequence. Initiation of replication at *oriC* is activated by one protein, DnaA, and inhibited by another, SeqA. Immediately after a cell has

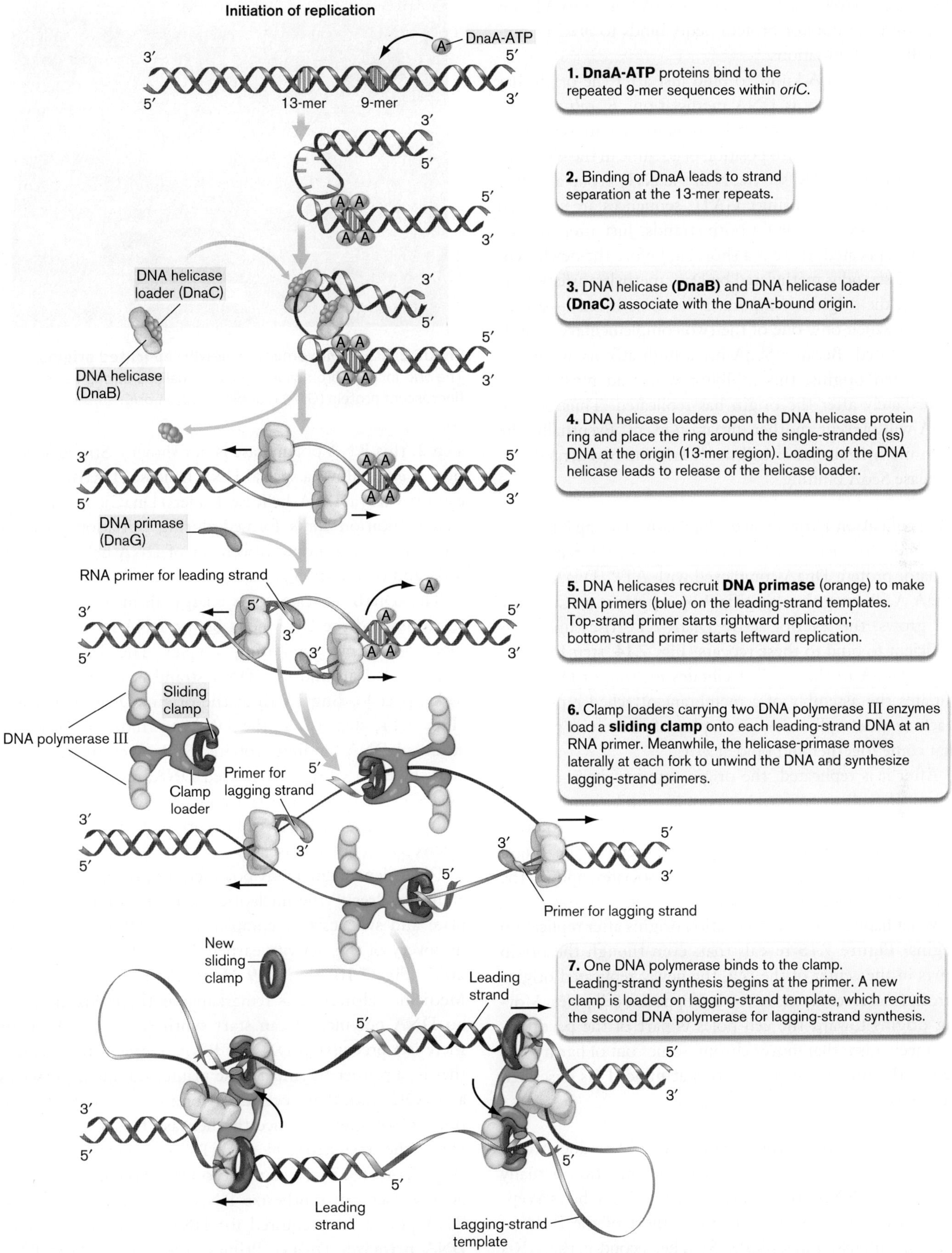

FIGURE 7.14 ■ **Initiation of DNA replication.**

divided, the level of active DnaA (DnaA bound to ATP) is low, and the inhibitor protein SeqA binds to *oriC* to prevent ill-timed initiations.

How does SeqA know to bind just after the origin has replicated? The key is DNA methylation. *E. coli* uses the enzyme DNA adenine methyltransferase (Dam) to attach a methyl group to the N-6 position of adenine in the sequence GATC (in **Fig. 7.4A**, see the N of the NH_2 group attached to the six-membered ring). GATC sequences are scattered along the chromosome on both strands. Just after the origin has replicated, there is a short lag before the newly synthesized strand is methylated by Dam methyltransferase. As a result, the origin is temporarily hemimethylated—a situation in which only one of the two complementary strands is methylated. Because SeqA has a high affinity for hemimethylated origins, this inhibitor will bind most tightly immediately after the origin has replicated. Thus bound, SeqA will prevent another initiation event. Eventually, the Dam methyltransferase will methylate the new strand and decrease SeqA binding.

The replication initiator protein, DnaA. Timing of initiation is determined by the concentration of the replication initiator protein DnaA complexed with ATP (DnaA-ATP). DnaA-ATP recognizes specific 9-bp repeats at *oriC*. As the cell grows, the level of active DnaA-ATP rises until it is sufficient to bind to these repeats (**Fig. 7.14**, step 1). Binding of DnaA to the origin facilitates melting of DNA and initiates the assembly of a membrane-attached replication machine (the replisome), a complex of numerous proteins that come together and bind at *oriC*.

After it is replicated, the origin cannot trigger another round of replication for two reasons: inhibition by SeqA and decreasing levels of unbound DnaA-ATP. Another round of replication can begin only after (1) the origin becomes fully methylated, (2) SeqA dissociates, and (3) the DnaA-ATP concentration rises.

What happens to new replication origins after replication begins? **Figure 7.15** reveals that, even though the origin starts in the center of the cell, the newly replicated origins (green fluorescence) move toward opposite cell poles. Moving origins toward the cell poles is part of the partitioning mechanism that moves chromosomes out of harm's way before the division septum forms at midcell (as described later in the chapter).

Initiation requires RNA polymerases. An unexpected feature of DNA replication is that its initiation actually requires two RNA polymerases. The first is the housekeeping RNA polymerase used to make most of the RNA in the cell (discussed in Chapter 8). The second is the DNA primase (discussed shortly).

The housekeeping RNA polymerase transcribes DNA at *oriC*, which helps separate the two DNA strands (**Fig. 7.14**, step 2; this RNA polymerase is not shown). Strand separation at *oriC* allows a special DNA helicase (DnaB), in association with a DNA helicase loader (DnaC), to bind the two replication forks formed during initiation (step 3). DnaC facilitates proper placement of DNA helicase at the fork, and then it disengages and leaves.

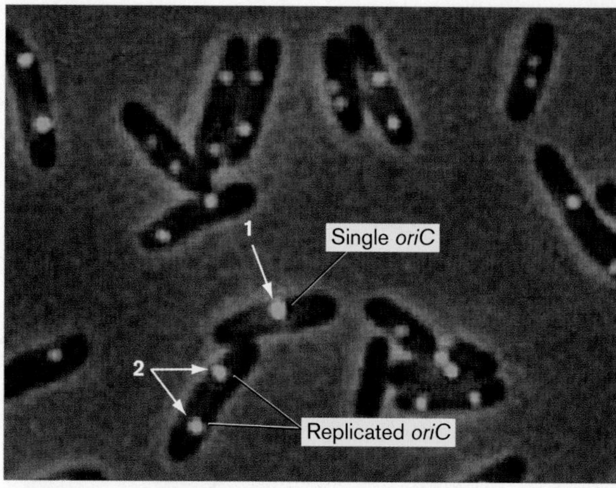

FIGURE 7.15 ■ Movement of newly replicated origins. The *oriC* loci on chromosomes were visualized through a green fluorescent protein (GFP) that binds only at *oriC*.

The DnaB helicase uses energy from ATP hydrolysis to unwind the DNA helix before DNA moves into the DNA polymerase replicating complex. The ringlike DnaB is assembled around one DNA strand at each replication fork. After loading DnaB at the origin, DnaC is released (**Fig. 7.14**, step 4). As the DNA unwinds, small single-stranded DNA-binding proteins (SSBs, seen in **Fig. 7.17**) coat the exposed single-stranded DNA, protecting it from nucleases patrolling the cell and preventing re-formation of double-stranded DNA. The origin is almost ready to receive DNA polymerase.

DNA-dependent DNA polymerases possess the unique ability to "read" the nucleotide sequence of a DNA template and synthesize a complementary DNA strand. The discovery of this activity earned Arthur Kornberg (1918–2007; **Fig. 7.16**) the 1959 Nobel Prize in Physiology or Medicine. However, as remarkable as these enzymes are, no DNA polymerase can start synthesizing DNA unless there is a preexisting DNA or RNA fragment to extend—that is, a primer fragment. The primer fragment possesses a 3′ OH end that receives incoming deoxyribonucleotides. Consequently, once the helicase DnaB is bound to DNA, the next step is to make RNA primers at each fork (**Fig. 7.14**, step 5). In contrast to DNA polymerases, RNA polymerases can synthesize RNA without a primer. The RNA polymerase required for DNA replication is called DNA **primase** (DnaG). Primase synthesizes short RNA primers (10–12 nucleotides) at the origin that can launch DNA replication. One primase is loaded at each of the two replication forks.

FIGURE 7.16 ■ Arthur Kornberg and Sylvy Kornberg, circa 1960. A biochemist in her own right, Sylvy Kornberg shared in the discoveries of DNA replication.

Why do DNA polymerases require RNA primers? RNA primers may be a holdover from the "RNA world" (an exciting model of molecular evolution, discussed in Chapter 17), when RNA served as the genetic material for life. In addition, the initial stages of nucleotide condensation involved in generating a primer are relatively inaccurate, and distinguishing these early polymers as RNA rather than DNA enables their targeted removal and replacement with DNA synthesized by a high-fidelity DNA polymerase. The mechanism of primer removal and replacement is described later in the chapter.

A sliding clamp tethers DNA polymerase to DNA. At this point, the DNA is almost ready for DNA polymerase. But first a **sliding clamp** protein (the beta subunit of DNA polymerase III) must be loaded to keep the DNA polymerase affixed to the DNA (**Fig. 7.14**, step 6). Without this clamp, DNA polymerase would frequently "fall off" the DNA molecule (see **eTopic 7.1**). A multisubunit complex (called the clamp-loading complex) places the beta clamp, along with an attached pair of DNA polymerase molecules, onto DNA. DNA polymerase (specifically DNA Pol III, discussed in the next section) then binds to the 3′ OH terminus of the primer RNA molecule and begins to synthesize new DNA (step 7). The molecular structure of the beta clamp loaded onto DNA is shown in **eTopic 7.1**.

Elongation of Replicating DNA

Escherichia coli contains five different DNA polymerase proteins, designated Pol I through Pol V. All DNA polymerases catalyze the synthesis of DNA in the 5′-to-3′ direction. However, only Pol III and Pol I participate directly in chromosome replication. The other polymerases conduct operations to rescue stalled replication forks and repair DNA damage.

DNA polymerase III. The main replication polymerase, Pol III, is a complex, multicomponent enzyme. Pol III was another discovery by Arthur Kornberg. The DNA synthesis activity of Pol III is held in the alpha subunit of the complex, while other subunits are used for improving fidelity (accuracy of replication) and processivity (a measure of how long the polymerase remains attached to, and replicates, a template). The Pol III epsilon subunit (DnaQ), for example, contains a **proofreading** activity that corrects mistakes and improves fidelity.

Proofreading activities within DNA polymerases scan for mispaired bases that have been mistakenly added to a growing chain. A mispaired base is more mobile than the correct base in a DNA molecule because a mispaired base does not properly hydrogen-bond to the template base. This motion halts DNA elongation by Pol III because the base is not properly positioned at the enzyme's active site. Stalling of Pol III activity triggers an intrinsic 3′-to-5′ **exonuclease** activity in the epsilon subunit. Exonucleases degrade DNA starting from either the 5′ end or the 3′ end, depending on the enzyme. The exonuclease activity of Pol III cleaves the phosphodiester bond, releasing the improperly paired base from the growing chain. Once the wayward base has been excised, Pol III can resume elongation.

Both DNA strands are elongated simultaneously. After initiation, each replication fork contains one elongating 5′-to-3′ strand, called the "leading strand" (look back at **Fig. 7.14**, step 7). But how is the opposite strand at each fork replicated? There are no known DNA polymerases capable of synthesizing DNA in the 3′-to-5′ direction, which would seem to be needed if both strands are to be synthesized simultaneously. DNA synthesis of one strand continuing all the way back to the origin is not a solution, because it would leave the unreplicated strand at each fork exposed to possible degradation for too long and would double the time needed to complete DNA replication.

The cell has solved this dilemma by coordinating the activity of two DNA Pol III enzymes in one complex—one for each strand. The two associated Pol III complexes, together with DNA primase and helicase, form the **replisome**. As the dsDNA unwinds at the fork, the problem strand loops out, and primase (DnaG) synthesizes a primer. The second Pol III enzyme binds to the primed section of the loop and synthesizes DNA in the 5′-to-3′ direction (imagine the lower template strand in **Figure 7.17** threading from left to right, through the polymerase ring). All the while, the second polymerase moves with the first polymerase (on the leading strand) toward the fork (**Fig. 7.17**, step 1).

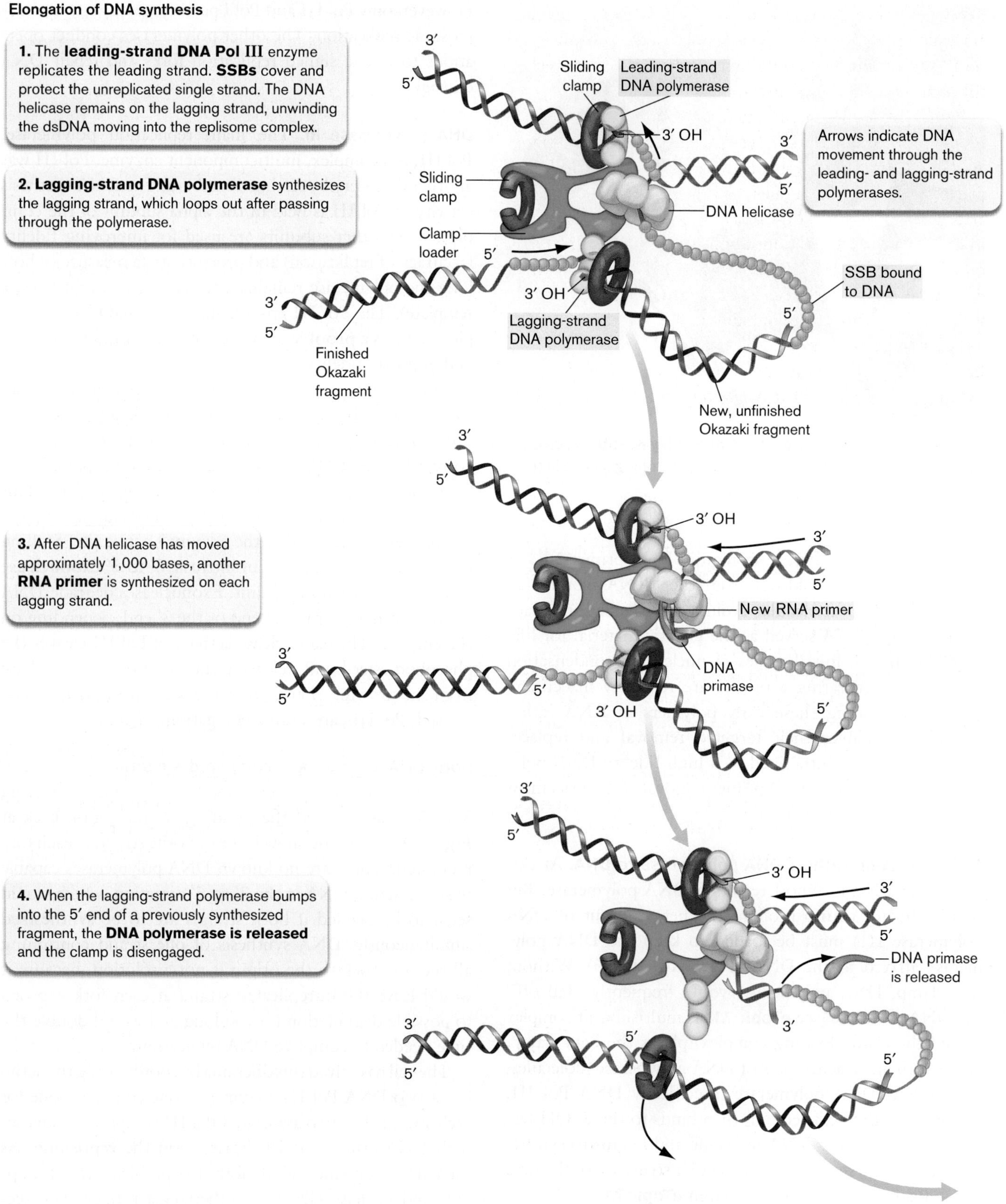

Elongation of DNA synthesis
1. The leading-strand DNA Pol III enzyme replicates the leading strand. SSBs cover and protect the unreplicated single strand. The DNA helicase remains on the lagging strand, unwinding the dsDNA moving into the replisome complex.
2. Lagging-strand DNA polymerase synthesizes the lagging strand, which loops out after passing through the polymerase.
3. After DNA helicase has moved approximately 1,000 bases, another RNA primer is synthesized on each lagging strand.
4. When the lagging-strand polymerase bumps into the 5′ end of a previously synthesized fragment, the DNA polymerase is released and the clamp is disengaged.
Arrows indicate DNA movement through the leading- and lagging-strand polymerases.
Sliding clamp
Leading-strand DNA polymerase
3′ OH
DNA helicase
Sliding clamp
Clamp loader
SSB bound to DNA
Lagging-strand DNA polymerase
Finished Okazaki fragment
New, unfinished Okazaki fragment
New RNA primer
DNA primase
DNA primase is released
3′
5′

FIGURE 7.17 ■ The DNA polymerase dimer acting at a replication fork. The leading and lagging strands are synthesized simultaneously in the 5′-to-3′ direction. For clarity, the beta clamp on the lagging strand is shown on the opposite side of Pol III as compared to its position on the leading strand.

Recent evidence indicates that the two *E. coli* replisomes move along DNA toward opposite poles of the cell, but generally stay within the middle third of the cell. They eventually meet again at the terminator located at midcell.

Note that simultaneous extension of the two strands at a single fork requires that synthesis of the looped strand must lag behind synthesis of the leading strand, and also that new RNA primers must be synthesized by DNA primase (DnaG) every 1,000 bases or so. Thus, the lagging strand is synthesized discontinuously, in pieces called Okazaki fragments, while the leading strand can be synthesized continuously. As the leading strand moves forward, advancing the fork, there remains a long stretch of lagging strand complementary to the already replicated leading strand. This lagging strand is single-stranded but protected by single-stranded DNA-binding proteins (SSBs; **Fig. 7.17**, step 2).

After about 1,000 bases, DNA primase reenters and synthesizes a new RNA primer in anticipation of lagging-strand DNA synthesis (step 3). At some point, the lagging-strand polymerase bumps into the 5′ end of the previously synthesized fragment. This interaction causes DNA polymerase to disengage from that strand (step 4), and the clamp loader loads a new clamp near the new RNA primer (step 5). The DNA polymerase binds to that clamp and begins synthesizing another Okazaki fragment (step 6).

The model just presented assumes that the replisome contains two DNA polymerase III molecules. However, the replisome actually consists of three polymerases—one on the leading strand and two on the lagging strand. The second polymerase on the lagging strand comes into play only when a large gap of unreplicated DNA remains on the lagging strand. For simplicity, the third polymerase is not included in the model shown in **Figure 7.17**.

DNA polymerase I. Discontinuous DNA synthesis results in a daughter strand containing long stretches of DNA punctuated by tiny patches of RNA primers. This RNA must be replaced with DNA to maintain chromosome integrity. To remove the RNA, cells typically use the 5′-to-3′ exonuclease activity of Pol I or an RNase enzyme specific for RNA-DNA hybrid molecules (called RNase H). A DNA Pol I enzyme then synthesizes a DNA patch using the 3′ OH end of the preexisting DNA fragment as a priming site (**Fig. 7.18**). When DNA Pol I reaches the next fragment, the enzyme removes the 5′ nucleotide and resynthesizes it. This process of replicating DNA increases accuracy and decreases mutations.

Once DNA Pol I stops synthesizing, it cannot join the 3′ OH of the last added nucleotide with the 5′ phosphate of the abutting fragment. The resulting nick in the phosphodiester backbone is repaired by **DNA ligase**, which in *E. coli* and many other bacteria uses energy gained by cleaving nicotinamide adenine dinucleotide (NAD) to form the phosphodiester bond (see **Fig. 7.18**). NAD is not used in its usual way, as a reductant that oxidizes substrates. Energy inherent in the diphosphate bond of NAD is captured upon cleavage by DNA ligase and used to re-join the 3′ OH and 5′ phosphate ends present at the nick. DNA ligase from eukaryotes and some other microbes uses ATP rather than NAD in this capacity.

DNA replication generates supercoils. As template DNA is threaded through the replisome, the helicase continually pulls apart the two strands of the DNA helix. As a result, the DNA ahead of the fork twists, introducing positive supercoils. (Try this yourself: Twist two pieces of string together, staple one end of the duplex to a piece of cardboard, and then pull the two strands apart from the free

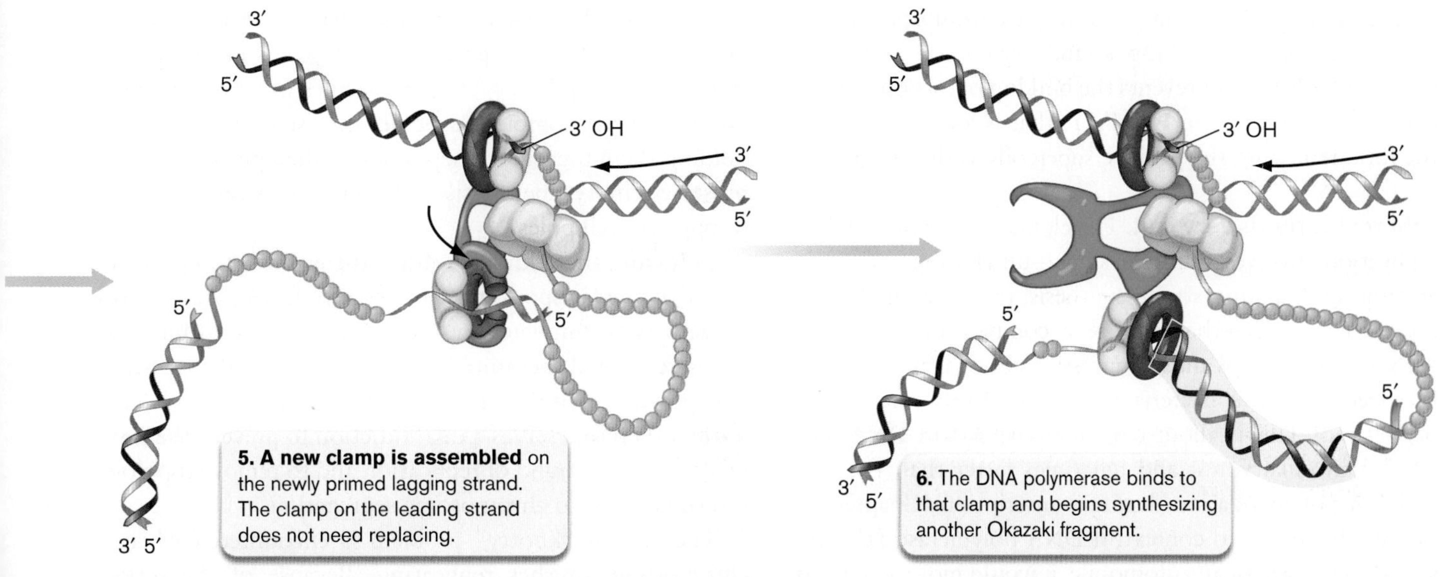
3′
5′
3′ OH
3′
5′
5′
5′
3′ 5′
5. A new clamp is assembled on the newly primed lagging strand. The clamp on the leading strand does not need replacing.
3′
5′
3′ OH
3′
5′
5′
5′
3′
5′
6. The DNA polymerase binds to that clamp and begins synthesizing another Okazaki fragment.

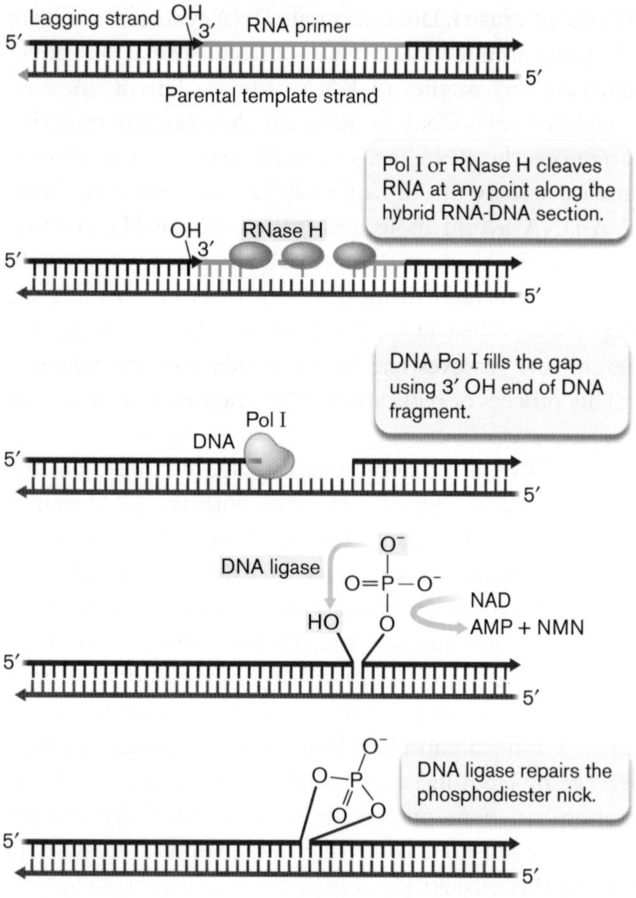

FIGURE 7.18 ▪ Removing the RNA primer. The 3′-to-5′ exonuclease activity of RNase H or the 5′-to-3′ exonuclease activity of Pol I cleaves the RNA primer (blue). In either case, DNA polymerase I uses the preexisting 3′ OH end of the DNA fragment to fill the gap. Finally, DNA ligase repairs the phosphodiester nick, using energy derived from the cleavage of NAD. AMP = adenosine monophosphate; NMN = nicotinamide monophosphate.

end. Notice the supercoiling that takes place beyond, or downstream of, the moving fork.)

The increasing torsional stress in the chromosome could stop replication by making strand separation more and more difficult. What prevents the buildup of torsional stress is the DNA gyrase (see **Fig. 7.11**) that is located ahead of the fork, removing the positive supercoils as they form.

DNA replication is very fast. The elongation phase of DNA replication involves multiple steps—unwinding, priming, leading- and lagging-strand synthesis, proofreading, relaxing of supercoils—that require a complex machinery to accomplish. Perhaps most impressive about this process is its speed, which in bacteria approaches 1,000 nucleotides per second! This is about ten times faster than DNA replication in eukaryotes, and ten times faster than the rate of RNA polymerization during transcription (Chapter 8). To put this rate into context, if DNA polymerase III were scaled to the size of an automobile, it would move at a speed of about 375 miles per hour. This is faster than the fastest recorded speed of a 10,000-horsepower dragster at the end of a quarter-mile race (330 miles per hour, as of 2017). The dragster's task is simply to move from point A to point B, whereas the polymerase must make a (near-perfect) copy of DNA at the same time it moves.

Terminating Replication and Segregating Sister Chromosomes

Bidirectional replication of a circular bacterial chromosome results in the two replication forks trying to replicate through the same DNA sequences 180° from the origin—that is, halfway around the chromosome. What tells the polymerases to stop? The *E. coli* chromosome has as many as ten terminator sequences (*ter*) that polymerases enter but rarely, if ever, leave (**Fig. 7.19A**). One set of terminators deals solely with the clockwise-replicating polymerase, while the other set halts DNA polymerases replicating counterclockwise relative to the origin. Which terminator site is used depends in part on whether replication of one fork has lagged behind replication of the other. A protein called Tus (terminus utilization substance) binds to these sequences and ensures that the polymerase complex does not escape and continue replicating DNA. The process by which the two ends of the circle are joined and the two replisomes are removed from the chromosome is still under investigation.

Thought Question

7.6 Reexamine **Figure 7.15**. If you GFP-tagged a protein bound to the *ter* region, where would you expect fluorescence to appear in the cell with two origins?

The structure of circular chromosomes poses "knotty" problems for their segregation after replication. These problems are solved by special enzymes that cut and re-join the DNA strands. The first problem develops as soon as replication begins. Replicated DNA molecules occasionally form knots between homologous genes on sister chromosomes, holding them together. These knots, called **pre-catenanes**, must be removed before sister chromosomes can segregate to opposite cell poles.

Resolving these knots requires the enzyme topoisomerase IV, a type II topoisomerase similar to DNA gyrase. Topo IV activity is stimulated by SeqA, the protein that binds to newly replicated, hemimethylated DNA. Recall that SeqA also prevents reinitiation at newly replicated origins. This protein therefore serves a dual function to prevent the start of a premature round of replication and to remove topological constraints to chromosome segregation.

The second "knotty" problem is encountered when a chromosome finishes replicating. Because of the topology of the chromosome, the two daughter molecules will

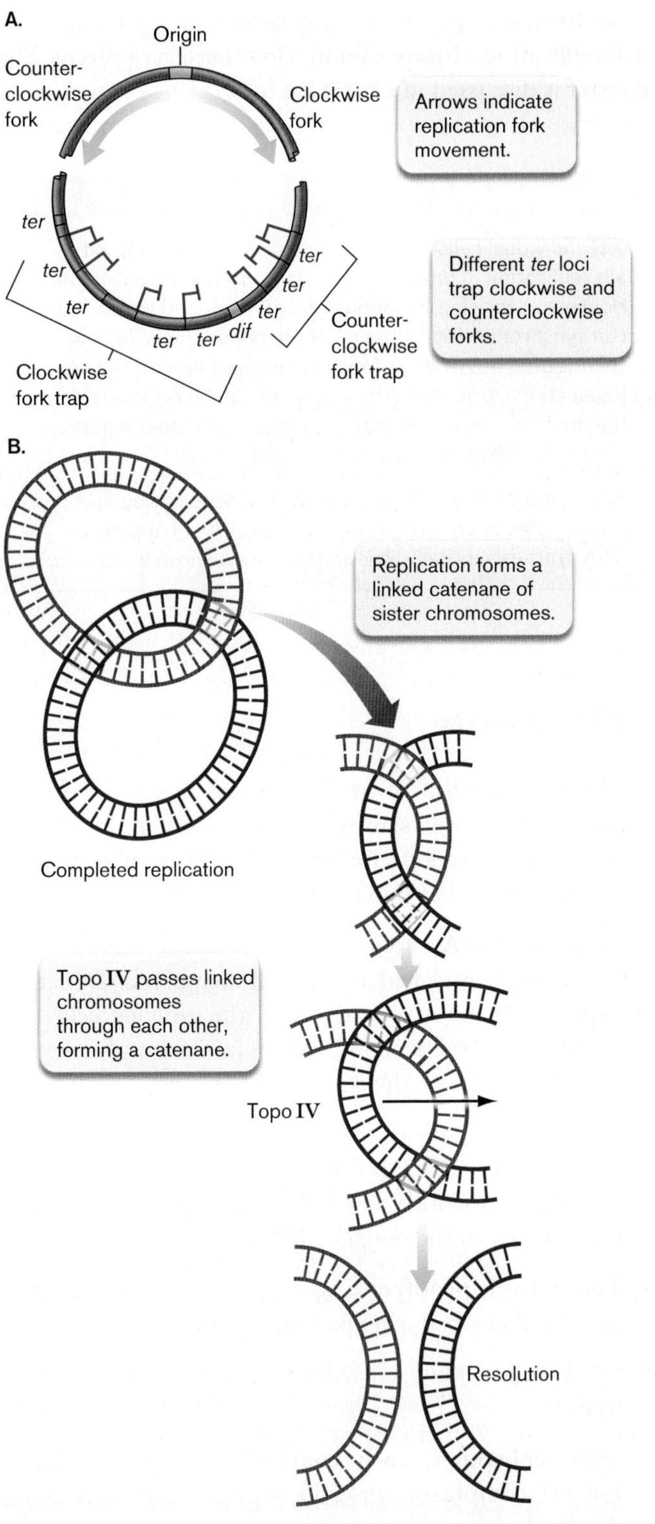

FIGURE 7.19 ■ Terminating replication of the chromosome. **A.** Terminator regions for DNA replication on the *E. coli* chromosome. The role of the *dif* site is described in Fig. 7.20. **B.** Resolution of DNA replication catenanes by Topo IV.

appear as a **catenane**, a pair of linked rings. The rings must be unlinked so that sister chromosomes can segregate at termination (**Fig. 7.19B**). The bacterial cell uses Topo IV at the site of fork convergence to resolve this final knot and release the sister chromosomes.

A third challenge for microbes with circular chromosomes is to resolve **homologous recombination** events that occur along the duplicated regions of the chromosome. Homologous, or general, recombination events (covered in detail in Chapter 9), involve the exchange of strands between two molecules of DNA; exchange occurs within extensive regions of identical or nearly identical sequence. Since duplicated chromosomes are essentially identical, crossovers between these DNA molecules are common.

If replication terminates with an odd number of recombination events between the sister chromosomes, they end up as an end-to-end two-chromosome dimer (**Fig. 7.20**, step 1). To resolve this dimer, a final recombination event is facilitated by proteins called XerC and XerD, which function at a specific 28-bp site, called *dif*, located in the terminus region of the chromosome. XerC and XerD are activated at the *dif* locus by the DNA translocase FtsK. FtsK monomers assemble as a ring structure around the DNA molecule, and this ring translocates along the DNA to the *dif* locus by following special "arrows," sequences known as KOPS (FtsK-orienting polar sequences) that are positioned along the chromosome (step 2). The polar orientations of the KOPS are opposite on the two replicating arms of the chromosome, such that the FtsK ring structure is directed to the *dif* site no matter which arm it assembles on.

Once at the *dif* site, FtsK activates XerC and XerD, which catalyze a second homologous recombination event (**Fig. 7.20**, step 3) that resolves the dimer into two separate chromosomes (step 4). Note that the two sister chromosomes have exchanged the DNA between the recombination sites, but unless mutation occurred during replication, these exchanged regions of DNA are identical.

FtsK is an important protein for the termination of chromosome replication: it activates not only XerC and XerD to resolve dimers but also Topo IV to resolve pre-catenanes. And using KOPS to act as a pump, FtsK can send DNA from midcell toward both poles at exceptionally high rates (1,700–16,000 base pairs per second!), thus ensuring that the sister chromosomes partition to the two halves of the cell before the septum forms at midcell.

Thought Question

7.7 Would you expect to find genes encoding Topo IV, XerC, and XerD in prokaryotes with exclusively linear chromosomes? Why or why not?

Once replicated, how do sister chromosomes move to opposite poles of a bacterial cell? In some species, such as *Bacillus subtilis*, a mechanical segregation system consisting of the proteins ParA and ParB and the DNA-binding

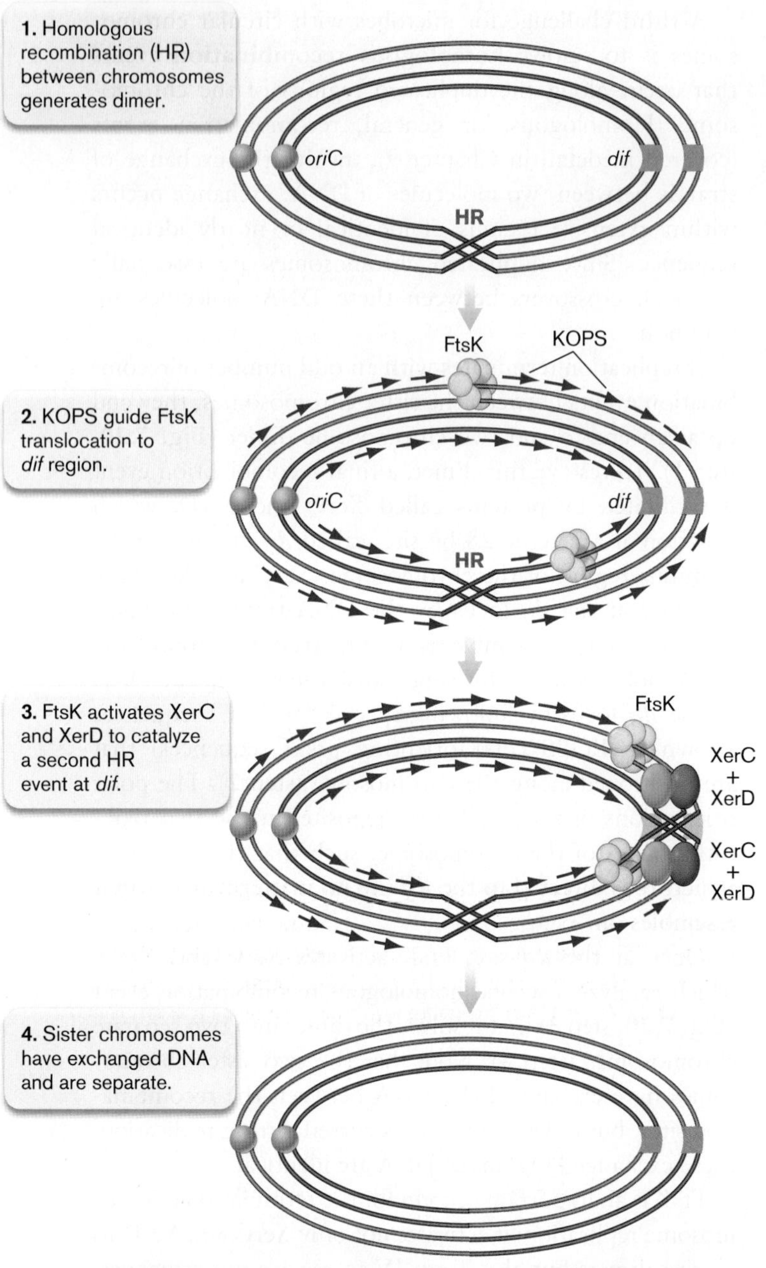

FIGURE 7.20 ▪ Resolution of chromosome dimers by XerC and XerD at the *dif* locus. The initial homologous recombination (HR) event results in the dimer, and resolution at the *dif* site results in separated chromosomes that have exchanged a segment of their DNA.

site *parS* pulls the chromosomes toward different ends of the cell. *E. coli*, however, does not possess a *parABS* system for segregating sister chromosomes. It seems that the process of replication itself, which takes place mostly in the middle third of the cell (see Fig. 3.28), propels *E. coli* chromosomes toward opposite cell poles. The terminator region remains in the center of the cell until after it is replicated. The replicating chromosome, then, takes on the appearance of a butterfly with its wings outstretched toward either cell pole. This configuration places the genome in a vulnerable position. A bacterial cell must avoid making a cell division septum too early or risk slicing through its still-replicating chromosomes. How bacteria prevent this disaster is discussed in **eTopic 7.2.**

Thought Questions

7.8 Individual cells in a population of *E. coli* typically initiate replication at different times (asynchronous replication). However, depriving the population of a required amino acid can synchronize reproduction of the population. Ongoing rounds of DNA synthesis finish, but new rounds do not begin. Replication stops until the amino acid is once again added to the medium—an action that triggers simultaneous initiation in all cells. Why is this replication synchronized?

7.9 The antibiotic rifampin inhibits transcription by RNA polymerase, but not by primase (DnaG). What happens to DNA synthesis if rifampin is added to a synchronous culture?

To Summarize

- **DNA replication of most genomes is semiconservative**, with newly synthesized strands lengthening in a 5′-to-3′ direction. Replication consists of initiation, elongation, and termination steps.
- **Bacterial DNA replication is initiated** from a fixed DNA origin attached to the cell membrane. Initiation depends on the mass and size of the growing cell. It is controlled by the accumulation of initiator and repressor proteins and by methylation at the origin.
- **During elongation**, primase (DnaG) lays down an RNA primer, DNA polymerase III acts as a dimer at each replication fork, and a sliding clamp keeps DNA Pol III attached to the template DNA molecule.
- **The 3′-to-5′ proofreading** activity of Pol III corrects accidental errors during polymerization.
- **DNA ligase** joins Okazaki fragments in the lagging strand.
- **Two replisomes, each containing a pair of DNA Pol III complexes**, move in opposite directions along the DNA.
- **Termination** involves stopping replication forks halfway around the chromosome at *ter* sites that inhibit helicase (DnaB) activity.
- **Ringed catenanes** formed at the completion of replication are separated by topoisomerase IV.
- **Chromosome dimers** are resolved by proteins XerC and XerD at the *dif* site.

7.4 Plasmids and Secondary Chromosomes

In this section we discuss genomes that are split into more than one replicating DNA molecule. The additional elements, plasmids and secondary chromosomes, complement the primary chromosome by adding various types of genes to the genome. Some microbes have multiple plasmids or secondary chromosomes that contribute to their genomes (**Table 7.1**). These DNA elements vary in size and genetic composition. While some coordination with the primary chromosome can exist, plasmids and secondary chromosomes control their own replication and copy number. Some plasmids can be lost during cell division, and some can be shared between distantly related species. This dynamic nature of plasmids challenges the notion that a microbial species is defined by a unique genome shared by all members of that species.

Plasmids Vary in Size, Copy Number, and Cargo

Plasmids are extrachromosomal elements found in bacteria, archaea, and eukaryotic microbes. Plasmids are usually circular and negatively supercoiled. They are typically much smaller than chromosomes (several thousand compared to several million base pairs) and may encode only a few genes (**Fig. 7.21**). Copy number varies widely among plasmid types, from a single copy to over 500 per cell. Genes found in high-copy-number plasmids can be dramatically overexpressed relative to genes found on chromosomes and low-copy-number plasmids, and this overexpression can be advantageous if a high concentration of the gene product is important for the gene's function in the cell.

The genes that plasmids carry are not essential for the basic functions involved in cell growth and metabolism, but they can play critical roles in certain situations. One important class of genes that some plasmids carry confers resistance to antibiotics (**Fig. 7.21B**; discussed in Chapter 27). Antibiotic resistance plasmids benefit bacteria, but they are a major problem for modern hospitals, where plasmids carrying multiple drug resistance genes are transmitted from harmless bacteria into pathogens. On the other hand, plasmids like pBR322 that contain drug resistance genes are the workhorses of genetic technology and have benefited society tremendously.

Other kinds of host survival genes carried by plasmids include genes providing resistance to toxic metals, genes encoding toxins that aid pathogenesis, and genes encoding proteins that enable symbiosis. These are discussed in later chapters. Most of the genes involved in the nitrogen-fixing symbiosis of *Rhizobium*, for example, are plasmid-borne.

A.

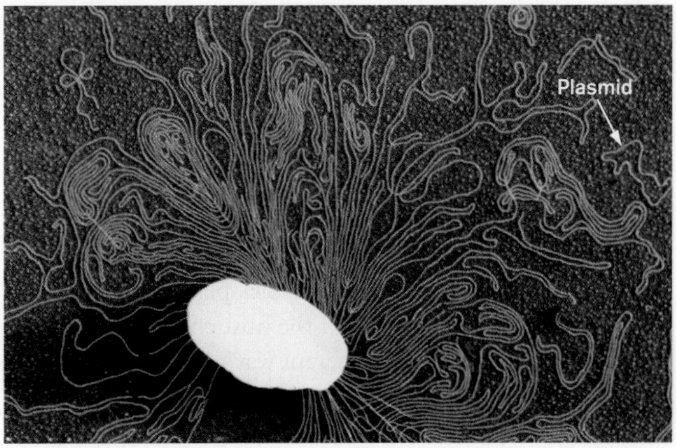

B.

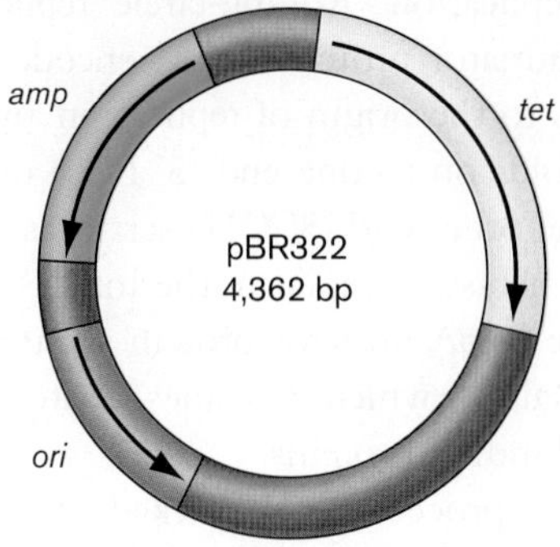

FIGURE 7.21 ▪ Plasmid map. A. Note the huge difference in size between a circular plasmid DNA molecule (arrow) and chromosomal DNA after both are gently released from a rod-shaped cell approx. 1 μm in length. **B.** Map of plasmid pBR322. This plasmid contains an origin of replication (*ori*) and genes encoding resistance to ampicillin (*amp*) and to tetracycline (*tet*).

Far from being freeloaders, plasmids often contribute significantly to the physiology of an organism.

Plasmids Are Transmitted between Cells

Plasmids have played a pivotal role in evolution because they are easily passed among bacterial species. This movement of plasmids is particularly evident today because of the rapid spread of antibiotic resistance, whose genetic basis is often a plasmid-encoded gene.

Some plasmids are self-transferable via conjugation, a process that requires cell-to-cell contact to move the plasmid from a donor cell to a recipient (as discussed in Section 9.3). Other plasmids are incapable of conjugation (nontransmissible). A third group can be transferred only if a self-transferable plasmid resides in the same cell. In this case, the conjugation mechanism produced by one plasmid will act on the other plasmid. Any plasmid released from dead cells can also be taken up by some bacteria in a process called transformation (see Section 9.3). Finally, plasmids can be transmitted in nature by accidentally being packaged into bacteriophage head coats—in other words, by bacteriophage transduction (also covered in Section 9.3).

Plasmids Regulate Their Replication

Plasmids may "borrow" the replication machinery of the host chromosome, but they have their own origin sequences and initiator proteins to control the frequency of replication. This autonomy enables plasmids to regulate copy number independently of the host chromosome. Plasmids can replicate in two different ways: by rolling-circle or bidirectional replication.

Used by some plasmids, rolling-circle replication (**Fig. 7.22**) is unidirectional, not bidirectional like chromosomal replication. Rolling-circle replication initiates when the initiator protein RepA, encoded by a plasmid gene, binds to the origin of replication and nicks one strand. RepA holds on to one end (5′ PO_4) of the nicked strand, while the other end (3′ OH) serves as a primer for host DNA polymerase to replicate the intact, complementary strand. The RepA initiator protein recruits a helicase that unwinds DNA, which becomes coated by single-stranded DNA-binding proteins.

As replication proceeds, the nicked strand progressively peels off without replicating, until the strand is completely displaced. Then the two ends of the displaced but nicked single strand are re-joined by the RepA protein and released. The single-stranded circular molecule is protected by host single-stranded DNA-binding proteins until host enzymes replicate a complementary strand and regenerate a double-stranded molecule.

Bidirectional replication of plasmids proceeds from a single origin, and both forks terminate at a single termination site, in a manner similar to initiation and termination of the bacterial chromosome. The molecular details of replication initiation of these plasmids has similarities to chromosomal replication, including the involvement of SeqA and DNA methylation. In most cases, binding of the chromosomal initiator protein DnaA near the origin is also involved; however, DnaA is not the master initiator, and its role in plasmid initiation is not completely understood. Instead, the initiation of replication is controlled by the plasmid-encoded Rep initiator protein, which binds near the origin and functions to melt the double helix at the origin and recruit the helicase to unwind the DNA and initiate replication. This action is different from RepA's action in rolling-circle replication. The sites at which the Rep proteins bind are called iterons (**Fig. 7.23**). Interons are direct repeats of 17–22 base pairs and vary from two to seven copies in plasmids analyzed to date.

Another distinction from DnaA-mediated initiation is that the Rep initiator can exist as both monomers and dimers. Both forms can bind iteron DNA, but only monomers initiate replication. Rep dimer formation prevents replication both by sequestering monomers and by forming an inactivating bridge between the iterons of two plasmids in a process called "handcuffing" (**Fig. 7.23**). Dimerization and handcuffing are thought to help control the rate of replication initiation to ensure that the plasmid does not overload the host cell with excess copies. Initiation occurs when the handcuffs are disrupted by excess monomers and dimer-targeting proteases. Monomers bound to the iteron then melt the adjacent DNA and recruit DnaB helicase to the plasmid to initiate replication.

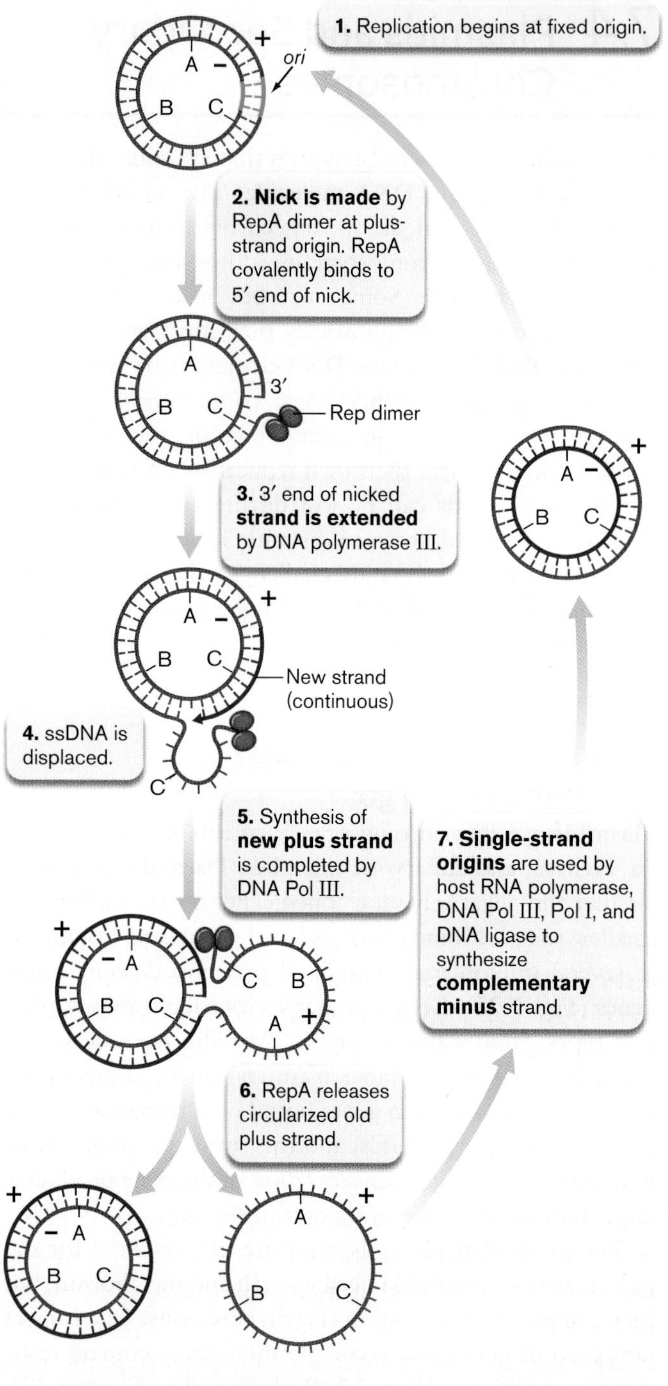

FIGURE 7.22 ■ **Plasmid replication: rolling-circle model.**

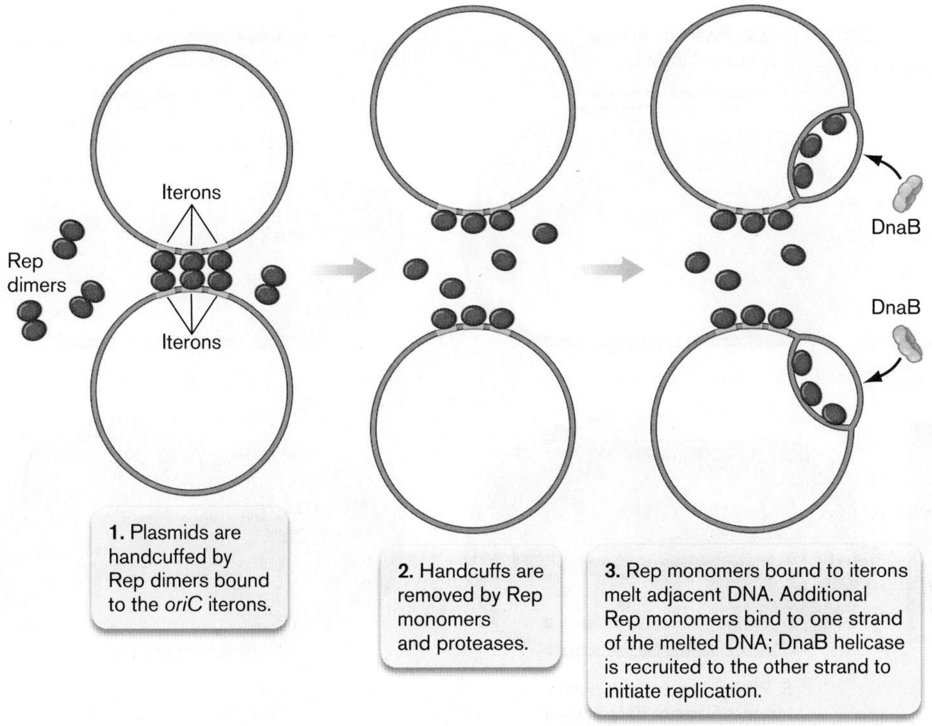

FIGURE 7.23 ■ Regulation of plasmid replication by handcuffing. Dimers of RepA handcuff two plasmids at the iteron loci. Removal of the dimers allows the Rep monomers to initiate replication.

Plasmid "Tricks" Ensure Inheritance

Given that plasmids do not carry essential genes, can cells easily lose their plasmids? When a plasmid-containing host cell divides, is it just chance that determines whether both daughter cells inherit the plasmid? In some instances the answer is yes. To maintain themselves in the host cell, plasmids can employ a number of different strategies. Some plasmids ensure their inheritance by carrying genes whose functions benefit the host bacterium under certain conditions. For instance, as discussed earlier, some bacterial plasmids confer resistance to antibiotics. As long as the antibiotic is present in the environment, any cell that loses the plasmid will be killed or stop growing. Still one more method of retention for the plasmid is to integrate into the host chromosome—a process described in detail in Chapter 9.

One of the surest means of plasmid retention is to flood the cytoplasm of the host with many copies. As the daughter cells split the cytoplasmic "inheritance" of the mother cells after cell division, there is a very high likelihood that both daughter cells will also inherit at least one copy of the plasmid.

Not all plasmids have a high number of copies, however. Low-copy-number plasmids limit how many copies they make to avoid draining the cell of energy. Cells forced to waste energy making many plasmid copies could be at a growth disadvantage when competing with plasmid-less cells in a natural environment. Low-copy-number plasmids evolved clever partitioning systems that ensure that both daughter cells will contain copies of the plasmid.

Figure 7.24 illustrates a model for how one type of partitioning works for plasmid R1, a *Salmonella* plasmid that imparts multidrug antibiotic resistance to host bacteria. The process employs three known plasmid-encoded genes: *parC*, *parM*, and *parR*. The *parC* DNA sequence is analogous to the centromere of eukaryotic chromosomes. The DNA-binding protein ParR binds to the *parC* sequence and forms a ParR-*parC* plasmid complex. ParM protein is an actin-like molecule that forms long filaments as it hydrolyzes ATP. The ParM filaments are dynamically unstable, constantly elongating and shortening. However, each end of a ParM actin-like filament can bind to a ParR-*parC* plasmid complex. When both ends of a ParM filament contact ParR-*parC* plasmid complexes, the filament stabilizes and elongates until the two plasmids hit the opposite poles of the cell. The plasmid is thus dislodged from the filament, and the filament quickly dissociates.

Secondary Chromosomes Carry Essential Genes

In contrast to *E. coli* and most bacteria studied so far, the genome of *Vibrio cholerae* is encoded on two chromosomes (**Fig. 7.2**). Typically, genomes with multiple chromosomes are composed of one primary chromosome and one or more secondary chromosomes. The secondary chromosomes are smaller than the primary chromosome. A **secondary chromosome** is distinguished from a plasmid because it carries at least one **essential gene**. An essential gene is required for cell viability under all environmental conditions. A secondary chromosome typically contains only a few of the essential genes in the genome.

Note: For some strains of *Vibrio cholerae*, the two chromosomes have fused into a single replicating unit. Both origins of replication are present, but it is unclear whether both are operational. The reason for the fusion event is unclear.

Thought Question

7.10 How would you demonstrate that a gene is essential?

Mechanisms of replication and segregation of the secondary chromosome may differ from that of the primary

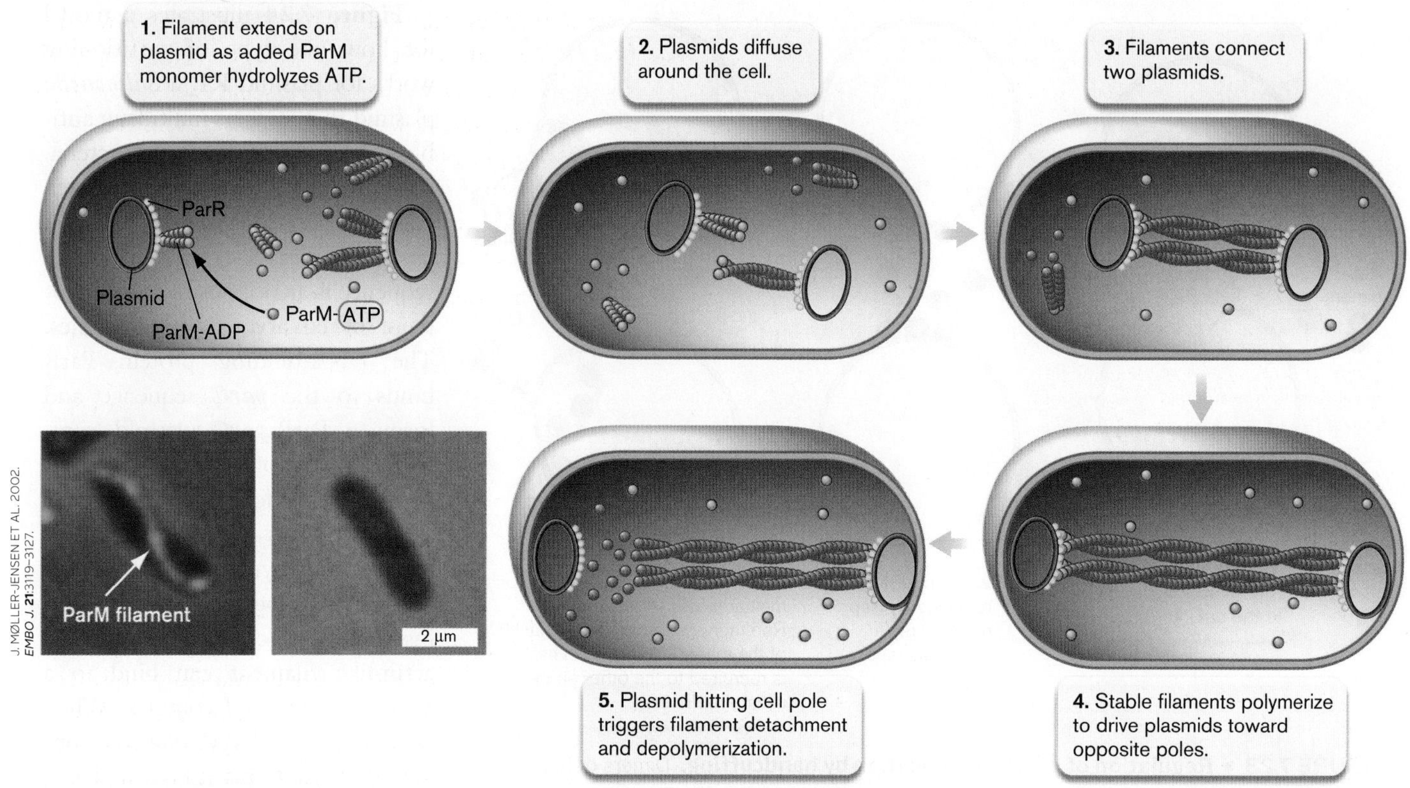

FIGURE 7.24 ■ Molecular model of plasmid segregation. The two micrographs at lower left show a cell with a *parM* mutant plasmid unable to make a ParM filament (right), and a cell with a *parM*$^+$ plasmid making a ParM actin-like filament (green; left) to partition daughter plasmids (step 4). ParM filament was seen by combined phase-contrast and immunofluorescence microscopy using rabbit anti-ParM antibodies and fluorescent Alexa 488–conjugated goat antirabbit IgG antibodies. *Source:* Modified from Christopher S. Campbell and R. Dyche Mullins. 2007. *J. Cell Biol.* **179**:1059–1066, fig. 4.

chromosome. Whereas the primary chromosome (Chr1) of *V. cholerae* (see **Fig. 7.2**) employs the canonical DnaA-dependent initiation machinery, the secondary chromosome (Chr2) uses machinery that is similar to that of bidirectionally replicated plasmids. The Chr2 version of the Rep initiator protein, RctB, can bind to two different types of iterons at the *oriC* region. When bound as a monomer to the 12-mer iteron (12 bp in length), RctB behaves as an initiator. When bound to the 39-mer region (39 bp), RctB behaves as an inhibitor by forming a handcuffing bridge to a 12-mer region of an adjacent copy of Chr2.

What controls whether RctB binds to the 12 mer (initiation) or the 39mer (inhibitor)? The critical factor is the copy number of the *crtS* gene on Chr1. As Chr1 is replicated, *crtS* is duplicated. When *crtS* is duplicated, RctB protein somehow loses affinity for the inhibitory 39mer site on Chr2 and switches to the initiator 12mer site. The distance of *crtS* from the Chr1 origin delays the start of replication for the smaller Chr2 such that both chromosomes complete replication at the same time. That the initiation of replication by Chr2 is controlled by the timing of Chr1 replication is a key distinction from plasmids, which control replication autonomously.

The two chromosomes of *V. cholerae* have distinct patterns of segregation during replication (**Fig. 7.25**). Both chromosomes segregate their copies with a plasmid-like Par system, but each chromosome uses a distinct Par system. The origin of Chr1 is found at the old pole in newborn cells (see Chapter 3 for a further discussion of polar asymmetry), and during replication one copy remains while the other copy moves to the new pole, pulling the duplicated region of the chromosome behind it. The terminus is at the new pole in newborn cells, and it migrates toward midcell before the cell replicates. In contrast, the origin of Chr2 is found at midcell in newborn cells, and the two replicated copies move poleward, toward the future midcell positions of the daughter cells. In the meantime, the terminus of Chr2 migrates toward midcell before the cell replicates. Replication is completed simultaneously at the Chr1 and Chr2 termini, and the dimers and catenanes of both chromosomes are resolved via Topo IV, XerC and XerD, and FtsK, as in *E. coli*. The terminus of Chr2 separates before that of Chr1, but both termini leave midcell before septum formation is complete. Septation results in daughter cells with their respective origins and termini at the same positions as in the parental (newborn) cell.

Secondary Chromosomes Evolve from Plasmids

How do secondary chromosomes form during bacterial evolution? One hypothesis is that the second chromosome

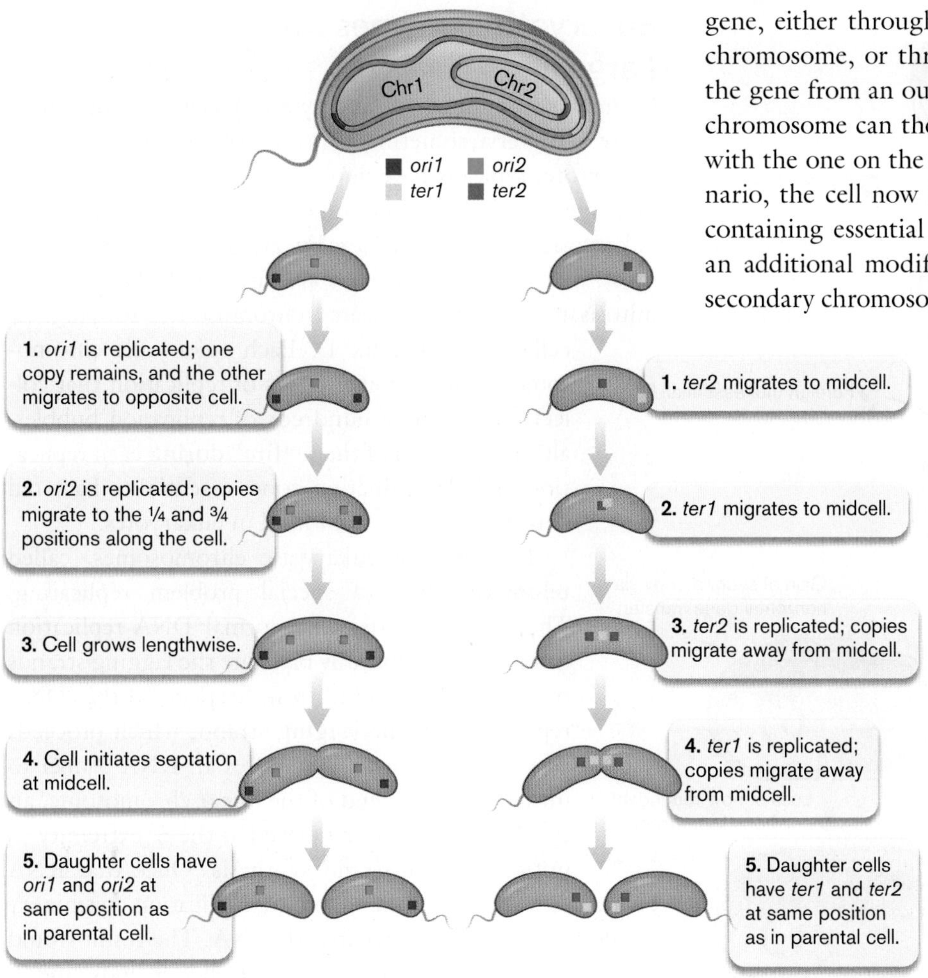

FIGURE 7.25 ■ Model for the replication and dynamic localization of the two chromosomes of *Vibrio cholerae*. Arrangement of Chr1 and Chr2 in a newborn cell. Chronology of duplications and migrations of the chromosomal origins (left) and termini (right). *Source:* Modified from Marie-Eve Val et al. 2016. *Curr. Opin. Microbiol.* **22**:120–126, fig. 2.

splits off from the primary chromosome to establish its own replicating unit. For this mechanism to work, both chromosomes must have an origin of replication at the time of the split or they would be lost during cell division. The competing hypothesis is that secondary chromosomes evolved from plasmids that captured one or more essential genes from the original chromosome. While the first scenario remains theoretically possible, for all studied secondary chromosomes the evidence supports the plasmid origin hypothesis. The strongest support for the plasmid origin model is the presence of plasmid-type replication initiation and segregation machinery for every secondary chromosome examined to date. No primary chromosome-type replication proteins have been found associated with secondary chromosomes.

Plasmids are thought to evolve into secondary chromosomes via a series of gene transfer events (**Fig. 7.26**). A foreign plasmid is acquired by a cell with a single chromosome, and through horizontal gene transfer the plasmid can acquire additional (nonessential) genes. At some point in its evolution, the plasmid acquires an essential gene, either through translocation of that gene from the chromosome, or through acquisition of a second copy of the gene from an outside source. The copy on the primary chromosome can then be lost because it is now redundant with the one on the secondary chromosome. In either scenario, the cell now has two independent DNA molecules containing essential genes. In the case of *Vibrio cholerae*, an additional modification is made to the newly formed secondary chromosome: transfer of the control of initiation from the secondary to the primary chromosome.

Regardless of the mechanism, the cells end up with the gain of an essential gene on the secondary chromosome and a corresponding loss of that gene on the primary chromosome. Is there a selective advantage to moving an essential gene onto another DNA element? One thought is that this translocation ensures retention of the second replicon: by marking the replicon with an essential function, it requires that the replicon, and the features it contains, be passed on to progeny. These other features may not be required for growth and reproduction, but they could provide *V. cholerae* with benefits worth maintaining. The nature of those benefits is under active investigation and could provide valuable clues as to why some microbes have multiple chromosomes and some do not.

One popular hypothesis is that the second chromosome is a test bed for evolutionary experimentation and innovation, which involve not only acquisition of new genes via horizontal gene transfer, but optimization of genes via adaptive mutation as well. Chr2 carries the integron island of *V. cholerae*, which is a gene capture system that has acquired several hundred genes, including those implicated in pathogenesis (see Section 9.1). Also, for reasons not well understood, mutations that change the functions of genes are more tolerable on Chr2 than Chr1.

To Summarize

- **Plasmids are autonomously replicating** circular or linear DNA molecules that are part of a cell's genome.
- **Plasmids can be transferred** between cells.
- **Plasmids replicate** by rolling-circle and/or bidirectional mechanisms.

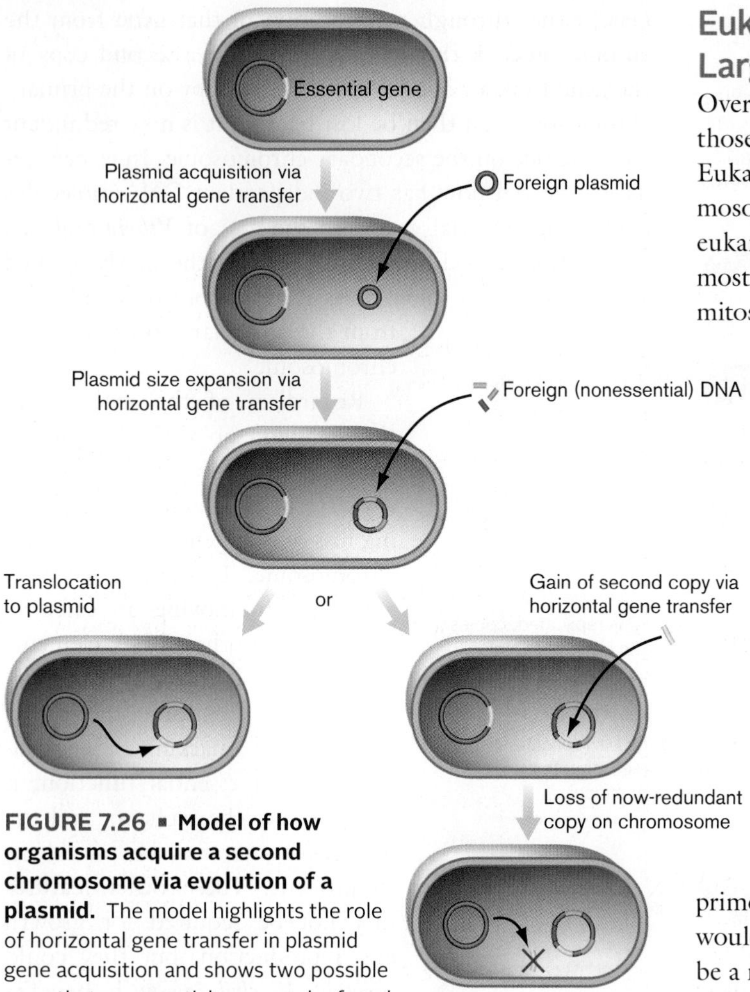

FIGURE 7.26 ■ Model of how organisms acquire a second chromosome via evolution of a plasmid. The model highlights the role of horizontal gene transfer in plasmid gene acquisition and shows two possible ways that an essential gene can be found exclusively on the new chromosome.

- **Secondary chromosomes** are distinguished from plasmids in that they carry essential genes.
- **Secondary chromosomes** evolve from plasmids by acquiring essential genes.

7.5 Eukaryotic and Archaeal Chromosomes

The chromosomes of eukaryotic microbes such as protists and algae have much in common with those of bacteria and archaea. All chromosomes consist of double-stranded DNA, for example, and are usually replicated bidirectionally. Nevertheless, important differences exist as well, particularly in genome structure. Eukaryotic chromosomes are linear and contained within a nucleus, and their replication involves mitosis (reviewed in eAppendix 2). Archaeal chromosomes, similar to bacterial chromosomes, are circular and lack a nuclear membrane, but archaeal proteins involved in transcription and replication appear more eukaryotic than bacterial.

Eukaryotic Genomes Are Large and Linear

Overall, the genomes of eukaryotic nuclei are larger than those of bacteria, sometimes by several orders of magnitude. Eukaryotes typically have duplicate copies of multiple chromosomes (diploid number ranges between 10 and 100). All eukaryotic chromosomes are linear, whereas many, if not most, bacterial chromosomes are circular. Eukaryotes use mitosis to segregate replicated chromosomes to daughter cells (see Appendix 1). Each eukaryotic chromosome has numerous origins of replication that collectively generate hundreds of replication bubbles, although not all of them "fire" during each replication cycle. Termination zones essential to bacterial chromosomes are not found in eukaryotes.

The ends of eukaryotic chromosomes, called **telomeres**, have a special problem replicating. The problem is that the normal DNA replication machinery cannot fully replicate the lagging strands of linear DNA. Recall from Section 7.3 that DNA replication on the lagging strand, which proceeds in a 5′-to-3′ direction, requires an RNA primer to initiate it. At the end of the linear chromosome, an RNA primer could be placed at the 5′ extremity to initiate lagging-strand synthesis. Once that RNA primer was removed, the standard replication machinery would be unable to replace it with DNA. The result would be a replicated chromosome with shortened 5′ ends (and 3′ overhangs). Each round of replication would shorten the chromosome a little more, until the loss of genetic information became catastrophic to the organism.

Eukaryotic microbes (and the germ cells of metazoa like humans) solve the problem of chromosomal shortening during replication with an enzyme complex called **telomerase**. Telomerase is actually a reverse transcriptase that reads RNA as a template to synthesize DNA (see Sections 6.3 and 11.3 for the role of reverse transcriptase in the replication of retroviruses such as HIV). Telomerase uses an <u>intrinsic</u> RNA (an RNA that is part of the enzyme) as a template to add numerous DNA repeat sequences to the 3′ ends of chromosomes. The purpose of this 3′ extension is to provide a new "upstream" location for the RNA primer to initiate lagging-strand synthesis. This RNA primer could then be used to replicate the 5′ end of the chromosome. In this situation, the only template DNA that is left unreplicated is the DNA placed by the telomerase. Telomere length is thus dynamic in replicating eukaryotic cells, with shrinkage due to RNA priming being countered by the synthesis activity of telomerases.

Telomerase may have evolved from the ancient progenitor cells that contained RNA rather than DNA genomes (see Section 17.2). The reverse transcriptase may also be the evolutionary source of retroviruses (see **eTopic 6.1**).

Note: Bacteria with linear chromosomes do not have telomerases. How do they replicate their 3′ ends? Some (for example, *Streptomyces*) cap the ends of their chromosomes with covalently bound terminal proteins that prime DNA replication. Others (for example, *Borrelia*) form covalently closed hairpin ends (essentially making one long, circular, single-stranded molecule). The ends appear to recombine after being replicated.

Unlike bacterial cells, eukaryotic cells pack their DNA within the confines of a nucleus where a series of proteins called **histones** compacts the DNA. Histones are rich in arginine and lysine, so they are positively charged, basic proteins that easily bind to the negatively charged DNA. The DNA becomes wrapped around the histones to form units called nucleosomes. Histones also play a regulatory role through methylation and acetylation. Bacteria, too, have DNA-packaging proteins, but they are less essential for function (see Chapter 3 and Section 7.2).

The detailed structures of the genomes of eukaryotes and bacteria reveal surprising differences (refer back to **Fig. 7.3**). The genome of the bacterium *E. coli* is packed with genes (green in the figure) that encode proteins or RNA molecules. These genes are separated by very little unused or noncoding DNA (purple), and by only an occasional mobile element, such as an insertion sequence, that can move from one DNA molecule to another. Most bacterial and archaeal genes are organized in coordinately regulated, multiple-gene operons. In contrast, 98% of the human genome consists of noncoding sequences (sequences that do not encode proteins).

Noncoding sequences include sequences that regulate protein-coding genes; genes that encode untranslated, small regulatory RNAs; and the fossil genomes of ancient viruses. Coding genes are separated by large stretches of noncoding sequences and usually are not clustered together in operons. Moreover, coding genes in the human genome are interrupted by **introns** (DNA within a gene that is not part of the coding sequence for a protein, shown as yellow in **Fig. 7.3B**) and ancient gene duplications that have decayed into nonfunctional, vestigial **pseudogenes**. Bacteria also have pseudogenes, but they are relatively rare, except in intracellular symbionts and pathogens, where gene functions are readily lost because the host cell provides the necessary resources.

Note: Pseudogenes differ from noncoding DNA because at least part of a pseudogene's sequence is similar to that of a gene with a known function. Also note that some pseudogenes express mRNA, but the protein products are typically truncated and nonfunctional.

By themselves, promoters of eukaryotic genes have very low activity and require enhancer DNA sequences to drive transcriptional activity. Enhancer sequences act at long distances from promoters, and scientists suspect that once enhancers became important, it was necessary to place enough DNA between them to reduce the activation of other, unrelated but adjacent promoters.

Archaeal Genomes Combine Features of Bacteria and Eukaryotes

Like bacteria, archaea are haploid and reproduce asexually via cell fission. Archaea are true prokaryotes because their cells lack a nuclear membrane, but the structures of their DNA-packing proteins, RNA polymerase, and ribosomal components more closely resemble those of eukaryotes. How are archaeal genomes organized and replicated?

Archaeal chromosome structure and replication share features with bacteria and eukaryotes. Archaea use several proteins to organize their chromosomes in the nucleoid. Many archaea use Alba proteins, unique to the archaeal domain, to form loops of DNA, and some also employ histones to further organize the chromosome. Like bacteria, most archaea have a single chromosome. Thus far, all known archaeal chromosomes are circular, and replication proceeds bidirectionally from the origin. However, most archaea have multiple origins of replication distributed around the chromosome, with some representatives, such as *Pyrobaculum calidifontis*, having as many as four.

The advantage of possessing multiple functional origins of replication is currently unknown, as genetic mutants with all but one origin removed have minimal to zero changes in growth rate. Notably, the additional origins seem to have been acquired by horizontal gene transfer, rather than through simple duplication of preexisting origin(s) on the chromosome. Each origin has its own specific initiator protein, a relative of the eukaryotic ORC (origin recognition complex) protein, sometimes called Cdc6. And even more intriguing, replication initiation at each origin is essentially synchronous: triggered by an unknown mechanism, all initiators are able to unwind the DNA at their respective origins and recruit DNA helicase and the remainder of the DNA polymerase machinery.

Note: The archaeal species *Haloferax volcanii* appears to initiate replication via homologous recombination, an unusual mechanism that may be similar to the initiation of bacteriophage T4 and the non-nuclear chromosomes of eukaryotes: mitochondria, chloroplasts, and kinetoplastids.

Archaea can possess one or two types of DNA polymerases: one that is related to the eukaryotic PolB enzyme, and another, PolD, that appears to be unique to the archaeal domain. Whether both participate in leading- and lagging-strand replication, or have alternative roles in the cell, is an open question, and the answer may be different depending on the archaeal lineage. The remaining components of the initiation and elongation machinery of archaea, including the helicase, primase, sliding clamp, and DNA ligase, are related to the eukaryotic proteins, rather than to the bacterial versions. Mechanisms of replication termination in archaea with multiple origins are still not well understood, but replication may be terminated as a consequence of replication fork collision. Chromosome dimers appear to be resolved by Xer proteins acting at *dif* sites, as in bacteria.

To Summarize

- **Eukaryotic chromosomes** are linear, double-stranded DNA molecules. After replication, the copies are segregated to daughter cells by mitosis.
- **A reverse transcriptase** called telomerase prevents net loss of DNA at the ends of eukaryotic chromosomes during replication.
- **Histones** (DNA-packing proteins) play a critical role in compacting chromosomes in eukaryotes and some archaea.
- **Introns and pseudogenes** are noncoding DNA sequences that make up a large portion of eukaryotic chromosomes.
- **Archaeal chromosomes** resemble those of bacteria in size and shape, but archaeal DNA replication machinery is more closely related to eukaryotic enzymes.

7.6 Microbiomes and Metagenomes

Microorganisms in nature do not generally exist as pure cultures. They grow instead as complex consortia containing numerous species, the sum total of which is called the **microbiome**. Most members of these microbiomes have never been grown in a laboratory, and thus cannot be investigated by traditional culture-based methods. However, we can use molecular-based methodologies such as high-throughput DNA sequencing and the polymerase chain reaction (PCR; see Chapter 12) to investigate the diversity of microbiomes without having to culture them. Like the explorers of old, scientists are finding new species never before seen and are beginning to understand the intricate ways in which bacterial species interact. The study of microbial communities is explored at length in Chapter 21.

The molecular investigation of the microbiome began with the work of Norman Pace and colleagues at Indiana University, who reasoned that while most microbes may not be readily cultured, their DNA could be isolated directly from an environmental sample and sequenced for analysis (Section 17.3). These investigators chose the 16S rRNA gene for sequencing, because it is highly conserved in microbes and phylogenic studies of 16S sequences could be used to determine whether members of the community belonged to known lineages or represented novel microorganisms (Section 17.3). Since then, environmental sequencing has revolutionized microbiology by shifting the focus even further away from cultivatable organisms toward the estimated 99% of microbial species that cannot be cultivated. Originally used to investigate exotic locations like the hot springs of Yellowstone, PCR-based environmental sequencing has been exploited to reveal the microbiomes present in a vast number of environments, including the human gut.

One recent molecular study revealed that human genes can influence the makeup of our gut microbiome. Approximately one hundred trillion microbes live in the human body, but we do not know the identity of most of them, because we cannot grow them in the laboratory. How we interact with our microorganisms and they with us is a major question in modern microbiology—one that has led to formation of the Human Microbiome Project, which examines gut, skin, and oral microbiomes. We already know that the composition of intestinal microbiota can influence blood chemistry and may contribute to irritable bowel syndrome (inflammation of the gastrointestinal tract), but microbiota are also suspected of influencing an individual's susceptibility to diabetes and obesity (see Section 23.2).

Ruth E. Ley (Cornell University; **Fig. 7.27A**) and collaborators cataloged the intestinal microbiomes of identical and fraternal human twins to determine whether any part of the microbiome might be heritable, that is, influenced by the genetic makeup of the human host. They used PCR to amplify 16S rRNA sequences from human fecal samples and sequenced the products. The 16S rRNA sequences were matched to a public database to identify bacterial species. Host genotype had the largest influence on *Christensenella minuta*, a recently discovered firmicute. Even more striking, this microbe was found more often in lean than in obese subjects, and it prevented weight gain when transplanted into germ-free mice (**Fig. 7.27B**). Knowing the identities of these microbes and their collective metabolic potential will undoubtedly lead to new advances in the treatment and prevention of human diseases.

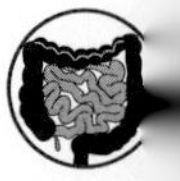

Analysis of an environmental sample using 16S rRNA sequences can provide valuable information on the composition of the microbiome but cannot reliably predict the activity of the community members. This is because, while the 16S rRNA gene is highly conserved, other genes in the genome can be lost and gained though horizontal gene transfer (Chapters 9 and 17). To gain a better understanding of the metabolic and physiological potential of a microbe, it is thus imperative to sequence every gene in every genome, even if you do not know which genes belong to which species. The result is called the **metagenome**. The objective of **metagenomics**, a term coined in 1998 by microbiologist Jo Handelsman (Yale University), is thus not only to identify the members of a microbial community, but also to reveal the potential activities and contributions to the ecosystem of these members. Metagenomic studies have radically accelerated with the development of high-throughput DNA sequencing technology (Chapter 21). Critically, the sheer scale of information in metagenomic studies can also identify patterns in microbial community function that would have been otherwise difficult or impossible to see in smaller data sets, as the next example will demonstrate.

In a comprehensive metagenomic study, Shinichi Sunagawa and colleagues sequenced 7.2 terabases (7.2×10^{12} bases) of data from 68 locations spanning the world's ocean (**Fig. 7.28**). The researchers first examined all of the 16S rRNA genes in the metagenome to determine how many microbes were in the microbiome. The level of sequence information could not resolve species versus genus classification, so they used a conservative term for each genotype called an operational taxonomic unit (OTU). The researchers identified 37,000 OTUs in their collection of ocean samples. When the individual OTUs were grouped into higher taxonomic ranks, such as the phylum Cyanobacteria, these taxa were found to contribute unequally to total taxonomic composition, and importantly, their contribution varied depending on the sampling location (**Fig. 7.28B**).

After identifying the microbiome members, Sunagawa and colleagues analyzed the rest of the metagenome. They found that a large fraction of the metagenome, 40%, consists of genes with no known function. That unknown 40% could be involved in processes critical for global nutrient cycling, and might include genes that can synthesize products of human interest, such as novel therapeutics and antibiotics. The remaining 60% of the metagenome consists of genes whose functions could be predicted by their sequence, and these could be categorized into gene families based on predicted function. Over 70% of these gene families were found at every station, and were classified as "core." The researchers noted that the distributions of these core gene families did not vary much at all between stations, in sharp contrast to the OTUs (**Fig. 7.28B**). This surprising finding suggests that certain core functions are important for the community, but which microbes perform those functions can vary from location to location. Studies like these have led microbial ecologists to consider that communities assemble as collections of gene functions, rather than as collections of taxonomic groups.

The Sunagawa study also compared the metagenomes from the ocean ecosystem to that of the human gut (**Fig. 7.29**). The researchers found that the ocean microbiome had a higher fraction of genes for photosynthesis and the uptake of amino acids, lipids, nucleotides, and secondary metabolites, while the gut microbiome had a higher fraction of genes for defense, signal transduction, and carbohydrate transport. Thus, at the community level the microbes behave quite differently in the human gut than they do in the ocean. Comparisons such as these provide new hypotheses and research directions for how microbial communities and functions are dictated by their environment, and provide clues as to which environmental factors may be most critical.

Thought Question

7.11 How might you interpret the discovery of genes for photosynthesis in the metagenome of the human gut? Could it indicate a possible error in the analysis, or could something else be going on?

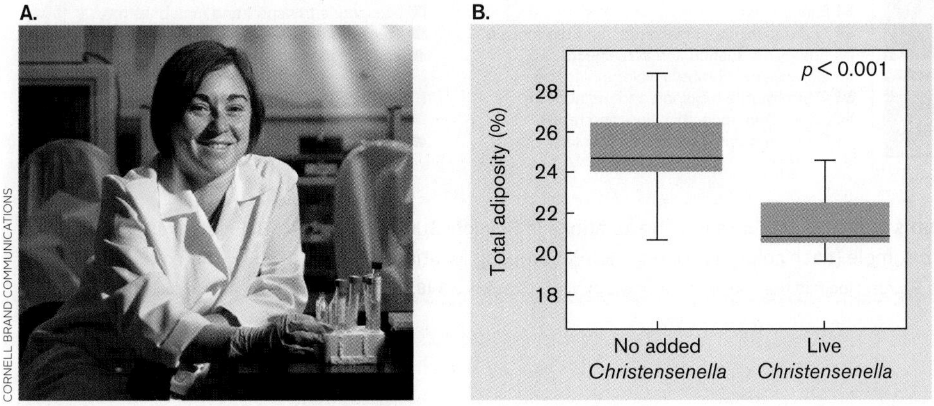

FIGURE 7.27 ▪ Heritable member of the human microbiome. **A.** Ruth Ley led a team that used metagenomics to discover that the abundance of some members of the intestinal microbiome are influenced by host genetics. **B.** One such organism, *Christensenella minuta*, also influenced weight gain and adiposity when orally "transplanted" into mouse intestines. The graph shows that adiposity 21 days after a fecal transplant was significantly lower in germ-free mice transplanted with *C. minuta*–containing fecal preparations (blue) than with preparations lacking this organism (orange). *Source:* Goodrich et al. 2014. *Cell* **159**:789–799, fig. 7B.

How are metagenomes such as the ones just described produced from microbiomes? The details of (meta)genomic analysis using high-throughput, "next-generation" sequencing are provided in Chapters 12 and 21; the essentials are provided here in brief. Current DNA sequencing technology cannot provide the sequence of an entire chromosome in a single read. Depending on the instrument, some reads may cover several genes, whereas others cover only a fraction of a gene. The strategy most often employed in metagenomics is to perform so-called shotgun sequencing, which provides vast quantities

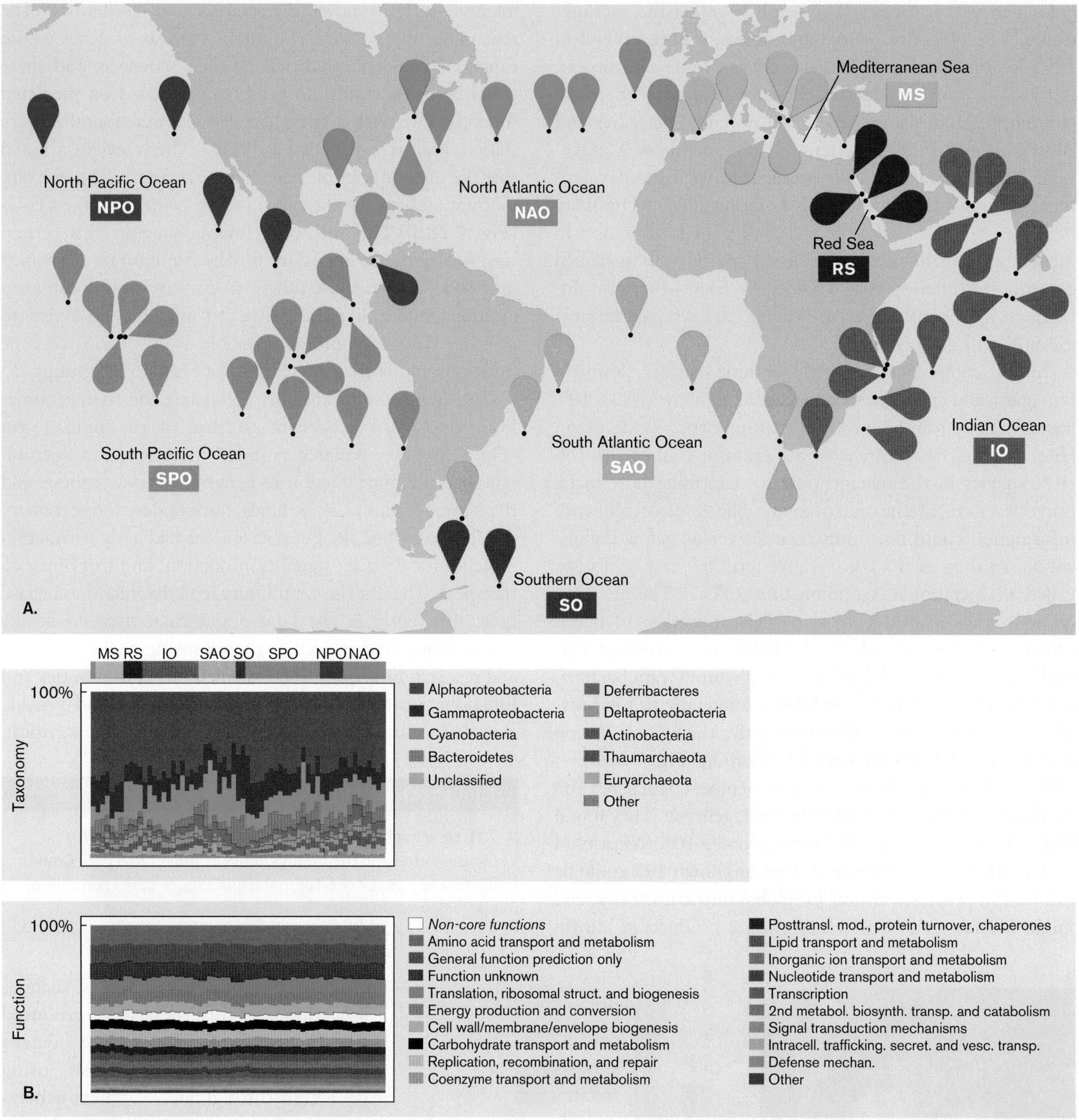

FIGURE 7.28 ■ Metagenome of the ocean's surface. **A.** The sampling locations in Shinichi Sunigawa's study. **B.** Fractional composition of taxa and functional gene categories for each sample (each column is one sample). Sampling locations are indicated by the color bar above the plots; refer to panel A for general location. *Source:* Modified from Shinichi Sunagawa et al. 2015. *Science* **348**:6237, figs. 1A (part A) and 8A (part B).

of information in the form of short DNA sequence reads. These reads are then subjected to various forms of bioinformatic analysis to extract information on the genetic potential of the community.

Metagenomic DNA is first purified from the microbial cells (**Fig. 7.30**). It is then fragmented into smaller lengths, and these fragments are prepared for shotgun sequencing according to the chemistry of the specific DNA sequencing instrument utilized. The sequences go through quality control steps, and then the reads are assembled by computational analysis of overlapping regions (see Chapter 21). These regions of overlap are used to stitch together

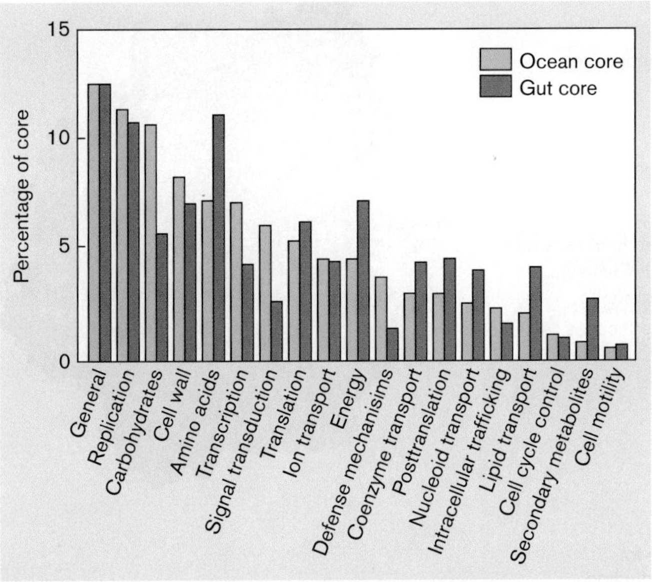

FIGURE 7.29 ▪ Comparison of the core set of genes found in the ocean and human gut microbiomes, arranged by functional category. *Source:* Modified from Shinichi Sunagawa et al. 2015. *Science* **348**:6237, fig. 7C.

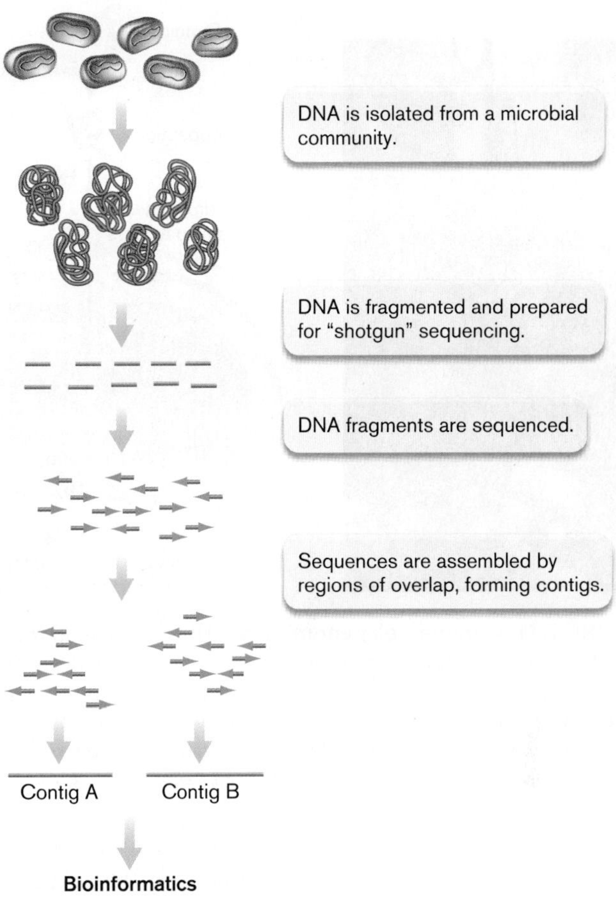

FIGURE 7.30 ▪ Steps in the metagenomic investigation of a microbial community. For simplicity, only a few cells, DNA molecules, and contigs are shown; practical uses have involved terabases of DNA encoding tens of millions of genes.

contiguous stretches of DNA, forming **contigs**, which may contain partial or complete genes. Finally, these genes are investigated with bioinformatics, which can provide the potential functions and taxonomies of the genes, as well as estimates of their abundances within the sampled microbial community.

The technology of DNA sequencing has progressed to a point where the genome of a single cell can be sequenced—enabling **single-cell genomics**. Single-cell genomics (SCG) has several advantages over shotgun sequencing–based metagenomics. First, it requires less sample DNA—only that found within a single cell—which may be important for study sites where microbial abundance is low. Second, because all reads come from one cell type, the assignment of the reads to a single taxon is more reliable. Hence, SCG has the power to identify more complex metabolic and physiological features of cells that may involve large suites of interacting genes. SCG requires special protocols, equipment, and reagents to isolate a single cell from an environment and, without ever growing it, extract and amplify its DNA. Once amplified, the genomic DNA is readily sequenced using the next-generation sequencing tools described in Chapter 12.

Using SCG, Karen Lloyd (**Fig. 7.31A**) and colleagues at the University of Tennessee, Aarhus University, and other institutions reconstructed partial genomes of archaea that are abundant in marine sediments but have thus far resisted all attempts at cultivation. Among the several metabolic pathways they were able to assemble, they discovered that these organisms have the potential for breakdown of extracellular proteins and peptides, an activity not previously recognized in sediment archaea (**Fig. 7.31B**). Bioinformatic analysis suggested that the cells secrete several classes of proteases into the environment. Consistent with this genomic evidence, sediments where these organisms were collected tested positive for the activity of these specific classes of proteases. Bioinformatics also identified candidate transporters for oligopeptide uptake and intracellular proteases that could, in concert with the secreted proteases, completely break down environmental proteins into amino acid monomers for use in both biosynthesis of new proteins and heterotrophic energy metabolism.

Environmental DNA sequence information in the form of metagenomes and single-cell genomes can provide valuable predictions of gene function but cannot test those predictions. How, then, can the functions be validated for genes discovered in microbes that cannot be cultured? One method that has proved successful is to clone and express the gene of interest in a surrogate host, such as *E. coli*. Gene cloning for expression is described in detail in Chapter 12, but in brief, *E. coli* can transcribe foreign DNA

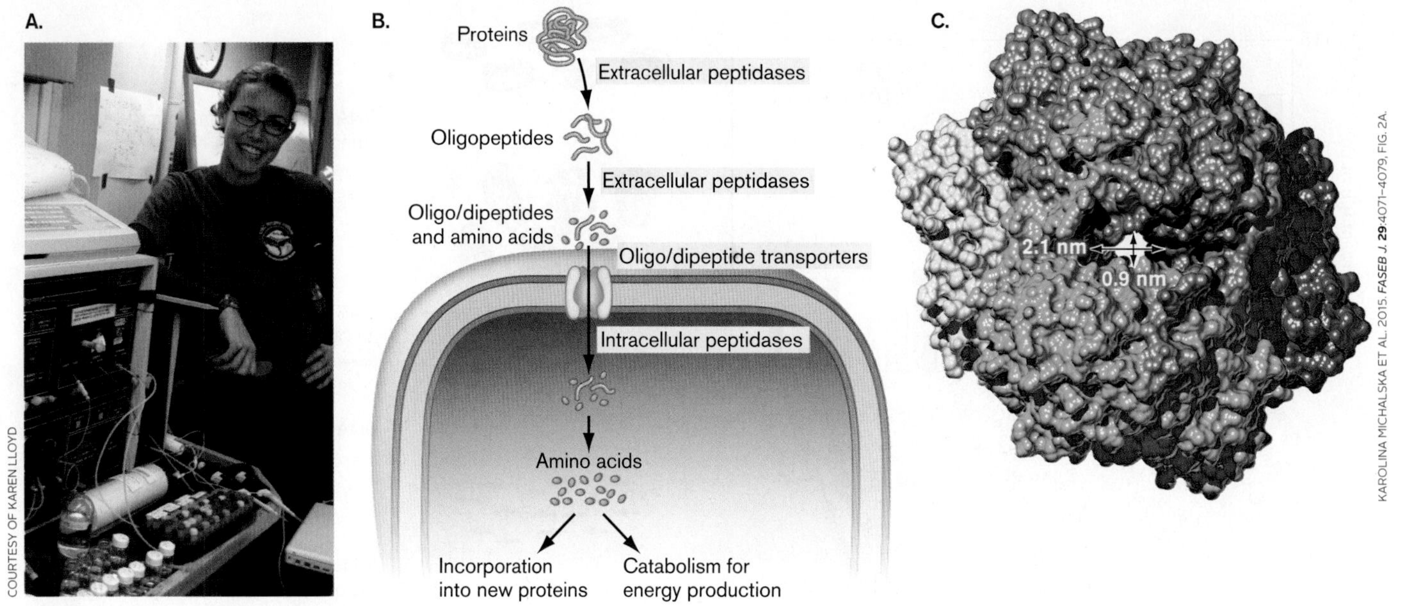

FIGURE 7.31 ■ Single-cell genomics identifies microbial functions in marine sediments. Karen Lloyd **(A)** and colleagues used SCG **(B)** to identify enzymes of an uncultured archaeon that degrades, imports, and processes extracellular proteins scavenged from marine sediments. *E. coli* was used as a surrogate host to produce proteins from this uncultured archaeon so that the proteins could be studied in the lab. **C.** One intracellular protein, S15, whose native state was a tetramer (each color is a monomeric subunit), was confirmed to have peptidase activity. S15 was the first protein of experimentally validated structure and function characterized for this uncultured microbe (PDB 2B9V).
Source: Part B modified from K. G. Lloyd et al. 2013. *Nature* **496**:7444, fig. 3.

into RNA, which can be translated into protein products that would be identical to those produced in the original organism.

Researchers at Argonne National Laboratory were interested in the potential new properties of the enzymes found in the uncultured archaea found by Lloyd's team, so they collaborated to clone and express these genes in *E. coli.* One protein of interest, S15, was predicted to function as an endopeptidase—that is, a peptidase that hydrolyzes the terminal amino acids of proteins. Purified preparations of S15 confirmed this prediction and showed that this enzyme, which assembles into a tetrameric complex (**Fig. 7.31C**), has a preference for cysteine and hydrophobic residues at the N terminus of proteins. Enzymes such as these, isolated from uncultured microbes in exotic, often extreme environments, may have novel substrates or products, and may have greater activity under extreme conditions (for example, high or low temperature or pH) that are optimal for industrial applications.

To Summarize

- **Microbiomes** are the complex microbial communities that live in environments such as the human gut, the ocean, and every other inhabited location on Earth.
- **Metagenomics** uses rapid DNA sequencing and other genomic techniques to study consortia of microbes directly in their natural environment.
- **Single-cell genomics** can analyze the partial genome of a single microbe without the need for cultivation.
- **Genes identified by metagenomics or SCG** can be cloned and expressed in other microbes as a form of bioprospecting.

Concluding Thoughts

We used to think of bacteria as simple, single-celled organisms. In fact, they are far from simple. Complexity is evident in the elegant mechanisms they use to organize and replicate their genomes. The next three chapters will expand on this idea and discuss the way bacteria make proteins, regulate genes, exchange DNA, and, in the process, evolve. Knowing how a bacterial cell replicates and genetically controls the repertoire of available proteins and enzymes provides a unique perspective from which to study the physiology of microbial growth, explored in Part 3 of this textbook. Furthermore, understanding the bacterial genome enables us to investigate the physiology of pathogenic, as well as nonculturable, organisms and gain greater insight into the evolutionary diversity of species, which we will discuss in Part 4.

CHAPTER REVIEW

Review Questions

1. Explain the structural types of bacterial genomes.
2. What are the differences between DNA and RNA?
3. Explain DNA supercoiling. Why is it important to microbial genomes?
4. Discuss the mechanisms of topoisomerases. What drugs target the topoisomerase enzyme DNA gyrase?
5. What are the basic mechanisms of DNA replication?
6. How does the bacterial cell regulate the initiation of chromosome replication?
7. What is the clamp loader? Primase? DNA helicase (DnaB)? Helicase loader (DnaC)? DNA proofreading?
8. How is the problem of replicating both strands at a replication fork solved?
9. What is a catenane? What does it have to do with DNA replication?
10. How are chromosome dimers that form by homologous recombination resolved to separate the chromosomes?
11. How does rolling-circle replication compare with bidirectional replication?
12. What is handcuffing?
13. What distinguishes a secondary chromosome from a primary chromosome or a plasmid?
14. What is a metagenome, and how can it be used for studies of microbiomes?

Thought Questions

1. Why might metagenomics or single-cell genomics be more successful at bioprospecting for novel microbial activities than traditional cultivation approaches are?
2. During rapid growth, why would a bacterial cell die if an antibiotic drug formed, as the text says, "a physical barrier in front of the DNA replication complex" (Section 7.2)?
3. If you were to synthesize a microbe from scratch, perhaps for industrial production of antibiotics, how would you design the genome? Single chromosome or multiple chromosomes? Plasmids included or not? Linear or circular DNA?

Key Terms

antiparallel (244)
catenane (259)
contig (271)
denature (245)
DNA ligase (257)
DNA replication (251)
enhancer (242)
essential gene (263)
exonuclease (255)
gene (240)
genome (240)
histone (267)
homologous recombination (259)
hybridization (245)
intron (267)
metagenome (269)
metagenomics (269)
microbiome (268)
nitrogenous base (243)
nucleobase (243)
nucleoid (246)
Okazaki fragments (252)
origin (*oriC*) (252)
phosphodiester bond (244)
plasmid (261)
pre-catenane (258)
primase (254)
promoter (242)
proofreading (255)
pseudogene (267)
purine (244)
pyrimidine (244)
quinolone (249)
replication fork (251)
replisome (255)
secondary chromosome (263)
semiconservative (251)
single-cell genomics (271)
sliding clamp (255)
telomerase (267)
telomere (266)
termination (*ter*) site (252)
topoisomerase (249)
transformation (240)

Recommended Reading

Anda, Mizue, Yoshiyuki Ohtsubo, Takashi Okubo, Masayuki Sugawara, Yuji Nagata, et al. 2015. Bacterial clade with a ribosomal RNA operon on a small plasmid rather than the chromosome. *Proceedings of the National Academy of Sciences USA* **112**:14343–14347.

Ausiannikava, Darya, and Thorsten Allers. 2017. Diversity of DNA replication in the Archaea. *Genes* **8**:56.

Bouet, Jean-Yves, Mathiu Stouf, Elise Lebailly, and Francois Cornet. 2014. Mechanisms for chromosome segregation. *Current Opinion in Microbiology* **22**:60–65.

Bury, Katarzyna, Katarzyna Wegrzyn, and Igor Konieczny. 2017. Handcuffing reversal is facilitated by proteases and replication initiator monomers. *Nucleic Acids Research* **45**:3953–3966.

Campbell, Christopher S., and R. Dyche Mullins. 2007. In vivo visualization of type II plasmid segregation: Bacterial actin filaments pushing plasmids. *Journal of Cell Biology* **179**:1059–1066.

Costa, Allesandro, Iris V. Hood, and James M. Berger. 2013. Mechanisms for initiating cellular DNA replication. *Annual Reviews of Biochemistry* **82**:25–54.

Dame, Remus T., Olga J. Kalmykowa, and David C. Grainger. 2011. Chromosomal macrodomains and associated proteins: Implications for DNA organization and replication in Gram-negative bacteria. *PLoS Genetics* **7**:e1002123.

Dewar, James M., and Johannes C. Walter. 2017. Mechanisms of DNA replication termination. *Nature Reviews. Molecular Cell Biology* **18**:507–516.

diCenzo, George C., and Turlough M. Finan. 2017. The divided bacterial genome: Structure, function, and evolution. *Microbiology and Molecular Biology Reviews* **81**:e00019-17.

Duggin, Ian G., R. Gerry Wake, Stephen D. Bell, and Thomas M. Hill. 2008. The replication fork trap and termination of chromosome replication. *Molecular Microbiology* **70**:1323–1333.

Falkow, Stanley. 2001. I'll have chopped liver please, or how I learned to love the clone. *ASM News* **67**:555.

Georgescu, Roxana E., Isabel Kurth, and Mike E. O'Donnell. 2011. Single-molecule studies reveal the function of a third polymerase in the replisome. *Nature Structural & Molecular Biology* **19**:113–116.

Goodrich, J. K., J. L. Waters, A. C. Poole, J. L. Sutter, O. Koren, et al. 2014. Human genetics shape the gut microbiome. *Cell* **159**:789–799.

Hedlund, Brian P., Jeremy A. Dodsworth, Senthil K. Murugapiran, Christian Rinke, and Tanja Woyke. 2014. Impact of single-cell genomics and metagenomics on the emerging view of extremophile "microbial dark matter." *Extremophiles* **18**:865–875.

Kuzminov, Andrei. 2014. The precarious prokaryotic chromosome. *Journal of Bacteriology* **196**:1793–1806.

Le, Tung Bk, and Michael T. Laub. 2014. New approaches to understanding the spatial organization of bacterial genomes. *Current Opinion in Microbiology* **22**:15–21.

Lindas, A. C., and R. Bernander. 2013. The cell cycle of archaea. *Nature Reviews. Microbiology* **11**:627–638.

Männik, Jaana, Matthew W. Bailey, Jordan C. O'Neill, and Jaan Männik. 2017. Kinetics of large-scale chromosomal movement during asymmetric cell division in *Escherichia coli*. *PLoS Genetics* **13**:e1006638.

McHenry, Charles S. 2011. Bacterial replicases and related polymerases. *Current Opinion in Chemical Biology* **15**:587–594.

Michalska, Karolina, Andrew D. Steen, Gekleng Chhor, Michael Endres, Austen T. Webber, et al. 2015. New aminopeptidase from "microbial dark matter" archaeon. *FASEB Journal* **29**:4071–4079.

Reyes-Lamothe, Rodrigo, and David J. Sherratt. 2019. The bacterial cell cycle, chromosome inheritance and cell growth. *Nature Reviews Microbiology* **17**:467–478.

Reyes-Lamothe, Rodrigo, Tung Tran, Diane Meas, Laura Lee, Alice M. Li, et al. 2014. High-copy bacterial plasmids diffuse in the nucleoid-free space, replicate stochastically and are randomly partitioned at cell division. *Nucleic Acids Research* **42**:1042–1051.

Robinson, Andrew, and Antoine M. van Oijen. 2013. Bacterial replication, transcription and translation: Mechanistic insights from single-molecule biochemical studies. *Nature Reviews. Microbiology* **11**:303–315.

Sanchez-Romero, Maria A., Ignatio Cota, and Josep Casadesus. 2015. DNA methylation in bacteria: From the methyl group to the methylome. *Current Opinion in Microbiology* **25**:9–16.

Song, Dan, and Joseph J. Loparo. 2015. Building bridges within the bacterial chromosome. *Trends in Genetics* **31**:164–173.

Sunagawa, Shinichi, Luis Pedro Coelho, Samuel Chaffron, Jens Roat Kultima, Karine Labadie, et al. 2015. Structure and function of the global ocean microbiome. *Science* **348**:6237.

Val, Marie-Eve, Martial Marbouty, Francisco de Lemos Martins, Sean P. Kennedy, Harry Kemble, et al. 2016. A checkpoint control orchestrates the replication of the two chromosomes of *Vibrio cholerae*. *Science Advances* **2**:e1501914.

Wegrzyn, Katarzyna E., Marta Gross, Urszula Uciechowska, and Igor Konieczny. 2016. Replisome assembly at bacterial chromosomes and iteron plasmids. *Frontiers in Molecular Biosciences* **3**:39.

Wolanski, Marcin, Rafal Donczew, Anna Zawilak-Pawlik, and Jolanta Zakrzewska-Czerwinska. 2014. *oriC*-encoded instructions for the initiation of bacterial chromosome replication. *Frontiers in Microbiology* **5**:735.

CHAPTER 8

Transcription, Translation, and Protein Processing

8.1 RNA Polymerases and Sigma Factors

8.2 Transcription of DNA to RNA

8.3 Translation of RNA to Protein

8.4 Protein Modification, Folding, and Degradation

8.5 Secretion: Protein Traffic Control

The cell accesses the vast store of data in its genome using tiny molecular machines. One machine (RNA polymerase) reads the DNA template to make RNA (transcription). Another machine (the ribosome) decodes the RNA to synthesize a protein (translation). Once made, each polypeptide must fold properly and somehow find its correct cellular or extracellular location. Proteins that have outlived their usefulness must be destroyed and their amino acids recycled. In Chapter 8 we explain how the cell makes RNA and proteins and guides their fate. What emerges is a picture of remarkable biomolecular integration, controlled to maintain balanced growth and ensure survival.

CURRENT RESEARCH highlight

Detecting gut bacteria with ultrasound: *Escherichia coli* cells can make gas vesicles in their cytoplasm (panel A) when they transcribe and translate genes from *Bacillus megaterium* and *Anabaena flos-aquae*. Gas vesicles are protein shells that contain gas but exclude liquid and can scatter sound waves. The vesicles generate an ultrasound contrast that can be detected even in the deep tissue of animals. When they were injected into the colon of mice, the *E. coli* cells expressing the gas-vesicle genes could be detected using ultrasound, as seen in panel B where, within the blue oval, color intensity represents ultrasound signal.

Source: Bourdeau et al. 2018. *Nature* **553**:86–89.

A.

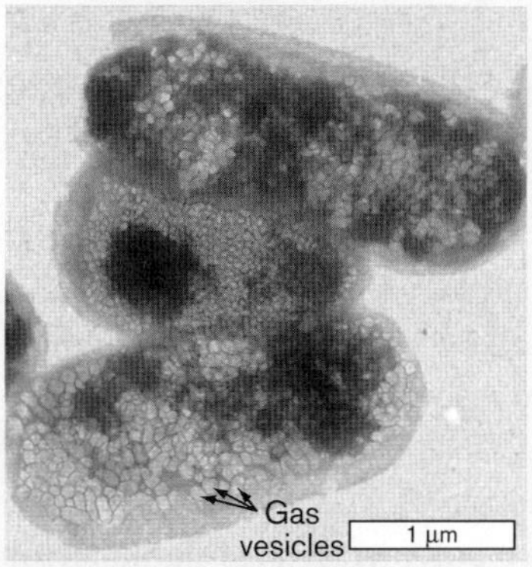

B.

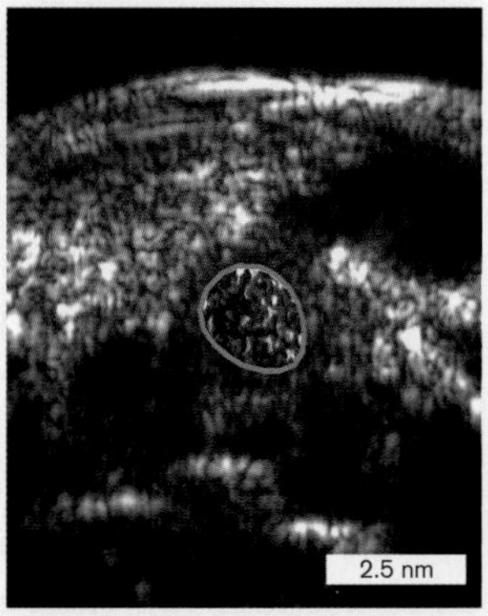

AN INTERVIEW WITH

DR. MIKHAIL SHAPIRO, BIOENGINEER, CALIFORNIA INSTITUTE OF TECHNOLOGY

How surprising was it that you could take genes from two foreign species of bacteria, express them in *E. coli*, and have them make functioning gas vesicles?

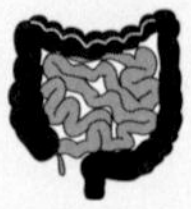

We were fairly certain that we would be able to eventually transfer gas-vesicle expression to other species, because these proteins are made naturally in relatively diverse organisms. However, getting all the genes to function properly, have correct stoichiometry, and assemble with another species always comes with some uncertainty!

Genomes of cellular microbes are made of very long stretches of DNA, usually millions of base pairs in length. These genomes can contain thousands of genes, whose function is to encode the RNAs and proteins of the cell. But what is the language of genes—the genetic code—and how is this code read by the cell to make RNA and protein? For all three domains of life, DNA is transcribed into RNA, and RNA is translated into protein. These universal processes—transcription and translation—are the means that cells use to read the instructions found in the genome and turn that information into functional proteins and RNA molecules. These same processes also can be exploited by researchers to express genes introduced from other organisms, as illustrated in the Current Research Highlight.

Where did our understanding of gene expression begin? By 1960 we knew that the DNA in a cell holds the genetic code, but the code itself and how the code produces proteins remained mysterious. RNA was a suspected, but unproven, intermediate until Marshall Nirenberg and his postdoctoral associate Heinrich Matthaei at the National Institutes of Health (**Fig. 8.1A**) designed a cell-free system (a cell lysate of *E. coli*) to test the RNA hypothesis. They used RNA molecules, synthesized by Maxine Singer (**Fig. 8.1B**), that contained simple, known, repeated sequences, such as poly-A (consisting only of adenylic acid), poly-U (polyuridylic acid), poly-AAU, and poly-ACAC.

These RNA molecules were tested to see whether the repetitive sequence might direct incorporation of a specific amino acid into a protein. Each synthetic polynucleotide was tested in the presence of a radiolabeled amino acid. If the radioactive amino acid was incorporated into a polypeptide, the polypeptide would also be radioactive. On the morning of May 27, 1960, the results of the experiment showed that the poly-U RNA specified the assembly of radioactive polyphenylalanine. It was the first breakthrough in cracking the genetic code. For his work, Nirenberg won the 1968 Nobel Prize in Physiology or Medicine.

In Chapter 8 we explore the way microbes, primarily bacteria, interpret the information held within a nucleotide sequence of DNA and convert that information into a string of amino acids—a protein. From there we look at what the cell does with those proteins once they are made. For instance, specific proteins are moved into the periplasm, while other proteins must be inserted into membranes. We also show how damaged proteins are selectively degraded.

A. B.

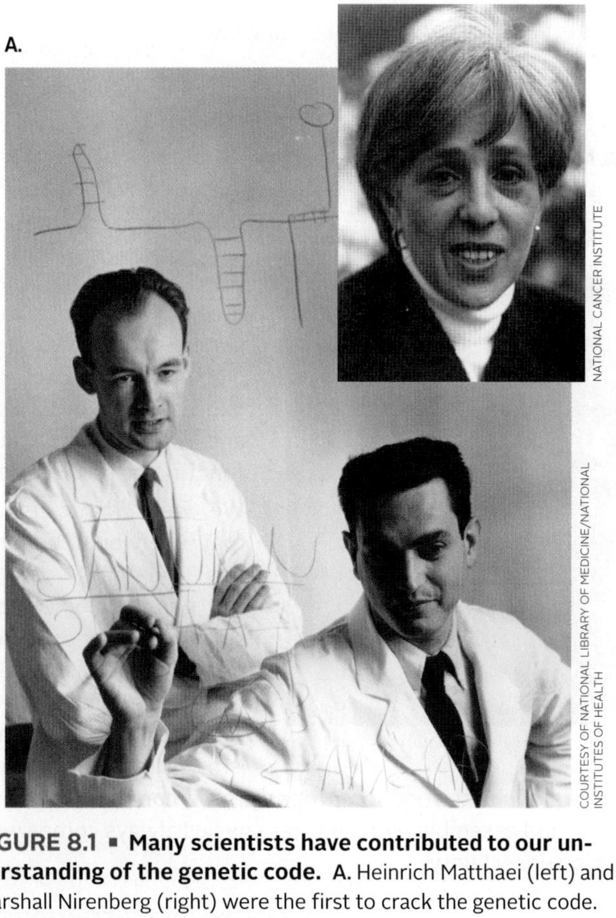

FIGURE 8.1 ■ Many scientists have contributed to our understanding of the genetic code. **A.** Heinrich Matthaei (left) and Marshall Nirenberg (right) were the first to crack the genetic code. An early hand-drawn model of what would eventually be known as translation is behind them. **B.** Maxine Singer, a key contributor to the genetic code experiments, also helped develop guidelines for recombinant DNA research.

8.1 RNA Polymerases and Sigma Factors

To survive and reproduce, every cell needs to access information encoded within DNA. How does this happen? Chromosomal DNA is large and cumbersome, so the first step in the process is to make multiple copies of the information as snippets of RNA that can move around the cell and, like disposable photocopies, can be destroyed once the encoded protein is no longer needed. This copying of DNA to RNA is called **transcription**.

RNA Polymerase Transcribes DNA to RNA

An enzyme complex called **RNA polymerase**, also known as **DNA-dependent RNA polymerase**, carries out transcription, making RNA copies (called **transcripts**) of a DNA template. The DNA **template strand** specifies the base sequence of the new complementary strand of RNA.

RNA polymerase in bacteria consists of a core polymerase and a sigma factor. Core polymerase contains the proteins required to elongate an RNA chain. **Sigma factor** is a protein needed only for initiation of RNA synthesis, not for its elongation. Together, core polymerase plus sigma factor are called the holoenzyme.

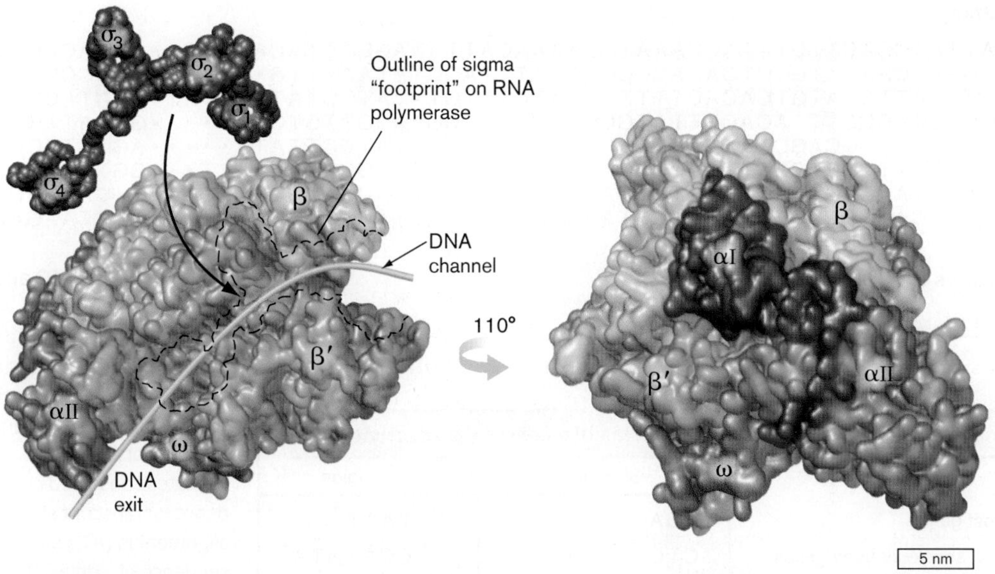

FIGURE 8.2 ■ **Subunit structure of RNA polymerase.** Two views of RNA polymerase. On the left, the channel for the DNA template is shown by the yellow line. Subunits (αI, αII, β, β′, and ω) are color-coded dark green, medium green, light green, cyan, and gold, respectively. The function of the omega (ω) subunit is currently unclear. Sigma factor (red), which recognizes promoters on DNA, is shown separate from core polymerase in the left-hand panel. Different functional areas of sigma are labeled sigma 1 through sigma 4 (σ_1–σ_4). Sigma factor interacts with the alpha (α), beta (β), and beta-prime (β′) subunits. The molecule on the left is rotated 110° to give the image on the right. 1L9Z in the RCSB Protein Data Bank. *Source*: Robert D. Finn et al. 2000. *EMBO J.* **19**:6833–6844.

A bacterial core RNA polymerase is a complex of four different subunits: two alpha (α) subunits, one beta (β) subunit, and one beta-prime (β′) subunit (**Fig. 8.2**). A fifth subunit, omega (ω), is shown in the figure but is not required for transcription. The beta-prime subunit houses the Mg^{2+}-containing catalytic site for RNA synthesis, as well as sites for the rNTP (ribonucleoside triphosphate, or ribonucleotide) substrates, the DNA substrates, and the RNA products. The 3D structure of RNA polymerase shows that DNA fits into a cleft formed by the beta and beta-prime subunits (**Fig. 8.2**). The alpha subunit assembles the other two subunits (beta and beta-prime) into a functional complex. It also communicates through physical "touch" with various regulatory proteins that can bind DNA. These protein-protein interactions inform RNA polymerase what to do after the enzyme binds DNA.

Sigma factors. Random transcription initiation occurs at low levels along the chromosome; transcription of the entire genome at high levels would be extremely wasteful and problematic to the cell. Consequently, bacteria use sigma (σ) factor proteins to guide RNA polymerase to the beginning of each gene whose expression is needed under certain conditions (**Fig. 8.2**). A sigma factor first binds to RNA polymerase through the beta and beta-prime subunits (see the dotted black outline in **Fig. 8.2**). The bound sigma factor recruits the core enzyme to a specific DNA sequence, called the **promoter**, marking the beginning of a gene. A single bacterial species can make several different sigma factors (see **Fig. 8.3C** for some examples). Each sigma factor helps core RNA polymerase find the start of a different subset of genes. However, a single core polymerase complex can bind only one sigma factor at a time. Hundreds of genes encoding sigma factors have been sequenced from numerous species of bacteria. Although they all encode somewhat different amino acid sequences, there are enough sequence and structural similarities to make them recognizable as sigma factors.

Promoters. Every cell has a "housekeeping" sigma factor that keeps essential genes and pathways operating. In the case of *E. coli* and other rod-shaped, Gram-negative bacteria, that factor is sigma-70 or σ^{70}, so named because it is a 70-kilodalton (kDa) protein (its gene designation is *rpoD*). Genes recognized by sigma-70 all contain similar promoter sequences that consist of two parts. The DNA base corresponding to the start of the RNA transcript is called nucleotide +1 (+1 nt). Relative to this landmark, promoter sequences are usually centered at −10 and −35 nt before the start of transcription (see **Fig. 8.3**). Other sigma factors typically recognize different consensus sequences at one or both of these positions (or at nearby locations in some cases).

The DNA promoter sequence recognized by a given sigma factor can be determined by comparison of known promoter sequences of different genes whose expression requires the same sigma. Similarities among these different promoter DNA sequences define a **consensus sequence** likely recognized by the sigma factor (**Fig. 8.3A** and **B**). A consensus sequence consists of the most likely base (or bases)

A. Strong *E. coli* promoters

Gene	Sequence (−35 … −10 … +1)
bioB	TGTCATAATCGACTTGTAAACCAAATTGAAAAGATT•TAGGTTTACAAGTC••••TACACCGAAT
cyoA	GTTCTCGATCAAATTGGCTGAAAGGCGGTAATTTAGCTATAAATTG••••••ATCACCGTCGAAA
galE	CTAATTTATTCCATGTCACACTTTTCGCATCTTTGT•TATGCTATGGTTATTTCATACCATAAG
lacP	AGGCACCCGAGGCTTTACACTTTATGCTTCCGGGTCGTATGTTGTGT••••••GGAATTGTGAGC
rpoD	TCGCCCTGTTCCGCAGCTAAAACGCACGACCATG••CGTATACTTAT••••••AGGGTTGCTGGT
rpsL	GGTGACGTTATCGGTGACTTGAGCCGTCGTCGTG••GTATGCTCAAA…•••••GGTCAGGAATC
trp	CCGGAAGAAAACCGTGACATTTTAACACGTTTG••••TTACAAGGTAAAGGC•••GACGCCGCCC
zwf	AGCGTTTACAGTTTTCGCAAGCTCGTAAAAGCA••••GTACAGTGCAC••••••CGTAAGAAAAT

B. Consensus sequences of σ^{70} promoters

−35 region — 17 ± 1 bp — −10 region

TTGACAT ——— TATAAT

C.

Sigma factor	Promoter recognized	Promoter consensus sequence	
		−35 region	−10 region
RpoD σ^{70}	Most genes	TTGACAT	TATAAT
RpoH σ^{32}	Heat shock–induced genes	TCTCNCCCTTGAA	CCCCATNTA
RpoF σ^{28}	Genes for motility and chemotaxis	CTAAA	CCGATAT
RpoS σ^{38}	Stationary-phase and stress response genes	TTGACA	TCTATACTT
		−24 region	−12 region
RpoN σ^{54}	Genes for nitrogen metabolism and other functions	CTGGNA	TTGCA

FIGURE 8.3 ■ −10 and −35 sequences of *E. coli* promoters. **A.** Alignment of the upstream region of a sub-sample of genes whose promoters are recognized by sigma-70 (σ^{70}). Dots do not represent bases; they help to align the sequences by accounting for short insertions and deletions. Yellow indicates conserved nucleotides; brown denotes transcript start sites (+1). **B.** The alignment in (A) generates a consensus sequence of sigma-70-dependent promoters (red-screened letters indicate nucleotide positions where different promoters show a high degree of variability). **C.** Some *E. coli* promoter sequences recognized by different sigma factors. "N" indicates that any of the four standard nucleotides can occupy the position.

at each position of the predicted promoter. Although promoters are double-stranded DNA (dsDNA) sequences, convention is to present the promoter as the single-stranded DNA (ssDNA) sequence of the sense (nontemplate) strand, which has the same sequence as the RNA product.

Some positions in a consensus sequence are highly conserved, meaning that the same base is found in that position in every promoter. Other, less conserved positions can be occupied by different bases. Few promoters actually have the most common base at every position. Even highly efficient promoters usually differ from the consensus at one or two positions.

Sigma factor recognition of promoters. How do sigma factors, or any other DNA-binding proteins for that matter, recognize specific DNA sequences when the DNA is a double helix? The phosphodiester backbone is quite uniform, and the interior of paired bases appears inaccessible. However, proteins can recognize side groups of the bases that protrude from the major and minor grooves of DNA (see Fig. 7.5). Portions of sigma-70 from *E. coli* wrap around DNA, allowing certain parts of the protein to fit into DNA grooves.

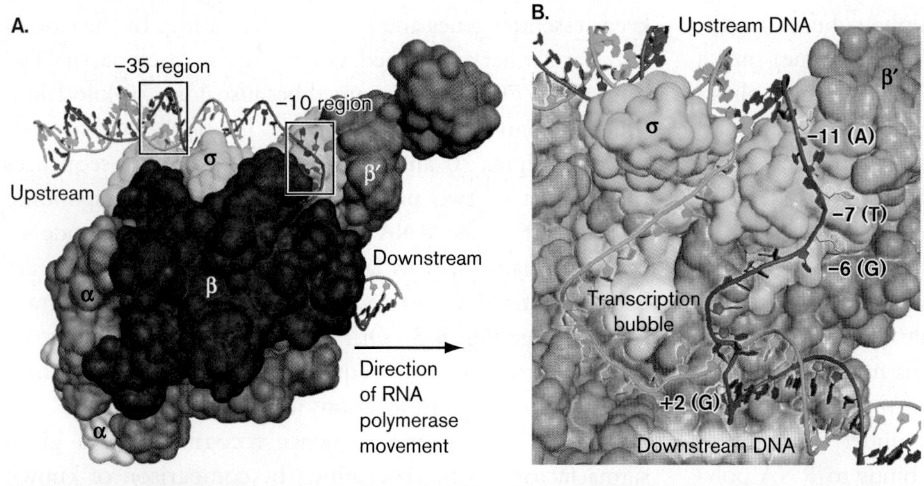

FIGURE 8.4 ■ RNA polymerase holoenzyme bound to a promoter. **A.** The initial open complex forms when holoenzyme binds to a promoter. DNA −10 and −35 contacts with sigma factor are shown. Nontemplate (coding) strand is color-coded magenta; template strand, green. **B.** Blowup of (A), with the beta subunit removed to view the transcription bubble. Some bases in the nontemplate strand are flipped outward (yellow) to interact with sigma factor or the beta subunit after the transcription bubble is formed. The base at position +1 is the first base transcribed.

Figure 8.4A shows the holoenzyme RNA polymerase complex positioned at a promoter and the points (−10 and −35 nt) where sigma factor contacts DNA. Sigma factors generally contain four highly conserved amino acid sequences, called regions (see **Fig. 8.2**, σ_1–σ_4). Part of region 2 of the sigma-70 family recognizes –10 sequences, whereas region 4 recognizes the –35 sites. Part of region 1 helps separate the DNA strands to make a transcription "bubble" in preparation for RNA synthesis (**Fig. 8.4B**).

Sigma factor control of complex physiological responses. A single cell will almost certainly experience numerous changes in its environment during its life. Each time the environment changes, the cell is challenged to readjust its physiology, sometimes very quickly to avoid death. Large readjustments can be made by sigma factors because they can simultaneously activate a large set of genes. Gene sets controlled by different sigma factors include those dealing with nitrogen metabolism, flagellar synthesis, heat stress, starvation, sporulation, and many other physiological responses.

In response to a change in a given environment, microbes will increase the abundance of the appropriate sigma factor (described in Chapter 10). The resulting high concentration of the specialty sigma factor will dislodge and replace other sigma factors, including sigma-70, from core polymerase. In this way, the cell redirects RNA polymerase to the promoters of genes best suited for growth or survival in the new environment. For instance, if the temperature suddenly rises, the cell produces sigma-32 (**Fig. 8.3C**), which displaces sigma-70 and redirects RNA polymerase to transcribe genes involved in the heat-shock response.

Thought Questions

8.1 If each sigma factor recognizes a different promoter, how does the cell manage to transcribe genes that respond to multiple stresses, each involving a different sigma factor?

8.2 Imagine two different sigma factors with different promoter recognition sequences. What would happen to the overall gene expression profile in the cell if one sigma factor were artificially overexpressed? Could there be a detrimental effect on growth?

8.3 Why might some genes contain multiple promoters, each one specific for a different sigma factor?

To Summarize

- **RNA polymerase holoenzyme**, consisting of core RNA polymerase and a sigma factor, initiates transcription of a DNA template strand.
- **Sigma factors** help core RNA polymerase locate consensus promoter sequences near the beginning of a gene. The sequences identified and bound by *E. coli* sigma-70 are located at –10 and –35 bp upstream of the transcription start site.
- **Dynamic changes in gene expression follow** changes in the relative levels of different sigma factors.

8.2 Transcription of DNA to RNA

Defining a Gene

Before we describe the mechanics of transcription and translation, it will help to illustrate the alignments between the DNA sequence of a structural gene (a gene encoding a protein), and the mRNA transcript containing translation signals and the protein-coding sequences. **Figure 8.5** shows the "sense" (coding) and "template" DNA strands of a two-gene operon. The sequence of the sense strand matches that of the mRNA transcript, but with Ts substituting for Us. The template strand is the strand actually "read" by RNA polymerase.

Note that +1 marks the DNA base where the mRNA transcript starts and that a single transcript is polygenic, containing both operon genes. In the mRNA transcript, an untranslated "leader" sequence precedes the gene *A* protein-coding region, located between translation start and stop signals (discussed later). The protein-coding region is also called the **open reading frame** (**ORF**). Downstream of the translation stop signal for gene *B* lies an untranslated "trailer." The leader and trailer sequences, also referred to as the 5′ and 3′ untranslated regions (UTRs), respectively, help regulate gene expression.

In bacteria, the RNA products of transcription can be monocistronic or polycistronic. Monocistronic RNA encodes the product of a single gene. This gene has its own promoter and transcription terminator. Polycistronic RNA encodes the products of one or more adjacent genes in one contiguous RNA molecule. **Figure 8.5** illustrates a polycistronic mRNA encoding two protein-coding genes. Transcription of these genes initiates at the promoter of the first gene, and terminates at the end of the last gene. The combination of the regulatory regions (promoter, terminator) and all the genes that are cotranscribed in the polycistronic message is referred to as an **operon**. As we will see in Chapter 10, placing multiple genes under the control of the same promoter is an efficient way to coordinate their regulation.

Thought Question

8.4 **Figure 8.5** illustrates an operon and its relationship to transcripts and protein products. Imagine that a mutation that stops translation (TAA, for example) was substituted for a normal amino acid about midway through the DNA sequence that encodes gene *A*. What would happen to the expression of the gene *A* and gene *B* proteins?

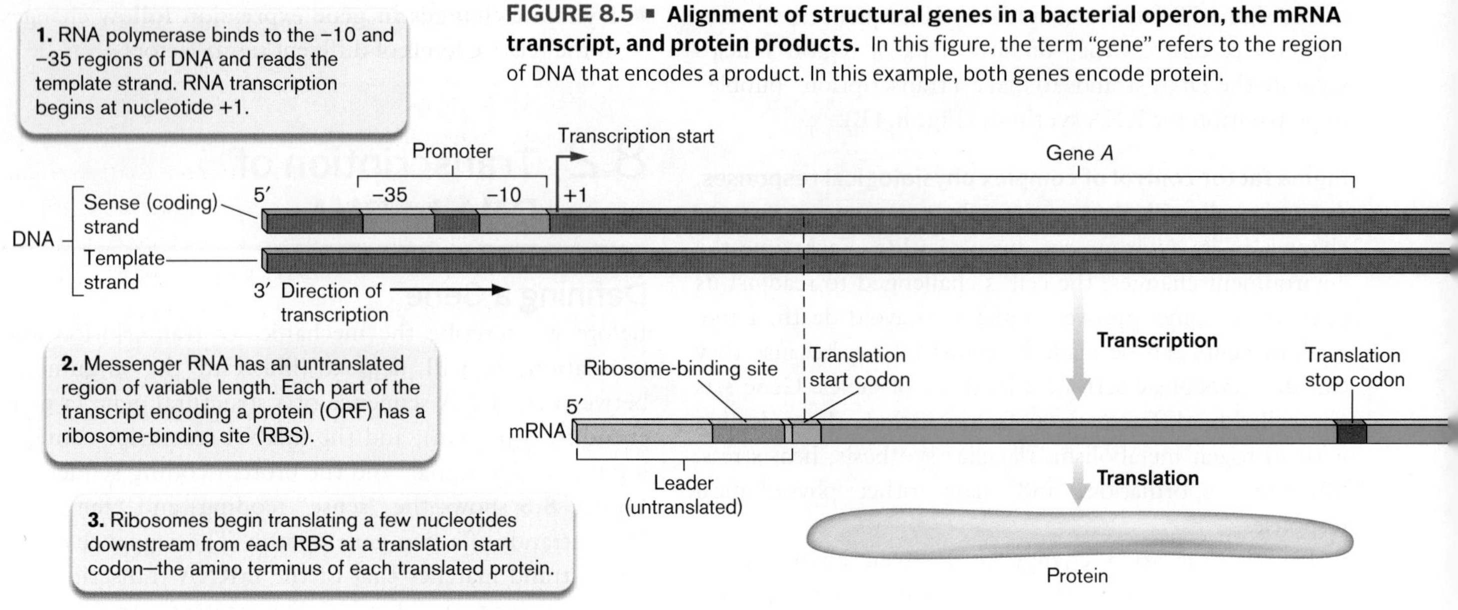
1. RNA polymerase binds to the −10 and −35 regions of DNA and reads the template strand. RNA transcription begins at nucleotide +1.
Promoter
Transcription start
Gene A
Sense (coding) strand
5′
−35
−10
+1
DNA
Template strand
3′ Direction of transcription
Transcription
2. Messenger RNA has an untranslated region of variable length. Each part of the transcript encoding a protein (ORF) has a ribosome-binding site (RBS).
Ribosome-binding site
Translation start codon
Translation stop codon
5′
mRNA
Leader (untranslated)
Translation
3. Ribosomes begin translating a few nucleotides downstream from each RBS at a translation start codon—the amino terminus of each translated protein.
Protein

FIGURE 8.5 ■ Alignment of structural genes in a bacterial operon, the mRNA transcript, and protein products. In this figure, the term "gene" refers to the region of DNA that encodes a product. In this example, both genes encode protein.

The Three Stages of Transcription

Like DNA replication, transcription of DNA to RNA occurs in three stages:

1. **Initiation**, in which RNA polymerase binds to the promoter, melts open the DNA helix, and catalyzes placement of the first RNA nucleotide
2. **Elongation**, the sequential addition of ribonucleotides to the 3′ OH end of a growing RNA chain
3. **Termination**, whereby sequences trigger release of the polymerase and the completed RNA molecule

The newly released RNA polymerase can then engage another sigma factor to seek a new promoter.

Transcription initiation. RNA polymerase constantly scans DNA for promoter sequences (**Fig. 8.6**, step 1). At the promoter, RNA polymerase holoenzyme forms a loosely bound, closed complex with DNA, which remains annealed and double-stranded—that is, unmelted (step 2). To successfully transcribe a gene, this closed complex must open through the unwinding of one helical turn, which causes DNA to become unpaired in this area (step 3; see also **Fig. 8.4B**). After promoter unwinding, RNA polymerase in the open complex becomes tightly bound to DNA.

The open-complex form of RNA polymerase begins transcription. The first ribonucleoside triphosphate (rNTP) of the new RNA chain is usually a purine (A or G). The purine base-pairs to the position designated +1 on the DNA template, which marks the start of the gene. As the enzyme complex moves along the template, subsequent rNTPs diffuse through a channel in the polymerase and into position at the DNA template. After the first base is in place, each

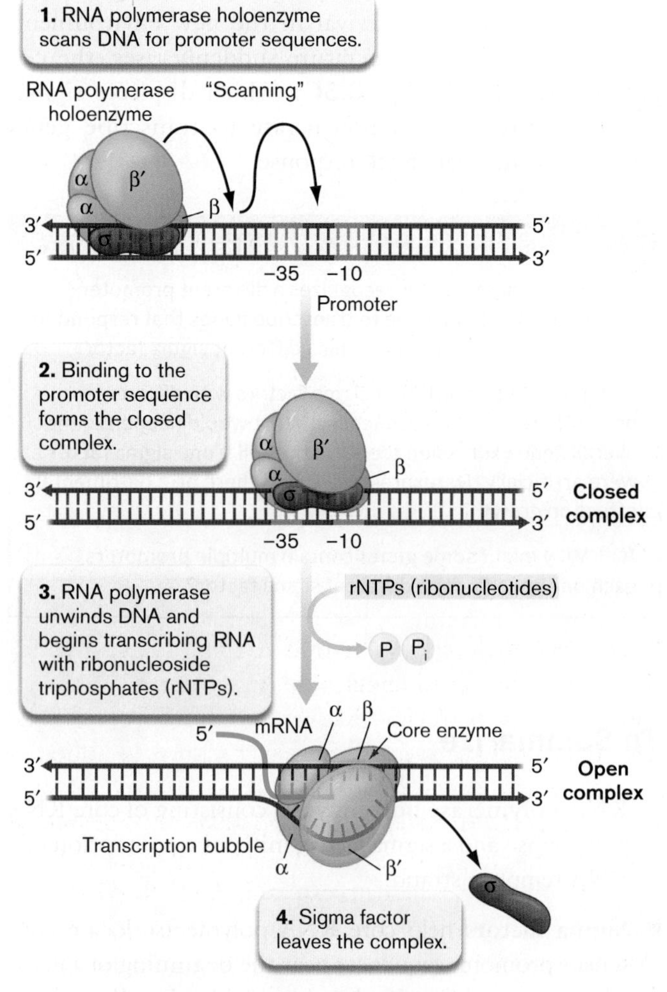
1. RNA polymerase holoenzyme scans DNA for promoter sequences.
RNA polymerase holoenzyme
"Scanning"
α
β′
α
β
σ
3′
5′
5′
3′
−35
−10
Promoter
2. Binding to the promoter sequence forms the closed complex.
α
β′
α
β
σ
3′
5′
5′
3′
Closed complex
−35
−10
3. RNA polymerase unwinds DNA and begins transcribing RNA with ribonucleoside triphosphates (rNTPs).
rNTPs (ribonucleotides)
P
Pi
5′
mRNA
α
β
Core enzyme
3′
5′
5′
3′
Open complex
Transcription bubble
α
β′
σ
4. Sigma factor leaves the complex.

FIGURE 8.6 ■ The initiation of transcription. Sigma factor helps RNA polymerase find promoters but is discarded after the first few RNA bases are polymerized. (Omega is not shown.)

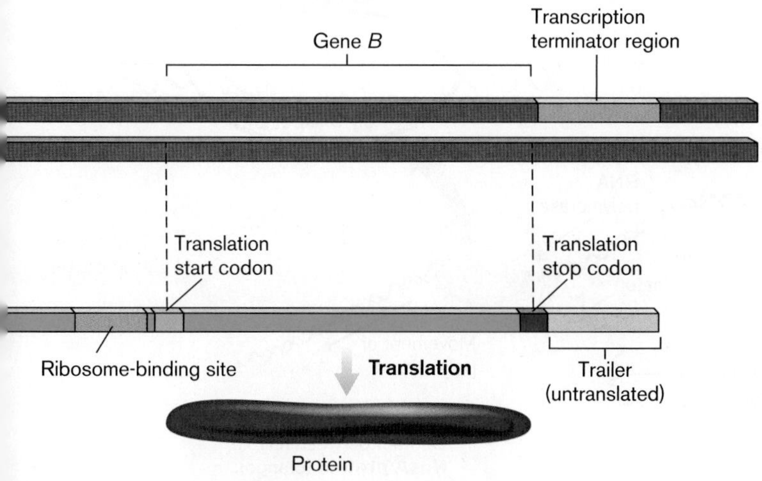

subsequent rNTP transfers a ribonucleoside monophosphate (rNMP) to the growing chain while releasing a pyrophosphate (PP_i):

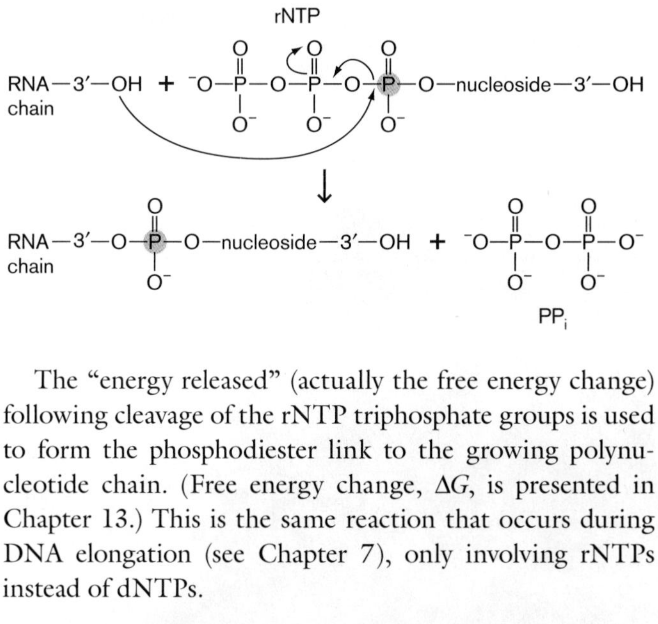

The "energy released" (actually the free energy change) following cleavage of the rNTP triphosphate groups is used to form the phosphodiester link to the growing polynucleotide chain. (Free energy change, ΔG, is presented in Chapter 13.) This is the same reaction that occurs during DNA elongation (see Chapter 7), only involving rNTPs instead of dNTPs.

Transcription elongation. A transcribing complex retains the sigma factor until about nine bases have been joined, and then the sigma factor dissociates (**Fig. 8.6**, step 4). The newly liberated sigma factor can recycle onto an unbound core RNA polymerase to direct another round of promoter binding. Meanwhile, the original RNA polymerase continues to move along the template, synthesizing RNA at approximately 45 bases per second. The NusG protein binds to RNA polymerase and helps to control the rate of elongation and the amount of pausing during transcription. As the DNA helix unwinds, a 17-bp transcription bubble forms and moves with the polymerase complex. DNA unwinding generates positive DNA supercoils ahead of the advancing bubble. The supercoils are removed by DNA topoisomerase enzymes (see Section 7.2).

Transcription termination. How does RNA polymerase know when to stop? Again, the secret is in the sequence. All bacterial genes use one of two types of known transcription termination signals: either Rho-dependent or Rho-independent.

Rho-dependent termination relies on a protein called Rho and an ill-defined sequence at the untranslated 3′ end of the gene that appears to be a strong pause site. **Rho factor** binds to an exposed region of RNA after the ORF at C-rich sequences that lack obvious secondary structure. This is the transcription terminator pause site. Rho monomers assemble as a hexamer around the RNA (**Fig. 8.7A**). Then, like a person pulling a raft to shore by a rope tied to a tree, Rho pulls itself to the paused RNA polymerase by threading downstream RNA through the ring via an intrinsic ATPase activity. Once Rho touches the polymerase, an RNA-DNA helicase activity also performed by Rho appears to unwind the RNA-DNA heteroduplex, which releases the completed RNA molecule and frees the RNA polymerase.

The second type of termination, called Rho-independent termination or intrinsic termination, occurs in the absence of Rho. Rho-independent termination requires a GC-rich region of RNA roughly 20 bp upstream from the 3′ terminus, as well as four to eight consecutive uridine residues (poly-U site) at the terminus (**Fig. 8.7B**). The GC-rich sequence contains complementary bases and forms a stem loop structure that contacts RNA polymerase. Contact halts nucleotide addition, causing RNA polymerase to pause. A protein called NusA stimulates transcription pause at these sites. While the polymerase is paused, the DNA-RNA duplex is weakened because the poly-U–poly-A base pairs at the 3′ terminus contain only two hydrogen bonds per pair, so they are easier to melt. Melting the hybrid molecule releases the transcript and halts transcription. The pause in polymerase movement, and thus transcription, is important to prevent tighter base-pairing downstream of the UA region.

Antibiotics Reveal RNA Synthesis Machines

To be useful in medicine, all antibiotics must meet two fundamental criteria: They must kill or inhibit the growth of a pathogen, and they must not harm the host. Antibiotics used against bacteria, for instance, must selectively attack features of bacterial targets not shared with eukaryotes. Many antibiotics possess highly specific modes of action, binding to and altering the activity of one specific "machine" of one type of cell.

One example is the antibiotic rifamycin B (**Fig. 8.8A**), produced by the actinomycete *Amycolatopsis rifamycinica* (**Fig. 8.8B**). Rifamycin B and its derivitives selectively target bacterial RNA polymerase and bind to the polymerase's beta subunit near the Mg^{2+} active site, thus blocking the

FIGURE 8.7 ▪ The termination of transcription.
A. Rho-dependent termination requires Rho factor but not NusA.
B. Rho-independent termination requires NusA but not Rho.

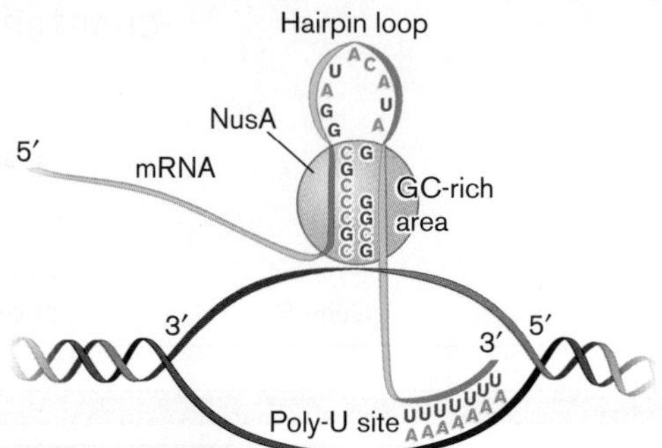

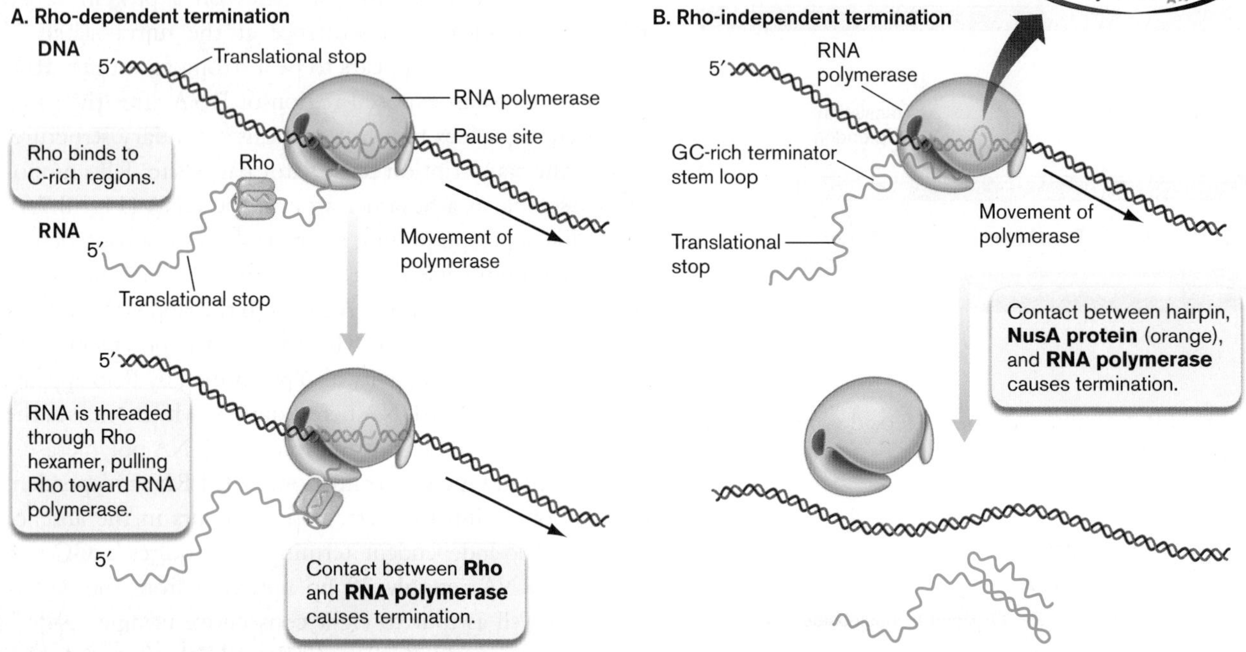

A.

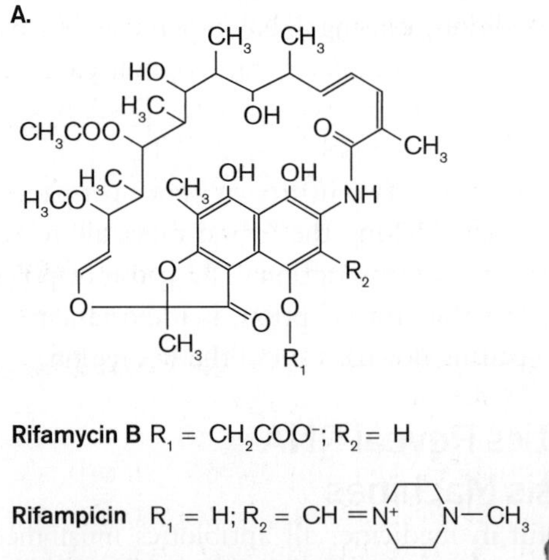

B.

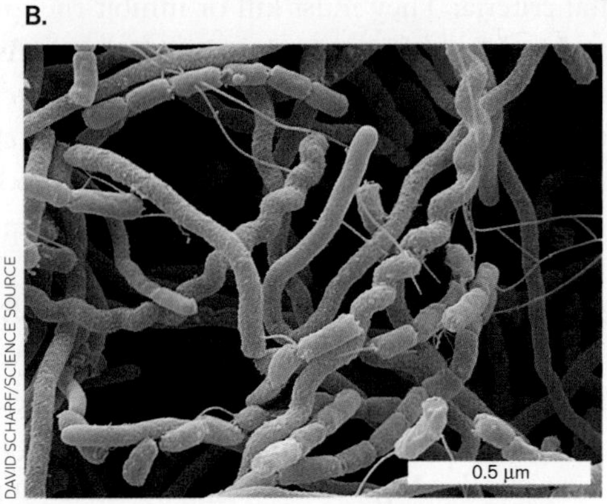

C.

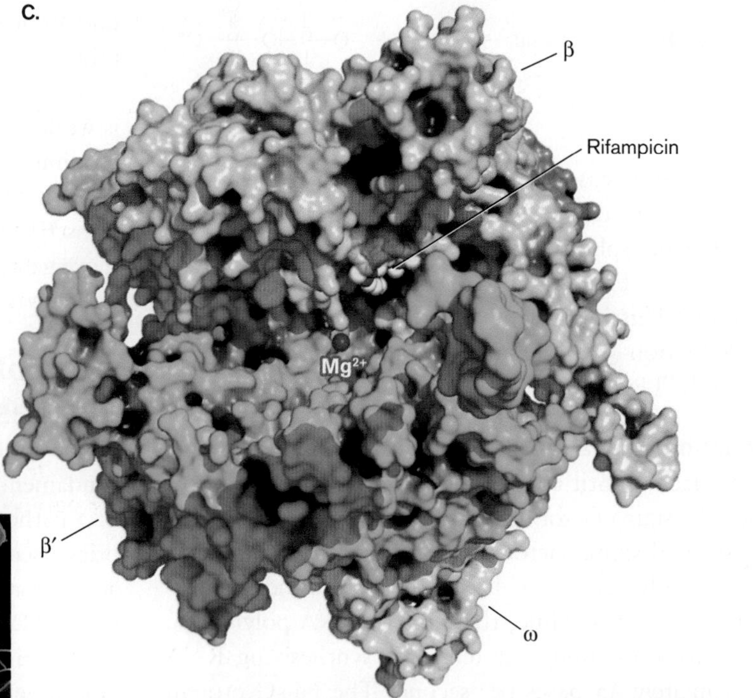

FIGURE 8.8 ▪ Structure and mode of action of rifamycin. A. Structure of rifamycin. The R groups indicated are added to alter the structure and pharmacology of the basic structure. **B.** Electron micrograph of *Amycolatopsis*. **C.** Rifampicin binds to the beta subunit and blocks the channel where elongating mRNA exits the RNA polymerase of *Thermus aquaticus* (PDB 1I6V). *Source:* Part C modified from Campbell, et al. 2001. *Cell* **104:** 910–912.

RNA exit channel (**Fig. 8.8C**). RNA polymerase can carry out two or three polymerization steps, but then it stops because the nascent RNA cannot exit. Rifamycin does not prevent RNA polymerase from binding to promoters, and because an mRNA molecule already passing through the exit channel masks the rifamycin-binding site, the drug cannot bind to RNA polymerases already transcribing DNA. The semisynthetic derivative of this drug, rifampicin (also called rifampin), is used for treating tuberculosis, leprosy, and bacterial meningitis caused by *Neisseria meningitidis.*

Thought Question

8.5 If rifamycin targets bacterial RNA polymerase, why doesn't it also kill its producer, the bacterium *Amycolatopsis mediterranei?*

Actinomycin D (**Fig. 8.9A**) is another antibiotic produced by an actinomycete. Its phenoxazone ring is a planar structure that inserts, or intercalates, between GC base pairs within DNA. The side chains then extend in opposite directions along the minor groove (**Fig. 8.9B**). Because it mimics a DNA base, actinomycin D blocks the elongation phase of transcription; but because it binds <u>any</u> DNA, it is <u>not</u> selective for bacteria. It can be used to treat human cancers because cancer cells replicate rapidly, but it has severe side effects because it inhibits DNA synthesis in normal cells too. (Antibiotics are discussed further in Chapter 27.)

Antibiotics helped scientists determine the details of RNA and protein synthesis. First, purified RNA polymerase and ribosomes were used to demonstrate transcription and translation in cell-free systems (in test tubes). Then, protein subunits of RNA polymerase extracted from antibiotic-resistant and antibiotic-sensitive strains of *E. coli* were mixed together to reconstruct chimeric molecules. To determine which RNA polymerase subunit was targeted by the antibiotic rifamycin, for instance, RNA polymerases from rifamycin-sensitive and rifamycin-resistant strains were purified from cell extracts. The component parts of the polymerases were separated and a chimeric

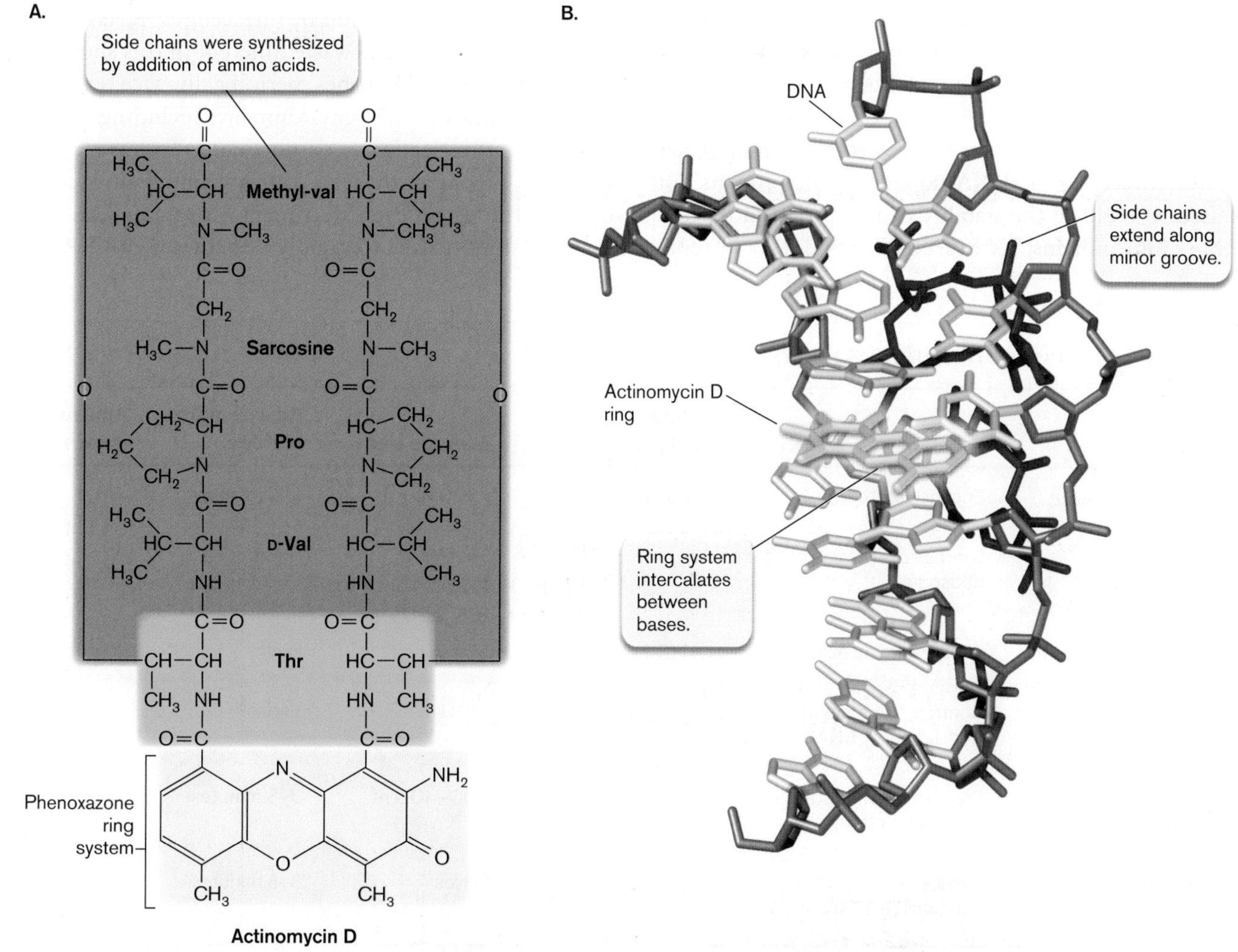

FIGURE 8.9 ▪ Structure and mode of action of actinomycin D. This antibiotic inserts its ring structure **(A)** between parallel DNA bases and wraps its side chains along the minor groove **(B)**. (PDB code for B: 1DSC)

RNA polymerase was reassembled (like a 3D jigsaw puzzle) by mixing subunits from antibiotic-sensitive and antibiotic-resistant polymerases. The reconstituted polymerase preparation was tested to see whether it could make RNA in the presence of rifamycin. Resistance was found to occur only when the subunit conveying resistance was added—in this case, the beta subunit. Investigators used this information in subsequent in vitro assays and X-ray crystallography to learn more about the enzymatic mechanism of the beta subunit.

Different Classes of RNA Have Different Functions

Not all RNA molecules are translated to protein. There are several classes of RNA in bacteria, each with a different function (**Table 8.1**). The class of RNA molecules that encode proteins is called **messenger RNA (mRNA)**. Molecules of mRNA average 1,000–1,500 bases in length but can be much longer or shorter, depending on the size of the protein they encode. Another class of RNA, called **ribosomal RNA (rRNA)**, forms the scaffolding on which ribosomes are built. As will be discussed later, rRNA also forms the catalytic center of the ribosome. **Transfer RNA (tRNA)** molecules ferry amino acids to the ribosome. A unique property of rRNA and tRNA is the presence of unusual modified bases not found in other types of RNA.

The fourth important class of RNA is called **small RNA (sRNA)**. Molecules of sRNA do not encode proteins but are used to regulate the stability or translation of specific mRNAs into proteins. As will be discussed later in more detail, all mRNA molecules contain untranslated leader sequences that precede the actual coding region. Complementary sequences within some of these RNA leader regions snap back on themselves to form double-stranded stems and stem loop structures that can obstruct ribosome access and limit translation. Regulatory sRNAs can base-pair with these regions in mRNA and either disrupt or stabilize intrastrand stem structures.

The final two classes of RNA are **tmRNA**, which has properties of both transfer RNA and messenger RNA (discussed in **eTopic 8.4**), and **catalytic RNA**. While catalytic RNA molecules, also referred to as **ribozymes**, are usually found associated with proteins, the enzymatic (catalytic) activity actually resides in the RNA portion of the complex rather than in the protein. All of these functional RNAs may represent remnants of the ancient "RNA world," where the earliest ancestral cells were built of RNA parts whose functions were later assumed by proteins. The RNA-world model is discussed in Chapter 17.

RNA Stability

Once released from RNA polymerase, most prokaryotic mRNA transcripts are doomed to a short existence, owing to their degradation by intracellular RNases. Why would a cell tolerate the waste inherent in this short-lived use for mRNA? Microbes periodically face extremely rapid changes in their environment, including changes in temperature, salinity, or nutrients. To survive, cells must be prepared to react quickly and halt synthesis of superfluous or even detrimental proteins. An effective way to do this is to rapidly destroy the mRNAs that

TABLE 8.1 Classes of RNA in Bacteria[a]

RNA class	Function	Number of types	Average size	Approximate half-life	Unusual bases
mRNA (messenger RNA)	Encodes protein	Thousands	1,500 nt	3–5 minutes	No
rRNA (ribosomal RNA)	Synthesizes protein as part of ribosome	3	5S: 120 nt; 16S: 1,542 nt; 23S: 2,905 nt	Hours	Yes
tRNA (transfer RNA)	Shuttles amino acids	41 (86 genes)	80 nt	Hours	Yes
sRNA[b] (small RNAs, or regulatory RNAs)	Controls transcription, translation, or RNA stability	20–30	<100 nt	Variable	No
tmRNA (properties of transfer and messenger RNA)	Frees ribosomes stuck on damaged mRNA	Roughly one per species	300–400 nt	3–5 minutes	No
Catalytic RNA	Carries out enzymatic reactions (e.g., RNase P)	?	Varies	3–5 minutes	No

[a] Six major classes of RNA exist in all bacteria. They differ in function, quantity, average size, and half-life, as well as whether they contain modified bases.

[b] Small RNAs include antisense RNA and micro-RNA.

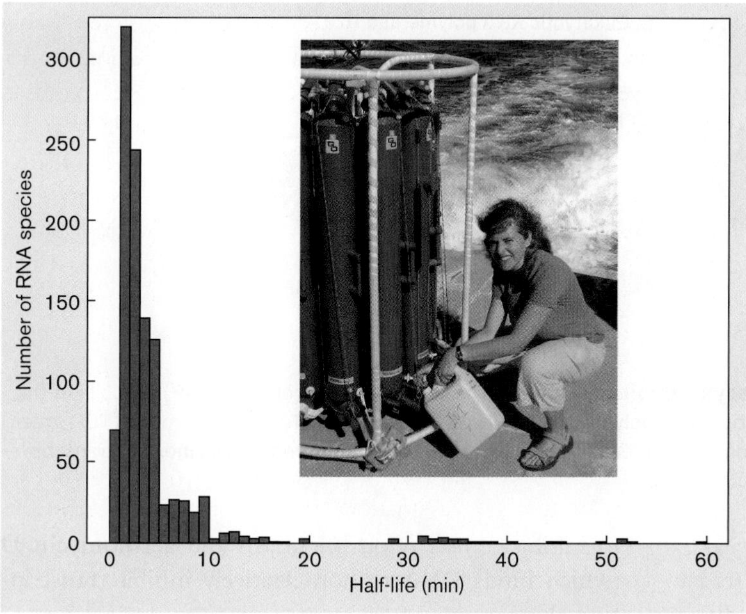

COURTESY OF CLAUDIA STEGLICH

FIGURE 8.10 ▪ Claudia Steglich (inset) collecting water samples for the study of *Prochlorococcus*. The half-lives of most *Prochlorococcus* transcripts are short, despite the long doubling time of this bacterium. *Source*: C. Steglich et al. 2010. *Genome Biol.* **11**:R54, fig. 1b.

encode those proteins. However, rapid mRNA degradation makes it necessary to continually transcribe those genes as long as their products are needed.

RNA stability is measured in terms of half-life, which is the length of time the cell needs to degrade half the molecules of a given RNA species. The modified bases found in rRNAs and tRNAs render them relatively resistant to RNase digestion. These more stable RNA molecules have half-lives on the order of hours. In contrast, the average half-life for mRNA is 1–3 minutes but can be as short as 15 seconds. Some mRNAs are far more stable, with half-lives of 10 minutes or longer.

The short half-lives of mRNA are found not only in fast-growing bacteria like *E. coli*, but also in slow-growing bacteria. The marine cyanobacterium *Prochlorococcus* doubles about once per day, but Claudia Steglich (**Fig. 8.10**) and colleagues at the Massachusetts Institute of Technology and the University of Freiburg showed that even in this slow-growing organism, mRNA is rapidly turned over, with a median mRNA half-life of 2.4 minutes. Thus, even slow growers rapidly recycle their mRNA, indicating that, like their fast-growing cousins, they must be primed for quick responses to environmental change.

Transcription in Archaea and Eukaryotes

Across all three domains—Bacteria, Archaea, and Eukarya—transcription of DNA into RNA proceeds in a similar manner, although there are important differences. Transcription is performed in all three domains by multisubunit DNA-dependent RNA polymerases. The catalytic cores of these polymerases (for example, the beta and beta-prime subunits in bacteria) are evolutionarily related, and may have their origin in the "RNA World" (see Chapter 17) as an RNA-dependent RNA polymerase. Homologs of the alpha homodimer of bacteria are also found in the RNA polymerases of Archaea and Eukarya, and all three domains share the elongation factor NusG. Given the conservation of these key components, it is not surprising that during the elongation phase, the mechanism of RNA polymerization in all three domains is similar. Where Archaea and Eukarya differ significantly from Bacteria is at the initiation and termination stages of transcription.

Archaea have a single RNA polymerase, while eukaryotes have three RNA polymerases. Eukaryotic RNA polymerases I and III synthesize stable, untranslated RNAs (rRNA, tRNA), while RNA polymerase II synthesizes mRNA. Structurally, the RNA polymerase of archaea is more similar to RNA polymerase II of eukaryotes than to the bacterial polymerase (**Fig. 8.11**; also see Section 19.1). Whereas the bacterial polymerase core is composed of only 5 subunits, the archaeal and eukaryotic versions contain 12 or more.

Rather than sigma factors, eukaryotes and archaea use other proteins to recruit the RNA polymerase to the promoter; hence, the sequence motifs within the promoter are also different. The TATA-binding protein (TBP) recognizes a motif in the promoter called the TATA box. TBP bends the DNA and recruits transcription factor B (TFB in archaea, TFIIB in eukaryotes) to the TFB recognition element of the promoter. The RNA polymerase binds TBP and TFB, and this complex is sufficient to generate the transcription bubble that initiates transcription. A third transcription factor, called TFE in archaea and TFIIE in eukaryotes, also plays a role in initiation when conditions are suboptimal. Eukaryotes employ additional initiator proteins, but these three are sufficient for initiation at strong promoters. The initiator proteins remain at the promoter (TBP, TFB/TFIIB) or are removed from the RNA polymerase shortly after chain elongation begins (TFE/TFIIE).

Note: Because the transcription machinery of archaea is different from that of bacteria, archaea are usually resistant to antibacterial antibiotics that target those processes.

Transcription is terminated in archaea by a mechanism that is similar to Rho-independent termination in

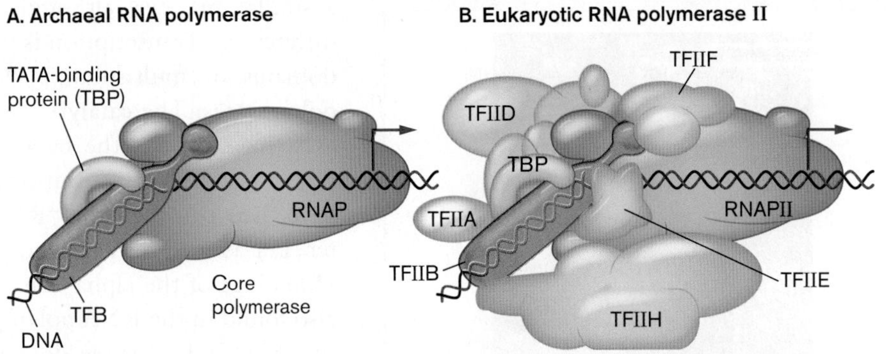

FIGURE 8.11 ■ RNA polymerases from Archaea and Eukarya exhibit homology. The core RNA polymerase (RNAP) subunits of the archaeon *Sulfolobus* **(A)** show homology to those of the eukaryotic RNA polymerase II (RNAP II) **(B)**. The TATA-binding protein (TBP, green) and transcription factor B (TFB, pink) of *Sulfolobus* also show homology to eukaryotic counterparts. The bent red arrows indicate transcription start sites.

bacteria: a string of Us at the 3′ UTR produces a weak RNA-DNA hybrid molecule that easily melts to release a completed mRNA (**Fig. 8.7**). Termination in archaea differs from termination in bacteria because it does not seem to require adjacent RNA stem loop structures. And unlike bacteria, archaea have many genes with multiple terminators, which can result in mRNAs with different 3′ UTR lengths depending on the terminator used. Termination in eukaryotes is less well understood, but it appears to involve specific terminator protein complexes for different types of RNA.

As in bacteria, many genes in archaea are found in operons. The mRNA transcripts from these operons will contain open reading frames for the translation of two or more genes. In contrast, most eukaryotic mRNAs are monocistronic. Like bacteria, archaea show much less posttranscriptional processing of mRNA than eukaryotes do. Archaeal mRNAs do not contain a 7-methlyguanosine cap on the 5′ ends of the mRNA and are not polyadenylated on their 3′ ends. While some archaeal genes contain introns, which are spliced out of the mRNA prior to translation, introns are not nearly as common as they are in eukaryotic genes. Thus, archaeal transcription shares features with both bacteria and eukaryotes.

To Summarize

- **RNA polymerase binds promoter DNA**, forming the closed complex.
- **The DNA strands separate** (melt) and form a "bubble" of DNA around the polymerase called the **open complex**.
- **Rho-dependent and Rho-independent** mechanisms mediate transcription termination.
- **Antibiotics that inhibit bacterial transcription include rifamycin B** (which binds to RNA polymerase to inhibit transcription initiation) and **actinomycin D** (which binds DNA to nonselectively inhibit transcription elongation).
- **Messenger RNA** (mRNA) molecules encode proteins.
- **Noncoding RNAs** include **ribosomal RNAs** (rRNAs), **transfer RNAs** (tRNAs), and **small RNAs**. Some noncoding RNAs can regulate gene expression (sRNA), have catalytic activity (catalytic RNA), or function as a combination of tRNA and mRNA (tmRNA).
- **The transcription machinery of archaea** is related more to eukaryotes than to bacteria, but the **organization of genes into operons** and a general **lack of posttranscriptional processing** is more similar to bacteria.

8.3 Translation of RNA to Protein

Once a gene has been copied into mRNA, the next stage is **translation**, the decoding of the RNA message to synthesize protein. An mRNA molecule can be thought of as a sentence in which triplets of nucleotides, called **codons**, represent individual words, or amino acids (**Fig. 8.12**). Ribosomes are the machines that read the language of mRNA and convert, or translate, it into protein. Translation involves numerous steps, the first of which is the search by ribosomes for the beginning of an mRNA protein-coding region (ORF). Because the code consists of triplet codons, a ribosome must start translating at precisely the right base (in the right frame), or the product will be gibberish. Before we discuss how the ribosome finds the right reading frame, let's review the code itself and the major players in translation.

2nd Base

1st Base	U: RNA codon	U: Amino acid	C: RNA codon	C: Amino acid	A: RNA codon	A: Amino acid	G: RNA codon	G: Amino acid	3rd Base
U	UUU	Phe	UCU	Ser	UAU	Tyr	UGU	Cys	U
U	UUC	Phe	UCC	Ser	UAC	Tyr	UGC	Cys	C
U	UUA	Leu	UCA	Ser	UAA	Stop	UGA	Stop	A
U	UUG	Leu	UCG	Ser	UAG	Stop	UGG	Trp	G
C	CUU	Leu	CCU	Pro	CAU	His	CGU	Arg	U
C	CUC	Leu	CCC	Pro	CAC	His	CGC	Arg	C
C	CUA	Leu	CCA	Pro	CAA	Gln	CGA	Arg	A
C	CUG	Leu	CCG	Pro	CAG	Gln	CGG	Arg	G
A	AUU	Ile	ACU	Thr	AAU	Asn	AGU	Ser	U
A	AUC	Ile	ACC	Thr	AAC	Asn	AGC	Ser	C
A	AUA	Ile	ACA	Thr	AAA	Lys	AGA	Arg	A
A	AUG	Met	ACG	Thr	AAG	Lys	AGG	Arg	G
G	GUU	Val	GCU	Ala	GAU	Asp	GGU	Gly	U
G	GUC	Val	GCC	Ala	GAC	Asp	GGC	Gly	C
G	GUA	Val	GCA	Ala	GAA	Glu	GGA	Gly	A
G	GUG	Val	GCG	Ala	GAG	Glu	GGG	Gly	G

Single-letter abbreviations for amino acids

Ala (A)
Arg (R)
Asn (N)
Asp (D)
Cys (C)
Gln (Q)
Glu (E)
Gly (G)
His (H)
Ile (I)
Leu (L)
Lys (K)
Met (M)
Phe (F)
Pro (P)
Ser (S)
Thr (T)
Trp (W)
Tyr (Y)
Val (V)

FIGURE 8.12 ■ The standard genetic code. Codons within a single box encode the same amino acid. Blue- and green-highlighted amino acids are encoded by codons in two boxes. Stop codons are highlighted red. Often, single-letter abbreviations for amino acids are used to convey protein sequences (see legend).

The Genetic Code and tRNA Molecules

When learning a new language, we need a dictionary that converts words from one language into the other. In the case of gene expression, how do we convert from the 4-base language of RNA to the protein language of 20 amino acids? Through painstaking effort, Marshall Nirenberg, Har Gobind Khorana, and colleagues cracked the molecular code and found that each codon (a triplet of nucleotides) represents an individual amino acid. Remarkably, and with few exceptions, the code operates universally across species.

In the standard genetic code (**Fig. 8.12**), we can see that most amino acids have multiple codon synonyms. Glycine (Gly), for example, is translated from four different codons, and leucine (Leu) from six, but methionine (Met) and tryptophan (Trp) each have only one codon. Because more than one codon can encode the same amino acid, the code is said to be degenerate or redundant. Notice that in most cases, synonymous codons differ only in the last base. How the cell handles this degeneracy (redundancy) in the code will be explained later.

Only 61 out of a possible 64 codons specify amino acids. The three triplets that do not encode amino acids (UAA, UAG, and UGA) are equally important, for they tell the ribosome when to stop reading a gene sentence. Called **stop codons**, they trigger a series of events (to be described later) that dismantle ribosomes from mRNA and release a completed protein.

Thought Question

8.6 How might the redundancy of the genetic code be used to establish evolutionary relationships between different species? *Hints:* 1. Genomes of different species have different overall GC content. 2. Within a given genome one can find segments of DNA sequence with a GC content distinctly different from that found in the rest of the genome.

Translation is the process of reading (decoding) the string of codons in mRNA to make a string of amino acids. Decoding mRNA for translation is biochemically different from decoding DNA for transcription. Specificity during transcription relies on the ability of incoming RNA nucleotides to form complementary base pairs with the bases along the DNA template strand. In contrast, during translation the mRNA template is not directly contacted by the amino acids it encodes. Rather, amino acids are attached to small adapter RNAs, called tRNAs, which have RNA sequences called **anticodons** that match and bind to specific codons on the mRNA being translated.

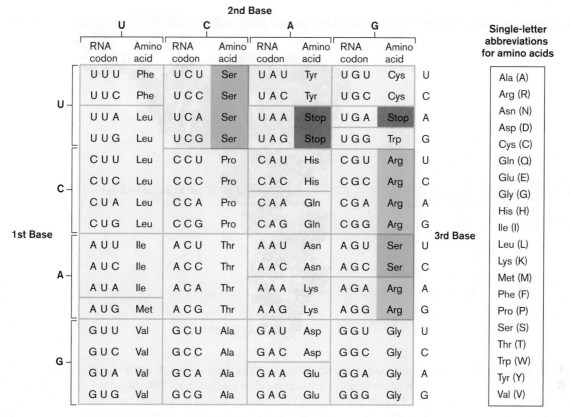

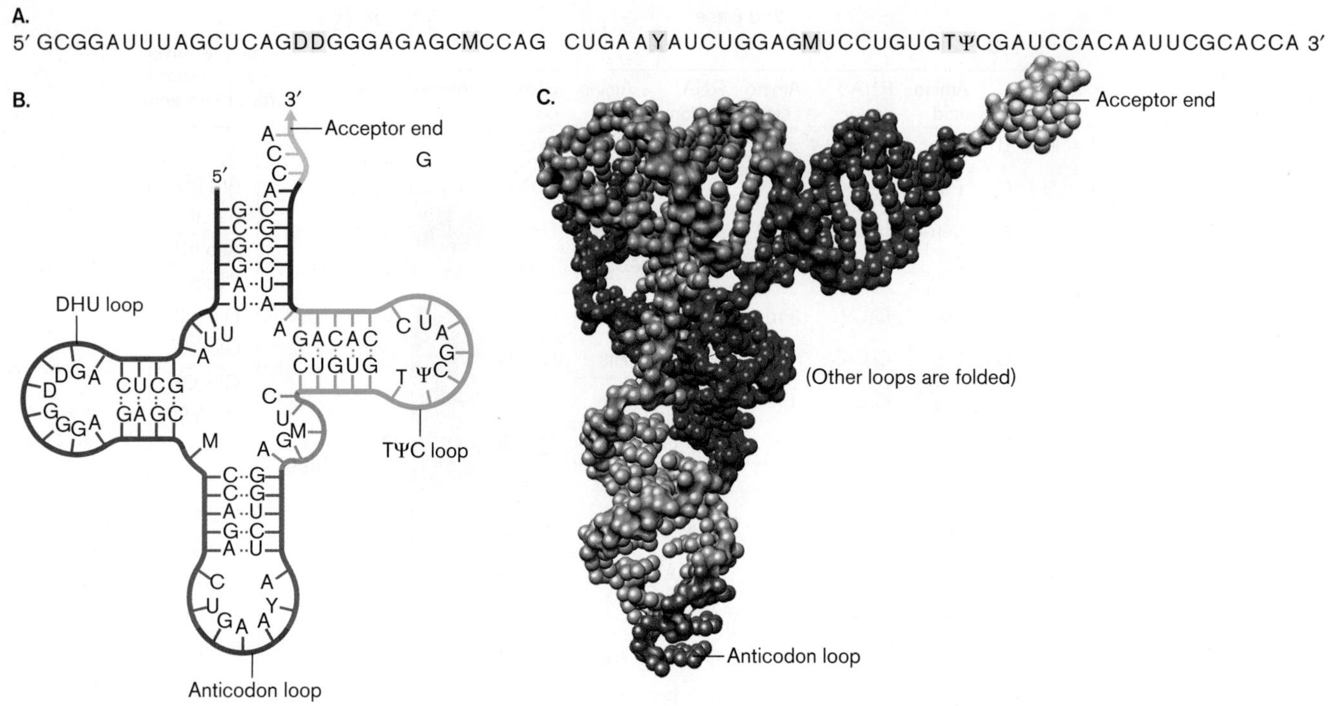

FIGURE 8.13 ■ **Transfer RNA.** **A.** Primary sequence of a tRNA molecule. The letters D, M, Y, T, and Ψ stand for modified bases found in tRNA. **B.** Cloverleaf structure. DHU (or D) is dihydrouracil, which occurs only in the DHU loop; TΨC consists of thymine, pseudouracil, and cytosine bases that occur as a triplet in the TΨC loop. The DHU and TΨC loops are named for the modified nucleotides that are characteristically found there. **C.** Three-dimensional structures. The anticodon loop binds to the codon, while the acceptor end binds to the amino acid. (PDB code: 1GIX)

Approximately 80 bases in length (**Fig. 8.13A**), a tRNA laid flat looks like a clover leaf with three loops (**Fig. 8.13B**). In nature, however, the molecule folds into the "boomerang-like" 3D structure shown in **Figure 8.13C**. The bottom loop of all tRNA molecules (as depicted in **Fig. 8.13B** and **C**) harbors the anticodon triplet that base-pairs with codons in mRNA (**Fig. 8.14**). As a result, this loop is called the anticodon loop. Notice that codon-anticodon pairings are aligned in an antiparallel manner. Most tRNA molecules begin with a 5′ G, and all end with a 3′ CCA, to which an amino acid attaches (see **Figs. 8.13** and **8.14**). Because the 3′ end of the tRNA accepts the amino acid, it is called the acceptor end.

As noted previously, in the genetic code many amino acids have codon synonyms. This redundancy is found primarily in the third position of the codon (for example, UU<u>U</u> and UU<u>C</u> both encode phenylalanine). A single tRNA for phenylalanine can recognize both codons because of "wobble" in the first position of the anticodon, which corresponds to the third position of the codon (remember, as with DNA, base-pairing between RNA strands is antiparallel). The wobble is due in part to the curvature of the anticodon loop and to the use of an unusual base (inosine) at this position in some tRNA molecules. The wobble structure allows one anticodon to pair with several codons differing only in the third position. Consequently, while 61 of the 64 codons encode amino acids, many microbes have economized with this wobble property and contain fewer than 61 tRNAs in their genome.

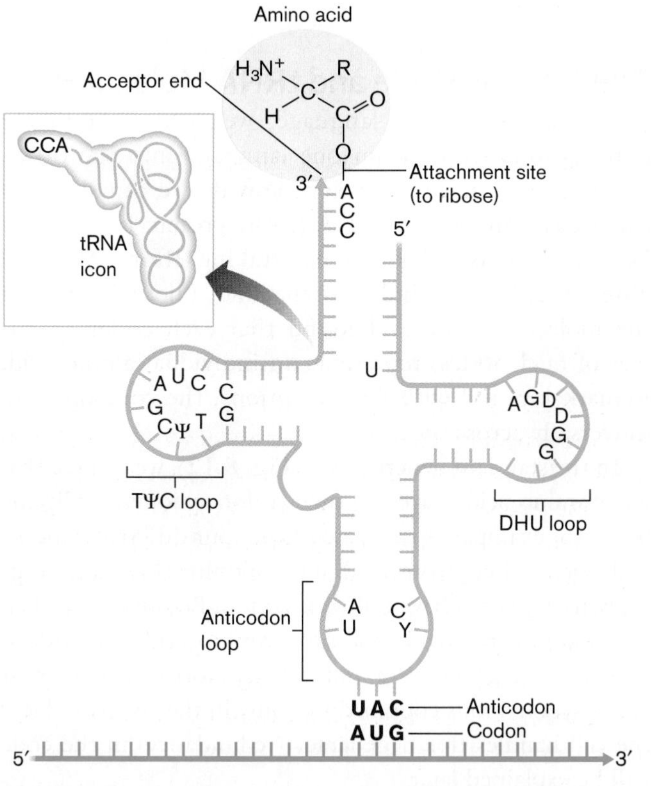

FIGURE 8.14 ■ **Codon-anticodon pairing.** The tRNA anticodon consists of three nucleotides at the base of the anticodon loop. The anticodon hydrogen-bonds with the mRNA codon in an antiparallel fashion. This tRNA is "charged" with an amino acid covalently attached to the 3′ end.

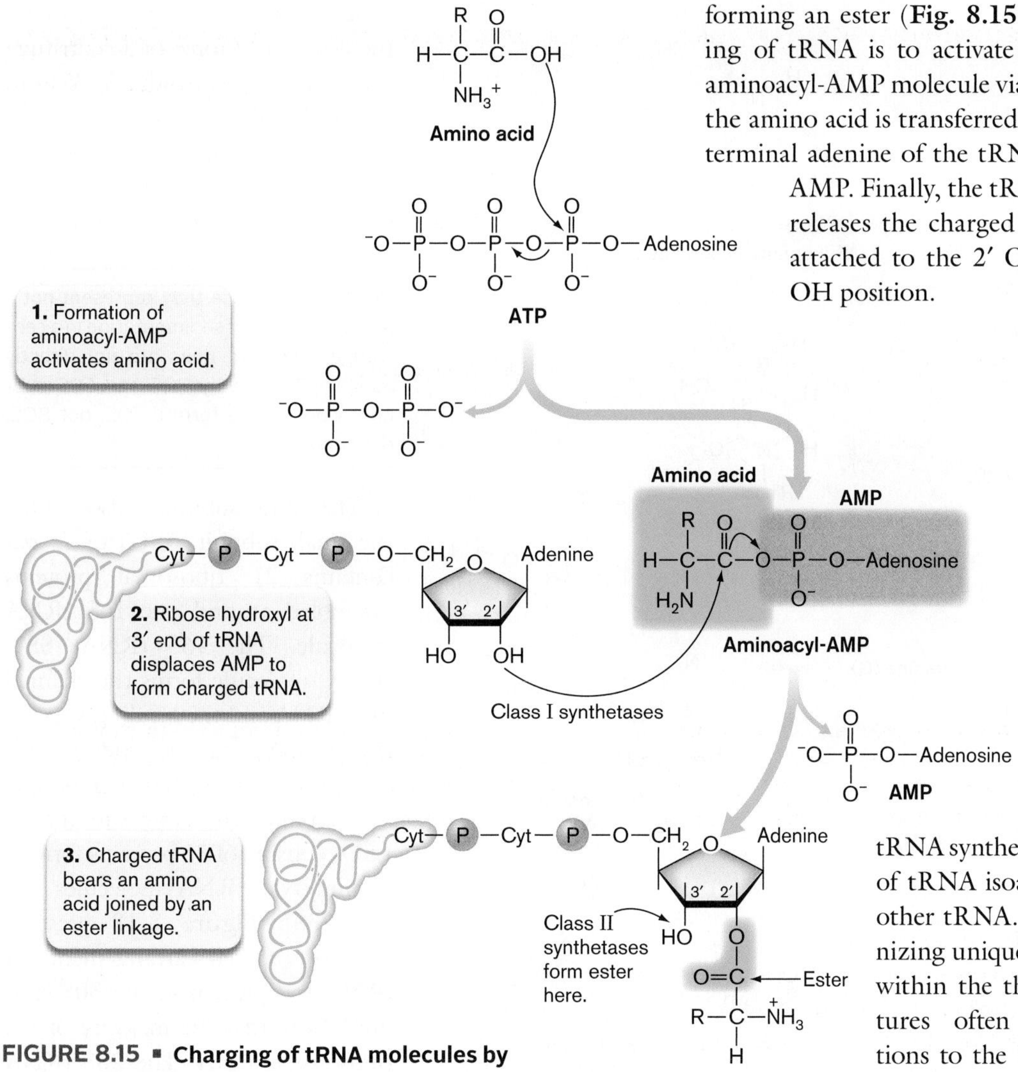

FIGURE 8.15 ■ Charging of tRNA molecules by aminoacyl-tRNA synthetases. At the end of this process, each amino acid is attached to the 3′ end of CCA on a specific tRNA molecule. Curved arrows indicate nucleophilic attack by electrons.

Aminoacyl-tRNA Synthetases Attach Amino Acids to tRNA

When a tRNA anticodon (for instance, GCG) pairs with its complementary codon (CGC in this case) during translation, the ribosome has no way of checking that the tRNA is attached (that is, charged) to the "correct" amino acid (here, arginine). Consequently, each tRNA must be charged with the proper amino acid before it encounters the ribosome. How are amino acids correctly matched to tRNA molecules and affixed to their 3′ ends?

Matching and attaching the correct amino acid to the correct tRNA is the function of enzymes called **aminoacyl-tRNA synthetases**. Every aminoacyl-tRNA synthetase has a specific binding site for its cognate (matched) amino acid. Each enzyme also has a site that recognizes the correct tRNA and an active site that joins the carboxyl group of the amino acid to the 3′ OH (class II synthetases) or 2′ OH (class I synthetases) of the tRNA by forming an ester (**Fig. 8.15**). The first step in the charging of tRNA is to activate the amino acid, forming an aminoacyl-AMP molecule via a reaction with ATP. Second, the amino acid is transferred to the hydroxyl residue of the terminal adenine of the tRNA, resulting in the release of AMP. Finally, the tRNA synthetase disengages and releases the charged tRNA. Amino acids initially attached to the 2′ OH are then moved to the 3′ OH position.

Most bacteria have 20 aminoacyltransferases—one for each amino acid. Some bacteria, however, have only 18, choosing instead to modify already-charged glutamyl-tRNA and aspartyl-tRNA by adding an amine to make glutamine and asparagine derivatives.

Transfer RNAs that have different RNA sequences but carry the same amino acid are referred to as isoacceptors. Each aminoacyl-tRNA synthetase must recognize its own set of tRNA isoacceptors but not bind to any other tRNA. Specificity is based on recognizing unique features of the tRNA located within the three loops. These unique features often involve unusual modifications to the bases, which accounts for the strange letter codes seen in **Figure 8.13A** (D and Ψ). Structures of some of the odd bases are shown in **Figure 8.16**. Wybutosine (yW), for example, has three rings instead of the two found in a normal purine. In the cloverleaf structure of tRNA, two of the loops are named after modifications that are invariantly present in those loops. One is called the TΨC loop because in every tRNA this loop has the nucleotide triplet thymidine–pseudouridine (Ψ)–cytidine. The other loop always contains dihydrouridine and is called the DHU loop (see **Figs. 8.13** and **8.14**).

How do these unusual bases end up in tRNA? During transcription of the tRNA genes, normal, unmodified bases are incorporated into the transcript. Some of these are modified later by specific enzymes to make inosine and other odd bases. The remarkable stability of tRNA molecules is explained, in part, by these unusual bases, because they are poor substrates for RNases.

The Ribosome, a Translation Machine

Ribosomes catalyze the linkage of amino acids during translation, using mRNA as the code and charged tRNAs as the source of amino acids. A ribosome is a massive

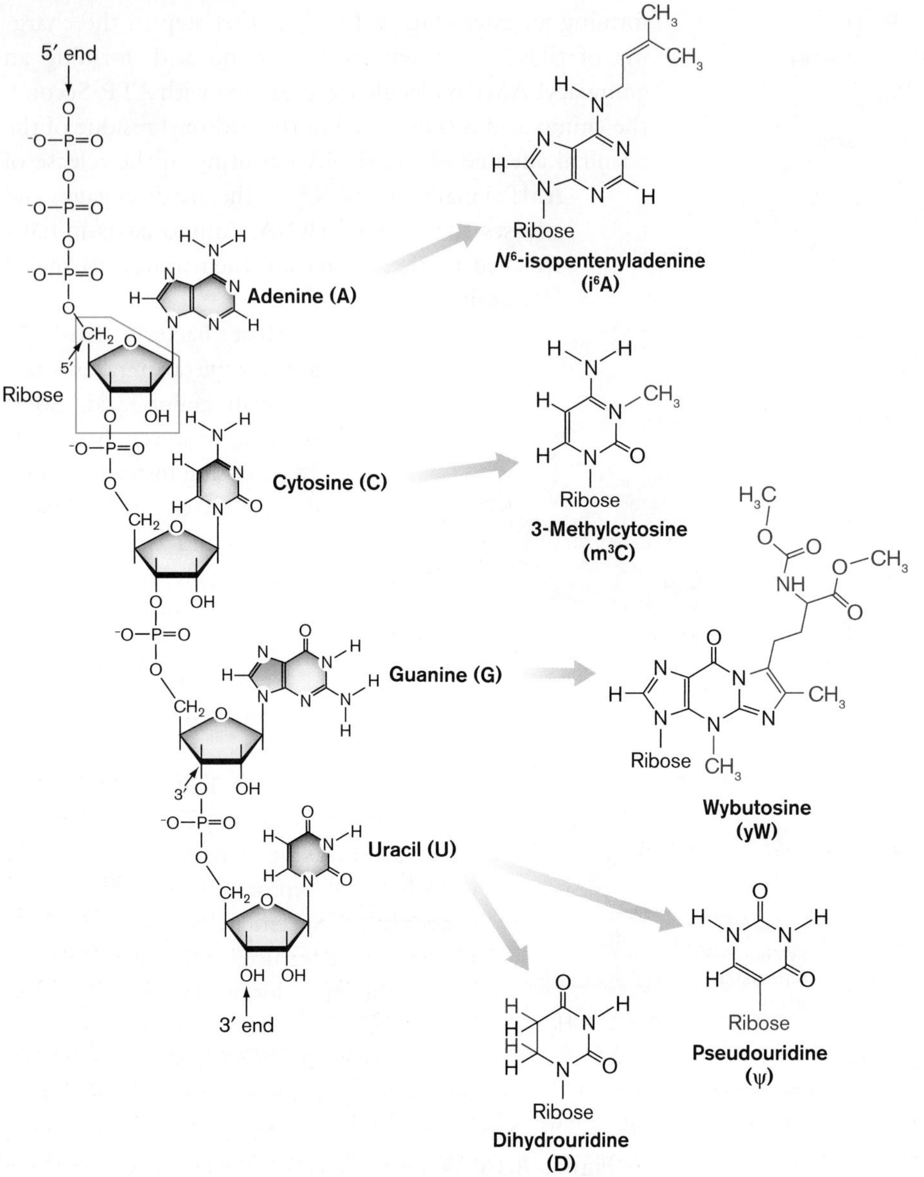

FIGURE 8.16 ▪ Modified bases in tRNA and rRNA molecules. Modifications to the canonical RNA bases are highlighted in red.

complex of protein and ribosomal RNA (rRNA). Whereas amino acid and nucleotide monomers average 110 and 325 daltons, respectively, a ribosome is over 1,000,000 daltons. The rRNA component forms the core structure of the enzyme, including the binding sites for tRNA, as well as the channels for mRNA and the elongating polypeptide. Remarkably, it is also the rRNA component, not the protein, that catalyzes the amino acid linkages (peptide bonds), through an enzyme called **peptidyltransferase**. Peptidyltransferase is a **ribozyme**, which is an RNA molecule that carries out catalytic activity. Proteins surrounding this active center offer structural assistance to ensure that the RNA is folded properly and to interact with tRNA substrates.

Ribosomes are composed of two complex subunits, each of which includes rRNA and protein components. In bacteria and archaea, the subunits are named 30S and 50S for their "size" in Svedberg units. Svedberg units represent the rate at which a molecule sediments under the centrifugal force of a centrifuge (discussed in eAppendix 3). Within the living cell, the two subunits exist separately but come together on mRNA to form the functional 70S ribosome (**Fig. 8.17**).

Note: Because they represent not a mass but a rate of sedimentation in a centrifuge, Svedberg units are not directly additive. That's why 30S and 50S subunits combine to form a 70S, not 80S, ribosome.

The 30S subunit (also called the small subunit, SSU) of *E. coli* contains 21 ribosomal proteins assembled around one 16S rRNA molecule. The 16S rRNA (SSU rRNA) molecule forms the channel for mRNA and the binding sites for tRNA, and it matches codons with anticodons during translation. The 50S subunit (the large subunit, or LSU) consists of 33 proteins formed around two rRNA molecules (5S and 23S). **Figure 8.18** presents the 3D spatial arrangement of rRNA and protein in the 50S subunit. Note that the majority of the ribosome is RNA. The 23S rRNA (LSU rRNA) mediates the peptidyltransferase activity of the ribosome (**Fig. 8.18** blowup, loop V).

How does such a complex molecular machine get built? As discussed in greater detail in **eTopic 8.1**, assembly begins with the transcription of the genes that encode ribosomal RNA (rRNA). These genes in the DNA are sometimes called rDNA.

The 16S, 23S, and 5S rRNAs are initially transcribed as one RNA molecule and posttranscriptionally processed into separate rRNA molecules. During rRNA transcription,

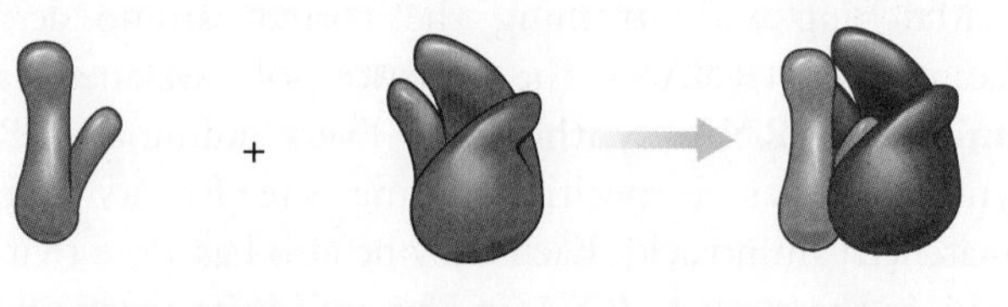

FIGURE 8.17 ▪ Bacterial ribosome structure. As this schematic illustrates, note that a section of the 30S subunit fits into the valley of the 50S subunit when forming the 70S ribosome.

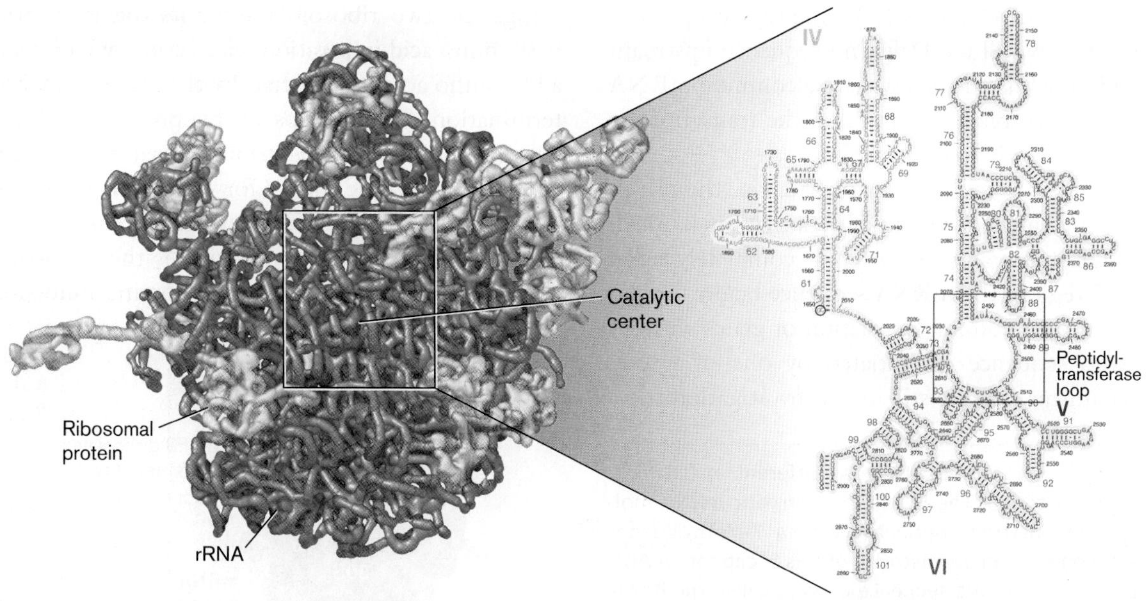

FIGURE 8.18 ■ RNA-protein interfaces in the large (50S) ribosome subunit. Blue = rRNA; gold = proteins. (PDB code: 1GIY) **Blowup:** Partial secondary structure of 23S rRNA, which includes the peptidyltransferase catalytic site. Note the many double-stranded hairpin structures that fold into domains (Roman numerals). Nucleotides are numbered in black; stem structures, in blue.

ribosomal proteins begin to assemble within the secondary rRNA structures. Thus, the ribosome is built by precise, timed molecular RNA-RNA, RNA-protein, and protein-protein interactions.

The ribosome is an ancient enzyme that is shared by all three domains of life. In size and structure, archaeal and bacterial ribosomes are similar, although some ribosomal proteins in archaea are not found in bacteria, and vice versa. Eukaryotic ribosomes are similar in overall structure, but larger and more complex than those of bacteria and archaea. Eukaryotic SSU and LSU subunits are 40S and 60S in size, respectively, and the holoenzyme is 80S. The eukaryotic SSU and LSU rRNAs are also larger (18S and 25S, respectively) because of their greater sequence length.

Despite differences in the length of rRNA and the sets of protein subunits involved, the ribosomes of bacteria, archaea, and eukaryotes have very similar mechanisms of translating mRNA into protein. The sequences of ribosomal RNAs from all microbes are very similar, in large part because rRNA plays an important catalytic role in the activities of ribosomes. But there are differences in rRNA sequences that increase in relation to the evolutionary distance between species. In fact, the close but not perfect similarity in the rRNA sequences allowed Carl Woese to argue successfully for the existence of the archaea as a domain distinct from bacteria and eukaryotes (see Chapter 17).

How Do Ribosomes Find Where to Start Translation?

The ribosome reads the mRNA as a string of consecutive codons, without overlap and without spacers between the codons. Because there are three bases per codon, each mRNA has three potential reading frames. For instance, in the sequence AUGCCAAA, one reading frame would include the codons AUG and CCA, the second reading frame would include UGC and CAA, and the third, GCC and AAA. How, then, is the right reading frame found by the ribosome, and how does the ribosome know where to begin translation along the mRNA?

Start codons mark where translation starts and set the correct reading frame. AUG is the most frequent start codon (90%), while GUG (8.9%), UUG (1%), and CUG (0.1%) are also used to lesser extents. Although these start codons are used to begin all proteins, they are not restricted to that role; they also encode amino acids found in the middle of coding sequences. Consequently, there must be something else about mRNA that signifies whether an AUG codon, for example, marks the start of a protein.

The key is a section of untranslated RNA sequence positioned upstream of the protein-coding segment of mRNA (see **Fig. 8.5**). This leader RNA contains a purine-rich consensus sequence (5′-AGGAGGU-3′) located four to eight bases upstream of the start codon. This upstream sequence is called the **ribosome-binding site** or **Shine-Dalgarno sequence**, after Lynn Dalgarno and her student John Shine at Australian National University, who discovered it in 1974.

In *E. coli*, the Shine-Dalgarno site is complementary to a sequence of 16S rRNA (5′-ACCUCCU-3′), found in the 30S ribosome subunit. Binding of mRNA to this site positions the start codon (such as AUG or GUG) precisely in the ribosome P site, ready to pair with its cognate tRNA and initiate translation. Elegant proof that the Shine-Dalgarno sequence actually binds to the 3′ end of 16S rRNA is described in **eTopic 8.2**. In multigene operons, each gene has its own Shine-Dalgarno sequence upstream of its start codon; while the genes share a common mRNA transcript, their open reading frames on the transcript are translated independently (see **Fig. 8.5**).

In archaea and in some lineages of bacteria, the use of Shine-Dalgarno sequences to initiate translation is much less common. In some cases, the mRNAs completely lack a 5′ untranslated region: the mRNA sequence begins immediately with the start codon. Translation of genes without Shine-Dalgarno sequences is initiated by one of several possible mechanisms that are still under investigation.

Note: Eukaryotic ribosomes have a different mechanism for finding the start codon. The 5′ end of eukaryotic mRNA molecules is modified posttranscriptionally with a 7-methylguanosine 5′ cap. Ribosomes scan downstream of this 5′ cap for an AUG located within a short specific sequence context called the Kozak sequence, named after its discoverer, Marilyn Kozak.

The Three Stages of Protein Synthesis

Once the ribosome has been properly positioned on the message, polypeptide synthesis can begin. The ribosome moves along the mRNA in a 5′-to-3′ direction, translating each codon of the message into an amino acid that it adds to the growing polypeptide chain. Like transcription, polypeptide synthesis has three stages: initiation, which brings the two ribosomal subunits together, placing the first amino acid in position; elongation, which sequentially adds amino acids as directed by the mRNA transcript; and termination. Several steps in this process can be inhibited by certain antibiotics (described later in this section).

As the ribosome moves forward, each incoming tRNA molecule successively occupies one of three binding sites (**Fig. 8.19**). The first position is the aminoacyl-tRNA **acceptor site (A site)**, where the incoming aminoacyl-tRNA

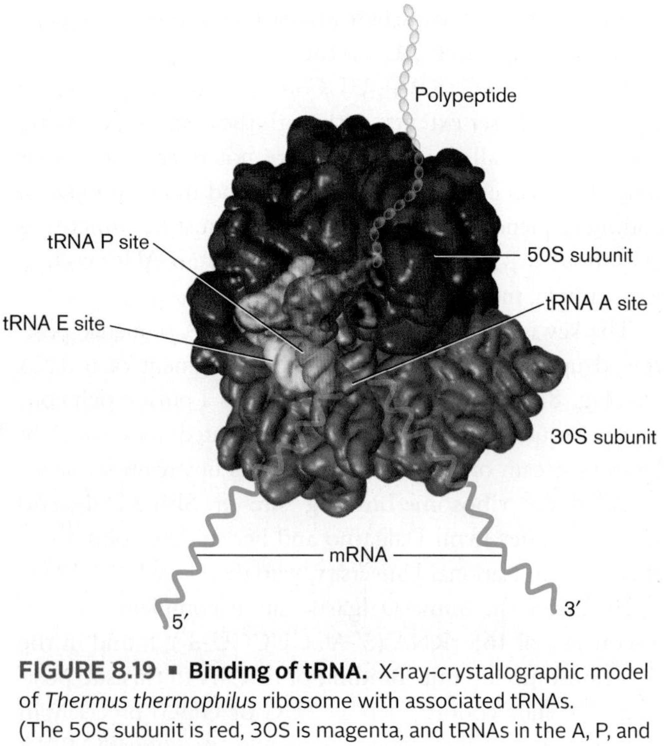

FIGURE 8.19 ■ Binding of tRNA. X-ray-crystallographic model of *Thermus thermophilus* ribosome with associated tRNAs. (The 50S subunit is red, 30S is magenta, and tRNAs in the A, P, and E sites are blue, green, and yellow, respectively.) The 30S subunit of the ribosome travels along the mRNA (light blue) in a 5′-to-3′ direction, and the growing peptide (yellow) exits from a channel formed in the 50S subunit. (PDB codes: 1GIX and 1GIY)

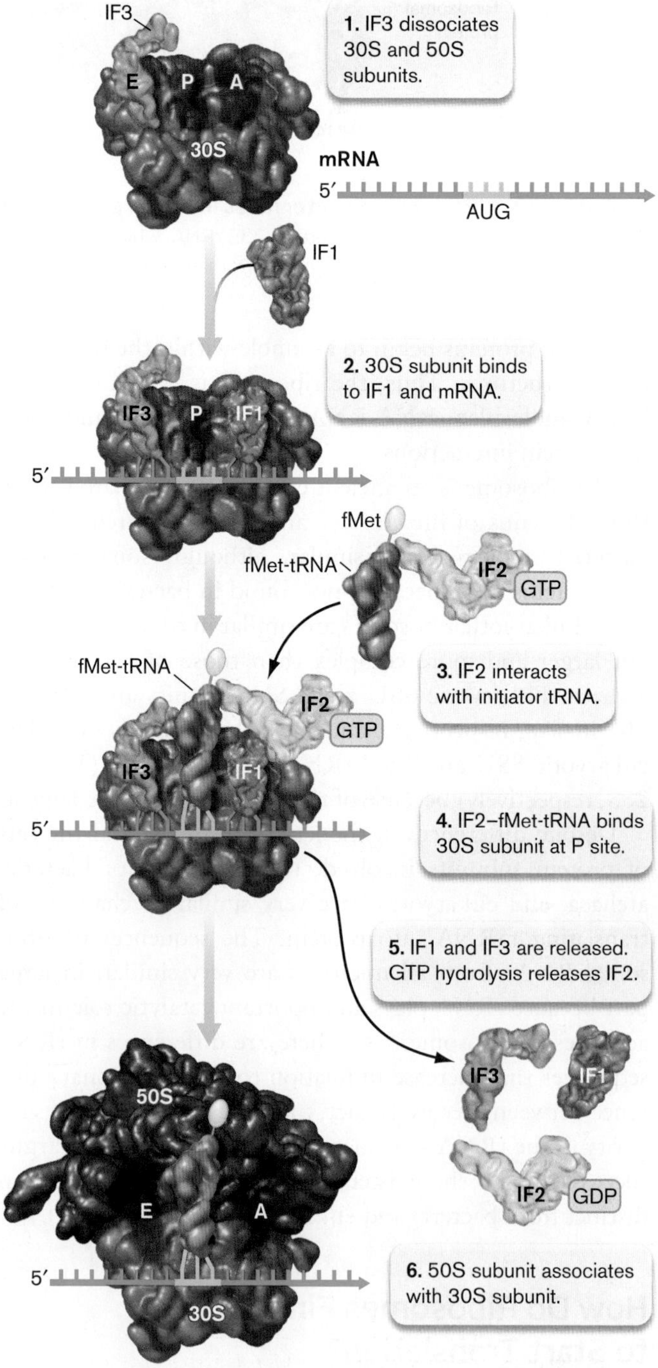

FIGURE 8.20 ■ Translation initiation. The end result of translation initiation is assembly of the 50S-30S-mRNA complex with the initiator tRNA-fMet set in the P site. See text for details.

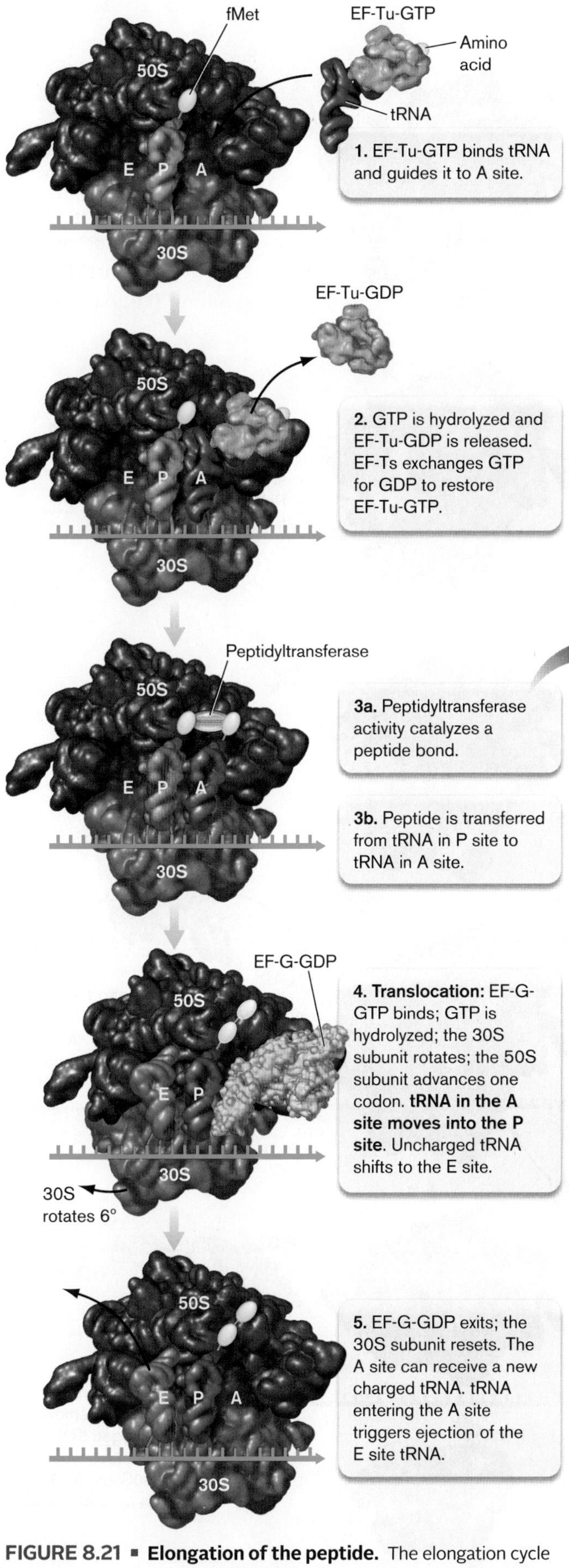

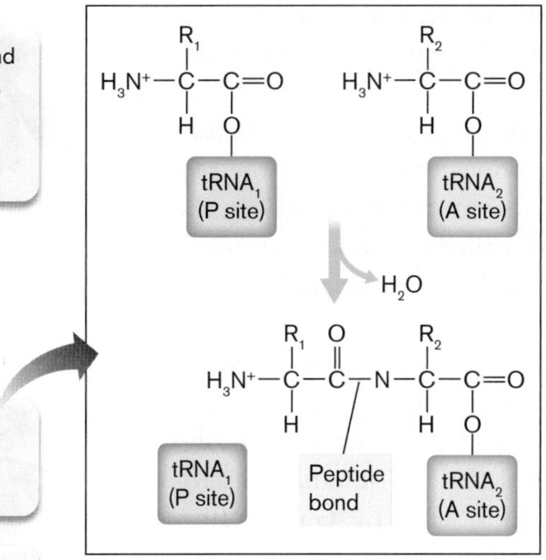

FIGURE 8.21 ▪ Elongation of the peptide. The elongation cycle moves the ribosome forward by one codon, lengthening the peptide chain by attaching it to the amino acid of the incoming tRNA, and ejecting the tRNA from which the peptide chain was transferred. The inset shows the peptidyltransferase reaction that generates the peptide bond.

enters the complex and its anticodon binds the codon of the mRNA. The second position is the **peptidyl-tRNA site** (**P site**), which binds the tRNA that is attached to the growing polypeptide. The third position is the **exit site** (**E site**), where the tRNA will be jettisoned from the ribosome after giving up its amino acid to the polypeptide chain. Next we discuss how these binding sites are used in the initiation, elongation, and termination of translation is discussed presently.

Initiation of translation. In bacteria such as *E. coli*, initiation of protein synthesis requires three small proteins called initiation factors (IF1, IF2, and IF3).

Initially, IF3 binds to the 30S subunit to separate the 30S and 50S subunits. Separating the subunits allows other initiation factors and mRNA access to the 30S subunit (**Fig. 8.20**, step 1). IF1 and mRNA join the 30S subunit. IF1 blocks the tRNA A site, and the mRNA sequence for the ribosome-binding site aligns with its complementary sequence on 16S rRNA (step 2).

Next, IF2 complexed to guanosine triphosphate (GTP) escorts the initiator *N*-formylmethionyl-tRNA (fMet-tRNA) to the start codon located at what will be the P site (steps 3 and 4). *N*-formylmethionyl-tRNA binds to all start codons and is the only aminoacyl-tRNA to bind directly to the P site. Once the initiator tRNA is in place, GTP hydrolysis releases IF2, along with IF1 and IF3 (step 5).

The 50S subunit then docks to the 30S subunit (step 6). The ribosome is now "locked and loaded," ready for elongation. Note that initiation is much more complex in archaea, involving as many as six different IF proteins.

Elongation of the polypeptide. In elongation, three basic steps are repeated: (1) A charged tRNA enters the A (acceptor) site; (2) the amino acid brought to the A site is covalently attached to the tRNA-bound amino acid in the P site, adding to the polypeptide chain; (3) the ribosome moves forward by one codon in a 5′-to-3′ direction (**Fig. 8.21**). First an elongation factor (EF-Tu) associates with GTP to form a complex (EF-Tu-GTP) that binds to most charged aminoacyl-tRNAs (except for the initiator tRNA). Sequences within the 50S rRNA recognize features of the EF-Tu-GTP–aminoacyl-tRNA complex, guiding it into the A site (**Fig. 8.21**, step 1). Correct selection of a tRNA is guided in large part by codon-anticodon pairing, but conformational changes sensed by various ribosomal proteins customize the fit. If the fit is not perfect, the tRNA is rejected.

Once an aminoacyl-tRNA is in the A site, a peptide bond is formed between the amino acid moored there and the terminal amino acid linked to tRNA in the P site. Simultaneously, a GTP is hydrolyzed and the resulting EF-Tu-GDP (EF-Tu complexed with guanosine diphosphate) is expelled. Peptide bond formation effectively transfers the polypeptide from the tRNA in the P site to the tRNA in the A site (**Fig. 8.21**, steps 2 and 3).

For protein synthesis to continue, the ribosome must advance by one codon, which moves the peptidyl-tRNA from the A site into the P site, leaving the A site vacant. The process, called **translocation**, involves another elongation factor, EF-G, associated with GTP. EF-G-GTP binds to the ribosome, GTP is hydrolyzed, and the 30S subunit rotates clockwise to ratchet the 50S subunit ahead on the message by one codon (step 4). Translocation is completed by the exit of EF-G-GDP and the rotation of the 30S subunit back to its initial conformation (step 5). How EF-G uses a finger-like motion to complete translocation is explained in **Special Topic 8.1**. The translocation maneuver opens up the A site, moves the peptidyl-tRNA into the P site, and slides uncharged tRNA into the E (exit) site. The next aminoacyl-tRNA that enters the A site stimulates a conformational change in the ribosome that telegraphs through to the E site and ejects the uncharged tRNA.

Notice that EF-Tu and EF-G recycle sequentially on and off the ribosome by binding to the same area of the ribosome. EF-G-GTP and the EF-Tu-GTP–aminoacyl-tRNA complex are structurally similar, an example of molecular mimicry. Because these factors bind to the same site, they must cycle on and off the ribosome sequentially. Although this process seems incredibly complex, the ribosomes of *E. coli* manage to link together 16 amino acids per second. Scientists have now visualized the staggered movements of a single ribosome as it translates an mRNA molecule (**eTopic 8.3**).

Termination of translation. Eventually, the ribosome arrives at the end of the coding region, but not the end of the RNA. As noted earlier, the end of the coding region is marked by one of three stop codons. Translocation after the last peptide bond forms brings the mRNA stop codon into the A site (**Fig. 8.22** ▶, step 1). No tRNA binds, but one of two **release factors** (RF1 or RF2) will enter (also

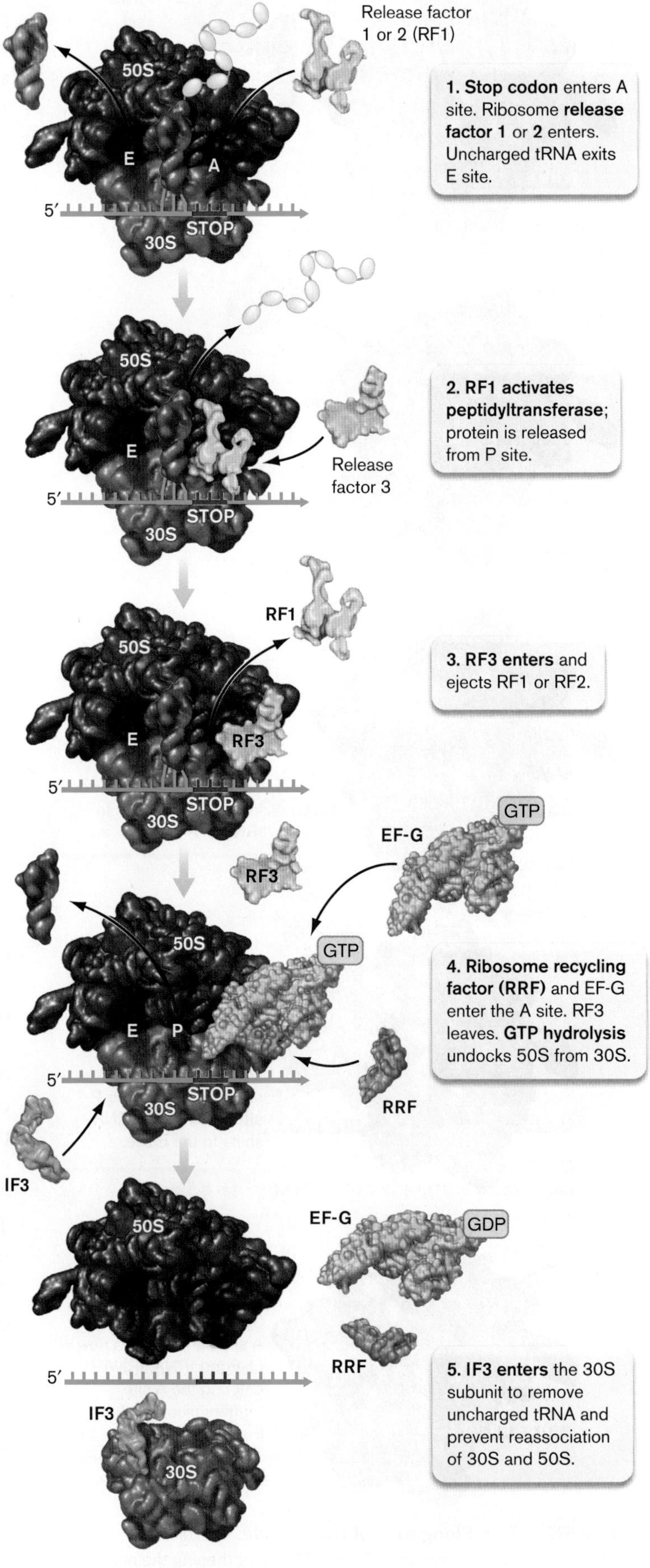

FIGURE 8.22 ■ Termination of translation. The completed protein is released, and the ribosome subunits are recycled. ▶

step 1). RF binding leads to ejection of tRNA in the E site and activates the peptidyltransferase, thereby cutting the bond that tethers the completed polypeptide to tRNA in the P site (step 2).

With the protein released, the ribosome disassembles. RF3 causes RF1 or RF2 to depart the ribosome (**Fig. 8.22**, step 3). Then, ribosome recycling factor (RRF), along with EF-G, binds at the A site, and accompanying hydrolysis of GTP undocks the two ribosomal subunits (step 4). IF3 then reenters the 30S subunit to eject the remaining uncharged tRNA and mRNA (step 5), thereby preventing the 30S and 50S subunits from redocking. The liberated ribosomal subunits are now free to diffuse through the cell, ready to bind yet another mRNA and begin the translation sequence anew.

Sometimes, premature transcription termination or cleavage of the mRNA leaves an mRNA without a translational stop codon. How a cell is able to "unstick" ribosomes bound to these truncated mRNAs is discussed in **eTopic 8.4**.

Thought Questions

8.7 How might one gene code for two proteins with different amino acid sequences?

8.8 Why involve RNA in protein synthesis? Why not translate directly from DNA?

8.9 Codon 45 of a 90-codon gene was changed into a translation stop codon, producing a shortened (truncated) protein. What kind of mutant cell could produce a full-length protein from the gene without removing the stop codon? *Hint:* What molecule recognizes a codon?

Antibiotics That Affect Translation

Most of our antibiotics were discovered as natural products of fungi or of bacteria such as actinomycetes (presented in Chapter 18). Streptomycin (**Fig. 8.23A**), a well-known member of the aminoglycoside family of antibiotics, is produced by the actinomycete *Streptomyces*. The drug targets bacterial small ribosomal subunits by binding to a region of 16S rRNA in the 30S subunit that forms part of the decoding A site and to protein S12, a protein critical for maintaining the specificity of codon-anticodon binding. Streptomycin bound to 16S rRNA makes decoding of mRNA at the A site "sloppy" by permitting illicit codon-anticodon matchups that result in a mistranslated protein sequence.

Bacteria can become resistant to streptomycin via spontaneous mutations in the S12 gene (*rpsL*) or 16S rRNA. The altered ribosomal protein or rRNA remains functional but does not bind streptomycin. Some bacteria gain resistance by acquiring an aminoglycoside phosphotransferase that modifies streptomycin so that it cannot bind to its target. Additional mechanisms are discussed in Chapter 27. Other therapeutically important aminoglycosides include gentamicin, tobramycin, and amikacin.

Tetracylines are polyketides, a class of antibiotics whose biosynthesis is presented in Chapter 15. Tetracycline and derivatives such as doxycycline (**Fig. 8.23B**) also target the 30S ribosomal subunit, where they bind to 16S rRNA near the A site. But instead of causing mistranslation, the tetracyclines prevent aminoacyl-tRNA from binding to the A site. Resistance to tetracycline can be conferred by an efflux transport system that effectively removes the antibiotic from the bacterial cell. Other resistance mechanisms are described in Chapter 27.

Another species of *Streptomyces* produces chloramphenicol (**Fig. 8.23C**), which attacks the 50S subunit. It binds to 23S rRNA at the peptidyl-tRNA site (that is, the P site) and inhibits peptide bond formation. Resistance to this drug comes from an ability to synthesize the enzyme chloramphenicol acetyltransferase, which modifies chloramphenicol in a way that destroys its activity.

Erythromycin, made by *Streptomyces erythraeus*, is one of a large group of related antibiotics called macrolides, whose hallmark is a large lactone ring of 12–22 carbon atoms (**Fig. 8.23D**). They all attack the 50S subunit by binding to 23S rRNA in the nascent peptide exit tunnel near the P site. Binding alters peptidyltransferase structure and interferes with peptide bond formation. Resistance to macrolides usually involves an efflux pump or methylation of the relevant area of 23S rRNA. A recently discovered resistance mechanism uses a protein to knock macrolides off of the ribosome (see the Current Research Highlight for Chapter 27.)

Other translation-targeting antibiotics interfere with mRNA binding to the ribosome (kasugamycin), prevent translocation by targeting EF-G (fusidic acid), or use structural similarity to tRNA (molecular mimicry) to trick peptidyltransferase into action without having a bona fide tRNA in the A site (puromycin). We chronicle the discovery and use of antibiotics more completely in Chapter 27.

Thought Question

8.10 While working as a member of a pharmaceutical company's drug discovery team, you find that a soil microbe snatched from the jungles of South America produces an antibiotic that will kill even the most deadly, drug-resistant form of *Enterococcus faecalis*, which causes bacterial endocarditis. Your experiments indicate that the compound stops protein synthesis. How could you more precisely determine the antibiotic's mode of action? *Hint:* Can you use mutants resistant to the antibiotic?

SPECIAL TOPIC 8.1 Translocation: EF-G Gets Physical

Translocation of the ribosome down an mRNA molecule is a critical part of translation. Ribosomes must move one codon along an mRNA molecule while tRNAs in the ribosome A (acceptor) site and P (peptidyl-tRNA) site shift to the P site and E (exit) site, respectively. Elongation factor G (EF-G) plays a major role in this process. When EF-G-GTP binds to a pretranslocation ribosome (see **Fig. 8.21**), it triggers the ratchet-like movement of the 30S subunit relative to the 50S subunit. During subunit ratcheting, the CCA acceptor ends of tRNAs move from the A and P sites to the P and E sites, leaving the anticodon ends in the original sites. Hydrolysis of GTP bound to EF-G then moves the anticodon ends of these tRNAs, along with mRNA, to the new sites. The result is the posttranslocation ribosome in which mRNA has moved one codon and tRNAs in the A and P sites have moved to the P and E sites, respectively (see **Fig. 8.21**).

One question, however, has troubled structural biologists for some time. Cryo-electron-microscopy and X-ray-crystallographic examinations of EF-G bound to posttranslocation ribosomes show an elongated form of EF-G with its domain IV projecting into the decoding center of the ribosome—a place where tRNA in the A site binds to its codon. The question is: How can EF-G bind to a pretranslocation ribosome without colliding with tRNA already in the A site? A team led by Nobel laureate Thomas Steitz (1940–2018) at Yale University (**Fig. 1**) discovered the answer by using a new way to trap EF-G bound to pretranslocation ribosomes from *Thermus thermophilus*.

Part of their strategy involved placing a nonhydrolyzable peptidyl-tRNA in the P site to prevent polypeptide transfer to tRNA in the A site. The result was a ribosome stalled in the pretranslocation state. Adding EF-G enabled them to crystallize the pretranslocation complex and perform X-ray crystallography to determine EF-G structure. The results, shown in **Figure 2A**, revealed that domain IV of EF-G was not extended before translocation, but was tucked away in a manner that did not impinge on tRNA in the A site. The authors propose that once EF-G-GTP binds to the ribosome, the GTP-binding site becomes ordered, triggering a conformational change in EF-G to an extended fingerlike form (**Fig. 2B**). The extended form reaches into the A site to complete translocation of the anticodon ends of tRNAs and moves the associated mRNA down the ribosome. The model is similar to a finger flicking a paper clip along a table.

The study by Steitz's team shows that EF-G is far more flexible than was previously thought, and suggests that controlling flexibility by GTP binding and stabilization is key to completing translocation. GTP hydrolysis, then, is required prior to the dissociation of EF-G from the posttranslocation ribosome.

FIGURE 1 ▪ Thomas Steitz (right) and associate research scientists Matthieu Gagnon (left) and Jinzhong Lin (center) examine the EF-G-binding site on a model of the 70S ribosome.

RESEARCH QUESTION

Propose three ways that an antibiotic might inhibit protein synthesis by interfering with EF-G.

Lin, Jinzhong, Matthieu G. Gagnon, David Bulkley, and Thomas A. Steitz. 2015. Conformational changes of elongation factor G on the ribosome during tRNA translocation. *Cell* **160**:219–227.

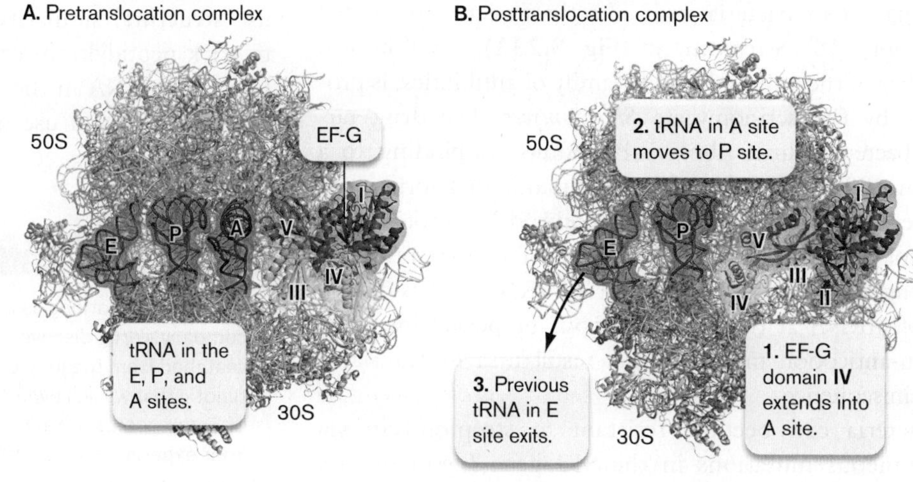

FIGURE 2 ▪ Structures of EF-G bound to the pretranslocation (A) and posttranslocation (B) ribosome.

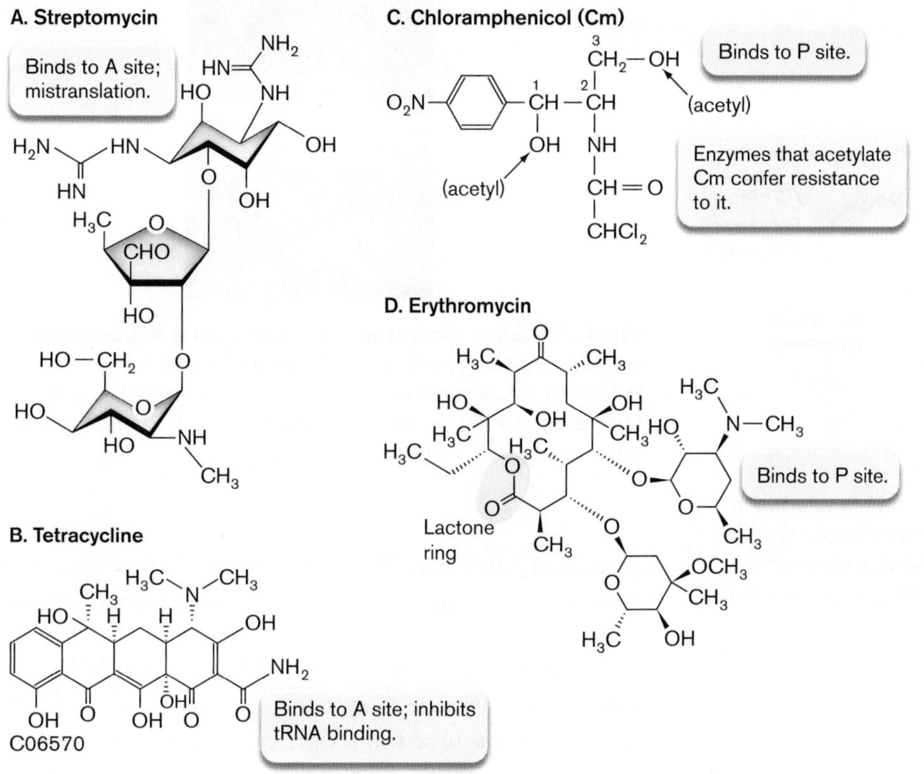

FIGURE 8.23 ■ **Antibiotics that inhibit protein synthesis in bacteria.** Streptomycin (A) and tetracycline (B) bind to the A site. Streptomycin causes mistranslation, tetracycline inhibits tRNA binding. Chloramphenicol (C) and erythromycin (D) bind to the peptidyl-tRNA site, thereby inhibiting peptide bond formation.

Polysomes and Coupled Transcription-Translation

Polysomes. Once a ribosome begins translating mRNA and moves beyond the ribosome-binding site, another ribosome can immediately jump onto that site. The result is an mRNA molecule with multiple ribosomes moving along its length at once. The multiribosome structure is known as a **polysome** (**Fig. 8.24**). Ribosomes in a polysome are closely packed along the mRNA, which helps protect the message from degradation by RNases and enables the speedy production of protein from a single mRNA molecule.

Coupled transcription and translation. Because bacteria and archaea lack nuclear membranes, ribosome subunits floating through the cytoplasm have an opportunity to bind to the 5' end of mRNA and begin making protein even before RNA polymerase has finished making the mRNA molecule. The simultaneous building of both mRNA and proteins is called coupled transcription and translation (**Fig. 8.25**). These coupled processes can also be coordinated with protein secretion across the membrane (see Chapter 3 for details). The coupling of transcription and translation in bacteria makes it possible for the cell to use translation as a means of regulating transcription. One such regulatory process, attenuation, is explained further in Chapter 10.

Once transcription is complete, the mature mRNA can diffuse out of the nucleoid. In rod-shaped cells, most ribosomes are located at the poles, being generally excluded from the nucleoid (**Fig. 8.26**), and the poles are where most translation occurs over the lifetime of the mRNA. Thus, although transcription-translation coupling happens when transcripts are first made in the nucleoid (**Fig. 8.27**), most translation occurs independent of transcription in nucleoid-free regions.

Transcription in eukaryotic microbes is not coupled to translation. In contrast to bacteria and archaea, eukaryotic microbes use distinct cell compartments to carry out most of their transcription and translation. They transcribe genes in the nucleus, where internal, noncoding parts of the mRNA (introns) are removed by a process known as RNA splicing. The processed transcripts are then exported to the cytoplasm and translated. Most eukaryotic DNA viruses do the same. However, a small amount of translation is performed in the eukaryotic nucleus, where it can be coupled to transcription.

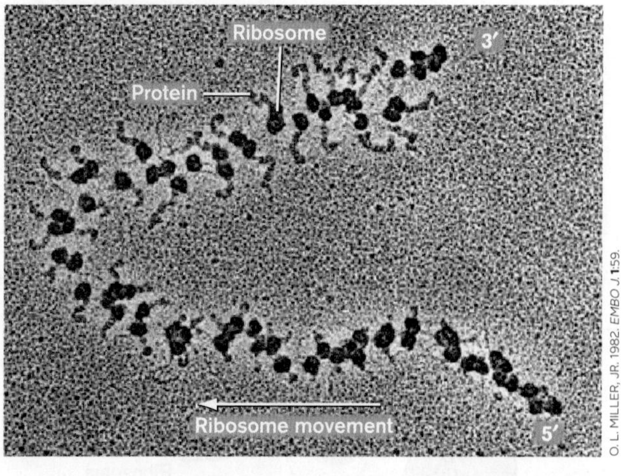

FIGURE 8.24 ■ **A polysome from a eukaryotic cell.** Several ribosomes may translate a single mRNA molecule at the same time. The beginning (5' end) of the mRNA is at lower right; the 3' end is at upper right. Note that the synthesized protein molecule grows longer and longer as the ribosome approaches the 3' end of the mRNA, where the protein molecule is most clearly seen. Polysomes also occur in prokaryotes.

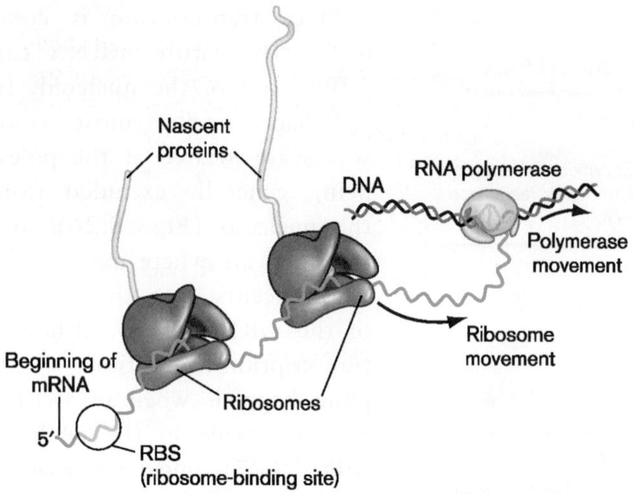

FIGURE 8.25 ▪ Coupled transcription and translation in bacteria. During coupled transcription and translation in prokaryotes, ribosomes attach at mRNA ribosome-binding sites and start synthesizing protein before transcription of the gene is complete.

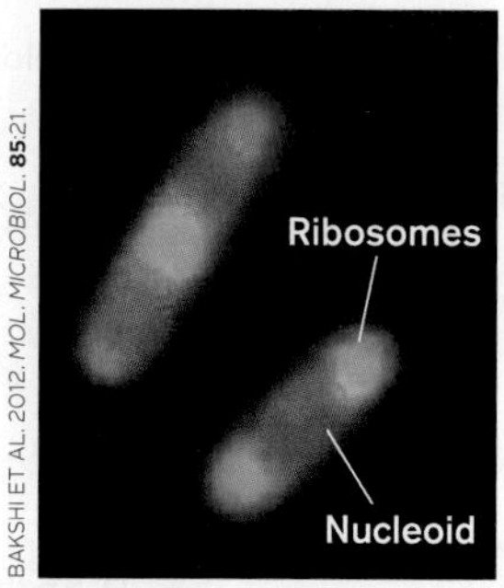

FIGURE 8.26 ▪ Most translation in *E. coli* is not coupled to transcription. Most ribosomes exist in nucleoid-free areas of the cell, primarily at the cell poles. DNA was visualized using a red fluorescent dye. Ribosomes contained a protein (S2) fused to yellow fluorescent protein, which here appears green.

Some Microbes Have Modified Genetic Codes

The standard genetic code (see **Fig. 8.12**) contains 61 sense codons and 3 nonsense (stop) codons. The code is ancient, and was likely established before the three domains of life emerged from the last universal common ancestor. A question worth asking is: Once the code is set, can it be changed? For example, if the codon UAU is reassigned from tyrosine to cysteine, then whenever the UAU codon appears in a transcript, tyrosine will be replaced with cysteine. Such a change could alter the sequence of hundreds or thousands of proteins at the same time, which could be lethal for the organism. Not surprisingly,

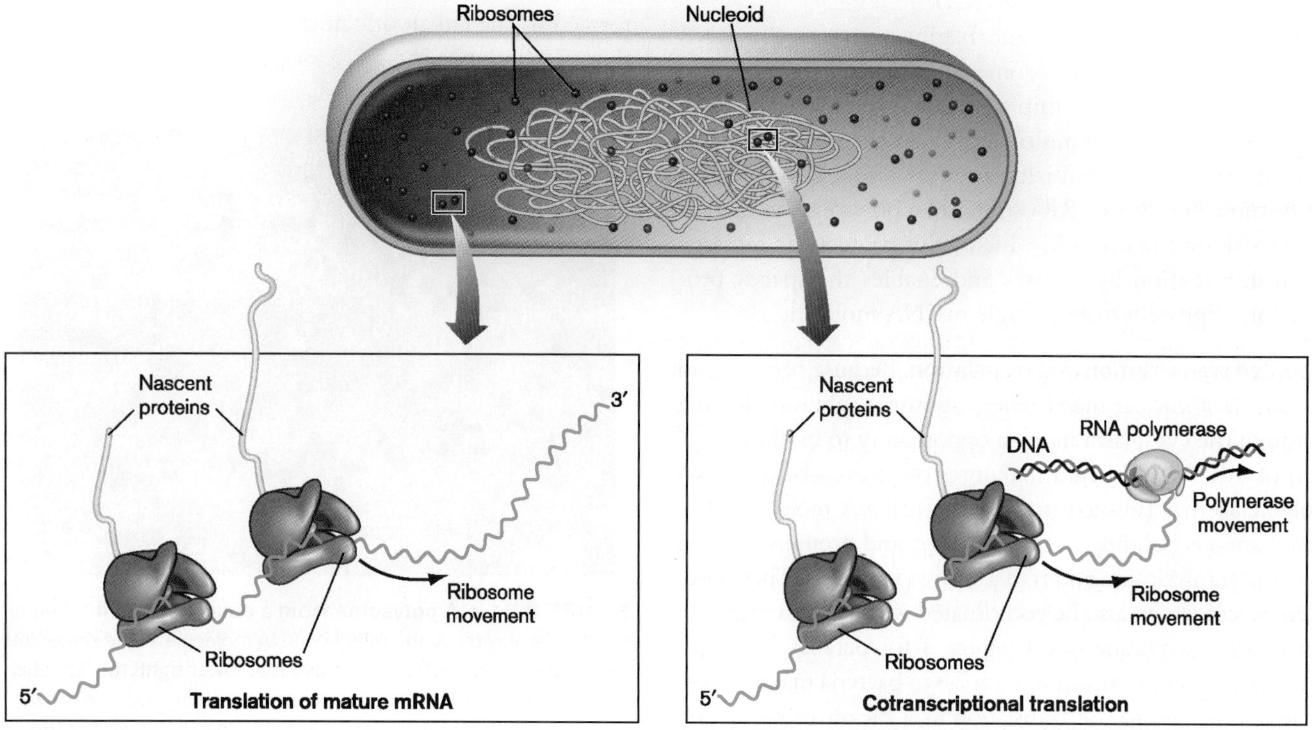

FIGURE 8.27 ▪ Spatial localization of transcription and translation in *E. coli*. Coupled transcription and translation occurs in or near the nucleoid (right blowup), whereas translation of mature, fully transcribed mRNA occurs at the cell poles (left blowup).

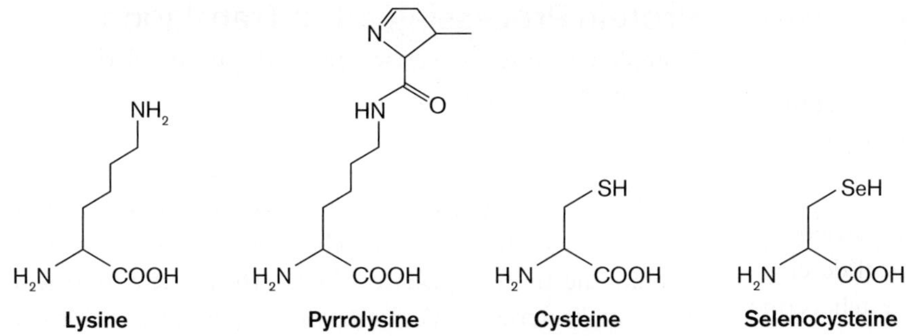

FIGURE 8.28 ■ The 21st and 22nd amino acids for translation. The structures of pyrrolysine and selenocysteine, compared to lysine and cysteine.

then, codon reassignments are not common in nature, and when found, they are usually limited to one or a few codon reassignments per organism.

How are codons reassigned? The usual way is to change the anticodon sequence in a tRNA so that it carries the same amino acid but now recognizes a different codon. In some cases, the codon was recognized as a stop codon, but after the change, the ribosome can "read through" the translation stop signal. For SR1, a lineage of bacteria associated with human periodontal disease, UGA has changed from a stop codon to one that encodes glycine. In the ciliate *Condylostoma magnum*, all three stop codons of the standard code (UGA, UAA, and UAG) encode amino acids. How does translation stop in this organism? The stop codons actually serve dual functions in this eukaryotic microbe: they encode amino acids when located at internal positions within the mRNA transcript, and they serve as stop codons when found at the end of the transcript. Just how the ribosome determines correctly which way to read these codons on the basis of location in the transcript is still unknown. In cases such as these, in which stop codons have been reassigned to sense codons, the adaptive benefit is not known, and it may be that they are simply tolerated evolutionary "accidents" that do no real help or harm to the organism.

In contrast, two other codon reassignments have more obvious benefit because they expand the genetic code and introduce novel amino acids—pyrrolysine and selenocysteine—into proteins. Pyrrolysine and selenocysteine have unique side chains that are related to lysine and cysteine, respectively (**Fig. 8.28**), and when incorporated into the polypeptide chain they confer new biochemical properties on the protein. Cells that utilize pyrrolysine and selenocysteine require special enzymes for their synthesis and ligation to their own dedicated tRNAs.

Pyrrolysine has been found in the proteins of some bacteria and archaea, but thus far not in eukaryotes. The marine bacterium *Acetohalobium arabaticum* is the only known organism that can expand its genetic code in response to environmental cues. *A. arabaticum* can grow on several carbon sources, including pyruvate and the methyl-containing molecule trimethylamine. When growing on pyruvate, it uses the canonical 20 amino acids to make proteins. However, growth on trimethylamine requires a pyrrolysine-containing enzyme. In response to this growth condition, *A. arabaticum* synthesizes the machinery to make pyrrolysine and ligate it onto its tRNA for incorporation into protein.

Selenocysteine has been found in proteins from all three domains of life. There are over 50 bacterial protein families that encode selenocysteine, many of which are enzymes involved in oxidation-reduction (redox) chemistry and protection from oxidative stress. In "selenoproteins" involved in redox chemistry, the selenocysteines have usually replaced cysteines at critical positions in the protein, and this change has been demonstrated to improve catalytic efficiency of the enzymes.

Thought Questions

18.11 Why do you think evolution by natural selection favors changes in codons, rather than in anticodons?

8.12 A major way that bacteria acquire new functions is through the acquisition and expression of genes from other microbes via horizontal gene transfer (Chapters 9 and 17). How would this mechanism of innovation be affected if the recipient bacterium changed its genetic code?

8.13 The incorporation of pyrrolysine and selenocysteine into the genetic code involved stop-to-sense changes, rather than sense-to-sense. Why do you think this was the case?

To Summarize

- **Triplet nucleotide codons** in mRNA encode specific amino acids. **Transfer RNA molecules** interpret the genetic code and bring specific amino acids to the A (acceptor) site in the ribosome.
- **Specific codons** mark the beginning and end of a gene. The Shine-Dalgarno sequence in mRNA, located before

the start codon, helps the ribosome find the correct reading frame in the mRNA.

- **Initiation of protein synthesis** in bacteria requires three initiation factors that bring the ribosomal subunits together on an mRNA molecule.
- **Peptide bond formation** by the ribosome is carried out by ribosomal RNA, not protein. The polypeptide elongates by one amino acid when the ribosome ratchets one codon length along the mRNA.
- **Translation terminates** when a stop codon is reached. A release factor enters the A site and triggers peptide bond formation, thus freeing the completed protein from tRNA in the P site.
- **Ribosome recycling factor** and EF-G bind to the A site to dissociate the two ribosomal subunits from the mRNA.
- **Antibiotics that affect translation** can cause ribosomes to misread mRNA (streptomycin), inhibit aminoacyl-tRNA binding to the A site (tetracycline), interfere with peptidyltransferase (chloramphenicol), trigger peptide bond formation prematurely (puromycin), cause translocation to abort (erythromycin), or prevent translocation (fusidic acid).
- **Transcription and translation** are coupled in bacteria and archaea. Translation of mature mRNA can occur after transcription is complete, but new mRNA is usually being translated while it's still being elongated.
- **Changes to the standard genetic code, though uncommon,** can turn stop codons into sense codons and can introduce novel amino acids such as pyrrolysine and selenocysteine into proteins.

8.4 Protein Modification, Folding, and Degradation

Once a protein is made, is it functional? For many proteins, translation is not the last step in producing a functional molecule. Often a protein must be modified after translation, either to achieve an appropriate 3D structure or to regulate its activity. Primary, secondary, and tertiary structures of proteins can be modified after the primary protein sequence has been assembled by the ribosomes. And what happens when a protein is damaged or is no longer needed? A healthy cell "cleans house" by degrading damaged or unneeded proteins. The precious amino acids are then recycled into making new proteins.

Protein Processing after Translation

Completed proteins released from the bacterial ribosome contain *N*-formylmethionine (fMet) at the N terminus, while archaeal and eukaryotic proteins have methionine (Met) in this position. In bacteria, archaea, and eukaryotes, this N-terminal amino acid is often removed or modified to produce the final polypeptide chain. For all three domains, methionine aminopeptidase removes the entire amino acid, while in bacteria another enzyme, methionine deformylase, provides the option to remove just the *N*-formyl group, leaving methionine. *N*-formylmethionine is important during the course of an infection because fMet peptides are produced only by bacteria and mitochondria, not by archaea or by the cytoplasmic ribosomes of eukaryotes. Our white blood cells can detect low concentrations of fMet peptides (about 10^{-12} M) as a sign of invading bacteria or of necrotic (dying) host cells releasing mitochondria.

A protein's function depends on its three-dimensional shape and its chemical properties. As we have seen, the standard genetic code provides 20 options for side-chain chemistry at each amino acid position in the protein polymer. Expanding the code to 21 or 22 with the inclusion of pyrrolysine or selenocysteine, as some microbes do, generates a moderate increase in options. An alternative way to vary protein chemistry is to modify proteins after translation. In fact, all three domains of life utilize protein modification by the covalent attachment of molecules, including sugars, lipids, and inorganic substrates, to specific amino acids after the protein is synthesized. Over 900 different types of posttranslational protein modifications have been reported thus far from in vivo and in vitro studies. These modifications serve various roles: stabilizing proteins, localizing proteins to specific regions of the cell, and regulating the activity of the proteins. For the latter, a key property of many posttranslational modifications is that they are reversible; thus, protein function can be regulated by reactions that add or remove the modifications.

As we will see in Chapter 10, the addition of phosphoryl or methyl groups (**Fig. 8.29**) can change the activity of signal transduction proteins in bacteria and archaea, such as the ones used in chemotactic motility. Adenylylation, the covalent attachment of adenosine 5′-monophosphate, can regulate the activity of enzymes such as glutamine synthetase. Attachment of acetyl groups via acetylation can serve multiple functions, including protein stabilization and the regulation of protein activity. Lipidation is the covalent attachment of lipids to proteins. Lipidation provides a hydrophobic tail that anchors these lipoproteins to the cytoplasmic membrane or to the outer membrane in Gram-negative organisms.

Glycosylation is the covalent addition of mono- or polysaccharides to generate glycoproteins (**Fig. 8.29** lower

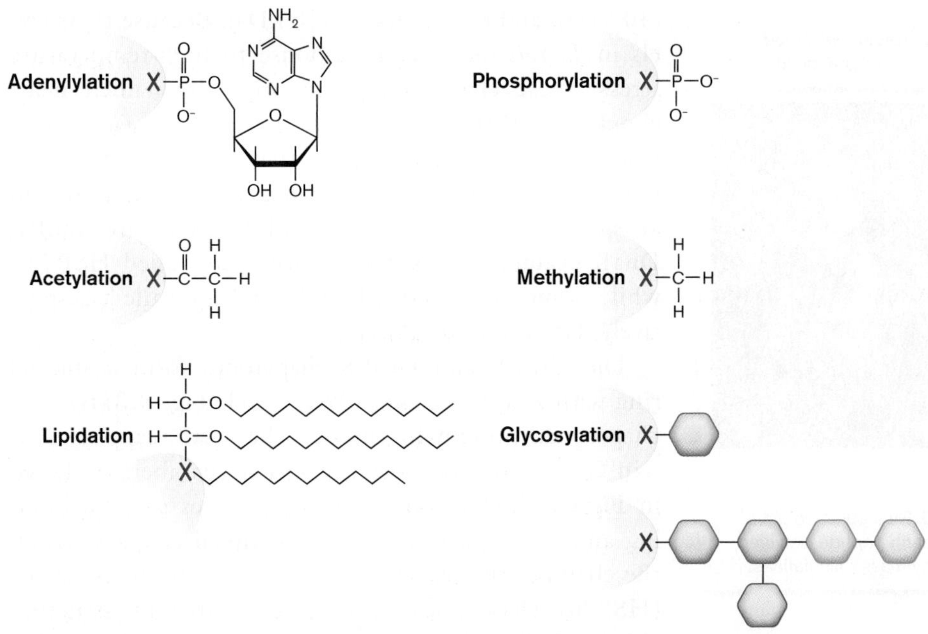

FIGURE 8.29 ■ Examples of posttranslational modifications to proteins. Molecules are typically covalently linked to an amino acid side chain (X). Shown at lower left is one of many variations of lipidation. Glycosylations can involve single sugars (represented by hexagons), or polysaccharide chains, often of varied sugar composition and degree of branching within the chain.

right). In bacteria and archaea, glycosylation can occur for a number of cell-surface proteins, including flagellar subunits, pili, adhesins, and the proteins that compose the S-layer. Glycoproteins can be involved in important microbial processes, such as biofilm formation, virulence, and colonization of the human gut. For example, glycosylation is important for *Bacteroides fragilis,* a member of the normal human gut microbiome. Critically, mutants unable to glycosylate proteins are outcompeted in the gut by wild-type *B. fragilis.* Glycosylation is also important for pathogens of the gut. *Clostridioides difficile* is a pathogen that can exploit the loss of normal gut microflora after antibiotic treatment (see Chapter 26). One of its secreted toxins, TcdA, is a glycosyltransferase that can glycosylate important regulatory proteins in human cells. These glycosylations inactivate the proteins, causing a disruption in normal regulation in the human cells, and ultimately making the host gut more susceptible to infection by *C. difficile.*

The application of mass spectrometry to the investigation of cell proteins (proteomics) has led to major advances in our understanding of posttranslational modifications, and has revealed that we have vastly underappreciated the extent to which bacterial proteins are modified after translation is complete. **Mass spectrometry** experimentally determines the exact mass of an unknown protein or peptide fragment. The experimentally determined mass can be compared to the predicted mass using genomic information for the organism, and by these comparisons the protein or protein fragment is identified.

To begin the analysis, proteins present in cell extracts are digested into distinct peptide fragments with a site-specific protease (trypsin; **Fig. 8.30**, steps 1 and 2). The fragment mixture is passed through an analytical column that separates peptides according to differences in hydrophobicity or some other parameter (step 3). The column effluent is then directly fed through tandem mass spectrometry (MS-MS) instrumentation (step 4). In MS-MS, each proteolytic fragment is subfragmented by ionization to produce progressively smaller secondary fragments missing one or more amino acids. Because the weight of each amino acid is distinct, MS-MS analysis can determine the amino acid sequence of the initial proteolytic fragment (step 5). Sophisticated computer programs identify the proteins by comparing the amino acid sequences of each protein fragment with the predicted sequences of proteolytic fragments from all ORFs in a genome (steps 6 and 7).

Peptide fragments that contain posttranslational modifications such as acetylations and phosphorylations can be isolated from unmodified fragments using special analytical columns. The samples are then subjected to mass spectrometry to identify the modified fragments on the basis of their amino acid sequence. From such analyses, hundreds of proteins have been determined to be acetylated in *E. coli.* In a study of a strain of the marine cyanobacterium *Synechococcus,* mass spectrometry data revealed 2,230 proteins that have one or more posttranslational modifications. This study discovered many previously unknown modifications to proteins involved in photosynthetic light harvesting. An important implication of this discovery is that fundamental physiological processes such as the photosynthetic conversion of light energy to chemical energy are still only partially understood at the molecular level.

Protein Folding: Assume the Position

As a new protein emerges from the ribosome, how does it fold into exactly the correct shape to do its job? Christian Anfinsen (1916–1995) won the 1972 Nobel Prize in Chemistry for demonstrating that, for some proteins, folding is governed solely by the protein itself. In other words,

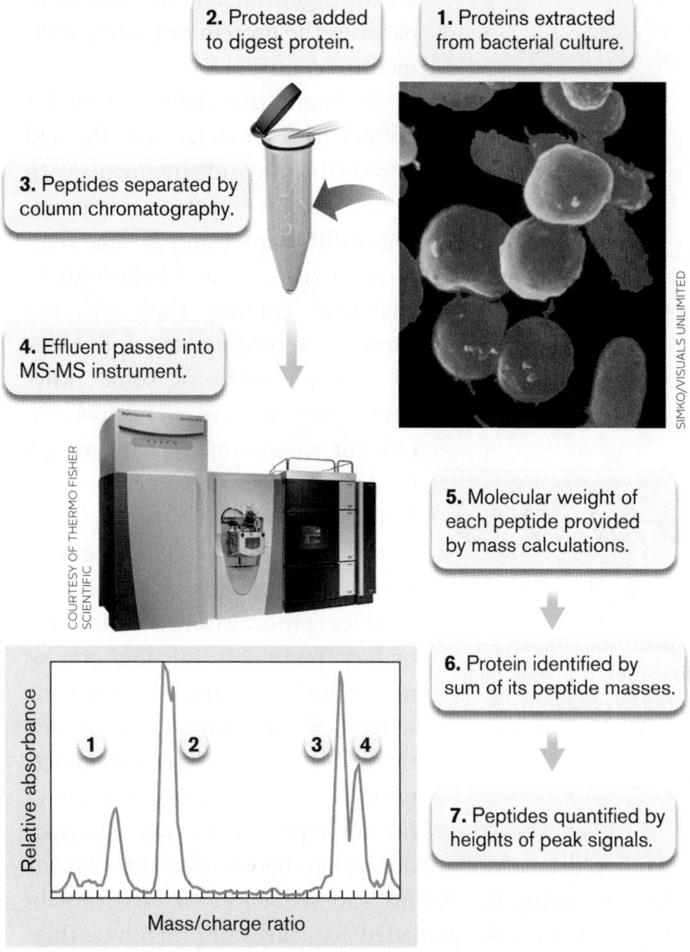

FIGURE 8.30 ■ Identifying proteins directly from whole-cell extracts by mass spectrometry. Proteins extracted from a bacterial culture are digested into peptides with trypsin. The peptides are separated by column chromatography and analyzed by mass spectrometry (here by the Thermo Scientific™ Q Exactive™ hybrid quadrupole-Orbitrap™ mass spectrometer). In tandem mass spectrometry (MS-MS), the mass of each peptide is determined first (peaks 1–4 in the graph), and then selected peptides are subjected to additional fragmentation by ion spray (not shown). Each resulting peptide fragment will differ in size by one or more amino acids. Knowing the mass of each amino acid and the masses of the different peptide fragments enables extrapolation of the original peptide's sequence.

the optimal 3D structure of a protein is determined solely by the linear sequence of amino acid residues. But three decades later, other scientists discovered that the folding of many proteins requires assistance from other proteins. These helper proteins are called **chaperones** (or chaperonins). Chaperones associate with target proteins during some phase of the folding process and then dissociate, usually after folding of the target protein is completed. Although chaperones exhibit some specificity, a given chaperone can help fold many different types of proteins.

The major chaperone family in most species includes GroEL (60 kDa), GroES (10 kDa), DnaK (70 kDa), DnaJ (40 kDa), and trigger factor (48 kDa). Because their levels in *E. coli* increase in response to high-temperature stress, these chaperones were originally named **heat-shock proteins** (**HSPs**), and they are, in fact, more resistant to heat denaturation than is the average protein. Representatives of these chaperones are found in all species. Because their molecular weights are similar, DnaK examples throughout nature are called HSP70s, while homologs of GroEL and DnaJ are called, respectively, HSP60s and HSP40s.

The GroEL and GroES chaperones form a stacked ring with a hollow center like a barrel (**Fig. 8.31A**). The chaperoned protein fits inside. The small, capping protein GroES controls entrance to the chamber, as shown in **Figure 8.31A** and **eTopic 8.5**. Cycles of ATP binding and hydrolysis cause conformational changes within the chamber that can reconfigure target proteins. DnaK (HSP70) chaperones have a very different structure (**Fig. 8.31B**). They do not form rings like the GroEL and GroES chaperones but can clamp down on a peptide to assist folding. Proteins emerging from a bacterial

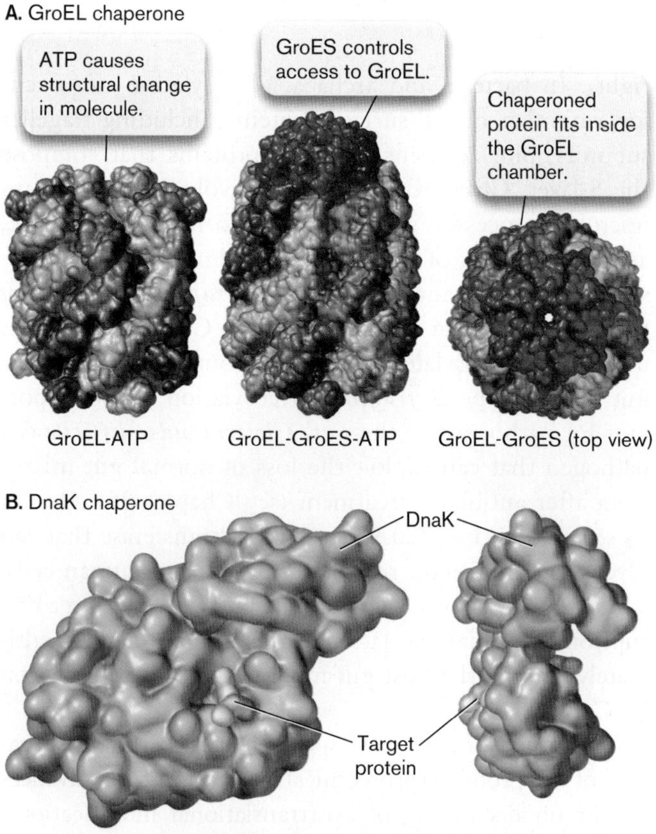

FIGURE 8.31 ■ *E. coli* GroEL, GroES, and DnaK structures. **A.** Three-dimensional reconstructions of GroEL-ATP, GroEL-GroES-ATP, and GroEL-GroES from cryo-EM. The first two panels are side views; the third panel is a top view. GroES is red. (PDB codes: 2C7E, 1PCQ) **B.** DnaK (HSP70) clamping down on a peptide (yellow). (PDB code: 1DKX)

ribosome enter a folding pathway that involves a hierarchy of these chaperones.

Protein Degradation: Cleaning House

What happens when a cell no longer needs a specific protein or when a cell synthesizes a protein with incorrect amino acids? Because the cell's needs constantly change, the presence of useless proteins can adversely affect the cell and they must be destroyed. This is particularly true of regulatory proteins, whose concentrations must change with time or in response to alterations in the cellular condition.

Many normal proteins contain degradation signals called degrons that dictate the stability of a protein. The **N-terminal rule** describes one type of degron. Recall that for some proteins, the N-terminal fMet is removed after translation, leaving a new amino acid at the N-terminus. The N-terminal rule states that the identity of the N-terminal amino acid correlates with the stability of the protein. For example, proteins beginning with leucine, phenylalanine, tryptophan, or tyrosine experience a short half-life (2 minutes or less), whereas proteins with aspartic acid, glutamic acid, or cysteine in the lead position have a longer half-life. A protein called ClpS facilitates degradation of these short-lived proteins. ClpS recognizes the destabilizing N-terminal amino acids and then presents the protein to the bacterial ClpAP protease.

Abnormally folded proteins are recognized by proteases in part because hydrophobic regions that are normally buried within the protein's 3D structure become exposed. The protein is progressively degraded into smaller and smaller pieces by a series of these proteases. Initial cuts, usually involving ATP-dependent endoproteases like Lon protein or ClpP, are followed by digestion with tripeptidases and dipeptidases. Endopeptidases cleave proteins somewhere within the sequence, but not from the ends of the sequence. Many peptidases use ATP hydrolysis to help unfold the target protein prior to digestion. Unfolding is necessary for the protein to slide into a barrel-shaped protease such as ClpAP, ClpXP, or ClpYQ in bacteria (**Fig. 8.32**).

Bacterial Clp proteases have a proteolytic core made of two homoheptameric protein rings of either ClpP or ClpY. The ClpP protease has interchangeable homohexameric ATPase caps made of ClpX, ClpA, ClpB, or ClpC, each of which recognizes different substrates. ClpY plays a similar capping role for ClpQ. The accessory proteins recognize and present different substrate proteins to the ClpP protease, thereby regulating which proteins are degraded. Protein-degrading enzymes are classified as serine, cysteine, or threonine proteases, depending on the key residue in their active sites.

Bacterial Clp proteases are structurally similar to eukaryotic proteasomes, which are even more complex protein-degrading machines. Proteasomes are found primarily in eukaryotes and archaea, although a few bacteria, such as the pathogen *Mycobacterium tuberculosis*, have them. Proteasomes are described in **eTopic 8.6**. Eukaryotic proteasomes recognize and then degrade proteins tagged by ubiquitin, a 76-amino-acid peptide. Some bacterial and viral pathogens exploit ubiquitination to reroute host metabolism.

What happens to proteins damaged by stress? Are they always degraded? Microbes are constantly exposed to environmental insults such as high temperature or pH extremes, which damage proteins and cause them to misfold. As an energy-saving device and to prevent interruption of protein function, injured proteins go through a kind of triage process that evaluates whether they are salvageable or must be destroyed before they can endanger the cell. Chaperones constantly hunt for misfolded (or otherwise damaged) proteins and attempt to refold them. But if the protein is released from a chaperone and remains misfolded, it can, by chance, either reengage the

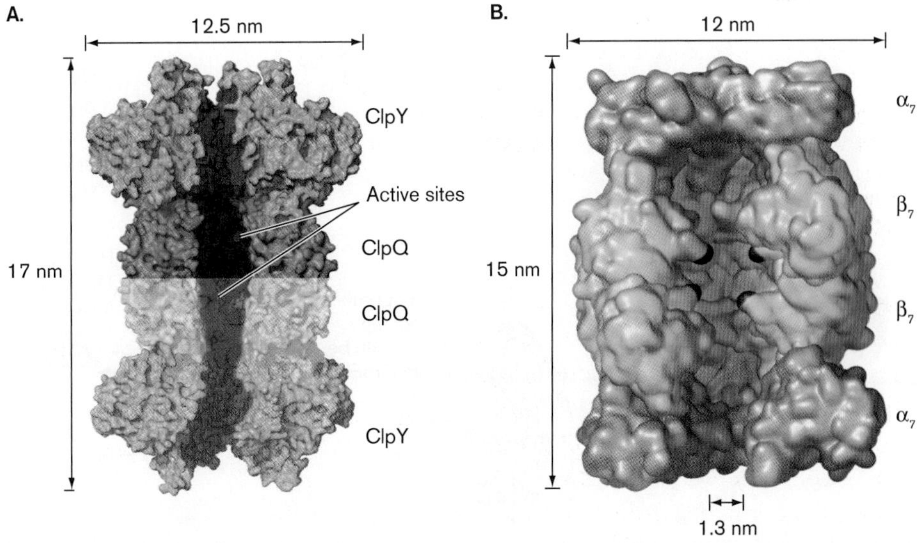

FIGURE 8.32 ■ Protein degradation machines. **A.** Bacterial ClpY ATPase and ClpQ protease (*Haemophilus influenzae*). (PDB code 1G3I) Two of the six subunits from each ring were removed to reveal the interior cavity. The active sites involved in peptide bond cleavage are indicated in pink. **B.** The 20S proteasome from the methanoarchaeon *Methanosarcina thermophila*. (PDB code: 1G0U)

chaperone or bind a protease that destroys it (**Fig. 8.33**). This fold-or-destroy triage system is essential if a microbe is to survive environmental stress.

To Summarize

- **Protein modifications** are made after translation is complete.
- **The N-terminal amino acid** (fMet or Met) can be removed by methionine aminopeptidase, or, in bacteria, just the formyl group can be removed by methionine deformylase.
- **Covalent additions** to proteins can enhance stability, add localization motifs, and regulate changes in activity.
- **Chaperone proteins** help translated proteins fold properly.
- **All proteins in all cells are eventually degraded** by specific devices such as proteases or proteasomes.
- **The N-terminal rule** describes one type of degradation signal (degron) that marks the half-life of a protein (that is, how long it takes 50% of the protein to degrade).
- **ATP-dependent proteases** such as Lon or ClpP usually initiate the degradation of a large protein.
- **Damaged proteins** enter chaperone-based refolding pathways or degradation pathways until the protein is repaired or destroyed.

8.5 Secretion: Protein Traffic Control

Microorganisms, especially Gram-negative bacteria, face a challenge in delivering proteins to different target locations of the cell. Recall that Gram-negative microbes are surrounded by two layers of membrane (the inner membrane, or cell membrane, and an outer membrane), between which lies a periplasmic space (see Section 3.3). Many proteins are specifically destined for one or another of these cell compartments. Other proteins are secreted completely out of the cell into the surrounding environment (for example, hemolysins that lyse red blood cells). But how do these diverse proteins know where to go? Protein traffic out of the cell is directed by an elaborate set of protein secretion systems. Each system selectively delivers a set of proteins originally made in the cytoplasm to various extracytoplasmic locations.

The term "secretion" is used to describe movement of a protein out of the cytoplasm. Some protein secretion systems move proteins out of the cytoplasm into the cytoplasmic membrane and across the membrane to the periplasm, others move proteins to the outer membrane, and still others deliver proteins across both of the membranes and into

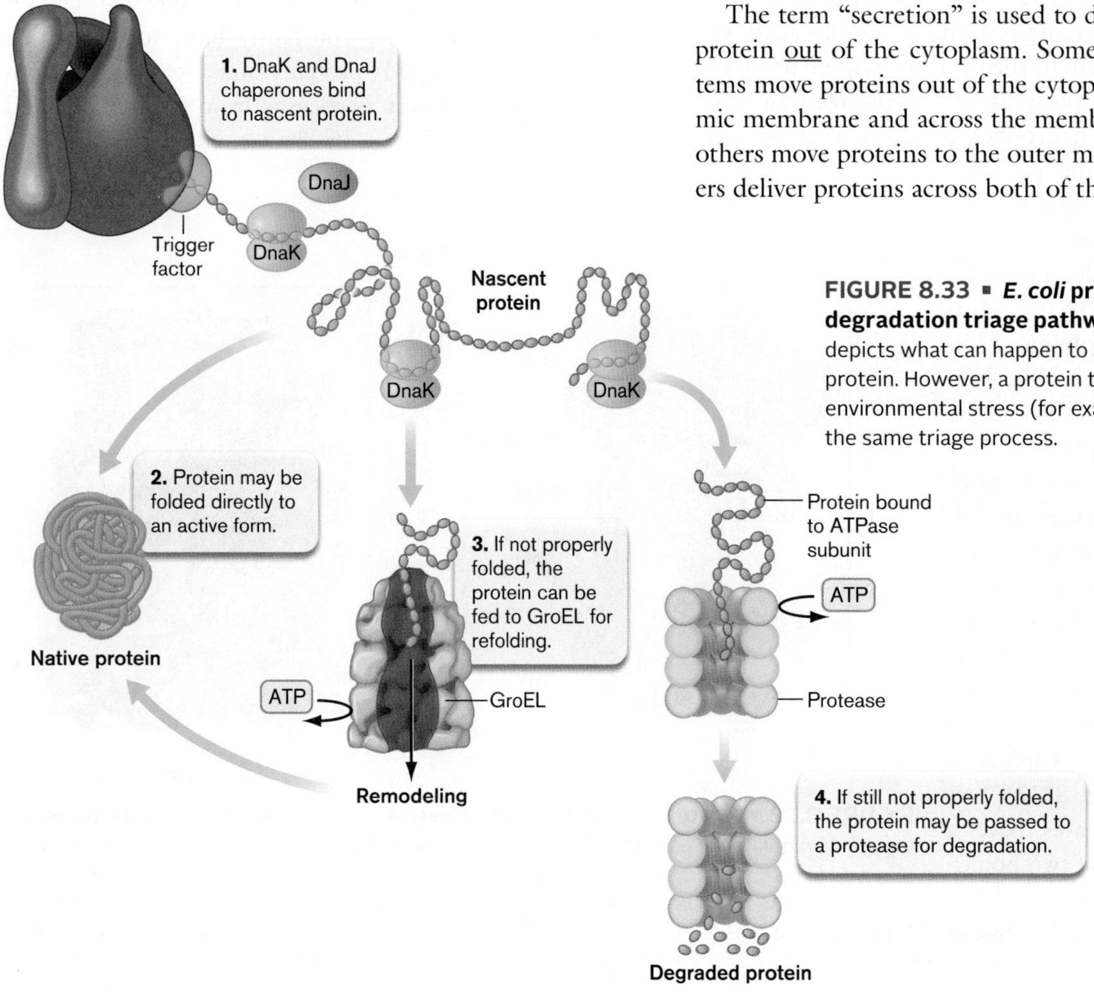

FIGURE 8.33 ▪ *E. coli* protein folding-versus-degradation triage pathways. The diagram depicts what can happen to a newly synthesized protein. However, a protein that unfolds in response to environmental stress (for example, heat) will undergo the same triage process.

the surrounding environment. An added complication of protein export is that periplasmic proteins are usually delivered unfolded into the periplasm and require another set of chaperones to fold properly in this cell compartment.

Protein Export Out of the Cytoplasm

Proteins destined for the bacterial cell membrane (such as membrane transport proteins), periplasm (binding proteins), outer membrane (porins), or extracellular spaces (proteases)—require special export systems. These systems manage to move hydrophilic proteins through one or more hydrophobic membrane barriers. Proteins meant for the inner membrane (for example, cytochromes) contain very hydrophobic N-terminal **signal sequences** of 15–30 amino acids. Signal sequences tether nascent proteins to the membrane and confer conformations that allow the proteins to melt into the fabric of the membrane. Inner membrane proteins also contain hydrophobic transmembrane regions (20–25 amino acids) that aid in this insertion process. These hydrophobic regions are important because they are compatible with the hydrophobicity of the membrane itself. A nutrient transport protein often has 12 such membrane-spanning regions, which weave back and forth across the membrane.

One special export system begins with a complex called the **signal recognition particle** (**SRP**), which targets proteins for inner membrane insertion. A second export mechanism uses a protein called trigger factor that assists proteins destined for the periplasm. (Trigger factor was mentioned earlier as a chaperone.) These two protein traffic pathways converge on a general secretion complex composed of three proteins, collectively called the SecYEG translocon, embedded in the cell (inner) membrane. Depending on the exported protein, SecYEG will assist export to the periplasm or insertion into the membrane.

Protein Export to the Cell Membrane

The pathway leading proteins to the inner (cell) membrane begins with an SRP. In *E. coli*, the SRP consists of a 54-kDa protein (Ffh) complexed with a small RNA molecule (*ffs*). SRP binds to the signal sequences of integral cell membrane proteins as they are being translated (**Fig. 8.34**) and halts further translation in the cytoplasm. The nascent protein with its paralyzed, nontranslating ribosome is delivered to the membrane-embedded protein FtsY, where translation resumes. The partially translated protein is now subject to one of two fates: It may be cotranslationally inserted directly into the cell membrane, meaning that the protein is inserted even as it is still being translated; or it may be completely synthesized, after which it is delivered to the SecYEG translocon for insertion. The route to membrane insertion depends on the protein to be inserted.

Protein Export to the Periplasm: The General Secretion Pathway

The periplasm contains important proteins that bind nutrients for transport into the cell and other proteins that carry out enzymatic reactions. For example, one form of superoxide dismutase (SOD), an enzyme that degrades superoxide, is a periplasmic protein in *Salmonella enterica* and other Gram-negative bacteria. Many periplasmic proteins, such as SOD and maltose-binding protein (which imports the sugar maltose), are delivered to the periplasm by a common pathway called the general secretion pathway.

The general secretion pathway has several steps. First, the peptide is completely translated in the cytoplasm

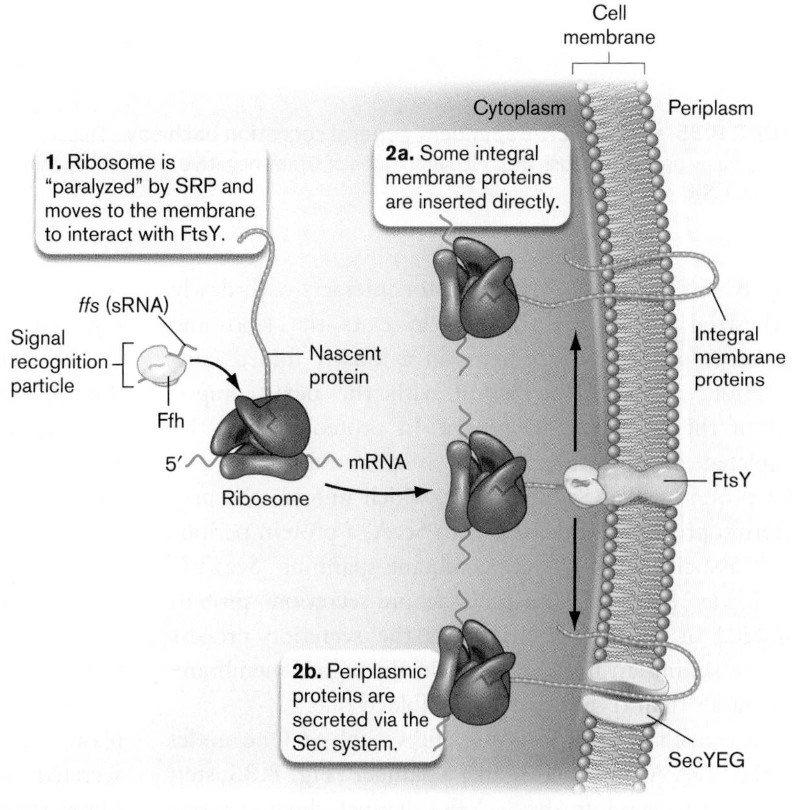

FIGURE 8.34 ■ SRP and cotranslational export in *E. coli*. A ribosome "paralyzed" by an SRP does not resume translating protein until it encounters FtsY in the membrane. Translation can then recommence. Some proteins destined for the membrane are inserted directly (step 2a). Other integral membrane proteins and proteins destined for the periplasm are inserted or secreted via the Sec system (step 2b).

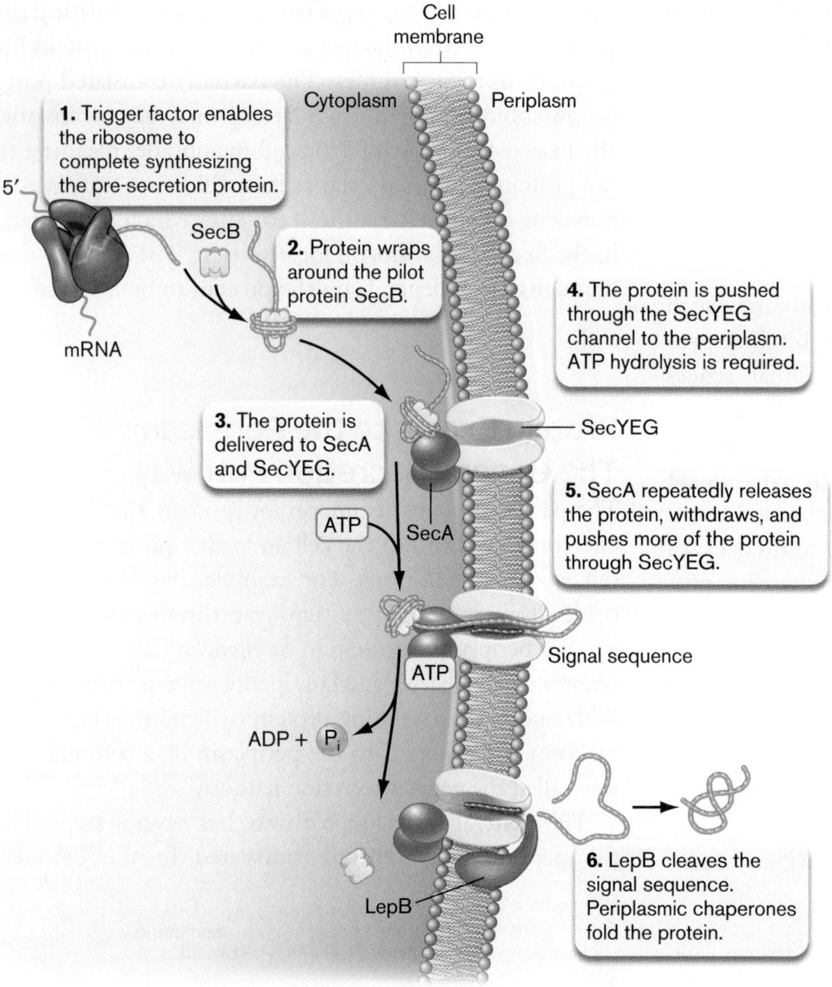

FIGURE 8.35 ▪ The SecA-dependent general secretion pathway. This pathway exports many proteins across the cell membranes of Gram-negative and Gram-positive bacteria. ▶

(**Fig. 8.35** ▶, step 1). Trigger factor interacts with newly synthesized protein as the protein exits the ribosome and keeps pre-secreted proteins in a loosely folded conformation, awaiting interaction with the next component of the secretion machinery. In proteobacteria, the completed pre-secretion protein is then captured by a pilot protein called SecB (step 2), which unfolds the pre-secretion protein and delivers it to SecA, a protein peripherally associated with the membrane-spanning SecYEG translocon (step 3). Keeping a pre-secretion protein unfolded in the cytoplasm assists the secretion process because sliding an unfolded protein through a membrane is far easier than trying to deliver a folded one.

The translocation process is not fully resolved. One model has the SecA ATPase acting like a plunger (**Fig. 8.35**, step 4). It inserts deep into the SecYEG channel, shoving about 20 amino acids of the target export protein into the channel. ATP hydrolysis causes SecA to release the protein and withdraw (step 5). At this point, SecA can bind fresh ATP, rebind the target protein, and reinsert, pushing another 20 amino acids through. The proton motive force at the cytoplasmic membrane (see Chapters 4 and 14) also contributes to the process, and is thought to help drive translocation of the protein after SecA release.

Proteins needed in the periplasm have cleavable signal sequences at their amino-terminal ends. Immediately following translocation of the amino-terminal sequence into the periplasm, the sequence is snipped off by periplasmic signal peptidases (LepB is one of several examples in *E. coli*). This cleaving completes translocation and releases the mature protein into the periplasm (**Fig. 8.35**, step 6). Signal peptidases, however, will not cleave signals from proteins destined to stay embedded within the membrane (integral membrane proteins).

Note: "Translocation" can refer to the movement of a ribosome along mRNA, or it can describe the movement of a protein from one cell compartment (cytoplasm) to another (periplasm).

Periplasmic proteins delivered by the Sec system arrive unfolded and inactive. Because the folding chaperones mentioned earlier are cytoplasmic, periplasmic proteins need a different set of dedicated chaperones to guide their tertiary folding. Another problem with periplasmic proteins is that the oxygen-rich environment of aerobic cells can oxidize cysteines within a protein and produce inappropriate cysteine disulfide bonds that destroy enzyme function. Special periplasmic disulfide reductases are required to reduce these S–S bonds back to two SH groups. Many periplasmic proteins, however, need certain disulfide bonds to be active, so the periplasm also contains a disulfide bond catalyst (DsbA) to make those bonds.

Gram-positive bacteria must also export proteins across the cell membrane and then fold and process them once they are secreted. However, Gram-positive bacteria lack the periplasmic space needed to facilitate interactions between newly secreted proteins and the accessory processing proteins. Many streptococci solve this problem by clustering their secretion systems and accessory factors at a microdomain of the cytoplasmic membrane called the ExPortal. The ExPortal is located near the cell septum and appears linked to peptidoglycan synthesis (**Fig. 8.36**). Proteins in the

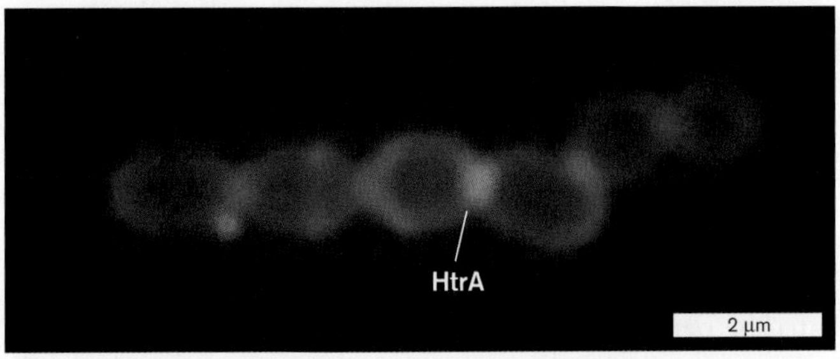

FIGURE 8.36 ▪ Location of the ExPortal in *Streptococcus pyogenes*. HtrA was identified using immunofluorescence. Note that HtrA is located at the septum.

ExPortal include HtrA (which assists in pili formation and covalent attachment of proteins to the cell wall) and sortase (which aids maturation of secreted proteins), as well as Sec system components and chaperones. Some proteins that pass through the ExPortal are truly secreted; others are not. The latter proteins have a transmembrane domain (missing in Gram-negative homologs) that anchors the proteins to the Gram-positive cytoplasmic membrane. The anchored proteins are extracellular but will not float away.

Eukaryotic microbes such as the yeast *Saccharomyces cerevisiae* also possess secretion systems that move proteins to the membrane and beyond. Most secreted proteins in eukaryotes are exported cotranslationally, using SRP to pause translation until the peptide is delivered to the SRP receptor and transferred to the Sec translocase. The Sec translocase of eukaryotes has homologs of the SecY and SecE proteins of bacteria, called Sec61α and Sec61γ, respectively. Their third component, Sec61β, however, is unrelated to the bacterial SecG. The translating ribosome provides the driving force for transport during cotranslational secretion (**Fig. 8.37A**). For posttranslational secretion, eukaryotes use the ATP-binding protein Bip, located in the lumen of the endoplasmic reticulum (ER), in coordination with a transmembrane complex of Sec62 and Sec63 to ratchet transport through the Sec complex (**Fig. 8.37B**).

Archaeal secretion systems are more similar to those of eukaryotes than to those of bacteria. However, archaea lack both SecA and Bip; thus, the driving force of protein secretion in archaea is still unknown.

Export of Prefolded Proteins to the Periplasm

In a dramatic departure from Sec-dependent transport systems, proteins can also be transported fully folded across the membrane to their periplasmic destination. Prefolding of periplasmic proteins in the cytoplasm enables cells to carefully control the insertion of cofactors. Cofactors such as flavins are important for proteins involved in respiration (see Chapter 14), and some of these cofactors are embedded within the protein structure as the protein folds after translation. Metalloproteins use metals as cofactors, but different metals can compete for the metal-binding sites within the protein. Prefolding of the metalloprotein in the cytoplasm ensures that the correct metal is inserted. In addition, other proteins without cofactors simply fold very quickly in the cytoplasm and need to be translocated in their fully folded state.

These proteins contain the amino acid motif RRXFXK within their N-terminal signal sequence (where R = arginine, F = phenylalanine, K = lysine, and X = any amino acid). This sequence, called the "twin arginine motif" targets the protein to the membrane-embedded twin arginine translocase (TAT), a transport complex that assembles on demand to ship fully folded proteins across the cell

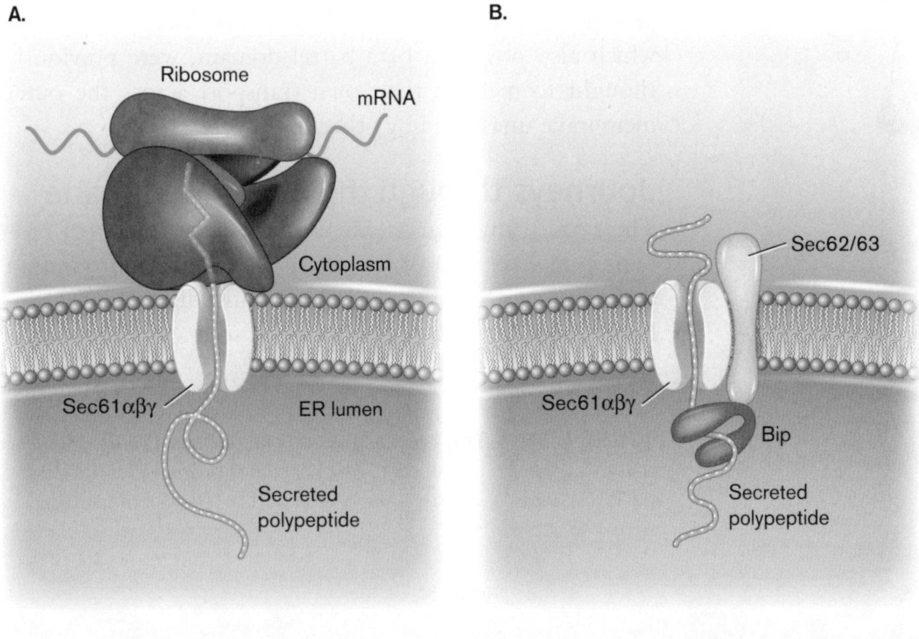

FIGURE 8.37 ▪ Protein secretion in eukaryotes. **(A)** Cotranslational secretion though the Sec61αβγ channel is driven by the ribosome. **(B)** Posttranslational secretion through the channel is driven by the ratcheting action of Bip and Sec62/63. *Source*: Modified from T. A. Rapoport et al. 2017. *Annu. Rev. Cell Dev. Biol.* **33**:369–390, fig. 8.

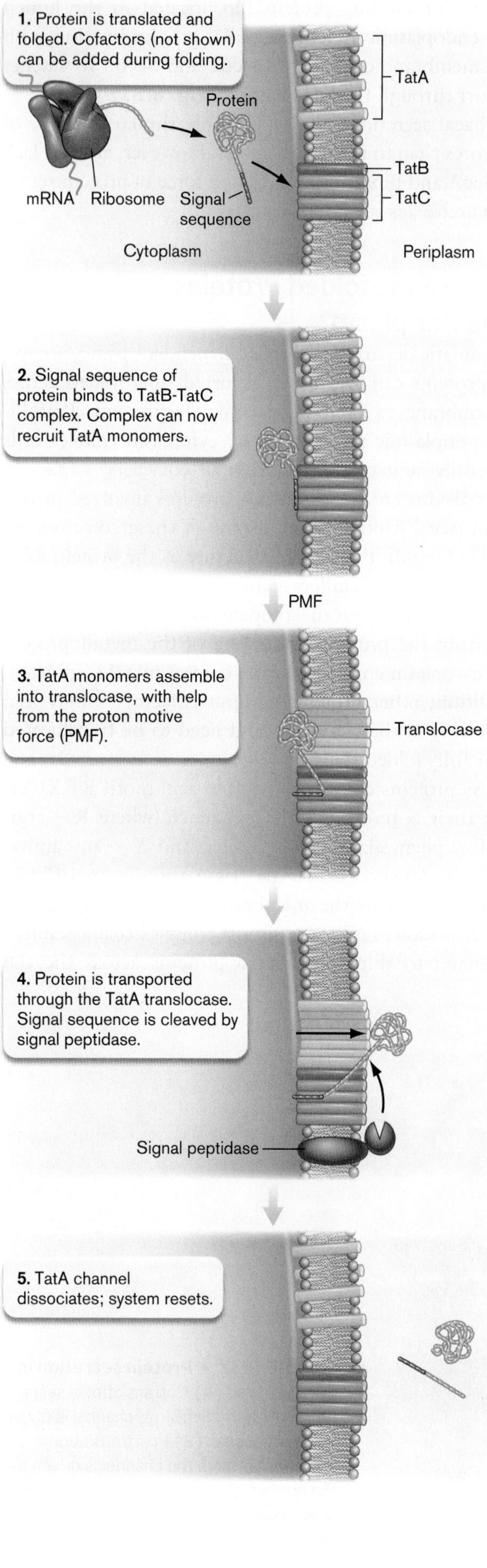

membrane to the periplasm (**Fig. 8.38**). Signal sequence binding triggers a conformational change in a complex of TatB and TatC proteins, which, with the assistance of the proton motive force, recruits and oligomerizes TatA, forming the translocase that enables fully folded protein substrates to pass through the membrane. After translocation the signal sequence is cleaved and the protein can diffuse into the periplasm. Once the protein leaves the translocase, the translocase dissociates into TatA monomers and the system resets to receive the next secreted protein.

Archaea also possess a TAT system, but eukaryotes do not. It is thought that a TAT system is unnecessary for eukaryotes, because translocation in eukaryotes occurs in the endoplasmic reticulum. In contrast to the extracytoplasmic space in bacteria and archaea, the lumen of the ER is ATP-rich, and is host to ATP-dependent enzymes that can provide transport assistance on the other side of the membrane.

Journeys to the Outer Membrane

Outer membrane proteins (OMPs) are made in the cytoplasm and exported to the periplasm by the SecA-dependent secretion system. Some OMPs have hydrophobic C-terminal signal sequences that facilitate insertion into the outer membrane, but all have a beta barrel structure that ultimately suits them for outer membrane placement; one example is TolC (see **Fig. 8.39** ▶). Periplasmic chaperones prevent aggregation of OMPs as they traverse the periplasm and deliver the proteins to a multisubunit, outer membrane machine called the BAM (beta barrel assembly machine) complex that facilitates OMP assembly in the outer membrane. Although the players seem to be known, the mechanism by which beta barrels are folded and inserted into the outer membrane bilayer remains unclear. BAM has also been implicated recently in the transport of so-called autotransporter proteins such as the attachment protein pertactin of *Bordetella pertussis.* Autotransporters, which also possess a beta barrel domain, were previously thought to mediate their own transport across the outer membrane unassisted by other transporters.

Journeys through the Outer Membrane

There are many reasons why bacteria need to export proteins completely out of the cell and into their surrounding environment. Some exported proteins digest extracellular peptides for carbon and nitrogen sources; others act as

FIGURE 8.38 ■ The twin arginine translocase (TAT). Model for the Tat protein translocase, which includes proteins TatA, TatB, and TatC. *Source:* Modified from Tracy Palmer and Ben C. Berks. 2012. *Nat. Rev. Microbiol.* **10**:483–496, fig. 2.

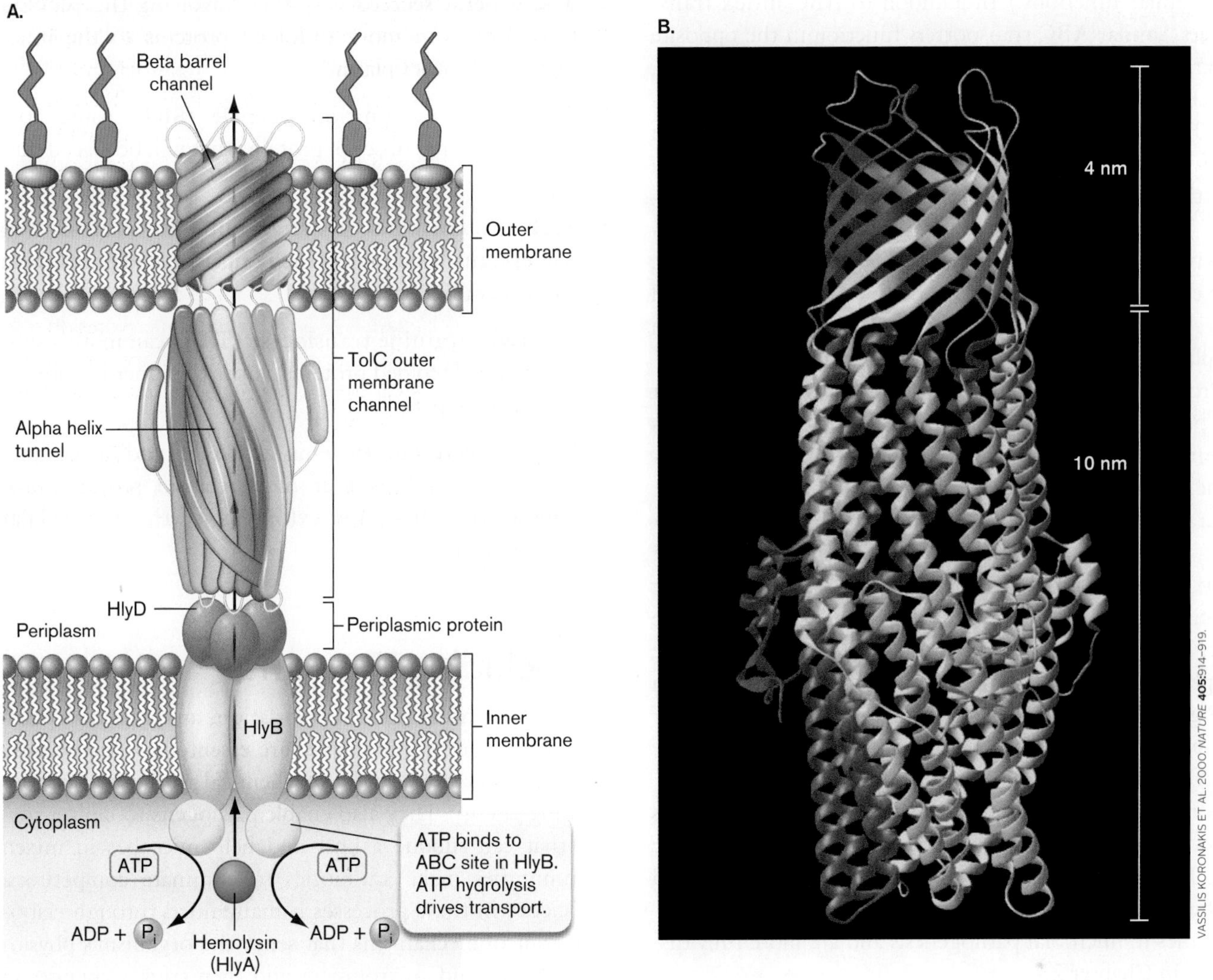

FIGURE 8.39 ■ Type I secretion: the Hly ABC transporter. **A.** Hemolysin (HlyA) is transported directly from the cytoplasm into the extracellular medium through a multicomponent ABC transport system. The HlyB and HlyD proteins are dedicated to HlyA transport. TolC is shared with other transport systems. Not drawn to scale. **B.** Molecular model of TolC. The beta barrel channel spans the outer membrane, and the alpha helix tunnel extends into the periplasm. Three monomers (red, yellow, and blue) make up the channel. *Source*: Part A modified from Moat et al. 2002. *Microbial Physiology,* 4th ed. Wiley-Liss.

free-floating toxins that bind and kill host cells. Still others are injected directly into eukaryotic cells by pathogenic or symbiotic microbes to commandeer host metabolic processes. Nine secretion systems, identified as type I through type IX, have been discovered thus far to ship proteins out of the cell. Most are exclusive to Gram-negative microbes, which have a periplasmic space and an outer membrane in addition to the cytoplasmic membrane. A few secretion systems start with the Sec system just to get the protein into the periplasm, where dedicated outer membrane systems take over and complete export. Other systems provide nonstop service, delivering the protein directly from the cytoplasm to the extracellular space.

The diversity of system architecture is impressive. It is the result, in some instances, of selective evolutionary pressures appropriating established cellular processes (for example, pilus assembly). New systems evolve through the accidental duplication of one set of genes followed by random mutations that provide the duplicated set with a new function. We know this because the footprints of genetic divergence have been left behind in the DNA sequence. Type I secretion is described here. Other systems will be covered during the discussion of pathogenesis in Chapter 25.

Type I Protein Secretion

Chapter 4 describes the family of ATP-binding cassette (ABC) influx transporters, whose signature is an amino acid motif that binds ATP. (The term "cassette" refers to a sequence of amino acids that is conserved in many proteins with similar functions.) In addition to ABC influx transporters, similar ABC transporters function in the opposite direction to export various toxins, proteases, and lipases, as well as antimicrobial drugs (multidrug efflux transporters). These ABC transporters are the simplest of the protein secretion systems and make up what is called type I protein secretion (for example, the Hly system that secretes hemolysin; **Fig. 8.39**). Type I systems all have three protein components, one of which contains an ATP-binding cassette. One component is an outer membrane channel, the second is an ABC protein at the inner membrane, and the third is a periplasmic protein lashed to the inner membrane.

Proteins secreted through type I systems never contact the periplasm, because they pass through a continuous channel that extends from the cytoplasm to the outer membrane. The inner membrane and periplasmic subunits are generally substrate specific, but numerous ABC export systems share the channel protein TolC, including multidrug efflux pumps that confer resistance. TolC is an intriguing protein composed of a beta barrel channel embedded in the outer membrane and an alpha helix tunnel spanning the periplasm (**Fig. 8.39B**). The type I transport system shown in **Figure 8.39A** exports a hemolysin (HlyA) from *E. coli* that lyses red blood cell membranes. HlyB and HlyD are the ABC and periplasmic components, respectively.

Some other protein secretion systems move proteins directly from the cytoplasm to the outside, similar to the type I system, whereas others pick up proteins deposited in the periplasm by the Sec system. They all play important roles in microbial pathogenesis and are more fully discussed in Chapter 25.

To Summarize

- **Special protein export mechanisms** are used to move proteins to the inner membrane, the periplasm, the outer membrane, and the extracellular surroundings.
- **N-terminal amino acid signal sequences** help target membrane proteins to the membrane.
- **The general secretory system**, involving the SecYEG translocon, can move unfolded proteins to the inner membrane or periplasm.
- **The signal recognition particle (SRP)** pauses the translation of a subset of proteins that will be placed into the membrane.
- **SecB protein** binds to certain unfolded proteins that will eventually end up in the periplasm, and pilots them to the SecYEG translocon.
- **The twin arginine translocase (TAT)** can move a subset of already folded proteins across the inner membrane and into the periplasm.
- **Type I secretion systems** are ATP-binding cassette (ABC) mechanisms that move certain secreted proteins directly from the cytoplasm to the extracellular environment.

Concluding Thoughts

The cellular transcription, translation, and secretion pathways described in this chapter are essential for life because they efficiently assemble biochemical pathways without wasting energy. They also enable pathogens to deliver toxins that can subdue a host and help microbes in mixed communities make antibiotics to eliminate competitors. Efficiency in these processes is maintained through elaborate control mechanisms that sense the organism's physiological state and environment and then trigger changes in replication, transcription, translation, and/or protein processing. How bacteria regulate gene expression in response to environmental stimuli, including threats to survival, will be discussed in Chapter 10.

But first, in Chapter 9, we explore how natural selection randomly redesigns genomes to adapt to ecological niches. Microbes use a variety of DNA exchange mechanisms, gene duplications, and gene alterations to evolve into forms better adapted to their environments and, in the process, may produce entirely new species.

CHAPTER REVIEW

Review Questions

1. What are some characteristics of an open reading frame?
2. What is a DNA sequence alignment, and what can it tell you?
3. What defines a promoter?
4. What are sigma factors, and what role do they play in gene expression?

5. Describe the three stages of transcription.
6. Explain the degeneracy of the genetic code. What is the wobble in codon-anticodon recognition?
7. Describe the stages of protein synthesis. Why is the ribosome called a ribozyme?
8. Discuss some antibiotics that affect transcription or translation.
9. What is coupled transcription and translation? Does it occur in eukaryotic cells?
10. How do bacterial cells release ribosomes that are stuck onto damaged mRNA molecules lacking termination codons?
11. What kinds of posttranslational modifications can be made to proteins, and how do those modifications affect the proteins?
12. What can happen to misfolded proteins?
13. Why are only certain proteins secreted from the bacterial cell? What are some secretion mechanisms?
14. In what major way do proteins transported by the twin arginine translocase (TAT) differ from other exported proteins?
15. Compare protein degradation in eukaryotes and bacteria.

Thought Questions

1. The process of transcription generates positive supercoils in front of the polymerase as it moves along a DNA template. Why doesn't the DNA in front of the polymerase become so knotted that the polymerase can no longer separate the DNA strands?
2. Why do cells secrete some proteins into their environments?
3. Type I protein secretion systems transport certain proteins from the cytoplasm of Gram-negative bacteria directly to the outside of the cell, across two membranes. How might the system "know" which proteins to transport?
4. Why might natural selection favor a system in which all DNA of the genome undergoes low levels of unregulated transcription?

Key Terms

acceptor site (A site) (294)
aminoacyl-tRNA synthetase (291)
anticodon (289)
catalytic RNA (286)
chaperone (304)
codon (288)
consensus sequence (279)
DNA-dependent RNA polymerase (278)
exit site (E site) (295)
heat-shock protein (HSP) (304)
mass spectrometry (303)
messenger RNA (mRNA) (286)
N-terminal rule (305)
open reading frame (ORF) (281)
operon (281)
peptidyltransferase (292)
peptidyl-tRNA site (P site) (295)
polysome (299)
promoter (279)
release factor (296)
Rho factor (283)
ribosomal RNA (rRNA) (286)
ribosome (291)
ribosome-binding site (293)
ribozyme (286, 292)
RNA polymerase (278)
Shine-Dalgarno sequence (293)
sigma factor (278)
signal recognition particle (SRP) (307)
signal sequence (307)
small RNA (sRNA) (286)
start codon (293)
stop codon (289)
template strand (278)
tmRNA (286)
transcript (278)
transcription (278)
transfer RNA (tRNA) (286)
translation (288)
translocation (296)

Recommended Reading

Bakshi, Somenath, Heejun Choi, Jagannath Mondal, and James C. Weisshaar. 2014. Time-dependent effects of transcription- and translation-halting drugs on the spatial distributions of the *Escherichia coli* chromosome and ribosomes. *Molecular Microbiology* **94**:871–887.

Bastos, P. A. D., J. P. da Costa, and R. Vitorino. 2017. A glimpse into the modulation of post-translational modifications of human-colonizing bacteria. *Journal of Proteomics* **152**:254–275.

Brandt, Florian, Stephanie A. Etchells, Julio O. Ortiz, Adrian H. Elcock, F. Ulrich Hartl, et al. 2009. The native 3D organization of bacterial polysomes. *Cell* **136**:261–271.

Burger, Adelle, Chris Whiteley, and Aileen Boshoff. 2011. Current perspectives of the *Escherichia coli* RNA degradosome. *Biotechnology Letters* **33**:2337–2350.

Costa, Tiago R., Catarina Felisberto-Rodrigues, Amit Meir, Marie S. Prevost, Adam Redzej, et al. 2015. Secretion systems in Gram-negative bacteria: Structural and mechanistic insights. *Nature Reviews. Microbiology* **13**:343–359.

Gehring, A. M., J. E. Walker, and T. J. Santangelo. 2016. Transcription regulation in Archaea. *Journal of Bacteriology* **198**:1906–1917.

Hui, Monica P., Patricia L. Foley, and Joel G. Belasco. 2014. Messenger RNA degradation in bacterial cells. *Annual Review of Genetics* **48**:537–559.

Keiler, Kenneth C. 2015. Mechanisms of ribosome rescue in bacteria. *Nature Reviews. Microbiology* **13**:285–297.

Ling, J., P. O'Donoghue, and D. Soll. 2015. Genetic code flexibility in microorganisms: Novel mechanisms and impact on physiology. *Nature Reviews. Microbiology* **13**:707–721.

Murakami, Katsuhiko S., Shoko Masuda, Elizabeth A. Campbell, Oriana Muzzin, and Seth Darst. 2002. Structural basis of transcription initiation: RNA polymerase holoenzyme-DNA complex. *Science* **296**:1285–1290.

Petrov, Anton S., Chad R. Bernier, Chiaolong Hsiao, Ashlyn M. Norris, Nicholas A. Kovacs, et al. 2014. Evolution of the ribosome at atomic resolution. *Proceedings of the National Academy of Sciences USA* **111**:10251–10256.

Preissler, Steffen, and Elke Deuerling. 2012. Ribosome-associated chaperones as key players in proteostasis. *Trends in Biochemical Sciences* **37**:274–283.

Proshkin, Sergey, A. Rachid Rahmouni, Alexander Mironov, and Evgeny Nudler. 2010. Cooperation between translating ribosomes and RNA polymerase in transcription elongation. *Science* **328**:504–508.

Saier, Milton H. 2019. Understanding the genetic code. *Journal of Bacteriology* **201**:e00091–19.

Schulze, Ryan J., Joanna Komar, Mathieu Botte, William J. Allen, Sarah Whitehouse, et al. 2014. Membrane protein insertion and proton-motive-force-dependent secretion through the bacterial holo-translocon SecYEG-SecDF-YajC-YidC. *Proceedings of the National Academy of Sciences USA* **111**:4844–4849.

Tsirigotaki, A., J. De Geyter, N. Sostaric, A. Economou, and S. Karamanou. 2017. Protein export through the bacterial Sec pathway. *Nature Reviews. Microbiology* **15**:21–36.

Vega, Luis A., Gary C. Port, and Michael G. Caparon. 2013. An association between peptidoglycan synthesis and organization of the *Streptococcus pyogenes* ExPortal. *mBio* **4**:e00485–13.

Washburn, Robert S., and Max E. Gottesman. 2015. Regulation of transcription elongation and termination. *Biomolecules* **5**:1063–1078.

CHAPTER 9

Genetic Change and Genome Evolution

9.1 Mutations

9.2 DNA Repair

9.3 Gene Transfer: Mechanisms and Barriers

9.4 Mobile Genetic Elements

9.5 Genome Evolution

DNA is a dynamic molecule. Genome sequences change over generations through mutations that include DNA rearrangements, and gene transfers between species. Bacterial and archaeal genomes can shuttle large clusters of genes between members of different taxonomic domains. In Chapter 9 we describe how the genome changes through gene mutation, gene gain, and gene loss, and we describe mechanisms that cells employ to limit change. The process of genetic change is the essential first step in evolution.

CURRENT RESEARCH highlight

Microfluidic "mother machines" reveal mutations in single cells. Within the microchannels of microfluidic chips, "mother cells" positioned at the closed-off end of the channel (bottom of figure) reproduce, and the progeny cells fill the channel as they grow and increase in number. With this device, Lydia Robert and colleagues could visualize replication-based mutations in individual cells. The mismatch repair machinery, stained green, localized to spots within the cell (stained red) where a mutation occurred (arrow; see inset for close-up).

Source: Robert et al. 2018. *Science* **359**:1283–1286.

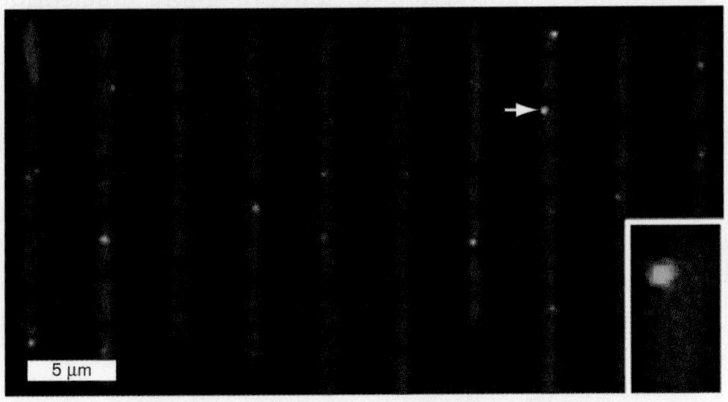

ROBERT ET AL. 2018. *SCIENCE* **359**:1283–1286, FIG. 1E.

AN INTERVIEW WITH

LYDIA ROBERT, MICROBIOLOGIST, UNIVERSITY OF PARIS

COURTESY OF LYDIA ROBERT

What additional kinds of discoveries can be made with single-cell microfluidics that might be missed by traditional culturing methods?

Microfluidics is the technique of choice to investigate responses to environmental perturbations at the single-cell level, potentially revealing subpopulations of cells with different behavior in a clonal population. Such heterogeneity would not be detected by traditional culturing methods that provide only average behaviors. It is also the best technique to investigate the behavior of a population in a heterogeneous environment, such as a chemical gradient.

Microbes have been evolving for millions of years, and they continue to evolve today. Evolutionary change can happen rapidly in the microbial world, with consequences that can directly impact us. Underlying evolutionary innovation is genetic change, which in microbes can result from gene mutation, as well as from the transfer of genes between microbes. One striking example of the latter involves the pathogen *Streptococcus pneumoniae*, which causes meningitis and serious lung infections in any age group, and also recurrent ear infections in infants.

A research team headed by Fen Hu and Garth Ehrlich (Drexel University) found evidence of evolution in a strain of *S. pneumoniae* isolated from an 8-month-old child stricken with recurrent ear infections over 7 months. Genome sequences of strains isolated during separate bouts revealed that 16 distinct gene transfer events had occurred from the earliest isolate to the latest. Over those 7 months the child had been colonized by multiple strains of *S. pneumoniae* that readily received DNA from other microbes. The resultant gene transfers enabled the species to quickly evolve and cause recurrent disease.

In another example of gene transfer, John Sullivan and Clive Ronson from New Zealand witnessed microbial evolution taking place in soil within a mere decade. These researchers were studying the microbe *Mesorhizobium loti*, a symbiotic bacterium that forms nitrogen-fixing nodules on plant roots. Genes encoding symbiosis are located as a group on the mesorhizobial chromosome. In a remarkable experiment, Sullivan and Ronson inoculated a single strain of *M. loti* into an area of land devoid of natural nodulating rhizobia. Seven years later, they discovered that the area contained many genetically diverse symbiotic mesorhizobia now able to nodulate the flowering plant *Lotus corniculatus*. These microbes did not exist before the experiment. The 500-kb genome segment encoding symbiosis had somehow made its way from *M. loti* into these other bacteria—essentially generating new species. How did this happen so quickly?

In Chapter 9 we explore the mechanisms of genetic change that facilitate evolution in microbes. First we look at how individual genes mutate, and how the cell repairs mutations. We then look at horizontal gene transfer and describe some of the cell defenses that limit transfer. Next we describe the role of transposable elements in both gene mutation and horizontal gene transfer. Finally, we describe how these mechanisms can add or remove genes in the genomes of microbes. Whereas this chapter explores the mechanisms of mutation and gene flow, the consequences of these processes during evolution will be discussed in Chapter 17.

9.1 Mutations

Any permanent, heritable alteration in a DNA sequence, whether harmful, beneficial, or neutral, is called a **mutation**. Mutations are the foundational events that facilitate evolutionary change. In addition, mutations provide the researcher with the ability to investigate the function of genes through manipulation of their DNA sequence. In this section we discuss the various types of mutations that can occur in microbes and how these mutations arise. Insertion of foreign DNA acquired by horizontal gene transfer, discussed in Section 9.3, is one form of mutation. In this section we discuss other classes of microbial mutations.

Mutations Can Change Genes in Many Ways

Changes in the DNA sequence of genes through mutation can have a variety of effects on those genes. Mutations can change the structure and function of the gene's products, be they protein or RNA, and can also change the regulation of the gene's expression. Gene mutations can have important consequences for the microbe, such as gaining resistance to antibiotics that normally target the gene's product. And as we will see in Section 9.5, mutation is an important mechanism that adds or subtracts genes from genomes.

Mutations that alter gene sequence fall into several different physical and structural classes:

- A **point mutation** is a change in a single nucleotide (**Fig. 9.1A** and **B**). Replacing a purine with a different purine or a pyrimidine with a different pyrimidine is called a **transition**. Swapping a purine for a pyrimidine (or vice versa) is a **transversion**.
- **Insertions** and **deletions** involve, respectively, the addition or subtraction of one or more nucleotides (**Fig. 9.1C** and **D**), making the sequence either longer or shorter than it was originally.
- An **inversion** results when a fragment of DNA is flipped in orientation relative to the flanking DNA on either side (**Fig. 9.1E**).
- A **duplication** produces a second copy of a sequence fragment on the DNA molecule, usually adjacent to the original copy (**Fig. 9.1F**).
- A **transposition** is the movement of a sequence fragment from one location to another. Transpositions are discussed in Section 9.4.
- A **reversion** restores a mutated sequence to its original sequence.

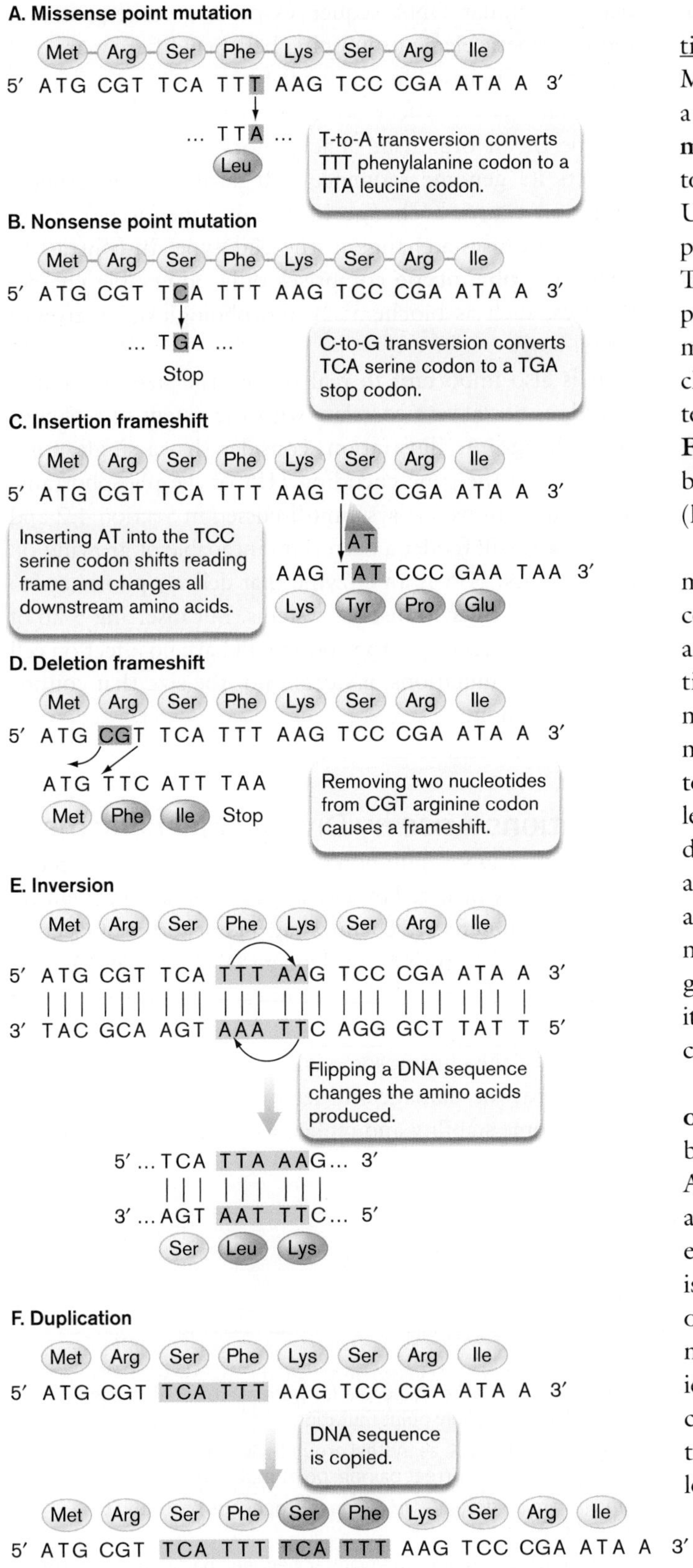

FIGURE 9.1 ■ Changes in a DNA sequence that result in different classes of mutations.

Mutations can be further categorized into informational classes based on how they impact the gene product. Mutations that do not change the amino acid sequence of a translated open reading frame (ORF) are called **silent mutations**. For example, a point mutation changing TTT to TTC in the sense DNA strand (corresponding to a UUU-to-UUC codon change in mRNA) still codes for phenylalanine (refer to Figure 8.12 for the genetic code). Thus, even though the DNA sequence has changed, the protein sequence remains the same (hence, it was a synonymous base substitution). However, if the UUU codon were changed to UUA (a U-to-A transversion), then the protein would have a leucine where a phenylalanine was (see **Fig. 9.1A**). This type of mutation is a **missense mutation** because it changes the amino acid sequence of the protein (hence, it was a nonsynonymous base substitution).

The amino acid substitution resulting from a missense mutation may or may not alter protein function. The outcome depends on the structural importance of the original amino acid and how close in structure and chemical properties the replacement amino acid is to the original. Missense mutations result in either conservative amino acid replacements, in which the new amino acid is structurally similar to the original (for example, leucine is substituted for isoleucine), or nonconservative replacements, in which a very different amino acid is substituted (for example, tyrosine for alanine). A missense change may decrease or eliminate the activity of the protein (a **loss-of-function mutation**), or it may make the protein more active. The protein could even gain a new activity, such as an expanded substrate specificity or a completely different substrate specificity (these are called **gain-of-function mutations**).

A mutation that eliminates function is known as a **knockout mutation**. Knockout mutations can include multiple-base insertions and deletions, as well as nonsense mutations. A **nonsense mutation** is a point mutation that changes an amino acid codon into a translation termination codon—for example, UCA (serine) to UAA (see **Fig. 9.1B**). The result is a truncated protein. Truncations can completely knock out the function of proteins, especially in cases where the mutation occurred early in the open reading frame. Typically, these defective, truncated proteins are degraded by cellular proteases (see Section 8.4). Nonsense mutations can turn genes into pseudogenes, which subsequently can be lost from the genome during evolution (Section 9.5).

Insertions and deletions can alter the reading frame of the DNA sequence (see **Fig. 9.1C** and **D**). Remarkable as it is, the ribosome simply reads RNA sequences one codon at a time, stringing amino acids together in the process. It translates each codon "word" but cannot understand the overall protein "sentence." It does not recognize when bases have been added or removed through mutation; instead, it keeps reading the sequence in triplets. If the number of bases inserted or deleted is not a multiple of three, the reading frame of translation changes. The result is a **frameshift mutation**, in which the ribosome produces a garbled protein product. Frameshift mutations often cause the ribosome to encounter a premature stop codon, originally in a different reading frame. Thus, like nonsense mutations, frameshifts can also turn genes into pseudogenes. However, if the insertion or deletion involves multiples of three bases, the reading frame is not changed, but one or more amino acids are added or removed.

Thought Question

9.1 How could frameshift mutations be used to confirm that codons consist of three bases, versus two or four? *Hint:* Think of how a series of "like" frameshifts (for example, single base-pair additions) along a gene would impact the reading frame.

An inversion mutation flips a DNA sequence (see **Fig. 9.1E**). Imagine the sequence highlighted in the figure rotating 180° while the adjacent sequences remained right where they were. The rotation would retain the 5′-to-3′ polarity in the new molecule. But if the inversion occurred within a gene, it would likely change the codons in the area and alter the resulting protein. **Figure 9.1E** shows a small inversion.

However, inversions often involve large tracts of DNA encompassing several genes. If an entire gene with its promoter inverts, the gene will likely remain functional, and its encoded protein may very well still be made. Inversions occur within a genome as a result of recombination events between similar DNA sequences or as a consequence of mobile genetic elements jumping between different areas of a genome (discussed in Section 9.4).

Mutations can affect both the genotype and the phenotype of an organism. The **genotype** of an organism reflects its genome sequences. Regardless of whether a mutation causes a change in a trait (phenotype), every mutation causes a change in the genotype. In contrast to genotype, **phenotype** comprises only observable characteristics, such as biochemical, morphological, or growth traits.

It is also important to realize that the size of a mutation does not always correlate with the extent of a phenotypic change. As illustration, consider that a single point mutation in the gene encoding HPr of the phosphotransferase sugar transport system (discussed in Section 4.2 and **eTopic 4.1**) will render a bacterium incapable of growing on many sugars. HPr is an enzyme that delivers phosphate to a number of sugar transport systems. But inserting 5 kb of DNA just past the *hpr* stop codon will have no effect on cell growth. For mutations, it's often not the size that counts; it's the location.

Mutations Arise by Diverse Mechanisms

What causes DNA mutation in microbes? DNA is susceptible to damage inflicted by a variety of physical and chemical agents. **Mutagens** are chemical agents or forms of electromagnetic radiation that can damage DNA (**Table 9.1**). For example, irradiation by X-rays can cause a massive number of double-stranded DNA breaks. When this happens, the integrity of the chromosome is lost, and if the DNA is left unrepaired, the damage will lead to severe problems with chromosome stability and replication and, ultimately, cause

TABLE 9.1 Mutagenic Agents and Their Effects

Mutagenic agent	Effects
Chemical agent	
Base analog *Examples:* caffeine, 5-bromouracil	Substitutes "look-alike" molecule for normal nitrogenous base during DNA replication: point mutation
Alkylating agent *Example:* nitrosoguanidine	Adds alkyl group, such as methyl group ($-CH_3$), to nitrogenous base, resulting in incorrect pairing: point mutation
Deaminating agent *Examples:* nitrous acid, nitrates, nitrites	Removes amino group ($-NH_2$) from nitrogenous base: point mutation
Acridine derivative *Examples:* acridine dyes, quinacrine	Inserts (intercalates) into DNA ladder between backbones to form a new rung, distorting the helix: can cause frameshift mutations
Electromagnetic radiation	
Ultraviolet rays	Link adjacent pyrimidines to each other, as in thymine dimer formation, thereby impairing replication; lethal if not repaired
X-rays and gamma rays	Ionize and break molecules in cells to form free radicals, which in turn break DNA; lethal if not repaired

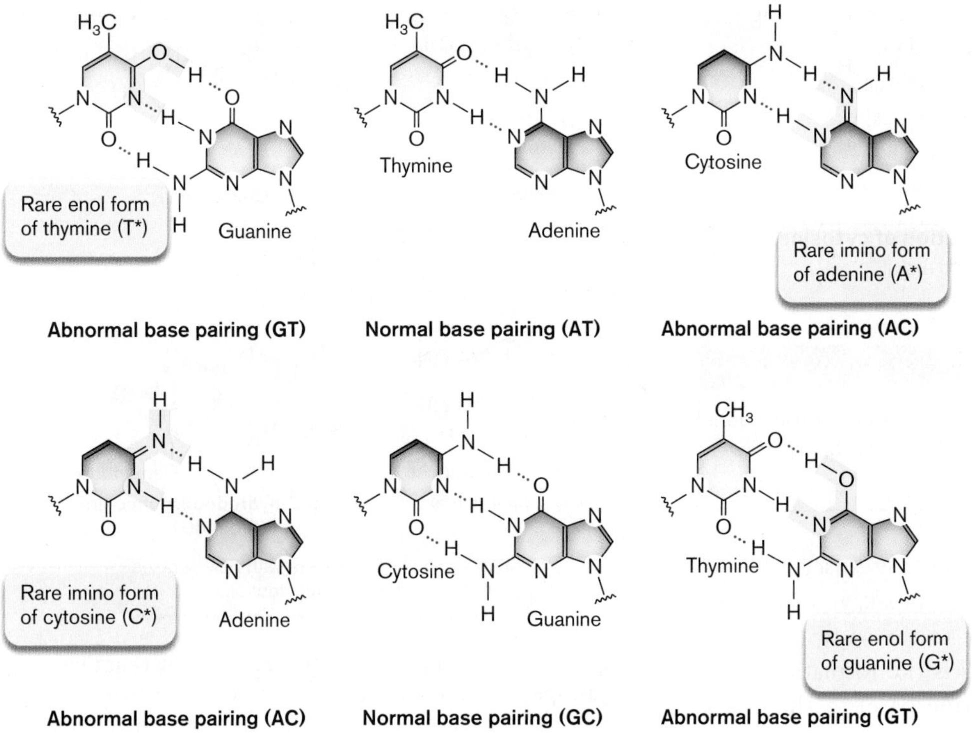

FIGURE 9.2 ■ Rare tautomeric forms of bases have altered base-pairing properties. Tautomeric transitions can lead to permanent mutations.

cell death. Other agents can directly modify the bases in DNA while leaving its overall structure intact. In this case, the modified bases have altered hydrogen bond base-pairing properties that result in the incorporation of an inappropriate base during replication. When that happens, a mutation results.

Mutations can arise even in the absence of a mutagen—that is, spontaneously. Despite the high accuracy of the replication apparatus (see Section 7.3), mistakes do occur, albeit at a very low rate. For example, spontaneous mutations in *E. coli* have a frequency of occurrence ranging from 10^{-8} to 10^{-6} per cell division in a given gene.

Spontaneous mutations in a genome arise for many reasons—for example, tautomeric shifts in the chemical structure of the bases (**Fig. 9.2**). Tautomeric shifts involve a change in the bonding properties of amino ($-NH_2$) and keto (C═O) groups. Normally, the amino and keto forms predominate (over 85%), but when an amino group shifts to an imino (═NH) group, for example, base pairing changes. A thymine that normally base-pairs with adenine will, in its rare enol form, base-pair with guanine. Tautomeric shifts that occur during DNA replication will increase the number of mutation events. **Figure 9.3** shows an example of how a tautomeric shift in thymine during replication results in an AT-to-GC transition mutation. Note that after the second round of replication, the mutation is "fixed" on both DNA strands in the mutant, and the mutation will therefore be passed on to all of the mutant's progeny.

Spontaneous mutations in DNA can also be caused by chemical reactions with water (hydrolysis). For example, cytosine spontaneously deaminates to yield uracil, which base-pairs with adenine instead of guanine (**Fig. 9.4**).

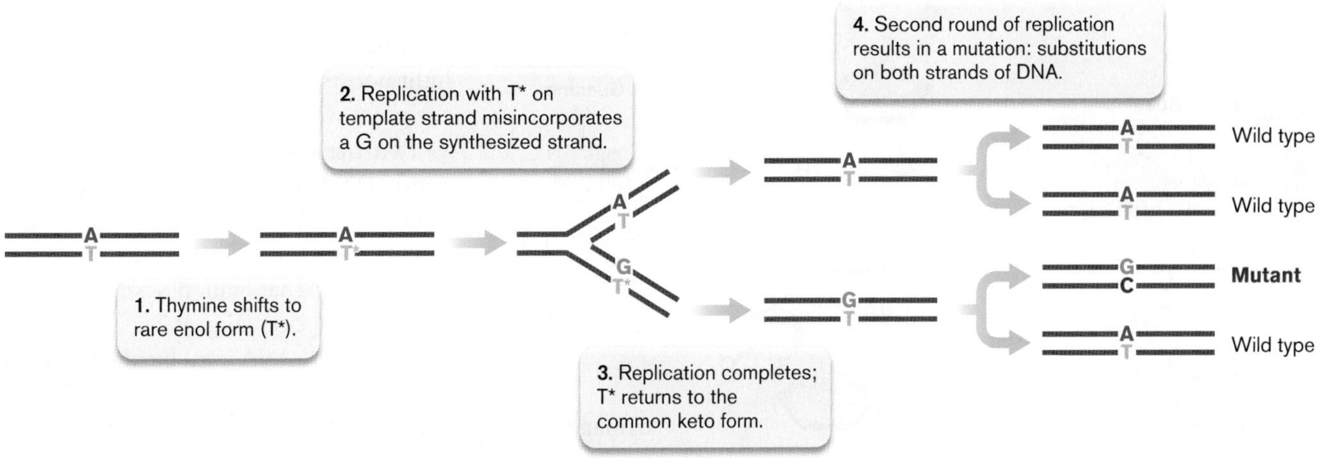

FIGURE 9.3 ■ Mutation arising from the tautomeric shift of thymine prior to DNA replication. This AT-to-GC transition mutation in one of the progeny occurs after the second round of replication.

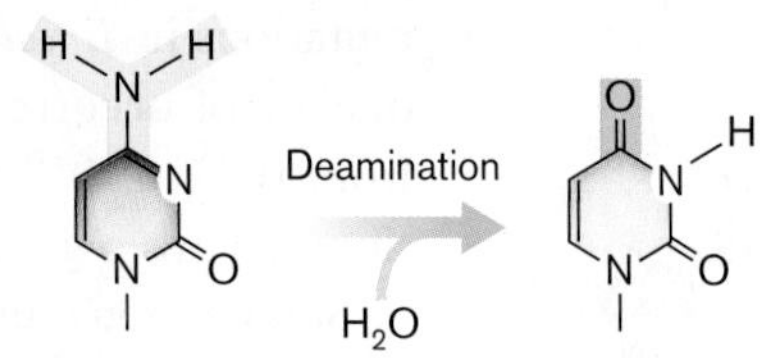

FIGURE 9.4 ■ Spontaneous deamination of cytosine. Oxidative deamination changes cytosine to uracil.

> **Thought Question**
>
> 9.2 Does deamination of cytosine to uracil lead to a transition mutation or a transversion mutation? Work this out in a drawing, using **Figure 9.3** as a guide.

In addition, purines are particularly susceptible to spontaneous ejection from DNA via breakage of the glycosidic bond connecting the base to the sugar backbone (**Fig. 9.5**). The result of this loss is the formation of an **apurinic site** (one missing a purine base) in the DNA. Lack of a purine would obviously hinder transcription and replication.

DNA can also be damaged by metabolic activities of the cell that produce reactive oxygen species, such as hydrogen peroxide (H_2O_2), superoxide radicals ($^{\bullet}O_2^-$), and hydroxyl radicals ($^{\bullet}OH$). Even though bacteria have biochemical mechanisms to detoxify reactive oxygen species, the systems can be overwhelmed. Oxidative damage causes the production of thymidine glycol or 8-oxo-7-hydrodeoxyguanosine in DNA (**Fig. 9.6**).

Naturally occurring intracellular methylation agents (for example, *S*-adenosylmethionine) can spontaneously

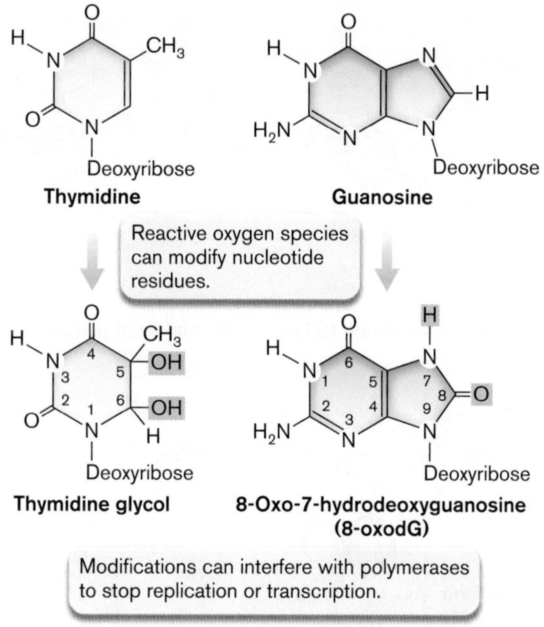

FIGURE 9.6 ■ Examples of damage caused by reactive oxygen species. The modifications can interfere with polymerase function and stop replication or interfere with the transcription of affected genes. The blue highlighting identifies modifications to thymidine and guanosine residues.

methylate DNA to produce a variety of altered bases. The spontaneous methylation of the N-7 position of guanine, for example, weakens the glycosidic bond and spontaneously releases the base (forming an apurinic site) or opens the imidazole ring (forming a methylformamide pyrimidine). In addition to mispairing, some of these spontaneous events can lead to major chromosomal rearrangements, such as duplications, inversions, and deletions. Mutagens can also increase the mutation rate by inducing repair pathways that themselves introduce mutations (error-prone DNA polymerases such as Pol IV and Pol V are discussed in Section 9.2).

Ultraviolet (UV) light will produce striking structural alterations in DNA molecules. Pyrimidines (more than purines) are highly susceptible to UV radiation. The energy absorbed by a pyrimidine hit with UV light boosts the energy of its electrons to the point where the molecule is unstable. If two pyrimidines are neighbors on a single DNA strand, their energized electrons can react to form a four-membered cyclobutane ring. The result is a pyrimidine dimer that will block replication and transcription (**Fig. 9.7**).

Although DNA can be damaged in numerous ways, the cell can repair that damage before it becomes fixed as a mutation. But the repair mechanisms are not perfect. Repair errors contribute heavily to the formation of heritable mutations and, thus, to evolution.

Depurination of DNA

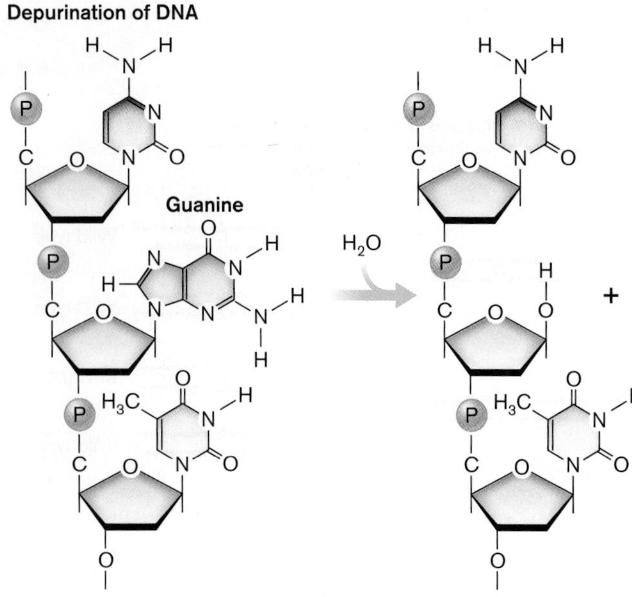

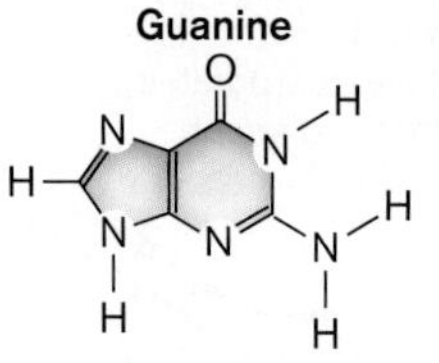

FIGURE 9.5 ■ Spontaneous formation of an apurinic site.

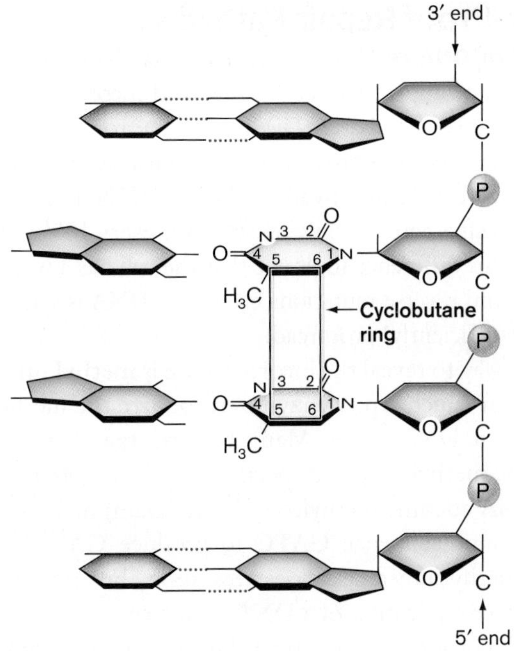

FIGURE 9.7 ▪ Production of a pyrimidine dimer. The energy from UV irradiation can be absorbed by pyrimidine molecules. The excited electrons of carbons 5 and 6 on adjacent pyrimidines can then be shared to form a four-membered cyclobutane ring between adjacent pyrimidines. The pyrimidine dimer blocks replication and transcription.

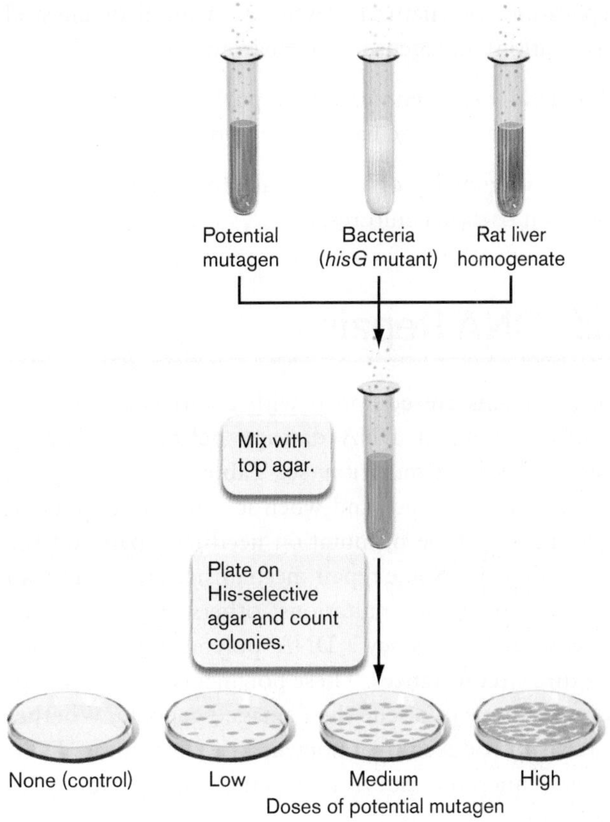

FIGURE 9.8 ▪ The modified Ames test to assay for the mutagenic properties of chemicals processed through the liver.

Identifying Mutagens Using Bacterial "Guinea Pigs"

In a world where we are continually exposed to new chemicals, it is important to determine which ones are potential mutagens. Bruce Ames and his colleagues invented a simple assay that uses bacteria as a rapid initial screen, for which Ames received the National Medal of Science in 1998. This method accelerates the process of drug discovery by providing an inexpensive preliminary screen for weeding out mutagenic chemicals before more expensive animal testing is undertaken. The method relies on a mutant of *Salmonella enterica* that is defective in the *hisG* gene, whose product is involved in histidine biosynthesis. The *hisG* mutant is **auxotrophic** for histidine and thus cannot grow on a defined minimal medium lacking histidine. However, if a reversion mutation occurs in the *hisG* gene and restores the gene to its original functional state, the new mutant cell will form a colony even in the absence of histidine. Generally, this method is called a reversion test, and the specific test that Ames developed is called the Ames test.

Today we use a modified form of the Ames test that takes into account that some chemicals become mutagenic only after they are processed by the liver. The liver is the chief organ for detoxifying the body—a task that liver enzymes accomplish by chemically modifying foreign substances. The potential mutagen, *hisG* mutant bacteria, and liver homogenate are combined and mixed with agar (**Fig. 9.8**). The combination is poured onto a Petri plate. If the liver extract enzymes act on the test compound and the metabolites produced are mutagenic, then increasing numbers of His^+ revertants will be observed with increasing doses of mutagen. If the compound is not mutagenic, the number of colonies will not exceed those found in a control group that was not exposed to the mutagen.

To Summarize

- **A mutation** is any heritable change in DNA sequence, regardless of whether a change in gene function results.
- **Classes of mutations** include point mutations, insertions, deletions, frameshift mutations, and transpositions.
- **Mutations can be further categorized by their impact on gene products**, and include silent, missense, and nonsense mutations.
- **Genotype** reflects the genetic makeup of an organism, whereas **phenotype** reflects its physical traits.
- **Spontaneous mutations** reflect tautomeric shifts in DNA nucleotides during replication, accidental incorporation of noncomplementary nucleotides during

replication, or "natural" levels of chemical or physical (irradiation) mutagens in the environment.

- **Chemical mutagens** can alter purine and pyrimidine structure and change base-pairing properties.
- **The mutagenicity** of a chemical can be assessed by its effect on bacterial cultures.

9.2 DNA Repair

Microorganisms are equipped with a variety of molecular tools that repair DNA damage before the damage becomes a heritable mutation (see **Table 9.2**). The type of repair mechanism used (and when it is used) depends on two things: the type of mutation needing repair and the extent of damage. Some repair mechanisms are error proof and do not introduce mutations; others are error prone and require "emergency" DNA polymerases expressed under dire circumstances. These polymerases sacrifice replication accuracy to rescue the damaged genome. Whether damage is introduced by mutagens or by inaccurate DNA synthesis, microbial survival depends on the ability to repair DNA.

We will first discuss the **error-proof repair** pathways that prevent mutations. These include photoreactivation, nucleotide excision repair, base excision repair, methyl mismatch repair, and recombinational repair. Then we will turn our attention to **error-prone repair** pathways. These pathways risk introducing mutations and operate only when damage is so severe that the cell has no other choice but to die. Note that many repair enzymes are in short supply. **Special Topic 9.1** describes how DNA repair enzymes can use an electric current in DNA to locate damage.

Error-Proof Repair Pathways

Repair of DNA replication errors. What happens if DNA polymerase simply makes a mistake and incorporates a normal but incorrect base? The inherent error rate of DNA polymerase III is approximately one mistake per 10^8 bases synthesized, after proofreading. However, the mutation rate in a live cell is actually only one mistake every 10^{10} replicated bases. This difference indicates that the cell has a way to recognize and reverse mutations even after DNA is synthesized and (insufficiently) proofread.

One way to reveal the incorrect base is **methyl mismatch repair**, in which repair enzymes recognize the methylation pattern in DNA bases. Many bacteria tag their parental DNA by methylating it at specific sites. In *E. coli*, for example, DNA adenine methyltransferase (Dam) methylates the palindromic sequence GATC to produce GA^{ME}TC. The Dam methyltransferase does this soon, but not immediately, after replication of a DNA sequence.

Misincorporation of a base during replication produces a mismatch between the incorrect base in the newly synthesized but unmethylated strand and the correct base residing in the parental, methylated strand (**Fig. 9.9**). Methyl-directed mismatch repair enzymes (MutS, MutL, and MutH) bind to the mismatch. MutS identifies the mismatch as a distortion in the usual base stacking in the DNA helix and recruits MutL and MutH (**Fig. 9.9**, steps 1 and 2). MutL recognizes the methylated strand (GA^{ME}TC) and brings it in a loop to meet MutS and MutH (steps 3 and 4). Then, MutH cleaves the unmethylated strand containing the mutation, near the GATC sequence (step 5). A DNA helicase called UvrD then unwinds the cleaved strand, exposing it to a variety of exonucleases (step 6). The result is a gap that is filled in by DNA polymerase I (Pol I) and sealed by DNA ligase.

TABLE 9.2 Types of DNA Repair

System	Proteins	Mutations recognized	Repair mechanism	Accuracy
Photoreactivation	PhrB	Pyrimidine dimers	Cyclobutane ring cleaved	Error proof
Nucleotide excision	UvrABCD	Helical destabilization (e.g., pyrimidine dimers)	Patch of nucleotides excised	Error proof
Base excision	Fpg, Ung, Tag, MutY, Nfo	Various modified bases	Glycosylases remove base from phosphodiester backbone; apurinic (AP) sites formed	Error proof
Methyl mismatch	MutHSL, Dam	Transitions, transversions	Nick on nonmethylated strand; excision of nucleotides	Error proof
Recombination	RecA	Single-strand gaps and double-strand breaks	Recombination	Error proof
Translesion bypass synthesis	UmuDC	Gaps	Part of SOS system	Error prone (generates mutations)
Nonhomologous end joining	Ku, LigD	Double-strand breaks	Ligation of strands after processing with exonuclease and polymerase	Error prone

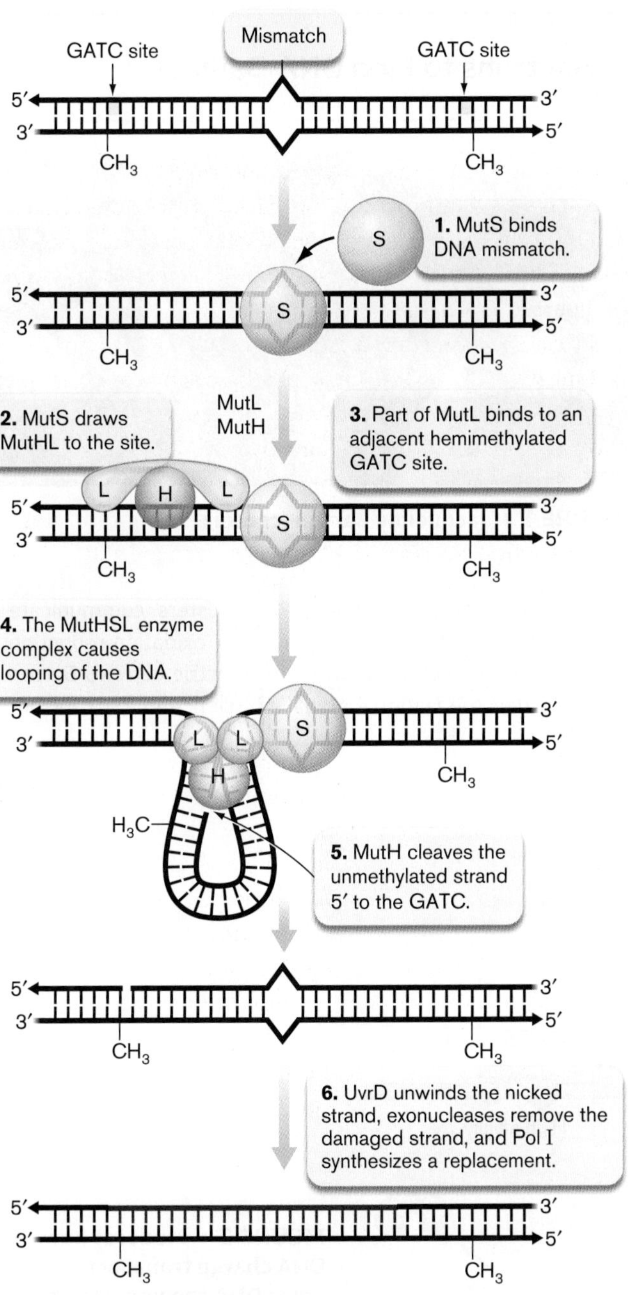

FIGURE 9.9 ■ Methyl mismatch repair. The bacterial cell, in this case *E. coli*, can use specific methylations on DNA to recognize parental DNA strands for preferential DNA repair. Newly replicated strands are not immediately methylated. So when a mismatch is found, the mismatch repair system views the newly synthesized strand as suspect and replaces the section of unmethylated DNA encompassing the mismatch. The steps following cleavage are similar to what is shown in Figure 1 of **eTopic 9.1** (steps 7 and 8).

The methyl-directed mismatch repair proteins (and genes) are called Mut (and *mut*) because a high mutation rate results when strains are defective in one of these proteins. A bacterial strain with a high mutation rate is called a **mutator strain**.

Note: Not all DNA methylations signal methyl mismatch repair. DNA methylation by restriction-modification systems, for example, prevents cleavage by DNA restriction endonucleases but is not used to designate parental DNA after replication.

Thought Questions

9.3 Would mutants that lack Dam have a mutator phenotype? Explain why or why not. What about mutants that overexpress Dam?

9.4 It has been reported that hypermutable bacterial strains are overrepresented in clinical isolates. Out of 500 isolates of *Haemophilus influenzae*, for example, 2%–3% were mutator strains having mutation rates 100–1,000 times higher than those of lab reference strains. Why might the mutator phenotype be beneficial to pathogens?

Repair of UV damage. Many microbes commonly encounter ultraviolet light in their natural habitat. The pyrimidine dimers that form as a result of ultraviolet irradiation can be repaired by several different mechanisms in bacteria. Recombinational repair (important during chromosome replication) and nucleotide excision repair are two mechanisms that remove pyrimidine dimers; they are described in detail in **eTopic 9.1**.

Photoreactivation is a third mechanism that repairs DNA damage from UV exposure. Interestingly, the photolyase enzymes involved in photoreactivation are activated by a lower-energy wavelength of light (in the visible range) and actually use this light to repair the damage from UV. In photoreactivation, photolyase binds to the dimer and uses light energy to cleave the cyclobutane ring linking the two adjacent, damaged nucleotides. The damage is repaired without any bases being excised.

DNA photolyases have been implicated in the evolution and range expansion of *Prochlorococcus*, the most abundant photosynthetic organism of the ocean. *Prochlorococcus* is found throughout the euphotic zone, which is the upper, sun-exposed layer of the ocean where photosynthesis can occur. The *Prochlorococcus* lineage is thought to have its origins in the deeper part of the euphotic zone, where UV light does not reach. Present-day lineages that live exclusively in the lower euphotic zone are most closely related to the ancestor of *Prochlorococcus*, and they lack DNA photolyases. In contrast, lineages found in the upper euphotic zone, where UV is present, have one or more photolyases. These lineages evolved more recently, and it is thought that the acquisition of photolyases by horizontal gene transfer (see Section 9.5) was instrumental in their ability to invade and colonize the upper euphotic zone.

SPECIAL TOPIC 9.1 DNA as a Live Wire: Using Electrons to Find DNA Damage

How do DNA repair proteins locate DNA damage? Do they randomly float in the cytoplasm until they happen to bump into an abasic site or a mispaired base? Or do they bind DNA and slide along its length until they run into something? Jacqueline Barton from the California Institute of Technology (**Fig. 1**) thinks that a different option is likely for repair proteins containing [4Fe-4S] iron-sulfur groups. The [4Fe-4S] group is a tetrahedral cluster (**Fig. 2A** blowup) with important roles in oxidation-reduction chemistry involving electron transfer. The positioning of the [4Fe-4S] clusters in proteins is coordinated by four cysteine residues that bind to the Fe molecules.

How might a redox center help repair proteins locate DNA damage? Many DNA repair enzymes are in very short supply. The DNA glycosylase MutY in *E. coli*, for instance, is present at only about 30 proteins per cell. It would take a long time for those few molecules to scan all of the DNA. Suppose different repair proteins could collaborate in some way to scan large stretches of DNA for defects. Barton and her collaborators think that is exactly what happens. Their research indicates that repair enzymes with [4Fe-4S] clusters communicate with each other through long-distance oxidation-reduction reactions using DNA as a "wire" for electric current. Barton was awarded the National Medal of Science in 2011 for her discoveries.

FIGURE 1 ■ Jacqueline Barton (center).

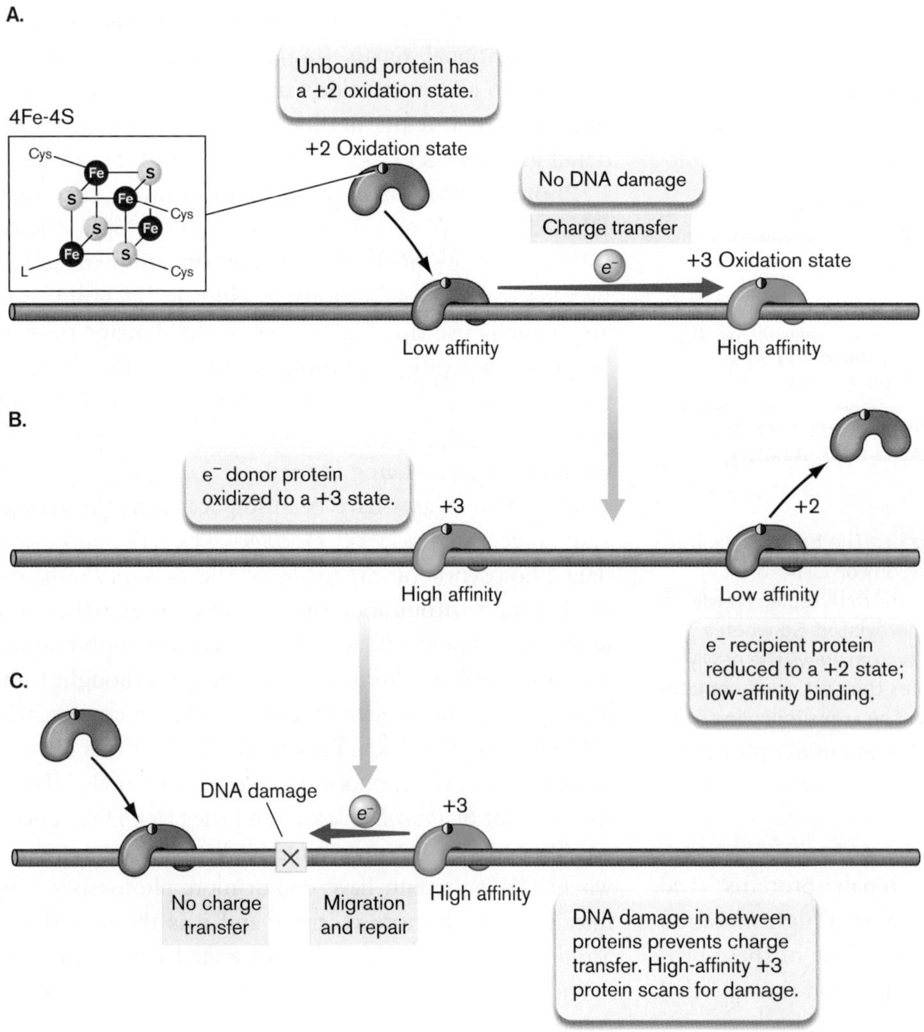

FIGURE 2 ■ Model of how DNA charge transport helps DNA repair proteins find DNA damage. A, B. Charge transport occurs between reduced (purple) and oxidized (turquoise) repair proteins when there is no DNA damage. Repair proteins cycle on and off undamaged DNA. C. Charge transport fails when there is damage to intervening DNA between repair proteins. High-affinity proteins (+3 oxidation state) accumulate near the site of damage. The redox-active [4Fe-4S] clusters in the proteins are signified by circles that are half yellow, half purple. **Blowup in (A):** [4Fe-4S] cluster.

The stacked core of aromatic bases that extends down the helical axis of DNA enables DNA to conduct electrons. But electrons flow only if the DNA is intact. DNA damage destroys the integrity of the wire, blocking electron flow. Repair proteins with [4Fe-4S] clusters that bind DNA can pass an electron along an undamaged DNA "wire" (a process called DNA charge transport, or CT) to another protein with a [4Fe-4S] cluster. Here is how the model works:

- The [4Fe-4S] cluster has a +2 oxidation state in repair proteins that are not bound to DNA (**Fig. 2A**; purple).
- When the repair protein is bound to DNA, an electron can be transferred through the DNA helix to another [4Fe-4S] repair protein bound to the DNA. Losing an electron means the [4Fe-4S] of the donor protein is oxidized from +2 to +3 (**Fig. 2A;** turquoise).
- The recipient protein is then reduced (gains an electron) from +3 back to a +2 oxidation state (**Fig. 2B**).
- The +2 recipient protein now has a lower DNA-binding affinity and dissociates, only to bind DNA again elsewhere. The process of scanning by charge transport then repeats from the new location.
- If the repair protein binds near damaged DNA (a "broken wire"), it cannot transfer an electron to a downstream +3 recipient protein (**Fig. 2C**). The bound, oxidized (+3) repair protein migrates two-dimensionally to find and repair the damage.

The [4Fe-4S] repair proteins use charge transport, or rather the lack of it, to accumulate in the vicinity of DNA damage.

Figure 3 provides evidence that charge transport through DNA works in living cells. The experiment starts with a strain of *E. coli* called InvA. InvA has a special mutation that makes the activity of a protein called DinG very important for growth. DinG is a DNA helicase that has a [4Fe-4S] cluster. This protein unwinds DNA-RNA hybrids called R loops that accumulate at stalled replication forks. The InvA mutant requires DinG to resolve these R loops; otherwise it has severe growth defects. The question that Barton's group asked was whether another [4Fe-4S]-containing repair enzyme, Endo III, helps DinG find R-loop damage.

In the experiment, growth (turbidity) means that DinG successfully found and removed R loops. Conversely, no growth means that DinG did not efficiently find R loops. If the hypothesis is correct and Endo III charge transport function helps DinG find R loops, then an InvA strain carrying a wild-type Endo III protein would be expected to grow (**Fig. 3**, tubes 1, 5, and 6). Any InvA strain that did not have Endo III or that carried a mutant Endo III protein lacking charge transport through DNA would not grow (tubes 2, 3, and 4). The endonuclease catalytic activity of Endo III, however, would not be important for localizing DinG, and mutants lacking this activity should still be able to grow (tube 5). The results in **Figure 3** match the prediction and indicate that Endo III charge transport helped DinG find R loops.

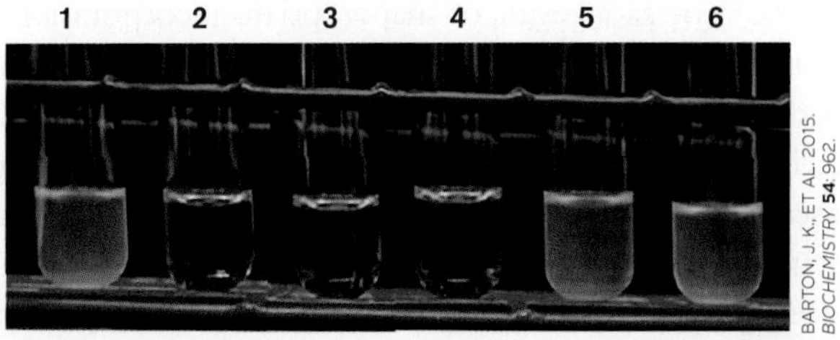

	1	2	3	4	5	6
Endo III on chromosome	+	–	–	–	–	–
Endo III on plasmid			Empty	Y82A	D138A	Wild type
				CT$^-$ Cat$^+$	CT$^+$ Cat$^-$	

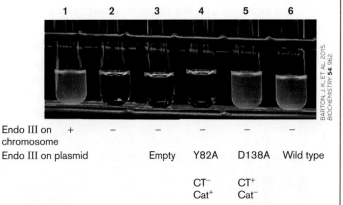

FIGURE 3 ▪ Charge transport (CT) works in vivo. All tubes contain an InvA mutant dependent on DinG for growth. Growth indicates that CT-proficient Endo III helps guide DinG to R-loop damage at stalled replication forks. Some strains (tubes 2–6) lack the chromosomal gene for Endo III, but certain Endo III mutants contain plasmids that encode replacement Endo III proteins. CT$^-$ or CT$^+$ = mutant Endo III that is, respectively, deficient or proficient in charge transport; Cat$^-$ or Cat$^+$ = mutant Endo III that is, respectively, defective or functional in catalytic activity.

DNA charge transport seems to be an efficient way to redistribute repair proteins to sites of DNA lesions for repair and could provide a general way for any DNA-bound [4Fe-4S] proteins to signal across the genome.

RESEARCH QUESTION

Atomic force microscopy can identify proteins bound to DNA molecules. Describe what you would observe, given the following: a mixture of DNA molecules of two different lengths. The larger DNA molecules are undamaged, while the smaller DNA molecules have a base-pair mismatch in the middle. Immediately after Endo III is added to the mix, the protein is seen bound to all molecules. What changes in Endo III binding would you observe for the large and small DNA molecules as the incubation proceeded?

Grodick, M. A., N. B. Muren, and J. K. Barton. 2015. DNA charge transport within the cell. *Biochemistry* **54**:962–973.

Replacement of damaged bases. Another error-proof process, known as **base excision repair (BER)**, detects and replaces damaged bases. Uracil can be found in DNA either as a spontaneous deamination product of cytosine (see **Fig. 9.4**) or as a result of inappropriate incorporation during replication. Spontaneous deamination of adenine residues produces hypoxanthine. Uracil and hypoxanthine have base-pairing properties very different from those of the original bases, so their formation can lead to mutations. Mutagens such as methyl methanesulfonate can alkylate adenine to produce 3-methyladenine. A 3-methyladenine residue will totally block DNA replication, making this a lethal form of damage. Consequently, repairing these mutations is vital for survival.

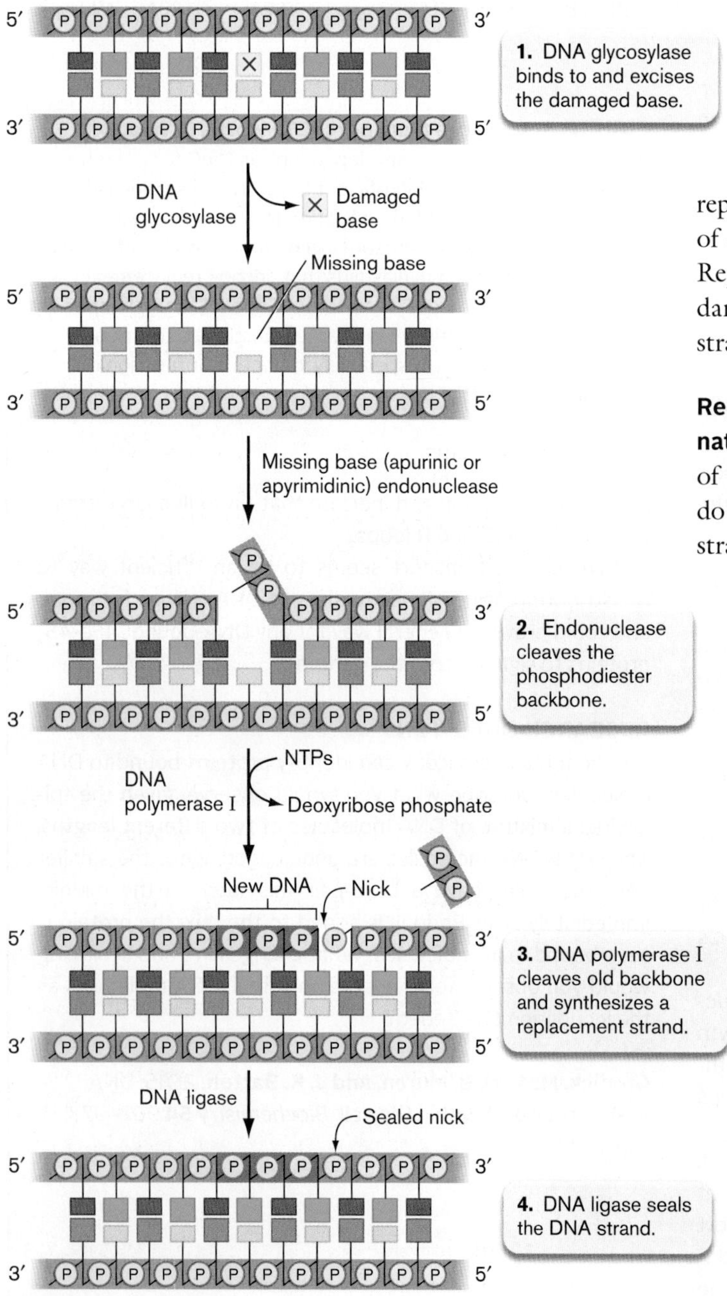

FIGURE 9.10 ■ Base excision repair. Replacement of a damaged base requires a succession of four enzymatic activities: glycosylase, endonuclease, polymerase, and ligase. NTPs = nucleoside triphosphates.

At the beginning of base excision repair, enzymes called glycosylases cleave the bond connecting the damaged base to the deoxyribose moiety in the phosphodiester backbone (**Fig. 9.10**, step 1). The result is an intact phosphodiester backbone missing a base. The site is called an **AP site** because it is either apurinic (missing a purine) or apyrimidinic (missing a pyrimidine). In step 2 of BER, AP endonucleases specifically cleave the phosphodiester backbone at AP sites. The 5′-to-3′ exonuclease activity of the gap-filling DNA polymerase I will degrade the cleaved strand downstream of the AP site and at the same time synthesize in its stead a replacement strand containing the proper base (step 3). DNA ligase seals the remaining nick, and the repair process is complete (step 4).

All of the repair processes described thus far (mismatch repair, photoreactivation, and base excision repair) are error-proof pathways that rely on the presence of an intact template strand opposite the damaged strand. Replication of the template strand is used to replace the damaged bases accurately. But what happens when both strands are damaged?

Repair of double-strand breaks by homologous recombination. Double-strand breaks that generate "blunt ends" of the chromosome are challenging to repair. Blunt ends do not have overlapping "sticky ends" that allow the DNA strands to reanneal and provide a stable substrate for DNA ligase to perform the repair. While possible, ligation of blunt ends is highly inefficient, because it relies on rare and short-lived conditions in which the two DNA ends have contacted each other through random collision. However, if the bacterial cell contains a second, intact chromosome copy in the cell, it can use this undamaged copy as a guide to repair the broken chromosome. This repair mechanism involves **homologous recombination**, which is a form of strand exchange that occurs at shared regions of homology between DNA molecules.

Figure 9.11 illustrates the mechanism of double-strand-break repair by homologous recombination. First, the two ends of the broken chromosome are trimmed back by nucleases to generate 3′ single-stranded ends (overhangs; **Fig. 9.11**, step 1). In *E. coli*, this trimming is performed by the RecBCD protein complex. The 3′ overhangs are recognized by the protein RecA, which binds in multiple copies to form a filament (step 2). The RecA filament mediates strand invasion of one 3′ overhang into the double helix of the second chromosome. This strand swapping occurs at the region of the second chromosome that is complementary to the sequence of the overhang. Using the undamaged complementary DNA as a template, the 3′

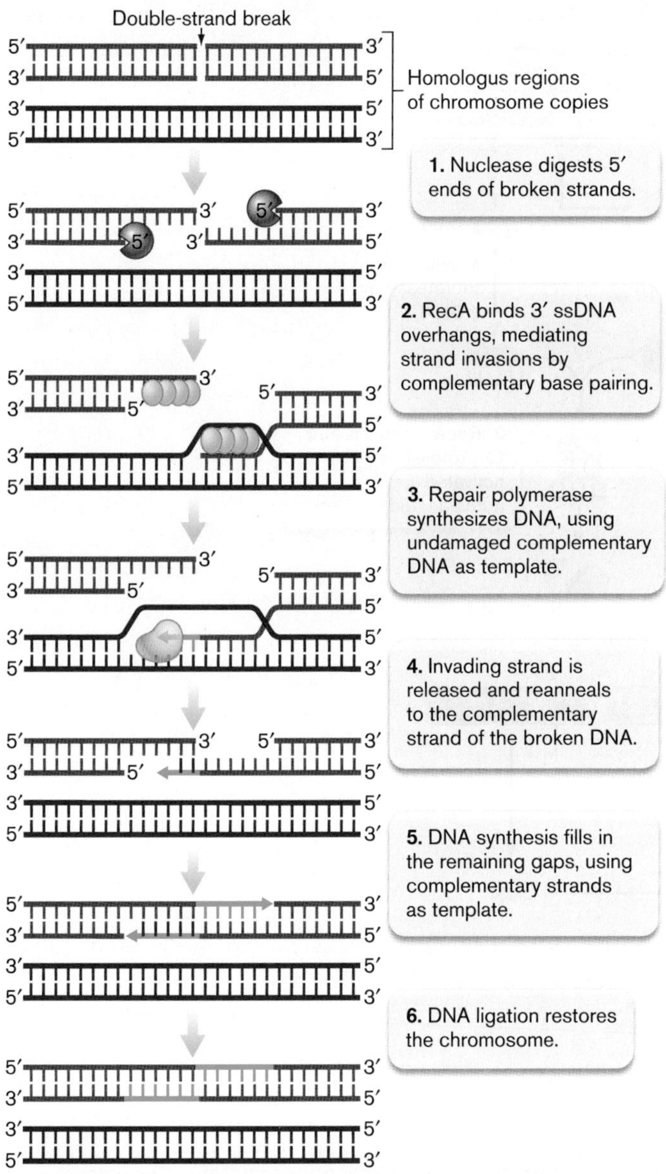

FIGURE 9.11 ▪ Repair of a double-strand break by homologous recombination involving a second, undamaged chromosome.

end is extended via a repair DNA polymerase (step 3). This extended DNA can now dissociate from the intact chromosome and anneal to the other 3′ overhang of the damaged chromosome by complementary base pairing (step 4). Repair polymerases can now complete the replacement of missing nucleotides on both strands of the damaged chromosome (step 5). Finally, DNA ligase seals the two nicks and restores the chromosome (step 6).

Error-Prone DNA Repair

SOS ("Save Our Ship") repair. When DNA damage is extensive, repair strategies that excise pieces of DNA or that require recombination will destroy the integrity of the chromosome and kill the cell. To save the chromosome, and itself, the cell must take more drastic measures and induce the **SOS response**, a system that introduces mutations into severely damaged DNA. The cell relaxes replication fidelity to maintain an intact chromosome even if incorrect bases are introduced. In the SOS system, the RecA protein, normally located at the cell poles, senses the extent of DNA damage by monitoring the level of single-stranded DNA produced. For example, excessive ultraviolet irradiation produces numerous ssDNA gaps because DNA polymerase cannot replicate through pyrimidine dimers. RecA interacts with the ssDNA and disengages from the poles.

Formation of RecA filaments (see **Fig. 9.11**, step 2) on ssDNA activates a second function of RecA, called coprotease activity. Coprotease activity stimulates autodigestion of LexA, a repressor protein that binds to the promoters of DNA repair genes and prevents their transcription (**Fig. 9.12A** and **B**). Autodigestion of LexA repressor unleashes the production of DNA repair enzymes. Among these enzymes are two "sloppy" DNA polymerases that lack proofreading activity: DinB (also called DNA polymerase IV, or Pol IV) and UmuDC (also called DNA polymerase V, or Pol V; **Fig. 9.12C**). The UmuDC enzyme is perfect for replicating through damaged bases (a process called translesion bypass replication) because it sacrifices accuracy for continuity. When the enzyme encounters an undecipherable damaged base, it will insert whatever nucleotide is available. The result, of course, will be numerous permanent mutations, but the benefit is that the cell has a chance to live if it can tolerate the mutations. The choice really is "mutate or die," because housekeeping DNA polymerases like Pol III cannot move through damaged DNA. (Note that the term "housekeeping" is applied to proteins or enzymes that keep the cell running at all times.)

When the cell is severely compromised by mutation, it temporarily halts cell division. Like a pit stop during an auto race, this pause in cell division allows time for repair enzymes to fix the damage. The product of another SOS-regulated gene, called *sulA*, causes this pause by binding to the FtsZ cell division protein, keeping it from initiating cell division (see Sections 3.1 and 3.4). Once the damage has been repaired, RecA coprotease is inactivated and LexA repressor accumulates, turning off all the SOS genes, including *sulA*. The lingering SulA protein is then degraded by a protease called Lon. The protein is called Lon because when it is missing, SulA inappropriately accumulates, inhibits cell division, and produces <u>lon</u>g filamentous cells.

The coprotease activity of RecA can also activate prophages by stimulating autocleavage of proteins that prevent phage replication. This can be considered a form of eavesdropping on the host cell by the prophage: In environmental conditions that trigger the SOS response, the virus may have a better chance of survival and replication as a viral particle rather than as a lysogen.

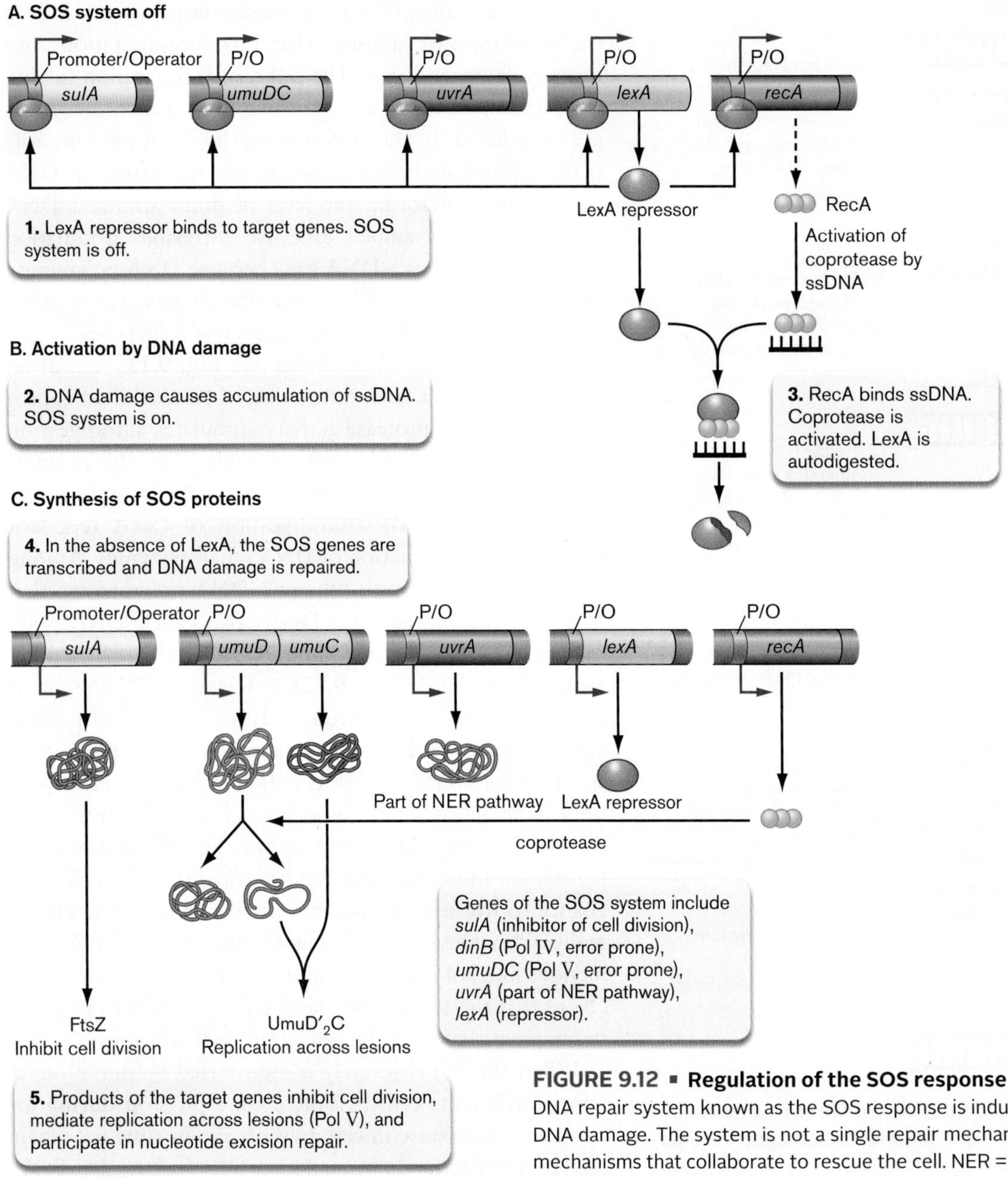

FIGURE 9.12 ■ Regulation of the SOS response system. The emergency DNA repair system known as the SOS response is induced when there is extensive DNA damage. The system is not a single repair mechanism but a set of different mechanisms that collaborate to rescue the cell. NER = nucleotide excision repair.

The SOS induction of viruses can be exploited by some organisms to displace competitors in an environmental niche. For instance, *Streptococcus pneumoniae*, a cause of pneumonia, displaces *Staphylococcus aureus* in the human nasopharynx by producing hydrogen peroxide. Hydrogen peroxide damages *S. aureus* DNA, which activates the SOS response and triggers replication of resident bacteriophages that lyse the cell. The loss of *Staphylococcus aureus* means more room for *Streptococcus pneumoniae*.

Nonhomologous end joining. Double-strand breaks in DNA are particularly dangerous to a cell because the loss of chromosome integrity is immediate. In rapidly growing bacteria such as *E. coli*, double-strand breaks can be repaired by homologous recombination as long as another copy of the chromosome is present to guide the repair. But in slow-growing bacteria such as *Mycobacterium tuberculosis*, a second chromosome copy is usually not available. These bacteria can use an intriguing repair mechanism called **nonhomologous end joining** (**NHEJ**), which is also used by mammals (**Fig. 9.13**).

Two bacterial proteins, Ku and LigD, carry out NHEJ repair. Ku protein binds to the ends of a double-strand break and recruits LigD, a protein with polymerase and 3′ exonuclease activities that fill in or remove single-strand overhangs, and a ligase activity that joins two double-strand breaks. Because the system does not require homology, it can be error prone, causing the loss or addition of a few nucleotides at the break site or even the joining of two previously unlinked DNA molecules.

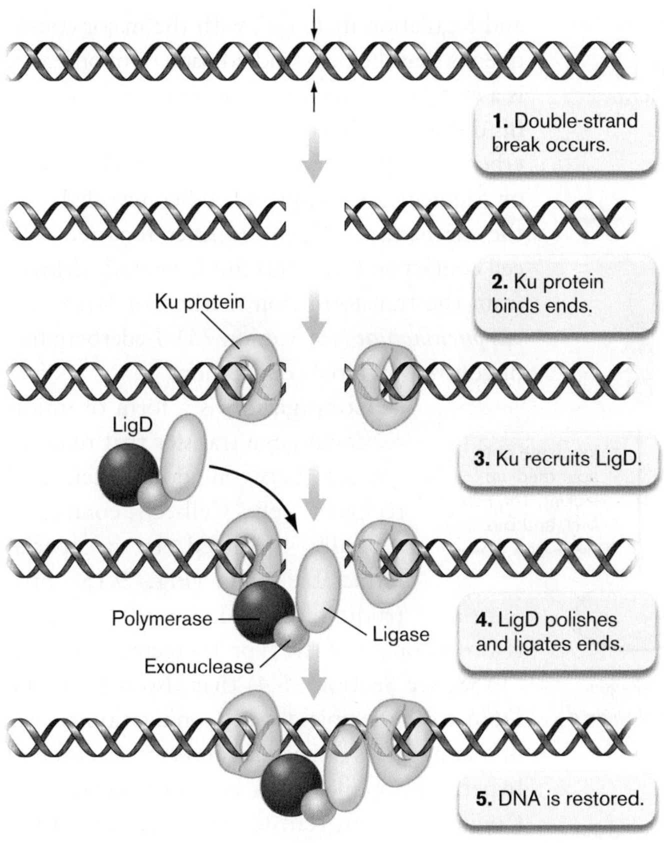

FIGURE 9.13 ▪ Nonhomologous end joining. NHEJ, a common repair mechanism in eukaryotes, has been documented in some bacteria, such as *Mycobacterium tuberculosis*. These pathogens can survive within macrophages that produce antibacterial compounds (nitric oxide and hydrogen peroxide) that generate double-strand DNA breaks. NHEJ mechanisms can repair those dsDNA breaks.

To Summarize

- **DNA repair pathways** in microorganisms include error-proof and error-prone mechanisms.
- **Methyl mismatch repair** uses methylation of the parental DNA strand to distinguish it from newly replicated DNA. The premise is that the parental strand will contain the proper DNA sequence.
- **Photoreactivation** cleaves the cyclobutane rings of pyrimidine dimers.
- **Base excision repair (BER)** excises structurally altered bases without cleaving the phosphodiester backbone. The resulting AP (apurinic or apyrimidinic) site is targeted by AP nucleases.
- **Recombinational repair** uses an intact copy of the DNA to repair a double-strand break in the damaged copy.
- **Extensive DNA damage leads to induction of the SOS response**, producing increased levels of the error-proof repair systems, as well as error-prone translesion bypass DNA polymerases that introduce mutations.
- **Nonhomologous end joining (NHEJ) mechanisms** repair double-strand DNA breaks in some, but not all, bacteria.

9.3 Gene Transfer: Mechanisms and Barriers

Genomic analysis shows that over millennia, microbes have undergone extensive gene loss and gain. Archaea, for example, arose from a common eukaryotic-archaeal phylogenetic branch and, as a result, possess many traits in common with eukaryotes, such as the structure and function of their DNA and RNA polymerases. However, many archaeal genes whose products are involved with intermediary metabolism are probably of bacterial origin. In fact, 37% of the proteins found in the archaeon *Methanocaldococcus jannaschii* are found in all three domains—Archaea, Eukarya, and Bacteria. Another 26% are otherwise found only among bacteria and archaea, while a mere 5% are confined to archaea and eukaryotes. All three domains, then, enjoy a mixed heritage.

Another surprise arising from bioinformatic studies is the mosaic nature of the *E. coli* genome and, indeed, of all microbial genomes. Though we have intensively studied *E. coli* for over 100 years, we now find that the organism's DNA is rife with pathogenicity islands and other genes with foreign origins. How does all this genomic blending happen during microbial evolution?

Before we describe the different mechanisms of gene transfer, it is worth stepping back and asking why it might be advantageous for bacteria and archaea to take up foreign DNA. Several advantages have been proposed:

- First, species that indiscriminately import DNA may use the DNA as <u>food</u>; it is an energy-rich molecule and a good source of carbon, nitrogen, and phosphorus.
- Second, imported DNA can be used to <u>repair</u> damaged chromosomes. This capability is especially relevant in instances when DNA from genetically similar microbes is taken up, because it can undergo homologous recombination to replace damaged segments of DNA.
- Finally, gene transfer can play a role in <u>evolution</u> by introducing new genes into the genome. Once expressed, these new genes can provide novel functions to the cell, which may enable them to compete at higher fitness in their current environment, and also allow them to invade new environments. Gene transfer may have enabled pathogens such as *Neisseria gonorrhoeae*, the cause of gonorrhea, to acquire genes whose protein products now help the organism evade the host immune system.

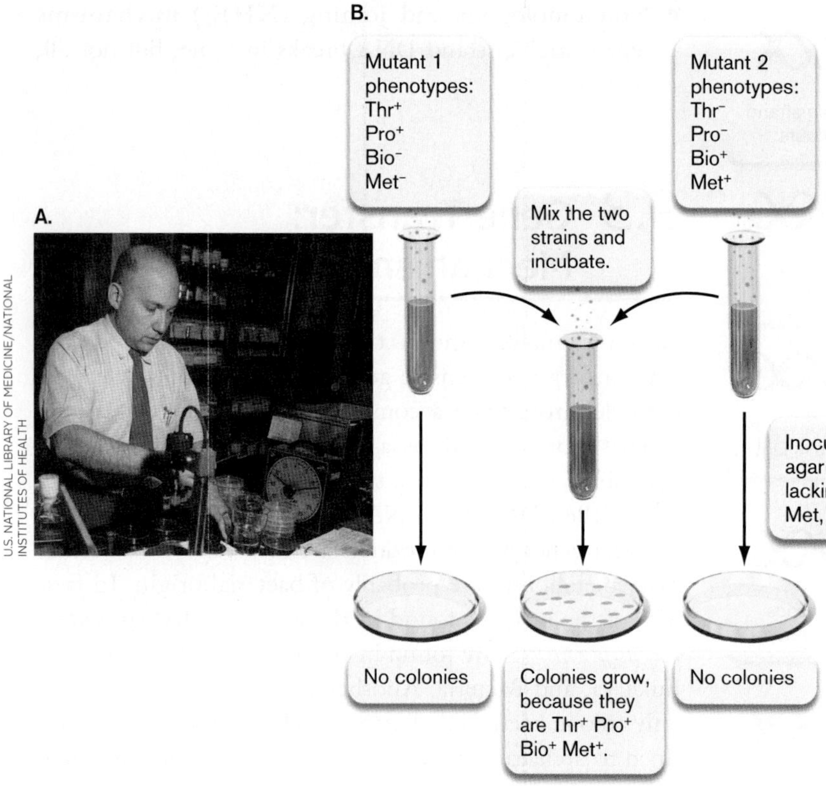

FIGURE 9.14 ■ The discovery of sexual recombination in *E. coli*. **A.** Joshua Lederberg in the lab (circa 1958). **B.** Mutants with two sets of auxotrophies, which prevent the cells from making either their own biotin and methionine (mutant 1) or threonine and proline (mutant 2), cannot grow on agar medium lacking these molecules. However, after these two strains are mixed and incubated, they generate recombinant offspring, some of which are prototrophic and are able to grow as colonies on plates where the parental auxotrophs could not.

Gene Transfer by Conjugation

In 1946, a 20-year-old student at Yale University named Joshua Lederberg (1925–2008; **Fig. 9.14A**) reported the first instance of recombination in prokaryotes. His adviser, Ed Tatum (of "one gene–one enzyme" fame), provided Lederberg with *E. coli* strains that each contained a different set of auxotrophic mutations (**Fig. 9.14B**). These mutants could not grow on agar plates unless the media were supplemented with the metabolites that their auxotrophic mutations prevented them from synthesizing. Lederberg hypothesized that if *E. coli* were capable of gene transfer and recombination, then when two different auxotrophs were mixed together, the alleles (auxotrophic and wild type) of the genes in question would sort out in various combinations in the progeny. Among these recombinant progeny would be a special class: those that received the wild-type alleles of all four genes. Such "prototrophic" cells could be found easily because they would be the only ones that could grow on agar plates without metabolite supplementation. This is exactly what Lederberg observed.

Lederberg's discovery that alleles can be recombined opened the gateway for the exploration of gene function and regulation in *E. coli*, with the major consequence that the best-studied bacterium on Earth is a member of the human gut microbiome. For his discovery of recombination in bacteria, Lederberg won the 1958 Nobel Prize in Physiology or Medicine. Subsequent studies revealed that the mechanism of gene transfer required cell-cell contact and was thus fundamentally distinct from the transformation method of *Streptococcus pneumoniae* (see Section 7.1). Lederberg had discovered bacterial conjugation.

Conjugation is a form of unidirectional gene transfer that requires contact between donor cell and recipient cell. Cell-cell contact is typically mediated by a special retractable pilus (**Fig. 9.15**) protruding from a donor cell. The pilus is a component of the type IV secretion system (T4SS; see Section 25.4) that also delivers the DNA to the recipient cell. Conjugation occurs in many species of bacteria and archaea, even in hyperthermophiles such as *Sulfolobus* species. Conjugation can transfer DNA to cells of the same species or different species, even different domains.

One striking example is *Agrobacterium tumefaciens*, which causes crown gall disease in plants (**Fig. 9.16A**). Using its type IV secretion machinery, *A. tumefaciens* transfers its tumor-inducing (Ti) plasmid via conjugation into plant cells. The bacteria detect and swim toward phenolic compounds released from the wound of a damaged plant. At the wound site, the bacteria attach to the plant cell (**Fig. 9.16B**). Subsequent transfer, integration, and expression of the Ti plasmid in the plant cell genome trigger the release of plant hormones that stimulate tumorous growth of the plant. Plant cells within the tumor release amino acid derivatives, called "opines," that the microbe can then use as a source of carbon and nitrogen. The unique mode of action of *A. tumefaciens* has made this bacterium an indispensable tool for plant breeding and enables entirely new (nonplant) genes to be engineered into crops (discussed in Chapter 16).

Note: Conjugation in bacteria and archaea is mechanistically different from conjugation among eukaryotic microbes. Some eukaryotic microbes exchange nuclei through a structure, called a conjugation bridge, that is very different from the T4SS of bacteria (see Chapter 20).

Bacterial conjugation typically involves special transferable plasmids that contain all the genes needed for pilus formation and DNA export. In some cases, plasmids can

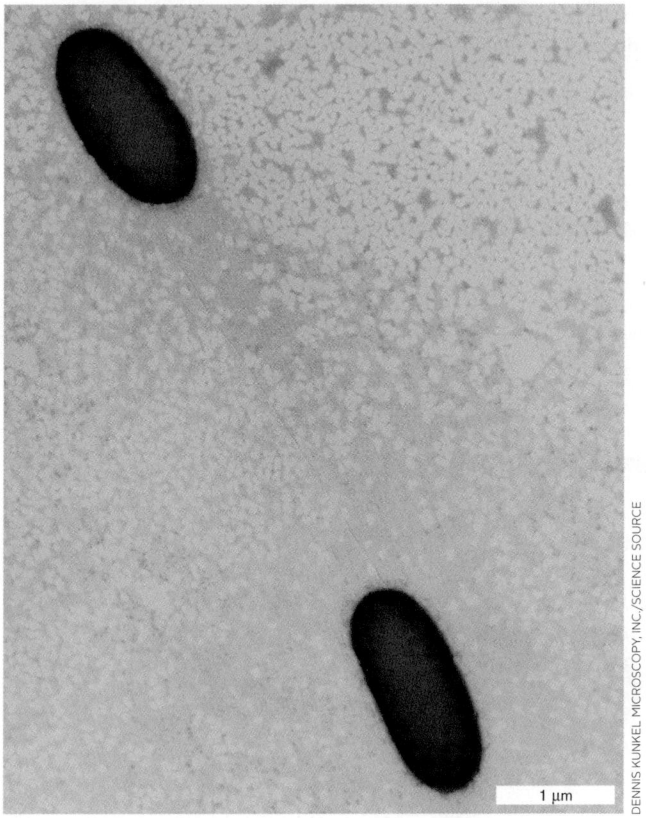

FIGURE 9.15 ■ **Sex pilus connecting two *E. coli* cells.** Pseudocolor added.

also contain genes encoding antibiotic resistance. Lederberg's discovery of recombination involved a well-studied, transferable plasmid in *E. coli* called **fertility factor**, or **F factor** (**Fig. 9.17**). To carry out plasmid replication and DNA transfer, F factor uses two replication origins, *oriV* and *oriT*, located at different positions on the plasmid. The origin *oriV* is used to replicate and maintain the plasmid in nonconjugating cells, whereas *oriT* is used only to replicate DNA during DNA transfer. F factor also contains several *tra* genes, whose protein products carry out the DNA transfer process.

Conjugation of the F factor begins with cell-cell contact between a donor **F⁺ cell** that carries the plasmid and a recipient **F⁻ cell** (**Fig. 9.17**, step 1). Contact causes pilus retraction and triggers activation of the plasmid-encoded relaxosome complex (step 2). The relaxosome is physically associated with the pilus, and before activation the relaxase subunit of the relaxosome binds the plasmid at *oriT*. Upon activation, the relaxosome produces a single-stranded DNA bubble in *oriT*, which recruits a second relaxase molecule. This relaxase unwinds the DNA to separate the strand that will be transferred from the strand that will be retained. At this point, the original relaxase nicks the phosphodiester backbone of one strand at the *nic* site in *oriT* (step 3). The relaxase remains covalently bound to the strand of DNA, and this complex is delivered to the T4SS

A.

B.

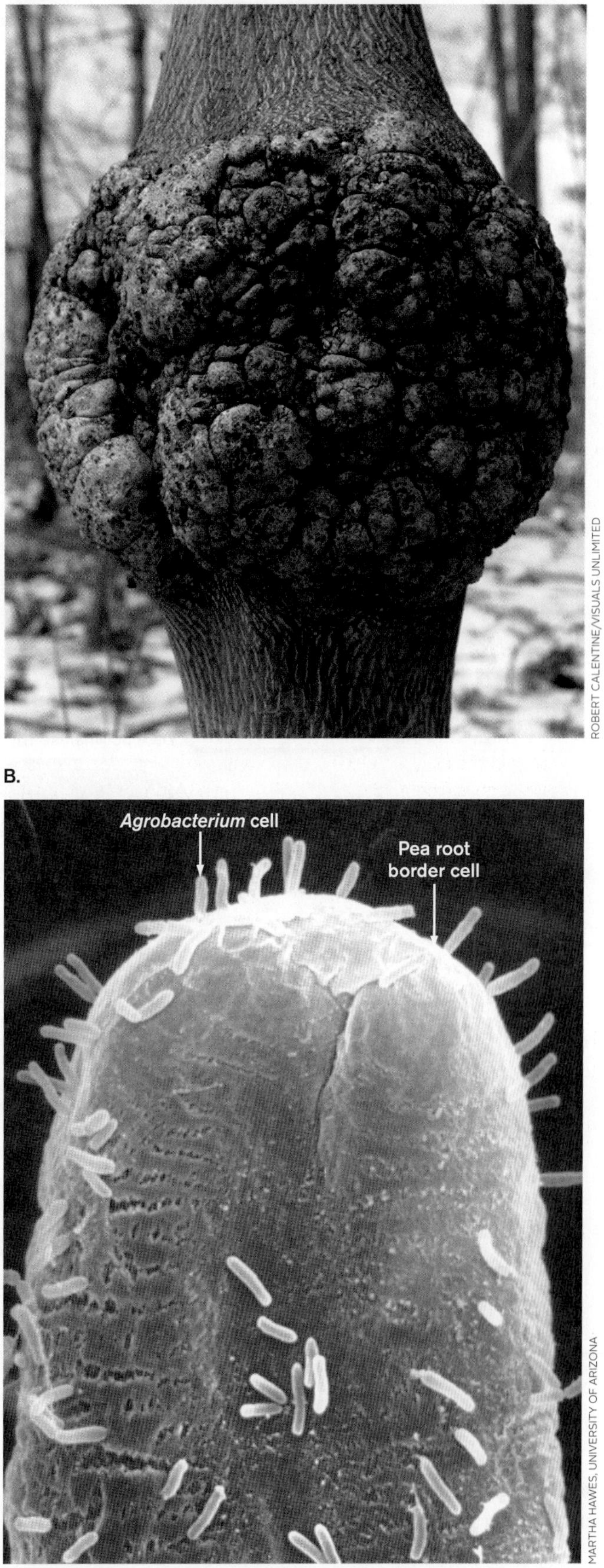

FIGURE 9.16 ■ **An example of gene transfer between bacteria and plants.** **A.** Crown gall disease tumor caused by the bacterium *Agrobacterium tumefaciens*. **B.** Electron micrograph of *A. tumefaciens* attached to a pea root border cell.

FIGURE 9.17 ■ The conjugation process. Some plasmids, such as F factor in *E. coli*, can mediate cell-to-cell transfer of DNA. Purple arrowheads mark the 3′ end of replicating DNA.

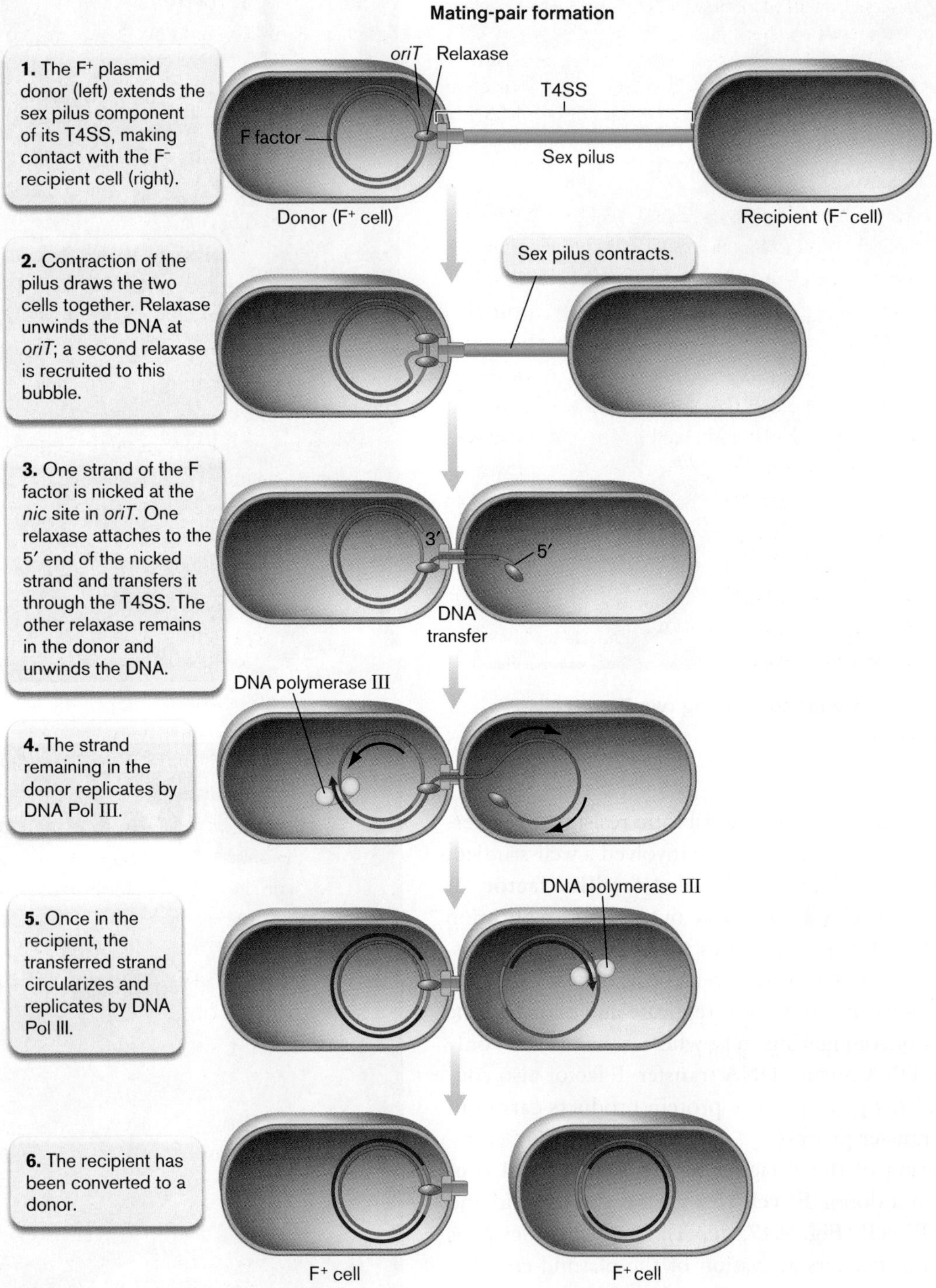

for transfer (see the blowup at step 3 in **Fig. 9.17**). The DNA-relaxase complex is then transferred to the recipient cell through the interior channel of the T4SS. The intact circular strand remains in the donor. DNA polymerase III (Pol III) is then recruited to *oriT* in the donor, where replication begins (step 4). In contrast to bidirectional replication, which is used to duplicate the chromosome, replication for conjugative transfer occurs unidirectionally, by rolling-circle replication (step 5; see Section 7.4).

Once the transfer is complete, the relaxase re-ligates the 5′ end it was holding to the 3′ tail of the transferred strand in the recipient. Thus, the last portion of F factor moved to the recipient is *oriT*. Once circularized, the plasmid is replicated by DNA polymerase III, and the F^+ recipient cell becomes a new F^+ donor cell (**Fig. 9.17**, step 6). Ultimately, the conjugation complex spontaneously comes apart, and the membranes seal. This transfer process is very quick, taking less than 5 minutes to transfer the entire 110-kb F factor.

Note: The sex pilus is hollow, and some evidence suggests it can serve as the channel for DNA transfer between cells. However, this role is not universally accepted in the field. Most evidence suggests DNA transfer occurs when there is close contact between cells, and may involve a T4SS whose pilus has been fully retracted (depolymerized into pilin subunits). As this is currently an unresolved question, for clarity we show DNA transfer steps with a T4SS that lacks the pilus component.

Is it possible for two donors to exchange DNA with each other? The answer is generally no; donors have mechanisms in place to prevent such exchange. For example, a membrane protein encoded by the F factor will inhibit formation of a conjugation complex with another donor that possesses the same protein. This mechanism prevents the pointless transfer of a plasmid to a cell that already has that plasmid.

Many different types of plasmids can be found in the microbial world. Most are not transferable by themselves. However, one group of plasmids that cannot transfer themselves can be mobilized if a transferable plasmid such as F factor is also present in the same cell. Mobilizable plasmids usually contain an *oriT*-like DNA replication origin recognized by the conjugation apparatus of the transferable plasmid. As a result, when the transferable plasmid begins conjugating, so does the mobilizable plasmid. This is one way in which antibiotic resistance genes on plasmids called R factors can be spread throughout a microbial population. Note that not all mobilizable plasmids contain antibiotic resistance genes.

The F-factor plasmid can transfer chromosomal genes. Thus far we have observed how the F factor of *E. coli* can transfer itself as a plasmid from cell to cell, but how was Lederberg able to use this system to recombine genes on the chromosome? This process required the F-factor plasmid to integrate into the chromosome (**Fig. 9.18**, step 1). Integration of the plasmid into the genome involves recombination between regions of homology between the plasmid and chromosome, such as the IS*3* element. A strain with the F factor integrated into the chromosome is called an **Hfr** (high-frequency recombination) **strain**. The strain is called "high-frequency" because every cell is capable of transferring chromosomal DNA to an F$^-$ cell. Hfr strains are rare in nature, and it was quite fortuitous that one such strain (mutant 1 in **Fig. 9.14B**) was in the hands of Lederberg.

The integrated F factor of the Hfr strain can transfer one strand of DNA through the T4SS as it would in plasmid form, except now it transfers chromosomal genes (see **Fig. 9.19**). Essentially, the entire circular chromosome becomes the F factor. Chromosomal genes adjacent to and behind *oriT* are transferred first (**Fig. 9.18**, step 2), and the remainder of the F factor, including the *tra* gene, is transferred last.

Figure 9.19 shows a reconstruction of the Hfr mating experiment that gave Lederberg the results of **Fig. 9.14B**, using current knowledge of the relative position of the genes and the F factor he used. The F factor transferred the *thr*$^+$, *pro*$^+$, and *bio*$^-$ alleles into the *thr*$^-$*pro*$^-$*bio*$^+$*met*$^+$ recipient (**Fig. 9.19**, step 4). The Hfr strain was unlikely to have transferred the *met*$^-$ allele, because it would

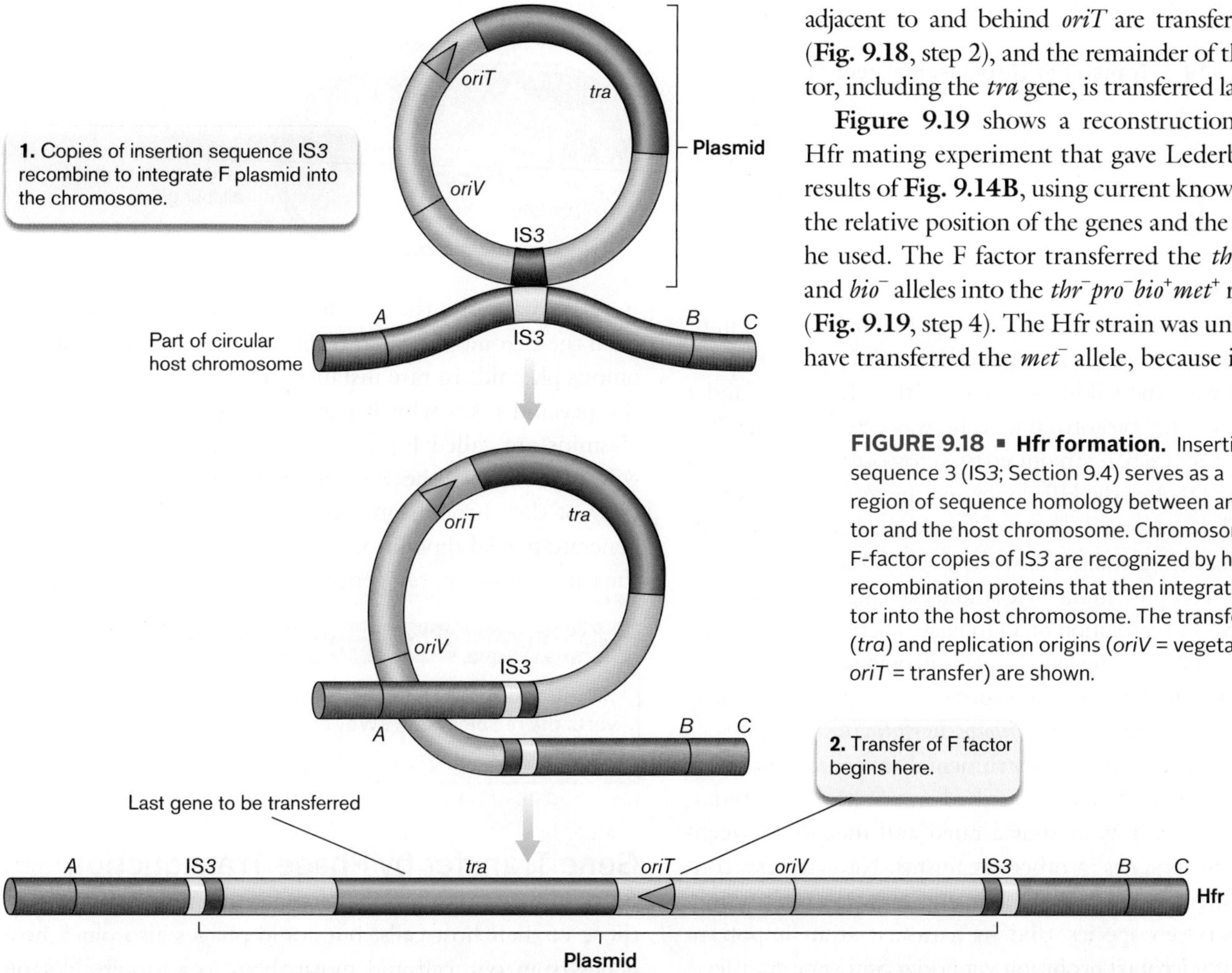

FIGURE 9.18 ▪ Hfr formation. Insertion sequence 3 (IS*3*; Section 9.4) serves as a region of sequence homology between an F factor and the host chromosome. Chromosome and F-factor copies of IS*3* are recognized by host cell recombination proteins that then integrate F factor into the host chromosome. The transfer genes (*tra*) and replication origins (*oriV* = vegetative; *oriT* = transfer) are shown.

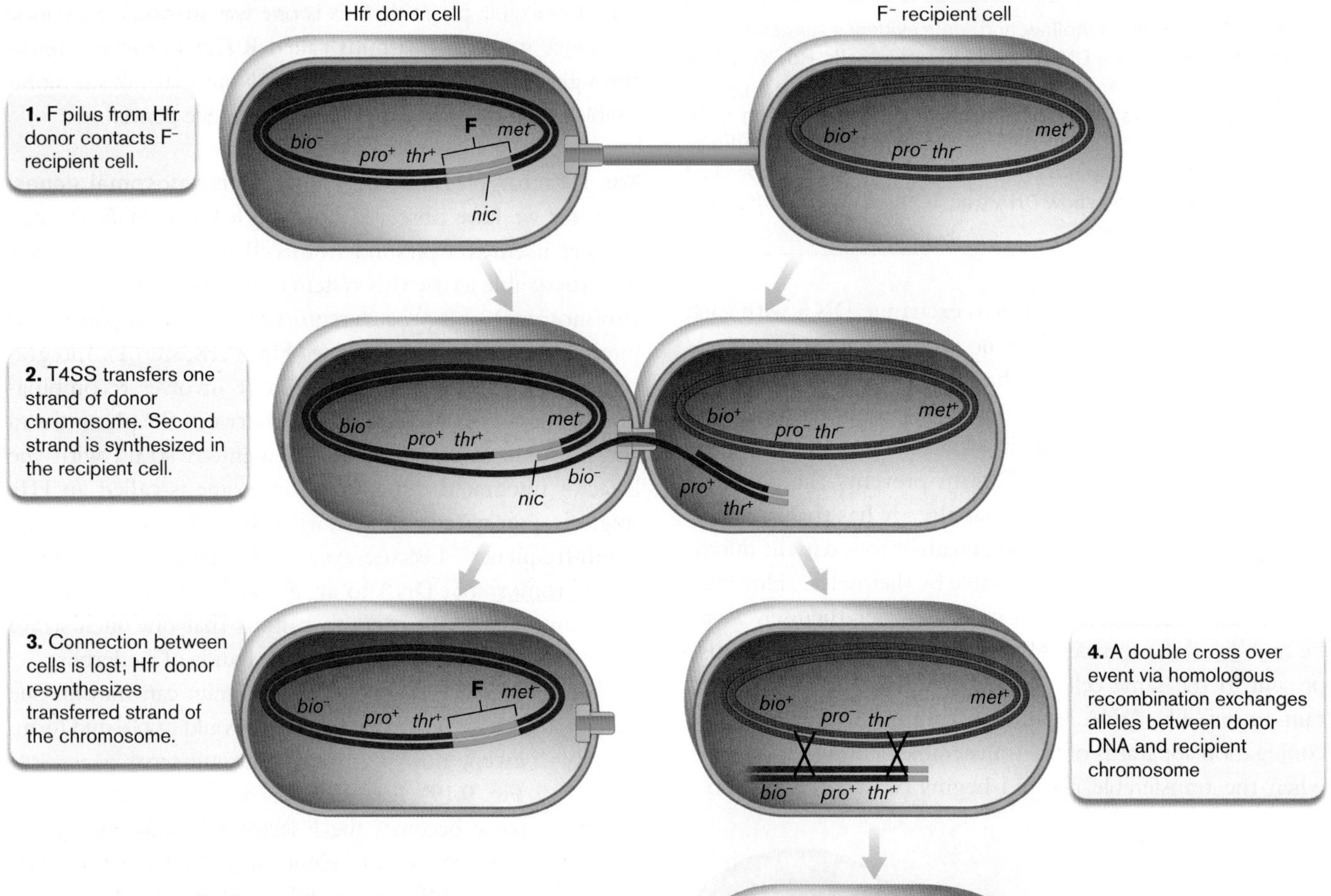

FIGURE 9.19 ■ Transfer of chromosomal genes of an Hfr strain.

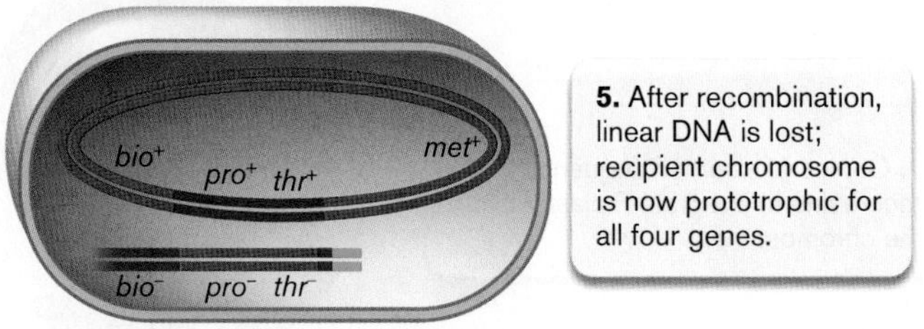

have been one of last genes transferred, and the pilus connection is usually broken long before this can occur. Once the donor DNA strand entered the cell, it was used as a template to synthesize the second strand. This double-stranded DNA could then recombine with the recipient chromosome, and alleles could be exchanged (step 4). For some matings, the auxotrophic alleles in the recipient (thr^- and pro^-) were replaced with the wild-type alleles of the Hfr donor, and a $thr^+pro^+bio^+met^+$ prototrophic strain was generated (step 5).

This allelic exchange required two crossover events between incoming DNA of the Hfr donor and the recipient's chromosome. The crossover events were performed by homologous recombination. Details of how the RecBCD enzyme complex and the RecA protein mediate strand exchange, including the formation and resolution of a Holliday junction, are provided in **eTopic 9.2**. Also included in this eTopic is a discussion of the special mode of recombination used by the extremely radiation-resistant species *Deinococcus radiodurans*.

Hfr conjugation was instrumental in constructing the genetic map of *E. coli* (described in **eTopic 9.3**). Today, DNA sequencing is used as a rapid and inexpensive technique to map genes in other organisms. Nevertheless, conjugation remains important as a tool for scientists to move genes between species, and in nature it is an important means of microbial evolution via horizontal gene transfer.

As a final note, the F factor in an Hfr strain can excise from the chromosome and replicate once again as an autonomous plasmid. In rare instances, excision is imperfect and the plasmid takes with it part of the chromosome. These plasmids are called F-primes (F′). If transferred to an F⁻ wild-type *E. coli*, the F-prime plasmid provides a second copy of the chromosomal genes it carries. This ability to generate partial diploids made F-primes highly valuable in the early days of *E. coli* genetics, as described in **eTopic 9.4**.

Thought Question

9.5 Transfer of an F factor from an F⁺ cell to an F⁻ cell converts the recipient to F⁺. Why doesn't transfer of an Hfr do the same?

Gene Transfer by Phage Transduction

Bacteriophages harbor their own genomes, distinct from those of their host cells, but could phages also pluck host genes from one cell and move them to another? Norton

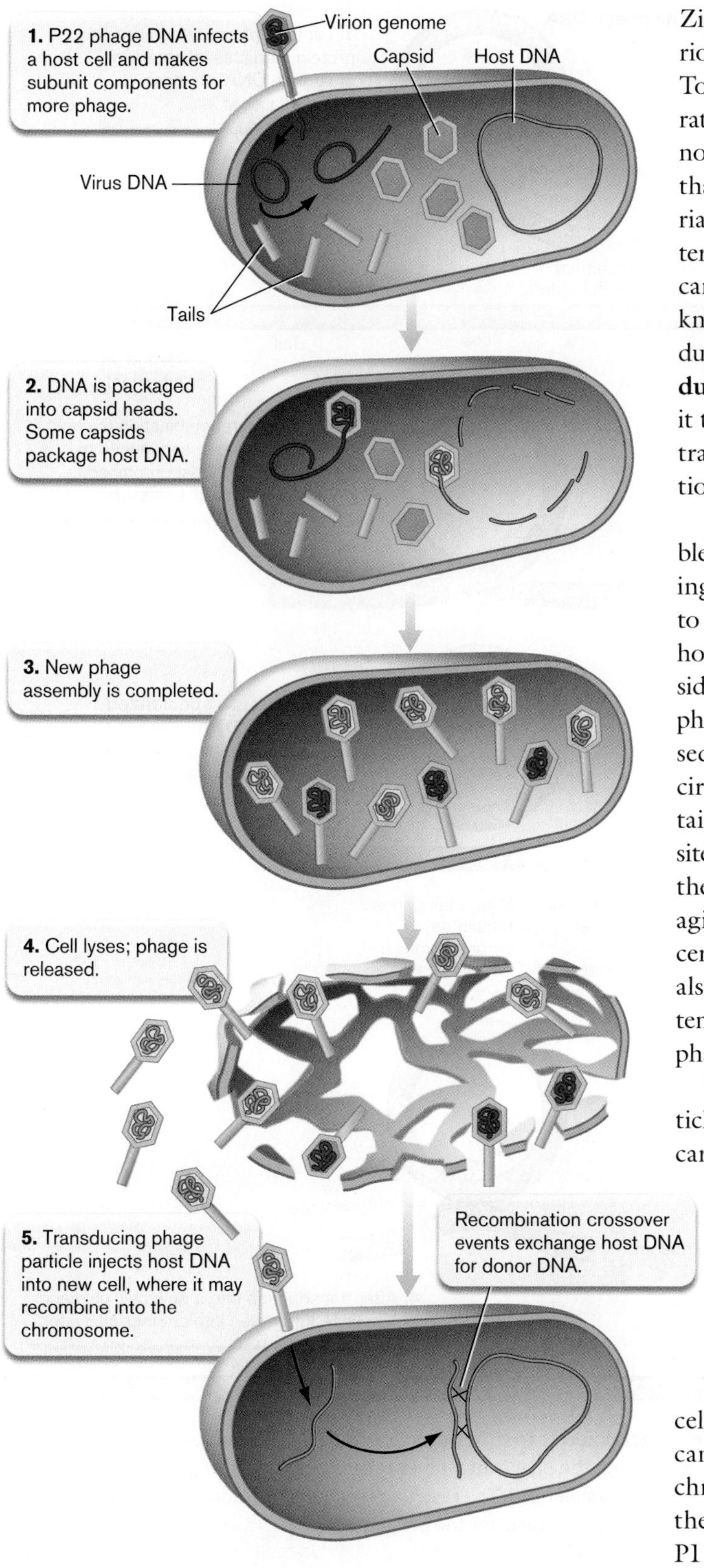

FIGURE 9.20 ■ Generalized transduction. Generalized transduction by phage vectors can move any segment of donor chromosome to a recipient cell. The number of genes transferred in any one phage capsid is limited, however, to what can fit in the phage head.

Zinder (1928–2012) was the first to suspect that bacteriophages could taxi chromosomal DNA between cells. To test this idea, he grew *Salmonella* in two tubes separated by a fine filter through which viruses could pass, but not bacteria. The experiments, reported in 1966, showed that a filterable agent, or virus, could carry genetic material between bacterial strains; direct contact between bacteria was unnecessary. The process in which bacteriophages carry payloads of host DNA from one cell to another is known as **transduction**. There are two basic types of transduction: generalized and specialized. **Generalized transduction** can take any gene from a donor cell and transfer it to a recipient cell, whereas **specialized transduction** can transfer only genes located near the chromosomal integration site of the bacteriophage.

How does transduction happen? Bacteriophages capable of generalized transduction have trouble distinguishing their own DNA from that of the host when attempting to package DNA into their capsids, so pieces of bacterial host DNA accidentally become packaged in the phage capsid instead of phage DNA. In the case of *Salmonella* P22 phage, the packaging system recognizes a certain DNA sequence on P22 DNA called a *pac* site. During rolling-circle replication, P22 DNA forms long concatemers containing many P22 genomes arranged in tandem. The *pac* site defines the ends of the phage genome, marking where the packaging system cuts the P22 DNA and starts packaging DNA into the next empty phage head. However, certain DNA sequences on the *Salmonella* chromosome also "look" like *pac* sites. As a result, the packaging system sometimes mistakenly packages host DNA instead of phage DNA.

The result of this mistake is that 1% of the phage particles in any population of P22 contain no phage DNA but carry host DNA plucked from around the chromosome, up to about 44 kb in length, or almost 1% of the genome (**Fig. 9.20**). The phages that carry host DNA are called transducing particles. Any single transducing particle will contain only one segment of host DNA, but different particles in the phage population will contain different segments of host DNA.

When a transducing particle injects its DNA into a cell, no new phages are made, but the hijacked host DNA can recombine, or exchange, with sequences in the host chromosome of the newly infected cell, thus changing the genetic makeup of the recipient. In *E. coli* the phage P1 is able to perform generalized transduction in a manner similar to that of P22 in *Salmonella*. The ability of these phages to transfer large sections of the genome (1%–2%) has been exploited extensively by microbiologists to construct strains of varying genotypes. And, in the days before high-throughput genome sequencing, they were also used to map the relative position of genes at very fine scales.

Thought Question

9.6 In a transductional cross between an $A^+B^+C^+$ genotype donor and an $A^-B^-C^-$ genotype recipient, 100 A^+ recombinants were selected. Of those 100, 15 were also B^+, while 75 were C^+. Is gene B or gene C closer to gene A?

Specialized (restricted) transduction is a phage-mediated gene transfer mechanism that requires alternating lysogenic and lytic events (see Section 6.4) to move genes from a donor cell to a recipient cell. In contrast to generalized transduction, specialized transduction can move only a limited number of donor genes. *E. coli* phage lambda (discussed in Chapter 11) provides the classic example of specialized transduction. Lambda phage DNA is linear when it first enters the cell; then it circularizes at cohesive ends called *cos* sites. Next, a 15-bp DNA sequence in lambda called *attP* can recombine with a similar host DNA sequence (called *attB*) located between the *gal* (galactose catabolism) and *bio* (biotin synthesis) genes on the *E. coli* chromosome, forming a lysogen (**Fig. 9.21**, step 1). This site-specific recombination event is carried out by the phage integrase protein.

The machinery used at this site is distinct from the RecA-dependent machinery used in generalized homologous recombination. Site-specific recombination produces a chromosome with an integrated phage genome, referred to as a prophage, flanked by chimeric *att* sites called *attL* and *attR*. These sites are called "chimeric" because each one is made half from the bacterial and half from the phage *att* sites. Prophage DNA remains latent until something happens to the cell to activate it.

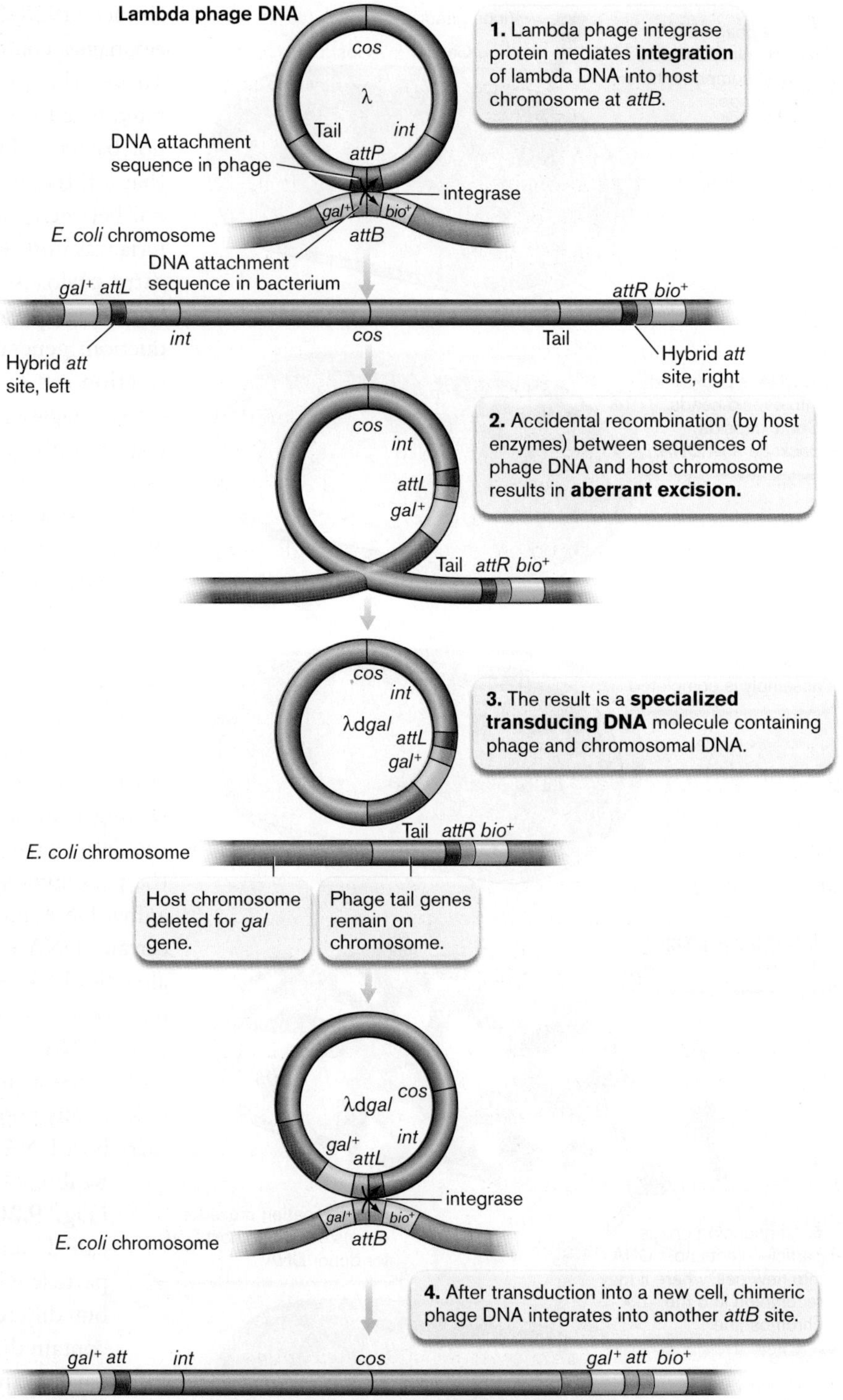

FIGURE 9.21 ▪ Specialized transduction. Specialized transduction is restricted to moving host genes flanking the phage attachment site. The number of genes transferred is limited by the size of the phage head. The resulting recipient, or transductant, chromosome becomes a partial diploid, in this case, for the *gal* gene.

Specialized transduction begins with the reactivation of this prophage DNA, which is often induced by DNA damage. Usually, the phage enzymes that excise the lambda DNA do so precisely. The chromosome and viral DNAs are restored to their native states. The viral DNA will then replicate and make more phage particles containing normal phage DNA. On rare occasions, however, improper excision (mediated by host recombination enzymes) can take place between host DNA sequences that lie adjacent to the phage insertion site (*attL* and *attR* in **Fig. 9.21**, step 2) and similar DNA sequences within the prophage. Improper excision yields a virus that will lack a few viral genes (tail genes remain on the chromosome in step 3) but will include host genes lying adjacent to the phage attachment site (the galactose utilization gene *gal* in step 3). With some phages, the specialized transducing particles can replicate unaided. However, in **Figure 9.21** the result is a defective, specialized transducing phage DNA (lambda d*gal*, or λd*gal*). Specialized transducing phage particles that are defective cannot replicate by themselves but require the presence of a second, helper phage that supplies missing gene products.

Once formed, specialized transducing phages can deliver the hybrid DNA molecule to a new recipient cell. This is the transduction process. Once in that cell, the phage DNA can integrate into the host *attB* site, carrying the donor host gene(s) with it and forming a lysogen once more (**Fig. 9.21**, step 4). The result will be a partial diploid situation in which the new recipient contains two copies of a host gene: one originally present on its chromosome, and one brought in by the transducing DNA.

Transformation of Naked DNA

In 1928, a perceptive English medical officer, Frederick Griffith (1879–1941), found that he could kill mice by injecting them with dead cells of a virulent pneumococcus (*Streptococcus pneumoniae*, a cause of pneumonia), together with live cells of a nonvirulent mutant. Even more extraordinary was that he recovered live, virulent bacteria from the dead mice. Were the dead bacteria brought back to life? Unfortunately, Griffith was killed by a German bomb during an air raid on London in 1941 and never learned the answer.

In a landmark series of experiments published in 1944, Oswald Avery (1877–1955), Colin MacLeod (1909–1972), and Maclyn McCarty (1911–2005) proved that Griffith's experiment was not a case of resurrecting dead cells. Instead, DNA released from dead cells of the virulent strain entered into the harmless living strain of *S. pneumoniae*—an event that transformed the live strain into a killer. The process of importing free DNA into bacterial or archaeal cells is now known as **transformation**.

Organisms in which transformation is a natural part of the growth cycle include Gram-positive bacteria like *Streptococcus* and *Bacillus*, as well as Gram-negative species such as *Haemophilus* and *Neisseria*. At least 82 species of bacteria have been shown to be capable of transformation. Transformation also happens in some archaea, although they do not have homologs to the bacterial transformasome genes and their mechanism of import is not yet known.

Taking up free DNA comes with several challenges. First, the cell needs a way to contact and recognize the DNA as a substrate for uptake. Second, the cell needs to pass the DNA through the cell envelope, which in Gram-negative organisms includes both the inner and outer membranes. The marine bacterium *Vibrio cholerae*, the causative agent of cholera, meets these challenges with a series of protein complexes at the cell surface, beginning with a retractable pilus.

Rather than rely on passive cell contact with the extracellular DNA, *V. cholerae* extends a type IV pilus beyond the outer membrane, which can grab double-stranded DNA (dsDNA) by its tip (**Fig. 9.22**). While some transformable species can take up only DNA that has species-specific sequence tags, the type IV pilus of *V. cholerae* will bind and take up any free dsDNA molecule. Ankur Dalia and colleagues at Indiana University discovered that once the pilus binds DNA, it can pull the DNA into the periplasm by retraction. Retraction occurs through the disassembly of the pilus filament at the base of the pilus. The PilA monomers that are removed from the filament are stored in the membrane for the next time the filament is assembled. Pilus retraction pulls the dsDNA molecule through the open channel that the pilus filament occupied (**Fig. 9.23**). Once in the periplasm, the dsDNA is bound by the ComEA protein. ComEA is too large to pass through the pilus channel, so each addition of ComEA

FIGURE 9.22 ▪ Time-lapse microscopy of dsDNA binding and retraction of the type IV pilus of *Vibrio cholerae*. Fluorescence microscopy of the same cell over time in seconds (s), showing the cell and its pilus (stained green) and the bound DNA (stained red).

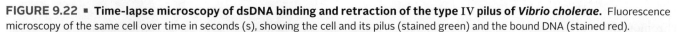

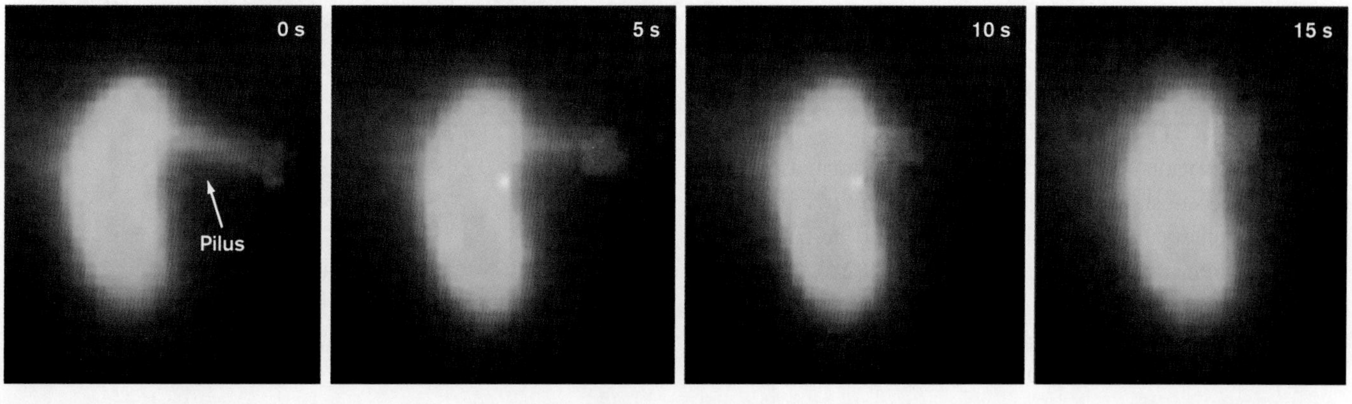

ELLISON, C. ET AL. 2018. *NATURE MICROBIOLOGY* DOI: 10.1038/s41564-018-0174-Y.

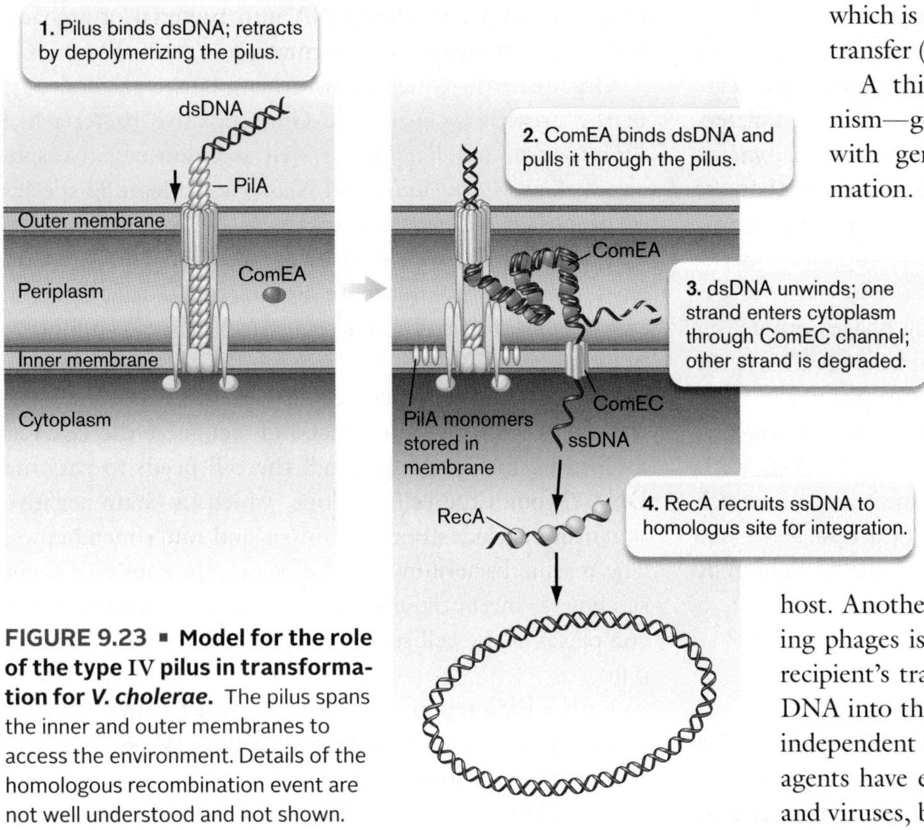

FIGURE 9.23 ■ Model for the role of the type IV pilus in transformation for *V. cholerae*. The pilus spans the inner and outer membranes to access the environment. Details of the homologous recombination event are not well understood and not shown.

protein to the incoming dsDNA serves to ratchet the dsDNA into the periplasm. ComEA-bound dsDNA is then delivered to the ComEC protein channel that spans the inner membrane.

The subsequent steps of transformation are not fully resolved for *V. cholerae*, but they are summarized in the model illustrated in **Figure 9.23**, on the basis of evidence from other transformable species. At the ComEC channel the dsDNA is unwound into two single strands (ssDNA). One strand is degraded in the periplasm, while the other is transported through the ComEC channel into the cytoplasm. Once in the cytoplasm, the ssDNA is bound by protective single-stranded DNA-binding proteins (SSBs). RecA proteins eventually replace the SSB proteins and mediate homologous recombination between the ssDNA and regions of homology in the chromosome.

Additional Modes of Gene Transfer between Bacteria

While conjugation, transduction, and transformation are probably the most prevalent means of transferring genes from one cell to another, some bacteria and archaea have developed alternative mechanisms (**Fig. 9.24**). Two of these mechanisms, membrane vesicles and nanotubes, are described in Chapters 3 and 5. Both membrane vesicles and nanotubes can transfer DNA between different species, which is an important aspect of horizontal gene transfer (discussed in Section 9.5).

A third alternative DNA delivery mechanism—gene transfer agents—shares properties with generalized transduction and transformation. Like P22 and P1 transducing viruses, gene transfer agents encode a phage-like capsid structures that packages random sections of chromosomal DNA. These phage-like particles are released when the host cell dies, and they deliver DNA to the recipient. Unlike phages, however, gene transfer agents do not package their own genes into the phage-like particle, and thus they cannot use the viral lifestyle to move from host to host. Another important difference from transducing phages is that gene transfer agents rely on the recipient's transformation machinery to release the DNA into the recipient's cytoplasm, rather than an independent mechanism of injection. Gene transfer agents have evolved multiple times within bacteria and viruses, but their contributions to fitness of the host and to the process of horizontal gene transfer are not currently known.

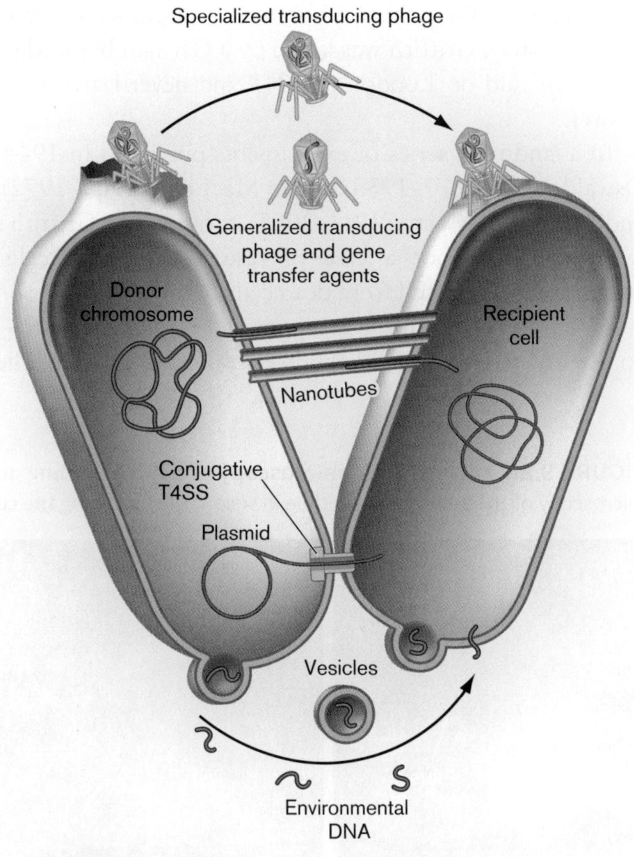

FIGURE 9.24 ■ Summary of mechanisms of gene transfer in bacteria and archaea.

Note: Current studies suggest that even more mechanisms of gene transfer may exist in archaea. Importantly, some of these may involve bidirectional DNA transfer, meaning that each cell involved in the mating sends as well as receives DNA. This is unlike the unidirectional transfers of the mechanisms described in this chapter. Bidirectional DNA transfer may happen in some archaea by cell fusion (cytoplasmic exchange), and in others by transfers at discrete intercellular bridge structures.

The Intestine: Cauldron of Horizontal Gene Transfer

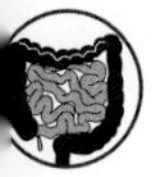

The human gastrointestinal tract is ideal for exchanging genetic information between widely diverse species (Section 23.1). The gut contains a high bacterial cell density and mixed-species biofilms that enable close and frequent contact between organisms, both living (a requisite for conjugation) and dead (a source of transforming DNA). The intestine is also rife with phages that can mediate transduction (see Chapter 6). Transduction is one of the key ways that antibiotic resistance genes can be transferred within the gut microbiome.

Perhaps the most intriguing and worrisome evidence of horizontal gene transfer in the gut involves antibiotic resistance. The gut microbiome can be considered a repository of many antibiotic resistance genes, usually within anaerobic members of the microbiome. Studies that examined the gut metagenomes of hundreds of individuals have identified upwards of 1,000 antibiotic resistance genes "lurking" among the microbial population. Their presence is not always evident until someone is treated with an antibiotic. For example, a plasmid encoding high-level resistance to carbapenem antibiotics was horizontally transferred from *Klebsiella pneumoniae* to *Escherichia coli* in the intestine of a 91-year-old patient. The resistance enzyme, carbapenemase, confers resistance to all cephalosporins, monobactams, and carbapenems used as first-line drugs for hospitalized patients (Chapter 27). The man was being treated for sepsis (a severe blood infection) with ertapenem. When the man's stool was tested after treatment, it contained carbapenem-resistant *K. pneumoniae*, but not *E. coli*. One month later, carbapenem-resistant *E. coli* containing the same plasmid was recovered from his stool.

Antibiotic exposure may even promote horizontal gene transfer among microbiome members. Human volunteers whose stools contained erythromycin-susceptible *Bacteroides* strains were given erythromycin. Within 7 days, erythromycin-resistant *Bacteroides* strains were found. These results implicate conjugation of erythromycin resistance plasmids from unknown members of the gut microbiome. **Figure 9.25** illustrates the mechanisms by which antibiotic resistance genes can be transferred from reservoir gut microbes to opportunistic pathogens that can escape the intestine, either by defecation or through intestinal lesion, to infect elsewhere.

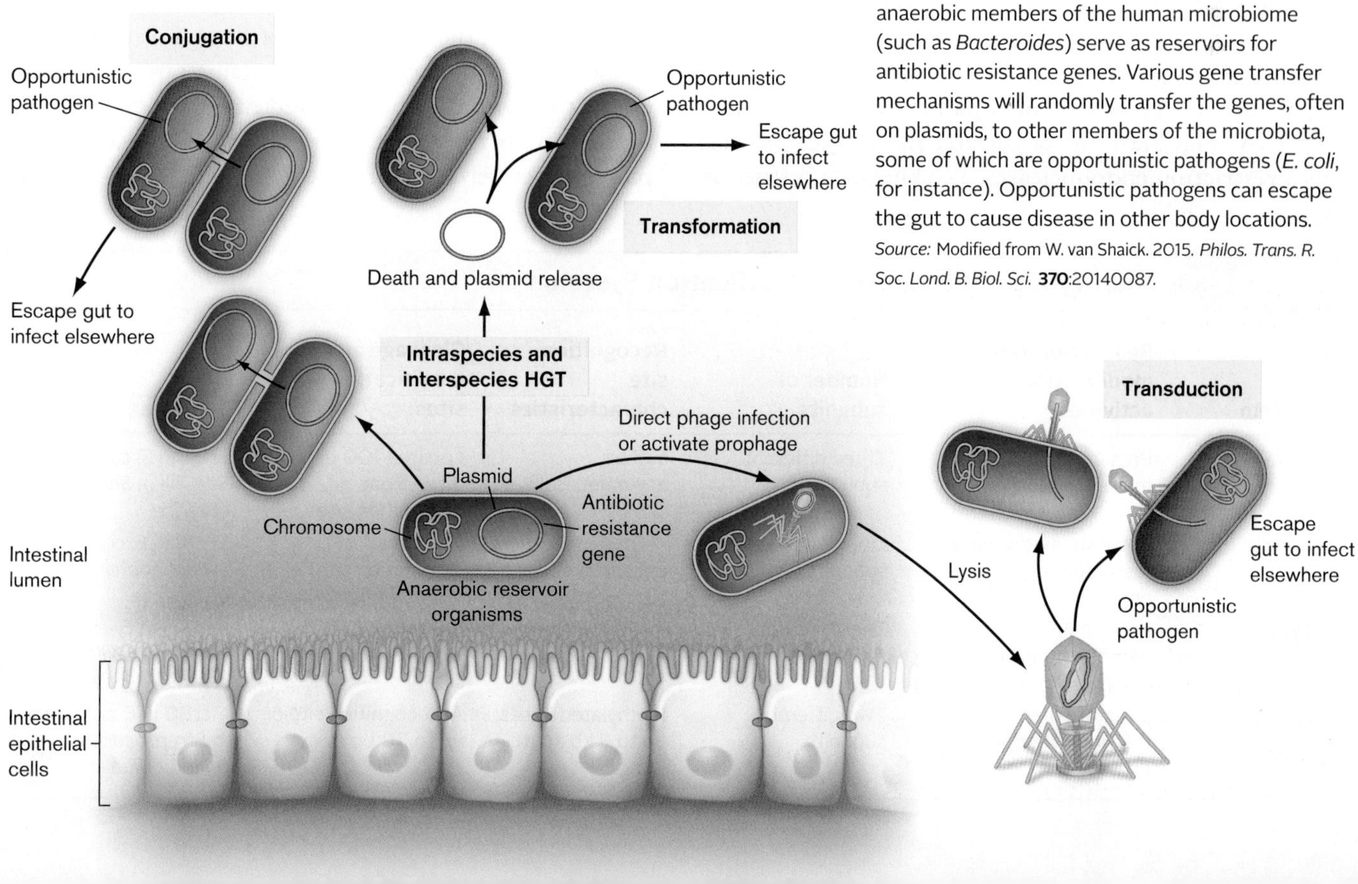

FIGURE 9.25 ■ Model for horizontal gene transfer (HGT) between gut microbiota. High cell density and biofilm formation in the intestine make the horizontal transfer of antibiotic resistance genes (or other genes) easier. Many anaerobic members of the human microbiome (such as *Bacteroides*) serve as reservoirs for antibiotic resistance genes. Various gene transfer mechanisms will randomly transfer the genes, often on plasmids, to other members of the microbiota, some of which are opportunistic pathogens (*E. coli*, for instance). Opportunistic pathogens can escape the gut to cause disease in other body locations. *Source:* Modified from W. van Shaick. 2015. *Philos. Trans. R. Soc. Lond. B. Biol. Sci.* **370**:20140087.

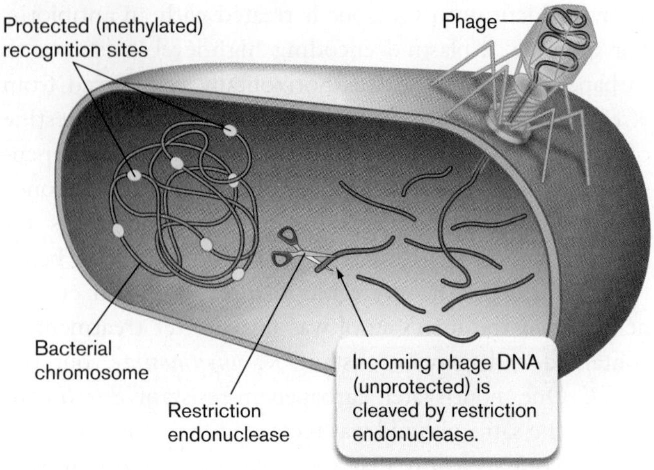

FIGURE 9.26 ■ Restriction of invading phage DNA. Phage DNA is injected into a host, where restriction endonucleases can digest it. Host DNA is protected because specific methylations of its own DNA prevent the enzymes from cutting it.

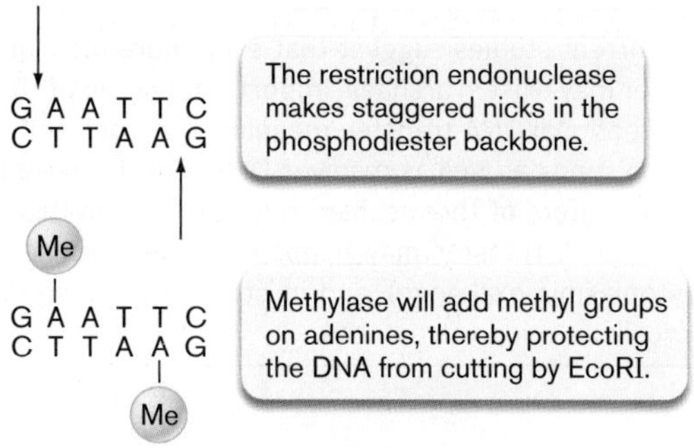

FIGURE 9.27 ■ EcoRI restriction site. EcoRI is an example of a class II restriction-modification system. Shown here are cleavage (top) and methyl modifications (bottom). The DNA sequence shown is specifically recognized by the endonuclease and methylase enzymes. Me = methyl group; $-CH_3$.

DNA Restriction and Modification

There are dangers to the recipient cell associated with the indiscriminate transfer of DNA between microbes. The most obvious risk involves phages whose goal is to replicate at the expense of target cells. A bacterium that can digest invading phage DNA while protecting its own chromosome has a far better chance of surviving in nature than do cells unable to make this distinction. As a result, bacteria have developed a kind of "Halt! Who goes there?" approach to gene exchange. It is an imperfect approach, however, that still leaves room for beneficial genetic exchanges.

This protection system, called "restriction and modification," involves the enzymatic cleavage (restriction) of alien DNA and the protective methylation (modification) of self DNA (**Fig. 9.26**). Most bacteria and archaea produce DNA **restriction endonucleases** (also known as restriction enzymes), enzymes that recognize specific short DNA sequences (known as recognition sites) and cleave DNA at or near those sequences (**Fig. 9.27**). This ability to cleave DNA at specific sequences has provided molecular biologists with a tremendously powerful tool for genetic engineering (see Chapter 12).

There are four types of restriction endonucleases (**Table 9.3**), called types I through IV. Type I and type III restriction endonucleases have their restriction and modification activities combined in one multifunctional protein and cleave DNA some distance away from the recognition site. Type II restriction endonucleases (which are used most often for cloning) possess only endonuclease activity; a separate type II modification protein methylates the same restriction site.

Importantly, the chromosome of an organism producing a restriction endonuclease also contains target sequences for that enzyme. So, how do bacteria avoid "committing suicide" with their own restriction endonucleases? They protect themselves with specific modification enzymes

TABLE 9.3 Main Types of Restriction-Modification Systems

System	Restriction and modification activities	Number of subunits	Recognition site characteristics	Cleavage and modification sites	Examples
Type I	Present in one multifunctional protein	Three different subunits	5–7 bp, asymmetrical	Located 100 bp or more from recognition site	EcoK in *E. coli*; StyLTIII in *Salmonella enterica*
Type II	Separate methylase and restriction endonuclease enzymes	One or two identical subunits per activity	4–6 bp, palindromic	At or near recognition site	EcoRI in *E. coli*; HindIII in *Haemophilus influenzae*
Type III	Present in one multifunctional protein	Two different subunits	5–7 bp, asymmetrical	Located 24–26 bp from recognition site	Eco571 in *E. coli*; BceSI in *Bacillus cereus*
Type IV	Separate methylase and restriction endonuclease enzymes	Two different subunits	Methylated bases, up to 3 kb apart	At recognition site or up to 30 bp away	McrBC in *E. coli*; DpnI in *Streptococcus pneumoniae*

that use *S*-adenosylmethionine to attach methyl groups to the restriction site sequences (see **Fig. 9.27**). Methylation makes the sequence invisible to the cognate (matched) restriction endonuclease (because the modified "A" no longer looks like adenine to the restriction endonuclease). Only one strand of the sequence needs to be methylated to protect the duplex from cleavage; thus, even newly replicated, and consequently hemimethylated (only one strand is methylated), DNA sequences are invisible to the restriction endonuclease.

Type IV restriction endonucleases are a curious class of enzymes. Whereas type I–III enzymes cleave only unmethylated DNA, type IV enzymes cleave only methylated DNA. As one might imagine, targeting methylated DNA, typically a signal of "self" in most microbes with methyltransferases, establishes an interesting dynamic with phage as well as foreign DNA that could serve as material for genomic innovation.

Note: Restriction endonucleases are named according to the species from which the enzyme was isolated. Thus, EcoRI is an enzyme from *E. coli*. Previously written as *Eco*RI, these enzymes no longer have the first three letters italicized.

Thought Questions

9.7 How do you think phage DNA containing restriction sites evades the restriction-modification screening systems of its host?

9.8 *E. coli* has several DNA methyltransferases, such as Dam (see Chapter 7), that methylate many bases in the genome. If a researcher wishes to transfer a plasmid from *E. coli* into a new species, the presence of which type(s) of restriction-modification systems in the new species might prompt a researcher to use a Dam-minus mutant of *E. coli*?

CRISPR Interference: Adaptive Immunity of Bacteria and Archaea

The restriction systems just described have been referred to as a primitive form of innate immunity in bacteria and archaea. They are innate because they are already in surveillance mode by the time foreign DNA enters the cell, and the DNA sequences they target do not change in response to infection. Many bacteria and archaea also employ a second system of defense, one that serves as a primitive form of adaptive immunity. This system, first introduced in Chapter 6, is called **CRISPR** (clustered regularly interspaced short palindromic repeats), and CRISPR works against incoming double-stranded DNA, such as that from bacteriophages or plasmids. If a CRISPR-containing organism manages to survive a phage attack, its CRISPR system can capture a piece of the invader's genome and wield it as a defense against future attack. There are at least six distinct types of CRISPR systems, although they share overall similar function in adaptive immunity. Nearly 50% of bacteria and 85% of archaea possess one or more of these CRISPR loci.

CRISPR anatomy. Emmanuelle Charpentier, currently at the Helmholtz Centre for Infection Research in Germany, and Jennifer Doudna at UC Berkeley (**Fig. 9.28**) played integral roles in deciphering the CRISPR mechanism. For their work, Charpentier, Doudna, and Virginijus Šikšnys, the person who originally discovered CRISPR, shared the Kavli prize in nanoscience (next only to the Nobel in prestige). A CRISPR locus on a bacterial or archaeal chromosome (**Fig. 9.29**) is composed of short direct-repeat sequences (averaging 32 bp) separated by spacers of uniform length (20–72 bp, depending on the species). Although the sequences of direct repeats are, by definition, nearly identical, the sequences of the spacers vary. Note that these repeats and spacers do not encode proteins. Near these sequence clusters lie CRISPR-associated

A.

B.

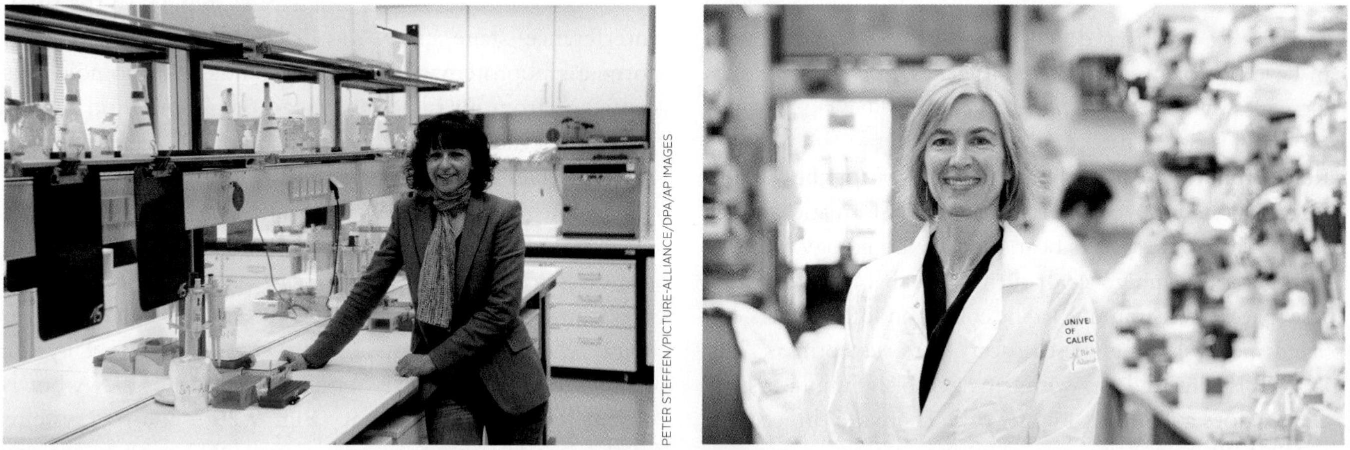

FIGURE 9.28 ■ **Emmanuelle Charpentier (A) and Jennifer Doudna (B) performed some of the seminal work on the CRISPR mechanism.**

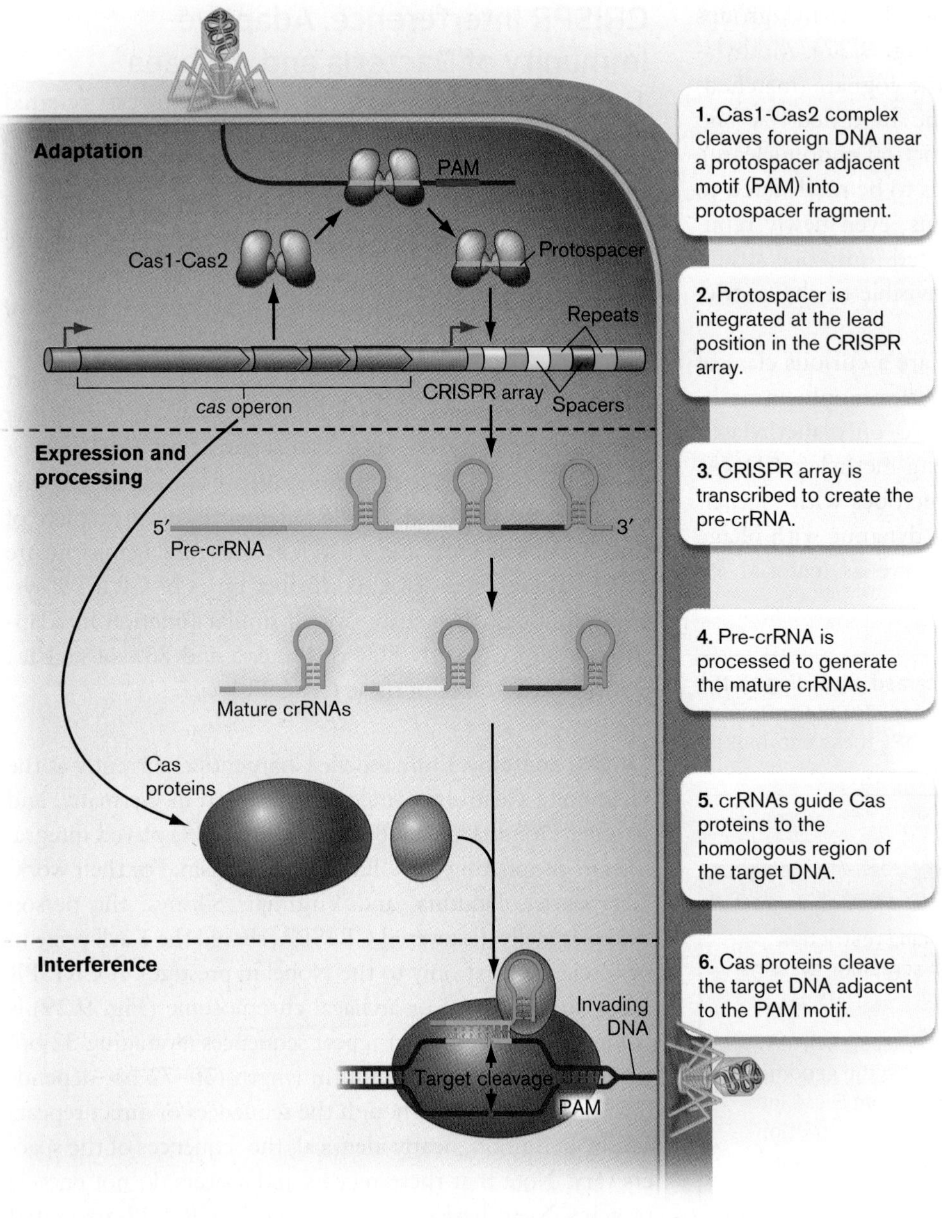

FIGURE 9.29 ■ The CRISPR-Cas adaptation and defense pathway. crRNA = *cis*-repressed mRNA. *Source:* Modified from Frank Hille et al. 2018. *Cell* **172**:1239–1259, fig. 1.

gene families (*cas*) that do encode proteins. A single species can have one or more of these *cas* genes, as well as *cas* subtype genes (which in *E. coli* are called *cse*).

CRISPR function. Clues about the function of the variable spacer regions were uncovered using bioinformatics, which revealed that some spacers bear sequence homology to bacteriophage or plasmid genes. It turned out that cells harboring these spacers were immune to the corresponding invaders, but related species lacking these spacers were susceptible. Thus, CRISPR is perceived as a primitive microbial immune system.

How does a CRISPR locus work? There are three stages of CRISPR-mediated immunity: adaptation, expression and processing, and interference (**Fig. 9.29**). During adaptation, bacteria with a CRISPR locus acquire new spacers by incorporating a piece of an invader's DNA. First, the Cas1 and Cas2 proteins recognize the foreign DNA and cleave it, generating a protospacer fragment (**Fig. 9.29**, step 1). In most CRISPR-Cas systems, part of the recognition involves a short stretch of DNA called the protospacer adjacent motif (PAM) located a few bases away from the cut site. For the Cas9 system of *Streptococcus pyogenes*, the PAM sequence is 5′-NGG-3′, where N could be any of the four bases. The PAM motif is not included in the protospacer, and this absence is important for distinguishing self from foreign DNA, as we will see. The Cas proteins integrate the protospacer into the lead position in the CRISPR region (step 2). With the protospacer in the CRISPR array, the cell "remembers" that this DNA sequence, now called a spacer, is foreign, and it will target the sequence for destruction.

The expression and processing stage begins with the transcription of the CRISPR locus, starting from the upstream leader sequence (**Fig. 9.29**, step 3). The RNA transcript is then cleaved and trimmed (processed) by some of the Cas products and host RNases into small RNAs composed of a single spacer sequence (crRNA, also called guide RNA; step 4). In the final, interference, stage, crRNA associates with Cas proteins to form a riboprotein complex that binds to a homologous sequence from an infecting phage or plasmid and directs cleavage of the foreign DNA (steps 5 and 6). As a result, the infected cell is spared destruction.

Given that the crRNA targets the Cas system to regions of homology in DNA, what prevents the Cas proteins from the disastrous event of cleaving the spacer that the crRNA was transcribed from in the CRISPR array on the chromosome? Most Cas systems identify the phage target by the PAM sequence, which is found in the target DNA but is absent in the CRISPR array (**Fig. 9.29**). The Cas systems of the interference stage only cleave DNA with homology to the crRNA if the homologous region is also adjacent to a PAM. The CRISPR array is protected because it lacks the PAM motif.

As part of the evolutionary arms race between predator and prey, some viruses have evolved so-called anti-CRISPR proteins that inactivate the host cell's CRISPR machinery. Researchers are currently investigating whether bacteria respond in kind by evolving possible anti-anti-CRISPR defenses.

Thought Question

9.9 Some bacteriophage genomes contain functional CRISPR-Cas arrays. Can you think of a reason why this might be advantageous to the bacteriophage?

The CRISPR system has been modified for in vivo genetic engineering purposes (see Chapter 12). Molecules of crRNA can be designed to direct Cas nucleases to cleave almost any in vivo DNA sequence. CRISPR-Cas technology has gained widespread use for modifying eukaryotic genes.

Novel CRISPR functions. A number of studies have now shown that CRISPR loci contribute to cell function beyond providing immunity from phage infections. One novel function of CRISPR loci, discovered from work in George O'Toole's laboratory at Dartmouth College, may be to banish a lysogenic cell from a biofilm, so that phages produced by the lysogen cannot kill the rest of the biofilm community. **Figure 9.30** demonstrates this phenomenon with *Pseudomonas aeruginosa*.

Figure 9.30A shows that a lysogenized (infected) cell of *P. aeruginosa* does not form a biofilm. If this lysogen were a single cell among many biofilm-producing nonlysogens, it would not associate with the biofilm. This self-exile saves the biofilm from infection, should the prophage in the lysogen become active. The CRISPR locus is essential for this loss, since disruption of the Cas-encoding gene *cys4* restores a lysogen's ability to form biofilms (**Fig. 9.30B**). Introducing a wild-type copy of *cys4* (**Fig. 9.30C**) into the mutant cell once again prevents biofilm formation. Thus, a mechanism in which an infected cell imposed self-exile would save the population.

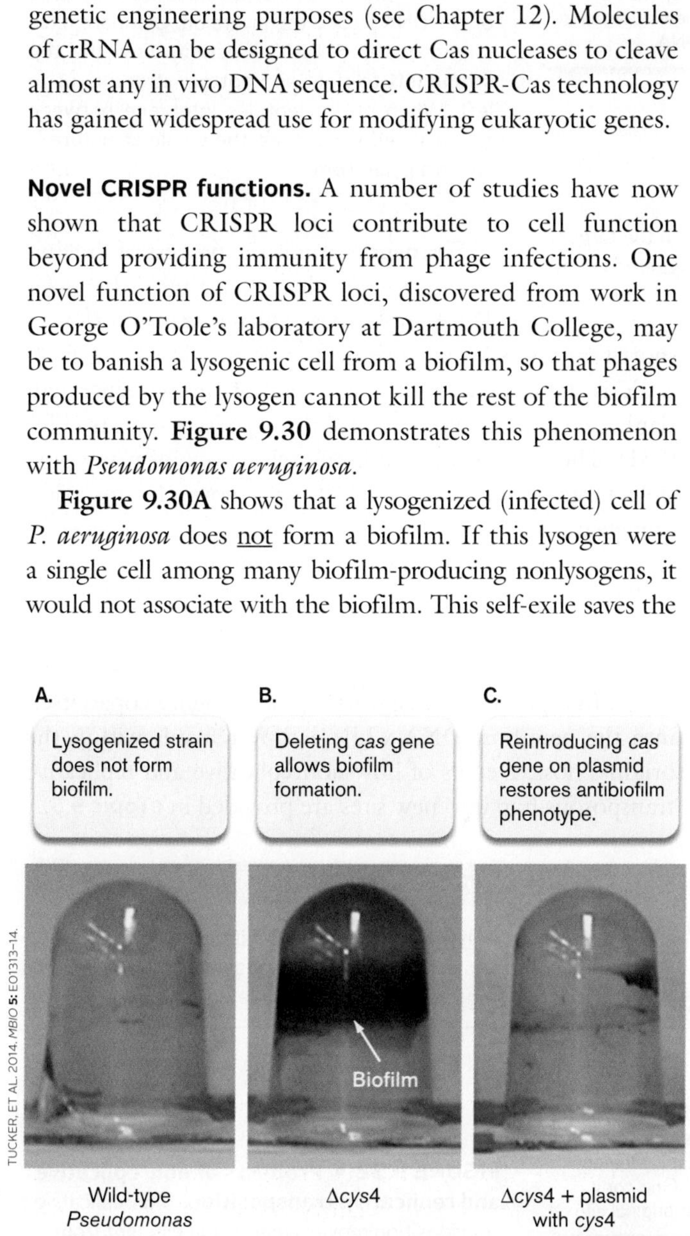

FIGURE 9.30 ■ The effect of CRISPR on *Pseudomonas aeruginosa* biofilm formation. Shown are upside-down tubes in which *P. aeruginosa* lysogenized with DMS3 phage were grown. Cells that form a biofilm stick to the side of the tube and are not dislodged by washing. The biofilm is revealed after staining with crystal violet. *cys4* is a Cas-encoding gene.

To Summarize

- **Conjugation** is a DNA transfer process mediated by a transferable plasmid that requires cell-cell contact and formation of a protein complex between mating cells.
- **Some bacteria can transfer DNA across phylogenetic domains.** For example, *Agrobacterium tumefaciens* conjugates with plant cells.
- **Integration of the F factor** into the chromosome creates an Hfr strain, which can transfer chromosomal genes via conjugation.
- **Transduction** is the process whereby bacteriophages transfer fragments of bacterial DNA from one bacterium to another. In **generalized transduction**, a phage can move any gene in a bacterial genome to another bacterium. In **specialized transduction**, a phage can move only a limited number of bacterial genes.
- **Transformation** is the uptake by living cells of free-floating DNA from dead, lysed cells.
- **Membrane vesicles, nanotubes, and gene transfer agents** contribute to horizontal gene transfer by mechanisms distinct from conjugation, transduction, or transformation.
- **Restriction endonucleases** protect bacteria from invasion by foreign DNA. Restriction-modification enzymes methylate restriction target sites in the host DNA to prevent self-digestion.
- **The CRISPR interference system** is a small, interfering RNA system in many bacteria and archaea that captures a piece of an invader's DNA and uses it to fend off future attacks, thus serving as a primitive form of adaptive immunity.

9.4 Mobile Genetic Elements

In 1948, Barbara McClintock (1902–1992) noticed that certain traits of maize defied the laws of Mendelian inheritance. The genes encoding these traits seemed to hop from one chromosome to another. Although McClintock's theories of these so-called jumping genes were provocative at the time, we now know that these types of genes, referred to as **transposable elements**, exist in virtually all life forms and can move both within and between chromosomes. These

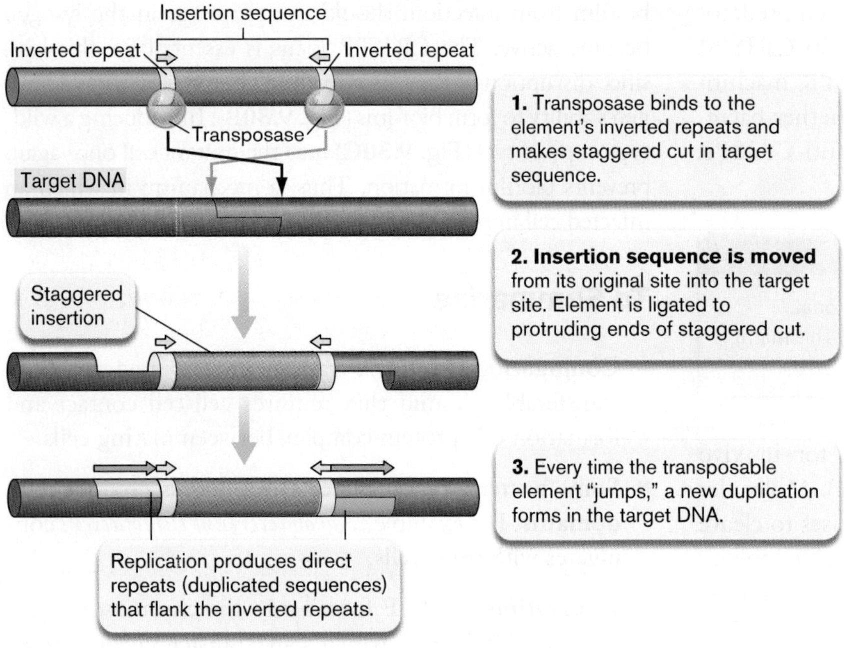

FIGURE 9.31 ▪ Basic transposition and the origin of target site duplication. Enzymes that catalyze transposition generate duplications in the target site by ligating the ends of the insertion element to the long ends of a staggered cut at the target DNA site.

mobile fragments of DNA have contributed greatly to genome rearrangements during the evolution of all species.

Transposons and Transposition

Transposable elements exist primarily as hitchhikers integrated into some other DNA molecule (**Fig. 9.31**). All transposable elements include a gene that encodes a **transposase**, which is an enzyme that catalyzes the transfer or copying of the element from one DNA molecule into another. Bacteria have two types of transposable elements: insertion sequences and transposons. An **insertion sequence (IS)** is a transposable element (typically 700–1,500 bp) consisting of a transposase gene flanked by short inverted-repeat sequences that are targets of the transposase (see **Fig. 9.31**). **Transposons** are mobile elements that carry other genes in addition to those required for transposition. Genes encoding antibiotic resistance proteins are one important class of "cargo" carried by transposons.

Note: An inverted repeat is a DNA sequence identical to another downstream sequence. The repeats have reversed sequences and are separated from each other by intervening sequences:

5′-AATCGAT............ATCGATT-3′
3′-TTAGCTA............TAGCTAA-5′

Recall that you must consider both strands of the DNA to see the inversion. Compare each strand in the 5′-to-3′ direction. View the top strand left to right and bottom strand right to left. When no nucleotides intervene between the inverted sequences, the whole structure is called a palindrome.

Transposition is the process of moving a transposable element within or between DNA molecules. During transposition, a short target DNA sequence on the destination DNA molecule is duplicated so that one copy of the sequence will flank each end of the element, forming direct repeats (**Fig. 9.31**). The transposase randomly selects one of many possible target sequences where it will move the insertion sequence.

Transposition occurs by two different mechanisms: nonreplicative or replicative transposition (**Fig. 9.32**). In nonreplicative transposition, the insertion sequence excises itself from one host DNA and integrates into the destination DNA. In replicative transposition, the sequence copies itself into the new host DNA while a copy remains within the original host. Details of how nonreplicative and replicative transposons hop into new sites are provided in **eTopic 9.5**.

Thought Question

9.10 Evolutionarily speaking, why might it be advantageous for a transposon to utilize replicative transposition? Why might nonreplicative transposition be advantageous?

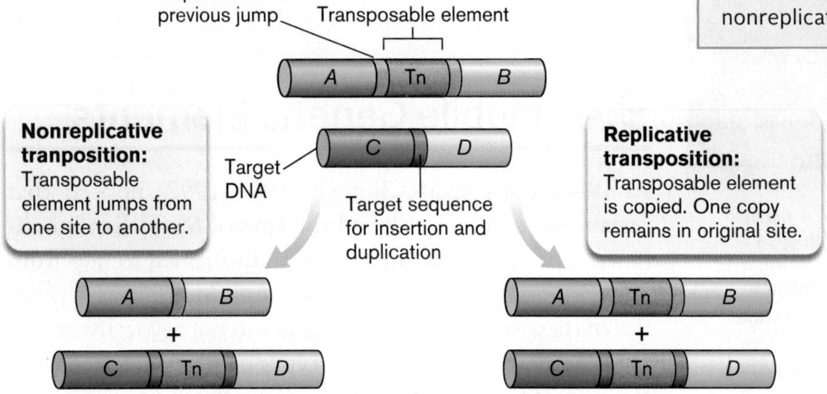

FIGURE 9.32 ▪ Products of nonreplicative and replicative transposition. Nonreplicative transposition moves an insertion element from one DNA site to another without leaving a copy of the element at the original site. Replicative transposition leaves the element at the original site and moves a replicated copy to the new site.

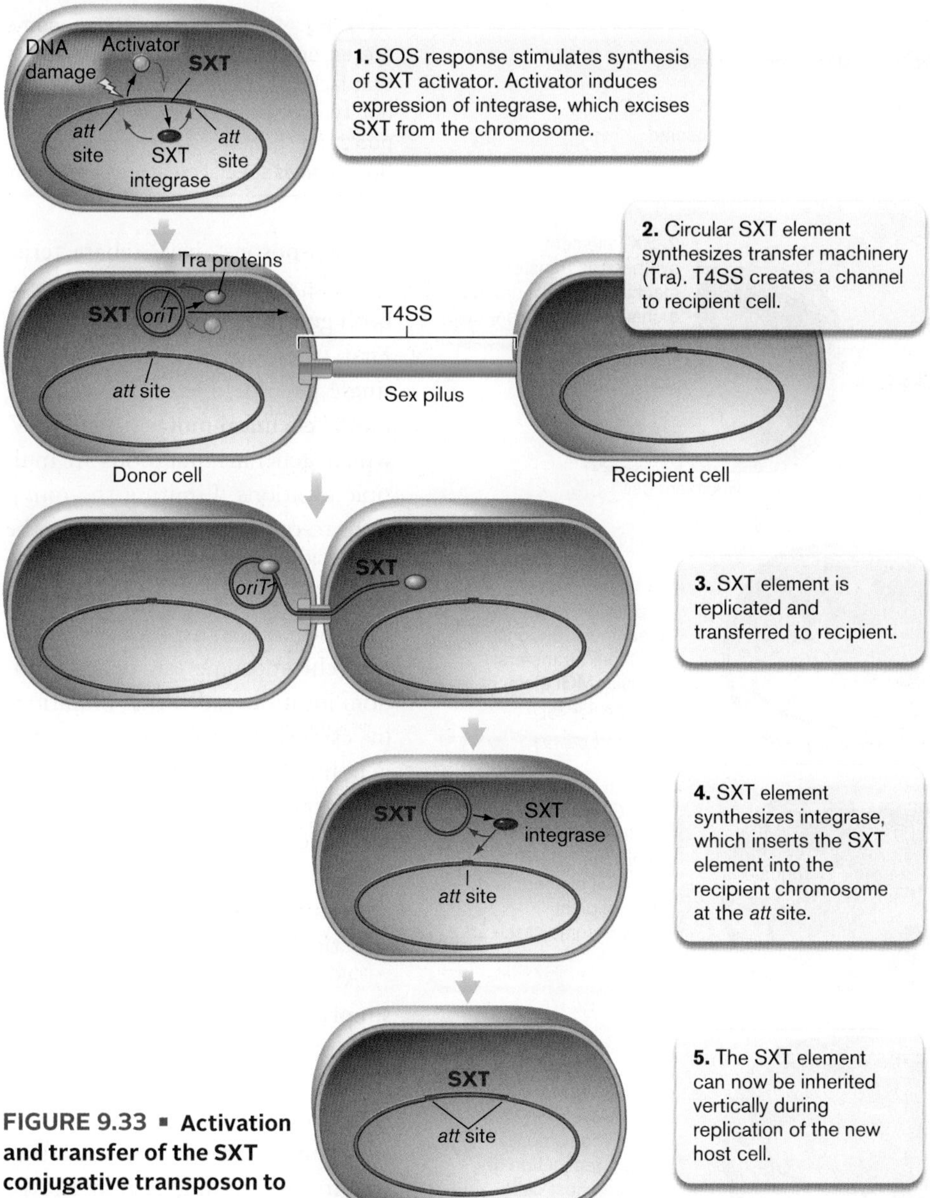

FIGURE 9.33 ■ Activation and transfer of the SXT conjugative transposon to a new cell.

Conjugative Transposons

The transposition events described thus far involve transfer between DNA molecules within the same cell. Some transposons, called **conjugative transposons**, are able to transfer from one cell to another by conjugation. These mobilizable transposons have been found in both bacteria and archaea. One example of a mobilizable transposon is the transposon SXT of the marine bacterium *Vibrio cholerae*. SXT carries genes that confer resistance to the antibiotics sulfamethoxazole and trimethoprim. Conjugative transposons like SXT therefore contribute to the spread of antibiotic resistance in the microbial world. Importantly, their mobilization can be triggered by antibiotic exposure.

Like many other conjugative transposons, SXT is induced by DNA-damaging agents, which include some antibiotics. DNA damage triggers the SOS response of the host, which induces expression of the SXT activator proteins (**Fig. 9.33**). In turn, these activator proteins induce expression of the SXT integrase, which excises the SXT element from donor DNA. Excision results in the formation of a circular intermediate. Like the F plasmid of *E. coli*, the SXT transposon encodes its own transfer machinery, including a T4SS used to contact the recipient cell and to provide the transfer channel. Once transferred into a new host, the conjugative transposon expresses its integrase, which recognizes the specific *att* site in the host chromosome and inserts the SXT transposon at that location.

In an unusual but intriguing process, some genomic islands can hijack the conjugation machinery of conjugative transposons such as SXT to move into new hosts. So-called **mobilizable genomic islands** (**MGIs**) lack the machinery for conjugation but have an *oriT* that mimics that of the conjugative transposon, and they use this *oriT* to trick the conjugative transposon into transferring the island. For example, the SXT transposon's transcription activator can induce expression of the integrase encoded by the MGI, which then excises and circularizes the MGI, much as the conjugative transposon does (**Fig. 9.34**). The SXT transfer machinery then recognizes the MGI's *oriT* and transfers the island to a recipient cell. Once transferred, the MGI uses its own integrase to recombine into the new host chromosome.

Like bacteriophage and conjugative plasmids and transposons, MGIs take a more active role in mediating horizontal gene transfer between microbes. MGIs were first discovered in *Vibrio* and have subsequently been found in an array of other marine bacteria. The cargo genes of MGIs can vary, but they often include restriction-modification systems, which could improve the viral defense of their host cells.

While some mobile elements, such as conjugative transposons, mediate the transfer of genetic material from a donor cell to a recipient, mobile elements of another interesting class, called integrons, have evolved to help the recipient cell capture DNA cassettes once they enter the cell (see **eTopic 9.6**). These mobile elements have facilitated the acquisition of antibiotic resistance genes in pathogens such as *Vibrio cholerae*.

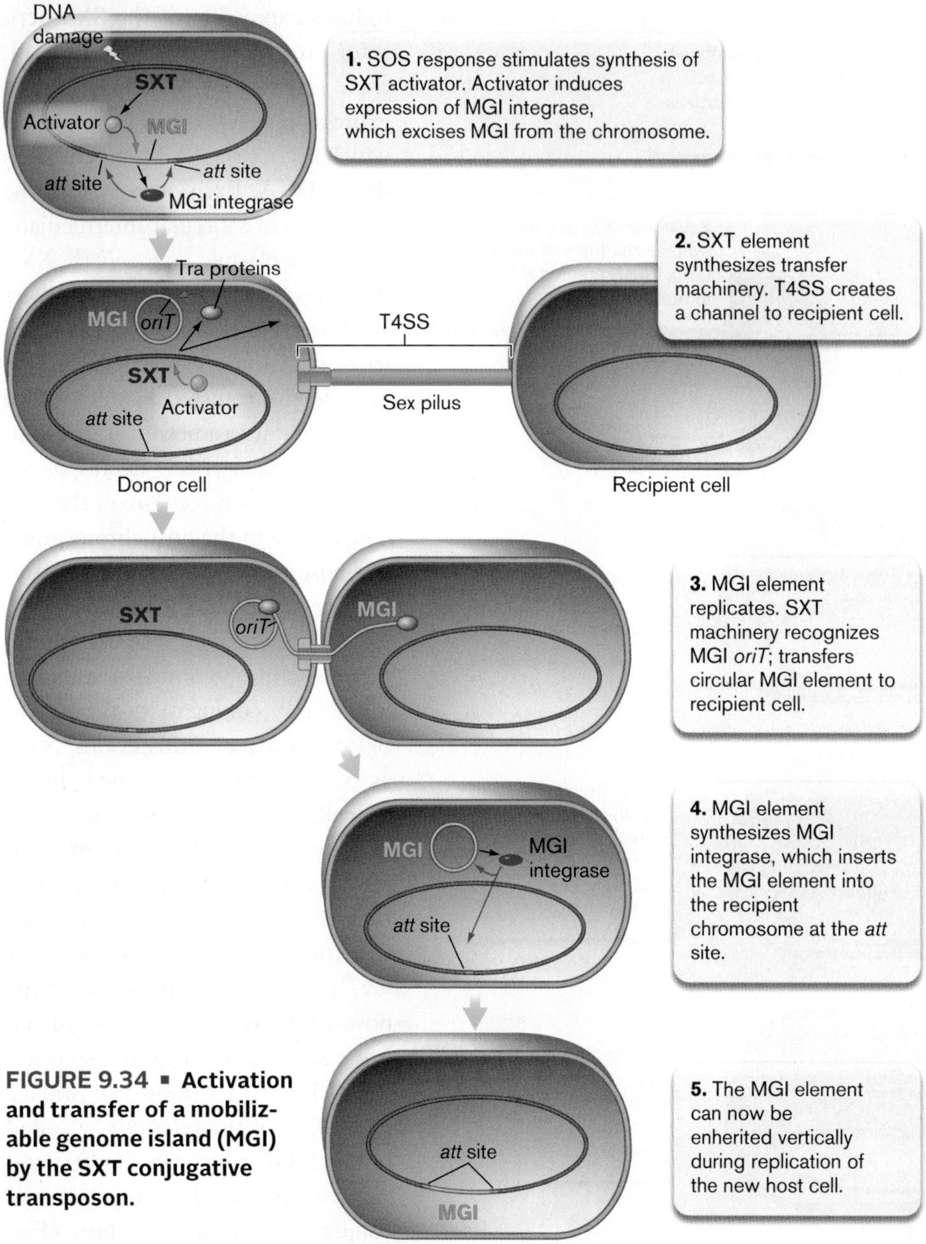

FIGURE 9.34 ■ Activation and transfer of a mobilizable genome island (MGI) by the SXT conjugative transposon.

Note: Many genes in the genome are essential for growth, and mutants that lack the activity of these genes do not typically contribute to the mutant pools for genetic screens and selections, unless special measures are taken.

Transposable Elements in Genetic Analysis

Transposable elements have been of immeasurable service as tools for gene discovery. To identify the genes involved in a microbial function, one genetic approach is to generate mutations in the microbe and assess which mutation(s) affect the function in question. Ideally, only a single gene or operon would be mutated for each mutant cell, and mutants of every gene in the genome would be present in the mutant pool.

Once the pool is generated, an assay is used to identify the mutants that lost the function in question. The assays come in two types: screen and selection. In a **screen assay**, all mutants can grow but only the mutants involved in the function of interest show a phenotype different from the wild type. In a **selection assay**, only the mutants involved in the function of interest can grow, while all other mutants and the wild-type parent die under the assay conditions.

Transposons have characteristics that make them great tools for both genetic screens and selections. First, they usually disrupt only a single gene or operon per mutant, unlike chemical mutagens and UV, which generate mutations in multiple locations. Limiting the mutations to one per mutant enables the researcher to assign the phenotypic change of the cell to a single mutation. Second, transposons hop into the chromosome at nearly random locations, such that mutations in essentially every gene in the genome can be generated. Third, transposons that carry antibiotic resistance markers can provide the researcher with a means to exclude all nonmutants from the screen or selection assay.

After mutagenesis, the cells are simply grown in the presence of the antibiotic, and all cells without the transposon are killed. Finally, transposons can generate loss-of-function mutations by inserting within the coding region of the gene. Transposon insertions can truncate the inserted gene's product if the transposon carries transcription and translation terminators. Thus, the region downstream (3′) of the insertion site is usually not transcribed or translated into protein.

For practical considerations, most screen and selection assays use modified transposons called minitransposons. Minitransposons have the transposase enzyme removed from the transposon (**Fig. 9.35**). Typically, the transposase gene is moved to another region of the plasmid or virus vector that delivers the transposon to the recipient. This way, transposase can still be expressed in the recipient cell but will not be included in the minitransposon that hops into the chromosome. The plasmid carrying the transposase cannot replicate in the recipient, so after cell division, transposase activity disappears and the new insertion is stable. Stability means that the minitransposon cannot hop to a second location, where it could further change the genotype and phenotype of the mutant.

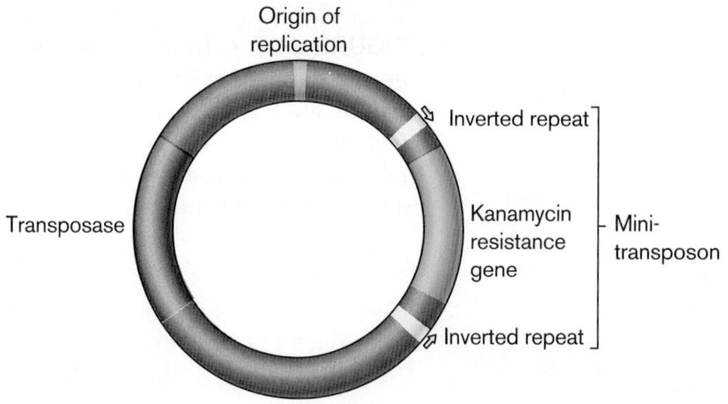

FIGURE 9.35 ■ Simplified model of a typical transposon delivery vector. Note that the origin of replication is usually engineered to function in the donor strain, but not the recipient strain.

In an application of transposon mutagenesis, Alison Buchan and colleagues at the University of Tennessee (**Fig. 9.36A**) performed a screen of marine bacteria called roseobacters to identify genes that are involved in the killing of other bacteria. Roseobacters are important decomposers of plant material in coastal salt marshes, habitats where competition for resources is stiff. One way that some microbes, such as roseobacters, can outcompete rival microbes is to eliminate them—an activity that can be observed on agar plates in a killing assay. When the (wild-type) roseobacter *Phaeobacter* strain Y4I is spotted onto a lawn of *Vibrio* (*Aliivibrio*) *fischeri*, a factor released by the roseobacter is able to lyse the *V. fischeri* cells, forming a halo around the patch of roseobacter cells (**Fig. 9.36B**).

Buchan screened a pool of transposon insertion mutants for those that lost the ability to form a halo and found a small subset of mutants with this loss-of-function phenotype. Using genomics to find the location of the transposon insertion, they found that some of these mutants were no longer able to make indigoidine, a molecule that they subsequently demonstrated has the antimicrobial activity. Interestingly, some other mutants in this screen showed a hyperkilling phenotype, with a larger halo and deeper pigmentation from indigoidine. As this example highlights, screens can also provide unexpected phenotypic outcomes that may spur new lines of investigation.

The development of high-throughput sequencing technology (see Chapters 7 and 12) has enabled researchers to screen for phenotypes among thousands of transposon mutants simultaneously. The TnSeq (also called InSeq) technique is essentially a test of fitness of the pool of mutants in a specific environmental condition. The relative abundances of tens of thousands of insertion mutants are compared before and after the environmental incubation. Those that decrease in abundance have insertions in genes whose activity contributes positively to fitness, and those that increase in abundance involve genes whose activity contributes negatively to fitness.

For example, in a study by Jeffrey Gordon and colleagues at Washington University in St. Louis, transposon mutants of the human gut bacterium *Bacteroides* were inoculated into the guts of mice to look for genes that were important for colonization and persistence. Over the course of several weeks the population of bacteria in the fecal pellets was monitored by TnSeq (**Fig. 9.37**). Of the mutants that decreased in abundance during incubation in the mice, many were involved in carbohydrate consumption and amino acid biosynthesis. These results indicate that the ability to acquire carbohydrates from the gut environment and convert them to amino acids for protein production contributes significantly to the growth and competitiveness of *Bacteroides* in the gut. The advantages of TnSeq for a

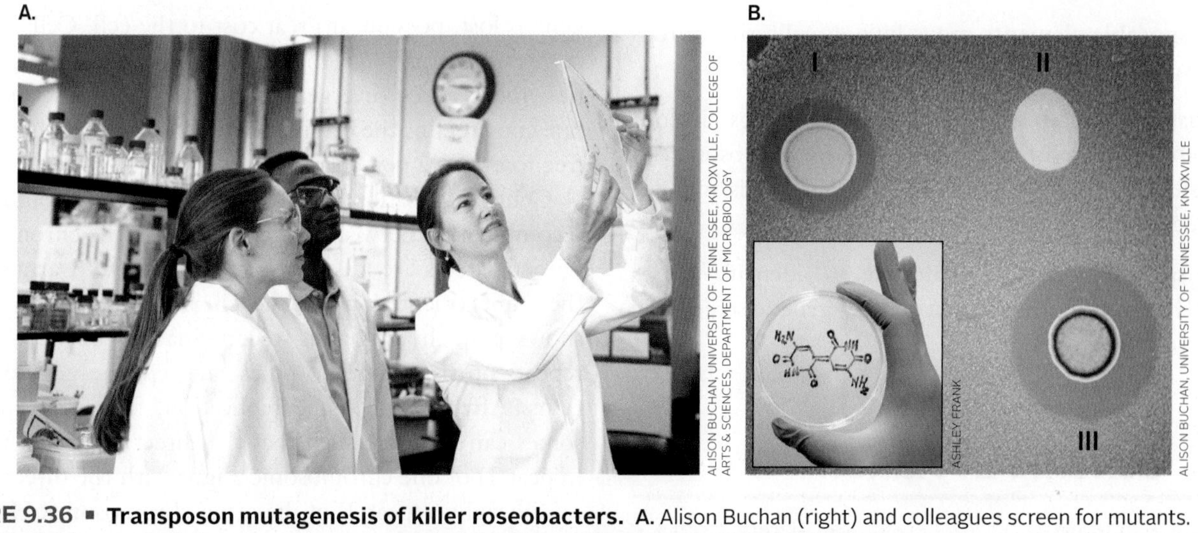

FIGURE 9.36 ■ Transposon mutagenesis of killer roseobacters. **A.** Alison Buchan (right) and colleagues screen for mutants. **B.** Individual cultures of *Phaeobacter* spp. are spotted onto lawns of *Vibrio* (*Aliivibrio*) *fischeri*. Wild-type cultures formed a zone of *V. fischeri* clearing when they grew (I), whereas one transposon mutant lost the ability to form a halo (II), and another one produced a larger halo than the wild type did (III). Note the band of increased pigment indigoidine in spot III. **Inset:** Indigoidine-producing roseobacters streaked on agar to demonstrate the structure of the pigment.

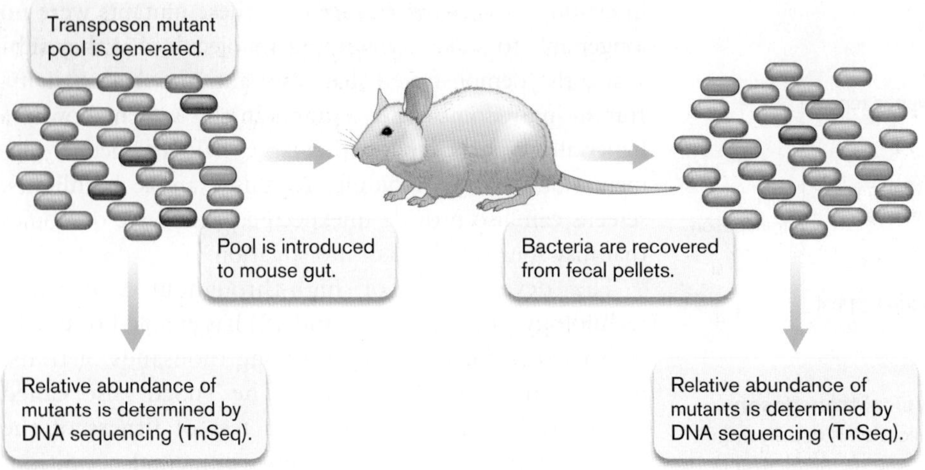

FIGURE 9.37 ■ TnSeq identifies genes involved in fitness in the mouse gut. Green and red cells are mutants that increase or decrease relative abundance, respectively, when introduced into the mouse.

study such as this should be apparent: imagine the time (and number of mice!) required to test each of the thousands of mutants individually for a fitness phenotype in the mouse gut.

To Summarize

- **Transposable elements** ("jumping genes") move from one DNA molecule to another, usually without replicating separately (that is, they are not plasmids).
- **Transposase** catalyzes the transfer or copying of transposable elements from one DNA molecule into another.
- **Insertion sequences** are simple transposable elements containing a transposase gene flanked by short inverted-repeat sequences.
- **Transposons** are complex transposable elements carrying additional genes (encoding, for example, drug resistance).
- **Transposable elements move** by nonreplicative or replicative mechanisms.
- **Transposons can carry a variety of genes**, including antibiotic resistance genes. Some transposons can transfer themselves or mobilizable genomic islands via conjugation.
- **Transposons can be exploited to identify genes involved in microbial functions** via genetic screen and selection assays.

9.5 Genome Evolution

Genomes evolve through the gain and loss of genes, with each gene having its own evolutionary history and origin. The genome of the best-studied organism on the planet, *Escherichia coli* strain K-12, consists of 4,489 genes. This was the collection of genes that was present when the organism was isolated in 1922 from a fecal sample of an unidentified human donor at a hospital in Palo Alto, California. How did this genome assemble to encode the gut-colonizing bacterium we call *E. coli* K-12? To put it another way, where did all of the *E. coli* genes come from?

In this section we revisit the topics of mutation and gene transfer and discuss them in the context of genomic change: how genes are gained and lost during evolution. The mutation events can be very rapid, such as the mutation from one base to another that gives a gene a new function. The impact that such changes have on microbial populations in longer evolutionary timescales is discussed in Chapter 17.

Homology, Duplications, and Divergence

Genes evolve through mutation and, given sufficient mutation, can acquire novel function. One problem with this mechanism of evolution is that the original function of the gene is lost, possibly at great cost to the cell. Cells can circumvent this problem through the mechanism of duplication followed by divergence: the gene is copied such that one copy retains the original function and the other copy is free to evolve a new function.

Gene duplication can be mediated by homologous recombination between direct repeats on the chromosome. Direct repeats are identical stretches of DNA arranged in the same orientation on the chromosome. **Figure 9.38** shows how homologous recombination between direct repeats on sister chromosomes can generate gene duplications. After chromosome replication, the sister chromosomes can associate such that the direct repeat on the left (repeat 1) of one chromosome aligns with the direct repeat on the right (repeat 2) of the other chromosome. Homologous recombination between these repeats results in two mutant chromosomes: one with a tandem duplication of the genes flanked by the repeats, the other with a deletion

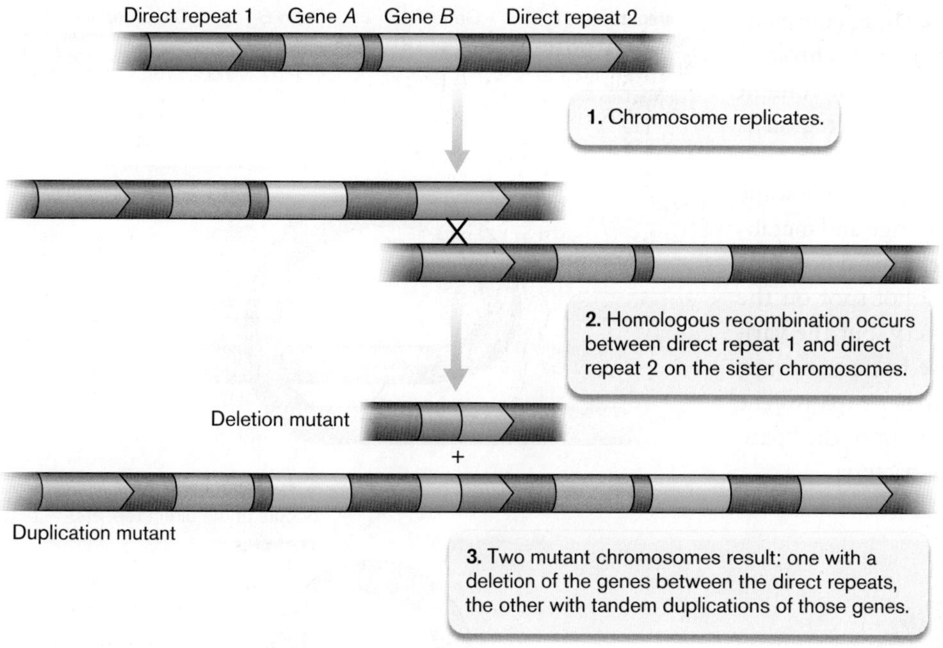

FIGURE 9.38 ▪ Gene loss and gene duplication resulting from homologous recombination between two sister chromosomes. Recombination at direct repeats flanking genes *A* and *B* leads to sister chromosomes with a deletion or tandem duplication of those genes. After chromosomes are partitioned and cell division completes, one daughter cell is a deletion mutant, and the other is a duplication mutant.

of those genes. If the chromosomes are circular, a second recombination event at *dif* or elsewhere along the chromosomes would be required to resolve the dimer that formed by the first recombination event (see Figure 7.20). Chromosome partitioning followed by cell division generates a daughter cell with the duplication and a daughter cell with the deletion. Gene deletion as an agent of genome reduction is discussed later in this section.

Once the gene is duplicated, one copy (or both copies) can acquire new function through mutation. Gene duplication followed by divergence is evident when the sequences of genes within and between genomes are compared. Genes with shared ancestry have sequence similarity, or homology, and are referred to as **homologs**. Homologs can be classified as either orthologs or paralogs. Orthologous genes (**orthologs**) are found in different species and usually have the same function. Paralogous genes (**paralogs**) are homologs found within a single genome and typically have related but distinct functions.

Paralogs arise by gene duplication (**Fig. 9.39**). Duplication facilitates divergence because while one copy maintains the preexisting function, the second copy is free to evolve into a paralogous gene. For example, genes encoding certain exporters of virulence factors in pathogenic bacteria evolved from paralogous genes whose products assemble flagella. The two sets of genes share a common ancestor but have since evolved completely different functions. Paralogous genes are maintained in a microbial genome because their distinct functions contribute to the organism's adaptive potential.

Genome Reduction

Evolution involves gene loss as well as gene acquisition. **Figure 9.38** illustrates how genes can be lost if two chromosomes recombine at direct repeats. Gene deletions can also occur at direct repeats of the same chromosome (**Fig. 9.40**). In this scenario, the DNA loops back to align the direct repeats. Homologous recombination between the repeats leads to a chromosome with the genes between the repeats lost as an extrachromosomal circular fragment. This fragment cannot replicate (it lacks the origin of replication *oriC*) and is eventually lost by degradation. If the deletion is not lethal, the chromosome with the deletion mutation is able to replicate and be passed on to the mutant's progeny.

Gene loss can be "neutral," neither hindering nor helping the microbe, but in some cases the loss of genes can actually benefit the microbe. For example, pathogenic shigellae (the cause of bacillary dysentery) exhibit chromosomal "black holes," regions lacking genes that occur in the closely related *E. coli*. The absence of these genes is required for a fully virulent phenotype. One black hole of shigellae removed *cadA*, which encodes lysine decarboxylase. Lysine decarboxylase produces cadaverine, which inhibits the activity of enterotoxin, an important virulence factor for shigellae.

The large-scale loss of genes through evolution is known as **genome reduction**. Genome reduction is fairly common in pathogens and symbionts associated with plants and animals, but it can also be found in free-living microbes. The evolutionary pressures leading to genome reduction are covered in more detail in Chapter 17; here we provide several interesting examples of genome reduction.

The most abundant organism in the ocean, *Pelagibacter* (also referred to as SAR11), has a highly streamlined genome (1,357–1,576 genes, compared to the 4,489 genes of *E. coli* K-12). Streamlining is thought to be a consequence of selection for efficient growth under the nutrient-poor conditions of the oceanic habitat in which *Pelagibacter* has evolved. Notably, many of the genes found in the genomes of different *Pelagibacter* strains have been found in every *Pelagibacter* genome examined to date. These universally shared (core) genes are found within each *Pelagibacter* chromosome at roughly the same position (**Fig. 9.41**). Therefore, much of the chromosome's gene content seems to diverge only minimally in the different strains.

However, all *Pelagibacter* strains have a large, conspicuous region at roughly the same position in their chromosomes where they show vast differences in gene content between strains (**Fig. 9.41**). This hypervariable region is a spot where rampant horizontal gene transfer has provided strains with unique complements of genes. For some strains, genes in this region are used to scavenge and metabolize sugars and sulfur molecules. Why extensive horizontal gene transfer is restricted to a defined hot spot on the *Pelagibacter* chromosome remains a mystery, but the variety within this hot spot likely helps these bacteria occupy vast regions of the ocean. Chapter 17 further explores the role of genetic variation, including the concept of the "pan-genome," in the evolution and ecology of bacteria.

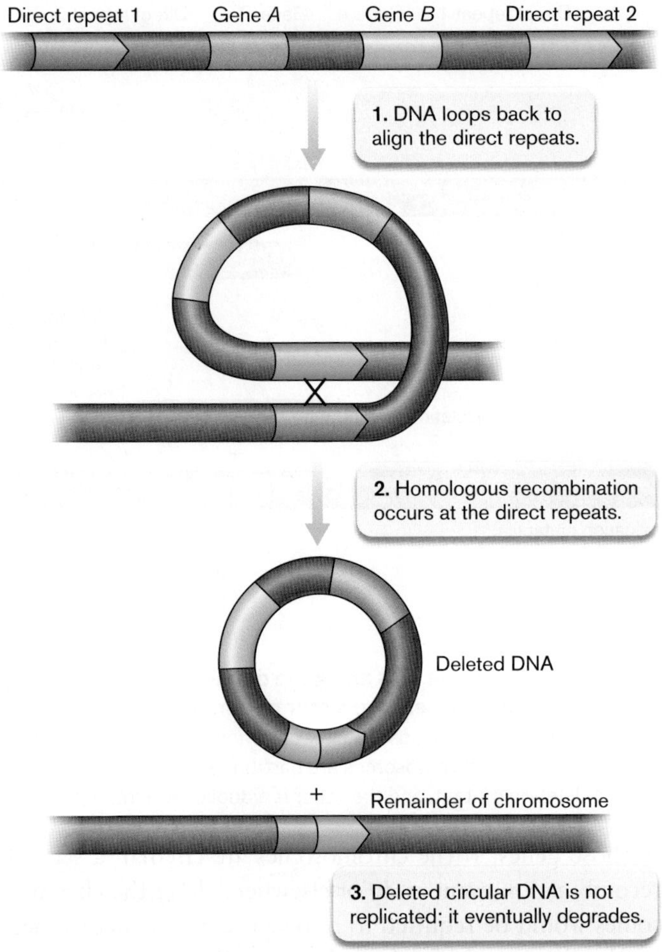

FIGURE 9.40 ■ Gene loss by homologous recombination at direct repeats of a single chromosome.

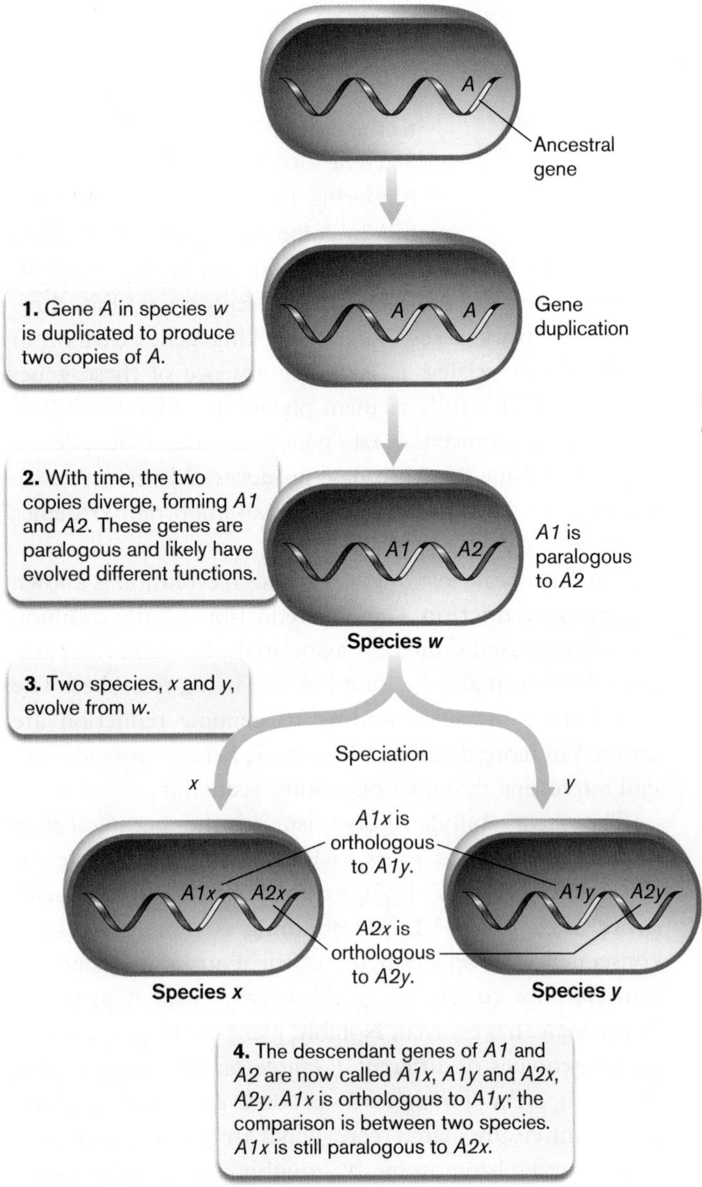

FIGURE 9.39 ■ Paralogous versus orthologous genes. An ancestral gene can undergo a duplication to evolve an orthologous or paralogous gene.

Sometimes we see evidence of genome reduction in the making. Many genomes contain **pseudogenes**, genes that by homology appear to encode an enzyme but are nonfunctional because a portion is missing as a result of mutation. A pseudogene is the remnant of a gene whose function became superfluous to the organism, and consequently the selective pressure to maintain the gene's functionality was lost. Cells with such pseudogenes are not eliminated by natural selection, because the mutated gene did not contribute to fitness. Pseudogenes can eventually be removed completely from the genome via deletion, and in this way they contribute to the process of genome reduction.

Genes become superfluous and subject to mutation into pseudogenes for multiple reasons. One copy of a duplicated gene can become a pseudogene if the second copy is "covering" the gene's function for the cell. Genes can also become superfluous if the organism changes its lifestyle or invades a new environment. One pathogen that has been "caught in the act" of losing genes because of its lifestyle is *Mycobacterium leprae*, the cause of Hansen's disease (leprosy). Over half of the *M. leprae* genome is made of pseudogenes, presumably formed from genes it no longer needs after becoming an obligate intracellular pathogen.

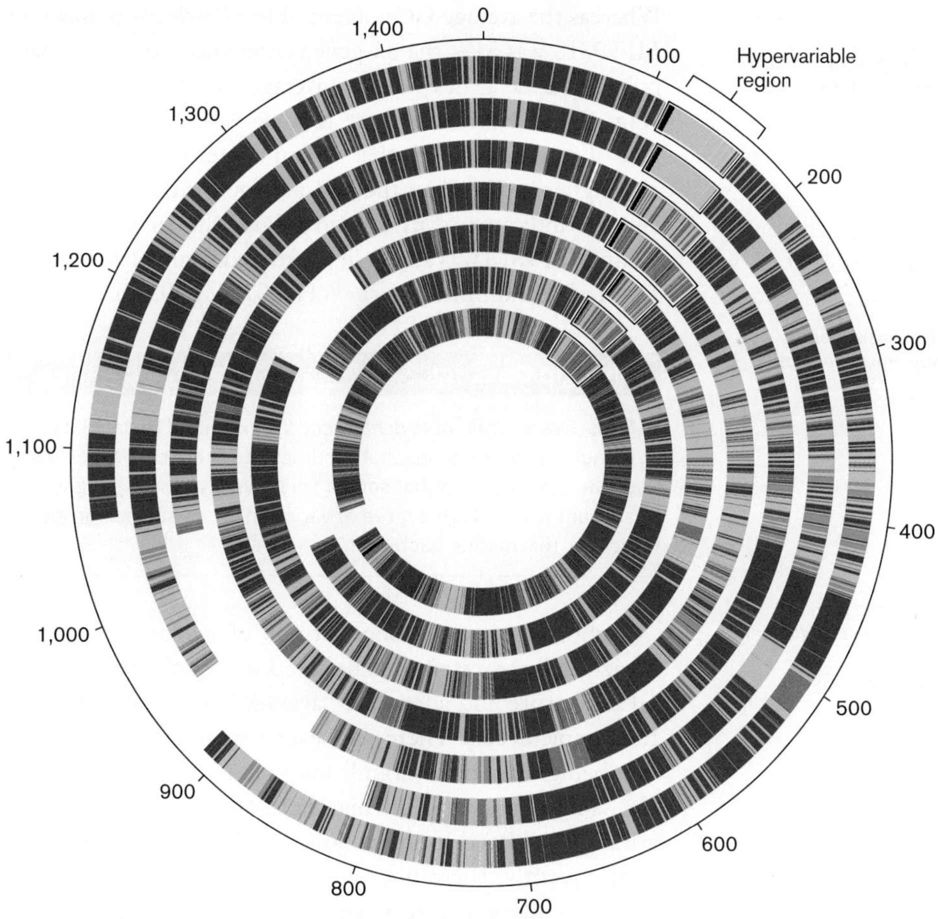

FIGURE 9.41 ■ Circular alignment of SAR11 genomes reveals a hypervariable region: a horizontal gene transfer hot spot. Each ring represents a SAR11 genome. Colors identify genes by category: blue = core SAR11 genes; bright green = additional core genes for SAR11 subclade Ia; orange = shared non-core genes; red = unique genes; black = rRNA genes. The numerical scale is measured in increments of 10,000 base pairs. Because of the disparity in genome sizes, large gaps were necessary to display the genomes in this circular manner. *Source:* Modified from J. Grote et al. 2012. *mBio* **3**:e00252-12.

Horizontal Gene Transfer

The natural movement of genes between organisms is called **horizontal gene transfer** or lateral gene transfer. Horizontal gene transfer between cells differs from **vertical gene transfer**, which is the generational passing of genes from parent to offspring (as in cell division). In contrast, horizontal gene transfer involves the import of genetic material from a foreign source into the cell and incorporation of that material into the genome. Sections 9.3 and 9.4 describe the major ways that genes are transferred between microbes. In this discussion we place these transfer events into the context of evolutionary gene acquisition.

Evolution can occur during both vertical and horizontal gene transfer, but the mechanisms and outcomes can be quite different. During vertical gene transfer, genomes can evolve through mutation of genes already present in the genome, via mechanisms such as gene divergence after duplication. Horizontal gene transfer allows genomes to evolve by introducing new genes from foreign sources. These genes can be entirely unrelated to anything in the original genome, and can provide brand-new properties to the organism once expressed.

Horizontal gene transfer is a prevalent mode of genetic innovation in the microbial world. For example, it has been estimated that nearly 20% of the *E. coli* genome may have originated in other microbes. *E. coli* strain O157:H7, the culprit in several fatal outbreaks of food-borne disease throughout the world, contains 1,387 genes that are not in strain K-12. These additional genes represent about 25% of the O157:H7 genome and encode virulence factors, metabolic pathways, and prophages (phage genes integrated into host chromosomes), all of which were acquired from other species.

Thought Question

9.11 Gene homologs of *dnaK* encoding the heat-shock chaperone HSP70 exist in all three domains of life. All bacteria contain HSP70, but only some species of archaea encode a *dnaK* homolog. The archaeal homologs are closely related to those of bacteria. Knowing this information, how do you suppose *dnaK* genes arose in archaea?

Genomic Islands Are Acquired by Horizontal Transfer

Many of the new functions that horizontal gene transfer can provide require the products of multiple genes. In such cases, these genes are usually all transferred together, forming a **genomic island** in the recipient chromosome (**Fig. 9.42**). Genomic islands serve a variety of functions, including virulence (**pathogenicity islands**), symbiosis (**symbiosis islands**), metabolism (**metabolic islands**), and antibiotic resistance (**resistance islands**). Pathogenicity and symbiosis islands can encode remarkable protein secretion systems (called type III and type IV secretion systems) capable of injecting effector proteins directly from the bacterium straight into a eukaryotic cell. Once inside the target cell, these effector proteins alter the function of eukaryotic cell proteins, making the host organism, in the case of

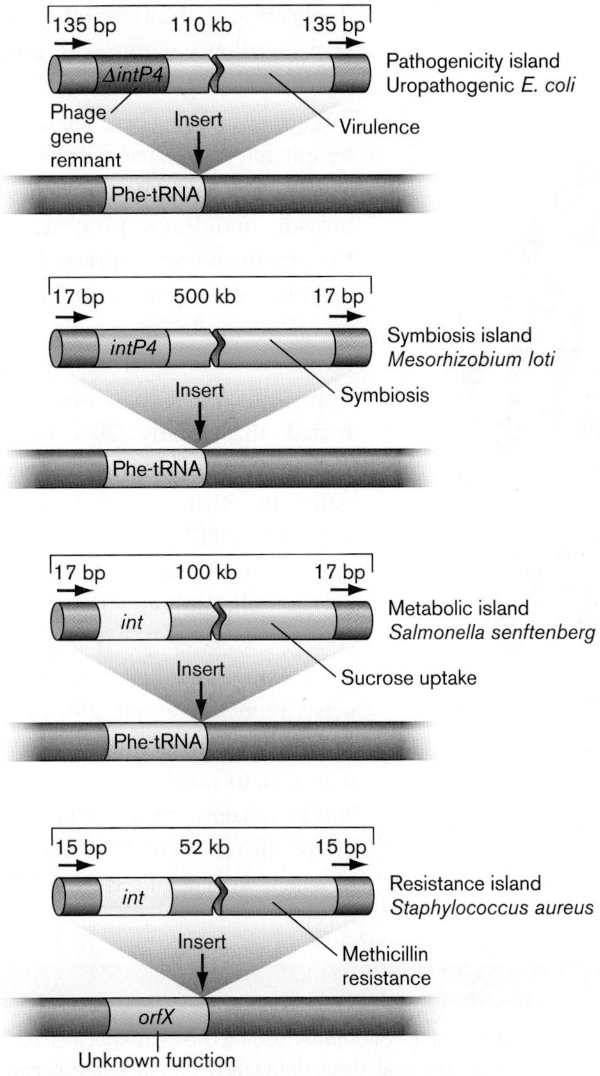

FIGURE 9.42 ■ Genomic islands from different microorganisms. Genomic islands are often found inserted adjacent to tRNA genes and have direct repeats at their ends (arrows). Functions for each island are identified.

symbiosis, more accepting of the symbiont. Type III and IV secretion systems are discussed further in Chapter 25.

Genomic islands are often flanked by boundary regions such as direct repeats or insertion elements, usually found near tRNA genes (**Fig. 9.42**). These flanking elements are hot spots for genetic exchange (recombination), and are the usual targets for insertion of genomic islands into the chromosome.

Genomic islands recently acquired by horizontal gene transfer can be identified through analysis of the frequency of GC base pairs (versus AT base pairs) along the chromosome (also known as GC content). A sudden change in the GC content in a discrete section of the chromosome is the equivalent of an evolutionary "footprint" marking a horizontal gene transfer. As an example, genomic analysis indicated that genes involved in synthesizing the cell surface of the marine photosynthetic bacterium *Prochlorococcus* are of foreign origin and arrived via horizontal gene transfer as a genomic island. Whereas the average GC content of the *Prochlorococcus* strain MIT9313 was 51%, the 33-gene cluster encoding the surface polysaccharide genes was, on average, much lower, at 42% (**Fig. 9.43**). Note that there is some variation in GC content of the cluster, such that several cluster regions are actually higher in GC than the 51% average for the genome. As another indicator of its foreign origin, this gene cluster of strain MIT9313 was absent in the related strains *Prochlorococcus* MED4 and *Synechococcus* WH8102 (**Fig. 9.43**).

Thought Question

9.12 Every strain of *Prochlorococcus* examined to date has a unique cluster of genes that modifies the chemical properties of the cell surface. What sort of selective pressure(s) might account for the high degree of variation in cell-surface properties for this marine bacterium?

Importantly, over long periods of evolutionary time, horizontally transferred genes tend to lose their distinctive GC signature and adopt the "average" GC composition of the recipient cell. There are several reasons for this shift, one being that even though multiple codons can code for the same amino acid during translation (Chapter 8), cells are usually biased in the codons they use for this purpose. Consequently, there is evolutionary pressure for acquired genes to adopt the recipient's predominant codon usage in their own DNA sequence, which will ultimately change the GC content of the genes. As such, many of the more ancient horizontal gene transfer events cannot be identified by examination of the GC content, and other methods must be used to discover them.

Horizontal gene transfers between Bacteria and Eukarya. In eukaryotes, the sexual exchange of genes is usually only within a single species. Bacteria and archaea, on the other hand, are more promiscuous. Matings between microbial genera are common and perhaps even desirable from an evolutionary viewpoint. By sharing genes, different species can sample genes from each other and keep genes (through natural selection) that increase fitness. But what about transfers between Bacteria and Eukarya? Do they happen? If so, are they useful? Work by Gos Micklem and colleagues at the University of Cambridge suggests that interdomain transfer of DNA over the course of evolution may be more common than was previously realized. These researchers report that DNA sequences from many eukaryotes, including humans, appear to include a variety of genes horizontally transferred from bacteria (**Fig. 9.44**). How the genes jumped domains is unclear.

In a bizarre example, Joseph Mougous (University of Washington) and colleagues described a eukaryotic gene called *dae* (domesticated amidase effector) that was transferred from bacteria to eukaryotic ticks and mites several times over millions of years. The bacterial version of the

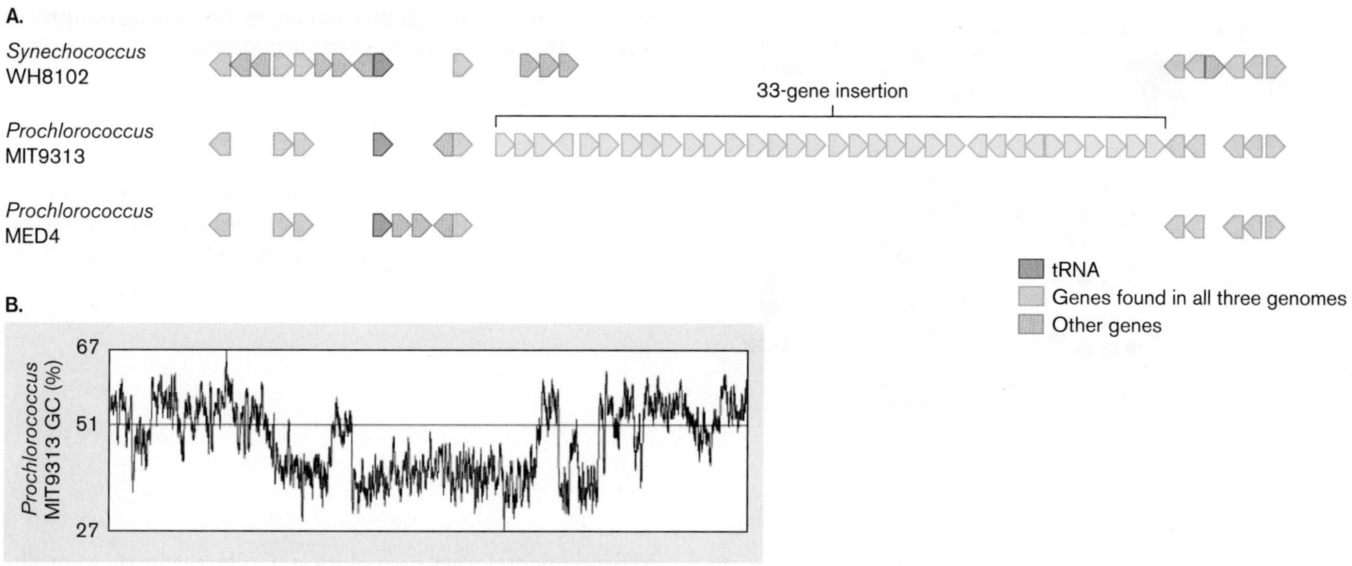

G. ROCAP ET AL. 2003. *NATURE* 424:1042–1047, FIG. 3B.

FIGURE 9.43 ■ Genomic island in *Prochlorococcus* strain MIT9313. **A.** The genomic region of the island, comparing the same genomic region in related strains *Synechococcus* WH8102 and *Prochlorococcus* MED4. Gaps between genes do not represent DNA; they are in place to align the genes shared between strains. Genes in purple are tRNAs, genes in orange are found in all three genomes, genes in gray are found in one or two strains, and genes in green are the genomic island of MIT9313. **B.** GC content for the MIT9313 genome in this region. *Source:* Modified from G. Rocap et al. 2003. *Nature* **424**:1042–1047, fig. 3b.

protein, called Tae, degrades peptidoglycan and is delivered from one bacterium to another via type VI secretion conducted by a protein complex like the tip of a spear (discussed in **eTopic 25.8**). Once injected into the "enemy" bacterium, Tae kills the recipient by disassembling the bacterial cell wall. Importantly, eukaryotes lack bacterial cell walls and so are unaffected by the toxin. Among the eukaryotic recipients of the *dae* gene is the deer tick *Ixodes scapularis*, which serves as the reservoir for the Lyme disease spirochete *Borrelia burgdorferi* (Lyme disease is described in Chapter 26). After their horizontal transfer, the *dae* genes became expressed and eventually acquired eukaryotic protein secretion signals.

Figure 9.45 illustrates that the *I. scapularis dae* gene helps the insect control the levels of *B. burgdorferi* colonizing the tick's midgut. The bacterial gene, therefore, became part of the *Ixodes* innate immune system against *B. burgdorferi* and probably other bacteria. Innate immunity is described in Chapter 23. Some endosymbiotic bacteria transfer enormous numbers of genes to their eukaryotic hosts—a topic that is discussed in **eTopic 9.7**.

To Summarize

- **Homologs** are genes with shared ancestry and have similar sequences. **Orthologs** are homologs found in different species; **paralogs** are homologs found within a single genome.
- **Gene loss is an important part of microbial evolution**, and some pathogens, symbionts, and free-living microbes have undergone extensive **genome reduction**.

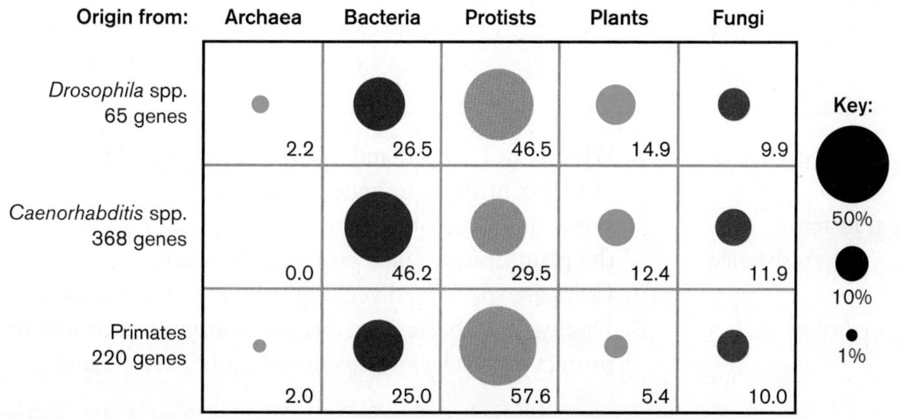

FIGURE 9.44 ■ Proportions of genes in eukaryotic genomes that are predicted to have been transferred horizontally from *Drosophila*, *Caenorhabditis*, and primates. Numbers show percent contribution.

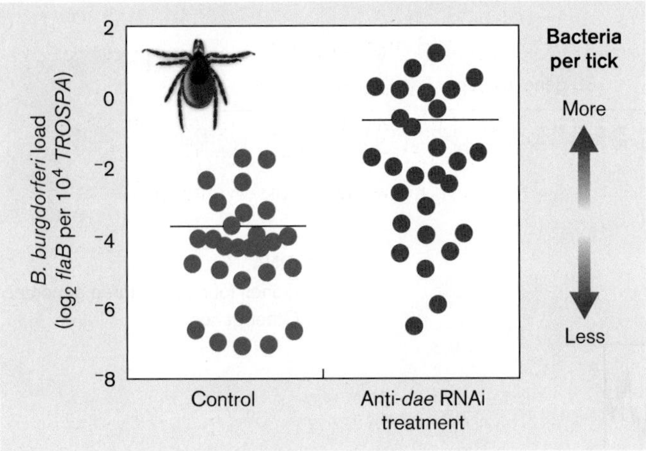

FIGURE 9.45 ■ The *dae* product helps the deer tick control levels of the spirochete *Borrelia burgdorferi*. Deer tick nymphs (inset) fed on spirochete-infected mice, and 2 weeks later *B. burgdorferi* levels were quantified to identify the number of *B. burgdorferi flaB* genes. To estimate relative numbers of bacteria per tick (*B. burgdorferi* load), the ratio of *B. burgdorferi*–specific *flaB* gene copies relative to copies of the tick-specific gene *TROSPA* were determined. Control nymphs received no treatment. Treated nymphs were administered an interfering RNA (RNAi) that binds to *dae* mRNA to reduce *dae* expression. Horizontal lines indicate mean values.

- **Horizontal gene transfers** occur between cells of the same species and also between different species, even members of different domains.
- **Genomic islands** result from horizontal gene transfers that expand the competitiveness of a recipient by contributing to its pathogenicity, symbiosis, or fitness in a hostile environment.
- A DNA sequence with a **GC content** different from that of flanking chromosomal DNA is one sign of horizontal gene transfer.

Concluding Thoughts

Fortuitous movements of genes between species and the divergence of genes following duplication have contributed in fundamental ways to evolution. Collectively, these mechanisms facilitated evolutionary innovation and enabled ancestral microbes to develop (and trade) the symbiotic and pathogenic mechanisms needed to inhabit new ecological niches. DNA repair mechanisms that prevent the establishment of deleterious mutations nevertheless allow some mutations to occur. An imperfect DNA repair mechanism offers the cell the opportunity to throw the genetic dice. The vast majority of mutations will be detrimental or neutral, but a key few mutations may help the variant strain outcompete its neighbors in its ecosystem. Given our new awareness of genetic mobility, how do we now define a species? For instance, numerous strains of *Helicobacter pylori* show large differences in gene sequence and organization. Should each strain be called a different species? For further discussion of the species concept and microbial evolution, see Chapter 17.

CHAPTER REVIEW

Review Questions

1. What are the basic ways microorganisms exchange DNA?
2. Discuss horizontal versus vertical gene transfer.
3. What are pathogenicity, metabolic, and resistance islands? What are their characteristics?
4. What is the difference between an ortholog and a paralog?
5. What is an F factor, and how does it (and other factors like it) contribute to gene exchange?
6. What does microbial gene exchange have to do with the plant disease called crown gall disease?
7. Compare specialized versus generalized transduction.
8. Discuss how bacteria use restriction endonucleases to protect themselves against invading bacteriophages.

9. Describe how CRISPR-Cas systems acquire foreign DNA and use it to protect a cell from infection.
10. Describe transformation and how it occurs.
11. What is the value of recombination to a species?
12. List and explain the different types of mutations.
13. How are genotype and phenotype different? How are they related?
14. Describe several different DNA repair mechanisms. Which ones contribute to mutations?
15. Explain the basic process of transposition. Why are insertion sequences always flanked by direct repeats of host DNA? How are transposons different from plasmids?

Thought Questions

1. How would you use transposon mutagenesis to identify genes involved in the repair of UV-damaged DNA? Would this be a genetic screen or a selection?
2. What genetic features would allow you to determine whether a chromosome region contains a conjugative transposon versus a mobilizable genomic island?
3. *Agrobacterium tumefaciens* Ti plasmid has been used as a tool to genetically modify plants. Why would a plant biologist use a tumor-causing plasmid to breed new plants? Wouldn't the genetically altered plant develop a tumor?
4. Though we can identify sets of genes that have been horizontally transferred from one species of microbe to another, rarely can we identify the source species. What might account for this failure?
5. You have just isolated a new temperate bacteriophage for *Salmonella enterica*. How can you determine whether this phage mediates generalized or specialized transduction?
6. Is it possible for a microbe to be too good at repairing mutations? What might the trade-off be?

Key Terms

AP site (326)
apurinic site (320)
auxotrophic (321)
base excision repair (BER) (326)
conjugation (330)
conjugative transposon (344)
CRISPR (341)
deletion (316)
duplication (316)
error-prone repair (322)
error-proof repair (322)
F^- cell (331)
F^+ cell (331)
fertility (F) factor (331)
frameshift mutation (318)
gain-of-function mutation (317)
generalized transduction (335)
genome reduction (350)
genomic island (351)
genotype (318)
Hfr strain (333)
homolog (349)
homologous recombination (326)
horizontal gene transfer (351)
insertion (316)
insertion sequence (IS) (344)
inversion (316)
knockout mutation (317)
loss-of-function mutation (317)
metabolic island (351)
methyl mismatch repair (322)
missense mutation (317)
mobilizable genomic island (MGI) (345)
mutagen (318)
mutation (316)
mutator strain (323)
nonhomologous end joining (NHEJ) (328)
nonsense mutation (317)
ortholog (349)
paralog (349)
pathogenicity island (351)
phenotype (318)
photoreactivation (323)
point mutation (316)
pseudogene (350)
resistance island (351)
restriction endonuclease (340)
reversion (316)
screen assay (346)
selection assay (346)
silent mutation (317)
SOS repair (327)
SOS response (327)
specialized transduction (335)
symbiosis island (351)
transduction (335)
transformation (337)
transition (316)
transposable element (343)
transposase (344)
transposition (316, 344)
transposon (344)
transversion (316)
vertical gene transfer (351)

Recommended Reading

Ambur, O. H., J. Engelstadter, P. J. Johnsen, E. L. Miller, and D. E. Rozen. 2016. Steady at the wheel: Conservative sex and the benefits of bacterial transformation. *Philosophical Transactions of the Royal Society of London. Series B, Biological Sciences* **371**:20150528.

Baidya, A. K., S. Bhattacharya, G. P. Dubey, G. Mamou, and S. Ben-Yehuda. 2017. Bacterial nanotubes: A conduit for intercellular molecular trade. *Current Opinion in Microbiology* **42**:1–6.

Bondy-Denomy, Joseph, and Alan R. Davidson. 2014. To acquire or resist: The complex biological effects of CRISPR–Cas systems. *Trends in Microbiology* **22**:218–225.

Cabezón, Elena, Jorge Ripoll-Rozada, Alejandro Pena, Fernando de la Cruz, and Ignacio Arechaga. 2015. Towards an integrated model of bacterial conjugation. *FEMS Microbiology Reviews* **39**:81–95.

Chou, Seemay, Matthew D. Daugherty, S. Brook Peterson, Jacob Biboy, Youyun Yang, et al. 2015. Transferred interbacterial antagonism genes augment eukaryotic innate immune function. *Nature* **518**:98–101.

Dini-Andreote, Francisco, Fernando D. Andreote, Welington L. Araujo, Jack T. Trevors, and Jan D. van Elsas. 2012. Bacterial genomes: Habitat specificity and uncharted organisms. *Microbial Ecology* **64**:1–7.

Hu, Bo, Pratick Khara, and Peter J. Christie. 2019. Structural bases for F plasmid conjugation and F pilus biogenesis in *Escherichia coli. Proceedings of the National Academy of Sciences USA* **116**:14222–14227.

Ilangovan, A., C. W. M. Kay, S. Roier, H. El Mkami, E. Salvadori, et al. 2017. Cryo-EM structure of a relaxase reveals the molecular basis of DNA unwinding during bacterial conjugation. *Cell* **169**:708–721.

Johnston, Calum, Bernard Martin, Gwennaele Fichant, Patrice Polard, and Jean-Pierre Claverys. 2014. Bacterial transformation: Distribution, shared mechanisms and divergent control. *Nature Reviews. Microbiology* **12**:181–196.

Lang, A. S., A. B. Westbye, and J. T. Beatty. 2017. The distribution, evolution, and roles of gene transfer agents in prokaryotic genetic exchange. *Annual Review of Virology* **4**:87–104.

Louwen, Rogier, Raymond H. J. Staals, Hubert P. Endtz, Peter van Baarlen, and John van der Oost. 2014. The role of CRISPR-Cas systems in virulence of pathogenic bacteria. *Microbiology and Molecular Biology Reviews* **78**:74–88.

Million-Weaver, Samuel, Ariana Nakta Samadpour, and Houra Merrikha. 2015. Replication restart after replication-transcription conflicts requires RecA in *Bacillus subtilis. Journal of Bacteriology* **197**:2374–2382.

Polz, Martin F., Eric J. Alm, and Willian P. Hanage. 2013. Horizontal gene transfer and the evolution of bacterial and archaeal population structure. *Trends in Genetics* **29**:170–175.

van der Veen, Stijn, and Christoph M. Tang. 2015. The BER necessities: The repair of DNA damage in human-adapted bacterial pathogens. *Nature Reviews. Microbiology* **13**:83–94.

Van Houten, Bennett, and Neil Kad. 2014. Investigation of bacterial nucleotide excision repair using single-molecule techniques. *DNA Repair* (Amsterdam) **20**:41–48.

Wagner, A., R. J. Whitaker, D. J. Krause, J. H. Heilers, M. van Wolferen, et al. 2017. Mechanisms of gene flow in archaea. *Nature Reviews. Microbiology* **15**:492–501.

CHAPTER 10
Molecular Regulation

10.1 Transcription Repressors and Activators

10.2 Alternative Sigma Factors and Anti-Sigma Factors

10.3 Regulation by RNA

10.4 Second Messengers

10.5 Clocks, Thermometers, and Switches

10.6 Chemotaxis: Posttranslational Regulation of Cell Behavior

A typical bacterial genome encodes thousands of different proteins, many of which are useful in only certain environments. To compete with other species, a microbe cannot waste energy making unneeded proteins. Cells achieve efficiency by using elegant control systems that selectively increase or decrease gene transcription, mRNA translation, and mRNA degradation, as well as by degrading, sequestering, or covalently modifying regulatory proteins. In Chapter 10 we describe how a cell detects or even anticipates when it needs to alter its physiology, and we explain the regulatory mechanisms used to effect that change.

CURRENT RESEARCH highlight

Melatonin activates circadian rhythm in gut bacteria. Vincent Cassone and colleagues examined the swarming behavior of the human gut bacterium *Enterobacter aerogenes* in the presence of melatonin, a hormone that regulates our night-day sleep patterns. Swarming produces concentric rings on agar plates as the bacterial population expands from a single point of inoculation. Rings on plates without melatonin had no periodicity (panel A), but cells grown on plates containing melatonin formed rings every 25 hours (panel B). The results indicate that members of the gut microbiome may possess a circadian clock that is turned on by, and synchronized with, its host.

Source: Paulose et al. 2016. *PLoS One* **11**: e0146643.

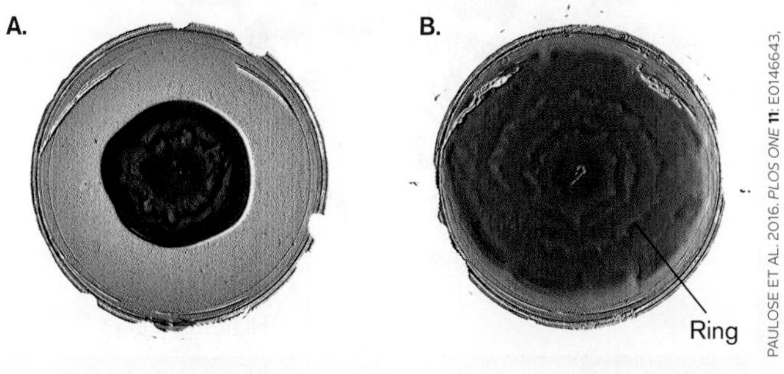

PAULOSE ET AL. 2016. *PLOS ONE* **11**: E0146643.

AN INTERVIEW WITH

VINCENT CASSONE, BIOLOGIST, UNIVERSITY OF KENTUCKY

COURTESY OF VINCENT CASSONE

What value would a circadian clock provide to a bacterium that lives in the gut?

The gastrointestinal system contains a robust clock that expresses circadian patterns of motility, secretion, excretion, body temperature, and other processes. Thus, it isn't surprising that the microbiome within the gut is profoundly rhythmic. Circadian clocks enable (micro)organisms to anticipate changes in the environment and to take advantage of predictable resources (for example, food intake) or to avoid dangers.

In Chapter 8 we discussed how genes are expressed as information passes from DNA to RNA to protein. In Chapter 10 we look at how gene expression and the activity of gene products are regulated. For a number of reasons, cells do not express every gene at maximal level under all conditions. For example, a cell has limited space to house all of those gene products. In addition, the building of a cell such as *E. coli* from a simple organic carbon resource requires an estimated 20–60 billion ATP equivalents of energy (that is, energy from ATP hydrolysis). Thus, wasting energy and resources on unnecessary gene products would be a significant disadvantage to a cell. Finally, the activities of some gene products can be counterproductive (or even harmful) to the cell if expressed under the wrong circumstance. Rapid growth of *Bacillus* in times of plenty would be difficult to achieve if the cells were always producing the machinery to differentiate into dormant spores. Natural selection has therefore favored the evolution of sophisticated mechanisms in microbes that control the expression of genes and the activity of gene products.

Microbes use numerous mechanisms to sense their internal and external environments. The information collected directs the synthesis of specific proteins whose concentrations change with changing environments. The cell's surface, for example, contains an array of sensory proteins that monitor osmolarity, pH, temperature, and the chemical content of the surroundings. The Current Research Highlight illustrates how a microbe may use chemical cues, in this case melatonin, to recognize that it (the microbe) is in the human gut, and thus to regulate cell activity in a day-night cycle. This regulation is quite elaborate: it begins with sensing the hormone at the cell surface, and ends with promoting periodic changes in the rotation of flagella that modulate the cell's motility. Once thought to be an uncontrolled bag of enzymes, a microbial cell, we now recognize, is exquisitely tuned to its dynamic environment.

In this chapter we discuss the fundamental principles of gene regulation in microbes and examine how these individual systems are woven into global regulatory networks that interconnect many processes throughout the cell. The expression of genes is regulated from the initial access of genetic information stored in DNA to the formation of functional proteins and RNA. The different sections of this chapter describe different classes of regulators and the steps in gene expression where they exert their control. Regulatory proteins and different types of sigma factors can control the initiation of transcription, the first step in gene expression. Transcription as well as translation can be controlled by RNA. Some proteins and RNAs can regulate expression by behaving as clocks or thermometers. Bacteria regulate some genes by modifying the DNA prior to transcription. Finally, some important cell behaviors are controlled at the posttranslational level, affecting the activity of proteins already synthesized. As each mode of control is introduced in this chapter, consider the advantages and disadvantages that it offers to the cell as a means of regulation. We begin our discussion of gene expression by describing the regulation of transcription initiation.

10.1 Transcription Repressors and Activators

As the first step in gene expression, initiation of transcription is a major place of regulatory control in bacteria. Bacteria use DNA-binding proteins to control initiation of transcription at the gene promoter. We begin this section with an overview of how changes in the environment within the cell or outside the cell can control the activity of regulatory proteins that control transcription initiation.

Transcriptional Control by Repressors and Activators

The initiation of transcription in bacteria is often controlled by **regulatory proteins**. These proteins bind DNA at or near the promoters of genes, where they either stimulate or prevent the binding of RNA polymerase to the promoter, the first step in transcription (Section 8.2). **Figure 10.1**

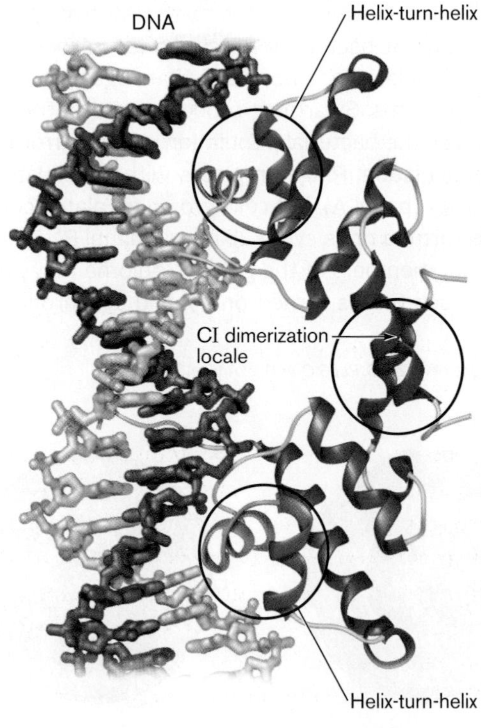

FIGURE 10.1 ■ Binding of a repressor protein to DNA. The dimer of lambda CI repressor binding to DNA. Note the helix-turn-helix motif located in two successive major grooves. Helices in this motif are brown and purple; the turns are green. (PDB code: 1LMB)

illustrates how one regulatory protein, CI from lambda phage, binds to DNA. The protein forms a dimer, and one part of each molecule, called the DNA-binding **domain**, interacts with DNA in the major groove. DNA sites that bind regulatory proteins typically exhibit a sequence symmetry that involves an inverted repeat (**Fig. 10.2**). Dimers of a regulatory protein bind to the DNA, with each subunit binding half of the symmetrical DNA sequence. Several techniques to demonstrate protein-DNA binding are described in Section 12.3.

The sequence of DNA affects the binding affinity of regulatory proteins. Regulatory proteins usually bind at high affinity to the upstream region of genes they regulate, and bind at low affinity elsewhere along the DNA. Regulatory proteins can use the low-affinity binding to remain in contact with the DNA as it scans for high-affinity targets.

Thought Questions

10.1 Why is the "scanning mode" that permits regulatory proteins to search for promoters along DNA more efficient than a random search within the cell's cytoplasm?

10.2 Knowing that the affinity of regulators for DNA depends on the DNA sequence, propose how evolution could result in the ability of a preexisting regulator to now regulate a newly acquired gene. Why might it be critical that the sequence encoding the regulator itself does not change during this evolutionary step?

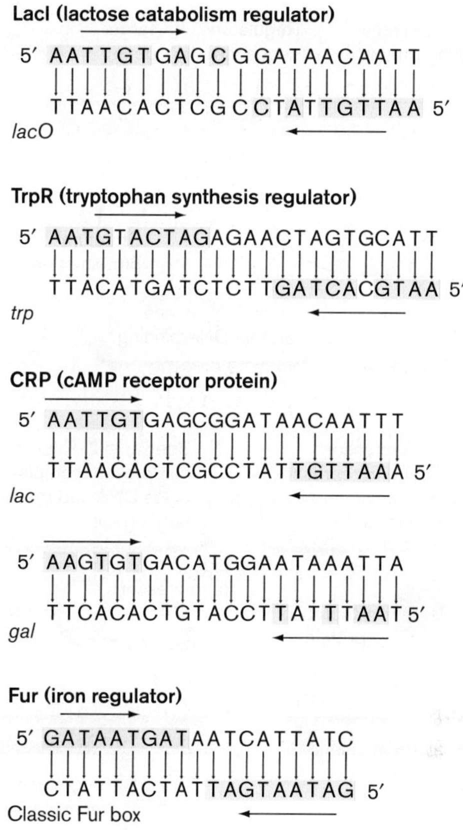

FIGURE 10.2 ▪ Examples of DNA regulatory sequences. The sequences shown are located upstream of the genes specified below each sequence. Inverted repeats are shown in yellow. Note that in some inverted repeats, occasionally a base is not repeated (only those bases that repeat are highlighted in the diagram). Arrows indicate the direction of symmetry.

Cells use different mechanisms to sense and respond to conditions within the cell and outside the cell membrane. Sensing conditions within the cell to regulate transcription is relatively straightforward. Many regulatory proteins bind specific low-molecular-weight compounds called ligands (**Fig. 10.3A**). Different regulatory proteins bind different ligands. For example, one regulator binds a carbohydrate ligand and alerts the cell that a new carbon source is available, while a different regulator senses whether enough of the amino acid tryptophan is present in the cytoplasm to carry out protein synthesis. The ligand, once bound, then alters the ability of the regulatory protein to latch onto specific DNA regulatory sequences located near the promoters of target genes.

Genes encoding regulatory proteins are usually transcribed separately from the target gene (**Fig. 10.3A**). Regulatory proteins come in two forms: **repressors** and **activators**. Repressor proteins bind to regulatory sequences, referred to as **operators**, and prevent the transcription of target genes—an event known as **repression**. Repression happens in one of two ways (scenarios 1 and 2 in **Fig. 10.3B**), depending on the repressor. In scenario 1, the repressor binds a specific operator and prevents transcription of a target gene. Relief from repression requires that a specific ligand, called an **inducer**, bind to the repressor protein, causing it to release from the operator. Because a small inducer molecule is required, the increased expression of the target gene is called **induction**. The lactose operon, discussed later in this section, is one example of an inducible system.

Note: DNA regulatory sequences are called "operators" when binding to them decreases expression of the target genes. But when binding to them increases expression, they are called "activator sequences."

Other repressor proteins (scenario 2) bind poorly to operators unless they first bind a small ligand called a **corepressor**. As the corepressor disappears from the cell, it is no longer available to bind to the repressor protein. Release of the corepressor triggers release of the repressor from the DNA, and the target gene is expressed. This process is called **derepression** rather than induction. The tryptophan operon, discussed later in this section, is an example of a repression-derepression system.

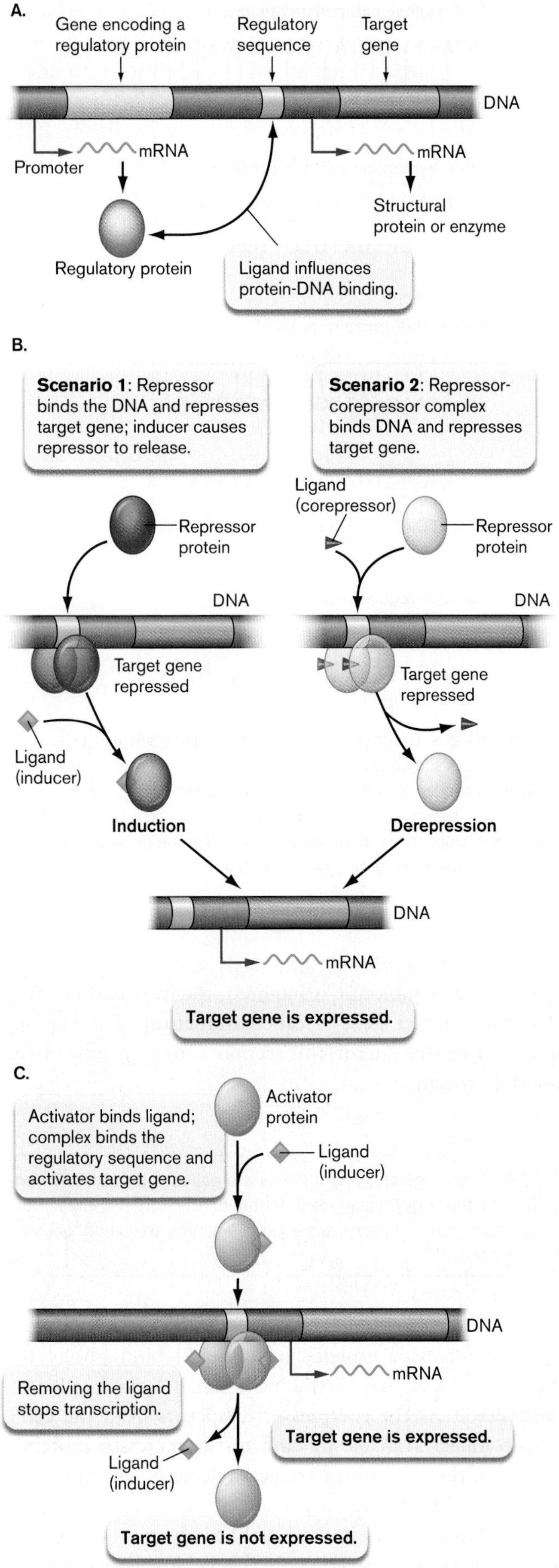

FIGURE 10.3 ■ General aspects of transcriptional regulation by repressor and activator proteins. **A.** Schematic of a regulatory system. The product of the regulatory gene (the regulatory protein) binds to DNA sequences near the promoter of the target gene and controls whether transcription occurs. Specific ligands influence binding. **B.** Repressor proteins bind to DNA sequences and prevent transcription. **C.** Activator proteins generally bind to specific chemical ligands in the cytoplasm before the protein can bind to DNA sequences near target genes. Activator proteins stimulate transcription.

Thought Question

10.3 Operators and repressors were discovered before promoters (the DNA sequences recognized by the sigma factor of RNA polymerase). Why do you think it took longer to discover the promoters? *Hint:* Finding elements of transcription initiation using genetics involves analysis of mutants; think about the phenotype of a repressor mutant compared to that of a promoter mutant.

Activator proteins also bind DNA, but they stimulate transcription by contacting an RNA polymerase positioned at a nearby promoter, spurring it to initiate transcription (**Fig. 10.3C**). Most activator proteins bind poorly to DNA sequences, unless their ligand is present. This ligand is called an inducer because it stimulates binding of the protein to the DNA. Note that both activators and repressors may have inducers, but the effect of inducer binding on the regulator's ability to bind DNA is opposite for the two classes of regulators (compare **Fig. 10.3B**, scenario 1, and **Fig. 10.3C**). When the intracellular concentration of inducer falls, the activator protein (without inducer) either leaves the DNA or moves to a nearby site from which it can no longer contact RNA polymerase. As a result, transcription of that target gene decreases or stops entirely.

Sensing the Extracellular Environment

Sensing what goes on outside the cell is more challenging than sensing intracellular conditions because intracellular regulatory proteins cannot reach through the membrane and interact with what is outside the cell. Instead, microbes rely on membrane-embedded signaling molecules to tell the cell what is happening outside. A common mechanism for collecting and transmitting information from outside the cell relies on two-member protein phosphorylation relay systems called **two-component signal transduction systems**. Each two-component system regulates a different set of genes. The first protein in each relay, the **sensor kinase**, spans the membrane (**Fig. 10.4**). A kinase transfers a phosphoryl group from ATP to a protein. The sensory domain of most sensor kinase proteins contacts the outside

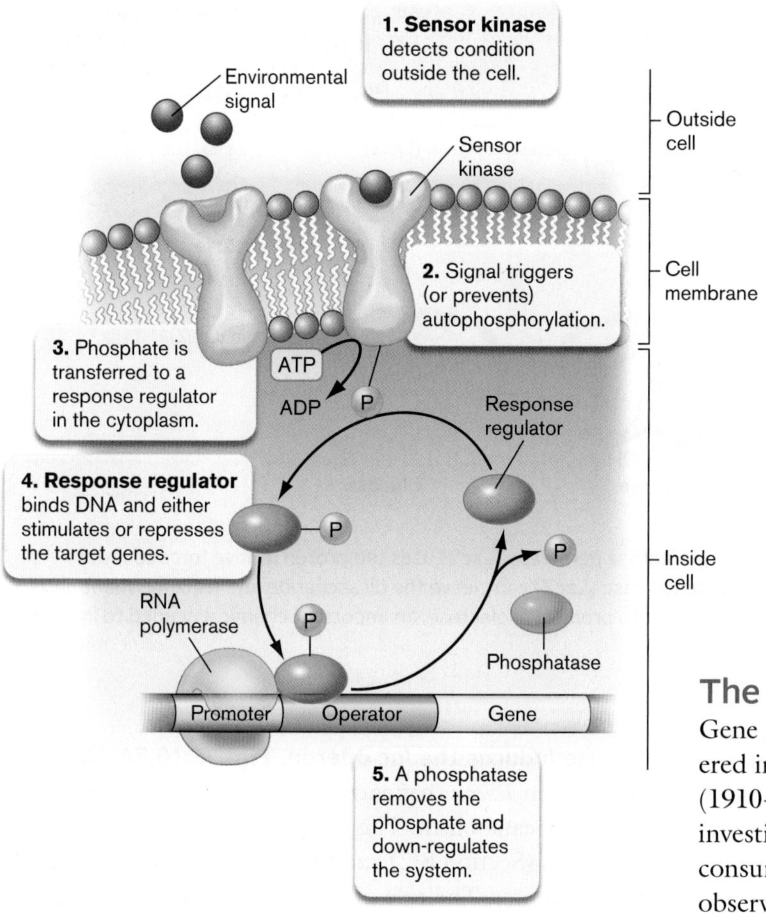

FIGURE 10.4 ▪ Two-component signal transduction systems sense the external environment. The first component of the system, a transmembrane sensor kinase protein, senses an environmental condition outside the cell (as in Gram-positive bacteria) or in the periplasm (as in Gram-negative bacteria). The sensor kinase sends a signal that travels to the second component, the response regulator, which then either stimulates or represses the target genes.

environment (or periplasm), while the other end (the kinase domain) protrudes into the cytoplasm.

Each sensor protein of a two-component system recognizes a different molecule or condition (for example, PhoQ in *Salmonella* senses magnesium). Once activated, the external sensory domain triggers a conformational change in the kinase domain that activates a self-phosphorylation reaction. Then, like two relay runners passing a baton, the phosphorylated sensor kinase protein passes the phosphate to activate a cognate (matched) cytoplasmic protein called a **response regulator**. The phosphorylated response regulator commonly binds to regulatory DNA sequences in front of one or more specific genes and activates or represses expression of those genes.

Note that response regulators are controlled by phosphorylation—a covalent modification. Compare this form of regulation with the earlier examples of repressors and activators that are controlled by noncovalent interactions with ligands.

The two-component system is down-regulated when a phosphatase cleaves the phosphate from the response regulator. For some two-component systems, the sensor kinase can serve as the phosphatase that resets the response regulator. Note that in these systems, the phosphate is not transferred back to the sensor kinase but is simply cleaved from the response regulator.

Thought Question

10.4 The transmembrane sensor kinase could have a direct role in gene expression, if its cytoplasmic domain has the ability to bind DNA. Why, then, might it be advantageous for the cell to use the response regulator as an intermediate in this signaling process? *Hint:* Consider the spatial organization of the cell.

The Lactose Operon

Gene regulation, a universal feature of all life, was discovered in bacteria. In 1961, French scientists Jacques Monod (1910–1976) and François Jacob (1920–2013; **Fig. 10.5**) investigated the enzymes that enable *Escherichia coli* to consume the carbohydrates glucose and lactose. They observed that the enzymes that metabolize glucose were always present (that is, constitutive) in the cell, even if glucose was absent. In contrast, the enzyme that metabolized lactose (beta-galactosidase) was produced only when the

KEYSTONE PRESS/ALAMY STOCK PHOTO

FIGURE 10.5 ▪ Discoverers of gene regulation. Jacques Monod (left), André Lwoff (center), and François Jacob (right) in 1965. This trio of scientists worked at the Pasteur Institute and won the 1965 Nobel Prize in Physiology or Medicine for their groundbreaking work on induction and gene regulation.

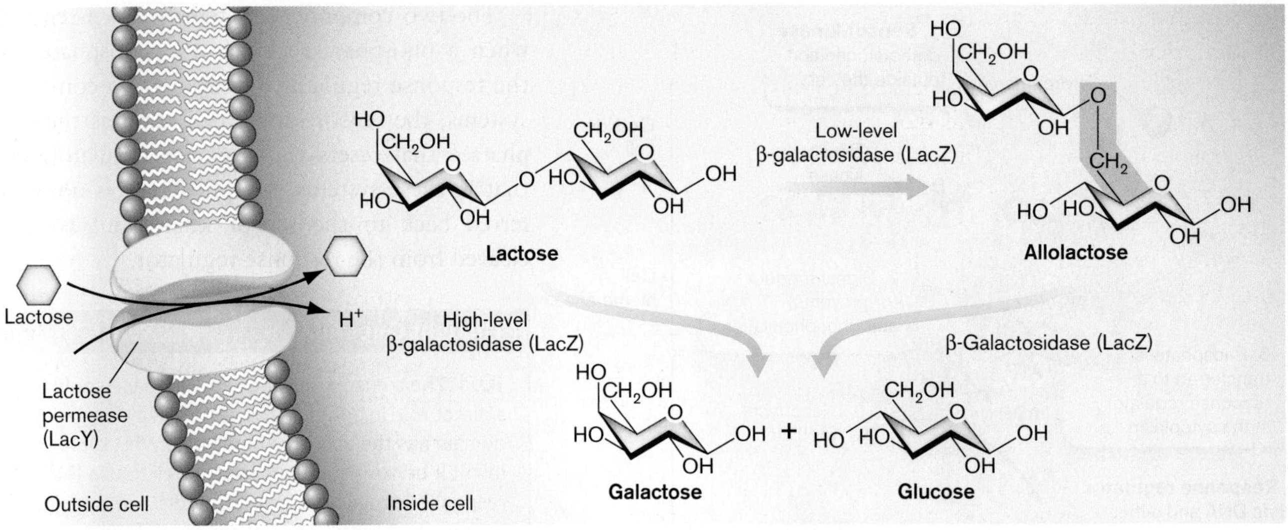

FIGURE 10.6 ▪ Lactose transport and catabolism. A dedicated lactose permease (LacY) uses the proton motive force to move lactose (and a proton) into the cell. Once there, the enzyme beta-galactosidase (LacZ) can cleave the disaccharide into its component parts (galactose and glucose) or alter the linkage between the monosaccharides to produce allolactose, an important chemical needed to induce the genes that encode this pathway.

researchers added lactose. Monod and Jacob coined the term "induction" to describe this phenomenon of beta-galactosidase regulation. This groundbreaking discovery of induction launched the field of gene regulation, a scientific realm where microbes continue to surprise us. For recognition of their pioneering research, Monod and Jacob were awarded a Nobel Prize, which they shared with André Lwoff (1902–1994; Fig. 10.5) for his study of phage lysogeny.

It took many years after Monod and Jacob's initial discovery to learn exactly how the expression of the lactose-degrading enzyme beta-galactosidase is regulated. As we will see, regulation involves both repressor and activator proteins that control gene expression, as well as protein interactions at the membrane that control lactose uptake. Many of the concepts presented here apply to numerous other bacterial gene regulatory systems, including some required for *Erwinia* infection of plants and *Streptococcus pneumoniae* infection of humans.

Lactose catabolism. Lactose is a disaccharide sugar made of glucose and galactose that can be used as a carbon and energy source (**Fig. 10.6**). The *E. coli* integral membrane protein that imports (transports) lactose from the extracellular environment is LacY. The enzyme that subsequently cleaves lactose into glucose and galactose is a beta-galactosidase (called LacZ). The products, glucose and galactose, are subsequently degraded by the enzymes of glycolysis to harness energy and capture carbon (see Section 13.4). Without LacY and LacZ, the catabolic energy of lactose is unavailable to the cell.

Lactose induces the *lac* operon. Figure 10.7A ▶ shows the genes in *E. coli* that encode the simple regulatory circuit for lactose catabolism. The genes *lacZ*, *lacY*, and *lacA* form an operon (Section 8.2) and are cotranscribed from a common promoter. The role of *lacA*, which encodes thiogalactoside transacetylase (LacA), is unclear. It is not needed to ferment lactose, but it may detoxify a harmful by-product of lactose metabolism.

Note: "Lactose operon," "*lac* operon," and "*lacZYA* operon" all refer to the same system.

In the absence of lactose, the *lac* operon is transcribed at extremely low levels (fewer than ten molecules of LacZ per cell). The reason for the low expression is that transcription of *lacZYA* is repressed by the protein product of the regulator gene *lacI* (**Fig. 10.7B**). The *lacI* gene is situated immediately upstream of *lacZYA* and is transcribed from a different promoter. A tetramer of LacI repressor protein forms in the cell and binds to two operator regions of DNA. One operator sequence is called *lacO*, which partially overlaps the *lacZYA* promoter (P_{lacZYA} or *lacP*; **Fig. 10.7A** and **B**). The second operator site is found within *lacI* and is called $lacO_I$ (**Fig. 10.7A**). The *lac* operators control whether the three structural genes are transcribed. The LacI tetramer simultaneously binds the *lacO* and $lacO_I$ operators (a dimer at each site). As a result, the intervening DNA loops out (**Figs. 10.7B** and **10.8**) and prevents RNA polymerase from continuing transcription into the structural *lacZYA* genes.

Once lactose is added, the *lac* operon is expressed at 100-fold higher levels. How does lactose get into the cell to induce the operon if the operon encoding its transport protein is repressed? As noted already, when repressed, the *lacZYA* operon is transcribed at a low level. (Most genes are expressed at low constitutive levels when uninduced.) This means that a small amount of the lactose transporter LacY and the beta-galactosidase LacZ are made. In the presence of lactose outside the cell, LacY transports the lactose into the cell, where LacZ catalyzes the hydrolysis of lactose. But LacZ also carries out a second reaction. Instead of all the lactose being hydrolyzed into glucose and galactose, a small amount of lactose is rearranged into a molecule called allolactose (whose structure is shown in **Fig. 10.6**). Interestingly, it is allolactose rather than lactose that activates the *lac* operon. Allolactose binds the repressor and "unlocks" the protein (by altering the conformation of LacI) so that it is released from the operator (**Fig. 10.7C**). Once this happens, RNA polymerase guided by sigma factor (not shown in **Fig. 10.7C**) can find the *lac* promoter sequences and initiate the transcription of large amounts of the *lacZYA* structural genes. A rapid increase in lactose metabolism results.

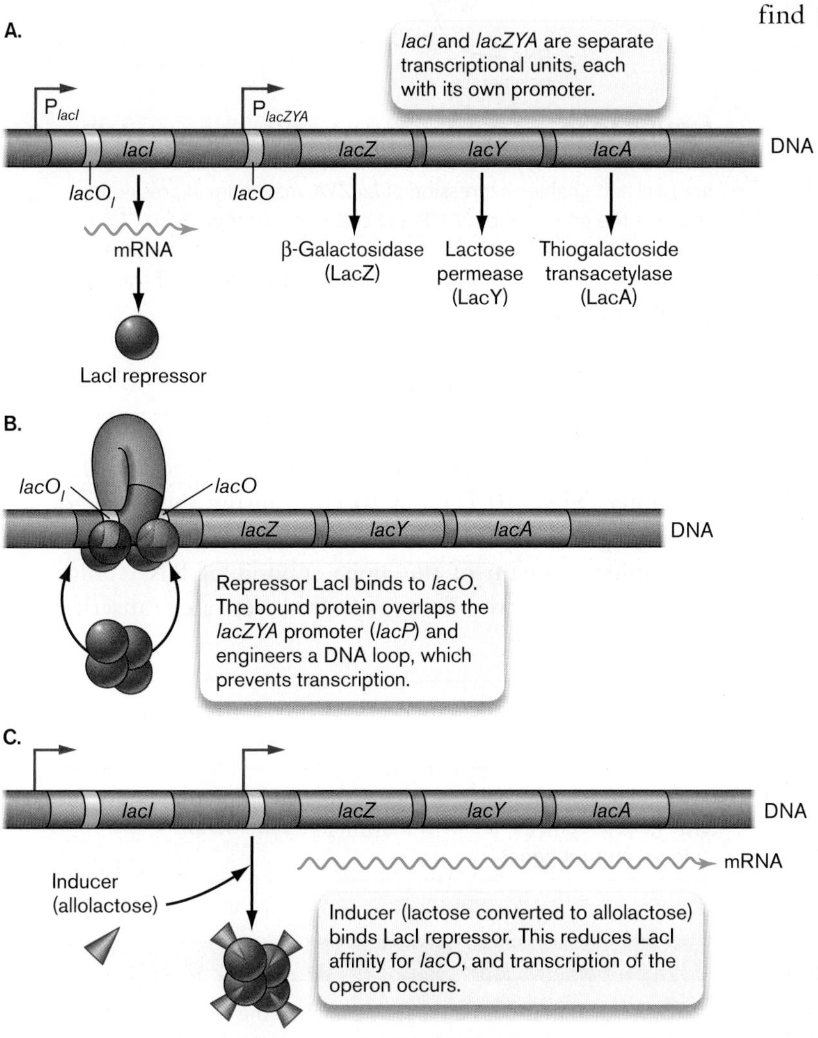

cAMP and cAMP receptor protein stimulate transcription. Another important mechanism that governs the level of *lacZYA* transcription in *E. coli* involves a small molecule called cyclic AMP (cAMP), which accumulates when a cell is starved for carbon. Cyclic AMP is a derivative of AMP (adenosine monophosphate) in which the 5′ phosphate is linked to the 3′ OH group of the ribose, making a cyclic structure (**Fig. 10.9**). Intracellular cAMP levels fluctuate with the activities of enzymes that synthesize (adenylyl cyclase), degrade (phosphodiesterases), or export this molecule. Various internal and external signals alter the balance of these processes (see **eTopic 10.1**).

As it accumulates, cyclic AMP controls the expression of many genes by combining with a dimeric regulatory protein called cAMP receptor protein (CRP). The cAMP-CRP complex binds to specific DNA sequences located near many bacterial genes and modifies their transcription, usually acting as an activator. This is the case for the *lacZYA* operon, which has a CRP-binding site just upstream of the promoter (**Fig. 10.7D**).

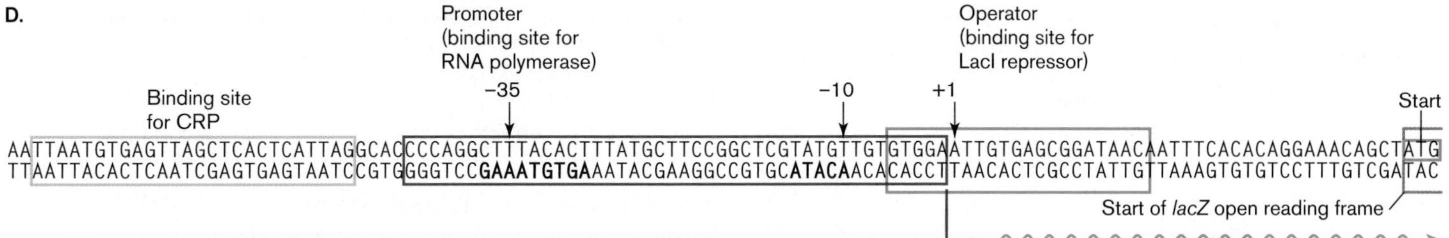

FIGURE 10.7 ▪ Transcriptional induction of the lactose operon. **A.** Organization of the operon. Bent arrows mark promoters. Green indicates LacI protein-binding sites on DNA. **B.** The LacI tetrameric repressor binds to specific DNA sites (the operator: *lacO*). **C.** The inducer allolactose (an altered form of lactose made by low levels of beta-galactosidase) removes the repressor LacI and allows expression of *lacZYA*. **D.** DNA sequence of the *lac* control region.

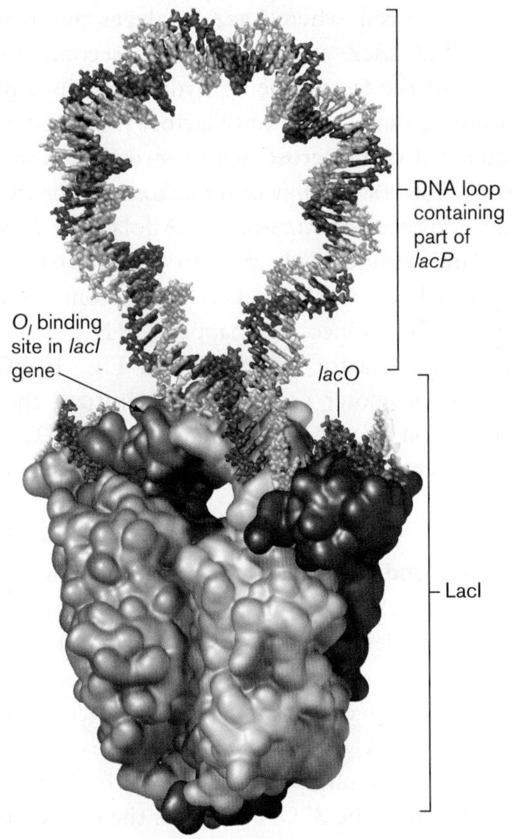

FIGURE 10.8 ▪ Regulatory protein interactions with DNA at the *lacZYA* control region. LacI repressor binds to two operator regions, *lacO* and $lacO_I$, so that the DNA forms a loop. (PDB codes: 1Z04, 1K8J)

Note: The original name for CRP was "catabolite activator protein" (CAP), which reflected its role in catabolite repression (described in the text). Though some investigators still use the "CAP" designation, the term "CRP" is generally preferred today. Thus, the gene encoding the protein is called *crp*.

How does the cAMP-CRP complex ultimately activate the expression of *lacZYA*? RNA polymerase cannot easily form an open complex (see Section 8.2) and is essentially stuck at *lacP*, even in the absence of a LacI repressor. This problem is overcome when the cAMP-CRP

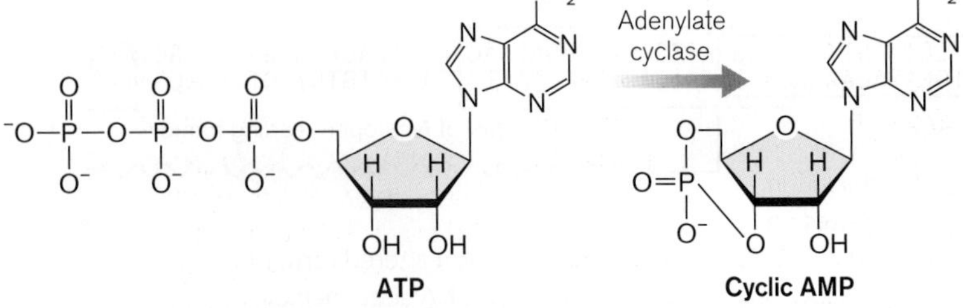

FIGURE 10.9 ▪ Cyclic AMP (cAMP) is derived from ATP and produced by adenylate cyclase.

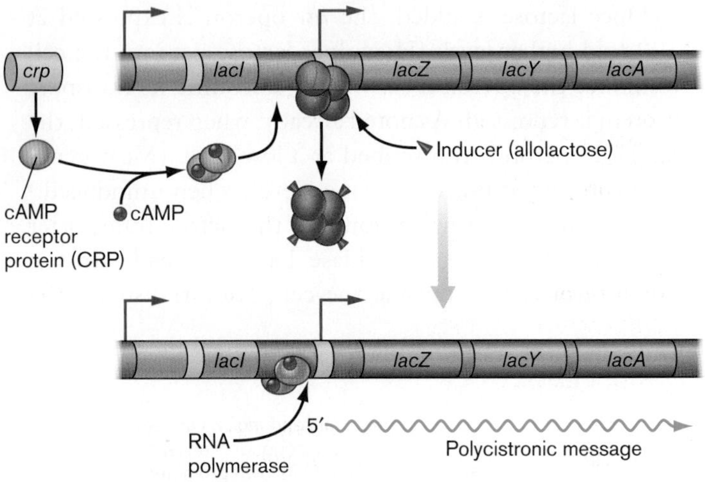

FIGURE 10.10 ▪ Activation of the *lacZYA* operon by cAMP and CRP. Although the inducer allolactose removes the repressor LacI and enables expression of *lacZYA*, maximum expression requires the presence of cAMP and cAMP receptor protein (CRP), which bind to a separate site at the *lacZYA* promoter. Bound CRP can interact with RNA polymerase and increase the rate of transcription initiation.

complex binds to the region upstream of the *lacZYA* promoter (**Figs. 10.7D** and **10.10**), causing the DNA to bend (**Fig. 10.11**). CRP then directly interacts with the C-terminal domain of the alpha subunit of RNA polymerase bound at the *lac* promoter to activate transcription (**Fig. 10.11**).

Note: Recall that the protein product of a gene is written without italics, and usually the first letter is capitalized. Thus, *lacZ* is the gene, but LacZ is the protein. In the presentation of a genotype, gene names written with a superscript "+" are considered wild-type genes. Gene names written without a superscript "+" are considered mutant genes.

Thought Questions

10.5 Null mutations completely eliminate the function of a given mutated gene. Predict the effects of the following null mutations on the induction of beta-galactosidase by lactose, and predict whether the *lacZ* gene is expressed at high or low levels in each case. The inactivated, mutant genes to consider are *lacI*, *lacO*, *lacP*, *crp*, and *cya* (the gene encoding adenylyl cyclase). What effects will those mutations have on catabolite repression?

10.6 Predict what will happen to the expression of *lacZ* when a second copy of the *lac* operon region containing various mutations is present on a plasmid. The genotypes of these partial diploid strains are presented as chromosomal genes/plasmid gene. (a) $lacI^- \ lacO^+P^+Z^+Y^+A^+$/plasmid $lacI^+$; (b) $lacO^- \ lacI^+P^+Z^+Y^+A^+$/plasmid $lacO^+$; (c) $crp^- \ lacI^+O^+P^+Z^+Y^+A^+$/plasmid crp^+.

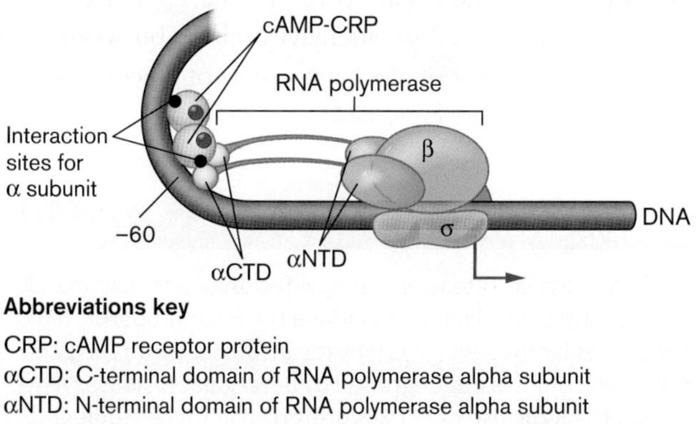

FIGURE 10.11 ■ CRP interactions with RNA polymerase. Promoters like the one at the *lacZYA* operon possess a CRP-binding site positioned about −60 bp from the transcriptional start. CRP on these promoters interacts with the alpha subunit C-terminal domain (αCTD) of RNA polymerase. *Source:* Modified from Moat et al. 2002. *Microbial Physiology*, 4th ed. Wiley-Liss.

Glucose represses the *lac* operon. What happens if, in addition to lactose, the medium contains an alternative carbon source, such as glucose? Enzymes for glucose catabolism (glycolysis) are always produced at high levels because glucose is the favored catabolite (carbon source), providing the quickest source of energy. Many carbohydrates, including lactose, must first be converted to glucose to be catabolized. So, in the interest of greater efficiency, *E. coli* avoids inducing the *lac* operon while glucose is present. The phenomenon is known as **catabolite repression**, in which the presence of a more favorable catabolite (commonly glucose) prevents the expression of operons that enable catabolism of a second carbohydrate.

When glucose and lactose are both present in the medium, cells initially grow by breaking down glucose until the glucose is depleted. Growth temporarily stops while lactose present in the medium induces the *lacZYA* operon. Once induced, cells can consume lactose and resume growth. The resultant biphasic growth curve is often called **diauxic growth** (**Fig. 10.12A**). But if lactose was present from the start, why was *lacZYA* turned off? It was turned off because glucose indirectly prevents the induction of *lacZYA*. **Figure 10.12B** illustrates that even when *lacZYA* is already induced, adding glucose stops (or represses) induction.

Catabolite repression via inducer exclusion. Failure of lactose to induce *lacZYA* during growth on glucose (**Fig. 10.12B**, red horizontal line) is due mainly to the fact that growth on glucose keeps lactose out of the cell. This phenomenon is known as **inducer exclusion**. If lactose cannot enter the cell, the *lacZYA* operon cannot be induced. The key to inducer exclusion is that a component of the glucose transport system (phosphotransferase system, or PTS; see **eTopic 4.1**), while transporting glucose, will bind to and inhibit LacY permease (**Fig. 10.13**).

The PTS transfers a phosphate from phosphoenolpyruvate (PEP) along a series of proteins to glucose during transport. When glucose is present (**Fig. 10.13A**), the glucose transport proteins continually transfer phosphate to glucose and, so, are usually left without phosphate. Unphosphorylated Enzyme IIA interacts with LacY in

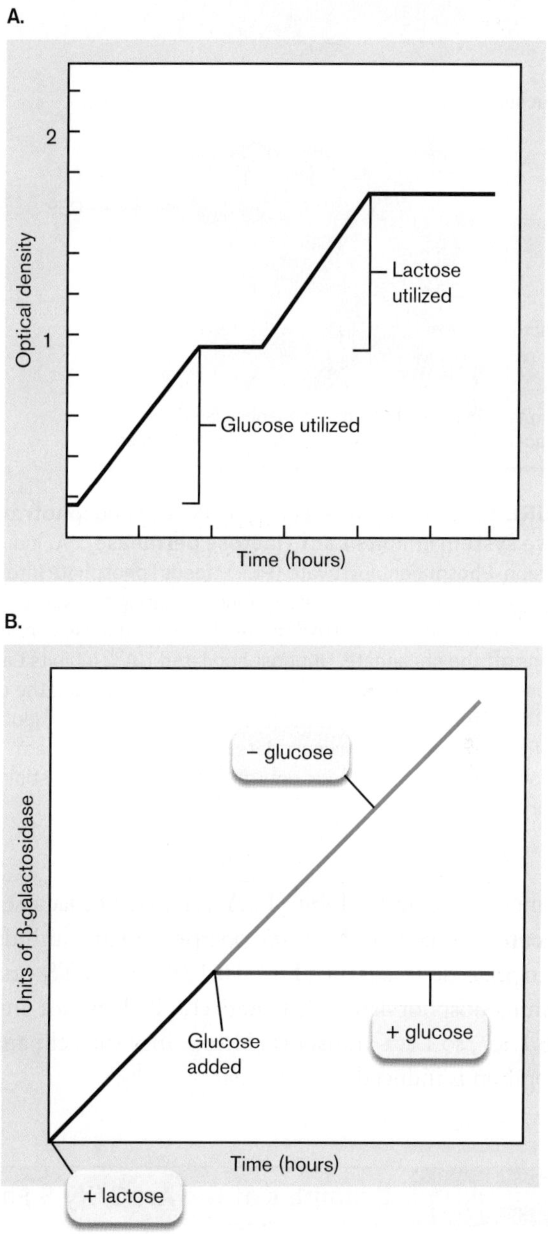

FIGURE 10.12 ■ Catabolite repression of the *lacZYA* operon. A. The diauxic growth curve of *E. coli* growing on a mixture of glucose and lactose. B. Glucose repression of LacZ (beta-galactosidase) production. Lactose was added at the beginning of the experiment to parallel cultures, and beta-galactosidase activity was measured. At the point indicated, glucose was added to one culture. From that point on, synthesis of beta-galactosidase continued to increase in the culture that lacked glucose but stopped in the culture that contained glucose.

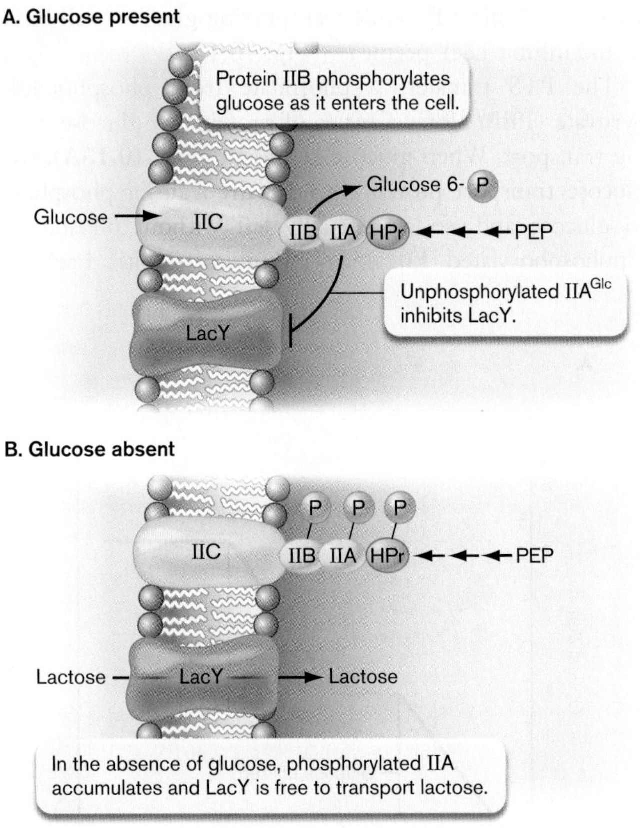

FIGURE 10.13 ■ Glucose transport via the phosphotransferase system inhibits LacY (lactose permease). **A.** Inducer exclusion. Phosphoenolpyruvate (PEP) "feeds" phosphate into the PTS, which relays the phosphate to glucose during transport. The level of unphosphorylated IIA^{Glc} is high because glucose continually siphons off the phosphate. Unphosphorylated IIA^{Glc} inhibits LacY (lactose permease) activity to keep lactose from entering the cell. **B.** In the absence of glucose, the phosphorylated forms of glucose-specific IIA^{Glc} and IIBC^{Glc} accumulate and cannot inhibit LacY. LacY transports lactose, and the *lac* operon is induced. HPr = histidine-rich protein.

the membrane and inhibits LacY activity. So, lactose cannot enter the cell, and the *lac* operon remains uninduced. When glucose is absent (**Fig. 10.13B**), the PTS proteins remain phosphorylated; Enzyme IIA-P does not interact with LacY, so LacY transports lactose into the cell; and the *lac* operon is induced.

Transport of some sugars through the PTS also affects the synthesis of cAMP by adenylyl cyclase, but control of cAMP production is not the major cause of glucose/lactose diauxic growth (see **eTopic 10.1**).

Thought Question

10.7 Researchers often use isopropyl-β-D-thiogalactopyranoside (IPTG) rather than lactose to induce the *lacZYA* operon. IPTG resembles lactose, which is why it can interact with the LacI repressor, but it is not degraded by beta-galactosidase. Why do you think the use of IPTG is preferred in these studies?

The AraC/XylS Family of Transcriptional Regulators

What could be better than having a regulatory protein that either represses (like LacI) or activates (like CRP) expression of an operon? How about a regulator that can repress and activate gene expression, depending on whether a carbohydrate substrate is available? A regulator this versatile provides very tight control over operon expression. One such regulator, called AraC, regulates the genes encoding arabinose catabolism. When arabinose is absent, AraC represses expression of the genes that break down arabinose, but when it is present, AraC activates these same genes. The products of these genes ultimately convert the five-carbon sugar L-arabinose to D-xylulose 5-phosphate, an intermediate in the pentose phosphate shunt, a pathway that provides reducing energy for biosynthesis (see Section 13.4).

Bioinformatic computer analysis of microbial genomes reveals a family of over 10,000 regulators with homology to AraC and the closely related XylS activator (xylose catabolism). Although first described for sugar metabolism of a harmless strain of *E. coli*, many of the AraC/XylS regulators are found in pathogens and control genes involved in virulence (**Table 10.1**). For example, the plague bacillus *Yersinia pestis* possesses an AraC-type regulator YbtA that governs uptake of iron. This regulator helps *Y. pestis* extract iron from blood plasma during infection.

TABLE 10.1 Examples of the AraC/XylS Family of Transcriptional Regulators

Regulator	Organism	Function
ExsA	*Pseudomonas aeruginosa* (human pathogen)	Controls type III secretion system in response to Ca^{2+}
NitR	*Rhodococcus rhodochrous* (plant pathogen)	Regulates synthesis of indole-3-acetic acid
ToxT	*Vibrio cholerae* (human pathogen)	Controls several virulence genes, including one encoding cholera toxin
TxtR	*Streptomyces scabies* (taproot pathogen)	Controls synthesis of thaxtomin, a plant toxin
YbtA	*Yersinia pestis* (human pathogen)	Controls iron uptake transporters

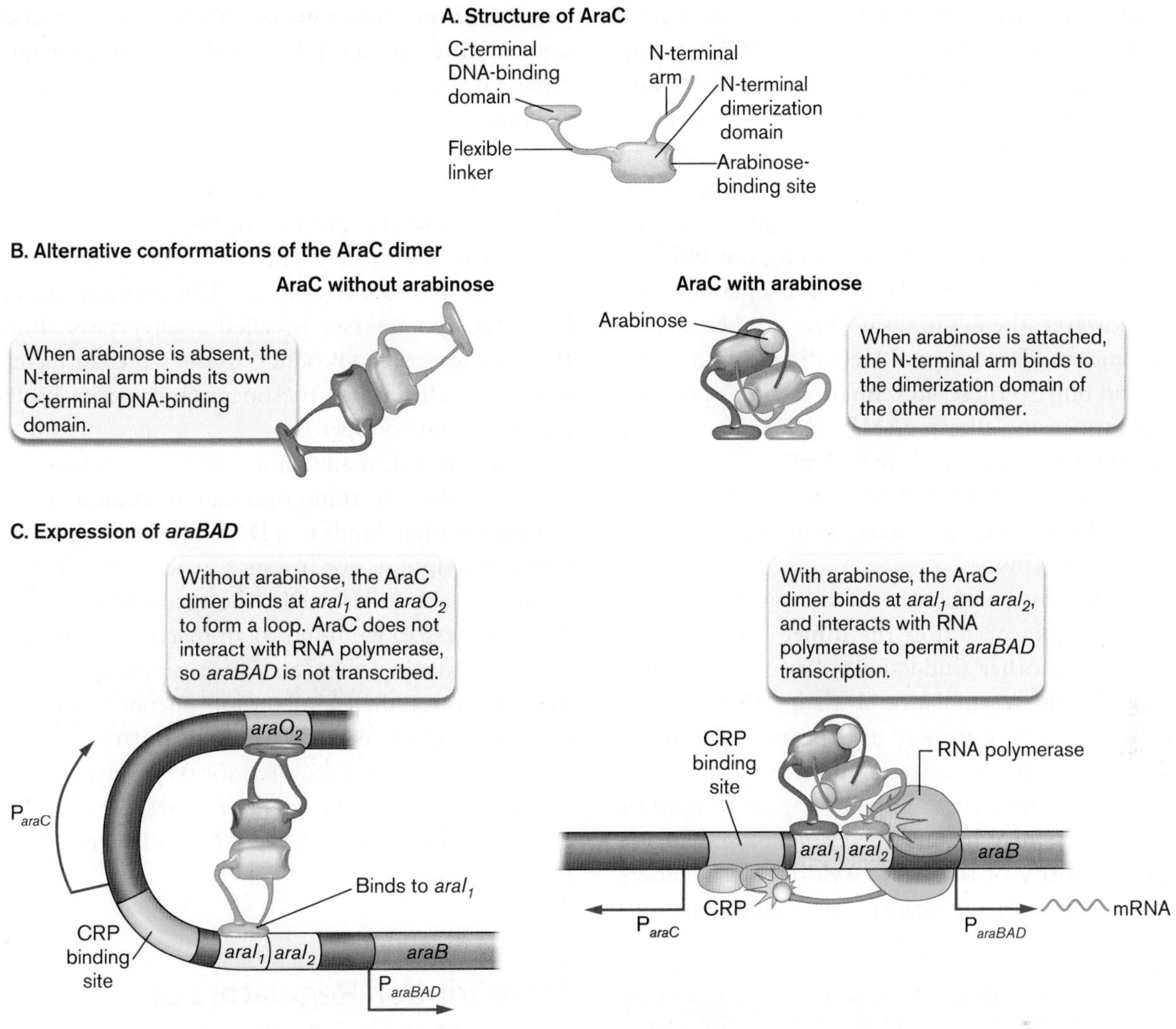

FIGURE 10.14 ■ Regulation of the *araBAD* operon by the AraC regulator. A. View of AraC showing dimerization and DNA-binding domains. B. Alternative conformations of the AraC dimer. C. The different conformations change the location of where the dimer can bind DNA. The region shown is about 400 bp. Note that O_2 is an operator, and I_1 and I_2 are other DNA-binding sites that, when occupied by AraC, induce expression. CRP can bind and stimulate expression when the dimer is bound to arabinose, but it is blocked by the DNA loop when the dimer lacks arabinose.

The AraC/XylS advantage. One advantage of AraC/XylS family regulators, as displayed by the AraC prototype, is that they remain affixed to their target genes. By contrast, the LacI repressor must fully dissociate from operator DNA during induction and disperse throughout the cell. Reestablishing repression requires slow, random diffusion of bulky LacI proteins back to the chromosome and the *lacO* DNA sequence. This takes some time. The AraC/XylS family strategy keeps the regulatory protein bound near target operons. The regulators simply shuttle back and forth between repressor and activator DNA-binding sites, shortening the delay between induction and repression. The known small chemical inducer molecules for these regulators quickly diffuse through the cytoplasm to find their cognate regulatory proteins already camped at target promoters.

The AraC strategy. The model for regulation by AraC/XylS regulators is based primarily on AraC itself, the most intensely studied family member. AraC is a 33-kDa protein with a DNA-binding C-terminal domain attached by a flexible linker peptide to the N-terminal dimerization domain (**Fig. 10.14A**). AraC forms a dimer in vivo that can assume one of two conformations, depending on whether arabinose is available. When arabinose is absent, the dimer exists in a rigid, elongated form because the N-terminal arm of each monomer binds to its own C-terminal domain. This form represses expression of the *araBAD* operon for arabinose degradation. When arabinose is present, however, the dimer assumes a more compact form because the N-terminal arm of one monomer binds to the N-terminal domain of the other monomer (**Fig. 10.14B**). This compact form activates expression of *araBAD* (**Fig. 10.14C**).

How does AraC act as both repressor and activator? AraC is able to bind to three important DNA-binding sites. Two of them, O_2 and I_1, are widely separated and flank a binding site for cAMP-CRP that, as discussed earlier in this section, is a global activator of many bacterial genes, including *araBAD*. The elongated AraC dimer (no arabinose) can bind these two sites. The result is a looped DNA structure that blocks the CRP-binding site and limits activation by cAMP-CRP. The *araBAD* operon is not expressed. However, the I_1 site sits next to another site, I_2, near the promoter. The compact AraC dimer (with arabinose) can bind only to these sites. Binding to I_1I_2 prevents DNA loop formation, allows cAMP-CRP access to the CRP DNA-binding site, and brings AraC close enough to physically contact an RNA polymerase already bound (but stuck) at the *araBAD* promoter. Transcription of the *araBAD* operon begins.

According to this model, the DNA region is rarely free of AraC. Arabinose moves the dimer from one site (repression) to the other (induction). AraC-like proteins show strong family resemblance at their DNA-binding domains, but homology usually disappears at the other end of the protein, the dimerization domain. The dimerization domain is where these proteins appear to bind or respond to a particular ligand (for example, arabinose). For the vast majority of the AraC/XylS family members in bacterial genomes, the identity of the ligand remains a mystery.

Thought Question

10.8 When the *lacI* gene of *E. coli* is missing because of mutation, the *lacZYA* operon is highly expressed regardless of whether lactose is present in the medium. Judging by the illustration of arabinose operon expression in **Figure 10.14**, what do you think would happen to *araBAD* expression if AraC were missing? Why?

Repression of Anabolic (Biosynthetic) Pathways

Regulation of biosynthetic pathways has a fundamentally different objective than regulation of catabolic pathways. Cells respond to a carbon and energy source by inducing the machinery to catabolize the substrate. In contrast, anabolic pathways are repressed by high concentrations of their products, and thus wasteful synthesis of these products is avoided. This difference means that gene expression regulators are used differently for catabolic versus anabolic pathways.

Repressor proteins that control catabolic pathways, such as lactose degradation, typically bind the initial substrate or a closely related product (for example, allolactose in the case of the *lac* operon). Binding the substrate decreases repressor protein affinity for operator DNA. Thus, increased concentration of the substrate or inducer actually removes the repressor from the operator and derepresses expression of the operon (**Fig. 10.3B**, scenario 1).

In contrast, genes encoding biosynthetic enzymes are regulated by repressors (called inactive aporepressors) that must bind the end product of the pathway (for example, tryptophan for the *trp* operon) to become active repressors (**Fig. 10.3B**, scenario 2). The pathway product that binds the aporepressor is called a corepressor. Binding of the corepressor (end product) to the repressor increases the repressor's affinity for the operator sequence upstream of the target gene or operon.

As just noted, many amino acid biosynthetic pathways are controlled by transcriptional repression, in which a repressor protein binds to a DNA operator sequence to prevent transcription. For instance, when internal tryptophan levels exceed cellular needs, the excess tryptophan (acting as a corepressor) will bind to an inactive repressor protein, TrpR, converting it to an active repressor (**Fig. 10.15**). TrpR repressor then binds to an operator DNA sequence positioned upstream of the tryptophan (*trp*) operon, which encodes the enzymes required for tryptophan biosynthesis. Repressor bound to the operator represses expression of the *trp* operon by preventing the RNA polymerase from binding the promoter.

Transcription Regulators of Eukaryotes and Archaea

As in bacteria, transcription in eukaryotes is regulated by activators and repressors. In eukaryotes, the sites where activator proteins bind DNA are called enhancers. In contrast to the situation in bacteria, these activator binding sites are usually positioned thousands of bases upstream or downstream of the transcription start site. How can an activator protein regulate transcription when it is bound to DNA so far away?

Recall from Chapter 8 that eukaryotes use RNA polymerase II to transcribe the DNA of protein-coding genes into mRNA. The transcription initiation complex (Fig. 8.6) forms at the promoter upstream of the transcription start site. The TATA-binding protein (TBP) binds to the TATA box of the promoter and recruits RNA polymerase II and additional transcription factors to the site. This initiation complex is unstable by default and, without assistance, falls off the promoter before transcription can initiate. One class of transcription activators stabilizes the complex long enough that the polymerase can eventually "escape" the complex and proceed with transcription elongation.

Transcription activators at enhancers stabilize the initiation complex through a multiprotein bridge called

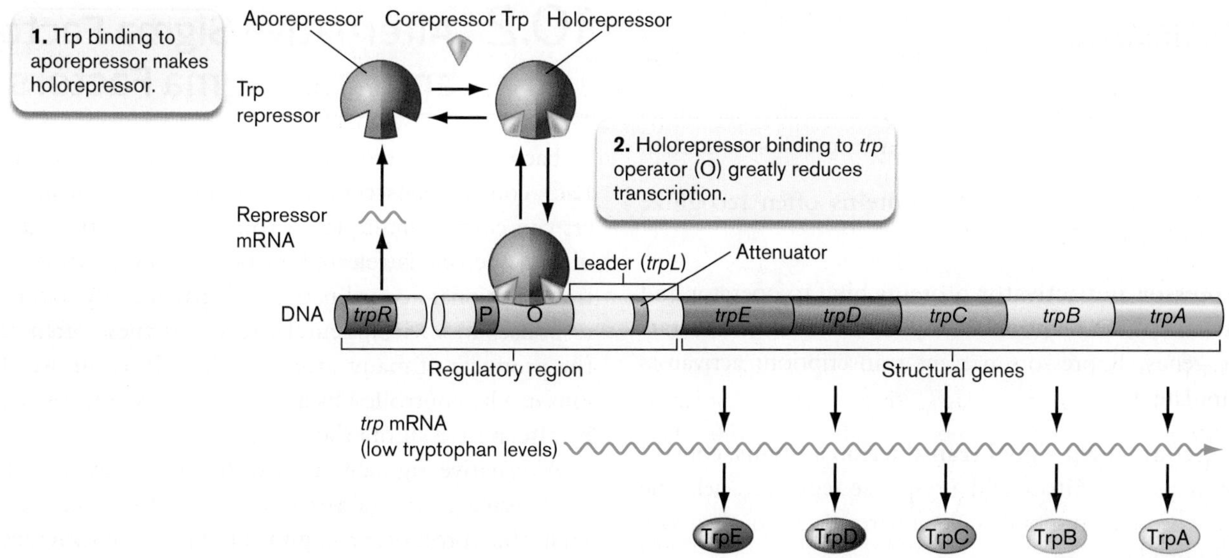

FIGURE 10.15 ▪ The tryptophan biosynthetic pathway in *E. coli* and repression of the *trp* operon. The tryptophan biosynthetic enzymes and their encoding genes are shown. TrpR aporepressor (inactive repressor) binds excess tryptophan when intracellular concentration exceeds need. The holorepressor (active repressor) then binds to the *trp* operator and greatly reduces transcription. Note the long polycistronic message in blue. Repression lowers expression about 100-fold.

Mediator. When the intervening DNA is looped out, Mediator can bind both the RNA polymerase II complex and the activator protein (**Fig. 10.16**). Mediator can bind a broad array of different activator proteins, and often binds to activators at multiple enhancer sites during transcription initiation.

Transcription regulation in eukaryotes has even greater complexity, through the dynamics of chromatin formation and remodeling. Nucleosomes are histone-DNA complexes that, when positioned at a promoter, can silence gene expression. Both activators and repressors can function in gene regulation by their ability to cause or prevent remodeling of these nucleosomes around the promoter region.

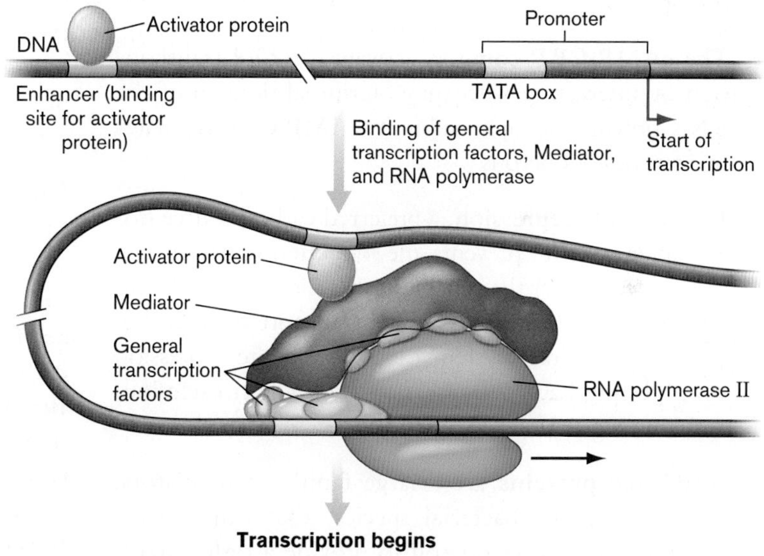

FIGURE 10.16 ▪ Eukaryotic gene activation at a distance through the Mediator connector. The gaps in the DNA indicate long stretches of intervening DNA between enhancer and promoter, often thousands of bases long. In this simplified example, a single activator binds an enhancer upstream of the promoter. In other cases, multiple activators are involved, binding enhancers located upstream as well as downstream of the gene. *Source:* Modified from Alberts et al. 2019. *Essential Cell Biology*, 5th ed., fig. 8–10.

How are genes regulated in archaea? Like the other two domains, archaea regulate their genes with both activators and repressor proteins. Recall from Chapter 8 that the RNA polymerase of archaea is structurally and functionally related to RNA polymerase II of eukaryotes (Fig. 8.11). Despite these similarities, regulation of transcription initiation in archaea lacks the complexity found in eukaryotes and in some aspects is more similar to regulation in bacteria. Archaea lack the Mediator complex of eukaryotes, and their activator proteins bind close to the promoter to help assemble the initiation complex. Their repressor proteins typically bind the promoter regions to prevent assembly of the initiation complex. Although archaea contain histones that package their DNA into nucleosomes, the involvement of nucleosome formation and remodeling in gene regulation appears to be less complex than in eukaryotes, and it may be restricted to only certain lineages of archaea.

To Summarize

- **Regulatory proteins** help a cell sense changes in its internal environment and alter gene expression to match.
- **DNA-binding domains of proteins** often recognize symmetrical DNA sequences.
- **Repressor and activator proteins** bind to operator and activator DNA sequences, respectively, in front of target genes. Repressors prevent transcription; activators stimulate it.
- **Two-component signal transduction systems**, consisting of a sensor kinase and a response regulator, help the cell sense and respond to its environment, both inside and outside.
- **The *lacZYA* operon of *E. coli* is controlled by repression and activation.** The **tetrameric LacI repressor** binds operator DNA sequences to prevent RNA polymerase from accessing the promoter. **Allolactose** (rearranged lactose) binds LacI, reduces repressor affinity for the operator, and allows induction of the operon.
- **The cAMP-CRP complex** activates *lacZYA* transcription by interacting with the C-terminal domain of the RNA polymerase alpha subunit. cAMP-CRP regulates many types of operons.
- **In catabolite repression**, a preferred carbon source (for example, glucose) prevents the induction of an operon (for example, *lac*) that enables catabolism of a different carbon source (lactose). Glucose transport through the phosphotransferase system causes catabolite repression by inhibiting LacY permease activity (inducer exclusion) and lowers cAMP levels.
- **AraC-like proteins** are a large family of regulators, present in many bacterial species, that can activate and repress a target operon to provide a tight on/off switch.
- **Many anabolic pathway genes** (for example, for tryptophan biosynthesis) are repressed by the end product of the pathway (for example, the amino acid), which binds to a corepressor that inhibits transcription.
- **Regulation of transcription in eukaryotes** is more complex than in bacteria, can involve activators that bind distant DNA sites called enhancers, and also involves extensive chromatin reorganization.
- **Archaeal gene regulation** involves activators and repressors that bind near the promoter, as in bacteria.

10.2 Alternative Sigma Factors and Anti-Sigma Factors

In bacteria, some regulatory proteins, such as the repressor LacI, control transcription of a single operon, whereas others can control multiple operons. The collection of coregulated operons is referred to as a **regulon**. Multi-operon transcriptional coordination is particularly useful for responses to environmental stresses, as these often require the products of many genes for the cell to survive. Regulons can be controlled by activators and repressors, but also by **alternative sigma factors**.

Alternative sigma factors differ from the "housekeeping" sigma factor (RpoD, or sigma-70) in several ways. First, their promoter recognition sequence is different from that of the housekeeping sigma factor (see Section 8.1), and when expressed, these sigma factors compete with the housekeeping sigma factor to deliver RNA polymerase to the promoters of their regulon. Second, whereas the housekeeping sigma factor is always expressed and functioning in the cell, the expression or activation of alternative sigma factors is usually triggered by an environmental change. The regulons controlled by these sigma factors typically help the cell survive this change. One example of a stress-induced alternative sigma factor, sigma S, is described next.

Sigma S and Its Regulon

Many Gram-negative bacteria such as *E. coli* have an alternative sigma factor (sigma S or RpoS) that is expressed in the stationary phase of growth. Its regulon protects the cell from starvation and other stresses that may occur during stationary phase, such as acid stress or oxidative stress. In large part, *E. coli* orchestrates the stationary phase–dependent accumulation of sigma S by modulating its degradation (protein degradation is discussed in Section 8.4). ClpXP protease degrades sigma S rapidly during exponential growth. When cells enter stationary phase or experience an environmental stress that slows growth, degradation of sigma S stops. Sigma S levels increase in these situations, and the increase triggers the expression of the stress survival regulon.

How are the genes of regulons, such as that of sigma S, identified in the lab? Historically, genes of regulons have been identified through genetic screens or gel electrophoresis of total cell proteins, which can be quite laborious to perform. Thanks to recent developments in nucleic acid techniques, regulons can now be identified more quickly. By comparing the expression of every gene in the genome, what is called the **transcriptome**, between the wild type and a mutant strain that lacks a regulator such as sigma S, one can identify regulon members.

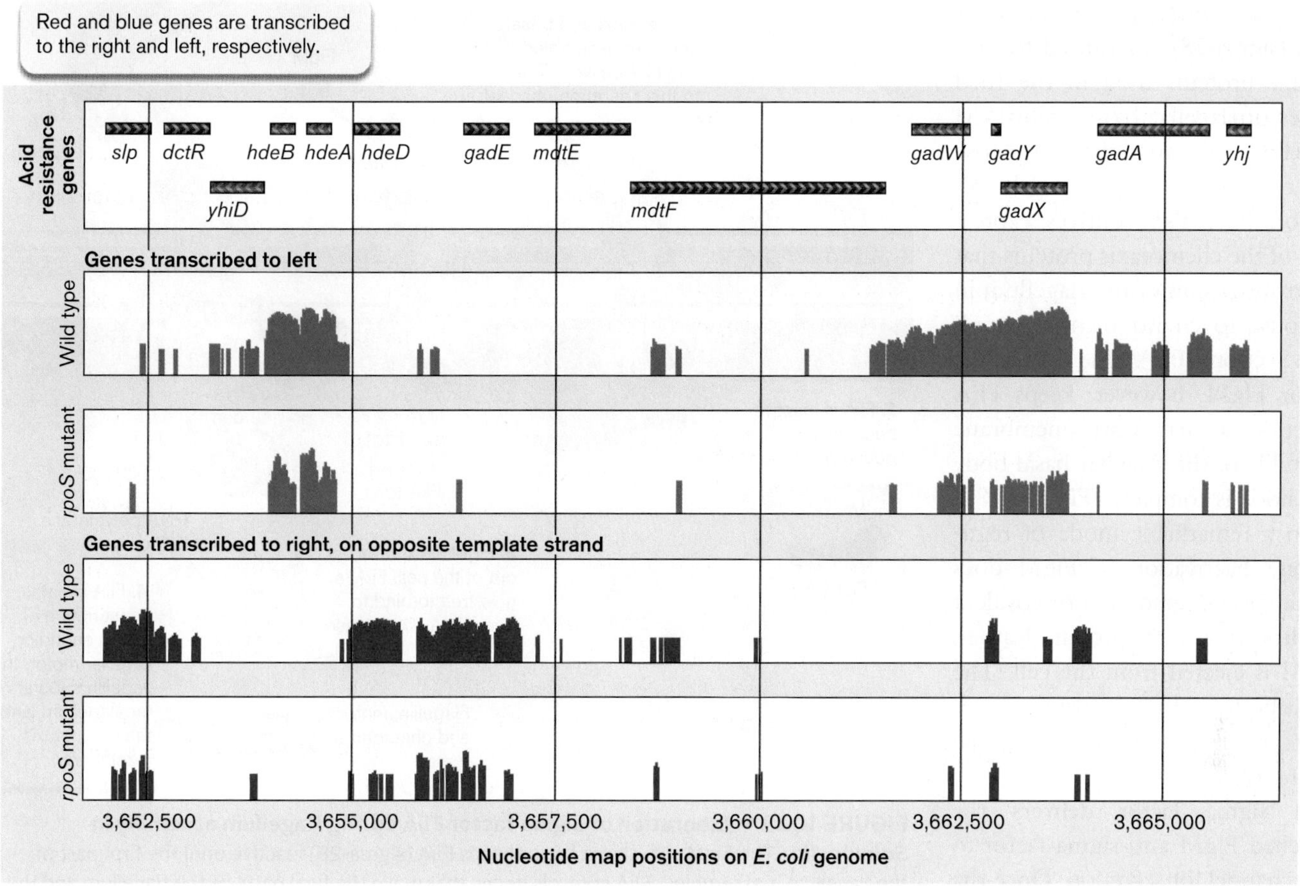

FIGURE 10.17 ■ Transcriptomics of *Escherichia coli* acid resistance genes. Deep sequencing was used to quantify the relative numbers of transcripts made from each gene in wild-type and *rpoS* mutant strains. Each vertical bar in the bottom four panels represents a short, partial read of a transcript. The experiments actually captured data from all genes expressed from the genome, but only some of these data are shown. *Source:* Modified from University of Oklahoma Gene Expression Database.

The state of the art in transcriptomic analysis is RNAseq, which is an RNA-based application of high-throughput DNA sequencing technology. Because most high-throughput sequencers can read only DNA (see Chapter 12), transcriptomes analyzed by RNAseq are first reverse-transcribed to DNA. Sequence reads are then compared to the reference genome in order to calculate transcript abundances for each gene.

Tyrrell Conway and colleagues at the University of Oklahoma used transcriptomics to identify acid resistance genes in *E. coli* that are members of the bacterium's sigma S regulon. Conway (now at Oklahoma State University) used RNAseq to quantify transcript levels of the known acid resistance genes in stationary-phase cells of wild-type and *rpoS* mutants of *E. coli* lacking sigma S. RNAseq provides only short reads of transcripts, but if all the individual reads are compiled together as in **Figure 10.17**, transcript abundances for each gene can be compared between the wild type and the *rpoS* mutant. The relative expression of the *dct*R, *gad*A, and *yhi*D acid resistance genes in the *rpoS* mutant strain (third and fifth panels in **Fig. 10.17**) was lower than the wild-type (second and fourth panels), indicating that these genes are under RpoS control.

Note: While the expression of a gene may change in the absence of a transcription regulator such as sigma S, one cannot automatically conclude that the regulator binds to the promoter region of that gene. The effect may be indirect if, for instance, the expression of the direct regulator of the gene is itself under the control of sigma S.

Regulation by Anti-Sigma Factors and Anti-Anti-Sigma Factors

Some sigma factors are controlled by **anti-sigma factor** proteins that inhibit sigma factor activity. Anti-sigma factor proteins bind specific sigma factors and block access to core RNA polymerase. The anti-sigma strategy prevents the expression of target genes until they are needed. Many mechanisms liberate sigma factors from anti-sigma factors. Liberation of sigma factors is an important part of the regulation of spore formation in *Bacillus subtilis*, as described in eTopic 10.2. Sigma factor release is also used to control the construction of flagella, complex multiprotein machines (Fig. 3.40). How the anti-sigma factor is eliminated in this latter example is quite impressive.

In *Salmonella*, the sigma factor FliA (sigma-28) is required to synthesize proteins used in the final stages of flagellar biosynthesis: the flagellin subunits that compose the flagellum and the flagellum's motor. FliA also controls expression of the chemotaxis proteins that control the spin of the flagellum in response to environmental stimuli (see Section 10.6). The anti-sigma factor FlgM, however, keeps FliA function at bay until membrane assembly of the flagellar basal body and hook is complete (**Fig. 10.18**).

In a remarkable mode of regulation, inactivation of FlgM does not involve degradation or covalent modification of the protein. Rather, FlgM is ejected from the cell. The basal body and hook form a secretory channel, and in an unusual mode of protein trafficking, the FliA sigma factor delivers the attached FlgM anti-sigma factor to this channel for secretion. Once the FlgM is ejected from the cell, FliA becomes free to direct transcription of the final set of flagellar assembly genes. The ability of the partially assembled flagellum to eject FlgM thus provides "proof" that flagellum assembly has progressed to the final phase of construction.

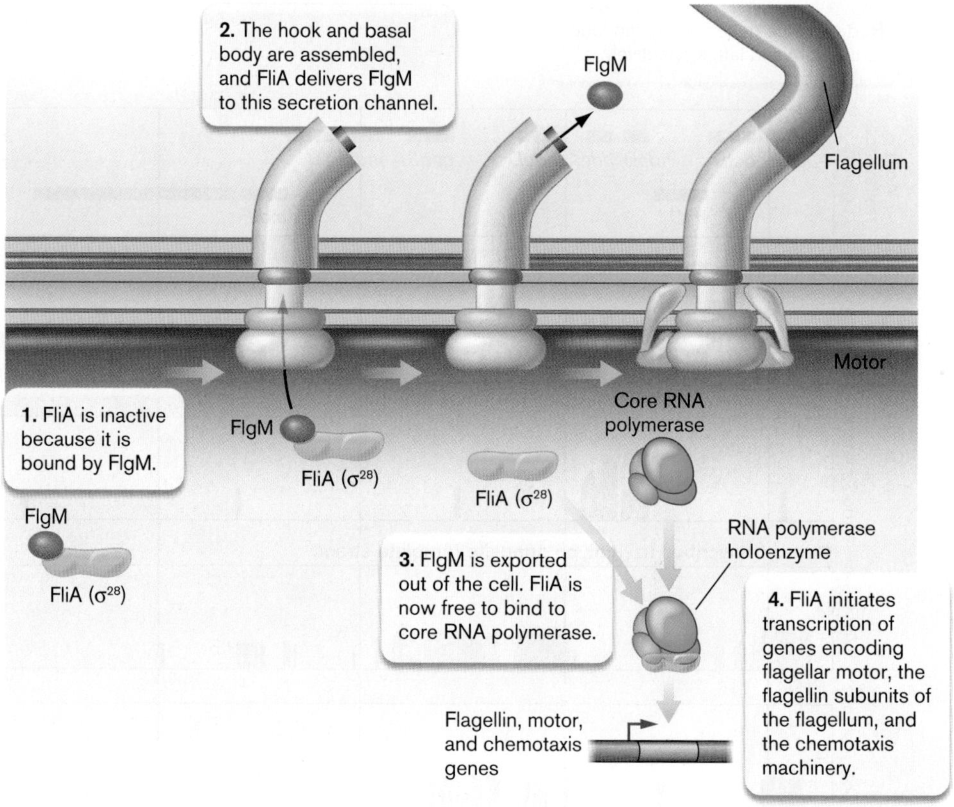

FIGURE 10.18 ■ Liberation of sigma factor FliA during flagellum assembly in *Salmonella*. The FlgM anti-sigma factor keeps FliA (sigma-28) inactive until the first part of the flagellum is assembled. FliA controls genes that make the final parts of the flagellum and the chemotaxis machinery. *Source:* Modified from Chevance and Hughes. 2008. *Nat. Rev. Microbiol.* **6**:455–465, fig. 3.

Anti-sigma factors can themselves be neutralized by **anti-anti-sigma factors** that bind the anti-sigma factor more tightly than sigma factor does. The anti-anti-sigma factor acts as a decoy to release the actual sigma factor from the anti-sigma factor. Freed sigma factor can then join core RNA polymerase and direct the transcription of target genes.

To Summarize

- **Changing the synthesis or activity of a sigma factor** will coordinately regulate a set of related genes termed the **regulon**.
- **The sigma S regulon protects cells** from starvation and other stresses during stationary phase.
- **Anti-sigma factors bind sigma factors** to prevent them from initiating transcription.
- **Anti-anti-sigma factors release sigma factors** from anti-sigma factors.

10.3 Regulation by RNA

Studies of the lactose and arabinose operons in *E. coli* established the paradigm of gene regulation at the step of transcription initiation. Subsequent studies have shown that bacteria have many ways to regulate gene expression after transcription begins. Regulatory sequences in the mRNA can cause premature termination of transcription or prevent translation of the mRNA into protein. In addition, regulatory RNAs transcribed from different promoters can influence the fate of the transcribed mRNA. We begin this section with a discussion of regulatory sequences within the transcribed mRNA.

Transcriptional Attenuation

In Section 10.1 we described how transcription of the *trp* operon is regulated by repressor binding to tryptophan, the biosynthetic product of the encoded proteins of the operon. Repression, however, is not the whole story in regulating the *trp* operon. Many amino acid biosynthetic operons, including the *trp* operon, have adopted a second strategy for down-regulating amino acid synthesis, which can be used alone or in conjunction with repression. This second mechanism is called **transcriptional attenuation**. Attenuation halts transcription in progress, before it even reaches the first gene. This mechanism affords the cell an even quicker response to changing amino acid levels than does simple repression.

Transcriptional attenuation was discovered by Charles Yanofsky (**Fig. 10.19A** ▶) and his colleagues at Stanford University. While examining the beginning of the *trp* operon in *E. coli*, they discovered an odd DNA region, called the leader sequence, located between the *trp* operator and *trpE*, the first structural gene of the operon. The leader sequence encodes a short peptide, but the peptide has no enzymatic function. However, the leader sequence includes a pair of tryptophan codons, and this information was key to discovering the mechanism of attenuation because it prompted the question: What happens at these codons when the cell has no tryptophan?

Recall from Chapter 8 that in bacteria, ribosomes can begin to translate mRNA before transcription is complete. When a ribosome latches on to a nascent mRNA from the *trp* operon and translates the leader sequence, it will stall at the tryptophan codons if the cell lacks tRNAs charged with tryptophan. If the ribosome stalls, it transmits a message to the RNA polymerase: the cell lacks tryptophan, so the operon needs to be transcribed so that more tryptophan can be synthesized. How does the ribosome transmit this message?

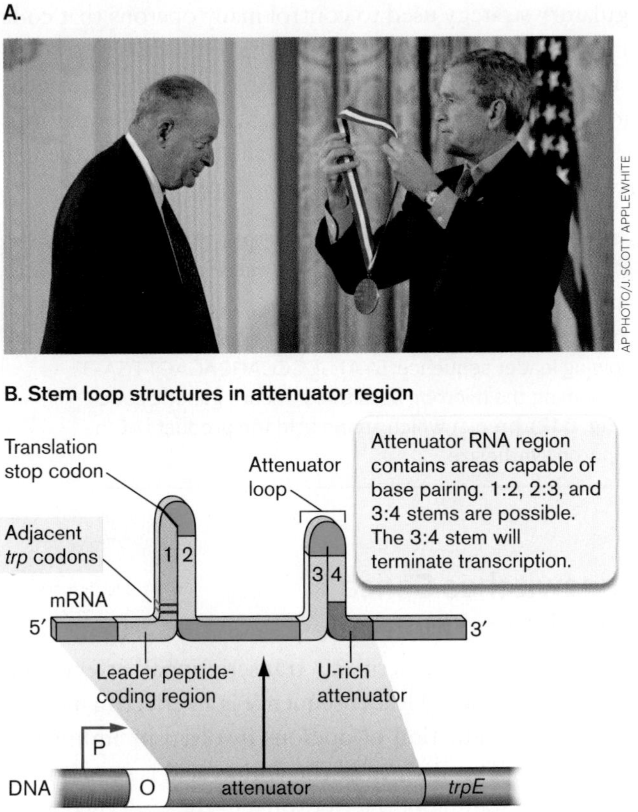

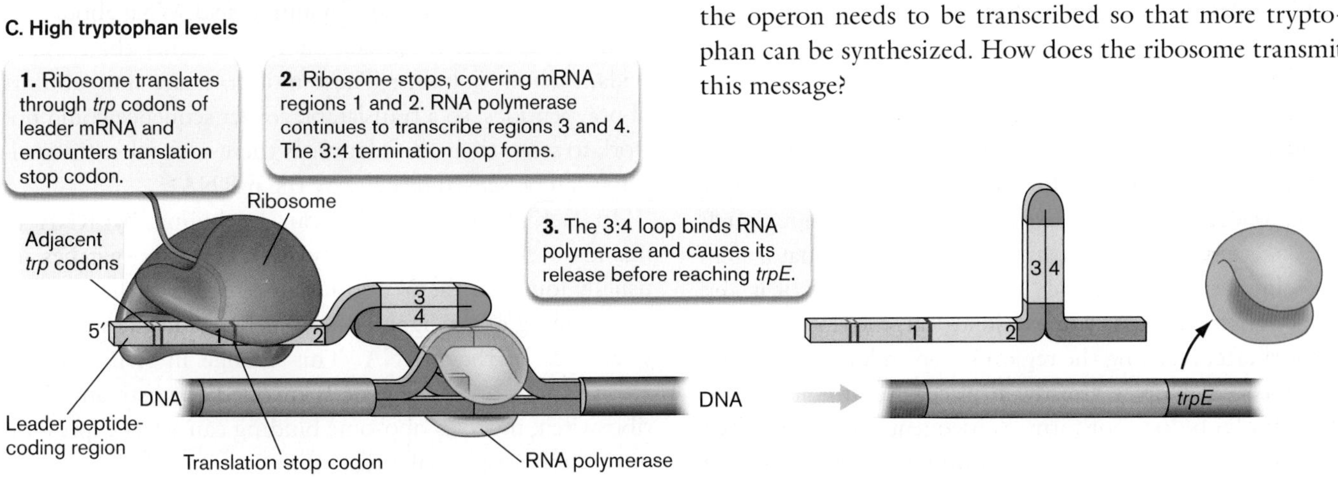

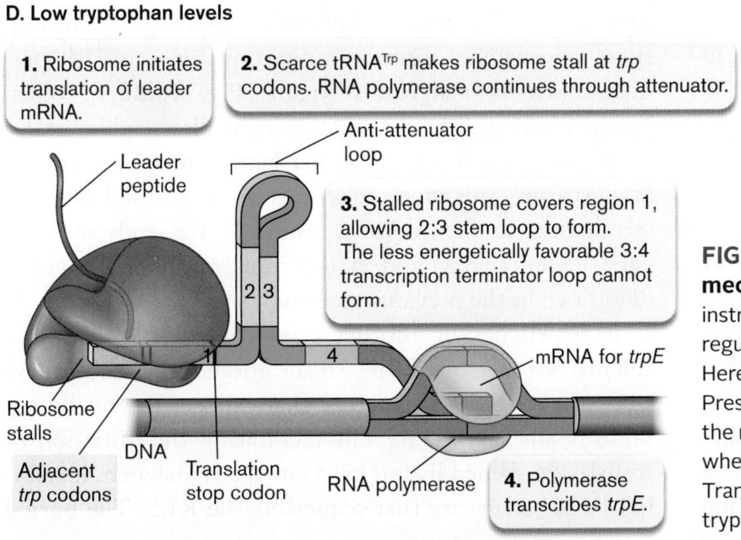

FIGURE 10.19 ■ The transcriptional attenuation mechanism of the *trp* operon. **A.** Charles Yanofsky was instrumental in discovering attenuation and several other gene regulatory mechanisms while studying tryptophan metabolism. Here he is seen receiving the National Medal of Science from President George W. Bush in 2003. **B.** Relationship between the mRNA attenuator region and encoding DNA. **C.** Attenuation when *E. coli* is growing in high tryptophan concentrations. **D.** Transcriptional read-through when *E. coli* is growing in low tryptophan concentrations. $tRNA^{Trp}$ = tryptophanyl-tRNA. ▶

Transmission of the message involves four complementary nucleotide stretches within the leader mRNA. These regions, numbered 1–4, can base-pair to form competing stem loop structures (**Fig. 10.19B**). Two of the stem loop structures are critical to the mechanism. These are the **anti-attenuator stem loop** formed by regions 2 and 3 and the **attenuator stem loop** (or terminator stem loop) formed by regions 3 and 4. Downstream of the attenuator stem loop is a stretch of U's (**Fig. 10.19B**). Recall from Chapter 8 that one type of transcription terminator (Rho-independent) is composed of a run of U's after a stem loop. Thus, if the 3:4 attenuator stem loop forms, the RNA polymerase is ejected and transcription stops before it reaches *trpE* (**Fig. 10.19C**). Formation of the 2:3 anti-attenuator stem loop, however, prevents formation of the 3:4 stem loop because the 2:3 stem is longer and more thermodynamically stable than the 3:4 stem. The anti-attenuator stem, if it forms, enables RNA polymerase to transcribe into *trpE* and the remainder of the operon (**Fig. 10.19D**). But what controls which stem loop forms?

High tryptophan levels. Because the ribosome is very large, it can barrel through RNA stem loop structures. When the cell is replete with charged tryptophanyl-tRNA and needs no more (**Fig. 10.19C**), the ribosome quickly translates through the key tryptophan codons in the leader sequence but runs into a translation stop codon between regions 1 and 2. The ribosome stops in this position, enveloping region 2 and preventing formation of the 2:3 stem. As a result, once RNA polymerase transcribes through region 4, the 3:4 attenuator stem snaps together. The attenuator stem then interacts with the RNA polymerase ahead of it and halts transcription. As you would expect, the ribosome dissociates after reaching the region 2 stop codon, but because the 3:4 stem loop is already in place, the anti-attenuator 2:3 stem loop does not form. Subsequent ribosome release leads to the formation of a 1:2 stem structure, precluding all possibility of regions 2 and 3 annealing.

Low tryptophan levels. If the level of charged tryptophanyl-tRNA is low (**Fig. 10.19D**), then the ribosome following behind RNA polymerase stalls over the tryptophan codons. Because these codons are right at the beginning of region 1, the ribosome does not cover region 2. So, as soon as RNA polymerase transcribes region 3, the 2:3 anti-attenuator stem loop forms and stops formation of the 3:4 attenuator stem loop. The result is that RNA polymerase can continue into the structural genes, and ultimately more tryptophan is made. (A new ribosome binds to a ribosome-binding site at the *trpE* message.)

Note that for the *trp* operon, attenuation is a fine-tuning mechanism. The repressor provides the majority of control. However, transcriptional attenuation is a common regulatory strategy used to control many operons that code for amino acid biosynthesis. Note that even though translation is part of the attenuation control mechanism, attenuation is not considered translational control. The reason is that RNA polymerase, rather than the ribosome, is the target of the control.

Thought Question

10.9 In a newly discovered bacterium, an operon suspected to synthesize an amino acid has what appears to be the following leader sequence: 5′-ATGCCCCAGCAGAGTTGA-3′. Assuming the microbe uses the standard genetic code (**Fig. 8.12**), predict which amino acid the products of the operon synthesize.

Riboswitches Sense Cytoplasmic Molecules

As we have just described, transcriptional attenuation involving a translated leader sequence is an effective method for feedback regulation of operons involved in amino acid biosynthesis. The absence of the amino acid is sensed indirectly by the stalling of the ribosome in the leader sequence at the codons that encode that amino acid. What about biosynthetic operons that make molecules other than amino acids, such as vitamins? These molecules are not subunits of polypeptides, so a translatable leader sequence could not work to sense their abundance. Is there a way that the molecule can be sensed directly by the mRNA?

In fact, many molecules can be bound by RNA as ligands, at sequences called **riboswitches**. Riboswitches are usually found in the 5′ untranslated region of the mRNA. The binding of the ligand causes a change in the stem loop structures of the mRNA. This change in structure can affect expression of the gene(s) encoded downstream of the riboswitch, much as ribosome binding can affect expression during transcriptional attenuation.

Some riboswitches control gene expression through early termination of transcription. In the absence of ligand, these riboswitches form an antiterminator stem that permits transcriptional read-through (**Fig. 10.20A**). When ligand binds, the antiterminator stem disassembles and a terminator stem forms. Together with a run of U's located downstream, this stem functions as a Rho-independent terminator, similar to those found at the ends of some genes (Section 8.2) and at the transcriptional attenuator (described in the previous section).

Riboswitches of another class control translation of the mRNA (**Fig. 10.20B**). In the absence of ligand, these riboswitches form a structure that exposes the ribosome-binding site (RBS), thereby facilitating the initiation of translation. When ligand binds to the riboswitch, it stabilizes a 3D structure that sequesters the RBS. The buried

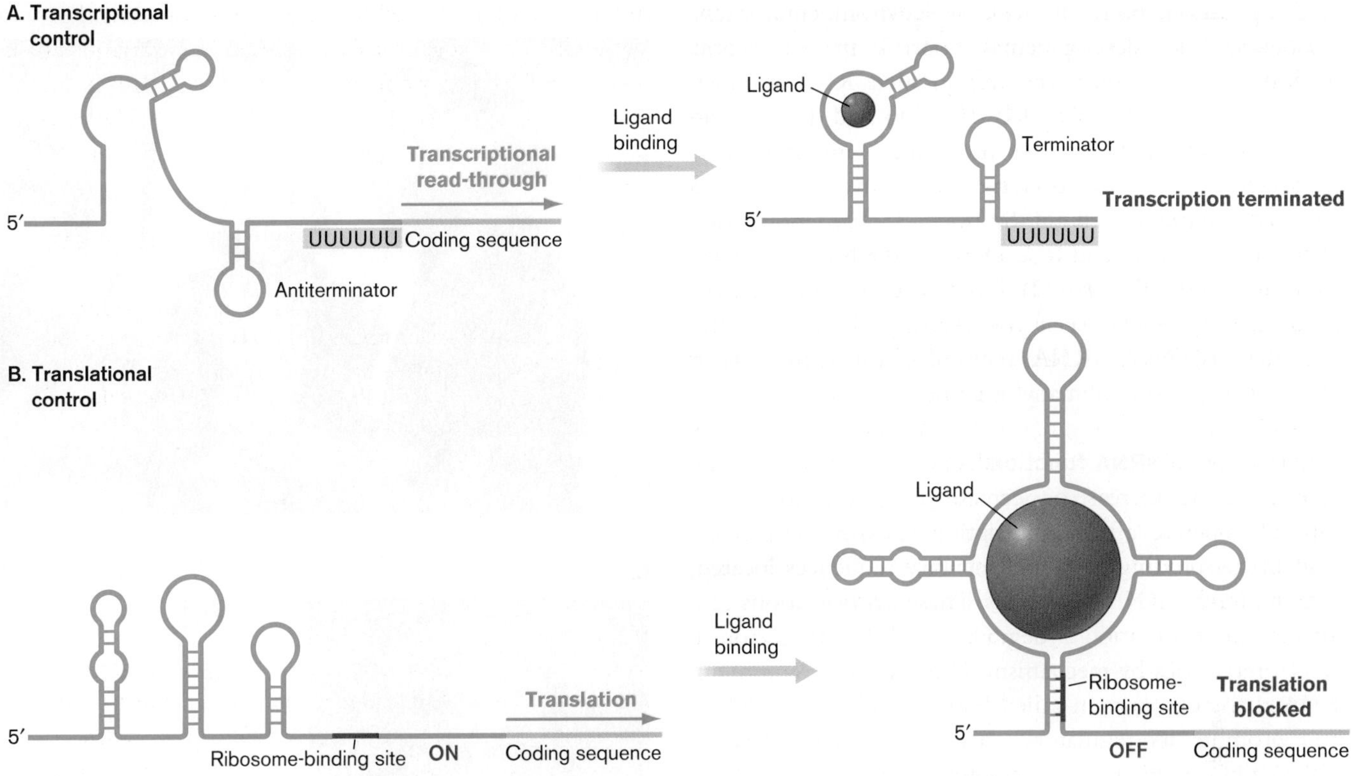

FIGURE 10.20 ■ Riboswitch regulation. **A.** Transcriptional control. Ligand binding stabilizes a secondary structure that forms a transcription termination stem. Transcription terminates before the coding sequence is transcribed. **B.** Translational control. Ligand binds to a riboswitch and stabilizes a structure that sequesters a ribosome-binding site. The coding region is not translated.

RBS prevents translation of the coding region. Riboswitches in this class do not interfere with transcription, and they can operate on mature, fully transcribed mRNA.

Riboswitches bind to a diverse set of ligands, including vitamins, metals, and regulatory molecules like cyclic di-GMP (discussed later). Ron Breaker's laboratory at Yale University discovered an unusual riboswitch in bacteria and archaea that specifically senses fluoride, which is not a cell metabolite (**Fig. 10.21**). Fluoride has been used for decades to inhibit dental caries, but it is also abundant in Earth's crust. The fluoride riboswitch was found attached to genes such as enolase that are inhibited by this halide. Fluoride buildup in bacteria will be detected by the fluoride riboswitch, which triggers increased translation of the enzyme being inhibited. The result is increased fluoride resistance in the microbe.

Untranslated Regulatory RNAs

Attenuators and riboswitches are part of the mRNA transcript that they control. Messenger RNA can also be regulated by RNAs transcribed from different promoters. A surprisingly large fraction of a bacterial chromosome does not encode mRNA, rRNA, or tRNA, but instead makes untranslated RNA with regulatory functions. Regulatory RNAs help control a variety of processes, such as plasmid replication, transposition, phage development,

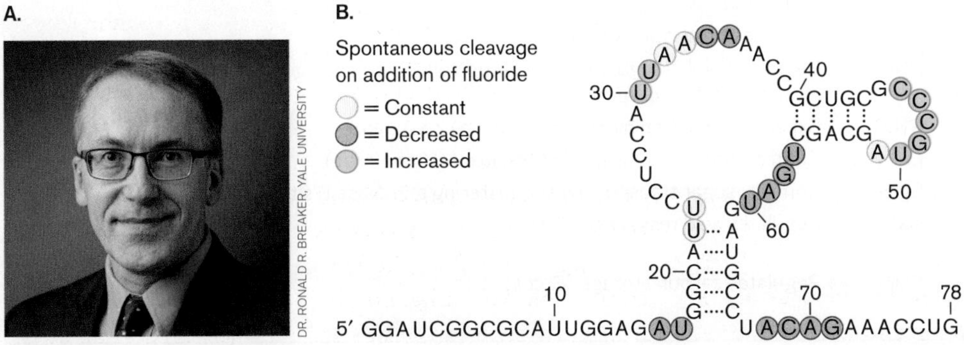

FIGURE 10.21 ■ A fluoride riboswitch. Ron Breaker (**A**) discovered the fluoride riboswitch called WT 78 Psy (**B**). Colored circles mark bases that become susceptible to spontaneous cleavage after the riboswitch binds fluoride—an indication that the structure of the RNA molecule changes. A pseudoknot contains at least two stem loop structures that form a knot-shaped 3D conformation.

viral replication, bacterial virulence, environmental stress responses, and developmental control in eukaryotic microbes. For instance, the untranslated RNA product GadY stimulates the expression of amino acid decarboxylases that confer acid resistance on *E. coli*, an important factor in the ability of this bacterium to survive the stomach and colonize the human gut (Section 5.3). The laboratories of Susan Gottesman and Gisela Storz at the National Institutes of Health (**Fig. 10.22**) were instrumental in discovering that regions between genes (intergenic regions) can encode **small RNA (sRNA)** molecules (100–200 nt) that affect the expression of many other genes.

A.

DALE LEWIS

B.

NIH MEDICAL ARTS

FIGURE 10.22 ■ Susan Gottesman and Gisela Storz. Gottesman (**A**, right) and Storz (**B**) played instrumental roles in establishing the importance of sRNA molecules in bacteria.

Mechanisms of sRNA function. Regulatory sRNA molecules can either increase or decrease gene expression. They typically operate after transcription (posttranscriptional control) by binding to complementary sequences located within target mRNA transcripts. These hybridizations can affect either translation or degradation. **Table 10.2** groups sRNA molecules by mechanism. Many sRNAs require an RNA chaperone protein called Hfq to stabilize the sRNA and promote its regulatory effects on target mRNAs. Hfq is a hexameric ring protein with sRNA- and mRNA-binding faces.

Most known sRNAs inhibit translation by base-pairing to a region of mRNA that overlaps the ribosome-binding site (RBS). Binding, therefore, prevents the ribosome from accessing the RBS (**Fig. 10.23A**). Other sRNA molecules can enhance translation by binding to part of a long, 5′, untranslated mRNA sequence located upstream of an RBS (**Fig. 10.23B**). Without the sRNA, the untranslated mRNA sequence folds in a way that occludes the RBS and

TABLE 10.2 Classes of sRNA Molecules

Mechanism	sRNA	Function
1. Inhibits translation by blocking ribosome-binding site (RBS).	OxyS	Regulates oxidative stress gene expression (*E. coli*).
	CyaR	Represses porin OmpX (*E. coli*).
	ChiX	Prevents transport of chitosugars by preventing the synthesis of ChiP porin (*E. coli*).
	SprD	Regulates Sbi (*Staphylococcus aureus* binder of IgG) immune evasion molecule.
	Qrr3	Controls quorum sensing by sequestering *luxO* (*Vibrio cholerae*).
	$sRNA_{162}$	Inhibits translation of regulator MM2241; affects methyltransferases (*Methanosarcina mazei*).
2. Permits translation by exposing RBS.	DsrA	Increases translation of sigma S mRNA (*E. coli*).
3. Promotes degradation of mRNA.	RNAIII	Regulates global regulator of *agr*-controlled virulence genes; also encodes a delta hemolysin (*S. aureus*).
	RyhB	Expands regulation by Fur repressor (*E. coli*).
	Qrr3	Controls quorum sensing via *luxR*, *luxM* (*V. cholerae*).
4. Inhibits degradation of mRNA.	SgrS	Controls sugar transport by sequestering an RNase E site (*E. coli*).
5. Stimulates processing of mRNA to make more stable transcripts.	GadY	Regulates acid resistance (*E. coli*).
6. Titrates regulatory proteins away from target mRNAs.	CsrB	Regulates carbon storage (*E. coli*).

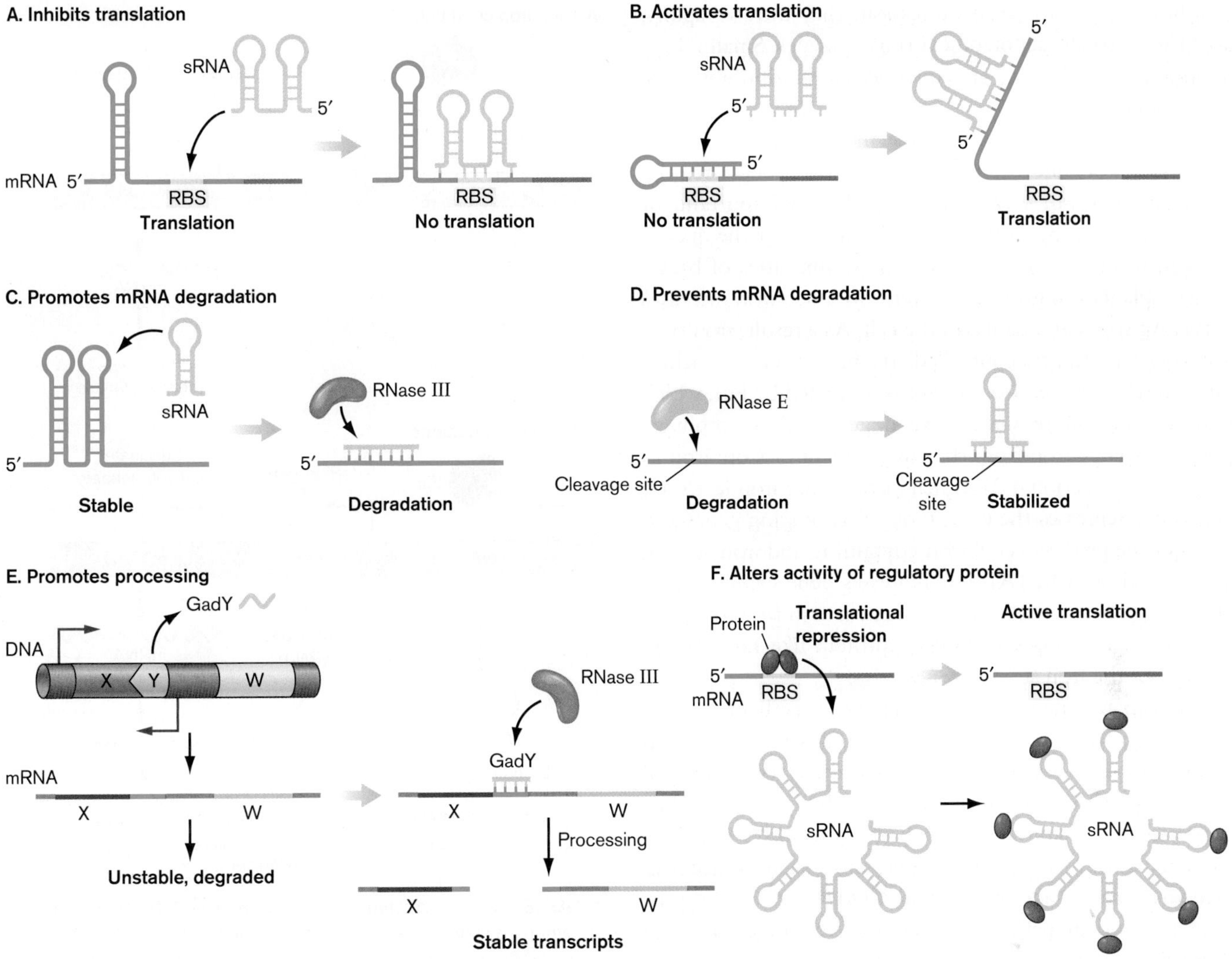

FIGURE 10.23 ▪ Mechanisms of regulatory sRNA function. Some sRNA molecules will **(A)** inhibit or **(B)** activate translation by blocking or exposing a ribosome-binding site (RBS). Other groups of sRNA molecules can **(C)** promote or **(D)** prevent the degradation of target mRNA; **(E)** mediate processing of long, unstable multigene mRNAs into more stable, shorter molecules; or **(F)** interfere with regulator protein activity.

prevents translation. However, when sRNA binds the target mRNA sequence, the untranslated region refolds to expose the RBS (an example is DsrA, a regulator of sigma S translation; **Table 10.2**). Ribosomes can now bind the mRNA and translate the message.

Some sRNA molecules control gene expression by affecting mRNA stability. These sRNAs operate by exposing or masking sites on the mRNA that are cleaved by different types of RNases (**Fig. 10.23C–E**). RNase III cleaves double-stranded RNA, while RNase E cleaves single-stranded RNA. Some sRNAs form a duplex with the mRNA to promote degradation by RNase III (**Fig. 10.23C**). The way this mode of regulation is used by the pathogen *Staphylococcus aureus* during infection is described in **eTopic 10.3**. Other sRNAs prevent degradation of the message by binding to and masking an RNase E–binding site (**Fig. 10.23D**).

Sometimes cleavage by RNases serves to stabilize rather than degrade the mRNA. Assisted by sRNA, a long, unstable polycistronic transcript can be processed to make shorter, more stable monocistronic mRNAs. For instance, the GadY sRNA of *E. coli* binds between two coding regions of a polycistronic message and generates a double-stranded target for cleavage by RNase III (**Fig. 10.23E**).

In a final class of regulatory sRNA molecules, sRNAs bind to proteins that control translation of certain mRNAs (**Fig. 10.23F**). When bound by the sRNA, the regulatory proteins are sequestered away from their mRNA target sequences and translation of the mRNA is affected.

Small RNA molecules can expand the reach of regulatory proteins. While large regulatory proteins (for example, LacI or TrpR) typically control only a few genes, their synthesis is energetically expensive. But if these proteins also control the expression of regulatory sRNAs that themselves control the expression of additional genes, then the return on investment for these regulatory proteins is improved. The benefits of sRNAs are that they do not require protein synthesis to control gene expression, they diffuse rapidly, and they typically act on preexisting messages. Small RNAs represent one of the most economical ways to inhibit gene expression.

One example of an sRNA molecule extending the reach of a regulatory protein involves the ferric uptake regulator (Fur) repressor of *E. coli.* Iron is hugely important for pathogens growing in the human body and for the quality of soil, where it determines which communities of bacteria and plants can grow. However, too much iron increases oxidative stress and damages the cell. As a result, iron content must be tightly controlled. In many bacteria, including intestinal *E. coli*, iron uptake is regulated by Fur, which senses iron and regulates several genes whose products either scavenge iron from the environment or store iron in the cell (see Section 4.2). When intracellular iron levels are high, Fur represses the expression of scavenging genes and induces the production of iron-containing and iron-storing proteins (**Fig. 10.24A**). Fur represses gene expression by directly binding DNA operator sequences in front of target genes. But how does a repressor protein like Fur activate genes like the iron storage genes?

Fur can activate iron storage genes indirectly by repressing the expression of an inhibitory sRNA called RyhB (**Fig. 10.24B**). RyhB sRNA is made under conditions of low iron (such as within the human body) because the Fur repressor is not active. RyhB hybridizes to mRNAs of several iron-storing and iron-using proteins (such as succinate dehydrogenase made from the *sucCDAB* operon) and promotes their degradation by generating a cleavage site for an RNase. As a result, dwindling iron reserves can be put to more productive use. When iron is plentiful, however, Fur will directly repress *ryhB*. The lack of RyhB sRNA stabilizes the expression of the iron storage genes. Thus, succinate dehydrogenase is made, binds intracellular iron, and enables the use of succinate as a carbon and energy source.

> **Thought Question**
>
> **10.10** The relationship between the small RNA RyhB, the iron regulatory protein Fur, and succinate dehydrogenase is shown in **Figure 10.24**. Given this regulatory circuit, will a *fur* mutant grow on succinate?

Small RNAs in archaea. Archaeal species also use several classes of sRNA molecules to fine-tune their physiology. One class, called small nucleolar RNAs (snoRNAs), is also prominent in eukaryotes. The snoRNA designation was retained for archaea even though archaea lack a nucleolus—or nucleus, for that matter. Some archaeal snoRNAs help guide the activity of enzymes that modify other RNAs. For example, one snoRNA (in *Pyrococcus* and *Sulfolobus*) guides a methylation enzyme to methylate rRNA at specific

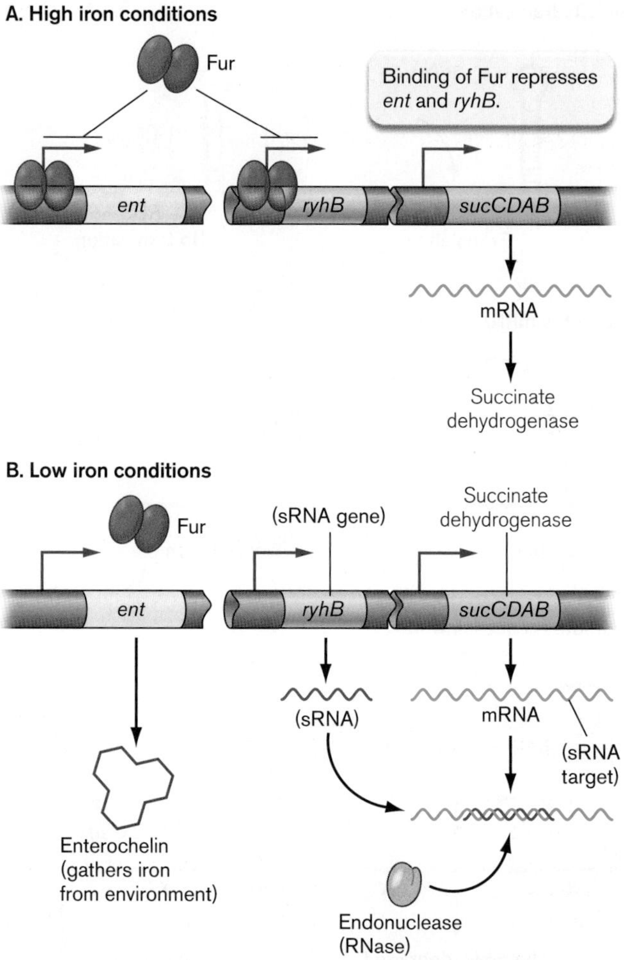

FIGURE 10.24 ■ Activity of a regulatory sRNA molecule. **A.** When iron levels are high, Fur repressor protein binds to the *ent* and *ryhB* Fur box DNA sequence (a short, specific DNA sequence in front of the genes regulated by Fur) and represses their expression. Enterochelin is no longer made, but the *sucCDAB* message encoding succinate dehydrogenase can be translated. **B.** Under low iron conditions, RyhB sRNA is expressed. RyhB sRNA binds to the *sucCDAB* message and renders it susceptible to an RNase.

sites. Another snoRNA guides the conversion of uridine to pseudouridine in rRNAs. A second class of archaeal sRNA is called tRF. One example is a tRF produced by the halophile *Haloferax volcanii* during alkali stress. This tRF binds to *Haloferax* ribosomes and inhibits translation, thereby helping the organism survive high pH.

***Cis*-antisense RNA.** Some operons can be read backward, generating an RNA that can regulate the mRNA transcript made in the forward direction. A ***cis*-antisense RNA** (**asRNA**) is transcribed from the nontemplate (coding) DNA strand that lies opposite an mRNA-encoding template strand (see Fig. 8.5). *Cis*-antisense regulatory transcripts can base-pair with cognate sense mRNAs and control their expression. Typically, asRNAs are 700–3,000 nt long, originate within a protein-coding gene, and affect only that gene. These features distinguish asRNA from sRNA. Deep

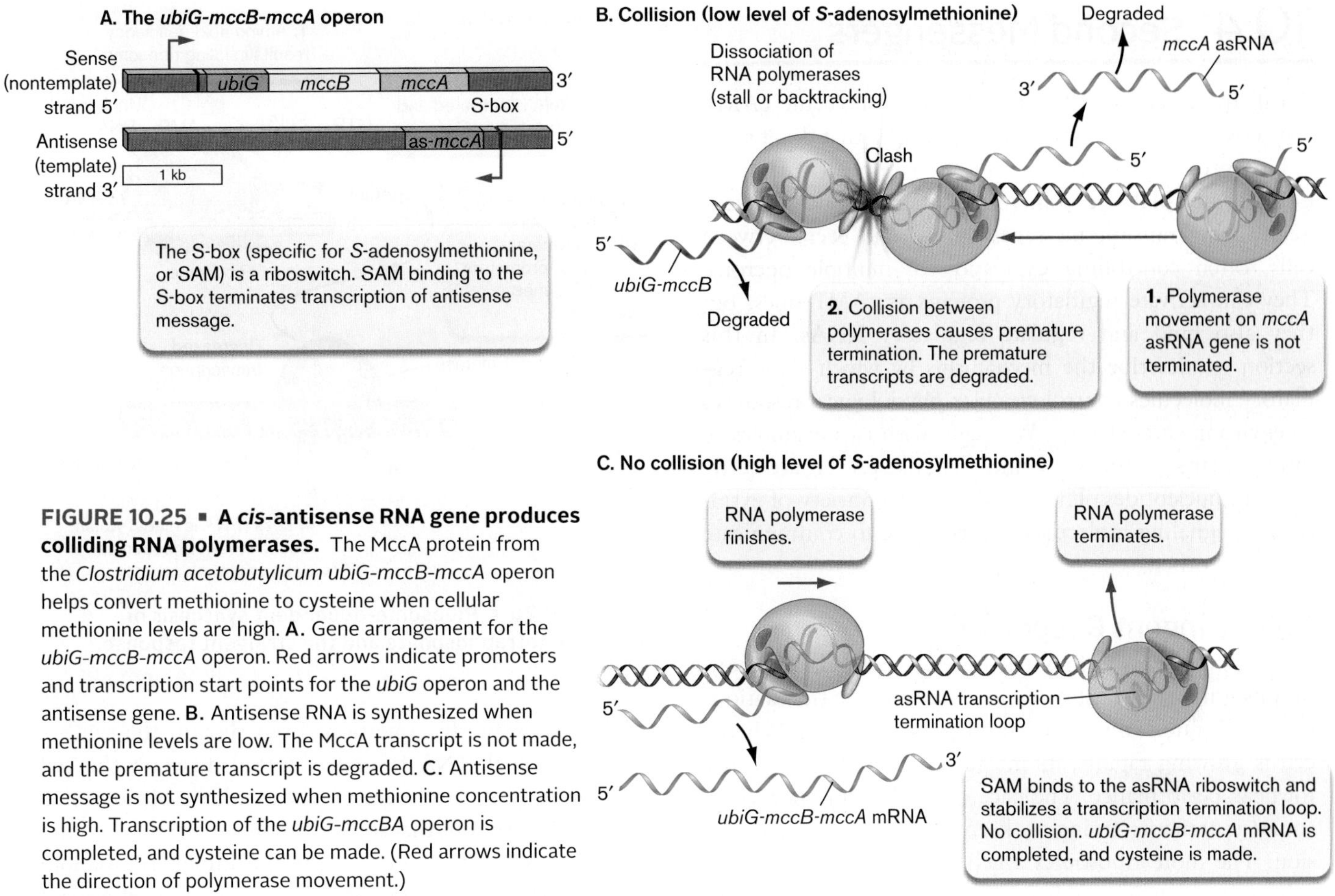

FIGURE 10.25 ■ A *cis*-antisense RNA gene produces colliding RNA polymerases. The MccA protein from the *Clostridium acetobutylicum ubiG-mccB-mccA* operon helps convert methionine to cysteine when cellular methionine levels are high. **A.** Gene arrangement for the *ubiG-mccB-mccA* operon. Red arrows indicate promoters and transcription start points for the *ubiG* operon and the antisense gene. **B.** Antisense RNA is synthesized when methionine levels are low. The MccA transcript is not made, and the premature transcript is degraded. **C.** Antisense message is not synthesized when methionine concentration is high. Transcription of the *ubiG-mccBA* operon is completed, and cysteine can be made. (Red arrows indicate the direction of polymerase movement.)

sequencing of the *E. coli* transcriptome revealed the presence of about 1,000 different *cis*-antisense RNA genes out of 4,290 protein-encoding genes. In fact, every microbe examined to date produces asRNA molecules.

Different asRNAs have different effects on their target genes. When bound to their sense mRNA counterparts, asRNAs can engineer attenuator loops that stop transcription, prevent mRNA translation, or trigger mRNA degradation. Even the simple act of asRNA transcription can produce collisions between converging RNA polymerase complexes that prematurely terminate transcription. An example of the collision mechanism was found for *Clostridium acetobutylicum*. The *ubiG-mccB-mccA* operon encodes proteins needed to convert methionine to cysteine (**Fig. 10.25A**), but it is expressed only when methionine levels are high. When cytoplasmic methionine concentration is low, the RNA polymerase that transcribes the antisense *mccA* gene will collide with, and stop, the RNA polymerase transcribing the *ubiG-mccB-mccA* operon (**Fig. 10.25B**). The sense MccA transcript is not made, and the premature transcript is degraded. However, when methionine concentration is high, *S*-adenosylmethionine (made from methionine) binds to the asRNA at the S-box (a riboswitch) and stabilizes a transcription termination loop (**Fig. 10.25C**). The *mccA* antisense RNA is not produced, and because the RNA polymerase making asRNA disengages, no polymerase collision happens. The *ubiG-mccB-mccA* operon mRNA is completed, MccA is made, and cysteine is synthesized.

To Summarize

- **Attenuation** is a transcriptional regulatory mechanism in which translation of a leader peptide affects mRNA structure to influence the transcription of a downstream structural gene.
- **Riboswitches** are secondary structures at the 5′ end of specific mRNA molecules that can obscure access to ribosome-binding sites or transcriptional terminator stems.
- **Regulatory sRNA molecules** found within bacterial intergenic regions regulate the transcription or stability of specific mRNA molecules and broaden the reach of protein regulators.
- ***Cis*-antisense RNAs** are produced from the sense DNA strand of a protein-encoding gene and affect the expression of only that gene. They bind to complementary target mRNA and either stabilize the mRNA or make it susceptible to degradation.

10.4 Second Messengers

Recall from Section 10.1 that the small molecule cAMP can activate operons such as *lac*. Molecules that don't serve as biosynthetic precursors but rather have regulatory function, like cAMP, are sometimes called **second messengers**. They transmit messages within and even between cells, often controlling expression of multiple operons. They can activate regulatory proteins as cAMP does, but they also bind and regulate regulatory RNAs. In this section we describe the mechanisms by which some regulatory molecules control complex physiological responses to environmental change. We begin with two examples of intracellular signaling molecules that, like cAMP, are derivatives of nucleotides. Then we discuss the variety of extracellular signaling molecules that cells use to communicate with each other.

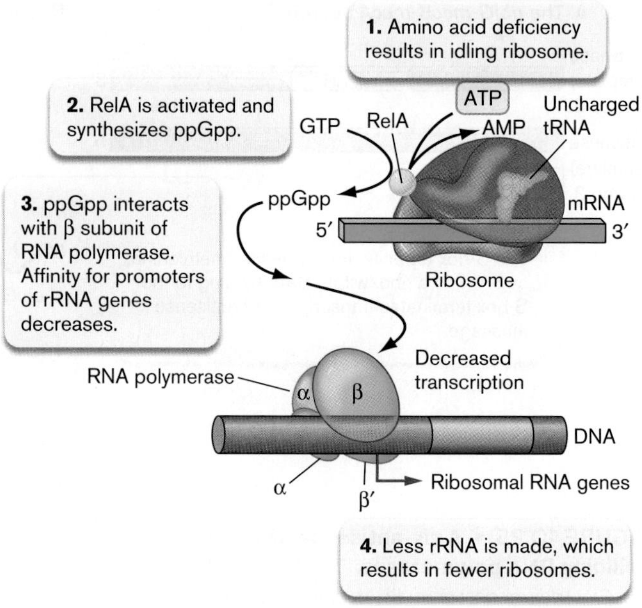

FIGURE 10.26 ▪ Ribosome-dependent synthesis of guanosine tetraphosphate and the stringent response.

The Stringent Response

During transitions from nutrient-rich to nutrient-poor conditions, microbes must contend with dramatic fluctuations in growth rate. This variation presents a problem. When a cell is growing rapidly, its molecular machinery is geared for peak performance. The pace of synthesizing new ribosomes is frenetic, trying to keep up with rapid cell division. The more ribosomes a cell contains, the faster that cell can make new proteins and the faster it can grow. But what happens when the party's over—when poor carbon and energy sources cannot supply enough energy to maintain rapid cell division? Without a way of curbing ribosome construction, cells would soon fill with idle ribosomes.

Under these conditions, bacteria undergo a process called the **stringent response**. The stringent response causes a decrease in the number of rRNA transcripts made for ribosome assembly and alters the expression of numerous other genes. Stringent response enables some bacteria to resist an antibiotic until they evolve more specific resistance (see Chapter 17).

In the stringent-response strategy, idling ribosomes trigger the synthesis of a small signaling molecule called guanosine tetraphosphate (ppGpp), which interacts with RNA polymerase and lowers the enzyme's ability to transcribe genes encoding ribosomal RNA (**Fig. 10.26**). How is ppGpp production regulated by stalled ribosomes? When an uncharged tRNA binds at the ribosome A site, which can happen during amino acid starvation, a ribosome-associated protein called RelA transfers phosphate from ATP to GTP to form ppGpp. This signal nucleotide interacts with the beta subunit of RNA polymerase and diminishes its recognition of promoters for operons producing rRNA and tRNA. The result is down-regulation of rRNA and tRNA synthesis. The less rRNA that is available for building ribosomes, the fewer ribosomes will be produced.

This raises another question. Even though the synthesis of ribosomal RNA has been curtailed, won't the cell continue to waste resources on the synthesis of ribosomal proteins? It turns out that some ribosomal proteins can bind to the mRNA that encodes them and inhibit translation. So, when rRNA levels in the cell are low, free ribosomal proteins accumulate in the cytoplasm, unassociated with ribosomes. These excess ribosomal proteins begin to bind to their own mRNA molecules and inhibit the translation of their own coding regions, as well as the coding regions of other ribosomal proteins residing on the same polycistronic mRNA. This process is called **translational control** because regulation affects the translation of an mRNA by ribosomes rather than transcription by RNA polymerase.

Cyclic Di-GMP and Biofilm Formation

E. coli cells transition between a motile, single-cell state (planktonic) and an adhesive multicellular biofilm. The transition can be seen clearly in batch cultures. The highly motile state appears during post-exponential growth when nutrient limitation forces *E. coli* to "forage" for food. When resources diminish further (stationary phase), the organism changes strategy: growth slows, motility decreases (a huge energy savings), and the synthesis of adhesins such as pili increases. After adhering to a surface, the cells start synthesizing exopolysaccharide matrix (see Chapter 4). The molecule cyclic di-GMP (c-di-GMP; **Fig. 10.27**) coordinates the transition by repressing flagellar synthesis genes and activating biofilm-promoting genes such as those encoding pili.

This coordinated shift in physiology is achieved by the balance between synthesis and degradation of c-di-GMP. Synthesis of c-di-GMP from GTP is carried out by many diguanylate

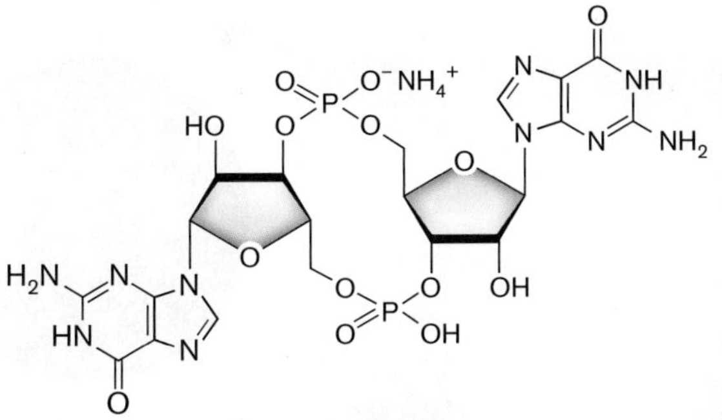

FIGURE 10.27 ■ Cyclic di-GMP [bis-(3′-5′)-cyclic dimeric guanosine monophosphate], or c-di-GMP.

cyclases (DGCs) in the cell, all of which contain the amino acid motif GGDEF (see Figure 8.12 for explanation of the single-letter amino acid abbreviations). Dedicated phosphodiesterases (PDEs) degrade c-di-GMP to GMP (**Fig. 10.28A**). Each DGC and PDE protein becomes activated by a different signal, thus increasing or decreasing c-di-GMP levels under different conditions. As shown in **Figure 10.28B**, PDEs predominate in post-exponential-phase cells, which keeps c-di-GMP level low. Low c-di-GMP concentration favors motility and scavenging. In contrast, DGCs predominate in stationary phase and increase c-di-GMP level. High c-di-GMP inhibits motility and scavenging. Thus, highly motile cells are made in post-exponential phase, while sessile, adherent cells able to form biofilms are produced in stationary phase. **Figure 10.28C** shows that c-di-GMP is required for *Salmonella* to make biofilms. Deleting all known GGDEF-motif proteins eliminates c-di-GMP synthesis and halts biofilm formation.

Cyclic di-GMP is a ubiquitous second messenger in bacteria and is not found in eukaryotes or archaea. But why does one cell possess so many DGC and PDE proteins? One hypothesis is that cognate pairs of these proteins may operate at separate locations within a single cell and generate localized changes in c-di-GMP concentrations. Multiple, focused c-di-GMP gradients could produce different outputs in different parts of a single cell.

Quorum Sensing and Cell-Cell Communication

Studies of the Hawaiian bobtailed squid (*Euprymna scolopes*; **Fig. 10.29**) led to a discovery that fundamentally changed the way we think about microbes. During the day, this tiny squid remains buried in the sand of shallow reef flats around Hawaii. After sunset, the animal emerges from its hiding place and begins its search for food. As it swims in the moonlit night, its light organ projects light downward in an apparent attempt to camouflage the squid from predatory fish swimming below. Looking up, the fish see only light (called counterillumination), not a squid's

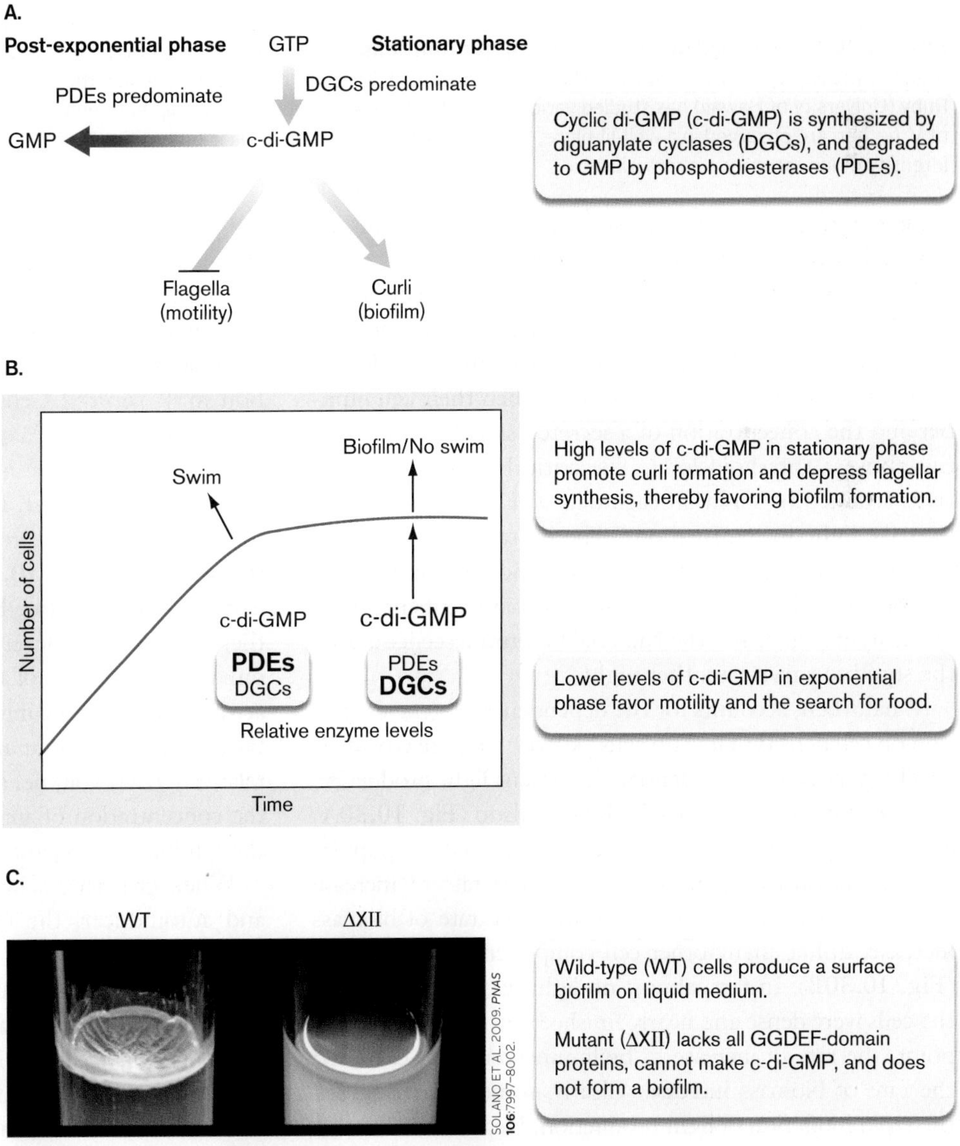

FIGURE 10.28 ■ Cyclic di-GMP coordinates the switch from planktonic growth to biofilm formation. **A, B.** The relative levels of PDEs and DGCs at different growth phases. **C.** Surface biofilm production by *Salmonella* requires c-di-GMP.

A. B. C. D.

M. J. MCFALL-NAGI AND E. G. RUBY, UNIVERSITY OF HAWAII

COURTESY OF EDWARD RUBY, UNIVERSITY OF HAWAII

COURTESY J. W. HASTINGS, HARVARD UNIVERSITY, THROUGH E. G. RUBY, UNIVERSITY OF HAWAII

FIGURE 10.29 ■ Visual demonstration of quorum sensing. **A.** The luminescent bacterium *Vibrio (Aliivibrio) fischeri* colonizes the light organ of the Hawaiian bobtailed squid (*Euprymna scolopes*). The light organ is deep inside the squid and therefore cannot be seen. **B.** Edward Ruby (University of Hawaii) has studied various aspects of the symbiotic relationship between *V. fischeri* and its squid host. **C.** When colonies of *V. fischeri* are observed in a well-lit place, the light emitted by the bacteria is not visible. **D.** If the same colonies are viewed in darkness, the intensity of luminescence is remarkable.

shadow moving against the surface-filtered light of the moon. The light, however, is not made by the squid.

Inside the squid's light organ are luminescent bacteria called *Vibrio* (or *Aliivibrio*) *fischeri.* Bacteria, not the squid, produce the light. However, these microbes do not glow all the time. They light up only when their cell number and the concentration of a secreted signaling molecule rise above a threshold level. The critical density of bacteria is attained by nightfall each day, and the genes needed to make light are "turned on." The bacteria and the squid have formed a symbiotic relationship known as mutualism (see Section 21.3): The bacteria feed on nutrients provided in the light organ, and the bacterial bioluminescence allows the squid to survive another night.

What, then, accounts for the dependence of gene expression on cell density? How do cells "know" they are crowded?

The phenomenon of density-dependent light production in *V. fischeri* was discovered by Ken Nealson (**Fig. 10.30A**) and colleagues in 1970. Nealson noted that as populations of *V. fischeri* grew in culture, their rate of increase in bioluminescence did not track with the rate of biomass increase, unlike many other cell components and activities (**Fig. 10.30B**). In fact, almost no light was produced until the cells were dense and nearly finished with growth. At this point, the rate of increase in bioluminescence far exceeded the rate of biomass increase. Clearly, something triggered the rapid induction of light production. Nealson named this phenomenon **autoinduction** to reflect the fact that cells need no external factors or stimuli to promote this rapid increase in luciferase-based light production. After this study the term **quorum sensing** was applied to this phenomenon because it seemed akin to parliamentary rules of order that require a minimum number of members (a quorum) to be present at a meeting in order to conduct business.

What triggers the eventual burst in luciferase production in *V. fischeri*? Cells do not count each other directly. Instead, induction of a quorum-sensing gene system involves the accumulation of a membrane-permeable small molecule called an **autoinducer**. In *V. fischeri*, the autoinducer is synthesized by the LuxI protein, the product of the first gene in the *lux* operon (**Fig. 10.31**). At low cell densities, the operon is transcribed at a low but constitutive rate. Consequently, these cells have low levels of LuxI protein and therefore make only small quantities of autoinducer. Autoinducer can diffuse across the membrane into the medium, such that the concentrations inside and out are the same. If growth proceeds in a relatively fixed volume, such as the light organ of the squid, the concentration of autoinducer in the environment and in the cell increase in parallel as a function of cell density.

When growth reaches a threshold concentration of cells and autoinducer, the intracellular concentration of autoinducer becomes high enough to bind and activate the regulatory molecule LuxR (**Fig. 10.31** ▶). The LuxR-autoinducer complex activates transcription of the *lux* operon, which has two important effects. First, more LuxI is produced from *luxI*, the first gene of the operon. More LuxI synthesizes more autoinducer, which in turn increases expression of the *lux* operon. This chain of steps establishes a positive feedback loop that rapidly generates high concentrations of LuxI and autoinducer. The second effect of

A.

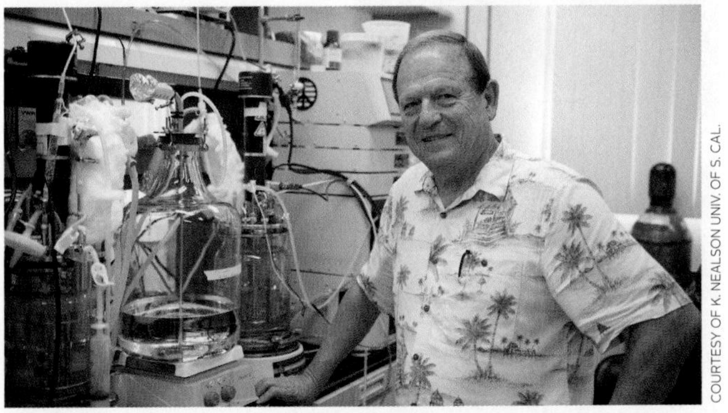

B. Biomass and bioluminescence of *Vibrio fischeri*

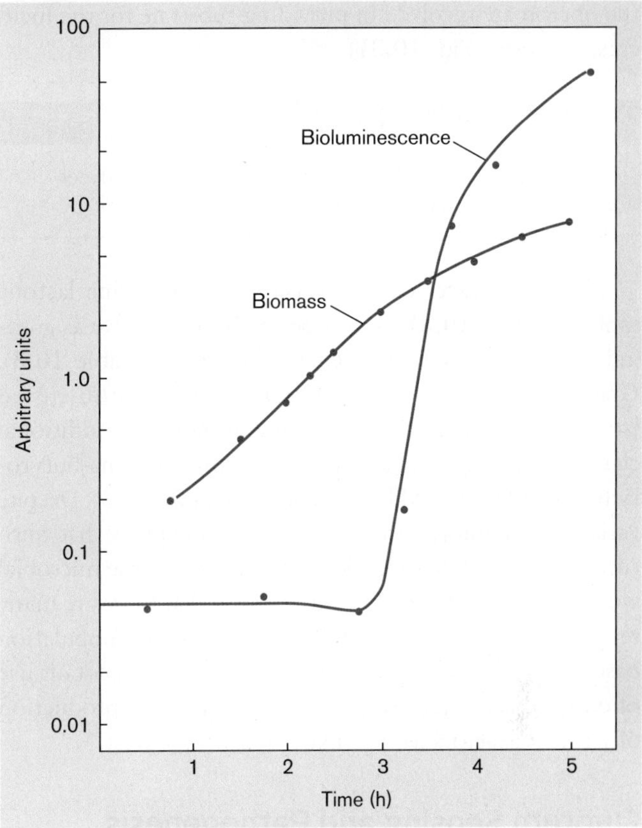

FIGURE 10.30 ■ Bioluminescence of *Vibrio* (*Allivibrio*) *fischeri* is controlled by quorum sensing. A. Ken Nealson. **B.** His experiment began with a dilute culture, whereby the biomass of the population increased at a constant rate until nutrient depletion. Bioluminescence of the growing population did not increase until 3 hours after inoculation, at a rate that far exceeded biomass increase. *Source:* Part B modified from K. H. Nealson et al. 1970. *J. Bacteriol.* **104**:313–322, fig. 4.

LuxR-autoinducer activation of the *lux* operon involves the genes downstream of *luxI*. These genes are cotranscribed with *luxI* and experience the same rapid elevation in expression. These downstream genes confer bioluminescence, and their rapid increase in expression can be witnessed by the rapid increase in luciferase expression and bioluminescence evident in **Figure 10.30B**. Luciferase is

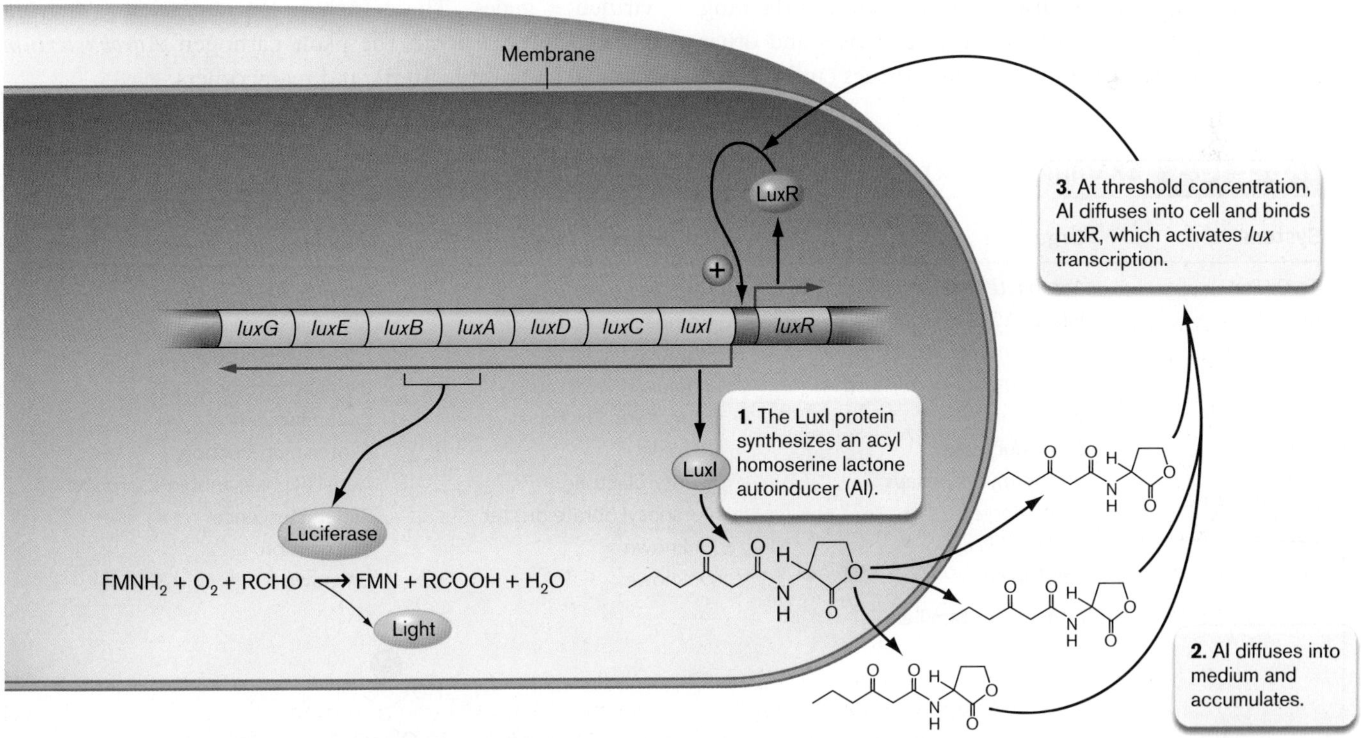

FIGURE 10.31 ■ Microbial communication through quorum sensing. The *lux* system of *Vibrio fischeri* mediates that organism's bioluminescence. Synthesis and accumulation of an autoinducer (AI) trigger expression of the *lux* operon. The greater the cell number and the smaller the container, the faster AI will accumulate. The resulting luciferase enzymes catalyze bioluminescence. The luciferase reaction, catalyzed by LuxA and LuxB, uses oxygen and reduced flavin mononucleotide (FMN) to oxidize a long-chain aldehyde (RCHO) and, in the process, produces blue-green light. Other *lux* gene products are involved in synthesis of the aldehyde.

encoded by *luxA* and *luxB*, while the remaining genes in the operon are involved in providing substrate for the luciferase reaction (**Fig. 10.31**).

Thought Question

10.11 What would be the outcome if purified autoinducer were experimentally provided to a low-density culture?

The autoinducer of *V. fischeri* is a homoserine lactone molecule (**Fig. 10.31**). Homoserine lactones serve as autoinducers for many Gram-negative microbes (**Table 10.3**). Gram-positives such as *Streptococcus pneumoniae* (discussed later) usually use short peptides as autoinducers. Additional classes of autoinducers also exist, including gamma-butyrolactones and furanosyl borate diesters (**Table 10.3**). Despite variations in autoinducer structure, it is evident that quorum-sensing systems are widely distributed in the microbial world. This broad distribution suggests that there are many reasons why the coordinated behavior of the population based on cell numbers is advantageous. One important use of quorum sensing is for pathogens to time the production of virulence factors for optimal effect on the host.

Quorum Sensing and Pathogenesis

Pseudomonas aeruginosa is a human pathogen that commonly infects patients who have cystic fibrosis, a genetic disease of the lung. The organism forms a biofilm in the lung and secretes virulence factors (such as proteases and other degradative enzymes) that destroy lung tissues (and thereby severely compromise lung function). These virulence proteins, however, are not made until cell density is fairly high—that is, at a point where the organism might have a chance of overwhelming its host. Made too early, virulence proteins would alert the host to launch an immune response.

E. PETER GREENBERG, UNIVERSITY OF WASHINGTON

FIGURE 10.32 ■ Peter Greenberg, one of the pioneers of cell-cell communication research. Greenberg has studied quorum sensing in *Vibrio* species and various other pathogenic bacteria, such as *Pseudomonas*.

Scientists such as Peter Greenberg at the University of Washington (**Fig. 10.32**) have discovered that the induction mechanism involves two interconnected quorum-sensing systems, called Las and Rhl, both composed of regulatory proteins homologous to LuxR and LuxI of *Vibrio* (*Aliivibrio*) *fischeri*. Many pathogens besides *Pseudomonas* appear to use chemical signaling to control virulence genes. These include *Staphylococcus*, *Yersinia pestis*, *Vibrio cholerae*, the plant pathogen *Agrobacterium tumefaciens* (**Table 10.3**), and many others.

TABLE 10.3 Examples of Microbial Quorum-Sensing Systems

System	Organism	Autoinducer family	Function
TraR/TraI	*Agrobacterium tumefaciens*	Homoserine lactone	Conjugation
LuxR/LuxI	*Vibrio (Aliivibrio) fischeri*	Homoserine lactone	Bioluminescence
LasR/LasI	*Pseudomonas aeruginosa*	Homoserine lactone	Exoenzyme production
Rhl	*Pseudomonas aeruginosa*	Homoserine lactone	Exoenzyme production
LuxN/LuxLM	*Vibrio harveyi*	Homoserine lactone	Bioluminescence
Agr	*Staphylococcus*	Peptide	Exotoxin production
StrR	*Streptomyces griseus*	γ-Butyrolactone	Aerial hyphae; antibiotic production
LuxQ/LuxS	*Vibrio harveyi*	Furanosyl borate diester	Bioluminescence
SdiA	*Escherichia coli*	Unknown	Cell division
YpeR/YpeI	*Yersinia pestis*	Unknown	Unknown

Structures of different autoinducer families:

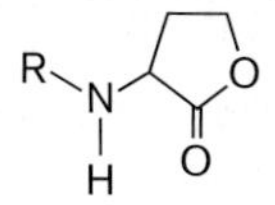

Homoserine lactone family

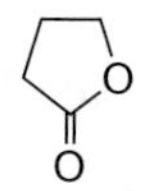

γ-Butyrolactone

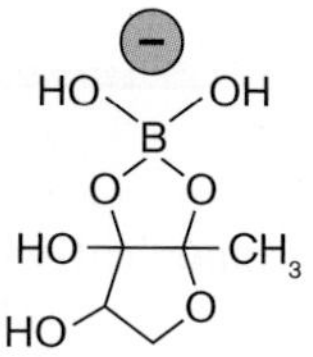

Furanosyl borate diester

Interspecies Communication

Some microbial species not only chemically talk among themselves but can communicate with other species. The marine bacterium *Vibrio harveyi*, for example, uses three different, but converging, quorum-sensing systems to coordinate control of its luciferase. These sensing pathways are very different from the *Vibrio fischeri* system. One utilizes an acyl homoserine lactone (AHL) as an autoinducer (AI-1) to communicate with other *V. harveyi* cells.

A second system produces a different autoinducer (AI-2), which contains borate. Bonnie Bassler (**Fig. 10.33**) and colleagues at Princeton University found that distantly related organisms such as *Salmonella* can activate the AI-2 pathway of *V. harveyi*, dramatically supporting the concept of cross-species communication. Because many bacterial species can produce this second signaling molecule, it is thought that mixed populations of microbes use it to "talk" to each other. Recently, Karina Xavier (NOVA University of Lisbon, Portugal) demonstrated that *E. coli* engineered to overproduce AI-2 could change the composition of gut microbiota when fed to mice.

The third quorum-sensing system of *V. harveyi* was discovered initially in *Vibrio cholerae* and was named cholera autoinducer 1 (CAI-1). Whereas AI-1 is species-specific and AI-2 allows for communication between distantly-related species, CAI-1 falls in between with its specificity, allowing different species of the *Vibrio* genus to communicate. Clearly, cell-cell communication is a highly sophisticated regulatory mechanism for light production by *V. harveyi*.

The three autoinducers of *V. harveyi* are synthesized by cytoplasmic enzymes and are recognized by specific membrane sensor kinase proteins (**Fig. 10.34**). Note that these systems detect autoinducers in the periplasm, not within the cytoplasm like the systems of *V. fischeri* (**Fig. 10.31**). At low cell densities (no autoinducer), all three sensor kinases initiate phosphorylation cascades that converge on a shared response regulator, LuxO, to produce phosphorylated LuxO. LuxO-P activates expression of small RNAs called Qrr that promote degradation of mRNA encoding the *lux* operon activator

FIGURE 10.33 ▪ **Bonnie Bassler.** Bassler was instrumental in characterizing interspecies communication in bacteria.

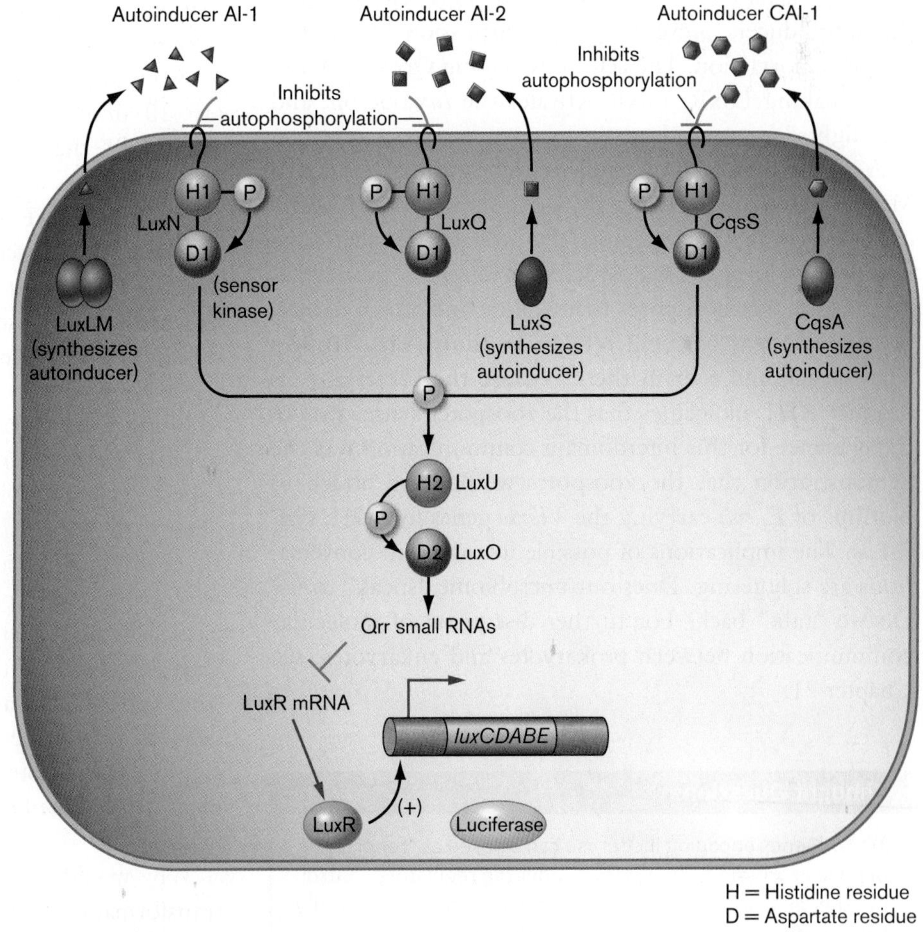

FIGURE 10.34 ▪ **Three quorum-sensing systems of *Vibrio harveyi*.** In the absence of autoinducers (AI-1, AI-2, and CAI-1), all three sensor kinases trigger converging phosphorylation cascades that end with the phosphorylation of LuxO. Phosphorylated LuxO (LuxO-P) activates the expression of Qrr small RNAs that promote the degradation of LuxR mRNA. Luciferase is not made. As autoinducer concentrations increase, they inhibit autophosphorylation of the sensor kinases and the phosphorylation cascade. As a result, Qrr levels decrease, allowing synthesis of the LuxR regulator, which activates the *lux* operon.

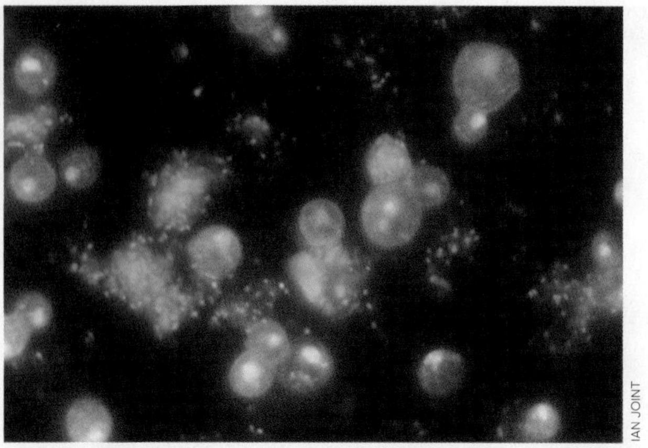

FIGURE 10.35 ▪ *Enteromorpha* zoospores. Zoospores of the alga *Enteromorpha* (red) attach to biofilm-producing bacteria (blue) in response to lactones produced by the bacteria.

LuxR (LuxR is not a homolog of the *V. fischeri* LuxR, but it plays a similar role). Thus, at low cell densities the culture does not display bioluminescence. At high cell densities, the autoinducers prevent signal transmission by inhibiting phosphorylation. The cell stops making Qrr sRNA and starts making LuxR. LuxR activates the *lux* operon, and the "lights" are turned on.

A report by Ian Joint and his colleagues at Plymouth Marine Laboratory, Plymouth, U.K., showed that bacteria can even communicate across the prokaryotic-eukaryotic boundary. The green seaweed *Enteromorpha* (a eukaryote) produces motile zoospores that explore and attach to *Vibrio anguillarum* bacterial cells in biofilms (**Fig. 10.35**). They attach and remain there because the bacterial cells produce AHL molecules that the zoospores sense. Part of the evidence for this interdomain communication was the demonstration that the zoospores would even attach to biofilms of *E. coli* carrying the *Vibrio* genes for AHL synthesis. The implications of possible interdomain conversations are staggering. Does our microbiome "speak" to us? Do we "talk" back? For further discussion of molecular communication between prokaryotes and eukaryotes, see Chapter 21.

Thought Questions

10.12 Genes encoding luciferase can be used as "reporters" of gene expression when placed under the regulatory control of other genes. Luminometers are machines that can quantify light production (luminescence) from luciferase. From the discussion in Section 9.2, propose an experiment to confirm that RecA is induced during the SOS response.

10.13 What would happen if a culture were coinoculated with *Vibrio (Aliivibrio) fischeri luxI* and *luxA* mutants, neither of which produces light?

Quorum Sensing in Gram-Positive Organisms, and the Activation of Natural Transformation

Natural transformation (see Section 9.3) in Gram-positive organisms typically involves the growth phase–dependent assembly of a transformasome complex across the cell membrane (**Fig. 10.36**). The transformasome is composed of a binding protein that captures extracellular DNA floating in the environment, plus proteins that form a transmembrane pore. A nuclease degrades one strand of a double-stranded DNA molecule while pulling the other strand intact through the pore and into the cell. Once inside, the strand can be incorporated into the chromosome by recombination, a process discussed in Section 9.2.

Once the transformasome is assembled, the cell is **competent**, meaning that it can import free DNA fragments and incorporate them into its genome by recombination. What triggers growth phase–dependent competence? For some Gram-positive bacteria, competence for transformation is generated by quorum sensing that takes place between members of the culture. Every individual in a growing population produces and secretes a small peptide (15–20 amino acids), called competence stimulation peptide (CSP), that progressively accumulates in the medium until it induces a genetic program that makes the population competent (**Fig. 10.36**, step 1). The CSP sequence is unique to each species, as are the specifics of the induction process. For *Streptococcus pneumoniae*, the level of CSP in the medium increases (step 2) as the population increases—that is, as the cell density increases.

Above a certain concentration threshold, CSP is able to bind to a sensory protein built into the cell membrane (ComD for *S. pneumoniae*). This binding begins what is called a phosphorylation cascade (the passing of a phosphate group from one protein to another; **eTopic 4.1**). In the competence phosphorylation cascade, the sensory protein phosphorylates itself using ATP and then passes the phosphate to a cytoplasmic regulatory protein, ComE, which stimulates expression of *comX* (**Fig. 10.36**, step 3). ComX is an alternative sigma factor specifically used to transcribe genes encoding the transformasome (step 4). The protein products of these genes are assembled at the membrane, and the cell becomes competent (step 5).

Why would organisms use quorum sensing to regulate transformation competence? One hypothesis holds that cells are unlikely to encounter stray DNA when growing in dilute natural environments such as ponds, where other bacteria are scarce. So, in this situation, why waste energy making the transformasome? When these same cells are growing at high density, as in a biofilm, they are more likely to encounter DNA released from dying neighbors.

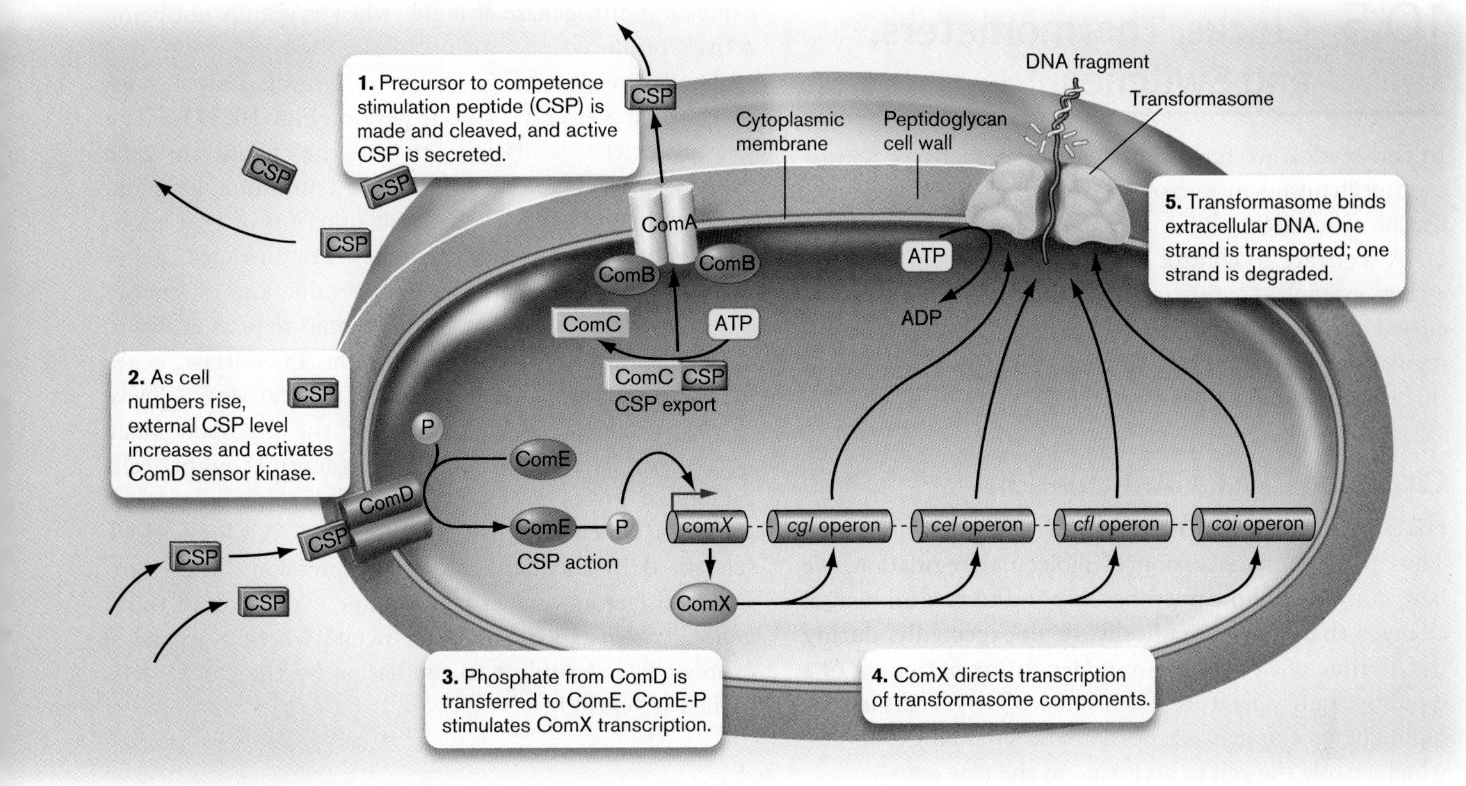

FIGURE 10.36 ■ Quorum-sensing regulation of transformation in *Streptococcus*. The process of transformation in *Streptococcus* begins with the synthesis of a signaling molecule (competence stimulation peptide, CSP) and concludes with the import of a single-stranded DNA strand through a transformasome complex.

This is DNA they could use to repair their own damaged genomes, to consume as food, or to sample for a new survival mechanism, should the DNA come from a different species present in a biofilm consortium. Regulation by quorum sensing would ensure that the transformasome would not form until there was a good chance that free foreign DNA was available.

Thought Question

10.14 **Figure 10.36** illustrates the process of transformation in *Streptococcus pneumoniae*. Would a mutant of *Streptococcus* lacking ComD be able to transform DNA?

To Summarize

- **The stringent response** is triggered during nutrient limitation when low cellular amino acid levels cause ribosomes to idle and synthesize the signaling molecule ppGpp. Binding of ppGpp to RNA polymerase decreases synthesis of rRNA, which slows the rate of new ribosome synthesis. The overall rate of translation, then, will match growth rate.
- **The second messenger cyclic di-GMP (c-di-GMP)** is made by numerous proteins containing a GGDEF amino acid motif. Many cell functions are influenced by c-di-GMP, including biofilm formation and motility.
- **Quorum sensing** involves the synthesis, secretion, and extracellular accumulation of small autoinducer signaling molecules. Cells within a population sense a threshold concentration of autoinducer and simultaneously respond by expressing a subset of genes.
- **Quorum sensing enables communication between cells** of a single species or between multiple species.
- **Pathogens use quorum sensing** to time the expression of virulence genes during growth within a host.
- **Quorum sensing can control natural transformation** in Gram-positive bacteria such as *Streptococcus pneumoniae*.

10.5 Clocks, Thermometers, and Switches

In this section we further explore the diversity of regulatory mechanisms in microbes. We will examine how cells use protein clocks, RNA thermometers, and changes in DNA sequence to control gene expression. Another interesting example—regulation by protein splicing—is discussed in **Special Topic 10.1**. These examples all show how regulation can be customized to fit the specific demands that the environment places on the cell.

Circadian Clocks: Anticipation Rather than Response

Thus far in our discussion of molecular regulation, we have considered how microbes respond to environmental changes that arise sporadically, or unexpectedly, during the lifetime of a microbe—a sudden influx of lactose, or a rapid rise in temperature, for instance. Such events induce rapid changes in gene expression and physiology, which enable the cell to acclimate to the new conditions. But what if the environmental change were predictable in the sense that it happened at regular time intervals? In this case, might evolution favor organisms that could anticipate this change and optimize its activities in preparation? The answer is yes, and we need not look any further than our own bodies, which have the ability to measure out time of day through the use of a circadian clock, which helps to set our sleep rhythms. As we will discuss, some microbes also possess clocks that help them regulate their activities in anticipation of the changes that occur over the day.

Circadian clocks maintain a period of approximately 24 hours because that is how long it takes Earth to rotate on its axis (**Fig. 10.37A**). Cells at a fixed position on the globe receive exposure to sunlight that varies as a function of this 24-hour period of rotation. Because the 24-hour light-dark cycles are highly reproducible, natural selection has generated biological clocks that can keep pace.

Photosynthetic bacteria of the phylum Cyanobacteria possess circadian clocks that help cells prepare for activities at specific times of day. For example, cyanobacteria perform photosynthesis during the day, converting light into chemical energy to fix carbon dioxide into biomass, producing oxygen as a waste product. Some cyanobacteria can also fix N_2 gas, but the key enzyme nitrogenase is inhibited by oxygen. Unicellular nitrogen-fixing cyanobacteria solve this dilemma by expressing nitrogenase at night, when oxygen is not being actively produced.

The expression of genes under circadian clock control oscillates over a 24-hour period (**Fig. 10.37B**). The phasing of the oscillation—that is, the changes in gene expression from a maximum level to a minimum level and back again—is often tied to the function of that regulated gene or set of related genes. For instance, genes involved in photosynthesis and production of energy storage molecules such as glycogen tend to peak at dawn, in anticipation of the coming daylight. In contrast, genes involved in "nighttime" processes peak at dusk. These genes include those that consume the glycogen made during the day. Deprived of light energy at night, cells catabolize glycogen as an energy source for nighttime activities, which can include cell division. While the clock sets the daily oscillation in gene expression, other transcription factors can set the baseline expression of those genes. For instance, they can dictate whether a gene is expressed (and subject to oscillation by the clock) or is turned off entirely.

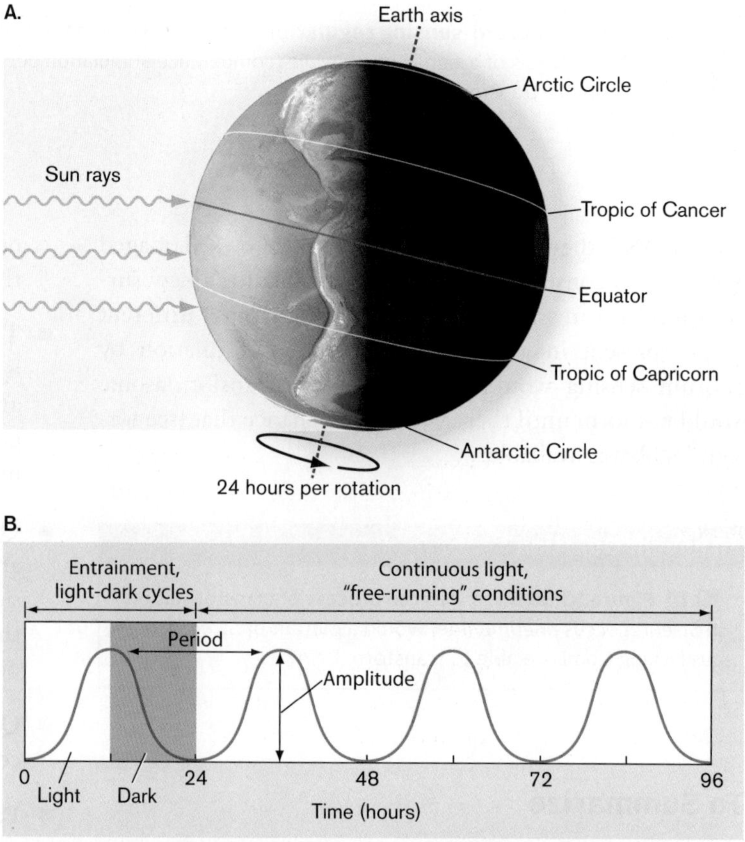

FIGURE 10.37 ▪ Circadian expression in cyanobacteria is defined by light-dark periodicity. A. The 24-hour circadian period is set by the speed of Earth's rotation on its axis. **B.** Oscillation of gene expression over the light-dark photoperiod entrains the clock. In continuous light, the clock maintains the 24-hour oscillations in output during this "free-running" condition. *Source:* Part A modified from Wikimedia.org; part B modified from Johnson et al. 2017. *Nat. Rev. Microbiol.* **15**:232–242, box 1 fig.

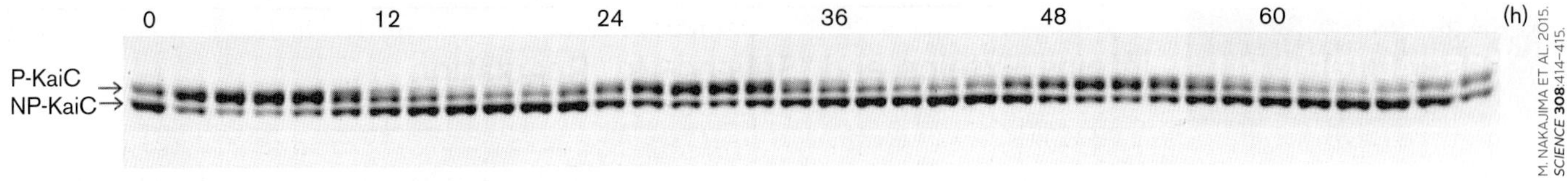

FIGURE 10.38 ■ Purified circadian clock proteins in a test tube maintain 24-hour periodicity in the phosphorylation state of KaiC. KaiA, KaiB, and KaiC proteins were incubated with ATP, and the phosphorylation state of KaiC was assayed every 2 hours for several days. P-KaiC and NP-KaiC refer to phosphorylated and nonphosphorylated states, respectively, which can be distinguished because they migrate as separate bands through a gel during electrophoresis. Band intensity is a function of protein concentration.

Scientists use three criteria to establish the presence of a circadian clock regulator of gene expression. First, the 24-hour periodicity of gene expression oscillation must persist if the cell is shifted to constant light (or dark) conditions (**Fig. 10.37B**). Second, it must be possible to entrain (reset) the period by manipulating the environmental cycle of light and dark, which can help the cell make subtle adjustments to the clock. Third, the oscillator must be temperature compensated, meaning that it will maintain a 24-hour period even when temperature is changed. This third criterion is important because temperature can affect the rate of enzymatic reactions and potentially alter the timing mechanism of the circadian clock, unless compensation is possible.

The proteins that make up the circadian clock, or oscillator, of cyanobacteria are KaiA, KaiB and KaiC. These proteins are not homologous to the clock proteins of animals and thus have a different evolutionary origin. The cyanobacterial clock operates by cycling between phosphorylated and dephosphorylated states of KaiC. Phosphorylation of KaiC changes its structure and consequently its ability to bind other proteins. KaiC performs autophosphorylation and autodephosphorylation of itself, but uses the activity of KaiA and KaiB to control the timing of (de)phosphorylation and output of the clock. This three-protein clock has a remarkable property (**Fig. 10.38**): If it is reconstituted in a test tube and supplied with ATP, it can keep 24-hour time for days or even weeks, even in constant darkness!

The clock controls global gene expression through the ability to modify the extent of chromosome compaction and the ability to regulate the activity of a two-component regulatory system. Toward the end of the day, KaiC is in a semiphosphorylated state and is able to bind and activate a two-component regulatory system that activates some genes and represses others. Some of these proteins encode sigma factors and anti-sigma factors that disseminate the clock signal to a wider set of genes in the genome. At dusk, the two-component system also turns on genes involved in catabolism, important for energy production at night, and turns off genes involved in photosynthesis.

What happens to cyanobacteria if their clock is broken? Mutants that lack a clock are still viable and grow relatively well under a normal light-dark photoperiod. Thus, clocks are not essential for life in these organisms, but perhaps their contributions to fitness are more subtle. To address this possibility, Carl Johnson (**Fig. 10.39A**) and colleagues at Vanderbilt University first generated mutants of *Synechococcus* that had different free-running clock periods. Some mutants had free-running periods (FRPs) that were shorter (22 hours) or longer (30 hours) than that of the wild type (25 hours; **Fig. 10.39B**). These strains were placed in cocultures and subjected to different lengths of day to determine whether clock period affected fitness. The mutants with 22-hour periods outgrew the wild type when a 22-hour day (11 hours light, 11 hours dark) was imposed (**Fig. 10.39C**). Likewise, mutants with 30-hour periods outgrew the wild type when a 30-hour day was imposed. Importantly, this advantage was specific to day length, since the 22-hour mutants were less fit than the wild type in 30-hour days, while the 30-hour mutants were less fit than the wild type in 22-hour days. This study did not identify the physiological basis of the fitness advantages but did demonstrate clearly that when clocks match the light-dark cycle, they provide a growth advantage to the organism.

Clocks are not restricted to cyanobacteria, as we saw in the Current Research Highlight. While gut bacteria such as *Enterobacter aerogenes* do not photosynthesize, their environment is impacted by the human circadian clock. *E. aerogenes* has homologs to the clock genes of cyanobacteria, and these may provide this organism with a selective advantage in the dynamic gut environment.

Sigma Factor Control by RNA Thermometers and Proteolysis

Excessive heat, above 42°C for *E. coli*, causes many proteins to denature and membrane structure to deteriorate. All cells subjected to heat above their optimal growth range will express a set of proteins called heat-shock proteins. These proteins include chaperones that refold damaged proteins (see Fig. 8.33), and a variety of other proteins that affect DNA and membrane integrity. The transcription of many *E. coli* heat-shock genes requires the heat-shock sigma factor sigma H (also called sigma-32, RpoH, σ^H, or σ^{32}). So, one of the first responses to exposure to

SPECIAL TOPIC 10.1 Inteins, Exteins, and "Spliced-Up" Regulation

Pyrococcus horikoshii is a marine, thermophilic member of the Archaea, isolated from hydrothermal vent fluid in the Okinawa Trough, located in the western Pacific Ocean. Like most microbes, it utilizes a recombinase system to repair DNA (see Section 9.2). Its RecA homolog (called RadA) has a very unusual feature: It contains a special polypeptide fragment, called an "intein," that keeps the protein in an inactive state. Remarkably, the substrate of the RadA protein, single-stranded DNA, triggers this intein to excise itself, splicing the two fragments of the RadA protein together, which enables this enzyme to perform its function in DNA recombination. This process was discovered by Marlene Belfort and Christopher Lennon at the University of Albany (**Fig. 1**).

Inteins are a special form of mobile genetic element found in bacteria, archaea, and single-celled eukaryotes (**Fig. 2A**). They have a unique and remarkable relationship with their host gene: They insert as DNA but can excise as protein. Their DNA sequence encodes endonucleases that insert a copy of the DNA of the mobile element into the coding region of genes by exploiting the cell's double-strand-break repair recombination mechanism. When the target gene is expressed, it translates the element's code as well, resulting in a hybrid protein, with the intein-encoded peptide between the split halves of the protein, called "exteins" (**Fig. 2A**). This hybrid protein is usually inactive, because the inteins bias their insertion to the active sites of enzymes. However, the intein peptide is catalytic and is capable of excising itself perfectly out of the host protein, even after the host protein folds into its 3D structure. When excision happens, the host protein's function is restored, because the intein splices together the extein fragments that the host leaves behind, without a trace of its presence remaining (**Fig. 2A**).

COURTESY OF CHRISTOPHER LENNON AND MARLENE BELFORT

FIGURE 1 ▪ Christopher Lennon (left) and Marlene Belfort.

Although inteins were originally viewed as "selfish" elements, some inteins have been recently observed to serve as sensors: they regulate their excision events in response to environmental stimuli, and by this excision they activate the host protein. Some of these proteins are involved in responding to environmental stress, such as oxidative stress, and remarkably, the stress triggers intein excision and protein activation. In this way, inteins serve as an unusual but nonetheless effective means of posttranslational regulation of gene expression.

Belfort and Lennon used an in vitro assay of purified RadA to analyze the intein excision event. Different stimuli were provided, and they ran the products on a polyacrylamide gel to separate the proteins by size. Proteins were visualized with the Coomassie blue stain (**Fig. 2B**). At time zero, before any stimulus was provided, the RadA protein consisted mostly of a 49-kDa "precursor" protein, indicating that the intein was present. A smaller amount of intein-free RadA (29.3 kDa) was also present, representing "baseline" intein excision in the absence of stimulus.

When incubated in TE buffer alone, the protein profile did not change. Exposure to ssDNA, however, increased intein excision and yielded more intein-free RadA and a new product consistent with the excised intein dimer (19.8 kDa × 2). This result demonstrated that ssDNA activates intein excision and extein splicing. Stimulation of excision was specific for single-stranded DNA, the specific substrate for RadA when it's a functional enzyme. When the investigators substituted

elevated temperature is an increase in the amount of sigma H protein. The concentration of sigma H protein is tightly regulated by two temperature-sensitive processes: RNA melting and protein degradation (**Fig. 10.40**).

The gene encoding sigma H is *rpoH*. At 30°C, *rpoH* mRNA adopts a secondary structure at the 5′ end that buries a ribosome-binding site, so *rpoH* mRNA is poorly translated. The 5′ region of *rpoH* mRNA is called the ROSE element for repression of heat-shock gene expression. A sudden rise in temperature melts this secondary structure and exposes the ribosome-binding site, enabling translation to initiate more frequently. Thus, heat shock increases sigma H synthesis, which in turn increases transcription of the heat-shock genes whose products include chaperones and proteases.

Proteolysis also controls sigma H accumulation. At 30°C, the *rpoH* mRNA is poorly translated, as previously described, but some sigma H protein is made. Inappropriate expression of heat-shock genes at 30°C is prevented by the ssDNA with deoxyribonucleotides (dNTPs), double-stranded DNA (dsDNA), or RNA, they saw no stimulation. Hence, excision and resulting activation of the enzyme is triggered by the presence of the enzyme's substrate.

Subsequent study of this system led Belfort and Lennon to propose a model for how ssDNA stimulates intein excision (**Fig. 2C**). In the absence of ssDNA, the intein-containing RadA precursor folds such that the intein component interacts with the C-terminal RadA extein to keep the intein's excision machinery inactive. When present, ssDNA disrupts the interaction between the intein and the C-terminal extein. This disruption activates the intein machinery to excise the intein and join the two exteins to make an active RadA recombinase. The RadA enzyme can now bind its ssDNA substrate and perform its function in DNA recombination for repair and/or horizontal gene transfer.

RESEARCH QUESTION

If you found that a protein became activated posttranslationally, during exposure to an environmental stimulus, what steps would you take to determine whether this change was due to the presence of an intein in the protein that could sense the stimulus?

Lennon, Christopher W., Matthew Stanger, and Marlene Belfort. 2016. Protein splicing of a recombinase intein induced by ssDNA and DNA damage. *Genes and Development* **30**:2663–2668.

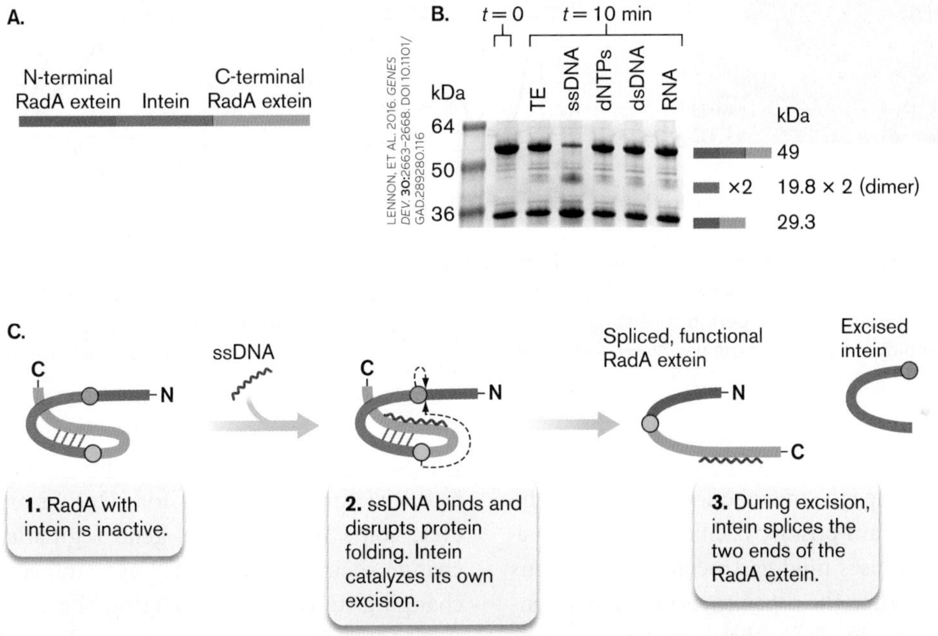

FIGURE 2 ■ Intein excision from RadA is induced by ssDNA. **A.** The intein (red) is situated between two "extein" fragments of RadA (blue and green). **B.** Coomassie blue–stained gel of the products of in vitro reactions at time zero and after 10 minutes of incubation with various substrates. The first lane is a set of size standards (units in kilodaltons, kDa). TE is a buffer-only negative control. **C.** Model for how ssDNA activates the excision machinery of the intein. *Sources:* Part B modified from Lennon et al. 2016. *Genes Dev.* **30**:2663–2668, fig. 1A; part C, from M. Belfort. 2017. *Curr. Opin. Microbiol.* **38**:51–58, fig. 4C.

DnaK-DnaJ-GrpE chaperone system, which interacts with sigma H and shuttles it to various proteases for digestion (see **Fig. 10.40**). At 42°C, however, proteolysis of sigma H decreases, and sigma H is allowed to accumulate. Sigma H degradation decreases because at the higher temperature, the chaperones are engaged in refolding and rescuing the heat-denatured proteins of the cell. This chaperone redeployment frees sigma H to transcribe the heat-shock genes, which include the chaperone genes *dnaK*, *dnaJ*, and *grpE*. These genes also have promoters that depend on other sigma factors to drive basal expression. Thus, as the temperature rises, the amount of sigma H is increased by two temperature-dependent mechanisms: One increases translation by exposing the ribosome-binding site (a so-called RNA thermometer), while the second redeploys the chaperones that direct its proteolysis.

Many other examples of RNA thermometers exist, some of which are involved in pathogenesis. Bacteria that infect

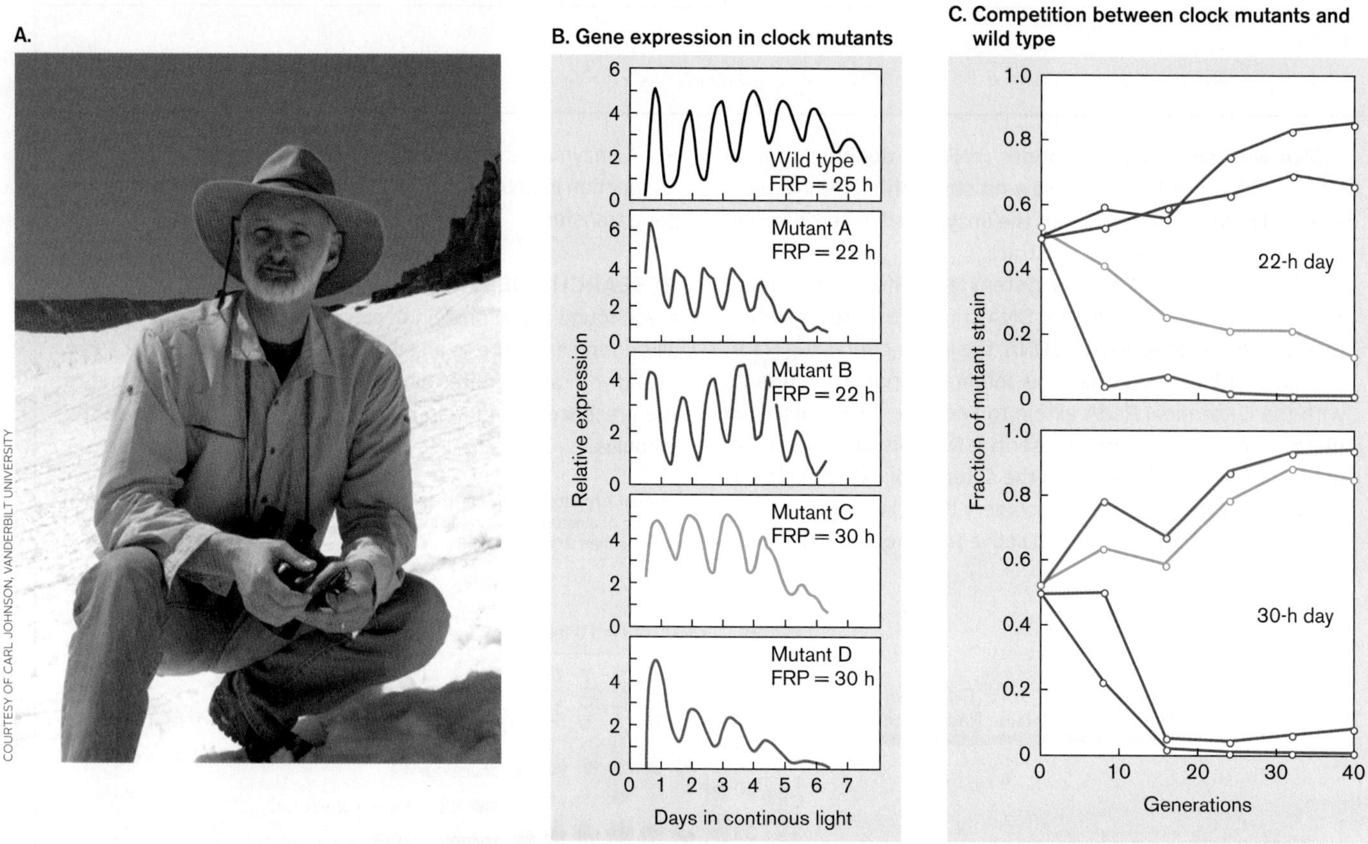

FIGURE 10.39 ▪ The clock confers a fitness advantage in cyanobacteria. **A.** Carl Johnson. **B.** Johnson and colleagues isolated mutants with free-running periods (FRPs) that were shorter (22 hours) or longer (30 hours) than that of the wild type (25 hours) when grown in continuous light. **C.** In competition under a period consisting of equal amounts of light and dark, the abundance of the clock mutants relative to the wild type is dictated by the length of the day. *Source:* Parts B and C modified from M. A. Woelfle et al. 2004. *Curr. Biol.* **16**:1481–1486, figs. 3A and 3C.

mammals have RNA thermometers set to body temperature, 37°C—a set point that helps the microbe sense entry into the mammalian host. The Gram-positive bacillus *Listeria monocytogenes*, for instance, causes mild gastroenteritis or serious meningitis in humans. When the organism is ingested, it synthesizes a regulatory protein, PrfA, that activates transcription of a number of virulence genes. Like *rpoH* in *E. coli*, the 5′ end of *prfA* mRNA contains a ROSE element that prevents translation until the temperature rises—after ingestion.

DNA Rearrangements That Alter Gene Expression

Most regulatory mechanisms that alter gene expression use interactions between proteins and DNA, proteins and RNA, or RNA and RNA. These control mechanisms are easily reversible. A more drastic means of control, however, involves altering the DNA sequence itself. A classic example of this strategy is phase variation. Phase variation helps microbial pathogens avoid the immune system.

Any infection of a host will trigger the production of antibodies specific to the invading microbe's component parts, such as pili, flagella, and lipopolysaccharides (discussed in Chapter 24). Antibodies that bind to these microbial surface structures are useful for clearing an infection. However, some microbes use gene regulation to periodically change their immunological appearance, like a chameleon changing its color, by changing the amino acid composition of a particular surface protein. This "shape-shifting" by the microbe, called **phase variation**, renders useless those antibodies specific for the old structure. The embattled immune system must start all over again making new antibodies, thus prolonging the course of infection. Two types of DNA rearrangement can be used to generate phase variations: gene inversions and slipped-strand mispairing.

Gene inversion: an on/off switch. Flagellar phase variation in the Gram-negative bacterium *Salmonella enterica* involves a DNA recombination event known as gene inversion that flips the orientation of a gene or DNA segment in the chromosome. *S. enterica* has two genes, widely separated on the chromosome, that encode different forms of flagellin, the protein from which flagella are made. A reversible DNA inversion turns off one gene while turning on the other. The invertible switch is a 993-bp DNA fragment (or cassette), called the H region. The H region is flanked by short

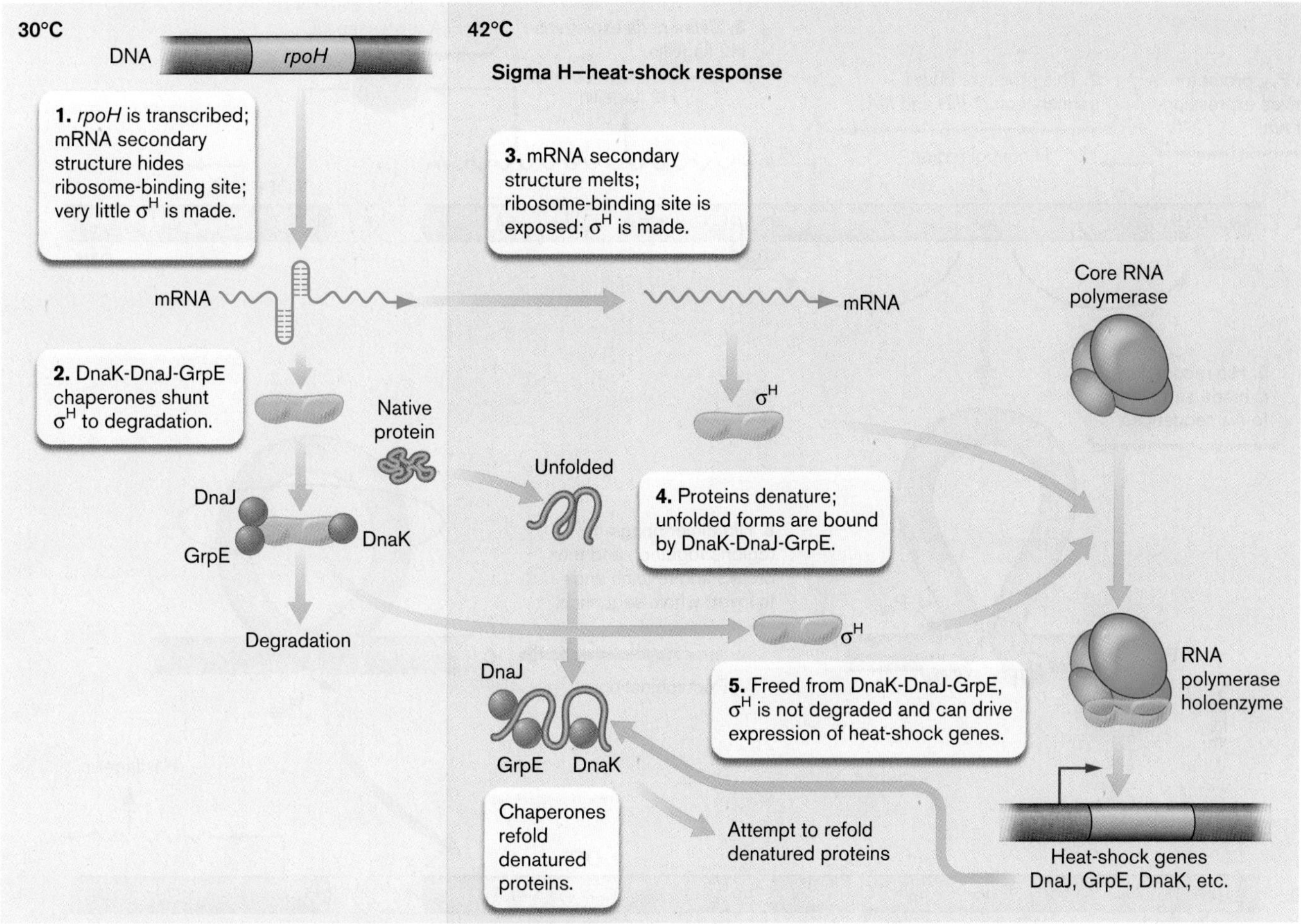

FIGURE 10.40 ■ **The heat-shock response of *E. coli*.** Two mechanisms control sigma H levels. The small amount of sigma H that can be made at 30°C (left) is met by the DnaK-DnaJ-GrpE chaperone system and shuttled toward degradation. At 42°C (right), however, misfolded cytoplasmic proteins siphon off the chaperone trio and release sigma H to direct transcription of the heat-shock genes.

(26-bp) inverted repeats called *hixL* (left) and *hixR* (right). These inverted repeats are where the inversion occurs.

Note: A **direct repeat** is a sequence found in identical form at two sites on the same double helix (for example, 5′-ATCGATCGnnnnnnATCGATCG-3′). A **tandem repeat** is a direct repeat without any intervening DNA sequence (for example, ATCGATCGATCGATCGATCGATCG). Like a direct repeat, an **inverted repeat** is a sequence found in identical forms at two sites on the same double helix; however, the sequences are inverted relative to one another (for example, 5′-ATCGATCGnnnnnnCGATCGAT-3′).

The H region contains the *hin* gene, controlled by the P_{hin} promoter. The *hin* product Hin recombinase mediates the inversion (**Fig. 10.41**). In antigenic terms, "H antigen" refers to flagella, so the acronym "Hin" stands for H inversion. The final important component of the H region is the P_{fljB} promoter, found at the 3′ end of the region.

Hin recombinase collaborates with other less specific DNA remodeling proteins to link the 26-bp left (*hixL*) and right (*hixR*) ends of the invertible DNA element. The two ends, each bound to a Hin monomer, are brought together by Hin-Hin protein interactions. DNA within the cassette then forms a loop. Hin cuts within the center of each *hix* site, producing staggered ends. An exchange of Hin subunits leads to strand inversion, so that the orientation of the DNA cassette is reversed relative to the flanking DNA on either side.

In one orientation, the P_{fljB} promoter of the H region directs expression of H2 flagellin (encoded by *fljB*) and a repressor (FljA) that prevents transcription of the other flagellin gene, *fliC* (see **Fig. 10.41A**). After the inversion, however, the P_{fljB} promoter points in the wrong direction, so there is no production of H2 flagellin or FljA, the repressor of *fliC* (see **Fig. 10.41D**). Once the existing repressor proteins degrade or become diluted during cell replication, the *fliC* flagellin gene can be expressed. Thus, H1 flagellin (present in phase 1 cells) is synthesized instead of H2 flagellin (present in phase 2 cells). The amino acid sequences, and thus the antigenicity, of the two flagellar proteins are different. Inversion of the H region enables *Salmonella* to change how it appears to a host immune

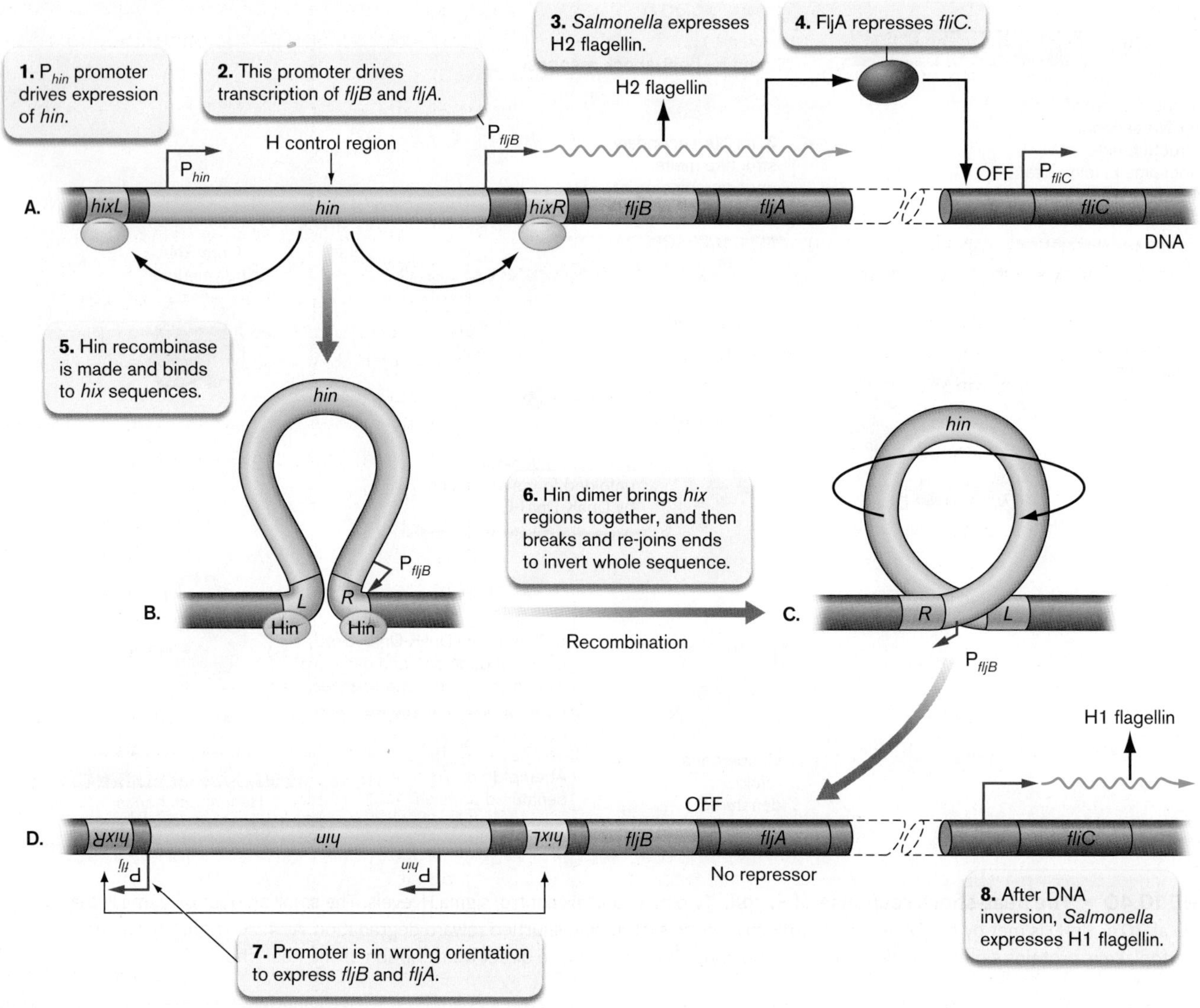

FIGURE 10.41 ■ Phase variation of flagellar proteins in *Salmonella enterica*. An invertible region containing a promoter controls the expression of two unlinked flagellar protein genes. In one orientation **(A)**, the *fljB* promoter drives synthesis of H2 flagellin (*fljB*) and a repressor (FljA) of the H1 flagellin gene (*fliC*). Action by Hin recombinase causes the segment to invert **(B, C)**, thereby reorienting the promoter. Because the repressor FljA is no longer synthesized, the gene for H1 can be expressed **(D)**.

system. In each generation, the rate of the reversible switch varies from about one cell in 10^3 to one in 10^5. Flagellar switching is especially important for the late stage of infection by *Salmonella*, once it has passed through the intestine and reaches the blood and spleen (see Chapter 25).

Thought Question

10.15 While viewing **Figure 10.41**, imagine the phenotype of a cell in which *fljB* has been deleted but *fljA* is still expressed. Would cells be motile? What type of flagella would be produced? Would the cells undergo phase variation? What would happen if *fliC* alone were deleted?

Slipped-strand mispairing. A different type of phase variation relies on multiple, short sequence repeats within a gene. The repeats "confuse" DNA polymerase as it replicates, causing it to slip occasionally during replication. Slippage either adds a repeat to the gene or deletes a repeat from it and alters the gene's translational reading frame. If the mRNA produced during transcription is out of frame, the protein is not made. This random process can alternately turn a gene off and then back on again in subsequent generations.

Like gene inversion in *Salmonella*, slipped-strand mispairing can contribute to virulence in pathogenic microbes. As discussed in **eTopic 10.4**, *Neisseria gonorrhoeae*, the causative agent of gonorrhea, uses slipped-strand mispairing to vary the types of outer membrane proteins expressed and exposed to the immune system. Eukaryotic microbes, especially pathogenic sporozoa, possess elaborate phase variation mechanisms. The trypanosome that causes "sleeping sickness" undergoes extensive genetic shuffling and mutation of its coat proteins over successive generations, essentially overwhelming the host immune system by presenting every possible form of antigen.

To Summarize

- **Circadian clocks** provide oscillations in gene expression that anticipate recurring changes in environment, such as the daily light-dark cycle on Earth.
- **RNA thermometers** are secondary structures at the 5′ end of specific mRNA molecules that can obscure access to ribosome-binding sites or transcriptional terminator stems. Cold conditions that stabilize the secondary structures will block translation of an mRNA.
- **Gene rearrangement controls** include invertible promoter switches or repetitive DNA sequences within a coding region that cause DNA polymerase to "slip" during DNA synthesis.

10.6 Chemotaxis: Posttranslational Regulation of Cell Behavior

In the prior sections of this chapter we have described the regulation of gene expression and given examples such as biofilm formation, whereby gene expression changes in response to environmental stimuli confer physiological changes on the microbe. Microbes can also control their behaviors through the posttranslational modification of regulatory proteins. Such modification often involves the phosphorylation cascades of two-component regulatory systems (see **Fig. 10.4**). In these cases, the response regulator controls the activity of another protein, rather than the initiation of transcription in target genes. One common example of such posttranslational regulation in bacteria involves the ability of cells to direct their movement through aqueous environments.

Chemotaxis, introduced in Chapter 3, is the ability of an organism to sense chemical gradients and modify the direction of its motility in response. Chemotaxis is important to bacteria for a number of reasons. It provides a useful survival strategy in nature, keeping bacteria moving toward nutrients and, with chemoreceptors that sense repellents, away from trouble (for example, toxic compounds). For commensal bacteria or pathogens, chemotaxis can also be used to move the organism toward a cell surface to which it can attach. For example, the intestinal lining can exude chemical attractants that, like a beacon, will lead the bacteria in the intestine toward the cell surface where they can attach. In sum, chemotactic sensory perception plays a major role in structuring microbial communities, in affecting microbial activities, and in influencing various microbial interactions with their surroundings.

Moving with a Purpose

Chemotactic cells can sense chemical gradients with receptor proteins in the cytoplasmic membrane and use a signal transduction mechanism to change the direction of spin of their flagella. In this section we connect the dots between the receptor and the flagellum, to understand how the cell controls flagellar spin and, ultimately, its swim direction.

Let's first talk about how bacteria can change the direction of their swim. Recall that the flagellar motor can rotate clockwise or counterclockwise (see Section 3.6, Fig. 3.42) and that, on any one bacterium, all motors coordinate their rotations. Thus, all flagellar motors on a single cell will rotate clockwise (or counterclockwise), and then suddenly switch to the opposite rotation. In the counterclockwise mode, all flagella sweep behind the cell, forming a rotating bundle that propels the organism forward in what is called smooth swimming or a "run." When flagellar rotors suddenly switch to a clockwise rotation, the bundle is disrupted and the bacterium "tumbles" in a random fashion. When the flagellar rotors switch back to counterclockwise rotation, swimming resumes, oriented in a new, random direction as a result of the tumble.

The key to chemotaxis is a mechanism that suppresses the number of tumbles an organism makes when it moves from lower to higher concentration of an attractant chemical (see Section 3.6, Fig. 3.42). For instance, an organism moving toward an attractant may tumble only twice in 5 seconds. In contrast, an organism moving in the wrong direction (that is, toward a lower concentration of attractant) may tumble eight times in 5 seconds. If, all of a sudden, the organism finds itself going in the right direction, the sensory transduction system will suppress the number of tumbles, and the cell will continue moving in the right direction.

How does the bacterium suppress tumble frequency? Let's start at the end of the system and work backward. The natural bias of the flagellar rotor is counterclockwise rotation (in other words, smooth swimming). In order to tumble, the cell phosphorylates a protein called CheY. CheY-P interacts with a rotor protein and flips the rotor in reverse to turn clockwise (**Fig. 10.42** ▶, step 1). CheA protein, the kinase that phosphorylates CheY, is the "key" to chemotaxis. The more CheA kinase activity there is, the more CheY-P is made, resulting in a higher frequency of tumbling (reorienting) events. CheA and CheY compose the two-component system that is embedded within the greater regulatory chemotaxis circuit. Another protein,

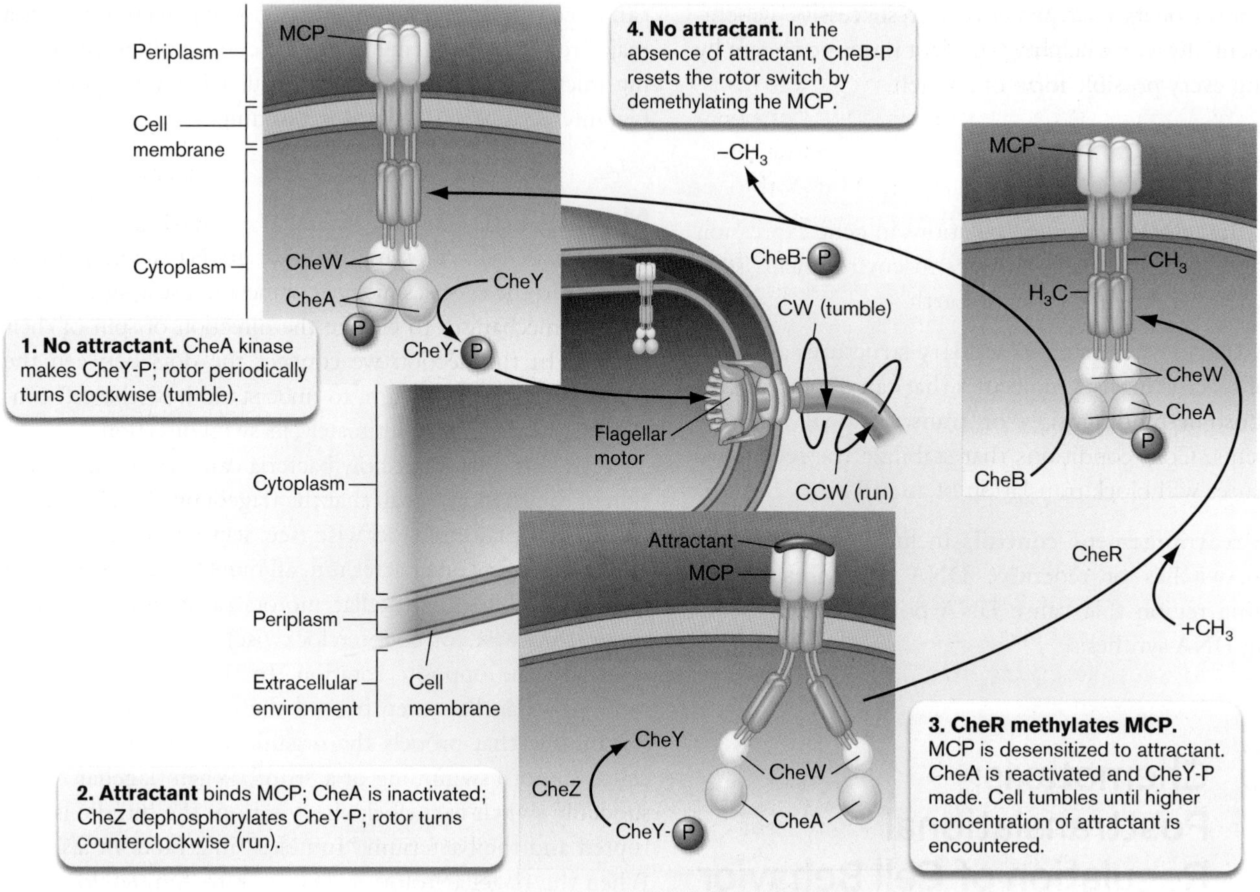

FIGURE 10.42 ■ **Chemotaxis.** Chemotaxis signaling pathway in *E. coli.*

CheZ, continually dephosphorylates CheY. When *E. coli* senses an attractant chemical like an amino acid at the cell surface, the result is decreased CheA kinase activity, which leads to lower CheY-P levels. The presence of fewer CheY-P molecules enables longer periods of counterclockwise rotation and, consequently, extended smooth swimming toward the attractant.

All this makes sense, but how does a cell recognize that it has moved into an area with attractant? The answer begins with clusters of special membrane-spanning proteins called **methyl-accepting chemotaxis proteins** (**MCPs**, or chemoreceptors) located at the poles of *E. coli* and other bacteria (**Fig. 10.42**, step 1). The periplasmic domains of different MCPs bind to different attractant molecules that flow into the periplasm as the cell moves. The cytoplasmic domain of each MCP binds to CheA (via an intermediary protein, CheW) and controls CheA activity. When a chemoattractant chemical, such as serine, binds to the periplasmic side of the MCP, the conformation of the cytoplasmic domain changes and inhibits CheA kinase activity (step 2). The CheY-P level then decreases (because CheZ continues to dephosphorylate CheY-P), and smooth swimming ensues.

Memory Helps Cells Move Up a Chemical Gradient

The chemotaxis signaling cascade just described makes sense when an organism first encounters a chemoattractant, but how does the cell know to keep moving into even higher concentrations? As it happens, chemotactic organisms possess a primitive form of memory. The conformational change in the MCP that inactivates CheA kinase activity also subjects the cytoplasmic side of the MCP to methylation (hence the name "methyl-accepting chemotaxis protein") by CheR methylase (**Fig. 10.42**, step 3). Methylation of glutamate residues in the cytoplasmic struts of the MCP reduces the binding affinity of MCP to the attractant, effectively desensitizing the system. This means that in order for MCP to maintain inhibition of CheA kinase and reward the cell with a continued run, an even higher concentration of the attractant must be present.

At the same time, CheR continues to reduce the affinity of MCP to the attractant by adding more methyl groups, and thus the cell needs to continue to move into higher concentrations of attractant to maintain the run. When the cell no longer moves into a higher concentration of attractant, MCP can no longer block phosphorylation of CheY, and the organism will tumble. If the bacterium moves to a lower concentration of attractant, the cell will keep tumbling at a high frequency until it moves back into a higher concentration, at which point tumbling is suppressed to enable a smooth run.

When the cell moves away from attractant, the methylation switch is reset by another protein, CheB-P, which removes methyl groups so that the system is resensitized to attractant (**Fig. 10.42**, step 4). The system is elegantly fine-tuned in that CheA kinase is also the protein that phosphorylates CheB. So, as the cell moves away from attractant, CheA kinase phosphorylates CheY to produce tumble, and it phosphorylates CheB to reset the sensitization switch.

Interestingly, while the basic chemotactic mechanism is evolutionarily conserved in bacteria, different genera use it differently. In the Gram-positive *Bacillus subtilis*, for example, ligand binding to an MCP stimulates CheA kinase, and CheY-P stimulates counterclockwise flagellar rotation and, thus, extended runs. This mechanism is the exact opposite of the one used in *E. coli*.

Thought Question

10.16 Using antibody to flagella, a cell of *E. coli* can be tethered to a glass slide via a single flagellum. Looking through a microscope, you will then see the bacterium rotate in opposite directions as the flagellar rotor switches from clockwise to counterclockwise rotation and back again (see **Fig. 10.42**). Which way will the bacillus rotate when an attractant is added? What would the interval between clockwise and counterclockwise switching be if you tethered the mutants *cheY*, *cheA*, *cheZ*, and *cheR* to the slide and then added attractant?

Diverse Mechanisms of Chemotaxis

Some chemotactic bacteria do not form flagellar bundles, and they perform runs and tumbles in different ways. For example, the rod-shaped *Rhodobacter sphaeroides* has a single flagellum, positioned curiously at the middle of the rod rather than at one of the poles. This flagellum provides a run when it spins at high speeds. If the spin slows down, the flagellar filament changes shape, forming a coil. This coil tumbles the cell as it spins. A return to higher spin speeds causes the flagella to revert to a shape capable of producing a run. Hence, this organism adjusts the speed, rather than direction, of flagellar rotation, to provide the runs and tumbles of chemotaxis.

To Summarize

- **Chemotaxis** is a behavior in which motile microbes swim toward favorable environments (chemoattractants) or away from unfavorable environments (chemorepellents).
- **The direction of flagellar motor rotation** determines the type of movement. Counterclockwise rotation results in smooth swimming; clockwise rotation results in tumbling.
- **Random movement toward an attractant** causes a drop in CheY-P levels, which enables counterclockwise rotation and smooth swimming.
- **Methyl-accepting chemotaxis proteins (MCPs)** clustered at cell poles bind chemoattractants and initiate a series of events that lowers CheY-P levels. Reversible methylation or demethylation of MCPs desensitizes or sensitizes MCPs, respectively.

Concluding Thoughts

Bacteria employ a vast array of regulatory mechanisms, both simple and complex, to make survival decisions. But do they possess cognitive ability? This is a controversial idea, but one worth pondering. Cognition in its broadest sense is the process of acquiring, processing, storing, and acting on information from the environment. Cyanobacteria use a clock to predict when the Sun will rise and fall. Two-component signal transduction systems, prominent in bacteria, contribute to cognition and can control the direction of motility in response to environmental gradients. Small RNA molecules also contribute to system cross talk by influencing disparate regulatory systems. Higher-order manifestations of cognition include communication and sociability. Bacteria communicate using autoinducer molecules and exhibit social behaviors like building biofilms. These examples, and many others, support the idea that bacteria have a type of cognitive ability. They lack self-awareness, certainly, but they can solve problems, communicate, and cooperate with each other. The next great task is to explore how individual bacteria integrate the massive amounts of information they collect from hundreds of signaling pathways to form coherent, adaptive responses.

CHAPTER REVIEW

Review Questions

1. List regulatory mechanisms discussed in this chapter that work at each of the following levels: DNA, transcription, translation, and posttranslation.
2. Describe a two-component signal transduction system.
3. How does lactose induce the *lacZYA* operon?
4. If *lacY* is induced only when lactose is present, how does external lactose induce the system?
5. How does tryptophan repress the tryptophan operon?
6. Discuss how glucose impacts the utilization of lactose as a carbon source.
7. Name a regulatory protein that can activate and repress an operon's expression. How does it do that?
8. Describe four ways that sigma factor production/activity can be regulated.
9. What is the regulatory mechanism that uses translation to control transcription? How does it work?
10. Compare the regulation mechanisms of transcriptional attenuators versus riboswitches.
11. Discuss how small regulatory RNA molecules can regulate gene expression by affecting translation or mRNA decay.
12. What are cAMP, ppGpp, and cyclic di-GMP, and how are they related in structure and function?
13. What is quorum sensing?
14. Describe how *Synechococcus* regulates its gene expression as a function of time of day.
15. Describe how phase variation can contribute to the virulence of pathogenic microbes.
16. Describe how *E. coli* senses a chemical gradient and changes its behavior in response.

Thought Questions

1. What would happen to the expression of the tryptophan operon if you replaced the key tryptophan codons in the attenuator region with tyrosine codons?
2. Adding tryptophan to *E. coli* will cause repression of the *trp* operon genes. Mutations in the *trpR* repressor gene and the *trp* operator will have the same phenotype; that is, adding tryptophan will no longer repress expression of the *trp* genes. What will happen to the phenotype if you transform each mutant with a plasmid carrying the wild-type *trpR* gene or the wild-type *trp* operator region?
3. When the cyanobacteria arose 3.5 billion years ago, Earth spun much faster, with a period of 6 hours! The moon's gravitational force has progressively slowed Earth's rotation and lengthened its day from 6 hours to the current 24 hours. Assuming the circadian clock is as old as the cyanobacterial lineage, what do you think all this implies about the ability of complex machines such as clocks to evolve over billions of years?
4. Phase variation occurs at low rates, such as one cell in 10^3 to one in 10^5 for flagellar switching in *Salmonella*. Why might a higher frequency of genetic switching be disadvantageous to the microbe, especially if it is a pathogen?
5. This chapter has outlined the many ways that bacteria sense their environment, their neighbors, and their location (rock, intestine, ocean). Can we then say that bacteria are conscious?

Key Terms

activator (359)
alternative sigma factor (370)
anti-anti-sigma factor (372)
anti-attenuator stem loop (374)
anti-sigma factor (371)
attenuator stem loop (374)
autoinducer (382)
autoinduction (382)
catabolite repression (365)
chemotaxis (395)
circadian clock (388)
cis-antisense RNA (asRNA) (378)
competent (386)
corepressor (359)
derepression (359)
diauxic growth (365)
direct repeat (393)
domain (359)
inducer (359)
inducer exclusion (365)
induction (359)
inverted repeat (393)
methyl-accepting chemotaxis protein (MCP) (396)

operator (359)
phase variation (392)
quorum sensing (382)
regulatory protein (358)
regulon (370)
repression (359)
repressor (359)
response regulator (361)
riboswitch (374)
second messenger (380)
sensor kinase (360)
small RNA (sRNA) (376)
stringent response (380)
tandem repeat (393)
transcriptional attenuation (373)
transcriptome (370)
translational control (380)
two-component signal transduction system (360)

Recommended Reading

Babski, Julia, Lisa-Katharina Maier, Ruth Heyer, Katharina Jaschinski, Daniela Prasse, et al. 2014. Small regulatory RNAs in Archaea. *RNA Biology* **11**:484–493.

Battesti, Aurelia, Nadim Majdalani, and Susan Gottesman. 2011. The RpoS-mediated general stress response in *Escherichia coli*. *Annual Review of Microbiology* **65**:189–213.

Commichau, Fabian M., Achim Dickmanns, Jan Gundlach, Ralf Ficner, and Jörg Stülke. 2015. A jack of all trades: The multiple roles of the unique essential second messenger cyclic di-AMP. *Molecular Microbiology* **97**:189–204.

Dalebroux, Zachary D., and Michelle S. Swanson. 2012. ppGpp: Magic beyond RNA polymerase. *Nature Reviews. Microbiology* **10**:203–212.

Fozo, Elizabeth M., Matthew R. Hemm, and Gisela Storz. 2008. Small toxic proteins and the antisense RNAs that repress them. *Microbiology and Molecular Biology Reviews* **72**:579–589.

Johnson, Carl H., Chi Zhao, Yao Xu, and Tetsuya Mori. 2017. Timing the day: What makes bacterial clocks tick? *Nature Reviews. Microbiology* **15**:232–242.

Jones, Christopher W., and Judith P. Armitage. 2015. Positioning of bacterial chemoreceptors. *Trends in Microbiology* **23**:247–256.

Lewis, Mitchell. 2005. The *lac* repressor. *Critical Reviews in Biology* **328**:521–548.

Lyon, Patricia. 2015. The cognitive cell: Bacterial behavior reconsidered. *Frontiers in Microbiology* **6**:264.

Mellin, J. R., and Pascale Cossart. 2015. Unexpected versatility in bacterial riboswitches. *Trends in Genetics* **31**:150–156.

Merino, Enrique, and Charles Yanofsky. 2005. Transcription attenuation: A highly conserved regulatory strategy used by bacteria. *Trends in Genetics* **21**:260–264.

Papenfort, Kai, and Bonnie L. Bassler. 2016. Quorum sensing signal-response systems in Gram-negative bacteria. *Nature Reviews. Microbiology* **14**:576–588.

Papenfort, Kai, and Carin K. Vanderpool. 2015. Target activation by regulatory RNAs in bacteria. *FEMS Microbiology Reviews* **39**:362–378.

Parker, Christopher T., and Vanessa Sperandio. 2009. Cell-to-cell signaling during pathogenesis. *Cellular Microbiology* **11**:363–369.

Paulose, Jiffin K., John M. Wright, Akruti G. Patel, and Vincent M. Cassone. 2016. Human gut bacteria are sensitive to melatonin and express endogenous circadian rhythmicity. *PLoS One* **11**:e0146643.

Schleif, Robert. 2010. AraC protein, regulation of the L-arabinose operon in *Escherichia coli*, and the light switch mechanism of AraC action. *FEMS Microbiology Reviews* **34**:779–796.

Sesto, Nina, Omri Wurtzel, Cristel Archambaud, Rotem Sorek, and Pascale Cossart. 2013. The excludon: A new concept in bacterial antisense RNA-mediated gene regulation. *Nature Reviews. Microbiology* **11**:75–82.

Staron, Anna, and Thorsten Mascher. 2010. Extracytoplasmic function sigma factors come of age. *Microbe* **5**:164–170.

Tseng, Roger, Nicolette F. Goularte, Archana Chavan, Jansen Luu, Susan E. Cohen, et al. 2017. Structural basis of the day-night transition in a bacterial circadian clock. *Science* **355**:1174–1180.

Turnbough, Charles L. 2019. Regulation of bacterial gene expression by transcription attenuation. *Microbiology and Molecular Biology Reviews* **83**: e00019–19.

Vink, Cornelis, Gloria Rudenko, and H. Steven Seifert. 2012. Microbial antigenic variation mediated by homologous DNA recombination. *FEMS Microbiology Reviews* **36**:917–948.

Yang, Ji, Marija Tauschek, and Roy M. Robins-Browne. 2011. Control of bacterial virulence by AraC-like regulators that respond to chemical signals. *Trends in Microbiology* **19**:128–135.

CHAPTER 11
Viral Molecular Biology

11.1 Phage Lambda: Enteric Bacteriophage

11.2 Influenza Virus: (–) Strand RNA Virus

11.3 Human Immunodeficiency Virus (HIV): Retrovirus

11.4 Endogenous Retroviruses and Gene Therapy

11.5 Herpes Simplex Virus: DNA Virus

Viruses such as influenza virus, hepatitis virus, and HIV sicken hundreds of millions of people worldwide. Yet many other viruses coexist with us without harm. What makes one virus deadly and another helpful? The answer lies in viral molecules. Endogenous viral genomes contribute molecular parts for our bodies, such as a protein for placental fusion. Even the viruses most deadly to humans can be converted to vectors that cure a human disease.

CURRENT RESEARCH highlight

Measles virus assembles. Measles is a dreaded disease found mainly in children and known for itchy red spots and high fevers. Elizabeth Wright's micrographs capture the measles virus in the act of assembly. The viral RNA is coated by nucleocapsid proteins (colorized blue or green). The protein-coated RNA snakes back and forth within the virion, which assembles at the host plasma membrane. The envelope is studded with glycoproteins that bind receptors on the next human cell. Understanding the details of virion formation may lead to new antiviral agents.

Source: Zunlong Ke et al. 2018. *Nat. Commun.* **9**:1736.

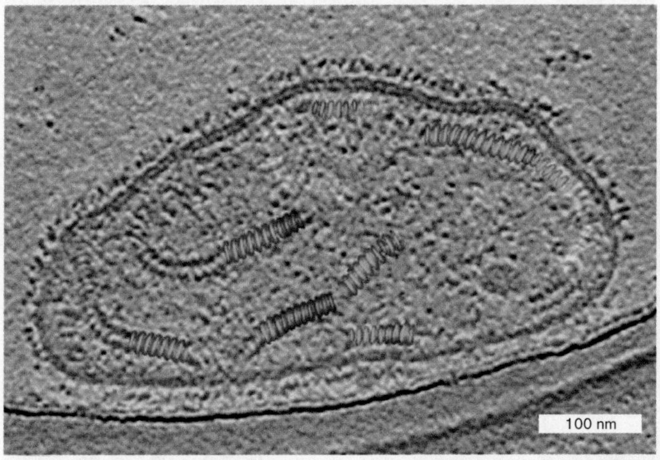

REPRINTED BY PERMISSION FROM SPRINGER NATURE: KE, Z., STRAUSS, J.D., HAMPTON, C.M. ET AL. *NAT COMMUN* **9:** 1736 (2018) DOI:10.1038/S41467-018-04058-2

AN INTERVIEW WITH

ELIZABETH WRIGHT, VIRAL MICROSCOPIST, UNIVERSITY OF WISCONSIN–MADISON

ROBIN DAVIES

What was exciting about the measles virus structure?

Our first view of the glycoprotein lattice and the nucleoprotein-coated RNA was exciting. By understanding the arrangement of the glycoproteins, we can consider how best to engineer antiviral agents (neutralizing antibodies) that target the actions of fusion and attachment. In addition, we can relate what we learn about measles virus to a number of other, more pathogenic paramyxoviruses, like Nipah and Hendra viruses—which, for biosafety considerations, we cannot study by cryo-EM.

One of the most infectious of all diseases is measles, caused by the virus shown in the Current Research Highlight. Nine of ten unvaccinated people who contact a measles patient will get sick. From the respiratory tract, the virus spreads to the lymph nodes, the blood vessels, internal organs, and skin, where the characteristic red blotches indicate the body's immune response. Throughout the world, children die of this disease. Yet this deadly virus is surprisingly simple in form: an envelope with glycoproteins, matrix proteins, and a single-stranded RNA genome that specifies seven types of proteins. How does such a simple reproductive entity propagate throughout the body?

Molecular interactions, such as a viral glycoprotein binding a cell-surface receptor, enable a simple virus to take over the host cell's machinery and thwart its defenses. For instance, consider another RNA virus, the influenza virus, whose seasonal strains kill up to half a million people each year. Influenza viruses show extraordinary diversity of envelope proteins. Each envelope variant binds to a different range of cell-surface receptor proteins found in our respiratory tissues (see Section 11.2). The structure of these cell-surface proteins determines whether a type of influenza virus can bind and succeed in infecting a given species, whether human or animal. Once inside a host cell, the viral genome and packaged molecules must collaborate with host enzymes to build virions, and to evade defense molecules of the immune system.

Molecular interactions also govern latency, the process by which a virus establishes dormant residence within a cell. In Chapter 10 we discussed how molecular mechanisms such as repressors and DNA inversions regulate the function of bacteria. Now, Chapter 11 presents the molecular mechanisms of viral infection and latent states. For background, we assume a fundamental understanding of the nature of viruses, presented in Chapter 6.

This chapter explores in depth four important viruses. First we consider the bacteriophage lambda, a type of phage found in your digestive tract and used by scientists for synthetic biology. Phages can actually be engineered to fight human pathogens; for example, in 2019 phage biologist Graham Hatfull at the University of Pittsburgh reported the first successful phage treatment of an opportunistic pathogen that threatened the life of a cystic fibrosis patient. The phages engineered for this therapy were originally discovered by undergraduates in Hatfull's SEA-PHAGE program.

We then examine in detail three viruses that infect humans: influenza (a negative-strand RNA virus), HIV (a retrovirus), and herpes simplex (a large DNA virus). We will see that all viral infection processes share common themes but also use molecular mechanisms unique to each virus. Section 11.4 presents our emerging awareness of endogenous retroviruses that express human proteins—and how we build retroviral vectors for lifesaving gene therapy.

Note: Online eTopics cover four additional viruses in depth:

- **eTopic 11.1** Phage T4: The Classic Molecular Model
- **eTopic 11.2** The Filamentous Phage M13: Vaccines and Nanowires
- **eTopic 11.3** Poliovirus: (+) Strand RNA Virus: A Research Model for Non-polio Enterovirus That May Cause Acute Flaccid Myelitis
- **eTopic 11.4** Hepatitis C: (+) Strand RNA Virus

11.1 Phage Lambda: Enteric Bacteriophage

Bacteriophage lambda infects *Escherichia coli* within the human gut, among the trillions of phages in our intestinal microbiome (**Fig. 11.1**). Historically, phage lambda's interaction with *E. coli* was the first living system simple enough to dissect at the molecular level. The phage yielded fundamental discoveries in gene regulation, most notably the control of lysogeny (introduced in Chapter 6). In microbial communities, lysogeny provides a way for many kinds of phages to transfer genes between bacterial genomes. When lambda lysogeny was first discovered, we had no idea that human genomes similarly carry genes of viral origin, such as those of endogenous retroviruses.

The "lambda switch" between lysis and lysogeny provided clues to the bacterial regulons presented in

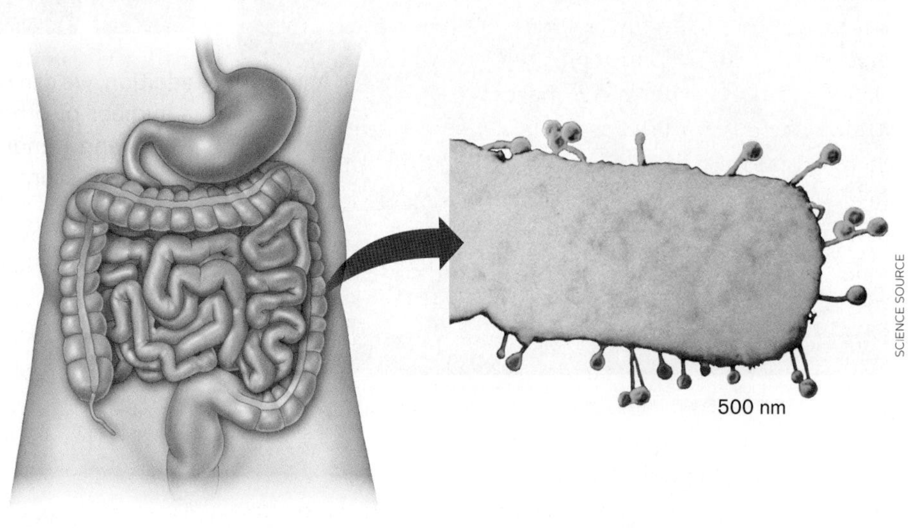

FIGURE 11.1 ■ Within the human intestine, *Escherichia coli* hosts phage lambda. Colorized SEM.

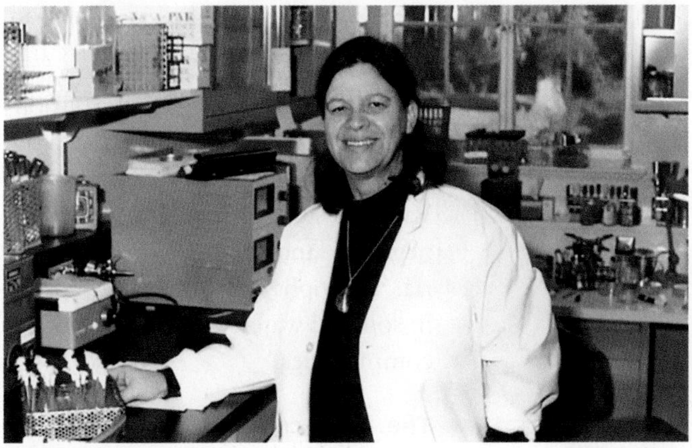

THE ESTHER M. ZIMMER LEDERBERG TRUST-WWW.ESTHERLEDERBERG.COM

FIGURE 11.2 ▪ Esther Lederberg discovered phage lambda. Lederberg in her laboratory at Stanford, holding phage stocks in cotton-stoppered tubes.

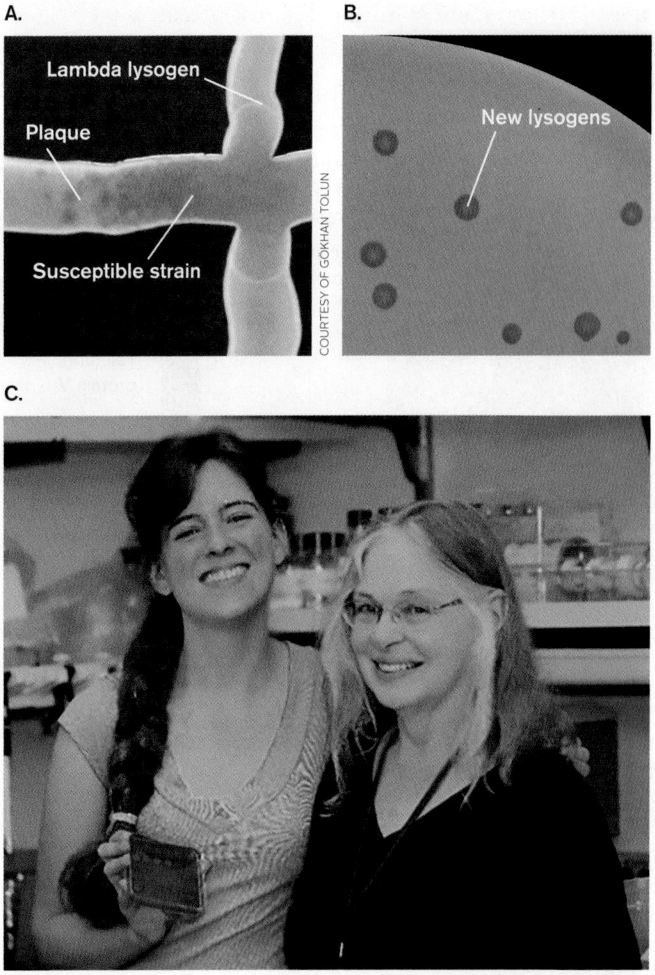

COURTESY OF GÖKHAN TOLUN

FIGURE 11.3 ▪ A lysogen reveals phage lambda. A. The lysogenic *E. coli* strain (vertical streak) releases phages that cause plaques in a susceptible *E. coli* strain (horizontal streak). B. Lambda plaques show a cloudy center where newly formed lysogens appear. C. Undergraduate Kathleen Morrill holds a plate of lambda *lacZ* reporter strains on MacConkey agar; with Lynn Thomason, at the National Cancer Institute.

Chapter 10, as well as molecular mechanisms of animals and plants. Today, the well-studied phage lambda provides tools for synthetic biosensors and even DNA chip devices.

Discovery of Phage Lambda

Bacteriophage lambda was discovered in 1950 by Esther Lederberg (1922–2006), pioneering bacterial geneticist at the University of Wisconsin–Madison, and later Stanford University (**Fig. 11.2**). At the time, it was known that some kinds of bacteria spontaneously lyse and release phages, but the mechanism, even the source of phage from within the cells, was a mystery. The strain *Escherichia coli* K-12 had been isolated in 1922 from a healthy patient's colon and was used for decades thereafter in teaching and research. In 1950, Lederberg was performing a genetic cross between two strains of *E. coli* K-12: a standard laboratory stock, and a strain that she had mutagenized with ultraviolet light. After the two strains were mixed, the mutant strain formed colonies that were "nibbled and plaqued" (**Fig. 11.3A**). The plaques resulted from phage particles lysing the mutant strain. The mutant had lost its lambda prophage and was therefore susceptible to infection. The infecting phage particles came from the lysogenic *E. coli* K-12, which Lederberg had streaked across.

Further experiments confirmed that at that time, all known stocks of K-12, the most widely used strain of *E. coli*, were lysogenic for phage lambda. **Lysogeny** means that the genome of a phage is incorporated into the genome of the host cell (discussed in Chapter 6). Phage particles then no longer exist as such, but the phage DNA replicates indefinitely within the host, as an integrated **prophage**. The presence of the prophage confers resistance to infection by the same type of phage. Phage lambda is thus called a "temperate phage," rather than a "virulent phage" such as T4, which always kills its host. But occasionally (about once in a million cells), a molecular signal tells the prophage genes to make progeny phages and lyse the cell. Thus, unknown to researchers at the time, a given tube of growing *E. coli* lysogen typically carried about a million lambda phages per milliliter. The phages released can form plaques on a strain that is not a lysogen (has no lambda prophage; **Fig. 11.3B**). The plaques have a cloudy center because some of the infected cells become lysogens, which start growing up where the infection began.

Lederberg and others learned to "cure" the K-12 lysogen to eliminate the phage from stock cultures, including most strains used today. Other notable discoverers of the nature of lysogeny include French microbiologists André Lwoff (1902–1994) and François Jacob (1920–2013). The molecular basis of the lysis/lysogeny "switch" was elucidated by Mark Ptashne and colleagues (discussed shortly). Full regulation of the switch includes numerous viral and host proteins, whose details continue to be discovered by researchers such as Lynn Thomason and her student Kathleen Morrill at the National Cancer Institute (**Fig. 11.3C**).

> **Thought Question**
>
> **11.1** Plaques from phage lambda quickly fill with resistant lysogens. Could there be a different way for the host cells to become resistant, without forming lysogens?

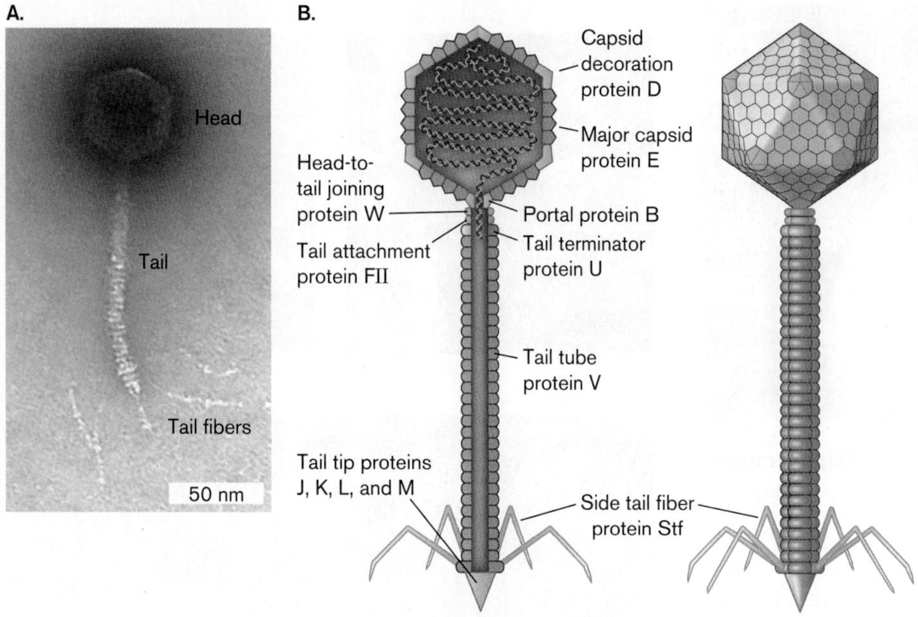

FIGURE 11.4 ■ **The lambda virion.** **A.** Phage particle visualized by heavy-atom negative stain (TEM). **B.** Diagram of components, colored to match coding genes in the genome (see Fig. 11.5).

Phage Structure and Genome Replication

The phage lambda particle (virion) is typical of tailed phages of the siphophage family (**Fig. 11.4**). Various siphophages infect Gram-negative and Gram-positive bacteria. Siphophages commonly appear in soil and water, as well as in enteric communities.

The head contains DNA. The part of a tailed phage that contains its genome is called the "head." Usually the head consists of an icosahedral protein complex, equivalent to the capsid of a tail-less virion. The head of phage lambda contains coiled double-stranded DNA. Thus, according to genome classification, phage lambda falls under Baltimore group I (see Chapter 6). The protein coat is composed of two types of subunits: the face subunit E and the "decoration" subunit D. The D subunits form a pentamer at each vertex of the icosahedron, except for the tail connector. The tail connector comprises four proteins encoded by different genes (B, FII, U, W). Each connector subunit is found in multiple copies that form a ring around the tail connector.

Tail tube. The tail itself consists of a long tube of 32 hexamer rings of subunit V. Remarkably, the exact length of the tail (the number of hexamers) depends on a "tape measure protein" (protein H). If we delete part of the gene that encodes the tape measure protein, the phage will assemble

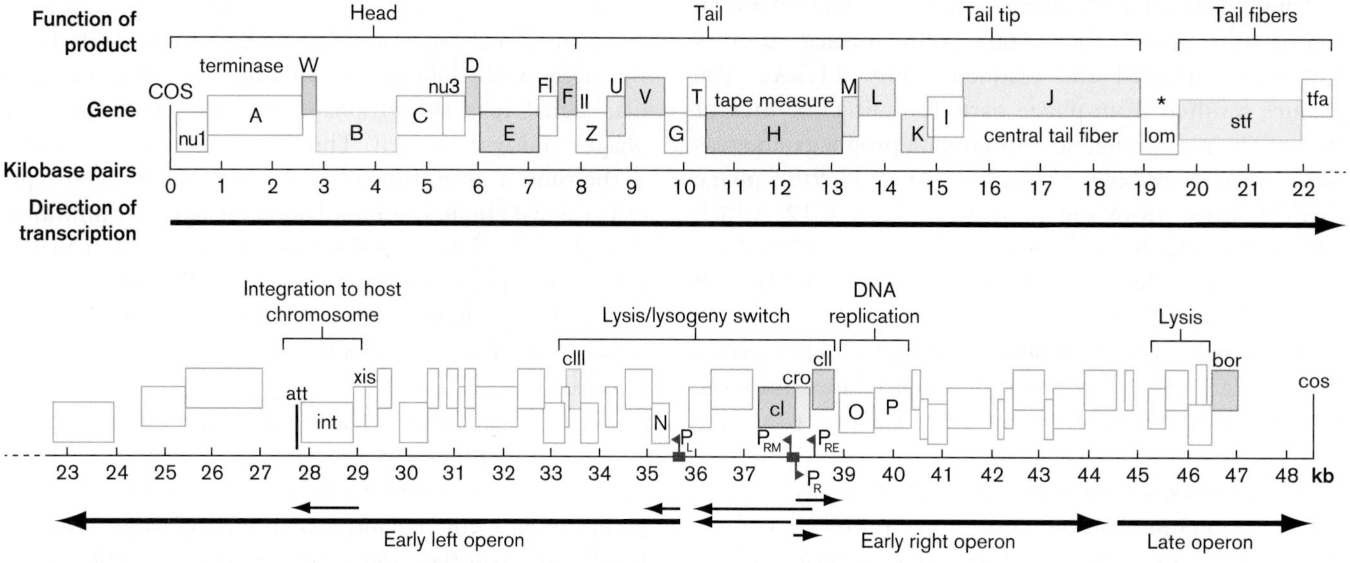

FIGURE 11.5 ■ **Genome of phage lambda.** For each gene, vertical offset indicates the reading frame.

a shorter tail. On the other hand, if the gene is lengthened by addition of nucleotides, the phage will make a longer tail. How the "tape measure" works is unclear.

With respect to function, the tail of all siphophages is noncontractile (unlike the contractile tail of myophages such as phage T4; **eTopic 11.1**). The tail tube has a tip and tail fibers, both of which help the phage attach to its host cell. Upon host attachment, the lambda DNA must uncoil and pass through the entire length of the tail to reach the host cytoplasm.

Genome. Phage lambda has a genome of 48.5 kilobase pairs (kb) that includes about 70 genes. In **Figure 11.5**, the left side of the genome encodes mainly components of the virion, including head, tail, and tail fibers. Most of these genes are color-coded to match their products, shown in **Figure 11.4**. The right side of the genome includes enzymes for key functions such as host integration, DNA replication and lysis, and the famous lysis/lysogeny switch (discussed shortly).

Within a phage head, the packaged genome is linear. The left and right ends each possess *cos*, a sequence of 200 bp that was cleaved by a terminase enzyme when the phage DNA was packaged. The cleavage generated a staggered cut, or "sticky ends" (**Fig. 11.6**). As the phage DNA now enters the new host cell, the *cos* sticky ends anneal together, and their backbones are sealed by DNA ligase (a replication enzyme discussed in Chapter 7). The genome is now circular, as the sealed ends form a long operon transcribed "rightward" from P_R, all the way through the virion structural components (**Fig. 11.5**). Another operon is transcribed "leftward," from P_L. Note that both the P_R and P_L operons use all three reading frames, and that some coding genes overlap. Long mRNA transcripts with gene overlap in three reading frames are a common feature of viral genomes.

Besides the P_R and P_L promoters, which function during the lytic cycle, other promoters are used for key events of the lysis/lysogeny switch.

During the lytic cycle, the circularized DNA molecule undergoes several rounds of bidirectional replication via host DNA polymerase (**Fig. 11.6**). The DNA is sometimes called a "theta form" because of the shape of the circle while replication is in progress. The completed circles are then nicked (cleaved on one strand), generating a 3′ OH end. The 3′ OH end of DNA serves as a primer for **rolling-circle replication**. Rolling-circle replication is commonly used by plasmids (see Chapter 7), as well as by circularized viral genomes such as that of herpesviruses (see Section 11.5).

In the rolling-circle process, one strand of DNA extends its 3′ end continually around the circular template, while the 5′ end "rolls away." The 5′ extension generates a long line of tandemly repeated genomes called a **concatemer**.

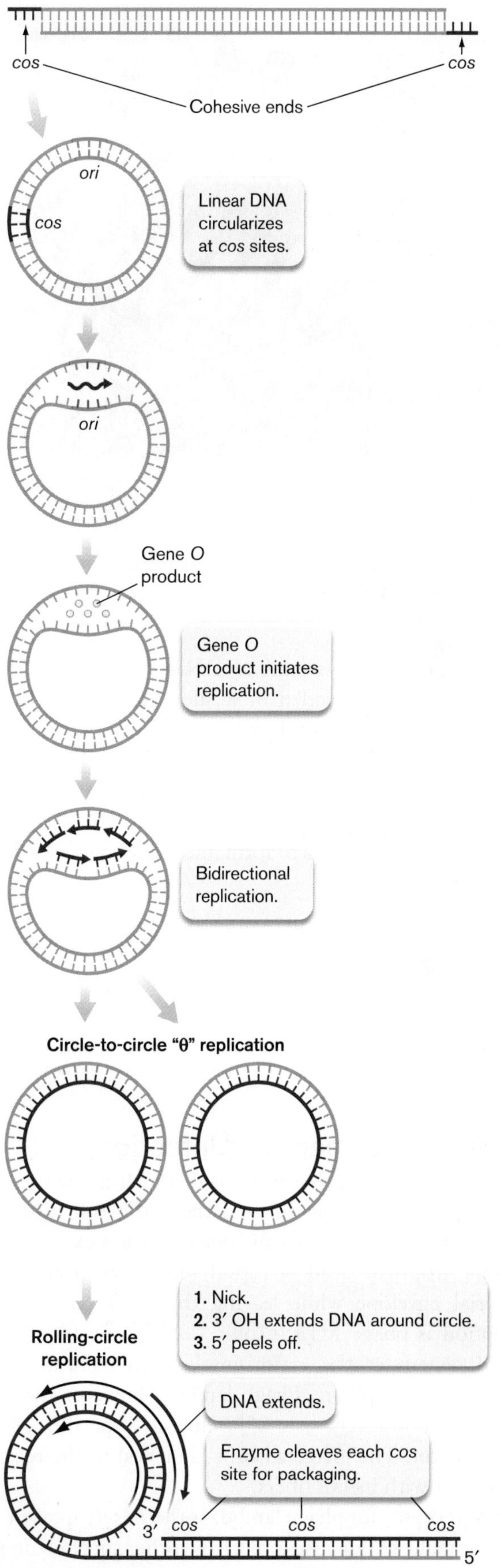

FIGURE 11.6 ■ Replication of phage lambda. Initially, the linear genome circularizes and replicates bidirectionally. Then it replicates multiple genomes end to tail, forming a concatemer.

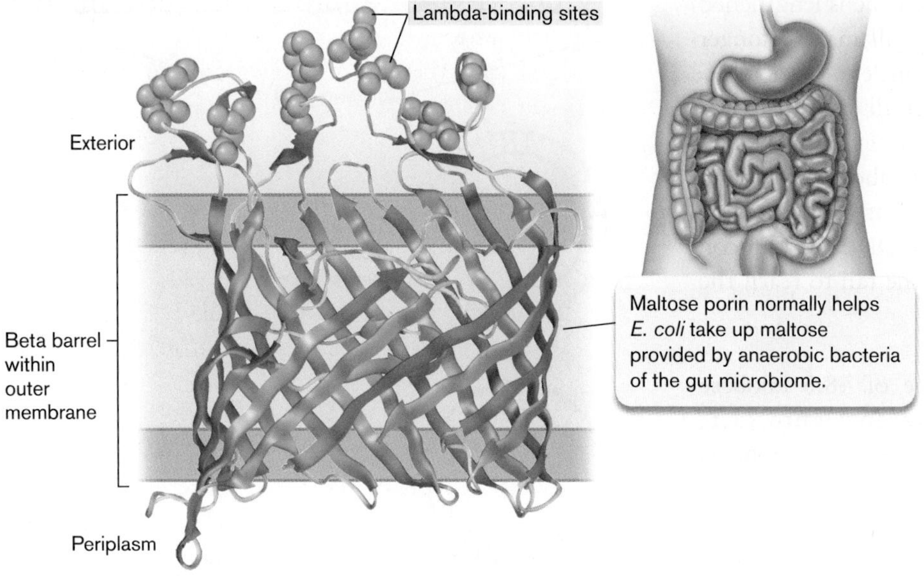

FIGURE 11.7 ■ Host receptor: maltose porin. The maltose porin is embedded in the outer membrane of *E. coli*. (PDB: 1MAL)

The complementary strand of DNA fills in later. Once the long, double-stranded molecule is complete, the terminase enzyme cleaves the concatemer into pieces that fit into phage head coats. For phage lambda, the genomes are cleaved at the *cos* site, which acts as a signal to package the progeny genome into the protein head coat. Cleavage restores the two *cos* ends to form linear genomes coiled into the phage head.

Thought Question

11.2 What advantages does rolling-circle replication offer a phage?

Phage Attachment and Infection

To infect a host cell, a virus needs to attach to the host surface and insert its genome into the cytoplasm (discussed in Chapter 6). Most types of bacteriophages extrude their genome from the head or capsid and thread it across the bacterial envelope while leaving the capsid outside. An exception is phage M13 (**eTopic 11.2**), whose filamentous capsid penetrates the entire envelope to replicate slowly within the cytoplasm. Phage lambda, however, uses the more common mechanism of binding to a specific cell-surface receptor. The phage adsorbs (attaches) to the receptor by contact with its tail fibers.

The receptor for phage lambda is the *E. coli* uptake complex for maltose (glucose dimer) and other short-chain sugars. In the colon, *E. coli* obtains these short glucose chains from anaerobes such as *Bacteroides* species that break down large, complex polysaccharides from plant material (discussed in Chapters 13 and 21). Sugar chain uptake is crucial for *E. coli*, so the bacteria are unlikely to lose the uptake complex by evolution, even though phage lambda takes advantage of it for infection.

The phage binds specifically to maltose porin, the outer membrane pore that transports maltose into the cell (**Fig. 11.7**). Porins are a large family of proteins that share a distinctive "beta barrel" structure (presented in Chapter 3). The beta barrel of maltose porin (blue in **Fig. 11.7**) is buried in the outer membrane. The phage-binding sites (green) were identified by amino acid substitution mutations that prevent phage binding and confer host resistance to lambda. The maltose porin was first called "lambda receptor," LamB, because it was discovered by its function as the lambda receptor. Of course, the protein actually evolved in the host as a way to obtain nutrients.

Attaching to the cell surface is just the first challenge that the phage faces to establish infection (**Fig. 11.8**, step 1). How does the phage get its DNA all the way across the periplasmic space and the inner membrane? Unlike phage T4 (whose tail contracts to expel DNA under pressure), phage lambda uses an extrusion mechanism that is not fully understood. Somehow the DNA threads across the periplasm and through the maltose transport complex of the inner membrane, to reach the cytoplasm (step 2). Within the cytoplasm, the staggered ends of the *cos* sites anneal and are ligated, circularizing the genome (step 3). Now, the host RNA polymerase can begin to transcribe the phage operons.

At first, though, the newly introduced phage genome supports transcription of just a few key proteins, most notably the "control proteins" Cro and CII (**Fig. 11.8**, step 4):

- **Cro protein leads to lysis.** Small amounts of Cro activate the lytic cycle.
- **CII protein leads to lysogeny.** Small amounts of CII block expression of Cro and lytic proteins, and CII induces expression of CI (known as "lambda repressor").

So, which protein wins—Cro or CII? The answer depends on numerous factors in the gut environment, whose signals combine to lead to one decision or the other. One such signal is a surge in nutrients, such as when you eat a meal, sending rich organic substrates

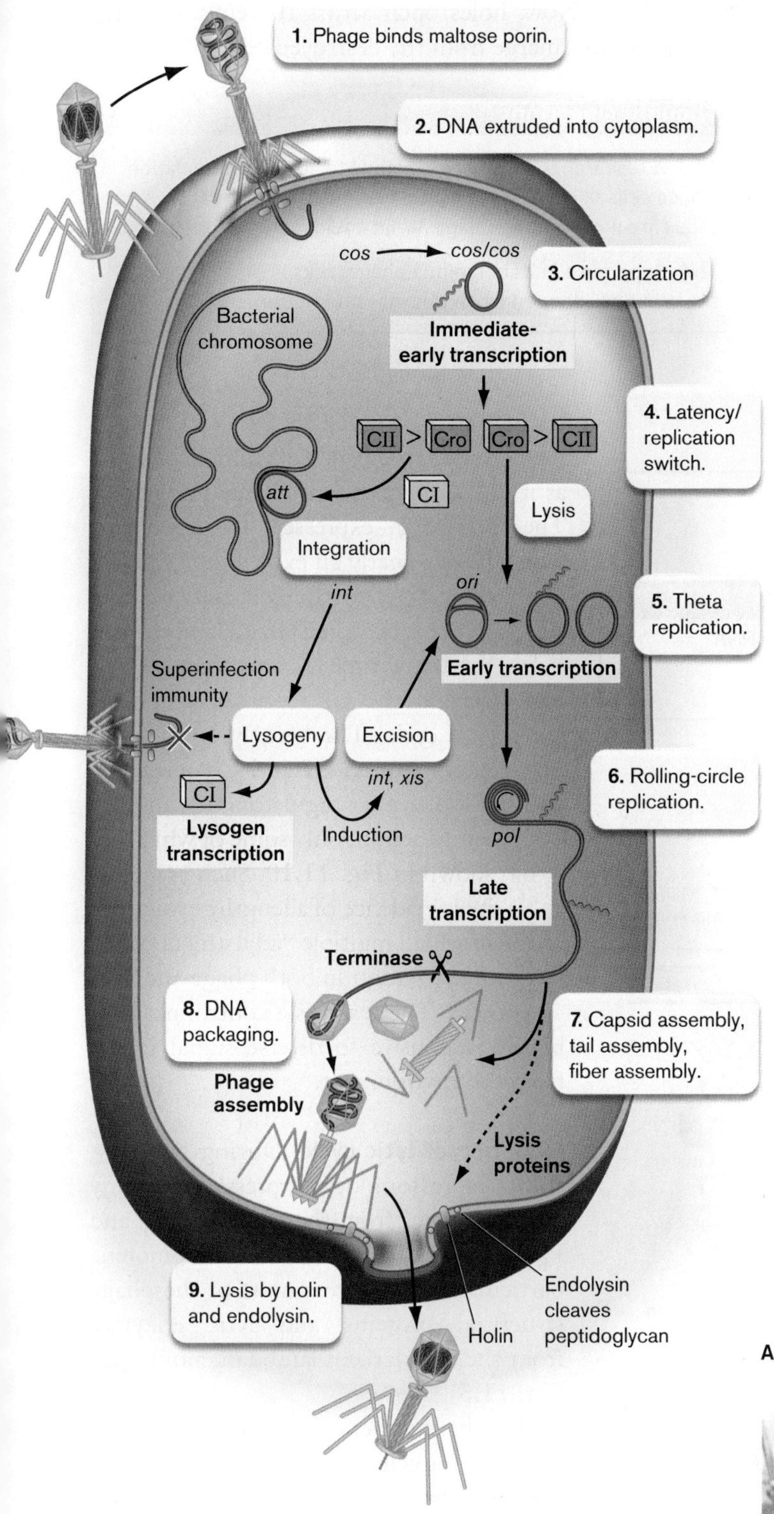

FIGURE 11.8 ■ Phage lambda infection of *E. coli* K-12.

down to your intestines. High nutrient concentrations cause the *E. coli* protease HflB to cleave the phage protein CII, leaving Cro to induce the lytic cycle. Low nutrient concentrations inhibit HflB, leaving CII available to block Cro. Alternatively, suppose multiple phage particles coinfect the cell at the same time. This event implies that the host population is outnumbered by phages, and thus progeny phages will have poor opportunities to find a host.

For this reason the phage has evolved a mechanism to avoid lytic reproduction: The multiple phages express a higher level of CII, some of which now evades the host protease and blocks Cro. When CII blocks Cro, the fortunate host cell becomes a lysogen instead of lysing. A phage genome integrates in the host genome as a prophage, at the *att* site. The phage DNA integrates with *att* by site-specific recombination—that is, recombination between the backbones of two DNAs that share a short sequence in common (discussed in Chapter 9).

The integrated prophage now expresses only a few proteins, including CI (lambda repressor). CI repressor prevents the lytic cycle by blocking transcription of lytic promoters. The CI repressor also prevents superinfection by other lambda phages. This is what happens to most of your intestinal bacteria between meals: the bacteria stay in stationary phase with their lysogenic phages repressed. The phage lambda CI repressor is famous because its binding to DNA was the first protein-DNA binding event to be described (**Fig. 11.9**). Mark Ptashne at Memorial Sloan Kettering Cancer Center (**Fig. 11.9A**) showed how CI protein forms a dimer that binds a specific DNA sequence. Each CI subunit binds DNA through an interaction between an alpha helix of the protein and a major-groove sequence of DNA (**Fig. 11.9B**). Similar alpha helix binding to the major groove mediates the function of many genetic regulators of animals and plants, as well as bacteria.

When phage lambda infects *E. coli*, what if the Cro control wins, instead of CI? Cro protein represses CI expression, thus promoting the lytic cycle. The lytic cycle leads to **lysis** (host cell destruction, and release of progeny phages). The circularized phage DNA replicates, first bidirectionally (**Fig. 11.8**, step 5, "theta replication"), and then by rolling-circle replication (step 6). Structural proteins are

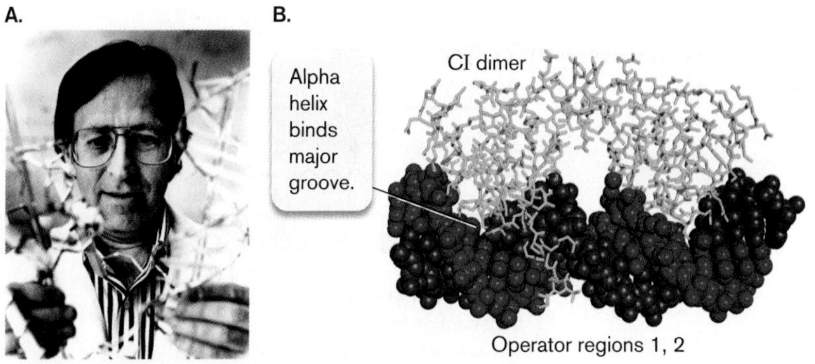

FIGURE 11.9 ■ CI repressor binds DNA. **A.** Mark Ptashne worked out the binding of CI repressor to DNA, and the mechanism of the lysis/lysogeny switch. **B.** The CI repressor dimer binds the operator sequence by fitting the DNA sequence at the major groove. (PDB: 1LMB)

synthesized and assembled to form the empty capsid (or head coat), tail tube, and tail fibers (step 7). The *cos* end of the DNA concatemer gets stuffed into a progeny head coat (step 8). The stuffed portion of the concatemer is then cleaved at the next *cos* site by a phage enzyme called terminase. Terminase also helps attach the filled head to the tail tube. When most of the progeny phages have assembled, a phage-encoded protein called holin punctures the inner cell membrane, providing a channel through which endolysin reaches the cytoplasm. Endolysin cleaves peptidoglycan (step 9, lysis). Now, holes open across the envelope, and phage particles emerge from the destroyed cell.

Thought Questions

11.3 A researcher adds phage lambda to an *E. coli* population whose cells fail to express maltose porin. After several days, the *E. coli* are now lysed by phage. What could be the explanation?

11.4 Suppose a mutant lambda phage lacks holin and endolysin. What will happen when this mutant infects *E. coli*?

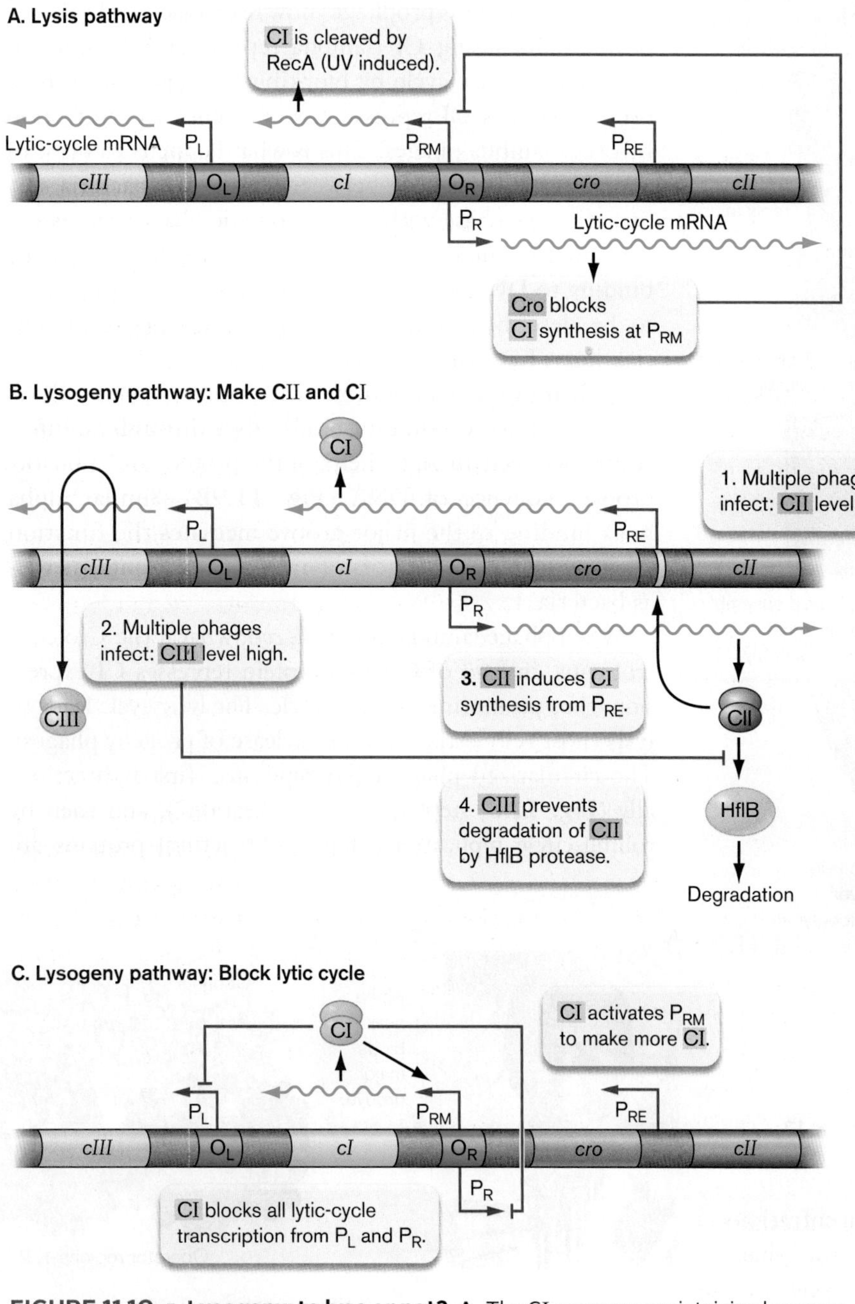

FIGURE 11.10 ■ Lysogeny: to lyse or not? A. The CI repressor maintaining lysogeny is cleaved by RecA (induced by UV exposure). Cro blocks CI synthesis, allowing expression of the lytic promoters P_L and P_R. **B.** If multiple phages infect simultaneously, the genome expresses enough CII protein to bind P_{RE}, activating expression of CI. **C.** The CI protein blocks promoters P_L and P_R (lytic cycle) and activates P_{RM} (to make more CI).

Lysogeny: To Lyse or Not?

As a lysogen, the host *E. coli* replicates just as it did without any prophage integrated. The prophage expresses CI repressor, which blocks nearly all expression of phage genes dangerous to the cell and prevents superinfection by other phage lambda particles. Yet at any time, a spontaneous event may override the CI repressor and trigger a lytic cycle. The lysis/lysogeny decision centers around CI and Cro proteins, with multiple other regulators that modify their function, only some of which are shown in **Fig. 11.10**. Such complexity is evidence of a lengthy evolution, in which multiple "adjustments" had time to occur in both phage and host genomes. The result is a control network highly responsive to diverse signal inputs from the environment.

Induction of lytic cycle. During lysogeny, the CI repressor blocks expression of nearly all other genes from the promoters P_L and P_R (**Fig. 11.10B**). These two promoters, particularly P_R, express most of the phage structural proteins and lytic enzymes from the circularized lambda genome (see **Fig. 11.5**).

But the number of repressor molecules present in a cell is small enough for fluctuation to lead to rare events. Occasionally, perhaps one in 10^7 cells, the CI concentration is lowered to a level at which P_R becomes exposed, enabling expression of Cro—which activates lysis. CI is further decreased by host RecA, which cleaves this protein along with many host repressors (**Fig. 11.10A**). RecA is activated by DNA damage during UV light exposure, so UV light can induce lysis in 100% of a lysogenic population.

When the Cro protein (lysis activator) is expressed, it binds to both operators (O_L and O_R) for the P_L and P_R operons, respectively. Cro up-regulates these lysis operons during the lytic cycle and, at the same time, blocks expression of CI from promoter P_{RM}. Thus, no further repressor can be made to block phage production, and the cell is committed to lysis.

Maintaining lysogeny. To avoid lysis requires continual expression of CI repressor (**Fig. 11.10B**). Early in lysogeny, CI is expressed from promoter P_{RE}. CI expression requires P_{RE} to bind CII protein. But CII protein is vulnerable to cleavage by host HflB (in a high-nutrient medium). HflB cleavage can be prevented by CIII, a protein expressed early from promoter P_L.

Once CI protein attains sufficient concentration, it activates its own expression from a different promoter, P_{RM}. P_{RM} activation requires the binding of CI dimers to O_R. Besides activating P_{RM}, the CI binding also blocks expression of P_R, the major lytic operon, as well as the leftward-transcribing operon from P_L. From then on, CI maintains its own expression while preventing induction of lysis. Lysis occurs only when stress such as DNA damage activates RecA to cleave CI.

The full repression of lysis requires eight molecules of CI in all: a pair of dimers at O_R, and another pair of dimers at O_L (**Fig. 11.10C**). The two pairs of dimers actually bind each other as an octamer, with DNA looped between them (not shown). The multiple molecules working together show "cooperativity," a property whereby the binding of two regulators is stronger than the sum of the individual regulators binding DNA. Cooperativity increases the on/off character of a molecular switch, lessening the occurrence of partial states in between. Other examples of cooperative molecular regulators are presented in Chapter 10.

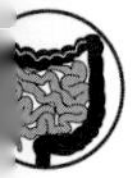

Synthetic Biology

Within natural microbiomes such as that of our intestine, lysogenic phages carry genes that offer useful functions to their host bacterium. An example is the *bor* gene (found at the right-hand end of the lambda genome; see **Fig. 11.5**). The *bor* gene is not needed for phage function; it encodes Bor lipoprotein, which resides in the outer membrane of the host bacterium. The presence of Bor protects a lysogen from destruction by the serum complement cascade (described in Chapter 23). The location of the *bor* gene near the end of the phage genome suggests that it could have been picked up from an ancestral host, by a process of specialized transduction (see Chapter 9). In specialized transduction, a prophage initiates lysis and picks up a small adjacent piece of host DNA while copying its genome out of the integration site. The host DNA is then copied into progeny phage and is transferred to the next host infected.

Today, we use genetic elements of the lambda switch for synthetic biology—in effect, imitating natural gene transfer mechanisms to construct bacteria with functions useful to us. For example, Pamela Silver's students used a CI/Cro switch to build a bacterial recorder that detects an environmental signal within the intestine. This kind of bacterial device could be developed as a detector of signals in the human gut, such as cancer molecules.

Silver's bacterial recorder possesses a stripped-down version of the CI/Cro switch (**Fig. 11.11A**). The *cro* gene is fused to *lacZ*, whose product generates blue colonies when expressed on indicator plates. (Gene fusion techniques are presented in Chapter 12.) But *cro* expression from P_R is blocked by CI repressor, which in turn is expressed from P_{RM}. Cro would block P_{RM}—if it were expressed. The net result is colonies that are white.

The same bacteria contain another fragment of the lambda switch, the "trigger element," in which Cro is expressed from a gene under control of the antibiotic-inducible promoter *tetP*. Now, suppose we add the inducing antibiotic (ATC) to the bacteria. The *tetP-cro* trigger element now expresses Cro, which represses expression of CI by binding P_{RM}. With CI expression turned off, the *cro-lacZ* fusion indefinitely expresses its protein, causing blue colonies (**Fig. 11.11B**). The blue phenotype recurs for many generations.

What happens when we put this bacterial recorder into the microbiome of a mouse? First, to colonize the mouse intestine, the bacterium needs to include a selective gene for competitive advantage within the microbial community—a gene conferring resistance to streptomycin. The mice are treated with streptomycin until the bacterial recorders are established (**Fig. 11.11C**). Then the researcher adds the test antibiotic ATC. After ATC is removed, bacteria are sampled from mice fecal pellets. For up to 7 days following ATC removal, blue colonies appear, showing that the bacteria record ATC exposure and report it long after the signal is gone.

Other applications of phage lambda sound more like science fiction. In 2015, Ido Yosef and colleagues at Tel Aviv University constructed a lambda phage that specifically kills antibiotic-resistant bacteria, while sparing bacteria that lack antibiotic resistance. The phage actually delivers a CRISPR-Cas defense system to its host, protecting the host from phage lysis. In a different kind of experiment, physicists at the Weizmann Institute of Science in Rehovot, Israel, used a CI/Cro switch to build an "artificial cell" on a DNA chip. This partly biological device is used to study biochemical networks relevant to medicine. There is no end in sight to future applications of the familiar phage lambda.

To Summarize

- **Bacteriophage lambda was discovered in an *E. coli* lysogen from a human colon.** An *E. coli* lysogen released phage that infected a sensitive strain. The human gut microbiome is full of phage lysogens.

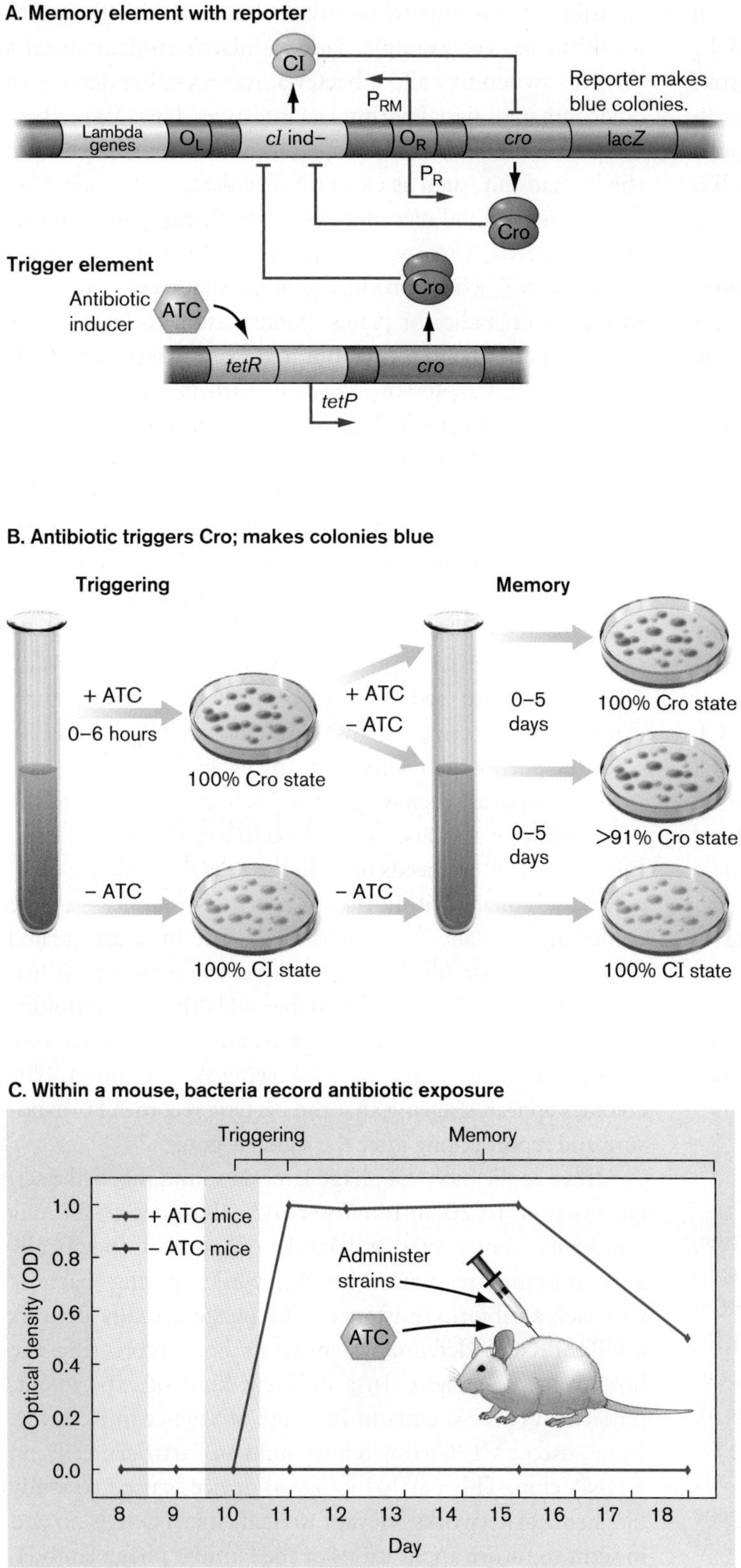

FIGURE 11.11 ■ Antibiotic reporter bacterium uses a CI/Cro switch. **A.** The CI/Cro switch was engineered to record exposure to an antibiotic (ATC). The Cro "memory" is triggered by an inducer that turns off CI expression over many cell generations. **B.** The ATC signal turns colonies blue. Blue colonies persist for 5 days. **C.** The bacterial detector works when administered to mice. Seven days after addition and removal of ATC, the engineered bacteria still make blue colonies (Cro state). *Source:* Jonathan Kotula et al. 2014. *PNAS* **111**:4838.

- **The phage lambda virion** consists of a head containing its DNA genome and accessory proteins, a tail composed of an internal tube, and tail fibers.
- **Phage lambda binds to the host maltose porin.** Phage DNA is inserted into the cytoplasm, where early phage genes are expressed.
- **Control proteins bind to DNA, leading to lysis or lysogeny.** Single-phage infection (Cro protein) leads to lysis, whereas multiple-phage infection (CII protein) more likely leads to lysogeny.
- **In a lytic cycle, rolling-circle replication generates progeny genomes.** The progeny genomes are packaged into head coats and then cleaved from the concatemer; the filled heads are attached to tails. Late-expressed proteins lead to lysis.
- **CII up-regulates expression of CI repressor, which maintains lysogeny.** CI then induces its own expression, while repressing expression of Cro and proteins of the lytic cycle. Human gut bacteria frequently switch between lysis and lysogeny.
- **The CI/Cro switch is used for synthetic biology.** The molecular switch is built into biomedical devices.

11.2 Influenza Virus: (–) Strand RNA Virus

We now turn to viruses that infect humans, focusing on three whose biology is well studied: influenza virus, a negative-strand RNA virus; human immunodeficiency virus (HIV), the retrovirus that causes AIDS; and herpes simplex, a double-stranded DNA virus causing oral and genital herpes. Viruses of humans show mechanisms that resemble those of bacteriophages. For example, retroviruses integrate the DNA copy of their genomes into the host cell genome, analogous to lysogeny by phage lambda. Other viruses, such as influenza, have no known latent form. The influenza virus is highly virulent and lyses infected cells with little or no latent state.

Influenza virus is a major human pathogen, causing up to half a million deaths per year worldwide. Besides the seasonal strains, influenza shows a cyclic appearance of strains that cause pandemic mortality, such as the famous pandemic of 1918, which infected

20% of the world's population and killed more people than those who died in World War I. The 1918 strain arose as a mutant form of an influenza strain infecting birds. In 2009, a highly transmissible strain related to swine influenzas ("swine flu") spread rapidly around the world but caused relatively mild illness. A future strain might emerge combining the high transmission seen in swine flu with the high human mortality seen in the avian strain.

Research focuses on the mechanisms of influenza infection as targets for new antiviral agents. But human viruses are challenging to study—more so than bacteriophages. Viruses infecting human cells require complex replication cycles among the compartments of a eukaryotic cell. Their progeny virions must navigate our organ systems to maintain infection and achieve transmission to new hosts; for example, avian influenza strains poorly transmit to humans because the cell-surface receptors they need are hidden deep within the human respiratory tract. And viruses face the defenses of our immune system (see Chapters 23 and 24). Even for the deadliest strains, such as the pandemic influenza of 1918, a large proportion of infected people eliminate the virus without symptoms (asymptomatic infection). Why do some people fight off the virus, whereas others who appear equally healthy succumb? Many mysteries remain to challenge the research virologist.

Virion Structure and Genome

The influenza virion has an asymmetrical structure (**Fig. 11.12**). Instead of an icosahedral capsid, the (–) strand RNA genome segments are individually coated by **nucleocapsid proteins** (**NPs**). The term "nucleocapsid" refers generally to proteins coating a viral genome and packaged within or as part of the virion. The NP-coated RNA segments are loosely contained by a shell of **matrix proteins** (M1). The matrix layer is further enclosed by the envelope. The envelope derives from the phospholipid membrane of the host cell, which incorporates the viral glycoproteins hemagglutinin (HA) and neuraminidase (NA). Upon host cell infection, the viral HA and NA bind specific carbohydrate chains on host cell-surface glycoproteins; the chains end in sialic acid. When a newly formed virion exits its host cell, neuraminidase acts as an enzyme to cleave a cell membrane glycoprotein, thus releasing the virion outside the cell. Neuraminidase can be blocked by the antiviral agent Tamiflu (oseltamivir), one of the main drugs available to treat influenza.

Segmented genome. Influenza virus has a **segmented genome** consisting of multiple separate nucleic acids, like the multiple chromosomes of a eukaryotic cell. A segmented genome has profound consequences for viral evolution. If two different strains of influenza virus infect a host simultaneously, their segments can reassort to generate a novel hybrid strain. Because influenza genomes are capable of **reassortment**, they can rapidly generate a new strain that our immune system fails to recognize, such as the pandemic H1N1 strain of 2009 (discussed shortly).

The influenza A genome includes eight segments, each a separate linear (–) strand of RNA (**Figs. 11.12** and **11.14**).

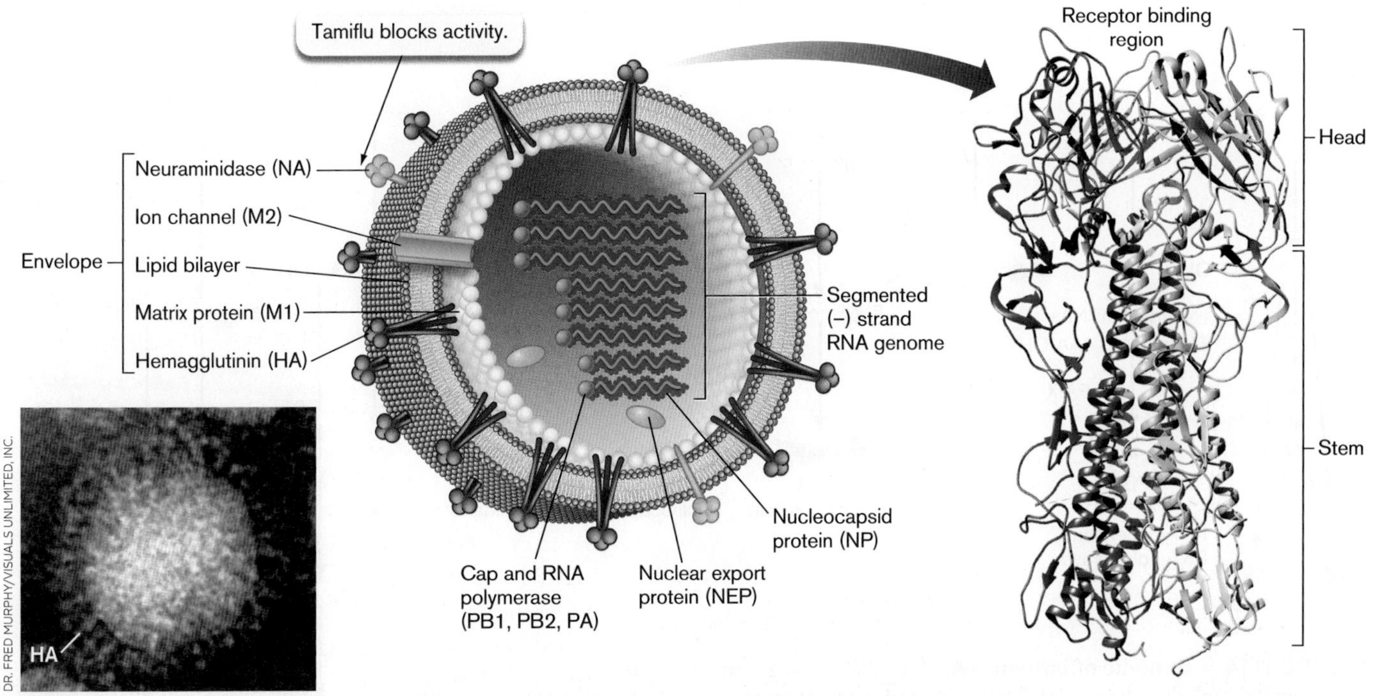

FIGURE 11.12 ▪ Structure of influenza A. Diagram of influenza A virion structure, showing envelope (colored tan), envelope proteins, matrix protein (yellow), RNA segments (blue) with attached polymerase, and the nuclear export protein NEP. **Inset:** Influenza A virion (TEM). The brush-like border coating the envelope consists of glycoproteins, hemagglutinin (HA), and neuraminidase (NA).

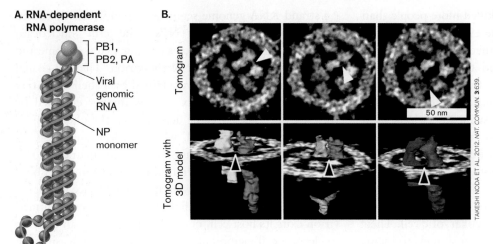

FIGURE 11.13 ■ **The eight ribonucleoprotein (RNP) complexes are linked and packed in the virion.** **A.** Structure of an RNP complex, including the RNA chromosome segment helically wrapped around NP monomers and complexed with the RNA-dependent RNA polymerase (PB1, PB2, PA). **B.** Influenza RNP complexes packed within a virion (colorized cryo-EM). Three tomography sections reveal links between specific RNPs (arrowheads). Source: Part A modified from Amie Eisfeld et al. 2015. *Nat. Rev. Microbiol.* **13**:28.

Within the infected cell, each (–) strand segment must be transcribed to a (+) strand mRNA with a host-derived 5′ cap and viral RNA polymerase, and a 3′ poly-A tail (AAA-$OH^{3'}$) typical of eukaryotic mRNA.

How does the virus express genes encoded by its (–) strand RNA genome? A complementary (+) strand RNA must be synthesized within the infected cell. Thus, within the virion each NP-coated RNA segment carries its own RNA-dependent RNA polymerase complex (proteins PB1, PB2, PA; **Figs. 11.12** and **11.13A**). The polymerase proteins were expressed in the previous host, before the virions formed and the cell lysed. Within the virion, each RNA segment forms a loop, complexed with NP subunits that condense the RNA in a helix. (The helix does not involve base pairs, and it differs from the standard helical forms of RNA.) The two ends of RNA are complexed with the RNA-dependent RNA polymerase, poised for RNA synthesis early in infection.

Once the mRNA molecules are synthesized, they are ready for processing (including 5′ cap and 3′ tail) and translation by host ribosomes. Certain segments (1, 2, 7, 8) encode more than one type of product from a common sequence (**Fig. 11.14**). The multiple products are enabled by alternate mRNA processing and by ribosome slippage and frameshifting. These nonstandard expression mechanisms are typical of the way RNA viruses economize to make diverse products from a small genome.

Note: Distinguish between **reassortment** (two different viruses contribute separate genome segments to a reassortant genome) and **recombination** (two different viruses contribute genetic material to a recombinant molecule.)

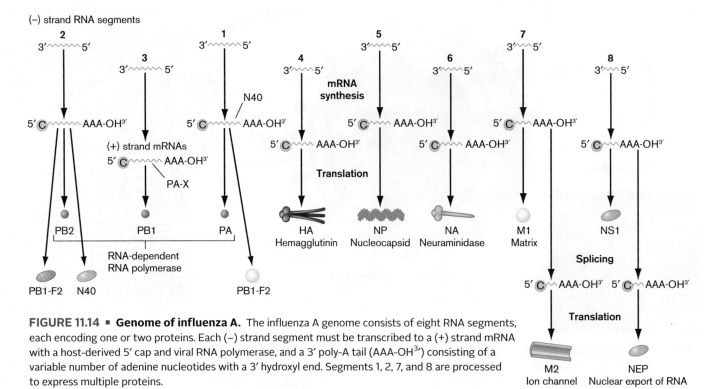

FIGURE 11.14 ■ **Genome of influenza A.** The influenza A genome consists of eight RNA segments, each encoding one or two proteins. Each (–) strand segment must be transcribed to a (+) strand mRNA with a host-derived 5′ cap and viral RNA polymerase, and a 3′ poly-A tail (AAA-$OH^{3'}$) consisting of a variable number of adenine nucleotides with a 3′ hydroxyl end. Segments 1, 2, 7, and 8 are processed to express multiple proteins.

> **Thought Question**
>
> **11.5** Could the influenza genome change by recombination of segments, rather than reassortment? What about the lambda phage genome?

Given that viral infection requires all eight segments, how does the assembly mechanism package exactly eight segments, one of each? Advanced electron microscopy imaging shows how the eight segments package precisely within the virion. The segments appear to link to each other in order as they arrange themselves. **Figure 11.13B** shows the evidence from cryo-electron tomography (a high-resolution EM technique described in Chapter 2). Cryo-EM combined images from frozen unstained virus particles, whose structure appeared highly consistent (unlike those imaged by earlier kinds of EM). The computed images show sections across the influenza virion, in which all eight RNA segments stand side by side, like a bundle of sticks. Sections taken from different depths through the particle reveal tiny molecular connections between adjacent segments. Further experiments with genetic constructs and fluorescence microscopy confirm that all eight unique segments link together in a defined, reproducible pattern.

Mutation and reassortment. Influenza viruses gradually accumulate mutations leading to newly virulent strains that go unrecognized by the host immune system. This gradual accumulation of mutations is known as "antigenic drift." Antigenic drift allows influenza viruses to continually acquire small mutations that can lead to new phenotypes with respect to drug resistance and host range.

A more radical way that strains change is reassortment. The key advantage of a segmented genome is that it enables reassortment of segments between two strains coinfecting the same cell. Reassortment leads to "antigenic shift," in which a new strain combines a mixture of components derived by two different strains. The new strain may enter a new host, where it evades the host immune system. Even strains that infect animals such as ducks or swine may reassort their segments with those of a coinfecting human virus.

Influx of genes from a distantly related strain can sharply increase virulence and mortality. For example, the 1968 Hong Kong flu strain, which killed over 33,000 people in the United States, derived three segments from avian strains. Major epidemics of exceptionally virulent influenza arise as a result of reassortment with genome segments from strains that evolved within ducks or swine, agricultural animals that live in close proximity to humans. In each genome, the "H" and "N" numbers designate alleles of the genes encoding envelope proteins <u>h</u>emagglutinin and <u>n</u>euraminidase, respectively. For example, the Hong Kong flu strain had alleles H3 and N2 (which have since become prevalent in "seasonal" human-flu strains).

In 1979, in Europe, an avian flu strain was found to have "jumped" into swine (the "avian-like" swine strain). In 1992, a triple-reassortant strain was identified that included segments PB2 and PA (encoding RNA-dependent RNA polymerase) from an avian virus; PB1 (polymerase subunit), NP (nucleocapsid), and M (matrix protein) from a swine virus; and PB, H, and N from a human seasonal strain of influenza A. Since the H and N envelope proteins came from a human strain, the triple reassortant could be transmitted readily between humans. Today, molecular surveillance reveals many emerging strains of human influenza A that combine avian and swine alleles. The "swine" strain in 2009 had alleles H1 and N1, similar to the pandemic 1918 strain (**Fig. 11.15B**). In 2009, prompt public health measures such as quarantine helped keep the disease rate low. Yet another reassortant avian strain, H7N9, emerged in China in 2013. So far, these avian strains have been contained—but the next time we may be less fortunate.

Transmission and Attachment to Host Cell

What determines which animals can be infected by a given flu strain? One factor is the requirement for a host cell protease to cleave the hemagglutinin protein on the virion envelope. Cleavage of hemagglutinin enables a small peptide called the **fusion peptide** to mediate viral entry into the host cell (discussed shortly; see **Fig. 11.17**). The presence of the protease is one **host factor** that determines which kind of host may be infected, and which tissues within the host support viral replication. Host factors of many kinds mediate viral infections.

Influenza receptor is a sialic acid glycoprotein. Another important host factor for influenza is cell-surface glycoproteins that contain a terminal sialic acid (**Fig. 11.16**). The sialic acid polysaccharide of the glycoprotein binds hemagglutinin, attaching the virion and enabling endocytosis. The precise structure of the sialic acid host receptor may determine whether a strain such as avian influenza H5N1 will spread directly between humans. For example, the sialic acid connection in the receptor polysaccharide can involve different OH groups of the sugar galactose: a linkage to the OH-3 (alpha-2,3) or to the OH-6 (alpha-2,6) bond (**Fig. 11.16**). The influenza strain H5N1 recognizes mainly the alpha-2,3-linked protein, found in birds. In humans, the upper respiratory tract contains mainly alpha-2,6-linked receptors; alpha-2,3-linked receptors are found only deeper within the lungs. But swine carry receptors of both types. For this reason, swine are believed to act as a "mixing bowl" for strains from birds and humans, as well as swine. Thus, swine incubated the avian strain in 1979, and then enabled later reassortment with human-flu genome segments, leading eventually to the

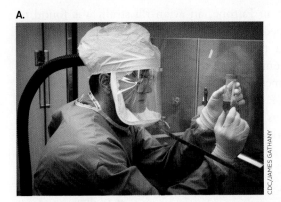

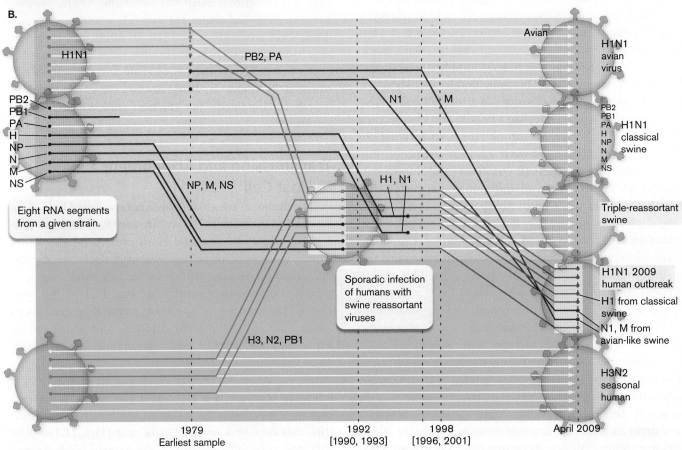

FIGURE 11.15 ▪ Reassortment between human, avian, and swine strains generates exceptionally virulent strains of influenza A. **A.** The reconstructed strain of the 1918 pandemic influenza virus is studied by Terrence Tumpey, a microbiologist at the Centers for Disease Control and Prevention (CDC), Atlanta. **B.** The 2009 strain of H1N1 influenza A arose from a series of reassortments of the eight RNA segments from avian, swine, and human influenza strains. The numbers following "H" and "N" refer to different alleles of the genes encoding hemagglutinin and neuraminidase, respectively. Source: Part B modified from Gavin J. D. Smith et al. 2009. Nature **459**:1122.

2009 strain—which sickened both pigs and humans. Today, large swine facilities are monitored for appearance of novel reassortant strains.

The avian influenza strain H5N1 causes exceptionally high mortality in humans, but it is rarely transmitted from one person to another. More rapid transmission could arise from antigenic drift, by accumulating mutations in the gene encoding avian hemagglutinin. Rapid person-to-person transmission of H5N1 might cause an influenza pandemic with high mortality. That is why public health organizations were so concerned when, in 2011, researchers announced that they had identified mutations conferring H5N1 transmission in the ferret model system, which closely resembles the human system (see **Special Topic 11.1**).

> **Thought Question**
>
> **11.6** How could swine play a role in generating a pandemic strain of influenza? How could avian influenza strain H7N9 become a pandemic strain endangering many people?

Host Entry and Replication Cycle

For replication, the influenza virus must gain entry to the host cell and evade its molecular defenses. Then the viral components travel in and out of the nucleus. Several viral enzymes and structural proteins travel along with its genetic material. As progeny envelope proteins are made, they require transport through the endoplasmic reticulum (ER) and the

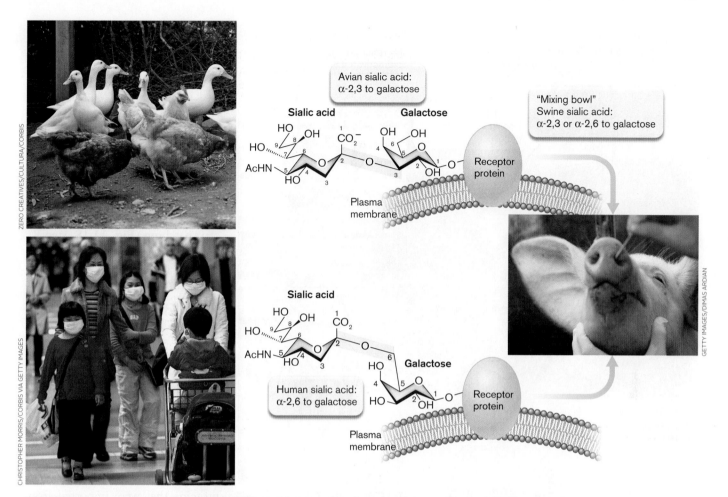

FIGURE 11.16 ■ Influenza receptors in different hosts. The avian host receptor polysaccharide contains sialic acid with an alpha-2,3 bond to galactose, whereas the human receptor in the upper respiratory tract has an alpha-2,6 bond. Swine receptors include both forms of sialic acid; thus, swine can be infected by both avian and human strains and may act as a "mixing bowl" for reassortment.

Golgi complex to the cell membrane. The overall replication cycle is highly complex—and the molecular details offer many opportunities to devise antiviral agents.

Endocytosis and membrane fusion. Endocytosis of the influenza virion involves a key step of acid-mediated membrane fusion, which offers a target for antiviral agents (**Fig. 11.17** ▶). As the influenza virion binds its sialic acid receptor (**Fig. 11.17**, step 1), a host protease cleaves each HA, forming a fusion peptide (step 2). The hemagglutinin trimer now contains three N-terminal fusion peptides. A fusion peptide is a portion of an envelope protein (cleaved from hemagglutinin in the case of influenza virus) that changes conformation so as to facilitate envelope fusion with the host cell membrane.

When the virion is taken up by endocytosis, the endocytic vesicle fuses with a lysosome and its interior acidifies (**Fig. 11.17**, step 3). The lowered pH (increased H^+ concentration) drives H^+ ions into the virion through the M2 ion channel in the matrix layer (see **Fig. 11.12** for virion structure). The influx of acid causes the matrix proteins to dissociate from the NP-coated RNA. The M2 ion channel is the target of amantadine, one of the first anti-influenza drugs; unfortunately, most strains today have evolved resistance to the drug. Low pH also induces a conformational change, shifting the C-terminal ends back and the N-terminal fusion peptides outward to face the vesicle membrane. The peptides extend into the membrane (step 4), where they mediate fusion between viral and host membranes. The fusion process expels the contents of the virion into the host cytoplasm (step 5).

Synthesis of (+) strand mRNA. After the influenza virion is taken up by endocytosis (**Fig. 11.18**, step 1), all the viral (−) RNA segments are released in the cytoplasm (step 2). Each RNA retains its coat of nucleocapsid proteins, as well as a prepackaged RNA-dependent RNA polymerase. The NP-coated RNA segments individually pass through a nuclear pore into the nucleus (step 3). Within the nucleus, each genomic (−) RNA segment with its prepackaged polymerase synthesizes (+) strand RNA for mRNA (step 4). Each mRNA synthesis initiates with a 7-methylguanosine-"capped" RNA fragment (portrayed as "C" in **Fig. 11.18**). The influenza polymerase obtains the cap fragments from

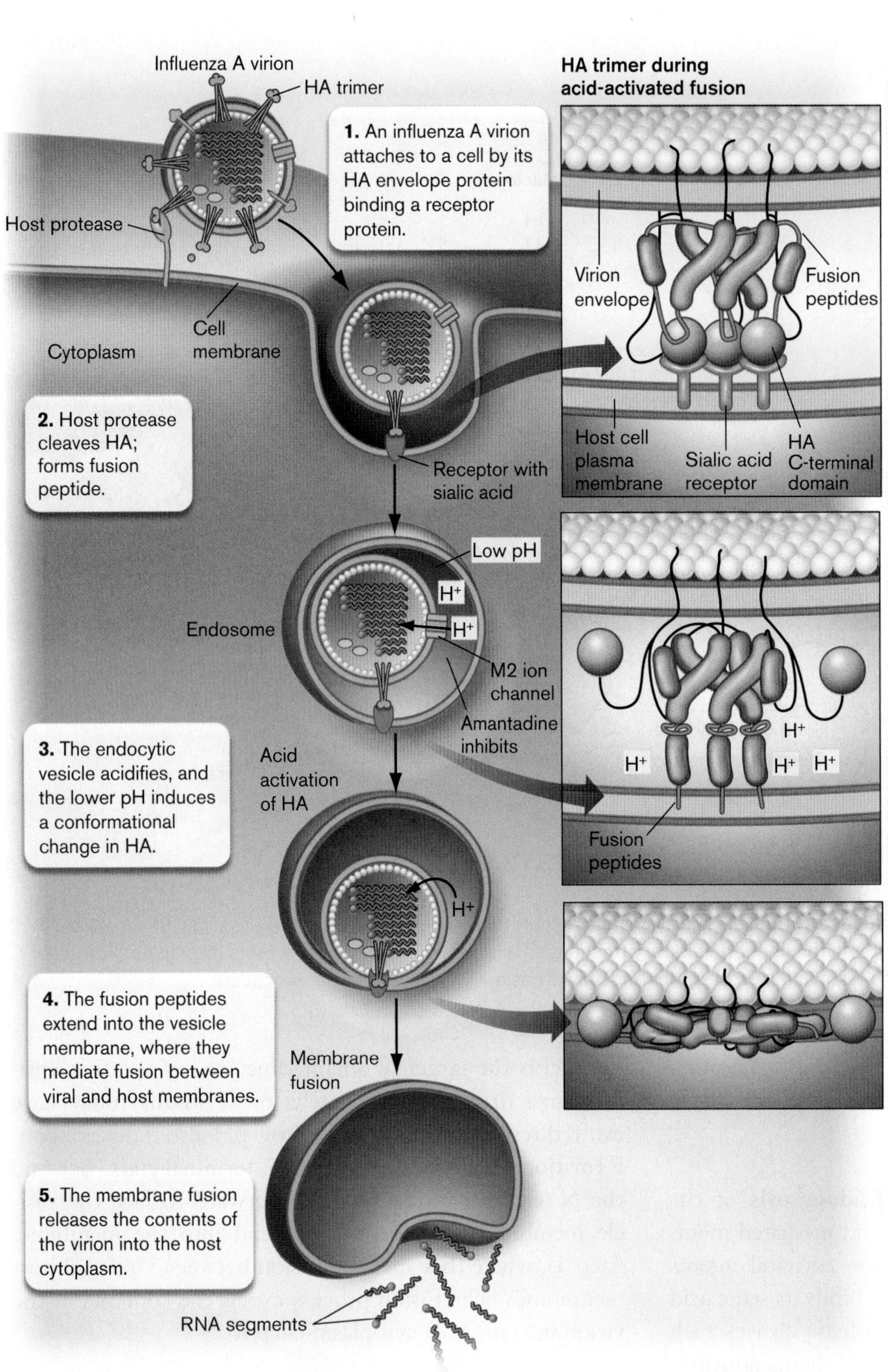

Influenza A virion
HA trimer
Host protease
1. An influenza A virion attaches to a cell by its HA envelope protein binding a receptor protein.
Cell membrane
Cytoplasm
2. Host protease cleaves HA; forms fusion peptide.
Receptor with sialic acid
HA trimer during acid-activated fusion
Virion envelope
Fusion peptides
Host cell plasma membrane
Sialic acid receptor
HA C-terminal domain
Low pH
H+
Endosome
M2 ion channel
Amantadine inhibits
3. The endocytic vesicle acidifies, and the lower pH induces a conformational change in HA.
Acid activation of HA
Fusion peptides
4. The fusion peptides extend into the vesicle membrane, where they mediate fusion between viral and host membranes.
Membrane fusion
5. The membrane fusion releases the contents of the virion into the host cytoplasm.
RNA segments

FIGURE 11.17 ▪ Influenza virion attachment to receptor, acid activation, and release of genome in the cytoplasm.

the host by cleaving them from host nuclear pre-mRNA—a process quaintly known as "cap snatching." The (+) strand mRNA molecules return to the cytoplasm for translation (step 5), using the snatched cap to bind the host ribosome. The RNA segments encoding envelope proteins attach to the ER for protein synthesis and transport to the host cell membrane (step 6). The newly synthesized nucleocapsid proteins (NPs), as well as RNA-dependent RNA polymerase components, subsequently return to the nucleus (step 7). Other genome-packaging proteins (M1 and NEP/NS2) also return to the nucleus.

Synthesis of (+) strand and (–) strand genomic RNA. Back in the nucleus, the original (–) strand RNA segments also serve as templates for RNA synthesis, without cap snatching (**Fig. 11.18**, step 8). The uncapped (+) strand RNA then becomes coated with the newly made NP subunits imported from the cytoplasm. The NP-coated (+) strand serves as template to synthesize (–) strand RNA (step 9), which also becomes coated with NP (step 10).

The NP-coated (–) RNA associates with a newly made polymerase for a future cycle of viral replication. The RNA is then complexed with matrix protein (M1) and nuclear export protein (NEP)—proteins that were imported from the cytoplasm earlier. At last, the fully packaged (–) RNA segments exit the nucleus to the cytoplasm (**Fig. 11.18**, step 11), where they approach the cell membrane for packaging into progeny virions (step 12).

Envelope synthesis and assembly. The envelope proteins synthesized at the ER include hemagglutinin (HA) and neuraminidase (NA). Within the ER lumen, these proteins are glycosylated by host enzymes and then transferred to the Golgi (**Fig. 11.18**, step 13) for export to the cell membrane (step 14). Within the cell membrane, the envelope proteins assemble around a group of (–) RNA segments complexed with their matrix and packaging proteins, completing the virion particle (step 15).

To exit the cell, the virion buds out (**Fig. 11.18**, step 16). Viral exit requires a final step of host release: Neuraminidase (the envelope protein N) cuts the sialic acid link of the host glycoproteins (step 17), releasing the virion out

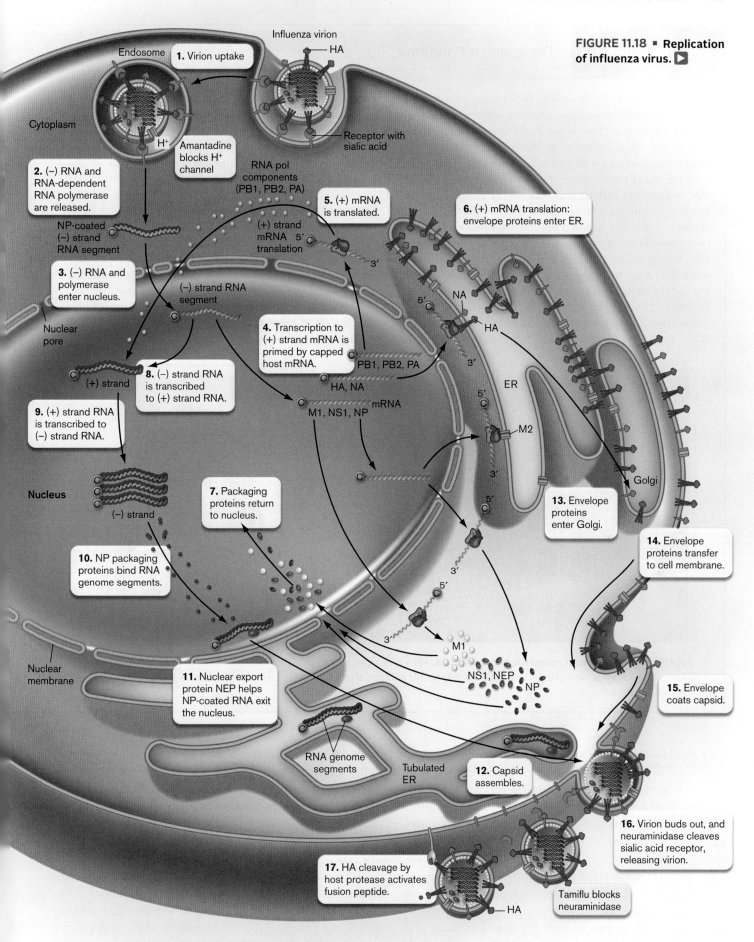

FIGURE 11.18 ▪ Replication of influenza virus.

SPECIAL TOPIC 11.1 Designing a Pandemic Flu

How could a deadly flu strain with poor transmission mutate to transmit readily? If we knew, could we prevent it—and thus prevent a pandemic? The answer might be yes. But how could we perform the experiments to find out? Could our experiments actually start the pandemic?

Some researchers believe the knowledge is worth the risk. Researchers such as Ron Fouchier, at Erasmus University Rotterdam, have investigated the question of which influenza mutations facilitate airborne transmission. In 2012, Fouchier and colleagues conducted experiments testing influenza A transmission in ferrets (**Fig. 1A**). Ferrets have an upper respiratory tract that shows many features in common with the human tract. Thus, ferrets are considered a good model system for human influenza transmission.

Fouchier (**Fig. 1B**) tested two variants of H5N1, the "avian" strain that shows high lethality but low transmission in humans. One H5N1 variant was engineered to contain four point mutations known to increase transmission of other strains. For example, two of the mutations increased the binding of hemagglutinin to human receptors. Initially, both variants showed no airborne transmission in ferrets. The ferrets were kept in separate cages, allowing no direct contact. But Fouchier passaged each H5N1 variant intranasally—that is, from nose to nose of infected ferrets (**Fig. 1C**). After ten intranasal passages (P1–P10 in **Fig. 1C**), the engineered variant suddenly showed airborne transmission. The new phenotype was associated with two extra mutations. The implication of this result was that a small number of specific mutations might convert avian influenza into a pandemic strain.

Fouchier's work generated enormous controversy, including a lawsuit to prevent publication of key details, and a halt to US government funding of certain projects. Despite the

A.

S. HERFST, ET AL. 2012. *SCIENCE* **336**: 1534. REPRINTED WITH PERMISSION FROM AAAS.

B.

DIRK-JAN VISSER/THE NEW YORK TIMES/REDUX

FIGURE 1 ■ Influenza transmission in ferrets. A. Ferrets in cages that allow airflow. B. Ron Fouchier. C. Intranasal passage of a mutant influenza virus selects for mutations that enable airborne transmission. *Source:* Part C modified from Herfst et al. 2012. *Science* **336**:1534–1521, fig. 2.

into the bloodstream. This host release activity of neuraminidase is inhibited by oseltamivir (Tamiflu), the main antiviral agent currently useful against influenza. Tamiflu remains useful today, but resistant strains of the 2009 H1N1 virus have emerged, so we urgently need new antivirals ahead of the next influenza pandemic.

Experimental evidence. How do researchers figure out all the steps of replication shown in **Figure 11.18**? A key technique is fluorescence microscopy, described in Chapters 2 and 3. Fluorophores such as green fluorescent protein (GFP), conjugated antibodies, and labeled nucleic acid probes reveal the presence of viral components within a cell, even tracking their motion over time. An example of fluorescence data is shown in **Figure 11.19**, from experiments conducted by Nadia Naffakh at the Pasteur Institute, France. Naffakh's experiments address the question: How is the assembly of progeny virions organized within the cell?

Transmission electron micrographs were obtained for host cells in tissue culture, both with and without influenza viral infection (**Fig. 11.19A** and **B**). The infected cell shows a marked difference in form of the endoplasmic reticulum. The infected ER shows intricate interconnecting

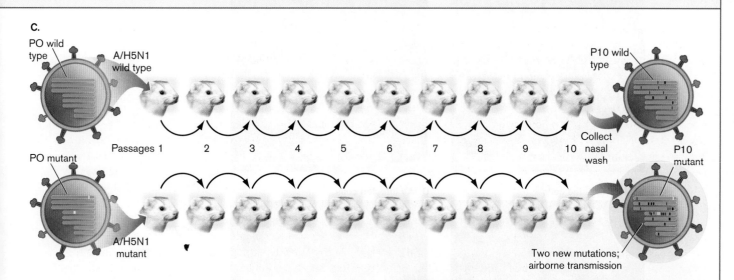

biosafety precautions (**Fig. 2**), what if such research could actually lead to escape of a pandemic strain? Does the knowledge gained actually aid prevention, or is it useless, given that other unknown mutations could still enhance transmission? Ultimately, a moratorium was declared on "gain of function" research (the engineering of strains with added functions that increase transmission) until stricter biosafety regulations were instituted. But the debate continues.

RESEARCH QUESTION

What questions could be tested about the evolution of pandemic influenza? Are these experiments worth the risk?

Sander Herfst, Eefje J. A. Schrauwen, Martin Linster, Salin Chutinimitkul, Emmie de Wit, et al. 2012. Airborne transmission of influenza A/H5N1 virus between ferrets. *Science* **336**:1534–1541.

Jessica A. Belser, Wendy Barclay, Ian Barr, Ron A. M. Fouchier, Ryota Matsuyama, et al. 2018. Ferrets as models for influenza virus transmission studies and pandemic risk assessments. *Emerging Infectious Diseases* **24**:965–971.

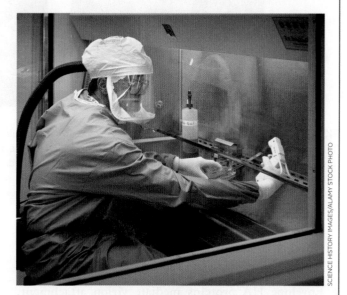

FIGURE 2 ▪ **Biosafety level 3 laboratory.** Study of influenza transmission requires special precautions.

tubes and vesicles that do not appear in the mock-infected cell (a cell inoculated with solvent but no virus). So, what is happening within these tubules? Fluorescence microscopy was performed in order to track the viral and cellular components (**Fig. 11.19C and D**). The ER was tracked with a fluorophore (red) conjugated to an antibody that binds a protein found within the ER lumen. The virus-infected cell shows a tubular pattern of the red fluorescence. There is also green fluorescence from an antibody fluorophore that binds the influenza viral protein NP. The tubules transport RNA segments, which later join the NP proteins forming progeny virions at the plasma membrane (not shown).

To Summarize

- **Influenza virus** causes periodic pandemics of respiratory disease. New virulent strains arise via antigenic drift (the gradual accumulation of mutations) and more suddenly by antigenic shift (through reassortment of gene segments from very different strains, such as human, avian, and/or swine strains.)

- **The influenza virus consists of segmented (−) strand RNA.** Each segment is packaged with nucleocapsid proteins. Segments from different strains reassort through coinfection of a shared host cell.

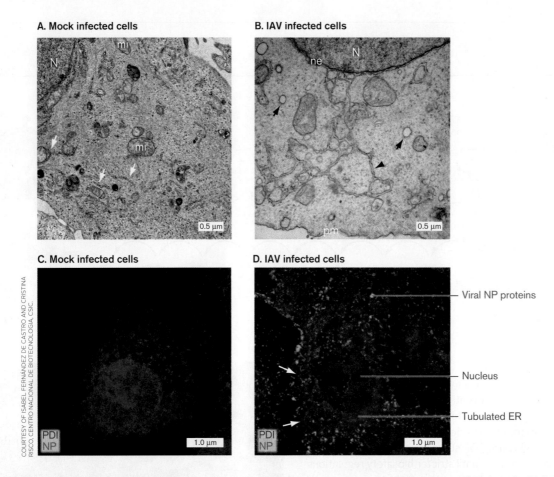

FIGURE 11.19 • Influenza virus tubulates ER. A. Lung tissue culture cells, without virus (TEM). **B.** Cells infected with influenza A virus (IAV) show tubulated ER. **C.** Cultured cells without virus (fluorescence microscopy) stained for nucleus (blue) and ER (red). No IAV protein (green) is present. **D.** Cultured cells with IAV. The ER (red) is tubulated, and the viral protein appears in the cytoplasm. *Source:* Parts A–D modified from I. F. de Castro Martin et al. 2017. *Nat. Commun.* **8**:1396, fig. 3a, c, e, f.

- **Nucleocapsid and matrix proteins** enclose the RNA segments of the influenza virus. The matrix is surrounded by an envelope containing the glycoproteins hemagglutinin (HA) and neuraminidase (NA).

- **Envelope HA proteins mediate virion attachment.** The HA protein includes a fusion peptide that undergoes conformational change to cause fusion between the viral envelope and host cell membrane. For influenza, the virion is internalized by endocytosis.

- **Lysosome fusion with endosomes** triggers viral envelope fusion with the endosome membrane. The viral genome and proteins are then released into the cytoplasm. Viral (−) strand RNA segments attached to RNA-dependent RNA polymerase enter the nucleus.

- **Influenza mRNA synthesis initiates with a capped RNA fragment** cleaved from host mRNA. The capped viral mRNAs return to the cytoplasm for translation.

- **Genomic RNA synthesis generates a (+) strand RNA as a template for (−) strand RNA segments.** Progeny RNA segments are then packaged in newly made nucleocapsid protein and exported to the cytoplasm, for coating with matrix, host cell membrane, and viral envelope proteins.

- **Neuraminidase cleaves the sialic acid connection to host glycoproteins.** This key step of virion release can be blocked by Tamiflu.

11.3 Human Immunodeficiency Virus (HIV): Retrovirus

Human immunodeficiency virus (HIV) is a **retrovirus** (Baltimore group VI). As discussed in Chapter 6, a retrovirus requires the enzyme **reverse transcriptase** to copy its RNA genome into DNA. Retroviruses are a large family of viruses known to infect all types of vertebrate and invertebrate animals; for examples, see **Table 11.1**. The "simple retroviruses" have genomes of just four genes, such as feline leukemia virus (FeLV), the number one killer of outdoor cats in the United States. Simple retroviruses generally cause cancer. The **lentiviruses**, or "slow viruses," cause diseases that progress slowly over many years. Lentiviruses

TABLE 11.1 Retroviruses of Animals (Examples)

Genus	Virus	Disease(s)	Hosts
Simple retroviruses			
Alpharetrovirus	Avian leukosis virus (ALV)	Leukemia	Birds
	Rous sarcoma virus (RSV)	Sarcoma (tumor)	Birds
Betaretrovirus	Mouse mammary tumor virus (MMTV)	Mammary tumor	Mice
Gammaretrovirus	Feline leukemia virus (FeLV)	Lymphoma, immunodeficiency	Cats
	Moloney murine leukemia virus (MMLV)	Leukemia	Mice
Deltaretrovirus	Bovine leukemia virus (BLV)	Leukemia	Cattle
	Primate T-lymphotrophic virus (PTLV-1) [formerly human T-cell leukemia virus (HTLV)]	Leukemia	Humans
Epsilonretrovirus	Walleye dermal sarcoma virus (WDSV)	Sarcoma	Fish
Lentiviruses			
Lentivirus	Human immunodeficiency virus (HIV-1, HIV-2)	AIDS	Humans
	Simian immunodeficiency virus (SIV)	Simian AIDS	Monkeys
	Equine infectious anemia virus (EIAV)	Anemia	Horses
	Maedi-Visna virus (MV)	Neurological disease	Sheep

possess additional regulator genes that modulate host interactions. In addition, most animal genomes show evidence of **endogenous retroviruses**, sequences from an ancient retrovirus whose genome integrated and became "fixed" by mutation, such as HERV-K (discussed in Section 11.4). Surprisingly, endogenous retroviruses can evolve into essential parts of host genomes (see Section 11.4).

The most famous lentivirus is **human immunodeficiency virus** (**HIV**), the cause of **acquired immunodeficiency syndrome** (**AIDS**). The virus HIV and its causative role in AIDS were discovered by French virologist Luc Montagnier, building on Robert Gallo's studies of retroviruses. In 2015, according to the United Nations, 34 million people globally were living with HIV, and AIDS-related infections claimed 1.1 million lives. But the rate of new infections has declined, thanks to research and new drugs that can save lives and prevent HIV transmission. And the virus has now been engineered to make "lentivectors," our most successful agents of gene therapy (see Section 11.4).

History of HIV and AIDS

HIV is a lentivirus that evolved from viruses infecting African monkeys. Two major types are recognized: HIV-1, the cause of most infections at present; and HIV-2, which appears to have evolved independently from a different strain infecting monkeys. The virus is transmitted through blood and through genital or oral-genital contact. HIV can hide in the host cell for many years, with only gradual buildup of virus particles, most of which are eliminated by the host. Eventually, however, the virus destroys the body's T lymphocytes, leaving the host defenseless against many organisms that normally would be harmless.

The first U.S. cases of AIDS were reported in 1981. A historical view of AIDS in the United States emerges in the book *And the Band Played On* by Randy Shilts, adapted as an award-winning film in 1993. The book and film show how American society failed for many years to grasp the significance of AIDS because the syndrome first appeared in societal groups considered marginal (homosexual men and certain ethnic immigrants), although it spread to all social classes. In addition, the virus proved extremely difficult to detect and grow in culture. The discovery of HIV sparked controversy because Gallo failed to acknowledge his use of a virus-producing cell line from Montagnier, the first to isolate HIV-1. Since that time, the two scientists (**Fig. 11.20A**) and many others have collaborated to develop a test for HIV-1 infection and to search for a vaccine. In 2008, the Nobel Prize in Physiology or Medicine was awarded to Montagnier and Françoise Barré-Sinoussi (**Fig. 11.20B**) for their discovery of HIV and its role in AIDS.

Thirty years later, HIV infects one in every 100 adults worldwide, equally among women and men. Treatment with antiretroviral therapy (ART), a mixture of antiretroviral drugs, can enable people carrying HIV to lead a normal life. The different antiretroviral drugs target different molecular mechanisms of HIV infection, as described in this chapter.

A surprising benefit of ART was discovered through humanitarian treatment programs in Africa, such as the President's Emergency Plan for AIDS Relief (PEPFAR), initiated in 2003 by President George W. Bush and continued under President Barack Obama. The PEPFAR program showed that, in communities that deliver ART to all members, regardless of infection status, the virus production

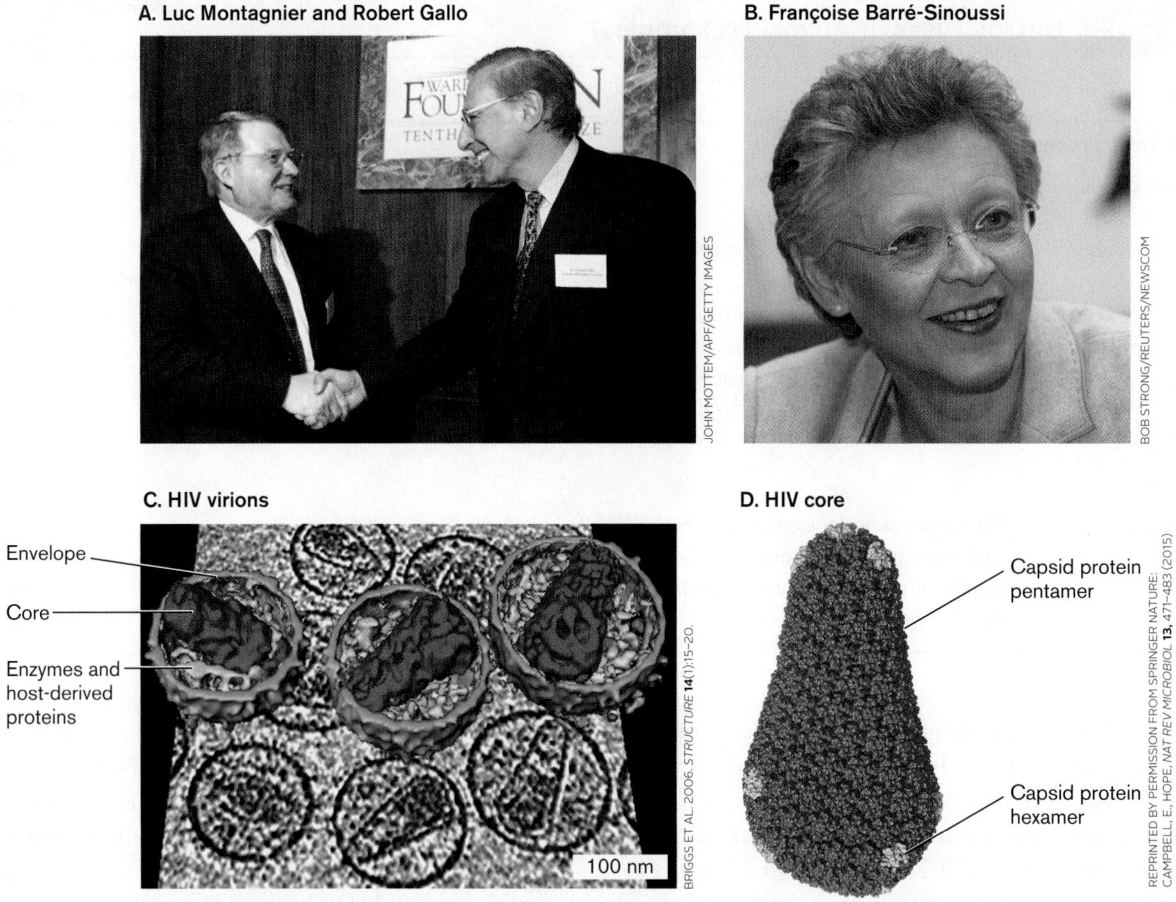

FIGURE 11.20 ▪ HIV discovery. **A.** Luc Montagnier (left) and Robert Gallo agree to collaborate on development of an AIDS vaccine, 2002. **B.** Françoise Barré-Sinoussi, at the Pasteur Institute, worked with Montagnier to discover the virus that causes AIDS. **C.** HIV virions (cryo-EM tomography). **D.** HIV core, model from cryo-EM and X-ray crystallography.

decreases so much that transmission declines. Thus, while HIV cannot be eliminated from infected individuals, new infections could be near zero.

In the United States, since 2012 the CDC has recommended routine HIV testing for all people 13 years of age and older. People at high risk for HIV infection can take a combination of antiviral agents called Pre-exposure Prophylaxis (PrEP), whose effectiveness approaches 99%. But still, infected individuals must take expensive drugs with side effects, and the HIV eventually acquires resistance.

We need a vaccine, but decades of research have failed. Why? The answers to this question are complex.

- **High mutation rate.** The mutation rate of HIV is among the highest known for any virus. Within one patient, the virus evolves into a quasispecies whose different strains attack different organs and predominate at different stages of the disease.
- **Integrated HIV genome hides within cells.** The maintenance of integrated lentiviral genomes involves a greater number of regulator proteins than do the "simple" retroviruses.

HIV Structure and Genome

The structure of HIV as visualized by TEM consists of an electron-dense **core particle** (or capsid) surrounded by a phospholipid envelope (**Fig. 11.20C**). The conical core is composed of capsid protein (CA) subunits whose arrangement is partly icosahedral (**Fig. 11.20D**). The membrane around the core contains **spike proteins**, which join the membrane to the matrix, as in influenza virus. The envelope forms around the core from host cell membrane, when a progeny virion is budding out. In HIV, budding out does not rapidly lyse the cell but has other devastating consequences for cell function (discussed shortly).

HIV core. The core contains two distinct single-stranded copies of the RNA genome (**Fig. 11.21A**). Unlike influenza segments, each of the two RNAs contains a complete "map" of HIV genes. However, the two RNAs of an HIV virion can have slightly different alleles arising from distinct replication events. Thus, the HIV virion is genetically "diploid." A nonfunctional mutant gene on one genome may be complemented by a functional gene on the other.

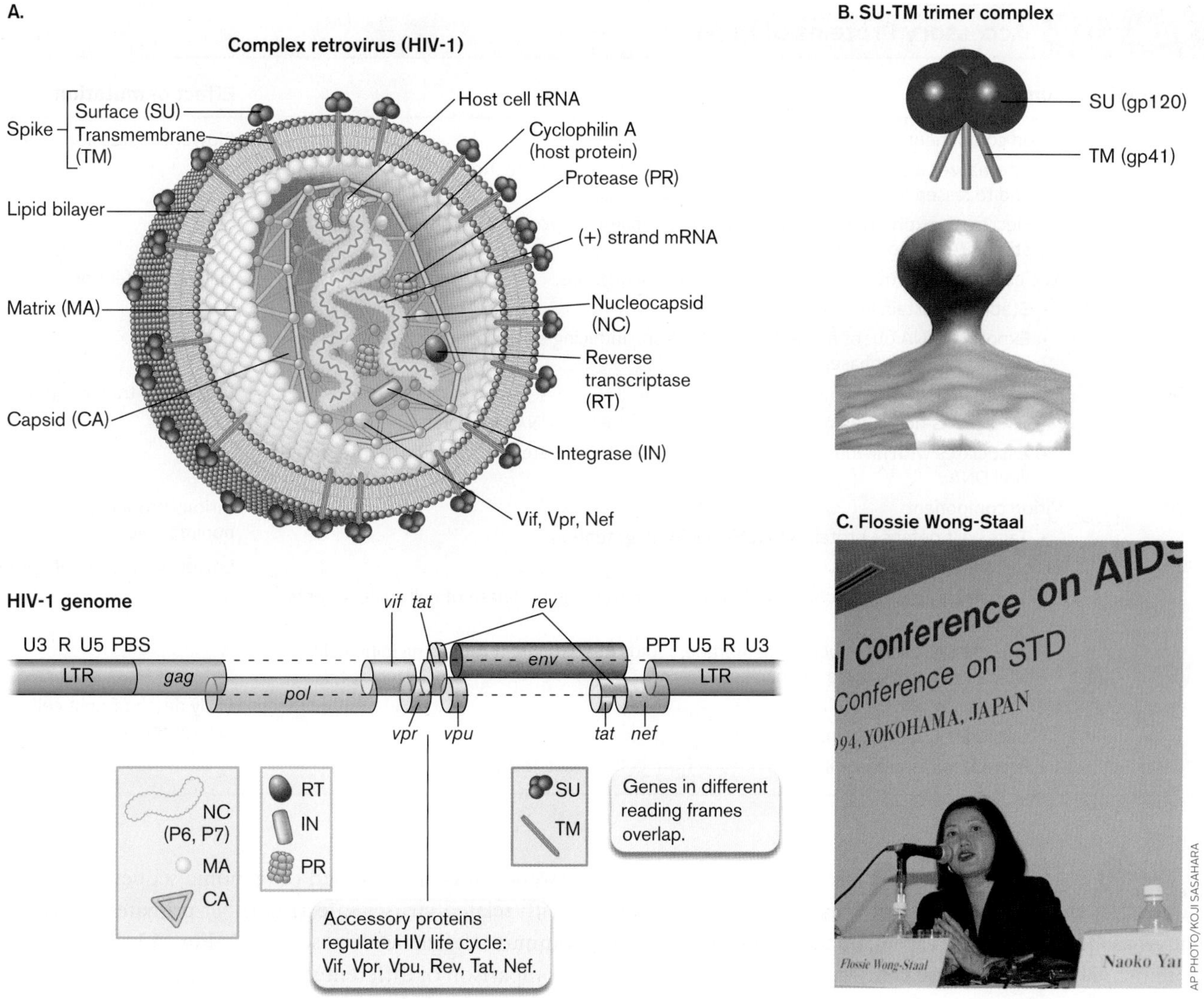

FIGURE 11.21 ■ HIV-1 structure and genome. **A.** Internal structure of the HIV-1 virion (top), color-coded to match the genome (bottom). In the genome sequence, the staggered levels indicate three different reading frames. LTR = long terminal repeat. **B.** Envelope spike complex (model based on cryo-EM tomography). **C.** Flossie Wong-Staal, pioneering AIDS researcher, was the first to clone the HIV genome. Wong-Staal now pursues gene therapy approaches to AIDS prevention and develops lentiviral gene vectors. *Source:* Part A modified from Briggs et al. 2006. *Structure* **14**:15–20; part B, from P. Zhu. 2008. *PLoS Pathog.* **11**:e1000203.

Each RNA genome is coated with nucleocapsid (NC) proteins similar in function to the NP proteins of influenza virus. Unlike influenza virus, each RNA of HIV requires a primer for DNA synthesis: a tRNA derived from the previously infected host cell. The host-derived tRNA is packaged in place on the RNA template, ready to go. The primed and packaged RNA is contained within the core composed of CA subunits. The core also contains about 50 copies of reverse transcriptase (RT) and protease (PR), as well as a DNA integration factor (IN). Unique to type HIV-1, subunits of a host chaperone named cyclophilin A are incorporated into the structure—about one for every ten core subunits. An HIV-1 mutant that fails to incorporate cyclophilin A can attach to a host cell and insert its capsid, but the core fails to come apart, and infection is halted.

The core is surrounded by a matrix (MA subunits), which reinforces the host-derived phospholipid membrane. The membrane is pegged to the matrix by spike proteins composed of the envelope subunits TM and SU (**Fig. 11.21B**). As in influenza virus, the spike proteins play crucial roles in host attachment and entry.

HIV genome. The genome of HIV (**Fig. 11.21A**) was first cloned for molecular study by Flossie Wong-Staal, now at UC San Diego (**Fig. 11.21C**). Born in China in 1947, Wong-Staal immigrated to the United States and then worked with Gallo on the early discoveries of HIV. Wong-Staal now pursues gene therapy approaches to combat HIV infection, and she is developing retroviral vectors for human gene therapy (discussed in Section 11.4).

TABLE 11.2 Accessory Proteins of HIV-1

Protein	Function	Effect of mutation
Nef	Virion component: • Internalizes and degrades CD4 receptors, to avoid superinfection by more HIV virions, and to lessen immune response to the infected cell. • Decreases expression of major histocompatibility complex (MHC) proteins that stimulate cytotoxic T cells.	Slower progression to AIDS
Rev	Nuclear phosphoprotein, combines with host cell proteins: • Stabilizes certain mRNAs in nucleus. • Exports mRNA out of nucleus into cytoplasm, inducing shift from latent phase to virion-producing phase.	Failure of infection
Tat	Transcription factor: • Binds trans-activation response (TAR) site on nascent RNA to activate transcription. • Associates with histone acetylases and kinases to activate transcription of integrated viral DNA.	Blocks HIV transcription
Vif	Virion component: • Tags host defense protein APOBEC3G for degradation.	Virions produced are noninfective
Vpr	Virion component: • Transcription factor; activates HIV transcription during G_2 phase of cell cycle; arrests T-cell growth. • Imports DNA across nuclear membrane; avoids need to infect rapidly dividing cells in which mitosis dissolves the nucleus.	Lower production of virions
Vpu	Membrane protein: • Degrades CD4, releasing bound spike proteins. • Promotes virion assembly and release from cell-surface tetherins.	Early death of host cell; lower production of virions

The HIV genome includes three main open reading frames that are found in all retroviruses: *gag*, *pol*, and *env*. The *gag* sequence encodes capsid, nucleocapsid, and matrix proteins; *pol* encodes reverse transcriptase (RT), integrase, and protease; and *env* encodes envelope proteins. The *gag* and *pol* sequences overlap, but because they are translated in different reading frames (different ways to align the triplet code), the ribosome expresses each independently of the other. During infection, each reading frame is transcribed and translated as a polyprotein; then, at subsequent stages, each polyprotein is cleaved by proteases to form the mature products.

In HIV-1, the *gag* and *pol* sequences overlap, and *env* overlaps with genes encoding **accessory proteins**, proteins that modify and regulate retroviral infection. Accessory proteins are unique to lentiviruses, mediating host response and maintaining the long-term slow progression of lentiviral disease. The accessory proteins are expressed within the infected host cell and regulate the replication cycle (**Table 11.2**). For example, Tat protein activates transcription of the viral genome. The HIV-1 genome encodes at least six accessory proteins—a greater number than in any other retrovirus. They are major targets for research and drug discovery aimed at preventing HIV proliferation.

Origin and evolution of HIV. Where did HIV come from? The origin of HIV has been traced back to the early twentieth century, based on genome sequence comparison with related viruses infecting other primates, called simian immunodeficiency viruses (SIV; **Fig. 11.22A**). Sequence comparison of different strains of HIV and SIV reveals that an immunodeficiency virus actually entered the human population more than once, from SIV strains derived from related primates in Africa. It is thought that human consumption of primates for meat may have introduced SIV strains that then adapted to human infection. Today, the vast majority of HIV-infected patients show the HIV-1 strain M, but some people have been infected by HIV-2, which derived independently from another SIV strain.

HIV is the most rapidly evolving pathogen known; its replication generates about one mutation per progeny virion. For this reason, physicians always prescribe a combination of antiretroviral drugs, with different molecular targets. The hope is that if any one mutation confers resistance to one drug, the mutant virus will still be blocked by another. This strategy of drug combination and continual testing for resistance enables many treated HIV carriers to remain free of AIDS for decades. But in some cases, recombination of different mutants can generate strains resistant to multiple antiviral agents (**Fig. 11.22B**).

Within a single infected patient, the high mutation rates generate multiple virus strains with differing properties of replication, tissue tropism, and resistance to antibiotics. This dynamic population of diverse mutant strains is called

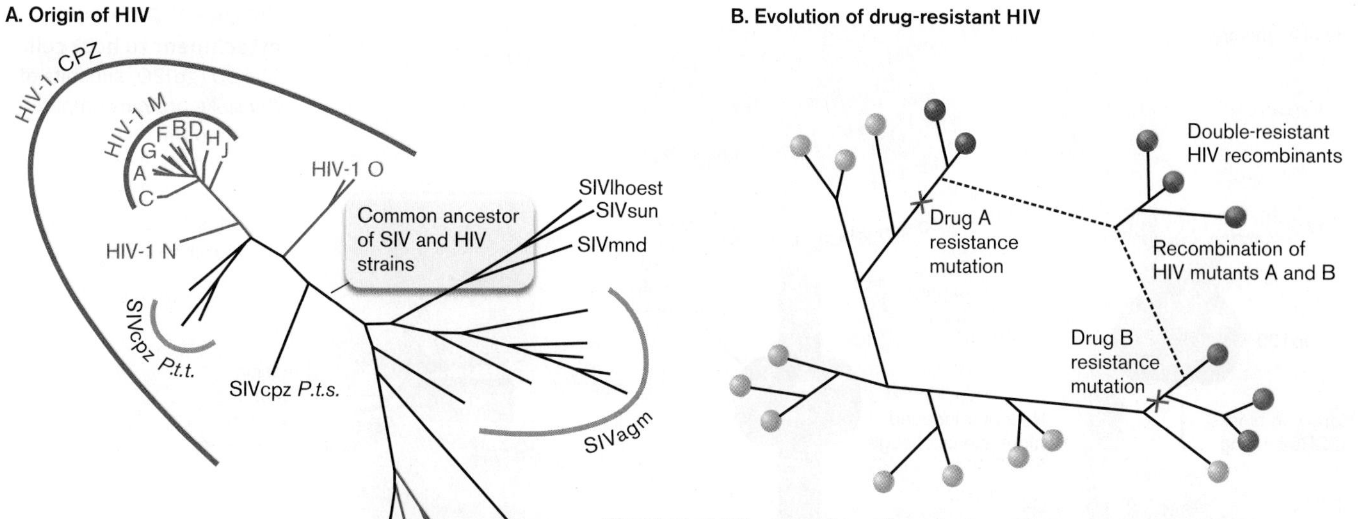

FIGURE 11.22 ▪ Origin and evolution of HIV. **A.** Strains of HIV and SIV arose independently multiple times over several decades, from a common origin in monkeys. *P.t.s.* = *Pan troglodytes schweinfurthii*; *P.t.t.* = *Pan troglodytes troglodytes*. **B.** When HIV infects a patient, different drugs may select strains with different resistance mutations. The mutant strains may then recombine to generate a double-resistant strain.

a **quasispecies** (**Fig. 11.23**). The quasispecies forms when an infective virion commences replication with rapid mutation and trait diversification. Many of the progeny virions have sequences and tropisms so different from those of their ancestor that, in isolation, they would be classified as different species. Different clonal variants of the virus colonize different tissues and organs. Virus types within a quasispecies may interact cooperatively on a functional level, by serving complementary roles in the disease state, and thus collectively define the traits of the viral population. But what happens when an infected "donor" introduces HIV to a recipient? Within the recipient, only one HIV type proliferates—close to the original type that generated the quasispecies. After this acute infection, the HIV population again diversifies into the quasispecies. It's as if a relatively narrow range of genotype carries the HIV "germ line," whereas the mutant types sustain the infected state and maintain immunosuppression.

HIV Attachment and Host Cell Entry

Like other viruses, HIV needs to recognize specific receptor molecules on the surface of its target cells. The primary receptor for HIV is the CD4 surface protein on CD4 T lymphocytes (T cells). The normal function of CD4 surface proteins is to connect the T cell with an antigen-presenting cell, which activates the T cell to turn on B-cell production of antibodies (discussed in Section 24.2). Disruption of this antibody production is the main cause of the AIDS-related susceptibility to opportunistic infections. Note, however, that CD4 proteins appear on many other cell types, such as

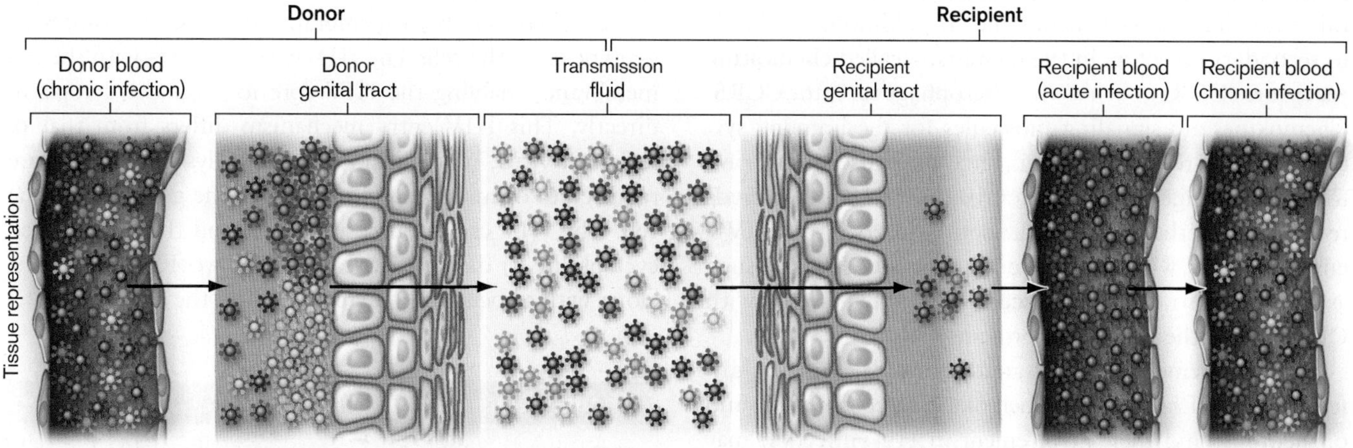

FIGURE 11.23 ▪ Quasispecies development. HIV rapidly generates mutant progeny (different colors) in a chronically infected "donor." The variants may colonize different tissues and organs. Only one of these types (green) is optimized to infect the next "recipient." Within the recipient, the new virions replicate and regenerate the quasispecies. *Source:* Modified from Sarah Joseph. 2015. *Nat. Rev. Microbiol.* **13**:414.

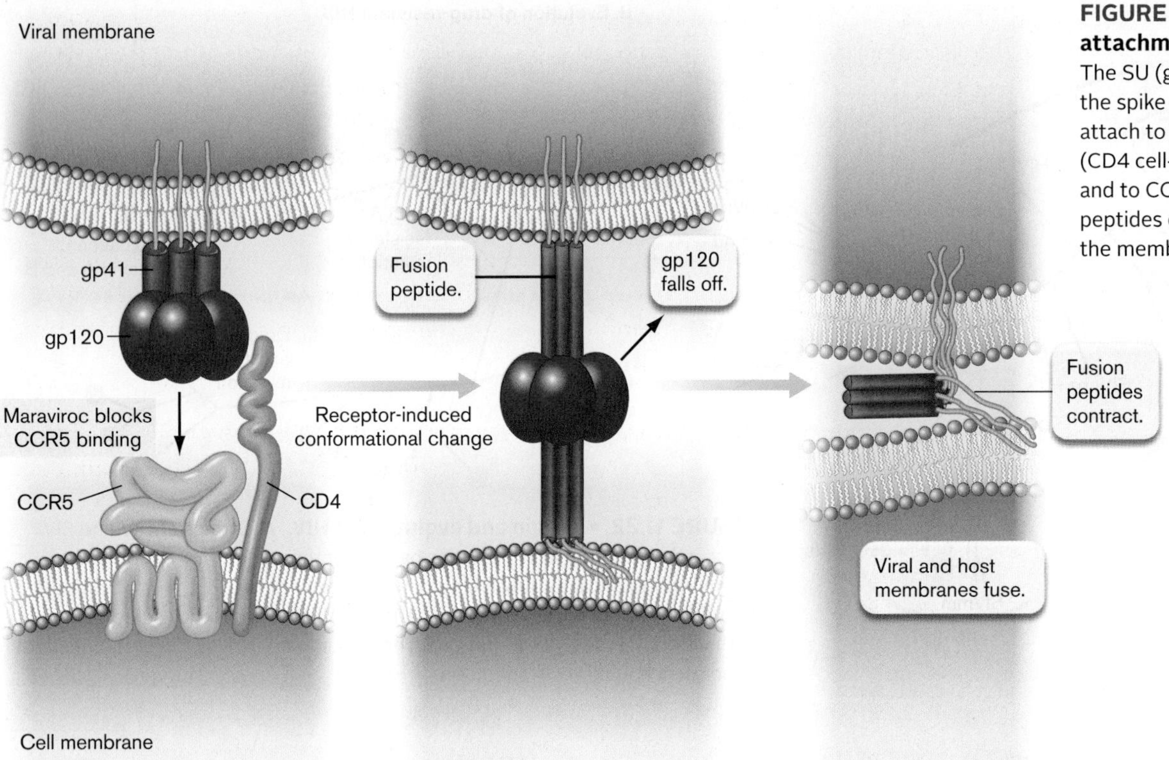

FIGURE 11.24 ■ HIV attachment to host cell. The SU (gp120) subunits of the spike protein complex attach to the receptor (CD4 cell-surface protein) and to CCR5. The fusion peptides contract, pulling the membranes together.

microglia (macrophage-like cells in the central nervous system) and Langerhans cells (immune cells of the epidermis). Their presence may make other cells susceptible to infection by HIV.

Spike proteins mediate membrane fusion. The binding of HIV to CD4 receptors involves the envelope spike protein SU (**Fig. 11.24**). Spike proteins are the main external proteins accessible to the host immune system.

HIV attachment to the cell membrane requires a fusion peptide rearrangement similar to that of influenza virus, except that it takes place at the cell surface (**Fig. 11.24**). When SU (gp120) binds to CD4, the spike transmembrane component TM (gp41) unfolds and extends its fusion peptide into the host cell membrane. In addition, SU binds to secondary receptors in the membrane called **chemokine receptors** (**CCRs**), such as the macrophage receptor CCR5. Chemokines are signaling molecules for the immune system, but their receptor proteins can bind viruses that evolve to take advantage of them. After the spike protein SU binds receptors and the TM fusion peptide inserts, the HIV-1 envelope fuses with the plasma membrane. A CCR such as CCR5 is also called a **coreceptor**, a protein acting with CD4 to bind the HIV spike proteins.

The requirement for CCR attachment varies among different types of HIV. Some chemokine receptors are found on neurons, and their involvement in HIV infection may mediate the neurological disorders seen in AIDS. Furthermore, the predominance of different viral envelope types with different receptor preferences varies over the course of infection. As HIV evolves a quasispecies, the virions target the CCR5 receptor early in infection, whereas later-evolved virions target a different host surface protein, called CXCR4. These "X4" virions can infect early-stage T cells that do not yet carry CCR5; thus the AIDS disease accelerates. The X4 virions are less infective when transmitted, so most new infections start with a CCR5 strain.

An exciting discovery was that individuals who lack the CCR5 protein because of a genetic defect show a high degree of resistance to HIV infection. This finding prompted the Pfizer company to develop an antiviral blocker of CCR5, maraviroc (see **Fig. 11.24**), which is now used for therapy.

After HIV binds to membrane receptors, how does its genome enter the cell? The HIV envelope fuses with the cell membrane, enabling the HIV core to enter the cytoplasm directly. This HIV entry mechanism differs from that of influenza virus, in which endocytosis and lysosome fusion are required to open the capsid and release the genome into the cell. The HIV core (composed of CA and the host-derived cyclophilin A) dissolves, releasing the two RNA genomes, along with associated viral enzymes, into the cytoplasm.

Thought Question

11.7 How do attachment and entry of HIV resemble attachment and entry of influenza virus? How do attachment and entry differ between these two viruses?

The two RNA genomes each possess a 5′ "cap" and a 3′ poly-A "tail" that enable them to mimic host nuclear mRNA. Each RNA is hybridized to a tRNA that serves as a primer for DNA synthesis. The primer is a lysine-specific tRNA from the previous infected cell. The primer might be expected to hybridize at the 3′ end of the template, where its 3′ OH "points" toward the opposite end, positioned to synthesize all the way down. Surprisingly, however, the 3′ OH end of the tRNA actually binds near the 5′ end, where initially it can generate only a brief sequence. These early sequences bind key regulatory factors for transcription and for DNA insertion into the host genome (discussed next).

Reverse Transcriptase Copies RNA to DNA

A retrovirus, unlike other RNA viruses, must integrate its entire genome into the host genome in order to replicate viral genomes and produce new progeny virions. Thus, the RNA genome needs to serve as a template to synthesize a DNA complement, but then the original RNA template must be degraded and replaced by a DNA strand, for host integration. All of these processes are accomplished by **reverse transcriptase** (**RT**), the defining enzyme of a retrovirus. Reverse transcriptase is the source of the high error rate of retroviral replication—on average, one or two errors per copy of HIV. This high error rate generates the quasispecies of different strains within an HIV-infected person (**Fig. 11.23**).

Reverse transcriptase is the target of the first clinically useful drug to treat HIV infection, the nucleotide analog **azidothymidine** (**AZT**). AZT is incorporated into the growing DNA chain in place of a thymidine, but because its 3′ OH is replaced by an azido group ($-N_3$), no further nucleotides can be added.

The reverse transcriptase complex actually possesses three different activities:

- **DNA synthesis from the RNA template.** Synthesis of DNA is first primed by the host tRNA, which was hybridized to the chromosome within the virion.
- **RNA degradation.** After DNA synthesis, the template RNA is gradually removed through an RNase H activity of the RT complex. Removal of RNA enables replacement of the entire original RNA template by DNA.
- **DNA-dependent DNA synthesis.** To make the DNA complementary strand replacing the RNA, the RT needs to use the newly made DNA as its template. Thus, RT has the rare ability to use either DNA or RNA as a template.

As shown in **Figure 11.25A**, the RNA template with its short RNA primer is threaded through the RT complex between the "thumb" and "fingers"—a configuration typical of other RNA polymerases (discussed in Chapter 8). The RT complex adds successive deoxynucleotides from deoxyribonucleoside triphosphates (dNTPs) starting at the 3′ OH end of the RNA primer. As DNA elongates, however, the RNA template is cleaved from behind by the RT complex. Thus, the new DNA actually replaces preexisting RNA sequence. This "destructive replication" is unique to retroviruses. The details are important, as they suggest possible targets for new antiviral drugs.

Reverse transcription of the HIV genome. Reverse transcription of the viral genome involves several unusual mechanisms (**Fig. 11.25B**). First, the host-derived tRNA primer initiates synthesis (by reverse transcriptase, RT) of a DNA strand complementary to the RNA chromosome. DNA is elongated toward the 5′ end of the HIV chromosome, generating a short segment (**Fig. 11.25B**, step 1). The original RNA template for this short segment is then degraded by the RNase H activity of RT, leaving only the DNA extension of the tRNA primer (step 2).

The new DNA primes the second template. The original RNA template had repeated ends (labeled "r," lowercase, for RNA in **Fig. 11.25B**), and the exposed DNA copy of the 5′ end has a complementary sequence ("R," uppercase, for DNA). The "R" DNA from the second tRNA extension hybridizes to the 3′ end of the original RNA (**Fig. 11.25B**, step 3). The hybridized DNA elongates along the rest of the chromosome (step 4), up to the primer-binding site (pbs) for mRNA transcription. DNA completion is followed by degradation of the remaining RNA template, except for occasional short fragments to serve as primers, such as the polypurine tract (ppt). A complementary DNA strand is then synthesized through the PPT primer, leaving a nick at U3 (step 5).

Interestingly, human host cells have evolved a protein, APOBEC3G, that interferes with reverse transcription by several mechanisms. Interference mechanisms include:

- Inhibition of tRNA priming of reverse transcription
- Deamination of cytosines to uridine in the HIV DNA product of reverse transcription, thus increasing the error rate of the HIV provirus (integrated DNA copy of the viral genome)
- Interference with removal of tRNA primer from completed DNA copy, thus inhibiting integration into the host genome

APOBEC3G can be packaged into progeny virions, and thus decreases the production of infective virions in the next host. However, HIV has evolved an accessory protein, Vif (see **Table 11.2**), that binds APOBEC3G and initiates its degradation.

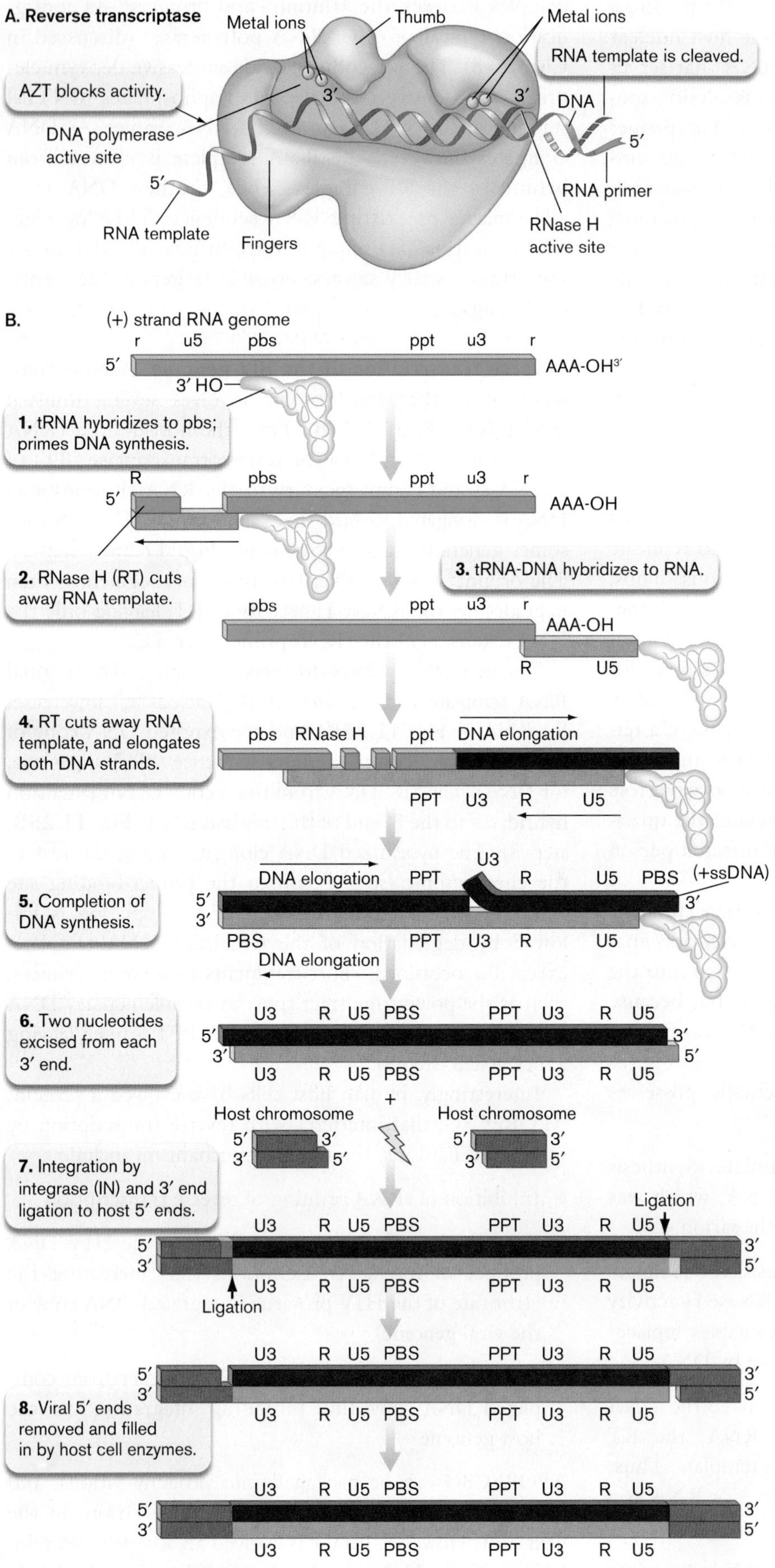

FIGURE 11.25 ▪ Reverse transcription of the HIV genome and integration into host DNA. **A.** The RNA template with its short RNA primer is threaded through the RT complex between the "thumb" and "fingers"—a configuration typical of other RNA polymerases. The RT complex adds successive dNTPs, as in regular DNA synthesis. As DNA elongates, the RNA template is cleaved from behind by the RNase H site of reverse transcriptase (RT). **B.** The tRNA primer initiates a short sequence of DNA complementary to the 5′ end of the HIV chromosome. The corresponding template is then degraded, and the DNA-primer complex is transferred to the opposite end, where it can complete synthesis of the genome. Remaining RNA is cleaved by RT and replaced by DNA. The double-stranded DNA circularizes at the U3-R-U5 sequence and then integrates somewhere in the host genome.

Integration into the host genome. The final phase of genome processing requires integration into host DNA. This complex process provides excellent targets for therapeutic inhibitors such as the drug raltegravir. Surprisingly, HIV integration requires no specific end homology, and the viral DNA generally inserts at positions with low host gene expression, such as introns. For this reason, HIV integration offers a compelling tool for gene therapy vectors (discussed in the next section).

The double-stranded DNA copy of the HIV genome undergoes integration catalyzed by the viral enzyme integrase (IN), which was packaged in the original virion (**Fig. 11.21**). First the integrase excises two nucleotides from each 3′ end of the duplex (**Fig. 11.25B**, step 6). This excision generates a 5′ overhang at each end. Integrase also nicks the host genome, generating 2-bp staggered ends. The HIV DNA then inserts at the nicked host site, with ligation of viral 3′ ends to the 5′ host ends (step 7). The viral 5′ ends are removed, and the gaps are filled in by host cell repair proteins (step 8). Overall, this mechanism forms an integrated viral genome, or **provirus**. The provirus includes two copies of the end sequence U3-R-U5, which is called a **long terminal repeat** (**LTR**). Proviral sequences can now be expressed, directing production of progeny virions.

An alternative to virion production is that the integrated HIV genome lies dormant, like the lambda prophage in *E. coli*. The integrated HIV genome is replicated passively within the genome of its host cell, hiding for many years with only infrequent production of virions. The few virions shed by the patient, however, can infect an unsuspecting individual who has sexual contact with or is exposed to the blood of the patient.

Note: The preceding description of HIV integration was revised from the previous edition of this book to reflect recent research discoveries. For further details, see Robert Craigie. 2012. The molecular biology of integrase. *Future Virology* **7**:679.

Replication Cycle of HIV

The steps of HIV replication are outlined in **Figure 11.26** ▶. The main points of viral entry and replication are typical of retroviruses. HIV, however, has an exceptionally large number of accessory proteins that govern the level of viral production and the duration of the quiescent phase, when the integrated chromosome replicates with the host cell.

Synthesis of HIV mRNA and progeny genomic RNA. After the HIV virion attaches to the host receptors, its envelope fuses with the host membrane (**Fig. 11.26**, step 1). Unlike influenza virus, the HIV core enters the cytoplasm directly, without endocytosis (step 2). The core partly uncoats, while the RNA chromosomes within are reverse-transcribed to make double-stranded DNA (step 3). The double-stranded DNA enters the nucleus through a nuclear pore (step 4)—a key step facilitated by Vpr accessory protein. Vpr enables infection of nondividing cells, which only lentiviruses can do; other retroviruses, such as those causing lymphoma, must infect dividing cells, in which the nuclear membrane dissolves during mitosis.

Upon entering the nucleus, the DNA copy of the HIV genome integrates its sequence as a provirus at a random position in a host chromosome (step 5). Integration is catalyzed by integrase (the IN protein; see **Fig. 11.21**). Integrase inhibitors such as raltegravir are an important class of anti-HIV drugs. Within the nucleus, full-length RNA transcripts are made by host RNA polymerase II, including a 5′ cap and a 3′ poly-A tail (step 6). Some of the RNAs exit the nucleus (step 7) to serve as mRNA for translation of polyproteins. Polyproteins are translated in alternative versions, such as Gag-Pol. Other full-length RNA transcripts exit the nucleus to form RNA dimers for progeny virions (step 8). Still other RNA transcripts within the nucleus are cut and spliced to complete the *env* gene sequence for translation of Env (envelope) proteins (step 9).

The Env proteins are made within the endoplasmic reticulum (**Fig. 11.26**, step 10). They pass through the Golgi for glycosylation and packaging (step 11) and are exported to the cell membrane (step 12). At the membrane, Env proteins plug into the core particle as it forms from the RNA dimers plus Gag-Pol peptides (step 13).

Virion assembly and exit. The core particles are packaged with envelope derived from host cell membrane containing Env spike proteins (**Fig. 11.26**, step 14). To escape the host cell, emerging virions require the accessory protein Vpu to bind a "tetherin," a host adhesion protein induced by interferon to cause reuptake and digestion of virions. Vpu causes proteasomal degradation of the tetherin. Emerging virions, some still tethered to the cell, are shown in **Figure 11.27A**.

As the virion buds off, the protease (PR) cleaves the Gag-Pol peptide to complete maturation of the core structure containing Gag subunits, as well as reverse transcriptase (RT; **Fig. 11.26**, step 15). The Gag subunits now form the conical core structure. Proteases that cleave Gag-Pol offer important drug targets, which have led to the development of anti-HIV drugs known as **protease inhibitors**.

However, HIV has alternative means of cell-to-cell transmission that avoid exposing virions to the immune system. One alternative is cell fusion, mediated by binding of Env in the membrane to CD4 receptors on a neighboring cell (see **Fig. 11.26**). The two cells then fuse, and HIV core particles can enter the new cell through their fused cytoplasm. The fusion of many cells can form a giant multinucleate cell called a syncytium. Cell fusion with formation of syncytia enables HIV to infect neighboring cells without

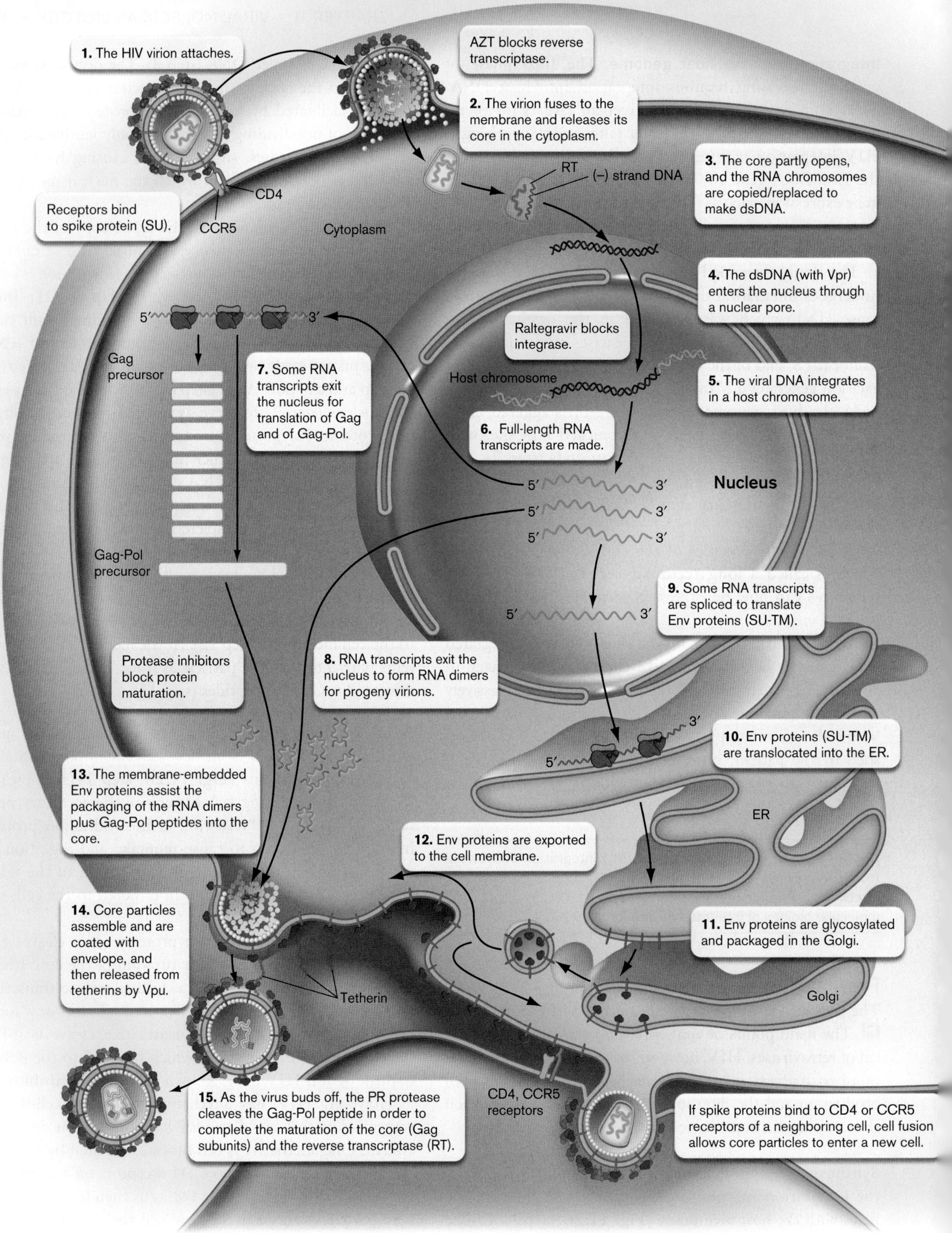

FIGURE 11.26 ▪ HIV replication cycle. The HIV virion attaches its receptor and fuses with the host cell membrane, releasing its contents in the cytoplasm to undergo a replication cycle.

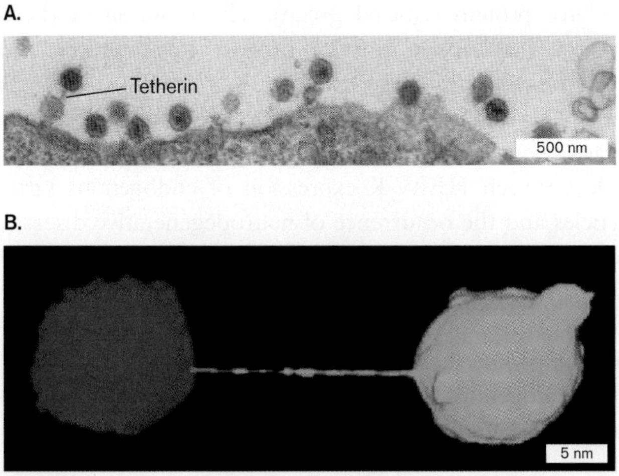

STUART NEIL ET AL. 2008. *NATURE* **451**:425

STEFANIE SOWINSKI ET AL. 2008. *NAT. CELL BIOL.* **10**:211.

FIGURE 11.27 ■ HIV exits from an infected cell. A. As virions emerge, they remain tethered to the cell surface by host tetherins, requiring a release step mediated by accessory protein Vpu (TEM). **B.** A T cell infected with HIV can transfer virions to an uninfected cell through a nanotubular connection. The two T cells are tagged here with different fluorescent labels (red versus green).

ever exiting a cell. Another means of cell-to-cell transmission is to travel through a "nanotube" connection between two T cells (**Fig. 11.27B**).

The intricate scheme in **Figure 11.26** actually omits many functions of HIV accessory proteins that enhance the virulence of HIV infection (see **Table 11.2**). Mutation of genes for accessory proteins often decreases virulence; thus, these proteins are potential targets for chemotherapy. Surprisingly, even a modest decrease of HIV infectivity can have major benefits for the patient, suggesting that HIV is so crippled by its high mutation rate that the slightest interference greatly decreases production of infective virions. Thus, numerous effective antiviral agents are now known—but all have side effects, and all select for resistant strains.

To Summarize

- **Human immunodeficiency virus (HIV)** is the cause of an ongoing pandemic of acquired immunodeficiency syndrome (AIDS). Molecular biology has led to drugs that control the infection.
- **HIV is a retrovirus** whose RNA genome is reverse-transcribed into double-stranded DNA, which integrates into the DNA of the host cell. HIV evolved from simian retroviruses.
- **The HIV core** contains two different copies of its RNA genome, each bound to a primer (host tRNA) and reverse transcriptase (RT). The core is surrounded by an envelope containing spike protein trimers.
- **HIV binds the CD4 receptor** of T lymphocytes together with the chemokine receptor CCR5. Following virion-receptor binding and envelope-membrane fusion, the HIV core particle is released into the cytoplasm, where it partly uncoats.
- **Reverse transcriptase synthesizes DNA from the HIV RNA template**, primed by the tRNA. **RNA degradation** enables formation of a double-stranded DNA. Entering the nucleus, the retroviral **DNA integrates** into the host genome.
- **Retroviral mRNAs are exported to the cytoplasm** for translation. Envelope proteins are translated at the endoplasmic reticulum and exported to the cell membrane.
- **Retroviruses are assembled at the cell membrane**, where virions are released slowly, without lysis. Alternative routes of cell-to-cell transmission involve cell fusion (forming syncytia) or travel through an intercellular nanotube.
- **Accessory proteins regulate virion formation** and the latent phase, in which double-stranded DNA persists without reproduction of progeny virions.

11.4 Endogenous Retroviruses and Gene Therapy

Suppose an integrated HIV genome mutated and lost the ability to produce progeny. What would happen to its genome? The integrated genome would be "trapped" within a cell, an endogenous retrovirus. If the cell entered the host germ line, over many host generations in its host the retroviral sequence would inevitably accumulate more mutations. It could even provide the material for evolution of a new trait. Such endogenous retroviruses inspired the idea that researchers could intentionally manipulate a retrovirus to impart a useful trait and use it for gene therapy.

Retroelements in the Human Genome

The human genome is riddled with remains of retroviral genomes, in various states of decay. These decaying genomes are collectively known as retroelements (**Fig. 11.28**). **Endogenous retroviruses** (in humans, HERVs) are retroelements that retain all the genomic elements of a retrovirus, including *gag*, *env*, and *pol* genes. Other endogenous elements retain part of the retroviral genome but have lost essential sequences through reductive evolution (discussed in Chapter 17).

Retrotransposons retain only partial retroviral elements but may maintain a reverse transcriptase to copy themselves

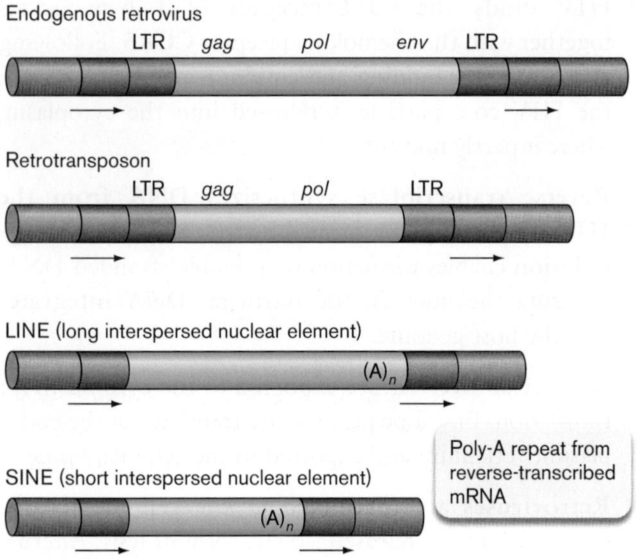

FIGURE 11.28 ▪ Retroelements in the human genome. Endogenous retroviruses and other retroelements in the human genome may arise from progressive degeneration of ancestral retroviruses, or they may be progenitors of new retroviruses.

into other genome locations. An example of a retrotransposon is the well-known Alu sequence, a short sequence found in about a million copies in the human genome. In some cases, a retrotransposon such as Alu can interrupt a key human gene, leading to a genetic defect such as a defective lipoprotein receptor associated with abnormally high cholesterol level and heart failure. Still other retroelements, known as LINEs (long interspersed nuclear elements) and SINEs (short interspersed nuclear elements), show more vestigial remnants of retroviral genomes. Amazingly, retroelements and transposons appear to have generated about half the sequence of the human genome.

Remarkably, many HERVs and other mammalian ERVs express viral proteins that have evolved into required parts of the normal host function. For example, the human embryo requires ERVW-1 to express a protein called syncytin, which enables fusion of placental and maternal cells. The gene encoding syncytin is derived from the retroviral gene for its envelope protein; in the original retrovirus, this protein causes fusion of infected host cells.

In another case, cells expressing a normal endogenous retrovirus form particles that appear to be actual virions. HERV-K proteins are expressed by the embryonic blastocyst (**Fig. 11.29**). At the blastocyst stage, cells express proteins of the endogenous retrovirus HERV-K and form virus-like particles (labeled red in the **Figure 11.29** EMs by HERV-K capsid antibody stain). The virus particle formation is not random, but highly regulated; HERV-K particles arise when cell nuclei express an embryonic regulator protein (labeled green). The function of these particles is unknown, but viral proteins might protect the embryo from infection by exogenous viruses.

On the downside, some HERV sequences may have negative consequences. An intriguing connection has been made between HERV-K expression of endogenous virus particles and the occurrence of neurodegenerative diseases such as amyotrophic lateral sclerosis (ALS). In one case, an ALS patient who was undergoing antiretroviral therapy for HIV actually experienced disappearance of his ALS symptoms, implying that the antiretroviral therapy had eliminated endogenous retroviruses. This speculative possibility has stimulated some exciting research.

Viral Gene Therapy

For millions of years, viruses have interacted with the genomes of humans and our prehuman ancestors—often contributing valuable genes whose products enhance our fitness. Could the genetic properties of such viruses be modified by our technology—to engineer therapy for medicine? Increasingly, the amazing answer is yes. Viruses are engineered as **gene transfer vectors** to deliver genes into our genome. A gene transfer vector is a DNA sequence that can express a recombinant gene within an animal or plant cell, either from a plasmid or from a sequence integrated into the host genome.

Gene transfer vectors are constructed from viruses whose replication cycles establish a viral genome within the host

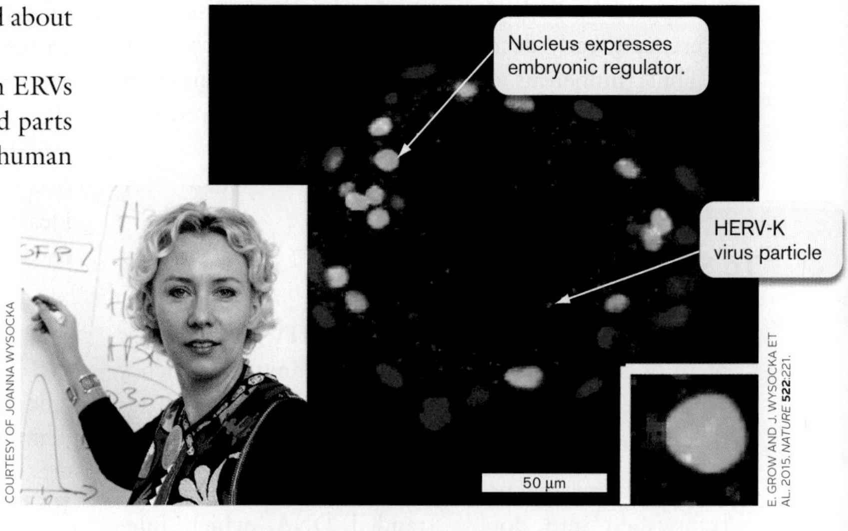

FIGURE 11.29 ▪ Human embryo makes retroviral particles. Human embryos at the blastocyst stage express proteins of the endogenous retrovirus HERV-K, including virus-like particles labeled by HERV-K capsid antibody stain (red). The blastocyst nuclei are labeled with DAPI fluorophore (blue). HERV-K virus-like particles arise when cell nuclei express an embryonic regulator protein (labeled green). **Left inset:** Joanna Wysocka, a stem cell biologist at Stanford University, studies the contribution of ERVs to the human genome. **Lower right inset:** Enlarged view of embryonic regulator protein.

TABLE 11.3 Gene Therapy with Viruses: Examples

Disease treated	Vector type	Reference
Spinal muscular atrophy type 1	Adeno-associated virus type 9	J. R. Mendel et al. 2017. *NEJM* **377**:1713.
Adenosine deaminase severe combined immunodeficiency (ADA-SCID)	Gammaretroviral vector (Strimvelis)	A. Aiuti et al. 2017. *EMBO Mol. Med.* **9**:737.
Acute lymphoblastic leukemia (ALL)	HIV-derived lentivector for chimeric antigen receptor (CAR) T-cell therapy	S. L. Maude et al. 2014. *NEJM* **371**:1507.
Sickle-cell disease	HIV-derived lentiviral vector (LentiGlobin BB305)	J.-A. Ribeil et al. 2017. *NEJM* **376**:848.

nucleus. Various kinds of viral vectors have produced exciting therapeutic results (**Table 11.3**). The first viral vectors were made from double-stranded DNA viruses such as adenoviruses, which cause mild respiratory illness. An adenoviral genome enters the cell nucleus, where it circularizes and replicates separately from the host chromosomes, in a cycle similar to that of herpesviruses (Section 11.5). Thus, adenoviral vectors avoid the long-term risks of inserting DNA permanently into the genome of the host cell. A promising example in 2017 was the use of an adenoviral vector to express protein SMN in infants with spinal muscular atrophy type 1.

A disadvantage of adenoviral vectors is that the adenoviral genes are eventually lost from the recipient, and thus the treatment generally must be repeated. Repeated exposure to the vector eventually stimulates an immune response that destroys it. Other challenges for adenoviral vectors involve the cellular trafficking of the virus and its gene product.

Another exciting class of gene transfer vectors derives from the lentivirus HIV; vectors in this class are called lentiviral vectors, or **lentivectors**. Lentivectors integrate genes into a host chromosome, providing longer-lasting therapy (**Fig. 11.30** ▶). Vectors derived from HIV are particularly useful for their ability to transfer genes into nonmitotic cells. And, surprisingly, the HIV integrase has a preference for avoiding integration into highly transcribed host genes; thus, lentivectors are less likely to interfere with host function. The basis of this property of integrase remains unknown, but it is highly useful for gene therapy.

For therapeutic use, the native lentiviral genome requires extensive modification. The lentivector is engineered to remove viral genes that cause disease, and to express an altered envelope protein from a different virus, such as vesicular stomatitis virus (VSV; **Fig. 11.30**). The VSV envelope protein increases the viral host range and tropism,

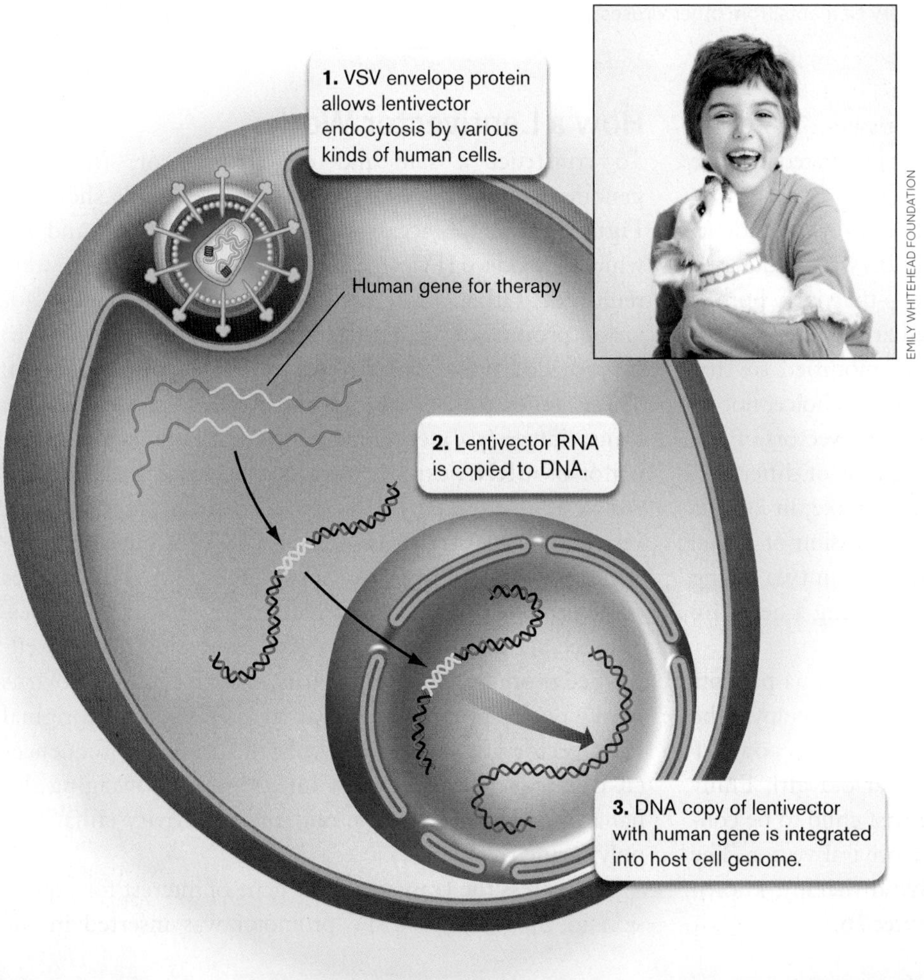
1. VSV envelope protein allows lentivector endocytosis by various kinds of human cells.
Human gene for therapy
2. Lentivector RNA is copied to DNA.
3. DNA copy of lentivector with human gene is integrated into host cell genome.
EMILY WHITEHEAD FOUNDATION

FIGURE 11.30 ▪ Lentiviral gene therapy. A lentiviral vector (lentivector) derived from HIV consists of virions coated with a VSV (vesicular stomatitis virus) envelope protein that enables uptake by various kinds of cells. The engineered viral genome lacks disease-causing genes but possesses a transgene needed by the patient. The lentivector RNA with the transgene is copied into DNA and integrated into a host chromosome. **Inset:** Emily Whitehead was the first child to be considered cured of an illness (B-cell leukemia) by a lentiviral vector. ▶

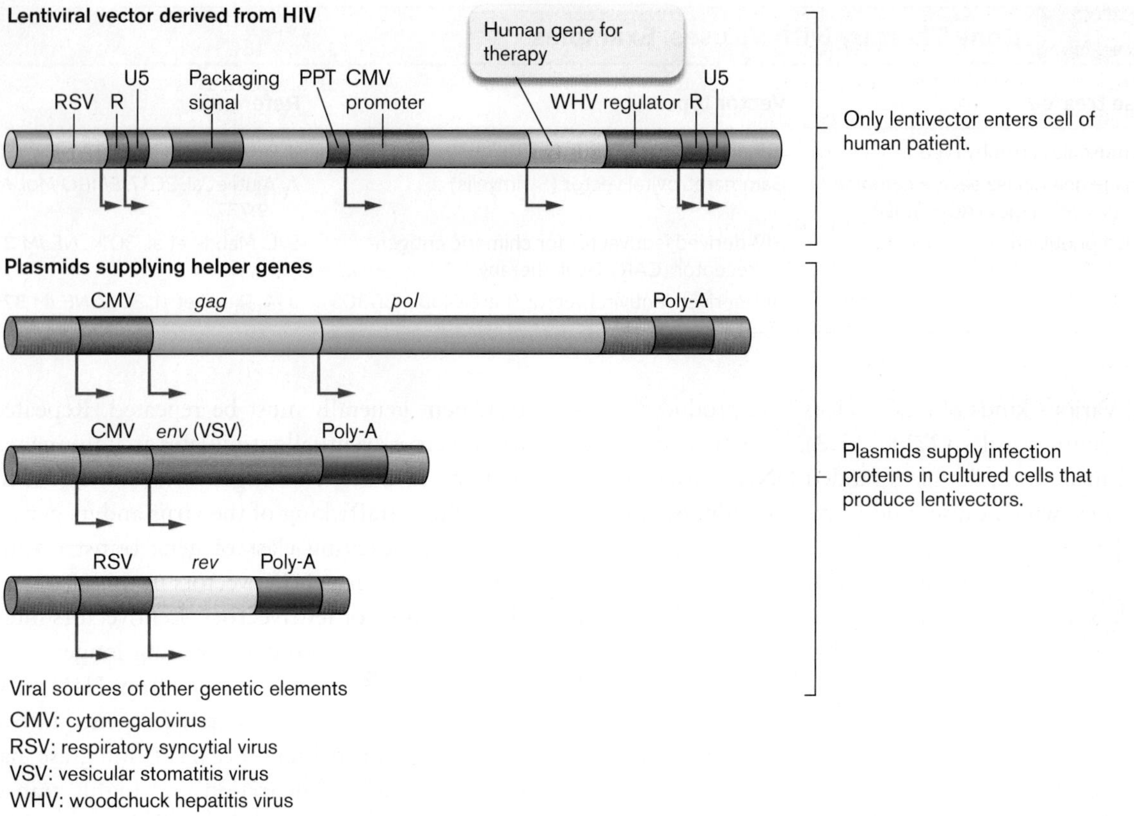

FIGURE 11.31 ▪ Lentivector with helper plasmids. The lentivector consists of an RNA sequence containing HIV signal elements required for genomic integration (dark blue), with promoter and regulator elements (red) derived from various other viruses, and the therapeutic human transgene (yellow). In order to produce the virions in cell culture, essential HIV genes are provided on DNA helper plasmids. Their expression is driven by regulatory elements from other viruses (red). *Source:* Modified from Blesch. 2003. *Methods* **33**:164.

allowing treatment of a wide range of tissues. The vector contains a human transgene that becomes integrated into the host genome.

Could lentiviral integration be dangerous for the patient? Lentiviral integration into host DNA poses the danger of activating an adjacent proto-oncogene—that is, a human gene that causes cancer when expressed at the wrong time. To avoid this problem, lentivectors are modified so that their promoters can activate only the gene of choice, not an adjacent cancer gene. Furthermore, lentiviral vectors offer a way to integrate DNA into nondividing cells of differentiated tissues such as brain neurons. For example, in 2009 a lentivector derived from HIV halted progression of a fatal brain disease, adrenoleukodystrophy (ALD), in two young boys. The lentivector inserted a gene, replacing a defective gene in each boy's blood stem cells.

In a different case, in 2012, the first leukemia patients were successfully treated by lentiviral gene therapy. The HIV-derived vector reprogrammed the patients' own T cells to attack cancerous B cells. A 7-year-old girl, Emily Whitehead (**Fig. 11.30** inset), was the first child to be considered fully "cured" of disease by a lentiviral vector. The commercial development of this lentiviral therapy, known as CAR-T therapy, is described in Chapter 16.

How a Lentivector Works

To construct a safe and effective vector, the HIV genome is modified extensively. In the example shown in **Figure 11.31**, accessory genes *vpr*, *vpu*, *nef*, and *vif*, which encode HIV virulence factors for disease, were removed. Other protein-encoding genes necessary for virion production were put into DNA helper plasmids, to be provided only in tissue culture for vector production. These genes provide the capsid monomer (*gag*), reverse transcriptase (*pol*), envelope glycoprotein (*env*), and a regulator of mRNA export from the nucleus (*rev*). The HIV *env* gene is replaced by an *env* gene from another virus, vesicular stomatitis virus (VSV). The VSV envelope protein has a broad tropism (host range) and thus enables the lentivector to infect a broad range of host cell types. Each helper plasmid drives its gene expression from a well-studied promoter of another virus, such as cytomegalovirus (CMV) or respiratory syncytial virus (RSV). The original HIV vector genome retains only the LTR end sequences (R-U5) required for genome integration, a packaging signal from the start of *gag*, and the infectivity-enhancing polypurine tract (PPT).

To express the **transgene** (the gene of interest for expression in the host), a CMV promoter was inserted in the

HIV-derived RNA vector. To further enhance transgene expression, a genetic enhancer sequence was added from woodchuck hepatitis virus (WHV). Note that a "lentiviral" gene transfer system, in fact, includes genetic elements from a diverse set of human and animal viruses, all found in previous studies to contribute specific properties to infection. Although these genes originate from different viruses, they nevertheless function together like parts of a machine.

To produce infective virions, the HIV-derived RNA vector plus the three helper plasmids are introduced into a special tissue culture line. The vector and plasmids enter the cells by a process, called **transfection**, in which calcium phosphate treatment promotes the uptake of nucleic acids across the cell membrane. The host tissue culture cells are derived from human embryonic kidney 293T cells containing a gene for a protein from simian virus 40 (SV40) that enables the replication of DNA plasmids containing an SV40 replication site. The 293T cells allow efficient expression of the viral genes on the helper plasmids, as well as a full virion production cycle in which the vector RNA is replicated and packaged into virions.

Safety of Lentivectors

The use of disease-causing viruses for human therapy raises important concerns about safety. The viruses might express toxins, induce cancer, or trigger a damaging immune response. These concerns need to be weighed against the risks of the conditions they are used to treat, such as severe combined immunodeficiency (SCID), cystic fibrosis, and cancer. In general, gene therapy is approved to treat only life-threatening conditions for which alternative therapies are inadequate.

Viruses used for gene therapy are engineered extensively to decrease risks, using techniques of genetic engineering presented in Chapter 12. Safety features of viral vectors include:

- **Deleting virulence genes.** Viral genes that promote disease and virion proliferation, but are not required for establishment of the viral DNA in the nucleus, are deleted from the vector genome. To produce the vector in tissue culture, the viral proliferation genes are provided on helper plasmids.
- **Avoiding genome insertion next to oncogenes.** Vectors engineered from adenoviruses are usually designed to avoid host chromosome integration altogether. The disadvantage of avoiding integration is that the separate viral DNA is soon lost from host tissues, and the therapy requires frequent repetition. In lentivectors, molecular modification avoids the activation of proto-oncogenes.
- **Altering tissue specificity.** The tropism, or tissue specificity, can be altered by replacing the gene for the viral envelope glycoprotein (spike protein) with the envelope gene from a different virus. For example, a rabies virus glycoprotein can be used to target the vector to brain cells. The alteration of viral tissue specificity by envelope gene replacement is called "pseudotyping." Pseudotyping can be used either to narrow the host range or to broaden it, depending on the needs of the vector.
- **Avoiding germ-line infection.** Current medical standards prohibit alteration of the germ line, the egg and sperm cells that transmit genes to the next generation. Because the long-term risks of gene therapy are unknown, only somatic gene therapy (gene insertion into somatic, or body, cells) is permitted.

Thought Question

11.8 What do you think are the arguments for or against lentivectors editing the germ line?

To Summarize

- **Ancient retroviral sequences persist within animal genomes**, including the human genome. Endogenous retroviruses (ERVs) retain the four main retroviral genes: *env*, *gag*, *pol*, *pro* (protease). Other sequences decay by mutation and are called retroelements. Retrotransposons can transcribe themselves into new locations in the host genome.
- **Gene transfer vectors** are made from viruses.
- **Virulence genes are deleted from a lentivector (lentiviral vector).** Efficient promoter sequences from other viruses are inserted.
- **Adenoviral vectors circularize and replicate separately from the host genome.**
- **Lentivectors integrate within the host genome.**

11.5 Herpes Simplex Virus: DNA Virus

Many important viruses of humans and other animals contain genomes of double-stranded DNA (**Table 11.4**). DNA viruses include the causative agents of well-known diseases such as smallpox, chickenpox, and infectious mononucleosis (mono). Most DNA viruses are considerably larger than RNA viruses and encode a wider range of viral enzymes; for example, the vaccinia genome encodes nearly 200 different proteins. The complexity of viruses such as vaccinia and herpes approaches that of small cells.

TABLE 11.4 DNA Viruses of Animals (Examples)

Virus	DNA replication	Disease(s)	Host(s)
Adenoviruses (many strains)	Viral DNA polymerase, single-strand binding protein, and protein primer	Enteritis or respiratory diseases	Humans, other mammals, birds
Papovavirus (simian virus 40, SV40)	Cellular DNA polymerases	Asymptomatic	Monkeys
Herpesviruses			
Herpes simplex virus 1 and 2	All viral components (DNA polymerase, primase, etc.)	Epithelial and genital lesions, latency in neurons	Humans
Varicella-zoster virus	Viral components	Chickenpox, shingles	Humans
Epstein-Barr virus	All cell components (DNA polymerase, etc.)	Infectious mononucleosis, Hodgkin's lymphoma	Humans
Other strains	Varies	Epithelial lesions, cancer	Monkeys, cattle, horses
Papillomaviruses	Viral DNA helicase; cellular polymerase		
Human papillomaviruses (many strains)		Genital warts, cervical and penile cancer, skin warts	Humans
Other papillomaviruses		Warts, cancer	Rabbits, cattle, sheep
Poxviruses	All viral components		
Variola major virus		Smallpox	Humans
Vaccinia virus		Cowpox	Cattle, humans
Other poxviruses		Monkeypox	Monkeys, camels, birds, humans

Herpesviruses have been associated with humans and our ape ancestors for hundreds of millions of years (see Chapter 6). Different herpesviruses cause diseases ranging from chickenpox (varicella-zoster virus) to birth defects (cytomegalovirus). Today, in a remarkable twist, we have engineered a "tumor-eating herpes" to treat metastatic tumors (discussed shortly).

Herpesvirus DNA replicates by mechanisms similar to those used in the replication of prokaryotic and phage genomes, either bidirectionally from an origin of replication (as in bacteria) or by the rolling-circle method (as in phages such as T4). As in the replication of cellular genomes, herpesvirus DNA replication requires more than a polymerase; enzymes such as helicase, primase, and single-strand binding proteins are also needed.

Herpes Simplex Virus Infects the Oral or Genital Mucosa

An important example of a DNA virus is herpes simplex virus (HSV). Strains HSV-1 and HSV-2 cause one of the most common infections in the United States. Approximately 60% of Americans acquire herpes simplex, usually HSV-1, in epithelial lesions commonly known as cold sores. About 30%–60% acquire genital herpes, usually HSV-2, through sexual contact (oral or vaginal). Genital herpes causes recurrent eruptions of infection in the reproductive tract (**Fig. 11.32**). Many of those infected are unaware of symptoms, but they can still transmit the disease to others.

Herpes simplex virus typically infects cells of the oral or genital mucosa, causing ulcerated sores. The primary infection is epithelial, followed by latent infection within neurons of the ganglia. A common site of infection is the trigeminal ganglion, which processes nerve impulses between the face and eyes and the brain stem.

The latent infection of the ganglia later leads to new outbreaks of virus, often triggered by stress such as menstruation, sunlight exposure, or depression of the immune system. Progeny virions travel back down the dendrites to the epithelia, causing lytic infection. In the trigeminal ganglion, herpes reactivation can lead to eye disease or lethal brain infection. In most cases, herpes symptoms can be controlled by antiviral agents such as acyclovir (discussed in Chapter 27). No cure or means of preventing future outbreaks exists. In pregnant women, HSV can be transmitted to the fetus, with serious complications for the child.

Herpes simplex virus is closely related to varicella-zoster virus, the cause of chickenpox, also an epithelial infection. Varicella, too, can hide in ganglial neurons, emerging decades later to cause painful skin lesions called shingles.

Herpes Simplex Virus Structure

The herpes virion comprises a double-stranded DNA chromosome packed within an icosahedral capsid (discussed in Chapter 6; **Fig. 11.33A** ▶). The capsid is surrounded by **tegument**, a collection of about 15 different kinds of virus-encoded proteins, as well as proteins from the previous host.

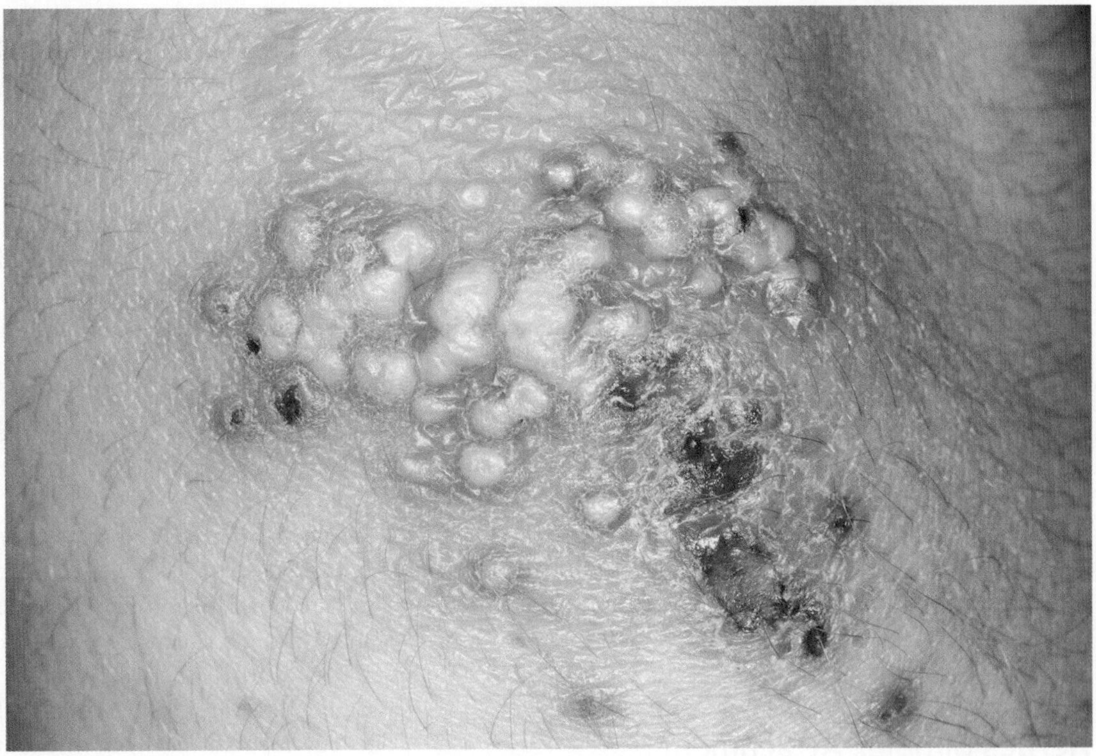

FIGURE 11.32 ■ Genital herpes infection. Lesions on the elbow of an 11-year-old, female patient, caused by HSV-2 infection.

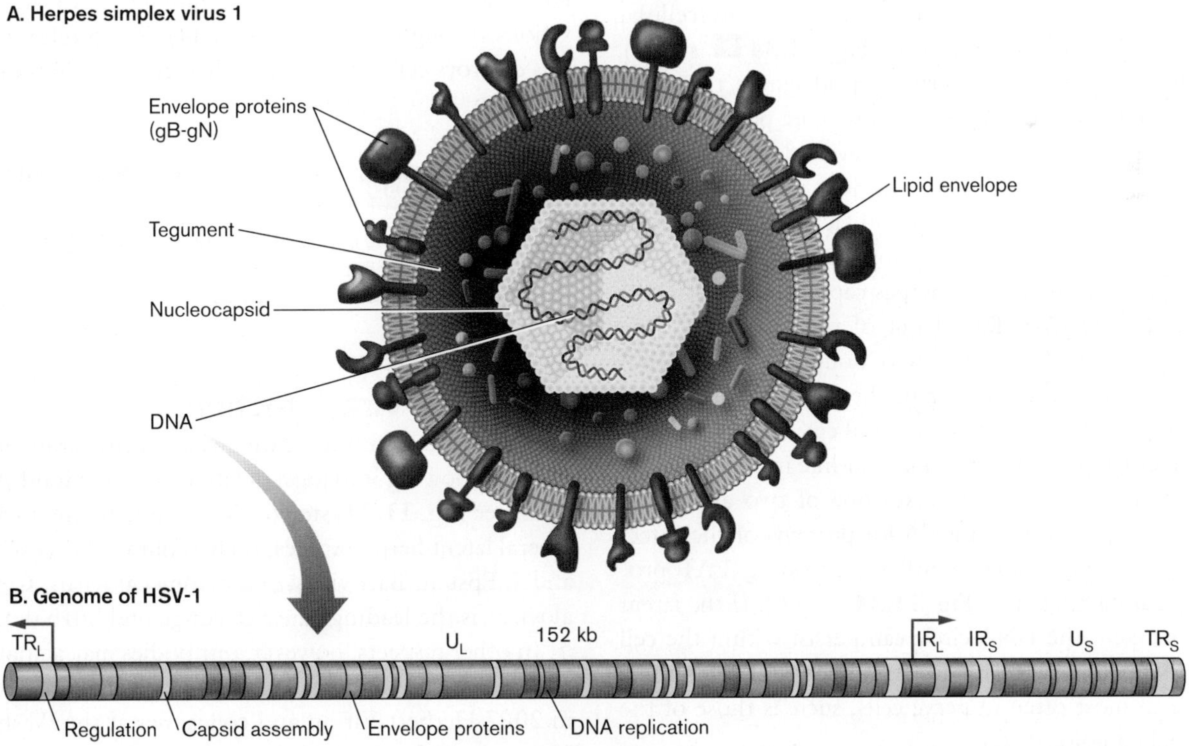

FIGURE 11.33 ■ Herpes simplex virus 1: virion and genome. **A.** The HSV-1 virion consists of a double-stranded DNA chromosome packaged within an icosahedral capsid. The capsid is surrounded by tegument, a collection of virus-encoded and host-derived proteins. The tegument is contained within a host-derived membrane envelope, including several kinds of envelope proteins. **B.** The genome of HSV-1 spans 152,000 base pairs, encoding more than 70 gene products. The HSV sequence consists of two segments, each containing a unique region (U_L or U_S) flanked by two inverted-repeat regions—terminal (TR_L or TR_S) and internal (IR_L or IR_S)—where the two segments meet. ▶

The tegument is contained within a host-derived membrane envelope with several kinds of spike proteins. The HSV-1 genome spans 152 kb, encoding more than 70 gene products (**Fig. 11.33B**). The sequence includes two unique segments, long (U_L) and short (U_S), each flanked by a terminal repeat (TR_L or TR_S) and an internal repeat (IR_L or IR_S). Within the host, the genome circularizes, so the genetic linkage map appears circular.

Herpes genes and gene regulatory elements resemble those of eukaryotic genes. Each gene has a promoter with eukaryotic control sequences such as the TATA box (Chapter 8). Genes are transcribed and translated individually; there is no polyprotein. Genes fall into temporal classes within the viral replication cycle: immediate, early, and late expressed genes. Each class has specific regulatory elements. Thus, the virus expresses only the gene products needed for a given phase of infection. An additional class, consisting of the LAT genes, is expressed for latent infection (discussed next).

Herpes Simplex Attachment and Host Cell Entry

Unlike HIV, the relatively large herpes virion has several envelope proteins that can bind to several alternative receptor molecules on the host cell surface, such as a homolog of tumor necrosis factor receptor called HveA or intercellular adhesion molecules called nectins (**Fig. 11.34** ▶, step 1). As with HIV, the entire herpes capsid enters the cytoplasm (step 2). But unlike HIV, whose core particle partly uncoats, the intact herpes capsid travels down a scaffold of microtubules (step 3) to the nuclear membrane. During this stage, the virion host shutoff factor (Vhs) degrades host mRNA, thus shutting off host protein synthesis. At a nuclear pore complex, the herpes capsid injects its DNA (step 4). The DNA is forced out of the capsid and into the nucleus by the high pressure of double-stranded DNA packing in the capsid, similar to the high-pressure injection of phage T4 DNA into a bacterial cell. The DNA then circularizes (step 5) to form a plasmid-like intermediate.

The herpes genome now takes one of two alternative directions: expression of mRNA for proteins of the infection cycle, or expression of mRNA encoding LAT proteins to maintain latency (**Fig. 11.34**, step 6). If the latent course is taken, the DNA circle can persist within the cell for decades before switching to lytic infection. Latent infection is seen most often in nerve cells, such as those of the trigeminal ganglion.

Replication of Herpes Simplex Virus

In the nucleus, herpes DNA is transcribed to mRNA by host RNA polymerase II (see **Fig. 11.34**, step 5). If mRNA for lytic infection is produced, it exits the nucleus to be translated by ribosomes. Many different mRNAs are produced and exported, including those required for "immediate" and "early" stages of infection (**Fig. 11.34**, step 7). The translated proteins return to the nucleus for packaging within capsids.

To generate progeny genomes, the circular DNA is replicated by viral enzymes, including DNA polymerase, single-strand binding protein, and a proofreading endonuclease (**Fig. 11.34**, step 8). Additional enzymes are provided by the host cell. DNA is replicated by the rolling-circle method, generating a concatemer similar to that of phage T4. Unlike T4, however, the herpes DNA is eventually cut into segments defined by the terminal repeat sequences.

The newly synthesized DNA expresses late-stage mRNA (**Fig. 11.34**, step 9), which exits the nucleus for translation. Translated envelope proteins are inserted into the ER membrane, through which they migrate to the nuclear membrane (step 10). Other late proteins reenter the nucleus for assembly into capsids containing DNA genomes (step 11). The envelope forms from the outer nuclear membrane (step 12). The virions are then transported through the ER, where they undergo secondary envelopment (step 13). The secondary-enveloped virions move to the Golgi, and ultimately to the cell membrane (step 14). The secondary envelope fuses with the cell membrane, releasing mature virions through exocytosis (step 14). Rapid release of virions destroys cells, causing the characteristic sores of herpes infection.

Thought Question

11.9 Compare and contrast the fate of the HSV genome with that of the HIV genome.

Persistent Viral Infections

Herpesviruses can infect humans and other animals indefinitely, following initiation of latency by the viral LAT proteins (see **Fig. 11.34**, step 6). Most humans are infected by several latent herpesviruses, such as human herpesviruses 6 and 7, Epstein-Barr virus, and cytomegalovirus. Cytomegalovirus is the leading cause of congenital birth defects.

In other respects, however, our bodies may actually benefit from the presence of certain herpesviruses. For example, in 2007 Herbert Virgin and colleagues at the Washington University School of Medicine in St. Louis showed that mice infected with a gamma herpesvirus similar to Epstein-Barr virus resist infection by the bacterial pathogens *Listeria monocytogenes* (the cause of listeriosis) and *Yersinia pestis* (the cause of bubonic plague). The mechanism of antibacterial resistance may involve stimulation of the immune

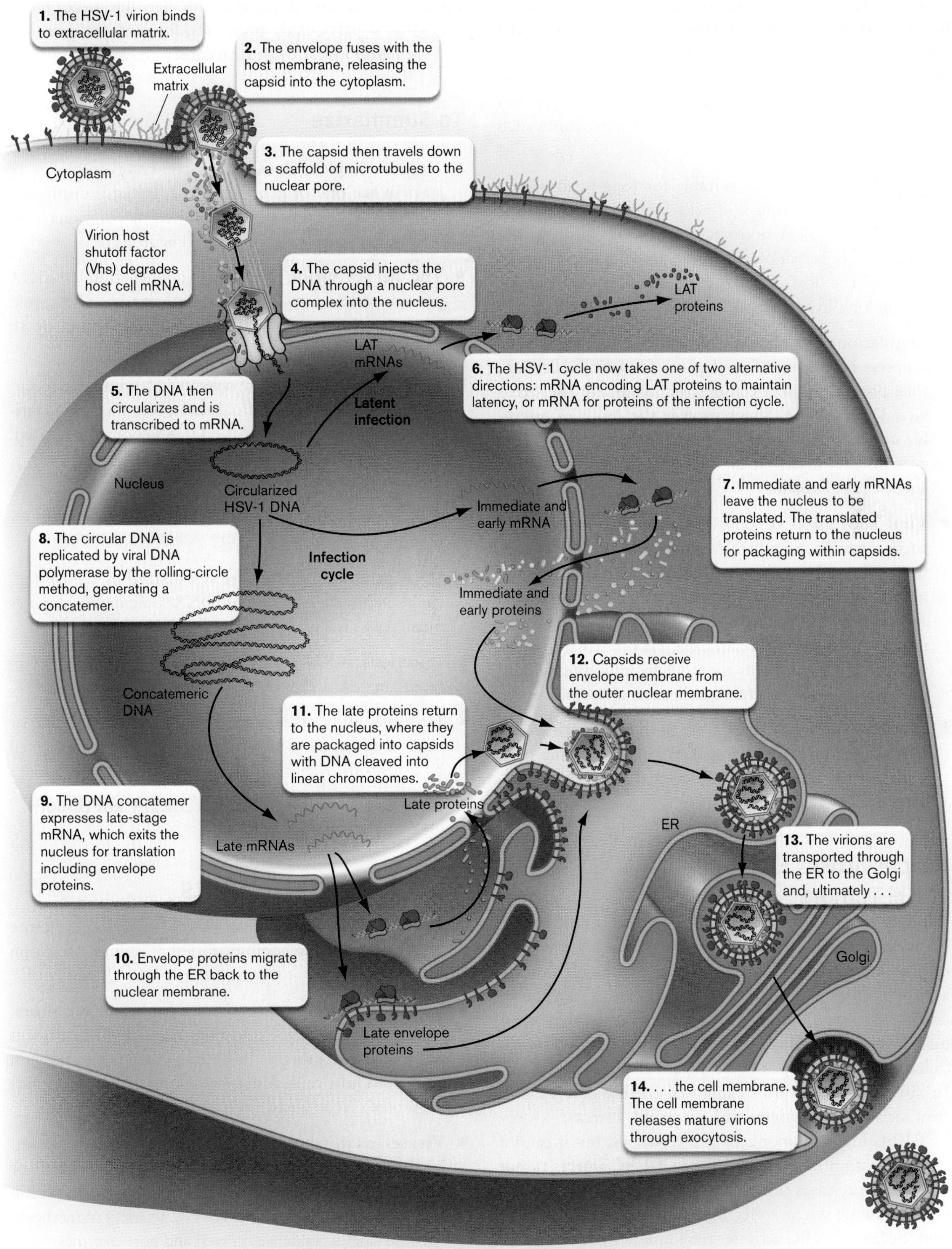

FIGURE 11.34 ■ Replication cycle of HSV-1. The HSV-1 virion binds to receptors on the host cell membrane and releases its capsid in the cytoplasm. The DNA chromosome is transferred into the host nucleus to conduct the replication cycle.

response. Thus, some of our silent herpesvirus "partners" may have coevolved with humans in a mutually beneficial relationship, or mutualism (discussed in Chapter 21).

The maintenance of persistent or latent viral infection involves several kinds of processes that are surprisingly similar in DNA viruses such as herpes and in retroviruses such as HIV. These processes include:

- **Infection of cell types suitable for long-term persistence.** After infecting skin cells for rapid viral replication, some HSV virions infect neurons. Neurons are long-lived cells that provide the virus with an everlasting home in the host. Similarly, Epstein-Barr virus persists within long-lived memory-B-cell lymphocytes.
- **Regulation of viral gene expression.** The LAT proteins suppress expression of viral genes for lytic replication, thus preventing host cell destruction where the latent viral DNA resides. Suppression is the result of assembly of "heterochromatin," chromosome-associated host proteins that inactivate all viral genes except those that encode LAT proteins.
- **Viral subversion of cellular apoptosis.** To maintain latent infection, a viral DNA must prevent host cell apoptosis, a form of programmed self-destruction in response to viral infection. For example, cytomegalovirus prevents apoptosis by mimicking one host apoptosis protein and inhibiting another. Similarly, the HIV retroviral accessory protein Tat prevents apoptosis by inhibiting protein p53, which suppresses tumors through apoptosis.
- **Evasion of immune responses.** Both herpesviruses and retroviruses express proteins that inhibit signaling molecules of the immune system, or that mimic immunosuppressive signals.

Herpesviruses Engineered to Kill Tumors

The long-standing adaptation of herpesviruses to the human body confers properties useful for virotherapy. Some kinds of viruses have a preference for infection of tumor cells. Such a virus is called "oncolytic." Oncolytic viruses are being developed to treat cancer. An example of an oncolytic virus for tumor therapy is the engineered form of HSV-1 virus called T-VEC, developed by a company acquired by Amgen. In 2015, the United States approved the use of T-VEC to treat melanoma in patients with inoperable tumors.

T-VEC infects tumor cells and replicates, but it cannot replicate within normal cells. When T-VEC infects tumor cells, it also releases tumor cell fragments that attract cells of the immune system to recognize the tumors for destruction. Because T-VEC activates the immune system, it helps destroy tumors throughout the body. This antitumor effect of a virus is particularly promising for metastatic tumors that are hard to treat by other means.

To Summarize

- **Herpes simplex virus** causes recurring eruptions of sores in the oral or genital mucosa. Initial transmission is by oral or genital contact, followed by eruptions from reactivated virus latent in ganglial neurons.
- **The herpes virion** contains a double-stranded DNA genome, packed in an icosahedral capsid. The capsid is surrounded by numerous matrix proteins and by an envelope.
- **HSV attachment** may involve several alternative receptors. A microtubular scaffold transports the herpes virions to the nucleus, where the DNA genome is inserted. The DNA circularizes for transcription.
- **LAT protein expression** leads to latent infection, usually in nerve cells, where the DNA persists silently for months or years.
- **DNA genomes of HSV** are synthesized by the rolling-circle method, using viral DNA polymerase supplemented by viral and host-generated components.
- **Infectious mRNA expression** leads to production of capsid, matrix, and envelope proteins for assembly of HSV.
- **HSV assembly** takes place at the nuclear membrane or other membranes. The virions are released from the cell by exocytosis. Rapid release leads to mucosal pathology.
- **HSV can be engineered to specifically kill tumors.**

Concluding Thoughts

In considering the mechanisms of viral infection, a picture emerges of interesting commonalities, as well as intriguing diversity:

- **Viral infection requires host membrane receptors.** Different viruses evolve to take advantage of different cell membrane proteins in order to recognize and gain entry into host cells. Mutation of host receptors confers resistance on the virus.
- **Viruses recruit host cells to replicate their genomes.** Viral genomes are remarkably diverse, including RNA or DNA, single- or double-stranded, linear or circular. Within the host cell, replication proceeds by a variety of methods, with varying dependence on viral and host enzymes.

- **Virus particles (and their genomes) range in size from a few components to assemblages approaching the complexity of cells.** The advantage of simplicity is the minimal requirement for resources. The advantage of complexity is the fine-tuning of infection mechanisms for evading host defenses.

The study of viral molecular biology raises intriguing questions about the nature of viral replication and about viral origins (see **eTopic 6.1**). Research in virology offers hope for new drugs and cures for humanity's worst plagues, as well as devastating diseases of agricultural plants and animals. At the same time, it is sobering to note that despite the enormous volumes we now know about viruses such as HIV and influenza, the AIDS pandemic continues, and we face the likely emergence of a new deadly flu strain. Molecular research can succeed only in partnership with epidemiology and public health (discussed in Chapter 28).

CHAPTER REVIEW

Review Questions

1. How does phage lambda attach to its correct host cell and insert its genome for replication?
2. How do the CI repressor and Cro protein modulate the switch between lysogeny and lysis?
3. How do influenza virions gain access to the host cytoplasm? Explain the role of fusion peptides.
4. How does influenza virus manage the replication and packaging of its segmented genome? What is the consequence of genome segmentation for virus evolution?
5. How does HIV provide ready-made components for replication of its genome? How does reverse transcriptase convert the single-stranded RNA genome into double-stranded DNA?
6. What is the role of protease in HIV replication? What is the significance of protease for AIDS therapy?
7. How did ancient retroviruses participate in evolution of the human genome?
8. For gene therapy, how can we construct a lentivector so as to avoid replication of the virus?
9. How does herpesvirus compartmentalize the expression and replication of its DNA genome?
10. Compare and contrast the needs of DNA genome replication with those of RNA genome replication.

Thought Questions

1. RNA viruses and DNA viruses represent fundamentally different reproductive strategies. How do their different strategies interact with the host response? How can our understanding of viral replication cycles help us develop new antiviral agents?
2. Discuss the roles of host-modulating viral proteins in HIV infection, and in herpesvirus infection. What various kinds of functions do these proteins serve for the virus, and what are their effects on the host cell?
3. Hemophilia is a life-threatening genetic blood disorder (the absence of a clotting factor) for which current therapies involve providing blood products and recombinant proteins. Discuss what might be the advantages and relative risks of using viral vectors to treat hemophilia.

Key Terms

accessory protein (424)
acquired immunodeficiency syndrome (AIDS) (421)
azidothymidine (AZT) (427)
chemokine receptor (CCR) (426)
concatemer (406)
core particle (422)
coreceptor (426)
endocytosis (415)
endogenous retrovirus (421, 431)
fusion peptide (413)
gene transfer vector (432)
host factor (413)
human immunodeficiency virus (HIV) (421)

lentivector (433)
lentivirus (420)
long terminal repeat (LTR) (429)
lysis (407)
lysogeny (403)
matrix protein (411)
nucleocapsid protein (NP) (411)
prophage (403)
protease inhibitor (429)
provirus (429)
quasispecies (425)
reassortment (411, 412)
recombination (412)
retrotransposon (431)
retrovirus (420)
reverse transcriptase (RT) (420, 427)
rolling-circle replication (405)
segmented genome (411)
spike protein (422)
tegument (436)
transfection (435)
transgene (434)

Recommended Reading

Barton, Erik S., Douglas W. White, Jason S. Cathelyn, Kelly A. Brett-McClellan, Michael Engle, et al. 2007. Herpesvirus latency confers symbiotic protection from bacterial infection. *Nature* **447**:326–329.

Cartier, Nathalie, Salima Hacein-Bey-Abina, Cynthia C. Bartholomae, Gabor Veres, Manfred Schmidt, et al. 2009. Hematopoietic stem cell gene therapy with a lentiviral vector in X-linked adrenoleukodystrophy. *Science* **326**:818–823.

Dedrick, Rebekah M., Carlos A. Guerrero-Bustamante, Rebecca A. Garlena, Daniel A. Russell, Katrina Ford, et al. 2019. Engineered bacteriophages for treatment of a patient with a disseminated drug-resistant *Mycobacterium abscessus. Nature Medicine* **25**:730–733.

Deng, Lei, Teena Mohan, Timothy Z. Chang, Gilbert X. Gonzalez, Ye Wang, et al. 2018. Double-layered protein nanoparticles induce broad protection against divergent influenza A viruses. *Nature Communications* **9**:539.

Engelman, Alan, and Peter Cherepanov. 2012. The structural biology of HIV-1: Mechanistic and therapeutic insights. *Nature Reviews. Microbiology* **10**:279–290.

Göke, Jonathan, Xinyi Lu, Yun-Shen Chan, Huck-Hui Ng, Lam-Ha Ly, et al. 2015. Dynamic transcription of distinct classes of endogenous retroviral elements marks specific populations of early human embryonic cells. *Cell Stem Cell* **16**:135–141.

Grow, Edward J., Ryan A. Flynn, Shawn L. Chavez, Nicholas L. Bayless, Mark Wossidlo, et al. 2015. Intrinsic retroviral reactivation in human preimplantation embryos and pluripotent cells. *Nature* **522**:221–225.

Heeney, Jonathan L., Angus G. Dalgleish, and Robin A. Weiss. 2006. Origins of HIV and the evolution of resistance to AIDS. *Science* **313**:462–466.

Joseph, Sarah B., Ronald Swanstrom, Angela D. M. Kashuba, and Myron S. Cohen. 2015. Bottlenecks in HIV-1 transmission: Insights from the study of founder viruses. *Nature Reviews. Microbiology* **13**:414–425.

Kane, Melissa, and Tatyana Golovkina. 2010. Common threads in persistent viral infections. *Journal of Virology* **84**:4116–4123.

Karzbrun, Eyal, Alexandra M. Tayar, Vincent Noireaux, and Roy H. Bar-Ziv. 2014. Programmable on-chip DNA compartments as artificial cells. *Science* **345**:829–832.

Kotula, Jonathan W., S. Jordan Kerns, Lev A. Shaket, Layla Siraj, James J. Collins, et al. 2014. Programmable bacteria detect and record an environmental signal in the mammalian gut. *Proceedings of the National Academy of Sciences USA* **111**:4838–4843.

Lakdawala, Seema S., Yicong Wu, Peter Wawrzusin, Juraj Kabat, Andrew J. Broadbent, et al. 2014. Influenza A virus assembly intermediates fuse in the cytoplasm. *PLoS Pathogens* **10**:e1003971.

Porter, David L., Bruce L. Levine, Michael Kalos, Adam Bagg, and Carl H. June 2011. Chimeric antigen receptor–modified T cells in chronic lymphoid leukemia. *New England Journal of Medicine* **365**:725–733.

Smith, Gavin J. D., Dhanasekaran Vijaykrishna, Justin Bahl, Samantha J. Lycett, Michael Worobey, et al. 2009. Origins and evolutionary genomics of the 2009 swine-origin H1N1 influenza A epidemic. *Nature* **459**:1122–1126.

Yosef, Ido, Miriam Manor, Ruth Kiro, and Udi Qimron. 2015. Temperate and lytic bacteriophages programmed to sensitize and kill antibiotic-resistant bacteria. *Proceedings of the National Academy of Sciences USA* **112**:7267–7272.

Zeng, Ming, Zeping Hu, Xiaolei Shi, Xiaohong Li, Xiaoming Zhan, et al. 2014. MAVS, cGAS, and endogenous retroviruses in T-independent B cell responses. *Science* **346**:1486–1492.

CHAPTER 12

Biotechniques and Synthetic Biology

12.1 DNA Amplification and Sequence Analysis

12.2 Genetic Manipulation of Microbes

12.3 Gene Expression Analysis

12.4 Applied Biotechnology

12.5 Synthetic Biology: Biology by Design

Microbial biotechnology uses living microbes or their products to improve human health or to perform specific industrial or manufacturing processes. Achieving these goals however, requires that we continually invent new technologies. Chapter 12 describes some classic and cutting-edge technologies and explores how they are used.

CURRENT RESEARCH highlight

Painting with motile bacteria. Like a crowd of pedestrians, swimming bacteria accumulate in spaces where their speed drops. Roberto Di Leonardo and colleagues at the University of Rome genetically engineered *E. coli* to convert light energy into the chemical energy that drives flagellum rotation. By projecting images such as Leonardo da Vinci's *Mona Lisa* onto a culture of this engineered *E. coli*, they achieved spatial variation in cell density via the relative swim speed of the cells: rapid swim in lit spaces, slow swim in dark spaces.

Source: Giacomo Frangipane et al. 2018. *eLife* **7**: e36608.

COURTESY OF ROBERTO DI LEONARDO

AN INTERVIEW WITH

ROBERTO DI LEONARDO, PHYSICIST, UNIVERSITY OF ROME

COURTESY OF ROBERTO DI LEONARDO

Are there potential scientific or industrial applications for your work?

We foresee several types of applications. Bacteria could provide the "living bricks" to fabricate new functional microstructures. Photokinetic bacteria could be engineered to produce extracellular cementing proteins to make these structures permanent when light is removed. In addition, remotely controlling millions of motile cells by means of low-power light patterns could provide new strategies for performing automated tasks such as the manipulation, transport, and sorting of single cells in miniaturized laboratories on a chip.

The ability to "paint" with motile bacteria is an amazing achievement in biotechnology. Although we think of biotechnology as a new field, the earliest biotechnologists actually lived about 10,000 years ago, when our ancestors unwittingly figured out how to use yeast to make alcohol, bread, and dairy products (discussed in Chapter 16). The field changed little over the millennia, until 1928, when the Scotsman Alexander Fleming discovered that microbes produce antibiotics (discussed in Chapter 1). That watershed event, along with unraveling the structure of DNA and deciphering the genetic code, ushered in the high-tech era of biotechnology. We have since learned a great deal about the molecular biology of the microbes our ancestors used. Now, through genetic engineering and synthetic biology, we can make bacteria produce hormones, engineer vaccines in plants, devise previously unimagined biochemical pathways, and design bacteria that detect explosives and flash a warning.

Biotechnology can also improve human existence. One example is the engineering of potatoes that resist devastation by a parasitic fungus. Today's potato farmers must spray huge amounts of antifungal agents on their crops to combat potato blight disease, the cause of the potato famine that wiped out 30% of the Irish population in the nineteenth century. Spraying plants as many as 25 times a season with chemicals, the typical U.S. potato farmer spends about $250 per acre to fight the disease. Considering that even the smallest farms are over 60 acres, there's a huge incentive to find less expensive means of prevention.

Biotechnologists have discovered a wild potato from Argentina that is highly resistant to *Phytophthora infestans*, the microbial eukaryote that causes potato blight disease (discussed in Chapter 20). Although the Argentine potato has no commercial value (it is inedible), the gene conveying resistance (*Rpi-vnt1*) was genetically inserted by recombination into commercial potatoes, making them resistant to the fungus-like pathogen. In 2017, these blight-resistant potatoes were cleared for agricultural use in North America by the FDA, EPA, and Health Canada.

We begin our discussion of biotechnology by explaining the techniques used to probe the inner workings of microbes. It is important to recognize that many of these techniques were "invented" by microbes, and we have learned how to harness them for our studies. First we examine ways to amplify and obtain the sequence information of DNA. Next we describe ways that genes can be manipulated experimentally. Then we discuss the methods used to analyze gene expression at both the RNA and protein levels. Many of these techniques, such as transcriptomics and proteomics, are described throughout the book; **Tables 12.2** and **12.3** give an overview. New techniques covered in this chapter include CRISPR-Cas gene engineering, as well as molecular techniques such as single-molecule DNA sequencing and chromatin immunoprecipitation. Finally, we describe some applications of microbial biotechnology and discuss the exciting new field of synthetic biology.

Rarely is one technique sufficient to characterize a gene's function in the cell. More often the results from the first technique prompt further exploration by a second technique. In this way, microbiologists commonly work through a progression of genetic and biochemical techniques to fully understand how genes and their proteins function and are regulated in the cell. A case study of such a progression is provided in a special extended eTopic, **eTopic 12.1**. This case study shows how acid resistance genes of *E. coli* were identified, how expression of these genes was discovered to be induced by acid, and how the transcriptional regulators of this response were found.

12.1 DNA Amplification and Sequence Analysis

Why do we study the nucleic acid sequences of genes and genomes? As discussed in Chapter 7, sequence information can be used to infer function for individual genes, as well as for whole genomes and community-wide metagenomes. In a cell, the DNA is transcribed into RNA and translated into protein (Chapter 8), and thus the DNA sequence can provide valuable clues about the structure and function of the gene product. In addition, DNA sequencing can identify mutations that change the phenotypes of the microbe (Chapter 9), and this information can provide additional clues about the functions of the mutated genes. In this section we describe methods to isolate and amplify DNA, the traditional and state-of-the-art methods for determining DNA sequence, and the bioinformatic methods used to annotate the DNA sequence to infer genome structure and gene function.

DNA Isolation and Purification

The chemical uniformity of DNA means that simple and reliable purification methods can be used to isolate it. A variety of techniques are used to extract DNA from microbial cells. Bacteria may be lysed by lysozyme, which degrades the peptidoglycan of the cell wall, followed by treatment with detergents to dissolve the cell membranes. Lysozyme is an enzyme that cleaves the bond linking residues of peptidoglycan. Once the cell contents are released, most proteins are precipitated in a high-salt solution. The precipitated proteins are removed by centrifugation, leaving a clear lysate containing DNA. This lysate is passed through a column containing a silica resin that specifically binds DNA. Any remaining proteins in the lysate are washed out of the column because they do not stick to the resin. Once the proteins are removed, the DNA is eluted (removed) from the resin with water.

Note: Lysozyme degrades the peptidoglycan of bacteria, but not the pseudopeptidoglycan or chitin cell walls of some archaea or eukaryotes, respectively (see Sections 19.1 and 20.2). Lysing cells of archaea and microbial eukaryotes requires alternative techniques.

Ethanol (or isopropanol) and salt are added to the DNA, which precipitates it from aqueous solution as ethanol removes water associated with DNA's phosphoryl groups (negatively charged), which then bind the Na^+ ions from the salt. The precipitated DNA is then dissolved in water or a very low-salt buffer. At this point, the extracted DNA can be examined with a variety of analytical tools.

There are other methods for isolating plasmid DNA, such as equilibrium density gradient centrifugation. (See **eTopic 12.2** for a discussion of equilibrium density gradient centrifugation and its use in the classic Meselson-Stahl experiment.)

PCR Amplifies Specific Genes from Complex Genomes

Imagine having the ability to amplify one copy of a gene into millions within 1 or 2 hours. This powerful technique, called the **polymerase chain reaction (PCR)**, has revolutionized biological research, medicine, and forensic analysis.

PCR is performed by heat-stable DNA polymerases that withstand the high temperatures required to denature double-stranded DNA into single-stranded DNA followed by cooling (annealing) of the denatured DNA with flanking sense and antisense primers that are extended by a thermostable DNA polymerase. This denature-anneal-extend cycle is repeated 25–40 times to amplify segments of DNA (as outlined here and in **Fig. 12.2**).

The mesophilic enzyme from *E. coli* originally used in this technique denatured at high temperature. Thus, more enzyme had to be added manually at each cycle, making the technique tedious and resource-costly. Fortunately, heat-resistant DNA polymerases can be found in microbes that live in hot places. In 1969, Thomas Brock and his undergraduate research assistant Hudson Freeze at Indiana University went to Yellowstone to study microbes that could grow in the hot springs found there. They isolated several hot-spring microbes by using growth media preconditioned to about 70°C. One such isolate, the bacterium *Thermus aquaticus* (**Fig. 12.1A**), provided the heat-resistant DNA polymerase, Taq, that made PCR cost-effective and revolutionary.

While Taq polymerase is still in use today, most applications use less error-prone DNA polymerases, such as Pfu. Pfu comes from the hyperthermophilic archaeon *Pyrococcus furiosus* (**Fig. 12.1B**), which has a temperature optimum of 100°C and lives at hydrothermal vents in the deep ocean (Chapter 21). It is amazing that the strange organisms found at a hot spring in Yellowstone and at a hydrothermal vent at the bottom of the ocean would lead to a revolution in DNA technology, medicine, and biology. For their National Science Foundation–funded discovery of *T. aquaticus*, Brock and Freeze won the 2013 Golden Goose Award, which recognizes "the tremendous human and economic benefits of federally funded research by highlighting examples of seemingly obscure studies that have led to major breakthroughs and resulted in significant societal impact."

A. B.

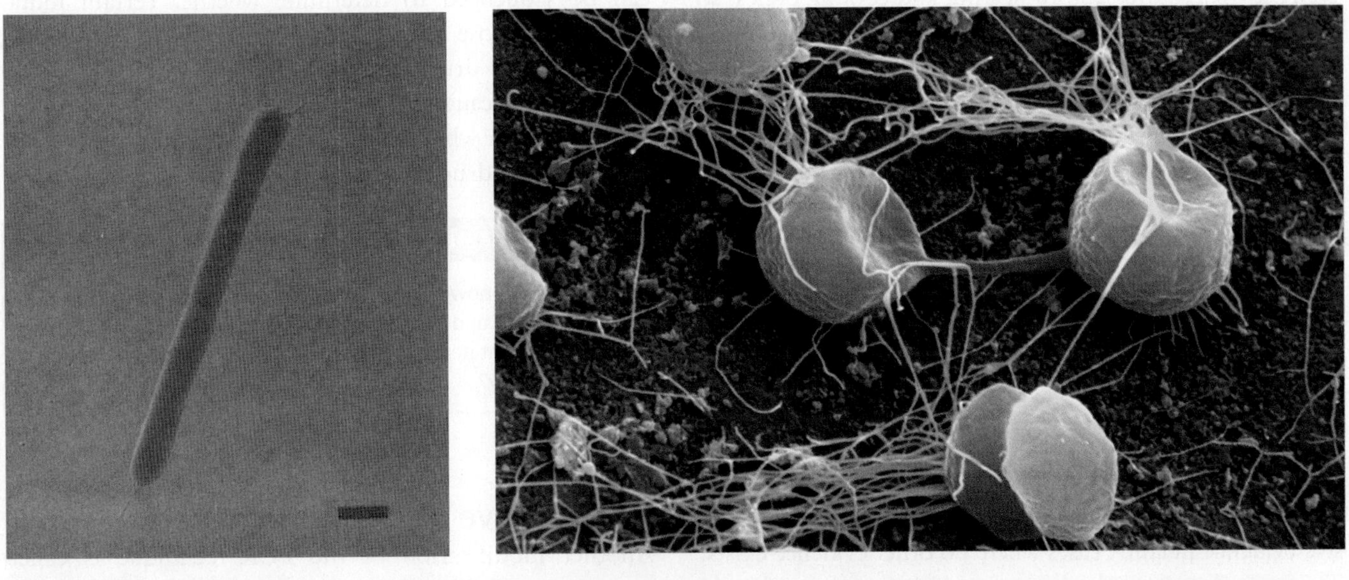

T. D. BROCK. 1969. *J. BACTERIOL.* **96**:289–297.

EYE OF SCIENCE/SCIENCE SOURCE

FIGURE 12.1 ▪ Microbial sources of DNA polymerase for PCR. **A.** *Thermus aquaticus*, the source of Taq. **B.** *Pyrococcus furiosus*, the source of Pfu.

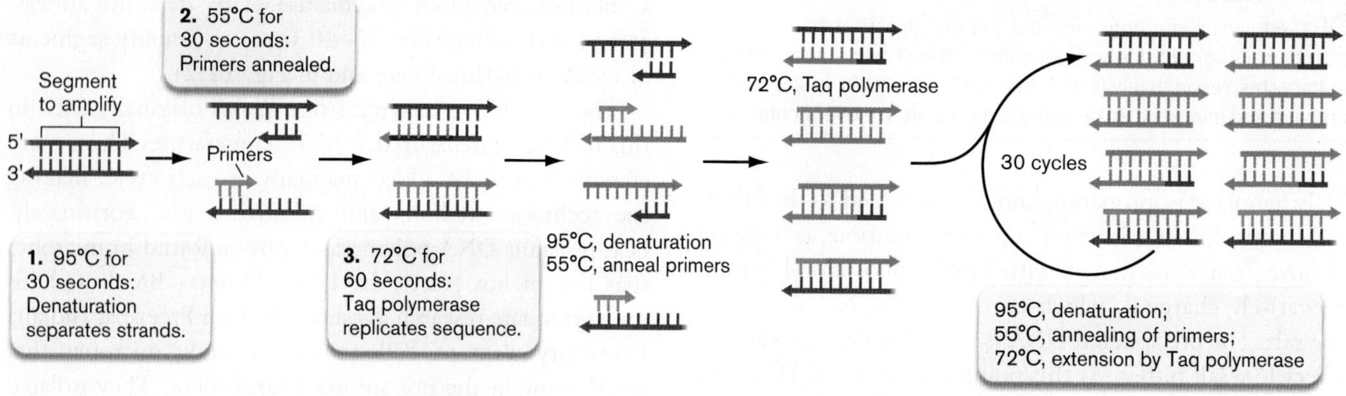

FIGURE 12.2 ■ The polymerase chain reaction. The cyclic PCR reaction makes a large number of copies of a small piece of DNA. Potentially 10^{30} copies of a fragment can be made from a single DNA molecule.

PCR, as well as other molecular techniques, such as gene cloning and reverse transcription, relies on the ability of complementary single strands of DNA or RNA to anneal and form double-stranded DNA or DNA-RNA complexes (for review, see Chapters 7 and 8). The PCR technique, outlined in **Figure 12.2**, requires the design of specific oligonucleotide **primers** (usually between 20 and 30 bp) that anneal to known DNA sequences flanking the DNA that will be amplified. The primers initiate replication of the target DNA by providing a "priming" substrate for DNA polymerase. The DNA polymerase synthesizes DNA by extending the primer molecule in a 5′-to-3′ direction, using the complementary DNA strand as the template. Heat-stable DNA polymerases are used because the PCR reaction mixture must undergo repeated cycles of heating to 95°C (to separate DNA strands so that they are available for primer annealing), cooling to 55°C (to enable primer annealing), and heating to 72°C (the optimal reaction temperature for the polymerase). To prepare enough target DNA, PCR requires 25–40 heating and cooling cycles, so a machine called a thermocycler is used to reproducibly and rapidly deliver these cycles.

This basic PCR technique has been modified to serve many purposes. Primers can be engineered to contain specific restriction sites that simplify subsequent cloning. If the primers used for PCR are highly specific for a gene that is present in only one microorganism, PCR can be used to detect the presence of that organism in a complex environment, such as the presence of the pathogen *E. coli* O157:H7 in ground beef. Multiplex PCR, involving several primer sets, can be used to detect multiple pathogens in a single reaction (see eAppendix 3).

PCR has profoundly changed our judicial system by making it possible to amplify the tiniest amounts of DNA present in a crime scene. The technique provides conclusive evidence in court cases where no other evidence exists. This technique has led to the successful reopening of "cold cases" such as the 2018 arrest of the Golden State Killer, whose unsolved crimes occurred between 1976 and 1986. And DNA evidence has also helped to exonerate innocent individuals convicted of crimes they did not commit.

Increasingly, PCR has facilitated the sequencing of individual human genomes—with profound ethical and societal implications. The hopes and fears raised by the invention of PCR-based genome sequencing inspired the 1997 science fiction film *Gattaca*, directed by Andrew Niccol, depicting an imaginary future in which everyone's destiny is determined by his/her DNA sequence. The film imagined that knowledge of DNA sequence could impact which jobs were available to someone, whether insurance coverage could be withheld, and even whether two individuals could marry.

Today, the Genetic Information Nondiscrimination Act passed by the U.S. Congress in 2008 protects individuals from "genetic discrimination" based on their DNA. In a more appropriate application, human sequence information is beginning to be used to predict which drugs will best treat individual patients. For example, tumor DNA can be sequenced to determine whether certain mutations make those tumors sensitive or resistant to existing anti-neoplastic drugs. This information is very helpful to oncologists because the patient can benefit from treatment with a drug for which his/her tumor is sensitive and not be treated with a drug for which that tumor is resistant.

Thought Question

12.1 PCR is a powerful technique, but a sample can be easily contaminated and the wrong DNA amplified—perhaps sending an innocent person to jail. What can be done to minimize this possibility?

Quantitative (Real-Time) PCR

Another modification of the PCR technique is called **quantitative PCR** (**qPCR**) or **real-time PCR**. Real-time PCR uses fluorescence to monitor the progress of PCR as it occurs (that is, in real time). Data are collected throughout the PCR process rather than just at the end of the reaction. Quantitative PCR can be used to quantify the level of DNA or RNA in a sample. DNA is quantified by how long it takes to first detect an amplified product while the polymerase chain reaction is still running. The higher the starting copy number of the nucleic acid target, the sooner a significant increase in fluorescence is observed. Thus, the time at which fluorescence first increases is a reflection of the amount of nucleic acid in the original sample.

How does qPCR work? Two techniques are commonly used. In one, a compound called SYBR Green is added to the reaction mix. This dye binds to double-stranded DNA and fluoresces. The fluorescence emitted increases with the amount of double-stranded PCR product produced.

The second quantitative PCR procedure, sometimes called **TaqMan** (**Fig. 12.3**) uses a reporter oligonucleotide probe containing a fluorescent dye on its 5′ end and a quencher dye on its 3′ end. As long as the probe remains intact, the quencher absorbs the energy emitted by the fluorescent dye—a process called **fluorescence resonance energy transfer** (**FRET**). The reporter probe does not itself prime DNA synthesis, but anneals to the target downstream of a priming oligonucleotide.

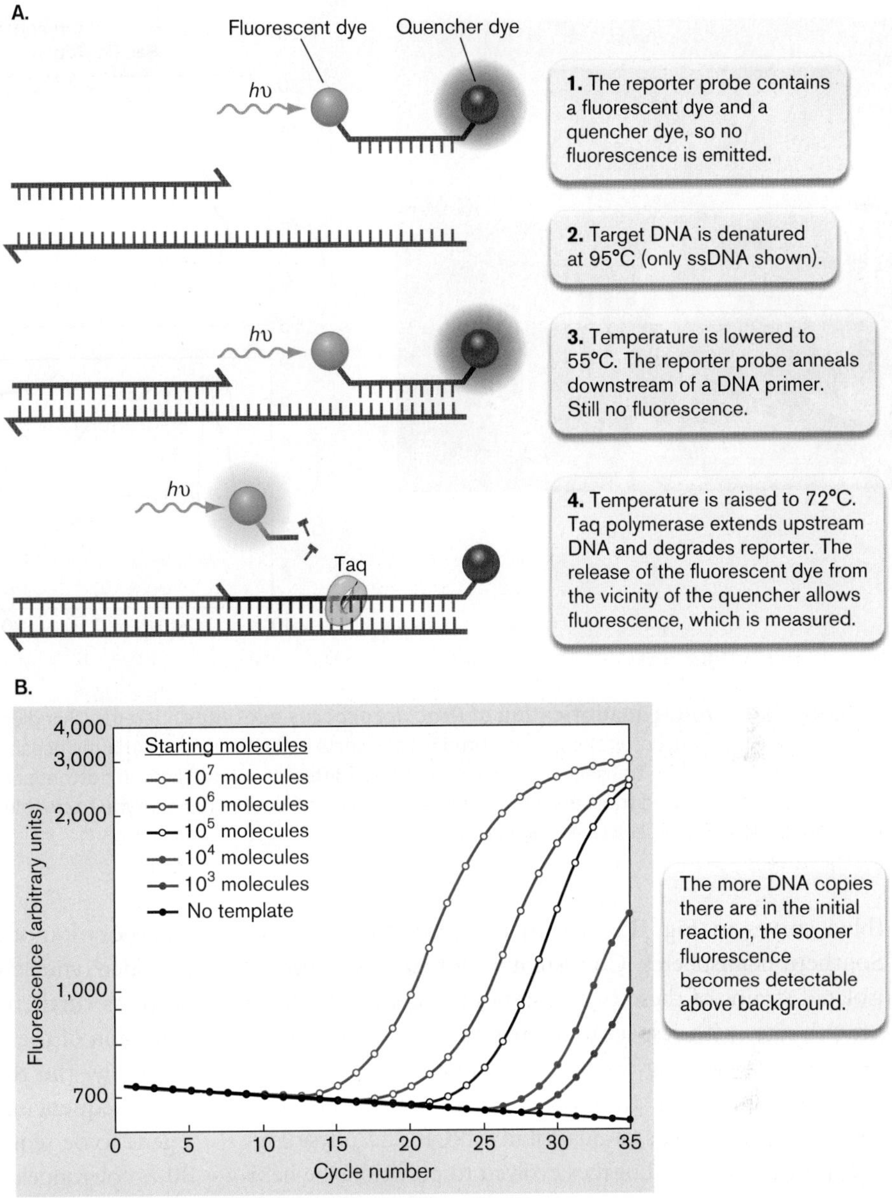

FIGURE 12.3 ■ Quantitative (real-time) PCR. A. The Taq DNA polymerase, extending an upstream primer, reaches the downstream reporter probe and degrades the probe, releasing the fluorescent dye from the vicinity of the quencher. **B.** Amplification plot of *Rhodococcus* with primer BPH4. Different numbers of DNA copies were used for each reaction mixture.

The priming oligonucleotide and the reporter probe both anneal to the target DNA sequence. Taq polymerase begins to synthesize DNA from the upstream primer (**Fig. 12.3**). However, Taq polymerase also has 5′-to-3′ exonuclease activity. So, Taq will run into and degrade the reporter probe, separating the fluorescent dye from the quencher dye. Fluorescence is emitted. Meanwhile, primer extension by Taq polymerase continues to the end of the template, and the template is amplified. After each annealing cycle, more reporter probe binds to the newly made templates and is cleaved by Taq polymerase during each round of polymerization. As a result, fluorescence continues to increase as the amount of amplified template increases. The more DNA copies there are in the initial reaction, the sooner fluorescence becomes detectable above background.

Quantitative PCR has been used extensively in the areas of medicine and ecology. As one example of the latter, Jeremy Chandler (**Fig. 12.4A**) and colleagues at the University of Tennessee investigated the abundances of the different genetic lineages of the marine cyanobacterium *Prochlorococcus* along long latitudinal transects in the Atlantic and Pacific oceans. Using primers specific to the different lineages, they found that each lineage has a unique distribution along these transects (**Fig. 12.4B**). One of these ecologically distinct lineages, or ecotypes, was the most abundant ecotype in the lower latitudes of both oceans (red curves in **Fig. 12.4B**), while another was most abundant in the high latitudes

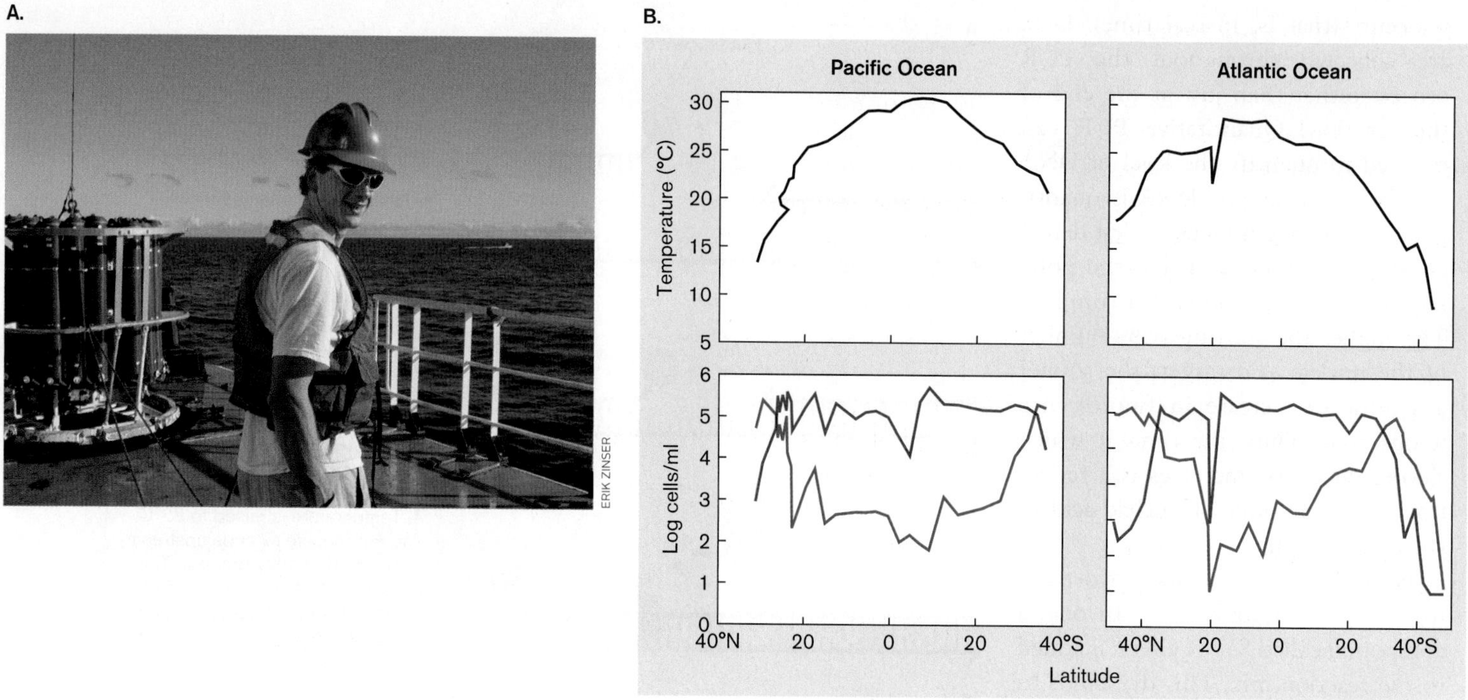

FIGURE 12.4 ■ **qPCR quantification of *Prochlorococcus* ecotypes.** Jeremy Chandler (**A**) and colleagues quantified the major ecotypes of *Prochlorococcus* along long latitudinal transects in the Pacific and Atlantic oceans. **B.** Temperature (top panels) varied greatly with latitude along these transects. The major ecotypes (different colored lines in bottom panels) were quantified with ecotype-specific PCR primers, and they showed transitions in numerical dominance from the red ecotype to the blue ecotype at higher latitudes. *Source:* Part B modified from Chandler et al. 2016. *Environ. Microbiol. Rep.* **8**:272–284, fig. 2.

(blue curves in **Fig. 12.4B**), in both the Northern and Southern hemispheres. Consistent with their distributions, isolated strains of these two ecotypes had different relative growth rates with respect to temperature, with the low-latitude strains growing faster at high temperature, and the high-latitude strains growing faster at low temperature. The combined results of quantitative PCR and physiology suggest that the two lineages evolved to partition the habitat along the temperature gradient that exists in the oceans with respect to latitude.

DNA Sequencing

Even with the development of gene cloning, our transition into the genomic age would have been impossible without the ability to rapidly sequence entire genomes. Previous sequencing methods used an ingeniously simple strategy but were not rapid. The dideoxy chain termination method developed by Fred Sanger in the 1970s relied on the fact that the 3′ hydroxyl group on 2′-deoxyribonucleotides (shown in Fig. 7.5A) is absolutely required for a DNA chain to grow. Thus, incorporation of a 2′,3′-dideoxyribonucleotide (also known as dideoxynucleotide; **Fig. 12.5**) into a growing chain prevents further elongation, because without the 3′ OH, extension of the phosphodiester backbone is impossible.

Briefly, the Sanger dideoxy method, or more generally called **sequencing by synthesis**, begins with the cloning of a gene to be sequenced into a plasmid vector. Next, a 20- to 30-bp oligonucleotide DNA primer molecule is designed to anneal at a site immediately adjacent to the DNA segment to be sequenced. The mixture is heated to 95°C to separate (denature) the DNA strands, and then cooled, enabling the primer to anneal to its complementary DNA. DNA polymerase can then use the 3′ OH end of this primer to synthesize DNA

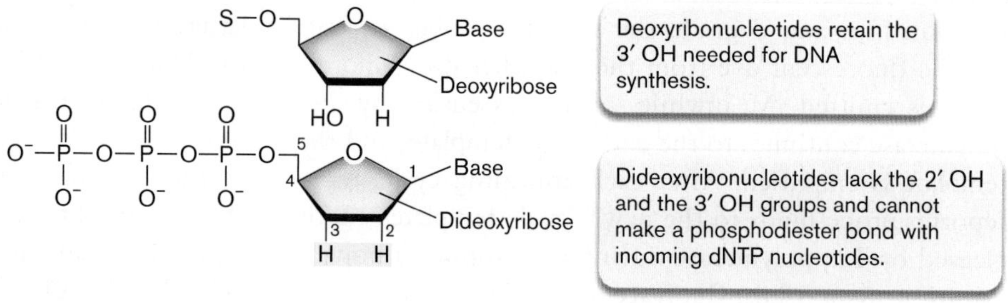

FIGURE 12.5 ■ **Dideoxyribonucleotide.** Review Figure 7.5 to see how lack of the 3′ OH group would terminate DNA synthesis. dNTP = deoxyribonucleoside triphosphate, or deoxyribonucleotide.

from the cloned-insert template. A small amount of dideoxyribonucleoside triphosphate (ddNTP—for example, dideoxy ATP) is included in the DNA synthesis reaction, which already contains all four normal deoxyribonucleoside triphosphates (dNTPs), also called deoxyribonucleotides.

Because normal 2′-deoxyadenosine is present at high concentration, chain elongation usually proceeds normally. However, the sequencing reaction contains just enough dideoxy ATP to make DNA polymerase occasionally substitute the dead-end base for the natural one, at which point the chain stops growing. The result is a population of DNA strands of varying length, each one truncated at a different adenine position. How the tagged fragments are used to determine sequence is described in eAppendix 3. For many applications, the Sanger method has been replaced by even more rapid techniques, all of which include an element of the Sanger chain termination approach.

Next-Generation Sequencing

The traditional way of sequencing genomes was to randomly clone DNA fragments into plasmid or phage vectors and then separately sequence each fragment. It took months, if not years, to sequence an entire genome. Today, new sequencing technologies, called "next-generation sequencing," have combined the power of robotics, computers, and fluidics such that an entire bacterial genome or community metagenome (Chapters 7 and 21) can be sequenced in a matter of days.

Perhaps the most used technique today is the sequencing-by-synthesis platform of Illumina (a technology developed by Solexa, a company now part of Illumina, Inc.). The Illumina process is outlined in **Figure 12.6**. Basically, the genome to be sequenced is sonically fragmented into segments of 100–300 bp, and different linker oligonucleotides are ligated to each end (**Fig. 12.6**, steps 1 and 2). Small fragments are used because the technique can sequence only 100–500 bp from any one fragment. Strands of each fragment are then separated and the mixture added to an optical flow cell. The fragments, millions of them, are randomly fixed to the solid surface (step 3). Each individual fragment sticks to a different area of the flow cell. There is also a dense lawn of oligonucleotides fixed at their 5′ ends to the glass surface. These oligonucleotides

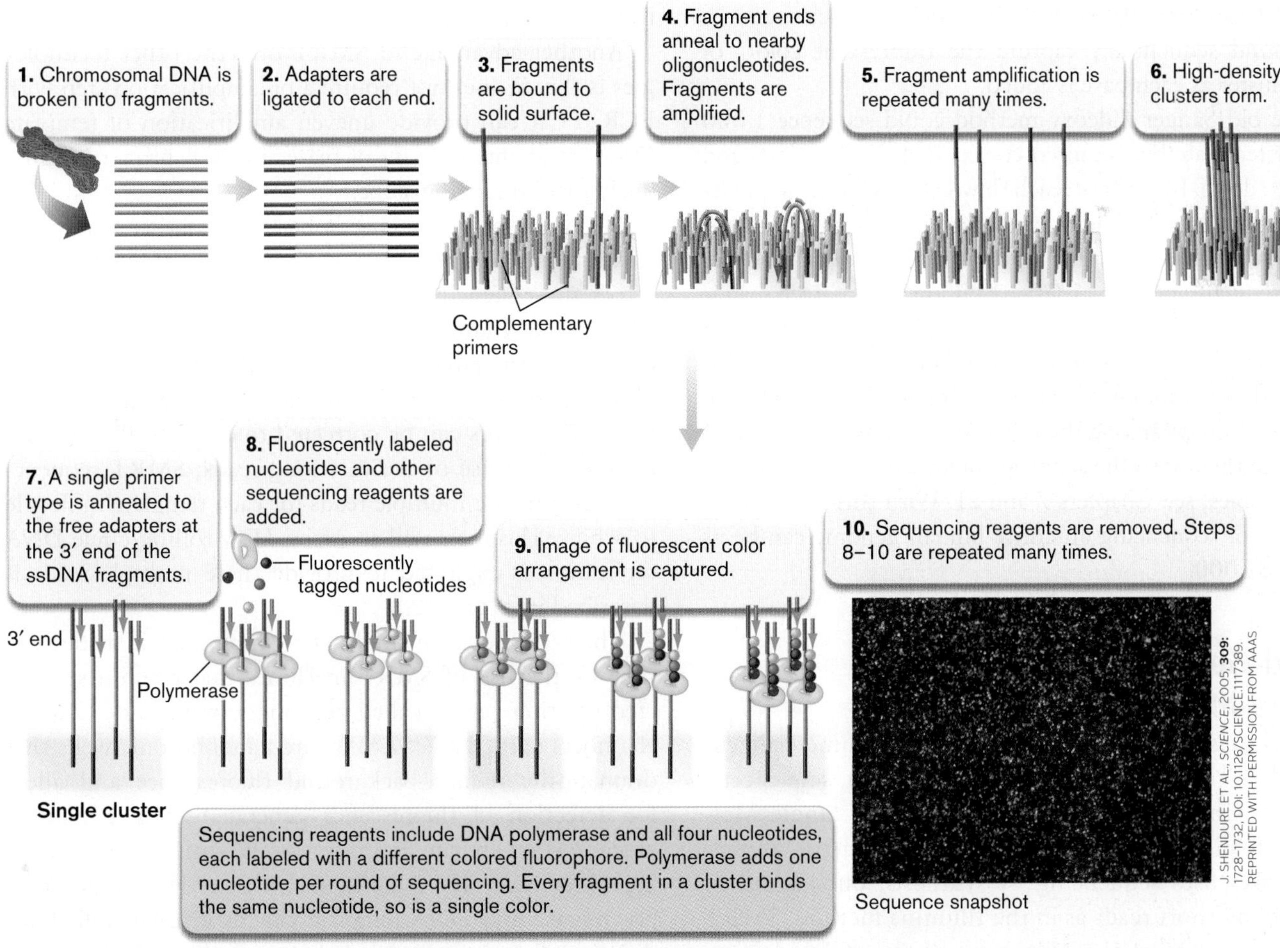

FIGURE 12.6 ▪ DNA sequencing: sequencing by synthesis. Steps 1–6: Generation of clusters by bridge amplification. **Steps 7–10:** Sequencing by synthesis using reversible fluorescent termination.

are complementary to the fragment linker ends but do not hybridize to them at this point.

Next, each single-stranded DNA (ssDNA) fragment is converted into a tight cluster of identical fragments through a series of so-called bridge amplifications (**Fig. 12.6**, steps 4 and 5). Ends of each fragment anneal to nearby matched oligonucleotides (fixed to the glass surface), which then serve as primers to amplify the fragments further. Multiple amplifications result in millions of different ssDNA fragment clusters dotting each lane of each slide (step 6).

All clusters are simultaneously sequenced in a repetitive series of single-step, reversible, chain termination reactions that attach fluorescently labeled nucleotides one at a time to the clusters (**Fig. 12.6**, steps 7 and 8). Because each individual cluster is a tuft of identical DNA molecules, all of the molecules in that cluster will have the same tagged base added and will fluoresce the same color. Note that the sequence of either strand can be determined by the addition of one or the other primer. After a base is added, a snapshot captures the colors, which reveal the bases added to each cluster (step 9). Next, the fluorescent marker (fluor) is removed from the growing chain, which also reverses chain termination, and the slide is again flooded with tagged bases (step 10). Snapshots taken after each sequencing round sequentially capture the fluorescent colors of each cluster as each base is added.

The old Sanger dideoxy method could sequence 1,000 bp per template but required considerable time, effort, and space to do so. In contrast, each flow-cell lane in the sequencing-by-synthesis technique produces 10–120 million DNA fragments, thus generating up to 1,800 gigabases (Gb) of DNA sequence per run (1 Gb = 1 billion bases). Each read is short (up to 300 bases per template), but because millions of templates are read, the massively parallel sequencing yields hundreds of millions of bases. Computer programs then find the overlaps among the millions of sequence results and assemble them into the sequence of an entire chromosome; for examples, see Chapters 7 and 21. With this technology, the cost of sequencing an entire human genome can be as low as $1,000.

Third-Generation Sequencing: Single-Molecule Sequencing

The latest sequencing platforms to be commercialized, sometimes referred to as third-generation sequencers, perform **single-molecule sequencing**. The Single Molecule, Real-Time (SMRT) system from Pacific Biosciences also uses sequencing by synthesis, but instead of assembling short reads as in the Illumina method, SMRT takes advantage of the high processivity of DNA polymerase to generate very long sequence reads. **Processivity** is the ability of a DNA polymerase to elongate a new DNA strand before falling off the template strand. SMRT sequencing uses the DNA polymerase from the bacteriophage phi29, which generates reads of up to 16 kb in length.

The very long reads from single molecules can be of great importance to the investigator because they limit the errors associated with assembling shorter reads such as the ones generated from Sanger and Illumina. Short sequencing reads can lead to assembly errors when a genome has stretches of repeated DNA, such as multiple copies of mobile elements. Because of the much longer reads produced, SMRT sequencing is considered advantageous in assembling genomes that have frequent repeats. For example, the genome of *Burkholderia pseudomallei*, an infectious bacterium that can cause fatal melioidosis, contains over 2,000 repeats. In a side-by-side comparison with Illumina, researchers led by Patrick Woo and Susanna Lau at the University of Hong Kong demonstrated that they could fully assemble the *B. pseudomallei* genome with the long reads of SMRT sequencing, but not with the short reads of the Illumina method. Complete, accurate, and rapid assembly of genomes for pathogens such as *B. pseudomallei* can improve our ability to understand and ultimately treat the diseases they cause.

Another advantage of SMRT over the other technologies is that it does not require a preamplification step with PCR. PCR can provide uneven amplification of template DNA fragments because of biases in PCR primer binding during the annealing stage.

No mode of sequencing is perfect, however, and a key disadvantage of the SMRT technique is that it generates errors in sequencing at very high frequencies relative to Sanger and Illumina (see **Table 12.1**). Fortunately, the errors are generated at random (different) locations each time. Therefore, if enough reads are generated and compared, the errors can be corrected to dramatically increase accuracy. Because of the very long reads, SMRT sequencing can generate multiple reads for each template molecule to achieve this increased accuracy. How rolling-circle DNA replication is exploited to provide these multiple reads is described in **eTopic 12.3**.

The sequencing of single DNA molecules is made possible by the use of SMRT machines that contain specialized reaction vessels called zero-mode waveguide (ZMW) chambers (**Fig. 12.7**). ZMWs are nanophotonic wells that dramatically reduce background fluorescence and allow for detection of fluorescence generated by single molecules. ZMWs are tiny reaction vessels, with volumes in the zeptoliter scale (10^{-21} liter!), such that only a single DNA polymerase and DNA template can fit in each well. Four different fluorescently labeled nucleotides (A, C, G, T), are added to this reaction, which can be excited by lasers located below the well. A camera also positioned below the

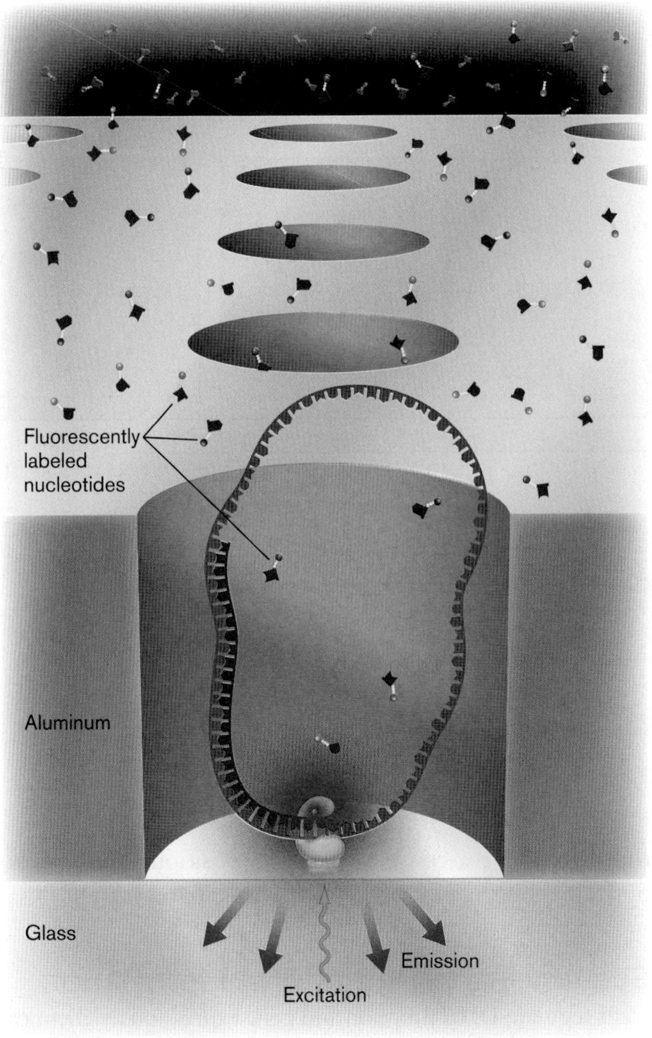

FIGURE 12.7 ■ Schematic of the zero-mode waveguide (ZMW) chamber for SMRT sequencing. Each ZMW contains a single complex of polymerase plus template DNA immobilized at the bottom. A laser below the well excites the fluorescent dye during replication of the DNA strand. A camera also positioned below the well is used to detect fluorescence emission from the dye.

well records the fluorescence emissions in movie form as the DNA is polymerized.

The dyes for each nucleotide are covalently linked via the nucleotide's triphosphate group, and when the nucleotides are free in solution they do not fluoresce. The first event in the sequencing reaction requires one of the labeled dNTPs to recognize and bind to its cognate base at the polymerase active site of the DNA template (**Fig. 12.8**, step 1). Once bound, the dye fluoresces its nucleotide-specific wavelength, and the camera records this increase in light intensity (step 2). The DNA polymerase then forms the phosphodiester bond between the new nucleotide and the growing chain (step 3). The dye released by this reaction loses fluorescence. The length of the fluorescence pulse thus depends on the rate of catalysis of the polymerization reaction. The next step is translocation of the polymerase to the next base in the template (step 4), and the cycle repeats (steps 5 and 6). The time between excitations is referred to as the interpulse duration. The length of the interpulse duration changes if the DNA is modified—a feature exploited to explore patterns of DNA methylation, as described shortly.

Sequencing at remote locations. The sequencing technologies discussed thus far work very well for their intended purposes but involve stationary pieces of machinery and often require extensive preparation of the DNA samples. What if a researcher wishes to analyze microbial DNA on location at a remote site, where time is short and there is no access to sequencing facilities? **Nanopore DNA strand sequencing** addresses that exact situation, as Sarah Johnson and colleagues at Georgetown University demonstrated in the McMurdo Dry Valleys region of Antarctica. The MinION instrument from Oxford Nanopore Technologies fits in the palm of the hand and can be powered by a laptop (**Fig. 12.9**). It also has the advantage of fast sample preparation and short run time (a few hours total). This device has been found to work in even more exotic locations, such as the zero-gravity International Space Station, suggesting that similar technologies could be exploited in the search for life on other planets. Closer to home, and no less important, nanopore sequencing has also been used for real-time surveillance of pathogens during epidemics, such as the 2015 Ebola outbreak in Guinea, West Africa.

Like SMRT, nanopore sequencing is another single-molecule sequencing technology. Unlike Sanger, Illumina, and SMRT sequencing, nanopore sequencing does not involve sequencing by synthesis. Rather, it relies on the changes in electric current that occur when a single strand of DNA passes through a nanopore structure (**Fig. 12.10**). DNA is first ligated to an adapter, and these adapters are recognized by an enzyme that ratchets one of the strands through the nanopore. The sequencer cannot recognize each base specifically, but it can recognize short strings of bases (three to six bases long) because they have signature changes in current that they produce as they pass through the nanopore. Machine-learning algorithms are then used to convert these current profiles into base sequences.

Because the system relies on changes in electric current rather than on synthesis, it can also be used to sequence RNA directly, without requiring that the RNA be reverse-transcribed to DNA (see Section 10.2). And finally, the sequencing is extremely processive, and can generate single reads that exceed 800 kb. This makes the instrument as good if not better than SMRT sequencing for providing long reads through complex, repetitive regions of chromosomal DNA.

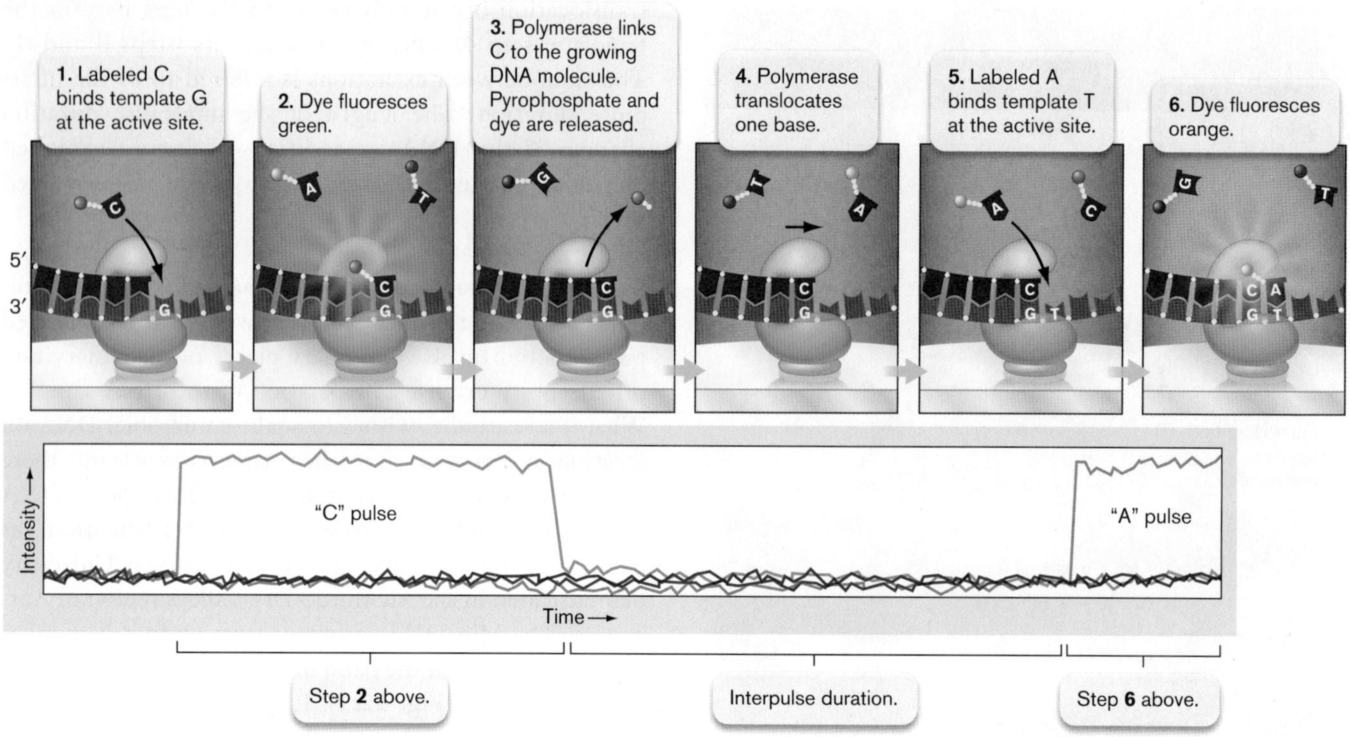

FIGURE 12.8 ■ SMRT sequencing.

Like SMRT sequencing, nanopore sequencing suffers from a high error rate, with single-read accuracy only as high as 92%. This drawback makes the technique inappropriate for applications in which sequencing accuracy is important, such as the identification of single-base-pair mutations. However, as with SMRT sequencing, accuracy can be increased by performing multiple reads and aligning those reads for a consensus at each position. Methods to optimize this process are currently in development.

SARAH JOHNSON

FIGURE 12.9 ■ Nanopore sequencing in Antarctica. The MinION sequencer is controlled and powered by a laptop computer, making it ideal for DNA sequencing at field sites.

Sequencing to Detect DNA Methylation

Cells regulate DNA function and gene expression by chemical modifications. One of the most common ways that microbes chemically modify their DNA is to attach methyl groups to specific bases. DNA methylation can serve multiple purposes for the cell. It helps to control the timing of DNA replication initiation (Section 7.3), to repair DNA using methyl-directed mismatch repair (Section 9.2), and to recognize the chromosome as "self" for restriction endonucleases involved in phage defense (Section 9.3). Identification of methylated bases has historically been time and resource intensive. Because methylation changes the physical properties of DNA bases, SMRT and nanopore sequencing technologies have provided scientists with high-throughput methods to identify a microbe's **methylome**, the bases in the genome that are chemically modified by methylation.

Methylated bases increase the period of nonfluorescence before a methylated base is sequenced (**Fig. 12.11**). This dark period, called the interpulse duration, can be used to identify several types of methylated bases found in microbes, including N^6-methyladenine, N^4-methylcytosine, and 5-methylcytosine.

In *E. coli*, the DNA adenine methyltransferase (Dam) enzyme produces N^6-methyladenine at the G<u>A</u>TC consensus sequences in the chromosome. Dam methylation at the origin of replication is important for restricting

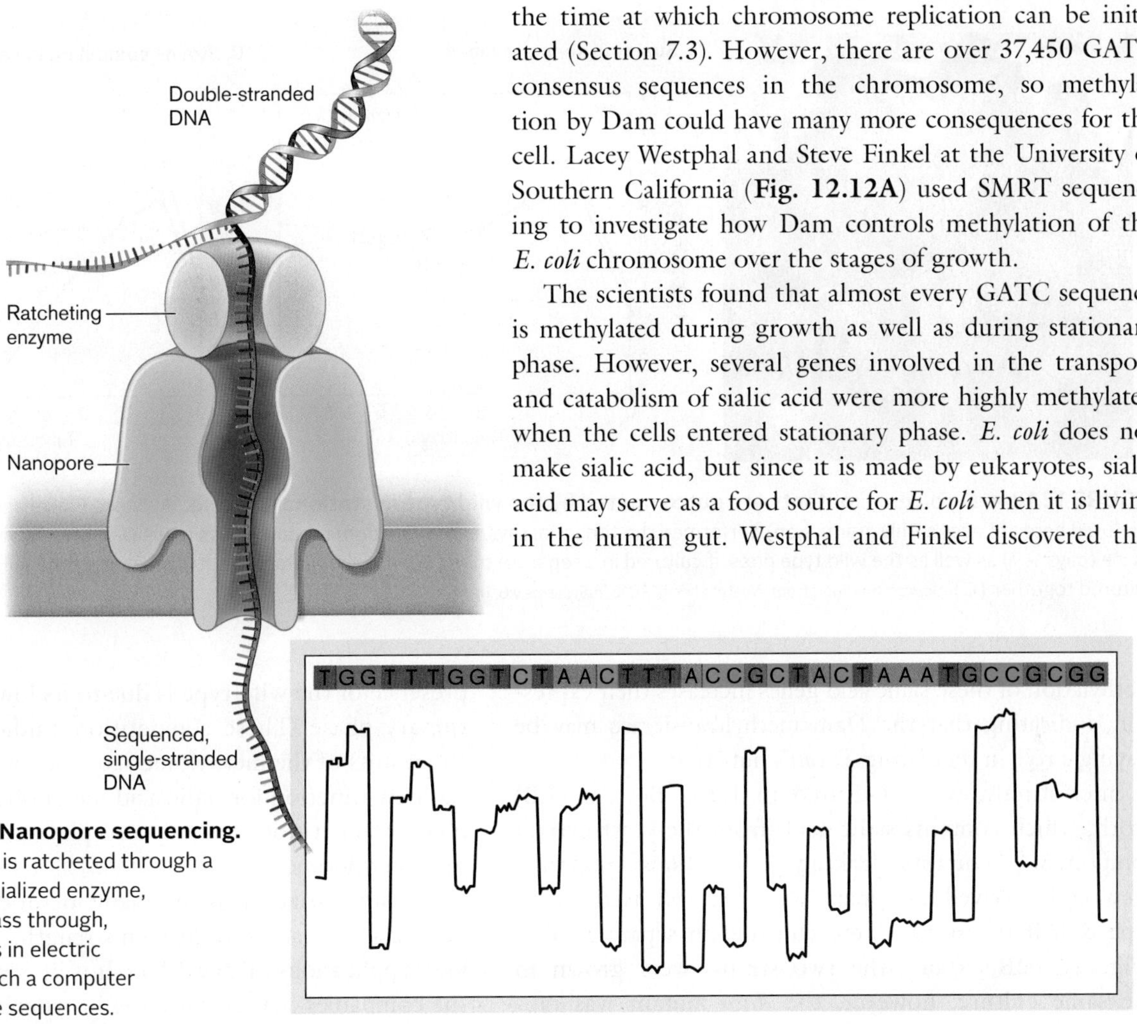

FIGURE 12.10 ▪ **Nanopore sequencing.** One strand of DNA is ratcheted through a nanopore by a specialized enzyme, and as the bases pass through, they cause changes in electric current (inset), which a computer translates into base sequences.

the time at which chromosome replication can be initiated (Section 7.3). However, there are over 37,450 GATC consensus sequences in the chromosome, so methylation by Dam could have many more consequences for the cell. Lacey Westphal and Steve Finkel at the University of Southern California (**Fig. 12.12A**) used SMRT sequencing to investigate how Dam controls methylation of the *E. coli* chromosome over the stages of growth.

The scientists found that almost every GATC sequence is methylated during growth as well as during stationary phase. However, several genes involved in the transport and catabolism of sialic acid were more highly methylated when the cells entered stationary phase. *E. coli* does not make sialic acid, but since it is made by eukaryotes, sialic acid may serve as a food source for *E. coli* when it is living in the human gut. Westphal and Finkel discovered that

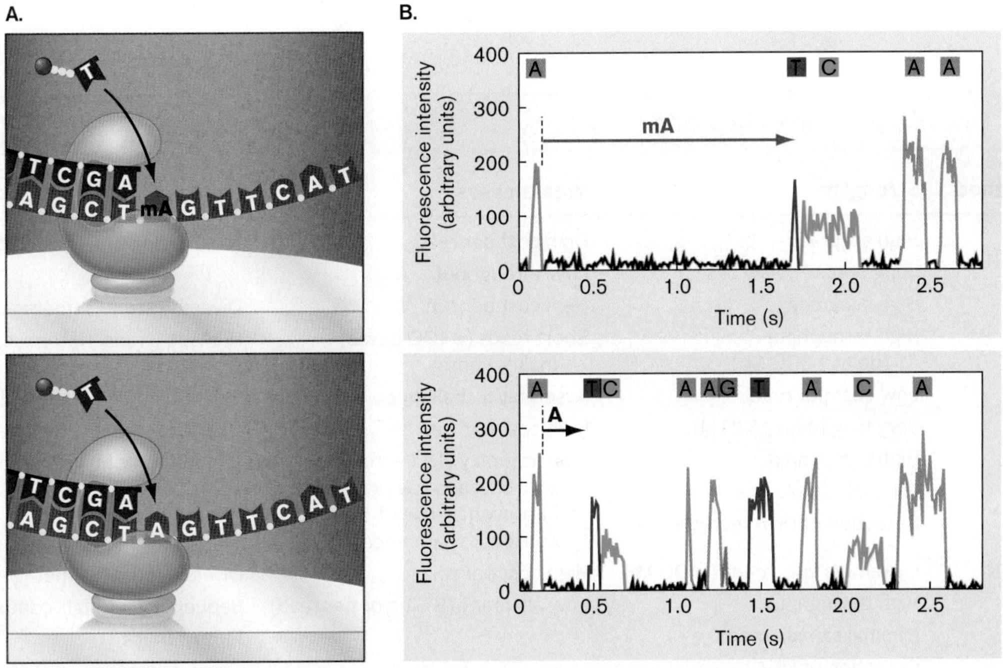

FIGURE 12.11 ▪ **SMRT sequencing can detect DNA methylation.** A. Loading a fluorescence-tagged thymine at a methylated adenine (mA; top) or nonmethylated adenine (bottom) in the template DNA. B. Methylation of the adenine causes a longer interpulse duration before the fluorophore on the thymine releases light. *Source:* Modified from Flusberg et al. 2010. *Nat. Methods* 7:461, fig. 1.

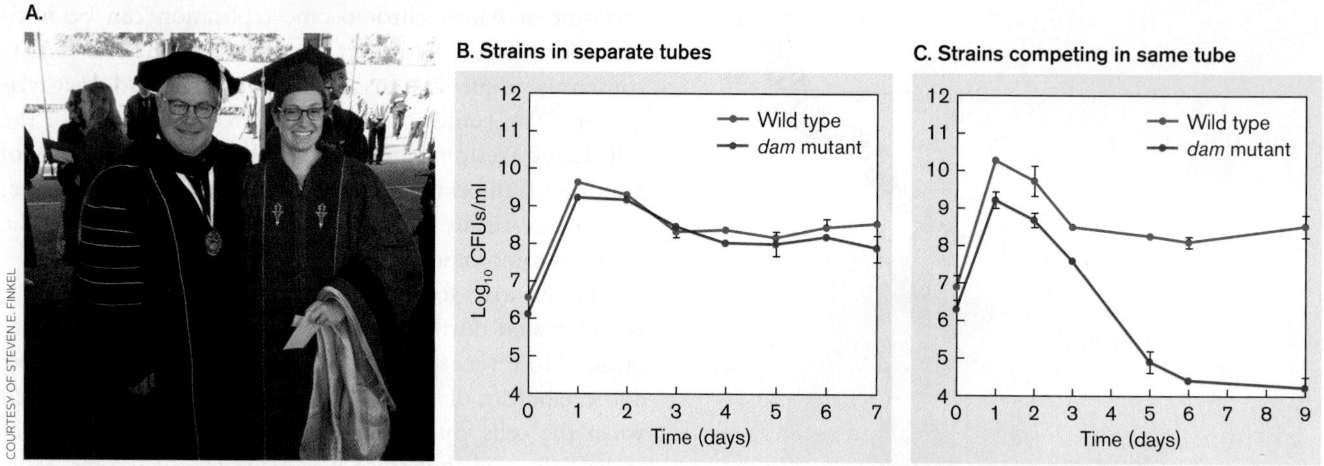

FIGURE 12.12 ▪ *E. coli dam* mutants are outcompeted by the wild type in stationary phase. **A.** Steve Finkel and Lacey Westphal (pictured here at Lacey's PhD graduation) performed the fitness assays. **B, C.** The *dam* mutant grows (days 0–1) and survives stationary phase (days 1–7) as well as the wild type does, if cultured in a separate tube (**B**), but the mutant is outcompeted by the wild type if they are cultured together (**C**). *Source:* Modified from Westphal et al. 2016. *mSystems* **1**:6, fig. 4.

methylation of these sialic acid genes increases their expression, indicating that the Dam methyltransferase may be playing a role in regulating *E. coli*'s nutrition.

Interestingly, when cultured in Luria Bertani (LB) broth, which contains sialic acid (from the yeast extract component), mutants lacking Dam (*dam* mutants) grew and survived stationary phase as well as the wild type did, if the strains were cultured in separate tubes (**Fig. 12.12B**). When the two strains were grown in the same culture, however, the *dam* mutant was rapidly outcompeted in stationary phase by wild-type cells (**Fig. 12.12C**), demonstrating that its death in the presence of the wild type is due to its lower fitness in stationary phase. These competition studies, prompted by the results of the methylome sequencing, hint at a potential new function for sialic acid metabolism in stationary-phase competition that may also play a role in the human gut microbiome.

To summarize, there are many methods for sequencing nucleic acids, each with its own strengths, weaknesses, and ideal applications (**Table 12.1**). It is important to note that the companies that produce the instruments used in these methods, as well as the scientific community that uses them, are continually improving performance and reducing

TABLE 12.1 DNA Sequencing Methods

Sequencing method	Strengths	Weaknesses	Ideal applications
Sanger	High accuracy Long reads (1,000 bases)	High cost per read Low throughput	Sequencing of a few PCR products or plasmids
Illumina	High accuracy High throughput (up to 1,800 Gb) Low cost per read	High cost per run Short reads (≤400 bases) makes contig Assembly a challenge	Diversity studies (metagenomics) RNAseq (transcriptomics, metatranscriptomics) Sequencing of low-complexity genomes
SMRT (Pacific Biosciences)	Very long reads (>20 kb) High throughput Minimal sample prep Detection of DNA methylation	High cost per run Low accuracy (12% error rate per read; accuracy greatly improved by use of circular consensus sequences, >99%)	Diversity studies (metagenomics) Sequencing of high-complexity genomes Methylomics
Nanopore (MinION)	Extremely long reads (>100 kb) High throughput Minimal sample prep Detection of DNA methylation Direct RNA sequencing Portable, fast	High cost per run Low accuracy (8% error per read)	Diversity studies (metagenomics) Sequencing of high-complexity genomes Methylomics Transcriptomics

cost, so some of the current limitations may be eliminated in the near future.

- Sanger is the gold standard for sequencing accuracy and is optimal for sequencing a few PCR products or genes cloned on plasmids.
- Illumina is optimal for sequencing large libraries of samples, where accuracy and high throughput are at a premium. Such applications include analysis of the diversity of a microbial community, using a single locus such as 16S rRNA (see Sections 7.6 and 21.1), the investigation of the transcriptome (Section 10.2) or metatranscriptome (Section 21.1) using RNAseq, and transposon mutant libraries using TnSeq (Section 9.5). The short reads generated by the Illumina sequencer make the analysis of long DNA fragments (for example, genomes, metagenomes) challenging because these reads need to be assembled (see Section 21.1).
- SMRT and nanopore sequencers yield much longer reads and hence have fewer assembly challenges. They can more readily assemble genomes with multiple repeat sequences. However, they are more error-prone and require additional measures to improve accuracy. One approach is to use these long-read methods to assemble the genome, and follow this with the more accurate Illumina method to correct for errors. Both the SMRT and nanopore single-molecule sequencers can detect methylated bases. Nanopore sequencers can also sequence RNA directly, and their small size and fast read times make them ideal for use in the field.

Annotating DNA Sequences

DNA sequence is the raw material for genetic investigation. Once the DNA sequence is acquired and assembled into the chromosomes and plasmids that constitute the genome, the next steps are to find the genes and then use the sequence information to infer function. These steps are performed using **bioinformatics**, an interdisciplinary field that combines computer science, mathematics, and statistics.

To illustrate the value of bioinformatics, **Figure 12.13** shows a circular display map of chromosome I of the Gram-negative pathogen *Vibrio vulnificus*. *V. vulnificus* is a marine bacterium that causes lethal blood or wound infections associated with eating, or catching,

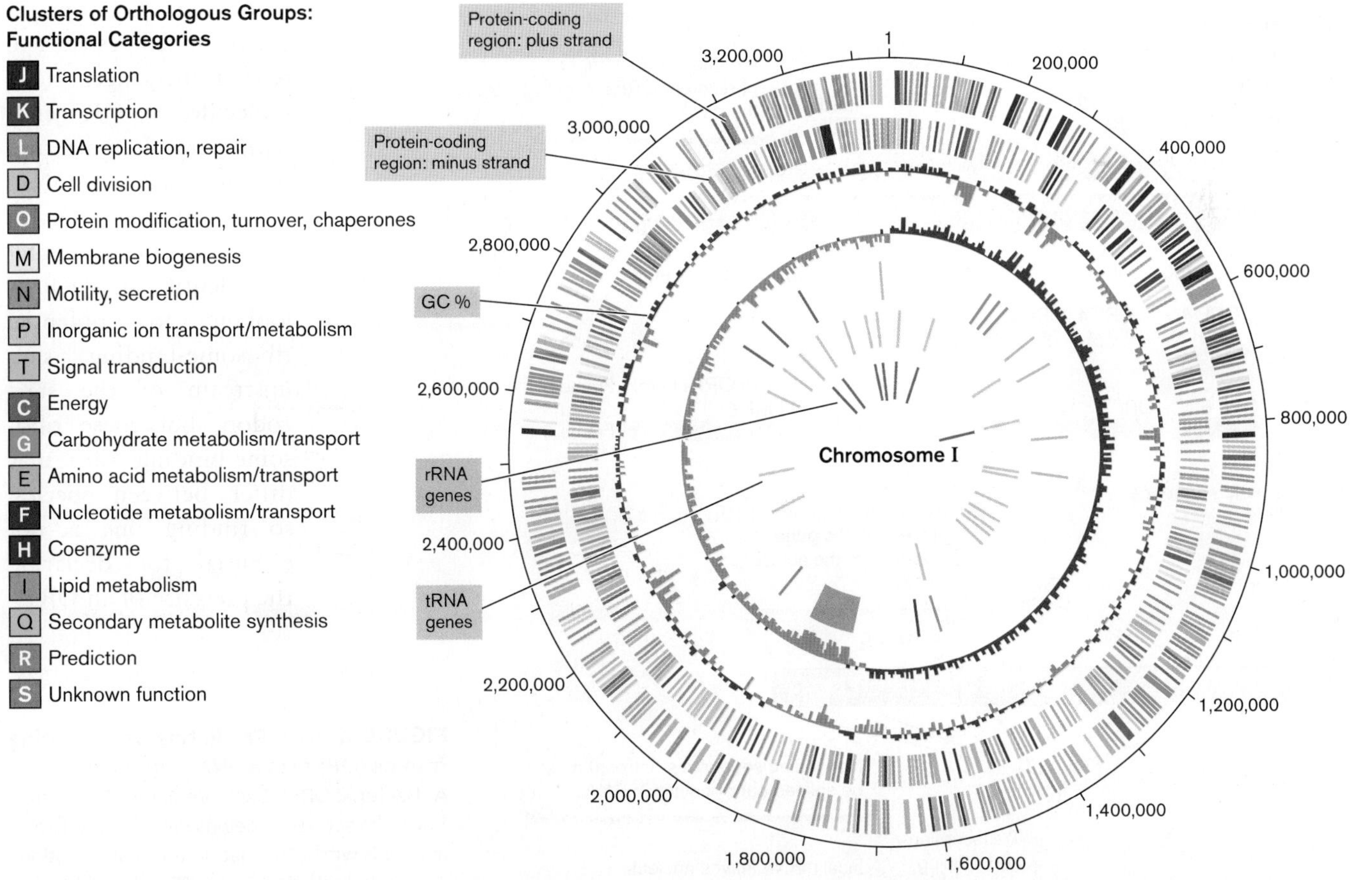

FIGURE 12.13 ■ Circular display map of the *Vibrio vulnificus* genome (chromosome I; 3,377 kb). The color code indicates gene clustering by function. The two outer rings represent genes transcribed from different DNA strands. The innermost circle uses peaks to show GC content levels above (red) or below (green) the genome average.

raw shellfish. One powerful approach to understanding how *V. vulnificus* causes disease is to probe its genome for genes associated with disease. Once such genes are located, they can be analyzed by bioinformatics to assign predicted functions based on their sequence. The genome-scale level of predicted microbial activity can be used to generate hypotheses about how this organism causes disease, and these hypotheses can be tested using other techniques, some of which are described in the sections that follow.

Bioinformatics has forever changed how the science of microbiology is conducted. Since 1998, over 140,000 complete microbial genomes have been sequenced, and a vast number of partial genomes from uncultured microbes is likewise available. Recall that fewer than 1% of microorganisms can be cultured. Bioinformatics, then, is critical to understanding the complexity of the uncultured microbial world.

In Chapter 7 we introduced the genome and the metagenome, the genes in a single microbe or a microbial community, respectively. But how do we learn which genes are encoded in a newly sequenced genome or metagenome? The first step is **annotation**. Annotation is analogous to identifying separate sentences and words in an unknown language. Computers annotate DNA sequences by using the known rules of transcription and translation to identify the start and stop sites of potential genes. DNA sequences are then translated *in silico* to determine amino acid sequences. When a sequence appears to encode a protein, the DNA sequence is called an open reading frame (ORF). Genes that encode transfer RNAs and ribosomal RNAs do not encode proteins but can be identified because of the conservation of these sequences across vast phylogenetic distances.

Seeking ORFs. Computer programs such as ORF finders use the universal genetic code to deduce protein sequences potentially formed in all reading frames on RNA molecules transcribed from either direction on the chromosome (**Fig. 12.14A**). An ORF, the equivalent of a sentence in our analogy, is defined as a DNA sequence that can potentially encode a string of amino acids of minimum length—say, 50 residues. The 50 residues of the ORF could encode a 5,500-Da (5.5-kDa) protein, since the average weight of an amino acid is 110 Da. Each ORF begins with a translation start codon (usually ATG or, more rarely, GTG or TTG). A translation start codon marks where ribosomes start to read a messenger RNA molecule. An ORF ends with a translation termination codon (in DNA: TAA, TAG, or TGA). In addition, the computer can identify an ORF by looking for potential ribosome-binding sites upstream of the start codon, but these ribosome-binding sites may differ between species, so finding one is not essential for declaring the presence of an ORF.

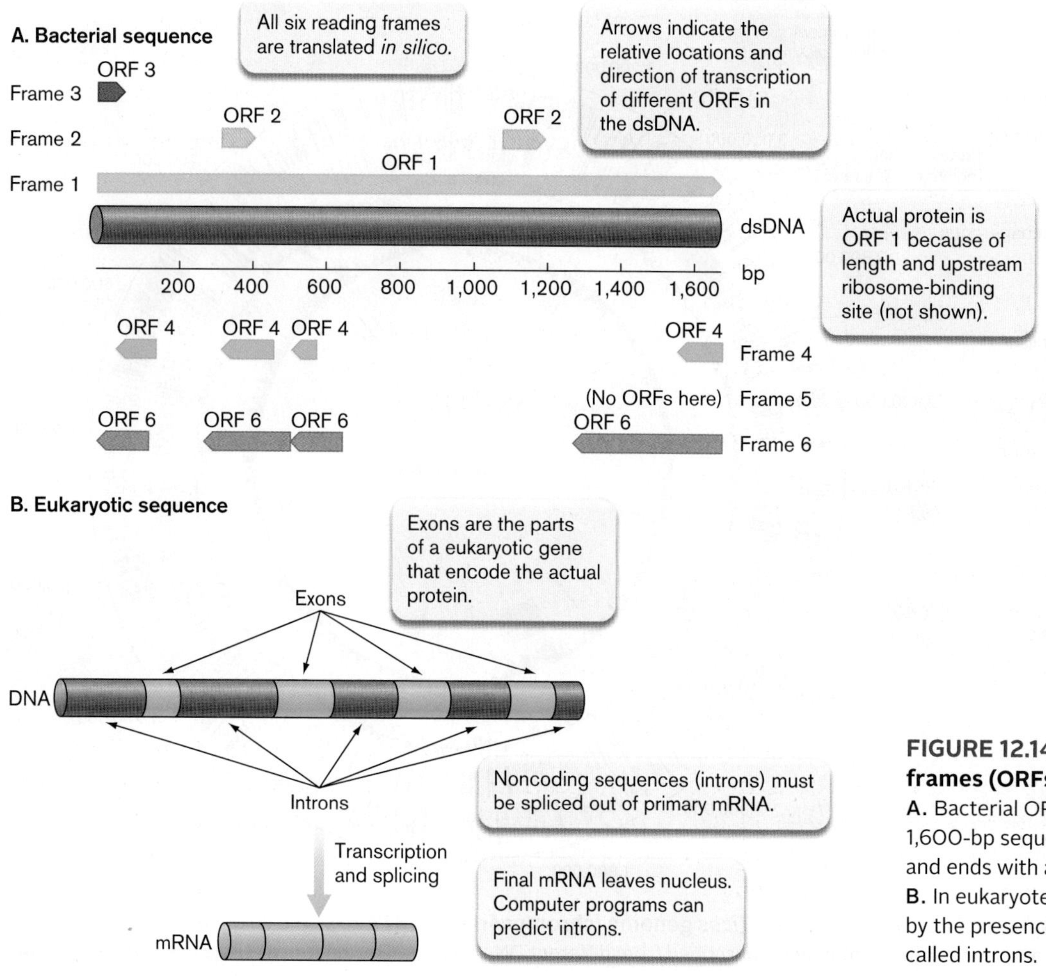

FIGURE 12.14 ▪ Predicting open reading frames (ORFs) in a DNA sequence. **A.** Bacterial ORFs. Each predicted ORF in this 1,600-bp sequence begins with AUG or GUG and ends with a translation terminator codon. **B.** In eukaryotes, finding ORFs is complicated by the presence of noncoding DNA sequences called introns.

Note: It is important to recognize that ORF predictions need to be verified experimentally. Additionally, it is important to be aware that arbitrary cutoffs for minimum ORF length, such as 50 amino acids, can exclude from the search smaller proteins that may play important roles in the cell.

While the preceding methods work for prokaryotes, identifying ORFs in eukaryotes is more difficult. In eukaryotes, most genes contain **introns** (long, noncoding sequences that occur in the middle of genes), as do the initial mRNA products made from those genes (**Fig. 12.14B**). The vast majority of known bacterial and archaeal genes do not contain introns. Introns serve several regulatory functions in eukaryotes that determine whether a protein product is made. However, the eukaryotic cell must use special splicing mechanisms to remove the intron sequences and re-join the protein-encoding sequences, called **exons**, of the mRNA prior to translation. (In eukaryotic organisms, the RNA transcribed directly from DNA is called the **preliminary mRNA transcript** or **pre-mRNA**; the "final" mRNA transcript is produced when the introns have been spliced out.) Therefore, to derive the ORF coding sequence of a eukaryotic gene, computer analysis must first identify and remove the intron sequences between the exons.

Assigning function. We know from considerable experimental precedent that proteins with the same function, regardless of species, usually have critical amino acid sequences in common because they evolved from the same ancestral sequence. Sequences common to two different protein or DNA molecules are called homologous sequences (see Section 9.2). During the annotation process, ORFs are compared to hundreds of thousands of known proteins to look for sequence homology. ORFs with sufficient homology to known proteins can then be assigned putative function. Powerful computer programs are available to carry out these analyses quickly (for examples, visit the Joint Genome Institute website at https://jgi.doe.gov).

If a query protein sequence does not show high levels of similarity with known proteins, "annotation by function" can be assigned when a protein domain exhibits characteristics or motifs similar to those possessed by proteins of known function. For example, periodic hydrophobic regions may indicate a membrane protein, or an amino acid sequence (or motif) matching those of known ATPases may suggest a similar function for the query protein. Annotation by function provides an extremely valuable starting point toward unraveling a protein's function, but biological validation is required for confirmation.

To Summarize

- **PCR** can rapidly amplify DNA, and quantitative (real-time) PCR is a specialized version in which the number of DNA template molecules can be accurately determined.
- **Sanger sequencing** provides accurate long reads of DNA suitable for the analysis of small numbers of DNA templates.
- **Illumina next-generation sequencing** is a high-throughput method to acquire accurate sequence information from many template DNA molecules, although the read lengths are shorter than with Sanger sequencing.
- **Single-molecule sequencing** can provide very long reads of single molecules of DNA or even RNA, but steps must be taken to improve read accuracy.
- **Sequencing by synthesis** is the method by which Sanger, Illumina, and SMRT sequencing generate DNA sequence information. In contrast, **nanopore sequencing** relies on the changes in electric current that occur when a single strand of DNA passes through a nanopore structure.
- **DNA methylomes** can be characterized by methods such as single-molecule sequencing that discriminate between methylated and unmethylated bases.
- **The field of bioinformatics** uses computer algorithms, mathematics, and statistics to annotate genes.
- **Annotation** requires computers that look for patterns in DNA sequences. Annotation predicts regulators, ORFs, and genes encoding untranslated RNAs such as tRNAs and rRNAs. Similarities in protein sequence (deduced from the DNA sequence) are used to predict protein structure and function.

12.2 Genetic Manipulation of Microbes

Key to the discovery of gene functions and gene regulation is our ability to manipulate an organism's DNA, and to characterize the impacts of these alterations on the gene and the microbe. We have discovered ways that nature cuts, copies, and pastes DNA with remarkable precision (Chapter 9) and have developed technologies that exploit these natural editing processes to provide us with sophisticated tools for genetic analysis.

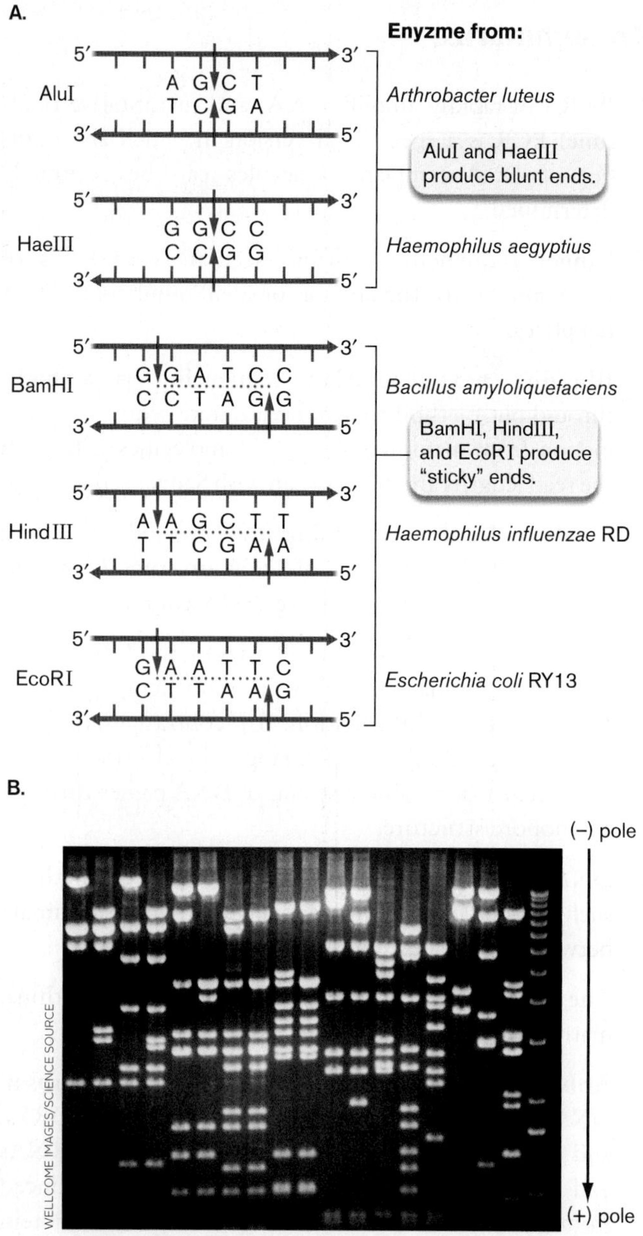

FIGURE 12.15 ▪ Bacterial DNA restriction endonucleases. **A.** Target sequences of sample enzymes. The names of these enzymes reflect the genus and species of the source organism. **B.** Agarose-gel size analysis of DNA fragments generated by restriction endonuclease digestion. Small fragments move toward the bottom faster than do large fragments. DNA bands are revealed by chemical stains that bind DNA and fluoresce under UV light.

Restriction Endonucleases and the Birth of Recombinant DNA

What seemed to be an esoteric line of research—exploring how bacteria cleave the "foreign" DNA of invading plasmids and viruses—ultimately led to a tool that revolutionized biology. One of the ways bacteria rid themselves of foreign DNA is with enzymes called **restriction endonucleases** that cut up unfamiliar DNA molecules at specific sequences called **restriction sites** (Section 9.3). Different species of bacteria possess unique restriction endonucleases, each of which recognizes a different DNA sequence. A few examples of restriction endonucleases and target sequences are shown in **Figure 12.15A**. The most useful restriction endonucleases for molecular biology recognize restriction sites that are four to six bases in length. Notice that the sequence of each of these restriction sites is a **palindrome**, because the top and bottom strands read the same in the 5′-to-3′ direction. A palindromic DNA restriction site enables a restriction endonuclease to attack both strands of a duplex by using the same substrate recognition site.

Restriction endonucleases cleave the phosphodiester backbones of opposite strands at locations either near or within the center of the restriction site. The cut produces fragments of DNA with either blunt ends or staggered ends. In staggered ends, the top strand is cut at one end of the site, and the bottom strand is cut at the other end (**Fig. 12.15A**). Staggered ends are also called "cohesive ends" because the protruding strand at one end can base-pair with a complementary protruding strand from any DNA fragment cut with the same restriction endonuclease, regardless of the source organism. The ability of cohesive ends from different organisms to base-pair is the property that first made recombinant DNA technology possible. For example, using restriction endonucleases along with other DNA editing processes made it possible to engineer bacterial cells to produce human insulin from the human insulin gene for the treatment of diabetes (as we will describe).

Agarose gels can be used to analyze DNA fragments generated by restriction endonuclease digestion (agarose-gel electrophoresis is explained in eAppendix 3). Agarose is a seaweed-based polysaccharide, which, when mixed with boiling water, cools to form a gel of cross-linked polysaccharide that possesses pores. Each lane of the gel shown in **Figure 12.15B** represents a different population of DNA molecules cut with the same restriction endonuclease. Because DNA is negatively charged, all DNA molecules travel to the positive pole during electrophoresis. Pore sizes in the agarose are such that they allow small molecules to speed through the gel, while larger molecules are slowed. Thus, DNA fragments in an agarose gel separate on the basis of size. The smaller the fragment, the farther it travels toward the positive pole in a given amount of time.

Thought Question

12.2 **Figure 12.15B** illustrates how agarose gels separate DNA molecules. If you ran a highly supercoiled plasmid in one lane of the gel and the same plasmid with one phosphodiester bond cut in another lane, where would the two molecules end up relative to each other in this type of gel?

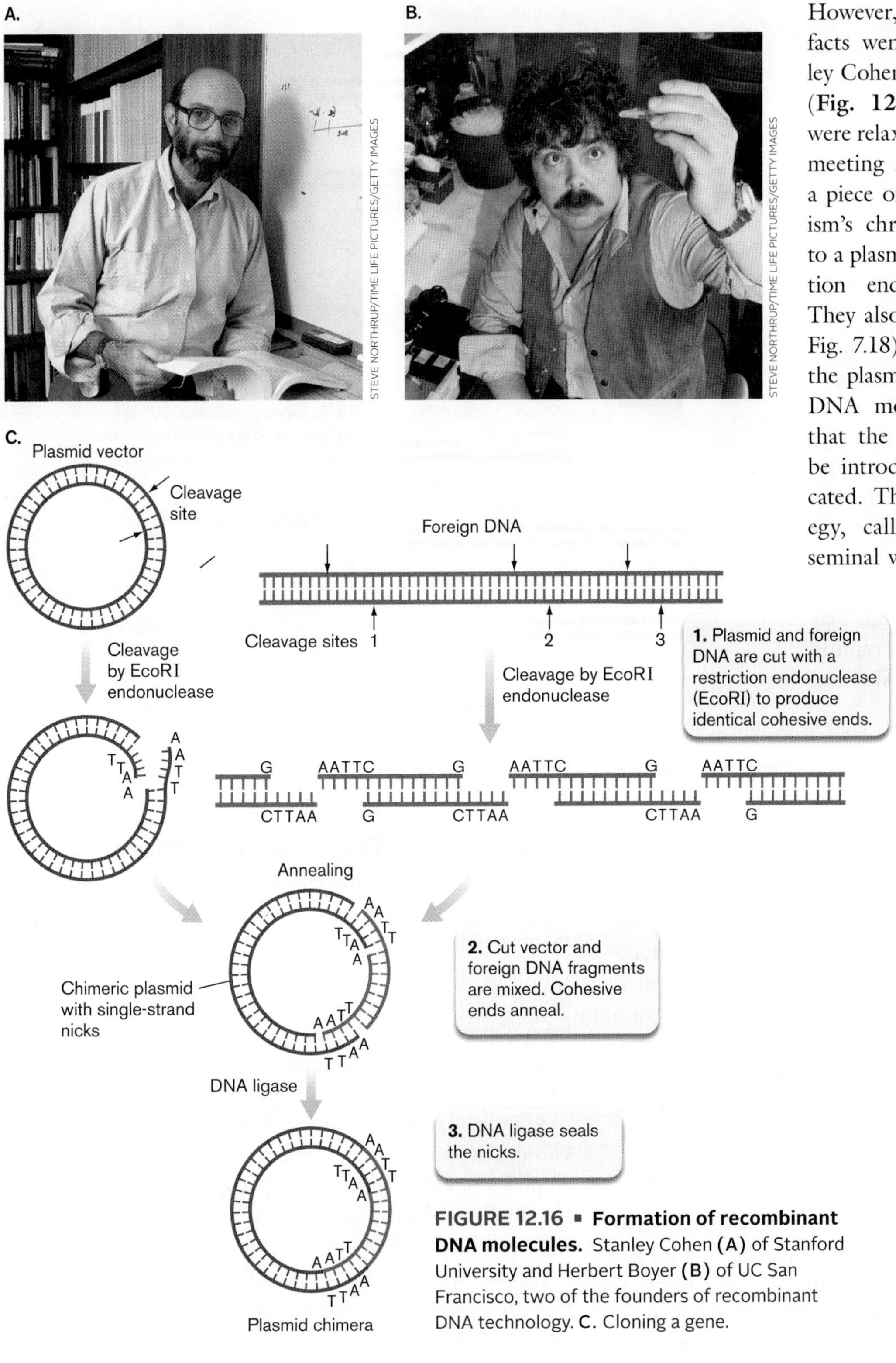

FIGURE 12.16 ■ Formation of recombinant DNA molecules. Stanley Cohen (**A**) of Stanford University and Herbert Boyer (**B**) of UC San Francisco, two of the founders of recombinant DNA technology. **C.** Cloning a gene.

Gene Cloning

How did scientists ever conceive of cloning a gene from one organism into the DNA of an unrelated organism? The beginning of the recombinant DNA revolution can be traced to 1972. Scientists knew that many restriction endonucleases generate cohesive ends and that some plasmids contain a single site for certain restriction endonucleases. However, the significance of these facts went unrecognized until Stanley Cohen (**Fig. 12.16A**), Herb Boyer (**Fig. 12.16B**), and Stanley Falkow were relaxing together after a scientific meeting in 1972. They realized that a piece of DNA cut from one organism's chromosome could be grafted to a plasmid cut with the same restriction endonuclease (**Fig. 12.16C**). They also knew that DNA ligase (see Fig. 7.18) would seal the fragment to the plasmid and form a new artificial DNA molecule. They hypothesized that the recombinant plasmid could be introduced into *E. coli* and replicated. Three months later, the strategy, called **cloning**, worked. This seminal work was published by Boyer and Cohen, along with students Annie Chang and Robert Helling.

Over the next decade, the technique became known as "gene cloning," an immensely powerful technology that thrust biology into the genomic era. Today, plasmid vectors are engineered with features specifically designed for cloning and expressing genes. One common element of plasmids is a gene for positive selection, such as a gene that confers antibiotic resistance to the host cell. Cells are unable to grow in media containing the antibiotic unless the plasmid is present. Growing the cells in the presence of antibiotics thus ensures that the plasmid is maintained.

Another common feature is a multicloning site, which is a region that is highly enriched for target recognition sequences of restriction endonucleases. This abundance of recognition sequences provides the researcher with many options for cloning by restriction digestion and ligation of the inserted DNA at the multicloning site. If the cloned insert is to be expressed, these multicloning sites may be downstream of a promoter that is recognized by the host cell (usually *E. coli*).

Some cloning methods utilize special enzymes for the restriction digestion or ligation steps. Golden Gate cloning uses a class of restriction endonucleases that cuts at a distance from the recognition sequence. This feature is exploited to eliminate the recognition sequence in the final product. TOPO TA cloning bypasses the restriction step and can clone PCR products directly into plasmids. TOPO TA cloning uses topoisomerase I rather than DNA ligase to ligate the insert into the plasmid. This cloning is made exceptionally efficient by the covalent attachment of the topoisomerase to the two ends of the linearized plasmid vector. TOPO TA cloning is especially useful in cloning DNA from environmental samples, where DNA concentrations may be limited and the goal is to capture a large fraction of the total microbial diversity.

In 1979, one of the first applications of cloning for gene expression turned *Escherichia coli* into a factory for human insulin. Before this achievement, treatment of diabetes relied on animal insulin, which was slow and costly to produce, and prone to inducing allergic reactions in patients. In collaboration with the biotech company Genentech, researchers at City of Hope National Medical Center synthesized the genes encoding the A and B subunits of insulin. Next they used restriction digestion and ligation to clone each gene in a plasmid, where expression was under the control of bacterial promoters. The synthesized protein subunits could then be purified and assembled into insulin that was chemically and functionally indistinguishable from insulin produced by humans. Importantly, the *E. coli* strain could grow quickly and densely in large fermentors to provide a massive, cost-efficient supply of much-needed insulin. Licensed by Eli Lilly and Company in 1982, human insulin became the first drug (Humulin) produced from recombinant DNA to be approved by the Food and Drug Administration and used in the treatment of disease.

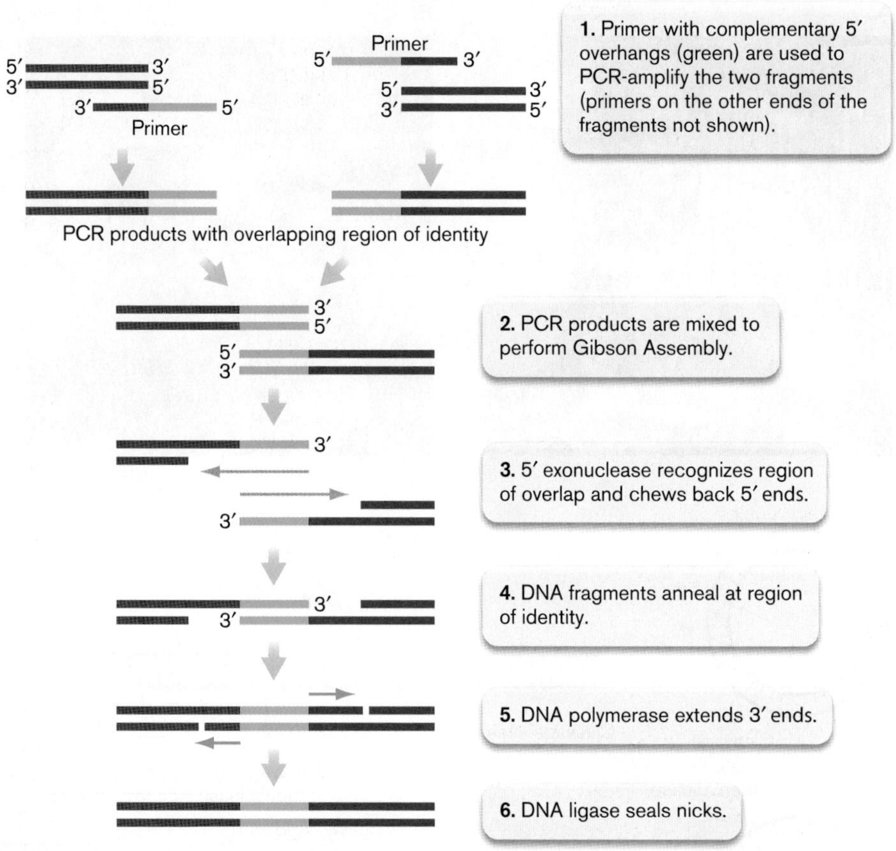

FIGURE 12.17 ▪ Formation of recombinant DNA by Gibson Assembly.

Cloning by PCR: Gibson Assembly

Restriction digestion and ligation works well for many cloning purposes but can be cumbersome if many DNA fragments need to be stitched together. Such was the case for the construction of the first chemically synthesized organism, JCVI-syn1.0. Its genome of 531,490 bp was assembled from oligonucleotides that were each less than 100 bp long. To perform this massive scale of assembly more efficiently, a restriction-independent method of cloning was utilized.

Gibson Assembly, named after its developer, Daniel Gibson, at the J. Craig Venter Institute, relies on PCR to amplify the DNA segments, and a special enzyme to splice the PCR products together. PCR primers are designed such that the products of PCR have overlapping regions of sequence identity at their ends (**Fig. 12.17**). When mixed together, the T5 exonuclease enzyme recognizes the region of overlap and "chews back" the 5′ end of one strand of each overlapping DNA. The complementary 3′ overhangs generated by this activity can then anneal, and with the addition of DNA polymerase and DNA ligase, the two strands of DNA become covalently attached. For simplicity, **Figure 12.17** shows a Gibson Assembly reaction at a single region of homology between two DNA fragments. If a PCR product has regions of homology to a plasmid at both of its ends, Gibson Assembly can be used to insert the product into the plasmid.

Thought Question

12.3 Given that *E. coli* possesses restriction endonucleases that cleave DNA from other species, how is it possible to clone a gene from one organism to another without the cloned gene sequence being degraded?

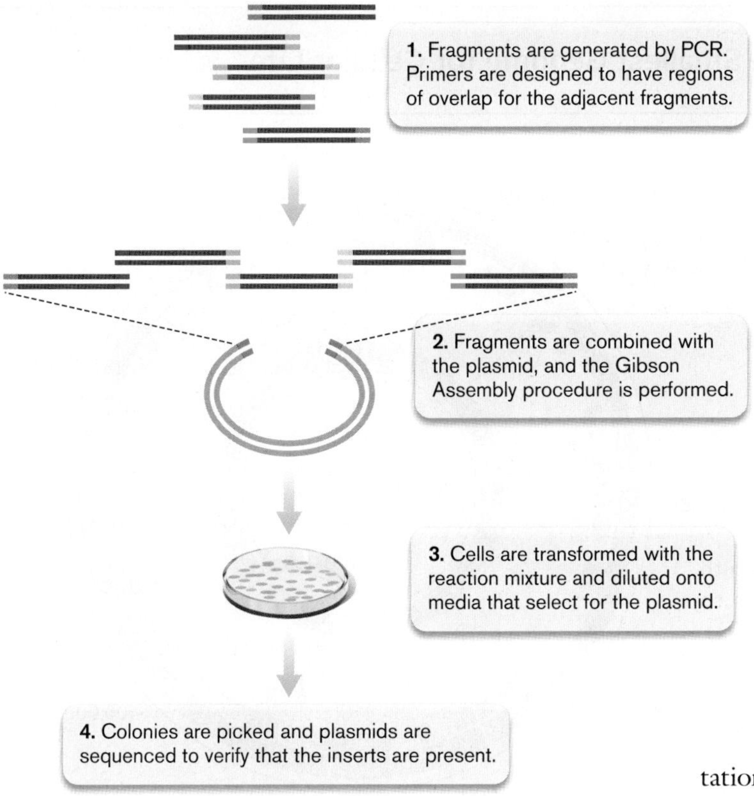

FIGURE 12.18 ▪ Gibson Assembly can clone multiple PCR fragments simultaneously.

Gibson Assembly has several advantages over classic restriction-and-ligation cloning. First, the site of splicing can be at any position along the DNA fragments, not just at restriction sites. Second, because the reaction occurs only at specific regions of homology, up to five fragments can be linked simultaneously, in the desired orientation and in a single reaction, if the regions of overlap are distinct (**Fig. 12.18**). This ability was fundamental to the creation of the first synthetic organism, JCVI-syn1.0. Additional details on how this microbe was built and streamlined to generate the smallest genome of a free-living organism is the subject of **Special Topic 12.1**.

Random Mutagenesis

Random mutagenesis is a valuable search technique to identify genes involved in microbial processes of interest. Its primary strength is its unbiased nature: a scientist doesn't have to guess which genes are involved; rather, all genes are examined simultaneously, and by the change in phenotype due to mutation, the ones involved reveal themselves. Random mutagenesis techniques are covered in greater detail in Chapter 9, and additional examples are provided in **eTopic 12.1** and **Special Topic 12.1**. In brief, a population of cells is exposed to a mutagen that alters the genetic code at random (or nearly random) locations in the genome. Chemical mutagens and UV irradiation were the traditional means of genetic alteration and are still in use today. The drawback to these methods is that they require a strong dose of mutagen, which will generate mutations at many loci in the genome. Thus, determining which of the mutations causes the phenotypic change can be quite challenging.

Transposable elements are currently the preferred means of performing random mutagenesis. In Chapter 9 we learned that transposable elements are mobile DNA sequences that can hop from one site to another. Transposition events are rare, and typically only a single transposon mutation occurs per cell. The inserted transposons can be located quickly via PCR, while the identity of the mutated gene can be determined by sequencing the DNA adjacent to the transposon. To confirm that the transposon insertion causes the phenotypic change, a wild-type copy of the gene without the insertion is cloned and expressed in a plasmid. If the wild-type phenotype is restored by the plasmid, the mutation is said to be complemented, and such complementation provides confidence that the transposon insertion caused the phenotype. Recently, the use of transposons for high-throughput phenotypic screening has been employed by a technique called TnSeq, discussed in Chapter 9.

Targeted (Site-Directed) Mutagenesis

When the gene of interest is known and the scientist wishes to mutate it specifically, site-directed mutagenesis is a warranted approach. Here we describe how site-directed mutagenesis can be applied to make a targeted deletion of the gene on the chromosome. **eTopic 12.4** describes how genes can be studied by site-directed mutagenesis when cloned on a plasmid.

In site-directed mutagenesis, the allelic replacement occurs by two crossover events between the host chromosome and a plasmid carrying the deletion allele (**Fig. 12.19**). The crossovers are mediated by the homologous recombination machinery of the cell. The plasmid is usually in the form of a "suicide vector" that cannot replicate without help from a gene present on the chromosome. This replication gene is present in the strain that is used to maintain the plasmid, but not in the strain to be mutated.

The vector contains several important genetic elements for the mutagenesis procedure (**Fig. 12.19**). The two regions on the chromosome that flank the target gene to be deleted are spliced together on the plasmid. These are the places where the two homologous recombination events will occur to generate the deletion. The plasmid also carries a gene for positive selection; in the example in **Figure 12.19** it is the *cat* gene that encodes resistance to the antibiotic chloramphenicol (see Section 8.3). Finally, the plasmid carries a gene for counterselection; in the

SPECIAL TOPIC 12.1 Constructing the Smallest Genome for Cellular Life

What is the minimum number of genes needed to make a self-replicating life form, and what are the essential functions that these genes encode? One way to address this question is to find all of the nonessential genes of an existing organism—particularly an organism with a small genome—then remove them all and see whether the organism can still replicate. Since there may be hundreds of genes to delete, deleting them one at a time would be slow and laborious. A better way to perform this large-scale genome reduction is to build the chromosome from scratch, adding only the genes thought to be essential. This strategy was employed by Clyde Hutchison III, Hamilton Smith (**Fig. 1**), and colleagues at the J. Craig Venter Institute to build a viable, replicating microbe named JCVI-syn3.0, whose genome, built entirely from chemically synthesized DNA, encodes a mere 473 genes.

Their approach began with strain JCVI-syn1.0, a synthetic reconstruction of the 1.08-Mb genome of *Mycoplasma mycoides* consisting of 901 genes (**Fig. 2**). Hypothesizing that not all genes in JCVI-syn1.0 were essential for growth, the scientists mutagenized the strain with a transposon and selected for viable transposon insertion mutants (Sections 9.5 and 12.2). The genes inactivated by the transposon insertions in these viable mutants were then identified and classified as "nonessential" for growth. Essential genes should not tolerate a transposon insertion, so they would be identified by their absence from the collection of viable mutants.

An initial genome built exclusively of the 240 identified essential genes did not grow, and thus some genes identified as nonessential were actually needed to establish a viable, replicating organism. What accounted for this need for "nonessential" genes? In some cases, the cell had two genes that could perform the same essential function; in other words, these genes had functional redundancy. Losing one gene through transposon insertion was nonlethal because the other gene was not disrupted. However, attempts to build a genome that lacked both genes were not successful; some genes had to be added back.

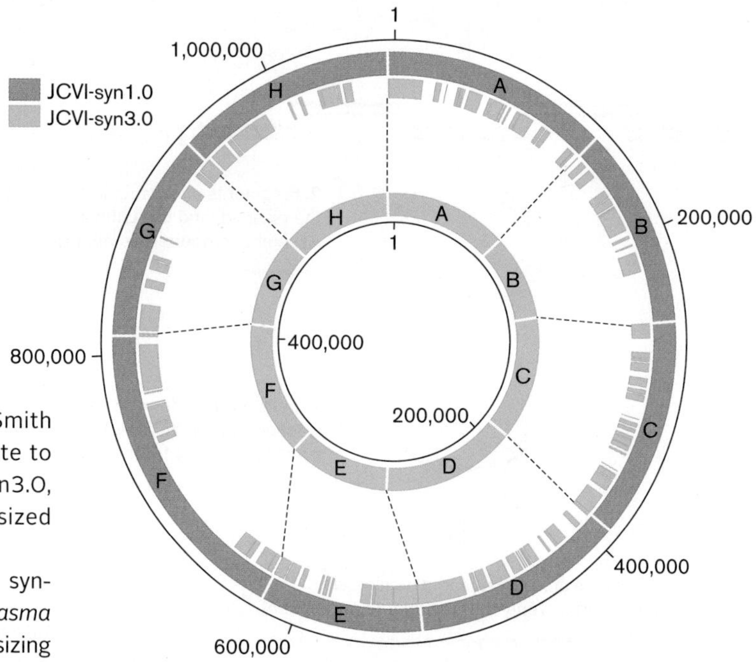

FIGURE 2 ▪ Derivation of JCVI-syn3.0 from JCVI-syn1.0. Comparison of the genomes, showing the locations of the genes (purple) in syn1.0 that were retained in syn3.0. *Source:* Modified from Hutchison et al. 2016. *Science* **351**:aad6253, unnum. fig., panel B.

J. CRAIG VENTER INSTITUTE

FIGURE 1 ▪ Clyde Hutchison III (left) and Hamilton Smith at the J. Craig Venter Institute.

Through trial and error, they arrived at version 3.0 of their synthetic genome: a viable, replicating life form with a genome of 531,490 bp and 473 genes (**Fig. 2**). Just how did they make their version 3.0 artificial chromosome, and get it to start the essential processes of life, such as gene expression and chromosome replication? Short oligonucleotides (less than 100 bases) were synthesized chemically, and then subsequent steps were performed to stitch all of these together to make the 531,490-bp whole genome (**Fig. 3**).

The first stitching step used PCR to assemble the oligonucleotides into 1,400-bp fragments. Five of these fragments were then stitched together using Gibson Assembly (described in the text) to make 7-kb fragments. The final assemblies into eight 66-kb fragments and subsequently to one 531-kb circular chromosome were performed inside yeast cells. When transformed by these fragments, the yeast could assemble the fragments into large replicating clones via their homologous recombination machinery. Once the whole genome was assembled, it was transformed into a related *Mycoplasma* species, *M. mycoides*. The transplanted syn3.0 chromosome used the existing machinery of the host cell to replicate and express its genes, eventually displacing the host *M. mycoides* chromosome and gene products for its own.

The JCVI-syn3.0 synthetic organism forms clusters of spherical cells of varying diameter (**Fig. 4A**). It is a heterotroph that requires extensive nutrient supplementation and grows about three times slower than the JCVI-syn1.0 strain. Its genome contains 438 protein-encoding genes and 35 untranslated RNA genes (rRNA, tRNA, and small RNAs), making it the smallest genome for an organism capable of growth in pure culture. The largest category of genes is for the expression of genomic information (**Fig. 4B**). Surprisingly, 79 genes (17%) included in the minimum genome have no known function. It thus remains unclear which cell functions are absolutely essential for survival and replication. But at least now we have a clearer picture of what we don't know.

RESEARCH QUESTION

What types of techniques found in this chapter might be applied to test the function of the unknown genes?

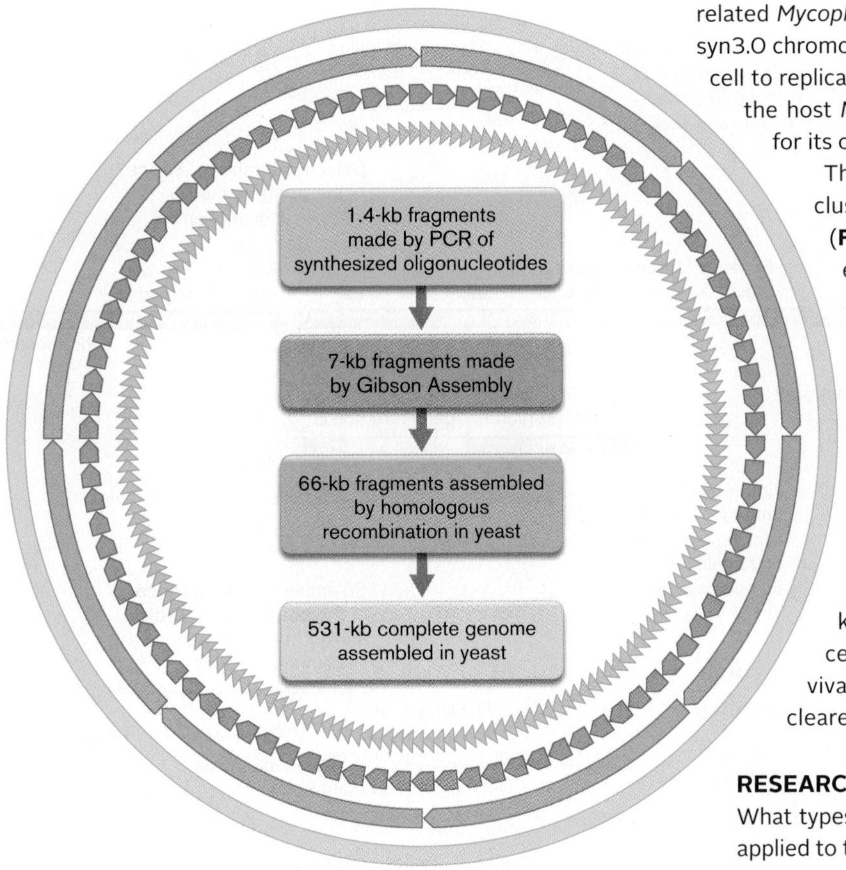

FIGURE 3 ■ Strategy for the synthesis of JCVI-syn3.0 from oligonucleotides. *Source:* Modified from Hutchison et al. 2016. *Science* **351**:aad6253, fig. 2.

Hutchison, C. A., III, R.-Y. Chuan, V. N. Noskov, N. Assad-Garcia, T. J. Deerinck, et al. 2016. Design and synthesis of a minimal bacterial genome. *Science* **351**:aad6253.

A.

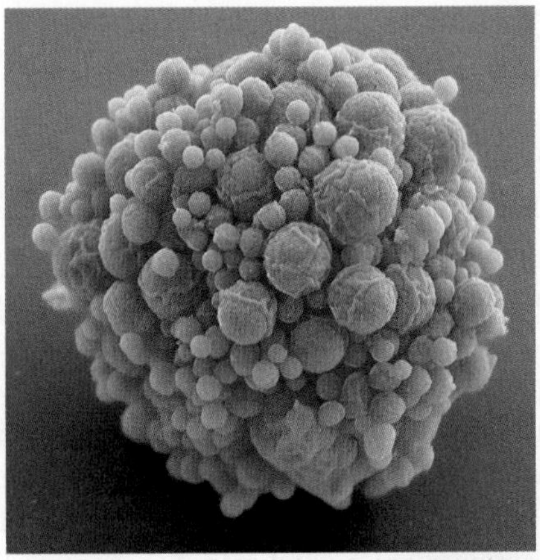

TOM DEERINCK AND MARK ELLISMAN OF THE NATIONAL CENTER FOR IMAGING AND MICROSCOPY RESEARCH AT THE UNIVERSITY OF CALIFORNIA AT SAN DIEGO

B.

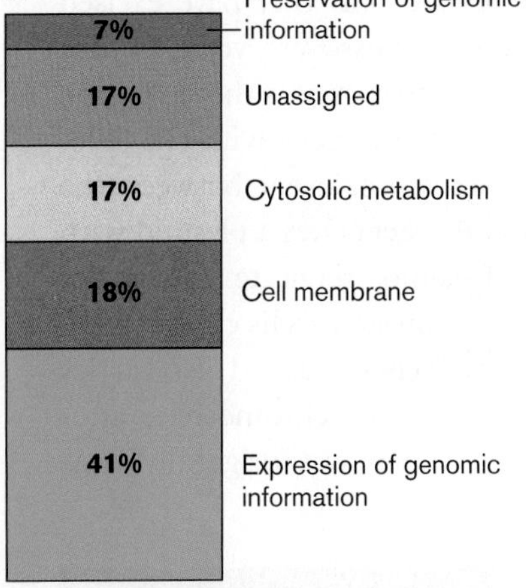

FIGURE 4 ■ Characterization of JCVI-syn3.0. A. A cell cluster (SEM). B. Functional group assignment of the genome. *Source:* Part B modified from Hutchison et al. 2016. *Science* **351**:aad6253, fig. 6.

example shown in **Figure 12.19**, this is the *sacB* gene encoding the enzyme levansucrase. In *Bacillus subtilis*, SacB catalyzes both the hydrolysis of sucrose (a disaccharide of glucose and fructose) and the synthesis of the fructose polymer levan. When expressed in *E. coli* and other microbes, SacB is toxic if sucrose is present. The exact reason for this toxicity remains unclear.

Construction of the deletion mutant requires two sequential selection steps. The first step (**Fig. 12.19A**) is to select for integration of the plasmid into the chromosome. After the suicide vector is delivered to the recipient cell, a crossover event occurs between DNA that flanks the target gene on the plasmid and its homologous region on the chromosome. **Figure 12.19A** shows this intermolecular crossover happening at flanking region B, but it could also occur at flanking region A. The crossover integrates the entire plasmid into the chromosome. A cell containing the integrated plasmid is now resistant to chloramphenicol. Cells with the integrated plasmid are selected on agar media plates containing chloramphenicol.

In the second selection step (**Fig. 12.19B**), the plasmid is excised, and along with it the target gene. Colonies of chloramphenicol-resistant cells are picked from the agar plates, inoculated in liquid medium, allowed to grow, then washed to remove the antibiotic and plated on agar media containing sucrose but not chloramphenicol. The only cells that can grow on these plates are cells that have excised the *sacB* gene through a second crossover event. **Figure 12.19B** shows how the crossover that excises the *sacB* gene also leads to replacement of the target gene with the deletion on the chromosome. A crossover event between the two copies of flanking region A regenerates a plasmid with the *sacB* gene. This plasmid cannot replicate and is not inherited by progeny cells. These progeny cells can grow on sucrose plates. Importantly, the excised plasmid also carries the target gene. What remains on the chromosome after the crossover is the deletion, which was originally constructed on the plasmid.

> **Thought Question**
>
> **12.4** Refer to **Figure 12.19**. What would be the outcome for the chromosome if the second crossover event occurred between the copies of flanking region B, instead of between the copies of flanking region A?

FIGURE 12.19 • Generating a deletion knockout strain using sucrose counterselection.

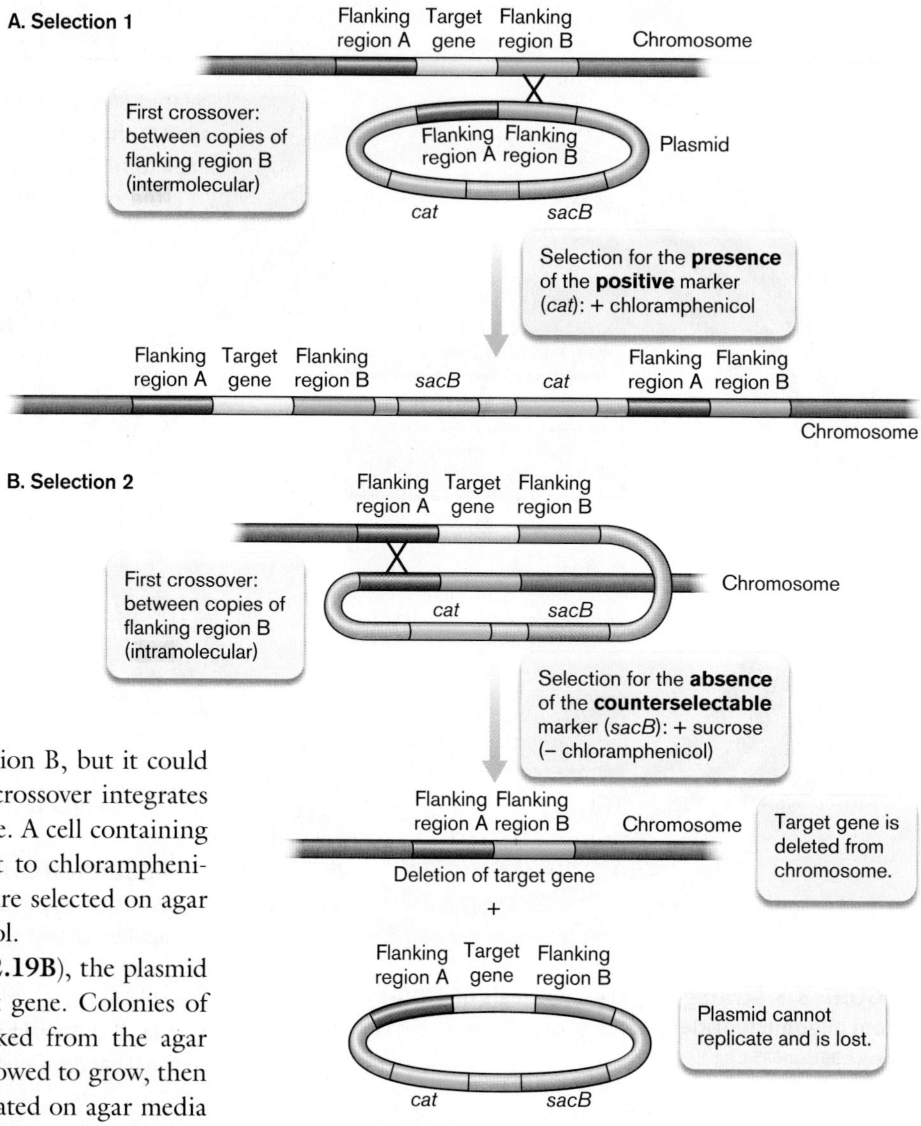

Targeted Gene Editing with CRISPR-Cas9 Technology

In Chapters 6 and 9 we described a form of bacterial and archaeal "immune system" called CRISPR (clustered regularly interspaced short palindromic repeats). Recall that when a bacterial cell survives a viral attack, a short DNA sequence from the virus is captured and inserted into the CRISPR locus as a kind of memory of the attack (see Fig. 9.13). The CRISPR locus includes sequences captured from many viruses, forming a kind of rogues' gallery of viral suspects. The virus-derived sequences are separated from each other by palindromic repeat sequences 32 base pairs in length. Once captured, transcripts of the virus memory sequences are periodically incorporated into a search-and-destroy endonuclease called Cas (Fig. 9.13). If the same virus reinvades the cell at a later date, the CRISPR transcript (called csRNA) guides Cas to the target viral DNA sequence and cleaves it, thereby destroying the virus. This is how the system works in nature.

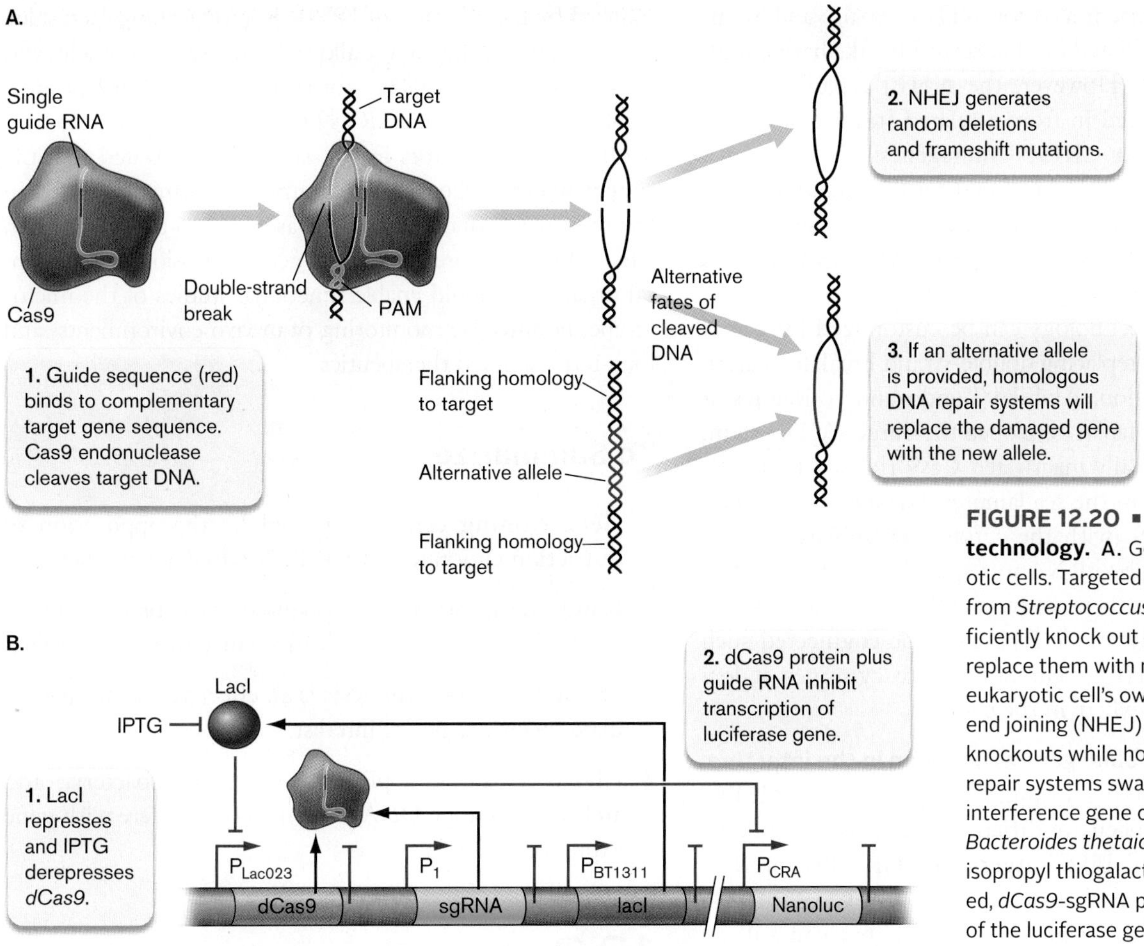

FIGURE 12.20 ▪ CRISPR-Cas9 technology. **A.** Gene editing in eukaryotic cells. Targeted endonuclease Cas9 from *Streptococcus pyogenes* can efficiently knock out eukaryotic genes or replace them with modified forms. The eukaryotic cell's own nonhomologous end joining (NHEJ) enzymes achieve knockouts while homologous DNA repair systems swap alleles. **B.** CRISPR interference gene circuit engineered in *Bacteroides thetaiotaomicron*. When isopropyl thiogalactoside (IPTG) is added, *dCas9*-sgRNA prevents transcription of the luciferase gene Nanoluc.

One CRISPR-Cas system has now been repurposed into a powerful molecular tool that can genetically edit any gene in a eukaryotic cell. The technology has enabled precise gene editing in species where this has never been possible. A striking example of this technology's potential was its use to disrupt latent HIV provirus in infected cells. As described in Chapter 11, HIV DNA made by reverse transcriptase integrates into the host genome. Current drug therapies can stop HIV replication and transmission but will not expunge the provirus from infected cells. Scientists from the Lewis Katz School of Medicine at Temple University used the CRISPR-Cas9 system to excise HIV provirus from the genomes of living animals, offering hope that this system may provide a tool for curing HIV infection in humans.

The system most often used is CRISPR-Cas9 from *Streptococcus pyogenes*. The key to this tool is in programming a synthetic csRNA (now called a guide RNA, or gRNA) to direct Cas9 endonuclease to a specific eukaryotic gene. The programmed gRNA is fused to what is called the tracer RNA, a molecule that binds to and activates Cas9 (**Fig. 12.20**). How are these genes introduced into eukaryotic cells? The gene that encodes Cas9 and the gene for the programmed guide RNA are inserted into plasmids that are introduced into eukaryotic cells via transformation, and the genes are transiently expressed in the nucleus.

The guide RNA binds to the Cas9 endonuclease and helps it recognize the target eukaryotic sequence. Recall from Chapter 9 that this target sequence needs to be adjacent to a three-nucleotide protospacer adjacent motif (PAM), which for Cas9 is 5′-NGG-3′. The CRISPR array lacks the PAM and hence is protected from Cas9 activity. The Cas9 endonuclease introduces a double-strand break in the target DNA—a critical feature of this gene-editing tool. To survive, the eukaryotic cell will try to re-join the blunt ends using nonhomologous end joining (NHEJ) repair (**Fig. 12.20A**; described in Section 9.2). However, the process often deletes a few bases, which may cripple the gene. Once mutated, the effect of the altered gene on cell physiology can be assessed.

The targeted gene can also be changed or replaced using CRISPR-Cas9 technology. In this case, a third plasmid is introduced that expresses a modified version (or allele) of the gene. Both ends of the replacement gene are homologous to DNA flanking the target gene, so homologous recombination systems will replace the broken gene with the modified allele. Any type of insertion, deletion, or change in sequence can be introduced.

CRISPR mutagenesis in bacteria and archaea. The application of CRISPR-Cas9 technology for genetic modification in bacteria and archaea has been more limited than for eukaryotes. One major reason is that many bacteria and archaea lack the machinery for NHEJ repair, and without it the Cas9-mediated double-strand breaks have a high degree of lethality. However, the NHEJ repair machinery can be transferred in from a related strain to improve survival and facilitate CRISPR mutagenesis. For example, CRISPR mutagenesis was improved for the archaeal methanogen *Methanosarcina acetivorans* when the NHEJ repair machinery of the related species *M. paludicola* was introduced.

CRISPR-Cas9 technology can be customized for bacterial gene editing by replacing double-strand-break formation with base substitution. Akihiko Kondo and colleagues at Kobe University in Japan developed the Target-AID system, in which a catalytically inactivated Cas9 is fused to a cytidine deaminase from the sea lamprey *Petromyzon marinus*. The defective Cas9 (with the cytidine deaminase in tow) can still be guided by gRNA to a specific target gene. The cytidine deaminase then converts cytidine residues to thymidine. In this way, guide RNAs can be engineered such that the Target-AID system can mutate cytosines at loci near PAM sites anywhere in the genome.

Manipulating microbial gene expression in the intestine. In addition to making changes to DNA, the CRISPR-Cas9 system has been modified to serve as a regulator of gene expression. Like Target-AID, CRISPR interference (CRISPRi) uses an inactive version of Cas9 (dCas9) that no longer works as an endonuclease but can still be guided to target DNA sequences. Once bound to its target, dCas9 blocks transcription. Timothy Lu's laboratory at the Massachusetts Institute of Technology recently used CRISPRi to control gene expression in *Bacteroides thetaiotaomicron* growing in a mouse intestine. *B. thetaiotaomicron* is an important bacterial member of mammalian gut microbiomes (discussed in Sections 13.3 and 23.2).

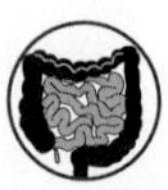

The scientists designed an sgRNA (single guide RNA) to target a luciferase gene that was part of the organism's genome. The sgRNA and the gene for dCas9 were integrated into the microbe's genome (**Fig. 12.20B**). The gene for dCas9 also included the *lacO* operator, which means that dCas9 can be induced by the addition of IPTG (isopropyl thiogalactoside). Without IPTG, the circuit allows luciferase to be made, and cells glow. Adding IPTG, however, induces dCas9 production. The sgRNA guides dCas9 to the luciferase gene to turn it off. The investigators in this study demonstrated that the system worked when the programmed organism resided in the mouse intestine and IPTG was added to drinking water. The ability to precisely modulate gene expression in commensal organisms should enable functional studies of the microbiome, noninvasive monitoring of in vivo environments, and long-term targeted therapeutics.

To Summarize

- **Gene cloning** can be achieved by the application of restriction endonucleases or PCR-based approaches.
- **Random mutagenesis** involving transposons is an effective way to identify genes involved in a microbial activity.
- **Site-directed mutagenesis** is an effective way to manipulate specific genes of interest.
- **CRISPR** can be applied to eukaryotes, bacteria, and archaea to change DNA sequences or interfere with gene expression.

12.3 Gene Expression Analysis

As described in Chapters 8 and 10, microbes can exert sophisticated control over gene expression at the transcriptional, translational, and posttranslational levels. The changes in a gene's expression in response to changes in environmental conditions can provide valuable clues about the potential function of both characterized and uncharacterized genes. **Tables 12.2** and **12.3** summarize some of the major techniques that microbiologists employ to study gene expression. Many of these techniques are described elsewhere in the book and will be only briefly summarized here. We begin with a description of the methods to study RNA and

TABLE 12.2 RNA and Transcription Analytical Techniques

Technique	Designed use	Location in book
Primer extension	Identify transcription start site	**eTopic 12.1**
Northern blot	Determine transcript size, abundance	**eTopic 12.1**
qRT-PCR	Rapidly determine transcript abundance	Section 28.2
DNA microarray	Transcriptomics	eAppendix 3.8
RNAseq	Transcriptomics, metatranscriptomics	Sections 10.2 and 21.5; **eTopic 12.5**
Transcriptional fusion	Large-scale gene expression studies, single-cell gene expression studies, genetic screens for regulators	Section 12.3; **eTopic 12.1**

transcription and will finish this section with a description of methods to study protein and posttranslational processing.

RNA and Transcription

Table 12.2 lists several common techniques currently employed to study RNA and transcription regulation. Targeted approaches including primer extension, northern blot, and quantitative reverse-transcription PCR (qRT-PCR) are designed to analyze individual genes and their RNA products. Primer extension (**eTopic 12.1**) is useful to characterize transcription start sites and the 5′ untranslated regions of operons. Northern blots (**eTopic 12.1** and eAppendix 3) can be used to quantify transcripts, and they are also useful in determining which genes are cotranscribed in the same operon. Quantitative reverse-transcription PCR (Section 28.2) is a fast and accurate method to quantify RNA transcripts, but it usually does not reveal information on transcript length.

In contrast to these targeted approaches, "omic" approaches are useful for analyzing transcripts of all genes in a single strain (transcriptomics) or a community (metatranscriptomics). DNA microarrays (eAppendix 3) were the first devices used in the analysis of transcriptomes, but because of the ever-shrinking price of DNA sequencing, this approach is being largely replaced with high-throughput sequencing of the RNA transcripts using RNAseq. RNAseq can provide genome-wide transcription information as a transcriptome and can even be used in mixed microbial communities to assess gene expression as metatranscriptomes (see Sections 10.2 and 21.5, respectively, for examples). Finally, as described in **eTopic 12.5**, a modification to the RNAseq technique can also be exploited to map transcription start sites.

All of the techniques just described require purification of RNA transcripts from the cells, a process that can be tedious and prone to contamination by ribonucleases, enzymes that hydrolyze RNA. These techniques perform well if the sample size is fairly small, but if transcription is to be monitored over a prolonged time series, or used for a genetic screen, a reporter fusion is often the preferred approach. This technique involves linking (or fusing) the promoter of the gene of interest to what is called a promoter-less **reporter gene**, such as *lacZ* (which encodes beta-galactosidase) or *gfp* (which encodes GFP, green fluorescent protein; see Chapter 2). Reporter genes encode proteins whose activities are easily assayed. In the case of *lacZ*, beta-galactosidase converts the substrate *o*-nitrophenyl galactoside (ONPG) to a yellow product that is easily measured with a spectrophotometer. Fusions to GFP are detected by a fluorometer, which detects fluorescence emitted by the GFP fusion protein.

A promoter-less reporter gene fused to the 3′ end of a target gene produces an artificial operon (**Fig. 12.21A**). Because there is only one promoter, only one mRNA transcript is made. The 5′ end of the newly engineered polycistronic mRNA will contain the target gene transcript, followed at the 3′ end by the reporter gene transcript. Now, any factor that controls expression of the target gene's promoter will also control production of the reporter. Because the reporter retains its own ribosome-binding site, it will not respond to translational controls over the target transcript. The fusion just described is known as a **transcriptional fusion** (**Fig. 12.21A**), which can monitor transcriptional control of the target gene.

Each type of fusion can be constructed in vitro using plasmids and then transferred into recipient cells in a way that promotes the exchange of the reporter fusion for the resident gene. Once constructed, these fusions enable the

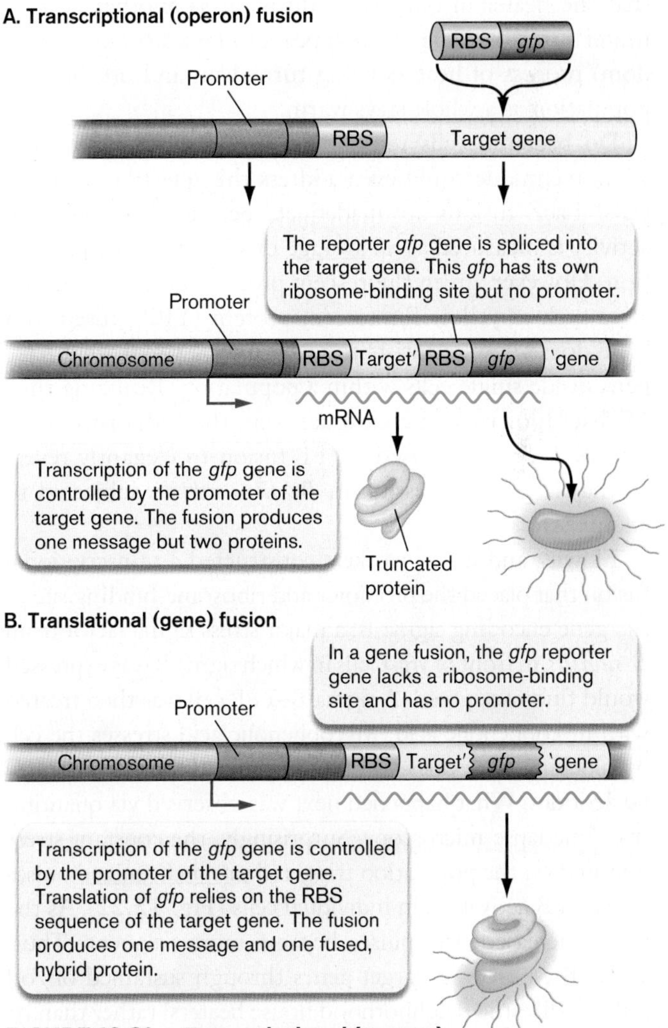

FIGURE 12.21 ▪ Transcriptional (operon) and translational (gene) fusions. Aspects of a gene's regulation can be monitored by fusing a second gene, one that produces an easily assayed protein, to the gene under study. **A.** Construction of a transcriptional fusion. **B.** Construction of a translational fusion. The *gfp* open reading frame must be fused in-frame to the amino-terminal end of the target gene.

researcher to monitor how different growth conditions influence the transcription or translation of the gene, or whether specific suspected regulators are involved. These fusions provide convenient phenotypes (for example, lactose fermentation or fluorescence) that can be exploited in mutant hunts to screen for new regulators. **eTopic 12.1** provides an example of how *lacZ* fusions can be used in genetic screens to find regulatory genes.

Gene expression in single cells. When a population of bacteria experiences a type of stress, do all cells simultaneously activate their stress response genes? And once activated, do the genes stay on until the stress dissipates? Think of your own house as a single cell. When it becomes cold outside (and then inside), the thermostat activates your heater. At some point, however, the heater turns off, and then it turns on again whenever the inside temperature dips. Now imagine looking down at a neighborhood of houses. You see that the heater in one house turns on as another house's heater turns off. The result appears to be a stochastic (random) process of heaters being turned on and off, but the population as a whole stays warm.

Do bacterial cells respond to stress in the same way? What technique could even address this question in single cells? LacZ fusions are inadequate because LacZ enzyme activity is measured as an average of all cells in the population. However, green fluorescent protein (GFP), or a derivative such as yellow fluorescent protein (YFP), fused to a gene does allow scientists to view (in real time) what happens inside single cells within a population. Realizing this, Michael Elowitz and colleagues from the California Institute of Technology used a YFP fusion to elegantly reveal stochastic gene expression in *Bacillus subtilis* subjected to energy stress.

Elowitz and his co-workers constructed a transcriptional fusion that placed the promoter and ribosome-binding site of the gene encoding sigma B, a major stress sigma factor from *B. subtilis*, in front of *yfp*. Cells in which sigma B was expressed would fluoresce green. A sigma B–YFP cell was then treated with mycophenolic acid. Mycophenolic acid stresses the cell by blocking the synthesis of guanosine triphosphate (GTP) nucleotides. What happened next was observed via quantitative time-lapse microscopy. Surprisingly, the constant stress imposed on the population triggered unsynchronized pulses of sigma B activation in individual cells (**Fig. 12.22**). As the stress increased, the pulses became more frequent. Thus, sigma B controls its target genes through sustained on/off pulsing (like the neighborhood house heaters) rather than by continuous activation.

As this example illustrates, GFP fusions are more versatile than LacZ fusions in many ways. GFP fusions can be used to track gene expression, to identify the locations of proteins inside cells (see below), and to follow the movement of pathogens within host animals.

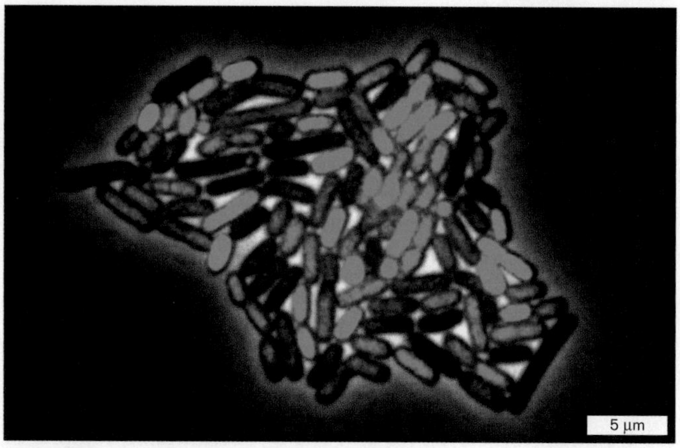

FIGURE 12.22 ■ Flashing bacteria. YFP fusion (manipulated to appear green) revealed the stochastic expression of the sigma B gene among single cells of *Bacillus subtilis* subjected to energy stress. Individual cells flash on and off much like a thermostat-controlled house heating system. A movie of the process can be seen at: http://science.sciencemag.org/highwire/filestream/592492/field_highwire_adjunct_files/0/1208144s1.mov.

Thought Question

12.5 How could you quickly separate cells in the population that express sigma B–YFP from those that do not express this fusion? And what could you learn by separating these subpopulations? *Hint:* A technique presented in Chapter 4 will help.

Protein Abundance and Function

As with RNA and transcription, analytical methods to study proteins, translation, and posttranslational processing (**Table 12.3**) fall into two categories: targeted approaches and "omic" approaches. One way to target specific proteins for expression analysis is to raise antibodies against them. Western blotting, which uses antibodies to detect proteins separated on polyacrylamide gels, is discussed in **eTopic 12.1**. Antibodies can also be covalently linked to fluorophores or gold particles, enabling them to reveal the intracellular locations of their target proteins by fluorescence or transmission electron microscopy (see Chapter 2).

Untargeted "omic" approaches for protein detection and quantification have traditionally relied on general protein stains such as Coomassie blue or the incorporation of ^{35}S-labeled methionine, followed by 1D or 2D gel electrophoresis for protein separation (see eAppendix A.3). Proteins of interest could then be cut out of the gel and identified by mass spectrometry. Current mass spectrometry technology enables researchers to bypass the gel electrophoresis step and obtain a proteome for the organism (Section 8.4), or even a metaproteome for a microbial community (Chapter 21). Coupled with specialized preparatory techniques, these technologies can also assess

TABLE 12.3 Protein and Translation Analytical Techniques

Technique	Designed use	Location in book
Western blot	Determining size, abundance of specific proteins	**eTopic 12.1**
Labeling with fluorophore- or gold particle–tagged antibody	Fluorescence or electron microscopy, localizing specific proteins in cells	Sections 2.4 and 2.6
Coomassie blue staining	Generic protein labeling for 1D and 2D gel electrophoresis of proteome	Section 12.3
^{35}S methionine labeling	Generic, translation-dependent labeling of proteins for 1D and 2D gel electrophoresis of proteome	eAppendix 3.3
Mass spectrometry	Protein sequencing, proteomics, metaproteomics, posttranslational modification detection	Sections 8.4 and 21.5
Translational fusion	GFP-mediated fluorescence tracking of protein abundance and location in living microbes	Section 12.3
Electrophoretic mobility shift assay (EMSA)	Assaying DNA-binding capacity of proteins such as transcription regulators	**eTopic 12.1**
DNase I footprinting	Determining the binding-site sequence of DNA-binding proteins	**eTopic 12.1**
ChIP-seq	Genome-wide screening for DNA-binding sites of a transcription regulator	Section 12.3
Two-hybrid analysis	Identifying proteins that bind other proteins of interest	Section 12.3

posttranslational modifications such as phosphorylation and acetylation at the proteomic scale (Section 8.4).

GFP fusions track proteins in live cells. The techniques that we have described thus far are very powerful, but they are unable to track proteins in real time in living organisms. A major innovation for the quantification and localization of proteins is the green fluorescent protein **translational fusion**. The GFP gene lacks both its promoter and its ribosome-binding site, and relies on the target gene to which it is fused for both transcription and translation. For GFP to be synthesized, the *gfp* sequence must be joined to the target gene sequence in the proper codon reading frame (**Fig. 12.21B**). Now, not only will the mRNA transcripts of the target and reporter genes be fused, but the peptides of the truncated target gene and the reporter gene will also be linked. In this case, anything controlling the transcription or translation of the target gene will also control GFP levels. Likewise, molecular cues that dictate the location of the target gene will also control the location of the GFP signal.

GFP technology can be used to target how microbial proteins move through bacterial cells (Section 2.4) and eukaryotic cells. For instance, the potato virus X (or potexvirus) is a positive-strand RNA virus (a positive-strand RNA is equivalent to mRNA). One viral protein, called TBGp3, is required for the virus to move between plant cells. Investigators fused GFP to the TBGp3 protein and asked where the protein ended up in the plant leaves. They delivered the genes to plant cells by bombarding leaves with the fusion plasmids. As shown in **Figure 12.23**, GFP

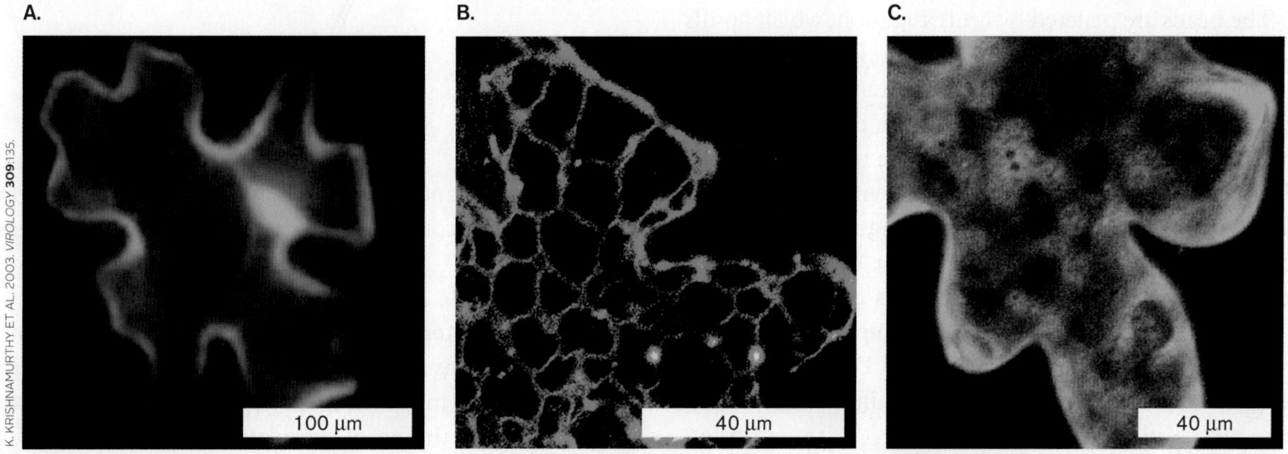

FIGURE 12.23 ■ Using GFP protein to track the potexvirus protein TBGp3 in a plant leaf. **A.** Plasmid containing GFP only. Fluorescence is distributed throughout the leaf. **B.** Plasmid containing GFP-TBGp3 fusion. Fluorescence is restricted to the reticulated network of the leaf. **C.** Plasmid containing mutant GFP-TBGp3 fusion gene. The mutant protein is not properly targeted to the reticulated network.

alone was diffusely located throughout the leaf. However, the GFP-TBGp3 fusion protein inserted only into the reticulated network of veins. Altering specific residues within the protein via mutation prevented this localization, causing the fusion protein to remain diffused throughout the plant cells. These results suggest that TBGp3 protein helps guide the virus through the plant's reticulated network.

This type of gene fusion technology has been used extensively to trace the movement and final location of numerous proteins delivered into host cells by pathogenic bacteria. GFP tagging has also been used to track the intermingling of different bacteria forming a biofilm (**eTopic 12.6**).

Thought Question

12.6 You have monitored expression of a gene fusion in which *gfp* is fused to *gadA*, the glutamate decarboxylase gene of *E. coli*. You find a 50-fold increase in expression of *gadA* (based on fluorescence level) when the cells carrying this fusion are grown in media at pH 5.5, compared to pH 8. What additional experiments must you do to determine whether the control is transcriptional or translational?

Whole-genome DNA-binding analysis: ChIP-seq. The techniques discussed in **eTopic 12.1** are great for identifying a site on DNA where you already suspect that a protein binds. But is there a way to determine all the sites in a genome to which a given protein binds? There is, and the basic process, called **ChIP sequencing** (**ChIP-seq**), is outlined in **Figure 12.24**. In the cell, the DNA-binding protein of interest, such as the stationary-phase sigma factor of *E. coli*, sigma S (Chapter 8), binds to all of its DNA target sites in the genome. The cells are treated with formaldehyde to covalently cross-link proteins to DNA. The DNA and its bound proteins are isolated and sheared into small fragments. Next, the transcription factor–DNA complexes are "fished" from the extract using tiny beads with antibodies attached that bind only to that particular TF protein. The beads are pelleted by centrifugation, which "pulls down" antibody bound to the TF protein–DNA complexes. Unbound DNA fragments are washed away. This process is called **chromatin immunoprecipitation**, or **ChIP**.

Next, the protein-DNA cross-links are removed (usually by heat), and the released DNA fragments are amplified by PCR. (To amplify the unknown fragments, the fragments are ligated at both ends to linker oligonucleotides that can be amplified by known primers containing fluorescent dye.) The fluorescently tagged, amplified fragments are then sequenced by massively parallel DNA sequencing (next-generation sequencing, discussed in Section 12.1). The sequence identifies precisely where the transcription factor bound to the genome and which genes it likely controls. In the case of sigma S, ChIP-seq identified 63 binding sites and discovered that several new genes in oxidative stress resistance and cell-surface chemistry are under sigma S control and could be important for stationary-phase survival.

FIGURE 12.24 ■ ChIP-seq technology. TF = transcription factor.

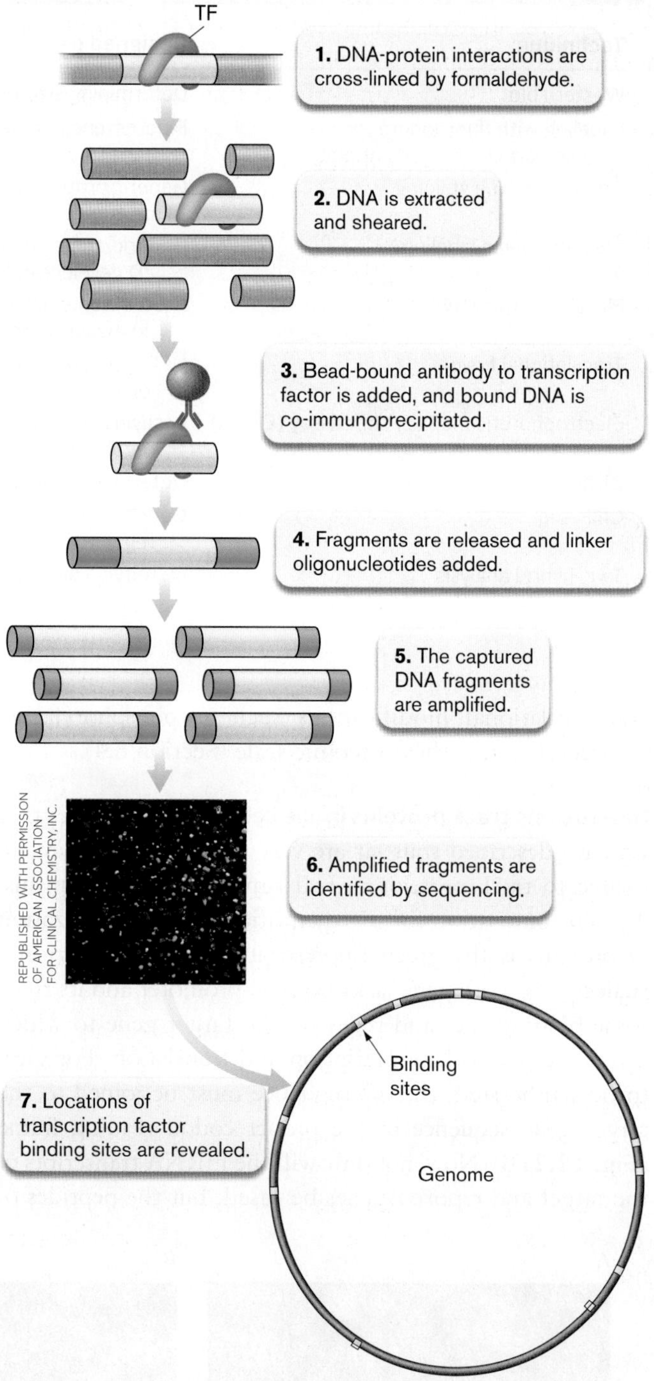

Mapping the interactome: protein-protein interactions. Many cell proteins interact with and influence the function of other proteins in the cell. For example, anti-sigma factors help control the timing of sporulation in *Bacillus* by interacting with sigma proteins. Many proteins, such as RNA polymerase and the ribosome, function as components of multisubunit complexes. In some intracellular situations, however, protein complexes can simplify physiological processes. A complex containing fatty acid biosynthesis enzymes, for instance, allows intermediates to pass quickly from one subunit to another without having to diffuse through the cell. Thus, all the protein-protein interactions of a living cell, which are collectively referred to as the **interactome**, are essential to the normal workings of the cell. Unraveling these interactions is an invaluable way of understanding protein function. How, then, does one map an interactome?

It is important to probe protein-protein interactions in vivo, where the proteins actually work. An ingenious tool to measure in vivo protein-protein interaction is **two-hybrid analysis**. Two-hybrid analysis can be used to mine for unknown "prey" proteins that interact with a known "bait" protein.

While there are many types of two-hybrid techniques, the classic example is the yeast two-hybrid system shown in **Figure 12.25**. In yeast two-hybrid analysis, genes encoding two potentially interacting proteins are fused to separated parts (domains) of the yeast GAL4 transcription factor, which is normally one protein (**Fig. 12.25**). One gene is fused to the GAL4 activation domain, while the other gene is fused to the GAL4 DNA-binding domain. If the two proteins of interest interact with each other, then the two parts of GAL4 are brought together. The DNA-binding domain binds to the yeast *GAL1* gene promoter region, and the activation domain (towed behind by the interacting proteins) is positioned to bind RNA polymerase and stimulate *GAL1* transcription.

The target reporter gene is typically a chromosomal *GAL1-lacZ* transcriptional fusion, although other reporter genes can be used. When the two hybrid proteins interact, the complex activates the transcription of *GAL1-lacZ*, which is visualized on agar plates containing X-Gal. X-Gal is a colorless substrate of beta-galactosidase that, when cleaved, produces a blue product.

The power of this technique is that it can also be used to find unknown proteins that interact with a known protein. A known protein fused to one of the GAL4 domains can be used as "bait" to find the "prey" proteins expressed by randomly cloned "prey" genes fused to the other GAL4 domain. Plasmids containing the bait fusion and prey fusion are transformed into yeast cells, and the transformants are plated onto a medium containing X-Gal. Colonies that express beta-galactosidase are the result of protein-protein interactions between the bait and prey fusion proteins. Sequencing the DNA insertion can then identify the gene encoding the prey protein.

Thought Question

12.7 You suspect that two proteins, A and C, can simultaneously bind to a third protein, B, and that this interaction then allows A to interact with C in the complex. How could you use two-hybrid analysis to determine whether these <u>three</u> proteins can interact?

Another approach to mapping interactomes uses individual bait proteins to pluck <u>all</u> of a particular protein's interacting partners out of the cell. A small-peptide affinity tag added to one end of the bait protein will enable the bait protein and its interacting proteins to be purified by affinity chromatography. The process is described in **eTopic 12.7**.

To Summarize

- **RNA properties** such as abundance and length can be determined through targeted or transcriptomic techniques, depending on the objective.
- **Protein properties** such as abundance, localization, and posttranslational processing can be determined through targeted or proteomic techniques, depending on the objective.
- **Transcriptional and translational fusions** use reporter genes like lacZ or GFP to monitor transcription and translation, respectively, of target genes.

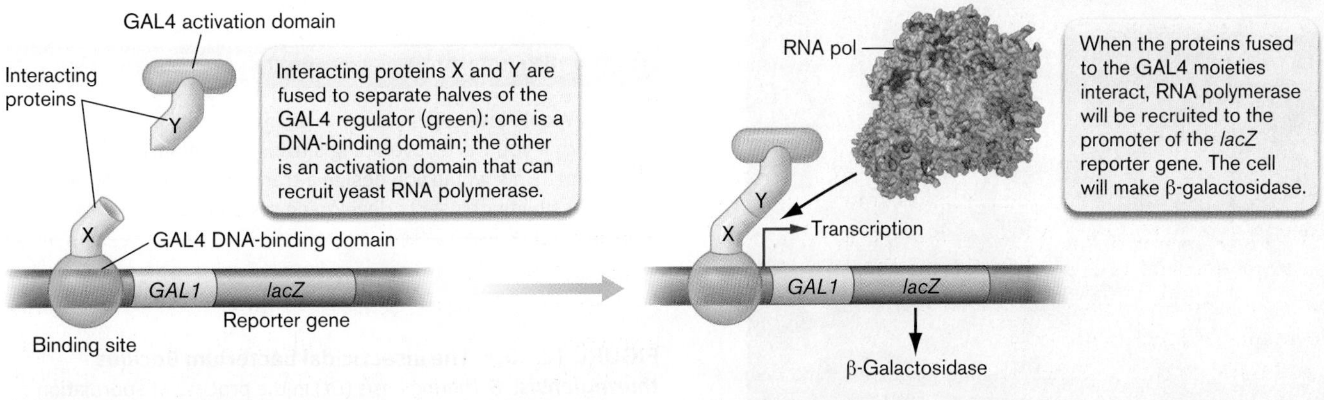

FIGURE 12.25 ▪ Detecting protein-protein interactions: the yeast two-hybrid system. Interacting proteins are fused to separate halves of the GAL4 regulator.

- **A protein's location or movement** can be visualized in live cells by fusing it to green fluorescent protein (GFP).
- **ChIP-seq technology** can identify all of the sequences in a genome to which a given protein will bind.
- **Two-hybrid analysis** uses hybrid proteins to detect protein-protein interactions, and it can be used to build protein interaction maps.

12.4 Applied Biotechnology

It is remarkable how many useful products are being developed with the tools of biotechnology (see Chapter 16). You may be surprised to learn that plants have been engineered to produce vaccines, and that vaccines can be delivered through not only the injection of a protein but also the injection of simply the DNA or RNA that encodes the antigenic protein (DNA vaccines are discussed in **eTopic 12.8**). We even have technology that uses a genetically engineered virus to repair a human genetic defect (**eTopic 12.9**). These are all remarkable additions to our scientific toolbox, and they are already helping to improve human life.

Bacterial Genes Save Crops

Protecting crops such as corn and cotton from hungry insects is an age-old problem. Insects such as moths or beetles can cause widespread damage to a crop in a short period of time. In the nineteenth and twentieth centuries, chemical pesticides were widely used to kill these pests. While generally effective, chemical pesticides can contaminate soil and groundwater and can negatively impact human health. Through bioengineering, we now have a better, safer way.

Bacillus thuringiensis (*Bt*), a close relative of *Bacillus anthracis*, sporulates on the surfaces of plants. In addition to the spore, the sporulating cell makes a separate parasporal body that cradles the spore (**Fig. 12.26A**). The parasporal body contains crystallized proteins that are toxic to insects feeding on the plant. A single subspecies of *B. thuringiensis* can produce multiple insecticidal proteins. After the insect or insect larva ingests the insecticidal protein crystals, the alkaline environment in the insect midgut dissolves the crystals, and insect proteases inadvertently activate the proteins. Activated insecticidal proteins insert into the membrane of the midgut cells and form pores that lead to a loss of membrane potential, cell lysis, and death of the insect through starvation. It is also proposed that death is due to septicemia by enteric bacteria that escape the damaged midgut (antibiotic treatment reduces death). Because of its effectiveness, farmers have taken to spreading *Bt* directly on their crops, where even dead cells are effective.

Bt is considered a highly beneficial pesticide with few downsides. Unlike most insecticides, *Bt* insecticides do not have a broad spectrum of activity, so they do not kill animals or beneficial insects—including the natural enemies of harmful insects (predators and parasites) and beneficial pollinators (such as honeybees). Therefore, *Bt* integrates well with other natural crop controls. Perhaps the major advantage is that *Bt* is nontoxic to people, pets, and wildlife, so it is safe to use on food crops or in other sensitive sites where chemical pesticides can have adverse effects.

Although spreading *Bt* on crops is helpful, a more direct means of delivering the toxin has been developed. The genes encoding insecticidal proteins have been spliced right into the genomes of agriculturally important plants such as cotton and corn. The gene is placed under the control of a plant promoter, so that the transgenic plant makes its own insecticide. As of 2013, over 75% of the cotton and corn planted in the United States contained *Bt* genes.

Thought Question

12.8 Would insect resistance to an insecticidal protein be a concern if you were developing a transgenic plant? Why or why not? How would you design a transgenic plant to limit the possibility of insects developing resistance?

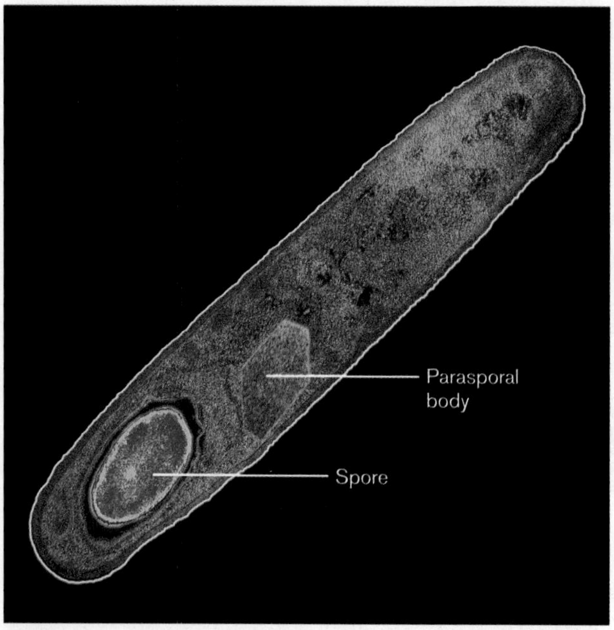

DR. GEORGE CHAPMAN/VISUALS UNLIMITED, INC.

FIGURE 12.26 ▪ The insecticidal bacterium *Bacillus thuringiensis*. *B. thuringiensis* (*Bt*) in the process of sporulation, and displaying a parasporal crystal, that can be harvested and used as insectiside.

TABLE 12.4 Potential Vaccine Antigens Expressed in Plants

Source of protein	Vaccine protein or peptide	Plant
Enterotoxigenic *E. coli*	Heat-labile enterotoxin B subunit (LT-B)	Tobacco, potato, tobacco chloroplast, maize kernels
Vibrio cholerae	Cholera toxin B subunit (CT-B)	Potato, tobacco chloroplast
Hepatitis B virus	Hepatitis B surface antigen	Tobacco, potato
Norwalk virus	Norwalk virus capsid protein (NVCP)	Tobacco, potato
Rabies virus	Glycoprotein	Tomato
Foot-and-mouth disease virus	Viral protein I	Alfalfa
Clostridium tetani	TetC	Tobacco chloroplast

Vaccine Antigens Produced in Plants

Starting with the war against smallpox in the late 1700s, vaccines have been used to inoculate humans against many terrible diseases (see Chapters 25 and 26). However, the cost of producing vaccines is enormous, and the price, of course, is passed on to those who are vaccinated or to governments that conduct vaccination programs. Because the cost is problematic for impoverished nations, 20% of the world's infants are not vaccinated properly, resulting in over 2 million preventable deaths annually.

One potential solution to this problem was to engineer plants to express vaccine proteins. The gene encoding a vaccine protein is inserted into the genome of a plant so that the plant makes the antigenic protein as part of itself. The vaccine gene must be placed under the control of a well-expressed plant promoter (recently, plastid genes have been used). Transgenic plants expressing bacterial or viral virulence proteins can be used in two ways: They can be direct vaccine delivery systems, as through edible fruits and vegetables (such as potatoes, tomatoes, and corn); or, as with transgenic tobacco crops, they can be a cheap way to grow large amounts of the antigenic protein that can be extracted, purified, and delivered in more conventional ways. Producing vaccine in edible fruits is an attractive idea because the fruit could be grown locally and distributed to residents of poor countries, even without refrigeration. But practical and ethical problems have, so far, prevented implementation. However, engineered plants are currently being used to mass-produce vaccine antigens, especially vaccines for use in animals. Transgenic plants and the vaccines they produce are listed in **Table 12.4**.

Phage Display Technology

Phage display is a technique in which DNA sequences encoding nonphage peptides are cloned into a phage capsid gene. These peptides are synthesized as part of the capsid protein and then "displayed" on the surface of the phage (**Fig. 12.27A**). The DNA sequences can code for random peptides, antibodies cloned from natural sources, or libraries (collections) of mutant enzymes that one wishes to study. Phage display enables scientists to screen for peptide variants with altered properties such as increased affinity for ligand.

The phage vectors used most often are the filamentous bacteriophages fd and M13 (discussed in **eTopic 11.2**). These phages infect *E. coli* strains containing the F plasmid (the phages attach to the sex pili) and replicate without killing the host cell. The phage particle consists of a single-stranded DNA molecule encapsulated in a long, cylindrical capsid made up of five proteins. The product of gene 3 (g3p)—present in three to five copies—confers phage infectivity. Most phage display libraries have been produced by cloning into gene 3. Frances H. Arnold, George P. Smith, and Sir Gregory P. Winter shared the 2018 Nobel Prize in Chemistry for their phage display work.

Peptides displayed on filamentous phages have been used to select peptides or antibodies with high affinities for receptors or antigens (see Section 24.2). Recombinant phages that display high-affinity binding peptides are isolated from libraries in a process called biopanning (**Fig. 12.27B**). In biopanning, the target ligand—for example, a human virus attachment protein—is attached to a plastic plate or other support, and phages displaying random peptides are added. The phages that do not stick to the human virus protein are washed away. Those that do bind are subsequently released by the addition of free ligand (for example, human virus protein). This process is analogous to the way miners would pan rivers for gold.

Several rounds of biopanning are required to obtain high-affinity clones. Following each round, the selected phages are enriched by repropagation in *E. coli*. Phage display was successfully used to identify high-affinity peptides that bind to attachment proteins of some eukaryotic viruses. These virus-binding peptides effectively block viral infection. More recently, phage display was used to identify specific single-chain antibodies that bind to a surface protein uniquely present on cancer cells. The antibody disrupts

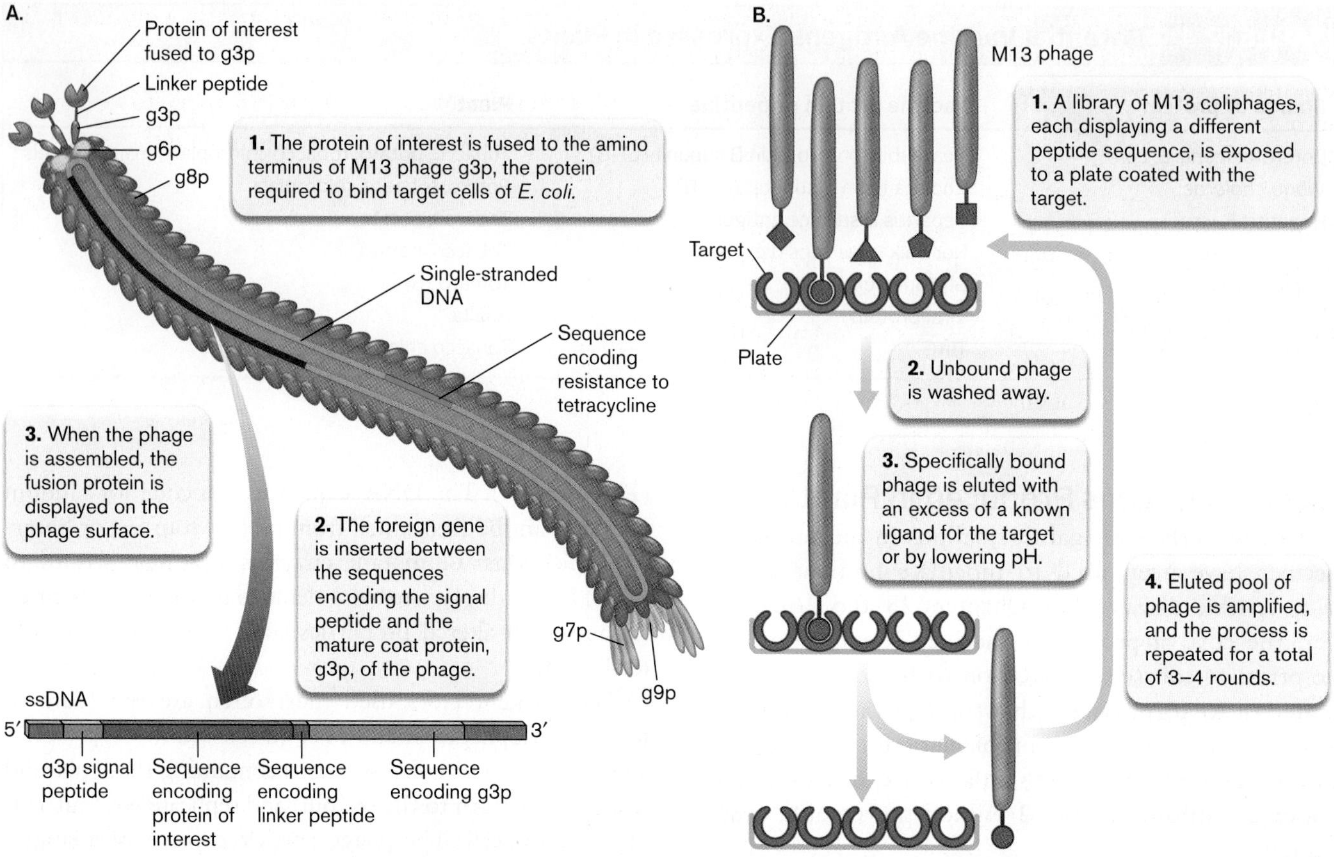

FIGURE 12.27 ▪ Phage display technology. Phage display is a tool for the directed evolution of peptides. The discovery of peptides that bind to specific target proteins is simplified by phage display techniques. **A.** Construction of the phage display protein. The product of gene 8 (g8p) is a small protein (5.2 kDa) that forms the cylinder of the capsid. The other coat proteins (g3p, g6p, g7p, and g9p) cap the ends of the cylinder. The g3p protein is an attachment protein. **B.** Biopanning a library of M13 display phage for a peptide that specifically binds to a target molecule.

the function of this protein and significantly decreases viability of the cancer cell.

Phage display can also be used for directed evolution projects that retool enzymes (directed evolution is discussed in Chapter 17). Natural enzymes, because they were selected by evolution to sustain life, do not necessarily have the stability or catalytic activity that would make them suitable for more harsh biotechnological applications. Phage display, with its ability to screen millions of mutants, can quickly find the variant with the desired trait. An example of how phage display was used to direct the evolution of a more effective beta-lactamase enzyme that cleaves penicillin is described in **eTopic 12.10**. Other in vitro evolution techniques—DNA shuffling and site-directed mutagenesis—are described in **eTopics 12.4 and 12.11**.

In addition to what we describe here, microbes have been genetically engineered for many other applications that we will discuss in later chapters. These applications include the bioremediation of organic wastes (Chapter 13) and inorganic wastes (Chapter 14); mine leaching (Chapter 14); engineering of producer microbes (Chapter 16); and oil-spill cleanup, mercury removal, and wastewater treatment (**eTopic 22.2**).

Biocontainment: "Who Let the Bugs Out?"

Since the advent of genetic engineering in the 1970s, scientists and ethicists have struggled with how to prevent the accidental escape of a genetically engineered microbial "Frankenstein" from the lab. Physical containment such as air locks at the entrances of some labs is necessary to prevent the escape of lethal pathogens (see Section 28.1), but what about genetically modified organisms (GMOs)? The potential consequences of releasing GMOs into an environment are largely unknown. A GMO designed for laboratory use but accidentally released into the natural environment might alter ecological balances through competition or by sharing modified genes with other organisms in the environment. A highly improbable, but not inconceivable, concern is that a GMO microbe could evolve to become pathogenic to plants, animals, or humans. With these unknowns in mind, might there be a way to efficiently contain these organisms without resorting to extreme physical containment strategies? For instance, could we engineer the organisms to quickly self-destruct if they escape the laboratory environment?

Existing biocontainment strategies employ natural auxotrophies (growth requirements) that are not easily met in nature, or conditional suicide switches introduced into the organism's genome (described in Section 12.5). But many of these mechanisms can be compromised through metabolic cross-feeding or genetic mutation. However, a promising new approach involving a bioengineered *synthetic* auxotrophy has emerged from the Yale University laboratory of Farren Isaacs. This team of scientists altered the genetic code of an *E. coli* strain to require an unnatural amino acid found only in laboratories. To recode the bacterium, they replaced all UAG stop codons with UAA stop codons. They then converted UAG into a sense codon by introducing an orthologous pair, consisting of aminoacyl-tRNA synthetase and tRNA, that is derived from the archaeon *Methanocaldococcus jannaschii*. The tRNA recognizes the UAG codon, and the amino acid–binding pocket of the tRNA synthetase was modified to accept the unnatural amino acid *p*-azido-L-phenylalanine (pAzF). The new tRNA synthetase will charge the UAG tRNA with pAzF.

Finally, UAG codons were introduced into three essential genes encoding DnaA (initiation of replication), MurG (peptidoglycan synthesis), and SerS (serine aminoacyl-tRNA synthetase). For this strain to grow, the unnatural amino acid pAzF must be incorporated by translation into those three essential proteins. Because pAzF is not found in the environment, any recoded microbe released from the lab will die and not pose a threat. Synthetic auxotrophies like the one just described will enable us to control when and where genetically engineered microbes of unknown risk can grow.

To Summarize

- **Genes encoding insecticidal proteins** from *Bacillus thuringiensis* (*Bt*) have been engineered into plant genomes to protect crops from insect devastation.
- **Proteins from pathogenic bacteria and viruses** cloned into plants may provide a cost-effective means to vaccinate large populations.
- **Phage display** is a form of in vitro evolution in which DNA sequences encoding random peptides are cloned into phage capsid protein genes. Phage particles displaying peptides with a desired binding property are pulled from the general phage population by biopanning.
- **Biocontainment** strategies are necessary to counteract an accidental release of a genetically engineered microbe into the natural environment.

12.5 Synthetic Biology: Biology by Design

What if we could, with all our knowledge of cells, genes, and gene circuits, make bacteria do things they wouldn't ordinarily do, such as sensing explosives in mines and then telling us about it? Or detecting pathogens? Or keeping time? Or even, as Special Topic 7.1 highlights, storing information—much like a computer? These are all goals of **synthetic biology**. Synthetic biology is the design and construction of new biological parts, devices, and systems for a desired purpose, using principles from electrical engineering.

How do we make these new parts or components? All living organisms already contain an instruction set encoded in DNA that determines what the creature looks like and what it does. Humans have been altering the genetic codes of plants and animals for millennia by selectively breeding individual plants and animals with desirable features. Today, scientists easily take pieces of genetic information from one organism and insert them directly into another, bypassing the need to breed. This is the basis of genetic engineering.

Recent technological advances now enable scientists to synthesize and manipulate DNA in ways never before possible. By applying engineering principles to these genetic manipulations, researchers can take components of genes from several different organisms, link them together, and design customized organisms that do new things. For instance, Jay Keasling's laboratory (UC Berkeley) used genes from several species to engineer a new pathway for artemisinin, an important antimalarial drug. By exploiting *E. coli*, Keasling's team eliminated the effort required to chemically synthesize a structurally complex molecule.

In another example, Chang Li and colleagues (Harvard Medical School) engineered an *E. coli* strain that prevented cancer in mice. The new strain contained the invasin gene from *Yersinia pseudotuberculosis* and the hemolysin gene from *Listeria monocytogenes*. The invasin gene enabled the modified *E. coli* to invade mouse cells. The hemolysin enabled delivery of an inhibitory small RNA molecule that decreased expression of a tumor-initiating gene. Amazingly, this novel *E. coli* prevented cancer in this mouse model.

Principles of Synthetic Biology

The key to synthetic biology is engineering. Engineering a life form from scratch is one of the most impressive feats that synthetic biology has accomplished to date (see **Special Topic 12.1**). On a smaller scale, a suite of genes is introduced into the host organism along with sophisticated mechanisms for regulation. For example, bioengineers turned *E. coli* into an alarm system that flashes light in the presence of toxic metals such as arsenic (**eTopic 12.12**).

Which engineering principles are employed by synthetic-biology scientists? Many of the principles are borrowed from the field of electrical engineering and involve "logic gates." An electronic logic gate (as in semiconductors) receives a tiny current as an input and produces voltage as an output. A genetic logic gate is a promoter and a gene. An input signal such as a regulatory protein affects the promoter, which drives the output signal (mRNA and protein). The output signal for one gate can be an input signal for another part of the logic circuit. These logic gates can be combined to produce biologically useful switches that control gene expression and cellular activity. Toggle, oscillator, and kill switches are examples that we will describe after introducing the types of genetic logic gates that compose them. As we will show, these switches can be used to make cells glow, keep time, or die if they are released accidentally into the environment.

Many of the components used in building an electronic circuit have biological counterparts that are used for building complex genetic circuits. **Figure 12.28** shows three common logic gates used in synthetic biology. The first is a "buffer gate" that amplifies signals (**Fig. 12.28A**). For a simple gene, the protein input activates a promoter; a message is then transcribed and a protein is made. With a "NOT gate," a protein input represses a promoter, and the output protein is not made (**Fig. 12.28B**). Finally, an "OR gate" involves several genes. For instance, two alternative gene output proteins can activate a third gene. So, in **Figure 12.28C**, the alternative input proteins I_1 and I_2 activate gene 1 or gene 2, respectively, to make product 1 or product 2 (Pr_1 or Pr_2). Either of those output signals can activate gene 3 to make product 3, Pr_3 (such as GFP). Once you understand how these gates work, you can make any kind of circuit.

Toggle Switches

Electrical systems rely heavily on toggle switches to control whether a system is turned on or off. Similarly, synthetic-biology circuits depend on biological toggle switches. **Figure 12.29** illustrates a basic, genetically engineered toggle switch designed to control whether a *gfp* gene is turned on or off. The circuit involves two repressor genes (called NOT gates) whose products can repress each other's transcription. The switch depends on whether one of two different inducer signals is present. Inducer 1 inactivates repressor 1, which means repressor 2 is produced. Repressor 2, in turn, stops transcription of the repressor 1 gene and the reporter gene. So, when inducer 1 is added, GFP is not made and the system is stably toggled off.

Alternatively, inducer 2 inactivates repressor 2, which means that the genes for repressor 1 and GFP are transcribed and the cell lights up. The system is stably toggled on because repressor 1 halts transcription of repressor 2. So, a genetic engineer can control the on/off switch of these cells by adding one or the other inducer. Of course, for this system to toggle, at least a small amount of both repressor proteins must be made at all times—just enough to bind to inducer molecules and have an effect on gene expression. Switches can also be engineered in bioreporter bacteria to record exposure to molecules such as antibiotics, as was shown in Figure 11.11.

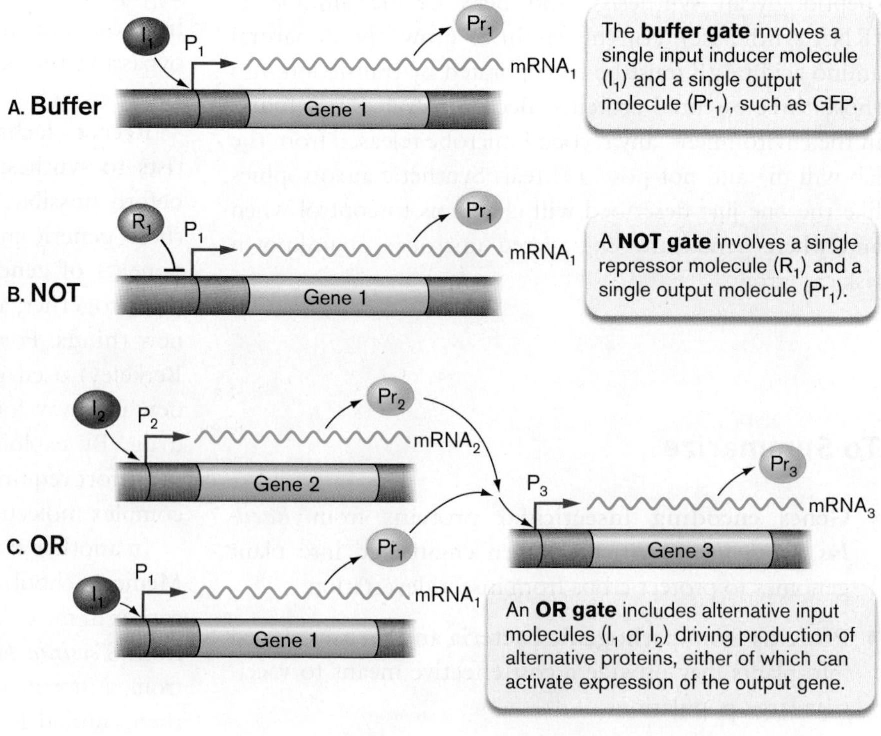

FIGURE 12.28 ▪ Examples of logic gates used in electronics, and their synthetic-biology equivalents. **A.** When inducer is added, the transcript for Pr_1 is made. Without inducer, Pr_1 is not made. **B.** When repressor is added, the transcript for Pr_1 is no longer made. **C.** When either I_1 or I_2 is added, the transcript for Pr_3 is made.

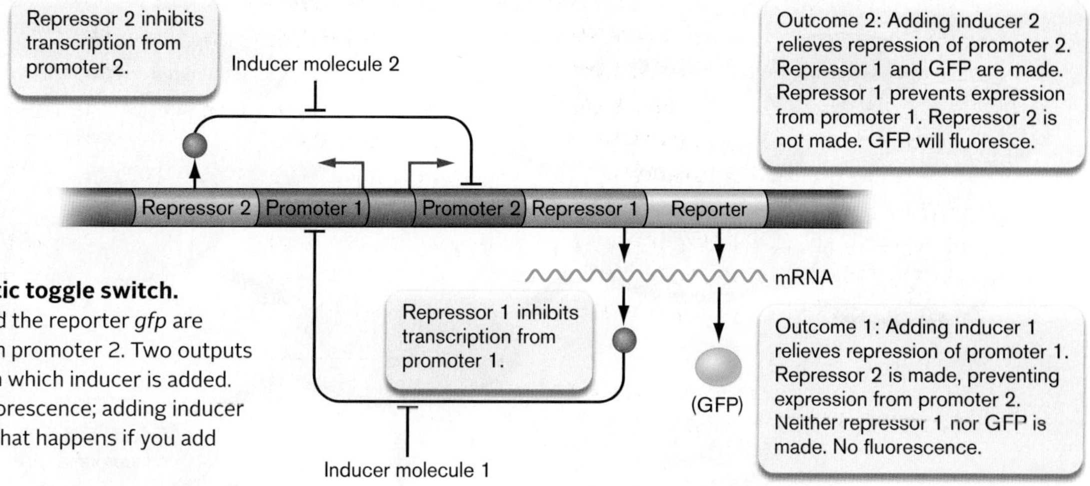

FIGURE 12.29 ■ Genetic toggle switch. Notice that repressor 1 and the reporter *gfp* are transcribed colinearly from promoter 2. Two outputs are possible, depending on which inducer is added. Adding inducer 1 stops fluorescence; adding inducer 2 triggers fluorescence. What happens if you add them both?

> **Thought Question**
>
> 12.9 What would happen with the toggle switch shown in Figure 12.29 if the genetic engineer could set repressor 1 and repressor 2 protein levels to be exactly equal? Imagine that this is done without either inducer present.

An Oscillator Switch

An oscillator is another important component of many electronic systems. In an electronic circuit, the oscillator produces a repetitive electronic signal viewed as a wave. In synthetic biology, scientists can also make an oscillating genetic circuit. **Figure 12.30A** shows a basic genetic oscillator switch designed by Jeff Hasty's laboratory at UC San Diego. The components come from some of the systems discussed in Chapter 10. One component is the gene for the AraC activator protein, which was linked by Hasty's group to a hybrid operator region that included an *ara* activator sequence and a LacI repressor control sequence. The next component is the gene encoding the repressor LacI, which was also spliced to the hybrid *ara* activator–LacI repressor control sequences. The resulting circuit contains negative and positive feedback loops. The *lac* and *ara* operons are discussed in Section 10.1.

To see how this oscillator works, IPTG (the chemical that inactivates LacI) and L-arabinose (the sugar that activates AraC) are added at the same time. The increase in AraC production drives expression of GFP but also increases LacI repressor, which eventually represses *araC*. However, IPTG inactivates LacI protein, which enables renewed *araC* and *lacI* expression. As shown in **Figure 12.30B**, the differential activity of the two feedback loops drives oscillation back and forth between fluorescence (on) and no fluorescence (off). Changing the concentrations of arabinose and IPTG will modulate the oscillation frequency, making the circuit tunable.

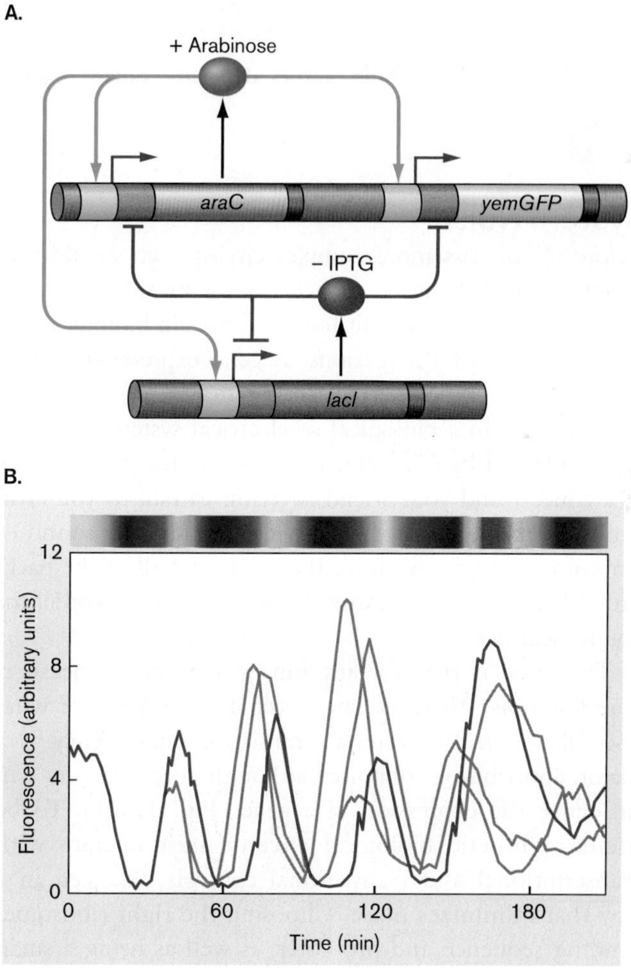

FIGURE 12.30 ■ A dual-feedback oscillator constructed in *Escherichia coli*. A. Network diagram. A hybrid promoter, $P_{lac/ara}$ (small pink and green boxes), drives the transcription of *araC* and *lacI*, forming positive and negative feedback loops. B. Single-cell fluorescence oscillations induced with 0.7% arabinose plus 2-mM IPTG (red line in the graph) or 1-mM IPTG (gray lines). The points represent experimental fluorescence values. The colored bar across the top represents the intensity of the fluorescence, from high (red) to low (blue). *Source:* Modified from J. Stricker et al. 2008. *Nature* **456**:516–519.

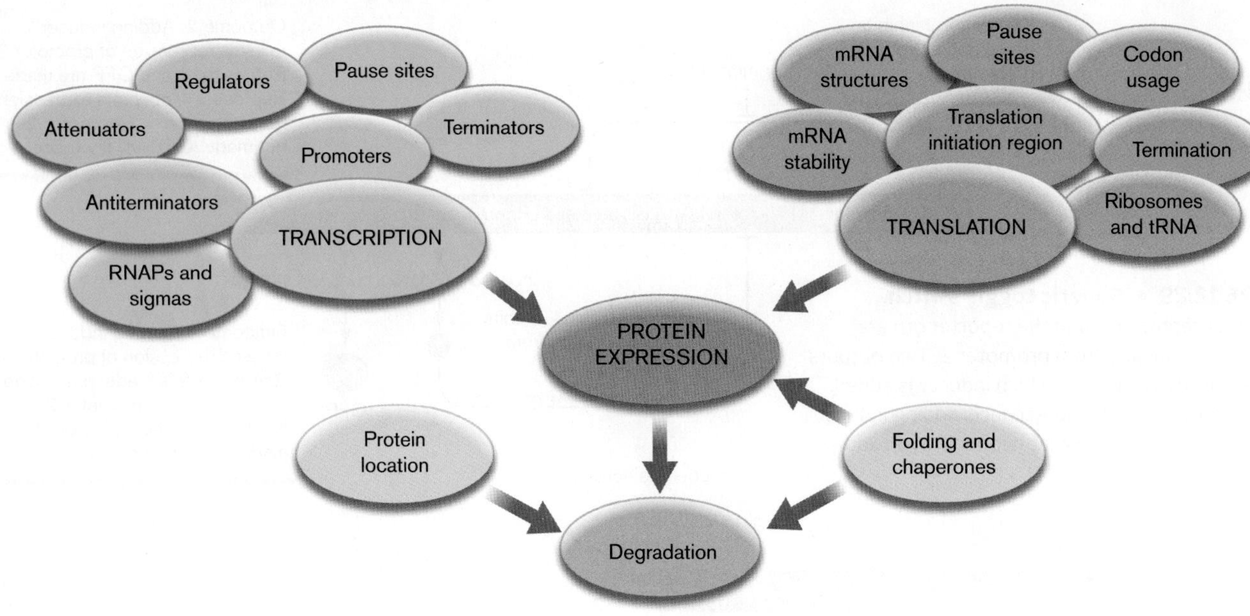

FIGURE 12.31 ■ Some factors affecting "noise" in a biological circuit.

System Noise

Before we discuss more complex circuits, we should talk about "noise." **Noise** (variance from a mean) is a problem in any electrical circuit but also arises in biological circuits, because of fluctuations in gene expression among single cells in a population (see Section 12.3). The concept of noise in a biological or electrical system might be best explained by the following analogy: Imagine you are at a concert and your friend is trying to talk to you. You can't understand what she is saying, because the sound of her voice is not rising above the music and all of the background noise. The band would have to go quiet in order for you to hear her.

The same is true for anything you want to measure. The less noise there is, the easier it is to measure what rises above it. In bacteria, translation inefficiency is a major contributor to noise, although many factors can influence the expression of a gene (**Fig. 12.31**). To be useful, synthetic biological circuits must operate with transcriptional and translational controls balanced in a way that minimizes noise. Choosing the right ribosome-binding sequence and promoter, as well as using a small RNA to control message stability or translation, can help to decrease noise.

Engineered Riboswitches and Switchboards

In Chapter 10 we described how the cell uses riboswitches to sense cell metabolites and control the translation of mRNA molecules. Synthetic-biology scientists seized upon this concept and have learned to tailor small RNAs, called synthetic riboswitches, to control the translation of nearly any gene they want. **Figure 12.32** reveals that a basic riboswitch consists of two parts: a *cis*-repressed mRNA (crRNA; **Fig. 12.32A**) and a *trans*-activating RNA (taRNA; **Fig. 12.32B**). The crRNA sequence is linked to an output gene and, by forming a hairpin, hides a ribosome-binding site to prevent translation. The taRNA, once it is made, will promote translation of the output gene by base-pairing with the crRNA (**Fig. 12.32C**). The base pairing releases the RBS for translation. Transcription and translation of the output gene can now be manipulated by whatever activates the promoters used to express the crRNA and taRNA genes. This process can reduce noise in a system.

Members of James Collins's laboratory (Harvard University) designed a series of matched crRNA-taRNA riboswitches and linked them to various promoters and output genes. The design of these circuits, and their responses to various environmental signals, are shown in **Figure 12.33**. Notice that each riboswitch responded to only one signal.

The Collins group then linked these riboswitches to genes encoding carbon metabolism enzymes and placed them all in the same cell, essentially making a metabolic switchboard that could channel carbon flow through alternative metabolic pathways depending on which riboswitch was activated. Metabolic switchboards would have important industrial applications. For instance, the device could simultaneously sense a variety of metabolic states in a large-batch fermentation system and maximize the efficiency of an industrially beneficial pathway.

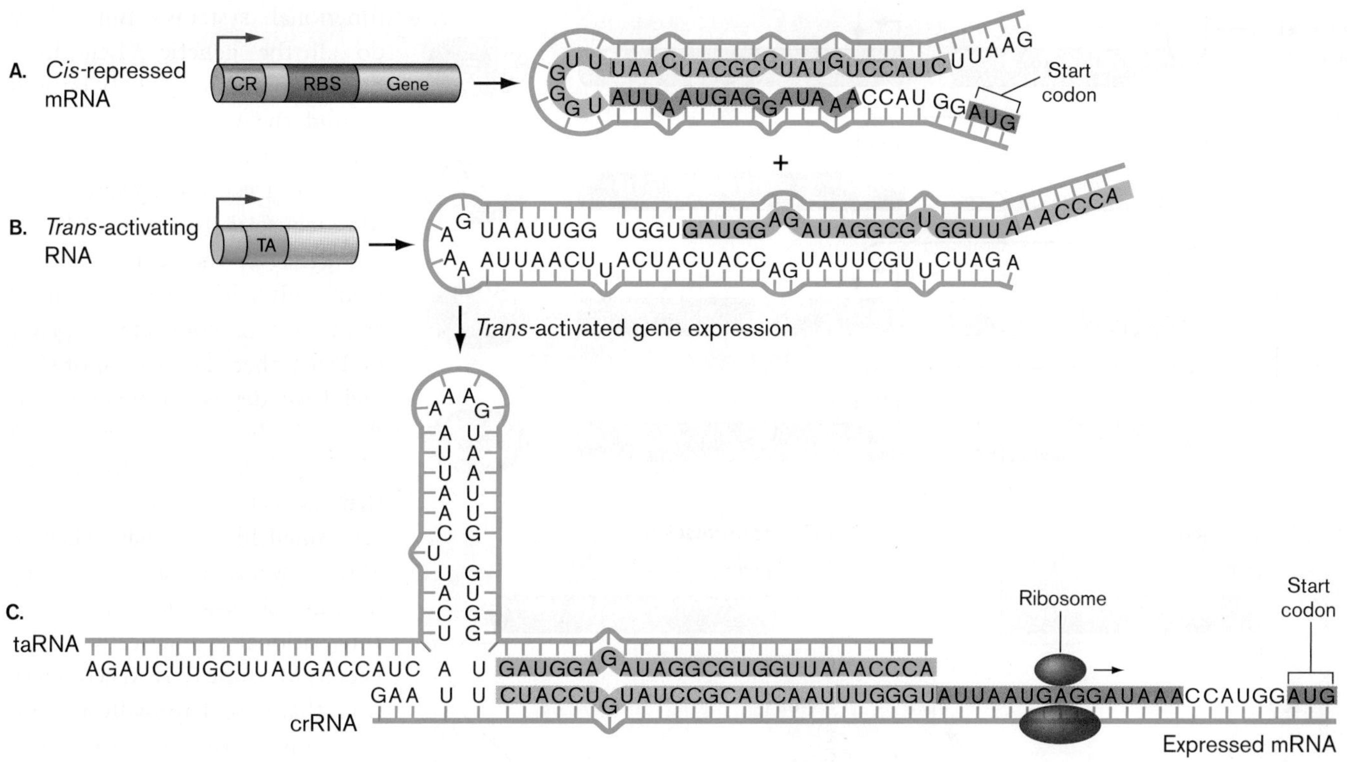

FIGURE 12.32 ■ The basic synthetic riboswitch. **A.** *Cis*-repressed mRNA (crRNA) folds to obscure a ribosome-binding site (RBS) needed to translate a downstream output gene. **B, C.** *Trans*-activating RNA (taRNA), driven by a different promoter **(B)**, can base-pair with crRNA **(C)**, thus opening up the hairpin to expose the RBS. Whatever activates the taRNA will activate translation of the output genes' mRNA.

Kill Switches

Most of what we have described about synthetic biology has little chance of dangerous unintended consequences, but what about the long term? Biologists would like to engineer new strains that do new tasks, such as gobble up toxins in the environment. Remediating toxin contamination in nature, however, would require the release of genetically modified organisms outside the laboratory. This raises concerns about potential, unknown havoc that these organisms might cause, as we discussed in Section 12.4.

To allay these concerns, scientists must engineer fail-safe mechanisms that kill the organism at a predetermined point. Hence the quest for effective "kill switches." A genetically modified organism equipped with a kill switch can be made to commit suicide once the bacterium's job is done. One proof-of-principle kill switch engineered by synthetic-biology techniques is shown in **Figure 12.34**.

The gene for CcdB, a potent DNA-damaging toxin, was linked to the promoter P_{LtetO}. This promoter is regulated by the TetR repressor protein. In the absence of tetracycline, TetR binds the promoter and prevents the transcription of *ccdB*. Cells with this switch in this off state are alive. Tetracycline controls the switch: when added, it binds and inactivates TetR, which then releases from the *ccd* promoter. As a result, the kill toxin gene *ccdB* is expressed and kills the cell.

Because transcription can be leaky, the engineers added a riboswitch mechanism (see **Fig. 12.32**) to prevent the organism from killing itself too early in this proof-of-concept experiment. They made sure that the translation of CcdB mRNA is blocked by a crRNA. A second component of the system, activated by AraC upon addition of arabinose, encodes the compensatory taRNA that can expose the RBS buried within the crRNA-CcdB message. So, both tetracycline and arabinose must be added for the cell to effectively produce CcdB and kill itself.

Although this system, as designed, is impractical for real-world use, its success shows that kill switches can be made. How could this kill switch be modified so that it could be used in the real world? What if repression of CcdB were tied to the presence of a toxic product found in the environment? Once the organism destroyed the toxin, the kill switch would be "thrown," and the bacterium would kill itself.

BioBricks and Do-It-Yourself Synthetic Biology

The science of synthetic biology makes new logic circuits by linking promoters from one system to genes from another system. The art of synthetic biology is to combine these new genetic logic circuits in ways that produce new

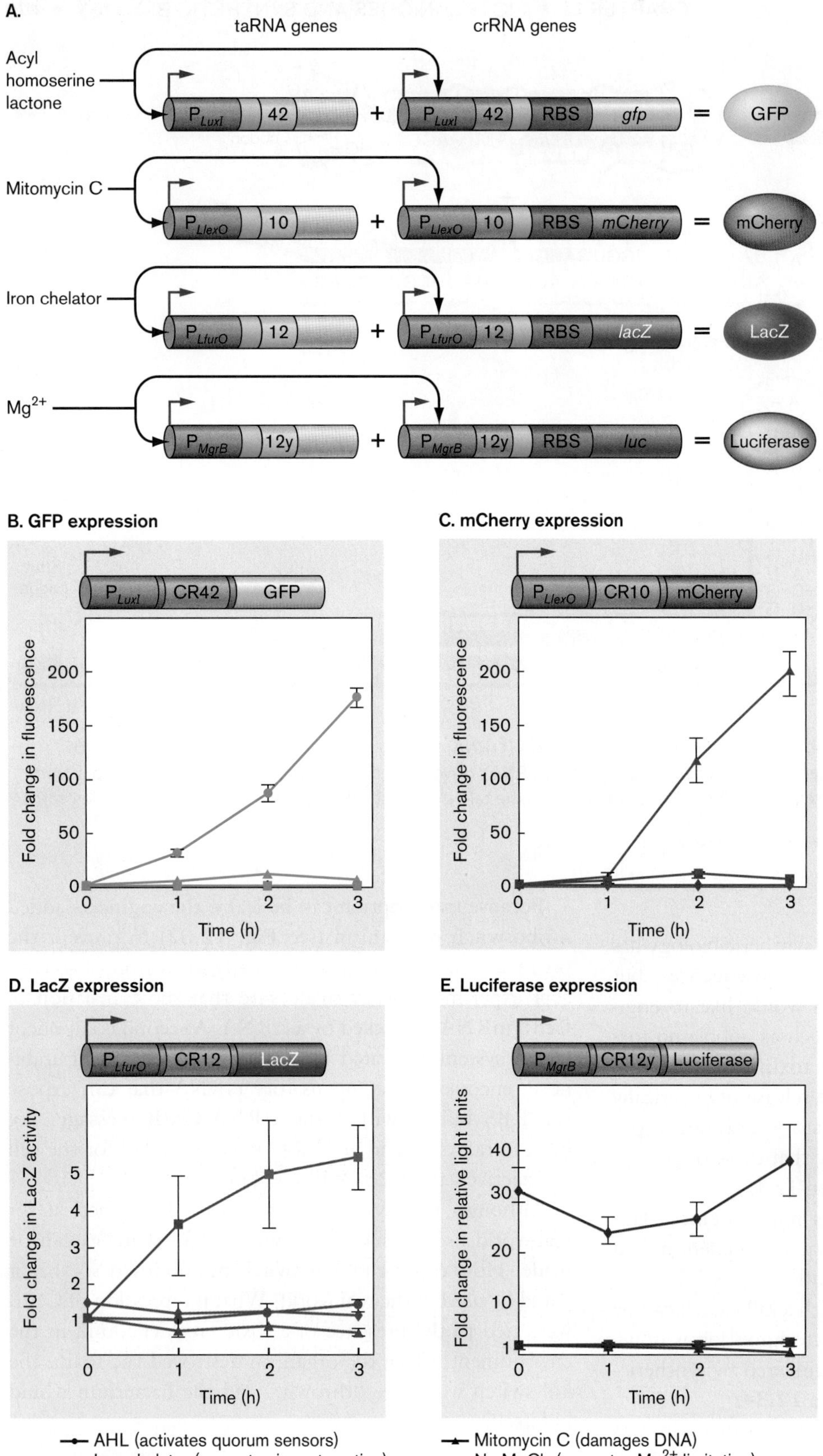

FIGURE 12.33 ▪ Riboswitchboard.
A. Four separate circuits were designed. Each circuit is controlled by a different promoter that senses a different environmental parameter: P_{LuxI} (acyl homoserine lactone), P_{LlexO} (mitomycin C), P_{LfurO} (iron), or P_{MgrB} (magnesium). Different taRNA-crRNA riboswitch pairs (labeled 42, 10, 12, and 12y) were used to control the translation of the output reporters for each circuit. **B–E.** These graphs show that each circuit responded to only a single environmental input signal. Inducers for GFP, mCherry, and LacZ **(B, C, D)** were added at time 0. The inoculum for the luciferase circuit **(E)** was grown in low Mg^{2+} and was active even at time 0.

functional systems. But where do all the genetic "Lego-like" blocks come from? Laboratories around the world construct and deposit their "building materials" into a central registry at the Massachusetts Institute of Technology. Over the years, scientists from universities have assembled thousands of connectable pieces of DNA that they call BioBricks and have deposited them in the BioBricks Foundation registry at MIT. BioBricks range from those that kill cells to one that makes cells smell like bananas. There is also a weekend-long synthetic-biology showdown, called the International Genetically Engineered Machine (iGEM) competition, that is held annually at MIT. The winner's trophy of the iGEM competition is a large, aluminum Lego (**Fig. 12.35**).

Despite its potential for good, synthetic biology raises the concern that anyone, theoretically, can do it. In fact, the lure of constructing a new organism, combined with the ever-lowering cost of equipment needed to carry out these experiments, has spawned a community of do-it-yourself (DIY) genetic engineers, including high school students and so-called garage scientists. Most of these DIY efforts are well intentioned and could yield useful products as a result of outside-the-box thinking, but we should be mindful, again, of unintended consequences.

To Summarize

- **Synthetic biology** applies engineering principles to designing and constructing new biological parts for a desired purpose.
- **Synthetic-biology circuits use genetic logic gates**, such as buffer gates, NOT gates, and OR gates.

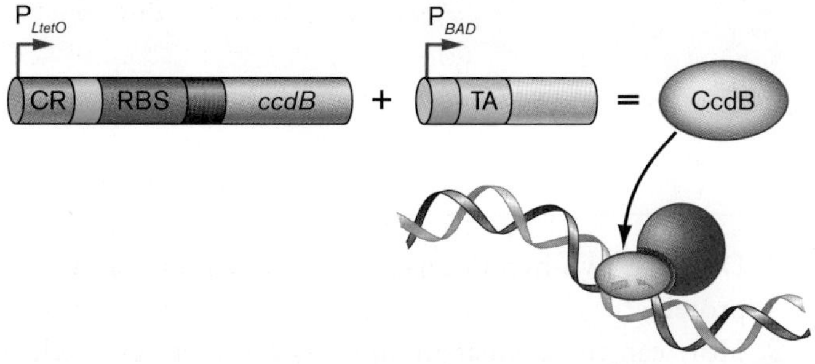

FIGURE 12.34 ■ A synthetic-biology "kill switch," or suicide module. The addition of tetracycline relieves TetR repression of the *ccdB* gene, but the gene is equipped with a *cis*-repressed RNA (CR) that prevents translation. The addition of L-arabinose enables AraC to activate the complementary *trans*-activating RNA gene (TA). The *ccdB* mRNA is translated, and CcdB causes DNA damage, which kills the cell.

- **A toggle switch**, built from two NOT gates, can turn a gene on or off by sensing two different chemical signals.
- **An oscillator switch** will repeatedly turn on and off in response to a signal.
- **System noise** is the fluctuation of gene expression among single cells of a population. System noise can dull the clarity of a response circuit. **Riboswitches** engineered into a genetic circuit can minimize noise.
- **Genetic kill switches** can eliminate a genetically modified organism when its job is done.
- **BioBricks** are connectable pieces of DNA that can be used in various combinations for synthetic biology.

THE IGEM FOUNDATION AND JUSTIN KNIGHT

THE IGEM FOUNDATION AND JUSTIN KNIGHT

FIGURE 12.35 ■ Winners of the 2018 iGEM competition Grand Prize trophy. Bottom: The team from the Polytechnic University of Valencia won Grand Prize at the 2018 competition for their design of Printeria, a combination of software and hardware that genetically engineers bacteria. **Top:** The BioBricks trophy.

Concluding Thoughts

In this chapter we have discussed some of the available molecular techniques and developing technologies (such as synthetic biology) used to probe biological processes and manipulate them for our benefit. Keep them in mind as we progress deeper into the physiology of microbial growth and the mechanisms of microbial pathogenesis. Many of the approaches described here and in eAppendix 3 were used to elucidate the concepts presented in later chapters. Realize, too, that the power of biotechnology brings with it a sobering responsibility. If nature has taught us anything, it's that tampering with it is risky. For instance, what would happen if an *Escherichia coli* that produced human growth hormone—or worse yet, one that had been engineered to make botulism toxin—managed to colonize the human gut? While these possibilities may seem unlikely, future advances in biotechnology must be guided by serious considerations of bioethics and an eye toward unintended consequences.

CHAPTER REVIEW

Review Questions

1. What DNA sequencing method would be appropriate for a small set of genes? A large set from an environmental sample? An entire genome with many repeats?
2. Once the DNA sequence of a gene is known, what specific methods can be used to gain clues as to the possible function of the gene product?
3. How can bioinformatics predict a metabolic pathway for an organism that cannot be grown in the laboratory?
4. What are the important considerations in designing a mutant selection strategy?
5. What are the key features of reporter genes used to measure transcriptional control? Translational control?
6. What techniques are used to determine whether a protein binds to a DNA sequence?
7. How can protein-protein interactions be determined in vitro?
8. Describe two-hybrid analysis. What are bait and prey proteins?
9. How can the cellular location of a protein be tracked in vivo?
10. Explain how PCR can be used in a single reaction to identify which of two species of pathogenic bacteria is/are present in a food sample.
11. Discuss how phage display might be used to produce a more toxic toxin. What are the ramifications of technologies such as this?
12. Discuss some ways that biotechnology has helped the agricultural industry. How has the field impacted vaccine delivery or our approaches to the treatment of metabolic diseases in humans?
13. Describe how BioBricks and logic gates are used in synthetic biology.

Thought Questions

1. You have identified two regulatory proteins, RegA and RegB, which affect the expression of gene *yxx*, a gene involved in quorum sensing. Mutants lacking RegA or RegB fail to express *yxx*. Describe two alternative models that would explain this effect. Next, describe how results from ChIP-seq analysis could help distinguish between those possibilities.
2. The Gram-negative bacillus *Shigella dysenteriae*, the agent that causes bacillary dysentery, will live and grow inside eukaryotic cells. Despite lacking flagella, the organism moves through the cytoplasm of the host cell by polymerizing host actin at one pole of the cell. This action pushes the organism through the cytoplasm with such force that the bacterium will poke into an adjacent cell. You have identified a *Shigella* protein that you think helps mediate actin tail formation and suspect that the protein might localize to one pole of the cell. What biotechnique would you use to test your hypothesis?
3. In Chapter 10 we discussed the *lac* operon and the *Salmonella* phase variation system in which DNA recombination controls a switch between two different flagellin proteins. You would like to understand more about the mechanism of regulation, so you have genetically engineered a strain of *E. coli* to contain the following genes: (1) the *hin* gene controlled by the *lacP* promoter, (2) a transcriptional activator gene flanked by *hix* sites, and (3) a green fluorescent protein gene (*gfp*) controlled by the transcriptional activator. Briefly describe what happens if (a) the strain is grown in lactose, (b) the strain is grown without lactose, (c) you delete the *lacI* gene, and (d) you eliminate the *lac* operator.

Key Terms

annotation (456)
bioinformatics (455)
ChIP sequencing (ChIP-seq) (470)
chromatin immunoprecipitation (ChIP) (470)
cloning (459)
exon (457)
fluorescence resonance energy transfer (FRET) (447)
interactome (471)
intron (457)
methylome (452)
nanopore DNA strand sequencing (451)
noise (478)
palindrome (458)
phage display (473)
polymerase chain reaction (PCR) (445)

preliminary mRNA transcript (pre-mRNA) (457)
primer (446)
processivity (450)
quantitative PCR (qPCR) (446)
real-time PCR (446)
reporter gene (467)
restriction endonuclease (458)
restriction site (458)
sequencing by synthesis (448)
single-molecule sequencing (450)
synthetic biology (475)
TaqMan (447)
transcriptional fusion (467)
translational fusion (469)
two-hybrid analysis (471)

Recommended Reading

Berens, Christian, Florian Groher, and Beatrix Suess. 2015. RNA aptamers as genetic control devices: The potential of riboswitches as synthetic elements for regulating gene expression. *Biotechnology Journal* **10**:246–257.

Bouveret, Emmanuelle, and Christine Brun. 2012. Bacterial interactomes: From interactions to networks. *Methods in Molecular Biology* **804**:15–33.

Clark, Tyson A., Ian A. Murray, Richard D. Morgan, Andrey O. Kislyuk, Kristi E. Spittle, et al. 2012. Characterization of DNA methyltransferase specificities using single-molecule, real-time DNA sequencing. *Nucleic Acids Research* **40**: e29.

Ebina, Hirotaka, Naoko Misawa, Yuka Kanemura, and Yoshio Koyanagi. 2013. Harnessing the CRISPR/Cas9 system to disrupt latent HIV-1 provirus. *Scientific Reports* **3**(art. 2510).

Eid, John, Adrian Fehr, Jeremy Gray, Khai Luong, John Lyle, et al. 2009. Real-time DNA sequencing from single polymerase molecules. *Science* **323**:133–138.

Huang, Johnny X., Sharon L. Bishop-Hurley, and Matthew A. Cooper. 2012. Development of anti-infectives using phage display: Biological agents against bacteria, viruses and parasites. *Antimicrobial Agents and Chemotherapy* **56**:4569–4582.

Jain, Miten, Hugh E. Olsen, Benedict Paten, and Mark Akeson. 2016. The Oxford Nanopore MinION: Delivery of nanopore sequencing to the genomics community. *Genome Biology* **17**:239.

Locke, James C., Jonathan W. Young, Michelle Fontes, Maria J. Hernandez Jimenez, and Michael B. Elowitz. 2011. Stochastic pulse regulation in bacterial stress response. *Science* **334**:366–369.

Loh, Belinda, Andreas Kuhn, and Sebastian Leptihn. 2019. The fascinating biology behind phage display: filamentous phage assembly. *Molecular Microbiology* **111**: 1132–1138.

Martínez-Garcia, Esteban, and Victor de Lorenzo. 2012. Transposon-based and plasmid-based genetic tools for editing genomes of Gram-negative bacteria. *Methods in Molecular Biology* **813**:267–283.

Mimee, Mark, Alex C. Tucker, Christopher A. Voigt, and Timothy K. Lu. 2015. Programming a human commensal bacterium, *Bacteroides thetaiotaomicron*, to sense and respond to stimuli in the murine gut microbiota. *Cell Systems* **1**:62–71.

Qi, Lei S., and Adam P. Arkin. 2014. A versatile framework for microbial engineering using synthetic noncoding RNAs. *Nature Reviews. Microbiology* **12**:341–354.

Rajagopala, Seesandra V., Patricia Sikorski, Ashwani Kumar, Roberto Mosca, James Vlasblom, et al. 2014. The binary protein-protein interaction landscape of *Escherichia coli*. *Nature Biotechnology* **32**:285–290.

Rovner, Alex J., Adrian D. Haimovich, Spencer R. Katz, Zhe Li, Michael W. Grome, et al. 2015. Recoded organisms engineered to depend on synthetic amino acids. *Nature* **518**:89–93.

Ruder, Warren C., Ting Lu, and James J. Collins. 2011. Synthetic biology moving into the clinic. *Science* **333**:1248–1252.

Saade, Fadi, and Nikolai Petrovsky. 2012. Technologies for enhanced efficacy of DNA vaccines. *Expert Review of Vaccines* **11**:189–209.

Selle, Kurt, and Rodolphe Barrangou. 2015. Harnessing CRISPR–Cas systems for bacterial genome editing. *Trends in Microbiology* **23**:225–232.

Uhlig, Christiane, Fabian Kilpert, Stephan Frickenhaus, Jessica U. Kegel, Andreas Krell, et al. 2015. In situ expression of eukaryotic ice-binding proteins in microbial communities of Arctic and Antarctic sea ice. *ISME Journal* **9**:2537–2540.

Zhang, Weiwen, and David R. Nielsen. 2014. Synthetic biology applications in industrial microbiology. *Frontiers in Microbiology* **5**(art. 451).

CHAPTER 13

Energetics and Catabolism

13.1 Energy for Life

13.2 Energy Carriers and Electron Transfer

13.3 Catabolism: The Microbial Buffet

13.4 Glucose Fermentation and Respiration

13.5 The Gut Microbiome: Friends with Benefits

13.6 Aromatic Catabolism and Syntrophy

To grow and multiply, all living cells need energy. Energy-yielding reactions such as photosynthesis and catabolism enable cells to build biomass and grow. Catabolism is the step-by-step process of breaking down complex molecules into smaller ones. Microbes catabolize food molecules within our own digestive tract and in the soil and water all around us. Our uses of microbial catabolism range from producing alcohol to remediating hazardous wastes, from our local wastewater plants to remote stations of Antarctica.

CURRENT RESEARCH highlight

Gut bacteria digest our food. Our gut bacteria digest hundreds of kinds of sugar chains called glycans. Julie Biteen watches *Bacteroides thetaiotaomicron* digest glycans using protein complexes called Sus. A Sus complex can access branched sugar chains too large to enter the cell. The glycan-binding protein SusE has a fixed position on the cell, as shown by time tracks of the fluorescent-tagged protein (left). Each SusE molecule anchors a glycan. Another protein, SusG, cleaves the glycan into short chains for transport across the membrane. SusG time tracks crisscross the cell surface (right).

Source: Hannah H. Tuson et al. 2018. *Biophys. J.* **115**:242–250.

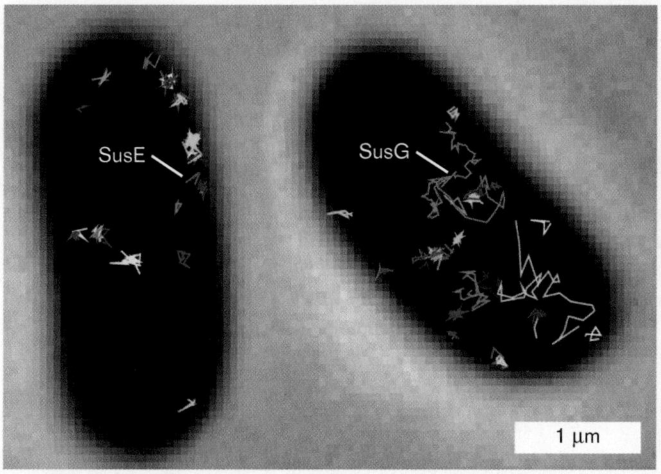

H. H. TUSON ET AL. 2018. *BIOPHYS. J.* **115**:242–250, FIGS. 2A AND 2C.

AN INTERVIEW WITH

JULIE BITEEN, MICROBIAL CELL BIOLOGIST, UNIVERSITY OF MICHIGAN

COURTESY OF THE UNIVERSITY OF MICHIGAN

What do the protein tracks tell us?

The proteins that collaborate to capture starch at the cell surface have distinctly different mobilities. SusE stays in one place, as part of a larger complex; SusG moves across the surface and stops only when it binds starch. Sus proteins may interact at immobile centers that form transient complexes. Human enzymes can digest only a limited range of sugars, but the human gut microbiome produces a far broader array of enzymes and releases nutrients from the remaining, host-indigestible food sources.

Our digestive organs, and those of other animals, support vast communities of microbes (our **gut microbiome**) that digest much of the food we eat. Gut bacteria digest a wide range of polysaccharides (or **glycans**) made by plants, as well as host-derived glycans such as those found in breast milk. Breaking down complex foods by reactions that yield energy is called **catabolism**. Many of our gut microbes catabolize food molecules using enzymes that our bodies don't have—thus extending the repertoire of our genome. In return for the breakdown of complex foods into small molecules that we can digest, our bodies provide a home for our gut microbiome.

Many foods we eat consist of long-chain polymers that cannot be taken up through the bacterial envelope for catabolism. So, our gut bacteria possess "starch utilization systems," Sus complexes, that capture and degrade glycans (**Fig. 13.1**; see also the Current Research Highlight). The Sus complex is named for starch, a simple polymer of glucose, but similar complexes digest more exotic glycans decorated with amino and sulfate groups. After initial chain cleavage, the short-chain breakdown products (oligosaccharides) then pass through the outer membrane via transporters such as SusC, into the periplasm. Within the periplasm, oligosaccharides are further degraded to monosaccharides, which then cross the inner membrane to the cytoplasm. Remarkably, some bacteria share their catabolic enzymes with the gut community, through vesicles that pinch off from the outer membrane. These vesicles can then be shared with other bacteria by fusing with the recipient outer membrane. Thus the gut contains a complex network of energy-yielding catabolic processes, with competition as well as sharing of enzymes and products.

Chapter 13 explains how energy-yielding reactions such as those of catabolism enable microbes to do work. The energy released from a reaction drives biosynthetic reactions that build simple molecules into a complex cell. The products of energy-yielding reactions have important consequences for the organism and environment. For example, cyanobacterial photosynthesis produces molecular oxygen that enables other organisms to respire, whereas anaerobic fermentation acidifies the environment. In humans, the products of bacterial catabolism regulate our immune system and our brain.

Next, Chapter 14 will present the mechanism of membrane-embedded electron transfer systems (ETS) and the transmembrane proton motive force (PMF). Electron transport systems play key roles in respiration, lithotrophy, and photosynthesis. The energy from all these pathways is used to build cells by anabolism (biosynthesis), which is the focus of Chapter 15. Finally, Chapter 16 explores commercial applications of microbial metabolism to produce food and beverages, as well as industrial products and pharmaceuticals.

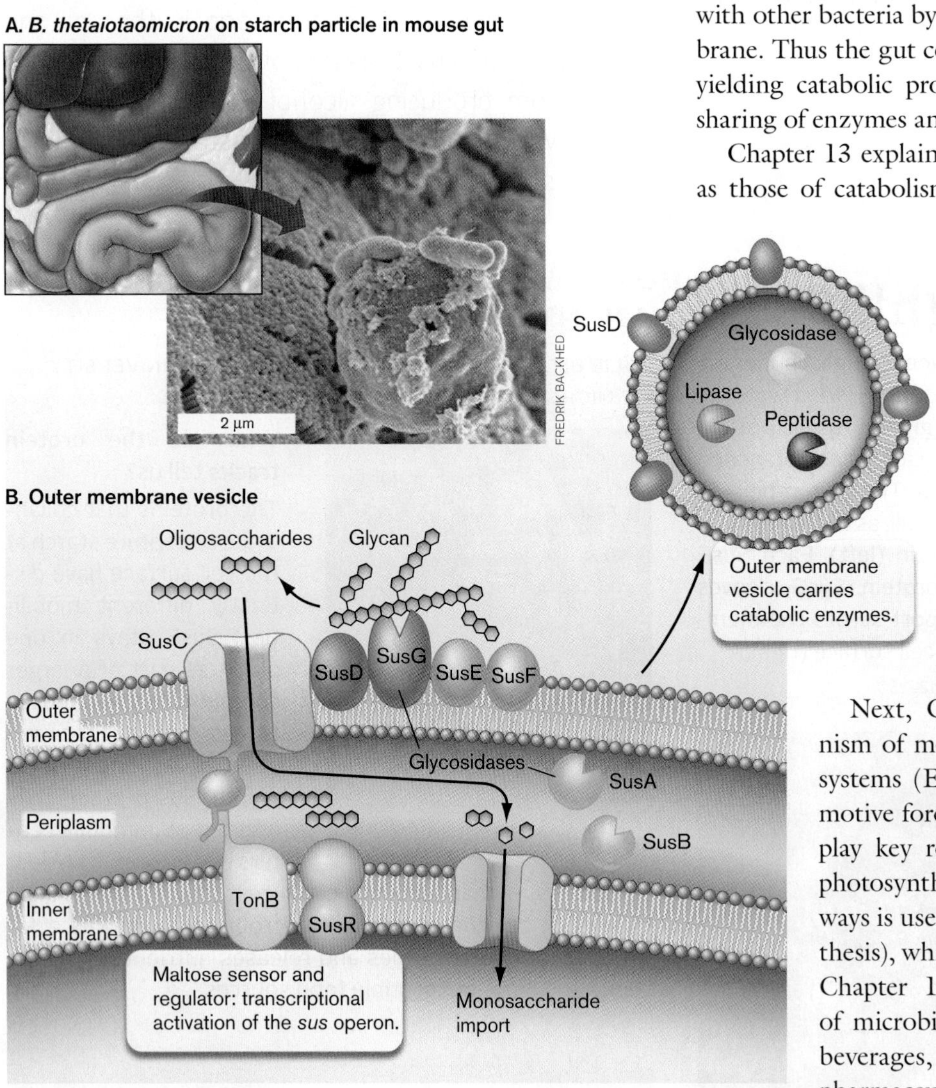

FIGURE 13.1 ■ Glycan catabolism by intestinal bacteria. **A.** *Bacteroides thetaiotaomicron* (colorized yellow) attached to a starch food particle within mouse intestine. **B.** Sus complex of glycan degradation enzymes and transporters in the *Bacteroides* envelope. Some enzymes and transporters become incorporated into outer membrane vesicles.

Note: For a review of molecular structure and biochemical reactions, see eAppendix 1.

13.1 Energy for Life

Every form of life, from a composting microbe to a human body, uses energy (as introduced in Chapter 4). **Energy** is the ability to do work, such as flagellar propulsion or cell growth. Energy is used to organize proteins, maintain ion gradients, and build biomass. To conduct all these processes, life needs to perform reactions that store chemical energy and control how the energy is spent.

Many Sources of Energy

Collectively, microbes use diverse energy sources, many of which multicellular organisms cannot use (**Table 13.1**). Recall from Chapter 4 that an organism may gain energy from chemical rearrangement of molecules (prefix "chemo-") or from light absorption (prefix "photo-"). **Chemotrophy** yields energy from electron transfer between chemicals, releasing products that are more stable. Chemotrophy in which organic compounds donate electrons to yield energy is known as organotrophy or chemoorganotrophy. Organic compounds include the foods we eat—which our intestinal bacteria help catabolize. Most organotrophs are also heterotrophs, organisms that use preformed organic compounds for biosynthesis (a process called **heterotrophy**). But chemotrophy also includes lithotrophy or chemolithotrophy (literally "rock eating"), obtaining energy from inorganic reactions. For example, the archaeon *Pyrodictium occultum* gains energy by oxidizing hydrogen gas with sulfur.

Phototrophy yields energy from light absorption. Phototrophs such as marine cyanobacteria are autotrophs (build their own carbon compounds). A photoautotroph gains energy solely from light and builds biomass solely from CO_2. Other bacteria, such as purple proteobacteria, practice photoheterotrophy, in which energy from light supplements chemotrophy. Phototrophy and lithotrophy are discussed in Chapter 14.

But all sources of energy pose the challenges of how to obtain the energy, how to avoid losing it, and how to convert it for cell growth and function.

TABLE 13.1 Energy Acquisition in Bacteria and Archaea

Energy source	Class of metabolism	Examples of energy-yielding reactions	Electron acceptor	Systems for energy acquisition
		CHEMICAL		
Chemoorganotrophy Organic compounds (at least one C–C bond) donate electrons	**Fermentation** Catabolism	$C_6H_{12}O_6 \rightarrow 2C_3H_6O_3$ (or other small molecules)	Organic	Glycolysis and other catabolism
	Organic respiration Catabolism with inorganic electron acceptor, or with small organic electron acceptor	$C_6H_{12}O_6 + 6H_2O + 6O_2 \rightarrow 6CO_2 + 12H_2O$	O_2	Glycolysis and other catabolism, TCA cycle, and electron transport systems
		$C_6H_{12}O_6 + 6H_2O + 12NO_3^- \rightarrow 6CO_2 + 12H_2O + 12NO_2^-$	NO_3^-, SO_4^{2-}, Fe^{3+}, or other	
Chemolithotrophy Inorganic compounds donate electrons	**Lithotrophy or chemolithoautotrophy** CO_2 fixation	Electron donor for respiration is H_2, Fe^{2+}, H_2S, NH_4^+	O_2, NO_3^-, or other	Electron transport system
	Methanogenesis	Electron donor is H_2: $CO_2 + 4H_2 \rightarrow CH_4 + 2H_2O$	CO_2	Methanogenesis
		LIGHT		
Phototrophy Light absorption provides electrons	**Photoautotrophy** Light absorption drives CO_2 fixation	Photolysis of H_2O: $6CO_2 + 12H_2O \rightarrow C_6H_{12}O_6 + 6H_2O + 6O_2$	CO_2	Photosystems I and II
		Photolysis of H_2S, HS^-, or Fe^{2+}: $6CO_2 + 12H_2S \rightarrow C_6H_{12}O_6 + 6H_2O + 12S$	CO_2	Photosystem I or II
	Photoheterotrophy Light absorption without CO_2 fixation	Photolysis of H_2S, HS^-, or light-driven H^+ pump. Usually supplements organotrophy.	Organic	Photosystem I or II; bacteriorhodopsin or proteorhodopsin

Note: Distinguish the following prefixes for "-trophy" terms.

Carbon source for biomass

Auto-: CO_2 is fixed and assembled into organic molecules.
Hetero-: Preformed organic molecules are acquired from outside and assembled into new organic molecules.

Energy source

Photo-: Light absorption captures energy.
Chemo-: Chemical reactions yield energy without absorbing light.

Electron source

Litho-: Inorganic molecules donate electrons.
Organo-: Organic molecules donate electrons.

Microbes Use Energy to Build Order

How can microbes use energy to assemble less-ordered molecules into a complex cell? As complexity increases, the cell is said to decrease in **entropy**, or disorder. But the decrease in entropy is local to the cell. Ultimately, the cell's energy must be spent as heat, which radiates away, causing entropy to increase overall. In other words, the local, temporary gain of energy enables a cell to grow. Continued growth requires continual gain of energy and continual radiation of heat. We see this release of heat, for example, in a compost pile, where heat is produced faster than it dissipates. The temperature of compost typically rises to 60°C.

Similarly, throughout Earth's biosphere, the total metabolism of all life must dissipate energy as heat. Biological heat production is not always obvious, because soil and water provide a tremendous heat sink. But overall, Earth's biosphere behaves as a giant thermal reactor (**Fig. 13.2**). As solar radiation reaches Earth, a small fraction is captured by photosynthetic microbes and plants. The fraction captured is largely in the range of visible light, the wavelengths at which photon energies can be absorbed for the controlled formation and dissociation of molecular bonds. At shorter wavelengths (X-rays), chemical bonds are broken indiscriminately; at longer wavelengths (microwaves and radio waves), the quantum energy is too low to drive chemical reactions.

Microbial and plant photosynthesis generates biomass, which is catabolized by consumers and decomposers. The consumers store a small fraction of their energy in biomass. At each successive trophic level, the majority of energy is lost, radiated from Earth as heat. Thus, despite the growth of living organisms on Earth, the universe as a whole becomes more disordered. The complex roles of microbial metabolism in global ecosystems are discussed in Chapters 21 and 22.

Note: The principles of energy change, discussed next, apply to all the reactions of **Table 13.1**. These reactions are covered in detail in Chapters 13 and 14.

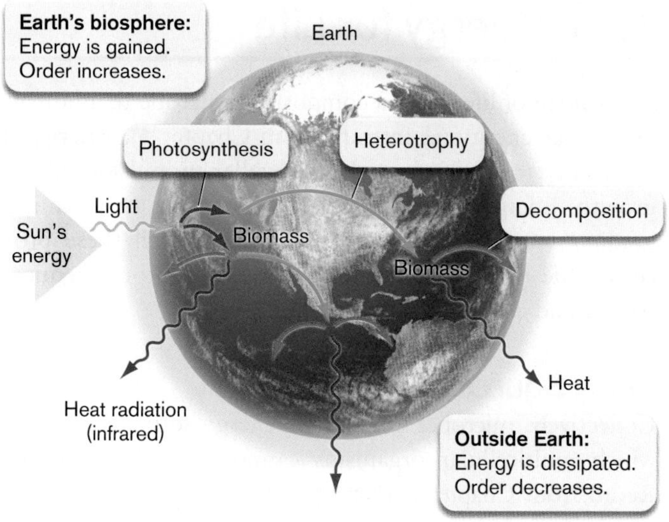

FIGURE 13.2 ■ Solar energy. Solar radiation reaches Earth, where a small fraction is captured by photosynthetic microbes and plants. The microbial and plant biomass enters heterotrophs and decomposers, which convert a small fraction to biomass at each successive level. At each level, the majority of energy is lost, radiated from Earth as heat.

Gibbs Free Energy Change

Table 13.1 shows that microbes can harness an enormous variety of chemical reactions for growth. But what determines whether a given reaction can support life?

Gibbs free energy change (ΔG). To provide energy to a cell, a biochemical reaction must go forward from reactants to products. The direction of a reaction can be predicted by a thermodynamic quantity known as the **Gibbs free energy change**, ΔG (also known as free energy change or Gibbs energy change). The ΔG value of a reaction determines how much energy is potentially available to do work, such as to drive rotary flagella, to build a cell wall, or to store accurate information in DNA. The sign of the free energy change, ΔG, determines whether a process may go forward. If ΔG is negative, the process may go forward, whereas positive values mean that the reaction will go in reverse. The sign of ΔG determines which "foods" a microbe can eat or, more precisely, which reactions between available molecules can be harnessed for microbial growth.

Under defined laboratory conditions, calculating the ΔG value of a reaction can predict how much biomass microbes will build (**Fig. 13.3**). **Figure 13.3A** lists various reactants for catabolic reactions performed by our gut bacteria, and by bacteria in soil. When oxygen is available, oxidative catabolism of sugars or of short-chain fatty acids yields relatively large amounts of energy. Once oxygen is used up, some bacteria can oxidize food with an alternative electron acceptor, such as nitrate (NO_3^-)—a process called **anaerobic respiration**. Alternatively, bacteria can ferment the carbon source with no mineral oxidant. Whatever the

A.

Carbon source	Oxidant	ΔG (kJ/mol)	Biomass (g/mol)
Glucose	O_2	−2,883.0	70.54
Propionate	O_2	−1,487.0	32.54
Lactate	O_2	−1,333.0	32.83
Ethanol	O_2	−1,308.0	26.32
Acetate	O_2	−847.0	18.72
Formate	O_2	−234.0	4.99
Glucose	NO_3^-	−2,774.0	60.00
Lactate	NO_3^-	−1,282.0	20.20
Ethanol	NO_3^-	−1,257.0	24.00
Acetate	NO_3^-	−813.0	16.95
Ethanol	None	−14.6	3.50

B.

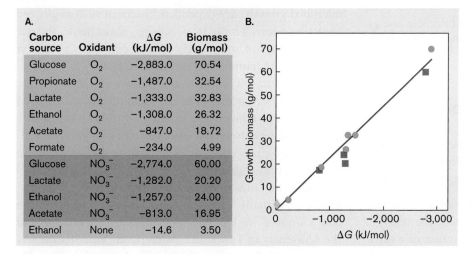

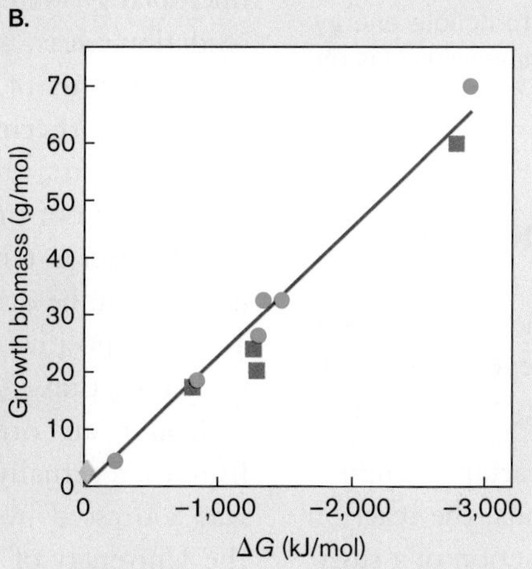

FIGURE 13.3 ▪ Bacterial-growth biomass depends on free energy change of metabolic reactions. A. Energy data used for panel B. Oxidant types (O_2, NO_3^-, none) are color-coded respectively. **B.** Observed biomass of cells grown on given pairs of carbon source and oxidant is plotted as a function of free energy change (ΔG). Product is CO_2, except for ethanol without oxidant (green diamond), whose products are acetate and hydrogen gas.
Source: Data from Eric E. Roden and Qusheng Jin. 2011. *Appl. Environ. Microbiol.* **77**:1907.

reaction, a linear relationship appears between ΔG and the biomass produced (**Fig. 13.3B**). In natural environments, however, the changing concentrations of substrates and products make the ΔG calculation more complicated.

ΔG includes enthalpy and entropy. The free energy change ΔG has two components:

- ΔH = change in **enthalpy**, the heat energy absorbed or released as reactants become products at constant pressure. When reactants absorb heat from their surroundings as they convert to products, ΔH is positive. When, instead, heat energy is released, ΔH is negative. Release of heat (negative value of ΔH) can yield energy for the cell to use. An example of a reaction with strongly negative ΔH is the oxidation of glucose by O_2.
- ΔS = change in **entropy**, or disorder. Entropy is based on the number of states of a system, such as the number of possible conformations of a molecule. If a cellular reaction splits one molecule into two, all else being equal, entropy increases; the system is more disordered, and ΔS is positive. Most catabolic reactions have a positive value of entropy change. A positive value of ΔS makes ΔG more negative and increases the potential energy yield of a reaction.

The relationship of the free energy change ΔG with ΔH and ΔS is given by:

$$\Delta G = \Delta H - T\Delta S$$

The overall sign of ΔG depends on its two components: ΔH (the absorption or release of heat energy) and $-T\Delta S$ [the negative product of entropy change (ΔS) and temperature (T)]. In living organisms, a sufficiently negative ΔH (energy lost as heat) often overrides $-T\Delta S$, the term for increase in order (negative value of ΔS, positive value of $-T\Delta S$). Thus, a living organism, whose development entails increasing order and decreasing ΔS, can grow as long as the sum of its metabolism has a sufficiently negative value of ΔH. The heat loss associated with ΔH is obvious in a compost pile, where temperature rises. Similarly, all living organisms and communities lose heat.

Negative ΔG Drives a Reaction Forward

An example of a thermodynamically favored reaction is the oxidation of hydrogen gas (H_2) to form water. Hydrogen is oxidized for energy by many kinds of bacteria in soil and water (a form of lithotrophy discussed in Chapter 14). For example, hydrogenotrophic bacteria of the genus *Ralstonia* have been isolated from ultrapure water used for nuclear fuel storage, where radioisotopes emit particles that ionize water, generating H_2. The chemical reaction of dissolved hydrogen and oxygen gases is:

$$2H_2 + O_2 \rightarrow 2H_2O$$

In this reaction, two molecules of hydrogen gas donate four electrons to oxygen, forming water. Under conditions of standard temperature (298 kelvins, or K) and pressure (sea level), $\Delta H = -572$ kilojoules per mole (kJ/mol). The ΔH is strongly negative (much heat is released) because the bonds of the product H_2O are much more stable than those of the substrates, H_2 and O_2.

However, entropy decreases because the three molecules are replaced by two—a more ordered state. Thus, ΔS is negative: -0.327 kJ/(mol · K). In the Gibbs equation the negative sign on the entropy term $-T\Delta S$ makes its contribution to ΔG positive—unfavorable for reaction. So which term wins: ΔH or $-T\Delta S$?

$$\begin{aligned}\Delta G &= \Delta H - T\Delta S \\ &= -572 \text{ kJ/mol} - (298 \text{ K}) \\ &\quad [-0.327 \text{ kJ/(mol} \cdot \text{K)}] \\ &= -572 \text{ kJ/mol} + 97 \text{ kJ/mol} \\ &= -475 \text{ kJ/mol}\end{aligned}$$

Overall, ΔG is negative. So bacteria with the appropriate enzyme pathways can use the reaction of hydrogen gas with oxygen to provide energy.

Note: The **joule** (**J**) is the standard SI unit to denote energy. 1 kilojoule (kJ) = 1,000 joules. Another unit commonly used is the kilocalorie (kcal). The conversion factor is: 1 kJ = 0.239 kcal.

Enthalpy and Entropy in Metabolism

How do ΔH and $-T\Delta S$ affect biochemical reactions? Each reaction in a living cell is associated with these two types of energy change, but to differing degrees, depending on the reaction. In general, a reaction yields energy for the cell if:

- **Molecular stability increases.** When reactants combine to form products with more stable bonds, the reaction has a negative ΔH. For example, the reaction of a sugar with oxygen has a negative ΔH because the relatively unstable oxygen molecules are reduced to H_2O.
- **Entropy increases.** Reactions in which a complex molecule is broken down to a greater number of smaller molecules increase entropy (positive ΔS, negative value of $-T\Delta S$). An important class of such product molecules is carboxylic acids, such as lactic acid, in which a H^+ dissociates from a carboxylate ion ($R\text{–}COO^-$). Entropy also increases with conversion of a solid reactant to a gas, such as CO_2. For example, glucose may be fermented to ethanol and CO_2, as in the production of alcoholic beverages.

The relative contributions of ΔH and $-T\Delta S$ are important because they determine the effect of temperature on microbial growth. ΔH-dependent reactions such as glucose oxidation release a lot of heat, causing, for example, the rise in temperature of an aerated compost pile. Such a reaction is called **exothermic**—that is, releasing heat.

By contrast, the entropy component of Gibbs energy, $-T\Delta S$, does not include heat loss. The magnitude of $-T\Delta S$ grows larger as temperature increases. Could an organism's metabolism be driven by a large entropy change $-T\Delta S$ with a smaller positive ΔH value? The positive ΔH would make such a reaction **endothermic**—that is, absorbing heat.

Could an organism conduct endothermic metabolism that actually cools its environment? This question was addressed by Urs von Stockar (**Fig. 13.4A**), now at the University of Lausanne, Switzerland. Von Stockar was a founder of the study of biological thermodynamics. He predicted the existence of entropy-driven metabolism, and he discovered an example, the conversion of acetate to methane and CO_2, which is conducted by the soil archaeon *Methanosarcina barkeri* (**Fig. 13.4B**):

$$CH_3COOH \rightarrow CH_4 + CO_2$$

The conversion of solid acetic acid into two gases incurs a tremendous increase in entropy, ΔS, which makes the free energy change more negative ($-T\Delta S$). But the sign of ΔH is positive; thus the organism actually absorbs heat from its soil environment. Despite the positive ΔH, the larger magnitude of the entropy change $-T\Delta S$ allows the reaction to proceed, yielding net free energy.

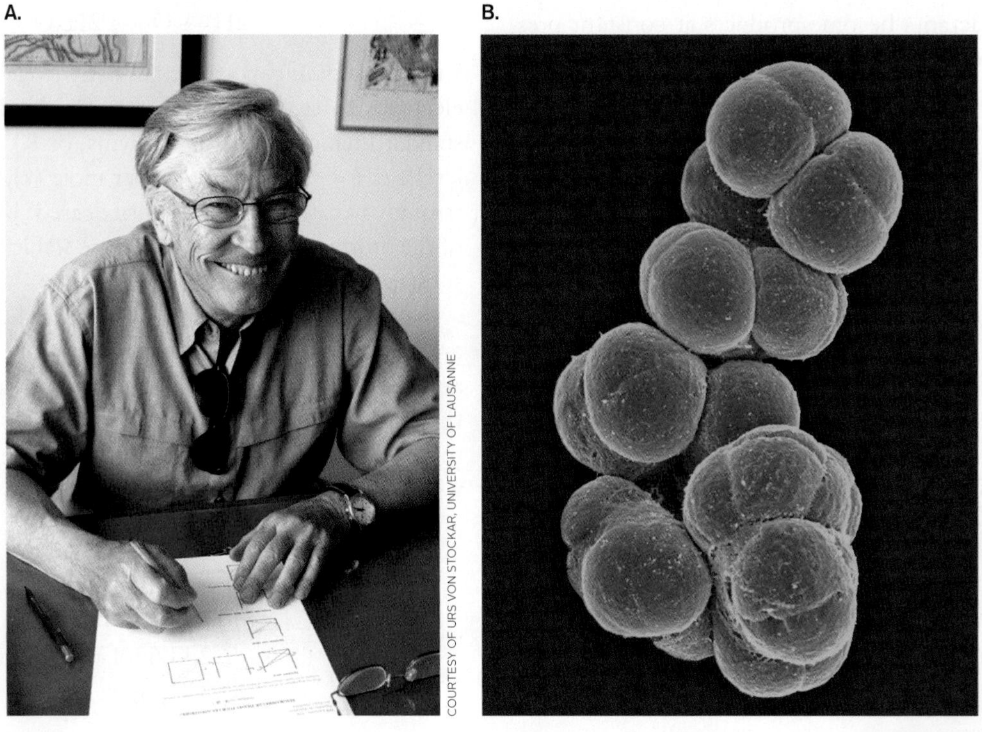

FIGURE 13.4 ■ *Methanosarcina barkeri* converts acetic acid to methane and CO_2. **A.** Urs von Stockar showed that *M. barkeri* metabolism cools its surroundings. **B.** *Methanosarcina barkeri*, a coccoid archaeon that converts acetic acid to CH_4 and CO_2.

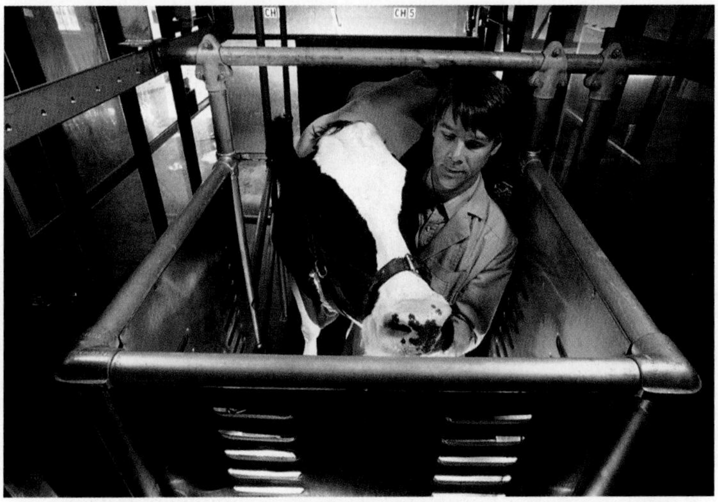

FIGURE 13.5 • Calorimeter for animal feed. A cow is prepared for a feeding test within a large-animal calorimeter at the U.S. Department of Agriculture (USDA) in Beltsville, Maryland. The test will show how much energy the cow and its digestive microbes obtain from feeding on ground corn.

Reaction Conditions

For a given reaction in a given environment, many factors determine ΔG. These factors fall into two classes—those intrinsic to the reaction, and those dependent on the environment:

- **Intrinsic properties of a reaction.** The intrinsic properties of a reaction are the changes of ΔH and ΔS contributing to ΔG. We can define standard values of these properties relative to arbitrary standard conditions such as concentration and temperature.

- **Concentrations and environmental factors.** The direction of the reaction depends on the concentrations of reactants and products. An excess of reactants over products makes ΔG more negative (forward reaction), whereas an excess of product makes ΔG more positive (reverse reaction). The direction of a reaction also depends on environmental factors such as temperature, pressure, and ionic strength (salt concentrations).

Standard Reaction Conditions

Scientists commonly present thermodynamic values under standard conditions for temperature, pressure, and concentration. The standard conditions make it possible to compare the intrinsic properties of reactions. The standard Gibbs free energy change is designated $\Delta G°$. The standard conditions for $\Delta G°$ are as follows:

- The temperature is 298 K (25°C).
- The pressure is 1 atm (standard atmospheric pressure).
- All concentrations of substrates and products are 1 molar (M).

For every chemical compound, a standard $\Delta G°$ of formation can be determined by chemical measurement. We obtain the $\Delta G°$ for a molecular reaction under standard conditions by summing the $\Delta G°$ values of the products, and subtracting the sum of $\Delta G°$ values of the reactants. A table of standard $\Delta G°$ values and a sample calculation are provided in Appendix 1.

But the $\Delta G°$ values reported in data tables hold only for isolated reactions under "standard reaction conditions." Standard conditions differ greatly from the actual conditions of living cells. These conditions include temperature, ionic strength, and gas pressure (in the case of gaseous components, such as CO_2), as well as the concentrations of reactants and products. To account for some of these differences, biochemists add special standard conditions: hydrogen ion concentration at pH 7, because living cells commonly maintain their cytoplasm within a unit of neutral pH; and water concentration of 55.5 M for dilute solutions. The free energy change in biochemistry is thus designated $\Delta G°'$.

How do we measure ΔG and its components, ΔH and $-T\Delta S$? The amount of energy released by a reaction is measured in an instrument called a **calorimeter**. A calorimeter may be designed to measure thermal energy (heat) released by a reaction of pure chemicals in a test tube, or of metabolizing bacteria—or even a large farm animal such as a cow (**Fig. 13.5**). The cow's thermal energy includes microbial digestion coupled to its own digestive processes, and to its own cellular biosynthesis; what is measured is residual energy lost as heat. In a calorimeter, the release of heat during a reaction (or collection of reactions) maintained at constant temperature gives a measure of ΔH. Measurement conducted at various temperatures (T) yields the temperature dependence of heat release, from which we calculate $-T\Delta S$. The sum of the two components yields ΔG.

> **Thought Questions**
>
> **13.1** Consider glucose catabolism in your blood, where the sugar is completely oxidized by O_2 and converted to CO_2:
>
> $$C_6H_{12}O_6 + 6O_2 \rightarrow 6CO_2 + 6H_2O$$
>
> Do you think this reaction releases greater energy as heat, or by change in entropy? Explain.
>
> **13.2** The bacterium *Lactococcus lactis* was voted the official state microbe of Wisconsin because of its importance for cheese production. During cheese production, *L. lactis* ferments milk sugars to lactic acid:
>
> $$C_6H_{12}O_6 \rightarrow 2C_3H_6O_3 \rightleftharpoons 2C_3H_5O_3^- + 2H^+$$
>
> Large quantities of lactic acid are formed, with relatively small increase in bacterial biomass. Why do you think biomass is limited? Cheese making usually runs more efficiently at high temperature; why?

The additivity of energy change is central to all living metabolism. Additivity makes it possible to do work by coupling an energy-yielding reaction to an energy-spending reaction (see Section 13.2). **Figure 13.6** shows an example, the initial reaction of glycolysis (see Section 13.4), in which ATP phosphorylates glucose to glucose 6-phosphate, catalyzed by the enzyme hexokinase. In this reaction, the loss of a phosphoryl group from ATP yields energy, some of which is captured by the enzyme to transfer the phosphoryl group onto glucose. The reactions and their $\Delta G^{\circ\prime}$ values are summed as follows:

Phosphorylation of glucose

$$C_6H_{12}O_6 + H_2PO_4^- \rightarrow C_6H_{12}O_6 - PO_3^- + H_2O$$

$$\Delta G_1^{\circ\prime} = +13.8 \text{ kJ/mol}$$

ATP hydrolysis

$$ATP + H_2O \rightarrow ADP + H_2PO_4^- + H^+$$

$$\Delta G_2^{\circ\prime} = -30.5 \text{ kJ/mol}$$

Sum:

ATP phosphorylation of glucose

$$C_6H_{12}O_6 + ATP \rightarrow C_6H_{12}O_6 - PO_3^- + ADP + H^+$$

$$\Delta G_3^{\circ\prime} = -16.7 \text{ kJ/mol}$$

Note: Distinguish these forms of the Gibbs free energy term:

Δ**G** = change in Gibbs free energy for a reaction under defined conditions.

Δ**G°** = ΔG at standard conditions of temperature (298 K) and pressure (1 atm, sea level), with all reactants and products at a concentration of 1 M. (Table of values in the Appendix.)

Δ**G°′** = ΔG at standard temperature, pressure, and concentrations for ΔG°, plus the biochemically relevant conditions of pH 7 (H^+ concentration of 10^{-7} M) and water (H_2O concentration of 55.5 M; activity = 1). Some biochemists standardize additional factors, such as magnesium ion concentration: $[Mg^{2+}]$ = 1 mM.

Concentrations of Reactants and Products

In living cells, the concentrations of reactants and products usually differ from 1 M; for example, in the *E. coli* cytoplasm, the concentrations of ATP and P_i are about 8 mM. Thus, the actual ΔG of reactions within a cell differs from ΔG° or $\Delta G^{\circ\prime}$.

Consider a reaction in which reactants A and B are reversibly converted to products C and D:

$$A + B \rightleftharpoons C + D$$

Higher concentration of reactants (A or B) drives the reaction forward, whereas higher concentration of products drives it in reverse. So, ΔG includes the ratio of products to reactants:

$$\Delta G = \Delta G^\circ + RT\ln\frac{[C][D]}{[A][B]}$$

$$= \Delta G^\circ + 2.303RT\log\frac{[C][D]}{[A][B]}$$

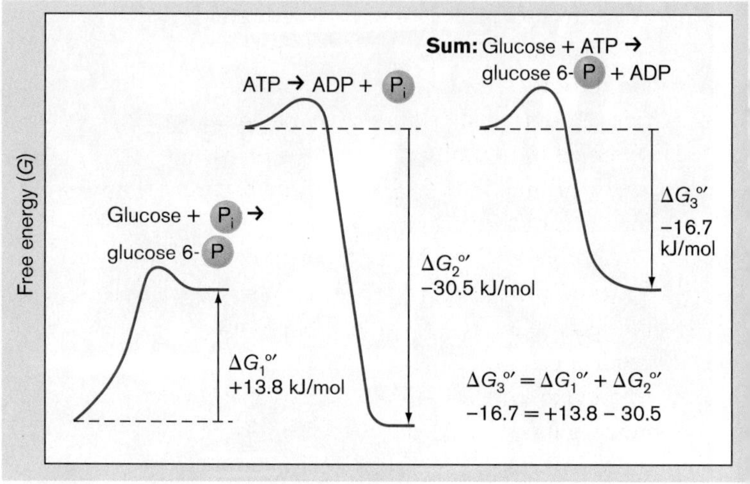

FIGURE 13.6 ■ Calculating the standard free energy of a reaction. The $\Delta G^{\circ\prime}$ value for the phosphorylation of glucose by ATP equals the sum of the values for the two coupled reactions: phosphorylation of glucose ($\Delta G_1^{\circ\prime}$) and ATP hydrolysis ($\Delta G_2^{\circ\prime}$).

where R is the gas constant [$R = 8.315 \times 10^{-3}$ kJ/(mol · K)] and T is the absolute temperature in kelvins (298 K, or 25°C). The factor 2.303 converts the logarithm of the ratio of products to reactants from base *e* (ln) to base 10 (log). Note that in $\Delta G^{\circ\prime}$ calculations, the water "activity" (concentration modified by a constant) is set at 1. The logarithm of 1 equals zero, so the water activity falls out of the equation.

Table 13.2 shows how the concentration ratio affects ΔG. In reactions at medium temperature (25°C–40°C), a 100-fold increase in the ratio of products to reactants adds about 11 kJ/mol to ΔG. ΔG is then less negative and the reaction less favorable. On the other hand, a 100-fold decrease in the concentration ratio (much fewer products than reactants) makes ΔG more negative by 11 kJ/mol.

In some environments, a highly negative concentration term can override a positive ΔG°, resulting in a reaction with negative ΔG that microbes can use for energy. For example, in an iron mine the high concentration of reduced iron

TABLE 13.2 Effect of the Concentration Ratio on ΔG

Initial ratio of products to reactants: $\frac{[C][D]}{[A][B]}$	Change in ΔG (kJ/mol) at standard temperature (298 K) and atmospheric pressure	Result of change from standard concentrations
10^{-4}	−23	Products increase
10^{-2}	−11	Products increase
1	0	$\Delta G = \Delta G^\circ$
10^2	+11	Reactants increase
10^4	+23	Reactants increase

favors iron-oxidizing microbes. Alternatively, a high temperature may increase the magnitude of the term $2.303RT \log$ [products]/[reactants] when it is negative, until it overrides a positive $\Delta G°$. This temperature dependence of ΔG is observed in thermophiles such as *Sulfolobus*, which metabolizes sulfur at 90°C; these sulfur reactions could not go forward at lower temperatures.

Another way the direction of reaction may change is for a second metabolic process to remove one of the reaction's products (C or D) as fast as it is produced. The decrease in product concentration could then change the ΔG of the first reaction to a negative value. For example, when bacteria break down glucose to pyruvate (see Section 13.4), many of the individual reactions have $\Delta G°'$ values of less than 5 kJ/mol. Their actual direction of reaction in the cell depends on concentrations of products and reactants. Glycolytic reactions with near-zero $\Delta G°'$ are reversible and can, in fact, participate in the biosynthetic pathway of gluconeogenesis (glucose biosynthesis, discussed in Chapter 15).

A surprising discovery has been that many bacteria and archaea in natural environments grow extremely slowly, using energy-yielding metabolism with values of ΔG approaching zero (that is, near thermodynamic equilibrium). When the actual ΔG (under actual reaction conditions) equals zero, a reaction proceeds equally forward and in reverse, and there is no net change in energy. At equilibrium, the ratio of product and reactant concentrations exactly cancels $\Delta G°$ or $\Delta G°'$:

$$\Delta G = 0 = \Delta G° + 2.303RT \log \frac{[C][D]}{[A][B]}$$

$$\Delta G° = -2.303RT \log \frac{[C][D]}{[A][B]}$$

Living cells can never grow exactly at equilibrium ($\Delta G = 0$). But most soil bacteria and archaea gain energy from anaerobic metabolism with near-zero values of ΔG. Such organisms must form biomass very slowly, but if no other metabolism is available, they may outgrow competitors. The discovery of low-ΔG energetics has opened new possibilities for environmental remediation previously thought impossible, such as the anaerobic digestion of complex organic pollutants in contaminated soil (see Section 13.6).

Some of the near-zero anaerobic pathways involve **syntrophy**, an intimate metabolic relationship between two species. For syntrophy, bacteria catabolize reactants by a reaction with a positive $\Delta G°'$ value, releasing a strong electron donor such as H_2. The hydrogen-releasing reaction is tightly coupled with that of H_2-oxidizing organisms such as methanogens, so the complete reaction has a net negative $\Delta G°'$. The bacterial catabolism provides H_2 gas, which the partner species consumes, thus keeping H_2 concentration low enough to drive the syntrophic reaction. Similar syntrophic partnerships may function in our gut (Section 13.5) and in anaerobic soil (Section 13.6).

To Summarize

- **Energy** enables cells to build ordered structures out of simple molecules from the environment. Transfer of energy is never perfectly efficient, so metabolism releases heat.
- **The free energy change (ΔG)** includes enthalpy (ΔH), the heat energy absorbed or released; and $-T\Delta S$, the negative product of temperature and entropy change (ΔS). Most reactions include both ΔH changes (such as oxidation-reduction) and ΔS changes (such as breakdown to a larger number of products).
- **Negative values of ΔG** show that a reaction can drive the cell's metabolism. The sign of ΔG depends on the relative magnitude of ΔH and $-T\Delta S$.
- **ΔH-driven reactions release heat.** Such reactions are called exothermic. By contrast, a reaction that absorbs heat (positive value of ΔH) is endothermic.
- **$T\Delta S$-driven reactions release less heat (or may absorb heat).** Reactions with a large $T\Delta S$ show a large temperature dependence of ΔG.
- **Intrinsic properties of the reaction determine the standard value of $\Delta G°$.** Properties include the molecular stability of reactants and products, and the entropy change associated with product conversion to reactants. By convention, standard conditions are set to define $\Delta G°$ (for biochemists, $\Delta G°'$).
- **Concentrations of reactants and products affect the actual value of ΔG.** The lower the concentration ratio of products to reactants, the more negative the value of ΔG. Environmental factors such as temperature and pressure also affect ΔG.
- **Within living cells, energy-yielding reactions are coupled with energy-spending reactions.** The measurement of energy flow under changing conditions requires calculations more complex than those shown here.

13.2 Energy Carriers and Electron Transfer

Our ΔG equations show only the total energy of a reaction such as glucose oxidation. If all the energy were released at once, however, it would dissipate as heat without building biomass. In living cells, glucose is never oxidized in one step. Instead, the energy yield is divided among a large number of stepwise reactions with smaller energy changes. In this way, the cell can be thought of as "making change"

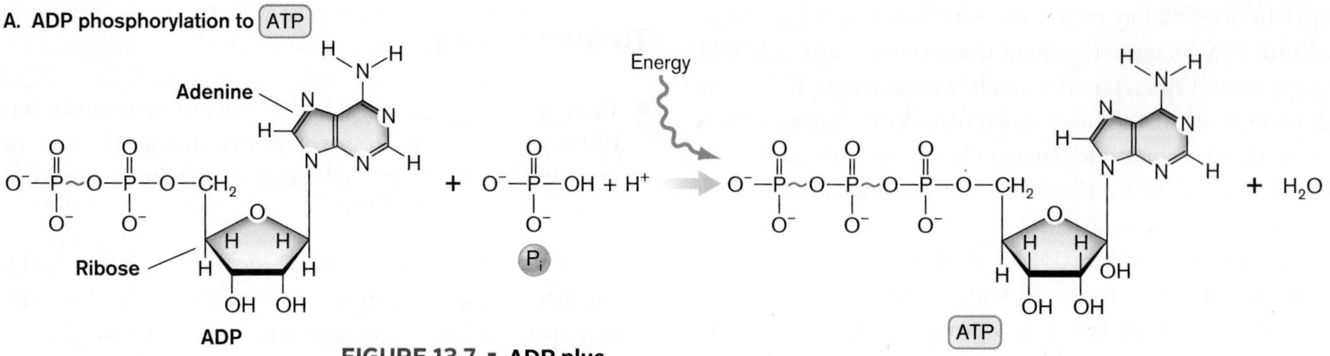

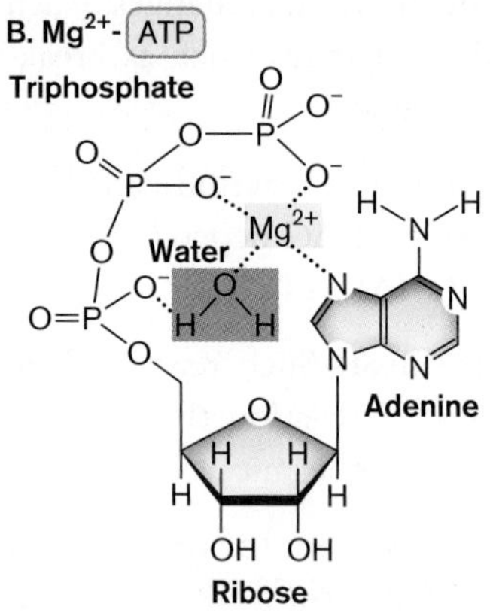

FIGURE 13.7 ■ ADP plus inorganic phosphate makes ATP. The reaction requires energy input (positive ΔG) because the negatively charged oxygens of the phosphoryl groups are forced to interact. **A.** The chemical reaction phosphorylating ADP (adenosine diphosphate) to ATP (adenosine triphosphate). **B.** Model of Mg^{2+}-ATP. The multiple negative charges of ATP are stabilized by binding a magnesium ion plus a water molecule.

by converting a large energy source to numerous smaller sources that can be "spent" conveniently for cell function and biosynthesis. The "spending" of energy is controlled by enzymes that couple all of the energy-yielding reactions to specific energy-spending reactions.

Energy Carriers Gain and Release Energy

Many of the cell's energy transfer reactions involve **energy carriers**. Examples of energy carriers are ATP (adenosine triphosphate) and NADH (the reduced form of nicotinamide adenine dinucleotide). Energy carriers are molecules that gain and release small amounts of energy in reversible reactions. Energy carriers are used to transfer energy in a wide range of biochemical reactions.

Some energy carriers, such as NADH, transfer energy associated with electrons received from a food molecule. A molecule that transfers, or "donates," electrons to another molecule is called an **electron donor** or a reducing agent; a molecule that receives, or "accepts," electrons is called an **electron acceptor**. For example, during glucose catabolism a molecule of glyceraldehyde 3-phosphate transfers a pair of electrons ($2e^-$) with a hydrogen ion (H^+) to NAD^+, forming NADH. NAD^+ is an electron acceptor that receives the electrons; it then becomes the electron donor NADH. Electron donors such as NADH transfer electrons from reduced food molecules to a terminal electron acceptor such as oxygen (see Chapter 14). Energy carriers that transfer electrons are needed for all energy-yielding pathways and for biosynthesis of cell components such as amino acids and lipids (see Chapter 15).

Note that in living cells, all energy transfer reactions are coupled by enzymes to specific biochemical processes. Without enzyme coupling, energy would dissipate and be lost from the living system.

ATP Carries Energy

Adenosine triphosphate, or **ATP** (**Fig. 13.7A**), is composed of a base (adenine), a sugar (ribose), and three phosphoryl groups. Note that adenine-ribose-phosphate (adenosine nucleotide) is equivalent to a nucleotide of RNA. The base adenine is a fundamental molecule of life, one that forms spontaneously from methane and ammonia in experiments simulating the origin of life on early Earth (discussed in Chapters 1 and 17). Like the sugar ribose, ATP is an ancient component of cells, found in all living organisms.

Under physiological conditions, ATP always forms a complex with Mg^{2+} (**Fig. 13.7B**). The magnesium cation partly neutralizes the negative charges of the ATP phosphates, stabilizing the structure in solution. Most enzyme-binding sites for ATP actually bind Mg^{2+}-ATP. This is one reason why magnesium is an essential nutrient for all living cells.

ADP phosphorylation to ATP. During cell metabolism, ATP is generated by **phosphorylation**, the condensation of inorganic phosphate with adenosine diphosphate (ADP):

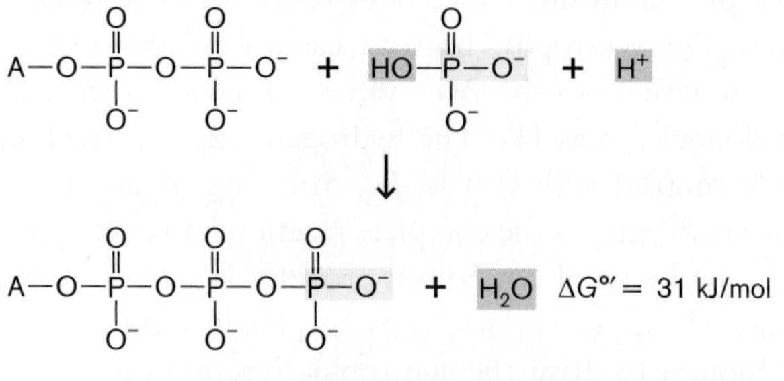

The phosphorylation of ADP to form ATP requires energy input (positive ΔG).

Why does ATP formation require energy? The inorganic phosphate molecule has four oxygen atoms that share a negative charge. When phosphate reacts with another phosphate to form a bond, the charged oxygens of adjacent phosphates are forced to interact despite charge repulsion. The charge repulsion in ATP limits the rotation of oxygens, and thus decreases entropy, resulting in a negative ΔG of hydrolysis. Hydrolysis of each phosphoryl group yields energy. The formation and hydrolysis of ATP can be shown as:

$$\text{A–P}{\sim}\text{P} + \text{H}^+ + \text{P}_i \rightleftharpoons \text{A–P}{\sim}\text{P}{\sim}\text{P} + \text{H}_2\text{O}$$

where ~ designates each energy-storing phosphoryl bond and P_i designates inorganic phosphate; or, in more concise shorthand:

$$\text{ADP} + \text{P}_i \rightleftharpoons \text{ATP} + \text{H}_2\text{O}$$

ATP is a "medium-sized" energy carrier, since the cell contains many phosphorylated molecules that yield greater energy upon hydrolysis. **Table 13.3** lists examples of free energy change associated with hydrolysis of various phosphoryl groups. As we will see, phosphoryl group hydrolysis with $\Delta G°'$ values larger than –31 kJ/mol can yield energy that is stored by ATP formation. ATP hydrolysis can yield energy to phosphorylate molecules at $\Delta G°'$ values smaller than –31 kJ/mol.

ATP transfers energy. ATP can transfer energy to cellular processes in three different ways: hydrolysis releasing phosphate, hydrolysis releasing pyrophosphate (diphosphate), and phosphorylation of an organic molecule. Each process serves different functions in the cell.

- **Hydrolysis releasing phosphate.** The **hydrolysis** of ATP at the terminal phosphate consumes H_2O to produce ADP and P_i, releasing energy. The energy released by ATP hydrolysis can be transferred to a coupled reaction of biosynthesis, such as building an amino acid. A "coupled reaction" means that a reaction that spends energy can go forward if it is connected to a reaction that yields energy. For a reaction driven by ATP hydrolysis, the two reactions are coupled by an enzyme with binding sites specific for ATP and the substrate of the energy-spending reaction.
- **Hydrolysis releasing pyrophosphate.** ATP can hydrolyze at the middle phosphate, releasing pyrophosphate (PP_i). The pyrophosphate usually hydrolyzes shortly afterward to make 2 P_i. Overall, the release of 2 P_i from ATP yields approximately twice as much energy as the release of 1 P_i. Pyrophosphate release and subsequent hydrolysis drives a reaction strongly forward, because twice as much energy would be required to reverse the reaction. Pyrophosphate is released in reactions that must avoid reversal—for example, the incorporation of nucleotides into growing chains of RNA.
- **Phosphorylation of an organic molecule.** ATP can transfer its phosphate to the hydroxyl group of a molecule such as glucose to activate the substrate for a subsequent rapid reaction. No inorganic phosphate appears, and no water molecule is consumed:

$$\text{ATP} + \text{glucose} \rightarrow \text{ADP} + \text{glucose 6-P}$$

Some enzymes catalyze ATP transfer of phosphate to activate sugar molecules for catabolism. Other enzymes couple the phosphorylation of a sugar to its transport across the cell membrane; consequently, these enzymes make up the phosphotransferase system (PTS). The PTS enzymes play a critical role in determining which nutrients from the environment some microbes can acquire and catabolize (discussed in Chapter 4).

TABLE 13.3 Hydrolysis of Phosphoryl Groups: Values of $\Delta G°'$ (pH 7.0, 25°C)

Reaction of hydrolysis	$\Delta G°'$ (kJ/mol)
Glucose 6-P + $H_2O \rightarrow$ glucose + P_i	–14
Fructose 1,6-bis-P + $H_2O \rightarrow$ fructose 6-P + P_i	–16
PP_i (pyrophosphate) + $H_2O \rightarrow 2\ P_i + H^+$	–19
ATP + $H_2O \rightarrow$ ADP + $P_i + H^+$	–31
ADP + $H_2O \rightarrow$ AMP + $P_i + H^+$	–33
ATP + $H_2O \rightarrow$ AMP + PP_i (pyrophosphate) + H^+	–46
1,3-Bis-P-glycerate + $H_2O \rightarrow$ 3-P-glycerate + P_i	–49
Phosphoenolpyruvate + $H_2O \rightarrow$ pyruvate + P_i	–62

Thought Questions

13.3 When ATP phosphorylates glucose to glucose 6-phosphate, what is the net value of $\Delta G°'$? What if ATP phosphorylates pyruvate? Can this latter reaction go forward without additional input of energy? (See **Table 13.3**.)

13.4 Linking an amino acid to its cognate tRNA is driven by ATP hydrolysis to AMP (adenosine monophosphate) plus pyrophosphate. Why release PP_i instead of P_i?

ATP produced by glucose catabolism. A large number of ATPs can be formed by coupling ATP synthesis to the step-by-step breakdown and oxidation of a food molecule such as glucose. In theory, complete oxidation of glucose through respiration can produce as many as 38 ATP molecules. The overall $\Delta G°'$ of the coupled reactions is:

$$\begin{aligned} &C_6H_{12}O_6 + 6H_2O + 6O_2 \rightarrow \\ &\qquad 12H_2O + 6CO_2 \quad \Delta G°' = -2{,}878\ \text{kJ/mol} \\ &38[\text{ADP} + \text{P}_i \rightarrow \text{ATP} + H_2O] \quad \Delta G°' = 38 \times (+31\ \text{kJ/mol}) \\ \hline &\qquad\qquad \text{Net}\ \Delta G°' = -1{,}700\ \text{kJ/mol} \end{aligned}$$

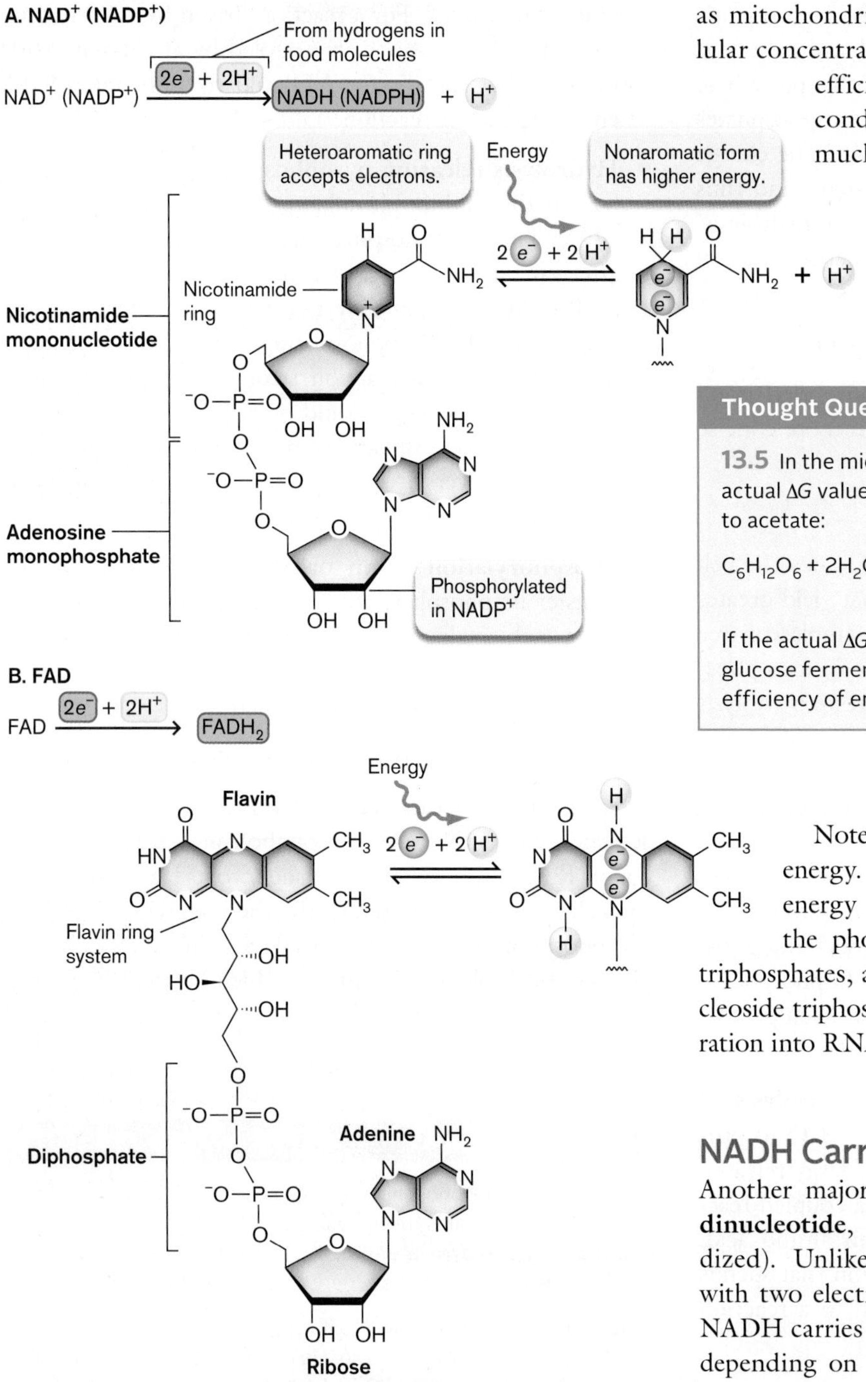

FIGURE 13.8 ■ Reduction of NAD⁺ and FAD. **A.** NAD^+ reduction: The nicotinamide ring (shaded pink) loses a double bond as two electrons are gained from an electron donor. Two hydrogen atoms are consumed; one bonds to NADH, while the other ionizes. **B.** FAD reduction: The flavin ring system gains two electrons associated with two hydrogens.

The difference in $\Delta G°'$ for the coupled reactions is the energy lost as heat and entropy—in this case –1,700 kJ/mol (–2,878 kJ/mol + 1,178 kJ/mol). Thus, the maximal efficiency of energy capture by ATP is about 40%, a level that may be approached by highly efficient systems such as mitochondria. When ΔG values are corrected for cellular concentrations of reactants and products, the actual efficiency may be greater than 50%. Under conditions such as low oxygen concentration, much smaller amounts of ATP are made per molecule of glucose. For comparison, the efficiency of a typical machine, such as an internal combustion engine, is about 20%.

> **Thought Question**
>
> **13.5** In the microbial community of the bovine rumen, the actual ΔG value has been calculated for glucose fermentation to acetate:
>
> $$C_6H_{12}O_6 + 2H_2O \rightarrow 2C_2H_3O_2^- + 2H^+ + 4H_2 + 2CO_2 \qquad \Delta G = -318\ \text{kJ/mol}$$
>
> If the actual ΔG for ATP formation is +44 kJ/mol and each glucose fermentation yields 4 ATP, what is the thermodynamic efficiency of energy gain? Where does the lost energy go?

Note that, besides ATP, other nucleotides carry energy. Guanosine triphosphate (GTP) provides energy for ribosome elongation of proteins. And the phosphodiester bonds of all four nucleotide triphosphates, as well as their corresponding deoxyribonucleoside triphosphates, carry energy for their own incorporation into RNA and DNA, respectively.

NADH Carries Energy and Electrons

Another major energy carrier is **nicotinamide adenine dinucleotide**, or **NAD** (NADH, reduced; NAD^+, oxidized). Unlike ATP, NADH carries energy associated with two electrons that reduce a substrate (**Fig. 13.8A**). NADH carries two or three times as much energy as ATP, depending on cellular conditions. During sugar catabolism, NADH carries electrons from breakdown products of glucose. Its oxidized form, NAD^+, receives two electrons ($2e^-$) plus a hydrogen ion (H^+) from a food molecule; a second H^+ from the food molecule enters the solution. Overall, reduction of NAD^+ consumes two hydrogen atoms to make NADH:

$$NAD^+ + 2H^+ + 2e^- \rightarrow NADH + H^+ \qquad \Delta G°' = +62\ \text{kJ/mol}$$

For this reaction, $\Delta G°'$ is positive; therefore, it requires input of energy from catabolism of the food molecule. The reduced energy carrier NADH can then reverse this reaction by donating two electrons ($2e^-$) to another molecule, regenerating NAD^+.

Note: A hydrogen ion (H^+), or "proton," does not exist free in solution. In water, a H^+ combines with H_2O to form a hydronium ion (H_3O^+), but for clarity we use H^+. An atom of hydrogen removed from a C–H bond consists of a proton (H^+) plus an electron (e^-). In a reaction, the proton and electron may be transferred to one molecule or to separate molecules.

NADH structure and function. NAD^+ consists of an ADP molecule attached to nicotinamide instead of a third phosphate. The nicotinamide mononucleotide, like a ribonucleotide, contains a nitrogenous base attached to a sugar phosphate. In NAD^+, the nicotinamide has a ring structure (shaded pink in **Fig. 13.8A**) that forms a stable cation.

NAD^+ is a relatively stable structure because the ring electrons are **aromatic**; that is, the bonding electrons delocalize around the ring, as in benzene. Aromatic rings that contain noncarbon atoms are said to be heteroaromatic. Many biologically active molecules are heteroaromatic, including adenine and other nucleotide bases. A heteroaromatic ring is stable, but its disruption requires less energy than the disruption of benzene. Thus, it is possible to disrupt the ring by adding two electrons with a H^+, eliminating one double bond. The donation of electrons eliminates the ring's aromaticity and thus stores energy. The reduced molecule NADH carries energy in an amount useful for cell reactions.

The electrons transferred to NADH eventually must be put somewhere else, onto the next electron acceptor. If NADH builds up in a cell, no NAD^+ remains to continue oxidizing food molecules. One way that the energy stored by NADH can be spent is to transfer $2H^+ + 2e^-$ onto a product of catabolism. For example, in ethanolic fermentation to make wine or beer, NADH reduces pyruvate to ethanol. In this case, however, the energy is lost to the cell. Alternatively, NADH can transfer its electrons to one of a series of electron carrier molecules known as the **electron transport system** (**ETS**), also called the electron transport chain (ETC). Electron transport within bacteria can actually be used to generate electricity for commercial power (discussed in Chapter 14). Examples of electron transfer reactions are shown in the "tower of power" in **Table 13.4**. For example, adding two electrons ($2e^-$) to NAD^+ to make NADH has a highly negative value of standard reduction potential $E°'$ (positive $\Delta G°'$), which means that the reaction requires energy input.

NADH oxidation and reduction participate in reactions of the ETS. The ETS includes a series of proteins and small organic molecules that can be reduced and cyclically reoxidized. The redox reactions store energy from electron transfer as ion gradients across the membrane of the cell or an organelle. At the end of the redox series, the electrons are transferred to a **terminal electron acceptor** whose product leaves the cell. For example, as a terminal electron acceptor, molecular oxygen (O_2) is reduced to H_2O. The reaction of O_2 reduction to H_2O may be coupled to oxidation of NADH:

$$NADH + H^+ \rightarrow NAD^+ + 2H^+ + 2e^- \quad \Delta G°' = -62 \text{ kJ/mol}$$
$$\tfrac{1}{2}O_2 + 2H^+ + 2e^- \rightarrow H_2O \quad \Delta G°' = -158 \text{ kJ/mol}$$
$$NADH + H^+ + \tfrac{1}{2}O_2 \rightarrow NAD^+ + H_2O \quad \Delta G°' = -220 \text{ kJ/mol}$$

Thus, the total energy released during NADH oxidation through the ETS is –220 kJ/mol (–62 kJ/mol – 158 kJ/mol). This energy is converted to transmembrane proton potential (composed of the H^+ concentration difference plus the charge difference across the membrane). Reduction potentials and the proton potential (see Chapter 14) drive nutrient transport, motility, and synthesis of ATP.

Note: Reduction potentials, electron transport, and proton motive force are discussed further in Chapter 14.

TABLE 13.4 Standard Reduction Potentials*

Electron acceptor[a]	→	Electron donor	$E°'$ (mV)	$\Delta G°'$ (kJ)
$2H^+ + 2e^-$	→	H_2	–420	+81
$NAD^+ + 2H^+ + 2e^-$	→	$NADH + H^+$	–320	+62
$FAD + 2H^+ + 2e^-$	→	$FADH_2$	–220	+42
$FMN + 2H^+ + 2e^-$	→	$FMNH_2$	–190	+37
Menaquinone + $2H^+ + 2e^-$	→	Menaquinol	–74	+14
Fumarate + $2H^+ + 2e^-$	→	Succinate	+33	–6
Ubiquinone + $2H^+ + 2e^-$	→	Ubiquinol	+110	–21
$NO_3^- + 2H^+ + 2e^-$	→	$NO_2^- + H_2O$	+420	–81
$\tfrac{1}{2}O_2 + 2H^+ + 2e^-$	→	H_2O	+820	–158

*For a more extensive list, see Table 14.1.

[a]FAD = flavin adenine dinucleotide; FMN = flavin mononucleotide.

Other energy carriers that transfer electrons. Different steps of metabolism utilize different but related energy carriers. For example, NADPH differs from NADH only in the extra phosphate attached to the 2′ carbon of adenine nucleotide; the amount of energy carried is the same. Some enzymes can use both NADPH and NADH, whereas other enzymes use only one or the other.

Another related energy carrier is **flavin adenine dinucleotide**, or **FAD** ($FADH_2$, reduced; FAD, oxidized), in which flavin substitutes for nicotinamide. The flavin nucleotide includes a ring structure whose aromaticity is eliminated by its receiving two electrons (**Fig. 13.8B**). The redox function of the flavin isoalloxazine ring system is similar to that of NADH:

$$FAD + 2H^+ + 2e^- \rightarrow FADH_2$$

Like NADH, $FADH_2$ donates $2e^-$ to an electron acceptor. $FADH_2$ is a weaker electron donor than NADH, but when $FADH_2$ is combined with a strong electron acceptor such as O_2, electrons are transferred and significant energy is released:

$$\Delta G^{\circ\prime} = -42 - 158 = -200 \text{ kJ/mol}$$

Why do different kinds of reactions use different energy carriers?

- **Different redox levels.** Food molecules may have more or fewer electrons (level of reduction/oxidation) than those associated with the cell structure. For example, lipids are more highly reduced than glucose. Thus, lipid catabolism requires a greater proportion of electron-accepting energy carriers (such as NAD^+ or $NADP^+$) than does glucose catabolism, and it makes relatively few ATP molecules directly. A combination of energy carriers with different redox states enables cells to balance their overall redox potential while transferring energy.
- **Different amounts of energy.** Biochemical reactions yield different amounts of energy—that is, different values of ΔG. Suppose a reaction can provide more than enough energy to generate ATP from ADP (31 kJ), but not quite enough to generate NADH from NAD^+ (62 kJ). An example is the conversion of succinate to fumarate in the TCA cycle (see Section 13.4). Succinate conversion provides the energy to reduce FAD to $FADH_2$, whose oxidation by O_2 can yield two molecules of ATP. Thus, the use of $FADH_2$ enables the cell to make more efficient use of its food than if generation of ATP or NADH were the only choices.
- **Regulation and specificity.** Specific energy carriers can direct metabolites into different pathways serving different functions. For example, in many bacteria NADH is directed into the ETS, whereas NADPH, the 2′-phosphorylated form of NADH, is directed into biosynthesis of cell components such as amino acids and lipids.

Concentration Gradients Store Energy

So far, our discussion of energy has assumed an isotropic system, in which concentrations are the same everywhere. But a living cell needs to obtain its molecules from outside, such as sugars, amino acids, and inorganic ions (discussed in Chapter 4). Suppose a blood pathogen needs to obtain a scarce substance such as iron (Fe^{2+}). How does the microbe move the iron "uphill" against its concentration gradient? In fact, a concentration gradient of any substance can store energy, just as an energy carrier molecule does.

A substance dissolved in water diffuses by random movements until its distribution has the same concentration throughout (**Fig. 13.9A**). The random distribution of molecules at uniform concentration represents the state of greatest entropy. Diffusion in the environment ultimately brings nutrients into contact with microbial cells, even cells that lack chemotactic motility to hunt for food. For example, sugars in the food we eat diffuse through our saliva to reach the bacteria growing in biofilms on our teeth.

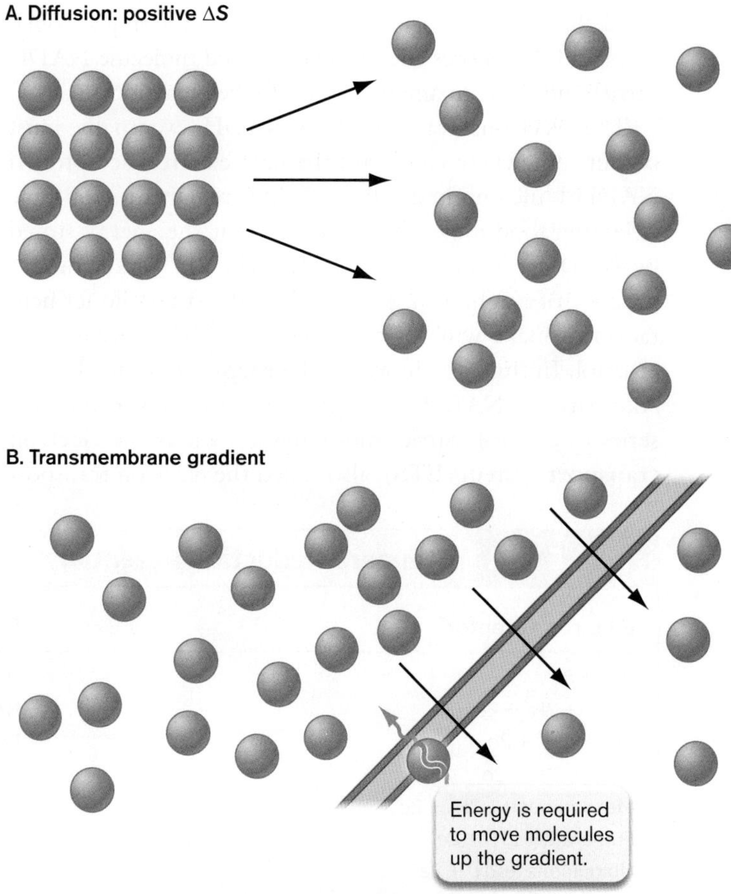

FIGURE 13.9 ▪ Diffusion and transport. A. Water-soluble molecules diffuse to uniform concentration throughout the solution: ΔS is positive, $-T\Delta S$ negative; the entropy term favors the process. B. If a membrane separating two compartments is permeable, molecules move from a compartment with high concentration to one with low concentration. Energy is required to move molecules up their concentration gradient.

The cell membrane contains transporters for useful molecules, allowing nutrient molecules to cross. Entropy favors their movement from higher to lower concentration (**Fig. 13.9B**). In most environments, however, the nutrients are at lower concentrations than inside the cell. To obtain these molecules from outside, the bacterial cell must transport them against their gradient—that is, from lower to higher concentration—increasing the concentration difference and thus decreasing entropy. Uptake against a concentration gradient requires an energy source to power transport proteins embedded in the membrane (as shown in Chapter 4). Alternatively, a transmembrane gradient of ions can store energy for the cell. The most important ion gradient is the H^+ gradient, or transmembrane pH difference, a component of the proton motive force (PMF).

Note: The proton motive force (PMF) is composed of the transmembrane concentration difference of hydrogen ions (ΔpH) and the electrical potential difference (–Δψ). The PMF is explained in detail in Chapter 14.

Enzymes Catalyze Metabolic Reactions

The Δ*G* values of energy change show whether a reaction goes forward, but not how long it will take. A few seconds, or a hundred years? The Δ*G* value does not specify the reaction rate—that is, the kinetics of reaction.

In living cells, each reaction must occur only as needed, in the right amount at the right time. The rate of a reaction is determined by the **activation energy** (E_a), the input energy needed to generate the high-energy transition state on the way to products (**Fig. 13.10**). Most biochemical reactions require an activation energy that exceeds the average kinetic energy of the reactant molecules colliding. Thus, no matter how negative the Δ*G* is, the reaction will proceed at a significant rate only when the activation energy is lowered by interaction with a catalyst, an agent that participates in a reaction without being consumed.

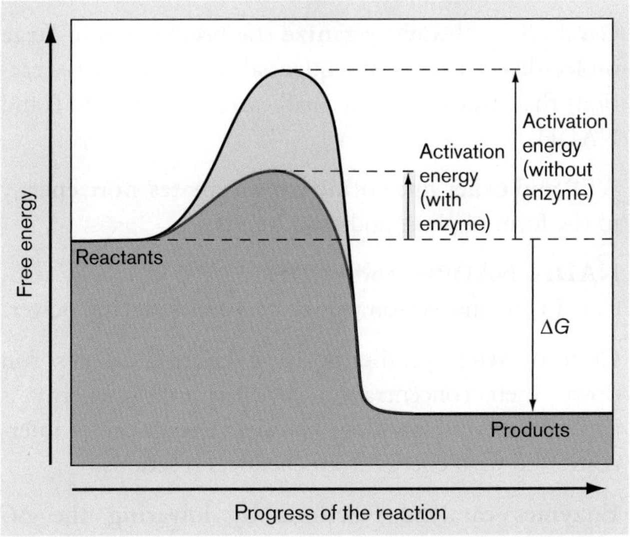

FIGURE 13.10 ■ Enzymes lower the activation energy of the transition state. In the presence of an enzyme, the activation energy of the reaction is decreased, allowing rapid conversion of reactants to products.

Biological reactions are catalyzed by **enzymes**, structures composed of protein (or, in some cases, RNA) that bind substrates of a specific reaction. The enzyme lowers the activation energy by bringing the substrates in proximity to one another and by correctly orienting them to react. In some cases, enzymes provide a reactive amino acid residue to participate in a transition state between reactants and products. Microbial enzymes have growing importance in industry; they are used for food production, fabric treatment, and drug therapies (discussed in Chapter 16).

Enzymes couple specific energy-yielding reactions (such as those of glucose breakdown) to the cell's energy-spending reactions (such as making ATP). An example of coupled reactions is shown in **Figure 13.11A**. The enzyme pyruvate kinase catalyzes removal of a phosphoryl group from the substrate phosphoenolpyruvate (PEP), a breakdown product of glucose. PEP is converted to pyruvate while the phosphoryl group is added to ADP, generating ATP:

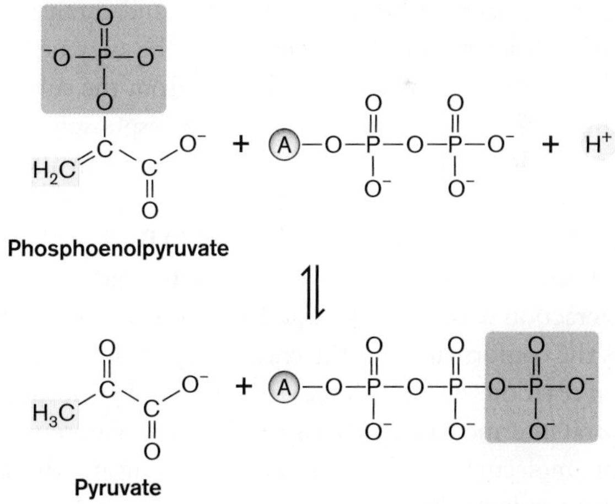

The $\Delta G°'$ of phosphate cleavage from PEP is –62 kJ/mol, whereas the $\Delta G°'$ of ATP formation is only +31 kJ/mol. Thus, the net $\Delta G°'$ is negative (–31 kJ/mol), and the reaction goes forward to pyruvate. But a high activation energy makes the reaction extremely slow; without help, it rarely goes forward on the timescale of life. The reaction proceeds only when the two reacting substrates (PEP and ADP) are coupled by the enzyme pyruvate kinase (**Fig. 13.11A**). Pyruvate kinase has specific binding sites for each of its substrates: PEP and ADP. ADP and PEP are brought together by the enzyme and positioned so as to lower the activation energy of phosphate transfer.

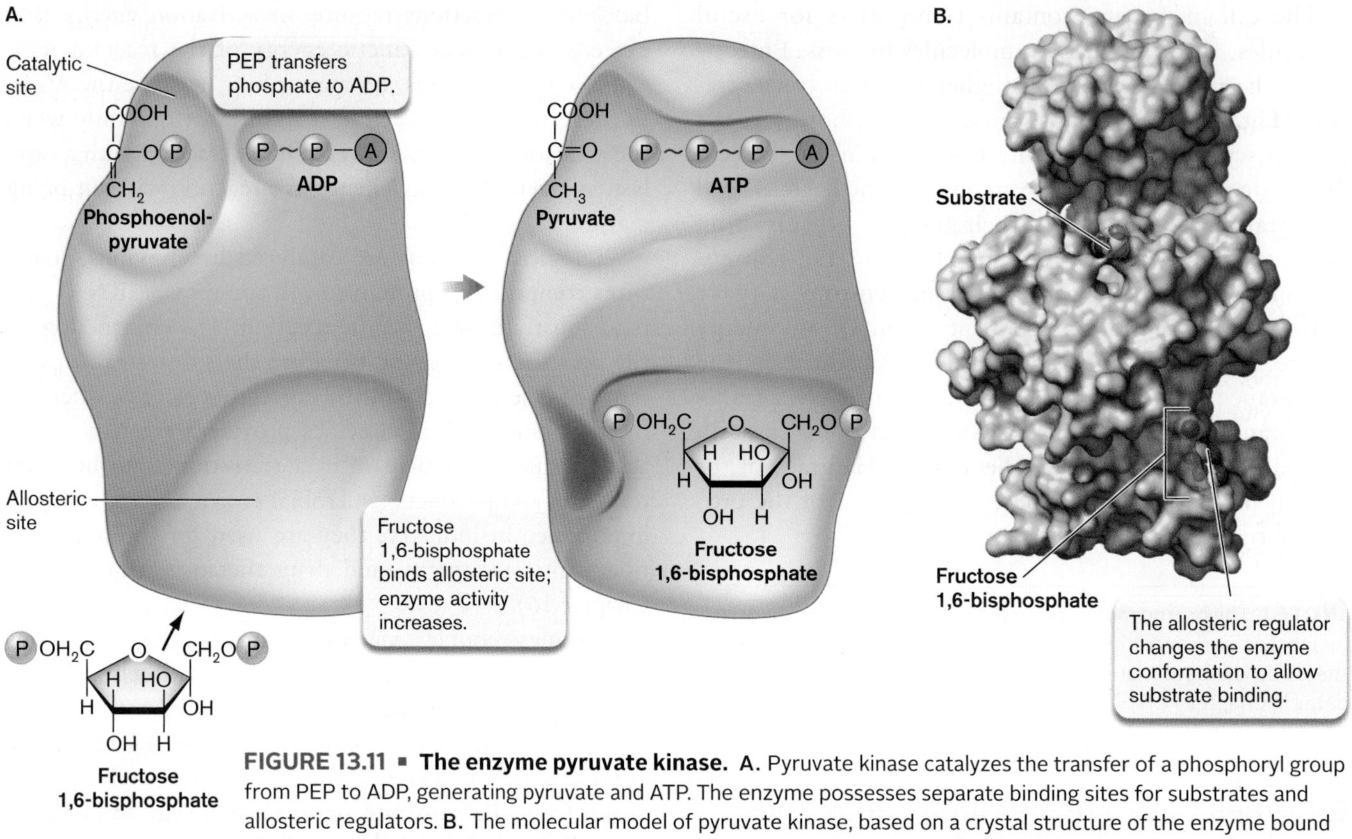

FIGURE 13.11 ■ The enzyme pyruvate kinase. A. Pyruvate kinase catalyzes the transfer of a phosphoryl group from PEP to ADP, generating pyruvate and ATP. The enzyme possesses separate binding sites for substrates and allosteric regulators. **B.** The molecular model of pyruvate kinase, based on a crystal structure of the enzyme bound to a substrate analog and an allosteric regulator, fructose 1,6-bisphosphate. (PDB code: 1A3W)

The **catalytic domain** or **catalytic site** is the part of the protein that actually binds the substrate and catalyzes the reaction. In addition to the catalytic site, the enzyme has an **allosteric site** (a regulatory site distinct from the substrate-binding site) for its activator, fructose 1,6-bisphosphate. The difference between a substrate-binding site and an allosteric site can be seen in the molecular model of pyruvate kinase based on X-ray crystallography (**Fig. 13.11B**). The allosteric site is found at a distance from the substrate-binding site, but its interaction with the regulator, fructose 1,6-bisphosphate, alters the conformation of the entire enzyme, increasing the rate of reaction. As we will see, fructose 1,6-bisphosphate is a central intermediate of glycolysis; thus, it makes sense that as this molecule builds up, it activates pyruvate kinase to remove products farther down the chain of catabolic reactions, maintaining the steady flow of metabolites.

Thought Question

13.6 What would happen to the cell if pyruvate kinase catalyzed PEP conversion to pyruvate but failed to couple this reaction to ATP production?

Note that pyruvate kinase is actually named for its reverse reaction: pyruvate phosphorylation to PEP. The enzyme may have been named for activity that was originally observed in the reverse direction, before the cell function was understood. Enzymes can catalyze both forward and reverse reactions. The predominant direction of catalysis depends on the concentrations of substrates and products (which determine ΔG) and on allosteric regulators.

To Summarize

- **Catabolic pathways organize the breakdown of large molecules** in a series of sequential steps coupled to reactions that store energy in small carriers such as ATP and NADH.
- **ATP and other nucleotide triphosphates store energy** in the form of phosphodiester bonds.
- **NADH, NADPH, and $FADH_2$** each store energy associated with an electron pair that carries reducing power.
- **Concentration gradients store energy.** Solutes run down their concentration gradient unless energy is applied to reverse the flow. Gradient energy can be interconverted with energy from chemical reactions.
- **Enzymes catalyze reactions by lowering the ΔG** required to reach the transition state. They couple energy transfer reactions to specific reactions of biosynthesis and cell function.

13.3 Catabolism: The Microbial Buffet

In the early twentieth century, it was thought that microbes could catabolize a limited subset of naturally occurring organic molecules, such as sugars. Molecules not known to be catabolized were termed "xenobiotics," especially if they were "synthetic" products of human industry. We have since found that virtually any organic molecule can be catabolized by a microbe that has evolved the appropriate enzymes. Catabolism by microbes in our own digestive system helps us digest complex food molecules. Catabolism also plays key roles in microbial disease; for example, the causative agent of acne, *Cutibacterium acnes*, degrades skin cell components such as lipids.

Note: Besides releasing energy, the breakdown of organic food molecules provides substrates for biosynthesis. Biosynthesis is covered in Chapter 15.

Substrates for Catabolism

Here we present several important classes of catabolic substrates (**Fig. 13.12**). While, in principle, virtually any organic constituent may be catabolized, certain kinds of substrates are used more rapidly because they require less activation energy or fewer types of enzymes to break them down. Many of these substrates form products important to our nutrition and technology. Others, such as aromatic components of petroleum, are environmental pollutants that only bacteria and fungi can degrade (see Section 13.6).

Carbohydrates. Carbohydrates include sugars and sugar polymers, which are generally called **polysaccharides** or **glycans** (**Fig. 13.12A**). Glycans are important as structural components of cells, and for their central role in human digestion. Glycans composed of glucose are also called glucans.

Microbial catabolism is central to our production of food and drink (presented in Chapter 16). Glucose as such is rarely available to microbes, except to pathogens growing within a host. But the pathways of glucose catabolism complete the digestion of a diverse range of food molecules found in the environment. The main glycan digested by human enzymes is **starch**, a polymer of glucose monomers in which an acetal (O–COH) condenses with a hydroxyl group, releasing H_2O. Starch is the main caloric component of foods such as bread, rice, and potatoes.

But intestinal bacteria such as *Bacteroides* species can digest a far greater range of polysaccharides or glycans, such as the pectin of fruit, and the multisugar xyloglucans of lettuce and tomatoes. Glycans may contain extra functional groups, such as the sulfates of porphyran, a glycan found in *Porphyra* seaweed. These glycans were originally defined as "fiber"—that is, polymers indigestible by humans. We now know that partial digestion of fiber contributes significant caloric content to the human diet.

Glycan sugar polymers are broken down by microbial enzymes, first to short chains (oligosaccharides), then to two-sugar units (disaccharides), and then to monosaccharides such as glucose or fructose. In some sugars, the aldehyde is replaced by a hydroxyl (sorbitol, mannitol) or a carboxylate (gluconate, glucuronate). The overall result is that glycans are hydrolyzed to products that enter central catabolic pathways such as glycolysis (**Fig. 13.13**).

Lipids. Many bacteria catabolize lipids (**Fig. 13.12B**) from sources such as milk, animal fats, and nuts. The oxidation of lipid catabolites causes the rancid odor of spoiled meat or butter (issues of food microbiology, discussed in Chapter 16). Microbes catabolize lipids by hydrolysis to glycerol and fatty acids (**Fig. 13.13**). Glycerol, a three-carbon triol (a compound with three –OH groups), can be considered a three-carbon sugar; it commonly enters catabolism as an intermediate of glycolysis. Alternatively, other pathways break down glycerol to acetate. Fatty acids, more highly reduced than glycerol, undergo oxidative breakdown by the fatty acid degradation pathway, forming acetyl groups. These acetyl groups enter the TCA cycle when a terminal electron acceptor is available; alternatively, they enter fermentation or anaerobic syntrophy.

Peptides. We commonly think of proteins as essential parts of a cell. When present in excess, however, proteins can be catabolized to provide energy. Initially, proteins are broken into peptides (**Fig. 13.12C**) by sequence-specific proteases. Peptides are then hydrolyzed to individual amino acids (**Fig. 13.13**).

Enzymes catalyze the early steps in the degradation of each amino acid until products are formed that can enter common pathways of carbohydrate catabolism or the TCA cycle. The initial step of amino acid degradation is one of two kinds: decarboxylation (removal of CO_2) to produce an amine, or deamination (removal of NH_3) to produce a carboxylic acid. Carboxylic acids are degraded through the TCA cycle. Amine products must be exported from the cell. For gut bacteria, these exported amines may serve host functions such as those of immunomodulators and neurotransmitters (discussed in Section 13.5). Other amines, such as cadaverine and putrescine, cause the noxious odor of decomposing flesh.

Aromatic molecules. Aromatic compounds are more difficult to digest than sugars because of the exceptional stability of aromatic ring structures. Yet many bacteria metabolize benzene derivatives, and even polycyclic aromatic molecules, either partly or all the way to CO_2. A particularly important aromatic substance found in nature is

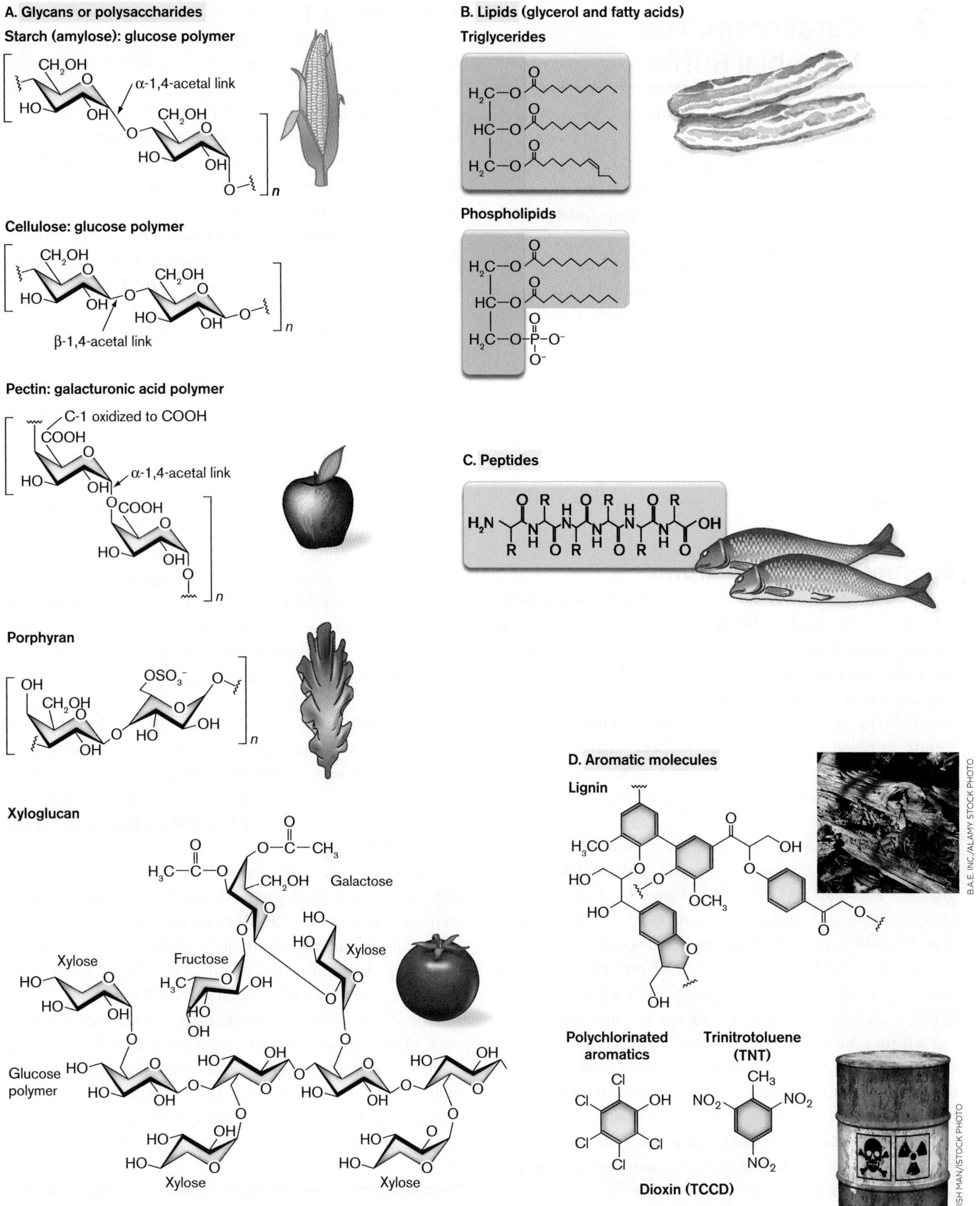

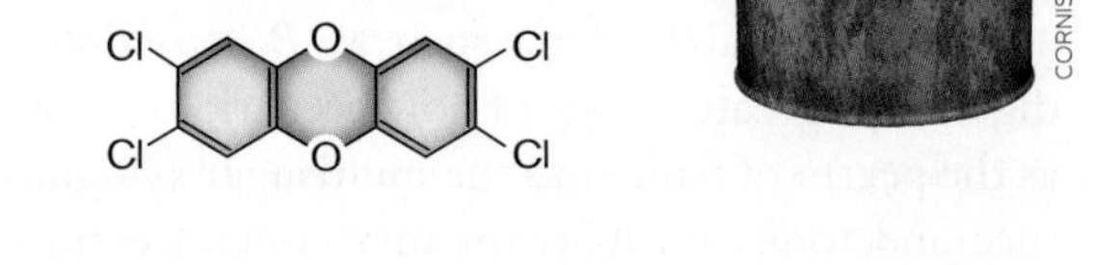

FIGURE 13.12 ■ Complex carbon sources for catabolism. **A.** Glycans or polysaccharides such as starch, cellulose, and pectin are hydrolyzed to six-carbon sugars. **B.** Lipids are broken down to acetate. **C.** Peptides are hydrolyzed to amino acids and then broken down to acetate, amines, and other molecules. **D.** Complex aromatic molecules such as lignins and halogenated aromatic pollutants are broken down to acetate and other molecules.

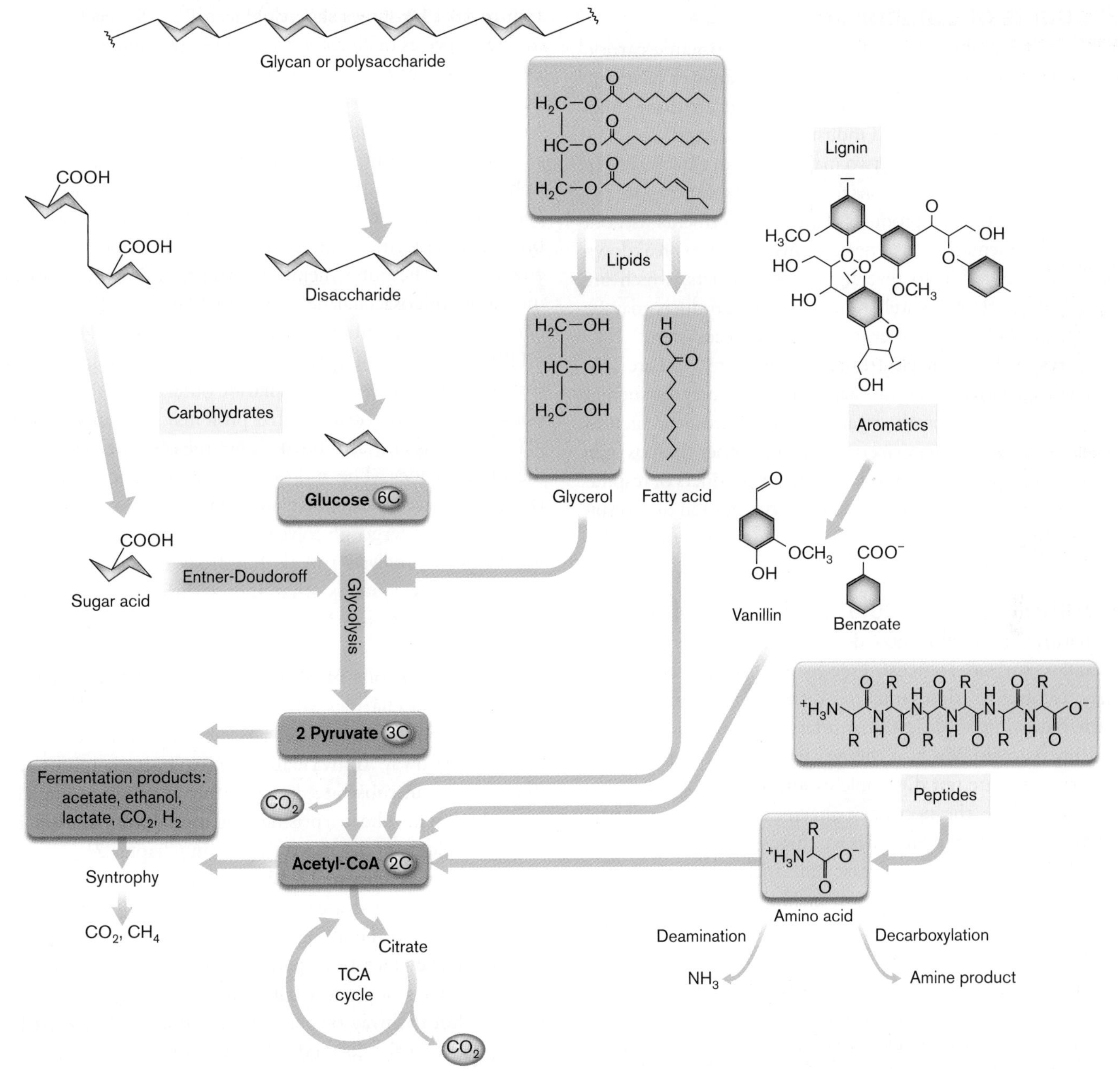

FIGURE 13.13 ■ Many carbon sources enter central pathways of catabolism. Glycans are broken down to disaccharides and then to monosaccharides such as glucose. Glucose and sugar acids are converted to pyruvate, which releases acetyl groups. Acetyl groups or acetate are also the breakdown products of fatty acids, amino acids, and lignin. Alternative routes of catabolism may form products that are excreted, such as organic amines from amino acid decarboxylation.

lignin (**Fig. 13.12D**), which forms the key structural support of trees and woody stems. A discouragingly complex molecule, lignin is made from six-carbon sugars converted to benzene rings, with ether connections that are difficult for enzymes to break down. Fungi and soil bacteria catabolize lignin to oxidized benzene derivatives such as benzoate and vanillin (**Fig. 13.13**). Further breakdown produces acetyl-CoA, which enters the TCA cycle.

Today, the environment contains increasing amounts of human-made aromatic compounds, produced for herbicides and other industrial uses, that are highly toxic pollutants. These include halogenated aromatics, such as polychlorinated biphenyls, the source of highly toxic dioxins. Halogenated aromatics turn out to be catabolized by a number of soil microbes, which are promising candidates for bioremediation of polluted environments (described in Section 13.6). Other types of aromatic molecules catabolized by microbes are polycyclic aromatic hydrocarbons (PAHs). Found in petroleum, the multiple fused rings of PAH compounds build up in natural environments—but with time, microbes can catabolize them and remediate the soil (see Section 13.6).

Products of Catabolism

What is the ultimate fate of all the bits of organic carbon chewed up by microbial catabolism? The answer has myriad consequences for human nutrition and environmental cycling, as well as food and industrial production.

From **Table 13.1**, recall two major forms of catabolism. In **fermentation**, all the electrons from organic substrates are put back onto the organic products. Thus, food ferments without oxygen. Fermentation has a negative ΔG, owing to breakdown of a large molecule to several smaller products, which are usually more stable as well. These organic products may be further catabolized by other members of an ecosystem.

In **respiration**, the electrons removed from food are ultimately transferred to an inorganic electron acceptor such as oxygen or nitrate, or an organic electron acceptor such as fumarate. The use of an electron acceptor other than oxygen is called anaerobic respiration. The major products of respiration are water and carbon dioxide. Fermentation and respiration are discussed in the next section, and in Chapter 14.

Catabolism: Molecular and Cell Biology

In natural ecosystems, how do cells manage to break down complex carbon sources? Free-living bacteria and fungi may catabolize thousands of different carbon sources, each requiring specific transporters and enzymes for initial breakdown. Most of these, particularly complex plant constituents, are not digestible by animals, whose genomes do not encode the necessary enzymes. The only polysaccharides that human enzymes can digest are starch, lactose, and sucrose. Yet human breast milk includes a mixture of complex milk oligosaccharides, including branched chains of *N*-acetylglucosamine, sialic acid, and fucose. Amazingly, our own breast milk has evolved to support bacteria with specific catabolic abilities to colonize the infant gut.

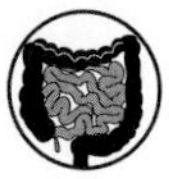

Many animals, including humans, contain anaerobes of the Gram-negative genus *Bacteroides* and the Gram-positive genus *Ruminococcus* (originally named for the bovine rumen). By evolving a symbiosis with microbes, animals avoid the need to acquire new catabolic genes in their own genome. The microbial genomes are functionally part of the human metagenome, the total sequence of genomes of a community of organisms.

The major glycan fibers of vegetables such as lettuce and tomatoes are xyloglucans, beta-linked glucose polymers with side chains containing xylose, galactose, and fructose or arabinose (see **Fig. 13.12A**). Digestion of each type of xyloglucan requires a slightly different set of genes, called a polysaccharide utilization locus (PUL). The PULs, or Sus-like systems, evolved from a common ancestor including the starch utilization system (SUS, shown in **Fig. 13.1B**). Most gut bacteria possess a number of PULs distributed around their genomes, showing evidence of horizontal gene transfer and evolution of genomic islands (discussed in Chapter 9).

Figure 13.14A shows the xyloglucan PUL for each of four different species of *Bacteroides* found in the human gut. The DNA sequences of these species show synteny—that is, sufficient similarity of sequence and map position to predict common ancestry (discussed in Chapter 17).

How do our gut bacteria actually handle complex food polymers? Since bacteria have solid cell walls and are incapable of phagocytosis, a bacterium cannot take up a large branched molecule, but it can secrete enzymes or place enzymes on the outer membrane to break down the fibers into short oligosaccharides (**Fig. 13.14B**). The PUL genes encode outer membrane–inserted enzymes (GH5A and GH9A in **Fig. 13.14B**) that specifically cleave xyloglucans. Other genes encode oligosaccharide outer membrane transporters (SusC and SusC-related proteins). Within the periplasm, various amylases break down the oligosaccharides to monosaccharides. The monosaccharide products are then transported across the inner membrane into the cytoplasm, where they undergo glycolysis (discussed in Section 13.4). Metagenome analysis reveals hundreds of Sus homologs, called Sus-like systems, that target different polysaccharides; these comprise, for example, 18% of the genome of *Bacteroides thetaiotaomicron*.

Different versions of SUS proteins catabolize different types of polysaccharides, such as those with sulfated sugars. Suppose two different species possess enzymes for different steps of a pathway. Individually, the two bacteria might not digest a certain substrate, but together they may complete its breakdown. The surprising nutritional cooperation of our microbiome is discussed further in Chapter 21.

As we saw in **Figure 13.13**, the products of catabolism of many diverse substrates ultimately funnel into a few common pathways of metabolism. The remainder of this chapter presents key metabolic pathways in detail: glycolysis and other pathways of glucose catabolism, the TCA cycle, and the catechol pathway of benzoate catabolism. These pathways play key roles in medical microbiology and in industrial fields such as bioremediation.

To Summarize

- **Carbohydrates such as polysaccharides (glycans)** are broken down to disaccharides, and then to monosaccharides. Sugars and sugar derivatives, such as amines and acids, are catabolized to pyruvate.
- **Pyruvate and other intermediary products of sugar catabolism** are fermented, or they are further catabolized to CO_2 and H_2O through the TCA cycle (in the presence of a terminal electron acceptor) or to CO_2, H_2, and CH_4 through fermentation and syntrophy.
- **Lipids, amino acids, and lignin** are catabolized to acetate and other metabolic intermediates.

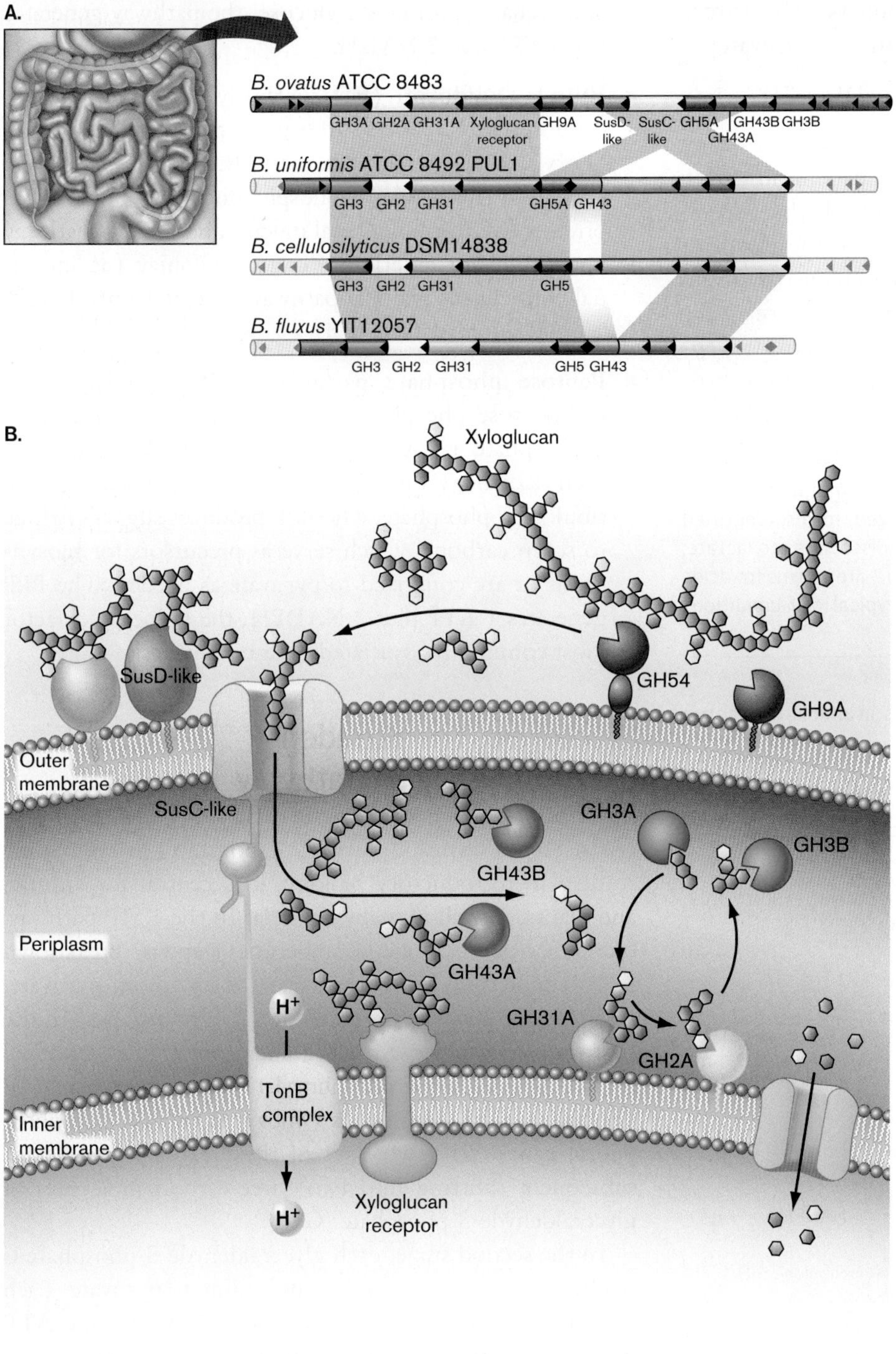

FIGURE 13.14 ■ Xyloglucan catabolism by gut bacteria. **A.** Some *Bacteroides* species have PUL sets of genes encoding xyloglucan degradation. The PULs show synteny, evidence of descent from a common ancestor. **B.** Xyloglucan degradation enzymes and transporters within a *Bacteroides* envelope. Xyloglucan cleavage by enzyme GH9A releases short sugar chains that cross the outer membrane via a SusC-like complex. Periplasmic enzymes further break down the chains. Protein colors are matched to their genes in (A).

- **Catabolism yields products that may be used by a host or other organisms in the ecosystem.**
- **Fermentation and respiration** complete the process of catabolism. In fermentation, the catabolite is broken down to smaller molecules without an inorganic electron acceptor. Respiration requires a terminal electron acceptor such as O_2.
- **Catabolism of complex substrates requires specialized enzymes and transporters.** Different bacteria contain homologous SUS or PUL complexes that fit different carbon sources.

13.4 Glucose Fermentation and Respiration

For microbes feasting on glycan, a major breakdown product is glucose. Glucose catabolism is important as a widespread source of energy, and also as a source of key substrates for biosynthesis, such as five-carbon sugars to build nucleic acids (discussed in Chapter 15). Blood plasma contains glucose, which pathogens can exploit.

Glucose and related sugars are catabolized through a series of phosphorylated sugar derivatives. A common theme in sugar catabolism is the splitting of a six-carbon substrate into two three-carbon products. The three-carbon products may form two molecules of pyruvate:

$$C_6H_{12}O_6 \rightarrow 2C_3H_4O_3 + 4H\ (2\ NADH + 2H^+)$$

Under anaerobiosis—the prevailing condition of many microbial habitats—the pyruvate obtained from sugar breakdown must be converted to compounds that receive electrons from NADH, in order to restore the electron-accepting form NAD^+. Different microbes reduce pyruvate to different end products of fermentation. Alternatively, through respiration, NADH may reduce an electron acceptor such as oxygen or nitrate, allowing pyruvate to be broken down via the TCA cycle (discussed in Section 13.6).

Note: The carboxylic acid intermediates of metabolism exist in equilibrium with their dissociated, or ionized, forms, identified by the suffix "-ate." For example, lactic acid dissociates to lactate; acetic acid dissociates to acetate. We use the "-ate" terms for acids whose ionized form predominates under typical cell conditions (around pH 7).

To catabolize glucose, bacteria and archaea use three main routes to pyruvate (**Fig. 13.15**):

- **Glycolysis**, or the **Embden-Meyerhof-Parnas (EMP) pathway**, in which glucose 6-phosphate isomerizes to fructose 6-phosphate, ultimately forming two molecules of pyruvate. EMP is used by many bacteria, eukaryotes, and archaea. From each glucose, the pathway generates net 2 ATP and 2 NADH.
- **Entner-Doudoroff (ED) pathway**, in which glucose 6-phosphate is oxidized to 6-phosphogluconate, a phosphorylated sugar acid. Alternatively, sugar acids may be converted directly to 6-phosphogluconate. Sugar acids often derive from intestinal mucus, and the ED pathway is essential for enteric bacteria to colonize the intestinal epithelium. The ED pathway generates only 1 ATP, 1 NADH, and 1 NADPH.
- **Pentose phosphate pathway (PPP)**, also known as the pentose phosphate shunt (PPS), in which glucose 6-phosphate is oxidized to 6-phosphogluconate, and then decarboxylated to a five-carbon sugar (a pentose), ribulose 5-phosphate. The PPP produces sugars of three to seven carbons, which serve as precursors for biosynthesis or are converted to pyruvate as needed. The PPP generates 1 ATP plus 2 NADPH, the reducing cofactor most commonly associated with biosynthesis.

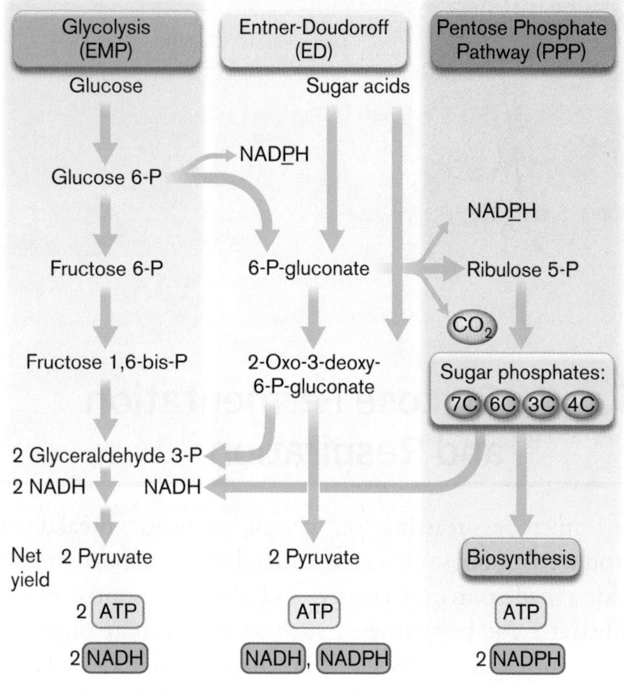

FIGURE 13.15 ■ From glucose to pyruvate: three pathways. The Embden-Meyerhof-Parnas (EMP) pathway of glycolysis, the Entner-Doudoroff (ED) pathway, and the pentose phosphate pathway (PPP) catabolize carbohydrates by related but different routes.

Glycolysis: The Embden-Meyerhof-Parnas Pathway

The **Embden-Meyerhof-Parnas (EMP) pathway**, or **glycolysis**, is the form of glucose catabolism most commonly studied in introductory biology; it is central for animals and plants, as well as many bacteria. In the EMP pathway, one molecule of D-glucose undergoes stepwise breakdown to two molecules of pyruvic acid (or its anion, pyruvate; **Fig. 13.16**). Glucose is broken down in two stages. In the first stage, the glucose molecule is primed for breakdown by two steps of sugar phosphorylation by ATP. Each ATP phosphotransfer step spends Gibbs free energy. The phosphoryl groups tag two sides of the glucose molecule for subsequent splitting into two three-carbon molecules of glyceraldehyde 3-phosphate (G3P).

In the second stage, each glyceraldehyde 3-phosphate is oxidized by NAD^+ through steps leading to pyruvate. Each conversion of G3P to pyruvate forms 2 ATP—one ATP molecule from dephosphorylation of the substrate, and one from addition of inorganic phosphate. Since 2 ATP were spent originally to phosphorylate glucose, the net gain of energy carriers from each glucose is 2 NADH plus 2 ATP.

Most of the conversion steps are associated with a small change in energy—so small that the sign of ΔG depends on the concentrations of substrates or products; thus, some individual steps are reversible. In the cytoplasm, however, as intermediate products form they are quickly consumed by the next step, so the pathway flows in one direction. The direction of flow is determined by the key irreversible reactions that consume ATP. These steps drive the pathway by spending energy.

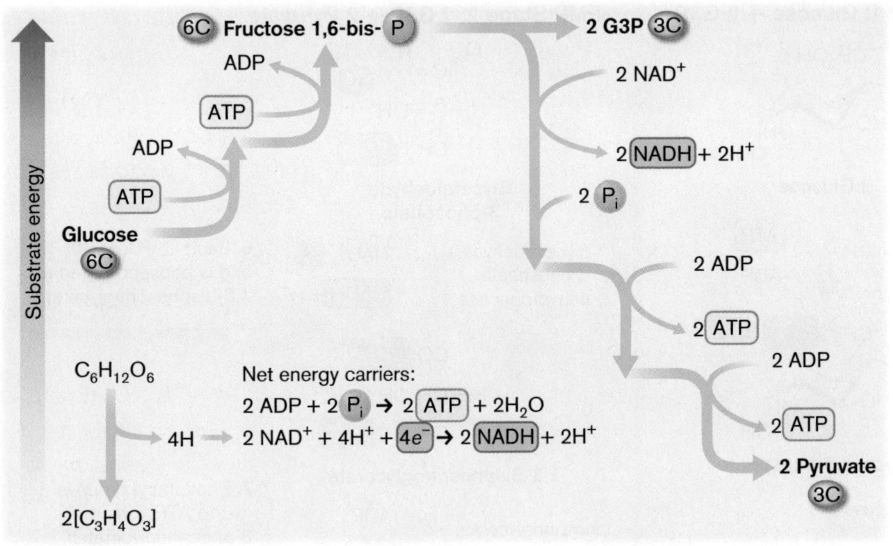

FIGURE 13.16 ■ Substrate energy changes during the Embden-Meyerhof-Parnas pathway (glycolysis). Glucose is activated through two substrate phosphorylations by ATP. The breakdown of glucose to two molecules of pyruvate is coupled to net production of 2 ATP and 2 NADH. Phosphoryl groups are shown as P.

Phosphorylation and splitting of glucose. In the first stage of the EMP pathway, the six-carbon sugar is activated by two phosphorylation steps (**Fig. 13.17**, left). The first phosphoryl group (phosphate) is added at carbon 6 of glucose. (In some species of bacteria, the first phosphoryl group is added by phosphoenolpyruvate instead of ATP, but the net effect is the same.) The next enzyme-catalyzed step—rearrangement of glucose 6-phosphate to fructose 6-phosphate—involves no significant change in energy but prepares the sugar to receive the second phosphoryl group, forming fructose 1,6-bisphosphate.

The sugar then splits into two three-carbon sugars (trioses), each tagged with one of the two phosphates. The splitting of this molecule has a favorable entropy change (ΔS), but its enthalpy change (ΔH) is unfavorable, largely canceling out the energy yield. The two triose phosphates—glyceraldehyde 3-phosphate (G3P) and dihydroxyacetone phosphate (DHAP)—have nearly the same energy, so an enzyme interconverts them reversibly. Interconversion is necessary because only G3P proceeds further in the pathway.

Note: In Chapters 13–16, every substrate conversion shown requires catalysis by an enzyme. For the EMP pathway, the enzymes are shown (see **Fig. 13.17**), but for other pathways, the enzyme names are omitted.

ATP generation. In the second stage of the EMP pathway, each three-carbon G3P is directed into an energy-yielding pathway to pyruvate (**Fig. 13.17**, right). First, the aldehyde (R–CHO) is converted to the carboxylate ion (R–COO$^-$) plus H$^+$. Conversion to a carboxylate releases a substantial amount of energy (negative ΔG). This oxidation of the aldehyde represents the major source of energy in glycolysis—the step at which the energy obtained is used to transfer a pair of electrons onto NAD$^+$, forming NADH (with an ionized H$^+$). In addition, sufficient energy is released to add a phosphoryl group (from inorganic phosphate, P_i) to the carboxylate, generating 1,3-bisphosphoglycerate.

In subsequent steps, the added phosphoryl group will be transferred to ADP, yielding net 1 ATP per pyruvate (2 ATP per glucose). The transfer of a phosphoryl group from an organic substrate to make ATP is called **substrate-level phosphorylation**. With subtraction of the initial two ATP molecules invested, the net energy carriers gained from each glucose molecule are as follows:

$$2\ NAD^+ + 4e^- + 4H^+ \rightarrow 2\ NADH + 2H^+$$
$$2\ ADP + 2\ P_i \rightarrow 2\ ATP + 2H_2O$$

Regulation of glycolysis. Enzymes of catabolism are regulated at the level of transcription of the enzyme. In addition, the activities of certain enzymes in long pathways require allosteric regulation. Allosteric regulation by enzyme substrates and products ensures that excess intermediates do not build up, and it avoids releasing more energy than the cell can use at a given time. Glycolysis is regulated so that its reactions go forward only when the cell needs energy, not when the cell is trying to synthesize glucose. The regulation occurs at steps where the products are consumed so rapidly that their forward reaction is effectively irreversible. Irreversible steps are shown as unidirectional arrows in **Figure 13.17**.

Thought Question

13.7 Some bacteria make an enzyme, dihydroxyacetone kinase, that phosphorylates dihydroxyacetone to dihydroxyacetone phosphate (DHAP). How could this enzyme help the cell yield energy?

Enzymes that catalyze irreversible steps are regulated so as to maintain consistent levels of intermediates in the pathway. The most important irreversible reaction in glycolysis is the phosphorylation of fructose 6-phosphate to fructose 1,6-bisphosphate, mediated by the enzyme

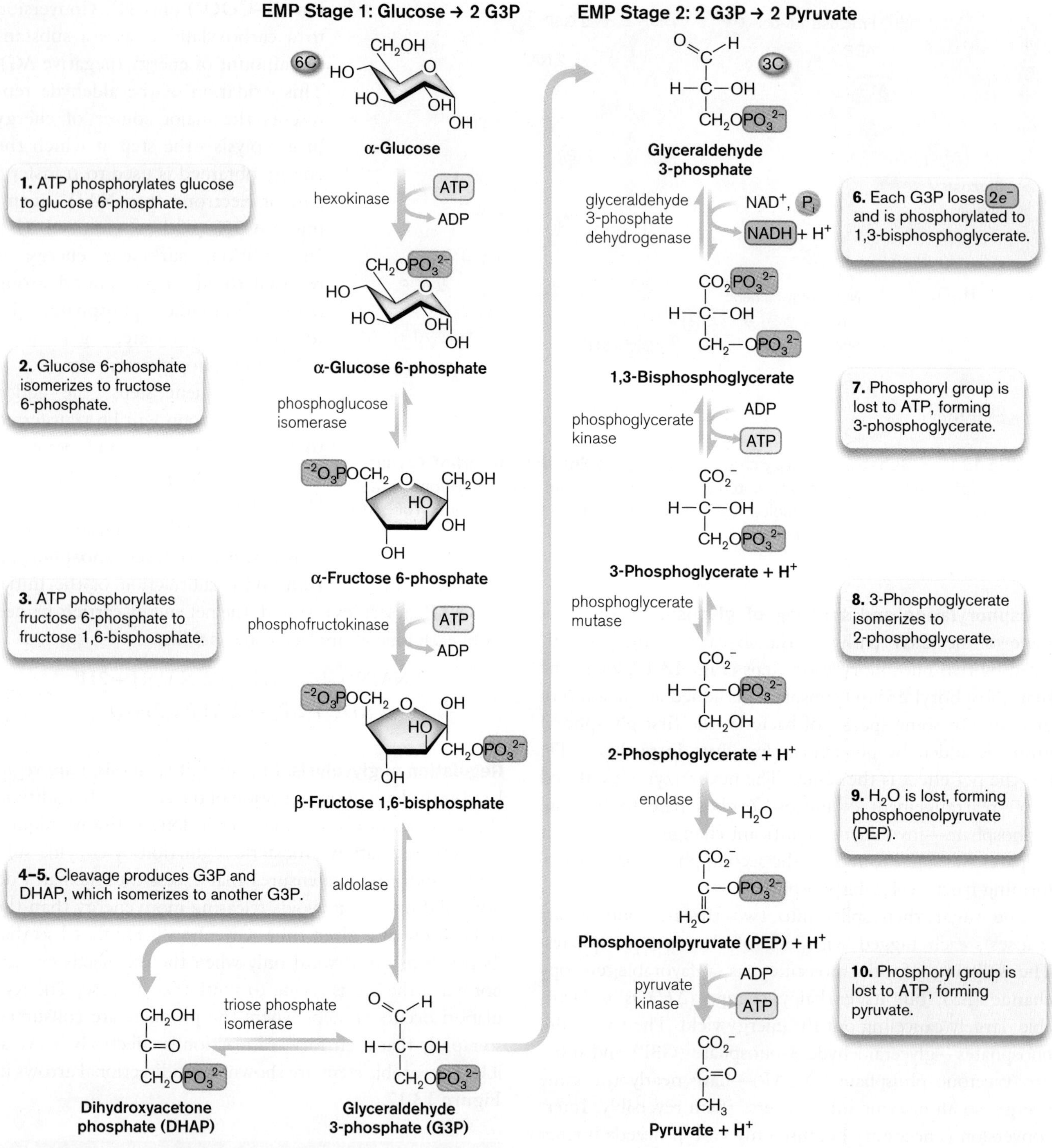

FIGURE 13.17 ■ Embden-Meyerhof-Parnas pathway (glycolysis).

phosphofructokinase. This enzyme is activated allosterically by ADP and inhibited by ATP or by the alternative phosphoryl donor, phosphoenolpyruvate.

What happens when the cell needs to reverse glycolysis in order to make glucose? Most of the intermediate reactions are reversible, so the same enzymes can be used for biosynthesis. Pathways that participate in both catabolism (breakdown) and anabolism (biosynthesis) are described as **amphibolic**.

The amphibolic pathway of glycolysis includes enzymes that can run in either direction, such as phosphoglucose isomerase. The direction of the pathway at a given time is determined by key enzymes that operate only in the catabolic direction or in the anabolic direction. For example, in glycolysis the ATP phosphorylation of fructose 6-phosphate is catalyzed by the enzyme phosphofructokinase, whereas in biosynthesis this step is reversed by a different enzyme, fructose bisphosphatase. Instead of regenerating ATP, fructose bisphosphatase removes the second phosphoryl group as inorganic phosphate—a step yielding energy and thus driving the whole pathway in reverse (toward biosynthesis of sugar). The two enzymes are regulated differently; the catabolic enzyme phosphofructokinase is activated by ADP, a signal of energy need, whereas the biosynthetic enzyme fructose bisphosphatase is inhibited by such signals.

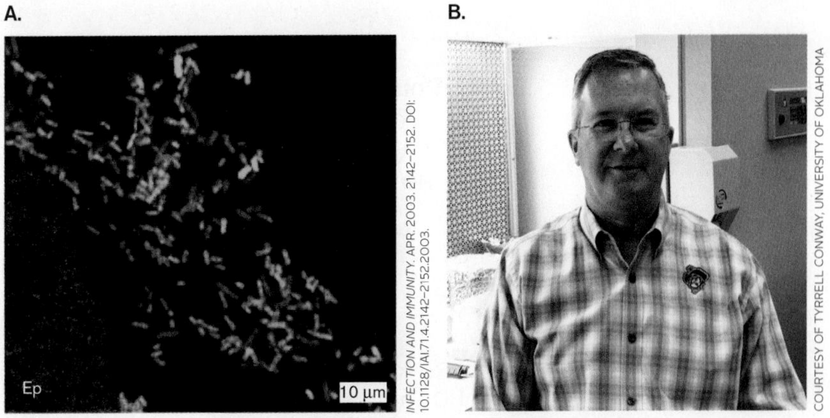

FIGURE 13.18 ■ Intestinal bacteria use the Entner-Doudoroff pathway. **A.** Intestinal *E. coli* (orange) feed primarily on gluconate from mucous secretions (fluorescence micrograph). **B.** Tyrrell Conway used genomics and genetic analysis to dissect the role of the ED pathway in the enteric bacterial catabolism of sugar acids from intestinal mucus.

The Entner-Doudoroff Pathway

The **Entner-Doudoroff (ED) pathway** offers a slightly different route to catabolize sugars, as well as sugar acids (sugars with acidic side chains). The ED pathway was originally studied for its role in production of the Mexican beverage *pulque*, or "cactus beer," by *Zymomonas* fermentation of the blue agave plant. Later, Tyrrell Conway and colleagues at the University of Oklahoma discovered genes encoding the Entner-Doudoroff enzymes in the genomes of many bacteria and archaea. In the human colon, the ED pathway enables *E. coli* and other enteric bacteria to feed on mucus secreted by the intestinal epithelium (**Fig. 13.18**). Some gut flora, such as *Bacteroides thetaiotaomicron*, actually induce colonic production of the mucus that they consume. These bacteria that "farm" intestinal mucus enhance human health by preventing pathogen colonization, and by stimulating the immune system.

The ED pathway appears to have evolved earlier than the EMP pathway, because it involves fewer substrate phosphorylation steps and produces less ATP, and it is found in a wider range of prokaryotes. As in the EMP pathway, glucose is phosphorylated to glucose 6-phosphate (**Fig. 13.19**). The next step, however, involves oxidation by $NADP^+$ at carbon 1, with loss of two hydrogens to form 6-phosphogluconate, a sugar acid. Gluconate is also found in intestinal mucus; the sugar acid can be phosphorylated to enter the ED pathway.

During ED, the hydrogens and electrons from glucose 6-phosphate are transferred to $NADP^+$ instead of NAD^+ as in the EMP pathway. This step differs from the EMP pathway in two respects: The carrier used is $NADP^+$ instead of NAD^+, and the electrons are transferred earlier, without a second ATP-consuming phosphorylation step. When the six-carbon substrate is eventually split into two three-carbon products, one of the three-carbon products is G3P, which enters the second stage of glycolysis (see **Fig. 13.17**). NADH is made, and 2 ATP are made by substrate-level phosphorylation. The remaining three-carbon product, however, is pyruvate. This one-step production of pyruvate short-circuits the catabolic pathway, missing the formation of an ATP. The unused potential energy is released as waste heat.

The net ATP gain from the Entner-Doudoroff pathway is only 1 ATP per molecule of glucose—half that of the EMP pathway (see **Fig. 13.19**, inset). The electrons transferred, however, are equivalent: Instead of 2 NADH, the Entner-Doudoroff pathway generates 1 NADH and 1 NADPH.

What is the significance of NADPH, compared to NADH? In most cases, NADH transfers electrons to the ETS to store energy, whereas NADPH is used for biosynthesis (see Chapter 15). Thus, enzymes for amino acid biosynthesis will use NADPH but not NADH. The ratio between NADH and NADPH enables cells to balance their need for energy with their need to build biomass.

Thought Question

13.8 Explain why the ED pathway generates only 1 ATP, whereas the EMP pathway generates 2 ATP. What is the consequence for cell metabolism?

The Pentose Phosphate Pathway

A third pathway of glucose catabolism is the **pentose phosphate pathway (PPP)**, which forms the key intermediate **ribulose 5-phosphate**, a five-carbon sugar. The pentose phosphate pathway generates 1 ATP with no NADH, but 2 NADPH for biosynthesis (**Fig. 13.20**). In addition, the PPP generates a complex series of intermediates that can be redirected as substrates for biosynthesis of diverse cell components such as amino acids and vitamins.

The pentose phosphate pathway starts like the Entner-Doudoroff pathway: Glucose 6-phosphate gives up two electrons to form NADPH and is oxidized to

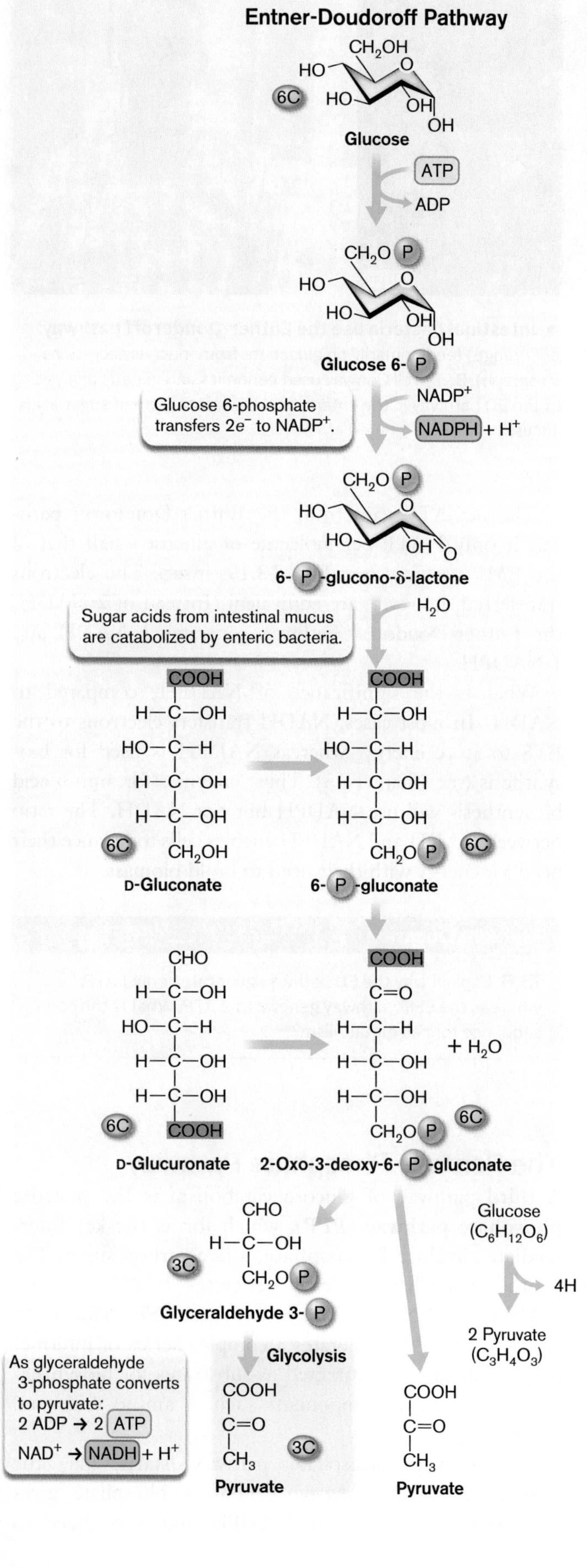

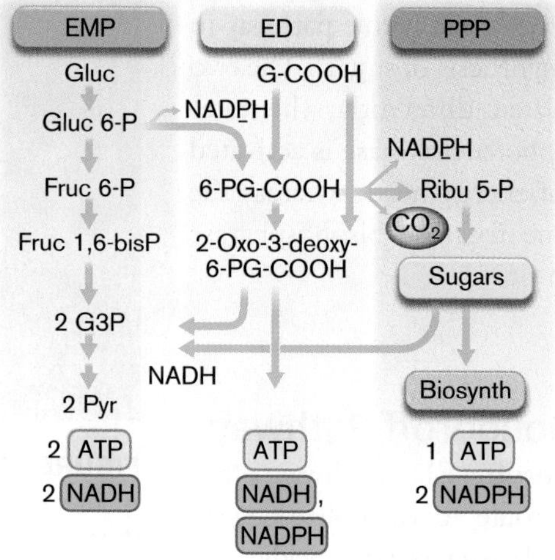

FIGURE 13.19 ■ Entner-Doudoroff pathway. Glucose 6-phosphate is oxidized to 6-phosphogluconate, with one pair of electrons transferred to NADPH. The 6-phosphogluconate is dehydrated and cleaved to form one pyruvate plus one glyceraldehyde 3-phosphate (G3P), which enters the EMP pathway to pyruvate.

6-phosphogluconate. The next step involves a second oxidation by $NADP^+$, with loss of a carbon as CO_2. The loss of CO_2 generates the five-carbon sugar ribulose 5-phosphate (hence the pathway name "pentose phosphate"). In succeeding steps, pairs of sugars, such as sedoheptulose 7-phosphate and glyceraldehyde 3-phosphate, exchange short carbon chains, giving rise to sugar phosphates of various lengths—for example, ribose 5-phosphate and erythrose 4-phosphate, which are precursors of purines and aromatic amino acids, respectively. Alternatively, if these routes to biosynthesis are not taken, the intermediates convert to fructose 6-phosphate and reenter the EMP pathway, where ATP and NADH are produced.

Fermentation Completes Catabolism

None of the pathways from glucose to pyruvate constitute completed pathways of catabolism, because NADH (and for some pathways, NADPH) remains to be recycled. In the absence of oxygen or other electron acceptors, heterotrophic cells must transfer the hydrogens from NADH + H^+ back onto pyruvate or its breakdown products, forming partly oxidized fermentation products with the same redox level (balance of O and H) as the original glucose had (see **Table 13.5**).

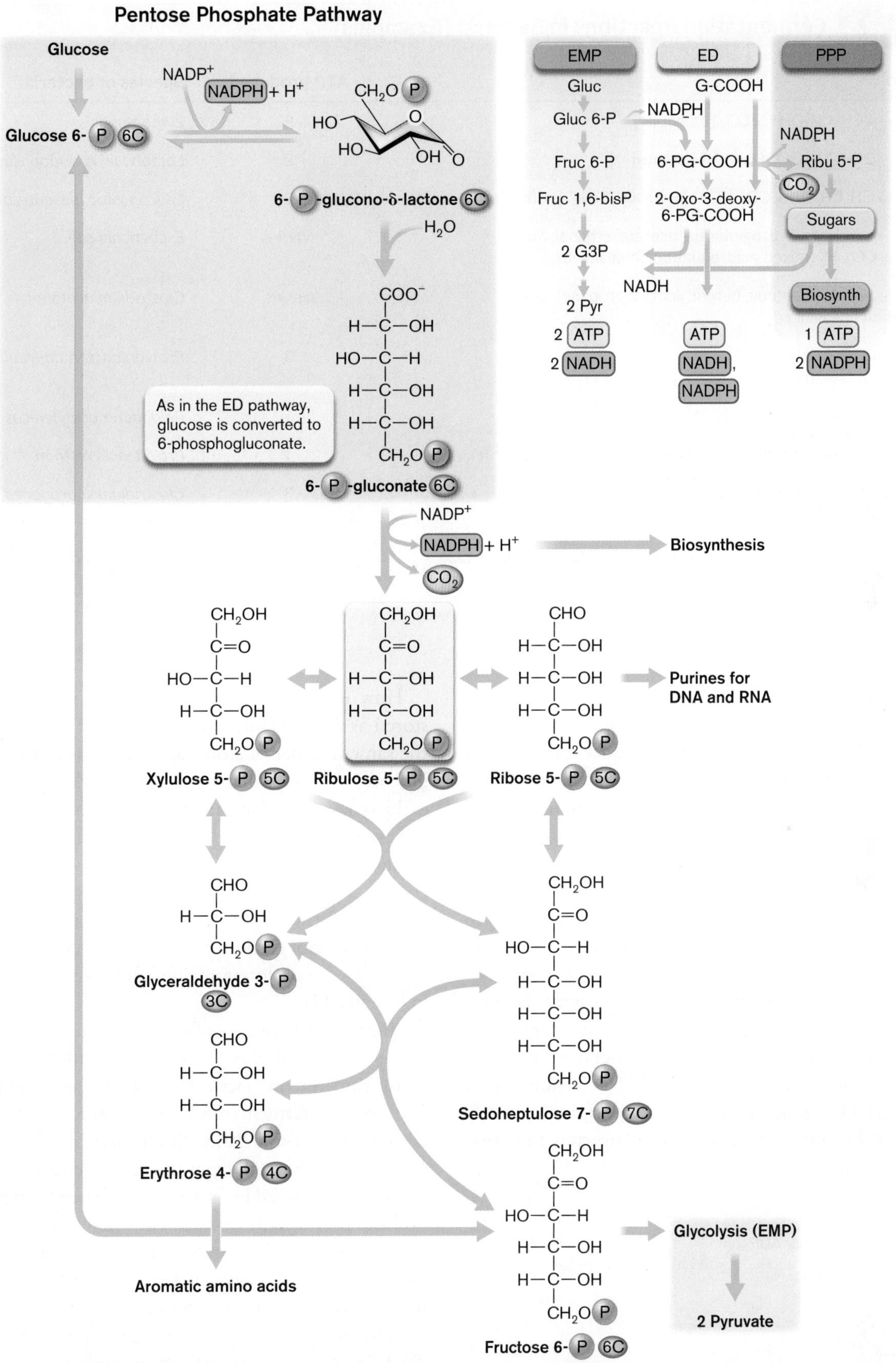

FIGURE 13.20 ▪ Pentose phosphate pathway. Like the Entner-Doudoroff pathway, the pentose phosphate pathway forms 6-phosphogluconate. One CO_2 is released, and 2 NADPH are produced for biosynthesis. The pathway can generate ribose 5-phosphate for purine synthesis or erythrose 4-phosphate to synthesize aromatic amino acids.

TABLE 13.5 Fermentation Reactions in Bacteria (Examples)

Reaction	ATP produced	Species of bacteria
$C_6H_{12}O_6 \rightarrow 2CH_3CH_2OH + 2CO_2$ [ethanolic]	2	*Zymomonas* sp.
$C_6H_{12}O_6 \rightarrow 2CH_3CH_2OCOO^- + 2H^+$ [lactate]	2	*Lactobacillus acidophilus*
$C_6H_{12}O_6 \rightarrow CH_3CH_2OH + CO_2 + CH_3CH_2OCOO^- + H^+$ [heterolactic]	2	*Leuconostoc mesenteroides*
$C_6H_{12}O_6 \rightarrow$ succinate, 2-oxoglutarate, acetate, ethanol, formate, lactate, CO_2, H_2 [mixed-acid; equation unbalanced]	Varies	*Escherichia coli*
$C_6H_{12}O_6 \rightarrow$ butanol, acetone, butyric acid, isopropanol, CO_2 [equation unbalanced]	Varies	*Clostridium acetobutylicum*
$2CH_3CH_2OCOO^-$ (lactate) $\rightarrow$ $CH_3(CH_2)COO^-$ (propionate) $+ CH_3COO^- + CO_2 + 2H^+$	3	*Propionibacterium freudenreichii*
$2C_2H_2$ (acetylene) $+ 3H_2O \rightarrow CH_3CH_2OH + CH_3COO^- + H^+$	1	*Pelobacter acetylenicus*
2 Citrate^{3-} $+ H^+ \rightarrow$ 2 succinate^{2-} $+ CH_3COO^- + 2CO_2$	1	*Providencia rettgeri*
CH_3CHNH_2COOH (alanine) $+ 2NH_2CH_2COOH$ (glycine) $+ 2H_2O \rightarrow$ $3CH_3COO^- + 3NH_4^+ + CO_2$ [Stickland reaction]	3	*Clostridium sporogenes*
$COOH(CH_2)_2COO^-$ (succinate) $\rightarrow CH_3(CH_2)COO^-$ (propionate) $+ CO_2$ [$Na^+_{in} \rightarrow Na^+_{out}$]	0	*Propionigenium modestum*

The "waste products" of fermentation retain much of their organic structure and food value. Their carbon-hydrogen bonds retain the ability to reduce oxygen, and their organic structures can be used for biosynthesis. Thus, fermentation products have proved extremely useful in human culture and technology (**Fig. 13.21**). For thousands of years, ethanolic fermentation by yeast has been used to produce wine and beer, while lactic acid fermentation has been used to produce yogurt and cheese (see Chapter 16). The minor product butyric acid (butyrate) lends taste to butter.

In **lactic acid fermentation**, the pyruvate is converted to lactate by the addition of two electrons (with two hydrogen atoms) at the ketone, generating an alcohol group. The product, lactic acid, has a number of atoms ($C_3H_6O_3$) equal to half of the original glucose ($C_6H_{12}O_6$). Overall, one glucose is converted into two molecules of lactic acid:

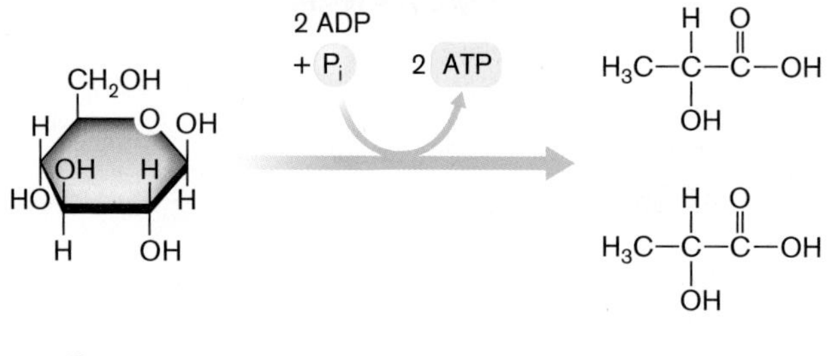

How does lactic acid fermentation yield energy to be stored as ATP? While two molecules of lactic acid contain the same number of atoms and electrons as those found in glucose, the molecular bonds are rearranged to form two carboxylic acids. The carboxylate/carboxylic acid has multiple states and increased entropy compared to the hydroxyl groups of glucose. Overall, glucose conversion releases free energy, and the reaction has a net negative value of ΔG. The ATP formation in such a reaction is called **substrate-level phosphorylation** because it involves only substrate reactions—no proton pumping across membranes (discussed in Chapter 14).

Alternative pathways of fermentation produce two molecules of ethanol plus two CO_2 molecules (**ethanolic fermentation**) or one lactic acid, one ethanol, and one CO_2 (heterolactic fermentation). Other kinds of fermentation are shown in **Table 13.5**. In all cases, fermentation products must be excreted from the cell. The large quantities of substrate consumed in fermentation generate large amounts of waste products to be excreted. These bacterial wastes are actually useful to human "fermentation industries" such as the production of alcoholic beverages (ethanolic fermentation) or cheese (lactic acid fermentation).

Why do fermenting bacteria give up such large quantities of waste products that retain usable energy? The reason is that in the absence of oxygen (or another electron acceptor), the fermentation products cannot yield energy. Most fermentation pathways do not generate ATP beyond that produced by substrate-level phosphorylation, the

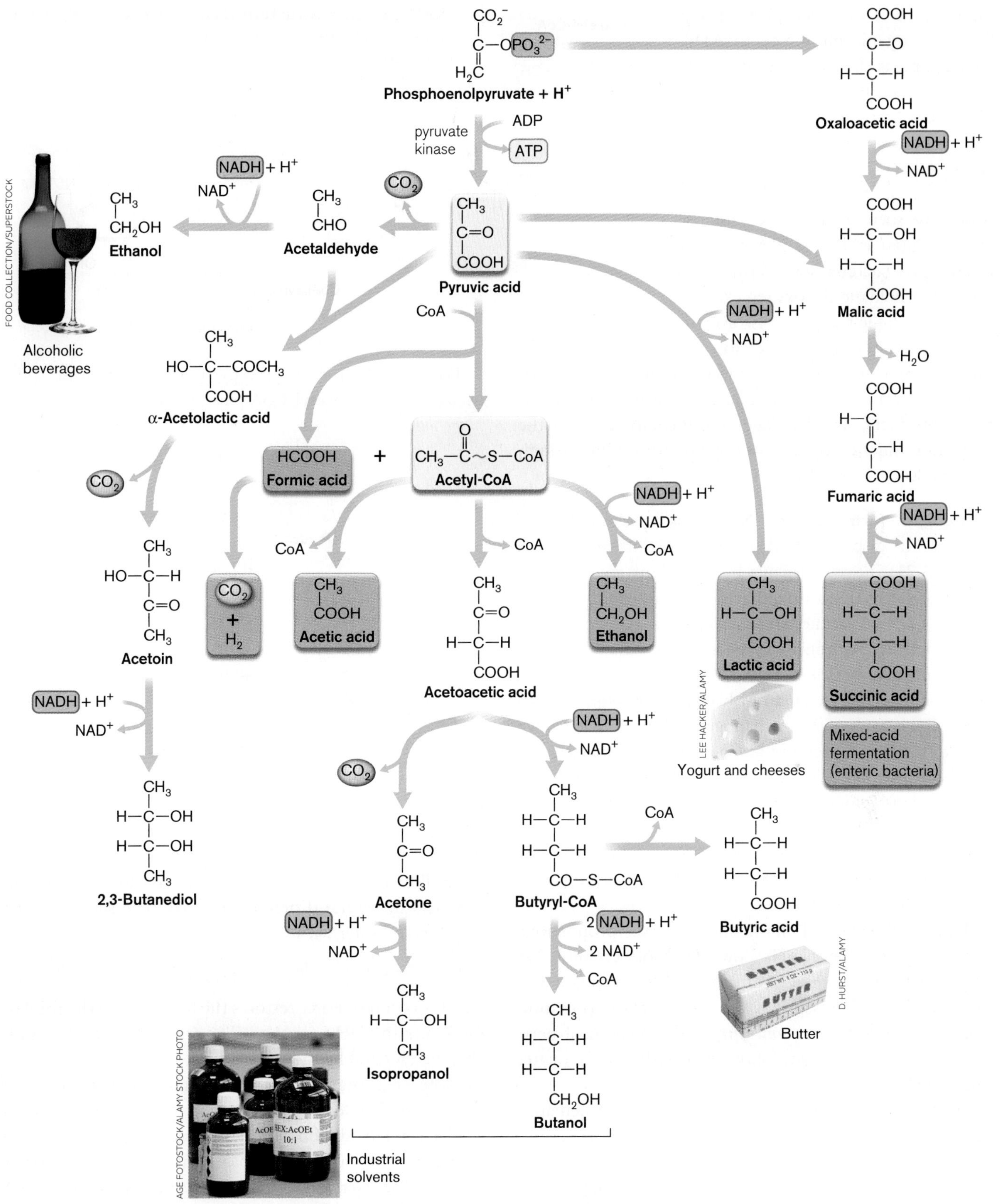

FIGURE 13.21 ■ Fermentation pathways. Alternative pathways from pyruvate and phosphoenolpyruvate to end products, many of which we use for food or industry. Different species conduct different portions of the pathways shown.

direct transfer of a phosphate group from an organic phosphate to ADP. An example is the final step of glycolysis, catalyzed by pyruvate kinase (see **Fig. 13.17**). Much of the energy available from glucose remains unspent or is lost as heat. Nevertheless, fermentation is essential for microbes in environments such as anaerobic soil or animal digestive tracts. Even aerated cultures of bacteria start fermenting once their demand for oxygen exceeds the rate at which oxygen dissolves in water. Microbes compensate for the low efficiency of fermentation by consuming large quantities of substrate and excreting large quantities of fermentation products. An advantage of fermentation is that the rapid accumulation of acids or ethanol can inhibit growth of competitors.

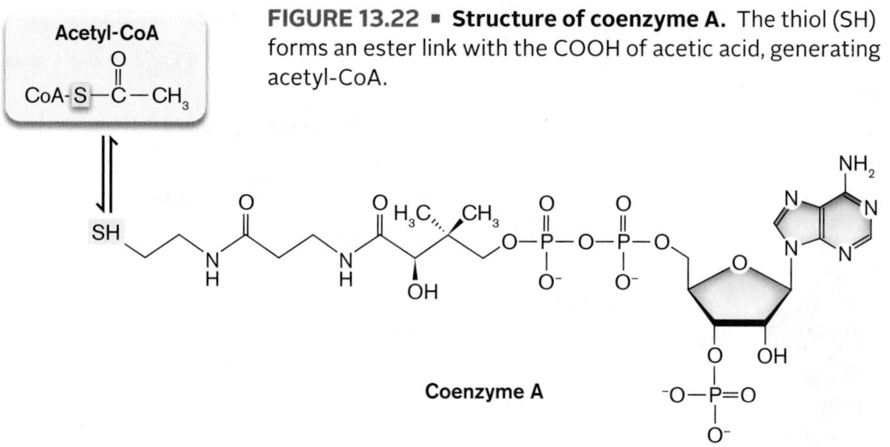

FIGURE 13.22 ■ Structure of coenzyme A. The thiol (SH) forms an ester link with the COOH of acetic acid, generating acetyl-CoA.

Mixed-acid fermentation. Numerous pathways have evolved to dispose of the waste in different forms (**Fig. 13.21**). *E. coli* ferments by a combination of routes known collectively as **mixed-acid fermentation**, forming acetate, formate, lactate, and succinate, as well as ethanol, H_2, and CO_2. Hydrogen (H_2) and carbon dioxide (CO_2) are the main gases passed by the human colon.

Colonic H_2 plus CO_2 from bacterial fermentation can yield further energy for methanogens, archaea that reduce CO_2 to methane (discussed in Chapter 14). Hydrogen and methane gases must be considered during medical procedures such as colonoscopy. When a polyp is removed from the colon by electrocautery (high-frequency electric current), the gas could ignite, causing an explosion. Colonic explosion is avoided by flushing out the gases before electrocautery.

During mixed-acid fermentation, the proportions of products vary with pH. Low pH favors ethanol and lactate (which minimize acidification) over formate and acetate. *Clostridium* species produce alcohols (butanol, isopropanol); *Porphyromonas gingivalis*, a cause of periodontal disease, produces short-chain acids (propionate, butyrate).

Many fermentation products share key intermediates, such as acetyl-CoA. Acetyl-CoA is a versatile two-carbon intermediate that serves as the "Lego block" of metabolism. The molecule consists of an acetyl group esterified to **coenzyme A** (**CoA**; **Fig. 13.22**), an important coenzyme whose discovery won Fritz Lipmann the 1953 Nobel Prize in Physiology or Medicine. CoA has a thiol (SH) that exchanges its hydrogen for an acyl group, thus activating the molecule for transfer in various metabolic pathways.

For example, the enzyme pyruvate formate lyase splits pyruvate to form acetyl-CoA plus formate:

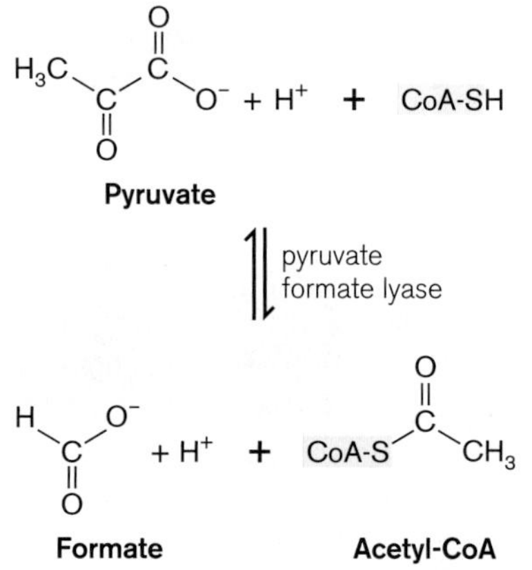

The SH group of CoA accepts the acetyl group to form acetyl-CoA. The hydrogen ($H^+ + e^-$) of the SH group is transferred onto the carboxyl carbon of pyruvate, generating formate.

Acetyl-CoA can be converted to various fermentation products by several different pathways. The simplest is an exchange of water with CoA, forming acetate:

$$CH_3CO\text{—S-CoA} + H_2O \rightleftharpoons CH_3COO^- + H^+ + \text{HS-CoA}$$

Incorporation of water restores the hydrogen to the thiol of CoA, and the OH to acetate. The acetate thus formed may then be excreted by the cell.

Note that the excreted acids and alcohols are readily recovered by cells when an electron acceptor becomes available for oxidation or when these fermentation products are needed as building blocks for biosynthesis. Alternatively, the excreted "wastes" may be utilized by other species capable of metabolizing them further. For example, the H_2 released by gut bacteria through mixed-acid fermentation can be oxidized to water by the gastric pathogen *Helicobacter pylori*. Alternatively, H_2 plus CO_2 can be used by gut methanogens to yield energy with release of methane.

In soil ecosystems, various other fermentations have evolved to utilize available substrates (see **Table 13.5**). *Clostridium* species generate butanol and other solvents of great industrial value. Other species ferment pairs of amino acids, in a type of mechanism called the **Stickland reaction**. Fermenting a pair of amino acids avoids generating H_2 (a loss that would waste reduction potential) and produces as much as 3 ATP. By contrast, the succinate fermentation of *Propionigenium modestum* is an anaerobic reaction that proceeds with too small a ΔG value to generate a single ATP. Instead, the decarboxylation step (catalyzed by methylmalonyl-CoA decarboxylase) is coupled to the pumping of sodium ions across the cell membrane. The sodium pump stores energy (discussed in Chapter 14).

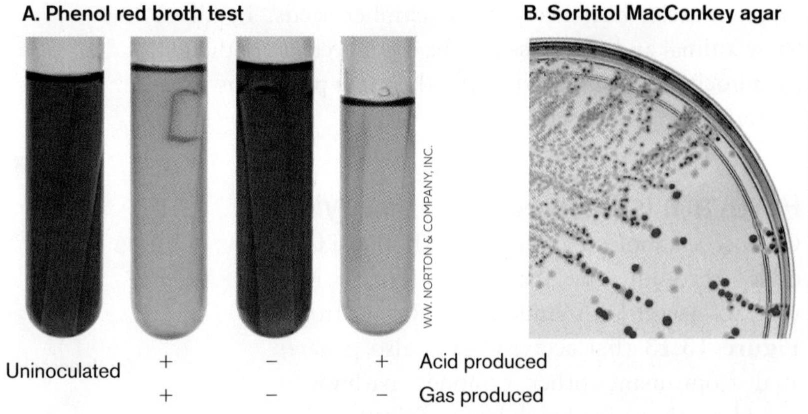

FIGURE 13.23 ■ Clinical tests based on fermentation. **A.** Phenol red broth test with Durham tube. From left to right: An inoculated control is red; *Escherichia coli* gives acidic fermentation products (yellow) and gas (CO_2 and H_2) in a Durham tube; a nonfermenter stays red; a fermenter without gas turns yellow, but forms no bubble. **B.** Sorbitol fermentation test for pathogen *E. coli* O157:H7. The pale colonies (strain O157:H7) failed to ferment sorbitol, unlike the red colonies (nonpathogenic *E. coli*).

Industrial and Clinical Applications

In the chemical industry, microbial fermentation produces industrial solvents such as butanol and acetone. Acetone production had historic impact during World War I, when Britain needed a source of acetone to manufacture gunpowder. Acetone and butanol were produced as fermentation products by *Clostridium acetobutylicum*, a bacterium identified by the biochemist Chaim Weizmann. Weizmann was a Russian-born Jew who helped the British government produce acetone via *C. acetobutylicum* fermentation. Weitzmann later became the first president of Israel. Today, the fermentation process discovered in *Clostridium* species is used to convert industrial waste into ethanol fuels and chemicals such as ethanol, butanol, and glycerol.

Another important application of fermentation lies in diagnostic microbiology. To quickly identify the microbe causing a disease and prescribe an effective antibiotic, hospitals use rapid and inexpensive biochemical tests (Chapter 28). A pH indicator added to growth media can detect the acid produced when a substrate is fermented, resulting in a color change (**Fig. 13.23A**). Phenol red is a pH indicator that is orange-red at neutral pH. It turns yellow in media acidified by fermentation acids (below pH 6.8) and red at higher pH (above pH 7.4). A culture of *Escherichia coli*, which ferments quickly, turns phenol red to a medium yellow after 24 hours.

More specific tests depend on the microbe's ability to ferment specific sugars. Different species possess different enzymes able to convert different sugars and sugar derivatives into glucose, which then enters the common fermentation pathway. An example is the use of sorbitol MacConkey agar to test for *E. coli* O157:H7, a lethal pathogen contaminating beef and vegetables. The O157:H7 strain of *E. coli* has a set of virulence genes absent from nonpathogenic *E. coli* strains that are present among our normal colon biota. But the pathogen also happens to lack genes present in normal *E. coli*, such as a gene encoding an enzyme to ferment sorbitol. Thus, failure to ferment sorbitol indicates a high probability that the strain is *E. coli* O157:H7. On sorbitol MacConkey agar, bacteria that ferment sorbitol during growth produce acids. The acidity causes a dye to turn red (the opposite of the phenol red test). Failure to ferment sorbitol (observed as pale colonies) indicates a high probability of *E. coli* O157:H7 (**Fig. 13.23B**).

The Tricarboxylic Acid Cycle

The products of sugar breakdown can be catabolized to CO_2 and H_2O through the **tricarboxylic acid (TCA) cycle**. The TCA cycle generates molecules of NADH and $FADH_2$, which donate electrons to an electron transport system (ETS) with a terminal electron acceptor such as O_2 (see Chapter 14). But sugars are not the cycle's only substrates. Lignins, fatty acids, and even polycyclic aromatic toxicants are degraded to acetyl-CoA and other products that enter the TCA cycle.

The TCA cycle is also known as the Krebs cycle, named for Hans Krebs (1900–1981), who shared the 1953 Nobel Prize in Physiology or Medicine with Fritz Lipmann. Krebs and his colleagues at Sheffield University, England, studied catabolism by observing the oxidizing activities of crude enzyme preparations from sources such as pigeon breast muscle, beef liver, and cucumber seeds. In all of these animal and plant tissues, the TCA cycle is conducted by mitochondria, using virtually the same process as their bacterial ancestors used.

Pyruvate Is Converted to Acetyl-CoA

Glucose catabolism connects with the TCA cycle through the breakdown of pyruvate to acetyl-CoA and CO_2. Recall from **Figure 13.13** that acetyl-CoA is also generated from many other catabolic pathways, including those for breakdown of fatty acids, amino acids, and lignin fragments in soil. Regardless of its source, acetyl-CoA enters the TCA cycle by condensing with the four-carbon intermediate oxaloacetate to form citrate (**Fig. 13.24**). Citrate undergoes two steps of oxidative decarboxylation, in which CO_2 is released and two hydrogens with electrons are transferred to make NADH + H^+ or $FADH_2$. The TCA cycle, in whole or in part, is found in all microbial species except for degenerately evolved pathogens that are dependent on host metabolism, such as *Treponema pallidum*, the cause of syphilis (**eTopic 13.1**).

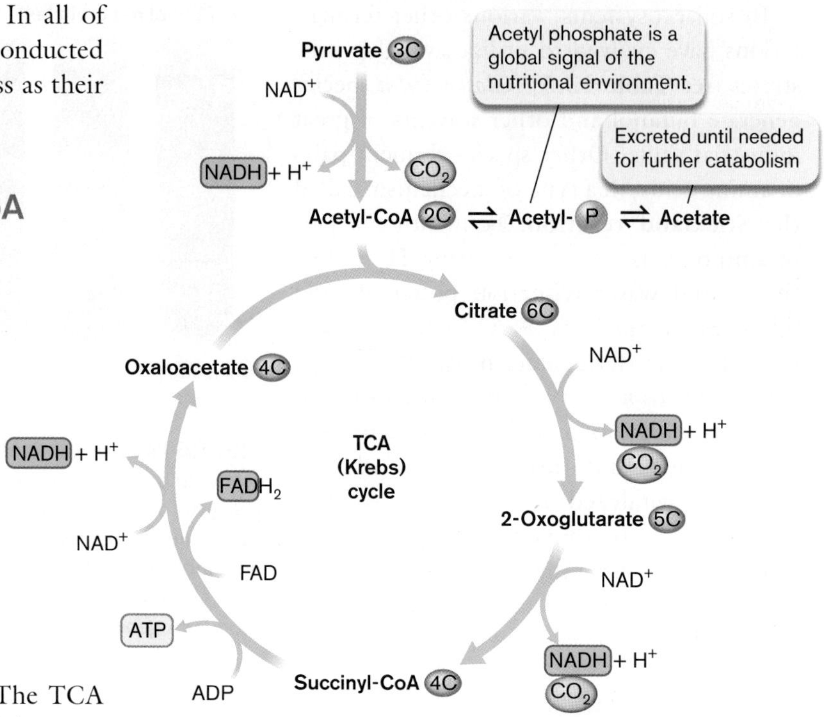

FIGURE 13.24 ■ Acetyl-CoA feeds into the TCA cycle. Pyruvate undergoes oxidative decarboxylation and incorporates CoA to form acetyl-CoA. Depending on the state of the cell, either acetyl-CoA is converted to acetate for excretion, or it enters the TCA cycle.

Thought Question

13.9 If a cell respiring on glucose runs out of oxygen and other electron acceptors, what happens to the electrons transferred from the catabolic substrates?

We present first the connecting step between pyruvate and acetyl-CoA, followed by the details of the TCA cycle. Bacteria and archaea use at least ten known variations on the TCA cycle, conducted by diverse species under various environmental conditions. You will no doubt be relieved to hear that we present only one: the Krebs pathway, which is found in most Gram-negative and Gram-positive bacteria, and in most eukaryotes. Other pathways are detailed in online resources such as the KEGG Pathway Database.

Pyruvate is converted to acetyl-CoA through removal of CO_2 and transfer of $2e^-$ onto NAD^+. The removal of CO_2 and transfer of two electrons is known as oxidative decarboxylation. The oxidative decarboxylation of pyruvate, coupled to CoA incorporation, is performed by an unusually large multisubunit enzyme called the **pyruvate dehydrogenase complex**, or **PDC**. PDC is a key component of metabolism in bacteria and mitochondria, the first molecular player to direct sugar catabolism into respiration. In human mitochondria, defects in PDC affect organs that have a high metabolic rate, such as the heart and brain, causing myocardial malfunction and heart failure, and neurodegeneration. A structural model for PDC, and the details of its reaction mechanism, are shown in **eTopic 13.2**.

The overall reactions catalyzed by PDC are:

$$CH_3COCOO^- + H^+ + HS\text{-}CoA \rightarrow CH_3CO\text{-}S\text{-}CoA + CO_2 + 2H^+ + 2e^-$$

$$NAD^+ + 2H^+ + 2e^- \rightarrow NADH + H^+$$

The removal of stable CO_2 yields energy for the electron transfer to NADH. The protons dissociated from pyruvate and from the thiol (SH) of CoA ($2H^+$ in total) are balanced by the net gain of protons by NADH + H^+.

The activity of PDC is increased by high concentrations of its substrates (CoA and NAD^+) and inhibited by its products (acetyl-CoA and NADH). The product acetyl-CoA may enter one of several pathways. In *E. coli*, when glucose is plentiful, acetyl-CoA is mostly converted to acetate via the intermediate acetyl phosphate (**Fig. 13.24**). Acetyl phosphate is a global signaling molecule that indicates to the cell the quantity and quality of carbon source available. As glucose decreases, the cell starts to reclaim acetate, converting it back to acetyl-CoA for entry into the TCA cycle. At the level of gene expression, PDC responds to environmental conditions. As would be expected, PDC gene expression is repressed by carbon starvation and by low levels of oxygen.

Thought Question

13.10 Compare the reactions catalyzed by pyruvate dehydrogenase and pyruvate formate lyase. What conditions favor each reaction, and why?

Acetyl-CoA Enters the TCA Cycle

Recall that acetyl-CoA can participate in various fermentations (see **Fig. 13.21**). But when a strong terminal electron acceptor is available (such as O_2), acetyl-CoA can enter the TCA cycle to transfer its electrons to electron carriers (**Figs. 13.24** and **13.25**). First, the acetyl group condenses with oxaloacetate, a four-carbon dicarboxylate (double acid). The condensation forms citrate, a six-carbon tricarboxylate. An advantage of intermediates with two or more acidic groups is that the concentration of the fully protonated form is extremely low; thus, the molecule is unlikely to be lost from the cell by diffusion across the membrane, as are monocarboxylic acids, such as acetate.

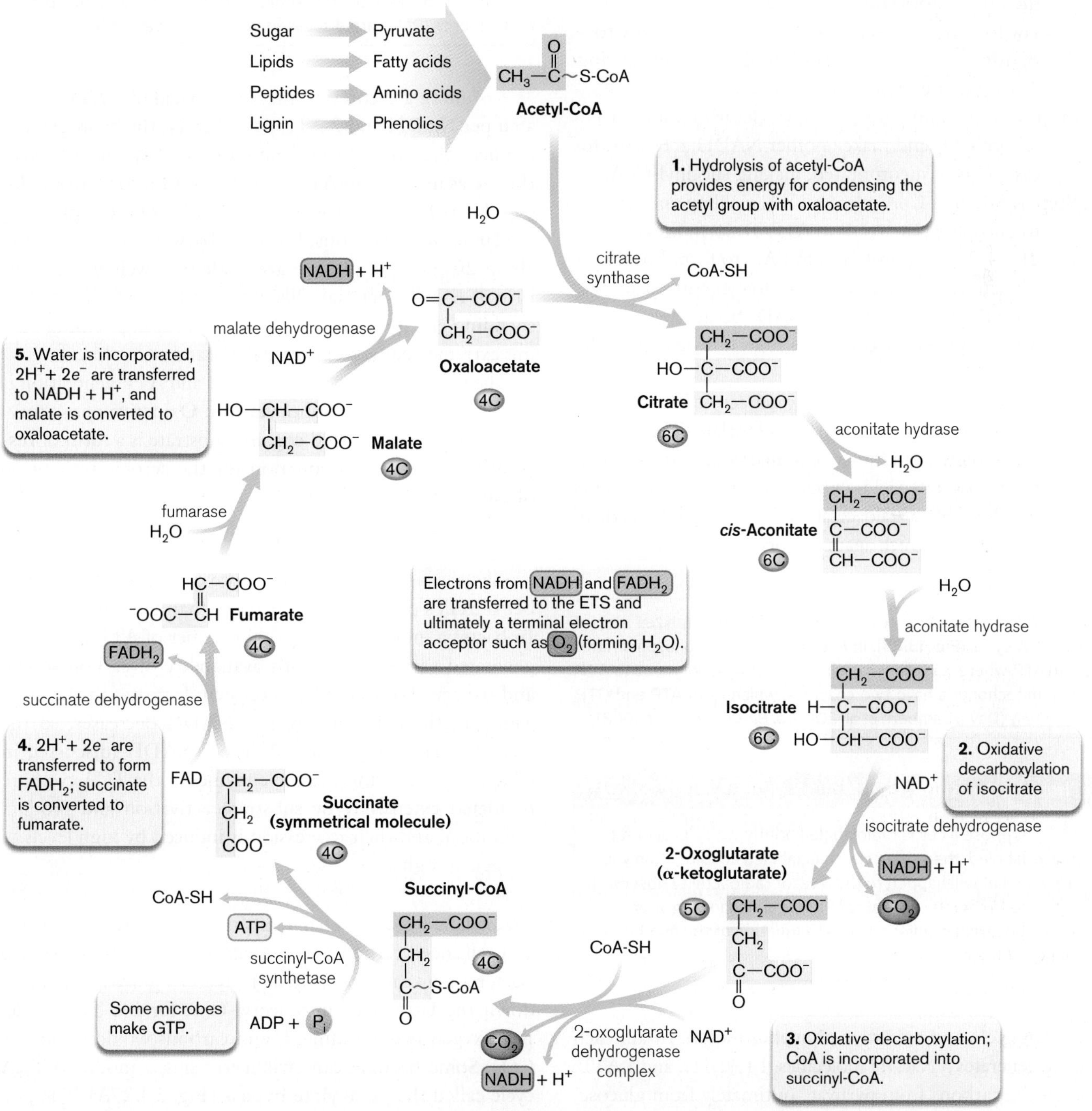

FIGURE 13.25 ■ The tricarboxylic acid (TCA) cycle. Acetyl-CoA derived from pyruvate and other catabolic pathways enters the TCA cycle. Green highlighting shows the fate of labeled acetate incorporated into the TCA cycle. Some forms of isocitrate dehydrogenase reduce $NADP^+$ instead of NAD^+.

Through the rest of the cycle, citrate loses two carbons as CO_2 by a series of reactions that transfer increments of energy to 3 NADH, 1 $FADH_2$, and 1 ATP. Each reaction step couples energy-yielding to energy-storing events (**Fig. 13.25**).

Step 1. As the acetyl group condenses with oxaloacetate, the hydrolysis of acetyl-CoA consumes a molecule of H_2O to restore HS-CoA. The removal of HS-CoA yields energy to condense acetate with oxaloacetate, forming citrate.

Step 2. Citrate undergoes two rearrangements (with little energy change) to form isocitrate. Isocitrate then undergoes oxidative decarboxylation. As we saw for pyruvate, removal of CO_2 yields energy to transfer $2H^+ + 2e^-$ to form NADH + H^+, producing 2-oxoglutarate (alpha-ketoglutarate).

Step 3. 2-Oxoglutarate undergoes decarboxylation to release CO_2 and make another NADH + H^+. In this case, CoA is incorporated, making succinyl-CoA.

Step 4. Succinyl-CoA releases CoA, yielding energy to phosphorylate ADP to ATP. To form fumarate, $2H^+ + 2e^-$ are transferred to FAD to form $FADH_2$—a reaction involving negligible free energy change. FAD is reduced instead of NAD^+ because electron donation from succinyl-CoA does not yield enough energy to reduce NAD^+.

Step 5. Fumarate incorporates water across its double bond, forming the hydroxy acid malate. The increasing stability from fumarate to malate and from malate to oxaloacetate yields enough energy to form the final NADH + H^+. Oxaloacetate is the original intermediate of the cycle, which again accepts the next acetyl-CoA.

Note: The enzyme succinyl-CoA synthetase catalyzes step 4 in the TCA cycle (**Fig. 13.25**). In *E. coli*, the enzyme phosphorylates ADP to ATP, whereas in other bacteria GDP is phosphorylated to GTP. Human mitochondria have two enzymes, which form ATP and GTP, respectively (David Lambeth et al. 2004. *J. Biol. Chem.* **279**:36621).

Thought Question

13.11 Suppose a cell is radiolabeled briefly with ^{14}C-acetate (pulse-labeled, then chased with unlabeled acetate). Can you predict what will happen to the level of radioactivity observed in isolated TCA intermediates? Plot a curve showing your predicted level of radioactivity as a function of the number of rounds of the cycle.

The TCA cycle and oxidative phosphorylation. In all, each acetate generates 3 NADH molecules, 1 $FADH_2$, and 1 ATP; and all the carbons from pyruvate (ultimately from glucose) are released as waste CO_2. From the standpoint of the carbon skeleton, the glucose breakdown is now complete. But do we have a completed metabolic pathway? No, because all of the NADH and $FADH_2$ need to be recycled by donating their electrons onto a terminal electron acceptor.

The process of electron transfer from NADH and $FADH_2$ is mediated by the electron transport system (ETS, discussed in Chapter 14). In the ETS, electrons are transferred from reduced proteins and cofactors to more oxidized proteins and cofactors, as in the examples in **Table 13.4**. Some of the membrane proteins use the energy of electron transfer to pump protons, generating a gradient of hydrogen ions across the membrane (**Fig. 13.26**).

Note: The electron transport system (ETS) and the proton motive force (PMF) are discussed in detail in Chapter 14.

Assuming a theoretical maximum yield of 3 ATP generated per NADH and 2 ATP per $FADH_2$, the hydrogen ion gradient then drives the membrane ATP synthase to synthesize as many as 34 ATP. Another 4 ATP come from glucose breakdown and the TCA cycle (38 total per glucose). Under actual conditions, however, bacteria make less ATP; about 20 ATP per glucose are made by a well-aerated culture of *E. coli*. Bacterial cells make trade-offs for flexibility, spending energy to maintain a stable proton potential during extreme changes in external pH and redox levels.

The process of electron transport and ATP generation is termed **oxidative phosphorylation**. Oxidative phosphorylation from a catabolized organic substrate is a form of **respiration**. The overall equation for the aerobic respiration of glucose is:

$$C_6H_{12}O_6 + 6H_2O + 6O_2 \rightarrow 12H_2O + 6CO_2$$

Glucose respiration can generate a relatively large number of ATPs per glucose—far more than fermentation can. In bacteria, however, the actual number of ATP molecules generated varies widely with availability of carbon source and oxygen. For example, as oxygen decreases in the environment, the ability to oxidize NADH decreases, so the cell may make only 1 or 2 ATP per NADH (discussed in Chapter 14). In addition, the enzymes of the TCA cycle are regulated extensively by substrate activation and product inhibition, and their expression is induced by high levels of oxygen and glucose.

Glyoxylate bypass. What happens when glucose is scarce and cells need carbon both for energy and for biosynthesis? Bacteria may catabolize lipids instead of glucose, breaking down the fatty acids into acetyl-CoA for the TCA cycle. But oxygen may be limited, and carbons would be lost as CO_2. Some bacteria can switch to using a modified TCA cycle called the **glyoxylate bypass** (**Fig. 13.27A**). The glyoxylate bypass enables lung pathogens such as *Pseudomonas aeruginosa* and *Mycobacterium tuberculosis* to catabolize fatty acids, via their breakdown to acetyl-CoA.

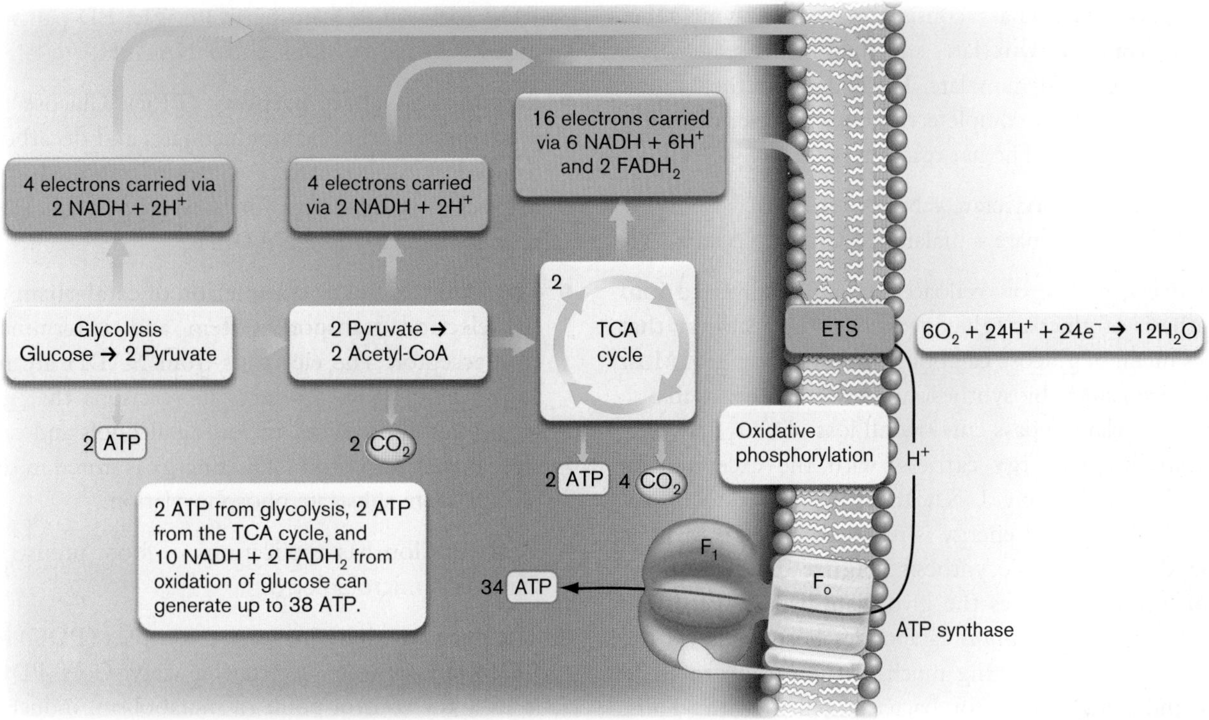

FIGURE 13.26 ■ Complete oxidation of glucose. Glucose catabolism generates ATP through substrate-level phosphorylation and through the electron transport system's pumping of H^+ ions to drive the ATP synthase. The complete oxidative breakdown of glucose to CO_2 and H_2O could theoretically generate up to 38 ATP. Under actual conditions, the number is smaller.

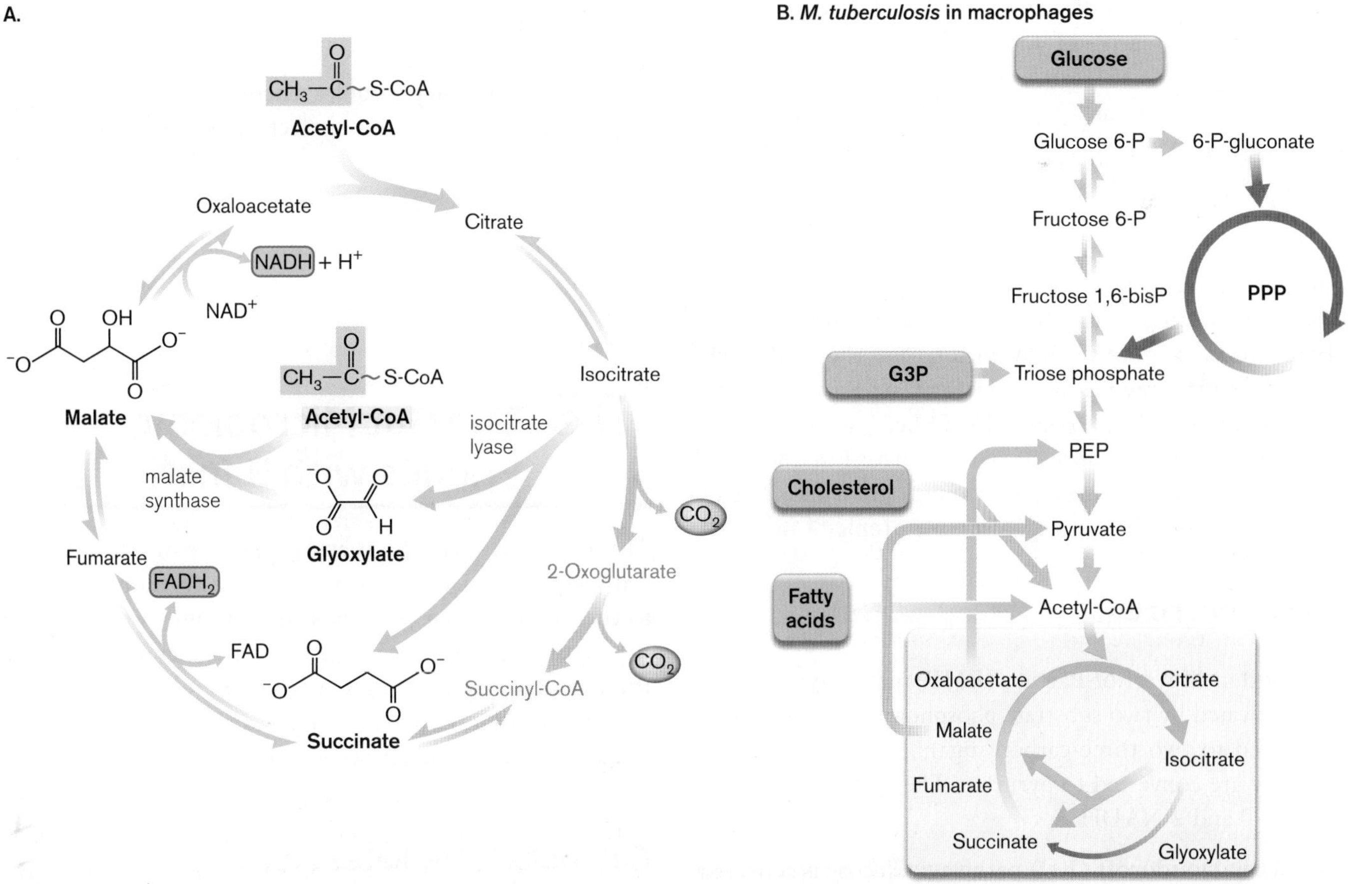

FIGURE 13.27 ■ The glyoxylate bypass. **A.** Instead of releasing two CO_2, the glyoxylate bypass incorporates a second molecule of acetyl-CoA, producing succinate plus malate. **B.** The glyoxylate bypass in the metabolism of *Mycobacterium tuberculosis* growing within mouse macrophages. For glucose biosynthesis, some glycolytic enzymes act in reverse.

The glyoxylate bypass consists of two enzymes that divert isocitrate to glyoxylate, and then incorporate a second acetyl-CoA to form malate. The malate can then regenerate oxaloacetate to complete the bypass cycle, donating $2e^-$ to form NADH. The net reaction is:

$$\text{2 Acetyl-CoA + oxaloacetate + NAD}^+ \rightarrow \text{succinate + malate + 2 CoA + NADH + H}^+$$

Alternatively, malate or oxaloacetate can be diverted into biosynthesis of glucose (gluconeogenesis)—a pathway that reverses much of glucose catabolism (see Chapter 15). Most bacteria need sugar biosynthesis to build their cell walls.

The glyoxylate bypass cuts out all loss of CO_2 and electron transfer to energy carriers, with the exception of 1 $FADH_2$ and possibly 1 NADH from malate to oxaloacetate. Thus, limited energy is released, but two carbons can be diverted to biosynthesis. **Figure 13.27B** shows how *M. tuberculosis* uses the glyoxylate bypass. Persistent intracellular *M. tuberculosis* catabolizes host lipids via the glyoxylate bypass, diverting much of the carbon to build sugars and amino acids for bacterial cell growth. Thus, key enzymes of the glyoxylate bypass, and their regulators, offer targets for new antibiotics.

The TCA cycle for amino acid biosynthesis. Analysis of pathway evolution indicates that the TCA cycle originally evolved to provide substrates for building amino acids (see Chapter 15). For example, the TCA cycle intermediate 2-oxoglutarate (alpha-ketoglutarate) is aminated to form glutamate, which leads to glutamine. The amine group comes from an ammonium ion. (The various sources of nitrogen for biosynthesis are discussed in Chapter 15). Oxaloacetate is aminated to form aspartate, which enters pathways to purines and pyrimidines. The TCA cycle, like glycolysis, is an amphibolic pathway that provides substrates for biosynthesis. Many bacteria use the TCA cycle and glycolytic enzymes to build their sugars and amino acids, as discussed in Chapter 15. Others, such as *Treponema pallidum*, the cause of syphilis, have lost the TCA cycle by reductive evolution. They must obtain amino acids synthesized by their host organism (see **eTopic 13.1**).

To Summarize

- **Embden-Meyerhof-Parnas (EMP) pathway:** Glucose is activated by two substrate phosphorylations, and then cleaved to two three-carbon sugars. Both sugars eventually are converted to pyruvate. The pathway produces 2 ATP and 2 NADH.
- **Entner-Doudoroff (ED) pathway:** Glucose is activated by one phosphorylation, and then dehydrogenated to 6-phosphogluconate. 6-Phosphogluconate is cleaved to pyruvate and a three-carbon sugar, which enters the EMP pathway to form pyruvate. The ED pathway produces 1 ATP, 1 NADH, and 1 NADPH.
- **Pentose phosphate pathway (PPP):** Glucose is dehydrogenated to 6-phosphogluconate, and decarboxylated to ribulose 5-phosphate. A series of intermediate sugars may serve as substrates for biosynthesis. The PPP may produce 1 ATP and 2 NADPH.
- **Fermentation is the completion of catabolism** without the electron transport system and a terminal electron acceptor. The electrons from NADH are restored to pyruvate or its products in reactions that generate fermentation products, including alcohols and carboxylates, as well as H_2 and CO_2. Energy is stored in the form of ATP from substrate phosphorylation.
- **Fermentation** has applications in food, industrial, and diagnostic microbiology.
- **The pyruvate dehydrogenase complex (PDC)** removes CO_2 from pyruvate, generating acetyl-CoA. PDC activity is a key control point of metabolism, induced when carbon sources are plentiful, and repressed under carbon starvation and low oxygen.
- **Acetyl-CoA enters the TCA cycle by condensing with oxaloacetate to form citrate.** A series of enzymes sequentially removes carbon dioxide and water molecules and generates 3 NADH, 1 $FADH_2$, and 1 ATP. Each reaction step couples energy-yielding to energy-storing events. A terminal electron acceptor such as O_2 must receive electrons.
- **The glyoxylate bypass** provides a way to gain limited energy from the TCA cycle while avoiding CO_2 loss; thus the bypass diverts intermediates to sugar biosynthesis.

13.5 The Gut Microbiome: Friends with Benefits

The focus of this chapter is energy and the variety of chemical conversions that microbes undergo to transfer energy to their own cellular processes. Increasingly, we recognize the broader significance of energy transfer reactions for the microbiome. The products of a microbial community such as our gut microbiome serve unexpected functions for other community members—including their host.

Catabolism Mediates New Niche Adaptation

As we saw in Section 13.3, different members of a host's digestive microbiome can catabolize different components

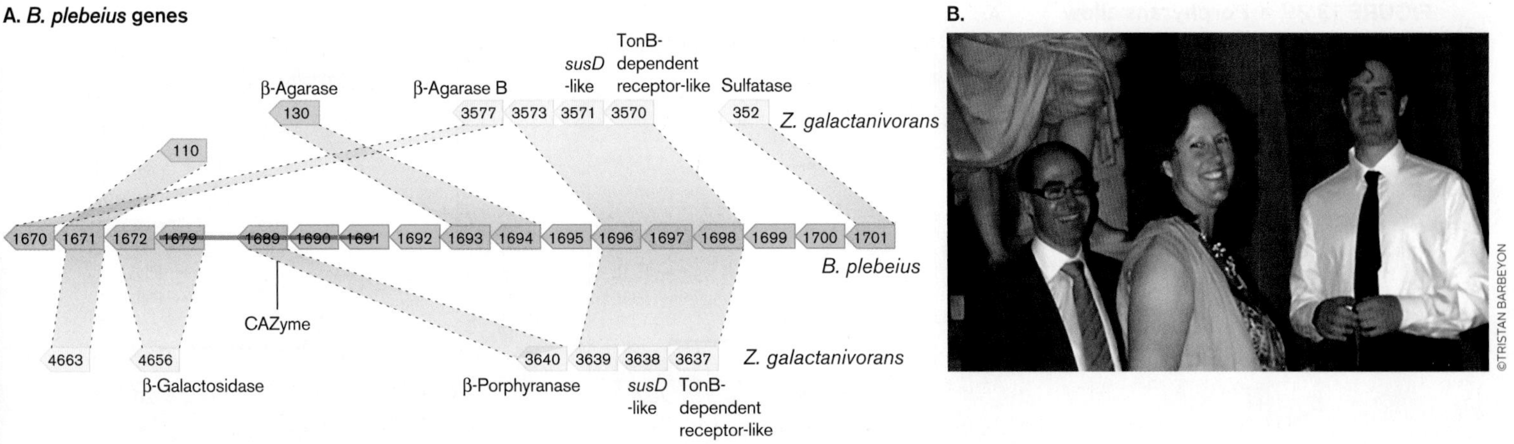

FIGURE 13.28 ■ Origin of *Bacteroides plebeius* CAZyme gene. **A.** Homology between *Bacteroides plebeius* gene sequences and genes horizontally transferred from *Zobellia galactanivorans* and *Microscilla*. **B.** Jan-Hendrik Hehemann (right), with colleagues Gurvan Michel (left) and Mirjam Czjzek. *Source:* Part A modified from Hehemann et al. 2010. *Nature* **464**:908–912.

of the host's diet. But how do the microbiome's digestive capabilities impact the host? One consequence of microbiome diversity is the capacity of the host to fill new nutritional niches. An example is that of koala populations adapted to feed on different varieties of eucalyptus. Certain koalas can consume only one variety of eucalyptus and starve to death if they attempt to consume a different variety. But when their gut microbiome is supplemented with feces transplanted from koalas that eat the new variety, the recipient koalas acquire the ability to digest it.

There is evidence for similar digestive diversity in humans. The ability to digest sulfonated glycans (porphyrans) is limited to populations in Japan, where people consume nori, a form of raw seaweed. The seaweed contains porphyrans and also harbors bacteria that catabolize them. The seaweed bacteria, such as *Zobellia galactanivorans*, show Sus-like (PUL) operons that encode the porphyran catabolism. Remarkably, similar operons appear in *Bacteroides plebeius*, a Gram-negative anaerobe found in the gut microbiome of Japanese who consume nori (**Fig. 13.28A**). The key porphyranase enzyme (CAZyme, gene 1689) was identified by Jan-Hendrik Hehemann and colleagues at the Pierre and Marie Curie University (**Fig. 13.28B**). Hehemann proposes that the gut microbial *Bacteroides* species acquired the porphyranase gene by horizontal transfer from a marine species, such as *Z. galactanivorans*, that entered the human digestive tract with ingested seaweed.

Could such a bacterium with augmented catabolism establish residence in the microbiome? Another research group, led by Elizabeth Shepherd and Justin Sonnenburg at Stanford University, conducted a bacterial transfer experiment in mice (**Fig. 13.29**). Mice with conventional gut microbiomes were fed a diet of diverse carbohydrates that *Bacteroides* species commonly catabolize. Their guts were then inoculated with a strain of *Bacteroides ovatus* that expresses a PUL system encoding porphyranase (designated PUL$^+$). The prevalence of *B. ovatus* was then measured by colony counts from fecal samples. When the mice were fed the standard carbohydrate diet (no porphyrans), *B. ovatus* soon disappeared from the gut (**Fig. 13.29A**). Mice fed a diet including porphyrans showed sustained colonization by the PUL$^+$ bacteria, but not by PUL$^-$ bacteria (**Fig. 13.29B**). This experiment demonstrates the possibility of manipulating the gut microbiome by providing a catabolic substrate.

A remarkable case of niche adaptation to catabolic substrates is that of glycans produced for mammalian breast milk. In humans, these glycans are called **human milk oligosaccharides** (**HMOs**). HMOs consist of branched chains of hexoses alternating galactose and *N*-acetylglucosamine (glucose with an added acetyl group, and an amine group). Some chains end in fucose (6-deoxy-L-galactose) or *N*-acetylneuraminic acid. These unique sugar chains form a large share of the nutritional content of milk, yet they cannot be digested by human enzymes. Instead, the HMOs select for gut colonization by microbes beneficial to the infant, particularly species of the Gram-positive anaerobe *Bifidobacterium*.

Bifidobacterium species have numerous benefits for infant digestion and development. These bacteria ferment by converting pyruvate to lactate (lactic acid fermentation) and thus avoid gas production, typical of *Escherichia coli* and related bacteria (see fermentation pathways in **Fig. 13.21**). Gas production causes infant colic (abdominal pain). Besides decreasing colic, bifidobacteria also produce anti-inflammatory molecules and immunomodulators. In effect, the maternal milk production uses catabolic substrates to manipulate the composition of the infant gut microbiome for improved health.

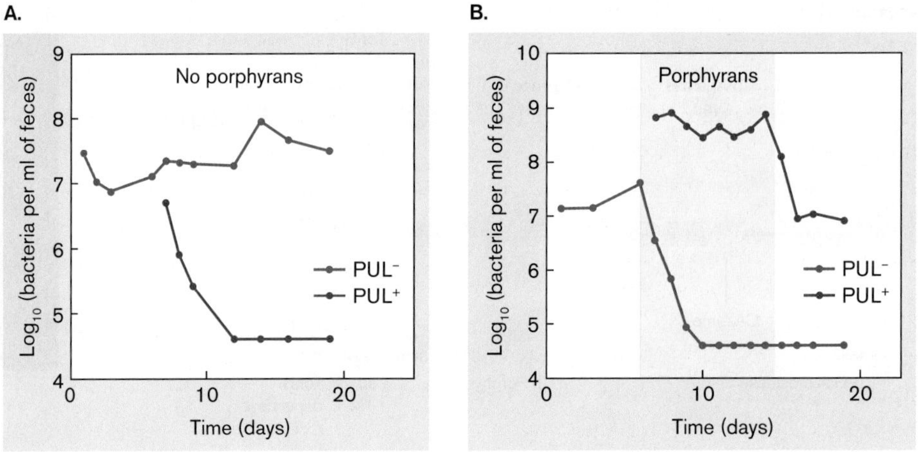

FIGURE 13.29 ■ Porphyrans allow mouse gut colonization by bacteria expressing porphyranase. A. Without porphyrans in the diet, bacteria expressing porphyranase (PUL^+) fail to compete in the mouse microbiome. **B.** PUL^+ bacteria successfully join the microbiome of mice that have been fed porphyrans. *Source:* Modified from Shepherd et al. 2018. *Nature* **557**:434–438.

Amino Acid Decarboxylation Produces Neurotransmitters

Peptides and amino acids in our diet are important for our own protein synthesis, but they also are catabolized by microbes. Amino acids are catabolized by fermentation processes that involve removal of either the amino group or the carboxylate. Decarboxylation of amino acids consumes a proton and produces an organic amine (**Fig. 13.30**). The amine is generally exported by an antiporter that exchanges the amino acid substrate for the exported amine. Thus, the overall reaction removes acid. Amino acid decarboxylation and amine export are induced by passage through acidic conditions such as in the stomach.

The exported amine "waste products" can be molecules of surprising importance to the host, including neurotransmitters and hormones such as dopamine, serotonin, and histamine. One example is the decarboxylation of glutamic acid to the neurotransmitter 4-aminobutanoic acid, or GABA (the abbreviation comes from its former name, "gamma-aminobutyric acid"). GABA is a neurotransmitter that binds inhibitory synapses in neurons of the intestine and other organs, as well as the brain. GABA modulates pain reception, brain development, and behavior. Another example is histamine, the product of decarboxylation of histidine. Histamine is an important anti-inflammatory molecule and an immunomodulator (discussed in Chapter 23).

Physicians have wondered whether bacteria that produce GABA serve as a probiotic to control pain. James Versalovic and colleagues tested this question in a rat model for abdominal pain. The rats, which show overstimulation of pain neurons, received a daily dosage of *Bifidobacterium dentium*, a GABA-producing

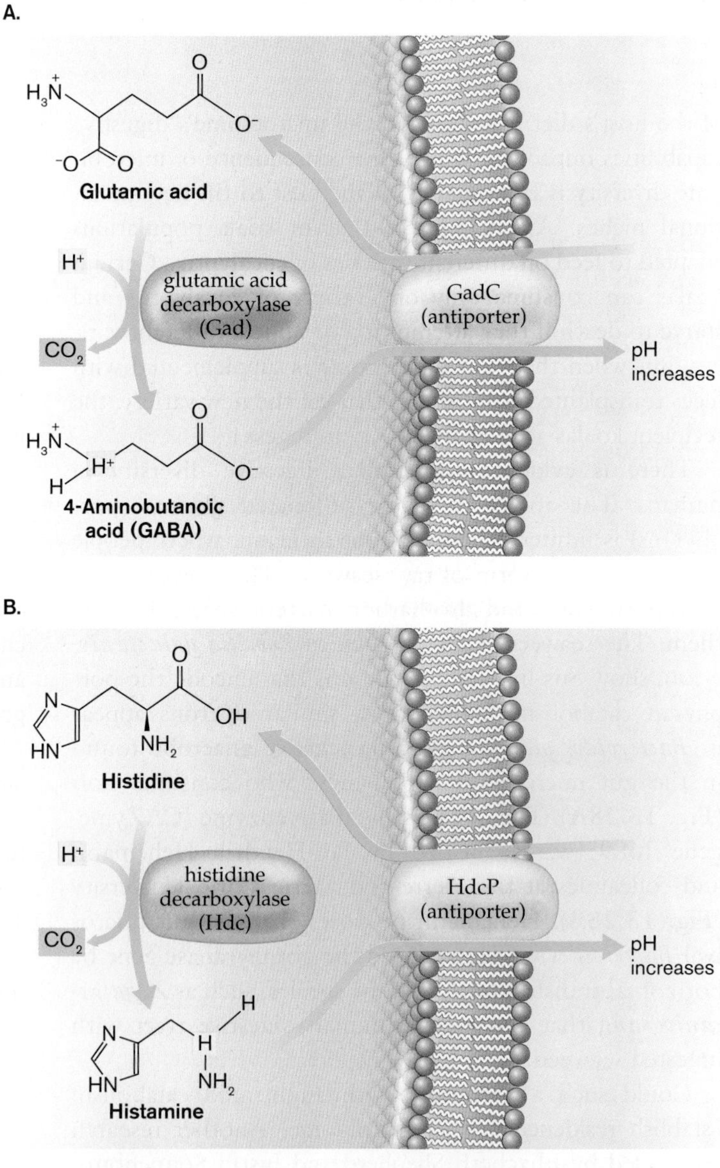

FIGURE 13.30 ■ Bacterial amino acid decarboxylases produce bioactive molecules. A. Glutamate decarboxylase produces GABA, which is exported by a glutamate/GABA antiporter. **B.** Histidine decarboxylase produces histamine, which is exported by a histidine/histamine antiporter.

species. The rats treated with *B. dentium* showed decreased activity of pain sensory neurons—but only if the bacteria administered had a functional *gadB* gene, encoding glutamate decarboxylase. Thus, GABA-producing bacteria might be developed for probiotic analgesia. Similar experiments show that histamine-producing bacteria, such as *Lactobacillus* species, may serve as probiotics to suppress chronic intestinal inflammation.

Could GABA-producing bacteria modulate our brain chemistry, and even our behavior? Fascinating experiments suggest roles for bacteria in the gut-brain axis (see **Special Topic 13.1**). In 2018, researchers found gut-type bacteria growing in tissue from normal human brains. Do bacteria in the brain have a functional role? For further exploration of the gut microbiome and its implications, see Chapters 21 and 23.

To Summarize

- **Gut microbial catabolism performs functions for the host.**
- **Bacterial catabolism of a novel substrate may enable host niche adaptation.** An example is the acquisition of porphyranase by gut bacteria in Japan, where people consume raw seaweed. Another example is that of human milk oligosaccharides (HMOs), which are catabolized by *Bifidobacterium* in the colon of human infants.
- **Amino acid decarboxylation produces neurotransmitters.** Bacterial production of GABA may regulate sensory neurons and may influence brain function and development. Bacterial production of histamine may modulate the immune system.

13.6 Aromatic Catabolism and Syntrophy

Catabolism of Benzene Derivatives

The TCA cycle plays a key role in the breakdown of aromatic carbon sources such as the benzene ring. Natural sources of aromatic carbon include lignin from wood, and polycyclic aromatic hydrocarbons (PAHs) from petroleum. Aromatic components of petroleum cause much of the environmental damage during oil spills such as the *Deepwater Horizon* spill in 2010. Soil is often polluted by industrial aromatic compounds, such as nitrate explosives, aniline dyes, and the solvent toluene. Even the most remote places on Earth, such as Antarctica, show traces of such pollutants.

To remove aromatic pollutants from water and soil, we depend on microbial catabolism. Aromatic molecules are notoriously difficult to catabolize because of the stability of the benzene ring. The benzene ring breaks down slowly, but over time, bacteria and fungi catabolize a wide range of aromatic molecules. **Figure 13.31** shows an example of research to study microbial digestion of phenanthrene (a tricyclic PAH) in the soil of Antarctica. Because the Antarctic Treaty forbids introduction of exogenous organisms, any bioremediation of pollutants must be accomplished by native microbes.

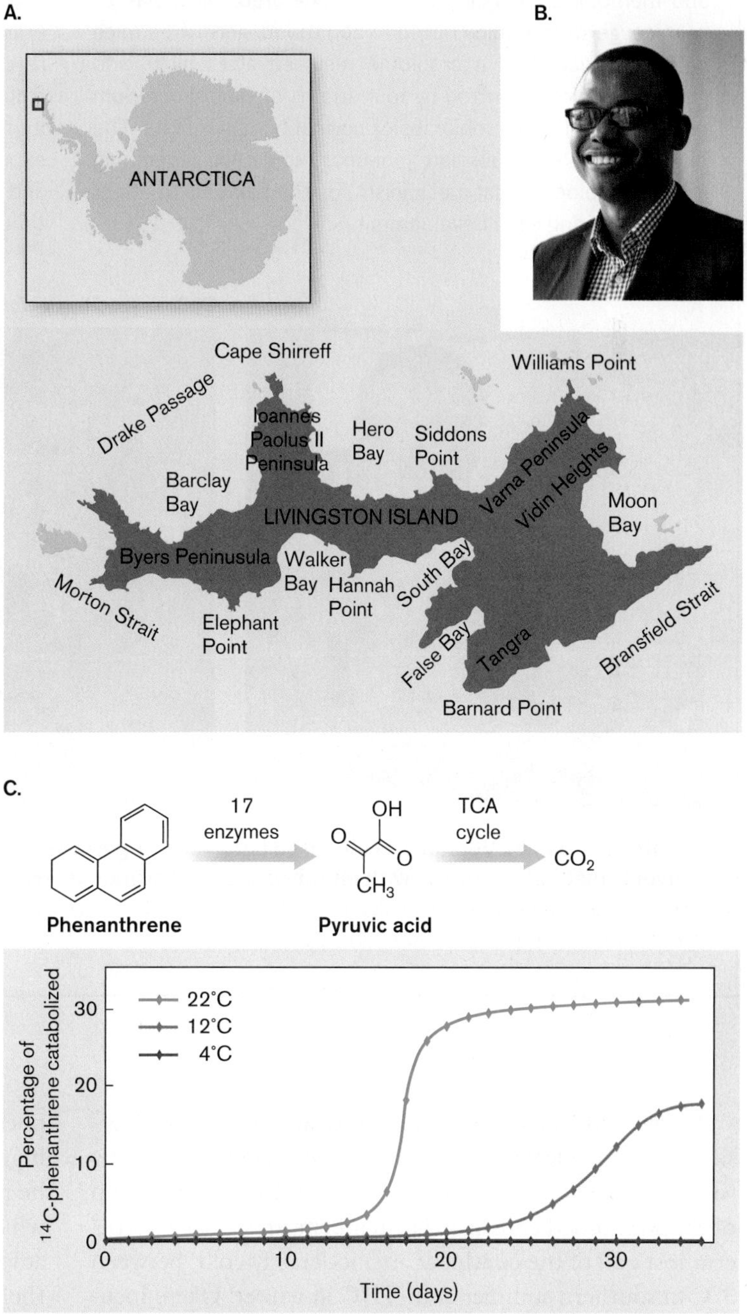

FIGURE 13.31 ■ Phenanthrene bioremediation by Antarctic psychrotrophs. **A.** Livingston Island, Antarctic Peninsula. **B.** Uchechukwu Okere, environmental microbiologist. **C.** Phenanthrene is broken down to CO_2 by indigenous psychrotrophs. *Source:* Uchechukwu Okere et al. 2012. *FEMS Microbiol. Lett.* **329**:69.

SPECIAL TOPIC 13.1 Gut Bacteria Rule Host Behavior

Bacteria export GABA as a fermentation product, one of the most important mammalian neurotransmitters. Could gut bacteria actually regulate development and function of the brain? James Versalovic and colleagues at Baylor University tested this question, starting with germ-free mice (**Fig. 1**). Mice reared without microbes are known to show many defects in behavior, including motor performance, anxiety, and memory. So Versalovic's team generated three classes of mice: germ-free mice raised to adulthood, germ-free mice with a conventional microbiome restored after birth, and germ-free mice colonized by four strains of *Bifidobacterium* prominent in the microbiomes of normal human infants. The bifidobacteria decarboxylate glutamate to GABA, suggesting a possible biochemical mechanism for the bacteria to affect the brain during early development.

Figure 1B shows sections of mouse colon colonized by bifidobacteria or by conventional mouse gut bacteria. Staining by FISH (fluorescence in situ hybridization) labels a DNA sequence specific to the genus *Bifidobacteria* (red) or to a general bacterial sequence of 16S rRNA (green).

The researchers found that postnatal introduction of either the human infant-associated bifidobacteria or a conventional mouse microbiome significantly improved several types of behavior, including motor function, anxiety response, and recognition memory. **Figure 2** shows the data from an experiment testing recognition memory (the ability to distinguish a novel object from a familiar one). In this experiment, each mouse is presented with a familiar object and a novel object. A camera then records the amount of time the mouse spends exploring each object; normally, mice

A. B. Colonized by *Bifidobacteria* Colonized by conventional microbiome

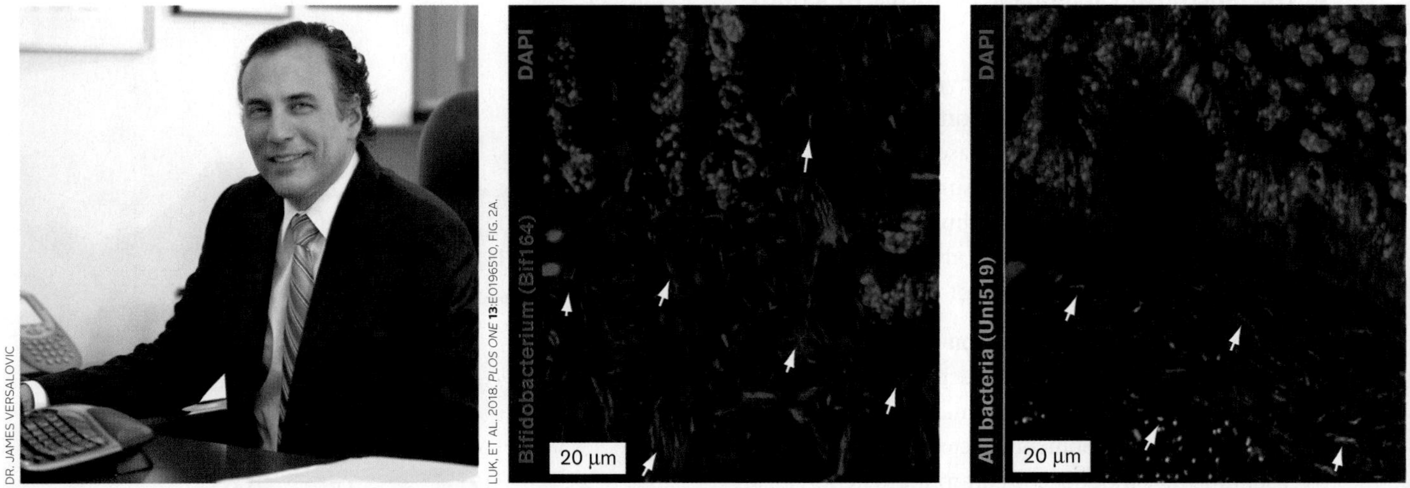

FIGURE 1 ■ James Versalovic compared behaviors of germ-free mice with those of mice colonized postnatally with conventional microbiota or with infant-associated bifidobacteria. A. James Versalovic. B. Fluorescence micrographs of murine colon sections labeled with *Bifidobacterium*-specific rRNA-hybridizing probe (red) or with a universal bacterial probe (green). Blue label (DAPI) indicates mouse chromosomes.

Uchechukwu Okere and colleagues at Lancaster University sought evidence for phenanthrene biodegradation by soil microbial communities at Livingston Island, off the tip of the Antarctic Peninsula. The temperatures at this northernmost end of the continent are moderately cold, between 3°C in southern summer and –11°C in winter. Okere incubated soil samples with phenanthrene substrates containing ^{14}C-radiolabeled carbons. The incubation was performed in a respirometer with a CO_2 trap, and the ^{14}C radioactivity was measured in the trapped CO_2. Okere found that over 10–30 days, microbes degraded as much as 30% of the phenanthrene to CO_2. The degradation rate was higher at higher temperatures; thus, the catabolic microbes appear to be psychrotrophs (cold tolerant) rather than psychrophiles, which require cold for growth. Psychrotrophs may take advantage of a temporary rise in temperature to outgrow their competitors.

How do microbes catabolize such tough carbon sources? As we saw earlier (**Fig. 13.13**), fungi and bacteria break down lignin and PAHs to form single-ring aromatic compounds such as benzoate and phenols. Some of these products have commercial uses: Benzoate is a food preservative;

spend more time exploring an object that is novel. **Figure 2B** shows the results for the three treatment groups: the germ-free mice, the mice with a conventional microbiome restored after birth, and germ-free mice colonized by infant-type *Bifidobacterium* species.

What were the results? The mice with conventional microbiomes, or with infant-type bifidobacteria-only microbiomes, spent more time exploring the novel object, whereas germ-free mice explored both objects equally, as if they had seen neither before. This result and results from experiments testing other behaviors are consistent with a role for gut bacterial catabolism in mammalian development. It remains to be seen how much of our brain function comes from our gut.

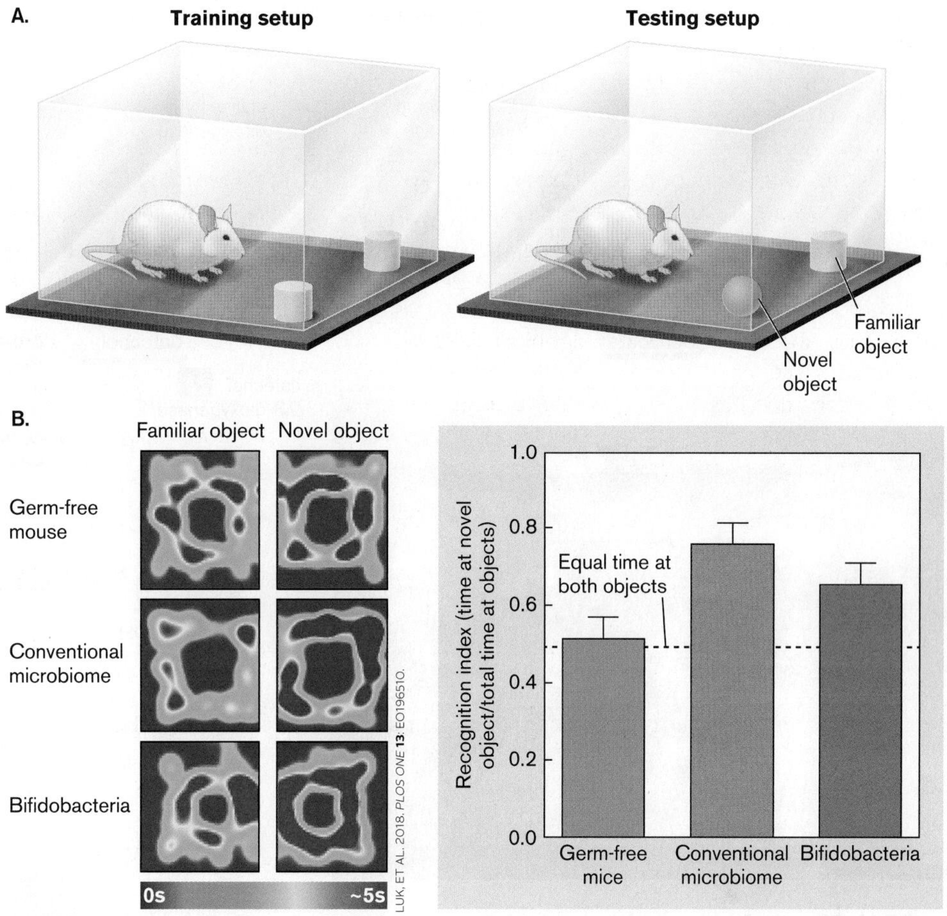

FIGURE 2 ▪ Bifidobacteria or conventional microbiome restores memory deficit in germ-free mice. A. Training setup: Each mouse is introduced to a pair of familiar objects, or a familiar object plus a novel object. **B.** Germ-free mice spend equal time around the familiar and novel objects, whereas mice with a conventional microbiome or with bifidobacteria spend more time exploring the novel object. Color scale indicates time spent in position, from zero (blue) to 5 seconds (red).

RESEARCH QUESTION

The authors speculate that GABA produced by metabolism mediates the effect of bifidobacteria on the behaviors of germ-free mice. Can you propose an experiment to test this hypothesis?

Berkley Luk, Surabi Veeraragavan, Melinda Engevik, Miriam Balderas, Angela Major, et al. 2018. Postnatal colonization with human "infant-type" *Bifidobacterium* species alters behavior of adult gnotobiotic mice. *PLoS One* **13**:e0196510.

vanillin, a flavor additive; phenol, an industrial solvent. In nature, the single-ring compounds are further catabolized to acetyl-CoA, which enters the TCA cycle. Benzoate, activated to benzoyl-CoA, has a central role in the catabolism of aromatic molecules, comparable to that of glucose in polysaccharide catabolism. Bacteria such as *Pseudomonas* and *Rhodococcus* species degrade benzoate and related molecules aerobically or anaerobically. Anaerobic degradation takes much longer, but it is critical because the volume of anaerobic habitat (such as soil) greatly exceeds that of oxygenated habitat.

Aerobic benzene catabolism. Benzene and related aromatic compounds, such as toluene (methylbenzene), chlorobenzoate, and nitrobenzene, can be catabolized via sequential oxidation steps, requiring the presence of an ETS that terminates with O_2 (**Fig. 13.32**). Early in the pathways, enzymes must remove substituents such as chlorides or nitrates. The methyl group of toluene is oxidized to carboxylate (R–COO^-), whose removal then drives a key breakdown step. Aerobic degradation commonly proceeds through the intermediate catechol, a benzene ring bearing two adjacent hydroxyl groups. Each benzene derivative is

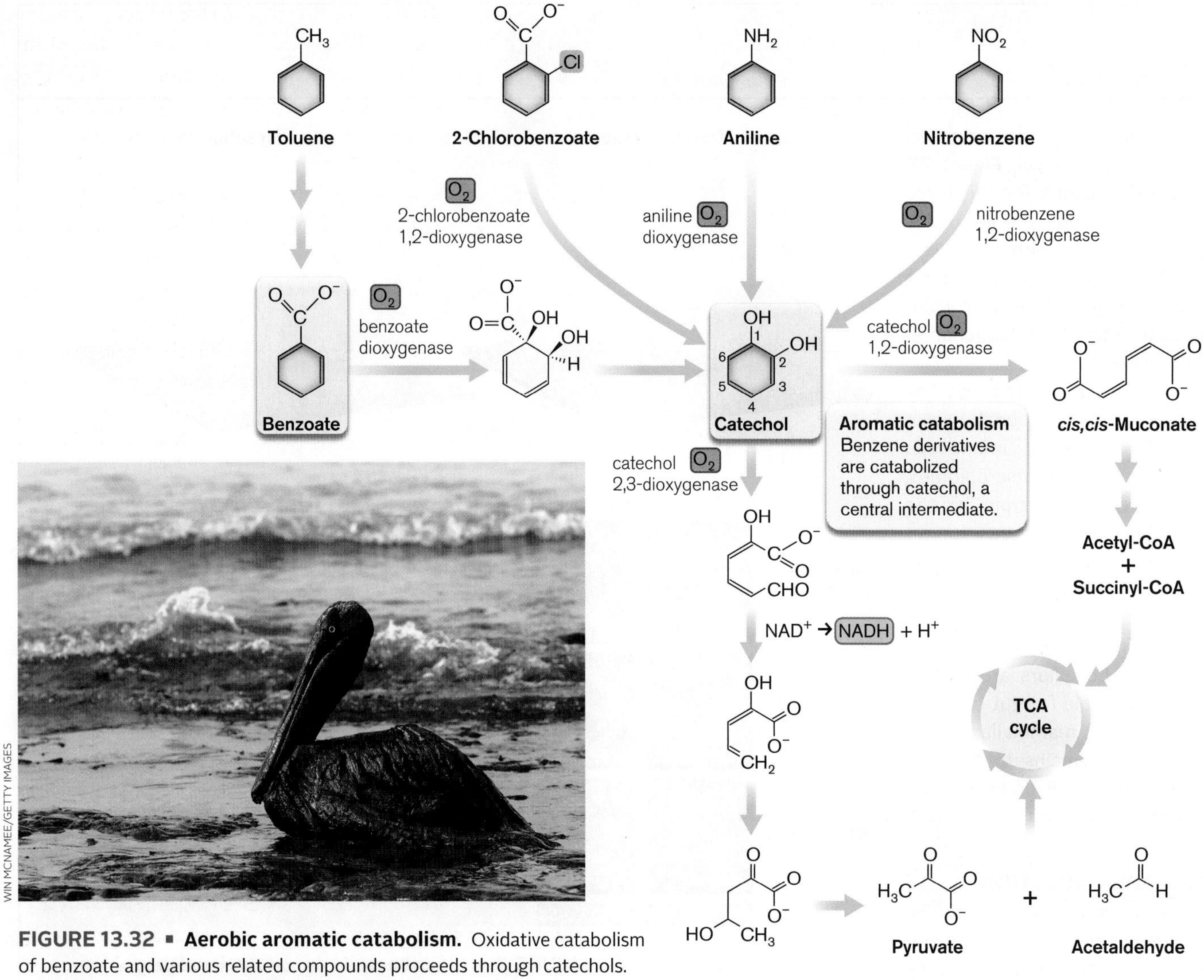

FIGURE 13.32 ■ Aerobic aromatic catabolism. Oxidative catabolism of benzoate and various related compounds proceeds through catechols. Catechols are degraded through several alternative pathways to the TCA cycle. Steps requiring oxidation are marked O_2. **Inset:** Bird contaminated by petroleum from an offshore wellhead.

converted to catechol by a specific dioxygenase, an enzyme that coordinately oxygenates two adjacent ring carbons. Next follows ring cleavage and breakdown to pyruvate (forming NADH) and to acetyl-CoA, which may enter the TCA cycle and respiration.

The intermediate catechol then undergoes another key oxidation by a catechol dioxygenase, which adds two more oxygens while cleaving the ring. In different bacterial species, the enzyme may oxidize catechol at either the 1 and 2 positions (as shown in **Fig. 13.32**) or the 2 and 3 positions. Typical products are succinyl-CoA and acetyl-CoA, which enter the TCA cycle, completing breakdown to CO_2. The details of benzene catabolism vary among bacterial species, and many diverse mechanisms continue to be discovered.

Anaerobic benzene catabolism. A challenge for biodegradation is that many pollution sources reach deep underground, where the soil is anoxic. Thus, oxygen is unavailable to conduct the conversions shown in **Figure 13.32**, particularly the formation of carboxylate (R–COO^-) and the introduction of hydroxyl groups. How can microbes catabolize benzene and benzoate derivatives without oxygen? In early stages of anaerobic catabolism (**Fig. 13.33**), benzene and naphthalene incorporate CO_2, forming the carboxylate group of benzoate. Because CO_2 is a very weak oxidant, these reactions require input of energy by hydrolysis of ATP. The carboxylate is then activated by HS-CoA, forming benzoyl-CoA, the same key intermediate as for aerobic benzene catabolism.

Anaerobically, the benzoyl-CoA must use reducing energy from NADH to hydrogenate its ring carbons, thus breaking the aromaticity. Some bacteria spend ATP as well, whereas others, such as the iron-reducing bacterium *Geobacter metallireducens*, can break the ring without spending ATP. The hydroxyl groups that enable shifting of the double-bond positions are introduced by incorporation

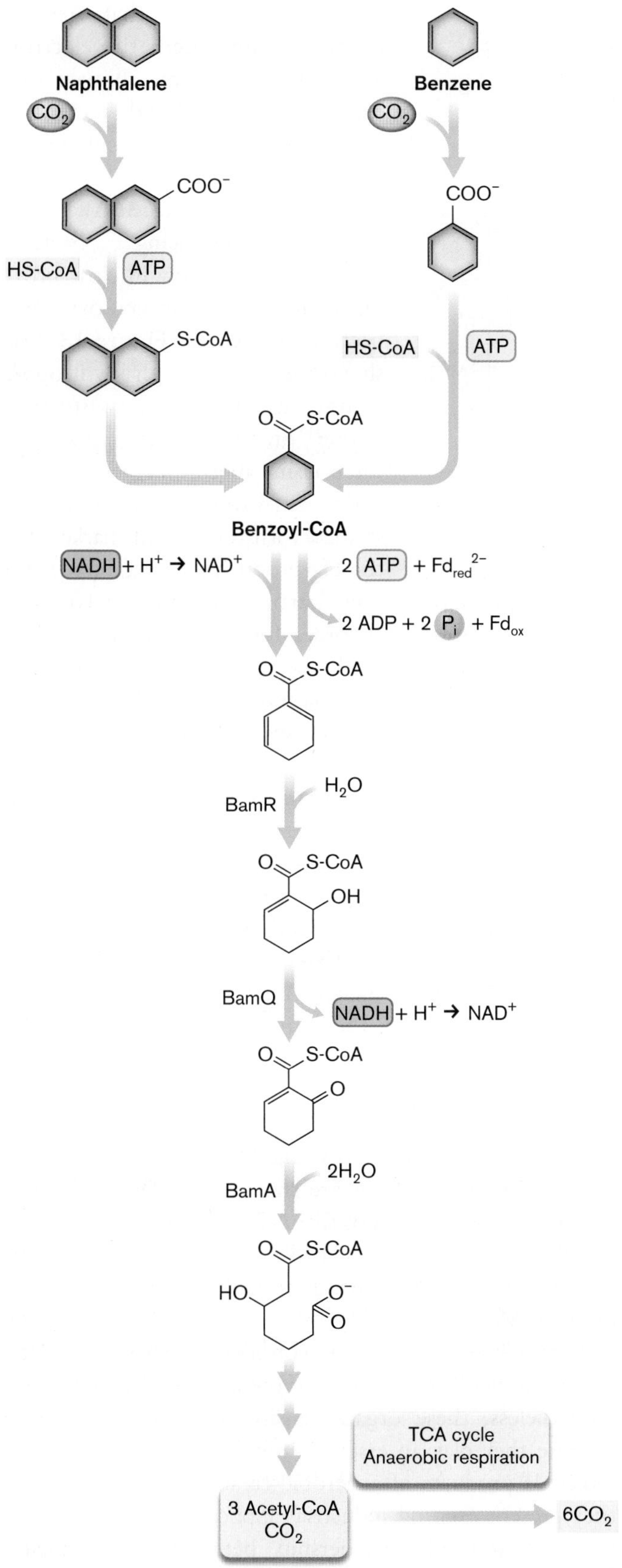

FIGURE 13.33 ■ Anaerobic benzoate catabolism. In the absence of oxygen, benzene and naphthalene incorporate CO_2 and are activated by HS-CoA to form benzoyl-CoA. Benzoyl-CoA breakdown requires hydrolysis (incorporating H_2O) and reduction by ferredoxin (Fd) and NADH. *Source:* Modified from Georg Fuchs et al. 2011. *Nat. Rev. Microbiol.* **9**:803.

of H_2O. Catabolism continues, forming three acetyl groups activated by HS-CoA, plus one CO_2. Aromatic catabolism requires high initial investment of reducing energy—one reason the process operates slowly. For soil bacteria, the energy invested must come from anaerobic phototrophy or from anaerobic respiration.

The discovery of effective benzene degraders is of great interest for bioremediation of sites such as the South Platte River, north of Denver, where a spill from the Suncor oil refinery in 2011 released benzene (**Fig. 13.34A**). To identify novel benzene-degrading microbes, Nidal Abu Laban and colleagues at the Helmholtz Centre in Munich, performed benzene enrichment culture of soil from a coal gasification site. From the enrichment culture, the predominant organism was identified as *Pelotomaculum* species of clostridia, Gram-positive endospore-forming bacteria. The genus *Pelotomaculum* was identified based on 16S rRNA gene sequence (for methods, see Chapter 17).

Pelotomaculum bacteria are related to other clostridia that use the anaerobic electron acceptor sulfate (SO_4^{2-}). So, Abu Laban hypothesized that sulfate could be used during benzene catabolism. In the experiment to test this hypothesis (**Fig. 13.34**), $^{13}CO_2$ was measured by gas chromatography–mass spectrometry (GC-MS), a technique that measures molecular masses and can thus distinguish "heavy isotopes" such as ^{13}C from the more common isotope, in this case ^{12}C. When benzene enrichment was performed in the presence of sulfate, the ^{13}C-labeled benzene showed conversion to $^{13}CO_2$, while at the same time the sulfate concentration decreased. The autoclaved controls showed no change in $^{13}CO_2$ or sulfate concentration. These data are consistent with anaerobic breakdown of benzene via the TCA cycle and ETS, with sulfate oxidation.

Microbial Syntrophy Cleans Up Oil

Aromatic and short-chain waste products of petroleum are often so stable that their breakdown reactions appear to incur positive values of ΔG. Michael McInerney at the University of Oklahoma, Norman, and his collaborator Jessica Sieber, now at the University of Illinois, think about waste cleanup from the point of view of microbial energetics (**Fig. 13.35**). They showed how key forms of anaerobic metabolism require **syntrophy**, the coupling of metabolism between two organisms such that the overall reaction yields energy ($-\Delta G$).

A fundamental challenge to degrading pollutants is the lack of oxygen. The molecules seep into the depths of soil or water, where any oxygen is quickly used up by microbes capable of respiration. As microbes use up the long-chain components of sewage or oil, the remaining short-chain molecules and aromatic rings can be broken down only

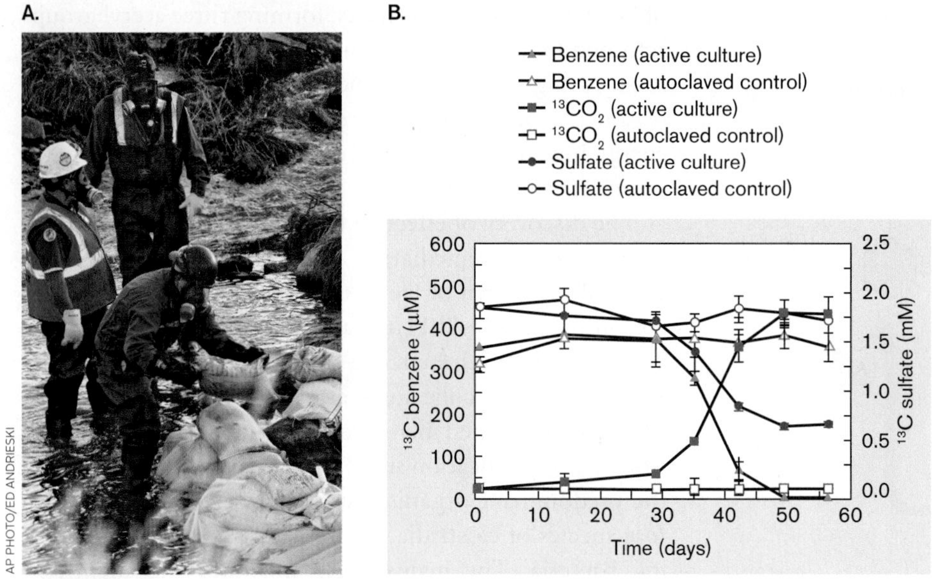

FIGURE 13.34 ▪ Benzene catabolism for bioremediation. **A.** Workers build a dam on the South Platte River in Denver, colorado in 2011, in their efforts to cleanup a contamination of benzene from the Suncor oil refinery. **B.** Anaerobic oxidation of benzene by sulfate. *Source:* Nidal Abu Laban et al. 2009. *FEMS Microbiol. Ecol.* **68**:300.

through the transfer of leftover electrons onto hydrogen ions, forming H_2. For example, *Syntrophus aciditrophicus* and *Syntrophomonas wolfei* catabolize benzoate into acetate and bicarbonate ions, releasing H_2 (**Table 13.6**).

But the H_2-releasing reactions of *Syntrophus* and *Syntrophomonas* have a positive value of $\Delta G°$ (see **Table 13.6**), and thus they cannot go forward to yield energy. So, how can these bacteria obtain energy? The ΔG can be shifted by rapidly removing H_2 gas, thus lowering the concentration of this key product. Once the equations are balanced, negative values of ΔG are obtained. H_2 can be removed by other microbes that oxidize hydrogen using alternative electron acceptors such as bicarbonate (for example, the archaeal methanogen *Methanospirillum*), sulfate, or nitrate (for example, the bacterium *Desulfovibrio*; **Fig. 13.36A**).

Hydrogen-producing bacteria have evolved to grow in close proximity to the hydrogen-oxidizing bacteria or archaea. **Figure 13.36B** shows a section through a syntrophic community within a wastewater sludge granule, where anaerobic catabolism must degrade sewage components to CO_2. The granule is stained with fluorescent markers for the H_2-producing bacteria (red) and their partner methanogen (yellow). In order for the syntrophy to proceed, the H_2 from the catabolizing bacteria must be transferred quickly to the methanogenic partner, by processes that are not yet understood. Microscopy reveals intercellular nanotubes that may transmit electrons from the H_2 producer to the partner.

FIGURE 13.35 ▪ Michael McInerney (a) and Jessica Sieber (b) collaborate on syntrophy in the pursuit of waste cleanup.

Under laboratory conditions, Sieber and McInerney demonstrated syntrophic growth in the presence of defined concentrations of reactants and products (**Table 13.6**). Their calculations showed that cells grew with negative values of free energy change. Yet even in the presence of the syntrophic partner, these anaerobic pathways have free energy changes that are remarkably low (about –20 kJ/mol), often below the theoretical minimum required to generate ATP (about –60 kJ/mol). Such extremely low energy levels require special arrangements called "reverse electron transport," in which electrons are transferred from one carrier to the next, despite a positive ΔG value (requiring energy input). The electron transfer must be coupled with a compensating reaction that spends enough energy to yield a net negative ΔG.

Syntrophic partners are considered "thermodynamic extremophiles" because they grow at such low ΔG values. Nonetheless, these organisms are of global importance because they play an essential role in recycling the small-molecule products of anaerobic catabolism. Furthermore, they present exciting industrial applications. For instance, the methanogenic partnership between H_2-producing bacteria and *Methanococcus maripaludis* might be used to attack unextractable oil residues and convert the carbon into methane for recovery as natural gas. Other syntrophic interactions may be used to increase the efficiency of sewage mineralization during water treatment.

TABLE 13.6 Syntrophy: Reactions Producing and Consuming H_2

Genus	Reaction	$\Delta G^{\circ\prime}$ (kJ/mol)[a]	$\Delta G'$ (kJ/mol)[b]		
$3H_2$ reduces an oxidant (HCO_3^-, SO_4^{2-}, or NO_3^-)					
Methanospirillum	$4H_2 + HCO_3^- + H^+ \rightarrow CH_4 + 3H_2O$	–136			
Desulfovibrio	$4H_2 + SO_4^{2-} + H^+ \rightarrow HS^- + 4H_2O$	–152			
Desulfovibrio	$4H_2 + NO_3^- + 2H^+ \rightarrow NH_4^+ + 3H_2O$	–600	**Oxidant for syntrophy:**		
Syntrophic catabolism producing H_2			HCO_3^-	SO_4^{2-}	NO_3^-
Syntrophus	Benzoate + $7H_2O \rightarrow$ 3 acetate + HCO_3^- + $3H^+$ + $3H_2$	+70	–25	–33	–45
Syntrophomonas	Butyrate + $2H_2O \rightarrow$ 2 acetate + H^+ + $2H_2$	+49	–5	–13	–18

[a] $\Delta G^{\circ\prime}$ is the standard free energy change at pH 7 when the concentrations of reactants and products are the same.

[b] $\Delta G'$ was the measured free energy change at pH 7 when syntrophic metabolism stopped.

Sources: Michael McInerney et al. 2007. *PNAS* **104**:7600; Ralf Cord-Ruwisch et al. 1988. *Arch. Microbiol.* **149**:350; Bradley E. Jackson and Michael J. McInerney. 2002. *Nature* **415**:454.

Thought Question

13.12 The thermophilic bacteria *Thermus* species grow in deep-sea hydrothermal vents at 80°C. It was proposed that they metabolize formate to bicarbonate ion and hydrogen gas:

$$HCOO^- + H_2O \rightarrow HCO_3^- + H_2$$

But the standard $\Delta G^{\circ\prime}$ is near zero (–2.6 kJ/mol). Under actual conditions, do you think the reaction yields energy? Assume concentrations of 150 mM for formate, 20 mM for bicarbonate ion, and 10 mM for hydrogen gas.

To Summarize

- **Catabolism of aromatic molecules** by bacteria and fungi recycles lignin and PAHs within ecosystems. Aromatic metabolism is used for bioremediation.
- **Benzoate undergoes aerobic catabolism to catechol.** The catechol ring is cleaved, generating acetyl-CoA, which enters the TCA cycle.

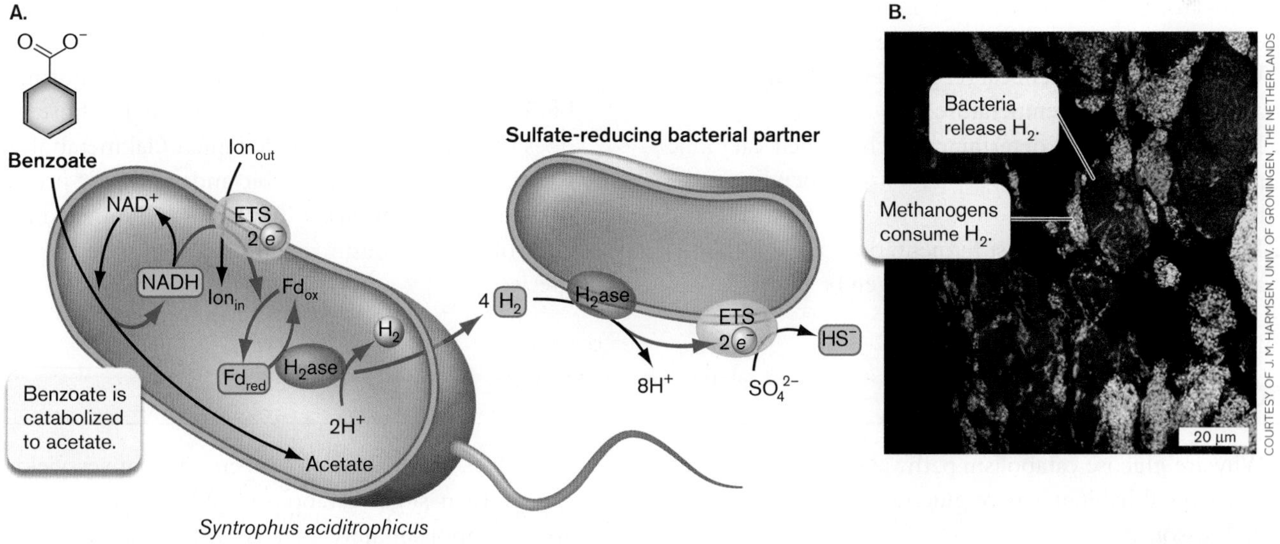

FIGURE 13.36 ■ Syntrophy between *Syntrophus aciditrophicus* and a sulfate-reducing bacterium, *Desulfovibrio*. **A.** *S. aciditrophicus* catabolizes benzoate to acetate, transferring electrons back onto the hydrogen ions removed from the substrate. The H_2 formed must donate electrons to an electron transport system (ETS), despite a positive ΔG value (reverse electron transfer). Rapid removal of H_2 by the sulfate-reducing partner (*Desulfovibrio* sp.) provides energy, making the net ΔG value negative. **B.** Syntrophic community labeled by fluorescence in situ hybridization (FISH). H_2ase = hydrogenase.

- **Anaerobic catabolism of benzoate** involves activation by HS-CoA and reduction by NADH. The energy for reduction comes from anaerobic respiration or phototrophy.
- **Benzene derivatives and short-chain acids can be catabolized by syntrophy.** In syntrophy, one partner species breaks down a molecule releasing H_2, while a tightly coupled partner species removes the H_2 to reduce an oxidant. The net reaction has a negative value of ΔG.

Concluding Thoughts

In this chapter we have seen how energy-yielding reactions enable living cells to generate order from disorder. In fact, the result of cell metabolism is to increase entropy (chaos) in the universe as a whole, even while complexity (order) increases in the living cell. Living organisms acquire energy by transferring it from reactions with negative ΔG to cell-building reactions with positive ΔG. The fundamental energy-yielding pathways of Earth's biosphere are those of photosynthesis, and the biomass generated through photosynthesis provides materials for catabolism. Much of human civilization has been built on harnessing microbial catabolism for waste treatment, food production, and biotechnology.

All these forms of energy-yielding metabolism involve chemical exchange of electrons through oxidation and reduction. In this chapter we have frequently mentioned oxidation-reduction reactions leading to an electron transport system. In Chapter 14 we focus on the mechanism of electron transport between donor and acceptor molecules, through membrane-embedded complexes that generate ion gradients across the membrane. This transport of electrons is fundamental to respiration, the oxidative completion of catabolism to CO_2 and water. It also underlies the inorganic sources of energy: the gain of energy from light (phototrophy) and from mineral oxidation (lithotrophy).

CHAPTER REVIEW

Review Questions

1. Why must the biosphere continually take up energy from outside? Why can't all the energy be recycled among organisms, like the fundamental elements of matter?
2. Explain how a biochemical reaction can be driven by a change in enthalpy, ΔH. Explain how a different reaction can be driven by a change in entropy, ΔS. In each case, explain the role of the free energy change, ΔG.
3. Why do some biochemical reactions release energy only above a threshold temperature?
4. How do organisms determine which of their catabolic pathways to use? How does catabolism depend on environmental factors?
5. Beer is produced by yeast fermentation of grain to ethanol. To make beer, why must oxygen be limited? Why are such large quantities of ethanol produced with a relatively small production of yeast biomass?
6. Explain the three different routes to catabolize glucose to pyruvate. Why is it necessary to start by spending one or two molecules of ATP?
7. Explain how the TCA cycle incorporates an acetyl group. How are the two CO_2 molecules removed?
8. How do bacterial catabolic reactions in the gut microbiome generate molecules that regulate the host? How does the human body regulate microbial metabolism?
9. Compare and contrast aerobic and anaerobic processes of benzoate catabolism. Explain why some kinds of anaerobic metabolism require syntrophy between two species.

Thought Questions

1. Why are glucose catabolism pathways ubiquitous, even in bacterial habitats where glucose is scarce? Give several reasons.
2. In glycolysis, explain why bacteria have to return the hydrogens from NADH back onto pyruvate to make fermentation products. Why can't NAD^+ serve as a terminal electron acceptor, like O_2?
3. Why does catabolism of benzene derivatives yield less energy than sugar catabolism? Why is benzene-derivative catabolism nevertheless widespread among soil bacteria?
4. Why do environmental factors regulate catabolism? For example, why are amino acids decarboxylated at low pH? Cite other examples.

Key Terms

activation energy (E_a) (499)
adenosine triphosphate (ATP) (494)
allosteric site (500)
amphibolic (508)
anaerobic respiration (488)
aromatic (497)
calorimeter (491)
catabolism (486)
catalytic domain (catalytic site) (500)
chemotrophy (487)
coenzyme A (CoA) (514)
electron acceptor (494)
electron donor (494)
electron transport system (ETS) (497)
Embden-Meyerhof-Parnas (EMP) pathway (506, 520)
endothermic (490)
energy (487)
energy carrier (494)
enthalpy (489)
Entner-Doudoroff (ED) pathway (506, 509)
entropy (488, 489)
enzyme (499)
ethanolic fermentation (512)
exothermic (490)
fermentation (504)
flavin adenine dinucleotide (FAD) (498)
Gibbs free energy change (ΔG) (488)
glycan (486, 501)
glycolysis (506)
glyoxylate bypass (518)
gut microbiome (486)
heterotrophy (487)
human milk oligosaccharide (HMO) (521)
hydrolysis (495)
joule (J) (490)
lactic acid fermentation (512)
lignin (503)
mixed-acid fermentation (514)
nicotinamide adenine dinucleotide (NAD) (496)
oxidative phosphorylation (518)
pentose phosphate pathway (PPP) (506, 509)
phosphorylation (494)
phototrophy (487)
polysaccharide (501)
pyruvate dehydrogenase complex (PDC) (516)
respiration (504, 518)
ribulose 5-phosphate (509)
starch (501)
Stickland reaction (515)
substrate-level phosphorylation (507, 512)
syntrophy (493, 527)
terminal electron acceptor (497)
tricarboxylic acid (TCA) cycle (515)

Recommended Reading

Eisenreich, Wolfgang, Thomas Dandekar, Jürgen Heesemann, and Werner Goebel. 2010. Carbon metabolism of intracellular bacterial pathogens and possible links to virulence. *Nature Reviews. Microbiology* **8**:401–412.

Fuchs, Georg, Matthias Boll, and Johann Heider. 2011. Microbial degradation of aromatic compounds—From one strategy to four. *Nature Reviews. Microbiology* **9**:803–816.

Head, Ian M., D. Martin Jones, and Wilfred F. M. Röling. 2006. Marine microorganisms make a meal of oil. *Nature Reviews. Microbiology* **4**:173–182.

Hickey, William J., Shicheng Chen, and Jiangchao Zhao. 2012. The *phn* island: A new genomic island encoding catabolism of polynuclear aromatic hydrocarbons. *Frontiers in Microbiology* **3**:125.

Jackson, Bradley E., and Michael J. McInerney. 2002. Anaerobic microbial metabolism can proceed close to thermodynamic limits. *Nature* **415**:454–456.

Koropatkin, Nicole M., Elizabeth A. Cameron, and Eric C. Martens. 2012. How glycan metabolism shapes the human gut microbiota. *Nature Reviews. Microbiology* **10**:323–335.

Ladas, Spiros D., George Karamanolis, and Emmanuel Ben-Soussan. 2007. Colonic gas explosion during therapeutic colonoscopy with electrocautery. *World Journal of Gastroenterology* **13**:5295–5298.

Larsbrink, Johan, Theresa E. Rogers, Glyn R. Hemsworth, Lauren S. McKee, Alexandra S. Tauzin, et al. 2014. A discrete genetic locus confers xyloglucan metabolism in select human gut Bacteroidetes. *Nature* **506**:498–502.

Martens, Eric C., Elisabeth C. Lowe, Herbert Chiang, Nicholas A. Pudlo, Meng Wu, et al. 2011. Recognition and degradation of plant cell wall polysaccharides by two human gut symbionts. *PLoS Biology* **9**:e1001221.

Mazzoli, Roberto, and Enrica Pessione. 2016. The neuro-endocrinological role of microbial glutamate and GABA signaling. *Frontiers in Microbiology* **7**:1934.

McInerney, Michael J., Jessica R. Sieber, and Robert P. Gunsalus. 2011. Microbial syntrophy: Ecosystem-level biochemical cooperation. *Microbe* **6**:479–485.

Okere, Uchechukwu V., Ana Cabrerizo, Jordi Dachs, Kevin C. Jones, and Kirk T. Semple. 2012. Biodegradation of phenanthrene by indigenous microorganisms in soils from Livingstone Island, Antarctica. *FEMS Microbiological Letters* **329**:69–77.

Shepherd, Elizabeth S., William C. DeLoache, Kali M. Pruss, Weston R. Whitaker, and Justin L. Sonnenburg. 2018. An exclusive metabolic niche enables strain engraftment in the gut. *Nature* **557**:434–438.

Sieber, Jessica R., Huynh M. Le, and Michael J. McInerney. 2014. The importance of hydrogen and formate transfer for syntrophic fatty, aromatic and alicyclic metabolism. *Environmental Microbiology* **16**:177–188.

Turroni, Francesca, Christian Milani, Sabrina Duranti, Chiara Ferrario, Gabriele Andrea Lugli, et al. 2018. Bifidobacteria and the infant gut: An example of co-evolution and natural selection. *Cell and Molecular Life Sciences* **75**:103–118.

Wolf, Patricia G., Ambarish Biswas, Sergio E. Morales, Chris Greening, and H. Rex Gaskins. 2016. H_2 metabolism is widespread and diverse among human colonic microbes. *Gut Microbes* **7**:235–245.

Yoshida, Shosuke, Kazumi Hiraga, Toshihiko Takehana, Ikuo Taniguchi, Hironao Yamaji, et al. 2016. A bacterium that degrades and assimilates poly(ethylene terephthalate). *Science* **351**:1196–1199.

CHAPTER 14

Electron Flow in Organotrophy, Lithotrophy, and Phototrophy

14.1 Electron Transport Systems and the Proton Motive Force

14.2 The Respiratory ETS and ATP Synthase

14.3 Anaerobic Respiration

14.4 Nanowires, Electron Shuttles, and Fuel Cells

14.5 Lithotrophy and Methanogenesis

14.6 Phototrophy

Microbes transfer energy by moving electrons—the equivalent of an electric current. Electrons move from reduced food molecules onto energy carriers, from energy carriers onto membrane proteins called cytochromes, and to organic electron carriers. The electrons ultimately reduce oxygen or other oxidized molecules. Electron flow drives protons across the membrane and generates a proton motive force. The proton motive force powers synthesis of ATP, which stores energy for later use. Some bacteria transfer electrons across a distance in biofilms. A fuel cell can harness these biofilms to generate electricity.

CURRENT RESEARCH highlight

Bacterial electric cables. A Dutch salt marsh is the last place one might expect to find a battery producing electric current. In fact, the marine sediment contains a vast expanse of voltage potential generated by cable bacteria. The cable bacteria multiply and extend their chains of cells from the reduced sulfides up toward the surface full of oxygen. Filip Meysman and colleagues showed that the bacterial envelopes contain filaments that conduct electrons—electricity—over chains of thousands of cells. These bacterial filaments convert the sediment into a source of electric power.

Source: Rob Cornelissen et al., 2018. *Front. Microbiol.* **9**:3044.

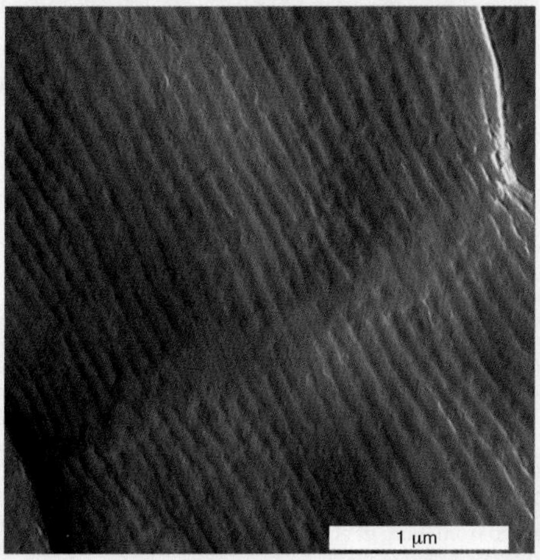

ROB CORNELISSEN, UNIVERSITY OF HASSELT

AN INTERVIEW WITH

FILIP MEYSMAN, MICROBIAL ECOLOGIST, UNIVERSITY OF ANTWERP, BELGIUM

COURTESY OF FILIP MEYSMAN

How did you show that bacteria conduct electricity?

We connected individual cable filaments to miniaturized electrodes and applied a voltage bias, just as you would add jumper cables to a car battery. We saw high currents running. The bacteria must contain an internal structure that can guide these electric currents. So, what is this conductive structure? The cell surface of the cable bacteria shows conspicuous ridges. Microscopy shows that fibers are located underneath the ridges, inside the cell envelope. These fibers channel electrons.

Electric current in living organisms has fascinated scientists ever since the eighteenth century, when Luigi Galvani (1737–1798) showed that a voltage caused a dead frog limb to flex. The idea of biological electricity inspired Mary Shelley's famous novel *Frankenstein: or, the Modern Prometheus*, in which a physician is imagined to "create life" by jolting body parts with an electric shock. Today we know that electric current flows in all cells, including those of bacteria.

Cable bacteria, as shown in the Current Research Highlight, form multicellular chains of exceptional length that can transfer electrons across several centimeters. Within the marine sediment, cells on the deep end of each bacterial "cable" collect electrons from reduced molecules such as sulfides (HS^-, H_2S; **Fig. 14.1**). Oxidation of sulfide (by removal of electrons) generates pure sulfur, deposited as precipitate. The electrons are captured by the bacteria, where they flow upward through the connected periplasm of the cell chain into the end cells exposed to oxygen gas. The upper-end bacteria then transfer the electrons to oxygen, which reacts with hydrogen ions to form water. Oxidation of sulfide, an inorganic electron donor, is an example of lithotrophy (Section 14.5). Marine cable bacteria are one of a growing number of microbial sources of electric power for devices, as discussed in Section 14.4.

How do cells make electricity? The energy-yielding reactions we saw in Table 13.1 involve transfer of electrons from one molecule to another. Much of the energy yield comes from successive redox steps within an **electron transport system** (**ETS**) of carriers in a membrane, whose energy-yielding redox reactions are coupled to transport of ions such as protons (H^+). Types of metabolism that use an ETS include **organotrophy** (organic electron donors), **lithotrophy** (inorganic electron donors), and **phototrophy** (light absorption excites electrons). Some bacteria and archaea are specialists, whereas others use more than one of these types of metabolism. Even organisms with very different forms of metabolism show common themes in their mechanisms of electron transfer. In fact, the protein complexes for processes as different as organotrophy and phototrophy show homology, indicating that they evolved from a common ancestral ETS.

Chapter 14 presents electron flow through the ETS in organotrophy, lithotrophy, and phototrophy. In each step of the ETS, a molecule becomes "reduced" (gains an electron) while the molecule donating the electron becomes "oxidized" (loses an electron). In various kinds of metabolism, some of the energy from electron transfer is stored in the form of an electrochemical potential (voltage) across a membrane. The potential is composed of the chemical concentration gradient of H^+ ions (or Na^+ ions, for some species) plus the charge difference across the membrane. The proton potential drives ATP synthesis. Ion potentials

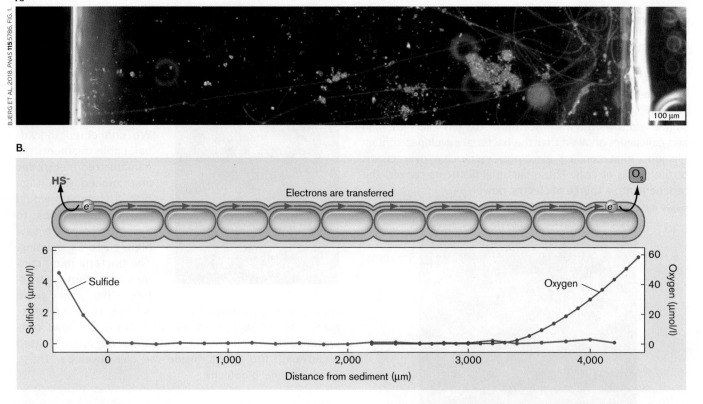

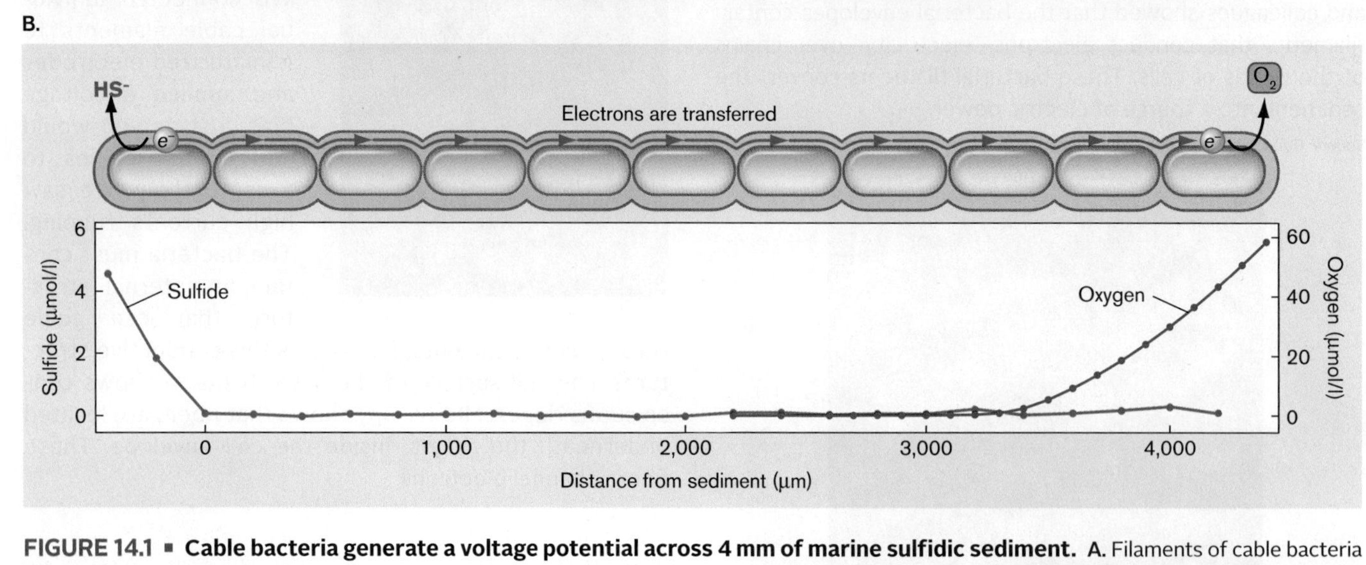

FIGURE 14.1 ■ Cable bacteria generate a voltage potential across 4 mm of marine sulfidic sediment. **A.** Filaments of cable bacteria extend from the reduced sulfidic zone (left) to the oxidized zone (right; dark-field microscopy). **B.** Concentration gradients of sulfide (blue line) and oxygen (red line) were recorded across the microscopic chamber. *Source:* Part B modified from Bjerg et al. 2018. *PNAS* **115**:5786, fig. 2D.

also drive nutrient uptake, mediate pathogenesis of infective microbes, and shape the chemistry surrounding a microbe. Microbial electron transfer reactions, and the ion potentials they generate, have profound consequences for ecosystems; they acidify or alkalinize soil and water, and they can precipitate minerals or dissolve them away.

14.1 Electron Transport Systems and the Proton Motive Force

As we learned in Chapter 13, energy to support life is obtained through reactions that convert substrates to products of lower energy—that is, reactions in which ΔG (free energy change) is negative. Many reactions involve transfer of electrons from a reduced **electron donor** to an oxidized **electron acceptor**. The simplest path of electron flow is found in fermentation, where electrons from the fermented substrate, such as glucose, are transferred onto NAD^+ to make NADH (reduction), and then returned to the glucose breakdown products, such as pyruvate. More complex kinds of metabolism, such as aerobic respiration, transfer electrons through a series of membrane-soluble carriers called an **electron transport system** (**ETS**), also known as an **electron transport chain** (**ETC**).

Unlike cytoplasmic redox reactions, a membrane-embedded ETS can convert its energy into an ion potential or electrochemical potential between two compartments separated by the membrane. The ion potential is most commonly a proton (H^+) potential, also known as **proton motive force** (**PMF**). The PMF drives essential cellular processes such as ATP synthesis and flagellar rotation.

Note: Recall the following prefixes.

Energy source

Photo-: Light absorption captures energy and excites electron.

Chemo-: Chemical electron donors are oxidized.

Electron source

Litho-: Inorganic molecules donate electrons.

Organo-: Organic molecules donate electrons.

Electron Donors and Acceptors

Different types of metabolism use electron transport systems with different components embedded in the bacterial membrane. For chemotrophy (oxidizing an electron donor to yield energy), each ETS must accept electrons from an initial electron donor. The electron donor may be an organic "food" molecule, such as glucose (organotrophy), or a reduced mineral, such as Fe^{2+} (lithotrophy). Next, the ETS proteins and cofactors act sequentially as electron donors and acceptors, whose oxidation-reduction reactions are coupled to H^+ transport across the membrane. Each ETS finally transfers its electrons to a terminal electron acceptor molecule.

Metabolism using an ETS is classified by the nature of the initial electron donors and terminal electron acceptors. In this chapter we cover three major classes of prokaryotic energy acquisition involving an ETS (their relationships are summarized in Table 13.1):

- **Organotrophy** (or **chemoorganotrophy**) is a form of metabolism in which organic molecules donate electrons. If the electrons are donated to a membrane-embedded ETS, and ultimately reduce a terminal electron acceptor, the overall pathway is called "respiration." The terminal electron acceptor may be O_2 for aerobic respiration, or an alternative electron acceptor such as nitrate (NO_3^-) for anaerobic respiration. Some terminal electron acceptors are organic (such as fumarate, $^-O_2C–CH{=}CH–CO_2^-$).
- **Lithotrophy** (or **chemolithotrophy**) is the oxidation of inorganic electron donors, such as reduced iron, or the sulfides for cable bacteria, described in the Current Research Highlight. The electron acceptor may be O_2 or an anaerobic alternative, such as NO_3^-. A strong electron donor, such as H_2, can use organic electron acceptors such as fumarate. Some chemoorganotrophs that catabolize organic or inorganic substrates are "facultative lithotrophs." But many lithotrophs are obligate autotrophs, fixing CO_2 for biosynthesis (discussed in Chapter 15). These specialist organisms are called chemolithoautotrophs. For example, the bacterium *Nitrosomonas europaea* obtains nearly all of its energy by oxidizing ammonia.
- **Phototrophy** consists of light capture by chlorophyll or another photopigment. The absorption of light energy is usually coupled to splitting of H_2S or H_2O (photolysis). Photoautotrophs couple photolysis to CO_2 fixation to form biomass (discussed in Chapter 15). Photoheterotrophs absorb light to generate ATP, while incorporating preformed organic compounds for biosynthesis. Photoheterotrophs may also capture light energy to supplement catabolism, which provides organic electron donors (photoorganotrophy).

Note: The term **respiration** refers to metabolism in which chemical electron donors yield energy through an ETS. In this book we reserve the term "respiration" for an ETS with organic electron donors (organotrophy). The use of inorganic electron donors for an ETS is called **lithotrophy**. Other fields use the term "respiration" differently. In medical physiology, "respiration" refers to catabolism with an ETS, or to the process of carbon dioxide exchange with oxygen through breathing. In ecology, "respiration" refers to carbon flux in an ecosystem, or throughout the atmosphere.

Energy Storage

In Chapter 13 we showed how the free energy change ΔG determines whether energy can be obtained from a given reaction of substrates to products. Similar considerations of ΔG govern the reactions of electron flow through an ETS and the storage of energy in transmembrane ion gradients.

The reduction potential. To obtain energy, biochemical reactions require a negative ΔG. In oxidation-reduction reactions (redox reactions), the ΔG values are proportional to the reduction potential (E) between the oxidized form of a molecule (electron acceptor) and its reduced form (electron donor). The reduction potential represents the tendency of a compound to accept electrons, measured in volts (V) or millivolts (mV). **A positive value of E has a negative ΔG,** so the gain of electrons yields energy. A negative value of E means that the reverse reaction (loss of electrons) yields energy.

The oxidized and reduced states of a compound are called a **redox couple**. An example of a redox couple is the electron acceptor O_2 and the electron donor H_2O. Because O_2 is a strong electron acceptor (readily gains $2e^-$), the redox couple $\frac{1}{2}O_2/H_2O$ has a high positive value of E. So when O_2 plus $2H^+ + 2e^-$ forms H_2O, the reaction yields a large amount of energy.

Standard values of E ($E°$) are known as the **standard reduction potential**. The standard reduction potential $E°'$ assumes a concentration of 1 M for all components at pH 7. For example, the value of $E°'$ for H_2 formation from $2H^+ + 2e^-$ is −420 mV. This large negative reduction potential means that it takes a lot of energy to add the two electrons to $2H^+$. Thus, the reverse reaction donating $2e^-$ from H_2 to an appropriate electron acceptor yields a lot of energy. An example of microbes that use this reaction for energy is hydrogen-oxidizing bacteria growing in oral biofilms, causing gum disease.

Reduction potentials represent a form of the standard free energy change, $\Delta G°'$. The value of $\Delta G°'$ (in kilojoules per mole, or kJ/mol) for an electron transfer reaction with a given reduction potential $E°'$ is given by

$$\Delta G°' = -nFE°'$$

where n is the number of electrons transferred, F is Faraday's constant (F = 96.5 kJ/V · mol), and $E°'$ is the reduction potential (in volts, V) at 1-M concentration, 25°C, and pH 7. Thus, $E°'$ represents the standard reduction potential per electron, whereas $\Delta G°'$ gives the overall free energy change. Note that the reaction is favored for positive values of E, which correspond to negative values of ΔG.

The standard reduction potential $E°'$ gives the potential difference between the oxidized and reduced states under standard conditions, when both oxidized and reduced forms are at equal concentrations—namely, 1 M. Assuming equal concentrations at 1 M, we may compare $E°'$ of various molecules available in the environment for microbial respiration. Such a comparison generates an "electron tower" representing the reduction potential $E°'$ of molecules that may act as electron acceptors or donors (**Table 14.1**). The oxidized state of each redox couple in the table is shaded red; the reduced state is shaded blue.

TABLE 14.1 "Electron Tower" of Standard Reduction Potentials

Electron acceptor	→	Electron donor	$E°'$ (mV)[a]	$\Delta G°'$ (kJ)
$CO_2 + 4H^+ + 4e^-$	→	$[CH_2O]$ glucose + H_2O	−430	+166
$2H^+ + 2e^-$	→	H_2	−420	+81
$NAD^+ + 2H^+ + 2e^-$	→	NADH + H^+	−320	+62
$S^0 + H^+ + 2e^-$	→	HS^-	−280	+27
$CO_2 + 2H^+ + 3H_2 + 2e^-$	→	$CH_4 + 2H_2O$	−240	+46
$SO_4^{2-} + 10H^+ + 8e^-$	→	$H_2S + 4H_2O$	−220	+170
FAD + $2H^+ + 2e^-$	→	$FADH_2$	−220[b]	+42
FMN + $2H^+ + 2e^-$	→	$FMNH_2$	−190	+37
Menaquinone + $2H^+ + 2e^-$	→	Menaquinol	−74	+14
Fumarate + $2H^+ + 2e^-$	→	Succinate	+33	−6
Ubiquinone + $2H^+ + 2e^-$	→	Ubiquinol	+110	−21
$Fe^{3+} + e^-$	→	Fe^{2+} (at pH 7)	+200	−19
$NO_3^- + 2H^+ + 2e^-$	→	$NO_2^- + H_2O$	+420	−81
$NO_2^- + 8H^+ + 6e^-$	→	$NH_4^+ + 2H_2O$	+440	−255
$MnO_2 + 4H^+ + 2e^-$	→	$Mn^{2+} + 2H_2O$	+460	−89
$NO_3^- + 6H^+ + 5e^-$	→	$\frac{1}{2}N_2 + 3H_2O$	+740	−357
$Fe^{3+} + e^-$	→	Fe^{2+} (at pH 2)	+770	−74
$\frac{1}{2}O_2 + 2H^+ + 2e^-$	→	H_2O	+820	−158

[a]Reduction potentials for redox couples when all concentrations are 1 M at 25°C and pH 7.

[b]This value is for free FAD; FAD bound to a specific flavoprotein has a different $E°'$ that depends on its protein environment.

Note: A more positive $E°'$ means that reducing the electron acceptor yields more energy. A more negative value of $E°'$ means that oxidizing the electron donor yields more energy. Either form of half reaction, in the favored direction, is associated with a negative value of $\Delta G°'$.

In the electron tower, the more negative values of $E°'$ represent couples (half reactions) with a stronger electron donor (H_2, NADH), whereas the more positive values represent stronger electron acceptors (O_2, NO_3^-). For example, in the redox couple NAD^+/NADH + H^+, the oxidized form is NAD^+ and the reduced form is NADH. The reduction potential is strongly negative (–320 mV), so NADH is a strong electron donor; that is, the reverse reaction, NADH + $H^+ \rightarrow NAD^+$, releases a lot of energy (+320 mV).

A complete redox reaction combines two redox couples: one accepting electrons (red column in **Table 14.1**), the other donating electrons (blue column, arrow reversed). Redox couples with more negative values of $E°'$ can provide electron donors (blue column) for electron acceptors with more positive values of $E°'$ (red column). The $E°'$ for the overall reaction is then given by adding the reversed reaction of the electron donor $E°'$ to the electron acceptor $E°'$. A positive value of $E°'$ means that the reaction may proceed to provide energy. For example, the oxidation of NADH by O_2 results from combining two couples: O_2/H_2O (forward) and NADH/NAD^+ (reversed from **Table 14.1**). For the reversed couple, the sign of $E°'$ changes:

	$E°'$	$\Delta G°'$
Electron acceptor is reduced:		
$\frac{1}{2}O_2 + 2e^- + 2H^+ \rightarrow H_2O$	+820 mV	–158 kJ/mol
Electron donor is oxidized:		
$NADH + H^+ \rightarrow NAD^+ + 2e^- + 2H^+$	–(–320 mV)	–(+62 kJ/mol)
$NADH + H^+ + \frac{1}{2}O_2 \rightarrow H_2O + NAD^+$	+1,140 mV	–220 kJ/mol

The aerobic oxidation of NADH pairs a strong electron donor (NADH) with a strong electron acceptor (O_2). Thus, NADH oxidation via the ETS provides the cell with a huge amount of potential energy (comparable to a 1-V battery cell) to make ATP and generate ion gradients. NADH is oxidized by many electron transport systems of bacteria and archaea, as well as mitochondria. The food we eat donates electrons to make NADH, for our mitochondria to oxidize.

Bacteria and archaea have evolved many alternative donor-acceptor systems; the one they use depends on what their environment provides. For example, in anaerobic marine sediment the proteobacterium *Shewanella* can donate electrons to many alternative oxidized minerals as terminal electron acceptors, including iron, lead, and nitrate. The use of terminal electron acceptors other than oxygen is called anaerobic respiration (discussed in Section 14.3).

Electron donors can also be oxidized by organic molecules. In the human colon, which has limited molecular oxygen, *Escherichia coli* can oxidize NADH by transferring $2e^-$ onto fumarate, which is reduced to succinate. **Table 14.1** shows that:

$$\text{NADH} + \text{H}^+ + \text{fumarate} \rightarrow \text{NAD}^+ + \text{succinate}$$
$$E°' = 320\text{ mV} + 33\text{ mV} = 353\text{ mV} \qquad \Delta G°' = -68\text{ kJ/mol}$$

The fumarate reduction potential is small but yields energy for growth. When oxygen reappears in the environment, the reverse reaction is favored—succinate can donate electrons to O_2, forming fumarate and water:

$$\tfrac{1}{2}\text{O}_2 + \text{succinate} \rightarrow \text{H}_2\text{O} + \text{fumarate}$$
$$E°' = 820\text{ mV} - 33\text{ mV} = 787\text{ mV} \qquad \Delta G°' = -152\text{ kJ/mol}$$

Thought Questions

14.1 *Pseudomonas aeruginosa*, a cause of pneumonia in cystic fibrosis patients, oxidizes NADH with nitrate (NO_3^-) to nitrite (NO_2^-) at neutral pH. What is the value of $E°'$?

14.2 Could a bacterium obtain energy from succinate as an electron donor with nitrate (NO_3^-) as an electron acceptor? Explain.

14.3 Use **Table 14.1** to write the chemical reaction used by cable bacteria. What is its standard reduction potential? Explain why electrons must travel such a long distance within the bacterial cable.

Concentrations of electron donors and acceptors. A reaction with a small value of $E°'$ can provide more energy if the concentration of an electron donor is high. A high donor concentration is particularly common for lithotrophic reactions such as iron oxidation. The standard reduction potential $E°'$ assumes 1-M concentrations of reactants and products at 25°C and pH 7, but actual values of E within cells depend on actual reactant concentrations.

In Chapter 13 we saw that ΔG includes a term incorporating the ratio of product concentrations (C and D) to reactant concentrations (A and B):

$$\Delta G = \Delta G°' + 2.303RT \log \frac{[\text{C}][\text{D}]}{[\text{A}][\text{B}]}$$

where R is the gas constant [8.315 joules per kelvin-mole, or J/(mol · K)], and T is the temperature in kelvins. The reduction potential E of a redox reaction depends on the log ratio of products to reactants:

$$E = \frac{-\Delta G}{nF}$$
$$= E°' - \frac{2.303RT}{nF} \times \log \frac{[\text{C}][\text{D}]}{[\text{A}][\text{B}]}$$
$$= E°' - 60\ (\text{mV}/n) \log \frac{[\text{C}][\text{D}]}{[\text{A}][\text{B}]}$$

where n is the number of electrons transferred and F is the Faraday constant. As the ratio of products to reactants increases, E decreases; in other words, there is less potential to do work. For a tenfold ratio of products to reactants, assuming moderate temperatures (25°C–40°C), the constants in the concentration term ($2.303RT/F$) combine to yield approximately 60 mV per electron. Thus, a tenfold ratio of products to reactants subtracts about 60 mV from E (decreases energy yield), whereas a tenfold excess of reactants adds 60 mV to E (increases energy yield).

An ETS Functions within a Membrane

An electron transport system transfers electrons onto membrane-embedded carriers in a series of increasing reduction potential (that is, lower on the electron tower) ending at a terminal electron acceptor such as O_2, Fe^{3+}, or NO_3^-. Some electron transfer steps yield energy to pump ions across the membrane. To store this energy, the ETS must maintain an ion gradient across a membrane that fully separates two aqueous compartments. In bacteria, the cytoplasmic membrane separates the cytoplasm and external medium.

Gram-negative bacteria such as *E. coli* have their ETS in the inner (cytoplasmic) membrane, which separates the cytoplasm from the periplasm (see Chapter 3). The Gram-negative outer membrane surrounding the periplasm is permeable to protons and other small molecules; thus, the outer membrane does not store energy. **Figure 14.2A** shows the inner membrane of *Helicobacter pylori*, a Gram-negative gastric pathogen. The inner membrane contains the ETS complexes; the cell wall and outer membrane do not participate in respiration. Some respiratory bacteria, such as the nitrite/ammonia oxidizer *Nitrospira*, pack their ETS within intracytoplasmic pockets called "lamellae."

Respiratory membranes similar to bacterial lamellae are found within our own mitochondria. Mitochondrial respiration uses only O_2 as a terminal electron acceptor. The ETS proteins are embedded in folds of the mitochondrial inner membrane called "cristae" (singular; crista; **Fig. 14.2B**). The inner membrane separates the inner mitochondrial space from the intermembrane space (between the inner and outer membranes). The mitochondrial inner membrane, including its electron transport proteins, evolved from the cell membrane of an endosymbiotic bacterial ancestor. Mitochondrial evolution is discussed further in Chapter 17.

Cytochromes in the membrane. Electron carrier molecules include proteins and small organic cofactors bound to the proteins. The protein components of an ETS were first discovered in the 1930s at Cambridge University by Russian entomologist David Keilin (1887–1963), who studied insect mitochondria. The mitochondrial inner membranes contain proteins called **cytochromes**, which were named for their deep colors, typically red to brown. The colors derive from absorption of visible light.

A.

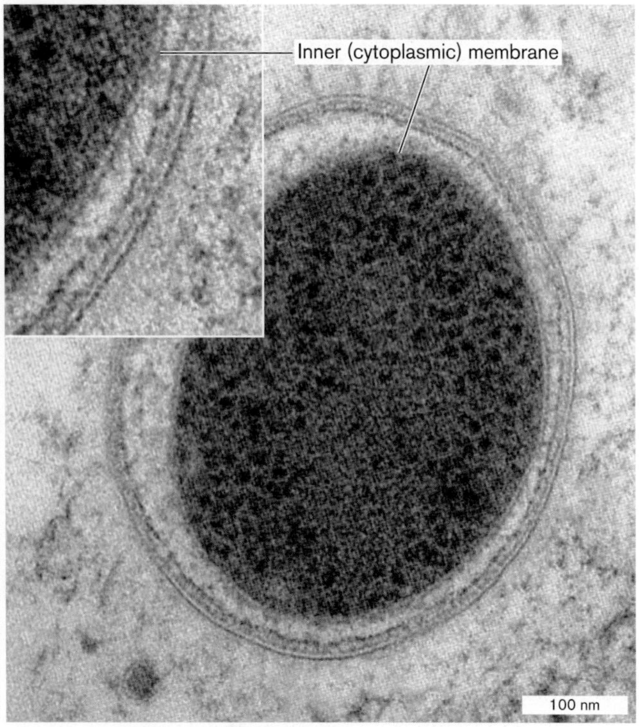

B.

FIGURE 14.2 ■ Respiratory membranes. A. Electron transport occurs in the inner (cytoplasmic) membrane of *Helicobacter pylori*. The cytoplasmic membrane and cell wall are surrounded by periplasm and outer membrane. **B.** Mitochondrion within the brain stem neuron of a cat, modeled by cryo-EM tomography, shown with one EM section through the cell. The mitochondrial outer membrane (blue) encloses inner membrane organized in pockets called cristae (other colors).

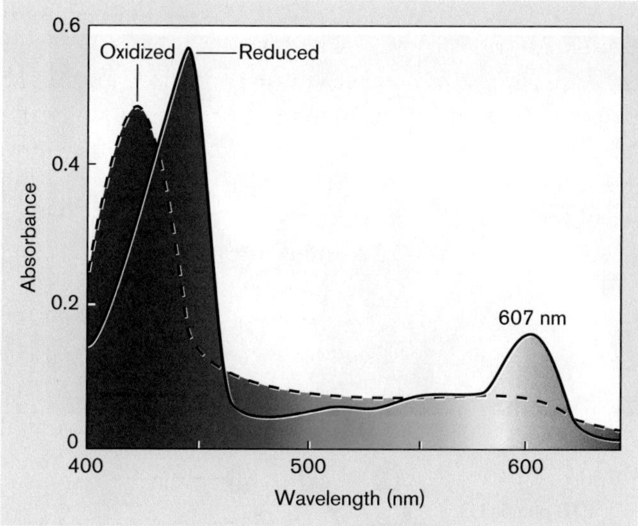

FIGURE 14.3 ▪ Light absorbance spectrum of a cytochrome. Absorption peaks shift between the oxidized and reduced forms of a cytochrome from the electron transport system of *Haloferax volcanii*, a halophilic archaeon.

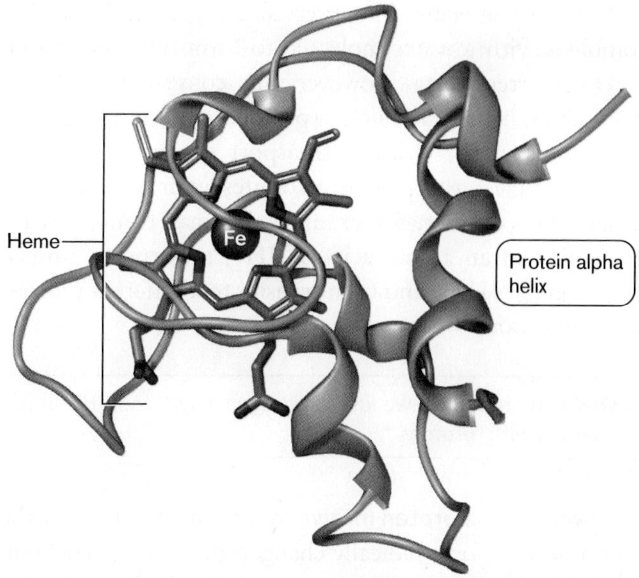

FIGURE 14.4 ▪ Cytochrome c. Portion of the protein structure of cytochrome *c* containing one heme cofactor, from the pathogen *Pseudomonas aeruginosa*. (PDB code: 2PAC)

Bacteria and archaea also have cytochromes. **Figure 14.3** shows the absorbance spectrum of a cytochrome from the cell membrane of *Haloferax volcanii*, a halophilic archaeon isolated from the Dead Sea. In the reduced cytochrome, light absorption peaks in the blue range (440 nm) and in the orange (607 nm). Upon oxidation, the cytochrome loses its absorption peak at 607 nm, and the 440-nm peak shifts to a shorter wavelength. Different species of bacteria make many different cytochromes, with different absorption peaks.

The changes in reduction state of cytochromes usually require a buried cofactor containing a metal ion, such as heme. The protein shown in **Figure 14.4**, cytochrome *c* of *Pseudomonas aeruginosa*, transfers electrons for nitrate respiration. Its reddish color derives from the **heme**, a ring of conjugated double bonds surrounding an iron ion (Fe^{2+} or Fe^{3+}). The heme plays a key role in acquiring and transferring electrons, with an Fe^{2+}/Fe^{3+} transition. A metal-reducing bacterium such as *Geobacter* may make more than a hundred different types of cytochromes in its envelope, where electrons accumulate until the bacterium finds an oxidant to accept them.

A membrane ETS typically includes several different cytochromes with different reduction potentials. Keilin proposed that the cytochromes pass electrons sequentially from each protein complex to the next-stronger electron acceptor, with each step providing a small amount of energy to the organism (**Fig. 14.5**). The electron transport proteins are called **oxidoreductases** because they oxidize one substrate (removing electrons) and reduce another (donating electrons). Thus, they couple different half reactions in the electron tower (see **Table 14.1**). Oxidoreductases consist of multiprotein complexes that include cytochromes as well as noncytochrome proteins. The structure and function of ETS oxidoreductase complexes are discussed in Section 14.3.

The Proton Motive Force

The sequential transfer of electrons from one ETS protein to the next yields energy to pump ions (in most cases H^+) across the membrane. Proton pumping generates a **proton motive force, PMF** (or **proton potential**) composed of the H^+ concentration difference plus the charge difference across the membrane. The proton motive force stores energy for all microbes, as well as the mitochondria of animals and plants.

The PMF (Δp) drives many different cellular processes, such as ATP synthesis and flagellar rotation. In pathogens, the PMF drives drug efflux pumps that confer resistance to antibiotics such as tetracycline.

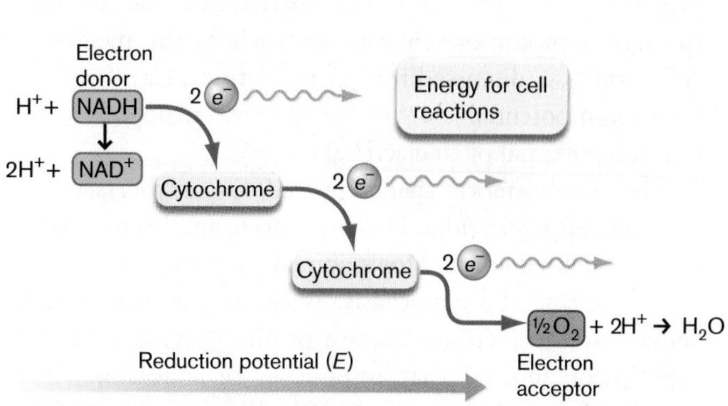

FIGURE 14.5 ▪ Electron transport system. David Keilin's model for electron transfer through cytochromes. Each cytochrome in sequence receives electrons from a stronger electron donor and transfers them to a stronger electron acceptor.

Note that in water, H^+ never occurs as such, because it combines with a water molecule to form hydronium ion, H_3O^+. Current data, however, are consistent with the passage of hydrogen nuclei (protons) through the proton pumps of the electron transport system. Within the pump complex, the proton associates with one chemical group after another; for example, H^+ may combine with the amine of an amino acid (RNH_2) to form an ammonium ion (RNH_3^+), and then transfer to a different proton-accepting group within the protein.

Note: In this book we use "hydrogen ions" and "H^+" interchangeably with "protons."

Discovery of the proton motive force. The discovery of the proton motive force radically changed the field of biochemistry. Early in the twentieth century, David Keilin and other scientists knew that the energy acquired by electron transport proteins was used to make ATP, but they did not know how. Most were convinced that electron transport was somehow directly coupled to ATP synthesis. The actual means of coupling was proposed and demonstrated by one of Keilin's students—Peter Mitchell (1920–1992)—and Mitchell's colleague Jennifer Moyle (1921–2016) (**Fig. 14.6A**). In 1961, Mitchell proposed an astonishing explanation, called the chemiosmotic theory, for the coupling of electron transport to ATP synthesis. The chemiosmotic theory states that the energy from electron transfer between membrane proteins is used to pump protons across the membrane, leading to a higher H^+ concentration outside the cell. The pump generates the proton motive force, Δp, which stores energy that can be used to make ATP (**Fig. 14.6B**).

Other biologists questioned how a proton potential could be coupled to ATP synthesis at an enzyme complex separate from the ETS, at a distant location on the membrane. They were skeptical that the H^+ concentration gradient and charge difference could exist everywhere on the membrane separating the two compartments. Because the membrane is impermeable to hydrogen ions, the H^+ current can flow back only through a proton-driven complex such as the membrane ATP synthase (discussed further in Section 14.2). In effect, the proton potential is a "proton battery," analogous to the electron potential of an electrical battery.

The chemiosmotic theory proved so controversial that Mitchell left Cambridge University to found his own laboratory, Glynn House. Moyle joined him to devise experiments testing the hypothesis. A key requirement was to show that the ETS generates a proton potential. Moyle's experiment showed that respiration of mitochondria is associated with proton efflux: Mitochondria isolated from rat livers were exposed to oxygen, causing efflux of hydrogen ions. The H^+ efflux occurred as electrons were transferred across the ETS and hydrogen ions were expelled

A.

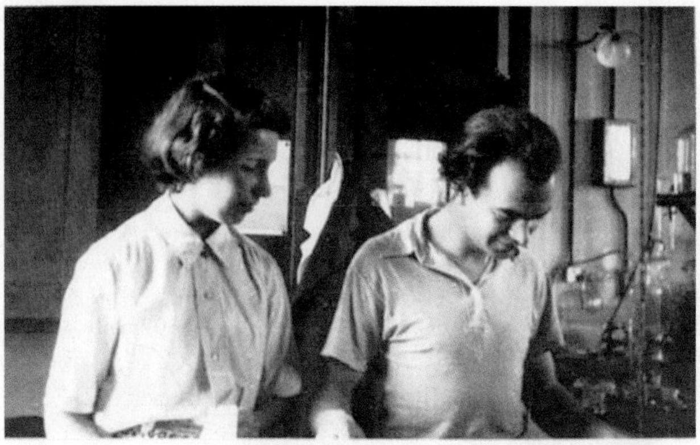

B.

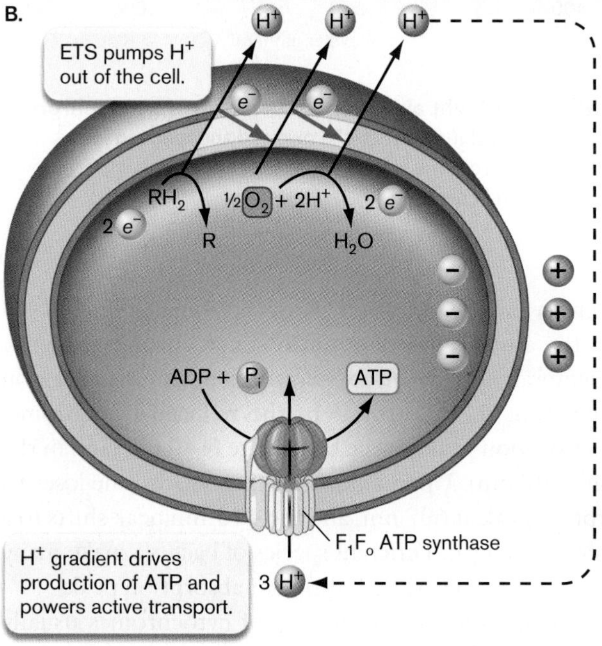

FIGURE 14.6 ■ Jennifer Moyle and Peter Mitchell discovered the proton motive force (Δp). **A.** Moyle and Mitchell proposed and tested the chemiosmotic theory. **B.** An electron transport system pumps protons out of the cell. The resulting electrochemical gradient of protons (proton motive force) drives conversion of ADP to ATP through ATP synthase.

by the proton pumps. In Moyle's experiments, the number of protons extruded per electron transferred down the ETS was consistent with the chemiosmotic model. Similar results were obtained with vesicles made first from mitochondrial membranes and later from the membranes of chloroplasts and bacteria.

A greater challenge was to demonstrate that a proton motive force could, in fact, drive the ATP synthase and other ion transporters without any undiscovered intermediate. A friend of Mitchell and Moyle, André Jagendorf, at Johns Hopkins University, tested the effect of a proton gradient imposed on spinach chloroplasts. Chloroplasts contain ATP synthase directed outward (that is, driven by proton flow from inside to outside; **Fig. 14.7A**). The chloroplasts were partly opened by osmotic shock and then

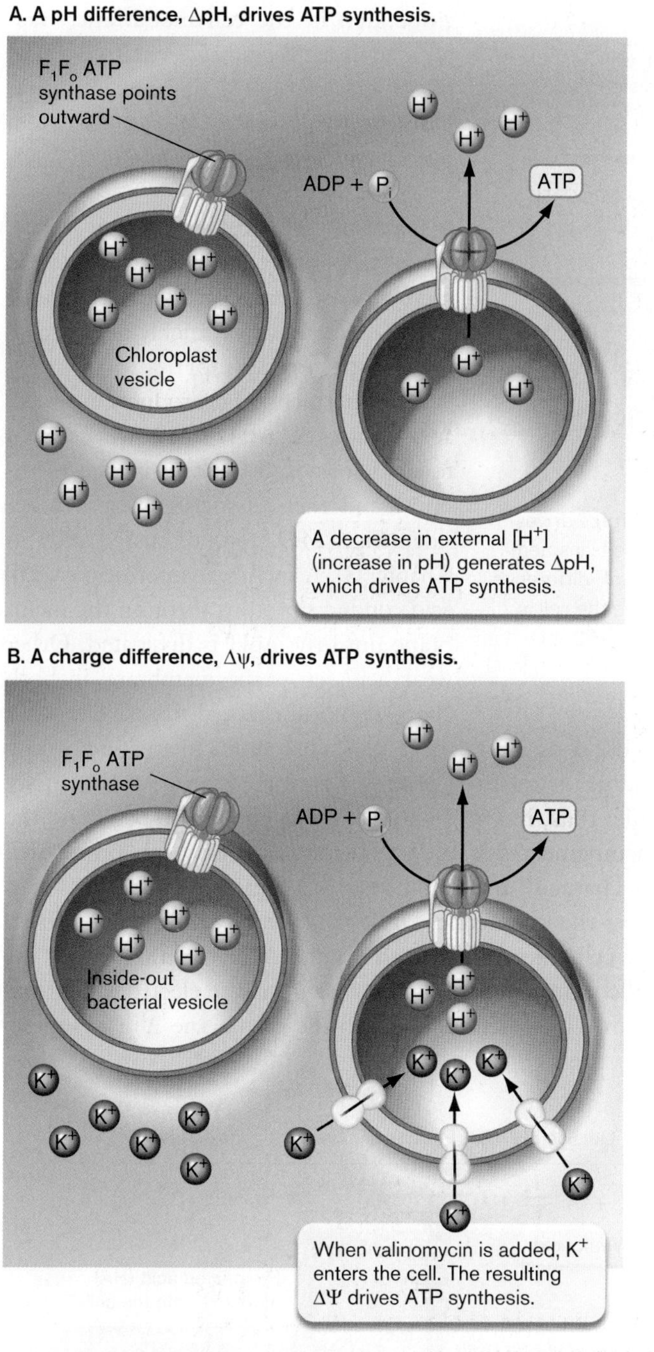

FIGURE 14.7 ■ Either ΔpH or Δψ drives ATP synthesis. A. A pH difference imposed across the chloroplast inner membrane drives a proton current through an outwardly directed ATP synthase. **B.** A charge difference generated by K^+ influx drives a proton current through ATP synthase.

suspended in medium containing concentrated hydrogen ions (pH 4). With the osmotic balance restored, the chloroplast membranes closed again, with increased [H^+] trapped inside. When the pH outside was raised to pH 8.3—that is, lower [H^+]—protons from the acidic interior of the vesicles flowed out through the ATP synthase, and ADP and inorganic phosphate were converted to ATP. Jagendorf interpreted this result as consistent with the chemiosmotic theory for the mechanism of phosphorylation.

Other researchers showed that ATP synthesis could be driven by a charge difference across a vesicle membrane (**Fig. 14.7B**). A charge difference (electrical potential, Δψ) was applied by loading vesicles with potassium ions (K^+), which add charge without affecting the chemical concentration of hydrogen ions. The potassium ions were conducted across the membrane by the ionophore valinomycin, a small hydrophobic peptide that binds K^+ and solubilizes it in the membrane. The potassium ions flow down their concentration gradient, along with their associated positive charge. In the experiment shown in **Figure 14.7B**, the vesicles are "inside-out" bacterial membrane vesicles, in which the ATP synthase points outward instead of inward. The K^+ influx adds positive charge, which drives H^+ out through ATP synthase, catalyzing formation of ATP. Thus, Δψ drives the formation of ATP.

Δ*p* includes Δψ and ΔpH. The transfer of H^+ through a proton pump generates a H^+ concentration difference across the membrane. Since H^+ carries a positive charge, the proton transfer also generates a charge difference across the membrane. The H^+ concentration difference (ΔpH) plus the charge difference (Δψ) make a proton potential (Δ*p*), also called a proton motive force (PMF).

Thus, when protons are pumped across the membrane, the proton motive force stores energy in two different forms—the separation of charge (electrical potential) and the gradient of H^+ concentration (pH difference)—as shown in **Figure 14.7**. Either form (or both) can drive Δ*p*-dependent cellular processes:

- **The electrical potential** (Δψ) arises from the separation of charge between the cytoplasm (more negative) and the solution outside the cell membrane (more positive). For many bacteria, this "battery" potential is about −50 to −150 mV.
- **The pH difference** (ΔpH) is the difference between internal and external pH ($pH_{int} - pH_{ext}$). For example, if the bacterial internal pH is 7.5 and the external pH is 6.5, the ΔpH is 1.0, and the ratio of $[H^+]_{ext}$ to $[H^+]_{int}$ is 10. A ΔpH of 1.0 corresponds to a proton potential of approximately −60 mV when the temperature is 25°C.

The relationship between electrical and chemical components of the proton potential Δ*p* (in millivolts) is given by:

$$\Delta p = \Delta\psi - (2.3RT/F)\Delta pH$$

or approximately:

$$\Delta p = \Delta\psi - 60\Delta pH$$

For cells grown at neutral pH, all three terms (Δ*p*, Δψ, and −60ΔpH) usually have a negative value, meaning that their force drives protons inward from outside.

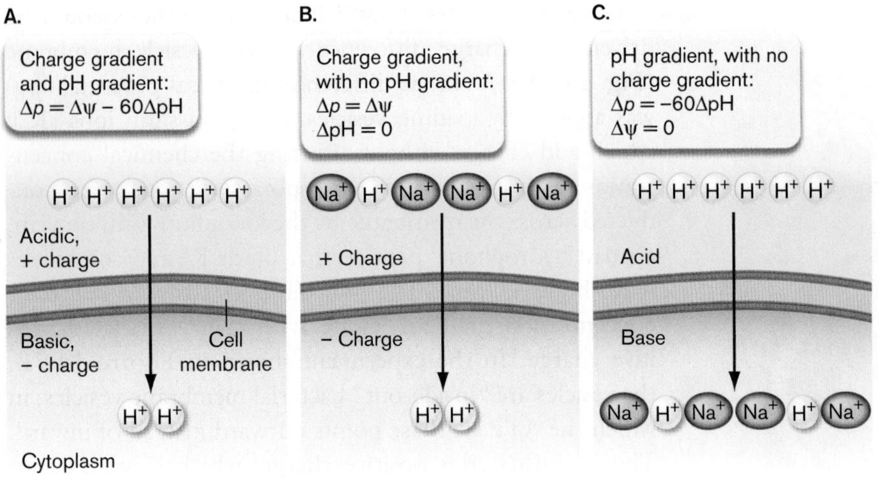

FIGURE 14.8 ■ Electrical potential and pH difference. Proton motive force drives protons into the cell (arrow). **A.** The proton motive force Δp is composed of the transmembrane electrical potential $\Delta\psi$ (the charge difference), plus the transmembrane pH difference ΔpH (the chemical concentration gradient of H^+). **B.** If the pH inside and outside the cell is equal, then $\Delta p = \Delta\psi$. **C.** If the electrical charge inside and out is equal, then $\Delta p = -60\Delta pH$.

In living cells, the relative contributions of $\Delta\psi$ and $-60\Delta pH$ vary, depending on other sources of charge difference and protons, as shown in **Figure 14.8**. Both the $\Delta\psi$ and ΔpH components of Δp are influenced by other factors besides ETS proton transport. For example:

- **The charge difference $\Delta\psi$** includes charges on other ions, such as K^+ and Na^+. These ions are pumped or exchanged by ion-specific membrane transport proteins.
- **The pH difference ΔpH** is affected by metabolic generation of acids, such as fermentation acids, or by pH changes outside the cell. Permeant acids can run down the gradient through the membrane, collapsing the ΔpH to zero.

Overall, the cell uses its various membrane pumps and metabolic pathways to adjust and maintain its proton motive force at a size sufficient to drive ATP synthesis but not so great as to disrupt the membrane.

Thought Questions

14.4 What do you think happens to $\Delta\psi$ as the cell's external pH increases or decreases? What could happen to the Δp of bacteria that are swallowed and enter the extremely acidic stomach?

14.5 How could a simple experiment provide evidence that a proton pump in the bacterial cell membrane drives efflux of antibiotics such as tetracycline?

What about other ions, such as the sodium ions in **Figure 14.8**? Can a sodium gradient drive cell processes? For some bacteria, a gradient of Na^+ concentration can drive transport of a nutrient through a transporter protein embedded in the cell membrane. Bacteria such as *Vibrio cholerae* (the cause of cholera) can use a Na^+ concentration gradient to power ATP synthesis by a Na^+-dependent ATP synthase.

Dissipation of proton motive force. What happens if something disrupts the membrane so that protons leak through? Even a small leak can quickly dissipate all the energy stored as ΔpH.

Many fermentation products are weak acids, such as acetic acid, whose protonated form can dissolve in the membrane and then dissociate on the other side (**Fig. 14.9**, top; discussed also in Chapter 3). A membrane-permeant weak acid conducts protons through the membrane until the ΔpH is dissipated. Other weak acids cross the membrane in both charged and uncharged forms. The weak acid 2,4-dinitrophenol (DNP) dissociates to an anion whose charge is relatively evenly distributed around the molecule; thus, it remains hydrophobic enough to penetrate the membrane (**Fig. 14.9**, bottom). Because both protonated (uncharged) and unprotonated (negatively charged) forms cross the membrane, DNP can cyclically bring protons into the cell and collapse both $\Delta\psi$ and ΔpH. Such molecules are called **uncouplers** because they uncouple electron transport from ATP synthesis through the membrane ATP synthase.

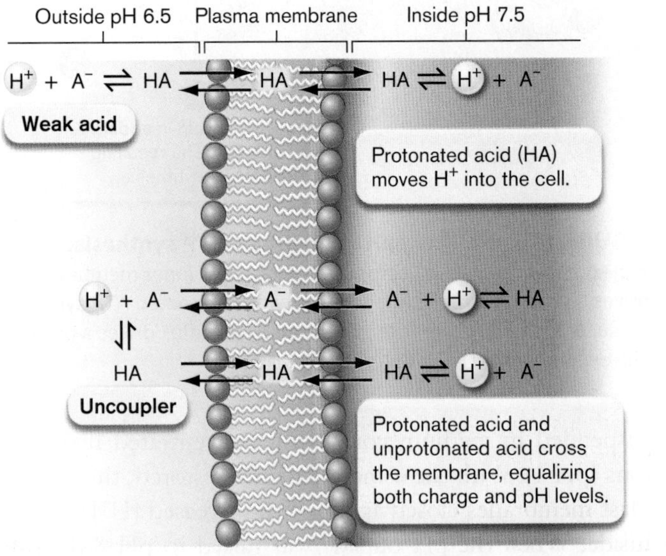

FIGURE 14.9 ■ Membrane-permeant weak acids and uncouplers. Top: A weak acid crosses the membrane in the protonated form and then dissociates, acidifying the cell. **Bottom:** An uncoupler crosses the membrane in both protonated and unprotonated forms, cyclically acidifying the cell and dissipating the charge difference (electrical potential, $\Delta\psi$).

With uncoupling, electron transport accelerates, but the energy dissipates as heat.

When the bacterial PMF is uncoupled from energy-spending reactions, the ETS pumps protons at an accelerated rate, while losing the energy as heat. Thus, uncouplers inhibit bacterial growth. They are highly toxic for human cells because the acceleration of electron transport causes overheating and brain damage.

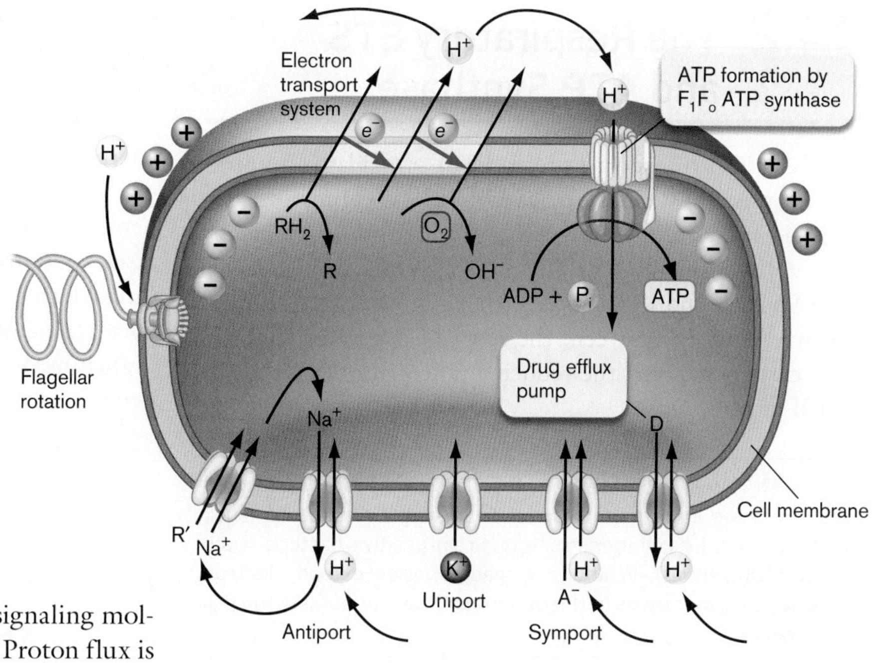

FIGURE 14.10 ■ Processes driven by the proton motive force. Processes powered by proton potential include ATP synthesis through the F_1F_o ATP synthase, flagellar rotation, uptake of nutrients, and efflux of toxic drugs. "R" represents an organic nutrient.

Δp Drives Many Cell Functions

Besides ATP synthesis, Δp drives many cellular processes directly (**Fig. 14.10**). Processes driven by the proton motive force include the rotation of flagellar motors (Chapter 3) and the secretion of toxins and signaling molecules across the cell membrane (Chapter 8). Proton flux is also coupled to transport of ions such as K^+ or Na^+ through parallel transport (symport) or oppositely directed transport (antiport), as discussed in Chapter 4. Ion flux then drives uptake of nutrients such as amino acids or efflux of molecules such as antibacterial drugs. Drug efflux pumps are a major problem in hospitals, where the pump proteins are encoded by genes carried on plasmids that spread among virulent strains (Chapter 9).

Thought Question

14.6 Suppose that de-energized cells of *E. coli* ($\Delta p = 0$) with an internal pH of 7.6 are placed in a solution at pH 6. What do you predict will happen to the cell's flagella? What does this effect demonstrate about the function of Δp?

To Summarize

- **An electron transport system (ETS)** consists of a series of electron carriers that sequentially transfer electrons to the carrier of next-higher reduction potential *E* (that is, the next-stronger electron acceptor). Electron flow through the ETS begins with an initial electron donor and ultimately transfers all electrons to a terminal electron acceptor.
- **The reduction potential *E*** for a complete redox reaction must be positive to yield energy for metabolism. The standard reduction potential $E^{\circ\prime}$ assumes that all reactant concentrations equal 1 M, at 25°C and pH 7.
- **Concentrations of electron donors and acceptors** in the environment influence the actual reduction potential *E* experienced by the cell.
- **The ETS is embedded in a membrane that separates two compartments.** Two aqueous compartments must be separate to maintain an ion gradient generated by the ETS.
- **The ETS includes protein complexes and cofactors.** Protein complexes called oxidoreductases include cytochromes and noncytochrome proteins. Cytochromes are colored proteins whose absorbance spectrum shifts with a change in redox state.
- **The chemiosmotic theory states that ETS complexes generate a proton motive force (Δp).** The force is usually directed inward, driving back into the cell protons that have accumulated outside the cell as a result of the ETS. The proton potential drives ATP synthase and other functions, such as ion transport and flagellar rotation.
- **The proton potential (Δp, measured in millivolts)** is composed of the electrical potential ($\Delta\psi$) and the hydrogen ion chemical gradient (ΔpH): $\Delta p = \Delta\psi - 60\Delta\text{pH}$.
- **Uncouplers are molecules taken up by cells in both protonated and unprotonated forms.** Uncouplers can collapse the entire proton potential, thus uncoupling respiration from ATP synthesis.

14.2 The Respiratory ETS and ATP Synthase

The electron transport system (ETS) of oxidative respiration consists of various carrier molecules and proteins. For aerobic bacteria, and for the mitochondrial inner membrane, the respiratory ETS receives electrons from NADH and $FADH_2$ and transfers electrons ultimately to O_2, producing H_2O. In between, the series of membrane-embedded carriers harvests the reducing potential of electrons in small steps.

Note: This section presents the classic respiratory ETS that resides in the inner membrane of Gram-negative bacteria (such as *E. coli*) or the cell membrane of Gram-positive bacteria (such as *Bacillus subtilis*). In addition, many species extend electron transfer to components in the outer membrane or outside the cell (Section 14.4).

Cofactors Allow Small Energy Transitions

ETS proteins such as cytochromes associate electron transfer with energy transitions that are small and reversible. The energy transitions are mediated by **cofactors**, small molecules that associate with the protein (**Fig. 14.11**). Cofactors such as heme allow small, reversible redox changes. The structure of each cofactor must allow transition of an electron between closely spaced energy levels to avoid "spending it all in one place." If all of the energy were spent in one transition, most of it would be lost as heat instead of being converted to several small processes, such as pumping H^+ across the membrane.

Small energy transitions typically involve these kinds of molecular structures:

- **Metal ions such as iron or copper**, coordinated (and hence held in place) with amino acid residues. Iron is often coordinated by sulfur atoms of cysteine residues in the protein; examples shown in **Figure 14.11B** are [2Fe-2S] and [4Fe-4S]. Transition metals make useful electron carriers because their outer electron shell has several closely spaced energy levels, facilitating small energy transitions.
- **Conjugated double bonds and heteroaromatic rings**, such as the nicotinamide ring of $NAD^+/NADH$, also provide narrowly spaced energy transitions. Membrane-soluble carriers such as **quinones** (reduced to **quinols**) allow even smaller energy transitions than does $NAD^+/NADH$.

The major protein complexes of electron transport each have one or more redox centers containing either metal ions or conjugated double bonds or both. In FMN, the conjugated double bonds allow a small energy transition (**Fig. 14.11A**). In iron-sulfur clusters—[2Fe-2S] and [4Fe-4S]—the metal atoms provide the site for a small energy transition, its size dependent on the cluster's connections within the associated protein (**Fig. 14.11B**). The heme group found in cytochromes and other oxidoreductases contains extensive conjugated double bonds, coordinated around the metal Fe^{3+} (**Fig. 14.11C**). Reduction by a transferred electron converts the iron to Fe^{2+}. The branching of side chains on the ring varies among different hemes, altering the magnitude of $E°'$ for the redox couple Fe^{3+}/Fe^{2+}.

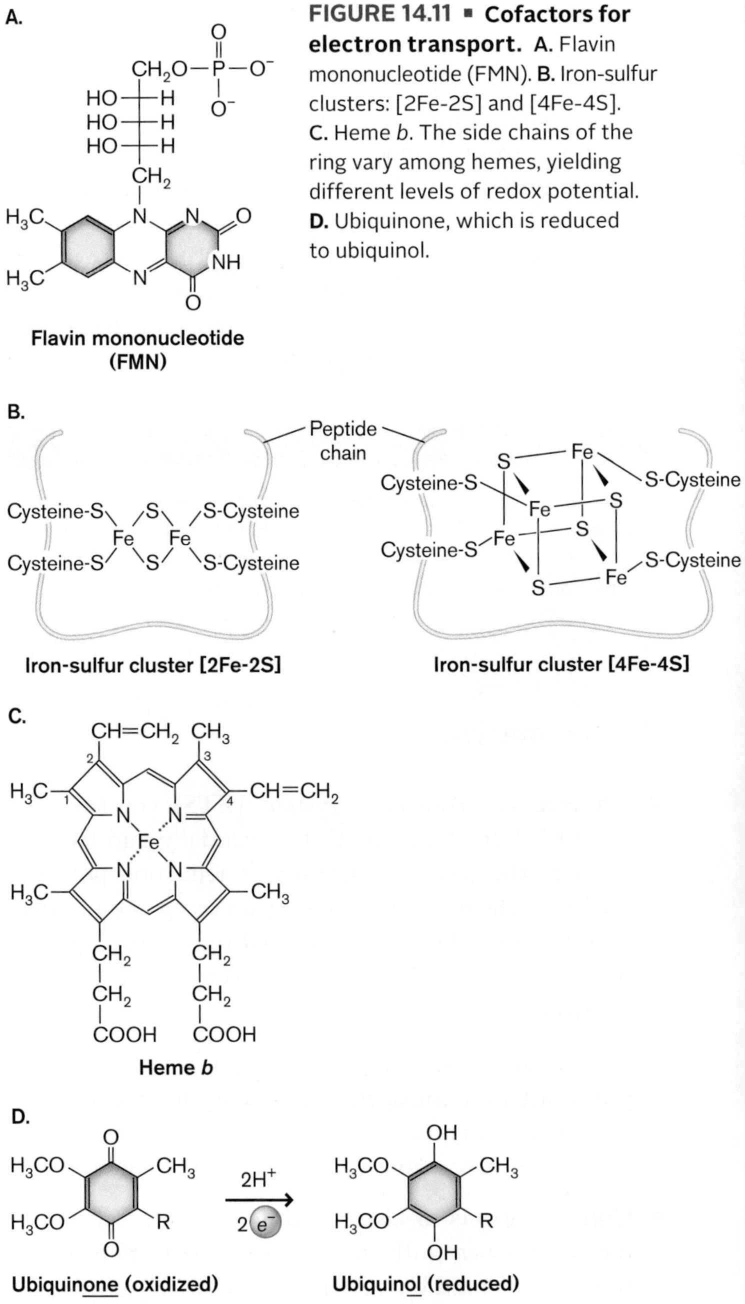

FIGURE 14.11 ■ Cofactors for electron transport. **A.** Flavin mononucleotide (FMN). **B.** Iron-sulfur clusters: [2Fe-2S] and [4Fe-4S]. **C.** Heme *b*. The side chains of the ring vary among hemes, yielding different levels of redox potential. **D.** Ubiquinone, which is reduced to ubiquinol.

Some electron carriers are small molecules that associate loosely with a protein complex and then come off to diffuse freely within the membrane. Mobile electron carriers include quinones (Q) such as ubiquinone (**Fig. 14.11D**), which can be reduced to quinols (QH_2):

$$Q + 2H^+ + 2e^- \rightarrow QH_2$$

Quinols carry electrons and protons laterally within the membrane between the proton-pumping protein complexes of the ETS. The hydrophobic quinols never leave the membrane; thus their electrons are kept in the membrane until transfer out of the ETS. After transferring their electrons to the next protein complex, quinols revert to quinones, capable of accepting electrons again.

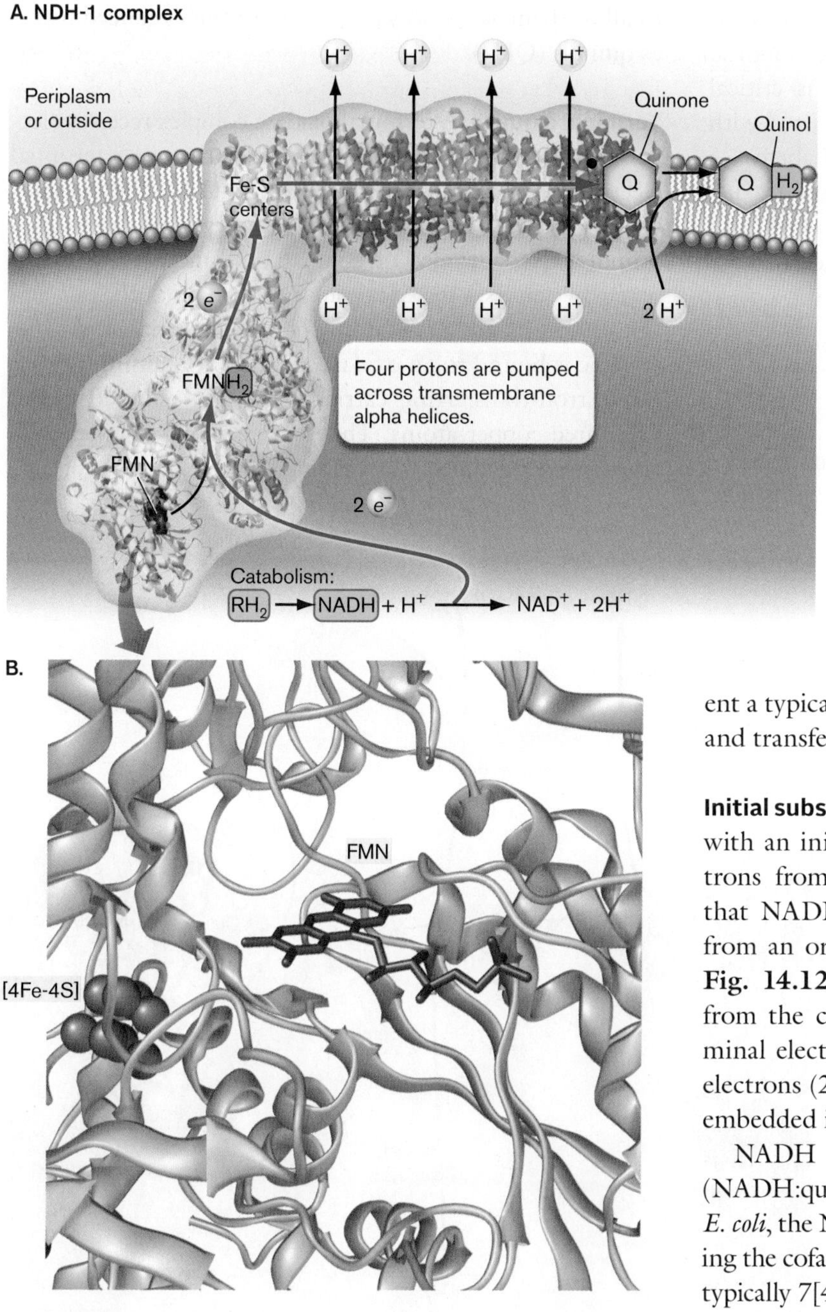

FIGURE 14.12 ■ NADH:quinone oxidoreductase complex (NDH-1). **A.** NDH-1 transfers two electrons from NADH onto NDH-1. The energy from oxidizing NADH is coupled to pumping $4H^+$ across the cell membrane. **B.** Within the NDH-1 complex, FMN lies adjacent to the first iron-sulfur center, [4Fe-4S]. (PDB code: 1NOX) *Source:* Part A modified from R. Efremov and L. Sazanov. 2010. *Nature* **476**:414; part B modified from L. A. Sazanov and P. Hinchliffe. 2006. *Science* **311**:1430–1436.

Note: Dehydrogenases, reductases, and oxidases are all **oxidoreductases**, which oxidize one substrate (remove electrons) and reduce another (donate electrons). Oxidoreductases that accept electrons from NADH or $FADH_2$ are also called dehydrogenases, because their reaction releases hydrogen ions.

Oxidoreductase Protein Complexes

A respiratory electron transport system includes at least three functional components: an initial substrate oxidoreductase (or dehydrogenase), a mobile electron carrier, and a terminal oxidase. Microbes make alternative versions of each component, using alternative electron-donating substrates and terminal electron acceptors, depending on what is available in the environment. Here we present a typical bacterial ETS receiving electrons from NADH and transferring them to oxygen.

Initial substrate oxidoreductase. A respiratory ETS begins with an initial oxidoreductase that receives a pair of electrons from an organic substrate such as NADH. Note that NADH forms by receiving two electrons plus $2H^+$ from an organic product of catabolism (designated RH_2; **Fig. 14.12A**). The $2H^+$ are ultimately balanced by $2H^+$ from the cytoplasm combining with O_2 (or another terminal electron acceptor) at the end of the ETS. The two electrons ($2e^-$) from NADH enter an ETS protein complex embedded in the membrane.

NADH donates electrons to NADH dehydrogenase (NADH:quinone oxidoreductase, NDH-1; **Fig. 14.12A**). In *E. coli*, the NDH-1 complex has 14 different subunits, including the cofactor FMN, as well as several iron-sulfur clusters—typically 7[4Fe-4S] and 2[2Fe-2S]. The cofactors "hand off" electrons to each other through adjacent connections; see, for example, the placement of FMN and the first [4Fe-4S] within the peptide coils of NDH-1 (**Fig. 14.12B**). Each electron from NADH travels through FMN and the iron-sulfur series. At the end of the chain, the electrons and $2H^+$ from solution are transferred to a quinone, which is thus reduced to a quinol. Quinones are designated Q; and quinols, QH_2.

Within the NDH-1 complex, the oxidation of NADH and reduction of Q to QH_2 yields energy to pump up to $4H^+$ across the membrane. A crystallographic model shows four apparent proton channels through transmembrane alpha helices of the protein subunits (**Fig. 14.12B**). Within each channel, a hydrogen ion hops along a series of amino acid residues. The H^+ translocation is driven by a conformational change in the alpha helices throughout the protein, arising from the initial two-electron reduction by NADH. The hydrogen ions pumped across the membrane contribute to the proton potential Δp. In human mitochondria, the NADH dehydrogenase (aka "complex I") is critical for health; genetic defects in complex I are associated with diseases such as Parkinson's and some forms of diabetes.

Note that the $4H^+$ pumped across the membrane are distinct from the $2H^+$ acquired by the quinone ($Q \rightarrow QH_2$). In some halophilic bacteria, $4Na^+$ are pumped instead of $4H^+$.

Note: In our figures, protons that cross the membrane by the end of the ETS (and thus contribute to Δp) are highlighted yellow.

Not all substrate oxidoreductases pump protons. For example, *E. coli* has an alternative NADH dehydrogenase (NDH-2) that transfers two electrons to Q without pumping additional protons across the membrane. (The unused energy is lost as heat.) NDH-2 functions during rapid growth, when the cell must limit its proton potential to avoid membrane breakdown. Other complexes, such as succinate dehydrogenase, transfer electrons from substrates in a reaction that lacks sufficient energy to pump extra protons. However, these electrons can be transferred to another ETS complex at lower voltage potential where redox reactions do pump protons.

Quinone pool. A quinone can receive $2e^-$ from the substrate oxidoreductase, along with $2H^+$ from solution, to balance the negative charges, yielding a quinol (**Fig. 14.12A**). The quinols diffuse within the membrane and carry reduction energy to other ETS components. After transferring $2e^-$ to the next protein complex, the quinol releases its $2H^+$. Usually the $2H^+$ released are on the opposite side of the membrane from where $2H^+$ were originally picked up (**Fig. 14.13**). Thus, besides transferring two electrons, a quinol may contribute two protons to the transmembrane proton potential. The reoxidized carriers then recycle back as quinones.

Each quinone can bind to a substrate dehydrogenase, pick up a pair of electrons and hydrogen ions, and then diffuse away and carry the electrons to a reductase. The quinones and quinols, referred to as the quinone pool, diffuse freely within the membrane. Thus, the quinones/quinols are able to transfer electrons between many different redox enzymes.

Different oxidoreductase complexes interact with slightly different quinones, such as ubiquinone and menaquinone. The reduction potentials of ubiquinone and menaquinone are given in **Table 14.1**. For clarity, in this chapter we refer to all of them as quinones (Q), and to their reduced forms as quinols (QH_2).

Terminal oxidase. A terminal oxidase complex receives electrons from a quinol (QH_2) and transfers them to a terminal electron acceptor, such as O_2 (**Fig. 14.13**). The complex usually includes a cytochrome that accepts electrons from quinols. Cytochromes of comparable function are designated by letters—for example, cytochrome *b* (*E. coli* and mitochondria) and cytochrome *c* (mitochondria). The cytochrome is bound to an oxidase complex containing a series of electron-transferring carriers: two iron-centered hemes and three copper atoms. This unique center couples electron transfer and proton pumping.

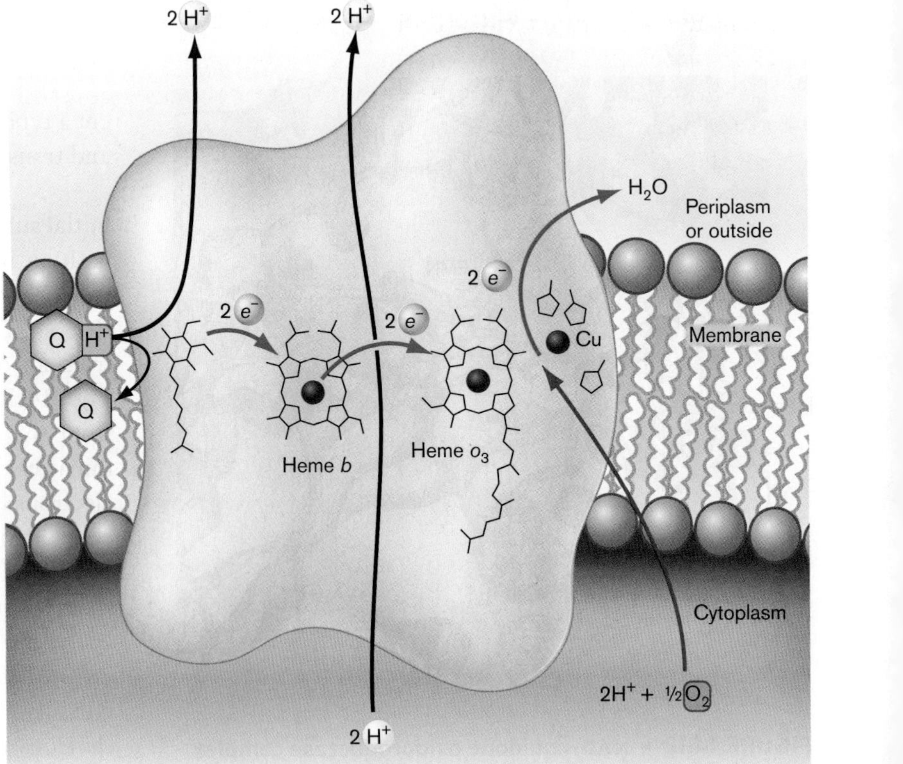

FIGURE 14.13 ■ Cytochrome *bo* quinol oxidase complex. Each quinol (QH_2) transfers two electrons to heme *b*. The two H^+ from each quinol are expelled to the periplasm (or outside the bacterial cell). The $2e^-$ from the quinols are transferred through the complex onto $2H^+$ plus an oxygen atom from O_2, forming water. *Source:* Based on the structure determined by Jeff Abramson et al. 2000. *Nat. Struct. Biol.* **7**:910–917.

The cytochrome *bo* quinol oxidase of *E. coli* consists of cytochrome *b* plus oxidase complex *o*. The cytochrome *b* subunit receives two electrons from a quinol ($QH_2 \rightarrow Q$) and releases the $2H^+$ out to the periplasm. Each electron from quinol travels through the two hemes of the oxidase complex. Because the two quinol hydrogens originated from $2H^+$ in the cytoplasm (see **Fig. 14.12**) and $2H^+$ were released outside by a quinol, there is net efflux of $2H^+$. In addition, the transfer of $2e^-$ between the two oxidase hemes is coupled to pumping $2H^+$ from the cytoplasm across to the periplasm (**Fig. 14.13**). The hydrogen ions then leak through the outer membrane, effectively outside the cell.

The second heme of the oxidase (heme o_3) acts to reduce an atom of oxygen from O_2. Each oxygen atom receives two electrons and combines with two protons ($2H^+$) from the cytoplasm to form H_2O. The $2H^+$ consumed balance the $2H^+$ released by catabolism to make NADH + H^+.

Note that the oxidase is conventionally shown as obtaining two electrons from the cytoplasm and donating them to an atom of an oxygen molecule ($\frac{1}{2}O_2$). A full reaction cycle of cytochrome *bo* quinol oxidase actually puts four electrons from two quinols (originally 2 NADH) onto O_2, taking up $4H^+$ from the cytoplasm to make two molecules of H_2O.

Besides cytochrome *bo* quinol oxidase, bacteria express different terminal oxidases that differ with respect to the ratio of cytoplasmic protons pumped to electrons transferred. For example, the alternative cytochrome *bd* quinol oxidase reduces O_2 to water but pumps no extra protons. Although it pumps no protons, cytochrome *bd* quinol oxidase can bind O_2 at much lower concentrations and donate electrons, completing the respiratory circuit. Thus, the *bd* oxidase enables *E. coli* to respire within low-oxygen habitats such as the mammalian intestine.

Figure 14.14 summarizes a complete ETS for oxidation of NADH by $\frac{1}{2}O_2$ in the inner membrane of *E. coli*. Overall, an ETS for respiration on organic substrates includes at least three phases of electron transfer, such as: (1) NADH from an organic substrate donates electrons to an initial oxidoreductase; (2) the electrons are transferred to a quinone, which is reduced to a quinol; (3) the quinol transfers electrons to a terminal oxidase and releases $2H^+$ outside the cell. Both enzymes and quinones show considerable complexity and diversity among different species in different environments.

The entering carrier NADH carries two electrons with protons obtained from catabolized food molecules. The two electrons ($2e^-$) from NDH-1 and two protons ($2H^+$) from solution are transferred onto Q (quinone), converting it to QH_2 (quinol). The transfer of $2e^-$ from NADH yields sufficient energy to pump as many as $4H^+$ across the membrane. The exact number depends on cellular conditions, such as the concentrations of NADH and the terminal electron acceptor.

The QH_2 diffuses within the membrane until it reaches a terminal oxidase complex, such as the cytochrome *bo* quinol oxidase. The $2H^+$ from QH_2 are released outside the cell, while the $2e^-$ enter the oxidase, reducing the two hemes. The two electrons join $2H^+$ from the cytoplasm, combining with an oxygen atom to make H_2O. The reaction is coupled to pumping of $2H^+$ across the membrane plus a net increase of $2H^+$ outside through redox reactions.

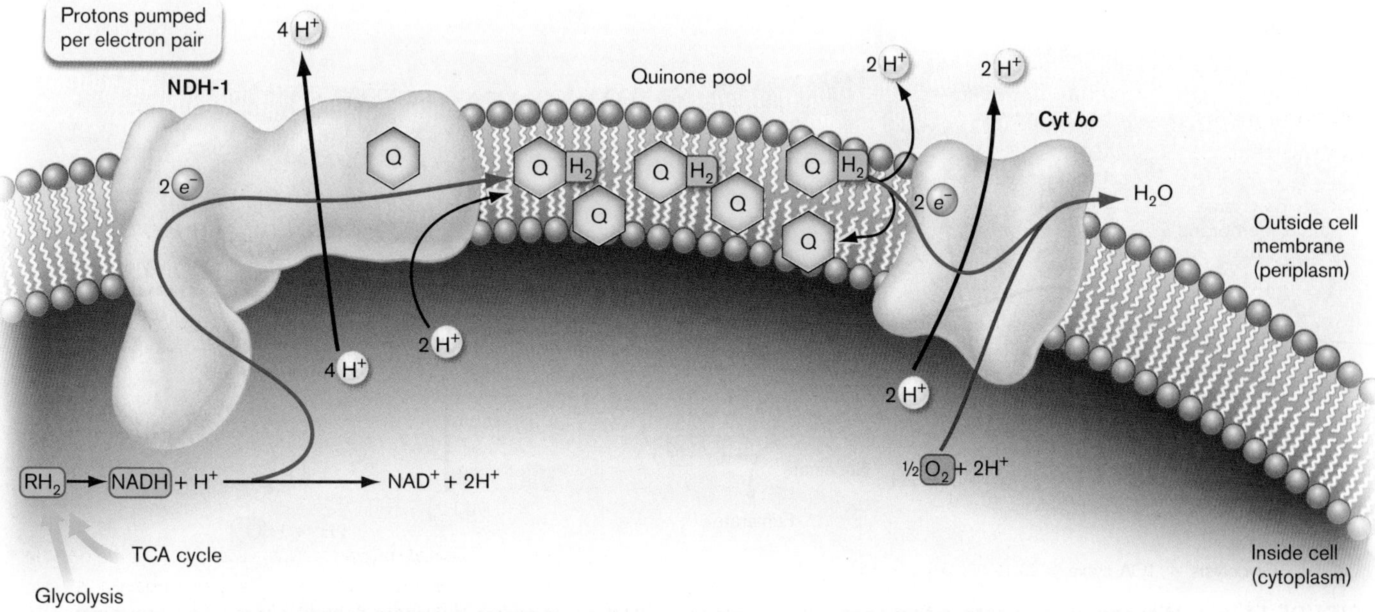

FIGURE 14.14 ■ A bacterial ETS for aerobic NADH oxidation. In *E. coli*, electrons from NDH-1 are transferred to quinones, generating quinols, which transfer electrons onto cytochrome *bo* (Cyt *bo*). For each NADH oxidized, up to $8H^+$ may be pumped across the membrane.

Overall, the oxidation of NADH exports about $8H^+$ per $2e^-$ transferred through the ETS to make H_2O. The export of protons generates a proton potential Δp.

Note: Protons (hydrogen ions, H^+) can have three different fates in the ETS:

1. Protons are pumped across the membrane (H^+) by an oxidoreductase complex. Contributes to Δp.
2. Protons are consumed from the cytoplasm by quinone/quinol, while other protons are released outside the membrane (H^+). Contributes to Δp.
3. Protons are consumed by combining with the terminal electron acceptor (O_2). If the loss balances protons released by catabolism, it does not affect Δp.

Thought Questions

14.7 In **Figure 14.14**, what is the advantage of the oxidoreductase transferring electrons to a pool of mobile quinones, which then reduce the terminal reductase (cytochrome complex)? Why does each oxidoreductase not interact directly with a cytochrome complex?

14.8 In **Figure 14.14**, why are most electron transport proteins fixed within the cell membrane? What would happen if they "got loose" in aqueous solution?

Environmental modulation of the ETS. The ETS just described represents optimal conditions, when food and oxygen are unlimited. What happens when food (that is, electron donors) or oxygen is scarce? When the environment changes, bacteria adjust the efficiency of their ETS by expressing alternative oxidoreductases. For example, at low concentrations of oxygen (microaerophilic conditions, discussed in Chapter 5), the reduction potential E is decreased. So the ETS may be unable to reduce O_2 to H_2O while pumping four protons. Instead, *E. coli* uses cytochrome *bd* quinol oxidase, which has higher affinity for oxygen but pumps no protons. (Protons are still pumped by the NADH oxidoreductase.) Thus, the bacteria will gain less energy, but they will still be able to grow. Environmental regulation of ETS is discussed further in **eTopic 14.1**.

Bacteria also have alternative oxidoreductases to serve different electron donors and acceptors. Some enzymes take electrons from donors lacking the potential of NADH; for example, succinate dehydrogenase catalyzes the one step of the TCA cycle that yields $FADH_2$—a step that provides not quite enough energy to produce NADH (discussed in Chapter 13). Succinate dehydrogenase is the only TCA enzyme embedded in the membrane as an ETS component. When O_2 concentration is so low as to be thermodynamically unavailable, some bacteria can use other electron acceptors, such as nitrate (anaerobic respiration, discussed in Section 14.3) or even a metal electrode (Section 14.4). Yet another alternative is to oxidize inorganic electron donors (lithotrophy; Section 14.5).

Mitochondrial respiration. In contrast to *E. coli*, mitochondria have only a single ETS, optimized for a relatively uniform intracellular environment (**Fig. 14.15**). Protected by the constant environment of the eukaryotic cytoplasm, mitochondria do not need to use alternative versions of their ETS to respire under different conditions. Instead, a set of just four electron-carrying complexes has evolved so

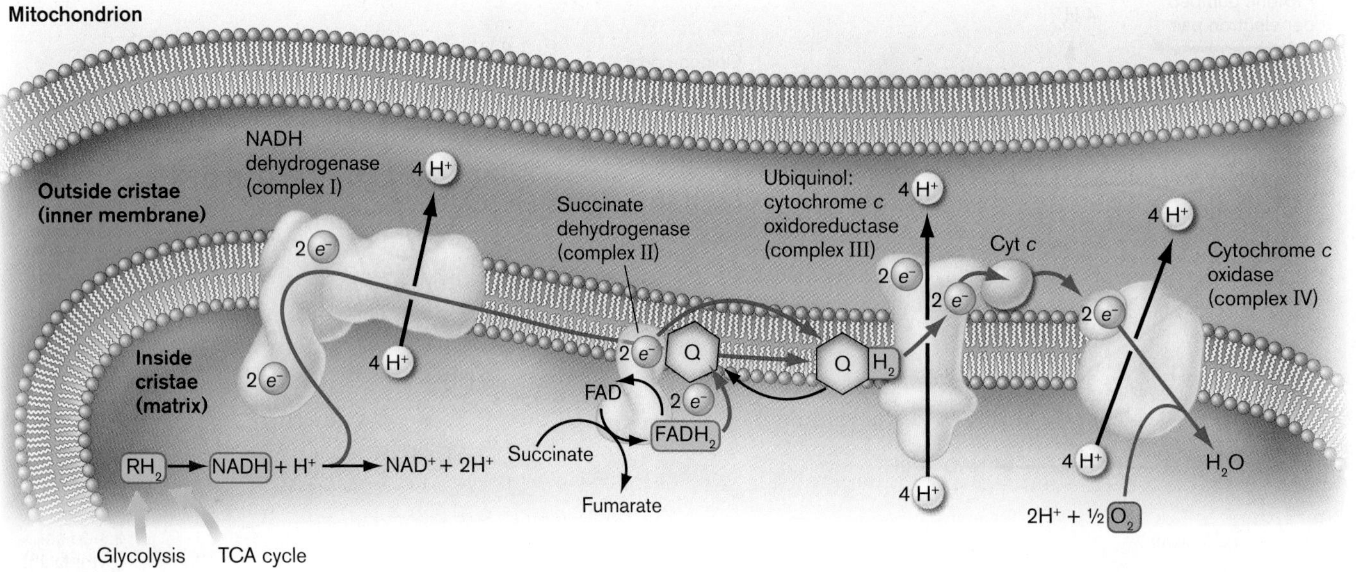

FIGURE 14.15 ■ Mitochondrial electron transport. In addition to NADH dehydrogenase, succinate dehydrogenase, and a terminal cytochrome oxidase, mitochondria possess ubiquinol:cytochrome c oxidoreductase, which provides an intermediate electron transfer step. As a result, mitochondrial membranes export 10–12 H^+ per NADH.

as to maximize the energy obtained from NADH, using oxygen as the terminal electron acceptor, and minimizing energy lost as heat. This mitochondrial ETS is nearly universal throughout the cells of animals, plants, and most eukaryotic microbes. In addition, some bacteria, such as *Paracoccus denitrificans*, have an ETS whose organization resembles that of mitochondria.

The mitochondrial ETS has homologs (proteins encoded by genes with a common ancestor) of bacterial ETS components, including NADH dehydrogenase, succinate dehydrogenase, and cytochrome *c* oxidase. However, the mitochondrial ETS differs from that of *E. coli* in the following respects:

- **An intermediate cytochrome oxidoreductase complex transfers electrons.** Besides NADH dehydrogenase (complex I) and cytochrome *c* oxidase (complex IV), mitochondria show an intermediate step of electron transfer to ubiquinol:cytochrome *c* oxidoreductase (complex III, shown in **Fig. 14.15**). The intermediate electron transfer step pumps an additional $2H^+$. Another $2e^-$ and $2H^+$ come from succinate dehydrogenase (complex II), which forms $FADH_2$ through the TCA cycle.
- **The mitochondrial ETS pumps more protons per NADH.** As many as 10–12 protons may be pumped per NADH, in contrast to 2–8 protons in *E. coli*.
- **Homologous complexes have numerous extra subunits.** Mitochondria have evolved additional nonhomologous subunits specific to eukaryotes. For example, the homologous subunits of cytochrome *c* oxidoreductase are enveloped by a series of eukaryotic proteins.

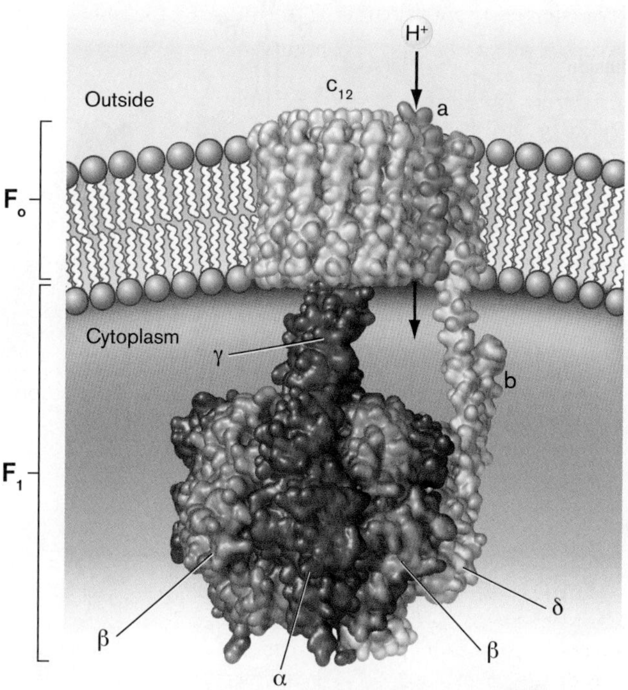

FIGURE 14.16 ■ Bacterial membrane-embedded ATP synthase (F_1F_o ATP synthase). The F_o complex is embedded in the bacterial plasma membrane, whereas the F_1 complex catalyzes ATP synthesis. (PDB codes: 1B9U, 1C17, 1E79, 2CLY)

The Proton Potential Drives ATP Synthase

The proton potential drives synthesis of ATP by the membrane ATP synthase, also known as the F_1F_o ATP synthase. The proton-driven synthesis of ATP completes the cycle of **oxidative phosphorylation**, in which hydrogen ions pumped by the ETS drive phosphorylation of ADP to ATP by the ATP synthase. The same ATP synthase can use the proton potential generated by lithotrophy (Section 14.5) or by phototrophy (Section 14.6). Surprisingly, despite the homology of ATP synthase across all forms of life, the complex is a target for antibiotics. For example, the ATP synthase of *Mycobacterium tuberculosis*, the cause of tuberculosis, is inhibited specifically by bedaquiline, an antibiotic approved in 2012 for patients whose disease resists all other drugs.

The F_1F_o ATP synthase is a protein complex highly conserved in the bacterial cell membrane, the mitochondrial inner membrane, and the chloroplast thylakoid membrane. An elegant molecular machine, the ATP synthase is composed of two complexes—F_o and F_1—that rotate relative to each other (**Fig. 14.16**). The F_o complex translocates protons across the membrane. Twelve c subunits form a cylinder embedded in the membrane, stabilized by subunits a and b.

The F_1 complex consists of six alternating subunits of types alpha and beta surrounding a gamma subunit that acts as a drive shaft. Each "third" of the F_1 complex (an alpha plus a beta subunit) interconverts ADP + P_i with ATP + H_2O. The gamma subunit connects the tripartite "knob" of F_1 to the membrane-embedded F_o. Proton transport through F_o drives ATP synthesis by F_1.

One proton at a time enters the subunit-a channel and moves into a c subunit of F_o (**Fig. 14.17** ▶). The proton potential directed inward ensures that protons more often enter from the outside than from the cytoplasm. Each entering proton causes a c subunit to rotate around the axle, with release of a bound proton to the cytoplasm. The flux of three protons through F_o is coupled to forming one molecule of ATP by one alpha-beta-gamma unit of F_1 (**Fig. 14.17A**). During each cycle generating ATP, the ring of c subunits of the F_o rotor rotates in the membrane one-third of one turn relative to the F_1 in the cytoplasm (**Fig. 14.17B**).

Note that the generation of ATP is completely reversible, so ATP hydrolysis by F_1 can pump protons back through F_o across the membrane. In the absence of a proton potential, a high ATP concentration can drive the F_o in reverse,

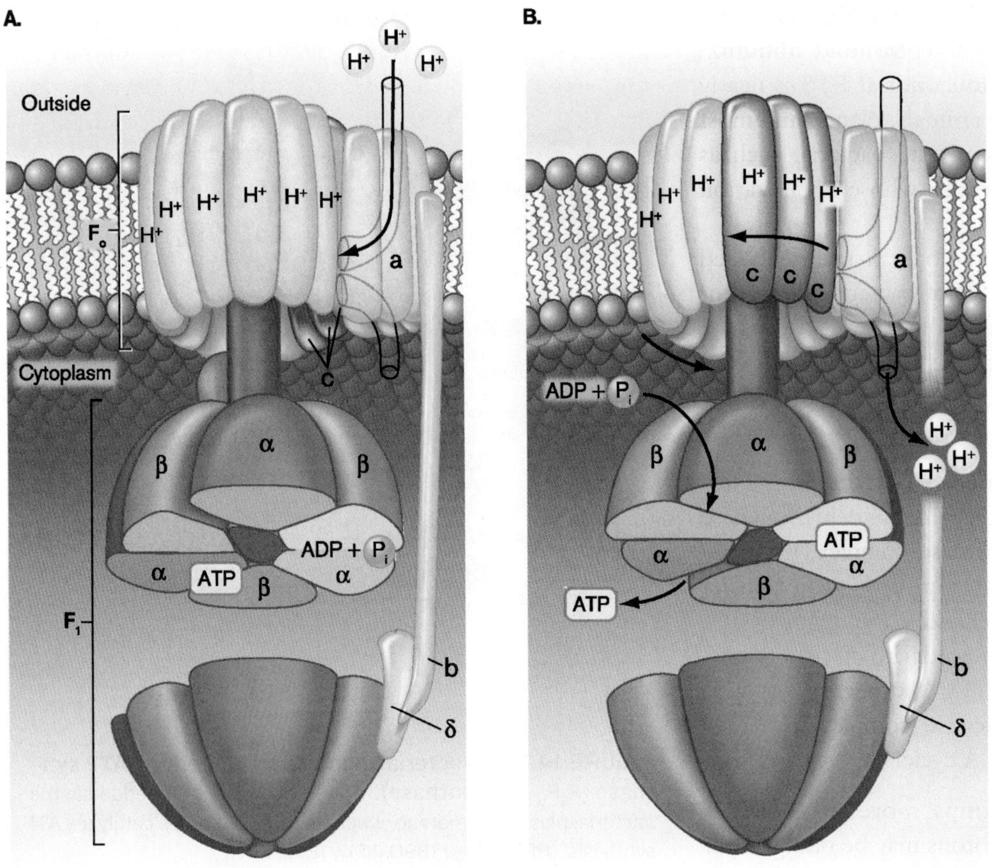

FIGURE 14.17 ■ H^+ flux drives ATP synthesis. **A.** Three protons enter c subunits of the F_o complex. The number of c subunits varies from 8 to 15 among bacterial species; 12 are shown here. **B.** The ring of c subunits rotates one-third turn relative to F_1. Flux of three protons through F_o is coupled to F_1 converting ADP + P_i to ATP.

actually pumping protons to generate Δp. This reversal of ATP synthase is used, for example, by *Enterococcus faecalis*, intestinal Gram-positive cocci that generate ATP mainly by fermentation. *E. faecalis* can operate the membrane-embedded F_1F_o ATP synthase in reverse, consuming ATP and thus generating a proton potential for nutrient uptake and ion transport.

Thought Question

14.9 Would *E. coli* be able to grow in the presence of an uncoupler that eliminates the proton potential supporting ATP synthesis?

Na^+ Pumps: An Alternative to H^+ Pumps

While a proton potential provides primary energy storage for most species, some bacteria generate an additional potential of sodium ions. A sodium motive force (ΔNa^+) is analogous to the proton motive force in that it includes the electrical potential $\Delta\psi$ plus the sodium ion concentration gradient (log ratio of the Na^+ concentration difference across the membrane). For extreme halophilic archaea, which grow in concentrated NaCl, the sodium potential entirely substitutes for the proton potential to drive ATP synthesis. These "haloarchaea" make use of the high external Na^+ concentration to store energy in the form of a sodium potential.

In some bacteria, an ETS oxidoreductase pumps Na^+ instead of H^+. For example, the proton-pumping NADH dehydrogenase can be supplemented by an NADH dehydrogenase that pumps Na^+ out of the cell. This primary sodium pump is found in many pathogens, including *Vibrio cholerae* (the cause of cholera) and *Yersinia pestis* (the cause of bubonic plague). These pathogens use the sodium-rich blood plasma to store energy in a sodium potential.

To Summarize

- **Cofactors allow small energy transitions. Electron carriers** containing metal ions and/or conjugated double-bonded ring structures are used for electron transfer. For example, cytochrome *bo* quinol oxidase has two hemes and three copper ions.
- **A substrate dehydrogenase** receives a pair of electrons from a particular reduced substrate, such as NADH. NADH dehydrogenase (NADH:quinone oxidoreductase) typically has an FMN carrier and nine iron-sulfur clusters.

- **Quinones receive electrons** from the substrate dehydrogenase and become reduced to quinols. Typically, a quinol receives $2H^+$ from the cytoplasm, and then releases $2H^+$ across the membrane upon transfer of $2e^-$ to an electron acceptor complex.
- **Protons are pumped** by substrate dehydrogenases (oxidoreductases) and terminal oxidases. The number of protons pumped by a bacterial ETS is determined by environmental conditions, such as the concentrations of substrate and terminal electron acceptor.
- **Protons are consumed** by combining with the terminal electron acceptor, such as combining with oxygen to make H_2O.
- **The proton potential drives ATP synthesis** through the membrane-embedded F_1F_o ATP synthase. Three protons drive each F_1F_o cycle, synthesizing one molecule of ATP. Some bacteria use a similar ATP synthase driven by Na^+.

14.3 Anaerobic Respiration

Most multicellular animals and plants must transfer electrons to oxygen. Likewise, some bacteria are called obligate aerobes because they grow only using O_2 as a terminal electron acceptor; examples include important nitrogen-fixing bacteria, such as *Sinorhizobium meliloti* and *Azotobacter vinelandii*. However, other organotrophic bacteria and archaea use a wide range of terminal electron acceptors, including metals, oxidized ions of nitrogen and sulfur, and chlorinated organic molecules. This **anaerobic respiration** generally takes place in environments where oxygen is scarce, such as wetland soil and water, and the human digestive tract.

E. coli possesses several different terminal oxidoreductases to reduce alternative electron acceptors (**Fig. 14.18**). These enzymes, although comparable to cytochrome quinol oxidases, are conventionally termed "reductases" to emphasize reduction of the alternative electron acceptor. Some of the electron acceptors are inorganic, such as nitrate (NO_3^-) reduced to nitrite (NO_2^-), or NO_2^- reduced to NO (nitric oxide). Because the electron acceptor stays outside the cytoplasm, even a toxic molecule may be used. For example, at the gut epithelium, *E. coli* can actually donate electrons to hydrogen peroxide (H_2O_2), a toxic molecule formed by the host immune system as a defensive response.

Other electron acceptors can be organic products of catabolism; for example, the TCA cycle intermediate fumarate can be reduced to succinate. Organic electron acceptors play important roles in food decomposition. For example, the substance trimethylamine oxide, used by fish as an osmoprotectant against sea salt, is reduced by bacteria to trimethylamine—the main cause of the "fishy" smell.

At the top of the ETS, alternative dehydrogenases (or substrate oxidoreductases) receive electrons from different organic electron donors, as well as from molecular hydrogen (H_2). The enzymes "connect" to various terminal oxidoreductases through the pool of quinones. Note that the various electron donors differ greatly in their reduction potential and hence in their capacity to generate proton potential (see **Table 14.1**). In a given environment, bacteria use the strongest electron donor and the strongest electron acceptor available. The best donor and best acceptor usually induce the expression of genes encoding their respective redox enzymes. For example, in the presence of nitrate, genes encoding nitrate reductase are expressed. At the same time, nitrate represses the expression of reductases for poorer electron acceptors, such as fumarate. (The mechanisms of gene induction and repression are discussed in Chapter 10.)

Oxidized Forms of Nitrogen

Respiration using oxidized forms of nitrogen is widespread among bacteria and archaea. Most eukaryotes must breathe oxygen, but some eukaryotic microbes respire on nitrate and nitrite. In anaerobic soil, many yeasts and filamentous fungi can reduce nitrate to nitrite, and nitrite to nitrous oxide.

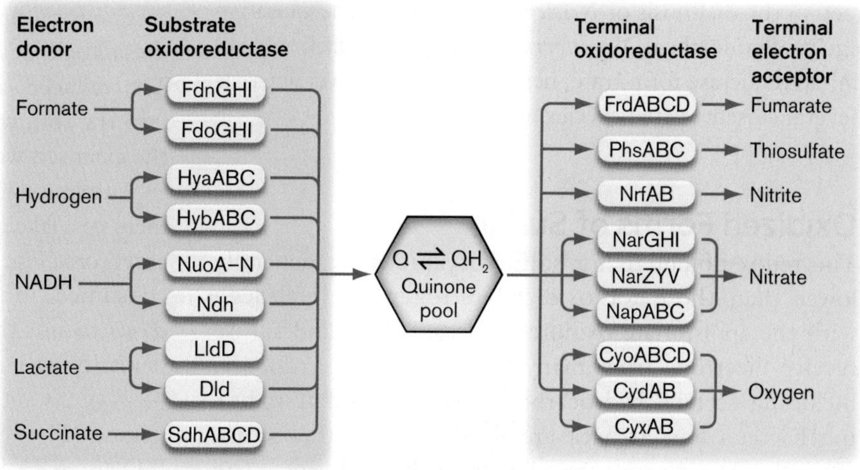

FIGURE 14.18 ■ Alternative electron donors and electron acceptors. *Escherichia coli* can oxidize various foods (electron donors) while reducing various electron acceptors. Each electron donor may use alternative substrate oxidoreductases, depending on the environmental conditions.

The nitrogen series offers an abundant source of strong electron acceptors. Reduction of oxidized states of nitrogen for energy yield is called **dissimilatory denitrification**. Dissimilatory nitrate reduction to ammonium (DNRA) contributes to respiration, whereas assimilatory reduction of nitrate generates ammonium ion for fixation into biomass (discussed in Chapters 15 and 22). Some pathogens are dissimilatory denitrifiers, such as *Neisseria meningitidis* (a cause of meningitis) and *Brucella* species that cause brucellosis in cattle, sheep, and dogs. In the dissimilatory nitrogen redox series, a given oxidation state can serve as an acceptor in one redox couple but as a donor in the next. The redox couples are summarized here:

$$\underset{\text{Nitrate}}{NO_3^-} \xrightarrow{2e^-} \underset{\text{Nitrite}}{NO_2^-} \xrightarrow{e^-} \underset{\text{Nitric oxide}}{NO} \xrightarrow{e^-} \underset{\text{Nitrous oxide}}{\tfrac{1}{2}N_2O} \xrightarrow{e^-} \underset{\text{Nitrogen gas}}{\tfrac{1}{2}N_2}$$

Each nitrogen state requires a specialized reductase, such as nitrate reductase or nitrite reductase, to receive electrons from the ETS. During the reactions, oxygen atoms are removed and combined with protons to form water. The full series of nitrate reduction to N_2 plays a crucial role in producing the nitrogen gas of Earth's atmosphere (discussed in Chapter 22).

An alternative option for many soil bacteria, such as *Bacillus* species, is to reduce nitrite to ammonium ion (NH_4^+):

$$NO_2^- + 8H^+ + 6e^- \rightarrow NH_4^+ + 2H_2O$$

Dissimilatory nitrate and nitrite reduction to ammonium ion can increase the soil pH. The high pH will precipitate metals such as iron.

The presence of a particular reductase can be used as a diagnostic indicator for clinical isolates of bacteria. For example, the chemical test for nitrate reduction is a key step in the diagnosis of *Neisseria gonorrhoeae*, the causative agent of gonorrhea. *N. gonorrhoeae* happens to lack the terminal reductase for nitrate; hence, it tests negative, whereas several closely related species test positive.

Oxidized Forms of Sulfur

The redox potentials for sulfur oxyanions are generally lower than those for oxidized nitrogen. Nevertheless, with the appropriate oxidoreductases, sulfate and sulfite receive electrons from many kinds of electron donors, including acetate, hydrocarbons, and H_2. Sulfate-reducing bacteria and archaea are widespread in the ocean, from the Arctic waters to submarine thermal vents. The ubiquity of sulfate reduction is related to the high sulfate content of seawater, in which SO_4^{2-} is the most common anion after chloride.

The major oxidized forms of sulfur that serve as bacterial electron acceptors include:

$$\underset{\text{Sulfate}}{SO_4^{2-}} \xrightarrow{2e^-} \underset{\text{Sulfite}}{SO_3^{2-}} \xrightarrow{2e^-} \underset{\text{Thiosulfate}}{\tfrac{1}{2}S_2O_3^{2-}} \xrightarrow{2e^-} \underset{\text{Elemental sulfur}}{S^0} \xrightarrow{2e^-} \underset{\text{Hydrogen sulfide}}{H_2S}$$

An important function of sulfate-reducing bacteria, such as the Desulfuromonadales group, is to help metabolize methane hydrates. The methane is oxidized by anaerobic methane-oxidizing archaea (ANME), which require sulfate-reducing bacteria (SRB) as symbiotic partners. This microbial partnership is vital for Earth's global climate because it dissipates large amounts of methane before the greenhouse gas reaches the atmosphere. The ANME-SRB microbial partnership is discussed further in Chapter 19.

Dissimilatory Metal Reduction

An important class of anaerobic respiration involves the reduction of metal cations, or **dissimilatory metal reduction**. The term "dissimilatory" indicates that the metal reduced as a terminal electron acceptor is excluded from the cell. This is in contrast to minerals reduced for the purpose of incorporation into cell components (assimilatory metal reduction).

The metals most commonly reduced through anaerobic respiration are iron ($Fe^{3+} \rightarrow Fe^{2+}$) and manganese ($Mn^{4+} \rightarrow Mn^{2+}$), but virtually any metal with multiple redox states can be reduced or oxidized by some bacteria. For example, *Geobacter* can reduce uranium ($U^{6+} \rightarrow U^{4+}$) to respire on acetate. Reduced uranium is insoluble in water, precipitating as uranite (UO_2). Thus, *Geobacter* respiration with uranium can remediate uranium-contaminated water. The U.S. Department of Energy uses uranium-reducing bacteria for remediation at Rifle, Colorado, where uranium contamination threatens the Colorado River. In this process, acetate is pumped into the water table, where *G. metallireducens* oxidizes the acetate to CO_2 by reducing U^{6+} to U^{4+}. The reduced uranium then precipitates out of the water into the soil, where it can be collected and removed, while the cleansed water flows through.

Anoxic or low-oxygen environments, such as the sediment of a lake or wetland, offer a series of different electron acceptors (**Fig. 14.19**). The stronger electron acceptors are consumed, in turn, by species that have the terminal oxidases to use them. For example, as oxygen grows scarce, denitrifying bacteria reduce nitrate to nitrogen gas ($NO_3^- \rightarrow N_2$). As nitrate is used up, other species reduce manganese ($Mn^{4+} \rightarrow Mn^{2+}$), iron ($Fe^{3+} \rightarrow Fe^{2+}$), and sulfate ($SO_4^{2-} \rightarrow H_2S$). At the bottom of the lake or sediment, methanogenic archaea reduce carbon dioxide to methane (see Section 14.5). Microbial reduction of minerals plays a critical role in every ecosystem and participates in the

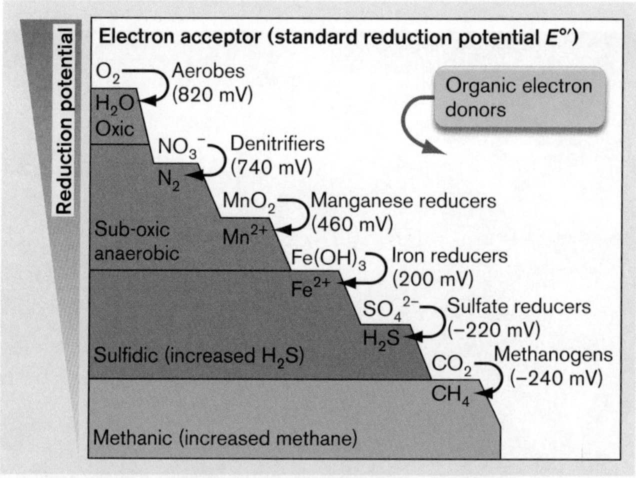

FIGURE 14.19 ▪ Anaerobic respiration in a lake water column. As each successive terminal electron acceptor is used up, its reduced form appears. The next-best electron acceptor is then used, generally by a different species of microbe.

geochemical cycling of elements throughout Earth's biosphere (discussed in Chapter 22).

Thought Question

14.10 The proposed scheme for uranium removal requires injection of acetate under highly anoxic conditions, with less than 1 part per million (ppm) dissolved oxygen. Why must the acetate be anoxic?

To Summarize

- **Anaerobic terminal electron acceptors**, such as nitrogen and sulfur oxyanions, oxidized metal cations, and oxidized organic substrates, accept electrons from a specific reductase complex of an ETS.
- **Nitrate** is successively reduced by bacteria to nitrite, nitric oxide, nitrous oxide, and ultimately nitrogen gas. Alternatively, nitrate and nitrite may be reduced to ammonium ion, a product that alkalinizes the environment.
- **Sulfate** is successively reduced by bacteria to sulfite, thiosulfate, elemental sulfur, and hydrogen sulfide. Sulfate reducers are especially prevalent in seawater.
- **Oxidized metal ions** such as Fe^{3+} and Mn^{4+} are reduced by bacteria in soil and aquatic habitats—a form of anaerobic respiration also known as dissimilatory metal reduction.
- **The geochemistry of natural environments** is shaped largely by anaerobic bacteria and archaea.

14.4 Nanowires, Electron Shuttles, and Fuel Cells

All electron transfer systems require a terminal electron acceptor to complete metabolism and yield energy. Soluble electron acceptors such as O_2 or NO_3^- can reach the cell by diffusion, but oxidized metals such as Fe^{3+} are often barely soluble in water. How does a bacterial membrane oxidoreductase interact with an insoluble metal particle outside the cell? It turns out that bacteria have evolved myriad ways to reach insoluble or distant electron acceptors. If bacteria can transfer electrons to metals, why not to a human-made electrode? Metal-reducing bacteria form **electrogenic biofilms**—biofilms that generate electricity in a human-made device.

Nanowires: Electrically Conductive Pili

Over the past decade, several laboratories have pursued the question of how bacteria reach outside their cytoplasm to deposit electrons. Ken Nealson and Moh El-Naggar, at the University of Southern California, have focused on the marine bacterium *Shewanella oneidensis*, which reduces iron, chromium, and uranium. Derek Lovley at the University of Massachusetts, Amherst, and Gemma Reguera at Michigan State University have investigated the cytochromes and conductive pili of *Geobacter* species that reduce numerous metals. Biophysical experiments reveal the molecular basis of electron transfer beyond the individual cell.

Lovley and colleagues observed how metal-reducing bacteria such as *G. sulfurreducens* grow adherent to particles of iron oxide or other metals. They developed conditions to culture *Geobacter* in a fuel cell, where the bacteria formed a biofilm over the anode (the electrode that receives electrons). The biofilm grew only when the anode was connected to a cathode in the fuel cell, allowing completion of an electrical circuit. When provided with an electron donor substrate, acetate, the biofilm generated electric power.

Lovley noticed that *Geobacter* formed long protein filaments called pili (**Fig. 14.20**). A pilus consists of a tube of helically stacked protein subunits (presented in Chapter 3). Lovley proposed that the pili of *Geobacter* conduct electrons, and that they connect the bacteria in a biofilm, allowing the relay of electric current across the biofilm to the anode (**Fig. 14.20B**). The pili thus serve as **nanowires**, completing an electrical circuit through the biofilm.

Note: The word "filament" may refer to a cellular filament (chain of cells), such as that of cable bacteria, or a protein filament (tubular complex of subunits) such as a pilus. The meaning must be inferred from context.

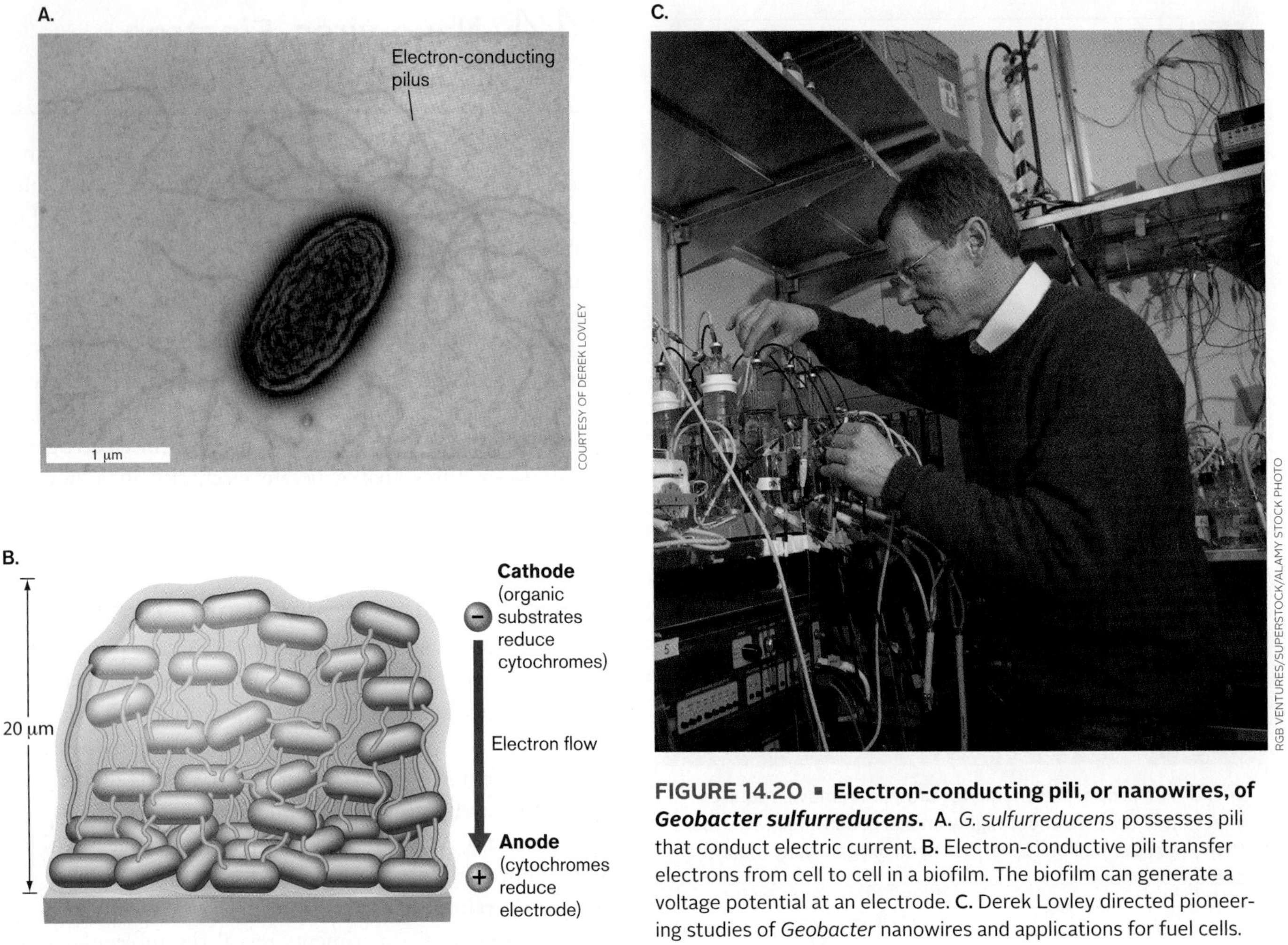

FIGURE 14.20 ■ Electron-conducting pili, or nanowires, of *Geobacter sulfurreducens*. **A.** *G. sulfurreducens* possesses pili that conduct electric current. **B.** Electron-conductive pili transfer electrons from cell to cell in a biofilm. The biofilm can generate a voltage potential at an electrode. **C.** Derek Lovley directed pioneering studies of *Geobacter* nanowires and applications for fuel cells.
Source: Part B adapted from G. Reguera. 2018. *FEMS Microbiol. Ecol.* **94**(7), fig. 5.

The model of pili as conducting current was highly controversial, in part because protein generally has the property of an insulator, resisting current, unlike metals, which conduct current. Nevertheless, Lovley's students and colleagues obtained strong evidence that pili act as nanowires:

- *G. sulfurreducens* that is provided with acetate (organic electron donor) generates electric current in a fuel cell. A strain genetically altered for increased number of pili generates greater electric current.
- With the electrical circuit open (no current possible), the bacteria grow a biofilm with few pili and low conductivity (the measured ability to conduct electric current).
- Pili isolated from *G. sulfurreducens* show conductivity measurements comparable to those of metal wires. Pili conductivity shows a temperature dependence (conductivity increases at lower temperature) comparable to that of metal wires.
- The conductivity of isolated *G. sulfurreducens* pili increases at low pH, similar to the pH dependence of organic metal compounds.

While the evidence for pili nanowires is strong, it raises the question: How do insulator-like proteins conduct electric current? Gemma Reguera addressed this question by investigating the amino acid residue content of *Geobacter* conductive pili (**Fig. 14.21**), which includes a high proportion of the aromatic amino acids tyrosine and phenylalanine. Aromatic rings contain conjugated pi bonds (bonds between p orbitals) that can readily transfer electrons via stacking of molecules whose orbitals overlap. Reguera showed that the tyrosines and phenylalanines are positioned at distances ideal for "electron hopping" from one aromatic ring to the next. Furthermore, a mutant was constructed in which one tyrosine was replaced by alanine, an amino acid with only a methyl R group. The loss of the one tyrosine broke the electrical connection. This mutant showed pili with normal molecular structure, but substantially decreased electrical conductivity.

Extracellular Cytochromes, Shuttles, and Cell Extensions

Further investigations reveal a virtual hardware store of diverse extracellular electrical devices (**Fig. 14.22A**). These devices are found on *Geobacter*, on the Gram-negative

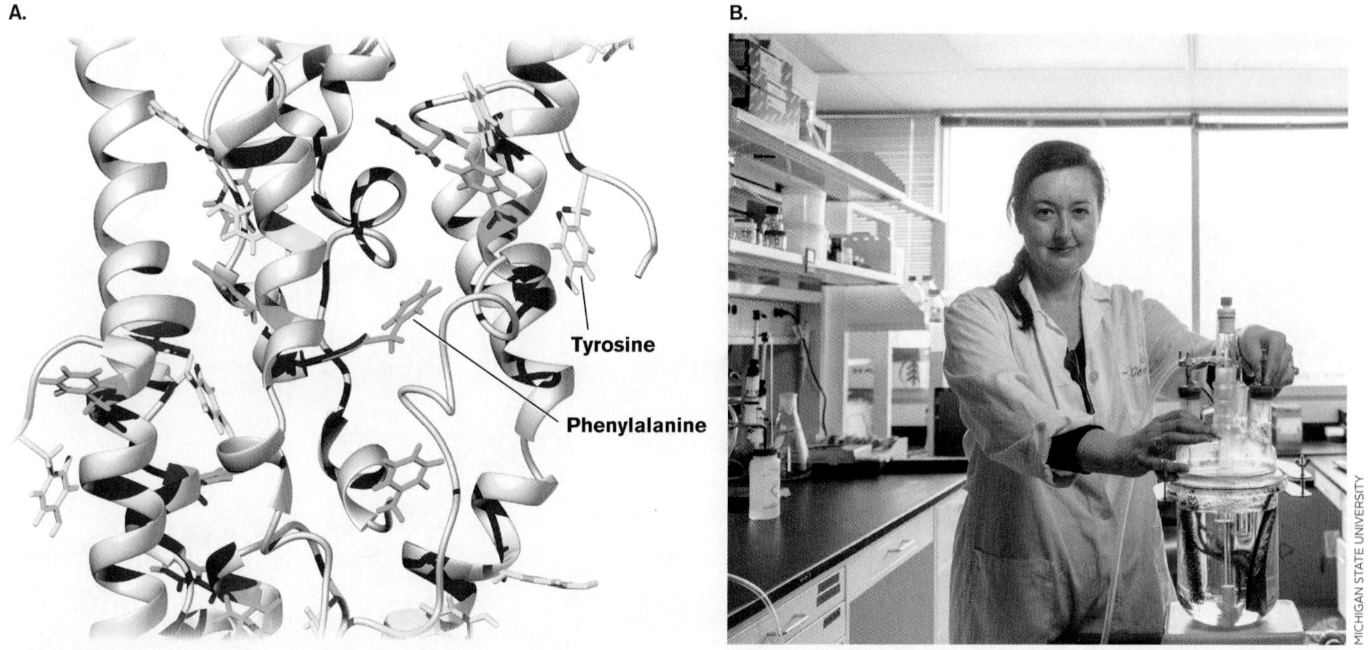

FIGURE 14.21 ■ Conductive pili transfer electrons via aromatic amino acid residues. **A.** Electron-conducting pilus protein includes aromatic amino acid residues (yellow = tyrosine; green = phenylalanine). The aromatic residues are spaced at distances that allow overlapping pi bond orbitals to transfer electrons. **B.** Gemma Reguera demonstrated electron conductivity in *Geobacter* pili. *Source:* Part A modified from G. Reguera. 2018. *FEMS Microbiol. Ecol.* **94**(7), fig. 4C.

pathogen *Pseudomonas*, and even on Gram-positive bacteria such as *Thermincola*, whose thick cell walls were once thought to preclude electron transfer. Electron-transferring cytochromes, once thought to reside only in the plasma membrane (or inner membrane) have now been found in bacterial outer membranes, as well as in the periplasm. The **outer membrane–associated proteins** (named Omc) contact extracellular oxidized minerals such as ferric oxide. Certain Omc proteins participate in porin-cytochrome complexes that connect via periplasmic cytochromes with the classic NADH-oxidizing ETS of the inner membrane (**Fig. 14.22A**). Other Omc proteins decorate the tips of conductive pili. The Omc contacts the electron acceptor, thus completing a circuit via the conductive pili all the way back to the inner membrane ETS.

The conductive pili are positioned by an extension-retraction apparatus embedded in the membrane. The apparatus is known as a type IV secretion system, similar to those found in pathogens for virulence attachment (discussed in Chapter 25). In the presence of electron acceptors, *Geobacter* bacteria build the pilus complex and extend it through the secretion apparatus. When electron acceptors are absent, pili are retracted.

The species *G. sulfurreducens* was shown to use extracellular **electron shuttles**. Electron shuttles are aromatic molecules similar to quinones that allow low-energy transitions for gaining or losing an electron. Unlike quinones, however, which remain embedded in the cell membrane, an electron shuttle dissolves in the aqueous medium and diffuses away from the cell. Thus, bacteria such as *G. sulfurreducens* can transfer electrons from their ETS to an Omc on the outer membrane, which then reduces the shuttle and allows it to diffuse away. The shuttle ultimately transfers its electron to a metal or some other electron acceptor at a distance. Do electron shuttles return to the bacteria that make them? The answer is unclear, but often electron shuttles are used in a biofilm where many cells share nutrients and a redox gradient.

Pathogens such as *Pseudomonas aeruginosa* use electron shuttles to transfer electrons to distant oxygen. Dianne Newman at the California Institute of Technology and Lars Dietrich at Columbia University have shown how electron shuttles enable *P. aeruginosa* to conduct aerobic respiration within biofilms buried by thick mucus. The shuttles used are polycyclic aromatics called phenazines. Phenazines enable bacteria under anaerobic conditions to pick up electrons from a cell-surface Omc and let them diffuse away to reach distant oxygen. Thus, phenazine electron shuttling favors *Pseudomonas* infections of cystic fibrosis patients.

In the soil, bacteria may shuttle electrons to extracellular metals using quinone-like degradation products of lignin, a complex aromatic substance that forms the bulk of wood and woody stems. Lignin degradation products are also called "humics" because of their presence in humus, the organic components of soil (discussed in Chapter 21). Humics accumulate in anaerobic environments, where decomposition is slow.

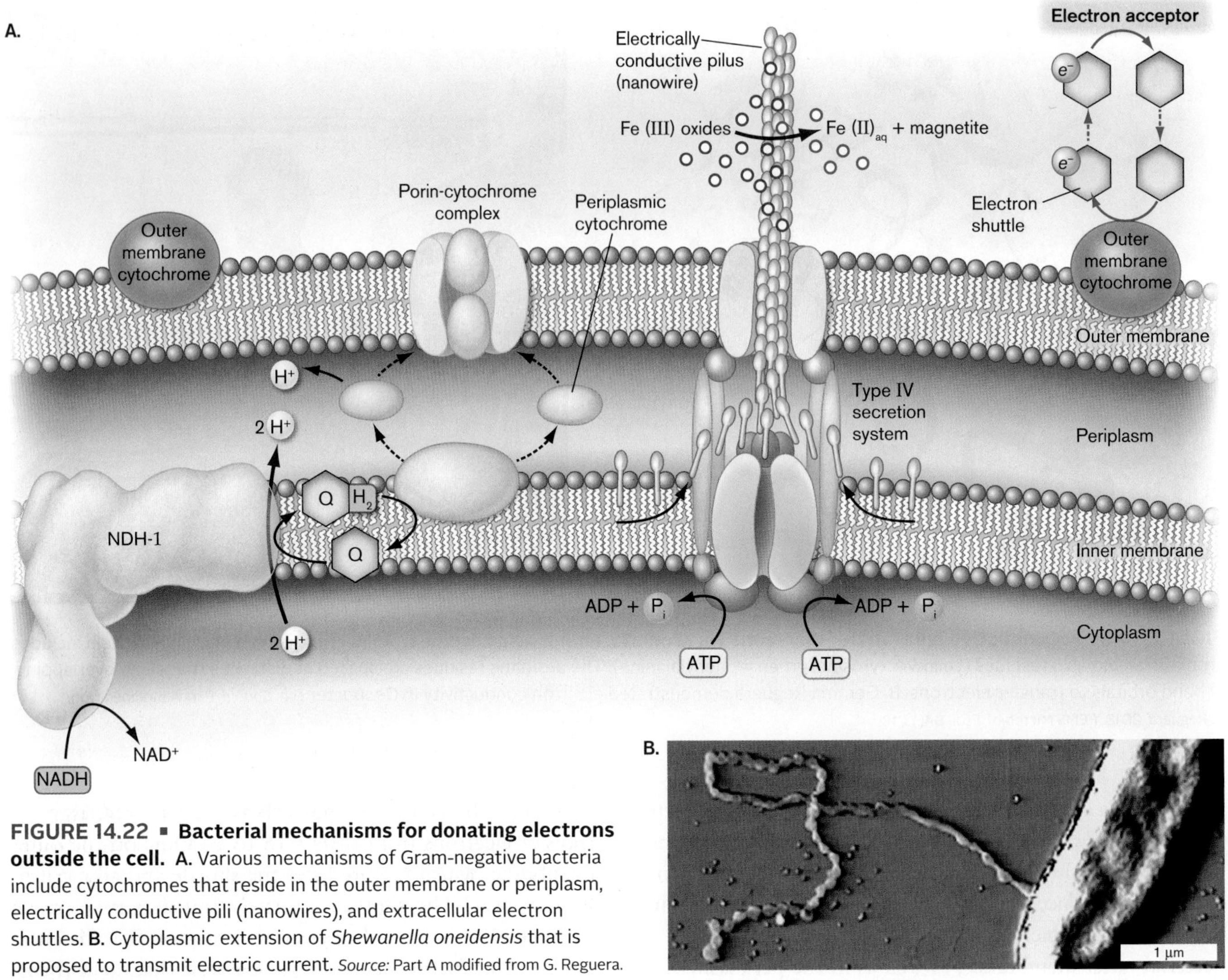

S. PIRBADIAN ET AL. 2014. *PNAS* **111**:12883–12888, FIG. 5B.

FIGURE 14.22 ■ Bacterial mechanisms for donating electrons outside the cell. **A.** Various mechanisms of Gram-negative bacteria include cytochromes that reside in the outer membrane or periplasm, electrically conductive pili (nanowires), and extracellular electron shuttles. **B.** Cytoplasmic extension of *Shewanella oneidensis* that is proposed to transmit electric current. *Source:* Part A modified from G. Reguera. 2018. *PNAS* **115**:5632–5634, fig. 1.

Yet another possible means of electron transfer is that of cytoplasmic extensions, studied by El-Naggar and Nealson in *Shewanella* (**Fig. 14.22B**). These cytoplasmic extensions are composed of chains of membrane vesicles whose properties are still under investigation. They may be the extracellular electrical conductors used by *Shewanella* to reduce iron and cobalt in clay sediments—and electrodes in a fuel cell.

Thought Question

14.11 What conditions might favor electron transfer by pili, versus by electron shuttles?

Fuel Cells: Bacterial Electric Power

How can we harness bacterial electric current to run human devices? In an electrolytic fuel cell, the bacteria form a biofilm on the anode (electron-attracting electrode; **Fig. 14.23**). As the bacteria oxidize organic substrates such as acetate, they transfer electrons to an anode. The result is charge separation between electrons and hydrogen ions. The electrons then pass through a circuit to the cathode, generating current that can run a device.

The "fuel" for the cell can be a mixture of organic substances derived from any kind of food waste or sewage. Organic waste includes small organic molecules such as lactate, acetate, and even formaldehyde. The biofilm bacteria on the anode can remove hydrogens from these organic molecules, separating the electrons and hydrogen ions (**Fig. 14.23B**). The hydrogen ions migrate through a polymer membrane, whereas the electrons enter the anode leading to an electrical wire. The remaining carbon and oxygen atoms of the fuel are released as CO_2. The process is similar to natural respiration, except that instead of molecular oxygen, the electron acceptor is an electrode made of graphite. To complete the circuit, the electrons from the wire current ultimately react with oxygen and hydrogen ions to form water, as in aerobic respiration.

So far, microbial fuel cells have been able to generate milliamps of current, enough to drive small devices, such as clocks and marine data sensors. **Special Topic 14.1** describes the use of a benthic fuel cell to power instruments analyzing deep ocean water over long periods.

Thought Question

14.12 An alternative mechanism for a fuel cell involves the use of a bacterium that receives electrons from an electrode, instead of donating electrons. How might this work?

To Summarize

- **Metal electron acceptors require extracellular devices to donate electrons.**
- **Electrically conductive pili transmit electrons** via pi bond orbital interactions between closely spaced aromatic amino acid residues in the pilus protein.
- **Outer membrane cytochromes and periplasmic cytochromes** extend the reach of the Gram-negative inner membrane–embedded ETS.
- **Electron shuttles reversibly transfer electrons** to electron acceptors at a distance from the cell.
- **Extracellular electron transmission devices enable formation of electrogenic biofilms.**
- **Electrogenic biofilms can power a fuel cell.**

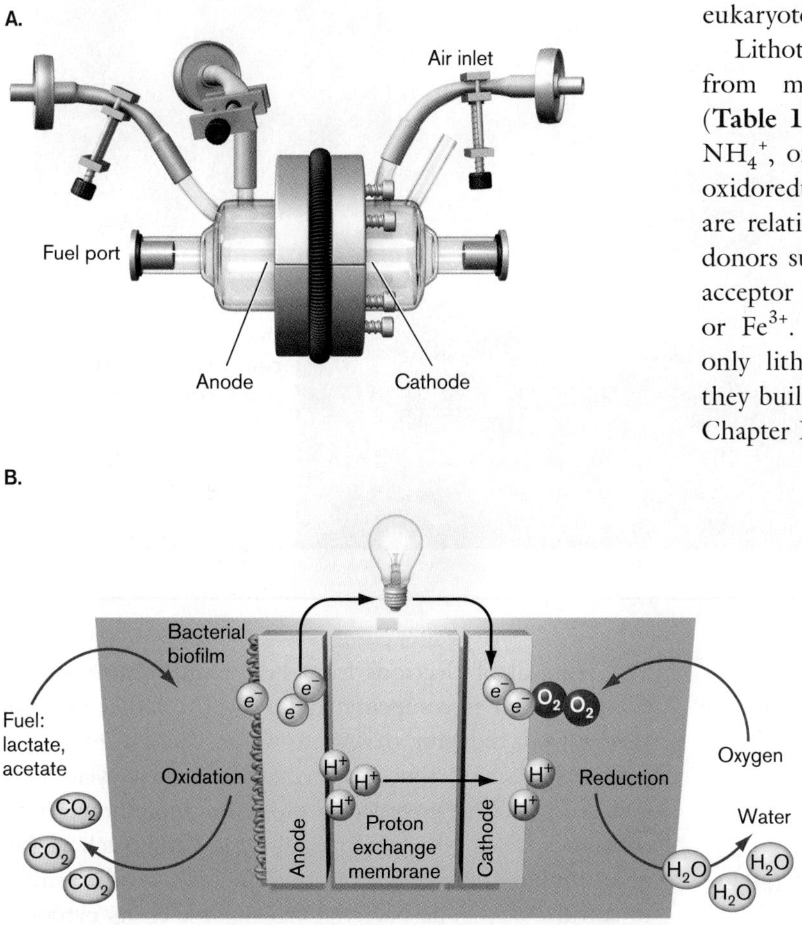

FIGURE 14.23 ■ A microbial fuel cell. A. A bacterial fuel cell. B. Reaction cycle of a bacterial fuel cell. *Source:* Part B modified from MURI Microbial Fuel Cell Project, University of Southern California (http://mfc-muri.usc.edu).

14.5 Lithotrophy and Methanogenesis

Many reduced minerals and single-carbon compounds can serve as electron donors for an ETS, in the energy-yielding form of metabolism known as **lithotrophy** (or **chemolithotrophy**). Some organotrophs have alternative oxidoreductases that conduct lithotrophy by oxidizing H_2 or Fe^{2+}; for example, the gastric pathogen *Helicobacter pylori* oxidizes hydrogen gas released by mixed-acid-fermenting bacteria in the colon. The cable bacteria, such as *Desulfobulbus*, oxidize sulfide or organic substrates. Other species are "obligate lithotrophs"; that is, they oxidize only inorganic molecules. All lithotrophs are bacteria or archaea. Thus, bacteria and archaea fill many key niches in ecosystems that eukaryotes cannot.

Lithotrophy includes many kinds of electron donors, from metals and anions to single-carbon groups (**Table 14.2**). Each type of electron donor, such as H_2, NH_4^+, or Fe^{2+}, requires a specialized electron-accepting oxidoreductase. Most inorganic substrates other than H_2 are relatively poor electron donors compared to organic donors such as glucose. Therefore, the terminal electron acceptor is usually a strong oxidant, such as O_2, NO_3^-, or Fe^{3+}. Obligate lithotrophs (organisms that conduct only lithotrophy) consume no organic carbon source; they build biomass by fixing CO_2 (a process discussed in Chapter 15).

Iron Oxidation

Reduced metal ions such as Fe^{2+} and Mn^{2+} provide energy through oxidation by O_2 or NO_3^-. Bacteria perform these lithotrophic reactions in soil where weathering exposes reduced minerals. They generate metal ions with higher oxidation states (such as Fe^{3+} or Mn^{4+}), which other bacteria use for anaerobic respiration. Environments such as ponds and wetlands that experience frequent shifts between oxygen availability and oxygen depletion are likely to host a variety of metal-oxidizing lithotrophs, as well as metal-reducing anaerobic heterotrophs. The roles of lithotrophy in ecology are discussed further in Chapters 21 and 22.

SPECIAL TOPIC 14.1 The Ocean Floor Is a Battery

The oceans are the heart of global ecosystems and climate, and their measurements are vital, including temperature, oxygenation, salt content, and many others. But the vast extent and depth of ocean water makes the cost of measurement prohibitive. Less than 15% of the Earth's oceans have sensory monitors reporting data. These monitors require power sources capable of running for months or years without refueling or maintenance.

An available source of power is the ocean itself. The sulfate from marine water enters the sediment, where anaerobic bacteria reduce it to sulfides. The sulfides accumulate, forming an anaerobic layer several centimeters below the oxygenated surface. Between the sulfides and the oxygenated surface water, the redox potential stores energy. The microbial community includes anaerobic respirers, as well as cable bacteria that oxidize the sulfides. In effect, the entire benthos (ocean floor) around the globe is an electrical battery, with the surface as anode and the subsurface as cathode.

Clare Reimers, an oceanographer at Oregon State University, developed a device to tap the ocean's energy to run instruments collecting benthic sensor data (**Fig. 1**). The device is a benthic microbial fuel-cell platform (**Fig. 2A**). The platform is deployed by being dropped on a line from a ship. Weights ensure that the device will sink into the sediment and tap the anaerobic layer.

COURTESY OF JAMES SWIFT

CLARE REIMERS

FIGURE 1 ▪ A benthic microbial fuel cell. Clare Reimers (inset) deployed the fuel cell at the ocean floor to draw current for instruments recording oxygen, temperature, and other properties of seawater.

An example of lithotrophy is iron oxidation by the bacterium *Acidithiobacillus ferrooxidans*. These bacteria oxidize Fe^{2+} (ferrous ion) using the reduction potential between Fe^{3+}/Fe^{2+} and O_2/H_2O (**Fig. 14.24**):

	$E°'$ (pH 2)	$G°'$ (pH 2)
$2Fe^{2+} \rightarrow 2Fe^{3+} + 2e^-$ (pH 2)	−(+770) mV	+149 kJ/mol
$\frac{1}{2}O_2 + 2H^+ + 2e^- \rightarrow H_2O$ (pH 2)	+1,100 mV	−212 kJ/mol
$2Fe^{2+} + \frac{1}{2}O_2 + 2H^+ \rightarrow 2Fe^{3+} + H_2O$	+330 mV	−63 kJ/mol

The removal of electrons from Fe^{2+} requires an input of energy, which is compensated for by the larger yield of energy from reducing oxygen to water. The net reduction potential is small; thus, *A. ferrooxidans* must cycle large quantities of iron in order to grow. Because the reaction consumes H^+, it goes forward only at low pH, such as that of acid mine drainage where other microbes oxidize sulfur to sulfuric acid. The bacterial cell must keep its cytoplasmic pH considerably higher (pH 6.5) by transmembrane exchange of H^+ with cations, and by inversion of the electrical potential $\Delta\psi$ (positive inside). Thus, the proton

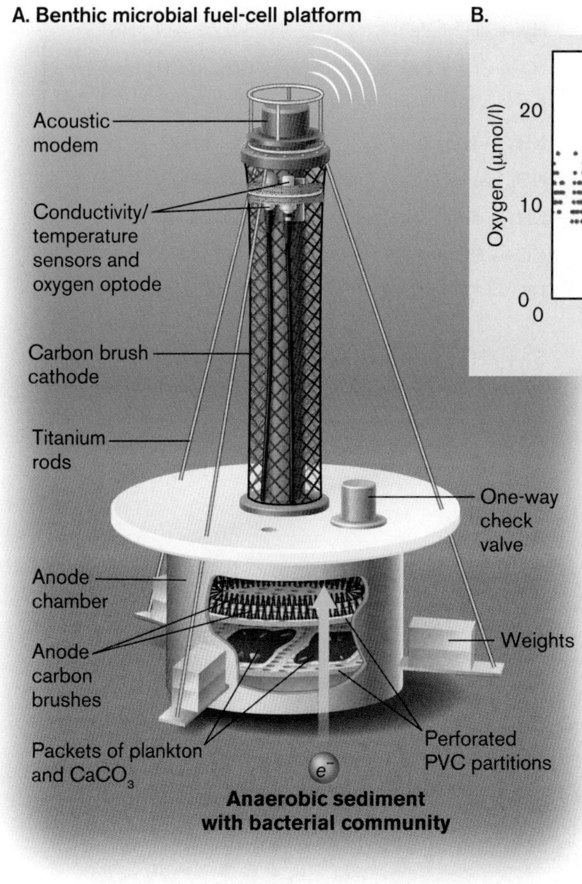

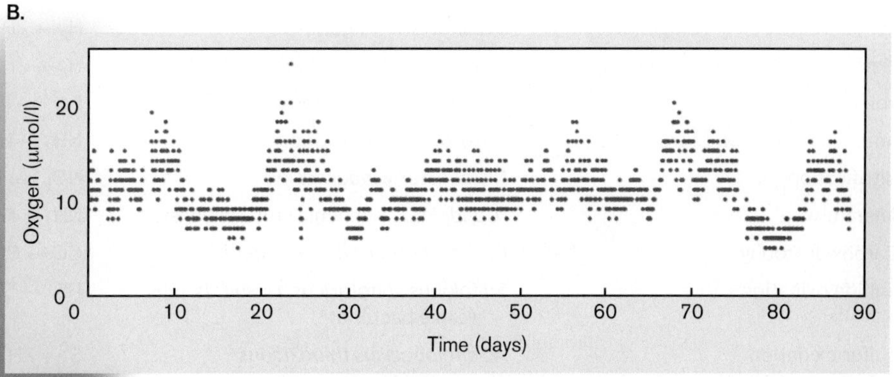

FIGURE 2 ■ Diagram of a benthic microbial fuel cell. **A.** The fuel cell contains a carbon anode to collect electrons from bacteria oxidizing electron donors supplemented by plankton flakes. **B.** Seawater oxygen concentrations recorded by a meter powered by the benthic fuel cell. *Source:* Modified from Clare Reimers and Michael Wolf. 2018. *Oceanography* **31**:98–103, figs. 1b (part A) and 3 (part B).

The device contains an anode consisting of carbon brushes where the anaerobic bacteria will form a biofilm. To promote biofilm formation, some extra organic electron donors are provided in the form of plankton flakes (fish food). A small power management device harvests energy and maintains storage in lithium ion batteries. The harvested electric current drives instruments such as sensors for oxygen and temperature. An example of the data obtained for oxygen levels is shown in **Figure 2B**. The trace shows daily records of oxygen concentration in benthic water over a 60-day period. The collected data then are transmitted to the research vessel via an acoustic modem, a device that converts digital signals into low-energy sound waves that can travel long distances through water. Reimers and her colleagues expect such fuel cells to operate continuously for several years, helping to build our understanding of the world's oceans.

RESEARCH QUESTION

How could you determine what kinds of bacteria in the sediment generate the electric current for the fuel cell?

Reimers, C. E., and M. Wolf. 2018. Power from benthic microbial fuel cells drives autonomous sensors and acoustic modems. *Oceanography* **31**:98–103.

potential Δp exists entirely in the form of ΔpH. The ΔpH drives ATP synthesis.

Unlike organic electron donors, metals such as iron must be accessed from insoluble particles outside the cell. In *A. ferrooxidans*, the electrons are collected from iron outside by an outer membrane cytochrome c_2. The cytochrome is associated with a periplasmic protein called rusticyanin. Rusticyanin collects electrons while excluding the potentially toxic metal from the cell. The electrons from rusticyanin can be transferred to an inner membrane cytochrome complex that reduces oxygen to water.

For iron oxidation (Fe^{3+}/Fe^{2+}), the reduction potential is too high to reduce $NADP^+$ to NADPH; that is, Fe^{2+} is a weaker electron donor than NADPH. So, how do cells obtain enough energy to form NADPH for biosynthesis? An alternative pathway directs electrons to an enzyme complex that uses proton influx (spends Δp) to convert $NADP^+$ to NADPH. This pathway is called **reverse electron flow** because it reverses the flux of electrons seen in an ETS that spends NADH (or NADPH) to form NAD^+ (or $NADP^+$). In reverse electron flow, a relatively poor electron donor (such as Fe^{2+}) reduces an ETS with an

TABLE 14.2 Lithotrophy: Electron Donors and Acceptors

Type of lithotrophy	Species example	Electron donor	Electron acceptor
Hydrogenotrophy	*Aquifex aeolicus*	$H_2 \rightarrow 2H^+ + 2e^-$	$O_2 \rightarrow H_2O$
Sulfate reduction (hydrogenotrophy)	*Desulfovibrio vulgaris*	$H_2 \rightarrow 2H^+ + 2e^-$	$SO_4^{2-} \rightarrow S^0, HS^-$
Methanogenesis (hydrogenotrophy)	*Methanocaldococcus jannaschii*	$H_2 \rightarrow 2H^+ + 2e^-$	$CO_2 \rightarrow CH_4$
Iron oxidation	*Acidithiobacillus ferrooxidans*	$Fe^{2+} \rightarrow Fe^{3+} + e^-$	$O_2 \rightarrow H_2O$
Ammonia oxidation (nitrosification)	*Nitrosomonas europaea*	$NH_3 \rightarrow NO_2^-$	$O_2 \rightarrow H_2O$
Nitrification	*Nitrobacter winogradskyi*	$NO_2^- \rightarrow NO_3^-$	$O_2 \rightarrow H_2O$
Anammox	"*Candidatus* Kuenenia stuttgartiensis"	$NH_3 + 4H^+ + 4e^- \rightarrow N_2$	$NO_2 \rightarrow N_2$
Carboxidotrophy	*Carboxydothermus hydrogenoformans*	$CO \rightarrow CO_2 + e^-$	$H_2O \rightarrow H_2$
Sulfide oxidation	*Sulfolobus solfataricus*, *Desulfobulbus* (cable bacteria)	$HS^- \rightarrow S^0 + H^+ + 2e^-$	$O_2 \rightarrow H_2O$
Sulfur oxidation	*Acidithiobacillus thiooxidans*	$S^0 + 2H^+ + 4e^- \rightarrow H_2SO_4$	$O_2 \rightarrow H_2O$

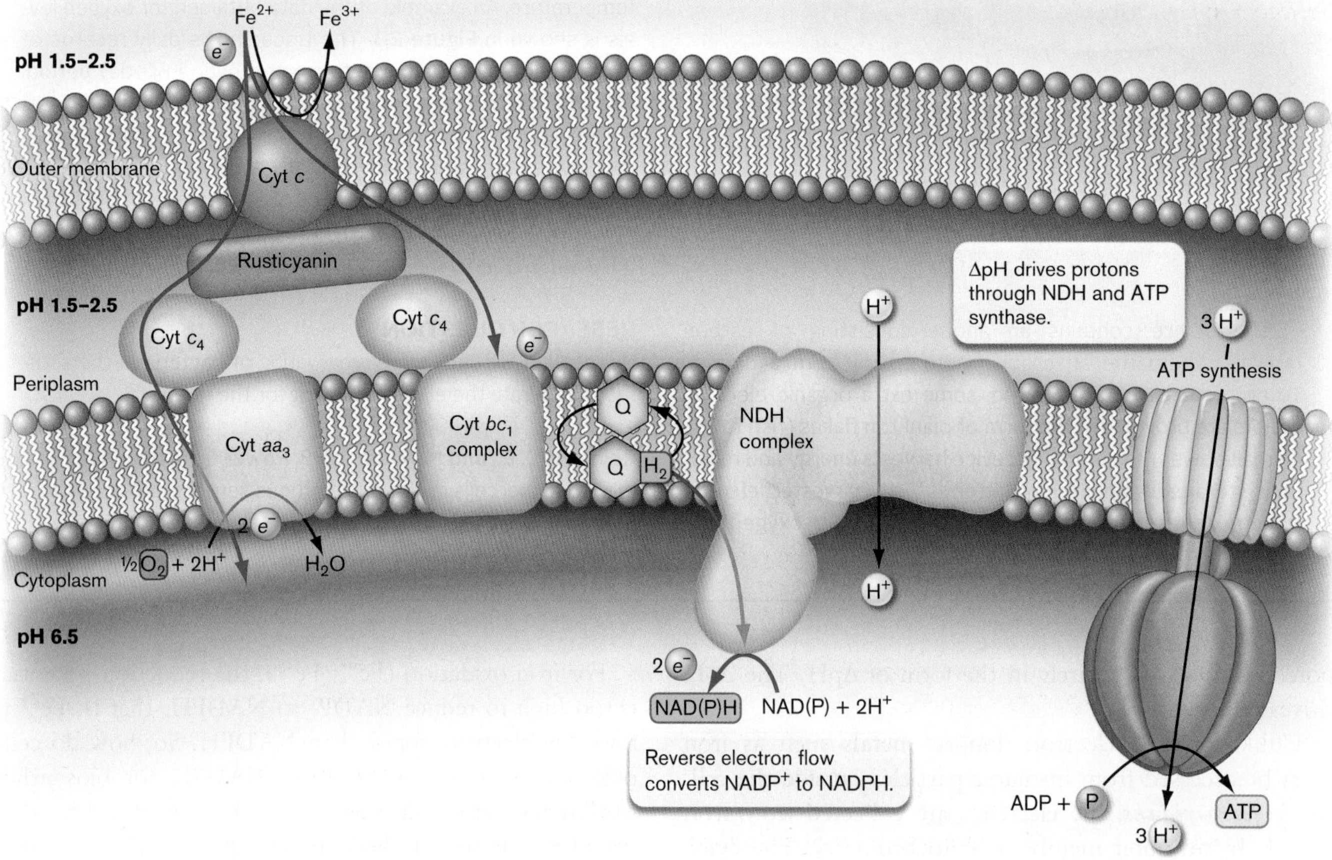

FIGURE 14.24 ■ Iron oxidation ETS. Fe^{2+} is oxidized to Fe^{3+} outside the cell, at about pH 2. The periplasm also is at about pH 2. Electrons from exogenous iron are collected by rusticyanin, which feeds electrons to the cytochrome aa_3 complex to reduce O_2 to H_2O. Alternatively, some electrons from rusticyanin reduce the cytochrome bc_1 complex. ΔpH drives reverse electron flow through NDH to form NADPH. *Source:* Modified from Raquel Quatrini et al. 2009. *BMC Genomics* **10**:394.

unfavorable reduction potential, requiring input of energy to generate NADH or NADPH. For iron oxidation, the energy input comes from the large ΔpH across the inner membrane (pH 6.5 inside, and pH 2 outside). Reverse electron flow is seen in some kinds of lithotrophy, in syntrophy (discussed in Chapter 13), and in phototrophy (discussed in Section 14.6).

Thought Question

14.13 Use **Figures 14.14** and **14.15** to propose a pathway for reverse electron flow in an organism that spends ATP from fermentation to form NADH. Draw a diagram of the pathway.

Nitrogen Oxidation

One kind of lithotrophy essential for the environment is oxidation of nitrogen compounds:

$$NH_4^+ \xrightarrow{\frac{1}{2}O_2} NH_2OH \xrightarrow{O_2} HNO_2 \xrightarrow{\frac{1}{2}O_2} HNO_3$$

Ammonium Hydroxylamine Nitr<u>ous</u> acid (nitrite) Nitr<u>ic</u> acid (nitrate)

Reduced forms of nitrogen, such as ammonium ion derived from fertilizers, support growth of **nitrifiers**, bacteria that generate nitrites or nitrates (forming nitrous acid or nitric acid, respectively). Acid production can degrade environmental quality. In soil treated with artificial fertilizers, nitrifiers decrease the ammonium ions obtained from such fertilizer and produce toxic concentrations of nitrites, which leach into groundwater. Still, nitrifiers can be useful in sewage treatment, where they eliminate ammonia that would harm aquatic life (discussed in Chapter 22).

Ammonia/ammonium and nitrite are relatively poor electron donors compared to organic molecules. Thus, their oxidation through the ETS pumps fewer protons, and the bacteria must cycle relatively large quantities of substrates to grow. As we saw for iron oxidation, reduced cofactors such as NADPH must be obtained through reverse electron flow involving redox carriers with different amounts of energy.

What happens to ammonium ion from detritus that accumulates in anaerobic regions, such as the bottom of a lake? Surprisingly, NH_4^+ can yield energy through oxidation by nitr<u>i</u>te (a product of nitr<u>a</u>te respiration):

$$NH_4^+ + NO_2^- \rightarrow N_2 + 2H_2O \qquad \Delta G^{\circ\prime} = -357 \text{ kJ/mol}$$

Under conditions of high ammonium and extremely low oxygen, nitrite oxidation of ammonium ion supports growth of bacteria. Known as the **anammox reaction**, anaerobic ammonium oxidation plays a major role in wastewater treatment, where it eliminates much of the ammonium ion from sewage breakdown. In the oceans, anammox bacteria cycle as much as half of all the nitrogen gas returned to the atmosphere.

Anammox is conducted by planctomycetes, irregularly shaped bacteria with unusual membranous organelles that fill much of the cell (**Fig. 14.25A**). The central compartment is called the anammoxosome. The anammoxosomal membrane is composed of unusual ladder-shaped lipids called ladderanes. An enzyme of the anammoxosomal membrane reduces nitrite to NO plus H_2O (**Fig. 14.25B**, step 1). Another membrane-embedded enzyme, hydrazine synthase, catalyzes NO reduction by NH_4^+ to form hydrazine (N_2H_4; step 2). Hydrazine is a high-energy compound that engineers use for rocket fuel; the substance is highly toxic, so the planctomycete keeps it sequestered within the specialized anammoxosomal membrane.

The hydrazine is further oxidized to N_2 (**Fig. 14.25B**, step 3), and its protons are released within the anammoxosome compartment. The protons then drive an ATP synthase embedded in the anammoxosomal membrane. The hydrazine oxidation enzyme obtains additional electrons from catabolism of organic substrates. Thus, anammox bacteria are examples of microbes whose metabolism combines lithotrophy and organotrophy.

Note: Distinguish between lithotrophy (<u>oxidation</u> of reduced minerals, usually by O_2, nitrate, or nitrite) and anaerobic respiration (<u>reduction</u> of oxidized minerals, usually by organic food molecules).

Sulfur and Metal Oxidation

Major sources of lithotrophic electron donors are minerals containing reduced sulfur, such as hydrogen sulfide and sulfides of iron and copper. As we saw for nitrogen compounds, each sulfur compound that undergoes partial oxidation may serve as an electron donor and be further oxidized.

$$H_2S \xrightarrow{\frac{1}{2}O_2} S^0 \xrightarrow{\frac{1}{2}O_2} \tfrac{1}{2}S_2O_3^{2-} \xrightarrow{O_2 + H_2O} H_2SO_4$$

Hydrogen sulfide Elemental sulfur Thiosulfate Sulfuric acid

Sulfur oxidation produces the strong acid sulfuric acid (H_2SO_4), which dissociates to produce an extremely high H^+ concentration. In the early twentieth century, no one would have believed that living organisms could grow in concentrated sulfuric acid, much less produce it. The *Star Trek* science fiction episode "The Devil in the Dark" portrayed an imaginary alien creature, the Horta, that produced corrosive acid in order to eat its way through solid rock (**Fig. 14.26A**). In actuality, no creatures beyond the size of a microbe are yet known to grow in sulfuric acid. But archaea such as *Sulfolobus* species oxidize hydrogen sulfide

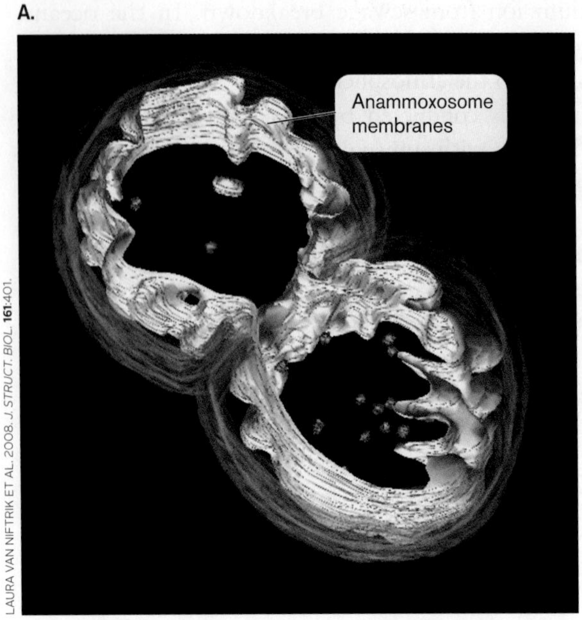

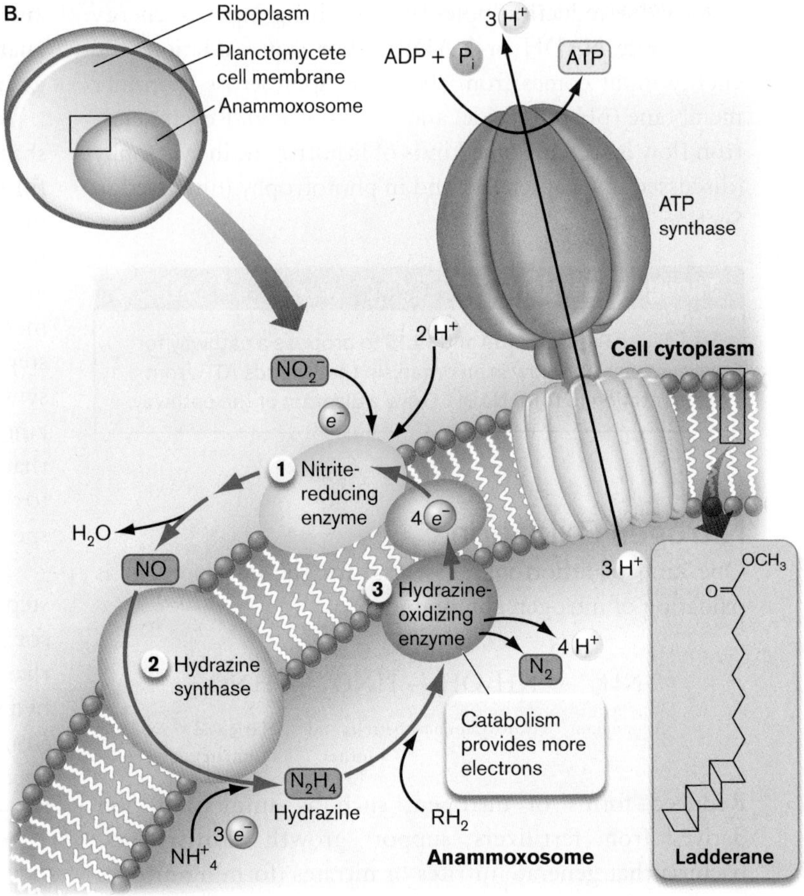

FIGURE 14.25 ■ Anammox ETS within a planctomycete. **A.** A planctomycete with anammoxosomal membranes (cryo-EM tomography). **B.** An enzyme of the anammoxosomal membrane reduces nitrite to NO plus H_2O. NO is reduced by NH_4^+ to form hydrazine (N_2H_4). Hydrazine is a toxic molecule that gets trapped in the anammoxosome by the membrane composed of ladderanes. Hydrazine donates electrons to reduce nitrite, and gives off N_2. Protons may be pumped by the hydrazine synthase complex.

to sulfuric acid and grow at pH 2, often in hot springs at near-boiling temperatures (**Fig. 14.26B** and **C**). *Sulfolobus* makes irregular Horta-shaped cells without even a cell wall to maintain shape; how it protects its cytoplasm from disintegration is not yet understood.

Microbial sulfur oxidation can cause severe environmental acidification, eroding concrete structures and stone monuments. The problem is compounded by sulfur oxidation in the presence of iron, such as in iron mine drainage. For example, the sulfur-oxidizing archaeon *Ferroplasma acidarmanus* was discovered in the abandoned mine Iron Mountain in northern California by Katrina Edwards (1968–2014; University of Southern California) and colleagues from the University of Wisconsin–Madison. *Ferroplasma* oxidizes ferrous disulfide (pyrite) with ferric iron (Fe^{3+}) and water:

$$FeS_2 + 14Fe^{3+} + 8H_2O \rightarrow 15Fe^{2+} + 2SO_4^{\ 2-} + 16H^+$$

This reaction generates large quantities of sulfuric acid. The acidity of the mine water is near pH 0—one of the most acidic environments found on Earth. As *Ferroplasma* grows, it forms biofilms of thick streamers in the mine drainage, which poison freshwater streams.

Anaerobic reactions between sulfur and iron cause hidden hazards for human technology, such as the corrosion of steel in underwater bridge supports. Anaerobic corrosion was long considered a mystery, since iron was known to rust by means of spontaneous oxidation by O_2. In anaerobic conditions, however, sulfur-reducing bacteria can corrode iron (**Fig. 14.27**). In one pathway, the bacteria reduce elemental sulfur (S^0) with H_2 to hydrogen sulfide (H_2S). H_2S then combines with iron metal (Fe^0), which gives up two electrons to form Fe^{2+}, precipitating as iron sulfide (FeS). The displaced $2H^+$ combine with the $2e^-$ from iron, regenerating H_2—now available to reduce sulfur once again. In an alternative mechanism, bacteria use Fe^0 to reduce sulfate directly to FeS. These damaging processes may resemble the iron-based metabolism of Earth's most ancient life forms.

Nevertheless, like the imaginary Horta that ended up helping miners with their excavations, acid-producing microbes are now used to supplement commercial mining—a process called "biomining." Lithotrophs such as *Acidithiobacillus ferrooxidans* oxidize sulfides of iron and copper found in minerals such as chalcopyrite ($CuFeS_2$), chalcocite (Cu_2S), and covellite (CuS). The oxidation of Cu^+ to Cu^{2+}, as well as the acidification resulting from production of sulfate, dissolves the metal from the rock. Cu^+ can be oxidized aerobically with O_2 or anaerobically with NO_3^- from the soil. Other metals oxidized

A. *Star Trek*: an imaginary creature produces H_2SO_4

B. A hot spring supports sulfur-oxidizing archaea

FIGURE 14.26 ■ **Organisms produce sulfuric acid: science and science fiction.** A. In the *Star Trek* episode "The Devil in the Dark," starship officers encounter an imaginary creature, called the Horta, that tunnels through rock by producing corrosive acid. B. Volcanic rocks and hot springs support growth of sulfur-oxidizing thermophilic archaea such as *Sulfolobus* species, whose growth at pH 2 colors the rocks in this photo. C. *Sulfolobus acidocaldarius* grows as irregular spheres (TEM).

C. *Sulfolobus acidocaldarius*

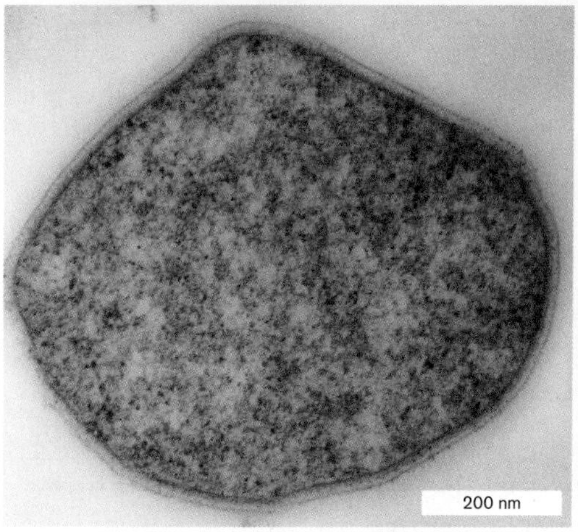

by *A. ferrooxidans* include selenium, antimony, molybdenum, and uranium. The process of metal dissolution from ores is called leaching. Leaching of minerals has been a part of mining since ancient times, long before the existence of microbes was known. Today, over 10% of the copper supply in the United States is provided by biomining ores too low in copper to smelt directly (**Fig. 14.28**). A similar process is being developed to biomine gold.

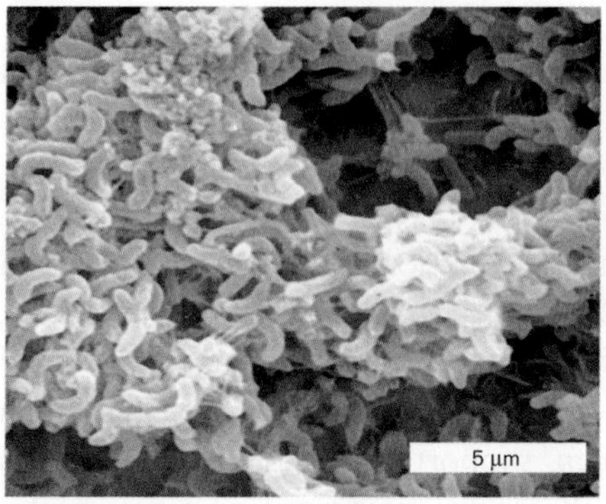

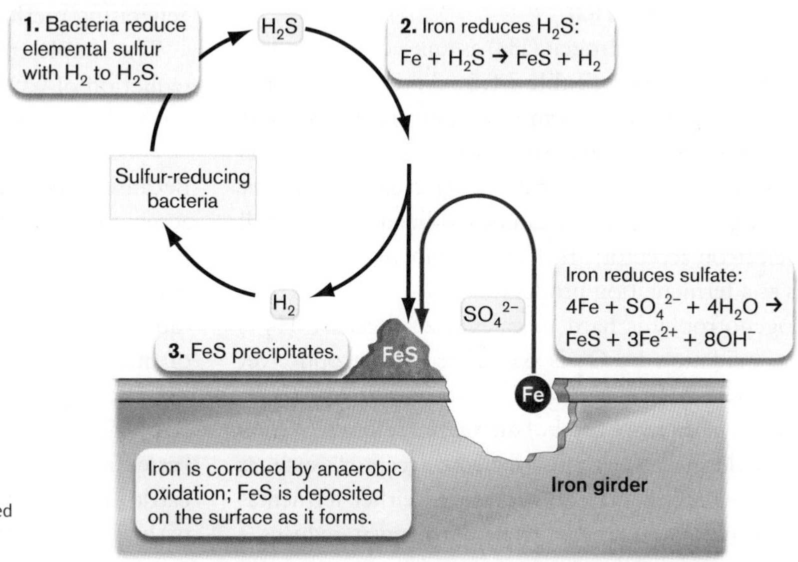

FIGURE 14.27 ■ **Anaerobic iron corrosion.** A. Sulfate-reducing bacteria corrode iron. B. Anaerobic corrosion of iron is accelerated by sulfur-reducing bacteria. Alternatively, in other bacteria, iron reduces sulfate. *Source:* Part B modified from Derek Lovley. 2000. *Environmental Microbe-Metal Interactions*, 163.

A. A copper mine

B. *Acidithiobacillus ferrooxidans*

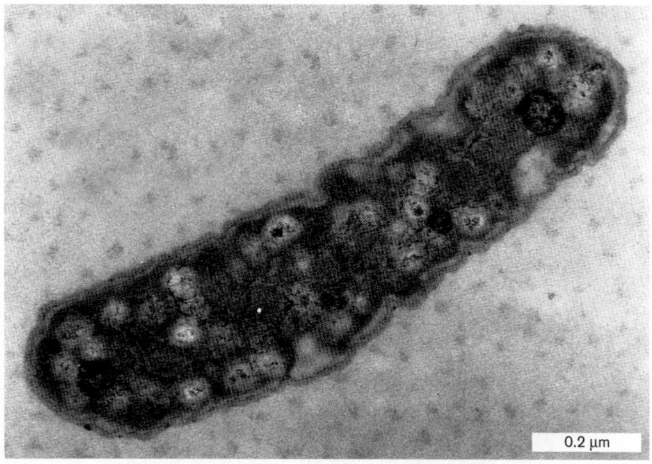

FIGURE 14.28 ▪ Copper mining. **A.** Bingham Canyon copper mine near Salt Lake City, Utah, where copper is leached from low-grade ores by *Acidithiobacillus ferrooxidans* **(B)**, a Gram-negative rod that oxidizes copper and iron sulfides (TEM).

Hydrogenotrophy Uses H_2 as an Electron Donor

The use of molecular hydrogen (H_2) as an electron donor is called **hydrogenotrophy** (**Table 14.3**). An example is oxidation of H_2 by sulfur to form H_2S, performed by *Pyrodictium brockii*, an archaeon that grows at thermal vents above 100°C. Hydrogen gas is a stronger electron donor than most organic foods, so it can be oxidized with the full range of electron acceptors. It is available in anaerobic communities as a fermentation product.

Hydrogenotrophy is hard to categorize. Because H_2 is inorganic, oxidation by O_2 is considered lithotrophy; yet many species that oxidize H_2 with molecular oxygen are organotrophs that also catabolize organic foods. When hydrogen reduces an organic electron acceptor, such as fumarate, the process may be considered either fermentation or anaerobic respiration. When hydrogen reduces a mineral such as sulfur or sulfate, the process is anaerobic lithotrophy.

TABLE 14.3 Hydrogenotrophy: Examples

Specific reaction	General description
$O_2 + 2H_2 \rightarrow 2H_2O$	Aerobic oxidation of H_2
Fumarate + $H_2 \rightarrow$ succinate	Organic + $H_2 \rightarrow$ organic
$2CO_2 + 4H_2 \rightarrow CH_3COOH + 2H_2O$	Mineral + $H_2 \rightarrow$ organic
$2H^+ + SO_4^{2-} + 4H_2 \rightarrow H_2S + 4H_2O$	Mineral + $H_2 \rightarrow$ mineral
$CO_2 + 4H_2 \rightarrow CH_4 + 2H_2O$	Methanogenesis

A form of hydrogenotrophy with enormous potential for bioremediation is "dehalorespiration," also known as organohalide respiration. In dehalorspiration halogenated organic molecules serve as electron acceptors for H_2 (**Fig. 14.29A**). Chlorinated molecules such as chlorobenzenes, perchloroethene, and polyvinyl chloride (known as PVC) are highly toxic environmental pollutants. In dehalorespiration, the chlorine is removed as chloride anion and replaced by hydrogen—a reaction requiring input of two electrons from H_2. Many soil bacteria respire on organohalides; a particularly unusual cell wall–less bacterium, *Dehalococcoides* (**Fig. 14.29B**), was discovered that dechlorinates chlorobenzene, a highly stable aromatic molecule.

Methanogenesis

Hydrogen is such a strong electron donor that it can even reduce the highly stable carbon dioxide to methane. Reduction of CO_2 and other single-carbon compounds, such as formate, to methane is called **methanogenesis**. Methanogenesis supports a major group of archaea known as methanogens, many of which grow solely by autotrophy, generating methane.

The simplest form of methanogenesis involves hydrogen reduction of CO_2:

$$CO_2 + 4H_2 \rightarrow CH_4 + 2H_2O \qquad E°' = 180 \text{ mV}$$

Because both sides of the equation contain a weak electron acceptor (CO_2, H_2O) and a strong electron donor (H_2, CH_4), it was surprising that such a reaction could yield energy for growth. But the presence of sufficient carbon dioxide and hydrogen supports enormous communities of methanogens. Such conditions prevail wherever bacteria grow by fermentation and their gaseous products are trapped, such as in landfills, where methane can be harvested as natural gas, as well as in the digestive systems of cattle and humans. Variants of methanogenesis also include pathways by which H_2 reduces various one-carbon and

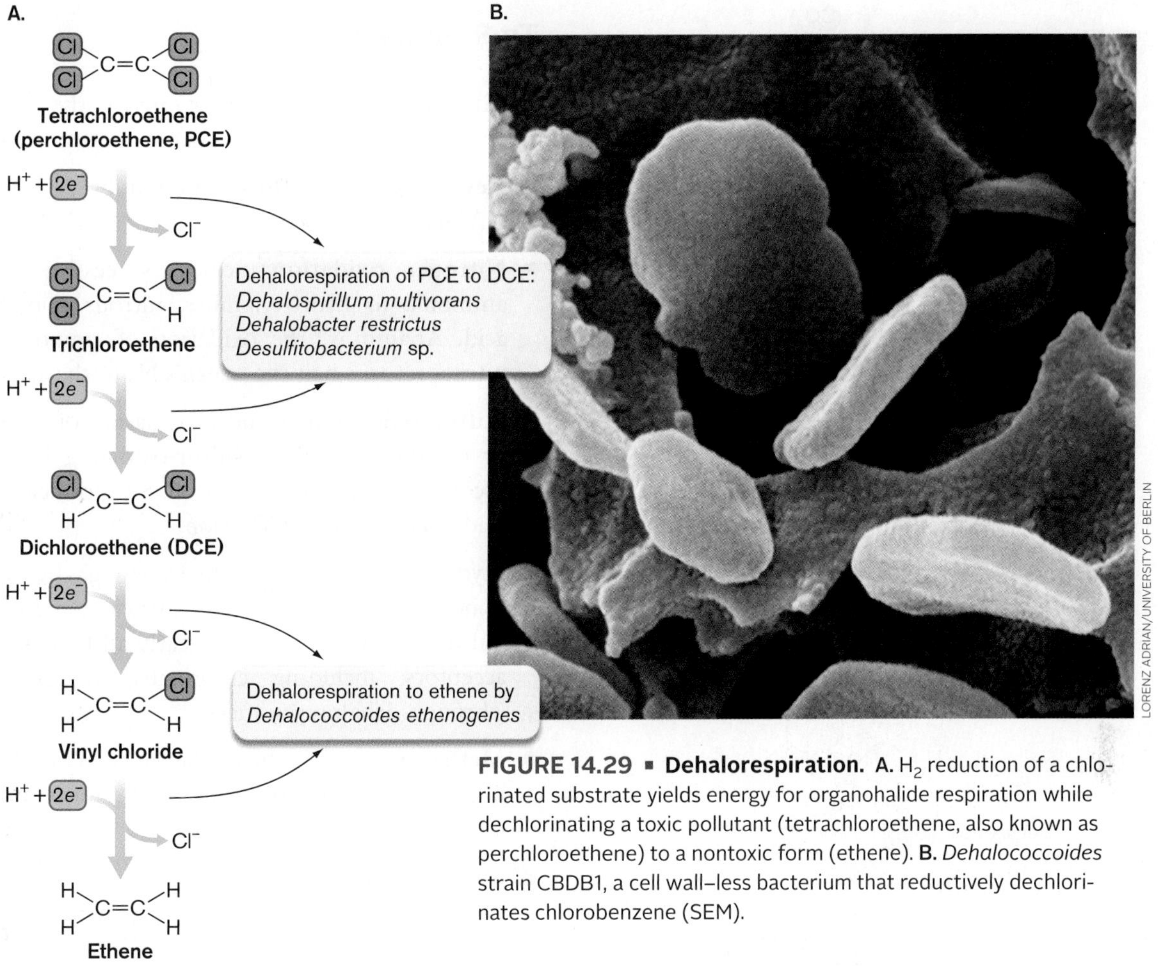

FIGURE 14.29 ▪ Dehalorespiration. **A.** H_2 reduction of a chlorinated substrate yields energy for organohalide respiration while dechlorinating a toxic pollutant (tetrachloroethene, also known as perchloroethene) to a nontoxic form (ethene). **B.** *Dehalococcoides* strain CBDB1, a cell wall–less bacterium that reductively dechlorinates chlorobenzene (SEM).

two-carbon molecules, such as methanol, methylamine, and acetic acid. Diverse methanogens are found in all kinds of environments (discussed in Chapter 19).

A simplified pathway of methanogenesis from CO_2 is shown in **Figure 14.30**. The CO_2 undergoes stepwise hydrogenation, and each oxygen is reduced to water. The increasingly reduced carbon is transferred through a series of unique cofactors (methanofuran, tetrahydromethanopterin, coenzyme M-SH). Three of the hydrogenation steps involve a membrane ETS complex that includes the carrier coenzyme F_{420}, whose reaction generates a proton potential. Note, however, that the final step of methane production generates a transmembrane sodium potential (ΔNa^+), which drives ATP synthesis by a sodium-powered ATP synthase embedded in the cell membrane. Further details of methanogenesis are presented in Chapter 19.

Thought Question

14.14 Hydrogen gas is so light that it rapidly escapes from Earth. Where does all the hydrogen come from to be used for hydrogenotrophy and methanogenesis?

Methylotrophy: Oxidation of Single-Carbon Substrates

Many forms of catabolism release reduced single-carbon molecules that other microbes can oxidize for energy, such as methanol (CH_3OH), methylamine (CH_3NH_2), methane (CH_4), and even carbon monoxide (CO). Oxidation of single-carbon molecules via an ETS is called **methylotrophy**, a form of metabolism outside the definitions of lithotrophy and organotrophy. The oxidant may be O_2 or anaerobic electron acceptors such as nitrite, sulfate, or metals. Methylotrophs are found in soil, marine, and freshwater sediments.

The methane released by methanogenesis provides a niche for **methanotrophy**, a form of methylotrophy in which bacteria and archaea oxidize methane. In deep-ocean sediment, the activity of methanogens is so high that enormous quantities of methane become trapped on the seafloor in the form of water-based crystals known as methane hydrates. If all the methane from these crystals were released at once, global warming would be greatly accelerated. Anaerobic methane oxidation may be critical for the global carbon cycle, since it suggests a mechanism for removal of deep-sea methane (discussed in Chapters 21 and 22).

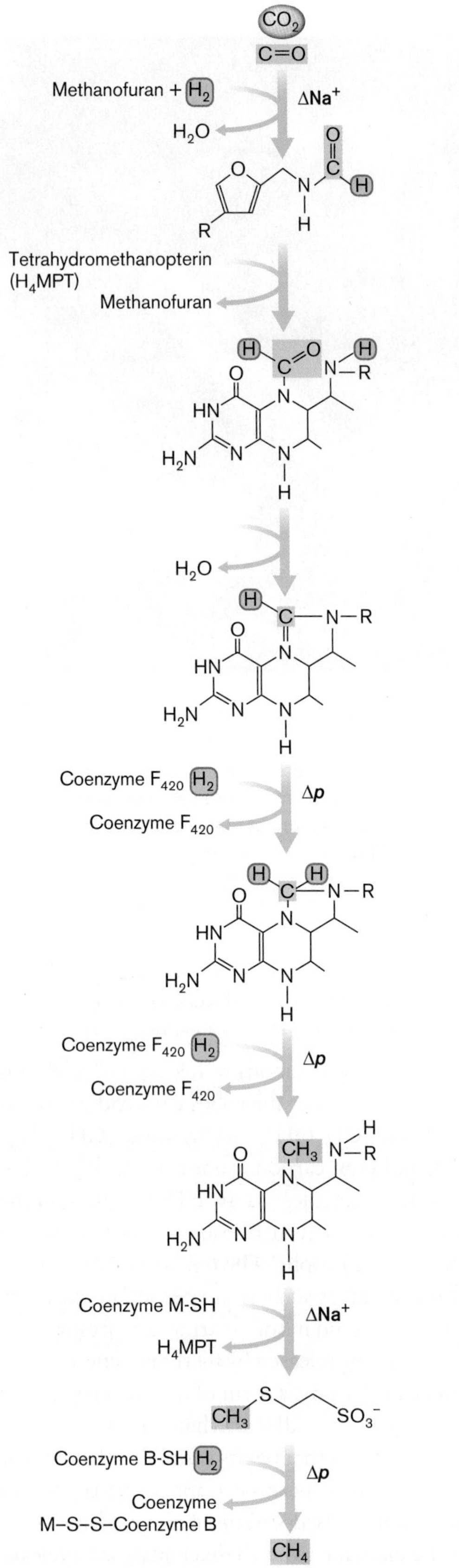

FIGURE 14.30 ■ Methanogenesis. A methanogen reduces carbon dioxide with hydrogen, generating methane (CH_4). The incorporation of hydrogen contributes to both a proton potential and a sodium potential (discussed in Chapter 19).

To Summarize

- **Lithotrophy (chemolithotrophy)** is the acquisition of energy by oxidation of inorganic electron donors.
- **Reverse electron flow** powered by Δp can generate NADH or NADPH.
- **Nitrogen oxidation** includes successive oxidation of ammonia to hydroxylamine, nitrous acid, and nitric acid. Anammox, the oxidation of ammonium ion by nitrite, returns half the ocean's N_2 to the atmosphere.
- **Sulfur oxidation** includes oxidation of H_2S to sulfur or to sulfuric acid by sulfur-oxidizing bacteria, often accompanied by iron oxidation. Sulfuric acid production leads to extreme acidification.
- **Hydrogenotrophy** uses hydrogen gas as an electron donor. Hydrogen (H_2) has sufficient reducing potential to donate electrons to nearly all biological electron acceptors, including chlorinated organic molecules (through dehalorespiration).
- **Methanogenesis** is the oxidation of H_2 by CO_2, releasing methane. Methanogenesis is performed only by the methanogen group of archaea.
- **Methylotrophs** use O_2, nitrite, or sulfate to oxidize single-carbon compounds such as methane, methanol, or methylamine. A class of methylotrophs called **methanotrophs** specifically oxidize methane.

14.6 Phototrophy

On Earth today, the ultimate source of electrons driving metabolism is phototrophy, the use of photoexcited electrons to power cell growth. Every year, photosynthesis converts more than 10% of atmospheric carbon dioxide to biomass, most of which then feeds microbial and animal heterotrophs. Most of Earth's photosynthetic production, especially in the oceans, comes from microbes (discussed in Chapter 21). Even the frigid seas of Antarctica support vast communities of phototrophic algae and bacteria, at −2°C beneath pack ice (**Fig. 14.31**).

In phototrophy, the energy of a photoexcited electron is used to pump protons. Different kinds of phototrophic systems include the bacteriorhodopsin proton pump; the single-cycle chlorophyll-based photosystems I and II; and the double-cycle Z pathway of oxygenic photosynthesis in cyanobacteria and chloroplasts.

Note: Oxygenic photosynthesis is coupled directly to fixation of CO_2 into biomass. The process of CO_2 fixation is discussed in Chapter 15.

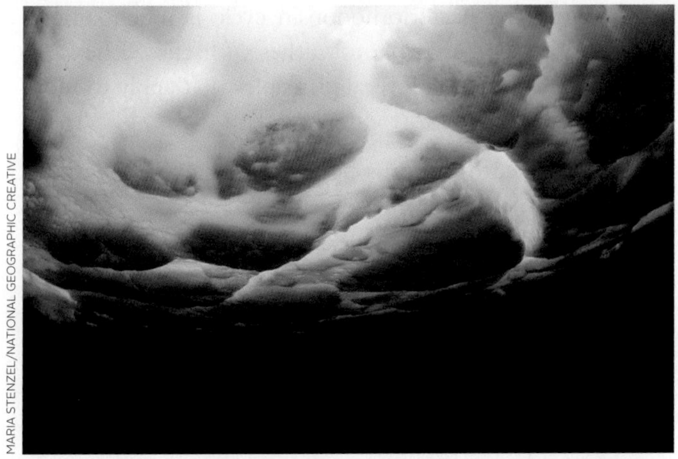

FIGURE 14.31 ■ Phototrophy beneath Antarctic pack ice. Underwater view of the underside of ice, showing yellow-green algae illuminated by sunlight.

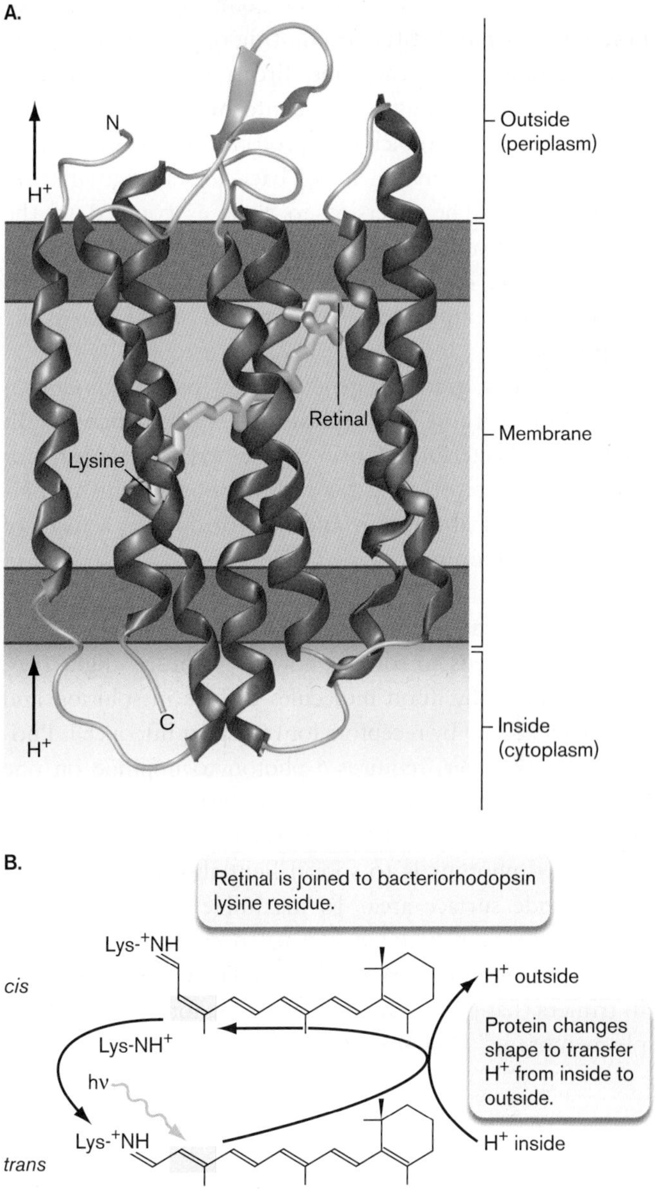

FIGURE 14.32 ■ The light-driven cycle of bacteriorhodopsin. **A.** Bacteriorhodopsin contains seven alpha helices that span the membrane in alternating directions and surround a molecule of retinal, which is linked to a lysine residue. (PDB code: 1FBB) **B.** A photon (hv) is absorbed by retinal, which shifts the configuration from *trans* to *cis*. The relaxation back to the *trans* form is coupled to pumping $1H^+$ across the membrane. (PDB code: 1MOL) *Source:* Purple Membrane: Theoretical Biophysics Group, VMD Image Gallery, NIH Resource for Macromolecular Modeling and Bioinformatics.

Retinal-Based Proton Pumps

In most ecosystems, the dominant source of carbon and energy is photosynthesis based on chlorophyll. At the same time, many halophilic archaea and marine bacteria supplement their metabolism with a simpler, more ancient form of phototrophy based on a single-protein, light-driven proton pump containing the pigment retinal. Many varieties of the retinal-based proton pump have evolved, known as **bacteriorhodopsin** in haloarchaea and as **proteorhodopsin** in bacteria. We present bacteriorhodopsin as a relatively simple form of phototrophy, followed by the more complex ETS-based photolysis in bacteria and chloroplasts.

Bacteriorhodopsin and proteorhodopsin. Bacteriorhodopsin is a small membrane protein commonly found in halophilic archaea (or haloarchaea) such as *Halobacterium salinarum*, a single-celled archaeon that grows in evaporating salt flats containing concentrated NaCl (discussed in Chapter 19). For many years, bacteriorhodopsin-like proton pumps were thought to be limited to extreme halophilic archaea. As bacterial genomes were sequenced, however, homologs of the protein appeared in several species of proteobacteria; the homologs were termed proteorhodopsin. The proteorhodopsin genes appear to have entered bacteria by horizontal transfer from halophilic archaea. In 2005, Oded Béjà and colleagues from Israel, Austria, Korea, and the United States surveyed the genomes of unculturable bacteria from the upper waters of the Mediterranean and Red seas. They found that 13% of the marine bacteria contain proteorhodopsins, accounting for a substantial—and previously unrecognized—fraction of marine phototrophy.

Bacteriorhodopsin absorbs light with a broad peak in the green range; thus, the organisms containing large amounts of bacteriorhodopsin reflect blue and red, appearing purple. The protein consists of seven hydrophobic alpha helices surrounding a molecule of **retinal**, the same cofactor bound to light-absorbing opsins in the vertebrate retina (**Fig. 14.32A**). In bacteriorhodopsin, the retinal is attached to the nitrogen end of a lysine residue (**Fig. 14.32B**).

Retinal has a series of conjugated double bonds that absorb visible light. Upon absorbing a photon, an electron in one of the double bonds is excited to a higher energy level. This process is called photoexcitation. As the electron falls back to the ground state, the double bond shifts position from *trans* (substituents pointing opposite) to *cis* (substituents pointing in the same direction). This change in shape of the retinal alters the conformation of the entire protein, causing it to pick up a proton from the cytoplasm. Eventually, the retinal switches back to its original *trans* configuration. The reversion to *trans* is coupled to the release of a proton from the opposite end of the protein facing outside of cell. Thus, photoexcitation of bacteriorhodopsin is coupled to the pumping of one H^+ across the membrane.

The proton gradient generated by bacteriorhodopsin drives ATP synthesis by a typical F_1F_o ATP synthase. Light capture by bacteriorhodopsin supplements catabolism for energy and heterotrophy for carbon source. This combination of light absorption and heterotrophy is a form of **photoheterotrophy**.

Purple membrane captures light rays. One problem every phototroph needs to solve is how to "capture" light rays. For chemotrophy, food molecules diffuse in solution and can be picked up by receptors for transport into a cell. Phototrophy, however, requires a photon to impinge on one point of the cell, where the photon either is absorbed or passes through. Thus, the only way to absorb a high percentage of photons is to spread light-absorbing pigments over a wide surface area. To maximize light absorption, *Halobacterium salinarum* archaea pack their entire cell membranes with bacteriorhodopsin. The protein assembles in trimers that pack in hexagonal arrays, forming the "purple membrane" (**Fig. 14.33**).

FIGURE 14.33 ■ Bacteriorhodopsin purple membrane. Trimers of bacteriorhodopsin (monomers shown red, blue, and green) are packed in hexagonal arrays, forming the "purple membrane." (PDB code: 1MOL) *Source:* Purple Membrane: Theoretical Biophysics Group, VMD Image Gallery, NIH Resource for Macromolecular Modeling and Bioinformatics.

Although the bacteriorhodopsin cycle is much simpler than chlorophyll-based photosynthesis (discussed next), it nevertheless illustrates several principles that apply to more complex forms of phototrophy:

- A photoreceptor absorbs light, causing excitation of an electron to a higher energy level, followed by return to the ground state.
- To maximize light collection, large numbers of photoreceptors are packed throughout a membrane.
- The photocycle (absorption and relaxation of the light-absorbing molecule) is coupled to energy storage in the form of a proton gradient.

Note: Distinguish among these terms of phototrophy:

- **Photo<u>excitation</u>** means light absorption that raises an electron to a higher energy state, as in bacteriorhodopsin.
- **Photo<u>ionization</u>** means light absorption that causes electron separation.
- **Photo<u>lysis</u>** means light absorption coupled to splitting a molecule.
- **Photo<u>synthesis</u>** means photolysis with CO_2 fixation and biosynthesis.

Chlorophyll Photoexcitation and Photolysis

Cyanobacteria and chloroplasts, as well as other kinds of bacteria, obtain energy by photoexcitation of chlorophylls. Figuring out their "light reactions"—the fundamental source of energy for Earth's biosphere—was one of the most exciting projects of the twentieth century. Among hundreds of important contributors, we note two major figures: the married couple Roger Stanier (1916–1982) and Germaine Cohen-Bazire (1920–2001; **Fig. 14.34A** and **B**). Stanier, a Canadian microbial physiologist at UC Berkeley, clarified the nature of cyanobacteria as phototrophic prokaryotes distinct from eukaryotic algae, and he helped distinguish the water-based photosynthesis of cyanobacteria from the sulfide metabolism of purple bacteria. Cohen-Bazire was a French bacterial geneticist who had studied *lac* operon regulation with Nobel laureate Jacques Monod at the Pasteur Institute in Paris. Cohen-Bazire applied her genetics skills to phototrophs, and she conducted the first genetic analysis of photosynthesis in purple bacteria and cyanobacteria.

Cyanobacteria are the only bacteria that split water to make oxygen. Most cyanobacterial species appear green, like algae or plants (**Fig. 14.34C**). Their green color arises from their chlorophyll, which absorbs blue and red but reflects green. Some species appear orange or brown because of secondary pigments. Cyanobacteria include a wide range of species, such as one of the ocean's major

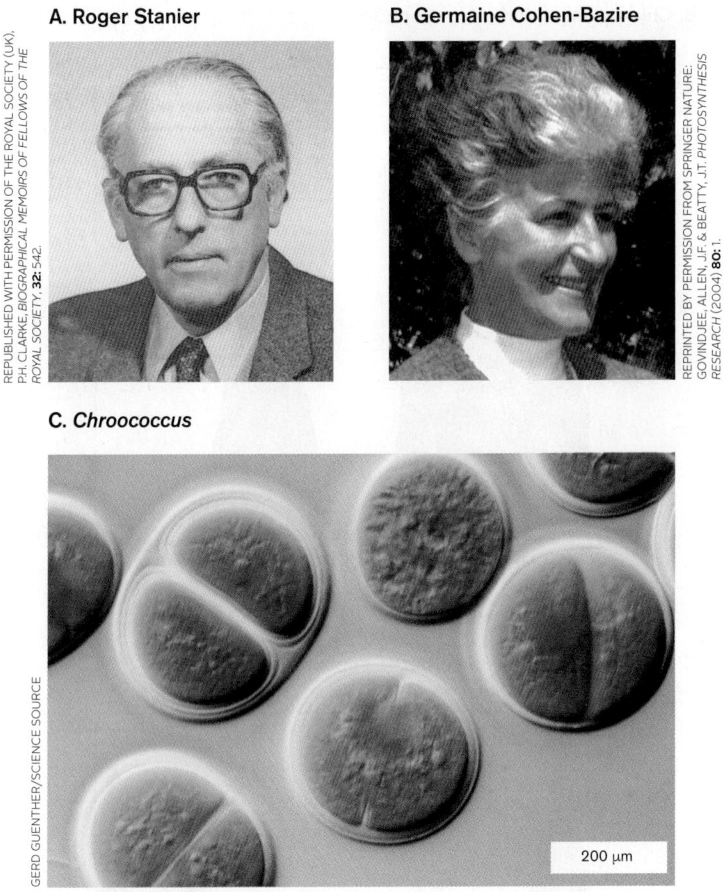

FIGURE 14.34 ■ Germaine Cohen-Bazire and Roger Stanier studied cyanobacterial photosynthesis. **A.** Stanier pioneered the study of cyanobacterial physiology. **B.** Cohen-Bazire performed the first studies of genetic regulation of bacterial photosynthesis. **C.** *Chroococcus*, a genus of cyanobacterium studied by Stanier.

producers, the tiny *Prochlorococcus marinus*, barely visible under a light microscope. Other cyanobacteria have cells as large as eukaryotic algae and form complex developmental structures with important symbiotic associations (see Chapters 18 and 21). Cyanobacteria are among the most successful and diverse groups of life on Earth. Along with the chloroplasts of algae and plants, cyanobacteria produce all the oxygen available for aerobic life.

Overview of photolysis. The energy for photosynthesis derives from the photoexcitation of a light-absorbing pigment. The light-absorbing molecule with an electron in an excited state becomes a strong electron donor. The photoexcited molecule can thus donate an electron to an electron acceptor molecule, which is coupled to an ETS. The components of the ETS are often homologous to those of respiratory electron transport, and they share common electron carriers, such as cytochromes.

In plant chloroplasts and in cyanobacteria, photolysis is known as the "light reactions," which are coupled to the "light-independent reactions" of carbon dioxide fixation. Note, however, that many of the sulfur- or organic-based bacterial phototrophs, such as *Rhodospirillum rubrum*, combine photolysis with heterotrophy instead of with CO_2 fixation. At the same time, lithotrophic bacteria such as *Nitrospira* and *Acidithiobacillus* fix CO_2 using energy from mineral oxidation instead of photolysis. In this chapter we focus on photolysis in cyanobacteria. We discuss CO_2 fixation along with other biosynthetic pathways in Chapter 15.

In ETS-based photosynthesis, photoexcitation leads to separation of an electron from a donor molecule such as H_2O or H_2S. Each electron is then transferred to an ETS, whose components show common ancestry with respiratory ETS proteins. The ETS generates a proton potential and the reduced cofactor NADPH. The proton potential drives ATP synthesis through an F_1F_o ATP synthase similar to the one for respiration.

Chlorophylls absorb light. The main light-absorbing pigments are **chlorophylls**. Each type of chlorophyll contains a characteristic **chromophore**, a light-absorbing electron carrier. The chlorophyll chromophore consists of a heteroaromatic ring complexed to a magnesium ion (Mg^{2+}; **Fig. 14.35A**). As we saw for ETS electron carriers, aromatic molecules offer electrons with relatively narrow energy transitions. The chromophore absorbs a photon through a reversible energy transition, such that the chlorophyll can alternate between excited and ground states.

Chlorophyll molecules differ slightly in their substituent groups around the ring; for example, chlorophyll *a* of chloroplasts has a methyl group in ring II, whereas chlorophyll *b* has an aldehyde. Both chlorophylls *a* and *b* are made by chloroplasts and by cyanobacteria, their nearest bacterial relatives. Because they absorb red and blue, they reflect the middle range of the spectrum and so appear green (**Fig. 14.35B**).

By contrast, the chlorophylls of anaerobic phototrophs, or "purple bacteria," such as *Rhodobacter* and *Rhodospirillum*, absorb most strongly in the far-red (infrared) and, in some cases, ultraviolet (**Fig. 14.35C**). Their chlorophylls are specifically named **bacteriochlorophylls**. The purple bacteria grow in pond water or sediment. Bacteriochlorophyll absorption over an extended range of wavelengths helps capture light missed by the cyanobacteria and algae at the water's surface. In purple bacteria, bacteriochlorophylls are supplemented by accessory pigments called **carotenoids**, which absorb light of green wavelengths and transfer the energy to bacteriochlorophyll. The combination of green-absorbing carotenoids and infrared-absorbing bacteriochlorophylls makes cultures appear deep purple or brown.

The infrared radiation absorbed by bacteriochlorophylls is too weak to permit splitting H_2O to produce oxygen.

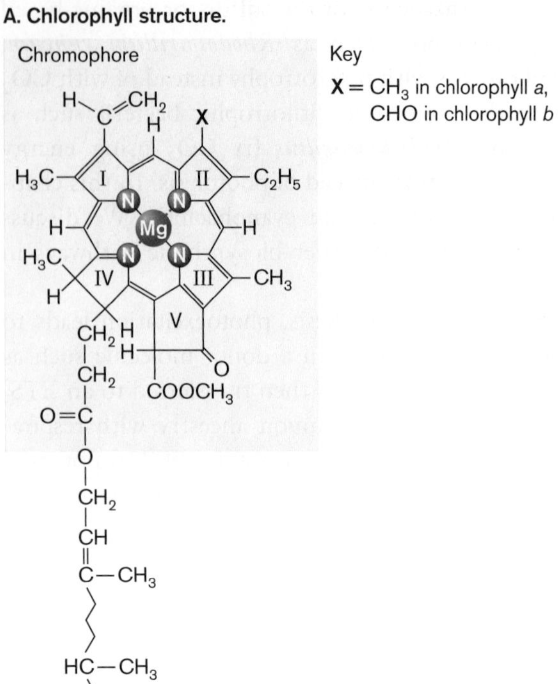

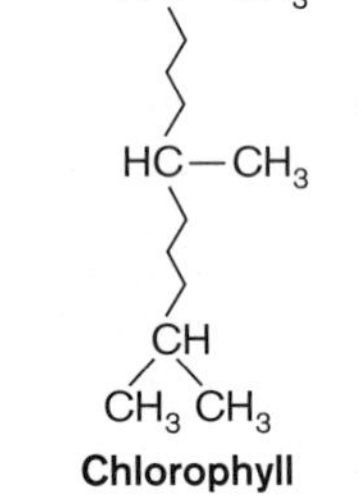

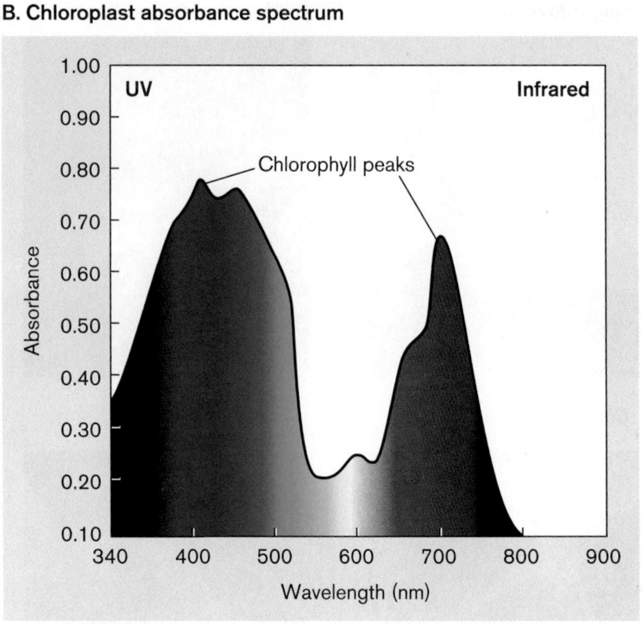

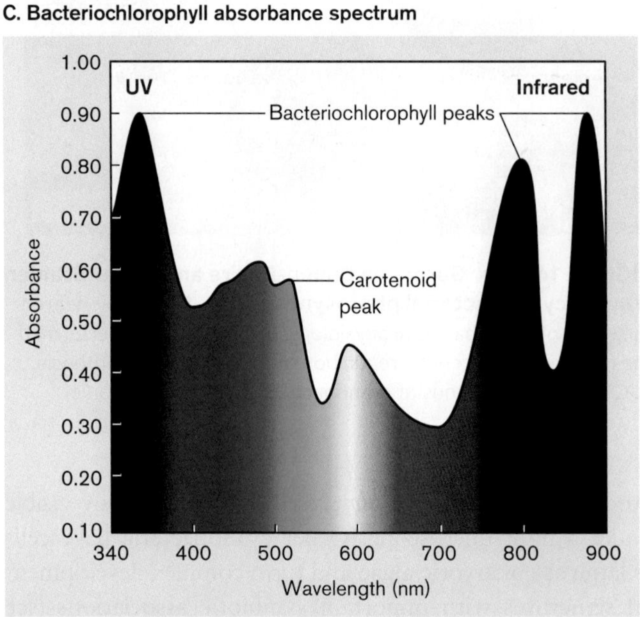

FIGURE 14.35 ■ Chlorophyll structure and absorbance. A. Chlorophyll molecular structure. **B.** Absorption by chloroplasts, including chlorophyll *a*, chlorophyll *b*, and carotenoid accessory pigments. The middle range (green) is reflected. **C.** Absorption by purple photosynthetic bacteria includes bacteriochlorophyll and carotenoids.

Thus, purple bacteria are limited to photolysis of H_2S and small organic molecules; many are photoheterotrophs. On the other hand, infrared rays are available in water below the oxygenic phototrophs absorbing red and blue. Thus, anaerobic phototrophs grow at depths where their more high-powered oxygenic relatives do not.

Note: Distinguish among these classes of photopigments:

- **Bacteriorhodopsin** is a retinal-containing proton pump.
- **Chlorophyll** is a charge-separating photopigment (usually referring to chloroplasts and cyanobacteria).
- **Bacteriochlorophyll** is a charge-separating chlorophyll of anaerobic purple and green bacteria (also known generically as chlorophyll).
- **Carotenoid** is an accessory pigment that absorbs the midrange of light wavelengths but does not directly conduct photolysis.

Antenna complex and reaction center. Light photons cannot be transported or concentrated in a compartment. Instead, they must be captured by absorption. The larger the array of absorptive molecules, the more photons will be captured.

To maximize light collection, many molecules of chlorophyll are grouped in an **antenna complex**. These antenna complexes are arranged like a satellite dish within the plane of the membrane in an elaborate cluster around accessory proteins. This cluster is called a light-harvesting complex (LH). Light-harvesting complex 2 (LH2) of *Blastochloris viridis* is shown in **Figure 14.36A**. *B. viridis* is a photoheterotroph, an alphaproteobacterium that absorbs an exceptionally broad range of light, into the infrared. The LH2 clusters of *B. viridis* then associate in a ring around light-harvesting complex 1 (LH1), shown in **Figure 14.36B**. In the center, surrounded by LH2 complexes, LH1 contains the **reaction center** (**RC**; **Fig. 14.36B**). The reaction center is the protein complex in which chlorophyll photoexcitation connects to the ETS.

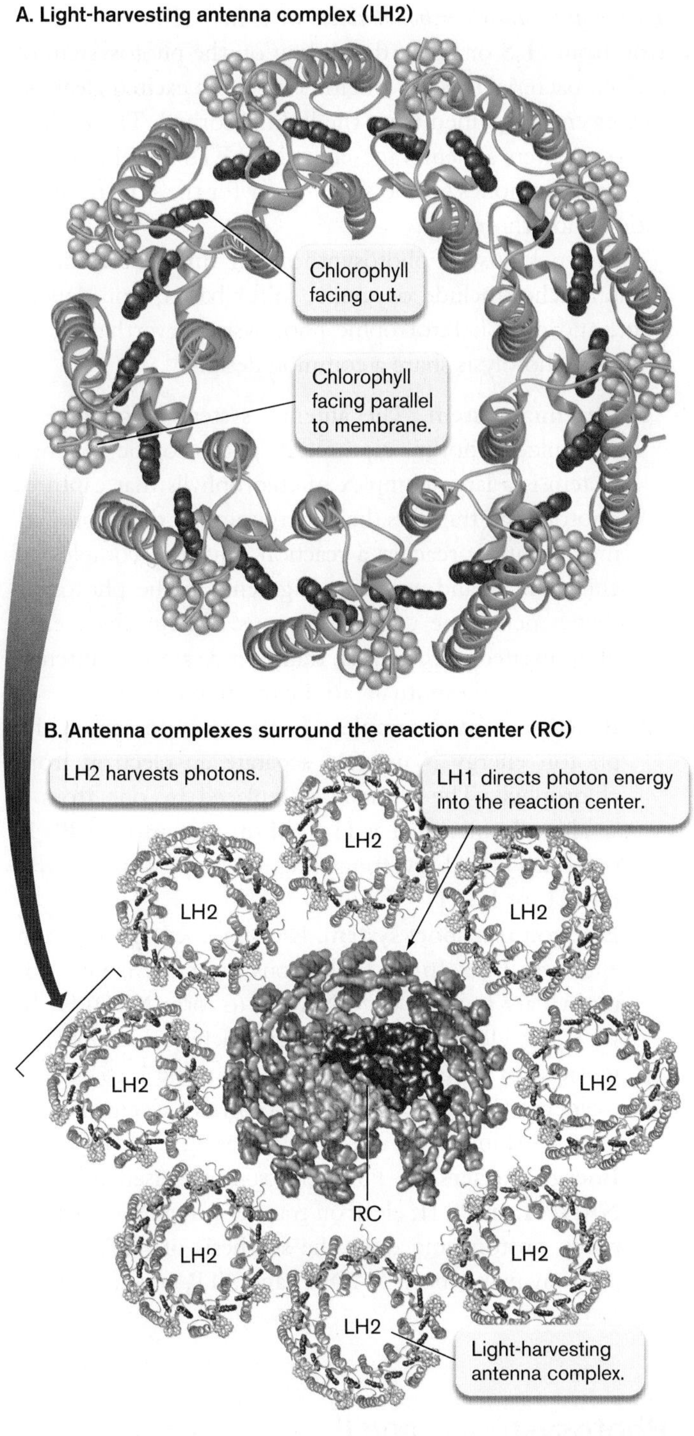

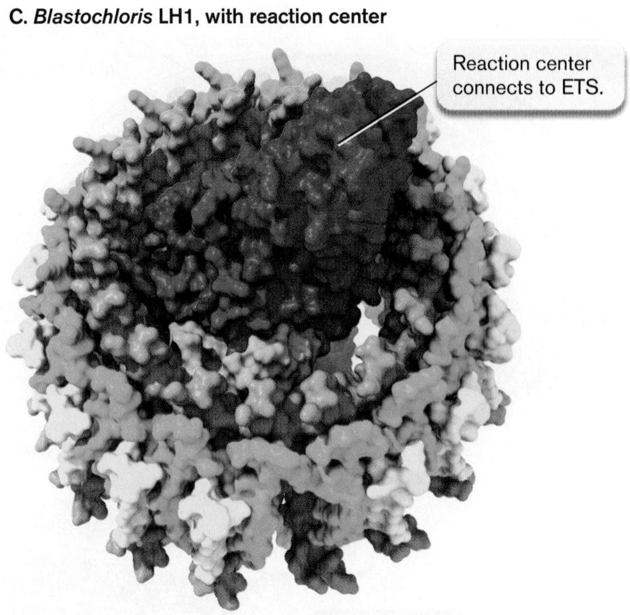

FIGURE 14.36 ■ Antenna complexes surround the reaction center. **A.** An antenna complex (LH2) of *Blastochloris viridis* contains 9 chlorophylls with chromophores facing parallel to the membrane (gold) and 18 with chromophores facing out from the ring (red). **B.** Multiple rings of chlorophyll-protein antenna complexes surround the reaction center (RC), like a funnel collecting photons. (PDB codes: 1PYH, 2FKW) **C.** Light-harvesting complex 1 (LH1), with reaction center. (PDB code: 6et5) *Sources:* Parts A and B modified from Quantum Biology of the PSU, NIH Resource for Macromolecular Modeling and Bioinformatics—BChl Antenna; part C modified from Qian et al. 2018. *Nature* **556**:203.

Throughout this antenna complex of bacteriochlorophylls and accessory pigments, whichever pigment molecule happens to be in the right place at the right time captures the photon. The energy from the photon then transfers from one chromophore to the next, until it arrives at the reaction center for electron transfer to the ETS. Other kinds of bacteria have other forms of antenna complexes, such as the "phycobilisome" of cyanobacteria (discussed shortly).

Purple bacteria and cyanobacteria increase their efficiency of photon uptake by extensive backfolding of the photosynthetic membranes in oval pockets stacked like pita breads (**Fig. 14.37**). These oval pockets are called **thylakoids**. The extensive packing of thylakoids gives an incident photon hundreds of chances to meet a chlorophyll at just the right angle for absorption. The hollow "thylakoid pockets" also store H^+ ions to generate a significant transmembrane Δp for ATP synthesis.

The thylakoids are connected by tubular extensions, so that there exists one interior space, the lumen, separated topologically from the regular cytoplasm, or stroma. Protons are pumped from the stroma across the thylakoid membrane into the lumen. The F_1F_o complex is embedded in the thylakoid, where it makes ATP using the proton current running through it into the cytoplasm. The F_1 knob of ATP synthase appears to face "outward" in photosynthetic organelles (as opposed to "inward" in respiratory chains). In each case, however, the proton current and ATP motor face in the same direction with respect to the cytoplasm (stroma). The proton potential is more negative in the cytoplasm (stroma), thus drawing protons through the ATP synthase to generate ATP.

The photolytic electron transport system. In photolysis, the absorption of light by chlorophyll or bacteriochlorophyll drives the separation of an electron. The chlorophyll may then gain an electron from the ETS, as in

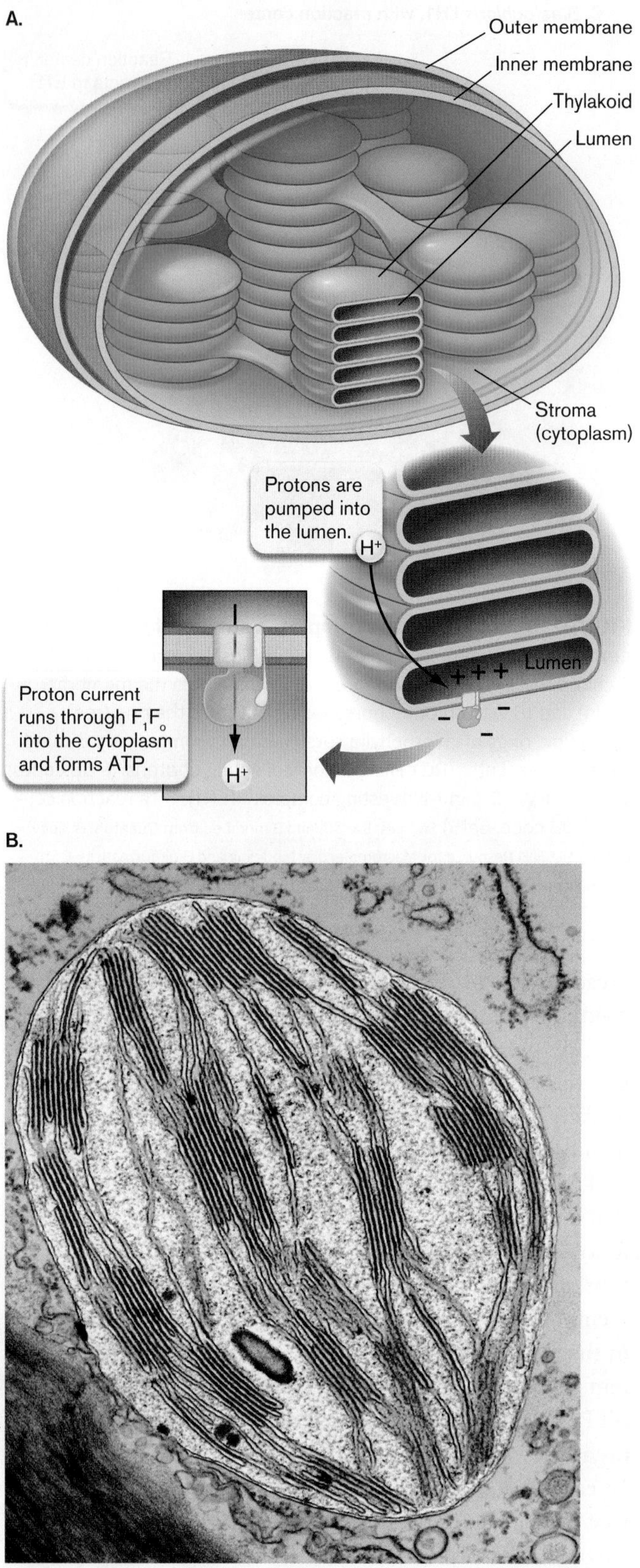

BIOPHOTO ASSOCIATES/SCIENCE SOURCE

FIGURE 14.37 ■ Photosynthetic membranes. **A.** The photosynthetic membranes of bacteria and chloroplasts appear as hollow disks with tubular interconnections. The disks are called thylakoids. Topologically, the membrane separates the cytoplasm (stroma) from the interior space (lumen). **B.** TEM section of a chloroplast, showing the stacked thylakoids.

Rhodospirillum or *Rhodobacter*, or it may remove an electron from H_2S or H_2O, depending on the photosystem of a given bacterial species. In either case, the excited electron carries energy gained from the light absorbed. The excited electron enters a membrane-embedded ETS of oxidoreductases and quinones/quinols, as we saw for the ETS of respiration and lithotrophy.

Diverse kinds of photosynthesis in different environmental niches include oxygenic, sulfur-based, iron-dependent, and even heterotrophic photolysis. Nevertheless, all forms of photolysis share a common design:

1. **Antenna system.** The antenna system (**Fig. 14.36**) maximizes photon capture. A phototrophic antenna system is a large complex of chlorophylls that captures photons and transfers their energy among the photopigments until it reaches a reaction center. A complex of chlorophylls and accessory pigments in the photosynthetic membrane collects photons. Energy from each photoexcited electron is transferred among antenna pigments and eventually to the reaction center.
2. **Reaction center complex.** In the reaction center, the photon energy is used to separate an electron from chlorophyll. The electron is replaced by one from a small molecule, such as H_2S (**photosystem I**, or **PS I**) or from a light-harvesting antenna complex (**photosystem II**, or **PS II**).
3. **Electron transport system.** Each photoexcited electron enters an ETS. In PS I, electrons separated from chlorophyll are transferred to $NADP^+$ to form NADPH. In PS II, the electron separated from bacteriochlorophyll is replaced by an electron returned from the ETS. In the **oxygenic Z pathway** (H_2O photolysis), electrons flow from PS II into PS I, ultimately releasing O_2 from H_2O.
4. **Energy carriers.** In PS I, electrons are used to make NADPH. In PS II, electron transfer provides energy to pump protons and drive the synthesis of ATP. The Z pathway makes both NADPH and ATP, which are used to fix CO_2.

Photosystems I and II

The steps of photolysis and electron transport occur in three different kinds of systems, in different classes of bacteria:

- **Anaerobic photosystem I** receives electrons associated with hydrogens from H_2S, HS^-, or H_2, or even from reduced iron (Fe^{2+}). Anaerobic PS I is found in chlorobia ("green sulfur" bacteria) and in chloroflexi (filamentous green bacteria).
- **Anaerobic photosystem II** returns an electron from the ETS to bacteriochlorophyll. Anaerobic PS II is found in alphaproteobacteria, "purple nonsulfur" bacteria, and other proteobacteria.

- **The oxygenic Z pathway** includes homologs of photosystems I and II. Two pairs of electrons are received from two water molecules to generate O_2. The Z pathway is found in cyanobacteria and in the chloroplasts of green plants.

The components of photosystems I and II (PS I and PS II) share common ancestry. Each system runs anaerobically, producing sulfur or oxidized organic by-products, but not O_2. Each photosystem shows more recent homology with the respective PS I and PS II components of the oxygenic Z pathway (so called because the electron path through a diagram of the two photosystems traces a Z). The Z pathway ultimately generates O_2—the source of nearly all the oxygen we breathe.

Note: In photolysis, each quantum of light excites a single electron. Some ETS components, such as the quinones/quinols ($Q \rightarrow QH_2$) actually process two of these electrons in completing their redox cycle. The single-electron intermediate states of quinones are not shown in **Figures 14.38, 14.39,** and **14.40**.

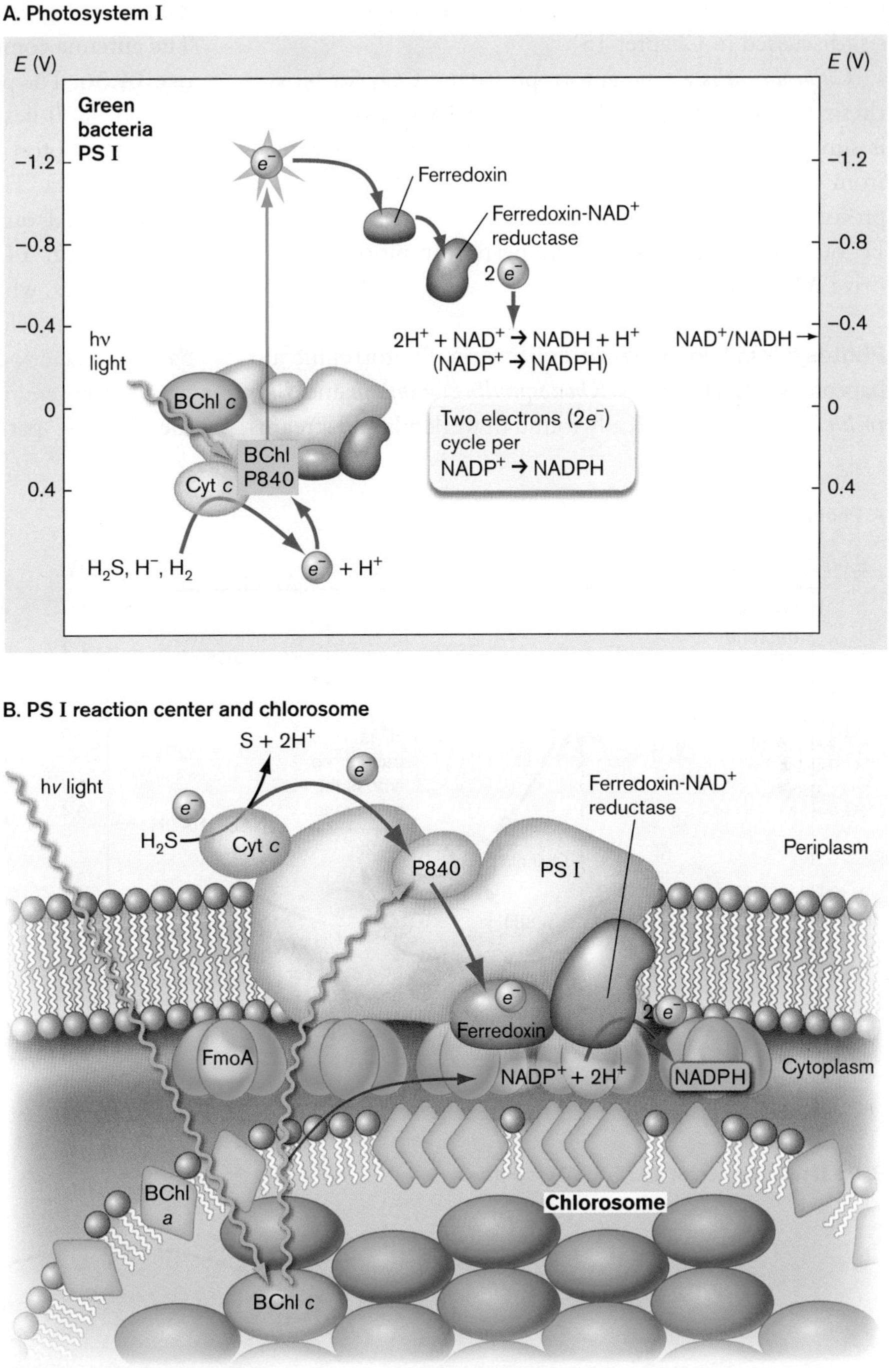

FIGURE 14.38 ■ Photosystem I separates electrons from sulfides and organic molecules. **A.** In green sulfur bacteria, photoexcitation of P840 transfers e^- to a quinone (phylloquinone, PQ), at high reduction potential E. From PQ the electron is transferred to ferredoxin (Fd). Ferredoxin is oxidized by ferredoxin-NAD$^+$ reductase (FNR), donating $2e^-$ to NAD$^+$ (or the energetically equivalent NADP$^+$) to form NADH (or NADPH). **B.** The chlorosome antenna complex transfers photon energy to the PS I reaction center.

Photosystem I in chlorobia. Bacteria such as *Chlorobium* species use PS I (**Fig. 14.38**). The reaction center (RC) contains bacteriochlorophyll P840, named for its peak absorption at 840 nm—that is, the near-infrared, actually beyond the range that humans can see. But P840 and the chlorophylls of the antenna complex also absorb light over shorter wavelengths, in the range of 400–550 nm.

The *Chlorobium* antenna complex consists of a membrane compartment called a chlorosome (**Fig. 14.38B**). A single chlorosome may contain 200,000 molecules of bacteriochlorophyll, harvesting photons with nearly 100% efficiency. The chlorosome is so sensitive that some chlorobia actually harvest thermal radiation from deep-sea thermal vents.

When any one bacteriochlorophyll absorbs a photon, the energy transfers among the photopigments until it reaches the PS I reaction center (**Fig. 14.38**). The photon yields sufficient energy to donate the high-potential electron to a high-potential quinone: phylloquinone/phylloquinol (PQ). Phylloquinol donates the electron to **ferredoxin**, an FeS protein. Ferredoxin transfers the electron to the enzyme ferredoxin-NAD$^+$ reductase; two of these electrons then reduce NAD$^+$ or the energetically equivalent NADP$^+$. The reduced carrier (NADH or NADPH) provides reductive energy for CO_2 fixation and biosynthesis (discussed in Chapter 15).

Chlorobium is a true autotroph, fixing CO_2 for biosynthesis (see Chapter 15). However, the phototrophic ETS is supplemented by lithotrophy, the donation of electrons from H_2S or H_2. The PS I electron flow generates a net proton gradient by consuming H^+ inside and generating H^+ outside the cell, thus providing proton motive force to drive ATP synthesis.

Photosystem II in alphaproteobacteria. Phototrophic alphaproteobacteria such as *Rhodospirillum rubrum* and *Blastochloris viridis* are typically found in wetlands and streams, where they capture light not used by other phototrophs. The antenna complex of *B. viridis* was shown earlier, in **Figure 14.36**. The peak wavelength absorbed by bacteriochlorophyll P870 lies so far into the infrared (800–1,100 nm) that the photon energy is insufficient to reduce NAD(P) to NAD(P)H. The electrons from P870 are transferred by low-potential quinols to a terminal cytochrome oxidoreductase (**Fig. 14.39**). The quinols pass their electrons to cytochromes, while moving $2H^+$ across the membrane.

When the electrons reach cytochrome *c*, they flow back to bacteriochlorophyll, where they can be reexcited by photon energy from the antenna complex. Because the electron path traces back to its source, the ETS of

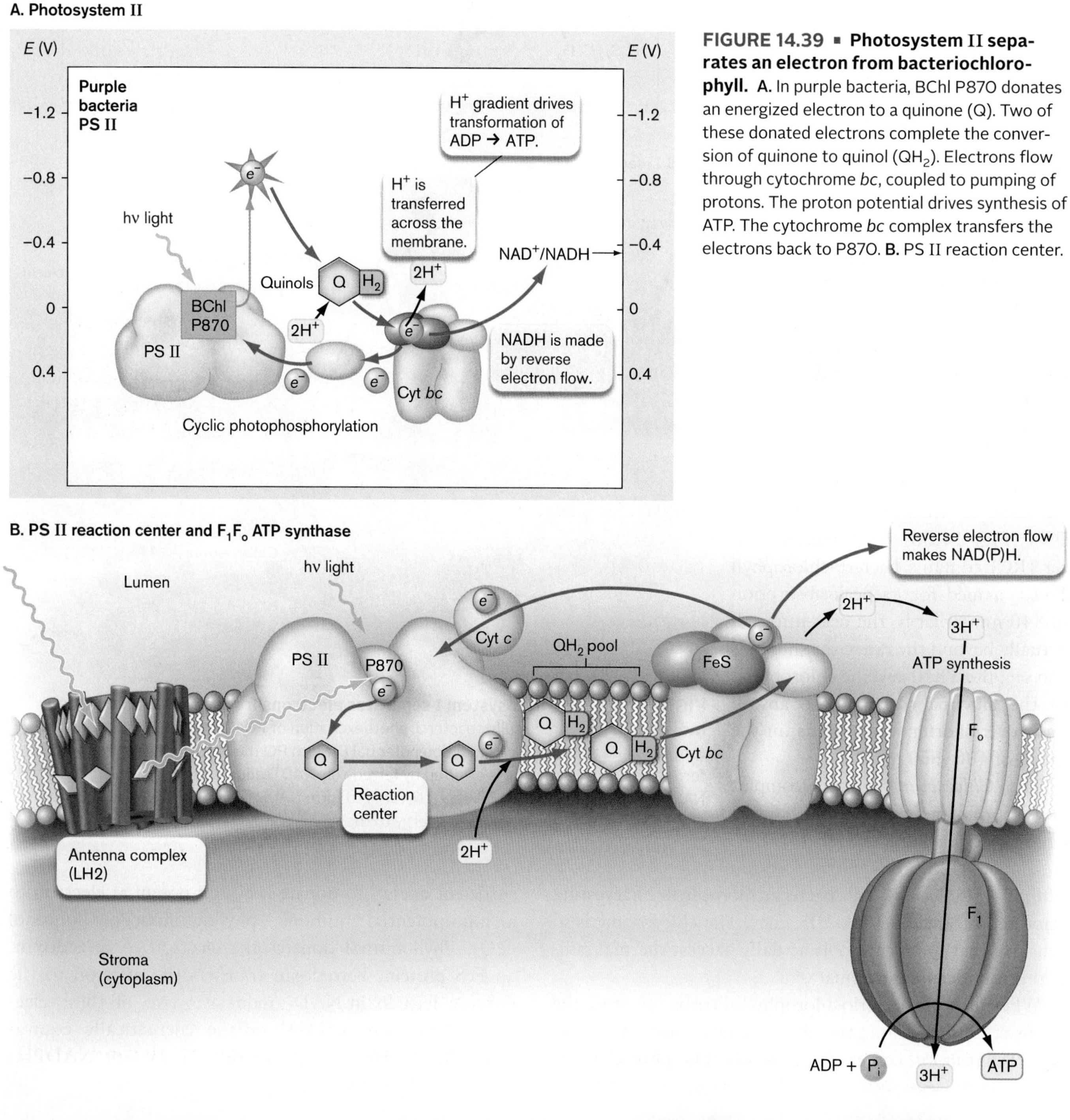

FIGURE 14.39 ■ Photosystem II separates an electron from bacteriochlorophyll. A. In purple bacteria, BChl P870 donates an energized electron to a quinone (Q). Two of these donated electrons complete the conversion of quinone to quinol (QH_2). Electrons flow through cytochrome *bc*, coupled to pumping of protons. The proton potential drives synthesis of ATP. The cytochrome *bc* complex transfers the electrons back to P870. **B.** PS II reaction center.

photosystem II leading to ATP synthesis is called **cyclic photophosphorylation**.

Since the reduction potential is too small to reduce $NADP^+$ to NADPH, photosystem II requires reverse electron flow. As we saw for iron oxidation (see Section 14.5), in reverse electron flow a low-potential electron donor reduces an ETS, requiring input of energy. Purple bacteria obtain this energy by spending ATP to increase the proton potential, or from a pathway outside photolysis, such as catabolism of organic compounds. The organic compounds also provide substrates for biosynthesis. Thus, most purple bacteria are photoheterotrophs.

Purple bacteria such as *Rhodopseudomonas palustris* are the ultimate generalists, often combining several major classes of metabolism. Caroline Harwood and colleagues at the University of Washington showed that *R. palustris* is a "photolithoheterotroph," capable of photosynthesis, catabolism on many substrates, and lithotrophy (**eTopic 14.2**). Photolithoheterotrophs are of interest for possible applications as solar-driven microbial fuel cells.

Thought Question

14.15 Suppose you discover bacteria that require a high concentration of Fe^{2+} for photosynthesis. Can you hypothesize what the role of Fe^{2+} might be? How would you test your hypothesis?

Oxygenic photolysis. The "Z" pathway of photolysis found in cyanobacteria and chloroplasts combines key features of both PS II and PS I (**Fig. 14.40**). Both reaction centers, however, contain chlorophylls that absorb at shorter wavelengths (higher energy) than those of the respective purple or green homologs: P680 instead of P870 (PS II), and P700 instead of P840 (PS I). Thus, the cyanobacterial reaction centers can split water—a highly stable molecule that cannot be photolyzed by anaerobic phototrophs. Photolysis of water requires greater energy input but ultimately yields greater energy overall, and produces molecular oxygen. The energy potentials are high enough to generate NADPH and fix CO_2 into biomass (presented in Chapter 15). Oxygenic phototrophs dominate the shallow water depths, whereas anaerobes grow at lower depths, using light at wavelengths unused by cyanobacteria and algae near the surface.

Cyanobacteria harvest light via an exceptionally efficient antenna complex called the phycobilisome (**Fig. 14.40B**). In the PS II reaction center, the photoexcitation of chlorophyll P680 yields enough energy to split H_2O. The entire cycle of water splitting involves $4H^+$ removed from $2H_2O$, $4e^-$ transferred to carriers, and the formation of O_2. The net reaction forming oxygen is:

$$2H_2O \rightarrow 4H^+ + 4e^- + O_2$$

Through the ETS, the electrons are transferred to quinones. As in respiration, each $2e^-$ reduction of quinone to quinol requires pickup of $2H^+$ from the stroma (equivalent to cytoplasm). Thus, the four electrons transferred generate a net change in the proton gradient of $4H^+$. Furthermore, the energy of electron transfer to cytochrome *bf* enables pumping of an additional $2H^+$ across the membrane. Thus, in all, for each conversion of $2H_2O$ to O_2 the net protons transferred across the thylakoid membrane include $4H^+$ (water photolysis) plus $4 \times 2H^+$ (quinones to quinols) through cytochrome *bf*, to yield a total of $12H^+$ for the proton gradient. The proton gradient drives the ATP synthase to make approximately 3 ATP per O_2 formed.

The electrons from PS II do not cycle back to the PS II reaction center, as they do in purple bacteria. Instead they are transferred to PS I by a protein called plastocyanin. The energy of the electron transferred by plastocyanin is augmented through absorption of a second photon by the chlorophyll of PS I. Subsequent electron flow through ferredoxin can now generate NADH or NADPH. Some of the electron flow instead cycles back to cytochrome *bf*, where it contributes to pumping protons.

The overall equation for energy yield of oxygenic photolysis can be represented as:

$$2H_2O + 2\ NADP^+ + 3\ [ADP + P_i] \rightarrow$$
$$O_2 + 2\ [NADPH + H^+] + 3\ ATP + 3H_2O$$

To release one O_2 and fix one CO_2 requires absorption of between 8 and 12 photons. The efficiency—that is, the proportion of photon energy converted to CO_2 fixation—is estimated to be 20%–30%. To generate one molecule of glucose by CO_2 fixation, we need six rounds of the photolysis equation—one per CO_2 molecule "fixed" into sugar ($C_6H_{12}O_6$; discussed in Chapter 15). The CO_2 fixation equation works out to:

$$12H_2O + 12\ NADP^+ + 18\ [ADP + P_i] \rightarrow$$
$$6O_2 + 12\ [NADPH + H^+] + 18\ ATP + 18H_2O$$

$$\mathbf{6CO_2} + 12\ [NADPH + H^+] + 18\ ATP + 18H_2O \rightarrow$$
$$\mathbf{C_6H_{12}O_6} + 6H_2O + 12\ NADP^+ + 18\ [ADP + P_i]$$

$$12H_2O + \mathbf{6CO_2} \rightarrow \mathbf{C_6H_{12}O_6} + 6H_2O + 6O_2$$

To Summarize

- **Bacteriorhodopsin** and **proteorhodopsin**—found in haloarchaea and in bacteria, respectively—are forms of a light-driven proton pump that contains retinal. The energy gained from light absorption supplements heterotrophy.
- **Chlorophylls and bacteriochlorophylls are photopigments that absorb light and transfer energy to an ETS.** Chlorophylls in cyanobacteria and in plant chloroplasts participate in oxygenic photosynthesis, whereas bacteriochlorophylls participate in phototrophy that does not produce oxygen.

A. Z pathway of oxygenic photosynthesis

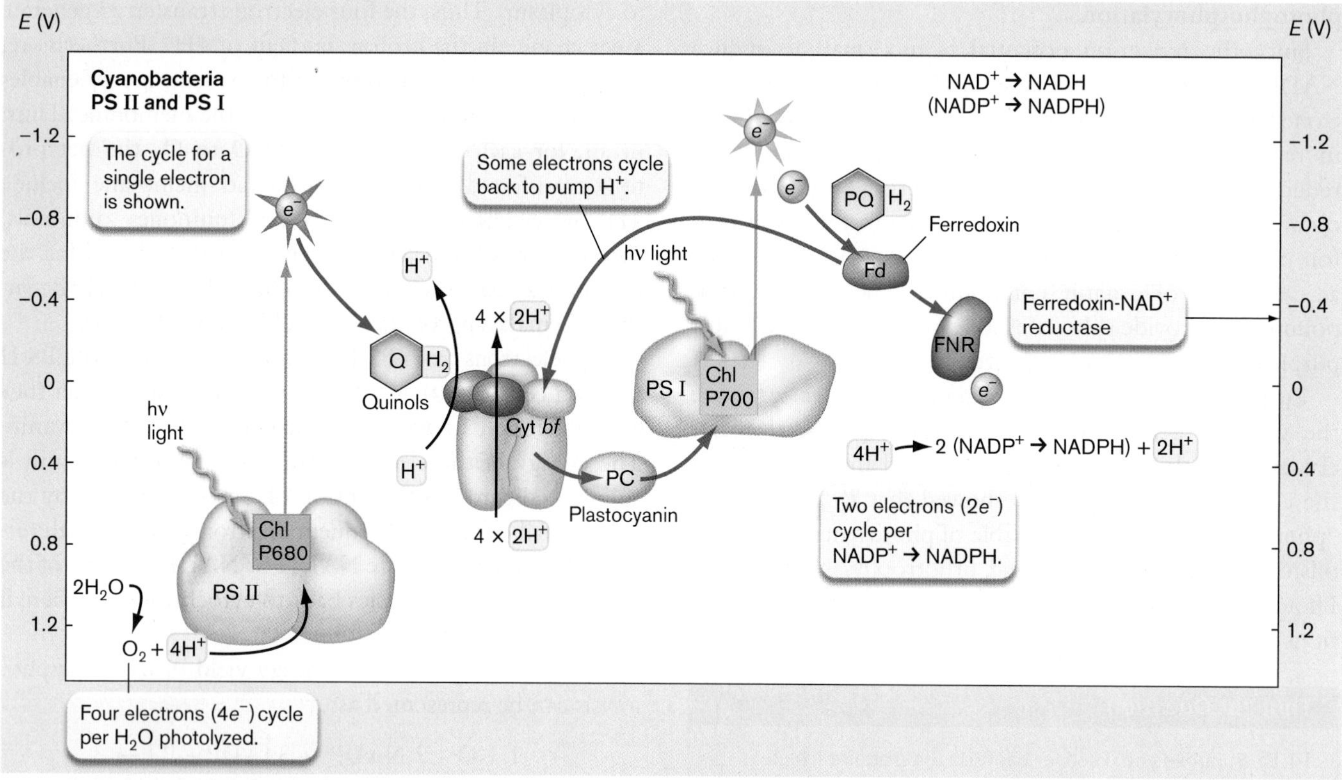

B. Z pathway reaction complexes and ATP synthase

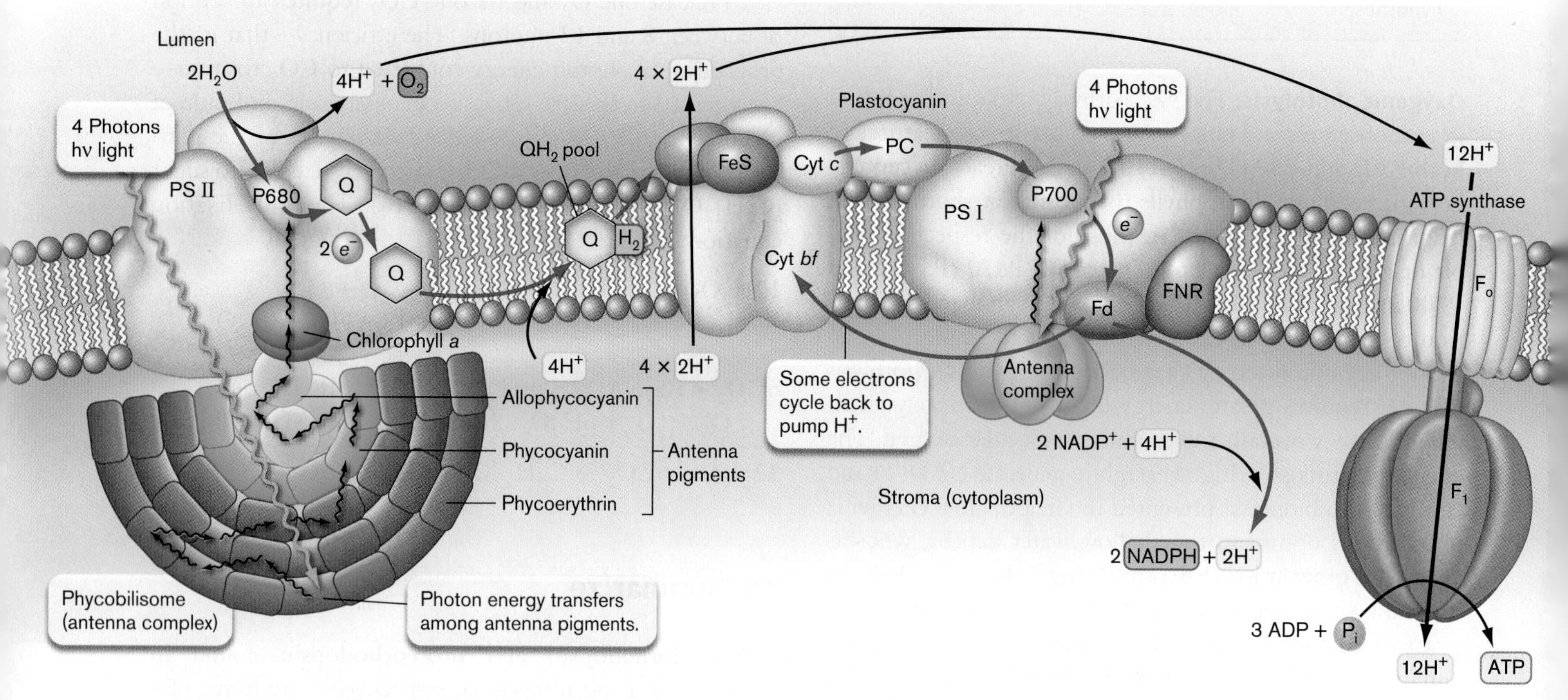

FIGURE 14.40 ■ Oxygenic photosynthesis in cyanobacteria and chloroplasts. **A.** Each H_2O is photolyzed via the Z pathway of PS II and PS I. The $2e^-$ from each water molecule ($4e^-$ in all) are transferred to the quinone pool, which transfers $4e^-$ to cytochrome *bf*. Cytochrome *bf* pumps $4 \times 2H^+$ across the membrane and transfers $4e^-$ via plastocyanin to a PS I containing chlorophyll P700. A second photon excites P700, enabling transfer of e^- to ferredoxin ($4e^-$ per O_2 formed), and from there to $NADP^+ \rightarrow NADPH$. **B.** The Z pathway within a photosynthetic membrane. *Source:* Part B based on crystallographic data from thermophilic cyanobacteria, modified from Genji Kurisu et al. 2003. *Science* **302**:1009.

- **The antenna complex** of chlorophylls and other photopigments captures light for transfer to the reaction center in chlorophyll-based photosynthesis.
- **Thylakoids** are folded membranes within phototrophic bacteria or chloroplasts. The membranes extend the area for chlorophyll light absorption, and they separate two compartments to form a proton gradient.
- **Photosystem I** obtains electrons from H_2S or HS^-. The electrons are transferred through an ETS to form NADH or NADPH.
- **Photosystem II** transfers an electron through an ETS and pumps H^+ to generate ATP. An electron ultimately returns to bacteriochlorophyll through cyclic photophosphorylation. Reverse electron flow generates NADPH or NADH.
- **The oxygenic Z pathway in cyanobacteria and chloroplasts** includes homologs of photosystems I and II. Eight photons are absorbed, and two electron pairs are removed from $2H_2O$, ultimately producing O_2.
- **Oxygenic photosynthesis generates 3 ATP + 2 NADPH** per $2H_2O$ photolyzed and O_2 produced. The ATP and NADPH are used to fix CO_2 into biomass.

Concluding Thoughts

Seemingly disparate biochemical means of nutrition, including organotrophy, lithotrophy, and photosynthesis, share common mechanisms of electron flow and proton transfer through an ETS. A remarkable consequence of these reactions is the oxygenation of our planet—the continual release of oxygen from photosynthesis by bacteria and plant chloroplasts—and the reduction of that oxygen by respiring heterotrophs. Disruption of these redox cycles by climate change will have profound consequences for our planet.

A recurring theme is that all forms of metabolism involve electron transfer reactions that yield energy for cell function. As molecules are rearranged by transfer of electrons from one substrate to another, energy is provided to form ion gradients and energy carriers. The energy carriers always need to be balanced between redox-neutral carriers, such as ATP, and reducing carriers, such as NADH or NADPH, for biosynthesis (as discussed in Chapter 15).

Whatever the means of obtaining energy, ultimately cells must spend energy for biosynthesis. Chapter 15 presents how microbes construct the fundamental "nuts and bolts" of their cells—including products surprisingly useful for biotechnology.

CHAPTER REVIEW

Review Questions

1. Explain the source of electrons and the sink for electrons (terminal electron acceptor) in respiration, lithotrophy, and photolysis.
2. How do bacteria combine redox couples for a metabolic reaction that yields energy? Cite examples, calculating the reduction potential.
3. How do environmental conditions affect the reduction potential of a metabolic reaction?
4. Explain the role of cytochromes and redox cofactors in electron transport systems. What features of a molecule make it useful for redox biochemistry?
5. Explain how a proton potential is composed of a chemical concentration difference plus a charge difference. Explain how each component of Δp can drive a cellular reaction.
6. Explain the role of the substrate dehydrogenase (oxidoreductase), the quinones, and the cytochrome oxidase (oxidoreductase) in the respiratory ETS.
7. Compare the ETS function in lithotrophy with that in respiration.
8. Summarize the inorganic redox couples that can be used in anaerobic respiration and those that can be used in lithotrophy. What constraints determine whether a given molecule can serve as electron acceptor or as electron donor?
9. How do diverse forms of anaerobic respiration and lithotrophy contribute to ecosystems?
10. Explain the differences and common features of bacteriorhodopsin phototrophy and chlorophyll phototrophy.
11. Explain the differences and common features of photosystems I and II. Explain how the two photosystems combine in the Z pathway. Why can the Z pathway generate oxygen, whereas PS I and PS II cannot?

Thought Questions

1. The lung pathogen *Pseudomonas aeruginosa*, which also grows in soil, can respire aerobically or else anaerobically using nitrate. Under what conditions would *P. aeruginosa* use each form of metabolism? What part of its ETS would need to change to accommodate the different forms?
2. What environments favor oxygenic photosynthesis, versus sulfur phototrophy? Explain.
3. In pathogens, which components of the ETS do you think would make good targets for new antibiotics, and why?
4. Devise a form of energy-yielding metabolism in which fumarate is converted to succinate; and a different form, in which succinate is converted to fumarate. Explain why the two reactions are reasonable and, on the Internet, try to find actual organisms that obtain energy through these reactions.

Key Terms

anaerobic respiration (551)
anammox reaction (561)
antenna complex (570)
bacteriochlorophyll (569, 570)
bacteriorhodopsin (567, 570)
carotenoid (569, 570)
chlorophyll (569, 570)
chromophore (569)
cofactor (544)
cyclic photophosphorylation (575)
cytochrome (538)
dissimilatory denitrification (552)
dissimilatory metal reduction (552)
electrogenic biofilm (553)
electron acceptor (535)
electron donor (535)
electron shuttle (555)
electron transport system (ETS) (electron transport chain, ETC) (534, 535)
ferredoxin (573)
heme (539)
hydrogenotrophy (564)
lithotrophy (chemolithotrophy) (534, 535, 557, 566)
methanogenesis (564)
methanotrophy (565)
methylotrophy (565)
nanowire (553)
nitrifier (561)
organotrophy (chemoorganotrophy) (534, 535)
outer membrane–associated protein (555)
oxidative phosphorylation (549)
oxidoreductase (539, 545)
oxygenic Z pathway (572)
photoexcitation (568)
photoheterotrophy (568)
photoionization (568)
photolysis (568)
photosynthesis (568)
photosystem I (PS I) (572)
photosystem II (PS II) (572)
phototrophy (534, 535)
proteorhodopsin (567)
proton motive force (proton potential) (535, 539)
quinol (544)
quinone (544)
reaction center (RC) (570)
redox couple (536)
respiration (535)
retinal (567)
reverse electron flow (559)
standard reduction potential ($E°'$) (536)
thylakoid (571)
uncoupler (542)

Recommended Reading

Andries, Koen, Peter Verhasselt, Jerome Guillemont, Hinrich W. H. Göhlmann, Jean-Marc Neefs, et al. 2005. A diarylquinoline drug active on the ATP synthase of *Mycobacterium tuberculosis. Science* **307**:223–227.

Beatty, J. Thomas, Jörg Overmann, Michael T. Lince, Ann K. Manske, Andrew S. Lang, et al. 2005. An obligately photosynthetic bacterial anaerobe from a deep-sea hydrothermal vent. *Proceedings of the National Academy of Sciences USA* **102**:9306–9310.

Bjerg, Jesper T., Henricus T. S. Boschker, Steffen Larsen, David Berry, Markus Schmid, et al. 2018. Long-distance electron transport in individual, living cable bacteria. *Proceedings of the National Academy of Sciences USA* **115**:5786–5791.

Farha, Maya A., Chris P. Verschoor, Dawn Bowdish, and Eric D. Brown. 2013. Collapsing the proton motive force to identify synergistic combinations against *Staphylococcus aureus. Chemistry & Biology* **20**:1168–1178.

Glasser, Nathaniel R., Scott H. Saunders, and Dianne K. Newman. 2017. The colorful world of extracellular electron shuttles. *Annual Reviews of Microbiology* **71**:731–751.

Jones, Shari A., Fatema Z. Chowdhury, Andrew J. Fabich, April Anderson, Darrel M. Schreiner, et al. 2009. Respiration of *Escherichia coli* in the mouse intestine. *Infection and Immunity* **75**:4891–4899.

Kracke, Frauke, Igor Vassilev, and Jens O. Krömer. 2015. Microbial electron transport and energy conservation—the foundation for optimizing bioelectrochemical systems. *Frontiers in Microbiology* **6**:575.

Kuenen, J. Gijs. 2008. Anammox bacteria: From discovery to application. *Nature Reviews. Microbiology* **6**:320–326.

Lobritz, Michael A., Peter Belenky, Caroline B. M. Porter, Arnaud Gutierrez, Jason H. Yang, et al. 2015. Antibiotic efficacy is linked to bacterial cellular respiration. *Proceedings of the National Academy of Sciences USA* **112**:8173–8180.

Lovley, Derek R. 2017. Happy together: Microbial communities that hook up to swap electrons. *ISME Journal* **11**:327–336.

Malvankar, Nikhil S., Madeline Vargas, Kelly P. Nevin, Ashley E. Franks, Ching Leang, et al. 2011. Tunable metallic-like conductivity in microbial nanowire networks. *Nature Nanotechnology* **6**:573–579.

Raghoebarsing, Ashna A., Arjan Pol, Katinka T. van de Pas-Schoonen, Alfons J. P. Smolders, Katharina F. Ettwig, et al. 2006. A microbial consortium couples anaerobic methane oxidation to denitrification. *Nature* **440**:918–921.

Reguera, Gemma. 2018. Harnessing the power of microbial nanowires (minireview). 2018. *Microbial Biotechnology* **11**:979–994.

Shrestha, Pravin M., and Amelia-Elena Rotaru. 2014. Plugging in or going wireless strategies for interspecies electron transfer. *Frontiers in Microbiology* **5**:237.

Steidl, Rebecca J., Sanela Lampa-Pastirk, and Gemma Reguera. 2016. Mechanistic stratification in electroactive biofilms of *Geobacter sulfurreducens* mediated by pilus nanowires. *Nature Communications* 7:12217.

CHAPTER 15

Biosynthesis

15.1 Overview of Biosynthesis

15.2 CO_2 Fixation: The Calvin Cycle and Other Pathways

15.3 Fatty Acids and Antibiotics

15.4 Nitrogen Fixation and Regulation

15.5 Amino Acids and Nitrogenous Bases

How do microbes build their cells? Some bacteria and archaea can build themselves entirely from carbon dioxide and nitrogen gas, plus a few salts. The cell assimilates carbon and nitrogen into small molecules, and then assembles more complex structures. How do cells organize their biosynthesis to build precisely the forms they need? How do bacteria avoid wasting energy on excess production? Microbial genomes reveal biosynthesis of antibiotics and antitumor agents that we can pursue using our biotechnology.

CURRENT RESEARCH highlight

Marine actinomycetes make lifesaving drugs. The oceans offer countless unknown microbes that synthesize molecules with antimicrobial properties. Paul Jensen discovered a new genus of actinomycetes, *Salinispora*, named for its culture requirement of seawater. A species such as *Salinispora arenicola* may produce ten different secondary products with potential properties as antibiotics or anticancer agents. Jensen sequenced the bacterial genomes, revealing new gene clusters for biosynthesis. "Orphan" gene clusters thought to be silent actually express enough enzymes to make products. Once identified, the products may be drugs with lifesaving potential.

Source: Gregory Amos et al. 2017. *PNAS* **114**:E11121.

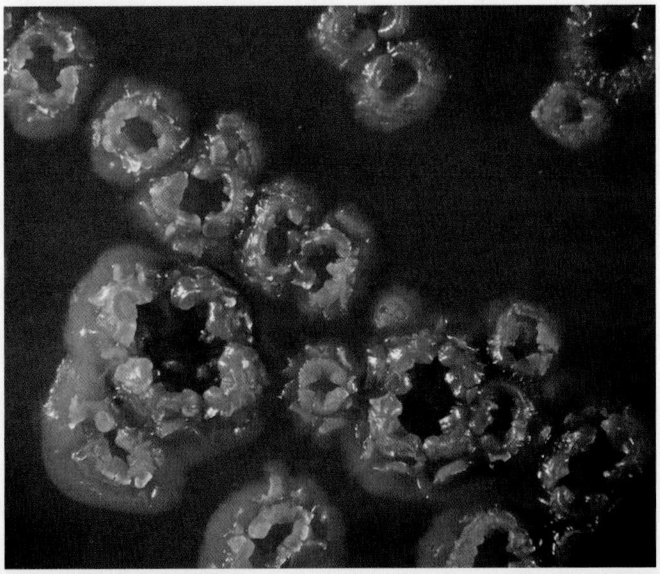

COURTESY OF PAUL JENSEN

AN INTERVIEW WITH

PAUL JENSEN, MICROBIOLOGIST, SCRIPPS INSTITUTION OF OCEANOGRAPHY, UC SAN DIEGO

COURTESY OF PAUL JENSEN

What is the most exciting thing about *Salinispora*?

Salinispora is an obligate marine actinomycete genus best known for the production of diverse natural products. Only a small percentage of the biosynthetic potential harbored by this genus has been realized. Accessing this potential has led to a new branch of research, called "genome mining," that aims to find the products of these "orphan" or "cryptic" gene clusters.

Natural product biosynthetic gene clusters have evolved to be exchanged among strains and provide an effective mechanism for bacteria to test the effects of small (and often biologically active) molecules on fitness. This adds to our growing understanding of the importance of horizontal gene transfer in bacterial evolution.

Every life form on Earth consists of biomass—a body built of carbon and other elements. Each atom of every molecule had to get there through biosynthesis, the process of an organism building itself.

To build a complex body from simple molecules, an organism must spend energy. In Chapters 13 and 14 we described how microbes gain energy from chemical reactions, to store the energy in ion gradients and in small molecules such as ATP and NADPH. Chapter 15 shows how the microbes spend this energy for biosynthesis.

Biosynthesis builds an extraordinary range of cell parts, from simple sugars and amino acids to exotic **secondary products** with bizarre forms that we cannot make in the laboratory. Secondary products are made by bacteria during stationary phase, such as antibiotics that enable the producer to outcompete other microbes for scarce nutrients. An example of a secondary product is salinipostin G, a bicyclic phosphodiester that inhibits eukaryotic parasites, such as *Plasmodium*, the cause of malaria (**Fig. 15.1**). This antimalarial agent was discovered as a product of the marine actinomycete *Salinispora pacifica*. Marine actinomycetes offer an enormous resource for biosynthesis of previously unknown products that might treat human diseases.

Chapter 15 presents the fundamental ways by which microbes build themselves. First, autotrophs must assimilate elements such as carbon and nitrogen to build useful carbon skeletons. We say that the organisms **fix** these elements—that is, combine the inorganic compounds into biomass, the organic molecules that form the cell. Microbial enzyme factories form simple organic building blocks such as acetyl groups and amino acids, and assemble these simple building blocks into remarkably complex biomolecules, including vitamins and antibiotics important for human health. Industrial applications of microbial biosynthesis are described in Chapter 16.

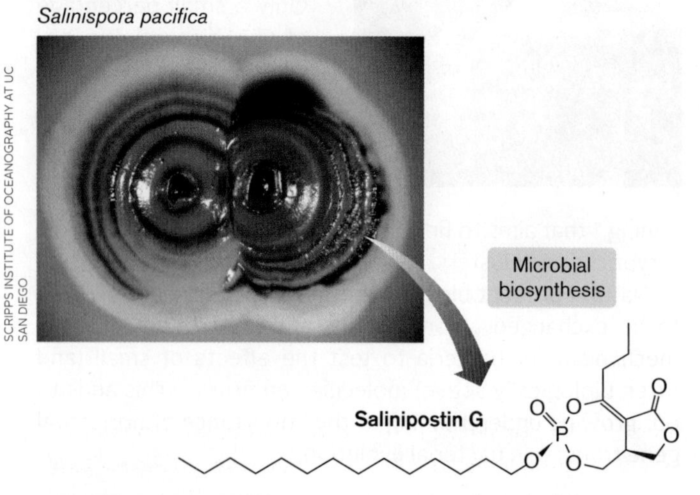

FIGURE 15.1 ■ Salinipostin G. This new antimalarial drug is synthesized by the actinomycete *Salinispora pacifica*.
Source: Salinipostin G structure modified from Gregory Amos et al. 2017. *PNAS* **114**:E11121–E11130, fig. 7.

15.1 Overview of Biosynthesis

Biosynthesis is the building of complex biomolecules, also known as **anabolism**, the reverse of catabolism. **Figure 15.2** presents an overview of biosynthesis, and how it relates to catabolic pathways we saw in Chapter 13. Enzyme pathways such as the TCA cycle are reversed to build up carbon skeletons, incorporating nitrogen to form amino acids. Ultimately, sugar monomers and amino acids are assembled to form polysaccharides, polypeptides, and the cell wall material peptidoglycan. Besides the biomass of the cell proper, cells synthesize and secrete products such as antibiotics, toxins for pathogenesis, quorum signals for cooperation, and matrix for biofilm (**eTopic 15.1**). All this biosynthesis requires substrates and energy.

Biosynthesis Requires Substrates

For their biosynthesis, where do microbes obtain their organic substrates, such as acetyl-CoA? Some microbes synthesize all their organic parts from minerals such as carbonate and nitrate (autotrophy). Two fundamental classes of autotrophy are photosynthesis and chemosynthesis. In **photosynthesis** or **photoautotrophy**, single carbon molecules (usually CO_2) are fixed into organic biomass, using energy from light absorption (discussed in Chapter 14). **Chemosynthesis** (or chemoautotrophy) fixes carbon dioxide via similar pathways, but without light. The energy for chemosynthesis usually comes from oxidation of minerals (lithotrophy; see Chapter 14). Other microorganisms must obtain organic substrates from their environment (heterotrophy). Many soil microbes are capable of both autotrophy and heterotrophy, depending on what their environment provides.

For all organisms, biosynthesis requires:

- **Essential elements.** Biosynthesis requires carbon, oxygen, hydrogen, nitrogen, and other essential elements. Carbon is obtained either through CO_2 fixation (autotrophy) or through acquisition of organic molecules made by other organisms (heterotrophy). Autotrophy assembles carbon and water into small molecules such as acetyl-CoA, which then serve as substrates or building blocks for the cell. By contrast, heterotrophy breaks down larger molecules such as carbohydrates and peptides to release acetyl-CoA and other small substrates. In addition to carbon, biosynthesis must assimilate nitrogen and sulfur for proteins, phosphorus for DNA, and metals for metal-containing enzymes.
- **Reduction.** Cell components such as lipids and amino acids are highly reduced. Their biosynthesis requires reducing a low-energy oxidized substrate, such as CO_2. A reducing agent such as NADPH hydrogenates the substrate and, in some cases, removes oxygen.

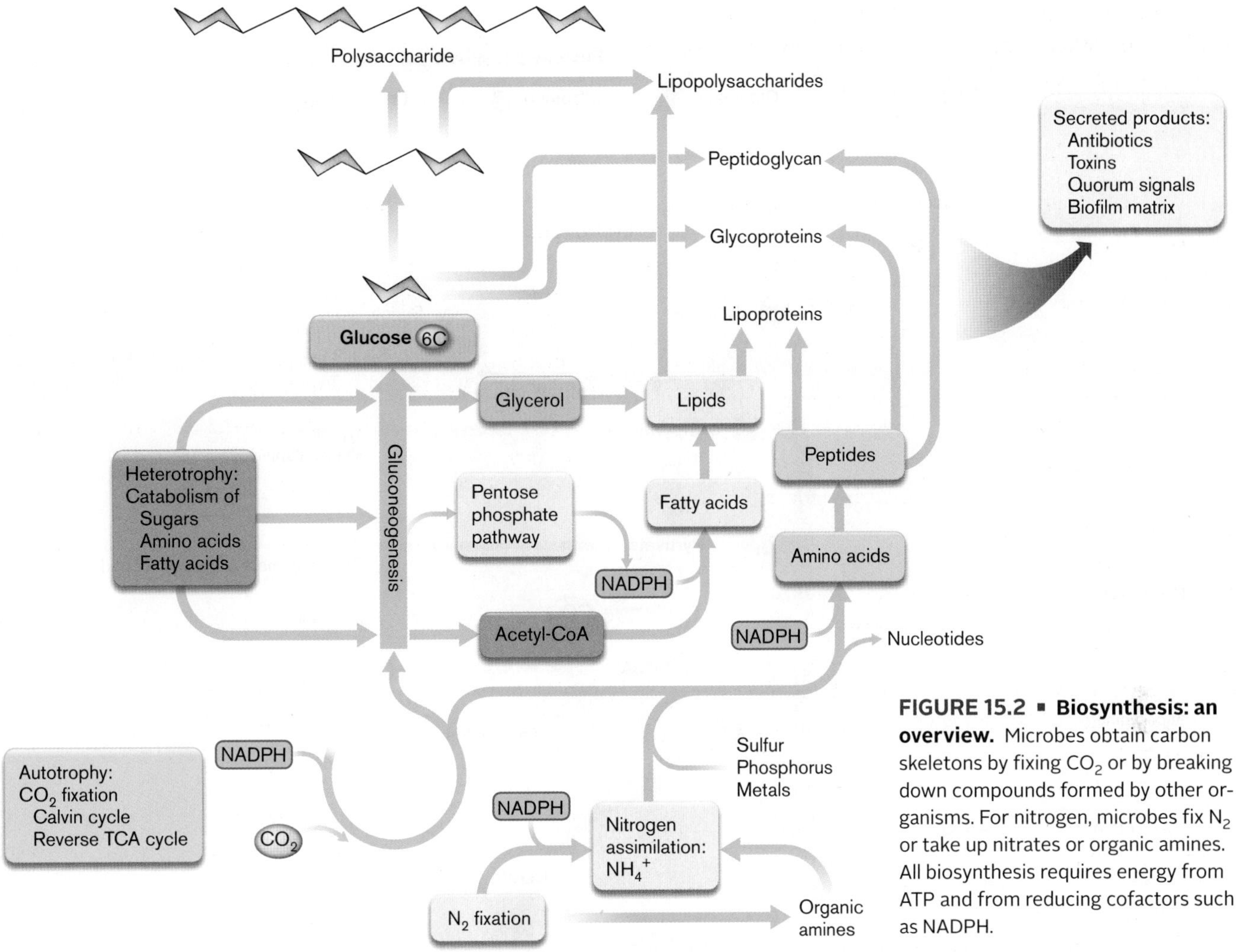

FIGURE 15.2 ■ Biosynthesis: an overview. Microbes obtain carbon skeletons by fixing CO_2 or by breaking down compounds formed by other organisms. For nitrogen, microbes fix N_2 or take up nitrates or organic amines. All biosynthesis requires energy from ATP and from reducing cofactors such as NADPH.

- **Energy.** Assembling small molecules into complex ones requires spending energy. Biosynthetic enzymes spend energy by coupling their reactions to the hydrolysis of ATP, the oxidation of NADPH, or the flux of ions down a transmembrane ion gradient.

Heterotrophs such as *E. coli* and actinomycetes obtain many substrates from glucose catabolism and the tricarboxylic acid (TCA) cycle, central catabolic pathways discussed in Chapter 13 (**Fig. 15.3**). For example, the succinyl-CoA molecules from the TCA cycle serve as the foundation of several kinds of amino acids, as well as vitamin B_{12}. Glycerol 3-phosphate provides the glyceride backbone of lipids. Pyruvate provides the backbone of several amino acids with aliphatic side chains. From the pentose phosphate pathway, erythrose 4-phosphate contributes to the ring structures of aromatic amino acids. Other amino acids derive their carbon skeleton from TCA cycle intermediates oxaloacetate and 2-oxoglutarate, which incorporate nitrogen in the form of ammonium ions (NH_4^+).

Note that glucose catabolism and the TCA cycle are both reversible. Some autotrophs can synthesize entire sugar molecules, starting with CO_2 and working up through the reverse TCA cycle to build glucose. The synthesis of glucose, called **gluconeogenesis**, reverses most of the enzymes of glycolysis. Thus, many common metabolites are available both to autotrophs and to heterotrophs, as well as to microbes capable of mixed metabolism.

Thought Question

15.1 To run the TCA cycle and glycolysis in reverse, does a cell use the same enzymes or different ones? Explain why some enzymes might be used in both directions, whereas other steps require different enzymes for catabolic and anabolic directions.

Biosynthesis Spends Energy

Biosynthesis costs energy in several ways. The organism's genome must maintain the DNA that encodes all the enzymes that catalyze all the steps of the pathway. The ribosomes spend energy to make enzymes. And the enzymes couple synthetic reactions to energy-releasing reactions. Collectively, these genomic and energetic costs generate enormous

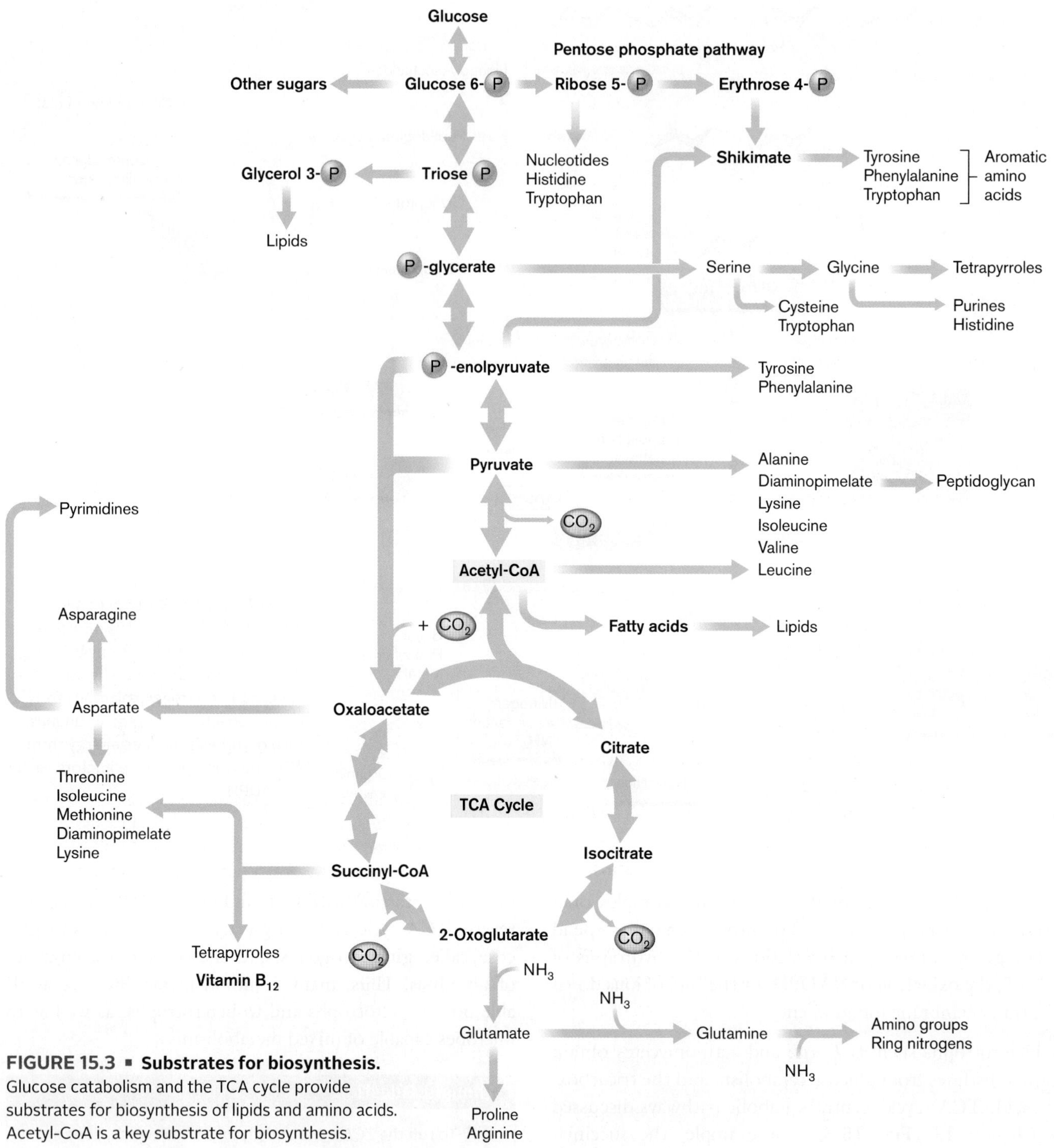

FIGURE 15.3 ■ Substrates for biosynthesis. Glucose catabolism and the TCA cycle provide substrates for biosynthesis of lipids and amino acids. Acetyl-CoA is a key substrate for biosynthesis.

selective pressure for microbes to evolve mechanisms that control these costs. The diversity of these cost-cutting mechanisms generates surprising ecological relationships.

- **Regulation.** Biosynthesis is regulated at multiple levels, such as gene transcription, protein synthesis, and allosteric enzyme control. In general, energy-expensive products block the expression or function of their biosynthetic enzymes. For example, as we saw in Chapter 10, if the cell's tryptophan concentration increases, the molecule binds a corepressor to block transcription of the genes for tryptophan biosynthesis, and the aminoacyl-tRNA stalls the ribosome midtranslation. In this way, microbes avoid making more than they need under given conditions. Microbes regulate their biosynthesis at the levels of transcription and translation of enzymes, as well as through feedback inhibition of enzyme activity. An important example of molecular regulation is nitrogen fixation, discussed in Section 15.4.

- **Competition and predation.** Free-living bacteria secrete antimicrobial agents that inhibit or kill competitors. The victors scavenge the spoils for raw materials. Actinomycetes, such as those discussed at the start of the chapter, possess large genomes that encode enzymes for biosynthesis of several antibiotics. High nutrient levels inhibit antibiotic production, but as the bacterial filaments grow more slowly in stationary phase, they start antibiotic production.
- **Genome loss and cooperation.** A drastic way to avoid the expense of biosynthesis is to lose the enzymes through reductive evolution (discussed in Chapter 17). But the species then depends on intricate relations with other community members. Parasites and mutualists that grow only within a host cell show the most extensive loss of biosynthetic genes. Each species takes an evolutionary path of maintaining the expense of a particular biosynthetic pathway, such as N_2 fixation, or of losing the pathway and becoming dependent on partner organisms.

Surprisingly, many marine phototrophs no longer encode certain biosynthetic pathways and depend on synthesis by partner heterotrophs. For example, on the sea ice of the Southern Ocean, Andrew Allen at the J. Craig Venter Institute discovered a fundamental mutualism of Antarctic diatoms and bacteria (**Fig. 15.4**). The Southern Ocean is highly productive for biomass and oxygen, because of the high solar irradiance and iron availability. Many of the sea ice phytoplankton (phototrophic algae) fail to synthesize vitamin B_{12}, a fundamental cofactor for enzymes but one of the most complicated biomolecules to synthesize. Allen's group found that the Antarctic diatoms obtain vitamin B_{12} from adherent bacteria. The major bacteria providing the vitamin are Oceanospirillales, an order of proteobacteria known for cleaning up oil spills (see Chapter 21).

Closer to home, our gut bacteria also evolve ingenious ways to share substrates and minimize biosynthesis. Samay Pande and co-workers at the Max Planck Institute for Chemical Biology, Jena, Germany, investigated resource sharing in *E. coli* by constructing a complementary pair of mutants in which each overproduced an amino acid (histidine or tryptophan). Pande found that when the growth medium lacks the two complementary biosynthesis pathways (for His or Trp), the two bacterial strains form nanotubes of phospholipids derived from the cell envelope (**Fig. 15.5**). But when the medium provides the amino acids needed, the nanotubes fail to form. Thus, even within the competitive gut microbial community, bacteria find ways to share excess carbon substrates for biosynthesis—and regulate their sharing to avoid helping a competitor when a ready source is available.

In this chapter we present the pathways by which autotrophs assimilate carbon and nitrogen into simple building blocks, such as acetyl-CoA. Then we present pathways of construction of the key parts of the cell, such as fatty acids and

A.

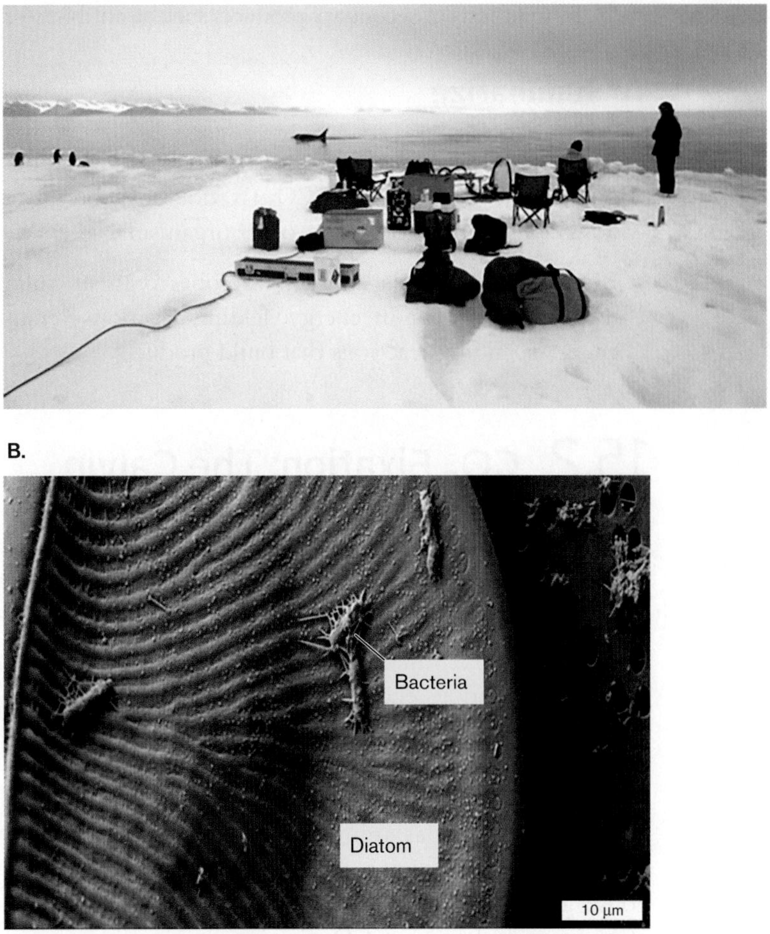

ANDREW ALLEN, SCRIPPS INSTITUTION OF OCEANOGRAPHY AND THE J. CRAIG VENTER INSTITUTE

GREG WANGER AND THE ALLEN LAB AT JCVI

FIGURE 15.4 ■ Bacteria provide vitamin B_{12} for Antarctic diatom. A. Explorers collect microbes on the ice edge, McMurdo Sound. Typical water temperature in December is –1.9°C. B. The diatom *Amphiprora* sp. hosts vitamin-producing bacteria, in sea ice of the Southern Ocean (SEM). *Source:* Eric Bertrand and Andrew Allen. 2012. *Front. Microbiol.* **3**:375.

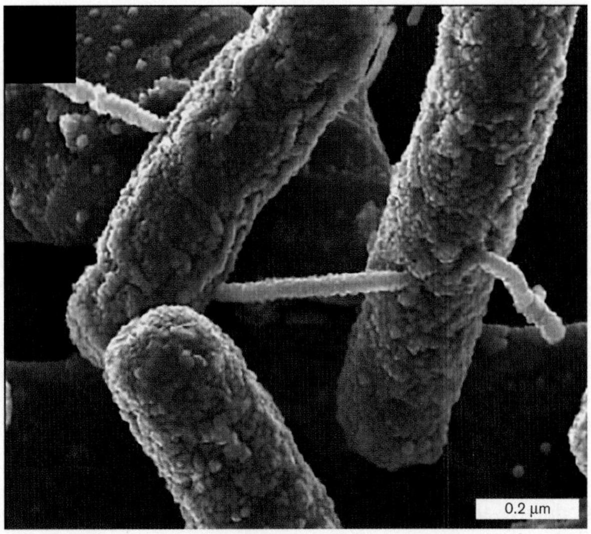

PANDE ET AL. 2015. *NAT. COMMUN.* **6**:6238, FIG. 5A.

FIGURE 15.5 ■ Sharing amino acids via nanotubes. *E. coli* strains connect via nanotubes to obtain the amino acid that each lacks but the other strain overproduces.

amino acids. Finally, we consider exciting recent discoveries in the biosynthesis of secondary products such as antibiotics.

To Summarize

- **Biosynthesis requires organic substrates.** Microbes obtain carbon skeletons by CO_2 fixation or by breaking down compounds formed by other organisms.
- **Biosynthesis spends energy.** Microbes regulate their relative investment in energy-yielding reactions versus energy-spending reactions that build products.

15.2 CO_2 Fixation: The Calvin Cycle and Other Pathways

The fundamental significance of **carbon dioxide fixation** by green plants and microbes was recognized in the early twentieth century. Fixing carbon into biomolecules requires tremendous energy input, as well as a large degree of reduction to incorporate hydrogen atoms. Bacteria have evolved a variety of cycles that fix CO_2 (**Table 15.1**). CO_2 fixation plays essential roles in soil, aquatic, and wetland ecosystems, as well as animal digestive systems. Globally, CO_2 fixation removes our atmosphere's most potent greenhouse gas—a major form of **carbon sequestration**. Our planet's survival requires CO_2 fixation.

The Calvin Cycle

The majority of the biomass on Earth consists of carbon fixed by chloroplasts and bacteria through the **reductive pentose phosphate cycle**, which recycles a pentose phosphate intermediate. This cycle is also known as the **Calvin cycle**, for which Melvin Calvin (1911–1997) was awarded the 1961 Nobel Prize in Chemistry. The full Calvin cycle is found only in bacteria and in chloroplasts, which evolved from bacteria. It has not been found in archaea or in the cytoplasm of eukaryotes. Archaea and some lithoautotrophic bacteria fix carbon by other pathways (**Table 15.1**). Thus, the Calvin cycle appears to have evolved after the divergence of the three domains of life.

The mechanism of the pentose phosphate cycle was solved by Calvin with colleagues Andrew Benson and James Bassham, at UC Berkeley. The importance of the Calvin cycle to the global ecosystem can scarcely be overestimated; it plays a major role in removing atmospheric CO_2. The Calvin cycle's responsiveness to CO_2, temperature, and other factors must be considered in all models of global warming.

Note: The Calvin cycle is also known as the **Calvin-Benson cycle**, the **Calvin-Benson-Bassham cycle**, or the **CBB cycle**. This book uses the terms "Calvin cycle" and "CBB cycle."

Several categories of bacteria conduct the Calvin cycle (**Fig. 15.6** and **Table 15.1**). Photoautotrophs such as cyanobacteria—and all plant chloroplasts—use the Calvin cycle to fix CO_2. Some photoheterotrophs, such as *Rhodobacter sphaeroides*, use the Calvin cycle only when light is available; in the dark, they use preformed substrates (heterotrophy). Still other bacteria fix CO_2 using energy from metal oxidation (lithoautotrophy; see Chapter 14). For example, *Acidithiobacillus ferrooxidans* (**Fig. 15.6C**) oxidizes sulfur and iron, generating acid that corrodes concrete.

- **Oxygenic phototrophic bacteria (cyanobacteria) and the chloroplasts of algae and multicellular plants fix CO_2** by the Calvin cycle coupled to oxygenic photosynthesis. The Calvin cycle is also called the "dark reactions,"

TABLE 15.1 Carbon Dioxide Fixation Pathways

	Organisms in which pathways occur		
Pathway	**Bacteria**	**Archaea**	**Eukaryotes**
	Photoautotrophs, photoheterotrophs, and lithoautotrophs		
Calvin cycle	Cyanobacteria; purple phototrophs; lithotrophs	Archaea have Rubisco homologs, but their function is unclear	Chloroplasts
	Lithoautotrophic bacteria and archaea		
Reductive (reverse) TCA cycle	Green sulfur phototrophs (*Chlorobium*); thermophilic epsilonproteobacteria	Hyperthermophilic sulfur oxidizers (*Thermoproteus* and *Pyrobaculum*)	Anaplerotic reactions fix CO_2 to regenerate TCA intermediates
Reductive acetyl-CoA pathway	Anaerobes: acetogenic bacteria and sulfate reducers	Methanogens; other anaerobes	None known
3-Hydroxypropionate cycle	Green phototrophs (*Chloroflexus*)	Aerobic sulfur oxidizers (*Sulfolobus*)	None known

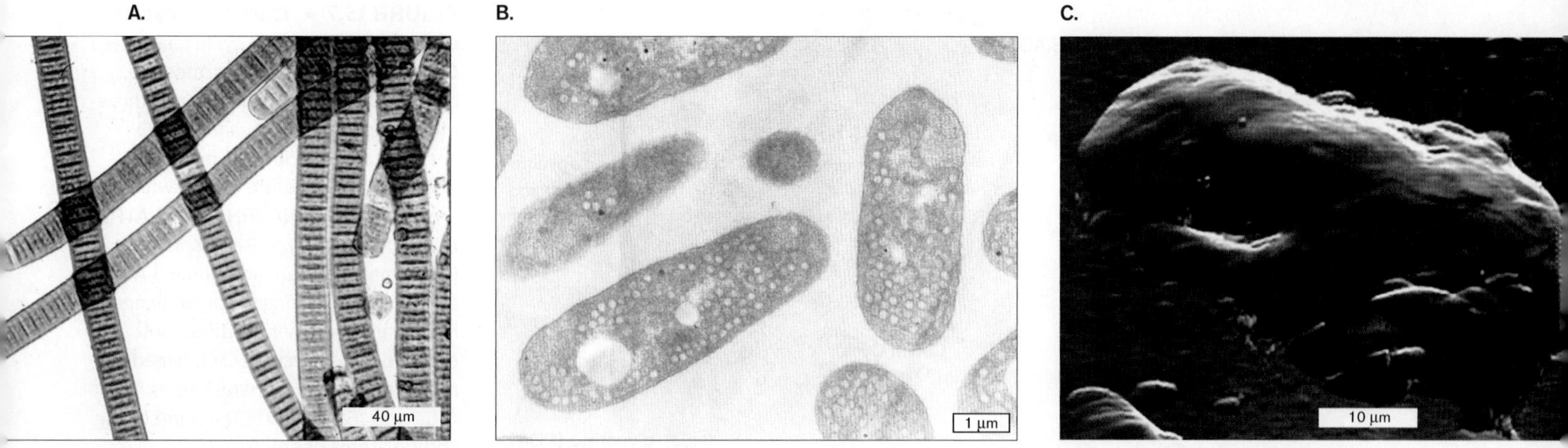

MICHAEL CLAYTON (EMERITUS), UNIVERSITY OF WISCONSIN–MADISON

COURTESY OF JILL ZEILSTRA-RYALLS

S. N. TAN AND M. CHEN. 2012. *HYDROMETALLURGY* **119–120**:87–94.

FIGURE 15.6 ■ Phototrophs and lithotrophs fix carbon via the Calvin cycle. **A.** *Oscillatoria*, a filamentous cyanobacterium, fixes CO_2 by oxygenic photosynthesis (LM). **B.** *Rhodobacter sphaeroides*, purple photoheterotrophs with spherical photosynthetic membranes (TEM). **C.** *Acidithiobacillus ferrooxidans* corrodes concrete while oxidizing sulfur and iron (AFM). This form of lithotrophy uses the Calvin cycle to fix CO_2.

or "light-independent reactions," of oxygenic photosynthesis. Cyanobacteria are believed to generate the majority of the oxygen gas in Earth's atmosphere.

- **Facultatively anaerobic purple bacteria**, including sulfur oxidizers and photoheterotrophs such as *Rhodospirillum* and *Rhodobacter*, use the Calvin cycle to fix CO_2. These bacteria also obtain carbon through catabolism.
- **Lithoautotrophic bacteria** fix CO_2 through the Calvin cycle, using NADPH and ATP provided by the oxidation of minerals. Mineral oxidation actually consumes oxygen.

Overview of the Calvin Cycle

The Calvin cycle is a process by which each CO_2 becomes one "corner" of a glucose molecule. Alternatively, the fixed carbon can enter biosynthetic pathways for amino acids, vitamins, and other essential components of cells.

Early in the twentieth century, biochemists tried to figure out the mechanism of CO_2 fixation, believing that agricultural photosynthesis could be made more efficient. With the tools then available, however, researchers had no hope of sorting out the intermediate products through which CO_2 was fixed. A fundamental breakthrough was the use of tracer radioisotopes, which are specific compounds labeled with radioactivity. The discovery of the carbon isotope ^{14}C by Martin Kamen (1913–2002) in 1940 revolutionized biochemistry, enabling the discovery of all kinds of cell metabolism (**eTopic 15.2**). Another key technique was paper chromatography, a means of separating labeled compounds based on differential migration in a solvent. This technique reveals short-lived intermediate compounds of a cycle. Calvin used paper chromatography for his Nobel-winning experiments to figure out the cycle that bears his name (**eTopic 15.3**).

Figure 15.7 shows an overview of the key steps of the Calvin cycle. The cycle begins with CO_2 fixation by ribulose 1,5-bisphosphate, a key intermediate that can derive from the pentose phosphate pathway (presented in Chapter 13). The addition of CO_2 is catalyzed by the enzyme **Rubisco** (ribulose 1,5-bisphosphate carboxylase/oxygenase) (**Fig. 15.8**). Rubisco is believed to be the most abundant protein on Earth.

In each "turn" of the Calvin cycle (**Fig. 15.7**), one molecule of CO_2 is condensed (combined, forming a new C–C bond) with the five-carbon sugar ribulose 1,5-bisphosphate. After CO_2 is fixed by Rubisco, the resulting six-carbon intermediate splits into two molecules of 3-phosphoglycerate (PGA). The fixed CO_2 ends up as a carbon of glyceraldehyde 3-phosphate (G3P). The conversion of PGA to G3P and sugars is detailed in **Figure 15.9**.

For every three turns of the cycle, fixing three molecules of CO_2, the cycle feeds one molecule of G3P ($C_3H_5O_3$–PO_3^{2-}) into biosynthesis:

$$3CO_2 + 6\ NADPH + 6H^+ + 9\ ATP + 9H_2O \rightarrow C_3H_5O_3\text{–}PO_3^{2-} + 6\ NADP^+ + 9\ ADP + 8\ P_i$$

The H_2O and the phosphoryl group of G3P are ultimately recycled during biosynthetic assimilation of G3P. Two molecules of G3P may condense (that is, form a new C–C bond) in a pathway to synthesize glucose. The overall condensation of $6CO_2 \rightarrow 2$ G3P $\rightarrow$ glucose looks like this:

$$6CO_2 + 12\ NADPH + 12H^+ + 18\ ATP + 18H_2O \rightarrow C_6H_{12}O_6 + 12\ NADP^+ + 18\ ADP + 18\ P_i + 6H_2O$$

Each carbon fixed requires reduction by $2H^+ + 2e^-$ from NADPH + H^+. Recall from Chapter 13 that NADPH is

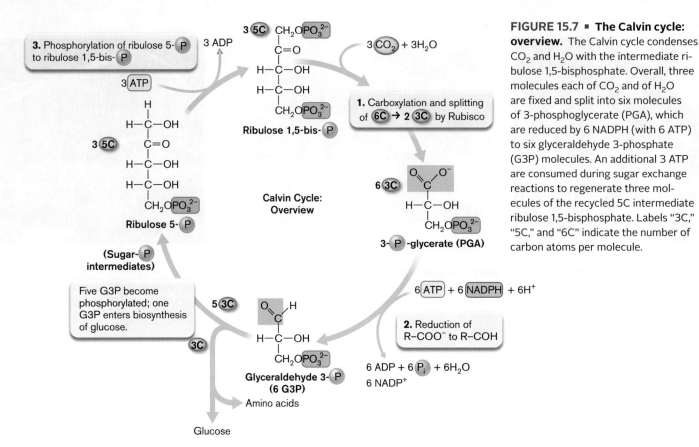

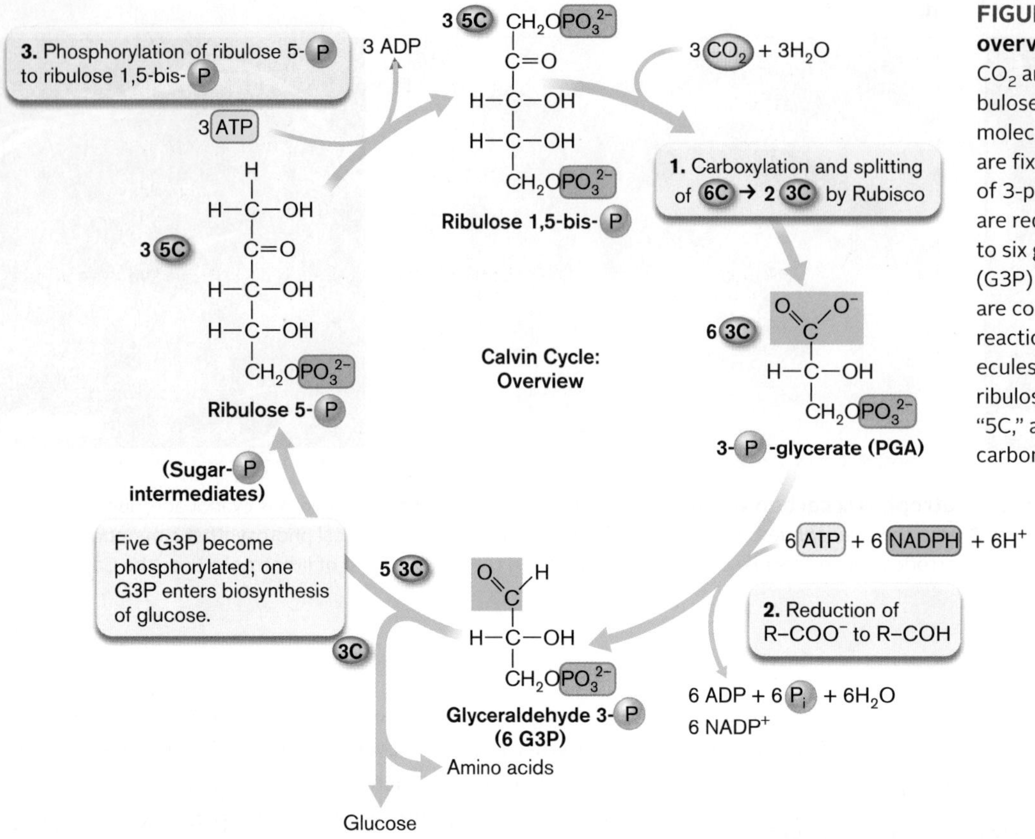

FIGURE 15.7 ■ The Calvin cycle: overview. The Calvin cycle condenses CO_2 and H_2O with the intermediate ribulose 1,5-bisphosphate. Overall, three molecules each of CO_2 and of H_2O are fixed and split into six molecules of 3-phosphoglycerate (PGA), which are reduced by 6 NADPH (with 6 ATP) to six glyceraldehyde 3-phosphate (G3P) molecules. An additional 3 ATP are consumed during sugar exchange reactions to regenerate three molecules of the recycled 5C intermediate ribulose 1,5-bisphosphate. Labels "3C," "5C," and "6C" indicate the number of carbon atoms per molecule.

a phosphorylated derivative of NADH commonly associated with biosynthesis. The phosphoryl group of NADPH, however, does not participate in energy transfer.

Glyceraldehyde 3-phosphate (G3P) is the fundamental unit of carbon assimilation into sugars or amino acids (**Fig. 15.9**).

Ribulose 1,5-Bisphosphate Incorporates CO_2

Each turn of the Calvin cycle has three main phases: carboxylation and splitting into two three-carbon (3C) intermediates, reduction of two molecules of PGA to two molecules of G3P, and regeneration of ribulose 1,5-bisphosphate (see **Fig. 15.7**):

1. **Carboxylation and splitting: 6C → 2[3C].** Ribulose 1,5-bisphosphate condenses with CO_2 and H_2O, mediated by Rubisco. Rubisco generates a six-carbon intermediate, which immediately hydrolyzes (splits into two parts by incorporating H_2O). The split produces two molecules of PGA, one of which contains the CO_2 fixed by this cycle (**Fig. 15.8**).
2. **Reduction of PGA to G3P.** The carboxyl group of each PGA molecule is phosphorylated by ATP. The phosphorylated carboxyl group is then hydrolyzed and reduced by NADPH, forming G3P.
3. **Regeneration of ribulose 1,5-bisphosphate.** Of every six G3P, resulting from three cycles of fixing CO_2, five G3P enter a complex series of reactions (including hydrolysis of 3 ATP) to regenerate three molecules of ribulose 1,5-bisphosphate (**Fig. 15.9**). The net conversion of five G3P molecules to three molecules of ribulose 1,5-bisphosphate releases $2H_2O$, restoring two of the $3H_2O$ fixed with $3CO_2$. The remaining sixth G3P exits the cycle, available to be used in the biosynthesis of sugars and amino acids. Thus, three fixed carbons lead to one three-carbon product.

Overall, each CO_2 fixed sends one carbon into biosynthesis and regenerates one ribulose 1,5-bisphosphate.

Rubisco catalyzes CO_2 incorporation. Rubisco is the key enzyme of the Calvin cycle. The CO_2-fixing enzyme is unique to bacteria and chloroplasts. Some archaeal genomes do show a homolog of Rubisco, but the archaeal enzyme does not fix CO_2. Instead, the archaeal Rubisco forms PGA out of adenine, through a "scavenging pathway" of catabolism.

The structure of Rubisco is highly conserved across bacterial and chloroplast domains. It consists of two types of subunits, designated small and large (**Fig. 15.8A**). The large (L) subunit contains the active (catalytic) site. The function of the small (S) subunit remains unclear; some bacteria, such as *Rhodospirillum rubrum*, have a Rubisco with no small subunits. Different species contain different multiples of the small and large subunits: eight

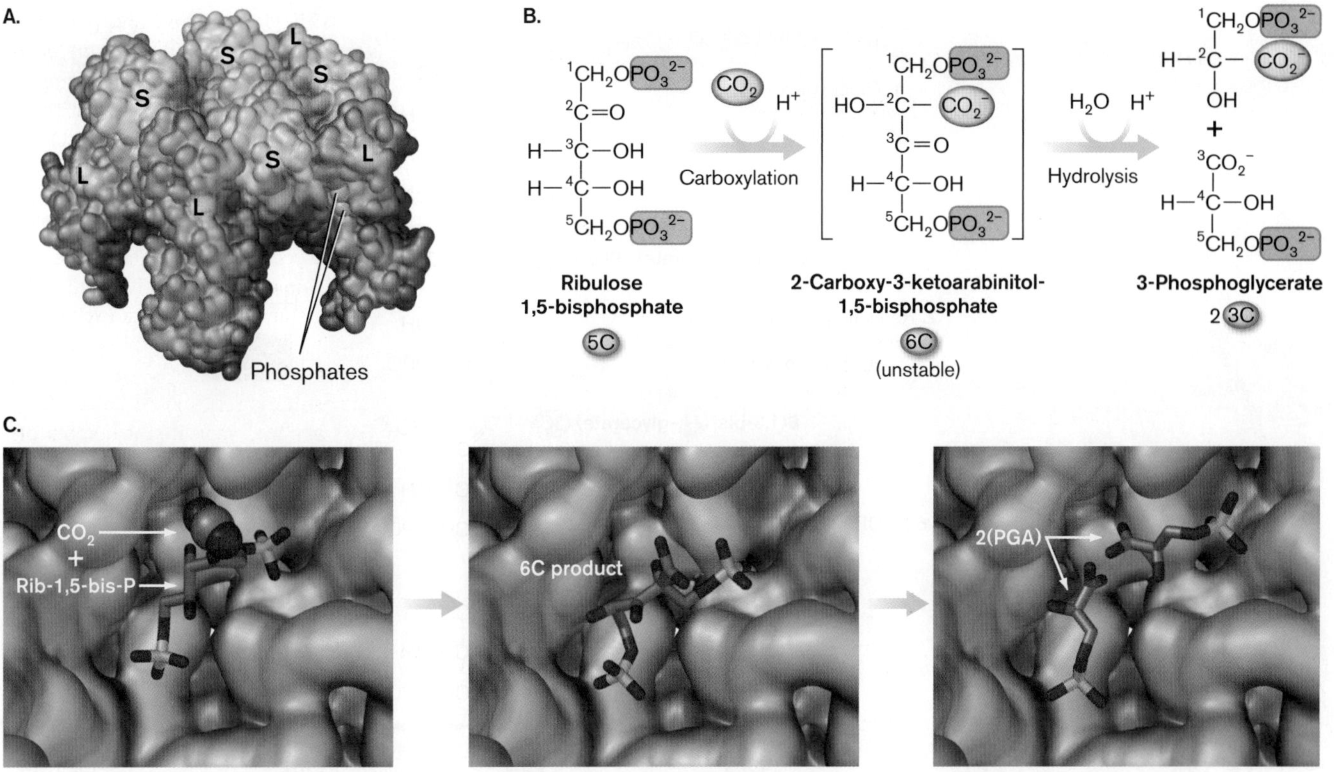

FIGURE 15.8 ■ The mechanism of Rubisco. **A.** Rubisco from *Alcaligenes eutrophus* consists of eight large subunits (L) crowned by eight small subunits (S). Only the upper four L and S subunits are shown. On each L subunit, a catalytic site contains two phosphates (orange), which compete with the substrates for binding. (PDB code: 1BXN) **B.** Rubisco adds CO_2 to ribulose 1,5-bisphosphate to give an unstable six-carbon intermediate, a ketone, bound to the enzyme. The bound intermediate hydrolyzes to form two molecules of 3-phosphoglycerate (PGA). **C.** The mechanism of CO_2 fixation at the catalytic site. CO_2 adds to the ketone, generating the 6C intermediate, which splits into two molecules of PGA. Rib-1,5-bis-P=ribulose 1,5-bisphosphate. (PDB code: 1RUS)

(in lithotrophs, green phototrophs, and some algae), six (in plant chloroplasts and some algae), or two (in purple phototrophs). Nevertheless, the fundamental mechanism of CO_2 fixation in all these organisms appears to be similar.

Initially, Rubisco adds CO_2 to ribulose 1,5-bisphosphate to give an unstable six-carbon intermediate that remains bound to the enzyme (**Fig. 15.8B** and **C**). The intermediate hydrolyzes into two molecules of PGA, each held at the active site by its phosphate. Only one PGA molecule contains a carbon from the newly fixed CO_2; nevertheless, as both molecules dissociate from the enzyme, they enter the cellular pool of PGA and behave equivalently in the rest of the cycle.

Rubisco has a high affinity for CO_2, and the typical concentration of Rubisco active sites (one per large subunit, six per complex) in plant chloroplasts is 4 mM—about 500 times greater than the concentration of CO_2. Thus, considerable energy is invested in CO_2 absorption. Yet the efficiency of carbon fixation by Rubisco is lowered by the existence of a competing reaction with O_2 that leads to 2-phosphoglycolate instead of 3-phosphoglycerate. This oxygenation reaction is called photorespiration. The function of photorespiration has been studied intensively but remains unclear.

Thought Question

15.2 Speculate on why Rubisco catalyzes a competing reaction with oxygen. Why might researchers be unsuccessful in attempting to engineer Rubisco without this reaction?

Regeneration of ribulose 1,5-bisphosphate. Overall, each glyceraldehyde 3-phosphate (G3P) arises from three rounds of CO_2 fixation by ribulose 1,5-bisphosphate. Thus, to maintain the cycle, for each G3P provided to biosynthesis, five other molecules of G3P must be recycled into three five-carbon molecules of ribulose 1,5-bisphosphate. Details of regenerating the intermediate are shown in the lower half of **Figure 15.9**.

The regeneration pathway is summarized here:

- Two molecules of G3P condense to form the six-carbon sugar fructose 6-phosphate.
- Fructose 6-phosphate condenses with a third G3P. The nine-carbon molecule splits to form a five-carbon sugar (xylulose 5-phosphate) and a four-carbon sugar (erythrose 4-phosphate). Xylulose 5-phosphate rearranges to ribulose 5-phosphate.

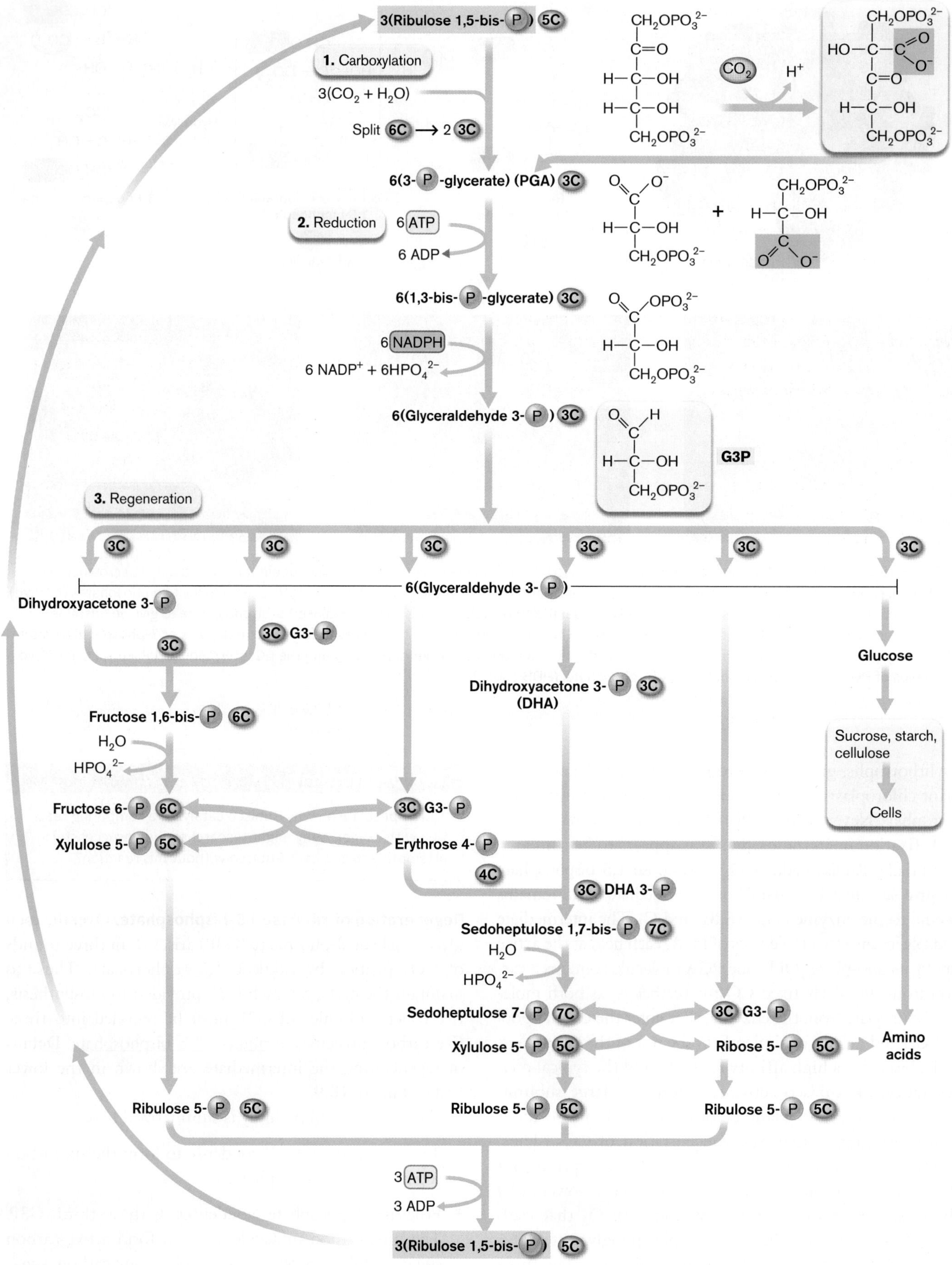

FIGURE 15.9 ■ **The Calvin cycle in detail.** The Calvin cycle assimilates three CO_2 molecules, forming six 3-phosphoglycerate (PGA) molecules reduced to glyceraldehyde 3-phosphate (G3P), and converts five G3P into three molecules of ribulose 1,5-bisphosphate. Labels "3C," "5C," "6C," and so on indicate the number of carbon atoms per molecule. Some sugars may enter alternative pathways to build amino acids.

- Erythrose 4-phosphate condenses with a fourth G3P (rearranged to dihydroxyacetone 3-phosphate) to make a seven-carbon sugar (sedoheptulose 7-phosphate).
- The seven-carbon sugar condenses with the fifth G3P and splits into two molecules of the ribulose 5-phosphate (via xylulose 5-phosphate and ribose 5-phosphate).
- Each of the three five-carbon sugars receives a second phosphoryl group from ATP, generating ribulose 1,5-bisphosphate. Each ribulose 1,5-bisphosphate is now ready for Rubisco to fix another CO_2.

Why would a cycle have evolved requiring so many enzymatic steps to so many different intermediates? First, a cycle with many steps breaks down the energy flow into numerous reversible conversions with near-zero values of ΔG. The nearer to equilibrium, the more energy is conserved in the conversion.

Second, the multiple different intermediates provide substrates for biosynthesis. For example, some molecules of erythrose 4-phosphate and ribose 5-phosphate are withdrawn from the cycle to build aromatic amino acids and nucleotides. These amino acids provide the plant proteins vital for human consumption of plant foods.

Thought Questions

15.3 Why does ribulose 1,5-bisphosphate have to contain two phosphoryl groups, whereas the other intermediates of the Calvin cycle contain only one?

15.4 Which catabolic pathway (see Chapter 13) includes some of the same sugar-phosphate intermediates that the Calvin cycle has? What might these intermediates in common suggest about the evolution of the two pathways?

CO_2 Uptake and Concentration

The concentration of CO_2 is a special problem because CO_2 diffuses readily through phospholipid membranes. Thus, cells cannot concentrate this substrate across the cell membrane to reach the level needed to drive Rubisco. Some carbon-fixing bacteria solve the gas concentration problem by enzymatic conversion of CO_2 to bicarbonate (HCO_3^-), which is trapped in the cytoplasm, unable to leak out of the cell membrane. This enzyme system is called the **carbon-concentrating mechanism** (**CCM**). Other bacteria use alternative CO_2-fixing systems adapted to different CO_2 concentrations.

Carboxysomes contain Rubisco. Many organisms that fix CO_2 contain the Rubisco complex within subcellular structures called **carboxysomes** (**Fig. 15.10**). Carboxysomes are found within CO_2-fixing lithotrophs, as well as within cyanobacteria and chloroplasts. A carboxysome consists of a polyhedral shell of protein subunits surrounding tightly packed molecules of Rubisco. The carboxysome takes up bicarbonate (converted from CO_2). Once inside the carboxysome, the bicarbonate is immediately converted to CO_2 by the enzyme carbonic anhydrase. The CO_2 is then fixed by Rubisco to PGA—the first step of CO_2 fixation (shown in **Fig. 15.9**). The PGA exits the carboxysome to complete the Calvin cycle in the cytoplasm. Mutant strains of bacteria lacking carboxysomes can fix CO_2 only at high concentration (5%), much higher than the atmospheric CO_2 concentration (0.037%).

Alternative concentration pathways. Instead of carboxysomes, some phototrophs possess alternative systems of CO_2 uptake and concentration adapted to different levels of CO_2. For example, the purple phototroph *Rhodobacter sphaeroides* has two unlinked operons, each encoding a different set of CO_2 fixation genes, called form I and form II. Form I and

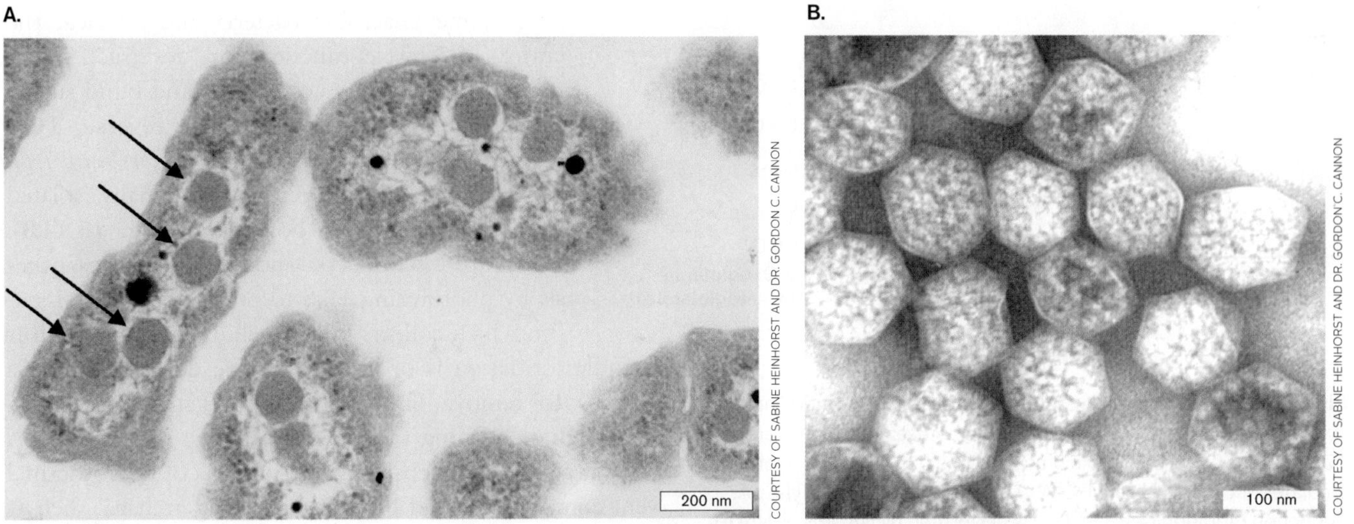

FIGURE 15.10 ■ **Carboxysomes. A.** Thin section of *Halothiobacillus neapolitanus*, a sulfur-oxidizing lithoautotroph (TEM), showing polyhedral carboxysomes (arrows). **B.** Isolated carboxysomes, packed with Rubisco complexes (TEM). *Source:* Modified from Tsai et al. 2007. *PLoS Biol.* **5**:E144.

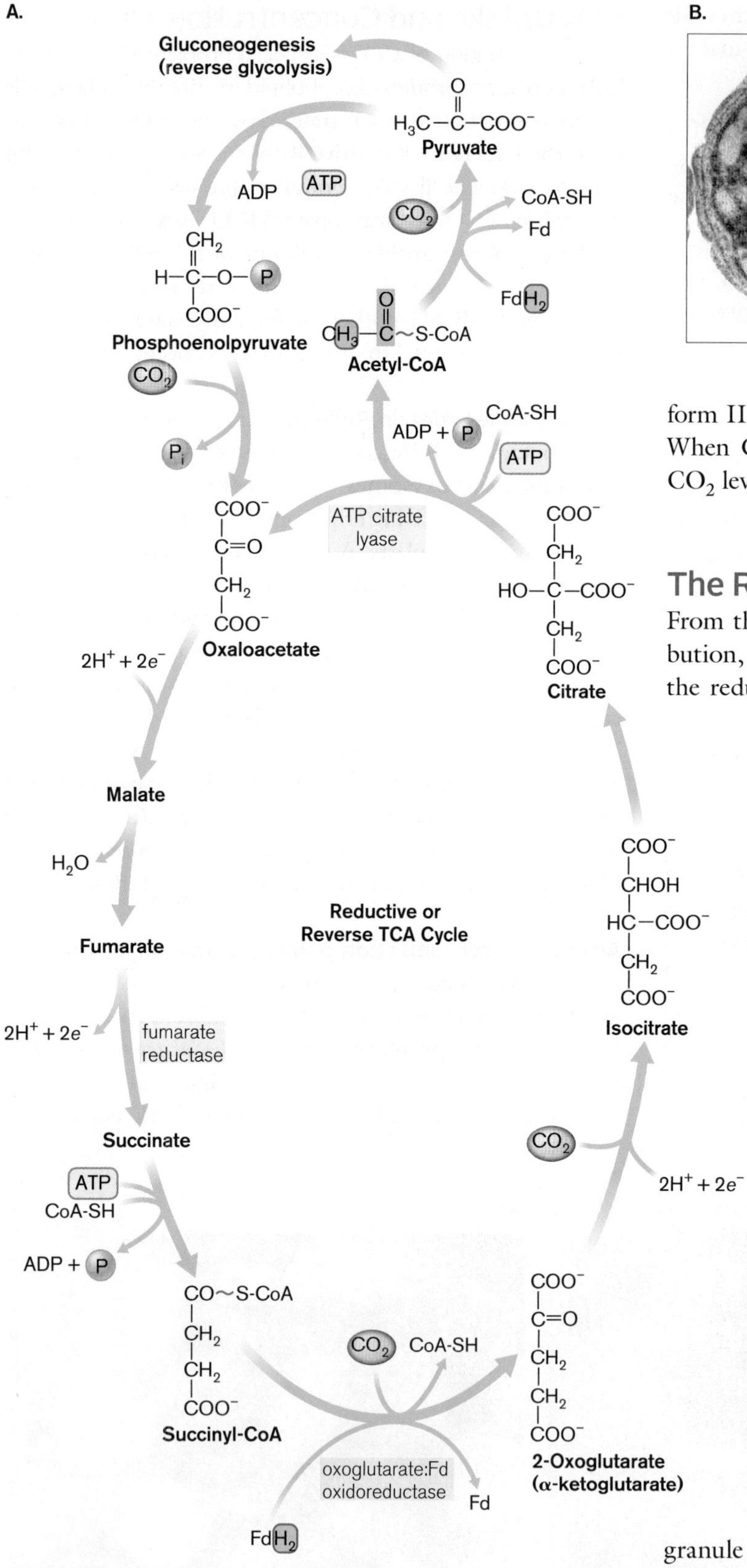

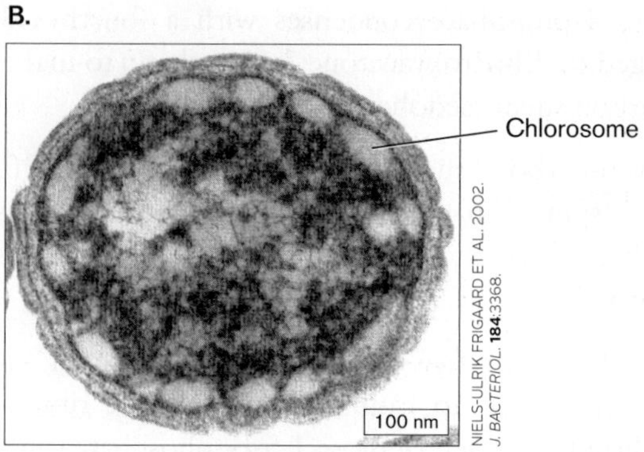

FIGURE 15.11 ■ The reductive, or reverse, TCA cycle for CO_2 fixation. **A.** Anaerobic phototrophs and archaea use the reverse TCA cycle to fix carbon into biomass. CO_2 is fixed by several intermediates, including succinyl-CoA, 2-oxoglutarate, and acetyl-CoA. Reduction (addition of $2H^+ + 2e^-$) is performed by NADPH or NADH and by reduced ferredoxin (FdH_2). **B.** *Chlorobium tepidum*, a green sulfur bacterium that fixes CO_2 by the reductive TCA cycle (TEM). Chlorosome membranes contain the reaction centers of photolysis.

form II each encode all the key enzymes, such as Rubisco. When CO_2 is limiting, form I enzymes work best; when CO_2 levels are saturating, form II enzymes work best.

The Reductive, or Reverse, TCA Cycle

From the standpoint of evolution and phylogenetic distribution, the most ancient pathway of metabolism may be the reductive, or reverse, TCA cycle. All major groups of organisms run the TCA cycle, or some part of it (discussed in Section 13.6). Most of the individual reaction steps of the TCA cycle are reversible because each has a relatively small ΔG of reaction. Thus, the steps that release CO_2 in the forward direction may assimilate small amounts of CO_2 in the reverse direction. Furthermore, all organisms, including humans, fix small amounts of CO_2 through special reactions that regenerate TCA cycle intermediates. These regeneration steps are called **anaplerotic reactions**. Common anaplerotic reactions are the formation of oxaloacetate from phosphoenolpyruvate (PEP), catalyzed by PEP carboxylase; and the formation of malate from pyruvate, catalyzed by malic enzyme.

In some anaerobic bacteria and archaea, the entire TCA cycle functions in "reverse," reducing CO_2 to generate acetyl-CoA and build sugars (**Fig. 15.11A**). The **reductive**, or **reverse, TCA cycle** is used by bacteria such as *Chlorobium tepidum*, a green sulfur bacterium originally isolated from a New Zealand hot spring (**Fig. 15.11B**). *Chlorobium* conducts anoxygenic photosynthesis by photolyzing H_2S to produce elemental sulfur, which collects in extracellular granules. Sulfur granule formation is of interest for the development of a process to remove H_2S from sulfide-generating industries such as refining of coal and oil. The reductive TCA cycle is also used by epsilonproteobacteria of hydrothermal vent communities and by sulfur-reducing archaea, such as *Thermoproteus* and *Pyrobaculum*.

Reversal of the TCA cycle uses four or five ATP to fix four molecules of CO_2 and generates one molecule of oxaloacetate. The enzymes used are the same as for the "forward" cycle, except for three key enzymes that spend ATP to drive the reaction in reverse: ATP citrate lyase, 2-oxoglutarate:ferredoxin oxidoreductase, and fumarate reductase. For example, ATP citrate lyase catalyzes the cleavage of citrate into acetyl-CoA and oxaloacetate. CO_2 assimilation requires reduction by NADPH. The reductive TCA cycle is believed to be the most ancient means of CO_2 fixation and amino acid biosynthesis, the original cycle of biomass generation in the ancestors of all three living domains.

From CO_2 to acetyl-CoA, the reverse TCA cycle essentially reverses the overall cycle of catabolism:

$$2CO_2 + 2\ ATP + 8H^+ + 8e^- + HS\text{-}CoA \rightarrow CH_3CO\text{-}S\text{-}CoA + 3H_2O + 2\ ADP + 2\ P_i$$

The coenzyme A is recycled as the acetyl group enters biosynthesis. For example, acetyl-CoA can assimilate another CO_2 with reduction to pyruvate. Pyruvate then builds up to glucose and related sugars by **gluconeogenesis** (reverse glycolysis), with expenditure of ATP and NADPH. Alternatively, acetyl-CoA can enter biosynthetic pathways to produce fatty acids or amino acids.

Unlike the Calvin cycle, the reverse TCA cycle provides several different molecular intermediates that can assimilate CO_2. Here are the key steps:

- Succinyl-CoA assimilates CO_2 to form 2-oxoglutarate (alpha-ketoglutarate).
- 2-Oxoglutarate assimilates CO_2 to form isocitrate.
- Acetyl-CoA (produced by the reverse TCA cycle) assimilates CO_2 to form pyruvate.

Each CO_2 assimilation requires one or more reduction steps. 2-Oxoglutarate is reduced by the protein ferredoxin (FdH_2). FdH_2 also mediates acetyl-CoA reduction to pyruvate. Other reduction steps may be accomplished by NADPH or NADH.

The Reductive Acetyl-CoA Pathway

Yet another ancient route for CO_2 assimilation into acetyl-CoA and pyruvate is the **reductive acetyl-CoA pathway** (**Fig. 15.12**). The acetyl-CoA pathway is fundamental for methanogenic archaea. In the acetyl-CoA pathway, two CO_2 molecules are condensed through converging pathways to form the acetyl group of acetyl-CoA. The acetyl-CoA pathway is used by anaerobic soil bacteria, such as *Clostridium thermoaceticum*, and by most autotrophic sulfate reducers, such as *Desulfobacterium autotrophicum*. It also provides the main source of biomass for methanogens. Most of the CO_2 absorbed by methanogens is used to yield energy by reducing CO_2 to methane (discussed in Chapter 14). However, 5% of the CO_2 absorbed by a methanogen enters the acetyl-CoA pathway for biosynthesis, generating the entire biomass of the organism.

The substrates and ultimate products of the acetyl-CoA pathway are the same as those of the reverse TCA cycle, except that the reducing agent is H_2 instead of NADPH. The reductive acetyl-CoA pathway is slightly more efficient than the reductive TCA cycle, requiring one less ATP:

$$2CO_2 + ATP + 4H_2 + HS\text{-}CoA \rightarrow CH_3CO\text{-}S\text{-}CoA + 3H_2O + ADP + P_i$$

Nevertheless, the order of the pathway and its intermediate compounds are completely different from those of the TCA cycle. The reductive acetyl-CoA pathway is linear, with no recycled intermediates.

Reduction of the first CO_2. The first CO_2 enters the linear pathway by reduction to formate (**Fig. 15.12A**). The

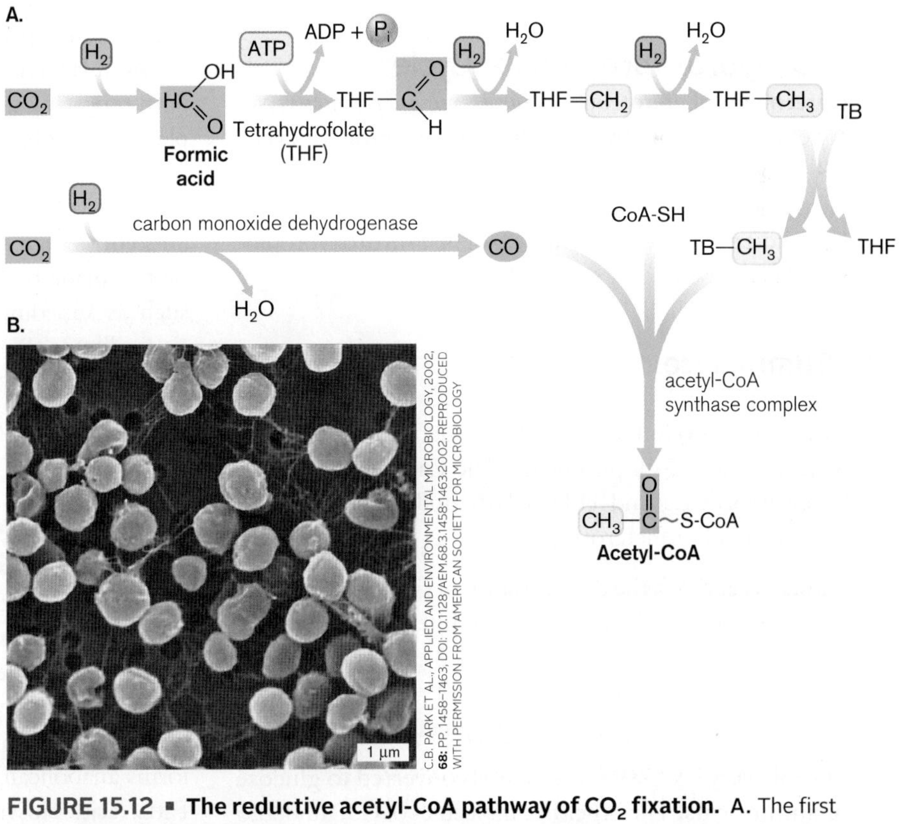

FIGURE 15.12 ■ The reductive acetyl-CoA pathway of CO_2 fixation. A. The first CO_2 is reduced to formate and is transferred onto the cofactor tetrahydrofolate (THF). After three reduction steps, the methyl group is transferred to a vitamin B_{12}–like cofactor (TB). The second CO_2 is reduced to carbon monoxide (CO) by the enzyme carbon monoxide dehydrogenase, and then incorporated into acetyl-CoA. **B.** *Methanocaldococcus jannaschii*, a thermophilic marine methanogen that fixes CO_2 by the reductive acetyl-CoA pathway (SEM).

formate is transferred onto a complex carrier cofactor. In bacteria, the cofactor is tetrahydrofolate (THF), a reduced form of **folate**. Folate (folic acid) is a heteroaromatic cofactor; it is an essential vitamin required by many organisms, including humans. A different cofactor, methanopterin (MPT), carries the formate in methanogenic archaea. In either case, the formate carbon is reduced in successive steps to a methyl group ($-CH_3$).

Reductive CO_2 incorporation into acetyl-CoA. The second CO_2 is reduced to carbon monoxide (CO) by the enzyme carbon monoxide dehydrogenase. This same enzyme is used in reverse reaction by some lithotrophs to gain energy from carbon monoxide (CO) as an electron donor. For fixation, the CO is condensed with the methyl group carried by the vitamin B_{12}–like cofactor TB to form acetyl-CoA. The acetyl-CoA then enters pathways of biosynthesis, as it does when formed by the reverse TCA cycle.

One place where methanogens fix CO_2 and release methane is the bovine rumen, or digestive fermentation chamber (see Chapter 21). Within the rumen, bacteria convert cattle feed into CO_2 and H_2, and methanogens divert a significant portion of these gases into methane. The methane then escapes through burping and flatulence. Beef microbial methane production is a major source of global warming, causing our climate change crisis (see Chapter 22).

The 3-Hydroxypropionate Cycle

The green photoheterotroph *Chloroflexus aurantiacus* fixes CO_2 yet another way: by the **3-hydroxypropionate cycle**. The 3-hydroxypropionate cycle is also used by archaea, such as the thermoacidophile *Acidianus brierleyi* and the aerobic sulfur oxidizer *Sulfolobus metallicus*. The 3-hydroxypropionate cycle is presented in **eTopic 15.4**.

To Summarize

- **The Calvin cycle** fixes CO_2 by reductive condensation with ribulose 1,5-bisphosphate. The Calvin cycle is used by cyanobacteria and chloroplasts and by some lithotrophs and photoheterotrophs.
- **Rubisco** catalyzes the condensation of CO_2 with ribulose 1,5-bisphosphate. The six-carbon intermediate immediately splits into two molecules of 3-phosphoglycerate (PGA), which are activated by ATP and reduced by NADPH to glyceraldehyde 3-phosphate (G3P).
- **One of every six G3P molecules is converted to glucose or amino acids.** The other five molecules of G3P undergo reactions to regenerate ribulose 1,5-bisphosphate.
- **Carboxysomes sequester and concentrate CO_2** for fixation by the Calvin cycle.
- **The reductive, or reverse, TCA cycle** fixes CO_2 in some anaerobes and archaea.
- **Anaplerotic reactions** in all organisms regenerate TCA cycle intermediates, sometimes by fixing CO_2.
- **The acetyl-CoA pathway** in anaerobic bacteria and methanogens fixes CO_2 by condensation to form acetyl-CoA.
- **The 3-hydroxypropionate cycle** in *Chloroflexus* and in some hyperthermophilic archaea fixes CO_2 in a cycle generating the intermediate 3-hydroxypropionate.

15.3 Fatty Acids and Antibiotics

Fatty acid production exemplifies the construction of cell components from repeating units (**Fig. 15.13**). The construction of molecules based on repeating units requires a cyclic pathway that feeds its products back repeatedly as substrates for further synthesis. The advantage of a cyclic process is that large polymers can be made with a limited number of enzymes; for example, the mycolic acids of *Mycobacterium tuberculosis* are extended to lengths of over 100 carbons. Such a process is also called **modular synthesis**. By contrast, other cell components, such as amino acids, have more complex, nonrepeating structures whose biosynthetic enzymes require long operons (presented in Section 15.5).

Modular synthesis also builds more complex products with repeating units that have various R groups. For example, **polyketides** consist of a hydrocarbon backbone equivalent to that of fatty acids, with a unique R group attached to each two-carbon unit. Polyketides include a major class of antibiotics such as erythromycin. Another vital class of antibiotics made by bacteria is that of **nonribosomal peptides** such as vancomycin. The mechanism of modular synthesis facilitates the evolution of an extraordinary diversity of antibiotics, which microbes use in natural environments for competition and predation. Researchers mine these environmental microbial communities to discover new antibiotics.

Fatty Acids Are Built from Repeating Units

The structure of fatty acids was presented in Chapter 3. Pathways for biosynthesis of fatty acids start with the small, versatile substrate acetyl-CoA. Acetyl-CoA is a product of catabolic pathways such as glycolysis and benzoate catabolism (see Sections 13.5 and 13.6). The acetyl group also forms anabolically from the reverse TCA cycle (**Fig. 15.11**). Fatty acid biosynthesis provides targets for antimicrobial drugs such as triclosan, an inhibitor of enoyl-ACP reductase.

The fatty acid synthase complex. The cyclic process of fatty acid synthesis is managed by the fatty acid synthase complex.

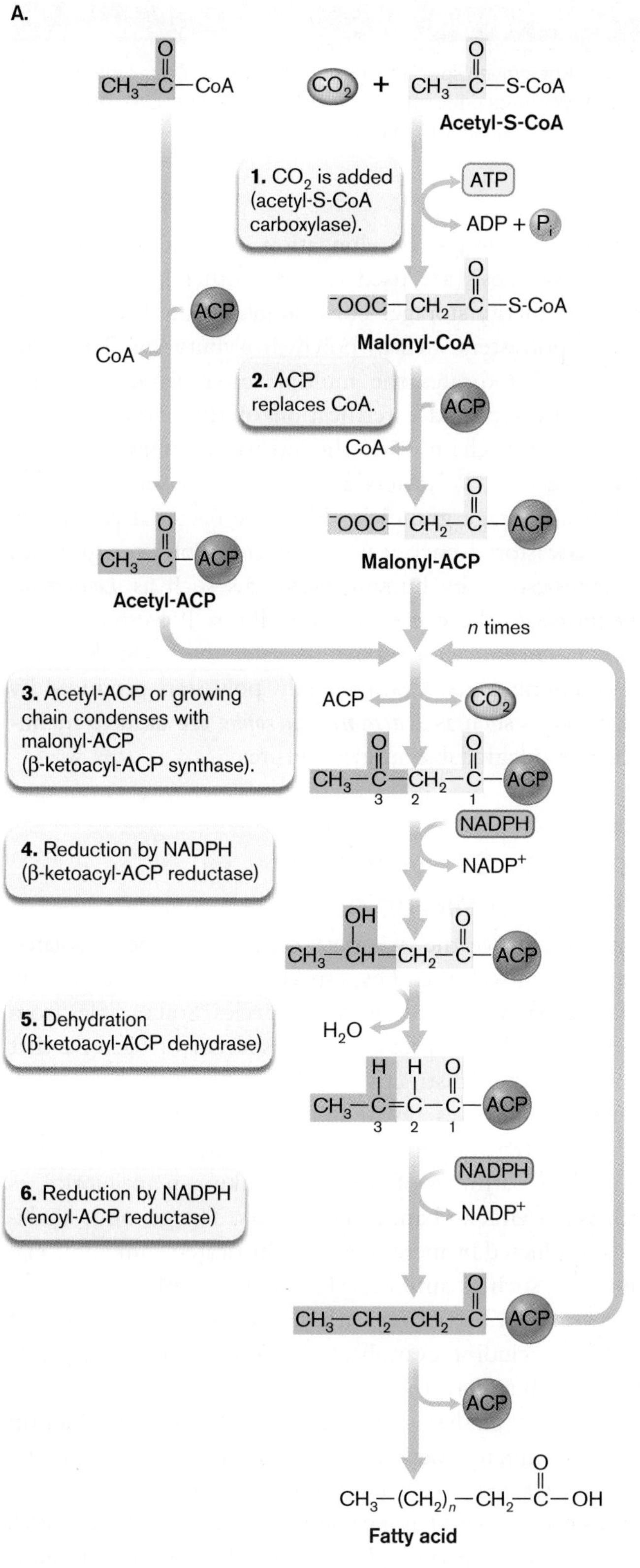

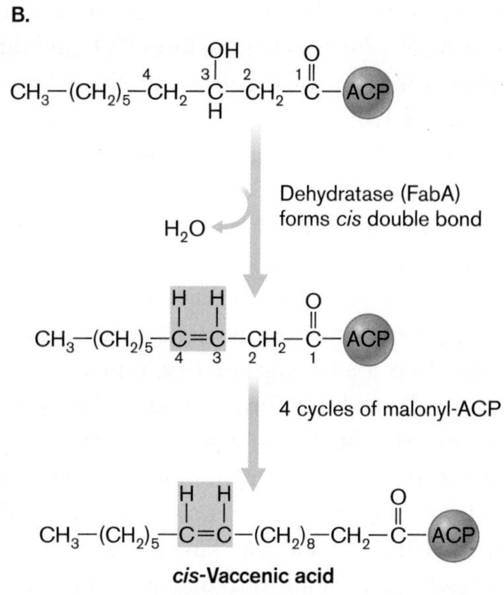

FIGURE 15.13 ▪ Fatty acid biosynthesis. **A.** Stepwise elongation of a saturated fatty acid. Units of acetyl-CoA are carboxylated to malonyl-CoA, and then successively condense to form the long chain of a fatty acid. Each two-carbon unit requires reduction by 2 NADPH. **B.** An alkene "kink" can be left unsaturated during chain extension. The double bond then forms between the third and fourth carbons, rather than the second and third. In this position, the double bond escapes reduction as the chain lengthens through subsequent cycles of malonyl addition. ACP = acyl carrier protein.

The complex contains all the enzymes and component binding proteins bound together in proximity, so that all steps proceed in one place without "losing" the unfinished molecule. The main bacterial version of this complex is designated FASII. All of the components are essential for viability and thus are potential drug targets. Analogous multienzyme complexes are used by actinomycete bacteria to synthesize other long-chain products, such as polyketide antibiotics.

Activation of acetyl-CoA. Most acetyl groups within the cell are "tagged" with coenzyme A (CoA), a cofactor that directs acetate into catabolic pathways such as the TCA cycle. To redirect acetyl groups into fatty acid biosynthesis, acetyl-CoA molecules are first tagged at the "back end" by condensation with CO_2 (**Fig. 15.13A**, step 1), catalyzed by acetyl-CoA carboxylase. The addition of CO_2 in this step does not count as carbon fixation, because the CO_2 added is not permanently incorporated into the carbon skeleton. Instead, like phosphate or CoA, the CO_2 exists to be displaced when its function is no longer required. The CO_2-tagged acetyl-CoA is called malonyl-CoA. The coenzyme A is replaced by **acyl carrier protein (ACP)**, making malonyl-ACP (step 2).

Malonyl-ACP condenses with the growing chain. In step 3 (see **Fig. 15.13A**), malonyl-ACP hooks onto the head of an acetyl-ACP or a longer-growing chain. The process of hooking onto the new unit involves displacement of CO_2 from the back end of malonyl-ACP, which replaces ACP from the front end of the growing chain.

The growing chain now contains a ketone (at carbon 3, from the former acetyl group). The ketone is reduced by NADPH and dehydrated, forming an alkene (C=C, unsaturated double bond between adjacent carbons 2 and 3). The alkene is further reduced by a second NADPH (steps 4–6). In all, the chain gains $4H^+ + 4e^-$ and is now fully hydrogenated (saturated).

Once hydrogenated, the chain is ready to take on the next malonyl-ACP, which, as before, loses CO_2 and replaces the ACP from the front of the growing chain (return to step 3 in **Fig. 15.13A**). Successive addition can continue many times to build a saturated fatty acid.

Unsaturation. Certain fatty acids require unsaturated "kinks" in the chain (alkenes) to serve a structural purpose, such as to increase the fluidity of the membrane. Unsaturation can be generated during a cycle of elongation (**Fig. 15.13A**, step 5). During the first four cycles of acetyl addition, an alkene bond forms between the second and third carbons of the fatty acid. After the fifth two-carbon addition, however, a special dehydratase enzyme generates the alkene double bond between the third and fourth carbons (**Fig. 15.13B**). This alkene fails to be hydrogenated further. Instead, several additional malonyl units are added, generating a long-chain fatty acid with a kink of a *cis* double bond.

Regulation of fatty acid synthesis. The synthesis of fatty acids consumes enormous quantities of reducing energy; thus, cells must regulate it to avoid waste. From a structural standpoint, production of fatty acids incorporated into membranes must be balanced with growth of the cytoplasm. Furthermore, the many different variants of fatty acids are regulated in response to particular environmental needs. Some mechanisms of regulation include the following:

- **Acetyl-CoA carboxylase represses its own transcription.** Transcription of an operon encoding two subunits of acetyl-CoA carboxylase (AccB, AccC) is repressed by one of its subunits, protein AccB. As AccB increases in concentration, it binds the promoter of the *accBC* operon, repressing further transcription. Thus, initiation of fatty acid biosynthesis (**Fig. 15.13A**, step 1) is always limited by the number of AccB and AccC enzyme subunits that are present.
- **Starvation blocks fatty acid biosynthesis.** Starvation for carbon sources blocks fatty acid biosynthesis through the "stringent response." Blockage is mediated by the polyphosphorylated nucleotide ppGpp (guanosine tetraphosphate), the global regulator of the stringent response (discussed in Section 10.4).
- **Temperature regulates fatty acid composition.** Bacterial fatty acid composition is regulated by environmental factors such as temperature. In *E. coli*, low temperature favors unsaturated fatty acids because they are less rigid and maintain membrane flexibility. Low temperature induces expression of the gene *fabA*, which encodes the dehydratase enzyme that desaturates the fatty acid bond. As the dehydratase activity is increased, more unsaturated fatty acids are made.

Thought Question

15.5 For a given species, uniform thickness of a cell membrane requires uniform chain length of its fatty acids. How do you think chain length may be regulated?

Cyclic pathways of elongation comparable to fatty acid biosynthesis are used to build other kinds of polymers for energy storage. For example, many bacteria synthesize polyesters such as polyhydroxybutyrate. The term "polyester" indicates the multiple ester groups that are formed by repeated esterification of the carboxylic acid group of the chain with the hydroxyl group of a new alkanoate unit. Polyesters are insoluble in water, so they collect as storage granules within the bacterial cell, ready to release stored energy when needed. Polyester granules are synthesized by human pathogens such as *Legionella pneumophila*, the cause of legionellosis. Polyester storage helps *L. pneumophila* survive in water sources such as air-conditioning units. Commercially, polyesters produced by soil bacteria such as *Ralstonia eutropha* are used to manufacture biodegradable surgical sutures.

Antibiotic Biosynthesis: Mining the Microbiomes

Antibiotics since the twentieth century have been isolated largely from soil microbes, especially fungi and actinomycete bacteria such as *Streptomyces* species. Soil samples from all over the world provide novel sources of bacteria that synthesize their distinctive agents of "microbial warfare." Genome sequences reveal homologs of known enzymes for antibiotic biosynthesis, and these homologs turn out to mediate pathways of previously unknown antibiotics, as well as antiviral and antitumor agents. The genomic analysis is conducted by increasingly sophisticated computational pipelines, such as antiSMASH. More recently, such analysis reveals antibiotic biosynthesis in a surprising range of bacteria, including cyanobacteria, as well as Gram-negative genera such as *Burkholderia*.

Genomic analysis also reveals antibiotic production in environments beyond the soil, most excitingly the oceans. Marine ecosystems yield an extraordinary diversity of bacteria and habitats, evolved under extreme conditions of temperature and pressure. Marine microbes also show complex interactions with host animals, particularly invertebrates such as corals and sponges. In these host relationships, the bacterial antibiotics may actually protect their host animals from predation or infection. An example of promising products obtained from sponges is the discodermins, nonribosomal peptides with anticancer activity.

Polyketide Antibiotics

The cyclic biosynthesis of fatty acids provides a model for biosynthesis of the polyketide antibiotics. As an example, we present here the well-known polyketide erythromycin, a broad-spectrum antibiotic prescribed for bacterial pneumonia and chlamydia infections (**Fig. 15.14**). Erythromycin blocks bacterial translation at the step of peptide elongation (see Chapter 8). The antibiotic was discovered in 1949 by scientists at the Eli Lilly company from a soil sample obtained by Filipino scientist Abelardo Aguilar. Aguilar's soil sample contained a previously unknown species of actinomycete, *Saccharopolyspora erythraea*, which produced a molecule that inhibits bacteria. The molecule, called erythromycin, showed bactericidal activity against a wide range of Gram-negative and Gram-positive bacteria.

The synthesis of a polyketide involves cyclic elongation by an enormous enzyme complex called a **modular enzyme**. A modular enzyme consists of multiple modules that add similar but nonidentical units to a growing chain (**Fig. 15.14**). Like fatty acids, polyketides are built by the successive condensation of malonyl-ACP units—but

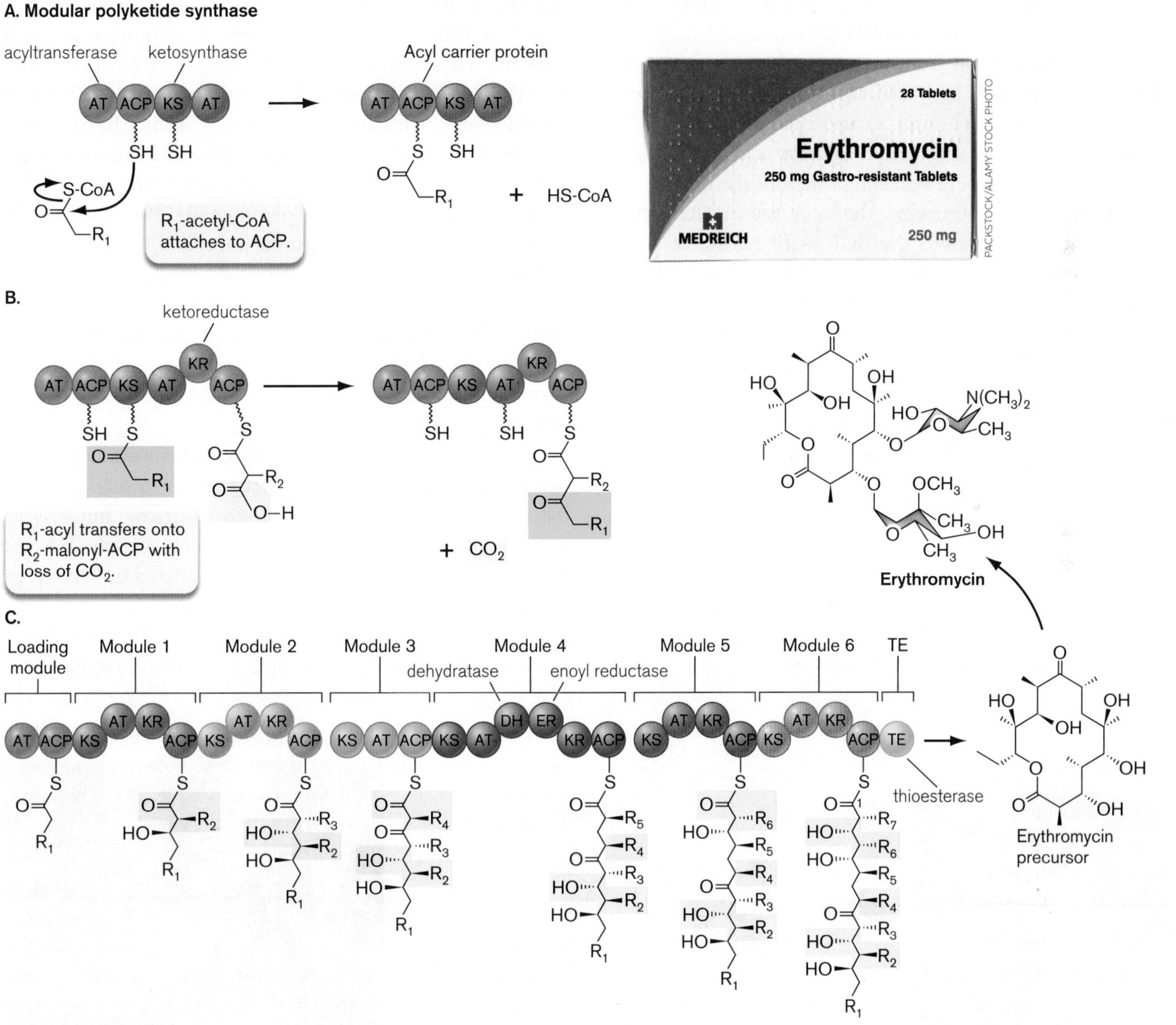

FIGURE 15.14 ■ Synthesis of the erythromycin ring. **A.** An acyltransferase (AT) transfers R_1-acetyl-CoA onto an ACP, with release of coenzyme A. **B.** The R_1-acetyl group is then transferred onto a ketosynthase (KS). The R_1-acetyl group condenses with R_2-malonyl-ACP, with release of CO_2. **C.** Modular subunits of the polyketide synthase complex elongate the polyketide to form the ring precursor of erythromycin. In this example, all R groups are methyl ($-CH_3$). Some modules include reducing enzymes such as ketoreductase (KR), dehydratase (DH), and enoyl reductase (ER). Elongation is terminated by thioesterase (TE). **Inset:** *Saccharopolyspora erythraea* synthesized the original erythromycin (now produced by commercially modified organisms). *Source:* Parts A–C modified from David E. Cane et al. 1998. *Science* **282**:63.

each malonyl group carries a unique extension, or R group, instead of a CO_2. In erythromycin, the R groups are all methyl groups. In other polyketides, the R groups range from hydroxyl groups and amides to aromatic rings, some of which undergo secondary reactions and interconnections.

How is the order of different R groups determined? The modular enzyme contains a series of active sites, or modules, each of which catalyzes the addition of one of a series of units. Each module contains all the enzyme activities needed to add the unit and recognizes a unit with one specific R group. The modular polyketide synthase may consist of a single protein with a series of domains, or it may consist of a complex of several protein subunits. Either the multidomain enzyme or the complex acts as an assembly line to generate the specific polyketide.

Figure 15.14A and **B** show the chain extension synthesis of a polyketide. The initial two domains of enzyme 1 are acyltransferase (AT) and an acyl carrier protein (ACP) domain, analogous to the ACP of fatty acid biosynthesis but specialized to accept one component of the polyketide. The acyltransferase transfers the acyl group (R_1-acetyl) from R_1-acetyl-CoA onto the ACP, with release of coenzyme A (HS-CoA). The R_1-acetyl is then transferred to a ketosynthase domain (KS).

Meanwhile, a malonyl group with its own R group (R_2-malonyl) has been transferred by a second AT onto the second ACP domain (**Fig. 15.14B**). The R_1-acetyl group then leaves KS and condenses with R_2-malonyl-ACP, releasing CO_2. The R_1R_2 acyl chain subsequently undergoes a series of extensions by R-malonyl groups (**Fig. 15.14C**). For erythromycin, all R groups are methyl; other polyketides have R groups that are more complex.

The initial AT and ACP domains constitute a "loading module" for the first acyl group, whereas subsequent sets of domains constitute "extender modules" that each add a different R-malonyl group. Extender modules can include secondary activities such as dehydratase (DH; removes OH) and ketoreductase (KR; hydrogenates a ketone to OH). In principle, the modular approach can construct a limitless range of products with diverse antibiotic properties.

Elongation of the polyketide is terminated by thioesterase (TE), which hydrolyzes the thioester bond to the final ACP. This polyketide chain is an erythromycin precursor. The precursor requires additional enzymes (not shown) to add extra components, including two sugars (glycosylation). Sugar addition completes the erythromycin.

Nonribosomal Peptide Antibiotics

Bacteria synthesize a vast array of peptide products that serve the functions of development, communication, and combat. These peptides have unique modifications that depart from the standard 20 amino acid residues. Some peptides are translated by ribosomes using the standard amino acids, and then undergo posttranslational modification. These are called ribosomally synthesized and posttranslationally modified peptides (RiPPs). Many RiPPs have antimicrobial activity; some are called "lantibiotics."

Other peptides, however, are synthesized without ribosomes. The **nonribosomal peptides** are synthesized entirely by modular enzymes comparable to those that make polyketides. Nonribosomal peptides include powerful antibiotics such as vancomycin and immunosuppressant drugs such as cyclosporine.

We present here the example of the nonribosomal peptide vancomycin (**Fig. 15.15**), the drug of last resort for combating life-threatening *Clostridioides difficile* and "flesh-eating" methicillin-resistant *Staphylococcus aureus* (MRSA) infections. Vancomycin was originally isolated in 1953 from a soil sample sent by a missionary in the jungle of Borneo to a friend at the Eli Lilly company. The Lilly biochemists isolated a new actinomycete, now called *Amycolatopsis*

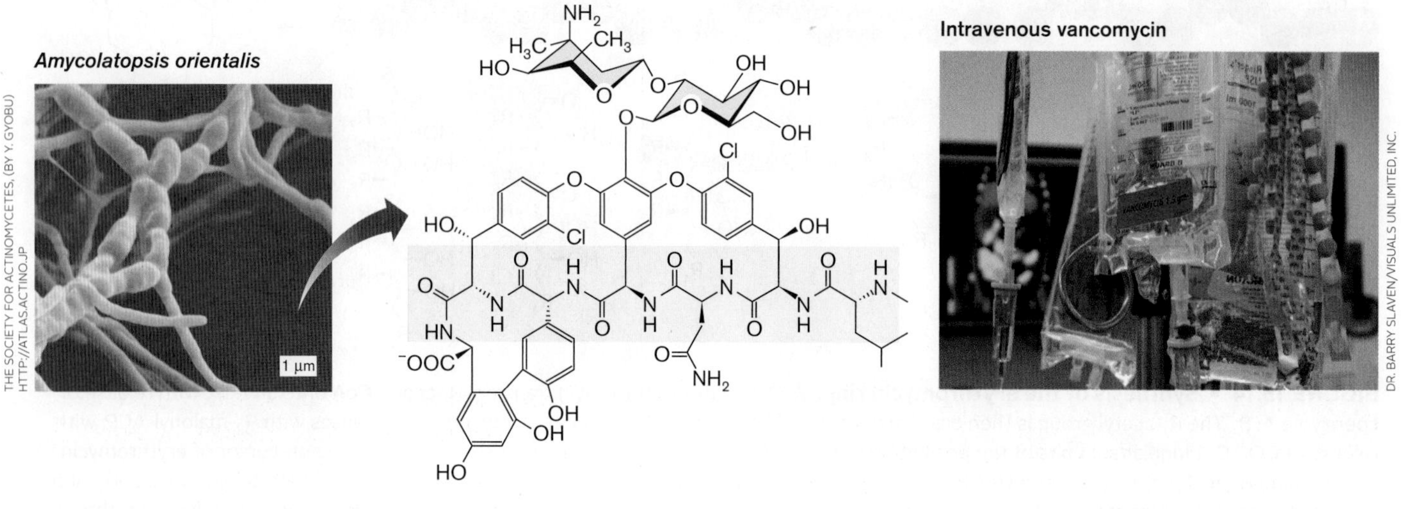

FIGURE 15.15 ■ Vancomycin, a nonribosomal peptide antibiotic. Vancomycin is produced by *Amycolatopsis orientalis* (left). The peptide backbone of the molecule is shaded. *Source:* Modified from Christopher Walsh. *2003. Antibiotics: Actions, Origins, Resistance.* ASM Press.

orientalis. This actinomycete produced a substance that killed Gram-positive pathogens. The active molecule was isolated and called vancomycin, for its ability to vanquish tough microbes.

Vancomycin has a peptide backbone (shaded in **Fig. 15.15**) equivalent to that of a ribosomal peptide, but its aminoacyl residues are nonstandard and show atypical secondary connections. Like polyketides, polypeptide antibiotics are synthesized by an enormous modular enzyme complex.

The complex is called a **nonribosomal peptide synthetase** (**NRPS**). The vancomycin peptide backbone is built by an NRPS that contains seven repeating modules. Each module includes the following key domains (**Fig. 15.16**):

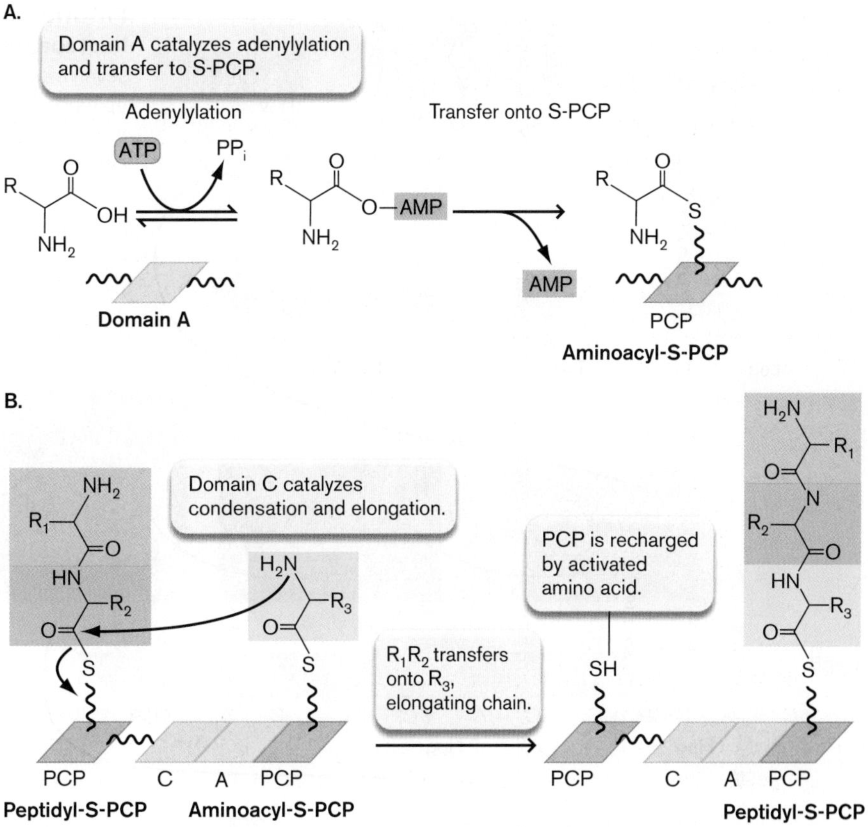

FIGURE 15.16 ■ Elongation of a nonribosomal peptide. **A.** Enzyme domain A adenylylates (activates) an amino acid and then transfers it onto S-PCP (peptidyl carrier protein). **B.** Domain C catalyzes transfer of the aminoacyl group to form a peptide bond. In subsequent rounds, the nascent peptide is transferred onto the next amino acid. *Source:* Modified from Christopher Walsh. 2003. *Antibiotics: Actions, Origins, Resistance.* ASM Press.

- **Domain A: adenylylation and transfer.** The A domain catalyzes adenylylation (addition of adenosine monophosphate, AMP) to the carboxylate of the amino acid (**Fig. 15.16A**). The adenylylation reaction is driven by ATP-releasing pyrophosphate (PP_i). Next, release of AMP drives transfer of the aminoacyl group onto a sulfur atom of the peptide carrier protein (S-PCP). Each S-PCP recognizes only one specific aminoacyl group.
- **Domain C: condensation and elongation.** The C domain catalyzes transfer of the peptidyl-S-PCP from one module onto the amine of the next aminoacyl-S-PCP (**Fig. 15.16B**). In each subsequent round, the nascent peptide is transferred onto the amino group of the next aminoacyl PCP. As each peptide travels down all seven modules, it accretes seven aminoacyl groups in all.
- **Domain E: epimerization.** Some (not all) modules include an E domain to epimerize the aminoacyl group (change its configuration from L to D).

The vancomycin NRPS complex includes three multidomain proteins: CepA, CepB, and CepC (**Fig. 15.17**). Each protein provides one to three modules for chain extension. The seven aminoacyl groups include leucine (Leu), asparagine (Asn), three tyrosines (Tyr), and two hydroxyphenylglycines (Hpg); hydroxyphenylglycine is a nonstandard derivative of glycine. Once the seven-member peptide is formed, additional enzymes (not shown) catalyze chlorination, cross-linking of hydroxyls, and sugar transfer to complete the structure shown in **Figure 15.15**. In all, vancomycin biosynthesis requires about 30 different genes, as predicted by bioinformatic analysis of the vancomycin gene cluster.

Modular enzyme biosynthesis of peptide antibiotics offers exciting prospects for the discovery and design of new antibiotics, anticancer agents, and other therapeutic molecules. Such research requires large-scale screening of microbial communities and mining of genomic data, as described in **Special Topic 15.1**.

To Summarize

- **Fatty acid biosynthesis** involves successive condensation of malonyl-ACP groups formed from acetyl groups tagged with acyl carrier protein (ACP) and a carboxylate. Each successive malonyl group is transferred onto the growing acyl chain, with release of CO_2.
- **The growing acyl chain is hydrogenated.** Each added unit is hydrogenated by two molecules of NADPH, unless an unsaturated kink is required.
- **Some fatty acids are partly unsaturated.** An unsaturated kink may be generated by a special dehydratase,

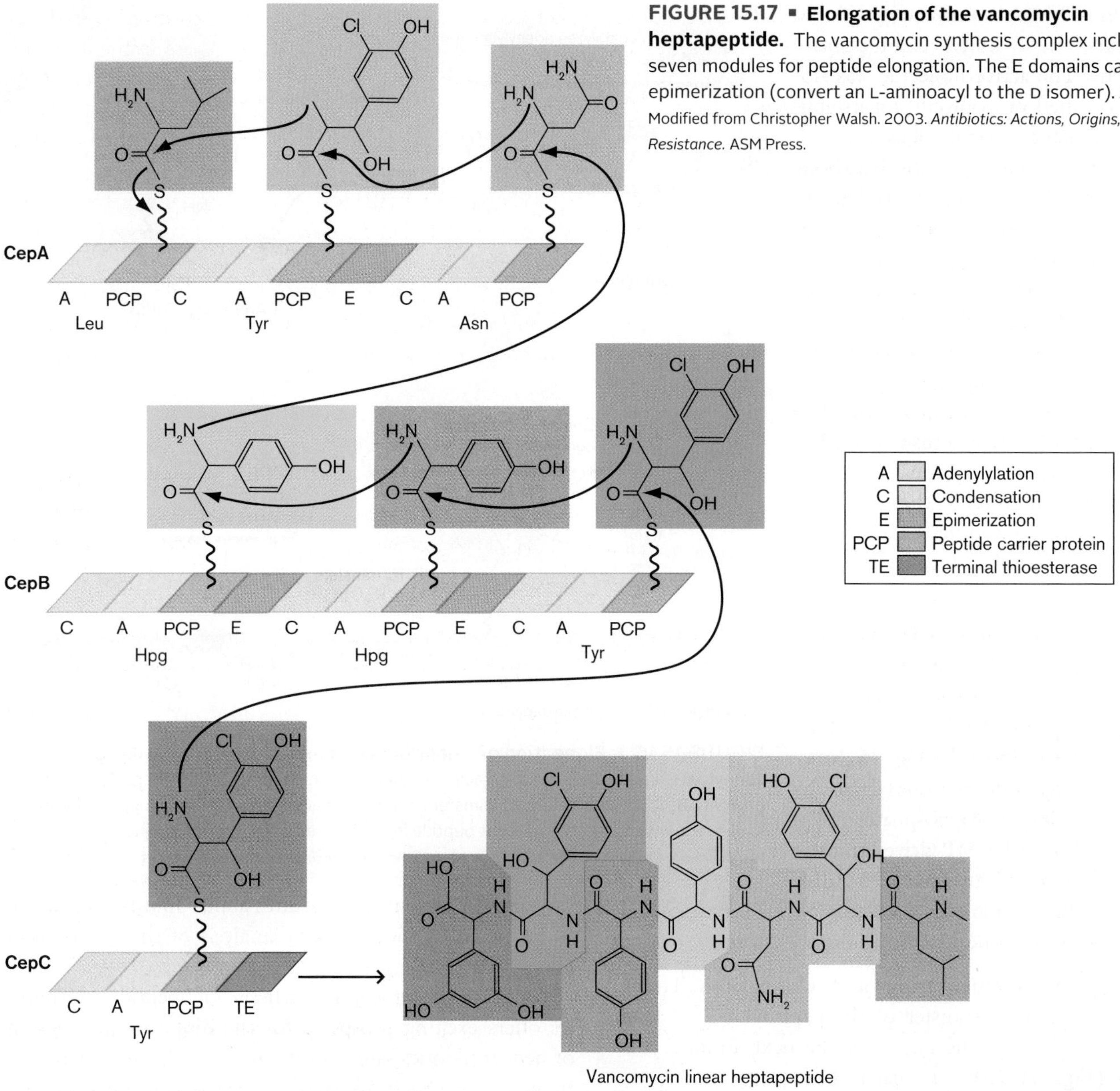

FIGURE 15.17 ▪ Elongation of the vancomycin heptapeptide. The vancomycin synthesis complex includes seven modules for peptide elongation. The E domains catalyze epimerization (convert an L-aminoacyl to the D isomer). *Source:* Modified from Christopher Walsh. 2003. *Antibiotics: Actions, Origins, Resistance.* ASM Press.

which generates the alkene double bond between the third and fourth carbons.

- **Fatty acid biosynthesis is regulated** by the levels of acetyl-CoA carboxylase and by the stringent response to carbon starvation. Bond saturation is regulated by temperature and other environmental factors.
- **Polyketide antibiotics are synthesized by modular enzymes.** The repeating units consist of acetyl groups with added functional groups.
- **Nonribosomal peptide antibiotics are also made by modular enzymes.** The repeating units consist of amino acids, many of which are outside the 20 ribosomal amino acids.

15.4 Nitrogen Fixation and Regulation

A cell contains DNA, amino acids, cell walls—think how much of this biomass is nitrogen. Although nitrogen gas makes up more than three-quarters of our atmosphere, our own bodies cannot assimilate or use any of it. The dinitrogen molecule (N_2) with its triple bond requires an enormous input of energy to split and reduce to two molecules of ammonia. Ammonia, protonated to ammonium ion (NH_4^+), is incorporated into carbon skeletons.

Early in evolution, all cells may have fixed their own N_2, but today only certain species of bacteria and archaea retain the ability. All other organisms depend on reduced or oxidized forms of nitrogen, which ultimately derive from N_2. Thus, all living organisms, directly or indirectly, depend on N_2-fixing prokaryotes within the biosphere (discussed in Chapters 21 and 22). Nitrogen limits microbes in many ecosystems, such as algae in freshwater lakes.

Given the limitations, nitrogen use by cells is subject to several levels of regulation. In this section we present some important molecular mechanisms of nitrogen regulation, which serves as an example of how cells regulate all biosynthesis and catabolism, depending on their relative needs for energy and materials.

Note: Bacteria and archaea play essential roles in global cycling of nitrogen, sulfur, and phosphorus. The geochemical cycling of these elements is discussed in Chapter 22.

Nitrogen Fixation

To fix nitrogen into biomass, bacteria or archaea must reduce the nitrogen completely to ammonia (NH_3). At pH 7, ammonia is mostly protonated to ammonium ion (NH_4^+). Unlike carbon, nitrogen rarely appears in oxidized form in complex biomolecules. Inorganic forms, such as nitric oxide (NO), are used as defense mechanisms against invading pathogens (see Chapter 23) or as signaling molecules. In macromolecules, however, virtually all the nitrogen is reduced; organic compounds containing oxidized nitrogen are generally toxic.

While N_2 is the ultimate source and sink of biospheric nitrogen, several oxidized or reduced forms are found in the environment, produced by living organisms (**Fig. 15.18**). Most free-living bacteria can acquire nitrate (NO_3^-) or nitrite (NO_2^-) for reduction to ammonium ion. Even nitrogen-fixing legume symbionts, such as *Rhizobium*, can use nitrate.

In natural environments, most potential sources of nitrogen are subject to competition from dissimilatory metabolism in which the molecule is oxidized or reduced for energy, as discussed in Chapter 14. For example, anaerobic respirers convert nitrate and nitrite to N_2 (denitrification), whereas lithotrophs oxidize NH_4^+ to nitrite and nitrate (nitrification). An important consequence of nitrification is that most of the commercial ammonia fertilizer spread on agricultural fields is soon oxidized by lithotrophs to nitrates and nitrites. High concentrations of nitrates in water are harmful because they combine with hemoglobin, generating a form that cannot take up oxygen. When infants drink water with high nitrite, they may become ill with "blue baby syndrome."

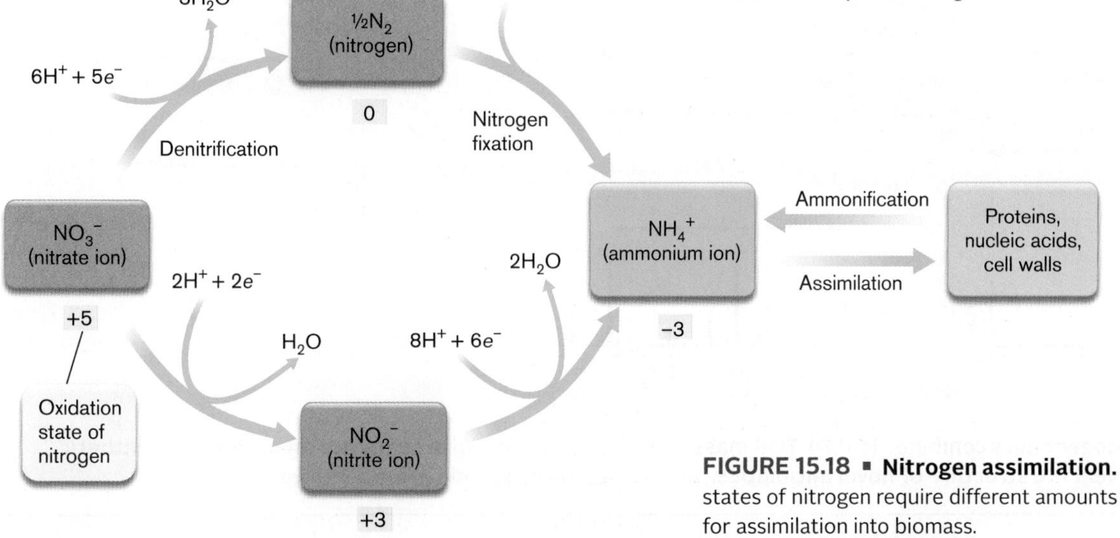

FIGURE 15.18 ■ Nitrogen assimilation. Different oxidation states of nitrogen require different amounts of reducing energy for assimilation into biomass.

Nitrogen Fixation: Early Discoveries

The first N_2 fixers, discovered by Martinus Beijerinck and colleagues in the late 1800s, were soil and wetland bacteria such as *Beggiatoa* and *Azotobacter*. Species of rhizobia, such as *Bradyrhizobium japonicum* (see **Fig. 15.20** inset) were discovered to fix nitrogen as endosymbionts of leguminous plants (see Chapter 21). For several decades, it was believed that N_2 was fixed only by a few special bacteria in the soil or in symbiotic plant bacteria. But in the 1940s, Martin Kamen and colleagues noticed that phototrophs such as *Rhodospirillum rubrum* produce hydrogen gas, a known by-product of nitrogen fixation. Nitrogen fixation was hard to demonstrate reliably in the laboratory because at that time researchers did not know that in *R. rubrum*, nitrogen fixation is repressed by the presence of alternative nitrogen sources, such as ammonia. When traces of ammonia and other nitrogen sources were eliminated, Kamen's student Herta Bregoff found that these photosynthetic bacteria actually fix nitrogen.

Nitrogen is now known to be fixed by most phototrophic bacteria (green and purple bacteria, as well as cyanobacteria) and by many archaea. Marine cyanobacteria fix a large proportion of both the nitrogen and carbon dioxide assimilated by our biosphere.

SPECIAL TOPIC 15.1 Mining Bacterial Genomes for Antibiotics

As resistance evolves, in our arms race with pathogens we always need more new antibiotics. Where do we find them? We can mine the genomes of the ever-evolving strains of bacteria in our environment, which engage in their own arms races with each other. The challenges are (1) to detect previously unknown antibiotics within cells—a chemical "needle in the haystack"—and (2) to reveal the genes encoding their biosynthetic enzymes. The genes can then be cloned in an engineered fermenter strain (discussed in Chapter 16).

So, which clues come first—the peptide or the biosynthetic genes? Either the peptide or the gene sequence may offer a clue. A multistep interactive approach between peptide and gene was devised by Pieter Dorrestein and colleagues at UC San Diego (**Fig. 1**). The approach is called peptidogenomics. **Figure 2** outlines the peptidogenomic approach that identified the structure and biosynthetic genes for stendomycin I (**Fig. 3**), an antifungal agent produced by the actinomycete *Streptomyces hygroscopicus*. The stendomycins are an emerging class of lipopeptides that target fungi; they selectively inhibit protein import into mitochondria. Lipopeptides are nonribosomal peptides attached to lipid. The nonribosomal peptide portion is synthesized by a modular enzyme, similar to that for vancomycin.

Stendomycin was the first antibiotic identified in *S. hygroscopicus* by a version of mass spectrometry called MALDI-TOF (matrix-associated laser desorption/ionization–time-of-flight). In the MALDI technique, a UV laser beam is trained on a sample embedded in a chemical matrix. A thin portion of sample plus matrix is "desorbed"—that is, ionized—and the ions are taken up into the mass spectrometer. The "time of flight" into the detector offers a measure of the size of the ionized molecules.

COURTESY OF PIETER DORRESTEIN/UNIVERSITY OF CALIFORNIA, SAN DIEGO

FIGURE 1 ■ Pieter Dorrestein discovers new antibiotic peptides by peptidogenomics.

Dorrestein aimed the MALDI laser at a matrix-embedded colony of *S. hygroscopicus*. Fragments of thousands of molecules were generated, among them components with sizes characteristic of certain uniquely modified amino acids that appear only in nonribosomal peptides, such as dehydroalanine.

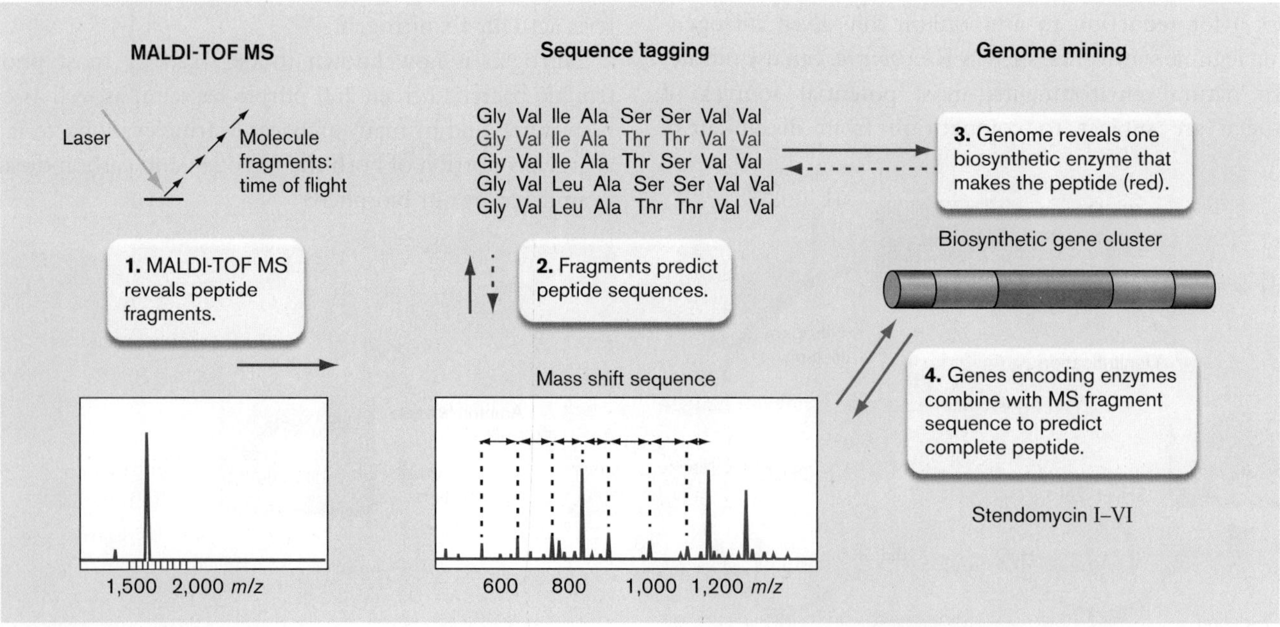

FIGURE 2 ■ Peptidogenomics combines MALDI-TOF mass spectrometry (MS) with sequence analysis of biosynthetic gene clusters to reveal the structure of novel antibiotics. *m/z* = mass-to-charge ratio (mass per charge).

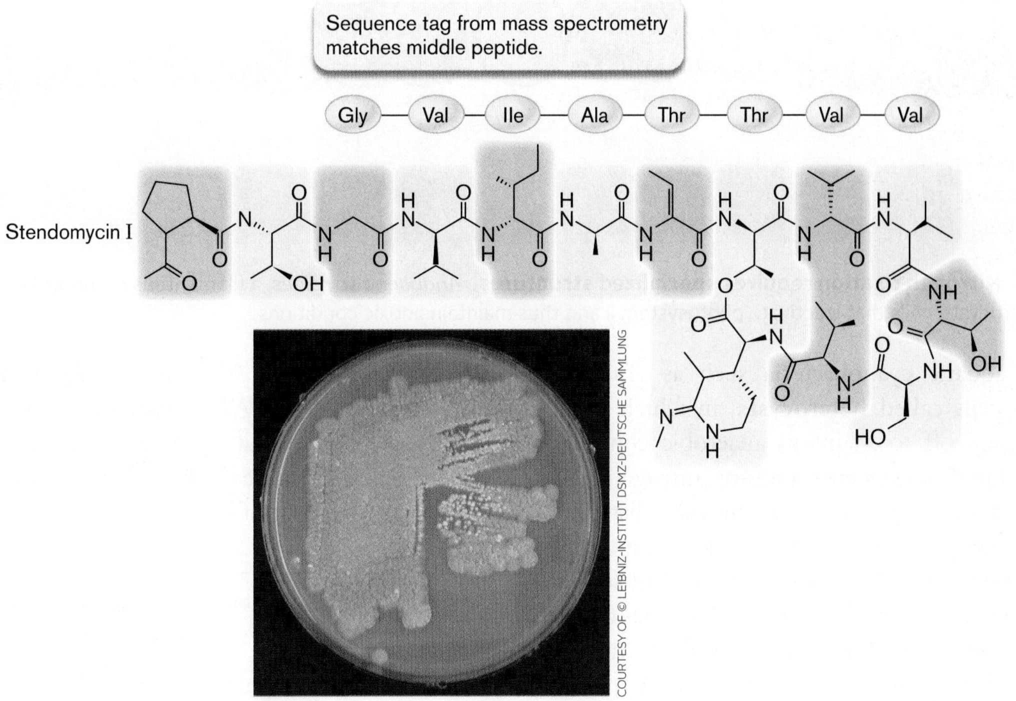

FIGURE 3 ■ A nonribosomal peptide antibiotic, stendomycin I, is synthesized by *Streptomyces hygroscopicus*.

In addition, the fragments predicted the existence of a peptide containing eight amino acids, whose sequence did not match any translated sequence within the *S. hygroscopicus* genome. The data were consistent with eight slightly different versions of the peptide, called a sequence tag (**Fig. 2**).

The peptide sequence tag was then used to query the genome for genes encoding modular enzymes that might catalyze the synthesis of the amino acids in the peptide. Previous bioinformatic studies characterized many families of modular enzymes that evolved from common ancestors. In the search for signature patterns for such genes, a cluster of genes for modular enzymes was identified that appeared likely to synthesize one version of the peptide: Gly-Val-Ile-Ala-Thr-Thr-Val-Val. This peptide proved to be the backbone of the lipopeptide stendomycin I (**Fig. 3**). Note that in the mature lipopeptide, other enzymes have catalyzed various modifications, such as an ester linkage between the carboxyl terminus and the hydroxyl group of a threonine residue. The full structure of stendomycin I was identified and confirmed by back-and-forth testing between models predicted by the mass spectrometry fragments and by gene cluster analysis.

Peptidogenomics is now used to identify novel peptides by screening against databases of mass spectra from millions of previously reported products. Such large data sets pose challenges, such as how to quickly identify and rule out peptides that have already been reported. To meet this challenge, Dorrestein and colleagues developed a computational pipeline called Dereplicator. The aim of Dereplicator is to identify "replicated" products already known—and to reveal novel products that are just slightly different from those known before. A small modification can lead to major new functions.

Impressive as our computations may be, our biosphere's microbes continually reproduce and evolve. We can only hope that our rate of drug discovery outpaces the rate of resistance arising in pathogens.

RESEARCH QUESTION

The genomic cluster shown in **Figure 2** encodes three modular enzymes for stendomycin biosynthesis, one of which synthesizes the tag peptide. What do you think the other two enzymes synthesize? How could you test your hypothesis using peptidogenomics?

Hosein Mohimani, Alexey Gurevich, Alla Mikheenko, Neha Garg, Louis-Felix Nothias, et al. 2017. Dereplication of peptidic natural products through database search of mass spectra. *Nature Chemical Biology* **13**:30–37.

Kersten, Roland D., Yu-Liang Yang, Yuquan Xu, Peter Cimermancic, Sang-Jip Nam, et al. 2011. A mass spectrometry–guided genome mining approach for natural product peptidogenomics. *Nature Chemical Biology* **7**:794–802.

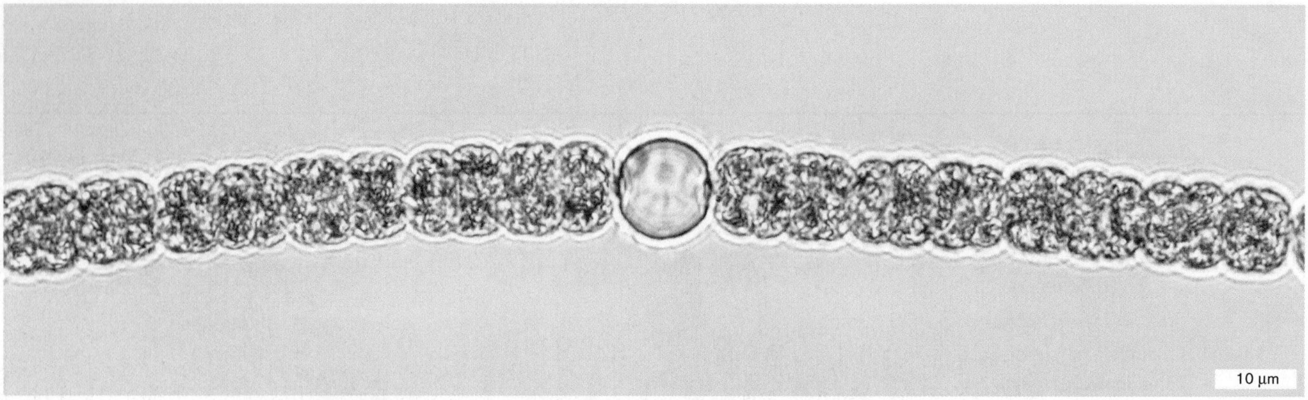

FIGURE 15.19 ■ Nitrogen fixation requires specialized structures. *Anabaena spiroides*, a filamentous cyanobacterium, segregates N_2 fixation in heterocysts, cells that inactivate photosystem II and thus maintain anoxic conditions.

To fix N_2, aquatic cyanobacteria such as *Anabaena* develop special cells called **heterocysts**, in which photosynthesis is turned off to maintain anaerobic conditions (**Fig. 15.19**). Land ecosystems require nitrogen-fixing bacteria and archaea in the soil, often in mutualistic relationships with plants. For example, the bacterium *Bradyrhizobium japonicum* associates with the roots of soybean plants, causing them to grow nodules. The nodules contain "bacteroids," forms of the bacteria that grow within the root cells. The bacteroids release extra reduced nitrogen into the soil; for this reason, farmers alternate soy crops with nitrogen-intensive crops such as corn. Nitrogen-fixing mutualism is discussed in Chapter 21.

Bacterial nitrogen fixation, however, has not sufficed to drive modern high-yield agriculture. Industrial nitrogen fixation uses the **Haber process**, in which nitrogen gas is hydrogenated by methane (natural gas) to form ammonia, under extreme heat and pressure. Scientists estimate that the Haber process reduces more atmospheric N_2 than all of Earth's other processes combined. While this amount of nitrogen represents a small fraction of the atmosphere, the environmental effects are huge, polluting waterways and causing marine areas of hypoxia called "dead zones" (discussed in Chapters 21 and 22). The anthropogenic (human-caused) influx of nitrogen into the biosphere is an astonishing example of the influence of human society on our planet.

The Mechanism of Nitrogen Fixation

In living cells, nitrogen fixation is an enormously energy-intensive process. The mechanism is largely conserved across species:

$$N_2 + 8H^+ + 8e^- + 16\ ATP \rightarrow 2NH_3 + H_2 + 16\ ADP + 16\ P_i$$

The electrons are donated by NADH, H_2, or pyruvate, or obtained through photosynthesis. The total energy investment includes approximately 3 ATP-equivalents per $2e^-$, plus 16 ATP molecules, as shown. That makes 12 + 16 = 28 ATP in all—a large part of the energy gained from oxidation of glucose. The production of H_2 is surprising, as it consumes extra ATP. Hydrogen loss results from initiating the cycle of nitrogen reduction (discussed shortly). Some bacteria have secondary reactions to reclaim the lost hydrogen with part of its lost energy.

Note: Although nitrogen fixation is commonly represented as producing ammonia (NH_3), under most conditions of living cells the predominant form is the protonated ammonium ion (NH_4^+).

Nitrogenase reaction mechanism. The overall conversion of nitrogen gas to two molecules of ammonia is catalyzed in four cycles by **nitrogenase**, an enzyme highly conserved in nearly all nitrogen-fixing species. Nitrogenase probably evolved once in a common ancestor of all nitrogen-fixing organisms.

The mechanism of nitrogenase is of intense interest to agricultural scientists because of the potential benefits of improving efficiency of plant growth and of extending nitrogen-fixing symbionts to nonleguminous plants such as corn. In 1960, scientists at the DuPont laboratory first isolated the nitrogenase complex from a bacterium, *Clostridium pasteurianum*. They measured its activity by incorporating heavy-isotope nitrogen gas, $^{15}N_2$, into $^{15}NH_3$. Since then, the detailed structure of nitrogenase has been solved (**Fig. 15.20**).

The active nitrogenase complex includes two kinds of subunits encoded by different genes: Fe protein (shaded green in **Fig. 15.20A**), containing a [4Fe-4S] center; and FeMo protein (shaded cyan), containing iron and molybdenum. The Fe protein contains a typical [4Fe-4S] structure to facilitate electron transfer. As we learned in Chapter 14, metal atoms help to transfer electrons because their orbitals are closely spaced and transitions between these orbitals involve relatively small amounts of energy. The electrons are funneled through a second Fe-S cluster (the P cluster) to the FeMo (iron-molybdenum) cluster. The FeMo cluster is an unusual structural characteristic of nitrogenase, in which trios of sulfur atoms alternate with trios of iron

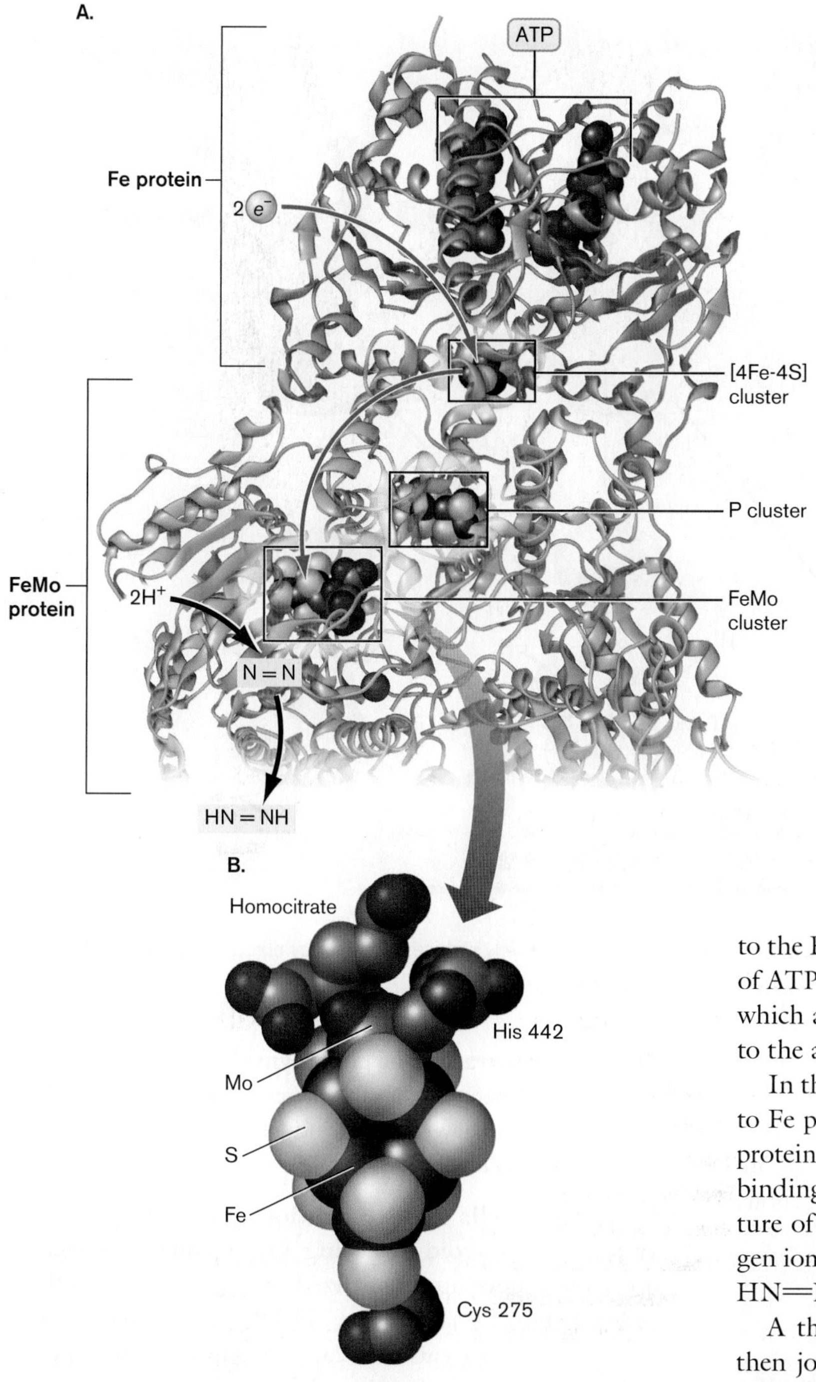

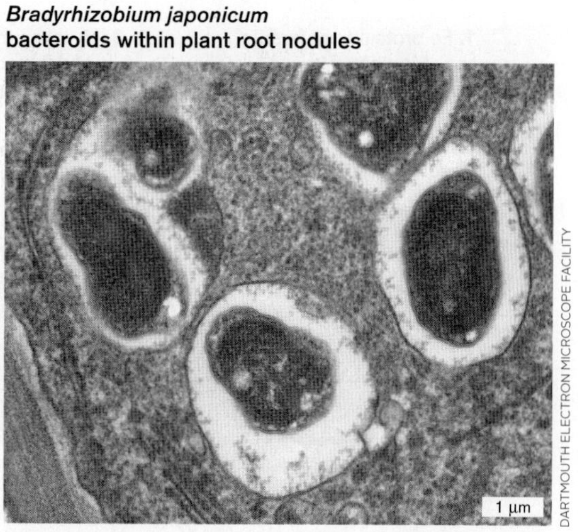

FIGURE 15.20 ▪ Structure of the nitrogenase complex. A. In each active site of the Fe protein–FeMo protein complex, the Fe protein binds ATP and receives electrons from the electron donors. The electrons are subsequently channeled down through the [4Fe-4S] cluster and the P cluster (Fe-S cluster) to the FeMo cluster, where they reduce the N_2. (PDB code: 1N2C) B. The metal cluster (Fe_7-S_9Mo) is held in place by coordination with a molecule of homocitrate plus two amino acid residues of nitrogenase: His 442 and Cys 275. **Inset:** Bacteroids of *Bradyrhizobium japonicum*, bacteria that fix nitrogen within plant root nodules.

atoms. One end of the cluster is capped with another iron, and the other end is capped with an atom of molybdenum (Mo). A consequence of the nitrogenase structure is that most nitrogen-fixing organisms (and their plant hosts, for leguminous symbionts) require the element molybdenum for growth. Some bacteria make an alternative nitrogenase that substitutes vanadium for molybdenum.

Four cycles of reduction. Nitrogen fixation requires four reduction cycles through nitrogenase (**Fig. 15.21**). To initiate the first cycle of N_2 reduction, $2e^-$ from an electron donor such as NADH or H_2 are transferred by a ferredoxin to Fe protein. The reduced Fe protein transfers each electron to the FeMo center, with energy supplied by four molecules of ATP (**Fig. 15.21**, step 1). The FeMo protein binds $2H^+$, which are reduced to H_2 (step 2). Only then does N_2 bind to the active site, by displacing the H_2 (step 3).

In the next reduction cycle, two electrons are transferred to Fe protein, where they reduce the iron. The reduced Fe protein transfers each electron to the FeMo center, near the binding site for N_2. The electron transfer requires expenditure of 2 ATP per electron, or 4 ATP per $2e^-$. Two hydrogen ions join the N_2 and receive the two electrons, forming HN=NH.

A third pair of electrons enters the Fe protein, which then joins another $2H^+$, reducing N_2 to H_2N–NH_2. The cycle again requires hydrolysis of 4 ATP. A final cycle of electron transfer and H^+ uptake reduces H_2N–NH_2 to $2NH_3$. At typical cytoplasmic pH values (near pH 7), NH_3 is protonated to NH_4^+.

The loss of H_2 during nitrogen fixation is puzzling because it represents lost energy in a highly energy-intensive process. The actual fate of the H_2 varies among species. *Klebsiella pneumoniae*, a Gram-negative bacterium, gives off the H_2 without further reaction. On the other hand, *Azotobacter* uses an irreversible hydrogenase enzyme to convert H_2 back to $2H^+$ and recover the transferred electrons. In leguminous rhizobial symbionts such as *Sinorhizobium* species, H_2 recovery varies surprisingly among strains and can affect the efficiency of plant growth.

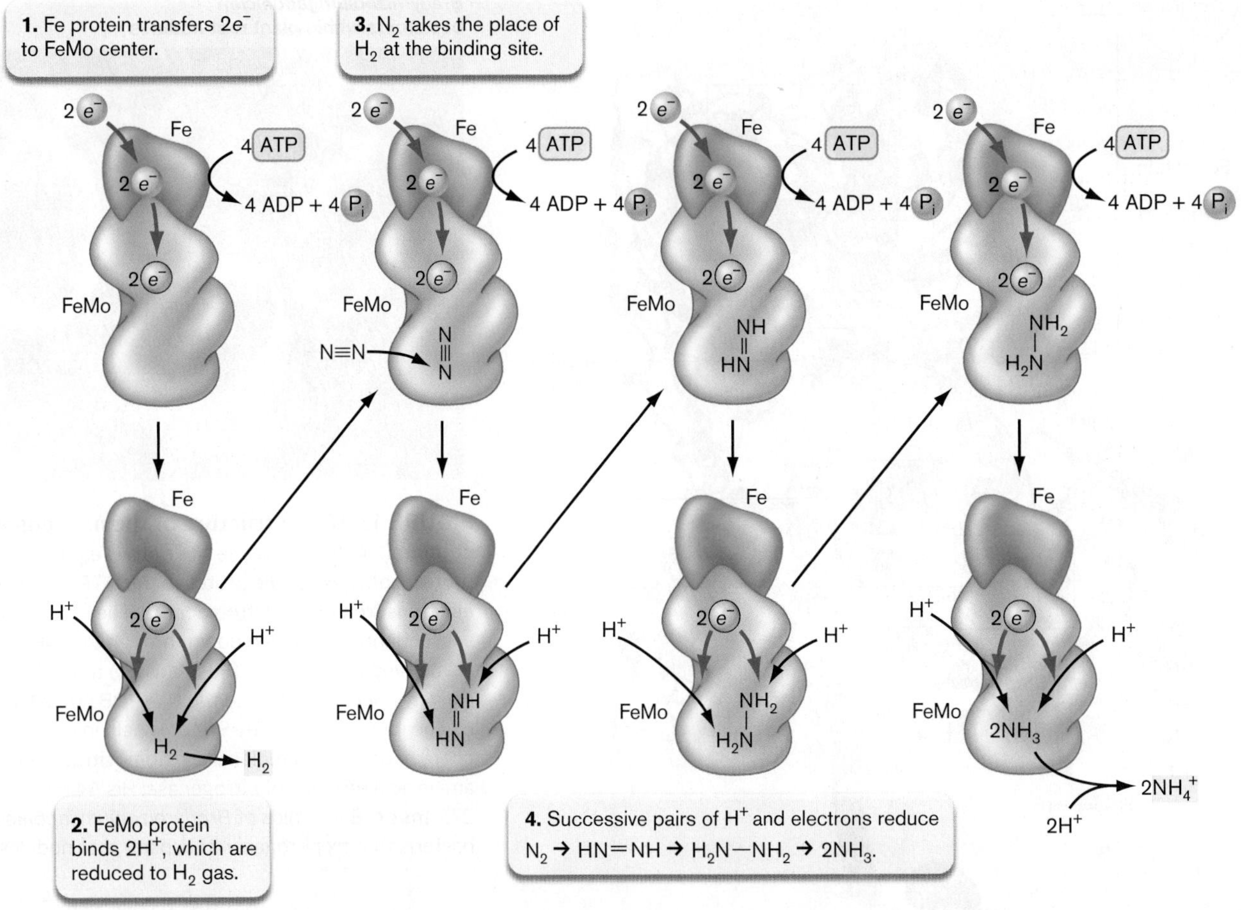

FIGURE 15.21 ■ Nitrogen fixation by nitrogenase. The enzyme nitrogenase successively reduces nitrogen by electron transfer, ATP hydrolysis, and H^+ incorporation.

Nitrogen fixation is expensive, both in terms of protein synthesis and in terms of reducing energy and ATP. Even bacteria capable of nitrogen fixation repress the process in the presence of other nitrogen sources, such as nitrate (NO_3^-), nitrite (NO_2^-), ammonia (NH_3), and nitrogenous organic molecules acquired from dead cells.

Anaerobiosis and N_2 Fixation

The reductive action of nitrogenase is extremely sensitive to oxygen because of the large reducing power needed to make NH_4^+. Thus, cells can fix nitrogen only in an anaerobic environment. This is no problem for anaerobes. But oxygenic phototrophs such as cyanobacteria face an obvious problem, as do bacterial symbionts of oxygenic plants. Aerobic and oxygenic organisms have developed several solutions to the problem of fixing nitrogen in aerobic environments:

- **Protective proteins.** Aerobic *Azotobacter* species synthesize protective proteins that stabilize nitrogenase and prevent attack by oxygen. For rhizobia, on the other hand, the plant hosts produce leghemoglobin (named for "legume" plants), a form of hemoglobin that sequesters oxygen away from the bacteria.
- **Temporal separation of photosynthesis and N_2 fixation.** Some species of cyanobacteria fix nitrogen only at night, when they do not conduct photosynthesis and thus release no oxygen.
- **Specialized cells for N_2 fixation inactivate photosystem II to avoid releasing O_2.** Filamentous cyanobacteria develop specialized nitrogen-fixing cells called heterocysts (see **Fig. 15.19**). The heterocysts lose their photosynthetic capacity entirely and specialize in nitrogen fixation. Heterocyst development is directed by a complex genetic program induced by nitrogen starvation. In natural environments, the heterocysts leak organic acids that attract heterotrophic bacteria, which use up all the oxygen around the heterocyst, generating ideal anaerobic conditions for nitrogen fixation.

Molecular Regulation of N_2 Fixation

Nitrogen fixation costs substantial energy, and therefore the process is regulated with extraordinary fine-tuning. Multiple factors determine the expression of genes encoding nitrogenase (*nifHDKTY*) and other nitrogen fixation proteins. Oxygen represses the expression of *nif* genes because nitrogenase is inactivated by oxygen; and high NH_4^+ depresses expression because enough fixed nitrogen is already available to the cell. Responses are mediated by several molecular regulators, including a nitrogen starvation sigma factor (sigma-54) and the NtrB-NtrC two-component signal transduction system. (Sigma factors and two-component regulators are discussed in Chapter 10.)

The NtrB-NtrC system was dissected in *Klebsiella pneumoniae* by Sydney Kustu and colleagues at UC Berkeley. When NH_4^+ is low (**Fig. 15.22**), the cell needs to fix nitrogen. The nitrogen sensor kinase NtrB autophosphorylates (obtains a phosphoryl group from ATP). NtrB-P then phosphorylates NtrC, forming NtrC-P, the nitrogenase activator. The NtrC-P phosphoprotein binds an upstream enhancer sequence to activate expression of *nifLA*, making NifL and NifA proteins. Expression of *nifLA* is coactivated by sigma-54. Sigma-54 and the NifA protein together activate expression of *nifHDKTY* to make nitrogenase. Thus, low NH_4^+ concentration turns on nitrogen fixation.

When cellular levels of NH_4^+ are high, NtrC remains unphosphorylated. The *nifLA* operon is not transcribed, no activators bind the *nifHDKTY* promoter, and the nitrogen fixation genes are not expressed.

Nitrogenase activity is stopped by oxygen, so high oxygen levels block expression of the gene. When oxygen is high, NifA binds NifL and is prevented from binding the nitrogenase promoter. Low oxygen allows NifA to bind the promoter and express nitrogenase to fix nitrogen. Overall, NtrC governs response to NH_4^+, whereas NifA governs response to oxygen. Being responsive to multiple environmental signals is typical of enzymes in energy-expensive biosynthetic pathways.

NtrB and NtrC also regulate nitrogen usage at the level of biosynthesis of the amino acids glutamate and glutamine (discussed in the next section).

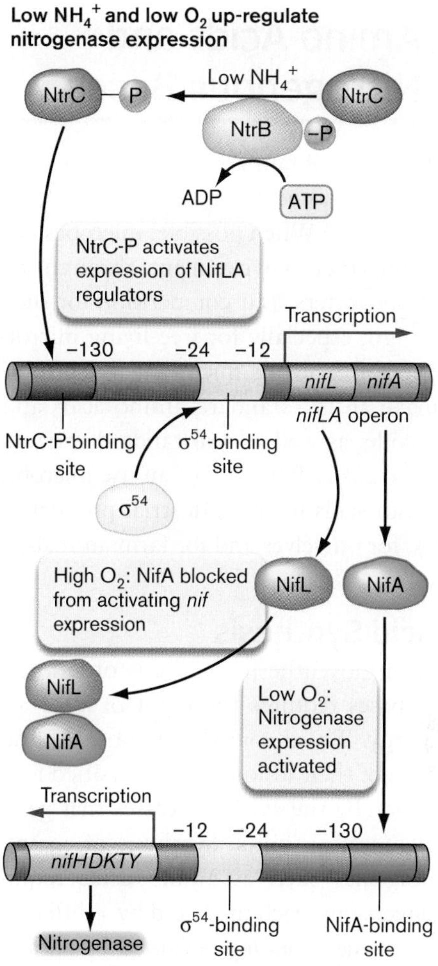

FIGURE 15.22 ■ Regulation of nitrogen fixation. When cellular levels of NH_4^+ are low, NtrC is phosphorylated and activates expression of *nifLA*. Expression is coactivated by the nitrogen starvation sigma factor, sigma-54. The NifA protein (and sigma-54) activates expression of *nif* genes encoding nitrogenase. When oxygen levels are high, however, nitrogen fixation cannot occur, so NifA is blocked by binding NifL. (Numbers refer to positions upstream of the translation site.)

To Summarize

- **Oxidized or reduced forms of nitrogen**, such as nitrate, nitrite, and ammonium ions, can be assimilated by bacteria and plants. Assimilation competes with dissimilatory reactions that obtain energy.
- **Nitrogen gas (N_2)** is fixed into ammonium ion (NH_4^+) only by some species of bacteria and archaea—never by eukaryotes.
- **Nitrogenase enzyme** includes a protein containing an iron-sulfur core (Fe protein) and a protein containing a complex of molybdenum, iron, and sulfur (FeMo protein). Electrons acquired by Fe protein (with energy from ATP) are transferred to FeMo protein to reduce nitrogen.
- **Four cycles of reduction by NADPH** or an equivalent reductant reduce one molecule of N_2 to two molecules of NH_3. At neutral pH, NH_3 is protonated to NH_4^+.
- **Oxygen inhibits nitrogen fixation.** Bacteria have various means of separating nitrogen fixation from aerobic respiration, such as heterocyst development or temporal separation.
- **Nitrogen and oxygen regulate transcription of nitrogenase.**

15.5 Amino Acids and Nitrogenous Bases

Where do microbes obtain amino acids to make their proteins and cell walls, as well as nitrogenous bases to synthesize DNA and RNA? When possible, microbes obtain these molecules from their environment through membrane-embedded transporters. But competition for such valuable nutrients is high, especially for free-living microbes in soil or water. Most free-living microbes and plants have the ability to make all the standard amino acids and bases of the genetic code, as well as nonstandard variants used for cell walls and transfer RNAs. We can use microbial biosynthesis of amino acids in the industrial production of food supplements, for ourselves and for farm animals.

Amino Acid Synthesis

Like fatty acid biosynthesis, synthesis of amino acids and nitrogenous bases requires the input of large amounts of reducing energy. These compounds pose additional challenges because of their unique and diversified forms, which cannot be made by the cyclic processes that generate molecules from repeating units. Synthesis of complex, asymmetrical molecules such as amino acids requires many different conversions, each mediated by a different enzyme. Nevertheless, some economy is gained by an arrangement of branched pathways in which early intermediates are utilized to form several products (**Fig. 15.23**). For example, oxaloacetate is converted to aspartate, which can be converted to four other amino acids.

The carbon skeletons of amino acids arise from diverse intermediates of metabolism (see **Fig. 15.23**). As in fatty acid biosynthesis, precursor molecules are channeled into amino acid biosynthesis by specialized cofactors and reducing energy carriers such as NADPH. Note that certain amino acids arise directly from key metabolic intermediates (for example, glutamate from 2-oxoglutarate), whereas others must be synthesized from preformed amino acids (for example, glutamine, proline, and arginine from glutamate). Some amino acids can arise from more than one source; for example, leucine and isoleucine can be made from succinate as well as from pyruvate.

It has been hypothesized that the amino acids arising in just one or two steps from central intermediates are more ancient in cell evolution than those requiring more complex pathways. Five of these "ancient" amino acids—glutamate, aspartate, valine, alanine, and glycine—are the same as those detected in meteorites, whose composition resembles that of prebiotic Earth (**Fig. 15.24**). These same five amino acids also appear in early-Earth simulation experiments in which methane, ammonia, and water are heated under reducing conditions and subjected to electrical discharge. Thus, we speculate that the first amino acids that early cells evolved to make were the same as those that arose spontaneously in the prebiotic chemistry of our planet.

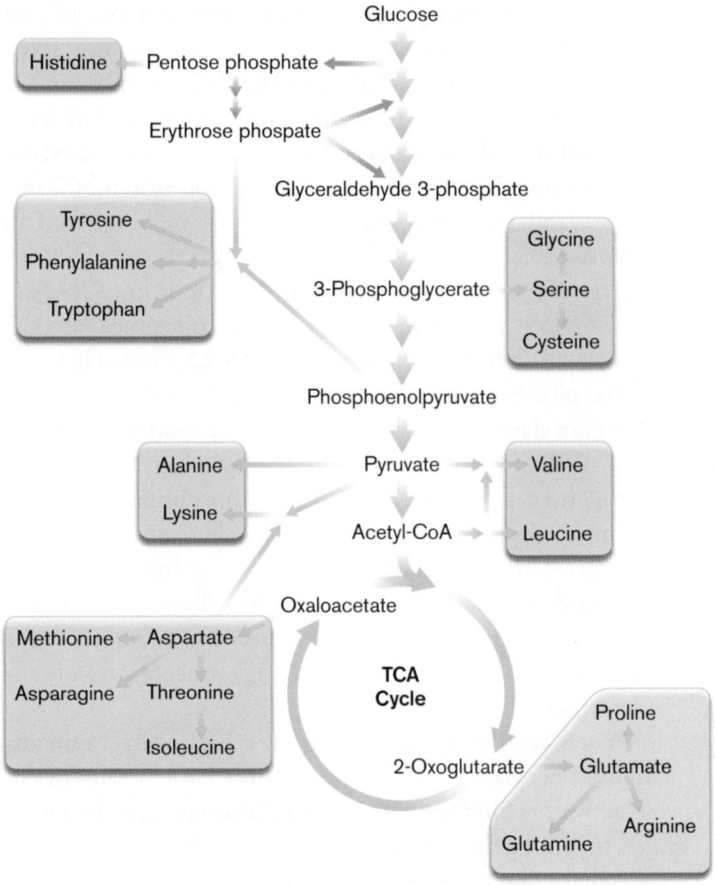

FIGURE 15.23 ▪ Major pathways of amino acid biosynthesis. Some amino acids arise from key metabolic intermediates, whereas others must be synthesized out of other amino acids.

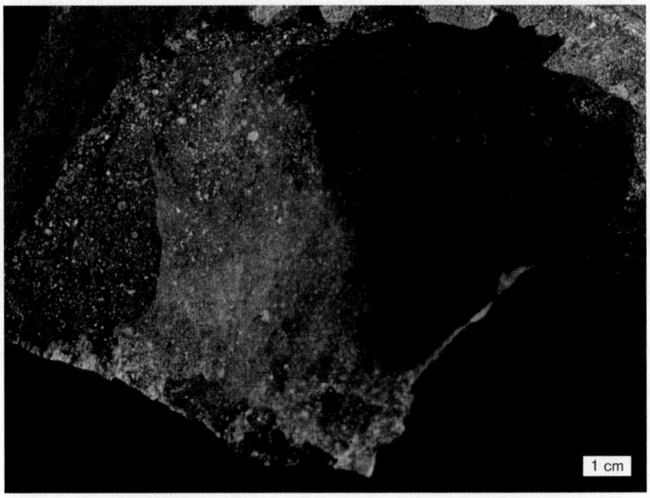

FIGURE 15.24 ▪ The Murchison meteorite. Fragments of a meteorite that fell in Murchison, Australia, in 1969 were shown to contain the five fundamental amino acids of biosynthetic pathways (glutamate, aspartate, valine, alanine, and glycine).

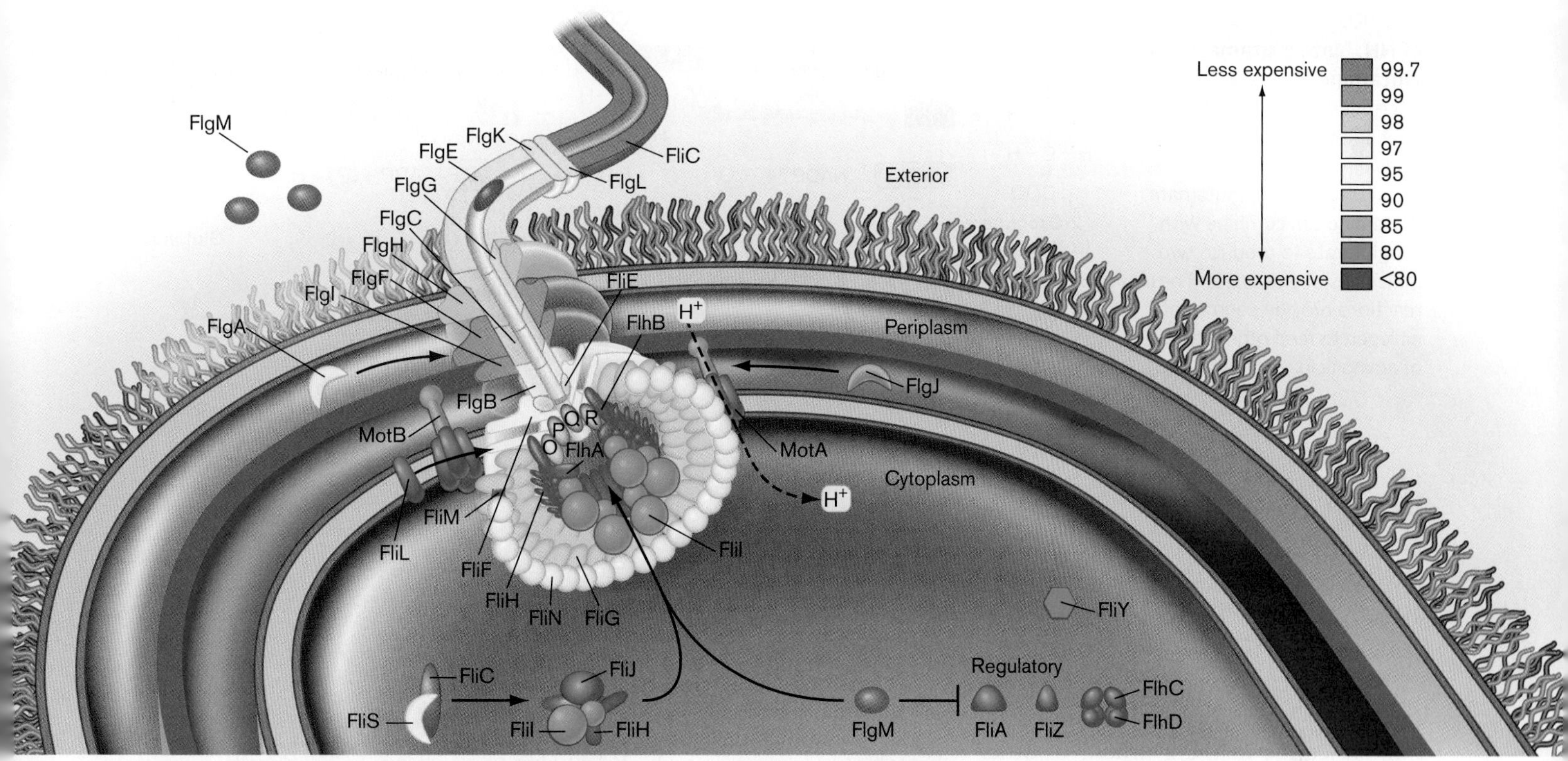

FIGURE 15.25 ■ **External proteins and proteins extending outside the cell use less expensive amino acids.** Diagram of flagellar proteins (FlgE, FlgK, FlgL, FlgG, FliC) attached to the motor complex. The external and secreted structures (FlgE, FliC, FlgM) are composed of amino acids that are less energy-expensive. *Source:* Modified from Daniel Smith and Matthew Chapman. 2010. *MBio* **1**:e00131.

A more subtle effect of evolution has been to adjust the amino acid composition of proteins based on the energetic cost of biosynthesis. Proteins that are secreted or that project outside the cell cannot be recycled; their amino acids are ultimately lost to the cell. Thus, secreted and externally projecting proteins have evolved to contain the "cheaper" amino acids—that is, the amino acids whose synthesis requires spending fewer molecules of ATP and NADPH. This effect can be seen in the diagram of a bacterial flagellum and its attached motor (**Fig. 15.25**), which contains extracellular as well as cytoplasmic components, colored in the figure on a scale based on biosynthetic "expense." The external and secreted components favor less expensive amino acids, compared to the cytoplasmic components.

Assimilation of NH_4^+

Unlike sugars and fatty acids, amino acids must assimilate another key ingredient: nitrogen. The NH_4^+ produced by N_2 fixation or by nitrate reduction is the key source of nitrogen for biosynthesis. But NH_4^+ always exists in equilibrium with NH_3, which raises pH to toxic levels. Moreover, NH_3 travels freely through membranes, making it difficult to store NH_4^+ within a cell. Deprotonation to NH_3 increases as pH rises; by pH 9.2, deprotonation reaches 50%. Even at neutral pH, a very small equilibrium concentration of NH_3 can drain NH_4^+ out of the cell. Thus, cells avoid storing high levels of NH_4^+; instead, the fixed nitrogen is incorporated immediately into organic products.

2-Oxoglutarate and glutamate condense with NH_4^+. In most bacteria, the main route for NH_4^+ assimilation is the condensation of NH_4^+ with 2-oxoglutarate to form glutamate, or with glutamate to form glutamine (**Fig. 15.26**). Three key enzymes interconvert these substrates:

- **Glutamate dehydrogenase** (**GDH**), actually named for its reverse activity, condenses NH_4^+ with 2-oxoglutarate to form glutamate. The condensation requires reduction by NADPH.
- **Glutamine synthetase** (**GS**), or GlnA, condenses a second NH_4^+ with glutamate to form glutamine—a process driven by spending one ATP.
- **Glutamate synthase** (**GOGAT**, **glutamine:2-oxoglutarate aminotransferase**) converts 2-oxoglutarate plus glutamine into two molecules of glutamate. Different variants of this enzyme use different reducing agents: NADPH, NADH, or ferredoxin. The NADPH variant is shown in **Figure 15.26**.

Through these reactions, all three substrates exchange amino groups readily. High nitrogen levels induce GDH

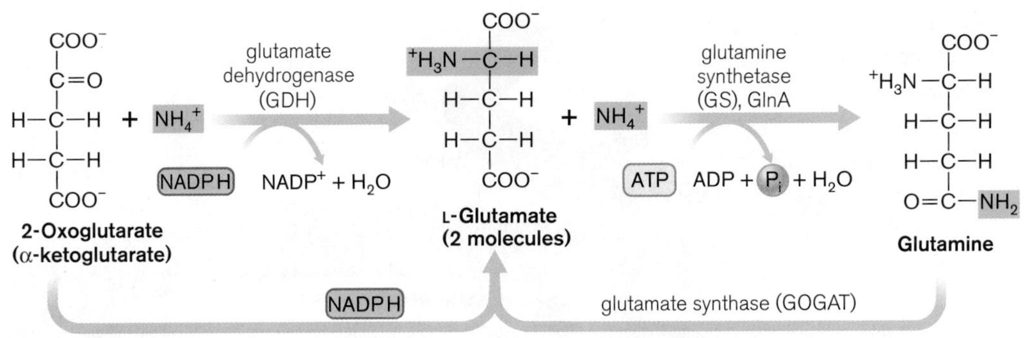

FIGURE 15.26 ■ Assimilation of NH_4^+ into glutamate and glutamine. The key TCA cycle intermediate 2-oxoglutarate (alpha-ketoglutarate) incorporates one molecule of NH_4^+ at the ketone to form glutamate. Glutamine can combine with oxoglutarate to produce two molecules of glutamate. These reactions provide sources of nitrogen to feed other pathways of amino acid synthesis.

to take up NH_4^+, but repress GS and GOGAT. GS has a higher affinity than GDH for NH_4^+. Low nitrogen levels induce GS (to make glutamine) and GOGAT (to convert some glutamine to glutamate as needed).

Both glutamate and glutamine contribute an amine, as well as their carbon skeletons, to the synthesis of other amino acids in the biosynthetic "tree." The transfer of ammonia between two metabolites such as glutamate and glutamine is called **transamination**. Many other pairs of amino acids and metabolic intermediates undergo transamination. For example, glutamate transfers NH_3 to oxaloacetate, making aspartate and 2-oxoglutarate; the reaction can also be reversed. In another example, valine transfers an amine group to pyruvate, generating alanine and 2-ketoisovalerate.

Thought Question

15.6 Suggest two reasons why transamination is advantageous to cells.

Glutamate and Glutamine Signal Nitrogen Availability

The cellular levels of glutamate and glutamine act as indicators of nitrogen availability. When nitrogen is scarce, NtrC is phosphorylated to NtrC-P as we saw earlier, in **Figure 15.22**. Along with nitrogenase, NtrC-P up-regulates the expression of glutamine synthetase (**Fig. 15.27**), as well as a high-affinity ammonia transporter, and transporters for organic sources of nitrogen such as amino acids, oligopeptides, and cell wall fragments containing amino sugars. All these molecules can be "scavenged" to obtain nitrogen for biosynthesis.

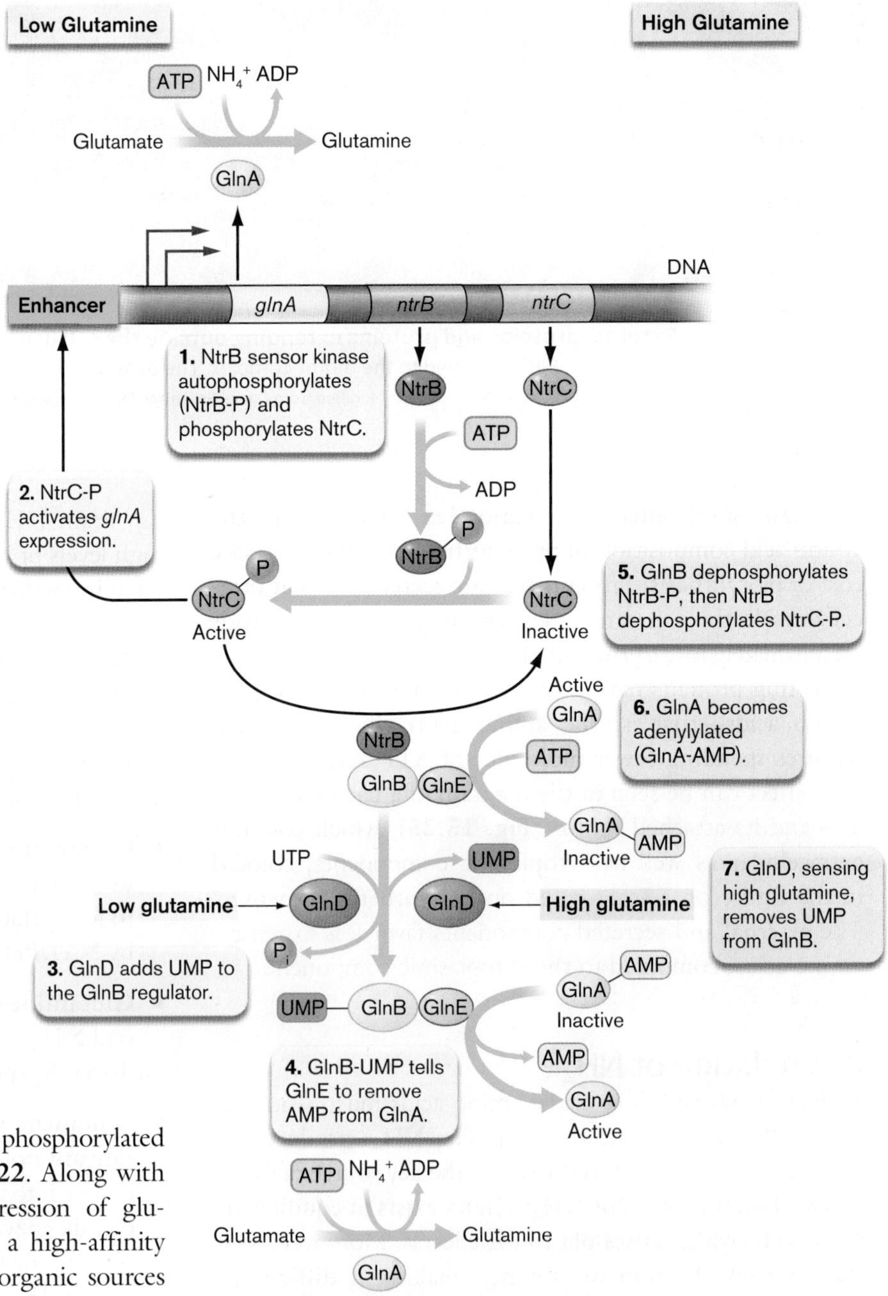

FIGURE 15.27 ■ Glutamine regulation. The NtrB-NtrC two-component system regulates glutamine synthetase (GlnA) expression. GlnB, D, and E regulate GlnA activity.

When ammonium ion (NH_4^+) is abundant, cells store the nitrogen safely as glutamine, via glutamine synthetase (GlnA; **Fig. 15.27**). GlnA uses the energy released from ATP hydrolysis to assimilate NH_4^+ into glutamic acid and produce glutamine. But what if the cell makes too much active GlnA? Then all of the cell's glutamate will be converted to glutamine, and not enough glutamate will remain to build proteins. To prevent a glutamate shortage, excess glutamine signals the cell to stop making GlnA and inactivate whatever GlnA is already present. To maintain the balance requires an intricate series of regulators responding to low glutamine or to high glutamine (**Fig. 15.27**).

Glutamine acts at the levels of transcriptional and post-translational control. Here we describe only a portion of this very complex regulatory system. As you study the system, imagine how such molecular controls interconnect all the enzymes and regulators throughout a bacterium.

Low glutamine. NtrB and NtrC form a two-component signal transduction system (see Chapter 10). When glutamine levels are low (**Fig. 15.27**, left), the sensor kinase NtrB autophosphorylates (NtrB-P) and then phosphorylates the response regulator NtrC (step 1). NtrC-P regulator binds to an enhancer sequence, which can influence target gene expression from great distances (1 or 2 kb; step 2). As a result of NtrC-P binding to its enhancer, the bacterium makes glutamine synthetase, assimilates nitrogen (in the form of ammonium ion), and incorporates the ammonium ion as glutamine. Maintaining GlnA requires active NtrC-P and active NtrB-P. When the nitrogen level increases, NtrB-P gets dephosphorylated by GlnB (see the next section).

Glutamine synthetase (GlnA) control is so important that after the enzyme is synthesized, its activity is also regulated. To regulate GlnA activity, GlnB is modified by another protein, GlnD, which adds uridine monophosphate (UMP). GlnB-UMP then activates GlnA by removing AMP from the deactivated form, GlnA-AMP (discussed next).

High glutamine. When glutamine levels are in excess (**Fig. 15.27**, right), the cell needs to stop making GlnA and inactivate whatever still remains. To halt *glnA* transcription, GlnB dephosphorylates NtrB-P, and then NtrB dephosphorylates NtrC-P (step 5). NtrC cannot bind the enhancer, and thus *glnA* transcription stops.

Meanwhile, GlnB also compels GlnE to add an AMP to GlnA (adenylylation, step 6). This modification inactivates whatever GlnA remains in the cell. The reaction is amplified by GlnD, which, under high glutamine, reverses its reaction, removing UMP from GlnB (step 7). The result is that ammonium ion no longer condenses with glutamate, and the glutamate/glutamine ratio is maintained.

Building Complex Amino Acids

Seven "fundamental" amino acids have relatively simple biosynthetic pathways. Others require longer pathways involving numerous enzymes. One amino acid produced by a complex pathway is arginine (**Fig. 15.28**). Bacterial arginine biosynthesis generally involves about a dozen different enzymes distributed among four to eight operons. Operon expression is regulated by an arginine repressor that binds the promoter of each operon to prevent transcription in the presence of sufficient arginine for cell needs. In some species, the arginine repressor also activates enzymes of arginine catabolism, making the excess amino acid available as a carbon source.

Arginine biosynthesis. Arginine synthesis begins with the condensation of glutamate with acetyl-CoA (**Fig. 15.28**, step 1), transferring an amino group ($-NH_2$) to the arginine precursor. Three more amino groups are transferred subsequently by glutamate, glutamine, and aspartate. A second glutamate transfers an amino group to the arginine precursor (step 2), while its own carbon skeleton cycles back as 2-oxoglutarate (alpha-ketoglutarate). This transfer of NH_4^+ between two organic intermediates is an example of transamination. The original acetyl group from acetyl-CoA is then hydrolyzed, producing ornithine. Ornithine is a central intermediate, used by cells to synthesize proline and various polyamines, as well as arginine.

Thought Question

15.7 Which energy carriers (and how many) are needed to make arginine from 2-oxoglutarate?

Ornithine receives a third amino group (**Fig. 15.28**, step 3) from glutamine via conversion of CO_2 to carbamoyl phosphate [$H_2N{-}CO(PO_4)^{2-}$]:

$$CO_2 + H_2O \rightarrow H^+ + HCO_3^-$$

$$HCO_3^- + ATP + \text{glutamine} \rightarrow H_2N{-}CO(PO_4)^{2-} + ADP + \text{glutamate}$$

Carbamoyl phosphate provides a way to assimilate one nitrogen plus one carbon into a structure.

The fourth nitrogen is acquired from aspartate in a two-step process requiring ATP release of pyrophosphate (PP_i; **Fig. 15.28**, step 4). The release of fumarate, a TCA cycle intermediate, yields arginine.

Aromatic amino acids. The complexity of aromatic amino acids requires particularly energy-expensive biosynthesis. The number of different enzymes required, however,

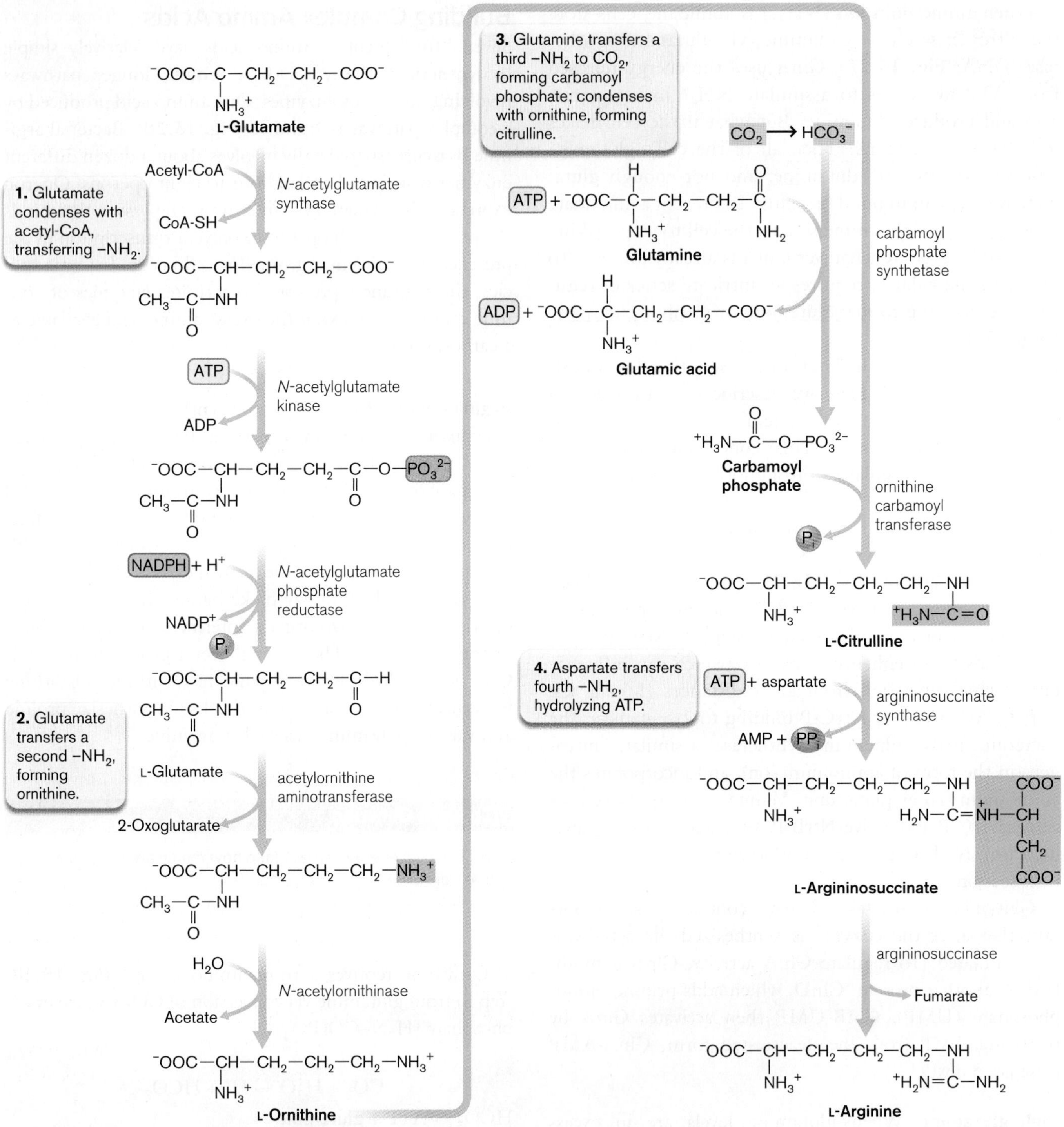

FIGURE 15.28 ■ Arginine biosynthesis in *E. coli*. In arginine biosynthesis, two glutamate molecules condense with acetyl-CoA, and nitrogen is donated by glutamate and glutamine.

is minimized by the presence of a common core pathway branching to several different amino acids (**Fig. 15.29**).

The three aromatic amino acids—phenylalanine, tyrosine, and tryptophan—each require a common precursor, chorismate. The pathway to chorismate starts with simple glycolytic intermediates (phosphoenolpyruvate and erythrose 4-phosphate), but its synthesis requires 10 enzymes encoded in a single operon, *aroABCDEFGHKL*. From chorismate to phenylalanine or tyrosine requires three additional enzymatic steps. From chorismate to tryptophan, the most complex amino acid specified by the genetic code, requires another five enzymes, expressed by *trpABCDE*. Thus, at least 15 enzymes encoded by different genes are required to make tryptophan out of simple substrates.

Not surprisingly, the presence of tryptophan in the cell represses expression of its own biosynthetic enzymes.

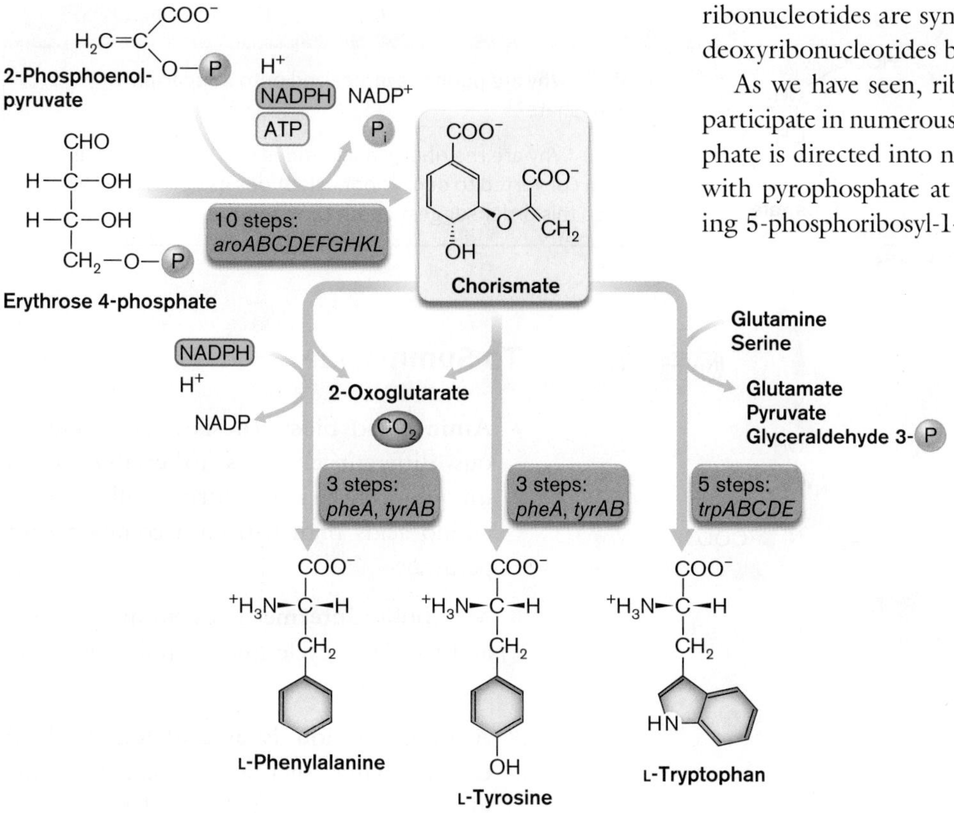

FIGURE 15.29 ▪ Biosynthesis of aromatic amino acids. Aromatic amino acids are assembled out of various carbon skeletons. In *E. coli*, the pathways to chorismate and tryptophan are encoded by operons of contiguous genes, *aro* and *trp*.

Repression of the *trp* operon includes two mechanisms: the Trp repressor, effective over moderate to high levels of tryptophan; and an RNA loop mechanism called attenuation, sensitive to lower levels of tryptophan (discussed in Chapter 10). Attenuation involves the destabilization of the transcription complex by formation of a specific stem loop on the nascent *trp* mRNA. The mechanism of attenuation in *E. coli* was discovered in 1981 by Charles Yanofsky, winner of a National Medal of Science in 2003. Several other amino acids commonly show attenuation in bacteria—particularly histidine, threonine, phenylalanine, valine, and leucine.

Purine and Pyrimidine Synthesis

Nitrogenous bases include purines and pyrimidines, the essential coding components of DNA and RNA. The accuracy of the entire genetic code requires accurate synthesis of these bases. Besides their role in nucleic acids, purine and pyrimidine nucleotides such as ATP serve as energy carriers.

Bases are built onto ribose 5-phosphate. The purine and pyrimidine bases are not built as isolated units; rather, they are constructed on a ribose 5-phosphate substrate, forming a nucleotide (sugar-base-phosphate). Note that ribonucleotides are synthesized first and then converted to deoxyribonucleotides by enzymatic removal of the 2′ OH.

As we have seen, ribose 5-phosphate and related sugars participate in numerous metabolic pathways. Ribose 5-phosphate is directed into nucleotide synthesis when it is tagged with pyrophosphate at the carbon 1 (C-1) position, forming 5-phosphoribosyl-1-pyrophosphate (PRPP; **Fig. 15.30**). PRPP is a major metabolic intermediate, the starting point for the synthesis of purine and pyrimidine nucleotides. PRPP is also produced by "scavenger" pathways, in which excess nitrogenous bases are broken down and recycled. Its synthesis is regulated by feedback inhibition to avoid overproduction of purines. In humans, overproduction of PRPP is one mechanism that leads to gout, a condition in which the purine breakdown product uric acid precipitates in the joints, causing painful swelling of wrists and feet.

Purine synthesis. The hydrolysis of PRPP releases pyrophosphate. This irreversible reaction drives forward the subsequent reactions to make the purine (**Fig. 15.30**, left). To construct the purine, the pyrophosphate at carbon 1 (C-1) is replaced by an amine from glutamine. The purine is built up by addition of a series of single-carbon groups (formyl, methyl, and CO_2) alternating with nitrogens from glutamine and carbamoyl phosphate. The number of single-carbon assimilations, including assimilation of CO_2, is striking. It may reflect the ancient origin of the purine synthetic pathway, which evolved in CO_2-fixing autotrophs.

The first purine constructed is inosine monophosphate. Inosine is the "wobble" purine found at the third position of tRNA anticodons. Inosine monophosphate is subsequently converted to adenosine monophosphate (AMP) or guanosine monophosphate (GMP).

Pyrimidine synthesis. The pyrimidine is built by a slightly different route (**Fig. 15.30**, right). First, the six-membered pyrimidine ring forms from aspartate plus carbamoyl phosphate. The pyrimidine ring then displaces the pyrophosphate of PRPP, attaching by a nitrogen to the ribosyl carbon 1. The first pyrimidine built is uracil (as UMP), which can be converted to cytosine or thymine—as cytidine monophosphate (CMP) or thymidine monophosphate (TMP), respectively.

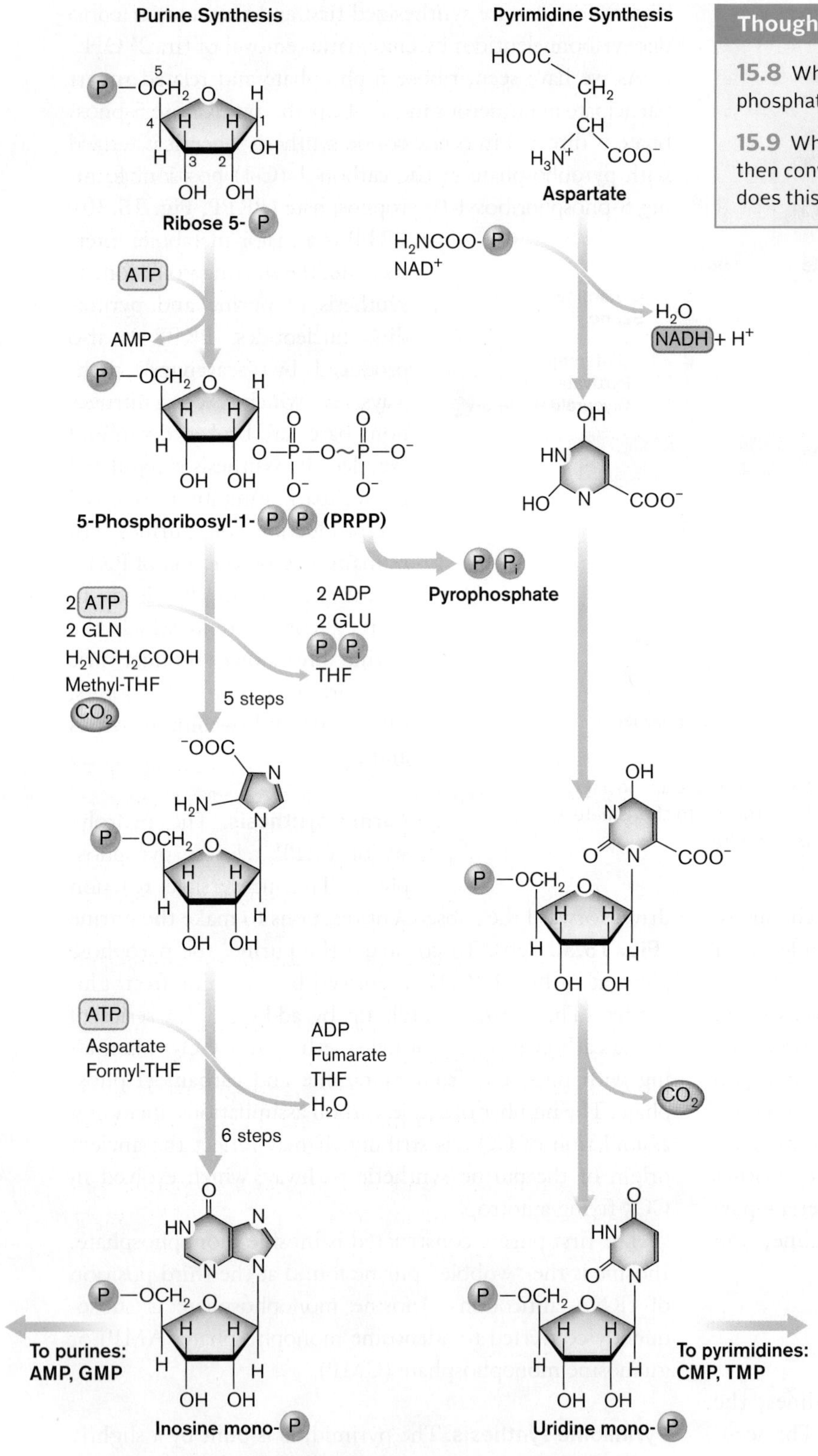

FIGURE 15.30 ■ Biosynthesis of purines and pyrimidines. Both purines and pyrimidines are built onto ribose 5-phosphate. Ribose 5-phosphate is directed into purine and pyrimidine biosynthesis by the activating step of ATP → AMP, converting the sugar to 5-phosphoribosyl-1-pyrophosphate (PRPP). The purine ring is built up out of successive additions of amines, one-carbon units, plus a formyl group from formyltetrahydrofolate (formyl-THF). The purine ring is modified to yield AMP and GMP. Likewise, the pyrimidine ring is modified to yield CMP or TMP.

Thought Questions

15.8 Why are purines synthesized onto a sugar-base-phosphate?

15.9 Why are the ribosyl nucleotides synthesized first and then converted to deoxyribonucleotides as necessary? What does this order suggest about the evolution of nucleic acids?

To Summarize

- **Amino acid biosynthesis** requires numerous different enzymes to catalyze many unique conversions. Structurally related amino acids branch from a common early pathway.
- **Metabolic intermediates** from glycolysis and the TCA cycle initiate amino acid biosynthetic pathways.
- **Ammonium ion is assimilated by TCA intermediates**, such as 2-oxoglutarate into glutamate. Glutamate assimilates ammonium ion to form glutamine. Transamination is the donation of NH_4^+ from one amino acid to another, such as the transfer of ammonia from glutamine to 2-oxoglutarate to make aspartate.
- **Arginine biosynthesis** requires multiple steps of NH_3 transfer and carbon skeleton condensation.
- **Aromatic amino acids** are built from a common pathway that branches out. Their biosynthesis is regulated tightly at both transcriptional and translational levels.
- **Purines are built as nucleotides** attached to a ribose phosphate. Several single-carbon groups are assimilated, including CO_2—a phenomenon suggesting an ancient pathway. **Pyrimidines are made from aspartate**, and then added onto PRPP.

Concluding Thoughts

In living cells, no enzymatic pathway occurs in isolation. All pathways of energy acquisition and biosynthesis occur together, sharing common substrates and products. For example, acetyl-CoA is produced by glycolysis or by fatty acid degradation; it is then utilized by pathways that synthesize fatty acids, amino acids, and nucleobases. Other precursors to these molecules arise from the TCA cycle. At the same time, cells need the TCA cycle to yield energy through catabolism. The microbial cell regulates these competing needs on an extremely rapid timescale: Within seconds of the disappearance or appearance of a nutrient, such as an amino acid, its biosynthesis is turned on or shut off.

A given microbe needs to build as much biomass as it can by the most efficient means available. For a pathogen growing within a host, this means using preformed host compounds for rapid growth and replication. Free-living microbes, however, face intense competition for organic nutrients; so in addition, they need to fix essential elements into readily accessible forms. Marine microbes build complex biological molecules out of single-carbon and single-nitrogen sources, but they also obtain molecules such as vitamins from their community. In the soil, actinomycetes build antibiotics to eliminate competitors and extract their nutrients. Those antibiotics offer powerful tools for human medicine.

Microbial metabolism has many applications in food preparation and preservation, and in the industrial production of pharmaceuticals. Food and industrial microbiology are discussed in Chapter 16. Later, in Chapters 21 and 22, we will show how microbial metabolism contributes to communities of organisms, and ultimately to Earth's global cycles.

CHAPTER REVIEW

Review Questions

1. What are the sources of substrates for biosynthesis? From what kinds of pathways do they arise?
2. How do microbial species economize by synthesizing only the products they need? Cite long-term as well as short-term mechanisms.
3. Compare and contrast the different cycles of carbon dioxide fixation. What classes of organisms conduct each type?
4. How is ribulose 1,5-bisphosphate consumed and re-formed through the Calvin cycle? What key products emerge to form sugars and amino acids?
5. How do oxygenic phototrophs maintain CO_2 at sufficient levels to conduct the Calvin cycle?
6. Explain the cyclic process of chain extension in fatty acid biosynthesis. Explain the generation of occasional unsaturated "kinks" in the chain.
7. Explain the different kinds of regulation of fatty acid biosynthesis.
8. Explain how modular synthesis generates a polyketide antibiotic, and how it generates a nonribosomal polypeptide antibiotic.
9. What are the different environmental sources of nitrogen? Explain how and why microbes use these different sources.
10. Explain the process by which nitrogenase converts N_2 to $2NH_4^+$. Why is H_2 formed?
11. Explain the different ways that microbes maintain anaerobic conditions for nitrogenase.
12. Explain the molecular basis for regulation of nitrogen fixation and nitrogen scavenging.
13. Compare and contrast the general scheme of the biosynthesis of amino acids with that of fatty acids.
14. Outline the interconversions of 2-oxoglutarate, glutamate, and glutamine that provide nitrogen for amino acid biosynthesis.
15. Compare and contrast the processes of purine and pyrimidine biosynthesis. What is the role of the sugar ribose in each case?

Thought Questions

1. Why do some soil microbes fix N_2, whereas others depend on available nitrate, ammonium ion, or organic nitrogen? What environmental conditions would favor each strategy?
2. Disease-causing bacteria vary widely in their ability to synthesize amino acids. What kinds of pathogens would be likely to make their own amino acids, and what kinds would not?
3. *Mycoplasma genitalium*, an organism growing in human skin, lacks the ability to synthesize fatty acids. How do you think it makes its cell membrane? How could you test this?

Key Terms

acyl carrier protein (ACP) (595)
anabolism (582)
anaplerotic reaction (592)
biosynthesis (582)
Calvin cycle (Calvin-Benson cycle, Calvin-Benson-Bassham cycle, CBB cycle) (586)
carbon-concentrating mechanism (CCM) (591)
carbon dioxide fixation (586)
carbon sequestration (586)
carboxysome (591)
chemosynthesis (chemoautotrophy) (582)
fix (582)
folate (594)
gluconeogenesis (583, 593)
glutamate dehydrogenase (GDH) (609)
glutamate synthase (glutamine: 2-oxoglutarate aminotransferase, GOGAT) (609)
glutamine synthetase (GS) (609)
Haber process (604)
heterocyst (604)
3-hydroxypropionate cycle (594)
modular enzyme (597)
modular synthesis (594)
nitrogenase (604)
nonribosomal peptide (594, 598)
nonribosomal peptide synthetase (NRPS) (599)
photoautotrophy (582)
photosynthesis (582)
polyketide (594)
reductive acetyl-CoA pathway (593)
reductive pentose phosphate cycle (586)
reductive (reverse) TCA cycle (592)
Rubisco (587)
secondary product (582)
transamination (610)

Recommended Reading

Bartholomae, Maike, Andrius Buivydas, Jakob H. Viel, Manuel Montalbán-López, and Oscar P. Kuipers. 2017. Major gene-regulatory mechanisms operating in ribosomally synthesized and post-translationally modified peptide (RiPP) biosynthesis. *Molecular Microbiology* **106**:186–206.

Gogineni, Vedanjali, and Mark T. Hamann. 2018. Marine natural product peptides with therapeutic potential: Chemistry, biosynthesis, and pharmacology. *Biochimica et Biophysica Acta* **1862**:81–196.

Iancu, Cristina V. H., Jane Ding, Dylan M. Morris, D. Prabha Dias, Arlene D. Gonzales, et al. 2007. The structure of isolated *Synechococcus* strain WH8102 carboxysomes as revealed by electron cryotomography. *Journal of Molecular Biology* **372**:764–773.

Kerfeld, Cheryl A., William B. Greenleaf, and James N. Kinney. 2010. The carboxysome and other bacterial microcompartments. *Microbe* **5**:257–263.

Pande, Samay, Shraddha Shitut, Lisa Freund, Martin Westermann, Felix Bertels, et al. 2015. Metabolic cross-feeding via intercellular nanotubes among bacteria. *Nature Communications* **6**:6238.

Rehm, Bernd H. A. 2010. Bacterial polymers: Biosynthesis, modifications and applications. *Nature Reviews. Microbiology* **8**:578–592.

Schirmer, Andreas, Rishali Gadkari, Christopher D. Reeves, Fadia Ibrahim, Edward F. DeLong, et al. 2005. Metagenomic analysis reveals diverse polyketide synthase gene clusters in microorganisms associated with the marine sponge *Discodermia dissoluta*. *Applied and Environmental Microbiology* **71**:4840–4849.

Singh, Rahul K., Shree P. Tiwari, Ashwani K. Rai, and Tribhuban M. Mohapatra. 2011. Cyanobacteria: An emerging source for drug discovery. *Journal of Antibiotics* **64**:401–412.

Xu, Ying, Roland D. Kersten, Sang-Jip Nam, Liang Lu, Abdulaziz M. Al-Suwailem, et al. 2012. Bacterial biosynthesis and maturation of the didemnin anticancer agents. *Journal of the American Chemical Society* **134**:8625–8632.

Yang, Yu-Liang, Yuquan Xu, Paul Straight, and Pieter C. Dorrestein. 2009. Translating metabolic exchange with imaging mass spectrometry. *Nature Chemical Biology* **5**:885–887.

Zimmer, Daniel P., Eric Soupene, Haidy L. Lee, Volker F. Wendisch, Arkady B. Khodursky, et al. 2000. Nitrogen regulatory protein C–controlled genes of *Escherichia coli*: Scavenging as a defense against nitrogen limitation. *Proceedings of the National Academy of Sciences USA* **97**:14674–14679.

CHAPTER 16

Food and Industrial Microbiology

16.1 Microbial Foods

16.2 Acid- and Alkali-Fermented Foods

16.3 Ethanolic Fermentation: Bread and Wine

16.4 Food Spoilage and Preservation

16.5 Industrial Microbiology

16.6 Microbial Gene Vectors for Plants and Human Gene Therapy

Microbes have nourished humans for centuries, generating cheese, bread, wine and beer, tempeh, and soy sauce. Yet from the moment of harvest, microbes on the food's surface or from the air colonize the food. The need for food preservation has led to drying, salting, smoking, and adding spices, all of which retard microbial growth. The principles of food microbiology extend to a growing field, industrial microbiology. Industrial microbiology includes the development of microbial antibiotics and enzymes. Giant fermentors grow transgenic microbes. And now we culture lentiviral vectors to deliver "miracle" gene therapies.

CURRENT RESEARCH highlight

Flavors in fermented cocoa. Chocolate production requires fermentation of cocoa beans by *Lactobacillus*, *Acetobacter*, and various yeasts. Rosane Freitas Schwan at the Federal University of Lavras, Brazil, analyzes the products of cocoa fermentation that contribute to the chocolate flavor. Lipids and polyphenols, as well as proteins, were extracted and identified from five different variants of cocoa beans. Ultimately, the chocolate produced from all the bean types was compared and subjected to consumer acceptance tests. High consumer acceptance rating was correlated with the presence of certain alcohols, such as 2,3-butanediol and 2-methyl-1-butanol.

Source: I. M. d. V. Moreira et al. 2018. *Food Res. Int.* **109**:196–203.

FREEDOMNARUK/SHUTTERSTOCK

AN INTERVIEW WITH

ROSANE FREITAS SCHWAN, MICROBIOLOGIST, FEDERAL UNIVERSITY OF LAVRAS, BRAZIL

COURTESY OF ROSANE FREITAS SCHWAN

Why is cocoa fermentation important?

Fermentation of cocoa is essential to form precursors of chocolate-specific flavor. Microbial activity in the pulp and proteolytic reactions inside the beans start after 3 days of fermentation. Small peptides and free amino acids are released. Peptides and reducing sugars in chocolate production promote the Maillard reaction, along with intermediate compounds produced from this reaction, such as furans, aldehydes, ketones, pyrroles, and other volatile compounds. These compounds greatly influence the aroma profile of cocoa and chocolate.

Chocolate, as introduced in the Current Research Highlight, is just one of many lucrative commercial products of microbial activity. The **fermentation industry** is the culturing of microbes to make products for market. It includes processes of fermentation (described in Section 13.4), as well as other kinds of metabolism. While today's industrial use of microbes includes sophisticated biotechnology, the roots of the fermentation industry reach back to 5000 BCE, the date of pottery jars excavated from a Neolithic mud-brick kitchen in the Zagros Mountains of modern Iran (**Fig. 16.1A**). The jars contain residue showing chemical traces of grapes fermented to wine. Besides making wine, Neolithic people leavened bread, another staple food requiring microbial growth. For thousands of years, microbial fermentation has been a daily part of human life and commerce.

Today, microbial products are made by bacteria grown in giant fermentation vessels (**Fig. 16.1B**; discussed in Section 16.5). Industrial "fermentation" generally uses respiratory metabolism to maximize microbial growth. Microbial growth drives companies such as Novozymes now earning hundreds of millions of dollars in annual revenues from engineering microbial enzymes for industrial use. Novozyme's microbial products range from contact lens cleansing agents to fashion finishes for denim at manufacturing centers around the world.

The fermentation industry originated from long-standing microbial associations with human food. Until the relatively recent invention of steam-pressure sterilization, all foods contained live microbes. Microbial metabolism could spoil food—or it could improve the food by adding flavor, preserving valuable nutrients, and preventing growth of pathogens. As we described in Chapter 1, one of the first great microbiologists, Louis Pasteur (1822–1895), began his career as a chemist investigating fermentation in winemaking. In Chapter 16 we discuss how the many kinds of microbial biochemistry presented in Chapters 13–15 give rise to the characteristics of food that are so familiar—from the taste of cheese and chocolate to the rising of bread dough and the physiological effects of alcoholic beverages. Industrial research continues to improve food through food biochemistry, discovering the molecular basis for the flavor and texture of microbial foods; food preservation, eliminating undesirable microbial decay by preservation methods; and food engineering, improving the quality, shelf life, and taste of microbial foods.

16.1 Microbial Foods

Certain kinds of microbial bodies are eaten as food, especially mushrooms and other fungal fruiting bodies, and some types of algae. These microbes can provide important sources of protein, vitamins, and minerals. Still other microbes, such as yeasts and lactic acid bacteria, ferment substrates to form valuable food products, such as bread, cheese, and alcoholic beverages.

Edible Fungi

Fungal fruiting bodies, multicellular reproductive structures that generate spores, are commonly known as mushrooms (see Chapter 20). Mushrooms offer a flavorful source of protein and minerals. The protein content of edible mushrooms can be as high as 25% dry weight, comparable to that of whole milk, and includes all essential dietary amino acids.

In the children's classic *Homer Price*, by Robert McCloskey, settlers on the Ohio frontier save themselves from starvation

A.

ALBUM/ART RESOURCE

B.

©NOVOZYMES, 2019.

FIGURE 16.1 ■ The fermentation industry: then and now. **A.** A Greek jar that contained wine, from the 5th to 3rd centuries BCE. Excavated in Catalonia, Spain. **B.** An industrial fermentation apparatus for growing microbes to produce enzyme products, at the world's largest enzyme manufacturing plant, Novozymes, Kalundborg, Denmark.

KELLY IVORS

FIGURE 16.2 ▪ Commercial mushroom production. Farming of *Agaricus bisporus* mushrooms on horse manure compost, by Penn State graduate student Kelly Ivors.

when they discover "forty-two pounds of edible fungus, in the wilderness a-growin'." Mushrooms aided the survival of preindustrial humans but killed those unlucky enough to consume varieties that were poisonous. Less than 1% of mushrooms are poisonous, but those few are deadly, such as the amanita, or "destroying angel," which produces toxins including the RNA polymerase inhibitor alpha-amanitin. People who ingest the amanita typically die of liver failure.

Other kinds of mushrooms actually invite consumption in order to disperse their spores. For example, the underground truffles prized in European cooking produce odorant molecules that attract animals to dig them up. Truffle odorants include sex pheromones such as androstenol, found in human perspiration.

Many varieties of edible mushrooms are farmed and marketed for human fare. **Figure 16.2** shows the culturing of *Agaricus bisporus*, the mushroom variety most commonly sold in the United States as button mushrooms and portobellos. Button mushrooms are harvested at an early stage, whereas portobellos are harvested later, when the gills are fully exposed and some of the moisture has evaporated. The decrease in moisture in portobellos concentrates the flavor and gives them a dense, meaty texture; the mushrooms are often served in gourmet sandwiches as a vegetarian alternative to hamburger. *Agaricus* culture was first developed around 1700 in France, where the mushrooms were grown in underground caves. In modern mushroom farms, *Agaricus* mushrooms are cultured on composted horse manure or chicken manure in chambers controlled for temperature and humidity.

Other mushroom varieties, from China and Japan, are grown on logs or wooden blocks. Wood-grown mushrooms include the strong-flavored *Lentinula edodes* (black forest mushrooms, or shiitake), *Pleurotus* (oyster mushrooms), and *Flammulina velutipes* (enoki mushrooms), with long, thin, white stalks and delicate flavor.

Yeasts (single-celled fungi) can provide a supplemental source of protein and vitamins, most importantly vitamin B_{12}. Historically, yeasts such as *Saccharomyces* species have been grown to high concentration in fermented milks and grain beverages. Traditional beers contained only a low percentage of alcohol with a thick suspension of nutrient-rich yeasts.

Edible Algae and Cyanobacteria

Several kinds of seaweed (marine algae) are cultivated, most notably in Japan. The red alga *Porphyra*, a eukaryotic "true alga" (discussed in Chapter 20), forms large, multicellular fronds cultured for **nori** (**Fig. 16.3**). Nori is best known for its use in wrapping rice, fish, and vegetables to form sushi. For nori production, the red algae are grown as "seeds," or starter cultures, in enclosed tanks. The starter cultures are then distributed on nets in a protected coastal area, usually an estuary. The cultures grow until they hang heavy from the nets, when they are harvested for processing into sheets. The sheets are toasted, turning dark green. Other edible seaweeds include wakame and kombu, forms of kelp.

Few bacteria are edible as isolated organisms, mainly because their small cells contain a relatively high proportion of DNA and RNA. The high nucleic acid content is a problem because nucleic acids contain purines, which the human digestive system converts to uric acid. Because we humans lack the enzyme urate oxidase, the uric acid cannot be metabolized. Consumption of more than 2 grams per day of nucleic acids causes uric acid to precipitate, resulting in painful conditions such as gout and kidney stones. For this reason, most bacteria can be consumed only as a minor component of food mass, as in fermented foods.

An exception is the cyanobacterium *Spirulina*, whose purine content is low enough to include as a modest part of the human diet. *Spirulina* consists of spiral-shaped cells that

A.

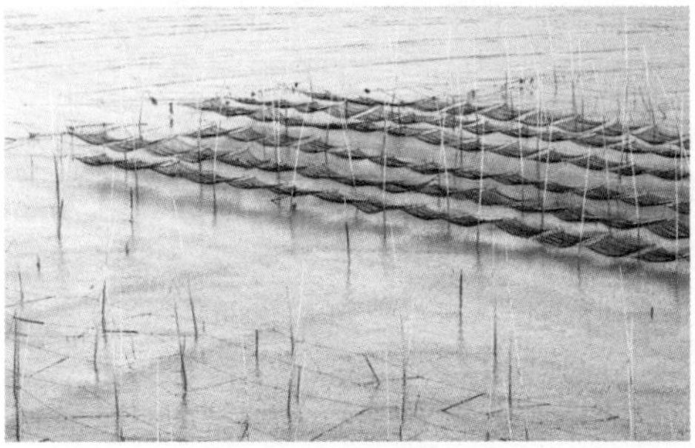

XVISION/GETTY IMAGES

B.

STOCKFOOD LTD/ALAMY STOCK PHOTO

FIGURE 16.3 ▪ Nori production for sushi. **A.** Nori grows from nets in seawater. **B.** Toasted sheets of nori are used to wrap rice, vegetables, and fish to make sushi.

grow photosynthetically in freshwater. *Spirulina* is sold as a food additive rich in protein, vitamin B_{12}, and minerals. It also contains antioxidant substances that may prevent cancer. *Spirulina* is grown with illumination in special ponds lined for food production. The final product is collected and vacuum-dried to form a dark green powder of flour-like consistency.

In the food industry, *Spirulina* is classified as a form of **single-celled protein**, a term for edible microbes of high food value. Other kinds of single-celled protein foods include eukaryotes such as yeast and algae.

Fermented Foods: An Overview

Virtually all human cultures have developed varieties of **fermented foods**, food products that are modified biochemically by microbial growth. In most cases, the modification involves true fermentation, a process of metabolism in which the electrons transferred from an electron donor are returned to the organic substrates, generating molecules that are rearranged. Recall from Chapter 13 how glucose catabolism leads to pyruvate, yielding energy in the form of ATP (Fig. 13.15). Electrons are transferred to NADH, but fermentation returns the electrons from NADH to pyruvate, generating various organic products (Fig. 13.21). The absence of a strong electron acceptor (O_2) limits the breakdown of substrate and thus preserves food value.

The purposes of food fermentation include the following:

- **Preserve food.** Certain microbes, particularly the lactobacilli, metabolize only a narrow range of nutrients before their waste products build up and inhibit further growth. Typically, the waste fermentation products that limit growth are carboxylic acids, ammonia (alkaline), or alcohol. Buildup of these substances renders the product stable for much longer than the original food substrate, for example, sauerkraut from cabbage.
- **Improve digestibility.** Microbial action breaks down fibrous macromolecules and makes the food easier for humans to digest. Meat and vegetable products are tenderized by fermentation.
- **Add nutrients and flavors.** Microbial metabolism generates vitamins, particularly vitamin B_{12} and riboflavin (vitamin B_2). Microbial action also generates flavor molecules that lead to the diverse tastes of foods such as cheeses and chocolate (see the Current Research Highlight).

Different societies have devised thousands of different kinds of fermented foods. Examples are given in **Table 16.1**. Fermented foods that are produced commercially include dairy products such as cheese and yogurt, soy products such as miso (from Japan) and tempeh (from Indonesia), vegetable products such as sauerkraut and kimchi, and various forms of cured meats and sausages. Alcoholic beverages are made from grapes and other fruits (wine), grains (beer and liquor), and cacti (tequila). Other kinds of foods require microbial treatment for special purposes, such as leavening by yeast (for bread) or cocoa bean fermentation (for chocolate). Besides commercial production, numerous fermented products are homemade by traditional methods thousands of years old. Such products are known as "traditional fermented foods." Occasionally, a traditional fermented food enters commercial production and becomes widespread. For example, soy sauce, a traditional Japanese product, was marketed by the Kikkoman company and achieved global distribution in the twentieth century.

The nature of fermented foods depends on the quality of the fermented substrate, as well as on the microbial species and the type of biochemistry performed. Traditional fermented foods usually depend on **indigenous microbiota**—that is, microbes found naturally in association with the food substrate; or on starter cultures derived from a previous fermentation, as in yogurt or sourdough fermentation. Commercial food-fermenting operations use highly engineered microbial strains to inoculate their cultures, although in some cases indigenous microbiota still participate. For example, wines and cheeses aged in the same caves over centuries often include fermenting organisms that persist in the air and the containers used.

Major classes of fermentation reactions are summarized in **Figure 16.4** (see also Figure 13.21). The most common conversions involve anaerobic fermentation of glucose.

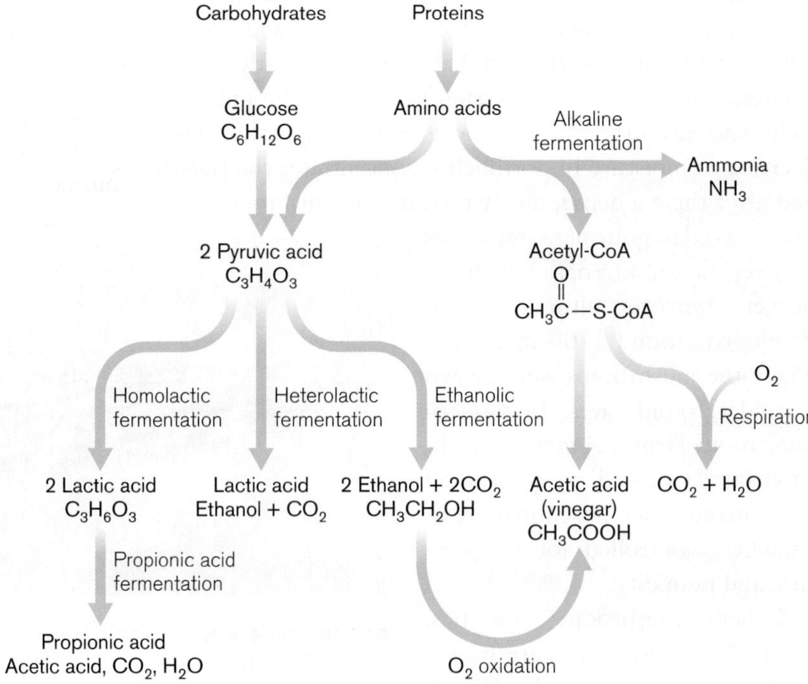

FIGURE 16.4 ■ Major chemical conversions in fermented foods.

TABLE 16.1 Fermented Foods and Beverages

Product (origin)	Description	Microbial genera
Acid fermentation of dairy products, meat, and fish		
Buttermilk (Asia, Europe)	Bovine milk; lactic fermented	*Lactococcus*
Yogurt (Asia, Europe)	Bovine milk; lactic fermented and coagulated	*Lactobacillus, Streptococcus*
Kefir (Russia)	Bovine or sheep's milk; mixed fermentation, acid with some alcoholic	*Lactobacillus, Streptococcus,* yeasts, others
Sour cream (Asia, Europe)	Bovine cream; lactic fermented	*Lactococcus*
Cheese (Asia, Europe)	Milk (bovine, sheep, or goat); lactic fermented, coagulated, and pressed; in some cases cooked; mold ripened (spiked or coated)	Acid fermentation: *Lactobacillus, Streptococcus, Propionibacterium* Mold ripening: *Penicillium*
Sausage (Asia, Europe)	Ground beef and/or pork encased with starter culture; lactic fermented, dried, or smoked	*Lactobacillus, Pediococcus, Staphylococcus,* others
Fermented fish (Africa, Asia)	Many kinds of fish; mixed fermentation, acid and amines produced	Halotolerant bacteria and haloarchaea
Acid fermentation of vegetables		
Tempeh (Indonesia)	Soybean cakes; fungal fermentation	*Rhizopus oligosporus*
Miso (Japan)	Soy and rice paste; fungal fermentation	*Aspergillus*
Soy sauce (China)	Extract of soy and wheat; fungal fermentation, brined, bacterial fermentation	*Aspergillus,* followed by halotolerant bacteria and haloarchaea
Kimchi (Korea)	Cabbage, peppers, and other vegetables, with fish paste, brined; container is buried	*Leuconostoc,* other bacteria
Sauerkraut (Europe)	Cabbage; fermented, making lactic and acetic acids, ethanol, and CO_2	*Leuconostoc, Pediococcus, Lactobacillus*
Pickled foods (Asia)	Cucumbers, carrots, fish; brined, then fermented	*Leuconostoc, Pediococcus, Lactobacillus*
Kenkey (western Africa)	Maize; fermented, wrapped in banana leaves and cooked	Unknown
Chocolate (South America)	Cocoa beans; soaked and fermented before processing to chocolate	*Lactobacillus, Bacillus, Saccharomyces*
Alkaline fermentation		
Pidan (China, Japan)	Duck eggs; coated in lime (CaO), aged, producing ammonia and sulfur odorants	*Bacillus*
Natto (China, Japan)	Whole soybeans; fermented	*Bacillus natto*
Dawadawa (Africa)	Locust beans; fermented	*Bacillus*
Ogiri (Africa)	Melon seed paste; fermented	*Bacillus*
Leavened bread dough		
Yeast breads (Asia, Europe)	Ground grain; dough leavened by yeast	*Saccharomyces*
Sourdough (Egypt)	Ground grain; dough leavened by starter culture from previous dough	*Saccharomyces, Torulopsis, Candida*
Injera (Ethiopia)	Ground teff grain; dough leavened and fermented 3 days by organisms from the grain	*Candida*
Alcoholic fermentation		
Wine (Asia, Europe)	Grape juice; yeast fermented, followed by malolactic fermentation	*Saccharomyces, Oenococcus*
Beer (Asia, Europe, Africa)	Barley and hops; yeast fermented	*Saccharomyces*
Sake (Japan)	Rice extract; yeast fermented	*Saccharomyces*
Tequila (Mexico)	Blue agave; yeast fermented and distilled	*Saccharomyces*
Whiskey (United Kingdom)	Barley or other grains or potatoes; fermented and distilled	*Saccharomyces*

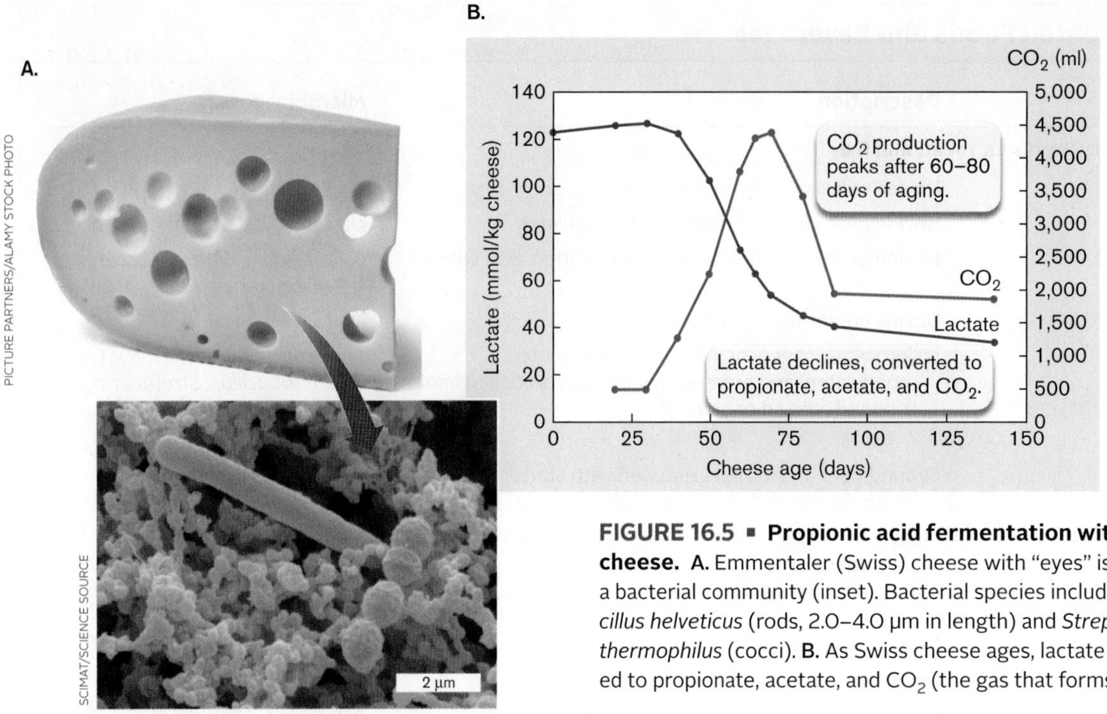

FIGURE 16.5 ■ Propionic acid fermentation within Swiss cheese. **A.** Emmentaler (Swiss) cheese with "eyes" is made by a bacterial community (inset). Bacterial species include *Lactobacillus helveticus* (rods, 2.0–4.0 μm in length) and *Streptococcus thermophilus* (cocci). **B.** As Swiss cheese ages, lactate is converted to propionate, acetate, and CO_2 (the gas that forms the eyes).

Recall from Chapter 13 how glucose is fermented to lactic acid (**lactic acid fermentation**). This fermentation occurs in cheeses and sausages, primarily by lactic acid bacteria such as *Lactobacillus.*

In Swiss cheese, a second stage of lactic acid fermentation produces propionic acid (**propionic acid fermentation**; **Fig. 16.5**). This fermentation, by *Propionibacterium freudenreichii*, generates the distinctive flavor of Swiss and related cheeses. As the cheese ages, first the lactic acid bacteria, such as *Lactobacillus helveticus*, convert sugars to lactic acid (**Fig. 16.5B**). Then *P. freudenreichii* starts to break down lactate further to acetate and CO_2—a reaction that yields ATP. The CO_2 gas forms the bubbles, or "eyes," that are characteristic of Swiss cheese (**Fig. 16.5A**).

Some kinds of vegetable fermentation, as in sauerkraut, involve production of lactic acid and CO_2, as well as small amounts of acetic acid and ethanol. This **heterolactic fermentation** is conducted primarily by *Leuconostoc*. Fermentation to ethanol plus carbon dioxide without lactic acid (**ethanolic fermentation**, also called alcoholic fermentation) is conducted by yeast during bread leavening and production of alcoholic beverages.

In some food products, particularly those fermented by *Bacillus* species, proteolysis and amino acid catabolism generate ammonia in amounts that raise pH (**alkaline fermentation**). For example, alkaline fermentation forms the soybean product natto. Other products require the growth of mold, such as the mold-spiked Roquefort cheese and the soy product tempeh. Mold growth requires some oxygen for aerobic respiration. Respiration must be limited, however, to avoid excessive decomposition of food substrate and loss of food value.

Note that the conversions cited here include only the major reactions in achieving the food product. In addition, thousands of minor or secondary reactions occur, some of which produce tiny amounts of potent odorants and flavors. While these flavor molecules have less nutritional consequence than the main fermentation products have, they provide the complex, "sophisticated" taste for which fine cheeses, wines, and soy products are known.

Thought Questions

16.1 Why do the lipid components of food experience relatively little breakdown during anaerobic fermentation?

16.2 Why does oxygen allow excessive breakdown of food, compared with anaerobic processes?

To Summarize

- **Edible fungi are protein-rich foods**, including yeasts, mushrooms, and truffles (fruiting bodies).
- **Edible algae** include nori (toasted red algae, used to wrap sushi), as well as wakame and kombu.
- ***Spirulina*** is an edible cyanobacterium, a source of single-celled protein. Most bacteria, however, are inedible in isolation because of their high concentration of nucleic acids.
- **Anaerobic fermentation of food** enhances preservation, digestibility, nutrient content, and flavor. Breakdown

of lipids and peptides is limited under anaerobic conditions of fermentation.

- **Acid fermentation of food** generates organic acid fermentation products that lower pH, such as lactate and propionate.
- **Alkaline fermentation** of food produces ammonia, which increases the food pH.
- **Ethanolic fermentation** produces ethanol and carbon dioxide.

16.2 Acid- and Alkali-Fermented Foods

Many food fermentations produce acids or bases. An acid or base serves as an effective preservative because the pH change is unlikely to be reversed, and because animal or plant bodies grown at near-neutral pH are unlikely to support growth of acidophiles or alkaliphiles, which grow at extreme pH conditions.

Acid Fermentation of Dairy Products

The major organic components of cow's milk are butterfat (about 4% unless skimmed), protein (3.3%), and the sugar lactose (4.7%). The conversion of milk to solid or semisolid fermented products dates far back in human civilization. The practice of milk fermentation arose among herders who collected the milk of their pack animals but had no way to prevent the rapid growth of bacteria. The milk had to be stored in a portable container such as the stomach of a slaughtered animal. After hours of travel, the combined action of lactic acid–producing bacteria and stomach enzymes caused the coagulation of milk proteins into **curd**. The curd naturally separated from the liquid portion, called **whey**. Both curds and whey can be eaten, as in the nursery rhyme "Little Miss Muffet." The curds, however, are particularly valuable for their concentrated protein content.

Curd formation. A **cheese** is any milk product from a mammal (usually cow, sheep, or goat) in which the milk protein coagulates to form a semisolid curd. Curd formation results from acidification, usually as a result of the microbial production of lactic acid; see **Figure 16.4**, and also Chapter 13, Figure 13.21.

Milk starts out at about pH 6.6, very slightly acidic. At this pH, the milk proteins are completely soluble in water; otherwise they would clog the animal's udder as the milk came out. Fermentation generally begins with bacteria such as *Lactobacillus* and *Streptococcus* (**Fig. 16.5A** inset). As bacteria ferment lactose to lactic acid, the pH starts to decline. The dissociation constant of lactic acid ($pK_a = 3.9$) allows greater deprotonation than with other fermentation products, such as acetate ($pK_a = 4.8$). Thus, lactic acid rapidly acidifies the milk product to levels that halt further growth of bacteria. Halting bacterial growth minimizes the oxidation of amino acids, thereby maintaining food quality.

Milk contains micelles (suspended droplets) of hydrophobic proteins called caseins. As the pH of milk declines below pH 5, the acidic amino acid residues of caseins become protonated, eventually destabilizing the tertiary structure. As the casein molecules unfold (or "denature"), they expose hydrophobic residues that regain stability by interacting with other hydrophobic molecules. The intermolecular interaction of caseins generates a gel-like network throughout the milk, trapping other substances, such as droplets of butterfat. This protein network generates the semisolid texture of **yogurt**, a simple product of milk acidified by lactic acid bacteria.

In most kinds of cheese formation, an additional step of casein coagulation is accomplished by proteases such as rennet. Rennet derives from the fourth stomach of a calf, although modern versions are made by genetically engineered bacteria. Calf rennet includes two proteolytic enzymes: chymosin and pepsin. Chymosin specifically cleaves casein into two parts, one of which is charged and water-soluble, the other hydrophobic. The hydrophobic portion forms a curd that is firmer than intact casein and results in the harder texture of solid cheeses. The water-soluble portion, about one-third of the total casein, enters the whey and is lost from the curd. Processing of some cheese varieties includes exposure to high temperature, which denatures even the whey protein, so it is retained in the curd.

Varieties of cheese. An extraordinary number of cheese varieties have been devised (**Fig. 16.6**). These fall into several categories based on particular steps in their production.

- **Soft, unripened cheeses**, such as cottage cheese and ricotta, are coagulated by bacterial action, without rennet. The curd is cooked slightly, and the whey is partly drained, but their water content is 55% or greater. These cheeses spoil easily; there are no steps of aging, or **ripening**.
- **Semihard, ripened cheeses**, such as Muenster and Roquefort, include rennet for firmer coagulation, and the curd is cooked down to a water content of 45%–55%. The cheese is aged for several months.
- **Hard cheeses**, such as Swiss cheese and cheddar, are concentrated to even lower water content. Extra-hard varieties, such as Parmesan and Romano, have a water content as low as 20%. These cheeses are aged for many months, even several years.

FIGURE 16.6 ■ Cheese varieties. **A.** Cottage cheese, an unripened perishable cheese. **B.** Emmentaler Swiss cheese, with eyes produced by carbon dioxide fermentation. **C.** Feta cheese, a soft cheese from goat's milk, preserved in brine. **D.** Roquefort, a medium-hard cheese ripened by spiking with *Penicillium roqueforti*.

- **Brined cheeses**, such as feta, are permeated with brine (concentrated salt), which limits further bacterial growth and develops flavor. Harder cheeses, such as Gouda, may be brined at the surface.
- **Mold-ripened cheeses** are inoculated with mold spores that germinate and grow during the ripening, or aging, process to contribute texture and flavor. The mold may be inoculated on the surface, to form a crust (as in Brie and Camembert), or it may be spiked deep into the cheese (as in blue cheese or Roquefort).

Cheese production. Commercial production of cheese involves a standard series of steps (**Fig. 16.7**). At each of these steps, choices of treatment lead to very different varieties. Key steps are illustrated in **Figure 16.8**.

In the first step, the milk is filtered to remove particulate objects, such as straw, and microfiltered or centrifuged to remove potentially pathogenic bacteria and spores. Most modern production includes flash pasteurization (brief heating to 72°C; discussed in Chapter 5), although some traditional cheeses continue to be made from unpasteurized milk. Unpasteurized milk in cheese has been linked to illness, particularly from *Listeria*, bacteria that grow at typical refrigeration temperatures.

The fermenting microbes are added as a **starter culture** (**Fig. 16.8A**). The starter was traditionally derived from a sample of the previous fermented product, in which case the flora is undefined. Traditional cheeses are defined by the precise location where they are made; for example, Emmentaler cheese is made only in the Emmen Valley of Switzerland. On a larger scale, commercial cheese production uses defined strains of bacteria, subject to government regulations on flavor, acid, and odor.

In all but the soft cheeses, bacterial coagulation and curd formation are supplemented by rennet or by genetically engineered proteases. The solid curd is then cut, or **cheddared** (hence the name "cheddar"

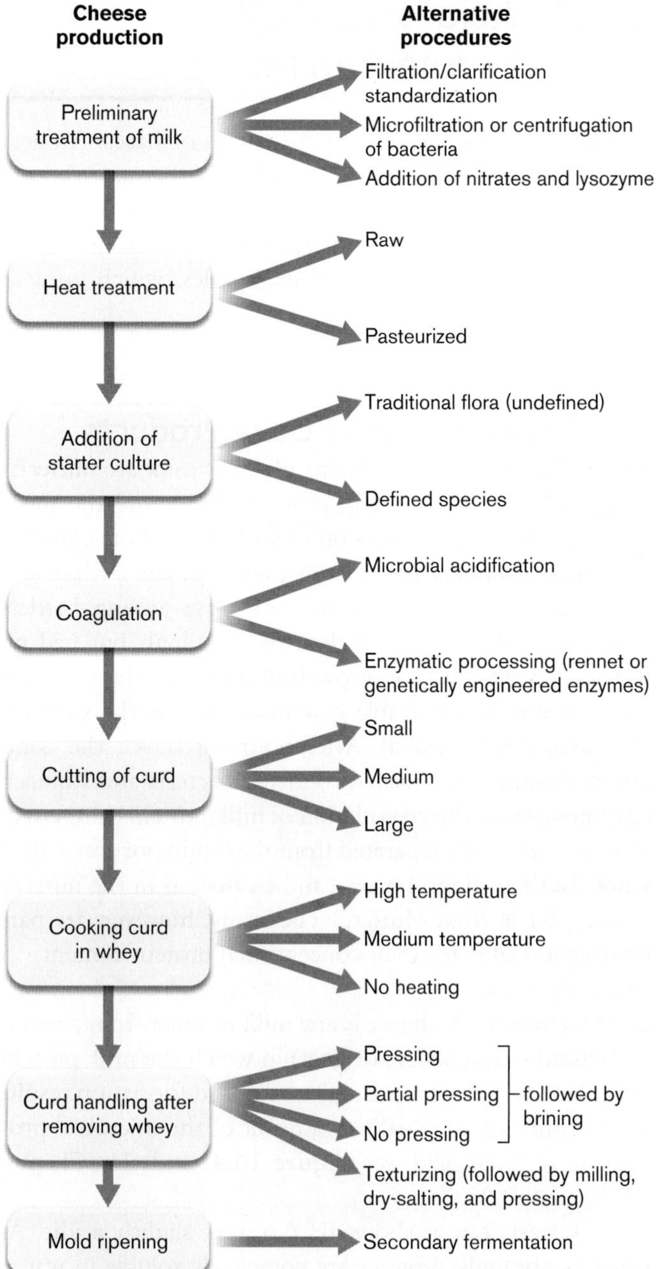

FIGURE 16.7 ■ Flowchart for cheese production. The alternative procedures shown on the right produce different varieties.

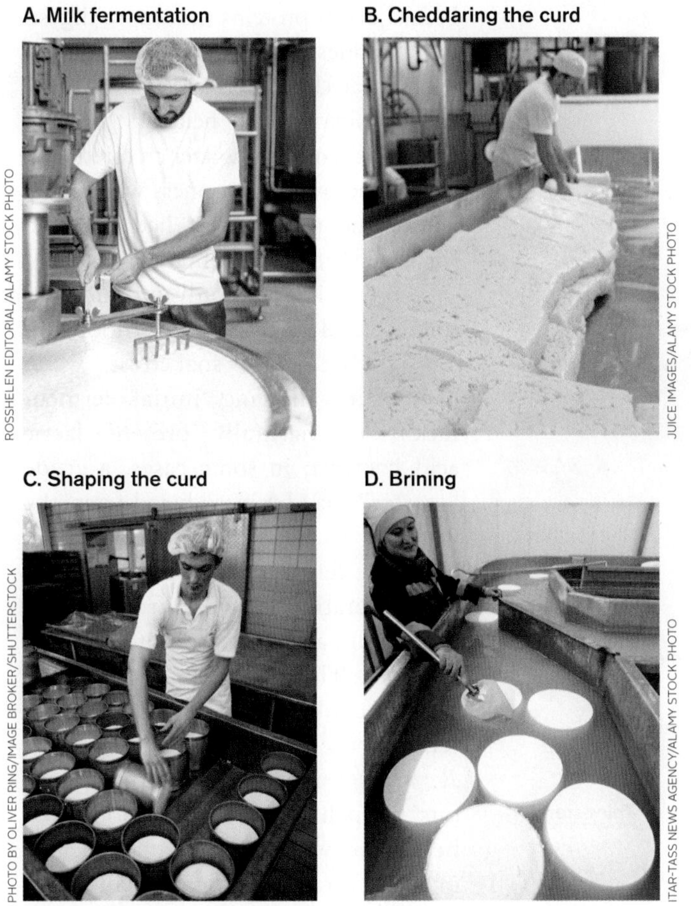

FIGURE 16.8 ■ Cheese production. **A.** Milk is poured into a fermentation tub with a bacterial starter culture and rennet. **B.** The milk curd is cut, or "cheddared." **C.** The curds are shaped into round molds and then pressed to remove whey. **D.** The solidified curds are floated in brine. The cheese then dries and ripens on the shelf.

cheese; **Fig. 16.8B**). The finer the pieces, the more whey that can be pressed out and the harder the cheese produced. Curd is then heat-treated, with or without the whey; if whey is included, more protein is retained. Brining at this stage leads to a salty cheese, such as feta.

The pressed curd is then shaped into a mold (**Fig. 16.8C**), which determines the ultimate shape of the cheese. Before ripening (or aging), the cheese may be floated in brine to generate a rind (**Fig. 16.8D**); or it may be coated or spiked with a *Penicillium* mold. The ripening period then allows flavor to develop. Texture also changes; for example, where fermentation has produced CO_2 the trapped gas forms "eyes," or holes.

Thought Questions

16.3 In an outbreak of listeriosis from unpasteurized cheese, only the refrigerated cheeses were found to cause disease. Why would this be the case?

16.4 Cow's milk contains 4% lipid (butterfat). What happens to the lipid during cheese production?

Flavor generation in cheese. In all fermented foods, microbial metabolism generates by-products that confer a characteristic aroma and flavor. In some cases, particular species confer distinctive flavors; for example, *Propionibacterium* ferments lactate or pyruvate to propionate, a flavor component of Swiss cheese. All bacteria generate a surprising range of side reactions, forming trace products that confer distinctive flavors. For example, while *Lactobacillus* converts most of the lactose to lactic acid, a small fraction of the pyruvate is converted to acetoin, acetaldehyde, or acetic acid, which contributes flavor. Most amino acids are retained intact, but traces are converted to flavorful alcohols, esters, and sulfur compounds (**Fig. 16.9**). For example, methanethiol (CH_3SH) contributes to the desirable flavor of cheddar cheese. Lipids are not significantly metabolized by *Lactobacillus*, but in mold-ripened cheeses such as Camembert, *Penicillium* oxidizes a small amount of the lipids to flavorful methylketones, alcohols, and lactones.

Acid Fermentation of Vegetables

Many kinds of vegetable products are based on microbial fermentation. Commercial products marketed globally include pickles, soy sauce, and sauerkraut. Other products provide staple foods for particular nations or regions, such as Indonesian tempeh and Korean kimchi.

Soy fermentation. Soybeans offer one of the best sources of vegetable protein and are indispensable for the diet of millions of people, particularly in Southeast Asia. In North America, soy products are important for vegetarian diets and as a milk substitute, as well as for animal feed. But soybeans also contain substances that decrease their nutritive value. Phytate, or inositol hexaphosphate, chelates minerals such as iron, inhibiting their absorption by the intestine. Lectins are proteins that bind to cell-surface glycoproteins within the human body. At high concentration, soybean lectins may upset digestion and induce autoimmune diseases. Soy protease inhibitors interfere with digestive enzymes chymotrypsin and trypsin, thus decreasing the amount of protein that can be obtained from soy-based food.

All of these drawbacks of soybeans are diminished by microbial fermentation, while the protein content remains comparable to that of the unfermented bean (40%). A variety of fermented soy foods have been developed. Most soy fermentation involves mold growth, supplemented by bacteria that contribute vitamins, including vitamin B_{12}.

A major fermented soy product is **tempeh** (**Fig. 16.10A**), a staple food of Indonesia, the world's fourth-most-populous country, as well as of other countries in Southeast Asia. Tempeh consists of soybeans fermented by *Rhizopus oligosporus*, a common bread mold. Besides decreasing the negative factors of soy, the mold growth

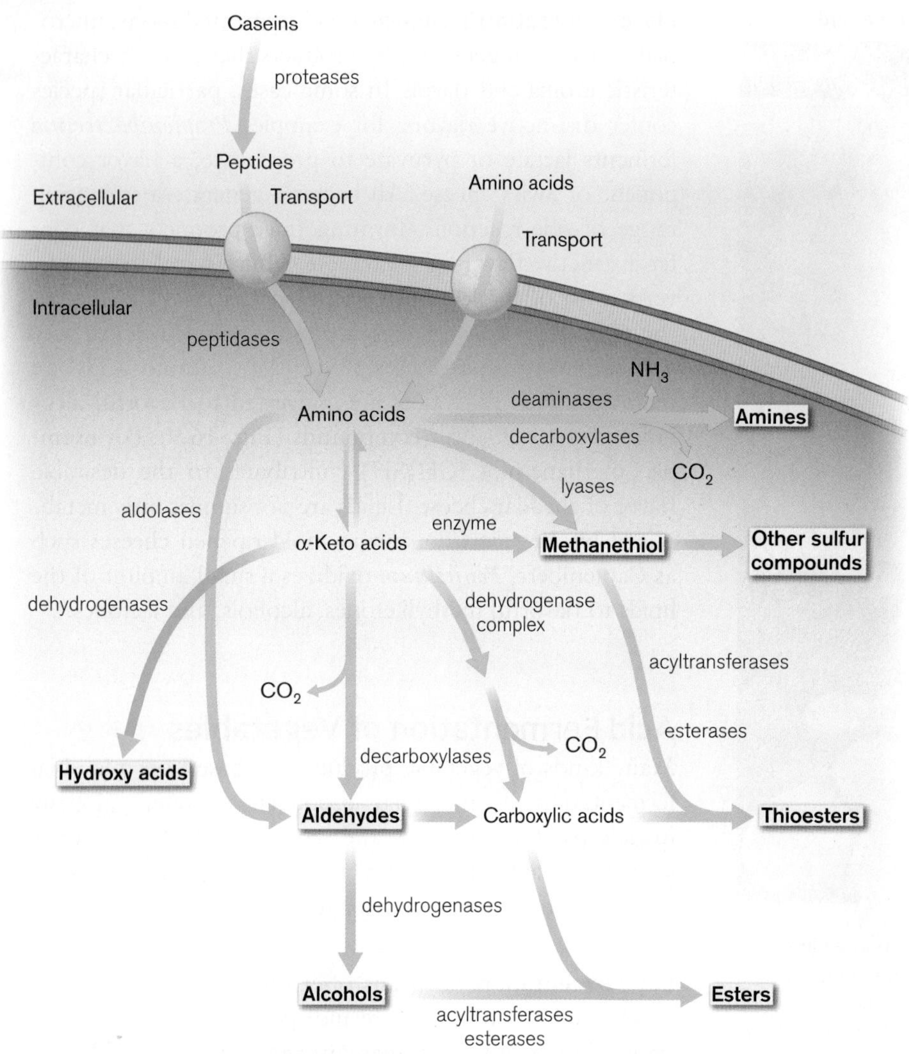

FIGURE 16.9 ■ Flavor generation from amino acid catabolism. Casein catabolism generates flavor molecules (highlighted). Extracellular enzymes break down casein into peptides and amino acids, which are taken into the bacterial cell by membrane transporters. The amino acids are fermented to volatile alcohols and esters. In some cases, they combine with sulfur to form methanethiol and other sulfur-containing odorants characteristic of cheese.

breaks down proteins into more digestible peptides and amino acids. During World War II, tempeh was fed to American prisoners of war held by the Japanese. The tempeh was later credited with saving the lives of prisoners whose dysentery and malnutrition had impaired their ability to absorb intact proteins.

Tempeh is commonly produced in home-based factories in Indonesia. The soybeans are soaked in water overnight, allowing initial fermentation by naturally present lactic acid bacteria; in some cases, a crude "starter" may be introduced from the water of soybeans soaked previously. This pre-fermentation allows bacterial generation of vitamins and produces mild acid that promotes growth of mold. The soaked beans are then hulled, cooked, and cooled to room temperature for inoculation with *R. oligosporus* spores or with a previous tempeh culture. The inoculated beans are wrapped in banana leaves or in perforated plastic bags and then allowed to incubate for 2 days. The mold grows as a white mycelium that permeates the beans (**Fig. 16.10B**), joining them into a solid cake. The final product has a mushroom-like taste and is served fried or grilled like a hamburger.

Other soy products undergo acid fermentation by the mold *Aspergillus oryzae*. The Japanese condiment **miso** is

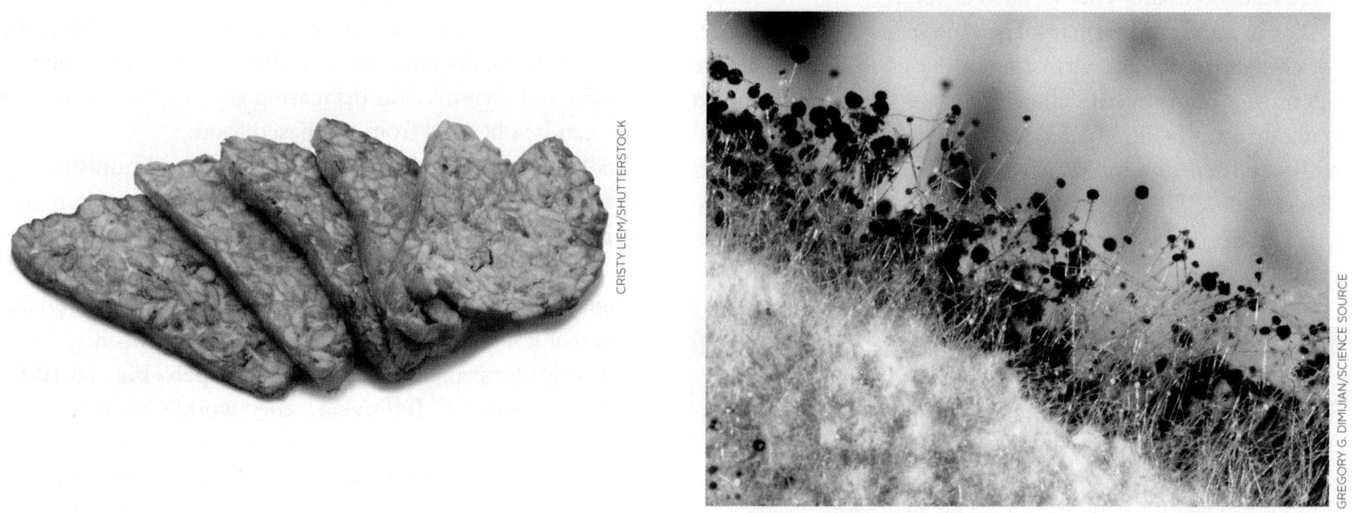

FIGURE 16.10 ■ Tempeh, a mold-fermented soy product. A. Fried tempeh. B. *Rhizopus oligosporus* mold, used to make tempeh.

made from ground soy and rice, salted and fermented for 2 months by *A. oryzae*. Soy sauce is made from *jiang*, a Chinese condiment similar to miso in which the rice starter culture is replaced by wheat. The fermentation generates glutamic acid, a flavor-enhancing compound known popularly in the form of its salt: monosodium glutamate, or MSG.

Fermentation of cabbage and other vegetables. Various leaf vegetables are fermented by traditional societies, originally as a means of storage over the winter months. In Europe and North America, the best-known fermented products include sauerkraut and pickles. Sauerkraut production involves heterolactic fermentation by *Leuconostoc mesenteroides*. In heterolactic fermentation, each fermented sugar molecule yields lactic acid, as well as ethanol and carbon dioxide. The culture is used to inoculate shredded cabbage, which is layered in alternation with salt. The salt helps limit the number of species and the extent of microbial growth. A similar brine-enhanced fermentation process is used to pickle cucumbers, olives, and other vegetables.

An important food based on brine-fermented cabbage is Korean **kimchi** (**Fig. 16.11**). Kimchi is prepared from Chinese cabbage, salted and layered with radishes, peppers, onions, and other vegetables. The vegetables are covered with a paste of fish, rice, and chili peppers. Pickled seafoods such as shrimp or oysters may be included. The entire mixture is stored in a pot, traditionally buried underground for several months. The main fermentation organism is *Leuconostoc mesenteroides*, although *Streptococcus* and *Lactobacillus* species participate.

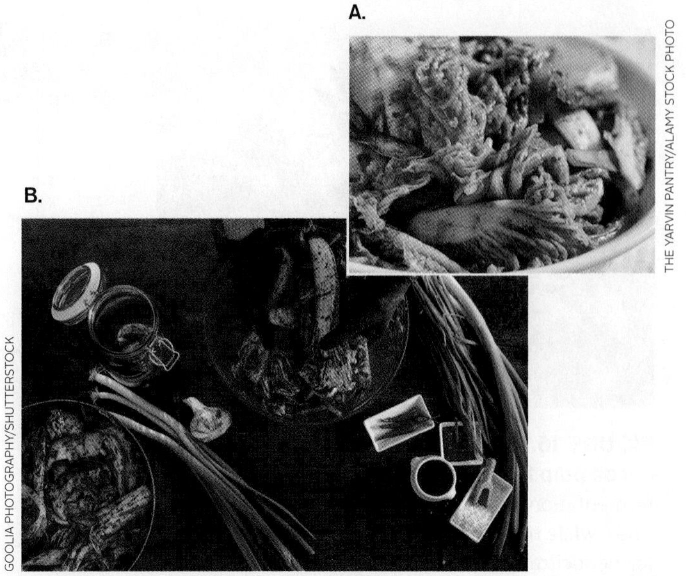

FIGURE 16.11 ▪ Kimchi. **A.** To make kimchi, cabbage leaves are layered alternating with chili paste containing salted fish and vegetables. **B.** The salted layers are packed, to be buried and aged for 2 months.

Chocolate from Cocoa Bean Fermentation

Cocoa and coffee beans both require fermentation within the juice of the fruit before the beans are dried and processed. Chocolate, the product of the cocoa bean, *Theobroma cacao* (or "food of the gods"), requires one of the most complex fermentations of any food. For all the commercialization of chocolate production, totaling 2.5 billion kilos per annum worldwide, no "starter culture" has yet been standardized to ferment the cocoa bean. The beans cannot be exported and fermented later; the fermentation must occur immediately where the beans are harvested. The cocoa beans harvested in Africa or South America are heaped in mounds upon plantain leaves for fermentation by indigenous microorganisms, essentially the same way cocoa has been processed for thousands of years (**Fig. 16.12A**).

The microbial fermentation actually takes place outside the cocoa bean, within the pulp that clings to the beans after they are removed from the cocoa fruit. The pulp contains approximately 15% sugars and pectin (a branched polysaccharide), 2% citric acid, plus a rich supply of amino acids and minerals. These nutrients support growth of many kinds of microbes, including three stages of succession: yeasts, lactic acid bacteria, and acetic acid bacteria (**Fig. 16.12B**).

Yeasts (anaerobic). Citric acid acidifies the pulp (pH 3.6). The acidity favors growth of yeasts, including *Candida*, *Kloeckera*, and *Saccharomyces*. The yeasts consume sugars and citric acid, increasing pH. They also degrade pectin into glucose and fructose, allowing the pulp to liquefy. As the liquefied pulp drains, yeast ferments the remaining sugars to ethanol, CO_2, and mixed acids, such as acetate. The reactions release heat, increasing the temperature and accelerating metabolism. But the acetate, a permeant acid (discussed in Chapter 13), crosses the yeast cell membrane, where it lowers cell pH and inhibits growth. The ethanol and acetate also penetrate the bean embryo, killing the cells and releasing enzymes that generate key flavor molecules of chocolate.

Lactic acid bacteria (anaerobic). The consumption of citric acid by yeast increases the bean pH to pH 4.2, encouraging growth of lactic acid bacteria, such as *Lactobacillus plantarum*. *Lactobacillus* species convert sugars to lactic acid, as well as acetate and CO_2. As fermentable substrates disappear, however, lactic acid bacteria are inhibited.

Acetic acid bacteria (aerobic). After 2 days, the beans are turned over and mixed periodically to permit access to oxygen. Aerobic acetic acid bacteria such as *Acetobacter* now

A.

B.

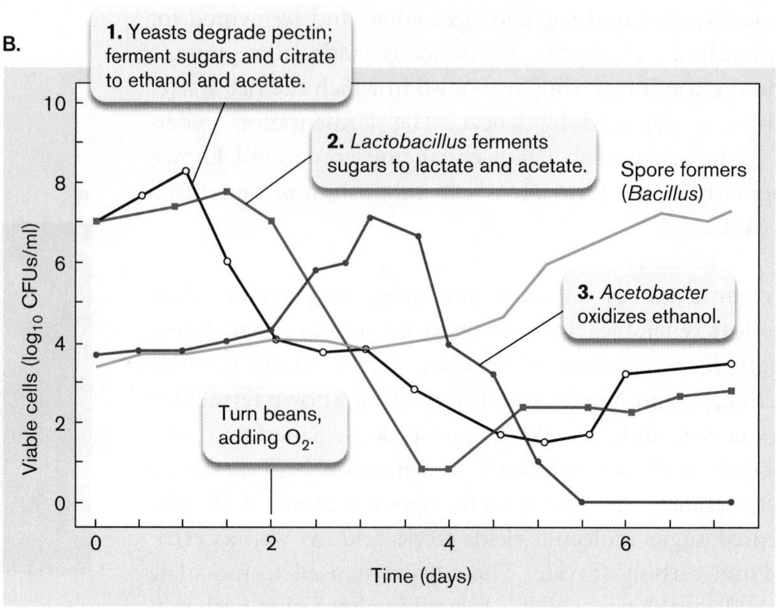

FIGURE 16.12 ■ Microbial succession during cocoa pulp fermentation. A. Cocoa beans undergo fermentation. Fruit pulp ferments, liquefies, and drains away, while the beans acidify and turn brown. B. Yeasts ferment citrate to alcohol; then, lactic acid bacteria convert sugars to lactate. With aeration, acetic acid bacteria oxidize ethanol to CO_2. CFUs = colony-forming units. *Source:* Data in part B from R. Schwan.

oxidize the ethanol and acids to CO_2. The consumption of acids neutralizes undesirable acidity. Heat is released, increasing the temperature to as high as 50°C. Oxygen penetrates the bean, oxidizing key components such as polyphenols. Polyphenol oxidation generates the brown color of cocoa and contributes flavor.

After fermentation and pulp drainage, the beans are dried and roasted. The roasting process completes the transformation of cocoa substances that contribute flavor. Cocoa liquor and cocoa butter are extracted from the beans and then recombined with sugar and other components to make "cocoa mass" (**Fig. 16.13**). The cocoa mass is stirred for several days to achieve a smooth texture; then it is molded into the decorative forms known as chocolate. But without the preceding fermentation process, no flavor would develop. Researchers are working to develop a defined starter culture, a community of microbes to produce a predictable high quality of flavor.

FIGURE 16.13 ■ Chocolate manufacture. Cocoa mass contains cocoa butter and liquor extracted from the cocoa beans, mixed with sugar and other ingredients.

Alkaline Fermentation: Natto and Pidan

In Western countries, food-associated fermentation is almost synonymous with acidification. In Africa and Southeast Asia, however, many food products involve increased pH. Such fermentations typically release small amounts of ammonia, which raises the pH to about pH 8, retarding growth of all but alkali-tolerant bacteria. The fermenting bacteria are usually *Bacillus*, aerobic species tolerant of moderate alkali and capable of extensive proteolysis and amino acid decomposition. The fermentation needs to be controlled to limit the loss of protein content, but the end result is a highly stable food product.

Natto. The Japanese soybean product **natto** is prepared by a process similar to that for tempeh. Soybeans are washed, pre-fermented, and cooked briefly before incubation with the starter organism, *Bacillus natto*. The fermenting beans are incubated in a shallow, ventilated container at a slightly raised temperature (40°C).

B. natto secretes numerous extracellular enzymes, including proteases, amylases, and phytases. These enzymes decrease the undesirable components of soy, such as phytates and lectins, while liberating more easily digestible peptides and amino acids. Some of the amino acids are deaminated, generating ammonia; a well-ventilated natto chamber allows most of this gas to escape. In addition, *B. natto* synthesizes extracellular polymers such as polyglutamate (a peptide chain consisting exclusively of glutamic acid residues). Polyglutamate generates long, elastic strings that bind the beans together. The stretching of these strings from chopsticks is considered a sign of a good natto (**Fig. 16.14A**).

Alkali-fermented vegetables. In Africa, numerous vegetable products are based on alkaline fermentation, predominantly by *Bacillus* species. An example is dawadawa, a paste of fermented locust beans common in western Africa. The locust beans are washed and supplemented with potash (potassium hydroxide) originally obtained from wood ashes. This addition of alkali retards growth of bacteria other than *Bacillus* species, which predominate at higher pH. The beans are fermented by indigenous bacteria (bacteria already present in the beans). The fermented beans are sun-dried, releasing most of the ammonia, and pounded into cakes for storage. Similar alkali-fermented vegetables include ogiri (from melon seeds) and ugba (from oil beans).

Pidan. An ancient means of preserving eggs led to the famous Chinese delicacy pidan, or "century egg," now a favorite at dim sum restaurants (**Fig. 16.14B**). To make pidan, duck eggs are covered with a mixture of brewed tea, lime (CaO), and sodium carbonate (Na_2CO_3). The lime and sodium carbonate react to form sodium hydroxide (NaOH), which penetrates the eggshells, raising pH and coagulating the egg white proteins. The eggs are buried in mud for several months, during which time the combined action of alkali and *Bacillus* fermentation generates dark colors and interesting flavors.

A.

AFLO CO. LTD/ALAMY STOCK PHOTO

B.

AGE FOTOSTOCK/ALAMY STOCK PHOTO

FIGURE 16.14 ■ Alkali-fermented foods. **A.** Natto consists of soybeans fermented by *Bacillus natto*. The fermentation generates long strings of polyglutamate. **B.** Pidan, or "century egg," consists of duck eggs coagulated by sodium hydroxide and fermented by *Bacillus* species. Eggs are cut open, revealing the transformed yolk, which develops a greenish color.

Thought Question

16.5 In traditional fermented foods, without a pure starter culture, how could someone control the kind of fermentation that occurs?

To Summarize

- **Milk curd** forms by lactic acid fermentation and rennet proteolysis, rendering casein insoluble. The cleaved peptides coagulate to form a semisolid curd.
- **Cheese varieties** include unripened cheeses, semihard and hard cheeses that are cooked down and ripened, brined cheeses, and mold-ripened cheeses.
- **Cheese flavors** are generated by minor side products of fermentation, such as alcohols, esters, and sulfur compounds.
- **Soy fermentation** to tempeh by *Rhizopus oligosporus* improves digestibility and decreases undesirable soy components such as phytates and lectins.
- **Vegetables** are fermented and brined to make sauerkraut, pickles, and kimchi.
- **Cocoa fermentation** for chocolate requires complex fermentation of cocoa beans within the fruit pulp, including anaerobic fermentation by yeast and lactic acid bacteria, and aerobic respiration by *Acetobacter* species.
- **Alkali-fermented vegetables** include the soy product natto, the egg product pidan, and the locust bean product dawadawa. The main fermenting organisms are *Bacillus* species.

16.3 Ethanolic Fermentation: Bread and Wine

Some of our most nutritionally significant and culturally important foods, including most bread and alcoholic beverages, require ethanolic fermentation by yeast fungi. Ethanolic fermentation converts pyruvic acid to ethanol and carbon dioxide:

$$C_3H_4O_3 \rightarrow CH_3CH_2OH + CO_2$$

The most prominent yeast used is *Saccharomyces cerevisiae*, known as baker's yeast or brewer's yeast (**Fig. 16.15**). A hardy organism, *S. cerevisiae* easily survives on a grocery

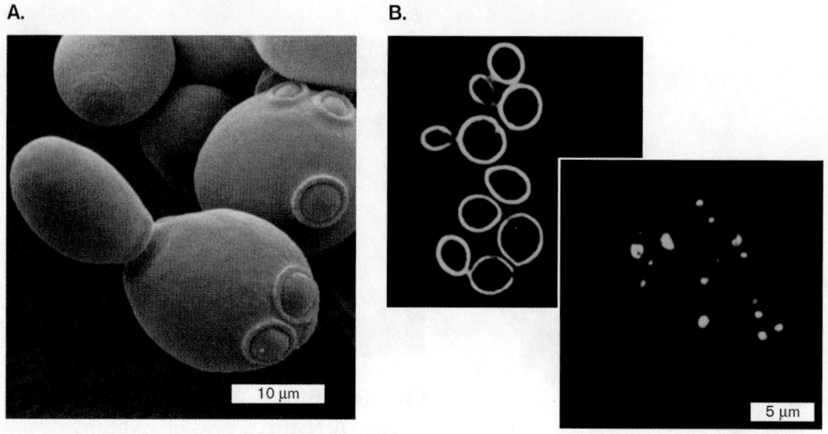

DAVID SCHARF/SCIENCE SOURCE

T.F. OUTEIRO AND S. LINDQUIST, *SCIENCE*, 2003, **302:** 1772. REPRINTED WITH PERMISSION FROM AAAS

FIGURE 16.15 ■ Baker's yeast, the "champion" fermenter. **A.** *Saccharomyces cerevisiae* cells budding; some show bud scars (SEM). **B.** *S. cerevisiae* is used to study the function of human proteins such as alpha-synuclein (green fluorescence), which plays a role in Parkinson's disease. Yeast cells engineered to express one copy of the gene (left panel) show the protein normally within their cell membrane. Two gene copies (right panel) cause the protein to clump and kill the cells.

shelf for home use and is genetically tractable for fundamental research. The yeast has been studied since the time of Pasteur, who used it to prove the biological basis of fermentation (presented in Chapter 1). *S. cerevisiae* today is a major model system of cell biology, yielding the molecular secrets of human cancer and other diseases.

Bread making depends on carbon dioxide to generate air spaces that **leaven** the dough, making its substance easier to chew and digest. The small amount of ethanol produced is eliminated during baking. For alcoholic beverages, however, ethanol is the key product, accompanied by carbon dioxide bubbles for "fizz," known as carbonation.

Bread Dough Is Leavened by Microbial CO_2

Bread is made in many different forms (**Fig. 16.16**) and from diverse kinds of flour, or ground grain. The earliest breads probably arose from grain mush naturally contaminated by yeast. Later, bread makers learned to include yeast left over from wine or beer production as a starter culture.

Yeast bread production. The preparation of all forms of yeast bread requires the same fundamental steps. A starter culture of yeast is included in the dough. The yeast can be commercial baker's yeast, or it can be **sourdough** starter, an undefined microbial population derived from a previous batch of dough. Analysis of sourdough shows mainly yeasts and lactobacilli, which release acids that favor the growth of the yeasts. The dough is kneaded to develop a fine network of air pockets and allowed to rise, expanding with production of the carbon dioxide gas (**Fig. 16.17**). The finest-textured breads are made from wheat flour that contains gluten. Gluten is a protein complex that forms a fine molecular network supporting the rising dough. In some individuals, gluten can trigger an autoimmune condition leading to gluten intolerance.

Thought Question

16.6 Compare and contrast the role of fermenting organisms in the production of cheese and bread.

Injera: extended fermentation. Most kinds of bread involve only a short fermentation period, just long enough to produce enough gas for leavening. A prolonged fermentation,

JOHN FOXX/GETTY IMAGES

FIGURE 16.16 ■ Yeast bread. Many varieties of bread are made.

SUSAN C. BOURGOIN/FOODPIX/JUPITERIMAGES

FIGURE 16.17 ■ Making bread. As yeast fermentation generates carbon dioxide gas, the dough rises.

A.

JIM SUGAR/GETTY IMAGES

B.

BERT DE RUITER ALAMY STOCK PHOTO

FIGURE 16.18 ■ Injera. A. After 3 days of fermentation, injera dough is baked in a ceramic pan upon a "Mirte" charcoal stove. B. Injera forms an edible tablecloth for a variety of Ethiopian foods.

with more extensive microbial activity, unfolds in the dough for an Ethiopian bread called **injera** (**Fig. 16.18**). The high microbial content provides a substantial source of vitamins not found in quick-rising breads.

Injera is made from teff (*Eragrostis tef*), a grain with small, round kernels that have high protein and lack gluten. Teff grows in arid regions and is now being cultivated in the United States for gluten-free products. Because teff lacks gluten, it cannot rise as much as wheat flour, but it makes a kind of flatbread. The dough is spread into a wide pancake, and the organisms present in the grain and air are allowed to ferment it for 3 days. The fermentation includes a succession of species, usually dominated by the yeast *Candida*. The extended fermentation generates a complex range of by-products that confer exceptional flavors in the baked product. In Ethiopia, injera forms the basis of an entire meal, served with other food items placed upon it as an edible tablecloth. Diners wrap samples of each food in a fold of injera and consume them together.

Alcoholic Beverages: Beer and Wine

Ethanolic fermentation of grain or fruit was important to early civilizations because it provided a drink free of waterborne pathogens. Traditional forms of beer also provided essential vitamins in the unfiltered yeasts.

Ethanol is unique among fermentation products in that it provides a significant source of caloric intake, but it is also a toxin that impairs mental function. A modest level of ethanol enters the human circulation naturally from intestinal flora, equivalent to a fraction of a drink per day. The human liver produces the enzyme alcohol dehydrogenase, which detoxifies ethanol. This enzyme in a healthy liver can metabolize small amounts of alcohol without harm. However, excess alcohol consumption can overload the liver's capacity for detoxification and permanently damage the liver and brain.

Beer: alcoholic fermentation of grain. Beer production is one of the most ancient fermentation practices and is depicted in the statuary of ancient Egyptian tombs dated to 5,000 years ago (**Fig. 16.19A**). The earliest Sumerian beers were made from bread soaked in water and fermented. Today, most beer is produced commercially by fermenting barley using giant vats (**Fig. 16.19B**). Production of high-quality beer involves complex processing with many steps, including germination of barley grains, mashing in water and cooking, and introduction of hops for flavor (**eTopic 16.1**).

A.

BORROMEO/ART RESOURCE, NY

B.

LIOR+LONE/STOCKSY

FIGURE 16.19 ■ Beer production: ancient and modern. A. Making beer in ancient Egypt, circa 3000 BCE. The mash was stirred in earthen jars. B. Fermentors in a modern brewery.

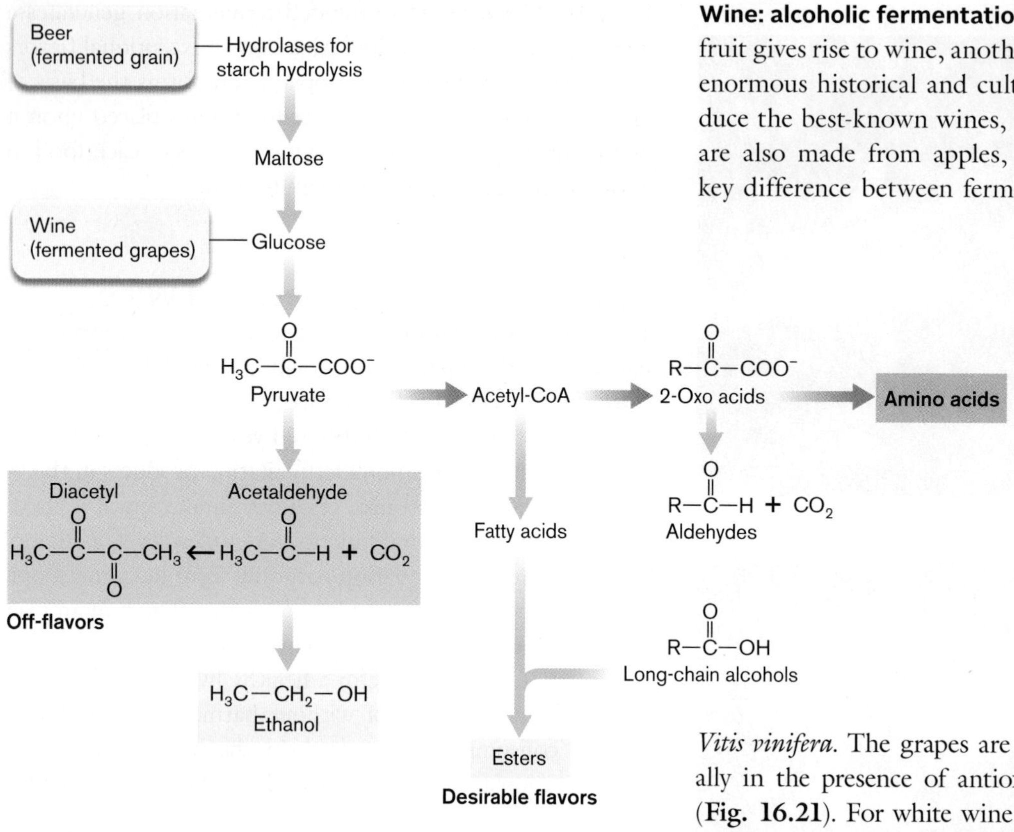

FIGURE 16.20 ■ Alcoholic fermentation in beer and wine. Yeast fermentation generates ethanol in substantial quantities. The biosynthesis of amino acids generates by-products that contribute both desirable flavors (long-chain alcohols) and off-flavors (acetaldehyde and diacetyl).

First the barley grains must germinate; that is, the seed embryos must start to grow. The germinating embryo makes enzymes needed to break down the barley starch to maltose (disaccharide) and glucose. Most of the sugars are fermented by yeast to ethanol and carbon dioxide. However, minor side products contribute flavors—or unpleasant off-flavors if present in too great an amount (**Fig. 16.20**). For example, off-flavors may result from the presence of small amounts of oxygen that oxidize some ethanol to acetaldehyde.

The yeast ferment most sugars to ethanol, but a small fraction is drawn off to make amino acids via TCA cycle intermediates, as discussed in Chapter 15. The 2-oxo acids of the TCA cycle are analogous to pyruvate, with the methyl group replaced by extended carbon chains (R group). A tiny amount of the 2-oxo acids is converted to long-chain alcohols, which add desirable flavor to beer.

Thought Question

16.7 Compare and contrast the role of low-concentration by-products in the production of cheese and beer.

Wine: alcoholic fermentation of fruit. The fermentation of fruit gives rise to wine, another class of alcoholic products of enormous historical and cultural significance. Grapes produce the best-known wines, but wines and distilled liquors are also made from apples, plums, and other fruits. The key difference between fermentation of fruits and fermentation of grains is the exceptionally high monosaccharide content in fruits. Grape juice, for example, can contain concentrations of glucose and fructose as high as 15%. The availability of simple sugars allows yeast to begin fermenting immediately, with no need for preliminary breakdown of long-chain carbohydrates, as in the malting and mashing of beer.

Most modern wine production uses strains of the grape *Vitis vinifera*. The grapes are crushed to release juices, usually in the presence of antioxidants such as sulfur dioxide (**Fig. 16.21**). For white wine, the skins are removed before juice is fermented. For red wine, the skins are included in early fermentation to extract the red and purple anthocyanin pigments, as well as phenolic flavor compounds. The first few days of fermentation are dominated by indigenous species of yeast naturally present on the grapes, such as *Kloeckera* and *Hanseniaspora* species. Commercial producers usually inoculate with standard *Saccharomyces cerevisiae*, whose population dominates the late stage of fermentation (6–20 days). Yeast growth ends once the ethanol level reaches about 15%; to achieve higher alcohol content, distillation is required.

After fermentation, the wine is drained, or "racked," from the sediment of grape and yeast material, the lees. The liquid may be further clarified by centrifugation. Then it is stored for 2–3 weeks in tanks or barrels. During storage, a second stage of fermentation may be performed, called **malolactic fermentation**. Malolactic fermentation is needed to decrease the acidity from malic acid (found in grapes). Malic acid is converted to lactic acid, with a higher dissociation constant (a weaker acid). The L-malate is decarboxylated to L- or D-lactate:

$$^{-}HOOC{-}CH_2{-}CHOH{-}COOH^{-} \rightarrow$$
$$CH_3{-}CHOH{-}COOH + CO_2$$

The wine is seeded with *Oenococcus oeni* bacteria, which ferment L-malate (deprotonated L-malic acid).

As in beer production, yeast fermentation of wine produces numerous minor products contributing flavor, such as long-chain alcohols and esters. At the same time, overgrowth of yeast or the growth of undesired species can produce excess amounts of these compounds, such as sulfides

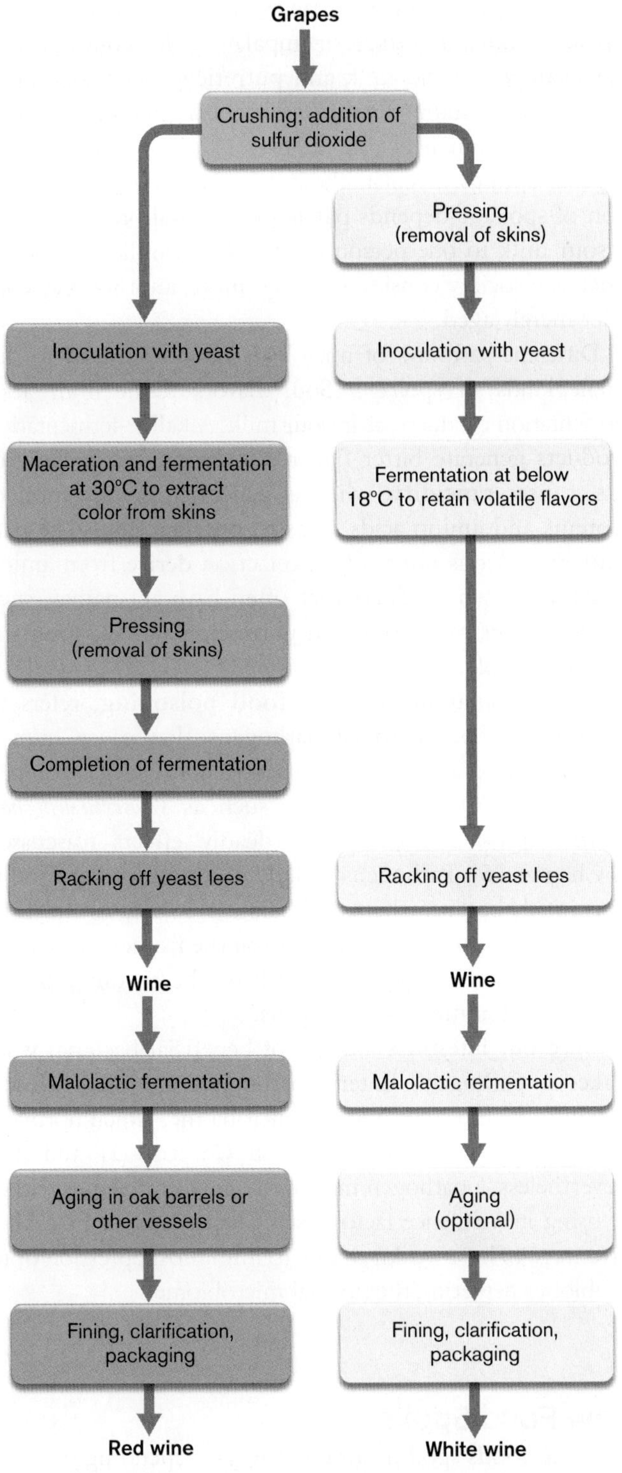

FIGURE 16.21 ▪ Production of red and white wines. Left: For red wine, the grapes are fermented with the skins at a temperature that increases extraction of color and tannins. **Right:** For white wine, the skins are removed before addition of yeast starter, and the temperature is kept low to retain volatile flavors. Both kinds of wine usually undergo malolactic fermentation by *Oenococcus oeni*, a species of lactic acid bacteria that consume malic acid.

and phenolics, giving rise to off-flavors. Some undesired species require oxygen exposure, whereas others can grow during storage and bottling. The balance of microbial populations is challenging to control and has a major role in determining the quality of a given wine vintage.

To Summarize

- **Bread is leavened** by yeasts conducting limited ethanolic fermentation, producing enough carbon dioxide gas to expand the dough.
- **Injera** bread dough undergoes more extensive fermentation by indigenous organisms and, as a result, generates multiple flavors.
- **Beer** requires alcoholic fermentation of grain. Barley grains are germinated, allowing enzymes to break down the starch to maltose for yeast fermentation.
- **Secondary products of grain fermentation**, such as long-chain alcohols and esters, generate the special flavors of beer.
- **Wine** derives from alcoholic fermentation of fruit, most commonly grapes. The grape sugar (glucose) is fermented by yeast to alcohol. A secondary product, malate, undergoes malolactic fermentation by *Oenococcus oeni* bacteria.

16.4 Food Spoilage and Preservation

We humans have always competed with microbes for our food. When early humans killed an animal, microbes commenced immediately to consume its flesh. Because meat perished so fast, it made economic sense to share the kill immediately and consume it all as soon as possible. Vegetables might last longer, but eventually they succumbed to mold and rot. Later societies developed preservation methods, such as drying, smoking, and canning, that enabled humans to survive winters and dry seasons on stored food.

Modern food preservation depends on **antimicrobial agents** (chemical substances that either kill microbes or slow their growth), as well as physical preservative measures, such as treatment with heat and pressure. These general principles of microbial control are described in Chapter 5. Here we focus on microbial contamination and food preservation from the perspective of the food industry.

Food Spoilage and Food Contamination

After food is harvested, several kinds of chemical changes occur. Some begin instantly, whereas others take several days to develop. Some changes, such as meat tenderizing, may be considered desirable; others, such as putrefaction, render food unfit for consumption. The major classes of food change include:

- **Enzymatic processes.** Following the death of an animal, its flesh undergoes proteolysis by its own enzymes. Limited proteolysis tenderizes meat. Plants after harvest undergo other changes; for example, in harvested corn the sugar rapidly converts to starch. That is why vegetables taste sweetest immediately after harvest.
- **Chemical reactions with the environment.** The most common abiotic chemical reactions involve oxidation by air—for example, lipid autooxidation, which generates rancid odors. Much research addresses oxidation, and the development of technologies to prevent it (**Fig. 16.22**). For example, produce may be packaged under an anaerobic atmosphere, wrapped in a film that prevents oxygen transmission. This process is called modified atmospheric processing.
- **Microbiological processes.** Microbes from the surface of the food begin to consume it—some immediately, others later in succession—generating a wide range of chemical products. In meat, internal organs of the digestive tract are an important source of microbial decay.

Microbial activity can aid food production, but it can also have various undesirable effects. Two different classes of microbial effects are distinguished: food spoilage and food contamination with pathogens.

FIGURE 16.22 ■ Research to prevent oxidation of produce. At the U.S. Department of Agriculture, microbiologist Arvind Bhagwat withdraws a gas sample from bagged lettuce, to test an anaerobic packaging method.

Food spoilage refers to microbial changes that render a product obviously unfit or unpalatable for consumption. For example, rancid milk and putrefied meats are unpalatable and contain metabolic products that may be deleterious to human health, such as oxidized fatty acids or organic amines. Even in these cases, however, the definition of spoilage depends partly on cultural practice. What is sour milk to one person may be buttermilk to another; what one society considers spoiled meat, another may consider merely aged.

Different pathways of microbial metabolism lead to different kinds of spoilage. Sour flavors result from acid-fermentation products, as in sour milk. Alkaline-fermentation products generate bitter flavor. Oxidation, particularly of fats, causes **rancidity**, whereas general decomposition of proteins and amino acids leads to **putrefaction**. The particularly noxious odors of putrefaction derive from amino acid breakdown products that often have apt names, such as the amines cadaverine and putrescine and the aromatic product skatole.

"Food contamination," or **food poisoning**, refers to the presence of microbial pathogens that cause human disease—for example, rotaviruses that cause gastrointestinal illness. Other pathogens, such as *Clostridium botulinum*, produce toxins with deadly effects (discussed in Chapter 25). Pathogens usually go unnoticed as food is consumed, because their numbers are very low, and they may not even grow in the food. Even the freshest-appearing food may cause serious illness if it has been contaminated with a small number of pathogens.

Since our intestines are full of beneficial bacteria, what makes a pathogen? Enteric pathogens are often closely related to members of our gut microbiome, which normally outcompete invaders (discussed in Chapters 21 and 23). Nevertheless, a pathogen may overcome our defenses, either by using its virulence factors (see Chapter 25) or by taking advantage of host weakness due to immunosuppression or to antibiotics depleting the normal microbiome.

How Food Spoils

Different foods spoil in different ways, depending on their nutrient content, the microbial species, and environmental factors such as temperature. **Table 16.2** summarizes common forms of spoilage.

Dairy products. Milk and other dairy products contain carbon sources, such as lactose, protein, and fat. In fresh milk, the nutrient most available for microbial catabolism is lactose, which commonly supports anaerobic fermentation to sour milk. Fermentation by the right mix of microbes, however, leads to yogurt and cheese production, as previously described.

TABLE 16.2 Food Spoilage (Examples)

Food product	Signs of spoilage	Microbial cause
Dairy products		
Milk	Sour flavor	Lactic acid bacteria produce lactic and acetic acids.
	Coagulation	Lactic acid bacteria produce proteases that destabilize casein and lower pH, causing coagulation.
	Bitter flavor	Psychrophilic bacteria degrade proteins and amino acids.
Cheese	Open texture, fissures	Lactic acid bacteria produce carbon dioxide.
	Discoloration and colonies	Molds such as *Penicillium* and *Aspergillus* grow on the cheese.
Meat, poultry, and eggs		
Meat and poultry	Rancid flavor	Psychrotrophic bacteria produce fatty acids that become oxidized.
	Putrefaction	*Pseudomonas* and other aerobes degrade amino acids, producing amines and sulfides.
	Discolored patches	Molds such as *Mucor* and *Penicillium* grow on the surface.
Eggs	Pink or greenish egg white	*Pseudomonas* and related bacteria grow on albumin, producing water-soluble pigments.
	Sulfurous odor	Bacterial growth on albumin releases hydrogen sulfide.
Seafood		
Fish	Fishy smell	Anaerobic psychrophiles such as *Photobacterium* convert trimethylamine oxide to trimethylamine.
	Odor of putrefaction	*Pseudomonas* and other Gram-negative species degrade amino acids, producing amines and sulfides.
Shellfish	Odor of putrefaction	*Vibrio* and other marine bacteria decompose the protein.
Fruits, vegetables, and grains		
Plants before harvest	Rotting or wilting	Plant pathogens, most commonly fungi such as *Alternaria*, *Aspergillus*, and *Penicillium*.
Stored plant foods	Rotting or wilting	Molds or bacteria produce degradative enzymes, such as pectinases and cellulases.
Apples, pears, cherries	Geosmin off-flavor	*Penicillium* mold.
Peeled oranges	Discoloration and off-flavor	*Enterobacter* and *Pseudomonas* spp.
Pasteurized fruit juices	Medicine-like phenolic off-flavor	Acid- and heat-tolerant spore former, *Alicyclobacillus* sp., produces 2-methoxyphenol (guaiacol).
Bread	Ropiness	*Bacillus* spp. grow, forming long filaments.
	Red discoloration	*Serratia marcescens*.

Under certain conditions, bitter off-flavors may be produced by bacterial degradation of proteins. The release of amines causes a rise in pH. Protein degradation is most commonly caused by psychrophiles, species that grow well at cold temperatures, such as those of refrigeration.

Cheeses are less susceptible than milk to general spoilage, because of their solid structure and lowered water activity. However, cheeses can grow mold on their surface. Historically, the surface growth of *Penicillium* strains led to the invention of new kinds of cheeses. But other kinds of mold, such as *Aspergillus*, produce toxins and undesirable flavors.

Meat and poultry. Meat in the slaughterhouse is easily contaminated with bacteria from hide, hooves, and intestinal contents. Muscle tissue offers high water content, which supports microbial growth, as well as rich nutrients, including glycogen, peptides, and amino acids. The breakdown of peptides and amino acids produces the undesirable odorants that define spoilage (for example, cadaverine and putrescine).

Meat also contains fat, or adipose tissue, but the lipids are largely unavailable to microbial action because they consist of insoluble fat (triacylglycerides). Instead, meat lipids commonly spoil abiotically by autooxidation (reaction with oxygen) of unsaturated fatty acids, independent of microbial activity. Thus, when meats are exposed to air during storage, they turn rancid—particularly meats such as pork, which contains highly unsaturated lipids. Autooxidation can be prevented by anaerobic storage, such as vacuum packing, which also prevents growth of aerobic microorganisms. The absence of the suppressed organisms, however, favors growth of lactic acid bacteria and facultative anaerobes such as *Brochothrix thermosphacta*. These organisms generate short-chain fatty acids, which taste sour.

In industrialized societies, the most significant factor determining microbial populations in meat spoilage is the practice of refrigeration. Refrigeration prolongs the shelf life of meat because contaminating microbes are predominantly mesophilic (grow at moderate temperatures, as discussed in Chapter 5). But ultimately, the few psychrotrophs initially present do grow; typically, these are *Pseudomonas* species. The pseudomonads are also favored by the low pH of meat (pH 5.5–7.0), which results from the accumulation of lactic acid in the muscle.

Seafood. Fish and other seafood contain substantial amounts of proteins and lipids, as well as amines such as trimethylamine oxide. Fish spoils more rapidly than meat and poultry for several reasons. First, fish do not thermoregulate, and they inhabit relatively low-temperature environments. Because fish grow in low-temperature environments, their surface microorganisms tend to be psychrotrophic and thus grow well under refrigeration. In addition, marine fish contain high levels of the osmoprotectant trimethylamine oxide, which bacteria reduce to trimethylamine, a volatile amine that gives seafood its "fishy" smell. Finally, the rapid microbial breakdown of proteins and amino acids leads to foul-smelling amines and sulfur compounds, such as hydrogen sulfide and dimethyl sulfide.

Thought Question

16.8 Why would bacteria convert trimethylamine oxide (TMAO) to trimethylamine? Would this kind of spoilage be prevented by exclusion of oxygen?

Plant foods. Fruits, vegetables, and grains spoil differently from animal foods because of their high carbohydrate content and their relatively low water content. The low water content of plant foods usually translates into considerably longer shelf life than animal-based foods have. Carbohydrates favor microbial fermentation to acids or alcohols that limit further decomposition, and this microbial action can be managed to produce fermented foods, as described in Section 16.2.

Plant pathogens rarely infect humans but may destroy the plant before harvest. Most plant pathogens are fungi, although some are bacteria, such as *Erwinia* species. Historically, plant pathogens have caused major agricultural catastrophes, such as the Irish potato famine, caused by a fungus-like pathogen. Plant pathogens continue to devastate local economies and cause shortages worldwide; for example, the witches'-broom fungus *Crinipellis perniciosa* causes a fungal disease of cocoa trees that has drastically cut Latin American cocoa production.

After harvest, various molds and bacteria can soften and wilt plant foods by producing enzymes that degrade the pectins and celluloses that give plants their structure. In general, the more processed the food, the greater the opportunities for spoilage. For example, citrus fruits generally last for several weeks, but peeled oranges are susceptible to spoilage by Gram-negative bacteria.

Baked bread usually resists spoilage, except for surface molds. In rare cases, however, improperly baked bread can show contamination. The appearance of red bread, caused by the red bacterium *Serratia marcescens*, is believed to have been the source of the "blood" observed in communion bread during a Catholic mass in the Italian town of Bolsena in 1263—an event that became known as the miracle of Bolsena.

Pathogens Contaminate Food

Intestinal pathogens spread readily because microbes can be transmitted through food without any outward sign that the food is spoiled. The U.S. Centers for Disease Control and Prevention (CDC) estimates that there are 76 million cases of gastrointestinal illness a year in this country, usually spread through water or food. Thus, one in four Americans experiences gastrointestinal illness in a given year. In 2013, under President Obama, the Food and Drug Administration implemented new controls and rules as part of the 2011 Food Safety Modernization Act, the largest reform of U.S. food safety laws in 70 years. The new laws require, for example, that crop irrigation water be free of pathogens and that agricultural workers have access to bathroom facilities.

An example of food contamination is the 2008 outbreak of *Salmonella enterica* from peanut products. Peanuts contaminated at one processing plant led to an epidemic that sickened 700 people across the United States (**Fig. 16.23A**). The first cases of *Salmonella* infection were reported to the CDC on September 1, 2008. Most infected individuals developed diarrhea, fever, and abdominal cramps 12–72 hours after infection, and symptoms lasted 4–7 days. Over the next 6 months, cases were reported from nearly all U.S. states. The curve of the outbreak (cases rising and then falling) followed the profile of a single-source epidemic, in which all infections are ultimately traced back to one source. (Epidemics are discussed in Chapter 28.)

What was the original source of the widespread outbreak? The CDC researchers compared the food intake histories of ill persons against matched controls. They found a statistical association between illness and intake of peanut butter, eventually narrowed to a specific brand of

A.

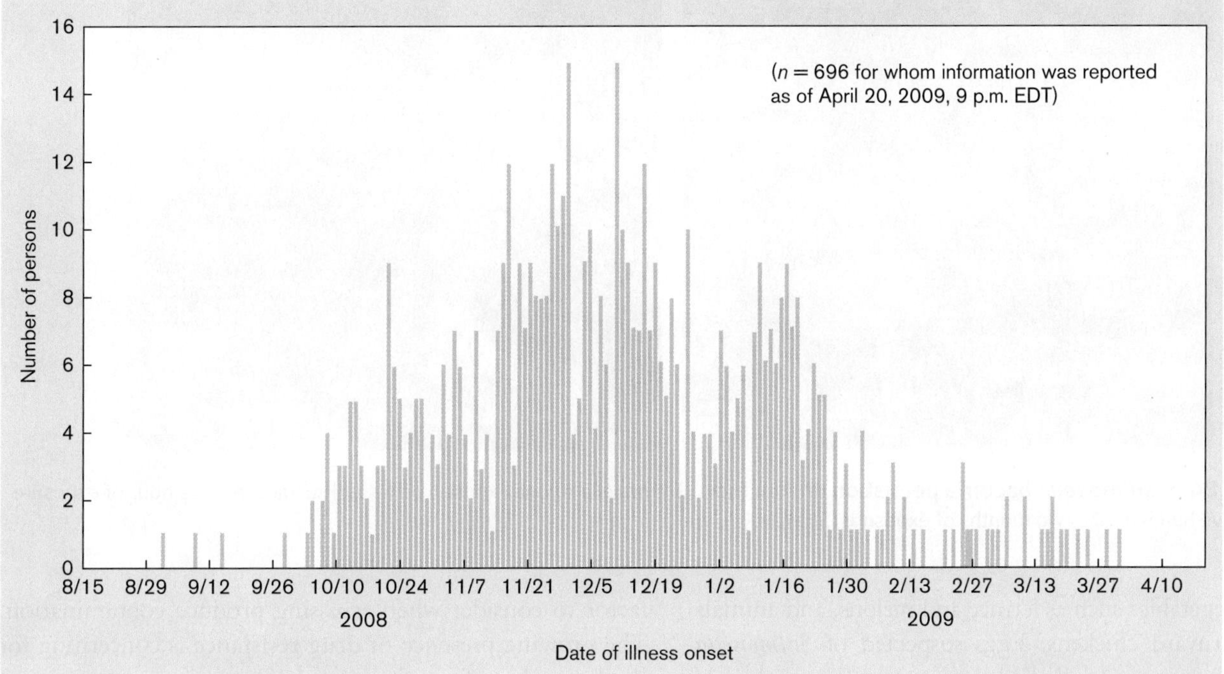

B.

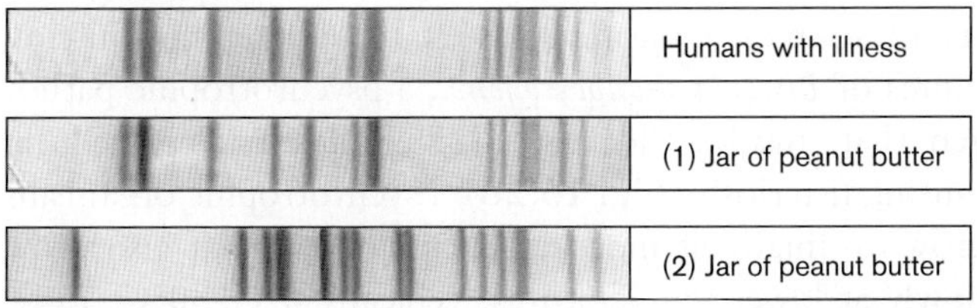

FIGURE 16.23 ■ *Salmonella enterica* outbreak from contaminated peanut butter. **A.** Number of infected individuals reported to the CDC from September 1, 2008, through April 1, 2009. **B.** Electrophoretic separation of restriction-digest DNA fragments from bacterial strains isolated from humans with illness. (1) Peanut butter containing the same strain; (2) peanut butter containing a different strain of *S. enterica*. *Source:* http://www.cdc.gov.

peanut butter sold to institutions. As the epidemic grew, cases emerged in which the contaminated food product was crackers filled with peanut butter cream. Ultimately, the peanut butter and cream were traced back to peanuts from a single factory in Georgia. At the food plant, the source of *Salmonella* contamination could not be identified, but the plant records showed that product samples had tested positive for *Salmonella*. Instead of discarding the product, the plant had retested the samples until they "tested negative." Numerous health violations were cited, including gaps in the walls and dirt buildup throughout the plant.

Salmonella bacteria are very common pathogens. Did all of the cases of illness result from a common strain? The CDC used DNA analysis to show that all patients carried a common strain of *S. enterica* serovar Typhimurium. (A serovar is a strain whose surface proteins elicit a distinctive immune response.) The strain was identified by analysis of its genomic DNA cleaved by restriction endonucleases (see Section 12.2). Each restriction endonuclease cleaves DNA at sequence-specific positions. Strains that differ at key restriction sites generate cleavage fragments of differing length, which are separated by pulsed-field electrophoresis (**Fig. 16.23B**).

In electrophoresis, applied voltage causes DNA fragments to migrate different distances according to size; the pulsed field optimizes separation of the largest sizes. The distance each fragment moves is visualized as a band in the gel. The band pattern, or "fingerprint," of *Salmonella* DNA from infected patients showed the same fragment lengths as *Salmonella* DNA from one peanut butter sample—the sample labeled "(1)" in **Figure 16.23B**. The band pattern from this peanut butter sample differed from that of another contaminated peanut butter sample, labeled "(2)"; the bacterium in the second sample proved unrelated to the *Salmonella* outbreak. The result of this simple test can be confirmed by whole-genome sequencing and comparison with known pathogenic strains.

This case illustrates several features of food contamination in modern society. It shows the consequence of a food production plant's failure to follow regulations, and the failure of health inspection to enforce them. The contaminated product shipped out to a diverse array of institutions such as schools and to secondary producers such as cookie manufacturers, which incorporated the peanut butter cream ingredient. The bacteria then remained viable in contaminated food products for many months, sickening people long after the contamination event occurred.

In recent years, an increasing range of products have been contaminated by *Salmonella*. During the year 2018, for example, in the United States contamination was reported for dry shelf products such as crackers and breakfast cereal,

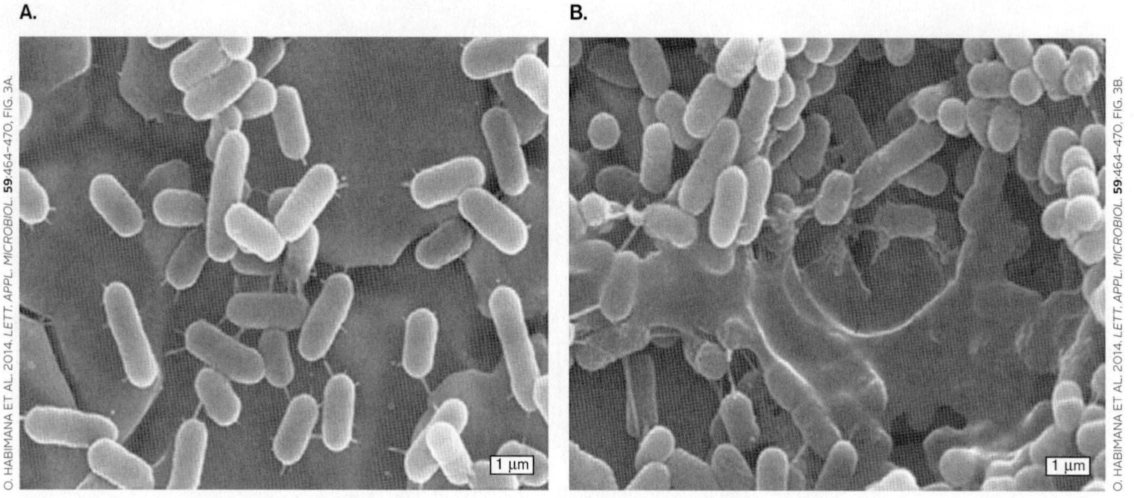

FIGURE 16.24 ■ *Salmonella* bacteria persist on dry surface. *Salmonella* Agona exposed on a steel surface. **A.** One hour of exposure at 35% relative humidity. **B.** Two months of exposure (SEM).

as well as vegetables such as lettuce and melons, and animals such as barnyard chickens. Eggs suspected of *Salmonella* contamination were recalled in seven states. The means of contamination could involve any stage of processing, such as water rinsing. *Salmonella* bacteria can persist on a dry surface for years. **Figure 16.24** compares the morphology of *Salmonella* bacteria exposed on a metal surface for 1 hour versus 2 months. The bacteria that persisted retained viability.

Food-Borne Pathogens Emerge from the Environment and Agriculture

Our environment and agriculture present constant sources of potential pathogens. Agricultural microbiologists continually assess the potential contamination of produce. For example, student Julia DeNiro worked with Douglas Doohan at the Ohio State University to study contamination of lettuce and tomatoes fertilized by manure (**Fig. 16.25**). DeNiro assessed the microbial contamination of produce as a function of many factors, such as distance from the manure, temperature, and rainfall. An example of her results is shown in **Figure 16.25B**. The surface of tomatoes was tested for colony-forming units of coliforms (intestinal bacteria) resistant to the antibiotics ampicillin, chloramphenicol, and streptomycin. With increasing rainfall, the tomatoes showed increasing numbers of drug-resistant colonies. Thus, rainfall is identified as a factor to consider when assessing produce contamination. The growing presence of drug resistance is concerning for the future biosafety of our food.

Food-borne pathogens can arise from a surprising variety of sources. Consider, for example, the transmission routes of *Listeria monocytogenes*, a psychrotrophic pathogen that invades the cells of the intestinal epithelium, causing listeriosis (**Fig. 16.26**). Psychrotrophic organisms grow optimally at moderate temperatures but also grow slowly at lower temperatures, typically 0°C–30°C. *Listeria* can be transferred from soil and feed to cattle, whose manure then cycles it back to soil. From cattle, the pathogen contaminates milk and meat, where it can eventually infect human consumers. Because *Listeria* is a psychrotroph, it outcompetes other food-borne bacteria under refrigeration.

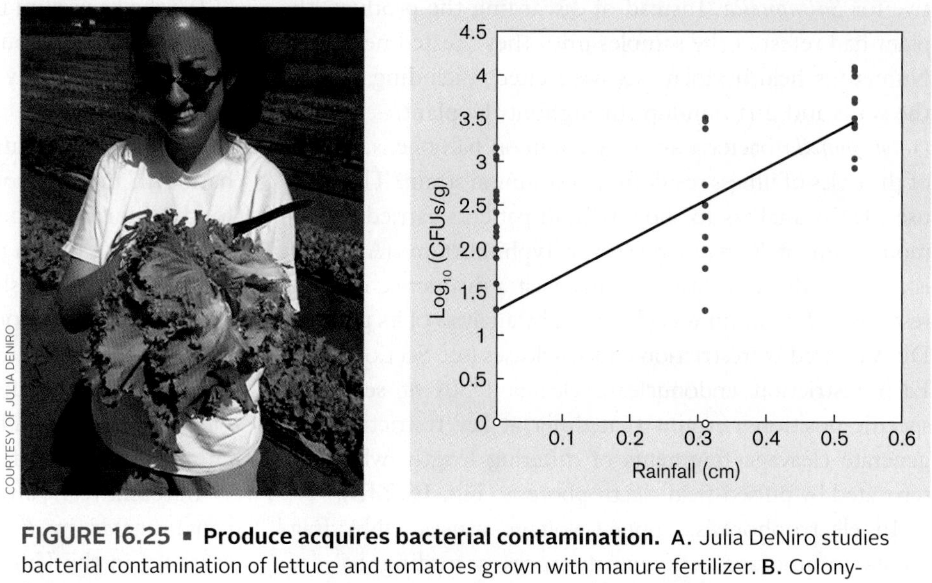

FIGURE 16.25 ■ Produce acquires bacterial contamination. **A.** Julia DeNiro studies bacterial contamination of lettuce and tomatoes grown with manure fertilizer. **B.** Colony-forming units (CFUs) from the surface of tomatoes increase with increasing rainfall.

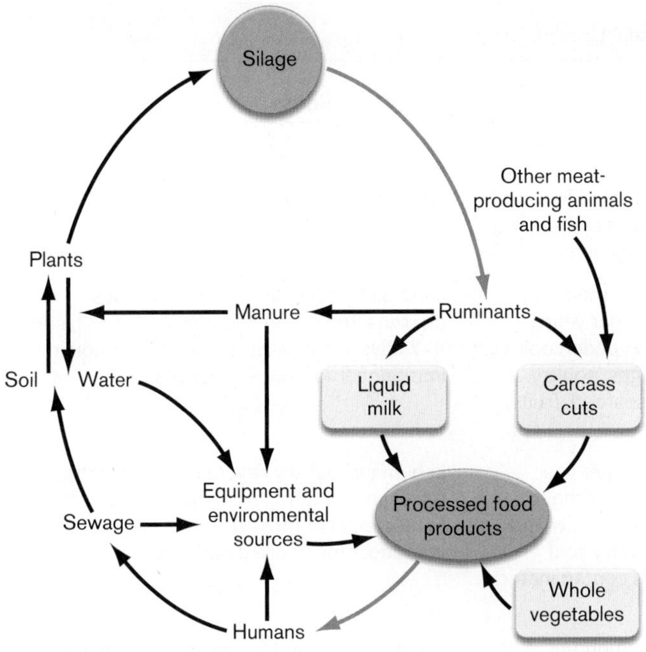

FIGURE 16.26 ▪ Transmission of *Listeria monocytogenes*. *Listeria* is transmitted through various routes, including passage through food products. Colored arrows indicate transmission of disease.

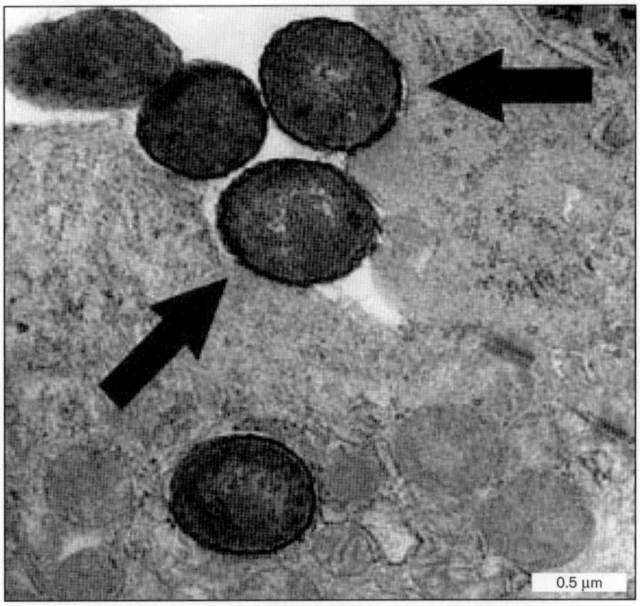

FIGURE 16.27 ▪ Intestinal crypt cells with adherent bacteria, *Escherichia coli* strain O157:H7. A gnotobiotic (germ-free) piglet was infected with the bacteria (arrows; TEM).

The U.S. Public Health Service judges the importance of food-borne pathogens by their incidence and/or the severity of the diseases they cause (**Table 16.3**). For example, the Norwalk-type viruses, or noroviruses, infect 180,000 Americans per year; their spread is very difficult to control, especially in close quarters, such as a cruise ship, where an outbreak can easily infect a large proportion of passengers. The course of illness is usually short, but it can lead to complications from dehydration. By contrast, the spore-forming pathogen *Clostridium botulinum* causes only about 100 cases of botulism per year. The incidence is relatively low, but if untreated, the fatality rate is 50%. In this case, the low incidence of botulism actually enhances its danger because the condition is likely to go undiagnosed.

What distinguishes a pathogen from a spoilage organism? Pathogens possess highly specific mechanisms for host colonization, as discussed in Chapter 25. **Figure 16.27** shows intestinal crypt cells covered with *Escherichia coli* O157:H7 bacteria, an emergent pathogen first recognized in 1982 in fast-food hamburgers. Since then, *E. coli* O157:H7 has also been found to contaminate spinach and other vegetables. The bacteria can actually grow as **endophytes** (plant endosymbionts) within the plant transport vessels. By 2010, six lesser-known strains of *E. coli* had sickened people through contaminated lettuce or beef.

Bacterial factors that contribute to disease are often encoded together in the genome in a region known as a **pathogenicity island**. A pathogenicity island consists of a set of genes and operons that function coordinately (as discussed in Section 9.5). The colocalization of the genes enables transfer of virulence capability to other species as pathogens evolve. **Figure 16.28** shows an example of a pathogenicity island in *Salmonella*. Four of its operons contribute to the type III secretion complex, which secretes toxins and host colonization factors (discussed in Chapter 25). Other genes encode outer membrane proteins that counteract host defenses, as well as regulators of virulence gene expression.

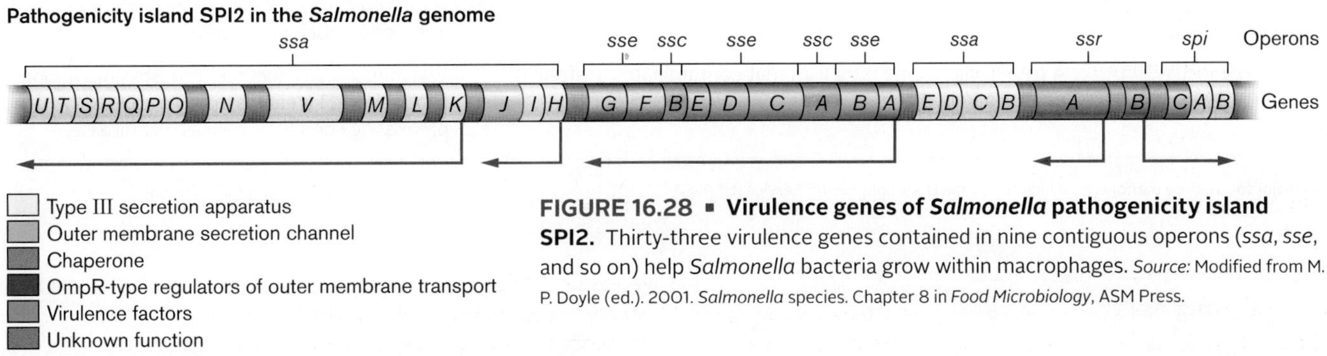

FIGURE 16.28 ▪ Virulence genes of *Salmonella* pathogenicity island SPI2. Thirty-three virulence genes contained in nine contiguous operons (*ssa*, *sse*, and so on) help *Salmonella* bacteria grow within macrophages. *Source:* Modified from M. P. Doyle (ed.). 2001. *Salmonella* species. Chapter 8 in *Food Microbiology*, ASM Press.

TABLE 16.3 Food-Borne Pathogens in the United States[a]

Pathogen	Incidence and transmission	Course of illness
Norovirus (Norwalk and Norwalk-like viruses)	Most common cause of diarrhea; also called "stomach flu" (no connection with influenza). 180,000 cases per year are estimated. Transmitted mainly by virus-contaminated food and water. Infection rates are highest under conditions of crowding in close quarters, such as inside a ship or a nursing home.	Disease lasts 1 or 2 days. Includes vomiting, diarrhea, and abdominal pain; headache and low-grade fever may occur.
Salmonella	Most common food-borne cause of death; more than 1 million cases per year; estimated 600 deaths per year. Transmission nearly always through food—raw, undercooked, or recontaminated after cooking, especially eggs, poultry, and meat; also contaminates dairy products, seafood, fruits, and vegetables.	Gastrointestinal disease that includes diarrhea, fever, and abdominal cramps lasting 4–7 days. Fatal cases are most common in immunocompromised patients.
Campylobacter	More than 1 million cases of campylobacteriosis per year; estimated 100 deaths per year. Grows in poultry without causing symptoms. Transmission is mainly through raw and undercooked poultry; contaminates half of poultry sold. Occurs less often in dairy products or in foods contaminated after cooking.	In humans, usually severe bloody diarrhea, fever, and abdominal cramps lasting 7 days. Fatal cases are most common in immunocompromised patients.
Escherichia coli O157:H7	An emerging pathogen, first recognized in a hamburger outbreak in 1982; now known to infect 73,000 people yearly, including 60 deaths per year. Grows in cattle without causing symptoms. Transmitted through ground beef; also through unpasteurized cider and from plant produce, where it grows as an endophyte.	In humans, usually severe bloody diarrhea and abdominal cramps lasting 5–10 days. About 5% of patients, especially children and elderly, develop hemolytic uremic syndrome, in which the red blood cells are destroyed and the kidneys fail.
Clostridium botulinum	Causes about 100 cases per year of botulism, with a 50% fatality rate if untreated. Grows in improperly home-canned foods, more rarely in commercially canned low-acid foods and improperly stored leftovers such as baked potatoes. Spores occur in honey, endangering infants under 2 years of age.	Botulinum toxin from growing bacteria causes progressive paralysis, with blurred vision, drooping eyelids, slurred speech, difficulty swallowing, and muscle weakness. Infant botulism causes lethargy and impaired muscle tone, leading to paralysis.
Listeria monocytogenes	*Listeria* bacteria grow in animals without causing symptoms. Animal feces may contaminate water, which is then used to wash vegetables. Transmission occurs mainly through vegetables washed in contaminated water and through soft cheeses. *Listeria* is psychrotrophic, growing at refrigeration temperatures.	Listeriosis involves fever, muscle aches, and sometimes gastrointestinal symptoms. In pregnant women, symptoms may be mild but lead to serious complications for the unborn child.
Shigella	Infects about 18,000 people a year in the United States; in developing countries, *Shigella* infections are endemic in most communities. Transmission occurs through fecal-oral contact or from foods washed in contaminated water.	Shigellosis involves gastrointestinal symptoms such as diarrhea, fever, and stomach cramps, usually lasting 7–10 days. Complications are rare.
Staphylococcus aureus	Best known as the cause of skin infections transmitted through open wounds. However, can also be transmitted through high-protein foods such as ham, dairy products, and cream pastries.	*S. aureus* causes toxic shock syndrome. Can also cause food poisoning via preformed toxins.
Toxoplasma gondii	A parasite believed to infect 60,000 people annually, most with no symptoms. In a few cases, serious disease results. Transmitted through contact with feces of infected animals, particularly cats, or through contaminated foods such as pork.	Toxoplasmosis causes mild flu-like symptoms; but in pregnant women, its transmission to the unborn child can lead to severe neurological defects, including death. Neurological complications also occur in immunocompromised patients.
Vibrio vulnificus	A free-living marine organism that contaminates seafood or open wounds. About 200 cases per year are reported.	*V. vulnificus* can infect the bloodstream, causing septic shock. Threatens mainly people with preexisting conditions such as liver disease.

[a]Ten major food-borne pathogens highlighted by the U.S. Public Health Service (USPHS).

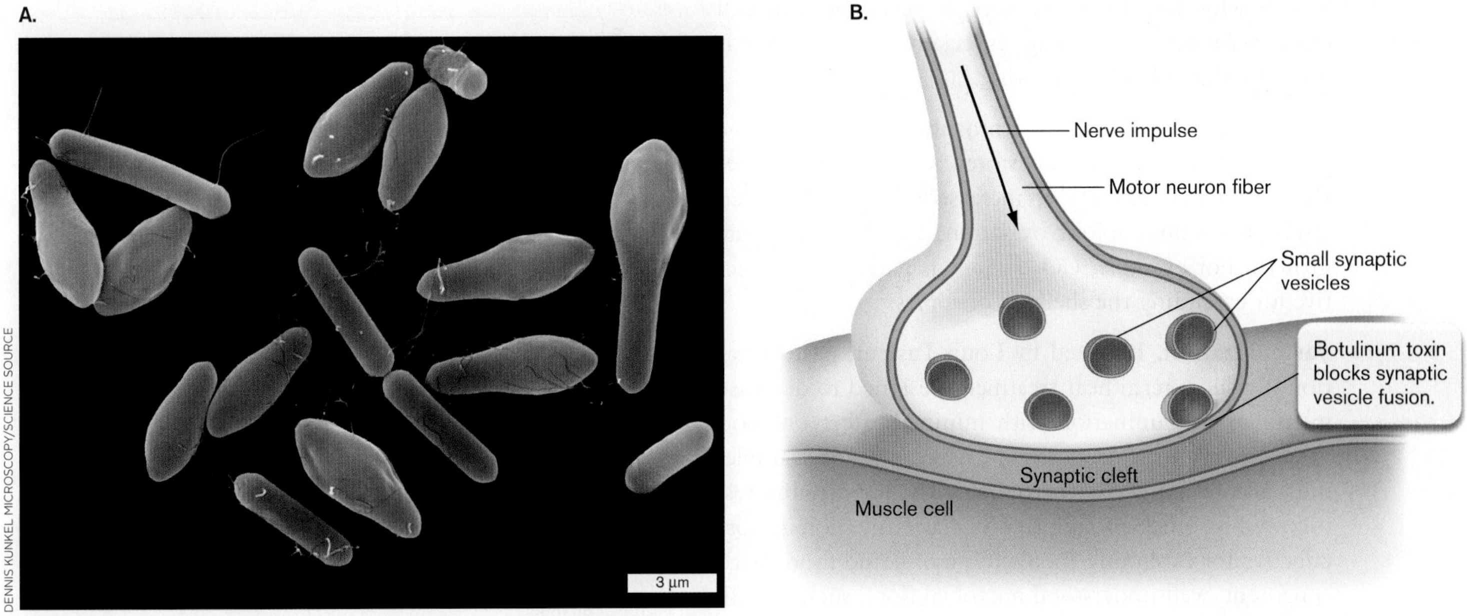

FIGURE 16.29 ■ *Clostridium botulinum* produces botulinum toxin. **A.** Club-shaped morphology of *C. botulinum* cells containing endospores. **B.** Botulinum toxin inhibits synaptic vesicle fusion in the terminal of a peripheral motor neuron, preventing activation of the muscle cell.

The most dangerous consequence of infection by foodborne pathogens is the production of a potentially fatal toxin. For example, *E. coli* O157:H7 infection of the intestines can be overcome, but the bacteria produce Shiga toxin, which can destroy the kidneys. In cases of adult botulism, *Clostridium botulinum* does not usually grow within the patient; the botulinum toxin comes from bacteria that grew previously in improperly sterilized food. The botulinum toxin has a highly specific effect, inhibiting synaptic vesicle fusion in the terminals of peripheral motor neurons (**Fig. 16.29**). Synaptic inhibition prevents activation of muscle cells, causing flaccid paralysis. Microbial toxins are discussed further in Chapter 25.

Food Preservation

Cultural practices and cuisines have long evolved to limit food spoilage. Such practices include cooking (heat treatment), addition of spices (chemical preservation), and fermentation (partial microbial digestion). In modern commercial food production, spoilage and contamination are prevented by numerous methods based on fundamental principles of physics and biochemistry that limit microbial growth (discussed in Chapter 5).

Physical means of preservation. Specific processes that preserve food based on temperature, pressure, or other physical factors include:

- **Dehydration and freeze-drying.** Removal of water prevents microbial growth. Water is removed either by application of heat or by freezing under vacuum (known as **freeze-drying**, or **lyophilization**). Drying is especially effective for vegetables and pasta. The disadvantage of drying is that some nutrients are broken down.
- **Refrigeration and freezing.** Refrigeration temperature (typically –2°C to 16°C) slows microbial growth, as shown in an experiment comparing bacterial growth in ground beef at different temperatures (**Fig. 16.30**). Nevertheless, refrigeration also selects for psychrotrophs, such as *Listeria*. Freezing halts the growth of most microbes, but preexisting contaminant strains often survive to grow again when the food is thawed.

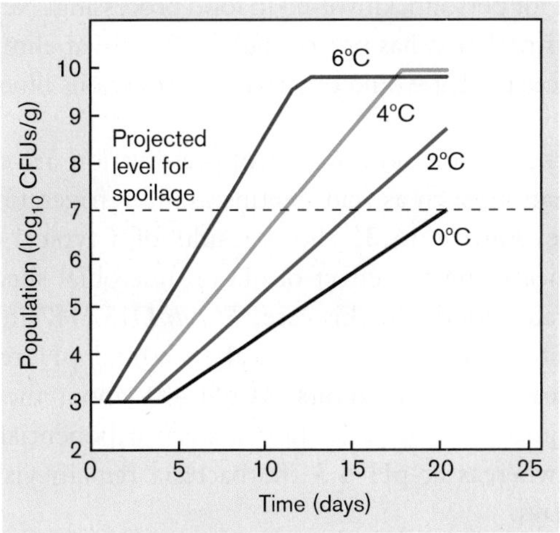

FIGURE 16.30 ■ Bacterial growth in ground beef. The growth rate of total aerobic bacteria in ground beef declines at lower storage temperatures. CFUs = colony-forming units.
Source: Modified from M. P. Doyle (ed.). 2001. Meat, poultry, and seafood. Chapter 5 in *Food Microbiology*, ASM Press.

This is why deep-frozen turkeys, for example, can still cause *Salmonella* poisoning, especially if the interior is not fully thawed before roasting.

- **Controlled or modified atmosphere.** Food can be packed under vacuum or stored under atmospheres with decreased oxygen or increased CO_2. Controlled atmospheres limit abiotic oxidation, as well as microbial growth. For example, CO_2 storage is particularly effective for extending the shelf life of apples.
- **Pasteurization.** Invented by Louis Pasteur, pasteurization is a short-term heat treatment designed to decrease microbial contamination with minimal effect on food value and texture (discussed in Chapter 5). For example, milk is commonly pasteurized at 63°C for 30 minutes, followed by quick cooling to 4°C. Pasteurization is most effective for extending the shelf life of liquid foods with consistent, well-understood microbial flora, such as milk and fruit juices.
- **Canning.** In canning, the most widespread and effective means of long-term food storage, food is cooked under pressure to attain a temperature high enough to destroy endospores (typically 121°C). Commercial canning effectively eliminates microbial contaminants, except in very rare cases. The main drawback of canning is that it incurs some loss of food value, particularly that of labile biochemicals such as vitamins, as well as loss of desirable food texture and taste. Ineffective canning can allow bacterial growth, causing toxin and gas production.
- **Ionizing radiation.** Exposure to ionizing radiation, known as food irradiation, effectively sterilizes many kinds of food for long-term storage. The main concerns about food irradiation are its potential for unknown effects on food chemistry and the hazards of the irradiation process itself for personnel involved in food processing. Nevertheless, irradiation has proved highly effective at eliminating pathogens that would otherwise cause serious illness.

Often, two or more means of preservation are used in combination, such as acid treatment and refrigeration. For example, **Figure 16.31** shows results of a typical experiment measuring the effect of pH on microbial survival in refrigerated food—in this case, *E. coli* O157:H7 in Greek eggplant salad. Note the critical threshold pH required to decrease bacterial counts. At pH 4.0, about the pH of lemon juice, the bacteria show a steep exponential death curve, whereas at pH 4.5 the bacteria remain viable for many days.

How does the food industry know how long to treat food for sterilization? Several measurements indicate the efficiency of heat killing. The D-value (decimal reduction time) was described in Chapter 5. An additional measure is 12D, the amount of time required to kill 10^{12} spores (or to decrease a population by 12 $\log_{10}$ units). A third measure is the z-value, the increase in degrees Celsius needed to lower the D-value to 1/10 of the time. If, for example, D_{100} (the D-value at 100°C) and D_{110} (the D-value at 110°C) for a given organism are 20 minutes and 2 minutes, respectively, then $12D_{100}$ equals 240 minutes (that is, 20 minutes × 12), and the z-value is 10°C (because a 10°C increase in temperature reduced the D-value to 1/10, from 20 minutes to 2 minutes). These measurements are determined empirically for each organism. The values are extremely important to the canning industry, which must ensure that canned goods do not contain spores of *Clostridium botulinum*, the anaerobic soil microbe that causes the paralyzing food-borne disease botulism (see **Fig. 16.29**).

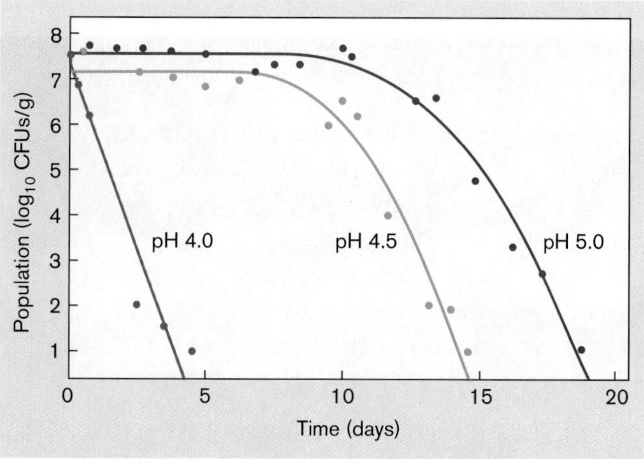

FIGURE 16.31 ▪ Bacteria die sooner at lower pH. Survival curves of *E. coli* O157:H7 in eggplant salad stored at 5°C at pH 4.0, pH 4.5, and pH 5.0. CFUs = colony-forming units. *Source:* Modified from Panagiotis N. Skandamis and George-John E. Nychas. 2000. *Appl. Environ. Microbiol.* **66**:1646.

Because the tastes of certain foods suffer if they are overheated, z-values and 12D-values are used to adjust heating times and temperatures to achieve the same sterilizing result. Consider an example in which D_{121} is 10 minutes and $12D_{121}$ is 120 minutes. Sterilizing at 121°C for 120 minutes might result in food with a repulsive taste, whereas decreasing the temperature and extending the heating time might yield a more palatable product. The D-values and z-values are used to adjust conditions for sterilization at a lower temperature. If D_{121} is 15 minutes (the time needed to kill 90% of cells) and the z-value is known to be 10°C (the temperature change needed to change D-value tenfold), then decreasing temperature by 10°C, to 111°C, will mean D_{111} is 150 minutes (10 × D_{121}). Therefore, the value of 12D (required to decrease a population by 12 logs) is $12D_{111}$ = 1,800 minutes. Sterilization may take longer, but food quality is likely to remain high, because the sterilizing temperature is lower.

Chemical means of preservation. Many kinds of chemicals are used to preserve foods. Major classes of chemical preservatives include:

- **Acids.** While microbial fermentation can preserve foods by acidification, an alternative approach is to add acids directly. Organic acids commonly used to preserve food include benzoic acid, sorbic acid, and propionic acid. The acids are generally added as salts: sodium benzoate, potassium sorbate, sodium propionate. These acids act by crossing the cell membrane in the protonated form and then releasing their protons at the higher intracellular pH. For this reason, they work best in foods that already have moderate acidity (pH 5–6), such as dried fruits and processed cheeses.
- **Esters.** The esters of organic acids often show antimicrobial activity whose basis is poorly understood. Examples include fatty acid esters and parabens (benzoic acid esters). They are used to preserve processed cheeses and vegetables.
- **Other organic compounds.** Numerous organic compounds, both traditional and synthetic, have antimicrobial properties. For example, cinnamon and cloves contain the benzene derivative eugenol, a potent antimicrobial agent.
- **Inorganic compounds.** Inorganic food preservatives include salts, such as phosphates, nitrites, and sulfites. Nitrites and sulfites inhibit aerobic respiration of bacteria, and their effectiveness is enhanced at low pH. These substances, however, may have harmful effects on humans; nitrites can be converted to toxic nitrosamines, and sulfites cause allergic reactions in some people.

Thought Question

16.9 Is it possible for physical or chemical preservation methods to completely eliminate microbes from food? Explain.

To Summarize

- **Food spoilage** refers to chemical changes that render food unfit for consumption. Food spoils through degradation by enzymes within the food, through spontaneous chemical reactions, and through microbial metabolism.
- **Food contamination, or food poisoning**, refers to the presence of microbial pathogens that cause human disease, or toxins produced by microbial growth. Food harvesting, food processing, and shared consumption are all activities that spread pathogens.
- **Dairy products** can be soured by excessive fermentation or made bitter by bacterial proteolysis.
- **Meat and poultry** are putrefied by decarboxylating bacteria, which produce amines with noxious odors.
- **Fish and other seafood** spoil rapidly because their unsaturated fatty acids rapidly oxidize, they harbor psychrotrophic bacteria that grow under refrigeration, and their trimethylamine oxide is reduced by bacteria to the fishy-smelling trimethylamine.
- **Vegetables** spoil by excess growth of bacteria and molds. Plant pathogens destroy food crops before harvest.
- **Food preservation** includes physical treatments, such as freezing and canning, as well as the addition of chemical preservatives, such as benzoates and nitrites.

16.5 Industrial Microbiology

The production and preservation of food is only one field of **industrial microbiology**, the commercial exploitation of microbes. For example, microbes can be cultured to produce complex molecules, such as vitamin B_{12}, more cheaply and easily than by abiotic chemistry (see **Special Topic 16.1**). The product may be generated by the original organism whose genome possesses the pathway; or alternatively, the genes encoding the pathway may be cloned into a **heterologous host** (different or distantly related species). A heterologous host may be optimized by genetic engineering for industrial control.

Industrial microbiology now includes a broad range of commercial products, among them microbial enzymes, vaccines and clinical devices, and genetically modified plants and animals using microbial vectors. Particularly exciting is the production of viral vectors for human gene therapy (Section 16.6). Other fields of industrial microbiology include wastewater treatment, bioremediation, and environmental management, which are covered in Chapters 21 and 22.

Industrial Microbiology Aims for Commercial Success

The practical application of a microbial product or device may arise out of an industrial laboratory, or it may be conceived by a research scientist with the aim of meeting a compelling need in society. In all cases, however, the key goal is to succeed in the marketplace—that is, to generate a product that customers adopt over alternative technologies. The product's sales must cover the costs of raw materials and production and (in a for-profit company) generate a profit for the shareholders. Success requires:

- **Identifying a useful product.** Possible products include small molecules such as antibiotics; human proteins from cloned genes; or proteins from microbes with

SPECIAL TOPIC 16.1 Microbial Vitamins for Sale

If you read the fine print of your vitamin supplement, did you ever wonder where all those molecules come from? Many of them were made by microbes. Vitamin B_{12} is an essential vitamin in our diet, required for red blood cell formation and brain function. This vitamin is one of the more complex molecules made by a living organism (**Fig. 1A**). Its synthesis requires more than 25 unique enzymes, with distinctive remodeling of a tetrapyrrole (uroporphyrinogen III) into the asymmetrical corrin ring. An unusual feature is the placement of the metal cobalt at the center of the heteroaromatic ring, held in place by a metal-carbon bond to a deoxyadenosyl group.

Since prehistoric times, vitamin B_{12} has been available to the human diet via consumption of animal products or of microbes such as yeasts. Today the vitamin is one of many sold in purified form, often added as a supplement to processed foods. Its industrial production requires a microbial host, most commonly a Gram-negative bacterium, either *Propionibacterium shermanii* or *Pseudomonas denitrificans*. The producer microbe must possess the entire biosynthetic pathway in its genome, starting from simple catabolic substrates such as sugars.

For industrial use, however, these native vitamin producers have drawbacks. Their fermentation takes longer than that of well-domesticated species such as *E. coli*. Their growth requires substrates that add expense. And their genetic tools for engineering are limited. A more economical alternative would be to transfer the biosynthetic genes into a heterologous strain that is optimized for commercial use. Commercial strains grow well in large fermentors and catabolize inexpensive carbon sources.

Dawei Zhang at the Tianjin Institute of Industrial Biotechnology, China, investigates biosynthetic pathways for vitamins and amino acids that offer potential success in commercial production. His lab is interested in developing a heterologous host strain for economical production of vitamin B_{12} (**Fig. 2**).

To construct a heterologous producer requires cloning all genes (more than 25) into the producer host.

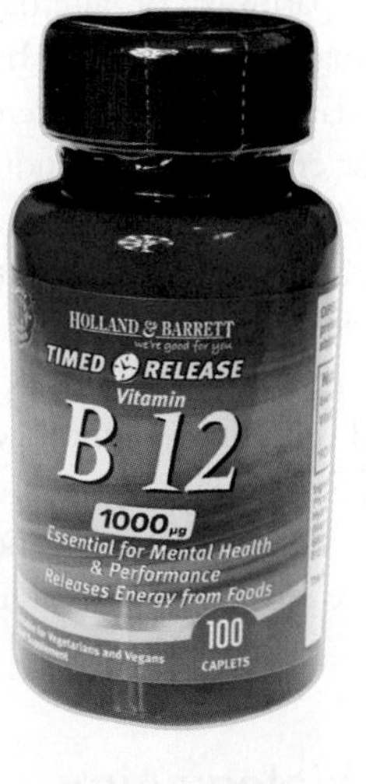

STEPHEN BARNES/MEDICAL/ALAMY STOCK PHOTO

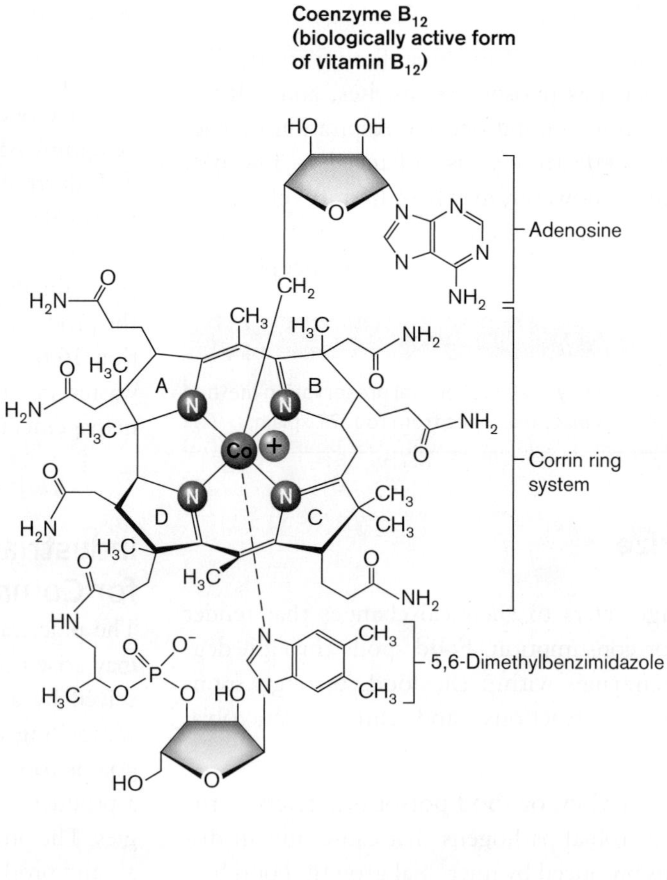

FIGURE 1 ■ Vitamin B_{12} structure. Vitamin B_{12} is a complex molecule that includes cobalt at the center of a corrin ring system.

FIGURE 2 ■ Developing a heterologous producer strain for vitamin B_{12}.

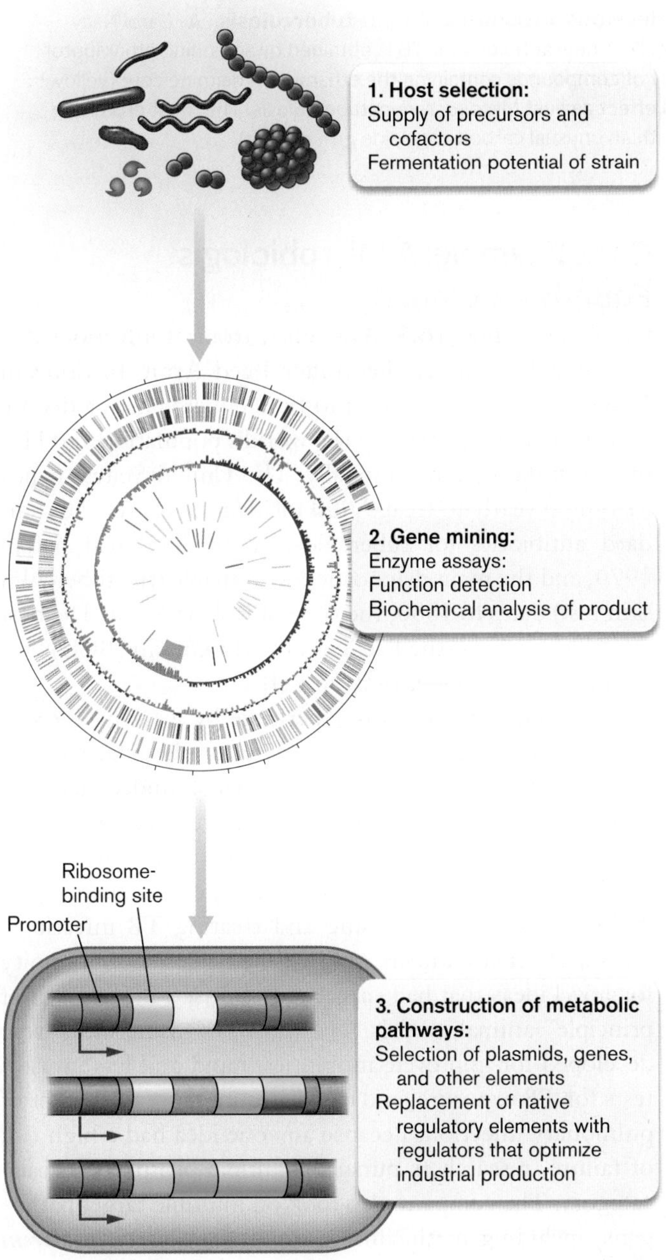

Genes must be mined from native producer organisms (**Fig. 2**, step 2). In the case of *E. coli*, the host already possesses genes for synthesis of the pyrrole precursor, delta-aminolevulinate. Genes encoding enzymes farther down in the pathway were transferred from the closely related species *Salmonella enterica*.

The biosynthetic pathway must be constructed for optimal overproduction, which differs from the native host's optimal needs for moderate production (step 3). Most important, all the native regulatory elements, such as promoters, activators, and ribosome-binding sites (RBS) must be deleted. For example, the B_{12} riboswitch (Chapter 10) detects the vitamin molecule at a low concentration and terminates transcription and translation. This feedback regulation makes sense for the native host but is undesirable for industrial production, which aims for as much product as possible. The native regulatory elements are then replaced by regulators that avoid bottlenecks (imbalance of substrate and product at intermediate steps) and optimize production per unit substrate. Finally, environmental variables such as media components and temperature are adjusted to optimize the rate of fermentation (step 4). Ultimately, the aim is to produce more of the vitamin at lower cost to consumers, while increasing profits for the company shareholders.

RESEARCH QUESTION

What kind of DNA regulatory elements might you select to replace the native ones? What regulatory behavior would you aim for in the producer strain?

Huan Fang, Jie Kang, and Dawei Zhang. 2017. Microbial production of vitamin B_{12}: A review and future perspectives. *Microbial Cell Factories* **16**:15.

A.

JOANNE LAWTON/WASHINGTON BUSINESS JOURNAL

B.

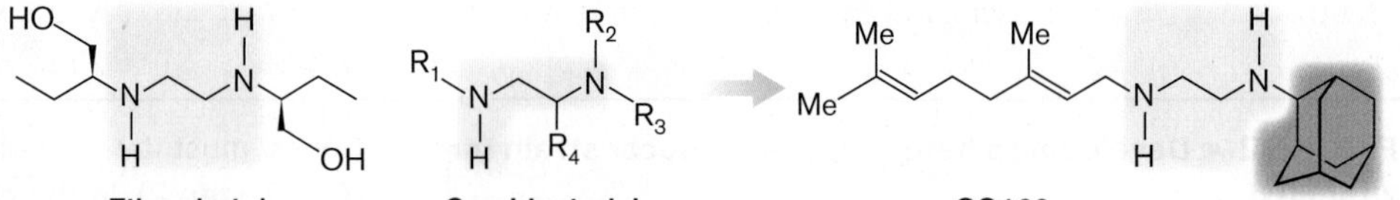

Ethambutol

Combinatorial library of 63,238 diamines

SQ109

FIGURE 16.32 ▪ Sequella develops antibiotics to fight tuberculosis. **A.** Carol Nacy founded the Sequella company. **B.** A new antibiotic for TB is obtained by screening ethambutol analogs. A combinatorial library of compounds containing the ethambutol diamine core (yellow) was screened for antibacterial effect against *Mycobacterium tuberculosis*. The most promising agent screened was SQ109, with an unusual carbon-cage side group (pink).

useful properties, such as thermostability or increased catalytic activity for a chemical process.

- **Isolating a microbe to produce the product.** A novel product, such as an antibiotic, is generally developed from a naturally occurring microbe. The genes encoding product biosynthesis may then be cloned into an industrial vector.
- **Scaling up production in quantity.** The producer microbial strain must be grown on an industrial scale, and the product must be isolated and purified.
- **Developing a business plan.** The scientist-entrepreneur must obtain partners skilled in industrial management, finance, and marketing. Patents must be filed to protect intellectual property rights.
- **Safety and efficacy testing.** Human consumption, environmental introduction, or consumer use requires many levels of testing prior to commercialization, including, in some cases, approval by government agencies.
- **Effective marketing.** The benefits of the new product must be communicated effectively to convince customers of its superiority over current products or processes.

Failure at any of these tasks spells doom for the product. Thus, a prudent business plan includes having multiple alternative products in development. Although the failure rate of new products is high, all the products we use had to overcome these risks. How does a bench scientist convert a microbial concept into a profitable product? Some scientists acquire partners to found their own company. Others join a mature company that offers many kinds of expertise in business and production.

Case Example: A Microbiologist Founds a Company

Carol Nacy (**Fig. 16.32A**) studied tropical infectious diseases for 17 years at the Walter Reed Army Institute of Research. She saw the need to fight tuberculosis, a disease that infects one-quarter of the world's population and kills over a million people annually. The United States spends $1 billion yearly to treat 9,000 incident cases; yet the standard antibiotics for tuberculosis were developed before 1970, and the main diagnostic test available (the tuberculin skin test) dates to 1880, the time of Robert Koch. The best available vaccine—the Bacille Calmette-Guérin (BCG) live, attenuated vaccine—is only 50% effective.

Nacy obtained business experience by working for several years as a chief scientific officer for the pharmaceutical company EntreMed, Inc. She then founded her own company, Sequella, to develop innovative drugs and treatment devices. Sequella targets innovative ideas with high risk but also high potential to improve performance, rapidity, and safety of diagnosing and treating TB infections. Nacy and her cofounders scanned the academic community for novel ideas that had succeeded against TB in "proof of principle" animal models. The most promising ideas were developed for improved antibiotics, rapid and less invasive tests for TB exposure, and devices to measure the extent of pulmonary infection. Because any one idea had a high risk of failure, researchers pursued multiple prospects in each category. The company also took on multiple target pathogens, including methicillin-resistant *Staphylococcus aureus* (MRSA) and vancomycin-resistant *Enterococcus* (VRE).

Sequella's most promising antibiotic, SQ109 was discovered via high-throughput screening of a chemical library, in collaboration with Clifton Barry at the National Institutes of Health. The chemical library consisted of over 60,000 analogs of a known TB antibiotic, ethambutol (**Fig. 16.32B**). Ethambutol is part of the current standard treatment for TB, whose 6-month time course has a poor compliance rate. It is hoped that improved drugs will shorten the time course and improve compliance, thereby decreasing the appearance of drug-resistant strains. The analog molecules were selected for their common diamine core, with different combinations of side chains. The 60,000 compounds in the library were subjected to combinatorial screening, a mathematically intensive analysis based on numerous tests. Of the compounds tested, 2,796 showed activity against *Mycobacterium tuberculosis* in the test tube. The 69 best compounds were tested for cytotoxicity in tissue culture, activity in TB-infected macrophages, and activity in infected animals. The compound with the greatest efficacy and fewest side effects was SQ109, a molecule with an unusual cage-like side group of three fused rings (**Fig. 16.32B**).

In 2018, SQ109 passed a phase IIb/III human clinical trial. The aim of the trial was to test the efficacy and safety of the molecule in human patients. The drug was tested in seven clinical centers in Russia, where the tuberculosis infection rate is high. A significantly greater number of patients treated with SQ109 showed clearing of tuberculosis bacteria in their sputum, compared with control patients on placebo. The drug also shows promise for treatment of *Helicobacter pylori*, as well as the parasites *Leishmania* and *Trypanosoma cruzi*.

Microbial Products

Companies such as Sequella seek agents to combat microbial pathogens, but increasingly, industry seeks beneficial microbes from natural sources to produce useful products. The microbe is cultured to synthesize a commercially valuable chemical substance such as the industrial enzyme lipase B, or a material for manufacturing.

Examples of commercial microbial products are listed in **Table 16.4**. Each product requires a gene or operon of genes encoding either the product itself or the enzymes for the product's biosynthesis. For example, the Danish company Novozymes markets over 700 microbial enzymes for products ranging from laundry detergents to agricultural inoculants (discussed in **eTopic 16.2**). Novozymes also produces numerous enzymes as biocatalysts for green chemistry. "Green chemistry" refers to environmentally friendly procedures for reactions in organic chemistry, typically using water solution in place of petroleum-derived organic solvents.

Many microbial products replace earlier chemical processes that were more costly, generated greater waste, and had more negative effects on environmental quality and global climate (**Table 16.4**). For example, phosphates in consumer clothes detergents have largely been replaced by bacterial proteases and yeast-derived lipases. The use of these enzymes improves the cleaning action, enables washing at lower temperatures (thus saving energy and carbon units), and avoids the use of phosphates, which cause eutrophication of lakes (see Chapter 22).

We may distinguish between two fundamentally different sources of products: cloned genes from human, animal, or plant sources; and native microbial products from newly discovered species in the environment, often from extreme conditions. Cloned human genes typically encode a protein of valuable function in the human body. For example, Therabron Therapeutics produces the recombinant human protein CC10, a lung development protein that is often deficient in the lungs of premature infants. The recombinant protein, produced and purified from a recombinant bacterium, can be used to reduce lung inflammation in premature infants, as well as in patients with chronic obstructive pulmonary disease (COPD). Cloning of gene products in recombinant organisms was discussed in Chapter 12.

To identify new microbial products, companies screen thousands of microbial strains from diverse ecosystems. The search for organisms with potential commercial applications is called **bioprospecting**. Bioprospecting can be done anywhere, from one's backyard to Yellowstone National Park. Unique ecosystems are the most promising sources of previously unknown microbial strains with valuable properties. Extreme environments such as the hot springs of Yellowstone are particularly promising because their products may tolerate higher temperatures that are required for industrial use. Psychrophiles from extremely cold environments such as Antarctica are a useful source of enzymes that are active in a wide range of temperatures, such as enzymes to clean clothing.

An important aspect of bioprospecting is "mining the genome." Once a promising source strain is obtained, its genome is sequenced to investigate the gene sequences that encode the useful product and regulate its expression. The operons encoding the product (or enzymes for its production) can then be cloned for optimal production (see Chapter 12). The cloned genes are transferred into an **industrial strain**, a strain whose growth characteristics are well studied and optimized for industrial production. An industrial strain must possess the following attributes:

- **Genetic stability and manipulation.** The industrial strain must reproduce reliably, without major DNA rearrangements. It must also have an efficient gene transfer system by which vectors can introduce genes of interest into its genome.
- **Inexpensive growth requirements.** Industrial strains must grow on low-cost carbon sources with minimal special needs, such as vitamins, and at easily maintained conditions of temperature and gases.
- **Safety.** Industrial strains must be nonpathogenic and must not produce toxic by-products.
- **High level of product expression.** The strain or recombinant vector must possess an efficient gene expression system to generate the desired product as a high proportion of its cell mass.

TABLE 16.4 Consumer Products Made with Industrial Microbiology

Product	Old manufacturing process	New industrial microbial process	Climate benefits	Consumer benefit
Bread 	Potassium bromate, a suspected cancer-causing agent, added as a preservative and a dough-strengthening agent	Genetically enhanced microorganisms produce baking enzymes to enhance rising, strengthen dough, prolong freshness.	Reduction of CO_2 emissions in grain production, milling and baking, and transportation	High-quality bread; longer shelf life; eliminates suspected carcinogen potassium bromate
Vitamin B_2 (Riboflavin) 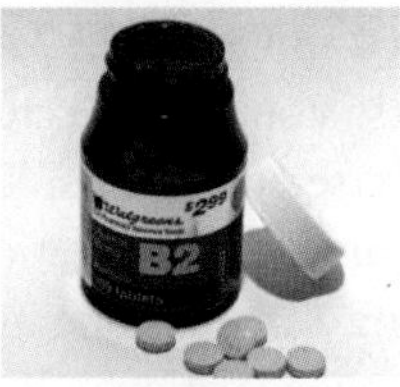	Toxic chemicals, such as aniline, used in a nine-step chemosynthesis process (hazardous waste generated)	*Lactobacillus* developed for one-step fermentation process uses vegetable oil as feedstock and sugar as nutrient.	Up to 33% reduction in energy use; 25%–33% reduction in CO_2 emissions	Greatly reduces hazardous waste generation and disposal
Personal care 	Chemicals such as propylene glycol and butylene glycol from petroleum used as solvents to mix ingredients	Genetically enhanced microbe produces 1,3-propanediol from renewable feedstocks, which can function as a solvent, humectant, emollient, or hand-feel modifier.	20% reduction of greenhouse gas emissions compared to 1,3-propanediol from petroleum	High purity; environmentally sustainable and renewable process; nonirritating for sensitive skin; enhanced clarity
Detergent 	Phosphates added as a brightening and cleaning agent	Microbes or fungi produce enzymes, which are added as brightening and cleaning agents. Protease enzymes remove protein stains; lipases remove grease; amylases remove starch.	Elimination of water pollution due to phosphates	Brighter, cleaner clothes with lower wash temperature; energy savings
Textiles 	New cotton textiles prepared with chlorine or chemical peroxide bleach	Microbial cellulase enzymes produce peroxides, allowing bleaching of textiles at low temperature (65°C) and in a neutral pH range.	25% decrease in greenhouse gases; 25% decrease in nonrenewable energy use	New fabrics have lower impact on the environment, better dyeing results, and a permanent soft and bulky handle
Paper 	Wood chips are boiled in a harsh chemical solution to yield pulp for papermaking	Wood-bleaching enzymes are produced by genetically enhanced microbes to selectively degrade lignin and break down wood cell walls during pulping.	Reduction in use of chlorine bleach, and in dioxins in the environment	Cost savings from lower energy and chemical use
Diapers 	Woven fabric coverings made from petroleum-based polyesters	*Bacillus* ferments corn sugar to lactic acid, which is heated to generate a biodegradable polymer for woven fabrics.	50%–70% decrease in CO_2 emissions	Biodegradable; disposal options include composting rather than landfills
Polyesters 	Polyester, a synthetic polymer fiber, produced chemically from petroleum feedstock	*Bacillus* ferments corn sugar to lactic acid, which is heated to generate a biodegradable polymer (such as NatureWorks' Ingeo).	75% decrease of CO_2, compared to PET; 90% reduction of CO_2 equivalent compared to nylon 6	Polylactic acid (PLA) polyester is biodegradable, does not harbor body odors, and does not give off toxic smoke if burned

TABLE 16.4 Consumer Products Made with Industrial Microbiology (*continued*)

Product	Old manufacturing process	New industrial microbial process	Climate benefits	Consumer benefit
Stonewashed jeans 	Open-pit mining of pumice; fabric washed with crushed pumice stone and/or acid	Fabric is washed with microbial enzymes (cellulases) to fade and soften jeans or khakis.	Less mining; decreased energy consumption	Softer fabric; lower cost
Enzymes 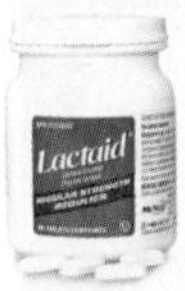	Chemical processes using materials generally derived from petroleum	Alpha-amylase—*Bacillus subtilis* Amyloglucosidase—*Aspergillus niger* (fungus) Lactase (beta-galactosidase)—*Kluyveromyces lactis* (fungus) Lipases—*Candida cylindraceae* (yeast) Alkaline protease—*Aspergillus oryzae* (fungus)	Decreased energy consumption; decreased pollution of environment	Lower cost
Food supplements 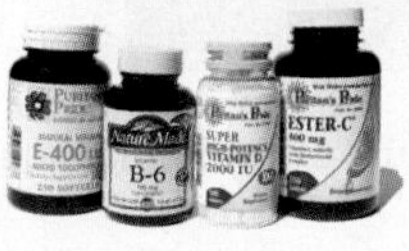	Derived from petroleum	L-Lysine—*Brevibacterium lactofermentum* L-Tryptophan—*Klebsiella aerogenes* Monosodium glutamate (MSG)—*Corynebacterium ammoniagenes* Vitamin B_{12}—*Pseudomonas denitrificans* Vitamin C (ascorbic acid)—*Acetobacter suboxidans*	Decreased energy consumption; decreased pollution of environment	Lower cost

Sources: Biotechnology Innovation Organization, 2018 (http://www.bio.org); and M. J. Waites. 2001. *Industrial Microbiology*. Blackwell Science.

- **Ready harvesting of product.** Either the product must be secreted by the cell or, if the product is intracellular, the cells must be easily breakable to liberate the product.

Common species for industrial strains include the bacteria *E. coli* and *Bacillus subtilis*, the yeast *Candida utilis*, and the filamentous fungus *Aspergillus niger*. Each of these species is safe; grows to high density on inexpensive carbon sources, such as molasses; and expresses desired products at high concentration.

Thought Question

16.10 Why would different industrial strains or species be used to express different kinds of cloned products?

Fermentation Systems

Commercial success requires optimizing every detail of the fermentation system. "Fermentation" in industrial terms refers not just to anaerobic metabolism, but to all means of growth of microbes on an industrial scale. In an **industrial fermentor**, the growth vessel and all its environmental supports, such as temperature control and oxygenation, must be scaled up to thousands of liters (**Fig. 16.33**). This increase in scale generates many problems of quality control, such as maintaining uniform temperature, pH, and oxygenation throughout the vessel and minimizing foaming of the culture liquid. A small change in any of the growth factors can impact production costs and profit margin. Another major concern is to avoid contamination by other organisms.

The fermentor is the core of the first half of industrial production, known as **upstream processing** (**Fig. 16.34**, top). Upstream processing is the culturing of the industrial microbe to produce large quantities of product or cell mass. All aspects of the process must be controlled to maximize the final concentration of product, which in most cases peaks at a specific time in the microbial growth cycle. Following microbial growth, the culture must be harvested and the product purified. These processes constitute **downstream processing** (**Fig. 16.34**, bottom). The first step of downstream processing is to separate the microbial cells from the culture fluid, by centrifugation or by filtration. Next, the **primary recovery** of product follows one of two different pathways,

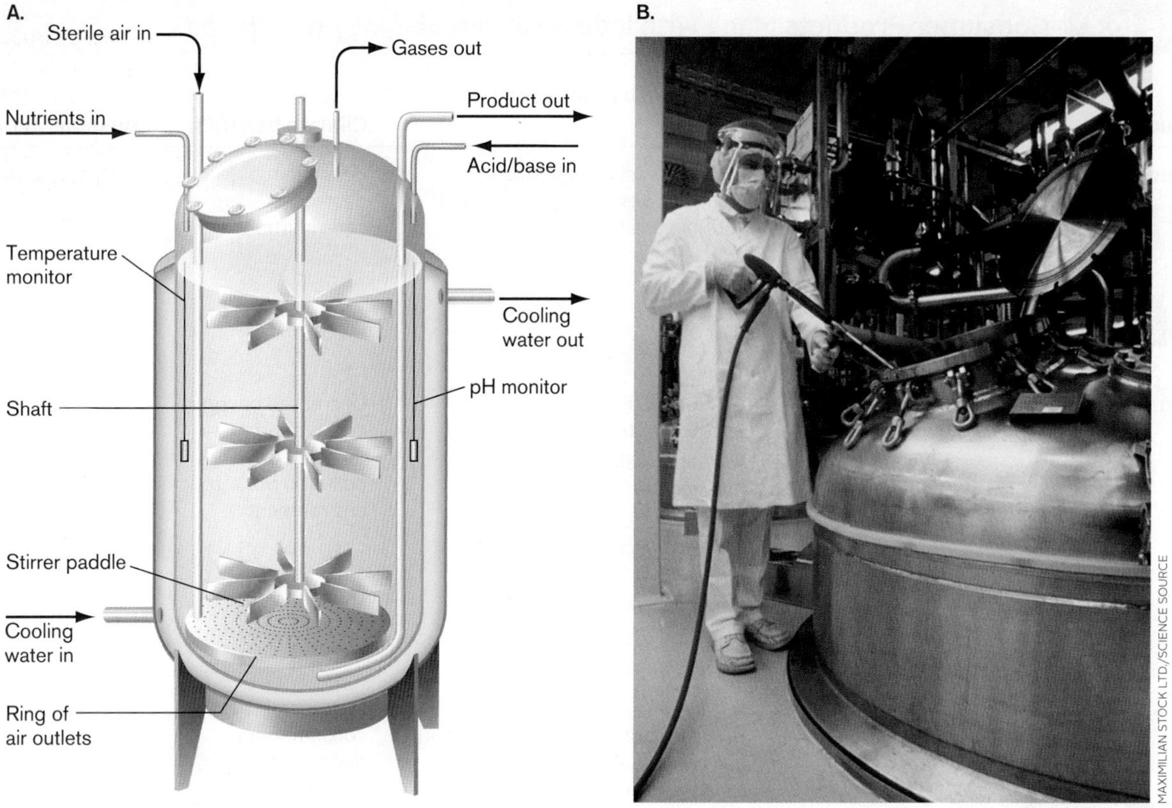

FIGURE 16.33 ■ Industrial fermentation. **A.** Industrial production of microbial products requires scaled-up culture of the production microorganism. **B.** An industrial fermentor.

depending on whether the product is maintained within the cells or secreted into the culture fluid. Many kinds of subsequent purification and finishing steps are necessary before the product has acceptable quality for its desired use. Again, failure of any detail can render the entire product unusable.

Products designed for human consumption or use in the environment face formidable hurdles in clinical testing, toxicological studies, and, finally, approval by the appropriate regulatory agency, such as the U.S. Food and Drug Administration (FDA). After millions of dollars are invested in process development, the product may still fail one of these late-stage hurdles and never come to market. Not surprisingly, a company must research thousands of potential products before achieving one that makes a profit. The consumer cost inevitably includes the development costs not only of the one successful product, such as recombinant insulin, but also of all the products that failed.

Despite all the hurdles to overcome, companies such as Novozymes sell hundreds of products and make billion-dollar profits. Examples of commercially successful products are described in **eTopic 16.2**.

Microbe as Product

Microbes as biological agents possess enormous potential for environmental use. The best-known example is the bacterial insecticide *Bacillus thuringiensis* (discussed in Chapter 12). Such microbes, derived from natural sources, may improve crop production and enhance agricultural sustainability. Most methods used by farmers to improve crop yields require large amounts of fertilizer and pest management, including herbicides and fungicides. To alleviate the financial and environmental impacts of large-scale farming, scientists are working to develop microbial solutions that can supplement chemical additives and pesticides, resulting in higher yields to farmers.

Scientists at Novozymes partner with Monsanto to conduct large-scale bioprospecting. The scientists screen thousands of microbes for positive impacts on agriculture. Some of the microbial products have resulted in higher corn yields across North America. One example is Jump-Start, which consists of the mold *Penicillium bilaii*. The mold solubilizes phosphate from the soil and makes it available to the roots of plants, whose exudates feed the mold. This phosphate-solubilizing inoculant improves growth for a variety of crops.

Another microbial product is Actinovate, a biological fungicide used for the suppression of root rot and damping-off fungi, as well as the suppression or control of leaf fungal pathogens. The active ingredient of Actinovate is the actinomycete *Streptomyces lydicus* (**Fig. 16.35B**). *S. lydicus* grows as a mutualist associated with the roots

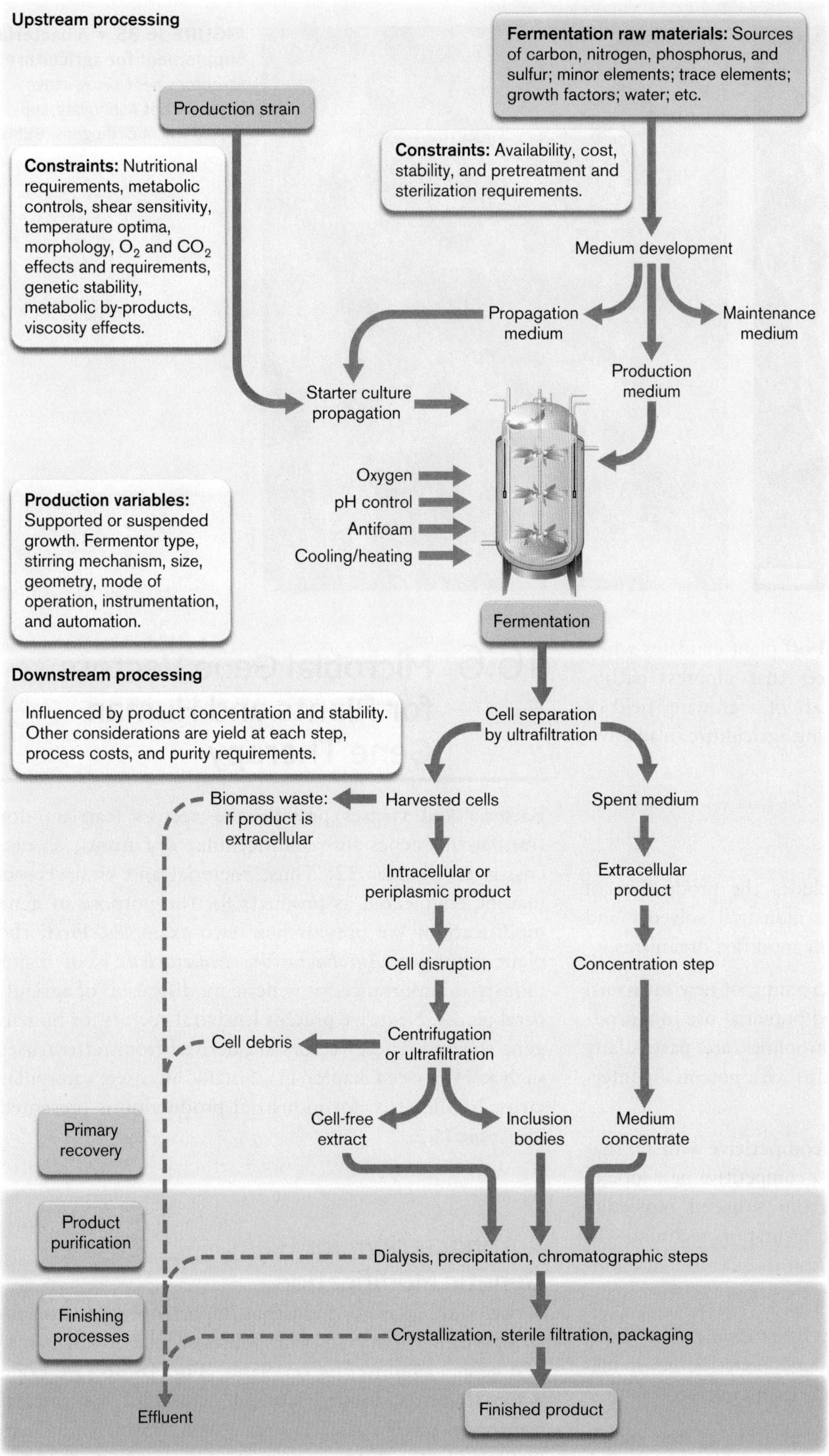

FIGURE 16.34 ■ Details of upstream and downstream processing. **Top:** Upstream processing consists of engineering the microbial strain and large-scale growth to generate the product. **Bottom:** Downstream processing consists of product concentration and purification.

A. B.

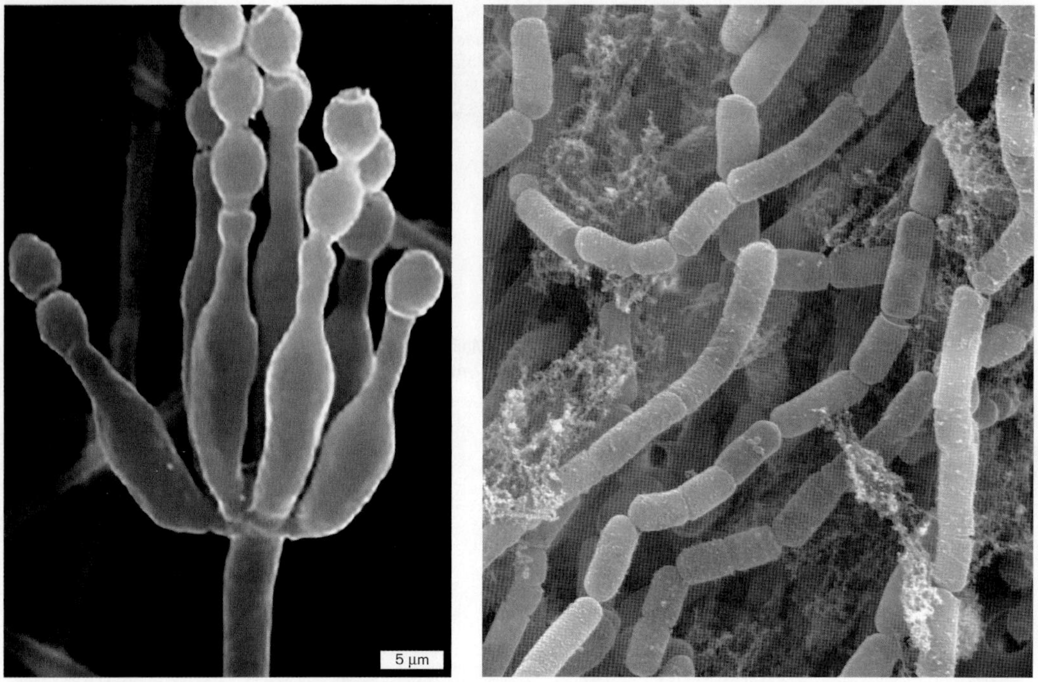

S.S. TZEAN, ET AL. 1995. TOXIC **33**:105. REPRINTED WITH PERMISSION FROM AAAS

DENNIS KUNKEL MICROSCOPY/SCIENCE SOURCE

FIGURE 16.35 ■ A bacterial supplement for agriculture. *Streptomyces lydicus*, active ingredient of Actinovate, suppresses plant pathogens (SEM).

of crop plants. The bacteria feed off plant exudates while secreting antimicrobial substances that suppress pathogens. These examples are the start of a growing field of microbial applications for improving agriculture in an environmentally sustainable manner.

To Summarize

- **Industrial microbiology** includes the production of vaccines and clinical devices, industrial solvents and pharmaceuticals, and genetically modified organisms.
- **Bioprospecting** is the mass screening of new microbial strains for promising traits and potential use in a product. Thermophiles and psychrophiles are particularly important sources of new strains with potentially interesting new properties.
- **Microbial products must be competitive** with alternative technologies. Developing a competitive new molecular product requires identifying a useful molecule, isolating and developing a fermentation technique to produce it, scaling up production in quantity, developing a business plan, and testing for safety.
- **Industrial strains**, commonly *Escherichia coli* or *Bacillus subtilis*, are often used to incorporate the newly discovered genes into an industrially useful microbe.
- **Upstream processing** is the culturing of the industrial microbe to produce large quantities of product. **Downstream processing** involves product recovery and purification.

16.6 Microbial Gene Vectors for Plants and Human Gene Therapy

Bacteria and viruses provide the vectors (carriers) for transferring genes into multicellular organisms, as discussed in Chapter 12. Thus, bacterial and viral vectors may be engineered as products for the purpose of gene modification. We present here two examples. First, the plant pathogen *Agrobacterium tumefaciens* is of major industrial importance for genetic modification of agricultural plants. Next, we present lentiviral vectors for human gene therapy. These vectors are derived from retroviruses such as HIV (see Chapter 11). Finally, an insect caterpillar virus, baculovirus, for industrial production is presented in **eTopic 16.3**.

Agrobacterium for Plant Genetic Modification

A bacterium of major industrial importance is *A. tumefaciens*, a tumor-inducing plant pathogen that conducts natural genetic engineering on plants (**Fig. 16.36**). Long before scientists invented "recombinant DNA," *A. tumefaciens* had evolved a gene transfer system by which it induces infected plant cells to generate food molecules to feed the pathogen. This highly efficient gene transfer system is readily modified to insert genes conferring traits of interest, such as herbicide resistance, into plant genomes.

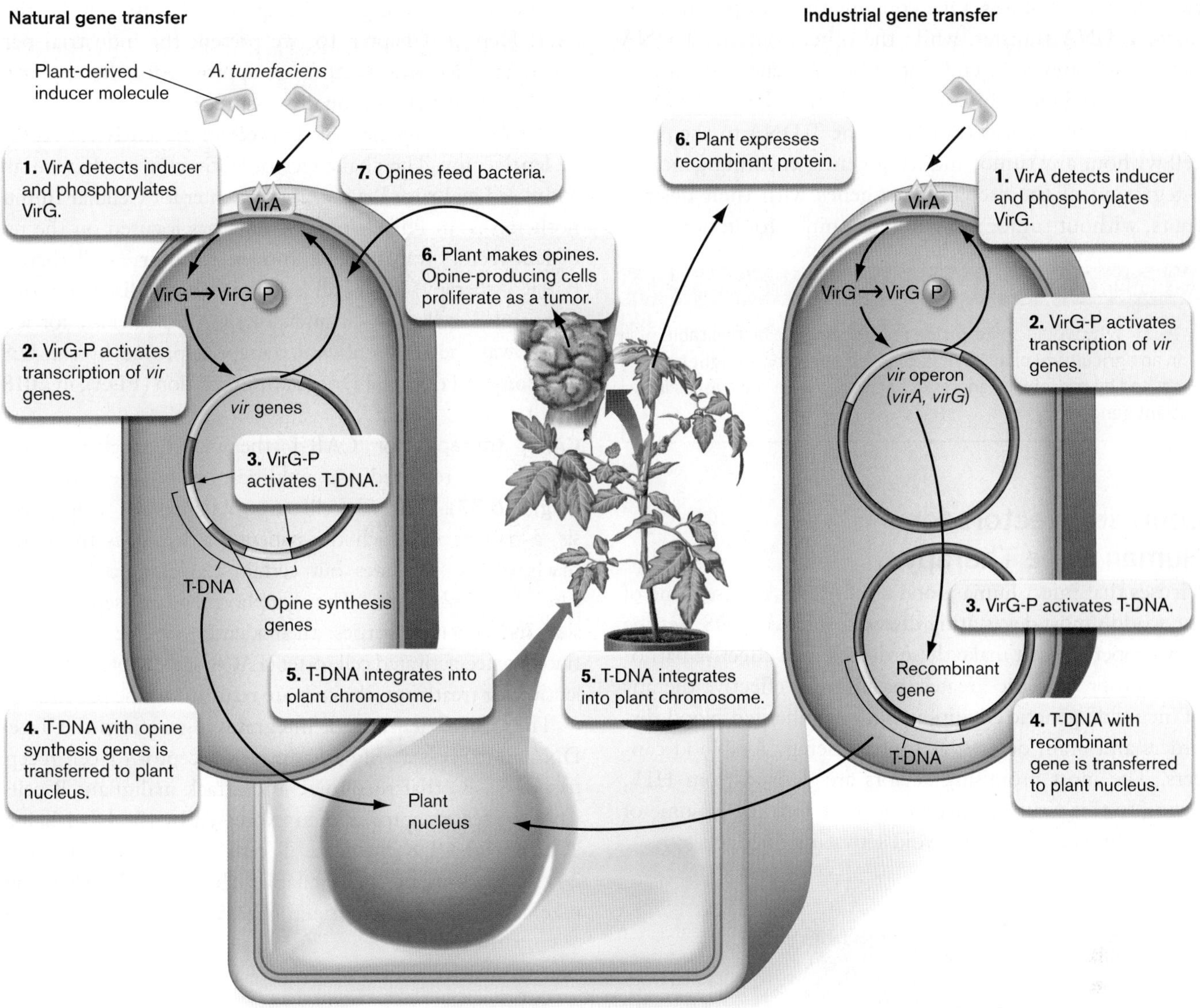

FIGURE 16.36 ■ *Agrobacterium tumefaciens* is a natural gene transfer vector for plants. Left: *A. tumefaciens* transfers T-DNA containing opine synthesis genes into a plant, which then produces opines to feed the bacteria. **Right:** *A. tumefaciens* can be engineered to transfer T-DNA containing a recombinant gene of interest into the plant genome. A recombinant strain of *A. tumefaciens* has the Ti plasmid divided into two separate plasmids: one containing the *vir* operon conducting DNA transfer, the other containing T-DNA with most of its genes substituted by a desired recombinant gene.

Tumorigenic strains of *A. tumefaciens* possess a special plasmid for engineering plant cells, called the **Ti plasmid** (tumor-inducing plasmid). The Ti plasmid is the source of the genetic material that gets transferred into a plant cell through a process mediated by bacterial proteins, similar to conjugation (see Section 9.3). The Ti plasmid includes the *vir* operons, which encode a virulence system, as well as the T-DNA (transferable DNA), a set of genes that will be transferred to the host plant and recombined into its genome. T-DNA encodes tumor induction genes, such as auxin synthesis genes, as well as enzymes for biosynthesis of a carbon and nitrogen source called an opine. Opines are specialized amino acids made by a one-step synthesis from arginine, typically by amination of a central metabolite such as pyruvate or 2-oxoglutarate. A given strain of *A. tumefaciens* typically provides one type of opine synthesis enzyme and has the ability to metabolize the corresponding opine.

The *Agrobacterium* vector transfers its DNA to the plant by means of the *vir* gene products. The *vir* gene products from the Ti plasmid detect the presence of a plant host and stimulate plasmid transfer (see **Fig. 16.36**, left). In the cell envelope, VirA protein detects a chemical signal from a wounded plant, which is capable of being infected. VirA then activates VirG to induce expression of other *vir* genes, encoding proteins that direct DNA transfer into the plant cell. Within the plant cell nucleus, the T-DNA becomes integrated into the plant genome, where it induces opine production. The opine-producing cells proliferate, forming a tumor.

For industrial use (**Fig. 16.36**, right), a recombinant strain of *A. tumefaciens* has the Ti plasmid divided into two separate plasmids. One contains the *vir* operons conducting DNA transfer, while the other contains T-DNA with its left and right ends intact but its genes substituted by the desired recombinant genes. Upon infection, the virulence system induces transfer of the T-DNA to the plant cell without any tumor-inducing genes, enabling genomic integration of the recombinant genes with their desired traits, without tumor induction or opine production.

Thought Question

16.11 Why would an herbicide resistance gene be desirable in an agricultural plant? What long-term problems might be caused by microbial transfer of herbicide resistance genes into plant genomes?

Lentiviral Vectors for Human Gene Therapy

Viruses that infect humans and modify DNA cause some of the world's most devastating diseases, such as AIDS. Yet the very properties that make these viruses such effective pathogens can provide the greatest promise for effective therapy of incurable genetic conditions such as inherited blood disorders and brain defects, as well as therapy for blood cancers. The most promising vectors are derived from HIV, owing to this virus's ability to integrate into the genome of nondividing cells—and to avoid activating oncogenes.

The viral basis of gene therapy was discussed in Section 11.4. Here in Chapter 16, we present the industrial perspective: Once viral vectors are approved for therapy, how are they produced and quality assured?

We focus on one class of viral vector: the **lentiviral vector**, or **lentivector**. The challenges of lentivector production are addressed by Bruce Levine, at the Center for Cellular Immunotherapies, in Philadelphia. Levine has focused on the use of lentivectors for chimeric antigen receptor T-cell therapy (CAR-T therapy) for B-cell leukemia. (For a discussion of B cells and T cells, see Chapter 24.) CAR-T therapy for leukemia was the first lentiviral therapy approved by the U.S. government Food and Drug Administration (FDA), in 2018.

CAR-T therapy. For CAR-T therapy, the patient's own T cells are removed for modification by the lentivector (**Fig. 16.37**). The T cells are obtained by leukapheresis, a process in which the patient's blood goes through a machine that separates out certain classes of cells, in this case T lymphocytes. The lymphocytes are separated by size and other properties; their identity can be tested by a fluorescence-activated cell sorter (FACS; see Chapter 24). After lentivector treatment, the cells are returned to the patient.

The CAR-T lentivector integrates a gene into the T-cell DNA to express a chimeric antigen receptor (T-cell receptor, or TCR) that recognize and attack malignant B cells. The antigen receptor is then expressed by the CD4 helper T cells and CD8 cytotoxic T cells (discussed in Chapter 24). The chimeric TCR enables those T cells to recognize a B cell–specific antigen (CD19) found on malignant

A.

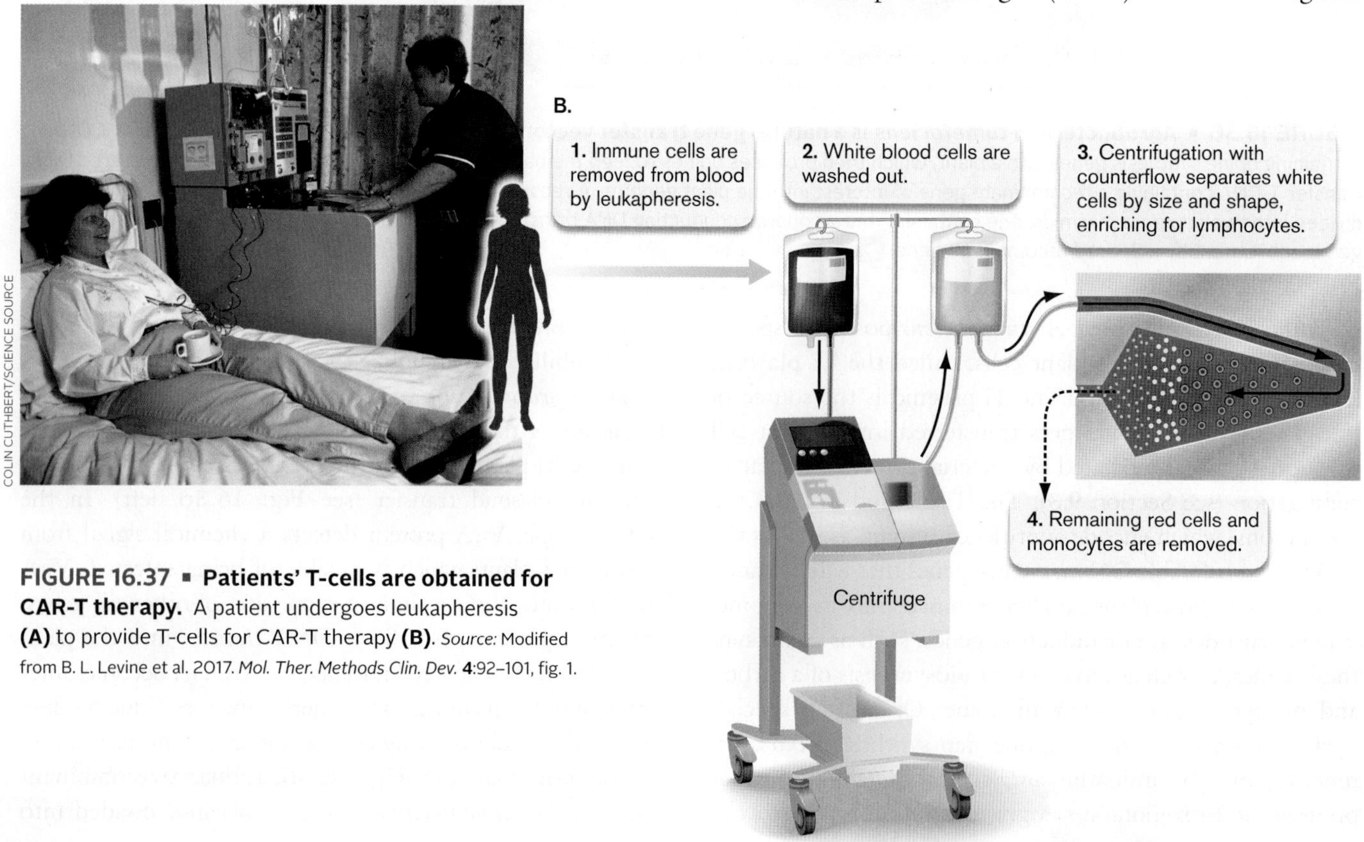

FIGURE 16.37 ■ Patients' T-cells are obtained for CAR-T therapy. A patient undergoes leukapheresis **(A)** to provide T-cells for CAR-T therapy **(B)**. *Source:* Modified from B. L. Levine et al. 2017. *Mol. Ther. Methods Clin. Dev.* **4**:92–101, fig. 1.

B cells and attack them. Normal B cells are destroyed also, but further immunotherapies can compensate.

The chimeric (fused gene) antigen receptor enables T cells to recognize the B cells growing out of control. To transfer the CAR gene into recipient T cells, lentivectors have several advantages. They express the HIV capsid, which enables the vector particle to enter the nuclear pore complex of human cell nuclei. Lentivectors are capable of integration at many different places in the human genome, but for unknown reasons they avoid integrating at sites that will start overexpression of oncogenes. The lentiviral cell envelope can undergo pseudotyping, the replacement of envelope proteins by those of another virus such as vesicular stomatitis virus (VSV-G). Pseudotyping can enable lentivectors to infect and integrate within a wide range of cell types.

Lentivector design and production. CAR-T therapy requires a dependable large supply of the lentivector. How do we generate large quantities of the lentivector in a process that avoids any chance of viral replication in humans? For safe production, the lentiviral genome is divided into four plasmids: a vector plasmid, two packaging plasmids, and an envelope plasmid (**Fig. 16.38**; see also Fig. 11.31). Only the vector plasmid has the long terminal repeats (LTR) that enable its RNA transcript to be packaged in virions (lentivector particles).

When the lentiviral vector infects a host cell from the patient, only the vector plasmid carrying the gene of interest (chimeric antigen receptor, CAR) integrates into the host cell genome. The other plasmids express proteins needed to accomplish integration of the desired gene but are not replicated, so they disappear from the host without any chance of future virus propagation.

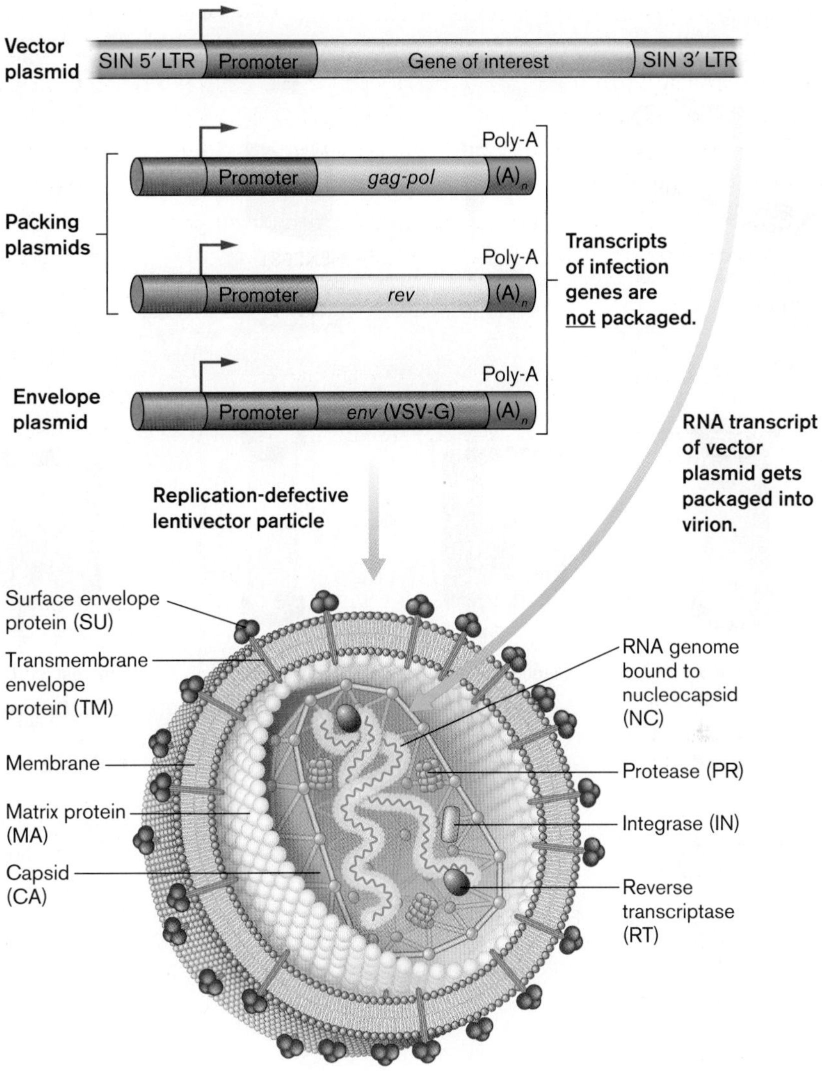

FIGURE 16.38 ▪ The lentiviral vector contains four plasmids. The vector plasmid encodes a gene of interest, flanked by lentiviral long terminal repeat (LTR) sequences that enable integration into the host cell genome. These sequences have been adjusted to be self-inactivating (SIN); that is, no further recombination occurs after genome integration. Packaging plasmids encode the Gag capsid protein and the Pol reverse transcriptase protein, needed to form infective virus particles. The envelope plasmid encodes envelope protein from vesicular stomatitis virus (VSV-G).
Source: Modified from Milone and O'Doherty. 2018. *Leukemia* **32**:1529–1541, fig. 2.

The vector plasmid encodes the CAR gene flanked by lentiviral long terminal repeat (LTR) sequences that enable integration into the host cell genome, as discussed in Section 11.4. The LTR sequences have been modified by deletion of a promoter sequence. This promoter deletion, called SIN (for "self-inactivating"), prevents future transcription of the full-length retrovirus from its position integrated in the host genome. Only the CAR gene is expressed from its own promoter. The packaging plasmids encode the Gag capsid protein and the Pol reverse transcriptase and integrase, which are needed to form infective virus particles. The envelope plasmid encodes envelope protein from vesicular stomatitis virus (VSV-G).

To produce the replication-defective vector particles, the four plasmids are obtained from frozen storage (**Fig. 16.39**) and transfected into a tissue culture line such as HEK293T cells. Unlike the patient's cells, these culture cells were derived from an epithelial cell line but now contain many modifications that enable rapid protein expression to form virions. Sterility and clean room conditions are important because the CAR-T vector particles cannot undergo complete sterilization. Over several days, the cell medium is exchanged from the transfected cells and the progeny vector particles are harvested. Filtration removes cell debris, and benzonase enzyme removes traces of DNA from the cell culture; only the RNA viral particles remain. The final lentivector is stored at –70°C.

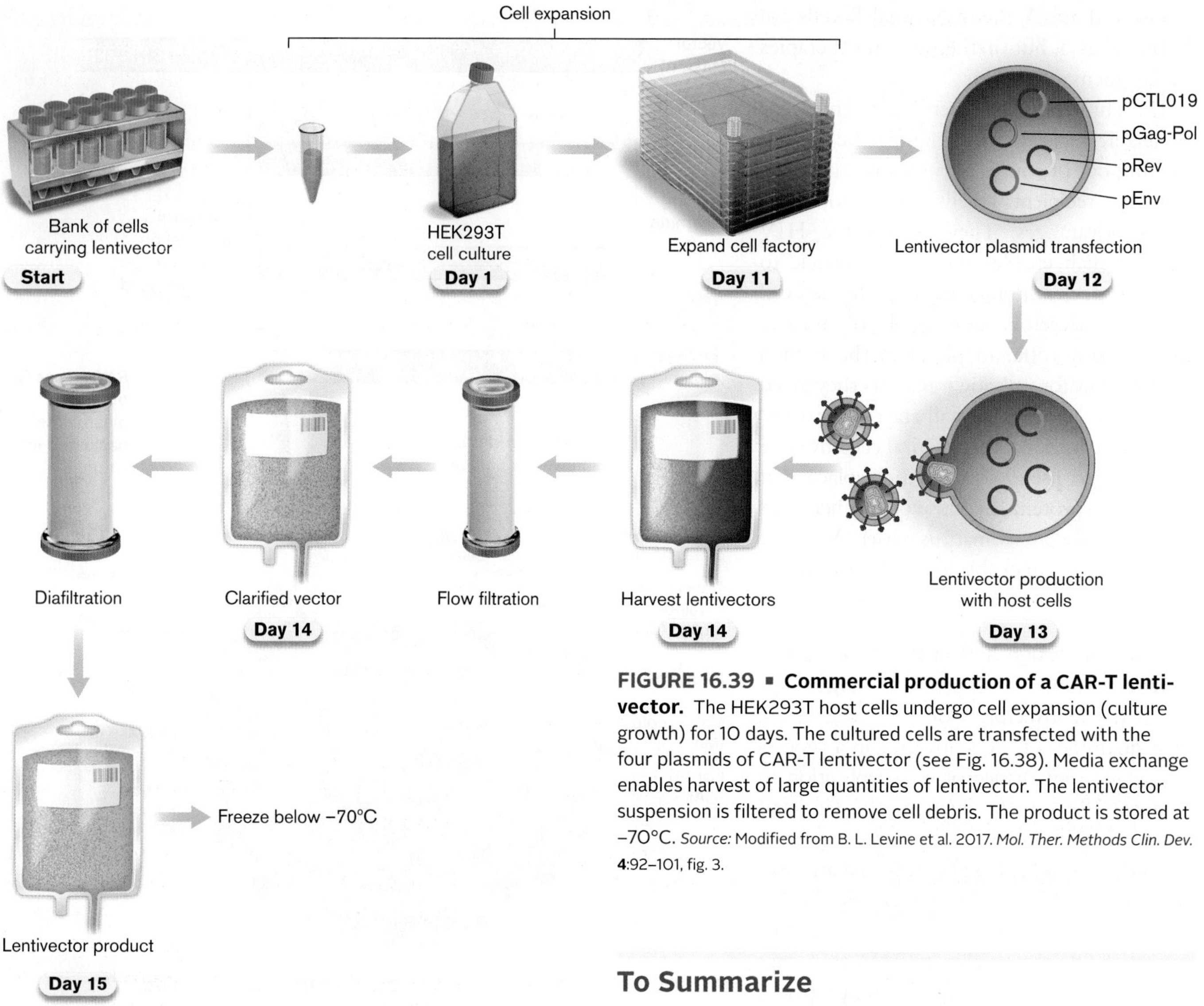

FIGURE 16.39 ■ Commercial production of a CAR-T lentivector. The HEK293T host cells undergo cell expansion (culture growth) for 10 days. The cultured cells are transfected with the four plasmids of CAR-T lentivector (see Fig. 16.38). Media exchange enables harvest of large quantities of lentivector. The lentivector suspension is filtered to remove cell debris. The product is stored at −70°C. *Source:* Modified from B. L. Levine et al. 2017. *Mol. Ther. Methods Clin. Dev.* **4**:92–101, fig. 3.

CAR-T production requires extreme quality control at all stages of production and use of the lentivector. The original cell culture must be kept free of contaminating viruses and cells such as mycoplasmas, a problem for all cell lines. During production of the lentiviral particles, filtration can remove some contaminants, but complete purity is challenging to obtain. At every stage of production, good manufacturing practice (GMP) must be maintained. GMP regulations mandate maintaining clean room conditions, avoiding open processing, and testing at every stage. The final stage of lentivector production must avoid the presence of any packaging cells (host cells) from the transfected culture.

Thought Question

16.12 Why do you think the SIN mutation is important in the lentivector, even though the integrated viral sequence lacks all the genes for virus replication (provided on the original plasmids)?

To Summarize

- ***Agrobacterium tumefaciens* is engineered to genetically modify agricultural plants.**
- **The Ti plasmid transfers *vir* genes to the plant.** The *vir* genes encode tumor induction and opine synthesis.
- **Modification of the *vir* plasmid allows gene transfer into the plant genome.**
- **Lentiviral vectors, or lentivectors, are derived from HIV.** The HIV virus can integrate in the genome of nondividing cells, and it avoids activating oncogenes.
- **CAR-T therapy uses a lentivector to modify the T cells of a cancer patient.** The CAR (chimeric antigen receptor) gene enables T cells to recognize malignant B cells and destroy them.
- **Production of lentivector particles starts with transfection by plasmids** that express genes needed for virus infection and genome integration. After genome integration, the plasmids are left behind and the vector genome is trapped in the host genome. SIN promoter

deletions ensure that the entire viral genome cannot be transcribed from the host.

- **CAR-T virus particles are produced in large quantities from tissue cultures.** Extremely clean rooms, ultrafiltration, and product testing are required at all stages.

Concluding Thoughts

In this chapter we have described how microbes from our environment contribute products for human use, from foods and vitamin supplements to industrial enzymes and highly specific protein therapeutics. These microbial applications incorporate much of the natural metabolism and biochemistry introduced in Chapters 13–15. Human inventiveness and industrial and financial management must meet many challenges to develop new microbial products.

What does the future hold for industrial microbiology? Entirely new microbes—viruses, bacteria, algae—are being engineered or "synthesized" to perform functions in an industrial system, or even within the human body. The field of synthetic biology is just beginning to develop such microbial partners for industry and for sustainable agriculture.

CHAPTER REVIEW

Review Questions

1. What kinds of microbes are consumed as food? Why are most bacteria inedible for humans?
2. What are the advantages of fermentation for a food product?
3. What are the differences between traditional fermented foods and commercial fermented foods?
4. How do acid fermentations contribute to the formation of different kinds of cheeses? What is the role of different kinds of metabolism performed by different microbial species?
5. Compare and contrast acid and alkaline fermentation processes. What different kinds of foods are produced?
6. Compare and contrast the role of ethanolic fermentation in bread making and winemaking.
7. How do relatively minor fermentation reactions contribute to the flavor of food?
8. Explain the differences between food spoilage and food poisoning.
9. What are the most important food-borne pathogens, based on infection rates? Based on mortality rates?
10. What are the major means of preserving food? Compare and contrast their strengths and limitations.
11. What tasks must be accomplished to develop a microbial product for commercial marketing?
12. What are the sources of potential new microbial products? What is the difference between a source strain and an industrial strain?
13. Explain upstream processing and downstream processing.
14. Explain why genes encoding industrially useful products may be transferred into plant or animal systems for production. What are the roles of microbes in these systems?
15. Explain the construction and use of lentiviral vectors. Explain how these vectors can be safe and effective.

Thought Questions

1. In cheese production, how do different kinds of fermenting microbes generate different flavors?
2. If you were a food safety regulator, which pathogen on the list in **Table 16.3** would you consider your top priority? Defend your answer by citing factors such as numerical incidence of infection, severity of disease, and economic losses due to illness.
3. Suppose you undertake industrial production of a recombinant glycoprotein for human therapy. Would you synthesize your product in bacteria, in yeast, or in caterpillars? Explain the advantages and limitations of your choice.

Key Terms

alkaline fermentation (622)
antimicrobial agent (633)
bioprospecting (647)
cheddared (624)
cheese (623)
curd (623)
downstream processing (649)
endophyte (639)
ethanolic fermentation (622)
fermentation industry (618)
fermented food (620)
food poisoning (634)
food spoilage (634)
freeze-drying (641)
heterolactic fermentation (622)
heterologous host (643)
indigenous microbiota (620)
industrial fermentor (649)
industrial microbiology (643)
industrial strain (647)
injera (631)
kimchi (627)
lactic acid fermentation (620)
leaven (630)
lentiviral vector (lentivector) (654)
lyophilization (641)
malolactic fermentation (632)
miso (626)
natto (628)
nori (619)
pathogenicity island (639)
pidan (629)
primary recovery (649)
propionic acid fermentation (622)
putrefaction (634)
rancidity (634)
ripening (623)
single-celled protein (620)
sourdough (630)
starter culture (624)
tempeh (625)
Ti plasmid (653)
upstream processing (649)
whey (623)
yogurt (623)

Recommended Reading

Centers for Disease Control and Prevention. 2008. Outbreak of *Listeria monocytogenes* infections associated with pasteurized milk from a local dairy—Massachusetts, 2007. *Morbidity and Mortality Weekly Report* **57**:1097–1100.

Centers for Disease Control and Prevention. 2009. Multistate outbreak of *Salmonella* infections associated with peanut butter and peanut butter–containing products—United States, 2008–2009. *Morbidity and Mortality Weekly Report* **58**:1–6.

Chang, Shu-Ting, and Philip G. Miles. 2004. *Mushrooms: Cultivation, Nutritional Value, Medicinal Effect, and Environmental Impact.* 2nd ed. CRC Press, New York.

Doyle, Michael P., and Robert L. Buchanan (eds.). 2013. *Food Microbiology: Fundamentals and Frontiers.* 4th ed. ASM Press, Washington, DC.

Giraffa, Giorgio. 2004. Studying the dynamics of microbial populations during food fermentation. *FEMS Microbiological Reviews* **28**:251–260.

Habimana, O., L. L. Nesse, T. Møretrø, K. Berg, E. Heir, et al. 2014. The persistence of *Salmonella* following desiccation under feed processing environmental conditions: A subject of relevance. *Letters in Applied Microbiology* **59**:464–470.

Hui, Y. H., Lisbeth Meunier-Goddick, Åse S. Hansen, Jytte Josephsen, Wai-Kit Nip, et al. (eds.). 2004. *Handbook of Food and Beverage Fermentation Technology.* Marcel Dekker, New York.

Leggett, Mary, Nathaniel K. Newlands, David Greenshields, Lee West, S. Inman, et al. 2014. Maize yield response to a phosphorus-solubilizing microbial inoculant in field trials. *Journal of Agricultural Science* **153**:1464–1478.

Levine, Bruce L., James Miskin, Keith Wonnacott, and Christopher Keir. 2017. Global manufacturing of CAR-T cell therapy. *Molecular Therapy: Methods & Clinical Development* **4**:92–101.

Marilley, L., and M. G. Casey. 2004. Flavours of cheese products: Metabolic pathways, analytical tools and identification of producing strains. *International Journal of Food Microbiology* **90**:139–159.

Mills, David A., Helen Rawsthorne, C. Parker, D. Tamir, and K. Makarova. 2005. Genomic analysis of *Oenococcus oeni* PSU-1 and its relevance to winemaking. *FEMS Microbiological Reviews* **29**:465–475.

Milone, Michael C., and Una O'Doherty. 2018. Clinical use of lentiviral vectors. *Leukemia* **32**:1529–1541.

Schwan, Rosane F., and Alan E. Wheals. 2004. The microbiology of cocoa fermentation and its role in chocolate quality. *Critical Reviews in Food Science and Nutrition* **44**:205–221.

CHAPTER 17

Origins and Evolution

17.1 Origins of Life

17.2 Forming the First Cells

17.3 Evolution: Phylogeny and Gene Transfer

17.4 Natural Selection and Adaptation

17.5 Microbial Species and Taxonomy

17.6 Symbiosis and the Origin of Mitochondria and Chloroplasts

Microbial life appeared as early as 3.8 billion years ago, soon after our planet, Earth, formed out of dust surrounding the young Sun. Since then, microbes have evolved into forms adapted to diverse ways of life, from psychrophiles beneath the Antarctic Ocean to anaerobes in the human colon. All living plants and animals, including ourselves, are descendants of those early microbes. How did microbes originate and evolve? How does their evolution continue to shape Earth's biosphere?

CURRENT RESEARCH highlight

Origin of multicellularity. How did multicellular animals and plants evolve from microbes that were single cells? A startling hypothesis is that ancient bacteria induced single-celled eukaryotes to develop multicellular forms. Roberto Nespolo, at the Austral University of Chile, tested how bacteria might cause the yeast *Saccharomyces cerevisiae* to evolve multicellularity. After 600 generations of coculture with environmental bacteria, the yeast formed aggregative clusters. These clusters may protect the yeast from bacterial wastes and toxins. The aggregation tendency was inherited via gene mutations, leading to a new strain of yeast.

Source: Julian F. Quintero-Galvis et al. 2018. *Ecol. Evol.* **8**:4619–4630.

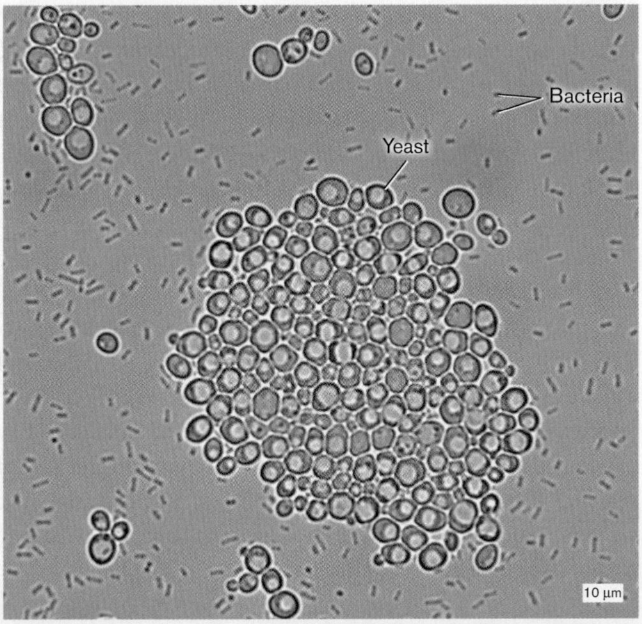

ROBERTO NESPOLO

AN INTERVIEW WITH

ROBERT NESPOLO, MOLECULAR MICROBIOLOGIST, AUSTRAL UNIVERSITY OF CHILE

ROBERTO NESPOLO

Why would bacteria induce evolution of multicellularity?

Bacteria have a profound influence on eukaryotic organisms: they compete for nutrients, they infect them or are mutualistic, and they generate toxic compounds. Thus, bacteria should have been an important selective factor during the transitions to multicellularity. Many unicellular organisms rapidly develop multicellular structures in the presence of bacteria, either for protection or because of mutualistic relationships. To me it seems obvious that an environment full of bacteria should have been crucial for the transition from unicellular to multicellular life.

Comparative evidence indicates that bacteria are involved in organ development in invertebrates or in the settling of larvae in some basal groups of animals. As humans, we are immersed in an environment saturated with bacteria, in our intestine, in our skin, symbiotic, parasitic, or just mutualistic.

Living organisms generate offspring of their own kind—but slightly different. Despite the high accuracy of DNA replication, a few genes mutate, yielding diverse progeny. Some of the mutants reproduce more than others, either by chance or because certain genetic traits work better in the given environment. The more successful variants may show greater fitness for the environment tested. This process of evolution generates all the different kinds of life on Earth. Amazingly, evolution continues today, in every living organism.

Microbes that rapidly generate large populations enable us to watch evolution as it happens. We can observe remarkable changes, such as the effect of bacteria on a single-celled eukaryote evolving a multicellular form (see the Current Research Highlight). The study of evolution also has broad-ranging implications for human medicine, from our use of model organisms to our awareness of drug resistance in pathogens.

Chapter 17 explores evidence for the origin of the earliest cells on Earth, as well as the challenges in interpreting data from so long ago. We show how molecular techniques reveal deep similarities among all life forms, such as the core macromolecular apparatus of DNA, RNA, and proteins. We also watch the mechanisms of ongoing microbial evolution emerge from laboratory experiments (discussed in Section 17.4). We present:

- The origin of life on Earth and the nature of the earliest cells.
- The divergence of microbes from common ancestors, modified by gene transfer and symbiosis.
- The mechanisms of microbial evolution, as it unfolds in nature and in the laboratory.

17.1 Origins of Life

For centuries, observers of the natural world have wondered where life came from. Medieval alchemists in Europe argued that life arose spontaneously from inert matter—a concept called spontaneous generation (discussed in Chapter 1). Others argued against spontaneous generation and devised experiments to show that even microbes have "parents." Today, genetic evidence overwhelmingly confirms that all life on Earth, including microbes, arises from preexisting life. But these experiments do not address the origin of the very first living cells—or how early life gave rise to multicellular plants and animals.

As early as 1802, the naturalist Erasmus Darwin, grandfather of Charles Darwin, wrote:

Organic life beneath the shoreless waves
Was born and nurs'd in ocean's pearly caves;
First forms minute, unseen by spheric glass,
Move on the mud, or pierce the watery mass;
These, as successive generations bloom,
New powers acquire and larger limbs assume.

Thus, nineteenth-century biologists developed the idea that "minute" life forms arose in the ocean—and that all organisms evolved from microbes, perhaps even from cells too small to be seen with the "spheric glass" of a microscope. Even without the tools of genetics, thoughtful observers recognized the commonalities among all living cells, such as the membrane-enclosed compartment of cytoplasm and common metabolic pathways, like sugar metabolism. Today, lines of evidence from geology, biochemistry, and genetics overwhelmingly support the microbial origin of life.

What might the first life forms have looked like?

Early Life

Microbiologists find clues to the nature of early life in geology—rock formations that preserve fossil evidence of the earliest life, from billions of years past. Fossil layers of microbes appear in granite from Pilbara Craton, Australia, where the rock is dated to 3.4 billion years ago (3.4 Gyr ago; **Fig. 17.1A**). Pilbara is a dry land, one of a few places on Earth where ancient rock was uplifted and remains exposed in cliffs and mountains. The Pilbara rock layers preserve the wavy form of **microbial mats**, thick masses of biofilm that probably included cyanobacteria. The cyanobacterial mats formed towering colonies called **stromatolites** (**Fig. 17.1B**).

A modern living stromatolite is a mass of layered limestone (calcium carbonate, $CaCO_3$) accreted by microbial mats. The mat layers can build over centuries, even more than a thousand years, reaching heights exceeding 2 meters. The outermost layers of the mat contain oxygenic phototrophs, such as diatoms and filamentous cyanobacteria, that exude bubbles of oxygen. A few millimeters below the surface, red light supports bacteria photolyzing H_2S to sulfate, which is then reduced by still lower layers of bacteria. Stromatolites today grow mainly in tidal pools whose high salt concentration excludes predators, as in Hamlin Pool, Shark Bay, Australia. But 3 billion years ago, stromatolites covered shallow seas all over Earth.

How do living microbes become fossilized? The stromatolite fossils formed as silicate grains sedimented in the mat and gradually replaced their organic structure. The sedimentary layers and wrinkled surface remain visible after billions of years. Remarkably, similar rock formations appear on Mars, suggesting that microbial life may have evolved there too. Of the planets in our solar system, Mars most closely resembles Earth in geology and distance from the Sun, and its crust might provide a habitat for life similar to Earth's (discussed in Chapter 22).

A. B.

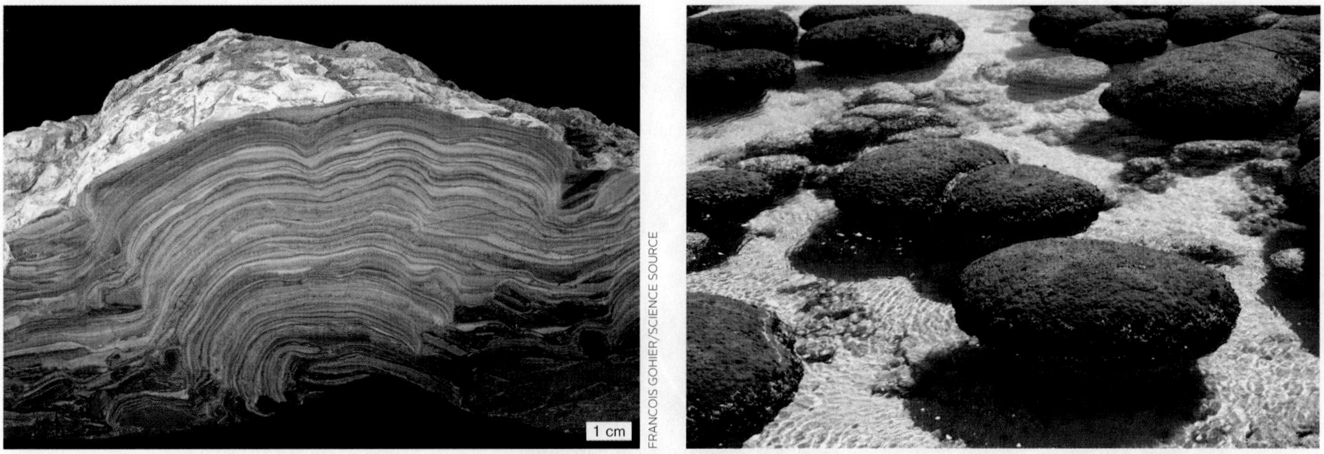

FIGURE 17.1 ■ Stromatolites: ancient life forms in modern seas. **A.** Cross section of a 3.4-billion-year-old fossil stromatolite from the Strelley Pool Chert, Pilbara Craton, Australia. **B.** Cyanobacterial stromatolites, present-day structures that resemble the earliest forms of life on Earth. Shark Bay, Western Australia.

Note: In geological description, a billion years (10^9) is a gigayear, or Gyr. A million years (10^6) is a megayear, or Myr.

Thought Question

17.1 Evolution by natural selection is based on competition, yet the earliest fossil life shows organized structures such as a stromatolite built by cooperating cells. How could this be explained?

Conditions for life. Fossils such as stromatolites reveal well-organized life forms, but little about what must have been their predecessors, the earliest rudimentary cells. How did Earth's very first life forms arise out of inert molecules? This process remains one of the great mysteries of science. But we know certain things that life required:

- **Essential elements.** The origin of life required fundamental elements that compose biological molecules, such as hydrogen, carbon, oxygen, nitrogen, sulfur, and phosphorus.
- **Continual source of energy.** The generation of life requires the continual input of energy, which ultimately is dissipated as heat. The main source of energy for life is solar radiation.
- **Temperature range permitting liquid water.** Above 150°C, life's macromolecules fall apart; below the freezing point of water, metabolic reactions cease. Maintaining the relatively narrow temperature range conducive to life depends on the nature of our Sun, our planet's distance from the Sun, and the heat-trapping capacity of our atmosphere.

Elements of Life

For life to arise and multiply, elements such as carbon and oxygen needed to be available on Earth. The planet Earth coalesced during formation of the solar system 4.5 Gyr ago. Central to the solar system is our Sun, a "yellow" star of medium size and surface temperature (5,770 K). The Sun's surface temperature generates electromagnetic radiation across the spectrum, peaking in the range of visible light. As we learned in Chapter 13, the photon energies of visible light are sufficient to drive photosynthesis but not so energetic that they destroy biomolecules. Thus, the stellar class of our Sun makes organic life possible.

The Sun's surface temperature and luminosity are generated by nuclear fusion reactions in which hydrogen nuclei fuse to form helium nuclei. (Be careful to distinguish nuclear reactions, involving nuclei, from chemical reactions, involving electrons.) Besides hydrogen and helium, 2% of the solar mass consists of heavier elements, such as carbon, nitrogen, and oxygen, as well as traces of iron and other metals—elements that compose Earth, including its living organisms. Where did these heavier elements come from? To answer this question, we must look to other stars in the universe at different stages of their development (**Fig. 17.2**).

Elements of life formed within stars. Throughout the universe, young stars such as our Sun fuse hydrogen to form helium. As stars age, they use up all their hydrogen. With hydrogen gone, the aging star contracts and its temperature rises, enabling helium nuclei to fuse, forming carbon (see **Fig. 17.2**). Carbon drives a cyclic nuclear reaction, the carbon-nitrogen-oxygen (CNO) cycle, to form isotopes of nitrogen and oxygen. Subsequent nuclear reactions generate heavier elements through iron (Fe). In this way, the

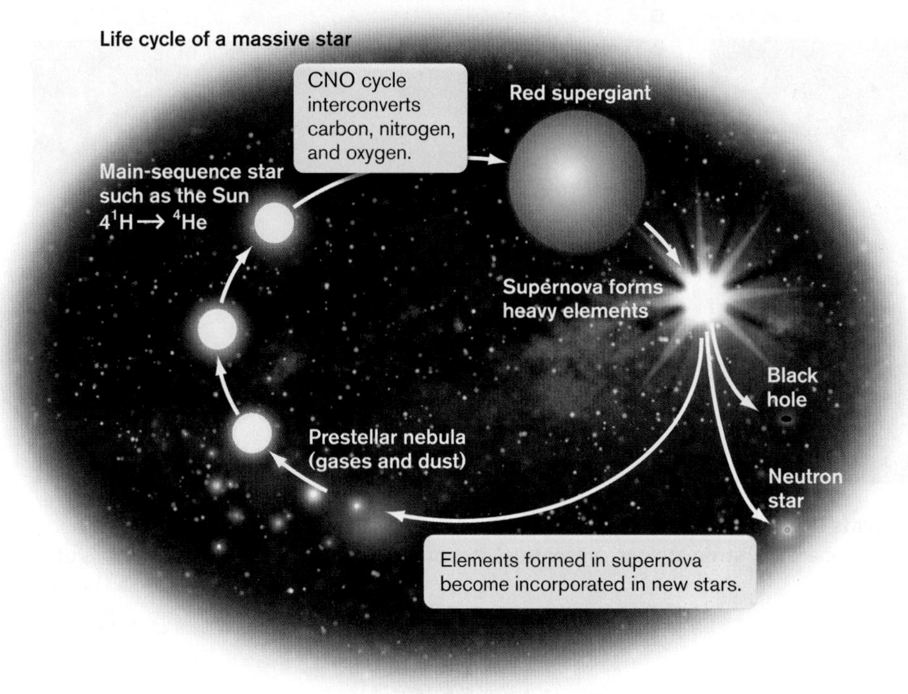

FIGURE 17.2 ■ Stellar origin of atomic nuclei that form living organisms. In young stars, hydrogen nuclei fuse to form helium. In older stars, fusion of helium forms carbon, nitrogen, oxygen, and all the heavier elements up through iron. Massive stars explode as supernovas, spreading all the elements of the periodic table across space. These elements are picked up by newly forming stars, such as our own Sun.

major elements of biomolecules were formed within stars that aged before our solar system was born.

The later nuclear reactions of aging stars generate heavier nuclei, as large as that of iron. The aging star expands, forming a red giant (see **Fig. 17.2**). When a star of sufficient mass expands (becoming a supergiant), it explodes as a supernova. The explosion of a supernova generates in a brief time all the heaviest elements and ejects the entire contents of the star at near light speed. Billions of years before our Sun was born, the first stars aged and died, spreading all the elements of the periodic table across the universe. Some of these elements coalesced with our Sun and formed the planets of our solar system. In effect, all life on Earth is made of stardust, the remains of stars long gone.

Thought Question

17.2 What would have happened to life on Earth if the Sun were of a different stellar class, substantially hotter or colder than it is?

Elemental composition of Earth. When our solar system formed, individual planets coalesced out of matter attracted by the force of gravity. Because of Earth's small size, most of the hydrogen gas escaped Earth's gravity very early. The most abundant dense component of Earth was iron (**Fig. 17.3**). Much of Earth's iron sank to the center to form the core. The core is surrounded by a mantle, composed primarily of iron combined with less dense crystalline minerals, such as silicates of iron and magnesium: $(Fe,Mg)_2SiO_4$. The mantle is coated by Earth's thin outer crust. The crust is composed primarily of silicon dioxide, SiO_2, also known as quartz or chert. Crustal rock contains smaller amounts of numerous minerals, including the carbonates and nitrates that provided the essential elements for life. Overall, the crust shows a redox gradient, reducing in the interior and oxidizing at the surface.

The crust provides a habitat for microbes, including **endoliths**, microbes that grow within the interstices of rock crystals. **Figure 17.3** (insets) shows endolithic algae that absorb light and photosynthesize, supporting diverse communities within the rock. Other endoliths are found at surprising depths, such as within gold mines excavated down to 3 kilometers (km). Some endolithic microbes obtain energy by oxidizing electron donors generated through decay of radioactive metals. The discovery of endolithic organisms deep in Earth's crust was of great interest to NASA scientists seeking life on Mars.

The outer surface of the crust and the atmosphere above it support the remainder of the **biosphere**, the sum total of all life on Earth. The biosphere generates oxidants (electron acceptors), most notably O_2. Oxygen-breathing organisms can live only on the outer surface, where O_2 is produced by photosynthesis.

Earth's atmosphere. From the crust and the mantle of early Earth, volcanic activity released gases such as carbon dioxide and nitrogen, which formed Earth's first atmosphere, while volcanic water vapor formed the ocean. The composition of this first atmosphere, before life evolved, looked much like that of Mars: thin, about 1% as dense as that of Earth today, and consisting primarily of CO_2. But unlike Mars, Earth developed living organisms that filled the atmosphere with gaseous N_2 and O_2 and that continue to produce these gases today. Organisms also produce CO_2, as well as fixing it into biomass. Some CO_2 and N_2 arise from geological sources such as volcanoes, but their contribution is small compared to that of biological cycles (discussed in Chapter 22). The overall composition of Earth's atmosphere is determined by living organisms, primarily microbes.

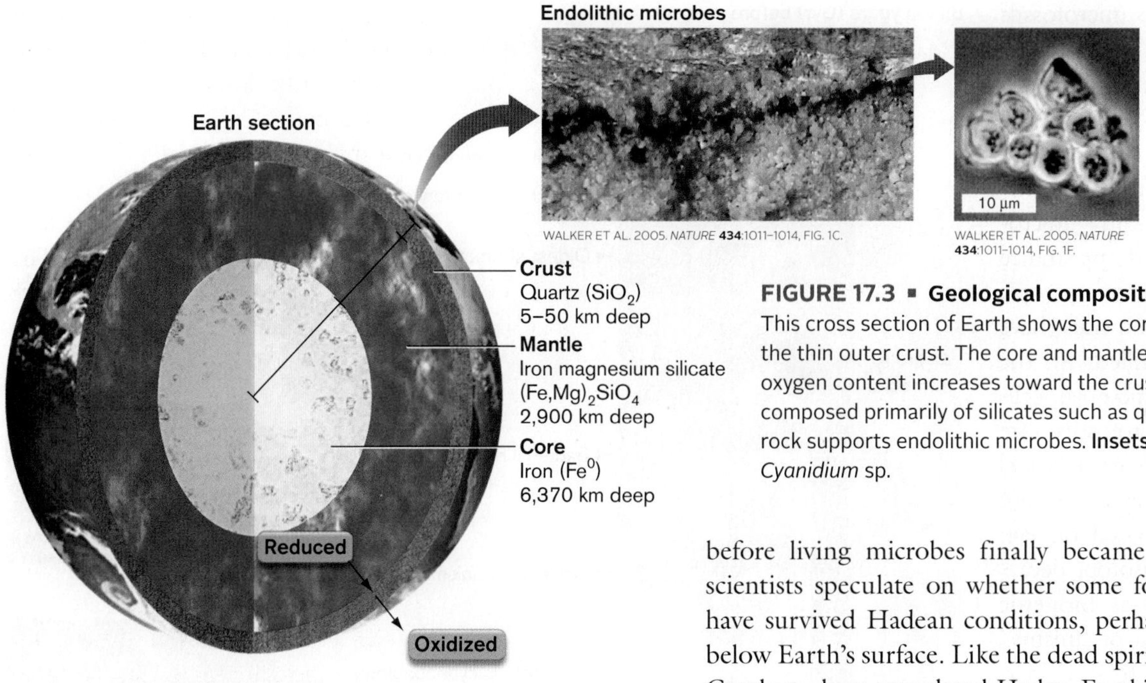

FIGURE 17.3 ▪ Geological composition of Earth. This cross section of Earth shows the core, the mantle, and the thin outer crust. The core and mantle are rich in iron; oxygen content increases toward the crust. The crust is composed primarily of silicates such as quartz (SiO_2). Crustal rock supports endolithic microbes. **Insets:** Endolithic algae, *Cyanidium* sp.

Temperature. Another important aspect of Earth's habitat, determined by the atmospheric density and composition, is temperature. Atmospheric gases absorb light and convert the energy to heat, raising the temperature of the surface and atmosphere. This rise in temperature is known as the **greenhouse effect**. Because carbon dioxide is an especially potent greenhouse gas, the CO_2-rich atmosphere of early Earth could have heated the planet to temperatures approaching those of Venus, eliminating the possibility of life. Instead, microbial consumption of CO_2 and generation of nitrogen and oxygen gases limited Earth's surface temperature to an average of 13°C. The cooling effect may have led to an ice age, possibly reversed by rising methane from methanogens. One way or another, the history of Earth's atmosphere is intimately related to the history of microbial evolution.

Geological Biosignatures for Early Life

Evidence for life in the geological record is called a **biosignature**, or **biological signature**. Biosignatures have been found that are even earlier than the oldest fossils. Their significance is limited, however, because it is hard to rule out nonbiogenic explanations, so researchers seek additional evidence based on independent principles (see **Table 17.1**).

When in Earth's geological record do the first biosignatures appear? Little evidence appears in the very earliest period of Earth's existence, ranging from 4.6 to 4.0 Gyr ago; it is called the **Hadean eon**, named for Hades, the ancient Greek world of the dead. During the Hadean eon, repeated bombardment by meteorites vaporized the oceans, which then cooled and recondensed. Meteor bombardment may have killed off incipient life more than once before living microbes finally became established. Still, scientists speculate on whether some forms of life might have survived Hadean conditions, perhaps growing 3 km below Earth's surface. Like the dead spirits imagined by the Greeks to have populated Hades, Earth's earliest cells may have reached deep enough within the crust that they were protected from the heat and vaporization at the surface.

The Archaean eon. The earliest geological evidence for life that is generally accepted dates to 4.0–2.5 Gyr ago, in the **Archaean eon** (**Fig. 17.4**). In the Archaean, meteor bombardment was less frequent, and Earth's crust had become solid. The Archaean marked the first period with stable oceans containing the key ingredient of life: liquid water. Water is a key medium for life because it remains liquid over a wide range of temperatures and because it dissolves a wide range of inorganic and organic chemicals. Rock strata dating to the Archaean eon reveal the first evidence of living organisms and their metabolic processes.

Note: The term "Archaean" refers to the earliest geological eon when life existed, whereas "archaeal" is the adjective referring to the taxonomic domain Archaea. The domain Archaea (originally "Archaebacteria") was named by Carl Woese, based on his theory that members of this domain most closely resembled the earliest life forms of the Archaean eon. In fact, early life may have encompassed diverse traits later associated with archaea, bacteria, and eukaryotes.

How and when did living cells arise out of inert materials? Without a time machine to take us back 4 Gyr, we must rely on evidence from Earth's geology. Interpreting geology is a challenge because most forms of evidence for early life are indirect and subject to multiple interpretations. The further back in time, the more the rock has changed, and the greater the difficulties are. One way to meet this challenge, however, is to compare the results from different kinds of biosignatures (**Table 17.1**). If two or more kinds of evidence (such as microfossils and isotope ratios) point to life in the same location, the conclusion is strengthened.

Microfossils. The most convincing evidence for early microbial life is the visual appearance of **microfossils**, microscopic fossils in which minerals have precipitated and filled in the form of ancient microbial cells (**Fig. 17.5**). Microfossils are dated by the age of the rock formation in which they are found, which in turn is based on evidence such as radioisotope decay.

To be accepted as **biogenic** (formed from living organisms), a microfossil needs to show regular 3D patterns of cells that resemble those of modern living cells, and that cannot be ascribed to abiotic (nonbiological) causes. The fossil needs to appear in nonmetamorphic rock—that is, rock whose fundamental form has not been reshaped by heat and pressure. The geochemistry must be consistent with rock deposition into organic materials. See, for example, fossil cells of filamentous algae, 1.2 Gyr old, from Arctic Canada (**Fig. 17.5A**). These cells appear comparable in form and size distribution to those of modern algae (**Fig. 17.5B**).

Older fossils, such as those of the Archaean eon, pose challenges. The rock is often metamorphic, and highly deformed. We can identify the macroscopic contours of a stromatolite, but identifying microfossils is more problematic. In the early 1990s, microfossils of cyanobacteria dated to 3.85 Gyr ago by William Schopf, a paleobiologist at UCLA, were accepted and described in many textbooks (**Fig. 17.6**). But these fossils were later reinterpreted by English paleobiologist Martin Brasier (1947–2014) as nonbiogenic artifacts (caused by abiotic processes). The form of the proposed Archaean microfossils is less regular and convincing than the form of later specimens, particularly when observed at different angles not shown in the original publication. Identifying microfossils remains a challenge because of the lack of quantitative criteria.

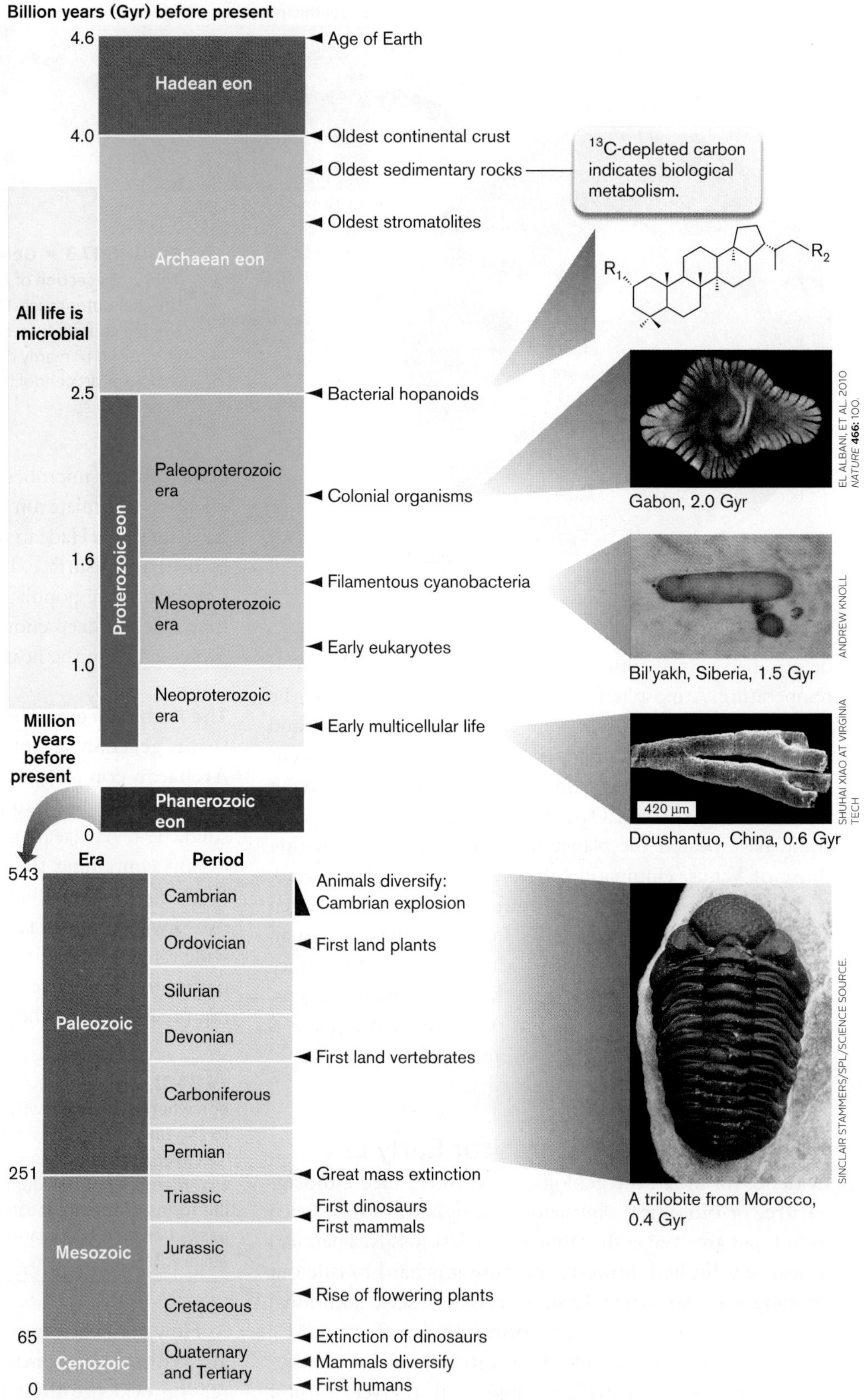

FIGURE 17.4 ▪ **Geological evidence for early life.** The geological record shows biosignatures of microbial life early in Earth's history, 3 Gyr before the first multicellular forms.

TABLE 17.1 Geological Evidence of Early Life

Type of evidence	Advantages	Limitations
Stromatolites		
Layers of phototrophic microbial communities grew and died, and their form was filled in by calcium carbonate or silica.	Fossil stromatolites appear in the oldest rock of the Archaean eon. Their distinctive shapes resemble those of modern living stromatolites.	Some layered formations attributed to stromatolites have been shown to be formed by abiotic processes.
Microfossils		
Early microbial cells decayed, and their form was filled in by calcium carbonate or silica. The size and shape of microfossils resemble those of modern cells.	Microfossils are visible and measurable, offering direct evidence of cell form. Elemental content can be analyzed by Raman spectroscopy, and by NanoSIMS for isotope ratios.	Microscopic rock formations require subjective interpretation. Some formations may result from abiotic processes.
Isotope ratios		
Microbes fix $^{12}CO_2$ more readily than $^{13}CO_2$. Thus, limestone depleted of ^{13}C must have come from living cells. Similarly, sulfate-respiring bacteria cause depletion of ^{34}S compared with ^{32}S.	Isotope ratios are a highly reproducible physical measurement. Isotope ratios generated by key biochemical reactions can calibrate the time lines of phylogenetic trees. NanoSIMS can reveal isotope ratios of microfossils.	We cannot prove absolutely that no abiotic process could generate a given isotope ratio. Isotope ratios tell us nothing about the shape of early cells or how they evolved.
Organic biosignatures		
Certain organic molecules found in sedimentary rock are known to be formed only by certain microbes. These molecules are used as biosignatures.	Biosignatures such as hopanoids are complex molecules specific to bacteria.	A molecule thought to be made only by living organisms may be discovered in abiotic reactions of organic chemistry. In the oldest rocks, organic biosignatures are eliminated by metamorphic processes.
Oxidation state		
The oxidation state of metals such as iron and uranium indicates the level of O_2 available when the rock formed. Banded iron formations suggest intermittent oxidation by microbial phototrophs.	Oxidized metals offer evidence of microbial processes even in highly deformed rock.	It is hard to rule out abiotic causes of oxidation. Even if the oxidation was biogenic, it does not reveal the kind of metabolism.
Raman spectroscopy		
Raman spectroscopy measures energy levels of laser light scattering. The levels are shifted by vibrational energy states characteristic of carbon molecular bonds.	Raman signals reveal the presence of organic molecules that compose living organisms. Measurement is nondestructive.	Raman spectroscopy requires high sensitivity and advanced equipment. It does not reveal details of complex molecules.

The best way to confirm the presence of microfossils is to analyze their chemistry and verify that organic material is present and consistent with biomass. Chemical analysis can be performed by Raman spectroscopy, which measures the effect of carbon bond vibrational energy on scattering of laser light. Recent technical advances enable the use of Raman spectroscopy to detect the presence of organic molecules in a microscopic sample.

Isotope ratios. An important physical biosignature is that of stable isotope ratios. An **isotope ratio** provides quantitative evidence (**Fig. 17.7**). The source is biogenic if the ratio between certain isotopes of a given element is altered by biological activity. Enzymatic reactions, unlike abiotic processes, are so selective for their substrates that their rates may differ for molecules containing different isotopes. For example, the carbon-fixing enzyme Rubisco, found in chloroplasts, preferentially fixes CO_2 containing ^{12}C rather than ^{13}C. The carbon dioxide fixed into microbial cells eventually is converted to calcium carbonate in sedimentary rock. The difference, $\delta^{13}C$, is defined by the fractional difference (in parts per thousand) between the $^{13}C/^{12}C$ ratios in a sample versus a standard inorganic rock:

$$\delta^{13}C = \frac{{}^{13}C/{}^{12}C\ (\text{experimental}) - {}^{13}C/{}^{12}C\ (\text{standard})}{{}^{13}C/{}^{12}C\ (\text{standard})} \times 1{,}000$$

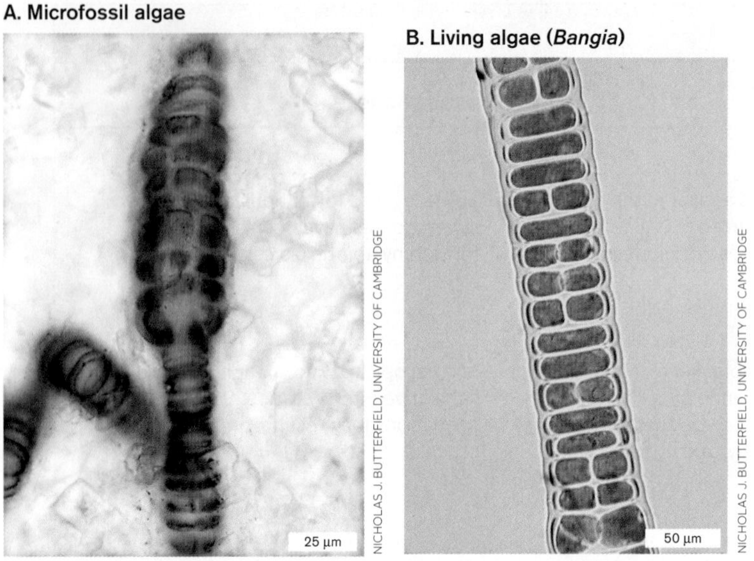

FIGURE 17.5 ▪ Microfossils compared with modern bacteria. **A.** Filamentous algae, 1.2 Gyr old, from Arctic Canada. **B.** Modern red algae, *Bangia* sp.

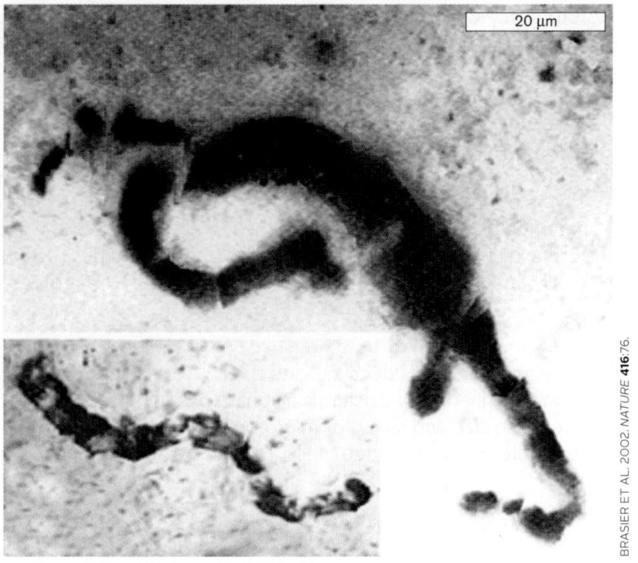

FIGURE 17.6 ▪ Microfossils or artifacts? Structures originally identified as cyanobacterial microfossils from Western Australia, dated to 3.85 Gyr ago. Further testing indicated that the structures are nonbiological artifacts.

The calcium carbonate deposited by CO_2-fixing autotrophs (such as cyanobacteria) shows lower ^{13}C content than does calcium carbonate deposited by abiotic processes. Typical $\delta^{13}C$ values are shown in **Figure 17.7A**. Organisms on land and sea, and the CO_2 derived from their breakdown, show $\delta^{13}C$ values of –10 to –30 parts per thousand. These negative $\delta^{13}C$ values represent a significant ^{13}C depletion through carbon fixation into the original biomass that entered the food web. A comparable $\delta^{13}C$ is observed in fossil fuels, which formed from plant and animal bodies decomposed by bacteria.

Thought Question

17.3 In **Figure 17.7A**, why does the shallow-ocean CO_2 show a positive value of $\delta^{13}C$?

The most ancient mineral samples showing a substantial $\delta^{13}C$ (about –18 parts per thousand) are graphite granules in the Isua rock bed of West Greenland, dated at 3.7 Gyr old. The graphite granules were analyzed by Minik Rosing, a native Greenlander at the Danish Lithosphere Centre (**Fig. 17.7B**). The graphite grains derive from microbial

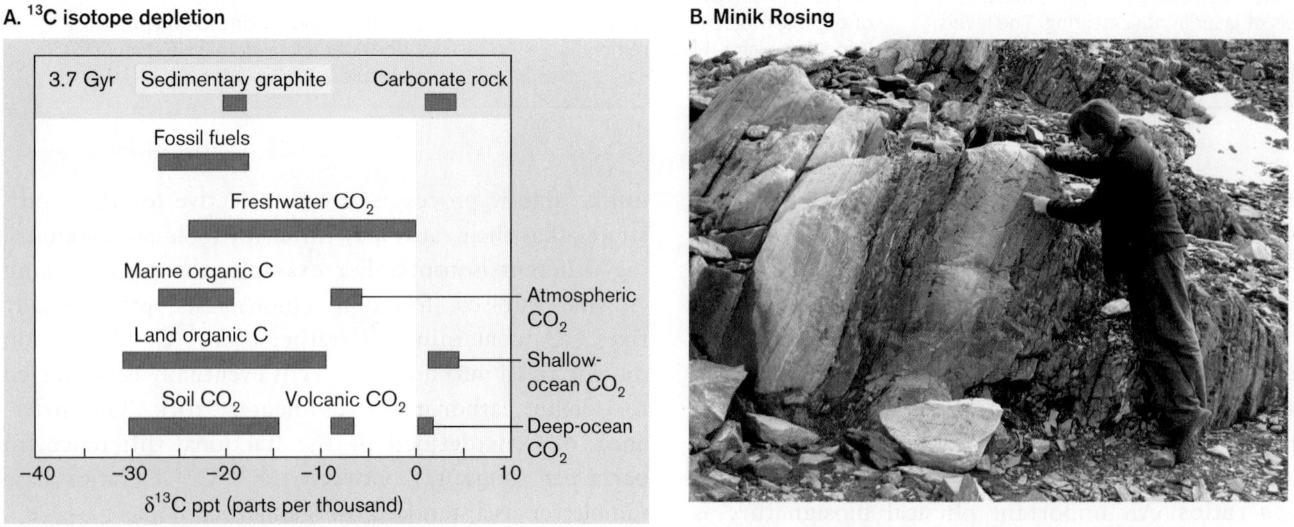

FIGURE 17.7 ▪ Carbon isotope depletion. **A.** ^{13}C isotope depletion (negative $\delta^{13}C$) occurs in biomass as a result of the Calvin cycle. Negative $\delta^{13}C$ is observed at 3.7 Gyr in sedimentary graphite, which may derive from sedimented phototrophs. Little or no isotope depletion is seen in carbonate rock, which has no biological origin. **B.** Minik Rosing (left), in Greenland, shows the Isua rocks whose carbon isotope ratios indicate photosynthesis at 3.8 Gyr ago.

remains buried within sediment that subsequently metamorphosed, driving out the water content but leaving behind the telltale carbon. By contrast, nonbiogenic carbonate rock from the same formation shows a $\delta^{13}C$ near zero.

Isotope ratios, as well as Raman spectroscopy, can confirm the content of ancient microfossils. **Figure 17.8A** shows a sample of a microfossil obtained from chert in a mountain outcrop dated at 1.5 Gyr ago, in Gaoyuzhuang, China. A thin section was taken for microscopy. In the micrograph, the regular formation of ovoid cells is consistent with the size and shape of modern cyanobacteria such as *Chroococcus* (discussed in Chapter 18). Raman spectroscopy shows concentrated organic biomass in the curved shape of each fossil cell (**Fig. 17.8B**). In addition, to measure $\delta^{13}C$, mass spectrometry was applied at the microscopic level—a technique known as NanoSIMS (described in Chapter 2). The NanoSIMS map of the sample (**Fig. 17.8C**) shows ^{13}C depletion located in the envelope of each putative cell. Thus, three types of experiments confirm the microbial identification of the Gaoyuzhuang microfossil. But the exact nature of these cells—their genetics and metabolism—remains a mystery.

How old are the oldest microfossils known? In 2017, geoscientist John Valley reported negative values of ^{13}C depletion in Schopf's microfossils in Western Australian rock dated to about 3.5 Gyr ago. This evidence supports the existence of microbial life that fixed carbon by a Rubisco-like mechanism more than 3 billion years ago. Researchers continue to pursue these questions.

Organic biosignatures. A different kind of biosignature is given by organic molecules specific to a particular life form. Certain organic molecules may last within rock for hundreds of millions of years. A particularly durable class of molecules consists of membrane lipids. Recall from Chapter 3 that some bacterial cell membranes contain steroid-like molecules called hopanoids. A hopanoid consists of four or five fused rings of hydrocarbon with side groups that vary depending on the kind of bacteria. The hopanoid derivative 2-methylhopane is found in sedimentary rock of the Hamersley Basin of Western Australia, dated to 2.5 Gyr ago. This biosignature offers evidence that some kind of bacteria existed by the end of the Archaean eon.

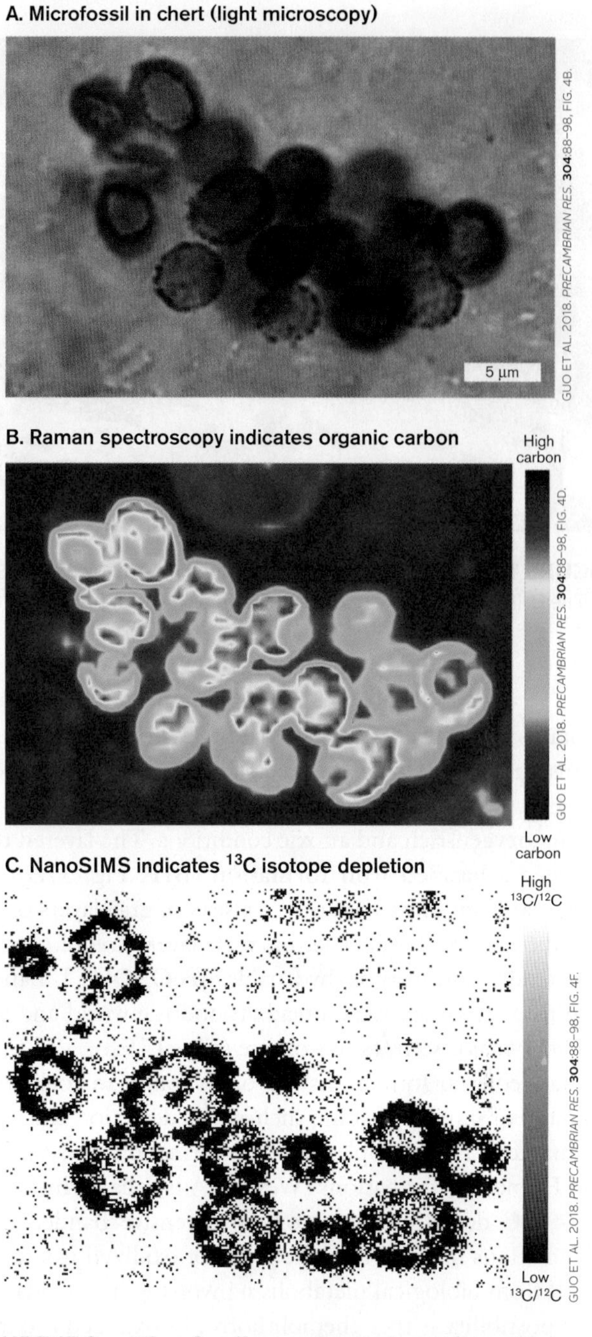

FIGURE 17.8 ■ Microfossil cyanobacteria confirmed by carbon isotope depletion. **A.** Microfossil cyanobacteria embedded in chert dated to 1.5 Gyr ago in Gaoyuzhuang, China, show size distribution consistent with living cells. **B.** Raman spectroscopy indicates organic carbon content of cells. **C.** NanoSIMS indicates ^{13}C isotope depletion levels consistent with cyanobacterial photosynthesis.

Banded Iron Formations Reveal Oxidation by O_2

An extraordinary event in the planet's history was the evolution of the first oxygenic phototrophs: cyanobacteria that split water to form O_2. The entry of O_2 into Earth's biosphere is often portrayed as a sudden event that would have been disastrous to microbial populations lacking defenses against its toxicity. In fact, geological evidence shows that oxygen arose gradually in the oceans, starting about 2 Gyr ago, and may have arisen and disappeared numerous times before reaching a high, steady-state level in our atmosphere. The mechanism of the oxygen fluctuation is unknown, but it caused cycles of aerobic and anaerobic microbial metabolism.

Evidence for oxygen in the biosphere comes from the oxidation state of minerals, particularly those containing iron. The bulk of crustal iron is in the reduced form (Fe^{2+}), which is soluble in water and reached high concentrations in the anoxic early oceans. Sedimentary rock, however,

FIGURE 17.9 ■ Banded iron formations. A. Jim Crowley, U.S. Geological Survey, studies a banded iron formation in Dales Gorge, Australia. B. The BHP Billiton Iron Ore mine at Newman, Western Australia.

contains many fine layers of oxidized iron (Fe^{3+}), which is insoluble and forms a precipitate, such as iron oxide (Fe_2O_3). The layers of iron oxide suggest periods of alternating oxygen-rich and anoxic conditions. The layered rock is called a **banded iron formation** (**BIF**; **Fig. 17.9A**). A common form of banded iron consists of gray layers of silicon dioxide (SiO_2) alternating with layers colored red by iron oxides and iron oxyhydroxides [$FeO_x(OH)_y$]. Banded iron formations are widespread around the world and provide our major sources of iron ore (**Fig. 17.9B**).

Banded iron formations are often found in rock strata containing signs of past life such as ^{13}C depletion and other biomarkers. For example, the Isua formations (Greenland) and Hamersley formations (Western Australia), which both show ^{13}C depletion, also contain extensive banded iron. Calculations indicate that the layers of oxidized iron could result from biological metabolism involving iron oxidation. One possibility is that chemolithotrophs oxidized the iron, using molecular oxygen produced by cyanobacteria. The Archaean and early Proterozoic eons experienced fluctuating levels of molecular oxygen in the atmosphere. These fluctuations could have led to oscillating levels of iron oxide, thus producing bands in the sediment, as microbes used up all the oxygen.

By 2.3 Gyr ago, the prevalence of oxidized iron and other minerals indicates the steady rise of oxygen from photosynthesis in Earth's atmosphere. Most of the dissolved Fe^{2+} from the ocean floor was oxidized, leaving the oceans in the iron-poor state that persists today. As the most efficient electron acceptor, molecular oxygen enabled the evolution of aerobic respiratory bacteria. Aerobic bacteria gave rise to mitochondria, which enabled the evolution of eukaryotes and, ultimately, multicellular organisms (**Fig. 17.10**). But even today, cells consist mainly of reduced molecules, highly reactive with oxygen—a relic of the time when our ancestral cells evolved in the absence of oxygen.

To Summarize

- **Elements of life** were formed through nuclear reactions within stars that exploded into supernovas before the birth of our own Sun.
- **Reduced molecules** compose Earth's interior. Oxidized minerals are found only near the surface. Early Earth had no molecular oxygen (O_2).
- **Archaean rocks show evidence for life** based on fossil stromatolites, isotope ratios, and chemical biosignatures. Fossil stromatolites appear in chert formations that arose 3.4 Gyr ago. Isotope ratios for carbon indicate photosynthesis at 3.7 Gyr ago. Bacterial hopanoids appear at 2.5 Gyr ago.
- **Microfossils of filamentous and colonial prokaryotes** date to 3.4 Gyr ago. At 1.2 Gyr ago, larger fossil cells resemble those of modern eukaryotes.
- **Microfossils are confirmed by Raman spectroscopy revealing organic carbon, and by NanoSIMS showing ^{13}C depletion.**
- **Banded iron formations** reflect the cyclic increase and decrease of oxygen produced by cyanobacteria and consumed through reaction with reduced iron. After most of the ocean's iron was oxidized, oxygen increased gradually in the atmosphere.

Origin and evolution of life on Earth

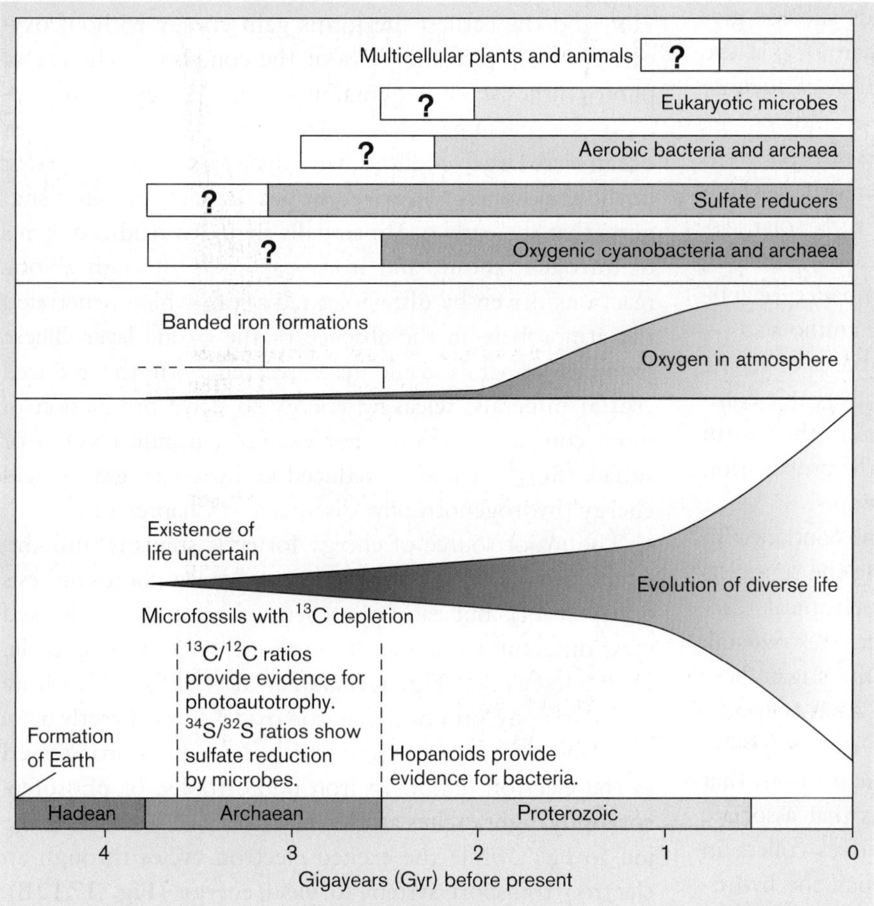

FIGURE 17.10 ▪ Proposed time line for the origin and evolution of life. The planet Earth formed during the Hadean eon (about 4.5 Gyr ago). The environment was largely reducing until cyanobacteria pumped O_2 into the atmosphere. When the O_2 level reached sufficient levels (about 0.6 Gyr ago), multicellular animals and plants evolved. Question marks designate periods when evidence for given life forms is uncertain.

17.2 Forming the First Cells

Various models have been proposed to explain how the first living cells originated from nonliving materials, and how they replicated and evolved. Models for early life attempt to address the following questions: In what kind of environment did the first cells form? What kind of metabolism did the first cells use to generate energy? What was their hereditary material?

The Prebiotic Soup

The **prebiotic soup** model proposes that **abiotic** (nonliving) reactions in the Archaean oceans could have assembled complex organic molecules of life. These reactions would have involved simple reduced chemicals such as ammonia and methane, which led to more complex macromolecules that acquired the apparatus needed for self-replication and membrane compartmentalization.

What evidence supports the prebiotic soup? In the mid-twentieth century, biochemists Aleksandr Oparin at Moscow University, and Stanley Miller and Harold Urey at the University of Chicago, showed that organic building blocks of life such as amino acids could arise abiotically out of a mixture of water and reduced chemicals, including CH_4, NH_3, and H_2 (**Fig. 17.11**). The mixture was subjected to an electrical discharge, similar to the lightning discharges that arise from volcanic eruptions, which would have been common

A. The prebiotic soup model

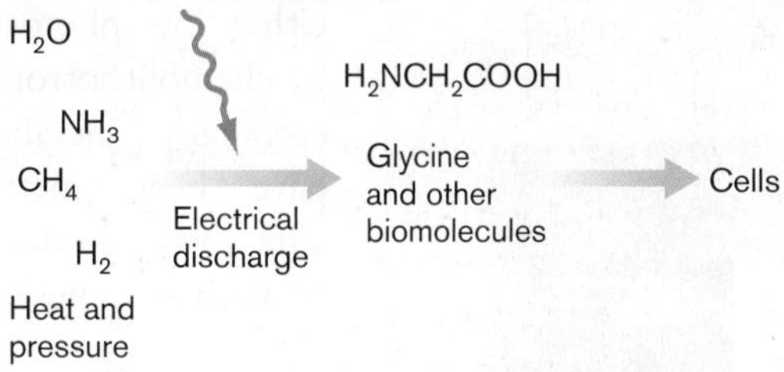

B.

FIGURE 17.11 ▪ The prebiotic soup model for the origin of life. A. In the prebiotic soup, inorganic molecules could have reacted to form complex macromolecules that eventually acquired the apparatus for self-replication and membrane compartmentalization. **B.** Lightning accompanies eruption of the Sakurajima Volcano in Japan, in 2013. The first biomolecules may have formed as a result of lightning triggered by volcanic eruption.

in the late Hadean or early Archaean eon. The chemical reaction produced fundamental amino acids such as glycine and alanine. Similar experiments by Juan Oró, at the University of Houston, showed the formation of adenine by condensation of ammonia and hydrogen cyanide.

The same amino acids and nucleobases arising from early-Earth simulation are also found in meteorites, which are believed to retain the chemistry of the early solar system "frozen" in time. But unlike the chemical experiments, the meteorites show a remarkable predominance of left-handed mirror forms (L enantiomers) of the amino acids—the same L form utilized in the protein synthesis of all life on Earth. Some researchers hypothesize that organic compounds formed in outer space were restricted to the L form by an unknown process and then "seeded" the propagation of L-form compounds in Earth's prebiotic soup.

How were the first biochemical reactions contained in a cell, the fundamental unit of life? Forming the first cells must have required enclosing the first biochemical reactants within a membrane-like compartment. Jack Szostak and colleagues at Harvard Medical School investigate how such compartments can arise spontaneously from fatty acid glycerol esters. The fatty acid derivatives are "amphipathic"; that is, they possess both hydrophobic portions that associate together and hydrophilic portions that associate with water. In water, the fatty acid derivatives collect in "micelles"—small, round aggregates in which the hydrophobic portions associate in the interior and the hydrophilic portions associate with water. Under certain conditions, micelles can aggregate to form hollow vesicles of membrane. Szostak showed how the vesicles can take up molecules such as RNA, suggesting a primitive cell-like form. These spontaneous processes of membrane formation offer models for how the first living cells may have arisen.

Early Oxidation-Reduction Reactions

How did the earliest life forms gain energy without oxygen gas to respire and without the complex machinery of photosynthesis? The original model conditions for the prebiotic soup assumed that molecules in the early Archaean ocean were largely reduced, with little or no oxygen present in the atmosphere. More recent geochemical evidence suggests that the early ocean actually included oxidized forms of nitrogen, sulfur, and iron that arose through abiotic reactions driven by ultraviolet radiation, which penetrated the atmosphere in the absence of the ozone layer. These oxidized minerals could have reacted with the reduced crustal minerals, releasing energy to drive production of more complex reactions. For example, nitrate (NO_3^-) or sulfate (SO_4^{2-}) could be reduced by hydrogen gas to yield energy (hydrogenotrophy; discussed in Chapter 14).

The major source of energy for ecosystems is sunlight, which drives oxidation-reduction cycles via photosynthesis (Chapter 14). But early photosynthesis could have looked very different from the Rubisco-mediated Calvin cycle. Dianne Newman (**Fig. 17.12A**), at the California Institute of Technology, proposes that iron oxides arose directly from anaerobic photosynthesis, in which the reduced iron served as the electron donor. In iron phototrophy, or **photoferrotrophy**, light excites an electron from Fe^{2+}, oxidizing the ion to Fe^{3+}, while the excited electron cycles through an electron transport system to yield energy (**Fig. 17.12B**). Newman discovered iron phototrophy in modern purple bacteria such as *Rhodopseudomonas palustris*, growing in the anoxic bottom layer of iron-rich lakes.

On the ancient Earth, photosynthetic oxidation of Fe^{2+} to Fe^{3+} could have occurred in cycles until the marine iron was all oxidized. The oxidized iron Fe^{3+} would then serve as an electron acceptor for benthic organisms. Alternating redox reactions generated the sedimentary layers of iron oxides and iron oxyhydroxides in banded iron formations.

Early ^{13}C depletion signals offer surprisingly early evidence for carbon fixation cycles based on Rubisco, powered either by photosynthesis or by chemolithotrophy. Another early redox metabolism would have been methanogenesis. The most fundamental form of methanogenesis involves the

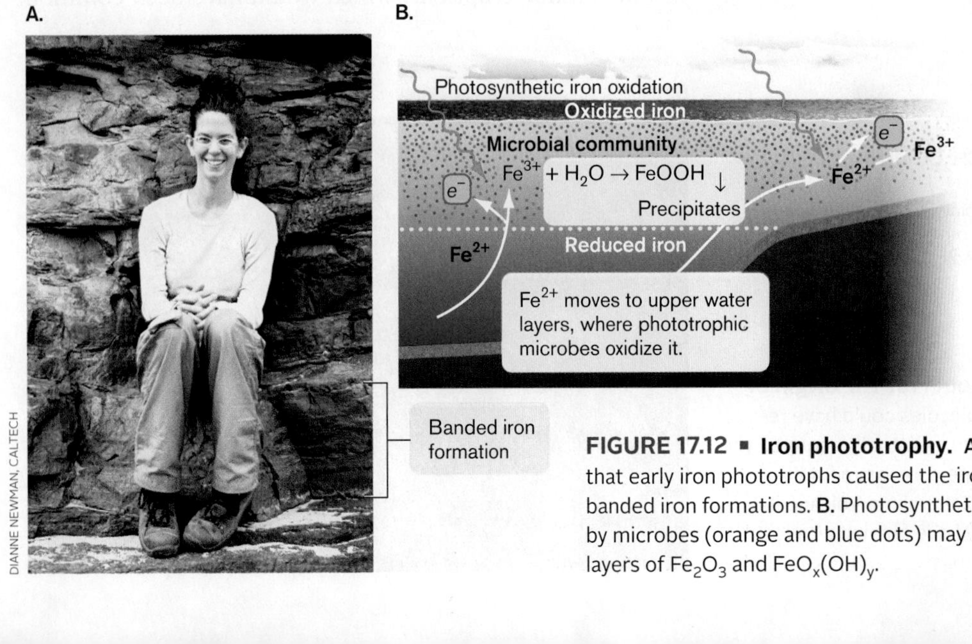

FIGURE 17.12 ■ Iron phototrophy. A. Dianne Newman proposes that early iron phototrophs caused the iron oxide deposition generating banded iron formations. **B.** Photosynthetic oxidation of Fe^{2+} to Fe^{3+} by microbes (orange and blue dots) may have generated sedimentary layers of Fe_2O_3 and $FeO_x(OH)_y$.

reaction of H_2 and CO_2 producing CH_4 and H_2O. Methanogens show highly divergent genomes—a finding that suggests early evolution of their common ancestor. Their biochemistry and evolution (discussed in Chapter 19) are consistent with proposed models of ancient life.

The RNA World

The prebiotic soup model does not account for the evolution of macromolecules that encode complex information, such as long chains of nucleic acids and proteins. A candidate for life's first "informational molecule" is RNA. The **RNA world** is a model of early life in which RNA performed all the informational and catalytic roles of today's DNA and proteins. The concept of an RNA world draws upon genome sequences, which reveal thousands of catalytic and structural RNAs. The concept has surprising relevance for medicine, as we investigate the function of RNA viruses such as influenza virus and human immunodeficiency virus (HIV; discussed in Chapters 6 and 11).

RNA is a relatively simple biomolecule, with only four different "letters," compared to the 20 standard amino acids of proteins. Its purine base adenine arises spontaneously from ammonia and carbon dioxide under conditions believed to resemble those of the Archaean eon. Its ribose sugar is a fundamental building block of living cells, with key roles in numerous biochemical pathways, such as the Calvin cycle. For several reasons, RNA is a better candidate than DNA for the earliest information molecule. Compared to DNA, RNA requires less energy to form and degrade. RNA's pyrimidine base uracil is formed early by biochemical pathways; only later is it converted to the thymine used by DNA.

Most important, RNA molecules have been shown to possess catalytic properties analogous to those of proteins. Catalytic RNA molecules are called **ribozymes**. The first ribozyme, discovered by Nobel laureate Thomas Cech in the protist *Tetrahymena*, can splice introns in mRNA. Other ribozymes actually catalyze synthesis of complementary strands of RNA, suggesting a model for early replication of RNA chromosomes. The most elaborate example of catalytic RNA is found in the ribosome. X-ray crystallography of the ribosome reveals that the key steps of protein synthesis, such as peptide bond formation, are actually catalyzed by the RNA components, not proteins (discussed in Chapter 8). The ribosomal proteins possess relatively little catalytic function; their main role seems to be protection and structural support of the RNA.

Could RNA molecules have composed the earliest cells? In 2009, Tracey Lincoln and Gerald Joyce at the Scripps Research Institute devised a system in which two RNA ribozymes catalyze each other's synthesis. The ribozyme model system suggests how, in the earliest cells, RNA might have fulfilled the key functions that are today filled

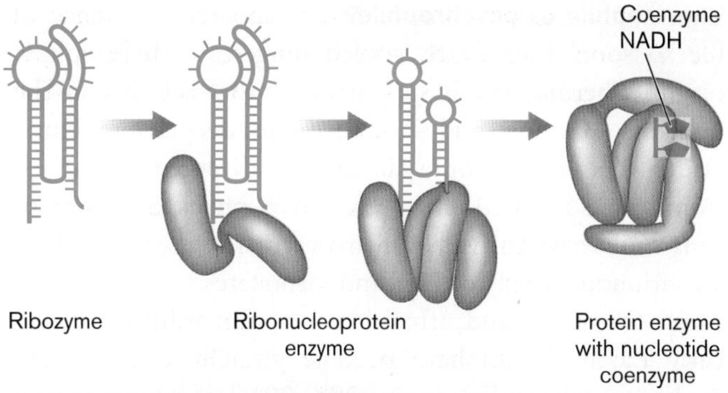

FIGURE 17.13 ▪ From the RNA world to proteins. The earliest cells may have been composed of RNA enzymes (ribozymes). As the RNA cells evolved, ribozymes acquired protein subunits that eventually assumed most of their catalytic functions. Remnants of the original RNA may persist as nucleotide cofactors such as NADH.

by DNA and proteins, including information storage, replication, and catalysis (**Fig. 17.13**). This model is known as the "RNA world."

The prominent function of RNA in the ribosome, one of life's most ancient and conserved molecular machines, suggests a model for the transition from an RNA world to the modern cell. In the ribosome, the actual steps of catalysis, such as forming the peptide bond, are conducted by RNA subunits acting as ribozymes. The ribozymes are stabilized by protein subunits. Thomas Cech, Sidney Altman, and colleagues propose that the earliest RNA components of cells evolved by acquiring proteins to enhance stability. The proteins helped prevent the tendency of RNA to hydrolyze (come apart in reaction with water). As cells evolved, their peptide components increased through natural selection, and the RNA subunits may have shrunk by reductive evolution (the evolutionary loss of unneeded parts). A few complexes, such as the ribosome, still maintain their ribozymes; for others, perhaps all that remains are one or two nucleotides. Dinucleotide cofactors such as NADH persist in enzymes today, perhaps representing vestigial remnants of the RNA world. For more on RNA evolution, and surprising applications for medicine, see **eTopic 17.1**.

Thought Question

17.4 Outline the strengths and limitations of the prebiotic soup model and the RNA world model of the origin of living cells. Which aspects of living cells does each model explain?

Unresolved Questions about Early Life

Overall, geology and biochemistry provide compelling evidence that organisms resembling today's cyanobacteria lived on Earth at least 2.5 Gyr ago, possibly 3.7 Gyr ago, and that bacteria or archaea with anaerobic metabolism evolved as early or earlier. Many intriguing questions remain.

Thermophile or psychrophile? The apparent existence of life so soon after Earth cooled suggests a thermophilic origin. Thermophily is supported by the fact that in the domains Bacteria and Archaea, the deepest-branching clades (that is, kinds of organisms that diverged the earliest from others in the domain) are thermophiles. Such organisms could have thrived at hydrothermal vents, which offer a continuous supply of H_2S and carbonates.

On the other hand, after meteoric bombardment abated, early Earth should have become glacially cold. In the Archaean, solar radiation was 20%–30% less intense than it is today, and the thin CO_2 atmosphere was insufficient to increase the temperature by a greenhouse effect. A colder habitat would support psychrophiles. Psychrophiles might have had an advantage in an RNA world, given the thermal instability of RNA compared to DNA and proteins.

A methane atmosphere? If the Archaean Earth was much cooler than today, what prevented Earth from freezing like Mars? Some researchers argue that the production of methane, an extremely potent greenhouse gas, could have greatly increased Earth's temperature during the Archaean eon. The methane produced would have been oxidized by methanotrophs (methane oxidizers), thus preventing the planet from overheating like Venus. The subsequent decline of methane and the rise of CO_2 then would have brought about relative thermal stability.

The debate over the temperature and climate of early Earth has interesting implications for Earth today, when we again face the prospect of massive global climate change. Human agriculture favors explosive growth of methanogens, which threaten to accelerate global warming faster than the biosphere can moderate it. Understanding the climate of early Earth may help us better understand and manage our own climate, discussed in Chapter 22.

Rapid evolution of first cells? Emerging evidence from fossils and geochemistry has inexorably pushed back the earliest known dates for several kinds of metabolism closer to 3.7 Gyr ago (see **Fig. 17.10**). This adjusted time frame implies that as soon as Earth cooled to a temperature suitable for life, all the fundamental components of cells evolved almost immediately. How could microbial life, with all its diverse kinds of metabolism, have arisen so quickly?

One possible explanation is that life evolved on early Earth faster than it does today. Today, RNA viruses such as influenza and HIV mutate and evolve much faster than modern cells (discussed in Chapter 11). If early cells with RNA genomes mutated as fast as viruses, they might have evolved and diverged faster than cellular organisms that we know today.

Some microbiologists propose an alternate explanation, that life forms originated elsewhere and "seeded" life on Earth. The concept that terrestrial life came from outside Earth is called **panspermia**. Theories of panspermia remain highly speculative. One hypothesis is that microbial cells originated on Mars and were then carried to Earth on meteorites. As the solar system formed, Mars would have cooled sooner than Earth, and because it is also smaller than Earth, Mars's weaker gravity would have generated less bombardment by meteorites. Martian rocks ejected into space by meteor impact have reached Earth, and calculations based on simulated space habitats show that microbes could survive such a journey. A Martian origin, however, gains us only about half a billion years; it does not really explain the origin of life's complexity and diversity. Did life forms come from still farther away, perhaps borne on interstellar dust from some other solar system? In that case, it would be hard to imagine how organic cells could survive light-years of travel subjected to cosmic radiation.

Thought Question

17.5 Suppose a NASA rover discovered living organisms on Mars. How might such a find shed light on the origin and evolution of life on Earth?

To Summarize

- **Prebiotic soup models** propose that the fundamental biochemicals of life arose spontaneously through condensation of reduced inorganic molecules.
- **Early metabolism** involved anaerobic oxidation-reduction reactions. Likely forms of early metabolism include sulfate respiration, light-driven ion pumps, photoferrotrophy, iron phototrophy, and methanogenesis.
- **The RNA-world model** proposes that in the first cells, RNA performed all the informational and catalytic roles of today's DNA and proteins.
- **Thermophile or psychrophile?** Classic models of early life assume thermophily, but Earth may actually have been cold when the first cells originated.
- **A world of methane?** If the first cells were methanogens, methane production could have led to the first greenhouse effect, warming Earth and enabling evolution of other kinds of life.
- **Origin on Earth or elsewhere?** Isotope ratios suggest the presence of complex metabolism by 3.7 Gyr ago, shortly after Earth cooled (4.0 Gyr ago). Simpler cells existing before 4.0 Gyr ago may have evolved much faster than life today does. A more speculative possibility is that life first evolved on another planet.

17.3 Evolution: Phylogeny and Gene Transfer

The unifying assumption of modern biology is genetic relatedness, or molecular phylogeny. Phylogeny generates a series of branching groups of related organisms called **clades**. Each clade is a **monophyletic group**—that is, a group of organisms that share a common ancestor not shared by any kind of organism outside the clade. Each monophyletic group then branches into smaller monophyletic groups, and ultimately **species**, the fundamental kind of organism (discussed in Section 17.5). Species inevitably diverge into strains or subspecies that have acquired different mutations.

The description of branching divergence of multiple species and strains is called **phylogeny**. Phylogeny, however, consists of more than branching divergence; the ancestry of all life forms includes **horizontal gene transfer** (discussed in Chapters 7 and 9). Both vertical gene flow (parent to offspring) and horizontal gene transfer shape all life forms, including humans.

Divergence through Mutation and Natural Selection

Populations of organisms diverge from each other through several fundamental mechanisms of evolution. These include:

- **Random mutation.** DNA sequences change through rare mistakes (in bacteria and archaea, typically one out of a million base pairs) as the chromosome replicates (discussed in Chapter 7). Replication errors result in mutation (discussed in Chapter 9). Most mutations are neutral; that is, they have no effect on gene function.
- **Natural selection and adaptation.** In a given environment, natural selection favors organisms that produce greater numbers of offspring in that environment (discussed in Section 17.4). Genes encoding traits under selection pressure may show mutation frequencies much higher or lower than those generated by the random mutation rate. Natural selection enables a population to adapt to a changing environment.
- **Reductive evolution (degenerative evolution).** In the absence of selection for a trait, the genes encoding the trait accumulate mutations without affecting the organism's reproductive success. Because mutations that decrease function are more common than mutations that improve function, accumulating mutations without selection pressure leads to decline and ultimately loss of the trait. The loss or mutation of DNA encoding unselected traits is called **reductive evolution** (or **degenerative evolution**).

Random mutations with neutral effects that are not subject to selection tend to accumulate at a steady rate over generations because the error rate of the DNA replication machinery stays about the same. After mutations accumulate, "genetic drift" can cause sequences in separate populations to diverge over time. The constancy of the mutation rate (within limits) provides a tool for us to measure the time of divergence of organisms based on their DNA sequences.

Thought Question

17.6 What kinds of DNA sequence changes have no effect on gene function? (*Hint:* Refer to the table of the genetic code, Figure 8.12.)

Molecular Clocks Are Based on Mutation Rate

An important conceptual advance of the twentieth century was that information contained in a macromolecule such as protein or DNA could measure the history of a species. The temporal information contained in a macromolecular sequence is called a **molecular clock**. Molecular clocks have revolutionized our understanding of the emergence of all living organisms, including human beings.

The first molecular clocks, based on protein sequencing and DNA hybridization, were developed in the 1960s. Subsequently, the sequences of ribosomal RNA (rRNA) were used by Carl Woese to reveal the divergence of three domains of life. The rRNA sequence is particularly useful because ribosome structure and function are highly similar across all organisms. Genome sequences now offer many other genes to measure divergence at different levels of classification.

A molecular clock is based on the acquisition of new random mutations in each round of DNA replication. In **Figure 17.14**, each offspring in generation 2 acquires two new mutations; their sequences now differ by 25%. In the next generation, each individual propagates the earlier mutant sequences while acquiring two more random mutations. Strain 3A now differs by 50% from strains 3B and 3C, which differ by 25% from each other. (Actual mutation frequencies, of course, are much lower—about one base per million per generation.) We assume that the chromosomes of offspring acquire a consistent number of random mutations from their parents, and therefore the number of sequence differences between two species should be proportional to the time of divergence between them.

Ideally, the molecular clock works best for a particular gene sequence with the following features:

- **The gene has the same function across all types of organisms compared.** That is, all versions are orthologous; they have not evolved to serve different functions. Functional difference may lead to different rates of change.

- **The generation time is the same for all organisms compared.** Shorter generation times (more frequent reproductive cycles) lead to overestimates of the overall time of divergence because of the increased opportunity for DNA mutation.

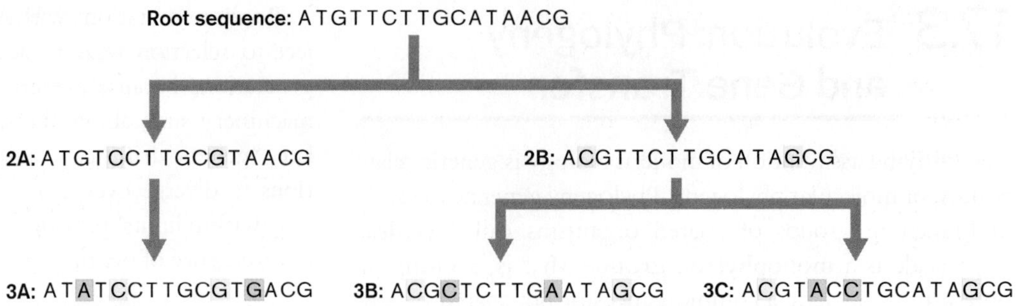

FIGURE 17.14 ▪ The molecular clock. As genetic molecules reproduce, the number of mutations (shaded in yellow and green) accumulated at random is proportional to the number of generations and thus the time since divergence. In each sequence designation (for example, "2A"), the number indicates the generation and the letter identifies a specific strain.

- **The average mutation rate remains constant among organisms and across generations.** If different kinds of life mutate at different rates, then organisms with more rapid rates of mutation will appear to have diverged over a longer time than is actually the case.

In practice, these requirements are never fulfilled exactly, but we do the best we can, and we remember the various ways that a tree may deviate from measuring time. Genes that show the most consistent measures of evolutionary time encode components of the transcription and translation apparatus, such as ribosomal RNA and proteins, tRNA, and RNA polymerase. The most widely used molecular clock is the gene encoding the **small-subunit rRNA (SSU rRNA)**. The SSU rRNA is also known by its sedimentation coefficient: 16S rRNA (bacteria and archaea) or 18S rRNA (eukaryotes). (Sedimentation coefficients are discussed in Chapter 3.)

The SSU rRNA is particularly useful because certain portions of its sequence are remarkably conserved across all forms of life. These portions can be used to define primers to amplify DNA of the gene encoding the rRNA, using the polymerase chain reaction (PCR; discussed in Section 12.1). The gene sequence lying between the pair of highly conserved rDNA sequences will show greater variation, allowing distinction between different clades. PCR can be used to amplify genes even from a mixture of uncultured organisms.

Use of a molecular clock requires the alignment of homologous sequences in divergent species or strains (**Fig. 17.15** ▶). Alignment is the correlation of portions of two gene sequences that diverged from a common ancestral sequence (homologous sequence). The process requires assumptions and decisions about base substitutions, as well as about insertions and deletions. For example, in **Figure 17.15** the best alignment requires us to assume that two bases were lost from the fourth sequence (or else inserted into the ancestor of the other sequences). The relative differences among the sequences can be used to propose a tree of divergence (**Fig. 17.15C**). In the tree, the length of each branch is proportional to the number of differences between two sequences. The number of differences, or divergence, is given by 100% minus the percentage of similarity between the aligned sequences.

In practice, a much larger amount of sequence with multiple differences is needed to calculate a phylogenetic tree. All trees are based on probability, with ambiguities depending on the assumptions of how sequences change. The data are calculated by computer programs, which may yield different results, depending on their assumptions. Standard assumptions include the following:

- **The minimum number of changes gives the best alignment.** The best alignment between two sequences is that which assumes the smallest number of mutational changes.

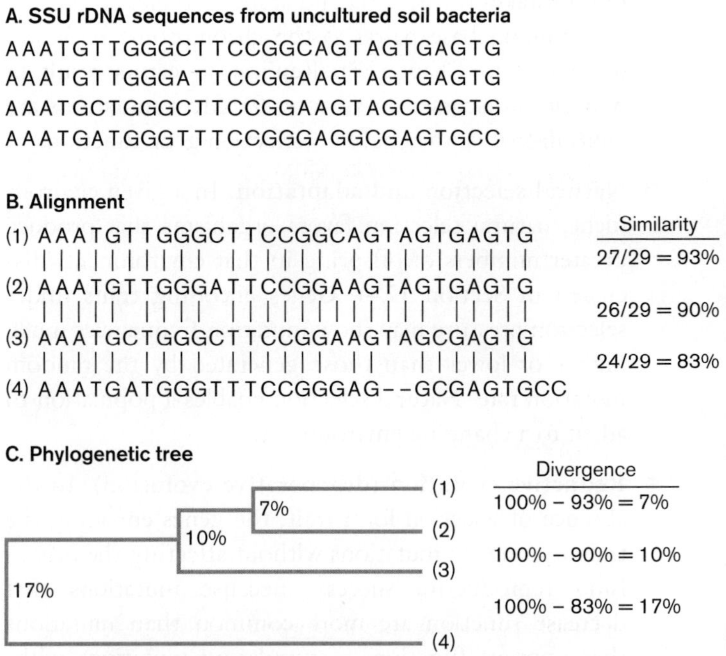

FIGURE 17.15 ▪ DNA sequence alignment. **A.** SSU rDNA sequences from different organisms can be aligned at homologous regions. **B.** The best alignment is that which minimizes mismatches. **C.** A possible phylogenetic tree of divergence of the four sequences. ▶

- **Functional sequences change more slowly.** Sequences that encode essential catalytic portions of the gene product are maintained by selection pressure and change more slowly than portions of the molecule without essential function.
- **Third-base codon positions show more random change.** In a protein-coding gene, many codons have multiple anticodons that differ at the third base, so third-base changes are least likely to change the amino acid. Thus, third-base nucleotides offer the best molecular clock information.

Phylogenetic Trees

Once homologous sequences are aligned, the frequency of differences between them can be used to generate a **phylogenetic tree** that estimates the relative amounts of evolutionary divergence between the sequences. If divergence rate over time, or mutation rate, is the same for all sequences compared, then divergence data can be used to infer the length of time since two organisms shared a common ancestor.

A phylogenetic tree is a model. All phylogenetic trees depend on complex mathematical analysis to measure degrees of divergence and propose a tree that most probably connects present sequences with their common ancestor. Because we can never know the exact tree with certainty, different types of calculations may lead to slightly different results. Two common approaches are:

- **Maximum parsimony.** Evolutionary distances are computed for all pairs of taxa based on the numbers of nucleotide or amino acid substitutions between them. A proposed common ancestor, or "ancestral state," is reconstructed. All possible trees comparing relative time of divergence from the common ancestor are computed. The "best fit" tree is defined as the one requiring the fewest mutations to fit the data (that is, the one that is most "parsimonious"). A limitation of parsimonious reconstruction is that more than one tree may produce results consistent with the data.
- **Maximum likelihood.** For each possible tree, one calculates the likelihood (probability) that such a tree would have produced the observed DNA sequences. The probability of given mutations is based on complex statistical calculations. Maximum-likelihood methods require large amounts of computation but obtain the most information from the data, usually generating one tree or a small set of probable trees.

A portion of a computed phylogenetic tree is shown in **Figure 17.16** ▶. The sequence data were obtained from PCR-amplified sequences of SSU rRNA from isolates in Obsidian Pool, a thermal spring at Yellowstone National Park. The sequences were analyzed by Susan Barns and colleagues at Indiana University in the laboratory of Norman Pace, a pioneering investigator of extreme habitats. Note that some of the Obsidian Pool samples are known species, such as the thermophile *Pyrodictium occultum*, whereas others are uncharacterized organisms given alphanumeric designations, their existence known only by the rRNA sequences reported here. In fact, any microbial habitat surveyed by PCR, including a human body, will yield previously unknown species (as in the shower curtain biofilm described in **eTopic 17.2**).

The Obsidian Pool rRNA sequences were aligned by the method illustrated in **Figure 17.15**, and divergence distances were analyzed by maximum likelihood. In each tree, the total distance (in terms of base-substitution frequency) between any two sequences is approximated by the distance from each sequence and its branch point, or **node** (**Fig. 17.16**). A node for a group of branches is called a **root**—that is, a common ancestor. Moving from the root to the tips of the branches means moving forward in time.

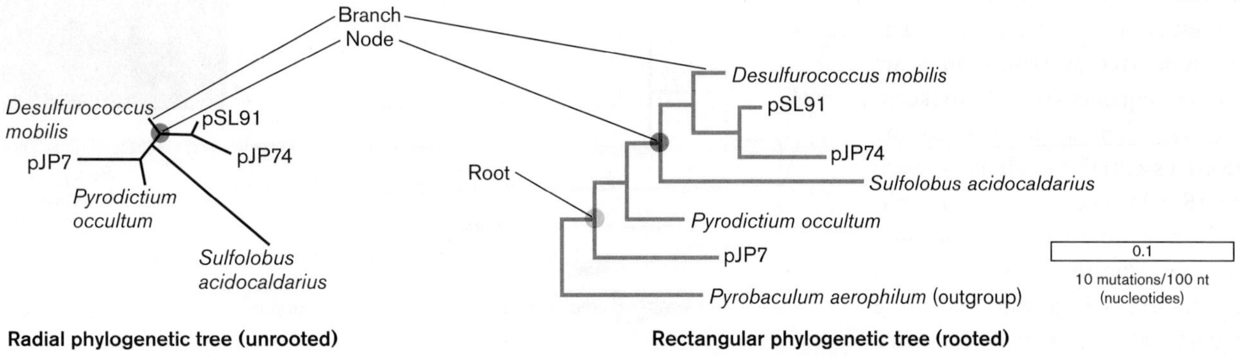

FIGURE 17.16 ■ Phylogenetic trees: rooted and unrooted. Comparison of rRNA sequences was used to generate a phylogenetic tree for thermophiles isolated from Obsidian Pool in Yellowstone National Park. **Left:** The radial phylogenetic tree shows the degrees of relatedness among taxa based on the number of molecular substitutions on each branch (see scale bar). **Right:** The rectangular tree shows the same divergence distances lined up in parallel. This rectangular tree includes a root, a common ancestor defined by divergence from an outgroup taxon. A rooted tree indicates the position of a common ancestor. *Source:* Susan M. Barnes et al. 1996. *PNAS* **93**:9188. ▶

Such a line of individuals, past and present, that descends from one ancestor is called a **lineage**. For example, in the tree on the right in **Figure 17.16** a lineage of earlier species connects the root of the tree (common ancestor) with the present-day archaeon *Desulfurococcus mobilis*.

Phylogenetic trees can be drawn in different ways. The tree at left in **Figure 17.16** is a radial tree, in which branches indicate distance outward from their nodes. The tree at right shows the same divergence distances from nodes, drawn as a rectangular tree. In a rectangular tree, all branches run in parallel. A rectangular tree helps compare divergent branches, although it takes up more space than a radial tree, especially when large numbers of organisms are compared.

The tree at right is rooted by an **outgroup**. An outgroup is a taxon that diverged from a lineage before the ancestor shared by all other taxa in a tree. The outgroup organism, in this case *Pyrobaculum aerophilum*, must be known to have diverged earlier, in order to define the **root**, or common ancestor, of all other branches.

If the length of each branch corresponds to a given length of time, we expect each of the lineages of branches to add up to the same total length. In some trees this condition is approximated, but in microbial trees the lengths differ greatly. In **Figure 17.16**, for example, *Sulfolobus acidocaldarius* appears to have evolved much more than *Desulfurococcus mobilis*. The reason is that *S. acidocaldarius* and its recent ancestors have accumulated mutations faster. In every tree, some lineages accumulate mutations faster than others. The difference in rate arises from differences in mutation rate and from differences in generation time between organisms whose sequences are compared. Thus, our molecular clocks are inevitably distorted, especially for distantly diverged organisms with disparate mutation rates.

A phylogenetic tree can compare any set of organisms, even from multiple habitats. **Figure 17.17** shows the phylogeny of selected bacteria from the human intestine, within a larger clade (the gammaproteobacteria, presented in Chapter 18). The phylogenetic tree is based on a set of protein sequences of "housekeeping" genes—that is, genes encoding functions essential for all cells (such as 16S rRNA), and usually transmitted vertically (parent to offspring). *Escherichia*, *Shigella*, and *Salmonella* are closely related genera, including pathogenic strains, as well as normal members of the gut microbiome. The genus *Klebsiella* includes species that grow outside the gut, causing pneumonia. *Photorhabdus luminescens* is a nematode bacterium that helps its host parasitize insects. *Erwinia carotovora* infects carrots and other plants; it is closely related to *Yersinia pestis*, the cause of bubonic plague. *Photobacterium profundum* is a marine barophile (high pressure) growing in a deep-sea trench.

Overall, the tree shown is rooted by the branch to an outgroup, *Shewanella*, a genus of metal reducers used to construct fuel cells. *Shewanella* defines the root because its branch diverged earlier than that of the common ancestor of the tree's other branches. Nevertheless, even *Shewanella* shares an ancestor with the other bacteria. The tree shows how bacteria of bewildering diversity, from human pathogens to deep-ocean dwellers, diverged relatively recently from a common ancestor.

One group, the genus *Buchnera*, appears to have diverged much faster than the others. *Buchnera* species are intracellular endosymbionts of aphids (**Fig. 17.17** inset). The intracellular symbionts have undergone reductive evolution, losing many functions supplied by their host. Intracellular endosymbionts typically mutate much faster than free-living bacteria. In addition, they undergo intense selection pressure for adaptation to their obligate host. And yet, other bacteria (*Regiella*) within the same host evolved the opposite way: They gained genes encoding parasitic factors. Endosymbiosis is discussed further in Section 17.6.

Buchnera and several other lineages show a widening branch. The widening branch indicates a "fan" or "bush" of strains equally distant from a common node (branch point). The cause of "bushy" branches is debated; some researchers argue that all branches become bushy once enough strains have been sequenced.

How do we calibrate a phylogenetic tree—that is, relate the number of mutations to the time since divergence? We

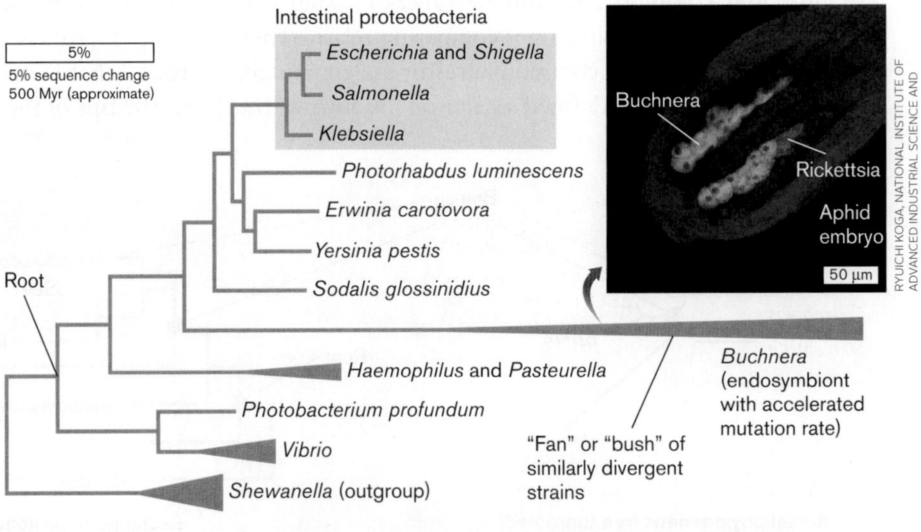

FIGURE 17.17 ■ Intestinal bacteria and related proteobacteria. The phylogenetic tree was derived from concatenated sequences of highly conserved "housekeeping" proteins. The scale bar corresponds to 5% amino acid sequence divergence. **Inset:** Aphid embryo contains bacterial symbionts, with DNA labeled by fluorescence in situ hybridization (FISH). Green = *Buchnera*; pink = *Regiella*; blue = aphid nuclei. *Source:* Phylogeny modified from Morgan Price et al. 2008. *Genome Biol.* **9**:R4; and from Fabia Battistuzzi et al. 2004. *BMC Evol. Biol.* **4**:44.

need an external measure of time. A tree can be calibrated if some kind of fossil evidence or geological record exists to confirm at least one branch point of the tree. But for microbes, such fossil calibrations remain speculative.

One convincing method of calibration is to correlate the divergence of microbial species growing only inside of particular host species with the divergences of their hosts, based on the host fossil record. For example, the exceedingly rapid divergence of *Buchnera* species (**Fig. 17.17**) can be calibrated on the basis of their host insects. Such calculations, however, reveal vastly divergent rates of evolutionary change in different bacterial taxa.

Could a life form mutate even faster than *Buchnera*? Some RNA viruses, such as HIV, mutate so fast that they form multiple strains within one host (discussed in Chapter 11).

Thought Question

17.7 What are the major sources of error and uncertainty in constructing phylogenetic trees?

Divergence of Three Domains of Life

Carl Woese first used SSU rRNA phylogeny to reveal the existence of a third kind of life, Archaea, roughly as distant from bacteria as from eukaryotes (**Fig. 17.18**). The three fundamental groups of life forms—Archaea, Bacteria, and Eukarya—are termed **domains**.

How was an entire domain of life missed in the past? Many archaea grow in habitats previously thought inhospitable for life; for example, *Thermoplasma* species grow at 60°C and pH 2 (**Fig. 17.19**). These thermoacidophiles surprisingly lack cell walls to protect their cells from osmotic stress, and their cells interconnect by extensive membrane nanotubes. Other archaea, such as methanogens and halophiles, were long known to microbial ecologists, but without tools for genetic analysis they were simply classified among bacteria. The fascinating diversity of Archaea is explored in Chapter 19.

Rooting the tree of life. Where is the root of the tree, the position of the last universal common ancestor of all life forms? Which of the three domains (Bacteria, Archaea, Eukarya) was the first to diverge from the other two? The question has profound importance for biology because the research community bases its "model systems" for study on their commonalities with organisms of importance to humans. For example, investigators of intron splicing in archaea argue that archaea represent a model system for related processes in complex eukaryotes such as humans.

The root, however, can be found only by measuring divergence relative to an outside group of organisms. Since no "outgroup" exists for the tree of all life, how is the tree rooted? One approach is to compare a pair of homologous genes within one organism—homologs that diverged from

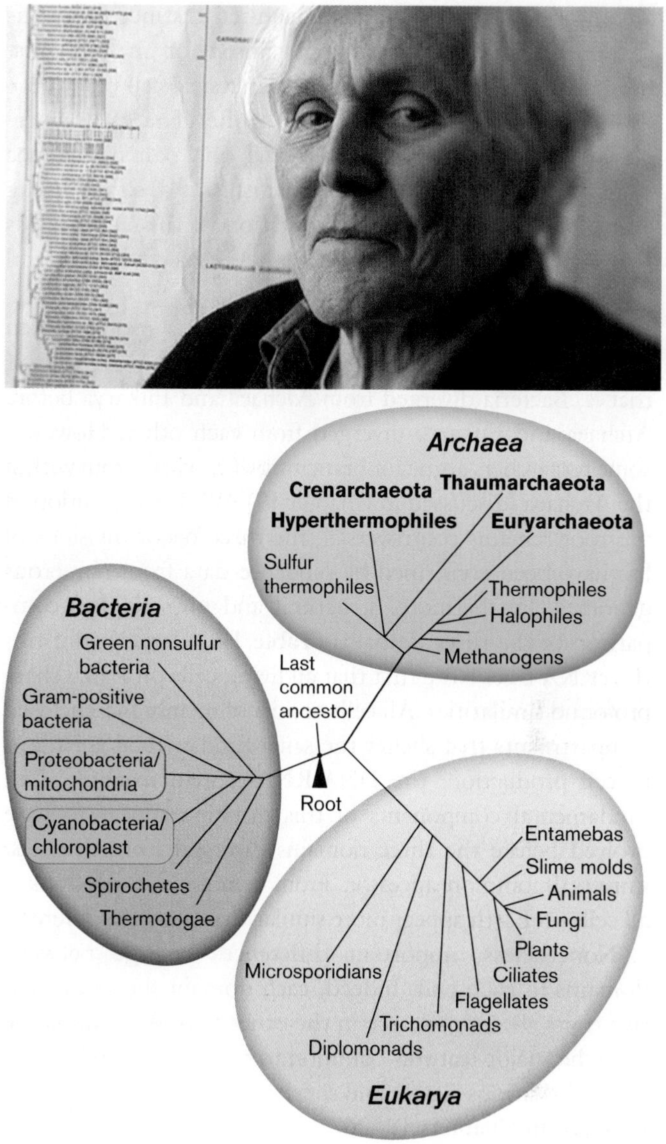

INSTITUTE FOR GENOMIC BIOLOGY UNIVERSITY OF ILLINOIS AT URBANA-CHAMPAIGN

FIGURE 17.18 ▪ Three domains of life. Carl Woese (1928–2012; inset) used SSU rRNA sequencing to reveal three equally distinct domains of life: Bacteria, Eukarya (eukaryotes), and Archaea.

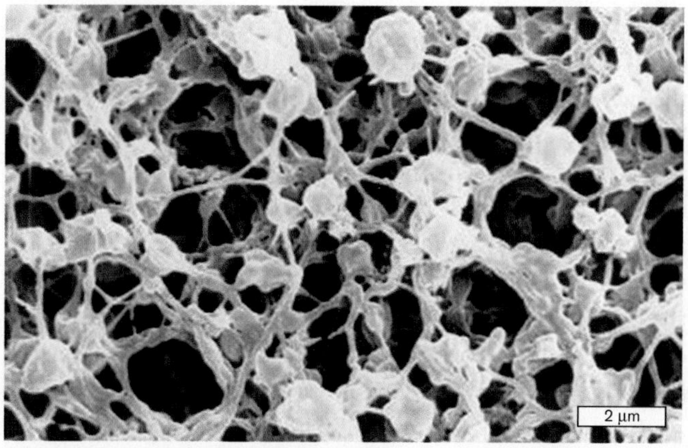

M. YASUDA ET AL. 1995. *APPL. ENVIRON. MICROBIOL.* **61**:3482–3485.

FIGURE 17.19 ▪ *Thermoplasma*. This archaeon lives at 60°C at pH 2—with no cell wall, only a cell membrane. Cells share cytoplasm by a network of membrane nanotubes.

a common ancestral gene and acquired distinct functions (paralogs). The pair of paralogs chosen for analysis must have diverged within the common ancestral cell before the divergence of the three domains of life. The early divergence is determined by comparing the pair's sequence similarity in many different organisms and constructing a tree of gene phylogeny. A limitation of this approach is that genes with different functions are subjected to natural selection in addition to random variation.

Most phylogenetic data so far indicate a root between Bacteria and the common ancestor of Archaea and Eukarya; that is, Bacteria diverged from Archaea and Eukarya before Archaea and Eukarya diverged from each other. However, some researchers argue for branching of Eukarya from within the Archaea (discussed in Chapter 19). While the position of the root remains controversial, the three major divisions of life have been confirmed by sequence data from numerous genomes. Furthermore, structural and physiological comparison of the three domains (**Table 17.2**) largely confirms the rRNA tree. Note first that all living cells on Earth share profound similarities. All cells consist of membrane-enclosed compartments that shelter the same fundamental apparatus of cell production: the DNA-RNA-protein machine. The fundamental components of this machine appear to have evolved before the three domains diverged from their last universal common ancestor. From a molecular standpoint, all cells on Earth appear more similar than they are different.

Nonetheless, important differences emerge between domains in each pair; indeed, each domain shows distinctive traits absent or scarce in the other two. We summarize here the major features common to most members of each domain. Various classes and species within each domain are explored in Chapters 18–20.

Archaea, bacteria, and eukaryotes. Of the three domains, the eukaryotes stand out as having a nucleus and other complex membranous organelles (see **Table 17.2**). Eukaryotic organelles include mitochondria and chloroplasts, which evolved from internalized bacteria. Bacteria and archaea possess no nucleus and have relatively simple intracellular membranes. Their size is limited by diffusion across the cell membrane, with occasional exceptions, such as the "giant bacterium" *Epulopiscium fishelsoni.* The larger size and complexity of eukaryotic cells mean that they generally require the most highly powered sources of energy, such as aerobic respiration and oxygenic photosynthesis, although some protists and fungi conduct fermentation. By contrast, the prokaryotes (bacteria and archaea) employ a wider range of metabolic alternatives, including lithotrophy and anaerobic respiration. Finally, eukaryotic plants and animals have attained a degree of multicellular complexity unknown in the prokaryotic domains.

On the other hand, eukaryotes share key traits with archaea that distinguish both from bacteria. The core information machinery of eukaryotes more closely resembles that of archaea. The two domains share closely related components of the central DNA-RNA-protein machine: RNA polymerase, ribosomes, and transcription factors. Even such hallmarks of eukaryotes as intragenic introns, the splicing machinery, and the "RNA interference" regulatory complexes are found in archaea. All this explains why rRNA trees place archaea closer to eukaryotes than to bacteria.

Nevertheless, eukaryotes share fundamental structures with bacteria that differ from those in archaea. Archaea possess unique cell membrane components, such as their ether-linked lipids (see Chapter 19). Outside the archaea, ether-linked membrane lipids are found only in deep-branching bacterial species that share habitat (and exchange genes) with hyperthermophilic archaea. Only the domain Archaea includes species capable of growth in the most "extreme" environments of temperature (above 110°C or below –20°C) and pH (below pH 1). At the same time, many other archaeal species grow well at mesophilic temperatures in soil or water.

Perhaps the most striking distinction of archaea is the absence of archaeal pathogens of animals or plants. Even the many methanogens that live within animal digestive tracts have not been shown to cause disease.

Overall, the three-domain phylogeny divides life usefully into three distinctive groups. Yet the tree also shows signs of gene flow unaccounted for by monophyletic descent. The eukaryotes contain mitochondria derived from assimilated bacteria whose genomes persist within the organelle. And pathogenic bacteria such as *Agrobacterium* species transfer DNA into the genomes of plants. Moreover, sequenced genomes reveal evidence of gene transfer between bacteria and archaea sharing high-temperature habitats. What if the tree of life is not strictly monophyletic?

Horizontal Gene Transfer

In retrospect, the very first demonstration of the molecular basis of heredity—the transformation of avirulent *Pneumococcus* to virulence in 1928—involved **horizontal gene transfer** (lateral gene transfer). Horizontal gene transfer is the acquisition of a piece of DNA from another cell, as distinguished from **vertical gene transfer**, the transmission of an entire genome from parent to offspring.

Among bacteria and archaea, DNA is transferred horizontally by plasmids, transposable elements, and bacteriophages, as well as through the process of transformation, as discussed in Chapter 9. For example, drug resistance genes are transferred from harmless human-associated bacteria to pathogens. Closely related taxa show evidence of numerous past transfer events. For example, the *Escherichia coli* genome acquired about 18% of its genes from closely related species

TABLE 17.2 Three Domains of Life

Characteristic	Traits of living organisms		
	All cells on Earth resemble each other in these traits:		
Chromosomal material	Double-stranded DNA		
RNA transcription	Common ancestral RNA polymerase		
Translation	Nearly universal genetic code; shared ancestral rRNAs and elongation factors		
Protein	Common ancestral functional domains		
Cell structure	Aqueous cell compartment enclosed by a membrane		
	COMPARISON OF DOMAINS		
	Bacteria	**Archaea**	**Eukarya**
	Archaea resemble bacteria in these traits:		
Cell volume	1–100 μm^3 (usually)		1–10^6 μm^3
DNA chromosome	Circular (usually)		Linear
DNA organization	Nucleoid		Nucleus with membrane
Gene organization	Multigene operons		Single genes
Metabolism	Denitrification, N_2 fixation, lithotrophy, respiration, and fermentation		Respiration and fermentation
Multicellularity	Simple		Simple or complex
		Archaea resemble eukaryotes in these traits:	
Intron splicing	Introns are rare	Introns are common	
RNA polymerase	Bacterial homologs	Eukaryotic homologs	
Transcription factors	Bacterial homologs	Eukaryotic homologs	
Ribosome sensitivity to chloramphenicol, kanamycin, and streptomycin	Sensitive	Resistant	
Translation initiator	Formylmethionine	Methionine (except mitochondria and chloroplasts use formylmethionine)	
	Bacteria resemble eukaryotes and differ from archaea in these traits:		
Methanogenesis	No	Yes	No
Thermophilic growth	Up to 95°C	Up to 120°C	Up to 80°C
Photosynthesis	Many species; bacteriochlorophyll (proteorhodopsin)	Haloarchaea only; bacteriorhodopsin (shares homology with proteorhodopsin)	Many species; chlorophyll (bacterial origin)
Light absorption by chlorophylls	Yes	No	Yes (chloroplasts of bacterial origin)
Membrane lipids (major)	Ester-linked fatty acids	Ether-linked isoprenoids	Ester-linked fatty acids
Pathogens infecting animals or plants	Many pathogens	No pathogens	Many pathogens

after its relatively recent divergence from the close relative *Salmonella enterica*. Some medically important genera, such as *Neisseria* (which causes gonorrhea and meningitis), are particularly "recombinogenic." Rapid gene exchange enables pathogens to evade the host immune system by expressing novel proteins not recognized by host antibodies.

How can we tell when a genome contains a DNA sequence "transferred" from a different species? One sign of past horizontal transfer is a DNA sequence whose GC/AT ratio (proportion of GC and AT base pairs) differs from that of the rest of the genome. A surprising proportion of genomic DNA can show "spikes" of GC content that differ from the

GC/AT ratio of neighboring sequences. These regions of anomalous GC/AT ratio, sometimes referred to as genomic islands (Chapter 9), indicate an origin elsewhere.

Genes are transferred most frequently between closely related strains or taxa. Yet, remarkably, some genes can be transferred across distant phyla, even from species of a different domain. A striking example is the transfer of genes encoding light-driven proton pumps (bacteriorhodopsin) from halophilic archaea into many species of marine bacteria (where they are called proteorhodopsins). Such transfer events are relatively rare, occurring perhaps once in a million generations. But over time, the number of such "rare" events can accumulate. In some archaea, particularly the hyperthermophiles *Pyrococcus* and *Aeropyrum*, 10%–20% of the genes appear to come from bacteria that share their high-temperature environment. Similarly, the thermophilic bacterium *Thermotoga maritima* shows many genes transferred from archaea.

How does horizontal gene transfer affect microbial phylogeny? In 1999, Ford Doolittle, at Dalhousie University, redrew the standard tree of life with a bewildering array of cross-cutting lineages to show how actual phylogeny combines horizontal and vertical transfer (**Fig. 17.20**). Doolittle's tree (**Fig. 17.20B**) acknowledges the ancestral transfer of entire bacterial genomes into eukaryotes, via the endosymbiotic ancestors of their mitochondria and chloroplasts (discussed further in Section 17.6). Then there are bacteria, such as the hyperthermophile *Aquifex pyrophilus* (**Fig. 17.20B** inset), that share genes with archaea—and with distantly related bacterial phyla, such as Epsilonproteobacteria and Actinobacteria (discussed in Chapter 18). For *Aquifex*, the very definition of domain, let alone species, is problematic. Further back, perhaps the "last common ancestor" of all life forms was actually a last common community of diverse life forms that contributed different parts of our genetic legacy.

Thought Question

17.8 What are the limits of evidence for horizontal gene transfer in ancestral genomes? What alternative interpretation might be offered?

A. Monophyletic tree: limited horizontal transfer

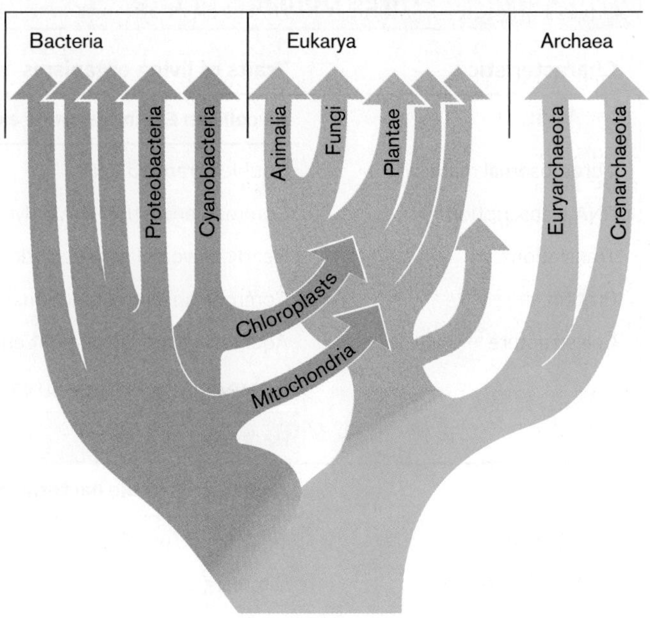

B. Horizontal transfer obscures phylogeny

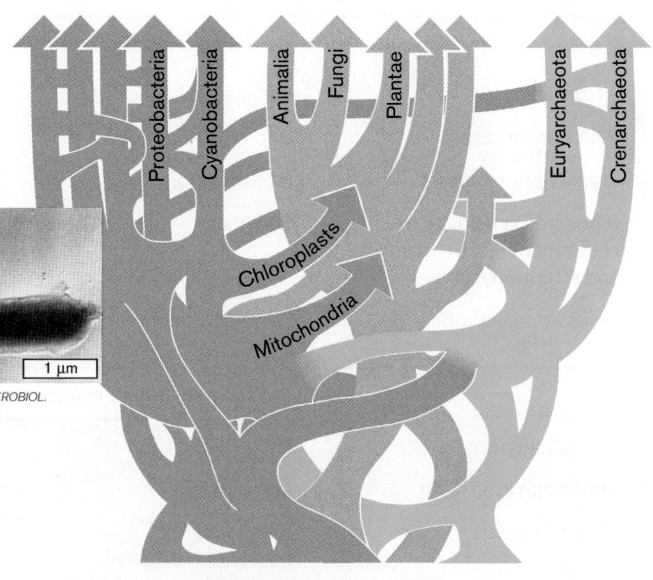

Aquifex pyrophilus

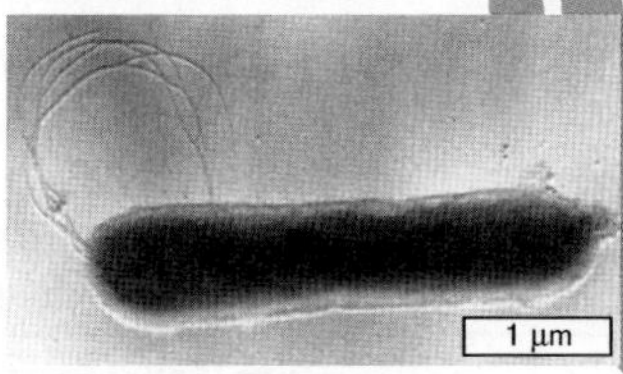

R. HUBER ET AL. 1992. *SYST. APPL. MICROBIOL.* **15**:340–351, FIG. 2.

FIGURE 17.20 ■ Vertical and horizontal gene transfer.
A. The traditional view of phylogeny. Most gene transfer is vertical, and horizontal transfer is limited to rare cases such as transfer of mitochondria and chloroplasts from bacteria into eukaryotes.
B. The modern view. Horizontal gene transfer occurs so often that it obscures monophyletic distinctions between taxa. *Source:* Modified from W. Ford Doolittle. 1999. *Science* **284**:2124.

Reconciling Vertical and Horizontal Gene Transfer

One approach to sorting out vertical and horizontal gene transfer was proposed by James Lake and colleagues at UCLA and developed further by Doolittle. Lake's approach assumes that some classes of genes nearly always transmit vertically—particularly "informational genes," which specify products essential for transcription and translation. Informational genes include those that encode RNA polymerase, as well as ribosomal RNAs such as SSU rRNA, and elongation factors. Informational genes need to interact directly in complex ways with large numbers of cell components; thus, their capacity for horizontal transfer is limited.

On the other hand, "operational genes" are those whose products govern metabolism, stress response, and pathogenicity. Operational genes function with relative independence from other cell components and consequently move more easily among distantly related organisms, particularly organisms that share the same habitat. An important category of such movable genes is virulence factors. Pathogens show extensive horizontal transfer of virulence genes and genes encoding resistance to host defenses and antibiotics (Chapter 25). Groups of such genes are often transferred together on plasmids or conjugative transposons (Chapter 9).

A balanced view of vertical and horizontal gene transfer is shown in **Figure 17.21**. The row of arrows designates vertical transfer of the bulk of the genomes of two closely related species—*E. coli* and *Salmonella enterica*—as they diverge from a common ancestor. Black lines represent vertically transferred genes, such as ribosomal RNA; gray lines indicate genes with more flexible function. At various levels, colored lines indicate horizontal transfer, where genes enter each lineage by processes such as conjugation or transformation. Gray lines peel out, indicating loss of a gene by mutation. Overall, the vertical lineage persists for most of the core informational genes (black lines), while horizontally acquired genes enter the genome. The persistence of genes showing vertical inheritance in nearly all life (such as SSU rRNA) may reflect the phylogeny of the organism as a whole, despite the other genes that are transferred horizontally. But to fully assess phylogeny, we must sequence entire genomes.

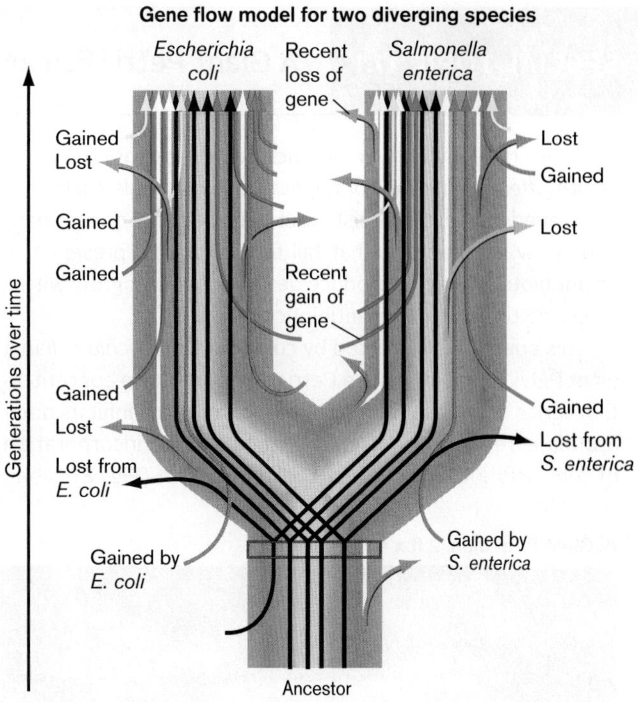

FIGURE 17.21 ▪ Genes enter and leave genomes of two closely related species. Both rRNA trees and whole-genome trees consistently reflect monophyletic descent (black lineages) down to the genus level. For closely related species, monophyletic descent may be obscured by high rates of horizontal transfer (colored lines entering, gray lines departing). White spaces indicate genes lost by reductive evolution.

To Summarize

- **Phylogeny is the divergence of related organisms.** Organisms diverge through random mutation, natural selection, and reductive evolution.
- **Molecular clocks are based on mutation rate.** Given a constant mutation rate and generation time, the degree of difference between two DNA sequences correlates with the time since the two sequences diverged from a common ancestor.
- **Different sequences diverge at different rates.** Under selection pressure, the actual divergence rates of sequences depend on structure and function of the RNA or protein products.
- **Phylogenetic trees are based on sequence analysis.** The more different the two sequences are, the longer is the branch representing time since divergence from the common ancestor. Rooting a tree requires comparison with an outgroup.
- **The tree of life diverges to three domains: Bacteria, Archaea, and Eukarya.** Eukaryotes are distinguished by the nucleus, which is lacking in archaea and bacteria. Archaea possess ether-linked isoprenoid lipids that are rare or absent in bacteria and eukaryotes, and they are never pathogens. The machinery of archaeal gene expression resembles that of eukaryotes.
- **Genes transfer between different species.** Horizontal transfer is most frequent between closely related species or between distantly related species that share a common habitat. Informational genes with many molecular interactions transfer vertically, whereas operational genes that function independently of other components are more likely to transfer horizontally.
- **Horizontal gene transfer is important for adaptation to new environments and for pathogenesis.** Gene transfer among pathogenic and nonpathogenic strains leads to emergence of new pathogens.

17.4 Natural Selection and Adaptation

In discussing phylogeny, we focused on the accumulation of random mutations that cause genome sequences to diverge at a steady rate. The rate of nonselected sequence changes enables us to measure evolutionary time. But how does evolution help life survive in a new environment?

SPECIAL TOPIC 17.1 A Giant Petri Dish and the Race to Resistance

Bacterial resistance to antibiotics is a huge threat to our health, affecting all kinds of medical care. Hospitals are breeding grounds for newly resistant strains. But how does it happen? How can bacteria that fail to grow in the presence of an antibiotic evolve to produce descendants that grow with a thousand times as much antibiotic present?

This question was tested by culturing *Escherichia coli* in a giant Petri dish (**Fig. 1**). The Petri dish contained a concentration series of the antibiotic trimethoprim, which inhibits dihydrofolate reductase, an enzyme needed for incorporation of folic acid into biosynthesis. Michael Baym (**Fig. 2**), along with colleagues in Roy Kishony's lab at Harvard School of Medicine, built the Petri dish 2 feet wide and 4 feet long. The dish was filled with a layer of agar media containing increasing amounts of trimethoprim as you move from the ends of the Petri dish toward the middle. The concentration of trimethoprim grew in four discrete steps. The first step (defined as one unit) contained three times the level of trimethoprim shown to kill the starting culture of *E. coli*. Each succeeding level contained a tenfold higher concentration than the preceding level. Atop the antibiotic agar, a thin layer of medium containing a lower agar concentration was provided through

A. Giant Petri dish, 2 ft x 4 ft **B.**

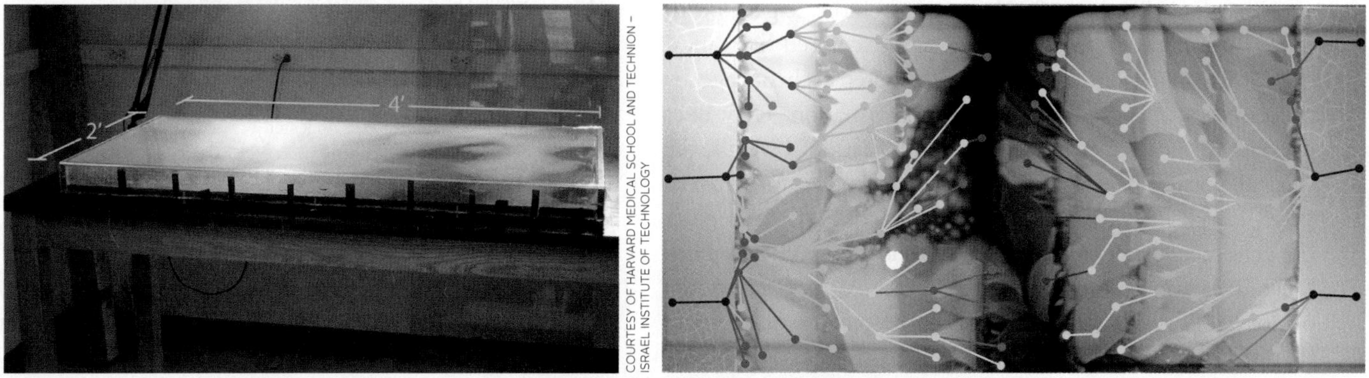

COURTESY OF HARVARD MEDICAL SCHOOL AND TECHNION – ISRAEL INSTITUTE OF TECHNOLOGY

COURTESY OF HARVARD MEDICAL SCHOOL AND TECHNION – ISRAEL INSTITUTE OF TECHNOLOGY

FIGURE 1 ■ Bacteria evolve antibiotic resistant strains in a giant Petri dish. **A.** Giant dish contains strips of agar with increasing concentrations of the antibiotic trimethoprim. **B.** *Escherichia coli* bacteria after 10 days growth. Bacteria were inoculated along each side (right and left), then allowed to swim across and grow in toward the center. Trimethoprim concentration starts at one selective unit, which is three times the minimum inhibitory concentration of the original inoculant. Each succeeding step of agar has ten-fold greater concentration. Colors indicate evolving clones that gain a mutation enabling growth into the higher antibiotic concentration.

The genetic variants that arise by random mutation differ in their chance of survival. Those variants that survive to leave more offspring undergo **natural selection**; that is, their traits are overrepresented in the next generation. A vivid demonstration of natural selection at work on drug-resistant bacteria is the "giant Petri dish" experiment of Michael Baym (**Special Topic 17.1**).

Note: Mutations occur randomly, without regard to selective pressure. Random mutations occur first, generating diversity in a population. Next, selection pressure acts on the population, altering the relative proportion of genetic variants.

How can we study the mechanisms of natural selection and adaptation? We have several kinds of evidence:

- **Genomic analysis.** Comparing gene sequences, both within and between genomes, enables us to track how organisms adapted in the past.
- **Strongly selective environments.** Environments under intensive selective pressure, such as exposure to antibiotics (see **Special Topic 17.1**) lead to rapid evolution that can easily be observed.
- **Experimental evolution.** Experimental strategies reveal evolution in the laboratory, enabling us to test predictive models.

Genomic Analysis

As discussed in Section 17.3, the sequence of genomes reveals descent over time based on steady accumulation of mutations. But other mutations undergo selection that provides a cell with new functions. A common way this happens is by **gene duplication**.

As cells replicate their DNA over many generations (discussed in Chapter 7), occasionally the DNA polymerase will make a duplicate copy of a gene. With duplicate copies

COURTESY OF HARVARD MEDICAL SCHOOL AND TECHNION - ISRAEL INSTITUTE OF TECHNOLOGY

FIGURE 2 ▪ Michael Baym studies the evolution of drug-resistant bacteria.

which the bacteria could swim—if they could survive the antibiotic diffusing from below.

Bacteria were inoculated at each end of the plate, and their growth was monitored for 10 days. A video recording was made: http://naturedocumentaries.org/13544/bacteria-evolution-mega-plate-michael-baym.

As the bacteria grew to fill one section of agar, their growth would slow to a near halt, until a single mutant clone was able to break through and grow into the next level. Other strains appeared with a different mutation that increased antibiotic resistance, and in some cases the different mutant strains competed with each other in the next level of antibiotic. Ultimately, bacteria from both sides of the plate gave rise to descendants that achieved 1,000-fold higher trimethoprim resistance than their ancestral strain.

What kinds of mutations were selected for resistance? The most frequent mutations were found in a gene encoding the enzyme dihydrofolate reductase—the target of the antibiotic trimethoprim. Secondary mutations were selected in general stress genes that confer resistance to multiple antibiotics. In some cases, a mutation led to drug resistance at the expense of growth rate; then a secondary mutation compensated, restoring growth rate. Yet another kind of mutation, called a "mutator," simply increased the rate of many different mutations.

While the setup of Baym's experiment differs from natural and clinical environments, nonetheless the general classes of mutation and the overall trajectory of evolution include mechanisms seen in nature, such as the appearance of secondary mutations that compensate for drawbacks of the early mutations. The more we learn, the more we realize the daunting challenge of antibiotic resistance for antimicrobial therapy.

RESEARCH QUESTION

What do you think would happen if the agar medium contained two kinds of antibiotics? How might the experimental setup require adjustment? What kind of bacterial genotypes and phenotypes do you predict?

Baym, Michael, Tami D. Lieberman, Eric D. Kelsic, Remy Chait, Rotem Gross, et al. 2016. Spatiotemporal microbial evolution on antibiotic landscapes. *Science* **353**:1147–1151.

of a gene, one copy may acquire mutations that change its function without detriment to the organism because the "backup" copy still functions. Now, suppose the mutated copy gains additional mutations that further alter its function, providing a new function to the cell. For example, a gene encoding a transporter for one sugar may now encode a transporter that better "fits" a different sugar. The two **paralogous genes**, or **paralogs**, have evolved to serve different functions.

Paralogs are a major source of raw material that contributes to new functions arising through evolution. We can detect paralogs in a genome, through sequence relatedness; for example, the genome of the hyperthermophilic bacterium *Thermotoga maritima* shows paralogous ABC transporters for lactose, cellobiose, mannose, and xylose, among others. Evolution of paralogs provides an important way for organisms to enhance fitness in a complex environment.

Does a population ever lose some of its paralogs? Certain environments select for loss of many genes—a phenomenon called **degenerative evolution** or **reductive evolution**. In cases of relatively small population size, such as intracellular symbionts, genes are lost via genetic drift. There can also be positive selection for gene loss, such that organisms save energy by avoiding replication and expression of unneeded genes. Intracellular pathogens and obligate symbionts often lose large chunks of their genome by processes such as intracellular recombination of the chromosome. The lost genes typically encoded functions supplied by the host, such as capture of nutrients and generation of energy. Lost genes can be identified by comparison with free-living species. For example, *Treponema pallidum*, the cause of syphilis, has a genome of 1.1 million base pairs, which is about a quarter the size of the *E. coli* genome. *T. pallidum* has lost all of its enzymes used in the TCA cycle, respiration, and amino acid biosynthesis. These genes remain in the genomes of free-living treponemes.

In other species, dilute, nutrient-poor natural environments select for gene reduction. Marine cyanobacteria, a major source of global photosynthesis, have lost numerous genes that confer little advantage in the open ocean. Missing genes encode transporters for sugars and amino acids (which are scarce in the ocean), as well as proteins for flagellar motility and pili. More surprising, *Prochlorococcus* species have lost catalase, an enzyme that destroys the hydrogen peroxide they produce. Their hydrogen peroxide is detoxified by catalase-positive bacteria that share their marine habitat. Erik Zinser at the University of Tennessee, Knoxville, showed that *Prochlorococcus* could be cultured on agar only in the presence of partner bacteria that supply catalase. Given that catalase is an expensive enzyme to produce, *Prochlorococcus* may have gained an advantage when this enzyme was deleted during its evolution in the nutrient-poor open ocean.

Strongly Selective Environments: Antibiotic Selection

Evolution requires many generations, but certain microbes may produce 40 generations in a day. When such microbes are under strong selection pressure, evolution may occur surprisingly fast. One of the strongest selective conditions we can study is the presence of an antibiotic (as seen in Baym's experiment, **Special Topic 17.1**). Antibiotics reveal evolution at work in real time, within hospital environments, and even within the body of one hospital patient.

Antibiotic resistance evolves within a patient. **Figure 17.22** depicts an actual clinical case of a hospitalized patient infected by MRSA, a deadly strain of *Staphylococcus aureus* that had already evolved resistance to drugs such as methicillin. The patient was an elderly man with a weakened immune system, unable to eliminate even small populations of an opportunistic pathogen such as MRSA. The original culture from the patient's blood showed bacteria sensitive to four antibiotics (different from methicillin). But prolonged exposure resulted in natural selection for resistance. Ultimately, following a total of 12 weeks' exposure to linezolid and other antibiotics, the bacteria isolated showed resistance to three other antibiotics plus partial resistance to linezolid.

In **Figure 17.22**, note that the latest drug-resistant strain actually made smaller colonies with or without the drug—the "small-colony variant." In other words, natural selection yielded a population that was the "fittest" under a particular environmental condition—the presence of linezolid in an immunocompromised host—where even a slow-growing strain could persist. In the absence of the drug, the original strain would outcompete the small-colony variant.

In the case just described, what was the molecular basis of the multidrug resistance? Researchers obtained DNA from the patient's original MRSA, as well as the later small-colony variant. They sequenced the two genomes and compared them. Just three point mutations in the small-colony variant accounted for the drug resistance, as well as the retarded growth. One of these mutations derepressed a stress regulon (group of genes under one regulator), causing accumulation of the stress signal ppGpp (guanosine tetraphosphate). The ppGpp stress regulon includes expression of many protective genes that enable a cell to survive in the presence of antibiotics. But the cost to the cell of expressing this regulon is a slower growth rate; like a community under "terror alert," the cell's normal everyday processes are slowed by the demands of the stress response. This cost, or downside, of a trait under natural selection is called a **fitness trade-off** or fitness cost.

The MRSA example illustrates two key points:

- **The "fittest" trait depends on the environment in which selection occurs.** The presence of an antibiotic selects individuals that are resistant, despite the fitness trade-off of slower growth (small-colony size). Without the antibiotic, the faster-growing individuals prevail.
- **Disabling regulation is an effective mechanism of adaptive evolution.** In this case, the ppGpp stress regulon was derepressed, enabling stress responses that normally would be turned off because they inhibit cell growth.

MRSA bacteria isolated from patient

Before antibiotics | After antibiotics

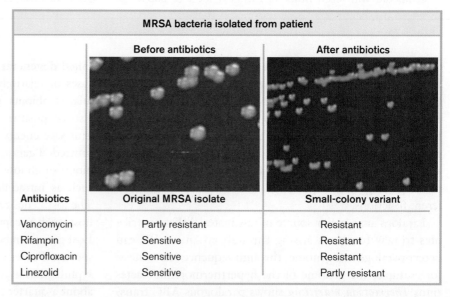

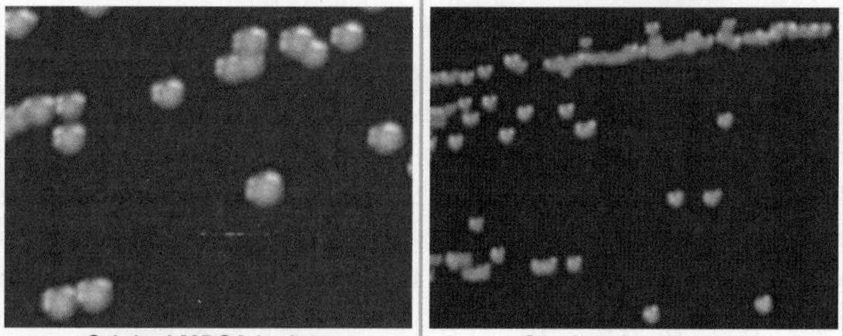

Antibiotics	Original MRSA isolate	Small-colony variant
Vancomycin	Partly resistant	Resistant
Rifampin	Sensitive	Resistant
Ciprofloxacin	Sensitive	Resistant
Linezolid	Sensitive	Partly resistant

FIGURE 17.22 ■ Evolution of antibiotic resistance in MRSA. **Left:** The initial MRSA isolate infecting the patient showed partial resistance to vancomycin but was sensitive to rifampin, ciprofloxacin, and linezolid. **Right:** Following exposure to these antibiotics, a new strain was isolated that grew more slowly (the small-colony variant) but was at least partly resistant to all the antibiotics. *Sources:* Wei Gao et al. 2010. *PLoS Pathog.* **6**:e1000944; images from Lin. 2016. *J. Antimicrob. Chemother.* **71**:1807.

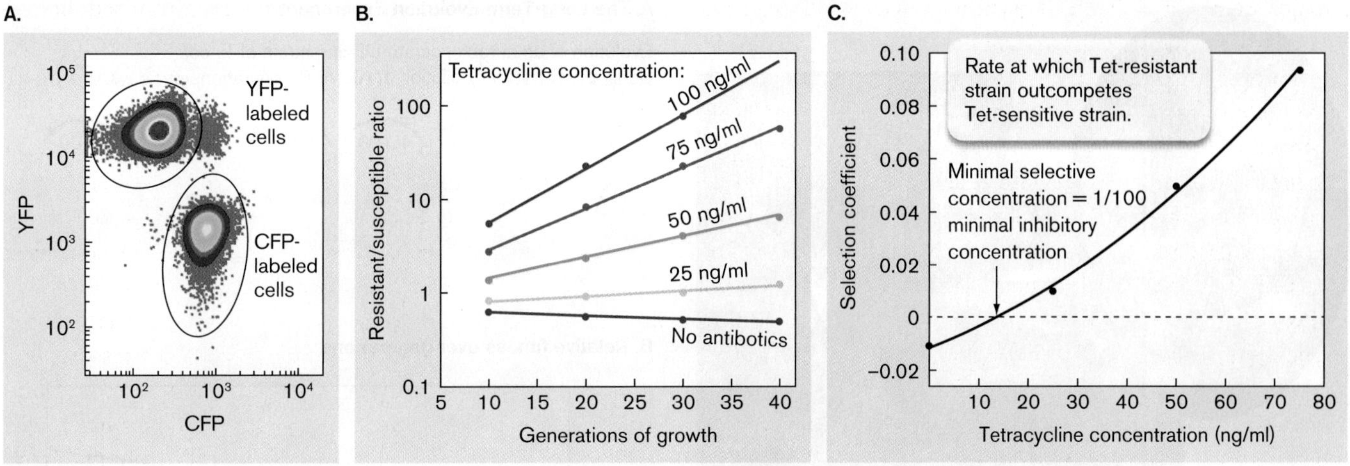

FIGURE 17.23 ■ Fitness competition measured by flow cytometry. **A.** Flow cytometry counts two members of a mixed population, labeled by YFP and CFP fluorophores. **B.** Relative fitness of *Escherichia coli* cells resistant or sensitive to an antibiotic (tetracycline). Strains were marked with genes encoding YFP or CFP, and differentially counted by flow cytometry. **C.** Rate of resistance selection (population sweep) plotted as a function of tetracycline (Tet) concentration. Positive selection for resistance is seen at 15 ng/ml, which is one-hundredth the concentration needed to prevent observable growth. *Sources:* Part A modified from S. Schaffner and J. Slonczewski; Parts B and C modified from E. Gullberg et al. 2011. *PLoS Pathog.* **7**:e1002158, figs. 2C and 2D, respectively.

Antibiotic resistance: selective concentrations. Quantitative questions about antibiotic resistance can be addressed in the laboratory. For example, what concentrations of an antibiotic are needed for natural selection? This question is important because every time a drug is prescribed, we provide potential conditions for selection of resistance. And low levels of medications continually enter sewage and reach our aquatic environments. Do these low levels constitute selective pressure? Until recently, it was thought that only drug levels sufficient to prevent microbial growth (the minimum inhibitory concentration, MIC) would have selective pressure. Most mutations conferring drug resistance have deleterious effects in the absence of the drug.

But small degrees of selective pressure can add up over time. Erik Gullberg and Dan Andersson at Uppsala University, Sweden, devised a method to measure very small population shifts arising from selective pressure. Their method uses flow cytometry to count individual bacteria that express one of two protein fluorophores: yellow fluorescent protein (YFP) or cyan fluorescent protein (CFP). Protein fluorophores are discussed in Chapter 2. Two different strains of bacteria (one sensitive to an antibiotic, the other resistant) each express YFP or CFP, respectively. These two strains are detected and counted by a laser (**Fig. 17.23A**). The technique can reveal growth rate differences as small as 0.3%.

Gullberg used flow cytometry to measure the ratios of resistant over sensitive cells during growth in the presence of various concentrations of an antibiotic, such as tetracycline (**Fig. 17.23B**). The slope of increase of resistant cells in the mixture was measured over generations, yielding a "selection coefficient" or measure of the degree of selection pressure. The selection coefficients were plotted as a function of tetracycline concentration (**Fig. 17.23C**). Surprisingly, selection pressure was detectable at a tetracycline concentration 100-fold lower than the concentration that prevents growth (MIC). Thus, a new term was defined, the minimum selective concentration (MSC). The MSC is defined as the threshold concentration at which resistance provides a positive selective value. The MSC for various antibiotics is comparable to the low levels entering our waterways and possibly our drinking supply.

Thought Question

17.9 In the fitness competition between cells labeled by YFP and CFP, how can we rule out fitness differences associated with the bacterial expression of the two different fluorescent proteins?

Experimental Evolution in the Laboratory

Natural selection for single genes is part of a much larger process of evolution that comprises entire genomes of individuals across populations over time. The study of overall evolution in the laboratory is called **experimental evolution** or **laboratory evolution**.

A landmark experiment on evolution in the laboratory was undertaken by Richard Lenski, Zachary Blount, and colleagues at Michigan State University (**Fig. 17.24**). In 1988, Lenski began the Long-Term Evolution Experiment (LTEE) by founding 12 populations of *E. coli* from a single clone. The 12 populations were cultured in a medium in which growth is limited by a small amount of glucose (**Fig. 17.25**). Every 24 hours, 1% of each population is

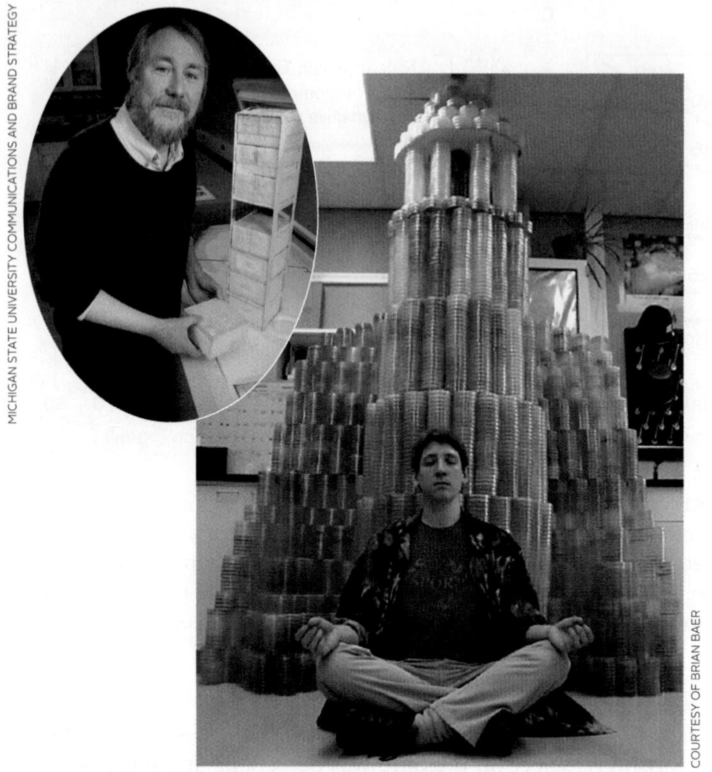

FIGURE 17.24 ■ Rich Lenski and Zack Blount conduct the Long-Term Evolution Experiment (LTEE). At Michigan State, Lenski (inset) opens a box of evolved clones stored in the –80°C freezer. Blount meditates before a tower of Petri plates that represent 1 year of competition assays for the relative fitness of evolved clones.

transferred to fresh medium and then grows at the rate of about 6.6 generations per day. Every 500 generations, samples of each population are frozen for later study. The populations have now evolved for more than 60,000 generations, and Lenski's students continue the daily dilutions.

The bacteria in Lenski's frozen populations remain alive and can be revived at any time for analysis and use in experiments. Experiments include the sequencing of genomic DNA of clones isolated from these frozen time points. With laboratory evolution experiments, we can test predictions of how new traits appear and use genome sequences to examine the mutations that led to the new traits. Evolution experiments can also include many replicate populations, making it possible to examine the repeatability of evolution.

Adaptation to the experimental environment. During the LTEE, the population has changed and adapted to the fixed conditions of the experimental environment, such as glucose as the sole available carbon source. Some bacteria have been able to grow faster than their ancestral strain. The researchers measure the **relative fitness**, a strain's ability to yield more progeny than a competitor. The relative fitness of each evolved isolate is compared to that of

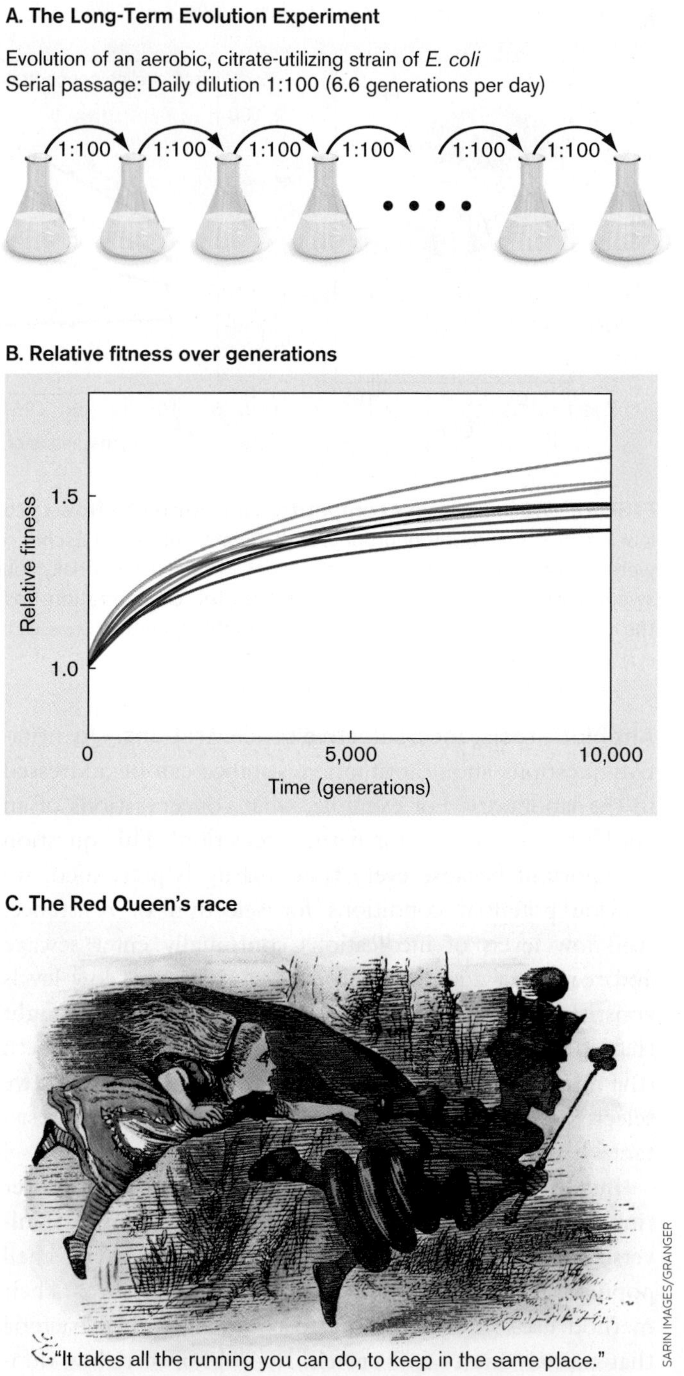

FIGURE 17.25 ■ Relative fitness increases over generations in a stable environment. **A.** Bacteria were diluted daily in a glucose-limited medium. The medium also contained citrate to enable the bacteria to take up iron. **B.** Relative fitness indicates the ratio of offspring produced compared to those of an ancestral strain in direct competition. Colored lines represent independent flasks undergoing evolution under the same conditions. In a constant experimental environment, fitness increases over generations. The rate of increase declines but never reaches zero. **C.** The Red Queen's race, from Lewis Carroll's *Through the Looking-Glass*. Even under constant abiotic conditions, individuals keep changing (by mutation) and thus change the environment experienced by their competitors. Thus, all population members must change in order to maintain relative fitness. *Source:* Part B modified from R. E. Lenski and M. Travisano. 1994. Dynamics of adaptation and diversification. Chapter 13 in *Tempo and Mode in Evolution*, National Academies Press.

the frozen original ancestor. Relative fitness is measured by direct competition assays, in which two strains are cultured together, then quantified by viable count assay on agar plates. **Figure 17.25B** shows the result of competition assays for nine different strains evolved for increasing numbers of generations (doublings) cultured together with the ancestral strain.

Several features of Lenski's relative fitness curve have proved surprisingly consistent over many evolution experiments conducted under various defined conditions. Most notably:

- Relative fitness increases at approximately equal rates for all replicate populations, even though different independent mutations are occurring.
- The rate of fitness increase declines over succeeding generations. But the fitness increase never goes to zero, even over 30 years and tens of thousands of generations.
- Genome sequences of persisting lineages show independent collections of distinct, independent mutations.

The continuing fitness increase of competing microbes, albeit at a decreasing rate, confirms a view of evolution formulated in the twentieth century as the "Red Queen's race" (**Fig. 17.25C**). The name of this hypothesis refers to the scene in Lewis Carroll's *Through the Looking-Glass* in which the Red Queen (an animated chess piece) tells Alice, "It takes all the running you can do, to keep in the same place." In other words, to sustain a given level of reproductive success, individuals must keep undergoing selection against their ever-evolving competitors. Even if the abiotic environmental conditions remain consistent, the individuals propagating within a population alter the environment inhabited by each other—for example, by producing or consuming molecules that are nutrients or toxins.

A radically new phenotype appears. Besides the general curve of fitness increase, we know that natural populations undergo radical shifts in phenotype, such as the appearance of the small-colony variant of MRSA (see **Fig. 17.22**). Such shifts have rarely been seen in evolution experiments—but one occurred in Lenski's LTEE. After generation 33,000, one of the 12 populations suddenly became much denser (**Fig. 17.26A**). In this one population, many of the bacteria had evolved a new phenotype: They were able to grow aerobically on citrate—a substance originally included in the medium as a buffer, not intended as a nutrient. Citrate catabolism is a trait found in other species but is rare in *E. coli*. The citrate-catabolizing *E. coli* are called citrate plus, or Cit^+.

Given that citrate has been available in the LTEE medium from the beginning, why did the Cit^+ mutation take so long to evolve? By testing *E. coli* from earlier generations that he had stored in a freezer, Blount was able to show that over the first 31,000 generations, no bacteria could metabolize citrate, but then a rare mutation occurred that enabled citrate uptake and catabolism. Over a few generations, the Cit^+ cells had increased their proportion of the population, until they became the predominant strain.

But was the overall phenotype really so sudden? Could any *E. coli* cell mutate to catabolize citrate? Lenski proposed that the earlier generations had acquired some kind of "potentiating" mutations that somehow enabled *E. coli* to gain the Cit^+ mutation. He called this the "historical contingency" hypothesis—that evolution of a new trait may require some other unknown trait to appear first. In support of the hypothesis, Lenski noted that the Cit^+ strains were extremely rare, arising in only one of the 12 populations. Therefore, mathematically it was likely that Cit^+ required the presence of some other rare mutation.

To test the hypothesis, Blount delved into the "fossil record" (frozen samples) of the evolving populations and isolated clones from various time points. He and Lenski then tested the ability of these clones, as well as of the ancestral strain, to mutate to Cit^+. They found that only clones isolated from time points after 20,000 generations reevolved the new Cit^+ phenotype; the original strain did not. This result supported the hypothesis that some other mutation was needed first.

To gain additional information about the early and late mutations, Blount sequenced the genomes of 29 clones isolated from the population's frozen fossil record at numerous time points through 40,000 generations (**Fig. 17.26B**). The strains were measured for citrate utilization, and for ability to evolve Cit^+. The data fit the following model of stages of evolving a new trait: (1) potentiation to achieve useful mutations; (2) actualization of a novel mutant phenotype; and (3) refinement, or increasing the degree of the phenotype.

Potentiating mutations. During the first 31,000 generations, early mutations occurred in some cells that did not confer citrate catabolism but somehow "potentiated" the ability of cells to gain a Cit^+ mutation later. These mutations led to the appearance of Cit^+ cells (**Fig. 17.26B**) and to the appearance of similar mutations in "replayed" evolution from late-generation strains. One potentiating mutation was shown to be *gltA*, a gene encoding citrate synthase, which synthesizes citrate in the TCA cycle. This mutation allowed growth on acetate, which is excreted during growth on glucose (discussed in Chapter 13).

Actualization of the novel phenotype. The Cit^+ mutation arose initially as a weak ability to transport citrate in the presence of oxygen, and therefore to respire on citrate. The genetic basis of this mutation was found to be a tandem duplication of a gene encoding a citrate/succinate antiporter, *citT* (**Fig. 17.26C**). Because the antiporter excretes succinate, the Cit^+ phenotype causes a metabolic

A.

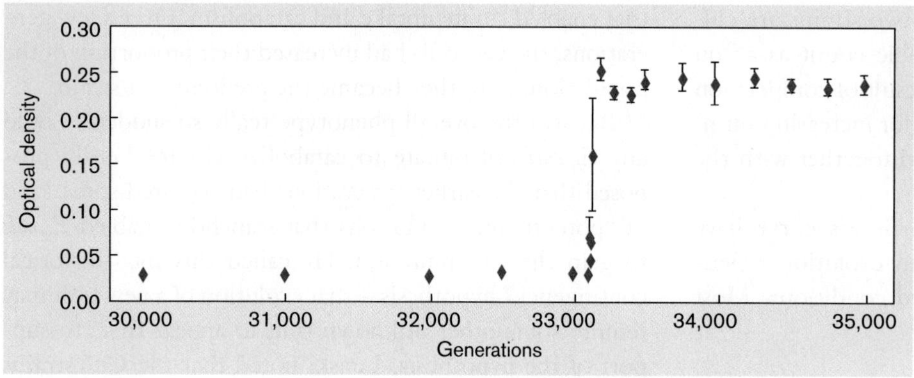

B. Evolution of Cit$^+$ (citrate catabolism)

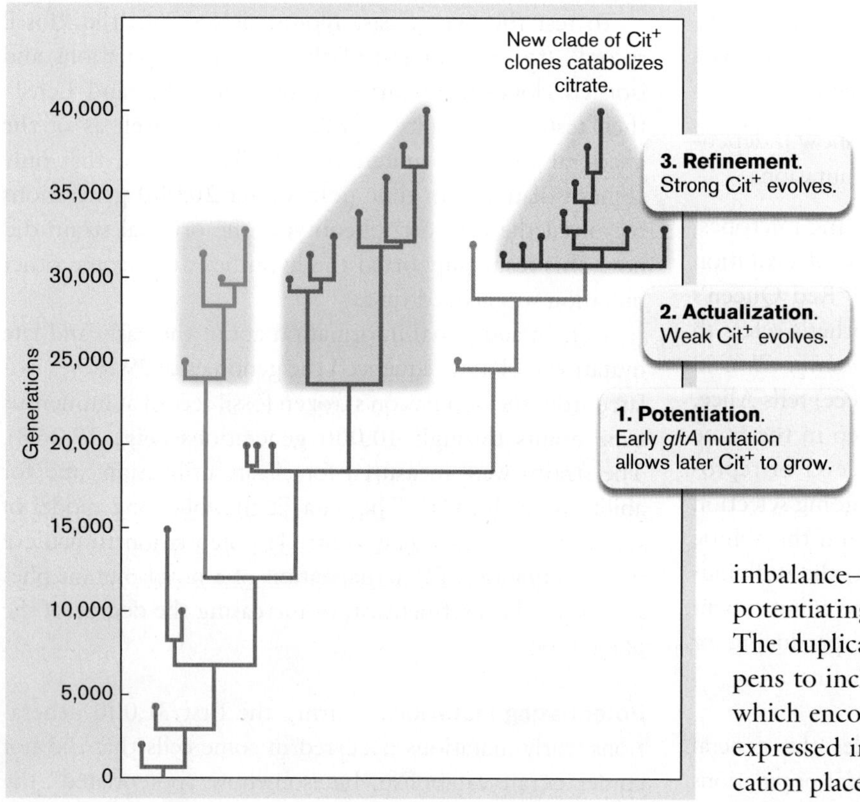

C. Tandem duplication fuses *rnk* promoter upstream of *citT*

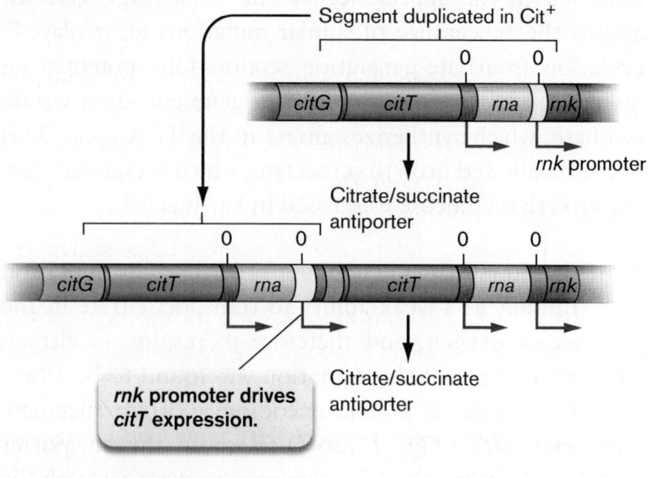

FIGURE 17.26 ■ Evolution of aerobic citrate catabolism in *E. coli*. A. The bacteria were diluted daily in a glucose-limited growth medium. The medium also contained citrate as a buffer. After 33,000 generations (estimated cell doublings), the bacterial population suddenly grew to a much greater density because of the rise to high frequency of a strain that had evolved the ability to catabolize the citrate. **B.** One or more of the mutations that accumulated over 30,000 generations potentiated evolution of the Cit$^+$ trait. The Cit$^+$ trait first appeared in about generation 31,000. Subsequent mutations refined the phenotype, increasing the rate of citrate utilization. Ball symbols indicate genomes from the population's history that were sequenced for analysis. Colors indicate evolving clades that share an ancestor. **C.** The Cit$^+$ mutation. After 31,000 generations, a tandem duplication event placed a copy of the *rnk* promoter, which directs expression when oxygen is present, upstream of the *citT gene* encoding a citrate/succinate antiporter that is normally repressed by oxygen. This new *rnk-citT* module now expresses CitT, which takes up citrate, enabling bacteria to respire on citrate from the medium.

Sources: Part A adapted from Zachary Blount et al. 2008. *PNAS* **105**:7899; part B, from Zachary Blount et al. 2012. *Nature* **489**:513.

imbalance—but this imbalance gets corrected by the potentiating mutation for acetate utilization, *gltA*. The duplicated *citT* (upstream of the first copy) happens to include a promoter for an adjacent gene, *rnk*, which encodes a regulator of nucleic acid metabolism, expressed in the presence of oxygen. Thus, the duplication places *citT* expression under control of the copied *rnk* promoter. The *rnk* promoter now expresses *citT* in the presence of oxygen, enabling the cell to respire on citrate.

The history of the initial Cit$^+$ clone thus required two rare events: the potentiating mutation *gltA*; and the tandem duplication of *citT-rnk*, which placed *citT* expression under control of a promoter active in the presence of oxygen.

Refinement of the phenotype. The earliest Cit$^+$ clones were very poor at growing on the citrate. But later clones evolved as a result of mutations that increased the rate of citrate utilization—that is, by evolutionary refinement of the Cit$^+$ trait. The researchers showed that this improvement was due to an increase in the copy number of the duplicated segment that produces the *rnk-citT* module, which increased the number of expressed copies of *citT*.

This improved growth on citrate eventually led the Cit^+ clones to dominate the population. Interestingly, though, even the "refined" Cit^+ cells never fully took over the cultures. Some cells always remained that had adapted to limiting glucose by other means.

Thought Question

17.10 The evolution of clones requiring nutrients from other clones may be more common than expected. How might the mechanisms of evolution favor the emergence of dependent clones?

The results of Lenski's Long-Term Evolution Experiment raise this question: If *E. coli* is evolving new traits, is it in the process of evolving a new species? We address the species concept next (Section 17.5). Meanwhile, beyond the philosophical debates, industrial laboratories now use advanced forms of experimental evolution. An example of industrial evolution is the development of hyperthermophilic forms of a xylanase enzyme used for paper production (discussed in **eTopic 17.3**). For an interview with Richard Lenski, see **eTopic 17.4**.

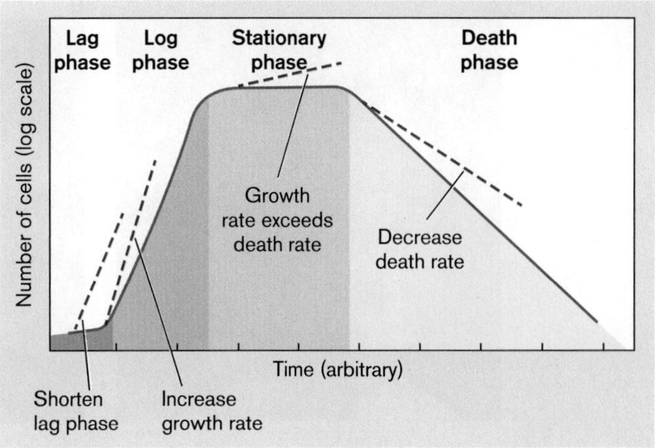

FIGURE 17.27 ■ Fitness competition during serial batch culture. A subpopulation may achieve higher fitness (more offspring compared to a competitor) by different mechanisms in different phases of the growth cycle (red lines). After dilution into fresh medium, an individual with a shortened lag phase may outcompete others in the population. During log phase, relative fitness increases by increasing growth rate. In stationary phase, higher fitness requires growth rate to exceed death rate. During death phase, higher fitness requires a lower relative death rate.

Evolution and Growth Cycles

Discussions of evolution tend to leave the impression that natural selection is all about producing large numbers of progeny. But actual growth cycles (discussed in Chapter 4) offer more than one way to outreproduce one's competitor.

Mechanisms of natural selection depend on the cell cycle. In natural environments, microbes experience all phases of batch culture (**Fig. 17.27**). When nongrowing cells experience a sudden influx of nutrients, they need a period of lag phase to gear up their metabolism for exponential growth. Forms of suspended animation such as endospores require an elaborate program of reversal by germination. Thus, within a population, those individuals requiring shorter times for lag phase or germination will get a head start and produce offspring before their neighbors do. So, a shorter lag phase is one mechanism that mediates natural selection.

Once exponential growth occurs, a lineage of microbes may outcompete others by having a faster doubling rate (*k* doublings per unit time, discussed in Chapter 4). But a comparable competition may be achieved by outsurviving one's competitors during death phase—that is, by dying more slowly than one's competitors (**Fig. 17.27**). A genetic variant that confers a fitness advantage in exponential phase may actually incur disadvantage during death phase. For example, *E. coli* bacteria cultured in acid (pH 4.8) show selection against genes encoding amino acid decarboxylases that degrade valuable nutrients. Yet the same bacteria exposed to more extreme acid (pH 2), where the cells start to die, show strong selection for the amino acid decarboxylases, which consume protons and protect the cytoplasm from acidification.

Stationary phase leads to a complex mix of selective effects, in which some subpopulations die off, whereas others show a net growth rate. In stationary phase, and under other kinds of stress, such as low oxygen or antibiotic exposure, highly expressed genes may show an increased rate of mutation. The higher mutation rate offers a temporary opportunity for genetic adaptation to conditions outside the optimal "window" of environment for the organism. Because the mutation increase is temporary, it incurs fewer long-term side effects than does the appearance of mutator gene variants that increase the mutation rate permanently.

Susan Rosenberg at Baylor College of Medicine (**Fig. 17.28A**) has conducted a series of experiments that reveal mechanisms by which starvation stress leads to mutation. Some of these mechanisms were first detected as the basis of sectored colonies because color phenotypes arise from late mutations that turn on *lac* fermentation within the colony (**Fig. 17.28B**). The exact basis of stress-associated means of mutation increase is controversial—and of great interest for medical fields such as cancer, where similar mechanisms may underlie the development of tumors. Examples found in bacteria include:

- Increased rate of errors during double-strand-break repair, which is mediated by homologous recombination (Chapter 9).

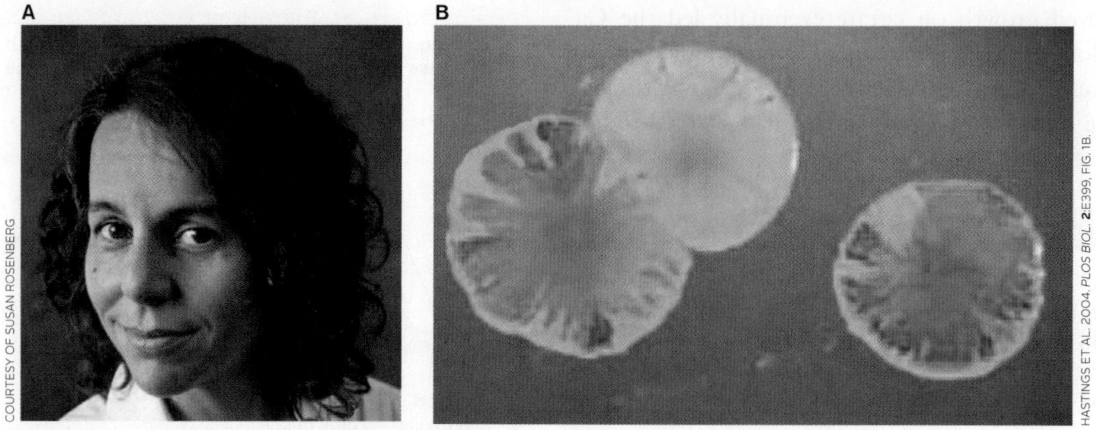

FIGURE 17.28 ▪ Stress conditions lead to increased rates of random mutation. **A.** Susan Rosenberg showed how bacteria under starvation stress increase their rate of random mutation and evolve novel variants faster. **B.** Sectored colonies form as a result of random mutations leading to Lac^+ phenotype (green indicator).

- Up-regulation of sigma S, induced by stresses such as starvation, acid, or oxygen radicals. Sigma S activates error-prone DNA polymerases, leading to base substitutions, as well as insertions and deletions.
- Mismatch repair (involving MutS, MutH, MutL) is transiently down-regulated, by unknown means.
- Double-strand breaks occur at genes highly transcribed during stress response. These double-strand breaks are repaired by error-prone means, leading to "hot spots" of mutations affecting those genes involved in stress response.

To Summarize

- **Natural selection** is a process by which a genetically defined subpopulation leaves a greater number of offspring than its competitors. The genome of the more successful competitor is said to show greater fitness for the given environment.
- **Strongly selective environments**, such as antibiotic exposure, lead to rapid evolution that is easy to observe.
- **Selective pressure depends on the particular environment.** A trait favored in one environment may be disadvantageous in another.
- **Genomic analysis** tracks how organisms adapted in the past. Gene duplications provide the opportunity to evolve paralogous genes with different functions.
- **Degenerative (reductive) evolution** happens when unneeded genes are lost from the genome. The organism saves energy by avoiding their replication and expression.
- **Experimental evolution in the laboratory enables** us to test hypotheses about natural selection. Richard Lenski's Long-Term Evolution Experiment shows potentiation to achieve useful mutations; actualization of a novel mutant phenotype; and refinement, or increasing the degree of the phenotype. Laboratory evolution is used widely in industrial development.
- **Mechanisms of natural selection depend on the cell cycle.** Microbes may outcompete their competitors by having a shorter lag phase, faster growth rate in log phase, increased survival in stationary phase, or slower death rate in death phase.
- **Stress conditions affect the mutation rate.** Under starvation, several processes increase the rate of mutation and thus increase the opportunity for genetic adaptation. The increased mutation rate may be higher at more highly transcribed genes.

17.5 Microbial Species and Taxonomy

What is a species? Among eukaryotes, a species is defined by the principle that members of different species do not normally interbreed. The failure to interbreed is the traditional property that distinguishes eukaryotic species. For bacteria and archaea, however, reproduction is primarily asexual; thus, species borders are harder to define. The definition of bacterial species is complex, subject to heated debate among microbiologists. Also much debated is the classifying of life forms into different kinds (**classification**) and the naming of species (**taxonomy**).

Note: Distinguish these terms:

Classification is the sorting of life forms into bins (categories) based on genetic relatedness and traits of the organisms.

Nomenclature is the naming of categories, including species.

Taxonomy is the classification and naming of organisms.

Defining a Species

As genome sequence data became available for microbes, scientists hoped that quantitative measures of divergence could provide a consistent basis for defining microbial species. But for some organisms, such as *Helicobacter pylori*, the genomes of different strains that cause the same disease (gastritis) differ by as much as 7%. On the other hand, strains of *Bacillus* with nearly identical genomes cause completely different diseases, such as anthrax (*B. anthracis*) and caterpillar infection (the biological pesticide *B. thuringiensis*). Even more puzzling, hyperthermophilic bacteria such as *Aquifex aeolicus* show a high proportion of genes from the domain Archaea. Some researchers argue that the species concept lacks meaning for microbes.

Amid the debates, microbiologists generally agree on the importance of two perspectives: phylogeny (based on DNA relatedness) and ecology (based on shared traits and ecological niche).

Phylogenetic relatedness. A species is a group of individuals that share relatedness of a key set of "housekeeping genes," typically informational genes such as ribosomal and transcriptional components. Ideally, these genes should all be orthologs (genes with a common origin and function), not paralogs (which diverged from a common ancestor but now differ in function). Within a genus, species that cannot be distinguished by SSU rRNA alone may be defined by analysis of multiple genes. For example, analysis of multiple gene loci effectively distinguishes *Neisseria meningitidis* (the cause of meningitis) from *Neisseria lactamica*, a harmless resident of the nasopharynx. The multigene approach has proved successful even in the case of highly "recombinogenic" organisms known to acquire and rearrange genes readily.

Ecological niche (ecotype). Besides a high degree of genomic relatedness, a species should include individuals that share common traits and an ecological niche, or "ecotype." Shared traits should include cell shape and nutritional requirements, and there should be a common habitat and life history (for example, causing the same disease). By these criteria, highly divergent strains of *Helicobacter pylori* causing gastritis make up one species, whereas *Bacillus anthracis* (anthrax) and *Bacillus thuringiensis* (caterpillar infection) are different species despite their highly similar genomes.

A working definition of species. While the debate goes on, many microbiologists accept the following criteria for a working definition of a microbial species:

- **SSU rRNA identity ≥95%.** Two organisms with 95% or greater similarity in SSU rRNA sequence generally are considered to share the same genus. Beyond the genus level, rRNA sequence lacks resolution.
- **Whole-genome similarity: average nucleotide identity (ANI) of orthologs ≥95%.** Within whole genomes, we can define all the orthologous genes (orthologs, genes of the same function) that two strains share. If the strains share 95% or greater ANI for their orthologs, they may be considered the same species.
- **Shared ecotype.** If two organisms with 95% or greater identity share a common habitat and metabolism, or cause the same disease, they are considered the same species.

This working definition classifies most known bacteria in a way that reconciles phylogeny with ecotype. For example, *Helicobacter pylori* genomes with a common gastric pathology show exceptional variation due to horizontal gene transfer, even including their SSU rRNA genes. Nevertheless, a set of orthologs can be defined that share 95% ANI.

At the same time, new questions arise from the availability of multiple sequenced genomes for a single species. Suppose that every time we sequence a new isolate from nature, or from a clinical specimen, we find that every new genome sequenced has a few new genes absent from previously sequenced isolates. How, then, do we define the gene map? This situation would be unheard of for animals and plants, in which the gene map is fixed by Mendelian recombination. It is common, however, for bacteria such as *Bacillus cereus*, a Gram-positive soil bacterium and food pathogen (**Fig. 17.29**). For each new genome sequenced, several hundred new genes are identified that are absent from all the

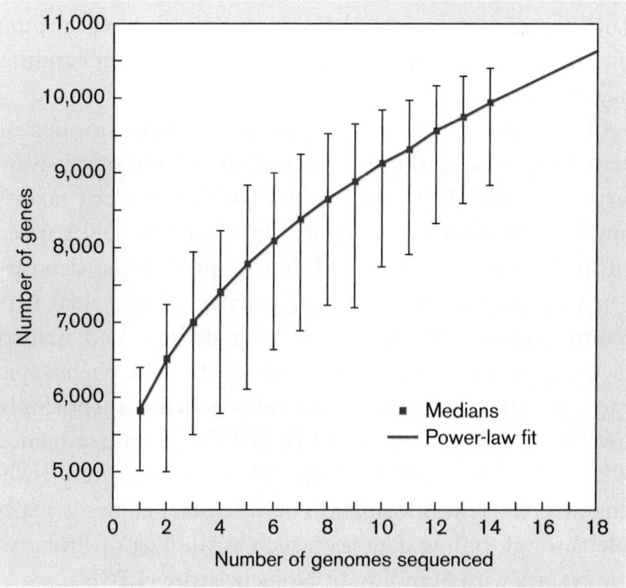

FIGURE 17.29 ■ Pangenome of *Bacillus cereus*. The median number of genes found is shown as a function of the number of *B. cereus* genomes sequenced. Error bars indicate the range in the number of genes found upon sequencing different combinations of genomes. *Source:* Modified from Hervé Tettelin et al. 2008. *Curr. Opin. Microbiol.* **12**:472.

other genomes so far. In a pathogen, some of these new genes could be involved in pathogenesis, with profound implications for medical therapy and vaccine development.

Should we now attempt to define bacterial and archaeal genomes not just by a single sequenced genome, but by the sum total of all expected genes in all possible isolates? This theoretical total is called the **pangenome**. The pangenome includes genes present in all sequenced genomes of a species (known as its **core genome**) plus "accessory genes" present in one or more sequenced isolates. But how do we estimate the size of a pangenome for a given species?

A statistical model can predict the chance of finding new genes every time we sequence another genome. A few species, such as the anthrax agent *Bacillus anthracis*, have a "closed" pangenome that appears to be defined by a relatively small number of natural isolates. But for *Bacillus cereus* (**Fig. 17.29**), after the first ten genomes, each new genome still reveals another 250 genes. Amazingly, the *B. cereus* pangenome appears to be infinite in size! This is called an "open" pangenome. Other species with open pangenomes include the pneumonia pathogen *Streptococcus pneumoniae*, the marine cyanobacterium *Prochlorococcus marinus*, and the halophilic archaeon *Haloquadratum walsbyi*. Open pangenomes may well predominate in nature; thus, most bacterial and archaeal species have access to their core genome plus an uncountable number of possible accessory genes.

Classification and Nomenclature

Defining a species is part of the task of **taxonomy**, the classifying of life forms into different categories with shared traits. Taxonomy is critical for every microbiological pursuit, from diagnosing a patient's illness to understanding microbial ecosystems.

Classification generates a hierarchy of **taxa** (groups of related organisms; singular, **taxon**) based on successively narrow criteria. The fundamental basis of modern taxonomy is DNA sequence similarity, but the use of DNA arises within a long historical tradition of phenotypic description that shapes the views and practice of microbial taxonomy. Historically, taxa have been defined and named on the basis of a combination of genetic and phenotypic traits. The nomenclature of microbes remains surprisingly fluid: As new traits are identified and the genetic sequence of a microbe is established, species are all too frequently renamed. Current information on microbial names is available through online databases such as the List of Prokaryotic Names with Standing in Nomenclature (LPSN).

Traditional classification designates levels of taxonomic hierarchy, or rank, such as phylum, class, order, and family (**Table 17.3**). Some levels of rank are designated by certain suffixes—for example, "-ales" (order) and "-aceae" (family). The ultimate designation of a type of organism is that of **species**, which includes the capitalized name of the genus (group of closely related species; plural **genera**) followed by the uncapitalized species name—for example, the well-known species *Streptomyces coelicolor*. *S. coelicolor* is a member of the Actinomycetales (informally called actinomycetes), filamentous Gram-positive bacteria that produce many kinds of antibiotics. Actinomycetes are subdivided into families and genera, such as the genus *Streptomyces*.

TABLE 17.3 Taxonomic Hierarchy of Classification

Taxon rank	A long-studied taxon
Domain	**Bacteria**
Division (phylum)	**Actinobacteria** High-GC; Gram-positive
Class	**Actinobacteria**
Subclass	**Actinobacteridae**
Order	**Actinomycetales** Filamentous; acid-fast stain
Family	**Streptomycetaceae** Hyphae produce spores
Genus	***Streptomyces***
Species (date first described)	***Streptomyces coelicolor*** **(1908)**

The definitions of taxonomic levels, however, are fluid, subject to change as microbiologists sequence new genomes. For example, the rhizobia are nitrogen-fixing soil bacteria that may develop intracellular mutualism within nodules of legume roots. The first rhizobial species characterized was *Rhizobium leguminosarum* in 1889. Since then, however, the "rhizobia" group was recognized to be **paraphyletic**; that is, it contains members of larger clades that do not share a unique common ancestor excluding members of other groups. Rhizobia are now classified under six families of Alphaproteobacteria, such as Rhizobiaceae and Bradyrhizobiaceae, as well as one family of Betaproteobacteria, the Burkholderiaceae.

Note: Taxonomic categories generally have two forms: formal and informal. The formal term is capitalized, with a latinized suffix: Actinomycetales, *Pseudomonas*, *Micrococcus*. The informal term is lowercase, in some cases with an anglicized ending, and informal references to genera are not italicized: actinomycetes, pseudomonads, micrococci.

Emerging Clades: Unclassified and Uncultured Bacteria

As we discover new microorganisms, how do we decide what to call them? The term **emerging** is used to refer to an organism recently discovered or described. If the new organism causes a disease, it is called an "emerging pathogen." An emerging organism that cannot be cultured may be known only by its habitat and its small-subunit rRNA sequence. Such organisms require a culture method and a description of essential cell structure and metabolism before designation as a new species. "Incompletely described" emerging organisms are designated as follows:

- **Unclassified/uncultured organism.** An unclassified organism, or uncultured organism, is assigned to a taxonomic rank based on SSU rRNA sequence but has not yet been grown in pure culture—for example, an "uncultured actinomycete."
- **Environmental sample.** An environmental sample is designated by its habitat and assigned a rank based on SSU rRNA. Examples of environmentally defined actinomycetes in the National Center for Biotechnology Information (NCBI) database include "oil-degrading bacterium AOB1" and "glacier bacterium FJS11."
- **Candidate species.** A cultured organism with some physiological characterization beyond DNA sequence may be published with a provisional status of **candidate species**, designated by the prefatory term "*Candidatus*." For example, in 2018 Philip Pope's lab at the Norwegian University of Life Sciences reported the discovery of "*Candidatus* Paraporphyromonas polyenzymogenes," a candidate species of the phylum Bacteroidetes that breaks down complex sugar chains in the rumen of cattle.

Nongenetic Categories for Medicine and Ecology

Genetic relatedness is the standard for classifying and naming organisms in all fields of biology. At the same time, several nongenetic systems of categorization serve a practical purpose in certain fields. These systems include:

- **Phenotypic categories for identification.** Categories such as pigmentation or cell shape (rod or coccus) may have minor genetic significance but are useful for practical identification of organisms isolated from field or clinical sources.
- **Ecological categories.** In ecology, the niche filled by an organism may be more important than its phylogeny. For example, cyanobacteria and sequoia trees both fill the role of photosynthetic producers of biomass. Both are primary producers—a trophic category of organisms that feed other organisms in the food web. Ecological categories are discussed further in Chapter 21.
- **Disease categories.** In medical microbiology, microorganisms are categorized according to the type of disease they cause or the host organ system they inhabit. For example, *Mycoplasma pneumoniae* (a cell wall–less bacterium) and influenza virus are both pulmonary pathogens, whereas *Escherichia coli* and *Bacteroides thetaiotaomicron* are both normal members of the gut microbiome. Disease categories are discussed further in Chapters 25 and 26.

Naming a Species

A commonly accepted set of names for species and higher taxa is essential for research and communication. The accepted rules for naming species and taxa have been determined by the International Committee on Systematics of Prokaryotes (ICSP). The ICSP establishes minimal criteria for designating species, genera, classes, and other taxa of bacteria and archaea. To establish a new species, a previously unknown form of microbe must be isolated and grown in pure culture; this cultured organism is known as an **isolate**. The isolate's unique genetic and phenotypic traits are published, with a proposed species name designated "*Candidatus*" for candidate species. The candidate species becomes accepted as an official species upon publication in the *International Journal of Systematic and Evolutionary Microbiology*, the official journal of record for novel prokaryotic taxa.

In recent years, the standard practice for accepting and naming species has been overwhelmed by the number of new isolates, as well as uncultured microbes, reported in the literature. Many uncultured microbes are known only by their SSU rRNA sequences. Increasingly, however, uncultured organisms have large portions of their genomes sequenced and assembled using **metagenomics** (presented in Chapters 7 and 21). Metagenomics is the sequencing of multiple genomes in an environmental community. The community can be sampled from any kind of habitat, such as the human digestive tract or the Sargasso Sea.

Identification

Once a species has been described and classified, we need a way to identify future members of the species isolated from natural environments. **Identification** requires recognition of the class of a given microbe isolated in pure culture. Identification poses special difficulties with microbes, which, by definition, are invisible to the unaided eye. Even under the electron microscope, thousands of divergent species may possess similar shape and form.

Because the definition of a species is based on its gene sequence, the most consistent way to identify an isolate is to sequence part or all of its genome and compare with known sequences in databases such as those of the National Center for Biotechnology Information (NCBI). In clinical practice, such DNA-based methods are used increasingly. Nevertheless, even in the clinical lab—as well as in field and environmental microbiology—it is convenient to narrow down the possibilities using various easily determined traits, such as cell shape, staining properties, and metabolic reactions. Thus, practical identification is based on a combination of phylogeny (relatedness based on DNA sequence divergence) and phenetic, or phenotypic, traits.

A traditional strategy of practical identification is the **dichotomous key**, in which a series of yes/no decisions successively narrows down the possible categories of species (**eTopic 17.5**). The dichotomous key approach is still used for some kinds of clinical diagnosis (Chapter 28). However, a disadvantage of the dichotomous key is that it requires a series of steps, each of which takes time. And one wrong step leads the clinician down a wrong path.

An alternative means of identification is the probabilistic indicator. A probabilistic indicator is a battery of biochemical tests performed simultaneously on an isolated strain (discussed in Chapter 28). The indicator requires a predefined database of known bacteria from a well-studied habitat, such as Gram-negative bacteria from the human intestinal tract. The database can be used to identify a specimen isolated from a patient, if the isolate coincides with a member of the predefined database.

To Summarize

- **Microbial species** are defined by sequence similarity of vertically transmitted genes such as SSU rRNA sequences and multiple orthologous genes. The species definition should be consistent with the ecological niche or pathogenicity.
- **A working definition of a species** includes shared identity of SSR rRNA above 95%, average nucleotide identity of orthologs greater than 95%, and a shared ecotype.
- A **pangenome** includes core genes possessed by all isolates of a species plus accessory genes found in some isolates but not others. A pangenome may be open (infinite number of genes) or closed (finite set of available genes).
- **Taxonomy** is the description and organization of life forms into classes (taxa). Taxonomy includes classification, nomenclature, and identification.
- **Classification** is traditionally based on a hierarchy of ranks. Groups of organisms long studied tend to have many ranks, whereas recent isolates have few.
- **DNA sequence relatedness** defines microbial taxa. Below genus level, however, the definition of bacterial species can be problematic.
- **Emerging taxa** are types of organisms recently discovered or described. They may be uncultured, and their phylogeny may be uncertain.
- **Practical identification** is based on phenotypic and genetic traits. Methods of identification include the dichotomous key and the probabilistic test battery. Both methods assume a predefined set of organisms.

17.6 Symbiosis and the Origin of Mitochondria and Chloroplasts

So far, we have largely considered single species in isolation. In fact, however, all organisms evolve in the presence of other kinds of species, with whom they share interactions, both positive and negative. A major engine of evolution is **symbiosis**, the intimate association of two unrelated species. The ecology and behavioral adaptations of microbial symbiosis are discussed in Chapter 21; here we focus on the role of symbiosis in the evolution of cells.

Evolution of Endosymbiosis

The word "symbiosis" is popularly understood to mean **mutualism**, a relationship in which both partners benefit and may absolutely require each other. Biologists, however, recognize **parasitism**, in which one partner is harmed, as a relationship equally as intimate as mutualism; both mutualism and parasitism are forms of symbiosis. Intimate relationships between species, either negative or positive, lead to **coevolution**, the evolution of two species in response to one another, showing parallel phylogeny.

An important bacterial mutualism is that of nitrogen fixation, in which rhizobia form intracellular "bacteroids" within legume plants that cannot fix nitrogen on their own. Both rhizobia and their plant hosts are highly evolved to respond to each other chemically and develop the nitrogen-fixing system. Another remarkable example of coevolution is that of leaf-cutter ants, which cultivate both fungal and bacterial partners (discussed in **eTopic 17.6**).

The most intimate kind of symbiosis is **endosymbiosis**, in which one partner population grows within the body of another organism. Endosymbiosis includes communities of microbes within the digestive tracts of animals, such as the human intestinal microbiome (discussed in Chapters 21 and 23). The internalized endosymbiont can also be intracellular, as in the case of rhizobial bacteroids within legume tissues. Rhizobia retain the genetic capacity for independence, growing readily in soil. Other intracellular endosymbionts, however, become wholly dependent on their host cells. Pathogenic endosymbionts, such as chlamydias, evolve specialized traits enabling their growth at the expense of the host and their evasion of the host immune system. But endosymbionts also undergo drastic reductive evolution, evolving ever-deeper interdependence with their host cells.

A simple example of intracellular endosymbiosis is that of the alga *Chlorella* growing within *Paramecium bursaria* (**Fig. 17.30**). The algae conduct photosynthesis and provide nutrients to the paramecium, which in turn shelters the algae from predators and viruses. The relationship is highly specific—only certain species of algae and paramecia

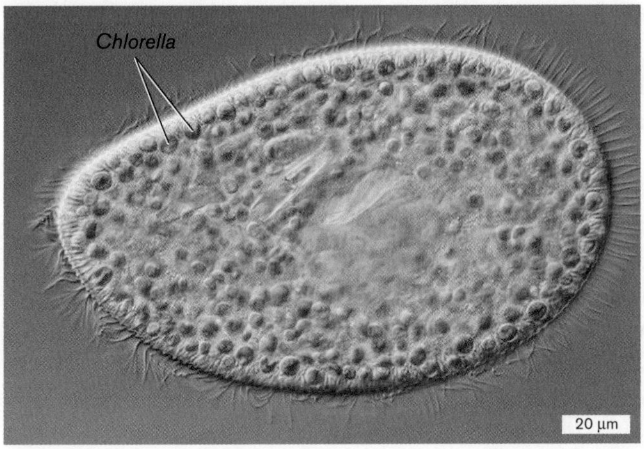

GERD GUENTHER/SCIENCE SOURCE

FIGURE 17.30 ▪ Endosymbiosis. *Paramecium bursaria*, a ciliate protist with endosymbiotic *Chlorella* algae.

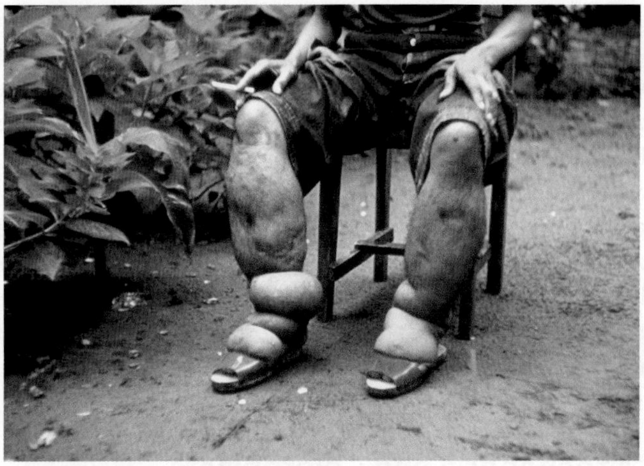

CDC

FIGURE 17.31 ▪ Filariasis. Patient suffering from a form of filariasis known as elephantiasis.

participate—and the algal growth is limited to a population that avoids harming the host. This relationship may give clues to how the bacterial ancestor of chloroplasts began its intracellular existence.

Despite this intimate relationship, *Chlorella* retains its ability to multiply outside the paramecium; thus this symbiosis is reversible. Moreover, under conditions of starvation in the absence of light, the paramecium may start to digest its endosymbionts as prey. Thus, the nature of the symbiosis (mutualistic or predatory) depends on the environment.

Many invertebrate animals, often themselves parasites of animals or plants, possess obligate bacterial endosymbionts. For example, the fruit fly, *Drosophila*, carries a parasitic endosymbiont, the bacterium *Wolbachia pipientis*. The *Wolbachia* strains that infect *Drosophila* cells are transmitted only through egg cells; they cannot exist outside the insect. Other invertebrate endosymbionts, however, are mutualists. In fact, 15% of insect species depend on intracellular bacteria to produce essential nutrients, such as certain amino acids or vitamins. In these mutualisms, both partners have lost essential traits by reductive evolution, and each now requires the partner species to provide the lost function.

A surprising number of human invertebrate parasites, such as filarial nematodes and *Anopheles* mosquitoes, carry bacterial endosymbionts required for host growth. This discovery has exciting implications for treatment of parasitic diseases. Invertebrate parasites such as filarial nematode worms invade human lymph nodes, causing forms of disease (filariasis) that are notoriously difficult to treat. Few anti-nematode compounds are sufficiently selective for worm metabolism versus human metabolism, since both are eukaryotic animals.

A form of filariasis is elephantiasis, in which a limb expands with edema because the lymph ducts are blocked by worms (**Fig. 17.31**). Filariasis afflicts more than 120 million people worldwide, largely in the Indian subcontinent and in Africa.

A filarial nematode, *Brugia malayi*, harbors endosymbiotic *Wolbachia* (a different strain from the one that infects *Drosophila*; **Fig. 17.32**). *Wolbachia* may have entered the nematode originally as a pathogen or parasite, and then persisted because of its metabolic contributions to the host. The nematode's *Wolbachia* inhabitants are mutualists; the nematode needs them for embryonic development. The bacteria are found within tissue layers beneath the nematodes' cuticle and within the uterine tubes of females, where they enter the developing offspring (**Fig. 17.32**). When human patients infected by the nematodes are treated with antibiotics such as tetracycline, the bacteria disappear from worm tissues. The worm burden gradually decreases, and no offspring are produced. Antibacterial antibiotics eliminate the worms sooner and more completely than does treatment with anti-nematode agents.

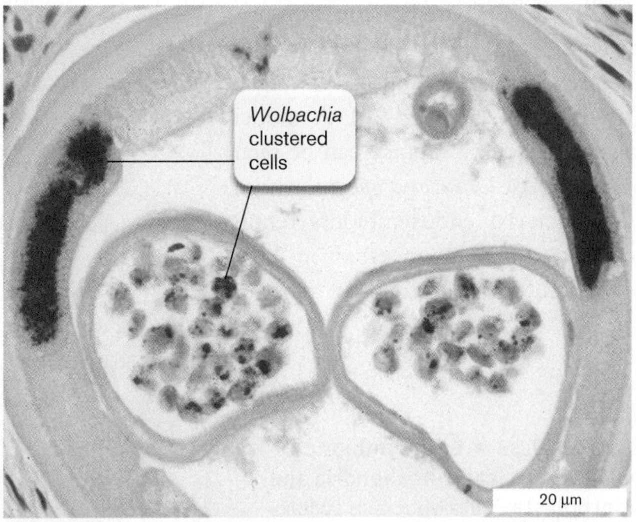

K. PFARR AND A. HOERAUF. 2005. *PLOS MED.* **2**:E110.

FIGURE 17.32 ▪ The filarial endosymbiont *Wolbachia*. Cross section of the nematode *Brugia malayi*, showing *Wolbachia* bacteria (stained pink) within the worm's dermis and uterine tubes.

The genome of a *Wolbachia* strain from a filarial nematode reveals extensive reductive evolution. With barely a million base pairs, the *Wolbachia* genome has lost many metabolic pathways. It retains glycolysis and the TCA cycle but has lost the pathways for biosynthesis of all amino acids and most vitamins. It nonetheless retains pathways to make purines, pyrimidines, and the coenzymes riboflavin and flavin adenine nucleotide (FAD)—essential pathways lost by its host nematode. Overall, *Wolbachia* appears to be evolving into an organelle of its host, like the ancestors of mitochondria and chloroplasts.

Mitochondria and Chloroplasts

As Lynn Margulis and colleagues showed, the assimilation of endosymbionts as mitochondria and chloroplasts played a central role in the evolution of eukaryotes (**Fig. 17.33**). Many similar endosymbioses are known today, such as the free-living alga *Chlorella* acquired by the protist predator *Paramecium bursaria* (see **Fig. 17.30**). Like *Wolbachia*, mitochondria evolved from a bacterium related to the rickettsias (intracellular pathogens). The mitochondrial ancestor must have entered the eukaryotic lineage as, or shortly after, the eukaryotes diverged from archaea, since all known eukaryotes retain mitochondria or vestigial remnants of mitochondrial genomes (discussed in Chapter 20).

Mitochondria provide the cell with the essential functions of electron transport and respiration. The electron transport system (ETS) is found in the mitochondrial inner membrane, believed to derive from the cell membrane of the ancestral bacterium. The outer membrane may derive from the invaginating membrane of the host cell that originally engulfed the endosymbiont.

Chloroplasts arose from cyanobacteria at some point before the divergence of red and green algae (discussed in Chapter 20). A model for cyanobacterial uptake can be seen in the protist *Glaucocystophyta*, whose cyanobacterial endosymbionts retain cell walls and some metabolism. Like mitochondria, chloroplasts possess inner and outer membranes, believed to derive from the ancestral endosymbiont and host, respectively. Photosynthetic complexes are located in the thylakoid membranes, similar to those of modern cyanobacteria.

The genomes of mitochondria and chloroplasts both show extreme reduction (**Fig. 17.34**)—even more extreme than that of any known endosymbiotic bacteria. The few genes that remain in the organellar genome include remnants of the central transcription-translation apparatus, such as rRNA and tRNAs, as well as a handful of genes whose products are essential for survival of the host cell: for respiration (mitochondria) or photosynthesis (chloroplasts).

Mitochondrial genome. In human mitochondria, the mitochondrial genome encodes key parts of the respiratory chain, including subunits of NADH dehydrogenase, cytochrome *c* oxidase, and ATP synthase (**Fig. 17.34A**). Mutations in these key genes lead to serious diseases; for example, damage to mitochondrial genes of respiration is associated with motor neuron disease, parkinsonism, and forms of ataxia.

But thousands of genes encoding ETS subunits, as well as other essential parts of mitochondria, have migrated from the mitochondrion to the nucleus. The nuclear acquisition probably occurred through accidental copying of mitochondrial genes into the nuclear genome. Reductive evolution then occurred, faster in the mitochondrial copy because of the faster mutation rate. Some of these nuclear-acquired

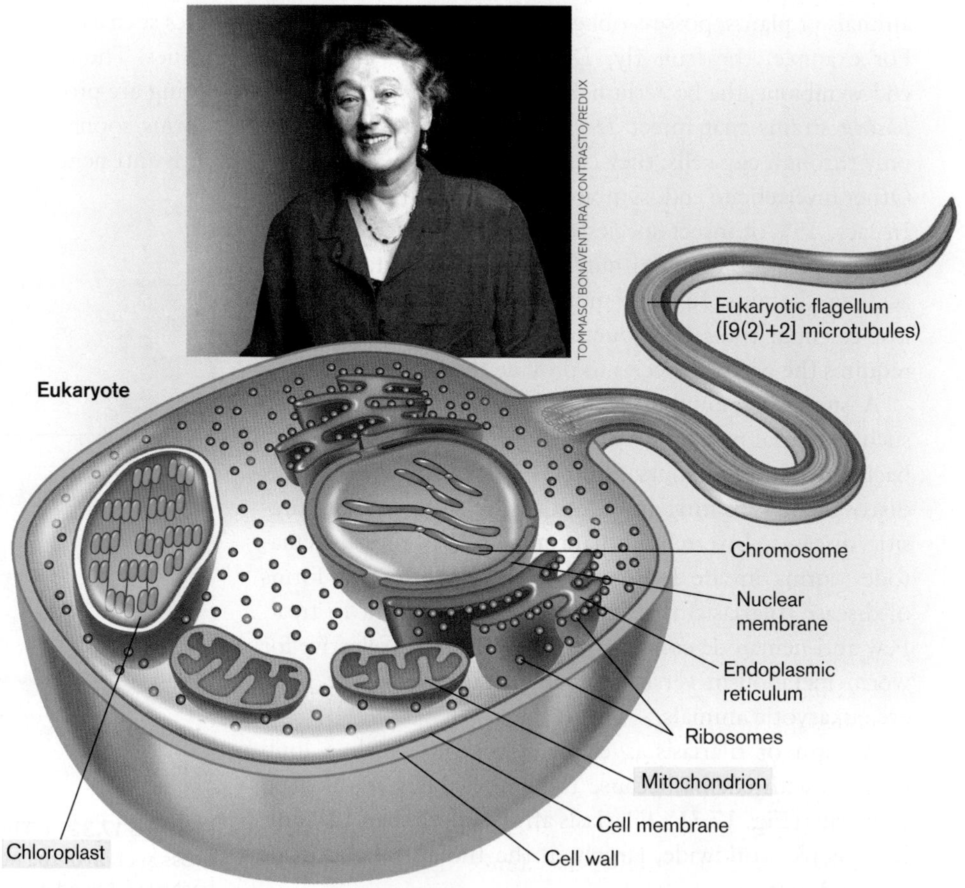

FIGURE 17.33 ■ Endosymbiotic cells evolved into mitochondria and chloroplasts. Eukaryotic cells contain mitochondria and chloroplasts, organellar remnants of ancient endosymbioses. **Inset:** Lynn Margulis (1938–2011).

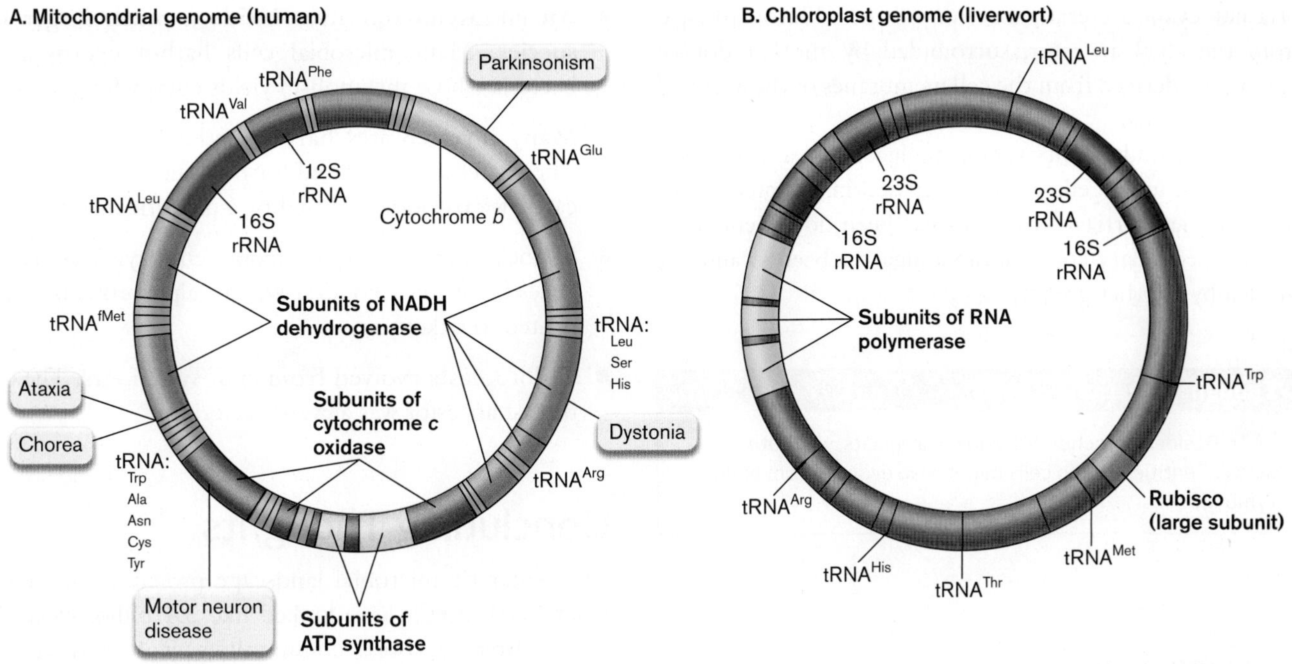

FIGURE 17.34 ▪ Genomes of mitochondria and chloroplasts. **A.** The mitochondrial genome retains large-subunit and small-subunit rRNAs, tRNA genes, plus subunits of the respiratory electron transport chain. Text bubbles indicate human diseases associated with mitochondrial defects. **B.** The chloroplast genome retains large-subunit and small-subunit rRNAs (23S and 16S), several tRNA and RNA polymerase genes, plus Rubisco (large subunit) and components of photosystems I and II (dispersed around the circle, not labeled in the figure).

mitochondrial genes show tissue-specific expression, resulting in different mitochondrial types associated with different tissues. Thus, some mitochondrial defects are actually inherited through the nuclear genome. The mitochondria have evolved as integral parts of the host cell.

Chloroplast genome. In chloroplasts, the organellar genome encodes essential products for photosynthesis, including photosystems I and II and the ATP synthase. The chloroplast genome shown in **Figure 17.34B** retains the large subunit of Rubisco, whereas the gene encoding the small subunit has migrated to the nucleus.

Remarkably, the process of symbiogenesis, the generation of new symbiotic associations, continues in many protists. Protist-algae, also known as "secondary symbiont" algae, result from symbiogenesis in which an alga (containing a chloroplast) was engulfed by an ancestral protist. The protist cell contains the degenerate remains of the algal endosymbiont (**Fig. 17.35**). The algal mitochondrion was lost through reductive evolution, and the nucleus shrank to a nucleomorph, the vestigial remains of a nucleus containing a small amount of the chromosomal DNA of the original algal nucleus. But the algal chloroplast was maintained, "enslaved" by its new host. The result is a new protist-alga species, *Guillardia theta*, capable of both phototrophy and heterotrophy. Its chloroplast has a double membrane derived from the

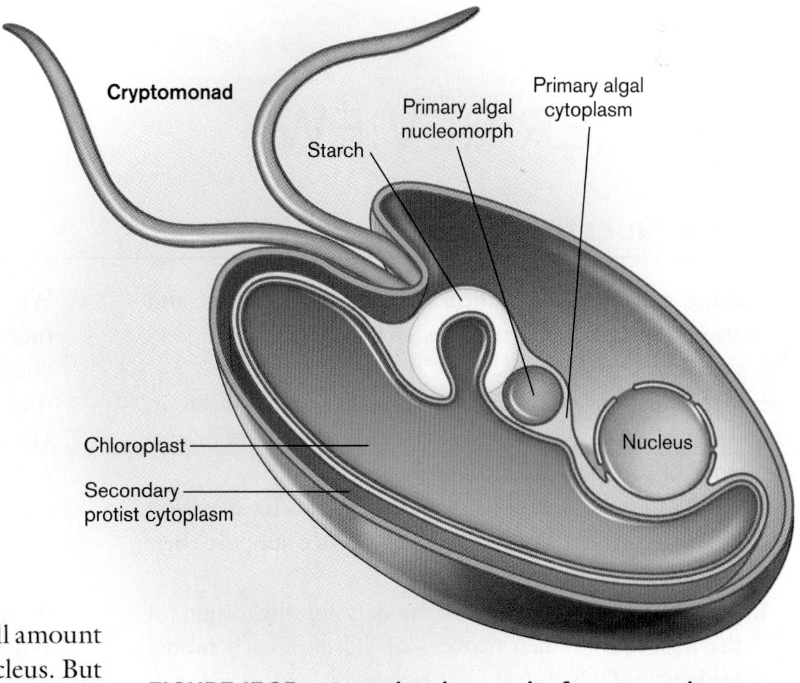

FIGURE 17.35 ▪ A protist-alga results from secondary endosymbiosis. The protist *Guillardia theta* contains the primary algal chromosome surrounded by the remains of the primary algal cytoplasm and nucleus, which has shrunk to a nucleomorph. *Source:* Modified from Paul R. Gilson. 2001. *Genome Biol.* **2**:1022.

original cyanobacterial ancestor of the chloroplast and from the algal ancestor, surrounded by another double membrane derived from the cell membranes of the alga and a secondary host.

Other secondary endosymbiont algae, such as kelps, diatoms, and dinoflagellates, are discussed in Chapter 20. In some species, tertiary symbiosis has been documented, in which a secondary-endosymbiont alga has been swallowed in turn by another protist.

Thought Question

17.11 Besides mitochondria and chloroplasts, what other kinds of entities within cells might have evolved from endosymbionts?

To Summarize

- **Symbiosis is the intimate association of two unrelated species.** A symbiosis in which both partners benefit is called **mutualism**. If one partner benefits while harming the other, the symbiosis is called **parasitism**.
- **Symbiotic partners undergo coevolution**, the evolution of two species in response to one another. Coevolution involves reductive (degenerative) evolution, in which each partner species loses some functions that the other partner provides.
- **An endosymbiont lives inside a much larger host species.** Many microbial cells harbor endosymbiotic bacteria whose metabolism yields energy for their hosts.
- **Many invertebrates harbor endosymbiotic bacteria.** The bacteria are required for host survival and, in some cases, for pathology caused by a parasitic invertebrate.
- **Mitochondria evolved from endosymbionts.** The ancestor of mitochondria was an alphaproteobacterium related to rickettsias.
- **Chloroplasts evolved from endosymbionts.** The chloroplast ancestor was a cyanobacterium.

Concluding Thoughts

The Antarctic microbial landscape provides a glimpse of what Earth might have looked like 3.4 billion years ago, as the first microbial communities evolved, their mats building rock layers that persist today. From those early microbes, all subsequent life evolved. It is hard to say which is more astonishing: the overall commonalities of all living cells, including membrane-enclosed support systems for genomes of shared ancestry, or the subsequent evolution of organisms with vastly different adaptations to exploit every possible niche of our planet. In the next three chapters we explore these diverse adaptations: Chapter 18, bacterial diversity; Chapter 19, archaeal diversity; and Chapter 20, diversity among microbial eukaryotes, including fungi, algae, and protozoa.

CHAPTER REVIEW

Review Questions

1. What was the composition of Earth's early crust and atmosphere? What processes changed their composition to that found today?
2. What kinds of evidence support the presence of life in the Archaean eon? What are the advantages and limitations of each kind of evidence?
3. What kinds of metabolism are believed to have existed in Archaean life? What kinds of evidence support their existence?
4. Compare and contrast two models for the origin of the first cells. Which features of life does each model explain, and which features are unexplained?
5. Explain the roles of classification, nomenclature, and identification for microbial taxonomy.
6. Why is the definition of species in bacteria and archaea more problematic than in eukaryotes? What is generally considered the present basis for defining prokaryotic species?
7. Discuss the roles of mutation, natural selection, and reductive evolution in the divergence of microbial species. Cite specific examples.
8. Explain the basis of a "molecular clock" for measuring microbial evolution. What fundamental properties must be met by a gene to function as a molecular clock? What are the limitations of a molecular clock?
9. Explain the basis of a phylogenetic tree. Why is the fundamental tree at the divergence of bacteria, archaea, and eukaryotes unrooted?

10. How does horizontal gene transfer determine genome content? What kinds of genes are likely to undergo horizontal transfer?
11. Explain three different ways that we can test questions about natural selection.
12. Explain how endosymbiosis can lead to obligate association. Explain how reductive evolution and gene transfer lead to the evolution of organelles that are inseparable from host cells.

Thought Questions

1. How convincing is the microfossil in **Figure 17.5A**? What criteria do you think would define a microfossil?
2. In the phylogeny shown here, where are the root and the outgroup? How does the outgroup organism differ from the others? Which two organisms are the most closely related? Which node represents the last common ancestor of *Neisseria* and *Haemophilus*? Which genome has evolved much faster than the others, and why?

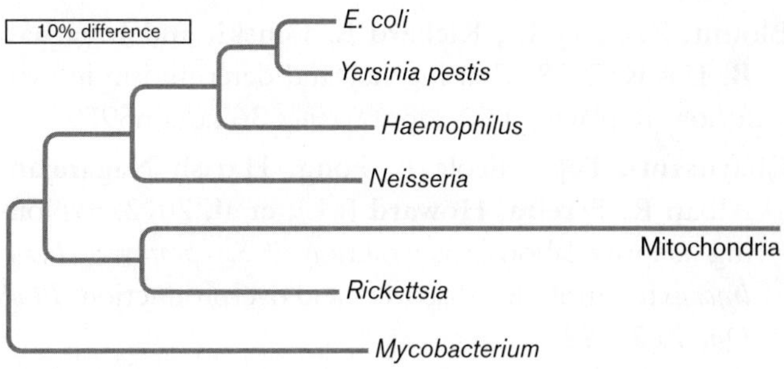

3. Design an evolution experiment in the laboratory to evolve a bacterium that breaks down a dangerous pollutant such as dioxin. How would you select the starting organism, and what experimental steps would you perform?

Key Terms

abiotic (669)
Archaean eon (663)
banded iron formation (BIF) (668)
biogenic (664)
biosignature (biological signature) (663)
biosphere (662)
candidate species (693)
clade (673)
classification (690)
coevolution (694)
core genome (692)
degenerative evolution (673, 683)
dichotomous key (694)
domain (677)
emerging (692)
endolith (662)
endosymbiosis (694)
experimental evolution (685)
fitness trade-off (684)
gene duplication (682)
genus (692)
greenhouse effect (663)
Hadean eon (663)
horizontal gene transfer (673, 678)
identification (693)
isolate (693)
isotope ratio (665)
laboratory evolution (685)
lineage (676)
metagenomics (693)
microbial mat (660)
microfossil (664)
molecular clock (673)
monophyletic group (673)
mutualism (694)
natural selection (682)
node (675)
nomenclature (690)
outgroup (676)
pangenome (692)
panspermia (672)
paralogous gene (paralog) (683)
paraphyletic group (692)
parasitism (694)
photoferrotrophy (670)
phylogenetic tree (675)
phylogeny (673)
prebiotic soup (669)
reductive evolution (673, 683)
relative fitness (686)
ribozyme (671)
RNA world (671)
root (675, 676)
small-subunit rRNA (SSU rRNA) (674)
species (673, 692)
stromatolite (660)
symbiosis (694)
taxon (692)
taxonomy (690, 692)
vertical gene transfer (678)

Recommended Reading

Attwater, James, Aditya Raguram, Alexey S. Morgunov, Edoardo Gianni, and Philipp Holliger. 2018. Ribozyme-catalysed RNA synthesis using triplet building blocks. *eLife* 7:e35255.

Blount, Zachary D., Jeffrey E. Barrick, Carla J. Davidson, and Richard E. Lenski. 2012. Genomic analysis of a key innovation in an experimental *Escherichia coli* population. *Nature* **489**:513–518.

Blount, Zachary D., Richard E. Lenski, and Jonathan B. Losos. 2018. Contingency and determinism in evolution: Replaying life's tape. *Science* **362**:eaam5979.

Charusanti, Pep, Nicole L. Fong, Harish Nagarajan, Alban R. Pereira, Howard J. Li, et al. 2012. Exploiting adaptive laboratory evolution of *Streptomyces clavuligerus* for antibiotic discovery and overproduction. *PLoS One* 7:e33722.

Didelot, Xavier, A. Sarah Walker, Tim E. Peto, Derrick W. Crook, and Daniel J. Wilson. 2016. Within-host evolution of bacterial pathogens. *Nature Reviews. Microbiology* **14**:150–162.

Fitzgerald, Devon M., P. J. Hastings, and Susan M. Rosenberg. 2017. Stress-induced mutagenesis: Implications in cancer and drug resistance. *Annual Review of Cancer Biology* **1**:119–140.

He, Amanda, Stephanie R. Penix, Preston J. Basting, Jessie M. Griffith, Kaitlin E. Creamer, et al. 2017. Acid evolution of *Escherichia coli* K-12 eliminates amino acid decarboxylases and reregulates catabolism. *Applied and Environmental Microbiology* **83**:e00442-17.

Jiao, Yongqin, Andreas Kappler, Laura R. Croal, and Dianne K. Newman. 2005. Isolation and characterization of a genetically-tractable photoautotrophic Fe(II)-oxidizing bacterium, *Rhodopseudomonas palustris* strain TIE-1. *Applied and Environmental Microbiology* **71**:4487–4496.

Johnson, Matthew D., James Bell, Kim Clarke, Rachel Chandler, Prachi Pathak, et al. 2014. Characterization of mutations in the PAS domain of the EvgS sensor kinase selected by laboratory evolution for acid resistance in *Escherichia coli. Molecular Microbiology* **93**:911–927.

Lapierre, Pascal, and J. Peter Gogarten. 2009. Estimating the size of the bacterial pan-genome. *Trends in Genetics* **25**:107–110.

Morris, J. Jeffrey, Richard E. Lenski, and Erik R. Zinser. 2012. The Black Queen hypothesis: Evolution of dependencies through adaptive gene loss. *mBio* **3**:e00036-12.

Noffke, Nora, Daniel Christian, David Wacey, and Robert M. Hazen. 2013. Microbially induced sedimentary structures recording an ancient ecosystem in the ca. 3.48 billion-year-old Dresser Formation, Pilbara, Western Australia. *Astrobiology* **13**:1103–1124.

Salzberg, Steven L., Julie C. Dunning Hotopp, Arthur L. Delcher, Mihai Pop, Douglas R. Smith, et al. 2005. Serendipitous discovery of *Wolbachia* genomes in multiple *Drosophila* species. *Genome Biology* **6**:R23.

Taylor, Tiffany B., Geraldine Mulley, Alexander H. Dills, Abdullah S. Alsohim, Liam J. McGuffin, et al. 2015. Evolutionary resurrection of flagellar motility via rewiring of the nitrogen regulation system. *Science* **347**:1014–1017.

Tettelin, Hervé, David Riley, Ciro Cattuto, and Duccio Medini. 2008. Comparative genomics: The bacterial pan-genome. *Current Opinion in Microbiology* **12**:472–477.

Yang, Tingting, Jun Zhong, Ju Zhang, Cuidan Li, Xia Yu, et al. 2018. Pan-genomic study of *Mycobacterium tuberculosis* reflecting the primary/secondary genes, generality/individuality, and the interconversion through copy number variations. *Frontiers in Microbiology* **9**:1886.

Yarza, Pablo, Pelin Yilmaz, Elmar Pruesse, Frank Oliver Glöckner, Wolfgang Ludwig, et al. 2014. Uniting the classification of cultured and uncultured bacteria and archaea using 16S rRNA gene sequences. *Nature Reviews. Microbiology* **12**:635–645.

CHAPTER 18

Bacterial Diversity

18.1 Bacterial Diversity at a Glance

18.2 Cyanobacteria: Oxygenic Phototrophs

18.3 Firmicutes, Tenericutes, and Actinobacteria (Gram-Positive)

18.4 Proteobacteria (Gram-Negative)

18.5 Spirochetes, Acidobacteria, Bacteroidetes, and Chlorobi (Deep-Branching Gram-Negative)

18.6 Planctomycetes, Verrucomicrobia, and Chlamydiae (PVC Superphylum)

Bacteria vary tremendously in their cell structure and metabolism. They include heterotrophs, phototrophs, and lithotrophs—and some species are all three. Bacterial cell shapes include rods, cocci, spirals, and budding forms. Ecologically, bacteria include mutualists, pathogens, and organisms that cannot be cultured in our laboratories. They range in size from half a millimeter down to cells that pass through a 0.2-μm filter. Every day we discover new species, ranging from sulfur metabolizers buried in the ocean benthos to mutualists clinging to our intestinal epithelium.

CURRENT RESEARCH highlight

The longest bacteria hang on to a worm. Bizarre bacteria live attached to nematode worms that crawl up and down through marine coastal sediment. Silvia Bulgheresi, at the University of Vienna, showed that the worms' hairlike bacteria encounter the reducing sulfides below and the oxygen-saturated water above. These "thioectosymbionts" oxidize toxic sulfides and give the worm organic nutrients. The bacteria shown here divide by Z-ring constriction—the longest known bacteria to do so. But other symbionts are short rods that widen and form a Z-ring down the middle.

Source: Nika Pende, et al. 2014. *Nature Communications* **5**:4803.

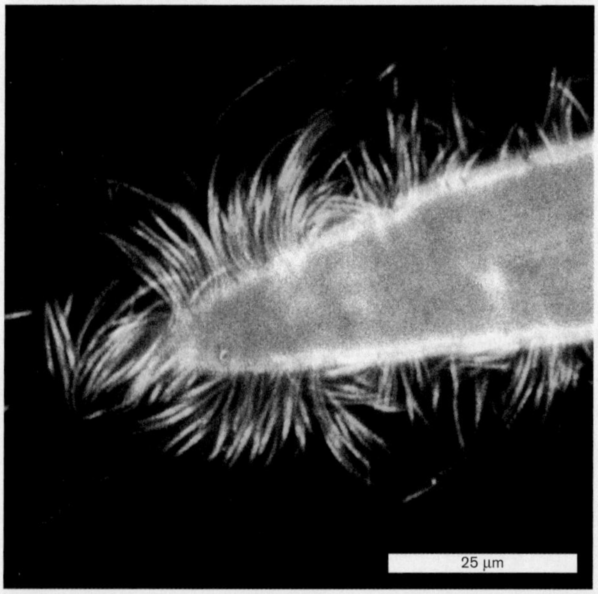

NIKA PENDE ET AL. 2014. *NAT. COMMUN.* **5**:4803, FIG. 3E. CREATIVE COMMONS LICENSE.

AN INTERVIEW WITH

SILVIA BULGHERESI, MICROBIOLOGIST, UNIVERSITY OF VIENNA, AUSTRIA

SILVIA BULGHERESI

Why do you think nematode ectosymbionts evolved such diverse cell forms?

It is still unknown why nematode ectosymbionts evolved such diverse cell shapes. One explanation might be that different ectosymbionts have different physiologies and/or are exposed to different environmental conditions (for example, sulfide or oxygen concentration). In such a scenario, specific shapes and spatial dispositions would have evolved to optimize nutrient exchange from or to the host and/or the environment.

Bacteria evolve a bewildering array of life forms that colonize every habitat on Earth, from the ocean depths to our own skin and digestive tract. You might think that, by now, we would know at least all the major phyla of bacteria, just as we know the major kinds of vertebrate animals and vascular plants. Yet we are constantly discovering new clades of bacteria never seen before, from marine sediment (see the Current Research Highlight) to the microbiome of our own intestines.

How do we begin to describe bacterial diversity, when even in familiar habitats the vast majority of species remain unknown? Chapter 18 surveys the bacteria that we do know something about. For instance, many kinds of bacteria inhabit the digestive tracts of animals, or grow in ponds by photosynthesis, or colonize legumes as nitrogen-fixing mutualists. Well-known bacteria are generally those we can culture in the laboratory and subject to experiments under controlled conditions. Sequencing their genomes shows how major groups of bacteria are related. But sequencing uncultured bacteria from metagenomes (Chapter 21) reveals entire new realms of bacteria that were previously overlooked in plain sight.

Chapter 18 emphasizes taxa that are of physiological, ecological, and medical importance. We organize our discussion by phylogeny (evolutionary relatedness of taxa), as well as key traits of a taxon, such as the Gram-positive cell wall of Firmicutes and the oxygenic photosynthesis of cyanobacteria. We show how diverse bacteria contribute to communities. Microbial communities and ecology are explored further in Chapter 21, and their roles in global biogeochemical cycles are discussed in Chapter 22.

For each major taxonomic group, we describe a few key species to represent the spectrum of diversity.

18.1 Bacterial Diversity at a Glance

To survey bacterial diversity in one chapter is like touring all the countries of a continent in a single day. Like countries, bacterial taxa have complex traits and histories, and often contested borders. But overall, bacteria share major traits in common. We review these common traits of bacteria and then go on to explore their differences.

Bacteria: Common Traits and Diverging Phylogeny

Chapter 17 summarized the differences and similarities of the three major domains—Bacteria, Archaea, and Eukarya (see Table 17.2). A common feature of bacteria is their central apparatus for gene expression, particularly their RNA polymerases, ribosomal RNAs, and translation factors. Bacterial gene expression complexes differ more from those of Archaea or Eukarya than the complexes of Archaea or Eukarya differ from each other. This subtle point of molecular biology has a profound consequence for human medicine and agriculture: It underlies the selective activity of many antibiotics, such as streptomycin, that attack only bacteria, without affecting animals or plants.

Another trait distinguishing bacteria from archaea and eukaryotes is that most bacterial cells possess a cell wall of peptidoglycan (discussed in Chapter 3). Peptidoglycan is composed of disaccharide-peptide chains that can cross-link in three dimensions; key enzymes that build the peptide links are blocked by antibiotics such as penicillin and vancomycin. Some archaea possess analogous sugar-peptide structures called "pseudopeptidoglycan" (discussed in Chapter 19), but their structural details and antibiotic sensitivity differ fundamentally from those of bacterial peptidoglycan. Eukaryotes such as fungi and plants have cell walls of polysaccharides such as cellulose and chitin (discussed in Chapter 20).

In bacteria, variant forms of peptidoglycan distinguish different species. For instance, the Gram-positive pathogen *Staphylococcus aureus* has cell wall peptides cross-linked by pentaglycine (a chain of five glycine residues). Some species, such as mycoplasmas, lack peptidoglycan altogether, although they arose by reductive evolution from bacteria that possess it.

A phylogeny of known bacteria is presented in **Figure 18.1**. The sequence data for this tree were obtained from a study of cultured and uncultured microorganisms from multiple environmental habitats that was conducted in 2015 by Jillian Banfield's group at UC Berkeley. Our understanding of bacterial phylogeny is surprisingly fluid, dependent on the choice of criteria and the use of molecular clock genes. The phylogenetic tree was computed by sequence comparison of the ribosomal RNA genes encoding rRNA for both 16S (small subunit) and 23S (large subunit), as well as concatenated sequences encoding ribosomal proteins. The inclusion of ribosomal protein sequences provides a higher degree of resolution than that of trees based solely on 16S rRNA (SSU rRNA; discussed in Chapter 17).

In **Figure 18.1**, the names of well-studied phyla, and a few recently identified phyla, are lettered in blue. A **phylum** (plural, **phyla**) is defined as a group of organisms sharing a common ancestor that diverged early from other bacteria. Increasingly, whole-genome data reveal new kinds of bacteria faster than we can figure out how to culture or characterize them. We continually discover very different new kinds of bacteria, known as emerging clades (see Chapter 17). Apart from their genome sequences, we know little about those newly discovered bacteria. Lineages whose genome sequences diverged early, at or before the well-known phyla, are called **deep-branching**

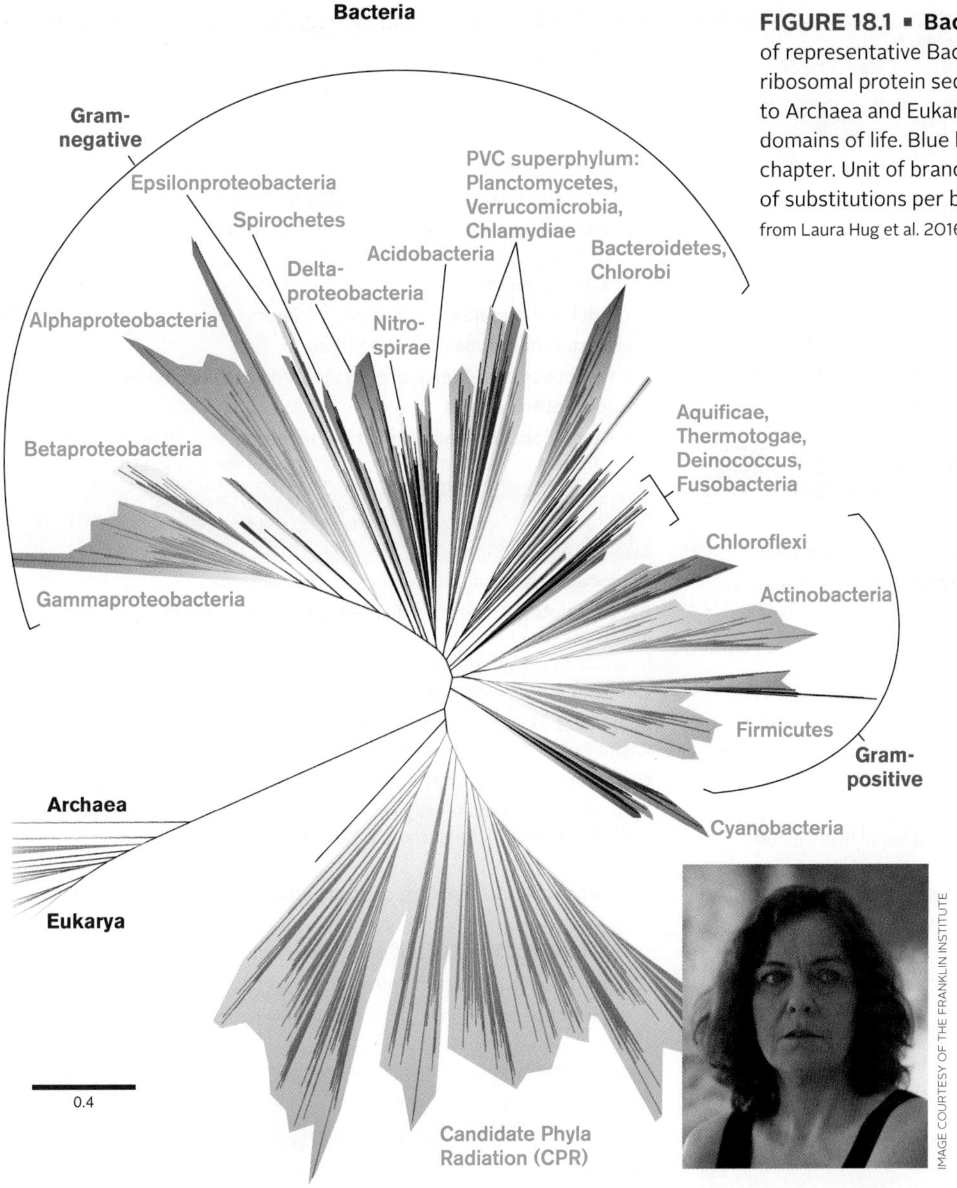

FIGURE 18.1 ▪ Bacterial phylogeny. A phylogenetic tree of representative Bacteria based on comparison of rRNA and ribosomal protein sequences. The tree is rooted with respect to Archaea and Eukarya. Black labels indicate the three domains of life. Blue labels indicate phyla discussed in this chapter. Unit of branch length represents the average number of substitutions per base. **Inset:** Jillian Banfield. *Source:* Modified from Laura Hug et al. 2016. *Nat. Microbiol.* **1**:16048, fig. 1.

taxa. In addition, other environmental DNA sequences of uncultivated bacteria (not shown) branch from all parts of this tree.

Phyla and other major divisions are also defined on the basis of historical convention and consensus of the research community. Before the advent of DNA sequencing, the Gram stain (Chapter 2) was considered a fundamental basis for dividing all bacterial species into two groups, those that retained the crystal violet stain (Gram-positive) and those that lost the stain in the classic procedure (Gram-negative). In general, bacteria that test Gram-positive possess a single membrane with a thick, multilayered peptidoglycan cell wall, whereas those that test Gram-negative usually have an outer membrane but a thin layer of peptidoglycan. The latter arrangement may be referred to as "diderm," meaning "two-skinned." DNA sequence analysis reveals that the Proteobacteria, as well as several distantly related phyla, are "diderm" and generally stain Gram-negative (**Fig. 18.1**), whereas the Actinobacteria and Firmicutes have thick cell walls and stain Gram-positive. Other phyla do not fit these patterns and are considered Gram-variable.

Generally, a well-studied bacterial phylum comprises species that share key traits as well as ancestry. While sharing key traits, the member species often show remarkable diversity in other ways. Some phyla, such as Cyanobacteria, share a unique form of metabolism (oxygenic photosynthesis), yet have evolved many diverse cell shapes and grow in diverse habitats. Other phyla, such as Spirochetes, share a unique cell structure while diverging in habitat and metabolism.

Table 18.1 introduces seven major groups of bacteria that appear commonly in the literature.

Bacterial Phyla

Certain groups of bacteria have been studied extensively, and the traits of many species are known in detail. Most of the groups presented here include cultured isolates.

Cyanobacteria. The phylum **Cyanobacteria** is a fundamental part of our biosphere, whose members conduct photosynthesis by splitting water and releasing oxygen (O_2). Marine cyanobacteria such as *Prochlorococcus* and *Synechococcus* are ubiquitous in Earth's oceans and produce a large portion of the oxygen we breathe (discussed in Chapter 21). Cyanobacteria share a common metabolism, yet the cell shape and physiology of cyanobacterial species show an immense range of different forms, including chains of cells, square arrays, and globular colonies (Section 18.2).

TABLE 18.1 Bacterial Diversity*

Cyanobacteria. Oxygenic photoautotrophs with thylakoid membranes. Share ancestry with chloroplasts.

- **Chroococcales.** Square colonies based on two division planes.
- **Gloeobacterales.** Lack thylakoids; conduct photosynthesis in cell membrane.
- **Nostocales.** Filamentous chains with N_2-fixing heterocysts. Some grow symbiotically with corals or plants. *Nostoc* sp.
- **Oscillatoriales.** Filamentous chains with motile hormogonia (short chains). *Oscillatoria* sp.
- **Pleurocapsales.** Globular colonies; reproduce through baeocytes.
- **Prochlorales.** Tiny single cells, elliptical or spherical (0.5 μm). *Prochlorococcus* sp.

Firmicutes, Tenericutes, and Actinobacteria (Gram-positive). Peptidoglycan multiple layers, cross-linked by teichoic acids.

Firmicutes. Low-GC, Gram-positive rods and cocci.

- **Bacillales.** Aerobic or facultative anaerobes. *Bacillus subtilis*.
- **Clostridia, order Clostridiales.** Anaerobic rods. *Clostridium botulinum* and *C. tetani*.
- **Lactobacillales.** Non–spore formers. Facultative anaerobes. Ferment, producing lactic acid. *Lactobacillus lactis*.

Tenericutes (Mollicutes). Lack cell wall; require animal host. *Mycoplasma pneumoniae*.

Actinobacteria. High-GC, Gram-positive bacteria with moderate salt tolerance.

- **Actinomycetales.**

  Actinomycetaceae. Filamentous, forming aerial hyphae and spores. *Streptomyces coelicolor*.

  Bifidobacteriaceae. Ferment without gas. *Bifidobacterium* sp. are important residents of the human infant microbiome.

  Corynebacteriaceae. Irregularly shaped rods. *Corynebacterium diphtheriae*.

  Micrococcaceae. Small, airborne cocci. *Micrococcus luteus*.

  Mycobacteriaceae. Exceptionally complex cell walls. *Mycobacterium tuberculosis*.

  Propionibacteriaceae. Propionic acid fermentation. *Propionibacterium shermanii*.

Proteobacteria. Gram-negative bacteria with diverse cell forms and metabolism.

Alphaproteobacteria

- **Caulobacterales.** Aquatic oligotrophs; alternate stalk and flagellum. *Caulobacter crescentus*.
- **Rhizobiales.** Plant mutualists and pathogens. *Sinorhizobium meliloti*.
- **Rhodobacterales, Rhodospirillales.** Flagellated photoheterotrophs.
- **Rickettsiales.** Intracellular parasites; related to mitochondria. *Rickettsia rickettsii* causes Rocky Mountain spotted fever.
- **Sphingomonadales.** Heterotrophs and photoheterotrophs. *Sphingomonas* sp.

Betaproteobacteria

- **Burkholderiales.** *Burkholderia pseudomallei* causes melioidosis.
- **Neisseriales.** Diplococci. *Neisseria gonorrhoeae* and *N. meningitidis*.

Gammaproteobacteria

- **Acidithiobacillales.** Lithotrophs. *Acidithiobacillus ferrooxidans* oxidizes iron and sulfur.
- **Aeromonadales.** Aquatic heterotrophs such as *Aeromonas hydrophila*.
- **Enterobacteriales.** Facultative anaerobes; colonize the human colon.
- **Legionellales.** *Legionella pneumophila* causes legionellosis pneumonia.
- **Pseudomonadales.** Rods; aerobic or respire on nitrate; catabolize aromatics. *Pseudomonas aeruginosa* infects lungs in cystic fibrosis patients.
- **Thiotrichales.** Lithotrophs and heterotrophs.
- **Vibrionales.** Marine heterotrophs. *Vibrio cholerae* causes cholera.

Deltaproteobacteria

- **Bdellovibrionales.** Periplasmic predators.
- **Desulfobacterales.** Reduce sulfate.
- **Myxococcales.** Gliding bacteria that form fruiting bodies.

Epsilonproteobacteria

- **Campylobacterales.** Spirillar pathogens. *Campylobacter jejuni* causes gastroenteritis. *Helicobacter pylori* causes gastritis.

Deep-branching Gram-negative phyla. Spirochetes, Bacteroidetes, and other phyla whose cells possess an outer membrane and usually stain Gram-negative.

Acidobacteria. Acidophiles and thermophiles.

Bacteroidetes. Anaerobes that feed on diverse carbon sources, in gut or soil (*Bacteroides thetaiotaomicron*); facultative aerobic soil heterotrophs (*Cytophaga*).

Chlorobi. Green sulfur-oxidizing phototrophs. *Chlorobium tepidum*.

Fusobacteria. Gram-negative anaerobic bacteria found in septicemia and in skin ulcers. *Fusobacterium nucleatum*.

Nitrospirae. Oxidize nitrite; aerobic or facultative. *Nitrospira marina*.

Spirochetes (Spirochaeta). Narrow, coiled cells with axial filaments, encased by sheath. Polar flagella beneath sheath double back around cell.

  Borrelia. *B. burgdorferi* causes Lyme disease, transmitted by ticks.

  Hollandina. Termite gut endosymbionts.

  Leptospira. L-shaped animal pathogens; cause leptospirosis.

  Spirochaeta. Aquatic, free-living heterotrophs.

  Treponema. *T. pallidum* causes syphilis.

Planctomycetes-Verrucomicrobia-Chlamydiae (PVC) superphylum. Irregular cells lacking peptidoglycan, with subcellular structures analogous to those of eukaryotes.

Chlamydiae. Intracellular cell wall–less pathogens of animals or protists. *Chlamydia trachomatis* causes sexually transmitted disease and trachoma (eye infection).

Planctomycetes. Some species have double membrane analogous to eukaryotic nuclear membrane. *Brocadia* species anaerobically oxidize ammonium and release N_2 (anammox reaction). *Pirellula* species inhabit marine sediment.

Verrucomicrobia. Stalk-like appendages contain actin. Aquatic oligotrophs. *Prosthecobacter* sp.

Deep-branching thermophiles. Thermophilic bacteria that diverged early from archaea and eukaryotes. Many genes transferred laterally from archaea.

Aquificae. Hyperthermophiles (70°C–95°C). Oxidize H_2. *Aquifex*.

Chloroflexi. Filamentous phototrophs, often with chlorosomes. *Chloroflexus aurantiacus*.

Deinococcus-Thermus. Radiation-resistant species and thermophiles. *Deinococcus radiodurans*.

Thermotogae. Thermophiles (55°C–100°C). Anaerobic heterotrophs. *Thermotoga* sp.

Candidate Phyla Radiation (CPR). Group of approximately 30–100 phyla known by DNA sequence, with no cultured isolates. Ultrasmall cells pass through a 0.2-μm filter. Anaerobic fermenters; have lost respiration and TCA cycle. Probably obligate symbionts.

*Bulleted terms are representative orders within the phylum (unless stated otherwise).

Firmicutes, Tenericutes, and Actinobacteria. The classic Gram-positive phylum is the **Firmicutes**, or "hard skin" bacteria. Firmicutes have an exceptionally thick cell wall, with several layers of peptidoglycan threaded by supporting molecules such as teichoic acids or mycolic acids. The thick, reinforced cell wall is what retains the Gram stain (discussed in Chapter 2). In addition, most Firmicutes possess a well-developed S-layer of glycoproteins (discussed in Chapter 3). Many Firmicutes form **endospores**, inert heat-resistant spores that can remain viable for thousands of years. Endospores are the most durable type of spore formed by bacteria.

Nevertheless, some bacteria related to Firmicutes fail to stain Gram-positive because they lack cell walls: the mycoplasmas. Now classified in the phylum **Tenericutes** (or **Mollicutes**), mycoplasmas lost their cell walls by reductive evolution as protected symbionts of host organisms.

Actinobacteria also have a thick peptidoglycan cell wall. Some species, however, possess a thick waxy coat that excludes the Gram stain. Actinobacteria differ genetically from Firmicutes in their high **GC content**—that is, the proportion of their genomes consisting of guanine-cytosine base pairs (as opposed to adenine-thymine). Actinobacteria were thus formerly known as the "high-GC Gram-positives." These bacteria include the **actinomycetes** (order Actinomycetales), which undergo complex life cycles, forming filamentous hyphae and arthrospores. Other groups closely related to actinomycetes grow as isolated rods and cocci, or with variable shape, such as the corynebacteria. Actinomycete relatives include the well-known causative agents of tuberculosis (*Mycobacterium tuberculosis*) and leprosy (*M. leprae*). Actinobacteria are ubiquitous in soil and water.

Proteobacteria. Proteobacteria comprise several phyla that are called "Gram-negative" because they possess an outer membrane of lipopolysaccharides (LPS) and their single layer of peptidoglycan fails to retain the Gram stain (see Section 18.4). The five main phyla are the **Alphaproteobacteria**, **Betaproteobacteria**, **Gammaproteobacteria**, **Deltaproteobacteria**, and **Epsilonproteobacteria**. In contrast to Cyanobacteria, the Proteobacteria show an immense range of diverse metabolism (heterotrophy, lithotrophy, and anaerobic phototrophy). Most species are aerobic or facultative anaerobes. Proteobacteria include famous model organisms and pathogens, such as *Escherichia coli*, *Salmonella enterica*, and *Yersinia pestis*. Others are human or animal symbionts, either as mutualists or as pathogens—including *Rickettsia*, the genus most closely related to mitochondria.

Note: Older literature designates the five main phyla of Proteobacteria as "classes," and separates the Greek letter from the name—for example, "Alpha Proteobacteria," "Beta Proteobacteria," and so on.

Deep-branching Gram-negatives. Several phyla with an outer membrane or sheath diverge distantly from the Proteobacteria, but stain Gram-negative. The **Spirochetes** have evolved a unique cell form of a flexible, extended spiral, resembling an old-style telephone cord. The cytoplasm and cell membrane are contained within an outer membrane called the sheath. Between the sheath and the cell membrane extend flagella doubled back from each pole. The rotation of these flagella is coordinated so as to twist and flex the helical body, generating motility and chemotaxis. Spirochetes include many free-living forms in aquatic systems, as well as digestive endosymbionts and pathogens.

Other Gram-negative clades show diverse metabolism and morphology (see Section 18.5). Members of widely different clades often coexist in multispecies biofilms, such as the biofilm growing on a colon tumor that is pictured in **Figure 18.2**. The biofilm includes **Bacteroidetes** bacteria, common residents of the gut and soil, as well as members of the **Fusobacteria**, which include human pathogens. Besides these two Gram-negative clades, the tumor biofilm includes Lachnospiraceae, which are Firmicutes (Gram-positive, introduced earlier). These biofilms, which differ in composition from the normal colon microbiome, are studied for their possible role in causing cancer.

Most members of the phyla Bacteroidetes and **Chlorobi** are obligate anaerobes. Within Bacteroidetes, *Bacteroides* species ferment complex carbohydrates, serving as the major mutualists of the human gut. By contrast, the closely related Chlorobi species are anaerobic "green sulfur" phototrophs that photolyze sulfides or hydrogen.

The **Nitrospirae** largely resemble proteobacteria in form. Most oxidize nitrite (NO_2^-) to nitrate (NO_3^-). Nitrite oxidation is a lithotrophic conversion essential for ecosystems (see Chapter 22). Another Gram-negative clade is the **Acidobacteria**, an important phylum of soil bacteria.

Planctomycetes-Verrucomicrobia-Chlamydiae (PVC) superphylum. The three related phyla of the PVC superphylum have unusual compartmentalized cells, with complex structural adaptations and development. These bacteria were thought to have lost their peptidoglycan cell walls, but peptidoglycan is now detected in some species.

Planctomycetes are free-living aquatic bacteria, with stalked cells that reproduce by budding. Each planctomycete cell contains an extra double membrane surrounding its nucleoid, analogous to a eukaryotic nuclear membrane, though it evolved independently. In the case of anammox bacteria (discussed in Chapter 14), the membrane compartment protects the cell from toxic intermediates of ammonia oxidation. **Verrucomicrobia**, also free-living, are bacteria with wart-like protruding structures containing actin. They are found in soil and water.

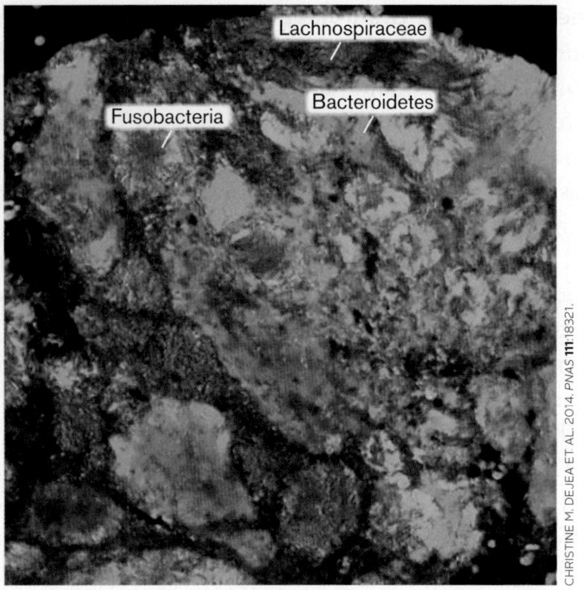

FIGURE 18.2 ▪ Multispecies biofilm on a colon tumor. Fluorescence in situ hybridization (FISH) of specific DNA probes shows surprisingly diverse bacteria growing on a tumor of the right ascending colon. Three different DNA probes hybridize with 16S rRNA of distinct bacterial taxa: Bacteroidetes (green), Lachnospiraceae (magenta), Fusobacteria (cyan).

The **Chlamydiae** (best-known genus *Chlamydia*) are intracellular parasites that lose most of their cell envelope during intracellular growth. The replicating parasites generate multiple spore-like "elementary bodies" that escape to infect the next host. *Chlamydia trachomatis* causes sexually transmitted infections, as well as the eye disease trachoma.

Deep-branching thermophiles. Deep-branching thermophiles show unusually large genetic divergence from other bacterial clades. This large divergence makes the thermophiles appear to have separated earlier from other kinds of bacteria. Other evidence, however, implicates high mutation rates and gene transfer between distant relatives as the source of genetic divergence of thermophiles.

The deep-branching thermophile taxa include extremophiles such as those in the phylum **Aquificae** (growing at up to 95°C at marine thermal vents). These organisms also show rapid growth and high mutation rates, which may have accelerated their molecular clock. These hyperthermophilic bacteria also share their high-temperature habitats with Archaea, and show surprising archaeal traits, such as archaeal ether-linked membrane lipids (discussed in Chapters 3 and 19). The genes encoding these "archaeal" lipids appear to have entered bacteria by horizontal transfer from archaea to bacteria sharing the high-temperature habitat. *Aquifex pyrophilus*, a flagellated rod, was first discovered by extremophile microbiologist Karl Stetter at the University of Regensburg in a submarine hydrothermal vent north of Iceland (for an interview, see **eTopic 18.1**). Most members of the Aquificae are hydrogenotrophs, oxidizing hydrogen gas with molecular oxygen to make water.

Another deep-branching phylum of thermophiles (growing at 50°C–80°C) is **Thermotogae**. *Thermotoga maritima* is a sulfur-reducing respirer, originally isolated from a geothermal vent in Vulcano, Italy. The cells have a loosely bound sheath, or "toga," for which the genus is named. As discussed in Chapter 17, both Aquificae and Thermotogae show remarkable mosaic genomes. Nearly a quarter of the *T. maritima* genome derives from archaea. This degree of mosaicism leads some researchers to argue that Aquificae and Thermotogae cannot be said to branch from one clade; or if they do, that we can never know, statistically, which clade that is.

Thought Question

18.1 What taxonomic questions are raised by the apparent high rate of gene transfer between archaea and thermophilic bacteria?

Chloroflexi are filamentous photoheterotrophs, supplementing heterotrophy with photosystem II (PS II) to generate ATP (discussed in Chapter 14). Together with other thermophiles, Chloroflexi species form massive microbial mats in the hot springs of Yellowstone (**Fig. 18.3**). Most species of *Chloroflexus* contain their photosynthetic apparatus within membranous organelles called **chlorosomes**. The process of photosynthesis by chlorosomes is presented in Chapter 14. Chloroflexi are informally called "green nonsulfur bacteria" to distinguish them from Chlorobi, a phylum of green phototrophs that are strict anaerobes (presented shortly, with the deep-branching Gram-negative phyla). However, *Chloroflexus* species may appear red or yellow, owing to accessory pigments.

The phylum **Deinococcus-Thermus** features a unique structural trait: the substitution of L-ornithine for diaminopimelic acid in the peptidoglycan cross-bridge. *Thermus* species (growing at 70°C–75°C) are heterotrophs commonly isolated from hot tap water. *Deinococcus* species, however, are not thermophilic. *D. radiodurans* bacteria resist extremely high doses of ionizing radiation (discussed in Chapters 5 and 9). A heterotroph, *D. radiodurans* was originally isolated from cans of meat supposedly sterilized with a radiation dose of several megarads. *Deinococcus* bacteria are found suspended in air, both indoors and high in the atmosphere; they resist drought.

Emerging Clades

Clades of microbes that are recently defined or characterized are referred to as **emerging**. Emerging clades are discovered by field microbiologists who devise ever-more-creative screens, finding new bacteria with unexpected traits.

For instance, in 2015 Jill Banfield's group at UC Berkeley discovered bacteria so small that they pass through a 0.2-μm filter—barely big enough for ribosomes. These ultrasmall bacteria were originally found in the Rifle, Colorado, aquifer; related bacteria have since been found throughout soil and water habitats. The DNA sequences of these tiny cells show deeply branching phylogeny—that is, very distant relatedness to all the bacteria known (see **Fig. 18.1**). The sequenced genomes "radiate"—that is, diverge from each other—so much that they may include more than 100 new candidate phyla. As a group, they are called the **Candidate Phyla Radiation (CPR)**.

Single-cell sequencing and TEM identify some CPR bacteria as cells that contain about 50 ribosomes, with a cell wall enclosed by an S-layer and long pili (**Fig. 18.4**). The tiny cells' genomes are smaller than those of most known bacteria. Most lack genes for respiration and biosynthesis, as well as certain ribosomal proteins that are found in all other bacteria. Bacteria in the CPR probably are obligate symbionts of other species. Some of their 16S rRNA sequences are so distant from those of other taxa that they cannot be amplified by so-called universal bacterial primers for PCR. Their extreme divergence may reflect an accelerated rate of evolution commonly found in dependent symbionts (such as *Buchnera*; Fig. 17.17). We know little about these bacteria—or others yet unknown, which future microbe hunters may discover.

A.

B.

C.

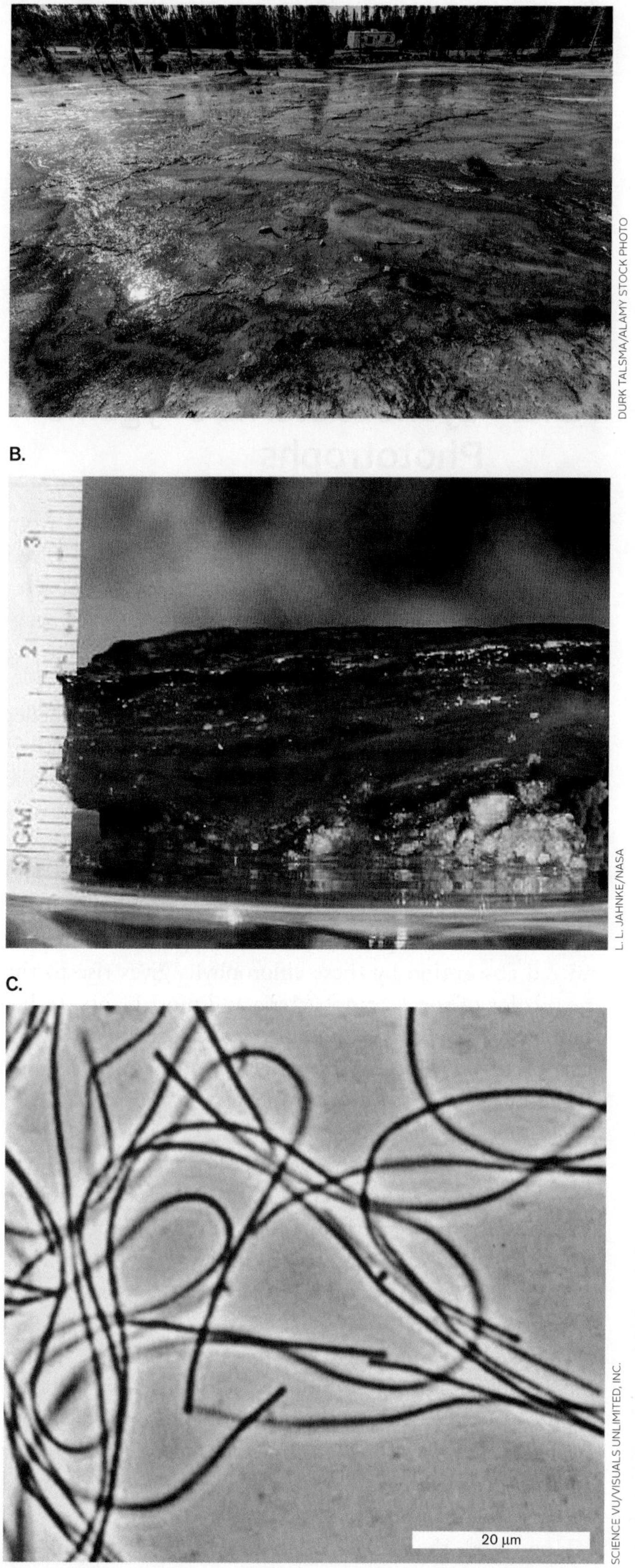

FIGURE 18.3 ■ A deep-branching thermophile: *Chloroflexus*. **A.** This hot-spring bacterial mat in Yellowstone National Park contains *Chloroflexus* species and other thermophilic bacteria. **B.** Bacterial mat section showing layers of *Chloroflexus*. **C.** *Chloroflexus* bacteria (LM).

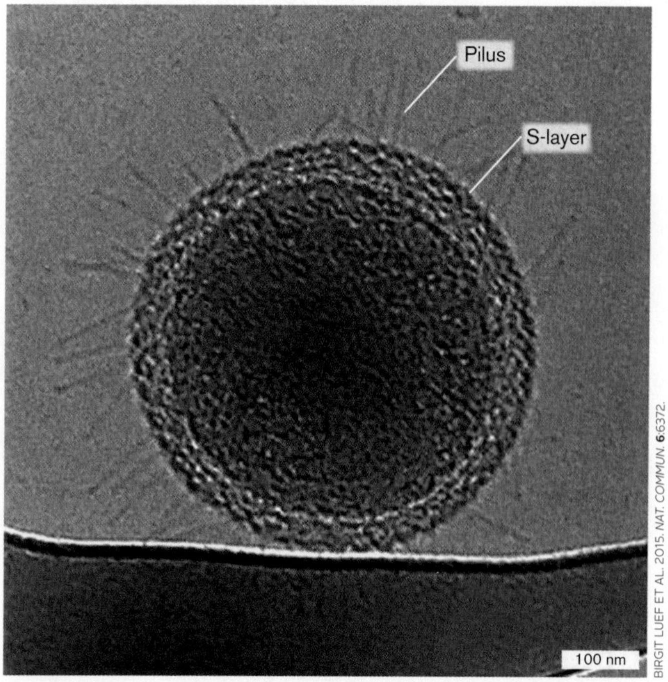

FIGURE 18.4 ■ Ultrasmall bacteria. Bacteria capable of passing through a 0.2-μm filter were discovered in 2015 by Jill Banfield and her colleagues in an aquifer in Rifle, Colorado. The tiny cells possess approximately 50 ribosomes, a cell wall surrounded by an S-layer, and long pili (cryo-TEM).

Note: The term "emerging" can refer to a newly discovered clade of organisms for which data are accumulating, or to an outbreak of disease in a location where the pathogen has not been seen recently. The meaning is inferred from the context.

Thought Questions

18.2 Which taxonomic groups in **Table 18.1** stain Gram-positive, and which stain Gram-negative? Which group contains both Gram-positive and Gram-negative species? For which groups is the Gram stain undefined, and why?

18.3 Which groups of bacterial species share common structure and physiology within the group? Which groups show extreme structural and physiological diversity?

Note: Formal names for taxa such as phyla, orders, and genera are capitalized (for example, phylum Cyanobacteria, genus *Streptococcus*). Informal forms are lowercase and roman (cyanobacteria, streptococci, and so on), as are adjectival forms (cyanobacterial, streptococcal).

Following our brief tour of the bacterial domain, we now explore some of the diverse species within well-studied phyla of **Table 18.1**.

To Summarize

- **Cyanobacteria** conduct oxygenic photosynthesis and interact closely with heterotrophs. Species vary widely in their cell shape and ecological niche.
- **Firmicutes and Actinobacteria** have a thick cell wall and generally stain Gram-positive. **Tenericutes** are related to Firmicutes but have lost their cell wall.
- **Proteobacteria** have a thin cell wall and an LPS-containing outer membrane. They show diverse metabolism and ecological adaptations.
- **Spirochetes** have flexible, spiral-shaped cells with complex intracellular architecture.
- **Other deep-branching Gram-negative phyla** include Acidobacteria, Bacteroidetes, Chlorobi, Fusobacteria, and Nitrospirae.
- **Chlamydiae, Planctomycetes, and Verrucomicrobia** have irregularly shaped cells with complex intracellular form and development.
- **Deep-branching thermophiles** such as *Aquifex* and *Thermotoga* species share traits and habitats with thermophilic archaea. Their genomes are highly mosaic, including many archaeal genes. *Deinococcus* species are highly resistant to ionizing radiation. *Chloroflexus* species are thermophilic photoheterotrophs.
- **Emerging organisms** continually reveal previously unknown clades of bacteria, such as the Candidate Phyla Radiation (CPR).

Note: The main organizing principle used in Chapters 18–20 is that of phylogeny based on DNA relatedness. Characteristic traits described for each branch apply to the majority of its known species, though many exceptions have evolved, such as members of Spirochetes that lack spiral form.

18.2 Cyanobacteria: Oxygenic Phototrophs

All of the oxygen gas in Earth's atmosphere comes from Cyanobacteria, and from plant chloroplasts that evolved from an ancient cyanobacterium. The phylum Cyanobacteria (also called Cyanophyta) is named for the blue phycocyanin accessory pigments possessed by some genera, giving them a bluish tint.

Cyanobacteria and chloroplasts (eukaryotic organelles that evolved from Cyanobacteria) are the only life forms that produce oxygen gas. These bacteria have a unique two-photosystem apparatus for oxygenic photosynthesis arranged in lamellar arrays of membranes called thylakoids (discussed in Chapter 14). Unique to Cyanobacteria and chloroplasts (discussed in Chapter 17) are the chlorophylls, most notably chlorophylls *a* and *b*; the predominant blue and red absorption by these chlorophylls gives rise to the green color of some cyanobacteria and most plants. Other cyanobacteria appear red because of the accessory pigment phycoerythrin, which absorbs blue-green light in a range missed by cyanobacterial chlorophylls.

Cyanobacteria are the only prokaryotes that use both photosystems I and II, photolyzing water to produce oxygen, as explained in Chapter 14. Under anoxic conditions, most cyanobacteria can also reduce sulfur compounds and organic compounds. Use of H_2S allows flexibility in habitats such as wetlands that alternate between aerobic and anoxic conditions.

Chlorophylls *a* and *b* are distinct from the bacteriochlorophylls used by non-oxygenic bacterial phototrophs (see **Table 18.2**). Most bacteriochlorophylls absorb lower-energy photons at longer wavelengths, which partly explains their inability to break down water, the most stable molecule that can be photolyzed.

How does the oxygenic photosynthesis of cyanobacteria compare with the non-oxygenic photosynthesis of thermophilic Chloroflexi, or of sulfur-based phototrophs among the Proteobacteria? **Table 18.2** summarizes key features of phototrophy in clades throughout the bacterial domain, including those presented in this chapter.

TABLE 18.2 Phyla That Include Phototrophic Bacteria

Taxon	Energy generation	Cell structure and photopigments	Absorption spectrum (whole cell)	Reaction center
Chloroflexi "Green nonsulfur" *Chloroflexus*	+O_2 Heterotrophy −O_2 Photoheterotrophy or use reduced sulfur	Chlorosome, BChl a, BChl c/d		PS II
Cyanobacteria "Blue-greens" *Anabaena*	+O_2 Oxygenic phototrophy −O_2 Photolithoautotrophy on reduced sulfur	Thylakoid, Chl a/b		PS I and PS II
Firmicutes "Sun bacteria" *Heliobacterium*	+O_2 −O_2 Photoheterotrophy	Forms endospore, BChl g		PS I
Proteobacteria "Purple nonsulfur" (Alpha and Beta phyla) BChl a *Rhodospirillum* BChl b *Blastochloris*	+O_2 Heterotrophy −O_2 Photoheterotrophy	BChl a / BChl b		PS II
"Purple sulfur" (Gamma phylum) *Chromatium*	+O_2 −O_2 Photolithoautotrophy on reduced sulfur	BChl a/b		PS II
"Proteorhodopsin" (Alpha and Gamma phyla) *Pelagibacter* (SAR11) SAR86	+O_2 Heterotrophy −O_2 Photoheterotrophy	Proteorhodopsin		PR
Chlorobi "Green sulfur" *Chlorobium*	−O_2 only Photolithoautotrophy on reduced sulfur	Chlorosome, BChl a, BChl $c/d/e$		PS I

Sources: J. Overman and F. Garda-Pichet. 2013. The phototrophic way of life. In *The Prokaryotes*. 4th ed. Springer; Oded Beja et al. 2001. *Nature* **411**:786–789.

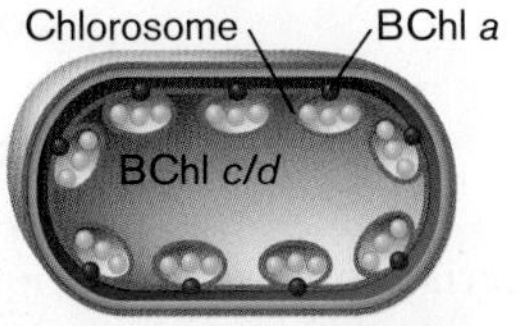

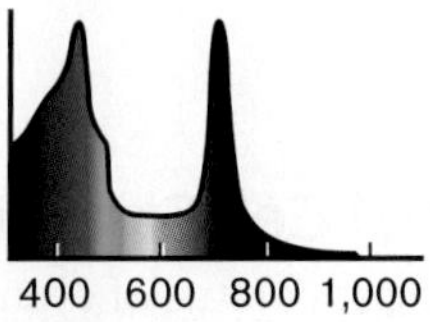

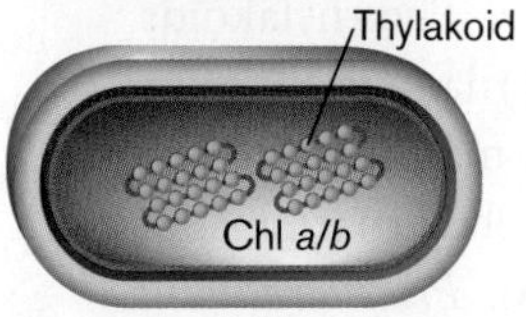

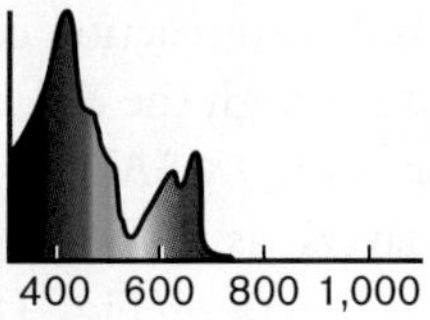

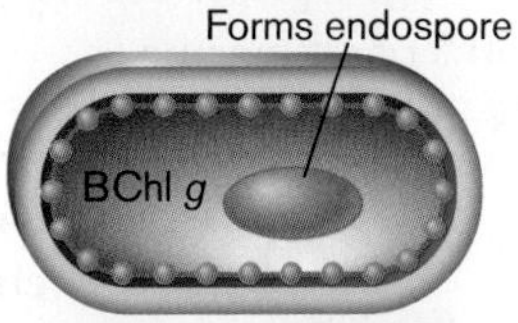

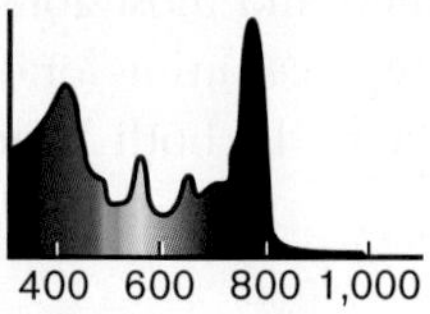

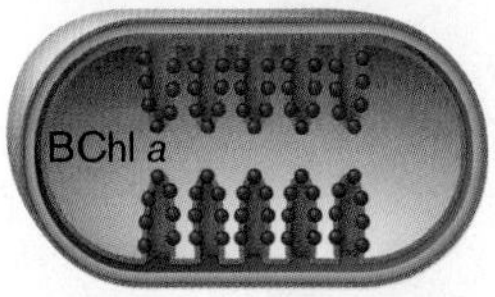

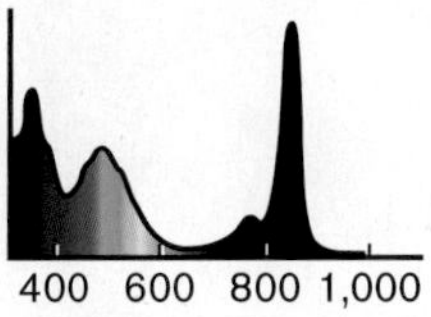

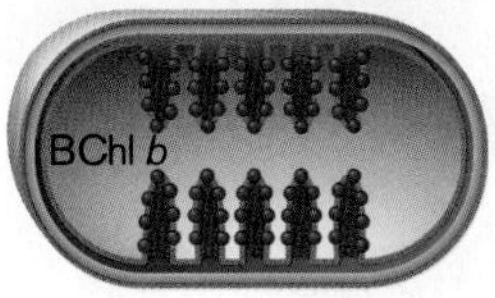

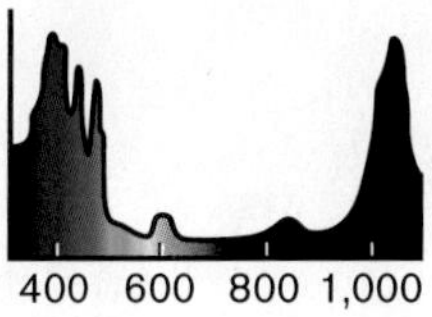

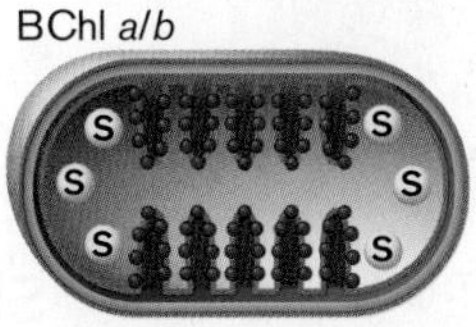

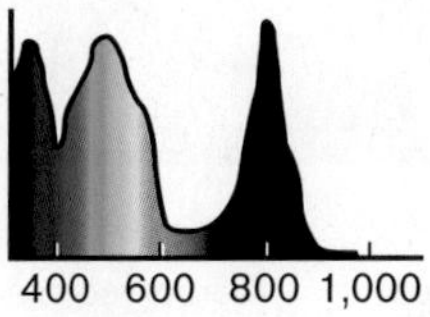

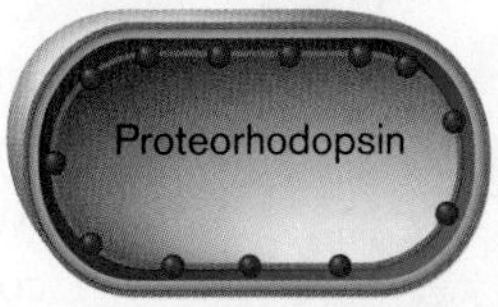

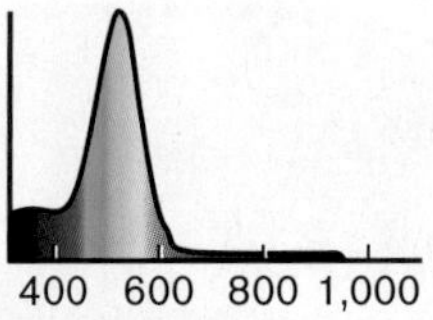

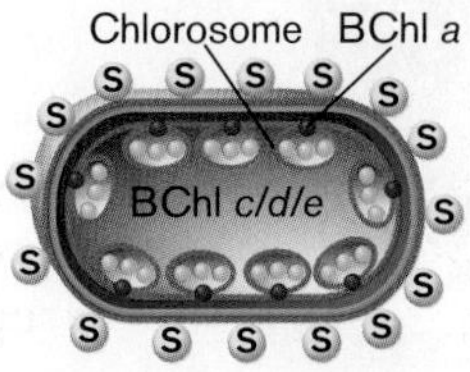

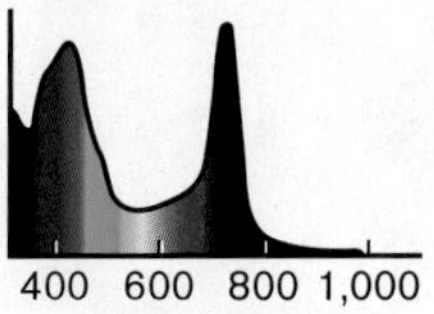

All light-harvesting complexes descend from one of two ancestral sources: the chlorophyll/bacteriochlorophyll with electron transport (PS I, PS II) or the proteorhodopsin proton pump. These photosystems have diverged through vertical inheritance, as well as by horizontal transfer between different clades.

Cyanobacterial Cell Structure

The photosynthetic apparatus of cyanobacteria is organized within thylakoids, pockets of membrane resembling flattened spheres packed with reaction centers. The thylakoids may be distributed through the cell, as in filamentous genera such as *Nostoc* (**Fig. 18.5A**), or they may encircle the cell in concentric layers, as in the single-celled marine species *Prochlorococcus marinus* (**Fig. 18.5B**). *Prochlorococcus* is one of the smallest and most abundant oxygen producers in the biosphere, accounting for 40%–50% of all marine phototrophic biomass. In both large cells and small, the thylakoids are completely separate from the plasma membrane, unlike the attached chlorosomes of Chloroflexi and the plasma membrane extensions of "purple" Proteobacteria (see **Table 18.2**). Cyanobacterial thylakoids resemble the thylakoids of eukaryotic chloroplasts; they are the most complex and specialized form of photosynthetic apparatus.

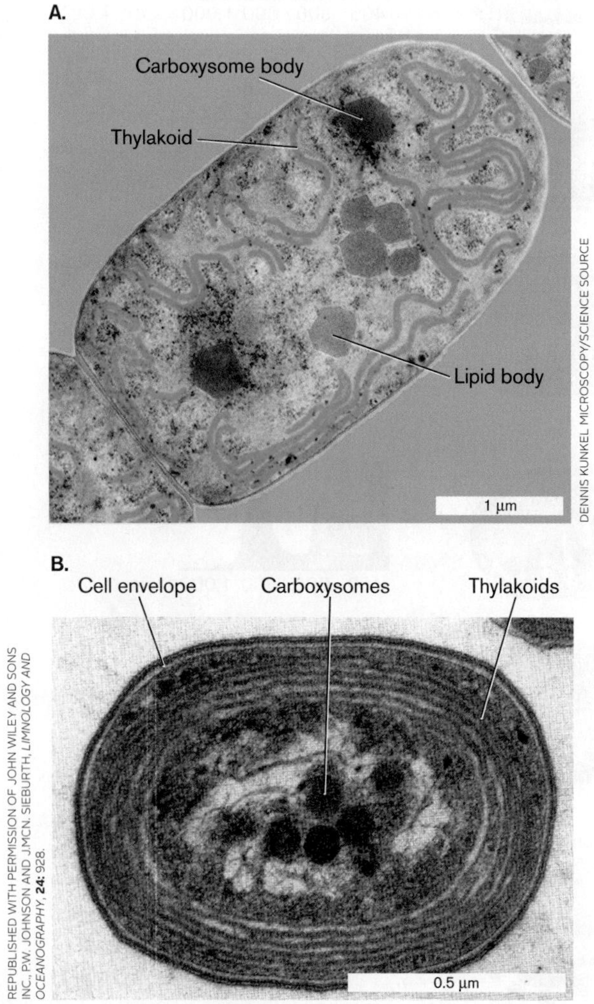

FIGURE 18.5 ■ Cyanobacterial cell structure. **A.** Intracellular organelles of *Nostoc*, a typical filamentous cyanobacterium (colorized TEM). **B.** Intracellular organelles of *Prochlorococcus*, a prochlorophyte cyanobacterium, the smallest known phototroph. *Prochlorococcus* accounts for 40%–50% of marine phototrophic biomass.

Cyanobacteria have several other subcellular structures (**Fig. 18.5**). **Carboxysomes** (also known as polyhedral bodies) are rich in the enzyme Rubisco, and they fix CO_2 (discussed in Chapter 15). Cyanobacteria store energy-rich compounds in lipid bodies. To maintain height in the water column and thus access to sunlight, cyanobacteria have **gas vesicles**, whose buoyancy enables cells to float. Their external structures include a thick peptidoglycan cell wall, similar to that of Gram-positive cells, plus several external layers that vary with different species. Many species move by "gliding," a form of motility whose mechanism is poorly understood.

Besides fixing CO_2, most cyanobacteria fix N_2. Because nitrogen fixation requires the absence of oxygen, cyanobacteria have to solve the problem of maintaining anaerobic biochemistry while producing huge quantities of highly toxic O_2. Different species solve this problem in different ways:

- Formation of specialized nitrogen-fixing cells called **heterocysts**. Heterocysts exclude oxygen, which associated heterotrophic bacteria consume.
- Temporal separation, alternating between photosynthesis during daylight and nitrogen fixation at night.
- Accumulation of large aggregates of cells in which the interior becomes sufficiently anaerobic for nitrogenase to function, while the exterior continues oxygenic photosynthesis.
- Symbiosis with fungi (as in lichens) or plants that consume oxygen or otherwise maintain anoxic conditions.

Single-Celled, Filamentous, and Colonial Cyanobacteria

Single-celled species include *Synechococcus* and *Prochlorococcus*, the most abundant phototrophs in the oceans. Another single-celled cyanobacterium is *Microcystis*, a colonial microbe found in freshwater that produces dangerous toxins (microcystins). Blooms of *Microcystis* form in the spring and summer when nitrogen and phosphorus runoff from agricultural fertilizer and animal waste enters lakes and rivers. A *Microcystis* bloom in Lake Erie in 2014 resulted in microcystin levels great enough that the water supply of Toledo was contaminated.

Since cyanobacterial species can have positive or negative impacts on their habitat, environmental scientists survey the cyanobacterial communities of lakes and ponds in order to assess their health. **Figure 18.6A** shows undergraduate Alex Schaal from Kenyon College sampling a pond in Ohio

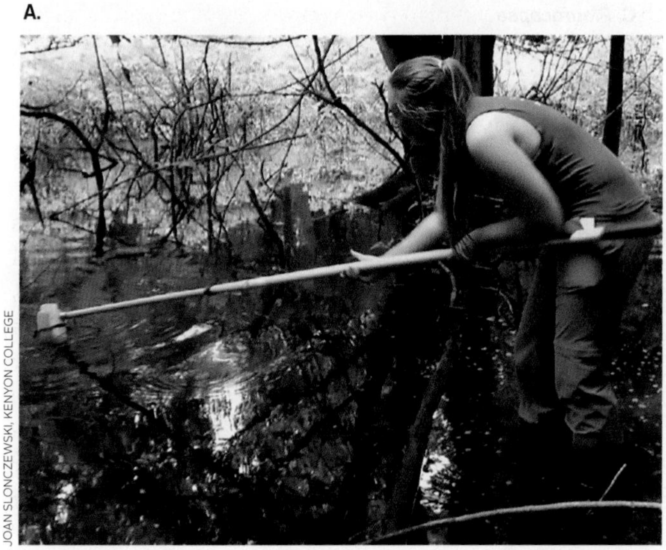

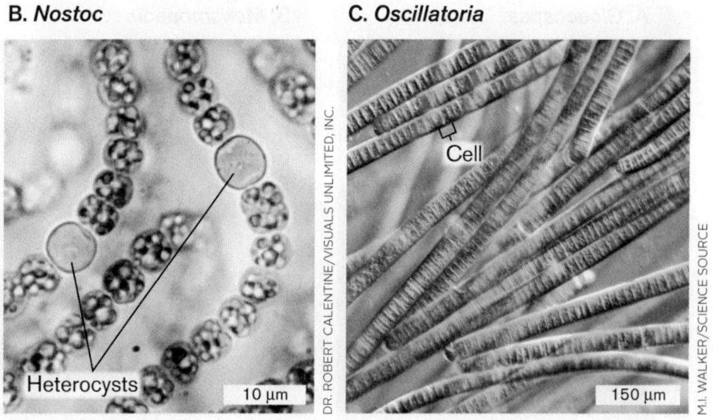

FIGURE 18.6 ■ Pond cyanobacteria. A. Alex Schaal samples water from a pond in Knox County, Ohio. Cyanobacteria and algae are filtered to obtain DNA for sequencing and identification. B. *Nostoc* filamentous cyanobacteria form heterocysts that fix nitrogen. C. *Oscillatoria* filaments consist of platelike cells.

to obtain cyanobacteria and eukaryotic algae. From these samples, she will sequence DNA to identify the species, and measure the concentration of microcystins to determine whether the water is safe for human use.

Other cyanobacterial genera, such as *Oscillatoria* and *Nostoc*, form multicellular filaments. Such filaments may contain hundreds or even thousands of cells (**Fig. 18.6B** and **C**). *Oscillatoria* cells are stacked like plates, wider than they are long. To disseminate their cells beyond the biofilm, the filaments produce **hormogonia** (singular, **hormogonium**), short motile chains of three to five cells. Many filamentous species, such as *Nostoc*, develop heterocysts to fix nitrogen (**Fig. 18.6B**; discussed in Chapter 15).

Under environmental stress, such as light limitation or phosphate starvation, filamentous cyanobacteria such as *Anabaena* form specialized spore cells called akinetes. An akinete forms as a long, oval cell adjacent to a heterocyst, where it stores nitrogen and develops a thickened envelope. Like other types of spores, akinetes resist desiccation and remain viable for long periods. **Table 18.3** compares

TABLE 18.3 Spore Types in Bacteria

Spore type	Bacteria that produce the spore	Initiation of spore formation	Formation of the spore	Properties of the spore
Akinete	Filamentous cyanobacteria	Light limitation Cold temperature Phosphate starvation	An akinete develops next to a heterocyst, as a large oval cell with a multilayered envelope.	Desiccation and cold resistant Viable for decades
Arthrospore	Actinomycetes	Carbon starvation Phosphate starvation	At the tip of an aerial mycelium, cells undergo vegetative division and pinch off as arthrospores.	Desiccation resistant Heat resistant
Elementary body	Chlamydias	Completion of intracellular life cycle	Intracellular chlamydia reticular bodies replicate and then develop into elementary bodies with cross-linked outer membrane proteins. Elementary bodies survive outside the host cell.	Survives outside host Desiccation resistant
Endospore	Firmicutes	Carbon starvation Nitrogen starvation Phosphate starvation Low pH Peptide antibiotics	Individual cell develops mother cell and forespore. The forespore develops into an endospore with a spore coat reinforced by keratin and calcium dipicolinate.	Highly heat and desiccation resistant Viable for centuries
Myxospore	Myxobacteria	Nutrient starvation Heat shock Glycerol Dimethyl sulfoxide	Myxobacteria aggregate to form a fruiting body in which vegetative cell division forms a mass of myxospores.	Desiccation resistant UV resistant

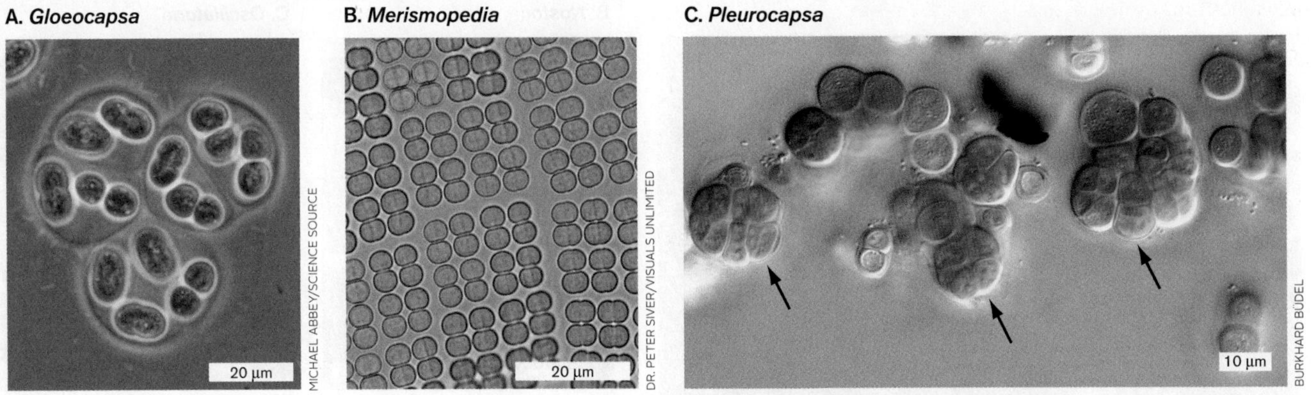

FIGURE 18.7 ■ Colonial cyanobacteria. **A.** *Gloeocapsa* is surrounded by mucus. Cells grow as single cells, doublets, or quartets. **B.** *Merismopedia* forms extended quartets, octets, and so on. **C.** Unicellular cyanobacterium *Chroococcidiopsis* sp. with baeocyte formation via successive multiple divisions.

the properties of akinetes with those of other spore types (which are discussed along with Firmicutes and other taxa). Akinetes lie dormant but viable until improved conditions permit germination and growth of new vegetative filaments. In lake water, akinete germination may cause toxic blooms of *Anabaena*.

Some filamentous cyanobacteria, such as *Lyngbya* and *Trichodesmium*, form algal blooms in the ocean. *Trichodesmium* can form giant blooms visible from outer space, covering many square kilometers of ocean surface. Such a bloom can be triggered by an influx of iron carried by wind from a dust storm blowing off the Sahara desert (discussed in Chapter 22).

Yet other species, called colonial cyanobacteria, divide to form small groups or larger colonies (**Fig. 18.7**). *Gloeocapsa* and *Chroococcus* form doublets or quartets, encased in a thick protective mucous slime (**Fig. 18.7A**). Others, such as *Merismopedia*, continue cell division in two planes, extending to form long, square sheets of attached cells (**Fig. 18.7B**). Colonial genera such as *Myxosarcina* and *Pleurocapsa* reproduce by multiple fission, forming large cell aggregates (**Fig. 18.7C**). As the aggregate matures, some cells continue to divide and release single cells called baeocytes. Each baeocyte reproduces and develops into a new cell aggregate. The aggregate group maintains anoxic conditions at the center for nitrogen fixation.

> **Thought Questions**
>
> **18.4** What are the relative advantages and disadvantages of propagation by hormogonia, as compared with akinetes?
>
> **18.5** What are the relative advantages and disadvantages of the different strategies for maintaining separation of nitrogen fixation and photosynthesis?

Multicellularity

Cyanobacterial filaments and colonies, in effect, represent multicellular organisms. Different bacteria have evolved diverse means of generating multicellularity (**Fig. 18.8**). As we have seen, filamentous cyanobacteria such as *Anabaena* or *Nostoc* form chains by serial cell division. At intervals, a cell differentiates into a heterocyst, which performs nitrogen fixation—a specific function for the filament as a whole. In other clades, we will see very different mechanisms of multicellularity (discussed in Sections 18.3 and 18.4).

Cyanobacterial Communities

Cyanobacteria share many kinds of mutualistic associations with animals, plants, fungi, and protists. Sponges growing on coral reefs may harbor communities of cyanobacteria that provide the sponges with nutrients from photosynthesis. The products of photosynthesis supplement the nutrients obtained by the sponges from their filter feeding, and these extra nutrients greatly augment the sponge growth rate within the competitive coral reef environment. Sponge symbionts produce many pharmaceutically active compounds.

In salt marshes and sand flats, cyanobacteria participate in multilayered microbial mats, such as the section shown in **Figure 18.9**, cut from the sand flats of Great Sippewissett Salt Marsh, on Cape Cod, Massachusetts. The high concentration of sulfides in the sediment supports growth of high populations of sulfur phototrophs. Typically, cyanobacteria and eukaryotic algae such as diatoms form the upper green layer. Below, the purple layer consists of "purple sulfur" proteobacteria, whose photopigments (primarily BChl *a*) absorb at longer wavelengths (discussed in Section 18.4). The pale-colored layer below the purple layer consists of proteobacteria with BChl *b*, which absorbs farther into the infrared.

FIGURE 18.8 ▪ Bacterial multicellularity. Multicellularity arises by various means. **A.** Filamentous cyanobacteria form chains by serial cell division. At intervals, a cell differentiates into a heterocyst. **B.** *Bacillus* species assemble by attachment to a substrate. They grow as a biofilm that produces antibiotics, then releases endospores. **C.** Actinomycetes form branching filaments that produce antibiotics, then undergo programmed cell death. Aerial filaments generate exospores. **D.** Myxobacteria assemble by chemotaxis of swarmer cells. They differentiate into a fruiting body and release myxospores.

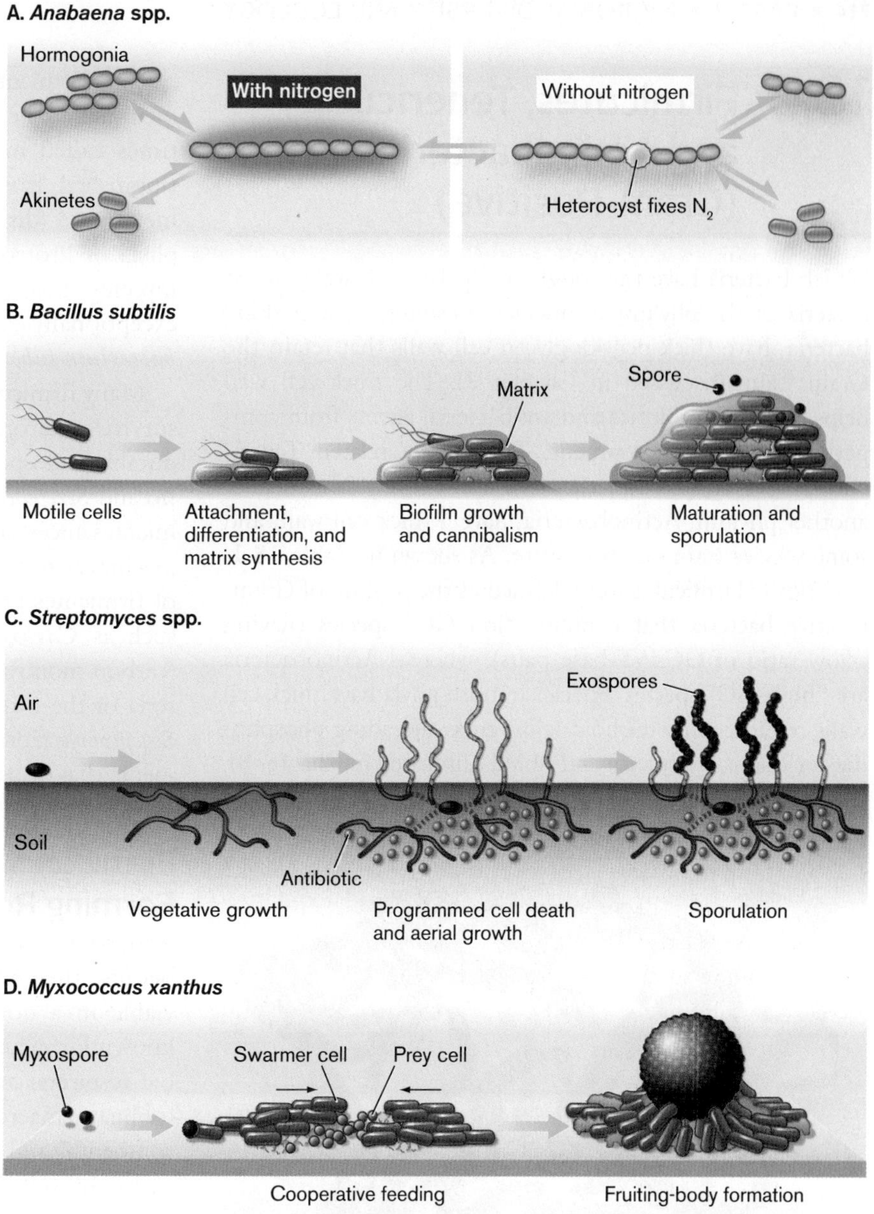

To Summarize

- **Cyanobacteria** are the only oxygenic prokaryotes. They conduct photosynthesis in thylakoids, fix CO_2 in carboxysomes, and maintain buoyancy using gas vesicles. They exhibit gliding motility.
- **Single-celled cyanobacteria** such as *Prochlorococcus* are among the smallest and most abundant phototrophic producers in the oceans. *Microcystis* species produce microcystins that poison lakes polluted by agricultural runoff.
- **Filamentous cyanobacteria** such as *Nostoc* and *Oscillatoria* are common in freshwater lakes. They form heterocysts to fix nitrogen, and reproduce by hormogonia or by akinetes.
- **Colonial cyanobacteria** such as *Myxosarcina* produce large cell aggregates with an anaerobic core for nitrogen fixation. The colonies reproduce through baeocytes.
- **Symbiotic associations** of cyanobacteria occur with animals, fungi, and plants.

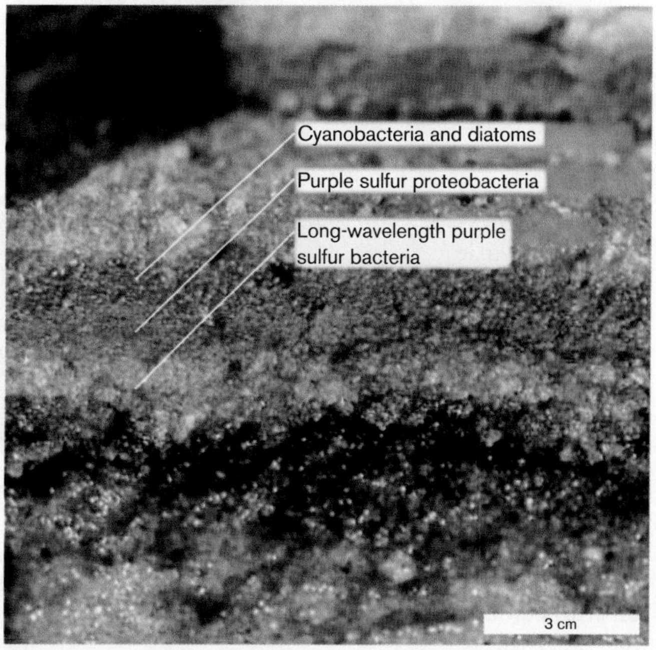

JORG OVERMANN AND FERRAU GARCIA-PICHEL

FIGURE 18.9 ▪ Cyanobacteria in microbial mats. Cutaway through a multilayered microbial mat from the sand flats of Great Sippewissett Salt Marsh (Cape Cod, Massachusetts). Cyanobacteria and diatoms form the upper green layer, above layers of purple sulfur proteobacteria.

18.3 Firmicutes, Tenericutes, and Actinobacteria (Gram-Positive)

Which bacteria have the toughest, thickest cell walls? Most bacteria of the phylum Firmicutes, meaning "tough skin" bacteria, have thick peptidoglycan cell walls that retain the Gram stain (discussed in Chapter 3). The thick cell wall helps exclude antibiotics and antibacterial agents from competitors in the environment. The closely related Tenericutes, however, have lost their cell wall entirely. Members of another phylum, Actinobacteria, have a thick cell wall, and some species stain Gram-positive. As shown in **Table 18.1**, historically Firmicutes were defined as the phylum of Gram-positive bacteria that contains "low-GC" species (having a low ratio of GC/AT base pairs), whereas Actinobacteria are "high-GC" species. Species in both phyla have thick cell walls reinforced by teichoic acids, cross-threading phosphodiester chains of glycerol and ribitol (discussed in Chapter 3).

The Firmicute cell is enclosed by a single phospholipid membrane (the cell membrane), so the bacteria are sometimes called **monoderm**. "Monoderm" bacteria are distinguished from **diderm** bacteria, those with an outer membrane (discussed in Chapter 3), such as the several phyla of Proteobacteria (discussed shortly). Most Actinomycetes, too, are monoderm—except for those with an exceptionally complex cell wall and envelope, such as *Mycobacterium tuberculosis* (discussed in Chapter 3).

Many firmicutes, such as *Bacillus* and *Clostridium* species, survive unfavorable environmental conditions by forming durable endospores. Non–spore formers such as *Lactobacillus* and *Streptococcus* may have evolved from a common Firmicutes ancestor that formed endospores. In many cases, the machinery to form endospores is discovered in the genomes of firmicutes previously thought to be non–spore formers, such as *Carboxydothermus*, soil bacteria that oxidize CO (carbon monoxide) to CO_2. On the other hand, actinobacteria of the order Actinomycetales (actinomycetes), such as *Streptomyces*, do not form endospores, but they develop filaments that disperse arthrospores (see **Table 18.3**).

Firmicutes Include Endospore-Forming Rods

Endospore-forming bacteria are common in soil and air because their spore forms resist desiccation and can remain viable in a dormant state for thousands of years. Well-known Firmicutes include the order Bacillales (mainly aerobic respirers) and the class Clostridia, order Clostridiales (obligate anaerobes). Both groups include species of environmental and economic importance, as well as causative agents of well-known diseases.

Bacillales. The genus *Bacillus* was one of the first bacterial genera to be classified, in the nineteenth century (**Fig. 18.10**). Colonies soon appear on a nutrient agar plate exposed to air. *Bacillus* species can be isolated from soil or food by suspending a sample in water and heating at 80°C

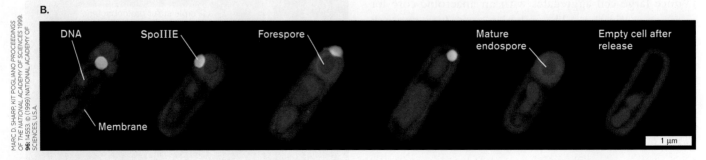

FIGURE 18.10 • *Bacillus* species: Gram-positive endospore formers. A. Gram-stained *Bacillus* sp., sporulating culture. Endospores stain green with malachite green. **B.** Correlated fluorescence imaging of membrane migration, protein translocation, and chromosome localization during *B. subtilis* sporulation. Membranes were stained with red fluorescent FM4-64. Chromosomes were localized with the blue fluorescent nuclear counterstain DAPI. The small, green fluorescent patches indicate the localization of a green fluorescent protein (GFP) gene fusion to SpoIIIE, a protein essential for both initial membrane fusion and forespore engulfment. Progression of the engulfment is shown from left to right.

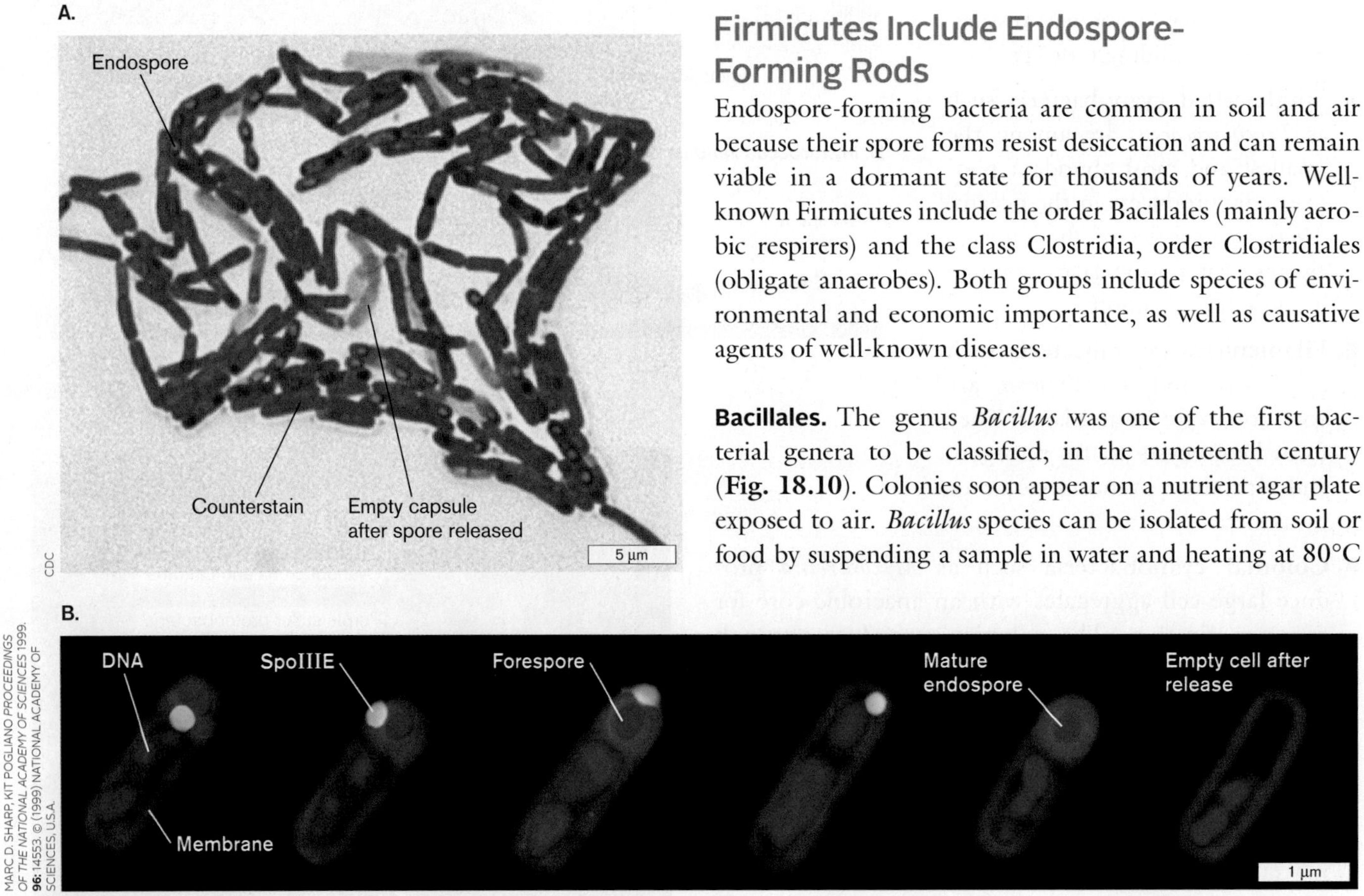

for half an hour. Vegetative cells (that is, cells undergoing binary fission) and non–spore formers are killed at that temperature. The remaining endospores will germinate and grow on a beef broth agar plate at 25°C–30°C. *Bacillus* species isolated in this way include over a thousand characterized strains, all but a few of them harmless to humans.

The large, rod-shaped **vegetative cells** (growing and replicating form) of *Bacillus* species are easily stained and visualized. A species of particular scientific importance is *Bacillus subtilis*, the best-studied Gram-positive organism, a "model system" for Firmicutes. The *B. subtilis* genome sequence reveals a large number of transporters for carbon sources and many secretory complexes for industrially important enzymes and drug resistance proteins. It also reveals several integrated prophages (integrated phage genomes). Phage genomes contribute to bacterial evolution by transferring genes between different strains and species, as discussed in Chapter 17.

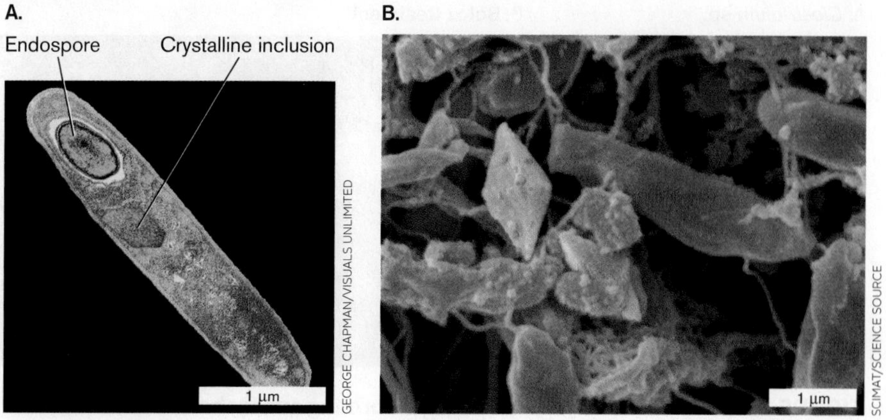

FIGURE 18.11 ■ *Bacillus thuringiensis*: the biological insecticide. **A.** Sporulating cell of *B. thuringiensis*, showing crystalline inclusion of insecticidal toxin (colorized TEM). **B.** Toxin crystals (colorized light purple) embedded in a *B. thuringiensis* colony.

B. subtilis is used as a model system to study stress response. Enormous changes in protein expression accompany starvation and general stress conditions. As the vegetative cells run short of nutrients on an agar plate, they begin a program to "sporulate"—that is, develop inert endospores (**Fig. 18.10A**). The life cycle of endospore production (see Chapter 4) involves a coordinated developmental plan, in which the cell divides near the pole instead of at the cell equator (**Fig. 18.10B**). The polar compartment develops as the **forespore**, directed by unique regulatory proteins such as SpoIIIE. The larger compartment, called the **mother cell**, provides DNA and nutrients to the growing forespore and disintegrates after release of the mature endospore. When the released endospore encounters favorable conditions of moisture and nutrients, it germinates and restarts vegetative growth. The sporulation cycle can proceed out of a biofilm (see **Fig. 18.8B**).

A *Bacillus* species of great economic importance is *B. thuringiensis*, the most successful biological control agent yet produced (**Fig. 18.11**). *B. thuringiensis* was discovered in 1901 by Japanese bacteriologist Shigetane Ishiwata, as a cause of disease in silkworms. The organism proved easy to culture on agar-based medium, and its spores are now applied as an insecticide against the gypsy moth caterpillar. During sporulation, *B. thuringiensis* generates a diamond-shaped crystal adjacent to its endospore (**Fig. 18.11A**). The crystal contains an insecticidal protein known as delta endotoxin. The crystals are ultimately released from the cells (**Fig. 18.11B**). The toxin is activated only at high pH in the digestive tracts of insect larvae; thus, it is safe for animals with acidic digestive tracts.

Many *Bacillus* species are extremophiles, growing at high pH (*B. alkalophilus*), at high temperature (*B. thermophilus*), or in high salt (*B. halodurans*). Some species combine alkaliphily with thermophily, as in the case of the "alkalithermophile" *B. alkalophilus.* Genomic research has focused on the surprisingly subtle differences that distinguish an extremophile from a closely related mesophile. For example, thermostability of proteins can be determined by comparing the protein sequences of a thermophile with those of a related mesophile. The thermophilic sequence shows specific patterns of amino acid residues that confer thermostability. Such amino acid substitutions provide useful information for industrial engineering of enzymes.

Clostridia. This diverse class includes anaerobic spore formers such as the genus *Clostridium* (**Fig. 18.12A**), species of which cause botulism (*C. botulinum*) and tetanus (*C. tetani*). The botulism toxin (botulinum, or "Botox") is famous for its therapeutic use to relax muscle spasms and smooth wrinkles in skin (**Fig. 18.12B**; discussed in Chapter 26). Other species, such as *C. acetobutylicum*, have economic importance as producers of industrial solvents such as butanol and acetone (for industrial microbiology, see Chapter 16). The butanol pathway is an example of the diverse fermentative strategies found among clostridia. Unlike *Bacillus* (a monophyletic clade with a common ancestor), the *Clostridium* group is polyphyletic, representing many clades that branch among different genera.

The best-known *Clostridium* species, such as *C. botulinum* and *C. tetani*, sporulate in a form distinct from that of *Bacillus* (see **Fig. 18.12A**). The growing *Clostridium* endospore swells the end of the cell, forming a "drumstick" appearance. Mature clostridial spores are generally less heat resistant than the spores of *Bacillus* and thus more difficult to isolate in pure culture from natural environments. Nevertheless, *Clostridium* spores are found in soil and

A. *Clostridium* sp. B. Botox treatment

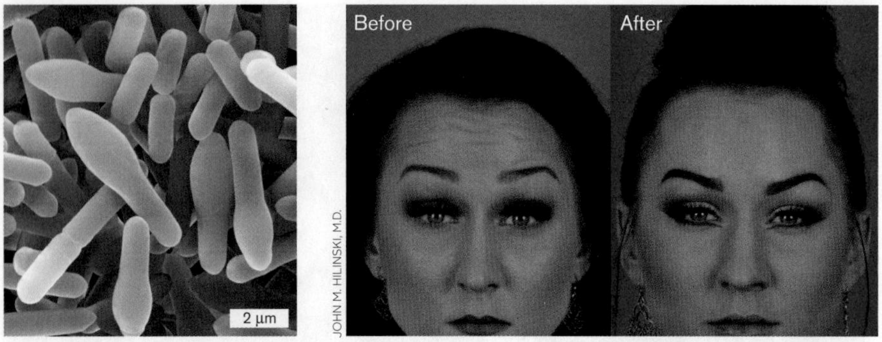

FIGURE 18.12 ■ *Clostridium* species: spore-forming anaerobes. A. As *Clostridium* cells sporulate, the endospore swells, forming a characteristic "drumstick" appearance. B. The deadly botulinum toxin (Botox) from *C. botulinum* is used to relax muscle spasms and reduce wrinkles. In the photo on the left, the woman is intentionally raising her eyebrows—creating unwanted wrinkles. On the right, she is unable to recreate the wrinkles (fire the same muscles) due to having the Botox injected.

water, ready to germinate and grow when the environment becomes anoxic. Many harmless species are found in the human colon, particularly in infants. Pathogenic *C. botulinum* can grow within the colon of very young infants, and infant botulism has been implicated in some cases of sudden infant death syndrome.

Another member of the clostridia, *Clostridioides difficile* (formerly classified *Clostridium difficile*), is a life-threatening intestinal pathogen, resistant to most antibiotics. *Clostridioides difficile* grows in patients treated with antibiotics that eliminate normal enteric bacteria, allowing growth of the pathogen. The pathogen can be so hard to eradicate that patients require fecal bacteriotherapy, or "fecal transplant" from a healthy person's colon. The healthy colon residents recolonize the gut and outcompete the pathogen.

A group of clostridia outside the order Clostridiales are the Heliobacteriaceae, or "Sun bacteria," the only known phototrophs in Firmicutes (see **Table 18.2**). Their name derives from their yellowish color, caused by the shift of their red-peak absorbance into the infrared, which allows transmission of red light plus green light (perceived as yellow). The heliobacteria are photoheterotrophs, with a PS I reaction center providing a modest boost of energy to supplement heterotrophic metabolism of simple organic compounds. Most heliobacterial cells are elongated rods with peritrichous flagella, and they form endospores—the only known endospore-forming phototrophs.

Note: Distinguish *Heliobacterium* from *Heli<u>c</u>obacter*, helical species of the Gammaproteobacteria.

Variant sporulation and "live birth." The order Clostridiales includes species of exceptionally large bacteria that grow only within the digestive tract of specific animal hosts. These species have evolved intriguing variations of the endospore former life cycle, as revealed by Esther Angert and colleagues at Cornell University (**Fig. 18.13A**). *Metabacterium polyspora* (size 15–20 μm) grows throughout the digestive tract of guinea pigs (**Fig. 18.13B**). Endospores ingested from feces germinate in the upper intestine but rarely undergo binary fission. Instead, the growing cell forms forespores at both poles (**Fig. 18.13C**). The forespores actually multiply within the mother cell to form several endospores, which are released in the colon before defecation.

An even larger enteric endosymbiont is *Epulopiscium fishelsoni*, found in the digestive tract of surgeonfish (**Fig. 18.14**). These bacteria are large enough to be seen by eye—about the size of the period at the end of this sentence. As in *M. polyspora*, Angert showed that *E. fishelsoni* reproduction is synchronized with the digestive cycle of its host, but it has gone even further in transformation of the sporulation cycle. Binary fission is eliminated; the cell must fission at both poles. Each polar fission generates an intracellular daughter cell that grows to nearly the full length of the mother cell. From two to seven intracellular offspring ultimately emerge in "live birth" from the mother cell, which then disintegrates.

An even more bizarre "live birth" species related to clostridia was discovered in the mammalian intestine. These bacteria, provisionally named "*Candidatus* Savagella," grow as filaments attached to the intestinal epithelial cells (**Special Topic 18.1**).

Non-Spore-Forming Firmicutes

Many Firmicute species do not form spores. An important order of non–spore formers is the Lactobacillales, or lactic acid bacteria, which are important for food production. In non-spore-forming firmicutes, endospore formation was probably lost by reductive evolution. Non–spore formers include human pathogens such as *Listeria* and *Streptococcus* species.

Listeria species are facultative anaerobic bacilli, named for the British surgeon Joseph Lister (1827–1912), who was the first to promote antisepsis during surgery. They include enteric pathogens, such as *L. monocytogenes*, that contaminate cheese and sauerkraut (discussed in Chapter 16). Unlike other food-associated organisms, *Listeria* grows at temperatures as low as 4°C. Under preindustrial conditions of food preparation, *L. monocytogenes* was generally outcompeted by other flora. The era of refrigeration led to the emergence of *Listeria* as the

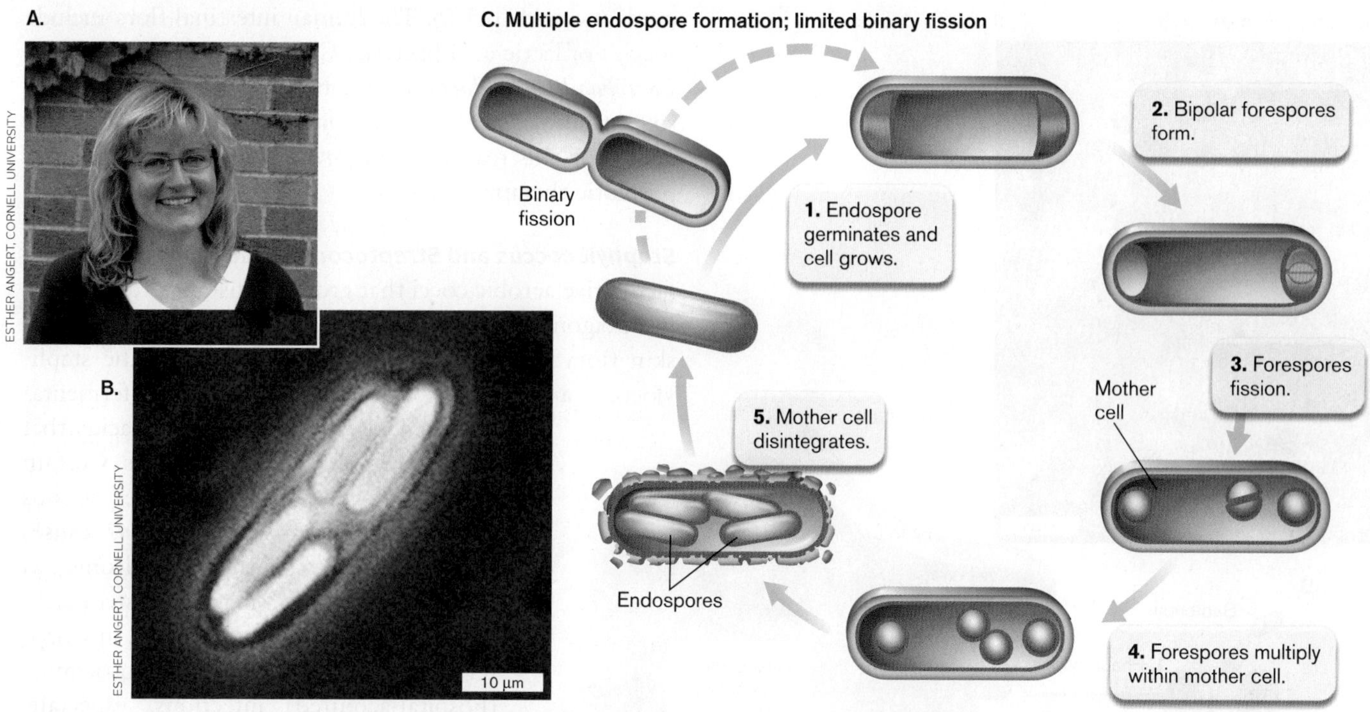

FIGURE 18.13 ■ Multiple endospore formation. A. Esther Angert characterized unusual forms of sporulation and reproduction in exceptionally large firmicute bacteria. B. *Metabacterium polyspora* forms multiple endospores (phase-contrast LM). C. A forespore forms at each pole. Forespores fission and multiply within the mother cell and then are released. Germinated cells undergo limited or no binary fission.

cause of listeriosis, a severe gastrointestinal illness that can progress to the nervous system. *L. monocytogenes* cells are taken up by macrophages into phagocytic vesicles, but they avoid digestion and escape the vesicles. The bacteria then multiply as they travel through the host cytoplasm, generating "tails" of actin (**Fig. 18.15** ▶). The actin tails eventually project the cells of *Listeria* out of the original host cell and enable it to penetrate a neighboring host cell.

Lactic acid bacteria. The lactic acid bacteria (order Lactobacillales) are aerotolerant (capable of growth in the presence of oxygen), though they do not use oxygen to respire. Most lactic acid bacteria are obligate fermenters; that is, they generate ATP by substrate-level phosphorylation (discussed in Chapter 13). They ferment primarily by converting sugars to lactic acid (a fermentation pathway discussed in Chapter 13). As the acid builds up, the pH decreases until it halts bacterial growth; thus, the carbon source retains much of its food value for human consumption. This is the basis of yogurt and cheese production. *Lactococcus* and *Lactobacillus* species are extremely important for the dairy industry (discussed in Chapter 16). Other common genera of lactic acid bacteria include *Leuconostoc*, which often spoils meat, and *Pediococcus*, found in

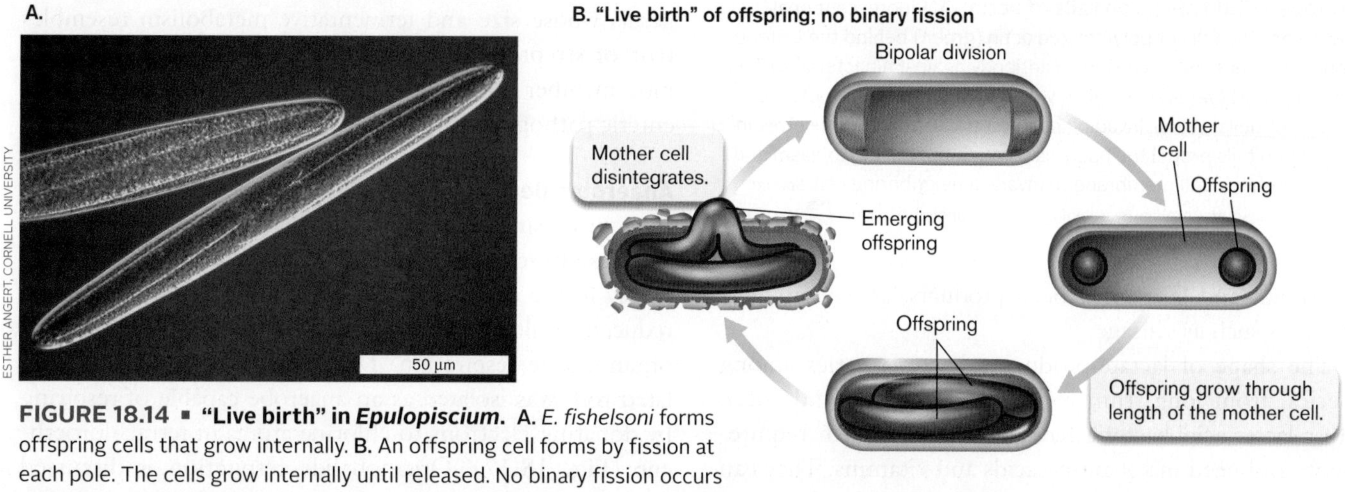

FIGURE 18.14 ■ "Live birth" in *Epulopiscium*. A. *E. fishelsoni* forms offspring cells that grow internally. B. An offspring cell forms by fission at each pole. The cells grow internally until released. No binary fission occurs outside the mother cell.

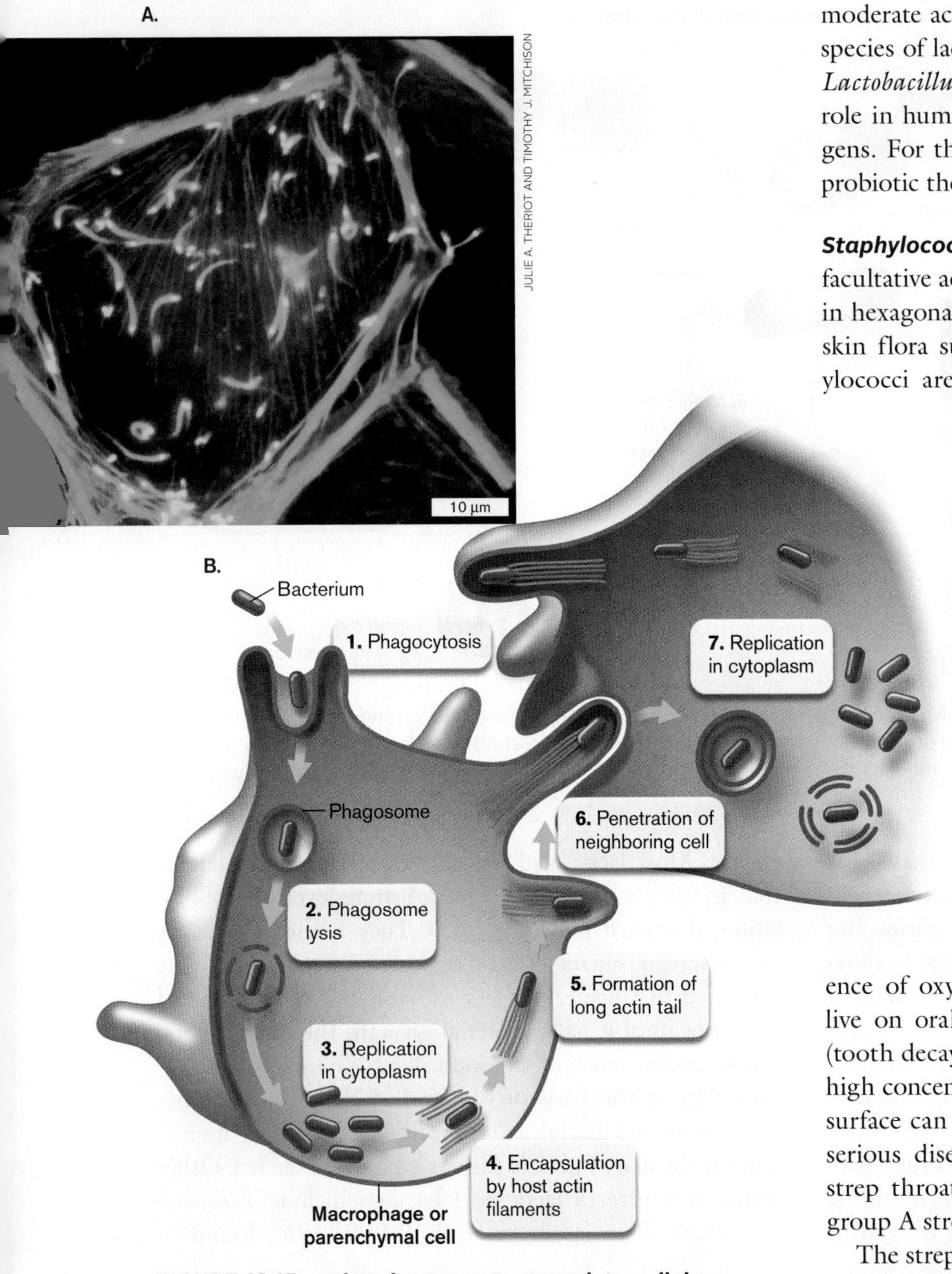

FIGURE 18.15 ■ *Listeria monocytogenes*: intracellular pathogen that travels on tails of actin. A. Fluorescent phalloidin mark the tails of polymerized actin (green) behind the *Listeria monocytogenes*, labeled using an antibody against a bacterial surface protein (ActA) (yellow), traveling within an infected PtK2 potoroo kidney epithelial cell. B. Invading bacteria encapsulate themselves in actin. Actin tails propel the bacteria through the host cytoplasm and out through the cell membrane to invade a neighboring cell. *Source:* Part B from U. South Carolina, Microbiology and Immunology On Line.

sauerkraut and fermented bean products, as well as meat products such as sausage.

The shape of lactate-producing bacteria varies among species, from long, thin rods to curved rods and cocci. Most lactic acid bacteria have fastidious growth requirements and need many amino acids and vitamins. They can be isolated from pasture grasses incubated anaerobically in moderate acid (pH 5). The human intestinal flora include species of lactic acid bacteria. Certain species, particularly *Lactobacillus acidophilus*, are believed to play a positive role in human health by inhibiting the growth of pathogens. For this reason *L. acidophilus* may be ingested as a probiotic therapy.

***Staphylococcus* and *Streptococcus*.** The staphylococci are facultative aerobic cocci that grow in clusters, often packed in hexagonal arrays (**Fig. 18.16A**). They include common skin flora such as *Staphylococcus epidermidis*. The staphylococci are generally salt tolerant, and their fermentation generates short-chain fatty acids that inhibit growth of skin pathogens. Certain species, however, are themselves serious pathogens. *Staphylococcus aureus* causes impetigo and toxic shock syndrome, as well as pneumonia, mastitis, osteomyelitis, and other diseases (discussed in Chapter 26). It is a major cause of nosocomial (hospital-acquired) infections, especially contamination of surgical wounds. The most dangerous strains, now resistant to most known antibiotics, are termed MRSA (methicillin-resistant *S. aureus*).

Streptococcus species generally form chains instead of clusters, because their cells divide in a single plane (**Fig. 18.16B**). They are aerotolerant (grow in the presence of oxygen) but metabolize by fermentation. Many live on oral or dental surfaces, where they cause caries (tooth decay). Their fermentation of sugars produces such high concentrations of lactic acid that the pH at the tooth surface can fall to pH 4. *Streptococcus* species cause many serious diseases, including pneumonia (*S. pneumoniae*), strep throat, erysipelas, and scarlet fever (*S. pyogenes* or group A streptococci).

The streptococci are less salt tolerant than *Staphylococcus* species, which tolerate as much as 5%–15% NaCl. Another genus whose size and fermentative metabolism resembles that of streptococci is *Enterococcus*. *E. faecalis* is a common member of the intestinal flora, and related strains are enteric pathogens.

Anaerobic dechlorinators. Some firmicutes from the soil show promising abilities to degrade chlorinated pollutants, such as dry-cleaning solvents that are biodegraded very slowly in the environment. The chlorinated molecules are reduced as alternative electron acceptors, a process called organohalide respiration. *Dehalobacter restrictus*, a flagellated rod, was isolated as an anaerobe capable of respiring by donating electrons to chlorine atoms in tetrachloroethene (**Fig. 18.17**). Organohalide respiration is discussed further in Section 14.5.

A. *Staphylococcus* (hexagonal clusters)

B. *Streptococcus* (chains)

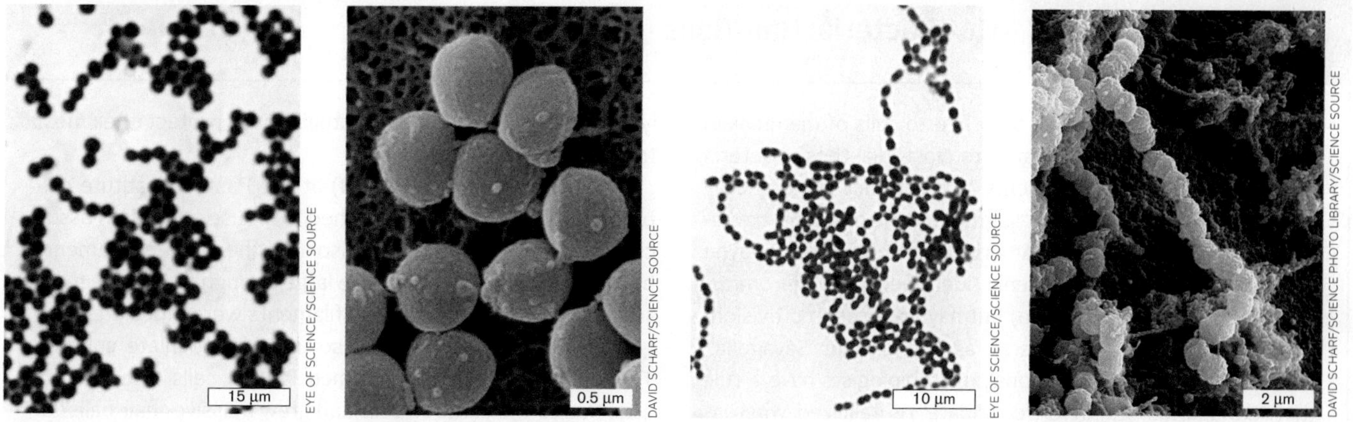

FIGURE 18.16 ■ Staphylococci and streptococci. A. *Staphylococcus* species: Gram stain (left); colorized SEM (right). B. *Streptococcus* species: Gram stain (left); colorized SEM (right).

While genetic analysis places *D. restrictus* among the clostridia, the bacterium actually stains Gram-negative, perhaps because its peptidoglycan layer is relatively thin. Nevertheless, the species shows no Gram-negative outer membrane. It does possess a thick S-layer of hexagonally tiled proteins, typical of Gram-positive bacteria.

Tenericutes: Mycoplasmas Lack a Cell Wall

The mycoplasmas are cell wall-less bacteria, classified as Tenericutes, or Mollicutes (Latin for "soft skin"). Mycoplasmas have completely lost their cell wall and S-layer through reductive evolution, retaining only their cell membrane. Presumably, the loss of these energy-expensive structures enhanced the reproductive rate of cells in a protected host environment. The Mollicutes comprise many genera of flexible wall-less cells that maintain a shape through some kind of cytoskeleton (**Fig. 18.18A** and **B**). On agar they form colonies that have a characteristic "fried-egg" appearance (**Fig. 18.18C**).

The best-known genus of Mollicutes is *Mycoplasma*, although its species are now known to include several distantly related branches. Mycoplasmas are found as parasites of every known class of multicellular organism, including vertebrates, insects, and vascular plants; in humans, they cause pneumonia and meningitis. Another medically important mycoplasma is *Ureaplasma urealyticum*, an opportunistic pathogen inhabiting the genital tracts of men and women. Most individuals are unaware that they harbor the organism, but *U. urealyticum* has been associated with urethritis, amniotic infections, and pulmonary infections.

Some *Mycoplasma* species remain adherent to the host cell, whereas others penetrate and grow intracellularly. Most mycoplasma cells have a rounded cell shape with one or two extended tips. In *M. penetrans*, an opportunistic pathogen infecting AIDS patients, the cell's attachment tip is coated with adhesion molecules that enable attachment to a host cell surface (**Fig. 18.18A**). The attachment tip penetrates deep into epithelial tissues. By contrast, the fish pathogen *M. mobile* has no attachment tip, but glides

A.

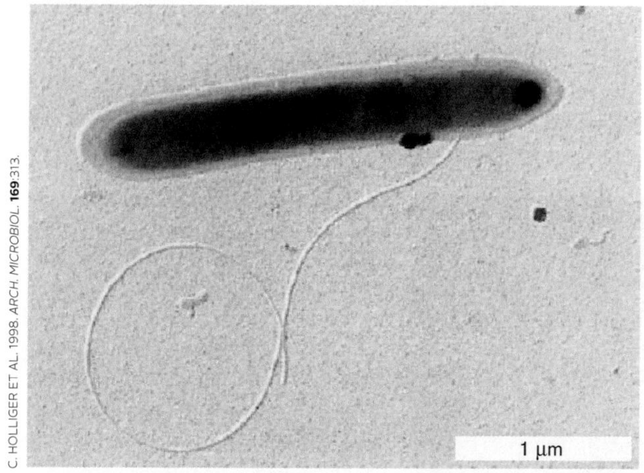

B. Dechlorination of tetrachloroethene by *Dehalobacter restrictus*

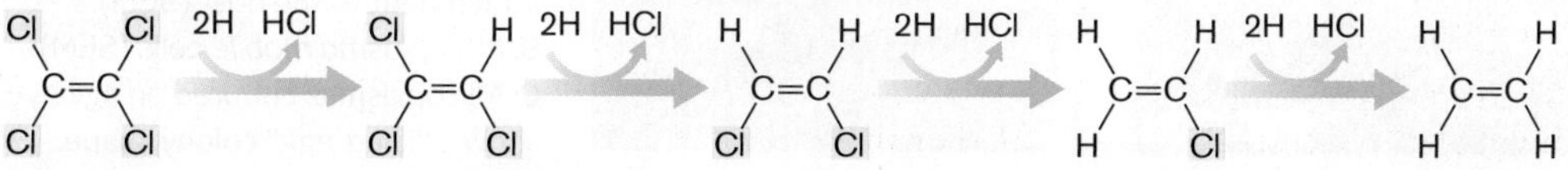

FIGURE 18.17 ■ *Dehalobacter restrictus* conducts anaerobic respiration by dechlorination. A. *Dehalobacter restrictus*, a flagellated rod related to clostridia. B. *D. restrictus* donates electrons to remove chlorine from tetrachloroethene, a major industrial pollutant.

SPECIAL TOPIC 18.1 Gut Bacterial Hair Balls

A bizarre kind of bacteria sit attached to cells of mammalian gut epithelium (**Fig. 1A**). Related to clostridia, these bacteria form long, segmented filaments that interact symbiotically with their host cell. These "segmented filamentous bacteria" were first described by American microbiologist Dwayne Savage. They were mistakenly identified as *Arthromitus* (a different kind of bacterium) but have now been provisionally named for their discoverer, as "*Candidatus* Savagella." Because the name is provisional, microbiologists have a rule that "*Candidatus*" (Latin for "candidate") is italicized, whereas the provisional name ("Savagella") is not.

For over 50 years, microscopists have seen these segmented filamentous bacteria in the gut of mice and humans. The bacteria require their host for growth, and in turn, they provide essential services by inducing development of the host immune system. Attached to the epithelium above gut lymph tissues, the bacteria actually help protect their host from infection.

Philippe Sansonetti (**Fig. 1B**) at the Pasteur Institute cultured Savagella filaments attached to mouse gut cells in tissue culture. To do this, Sansonetti isolated the multicell filaments by filtration at 5 µm, a pore size large enough to exclude most single-celled gut bacteria. The filaments were inoculated into germ-free mice. They were also used to inoculate cultured TC7 cells, a cell line of gut epithelial tumor cells. The bacteria attached to the cultured cells and grew massive "hair balls" of filaments.

The in vitro system enabled Sansonetti to observe the bacteria's unique process of differentiation and production of intracellular offspring (**Fig. 2**). Electron microscopy showed that each filament begins with a single "newborn" cell whose holdfast attaches to the surface of the host

A.

B.

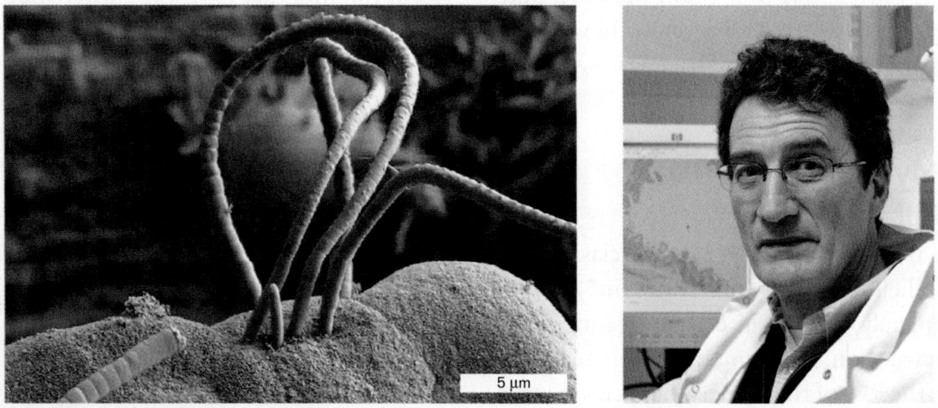

IVANOV ET AL. 2011. *CELL* **139**(3):485–498; FROM IVAYLO IVANOV (COLUMBIA UNIV. MEDICAL CENTER), DAN LITTMAN (NYU LANGONE MEDICAL CENTER), AND DOUG WEI (CARL ZEISS SMT, INC.)

PATRICK ALLARD/REA/REDUX

FIGURE 1 ▪ Segmented filamentous bacteria (Savagella) adhere to mouse gut epithelium. **A.** Savagella filaments adhere to mouse gut epithelium (SEM). **B.** Philippe Sansonetti, at the Pasteur Institute, Paris.

A. B. C.

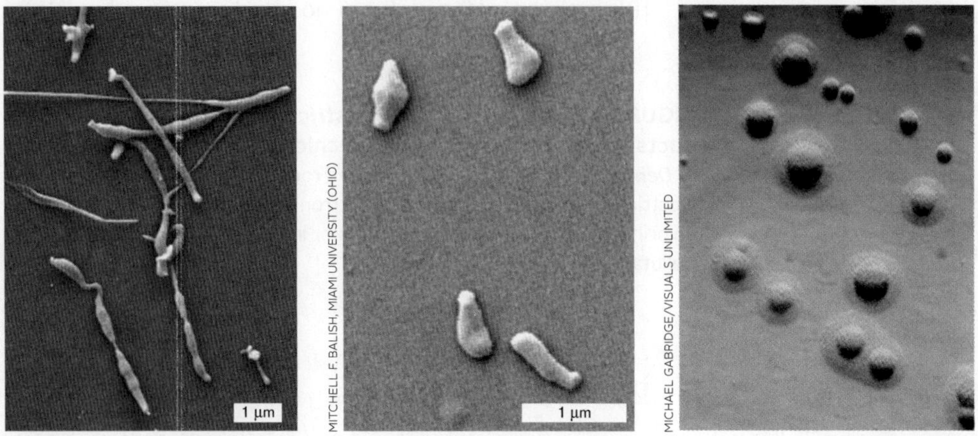

MITCHELL F. BALISH, MIAMI UNIVERSITY (OHIO)

MITCHELL F. BALISH, MIAMI UNIVERSITY (OHIO)

MICHAEL GABRIDGE/VISUALS UNLIMITED

FIGURE 18.18 ▪ Mycoplasmas: parasites without cell walls. **A.** *Mycoplasma penetrans* cells have an elongated tip used for attachment to the host (SEM). **B.** *Mycoplasma mobile* cells (SEM). **C.** Mycoplasmas cultured on agar show a "fried-egg" colony shape.

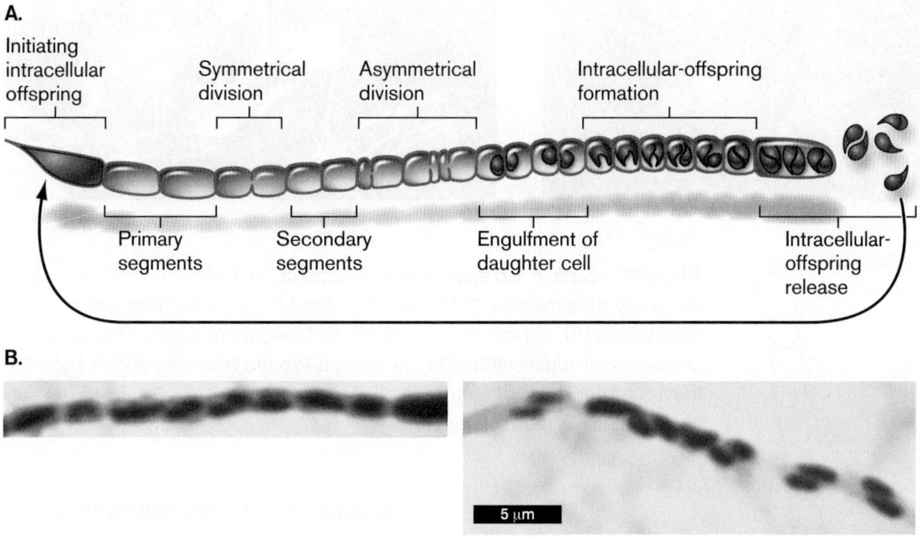

PAMELA SCHNUPF ET AL. 2015. *NATURE* **520**:99–103.

FIGURE 2 ■ Savagella growth and reproduction. **A.** Primary segments divide and differentiate until offspring bacteria develop within each cell. Offspring are released, to attach to an epithelial cell at a new position and grow a new filament. **B.** Intracellular offspring develop within cells of the filament (Gram stain).

epithelial cell. The cell then grows and divides, forming a multicellular filament (primary segment cells). When the filament length exceeds 50 μm, the distal-segment cells start to differentiate into mother and daughter cells. Each daughter cell becomes engulfed by a mother cell. The engulfed daughter cell then divides and differentiates into two intracellular offspring. Finally, the mother cell breaks open, releasing the two offspring. Each offspring cell possesses a holdfast to attach to a new site on the gut epithelium.

How do the attached bacterial filaments influence our gut epithelium? Schnupf has shown that the filament-attached epithelial cells have elevated transcription of genes that govern innate immune responses (discussed in Chapter 23). A more recent study by Bo Chen and colleagues finds that segmented filamentous bacteria specifically colonize human children. In children, the bacteria may modulate T cells regulating adaptive immunity. Future research may reveal how this bacterium's unusual life cycle plays a key role in the function of its mammalian host.

RESEARCH QUESTION

How do you think Savagella filaments communicate with the host immune system? Can you design an experiment to test your hypothesis?

Chen, Bo, Huahai Chen, Xiaoli Shu, Yeshi Yin, Jia Li, et al. 2018. Presence of segmented filamentous bacteria in human children and its potential role in the modulation of human gut immunity. *Frontiers in Microbiology* **9**:1403.

Schnupf, Pamela, Valérie Gaboriau-Routhiau, Marine Gros, Robin Friedman, Maryse Moya-Nilges, et al. 2015. Growth and host interaction of mouse segmented filamentous bacteria in vitro. *Nature* **520**:99–103.

along a surface at a rate of seven cell lengths per second (**Fig. 18.18B**). The basis of mycoplasma motility is not understood, although it may involve gliding or cytoskeletal contraction. In some species, motility involves a specialized cell tip called the "terminal organelle," but species that lack this organelle are equally motile. Species of *Spiroplasma* have a spiral shape and undergo corkscrew motion; the basis for this motion is also unknown.

Mycoplasma genomes are among the smallest in known cellular organisms. They also lack biosynthetic pathways for amino acids and phospholipids, which instead must be acquired from the host. Mycoplasmas have unique nutritional requirements, such as cholesterol, a membrane component typical of eukaryotes but rare for prokaryotes. For these reasons, mycoplasmas are difficult to grow in pure culture, although they readily infect tissue cultures. In fact, mycoplasma contamination of tissue culture is so prevalent that it has compromised major studies of cancer and AIDS.

Actinomycetes Form Multicellular Filaments

The phylum Actinobacteria, the "high-GC Gram-positives," comprises several orders, including Actinomycetales and Bifidobacteriales. Members of the order Actinomycetales are "actinomycetes." Actinomycetes include filamentous spore formers such as *Streptomyces* and *Frankia*, as well as short-chain organisms such as *Mycobacterium*, and irregularly shaped cells such as *Corynebacterium*. Members of an emerging group of marine actinomycetes, such as *Salinispora*, are isolated from sediment and from sponges.

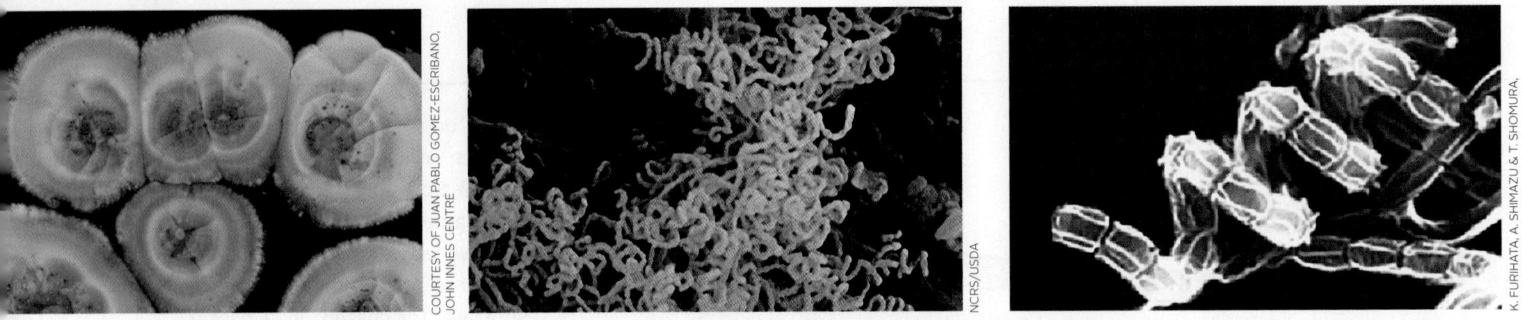

D. Telomere (end of chromosome)

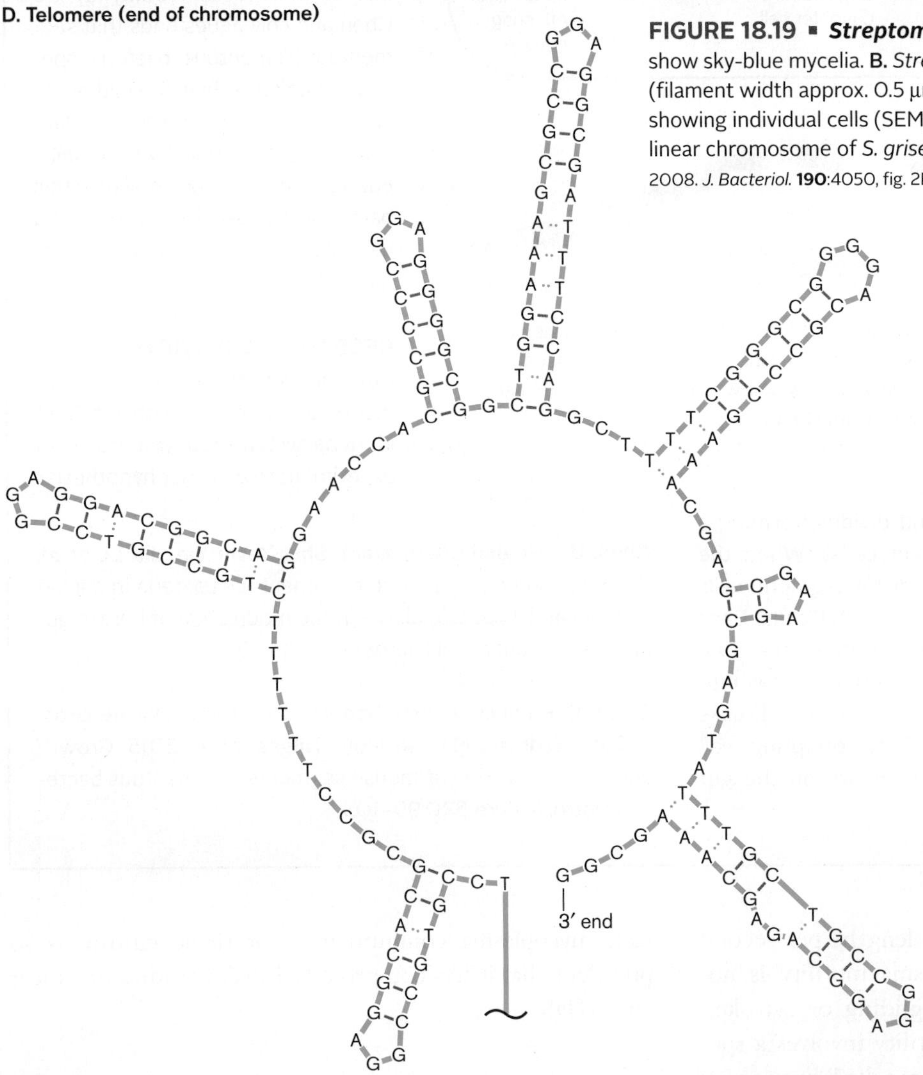

FIGURE 18.19 ▪ *Streptomyces* **bacteria. A.** Colonies of *S. coelicolor* show sky-blue mycelia. **B.** *Streptomyces* cells form coiled filaments (filament width approx. 0.5 μm; SEM). **C.** Close-up of a coiled filament, showing individual cells (SEM). **D.** Hairpin-looped telomere end of the linear chromosome of *S. griseus*. *Source:* Part D modified from Yasuo Ohnishi et al. 2008. *J. Bacteriol.* **190**:4050, fig. 2B.

***Streptomyces*: filamentous spore formers.** The best-studied actinomycetes are *Streptomyces* bacteria, which form multicellular filaments that generate dispersible spores (see streptomycete life cycle, Fig. 4.35). Streptomycetes play a major role in the ecosystems of soil (discussed in Chapter 21). Decaying *Streptomyces* cells produce the compound geosmin, which causes the characteristic odor of soil and can affect the taste of drinking water. In culture, the best-known species, *S. coelicolor* (Latin for "sky color"), forms strikingly blue colonies (**Fig. 18.19A**). The blue color derives from several pigments, including actinorhodin, a polyketide antibiotic. Other species produce filaments that are red, orange, green, or gray, depending on their distinctive products, many of which are antibiotics.

Thought Question

18.6 Why would *Streptomyces* produce antibiotics targeting other bacteria?

Streptomyces species are obligate aerobes, requiring access to air to complete their multicellular life cycle (see earlier, **Fig. 18.8C**). When a *Streptomyces* spore germinates, it extends **vegetative mycelia**, branched filaments that grow into the substrate. Some of the filaments then grow upward into the air, where they develop into **aerial mycelia**.

Salinispora species form numerous exotic secondary products that show promise as pharmaceutical agents.

The actinomycetes form complex multicellular filaments superficially resembling the branched "fuzzy" form of fungi (fungi are discussed in Chapter 20). Profoundly important for medicine, actinomycetes and fungi produce most of our antibiotics. Researchers mine the soil of remote environments to find actinomycetes producing antibiotics for which resistance genes are not yet widespread.

The aerial mycelia in some species grow in tightly coiled spirals (**Fig. 18.19B** and **C**). As the mycelial colony runs out of nutrients, older cells of the filament age and lyse, releasing nutrients that are absorbed by the younger cells. The nutrients also attract other scavengers, which may be killed by antibiotics produced by the aging streptomycete cells. The dead scavengers, too, release nutrients that feed the growing tip of the mycelium.

As mycelia mature, they fragment into smaller cells called exospores (also called arthrospores). Exospores are vegetative cells, not dormant like *Bacillus* endospores (see **Table 18.3**). The exospores separate and are dispersed by the wind, enabling them to colonize a new location. Streptomycete mycelia can be obtained from natural habitats by burying a glass slide in soil and then waiting several days for spores to germinate, covering the slide with mycelia. They are challenging to isolate in pure culture, however, because their coiled filaments trap cells of other bacteria.

The *S. coelicolor* genome is one of the largest prokaryotic genomes, containing over 8 million base pairs. Streptomycete chromosomes are linear with special "telomeres," single-stranded end sequences that double back to form hairpin loops (**Fig. 18.19D**). Much of the lengthy genome of a streptomycete encodes catabolism of a rich array of diverse organic components of decaying plant and animal matter, including even lignin. Other genes encode extensive operons for production of diverse secondary products (see Chapter 15), including antibiotics. More than half of the antibiotics currently used in medicine derive from *Streptomyces* species.

Note: Distinguish *Streptomyces* species, filamentous rod-shaped actinomycetes, from nonactinomycete *Streptococcus* species, Gram-positive cocci that form short, unbranched chains.

Actinobacteria associated with animals and plants. Many marine actinobacteria associate with invertebrate animals such as sponges, corals, mollusks, and ascidians (**Fig. 18.20**). In some cases the animal hosts a specific bacterium, such as the sponge *Hymeniacidon perleve*, which harbors *Actinoalloteichus* bacteria. More commonly,

A. ***Hymeniacidon*** **sponge harbors** ***Actinoalloteichus*** **bacteria.**

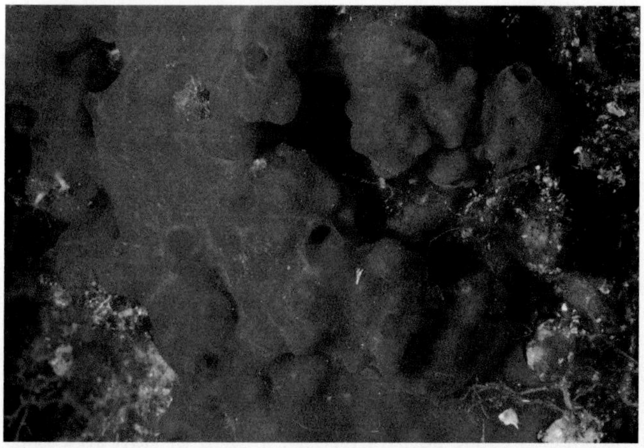

B. Gorgonian corals harbor many diverse Actinobacteria.

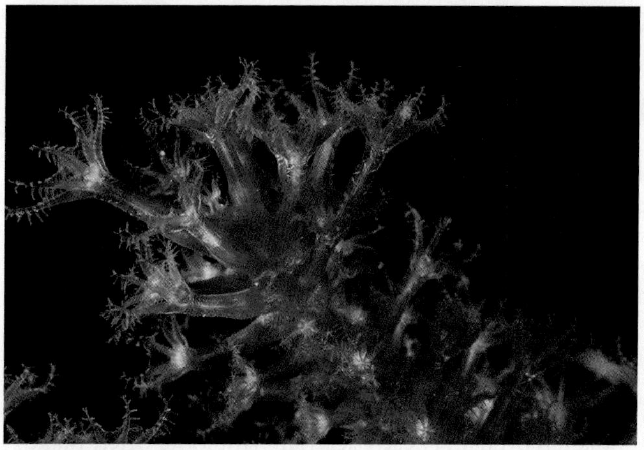

C. Cone snail ***Conus pulicarius*** **harbors diverse Actinobacteria**

D. Ascidian ***Eudistoma*** **harbors** ***Micromonospora***

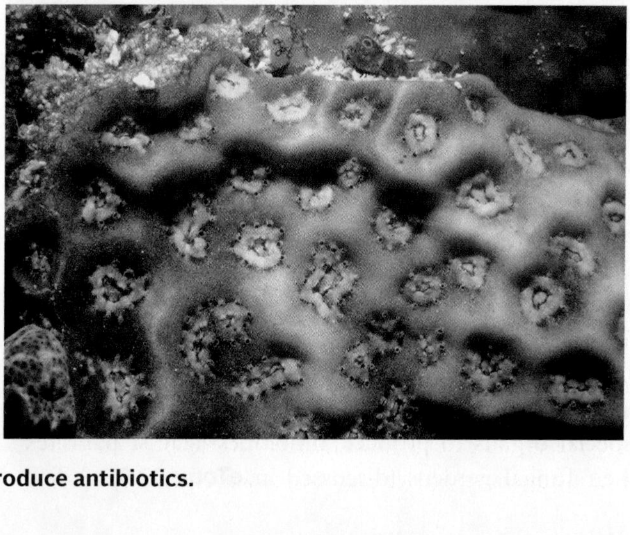

FIGURE 18.20 ■ Marine animals harbor actinobacteria that produce antibiotics.

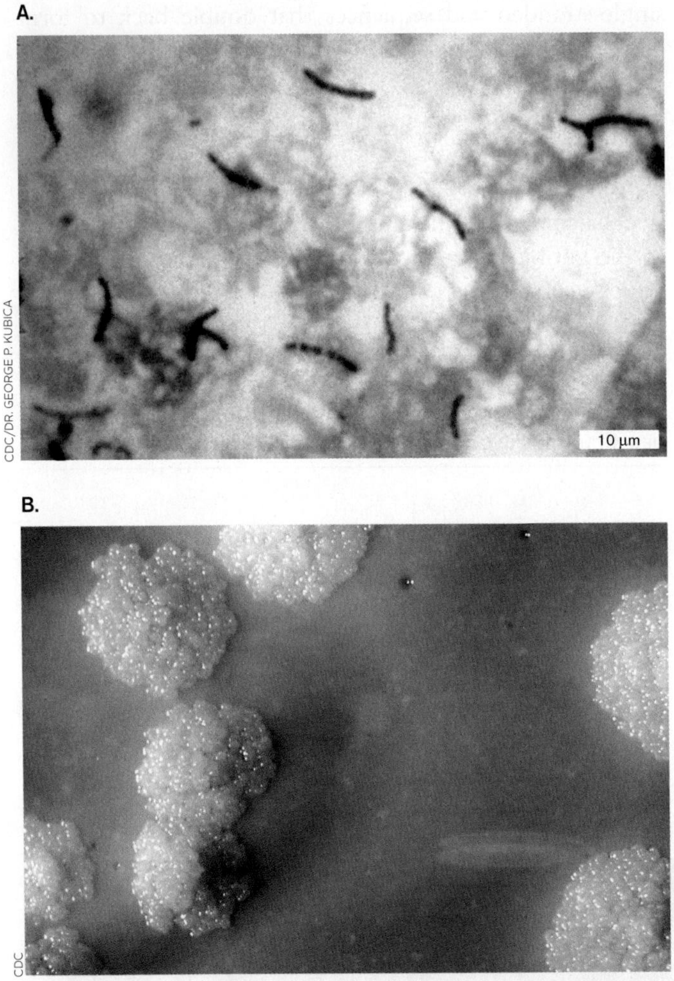

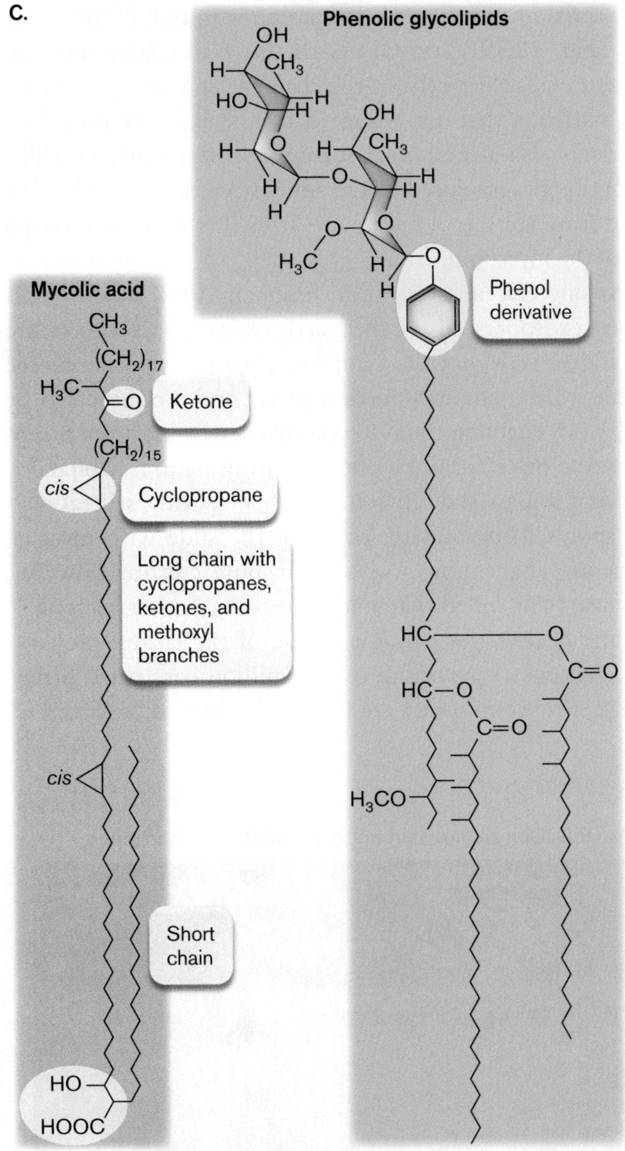

FIGURE 18.21 ■ *Mycobacterium tuberculosis* causes tuberculosis. **A.** Acid-fast stain of tissue sample containing *M. tuberculosis* (chains of pink rods). **B.** Crinkled appearance of *M. tuberculosis* colonies. **C.** Mycolic acids and phenolic glycolipids coat the cell wall of *M. tuberculosis*.

a sponge or coral supports a diverse community of many actinobacterial taxa. These bacterial symbionts often make up a substantial portion of the biomass of their host animal. The bacteria produce antibiotics and other secondary products that may help the animal resist pathogens or predators. They represent a vast, untapped potential source of novel therapeutic agents—antibiotics, antiparasitic agents, antitumor agents, immunomodulators, and agricultural agents. Mining the marine microbiomes is a growing aspect of the pharmaceutical industry—an aspect that depends on the global health and quality of our oceans.

On land, actinobacteria maintain symbiotic relationships with animals and plants. For example, *Frankia* bacteria form nitrogen-fixing nodules on the alder tree. Certain *Streptomyces* species have evolved a highly structured mutualism with leaf-cutter ants. The ants culture the bacteria on special organs to produce antibiotics against parasites of their fungal gardens (discussed in **eTopic 17.6**). A few actinomycetes are animal pathogens; for example, *Actinomyces* species cause actinomycosis, a form of skin abscesses in humans and cattle.

Nonmycelial Actinobacteria: *Mycobacterium* and *Corynebacterium*. These actinobacteria share the thick cell wall of *Streptomyces* but lack the mycelial lifestyle. Two genera that cause dreaded diseases are *Mycobacterium* (**Fig. 18.21**) and *Corynebacterium*. Both genera have thick cell envelopes containing **mycolic acids** and phenolic glycolipids. The mycolic acids of *M. tuberculosis* are extremely diverse and include some of the longest-chain acids known, up to 90 carbons (**Fig. 18.21C**). The mycolic acids are linked to arabinogalactan, a polymer of arabinose and galactose built on the peptidoglycan (discussed in Chapter 3). The mycolyl-arabinogalactan-peptidoglycan complex forms a waxy coat that impedes the entry of nutrients through porins and thus limits growth rate, but it also protects the bacterium extremities (hands and feet) whose temperature is lower than that of the body core. Culture on artificial media is impossible; the bacteria can be grown only within low-temperature animals, such as armadillos, or within genetically immunodeficient mice.

The genomes have been sequenced for both *M. tuberculosis* and *M. leprae*. *M. tuberculosis* has surprisingly few recognizable pathogenicity genes but a large number of environmental stress components, including 16 environmental sigma factors (discussed in Chapter 8), as well as 250 genes for its complex lipid metabolism. Over half of the *M. leprae* genome consists of pseudogenes, homologs of *M. tuberculosis* genes undergoing reductive evolution (**Fig. 18.22B**). Thus, *M. leprae* appears to be an evolving pathogen "caught in the act" of losing many genes no longer needed in its sheltered host environment. How it lost the need for so many genes preserved in *M. tuberculosis* remains a mystery. *M. leprae* causes disease worldwide, including 250 cases of leprosy annually in the United States. In 2013, a new test was approved that reveals leprosy infection a year before symptoms appear—enabling it to be cured with antibiotics before nerve damage is irreversible.

Mycobacteria also include a much larger number of harmless commensals, such as *M. smegmatis*, isolated from human skin. Species of mycobacteria can be isolated from soil and water, as well as from various animal sources. Their culture is difficult because of their slow growth rates, but isolation can be enhanced by treatment with a base (NaOH or KOH) at concentrations that kill most other bacteria.

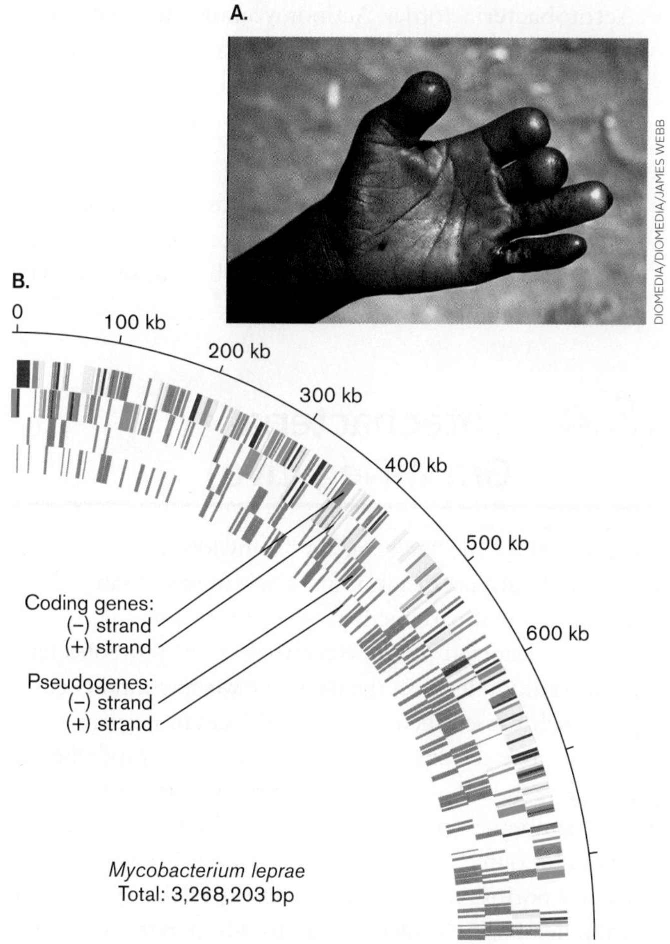

FIGURE 18.22 ▪ *Mycobacterium leprae* causes leprosy.
A. Hand disfigured by leprosy. **B.** The genome of *M. leprae* shows a high content of decaying pseudogenes (gray bars), most of which correspond to functional genes in the genome of *M. tuberculosis*.

Note: Distinguish *Mycobacterium* species, rods whose cell walls contain mycolic acids, from *Mycoplasma* species, bacteria of the phylum Tenericutes that lack cell walls.

from host defenses and antibiotics. For this reason, to cure tuberculosis requires an exceptionally long course of antibiotic therapy.

Mycobacterium includes the species *M. tuberculosis* and *M. leprae*, as well as lesser-known pathogens such as *M. ulcerans*. Cells of *M. tuberculosis* can be detected by the **acid-fast stain** as tiny rods associated with sloughed cells in sputum (**Fig. 18.21A**). In the acid-fast stain, cells are penetrated with a dye that is retained under treatment with acid alcohol (discussed in Chapter 2). The acid-fast property is associated with unusual cell wall lipids, such as mycolic acids. *M. tuberculosis* bacteria are challenging to culture; they form crinkled colonies after 2 weeks of growth on agar-based media (**Fig. 18.21B**).

The closely related species *M. leprae* causes the disfiguring disease leprosy (**Fig. 18.22A**). The species has one of the longest known doubling times of any pathogen (about 14 days), and it can take a year to grow enough cells in the laboratory for observation. Growth of *M. leprae* requires lower temperature; for this reason, leprosy attacks the

Irregularly shaped actinomycetes. Several nonmycelial actinomycetes show unusual cell shapes. Members of the genus *Corynebacterium* include soil bacteria, as well as pathogens such as *C. diphtheriae*, the cause of the lung disease diphtheria. *Corynebacterium* species grow as irregularly shaped rods, which may divide by a "half-snapping" mechanism in which one side of the cell remains attached like a hinge (**Fig. 18.23A**). Related soil bacteria include the genera *Nocardia* and *Rhodococcus*.

Soil bacteria of the genus *Arthrobacter* exhibit an unusual cell cycle in which coccoid stationary-phase cells sprout into rods, which eventually run out of nutrients and revert to the coccoid form. The growing rods form irregular branched filaments (**Fig. 18.23B**). An *Arthrobacter* species was discovered conducting anaerobic respiration by reduction of hexavalent chromium (Cr^{6+}), a toxic metal pollutant, to a less toxic oxidation state. *Arthrobacter* now shows potential as an agent of bioremediation of hexavalent chromium.

A. *Corynebacterium diphtheriae* B. *Arthrobacter globiformis*

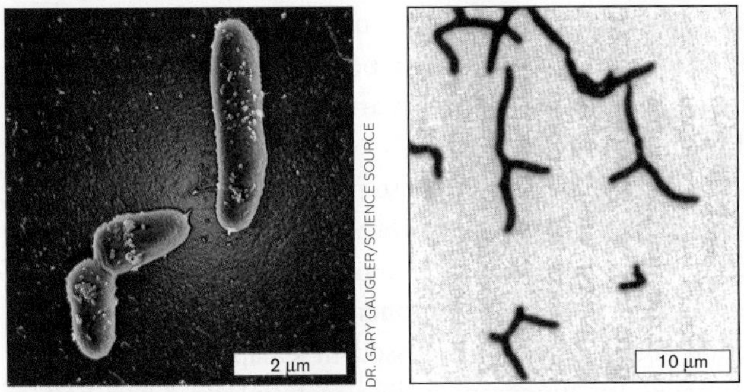

DR. GARY GAUGLER/SCIENCE SOURCE

J. J. GERMIDA AND L. E. CASIDA, JR., DEPARTMENT OF MICROBIOLOGY, PENNSYLVANIA STATE UNIVERSITY

FIGURE 18.23 ■ Irregularly shaped actinomycetes: *Corynebacterium* and *Arthrobacter*. **A.** *Corynebacterium diphtheriae* divides by snapping off one side while remaining attached at the other; the result is a typical V shape or "Chinese letter" arrangement (colorized SEM). **B.** *Arthrobacter globiformis* cultures form coccoid cells in stationary phase. With added nutrients, the coccoid cells grow out as rods.

Micrococcaceae. Relatives of *Arthrobacter* are nonfilamentous cocci such as *Micrococcus. M. luteus* is one of the most widespread of soil bacteria, appearing readily as yellow colonies on agar plates exposed to air. Historically, the genus *Micrococcus* in the family Micrococcaceae was classified with *Staphylococcus*, but its DNA sequence data now place *Micrococcus* and most of the Micrococcaceae within the actinomycetes. Micrococci are aerobic heterotrophs, commonly isolated from air and dust, although their habitat of choice is human skin. Micrococci grow in square or cuboid formations, dividing in two or three planes; cuboid clusters are known as **sarcinae** (singular, **sarcina**). *M. luteus* is harmless to humans and grows well at room temperature, so it makes an excellent laboratory organism for observation by students.

To Summarize

- **Firmicutes**, the low-GC Gram-positive bacteria, include endospore-forming genera such as *Bacillus* and *Clostridium*. The cycle of endospore formation was probably present in the common ancestor of this phylum.
- **Nonsporulating firmicutes** include pathogenic rods such as *Listeria*, as well as food-producing bacteria such as *Lactobacillus* and *Lactococcus*.
- ***Staphylococcus* and *Streptococcus*** are Gram-positive cocci that include normal human flora, as well as serious pathogens causing toxic shock syndrome, pneumonia, and scarlet fever.
- **Mycoplasmas** (Tenericutes) are related to Firmicutes but lack the cell wall and S-layer. They have flexible cytoskeletons and show ameboid motility. Mycoplasma species cause diseases such as meningitis and pneumonia.
- **Actinobacteria** (order Actinomycetales) include mycelial spore-forming soil bacteria, such as the actinomycete *Streptomyces.* Other actinobacteria have irregularly shaped cells, such as *Corynebacterium* species that cause diphtheria.
- **Mycobacteria** are actinobacterial rods whose cell envelope contains a diverse assemblage of complex mycolic acids. Mycobacterial species cause tuberculosis and leprosy. They stain acid-fast.

18.4 Proteobacteria (Gram-Negative)

Proteobacteria—the "protean," or "many-formed," bacteria—show highly diverse metabolism. Their lifestyles range from plant and animal pathogens to soil lithotrophs that reduce nearly any oxidized metal. Nevertheless, all proteobacteria share a common feature, the diderm (two-membrane) envelope, which causes their cells to stain Gram-negative. The Gram-negative cell envelope consists of an outer membrane, peptidoglycan cell wall permeated by the periplasm, and inner membrane (plasma membrane; discussed in Chapter 3). The outer membrane is packed with receptor proteins and porins, comprising two-thirds of the mass of the membrane. Porins evolved so as to admit nutrients while excluding antibiotics. The outer membrane lipids contain long sugar polymer extensions (lipopolysaccharide, or LPS). In pathogens, LPS repels phagocytosis and has toxic effects when released by dying cells.

The five phyla (previously classes) of Proteobacteria are designated Alpha through Epsilon (see **Table 18.1**). Even within each phylum, we see nearly as wide a range of cell shape and metabolism as we see in Proteobacteria as a whole.

The Protean Metabolism of Proteobacteria

A closer look at proteobacterial metabolism shows that processes that at first seem very different actually connect through linked biochemical modules (**Fig. 18.24**; metabolism is discussed in detail in Chapter 14). Many proteobacteria can oxidize H_2 (hydrogenotrophy) and small organic acids, as well as complex organic molecules with aromatic rings. Some "photoheterolithotrophs" carry out nearly all the fundamental classes of metabolism, depending on environmental conditions such as availability of light, oxygen, and nutrients. An example is *Rhodopseudomonas palustris*, a species of Alphaproteobacteria characterized and sequenced by Caroline Harwood at the University of Washington (**eTopic 14.2**). In such a "protean" organism, the core of

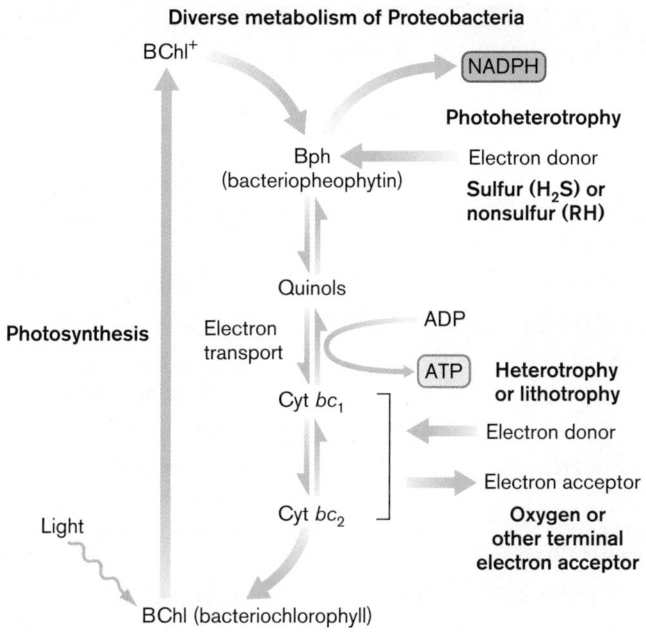

FIGURE 18.24 ■ The protean metabolism of Proteobacteria. In many Gram-negative species, metabolic diversity arises through minor "add-ons" of biochemical modules such as light absorption by bacteriochlorophyll, use of sulfide or organic electron donors, and use of oxygen or alternative (anaerobic) electron acceptors.

all proteobacterial energy acquisition is a respiratory chain of electron donors and acceptors. Electrons may enter the chain from a photoexcited chlorophyll, from organic electron donors such as sugars or benzoates, or from a mineral electron donor such as reduced sulfur or iron (lithotrophy). In facultative anaerobes, the electron acceptor may be a mineral such as nitrate or sulfate (anaerobic respiration). The capabilities of a given organism depend on which oxidoreductases its genome encodes.

Photoheterotrophy: light-supplemented heterotrophy. In the Alpha-, Beta-, and Gammaproteobacteria, diverse forms of light absorption have evolved from a common ancestor of photosystems I and II (see **Table 18.2**). The various bacteriochlorophylls of Proteobacteria peak in two ranges—in the blue and in the red or infrared. Bacteriochlorophyll *b*, in *Blastochloris viridis*, peaks well beyond 1,000 nm. Species with different photopigments often grow together in stratified layers of sediment or wetland—the infrared absorbers below, where they capture the longer wavelengths "left over" from the shorter-wavelength absorbers above.

The prevalence of homologous photosystems suggests that the common ancestor of the Gammaproteobacteria was a photoheterotroph. At the same time, other evidence supports horizontal transfer of photosystems among proteobacterial branches. The evidence is particularly strong in the case of **proteorhodopsin**, a homolog of the retinal-protein light-driven proton pump first characterized in halophilic archaea (discussed in Chapters 14 and 19). The proteorhodopsin light pump in *Pelagibacter*, a marine relative of *Rickettsia*, absorbs green and yellow (wavelength 500–600 nm), the midrange of solar radiation reaching Earth's surface. Different species have proteorhodopsins with very different absorption ranges, apparently adapted to different niches in the marine ecosystem.

In the literature, proteobacterial phototrophs are historically called "purple bacteria." The actual colors of all the phototrophs range from purple-red through yellow and brown. "Purple sulfur bacteria" are species that photolyze reduced forms of sulfur, such as H_2S, HS^-, S^0, or $S_2O_3^{2-}$ (thiosulfate), whereas "purple nonsulfur bacteria" photolyze H_2 or use photosystem II for cyclic photophosphorylation (discussed in Chapter 14). However, it turns out that many purple bacteria also use photosystem II. Most proteobacterial phototrophs conduct photosynthesis only in the absence of oxygen, reverting to aerobic heterotrophy when oxygen is present. Some, however, are obligate anaerobes. Still others, found in marine environments, surprisingly photolyze sulfide or organic compounds only in the presence of oxygen. Making sense of this extraordinary diversity is the job of marine and freshwater microbiologists, particularly those calculating global cycles of carbon and oxygen (discussed in Chapters 21 and 22).

Thought Question

18.7 Why might genes for the proteorhodopsin light-powered proton pump be more likely to transfer horizontally than the genes for bacteriochlorophyll-based photosystems PS I and PS II?

Lithotrophy: inorganic electron donors. Many different Proteobacteria oxidize inorganic electron donors, obtaining energy from the redox reaction without light absorption (photosynthesis or photolithotrophy, as discussed in the previous section). Examples of lithotrophy are given in **Table 18.4**. The ability to oxidize or reduce minerals evolved multiple times in different clades by modification of the electron transport chain. Many lithotrophic reactions generate acid, which conveniently breaks down rock containing additional reduced minerals. Bacterial and archaeal lithotrophy has tremendous significance for global cycling of elements in the biosphere (discussed in Chapter 22). Some lithotrophs are autotrophs, fixing carbon dioxide into biomolecules by the Calvin cycle (in carboxysomes) or the reverse TCA cycle. Others have alternative pathways of heterotrophy when organic foods are available. An interesting kind of lithotrophy performed by many Proteobacteria is the oxidation of carbon monoxide (CO; **eTopic 18.2**). Oxidizing CO is important because this toxic substance is a by-product of animal and plant metabolism.

TABLE 18.4 Lithotrophy and Methylotrophy (Examples)

Class/phylum	Example species	Reaction	$\Delta G°'/2e^-$ (kJ)	Class of reaction
Alphaproteobacteria	*Nitrobacter winogradskyi*	$NO_2^- + ½O_2 \rightarrow NO_3^-$	–54	Nitrite oxidation (nitrification to nitrate)
	Paracoccus pantotrophus	$H_2S + 2O_2 \rightarrow 2H^+ + SO_4^{2-}$		Sulfide oxidation to sulfate
	Roseobacter litoralis	$CO + H_2O \rightarrow CO_2 + 2H^+ + 2e^-$		CO oxidation
	Hyphomicrobium	$CH_3OH + O_2 \rightarrow CO_2 + H_2O + 2H^+ + 2e^-$		Methylotrophy
Betaproteobacteria	*Nitrosomonas europaea*	$NH_3 + O_2 \rightarrow HNO_2^- + 3H^+ + 2e^-$		Ammonia oxidation (nitrification to nitrite)
	Ralstonia eutropha	$2H_2 + ½O_2 \rightarrow 2H_2O$	–237	Hydrogen oxidation
	Thiobacillus denitrificans	$3S^0 + 4NO_3^- \rightarrow 3SO_4^{2-} + 2N_2$		Anaerobic sulfur oxidation
Gammaproteobacteria	*Acidithiobacillus ferrooxidans*	$4FeS_2 + 15O_2 + 14H_2O \rightarrow 4Fe(OH)_3 + 16H^+ + 8SO_4^{2-}$	–164	Iron-sulfur oxidation
	Acidithiobacillus thiooxidans	$2S^0 + 3O_2 + 2H_2O \rightarrow 2SO_4^{2-} + 4H^+$	–196	Sulfur oxidation
	Methylococcus capsulatus	$CH_4 + 2O_2 \rightarrow HCO_3^- + H^+ + H_2O$	–203	Methanotrophy
	Beggiatoa alba	$2H_2S + O_2 \rightarrow 2S^0 + 2H_2O$	–210	Sulfide oxidation
	Chromatium	$4Fe^{2+} + CO_2 + 11H_2O + h\nu \rightarrow 4Fe(OH)_3 + [CH_2O] + 8H^+$		Iron phototrophy (photoferrotrophy)
Deltaproteobacteria	*Desulfovibrio*	$4S^0 + 4H_2O \rightarrow SO_4^{2-} + 3HS^- + 5H^+$	–11.3 (at pH 8)	Sulfur oxidation
Nitrospirae	*Nitrospira*	$NO_2^- + ½O_2 \rightarrow NO_3^-$	–54	Nitrite oxidation (nitrification to nitrate)
Planctomycetes	*Brocadia anammoxidans*	$NH_4^+ + NO_2^- \rightarrow N_2 + 2H_2O$	–238	Anaerobic ammonium oxidation (anammox)

Alphaproteobacteria: Photoheterotrophs, Methylotrophs, and Endosymbionts

The Alphaproteobacteria include photoheterotrophs, heterotrophs, and bacteria that metabolize single-carbon compounds, as well as intracellular mutualists and pathogens.

Photoheterotrophs. Most alphaproteobacterial photoheterotrophs are unicellular. Their cell shapes range from flagellated spirilla to rounded rods (*Rhodobacter sphaeroides*; **Fig. 18.25A**), wide spirals (*Rhodospirillum rubrum*), and stalked cells (*Rhodomicrobium*; **Fig. 18.25B**). The cryo-EM model of *Rhodobacter sphaeroides* in **Figure 18.25A** shows that the cell is packed with photomembranes (membranes containing the photosynthetic complex) arranged in vesicles invaginated from the cytoplasmic membrane. These intracellular photomembranes expand the surface area for photon capture, but during growth with oxygen, the photomembranes disappear. The outer surface of the cell shows LPS filaments extending from the outer membrane.

The metabolism of *Rhodospirillum rubrum* shifts drastically with oxygen. Anaerobically, *R. rubrum* bacteria grow dark red because they synthesize membrane containing BChl *a*, which absorbs green and infrared and reflects primarily red. With oxygen, however, *R. rubrum* fails to synthesize BChl *a*, and the cells grow white as their metabolism switches to straight heterotrophy. Heterotrophy is supplemented by the oxidation of small molecules such as carbon monoxide.

Alphaproteobacterial photoheterotrophs that require oxygen (but do not photolyze water) were found unexpectedly in genomic surveys of marine bacteria. Originally thought to be straight heterotrophs, these organisms revealed genes encoding bacteriochlorophyll. Their aerobic heterotrophy was shown to be driven by light absorption. The reason for the oxygen requirement is unclear, since their photosystems do not involve oxygen. Examples of O_2-requiring phototrophs have since been found in most of the proteobacterial clades, and species have been isolated from all major habitats, including freshwater, marine, and soil ecosystems. Some show unusual cell morphology, such as the Y shape of *Citromicrobium* (**Fig. 18.25C**).

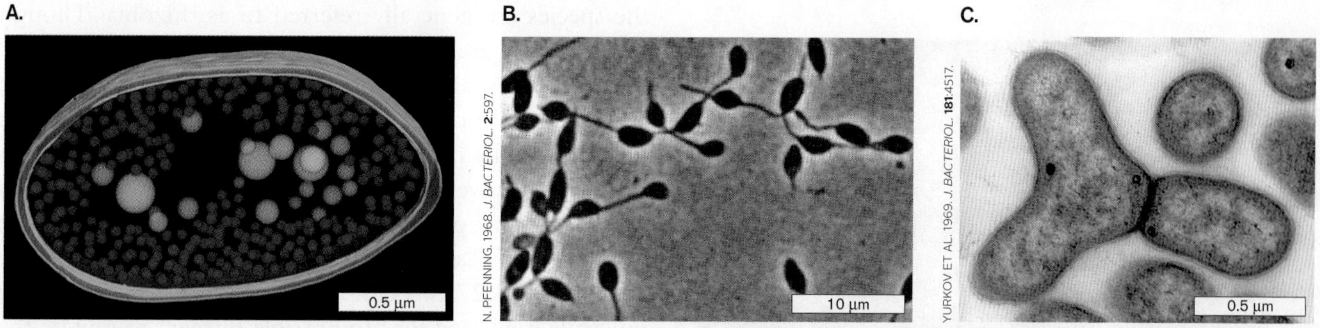

JADE M. NOBLE, ET AL. 2018. *MOL MICROBIOL* **109:** 812–825.
N. PFENNING. 1968. *J. BACTERIOL.* **2:** 597.
YURKOV ET AL. 1969. *J. BACTERIOL.* **181:** 4517.

FIGURE 18.25 ▪ Alphaproteobacterial photoheterotrophs. A. *Rhodobacter sphaeroides*, showing intracellular photosynthetic vesicles (cryo-EM tomography). B. *Rhodomicrobium vannielii* with stalked cells (phase contrast). C. *Citromicrobium* species, an aerobic photoheterotroph, forms highly pleomorphic shapes, including this Y shape (TEM).

The aerobic photoheterotroph *Erythromicrobium ramosum* can reduce toxic metal compounds such as tellurite ion (TeO_3^{2-}). The cells generate tellurium crystals that take up 30% of their cell weight—a trait that we might use to remove tellurite from liquid waste. Other toxic metals reduced by aerobic photoheterotrophs include selenium and arsenic. Thus, these obscure bacteria now have a promising future in remediation of metal-contaminated industrial wastes.

Aquatic and soil oligotrophs. Alphaproteobacteria also include many nonphototrophic heterotrophs of soil and water. Many aquatic and soil heterotrophs are oligotrophs adapted to extremely low nutrient concentrations; an example is *Caulobacter crescentus*, the stalk-to-flagellum organism discussed in Chapters 3 and 4. Oligotrophic bacteria often have unusual extended shapes enhancing nutrient uptake, such as the starlike cell aggregates of *Seliberia stellata*. In *Seliberia* species, the individual tightly coiled rods generate oval or spherical reproductive cells by a budding process. The budding reproductive cells germinate into rods, which then form new aggregates.

Some heterotrophs common in soil are pathogens. *Brucella* species are intracellular pathogens of animals that can also infect humans. The soil pathogen *Granulibacter bethesdensis* is associated with chronic granulomatous disease, an inherited disorder of the phagocyte oxidase system that leaves patients susceptible to infection. Other pathogens are carried by insects or animal hosts; for example, *Bartonella henselae* causes cat scratch disease.

Methylotrophy and methanotrophy. Methylotrophy is the ability of an organism to oxidize reduced single-carbon compounds such as methanol, methylamine, or methane. The Alphaproteobacteria include several genera of methylotrophs, which are found in all environments, including soil, freshwater, and the ocean. Most methylotrophs can grow on both single-carbon and organic compounds. One such versatile genus, *Methylobacterium*, is equally at home in soil and water, on plant surfaces, and as a contaminant of facial creams and purified water for silicon chip manufacture.

Other species are restricted to single-carbon compounds, incapable of metabolizing organic compounds with carbon-carbon bonds. Methylotrophs that grow solely on methane (CH_4) are called **methanotrophs**. Methane-oxidizing species of Alpha- and Gammaproteobacteria, and Verrucomicrobia, contribute to aquatic ecosystems, serving as major food sources for zooplankton. They eliminate much of the methane produced by methanogens before it reaches the atmosphere, where it has a potent greenhouse effect (see Chapter 22).

An interesting methylotroph is *Hyphomicrobium*, a bacterium with an unusual stalk-to-flagellum transition similar to that of the alphaproteobacterium *Caulobacter crescentus*, whose life cycle is described in Chapter 3. *Hyphomicrobium* species are found in environments as diverse as wastewater sludge and Antarctic island soil. For example, the Antarctic soil species *H. sulfonivorans* metabolizes sulfur compounds such as dimethyl sulfone (**Fig. 18.26**). Like *Caulobacter*, a flagellated cell of *Hyphomicrobium* species may lose its flagellum to form a stalk (also called "hypha"). But unlike *Caulobacter*, *Hyphomicrobium* then forms a daughter cell from the opposite end of the stalk! As the DNA replicates, one daughter nucleoid must migrate all the way through the stalk to reach the daughter cell. The daughter cell forms a flagellum and septates, separating from the parent (**Fig. 18.26B**).

Thought Question

18.8 Can you hypothesize a mechanism for migration of the daughter nucleoid of *Hyphomicrobium* through the stalk to the daughter cell? For possibilities, see Chapter 3 and consider the various molecular mechanisms of cell division and shape formation.

A.

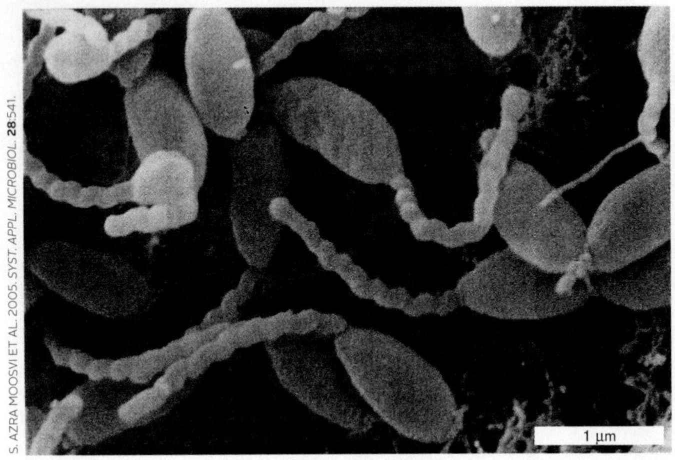

B.

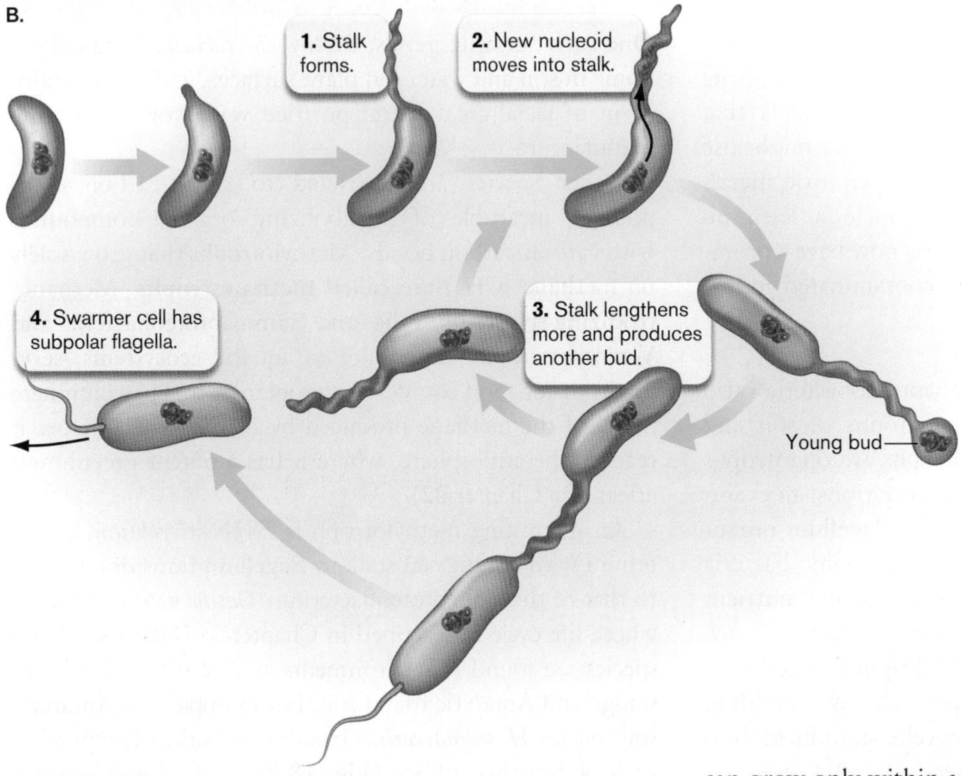

FIGURE 18.26 ▪ *Hyphomicrobium sulfonivorans* from Signy Island, Antarctica. A. These bacteria with coiled stalks catabolize dimethyl sulfone. B. Life cycle of *Hyphomicrobium* species.

Endosymbionts: mutualists and parasites. The Alphaproteobacteria include many highly evolved intracellular symbionts—some mutualists, others parasites. As isolated bacteria, they are generally rod-shaped and have aerobic metabolism, but their shape is transformed within the host cell. Intracellularly, they need to solve special problems, such as the exclusion of oxygen from nitrogen fixation (a process requiring anaerobiosis) or, in the case of pathogens, the need to resist host defenses.

Nitrogen-fixing endosymbionts of plants include genera such as *Rhizobium*, *Bradyrhizobium*, and *Sinorhizobium*. The nomenclature has undergone many changes, but the species are generally referred to as rhizobia. Though they can live freely in the soil, rhizobia prefer to colonize plants, usually legumes such as peas or alfalfa, where they form distinctive nodule structures (**Fig. 18.27**). The host plant cells provide the bacteroids with nutrients, as well as protective components such as **leghemoglobin**, an oxygen-binding protein that maintains anaerobiosis within infected cells. Leghemoglobin turns the nodule interior pink (**Fig. 18.27A**). Leghemoglobin (**Fig. 18.27A**) is the molecule now used for "Impossible Burger" to imitate the color and taste of ground beef (**Fig. 18.27B**).

Each bacterial species colonizes and infects a particular host range. Complex chemosensory processes attract the bacteria to the surface of the host root, where their presence induces root hairs to form curls around the bacteria (**Fig. 18.27C**). The curling of the root hairs enables the bacteria to form infection threads, chains of rod-shaped bacteria that invade the root cells. Within the host cells, the bacteria lose their cell wall and become rounded **bacteroids**, specialized for nitrogen fixation.

The genus *Agrobacterium* includes plant pathogens closely related to the rhizobia. *A. tumefaciens* is known for its ability to convert host cells to a form that produces tumors called galls (**Fig. 18.28**). The ability of *A. tumefaciens* to insert its DNA into plant genomes has made it a major tool for plant biotechnology (see Chapter 16).

Rickettsias are famous intracellular pathogens. Short coccoid rods, rickettsias lack flagella and can grow only within a host cell. The best-known rickettsia is *Rickettsia rickettsii*, the cause of Rocky Mountain spotted fever, a disease spread by ticks throughout the United States (**Fig. 18.29A**). *Rickettsia* species parasitize human endothelial cells (**Fig. 18.29B**). The bacteria induce phagocytosis and then dissolve the phagocytic vesicle and escape into the cytoplasm. Some rickettsias propel themselves through the host cell by polymerizing cytoplasmic actin behind them (**Fig. 18.29C**)—a process similar to that of *Listeria* (described earlier, along with Firmicutes). The actin tails eventually project outward as filopodia, extensions of host cytoplasm and membrane that protect the bacteria from host defenses while enabling them to invade adjacent cells.

The rickettsias show genetic relatedness to mitochondria. All eukaryotic mitochondria appear to be descendants of an ancient rickettsial parasite whose respiratory apparatus ultimately became essential to power eukaryotic cells.

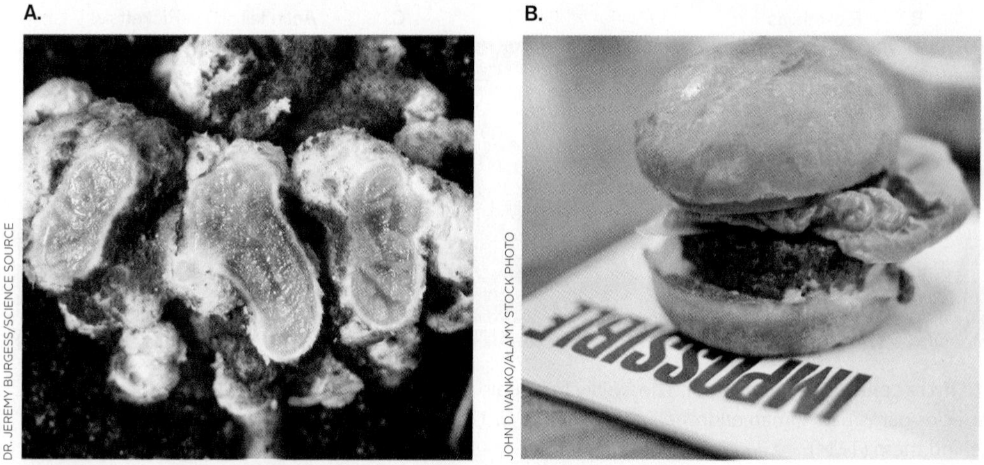

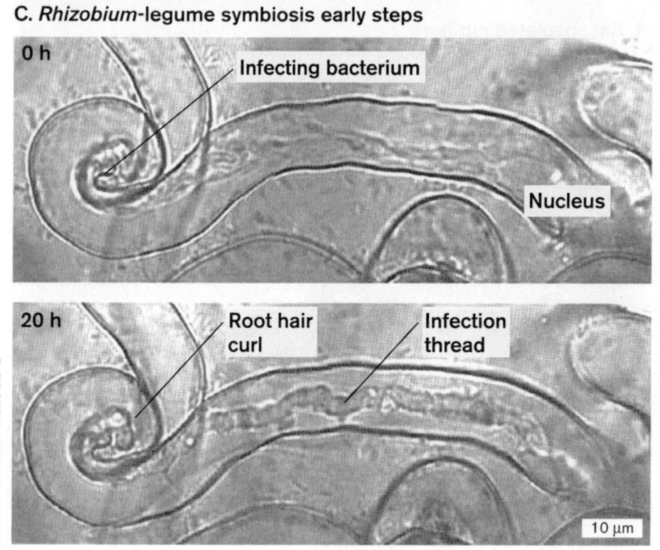

FIGURE 18.27 ■ Rhizobia: legume endosymbionts. **A.** Legume nodules cut open to show pink regions where the plant cells produce leghemoglobin to maintain anaerobic conditions for bacteroid nitrogen fixation. **B.** Leghemoglobin is the source of beef-like color and flavor of the "Impossible Burger." **C.** Clover root hair curls around infecting *S. meliloti*. The bacteria enter the curl and grow down the root hair as an infection thread that penetrates the legume cells, enabling the bacteria to colonize in the form of bacteroids.

Betaproteobacteria: Photoheterotrophs, Lithotrophs, and Pathogens

The Betaproteobacteria include photoheterotrophs such as *Rhodocyclus*, as well as a diverse range of lithotrophs (see **Table 18.4**). Several are important pathogens that infect humans and other animals.

Lithotrophs: nitrifiers and sulfur oxidizers. An important group of nitrogen lithotrophs consists of the nitrifiers. Nitrifiers oxidize ammonia (NH_3) to nitrite (NO_2^-), or nitrite to nitrate (NO_3^-). Typically, different species conduct the two reactions separately while coexisting in soil and water. Nitrifiers are of enormous economic and practical importance for wastewater treatment because they decrease the reduced nitrogen content of sewage. Special systems have been developed to retain nitrifier bacteria behind filters as one stage of water treatment. In the system shown in **Figure 18.30A**, nitrifying bacteria are encapsulated in pellets to retain them within the bioreactor while the treated water flows through a filter.

Commonly isolated ammonia oxidizers are species of *Nitrosomonas*, *Nitrosolobus*, and *Nitrosovibrio*. *Nitrosomonas* cells conduct electron transport through extensive internal membranes that are either stacked or invaginated (**Fig. 18.30B**). One genus of Gammaproteobacteria has been found to oxidize ammonia: *Nitrosococcus*. Nitrite oxidizers include the nonproteobacterial phylum Nitrospirae. Although genetically outside the Proteobacteria, the unusual spiral-shaped cells of Nitrospirae have a Gram-negative cell envelope.

Pathogens. The Betaproteobacteria include aerobic heterotrophic cocci such as *Neisseria*. The cocci of *Neisseria* species form distinctive pairs known as **diplococci**. Most *Neisseria* species are harmless commensals of the nasal or oral mucosa, but *N. gonorrhoeae* causes the sexually transmitted disease gonorrhea. *N. gonorrhoeae* is actually a microaerophile, requiring a narrow range of oxygen concentration; it has fastidious growth requirements, necessitating cultivation on a special blood-based medium. A related organism, *N. meningitidis*, may be carried asymptomatically by as much as a quarter of the human population,

FIGURE 18.28 ■ Plant gall induced by *Agrobacterium tumefaciens*. Crown gall on chrysanthemum plant.

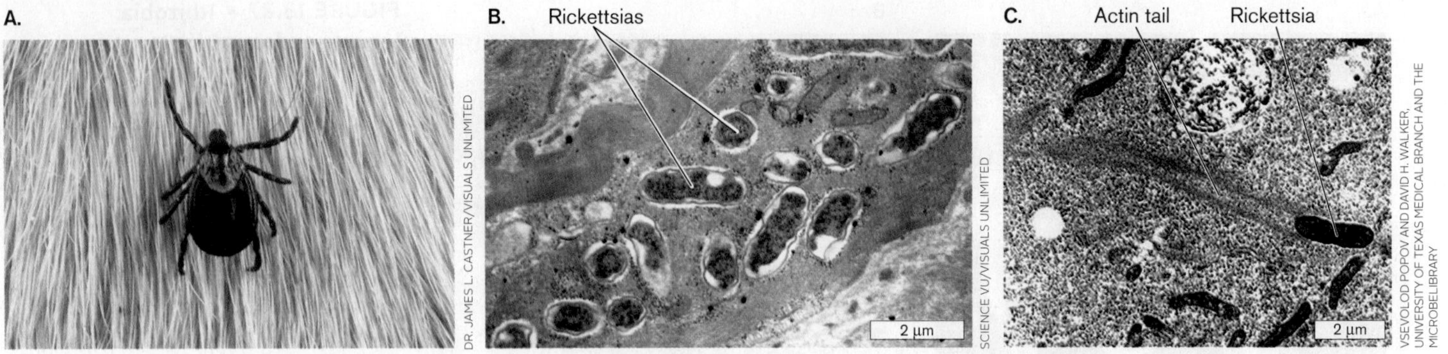

FIGURE 18.29 ■ Rickettsias are obligate intracellular parasites. **A.** The Rocky Mountain tick carries *Rickettsia rickettsii*, the cause of Rocky Mountain spotted fever. **B.** *Rickettsia* species parasitize human endothelial cells (TEM). **C.** The rickettsias propel themselves through the host cell by polymerizing cytoplasmic actin behind them (TEM).

but occasionally it causes meningitis, which can be fatal. Other neisserias, such as *N. sicca*, are easily isolated from skin and often presented to undergraduates as an unknown for identification.

Other members of the Beta class are important animal and plant pathogens, such as *Burkholderia*. *B. cepacia* was originally isolated from onions as a cause of bulb rot; it is now known to be a major opportunistic invader of the lungs of cystic fibrosis patients.

Gammaproteobacteria: Photolithotrophs, Enteric Flora, and Pathogens

The Gammaproteobacteria include vast numbers of marine organisms that catabolize complex organic pollutants such as petroleum hydrocarbons. However, the best-known Gammaproteobacteria are the Enterobacteriaceae family of facultative anaerobes found in the human colon—the family that includes *Escherichia coli*. Gammaproteobacteria also include unusual phototrophs that oxidize iron and nitrite.

Sulfur lithotrophs. The sulfur-oxidizing genus *Beggiatoa* was one of the first kinds of lithotrophs described by pioneer microbial ecologist Sergei Winogradsky (discussed in Chapter 1). *Beggiatoa* species oxidize H_2S to elemental sulfur, which collects as sulfur granules within the periplasm. *Beggiatoa* also stores carbon in cytoplasmic granules of polyhydroxybutyrate—a common strategy among proteobacteria. The cells of *Beggiatoa* grow as extended filaments with sulfur granules, forming biofilms on sulfide-rich sediment (**Fig. 18.31**). The hairlike thioectosymbionts shown in the Current Research Highlight are also sulfur-oxidizing Gammaproteobacteria.

Bacteria of the genus *Acidithiobacillus* oxidize iron or sulfur (**Fig. 18.32A**). Iron-oxidizing bacteria commonly

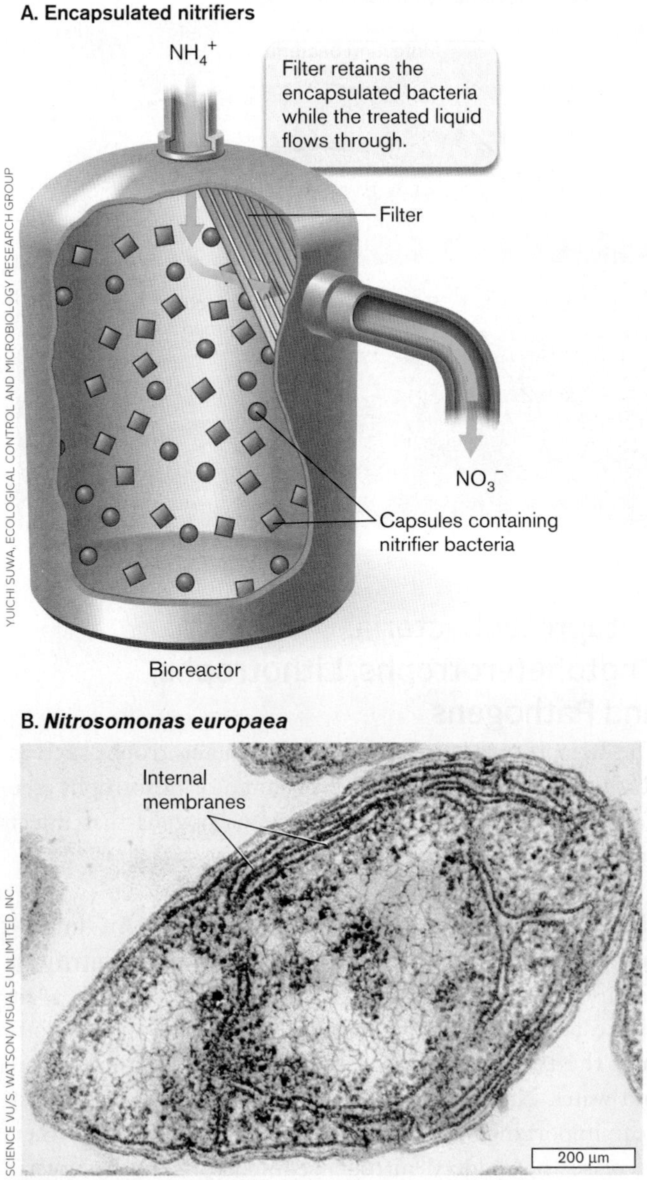

FIGURE 18.30 ■ Nitrifiers. **A.** Wastewater treatment uses nitrifier bacteria to remove ammonia. **B.** *Nitrosomonas europaea*, of the Betaproteobacteria, oxidizes ammonia to nitrite (TEM). Internal membranes contain the electron transport complexes.

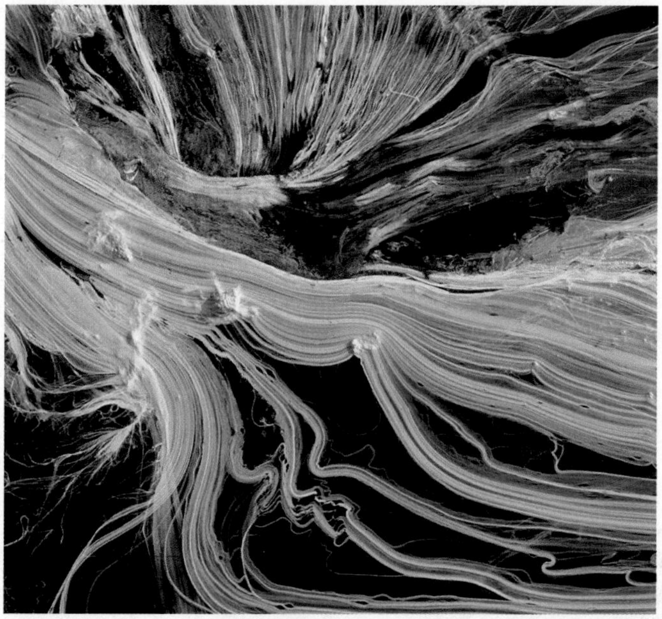

JANE E. ROCHE/NATURE PL.COM

FIGURE 18.31 ■ *Beggiatoa* species oxidize sulfur in marine sediment.

A.

ROMANO, P., BLÁZQUEZ, M. L., ET AL. (2001). *JOURNAL OF CHEMICAL TECHNOLOGY AND BIOTECHNOLOGY*, **76:** 723.

B.

ROYCE BAIR/GETTY IMAGES

FIGURE 18.32 ■ *Acidithiobacillus*: iron oxidizers. **A.** *A. ferrooxidans* leaches copper and iron from molybdenite ore (SEM). The hexagonal object is a molybdenite crystal. **B.** A copper mine in Utah, where *Acidithiobacillus* species can oxidize copper ores, leaching the copper into solution for retrieval.

form the brown stains found inside plumbing. Most species are short rods or vibrios (comma-shaped). *Acidithiobacillus* and other sulfur-oxidizing genera can undergo a number of different reactions oxidizing H_2S to S^0 and S^0 to SO_4^{2-} (see **Table 18.4**). Sulfate production makes an environment acidic enough to erode stone monuments and the interior surface of concrete sewer pipes. Sulfur oxidation is often coupled to oxidation of iron, $Fe^{2+} \rightarrow Fe^{3+}$. The bacterium *A. ferrooxidans* is known for its role in acidification of mine water, which contributes to leaching of iron, copper, and other minerals (**Fig. 18.32B**).

Sulfur and iron phototrophs. The gammaproteobacterial phototrophs, such as *Chromatium* species (**Fig. 18.33A**), mainly utilize sulfide and produce sulfur, which is deposited as intracellular granules visible within the cytoplasm. Their phototrophy is entirely anaerobic. Some of the Gamma group are true autotrophs and do not use organic substrates. Some of these actually conduct phototrophy using iron (Fe^{2+}) to donate electrons. Iron phototrophy, or "photoferrotrophy" (**Fig. 18.33B**), is considered an intriguing possibility for the metabolism of early life (discussed in Chapter 17). *Thiocapsa* uses NO_2^- and was the first nitrogen-based phototroph discovered.

A.

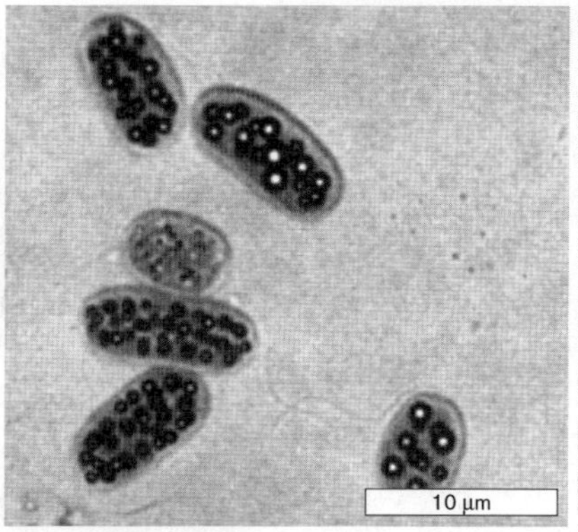

H. G. SCHLEGEL AND N. PFENNIG. 1961. *ARCH. MICROBIOL.* **38:**1–39.

B.

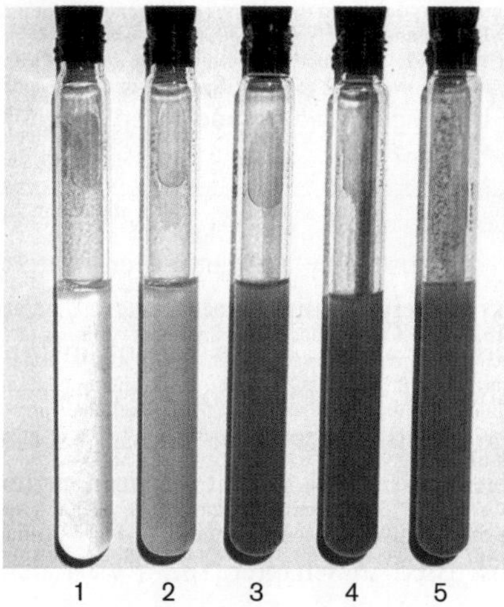

A. EHRENREICH ET AL. *APPLIED AND ENVIRONMENTAL MICROBIOLOGY* **60**:12 (1994): 4517–26. REPRODUCED WITH PERMISSION FROM AMERICAN SOCIETY FOR MICROBIOLOGY

FIGURE 18.33 ■ *Chromatium*: sulfur and iron phototrophs. **A.** *Chromatium* forms single-flagellated rods full of sulfur granules. **B.** Photoferrotrophy. Under illumination, color develops over time (tubes 1–5) as a *Chromatium* isolate oxidizes Fe^{2+} to Fe^{3+}.

Enterobacteriaceae: intestinal fermenters and respirers. The family Enterobacteriaceae, facultative anaerobes of the Gammaproteobacteria, include some of the most intensively studied species of all bacteria. Species are readily isolated from the contents of the human digestive tract and easily grown on laboratory media based on human food. The best-known species of Enterobacteriaceae—indeed, the most studied of all bacterial species—is the model organism *Escherichia coli*. Some strains of *E. coli* grow normally in the human intestine, feeding on our mucous secretions and producing vitamins, such as vitamin K. They may grow in symbiosis with anaerobic fermenters such as *Bacteroides* species, which release short-chain sugars that *E. coli* digests (discussed in Chapter 21). But other strains, such as *E. coli* O157:H7, cause serious illness. A large proportion of the world's children die of *E. coli*–related intestinal illness before the age of 5. Related pathogens include *Salmonella enterica* and *Shigella flexneri*.

The Enterobacteriaceae are Gram-negative rods, although as nutrients diminish, their size dwindles almost to a coccoid form. They grow singly, in chains, or in biofilms. Many species are motile, with numerous flagella. Most strains grow well with or without oxygen, by either respiration (aerobic or anaerobic) or fermentation. They ferment rapidly on carbohydrates, generating fermentation acids, ethanol, and gases (CO_2 plus H_2) in varying proportions, depending on the species. Their presence in the intestine supports the growth of organisms utilizing these gases, including methanogens (discussed in Chapter 19). Many strains form biofilms. Biofilm formation explains the persistence of drug-resistant infections, such as those associated with urinary catheters in long-term hospital patients.

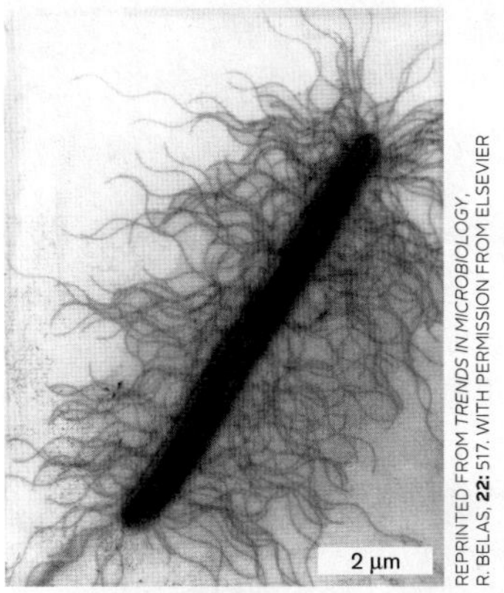

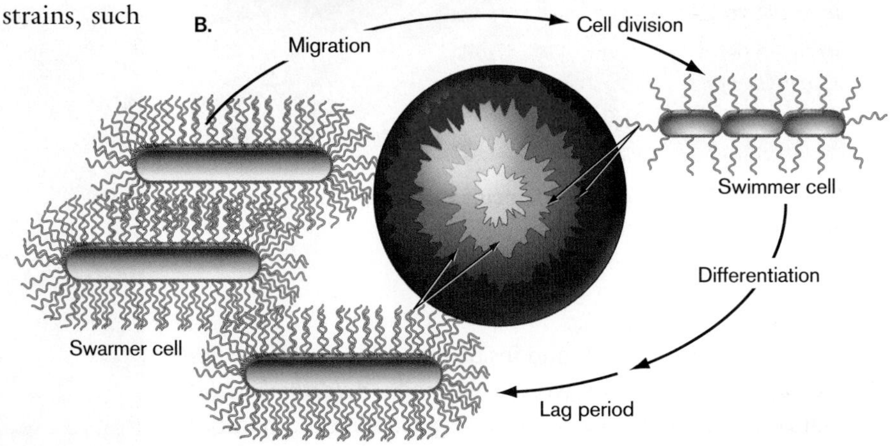

FIGURE 18.34 ■ The enteric rod *Proteus mirabilis*: isolated swimmer or cooperative swarmer. **A.** A thickly flagellated swarmer cell (TEM). Cell length can reach 20 μm. **B.** Swarmer rafts of *P. mirabilis* migrate through blood agar.

Thought Question

18.9 Why do you think it took many years of study to realize that *Escherichia coli* and other Proteobacteria can grow as a biofilm?

Because they are easy to cultivate and have been studied extensively in clinical laboratories, many genera of Enterobacteriaceae are familiar to students in introductory microbiology laboratory courses. A common laboratory exercise is to distinguish *Enterobacter* and *Klebsiella* species from *E. coli* by their fermentation to the pH-neutral product butanediol, which tests positive in the Voges-Proskauer test. (Fermentation is discussed in Chapter 13.) *Enterobacter* species occur more frequently in freshwater streams than in the human body, although a few species colonize the intestine and cause illness.

Proteus mirabilis and *P. vulgaris* cause bladder and kidney infections, particularly as a complication of surgical catheterization. *Proteus* species are heavily flagellated and display a remarkable **swarming** behavior (**Fig. 18.34**). In response to an environmental signal, the flagellated rods grow into long-chain swarmer cells. The swarmers gather together, forming "rafts" that swim together and grow into a complex biofilm.

Besides symbionts of animals, Enterobacteriaceae include plant pathogens, such as *Erwinia carotovora* and

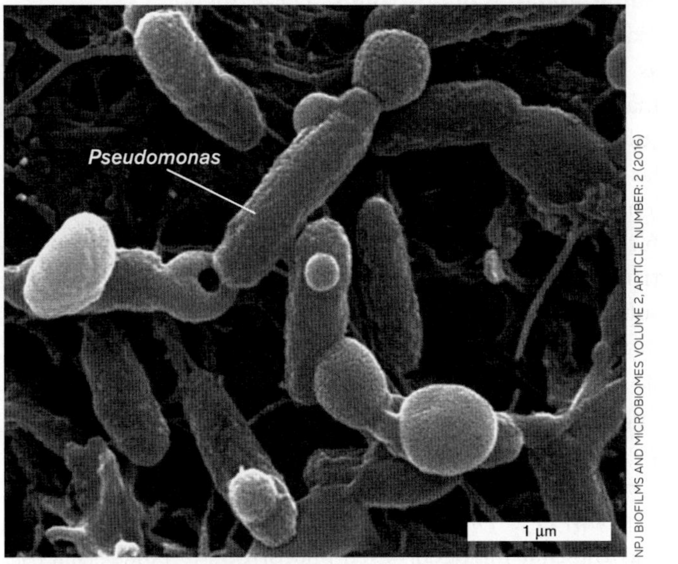

FIGURE 18.35 ■ *Pseudomonas* species form biofilms. *P. fluorescens* biofilm on a plant surface.

related species. *Erwinia* species cause wilts, galls, and necrosis of a wide variety of plants, including bananas, tomatoes, and orchids.

Related facultative rods in soil and water conduct anaerobic respiration by donating electrons from organic substrates to a variety of metals, such as iron and magnesium. *Shewanella oneidensis* and other metal reducers are used to make electricity in fuel cells (discussed in Chapter 14).

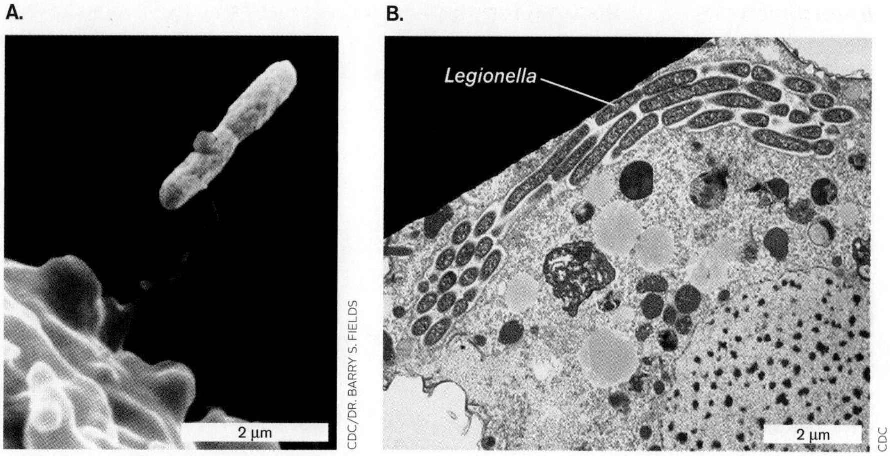

FIGURE 18.36 ■ *Legionella pneumophila* colonizes an ameba. **A.** *L. pneumophila* cell caught by an ameba's pseudopod (colorized SEM). **B.** *L. pneumophila* cells have colonized the ameba (TEM).

Aerobic rods. Closely related to Enterobacteriaceae are several genera of rod-shaped bacteria that are obligate respirers. Many are obligate aerobic respirers (requiring O_2 for growth), although some can use alternative electron acceptors, such as nitrate. These genera catabolize an extraordinary range of natural compounds, including aromatic derivatives of lignin; thus, they have important roles in natural recycling and soil turnover.

The Pseudomonadaceae are a large and amorphous group of Gammaproteobacteria. As the "pseudo-" prefix suggests, their taxonomic unit is poorly defined and includes species whose DNA sequence has necessitated their reassignment to other groups. The pseudomonads, such as *Pseudomonas aeruginosa* and *P. fluorescens*, respire on oxygen or nitrate and are vigorous swimmers with single or multiple polar flagella. *P. aeruginosa* can swim throughout a standard agar plate (much to the chagrin of students attempting to isolate colonies). Nevertheless, under appropriate environmental conditions, pseudomonad cells give up their motility and develop biofilms (**Fig. 18.35**; discussed in Chapter 4). Biofilms of *P. aeruginosa* cause lethal infections of the pulmonary lining in cystic fibrosis patients.

Some pseudomonad pathogens have the unusual ability to infect both plants and animals. For example, *P. aeruginosa* commonly infects plants as well as humans. *P. fluorescens* infects seedlings and causes rotting of citrus fruit, while it also appears as an opportunistic pathogen of immunocompromised cancer patients. Other pseudomonad species, however, are harmless residents of soil or sewage.

Legionella pneumophila is a well-publicized pathogen related to the pseudomonads. Incapable of growth on sugars, *L. pneumophila* requires oxygen to respire on amino acids. The organism exhibits an unusual dual lifestyle, alternating between intracellular growth within human macrophages and intracellular growth within freshwater amebas (**Fig. 18.36**). Growth within amebas facilitates transmission through aerosols into the human lung. *L. pneumophila* is an environmental pathogen that takes advantage of our lifestyle (the prevalence of large-scale air-conditioning units that contain unfiltered water).

A related bacterium, *Coxiella burnetii*, causes Q fever, a respiratory illness of livestock and humans. *C. burnetii* converts to a spore-like form that persists in soil. *C. burnetii* resembles a rickettsia in some of its strategy of pathogenesis, but it is genetically closer to *Legionella*.

Gammaproteobacteria include important plant pathogens, such as *Xanthomonas* species. *Xanthomonas* is a flagellate rod that colonizes a wide range of agricultural plants, such as tomatoes, potatoes, onions, broccoli, and citrus fruits (**Fig. 18.37**). The disease it causes may mottle the leaves and fruit, and it may spread throughout the plant.

Deltaproteobacteria: Lithotrophs and Multicellular Communities

The Deltaproteobacteria include important sulfur and iron reducers, such as the fuel-cell bacterium *Geobacter metallireducens* (discussed in Chapter 14). Other species have complex life cycles that include multicellular developmental

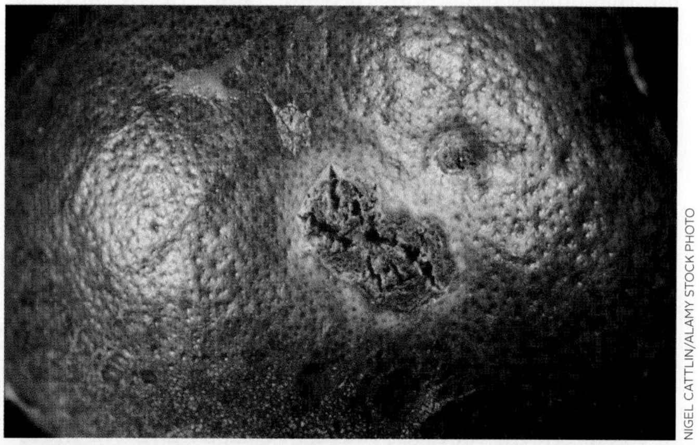

FIGURE 18.37 ■ Bacterial spot disease of orange fruit, caused by *Xanthomonas* species.

A.

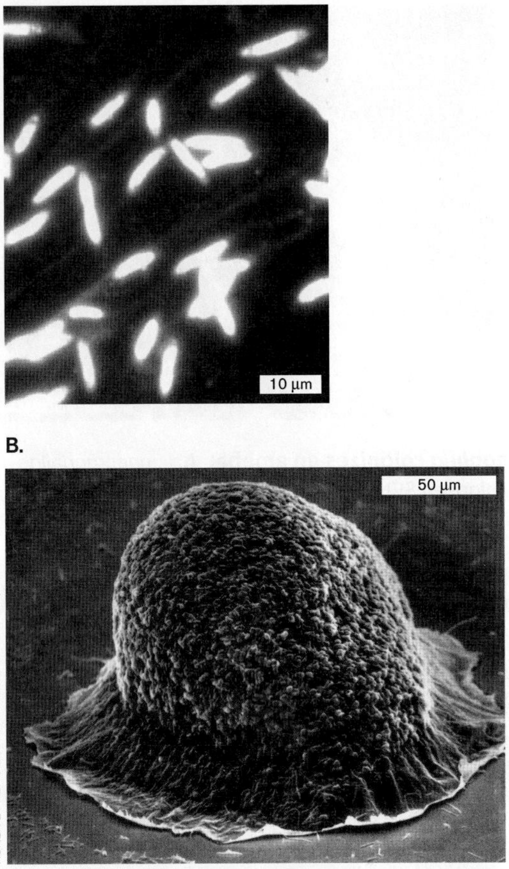

WOLGEMUTH ET AL. 2002. *CURR. BIOL.* **12**:369–377, FIG. 3A.

KUNER ET AL. 1982. *J. BACTERIOL.* **151**:458–461, FIG. 1.

FIGURE 18.38 ■ Myxobacteria. A. *Myxococcus xanthus* cells glide while producing trails of slime (dark-field LM). **B.** Upon starvation, cells come together to generate a fruiting body packed with spherical myxospores (SEM).

forms. The best-studied example is the myxobacteria. Myxobacteria have exceptionally large genomes, such as that of *Sorangium cellulosum* (12.6 Mb).

Myxobacteria. The myxobacteria, such as *Myxococcus xanthus*, are free-living soil bacteria that can grow as isolated cells but come together to form a multicellular structure for the purpose of spore dispersal (**Fig. 18.38**; also see **Fig. 18.8D**). When nutrients are plentiful, the myxobacteria grow and divide as individual cells. As nutrients run out, the cells begin to attract each other by quorum sensing, and they move into parallel formations. Myxobacteria have no flagella, but they move along a surface by using a form of motility called gliding. The aggregating cells coalesce to develop a **fruiting body**. For *Myxococcus*, the fruiting body rises from the aggregating cells to form a globular mass (**Fig. 18.38B**). Other taxa, such as *Stigmatella*, form a fruiting body that is branched with bulbous ends (**Fig. 18.39**).

Analogous fruiting bodies also form in slime molds, a class of eukaryotic microbes (discussed in Chapter 20). The myxobacteria, however, have evolved their process independently. Within the myxobacterial fruiting body, durable spherical

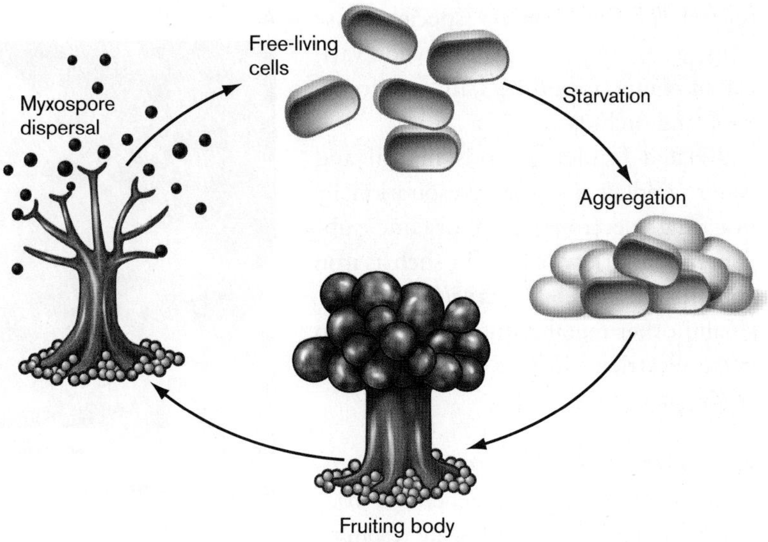

FIGURE 18.39 ■ Life cycle of a myxobacterium (*Stigmatella*). Starving *Stigmatella* cells glide toward each other to aggregate. The aggregation generates a fruiting body with bulges packed with small, spherical myxospores.

cells called **myxospores** develop (**Fig. 18.39**). Ultimately, the fruiting body releases the myxospores to be carried on the wind or by insects to a more favorable location.

Thought Question

18.10 Compare and contrast the formation of cyanobacterial akinetes, firmicute endospores, actinomycete arthrospores, and myxococcal myxospores.

Note: Distinguish myxobacteria from Mycobacterium, the Gram-positive genus that includes the causative agents of tuberculosis and leprosy.

Bdellovibrios parasitize bacteria. Bacteria can be parasitized or preyed on by smaller bacteria. The Deltaproteobacteria include *Bdellovibrio* species, which attack proteobacterial host cells. The structure of the "attack cell" is a small, comma-shaped rod with a single flagellum. The attack cell attaches to the envelope of its host and then penetrates into the periplasm, where it uses host resources to grow (**Fig. 18.40**). The growing cell produces enzymes that cross the inner membrane to degrade host macromolecules and make their components available to the bdellovibrio. The entire host cell loses its shape and becomes a protective incubator for the predator. This stage is called the bdelloplast.

Within the periplasm, the invading bdellovibrio elongates as a spiral filament while replicating several copies of its DNA. When most nutrients have been exhausted, the filament septates into multiple short cells. The cells develop flagella, and the bdelloplast bursts, releasing the newly formed attack cells (**Fig. 18.40B**).

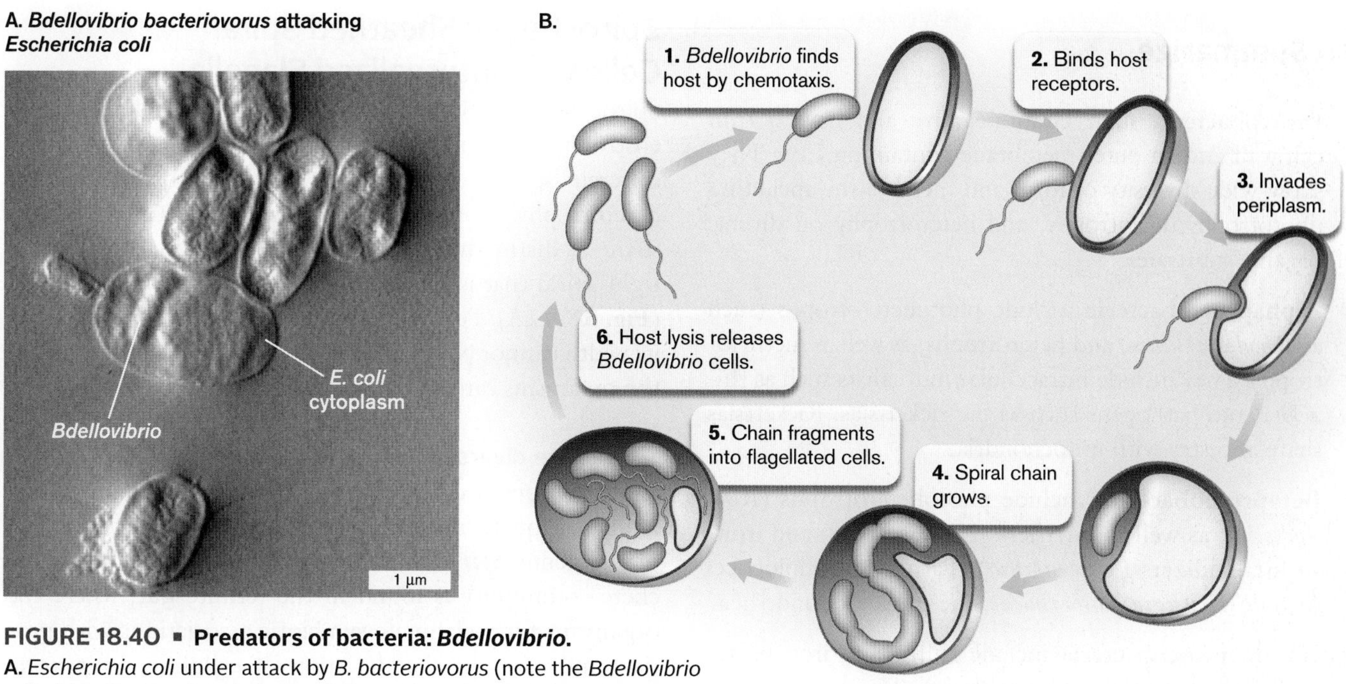

FIGURE 18.40 ▪ Predators of bacteria: *Bdellovibrio*. **A.** *Escherichia coli* under attack by *B. bacteriovorus* (note the *Bdellovibrio* cell within the *E. coli* periplasm; AFM). **B.** Life cycle of a bdellovibrio.

Bdellovibrios can be isolated from sewage, soil, or marine water—any environmental source of Gram-negative prey bacteria. Different species infect *E. coli* and *Pseudomonas*, as well as *Agrobacterium* and *Rhizobium* in their free-living state. They can be cultured as plaques on a top-agar plate containing host bacteria—the same procedure used to isolate bacteriophages (discussed in Chapter 6).

Epsilonproteobacteria: Helical Pathogens and Marine Sulfur Bacteria

Epsilonproteobacteria include the genera *Campylobacter* and *Helicobacter*. *Helicobacter pylori* is well known today as the causative agent of gastritis and stomach ulcers (**Fig. 18.41**). Yet, until recently, most microbiologists believed that bacteria could not live in the acidic stomach. The discovery of *H. pylori* as the cause of gastritis earned Barry Marshall and J. Robin Warren, of the University of Western Australia, the 2005 Nobel Prize in Physiology or Medicine. Marshall famously swallowed a *Helicobacter* culture to prove that these bacteria cause gastritis.

H. pylori and related species grow primarily on the stomach epithelium, at about pH 6, which is less acidic than the gastric contents (pH 2–4). The bacteria bury themselves in the epithelial layer and neutralize their acidic surroundings by making urease enzyme, which converts urea to ammonia and carbon dioxide. *Helicobacter* species form wide spiral cells (spirilla) with an unusual grouping of flagella at one end. The metabolism of *H. pylori* is microaerophilic, requiring a low level of oxygen. The bacteria can be isolated through biopsy of the gastric mucosa.

Several groups of related Epsilonproteobacteria are sulfur oxidizers and sulfur reducers, found in marine and freshwater habitats. *Thiovulum* species oxidize sulfides aerobically, using oxygen available in marine water. In deep-ocean sediment, near hydrothermal vents, hydrogenotrophs *Nautilia* and *Hydrogenimonas* oxidize H_2 using sulfur or nitrate. These bacteria enrich the marine habitat by cycling carbon, nitrogen, and sulfur. Surprisingly, genes from Epsilonproteobacteria are found in the sequenced genomes of deep-branching taxa, the *Aquifex* species (see Section 18.1). Like other hyperthermophiles, the Epsilons "get around," with their genes migrating into the genomes of distantly related organisms.

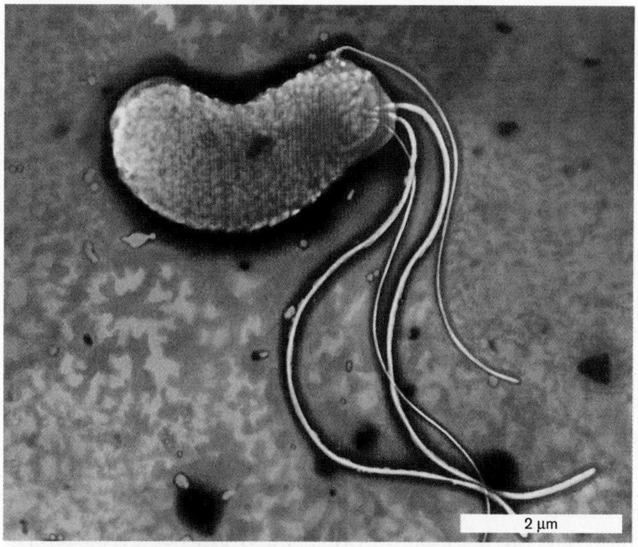

FIGURE 18.41 ▪ *Helicobacter*, a neutralophile growing within the acidic stomach. *H. pylori* is a short spirillum with unusual knobbed flagella projecting from one end.

To Summarize

- **Proteobacteria** stain Gram-negative and have a thin cell wall and an outer membrane containing LPS. They show wide diversity of form and metabolism, including phototrophy, lithotrophy, and heterotrophy on diverse organic substrates.
- **Alphaproteobacteria** include photoheterotrophs (such as *Rhodospirillum*) and heterotrophs, as well as methylotrophs. They include intracellular mutualists such as rhizobia, and pathogens such as the rickettsias. Rickettsias share ancestry with mitochondria.
- **Betaproteobacteria** include photoheterotrophs (*Rhodocyclus*), as well as nitrifiers (*Nitrosomonas*) and iron-sulfur oxidizers (*Thiobacillus*). Pathogenic diplococci include *Neisseria gonorrhoeae*, the cause of gonorrhea.
- **Gammaproteobacteria** include sulfur and iron bacteria (*Acidithiobacillus*, *Chromatium*). The most famous species are the Enterobacteriaceae found in the human colon. Intracellular pathogens include *Salmonella* and *Legionella*. The pseudomonads, aerobic rods, can respire on a wide range of complex organic substrates.
- **Deltaproteobacteria** include sulfur and iron reducers (*Geobacter*), fruiting-body bacteria (*Myxococcus*), and bacterial predators (*Bdellovibrio*).
- **Epsilonproteobacteria** include spirillar pathogens such as *Helicobacter pylori*, the cause of gastritis. They also include marine sulfur oxidizers (*Thiovulum*) and reducers (*Hydrogenimonas*).

18.5 Spirochetes, Acidobacteria, Bacteroidetes, and Chlorobi (Deep-Branching Gram-Negative)

Besides Proteobacteria, members of several other phyla possess an outer membrane and generally stain Gram-negative. They have diverse lifestyles and habitats, ranging from aquatic phototrophs to human pathogens. The spirochete cell has a distinctive form of a tightly coiled spiral, with internalized flagella. Spirochetes include famous human pathogens that cause Lyme disease and syphilis. Other important phyla of deep-branching Gram-negative bacteria include Acidobacteria, Bacteroidetes, Chlorobi, Fusobacteria, and Nitrospirae.

Spirochetes: Sheathed Spiral Cells with Internalized Flagella

Spirochetes (or Spirochaeta) is a unique phylum of heterotrophic bacteria that form tightly coiled spirals. While different species of spirochetes conduct a broad range of heterotrophy, from aerobic to anaerobic, all spirochetes share a distinctive cell structure consisting of a long, tight spiral that is flexible like an old-style telephone cord (**Fig. 18.42A**). For many species, the spiral is so thin that its width cannot be resolved by bright-field microscopy, and the organisms can pass through a filter of pore size 0.2 µm.

Spirochete diversity. Spirochetes grow in a wide range of habitats, from ponds and streams to the digestive tracts of animals. Aquatic systems carry free-living sugar fermenters of the genus *Spirochaeta*. A particularly interesting spirochete community is found in the termite gut, where the organisms form elaborate symbiotic associations with protists and assist in the digestion of cellulose. *Treponema azotonutricium* actually fixes nitrogen for its host termite (**Fig. 18.42A** and **B**). Other members of the genus *Treponema* are normal residents of the human and animal oral, intestinal, and genital regions.

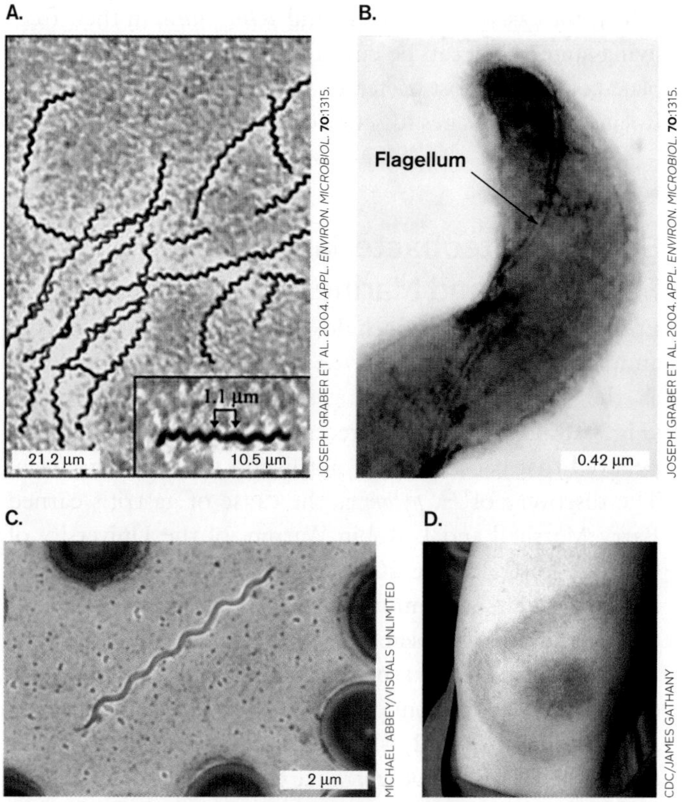

FIGURE 18.42 ■ Spirochetes. **A.** *Treponema azotonutricium* fixes nitrogen in the termite gut (TEM). **B.** Periplasmic flagellum of *T. azotonutricium*. **C.** *Borrelia recurrentis*, shown here in a blood film, is the cause of relapsing fever (stained LM). **D.** Lyme disease rash, caused by tick-borne *Borrelia burgdorferi*.

The best-known spirochetes are those that cause human and animal diseases. The causative agent of the sexually transmitted disease syphilis is the spirochete *Treponema pallidum*. The cell of *T. pallidum* is too narrow (0.2 μm) to visualize by bright-field microscopy; instead, dark-field, fluorescence, or electron microscopy must be used. *T. pallidum* is not culturable in the laboratory, but its genome sequence reveals much about its physiology. The genome of 1.1 million base pairs (presented in **eTopic 13.1**) is highly degenerate, lacking nearly all components of biosynthesis and of the TCA cycle; its only system for ATP production is glycolysis.

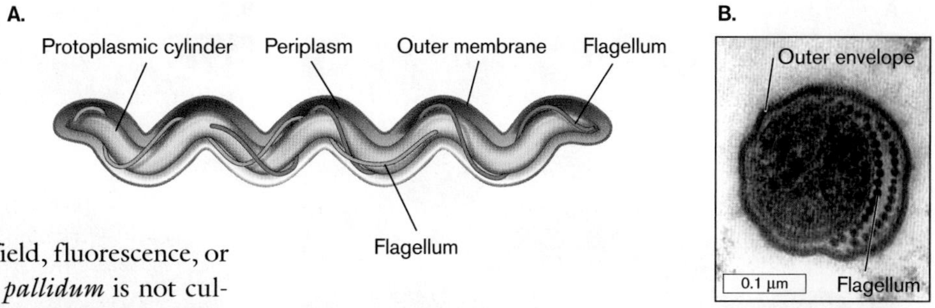

M. A. LISTGARTEN AND S. S. SOCRANSKY. 1964. *J. BACTERIOL.* **88**:1087.

FIGURE 18.43 ■ Spirochete structure. **A.** Spirochete cell structure, showing the arrangement of periplasmic flagella. **B.** Cross section through a human gingival spirochete, showing outer envelope and flagella (axial fibrils; TEM).

The spirochete genus *Borrelia* includes two pathogens that cause serious tick-borne diseases in the United States. *B. recurrentis* causes relapsing fever (**Fig. 18.42C**). The related species *B. burgdorferi* causes Lyme disease, known for the distinctive bull's-eye rash (**Fig. 18.42D**). A unique feature of *Borrelia* species is their multipartite genome. Each species possesses a linear main chromosome of less than a million base pairs, plus a number of linear and circular plasmids. *B. burgdorferi*, for example, has a linear chromosome of 910,725 base pairs, with at least 17 linear and circular plasmids that total an additional 533,000 base pairs. The reason for this unusual fragmentation of *Borrelia* genomes is unknown.

Spirochetes include important pathogens of animals, such as *Leptospira*, the cause of leptospirosis, a form of nephritis (kidney inflammation) with complications in the liver and other organs. *Leptospira* cells are known for the peculiar L shape of the cell, as each end of the spirochete turns out at an angle. Members of the genus *Treponema* also cause digital dermatitis in cattle and sheep. As in other bacterial phyla, however, the pathogenic species are far outnumbered by harmless organisms.

Note: Distinguish the spirochete *Leptospira* from the nitrite-oxidizing proteobacterium *Leptospirillum*.

Cell structure of Spirochetes. The spirochete cell is surrounded by a thick outer sheath of lipopolysaccharides and proteins. The spirochete sheath is similar to a proteobacterial outer membrane, except that the periplasmic space completely separates the sheath from the plasma membrane. At each end of the cell, one or more polar flagella extend and double back around the cell body within the periplasmic space (**Fig. 18.43**). The periplasmic flagella (axial fibrils) rotate on proton-driven motors, as do regular flagella; but because they twine back around the cell body, their rotation forces the entire cell to twist around, corkscrewing through the medium.

This corkscrew motion of the cell body turns out to have a physical advantage in highly viscous environments, such as human mucous secretions or agar culture medium. Few nonspirochetes can swim through agar at standard culture concentrations (1.5% agar); thus, growth within agar provides a way to isolate anaerobic spirochetes from environmental sources. When an environmental sample is inoculated into agar, other bacteria grow and concentrate at the injection point, whereas spirochetes migrate outward in a "veil" through the agar.

Acidobacteria

Acidobacteria are abundant in soils, where they metabolize a wide range of organic substrates using diverse electron acceptors. Many species grow in extreme conditions, such as in the presence of acid and metals—for example, in soils contaminated by uranium mining. Others grow at high temperature, such as isolate K22 (**Fig. 18.44**) obtained from the Taupo Volcanic Zone, New Zealand, by GNS Science, a geoscience prospecting company. The thin section of isolate K22 in **Figure 18.44B** shows its extensive outer membrane.

The candidate species *Chloracidobacterium thermophilum* is a thermophilic aerobic phototroph, isolated from Octopus Spring, at Yellowstone National Park. This acidobacterium is a photoheterotroph that supplements catabolism with light absorption by photosystem I (PS I). It grows chlorosomes similar to those of Chloroflexi (see Section 18.1). These distantly related organisms may have shared photosynthesis genes via horizontal gene transfer.

Bacteroidetes

The phylum Bacteroidetes includes genera such as *Bacteroides* and *Flavobacterium*. *Bacteroides* species, such as *B. fragilis* and *B. thetaiotaomicron*, are the major inhabitants of

A. B.

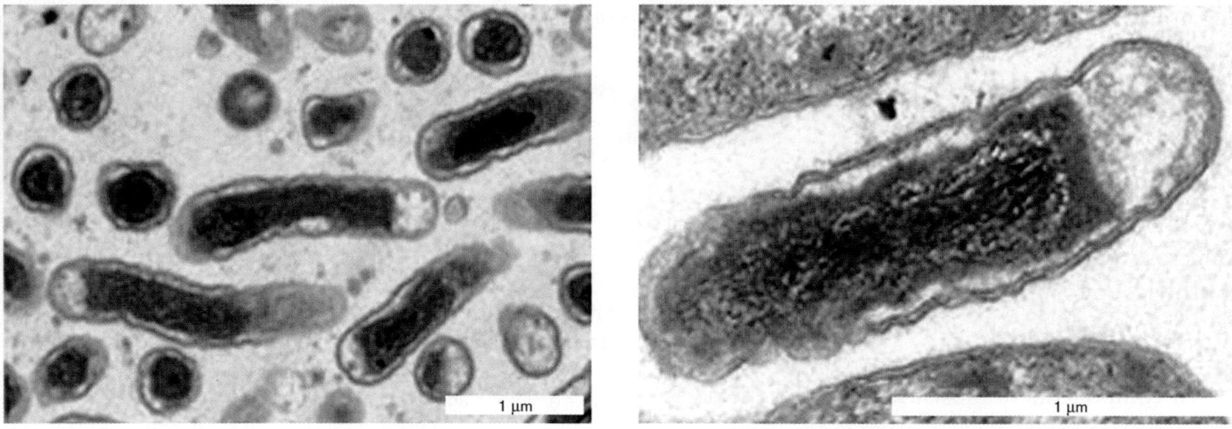

REPUBLISHED WITH PERMISSION OF JOHN WILEY AND SONS INC., P.F. DUNFIELD ET AL., *ENVIRONMENTAL MICROBIOLOGY*, 2008. **10:** 2030.

FIGURE 18.44 ■ Acidobacteria isolate K22. **A.** A thermophilic acidophile, showing a Gram-negative outer membrane. **B.** Thin section of isolate K22 (TEM).

the human colon (**Fig. 18.45**). Their envelope polysaccharides help the bacteria evade the immune system. *Bacteroides* species grow anaerobically under extremely low oxygen concentrations, such as those found in the human intestine (1 ppm), yet they can actually use the oxygen they find; thus they have been called "nanoaerobes."

Their main source of energy is fermentation of a wide range of sugar derivatives from plant material, compounds that are indigestible by humans and potentially toxic (discussed in Chapter 13). *Bacteroides* bacteria convert these substances into simple sugars and fermentation acids, some of which are absorbed by the intestinal epithelium; others feed associated gut bacteria such as *E. coli* (discussed in Chapter 21). Thus, *Bacteroides* species serve important functions for their host: They break down potential toxins in plant foods, and their fermentation products make up as much as 15% of the caloric value we obtain from food.

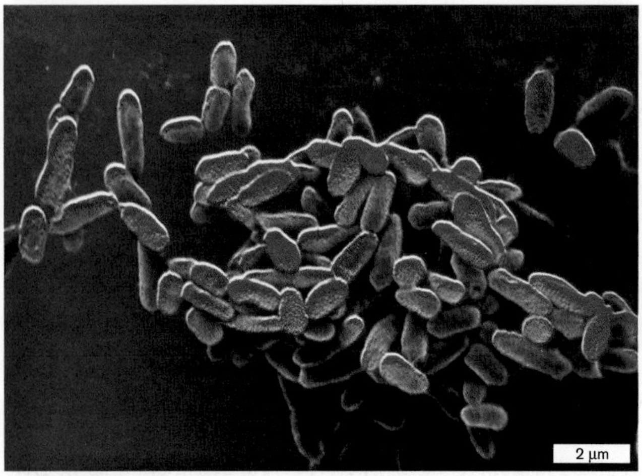

FIGURE 18.45 ■ *Bacteroides fragilis* cells colonize the human colon. *B. fragilis* causes intestinal infections (colorized SEM).

Yet another benefit of *Bacteroides* is their ability to remove side chains from bile acids, enabling the return of bile acids to the hepatic circulation. In effect, our gut bacteria, including *Bacteroides*, constitute a functional organ of the human body.

Bacteroides species cause trouble, however, when they reach parts of the body not designed to host them. During abdominal surgery, bacteria can escape the colon and invade the surrounding tissues. The displaced bacteria can form an abscess, a localized mass of bacteria and pus contained in a cavity of dead tissue. The interior of the abscess is anaerobic and often impenetrable to antibiotics.

Chlorobi

The Chlorobi (mainly *Chlorobium* species) are known informally as green sulfur bacteria. While they are genetically close to *Bacteroides*, their metabolism is surprisingly different. Chlorobi are strict photolithotrophs, using PS I to split electrons from H_2, H_2S, or other reduced sulfur compounds (**Fig. 18.46**; see also **Table 18.2**). During the oxidation of sulfide, *Chlorobium* species deposit elemental sulfur extracellularly, forming attached sulfur globules (**Fig. 18.46A**). In contrast, the sulfur granules of Gammaproteobacteria such as *Chromatium* are deposited internally.

Chlorobium species require extensive membrane systems full of photopigments for light absorption. The chlorophyll reaction centers of *Chlorobium* are contained within chlorosomes associated with the cytoplasmic membrane (**Fig. 18.46B**), similar to chlorosomes of the deep-branching phylum Chloroflexi. The photopigment of *Chlorobium* is predominantly BChl *c*, which absorbs in the blue (460 nm) and near-infrared (750 nm), thus reflecting the middle range, brownish green.

A.

B.

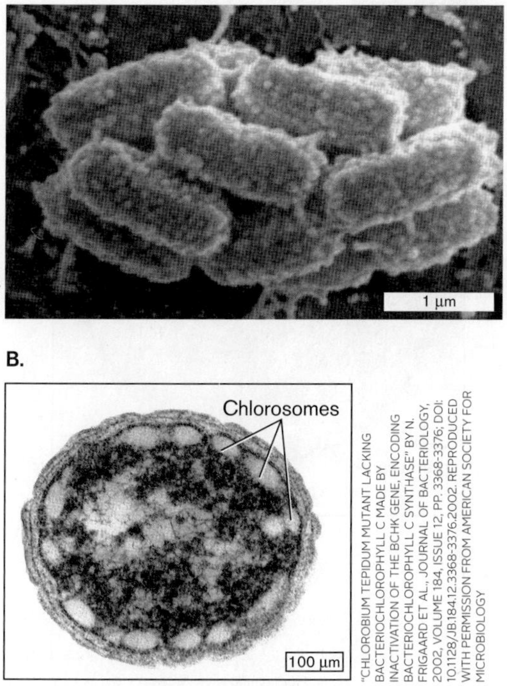

FIGURE 18.46 ▪ *Chlorobium*: Gram-negative green sulfur bacteria. A. *Chlorobium* sp. cells covered with sulfur globules; cells form a cluster surrounding a nonphototrophic symbiotic bacterium. B. *C. tepidum* cell containing chlorosomes.

Note: Distinguish phylum Chlorobi (sulfur photoautotrophs) from phylum Chloroflexi (thermophilic filamentous photoheterotrophs). Both taxa are phototrophs containing green photopigment complexes arranged in chlorosomes, but they are deeply divergent genetically.

Fusobacteria

Fusobacteria is an emerging phylum of surprisingly virulent pathogens. *Fusobacterium nucleatum* was identified from dental plaque; this bacterium is the second most frequent cause of human abscesses (after *Bacteroides*). Within biofilms such as dental plaque, *Fusobacterium* species form carbohydrate bridges with numerous other kinds of bacteria, such as spirochetes, proteobacteria, and firmicutes, as well as eukaryotic pathogens such as fungi. These distantly related biofilm partners have transferred exceptional numbers of genes to the *F. nucleatum* genome.

Nitrospirae

The phylum Nitrospirae consists of Gram-negative spiral bacteria that oxidize nitrite ion to nitrate (NO_2^- to NO_3^-). Their cell structure resembles that of the Proteobacteria, although their phylogenetic branch is deep enough for assignment to a separate phylum. Most species, such as *Nitrospira* species (see **Tables 18.1** and **18.4**), are lithotrophs that fix carbon in the form of carbon dioxide or carbonate using carboxysomes. *Nitrospira* species are generally found in freshwater or salt water. Their removal of excess nitrite makes a key contribution to aquatic ecosystems.

Another important genus is *Leptospirillum*, which includes acidophilic iron oxidizers. *Leptospirillum* species are also strict autotrophs, fixing carbon by using Fe^{2+} as their electron donor and O_2 as the electron acceptor. Their metabolism generates acid, contributing to acid mine drainage in iron mines at Iron Mountain, California, where they grow in massive pink biofilms.

To Summarize

- **The spirochete cell** is a tight coil, enclosed by a sheath and periplasmic space containing periplasmic flagella.
- **Spirochete motility** is driven by a flexing motion caused by rotation of the periplasmic flagella, propagated along the length of the coil.
- **Spirochetes grow in diverse habitats.** Some are free-living fermenters in water or soil. Others are pathogens, such as *Treponema pallidum*, the cause of syphilis. Still others are endosymbionts of an animal digestive tract, such as the termite gut.
- **Acidobacteria** are Gram-negative soil bacteria, including many acidophiles.
- **Bacteroidetes** are anaerobes that ferment complex plant materials in the human colon. They may enter body tissues through wounds and cause abscesses.
- **Chlorobi** are green sulfur phototrophs, obligate anaerobes incapable of heterotrophy.
- **Fusobacteria** are pathogens that cause septicemia and skin ulcers.
- **Nitrospirae** are Gram-negative spiral bacteria that oxidize nitrite to nitrate (*Nitrospira*).

18.6 Planctomycetes, Verrucomicrobia, and Chlamydiae (PVC Superphylum)

Several related phyla of bacteria—the Planctomycetes, Verrucomicrobia, and Chlamydiae—possess compartmentalized cells with relatively diminished cell walls. Genomic evidence suggests they evolved from a common compartmentalized ancestor, yet their diverse cell form and metabolism show remarkable environmental adaptations.

Planctomycetes: A Nucleus-like Compartment

The Planctomycetes evolved largely as free-living organisms. Planctomycetes are oligotrophs, heterotrophs requiring nutrients at extremely low concentrations. They grow in freshwater, marine, and saline environments. Their mechanism of osmoregulation remains poorly understood.

Some planctomycete cells possess multiple internal membrane compartments of unknown function (**Fig. 18.47A**). In these species, an internal membrane just inside the cell membrane divides the cytoplasm into concentric portions. A double membrane surrounds the entire nucleoid, analogous to the double membrane surrounding the eukaryotic nucleus—a remarkable example of independent analogous evolution between bacteria and the eukaryotes. The actual function of the nucleoid-surrounding membrane varies with different species; for example, in anammox planctomycetes, a single membrane enclosing the nucleoid contains the complexes that conduct the anammox reaction (anaerobic ammonium oxidation; discussed in Chapter 14).

Like eukaryotic protists (discussed in Chapter 20), the planctomycetes have flexible cell bodies that can assume diverse forms. Remarkably, in 2019 planctomycetes were found to engulf smaller microbes, analogous to the eukaryotic process of phagocytosis. Some *Planctomyces* species have rotary flagella (**Fig. 18.47B**), whereas *P. bekefii* cells have stalks that attach to each other to generate a starlike aggregate. Most planctomycetes reproduce by budding, a strategy typical of eukaryotic yeasts.

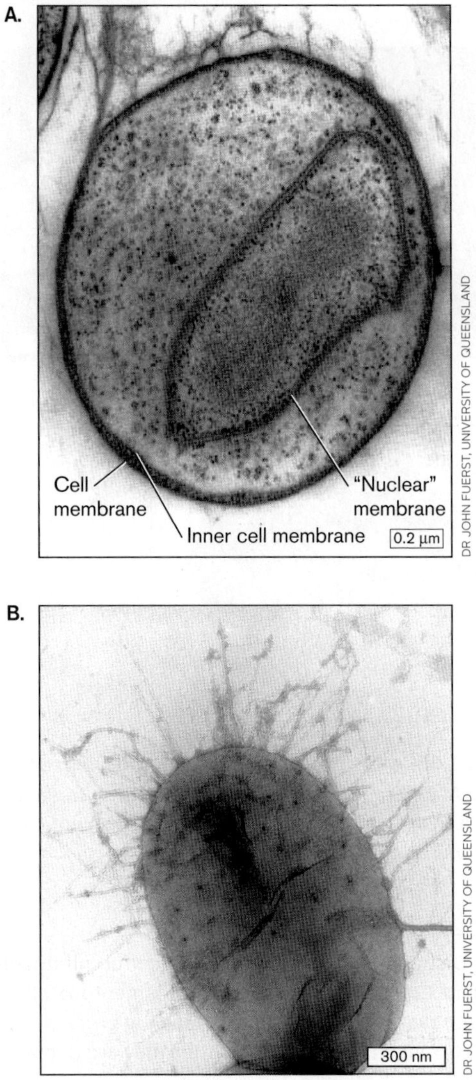

FIGURE 18.47 ▪ *Planctomyces*: bacteria with a "nuclear membrane." **A.** Section through the planctomycete *Gemmata obscuriglobus*, showing membrane compartmentalization (TEM). The DNA is contained within a double membrane analogous to a eukaryotic nuclear membrane. **B.** A swarmer cell of *Planctomyces* sp. with multiple flagella (TEM).

Verrucomicrobia: Wrinkled Microbes

Verrucomicrobia, or "wrinkled microbes," are irregularly shaped bacteria found in a wide variety of aquatic and terrestrial environments, as well as in the mammalian gastrointestinal tract (**Fig. 18.48**). Most Verrucomicrobia are oligotrophs, growing heterotrophically in low-salt habitats. Some are ectosymbionts of protists, attached to the cell surface of the eukaryote, where they eject harpoon-like objects. They are rarely cultured, and until recently they failed to show up in PCR amplification of natural isolates, because their rDNA sequences are poorly amplified by the standard primer sequences. When appropriate primer pairs are used, Verrucomicrobia comprise 5% of all surveyed microbial sequences in some natural environments.

Verrucomicrobia have peptidoglycan cell walls, but their shape is dominated by wart-like projections. The wart-like cytoskeleton appears to contain tubulin, a cytoskeletal protein previously believed to exist only in eukaryotes. In 2002, genes encoding tubulin were found in the partly sequenced genome of *Prosthecobacter dejongeii*, a free-living member of Verrucomicrobia. The genes appear so similar to those of eukaryotes that they must have undergone horizontal transfer from a eukaryotic genome. This horizontal transfer of a eukaryotic trait may be contrasted with the independent development of a nucleus-like structure in planctomycetes.

Chlamydiae: Intracellular Parasites

In the phylum Chlamydiae, the genera *Chlamydia* and *Chlamydophila* evolved complex developmental life cycles

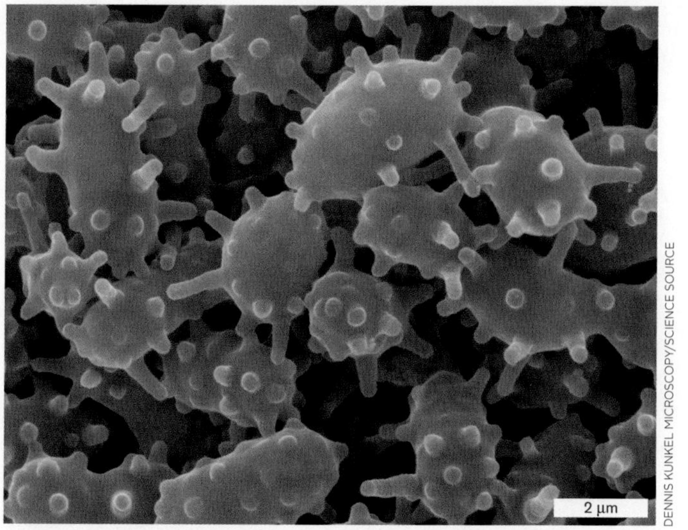

FIGURE 18.48 ■ Verrucomicrobia: the "warty" bacteria. *Verrucomicrobium spinosum* with wart-like cell protrusions (colorized SEM).

of parasitizing host cells. *Chlamydia trachomatis* causes the most prevalent sexually transmitted infection among young people in the United States. The same pathogen also causes trachoma, an eye disease dating back to records in ancient Egypt. The related species *Chlamydophila pneumoniae* causes pneumonia and has been implicated in cardiovascular disease.

Chlamydiae, or chlamydia, alternate between two developmental stages with different functions: elementary bodies and reticulate bodies (**Fig. 18.49A**). The form of chlamydia transmitted outside host cells is called an **elementary body**. Like endospores, elementary bodies are metabolically inert, with a compacted chromosome. While lacking a cell wall, they possess an outer membrane whose proteins are cross-linked by disulfide bonds, making a tough coat that provides osmotic stability. The elementary body adheres to a host cell surface and is endocytosed (**Fig. 18.49B**).

To reproduce, the elementary body must transform itself into a **reticulate body**, named for the netlike appearance of its uncondensed DNA. The reticulate body has an active metabolism and divides rapidly, but outside the cell it is incapable of infection and vulnerable to osmotic shock. To complete the infection cycle, therefore, reticulate bodies must develop into new elementary bodies before exiting the host. When the host cell lyses, the elementary bodies are released to infect new cells. Chlamydiae infect a wide range of host cell types, from respiratory epithelium to macrophages.

To Summarize

- **Planctomycetes** include species that have evolved a membrane enclosing the nucleoid, analogous to the eukaryotic nuclear membrane.
- **Verrucomicrobia** have cell projections containing tubulin. Their tubulin genes are thought to have arisen by horizontal transfer from a eukaryote.

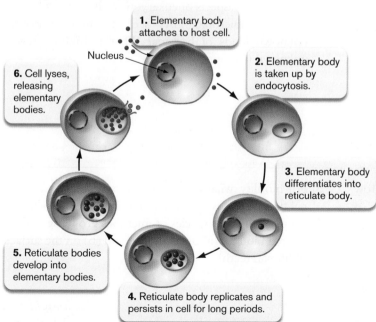

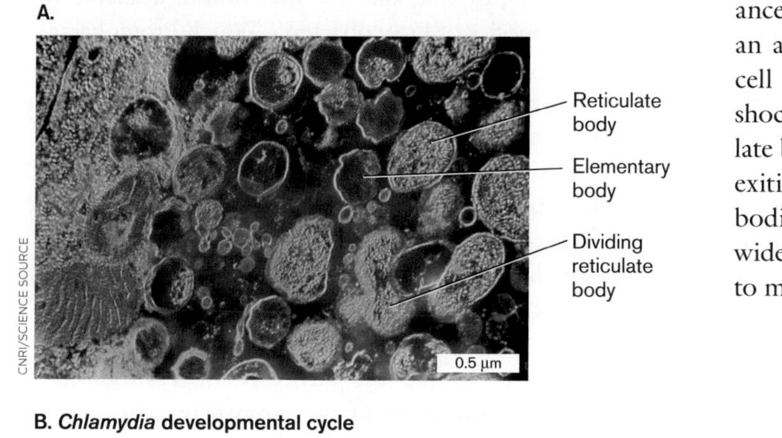

FIGURE 18.49 ■ *Chlamydia* life cycle. **A.** *C. trachomatis*, multiplying within a human cell (colorized TEM). Infected cell, equivalent to step 5 in part (B), contains reticulate bodies (yellow) growing and dividing, as well as newly formed elementary bodies (red). **B.** *Chlamydia* spp. persist outside the host as a spore-like "elementary body." Upon endocytosis, the elementary body avoids lysosomal fusion and develops into a "reticulate body" in which the DNA, now uncondensed, has a reticular (netlike) appearance. The reticulate body replicates within the host cytoplasm and then develops into new elementary bodies that are released when the cell lyses.

- **Chlamydiae** are obligate intracellular parasites that undergo a complex developmental progression, culminating in a spore-like form called an elementary body that can be transmitted outside the host cell.

Concluding Thoughts

Several themes emerge in the diversity of the domain Bacteria. Some phyla, such as Alphaproteobacteria and Actinobacteria, have evolved highly diverse metabolism and cell form. Others show remarkable uniformity in cell structure (Spirochetes) or in metabolism (Cyanobacteria). At the same time, our attempts to generalize about any given clade inevitably run up against the unexpected appearance of exceptions, such as the heliobacteria, photosynthetic endospore formers that unaccountably branch from clostridia. Given the range of bacterial phenotypes, it is hard to imagine that yet more diverse forms of microbial life exist, but they do, as we will find among the Archaea (Chapter 19) and the microbial Eukarya (Chapter 20).

CHAPTER REVIEW

Review Questions

1. Which deep-branching phyla include hyperthermophiles? Why is the actual branch position of these groups controversial?
2. Compare and contrast Chloroflexi and Cyanobacteria with respect to habitat, cell structure, and means of photosynthesis.
3. Compare and contrast the colonial and filamentous cyanobacteria with respect to life cycle and means of nitrogen fixation.
4. Name and describe three genera of Firmicutes that form endospores. Discuss the difference between single and multiple spore formers. Explain the existence of non-spore-forming species of Firmicutes.
5. Compare and contrast firmicute endospores and actinomycete arthrospores with respect to their means of production, their resistance properties, and their dispersal mechanisms.
6. Describe the diverse kinds of metabolism available to different species of Proteobacteria. Which Proteobacteria conduct lithotrophy? Anaerobic respiration? Which species are pathogens?
7. What do species of Bacteroidetes and Chlorobi have in common, and how do they differ?
8. Explain the structure and mechanism of motility typical in species of Spirochetes.
9. Discuss the properties of as many intracellular bacteria as you recall from various phyla. What do they have in common, and how do they differ?
10. Describe the unique cell structure of *Planctomyces.* Explain how the traits of planctomycete cells appear analogous to aspects of eukaryotic cells.
11. How do microbiologists seek to find previously unknown kinds of bacteria? What techniques reveal unknown microorganisms?

Thought Questions

1. Why are the Proteobacteria metabolically diverse? How is it possible that species of Alphaproteobacteria use so many kinds of molecules to yield energy?
2. How do you think *Mycobacterium tuberculosis* manages to grow, despite its thick envelope screening out most nutrients?
3. For motility, what are the relative advantages of external flagella versus the flexible spiral cells of spirochetes?
4. Why do different species of microbes grow together in layered biofilms (a) on a sand flat and (b) on the surface of human teeth?

Key Terms

acid-fast stain (725)
Acidobacteria (705)
Actinobacteria (705)
actinomycete (705)
aerial mycelium (722)
Alphaproteobacteria (705)
Aquificae (706)
bacteroid (730)
Bacteroidetes (705)
Betaproteobacteria (705)
Candidate Phyla Radiation (CPR) (707)
carboxysome (710)
Chlamydia (743)
Chlamydiae (706)
Chlorobi (705)
Chloroflexi (706)
chlorosome (706)
Cyanobacteria (703)
deep-branching taxa (702)
Deinococcus-Thermus (706)
Deltaproteobacteria (705)
diderm (714)
diplococcus (731)
elementary body (743)
emerging (706)
endospore (705)
Epsilonproteobacteria (705)
Firmicutes (705)
forespore (715)
fruiting body (736)
Fusobacteria (705)
Gammaproteobacteria (705)
gas vesicle (710)
GC content (705)
heterocyst (710)
hormogonium (711)
leghemoglobin (730)
methanotroph (729)
methylotrophy (729)
Mollicutes (705)
monoderm (714)
mother cell (715)
mycolic acid (724)
myxospore (736)
Nitrospirae (705)
phylum (702)
Planctomycetes (705)
Proteobacteria (705)
proteorhodopsin (727)
reticulate body (743)
sarcina (726)
Spirochetes (705)
swarming (734)
Tenericutes (705)
Thermotogae (706)
vegetative cell (715)
vegetative mycelium (722)
Verrucomicrobia (705)

Recommended Reading

Angert, Esther. 2006. Beyond binary fission: Some bacteria reproduce by alternative means. *Microbe* **1**:127–131.

Beatty, J. Thomas, Jörg Overmann, Michael T. Lince, Ann K. Manske, Andrew S. Lang, et al. 2005. An obligately photosynthetic bacterial anaerobe from a deep-sea hydrothermal vent. *Proceedings of the National Academy of Sciences USA* **102**:9306–9310.

Brown, Christopher T., Laura A. Hug, Brian C. Thomas, Itai Sharon, Cindy J. Castelle, et al. 2015. Unusual biology across a group comprising more than 15% of domain Bacteria. *Nature* **523**:208–211.

Campbell, Barbara J., Annette Summers Engel, Megan L. Porter, and Ken Takai. 2006. The versatile epsilonproteobacteria: Key players in sulphidic habitats. *Nature Reviews. Microbiology* **4**:458–468.

Claessen, Dennis, Daniel E. Rozen, Oscar P. Kuipers, Lotte Søgaard-Andersen, and Gilles P. van Wezel. 2014. Bacterial solutions to multicellularity: A tale of biofilms, filaments and fruiting bodies. *Nature Reviews. Microbiology* **12**:115–124.

Geissinger, Oliver, Daniel P. R. Herlemann, Erhard Mörschel, Uwe G. Maier, and Andreas Brune. 2009. The ultramicrobacterium "*Elusimicrobium minutum*" gen. nov., sp. nov., the first cultivated representative of the termite group 1 phylum. *Applied and Environmental Microbiology* **75**:2831–2840.

Human Microbiome Project Consortium. 2012. Structure, function and diversity of the healthy human microbiome. *Nature* **486**:207–214.

Kindaichi, Tomonori, Tsukasa Ito, and Satoshi Okabe. 2004. Ecophysiological interaction between nitrifying bacteria and heterotrophic bacteria in autotrophic nitrifying biofilms as determined by microautoradiography-fluorescence in situ hybridization. *Applied and Environmental Microbiology* **70**:1641–1650.

Luef, Birgit, Kyle R. Frischkorn, Kelly C. Wrighton, Hoi-Ying N. Holman, Giovanni Birarda, et al. 2015. Diverse uncultivated ultra-small bacterial cells in groundwater. *Nature Communications* **6**:6372.

Masaki, Toshihiro, Jinrong Qu, Justyna Cholewa-Waclaw, Karen Burr, Ryan Raaum, et al. 2013. Reprogramming adult Schwann cells to stem cell-like cells by leprosy bacilli promotes dissemination of infection. *Cell* **152**:51–67.

Schnupf, Pamela, Valérie Gaboriau-Routhiau, Marine Gros, Robin Friedman, Maryse Moya-Nilges, et al. 2015. Growth and host interaction of mouse segmented filamentous bacteria in vitro. *Nature* **520**:99–103.

Shams ul Hassan, Syed, Komal Anjuma, Syed Qamar Abbas, Najeeb Akhter, Bibi Ibtesam Shagufta, et al. 2017. Emerging biopharmaceuticals from marine actinobacteria. *Environmental Toxicology and Pharmacology* **49**:34–47.

Shiratori, Takashi, Shigekatsu, Suzuki, Yukako Kakizawa, and Ken-ichiro Ishida. 2019. Phagocytosis-like cell engulfment by a planctomycete bacterium. *Nature communications* **10**:5529.

Soo, Rochelle M., James Hemp, Donovan H. Parks, Woodward W. Fischer, and Philip Hugenholtz. 2017. On the origins of oxygenic photosynthesis and aerobic respiration in Cyanobacteria. *Science* **355**:1436–1440.

Tolli, John D., S. M. Sievert, and C. D. Taylor. 2006. Unexpected diversity of bacteria capable of carbon monoxide oxidation in a coastal marine environment, and contribution of the *Roseobacter*-associated clade to total CO oxidation. *Applied and Environmental Microbiology* **72**:1966–1973.

Valliappan, Karuppiah, Wei Sun, and Zhiyong Li. 2014. Marine actinobacteria associated with marine organisms and their potentials in producing pharmaceutical natural products. *Applied Microbiology and Biotechnology* **98**:7365–7377.

Wiegand, Sandra, Mareike Jogler, and Christian Jogler. 2018. On the maverick Planctomycetes. *FEMS Microbiological Reviews* **42**:739–760. http://doi.org/10.1093/femsre/fuy029.

Wu, Martin, Qinghu Ren, A. Scott Durkin, Sean C. Daugherty, Lauren M. Brinkac, et al. 2005. Life in hot carbon monoxide: The complete genome sequence of *Carboxydothermus hydrogenoformans* Z-2901. *PLoS Genetics* **1**:e65.

CHAPTER 19

Archaeal Diversity

19.1 Archaeal Diversity at a Glance

19.2 TACK Hyperthermophiles Eat Sulfur

19.3 Thaumarchaeota: Ammonia Oxidizers and Animal Symbionts

19.4 Euryarchaeota: Methanogens from Gut to Globe

19.5 Haloarchaea and Other Euryarchaeotes: Underground and Under Ocean

19.6 DPANN Symbionts, Altiarchaeales, and Asgard: Branch to Eukaryotes?

Archaea are found in all soil and water habitats, in symbiosis with animals and plants, and in extreme environments that exclude bacteria and eukaryotes. Archaea include hyperthermophiles inhabiting Earth's hottest habitats, as well as Arctic and Antarctic psychrophiles. Methanogens inhabit the digestive tracts of humans and other animals, as well as anaerobic soil and water, with consequences for global climate. Many archaea collaborate with bacteria in multispecies biofilms. No archaeon is yet known to cause disease, but archaea do interact with our immune system.

CURRENT RESEARCH highlight

A spaghetti-tubed hyperthermophile that looks like a eukaryote. In the ocean off the coast of Iceland, 600 meters below sea level, rises the Kolbeinsey Ridge volcano. From a thermal vent, Karl Stetter isolated the archaeon *Ignicoccus*, a hyperthermophile that grows best at 90°C. Microscopists Thomas Heimerl and Reinhard Rachel investigated the cell form of *Ignicoccus hospitalis*, which lacks any cell wall yet possesses an extensive endomembrane system. They used low-temperature processing and electron tomography on serial sections to map the extensive protrusions and tubules of the *Ignicoccus* endomembrane, which surprisingly resembles that of eukaryotes.

Source: Thomas Heimerl, et al. A complex endomembrane system in the Archaeon *Ignicoccus hospitalis* tapped by *Nanoarchaeum equitans*. 2017. *Front. Microbiol.* **8**:1072.

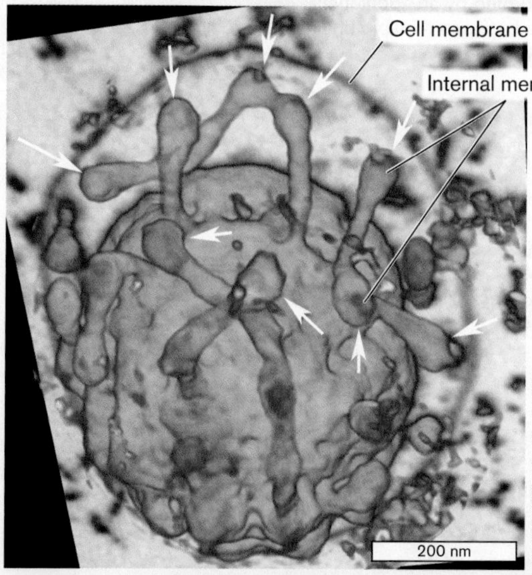

AN INTERVIEW WITH

REINHARD RACHEL, MICROBIOLOGIST, UNIVERSITY OF REGENSBURG, GERMANY

What is the most amazing thing about *Ignicoccus* endomembranes?

There are molecules found in *Ignicoccus* that are supposed to be involved in shaping membranes, and they are distantly related to molecules which are involved in shaping membranes in eukaryotic cells. Being more specific is tricky, as we cannot knock out these genes in *Ignicoccus*, nor are we able to manipulate any gene in this archaeon, and expressing this archaeal gene in other bacterial cells or eukaryotes is tricky, and so far not impossible. So the real function in *Ignicoccus* is a matter of speculation. Do *Ignicoccus* membranes look like those of Eukaryotes? We assume that today's archaea exhibit features, genes, molecules, and proteins that were basically invented much earlier in evolution, e.g. in archaeal cells, and then transferred to and further adapted in eukaryal cells. But, we do not *know* this for sure, as we cannot go back in time.

In 1977, Carl Woese revealed the existence of a third kind of life: the Archaea, a domain of life forms very different from the plants, animals, and bacteria that had long been known. Archaea are known for the extremophiles—hyperthermophiles such as *Ignicoccus* (see the Current Research Highlight) and hyperacidophiles, such as *Ferroplasma*, that grow in sulfuric acid below pH 1. Extremophiles are of interest to biotechnology because their enzymes may catalyze reactions under industrially useful conditions, such as high salt or acidity.

But we also find archaea throughout mesophilic soil and water, and within the digestive tracts of humans and other animals. Even human skin microbiota include more than 4% archaea. How do archaea in "moderate" habitats interact with bacteria and eukaryotes? Many archaea share intimate associations with bacteria. For example, soil methanogens conduct syntrophy with Deltaproteobacteria, obtaining H_2 from the bacterial fermentation (discussed in Chapter 13). Other archaea cooperatively colonize the surface of plant roots. And our intestines support methanogens making a living off the H_2 and CO_2 from bacterial mixed-acid fermentation (Chapter 13). While no archaea are yet known to cause human disease, gut methanogens are associated with flatulence and possibly some metabolic disorders (discussed in Section 19.4).

In Chapter 19 we explore the diversity of archaea. We emphasize the unique structures and metabolic pathways of archaeal cells. While surveying their major taxonomic categories, we introduce research techniques used to study organisms in extreme environments and those with unique forms of metabolism, such as methanogenesis.

19.1 Archaeal Diversity at a Glance

Archaea share many metabolic traits with bacteria, such as the enzymes of redox metabolism. But archaea share some core traits of DNA-RNA machinery and transcription factors with eukaryotes. Key traits such as ether-linked membrane lipids are found only in archaea, or in some bacteria via horizontal transfer. Such distinctive features of archaea are called "archaeal signatures" (**Table 19.1**).

Ether-Linked Isoprenoid Membranes

The most distinctive structure of archaea is their ether-linked membrane (**Fig. 19.1**). The membrane lipids of archaea differ profoundly from those of bacteria and eukaryotes, with the exception of a few thermophilic bacteria that obtained ether lipids horizontally from archaea in their shared high-temperature environment. Most features of archaeal lipids increase lipid stability in extreme environments such as high temperature or extreme acidity. Nevertheless, ether lipids are also widespread in mesophilic archaea that grow at moderate temperatures. Some researchers argue that the original archaea were hyperthermophiles but later evolved as mesophiles by acquiring genes from bacteria.

Distinctive features of archaeal lipids include:

- **L-Glycerol.** Archaeal membrane lipids incorporate L-glycerol (**Fig. 19.1A**), rather than the mirror-symmetrical form D-glycerol, which is used by bacteria and eukaryotes. The two chiral forms show similar thermal stability, but their biochemistry requires different enzymes, and thus they represent a deep divergence in

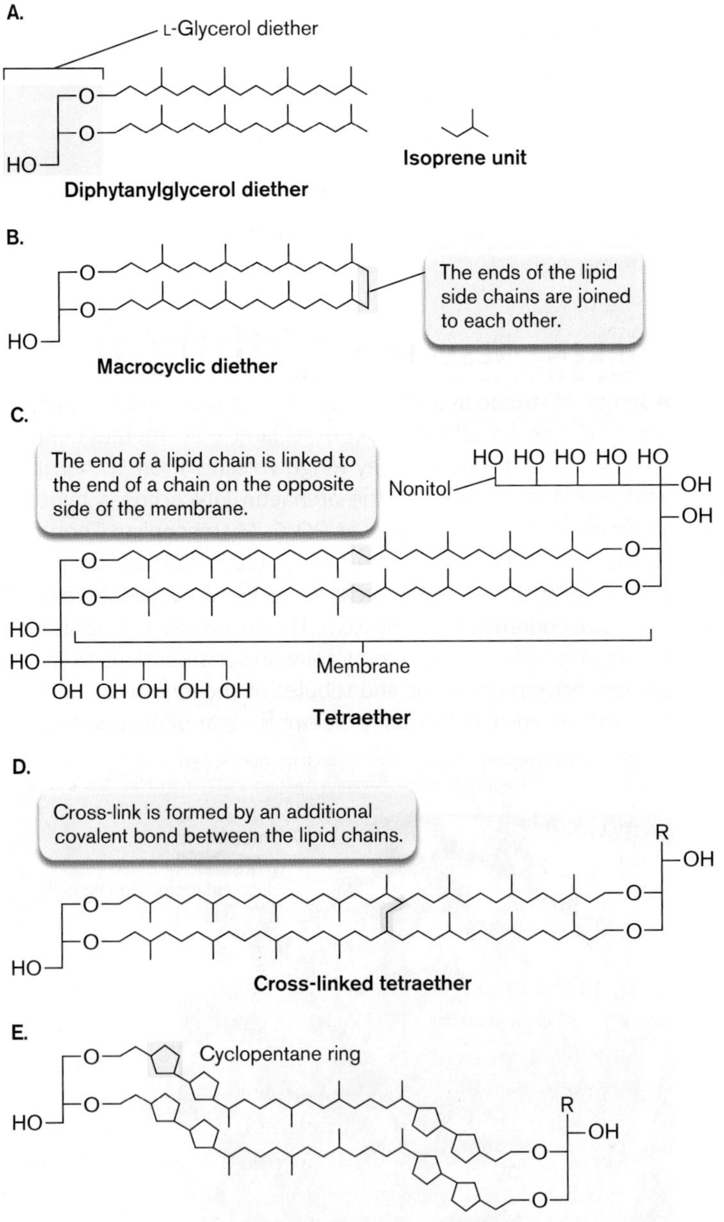

FIGURE 19.1 ■ Branched-chain ether lipids are characteristic of archaeal membranes. *Source:* Modified from R. M. Daniel and D. A. Cowan. 2000. *Cell. Mol. Life Sci.* **57**:250–264.

TABLE 19.1 Archaeal traits distinct from bacteria and eukaryotes

Trait	Archaea showing the trait	Alternative traits of bacteria and/or eukaryotes
Cell envelope		
Membrane lipids: isoprenoid L-glycerol ethers or diethers	All archaea	Membrane lipids: D-glycerol hydrocarbon diesters
Membrane lipid chains stiffened by covalent cross-links or by pentacyclic rings	Most archaea	Chains stiffened by saturation
S-layer of glycoprotein	Crenarchaeota and Thaumarchaeota	Peptidoglycan (bacteria); cellulose and other polysaccharides (eukaryotes)
S-layer of protein, methanochondroitin, or sulfated polysaccharide	Methanogens and Haloarchaea	Peptidoglycan (bacteria); cellulose and other polysaccharides (eukaryotes)
Pseudopeptidoglycan sacculus contains talosaminuronic acid; peptide bridges contain only L-amino acids	Methanobacteriales, Methanopyrales	Peptidoglycan (bacteria); cellulose and other polysaccharides (eukaryotes)
Metabolism		
Nonphosphorylated intermediates of sugar catabolism and synthesis (EMP glycolysis in some cases)	All archaea	EMP pathway of glycolysis, with phosphorylation mediated by NAD or NADP (or Entner-Doudoroff pathway)
Methanogenesis from H_2 and CO_2; or from CO, methanol, methyl sulfides, formate, or acetate	Methanogens	Anaerobic metabolism such as fermentation; no methane production
Archaeal coenzymes: cofactors F_{420} and F_{430}, and coenzyme M	Most methanogens	Flavin mononucleotide, coenzyme A, others
Retinal-associated light-driven membrane pumps for H^+ or Na^+	Haloarchaea	Chlorophyll-based photosynthesis (bacteria and eukaryotic chloroplasts)
Nucleic acid structure and function		
Positive superturns generated by reverse gyrase, which protect DNA from extreme acid	Hyperthermophilic archaea	Negative superturns generated by gyrase
Unique base structures in tRNA, such as the guanine analog archaeosine	Most archaea	tRNA bases, such as queuosine, found only in bacteria and eukaryotes

Sources: O. Kandler and H. König. 1998. *Cell. Mol. Life Sci.* **54**:305–308; C. Bullock. 2000. *Biochem. Mol. Biol. Ed.* **28**:186–191.

ancestry. In some archaea the glycerol is extended by six carbons, forming nonitol (nine OH groups).

- **Ether linkage.** The glycerol units are linked to side chains by ether links (R–O–R) instead of the ester links (R–COO–R) found in bacteria and eukaryotes (**Fig. 19.1A**). Ether links are much more stable than ester links; in other words, breaking them requires more energy.
- **Isoprenoid chains.** The side chains of archaeal lipids are branched at every fourth carbon. The methyl branches arise by condensation (C–C bond formation) of units of isoprene (see **Fig. 19.1A**). Condensed isoprene chains are called **isoprenoid** or diphytanyl chains; thus, the overall lipid is diphytanylglycerol diether. Isoprenoid branched chains increase membrane stability by hooking each other in place.
- **Cross-linked lipids.** In some hyperthermophiles, the ends of side chains are linked covalently, either to each other (**Fig. 19.1B**) or to a lipid on the opposite side of the membrane (**Fig. 19.1C**). Two pairs of lipid chains cross-linked across the membrane form a **tetraether**, so called because the complex contains four ether links in all. In some cases, an additional covalent bond links the two linked pairs of side chains across the middle (**Fig. 19.1D**).
- **Cyclopentane rings.** In some archaea, the lipid's methyl branches cyclize, forming cyclopentane rings (**Fig. 19.1E**). Cyclopentane rings strengthen membranes at high temperature.

Most archaea possess a single membrane, without any outer membrane such as that of Gram-negative bacteria (shown in Chapter 3). Many, such as the thermophilic crenarchaeotes, possess no cell wall at all, but only an S-layer of proteins plugged into the tetraether membrane (discussed in Section 19.2). Some archaea, such as methanogens and haloarchaea, do have a cell wall, but the structure differs fundamentally from the bacterial peptidoglycan (Sections 19.4 and 19.5).

Similarly, many archaea possess filamentous protein structures for attachment and motility that are analogous to bacterial pili and flagella. Despite functional similarity, the archaeal protein structures are very different from those of bacteria. Archaeal flagella are called **archaella** (singular, **archaellum**). Archaella have rotary motors, but they evolved independently from bacterial flagella, showing greater similarity to type IV pili.

Archaeal Gene Structure and Regulation

How do archaeal genomes compare with those of bacteria and eukaryotes? The genomes of archaea generally resemble those of bacteria in size and gene density, and genes of related function are often arranged in operons, like those of bacteria. Nevertheless, certain features of genome structure more closely resemble those of eukaryotes than those of bacteria.

The genes encoding most archaeal proteins contain uninterrupted coding sequences, as in bacteria, but certain tRNA gene sequences are interrupted by introns (nontranslated sequences), similar to the tRNA introns found in eukaryotes. Furthermore, the archaeal apparatus for DNA and RNA polymerases, transcription factors, and protein synthesis show remarkable similarity to those of eukaryotes. **Figure 19.2A** compares the components of an archaeal RNA polymerase (from *Sulfolobus*) with those of eukaryotic RNA polymerase II, the enzyme that catalyzes the synthesis of messenger RNA. The archaeal polymerase

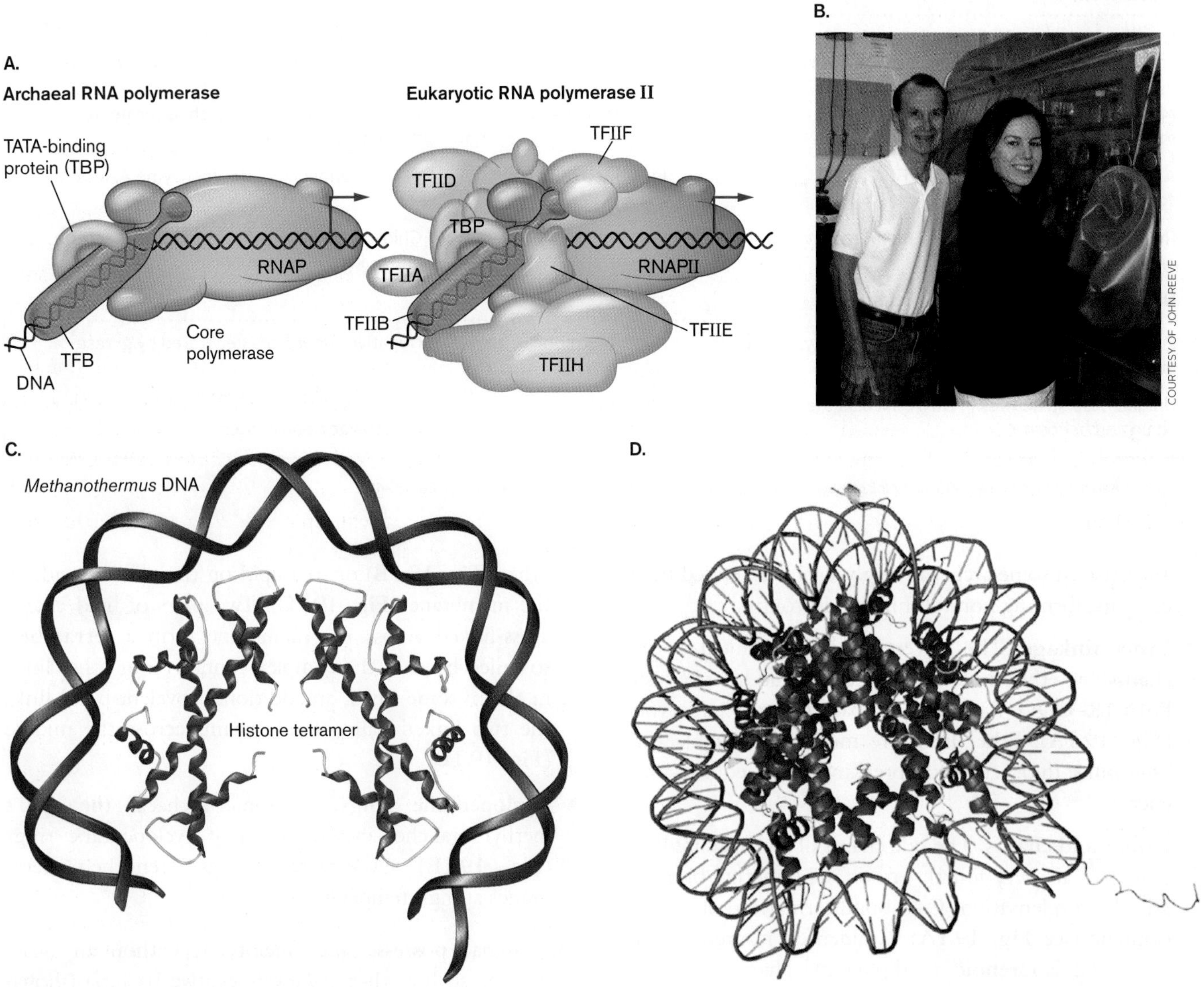

FIGURE 19.2 ■ Central genetic molecules of Archaea resemble those of Eukarya. A. The core RNA polymerase (RNAP) subunits of the archaeon *Sulfolobus* show homology to those of the eukaryotic RNA polymerase II (RNAP II). The TATA-binding protein (TBP, green) and transcription factor B (TFB, pink) also show homology to eukaryotic counterparts. **B.** John Reeve, pictured here with a graduate student, studies the molecular biology of methanogens using an anaerobic glove box. **C.** *Methanothermus fervidus* DNA binds to histone tetramers that show homology to the histones of eukaryotes. (PDB code: 1A7W) **D.** Eukaryotic histone octamer. (PDB code: 1AOI) *Sources:* Part C modified from Kathryn A. Bailey et al. 2002. *J. Biol. Chem.* **277**:9293; part D modified from K. Luger et al. 1997. *Nature* **389**:251.

possesses two transcription factors (regulatory protein components) that are found in eukaryotes: TATA-binding protein (TBP, a subunit of transcription factor II D, TFIID) and transcription factor II B (TFIIB, which in archaea is designated TFB). By contrast, bacteria have no homologs of these factors. A consequence of the eukaryotic-like transcription and translation in archaea is that archaea are resistant to antibacterial antibiotics that target transcription and translation.

Another eukaryotic structure for which archaeal homologs were discovered is that of **histones**, the fundamental packaging proteins of DNA. The histone complex found in eukaryotic chromosomes contains a histone $(H3 + H4)_2$ tetramer flanked by two histone (H2A + H2B) dimers. The archaeal homologs form an $(H3 + H4)_2$ tetramer, with no (H2A + H2B) homologs. John Reeve and colleagues at the Ohio State University (**Fig. 19.2B**) showed that isolated DNA of specific sequences could be bound and curved around a histone tetramer from the methanogen *Methanothermus fervidus* (**Fig. 19.2C**). The DNA has AT-rich sequences that specifically fit the histone complex. The key sequences of the histones (the ones that bind to DNA) show homology to eukaryotic histones. Histones have since been found in many species of archaea.

Other molecular genetic traits are unique to the domain Archaea. A distinctive feature of archaeal chromosome function is the "reverse gyrase" enzyme found in hyperthermophiles such as *Sulfolobus* and *Methanopyrus*. For comparison, all bacteria and eukaryotes, as well as mesophilic archaea, have gyrase to maintain their DNA in an "underwound" negatively supercoiled state (presented in Chapter 7). But hyperthermophilic archaea with reverse gyrase maintain positive supercoiling to stabilize their DNA. The positive superturns "overwind" the DNA, preventing the helix from melting into separate strands at high temperature. Reverse gyrase evolved only in Archaea, though it was transferred horizontally to some bacterial hyperthermophiles, such as *Thermotoga* (discussed in Chapter 18).

A unique feature of archaeal RNA is the distinctive modified bases in tRNA molecules. In particular, the guanosine analog archaeosine (7-formamidino-7-deazaguanosine) is used by nearly all archaea, but not by any bacteria or eukaryotes. Other unusual tRNA bases, such as queuosine, are found only in bacteria and eukaryotes, not in archaea.

Phylogeny of Archaea

What are the major kinds of Archaea? The answer to this question keeps changing, as DNA sequencing of uncultured organisms reveals a growing number of previously unknown phyla (**Fig. 19.3**). Two major advances in genomic analysis have enabled this remarkable pace of discovery, as described by Thijs Ettema (Uppsala University, Sweden) and colleagues in their 2017 review article of archaeal phylogeny.

First, the assembly of metagenomes (the genome sequences of a microbial community) from mixed environmental samples has enabled construction of "genomic bins" that approximate the genomes of uncultured organisms (Chapter 21). These genomic bins enable comparison of entire small-subunit (SSU) rRNA genes and ribosomal operons encoding ribosomal proteins and transfer RNAs, as pioneered by Jillian Banfield's lab at UC Berkeley. Ribosomal protein operons enlarge the volume of data available from 16S rRNA comparison, and improve the resolution of taxonomic distances.

Second is the sequencing of single-cell genomes, most prominently developed at the Bigelow lab in Maine, and by Tanja Woyke at the Joint Genome Institute of the U.S. Department of Energy. When a sequence is obtained from a single cell, it may be more fragmented than a genomic bin (Chapter 21), but we know that the entire sequence comes from one individual, not an artifactual hybrid. Thus, for example, we can see when an organism possesses multiple metabolic capabilities, such as the ability of some marine archaea to combine autotrophic and heterotrophic lifestyles.

With all the new data appearing, note that fundamental features of ancient divergence remain in dispute, such as the relative divergence times of domains Bacteria and Eukarya from Archaea. Recent trees from some laboratories show unexpectedly high similarity between Eukarya and the archaeal phylum Lokiarchaeota (**Fig. 19.3**). These data imply that the eukaryotic ancestor branched relatively late from one line of Archaea. Other research groups argue that better computation methods show a different branch point nearer the ancient Archaea-Bacteria divide. For further discussion, see Section 19.6.

Major archaeal taxa are outlined in **Table 19.2**. As of this writing, the clades labeled in blue in **Figure 19.3** are the most commonly studied.

For several reasons, the phylogeny of Archaea is a challenge to define. Most archaea are uncultured and are known solely through metagenomic bins. Many of their genomes are highly "recombinogenic." For example, two different samples of *Ferroplasma acidarmanus* show 99% identical SSU rRNA, yet their overall genomes differ by 22%, implying extensive horizontal gene exchange. And many archaeal genomes, particularly those of mesophiles, include large portions of DNA transferred from bacteria. Microbial ecologist Purificación López-García, at Paris-Sud University, proposes that hyperthermophilic archaeal species evolved into mesophiles by acquiring genes from bacteria. According to her model, the bacterial genes provided key metabolic pathways and cell structures optimized for growth at moderate or cold temperatures.

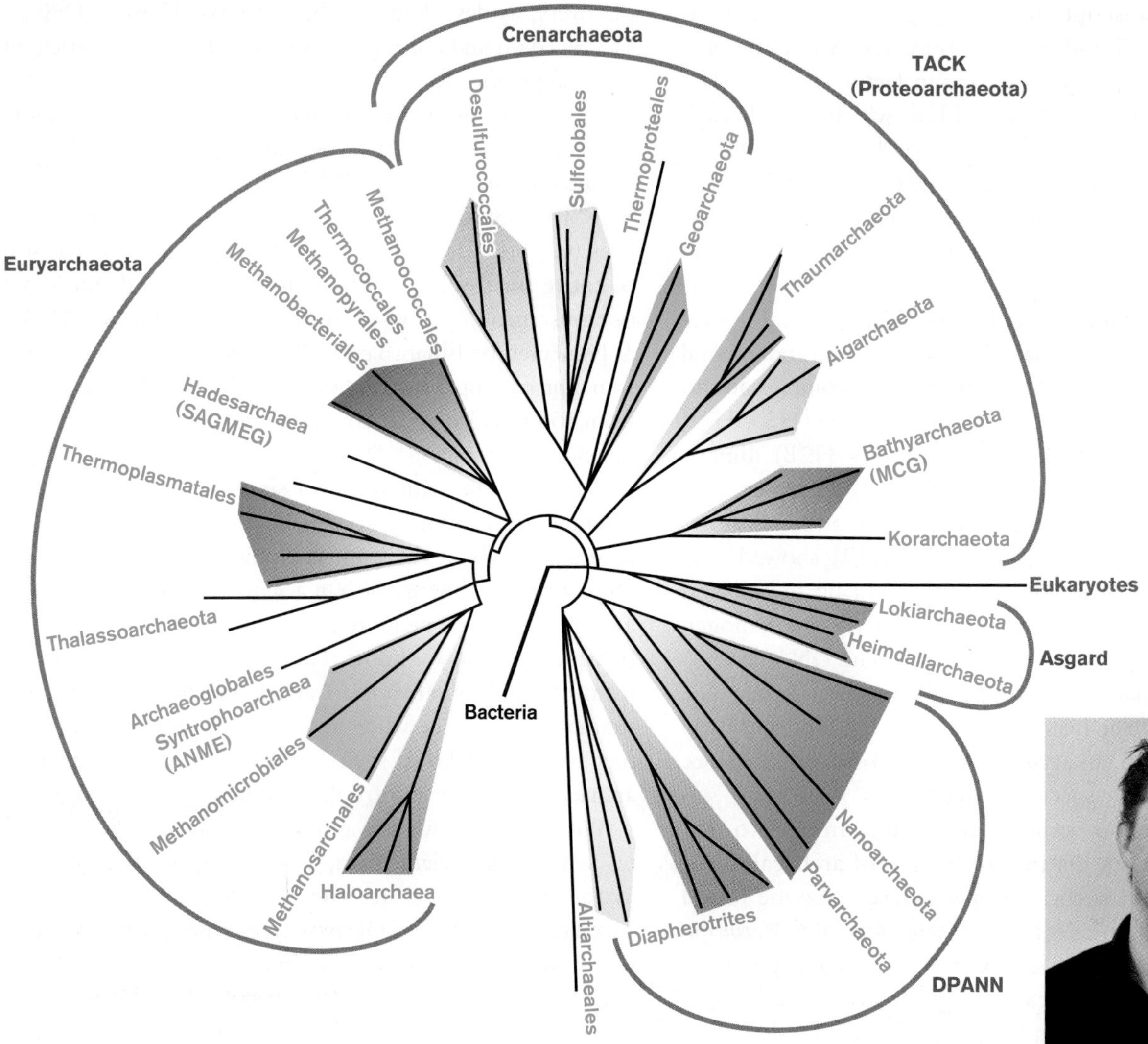

COURTESY OF THIJS ETTEMA

FIGURE 19.3 ■ Archaeal phylogeny. Major superphyla, phyla, and orders of Archaea. Divergence is based approximately on SSU rRNA and genome sequences. **Inset:** Thijs Ettema.

TACK Superphylum: Thaumarchaeota, Aigarchaeota, Crenarchaeota, Korarchaeota

Marine and geothermal hyperthermophiles. The TACK superphylum includes major clades of Archaea that grow at temperatures above 90°C, originally designated as the Crenarchaeota. Most TACK hyperthermophiles were discovered at deep-sea marine hydrothermal vents, or at hot springs such as those of Yellowstone National Park. Major clades (originally orders, now designated phyla) include **Desulfurococcales**, sulfur-reducing anaerobes at marine hydrothermal vents; **Sulfolobales**, sulfide-oxidizing acidophiles at hot springs; **Thermoproteales**, vent thermophiles; and **Geoarchaeota**, found in acidic iron mats.

Many TACK thermophiles metabolize sulfur, either by anaerobic reduction (such as by H_2 to form H_2S) or by aerobic oxidation (by O_2 to form sulfuric acid). Anaerobic sulfur metabolizers include moderate thermophiles (growth range about 60°C–80°C), as well as hyperthermophiles (90°C–120°C). Many of the hyperthermophiles are also barophiles, growing under high pressure at hydrothermal vents on the ocean floor; an example is *Pyrodictium abyssi*.

Besides thermophiles, the TACK clades include a substantial proportion of mesophilic soil, marine, and benthic (marine sediment) microbial communities. Many are found in the microbiomes of plants and animals. Understanding these organisms is important for assessing their contribution to the global carbon cycle. The **Bathyarchaeota** (formerly Miscellaneous Crenarchaeote Group, MCG) include vast populations throughout the open ocean. Their genomes reveal the capacity for methanogenesis. The Bathyarchaeota are the only known methanogens outside of the Euryarchaeota.

The TACK superphylum even includes psychrophiles (growing at temperatures below 20°C) found in Antarctic lakes.

TABLE 19.2 Archaeal diversity

TACK (Thaumarchaeota, Aigarchaeota, Crenarchaeota, Korarchaeota).

Aigarchaeota. Aerobic hyperthermophilic filaments in hot springs.

Bathyarchaeota (formerly Miscellaneous Crenarchaeote Group, MCG). Widely divergent autotroph-heterotrophs in diverse soil and water.

Korarchaeota. Anaerobic hyperthermophiles in Yellowstone hot springs and in deep-sea thermal vents.

Thaumarchaeota (Marine Group I). Tetraether membranes include crenarchaeol. Marine and soil archaea that oxidize NH_3 with O_2. Important marine sources of nitrates for phytoplankton, and of methylphosphonate ($CH_3–PO_3^{2-}$), which bacteria convert to methane.

- **Cenarchaeales.** Sponge symbionts; grow at 10°C. *Cenarchaeum symbiosum.*
- **Nitrosopumilales.** Ammonia-oxidizing archaea (AOA), marine and soil. *Nitrosopumilus maritimus, Nitrososphaera gargensis.*
- **Psychrophilic marine thaumarchaeotes (uncharacterized).** Marine water, deep sea, and Antarctica. Anaerobic heterotrophs, sulfate reducers, nitrite-reducing methanotrophs.

Thermoprotei. Most have tetraether membranes surrounded by S-layer. Includes thermophiles in hot springs and marine vents; also marine mesophiles and psychrophiles. **Crenarchaeoata** is an alternative name encompassing Desulfurococcales, Sulfolobales, Thermoproteales, Geoarchaeota.

- **Caldisphaerales.** Thermoacidophilic heterotrophs grow in hot springs. *Caldisphaera* spp.
- **Desulfurococcales.** Anaerobic sulfur reduction with organic electron donors. Irregularly shaped cells with glycoprotein S-layer; no cell wall (*Aeropyrum pernix, Desulfurococcus fermentans*). *Ignicoccus islandicus. Pyrodictium abyssi, Pyrodictium occultum, Pyrolobus fumarii* grow at marine thermal vents, up to 110°C.
- **Geoarchaeota.** Acidic thermal mats.
- **Sulfolobales.** Aerobic acidophiles; moderate thermophiles. Oxidize H_2S to H_2SO_4. *Sulfolobus, Sulfurisphaera, Acidianus.*
- **Thermoproteales.** *Pyrobaculum, Thermoproteus, Vulcanisaeta.*

Euryarchaeota. Metabolism includes methanogenesis, halophilic photoheterotrophy, and sulfur and hydrogen oxidation; acidophiles and alkaliphiles. Methanogens and halophiles have rigid cell walls.

Anaerobic Methane-Oxidizing Euryarchaeota (ANME). Anoxic marine sediments. Oxidize methane from methanogens, in syntrophy with sulfate-reducing bacteria.

Archaeoglobales. Hyperthermophiles; sulfate oxidation of H_2 or organic hydrogen donors; reverse methanogenesis. *Archaeoglobus fulgidus.*

Hadesarchaea. Formerly South African Gold Mine Miscellaneous Euryarchaeotic Group (SAGMEG). Found in water from gold mines deep underground.

Haloarchaea. Halophiles; grow in brine (concentrated NaCl). Conduct photoheterotrophy, with light-driven H^+ pump and Cl^- pump. *Haloarcula, Halobacterium, Haloferax* grow in salterns and salt lakes. *Haloquadra* (*Haloquadratum*) are square-shaped. *Halorubrum lacusprofundi* is an Antarctic psychrophile. *Natronococcus* spp. are alkaliphiles in soda lakes.

Methanogens (many classes). Generate methane from CO_2 and H_2, formate, acetate, other small molecules; strict anaerobes. Pseudopeptidoglycan or sulfated chondroitin cell walls. Grow in anaerobic soil, water, or animal digestive tracts. Deep-sea psychrophiles generate methane hydrates.

- **Methanobacteriales.** Lack cytochromes; reduce CO_2, formate, or methanol with H_2 by electron bifurcation. *Methanobrevibacter smithii* and *Methanosphaera stadtmanae* inhabit human digestive tract.
- **Methanomicrobiales, Methanococcales, and Methanopyrales.** Lack cytochromes; reduce CO_2, formate, or methanol with H_2 by electron bifurcation. *Methanocaldococcus jannaschii*, vent thermophile.
- **Methanosarcinales.** Includes acetoclastic methanogens.

Thermococcales. Hyperthermophiles (grow above 100°C) and barophiles (up to 200 atm pressure). Anaerobes; reduce sulfur. *Thermococcus, Pyrococcus abyssi, P. furiosus.*

Thermoplasmata. Extreme acidophiles; oxidize sulfur from pyrite (FeS_2), generating sulfuric acid. Mesophiles or moderate thermophiles.

- **Thermoplasmatales.** *Ferroplasma acidiphilum* and *F. acidarmanus* grow at 37°C–50°C. Oxidize sulfur from FeS_2, generating ambient pH as low as pH 0. No cell wall. *Thermoplasma acidophilum* grows at 59°C and pH 2.

Asgard. Vent thermophiles named for Norse gods.

Heimdallarchaeota. Formerly Ancient Archaeal Group (AAG).

Lokiarchaeota. Isolated from Loki's Castle hydrothermal vent, off coast of Norway. Formerly called Deep Sea Archaeal Group (DSAG) or Marine Benthic Group B (MBGB).

DPANN (Diapherotrites, Parvarchaeota, Aenigmarchaeota, Nanoarchaeota, Nanohaloarchaeota). Diminished genomes and metabolic capabilities suggest obligate symbionts.

Nanoarchaeota. Vent hyperthermophiles; obligate symbionts attached to *Ignicoccus. Nanoarchaeum equitans.*

Altiarchaeales. Form grappling-hook biofilms in cold sulfidic water. *Altiarchaeum*. The position of Altiarchaeales within DPANN is uncertain.

Overall, the TACK superphylum spans the widest range of growth temperatures of any division of life.

Thaumarchaeota. The TACK phylum **Thaumarchaeota** includes mesophiles originally classified with Marine Group I Crenarchaeota but now recognized as a distinct clade whose members are very important in marine environments. Thaumarchaeota include ammonia-oxidizing archaea (AOA), which oxidize ammonia to nitrite. These ammonia oxidizers play a major role in the nitrogen cycle (discussed in Chapters 21 and 22). Other Thaumarchaeota include mesophilic heterotrophs and sulfur oxidizers in soil and water. Some are key symbionts of invertebrate animals such as deep-sea sponges.

Unique metabolic pathways. Some archaea use distinctive metabolic pathways. TACK archaea catabolize glucose via several variants of the Entner-Doudoroff (ED) and Embden-Meyerhof-Parnas (EMP) pathways that rarely occur in bacteria (**Fig. 19.4**). For example, the sulfur thermophiles *Sulfolobus* and *Thermoplasma* convert glucose to gluconate without phosphorylation, ultimately generating pyruvate with no net production of ATP. (ATP is still produced by further breakdown of pyruvate.)

Among the Euryarchaeota halophilic archaea such as *Halobacterium* phosphorylate the dehydrated product of gluconate (2-oxo-3-deoxygluconate), which enters the "standard" ED pathway. The ED pathway generates one molecule of pyruvate and one molecule of 3-phosphoglycerate, which produces one net ATP through the second stage of the EMP pathway. This variant may be compared with different variants of the EMP pathway seen outside the TACK group, in the euryarchaeotic vent thermophile *Pyrococcus furiosus*. *P. furiosus* oxidizes glyceraldehyde 3-phosphate using ferredoxin instead of NAD^+ and avoids phosphorylation. The reduced ferredoxin is then used to reduce $2H^+$ to H_2 in an energy-yielding reaction.

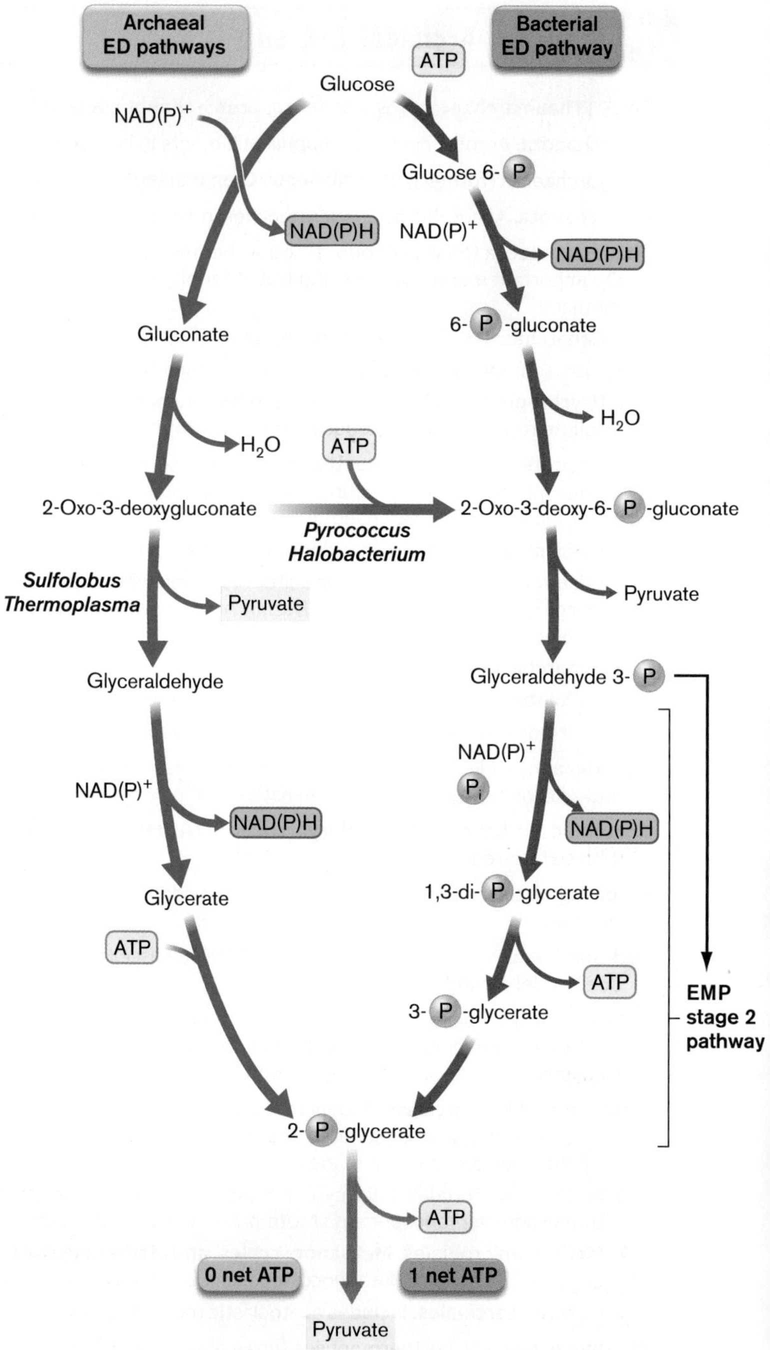

FIGURE 19.4 ■ Glucose catabolism in archaea. Left: *Sulfolobus* and *Thermoplasma* species catabolize glucose to pyruvate via a modified ED pathway without phosphorylating glucose and produce no net ATP. **Right:** *Halobacterium* species phosphorylate 2-oxo-3-deoxygluconate and produce one net ATP via the EMP stage 2 pathway. *Pyrococcus furiosus* oxidizes glyceraldehyde 3-phosphate using ferredoxin instead of NAD^+ and avoids phosphorylation.

Euryarchaeota: Methanogens, Halophiles, and Thermophiles

The superphylum **Euryarchaeota** also includes members throughout soil and water, and associated with plants and animals. The most highly divergent group of Euryarchaeota is methanogens, including several polyphyletic clades (clades not sharing a single common ancestor). Methanogens serve a key energetic role in ecosystems by offering an anaerobic mechanism for removing excess H_2 and other small-molecule reductants. But their metabolism releases methane—a potent greenhouse gas. The accelerating release of methane from Arctic tundra (**Fig. 19.5**) and from marine benthic sources has drastic consequences for our global climate.

The energy-yielding process of **methanogenesis** occurs only in archaea (discussed in Section 19.4). Despite their common energetic pathway, methanogens branch deeply among other euryarchaeotic groups, and they show a wide range of cell shapes and environmental adaptations.

How do methanogens and other chemoautotrophs form biomass? Many euryarchaeotes fix CO_2 into biomass by

FIGURE 19.5 ■ Polar methane release. In Fairbanks, Alaska, bubbles of methane from methanogens beneath the ice burst into the atmosphere, where the methane ignites into flame.

the Wood-Ljungdahl pathway of acetogenesis (acetate formation), also known as the reductive acetyl-CoA pathway (presented in Chapter 15). Biochemical and genomic comparisons suggest that two-carbon assimilation by acetogenesis may be the oldest means of carbon fixation, used by the ancient ancestral archaea.

Another branch of euryarchaeotes is the Haloarchaea, extreme halophiles that are the only form of life to grow in concentrated brine (NaCl). Most Haloarchaea, such as *Halobacterium* species NRC-1, are photoheterotrophs that can supplement their metabolism with retinal-containing light-driven ion pumps, called **bacteriorhodopsin**. These ion pumps (for protons or for sodium ions) are the only known form of phototrophy in archaea (discussed in Chapter 14). Bacteriorhodopsins are found in Haloarchaea such as *Halobacterium halobium*; the species was named before it was known to be an archaeon (discussed in Section 19.5). It turns out that Haloarchaea have transferred genes encoding light-absorbing proton pumps into many marine bacteria. By contrast, the chlorophyll-based phototrophy found in bacteria and plants is completely unknown in archaea.

Note: In **Table 19.2**, certain archaeal names contain a "bacteria" component—for example, the genus *Halo<u>bacterium</u>*. Such organisms were known and named as bacteria before 1977, when the category "archaea" was first defined.

Euryarchaeotes include extreme acidophiles, as well as extreme alkaliphiles—from *Ferroplasma* (Thermococcales), growing at pH 0, to *Natronococcus*, growing at pH 10. With respect to pH, the euryarchaeotes show the widest range of any taxonomic group.

Asgard and DPANN

A recently characterized clade of deep-sea archaea is the **Lokiarchaeota**, branching from a marine benthic microbial community originally known as the Deep-Sea Archaeal Group/Marine Benthic Group B (DSAG/MBGB). First identified at a thermal vent named Loki's Castle, the Lokiarchaeota genomes share surprising traits with eukaryotes. Several related phyla have been described and given similar Norse mythological names, such as the Heimdallarchaeota (formerly Ancient Archaeal Group, AAG). These clades are collectively known as the Asgard superphylum.

Eukaryotic-like genes found in Lokiarchaeota include those associated with an actin cytoskeleton and phagocytosis. These features, and phylogenetic trees, suggest that Lokiarchaeota may descend from a common ancestor of a cell that contributed to the ancestral eukaryote.

Another vast superphylum revealed by metagenomics is DPANN (named for the phyla Diapherotrites, Parvarchaeota, Aenigmarchaeota, Nanoarchaeota, and Nanohaloarchaeota). The genomes of these organisms all show evolutionary loss of many genes, typical of obligate symbionts. The most studied example is the **Nanoarchaeota** species *Nanoarchaeum equitans*, a cellular parasite of *Ignicoccus hospitalis* (see the Current Research Highlight and Section 19.6).

Meanwhile, sequencing of metagenomes and single-cell genomes continues to reveal new deeply branching phyla with unexpected traits. For example, the Altiarchaeales clade includes *Altiarchaeum hamiconexum*, a soil archaeon that forms networked biofilms connected by harpoon-like appendages. As of this writing, the taxonomic position of Altiarchaeales (within DPANN or outside) remains in dispute.

Thought Question

19.1 Suppose that two deeply diverging clades each show a wide range of growth temperature. What does this suggest about the evolution of thermophily or psychrophily?

To Summarize

- **Archaeal membranes are composed of L-glycerol diether or tetraether lipids**, with isoprenoid side chains that may include cross-links or pentacyclic rings. The membrane may be covered by a protein S-layer, but no cell wall. A few clades of Archaea possess a cell wall of pseudopeptidoglycan.

- **Glucose is catabolized by variants of the Entner-Doudoroff pathway.** Other metabolic pathways found in archaea include methanogenesis and retinal-associated light-driven ion pumps such as bacteriorhodopsin.
- **Central gene functions of archaea resemble those of eukaryotes**, as seen in the structure of DNA and RNA polymerases and of histone-like DNA-binding proteins.
- **Major groups of archaeal phyla are TACK, Euryarchaeota, DPANN, and Asgard.** TACK hyperthermophiles include sulfur reducers and sulfide oxidizers. TACK Thaumarchaeota include ammonia oxidizers, marine invertebrate symbionts, and others. Euryarchaeota include methanogens, halophiles, and acidophiles.
- **The deep-sea clade Asgard includes the phylum Lokiarchaeota**, which shows genomic evidence of relatedness to the eukaryotes.
- **Metagenomics reveals uncultured organisms.** We continue to discover deeply branching clades of archaea by sequencing metagenomes.

19.2 TACK Hyperthermophiles Eat Sulfur

Hyperthermophiles are most prevalent in two kinds of habitats: hot springs, where crustal fissures release steam; and undersea hydrothermal vents, found in the marine benthos.

FIGURE 19.6 ▪ Thermophiles colonize a hot spring. Morning Glory Pool, a hot spring in Yellowstone National Park, supports thermophilic archaea and bacteria.

Hot Springs

Thermophilic archaea commonly grow in hot springs and geysers, such as those of Yellowstone National Park (**Fig. 19.6**) or the Solfatara volcanic area near Naples, Italy. A hot spring occurs where water seeps underground above a magma chamber, which heats the water to near boiling. The heated water expands and is forced upward through fissures, coming out in a heated spring. In a geyser, the water is heated under pressure. As the water escapes upward, it turns into steam, which expands and jets upward, falling into a heated pool. These heated pools and their surrounding edges generate extreme ranges of temperature, mineral content, and acidity. They support a diverse range of microbial life, including thermophilic cyanobacteria and firmicutes, as well as archaea.

Several features of hot springs and geysers are important for thermophiles:

- **Reduced minerals.** The heated water dissolves high concentrations of sulfides and other reduced minerals. When the water emerges and cools, the minerals precipitate. These reduced minerals serve as rich energy sources for autotrophs.
- **Low oxygen content.** At higher temperatures, the oxygen concentration of water declines. Therefore, hyperthermophiles tend to be anaerobic, although there are important exceptions, such as the aerobic sulfur oxidizer *Sulfolobus*.
- **Steep temperature gradients.** The temperature of the water falls dramatically within a short distance from the source, forming a steep gradient. Different species of thermophiles are adapted to different temperatures and grow in separate patches at the different temperatures, causing a variegated pattern.
- **Acidity.** Some hot-spring environments show extreme acidity. The acidity results from oxidation of sulfur or iron in reactions that generate strong inorganic acids, such as sulfuric acid (H_2SO_4).

A special subcategory of volcanic hot-spring habitats is that of submarine hydrothermal vents on the ocean floor. The vent thermophiles must evolve to grow at high pressure under several kilometers of ocean. Pressure increases by approximately 100 atm per kilometer of ocean depth. Organisms that grow only at high pressure are called **barophiles** (discussed in Chapter 5).

Desulfurococcales: Reducing Sulfur

The **Desulfurococcales** archaea show distinctive cell structures and forms of metabolism (**Table 19.3**). All possess a membrane with a combination of diethers and tetraethers, surrounded by an elaborate S-layer. Many take advantage

TABLE 19.3 TACK Hyperthermophiles

Representative species	Growth temperature (°C)	Growth pH	Cell shape	Metabolism
Aeropyrum pernix	70–100°	pH 5–9	Cocci	O_2 respiration
Desulfurococcus fermentans	78–87°	pH 6	Motile cocci with archaella	Anaerobic S^0 respiration or fermentation
Ignicoccus islandicus	70–98°	pH 5–7	Cocci with periplasmic space	Anaerobic lithotrophy, S^0 oxidation of H_2
Pyrodictium abyssi	80–110°	pH 5–7	Disks linked by cannulae	Anaerobic oxidation of H_2 by S^0 or $S_2O_3^{2-}$ or fermentation
Sulfolobus solfataricus	50–87°	pH 2–4	Irregular cocci	O_2 respiration on S^0, producing H_2SO_4
Thermosphaera aggregans	65–90°	pH 5–7	Motile cocci in aggregates	Anaerobic fermentation

of the high temperatures that increase the thermodynamic favorability of sulfur redox reactions. An example is *Desulfurococcus fermentans* (**Fig. 19.7**), a motile coccoid cell with archaella, isolated from hot springs. *D. fermentans* grows optimally at 85°C. The species respires anaerobically by reducing elemental sulfur (S^0) to sulfide (HS^-). (Anaerobic respiration was discussed in Chapter 14.) Sulfur reduction is coupled to oxidation of small organic molecules such as sugars.

Another motile coccus, *Ignicoccus islandicus*, has an unusual cell architecture (**Fig. 19.8**; see the Current Research Highlight). *Ignicoccus* has an outer membrane surrounding its cytoplasmic membrane, with a large aqueous compartment between them. This outer compartment contains membrane-enclosed vesicles of unknown function. The evolution of this compartment, similar to the extra membranes of the bacterial genus *Planctomyces* (see Chapter 18), suggests a model for an intermediate stage of evolution of the eukaryotic nucleus. *I. islandicus*, unlike *Desulfurococcus*, is a marine organism, growing at temperatures as high as 98°C. It is a lithotroph, oxidizing hydrogen with sulfur:

$$H_2 + S^0 \longrightarrow H_2S$$

Most of the cultured species of Desulfurococcales are obligate anaerobes. An exception is *Aeropyrum pernix*, one of the first archaea to have its genome sequenced. *A. pernix* is an aerobic heterotroph, respiring with O_2 on complex compounds during growth at 70°C–100°C.

While *Desulfurococcus* species are motile, other TACK hyperthermophiles form dense biofilms. *Thermosphaera aggregans* forms colonies so tightly bound that they cannot be dissociated by protease treatment or sonication.

Thought Question

19.2 What might be the advantages of archaellar motility for a hyperthermophile living in a thermal spring or in a black smoker vent? What would be the advantages of growth in a biofilm?

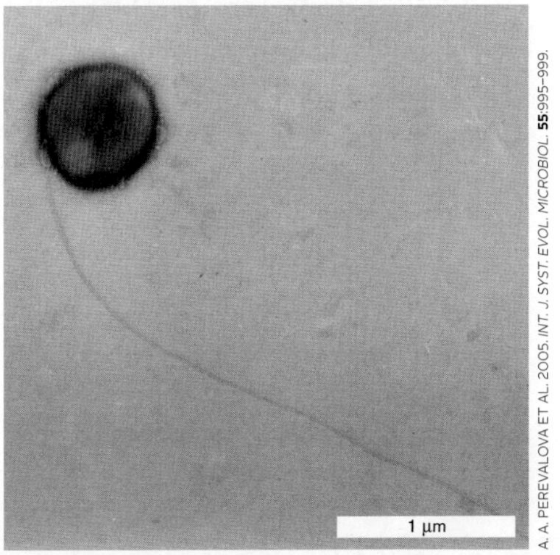

FIGURE 19.7 ▪ TACK hyperthermophilie. *Desulfurococcus fermentans* (shadow EM).

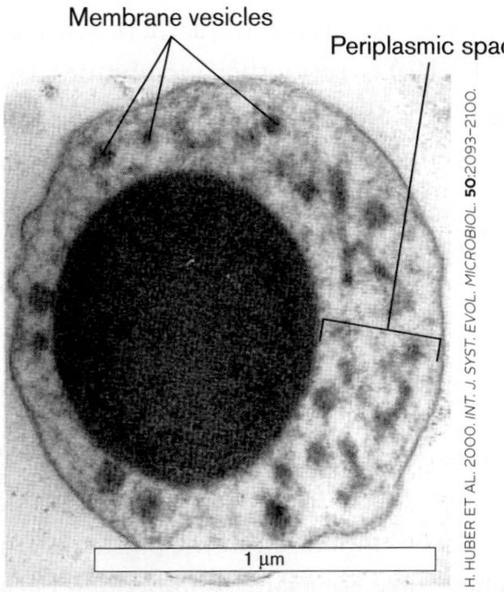

FIGURE 19.8 ▪ *Ignicoccus islandicus*. The unique periplasmic space of this archeaon contains membrane vesicles (TEM).

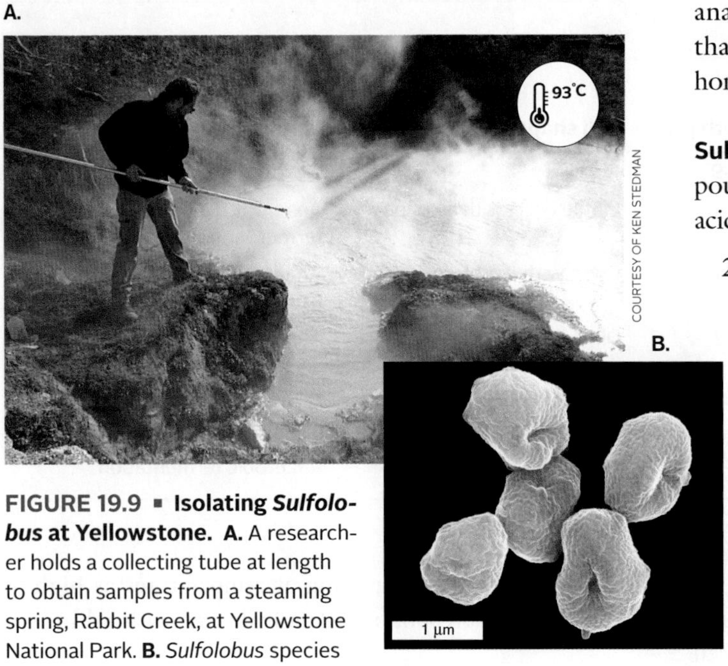

FIGURE 19.9 ▪ **Isolating *Sulfolobus* at Yellowstone.** **A.** A researcher holds a collecting tube at length to obtain samples from a steaming spring, Rabbit Creek, at Yellowstone National Park. **B.** *Sulfolobus* species grow at 80°C at pH 2–3.

Sulfolobales: High Temperature and Extreme Acid

Some archaea grow in extreme acid, as well as high temperature. Organisms that grow under multiple conditions considered "extreme" are called "polyextremophiles." The phylum **Sulfolobales** includes species that respire by oxidizing sulfur (instead of reducing it as *Desulfurococcus* does). These organisms, such as *Sulfolobus* species, grow at 80°C–90°C within hot springs and solfataras (volcanic vents that emit only gases). Ken Stedman and colleagues at Portland State University study *Sulfolobus solfataricus*, a species that grows at 80°C and pH 2 (**Fig. 19.9**). Unlike most archaea, *S. solfataricus* is readily cultured in the laboratory, and thus many interesting traits have been documented.

Cell membrane and S-layer. *Sulfolobus* cells have a membrane composed mainly of tetraethers with cyclopentane rings (**Fig. 19.10A**). Tetraether membranes are commonly seen in acidophilic thermophiles, probably because they are exceptionally impermeable to protons. Remarkably, *Sulfolobus* has no cell wall, but only an S-layer of glycoprotein (**Fig. 19.10B**). The S-layer proteins of *Sulfolobus* species lock together to form a sturdy array. While helping to keep the cell intact, the S-layer protein array nonetheless allows a flexible cell structure, in contrast to the more rigid structures of most bacteria.

Sulfolobus species possess archaella (**Fig. 19.11**). Like bacterial flagella, *Sulfolobus* archaella are helical filaments driven by a motor embedded in the plasma membrane (**Fig. 19.11B**). The motor structure is embedded between S-layer proteins. Unlike bacterial flagella, the archaeal motor is driven by ATP hydrolysis catalyzed by FlaI subunits, analogous to the type IV pilus assembly motor. Note that the archaeal proteins were named before their lack of homology to bacterial flagellar components was known.

Sulfide oxidation. *Sulfolobus* species oxidize organic compounds with oxygen, or they oxidize S^0 or H_2S to sulfuric acid:

$$2S^0 + 3O_2 + 2H_2O \longrightarrow 2H_2SO_4 \longrightarrow 4H^+ + 2SO_4^{2-}$$

As a result of sulfuric acid production, the pH of the organism's surroundings falls to pH 2–3, effectively excluding all but acidophiles. While other archaea grow at higher temperatures (up to 120°C) or lower pH (below pH 0), *Sulfolobus* is of interest as a "double extremophile," requiring both high temperature and extreme acidity simultaneously.

Sulfolobus species can grow heterotrophically on sugars or amino acids. In fact, many species are easily cultured in tryptone broth at 80°C, pH 3. Their internal pH is typically pH 6.5; thus, they maintain more than three units of pH difference across their membrane. The full metabolic potential of this organism is revealed by annotation of its genome, which contains

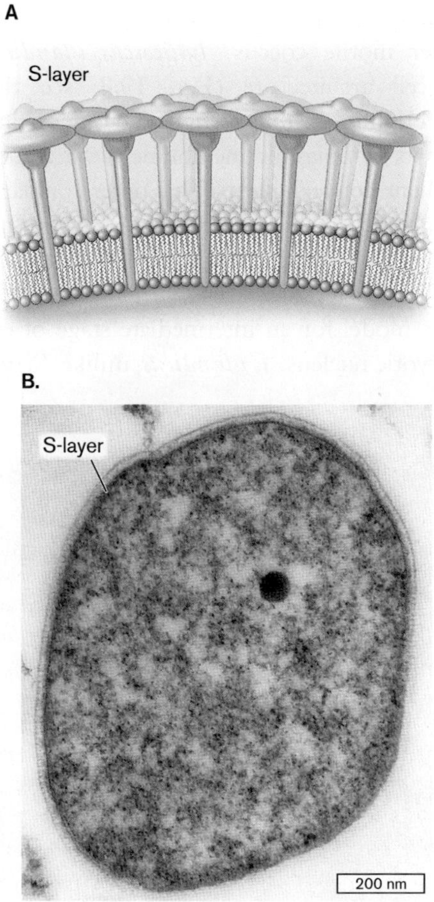

FIGURE 19.10 ▪ ***Sulfolobus.*** **A.** *Sulfolobus* species have S-layer proteins plugged into a tetraether membrane. **B.** *Sulfolobus* cell with thick S-layer.

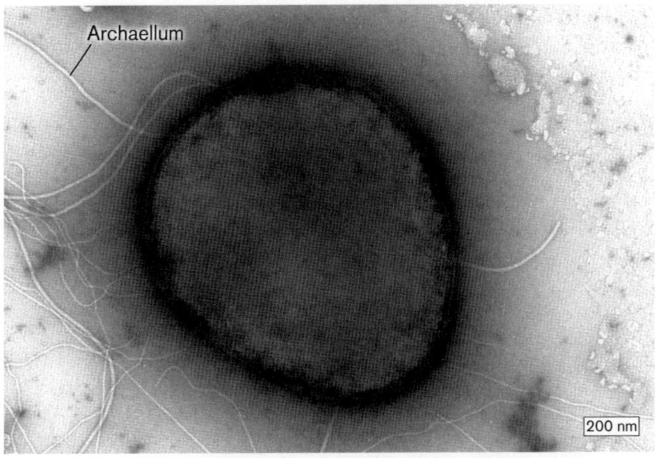

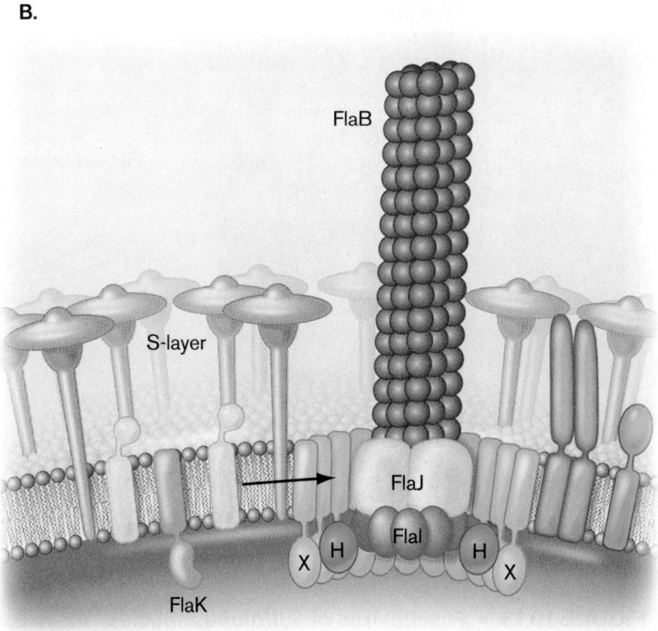

FIGURE 19.11 ▪ *Sulfolobus* sp. A20 possesses archaella. A. Motile *Sulfolobus* cell with archaella (TEM). **B.** Archaellum structure is embedded between S-layer subunits. *Source:* Part B modified from Sonja-Verena Albers and Ken F. Jarrell. 2015. *Front. Microbiol.* **6**:23, fig. 1B.

homologs of sugar and amino acid transporters, as well as the non-ATP-forming Entner-Doudoroff pathway of glucose catabolism. Genes are present for enzymes to oxidize HS^+ and $S_2O_3^{2-}$, as well as S^0. The main redox carrier for respiration appears to be ferredoxin (instead of NADH, which is relatively unstable at high temperature).

Thought Question

19.3 What problem with cell biochemistry is faced by acidophiles that conduct heterotrophic metabolism?

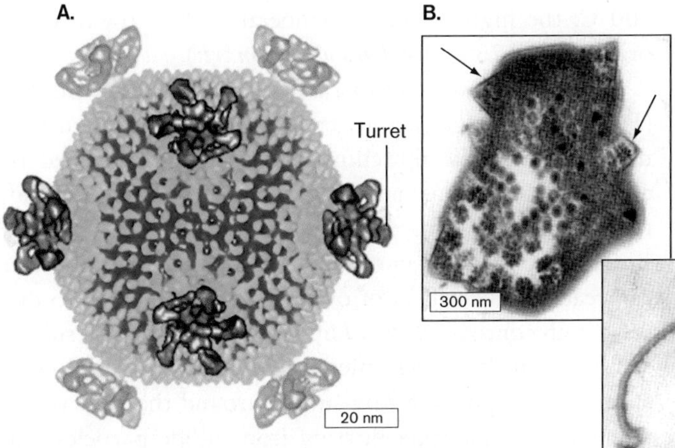

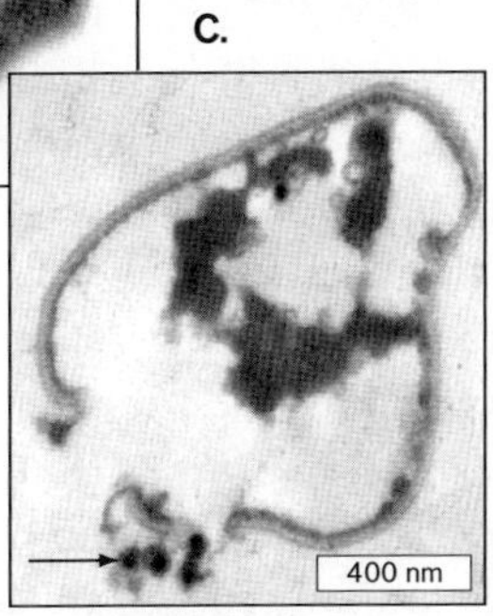

FIGURE 19.12 ▪ *Sulfolobus* turreted icosahedral virus (STIV) infects *Sulfolobus solfataricus*. A. STIV capsid with "turrets" (cryo-EM). Capsid diameter 60 nm. **B.** A cell of *S. solfataricus* packed with a hexagonal array of STIV particles. Arrows point to pyramidal bulges where virus arrays poke through a breach in the S-layer. **C.** Empty cell membrane and S-layer following lysis and viral release. Arrow points to released virus particles.

Viruses of *Sulfolobus*. A fascinating discovery in *Sulfolobus* was that of archaeal viruses. *Sulfolobus* species are attacked by a number of viruses, including turreted icosahedral virus. The process of a viral infection has been observed in *S. solfataricus* (**Fig. 19.12**). Cells were infected with *Sulfolobus* turreted icosahedral virus (STIV), a virus isolated from a boiling-acid hot spring in Yellowstone National Park. Each icosahedral particle of STIV has 12 turret-like projections, as shown in the cryo-EM image reconstruction in **Figure 19.12A**. Mature virions contain a double-stranded DNA genome coated with lipid within the capsid. The thin section of an infected cell (**Fig. 19.12B**) shows progeny virions packed into a hexagonal array, portions of which poke through the host cell's S-layer in pyramidal forms. After lysis (**Fig. 19.12C**), the S-layer complex is all that remains of the empty cell. The S-layer appears surprisingly intact, showing that it provides a sturdy cell covering perhaps comparable in strength to a cell wall.

This lytic cycle resembles lytic and fast-release cycles of bacterial and eukaryotic viruses, although the capsid "turrets" and the pyramidal bulges of the lysing cell are unique to archaea. The structure of a capsid protein component shows surprising homology to both bacterial and eukaryotic viral proteins.

Another unusual virus infecting *Sulfolobus* species is the fusellovirus (**Fig. 19.13**). The spindle shape of fusellovirus is found only in viruses of archaea. The DNA genome of this virus—which must retain its structure at high temperatures—is

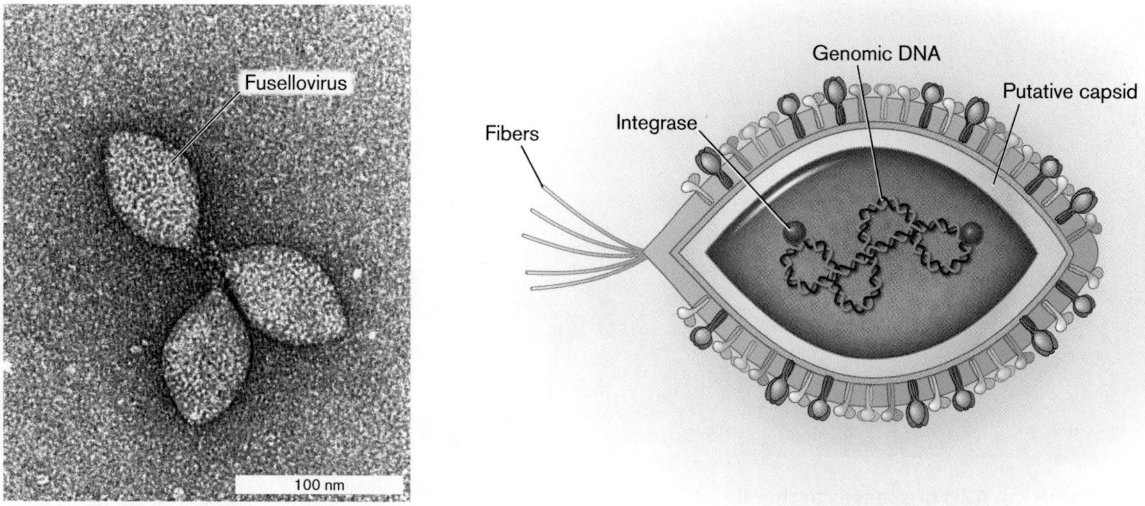

FIGURE 19.13 ■ Fusellovirus of *Sulfolobus* species. A. Spindle-shaped fusellovirus particles. **B.** Virion structure, including positively supercoiled DNA genome with integrase and capsid and envelope proteins.

positively supercoiled. The structure and genetics of fusellovirus have been studied, including its ability to integrate in the host genome at a tRNA-encoding gene.

Numerous other archaeal viruses have been discovered in high-temperature environments, including icosahedral, tailed, and filamentous forms. All known archaeal viruses have genomes consisting of double-stranded DNA, suggesting that only double-stranded DNA is stable enough for virus particles to persist at high temperature. In 2012, however, a possible archaeal RNA-dependent RNA transcriptase (discussed in Chapter 11) was identified from a Yellowstone hot spring. This finding suggests that there may exist positive-strand RNA viruses that infect hyperthermophilic archaea.

Thought Question

19.4 What hypotheses might be proposed about archaeal evolution if viruses of mesophilic archaea are found to have RNA genomes? What if, instead, all archaeal viruses have DNA genomes only?

Besides Sulfolobales, the group Thermoprotei includes other clades of thermoacidophiles. Caldisphaerales is a clade of thermoacidophiles first isolated from a Philippine hot spring at Mount Makiling. Members of Caldisphaerales, such as *Caldisphaera*, typically grow at pH 3 up to 80°C. Unlike *Sulfolobus* species, *Caldisphaera* species are anaerobes or microaerophiles, tolerating only low concentrations of oxygen. They grow by fermentation or anaerobic respiration. Another clade, Thermoproteales, includes hyperthermoacidophiles isolated from marine vents; thus, they survive extreme pressure, as well as heat and acid! *Thermoproteus* species grow at temperatures up to 97°C and at pH values lower than pH 3. Their rod-shaped cells are less than 0.3 μm in length—one of the smallest cell types known. They have autotrophic metabolism, gaining energy by reducing sulfur with H_2 to H_2S.

Barophilic Vent Hyperthermophiles

The most extreme hyperthermophiles are barophiles adapted to grow near hydrothermal vents at the ocean floor. The high pressure beneath several kilometers of ocean allows water to remain liquid at temperatures above 100°C; the highest known temperature for growth of an organism is 125°C (for *Pyrolobus fumarii*).

A common feature of thermal vents is the **black smoker** (**Fig. 19.14**). A black smoker is a chimney-like structure resulting from the upwelling of seawater superheated by an undersea magma chamber. As in a geyser aboveground, the heated water is forced upward through a small opening. Because the thermal vent is under steam pressure, the water can reach temperatures of over 400°C, enabling it to dissolve high concentrations of minerals such as iron II sulfide (FeS). When the rising water escapes, however, it immediately cools, depositing iron sulfide around the edge of the vent chimney and precipitating iron sulfide particles that cloud the water—hence the term "black smoker." While no organism can grow at 400°C, various species of archaea are adapted to grow in the range of 100°C–120°C, where the vent stream meets the seawater and minerals precipitate (**Fig. 19.14B**).

How do we study organisms under such extreme conditions? To study hyperthermophiles from black smoker

A.

DR. MICHAEL PERFIT, UNIVERSITY OF FLORIDA, AND NOAA VENTS PROGRAM

B.

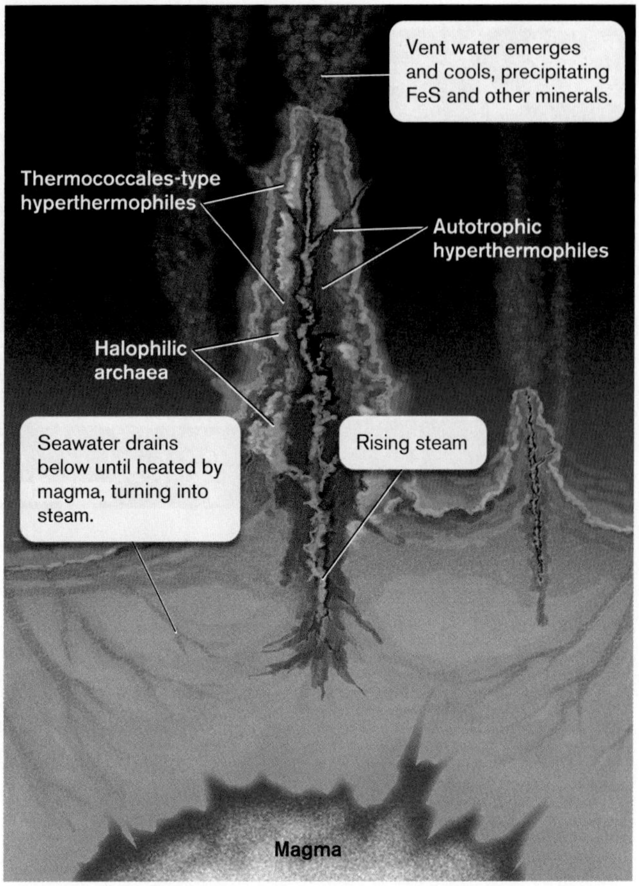

FIGURE 19.14 ■ Extreme temperature and pressure: black smoker vents. A. Black smoker vent with fluid escaping from "chimneys" of sulfide minerals that crystallize as the 350°C fluid hits the cold, 2°C ocean seawater. The photo was taken by the submersible *Alvin* at a depth of 2,250 m, at the Juan de Fuca Ridge, off the coast of Oregon. **B.** Different parts of the smoker vent system support different classes of archaea.

vents requires specialized equipment. The isolation of such organisms is a challenge because their habitats endanger our own survival. Undersea vent systems must be approached by a special submersible device with a robotic arm. An example is the Environmental Sample Processor from the Monterey Bay Aquarium Research Institute (**Fig. 19.15A**). The robotic system samples temperature and other properties of fluid emerging from a black smoker hydrothermal vent. It can then sample organisms for study. An advanced version of this device can actually process the organism's DNA. Thus, the DNA can be obtained from vent-adapted microbes that could not survive transfer to a laboratory at sea level. The robotic sample processor is supported by NASA as a model for a future space probe to explore one of Jupiter's moons, Europa, considered a possible source of extraterrestrial life.

Organisms that do survive transport to sea level must nonetheless be maintained at high pressure and temperature to ensure viability. In the laboratory, all devices for microscopy and cultivation must be kept under pressure and at high temperature (**Fig. 19.15B**). The culture must be provided with reduced minerals and gases needed for the growth of vent microbes.

Vent-adapted members of Desulfurococcales include *Pyrodictium abyssi* (**Fig. 19.16**), *P. occultum*, and *P. brockii*; the latter is named for Thomas Brock of the University of Wisconsin–Madison, a pioneering researcher of hyperthermophiles. For energy, *Pyrodictium* species reduce sulfur to H_2S, either with molecular hydrogen or with organic compounds. A membrane-embedded sulfur-reducing complex and a proton-translocating ATP synthase have been isolated from *P. abyssi*. The complexes are extremely heat stable, exhibiting a temperature optimum of 100°C.

A.

© 2006 MBARI

B.

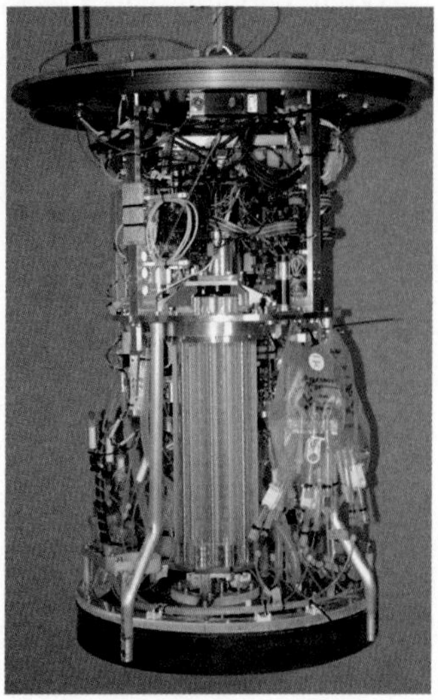

TODD WALSH © 2006 MBARI

FIGURE 19.15 ■ Robotic sampling from a black smoker vent. A. Engineer Gene Massion, from the Monterey Bay Aquarium Research Institute, deploys the submersible Environmental Sample Processor with robotic collection arm at an ocean site off the coast of Maine. **B.** Pressurized device for sampling and cultivation of vent organisms.

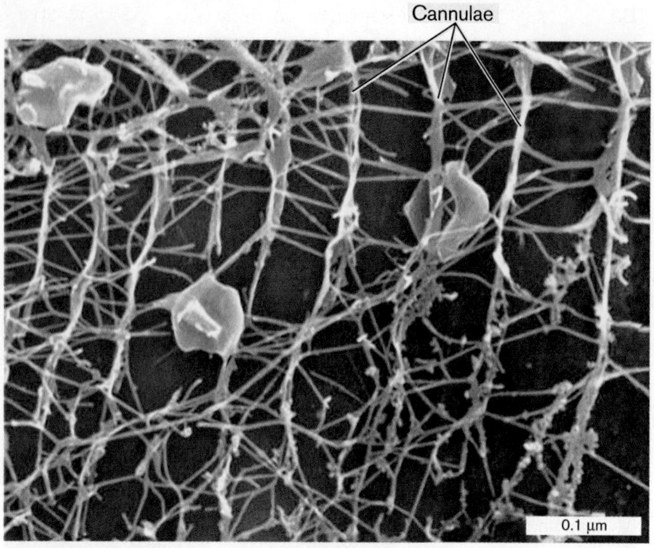

FIGURE 19.16 ▪ *Pyrodictium abyssi*, growing as networks of cells linked by cannulae. SEM.

Pyrodictium species grow as flat, disk-shaped cells that can be as thin as 0.1 µm. The cells contain a periplasm and outer membrane with an S-layer that is coated with zinc sulfide, a mineral that precipitates from the vent. The cell disks are interconnected by glycoprotein tubules called **cannulae** (singular, **cannula**). The cannulae can extend to more than 0.1 mm, forming complex networks of connections (**Fig. 19.16**). In liquid culture, the networks grow into white balls up to 10 mm in diameter. Cryo-electron tomography of a *Pyrodictium* cell shows that the cannulae bridge the periplasm between cells, but not the cytoplasm. The cannulae may enable *Pyrodictium* cells to share nutrients and maintain a biofilm, while keeping their cellular identity distinct with separated cytoplasm.

What happens when a *Pyrodictium* cell divides? Cells of *P. abyssi* generate new cannulae as they undergo fission (**Fig. 19.17**). Some of the new cannulae form as loops connecting the two daughter cells while simultaneously pushing the two cells apart. In this fashion, the cell division process expands the cell network.

Note: Two genera of vent thermophiles have similar names but only distant genetic relatedness: *Pyrodictium abyssi*, from TACK Desulfurococcales; and *Pyrococcus abyssi*, from Euryarchaeota.

Korarchaeota and Aigarchaeota: Cryptic Filaments of the Deep

Metagenomic analysis continues to reveal previously unknown branches of life. An example of a deeply branching clade is the **Korarchaeota** (the "K" in TACK). Korarchaeota are hydrothermal vent hyperthermophiles isolated by Susan Barns and Norman Pace. The organisms have not been isolated in pure culture, but in 2008 the genome of one korarchaeote was sequenced from an enriched mixed culture at 80°C–90°C at Obsidian Pool, Yellowstone National Park (**Fig. 19.18**). The organism, provisionally named "*Candidatus* Korarchaeum cryptofilum," grows in long, thin filaments less than 200 nm wide. Its genes suggest that it gains energy mainly from anaerobic peptide fermentation, yet we still do not know enough to culture this organism in the laboratory.

Another recent discovery, from a terrestrial subsurface hydrothermal ecosystem is the phylum **Aigarchaeota**. The Aigarchaeota include filamentous heterotrophs that are aerobic, using oxygen as their electron acceptor. Aigarchaeota is the "A" in the TACK superphylum name.

To Summarize

- **Habitats for hyperthermophiles** include hot springs and submarine hydrothermal vents. Vent organisms are barophiles as well as thermophiles. Anaerobic hyperthermophilic acidophiles include Caldisphaerales and Thermoproteales.
- **Desulfurococcales includes diverse thermophiles.** Most are anaerobes that use sulfur to oxidize hydrogen or organic molecules.
- ***Sulfolobus*** species oxidize sulfur or H_2S to sulfuric acid, and they catabolize organic compounds.

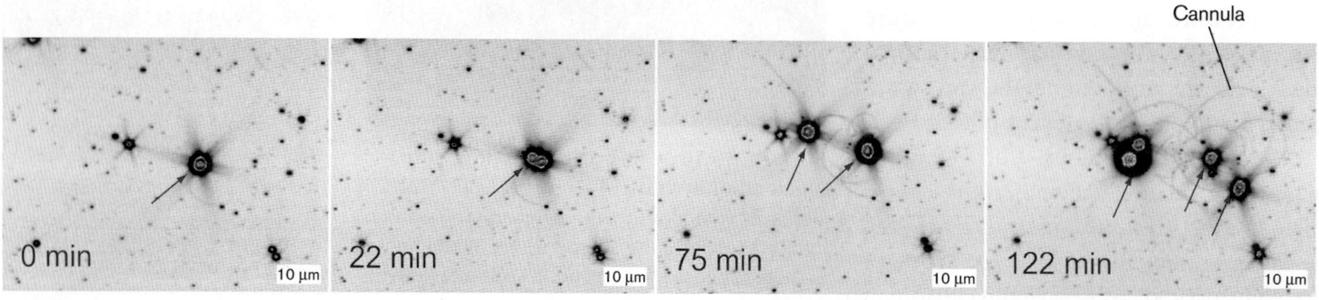

FIGURE 19.17 ▪ *Pyrodictium abyssi*, undergoing cell division. Cells of *P. abyssi* (arrows) generate new interconnecting cannulae as they divide.

A. B.

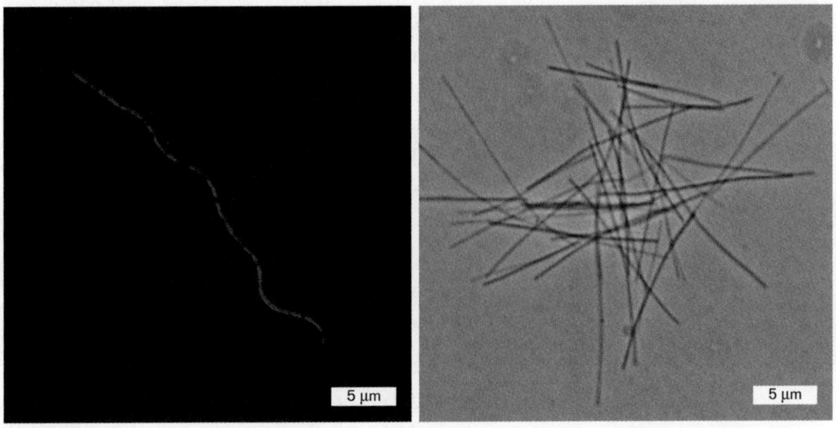

ELKINS ET AL. 2008. *PNAS* **105**:8102, FIG. 1A.

ELKINS ET AL. 2008. *PNAS* **105**:8102, FIG. 1B.

FIGURE 19.18 ■ Korarchaeota: a filamentous vent hyperthermophile. **A.** Filamentous chain of a candidate species: *Korarchaeum cryptofilum* (fluorescence microscopy, FISH with fluorescent DNA probe). **B.** *K. cryptofilum* under phase contrast.

- **Archaella are driven by rotary motors.** Rotation is driven by ATP hydrolysis. The archaellum may share ancestry with type IV pili.
- **Viruses infect *Sulfolobus*.** Archaeal viruses show unique shapes and positively supercoiled DNA.
- **Marine hydrothermal vents support archaea that are barophiles as well as hyperthermophiles.** *Pyrodictium* species are disk-shaped cells interconnected by cytoplasmic bridges called cannulae.
- **Korarchaeota and Aigarchaeota are filamentous hyperthermophiles.**

19.3 Thaumarchaeota: Ammonia Oxidizers and Animal Symbionts

Because archaea were first isolated from extreme habitats, it came as a surprise when SSU rRNA probes revealed mesophilic TACK archaea (then identified as Crenarchaeota) in moderate habitats throughout the biosphere. The first marine mesophilic archaea were found in 1992 by Jed Fuhrman, University of Southern California (**Fig. 19.19A**). Edward DeLong (**Fig. 19.19B**), Massachusetts Institute of Technology, further sampled the Pacific Ocean at the Hawaii Ocean Time-series station, where he found high numbers of mesophilic archaea, then called Crenarchaeota, now classified under TACK (**Fig. 19.19C**). The abundance of archaea, predominantly Thaumarchaeota, varied according to season and increased with depth, typically comprising 40% of the total microbial population at depths of 1,000 meters, where temperatures are cold.

Mesophilic archaea may have evolved from hyperthermophiles by gene acquisition from mesophilic bacteria. Many—perhaps the majority of archaea overall—grow in water or soil, often in association with plants and animals.

Ammonia-Oxidizing Thaumarchaeotes Cycle Global Nitrogen

From a global standpoint, the most significant thaumarchaeotes discovered are the ammonia-oxidizing archaea (AOA), such as those of the order Nitrosopumilales. The best-studied ammonia-oxidizing thaumarchaeote, *Nitrosopumilus maritimus*, was isolated from a marine water tank at the Seattle Aquarium by David Stahl and colleagues at the University of Washington (**Fig. 19.20**).

Ammonia-oxidizing archaea gain energy by aerobically oxidizing ammonia to nitrite:

$$2NH_3 + 3O_2 \longrightarrow 2NO_2^- + 2H_2O + 2H^+$$

This lithotrophic reaction yields redox energy, enabling the microbe to fix CO_2 for biomass (discussed in Chapter 14). In marine environments, excreted ammonia can build up to high levels, until it is oxidized by archaea and bacteria. In some habitats, the Thaumarchaeota perform most of the recycling of ammonia, such as the ammonia excreted by fish in your aquarium. The thaumarchaeotes thus help balance the ecology of intertidal pools and beaches. The reaction plays a key role in the global nitrogen cycle, as the first step of returning organic nitrogen to atmospheric N_2 (discussed in Chapters 21 and 22). It also provides a major source of nitrite for marine phytoplankton.

Ammonia-oxidizing archaea are ubiquitous in marine environments. Tatsunori Nakagawa and colleagues at Nihon University, Japan, used enrichment culture (described in Chapter 4) to hunt for new kinds of AOA (**Fig. 19.21**). Samples of sand from the seafloor at Tanoura Bay, Japan, were serially diluted in medium containing ammonium sulfate and adjusted to pH 8, a pH high enough for significant deprotonation of ammonium ion (NH_4^+) to ammonia (NH_3). During the enrichment culture, ammonia was progressively consumed and converted to nitrite (NO_2^-). In the culture, the gene encoding ammonia oxidase (*amoA*) was identified by PCR amplification. The AOA were identified by fluorescence in situ hybridization (FISH) using a fluorophore attached to DNA that hybridizes to rRNA of *Nitrosopumilus* species.

Nitrosopumilus and related AOA oxidize ammonia at extremely low concentrations. Thus, AOA quickly remove

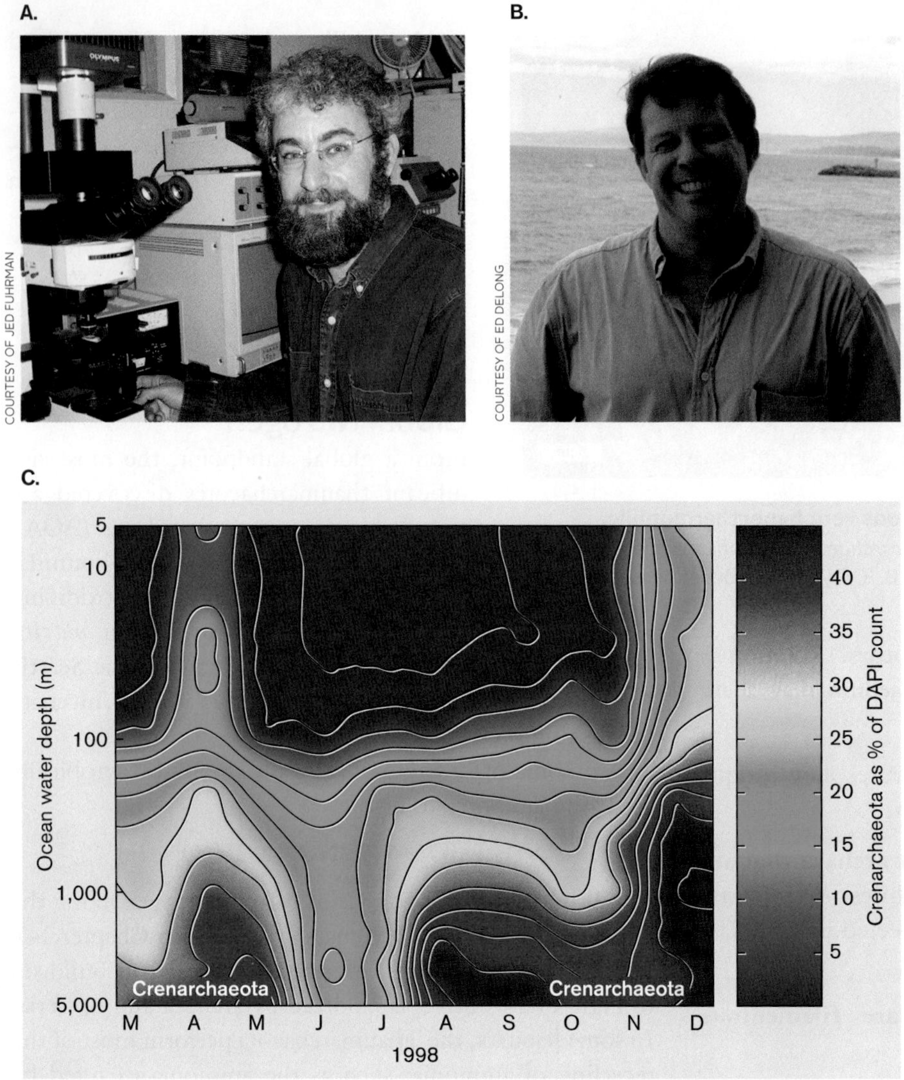

FIGURE 19.19 ■ Crenarchaeota in the Pacific Ocean. A. Jed Fuhrman samples mesophilic archaea in the Pacific Ocean. **B.** Ed DeLong mapped the abundance of marine mesophilic archaea. **C.** The proportion of TACK mesophilic archaea (color profile) measured as a function of depth and season, in the Hawaii Ocean Time-series station. Archaea were identified by FISH. A fluorescein-labeled DNA probe specific to TACK archaea was hybridized to a cellular rRNA sequence; DNA was detected using the DNA-binding fluorophore DAPI. *Source:* Part C modified from Markus Karner et al. 2001. *Nature* **409**:507.

a substance toxic to fish. AOA have since been found throughout marine and freshwater communities, as well as in soil, where they perform the important function of ammonia removal. They appear in industrial waste sludge, revealed by fluorescent DNA probes (**Fig. 19.22**). While bacteria also oxidize ammonia, in many habitats the AOA are the dominant oxidizers. Ammonia-oxidizing thaumarchaeotes continue to be discovered in other habitats, including Antarctic lakes down to the temperature of –20°C.

Symbiotic ammonia-oxidizing archaea. Human sweat glands release nitrogenous materials that bacteria break down to ammonia. Are ammonia-oxidizing archaea found on humans? Christine Moissl-Eichinger, at the Medical University of Graz, Austria, discovered AOA on human skin, where they are the major archaeal component of the skin microbiome. The human body continually emits ammonia, so AOA may help metabolize this toxic substance while controlling skin pH for the microbial community. The finding of ammonia-oxidizing thaumarchaeotes on the skin, as well as methanogens in the gut, may help explain another discovery made by Moissl-Eichinger—namely, the DNA

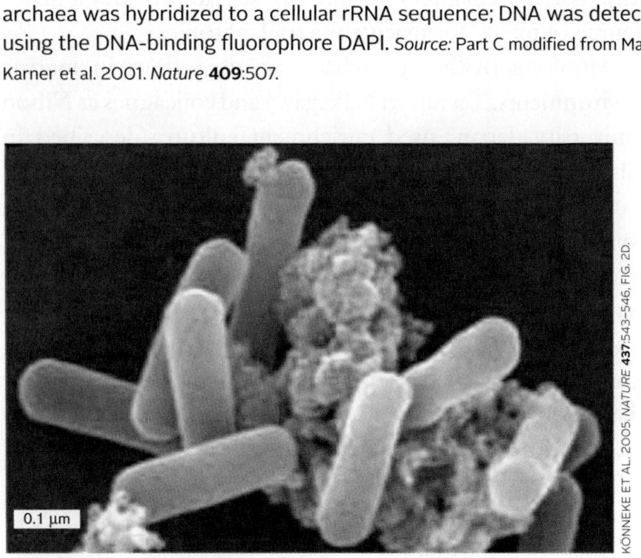

FIGURE 19.20 ■ *Nitrosopumilus*, an ammonia-oxidizing archaeon. SEM of cells clustered on sediment.

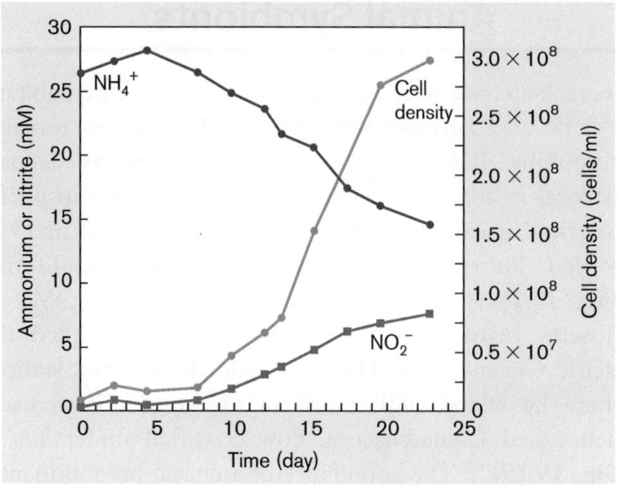

FIGURE 19.21 ■ Enrichment culture for ammonia oxidizers. Ammonia-oxidizing microbes from eelgrass seafloor samples convert ammonia to nitrite. *Source:* Modified from Naoki Matsutani et al. 2011. *Microbes Environ.* **26**:23–29.

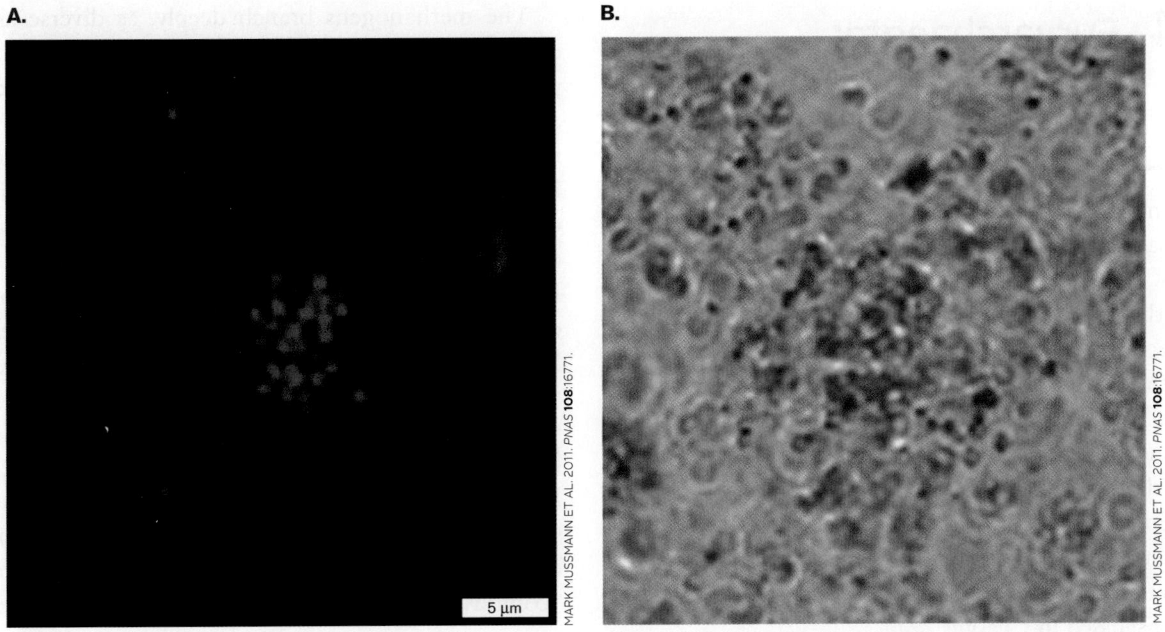

FIGURE 19.22 ■ Ammonia-oxidizing Thaumarchaeota in nitrogen-rich industrial sludge. A. FISH probe hybridizes with rRNA sequence specific to thaumarchaeotes. **B.** Ammonia-oxidizing archaea within a particle of sewage sludge (LM).

signatures of thaumarchaeotes and methanogens in hospital intensive care units and industrial clean-room facilities.

In the deep ocean, there are thaumarchaeote psychrophiles (cold-adapted microorganisms). Some psychrophilic thaumarchaeotes live as endosymbionts of marine animals such as sponges. The thaumarchaeote *Cenarchaeum symbiosum* inhabits the sponge *Axinella mexicana* (**Fig. 19.23A**). *C. symbiosum* has yet to be grown in culture, but the presence of the microbes is shown by the fluorescence of a DNA probe that hybridizes to sequences specific to *C. symbiosum* (**Fig. 19.23B**). How the microbe benefits its sponge host is unknown, but the sponge and its endosymbionts can be cocultured in an aquarium for many years. A hypothesis investigated by researchers is that *C. symbiosum* produces antimicrobial agents that protect the sponge from pathogens or predators (see Chapter 15). Some of the products of *C. symbiosum* are being tested for their pharmacological properties.

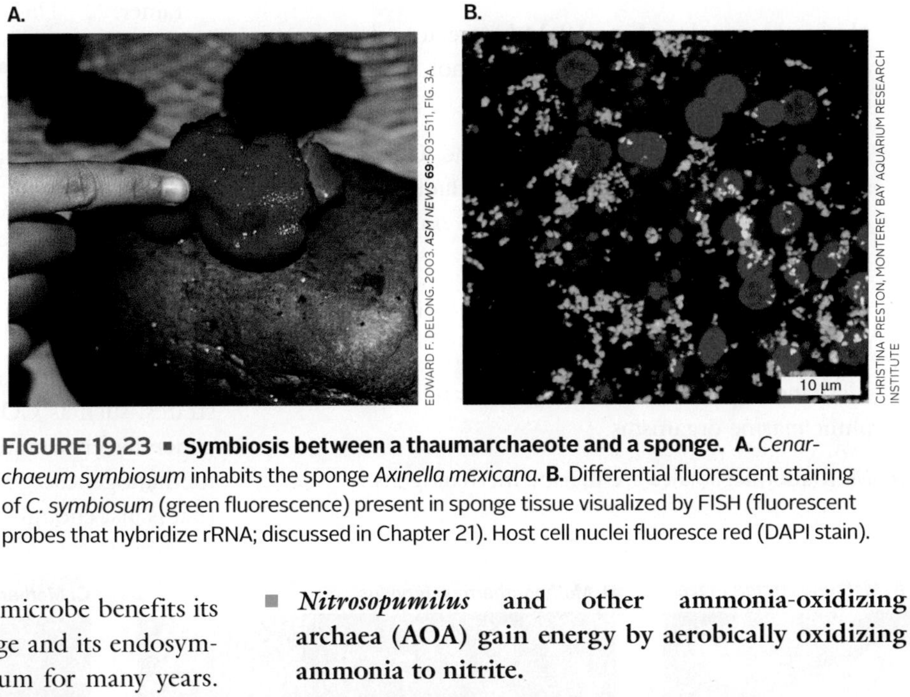

FIGURE 19.23 ■ Symbiosis between a thaumarchaeote and a sponge. A. *Cenarchaeum symbiosum* inhabits the sponge *Axinella mexicana*. **B.** Differential fluorescent staining of *C. symbiosum* (green fluorescence) present in sponge tissue visualized by FISH (fluorescent probes that hybridize rRNA; discussed in Chapter 21). Host cell nuclei fluoresce red (DAPI stain).

To Summarize

- **Oceans, soil, plant roots, and animals** provide habitats for mesophilic and psychrophilic TACK archaea.
- ***Nitrosopumilus*** **and other ammonia-oxidizing archaea (AOA) gain energy by aerobically oxidizing ammonia to nitrite.**
- **Ammonia-oxidizing archaea** make essential contributions to local and global nitrogen cycling in water and soil.
- **Ammonia-oxidizing archaea comprise a significant part of the human skin microbiome.**
- **Thaumarchaeotes include symbionts of marine sponges.** Emerging thaumarchaeotes include mesophiles, psychrophiles, and thermophiles in many environments.

19.4 Euryarchaeota: Methanogens from Gut to Globe

A third major branch of known archaea is Euryarchaeota, the "broad-ranging archaea" (see **Table 19.2**). The euryarchaeotes include multiple branches of **methanogens**, species that derive energy through reactions producing methane. Most are strict anaerobes; while their metabolism is chemolithotrophic, they depend on bacterial communities releasing CO_2, H_2, and other simple carbon compounds. Methanogens release methane from soil and animal digestive habitats found nearly everywhere on Earth. For the methanogen, methane is just a by-product of its energy-yielding metabolism, but this by-product is a greenhouse gas with profound consequences for our biosphere (discussed in Chapter 22).

Figure 19.3 shows the approximate branch points of five major clades of methanogens. The branches are polyphyletic—that is, they lack a distinctive shared ancestor—and they are interspersed by nonmethanogens.

- Methanosarcinales (**Fig. 19.24A**) are found in ocean and soil, breaking down acetate, alcohol, and amines to methane.
- Methanobacteriales (**Fig. 19.24B** and **C**) inhabit soil and animal digestive tracts. They include main human gut methanogens *Methanosphaera stadtmanae* and *Methanobrevibacter smithii*.
- Methanopyrales include hyperthermophilic methanogens growing at up to 122°C.
- Methanococcales include mesophilic and hyperthermophilic marine organisms.
- Methanomicrobiales are marine methanogens.

The methanogens branch deeply, as diverse as all the euryarchaeotes together. A possible explanation of this phylogeny would be that the ancestral euryarchaeote was a methanogen, but many descendant lineages lost the metabolism. Alternatively, divergent euryarchaeotes could have acquired the genes for methanogenesis later, independently by horizontal gene transfer. As yet, we lack sufficient genomic data to determine which model is correct. Furthermore, genes encoding enzymes of methanogenesis were found in one deep-branching clade outside the Euryarchaeota, the TACK phylum Bathyarchaeota.

Biogenic Paths to Methane

Biogenic methane formation has actually been observed since 1776, when the Italian priest Carlo Campi and physicist Alessandro Volta (1745–1827) investigated bubbles of "combustible air" from a wetland lake. Volta wrote:

> Being in a little boat on Lake Maggiore, and passing close to an area covered with reeds, I started to poke and stir the bottom with my cane. So much air emerged that I decided to collect a quantity in a large glass container. . . . This air burns with a beautiful blue flame.

Volta noted that methane arose from wetlands containing water-saturated decaying plant material. In 1882, the German medical researcher Hermann von Tappeiner (1847–1927) combined plant materials with ruminant stomach contents (a source of methanogens) and showed that both components were essential to produce methane. During the late nineteenth century, in England, methane was collected from manure and sewage and used as a fuel for street lamps.

Different species generate methane from different substrates, such as CO_2 and H_2 or small organic compounds, generally fermentation products of bacteria. Each pathway of methanogenesis generates a small free energy change ($\Delta G^{\circ\prime}$) that is just enough to drive processes of carbon fixation and

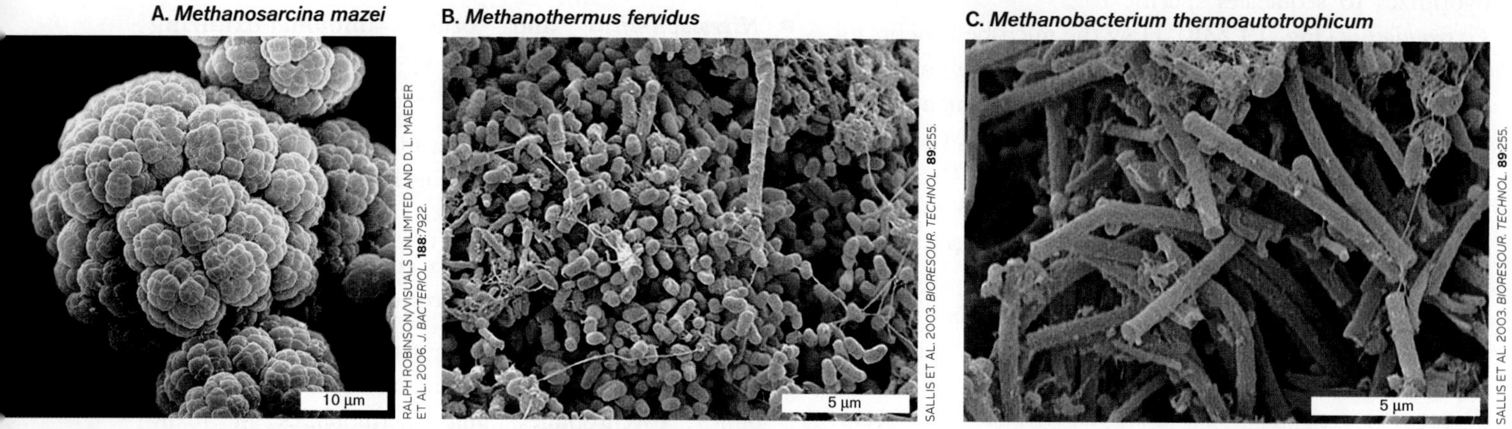

FIGURE 19.24 ■ Methanogens show a wide range of shapes. **A.** *Methanosarcina mazei*, a lobed coccus form lacking flagella (SEM). **B.** *Methanothermus fervidus*, a short bacillus (SEM). **C.** *Methanobacterium thermoautotrophicum*, an elongated bacillus (SEM).

biosynthesis. All types of methanogenesis are poisoned by molecular oxygen and therefore require extreme anaerobiosis. Major substrates and reactions include:

Carbon dioxide: $CO_2 + 4H_2 \longrightarrow CH_4 + 2H_2O$
Formic acid: $4CHOOH \longrightarrow CH_4 + 3CO_2 + 2H_2O$
Acetic acid: $CH_3COOH \longrightarrow CH_4 + CO_2$
Methanol: $4CH_3OH \longrightarrow 3CH_4 + CO_2 + 2H_2O$
Methylamine: $4CH_3NH_2 + 2H_2O \longrightarrow 3CH_4 + CO_2 + 4NH_3$
Dimethyl sulfide: $2(CH_3)_2S + 2H_2O \longrightarrow 3CH_4 + CO_2 + 2H_2S$

In CO_2 reduction, the most common form of methanogenesis, carbon dioxide plus molecular hydrogen combine to form water and methane. CO_2-reducing methanogens are autotrophs, growing solely on CO_2 and H_2 with a source of nitrogen and other minerals. Methanogenesis from other carbon sources, such as formate, acetate, or methanol, is heterotrophic and generates CO_2 as a product in addition to methane. The nitrogen-containing substrate methylamine generates ammonia in addition to methane and CO_2, whereas the sulfur-containing substrate dimethyl sulfide generates hydrogen sulfide.

Note that only a narrow range of substrates supports methanogenesis. For unknown reasons, most methanogens lack the vast array of energy-yielding pathways found in soil bacteria such as *Rhodopseudomonas* or *Streptomyces.* Thus, methanogens generally require close association with bacterial partners to provide their substrates—a relationship called **syntrophy** (discussed in Chapter 14). Syntrophic methanogens usually grow in habitats with minimal resource flux, where hydrogen and carbon dioxide gases can be trapped for their use. Removal of these gases then enhances bacterial metabolism.

Thought Question

19.5 Anoxic soil contains bacteria and methanogens. What will happen to the microbial populations when the soil is tilled and aerated?

Many kinds of methanogens can be cultured in the laboratory. Their culture poses special challenges because the reactions of methanogenesis are halted by oxygen. Thus, most methanogens are strict anaerobes, although a few that tolerate oxygen have been found. To exclude oxygen, microbes are cultured and manipulated within an anaerobic chamber. Furthermore, methanogens that use CO_2 and H_2 as substrates must receive a steady supply of these gases. Even more challenging is the culture of hyperthermophilic methanogens, such as *Methanopyrus*, a deep-ocean vent hyperthermophile that grows at scorching temperatures up to 122°C. For these organisms, we must maintain both high temperature and high pressure.

A striking aspect is the wide range of growth temperatures among closely related species. Thermophiles and even hyperthermophiles branch from closely related mesophiles; for example, the order Methanobacteriales includes *Methanobrevibacter ruminantium* (37°C–39°C), *Methanobacterium thermoautotrophicum* (50°C–75°C), and *Methanothermus fervidus* (60°C–97°C). **Table 19.4** summarizes five major orders of methanogens.

Methanogens Show Diverse Cell Forms

Despite their metabolic similarity, methanogens display an astonishing diversity of form, perhaps as diverse as the entire domain of bacteria (**Fig. 19.24**). For example, *Methanocaldococcus jannaschii* cells grow as cocci with numerous archaella attached to one side, while *Methanosarcina mazei* forms peach-shaped cocci lacking archaella. *Methanothermus fervidus* cells are short, fat rods without archaella, and *Methanobacterium thermoautotrophicum* grows as elongated rods reminiscent of the bacterial genus *Bacillus.* Still others, such as *Methanospirillum hungatei*, form wide spirals.

The morphological diversity of methanogens may be explained in part by their rigid cell walls, which can maintain a distinctive shape. The composition of methanogen cell walls is much more diverse than that of bacteria. *Methanobacterium* species have a cell wall composed of **pseudopeptidoglycan**, or **pseudomurein**, a structure in which chains of alternating amino sugars are linked by peptide cross-bridges analogous to those of peptidoglycan (**Fig. 19.25**). The bacterial *N*-acetylmuramic acid, however, is replaced by a related sugar, *N*-acetyltalosaminuronic acid; and the sugar linkage is β(1,3) instead of β(1,4) as in bacteria. As a result, these archaea are resistant to lysozyme, which degrades bacterial cell walls at the β(1,4) sugar linkage. The peptide cross-bridges of pseudopeptidoglycan differ as well, causing resistance to penicillin (discussed in Chapter 3).

By contrast, *Methanosarcina* species have a cell wall composed of sulfated polysaccharides. The genera

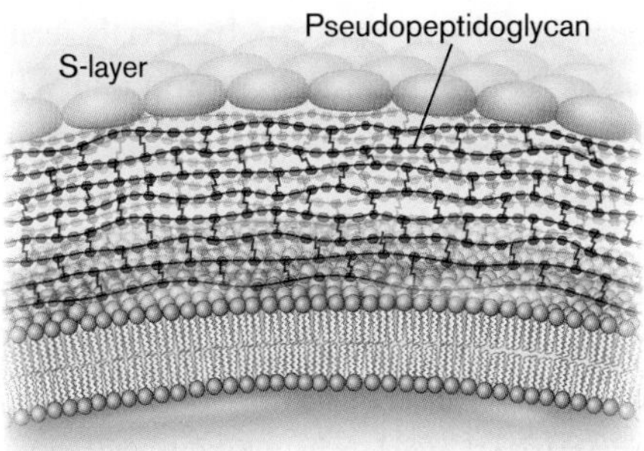

FIGURE 19.25 ■ Methanobacteriales and other methanogenic species have a cell wall of pseudopeptidoglycan between their membrane and S-layer.

TABLE 19.4 Methanogens

Representative species	Growth temperature (°C)	Growth pH	Cell shape	Substrates for methanogenesis
Methanobacteriales				
Methanobrevibacter ruminantium	37–39°	pH 6–9	Chains of short rods	H_2 and CO_2
Methanobacterium thermoautotrophicum	50–75°	pH 7–8	Filaments of long rods	H_2 and CO_2, formate
Methanothermus fervidus	60–97°	pH 6–7	Rods	H_2 and CO_2
Methanococcales				
Methanococcus vannielii	20–40°	pH 7–9	Cocci with archaella	H_2 and CO_2
Methanocaldococcus jannaschii	48–94°	pH 6–7	Cocci with archaella	H_2 and CO_2
Methanomicrobiales				
Methanoculleus olentangii	30–50°	pH 6–8	Cocci	Acetate, complex nutrients
Methanospirillum hungatei	20–45°	pH 6–7	Spirilla	Acetate
Methanopyrales				
Methanopyrus kandleri	84–122°	pH 6–8	Long rods (2–14 μm)	H_2 and CO_2
Methanosarcinales				
Methanosarcina barkeri	20–50°	pH 5–7	Aggregates of cocci	H_2 and CO_2, methanol, methylamine, acetate
Methanosaeta concilii	10–45°	pH 6–8	Filaments of rods	Acetate
Methanohalophilus zhilinae	45°	pH 8–10	Cocci	Methanol, methylamine, dimethyl sulfide

Source: Harald Huber and Karl O. Stetter. 2002. *The Prokaryotes*. Springer.

Methanomicrobium and *Methanococcus* have protein-derived cell walls.

Filamentous methanogens form chains of large cells similar to those of filamentous cyanobacteria. Filaments of *Methanosaeta* perform a key function in the treatment of sewage waste (**Fig. 19.26**). In waste treatment, the raw sewage first undergoes aerobic respiration by bacteria, in which the organic materials are converted into small molecules such as CO_2 and acetate, and then the remainder is digested anaerobically. The bacteria performing anaerobic decomposition become trapped in filaments of *Methanosaeta* species, which convert bacterial fermentation products such as acetate into methane and CO_2. The methanogenic filaments serve a key function by trapping bacteria into granules that settle out from the liquid.

Methanogenesis in soil and landfills. A major methanogenic environment is the anaerobic soil of wetlands. The wetlands that generate the most methane are typically disturbed or artificial wetlands, particularly rice paddies, which contain high levels of added fertilizer that bacteria convert to the substrates used by methanogens. From the standpoint of the organism, methane is an incidental by-product, but this product has great significance for our biosphere as a greenhouse gas (discussed in Chapter 22).

Chinese farmers have found that one way to decrease methane production is to drain the soils used for rice production. Changsheng Li and colleagues at the University of New Hampshire showed in experimental plots that drained soil produces less methane. The drained soil receives more oxygen, which blocks methanogenesis. Draining soil also stimulates rice root development and accelerates decomposition of organic matter in the soil to release nitrogen. Sufficiently dry rice paddy soil can actually become a sink for atmospheric methane, thanks to methane-oxidizing bacteria.

Another major source of methanogens is landfills. Landfills such as those outside New York City are among the largest human-made structures on Earth. They are rich in organic wastes, which bacteria ferment to CO_2, H_2, and short-chain organic molecules that methanogens convert to methane. As a result, large amounts of gas can build up and spontaneously combust, causing explosions. To avoid explosions, the methane needs to be piped out. In some cases, the gas can be collected and used to generate electricity.

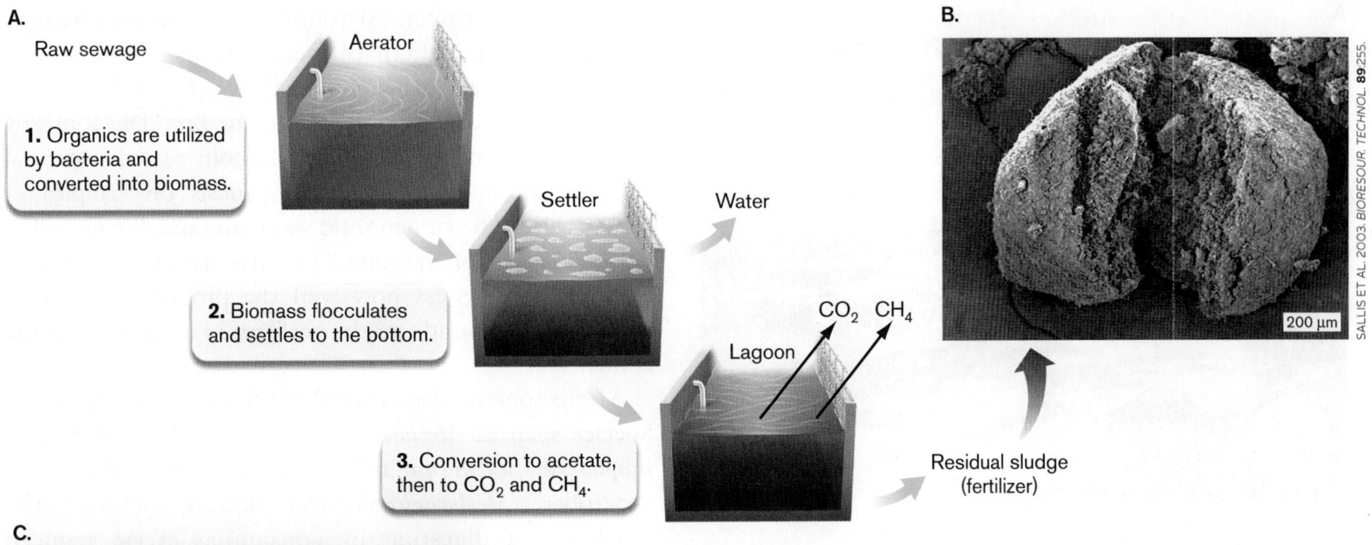

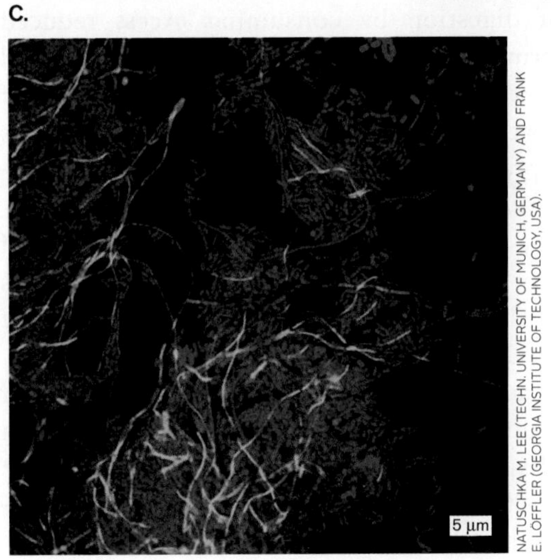

FIGURE 19.26 ▪ Filamentous methanogens bind bacterial communities in waste treatment. A. Raw sewage is aerated and decomposed by bacteria, followed by anaerobic incubation. Under anaerobiosis, the bacterial waste products are converted to methane and CO_2. The bacteria are packed together by filamentous methanogens to form sludge. **B.** Sludge (wastewater sediment) forms granules packed with bacteria and methanogens (discussed in Chapter 22). **C.** Filamentous methanogens entangle bacteria, forming "flocs" that settle in the waste treatment tank. Methanogens fluoresce green with a DNA probe, while other wastewater microbes fluoresce blue (DAPI stain).

In Antarctica, scientists such as Ricardo Cavicchioli from the University of New South Wales, Australia, have discovered psychrophilic methanogens (**Fig. 19.27**). The near-freezing water of Ace Lake is saturated with methane from methanogens present in high numbers. Cavicchioli showed that the Ace Lake methanogens have membrane lipids with many double bonds; the unsaturated form allows more fluidity at low temperature (see Chapter 3).

Rising temperatures accelerate methanogenesis, and the methane bubbles up so fast that it melts holes in the ice sheets covering Arctic lakes. In 2012, Andrew McDougall and colleagues at the University of Victoria, British Columbia, modeled the permafrost carbon release and its feedback effect on global warming. They project that permafrost could release more than a quarter of its carbon stores by the year 2100, and that this amount could add another 1.5°C to the global temperature by the year 2300.

Methane hydrates and Syntrophoarchaeota. Psychrophilic marine methanogens grow at or beneath the seafloor. These seabed methanogens generate large volumes of methane that seep up slowly from the sediment. Under the great pressure of the deep ocean, the methane becomes trapped as **methane gas hydrates**, which are crystalline materials in which methane molecules are surrounded by a cage of water molecules. The methane hydrates accumulate in vast quantities. Methane hydrates are of interest as a potential source of natural gas. But if a large part of this methane were to be released, it could greatly accelerate global warming (discussed in Chapter 22).

Fortunately, much of the methane produced by seafloor methanogens is oxidized to CO_2 by anaerobic methane oxidizers, classified as **Anaerobic Methane-Oxidizing Euryarchaeota** (**ANME**), also known as Syntrophoarchaeota. The ANME euryarchaeotes oxidize methane in syntrophy with sulfur-reducing bacteria such as *Desulfurococcus* (discussed in Chapter 18). The bacteria reduce sulfate (abundant in seawater) to sulfide. Coupled together, the reactions of methane oxidation and sulfate reduction have the negative value of ΔG needed to drive metabolism for both kinds of organisms.

A.

RICARDO CAVICCHIOLI. 2006. NAT. REV. MICROBIOL. 4:331-343.

B.

COURTESY RICK CAVICCHIOLI

FIGURE 19.27 ■ Antarctic lakes support psychrophilic methanogens. **A.** Ace Lake, which contains anoxic sulfidic layers, opens to the sea. **B.** Ric Cavicchioli (left) and Torsten Thomas process microbial samples collected at Ace Lake.

Digestive methanogenic symbionts. Numerous methanogens also grow within the digestive fermentation chambers of animals such as termites and cattle (**Fig. 19.28**; discussed in Chapter 21). Termites have to aerate their mounds continually in order to remove methane; when rainfall temporarily clogs the mound, the mound can be ignited by lightning and explode. Cattle support methanogenesis within their rumen and reticulum (a common veterinary trick is to insert a tube into the rumen and ignite the escaping methane gas). Bovine methanogenesis diverts carbon from meat production, and it makes a significant contribution to global methane.

Research on rumen microbiology focuses on attempts to suppress the growth of methanogens. For example, Stephen Ragsdale (now at the University of Michigan, Ann Arbor) and his graduate student Bree DeMontigny at the University of Nebraska–Lincoln tested potential chemical inhibitors of methanogenesis on samples of rumen fluid. The rumen fluid was maintained in vials that serve as artificial rumens. The artificial rumens enable researchers to assess how well the proposed methane-blocking compounds might perform in an actual bovine rumen.

Methanogens also contribute to human digestion. Species such as *Methanobrevibacter smithii* and *Methanosphaera stadtmanae* may constitute about 10% of gut anaerobes. *Methanobrevibacter smithii* increases the efficiency of digestion by consuming excess reduced products (formate and H_2) from bacterial mixed-acid fermentation (discussed in Chapter 13). High levels of H_2 inhibit bacterial NADH dehydrogenases and thus decrease the proton potential for ATP production. *Methanosphaera stadtmanae* consumes methanol, a potentially toxic by-product of the bacterial degradation of pectin, a major fruit polysaccharide. Thus, methanogens may enhance the growth of human enteric bacteria (see **Special Topic 19.1**).

If methanogens influence the fermentation efficiency of gut bacteria, do they affect the caloric content we obtain from food? Experiments with gnotobiotic mice test the interactions between methanogens and *Bacteroides thetaiotaomicron*, an anaerobic bacterium that ferments complex polysaccharides from plant foods. The mice were colonized in the presence or absence of *Methanobrevibacter smithii*, which consumes H_2 through methanogenesis. Mice colonized with both *B. thetaiotaomicron* and *M. smithii* were found to make more fat after consuming the same quantity of food than did mice colonized solely

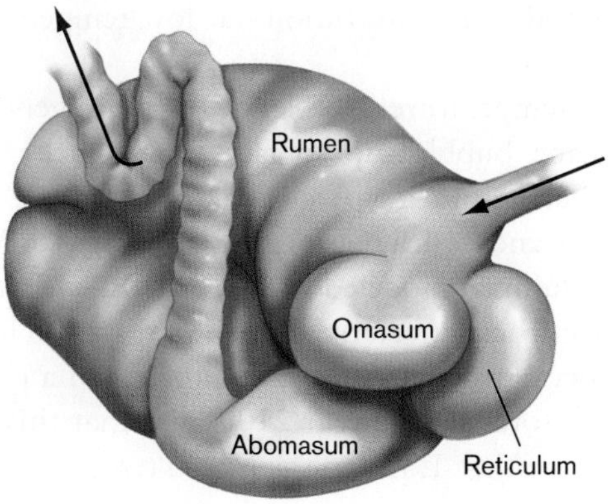

FIGURE 19.28 ■ Methane production in the bovine rumen. Methanogens grow with the community of fermentative bacteria in the rumen and reticulum of ruminating animals.

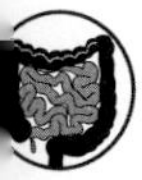

with *B. thetaiotaomicron*. Following up on this result, a study of humans found a correlation between methane breath release and obesity. Thus, methanogenesis may somehow interact with human gut bacterial digestion to increase the efficiency of fat storage. How the composition of intestinal microbiota may influence obesity is discussed in Section 23.2.

Biochemistry of Methanogenesis

Methanogenesis is extremely important for our environment because the process releases a potent greenhouse gas. Thus, much effort goes into better understanding how methanogenesis works. Knowledge of methanogenic biochemistry may help us develop controls, such as methanogen-specific antibiotics to minimize methane output from cattle.

Cofactors for methanogenesis. The process of methanogenesis uses a series of specific cofactors to carry each carbon from CO_2 (or other substrates) as it becomes progressively reduced by hydrogen. The hydrogen atoms also require redox carriers. Most of the cofactors are unique to methanogens, although the general structural types resemble those of redox cofactors that we saw in Chapter 14 (**Fig. 19.29**). For example, cofactor F_{420} is a heteroaromatic molecule (containing nitrogens in the aromatic rings) that undergoes a redox transition by acquiring or releasing two hydrogens (**Fig. 19.29C**)—a transition similar to that of the nicotinamide ring of NADH.

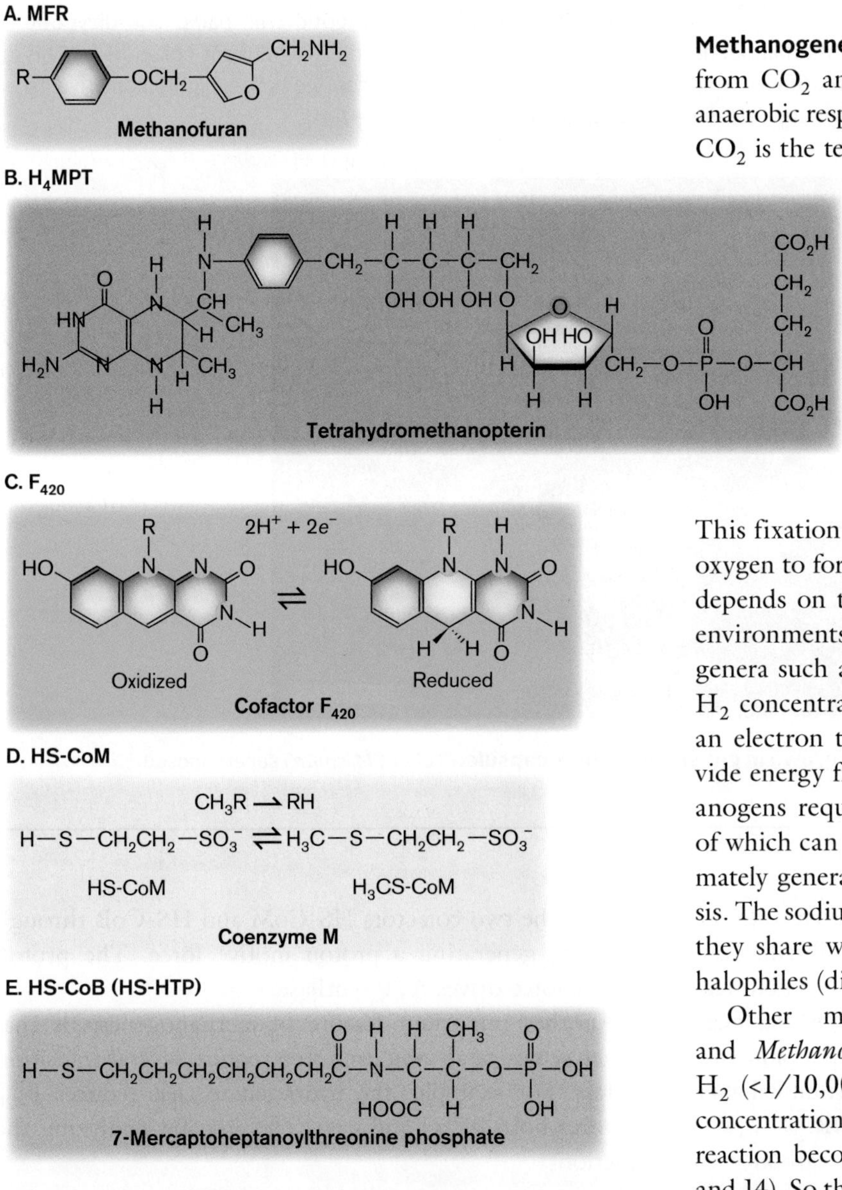

FIGURE 19.29 ■ Cofactors for methanogenesis. Cofactors specific for methanogenesis transfer the hydrogens and the increasingly reduced carbon to each enzyme in the pathway.

Methanogenesis from CO_2. The formation of methane from CO_2 and H_2 (**Fig. 19.30**) is technically a form of anaerobic respiration in which H_2 is the electron donor and CO_2 is the terminal electron acceptor (discussed in Chapter 14). The process fixes CO_2 onto the cofactor methanofuran (MFR) and then passes the carbon stepwise from one cofactor to the next, each time losing an oxygen to form water or gaining a hydrogen carried by another cofactor.

The first step in the conversion of CO_2 and H_2 to methane is fixing the carbon onto methanofuran (for its chemical formula, see **Fig. 19.29A**). This fixation step requires placement of protons onto the oxygen to form water. The mechanism of methanogenesis depends on the available concentration of H_2. High-H_2 environments, such as oil wells and sewage sludge, favor genera such as *Methanosarcina* (**Fig. 19.30A**). The high H_2 concentration allows electron donation to CO_2 from an electron transport system (ETS). The ETS may provide energy from a sodium potential (ΔNa^+). Most methanogens require Na^+ for growth, unlike bacteria, many of which can grow without sodium. Methanogenesis ultimately generates a Na^+ potential that drives ATP synthesis. The sodium requirement of methanogens is something they share with another division of Euryarchaeota, the halophiles (discussed in Section 19.5).

Other methanogens, such as *Methanobacterium* and *Methanococcus*, use much lower concentrations of H_2 (<1/10,000 atm) common in soil or water. At lower concentrations of substrate, the free energy change (ΔG) of reaction becomes less favorable (discussed in Chapters 13 and 14). So the methanogen needs to couple CO_2 reduction to an energy-spending reaction, the reduction of CoM-S-S-CoB (**Fig. 19.30B**). This coupling of an energy-spending

SPECIAL TOPIC 19.1 Methanogens for Dinner

If our gut is full of methanogens, what are they doing there? Do they just pass through, picking up hydrogen and CO_2 and producing gas that we pass? Or do they interact specifically with their host, and with their neighboring microbes in the gut microbiome?

Jeff Gordon and colleagues at Washington University in St. Louis set out to address these questions. They selected a well-studied gut methanogen, *Methanobrevibacter smithii* (**Fig. 1**). *M. smithii* can be cultured inside the cecum (entrance to the large intestine) of a germ-free mouse. When cultured in the mouse, the archaeal cell forms a thick capsule of polysaccharide. By contrast, *M. smithii* cultured in a batch fermentor forms cells whose capsule is very thin. This difference suggests that capsule formation is a response to the animal host.

Gordon asked, Does the existence of methanogens depend on the genetic identity of the specific host? This question is of interest because only some humans release methane, while others do not; thus, we infer that only some of us are colonized by methanogens. We do not know whether this difference arises from host genetics or from environmental factors. So, Gordon conducted a study comparing the methanogen prevalence between pairs of monozygotic twins (identical twins, from the same fertilized egg) and dizygotic twins (fraternal twins, from two different zygotes; **Fig. 2**). Methanogen prevalence was measured by quantitative PCR (qPCR) amplification of the signature gene *mcrA*. The gene encodes methyl-coenzyme M reductase, the enzyme that converts methyl-coenzyme M (CH_3–S-CoM) and thiol-coenzyme B (HS-CoB) to methane and the CoM-CoB disulfide (CoM-S-S-CoB, shown in **Fig. 19.30B**). A remarkably high correlation was found between the methanogen levels of the monozygotic twin pairs. The dizygotic twin pairs, however, showed no significant correlation. This finding suggests that methanogens are highly sensitive to the molecular biology of the host.

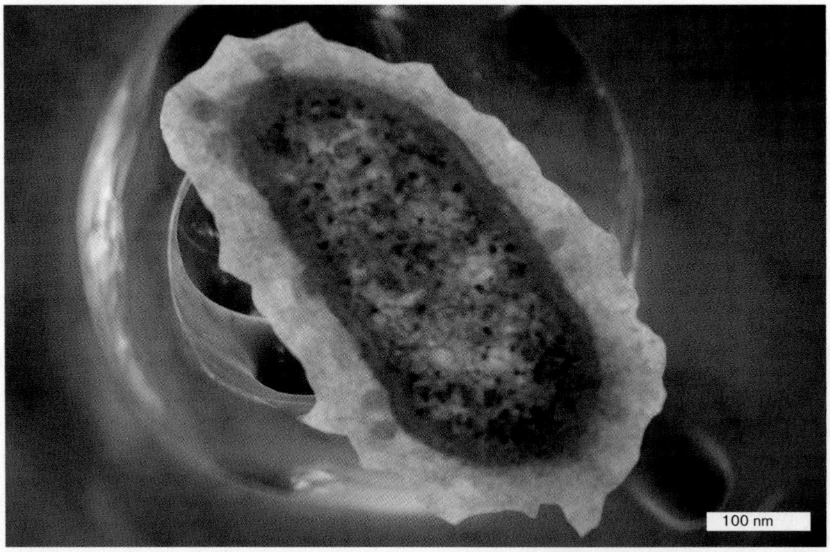

SEBASTIAN KAULITZKI/123RF; MICROGRAPH FROM BUCK SAMUEL ET AL. 2007. *PNAS* **104**:10643, FIG. 1A.

FIGURE 1 ■ *Methanobrevibacter smithii* grown in gut shows a thick capsule. TEM of *M. smithii* superimposed upon the image of a colon.

electron transfer (CO_2 reduction via ferredoxin) to an energy-yielding electron transfer (CoM-S-S-CoB reduction) is called **electron bifurcation**. The cost of electron bifurcation (compared to the ETS used with high H_2) is that fewer ATPs can be made, but it allows many more types of methanogens to grow in a much wider range of natural habitats. Electron bifurcation is shown in **Figure 19.30B**.

Later steps in methanogenesis reduce the carbon with H_2 to methane. In the final step, CoM-S-S-CoB serves as an anaerobic terminal electron acceptor for an ETS accepting electrons from H_2. Overall, H_2 reduces CoM-S-S-CoB back to the two cofactors HS-CoM and HS-CoB through an ETS, generating a proton motive force. The proton motive force drives ATP synthase.

Another important feature of methanogenesis is that the enzymes catalyzing each step require several transition metals. For example, the hydrogenase that reduces F_{420} requires both nickel and iron. The enzyme catalyzing the reaction

$$CO_2 + \text{methanofuran} + 3H \longrightarrow \text{CHO-methanofuran} + H_2O$$

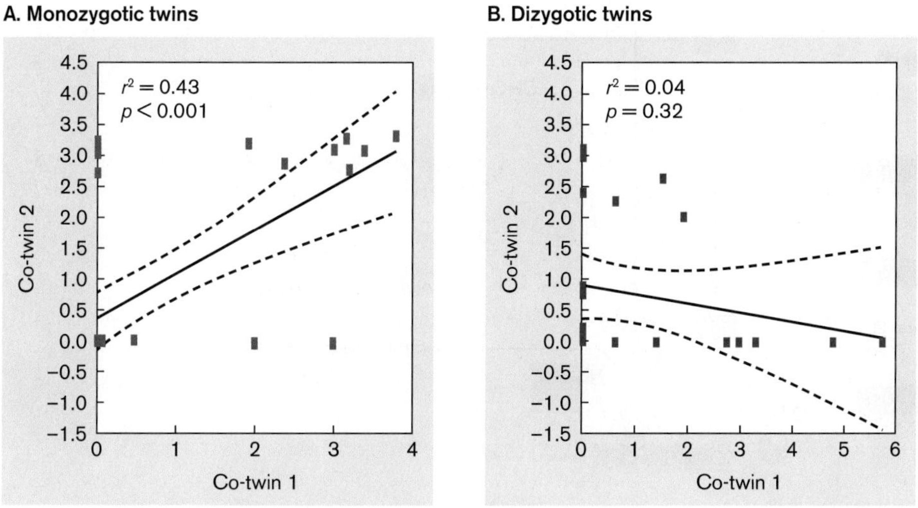

FIGURE 2 ▪ Correlation of intestinal methanogen levels between twins. The number of methanogens in fecal samples was measured by qPCR amplification of *mcrA*. **A.** Monozygotic twins show similar methanogen levels. **B.** Dizygotic twins show no correlation. *Source:* Modified from Elizabeth Hansen et al. 2011. *PNAS* **108**:4599.

Gordon then asked whether the presence of methanogens depends on specific bacterial members of the gut microbiome. The researchers sequenced metagenomes (sets of partial genomes from a microbial community; presented in Chapter 21). From the metagenomes, 22 partial bacterial genomes were assembled that showed significant correlation with the presence of *M. smithii*. Remarkably, 20 of the 22 partial genomes were assigned to a particular taxon, the bacterial order Clostridiales. This is the first such specific methanogen-bacterium correlation to be shown across multiple human host metagenomes. The basis of the connection is not known, but Gordon speculates that specific methanogens couple with specific clostridia-like bacteria to provide the hydrogen gas needed for methanogenesis. In effect, methanogens may bring preferred "dinner guests" to the intestinal buffet.

If methanogens are intimately involved with our digestion, could they be manipulated to improve gastrointestinal health or comfort? A study by Heeball Kim and colleagues at Seoul National University, South Korea, found that administration of probiotic bacteria to humans caused a decrease in abundance of *Methanobrevibacter* species, whose methane production generates intestinal gas. Participants who experienced a decrease in *Methanobrevibacter* reported a decrease in flatulence. Thus, adjustment of our gut methanogen community could improve one measure of gastrointestinal health.

RESEARCH QUESTION

If methanogens are associated with flatulence, how might the mechanism of this association be tested, and what therapies might be proposed?

Hansen, Elizabeth E., Catherine A. Lozupone, Federico E. Rey, Meng Wu, Janaki L. Guruge, et al. 2011. Pan-genome of the dominant human gut-associated archaeon, *Methanobrevibacter smithii*, studied in twins. *Proceedings of the National Academy of Sciences USA* **108**:4599–4606.

Seo, Minseok, Jaeyoung Heo, Joon Yoon, Se-Young Kim, Yoon-Mo Kang, et al. 2017. *Methanobrevibacter* attenuation via probiotic intervention reduces flatulence in adult human: A non-randomized paired-design clinical trial of efficacy. *PLoS One* **12**:e0184547.

requires either molybdenum or tungsten, depending on the species; some species have two alternative enzymes, depending on which metal is available. Another metal, cobalt, is required for a B_{12}-related cofactor that participates in methane production from methanol and methylamines.

Thought Question

19.6 What do the multiple metal requirements suggest about how and where the early methanogens evolved?

Methanogenesis from acetate. Methane production from acetate is particularly important for wastewater treatment, where most of the substrate consists of short-chain bacterial fermentation products. The acetate methanogenesis pathway is not fully understood, but the initial incorporation of H_2 probably requires a coupled gradient of sodium ion, as it does for CO_2 methanogenesis. The two carbons from acetate enter different pathways, which eventually converge:

$$CH_3\text{-S-CoM} + \text{HS-CoB} \longrightarrow CH_4 + \text{CoM-S-S-CoB}$$

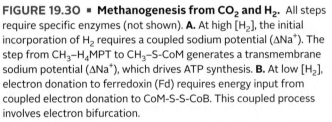

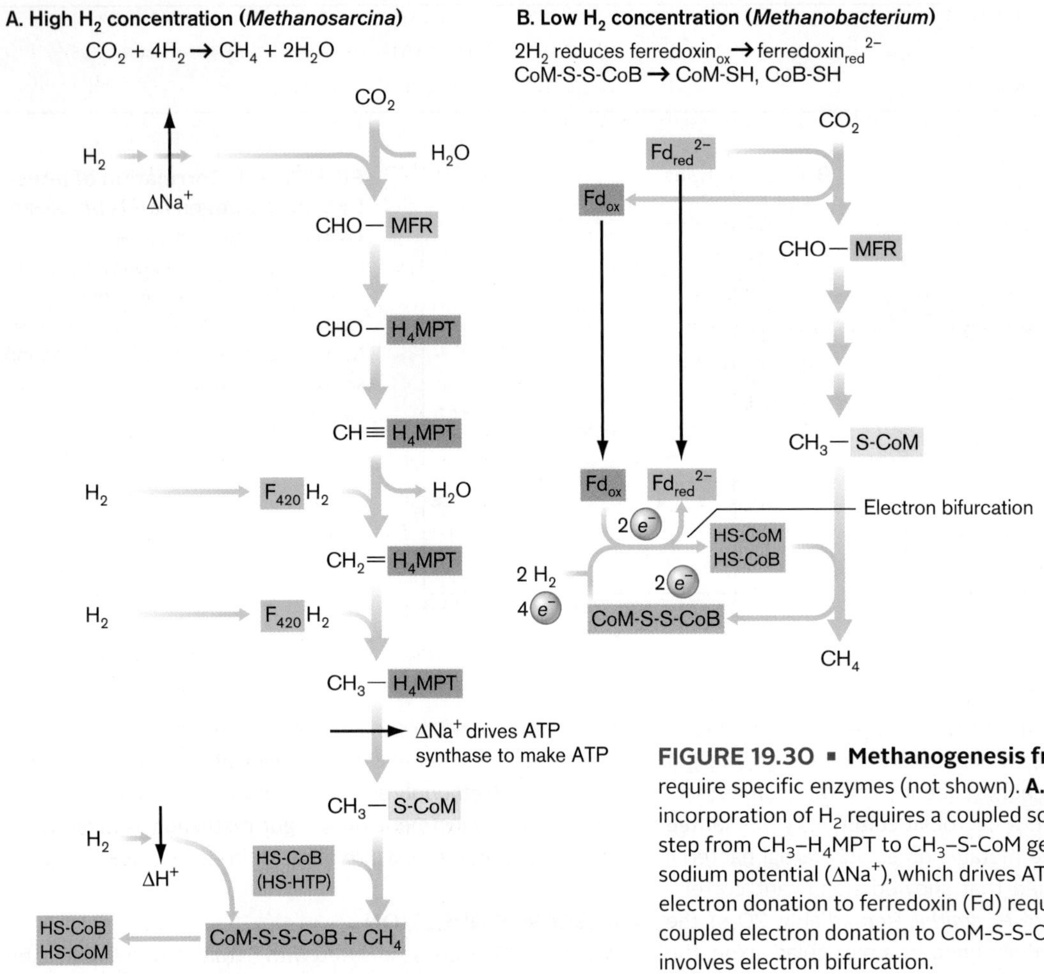

FIGURE 19.30 ■ Methanogenesis from CO_2 and H_2. All steps require specific enzymes (not shown). **A.** At high [H_2], the initial incorporation of H_2 requires a coupled sodium potential (ΔNa^+). The step from CH_3–H_4MPT to CH_3–S-CoM generates a transmembrane sodium potential (ΔNa^+), which drives ATP synthesis. **B.** At low [H_2], electron donation to ferredoxin (Fd) requires energy input from coupled electron donation to CoM-S-S-CoB. This coupled process involves electron bifurcation.

To Summarize

- **Methanogens gain energy through redox reactions that generate methane** by using H_2 to reduce CO_2, formate, acetate, and other small molecules. Methanogens require association with bacteria whose fermentation generates the needed substrates.
- **Methanogens have rigid cell walls of diverse composition in different species**, including pseudopeptidoglycan, protein, and sulfated polysaccharides.
- **Species of methanogens show a wide range of different shapes**, including rods (single or filamentous), cocci (single or clumped), and spirals.
- **Methanogens inhabit anaerobic environments** such as wetland soil, marine benthic sediment, and animal digestive organs.
- ***Methanobrevibacter smithii* and *Methanosphaera stadtmanae*** are the main methanogens of the human colon. They show possible involvement with caloric efficiency of digestion.
- **Biochemical pathways of methanogenesis** involve transfer of the increasingly reduced carbon to cofactors that are unique to methanogens.

19.5 Haloarchaea and Other Euryarchaeotes: Underground and Under Ocean

In the saturated brine of Utah's Great Salt Lake or Israel's Dead Sea, few living things can grow except for halophilic euryarchaeotes called **Haloarchaea**. Elsewhere, still other euryarchaeotes include sulfur hyperthermophiles. Some are also acidophiles—hyperthermoacidophiles that grow under the most extreme acid as well as high temperatures.

Haloarchaea: Life in Salt

The halophilic archaea require at least 1.5-M NaCl or equivalent ionic strength, and most grow optimally at near saturation (about 4.3 M, seven times the concentration of seawater). Salted foods such as meat and fish and salt-cured hides can be spoiled by haloarchaea.

Most haloarchaea belong to a monophyletic clade that was formerly named Halobacteria, before the archaea were classified as distinct from bacteria. Haloarchaea show relatedness to methanogens (see **Fig. 19.3**) but they do not conduct methanogenesis. Most haloarchaea grow as photoheterotrophs, using light energy to drive a retinal-based ion pump to establish a proton potential. Their photopigments color salterns, brine pools that are evaporated to mine salt (**Fig. 19.31A**). The red pigment bacterioruberin protects cells from damage by light (**Fig. 19.31B**).

Note: The terms "Halobacteria" and "Haloarchaea" are synonymous; all members of this clade are indeed archaea. Outside the Haloarchaea, true bacteria that grow in elevated NaCl are called halophilic bacteria or bacterial halophiles.

Haloarchaeal form and physiology. Halophilic microbes need a way to maintain turgor pressure—that is, to avoid cell shrinkage as cytoplasmic water runs down the osmotic gradient toward external high salt. Most bacterial halophiles compensate for high external salt by uptake or synthesis of other kinds of osmolytes, such as small organic molecules. Haloarchaea, however, adapt to high external NaCl by maintaining a high intracellular concentration of potassium chloride (about 4-M KCl). Potassium ion

A.

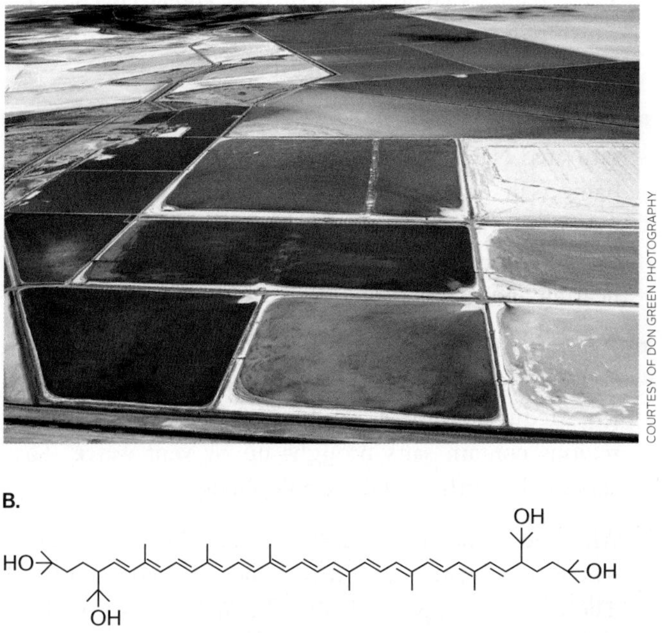

FIGURE 19.31 ▪ A saltern for salt production. A. Aerial view of a solar saltern facility in Grantsville, Utah. **B.** The red pigment bacterioruberin protects Haloarchaea from damage by light.

concentrations are moderately high in most microbial cells (commonly 200 mM), but the exceptionally high KCl concentration within haloarchaea requires major physiological adaptations:

- **High GC content of DNA.** High salt concentration decreases the fidelity in base pairing of DNA. Because the triple hydrogen bonds of GC pairs hold more strongly than those of AT pairs, the exceptionally high GC content of haloarchaea (above 60% for most species) may protect their DNA from denaturation in high salt.
- **Acidic proteins.** Most haloarchaeal proteins are highly acidic, with an exceptional density of negative charges at their surface. The high negative charge maintains a layer of water in the form of hydrated potassium ions (K^+). This unusual hydration layer keeps the acidic proteins soluble in the high salt within the cytoplasm.

The high salt content of haloarchaea provides a convenient way to lyse cell contents for analysis: Simply transfer cells into low-salt buffer, and they fall apart owing to osmotic shock. This technique provides a quick way for beginning students to isolate DNA. This and other techniques have made the organism *Halobacterium* sp. NRC-1 a model system for molecular biology education (**eTopic 19.1**).

The properties of diverse halophiles are presented in **Table 19.5**. The haloarchaea are generally uniform with respect to temperature range (mesophilic). With respect to pH, some halophiles grow at neutral pH, whereas others are alkaliphiles, growing in soda lakes above pH 9, such as Lonar Lake, Maharashtra State, India. The cell envelopes of most haloarchaea contain rigid cell walls of glycoprotein, as do some of the methanogens. Most species possess archaella for phototaxis and **gas vesicles** for maintaining their buoyancy in the upper layer of the water column. Unlike phospholipid vesicles, gas vesicles are made entirely of protein, and they are filled with air.

Haloarchaeal gas vesicles earned an unusual grant from the Bill & Melinda Gates Foundation. Shiladitya DasSarma at the University of Maryland School of Medicine proposes to use recombinant gas vesicles of *Halobacterium* NRC-1 as a delivery vehicle for recombinant vaccines against typhoid bacteria. The recombinant haloarchaeal vaccine can be embedded in salt crystals (**Fig. 19.32**)—a cheap and convenient delivery method for developing countries where typhoid is widespread.

With respect to cell shape, Haloarchaea display considerable diversity. Some species form symmetrical rods, as in *Halobacterium* NRC-1. Other species form pleomorphic cells, flattened like pancakes, and still others form regular cocci (*Haloferax mediterranei*; **Fig. 19.33A**). A few species grow as flattened squares—the only microbial cells

TABLE 19.5 Halophilic archaea

Representative species	NaCl range (M)	Growth temperature (°C)	Growth pH	Cell shape	Location of isolate
Haloarcula quadrata	2.7–4.3	53°	pH 7	Square flat; pleomorphic	Sabkha, Sinai, Egypt
Haloarcula valismortis	3.5–4.3	40°	pH 7.5	Pleomorphic rods	Salt pools, Death Valley, CA
Halobacterium salinarum	3.0–5.2	35–50°	pH 7	Rods	Salted cowhide
Halococcus morrhuae	2.5–5.2	30–45°	pH 7	Cocci	Dead Sea, Israel
Haloferax volcanii	1.5–5.2	40°	pH 7	Pleomorphic; dish-shaped	Dead Sea, Israel
Halorubrum lacusprofundi	1.5–5.2	1–44°	pH 7	Long rods (12 μm)	Deep Lake, Antarctica
Natronococcus occultus	1.4–5.2	30–45°	pH 9.5	Cocci	Lake Magadi, Kenya
Natronomonas pharaonis	2.0–5.2	45°	pH 9–10	Rods	Wadi El Natrun, Egypt

Sources: Aharon Oren. The order Halobacteriales. In Huber and Stetter. 2002. *The Prokaryotes*, Springer. The *Halorubrum lacusprofundi* growth temperature range comes from Ricardo Cavicchioli. 2006. *Nat. Rev. Microbiol.* **4**:331.

known to be square (*Haloquadratum walsbyi*; **Fig. 19.33B** and **C**). The mechanism that maintains the various shapes is unknown. In oligotrophic (low-nutrient) environments, it is likely that the greater surface-to-volume ratio gives flattened cells a competitive edge in obtaining nutrients.

Hypersaline habitats. Hypersaline (high-salt) habitats differ with respect to pH, temperature, and the presence of other minerals, such as magnesium ion. Different kinds of hypersaline habitats support different species of haloarchaea (see **Table 19.5**), as well as halophilic bacteria. Major types of habitats include:

- **Thalassic lakes.** Thalassic lakes (from the Greek *thalassa*, meaning "ocean"), such as the Great Salt Lake in Utah, contain saturated salts with essentially the same ionic proportions as the ocean: Na^+ and Cl^- (NaCl) predominate, followed by Mg^{2+}, K^+, and SO_4^{2-}. Thalassic lakes (also called brine lakes) support genera such as *Halobacterium*. Antarctic brine lakes support growth of cold-adapted halophiles, such as *Halorubrum lacusprofundi*, at temperatures as low as –18°C.

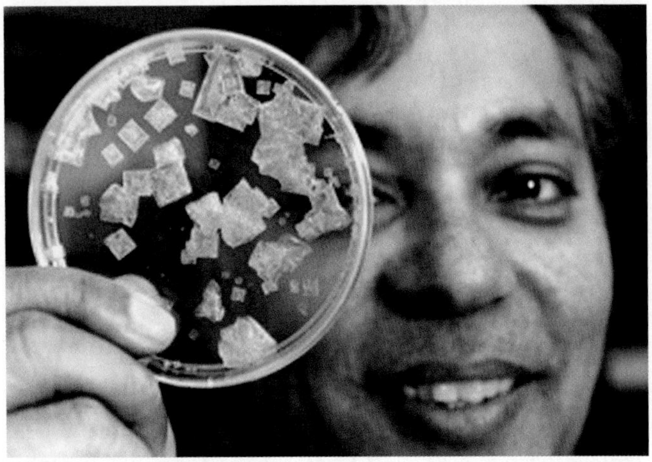

FIGURE 19.32 ▪ Pink-pigmented salt crystals deliver haloarchaeal vaccine. *Halobacterium* NRC-1 embedded in salt crystals contains gas vesicles formed of a recombinant protein vaccine. Crystals are held here by Shiladitya DasSarma.

- **Athalassic lakes.** Athalassic lakes, such as the Dead Sea in Israel, contain higher proportions of magnesium ions. These habitats favor genera such as *Haloarcula* species, which require 100 mM Mg^{2+} for growth and grow best at magnesium concentrations above 1 M.
- **Solar salterns.** These are artificial pools of brine, or saturated NaCl, that evaporate in sunlight, precipitating halite (salt crystals) for commercial production. Commercial evaporation pools for salt actually benefit from the red microbes, whose light absorption accelerates heating and evaporation.
- **Brine pools beneath the ocean.** Undersea brine pools collect near geothermal vents, as in the Gulf of Mexico, Mediterranean Sea, and Red Sea. These hypersaline regions contain salts brought up by vent water. They support hyperthermophilic halophiles.
- **Alkaline soda lakes**, such as Lake Magadi in Kenya, where carbonate salts drive the pH above pH 9. These lakes support alkaliphilic haloarchaea such as *Halobacterium*.
- **Underground salt deposits** contain micropockets of salt-saturated water where halophiles may survive for thousands of years.

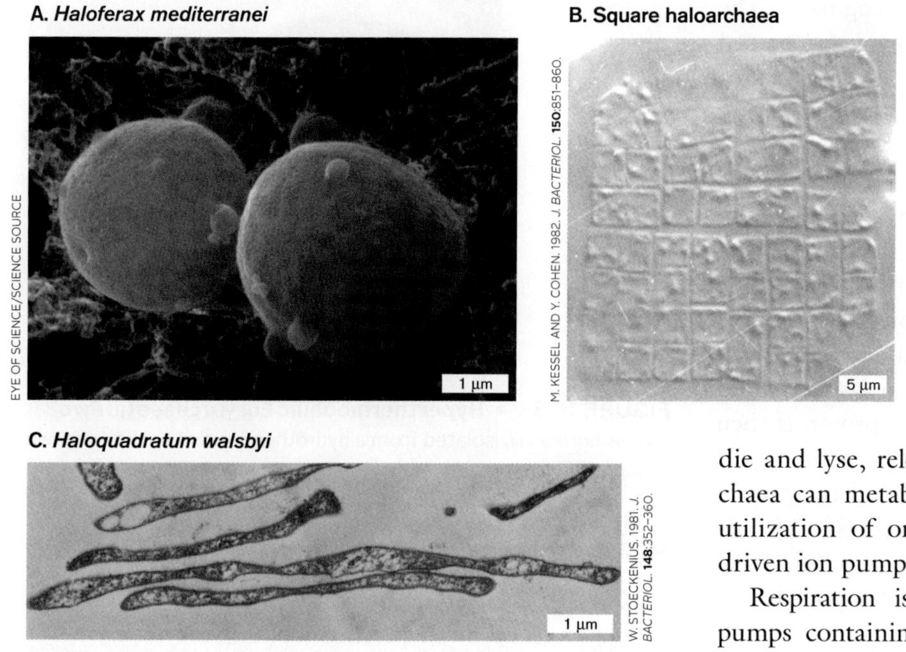

FIGURE 19.33 ■ **Diverse haloarchaea.** **A.** *Haloferax mediterranei* (SEM). **B.** Square haloarchaea (*Haloquadratum walsbyi*) during their sixth round of cell division, which alternates in vertical and horizontal directions (photomicrograph, Nomarski optics). **C.** Cross-sectioned cells of square haloarchaea show their thinness (TEM).

Retinal-based photoheterotrophy. Most haloarchaea are photoheterotrophs, in which light-directed energy acquisition supplements respiration on complex carbon sources.

Haloarchaea respire with oxygen, or anaerobically with nitrate. A typical habitat for haloarchaea starts out as a pool with moderate salt content, containing a range of bacterial and archaeal species with varied tolerance to salt. As the pool evaporates, the salt concentrates, and the various bacteria die and lyse, releasing cell components that the haloarchaea can metabolize. The halophiles supplement their utilization of organic substrate energy by using light-driven ion pumps.

Respiration is supplemented by light-driven proton pumps containing retinal. Light is captured by retinal-containing proton pumps called **bacteriorhodopsin** and the chloride pump **halorhodopsin** (**Fig. 19.34** ▶). (The role of light-driven ion pumps in phototrophy is discussed in Chapter 14.) The proton pump bacteriorhodopsin was named to indicate that it is a prokaryotic version of the better-known eukaryotic retinal rhodopsins; the name was

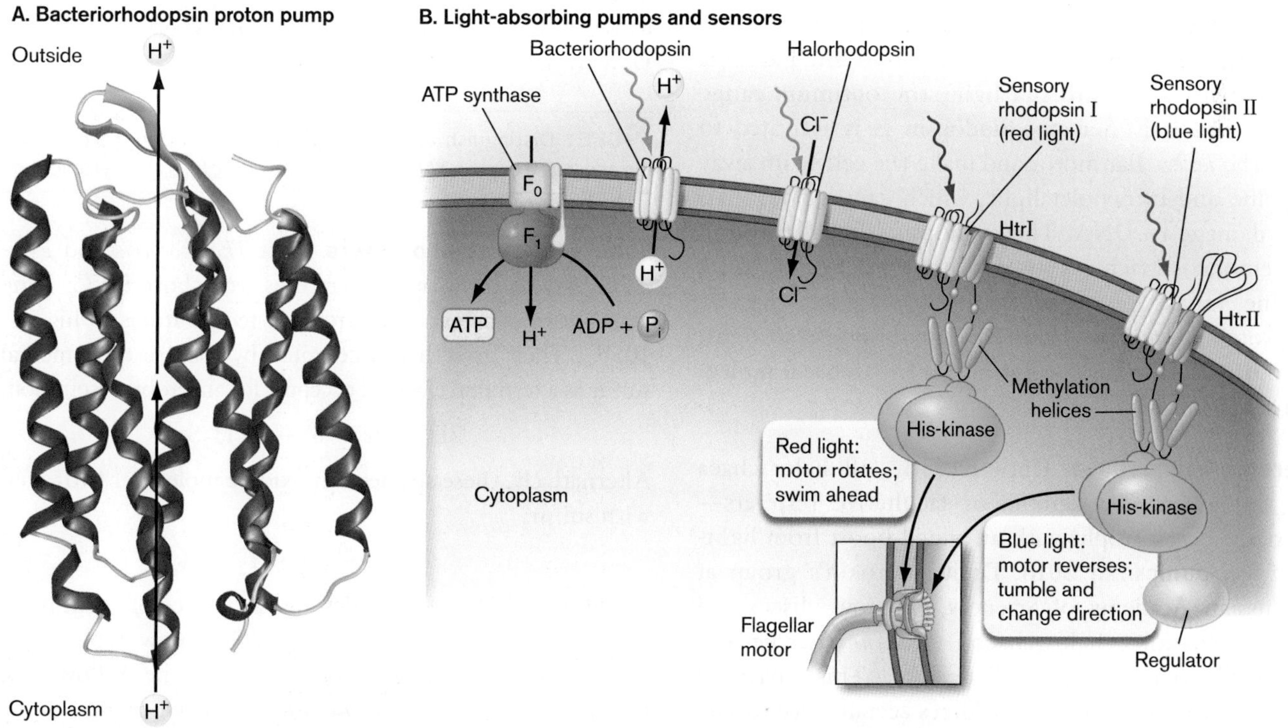

FIGURE 19.34 ■ **Light-driven ion pumps and sensors.** **A.** Bacteriorhodopsin absorbs light and pumps H^+ out of the cell (PDB code: 1QKO), whereas light-activated halorhodopsin pumps chloride into the cell (not shown). **B.** Rhodopsin family molecules in the membrane of *Halobacterium*: bacteriorhodopsin, light-driven proton extrusion; halorhodopsin, light-driven chloride intake; sensory rhodopsins I and II, with their signal transduction proteins Htr I and Htr II. The sensory transduction proteins phosphorylate and dephosphorylate a protein that regulates the direction of rotation of the flagellar motor. ▶

chosen before the domain Archaea was recognized. The chloride pump halorhodopsin was discovered and named in reference to the chloride ion (a halide). The two proteins are homologs with similar structure and mechanism. In the cell membrane, they form complexes that aggregate in patches called purple membrane.

Each bacteriorhodopsin proton pump contains seven alpha helices that traverse the membrane (**Fig. 19.34A**), surrounding a buried molecule of retinal, the same light-absorbing molecule found in photoreceptors of the human retina. Light absorption triggers a conformational change that enables a proton from the cytoplasmic face to be picked up by an aspartate residue. The proton is then transferred stepwise through several other amino acid residues in the protein, leading to release of the proton outside the cell. The net result of proton transfer by bacteriorhodopsin is generation of a proton motive force that can run a proton-driven ATP synthase, storing energy as ATP (**Fig. 19.34B**).

Halorhodopsin has a similar structure, in which chloride (instead of H^+) is pumped into the cell instead of outward. Because chloride is negatively charged, this chloride transport contributes to the proton motive force. Light-driven chloride pumps are unique to haloarchaea.

Haloarchaea also possess homologs of bacteriorhodopsin that serve as sensory devices: the sensory rhodopsins I and II. These proteins, too, each have seven alpha helices containing retinal. When the sensory rhodopsins absorb light, they signal the cell to swim using its archaella. The activated sensory rhodopsin I directs the cell to swim toward red light, the optimum range for photosynthesis. Sensory rhodopsin II is activated to reverse the archaellar motor and make the cell swim away from blue and ultraviolet light, which causes photooxidative damage to DNA. The rhodopsins signal through the chemotaxis machinery to the archaellar motor, using histidine kinase enzymes that phosphorylate regulatory proteins (see **Fig. 19.34B**). This response to light, or phototaxis, is an important model for archaeal molecular regulation.

Anaerobic haloarchaea. Until recently, all haloarchaea were thought to be aerobic or facultative respirers—essentially, heterotrophs with an added boost from light-driven ion pumps. In 2016, Dmitri Sorokin's group at the Russian Academy of Sciences, Moscow, discovered a strictly anaerobic haloarchaeon (*Halanaeroarchaeum*) from a Siberian chloride-sulfate lake. This anaerobe requires 3- to 5-M NaCl and oxidizes acetate with sulfate (a form of sulfate respiration, discussed in Chapter 14). A related genus from anaerobic sediment of a brine lake was shown to oxidize H_2 or formate with sulfur. Thus, the known metabolic capacity of Haloarchaea has been expanded to include anaerobes with sulfur reduction.

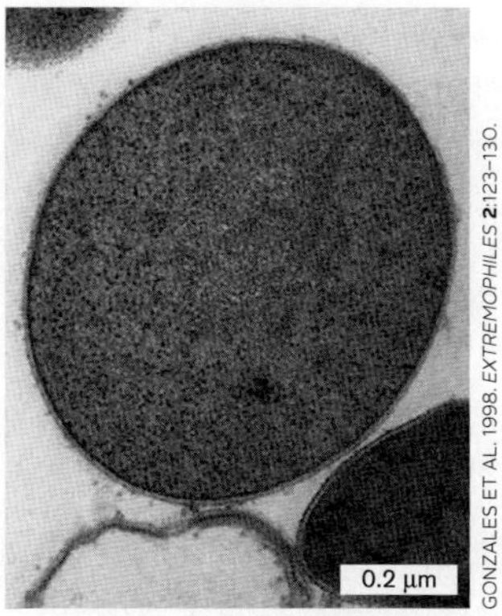

FIGURE 19.35 ■ Hyperthermophilic Euryarchaeota *Pyrococcus horikoshii*, isolated from a hydrothermal vent at the Okinawa Trough (TEM).

Hot and Hotter: Thermococcales and Archaeoglobales

A major group of hyperthermophilic euryarchaeotes is the order Thermococcales, including genera such as *Thermococcus* and *Pyrococcus* (**Fig. 19.35**). These genera are most commonly isolated from thermal vents at the ocean floor, and from submarine solfataras, volcanic vents that emit only gases. Besides high temperature, most of these microbes experience high pressure (they are barophiles). Most Thermococcales are anaerobes that ferment complex carbon sources such as peptides or carbohydrates. Despite deep genetic divergence, their traits superficially resemble those of TACK (crenarchaeote) sulfur hyperthermophiles. Unfortunately, their names also are similar.

Note: Distinguish the euryarchaeotes *Thermococcus* and *Pyrococcus* from the TACK genera *Thermoproteus* and *Pyrodictium*.

Sulfur reducers and oxidizers. Most *Thermococcus* and *Pyrococcus* species grow at temperatures well above 90°C; *Pyrococcus woesei*, for example, grows at temperatures as high as 105°C. Their growth is accelerated by the use of elemental sulfur as a terminal electron acceptor for anaerobic respiration:

$$2H^+ + 2e^- + S^0 \longrightarrow H_2S$$

Alternatively, these species can oxidize molecular hydrogen with sulfur:

$$H_2 + S^0 \longrightarrow H_2S$$

Species of Thermococcales are the source of "vent polymerases" for PCR amplification. They now replace the enzyme Taq polymerase obtained from the Yellowstone hot-spring bacterium *Thermus aquaticus*, which grows only at temperatures up to 90°C. The Taq enzyme has only limited stability at or above 95°C, but DNA polymerases with greater stability at higher temperatures and increased accuracy in replication have been produced from vent-dwelling archaea such as *Pyrococcus furiosus* and *Thermococcus litoralis*. These vent polymerases now allow higher-temperature denaturation and synthesis of GC-rich sequences that were difficult to amplify with the original thermostable enzyme.

The first member of Thermococcales whose genome was sequenced was *Pyrococcus abyssi*. (Remember to distinguish *Pyrococcus* abyssi from the TACK species *Pyrodictium* abyssi.) The genome of *Pyrococcus abyssi* reveals an especially high number of eukaryotic homologs for DNA replication, transcription, and protein translation. Examples include the eukaryotic-like primase, helicase, and endonuclease for generation of Okazaki fragments in the lagging strand of DNA synthesis. At the same time, the genome also shows bacterial homologs for cell division and DNA repair. Thus, *Pyrococcus abyssi* offers a striking view of mixed heritage from the common ancestor of the three domains.

A.

B.

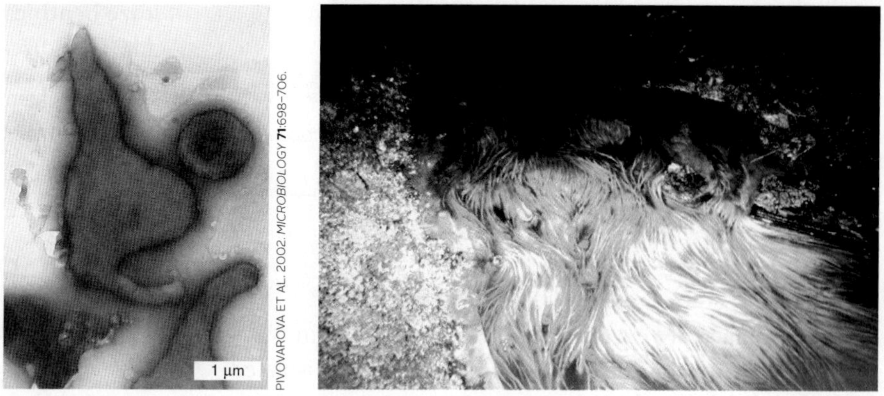

PIVOVAROVA ET AL. 2002. *MICROBIOLOGY* **71**:698–706.

K. J. EDWARDS ET AL. 2000. *SCIENCE* **287**:1796.

FIGURE 19.36 ■ The extreme acidophile *Ferroplasma*. A. *F. acidiphilum* grows at pH 0 (TEM). **B.** Streamers of *F. acidarmanus* anchored to deposits of pyrite within the Iron Mountain mine in California. The stream is about a meter across, its water about pH 0. This level of acidity will dissolve a metal shovel in a day.

Pyrococcus species possess enzymes conducting most of the classic conversions of the EMP pathway of glycolysis, but several steps use enzymes unrelated to those in bacteria. Examples include two ADP-dependent enzymes—glucokinase and phosphofructokinase—as well as the phosphate-independent enzyme glyceraldehyde 3-phosphate ferredoxin oxidoreductase. The glyceraldehyde 3-phosphate conversion is unique in that it sidesteps the use of inorganic phosphate, which is required at this step by bacterial glycolysis (shown earlier, in **Fig. 19.4**).

Pyrococcus and other members of Thermococcales have enzymes requiring tungsten, a metal rarely required outside the archaea. (Tungsten is present in elevated concentrations at hydrothermal vents.) Another unusual feature of *Pyrococcus* is that both proton motive force generation and the ATP synthase appear to involve a Na^+/H^+ antiport system. This dependence on sodium is a trait shared with the methanogens and the halophiles.

Thought Question

19.7 Compare and contrast the metabolic options available for *Pyrococcus* and for the TACK organism *Sulfolobus*.

Archaeoglobales reduce sulfate and reverse methanogenesis. While many benthic archaea reduce sulfur (S^0), only one clade, Archaeoglobales, is known to reduce sulfate ion (SO_4^{2-}) without a bacterial partner. *Archaeoglobus fulgidus* reduces sulfate by way of an acetyl-CoA degradation pathway that reverses part of methanogenesis. This process may be important for eliminating methane from methane hydrates before they release gas to the atmosphere (presented in **eTopic 19.2**).

Life in Hot Acid: Thermoplasmatales

A clade whose habitat is extreme even by archaeal standards is **Thermoplasmatales**. This clade includes thermophilic acidophiles with no cell wall and no S-layer, but only a plasma membrane. How they maintain their cells against the most extreme conditions known on Earth is poorly understood.

An example is a thermophile isolated from self-heating coal refuse piles, *Thermoplasma acidophilum*, growing at 59°C and pH 2. *Thermoplasma* cells are motile, but it is unclear how their archaella maintain a torque against the membrane without a rigid envelope or S-layer for support. The metabolism of *T. acidophilum*, like that of *Pyrococcus*, is sulfur (S^0) respiration of organic molecules.

The genome sequence of *T. acidophilum* contains just 1.5 million base pairs—one of the smallest known for a free-living organism. It shows substantial evidence of horizontal transfer from *Sulfolobus*, a TACK hyperthermophile that shares the same range of habitats. Horizontal transfer between distant relatives turns out to be common among the hyperthermophiles. For example, *T. acidophilum* has genes encoding an entire protein degradation pathway that was transferred horizontally from an ancestor of *Sulfolobus*.

Another acidophilic genus of the order Thermoplasmatales is *Ferroplasma* (**Fig. 19.36**). *Ferroplasma* species are mesophiles or thermophiles, found in mines containing iron pyrite ore (FeS_2). Using dissolved Fe^{3+} as an oxidizing agent in the presence of H_2O, they oxidize the sulfur to sulfuric acid. The chemical equation is:

$$FeS_2 + 14Fe^{3+} + 8H_2O \longrightarrow 15Fe^{2+} + 2SO_4^{2-} + 16H^+$$

The reaction generates pH values below pH 0 (1-M H^+). This degree of acidity can dissolve a metal shovel within a day.

The amorphous cells of *Ferroplasma acidarmanus* grow in biofilms that form long streamers into water draining from the mine (**Fig. 19.36B**). The oxidative disintegration of iron-bearing ores can be useful for leaching of minerals, but it also causes acid mine drainage into aquatic systems (discussed in Chapter 22).

Thought Question

19.8 Compare and contrast sulfur metabolism in *Pyrococcus* and in *Ferroplasma*.

To Summarize

- **Haloarchaea are extreme halophiles**, growing in NaCl at a concentration of at least 1.5 M. They are isolated from salt lakes, solar salterns, underground salt micropockets, and salted foods.
- **Haloarchaea show diverse cell shapes**, including slender rods (*Halobacterium*), cocci (*Halococcus*), and flat squares (*Haloquadratum*). The cell envelopes of most haloarchaea contain rigid cell walls of glycoprotein. Gas vesicles enable haloarchaea to remain near the top of the water column.
- **Molecular adaptations to high salt** include DNA of high GC content and acidic proteins (proteins with a high number of negatively charged residues).
- **Retinal-based photoheterotrophy** involves the proton pump bacteriorhodopsin or the chloride pump halorhodopsin. Both pumps contain retinal for light absorption.
- **Thermococcales are hyperthermophiles.** Most species use sulfur to oxidize complex organic substrates.
- ***Pyrococcus* and *Thermococcus* are the source of vent polymerases** for PCR. Thermococcal enzymes notably use tungsten at their active site.
- ***Archaeoglobus* hyperthermophiles reduce sulfate and oxidize benthic methane hydrates.** The methyl group of acetate is oxidized to CO_2 by partly reversing methanogenesis.
- **Thermoplasmatales includes extreme acidophiles.** *Ferroplasma* oxidizes iron pyrite ore (FeS_2) in a process that generates concentrated sulfuric acid, causing acid mine drainage.

19.6 DPANN Symbionts, Altiarchaeales, and Asgard: Branch to Eukaryotes?

The DPANN superphylum includes several deeply branching clades of ultrasmall cells with shrunken genomes that have lost major metabolic pathways and appear to require a symbiotic partner. They are most commonly found in freshwater and marine environments. A related group, Altiarchaeales, is free-living and shows unusual harpoon-like appendages called "hami." Finally, the deeply branching Asgard group includes marine thermophiles that may be the most closely related to the domain Eukarya.

Nanoarchaeota: Parasites on a Hyperthermophile

The best-studied clade of the DPANN superphylum is the Nanoarchaeota. The cultured species *Nanoarchaeum equitans* consists of exceptionally small cells that are obligate symbionts of the TACK hyperthermophile *Ignicoccus* (**Fig. 19.37**). The *Ignicoccus-Nanoarchaeum* isolates were obtained from Kolbeinsey Ridge hydrothermal vents.

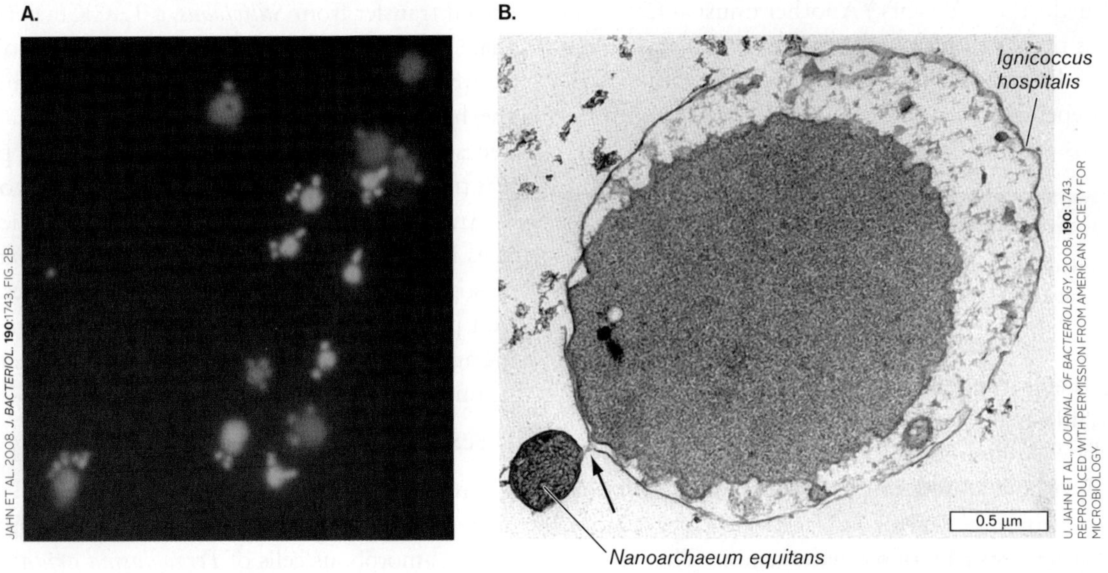

FIGURE 19.37 ■ *Nanoarchaeum equitans*, a small obligate symbiont attached to *Ignicoccus hospitalis*. A. Each *I. hospitalis* cell here has one to four attached cells of *N. equitans*. Fluorescence microscopy. Green (live cells); orange (dead cells). **B.** *Ignicoccus*, showing intracellular compartment and one cell of *N. equitans* attached by a membrane bridge (arrow; TEM).

The host cell may harbor up to four of the smaller cells (**Fig. 19.37A**). The *N. equitans* cell in **Figure 19.37B** is attached to *I. hospitalis* by a membrane bridge, which has been shown to connect the intracellular membrane tubules of *Ignicoccus* (see the Current Research Highlight). How the *Nanoarchaeum* symbiont maintains this intimate connection is unknown.

Both host and symbiont genomes have been sequenced, revealing extensive coevolution of the two. The *N. equitans* genome is exceptionally small (less than 500 kb) and shows evidence of rapid degenerative evolution. The diminished genome is typical of a dependent organism that has lost numerous genes for functions now provided by a host. Growth rate comparisons show that the host *I. hospitalis* grows faster in the absence of *N. equitans*; thus, *N. equitans* is inferred to be a parasite. It remains unclear, however, whether *N. equitans* makes any metabolic contribution to its host.

Members of other DPANN phyla, such as Diapherotrites, are identified from marine and underground sources. They show evidence of evolutionary gene loss as symbionts, but also the gain of some bacterial genes that may offer free-living alternatives. Their genes encoding SSU rRNA show multiple differences from the sequences generally used as primers for PCR amplification. Thus, it is hard to identify such organisms, much less study them.

Altiarchaeales: Archaea with Grappling Hooks

A newly discovered archaeon, *Altiarchaeum hamiconexum*, was found in marsh water rich in sulfides, from the Sippenauer Moor, Germany. First known as the SM1 euryarchaeon, the microbe was discovered in 2004 by Rudolph Huber and his students at the University of Regensburg, Germany. These archaea fix CO_2 by an unusual pathway related to methanogenesis.

A surprising trait of *Altiarchaeum* was discovered by Christine Moissl-Eichinger. Her students showed how this organism forms netlike biofilms by use of pilus-like grappling-hook appendages called **hami** (singular, **hamus**; **Fig. 19.38**). Each grappling hook contains paired barbs along an extended protein filament many times the length of the cell. The filament ends with a triple fishhook that enables filaments to clasp neighboring cells in a biofilm matrix (**Fig. 19.39**). No other kind of cell is known to make this type of appendage.

The matrix of grappling hooks also coats filaments of sulfide-oxidizing bacteria that may cooperate with *Altiarchaeum*. Thus, this archaeon represents yet another type of intimate community involving archaea.

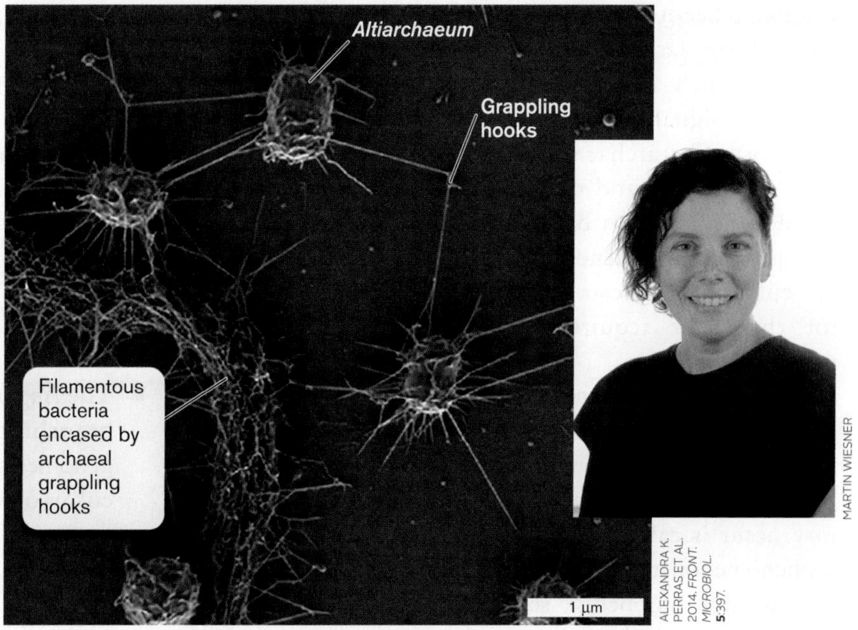

FIGURE 19.38 ▪ ***Altiarchaeum* forms chains with grappling hooks (hami).** **Inset:** Christine Moissl-Eichinger.

Asgard, the Norse Gods Superphylum

The superphylum Asgard (meaning "world of the gods") was named for its inclusion of the Lokiarchaeota, originally identified from a hydrothermal ridge known as Loki's Castle. Related deeply branching phyla were fancifully named for other Norse gods, such as Heimdallarchaeota. Asgard genomes have been sequenced from various marine and freshwater sediments. All show such deeply branching divergence that standard SSU rRNA primers may fail to amplify their DNA.

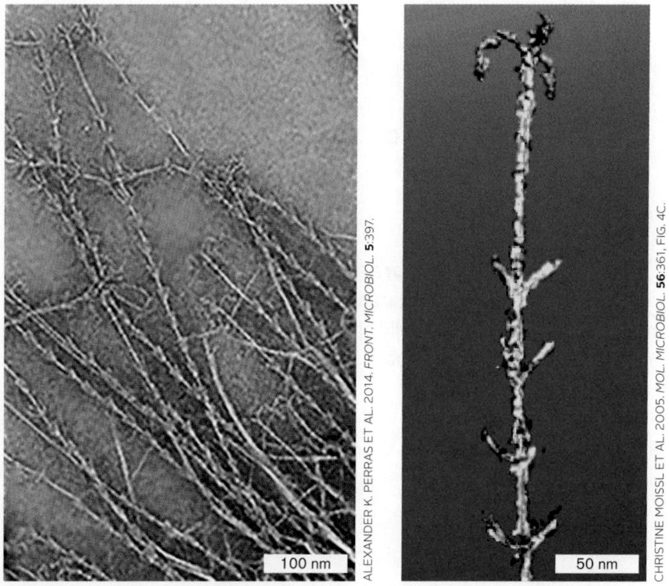

FIGURE 19.39 ▪ **Grappling hooks (hami) of *Altiarchaeum* species.** **A.** Hooked appendages extending from a cell (TEM). **B.** Cryo-EM tomography model of a grappling hook (hamus).

Lokiarchaeota genome sequences show genes with unusual similarity to signature genes of the domain Eukarya, such as those encoding components of cytoskeleton, signal transduction, and nucleocytoplasmic transport. The archaeal phylogenetic tree, as computed by Thijs Ettema and co-workers (see **Fig. 19.3**) shows the domain Eukarya branching from the Lokiarchaeota lineage. These sequence data are evidence that the original eukaryotic ancestor evolved from an Asgard ancestor that later acquired bacterial symbionts such as mitochondria.

However, the placement of the Eukarya branch remains controversial. The sequences diverge so deeply that small adjustments in tree computation may yield very different branch points for distant clades. One artifactual effect that may occur is called long branch attraction (**Fig. 19.40**), a phenomenon in which two fast-evolving lineages show such a large number of sequence mismatches that their sequences actually appear more similar to each other than to more closely related lineages that evolve more slowly. Long branch attraction occurs because each sequence position has only four possible base pairs, so as the rate of DNA mutation increases, the chance increases that two distant lineages might independently change to the same base at a given position. Thus, two distantly evolved lineages may branch together because they share dissimilarity to more slowly evolving organisms. Further study of Lokiarchaeota genomes, and the culturing of live microbes, may help resolve the fascinating question of the origin of eukaryotes.

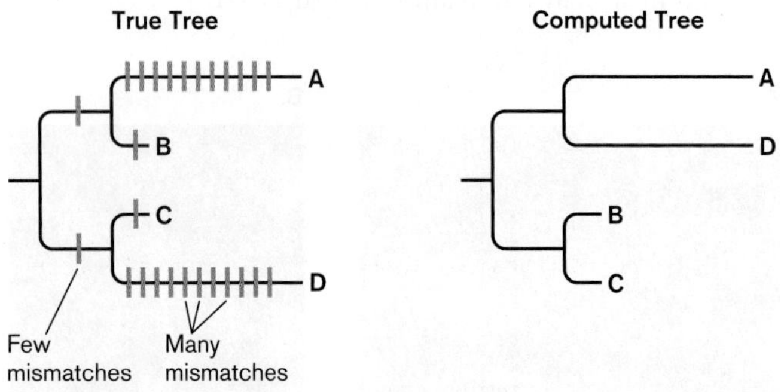

FIGURE 19.40 ■ Long branch attraction. Two fast-evolving lineages may show such a large number of sequence mismatches that their sequences actually appear more similar to each other than to more closely related lineages that evolve more slowly.

To Summarize

- **The DPANN phylum Nanoarchaeota includes *Nanoarchaeum equitans*, a tiny obligate symbiont of the marine hyperthermophile *Ignicoccus hospitalis*.** *N. equitans* is a parasite, impacting growth of its host.
- **Altiarchaeales includes *Altiarchaeum hamiconexum*, an archaeon with unique grappling-hook appendages.** *A. hamiconexum* forms netlike biofilms including a relationship with sulfide-oxidizing bacteria.
- **The Asgard phylum Lokiarchaeota shows evidence of branching with the domain Eukarya.** Lokiarchaeota genomes show a number of signature proteins found in eukaryotes.
- **Long branch attraction is a computational artifact that can lead distantly branching lineages to appear more similar than, in fact, they are.**

Concluding Thoughts

Overall, species of archaea inhabit a wider range of environments than either bacteria or eukaryotes do—from extreme heat to extreme cold, from high pH to extreme acid, as well as temperate environments such as the oceans. Archaea are found in all soil and water habitats, and as endosymbionts of marine animals. Archaea show unique cell structures, such as grappling-hook appendages, and unique forms of metabolism, such as methanogenesis. Methanogens colonize all soil and water environments, as well as the human digestive tract. Yet we probably know less about the actual scope of Archaea than we do about the other two domains because so many members remain uncultured. The TACK superphylum includes sulfur-metabolizing hyperthermophiles, as well as marine mesophiles and psychrophiles. The Thaumarchaeota include ammonia oxidizers and marine endosymbionts, and the Euryarchaeota include methanogens, haloarchaea, and sulfur-metabolizing extremophiles. All the while, our DNA sequencing of microbial communities reveals an expanding realm of previously unknown archaea, such as the Asgard superphylum of marine hyperthermophiles. Much of what we call Archaea remains to be explored.

CHAPTER REVIEW

Review Questions

1. What distinctive structures are seen in the archaeal cell membrane and envelope?
2. Which aspects of archaeal genetics resemble the genetics of bacteria, and which aspects have more in common with eukaryotes?
3. Compare and contrast diverse members of the TACK superphylum.
4. Outline the genetic phylogeny and key traits of these groups of archaea: Haloarchaea, methanogens, Thermococcales, Thermoplasmatales.
5. What are some specific physiological adaptations found in hyperthermophiles? Halophiles? Extreme acidophiles?
6. Outline three different specific types of mutualism involving an archaeal symbiont.
7. Explain how different archaea contribute to cycling of nitrogen and sulfur in ecosystems.
8. Explain what is known and what is unknown about the following groups of archaea: marine TACK archaea; marine benthic anaerobes; soil and plant root–associated archaea. What kinds of experiments may reveal additional traits of these organisms?
9. How do methanogens interact metabolically with communities of bacteria, within a host animal or within a soil environment?

Thought Questions

1. What approaches can you use to discover previously unknown deeply branching groups of archaea? Explain the strengths and limitations of each method.
2. Why do you think we have found no archaea that are pathogens of animals or plants?
3. Why do you think methanogens appear in many branches among different groups, whereas Haloarchaea branch as a single group?

Key Terms

Aigarchaeota (762)
Anaerobic Methane-Oxidizing Euryarchaeota (ANME) (769)
archaellum (750)
bacteriorhodopsin (755, 777)
barophile (756)
Bathyarchaeota (752)
black smoker (760)
cannula (762)
Desulfurococcales (752, 756)
electron bifurcation (772)
Euryarchaeota (754)
gas vesicle (775)
Geoarchaeota (752)
Haloarchaea (774)
halorhodopsin (777)
hamus (781)
histone (751)
isoprenoid (749)
Korarchaeota (762)
Lokiarchaeota (755)
methane gas hydrate (769)
methanogen (766)
methanogenesis (754)
Nanoarchaeota (755)
pseudopeptidoglycan (pseudomurein) (767)
Sulfolobales (752, 758)
syntrophy (767)
tetraether (749)
Thaumarchaeota (754)
Thermoplasmatales (779)
Thermoproteales (752)

Recommended Reading

Auernik, Kathryne S., Charlotte R. Cooper, and Robert M. Kelly. 2008. Life in hot acid: Pathway analyses in extremely thermoacidophilic archaea. *Current Opinion in Biotechnology* **19**:445–453.

Bang, Corinna, Katrin Weidenbach, Thomas Gutsmann, Holger Heine, and Ruth A. Schmitz. 2014. The intestinal archaea *Methanosphaera stadtmanae* and *Methanobrevibacter smithii* activate human dendritic cells. *PLoS One* **9**:e99411.

Bolduc, Benjamin, Daniel P. Shaughnessy, Yuri I. Wolf, Eugene Koonin, Francisco F. Roberto, et al. 2012. Identification of novel positive-strand RNA viruses by metagenomic analysis of archaea-dominated Yellowstone hot springs. *Journal of Virology* **86**:5562–5573.

Brumfield, Susan K., Alice C. Ortmann, Vincent Ruigrok, Peter Suci, Trevor Douglas, et al. 2009. Particle assembly and ultrastructural features associated with replication of the lytic archaeal virus *Sulfolobus* turreted icosahedral virus. *Journal of Virology* **83**:5964–5970.

Da Cunha, Violette, Morgan Gaia, Arshan Nasir, and Patrick Forterre. 2018. Asgard archaea do not close the debate about the universal tree of life topology. *PLoS Genetics* **14**:e1007215.

DasSarma, Priya, Regie C. Zamora, Jochen A. Müller, and Shiladitya DasSarma. 2012. Genome-wide responses of the archaeon *Halobacterium* sp. strain NRC-1 to oxygen limitation. *Journal of Bacteriology* **194**:5530. https://doi.org/10.1128/JB.01153-12.

DeMaere, Matthew Z., Timothy J. Williams, Michelle A. Allen, Mark V. Brown, John A. E. Gibson, et al. 2013. High level of intergenera gene exchange shapes the evolution of haloarchaea in an isolated Antarctic lake. *Proceedings of the National Academy of Sciences USA* **110**:16939–16944.

Golyshina, Olga V., and Kenneth N. Timmis. 2005. *Ferroplasma* and relatives, recently discovered cell wall-lacking archaea making a living in extremely acid heavy metal-rich environments. *Environmental Microbiology* 7:1277–1288.

Könneke, Martin, Anne E. Bernhard, José R. de la Torre, Christopher B. Walker, John B. Waterbury, et al. 2005. Isolation of an autotrophic ammonia-oxidizing marine archaeon. *Nature* **437**:543–546.

López-García, Purificación, Yvan Zivanovic, Philippe Deschamps, and David Moreira. 2015. Bacterial gene import and mesophilic adaptation in archaea. *Nature Reviews. Microbiology* **13**:447–456.

Pernthaler, Annelie, Anne E. Dekas, C. Titus Brown, Shana K. Goffredi, Tsegereda Embaye, et al. 2008. Diverse syntrophic partnerships from deep-sea methane vents revealed by direct cell capture and metagenomics. *Proceedings of the National Academy of Sciences USA* **105**:7052–7057.

Perras, Alexandra K., Gerhard Wanner, Andreas Klingl, Maximilian Mora, Anna K. Auerbach, et al. 2014. Grappling archaea: Ultrastructural analyses of an uncultivated, cold-loving archaeon, and its biofilm. *Frontiers in Microbiology* **5**:397.

Pester, Michael, Christa Schleper, and Michael Wagner. 2011. The Thaumarchaeota: An emerging view of their phylogeny and ecophysiology. *Current Opinion in Microbiology* **14**:300–306.

Ruepp, Andreas, Werner Graml, Martha-Leticia Santos-Martinez, Kristin K. Koretke, Craig Volker, et al. 2000. The genome sequence of the thermoacidophilic scavenger *Thermoplasma acidophilum*. *Nature* **407**:508.

Sauder, Laura A., Katja Engel, Jennifer C. Stearns, Andre P. Masella, Richard Pawliszyn, et al. 2011. Aquarium nitrification revisited: Thaumarchaeota are the dominant ammonia oxidizers in freshwater. *PLoS One* **6**:e23281.

Spang, Anja, Eva F. Caceres, and Thijs J. G. Ettema. 2017. Genomic exploration of the diversity, ecology, and evolution of the archaeal domain of life. *Science* **357**:eaaf3883.

Vanderhaeghen, Sonja, Christophe Lacroix, and Clarissa Schwab. 2015. Methanogen communities in stools of humans of different age and health status and co-occurrence with bacteria. *FEMS Microbiology Letters* **362**:fnv092.

Zaremba-Niedzwiedzka, Katarzyna, Eva F. Caceres, Jimmy H. Saw, Disa Bäckström, Lina Juzokaite, et al. 2017. Asgard archaea illuminate the origin of eukaryotic cellular complexity. *Nature* **541**:353–358.

CHAPTER 20

Eukaryotic Diversity

20.1 Phylogeny of Eukaryotes

20.2 Fungi

20.3 Amebas and Slime Molds

20.4 Algae

20.5 Alveolates: Ciliates, Dinoflagellates, and Apicomplexans

20.6 Parasitic Protozoa

The domain Eukarya encompasses a breathtaking range of size and shape, from giant whales and sequoias to microbial fungi, algae, and protozoa. Fungi include multicellular forms such as mushrooms, as well as unicellular yeasts and filamentous *Penicillium*. Algae conduct photosynthesis using chloroplasts; some form long sheets of seaweed, while others are single-celled phytoplankton. Protozoa include amebas and stalked vorticellae. Most are free-living predators that use cilia to catch their prey, but some are parasites that cause diseases such as malaria and sleeping sickness.

CURRENT RESEARCH highlight

Photosynthetic coral symbionts take the heat. Corals throughout the ocean provide a home for *Symbiodinium* species, dinoflagellates whose photosynthesis feeds the coral. But their mutualism is highly sensitive to environmental change. Mónica Medina at Penn State University studies how coral endosymbionts respond to acidification and temperature rise. She found that the thylakoids (photosynthetic membranes) of *Symbiodinium* melt at higher temperatures, and that different species show widely different tolerances to heat. These observations add to our understanding of "coral bleaching," the expulsion of *Symbiodinium* followed by death of the coral. The micrograph shows *Effrenium voratum*, a genus and species in the family *Symbiodiniaceae*.

Source: Erika M. Díaz-Almeyda et al. 2017. *Proc. R. Soc. Lond. B. Biol. Sci.* **284**: 20171767.

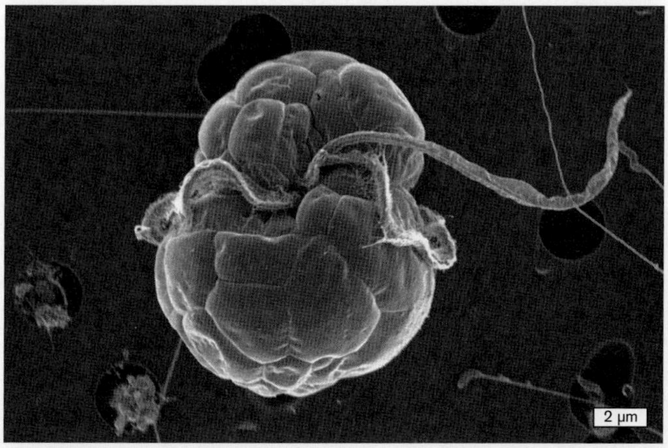

COURTESY OF TODD LAJEUNESSE, PENN STATE UNIVERSITY

AN INTERVIEW WITH

MÓNICA MEDINA, PROFESSOR OF BIOLOGY, PENN STATE UNIVERSITY

COURTESY OF MÓNICA MEDINA

What is the most important thing to know about how *Symbiodinium* responds to climate change?

Light and temperature are major drivers in the ecology and biogeography of symbiotic dinoflagellates living in corals and other cnidarians. There are multiple new genera [of dinoflagellate symbionts] besides *Symbiodinium,* and the group is now called the family Symbiodiniaceae. Many of the symbionts have local adaptations, and the host-algal combination is a key aspect that will determine the survival of a coral species in a warming climate. There is a lot of genetic variation even within algal species, and their physiological thermal tolerance ranges are quite diverse as well.

When we think of eukaryotes, we think first of plants and animals consisting of complex, multicellular bodies. But eukaryotes also include vast numbers of microbes, including fungi and protists (algae and protozoa). Many protists play vital roles in our oceans. A particularly compelling example is the dinoflagellate *Symbiodinium* species, which form endosymbioses with corals. *Symbiodinium* dinoflagellates take up residence within the coral tissues, where their photosynthesis drives the growth of reefs that support vast ecosystems of marine life. Yet the existence of corals worldwide is threatened by global climate change. Rising temperatures and acidity upset the delicate mutualism of coral and its dinoflagellate partners (see the Current Research Highlight).

In Chapter 20 we present the phylogeny of major groups of microbial eukaryotes, along with the branching of animals and plants from the microbial family tree. We explore the form and function of algae, such as the calcium carbonate–plated coccolithophores that bloom over vast stretches of ocean (**Fig. 20.1A**). Fungi range from yeast to mushrooms to common molds such as *Penicillium*, our source of penicillin (**Fig. 20.1B**). Protozoa include a dozen deep-branching clades, from free-living amebas to diarrhea-causing parasites such as *Giardia* (**Fig. 20.1C**). We get to know the eukaryotic microbes as essential partners in ecosystems and as infectious agents that cause some devastating diseases.

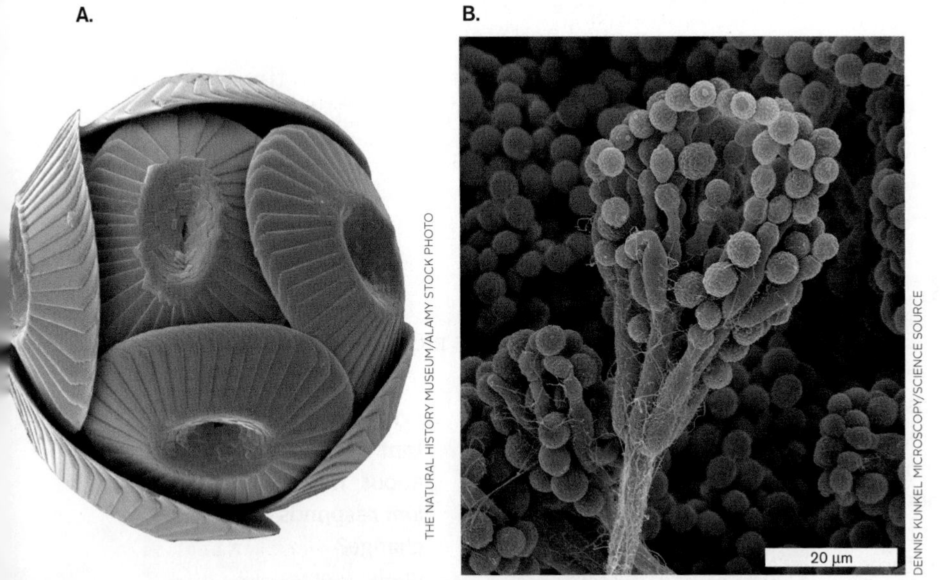

THE NATURAL HISTORY MUSEUM/ALAMY STOCK PHOTO

DENNIS KUNKEL MICROSCOPY/SCIENCE SOURCE

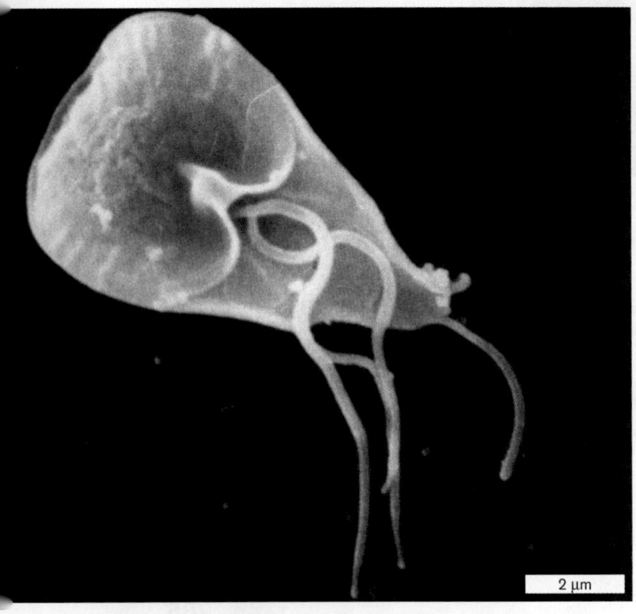

CDC/JANICE HANEY CARR

FIGURE 20.1 ▪ Microbial eukaryotes. **A.** *Coccolithus pelagicus*, a marine coccolithophore, single-celled alga that forms scales of calcium carbonate (colorized SEM). **B.** Conidiophores of *Penicillium notatum*, an airborne fungus (colorized SEM). **C.** *Giardia intestinalis*, a waterborne parasite (SEM).

20.1 Phylogeny of Eukaryotes

With all their diversity, eukaryotic cells share a common structure defined by the presence of the nucleus and other membrane-enclosed organelles (for review, see eAppendix 2). The extensive compartmentalization of the eukaryotic cell, including the nucleus and endomembrane system, enables eukaryotic cells to grow a thousandfold larger than prokaryotic cells. Yet these organelles are found within even the tiniest of eukaryotes, such as *Ostreococcus tauri*, a green alga less than 2 µm across (**Fig. 20.2**). This alga consists of a flat, disk-shaped cell containing one or two of each major type of organelle—a mitochondrion, a chloroplast, and a stack of Golgi—all packed within the cell's small volume. The genome of *O. tauri* is also downsized; at only 8 Mb, the genome is barely twice as long as that of the bacterium *Escherichia coli*. Though poorly understood, such tiny, bacteria-sized eukaryotes are believed to be the most numerous and ubiquitous forms of eukaryotic life.

Note that despite their extraordinary range of form, the metabolism of eukaryotes is less diverse than that of either bacteria or archaea. Most eukaryotes conduct either oxygenic photosynthesis or heterotrophy—or both. All have descended from an ancestral cell that engulfed a bacterial endosymbiont, giving rise to mitochondria, the source of aerobic respiration. And all eukaryotic phototrophs descended from a cell that engulfed the bacterial ancestor of chloroplasts.

Historical Overview of Eukaryotes

For most of human history, life was understood in terms of macroscopic, multicellular eukaryotes: animals (creatures that move to obtain food) and plants (rooted organisms that grow in sunlight). **Fungi** (singular, **fungus**),

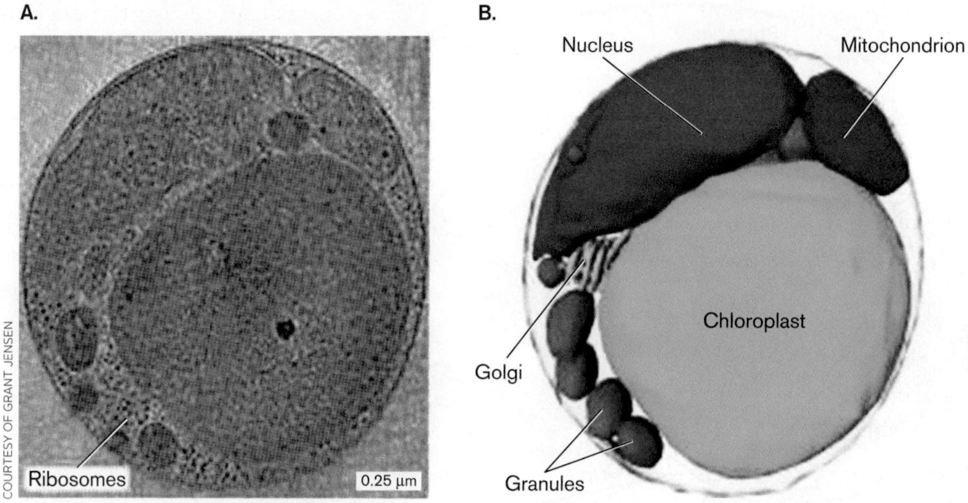

FIGURE 20.2 ▪ A tiny eukaryote still has organelles. *Ostreococcus tauri*, visualized by cryo-EM **(A)** and in 3D by tomography **(B)**.

which lack photosynthesis, were nonetheless considered a form of plant because they grow on the soil or other substrate. Thus, **mycology**, the study of fungi, was often included with botany, the study of plants. But the basis of fungal growth was poorly understood, and its mystery was often associated with magic. For example, people were mystified by the sudden growth of mushrooms in a ring, which they called a "fairy ring" (**Fig. 20.3A**). The mushrooms actually arise as fruiting bodies from the tips of fungal hyphae (filaments of cells) that propagate from a single spore and extend radially underground.

In the eighteenth and nineteenth centuries, microscopists came to recognize microscopic forms of fungi such as filamentous hyphae and unicellular yeasts. Other unicellular life forms, such as amebas and paramecia, were motile and appeared more like microscopic animals. These animal-like organisms were called **protozoa** (singular, **protozoan**; **Fig. 20.3B**). The cellular dimensions of protozoa were typically ten- to a hundredfold larger than those of bacteria, and their form and motility offered intriguing subjects for observation. So did single-celled phototrophs such as diatoms and dinoflagellates, which were called **algae** (singular, **alga**). Algae were thought of as unicellular plants, although simple multicellular algae were known. The unicellular and microscopic forms of fungi, protozoa, and algae came to be included in the subject of microbiology.

Discoveries in physiology led us to redefine these organisms. For example, the motile organisms defined as protozoa often contain chloroplasts and fit the classification of algae. On the other hand, slime molds, originally classified with fungi, show form and motility more typical of protozoa. By the mid-twentieth century, naturalists classified protozoa, unicellular algae, and undifferentiated colonial forms as **protists**. Researchers including Herbert Copeland, Robert Whittaker, and Lynn Margulis attempted to refine the definition of "protist" to better distinguish microbial life forms.

Today, molecular phylogeny shows that protists comprise several clades equally distant from each other as they are from animals and plants (**Fig. 20.4**). In the terminology used today:

- **Protist** refers to single-celled and colonial eukaryotes other than fungi. Protists include many diverse clades of algae and protozoa.

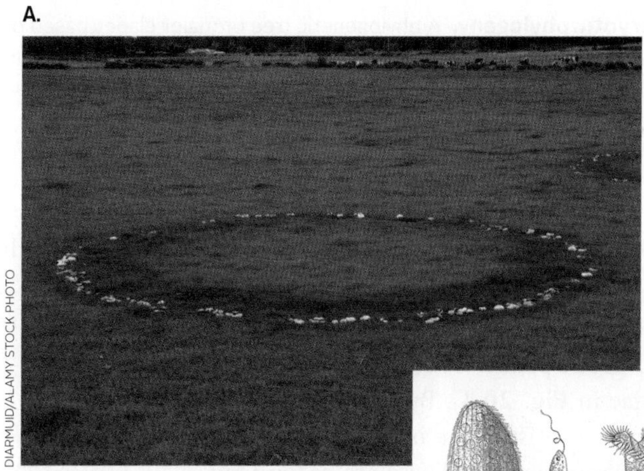

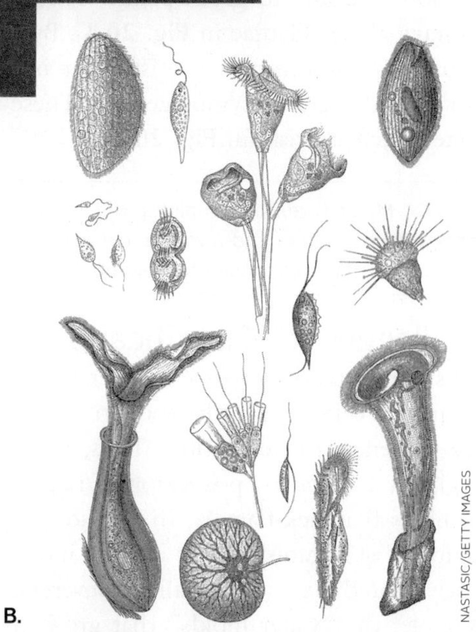

FIGURE 20.3 ▪ Traditional views of fungi and protozoa (protists). **A.** Basidiomycete mushrooms growing in a "fairy ring." **B.** A nineteenth-century depiction of ciliated protozoa, by Rudolf Leuckart.

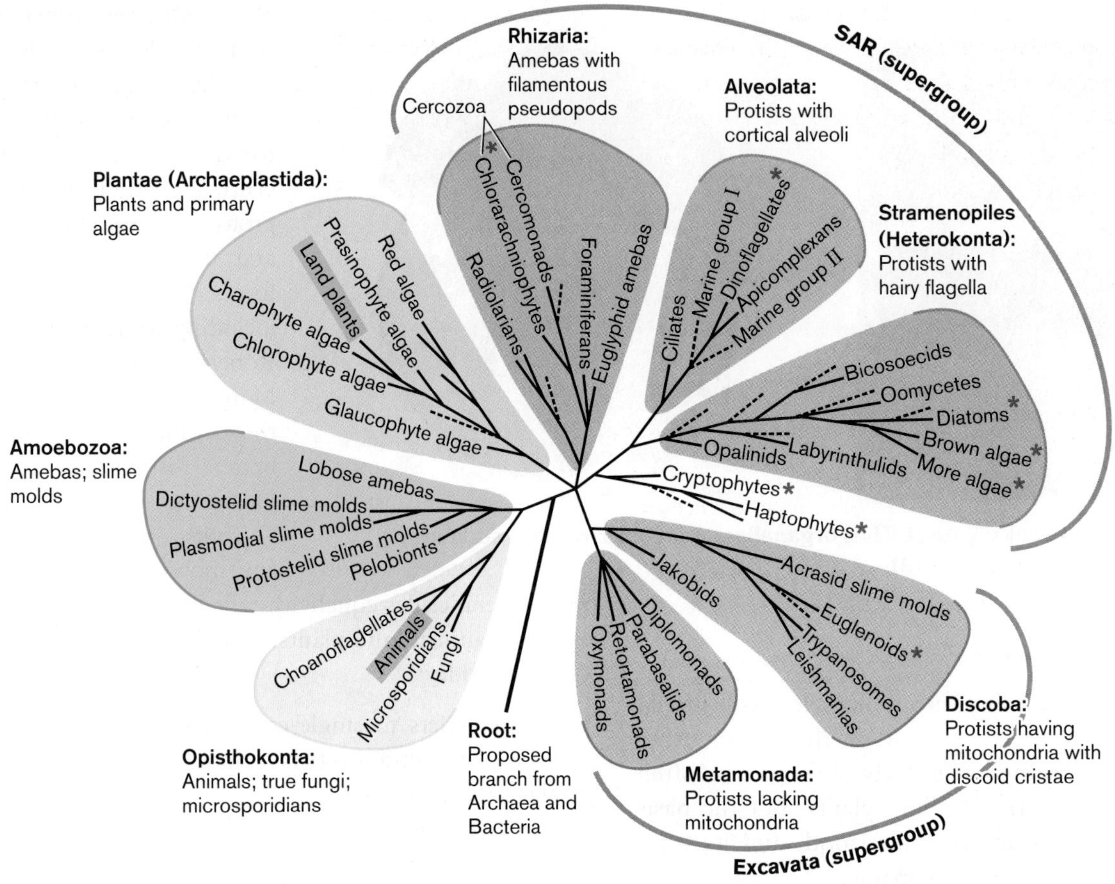

FIGURE 20.4 ■ **Eukaryotic phylogeny.** A phylogenetic tree of major clades based on DNA sequence data. Dotted lines denote emerging, little-known taxa. Green asterisks denote secondary algae whose ancestor engulfed a primary alga with chloroplast. Gray branches indicate less certain position. *Source:* Modified from Sandra L. Baldauf. 2003. Science **300**:1703; and Fabien Burki. 2014. *Cold Spring Harb. Perspect. Biol.* **6:**a016147.

- **Protozoa** are protists that are single-celled heterotrophs. They include environmental consumers, as well as medically important parasites such as *Giardia*.

The algae include two major kinds. Those of the first kind are derived from a single endosymbiotic event and are closely related to green plants. These algae are called **primary algae** (Plantae in **Fig. 20.4**). By contrast, various groups of heterotrophic protists later incorporated algae in a second event of symbiogenesis. These are **secondary algae** (green asterisks in **Fig. 20.4**).

Note: Medical textbooks also cover invertebrate animal parasites such as worms and mites as eukaryotic agents of disease, although they are not considered microbes.

Challenges for Classification

Classifying eukaryotes presents several challenges. Complex eukaryotic cells frequently lose structures through reductive (degenerative) evolution. Thus, for example, a clade originally defined by possession of flagella often includes members that lack flagella. In addition, superficially similar forms of organisms have evolved independently in distantly related taxa; this is called convergent evolution. For example, the "water molds" that grow on aquarium fish superficially resemble fungi, but they actually evolved in a clade that includes brown algae and diatoms.

Another challenge for classification is the size and complexity of eukaryotic genomes. Eukaryotic genomes are typically severalfold larger than those of bacteria and archaea, and 50%–90% of their DNA consists of noncoding sequences. Thus, eukaryotic genomes take longer to sequence and are more challenging to annotate than those of prokaryotes.

Furthermore, the evolution of eukaryotes includes multiple events of **endosymbiosis**, in which an engulfed cell evolved into an essential organelle. An endosymbiotic incorporation of a proteobacterium by the ancestor of all eukaryotes gave rise to mitochondria. Later, incorporation of a cyanobacterium by the ancestor of plants and algae gave rise to chloroplasts. Much later, several lineages of protists took up chloroplast-bearing algae, which now show varying stages of evolution as organelles.

The names and groupings of eukaryotic clades are changed as new evidence emerges. A current consensus view of eukaryotic phylogeny (**Fig. 20.4**) is based on DNA sequence comparison, protein trees, and the appearance of specific gene fusions and deletions. In current practice, the clades of protists including lobed amebas and slime molds (Amoebozoa) share relatively close ancestry with Opisthokonta.

More deeply branching protist taxa include the amebas with needlelike pseudopods (Rhizaria); ciliates and dinoflagellates (Alveolata); oomycetes, brown algae, and diatoms (Stramenopiles, or Heterokonta); and groups of parasitic protozoa such as Metamonada and Discoba. The plants and primary algae (algae descended from a single plastid evolution event) are called Plantae (also called Archaeplastida).

We will now look at key traits of each major group. **Table 20.1** summarizes representative clades of microbial eukaryotes.

TABLE 20.1 Eukaryotic Microbial Diversity

Opisthokonta (fungi and metazoan animals). Single flagellum on reproductive cells. Includes multicellular animals.

Metazoa (animals). Multicellular organisms with motile cells and body parts. Includes colonial animals, both invertebrates and vertebrates. *Homo sapiens.*

Choanoflagellata. Single flagellum and collar of microvilli. Resemble sponge choanocytes. Possible link to common ancestor of multicellular animals.

Fungi (Eumycota). Cells form hyphae with cell walls of chitin.

- **Ascomycota.** Fruiting bodies form asci containing haploid ascospores. Includes opportunistic pathogens and bread-making yeasts. *Penicillium, Aspergillus, Saccharomyces cerevisiae, Stachybotrys*. Lichens are a mutualism between an ascomycete and green algae (*Trebouxia*) or cyanobacteria (*Nostoc*).
- **Basidiomycota.** Basidiospores form primary and secondary mycelia; some generate mushrooms. May be edible (*Lycoperdon*) or toxic (*Amanita*). Plant pathogens (*Ustilago maydis* causes corn smut). Human pathogens (*Cryptococcus neoformans*).
- **Chytridiomycota.** Motile zoospores with a single flagellum. Saprophytes or anaerobic rumen fungi. *Allomyces*. Frog pathogens (*Batrachochytrium dendrobatidis*). Bovine rumen digestive endosymbionts (*Neocallimastix*).
- **Glomeromycota.** Mutualists of plant roots, forming arbuscular mycorrhizae, filamentous networks that share nutrients with and among diverse plants.
- **Zygomycota.** Sexual hyphae (1*n*) grow toward each other and fuse to form the zygote (zygospore). Saprophytes or insect parasites. Some form mycorrhizae.

Microsporidia. Single-celled parasites that inject a spore through a tube into a host cell, causing microsporidiosis. *Encephalitozoon* species. Commonly infect AIDS patients.

Amoebozoa (amebas and slime molds). Lobe-shaped (lobose) pseudopods driven by sol-gel transition of actin filaments. Share branch with Opisthokonta.

Amebas. Unicellular. No microtubules to define shape. Life cycle is primarily asexual. Predators in soil or water. Giant free-living amebas (*Amoeba proteus*); parasites (*Entamoeba histolytica*).

Mycetozoa. Slime molds. Cellular slime molds (*Dictyostelium*). Upon starvation, amebas aggregate to form a fruiting body, which produces spores. Plasmodial slime molds (*Physarum polycephalum*) undergo meiosis, producing spores that germinate to ameboid gametes.

Plantae or Archaeplastida (primary endosymbiotic algae and plants). Includes green algae and multicellular land plants. Chloroplasts all arose from a single cyanobacterial endosymbiont.

Charophyta. Multicellular algae with rhizoids that adhere to sediment. *Spirogyra.*

Chlorophyta (green algae). Chlorophyll *a* confers green color. Inhabit upper layer of water.

- **Unicellular with paired flagella.** *Chlamydomonas, Volvox* (colonial).
- **Multicellular.** *Ulva* grows in sheets; *Cymopolia* forms calcified stalks with filaments.

Glaucophyta. Unicellular algae whose chloroplasts have peptidoglycan.

- **Picoeukaryotes.** *Ostreococcus* and *Micromonas* are unicellular algae.
- **Siphonous algae.** *Caulerpa* species consist of a single cell with multiple nuclei, growing to indefinite size.

Rhodophyta (red algae). Phycoerythrin obscures chlorophyll, colors the algae red. Absorption of blue-green light enables colonization of deeper waters. *Porphyra* forms sheets edible by humans; *Mesophyllum* is a coralline alga, hardened by calcium carbonate crust; resembles coral.

SAR (Stramenopiles, Alveolata, Rhizaria)

Stramenopiles (Heterokonta). Paired flagella of dissimilar form, one shorter with hairs.

- **Bacillariophyceae.** Diatoms.
- **Chrysophyceae.** Golden algae.
- **Oomycetes.** Water molds.
- **Phaeophyceae.** Kelps.
- **Prymnesiophyceae.** Coccolithophores.

Alveolata (having cortical alveoli). Cortex contains flattened vesicles called alveoli, reinforced below by lateral microtubules.

- **Apicomplexa (formerly Sporozoa).** Parasites with complex life cycles. Lack flagella or cilia; possess apical complex for invasion of host cells. Vestigial chloroplasts. *Plasmodium falciparum* causes malaria; *Toxoplasma gondii* causes feline-transmitted toxoplasmosis; *Cryptosporidium parvum* is a waterborne opportunistic parasite.
- **Ciliophora.** Common aquatic predators. Undergo sexual exchange by conjugation, in which micronuclei are exchanged, then regenerate macronuclei. Covered with cilia (*Paramecium*); mouth ringed with cilia (*Vorticella*); suctorians (*Acineta*).
- **Dinoflagellata.** Secondary or tertiary endosymbiont algae, from engulfment of primary or secondary algae. Cortical alveoli contain stiff plates. Pair of flagella, one wrapped around the cell. Free-living aquatic (*Peridinium*); zooxanthellae, endosymbionts of coral (*Symbiodinium*).

Rhizaria (amebas with filament-shaped pseudopods). Filament-shaped (filose) pseudopods. Some species have a test (shell) of silica or other inorganic materials. Possess flagella or pseudopods (Cercozoa); form spiral tests (Foraminifera); form thin pseudopods called filopodia (Radiolaria).

Excavata. Reduced genomes, modified mitochondria, parasitic.

Discoba (having disk-shaped cristae). Disk-shaped cristae of mitochondria. Parasitic or symbiotic flagellates; some alternate with ameboid forms. *Euglena, Trypanosoma, Naegleria.*

Metamonada (vestigial mitochondria). Parasitic or symbiotic flagellates. Mitochondria and Golgi degenerated through evolution. Human intestinal parasites (*Giardia intestinalis*, or *G. lamblia*); symbionts of termite gut (*Pyrsonympha*).

Opisthokonts: Animals and Fungi

Where do humans and other multicellular animals fit into this taxonomy? The position of animals (Metazoa) among the eukaryotic microbial clades is of interest because it suggests which contemporary microbes most closely resemble our own cells. The degree of relatedness can help define microbial model systems for probing key questions of human cell biology. For example, the baker's yeast *Saccharomyces cerevisiae* shares so much of its genetic machinery with humans that it provides a model for cancer, defects in cellular trafficking, and degenerative brain disorders.

The yeast *S. cerevisiae* is a fungus—and in fact, genomic analysis relates animals more closely to fungi (such as yeasts) than to motile microbial eukaryotes (such as paramecia or amebas). The **true fungi**, or **Eumycota**, are heterotrophs, either single-celled or growing in nonmotile filaments of cells called hyphae. Animal and fungal cells also share a structural feature distinguishing them from protists: the presence of an unpaired flagellum. Both animals and fungi include species whose life cycle has a uniflagellar stage, in contrast to other microbial eukaryotes whose flagella are paired (for example, euglenas). In the case of humans, the uniflagellar stage is the spermatozoan. Similarly, some species of fungi generate uniflagellar reproductive cells called **zoospores**. As a whole, the clade including single-flagellum members is termed "opisthokont," based on the Greek words meaning "backward-pointing pole," because the flagellum points backward like an oar.

Among opisthokonts, the microbes that diverged most recently from animals (600 million years ago) appear to be the choanoflagellates. Genetic studies of choanoflagellates reveal several genes found only in animals. The prefix "choano-" (meaning "funnel") refers to the collar of filaments surrounding the flagellum. The collared cells of choanoflagellates closely resemble the choanocyte cells of colonial sponges, an ancient form of animal (**Fig. 20.5**). In 2019, Nicole King and colleagues at the University of California-Berkeley showed that some choanoflagellates can assemble in a cup-like colony that undergoes coordinated movement, like an animal tissue. Thus, choanoflagellates may represent a "missing link" between animals and the microbial eukaryotes.

Note: Eukaryotic flagella are whiplike organelles composed of microtubules and surrounded by a membrane; their action is powered by ATP along the entire filament. Distinguish them from bacterial and archaeal flagella, which are rotary, helical filaments composed entirely of protein subunits; their rotation is powered at the base by proton motive force.

Thought Question

20.1 How could you demonstrate that eukaryotic flagella move with a whiplike motion, instead of rotary motion? What experiment might you conduct?

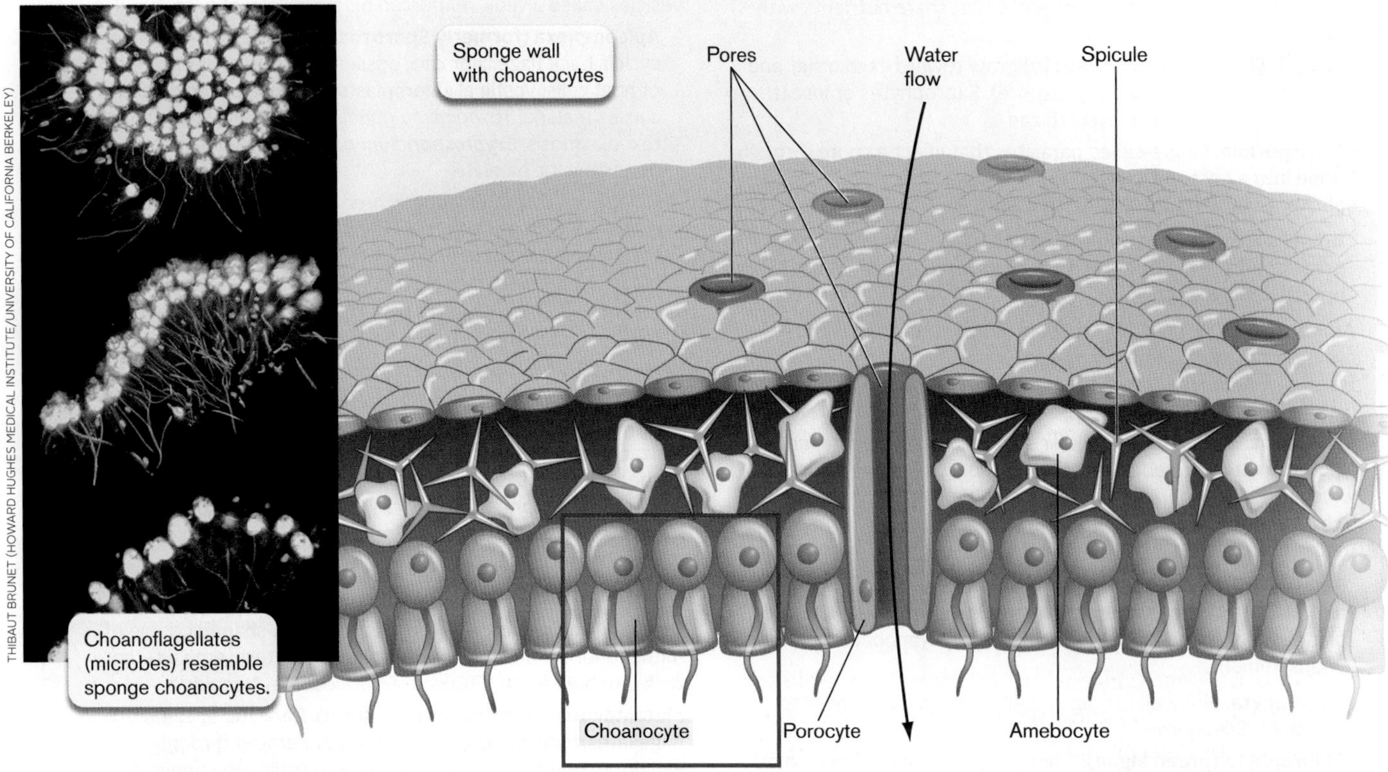

FIGURE 20.5 ■ Choanoflagellates resemble sponge choanocytes. Sponge choanocytes resemble choanoflagellates. Within the sponge, choanocytes assist the circulation of water and the uptake of nutrients. Inset: Choanoflagellates, single-celled microbes that form colonies resembling an animal cell layer. *Source*: Nicole King.

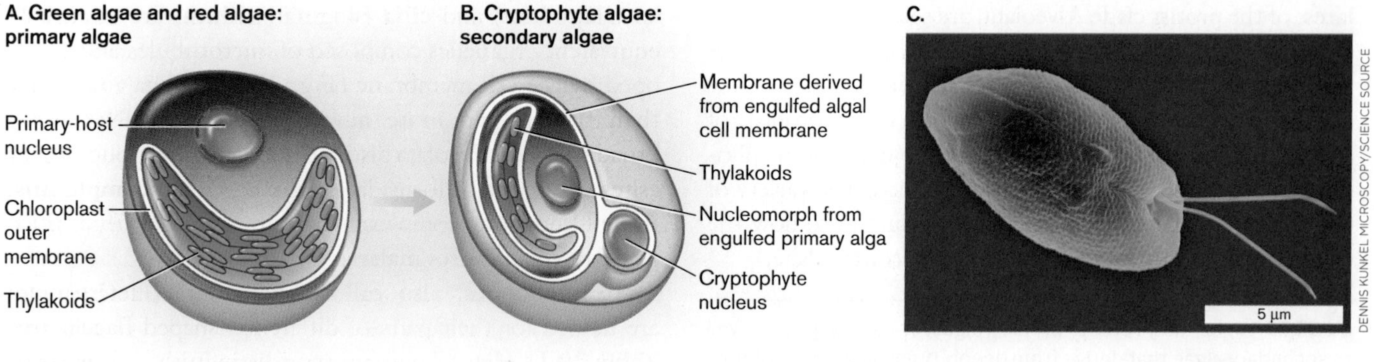

FIGURE 20.6 ▪ Chloroplast evolution: primary and secondary endosymbiosis. **A.** Green algae (Chlorophyta) and red algae (Rhodophyta) contain chloroplasts (green) that evolved from engulfed cyanobacteria. **B.** Cryptophyte (cryptomonad) algae contain chloroplasts (green); vestigial primary-host cytoplasm (yellow); and a vestigial nucleus, or nucleomorph (purple), from the engulfed primary endosymbiont. **C.** Marine cryptomonad, *Rhodomonas salina* (colorized SEM), shows double flagella.

Fungi (Eumycota) consist of cells with chitinous cell walls that may grow in chains called **hyphae** (singular, **hypha**). Fungi range from single-celled organisms, such as yeasts, to complex multicellular forms such as mushrooms. The deepest-branching clade of fungi, Chytridiomycota, produces the uniflagellar zoospores that define opisthokonts. Other fungi, however, generate nonmotile reproductive cells—a result of reductive evolution.

Several taxa that historically were grouped with fungi (for example, slime molds) are now classified in more deeply branching clades. Slime molds generate populations of cells that migrate into a unified structure called a **fruiting body** to make reproductive cells. Slime molds are now grouped with amebas (Amoebozoa, discussed in Section 20.3). Water molds, which are plant and animal pathogens of the class Oomycetes (Oomycota), are now recognized as heterokont protists.

Amoebozoa Branch near Opisthokonts

Amebas (also spelled **amoebas**) are unicellular organisms of highly variable shape that form **pseudopods**, locomotory extensions of cytoplasm enclosed by the cell membrane. Their size can reach several millimeters, and they can eat small invertebrate animals. The Amoebozoa, the most familiar kind of amebas, have lobed pseudopods, pseudopods that extend lobes of cytoplasm through cytoplasmic streaming. Most lobed amebas are free-living in aquatic habitats, but some cause human diseases such as amebic dysentery. Still other kinds of lobed amebas are cellular slime molds, in which individual amebas converge to form a fruiting body.

A more deeply branching group of protists that are also called "amebas" is the **Rhizaria** (see **Table 20.1**). The Rhizaria have thin, filamentous pseudopods, often radially arranged like a star, as in heliozoa. Despite the term "amebas," the genetics and physiology of Rhizaria differ greatly from those of Amoebozoa. Some Rhizaria, such as the foraminiferans, form inorganic shells called tests. Fossil foraminiferan tests are common in rock formations derived from ancient seas. Foraminiferan shells formed the White Cliffs of Dover in Britain and the stone used to build the Egyptian pyramids.

Algae Evolved by Engulfing Phototrophs

Algae are commonly defined as single-celled plants and simple multicellular plants lacking true stems, roots, and leaves. Algal cells contain **chloroplasts**, membrane-enclosed organelles of photosynthesis that evolved from a cyanobacterium (**Fig. 20.6**). In Earth's biosphere, algae plus bacterial phototrophs feed all marine and freshwater ecosystems, producing the majority of oxygen and biomass available for Earth's consumers.

How did ancestral algae get their chloroplasts? The **Plantae** include multicellular plants as well as **primary algae**, all descended from a common ancestor containing a chloroplast. The chloroplast evolved by endosymbiosis between an ancient protist and a cyanobacterium (discussed in Chapter 17). In primary algae, the chloroplast is enclosed by two membranes (**Fig. 20.6A**): the inner membrane (from the ancestral phototroph's cell membrane) and the outer membrane (from the host cell membrane as it enclosed its prey). Both **green algae** (**chlorophytes**) and **red algae** (**rhodophytes**) are primary algae. Their chloroplasts diverged from their common ancestor to use pigments absorbing different ranges of the light spectrum.

Surprisingly, several other taxa traditionally called algae evolved through a secondary endosymbiosis with a second protist host. The symbiotic history of these **secondary algae** is most evident in the cryptophytes (**Fig. 20.6B**), which still retain a vestigial nucleus, or **nucleomorph**, derived from the engulfed alga. In secondary algae, the chloroplast is surrounded by two extra membranes—one from the primary alga and one from the secondary host. An example of a secondary alga, the cryptomonad *Rhodomonas salina*, is shown in **Figure 20.6C**.

Cryptomonads and other secondary algae are abundant in the oceans. Other secondary algae include the

chrysophytes, such as kelps and diatoms. The **dinoflagellates**, of the protist clade Alveolata, are secondary or tertiary algae, descended from a flagellate that consumed one or more types of algae. Dinoflagellates also engage in "kleptoplasty," or "chloroplast stealing," in which the chloroplast of a digested prey is retained long enough to derive some photosynthetic energy, but ultimately consumed. The variety of endosymbiosis among protists provides clues as to how the original chloroplast evolved within the ancestral algae.

Note: "Algae" may refer to primary algae, or "true algae," as well as secondary algae that derive from deeply branching clades of protists. Fungi (the "true fungi," or Eumycota) are opisthokonts. Several fungus-like organisms have been reclassified within protist clades.

Other Protist Clades

Protists include multiple distantly related categories of eukaryotes (see **Table 20.1**). Aside from primary algae, all protists are heterotrophs, commonly predators or parasites, although many also conduct photosynthesis as secondary algae. Protists are important producers and consumers in marine, freshwater, and soil food webs. In ecology, phototrophic protists are termed "phytoplankton" and heterotrophs are termed "zooplankton," although many, in fact, are "mixotrophs" that act as both producers and consumers.

Alveolates (Alveolata) include ciliated protists (ciliates), dinoflagellates, and apicomplexans. An example is the ciliated protist *Paramecium* (**Fig. 20.7A**). Alveolates are known for their complex outer covering, or cortex. The cortex contains networks of vesicles called **cortical alveoli** (singular, **alveolus**) (**Fig. 20.7B**). Alveoli store calcium ions, and in some species protective plates form from them. Organisms equipped with paired flagella or cilia are known as **flagellates** and **ciliates**, respectively. **Flagella** (singular, **flagellum**) and **cilia** (singular, **cilium**) are essentially equivalent organelles composed of microtubules and enveloped by the cell membrane (**Fig. 20.7C**). Cilia are shorter than flagella and more numerous, and usually cover a broad surface. Alveolata also includes a major group of parasites whose major forms lack flagella, the **apicomplexans**. A well-known apicomplexan parasite is *Plasmodium falciparum*, which causes malaria.

Stramenopiles, also called heterokonts (Heterokonta) are named for their pairs of differently shaped flagella (see **Table 20.1**). Heterokonts are thus distinguished from opisthokonts, which possess a single unpaired flagellum (if any). Flagellated heterokonts often possess two flagella of unequal length, one of which has hairs. The heterokonts include voracious zooplankton, common in marine and freshwater environments. Other heterokonts include the oomycetes, or water molds (formerly classified with fungi), as well as diatoms and kelps (secondary endosymbiotic algae with chloroplasts). **Diatoms** are single cells with unique bipartite shells that fit together like a Petri dish. The shells of diatoms form an infinite variety of different patterns for different species. Kelps, also known as brown algae, extend multicellular sheets floating at the water's surface. Some taxonomists group the Stramenopiles, Alveolata, and Rhizaria together as the "SAR" clade (see **Fig. 20.4** and **Table 20.1**).

The Discoba include free-living protists such as euglenas, as well as parasites showing extensive evolutionary reduction. Most Discoba have mitochondria with distinctive disk-shaped cristae (membrane pockets). Parasitic Discoba include the trypanosomes that cause sleeping sickness and Chagas' disease. A related clade, Metamonada, includes the waterborne parasite *Giardia intestinalis* (also known as *G. lamblia*). In

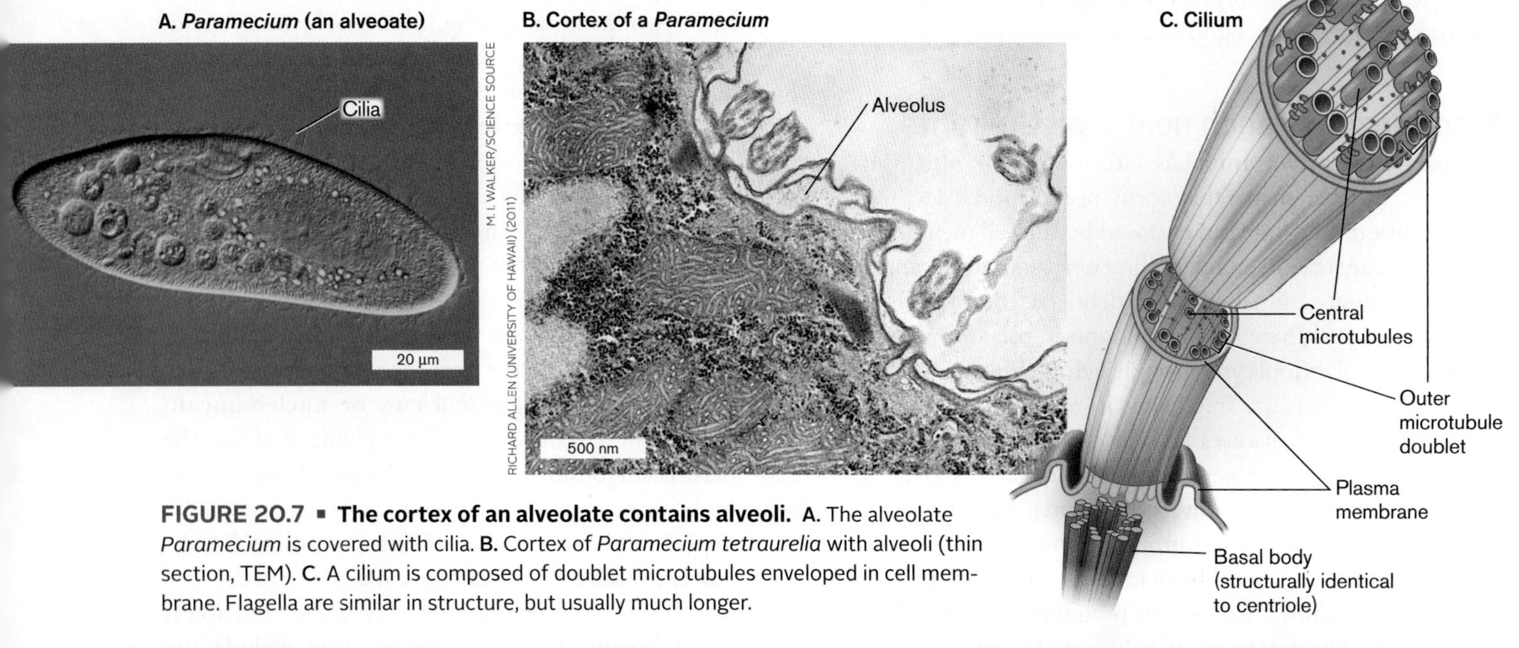

FIGURE 20.7 ■ The cortex of an alveolate contains alveoli. **A.** The alveolate *Paramecium* is covered with cilia. **B.** Cortex of *Paramecium tetraurelia* with alveoli (thin section, TEM). **C.** A cilium is composed of doublet microtubules enveloped in cell membrane. Flagella are similar in structure, but usually much longer.

metamonads the mitochondria have lost their genomes and degenerated into "mitosomes" (discussed in Section 20.6).

Emerging Eukaryotes

Are there still eukaryotes that we have yet to discover? Genes from natural communities continually reveal new species of microbial eukaryotes in previously unknown divisions. Many of the new isolates are single cells as small as bacteria, designated nanoeukaryotes (3–10 μm) and picoeukaryotes (0.5–3 μm).

Genetic analysis shows that similar miniaturized eukaryotes branch deeply from all groups in the phylogenetic tree, potentially doubling the known number of major eukaryotic taxa. Furthermore, our metagenomic analysis reveals eukaryotes in environments previously believed to be restricted to bacteria and archaea, such as anaerobic submarine sediments, and the hyperacidic Tinto River in Spain. In every new habitat tested, new deep-branching clades of eukaryotes emerge, with new implications for ecology and global cycling of elements (discussed in Chapters 21 and 22). The wealth of new genomic data is reshaping our understanding of the domain Eukarya.

To Summarize

- **Opisthokonta** includes true fungi (Eumycota) and multicellular animals (Metazoa), as well as related clades such as Microsporidia.
- **Plantae** (also called Archaeplastida) includes green plants and primary algae.
- **Other protist clades** include Amoebozoa, the unshelled amebas with lobe-shaped pseudopods; Rhizaria, the filopodial amebas; Alveolata, ciliates and flagellates with complex cortical structure; Stramenopiles, the kelps, diatoms, and flagellates with nonequivalent paired flagella; and Discoba and Metamonada, primarily parasites.
- **Many protists are phototrophs as well as heterotrophs**, based on secondary or tertiary endosymbiosis derived from engulfed algae.
- **Metagenomic analysis reveals new clades of microbial eukaryotes.** New kinds of microbial eukaryotes emerge from all habitats.

20.2 Fungi

Fungi provide essential support for all communities of multicellular organisms. Fungi recycle the biomass of wood and leaves, including substances such as lignin, which other organisms may be unable to digest. Underground fungal filaments called mycorrhizae extend the root systems of most plants, forming a nutritional "internet" that interconnects the plant community (discussed in Chapter 21). Mycorrhizae may have inspired the fictional underground tree network depicted in the film *Avatar* (2009).

Within the ruminant digestive tract, fungi ferment plant materials. On the other hand, pathogenic fungi infect plants and animals, and they contribute to the death of immunocompromised human patients. Still other fungi produce antibiotics such as penicillin, as well as food products such as wine and cheeses (discussed in Chapter 16).

Shared Traits of Fungi

Most fungi share these distinctive traits:

- **Absorptive nutrition.** Most fungi cannot ingest particulate food, as do protists, because their cell walls cannot part and re-form, as do the flexible pellicles of amebas and ciliates. Instead they secrete digestive enzymes and then absorb the broken-down molecules from their environment.
- **Hyphae.** Most fungi grow by extending multinucleate cellular filaments called hyphae (singular, hypha; **Fig. 20.8A**).

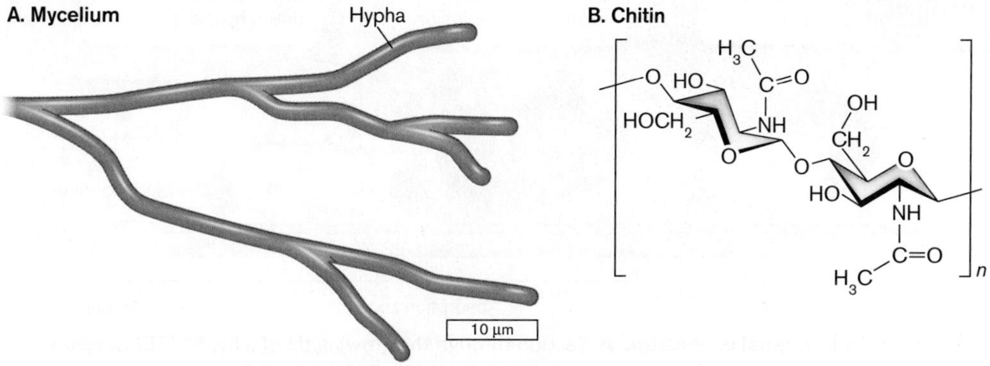

FIGURE 20.8 ■ Fungi grow hyphae with cell walls of chitin. A. Fungal hyphae extend and form branches, generating a mycelium. B. Chitin consists of beta-linked polymers of *N*-acetylglucosamine.

As a hypha extends, its nuclei divide mitotically without cell division, generating a multinucleate cell. Later, septa may form to partition the hypha into cells. Hyphae grow by cytoplasmic extension and branching. A branched mass of extending hyphae is called a **mycelium** (plural, **mycelia**).

- **Cell walls contain chitin.** Chitin is an acetylated amino-polysaccharide of immense tensile strength, stronger than steel (**Fig. 20.8B**). Its strength derives from multiple hydrogen bonds between fibers. Chitinous cell walls enable fungi to penetrate plant or animal cells, including tough materials such as wood. Inhibitors of chitin synthesis, such as the polyoxins and nikkomycins, are used as antibiotics against fungal infections.
- **Membranes contain ergosterol.** Ergosterol is an analog of cholesterol not found in animals or plants. Ergosterol is so distinctive to fungi that its presence can be used as a measure of fungal content in plant food products such as grains. Inhibitors of ergosterol biosynthesis, such as the triazoles, are used to treat fungal infections. Another antifungal agent, nystatin, specifically binds ergosterol and forms membrane pores that leak K^+ ions.

Fungal Hyphae Absorb Nutrients

How do fungal hyphae grow and form colonies of "mold"? A fungal hypha expands at the tip. Cytoplasmic expansion is driven by turgor pressure against the chitin cell wall—the force that enables fungi to penetrate tough materials such as wood. Fungal hyphae can extend as fast as half a centimeter per hour.

The cytoplasmic turgor pressure is regulated by uptake of hydrogen ions in exchange for potassium ions (**Fig. 20.9**). Loss of turgor pressure—for example, by puncture of the hypha—leads to accelerated K^+ uptake and water influx, restoring turgor. Molecules that cause loss of K^+, such as nystatin, serve as antifungal agents.

At the hypha's growing tip, turgor pressure pushes the cell membrane forward, and the membrane expands by incorporating vesicles generated from the endoplasmic reticulum (as seen in **Fig. 20.9A**). The ER and mitochondria store Ca^{2+}, whose release triggers vesicle fusion with the plasma membrane. The fused vesicles provide phospholipids and proteins to extend the membrane surface area as the cytoplasm expands.

A.

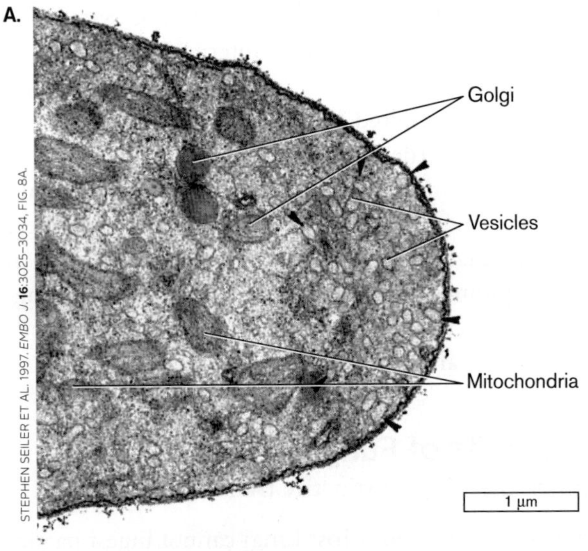

B.

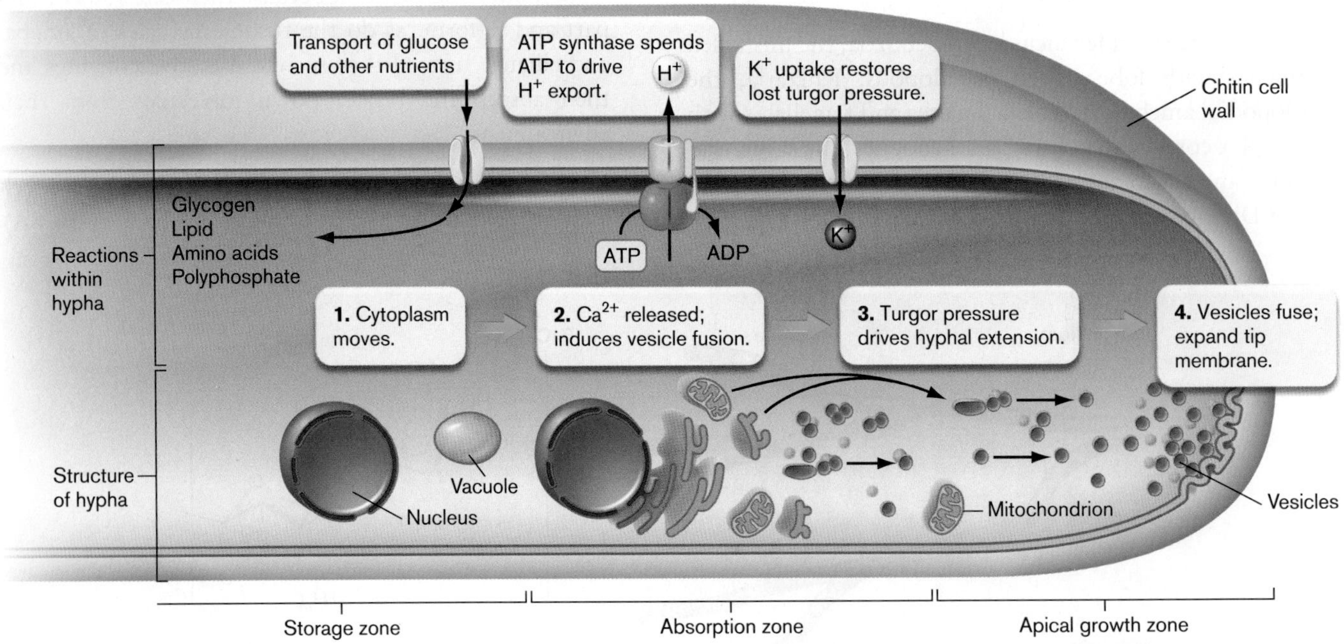

FIGURE 20.9 ■ Cellular basis of hyphal extension. **A.** Section through the growing tip of a hypha (TEM). Vesicles collect at the tip, where they fuse into the cell membrane, enabling extension. **B.** The absorption zone takes in nutrients. Cytoplasm moves toward the tip of the apical growth zone, driven by turgor pressure. Turgor pressure is regulated by H^+ export and K^+ uptake. Ca^{2+} released by the ER and mitochondria induces vesicles to fuse and to expand the plasma membrane at the growing tip.

Just behind the hypha's growing tip lies its absorption zone (**Fig. 20.9B**). The absorption zone takes in nutrients from the surrounding medium, such as the cytoplasm of an invaded animal cell. Behind the absorption zone, the older part of the hypha collects and stores nutrients. As the storage zone expands, the nucleus divides multiple times. Septa form across the hypha, partly compartmentalizing the cytoplasm. As the older part of the hypha ages, its tubular form begins to lyse, releasing cell contents. This aging part of the hypha is called the senescence zone (not shown in the figure).

As hyphae grow, branches extend from their sides. The hyphae branch and extend radially, forming the mycelium. The mycelium forms the characteristic round, fuzzy colony of a fungus or "mold." On a substrate such as wood or agar, mycelia grow in two forms: aerial mycelium, which extends out into the air; and surface mycelium, which grows into and along the surface of the substrate.

Unicellular Fungi

Despite the advantages of multinucleate hyphae, some fungi are unicellular, known as **yeasts**. Yeast forms evolved in many different fungal taxa. The yeast *Saccharomyces cerevisiae* (**Fig. 20.10A**) is used to leaven bread and to brew wine

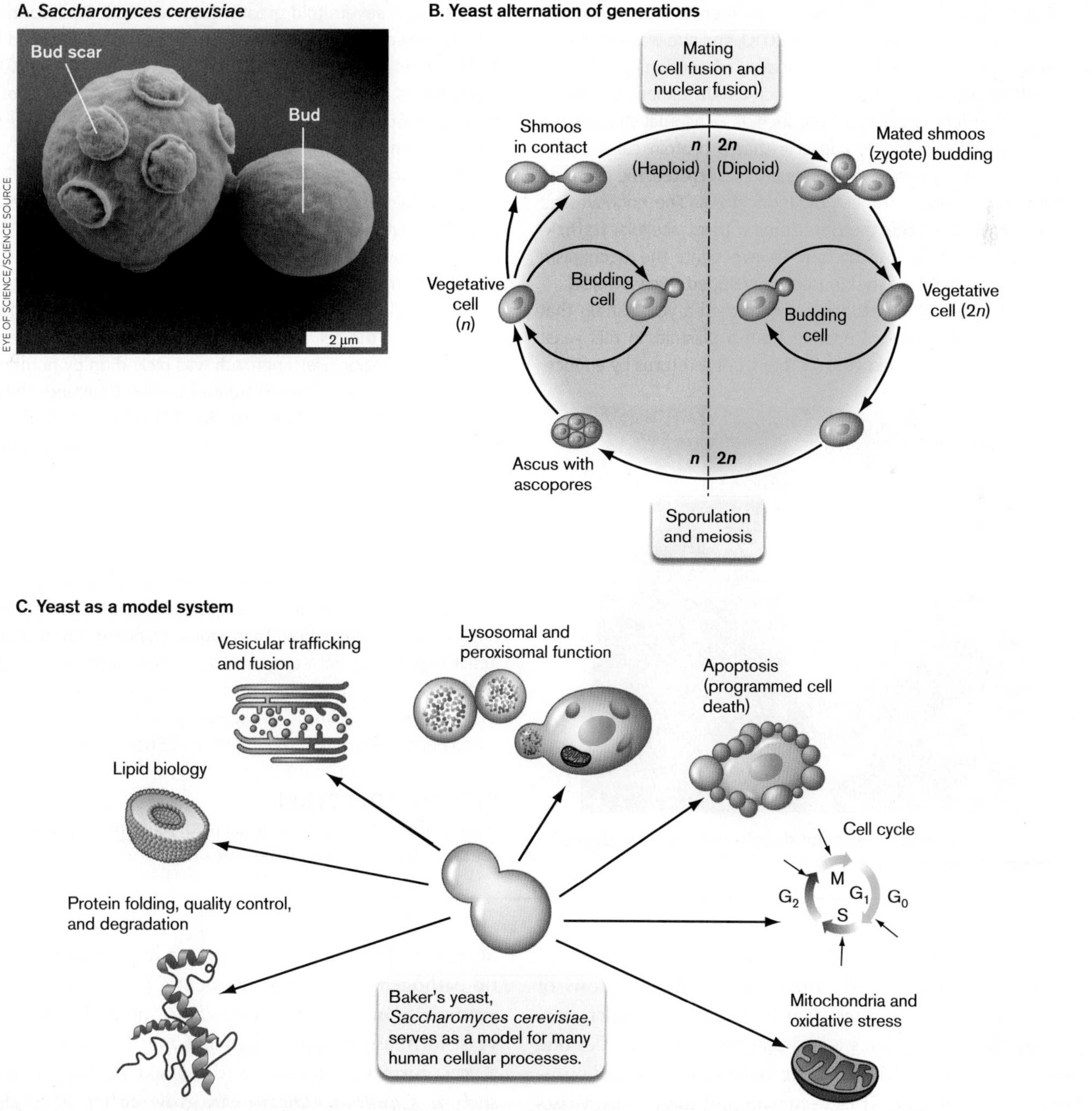

FIGURE 20.10 ▪ Yeasts are nonmycelial fungi. A. *Saccharomyces cerevisiae*, or baker's yeast, reproduces by budding (SEM). Upper cell shows six bud scars. **B.** In the life cycle of *S. cerevisiae*, haploid cells reproduce many generations by budding. **C.** *S. cerevisiae* serves as a model for human cellular processes. *Source:* Part C modified from Vikram Khurana and Susan Lindquist. 2010. *Nat. Rev. Neurosci.* **11**:436–449.

SPECIAL TOPIC 20.1 Yeast: A Single-Celled Human Brain

The yeast *Saccharomyces cerevisiae* provides a model for human diseases such as cancer because it shares so many homologs of our genes. Most of these genes encode fundamental parts of cells, such as actin and cell growth regulators. But, of course, the single-celled fungus lacks the differentiated parts and connectors of human cells such as neurons. So, yeast could not serve as a model for complex diseases of the brain. Or could it? What sounds impossible was accomplished by the pioneering yeast molecular biologist Susan Lindquist (1949–2016) at the Whitehead Institute for Biomedical Research (**Fig. 1**). An extraordinary molecular biologist, Lindquist earned the National Medal of Science in 2010, and she was elected as a Foreign Member of the Royal Society of London in 2015.

Neurodegenerative diseases such as Alzheimer's can result from disorders of cell function, such as lysosomal storage and degradation. A possible cause is that a protein called beta-amyloid gets cleaved to a peptide that is secreted; the secreted peptide returns to the cell by endocytosis. The endocytosed beta-amyloid peptide then interferes with intracellular trafficking by an unknown mechanism. To investigate the mechanism of beta-amyloid toxicity, Lindquist developed a yeast model.

First, she and her students constructed a yeast strain that expresses beta-amyloid peptide from a plasmid. In this yeast strain the beta-amyloid gets secreted, but it returns by endocytosis to the endoplasmic reticulum (ER) for trafficking—just as it does in a human cell. So, what happens to the yeast? The yeast grows more slowly because beta-amyloid disrupts ER trafficking.

FIGURE 1 ■ Susan Lindquist developed yeast models of human disease.

The disruption of trafficking by beta-amyloid was consistent with its proposed role in Alzheimer's disease. But to strengthen the connection—and to reveal other parts of the process—Lindquist used her yeast model to screen for yeast proteins that could overcome beta-amyloid interference. So she transformed her yeast model strain with a library of yeast genes that were "overexpressed" (that is, expressed on a plasmid at levels severalfold greater than normal). She identified 23 suppressors, genes whose overexpression suppressed the effect of beta-amyloid and restored normal yeast growth. The suppressors also restored normal trafficking of a reporter protein, a fluorescent YFP fusion protein expressed by the yeast genome (**Fig. 2**). The control cells showed reporter protein fluorescence localized normally to the vacuole (**Fig. 2A**). Beta-amyloid expression prevented trafficking to the vacuole (**Fig. 2B**), but expression of the suppressor protein YAP1802 restored movement to the vacuole (**Fig. 2C**).

At the time of her death in 2016, Lindquist was using the yeast model to explore the mechanism of beta-amyloid's effects within the cell and test possible therapeutic agents for neurodegeneration. Her approach was picked up by numerous colleagues, including Dina Petranovic at the Chalmers University of Technology, Gothenburg, Sweden (**Fig. 3A**). Petranovic developed a "humanized" yeast beta-amyloid model that more closely resembles human cells in the timing of expression of the toxic protein. Her model reveals several processes that impact cell metabolism and morphology. **Figure 3B** and **C** shows the visualization of yeast mitochondria, using two-photon excitation fluorescence (TPEF), a method in which two infrared photons simultaneously excite the fluorophore; this method achieves higher spatial resolution than does standard fluorescence. Yeast expressing beta-amyloids show fragmented mitochondria, while respiration is decreased. This pathology resembles that seen in the brains of Alzheimer's patients. Petranovic dedicated her publication to the memory of Susan Lindquist.

RESEARCH QUESTION

How would you use the humanized yeast model to identify therapeutic agents for Alzheimer's disease?

and beer (for more on food microbiology, see Chapter 16). *S. cerevisiae* reproduces by **budding**, in which mitosis of the mother cell generates daughter cells of smaller size. The mother cell acquires a bud scar where the smaller one pinched off (**Fig. 20.10A**). After generating a limited number of buds, the mother cell senesces and dies. Thus, yeasts provide a unicellular model system for the process of aging.

Other yeasts, such as *Candida albicans*, are important members of the human vaginal microbiome but can cause opportunistic infections. Some are important opportunistic pathogens, occurring frequently in AIDS patients; for example, *Pneumocystis jirovecii* (formerly *P. carinii*) is a yeast-form ascomycete, whereas *Cryptococcus neoformans* is a yeast-form basidiomycete (discussed shortly). Pathogens such as *Candida albicans* can grow either as single cells (yeast form) or as mycelia; because they exist in two different forms, these are known as "dimorphic" fungi. The yeast form grows normally in the mucosa, but germination

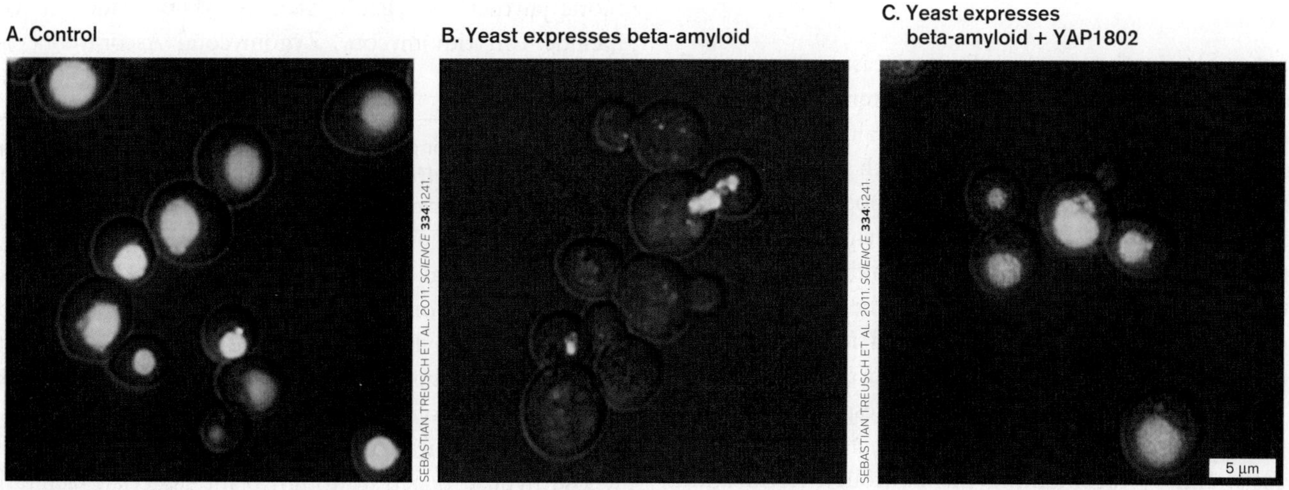

SEBASTIAN TREUSCH ET AL. 2011. *SCIENCE* **334**:1241.

FIGURE 2 ■ Beta-amyloid disrupts traffic of yeast protein. **A.** A yeast protein fused to yellow fluorescent protein (YFP) is trafficked normally to the vacuole (fluorescence microscopy). **B.** Beta-amyloid expression interferes with transport. **C.** YAP1802 expression restores transport.

Treusch, Sebastian, Shusei Hamamichi, Jessica L. Goodman, Kent E. S. Matlack, Chee Yeun Chung, et al. 2011. Functional links between Ab toxicity, endocytic trafficking, and Alzheimer's disease risk factors in yeast. *Science* **334**:1241–1245.

Chen, Xin, Markus M. M. Bisschops, Nisha R. Agarwal, Boyang Ji, Kumaravel P. Shanmugavel, et al. 2017. Interplay of energetics and ER stress exacerbates Alzheimer's amyloid-β (Aβ) toxicity in yeast. *Frontiers in Molecular Neuroscience* **10**:232.

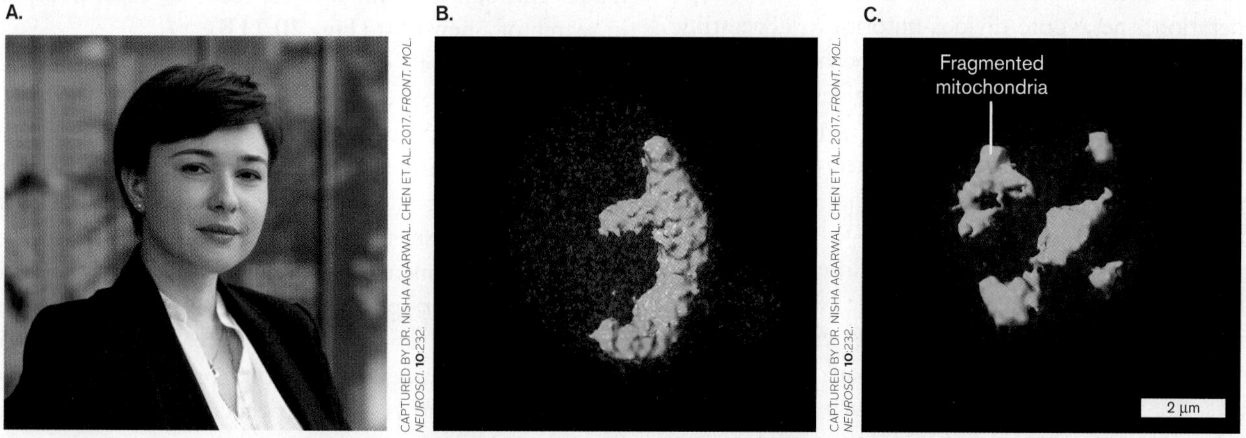

PHOTO BY JAN-OLOF YXELL, CHALMERS UNIVERSITY OF TECHNOLOGY

CAPTURED BY DR. NISHA AGARWAL. CHEN ET AL. 2017. *FRONT. MOL. NEUROSCI.* **10**:232.

FIGURE 3 ■ Human beta-amyloid expression alters yeast mitochondria. **A.** Dina Petranovic. **B.** Yeast control with mitochondria stained green by rhodamine 123, viewed by two-photon excitation fluorescence (TPEF) microscopy. **C.** Yeast expressing beta-amyloid oligomer show fragmented mitochondria.

of mycelia leads to disease. A very different dimorphic fungal pathogen is *Blastomyces dermatitidis*, the cause of blastomycosis, a type of pneumonia. *B. dermatitidis* forms a mycelium in culture and in soil environments, but it grows as a yeast within the infected lung.

Yeast as a model organism for research. The yeast *Saccharomyces cerevisiae* is a model organism for research in eukaryotic biology (**Fig. 20.10C**). Its cells grow rapidly, in haploid and diploid forms, and are amenable to genetic recombination and transformation. The yeast genome of 6,000 genes includes many human homologs, such as the *ras* proto-oncogene (a gene involved in cancer). Often, entire networks of proteins interacting in yeast have human homologs; thus, *S. cerevisiae* has been called a "single-celled human." Susan Lindquist (1949–2016) at MIT's Whitehead Institute for Biomedical Research developed yeast as a model for brain neuronal diseases such as Alzheimer's (**Special Topic 20.1**).

Thought Question

20.2 Why would yeasts remain unicellular? What are the relative advantages and limitations of hyphae?

Yeast reproductive cycles. Some yeasts are asexual, whereas others can undergo sexual **alternation of generations** (**Fig. 20.10B**). This life cycle alternates between generation of a haploid population, with a single copy of each chromosome (n), and a diploid population, with a diploid chromosome number ($2n$). The haploid form develops gametes to fertilize each other, making a $2n$ zygote. After vegetative (nonsexual) divisions, the $2n$ form undergoes meiosis, regenerating the haploid form. (The process of meiosis is reviewed in eAppendix 2.) Alternation of generations allows an organism to respond genetically to environmental change by reassorting its genes through meiosis, and by recombining them through fertilization. Gene reassortment and recombination provide new genotypes, some of which may increase survival in the changed environment.

In baker's yeast (*Saccharomyces cerevisiae*), haploid spores divide and proliferate by mitosis. Under environmental signals such as starvation, mating factors induce the haploid cells to differentiate into gamete forms called "shmoos" (see **Fig. 20.10B**). Gametes of two different mating types fuse, and their nuclei combine to form a zygote. In the diploid generation, the zygote divides mitotically, generating a population of diploids that appear superficially similar to haploid cells. Under stress, particularly desiccation, the diploids undergo meiosis to reassort their genes for combinations that may better survive the changed environment. Meiosis generates an **ascus** (plural, **asci**) that contains four haploid spores.

Many fungi and protists undergo modified versions of alternation of generations, utilizing a wide variety of haploid and diploid structures to accomplish essentially the same genetic tasks. In many fungi, the haploid form predominates; for example, ascomycetes such as *Aspergillus* and *Neurospora* form mainly haploid mycelia. In some fungi, no sexual reproduction has been observed, probably because the inducing conditions are unknown. Species that lack a known sexual cycle are called **mitosporic fungi**, also known as "imperfect fungi." Mitosporic species are found in many different clades. An example is the famous *Penicillium* mold, an ascomycete, from which we discovered the antibiotic penicillin.

Thought Question

20.3 Why would some fungi avoid sexual reproduction? What are the advantages and limitations of sexual reproduction?

Mycelia, Mushrooms, and Mycorrhizae

Different species of fungi show vastly different forms, from the familiar mushrooms (fruiting bodies that can weigh several pounds) to the mycelia of pathogens and the symbiotic partners of algae in lichens. Major clades of fungi include Chytridiomycota, Zygomycota, Ascomycota, and Basidiomycota.

Note: The major groups of fungi are also known by names with the alternative suffix "-etes": Chytridiomycetes, Zygomycetes, Ascomycetes, Basidiomycetes.

Chytridiomycota: motile zoospores. The deepest-branching clade of fungi is Chytridiomycota (the chytridiomycetes, or chytrids), which possess motile, flagellated reproductive forms called zoospores. The zoospore form has been lost by other fungi.

Chytrid species include bovine rumen inhabitants whose hyphae penetrate tough plant material, facilitating digestion. Other chytrids are aerobic animal pathogens. **Figure 20.11A** shows the skin of a frog infected by the chytridiomycete *Batrachochytrium dendrobatidis*. *B. dendrobatidis* has caused a widespread die-off of frogs in Central and South America, in an epidemic associated with global warming. The mycelium of *B. dendrobatidis* grows within the frog skin, producing capsules full of diploid zoospores called zoosporangia. Each zoosporangium protrudes through the skin surface, ready to expel zoospores in search of a new host (**Fig. 20.11B**).

The life cycle of a chytrid includes mycelia that are haploid (gametophyte) or diploid (sporophyte; **Fig. 20.11C**). Haploid mycelia produce motile gametes that detect each other by sex-specific attractants. The gametes fuse to produce a motile zygote. The zygote forms a cyst, a cell with arrested metabolism that can persist for long periods. In a favorable environment, the cyst germinates to form a diploid mycelium, or sporophyte. The sporophyte generates zoosporangia full of zoospores. There are two alternative forms of zoosporangia: those that produce diploid zoospores, which form cysts and regenerate the diploid mycelium; and those that undergo meiosis to produce haploid zoospores. The haploid zoospores generate a haploid mycelium (gametophyte) capable of producing haploid gametes.

Zygomycota: nonmotile sporangia. The zygomycetes and other nonchytridiomycete fungi generate nonmotile spores. Nonmotile spores require transport by air or water, or ballistic expulsion (expulsion under pressure) from a spore-bearing organ, called the **sporangium** (plural, **sporangia**). A common zygomycete is the bread mold *Rhizopus* (**Fig. 20.12A**). Most zygomycetes, such as *Mucor* species, are soil molds that decompose plant material or other fungi or the droppings of animals (**Fig. 20.12B**). These modest molds fill important niches in all terrestrial ecosystems.

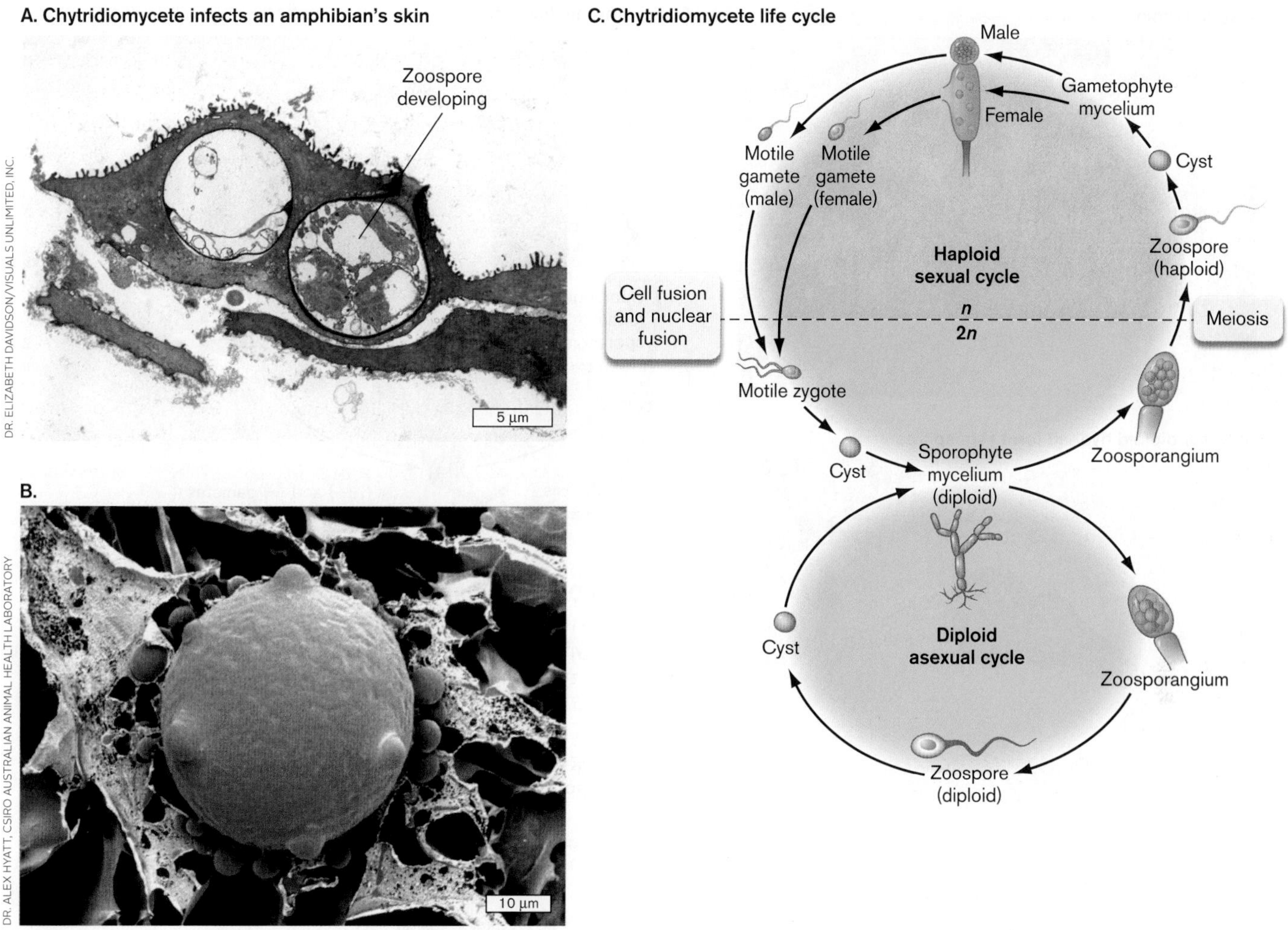

FIGURE 20.11 ■ Chytridiomycete form and life cycle. **A.** A pathogenic chytrid, *Batrachochytrium dendrobatidis*, infects the skin of an amphibian. Amphibian skin cells are penetrated by discharge tubes of diploid zoosporangia about to release zoospores. **B.** Zoosporangium releasing zoospores. **C.** Life cycle of a chytrid. The diploid mycelium produces motile zoospores that form cysts in a poor environment. Alternatively, the diploid mycelium undergoes meiosis to form a haploid mycelium (gametophyte) that produces motile gametes.

As with chytrids, the life cycle of a zygomycete alternates between haploid (n) and diploid ($2n$) forms (**Fig. 20.12C**). The mechanics differ, however, owing to the lack of motile gametes. The haploid spore (sporangiospore) is disseminated through air currents. A sporangiospore does not directly undergo sexual reproduction; it grows into a haploid mycelium. The haploid mycelium then forms special hyphae whose tips differentiate into gamete cells. The gametes cannot separate from the filament; instead, two gamete-bearing hyphae must grow toward each other in order to fuse and form a **zygospore**. The zygospore undergoes meiosis and generates the sporangium, a haploid structure that releases **sporangiospores**.

Thought Question

20.4 What are the advantages and limitations of motile gametes, as compared to nonmotile spores?

Ascomycota: mycelia with paired nuclei. The **ascomycete** fungi are famous in the history of science, as well as in the culinary arts (**Fig. 20.13**). The bread mold *Neurospora* was used by George Beadle and Edward Tatum in the 1940s to formulate the one gene–one protein theory. In *Neurospora*, meiosis produces pods (asci) of **ascospores** aligned in rows that reflect the ordered tetrads of meiotic division (**Fig. 20.13A**). The tetrad patterns were used by geneticists to demonstrate the segregation and independent assortment of chromosomes. In other species, by contrast, the asci are packed in large mushroom-like fruiting bodies known as morels (*Morchella hortensis*; **Fig. 20.13B**) and truffles (*Tuber aestivum*). The ascospores of such fruiting bodies are spread by animals attracted by their delicious flavor. Human collectors traditionally use muzzled pigs to detect and unearth the famous underground truffles.

The ascomycete life cycle (**Fig. 20.13C**) includes a phase in which each cell possesses a pair of separate nuclei, one from each parent (chromosome number is designated $n + n$).

A. Bread mold, *Rhizopus*

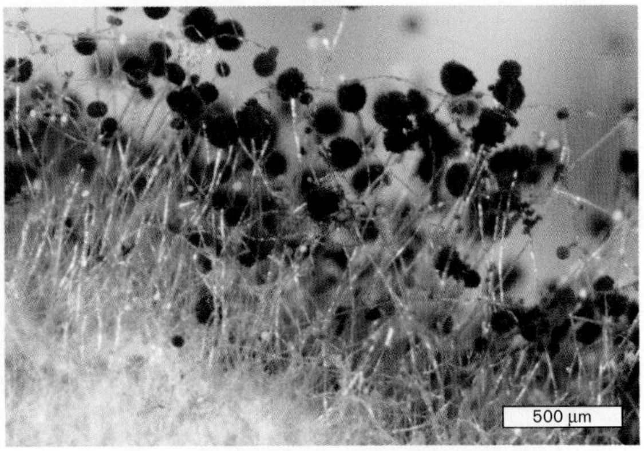

B. *Mucor* diploid hyphae form zygospores

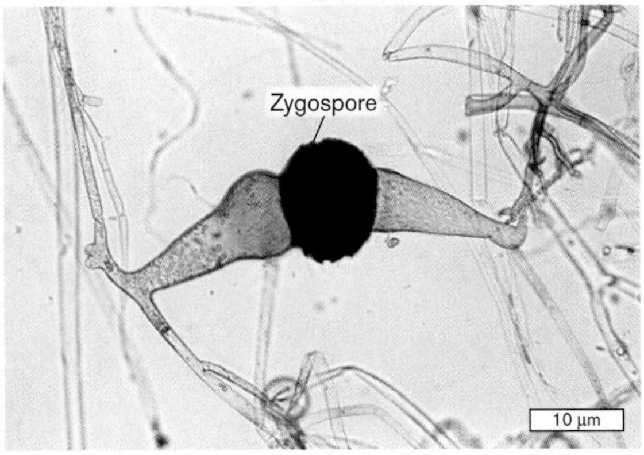

C. Zygomycete life cycle

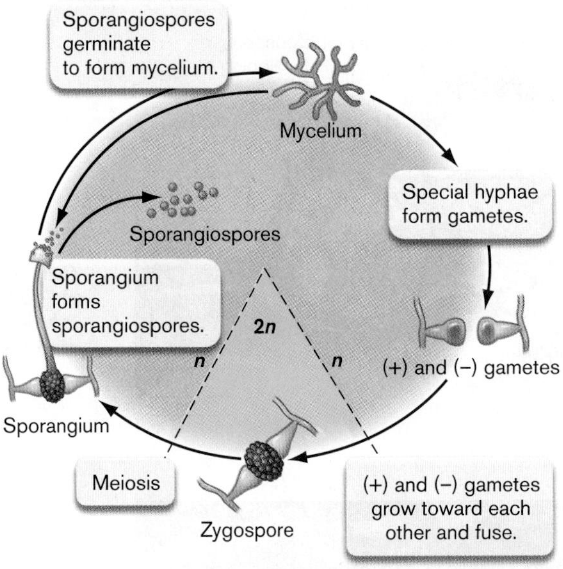

FIGURE 20.12 ▪ Zygomycete fungi form nonmotile sporangia. **A.** *Rhizopus* (bread mold) haploid sporangia contain sporangiospores. **B.** Diploid hyphae of Mucor species terminate in zygospores. **C.** The life cycle of zygomycetes involves primarily haploid mycelia. Special hyphae form gametes at their tips. Gametes of different mating types fuse to form the diploid zygospore. The zygospore undergoes meiosis, regenerating haploid cells that form sporangia. The sporangia release sporangiospores, which germinate to form new mycelia.

A. Asci containing ascospores

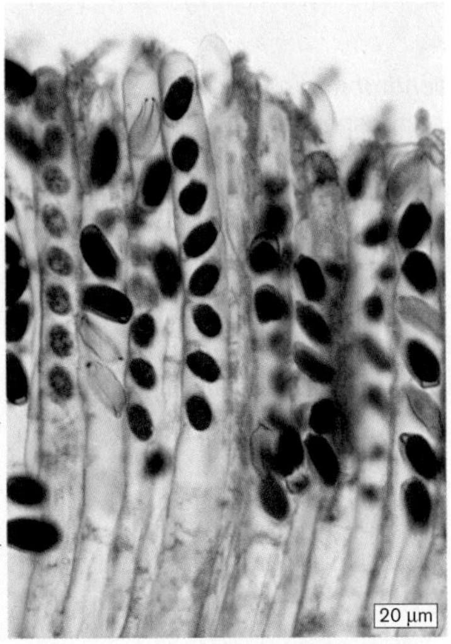

B. Morel (an ascomycete fruiting body)

C. Ascomycete life cycle

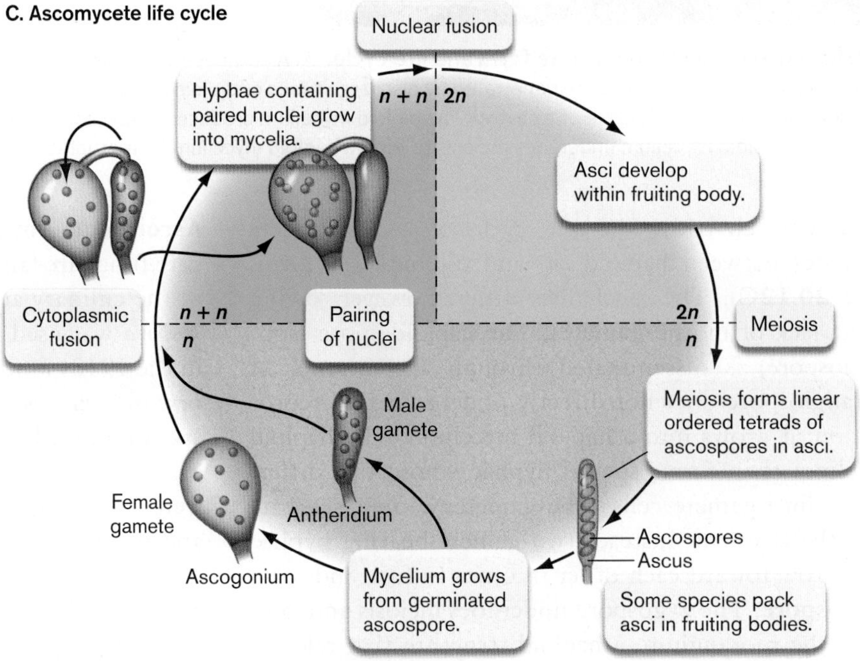

FIGURE 20.13 ▪ Ascomycetes produce large fruiting bodies. **A.** Ascomycete asci containing ascospores (stained red). **B.** The culinary delicacies known as morels are fruiting bodies of the species *Morchella hortensis*. The dark pits of the morel are lined with asci. **C.** The life cycle of an ascomycete alternates between the diploid and haploid forms. The diploid mycelium produces asci, within which the haploid ascospores are formed.

The "dikaryotic" (paired-nuclei) phase is generated by haploid mycelia in which male and female reproductive structures fuse, followed by migration of all the male nuclei into the female structure. The paired nuclei then undergo several rounds of mitotic division while migrating into the growing mycelium. In the mycelial tips, the paired nuclei finally fuse (becoming $2n$), and the mycelial tips develop into asci. Each ascus then undergoes meiosis in which the haploid products segregate in the same order that the meiotic chromosomes separated.

Some ascomycetes, such as *Aspergillus* and *Penicillium* species, form small asexual fruiting bodies called conidiophores for airborne spore dispersal (**Fig. 20.14A**). *Penicillium* is known for producing penicillin, the first antibiotic in widespread use; forms of penicillin are still used today (discussed in Chapter 1). *Aspergillus* (**Fig. 20.14A** and **B**) is a growing medical problem as an opportunistic pathogen of immunocompromised patients. *Aspergillus* can produce toxins (called mycotoxins) such as aflatoxin. Aflatoxin poisoning commonly affects livestock and, in some cases, agricultural workers; the toxin causes liver damage, immunosuppression, and cancer.

Conidiophore-forming ascomycetes such as *Aspergillus* and *Stachybotrys* are the major forms of mold associated with dampness in human dwellings; for example, they caused massive damage to homes flooded in the wake of Hurricane Katrina in 2005 (**Fig. 20.14C**). The flooding of homes full of drywall made ideal conditions for the growth of mold, which commonly consists of airborne ascomycete mycelia. Mold grew not only on materials submerged, but also on the surface above exposed to water-saturated air, up to 3 feet above the flood line (the highest level submerged). Unfortunately, most homeowner insurance policies covered damage only "up to the flood line."

Many ascomycetes are pathogens of animals or plants. For example, *Microsporum* and *Trichophyton* species cause ringworm skin infection, whereas *Magnaporthe oryzae* causes rice blast, the most serious disease of cultivated rice. Other species, however, are beneficial symbionts of plants, including crop plants such as beans, cucumbers, and cotton. *Trichoderma* species grow on the roots, or in some cases within the vascular tissue, of the plant. The fungi share nutrients with the plant, and they induce plant defenses against pathogens. *Trichoderma* even has a commercial use in cloth processing; the fungus is used to make "stonewashed jeans," as its cellulase enzymes partly digest the cotton.

Thought Question

20.5 Compare the life cycle of an ascomycete (**Fig. 20.13C**) with that of a chytridiomycete (**Fig. 20.11C**). How are they similar, and how do they differ?

Basidiomycota: cells with paired nuclei form mushrooms. The **basidiomycetes** form large, intricate fruiting bodies known as "true" mushrooms (**Fig. 20.15**). Mushrooms produce some of the world's deadliest poisons, such as alpha-amanitin, which inhibits RNA polymerase II. Alpha-amanitin is produced by the amanita, or "destroying angel" (**Fig. 20.15A**), a taste of which is usually fatal. (The amanita's own RNA polymerase is insensitive to the toxin.)

FIGURE 20.14 ▪ Ascomycete molds. **A.** *Aspergillus* forms a microscopic asexual fruiting structure called a conidiophore, containing spores in its spherical tip. **B.** Colony of *Aspergillus nidulans* on an agar plate. **C.** Black mold (*such as Stachybotrys*) grows above the flood line on a kitchen wall. An undergraduate volunteer points out the presence of mold in New Orleans, 6 months after flooding caused by Hurricane Katrina.

FIGURE 20.15 ▪ Mushrooms and other basidiomycetes form large, complex fruiting bodies. A. *Amanita phalloides* makes one of the most dangerous toxins known: alpha-amanitin, an inhibitor of RNA polymerase II. B. *Aleuria aurantia*, "orange fungus."

Other mushroom species include some of the world's most prized culinary delights, such as the portobello. Many grow in soil, while others, such as *Piptoporus*, grow on tree bark. Some mushrooms have evolved elaborate insect-attracting structures and odors, such as the "starfish stinkhorn," with its ring of bright red horns.

The mushroom itself is only the fruiting body of the basidiomycete, whose life cycle involves transitions among n, $n + n$, and $2n$ (**Fig. 20.16A**) similar to those of ascomycetes (see **Fig. 20.13C**). In the basidiomycete, however, the fruiting body consists largely of cells with paired nuclei ($n + n$). A few of the paired nuclei fuse to form diploid cells ($2n$) called basidia (singular, basidium), which line the gills of the mushroom. The basidia undergo meiosis to form haploid **basidiospores** (n). Some types of basidia can release basidiospores under pressure, whereas other basidiospores are dispersed by wind. Some fungi that resemble mushrooms, such as *Aleuria*, are actually classified as ascomycetes (**Fig. 20.15B**).

The basidiospores germinate to form underground mycelium. This haploid "primary mycelium" generates gametes that ultimately fuse to form $n + n$ "secondary mycelium." The primary and secondary mycelia may radiate

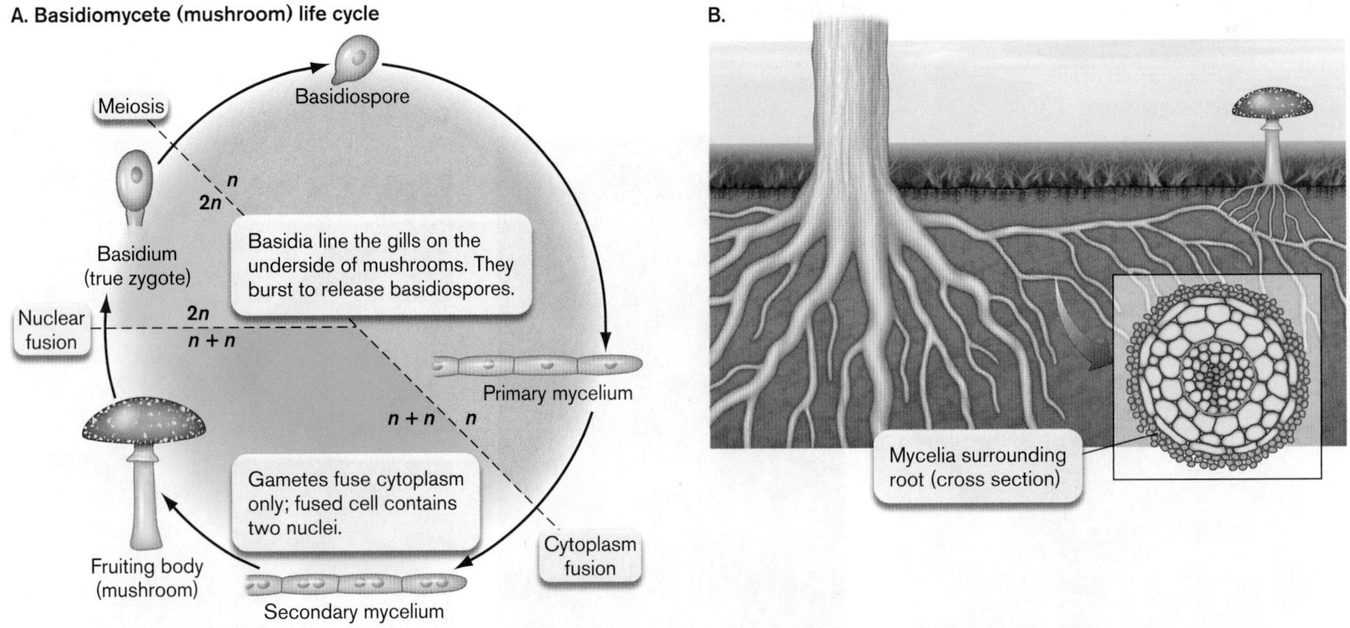

FIGURE 20.16 ▪ Mushroom life cycle. A. Haploid basidiospores generate primary mycelium underground, where they form gametes. Gametes of opposite mating types fuse their cytoplasm only, forming secondary mycelium. The parental nuclei remain separate throughout many generations of mitosis during development of the fruiting body (mushroom). As the mushroom matures, the basidia undergo nuclear fusion and meiosis, forming progeny basidiospores. B. The underground secondary mycelia of some mushrooms form mycorrhizae with tree roots. Mycorrhizae enhance and extend the absorptive power of the tree roots, while obtaining plant sugars for the fungus.

underground, unseen, until their tips generate mushrooms aboveground, at points approximately equidistant from the origin. The result is a mysterious "fairy ring" of mushrooms (see **Fig. 20.3A**). In a forest, these invisible underground hyphae of basidiomycetes and glomeromycota contribute **mycorrhizae** (singular, **mycorrhiza**) that gain sugars from trees while extending the tree root systems (**Fig. 20.16B**).

Glomeromycota form arbuscular mycorrhizae. A remarkable group of fungi is the Glomeromycota, all of which are obligate mutualists of plants. These fungi, such as *Glomus* species, form extensive networks of filamentous connections with plant roots similar to the mycorrhizae formed by basidiomycetes; but unlike the basidiomycete mycorrhizae, the mycorrhizae formed by glomeromycota are **arbuscular mycorrhizae**, in which the fungal filaments actually penetrate plant cells in a most intimate symbiosis (**Fig. 20.17**).

Mycorrhizae expand the roots' absorptive capacity, while obtaining plant sugars for the fungus. More than 90% of all land plants, including trees, depend on these fungal interconnections, which share nutrients among many unrelated plants, as well as the fungi (discussed in Chapter 21).

In arbuscular mycorrhizae, a fungal hypha first grows between plant cells without breaching the plant cell wall (**Fig. 20.17B**). Next, the hypha branches into the plant cytoplasm; the plant cell membrane invaginates to accommodate the branch, while maintaining a "periarbuscular space" between the plant cell membrane and the fungal plasma membrane. This highly regulated invaginating branch is called an "arbuscule"—hence the term "arbuscular mycorrhiza." Arbuscule formation is regulated by plant hormones called strigolactones. The arbuscule expands the surface area to exchange sugars from the plant for ammonium and phosphate from the fungus.

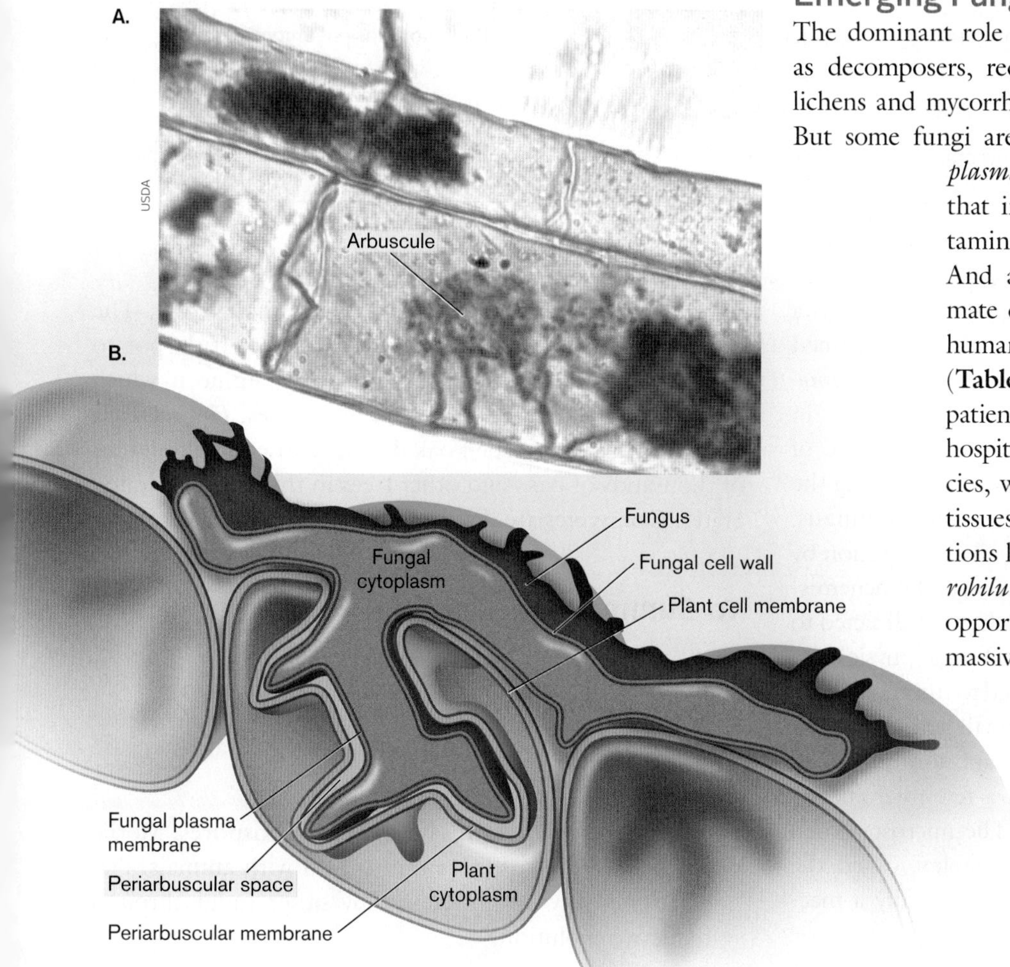

FIGURE 20.17 ■ Arbuscular mycorrhiza. A. Glomeromycota fungi invade corn root cells, as part of a mutualistic symbiosis to exchange nutrients. **B.** The fungal hypha grows between the plant cells, and then into a plant cell to form an arbuscule. The arbuscule expands surface area for exchange while maintaining the plant cell wall intact. *Source:* Part B modified from Martin Parniske. 2008. *Nat. Rev. Microbiol.* **6**:763.

Emerging Fungal Pathogens

The dominant role of fungi in our biosphere is positive—as decomposers, recyclers, and symbiotic partners within lichens and mycorrhizae (discussed further in Chapter 21). But some fungi are important pathogens, such as *Histoplasma capsulatum*, an ascomycete fungus that infects healthy people who inhale contaminated dust, causing a deadly pneumonia. And as human demographics shift and climate change increases, a growing number of human, animal, and plant pathogens emerge (**Table 20.2**). For immunocompromised patients, especially the elderly confined to hospitals, a growing threat is *Aspergillus* species, which can colonize the lungs and other tissues. In 2012, contaminated steroid injections led to an outbreak of infections by *Exserohilum rostratum* and other previously rare opportunists. Other fungal pathogens cause massive mortality of animals and plants.

Several parasitic organisms originally classified as protozoa because of their superficial appearance have now been shown to be fungi or fungus-related, based on their genome sequence and biochemistry. The reclassification has important consequences for research, taxonomy, and therapy. An example is the ascomycete *Pneumocystis jirovecii*. The organism was first described in 1909, when it was thought to be a life stage of a trypanosome causing Chagas' disease (discussed

TABLE 20.2 Emerging Fungal and Microsporidian Pathogens

Species	Phylum	Host	Disease
Aspergillus fumigatus	Ascomycota	Humans (immunocompromised)	Aspergillosis of lung, and elsewhere in the body
Batrachochytrium dendrobatidis	Chytridiomycota	Amphibians (frogs and toads)	Chytridiomycosis
Cryptococcus neoformans	Basidiomycota	Humans (immunocompromised)	Meningitis and meningoencephalitis
Encephalitozoon intestinalis	Microsporidia	Humans with AIDS	Microsporidiosis (intestinal)
Exserohilum rostratum	Ascomycota	Humans	Wound and skin infections; contaminated injections
Fusarium solani	Ascomycota	Sea turtles (loggerheads)	Hatch failure
Geomyces destructans	Ascomycota	Brown bats	White nose disease
Histoplasma capsulatum	Ascomycota	Humans, dogs, cats	Histoplasmosis (lung)
Magnaporthe oryzae	Ascomycota	Rice	Rice blast disease
Nosema species	Microsporidia	Honeybees	Colony collapse disorder
Pneumocystis jirovecii	Ascomycota	Humans (immunocompromised)	Pneumocystis pneumonia
Puccinia graminis	Basidiomycota	Wheat	Wheat stem rust
Stachybotrys chartarum	Ascomycota	Humans	Black mold disease; respiratory damage

in Section 20.6). When the organism was recognized as a distinct species of protists infecting animals, it was named *Pneumocystis carinii.* In the 1970s, this strain of *Pneumocystis* was renamed *P. jirovecii* for causing pneumonia in immunocompromised humans. In the 1980s, the rise of AIDS led to a sudden increase in infections. Sequencing the organism's genome revealed it to be an ascomycete fungus. As a result, the organism's name was called into question by fungus researchers; its developmental forms and biochemistry were reevaluated, and medical research was redirected to culture and treat the organism based on fungal physiology.

A major clade of parasites closely related to fungi is Microsporidia. Microsporidia have relatively small genomes; their mitochondria have lost their DNA and are nonfunctional. Microsporidia form spores as small as a few micrometers in size, which can infect animal cells. The microsporidian spore extrudes a specialized invasion complex, called the polar tube, that penetrates the host cell, typically a macrophage. In humans, microsporidians are opportunistic pathogens, such as the intestinal parasite *Encephalitozoon intestinalis*, emerging with the rise of AIDS and the growth of elderly and immunocompromised populations.

Other taxa originally assigned as fungi because of superficial appearance, are now classified as protists because of genetic similarity. The **oomycetes**, or "water molds," were originally classified as fungi because of their fungus-like filaments, which infect plants and animals. The oomycete *Phytophthora infestans* devastated Irish potato crops in the 1840s, causing the Great Irish Famine, in which a million people died. Today, a related species, *Phytophthora ramosum*, causes "sudden oak death," a disease killing tens of thousands of oaks and other trees in the western United States. Oomycetes are discussed further in **eTopic 20.1**.

To Summarize

- **Fungi form hyphae with cell walls of chitin.** Hyphae absorb nutrients from decaying organisms or from infected hosts. Some fungi remain unicellular; these are called yeasts or mitosporic fungi.
- **Chytridiomycete fungi have motile zoospores.** Motile reproductive forms are a trait shared with animals. Flagellar motility has been lost by other fungi through reductive evolution.
- **Zygomycete fungi form haploid mycelia.** Hyphal tips differentiate into gametes and grow toward each other to undergo sexual reproduction. Some zygomycetes form mycorrhizae that connect the roots of plants.
- **Ascomycete fungal mycelia form paired nuclei.** Within the hyphal cells, the paired nuclei fuse, followed

by meiosis and development of ascospores. Some ascomycetes form fruiting bodies called conidiophores.

- **Basidiomycete fungi form mushrooms.** Cells with paired nuclei (secondary mycelium) form large fruiting bodies called mushrooms. The paired nuclei fuse to form the diploid basidium, which generates haploid basidiospores. The basidiospores develop underground hyphae or mycorrhizae that interconnect plant roots.
- **Glomeromycota form arbuscular mycorrhizae.** These fungi form intimate mutualistic networks of connections with plant roots, exchanging minerals for plant sugars.
- **Emerging fungal pathogens threaten humans, plants, and animals.**

20.3 Amebas and Slime Molds

The ameba (alternative spelling "amoeba") is familiar to most of us as an apparently amorphous form of microscopic life, capable of engulfing and consuming prey in a dramatic fashion. Amebas are a polyphyletic group; that is, they lack a common ancestor that share the ameba form but include members that branch from distantly related clades. The taxonomy of amebas and slime molds remains problematic, with diverse views as to the number of clades, their relatedness, and their degree of divergence.

An ameba's shape is exceptionally variable, but the pseudopods, or "false feet" (**Fig. 20.18**), that it extends are complex structures that undertake highly controlled movements. The "classic" amebas of the Amoebozoa are free-living predators in soil or water, engulfing prey by **phagocytosis**. They range in size up to 5 mm, large enough to phagocytose bacteria, algae, ciliates, smaller amebas, and even invertebrates such as rotifers. A few are dangerous parasites of humans or animals. Furthermore, free-living amebas can harbor bacterial pathogens such as *Legionella pneumophila*, which contaminates water supplies and air ducts. The bacteria cause legionellosis, an often fatal form of pneumonia. The host ameba enables the pathogen's persistence and transmission to human hosts.

Free-living amebas such as *Acanthamoeba* species are common predators in the soil microbial community. They cause problems when they contaminate contact lens cleaning solutions, causing keratitis (infection of the cornea). Wearers of contact lenses have an increased risk of *Acanthamoeba* keratitis.

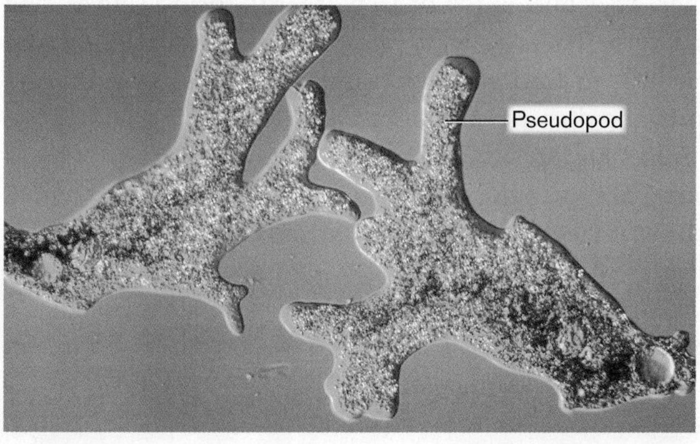

FIGURE 20.18 ■ *Amoeba proteus* moves by extending its pseudopods.

Thought Question

20.6 What cellular interactions can happen when an ameba phagocytoses algae?

Pseudopod Motility

Species from diverse clades form ameba-like cells with pseudopods. The Amoebozoa persist as an ameba throughout all or most of the life cycle. Other protists can convert to flagellated forms, particularly when the habitat fills with water—a low-viscosity condition favoring flagellar motility. For example, *Naegleria fowleri*, the "brain-eating ameba," is a parasite of the Discoba group (Section 20.6). A pond organism that causes amebic meningoencephalitis, *N. fowleri* exists as a flagellate and as an ameboid form with pseudopods. Other species such as dinoflagellates never become fully ameboid, but they can extend a pseudopod to engulf prey.

Different kinds of amebas have different kinds of pseudopods. The Amoebozoa have lobe-shaped pseudopods (**Fig. 20.18**). Lobe-shaped pseudopods have the most variable shape. A different form is the sheetlike pseudopod, or lamellar pseudopod, used by dinoflagellates. Similar lamellar pseudopods are generated by human white blood cells such as leukocytes. Finally, needlelike pseudopods, or filopodia, are thin extensions reinforced by parallel actin filaments. These are typical of Rhizaria amebas, although some Rhizaria instead make long, thin pseudopods supported by microtubules. The most famous protists with microtubule-supported pseudopods also have mineralized supporting structures, such as Foraminifera (spiral shells) and Radiolaria (radial-form "skeleton").

The extension of lobe-shaped and lamellar pseudopods has been studied closely for its relevance to human white blood cells (for more on white blood cells as host defenses, see Chapter 23). The mechanism of pseudopod motility remains poorly understood, but it is known to involve a sol-gel transition between cortical cytoplasm (just beneath the cell surface) and the cytoplasm of the deeper interior (**Fig. 20.19**). The tip of a pseudopod contains a gel of polymerized actin beneath its cell membrane. From the center of the ameba, liquid cytoplasm (sol) containing actin subunits streams forward along

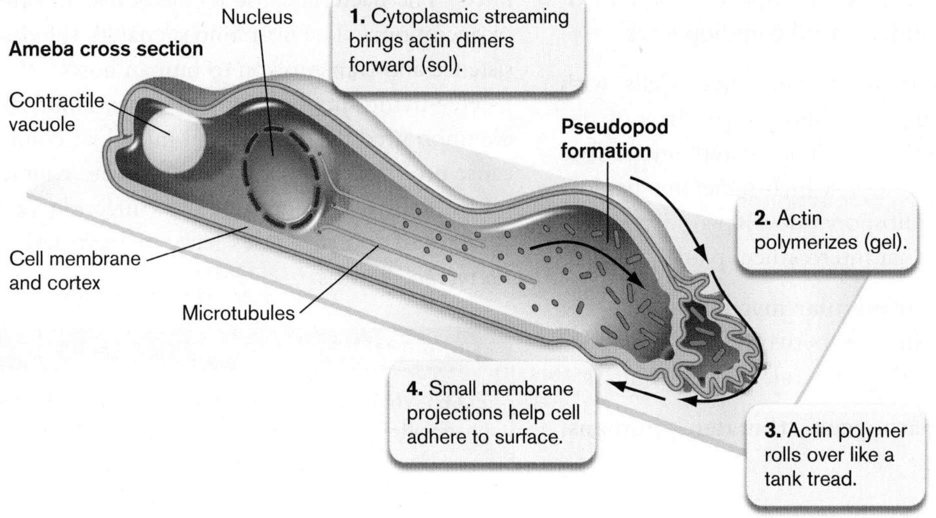

FIGURE 20.19 ■ Pseudopod motility. A pseudopod extends by flow of liquid cytoplasm (sol state) followed by actin polymerization (gel state). As actin polymerizes, the cell rotates down toward the substrate like a tank tread.

microtubular "tracks," powered by ATP hydrolysis. The actin subunits stream into the pseudopod, where they polymerize, forming a gel. The gel region grows, pushing the membrane forward and extending the pseudopod. As the gel is pushed backward, it resolubilizes to continue the cycle.

Amebas can have one nucleus or multiple nuclei. They are usually haploid and reproduce asexually by nuclear mitosis, without dissolution of the nuclear membrane, followed by fission of the cytoplasm. Some species do have developmental alternatives, such as cyst formation, gamete fusion and meiosis, and even growth of flagella in a favorable habitat.

Thought Question

20.7 What kind of habitat would favor a flagellated ameba?

Ameba genetics is poorly understood, but at least one ameba genome has been sequenced—that of the intestinal parasite *Entamoeba histolytica*. The sequence contains 20 Mb of DNA in 14 chromosomes; some of these are linear, whereas others are circular. Closely related strains show considerable variation in organization, suggesting that ameba genomes undergo extensive rearrangement.

Slime Molds

Some amebas conduct a life cycle in which thousands of individuals (all members of one species) aggregate into a complex, differentiated fruiting body (**Fig. 20.20A**). Such an organism is called a **cellular slime mold**. A cellular slime mold forms from ameboid cells that aggregate into a multicellular "slug." Slime molds, as the name implies, were originally classified with fungi because their fruiting bodies superficially resemble fungal reproductive forms.

A well-studied example of a cellular slime mold is *Dictyostelium discoideum*, historically an important model system for multicellular development (**Fig. 20.20**). *D. discoideum* amebas are relatively small, about 10 μm, but large enough to consume bacteria. They can be cocultured on a plate with *Escherichia coli*. As the haploid amebas consume bacteria, they divide asexually until their food runs out. At this point, a few amebas begin to emit the aggregation signaling molecule cyclic AMP (cAMP). An ameba emitting cyclic AMP attracts other amebas nearby, which move toward the center and begin emitting cyclic AMP as well. Successive waves of cyclic AMP continue to attract thousands of amebas to the center, where they pile on top of each other to form a slug as long as 1 mm. The slug then migrates, attracted by light and warmth, to find an appropriate place to form a fruiting body and disperse its spores.

The slug at last differentiates into a fruiting body, a spherical sporangium supported on a stalk of largely empty cells that emerges from a basal disk. The sporangium then releases spores (also called cysts), which are dispersed on air currents and can remain viable for several years. When a spore detects chemical signals from bacteria, it germinates an ameba to feed on them.

Note that the entire reproductive cycle just described is asexual; the amebas and their differentiated structures remain haploid throughout. *D. discoideum* amebas do

A.

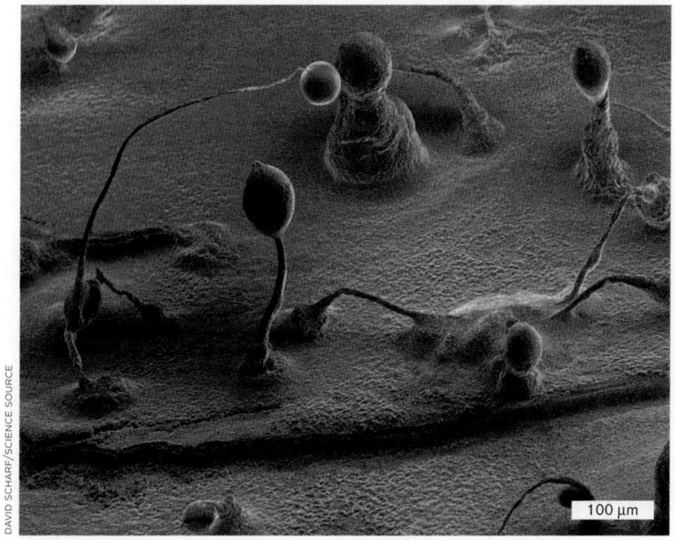

B.

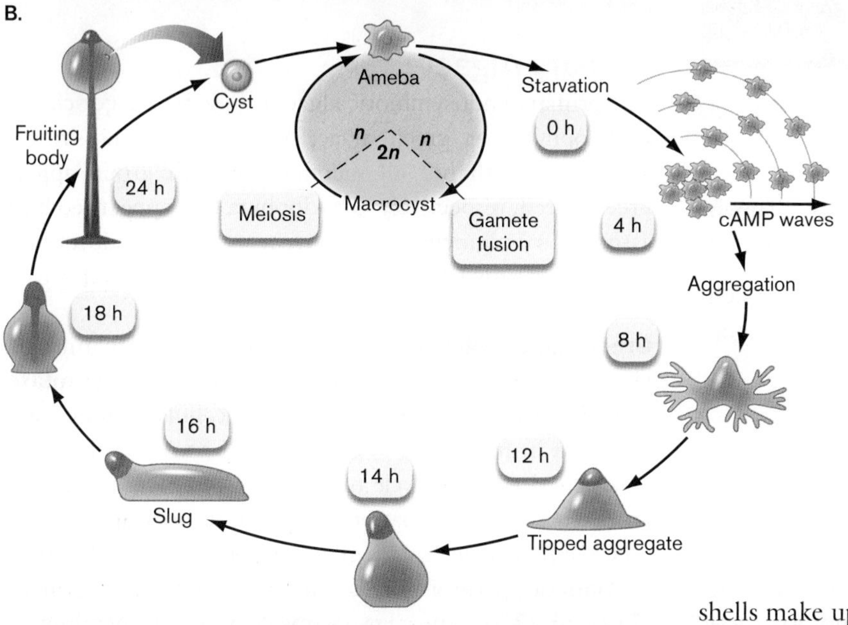

FIGURE 20.20 ■ A cellular slime mold: *Dictyostelium discoideum*. A. Fruiting bodies of *D. discoideum* (composite SEM). B. Life cycle of *D. discoideum*.

have a sexual alternative (illustrated in **Fig. 20.20B**), in which cells of opposite mating type can fuse to form a diploid zygote and then undergo meiosis, restoring haploid amebas.

A different kind of slime mold is the **plasmodial slime mold**, in which an ameba undergoes mitosis without cell division, forming a multinucleate single cell. A plasmodial slime mold such as *Physarum polycephalum* develops from a single diploid ameba. As the ameba grows, its nuclei multiply, forming a **plasmodium** (plural, **plasmodia**), a giant multinucleate cell that can spread over an area of many square centimeters. Out of the plasmodium arise fruiting bodies whose sporangia undergo meiosis, producing haploid spores. Two spores then fertilize each other to form a diploid ameba that can develop again into a plasmodium. A large plasmodium can occasionally be seen as a yellow mass of slime spreading over decaying wood.

Note: Distinguish the term "plasmodium" (a large multinucleate cell) from the genus *Plasmodium* (an apicomplexan parasite, such as *Plasmodium falciparum*, which causes malaria).

Filamentous and Shelled Amebas

The Rhizaria form needlelike pseudopods. Despite the term "filamentous amebas," the Rhizaria form a clade distant from Amoebozoa and more closely related to Alveolata (Section 20.5). Most Rhizaria are encased by mineral shells called tests. A major group of "shelled amebas" is the **radiolarians**, whose skeleton-like tests are made of silica perforated with numerous holes through which pseudopods appear to radiate in all directions (**Fig. 20.21A**). Their shells make important contributions to marine sediment and sand. Many different kinds of radiolarians exist today, and many can be recognized in fossil rock. A second group of shelled amebas is the **foraminiferans** (**Fig. 20.21B**). The foraminiferans, or forams, generate shells of calcium carbonate as chambers laid down in helical succession. Their pseudopods all extend from one opening in the most recent chamber.

Radiolarians and shelled forams grow in marine and freshwater habitats. Their shells make up a large part of reef formations, sedimentary rock, and beach sand. Forams, in particular, are used in geological surveys as indicators of petroleum deposits.

To Summarize

- **Amebas** move using pseudopods. In different species, pseudopods are lobe-shaped, lamellar, or filamentous (filopodia).
- **Cytoplasmic streaming** through cycles of actin polymerization and depolymerization drives the extension and retraction of pseudopods.
- **Slime molds** show an asexual reproductive cycle in which a fruiting body produces spores. In cellular slime molds, amebas aggregate to form a slug. In plasmodial slime molds, a single ameba develops into a multinucleate cell.

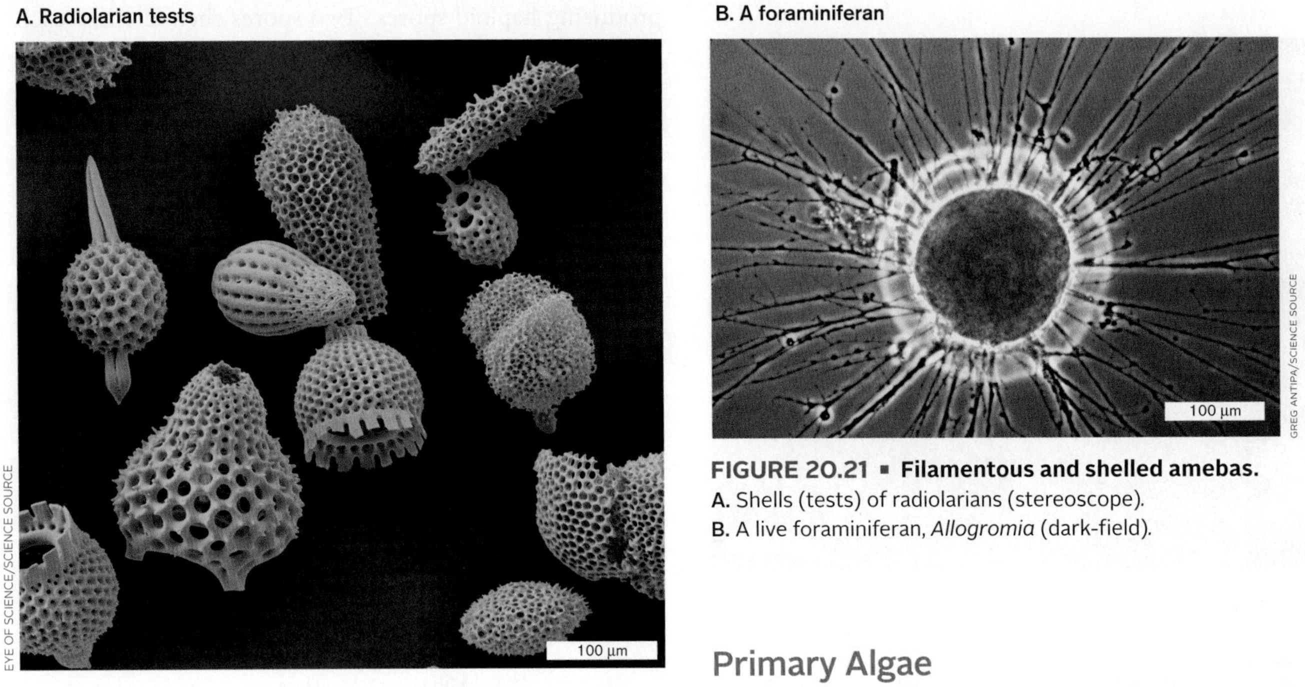

FIGURE 20.21 ■ Filamentous and shelled amebas. A. Shells (tests) of radiolarians (stereoscope). B. A live foraminiferan, *Allogromia* (dark-field).

- **Radiolarians** have silicate shells penetrated by filamentous pseudopods.
- **Foraminiferans** have calcium carbonate shells with helical arrangement of chambers. The most recent chamber opens to extend filamentous pseudopods.

20.4 Algae

Algae are CO_2-fixing producers in all ecosystems, most crucially freshwater and marine habitats. In freshwater and marine ecology, the algae, together with photosynthetic bacteria, are known as **phytoplankton** (see Chapter 21). All algae possess chloroplasts.

The primary algae are products of a single ancestral endosymbiosis that also gave rise to land plants. The biochemistry and cell structures of true algae and land plants are similar; thus, the alga *Chlorella* was the model organism of choice for pioneers of photosynthesis research, including Martin Kamen, Melvin Calvin, and others (discussed in Chapter 15).

Other photosynthetic eukaryotes, or secondary endosymbiotic algae, arose from protists that engulfed a primary or secondary alga. Secondary algae often show "mixotrophic" nutrition, involving both phototrophy and heterotrophy. For example, dinoflagellate "algae" are voracious predators of smaller protists. The heterokont algae (diatoms, coccolithophores, and kelps) are covered in this section. Dinoflagellates are covered together with alveolates (Section 20.5).

Primary Algae

The primary endosymbiotic algae include two major clades: Chlorophyta, or green algae; and Rhodophyta, or red algae—although not all members of each group appear green or red, respectively. Rhodophyta that appear red have a secondary pigment called phycoerythrin, in addition to green chlorophyll.

Green algae (Chlorophyta). Many green algae are unicellular. An important model system for genetics and phototaxis is *Chlamydomonas reinhardtii*, a unicellular chlorophyte common in freshwater systems as well as Antarctic pools. The genetics of cell cycle regulation in *C. reinhardtii* provides clues to the formation of human tumors.

C. reinhardtii has a symmetrical pair of flagella—a common pattern for green algae and their gametes (**Fig. 20.22A**). The alga swims forward by bending its flagella back toward the cell, like a breaststroke. *Chlamydomonas* cells are mostly haploid, reproducing by asexual cell division (**Fig. 20.22B**). For sexual reproduction, opposite mating types fuse to form a zygote, which loses flagella and grows a spiny protective coat. The zygote undergoes meiosis to regenerate haploid cells.

The cell ultrastructure of *Chlamydomonas* is typical of algal cells (**Fig. 20.23**). The nucleus is cupped by a single chloroplast, which is surrounded by a double membrane. The double membrane indicates primary algae; the inner membrane derives from the ancestral bacterium, and the outer membrane derives from the engulfing host. Within the chloroplast is a pyrenoid, an organelle that in some species concentrates bicarbonate (HCO_3^-) and converts it to CO_2 for fixation. The pyrenoid is surrounded by one or more starch bodies, which are used for energy storage. The starch is broken down to sugars as needed, followed by glycolysis and respiration in the mitochondria. Osmolarity is

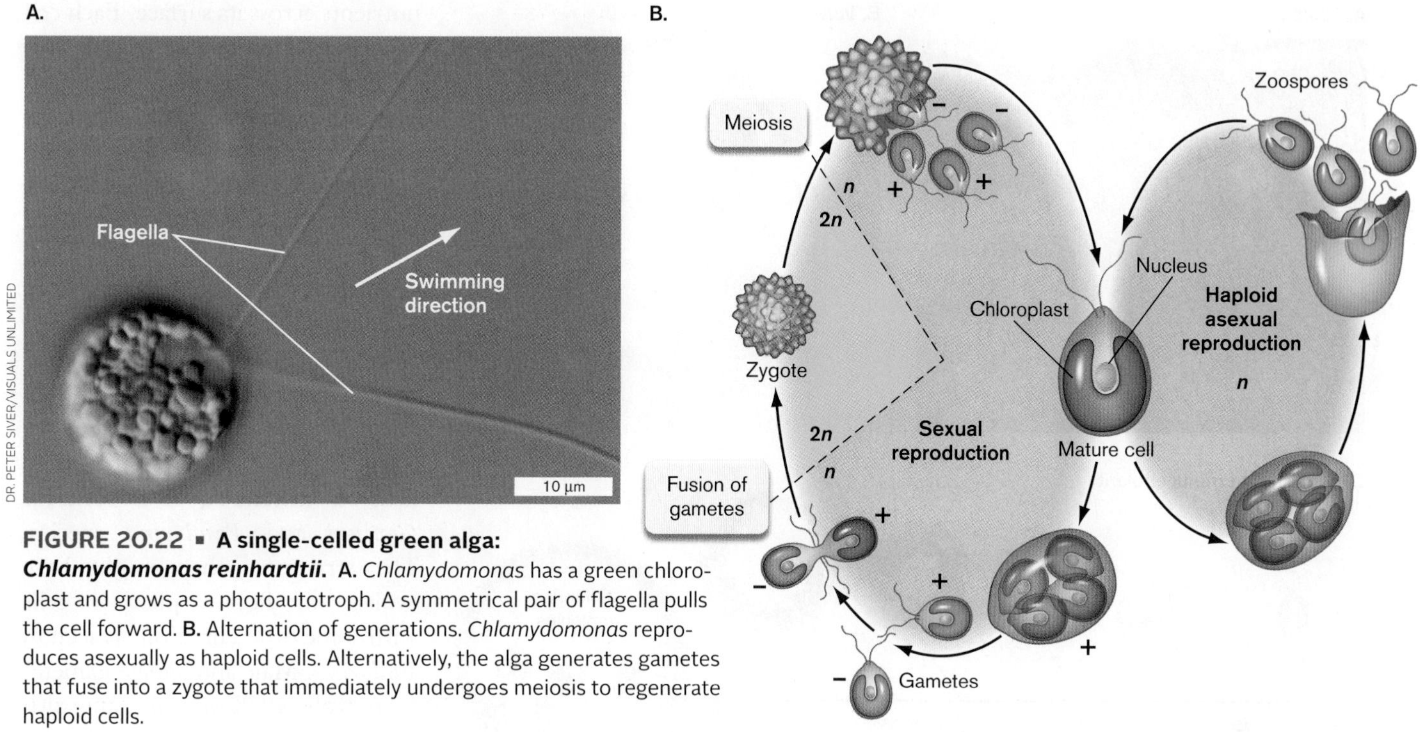

FIGURE 20.22 ▪ A single-celled green alga: *Chlamydomonas reinhardtii*. **A.** *Chlamydomonas* has a green chloroplast and grows as a photoautotroph. A symmetrical pair of flagella pulls the cell forward. **B.** Alternation of generations. *Chlamydomonas* reproduces asexually as haploid cells. Alternatively, the alga generates gametes that fuse into a zygote that immediately undergoes meiosis to regenerate haploid cells.

maintained by the contractile vacuole. The *Chlamydomonas* cell is encased in a cell wall composed predominantly of glycoprotein. Other green algae have cellulose cell walls similar to those of plants.

While *C. reinhardtii* is unicellular in nature, an exciting experiment showed that this alga can quickly evolve into a multicellular form. In 2013, William Ratcliff at the Georgia Institute of Technology (**Fig. 20.24A**) and Michael Travisano at the University of Minnesota showed that *C. reinhardtii* could undergo selection by repeated subculturing in a standing tube. In each culture, a few cells would clump at the bottom, and the clumps were selectively cultured further. Eventually, a strain was isolated that grew regularly in connected cell clusters that alternated with dispersal.

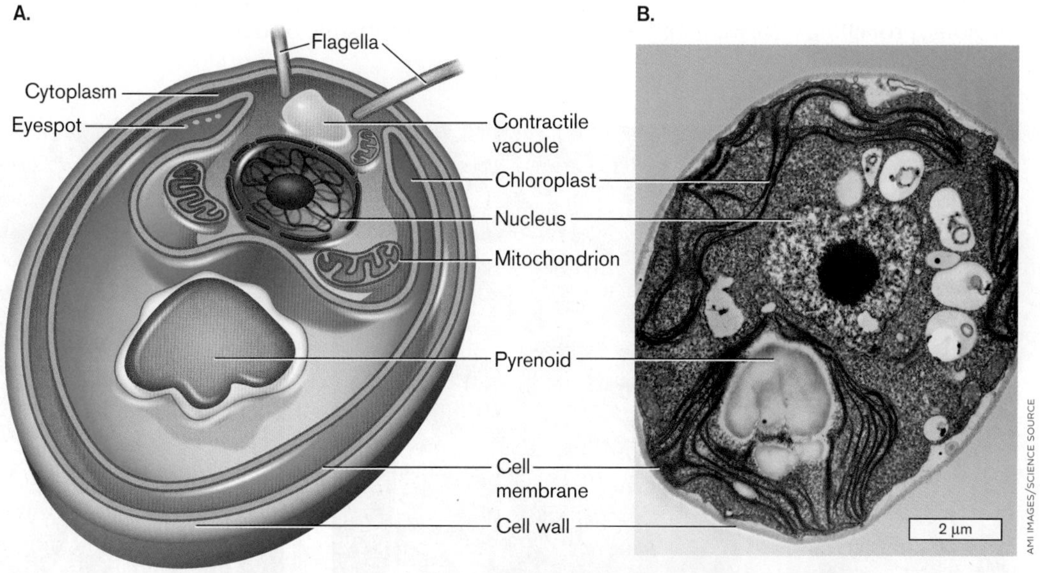

FIGURE 20.23 ▪ Cell structure of *Chlamydomonas reinhardtii*. **A.** The cell membrane is surrounded by a cell wall of cellulose and glycoproteins. The single chloroplast fills much of the cell and wraps around the nucleus. Within the chloroplast lies a pyrenoid, a structure for concentrating bicarbonate ion for conversion to CO_2. The pyrenoid is surrounded by starch bodies that store high-energy compounds. A contractile vacuole maintains constant osmotic pressure. **B.** Electron micrograph of C. *reinhardtii* shows the nucleus, chloroplast, pyrenoid, and other organelles.

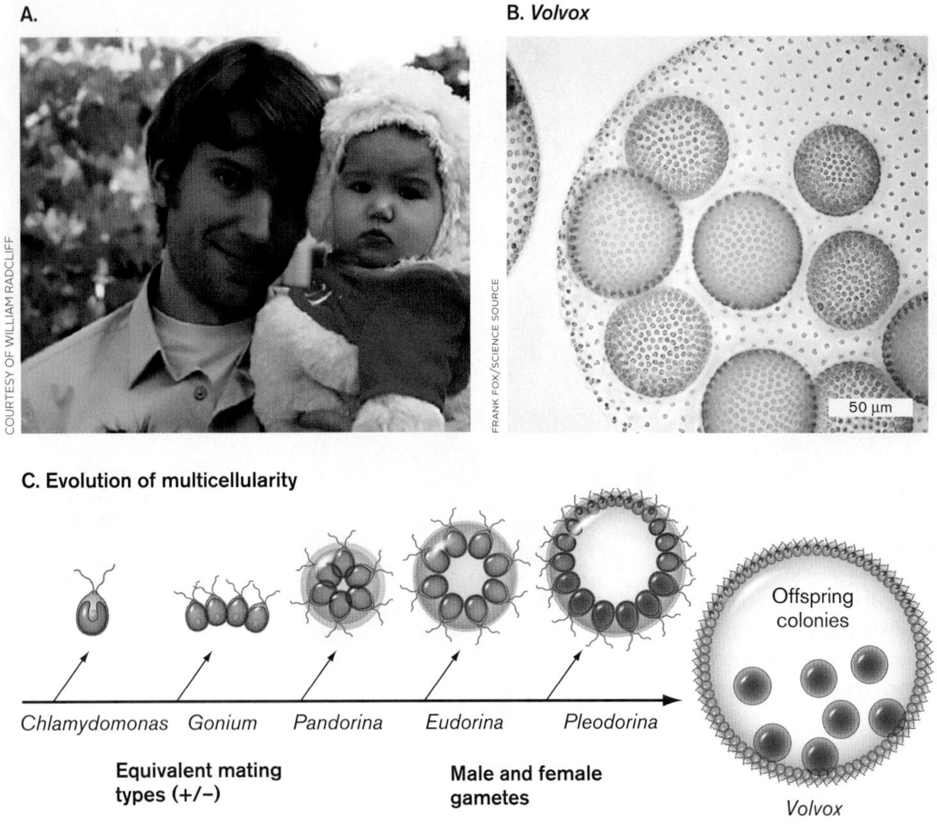

FIGURE 20.24 ■ *Volvox* is multicellular. **A.** William Ratcliff, who studies origins of multicellularity, with his own multicellular offspring. **B.** *Volvox* forms a spherical colony of cells (1–3 mm) that generates offspring colonies within. **C.** Different types of algae suggest a model for the evolution of multicellularity from a unicellular ancestor (like *Chlamydomonas*) to a multicellular colony with female and male gametes (*Volvox*).

Some marine and freshwater algae that resemble *Chlamydomonas* grow naturally in intricate multicellular colonies. Colonial algae such as *Volvox* (**Fig. 20.24B**) generate geodesic spheres of biflagellate cells. Their flagella point outward from the colony, propelling it forward and drawing nutrients across its surface. Each cell of *Volvox* connects by cytoplasmic bridges to five or six of its neighbors. The colony reproduces by generating daughter colonies within the sphere, which grow until the outer sphere falls apart, liberating the daughters.

A survey of algal diversity from *Chlamydomonas* through *Volvox* suggests a model for the progression of evolving multicellularity (**Fig. 20.24C**). First, cells might have evolved to grow in small, ordered structures, as is seen for *Gonium* and *Pandorina* species. The structures then might have evolved into hollow spheres of cells with some differentiation, as seen in *Eudorina*. Male and female gametes then evolved (*Pleodorina*), and ultimately colonies that generate colonial offspring (such as *Volvox*).

Other species of algae grow as multicellular filaments. An example is *Spirogyra*, a common pond dweller known for its spiral chloroplasts (**Fig. 20.25A**). As with *Chlamydomonas*, the haploid form predominates; but unlike the unicellular alga, *Spirogyra* forms neither flagellated gametes nor zoospores. Instead, its sexual reproduction requires alignment of two filaments of opposite mating type (**Fig. 20.25B**). Cells conjugate (form cytoplasmic bridges) between the two filaments. The

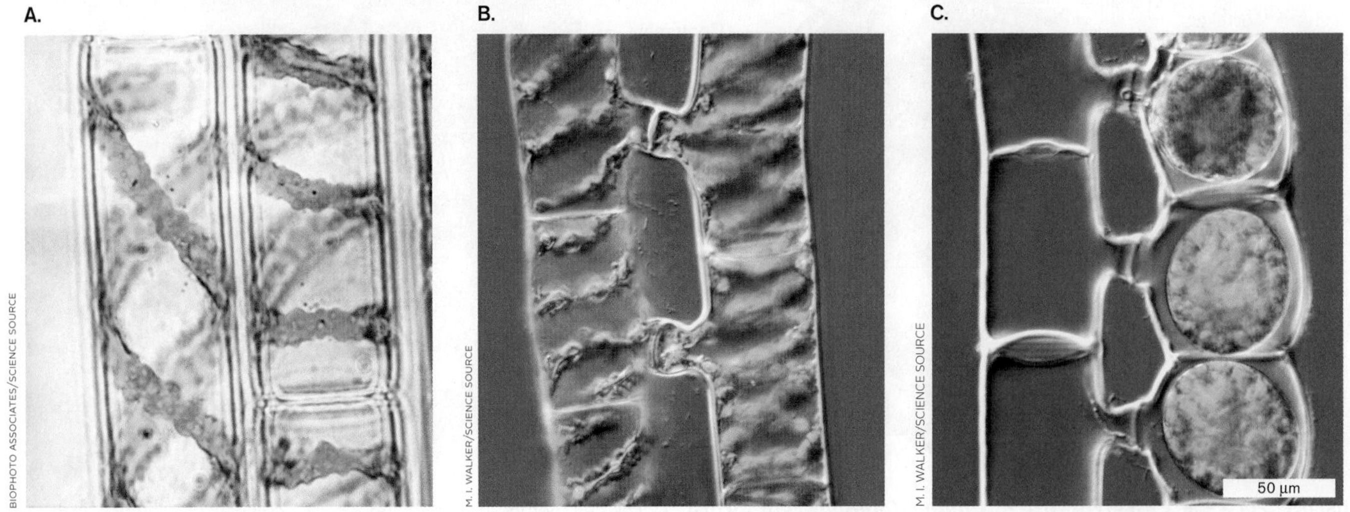

FIGURE 20.25 ■ Filaments with spiral chloroplasts: *Spirogyra*. **A.** *Spirogyra* species grow in long, multicellular filaments, typically 25 mm wide and several centimeters long. Each cell contains one or more chloroplasts that spiral around the cytoplasm. **B.** Sexual reproduction involves conjugation between cells of two mating types. The cytoplasm from each male cell exits its cell wall and enters the female cell. **C.** Gamete fusion is complete, generating zygotic spores.

A.

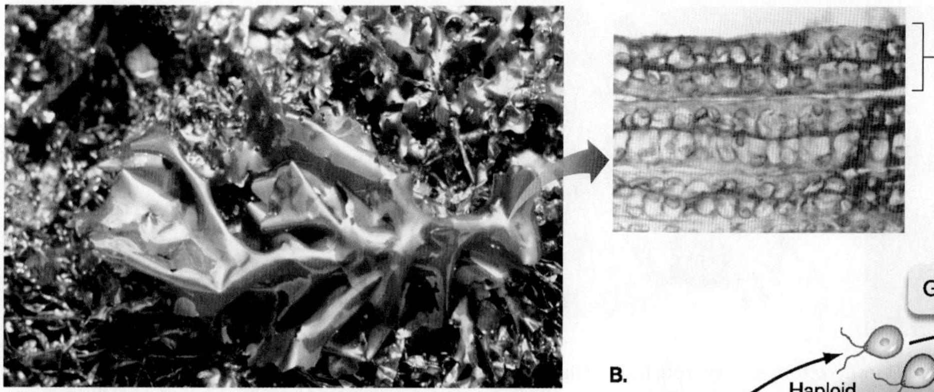

BIOPHOTO ASSOCIATES/SCIENCE SOURCE; INSET: BOBIN-DUBIGEON ET AL. 1997. *J. SCI. FOOD AGRIC.* **75**:341–351.

B.

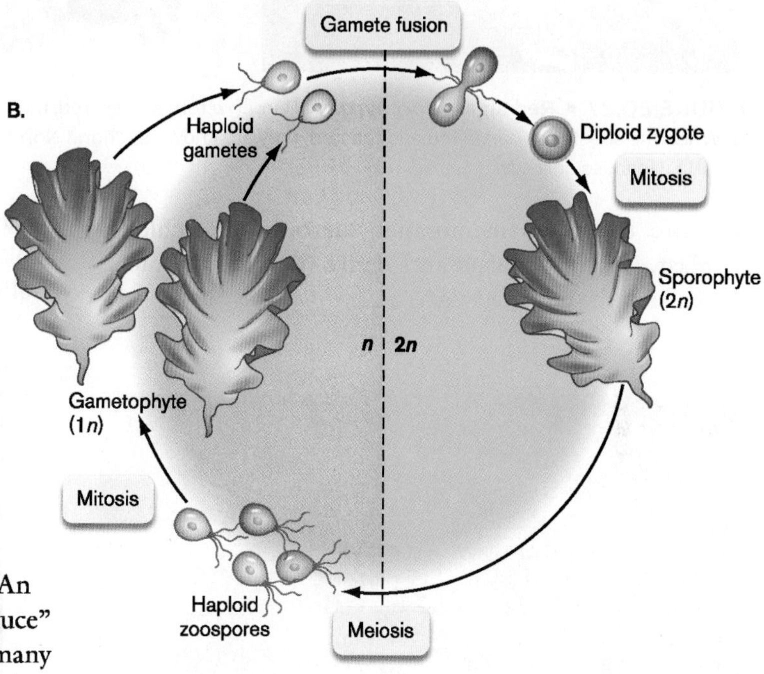

FIGURE 20.26 ■ Multicellular algae: *Ulva*. A. *Ulva* species generate large, undulating sheets of cells in double layers (inset). B. *Ulva* undergoes symmetrical alternation of generations. The haploid and diploid forms appear very similar. *Source:* Part B modified from Christine Bobin-Dubigeon et al. 1997. *J. Sci. Food Agric.* **75**:341–351.

"male" gametes are those whose cytoplasm inserts through the conjugation bridge to join that of the "female" gamete. As the gametes fuse, a row of empty cell walls is left behind (**Fig. 20.25C**). **The zygote eventually hatches and germinates a new chain** of cells.

Some marine algae grow in undulating sheets. An example familiar to beach bathers is the "sea lettuce" *Ulva* (**Fig. 20.26A**). Sheets of *Ulva* **can extend over many square meters, although they are only two cells thick (see Fig. 20.26A** inset). The *Ulva* life cycle shows classic alternation of generations between haploid and diploid forms (**Fig. 20.26B**). Haploid sheets of cells (the gametophyte) produce symmetrically biflagellate gametes, similar to unicellular *Chlamydomonas*. But when *Ulva* gametes fuse, the zygote grows into an immense diploid sheet of cells. This sporophyte (diploid multicellular body) appears similar in form to the gametophyte. The sporophyte eventually undergoes meiosis, releasing haploid zoospores with two pairs of flagella. The zoospores undergo mitosis and regenerate the gametophyte.

Algae serve as symbiotic partners for many important living systems. For example, certain algae grow in intimate association with fungi to form unified structures called **lichens**, important colonizers of dry and cold habitats (discussed in Chapter 21). A form of algal-fungal ground cover similar to lichens is **cryptogamic crust**, common on desert soil. Still other algae grow within the cells of paramecia and hydras, providing photosynthetic nutrition in exchange for protection.

Thought Question

20.8 What are the relative advantages of being unicellular or multicellular?

Red algae (Rhodophyta). Red algae, or rhodophytes, are colored red by the photopigment phycoerythrin. Phycoerythrin absorbs efficiently in the blue and green range, which green algae fail to absorb. Because the red algae absorb wavelengths missed by the green algae, they can colonize deeper marine habitats below the green algal populations.

Rhodophytes include unicellular, filamentous, and multicellular forms. Several kinds are human food sources. *Porphyra* forms large sheets that are harvested in Japan for use as nori for wrapping sushi, a delicacy that includes rice, vegetables, and uncooked fish (**Fig. 20.27**). Red algae contain valuable polymers called sulfated polygalactans (sugar polymers with sulfate side chains). Sulfated polygalactans include agar, used to solidify microbial growth media; agarose, a processed sugar derivative used to form electrophoretic gels; and carrageenan, an additive used in processed foods.

Secondary Algae: Diatoms and Kelps

Many kinds of secondary algae arose from ancestral protists that once engulfed a primary endosymbiotic alga (see **Fig. 20.6**). Secondary algae show two traits that distinguish them from primary algae:

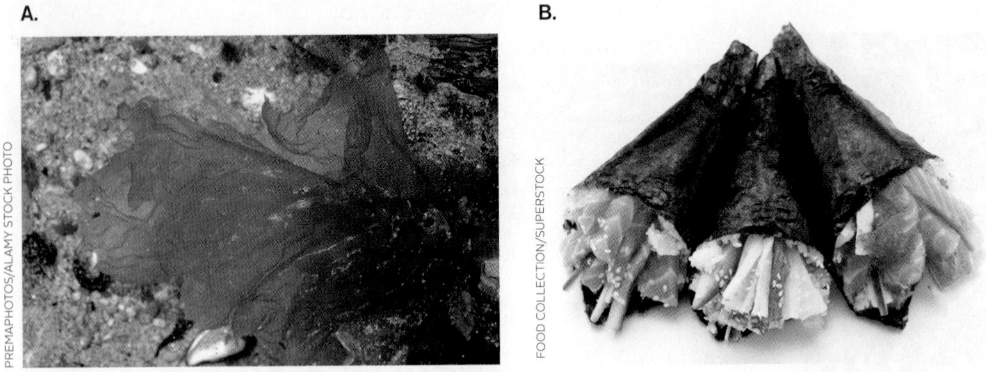

FIGURE 20.27 ▪ Red algae: *Porphyra*. A. *Porphyra* forms large, red, multicellular sheets. B. When the sheets are harvested and toasted, they are known as "nori." Nori are used to wrap sushi, a Japanese delicacy.

- **More than two membranes surround the chloroplast.** The extra membranes derive from the cell membrane of the engulfed alga.
- **Metabolism of secondary algae includes heterotrophy.** By contrast, the primary algae are near-obligate autotrophs, catabolizing only the simplest substrates, such as acetate.

Several major groups of secondary algae are heterokonts (stramenopiles). These include the diatoms (Bacillariophyceae); the brown algae (Phaeophyceae), such as kelps; the golden algae (Chrysophyceae), mainly flagellates; the yellow-green algae (Xanthophyceae); and other less studied forms of phytoplankton. Flagellated species of heterokont algae generally show a pair of differently shaped flagella. Typically,

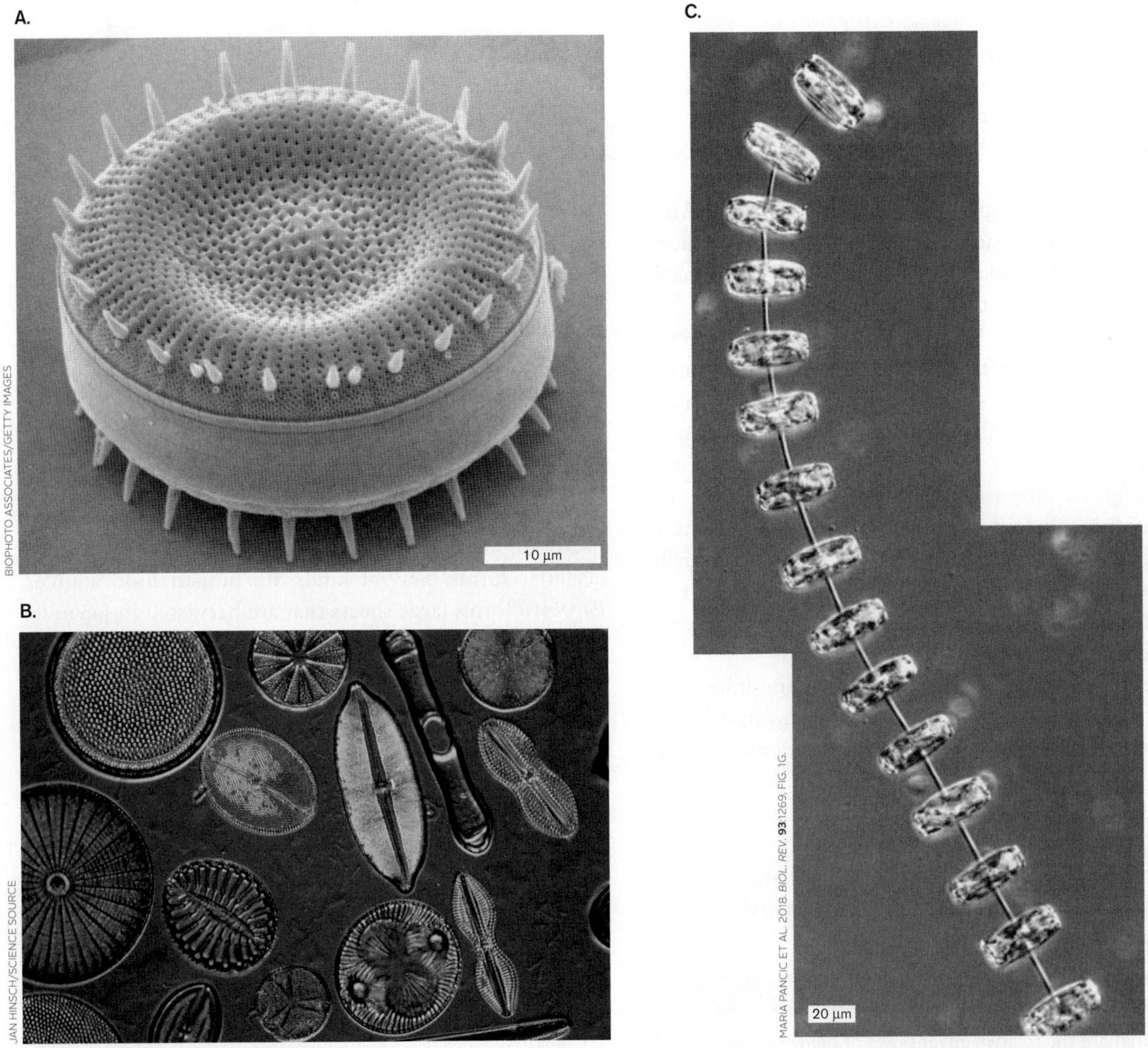

FIGURE 20.28 ▪ Diatoms of various species. A. *Stephanodiscus astraea*, a centric diatom. B. Diverse centric and pennate diatoms. C. Chain of disk-shaped centric diatoms, *Thalassiosira rotula*.

the flagellum that drives the cell through the environment is brush-like with lateral hairs, while the other is usually shorter and sometimes missing altogether. In some cases it functions like a rudder.

Diatoms (Bacillariophyceae). Diatoms are unicellular algae found ubiquitously in both fresh and marine waters (**Fig. 20.28**). They conduct a fifth of all photosynthesis on Earth, and they fix as much biomass as all the terrestrial rain forests. A diatom grows a unique kind of bipartite shell called a **frustule**. The frustule is composed of silica (cross-linked silicon dioxide, SiO_2). The silicate frustules protect diatoms from many kinds of predators. Diatoms are nonetheless consumed by flagellates and amphipods (shrimplike invertebrates) and are infected by viruses.

Frustules of different species form an extraordinary range of shapes with intricate pore formations (**Fig. 20.28A**). The shapes fall into two classes: centric, with radial symmetry; or pennate, with bilateral symmetry (**Fig. 20.28B**). Spectacular chains of diatoms can be found, such as the disk-shaped diatoms shown in **Figure 20.28C**, which are held together by threads. Frustules of decomposed diatoms eventually sediment on the ocean floor, where they build sedimentary rock strata more than a kilometer thick. This "diatomaceous earth" is used in insulation material and in toothpaste. Diverse species of diatoms are highly sensitive to environmental factors such as pH, and the frequency of their shells in sediment can be used to track a lake's environmental history.

Diatoms pose a unique challenge to cell division (**Fig. 20.29**). As the diatom grows and fissions, each daughter cell receives one parental half of the frustule while forming a new half fitting within the parental half, like the bottom dish of a Petri plate. Thus, each generation results in an inexorable decline in size of the organism. As its cell size reaches a critical point, the diatom must undergo meiosis to generate gametes. Maria Vernet at the Scripps Institution of Oceanography studies diatoms of the polar oceans, such as the Antarctic diatom *Corethron criophilum*. Centric diatoms such as *Corethron* form egg cells and flagellated sperm. When the gametes fuse, they form a special kind of zygote called an auxospore. The auxospore generates a frustule of the same size as the original diatom.

Coccolithophores (Prymnesiophyceae). Coccolithophores are secondary algae of the haptophytes (see **Fig. 20.4**). Coccolithophores superficially resemble diatoms in that their cells have a solid mineral exoskeleton (**Fig. 20.30**). Instead of silicate, however, their exoskeleton is composed of calcium carbonate ($CaCO_3$). The calcium carbonate grows in multiple scales (unlike the unitary exoskeleton of a diatom). The plates may extend radially in all directions, such as the "trumpet" shapes of *Discosphaera*

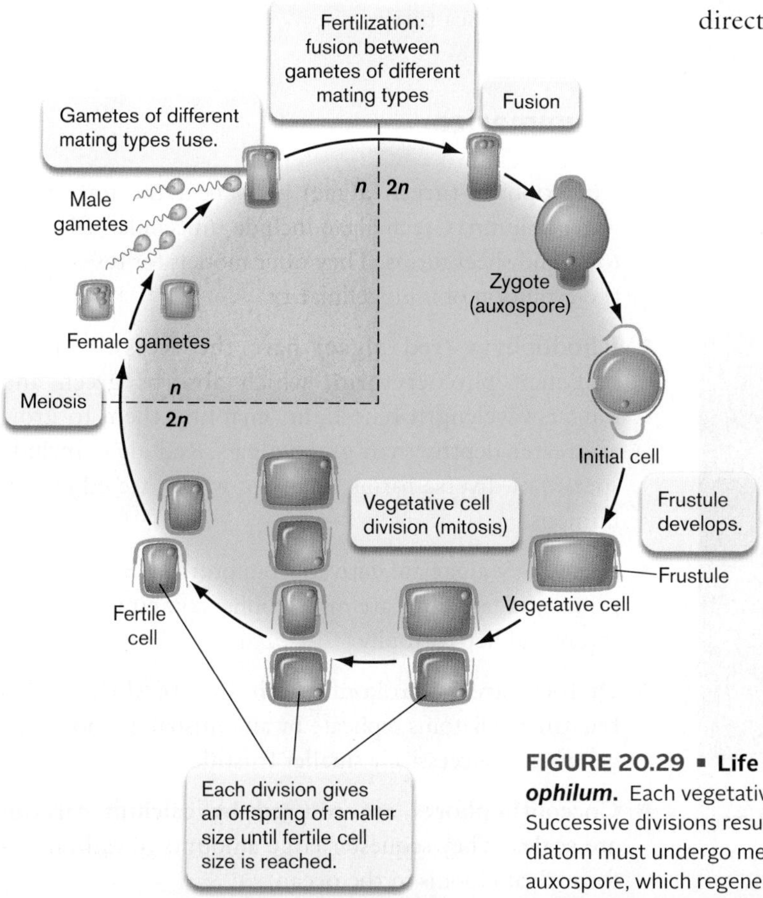

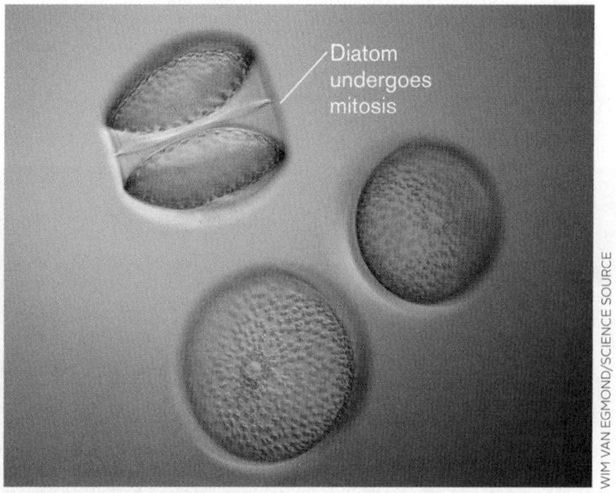

FIGURE 20.29 ■ Life cycle of a centric Antarctic diatom, *Corethron criophilum*. Each vegetative cell division requires formation of an in-fitting frustule half. Successive divisions result in progressive decrease in size. At a critical size limit, the diatom must undergo meiosis to form eggs and sperm. As gametes fuse, they form an auxospore, which regenerates a frustule of the original size.

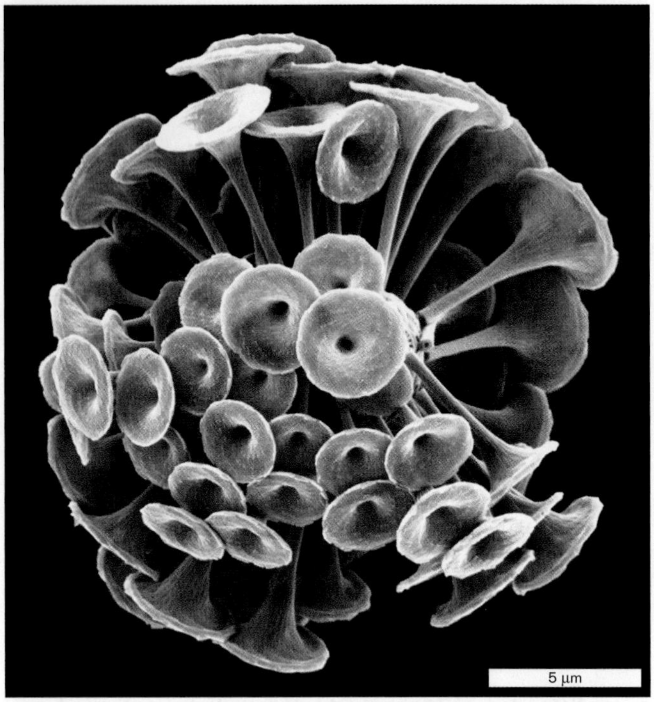

FIGURE 20.30 ■ A coccolithophore, *Discosphaera tubifera*. The tiny cell is surrounded by a much larger volume of trumpet-shaped coccoliths. From the Alboran Sea, western Mediterranean.

tubifera (**Fig. 20.30**), or form multiple layers of flat, oval shapes encasing the cell, as in *Emiliana huxleyi*. These scales are the "coccoliths" that give coccolithophores their name. The calcium carbonate scales protect the tiny cell from some predators. Coccolithophores such as *E. huxleyi* generate huge, milky blooms that appear in NASA satellite images. The blooms are dissipated by predation or by virus infection.

Sargassum weed

FIGURE 20.31 ■ Sargassum forests. *Sargassum natans* is the basis of the Sargasso Sea. The brown alga forms stalks with leaflike blades and round gas bladders to keep the alga afloat. Sargassum forests support animals such as the sea turtle.

Coccolithophores are increasingly recognized as major players in the ocean's carbon cycle (discussed in Chapter 22). They sequester large amounts of carbon from CO_2 into their carbonate shells. Unfortunately, their calcium carbonate is sensitive to acidification caused by the global rise in CO_2. For this reason, intensive research is focused on understanding the response of coccolithophores such as *E. huxleyi* to pH change. In 2011, a large-scale study conducted by several European universities concluded that increasing CO_2 levels correlate with a decline in the overall mass of marine coccoliths. At the same time, however, certain strains of *Emiliana* were shown to grow despite low pH, so many uncertainties remain.

Brown algae (Phaeophyceae). Brown algae, such as kelps, possess vacuoles of leucosin, a polysaccharide for energy storage whose color gives the organism a brown or yellow tint. Kelps are familiar to ocean bathers as the long, dark brown blades that root near the beach until the surf rips the blades off and tosses them ashore. Kelps support important communities of multicellular organisms known as kelp forests. Another type of marine "forest" consists of the unrooted **sargassum weeds** that float on the Sargasso Sea. Sargassum consists of stalks with photosynthetic blades and round gas bladders to keep the organism afloat (**Fig. 20.31**). Sargassum supports a complex food web of invertebrate and vertebrate animals, including worms, crabs, fish, and sea turtles.

To Summarize

- **Chlorophyta (green algae)** grow near the top of the water column. Green algae include unicellular, filamentous, and sheet forms. They offer models for research on the evolution of multicellularity.
- **Rhodophyta (red algae)** have the accessory photopigment phycoerythrin, which absorbs green and longer-wavelength blue light, enabling them to grow at greater depths than green algae. Red algae include species of diverse forms, many of which are edible for humans.
- **Secondary algae** are derived from protists that engulfed primary algae. They are mixotrophs, combining phototrophy and heterotrophy.
- **Diatoms are heterokonts with silicate shells called frustules.** Diatoms replicate by an unusual division cycle generating successively smaller frustules.
- **Coccolithophores are covered by calcium carbonate scales.** They sequester large amounts of carbon and form algal blooms in the ocean.

- **Kelps are heterokonts that grow in long, sheetlike fronds.** Kelps play an important role in the ecology of the coastal ocean, as well as the ecology of marine beaches.

20.5 Alveolates: Ciliates, Dinoflagellates, and Apicomplexans

The Alveolata include voracious predators such as the ciliated protists (**Fig. 20.32A**), as well as a major group of algae, and perhaps the most successful group of parasitic microbial eukaryotes. Alveolates are named for the flattened vacuoles called "alveoli" (singular, alveolus) within their outer cortex (**Fig. 20.32B**; see also **Fig. 20.7A**). Some alveoli contain plates of stiff material, such as protein, polysaccharide, or minerals. Besides alveoli, most alveolate protists possess other kinds of cortical organelles, such as extrusomes for delivery of enzymes or toxins, bands of microtubules for reinforcement, and whiplike cilia or flagella. The alveolate cell form is highly structured, in contrast to the amorphous shape of amebas. Major groups of alveolates include ciliates, dinoflagellates, and apicomplexans.

Ciliates

Alveolates of a diverse group known as Ciliophora, or ciliates, possess large numbers of cilia, short projections containing [9(2)+2] microtubules. Their whiplike action is driven by ATP (for review, see eAppendix 2). The cilia beat in coordinated waves that maximize the efficiency of motility. Cilia serve two functions:

- **Cell propulsion.** Coordinated waves of beating cilia, usually covering the cell surface, propel the cell forward.
- **Food acquisition.** By generating water currents into the mouth of the cell, a ring of cilia around the mouth brings food into the cell.

Ciliate cell structure. *Paramecium* is one of the most studied ciliates. Paramecia feed on bacteria and in turn are consumed by other ciliates, such as *Didinium* (**Fig. 20.32A**). They can also take up smaller particles through endocytosis by specialized pores in their cortex, called parasomal sacs (**Fig. 20.32B**).

The cell structure of *Paramecium* includes an **oral groove** for uptake of food driven by the beating cilia (**Fig. 20.32C**). Once ingested through the oral groove, a food particle travels within the digestive vacuole in a circuit around the cell. The digestive vacuole ultimately empties into the cytoproct, a specialized vacuole for the discharge of waste outside the cell. Paramecia maintain osmotic balance by means of a **contractile vacuole**, a vacuole that withdraws water from the cytoplasm to shrink it or contracts to expand it. Contractile vacuoles are widespread among protists and algae, but their mode of action has been most studied in paramecia.

Genetics and reproduction. Most ciliates have a complex genetic system involving one or more **micronuclei** and **macronuclei**. The micronucleus contains a diploid set of chromosomes that undergoes meiosis for sexual exchange (a process reviewed in eAppendix 2). For gene expression, however, one of the micronuclei develops into a macronucleus that forms hundreds of copies of its DNA. The DNA copies in the macronucleus are rearranged and fragmented to small segments. The small DNA segments generate a large number of "telomeres," chromosome ends. In humans, telomere shortening is associated with aging; thus, ciliate macronucleus formation provides a model system for study of human aging (**eTopic 20.2**).

In ciliates, only macronuclear genes are transcribed to RNA and translated to protein. When a ciliate reproduces asexually, the micronucleus undergoes mitosis, whereas the macronucleus divides by a different mechanism that is poorly understood. Cell division occurs across the long axis, necessitating generation of a new oral groove for the posterior daughter cell and a new cytoproct for the anterior daughter cell—again, a process poorly understood.

Most ciliates are diploid and never produce haploid gamete cells. Instead, their sexual reproduction involves exchange of micronuclei. The two ciliates of a mating pair exchange haploid micronuclei by **conjugation**. In conjugation, two cells of opposite mating type form a cytoplasmic bridge and exchange their nuclear products of meiosis (**Fig. 20.33**). While the cells are connected, the micronucleus of each cell undergoes meiosis to form four haploid nuclei. Three out of four of the haploid nuclei disintegrate, as does the entire macronucleus. The haploid micronuclei then undergo mitosis, and one daughter nucleus from each of the two conjugating cells is exchanged across the cytoplasmic bridge. Each transferred nucleus then fuses with its haploid counterpart, restoring diploidy. The two cells come apart, and each recombined micronucleus fissions several times. One of the daughter micronuclei then transforms into the new macronucleus.

Thought Questions

20.9 Compare and contrast the process of conjugation in ciliates and bacteria (see Chapter 9).

20.10 For ciliates, what are the advantages and limitations of conjugation, as compared with gamete production?

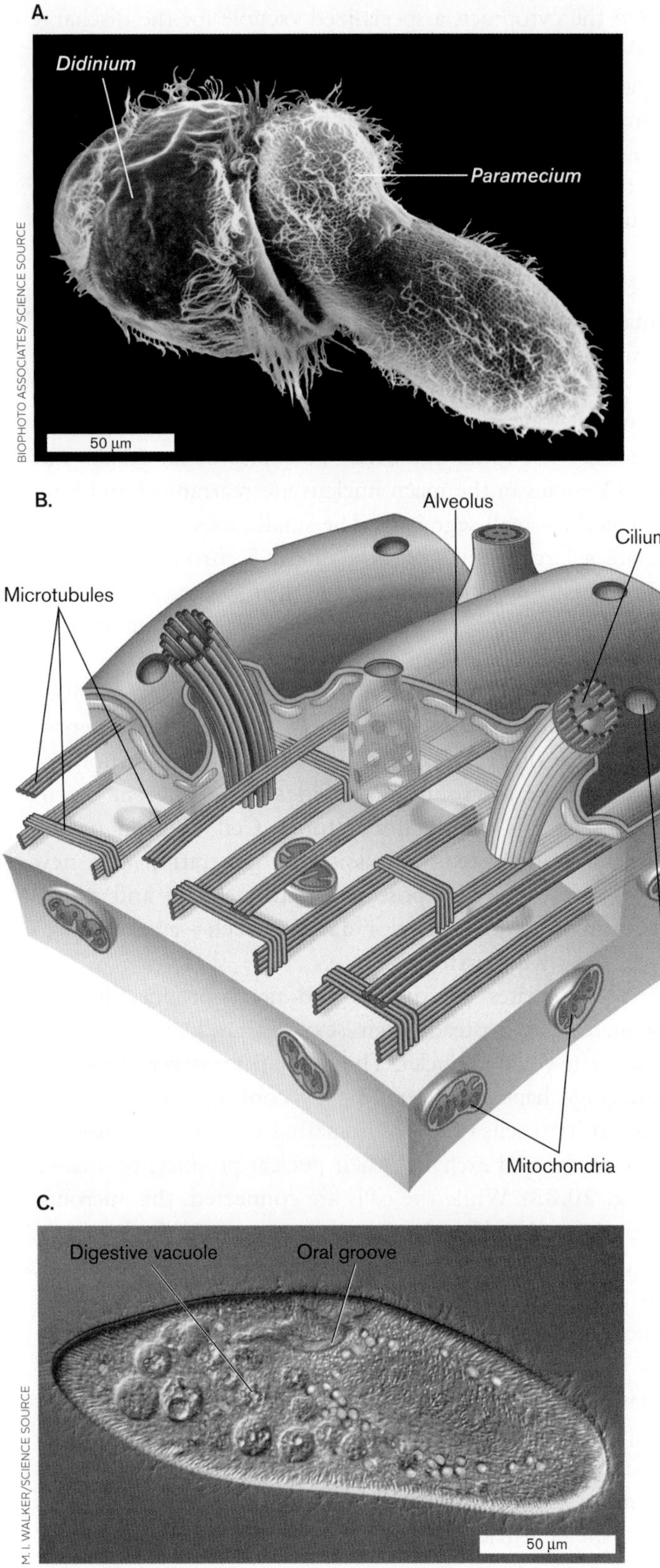

FIGURE 20.32 ■ Ciliated protists. **A.** *Didinium*, consuming *Paramecium* (SEM). **B.** Cortical structure of a ciliate. Beneath the outer cell membrane lie flattened sacs called alveoli. The cilia, composed of [9(2)+2] microtubules, are rooted in a complex network of microtubules. Parasomal sacs take up nutrients and form endocytic vesicles. **C.** A *Paramecium* has digestive vacuoles and an oral groove for ingestion (differential interference contrast, colorized).

Stalked ciliates. Some ciliates adhere to a substrate and use their cilia primarily to obtain prey. **Stalked ciliates** such as *Stentor* and *Vorticella* have a ring of cilia surrounding a large mouth (**Fig. 20.34A**). The ciliary beat is specialized to draw large currents of water and whatever prey it carries. Stalked ciliates are commonly found in pond sediment and in wastewater during biological treatment by microbial digestion, where they are attached to flocs of filamentous bacteria.

Another group of stalked ciliates, the suctorians (**Fig. 20.34B**), possess cilia for only a short period after a daughter cell is released by the stalked cell. The daughter cell swims by ciliary motion until it finds a good habitat in which to settle, whereupon its cilia are replaced by knobbed tentacles similar to the filopodia of shelled amebas. Suctorians prey on swimming ciliates such as paramecia.

Dinoflagellates Are Phototrophs and Predators

The dinoflagellates (Dinoflagellata) are a major group of marine phytoplankton, essential to marine food webs. Like ciliates, they are highly motile, but instead of numerous short cilia, dinoflagellates possess just two long flagella, one of which wraps along a crevice encircling the cell (**Fig. 20.35A**). Some dinoflagellates possess elaborate hornlike extensions (**Fig. 20.35B**). The cell extensions increase the range of nutrient uptake, and they may deter predation.

Dinoflagellates are secondary or tertiary algae. They have a chloroplast derived from a red alga, which in some species was later replaced by a heterokont alga, itself a secondary alga (**Fig. 20.35C**). Some dinoflagellates possess carotenoid pigments that confer a red color. Blooms of red dinoflagellates cause the famous red tide, which may have inspired the biblical story of the plague in which water turns to blood (**Fig. 20.35D**). Dinoflagellates release toxins that can be absorbed by shellfish, poisoning consumers.

The armor-plated appearance of a dinoflagellate results from its stiff alveolar plates, composed of cellulose (**Fig. 20.35C**). The complex outer cortex includes various extrusomes (organelles that extrude a defensive substance) and endocytic pores, as well as a species-specific pattern of alveolar plates. Dinoflagellates supplement their photosynthesis by predation, extending a special type of pseudopod to engulf prey. Many dinoflagellates have evolved to lose their chloroplasts altogether, becoming obligate predators or parasites.

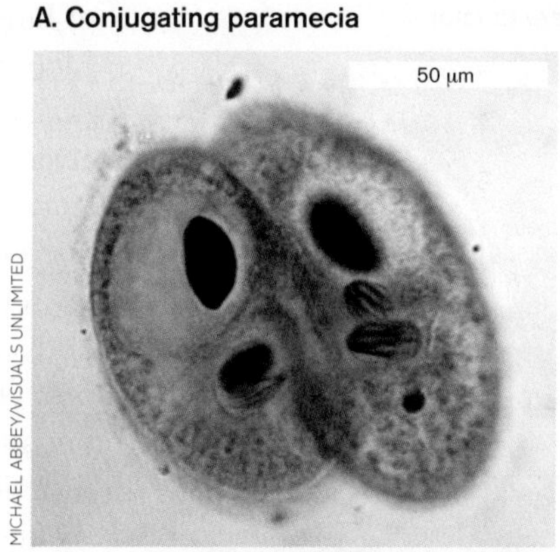

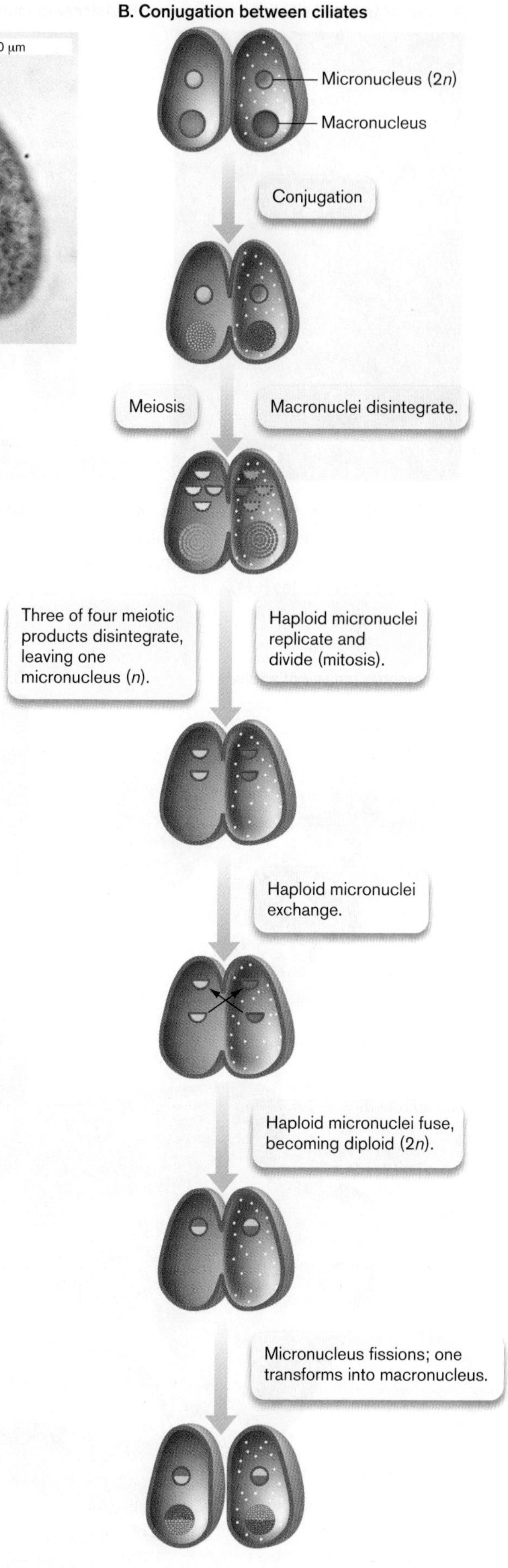

FIGURE 20.33 ■ Conjugation. **A.** Two paramecia conjugating (LM). **B.** In conjugation, two paramecia of opposite mating type form a cytoplasmic bridge. The 2*n* micronucleus of each cell undergoes meiosis. Each macronucleus, as well as three out of four meiotic products, disintegrates. The haploid micronuclei undergo mitosis, forming two daughter micronuclei. Daughter nuclei from each cell are exchanged across the cytoplasmic bridge and then fuse with their respective counterparts, restoring 2*n* micronuclei. The micronuclei fission several times, and one transforms into a new macronucleus.

Some dinoflagellates inhabit other organisms as endosymbionts, providing sugars from photosynthesis in exchange for a protected habitat. Their hosts include shelled amebas, as well as cnidarian animals: sponges, sea anemones, and, most important, reef-building corals. Coral endosymbionts, such as zooxanthellae, are vital to reef growth. The major genus of zooxanthellae is *Symbiodinium*. *Symbiodinium* species are taken up by the coral's gastrodermal cells (**Fig. 20.36**; see also the Current Research Highlight). Each dinoflagellate is enclosed by an intracellular vacuole, called a symbiosome (**Fig. 20.36A**). The vacuole enables the dinoflagellate to transfer products of photosynthesis to the host coral cell. In return, the host cell provides ammonium excreted from catabolism. Ammonium and nitrate are then used by the dinoflagellate to synthesize amino acids—something the host coral cannot do. Thus, *Symbiodinium* provides essential nutrients while receiving a home protected from predators.

The coral symbiosis with *Symbiodinium* is highly sensitive to temperature. Rising temperatures in the ocean lead to coral bleaching (the expulsion of zooxanthellae), after which the coral dies. Thus, the health of coral reefs, and of large-scale ecosystems such as Australia's Great Barrier Reef, is endangered by global climate change. Todd LaJeunesse (**Fig. 20.36B**), at Penn State University, studies the ecology of the coral-*Symbiodinium* mutualism, and the effects of climate warming.

Apicomplexans Are Specialized Parasites

Apicomplexans include many human parasites, such as the intestinal parasite *Cryptosporidium*, which infected hundreds of people in the United States in 2013. Apicomplexan cells have an apical complex, a highly specialized structure that facilitates entry of the parasite into a host cell. Another important apicomplexan is *Toxoplasma gondii*, a parasite commonly carried by cats and transmissible

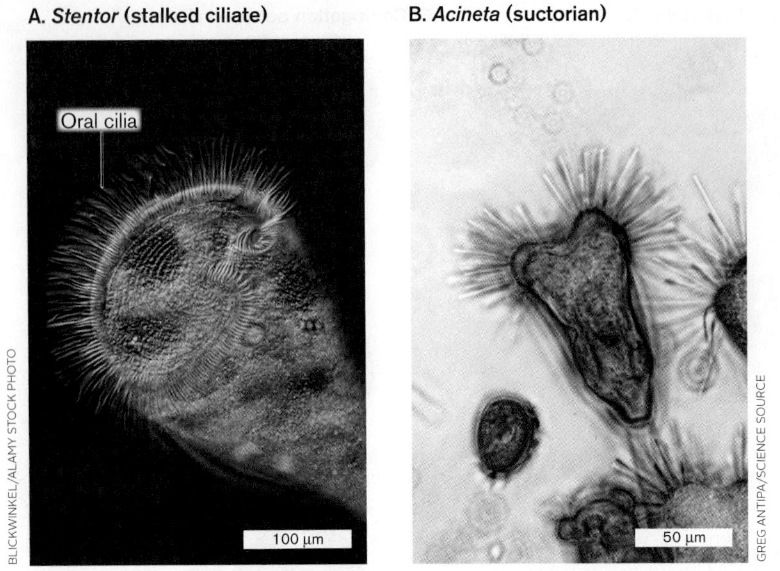

FIGURE 20.34 ▪ Stalked ciliates. **A.** *Stentor*, a ciliate with a flexible stalk, 1.5–2.0 mm in length (phase contrast). The oral ring of cilia generates currents drawing food into the mouth. **B.** The suctorian *Acineta* replaces cilia with knobbed tentacles (LM).

to humans, where it can harm a developing fetus. Like the ciliates and dinoflagellates, apicomplexans possess an elaborate cortex composed of alveoli, pores, and microtubules. But as parasites, apicomplexans have undergone extensive reductive evolution, losing their flagella or cilia. They possess a unique organelle called the apicoplast, derived by genetic reduction from an endosymbiotic chloroplast. No capacity for photosynthesis remains, but the apicoplast provides one essential function in fatty acid metabolism.

The best-known apicomplexan is *Plasmodium falciparum*, the main causative agent of **malaria**, the most important parasitic disease of humans worldwide. **Figure 20.37** shows red blood cells in the early "ring stage" of infection, and in the "schizont" stage, which eventually bursts, releasing progeny parasites. *P. falciparum* is carried by mosquitoes, which transmit the parasite to humans when the insect's proboscis penetrates the skin. The disease is endemic in areas

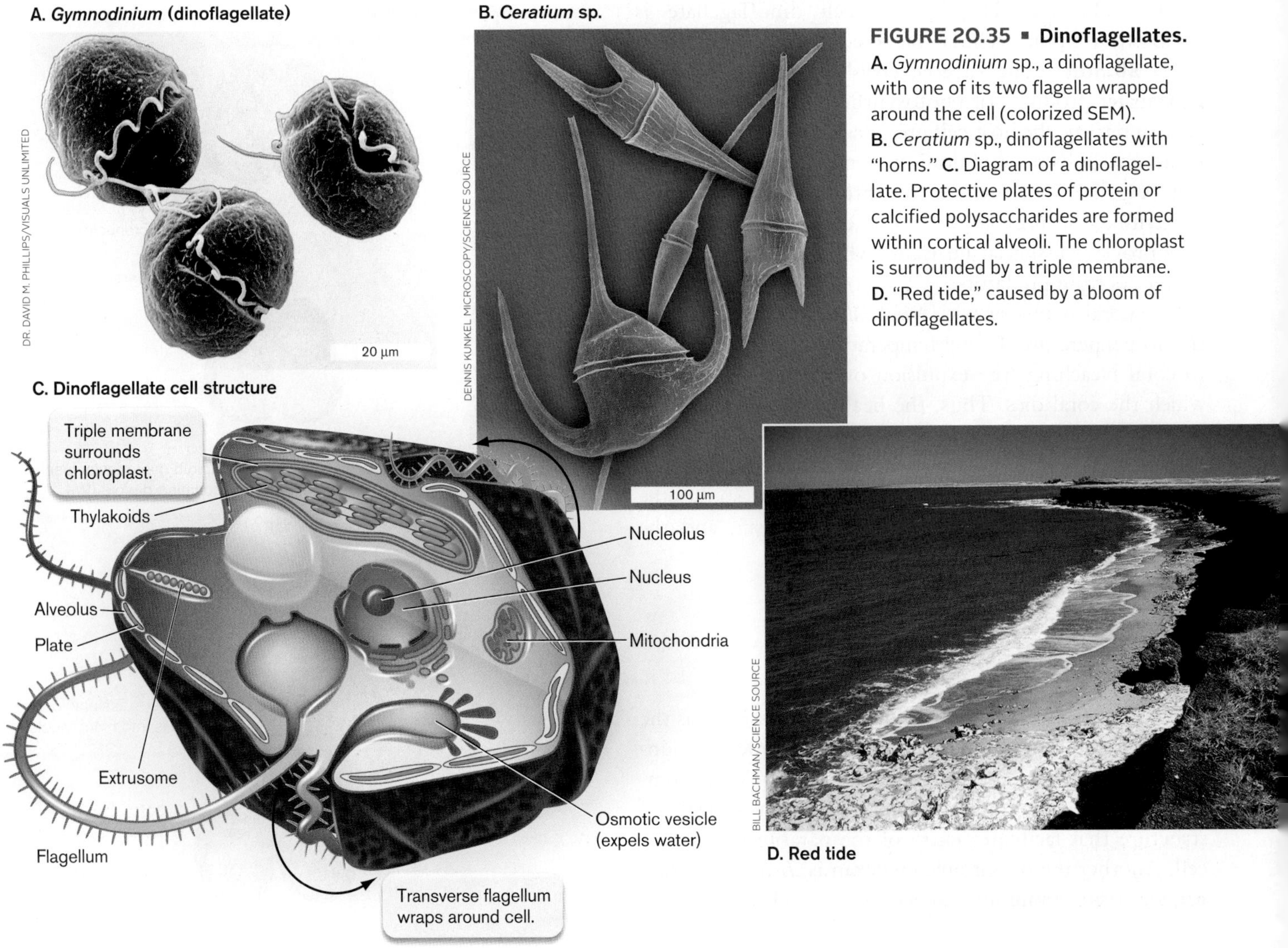

FIGURE 20.35 ▪ Dinoflagellates. **A.** *Gymnodinium* sp., a dinoflagellate, with one of its two flagella wrapped around the cell (colorized SEM). **B.** *Ceratium* sp., dinoflagellates with "horns." **C.** Diagram of a dinoflagellate. Protective plates of protein or calcified polysaccharides are formed within cortical alveoli. The chloroplast is surrounded by a triple membrane. **D.** "Red tide," caused by a bloom of dinoflagellates.

A.

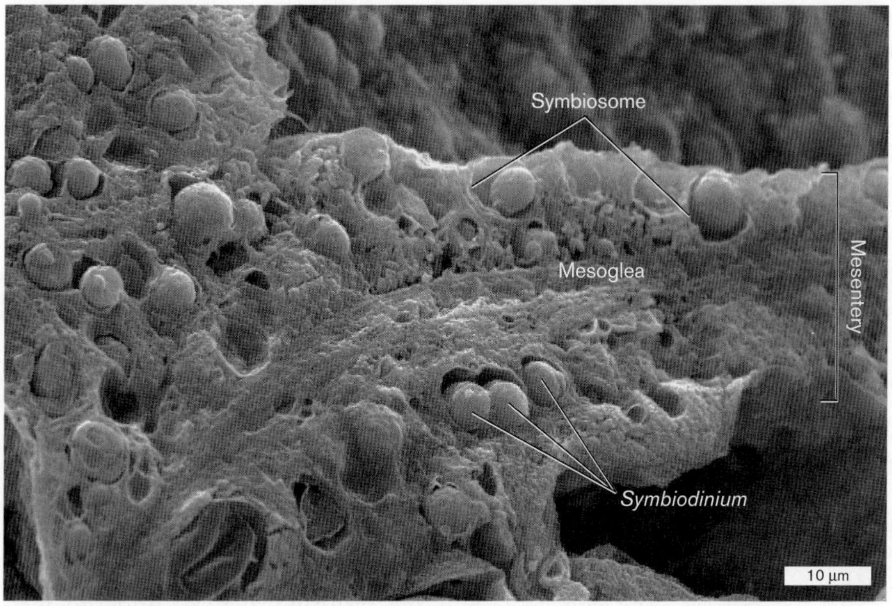

B.

FIGURE 20.36 ■ *Symbiodinium*: endosymbiotic dinoflagellate partners of coral. **A.** SEM micrograph of freeze-fractured internal mesentery from a reef coral polyp (Porites Porites) that shows the distribution and density of symbiont cells in host cell vacuoles. Symbiosomes are vacuoles within a gastrodermal cell. Each symbiosome contains one cell of *Symbiodinium*, which exchanges products of photosynthesis with organic nutrients from the host cell. **B.** Todd LaJeunesse studies coral ecology and the effects of global climate change.

inhabited by 40% of the world's population; it infects hundreds of millions of people and kills half a million African children each year.

The transmitted parasites invade the liver and then develop into the **merozoite** form that invades red blood cells (**Fig. 20.38**). The merozoite first contacts a red blood cell with its apical complex. The apical complex contains secretory organelles called rhoptries that inject enzymes to aid entry by the parasite. The cone-like tip of the apical complex penetrates the host cell, enabling secretion of lipids and proteins that facilitate invasion. Eventually, the entire merozoite enters the host cell, leaving no traces of the parasite on the host cell surface. Thus, the internalized parasite becomes invisible to the immune system until its progeny burst out.

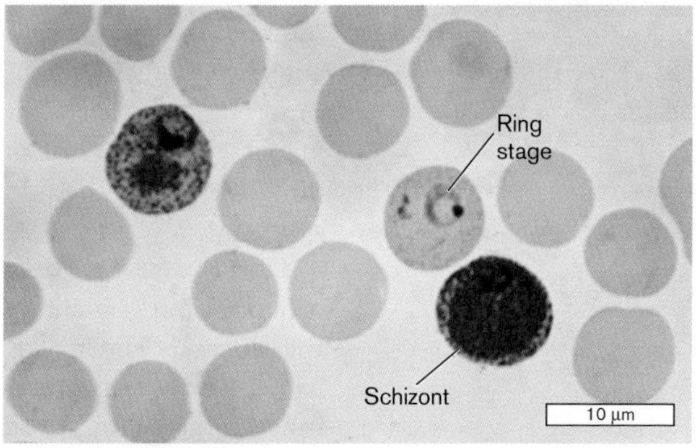

FIGURE 20.37 ■ *Plasmodium falciparum*, a cause of malaria. Red blood cells infected with *P. falciparum*, which is stained purple with a dye that interacts with DNA (LM). One late-stage infected blood cell (schizont) can be seen.

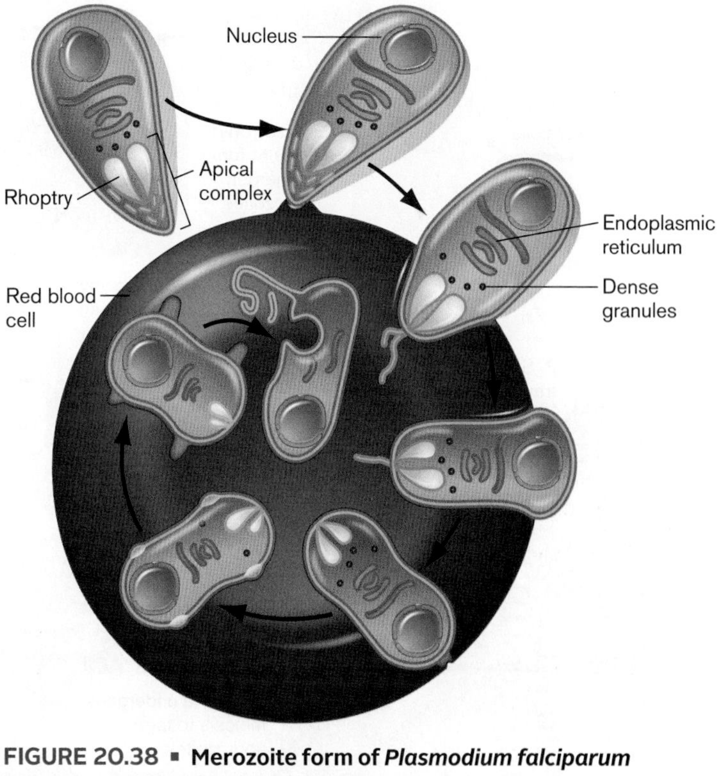

FIGURE 20.38 ■ Merozoite form of *Plasmodium falciparum* invades a red blood cell. The apical complex facilitates invasion and then dissolves as the merozoite transforms into an intracellular form.

P. falciparum acquires resistance rapidly, and many strains no longer respond to drugs, such as quinine, that nearly eliminated the disease half a century ago. The life cycle and molecular properties of *P. falciparum* have been studied extensively for clues to aid in the development of new antimalarial drugs and vaccines. The elaborate life cycle of *P. falciparum* and other apicomplexan parasites involves several common features:

- **Schizogony**, mitotic reproduction of a haploid form (in the mammalian host) to achieve a large population within a host tissue. Usually, the nuclei multiply first, followed by separation of individual nucleated cells.
- **Gamogony**, the differentiation of haploid cells into male and female gametes capable of fertilization.
- **Mitosis** and meiosis of the diploid zygote (within the insect) turns it into a haploid, spore-like form transmissible to the next host.

In the case of malaria (**Fig. 20.39**), whip-shaped sporozoites injected by the mosquito invade the liver, where they undergo **schizogony** (nuclear multiplication followed by cell separation). The cell products of schizogony are called merozoites. The merozoites from the liver then invade red blood cells, where they feed on hemoglobin. An early infected blood cell appears as a "ring stage" (**Fig. 20.39**). The parasite multiplies, filling the host cell, now called a "schizont." The schizont bursts, liberating progeny merozoites that invade another round of red cells. The bursting of red blood cells also releases cell fragments that trigger the cyclic fevers characteristic of malaria.

Some of the merozoites in the bloodstream develop into pre-gamete cells, or gametocytes. The gametocytes are acquired by bloodsucking mosquitoes and then multiply and mature (gamogony) in the mosquito's midgut. The gametocytes develop into female eggs and thin male cells with flagella (this is the only stage in the apicomplexan life cycle that has flagella). The male cells fertilize the egg cells, and the resulting zygotes undergo meiosis and differentiate into sporozoites, which enter the mosquito's salivary gland for transmission to the next human host.

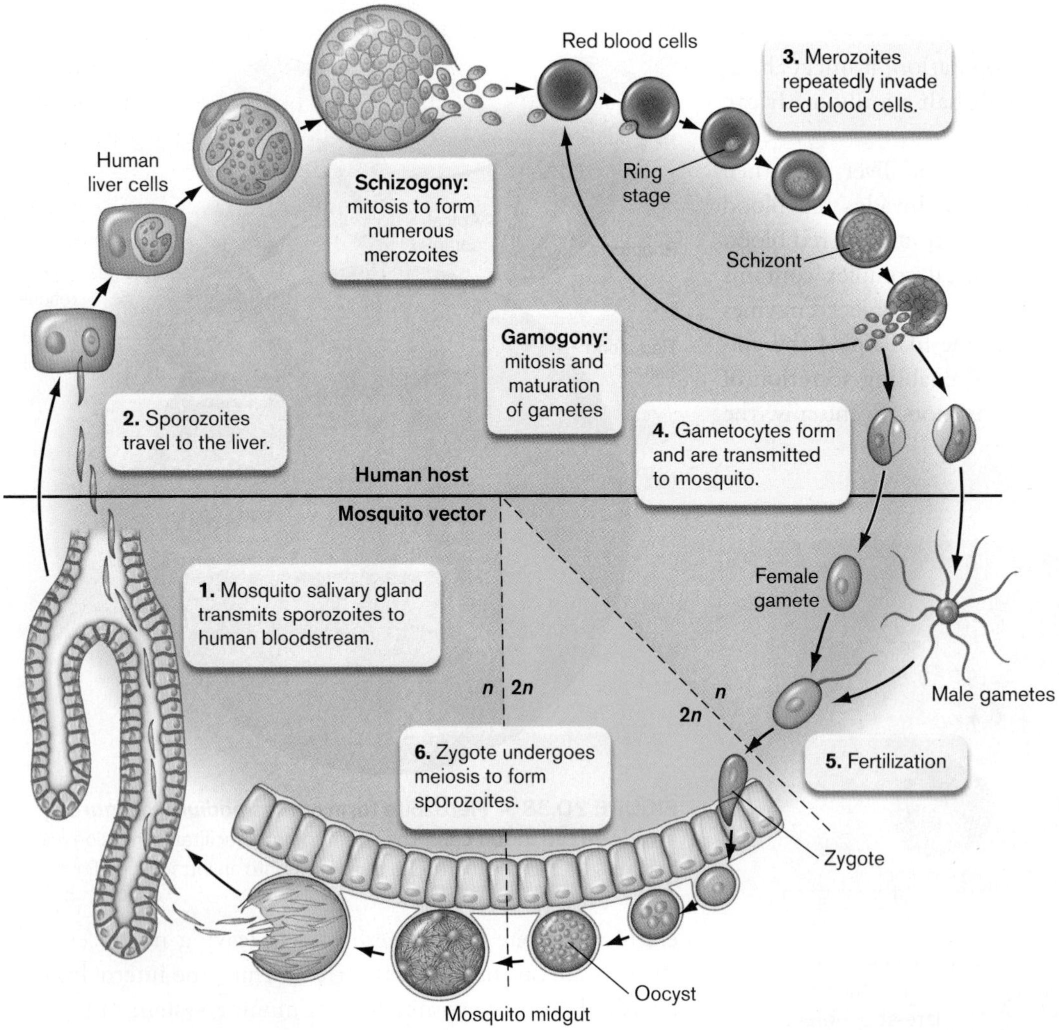

FIGURE 20.39 ■ Malaria: cycle of *Plasmodium falciparum* transmission between mosquito and human.

The nuclear genome of *P. falciparum* consists of 24 Mb contained in 14 chromosomes. Sequence annotation and expression studies predict 5,300 protein-encoding open reading frames (ORFs), comparable to the number in a yeast genome. The parasite has lost many genes encoding enzymes and transporters while expanding its repertoire of proteins involved in antigenic diversity. In addition, the parasite contains two smaller non-nuclear genomes: that of its mitochondria and that of the chloroplast-derived apicoplast.

How can we treat malaria and eradicate the disease? The malarial genome reveals promising targets for drug design. For example, the fatty acid biosynthesis within the apicoplast is targeted by triclosan and other antimicrobials.

Other promising targets for antimalarial drugs are the unique proteases required to digest hemoglobin within the *P. falciparum* food vacuole.

To Summarize

- **Ciliates are covered with numerous cilia.** Cilia provide motility and help capture prey. Ciliates undergo complex reproductive cycles involving exchange of micronuclei through conjugation.
- **Dinoflagellates are phototrophic predators.** Dinoflagellates are tertiary endosymbiotic algae. Their alveoli contain calcified plates; they have paired flagella, one of which is used for propulsion. Predation occurs by extension of a lamellar pseudopod.
- ***Symbiodinium* dinoflagellates form endosymbioses with corals.** Each dinoflagellate inhabits a vacuole within a coral host cell and provides products of photosynthesis. Coral bleaching (the expulsion of endosymbionts) is induced by temperature increase, a growing problem with global climate change.
- **Apicomplexans are parasites that penetrate host cells.** The apicoplast is a specialized organ for cell penetration. Apicomplexans such as *Plasmodium falciparum* conduct complex life cycles within mammalian and arthropod hosts.

20.6 Parasitic Protozoa

Surprisingly, many protists harmlessly inhabit the human gut as commensals. Such commensals include members of diverse clades, such as *Entamoeba coli* (an ameba), *Blastocystis* (a heterokont), and *Retortamonas intestinalis* (a metamonad). Such protists might provide positive benefits to their host. Nevertheless, closely related organisms, such as *Entamoeba histolytica*, may cause deadly disease. The previous section introduced apicomplexan parasites of major importance, especially those that cause malaria. Other major clades of parasites (and related harmless protozoa) are the trypanosomes and metamonads.

Trypanosomes

The group Euglenida includes flagellated protists such as *Euglena*, with chloroplasts arising from secondary endosymbiosis. Like other algal protists, euglenas combine photosynthesis and heterotrophic nutrition. Euglenida, however, also includes a group of obligate parasites called **trypanosomes**. Trypanosomes consist of an elongated cell with a single flagellum. The trypanosome has a unique organelle called the "kinetoplast," consisting of a mitochondrion containing a bundle of multiple copies of its circular genome, usually placed near the base of the flagellum.

Trypanosomes cause some of the most gruesome and debilitating conditions known to humanity, such as leishmaniasis (**Fig. 20.40A**). *Leishmania major* (**Fig. 20.40B**) causes skin infections that may enter the internal organs. If untreated, leishmaniasis can lead to swelling and decay of the extremities and eventually death. Carried by sand flies, *Leishmania* infects 1.5 million people annually, in South America, Africa and the Middle East, and southern Europe. *Leishmania* often infects Americans serving in Iraq; for this reason, returning veterans from Iraq are permanently restricted from donating blood.

Another major disease caused by trypanosomes is trypanosomiasis, also known as African sleeping sickness. The parasite, *Trypanosoma brucei* (**Fig. 20.40C**), is carried by the tsetse fly. *T. brucei* multiplies in the bloodstream of the host animal, causing repeated cycles of proliferation

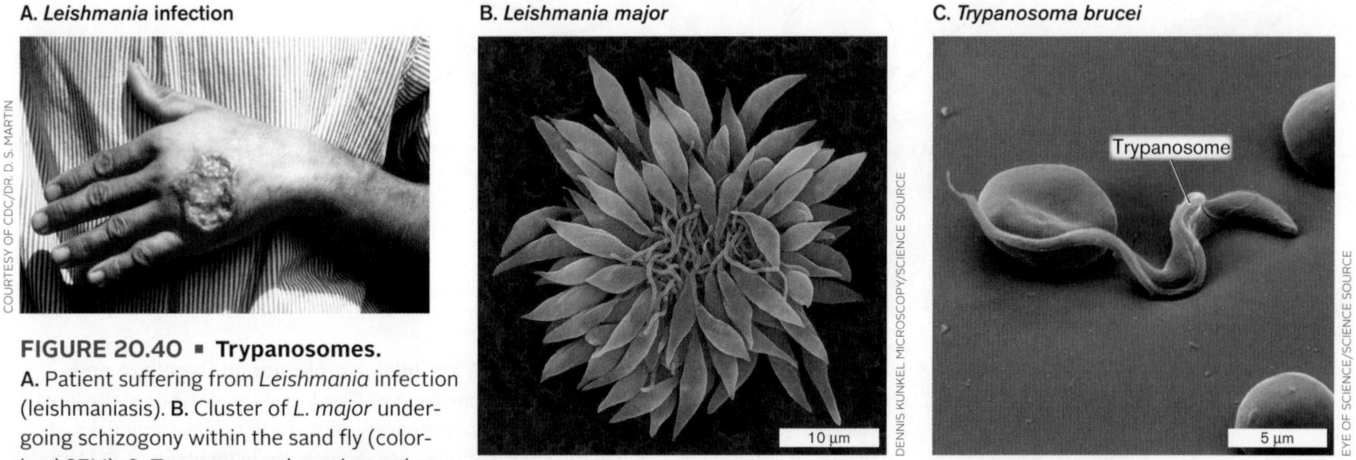

FIGURE 20.40 ■ Trypanosomes.
A. Patient suffering from *Leishmania* infection (leishmaniasis). **B.** Cluster of *L. major* undergoing schizogony within the sand fly (colorized SEM). **C.** *Trypanosoma brucei*, seen here among red blood cells, is the cause of African sleeping sickness (colorized SEM).

and fever that ultimately lead to death, if untreated. This trypanosome is known for its extraordinary degree of antigenic variation. Its genome includes 200 different active versions of its variant surface glycoprotein (VSG), the antigen inducing the immune response, as well as 1,600 different "silent" versions that can recombine with "active" VSG to make further variations. In effect, the trypanosome overwhelms the host immune system by continually generating new antigenic forms until the host repertoire of antibodies is exhausted. In order to infect its human host, the trypanosome needs to interconvert among several different forms of its life cycle. The molecular basis of conversion offers targets for drug therapy. An example of research on the trypanosome life cycle is described in **eTopic 20.3**.

A related trypanosome, *T. cruzi*, is carried by reduviid bugs, a kind of blood-feeding insect. *T. cruzi* causes Chagas' disease, a debilitating infection of the heart and other internal organs. Chagas' disease is prevalent in South and Central America, and global climate change is expected to expand its range north.

Metamonads

Metamonada is another major group of parasites and symbionts. The metamonad parasites include the diplomonads (named for their double nuclei), such as *Giardia intestinalis* (*G. lamblia*), and *G. duodenalis,* a common intestinal parasite (**Fig. 20.41**). *Giardia* is a frequent nemesis of day-care centers, but it also occurs in freshwater streams visited by bears and other wildlife. *Giardia* occasionally contaminates community water supplies in the United States and is endemic in major cities of Russia. *Giardia* and other metamonads are noted for their anaerobic metabolism and their degenerated organelles, reflecting their adaptation to the anaerobic intestinal environment.

The *Giardia* life cycle alternates between two major forms: the trophozoite and the dormant cyst. The trophozoite (**Fig. 20.41B**) has two nuclei (and, therefore, $4n$ chromosomes). There are four pairs of flagella, and an "adhesive disk" enabling the parasite to adhere to the intestinal epithelium. The cell body contains no Golgi, and its mitochondria have degenerated to "mitosomes." Mitosomes lack mitochondrial genomes. When the trophozoite experiences stress conditions, such as high levels of bile and a high pH, the organism encysts (**Fig. 20.41C**). The cyst detaches from the intestine and is expelled from the host. It remains dormant until ingestion by a new host, where

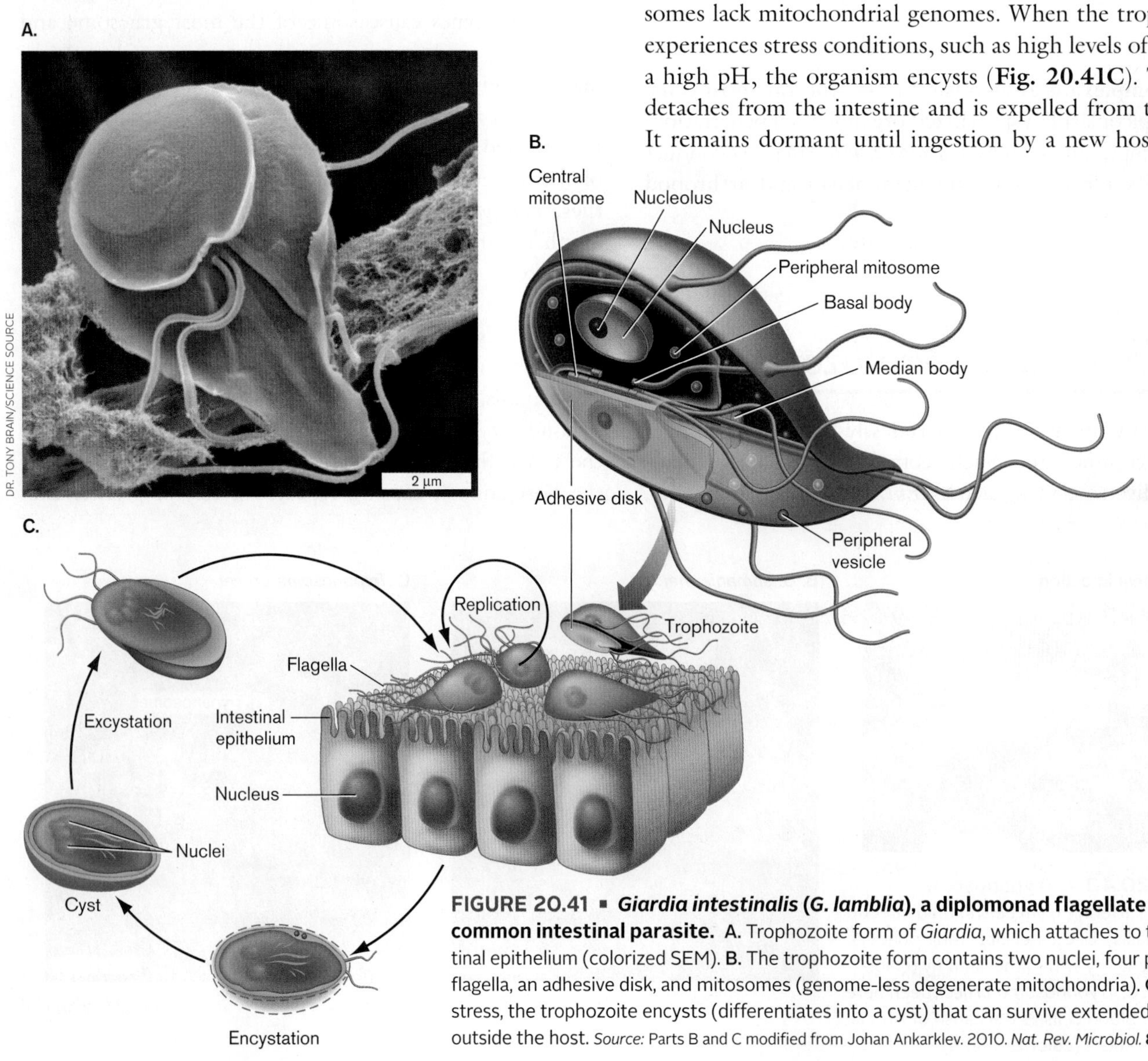

FIGURE 20.41 ■ ***Giardia intestinalis* (*G. lamblia*), a diplomonad flagellate and common intestinal parasite.** **A.** Trophozoite form of *Giardia*, which attaches to the intestinal epithelium (colorized SEM). **B.** The trophozoite form contains two nuclei, four pairs of flagella, an adhesive disk, and mitosomes (genome-less degenerate mitochondria). **C.** Under stress, the trophozoite encysts (differentiates into a cyst) that can survive extended periods outside the host. *Source:* Parts B and C modified from Johan Ankarklev. 2010. *Nat. Rev. Microbiol.* **8**:413.

stomach acid triggers differentiation into a trophozoite.

Intestinal Parasites

Giardia is just one of many unpleasant intestinal visitors acquired by humans and animals. The ameba *Entamoeba histolytica* grows in the human colon, causing amebiasis (**Fig. 20.42A**). The disease includes diarrhea and possible damage to the intestinal wall; in some cases, the parasite can invade the blood and internal organs. Worldwide, *E. histolytica* kills tens of thousands of people per year. The organism is challenging to diagnose because it appears very similar to a harmless ameba, *E. dispar*, which grows normally in the intestine.

The apicomplexan *Cryptosporidium parvum* (**Fig. 20.42B**) commonly contaminates water supplies in the United States; it caused the nation's largest waterborne disease outbreak to date, sickening more than 400,000 people in Milwaukee in 1993. *Cryptosporidium* is especially dangerous to immunocompromised patients.

An important ciliate parasite is *Balantidium coli* (**Fig. 20.42C**). *Balantidium* is transmitted by a fecal-oral route, most commonly in malnourished individuals whose stomach acid is low, thus failing to kill the pathogen. Infection may be without symptoms, or it can lead to diarrhea and damage the colon.

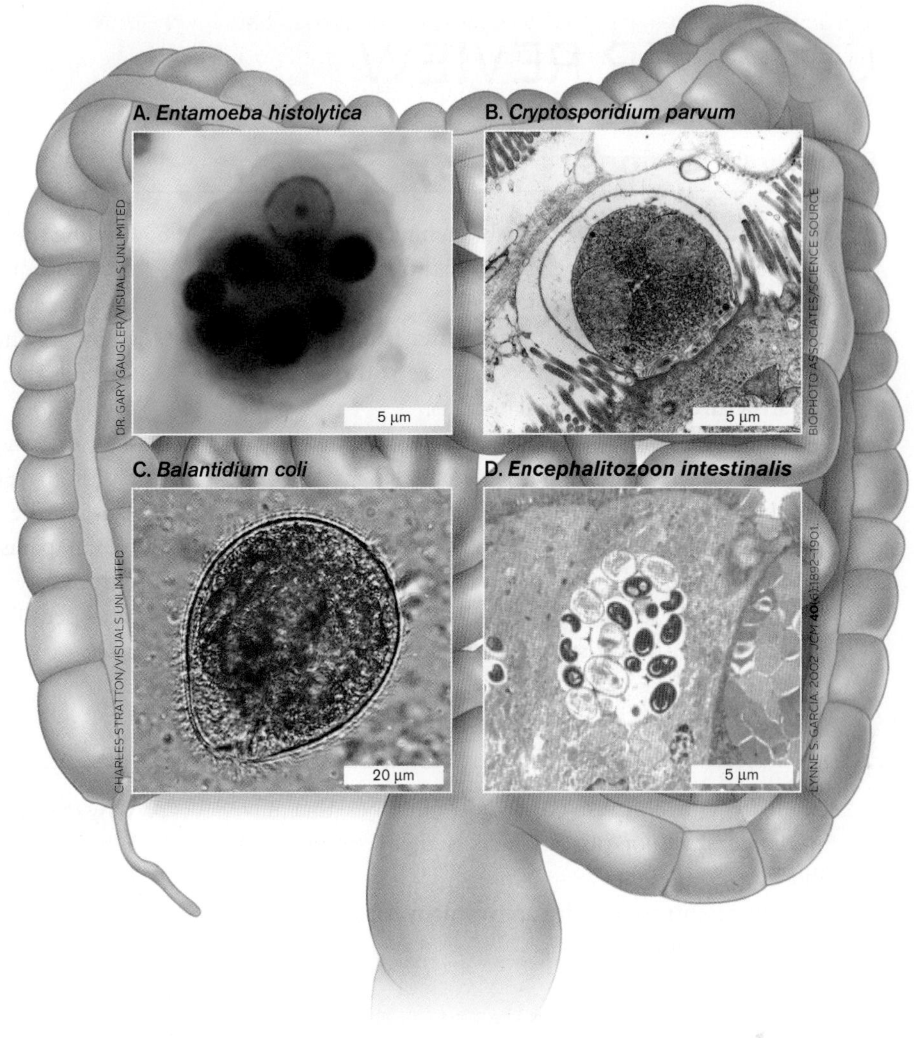

FIGURE 20.42 ■ Human intestinal protozoa.

Microsporidians (**Fig. 20.42D**) were once thought to be protozoa, but genetically and physiologically they are closely related to fungi (see Section 20.2). *Encephalitozoon intestinalis* is an obligate parasite of the intestine, causing problems especially for immunocompromised patients. More discussion of intestinal pathogens is found in Chapters 23 and 26.

To Summarize

- **Trypanosomes** include free-living euglenas, as well as the important parasites *Leishmania* (cause of leishmaniasis), *Trypanosoma brucei* (cause of African sleeping sickness), and *Trypanosoma cruzi* (cause of Chagas' disease).
- **Metamonads** include parasites such as *Giardia intestinalis*, a frequent contaminant of natural freshwater environments, and frequently transmitted among children.
- **Other intestinal parasites** include the ameba *Entamoeba histolytica*, the apicomplexan *Cryptosporidium parvum*, the ciliate *Balantidium coli*, and the microsporidian *Encephalitozoon intestinalis.*

Concluding Thoughts

The microbial eukaryotes, including fungi, algae, and many kinds of protozoa, serve many diverse roles in our ecosystems. A survey of protozoa may leave an impression that most exist primarily to parasitize humans. However, vast numbers of unknown clades of eukaryotic microbes have no direct connection with humans, yet they fill crucial niches in the ecosystems on which human existence depends. Marine ecosystems have many trophic levels of protists that feed invertebrate and vertebrate animals. Chapters 21 and 22 emphasize the interconnections among the myriad kinds of microbes that form the foundations of Earth's biosphere.

CHAPTER REVIEW

Review Questions

1. Discuss the evidence for the branching of fungi and animals within one clade, the Opisthokonta, which is distinct from algae and protists.
2. How do primary symbiont algae differ from secondary and tertiary symbiont algae? Compare with respect to cell structure and nutritional options.
3. Compare and contrast the molecular basis of motility in amebas and ciliates. Cite particular species.
4. Compare and contrast kelps, diatoms, and dinoflagellates in terms of cell structure, colony organization, and nutritional options.
5. Summarize the key traits of fungi. What do fungi have in common with protists, and how do they differ?
6. Outline the life cycles of the major phyla of fungi: Chytridiomycota, Zygomycota, Ascomycota, and Basidiomycota. Explain their ecological significance.
7. Compare and contrast the traits of green and red algae.
8. Outline the life cycle of the slime mold *Dictyostelium discoideum*. Compare and contrast its features with that of fungi that produce fruiting bodies, such as basidiomycetes.
9. Outline the complex parasitic life cycles of an apicomplexan parasite and a trypanosome. Cite evidence of reductive evolution, as well as of evolution of elaborate specialized structures to facilitate the parasite's life cycle.

Thought Questions

1. Compare and contrast eukaryotic microbes that have inorganic shells or plates. What is their composition, and how do they grow?
2. Explain mixotrophy. Why are so many marine eukaryotes mixotrophs?
3. Why do eukaryotes show such a wide range of cell size? Which selective forces favor large cell size, and what favors small cell size?
4. Do eukaryotic parasites have genomes that are larger or smaller than those of free-living organisms? Explain.

Key Terms

alga (787, 791)
alternation of generations (798)
alveolate (792)
ameba (791)
apicomplexan (792)
arbuscular mycorrhizae (803)
ascomycete (799)
ascospore (799)
ascus (798)
basidiomycete (801)
basidiospore (802)
budding (796)
cellular slime mold (806)
chlorophyte (791)
chloroplast (791)
chrysophyte (792)
ciliate (792)
cilium (792)
conjugation (815)
contractile vacuole (815)
cortical alveolus (792)
cryptogamic crust (811)
diatom (792)
dinoflagellate (792)
endosymbiosis (788)
Eumycota (790)
flagellate (792)
flagellum (792)
foraminiferan (807)
fruiting body (791)
frustule (813)
fungus (786)
gamogony (820)
green alga (791)
hypha (791)
lichen (811)
macronucleus (815)
malaria (818)
merozoite (819)
micronucleus (815)
mitosis (820)
mitosporic fungus (798)
mycelium (794)
mycology (787)
mycorrhizae (803)
nucleomorph (791)
oomycete (804)
oral groove (815)
phagocytosis (805)
phytoplankton (808)
Plantae (791)
plasmodial slime mold (807)
plasmodium (807)
primary algae (788, 791)
protist (787)
protozoan (787, 788)
pseudopod (791)
radiolarian (807)
red alga (791)
Rhizaria (791)
rhodophyte (791)
sargassum weed (814)
schizogony (820)

secondary algae (788, 791)
sporangiospore (799)
sporangium (798)
stalked ciliate (816)
Stramenopiles (792)
true fungi (790)
trypanosome (821)
yeast (795)
zoospore (790)
zygomycete (798)
zygospore (799)

Recommended Reading

Ankarklev, Johan, Jon Jerlström-Hultqvist, Emma Ringqvist, Karin Troell, and Staffan G. Svärd. 2010. Behind the smile: Cell biology and disease mechanisms of *Giardia* species. *Nature Reviews. Microbiology* **8**:413–422.

Armbrust, E. Virginia. 2009. The life of diatoms in the world's oceans. *Nature* **459**:185–192.

Armbrust, E. Virginia, John A. Berges, Chris Bowler, Beverley R. Green, Diego Martinez, et al. 2004. The genome of the diatom *Thalassiosira pseudonana*: Ecology, evolution, and metabolism science. *Science* **306**:79–86.

Beaufort, L., I. Probert, T. de Garidel-Thoron, E. M. Bendif, D. Ruiz-Pino, et al. 2011. Sensitivity of coccolithophores to carbonate chemistry and ocean acidification. *Nature* **476**:80–83.

Davy, Simon K., Denis Allemand, and Virginia M. Weis. 2012. Cell biology of cnidarian-dinoflagellate symbiosis. *Microbiology and Molecular Biology Reviews* **76**:229–261.

Fisher, Matthew C., Daniel A. Henk, Cheryl J. Briggs, John S. Brownstein, Lawrence C. Madoff, et al. 2012. Emerging fungal threats to animal, plant and ecosystem health. *Nature* **484**:186–194.

Henderson, Gregory P., Lu Gan, and Grant J. Jensen. 2007. 3-D ultrastructure of *O. tauri*: Electron cryotomography of an entire eukaryotic cell. *PLoS One* **8**:e749.

Keeling, Patrick J., Gertraud Burger, Dion G. Durnford, B. Franz Lang, Robert W. Lee, et al. 2005. The tree of eukaryotes. *Trends in Ecology and Evolution* **20**:670–676.

Kronstad, James W., Rodgoun Attarian, Brigitte Cadieux, Jaehyuk Choi, Cletus A. D'Souza, et al. 2011. Expanding fungal pathogenesis: *Cryptococcus* breaks out of the opportunistic box. *Nature Reviews. Microbiology* **9**:193–203.

LaJeunesse, Todd, John Everett Parkinson, Paul W. Gabrielson, Hae Jin Jeong, James Davis Reimer, et al. 2018. Systematic revision of Symbiodiniaceae highlights the antiquity and diversity of coral endosymbionts. *Current Biology* **28**:P2570–2580.E6.

Lew, Roger R. 2011. How does a hypha grow? The biophysics of pressurized growth in fungi. *Nature Reviews. Microbiology* **9**:509–518.

Lukeš, Jules, Christen R. Stensvold, Kateˇrina Jirkˇu-Pomajbíková, and Laura W. Parfrey. 2015. Are human intestinal eukaryotes beneficial or commensals? *PLoS Pathogens* **11**:e1005039. https://doi.org/10.1371/journal.ppat.1005039.

Martin, Francis, Annegret Kohler, Claude Murat, Claire Veneault-Fourrey, and David S. Hibbett. 2016. Unearthing the roots of ectomycorrhizal symbioses. *Nature Reviews. Microbiology* **14**:760–773.

Parfrey, Laura W., and Laura A. Katz. 2010. Dynamic genomes of eukaryotes and the maintenance of genomic integrity. *Microbe* **5**:156–163.

Parniske, Martin. 2008. Arbuscular mycorrhiza: The mother of plant root endosymbioses. *Nature Reviews. Microbiology* **8**:763–775.

Pounds, J. Alan, Martin R. Bustamante, Luis A. Coloma, Jamie A. Consuegra, Michael P. L. Fogden, et al. 2007. Widespread amphibian extinctions from epidemic disease driven by global warming. *Nature* **439**:161–167.

Ratcliff, William C., Matthew D. Herron, Kathryn Howell, Jennifer T. Pentz, Frank Rosenzweig, et al. 2013. Experimental evolution of an alternating uni- and multicellular life cycle in *Chlamydomonas reinhardtii*. *Nature Communications* **4**:2742.

Wahlgren, Mats, Suchi Goel, and Reetesh R. Akhouri. 2017. Variant surface antigens of *Plasmodium falciparum* and their roles in severe malaria. *Nature Reviews. Microbiology* **15**:479–491.

CHAPTER 21
Microbial Ecology

- **21.1** Microbial Communities: Metagenomes and Single-Cell Sequencing
- **21.2** Functional Ecology
- **21.3** Symbiosis
- **21.4** Animal Digestive Microbiomes
- **21.5** Marine and Freshwater Microbes
- **21.6** Soil and Plant Microbial Communities

Microbes dominate all habitats on Earth, from the Antarctic Southern Ocean to the human intestine. Microbial communities form functional components of all plants and animals, including human beings. While most microbes cannot be cultured, the sequences of their metagenomes (community DNA) and metatranscriptomes (community RNA) reveal their secrets. Succession of microbial communities transforms entire ecosystems. In Chapter 21 we explore how microbes interact with each other and with their diverse habitats on Earth.

CURRENT RESEARCH highlight

Symbiosis in a stinkbug. A common visitor to human habitations, the brown marmorated stinkbug hosts a surprising endosymbiont: a bacterium that Zakee Sabree identified and named *Pantoea carbekii*. The bacteria were detected by FISH probe hybridization to the bug's gut crypt cells (shown pink). The bug requires its symbiont to live; and the bacterium's genome shows the extensive loss of essential genes, as well as gain of genes useful to its host, indicating a true mutualism.

Sources: Raman Bansal et al. 2014. *Environ. Entomol.* **43**:617–625; Laura J. Kenyon et al. 2015. *Genome Biol. Evol.* **7**:620–635.

AN INTERVIEW WITH

ZAKEE SABREE, MICROBIOLOGIST, THE OHIO STATE UNIVERSITY

What's so interesting about a bug's mutualistic bacteria?

The brown marmorated stinkbug is an invasive stinkbug that attacks lots of agricultural and ornamental crops, and it arrived in the United States in the 1990s. It has an obligate bacterial symbiont that we discovered, named *Pantoea carbekii*. Although the *P. carbekii* genome is reduced relative to free-living gammaproteobacteria, it encodes enzymes that can generate essential nutrients potentially limited in the host's diet. *P. carbekii* may support the stinkbug's ability to exploit a wide range of host plants.

Microbes have evolved to colonize every habitat of Earth's biosphere, including soil, water, air, and the bodies of plants and animals. Within an animal, a microbial community can provide essential components for its host, such as the stinkbug studied by Zakee Sabree and his students at the Ohio State University (see the Current Research Highlight). The intracellular bacteria synthesize amino acids and vitamins needed by the insect. More broadly, the bug and its endosymbionts exist within a much larger ecosystem that includes trees and their fungal root mutualists, the mycorrhizae (discussed in Section 21.6). Microbes, in diverse roles, form the foundation of all ecosystems.

In Chapter 21 we explore the unique roles of microbes in their **ecosystems**. "No man is an island," nor is any microbe. All organisms evolve within an ecosystem, which consists of populations of species plus their habitat or environment. A **population** is a group of individuals of one species living in a common location. The sum of all the populations of different species constitutes a **community**. Microbial communities critically impact other organisms in all habitats, from oceans and forests to the interstices of rock. Microbes recycle organic material in aquatic and terrestrial ecosystems, providing resources for plants and animals, and deep below Earth's surface, microbes shape the rock of Earth's crust.

Extreme environments offer attractive targets for study because the native microbes are likely to show novel traits. Such environments may pose safety threats for researchers, however; for example, the acid mine drainage site sampled by Jillian Banfield of UC Berkeley (**Fig. 21.1A**) is acidic enough to dissolve an iron shovel. Nonetheless, Banfield sequenced community DNA, including the genome of the hyperacidophile *Leptospirillum* (**Fig. 21.1B**). Other target communities require interaction with a host animal, such as the rumen (fermentation organ) of a cow (**Fig. 21.1C** and **D**). The microbes associated with a host animal or plant are called the **microbiome**. Every multicellular organism possesses a microbiome. The term "microbiome" has become synonymous with the "microbial community" of a habitat, such as the soil microbiome.

The first problem of microbial ecology is how to find and identify a habitat's microorganisms. The great eighteenth-century taxonomist Carl von Linné (known also as Carolus Linnaeus) called microbes "chaos" because he thought it would never be possible to distinguish one from another (see Chapter 1). In the nineteenth century, Robert Koch developed techniques of pure culture, and Sergei Winogradsky developed enrichment culture methods that isolate microbial species and reveal their metabolic traits (discussed in Chapters 1 and 4). Yet the majority of microbes remain uncultured, and perhaps undiscovered. Today, researchers rely on increasingly sophisticated tools that probe community samples at the molecular level (metagenomes, metatranscriptomes, and single-cell genomes), as well as whole-cell protein populations (proteomes). These tools, and others, are known informally as **omics**.

In Section 21.1 we begin with the analysis of metagenomes and other "omics" to dissect microbial ecology. In subsequent sections we evaluate the functions of those microbes in their habitats and microbiomes, including symbioses with partner animals and plants, as well as the larger communities found in the oceans and Earth's crust.

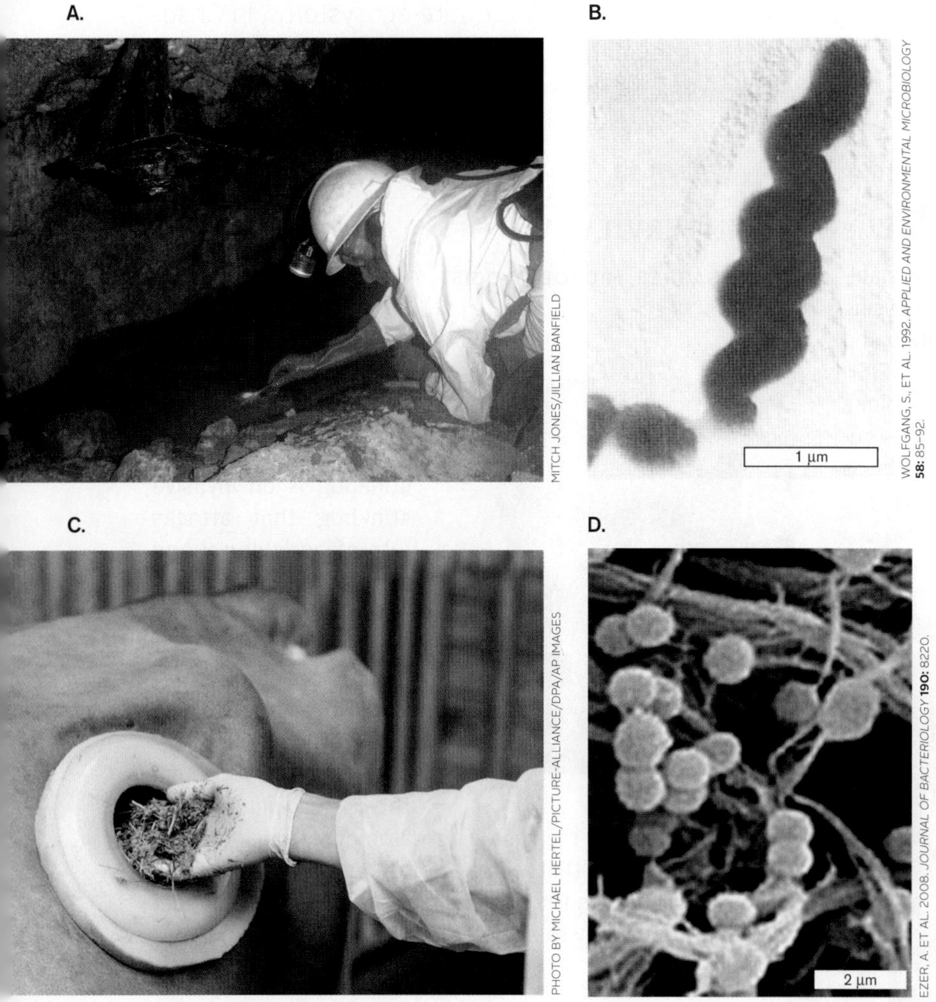

FIGURE 21.1 ▪ Sampling a target community of microbes.
A. Jillian Banfield samples the "pink slime" of archaea growing in acid mine drainage at Iron Mountain, California, one of the largest Superfund cleanup sites. **B.** *Leptospirillum* sp., a bacterium whose genome emerged out of Banfield's metagenomic analysis. **C.** A researcher samples rumen contents from a fistulated (cannulated) cow. The closable opening does not harm the animal. **D.** *Ruminococcus albus* bacteria digest plant fibers within the rumen of a cow.

Note: Chapter 21 presents analysis of microbial communities, the interactions between microbes and partner organisms, and the functional ecology of microbes.

Chapter 22 presents the role of microbes in global nutrient cycles and climate change, and the microbiology of human-built environments.

21.1 Microbial Communities: Metagenomes and Single-Cell Sequencing

A major breakthrough in microbial ecology was the discovery that we can identify uncultured microbes by their DNA. In 1991, one of the first to sequence genes from environmental samples was Norman Pace, then at Indiana University, who cloned ribosomal RNA genes from plankton (floating microbes) filtered from the Pacific Ocean. The gene encoding small-subunit (SSU) rRNA became the standard for identifying environmental taxa (discussed in Chapter 17). But SSU rRNA is just one highly conserved molecule, which tells us little about the rest of the organism.

In 1998, Jo Handelsman and colleagues, then at the University of Wisconsin–Madison, used shotgun sequencing to analyze large portions of genomes from a soil microbial community—a mixture of species that was previously thought impossible to interpret. Handelsman coined the term "metagenome" to refer to the DNA sequence obtained directly from a mixture of genomes. Today, metagenomes are sequenced from many kinds of communities using next-generation sequencing (NGS) technologies such as the Illumina sequencing platform (discussed in Chapter 12). With these methods we are addressing exciting questions:

- **Who is there?** What microbes inhabit a given environment or human body part?
- **What are they doing?** What function do the microbes contribute? Do they fix carbon, break down toxic waste, or produce antibiotics?
- **How do they vary under different conditions?** How do microbial communities change—that is, undergo succession? As the climate changes, what happens to the microbial community structure of forests and soil?

Analyzing a Metagenome

Examples of metagenomic studies are listed in **Table 21.1**. For instance, in 2019 an international consortium sequenced the metagenome of human intestinal microbiomes from 11,850 individuals distributed across six continents. They found that individual gut microbiomes show distinctive profiles of species and subspecies. This finding suggests that an understanding of individual microbiomes could help fine-tune drug therapies for different patients. Another

TABLE 21.1 Examples of Metagenome Analysis

Microbial target community	Discovery	Reference
Human gut microbiomes	More than 11,000 human gut microbiomes reveal 1,900 new candidate bacterial species. May lead to personalized chemotherapies.	Alexandre Almeida et al. 2019. *Nature* **568**:499–504.
Crystal Geyser underground environment saturated with CO_2	Single-cell genomes and metagenomes reveal bacteria and archaea that fix CO_2.	Alexander J. Probst et al. 2018. *Nat. Rev. Microbiol.* **3**:328.
Coal-based methane wells, Queensland, Australia	Archaeal phylum Bathyarchaeota contains previously unknown methanogens.	Paul Evans et al. 2015. *Science* **350**:435.
River sediments in Rifle, Colorado	Rare organisms (<0.1%) in new candidate phyla detected from diverse communities.	Itai Sharon et al. 2015. *Genome Res.* **25**:534.
Global ocean microbiome sampled at 68 locations	Ocean microbial catalog compiled; core genes vary with depth and temperature.	Shinichi Sunagawa et al. 2015. *Science* **348**:1261359.
Soils sampled from Antarctic desert, hot desert, and temperate regions	Desert soil genomes show more osmotic stress genes but fewer antibiotic resistance genes than nondesert genomes have.	Noah Fierer et al. 2012. *PNAS* **209**:21390.
Gulf of Mexico marine water contaminated by *Deepwater Horizon* oil spill	Metagenomes, metatranscriptomes, and single-cell genomes reveal microbial response to petroleum.	Olivia Mason et al. 2012. *ISME J.* **6**:1715–1727.
Rumen microbiome of fistulated cows fed switchgrass	Assembly of 15 bacterial genomes with 27,700 carbohydrate-digesting enzymes.	Matthias Hess et al. 2011. *Science* **331**:463.

landmark study was conducted by the U.S. Department of Energy on the response of marine microbial communities to petroleum contamination following the *Deepwater Horizon* oil rig blowout off the coast of Louisiana in 2010.

Metagenomic sequencing poses challenges far beyond those of sequencing a single intact genome. To sequence a metagenome requires a series of steps, each of which presents important choices (**Fig. 21.2**).

Sampling the target community. The first decision is to define a **target community** from which to obtain DNA. The cells of the target community must be separated from their surroundings without loss of DNA (**Fig. 21.2**, step 1). For example, sampling a soil community requires removal of humic acids (wood breakdown products), which inhibit DNA polymerases. Suppose the target community inhabits a host plant or animal; what additional separation is required? Before DNA extraction, we need to dislodge host-associated microbes from their host. Otherwise, the host DNA could contaminate the microbial DNA pool.

Isolating DNA. Once separated from the physical habitat, the cells of the target community must be opened in such a way that all of the DNA is released with minimal breakage of the strands (**Fig. 21.2**, step 2). We can lyse the cells by "bead beating" or by sonication (methods discussed in Chapter 3). The DNA can then be purified by phenol extraction and precipitation with ethanol, or by binding to special filters. But—unlike the analysis of a single-species genome—analysis of a metagenome requires accounting for different species that possess different kinds of enzyme inhibitors, as well as envelope, sheath, and S-layers of diverse composition.

How can we ensure that our protocol will be optimal for all the thousands of species in the community? In fact, no single best way exists to extract metagenomic DNA. Different researchers argue for one of two main approaches:

- **Use a single, universally applied method of DNA extraction for all target communities.** If all research groups sample metagenomes using a common DNA extraction method, then results may be compared across all projects.
- **Use multiple DNA extraction methods for each target.** If a research group uses multiple DNA extraction methods to sample one target community, then they have the best chance to maximize coverage of all the microbial genomes in the sample.

Thought Question

21.1 Suppose you are conducting a metagenomic analysis of soil sampled from different parts of a wetland. Would you use one DNA extraction method or multiple methods?

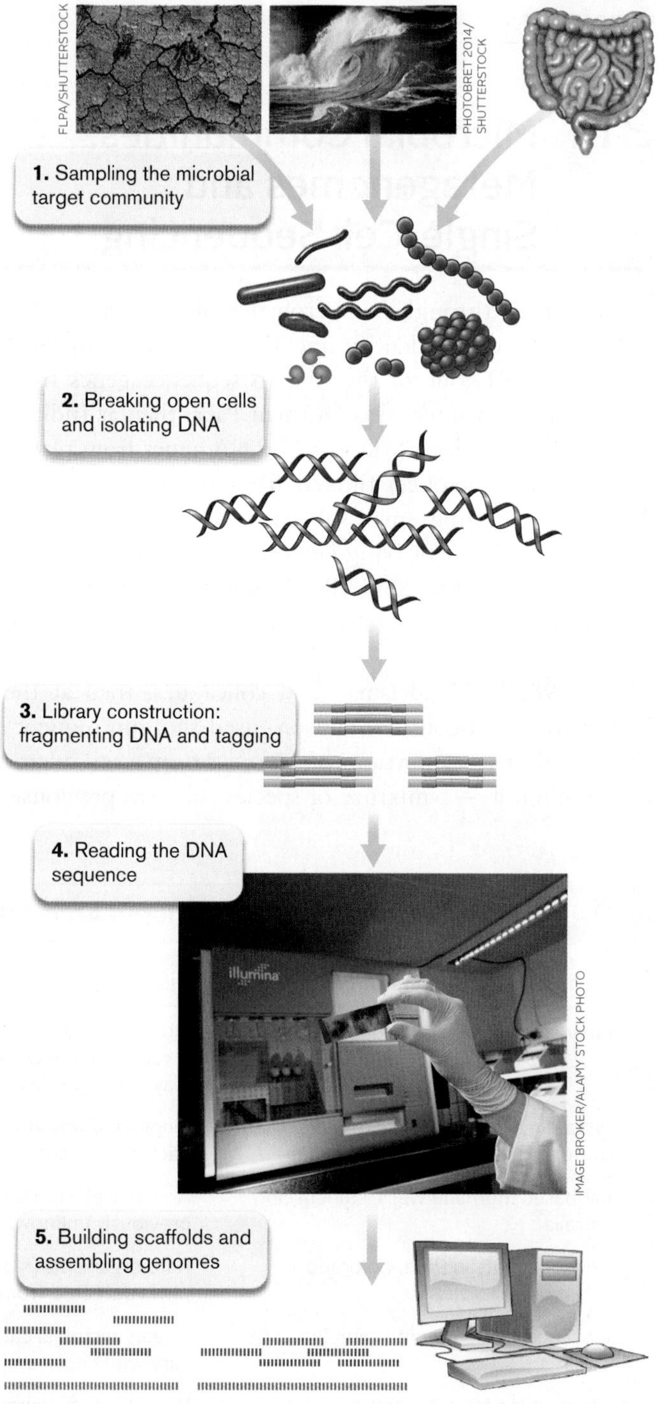

FIGURE 21.2 ▪ Sequencing a metagenome. First select a target community to sample, from a habitat such as soil, water, or host plant or animal. Remove the sampled microbes from their environment (step 1) and stabilize the content. Lyse the cells and isolate pure, intact DNA (step 2). Amplify the DNA library by constructing fragments with tagged ends (step 3). Read the DNA sequence using an NGS sequencer (step 4). Use a computational software pipeline to build scaffolds and assemble genomes (step 5). *Source:* John Wooley et al. 2010. *PLoS Comput. Biol.* **6**:e1000667.

Preliminary screen for diversity. Before investing in large-scale DNA sequencing, we can perform a preliminary screen for sequence diversity. The preliminary screen may answer questions such as: Since species in the community may differ in their fractional representation by several orders of magnitude, how important are the community's rarest members? Is our aim a complete description of the target community, such as the microbiota inhabiting the human stomach? Or do we focus on a narrower goal, such as identifying soil actinomycete genes that produce novel antibiotics?

A preliminary screen of the sample involves sequencing the SSU rRNA genes (discussed in Chapter 17). The genes specifying 16S rRNA (bacteria and archaea) or 18S rRNA (eukaryotes) may be amplified by PCR using universal primers known to detect a wide range of taxa. The classical approach of cloning has largely been replaced by amplification of the rRNA genes from NGS Illumina sequencing of DNA (presented in Chapter 12). This use of Illumina to obtain SSU rRNA sequences from a metagenome is called iTAG analysis or **metabarcoding** (as compared with **barcoding**, the SSU rRNA identification of single taxa). Compared to earlier methods, Illumina sequences provide a larger number of genes at lower cost.

Metabarcoding quickly identifies many taxa present in a community, including relatively rare members that do not yield full genomes when large-scale sequencing is performed. SSU rRNA gene similarity is used to define **operational taxonomic units (OTUs)**, a working metagenomic definition of "species." SSU rRNA gene sequencing reveals the general categories of microbes found in a community, their relative abundance in a community, and the overall diversity of the sample.

Library construction for next-generation sequencing. Samples, such as a human tissue biopsy or marine water, will contain DNA in such small quantities that it must be amplified by a form of PCR, and processed for the next-generation sequencing (NGS) reactions. The amplification and processing are called **library construction** (**Fig. 21.2**, step 3). The library construction depends on several factors.

- **DNA concentration.** Lower concentration requires greater amplification. However, amplification can introduce sequence errors and loss of sequences that fail to be amplified.
- **Fragmentation for size.** The DNA must be fragmented to the size required by NGS, typically the Illumina platform (see Chapter 12). Small fragment sizes may increase errors in alignment, but larger sizes may involve other kinds of errors.
- **Addition of adapters.** Specific adapter sequences called tags are added at the ends of the fragments, to allow the sequencing reaction to proceed.

Metagenome sequencing and assembly. The most common type of NGS metagenome sequencing is Illumina sequencing by synthesis (**Fig. 21.2**, step 4). The process of Illumina sequencing is described in Chapter 12. Illumina sequencing generates millions of short DNA sequences of defined length; these short sequences are called **reads**. To sequence the genome requires **assembly** of reads whose sequences of base pairs overlap. The overlapping reads are assembled into **contigs**, regions of contiguous DNA sequence without gaps (**Fig. 21.3**). Inevitably, though, gaps remain between contigs.

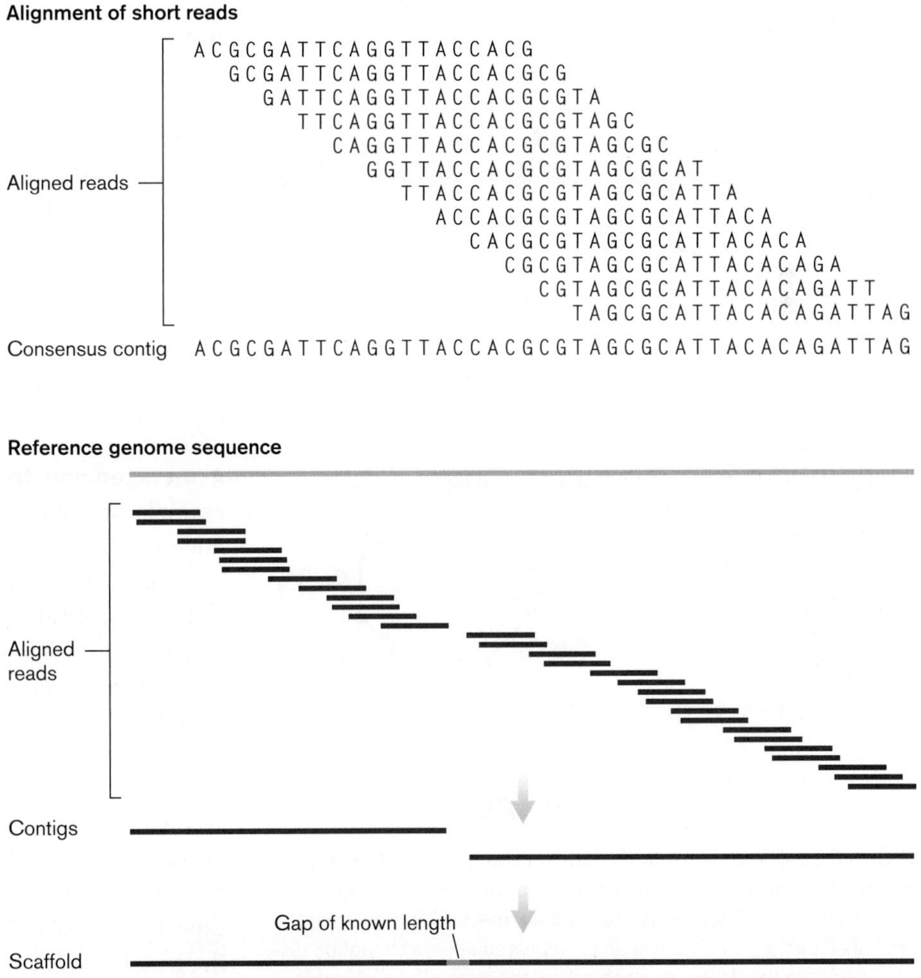

FIGURE 21.3 ■ **Assembly of reads into contigs and scaffolds.** Overlapping reads generate a contig. Contigs matched to a reference genome generate a scaffold. Scaffolds may still contain gaps of unknown sequence.

If the contigs show high relatedness to the genome of a known organism, this reference genome may be used to match contigs together in **scaffolds** containing gaps of presumed length. In effect, the assembly of contigs resembles a jigsaw puzzle with thousands of pieces. But a metagenome requires assembly of thousands of jigsaw puzzles, whose pieces are all jumbled together. In practice, metagenomes rarely yield complete genomes of individual species, but they can yield partial genomes of previously unknown organisms with interesting properties. These partial genomes are called metagenome-assembled genomes (MAGs). A metagenome-assembled genome may define an operational taxonomic unit (OTU) with greater resolution than those defined by a single SSU rRNA sequence.

Genome assembly requires a **computational pipeline**, a linear series of programs that combine mathematical tools with biological assumptions to propose assembled genomes. Mathematical tools of assembly are based on a formula known as the de Bruijn graph (**Fig. 21.4**). The de Bruijn graph compiles overlaps between sequences of base pairs, in a manner that predicts the identity of the original intact sequence. The short sequences are called k-mers, where "k" refers to a defined length. The k-value is chosen generally as an odd number, shorter than the Illumina read length, such that sequence errors are minimized. The k-mers are assumed to start from all possible positions in the overall sequence; thus, they overlap neighboring k-mers, in all possible ways. The computational algorithm links all k-mers by their overlapping ends. Repeated k-mers cannot be distinguished, so the algorithm collapses them with overlapping connections, generating a diagram of possible sequence connections with loops as shown. An actual de Bruijn graph from a set of reads may have numerous loops of this type.

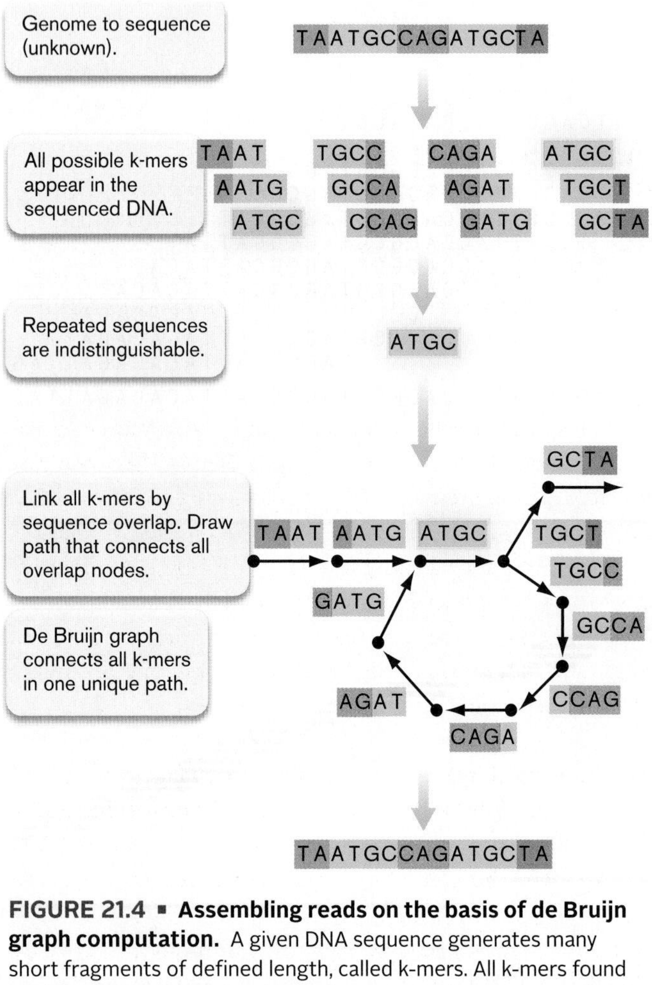

FIGURE 21.4 ▪ Assembling reads on the basis of de Bruijn graph computation. A given DNA sequence generates many short fragments of defined length, called k-mers. All k-mers found are linked by their overlapping ends. Repeated k-mers cannot be distinguished, so the de Bruijn graph collapses them with overlapping connections. To reveal the sequence that produced the fragments, find a path that passes through every k-mer exactly once. That path generates the original sequence of base pairs.

We can predict the sequence that produced the fragments by taking a path through the de Bruijn graph that passes through every k-mer exactly once. This path generates the original sequence of base pairs. In principle, the k-mer size is selected such that most reads have one unique position in the genome, and in this case, computation based on the de Bruijn graph predicts the exact sequence of the original DNA. In practice, of course, there are ambiguities and "unfinished" regions caused by errors in sequence, and by multicopy repeated regions that are larger than the k-mer size.

Genome assembly can be of two types: mapping to a reference and de novo assembly. Mapping to a reference, or "resequencing," involves aligning the contigs and scaffolds with known reference genomes. Reference genomes can be effective at "pulling out" genomes of organisms present in low abundance. Without a reference, de novo assembly takes much longer, although it offers greater possibility of identifying a previously unknown type of organism.

A metagenome from a bovine rumen microbiome. An example of a metagenome study is offered by the rumen microbiome of a cow, a highly diverse community sequenced by Matthias Hess and colleagues at the Joint Genome Institute (see **Table 21.1**). Hess's study yielded 15 partial genomes (OTUs) ranging from 60% to 93% estimated completeness; 12 are listed in **Figure 21.5**. The completeness of each genome was estimated from the number of "core genes" identified. Core genes are defined as those genes found in nearly all members of a given taxonomic clade, such as the order Clostridiales for the group of genomes (genome bin) *APb* (highlighted row in **Fig. 21.5**). The fraction of the core genes found, divided by those expected for the order, gives an estimate of completeness of the genome sequence.

The OTUs that Hess found, however, account for only a tiny fraction of the species present, out of 268 gigabases (billions of base pairs) sequenced. Each partial genome assembled from the rumen community represents a **bin**—that is, a set of sequence reads from closely related members of one taxonomic unit, showing a given level of similarity. Since no two individual organisms possess exactly the same sequence, the investigator must decide how much difference to allow in defining a taxon. **Binning**, the sorting of sequences into taxonomic bins, requires further computational analysis. Factors for computation may include, for example, similarity of base composition and similarity to reference database sequences.

Figure 21.5 shows an example of a binned partial genome, *APb*, as a ring composed of scaffolds matched to a Clostridiales reference genome. Between the scaffolds, there remain unsequenced gaps. In the wheel, the jagged trace represents the degree of coverage—that is, the number of sequence reads that cover a given region. Multiple copies of overlapped sequence represent a reliable assembly; typically, 30-fold coverage is considered good. But the remainder of the rumen sequences are unattached scaffolds and single-copy reads out of thousands of unknown genomes. These unattached sequences encode tens of thousands of novel enzymes for carbohydrate digestion, and they are of great interest to the biofuel industry.

Note that metagenome sequencing is limited by the challenge of assigning DNA sequence reads correctly to their shared genomes. All computational pipelines must choose assumptions about binning, sequence gaps, and other sequence characteristics. For this reason, different pipelines often predict different partial genomes.

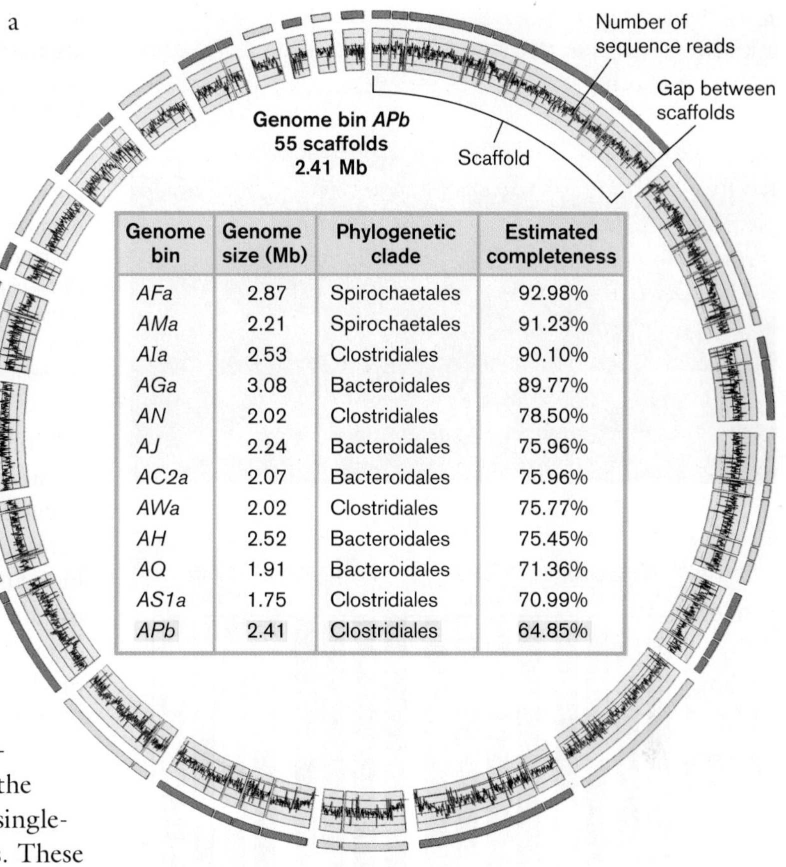

Genome bin	Genome size (Mb)	Phylogenetic clade	Estimated completeness
AFa	2.87	Spirochaetales	92.98%
AMa	2.21	Spirochaetales	91.23%
AIa	2.53	Clostridiales	90.10%
AGa	3.08	Bacteroidales	89.77%
AN	2.02	Clostridiales	78.50%
AJ	2.24	Bacteroidales	75.96%
AC2a	2.07	Bacteroidales	75.96%
AWa	2.02	Clostridiales	75.77%
AH	2.52	Bacteroidales	75.45%
AQ	1.91	Bacteroidales	71.36%
AS1a	1.75	Clostridiales	70.99%
APb	2.41	Clostridiales	64.85%

FIGURE 21.5 ■ Partial genomes assembled from a cow rumen microbiome. Map of genome bin *APb* (order Clostridiales) showing assembled scaffolds. Inner rings indicate fold coverage (number of sequence reads) backward and forward; red lines indicate 25-fold coverage. **Center:** Genome bins were matched to clades by comparison with reference sequences. *Source:* Modified from Matthias Hess et al. 2011. *Science* **331**:463.

Thought Question

21.2 Suppose you plan to sequence a marine metagenome for the purpose of understanding carbon dioxide fixation and release, to improve our model for global climate change. Do you focus your resources on assembling as many complete genomes as possible, or do you focus on identifying all the community's enzymes of carbon metabolism?

The *Deepwater Horizon* Spill: Metagenomes, Metatranscriptomes, and Single-Cell Analysis

An inadvertent "experiment" on microbial community dynamics occurred in 2010 with the *Deepwater Horizon* oil well blowout, which released 4 million barrels of oil into the Gulf of Mexico (**Fig. 21.6A**). Over the months that followed, several research groups studied the **succession**, or change in community structure, of the marine microbes following this amendment (addition of an organic nutrient, petroleum). How would the petroleum affect the distribution of species? Would the species richness (number of different species) and diversity (number and evenness of species) be altered? Most interestingly, would consumers of petroleum components experience positive selection?

To address these questions, researchers analyzed the microbial community structure through metagenome, metatranscriptome, and single-cell analyses.

Preliminary assessment of diversity. During several months that followed the *Deepwater Horizon* spill, Molly Redmond and David Valentine from UC Santa Barbara investigated the microbial community response to the influx of petroleum from the wellhead. They conducted a preliminary survey of the microbial community, comparing 16S rRNA gene sequences from water contaminated by the plume of oil rising at the site of the leak and from uncontaminated Gulf seawater. The **amplicons** (amplified DNA sequences) were assigned taxa by sequence homology (discussed in Chapter 17).

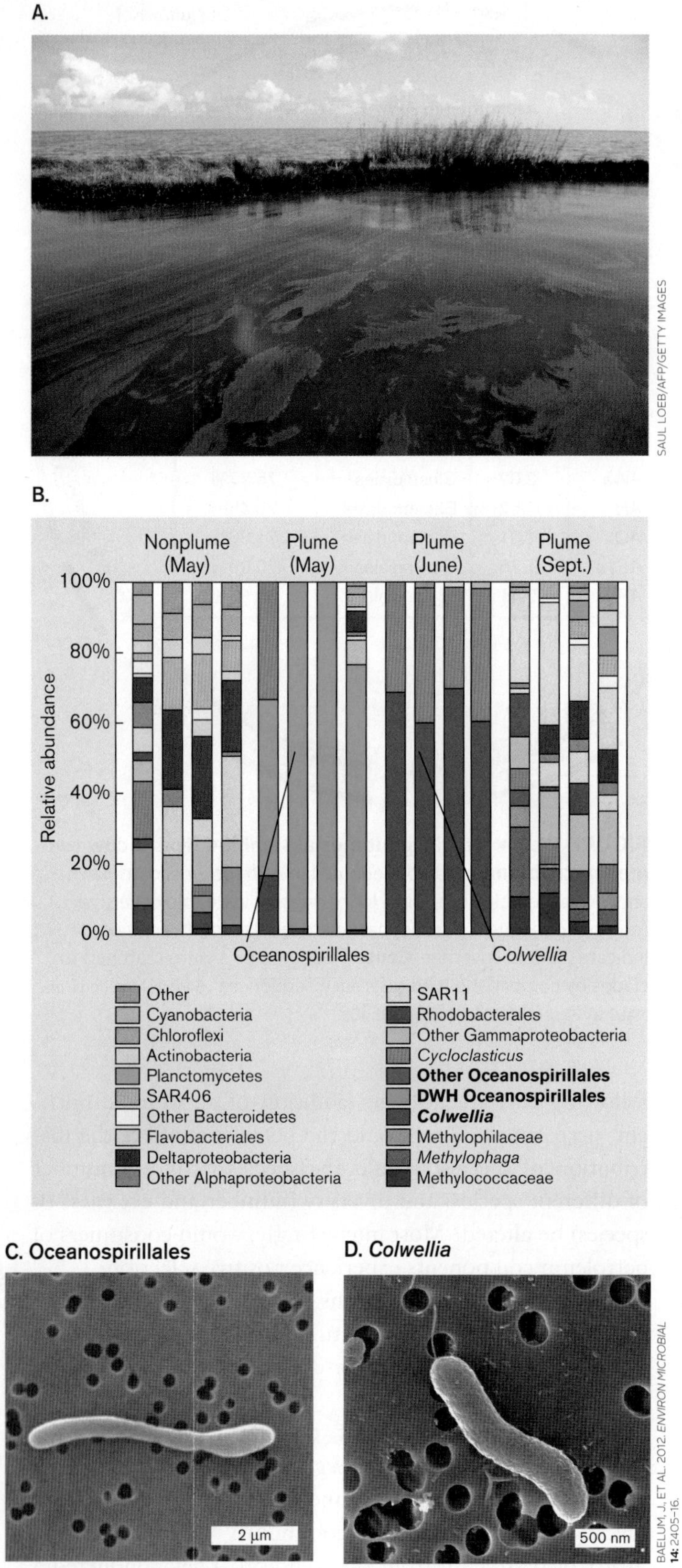

FIGURE 21.6 ■ Succession of bacterial community following the *Deepwater Horizon* oil spill. A. Petroleum from the *Deepwater Horizon* oil well blowout in April 2010 contaminated the Louisiana coast. **B.** Bacterial relative abundance in 16S rRNA genome libraries. **C.** Oceanospirillales, an order of bacteria whose DNA was found in oil-contaminated water in May (SEM). **D.** *Colwellia* bacteria dominated the oil-contaminated water in June (SEM). Samples were obtained from the plume of oil spreading through seawater, and from nonplume seawater. During May and June, major taxa shifted to oil-consuming Oceanospirillales and *Colwellia*. By September, after bacteria had consumed much of the oil, the taxon distribution appeared more similar to that before the oil spill. *Source:* Part B modified from Molly C. Redmond and David L. Valentine. 2012. *PNAS* **109**:20292–20297.

The relative abundance of various marine taxa at different times after the spill is shown in **Figure 21.6B**. A sample of nonplume (uncontaminated) water, taken in May, showed a broad range of diverse taxa, including Proteobacteria, Cyanobacteria, and Bacteroidetes. By contrast, samples of oil-contaminated water taken in May and again in June showed a marked shift to particular taxa, such as the genus *Colwellia* (**Fig. 21.6D**; blue-gray bars in **Fig. 21.6B**), a benthic gammaproteobacterium named for marine microbiologist Rita Colwell (Section 21.5). *Colwellia* is known to catabolize propane and benzene, which are components of petroleum and natural gas. Also found was a novel clade of Oceanospirillales (**Fig. 21.6C**; bright blue bars in **Fig. 21.6B**) with unknown metabolic potential.

Where did the oil-metabolizing bacteria come from? They were minority components of the community, present because of the natural seepage of petroleum from marine sediment throughout the Gulf. The sudden influx of high levels of petroleum initiated a bloom of microbes capable of catabolizing certain components, particularly those easiest to degrade, such as alkanes (saturated hydrocarbons). By September, when the more rapidly degradable molecules had dissipated, the range of diversity had been partly restored toward that of the uncontaminated Gulf seawater.

Thought Question

21.3 Could you design a metagenome experiment analogous to the oil plume experiment to test which kinds of human gut bacteria digest a certain food, such as hamburger meat? What follow-up experiments would be needed?

Metagenomes. Which of the marine microbes in the oil-contaminated community actually possessed the ability to degrade petroleum? The results are expected to be complex, since the community comprises numerous species and petroleum includes many kinds of molecules, ranging from short-chain and long-chain alkanes to aromatic rings, including polycyclic molecules.

The marine metagenomes were analyzed by Olivia Mason and Jerry Hazen and their colleagues at The Lawrence Berkeley National Laboratory (LBNL). Mason compared Illumina DNA sequencing reads from marine DNA samples taken from water at three different sites: near the oil well blowout, distant from the blowout but still containing measurable contamination, and untouched by the spill (uncontaminated). Mason's team performed functional analysis to identify genes encoding enzymes known to participate in hydrocarbon degradation (**Fig. 21.7B**).

The DNA reads from metagenome sequencing were compared for similarity to sequences from the GeoChip database of proteins involved in hydrocarbon degradation. Of particular interest were enzymes called alkane monooxygenases—that is, enzymes that incorporate oxygen to yield alcohols, which are later degraded to aldehydes and fatty acids. Aromatic-degrading enzymes were also detected in the metagenomes. These enzymes are of interest because they lead to pathways ultimately degrading alkanes and aromatics into acetyl-CoA that enters the TCA cycle, ultimately releasing CO_2 (discussed in Chapter 13).

Figure 21.7 shows that naturally occurring bacteria in the Gulf water possess DNA encoding enzymes that degrade and remove many petroleum constituents. Note, however, that the aromatic-degrading enzymes showed less enrichment than those for alkanes; for example, the enzymes for degradation of toluene and polycyclic aromatic hydrocarbons (PAHs) actually showed lower prevalence in the contaminated samples (red and blue bars in **Fig. 21.7B**) than in the samples from uncontaminated water (black bars). This finding is important because aromatic contaminants are known to decay more slowly than alkanes and to persist for longer times in marine sediment.

Functional analysis requires the existence of a database of genes known to encode products of given function in other studied organisms. No single method works best to ensure that we recognize all the actual genes encoding functional products—or that we don't mistakenly define some noncoding sequences as genes (false positives). Many bioinformatic tools are used to "call" genes, on the basis of gene structure and homology—a process known as annotation (discussed in Chapter 12).

Note that all gene-calling (annotation) approaches are incomplete because they miss truly novel genes for which no homologs or motifs exist in the databases. Furthermore, the presence of a sequence in a genome does not prove that the organism actually performs a given function. How do we know which of the genes present in metagenomes are actually expressed by the target community?

Metatranscriptomes. Beyond the DNA, we can sample the RNA pools of a community. **Metatranscriptomics** is the study of the RNA transcripts (using a high-throughput sequencing method known as RNAseq) obtained from an environmental community. The "metatranscriptome" gives a snapshot of gene expression activity of a community at a given point in time.

For the *Deepwater* samples, Mason's team observed the metatranscriptomes of samples taken near to or distant from the contamination site (**Fig. 21.7B**). The

A.

B. Gene functions

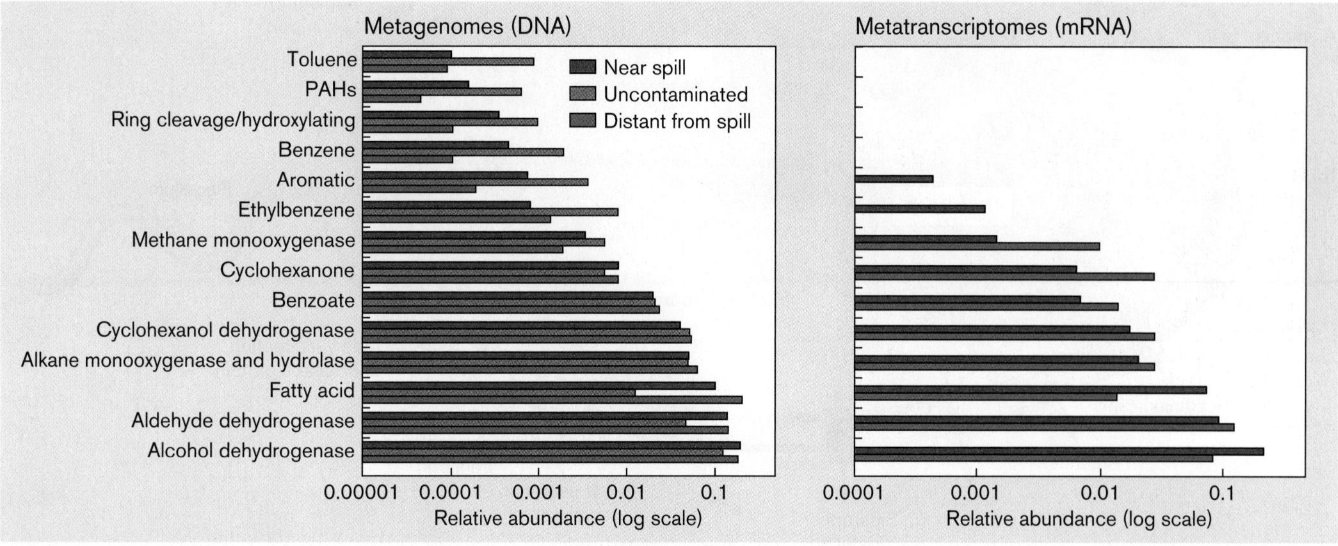

FIGURE 21.7 ■ *Deepwater Horizon*: bacteria show hydrocarbon catabolism enzymes. **A.** Olivia Mason. **B.** Metagenomes from oil-contaminated plume reveal genes associated with hydrocarbon degradation. Metatranscriptomes reveal mRNA transcripts encoding enzymes for hydrocarbon degradation. PAHs = polycyclic aromatic hydrocarbons. *Source:* Part B modified from Mason et al. 2012. *ISME J.* **6**:1715, figs. 3 and 4.

metatranscriptomes show the presence of ample RNA-encoding enzymes that degrade alkanes. Yet there is little sign of the monooxygenases needed for degradation of aromatic substrates such as toluene or benzene. This finding is of concern for the marine environment, indicating that, without biodegradation, aromatic contaminants from the oil spill are likely to persist for extended times.

Which kinds of bacteria or archaea are conducting most of the hydrocarbon degradation? Mason's metagenomes from contaminated water showed a high prevalence of Oceanospirillales, one of the taxa reported also by Redmond and Valentine. The metatranscriptomes also showed a high prevalence of 16S rRNA from Oceanospirillales. So, Mason sought to identify individual cells representing this taxon and investigate whether the organism actually possesses full pathways for hydrocarbon biodegradation.

Single-cell genome sequencing. A modification of metagenomic analysis is that of single-cell genome sequencing (described in Chapter 7). Single-cell analysis was first developed by the Bigelow Laboratory for Ocean Sciences, Maine. For single-cell analysis, we must isolate single cells from a microbial community (**Fig. 21.8**). This remarkable feat is most commonly performed by fluorescence-activated cell sorting (FACS), using a DNA-binding dye such as SYBR Green (**Fig. 21.9**). The DNA from a single cell is then amplified by a special non-PCR process called multiple displacement amplification (MDA). In MDA, unlike PCR, only the template is amplified, without multiple rounds. Amplification is performed with a highly accurate enzyme that requires no species-specific primer.

The MDA-amplified genome is then sequenced by Illumina and assembled as a single genome. Thus, in single-cell sequencing, multiple genomes from a community can be isolated and assembled separately, without needing to sort out a polymicrobial mixture of DNA or to obtain a pure culture in the laboratory

Mason's team used FACS to isolate two independent cells of the Oceanospirillales clade. The genomes of each cell were assembled, revealing genes known to encode enzymes of metabolic pathways and substrate transporters (**Fig. 21.8**). For example, a full degradation pathway was found for cyclohexane (a cyclic alkane). Alkanes are ultimately degraded to CO_2 via the TCA cycle (discussed in Chapter 13). Most of the genes found are also present in the metatranscriptome, confirming the hypothesis that Oceanospirillales is a dominant clade of bacteria capable of conducting early degradation of petroleum components in the Gulf seawater samples.

Note, however, that DNA analysis never shows that the organisms actually perform the metabolism or other functions predicted by their genes. Genomic predictions of a phenotype must be confirmed by physiology and

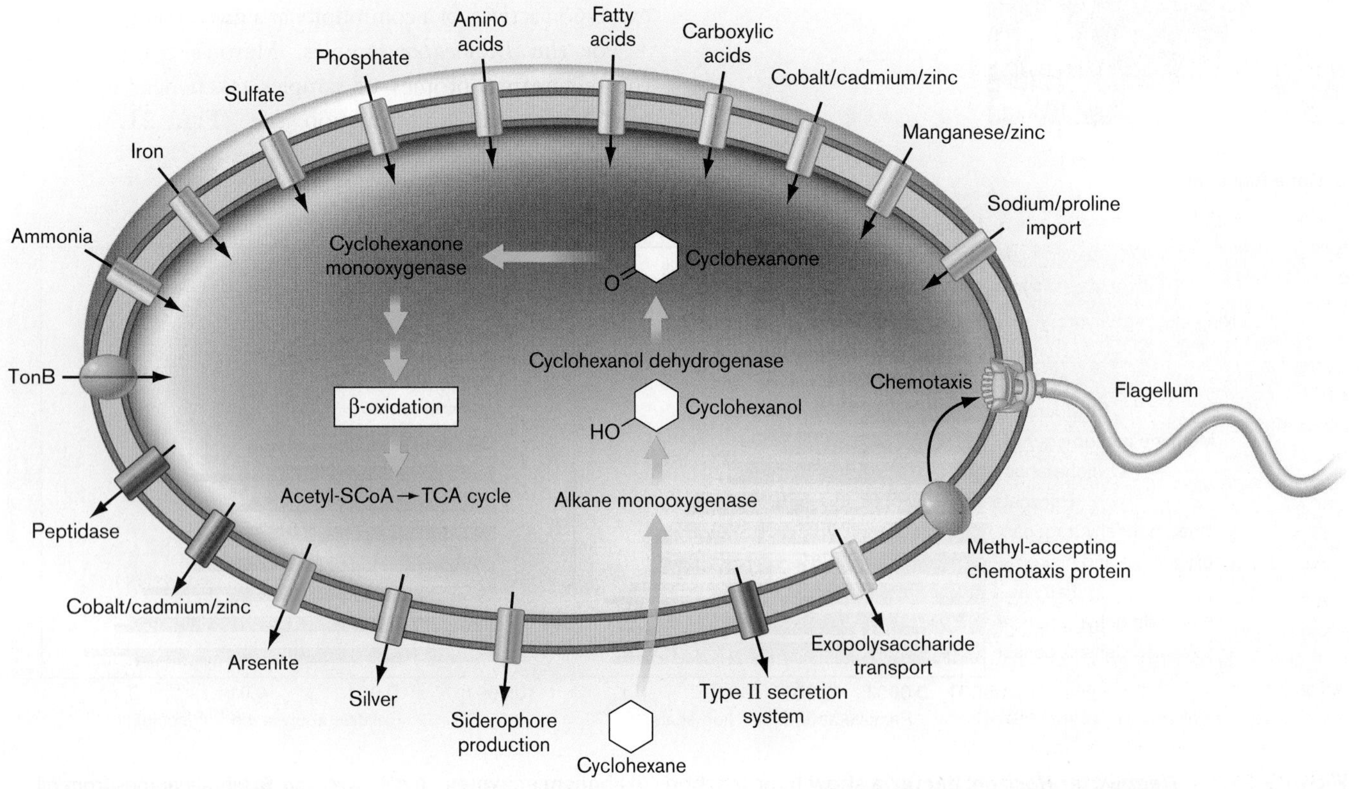

FIGURE 21.8 ■ Single-cell sequence of Oceanospirillales allows reconstruction of metabolic map. Gene functions were identified by comparison with annotated functional genes in public databases. *Source:* Modified from Mason et al. 2012. *ISME J.* **6**:1715, fig. 5.

biochemistry—for example, by isotope labeling. Additional examples are discussed in Section 21.2.

Thought Question

21.4 How could you determine whether an organism actually performs the functions predicted by your analysis of its genome and transcriptome, such as metabolizing petroleum components?

Beyond Omics: Cell Sorting and Culturing the Uncultured

The growing importance of single-cell analysis relies increasingly on FACS technology. At the same time, cell sorting enables additional dimensions of community analysis based on phenotypes. Furthermore, creative approaches to cell culture can actually reveal community members missed by the metagenomes.

Cell sorting and flow cytometry. A powerful tool to analyze community structure is **flow cytometry**, in which cells of multiple types are detected individually and distinguished using a set of phenotypic traits. The individual cells can further be collected into sorted samples by **cell sorting**. Flow cytometry, with or without cell sorting, provides intriguing data on marine and freshwater microbes. The phytoplankton (microbial phototrophs) include a wide range of bacterial and eukaryotic microbes that conduct photosynthesis. Their photosynthesis uses various chlorophylls and accessory light-harvesting pigments such as phycoerythrin (discussed in Section 14.6). These pigments are naturally fluorescent—a property called **autofluorescence**. Each subpopulation of the community can thus be defined by a combination of signals that include the intensity and wavelength of autofluorescence, and the intensity of scattered light (a property described in Chapter 2).

Figure 21.9A diagrams the inner workings of a cell sorter, an instrument that conducts both flow cytometry

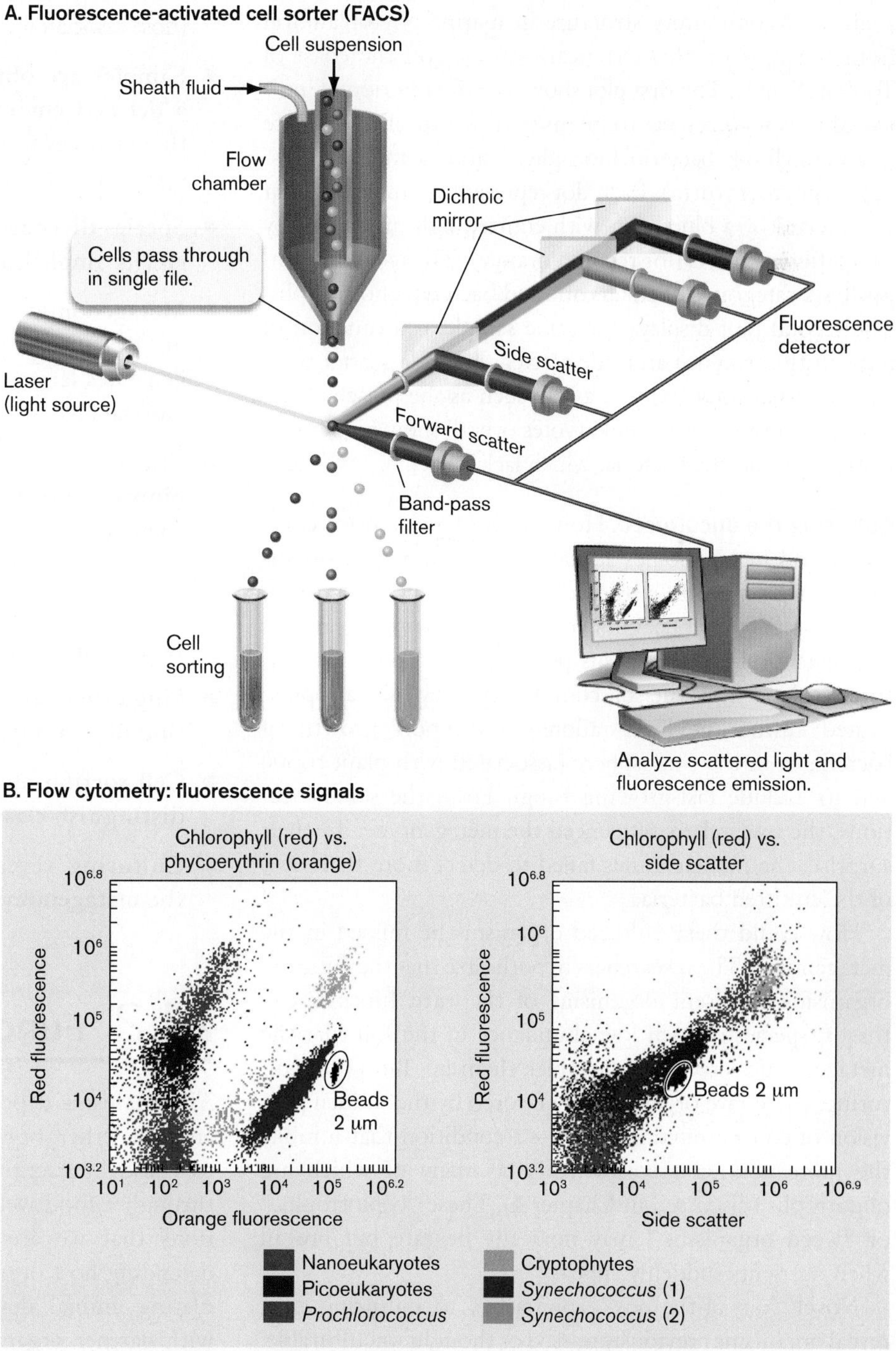

FIGURE 21.9 ■ Flow cytometry and FACS. A. In a fluorescence-activated cell sorter (FACS), a cell suspension is inoculated into flowing sheath fluid, which forms a stream of droplets. Each droplet carries one cell or none. Light from a laser interacts with each cell, generating forward scatter, side scatter, and fluorescence modulated by various filters. The pattern of scatter and fluorescence is analyzed (flow cytometry). A computed "gate" determines a combination of scatter and fluorescence intensities that activates deflection of the droplet into a collection tube (cell sorting). **B.** Flow cytometry reveals subpopulations of microbes from a sample of Mediterranean seawater off the coast of Toulon, France. Each subpopulation is defined by the intensities of red autofluorescence (chlorophyll) and orange autofluorescence (phycoerythrin; left graph), or of red autofluorescence (chlorophyll) and side scatter (right graph).

Source: Modified from Delpy et al. 2018, *Estuaries Coast* **41**:2039, fig. 3a and 3b.

and cell sorting. Within the instrument, a continuous flow of "sheath fluid" under pressure generates a stream of droplets. A cell suspension is inoculated into flowing sheath fluid, at a concentration adjusted such that each droplet carries one cell or none.

Light from a laser interacts with each cell to generate signals of forward scatter, side scatter, and fluorescence modulated by filters. The intensity of forward scattered light varies with particle size, whereas side scatter varies with the particle's internal complexity or granularity. For sorting and collection of cells, a computed "gate" determines a combination of scatter and fluorescence intensities that activates deflection of the droplet into a tube or microtiter plate.

Figure 21.9B shows an example of flow cytometry: the analysis of community structure in marine phytoplankton from a sample of Mediterranean seawater off the coast of Toulon, France. The first plot shows the distribution of individual cells with respect to intensity of red autofluorescence (chlorophyll or bacteriochlorophyll) and orange fluorescence (phycoerythrin). Each dot represents a single cell that was excited by a blue laser, with emitted light measured by two different filters (for red and orange). The results distinguish six categories of eukaryotic and bacterial phototrophs. The second plot displays the same sample as a function of red autofluorescence and side scatter. The side scatter indicates internal complexity of a cell, such as the presence of a nucleus. In this plot the eukaryotes overall show higher side scatter than do the bacteria, which lack nuclei.

Culturing the uncultured. How much of a microbial community do metagenomes actually catch—and what organisms do they miss? In a study of soil in an apple orchard, Jo Handelsman and colleagues compared the sampling rate of a metagenomic screen versus plate culture. To obtain cultured isolates from the soil community, they used a sophisticated culture medium tailored to support growth of bacteria from the rhizosphere (associated with plant roots) and to exclude fast-growing fungi. From the same locations, the researchers sequenced the metagenome. To their surprise, the metagenomes failed to detect more than 60% of the cultured bacteria.

How could these cultured organisms be missed in the metagenome? The researchers hypothesize that the cultured organisms represent organisms of the **rare biosphere**—that is, species of such low abundance in the soil that the metagenomic screen does not pick them up. But upon culturing, these rare organisms are favored by the sudden provision of concentrated nutrients—a condition that inhibits the more abundant soil organisms, many of which are oligotrophs (discussed in Chapter 4). These "copiotrophs," or "weed organisms," may normally be rare but prevail when nutrients suddenly appear.

Novel "out-of-the-box" approaches to culturing now reveal organisms previously missed or thought unculturable. An example is the teixobactin-producing soil microbe isolated by Kim Lewis (described in Chapter 4). Additional examples are presented in Sections 21.4 (intestinal biota) and 21.5 (*Prochlorococcus*).

To Summarize

- **A microbial community is the sum of populations of all taxa in a given ecosystem.** Change over time in a community or microbiome is called **succession**.
- **A metagenome is the sum total of all DNA sequenced from a microbial community.** The community may be any natural or human-made environment, or a **microbiome** (host-associated microbes) of a plant or animal.
- **Samples are obtained from a target community of a defined environment.** Sampling requires separating the microbes from their physical environment, breaking open the cells, and purifying the DNA.
- **Species diversity of a community is assessed by SSU rRNA amplification.**
- **Metagenomic DNA is sequenced.** Illumina sequencing by synthesis (next-generation sequencing, or NGS) generates large amounts of data, which require computational analysis.
- **The sequence reads are assembled into scaffolds and binned into partial genomes.** Assembly requires a computational pipeline that incorporates mathematical tools and biological assumptions.
- **Functional annotation offers clues as to the ecological functions contributed by the microbes.**
- **Single-cell genome analyses reveal coordinated function of genes within an organism.**
- **Cell sorting and flow cytometry offer new ways to distinguish classes of microbes within a community.**
- **Culturing reveals community members missed by the metagenomes.**

21.2 Functional Ecology

All organisms depend, directly or indirectly, on the presence of other organisms. How do microbes contribute to these interactions? Microbes cycle essential nutrients through a food web. They also serve more complex functions that we are just beginning to discover, such as defending host organisms from pathogens, and even modulating animal development and behavior. Cooperation with partner organisms may be incidental, as in the case of hydrogen-oxidizing bacteria using H_2 from fermenters; or it may involve mutualism, a highly developed partnership in which two or more species coevolve to support each other (discussed in Sections 21.3 and 21.4).

The Niche Concept

Within a community, each population of organisms fills a specific **niche**. The niche is a set of conditions, including its habitat, resources, and relations with other species of the ecosystem, that enable an organism to grow and reproduce. For example, the niche of *Anabaena*, a cyanobacterium, is that of a filamentous or mat-forming marine organism that fixes CO_2 into biomass of its vegetative cells while fixing nitrogen via specialized cells called heterocysts (see Chapter 18). *Anabaena*'s photosynthesis releases molecular oxygen that is used by swarms of respiring bacteria. The habitat of *Anabaena* is fresh or brackish water; its biomass provides food for invertebrates and fish. Despite being autotrophic, *Anabaena* needs the other organisms too. The cyanobacteria grow best in the presence of heterotrophic Proteobacteria, whose respiration depletes oxygen near the *Anabaena* heterocysts, which need anoxic conditions to fix nitrogen. Thus, organisms do more than fill a niche; they construct niches for other kinds of organisms. Organisms perform **niche construction** by shaping the biochemical dimensions of their habitat.

Spatial Organization of Microbes in a Habitat

A question unanswered by metagenomes is the spatial organization and interaction of microbes within a habitat. However, genome sequence data can be used to construct probes to answer such questions. A key technique that shows the spatial location of microbial taxa is **fluorescence in situ hybridization**, or **FISH** (introduced in Chapter 2).

The FISH technique combines probes derived from sequencing with fluorescence microscopy (**Fig. 21.10A**). This method makes use of a fluorophore-labeled oligonucleotide probe (usually a short DNA sequence) that hybridizes to microbial DNA or rRNA. Hybridizing to rRNA increases sensitivity because rRNA is present in approximately 100-fold to 10,000-fold excess over DNA. In a typical procedure, the cells of a sample are fixed to a slide (**Fig. 21.10A**, step 1) by a chemical treatment that maintains cell integrity while permeabilizing the cell so that the fluorophore-labeled DNA probe can enter (step 2). Next the fixed cells are incubated in a hybridization buffer containing the probe, at a temperature designed to maximize specificity of binding to the sequence of the desired taxa (step 3). A probe with broad specificity might hybridize to all bacterial rRNA, but not to archaeal or eukaryotic rRNAs. For greater specificity, a probe may have a sequence complementary to a sequence found only in rRNA of a given bacterial taxon. Following hybridization and a wash (step 4), the cells containing hybridized probes are observed by fluorescence microscopy (step 5).

Note: For microbial ecology, FISH usually involves probe hybridization to ribosomal RNA molecules, whereas for eukaryotic cells the probe is hybridized to a gene on chromosomal DNA.

Figure 21.10B shows an example of FISH used to reveal the deep-sea methanotrophic consortium that oxidizes 90% of the methane emitted by methanogens—a major contribution to the global climate (discussed in Chapter 22). The target community consists of a mixed biofilm of anaerobic methane-oxidizing archaea (ANME) and sulfate-reducing bacteria (SRB), obtained from a methane seep at the Guaymas Basin in the Gulf of California, located 2 km below sea level. The biofilm was labeled with oligonucleotide fluorescent probes specific for ANME (red) and for bacteria (green).

The FISH image shows the two kinds of cells grouped together in a remarkably regular pattern. The ANME archaea, which extract electrons from methane, thereby releasing CO_2, tend to be buried within the sample, surrounded by the bacteria. The bacteria receive electrons from the ANME and transfer electrons to sulfate, which is reduced to sulfide. Sulfate is a relatively poor electron accepter (as discussed in Chapter 14), but its high concentration in seawater increases its free-energy contribution (discussed in Chapter 13). The spatial organization of ANME and SRB cells thus facilitates their syntrophy, a form of metabolism in which each partner completes half of a reaction with an overall negative value of ΔG (Chapter 13). Tight association of the ANME archaea with the sulfate-reducing bacteria enables ANME to transfer electrons from methane directly to the bacteria.

All Ecosystems Require Microbes

Within communities, various microbial taxa depend on each other—and more broadly, entire ecosystems rely on essential microbes. The role of microorganisms in all ecosystems was originally formulated by the Dutch microbiologist Cornelis B. van Niel (1897–1985). Van Niel formalized the concept that bacteria in the soil and water can conduct photosynthesis without producing oxygen, using electron donors such as H_2S instead of H_2O. This surprising discovery revealed one of many kinds of metabolism unique to microbes, and unknown in plants or animals. Other unique forms of microbial metabolism (discussed in Chapters 13–15) include nitrogen fixation by bacteria and archaea, and the degradation of lignin by bacteria and fungi. Moreover, microbial metabolism provides ecosystems with their sole source of key elements such as sulfur, phosphorus, and iron.

A.

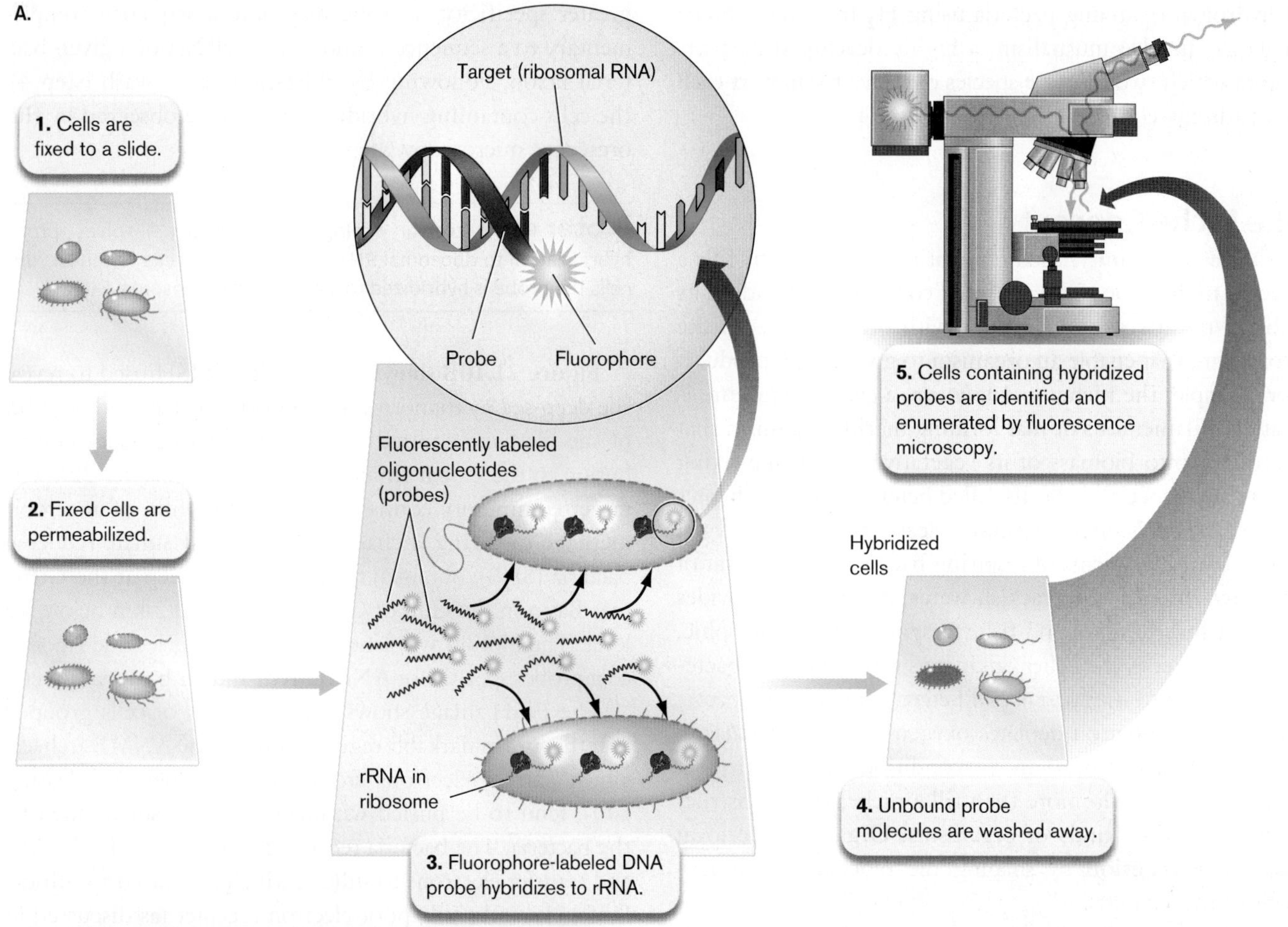

B.

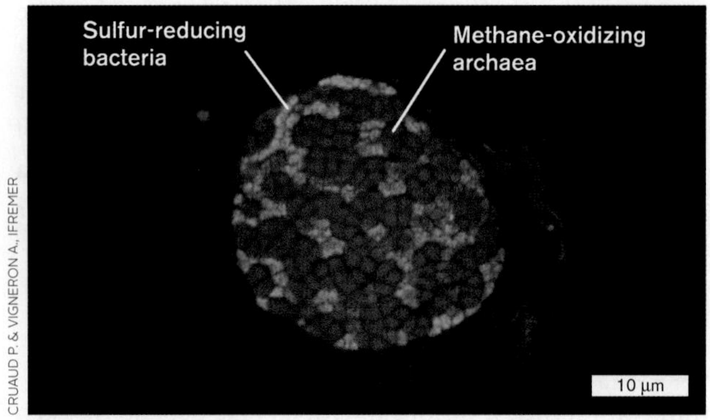

FIGURE 21.10 ■ Fluorescence in situ hybridization (FISH) of bacteria and archaea. A. Fluorophore-labeled DNA oligonucleotide hybridizes to a taxon-specific sequence of rRNA molecules within the cells that are fixed and permeabilized on a microscope slide. **B.** Syntrophy between anaerobic methane-oxidizing archaea (red FISH) and sulfate-reducing bacteria (green FISH) from a deep-sea cold seep at Guaymas Basin in the Gulf of California. *Source:* Part A modified from Rudolf Amann and Bernhard M. Fuchs. 2008. *Nat. Rev. Microbiol.* **6**:339, fig. 1.

These examples of microbial metabolism provide unique roles for microbes in ecosystems. Van Niel expressed this principle in two bold hypotheses of microbial ecology:

- **Every molecule existing in nature can be used as a source of carbon or energy by a microorganism somewhere.** Any molecule found in the environment can participate in some kind of energy-yielding reaction. If an energy-yielding reaction exists, some microbe will evolve to use it.
- **Microbes are found in every environment on Earth.** Every habitat where life is possible includes microbial life. In fact, the largest part of the biosphere (below Earth's surface) is inhabited solely by microbes.

The van Niel hypotheses imply that a limitless variety of species carry out different energy-yielding reactions, depending on what their environment has to offer. For example, oil-contaminated water from the *Deepwater Horizon* oil spill region shows enrichment for bacteria that catabolize molecules found in petroleum, small amounts of which seep naturally into Gulf sediment. In effect, a pipeline spill acts as a kind of **enrichment culture**, in which the addition of a particular class of nutrient favors the growth of microbes that can use that nutrient (discussed in Chapter 4). The catabolic enzymes were revealed by functional annotation of metagenomes and metatranscriptomes. The energy gained by microbial metabolism, as well as the elements that microbes assimilate into biomass, eventually circulate throughout the ecosystem.

What about new types of compounds, entirely novel in the environment, for which no catabolic enzymes exist? Compounds such as synthetic plastics were once considered nondegradable "xenobiotic" molecules. But evolution may generate microbes that can digest even the indigestible.

An example is the compound polyethylene terephthalate (PET). This polymer of esterified polyethylene monomers, commonly known as "polyester," was patented in 1941 and has been used to make bottles for beverages since the 1970s (**Fig. 21.11A**). Today, the worldwide production of PET is more than 50 million tons per year. Much of the material ends up in our environment as fragments of various sizes, including tiny microscopic fragments ingested by small invertebrates and consumed up the food chain. Until recently, no decomposition of PET was reported. The material appears to persist permanently in our environment.

A. B.

DUNCAN PHILLIPS/ALAMY STOCK PHOTO
COURTESY OF SHOSUKE YOSHIDA

FIGURE 21.11 ■ Discovery of a bacterium that degrades polyethylene terephthalate (PET). **A.** Bottles made of PET accumulate indefinitely in the environment. **B.** Shosuke Yoshida discovered a bacterium, *Ideonella sakaiensis*, that catabolizes PET.

In 2016, Shosuke Yoshida (**Fig. 21.11B**) and colleagues at the Kyoto Institute of Technology reported the discovery of a bacterium, *Ideonella sakaiensis*, that degrades PET film (**Fig. 21.12**). Yoshida's team discovered the bacteria by screening hundreds of soil samples from the PET-contaminated

A.

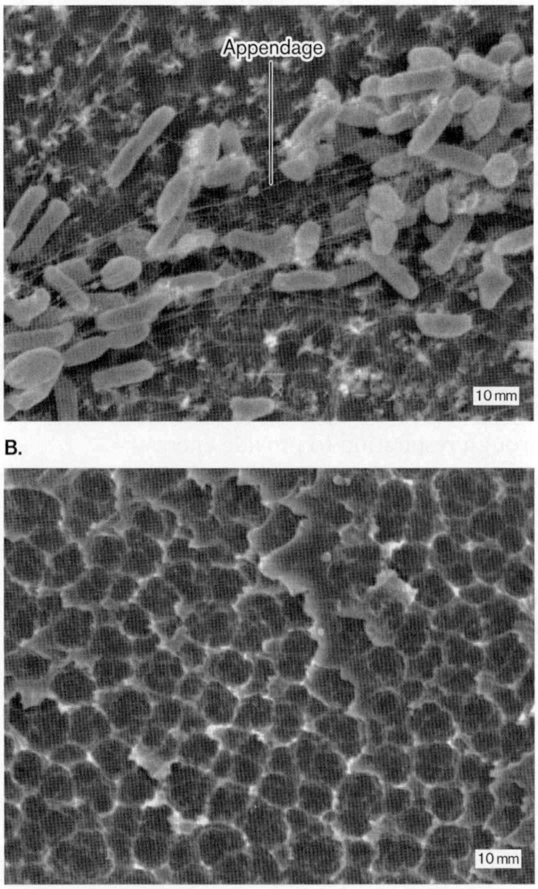

YOSHIDA ET AL. 2016. *SCIENCE* **351**:1199, FIG. 1D. REPRINTED WITH PERMISSION FROM AAAS.
YOSHIDA ET AL. 2016. *SCIENCE* **351**:1199, FIG. 1G. REPRINTED WITH PERMISSION FROM AAAS.

C.

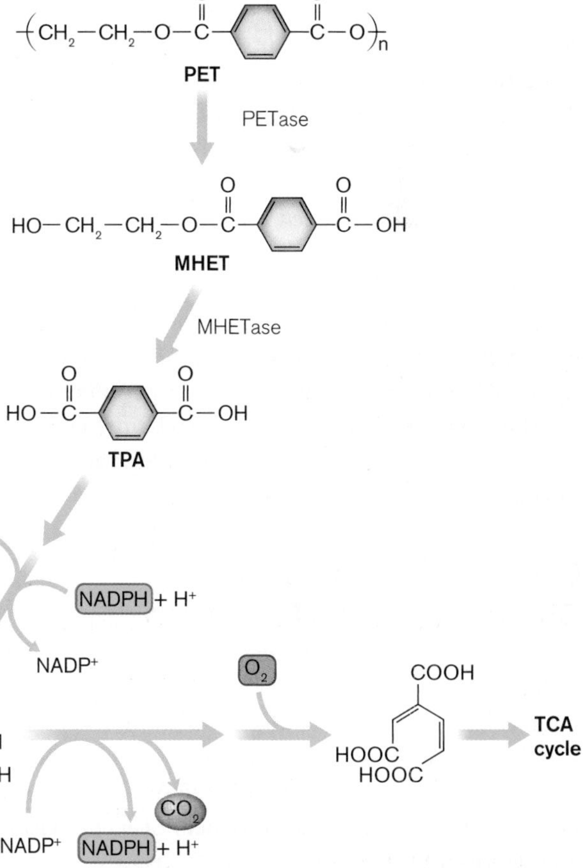

FIGURE 21.12 ■ Soil bacteria degrade PET. **A.** *Ideonella sakaiensis* cultured on a film of polyethylene terephthalate (PET). Cell appendages adhere to the film. **B.** After removal of bacteria, the film shows degradation (compare intact film, inset). **C.** The PET degradation pathway includes newly discovered catabolic enzymes: PETase and MHETase. TPA = terephthalic acid. *Source:* Part C adapted from Yoshida et al. 2016. *Science* **351**:1199, fig. 3B.

yard of a bottle-recycling factory. They cultured their samples in a lettuce-and-egg medium supplemented with a film of PET (**Fig. 21.12A**). One sample, number 46, had bacteria that colonized the film. The bacteria extended appendages to aid adherence and possibly contribute to breakdown of the material. **Figure 21.12B** shows the pockmarked appearance of the film after the colonizing bacteria were removed.

How do the bacteria manage to catabolize PET where others fail? Yoshida sequenced the organism's genome and found a gene showing similarity to the gene that encodes a hydrolase (an enzyme that hydrolyzes polymers). Hydrolases are common in soil microbes and in the human gut biota, where they aid human digestion of complex plant glycans (discussed in Chapter 13). Yoshida ultimately isolated the enzyme PETase and showed that it, indeed, hydrolyzes PET, releasing units of the monomer monoethylene terephthalate (MHET; **Fig. 21.12C**). In addition, a second enzyme predicted by the genome, MHETase, was shown to break down the monomer released by PETase. The monomer breakdown leads to a common pathway of aromatic catabolism whose products can enter the TCA cycle, as described in Chapter 13.

Where did this novel PETase come from? Microbiologists hypothesize that it evolved as a mutant form of a hydrolase that degrades some other kind of polymer. Subsequent research on the crystal structure of PETase confirms its evolutionary relationship with other hydrolases that break down cutin (a waxy polymer of plants) and lipids. Altered versions of PETase were made that have even higher activity, which show promise for bioremediation of plastic pollutants.

Overall, Yoshida's discovery is consistent with van Niel's prediction that any molecule can be catabolized by a microbe with the needed enzyme. Given that microbes continually undergo mutation and evolution, the presence of an energetically favorable food source—one whose oxidation or reduction has a negative value of free energy, ΔG—will eventually select for a lineage in which enzymes can perform the metabolism. For an experimental example, see the Chapter 17 discussion of Richard Lenski's Long-Term Evolution Experiment, which yielded an *E. coli* strain capable of catabolizing citrate.

Carbon Assimilation and Dissimilation: The Food Web

The interactions between microbes and their ecosystems include two common roles of metabolic input and output, often called assimilation and dissimilation, respectively. We discussed most of these metabolic processes in Chapters 13, 14, and 15, but ecology offers a community perspective: How do the substrates and products of metabolism affect populations of other organisms?

Assimilation refers to processes by which organisms acquire an element, such as carbon from CO_2, to build into cells. When the environment lacks organic compounds containing an element such as nitrogen or phosphorus, microbes may assimilate the element from mineral sources. Common kinds of assimilation include carbon dioxide fixation and nitrogen fixation. Organisms that produce biomass from inorganic carbon (usually CO_2 or bicarbonate ions) are called **primary producers**. Producers are a key determinant of productivity for other members of the ecosystem.

Dissimilation is the process of breaking down organic nutrients to inorganic minerals such as CO_2 and NO_3^-, usually through oxidation. Microbial dissimilation releases minerals for uptake by plants and other microbes, and it provides the basis of wastewater treatment (discussed in Chapter 22). But microbial dissimilation can decrease habitat quality by removing organic nitrogen. When soil bacteria break down amines (RNH_2) to ammonium ion (NH_4^+), nitrifying bacteria such as *Nitrosomonas* oxidize the ammonium to nitrite (NO_2^-) and nitrate (NO_3^-). These highly soluble anions are then washed out of soil into the groundwater.

This chapter covers microbial assimilation and dissimilation of carbon and nitrogen in association with the plants and animals of an ecosystem. The cycles of other key elements, and their effects on the global biosphere, are explored in Chapter 22.

Thought Question

21.5 From Chapters 13–15, give examples of microbial metabolism that fit patterns of assimilation and dissimilation.

The major interactions among organisms in the biosphere are dominated by the production and transformation of **biomass**, the bodies of living organisms. To obtain energy and materials for biomass, all organisms participate in **food webs** (**Fig. 21.13**). A food web describes the ways in which various organisms produce and consume biomass. Levels of consumption are called **trophic levels**. Organisms at each trophic level consume biomass of organisms from another level. At each trophic level, the fraction of biomass retained by the consumer is small; most is released as CO_2, through respiration to provide energy.

Every food web depends on primary producers for two things:

- **Absorbing energy from outside the ecosystem.** A key source of energy is sunlight, which drives production by photoautotrophy.
- **Assimilating minerals into biomass.** The biomass of producers is then passed on to subsequent trophic levels.

The majority of carbon in Earth's biosphere is assimilated by oxygen-producing phototrophs such as cyanobacteria, algae, and plants. Certain important ecosystems are founded on lithoautotrophs—for example, the hydrothermal vent communities, in which bacteria oxidize hydrogen sulfide to fix CO_2, capturing both gases as they well up from Earth's crust. The vent communities use oxygen generated by phototrophs living in the euphotic zone (see Section 21.5).

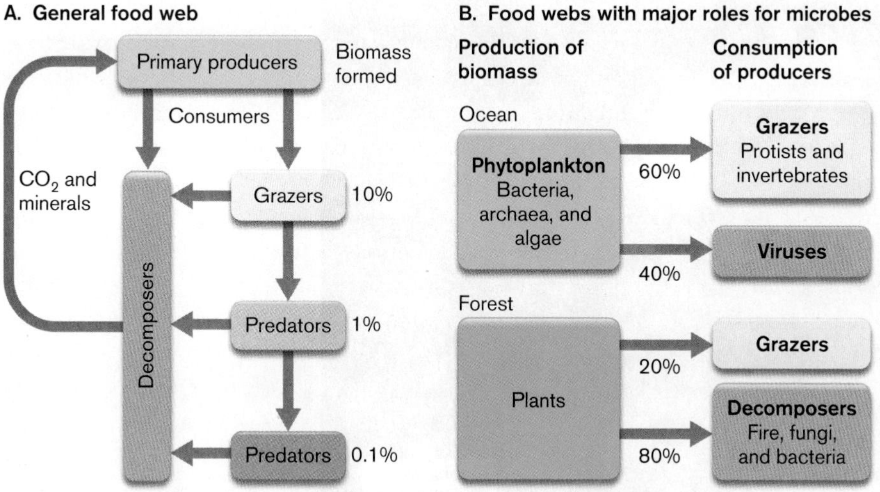

FIGURE 21.13 ▪ Microbes within food webs. A. Biomass production and carbon recycling. Percentages indicate the fraction of original CO_2 converted to biomass at each trophic level. **B.** In marine ecosystems, the primary producers are bacteria, archaea, and algae. Viruses break down both producer and consumer microbes. In a forest, the major producers are trees, while the main decomposers are fungi and bacteria.

In addition to producers, all ecosystems include **consumers**, which acquire nutrients from producers and ultimately dissimilate biomass by catabolism, returning carbon back to the atmosphere (see **Fig. 21.13A**). Consumers constitute several trophic levels, based on their distance from the primary producers. The first level of consumers, generally called **grazers**, directly feed on producers. Grazers usually convert 90% of the producer carbon back to CO_2 through respiratory metabolism, yielding energy. The next level of consumers, often called **predators**, feed on the grazers, again converting 90% back to atmospheric CO_2. In microbial ecosystems, the trophic relationships are often highly complex, because a given species may act as both producer and consumer.

At each trophic level, some of the organisms die, and their bodies are consumed by **decomposers**, returning carbon and minerals back to the environment for use by producers. All decomposers are microbes (fungi or bacteria). Decomposers have particularly versatile digestive enzymes capable of breaking down complex molecules such as lignin. Without decomposers, carbon and minerals needed by phototrophs would be locked away by ever-increasing mounds of dead biomass. Instead, all biomass is recycled somewhere in the biosphere. As we learned in Chapter 13, the energy gained by ecosystems can be cycled in part, but all is eventually lost as heat.

In the function of ecosystems, phylogenetic distinctions among the domains Bacteria, Archaea, and Eukarya have less importance than the biological and biochemical consequences of an organism's presence in the community—that is, what the organism produces or consumes. Thus, in this chapter we place greater emphasis on trophic roles than on phylogenetic distinctions.

The relative impact of microbial and multicellular producers and consumers varies considerably among different habitats. A major difference appears between marine and terrestrial ecosystems (**Fig. 21.13B**). In the oceans, the smallest inhabitants, cyanobacteria such as *Prochlorococcus*, perform most of the CO_2 fixation and biomass production. The main consumers are protists and viruses. Viruses are the most numerous replicating forms in the ocean—and they lyse most marine cells before any multicellular predators have a chance to consume them. In terrestrial ecosystems, by contrast, the major primary producers and fixers of CO_2 are multicellular plants. Plants generate **detritus**, discarded biomass such as leaves and stems, that requires decomposition by fungi and bacteria. While viruses are important, multicellular consumers such as worms and insects play a greater role in decomposition.

The differences between the food webs of ocean and dry land explain why most of the food we harvest from the ocean consists of predators at the higher trophic levels (fish), whereas most food harvested on land consists of producers and first-level consumers (plants and herbivores). Fish depend on a large number of trophic levels—including a vast base of microbes. Thus, the numbers of fish remain limited, despite the seemingly huge volume of ocean.

Oxygen and Other Electron Acceptors

The availability of oxygen and other electron acceptors is the most important factor that determines how nutrients containing carbon, nitrogen, and sulfur are assimilated and dissimilated. Examples of aerobic and anaerobic metabolism are presented in **Table 21.2**. In aerobic environments, microbes use molecular oxygen as an electron acceptor to respire on organic compounds (abbreviated CHO) produced by other organisms (discussed in Chapters 13 and 14). Aerobic respiration on organic compounds is highly dissimilatory in that it tends to break compounds down to CO_2. Microbes

TABLE 21.2 Aerobic and Anaerobic Metabolism

Element	Oxidized by O_2 (lithotrophy)	Reduced by CHO* or H_2 (anaerobic electron transport)
Nitrogen	$NH_3 + O_2 \rightarrow NO_3^-$	$NO_3^- + CHO \rightarrow N_2$
Manganese	$Mn^{2+} + O_2 \rightarrow Mn^{4+}$	$Mn^{4+} + CHO \rightarrow Mn^{2+}$
Iron	$Fe^{2+} + O_2 \rightarrow Fe^{3+}$	$Fe^{3+} + CHO \rightarrow Fe^{2+}$
Sulfur	$H_2S + O_2 \rightarrow SO_4^{2-}$	$SO_4^{2-} + CHO \rightarrow H_2S$
Carbon	$CH_4 + O_2 \rightarrow CO_2$	$CO_2 + H_2 \rightarrow CH_4$

*CHO = organic material.

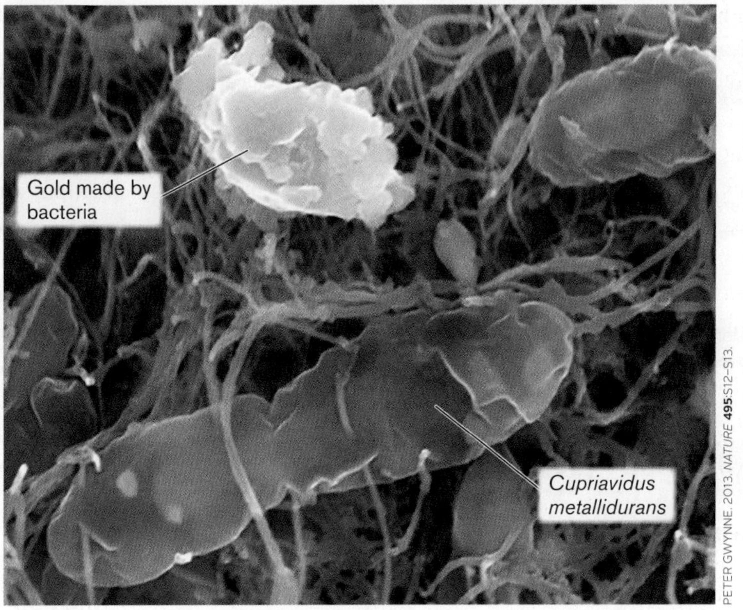

FIGURE 21.14 ▪ Bacteria make gold. *C. metallidurans* reduces Au^{3+} to gold metal (Au^0; SEM).

also use oxygen to respire on reduced minerals such as NH_3, H_2S, and Fe^{2+} (lithotrophy) and to oxidize hydrogen with CO_2 (methanogenesis). In most cases, lithotrophy is coupled to CO_2 fixation and is therefore assimilatory metabolism.

In anoxic environments, such as deep soil, microbes use minerals such as Fe^{3+} and NO_2^- to oxidize organic compounds supplied by other organisms (anaerobic respiration) or reduced minerals (anaerobic lithotrophy; discussed in Chapter 14). Some metals may be reduced to counteract their toxicity; for example, *Cupriavidus metallidurans* reduces gold ion (Au^{3+}) to form particles of gold metal (**Fig. 21.14**). Frank Reith, at the University of Adelaide, proposes that most of Earth's gold ore was originally generated by bacterial reduction. Anoxic environments require much slower rates of assimilation and dissimilation than occur in the presence of oxygen. The total biomass of anaerobic microbial communities, however, far exceeds that of our oxygenated biosphere.

Temperature, Salinity, and pH

Abiotic factors can profoundly affect the environment for microbes and other members of the food chain, either directly or by impacting other factors. Temperature limits the rate of metabolism (discussed in Chapter 4). Higher temperatures found in hot springs (80°C–100°C) enable some of the fastest growth rates measured (doubling times as short as 10 minutes for some hyperthermophiles). On the other hand, temperature limits the oxygen concentration in water, so hyperthermophiles often respire with sulfur instead. Extreme cold such as that of polar regions may limit diversity and exclude all but microorganisms and microscopic invertebrates. An example of a cold-adapted ecosystem is the Antarctic lake described in **Special Topic 21.1**.

The effects of abiotic factors are most obvious for **extremophiles**, species that grow in environments considered extreme by human standards (**Table 21.3**), discussed in Chapter 5. Extreme high temperature and pressure selects for hyperthermophilic archaea, whose membranes contain ether-linked acids and whose DNA is positively supercoiled by reverse gyrase (Chapter 19).

High salt concentration limits the growth of microbes adapted to freshwater conditions. By contrast, many microbial

TABLE 21.3 Extremophiles

Class of extremophile	Typical environmental conditions for growth
Acidophile	Acidic environments at or below pH 3
Alkaliphile	Alkaline environments at a range of pH 9–14
Piezophile (barophile)	High pressure, usually at ocean floor, from 200 to 1,000 atm
Endolith	Within rock crystals down to a depth of 3 km
Halophile	High salt, typically above 2-M NaCl
Hyperthermophile	Extreme high temperature, above 80°C
Oligotroph	Low carbon concentration, below 1 ppm
Psychrophile or cryophile	Low temperature, below 15°C
Thermophile	Moderately high temperature, 50°C–80°C
Xerophile	Desiccation, water activity below 0.8

species have adapted to high salinity (they are called halophiles). The haloarchaea, for instance, bloom in population as a body of water shrinks and becomes hypersaline.

Acidity is important geologically because a high concentration of hydronium ions accelerates the release of reduced minerals from exposed rock. Extreme acidity is often produced by lithotrophs (discussed in Section 14.5). The increasing acidity releases minerals to be oxidized, while excluding acid-sensitive competitors. Other kinds of habitats, such as soda lakes, show extreme alkalinity resulting from high sodium carbonate.

Deeper understanding of ecosystems leads to the discovery of other remarkable ways that organisms benefit each other. For example, plant-associated bacteria and fungi enhance the uptake of nutrients and protect the host from pathogens. Animal-associated microbes protect the host from pathogens, enhance digestion, and modulate the immune system. These processes are detailed in the next sections.

To Summarize

- **Microbial populations fill unique niches in ecosystems.** Every chemical reaction that may yield free energy can be utilized by some kind of microbe.
- **Microbial enzymes may evolve to consume a newly available energy source.**
- **FISH reveals spatial organization of community partners.**
- **Microbes fix or assimilate essential elements into biomass**, which recycles within ecosystems. Important elements, such as nitrogen, are fixed solely by bacteria and archaea.
- **Consumers and viruses break down the bodies of producers**, generating CO_2 and releasing heat energy. Dissimilation is the process of breaking down nutrients to inorganic minerals such as CO_2 and NO_3^-, usually through oxidation.
- **Primary producers fix single-carbon units, usually CO_2.** Microbial primary producers include algae, cyanobacteria, and lithotrophs.
- **Decomposers such as fungi and bacteria release nutrients from dead organisms.**
- **Microbial activity depends on levels of oxygen, carbon, nitrogen, and other essential elements**, as well as environmental factors such as temperature, salinity, and pH. The largest volume of our biosphere contains anaerobic bacteria and archaea. Beneath Earth's surface, most metabolism is anaerobic.

21.3 Symbiosis

One of the most fascinating features of evolution is how organisms adapt to the presence of others. Some relationships of microbes occur at a distance; for example, the oxygen gas released by marine cyanobacteria is breathed by organisms around the globe. Other relationships require intimate association between two or more specific partners. An intimate association between organisms of different species is called **symbiosis** (plural, **symbioses**). Symbiotic associations include a full range of both positive and negative relationships (**Table 21.4**). Whether the relationship is positive or negative, both partners evolve in response to each other. Symbiosis may involve two or more partner

TABLE 21.4 Types of Symbiotic Associations Involving Microbial Species

Type of interaction	Effects of interaction	Example
Mutualism	Two organisms grow in an intimate species-specific relationship in which both partner species benefit and may fail to grow independently.	Lichens consist of fungi and algae (in some cases, cyanobacteria) growing together in a complex layered structure. Each species requires the presence of the other.
Synergism	Both species benefit through growth, but the partners are easily separated and either partner can grow independently of the other.	Human colonic bacteria ferment, releasing H_2 and CO_2, which methanogens convert to methane. The methanogens gain energy, and the bacteria benefit energetically from the removal of their fermentation products.
Commensalism	One species benefits, while the partner species neither benefits nor is harmed.	In wetlands, *Beggiatoa* bacteria oxidize H_2S for energy. Removal of H_2S enables growth of other microbes for whom H_2S is toxic. The other microbes are not known to benefit *Beggiatoa*.
Amensalism	One species benefits by harming another. The relationship is nonspecific.	In the soil, *Streptomyces* bacteria secrete antibiotics that lyse other species, releasing their cell contents for *Streptomyces* to consume.
Parasitism	One species (the parasite) benefits at the expense of the other, a specific host. The relationship is usually obligatory for the parasite.	*Legionella pneumophila*, the cause of legionellosis, parasitizes amebas in natural aquatic habitats. Within the human lung, *L. pneumophila* parasitizes macrophages.

SPECIAL TOPIC 21.1 Antarctic Lake Mats: Have Ecosystem, Will Travel

What kind of life grows in an ice-covered Antarctic lake—and how does it escape? To investigate these questions, Rachael Morgan-Kiss at Miami University of Ohio leads expeditions to the Taylor Valley, Antarctica, one of Earth's coldest dry deserts. To reach Antarctica from New Zealand, scientists must take an 8-hour flight in an air force transport plane. The plane typically carries all kinds of workers and equipment, prefabricated buildings, and all-terrain vehicles—in effect, a sample of the human "ecosystem" needed to sustain science at McMurdo Station and other research locations.

It turns out that lake microbes have invented a similar system. The cyanobacterial mats of many Antarctic lakes support entire ecosystems limited to microbes and meiofauna (microscopic invertebrates). But within these lakes, the mats have evolved a surprising cycle that flies them out to colonize new locations—along with multiple components of their lake ecosystem.

The Taylor Valley lakes are isolated, without outlet, and their main source of water is glacier melt in the summer. Thus, each lake has little circulation or turnover, and the water is covered by about 5 meters of ice. The ice surface is sculpted by winds, from cold air masses that fall down the mountain slope reaching speeds of 200 km per hour.

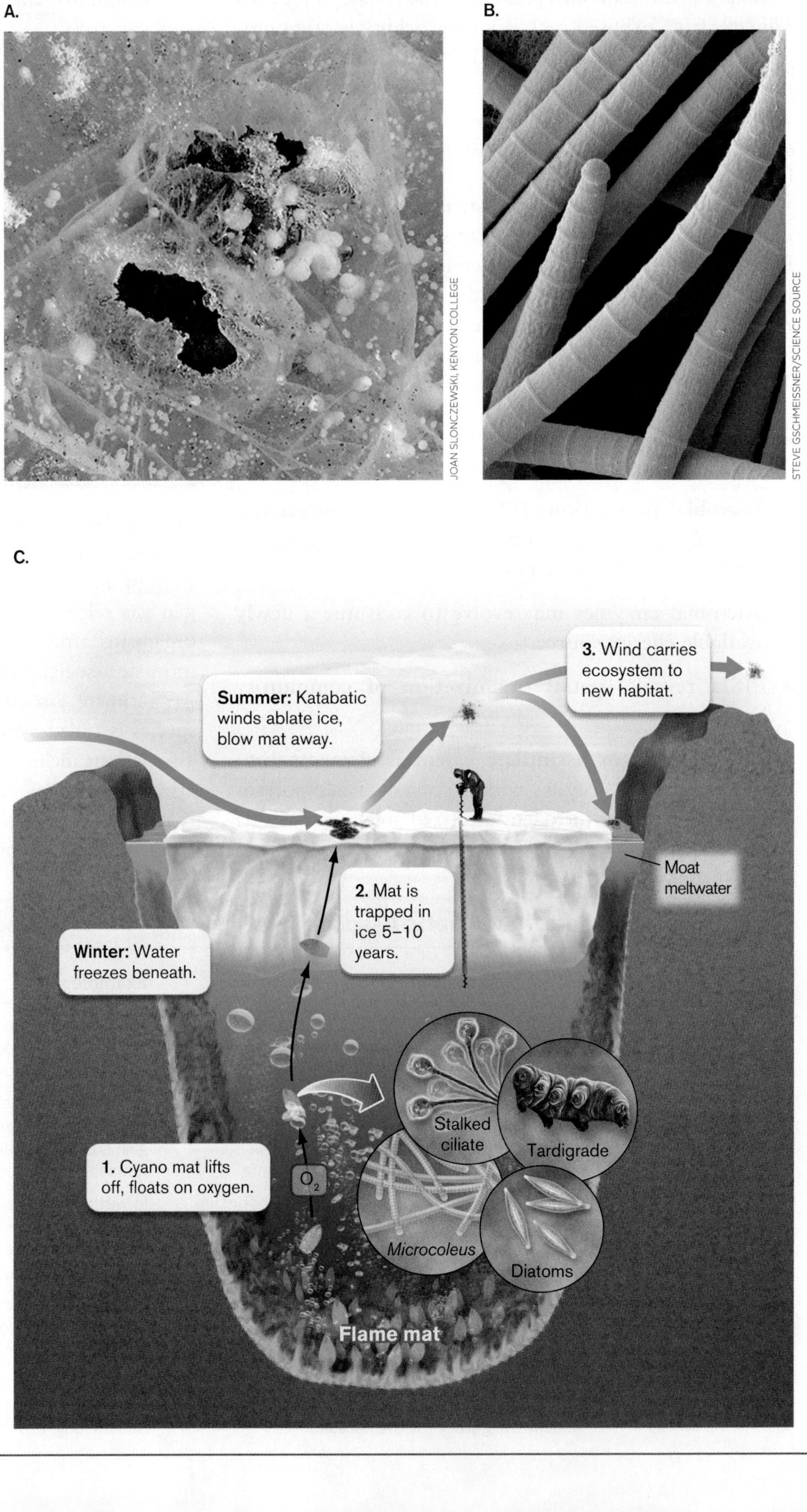

FIGURE 1 ■ Cyanobacterial mats carry an ecosystem. A. A cyanobacterial mat emerges from the ice. **B.** *Microcoleus* grows in "flame mats" at the bottom of an Antarctic lake. **C.** The cyanobacterial mats lift off upon bubbles of oxygen, carrying protists, heterotrophic bacteria, and meiofauna. In winter, water freezes the mat into the ice. Over several years, the mat works its way through the ice until winds scrape it out and carry the freeze-dried ecosystem to a new habitat.

FIGURE 2 ■ Bacterial taxa in mat and water samples. Taxa were quantified by marker genes found in Illumina reads of DNA samples from Lake Fryxell, Taylor Valley, Antarctica. **A.** Liftoff mat samples emerging from ice surface. **B.** Water column samples. **Inset:** Cultured organism identified as *Rhodoferax* sp.
Source: Data for A and B, Daniel Barich and Joan Slonczewski.

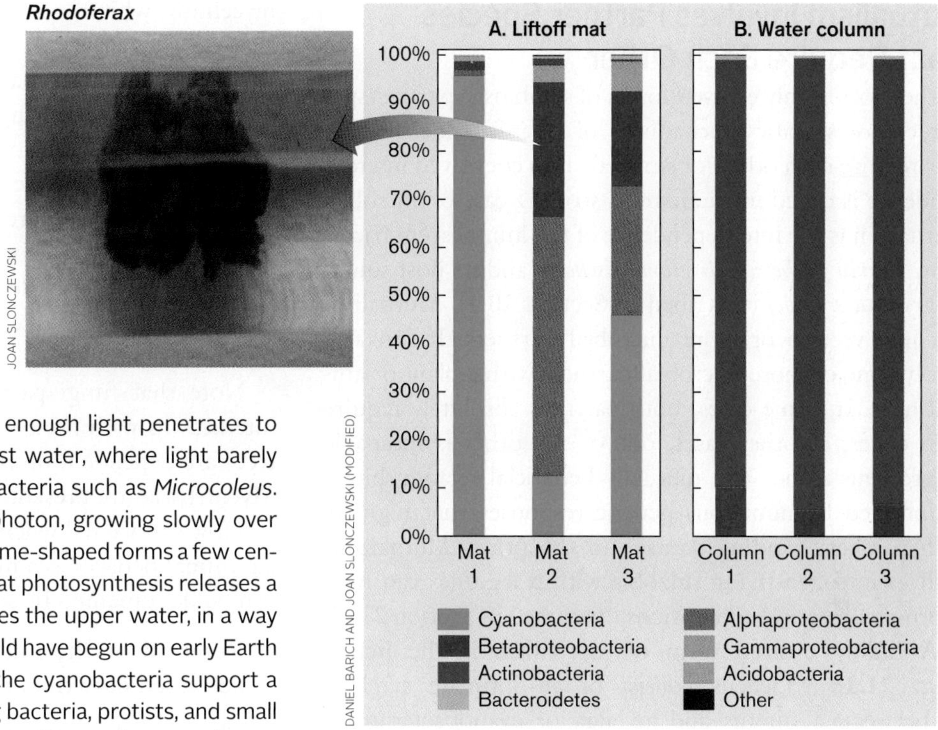

But through the meters of ice, enough light penetrates to power photosynthesis. The deepest water, where light barely appears, supports mats of cyanobacteria such as *Microcoleus*. The cyanobacteria absorb every photon, growing slowly over the years, in some cases building flame-shaped forms a few centimeters tall (**Fig. 1A** and **B**). The mat photosynthesis releases a steady supply of oxygen that aerates the upper water, in a way that suggests how oxygenation could have begun on early Earth 2 billion years ago. As producers, the cyanobacteria support a community of consumers including bacteria, protists, and small invertebrates. Invertebrates such as rotifers and nematodes are the top predators of the lake, where no fish can live.

All Antarctic lake organisms must survive 6 months of frozen dark, when water is below zero and surface temperatures reach −40°C. In the summer, with 24-hour sunlight, the phototrophs and consumers come back to life. The ice cover persists, except around the edge, where a "moat" melts through. Meanwhile, 10 meters below the ice, cyanobacteria are bubbling oxygen. The oxygen bubbles lift the mat, until a scrap breaks off and floats toward the surface (**Fig. 1C**). The scrap of cyanobacterial filaments carries with it all kinds of associated bacteria and protists, nematodes and water bears (tardigrades)—in effect, a sample of the whole lake ecosystem. The scrap halts beneath the ice, where it freezes, and come winter, new ice freezes beneath. But above, wind ablates the ice, wearing it down, as annual winter freezes more below. Thus, over the years the frozen scrap of life travels up the "ice elevator" to the surface. There, as the ice breaks open, the wind carries the scrap away. The scrap may fall into the moat, ready to colonize anew. Amazingly, even the nematodes and tardigrades come to life again with all their phototrophs ready for photosynthesis. And some lucky scraps are borne away to a moat in some other lake, ready to colonize a new home.

The "liftoff" mat microbial communities of Antarctica's Lake Fryxell were investigated by author Joan Slonczewski at Kenyon College and Zhong Wang at the Joint Genome Institute. Wang developed an advanced computational pipeline, called Constellation, in order to analyze complex metagenomes such as those of cyanobacterial mats. Wang compared the taxonomic diversity of mats emerging from the ice with those of bacteria from the water column below the ice (**Fig. 2**). Mat samples showed a high proportion of filamentous cyanobacteria such as *Microcoleus*. But some species showed a surprising proportion of Betaproteobacteria and Alphaproteobacteria. From one sample, Slonczewski cultured the phototroph *Rhodoferax*. This form of *Rhodoferax* grew under illumination at 10°C, producing few floating cells, but red globular clumps crept along the inside of the tube. Presumably, these red clumps form part of the liftoff mat at the bottom of the lake.

By contrast, the samples from the water column showed a predominance of Actinobacteria, a phylum known for producing many antibiotics. New species of actinobacteria could produce new antibiotics—perhaps forms that have never encountered human pathogens. Thus, the remote Antarctic ecosystem could provide unexpected benefits for medicine.

RESEARCH QUESTION

How do Antarctic lake microbes and meiofauna adapt to the seasonal extremes of oxygen and light, followed by freezing anaerobiosis and dark?

Wang, Zhong, Harrison Ho, Rob Egan, Shijie Yao, Dongwan Kang, et al. 2019. A new method for rapid genome classification, clustering, visualization, and novel taxa discovery from metagenome. *BioRxiv* doi: https://doi.org/10.1101/812917.

Slonczewski, Joan L., and Rachael Morgan-Kiss. 2015. Cyanobacterial communities of Antarctic Lake Fryxell lift-off mats and glacier meltwater. Award no. 1936, Genome Portal of the Department of Energy Joint Genome Institute.

species, even thousands of partners, as in animal digestive communities (see Section 21.4).

Mutualism Involves Partner Species That Require Each Other

In the most highly evolved forms of symbiosis, partner species evolve specific mechanisms of interdependence, and may require each other for survival. This coevolved interdependence is called **mutualism**. A striking case of microbial mutualism is the interdependence of the luminescent bacterium *Vibrio fischeri* (*Aliivibrio fischeri*) and its host squid, *Euprymna scolopes* (described in Section 10.4). Mutualism can involve two or more microbial partners. It can also involve one or more microbial partners with a plant or animal host. In some cases, both partners absolutely require each other; in other cases, one or the other is incapable of growing alone. The mutually beneficial relationship is maintained by numerous genetic responses that regulate each partner, avoiding damage to the other. Mutualisms such as nitrogen-fixing rhizobia within legumes can have enormous practical applications (discussed in Section 21.6).

A highly evolved form of mutualism is the **lichen** (**Fig. 21.15**). Lichens consist of an intimate symbiosis between a fungus and an alga or cyanobacterium—sometimes both. The symbiosis requires compatible partner species. The alga or bacterium provides photosynthetic nutrition, while the fungus provides minerals and protection. Lichens grow very slowly, but they tolerate extreme desiccation.

Lichens show a surprising variety of form. Different species may form a flat crust, branched filaments, or leaflike lobes. A cross section of the leaflike lichen *Lobaria pulmonaria* reveals a layer of algae (green) covered by fungal mycelium (white), which protects the algae from ultraviolet light damage (**Fig. 21.15C**). In addition to the algae, this lichen includes patches of cyanobacteria, which fix nitrogen. Thus, the fungus, algae, and cyanobacteria form a three-way mutualism. For dispersal, the lichen forms asexual clumps of algae wrapped in fungal mycelium. The clumps flake off and are carried by wind to new locations. In boreal (northern) forests, lichens cover the majority of the ground and provide food for grazing animals. Lichens are a winter food source for caribou, which dig beneath the snow to obtain them.

Note that fungi participate in widespread mutualistic interactions with the roots of plants, especially trees. Most trees are dependent on fungal partners called mycorrhizae. Mycorrhizae are presented in Section 21.6.

How can we tell when a microbe present in a multicellular animal behaves as a mutualistic partner? The mutualism can be documented by several kinds of evidence:

- **Removal of the microbial partner** leads to death or decreased growth of the host.
- **The microbial genome shows extensive degeneration (reduction)** of normally essential genes for metabolism and protective structures.
- **Isotope labeling** shows incorporation of products synthesized by one partner and used by the other.

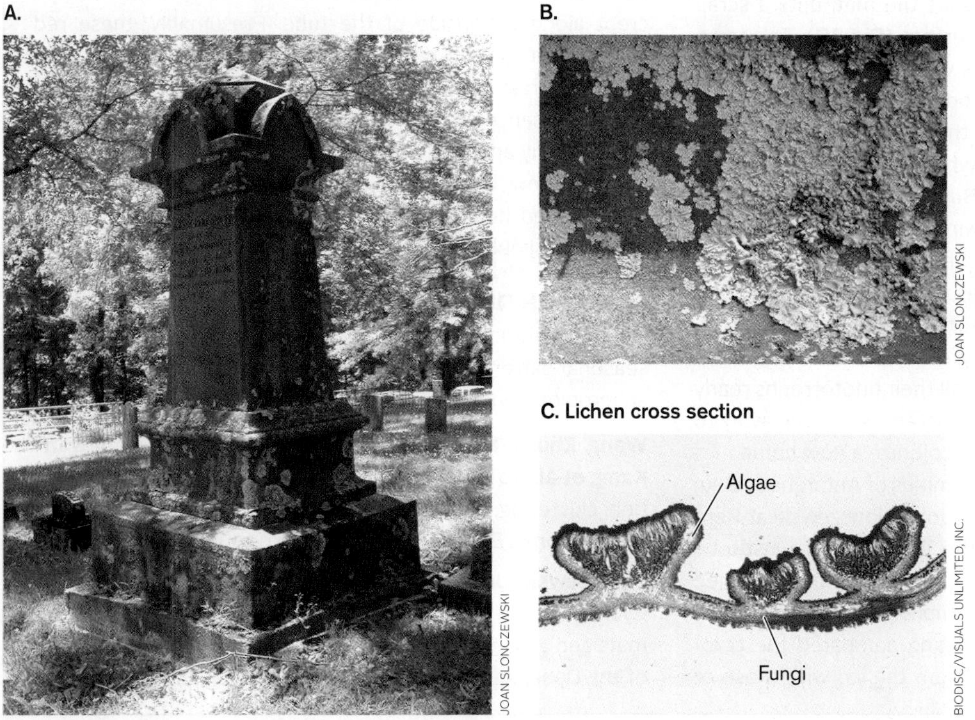

FIGURE 21.15 ■ Lichens. **A.** A tombstone at Kenyon College Cemetery in Ohio, encrusted with lichens (pale green) and mosses (dark green), a nonvascular plant. **B.** Close-up of the lichens in part (A). **C.** Section through a lichen (*Lobaria pulmonaria*) shows fungal and algal symbionts (stained LM).

Note, however, that genome reduction and partner product transfer also are traits of parasites. It can be surprisingly tricky to distinguish between a parasite and a mutualist.

Photosynthetic Endosymbionts of Animals

Photosynthetic animals were once thought to be the stuff of science fiction. Microbial ecologists, however, have found photosynthesis by elaborate mutualisms involving various kinds of protists, bacteria, and archaea with host invertebrate animals. Many invertebrate animals possess symbiotic bacteria and archaea that provide antimicrobial activity or defense against predation (discussed in the Current Research Highlight in Chapter 1, and in Chapters 18, 19, and 20).

A mutualism of major importance for marine ecosystems is that of corals with dinoflagellates, which is essential for the growth and sustenance of coral reefs (**Fig. 21.16A**). Besides corals, other cnidarians, such as jellyfish, anemones, and hydras, possess endosymbiotic algae. The most common algal partners are dinoflagellates of the genus *Symbiodinium* (discussed in Chapter 20). The algae receive protection from predators, while the animal receives photosynthetic products. Coral endosymbionts are extremely important to the biosphere because healthy coral is required for reef formation and much of the biological productivity of coastal shelf ecosystems. The unprecedented rise in temperature caused by human-made CO_2 emissions has already led to severe problems with **coral bleaching**, in which the algal symbionts die or are expelled. The coral turns white and soon dies, unless its symbionts return.

A given coral or anemone may harbor several different species of *Symbiodinium*, which show different preferences for light or shade and different tolerances for temperature change. Studies of coral bleaching due to temperature increase suggest that corals containing diverse species of symbionts are more likely to survive, because one of their species may happen to be resistant to a rise in temperature.

Photosynthetic mutualism was thought to be limited to invertebrates, but recently a vertebrate example was discovered: the colonization of salamander embryos (*Ambystoma maculatum*) by green algae (*Oophila amblystomatis*; **Fig. 21.16B**). These algae are chlorophytes, primary or "true" algae with chloroplasts similar to those of green plants (discussed in Chapter 20). Initially flagellated, they lose their flagella after they have invaded the salamander embryo. Within the embryo, the algae multiply to a limited population, producing oxygen and photosynthetic carbon products that are used by the growing embryo. Embryos lacking partner algae grow more slowly and show lower rates of survival.

Insects Possess Intracellular Bacteria

A remarkable form of endosymbiosis is the possession of intracellular bacteria by most species of insects (**Table 21.5**). In most insects the bacteria are inherited from mother to offspring, and their relationship is obligate: Bacteria and insect require specific functions from each other. The bacteria provide essential nutrients whose biosynthetic pathways the insect cells have lost through evolution, such as amino acids and vitamins. In turn, the insect cell provides numerous functions lost from the bacterial genomes by degenerative evolution. Endosymbionts may also defend the host insect from parasites and viruses.

A complex case is that of the tsetse fly, *Glossina morsitans*, the vector for the trypanosome of sleeping sickness (described in Chapter 20). The tsetse fly carries several

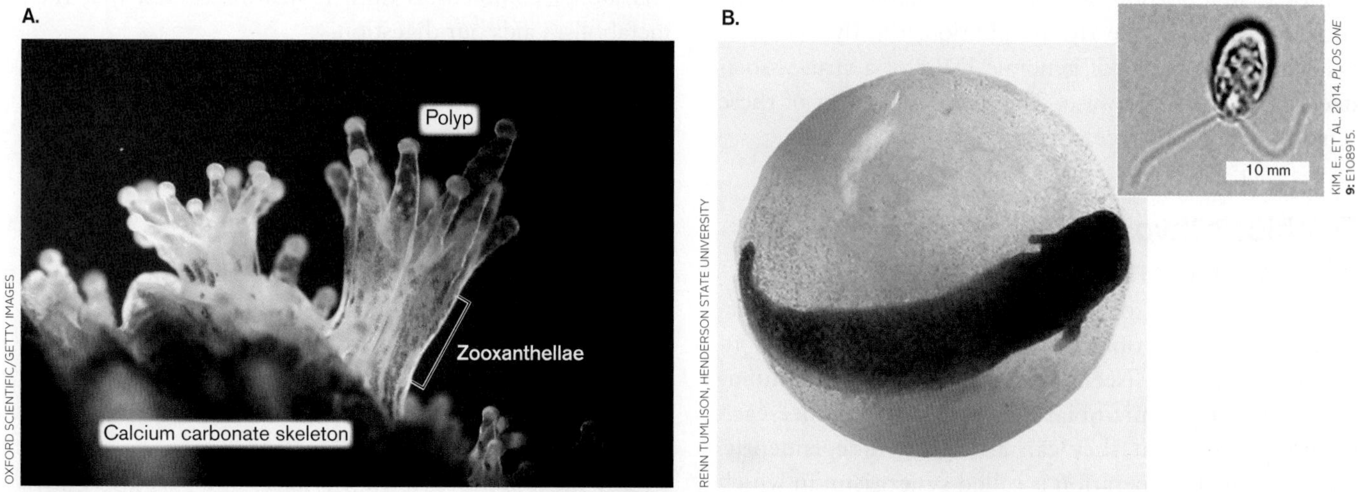

FIGURE 21.16 ▪ Animals harbor endosymbiotic algae. **A.** Coral polyps carry mutualistic dinoflagellates, of the genus *Symbiodinium*. **B.** Salamander embryo (*Ambystoma maculatum*) colonized by green algae (*Oophila amblystomatis*), which provide oxygen and fixed carbon, accelerating embryonic growth. **Inset:** *Oophila* algae prior to colonization.

TABLE 21.5 Intracellular Endosymbionts of Insects—Examples

Bacterial endosymbionts	Insect hosts	Relationship	Reference(s)
Wolbachia and *Spiroplasma* species	Fruit fly *Drosophila* species, ladybird beetles, and butterflies	Bacteria confer resistance to parasites and viruses; alter insect sex ratio, favoring females.	Julien Martinez et al. 2017. *Mol. Ecol.* **26**:4072. Tamara S. Haselkorn and John Jaenike. 2015. *Mol. Ecol.* **24**:3752.
Wigglesworthia, Sodalis, Wolbachia	Tsetse fly, *Glossina morsitans*	Bacteria provide essential vitamins to fly; provide resistance to bacterial and viral infection.	Jingwen Wang et al. 2013. *Front. Cell. Infect. Microbiol.* **3**:69.
Buchnera	Pea aphid, *Acyrthosiphon pisum*	Bacteria synthesize essential amino acids for aphid host.	Rebecca A. Chang and Nancy A. Moran. 2018. *ISME J.* **12**:898.
Blattabacterium	Cockroach, *Periplaneta americana*	Bacteria recycle nitrogen into essential amino acids and vitamins.	Zakee L. Sabree et al. 2009. *PNAS* **106**:19521.

kinds of intracellular endosymbionts (**Fig. 21.17**). *Wigglesworthia* is a gammaproteobacterium, an obligate endosymbiont with its genome decreased to only 700,000 bp. The bacteria are found in various parts of the fly but are concentrated in a specialized organ around the midgut, called the bacteriome. They produce B vitamins essential for fly larvae to develop. Another bacterium, *Wolbachia*, is related to the rickettsias, intracellular infectious agents of humans. *Wolbachia* resides in the fly germ cells, where it enhances the fertility of infected female embryos while decreasing the growth of uninfected embryos. *Sodalis* enterobacteria are considered secondary endosymbionts with a lesser relationship to the host. These bacteria may increase the tsetse fly's susceptibility to infection by the trypanosome.

An amazing feature of the tsetse fly's long evolutionary association with bacterial endosymbionts is the acquisition of a large portion of a *Wolbachia* genome within the nuclear genome of the fly. In addition, the fly's genome has acquired portions of genomic DNA of a virus associated with a parasitoid wasp. The functional effect of these genomic acquisitions remains unclear.

Symbiosis Involves Varying Degrees of Cooperation and Parasitism

Symbiosis between organisms involves a range of interdependence, from obligate mutualism (cooperation) to obligate parasitism (see **Table 21.4**). In a gut community, some of the microbial members may enhance each other's growth, but they can also grow independently. Their optional cooperation is called **synergism**, in which both species benefit but can grow independently and show less specific cell communication. For example, human colonic bacteria produce fermentation products that colonic methanogens metabolize to methane. The methanogens gain energy, and the fermenting bacteria benefit energetically through the steady-state removal of their end products. The bacteria found within marine sponges may offer an example of synergism. These bacteria fix carbon or secrete defense chemicals that protect the host sponge.

In other cases, one species derives benefit from another without return; for example, some wetland bacteria derive benefit from *Beggiatoa* because *Beggiatoa* bacteria oxidize H_2S, which inhibits growth of other species. An interaction that benefits one partner only is called **commensalism**. Commensalism is difficult to define in practice, since "commensal" microbes often provide a hidden benefit to their host. *Beggiatoa*, for example, requires a source of H_2S, probably produced by other community microbes. In the human gut, bacteria such as *Bacteroides* species were considered commensals until it was discovered that their metabolism aids our digestion.

An interaction that harms one partner nonspecifically, without an intimate symbiosis, is called **amensalism**. An example of amensalism is actinomycete production of antimicrobial peptides that kill surrounding bacteria. The dead bacterial components are then catabolized by the actinomycete.

Finally, **parasitism** is an intimate relationship in which one member (the parasite) benefits while harming a specific host. Many microbes have evolved specialized relationships as parasites, including intracellular parasitic bacteria such as the rickettsias, which cause diseases such as Rocky Mountain spotted fever.

The distinction between mutualist and parasite is often subtle. Lichens consist of a mutualistic association between fungus and algae, but environmental change can convert the fungus to a parasite. On the other hand, parasitic

Wolbachia
- Multiple natural *Glossina* populations harbor *Wolbachia* infections.
- Infections are localized to the ovarian germ cells and testes.
- Multiple horizontal transfer events have resulted in transfer of >1 Mb of the *Wolbachia* genome into the tsetse genome with unknown functional implications.
- Infection causes cytoplasmic incompatibility phenomena in tsetse, which gives infected females a reproductive advantage over uninfected flies.

Wigglesworthia
- Localized to bacteriome, milk gland tubules, and larval gut.
- Provides nutritional supplementation required for female fecundity and larval development, including B vitamins, which may be critical to supplement tsetse's diet.
- Provides immune stimulation required for larval immune system development.

Sodalis
- Found intra- and extracellularly throughout the fly.
- May contribute toward tsetse's susceptibility to trypanosomes.

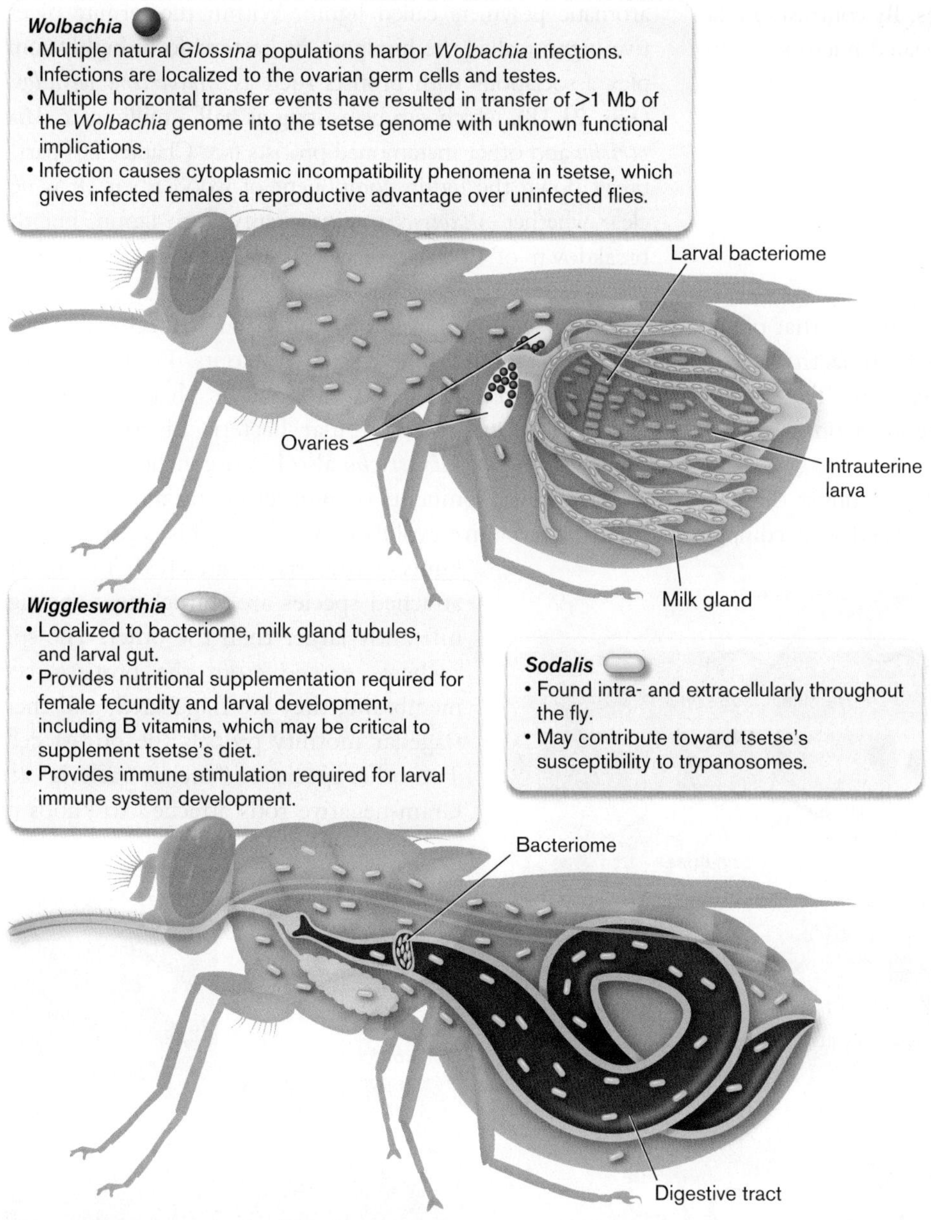

FIGURE 21.17 • Intracellular endosymbionts of the tsetse fly. *Wolbachia* infects germ cells and prevents male development, favoring females. *Wigglesworthia* provides vitamins and enhances female fertility. *Sodalis* may increase the tsetse fly's susceptibility to trypanosomes. Source: Modified from International Glossina Genome Initiative. 2014. *Science* **344**:380, fig. 2.

To Summarize

- **Symbiosis** is an intimate association between organisms of different species.
- **Mutualism** is a form of symbiosis in which each partner species benefits from the other. The relationship may be obligatory for growth of one or both partners.
- **Lichens** are a mutualistic community of algae and/or cyanobacteria with fungi. Lichens are essential producers for dry soil habitats.
- **Some animals harbor endosymbiotic algae that provide products of photosynthesis.** Corals and other cnidarians harbor mutualistic dinoflagellates. Embryos of one salamander species carry photosynthetic green algae.
- **Insects harbor intracellular obligate endosymbiotic bacteria with highly degenerate genomes.** Intracellular endosymbionts have evolved ancient relationships with their host, resulting in genome degradation and essential contributions to the host physiology.
- **Parasitism** is a form of symbiosis in which one species grows at the expense of another, usually much larger, host organism.
- **Interactions of multiple species can include both mutualism and parasitism.**

microbes may coevolve with a host to the point that each depends on the other for optimal health. For example, the high incidence of human allergies is proposed to correlate with lack of exposure to parasites that stimulate development of the immune system.

Multiple partner species can form a complex web of positive and negative dependence. An example is the leaf-cutter-ant symbiosis with fungi, which includes a fungal mutualist, a fungal parasite, and a bacterial mutualist that counteracts the parasite (presented in **eTopic 17.6**). Additional cases of complex multipartner systems are described in Sections 21.4 and 21.6, under microbial associations with animals and plants.

21.4 Animal Digestive Microbiomes

All animals have microbial communities on their surfaces or within organs that serve as digestive chambers, such as the bovine rumen or the human intestines. Many host-associated microbes have beneficial effects, such as enhancing digestion or generating protective substances in the skin. The beneficial properties are so essential that an animal is now considered a **holobiont**, an entity composed of multiple types of organisms, including microbes. By contrast, a relatively small proportion of animal-associated microbes cause disease. Human microbial interactions with the immune system are discussed in Chapter 23. Pathogenesis and disease in humans are discussed in Chapters 25 and 26.

The Termite Wood-Digesting Microbiome

A particularly complex metabolic mutualism is that of termites, whose digestive tract contains bacteria that catabolize wood polysaccharides such as cellulose. The termite feeds on wood and is completely dependent on its symbiotic bacteria and protists.

Wood particles ingested by the termite consist of cellulose and hemicellulose sugar chains entwined with complex aromatic polymers called lignin. Within the termite digestive organ (called the hindgut), its bacteria form highly complex associations with protists such as *Mixotricha paradoxa* (**Fig. 21.18**), which can be as long as half a millimeter. *Mixotricha* and other metamonad protists (see Chapter 20) partly break down the lignin component of wood fibers. It is not clear whether *Mixotricha* gains energy from lignin, but the breakdown of fibers makes cellulose available for bacterial catabolism.

M. paradoxa is covered with cilia and flagella, and possesses several kinds of bacterial symbionts. The protist takes up termite-ingested wood particles by phagocytosis, and then the protist's intracellular bacteria digest the wood polysaccharides. *Mixotricha* also has organelles that appear to be vestigial remnants of another endosymbiont, diminished by reductive evolution. On the protist's surface, four kinds of bacteria are attached. Two of the attached species are spirochetes, one significantly larger than the other. The spirochetes extend from the protist's cell membrane; they are flagellated, and their flagellar motility propels the protist cell. Two other species of "anchor bacteria" are Gram-negative rods attached to knobs of

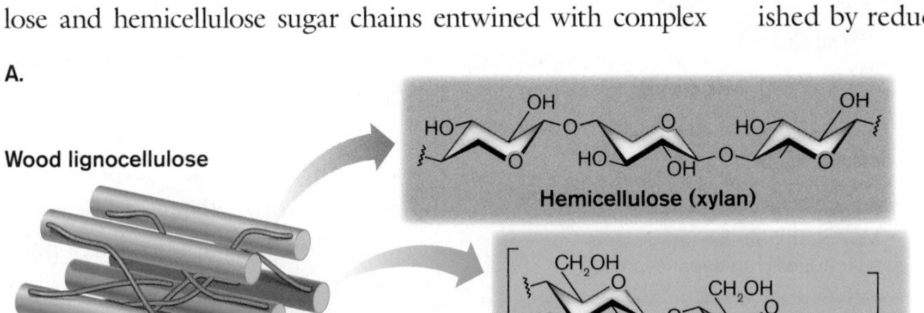

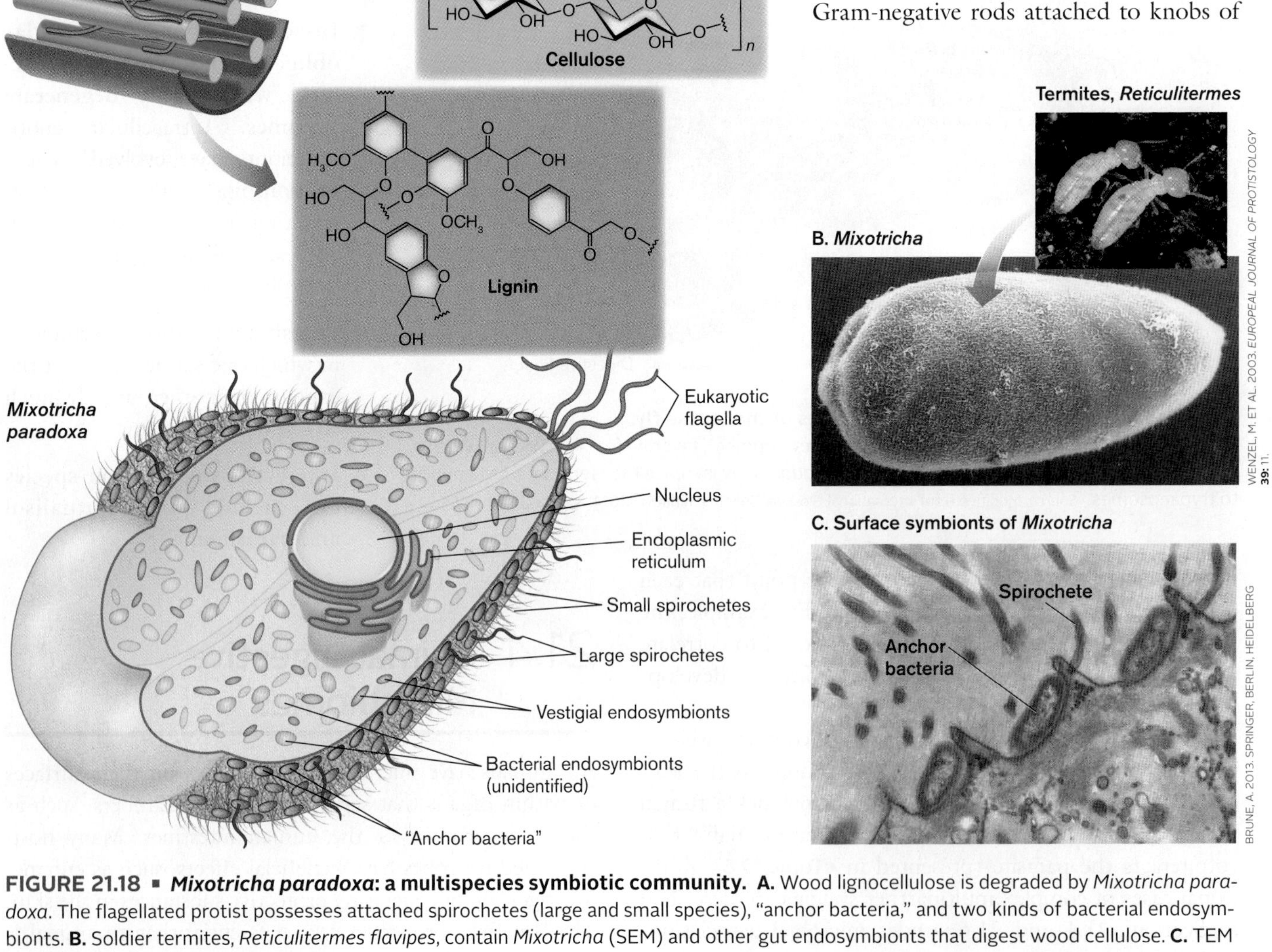

FIGURE 21.18 ■ *Mixotricha paradoxa*: a multispecies symbiotic community. **A.** Wood lignocellulose is degraded by *Mixotricha paradoxa*. The flagellated protist possesses attached spirochetes (large and small species), "anchor bacteria," and two kinds of bacterial endosymbionts. **B.** Soldier termites, *Reticulitermes flavipes*, contain *Mixotricha* (SEM) and other gut endosymbionts that digest wood cellulose. **C.** TEM section through *Mixotricha*'s pellicle, including anchor bacteria and attached spirochetes.

the protist surface. All members of the partnership have evolved an obligate relationship.

The relationship among microbial symbionts within the termite gut community generates a complex series of metabolic fluxes (**Fig. 21.19**). This kind of metabolic cooperation is called **syntrophy**, which means, "feeding together." For syntrophy, the fluxes within the community must balance energetically with a negative value of ΔG, as they would for a single free-living organism (discussed in Chapter 13). Most commonly, one species produces a substance that is consumed by the second species, which, if left to build up in high concentration, would result in an unfavorable ΔG for its continued production.

In the simplified model shown in **Figure 21.19**, the wood polysaccharides are hydrolyzed by bacteria and fermented to short-chain fatty acids, such as acetic acid, which is absorbed by the termite. (The termite's metabolism then oxidizes the acetate to CO_2.) Other products of the termite bacterial fermentation include CO_2 and H_2, which can be converted to methane by methanogenic archaea. In some termites the H_2 builds up to levels as high as 30%. The termite microbial mutualism is being studied as a model system for production of hydrogen biofuel.

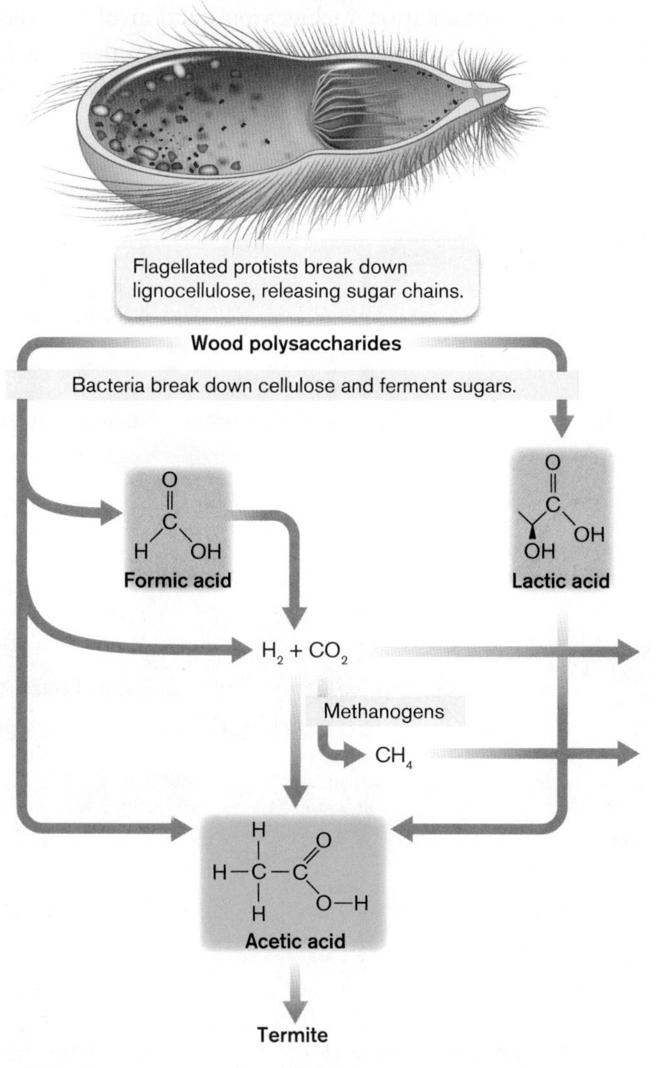

FIGURE 21.19 ■ Metabolic fluxes within the syntrophic community of the termite hindgut. Bacteria and their protist symbionts ferment wood polysaccharides to lactic acid, formic acid, acetic acid, H_2, and CO_2 within the hindgut of a termite, *Reticulitermes santonensis*. Some hydrogen is lost from the gut, some is converted to methane, and some is converted to acetic acid. Acetic acid is absorbed through the outer lining and feeds the termite.

The Bovine Rumen Fermenter

All vertebrate animals possess digestive microbiomes. The best-known microbiomes are those of ruminants, such as cattle, sheep, and caribou. Throughout most of human civilization, ruminants have provided protein-rich food, textile fibers, and mechanical work. A historical reference is the biblical injunction against consuming an animal that "is cleft-footed and chews the cud"—that is, "ruminates," or redigests its food in the fermentation chamber known as the **rumen** (**Fig. 21.20**).

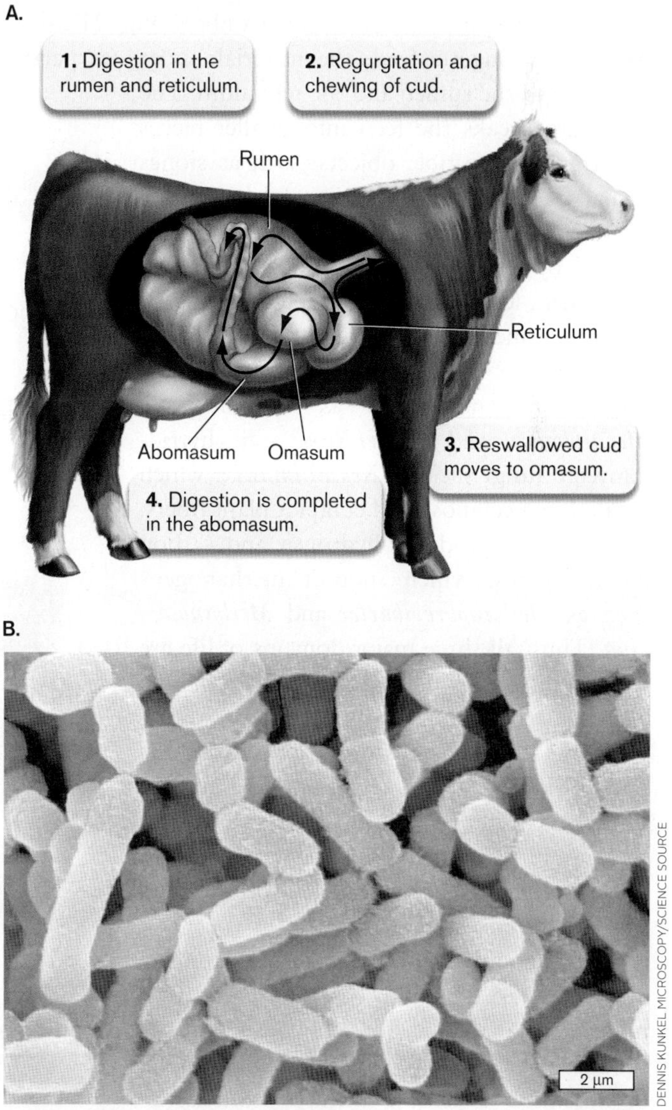

FIGURE 21.20 ■ The bovine rumen. A. The rumen is the largest of four chambers in the bovine stomach. **B.** *Prevotella ruminicola* bacteria from the rumen microbial community (SEM).

The microbial community of the rumen enables herbivores to acquire nutrition from complex plant fibers. From a genomic standpoint, such an arrangement makes evolutionary sense. If the animal had to digest all the diverse polysaccharide chains encountered in nature, its own genome would have to encode a wide array of different enzyme systems. Instead, the ruminant relies on diverse microbial species to conduct various kinds of digestion. As we saw for termites, microbes that partly digest a substrate provide short-chain fatty acids (SCFAs) that the host animal can absorb and digest to completion by aerobic respiration.

Thought Question

21.6 How does ruminant microbial fermentation provide food molecules that the animal host can use? How is the animal able to obtain nourishment from waste products that the microbes could not use?

The bovine gut system has four chambers (**Fig. 21.20A**). As the cow ingests feed, the material undergoes partial digestion in the rumen and the reticulum. The reticulum breaks the feed into smaller pieces and traps indigestible objects, such as stones or nails. After initial digestion, feed is regurgitated for rechewing and then returned to the rumen, by far the largest of the chambers. In the rumen, feed is broken down to small particles and fermented slowly by thousands of species of microbes. Rumen inhabitants include Firmicutes genera such as *Ruminococcus*, *Megasphaera*, and *Clostridium*, and chytridiomycete fungi such as *Neocallimastix*, which break down cellulose and complex plant fibers. Fermentation produces hydrogen and carbon dioxide gases, which support methanogens such as *Methanobrevibacter* and *Methanosarcina*. Thus, all three major domains of life are represented by the normal rumen microbiome.

The partially digested feed passes to the omasum, which absorbs water and short-chain acids produced by fermentation. The abomasum then decreases pH and secretes enzymes to digest proteins before sending its contents to the colon for further nutrient absorption and waste excretion.

In the twentieth century, Robert Hungate (1906–2004) at UC Davis pioneered techniques of anaerobic microbiology. One of Hungate's methods still in use today is that of obtaining anaerobic cultures from a **fistulated**, or **cannulated, cow**—that is, a cow in which an artificial connection is made between the rumen and the animal's exterior (see **Fig. 21.1C**). The cow is unharmed by the fistula and rumen sampling.

Metagenomics coupled with bioenergetic studies shows how different microbes fill different niches in ruminal metabolism (**Fig. 21.21**). Cattle grown on relatively poor forage (that is, forage high in complex plant content) show a high proportion of ruminal fungi, the chytridiomycetes (discussed in Chapter 20). Chytridiomycete mycelia appear on ruminal food particles, and their motile zoospores—formerly mistaken for protists—swim through rumen fluid. By contrast, cattle fed a high-cellulose diet, such as hay, grow faster and show cellulolytic bacteria such as *Ruminococcus albus* and *Fibrobacter flavefaciens.* The cellulolytic bacterial metabolism also requires the presence of *Megasphaera* and *Peptostreptococcus* species, which release small amounts of branched-chain fatty acids. Thus, while bacteria compete for food, they also share in a complex web of syntrophy.

Ruminal fermentation includes production of H_2 and CO_2, which support methanogens. From a farmer's point of view, methanogenesis wastes valuable carbon from feed

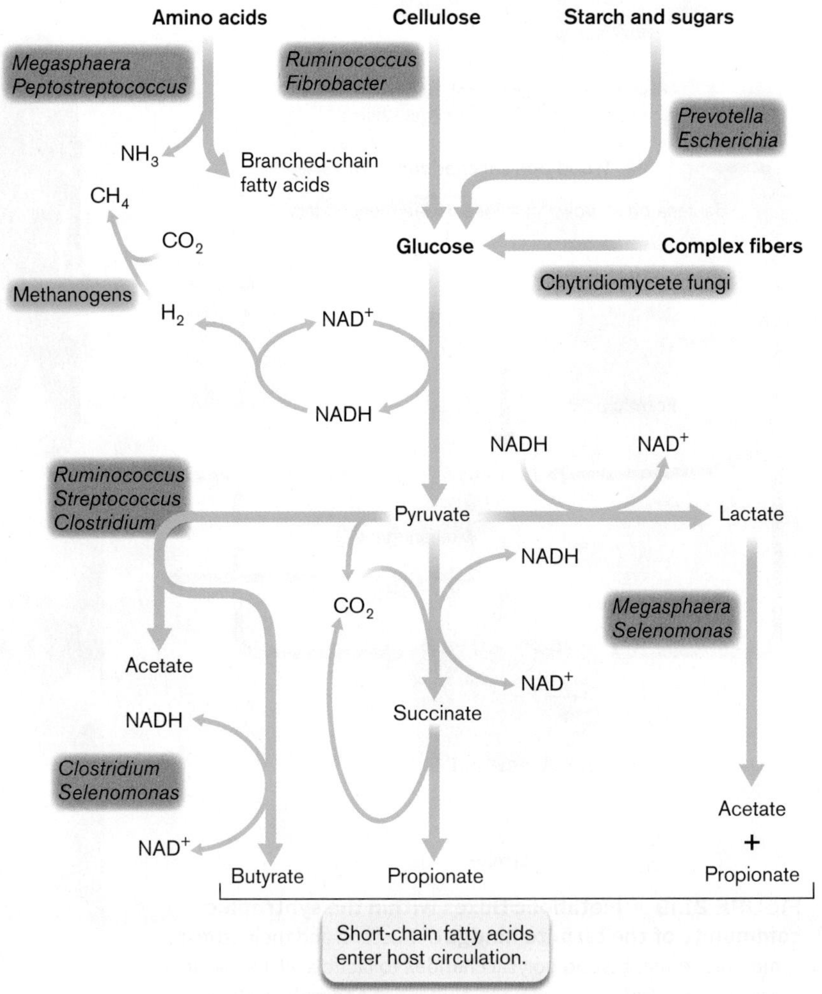

FIGURE 21.21 ■ **Ruminal metabolism.** Various microbes participate in digesting food, ultimately producing short-chain fatty acids that are absorbed by the bovine gut epithelium. *Source:* Modified from J. B. Russell and J. L. Rychlik. 2001. *Science* **292**:1119–1122.

and emits the greenhouse gas methane. So much methane forms that a cannula inserted into the rumen liberates enough of the gas to light a flame. Other vertebrate digestive tracts produce less methane than cattle do. For example, kangaroos have a shorter gut retention time, so there is less time for the action of slow-growing microbes such as methanogens. Switching from cattle to kangaroo meat has been proposed in Australia as a way to decrease greenhouse gas emissions.

Thought Question

21.7 How do you think cattle feed might be altered or supplemented to decrease methane production?

The Human Colon

In contrast to ruminants, humans (and other vertebrates) conduct much of their microbial fermentation at a much later stage of digestion, in the colon, which (unlike the rumen) resides near the end of the digestive tract (**Fig. 21.22**). Nonetheless, some prevalent genera of the rumen, such as *Ruminococcus* and *Clostridium*, are also represented in the human colon. The genomic coding capacity of human gut microbes may exceed that of the human genome by a factor of 100.

Colonic fermentation favors bacteria capable of digesting complex plant materials that pass undigested through the small intestine. For example, Bacteroidetes genera such as *Bacteroides* and *Prevotella* ferment mucopolysaccharides, pectin, and arabinogalactan, among many others. Because the colonic oxygen pressure is low, most bacterial digestion is fermentative, releasing acetate and other short chain fatty acids (SCFAs) that are absorbed by the intestinal epithelium, thus providing up to 15% of our caloric intake. Amazingly, our bodies also use amino acids synthesized by gut bacteria.

Human gut taxa. Human enteric metagenomes reveal an extraordinary range of bacterial and archaeal taxa, which

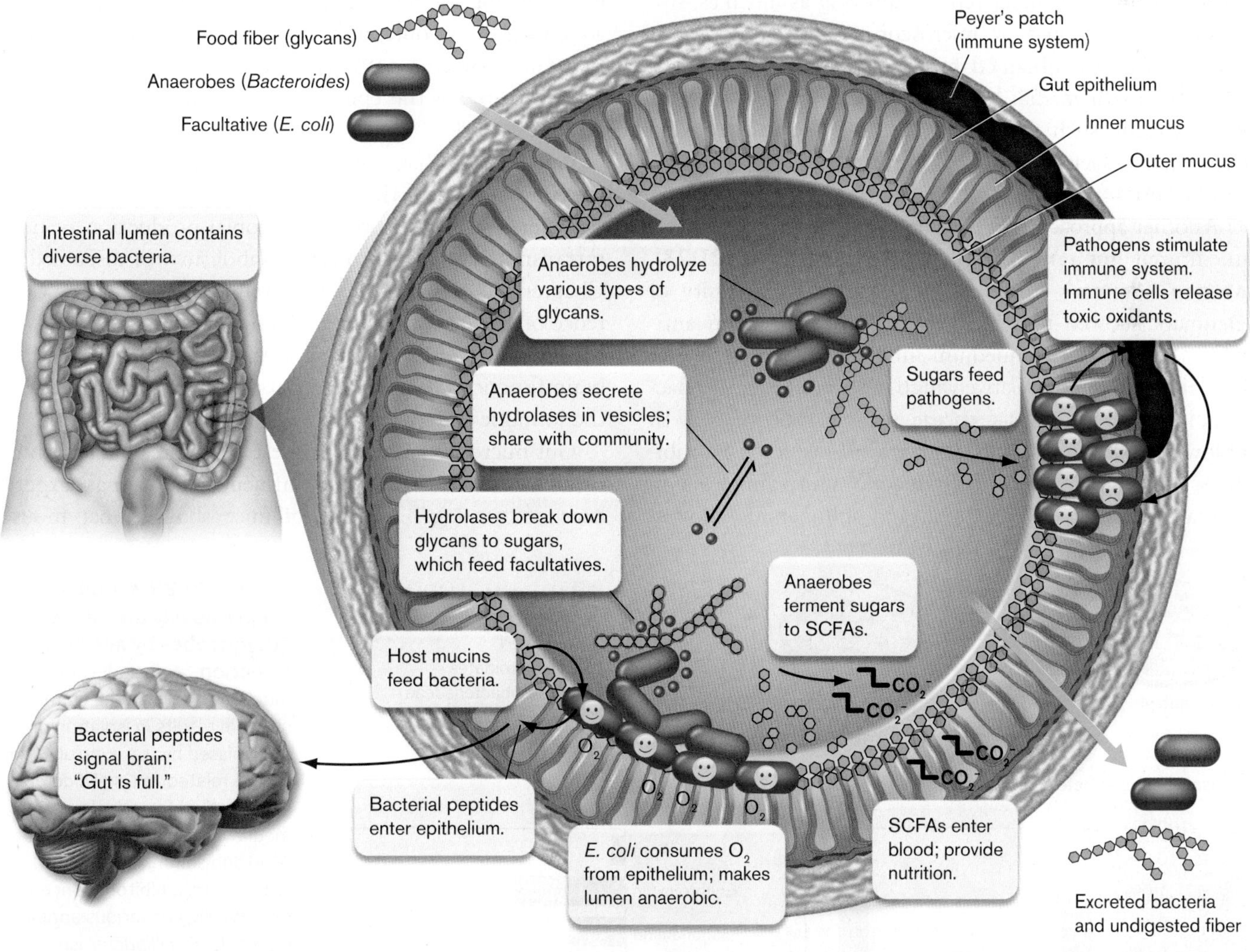

FIGURE 21.22 ■ Human gut microbiome digests our food. In a "restaurant" mixed-species biofilm, *Bacteroides* bacteria break down glycans into sugars that *E. coli* catabolizes, consuming oxygen from the host blood supply. The sugars may also feed pathogens, which stimulate an immune response.

vary by individual, diet, age, and health conditions. Besides *Bacteroides*, other major taxa include Firmicutes such as *Clostridium* and *Lactococcus*, Actinobacteria such as *Bifidobacterium*, and Verrucomicrobia such as *Akkermansia* (bacterial diversity is surveyed in Chapter 18). The ratio of Firmicutes to Bacteroidetes evolves over the human lifetime, with Firmicutes predominating in adults while Bacteroidetes dominate in infants and the elderly.

As in the bovine rumen, diverse colonic bacteria both compete with each other and collaborate through syntrophy. *Bacteroides* species release outer membrane vesicles of hydrolases that break down various types of glycans (polysaccharides) to release oligosaccharides (short-chain sugar polymers). These oligosaccharides may be picked up by other species of *Bacteroides* and *Bifidobacterium* that break them down further, to SCFAs. Some fermentation products support methanogens such as *Methanobrevibacter smithii*.

In assessing the functional importance of various taxa, a major challenge is the lack of cultured isolates. Until recently, few gut organisms could be cultured in the laboratory; but recent studies report culturing as much as 40% of microbes counted by microscopy. Culture of various isolates has been obtained by inclusion of specific nutrients in growth media; by replacing agar with a different gelling agent, such as the acetylated polysaccharide gellan (trade name Gelrite); and by culture under strict anoxic conditions.

Another approach that has enabled culture of previously uncultured gut taxa is antibiotic selection (**Fig. 21.23**). Morten Sommer's group at the Technical University of Denmark showed that inclusion of broad-spectrum antibiotics in the culture medium inhibits antibiotic-sensitive organisms present in relatively high proportion, such as Bacteroidetes and Enterobacteriaceae, while permitting growth of low-abundance taxa with higher tolerance for the antibiotics. For example, culture of gut community samples with media containing erythromycin and sulfonamide allowed the growth of *Oscillibacter*, a bacterium previously predicted from metagenome sequences. The graph in **Figure 21.23B** shows the percentage of colonies from a gut sample that were shown to be *Oscillibacter*, depending on the antibiotics included in the medium. *Oscillibacter* is of particular interest as an organism that may be associated with avoidance of Crohn's disease. Such health-associated organisms are cited on the Human Microbiome Project's (HMP) "Most Wanted" list—that is, metagenome-predicted organisms most desired for isolation and culture.

Gut bacteria-host interactions. A problem for the colon is the relatively short retention time of the human digestive tract. To remain in the colon with continual resupply of nutrients, *Bacteroides* and other anaerobes may form mixed biofilms with *E. coli* that adhere to the epithelium in the outer mucus layer (**Fig. 21.22**). The mucus layer turns over every 2 hours, but *E. coli* growth can outpace the rate of shedding. The mixed biofilm benefits both bacterial partners, as *E. coli* gains access to sugars while reducing oxygen that leaks in from the epithelium, sustaining the anoxic environment for fermenters. The mixed biofilm has the added benefit of outcompeting pathogens. When competition fails, however, pathogens can get through the mucus layer and invade the epithelium, stimulating the immune system.

Gut bacteria have many astonishing interactions with the human host, to the point where the gut microbiome is considered a human organ. Microbial production of neurotransmitters via amino acid catabolism was discussed in Chapter 13. Researchers are actively investigating the existence of a "gut-brain axis" whereby microbial peptides and neurotransmitters modulate anxiety, hunger, and other brain functions. Gut bacteria participate in development of the immune system, as discussed in Chapter 23.

Gut microbiomes even help establish circadian rhythms in mammals, and they may influence the human experience of "jet lag." A study of mice showed that mouse

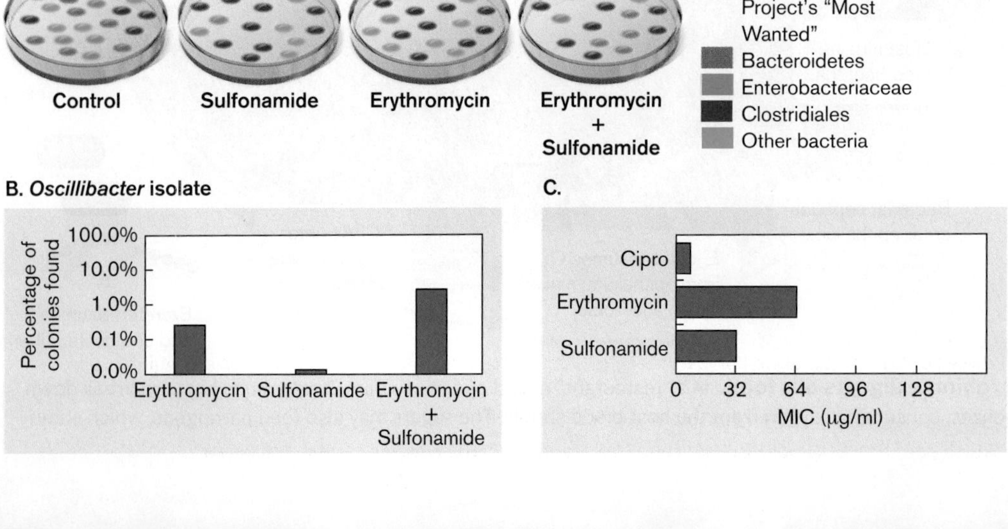

FIGURE 21.23 ■ Cultivating previously uncultured gut microbes by antibiotic selection. A. Including one or more antibiotics in the culture medium results in isolation of taxa missed by the control. **B.** An isolate related to *Oscillibacter ruminantium* forms colonies on media containing erythromycin (Ery) and sulfonamide (Sulfa). **C.** Minimum inhibitory concentration (MIC) of various antibiotics for the *Oscillibacter* isolate. *Source:* Modified from Rettedal et al. 2014. *Nat. Commun.* **5**:4714, figs. 3a (part A) and 3e (parts B and C).

microbial populations vary over the daily cycle, and that restoring normal gut populations can overcome a genetic defect in the molecular clock that governs animal activity. In humans, jet lag induces abnormal fluctuations in the gut microbiome. Perhaps someday we will develop probiotic therapies for jet lag. Other kinds of microbial circadian rhythms are discussed in Chapter 10.

Thought Question

21.8 How could you design an experiment to test the hypothesis that an animal makes use of amino acids synthesized by its gut bacteria?

To Summarize

- **Animals harbor digestive microbial communities.** The microbes possess numerous digestive enzymes absent in the host genome.
- **Termite gut mutualists** include protists that break down lignin, as well as bacteria that catabolize cellulose. A major by-product is hydrogen gas.
- **Syntrophy** is a metabolic association between (at least) two species, requiring both partners in order to complete the metabolism with a negative value of ΔG.
- **The rumen of ruminant animals is a complex microbial digestive chamber.** Rumen microbes, including bacteria, protists, and fungi, digest complex plant materials. The microbial digestion generates short-chain fatty acids that are absorbed by the intestinal epithelium.
- **The human gut microbiome contributes to our digestion.** Anaerobes such as *Bacteroides* and facultative respirers such as *E. coli* form mixed biofilms.
- **Modified culture methods reveal important members of the gut microbiome.**
- **Commensal members of the gut microbiome normally outcompete pathogens.** Pathogenic bacteria can invade the inner mucus layer and harm the gut epithelium.
- **Human gut microbes communicate with their host.** Commensal bacteria send chemical signals to our immune tissues and to our brain.

21.5 Marine and Freshwater Microbes

What microbial communities inhabit the oceans? Oceans cover more than two-thirds of Earth's surface, reaching depths of several kilometers and forming an immense habitat. Both oceans and freshwater support huge quantities of bacteria and algae, which drive vast ecosystems. Marine microbes contribute half the planet's productivity (production of biomass). Marine and freshwater microbes form the base of the food chain for seafood, with enormous impact on humans.

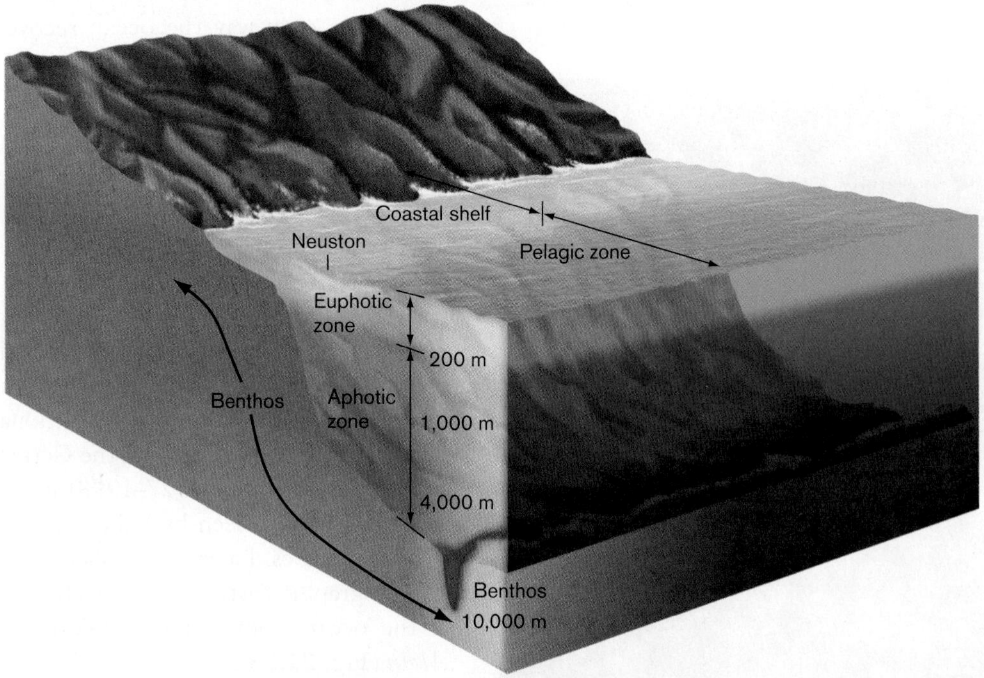

FIGURE 21.24 ▪ Regions of marine habitat. The marine habitat subdivides into several categories. The coastal shelf region is defined as the water extending from the shoreline out to a depth of 200 meters. The pelagic zone (open ocean) includes several depth regions: the neuston, the microscopic interface between water and air; the euphotic zone of light penetration, where phototrophs can grow, down to 100–200 meters; the aphotic zone, which supports only heterotrophs and lithotrophs; and the benthos, at the ocean floor.

Marine Habitats

Marine water has a salt concentration averaging 3.5% by weight. The major ions are Na^+ and Cl^-, with substantial amounts of sulfate and iodide. The salt concentration is high enough to prevent growth of many aquatic and terrestrial bacteria, such as *E. coli*. Nevertheless, salt-tolerant organisms such as *Vibrio cholerae* grow well over a broad range of salt concentrations.

The ocean varies considerably with respect to temperature, pressure, light penetration, and concentration of organic matter. In the open ocean (known as the **pelagic zone**), the water column is subdivided into distinct regions (see **Fig. 21.24**):

- **Neuston (about 10 μm).** The neuston is the air-water interface. Although extremely thin, the neuston layer contains the highest concentration of microbes. Many algae and protists have evolved so as to "hang" from the layer of surface tension that forms at the air-water interface.
- **Euphotic zone (100–200 m).** The **euphotic zone**, or **photic zone**, is the upper part of the water column, which receives light for phototrophs. In the open ocean, the euphotic zone extends down a couple hundred meters, whereas at the **coastal shelf** (<200 meters to the ocean floor), a higher concentration of silt and organisms decreases the photic zone to as little as 1 meter.
- **Aphotic zone.** Below the reach of sunlight, in the aphotic zone, only heterotrophs and lithotrophs can grow.
- **Benthos.** The benthos includes the region where the water column meets the ocean floor, as well as sediment below the surface. Organisms that live in the benthos, such as those of thermal vent communities, are called **benthic organisms**.

FIGURE 21.25 ■ Holger Jannasch discovered unculturable marine bacteria. The submersible *Alvin* used by researchers from WHOI for sampling deep marine organisms.

Another important determinant of marine habitat is the **thermocline**, a depth at which temperature decreases steeply and water density increases. A thermocline typically exists in an unmixed region. At the thermocline, a population of heterotrophs will peak, feeding on organic matter that settles from above. An experiment showing the thermocline, conducted by undergraduates, is described in **eTopic 21.1**.

Coastal regions show the highest concentration of nutrients and living organisms, and the least light penetration. By contrast, the open ocean is largely oligotrophic (having an extremely low concentration of nutrients and organisms). The concentration of heterotrophic microorganisms determines the **biochemical oxygen demand** (**BOD**; also called **biological oxygen demand**), the amount of oxygen removed from the water by aerobic respiration. Normally, the open ocean has such a low concentration of organisms that the BOD is extremely low; therefore, the dissolved oxygen content is high. This explains why enough oxygen reaches the ocean floor to serve chemolithoautotrophs such as sulfide-oxidizing bacteria. The BOD rises, however, when excess sewage or petroleum is present, such as that spilled by the *Deepwater Horizon* oil rig in 2010. Microbial consumption of these wastes at first deprives fish of oxygen, but the process is the only way the ocean recovers (see **Figs. 21.6** and **21.7**).

Marine Ecosystems: Where Are the Microbes?

To nineteenth-century microbiologists, the oceans appeared virtually free of bacteria. Trained in the tradition of Robert Koch (discussed in Chapter 1), microbiologists attempted to isolate marine bacteria by plate culture, considered the definitive way to study a microbial species. But the colonies that grew on traditional plate media were extremely few. Then, in 1959, the German microbial ecologist Holger Jannasch (1927–1998) showed that many more bacteria could be seen by light microscopy than could be grown on plates. Later, with colleagues at the Woods Hole Oceanographic Institution (WHOI), Jannasch studied life at the ocean floor, using the famous submersible vessel *Alvin* (**Fig. 21.25**).

Today, emerging culture methods enable many uncultured organisms to grow in the laboratory. Often, microbial growth requires hidden synergy or mutualism with other organisms in the natural environment. For example, the tiny cyanobacterium *Prochlorococcus* is the ocean's most abundant oxygenic phototroph, accounting for half the ocean's photosynthesis. The abundance map in **Figure 21.26**, obtained from marine samples worldwide, shows that in most regions, *Prochlorococcus* outnumbers its relative *Synechococcus* by tenfold; the dominance of the tiny phototroph is due in part to its unusual chlorophylls, which absorb more of the blue light from the solar spectrum. Nevertheless, *Prochlorococcus* species are very difficult to grow in pure culture.

Jeff Morris (now at the University of Alabama at Birmingham) and colleagues at the University of Tennessee, Knoxville, showed that culturing *Prochlorococcus* in the laboratory requires the presence of a "helper bacterium" that catalyzes the breakdown of hydrogen peroxide, a byproduct of oxygenic photosynthesis. The helpers produce catalase, which *Prochlorococcus* has lost through degenerative evolution, enabling more efficient growth when heterotrophs are present. Thus, in the ocean—as in gut microbiomes—microbes require surprising synergisms and symbioses, in the midst of competition. Synergism, rather than single-species growth, may be the more common condition for microbes in nature.

The complexity of marine microbial ecosystems includes surprising connections with human medicine. A pioneering investigator of marine-medical connections is Rita Colwell at the University of Maryland (**Fig. 21.27**; interviewed in **eTopic 1.1**) Colwell and her Bangladeshi associate Anwar Huq showed that the human cholera pathogen *Vibrio cholerae* is actually a mutualist of marine copepods (**Fig. 21.28**).

In natural water systems, most *V. cholerae* cells do not swim freely but colonize the surfaces of copepods (**Fig. 21.28A**). The copepods actually depend on these bacteria to eat through the chitin of their egg cases, releasing their young. Bacterial mutualists of copepods are virulent pathogens of humans, whose cholera diarrhea returns the organism to the water full of copepods. To decrease the incidence of cholera, Colwell and Huq showed that drinking water contaminated by *V. cholerae* could be partly decontaminated by filtering out the copepods through several layers of sari cloth (**Fig. 21.28B**). Sari cloth filtration is now a common practice in Bangladesh. Thus, the elucidation of this complex microbial partnership reaped benefits for people threatened by cholera.

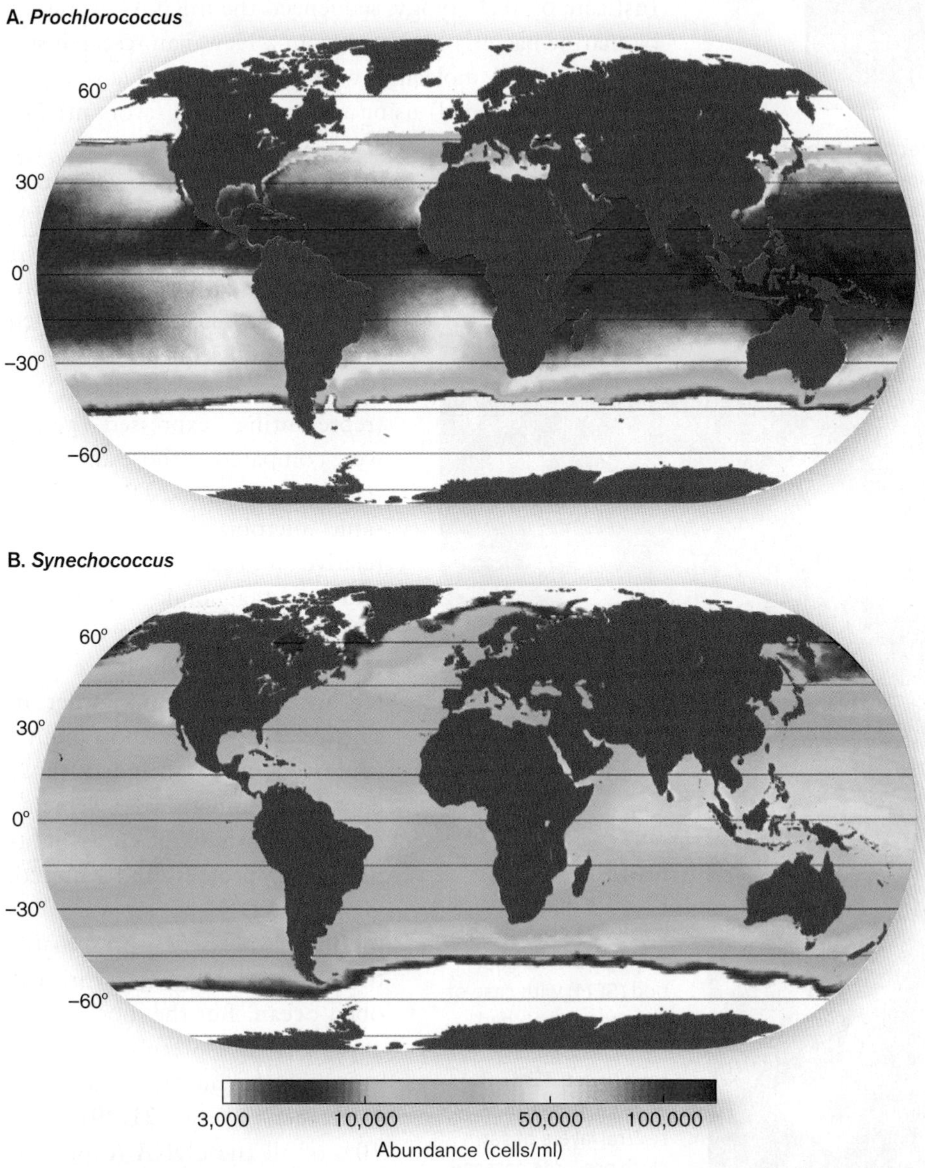

FIGURE 21.26 ■ Global distribution of marine phototrophs at the sea surface. A. *Prochlorococcus* abundance peaks in warmer water. **B.** *Synechococcus* is less abundant overall but reaches colder regions. *Source:* Pedro Flombaum et al. 2013. *PNAS* **110**:9824.

Marine Metagenomes

Marine microbial communities are now emerging via metagenomes. An example featured in Chapter 7 is Shinichi Sunagawa's international study that sequenced seven terabases (7×10^{12} bp), including samples of all the world's oceans. Another interesting study, conducted in 2016 by Jessica Bryant, Edward DeLong, and colleagues at

FIGURE 21.27 ▪ Rita Colwell, former director of the National Science Foundation. Colwell conducts pioneering studies of the complex marine ecology of *Vibrio cholerae*.

the University of Hawaii at Manōa, focused on microbial communities associated with plastic debris in the North Pacific Ocean, known as the "great Pacific garbage patch." The metagenomes showed a surprisingly consistent presence of Cyanobacteria and Alphaproteobacteria, as well as microscopic eukaryotes such as bryozoan invertebrates. The bacterial genomes showed a high prevalence of genes for chemotaxis, secretion systems, and nitrogen fixation. These findings will help us understand the fate of human wastes that flow into and degrade marine ecosystems.

As discussed in Section 21.1, even metagenomes miss community members that fill important niches. To find rare organisms with functional importance, researchers set out to explore expressed genes—a "metatranscriptome." Sallie Chisholm and Ed DeLong, then at the Massachusetts Institute of Technology, sequenced the mRNAs expressed by marine bacteria obtained from a Hawaiian research station. The RNA molecules were polyadenylated (a string of adenines was added) using an enzyme that favors mRNA and excludes ribosomal RNAs because of their high degree of secondary structure (intramolecular folding). The resulting RNA pool, enriched for expressed mRNA, was reverse-transcribed to DNA (called cDNA for "complementary DNA"). The cDNA was then amplified and sequenced.

The sequences of the cDNA (representing expressed genes) were compared to those amplified from metagenomic DNA of the same microbial community. The relative abundance of expressed and metagenomic sequences gives a measure of expression levels of all genes, which are plotted in declining rank order in **Figure 21.29**. The most highly expressed genes included those encoding functional elements of photosynthesis, such as light-harvesting proteins and Rubisco. Genes for DNA repair were also highly expressed, presumably to correct UV damage in the open ocean. But the most highly expressed genes were some of the rarest in the metagenomes (colored red in **Fig. 21.29**). And 40% of all the cDNA (expressed genes) had no counterpart in the genomic DNA (not shown in the figure). Thus, expression analysis

A.

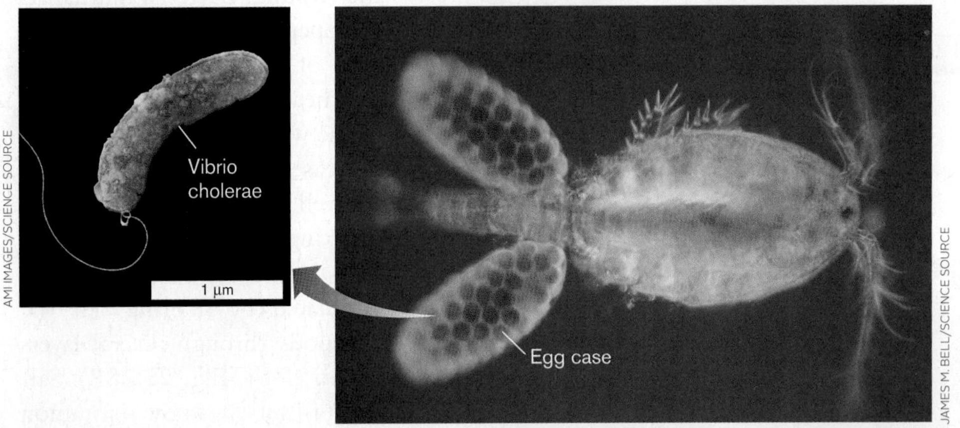

B.

FIGURE 21.28 ▪ *Vibrio cholerae* colonizes copepods. A. Copepod (SEM) with case of eggs to be "hatched" by *V. cholerae* bacteria (inset; TEM). **B.** Filtering water through several layers of sari cloth prevents passage of contaminated copepods and thus prevents transmission of cholera.

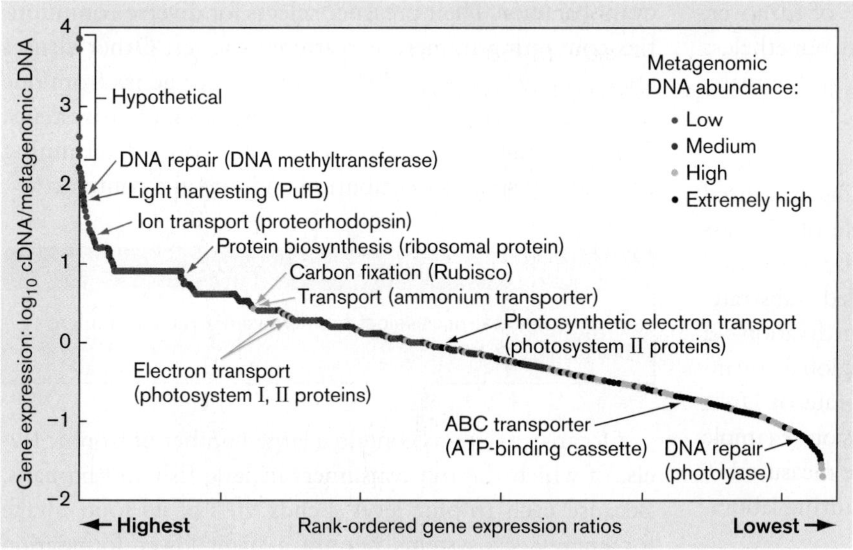

FIGURE 21.29 ▪ Genes expressed in a marine microbial community. Expression of each gene is measured as RNA (copied to cDNA), divided by the gene's relative abundance in the metagenome. The *y*-axis represents the ratio of cDNA to metagenomic DNA for all expressed genes, plotted in rank order (from highest to lowest). The relative abundance of the DNA sequence within the metagenome is color-coded. The genes most highly expressed tend to be rarest in the genome (colored red).

does indeed reveal genetic sequences of organisms undetected in the metagenome, presumably too rare for detection by our current sequencing methods. It also suggests that vast ranges of organisms may remain unreached by either genome or transcriptome sequencing.

Beyond DNA, microbial ecology addresses organisms in their relationships with their habitat and with each other. We will now introduce approaches to measuring microbial populations and their interactions within marine communities. These approaches have enormous practical applications, from the management of fisheries to the assessment of greenhouse gas fluxes (a topic pursued in Chapter 22).

Measuring Planktonic Communities

In marine science, the term **plankton** refers to organisms that float passively in water. Microbiologists use the term loosely, as microbial "plankton" include motile bacteria and protists. Marine phytoplankton (microbial phototrophs) produce a substantial part of the world's oxygen and consume much of the atmospheric CO_2.

Microbial plankton, or **microplankton**, include numerous members of the three domains discussed in Chapters 18, 19, and 20: Bacteria, Archaea, and Eukarya, respectively. Nanoplankton (about 2–20 μm) arbitrarily include smaller algae and flagellated protists, as well as filamentous cyanobacteria. Picoplankton (about 0.2–2 μm) consist of bacteria and the smaller eukaryotes, including the smallest living cells known. Besides these size classes of cells, the term "femtoplankton" may refer to marine viruses, the smallest detectable particles capable of reproduction. Note, however, that the use of these prefixes ("nano-," "pico-," "femto-") is approximate and does not refer to size units such as nanometers.

Despite the definition of "plankton," not all marine microbes float independently. Many form biofilms on colonial algae such as kelps, or on suspended inorganic particles known as **marine snow**. As many as half of all marine bacteria are associated with particulate substrates from broken-down organisms. As observed by Farooq Azam at the Scripps Institution of Oceanography (see **Fig. 21.30C**): "Seawater is an organic matter continuum, a gel of tangled polymers with embedded strings, sheets, and bundles of fibrils and particles, including living organisms, as 'hot spots.'" Thus, while the ocean as a whole has a low average nutrient concentration, marine waters include a suspension of concentrated tangles of nutrient-rich substrates. Collectively, their global volume is so large that their sedimentation rate influences calculations of Earth's carbon cycle.

Population size can be estimated in different ways: the number of individual reproductive units, the total organic biomass available to consumers, or the rate of productivity or assimilation of key nutrients, such as carbon dioxide. All ways of measuring these quantities present challenges; there is no single right method, but different approaches answer different questions.

Fluorescence microscopy. Marine microbes of all sizes, even viruses, can be detected and counted under the microscope using a DNA-intercalating fluorescent dye. Recall that fluorescence enables detection even of particles whose size is below the resolution limit defined by the wavelength of light (discussed in Chapter 2). Fluorescence microscopy with a DNA-binding fluorophore such as DAPI is used to detect planktonic microbes. Specific taxa can be detected by FISH (see **Fig. 21.10**). FISH uses a fluorophore attached to a DNA probe, a short sequence that can hybridize only to DNA or RNA of a particular taxon.

Biomass. The amount of biomass can be determined by standard chemical assays of protein and other forms of organic matter. Net biomass of a population, however, does not indicate productivity within an ecosystem, because it misses the amount of carbon cycled through respiration.

Marine microbes have an extremely rapid rate of turnover, but what appears to be a small population may nonetheless conduct tremendous rates of carbon fixation into biomass, which is rapidly consumed by the next trophic level.

Incorporation of radiolabeled substrates. The measured amount of biomass does not indicate the rate of biomass production. The rate of production can be estimated by the cells' incorporation of a radiolabeled substrate. Uptake of $^{14}CO_2$ indicates the rate of carbon fixation—a highly important property for the study of global warming. Uptake of ^{14}C-thymidine measures the rate of DNA synthesis, an indicator of the rate of cell division (example shown in **eTopic 21.1**). These properties can be measured in seawater under native conditions, without requiring laboratory cultivation.

A limitation of these methods is that addition of the radiolabeled substrate may raise a nutrient concentration to artificially high levels that distort the naturally occurring rates of activity. In the case of labeled thymidine, another limitation is that not all growing cells incorporate exogenous thymidine into their DNA.

Planktonic Food Webs

Analysis of microbial food webs was pioneered by Farooq Azam, at the Scripps Institution of Oceanography, at the University of California-San Diego (**Fig. 21.30A**). A simplified outline of the food web in the open ocean is shown in **Figure 21.30B**. The diagram of a food web is often called a "spaghetti diagram" because so many trophic interactions cross each other in various directions.

Phytoplankton. The phototrophic producers of the open ocean are known as **phytoplankton**. The phytoplankton include cyanobacteria such as *Prochlorococcus*, *Synechococcus*, and *Trichodesmium*. Some cyanobacteria, such as *Trichodesmium*, fix nitrogen as well as carbon. The phytoplankton also include microbial eukaryotes such as algae, diatoms, and dinoflagellates. In the food web, bacteria and protists are consumed by larger protists, which feed small invertebrates, which in turn feed larger invertebrates and, ultimately, vertebrates such as fish.

All levels of microbial plankton undergo intense predation by viruses. The degree of viral predation is difficult to measure, but recent studies indicate that cell lysis by viruses breaks down about half of microbial biomass (discussed in Chapter 6). Virus particles represent a major sink for carbon and nitrogen. They accelerate the return of minerals to producers, a process called the "viral shunt." The viral shunt necessitates a larger base of producers to sustain the ecosystem. Some marine viruses are highly host specific, infecting only certain species of dinoflagellates or cyanobacteria. Their presence selects for diverse communities containing numerous scattered species. Other viruses attack many hosts—and they can transfer genes from one host to another, such as the genes encoding photosystems. Thus, marine viruses are a dominant force determining community species distribution and genome content.

Thought Question

21.9 How do viruses select for increased diversity of microbial plankton?

Ocean ecosystems contain a large number of trophic levels, of which the top consumers include fish and humans. Because each trophic level spends 90% of its food intake for energy, ecosystems require a huge lower foundation to sustain the highest-level consumers. This is one reason why fisheries worldwide are now in danger of running out of fish for human consumption: Fish are being harvested faster than the ecosystem can replace them.

Note that actual food webs are far more complex than this simplified model; see **Figure 21.30C**. One source of complexity is that many algal protists, such as chrysophytes (golden algae) and dinoflagellates, are actually **mixotrophs**, organisms that both fix CO_2 through photosynthesis and consume microbial prey. Mixotrophy offers the opportunity to grow at night, without light, and to acquire scarce minerals, such as iron, from prey. Mixotrophs include secondary endosymbiont algae (discussed in Chapters 17 and 20), such as kelp and sargassum weed, as well as protists containing cyanobacterial or algal endosymbionts.

Another complicating factor is the vast number of diverse taxa associated with marine animals, particularly actinobacteria that produce antimicrobial molecules (see Chapter 18). The microbiomes of marine animals provide a rich source of novel compounds for therapeutic and biotechnological applications.

The Ocean Floor

The ocean floor (benthos) experiences extreme pressure beneath several kilometers of water. Most organisms that live there are pressure-dependent species, known as **piezophiles** or **barophiles** (discussed in Chapter 5). Barophiles require high pressure for growth (200–1,000 atm), failing to grow when cultured at sea level. Cold temperatures (about 2°C) select for **psychrophiles** (cold-adapted species), whose rate of growth is relatively slow. In the absence of light, heterotrophic bacteria depend on detritus from above as their carbon source, but by the time any material reaches the benthos, much of the organic carbon has been depleted by prior consumers. Metagenomes of the benthos (deep ocean) show primarily Proteobacteria, Bacteroidetes, and

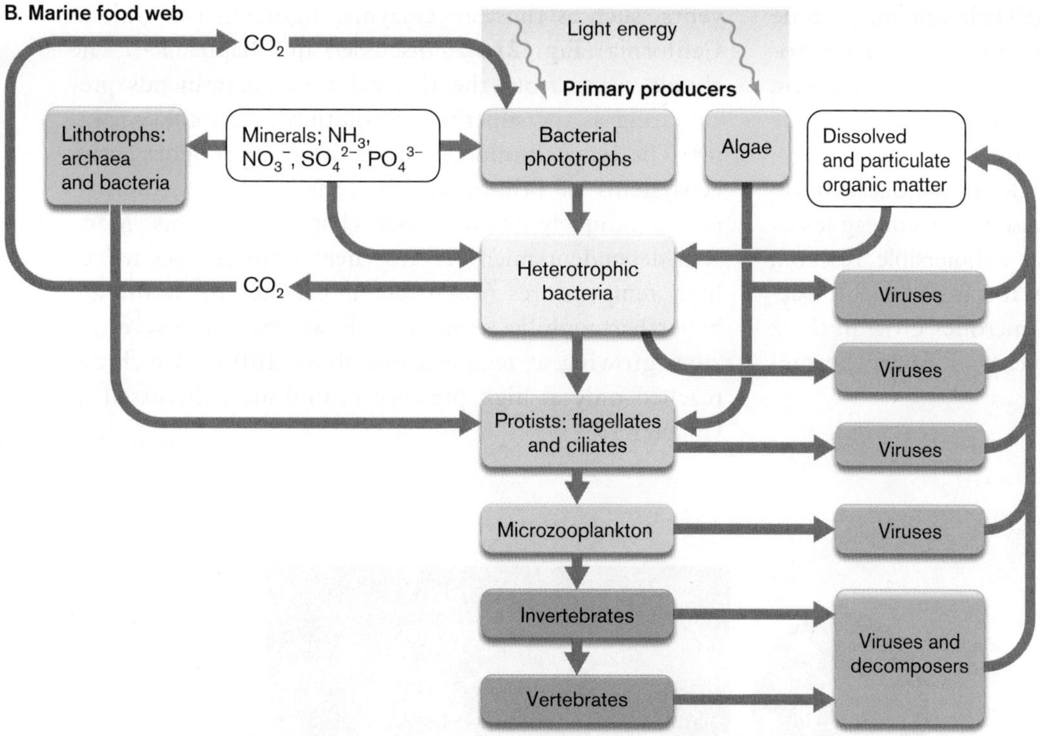

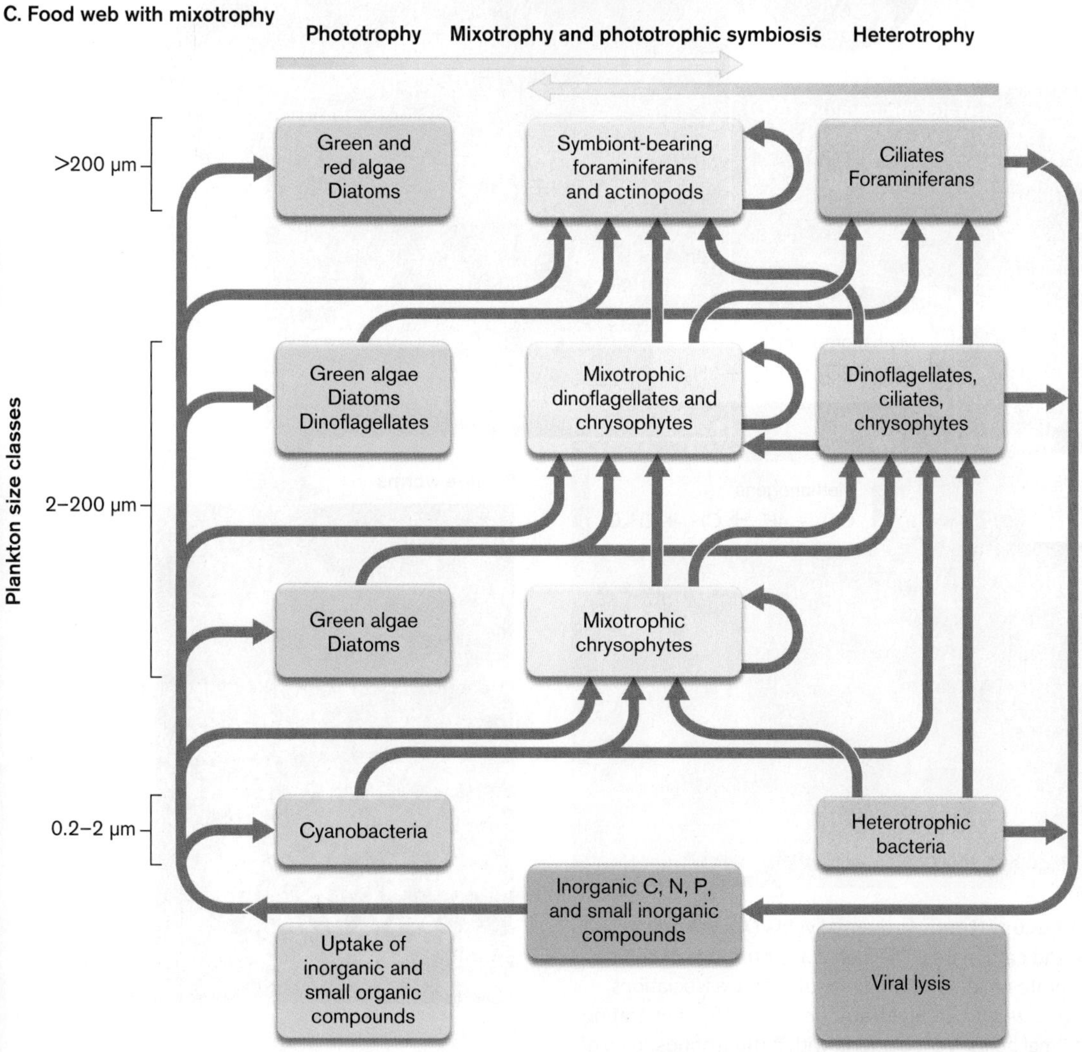

FIGURE 21.30 ■ The pelagic marine food web. A. Farooq Azam was elected to the American Academy of Arts and Sciences in 2016 for his fundamental contributions to the study of marine microbiomes. **B.** In the marine water column, the vast majority of carbon transfer occurs among microbes. Arrows lead from organisms consumed and point to their consumers. The primary producers are phototrophic bacteria and algae, with a smaller contribution from lithotrophic bacteria and archaea. Grazers include heterotrophic bacteria and protists. Predators include protist flagellates and ciliates. About 50% of bacterial and protist biomass is degraded by viral lysis. Multicellular organisms cycle a relatively small fraction of biomass. **C.** A more complex view of the marine food web includes mixotrophs (organisms that combine phototrophic and heterotrophic nutrition) and symbiotic associations between protists and phototrophic bacteria or algae.

Planctomycetes, as well as Archaea. Their genomes encode a large number of heavy-metal transport and resistance proteins, which help cells survive toxic metals upwelling from the ocean floor. These metal resistance genes may have practical applications for bioremediation.

Among the first to investigate life on the ocean floor, in the 1960s, were Holger Jannasch and colleagues at WHOI. Using the two-passenger submersible research vessel *Alvin*, they discovered new forms of benthic bacteria and archaea. Some of these microbes grow in thick biofilms at black smoker **thermal vents**, or **hydrothermal vents**, such as those of Guaymas Basin, in the Gulf of California (**Fig. 21.31**; discussed in Chapter 19). The clouds rising from the thermal vent are minerals precipitating as the superheated solution meets cold seawater. The reduced minerals from the vents support entire ecosystems of lithotrophic bacteria, including mutualists of uniquely evolved giant clams and worms. Many vent-dependent microbes are **thermophiles**, adapted to high temperatures (discussed in Chapter 5). Many are hyperthermophilic archaea such as *Pyrodictium occultum*, growing at temperatures above 100°C, which are reached only at high pressure (1,000 atm; discussed in Chapter 19).

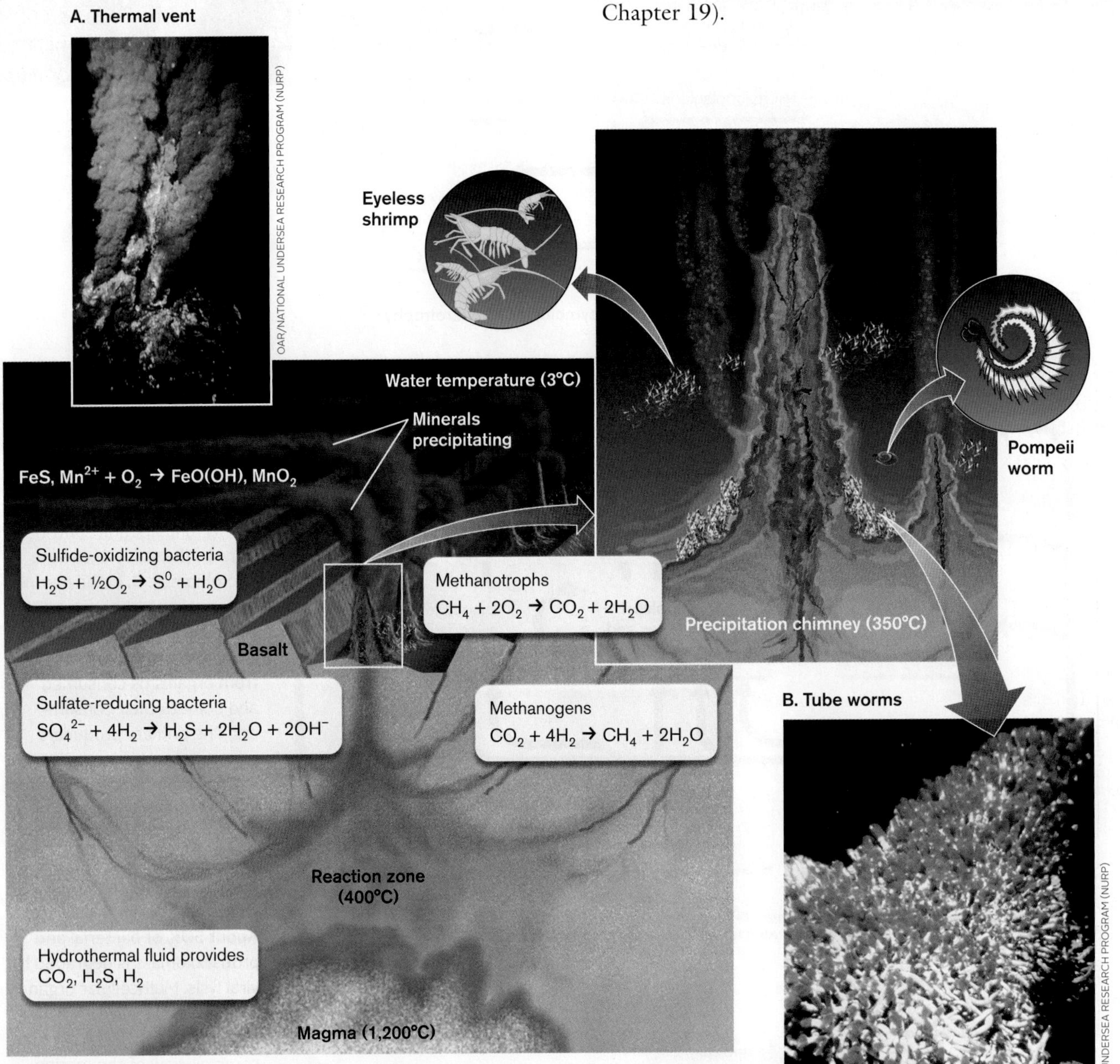

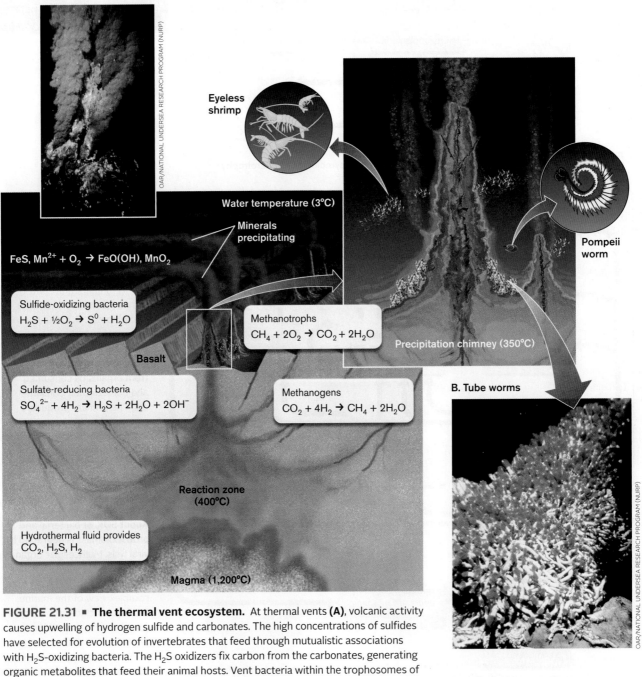

FIGURE 21.31 ■ The thermal vent ecosystem. At thermal vents **(A)**, volcanic activity causes upwelling of hydrogen sulfide and carbonates. The high concentrations of sulfides have selected for evolution of invertebrates that feed through mutualistic associations with H_2S-oxidizing bacteria. The H_2S oxidizers fix carbon from the carbonates, generating organic metabolites that feed their animal hosts. Vent bacteria within the trophosomes of tube worms **(B)** provide energy from oxidation of hydrogen sulfide. *Source:* Part A from Raina Maier et al. 2000. *Environmental Microbiology*. Academic Press. Reprinted by permission of Elsevier, Ltd.

The ocean floor provides reduced inorganic minerals, such as iron sulfide (FeS) and manganese (Mn^{2+}), that can combine with dissolved oxygen to drive chemolithoautotrophy (discussed in Chapter 14). Chemolithoautotrophy by microbes in the sediment generates a permanent voltage potential between the reduced sediment and the oxidizing water. As a result, the entire benthic interface between floor and water acts as a charged battery that can actually generate electricity.

The benthic redox gradient is enhanced dramatically at hydrothermal vents (**Fig. 21.31A**), where volcanic activity causes upwelling of hydrogen sulfide (H_2S), H_2, and carbonates. At a hydrothermal vent, these reduced minerals are brought up by seawater that seeps through the sediment until it reaches a magma pool, where it becomes superheated and rises to the surface as steam. Sulfate-reducing bacteria reduce sulfate from seawater with H_2 upwelling from the vent fluids, to form H_2S. As the H_2S rises, it is oxidized by sulfur-oxidizing bacteria such as *Thiomicrospira*. Nearby anoxic sediment supports methanogens, and the methane they produce seeps up into the oxygenated water, where it is oxidized by methanotrophs.

The high concentrations of sulfides near thermal vents have selected for a remarkable evolution of invertebrate species that feed through mutualistic associations with H_2S-oxidizing bacteria, such as the yeti crab. The H_2S oxidizers fix carbon from the carbonates, generating organic metabolites that feed their animal hosts—species of worms, anemones, and giant clams, all closely related to surface-dwelling species that would be poisoned by H_2S. The tube worm *Riftia* is colored bright red by a pigment carrying H_2S and O_2 in its circulatory fluid (**Fig. 21.31B**). The worm has evolved such a complete dependence on its symbionts that it has lost its own digestive tract.

Microbial communities related to those at thermal vents are also found in unheated benthic habitats known as cold seeps, where hydrogen and methane seep slowly through cracks in the rock. Cold-seep endosymbiosis within animals is presented in **eTopic 21.2**.

Freshwater Microbial Communities

Many features of marine habitats also apply to freshwater systems, such as lakes and rivers. Freshwater habitats, of course, contain much lower salt, usually less than 0.1%. Large, undisturbed lakes are usually **oligotrophic**, with dilute concentrations of nutrients and microbes. The warm upper water layer above the thermocline supports oxygenic phototrophs such as algae and cyanobacteria. The thermocline is a steep transition zone to colder, denser water below, a region that becomes anoxic.

A lake that receives large concentrations of nutrients, such as runoff from agricultural fertilizer or septic systems, becomes **eutrophic**. In a eutrophic lake, the nutrients support growth of algae to high densities, causing an **algal bloom**. The bloom may actually consist of cyanobacteria, *Microcystis aeruginosa* (**Fig. 21.32**). Such blooms regularly affect Lake Erie in the summers, where *Microcystis* produces a liver toxin, microcystin, that contaminates water supplies. In August 2014, microcystin from Lake Erie blooms caused the city of Toledo to issue a "Do Not Drink or Boil" order for its water supply. Boiling does not remove the microcystin.

As the bloom microbes die, they are consumed by heterotrophic bacteria, whose respiration removes all the oxygen. This oxygen loss causes the deep, anoxic region to reach nearly up to the surface of the lake. In a eutrophic lake, fish die off owing to lack of oxygen, which heterotrophic microbes have consumed. Such a region is called a **dead zone**. With increasing human pollution, dead zones now occur in the oceans as well as in lakes; marine dead zones are discussed in Chapter 22.

A.

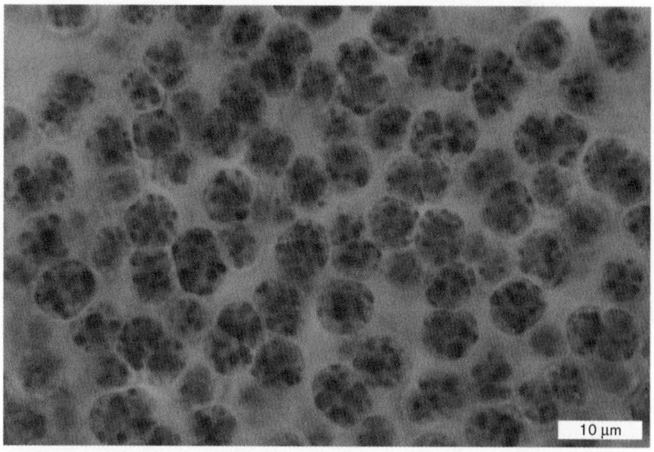

BIOPHOTO ASSOCIATES/SCIENCE SOURCE

B.

THE WASHINGTON POST/GETTY IMAGES

FIGURE 21.32 ■ *Microcystis* bloom on a freshwater lake kills fish. **A.** *Microcystis aeruginosa* cyanobacteria produce the toxin microcystin (stained LM). **B.** Shore of Lake Erie during a *Microcystis* bloom.

Common causes of eutrophication include:

- **Phosphates.** Because phosphorus is commonly a **limiting nutrient** (nutrient in shortest supply) for algae, addition of phosphates from detergents and fertilizers can lead to an algal bloom.
- **Nitrogen** from sewage effluents and agricultural fertilizer runoff can lead to algal blooms by relieving nitrogen limitation.
- **Organic pollutants** from sewage effluents overfeed heterotrophic bacteria, depleting the water of oxygen.

In a eutrophic lake, the lower layers have become depleted of oxygen as a result of overgrowth of microbial producers and consumers. Thermal stratification may break up, and the lake may mix one to several times per year, thus reoxygenating the entire lake. A permanently eutrophic lake typically supports ten times the microbial concentrations of an oligotrophic lake but shows greatly decreased animal life.

Anoxic water supports only anaerobic microbes. These include anaerobic phototrophs that do not produce O_2. Enough light may penetrate to support anaerobic H_2S-oxidizing phototrophs such as *Chlorobium* and *Rhodopseudomonas*. Although H_2S photolysis provides less energy than oxygenic H_2O photolysis, these bacteria have evolved to use chlorophylls and accessory pigments whose spectrum extends into the infrared (**Fig. 21.33**; see also Chapter 14). Light in the infrared portions of the spectrum cannot be used for oxygenic photosynthesis, because the photon energy captured by bacteriochlorophylls is insufficient for the complex to split water. The less efficient H_2S photolysis, however, can harness the energy of red and infrared radiation (discussed in Chapter 14).

H_2S-oxidizing phototrophs include some cyanobacteria and purple proteobacteria, which overlap metabolically with anaerobic heterotrophs and lithotrophs that reduce oxidized minerals. Some proteobacteria, such as *Rhodopseudomonas palustris* or *Rhodospirillum rubrum*, can grow anaerobically with or without light; others grow only by anaerobic metabolism, unassisted by light. These low-oxygen bacteria form a flourishing community, but they cannot support oxygen-breathing consumers such as fish.

At the bottom of the water column, the water meets the sediment (benthos). In the benthic sediment, gradients develop in which successive electron acceptors are reduced by anaerobic respirers and by lithotrophs (**Fig. 21.34**). Electron acceptors that yield the most energy are consumed first; as each in turn is depleted, the electron acceptor with the most energy is consumed next. First, molecular oxygen is used to oxidize organic material and reduced minerals such as NH_4^+. Below, as molecular oxygen falls off, bacteria use nitrate (NO_3^-) from oxidized ammonium ion as an electron acceptor to respire on remaining organic material. As the nitrate is used up, still other bacteria use manganese (Mn^{4+}) as an electron acceptor, followed by iron (Fe^{3+}) and sulfate (SO_4^{2-}). Reduction of sulfate leads to H_2S, which eventually returns to the upper layers supporting anaerobic photolysis. Reduction of CO_2 by H_2 produces methane (CH_4). Methane collects below and sometimes ignites when it escapes to the surface. Methane from freshwater lakes and streams is emerging as a major contributor to global warming (discussed in Chapter 22).

Note: Certain minerals and organic molecules occur in equilibrium between ionized and un-ionized states over the range of pH typical of most common habitats (pH 4–9). Examples include ammonia (NH_3), which protonates to ammonium ion (NH_4^+), and organic acids such as acetic acid (CH_3COOH), which deprotonates to acetate (CH_3COO^-). In this chapter we refer to the form most prevalent at pH 7 unless stated otherwise.

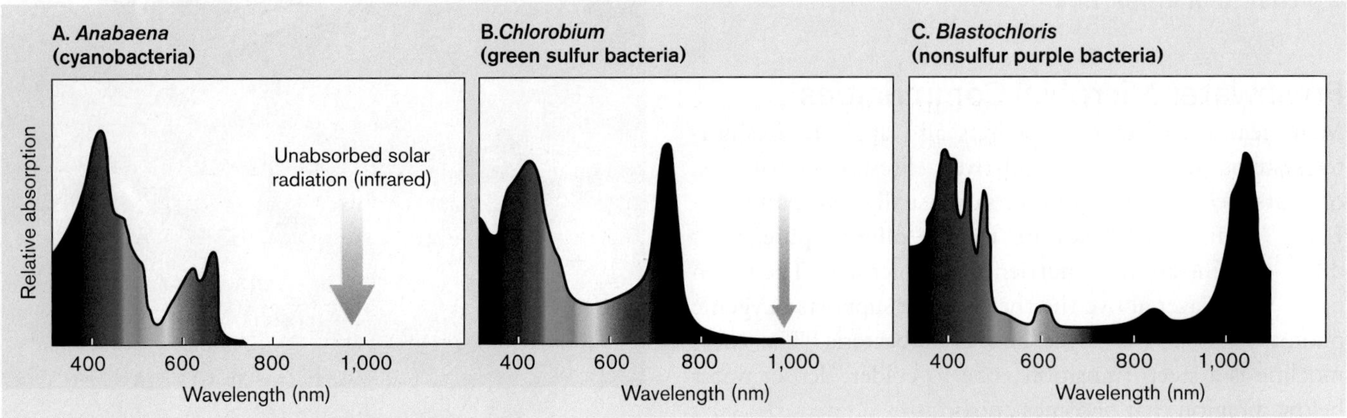

FIGURE 21.33 ■ Absorbance spectra of lake phototrophs, including chlorophylls and accessory pigments. A, B. In the upper waters, algae and cyanobacteria absorb primarily blue and red. **C.** Below, where red has been absorbed by microorganisms above, anaerobic phototrophs such as *Blastochloris* absorb infrared (wavelengths beyond 750 nm).

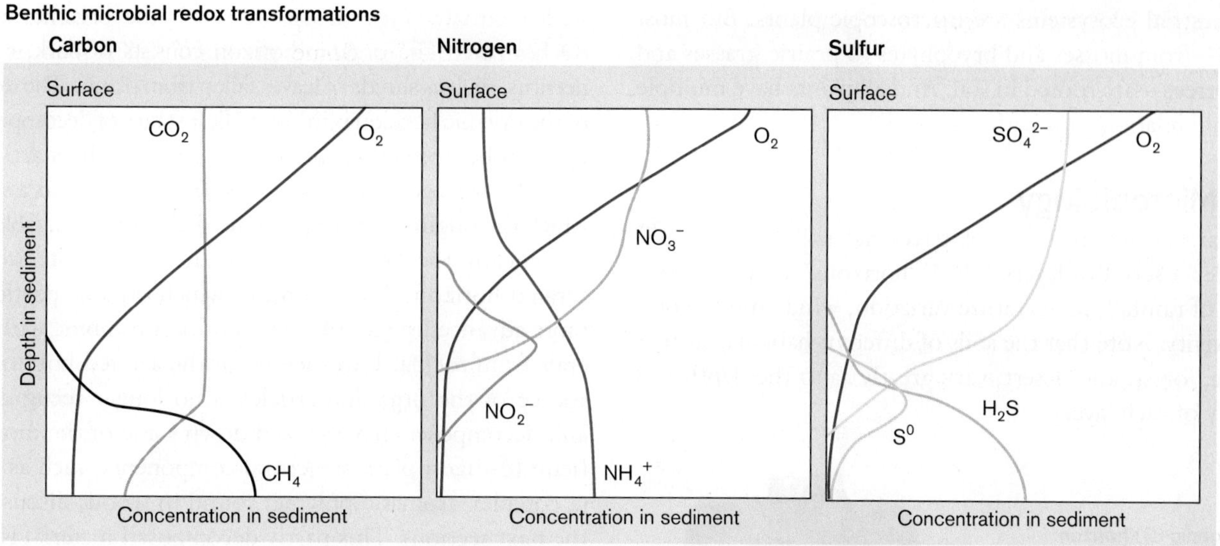

FIGURE 21.34 ▪ Redox gradients in the benthic sediment. At the top of the sediment interface with the water column, minerals are first oxidized lithotrophically by O_2; then, as O_2 declines, the oxidized minerals are used as alternative electron acceptors for anaerobic respiration.

Throughout the water column, protists consume algae and bacteria, while fungi decompose detritus. Protists and fungi also interact with invertebrates and fish as parasites. Other important consumers are the viruses, which lyse about half of microbial populations in lakes, as they do in the ocean. Viruses limit the number of microbes and may keep the water clear enough for light to penetrate.

To Summarize

- **The euphotic zone of the ocean is the upper part of the water column, the part that receives light for phototrophs.** Below, in the aphotic zone, only heterotrophs and lithotrophs can grow. The benthos includes the region where the water column meets the ocean floor, as well as sediment below the surface.
- **Most marine microbes either cannot be cultured or require special culturing methods.** Some can be cultured using conditions that mimic their natural habitat, and allow intimate association with other species that provide growth factors such as siderophores.
- **Marine microbes can include human pathogens.** *Vibrio cholerae* colonizes copepods as symbiotic partners, but it causes the disease cholera in humans.
- **Plankton are small floating organisms, including swimming microbes.** Phytoplankton are phototrophs such as cyanobacteria and algae. Microbial consumers include protists and viruses. Many marine protists are mixotrophs (producers and consumers in one).
- **Picoplankton include bacteria and small microbial eukaryotes.** They are measured by fluorescence microscopy. Their biochemical rate of production is estimated by uptake of radiolabeled nutrients.
- **Marine microbes may grow as biofilms on particles of detritus called marine snow.**
- **Benthic microbes are piezophiles.** The seafloor supports psychrophiles, whereas hydrothermal vents support thermophiles. Vents and cold seeps support sulfur- and metal-oxidizing bacteria, sulfur-reducing bacteria, methanogens, and methanotrophs. Bacteria that oxidize H_2S and methane feed symbiotic animals such as tube worms.
- **Freshwater lakes have stratified water columns.** As depth increases, minerals become increasingly reduced. Anaerobic forms of metabolism predominate, with the more favorable alternative electron acceptors used in turn.
- **Lakes may be oligotrophic or eutrophic.** Eutrophic lakes may show such high biochemical oxygen demand (BOD) that the oxygen concentration falls to levels too low to support vertebrate life.

21.6 Soil and Plant Microbial Communities

Soil is a complex mixture of decaying organic and mineral matter that feeds vast communities of microbes. As such, it is arguably the most complex microbial ecosystem on Earth. And soil-based agriculture is the major source of food for our planet's human inhabitants. The qualities of a given soil—oxygenated or water saturated, acidic or alkaline, salty or fresh, nutrient-rich or nutrient-poor—define what food can be grown and whether the human community will eat or starve.

In contrast to the ocean, where the photosynthetic producers are almost entirely microbial, the major producers of terrestrial ecosystems are macroscopic plants. But most plants—from mosses and bryophytes to prairie grasses and forest trees—are rooted in soil. And all plants have multiple intimate relationships with microbes.

Soil Microbiology

The large-scale structure of terrestrial soil (**Fig. 21.35**) includes a series of layers called "horizons" that arise as a result of rainfall, temperature variation, wind, and biological activity. Note that the soils of different habitats, such as prairie, forest, and desert, vary greatly as to the depth and quality of each layer.

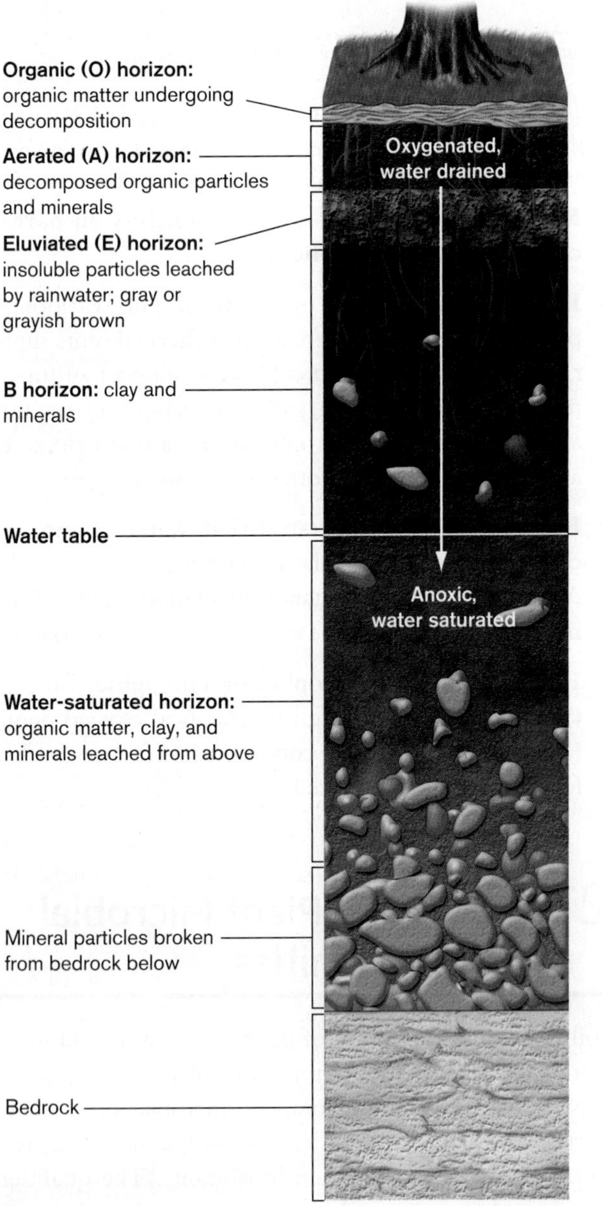

FIGURE 21.35 ■ The soil profile. Soil forms layers in which decomposing organic material predominates at the top, and minerals predominate toward the bottom, at bedrock. The top layers are aerated, providing heterotrophs with access to O_2, whereas the bottom layers are water saturated and anaerobic.

The surface layer of soil we see is the organic horizon (O horizon). The organic horizon consists of dark, organic detritus, such as shreds of leaves fallen from plants. The detritus of the organic horizon is in the earliest stages of decomposition by microbes, primarily fungi and bacteria such as actinomycetes. Early-stage decomposition is defined loosely as a state in which the origin of the detritus may be still recognizable.

Beneath the organic horizon lies the lighter-colored aerated horizon (A horizon), in which organic particles in more advanced stages of decomposition combine with minerals from rock at lower levels. In the aerated horizon, the source of the organic particles is no longer recognizable, and decomposers have broken down some of the more difficult-to-digest plant structural components, such as lignin (a complex aromatic polymer found in wood, discussed in the next section). This partly decomposed material is often sold by garden stores as peat or topsoil.

In well-drained soil, both the organic and aerated horizons are full of oxygen, as well as nutrients liberated by the decomposers and used by plants. Soil consists of a complex assemblage of organic and inorganic particles (**Fig. 21.35**). Between the soil particles are air spaces that provide access to oxygen, allowing aerobic respiration. Each particle of soil supports miniature colonies, biofilms, and filaments of bacteria and fungi that interact with each other and with the roots of plants (**Fig. 21.36**). Even within a single soil

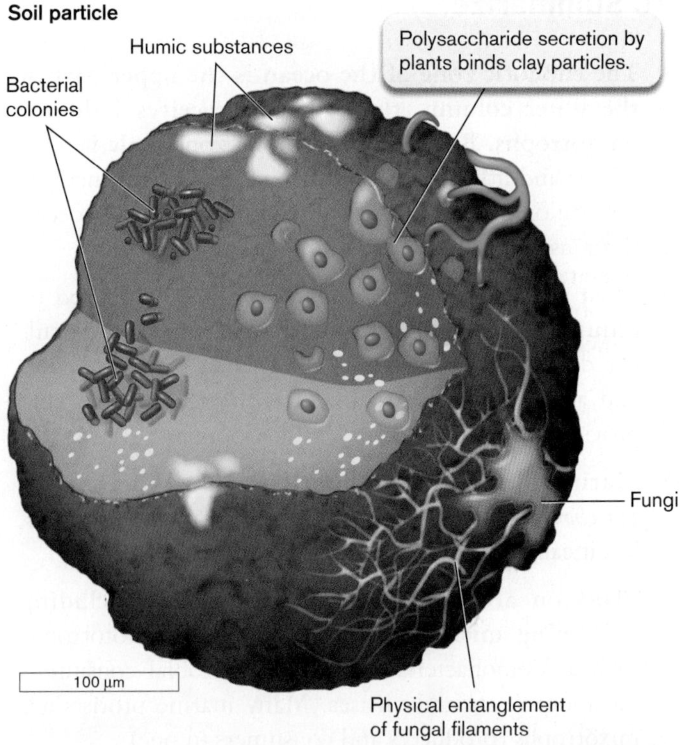

FIGURE 21.36 ■ Microbes in soil and rock. Soil particles support the growth of complex assemblages of microbes. A soil particle contains bacterial colonies, biofilm associations, and microbes associated with fungi and plant roots.

particle, there may be numerous microhabitats—some aerated, some anoxic enough to support methanogens. One side of a particle may be heavily influenced by a plant root, while the other side is separated by a piece of impenetrable detritus.

Below the aerated horizon, the eluviated horizon (E horizon) experiences periods of water saturation from rain (see **Fig. 21.35**). Rainwater leaches (dissolves and removes) some of the organic and mineral nutrients from the upper layers. Below the eluviated horizon lie increasing proportions of minerals and rock fragments broken off from bedrock below. These lower water-saturated layers form the **water table**. This anoxic, water-saturated region contains mainly lithotrophs and anaerobic heterotrophs.

The soil layers finally end at bedrock, a source of mineral nutrients such as carbonates and iron. Interestingly, bedrock is permeated with microbes. Core samples show that crustal rock as deep as 3 km down contains **endoliths**, bacteria growing between crystals of solid rock. What energy source feeds microbes trapped within rock? For some endoliths, a surprising answer may be the radioactive decay of uranium. Uranium-238 decay generates hydrogen radicals that combine to form hydrogen gas. The hydrogen gas combines with CO_2 from carbonate rock, providing an electron donor and a carbon source for methanogens and other endolithic lithotrophs.

The Soil Food Web

The top horizons of soil feature a food web of extraordinary complexity (**Fig. 21.37**). The major producers are green plants, whose leaves generate detritus and whose root systems feed predators, scavengers, and mutualists. Some carbon is also fixed by lithotrophs oxidizing reduced nitrogen (NH_3), hydrogen sulfide (H_2S), and iron (Fe^{2+}). In well-aerated soil, however, the proportion of carbon fixed by lithotrophs is small compared to that fixed by plants.

Which taxa are found in soil? Metagenomes from soil samples reveal the most diverse assemblage of microbes in any known habitat (**Fig. 21.38**). As in the ocean, the vast majority of soil microbes are uncultured, but recent research reveals methods to culture soil bacteria that require signals from their community. See, for example, the discovery of a novel antibiotic producer by Kim Lewis (Chapter 4).

Fungi and bacteria play key roles in decomposition, the process that generates soil. Plant material such as leaf litter (fallen leaves) is decomposed by fungal genera such as *Mycena*, and by bacteria such as the actinomycetes. The actinomycetes include *Streptomyces*, a genus famous for the production of antibiotics and for generating chemicals whose odors give soil its characteristic smell (**Fig. 21.39A**). The fungal and bacterial species composition of leaf litter and soil is highly diverse and depends on associated plant species, especially trees. For example, in one forest study the ascomycete fungal taxon Capnodiales (sooty mold fungi) was prevalent in all soil samples, whereas another ascomycete taxon, Jahnulales, was found only associated with basswood trees.

Besides leaf litter, another source of organic matter from plants is the **rhizosphere**, the region of soil surrounding plant roots. The rhizosphere contains proteins and sugars released by roots, as well as sloughed-off plant cells. These materials feed large numbers of bacteria, which then cycle minerals back to the plant. Bacteria in the rhizosphere may also discourage growth of plant pathogens.

In the aerated zone, heterotrophic bacteria feed on leaf detritus and root exudates; the various taxa of Proteobacteria are most common. Proteobacteria commonly possess the ability to interact with multiple electron acceptors or donors, both organic and mineral. Bacteria are then consumed by protists and nematodes. The nematode *Heterorhabditis bacteriophora* (**Fig. 21.39B**) carries symbiotic *Photorhabdus luminescens* bacteria, which it alternately consumes and transmits as a mutualist when it infects an insect. *Vampirella* protists drill holes in fungal hyphae to suck out their nutrients. Parasitic fungi prey on plants or invertebrates; some actually capture and strangle nematodes.

Mycorrhizal fungi extend the absorptive surface area of plant roots. The nutrient-sharing hyphae of mycorrhizae connect most tree roots in a vast underground "fungal internet" (discussed in the next section). Microbes ultimately feed invertebrates, which then feed larger invertebrates and vertebrate predators. Some predators, such as earthworms and burrowing animals, enhance the soil quality by turning over the matter, thus aerating the soil particles and helping to mix the organic matter from above with the mineral particles from below.

A critical role of fungal decomposers (also known as saprophytes) is the breakdown of extremely complex structural components of vascular plants such as grasses and trees (**Fig. 21.40**). Trees, in particular, accumulate vast stores of biomass in forms that are difficult to digest, such as **lignin**. Lignin is a highly complex and diverse covalent polymer composed of interlinked phenolic groups (benzene rings with OH or related oxygen-bearing side groups; **Fig. 21.40A**). We saw earlier that termites host protozoa that help degrade the lignin portion of wood particles. In soil, fungal and bacterial decomposers possess enzyme systems to degrade lignin and other complex components of plants. Examples of decomposers include white rot fungi (**Fig. 21.40B**) and actinomycete soil bacteria. The prevalence of lignin is one reason that decomposition by fungi plays a much larger role in terrestrial ecosystems than in marine ecosystems.

The first phase of microbial degradation (about 50% of the carbon) is complete within a year of deposition

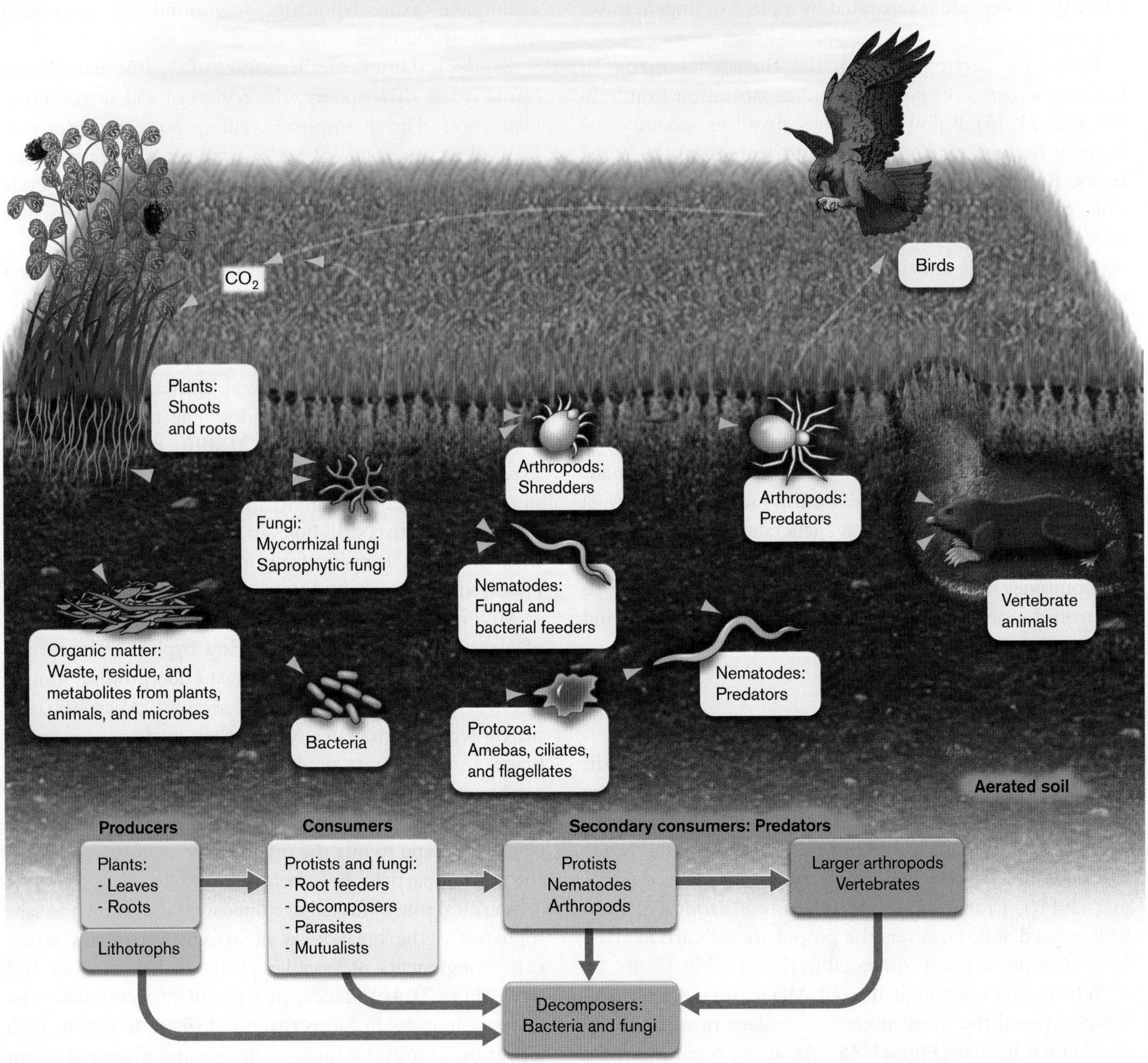

FIGURE 21.37 ■ The soil food web (aerated zone). Plants are the major producers, although some production also occurs from lithotrophs such as ammonia oxidizers. Detritus from plants is decomposed by fungi and bacteria, which feed protists and small invertebrates such as nematodes. Protists and small invertebrates are consumed by larger invertebrates and vertebrate animals.

in the soil. The remaining phenolics, however, may be degraded at a rate of less than 5% per year, and some samples dated by ^{14}C isotope ratios have been shown to last 2,000 years. These phenolic molecules are called **humic material** or **humus**. Because of its slow degradation, humic material provides a steady slow-release supply of nutrients for plant growth. But forests whose rate of microbial decomposition is particularly low—for example, the New Jersey Pine Barrens—depend on fire to clear the mounting layers of humus and return its minerals to the ecosystem.

Note that soil microbial communities also include a wide range of mesophilic archaea, particularly the thaumarchaeotes (see **Fig. 21.38**; taxa described in Chapter 19). The thaumarchaeotes include ammonia oxidizers, which provide an important link in the nitrogen cycle.

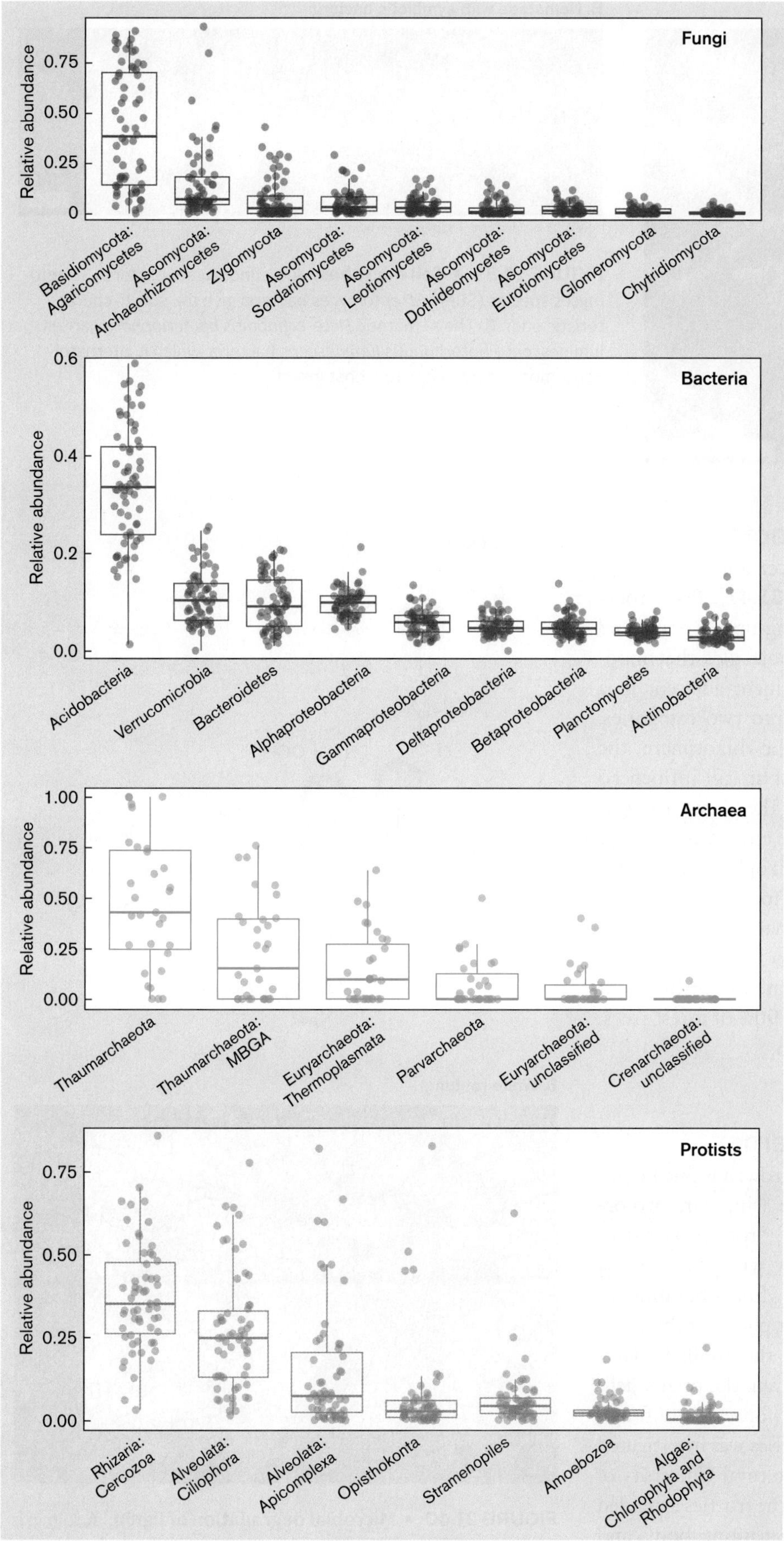

FIGURE 21.38 ■ Microbial community structure of soil. Fungi, bacteria, archaea, and protists of various taxa were identified by SSU rRNA gene sequencing of soil samples from around the world. MBGA = Marine Benthic Group A. *Source:* Modified from Noah Fierer. 2017. *Nat. Rev. Microbiol.* **15**:579–590, fig. 3.

A. *Streptomyces* bacteria

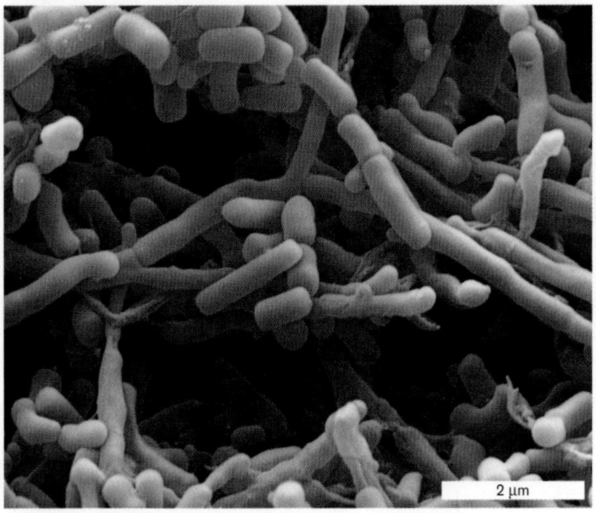

EYE OF SCIENCE/SCIENCE SOURCE

B. Nematode with symbiotic bacteria

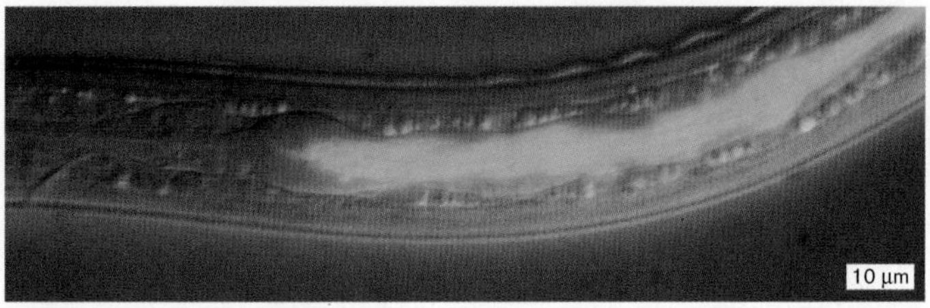

CICHE, T., ET AL. 2008. *APPL. AND ENVIRON. MICROBIO* **74:** 2275.

FIGURE 21.39 ■ Soil microbes. A. Actinomycete bacteria, *Streptomyces* species (SEM). *Streptomyces* bacteria give the soil its characteristic odor. **B.** The nematode *Heterorhabditis bacteriophora* carries luminescent *Photorhabdus luminescens* bacteria, which it alternately consumes and transmits to a host insect.

Microbes Associated with Roots

The presence of plant roots provides yet another level of complexity to soil communities (**Fig. 21.41**). Plant roots influence the surrounding soil by taking up nutrients and by secreting organic substances and molecules that modulate their surroundings. The environment adjacent to a plant root can be further subdivided into two categories: the rhizoplane, the root surface; and the rhizosphere, the region of soil outside the root surface but still influenced by plant exudates (materials secreted by the plant). Particular bacterial species are adapted to these environments. For example, in anoxic wetland soil, the rhizoplane and rhizosphere of plant roots provide oxygen for methanotrophs that oxidize methane produced by methanogens.

The rhizoplane and rhizosphere also provide the environment for symbiotic fungi that generate mycorrhizae. At least 80% of plants in nature, including 90% of forest trees, require mycorrhizae for optimal growth.

A. Lignin

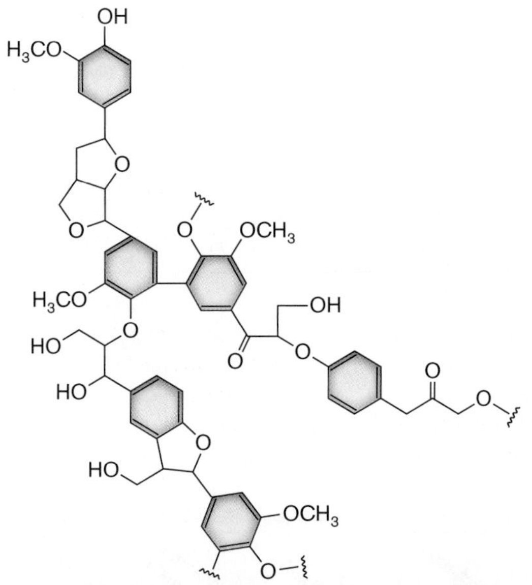

B. White rot fungi

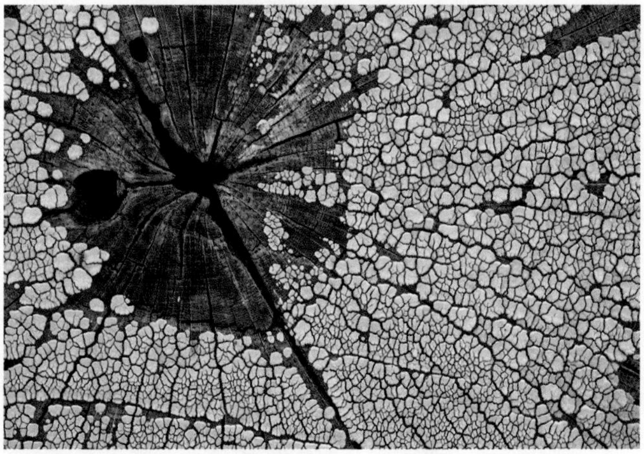

GAIL JANKUS/SCIENCE SOURCE

FIGURE 21.40 ■ Microbial degradation of lignin. A. Lignin is a complex organic polymer that is a component of wood and bark. **B.** *Xylobolus frustulatus*, a white rot fungus, growing on a willow log. The fungus degrades lignin.

Mycorrhizae: The Fungal Internet

The function of mycorrhizae for plant growth is just beginning to be understood. **Mycorrhizae** (singular, **mycorrhiza**; from *myco*, "fungal," and *rhiza*, "root") consist of fungal mycelia that associate intimately with the roots of plants, extending access to minerals while obtaining in return the energy-rich products of plant photosynthesis.

Mycorrhizae were first discovered in the 1880s by German truffle hunters who sought to cultivate the prized delicacy, the fruiting body of an ascomycete (for review of fungi, see Chapter 20). The propagation of truffles was investigated by mycologist Albert Frank at the Agricultural University of Berlin. To his surprise, Frank found that the truffles extended their mycelia far beyond the site of the fruiting body, and that the mycelia formed an impenetrable tangle with plant roots. Frank called the tangled mycelia "fungus-roots" or

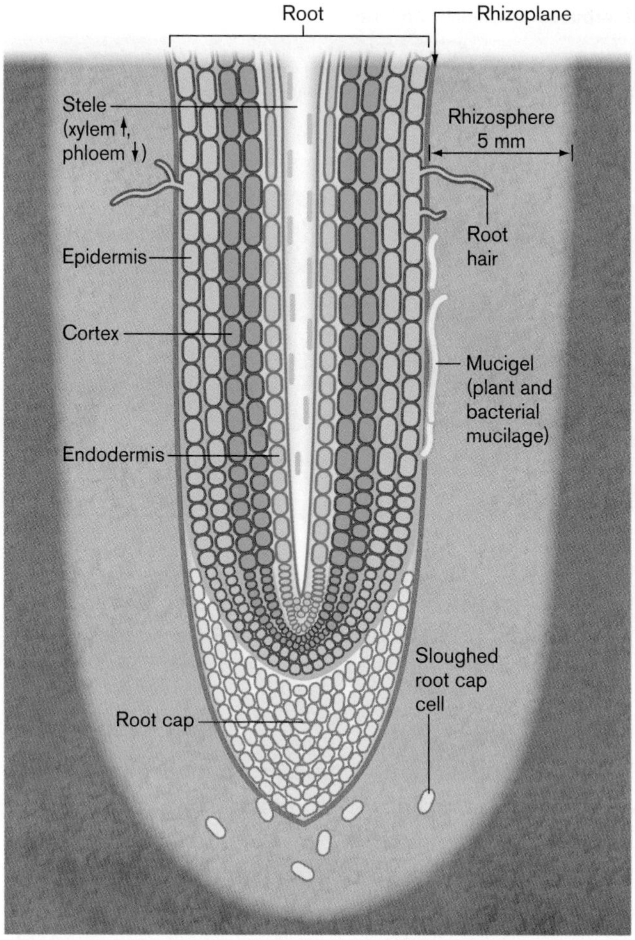

FIGURE 21.41 ■ Plant roots offer special habitats for microbes. The rhizoplane is the region of soil directly contacting the plant root surface. The rhizosphere is the soil outside the rhizoplane that receives substances from the root, such as mucilage, sloughed cells, and exudates.

mycorrhizae. A century later, we are beginning to appreciate that these mysterious fungus-root tangles offer a vast interconnected network for exchange of nutrients among fungi and many different plants, like an internet connecting countless sites. In temperate forests, 40% of the carbon in tree roots may be derived from photosynthesis by other trees of different species, connected by mycorrhizae.

Two different kinds of mycorrhizae are observed: ectomycorrhizae and endomycorrhizae. **Ectomycorrhizae** colonize the rhizoplane, the surface of plant rootlets—the most distal part of plant roots (**Fig. 21.42**). The fungal mycelia never penetrate the root cells. They form a thick mantle surrounding the root and growing between the root cells, and then extend long mycelia away from the root to absorb nutrients. Numerous kinds of fungi form ectomycorrhizae, including ascomycetes (such as truffles) and basidiomycetes, known by their mushrooms (such as stinkhorns). Plants grown with ectomycorrhizae invest less of their body mass in roots and more in the aboveground stems and leaves—an important consideration for agriculture, in which the aboveground plant is usually the part harvested.

Endomycorrhizae form a more intimate association, in which the fungal hyphae penetrate plant cells deep within the cortex (**Fig. 21.43**). The penetrating hyphae form knobbed branches that resemble microscopic "trees," or arbuscules, within the root cells. Some of the hyphae form specialized vesicles within the plant that store nutrients. Another name for this kind of mycorrhizae is **vesicular-arbuscular mycorrhizae** (**VAM**).

Endomycorrhizae are more specialized than ectomycorrhizae. They comprise a relatively small number of fungal species, such as members of the Glomeromycota genus

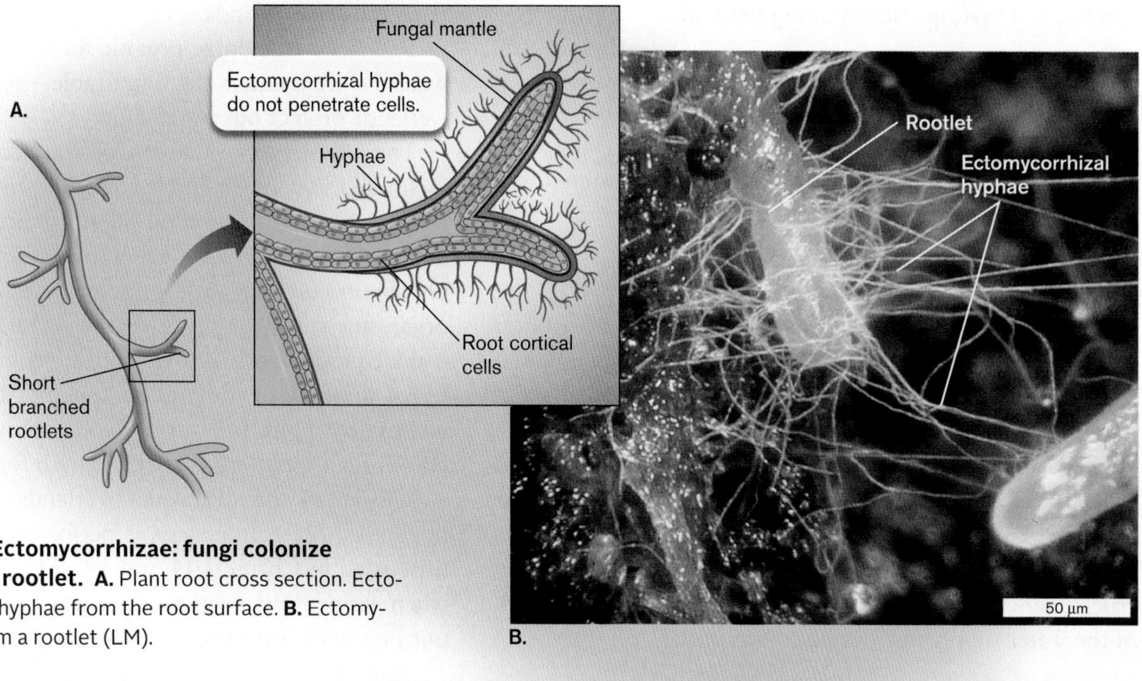

FIGURE 21.42 ■ Ectomycorrhizae: fungi colonize the surface of the rootlet. A. Plant root cross section. Ectomycorrhizae extend hyphae from the root surface. **B.** Ectomycorrhizal hyphae from a rootlet (LM).

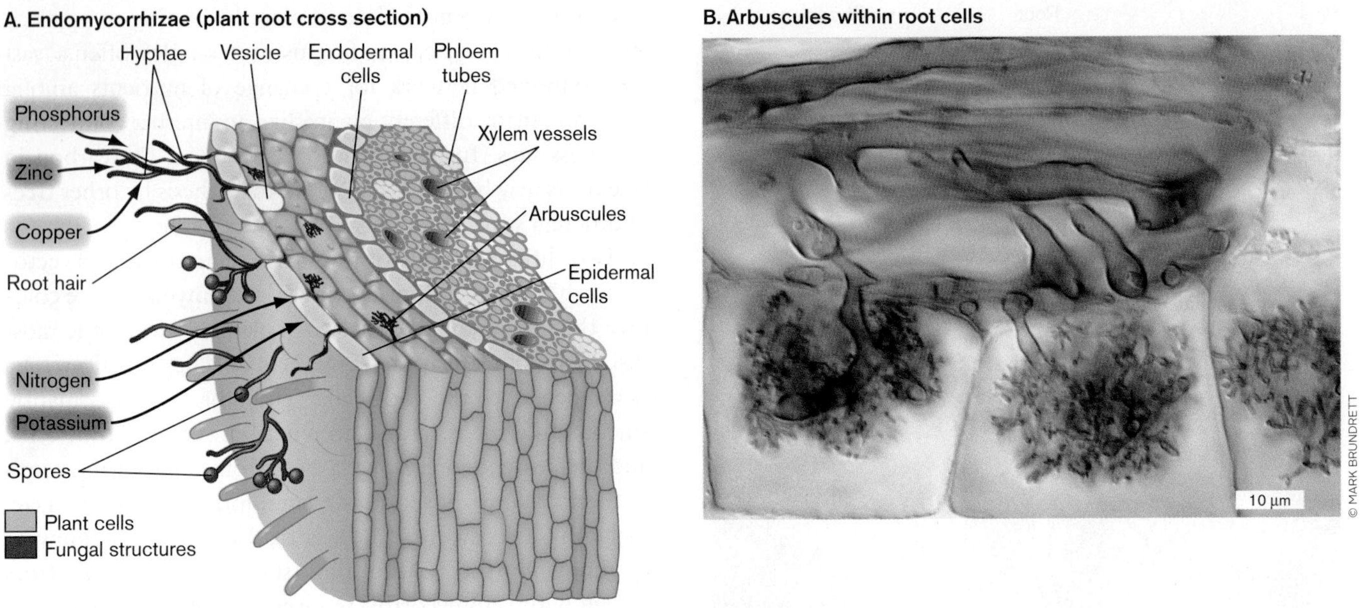

FIGURE 21.43 ■ Endomycorrhizae: fungi invade root cells, forming arbuscules. **A.** Plant root cross section. Endomycorrhizal hyphae penetrate cells deep within the root cortex. **B.** The penetrating hypha forms an arbuscule within a root cell.

Glomus, and they show obligate dependence on their host plants. Their presence in nature and their importance in the ecosystem, however, are actually greater. Endomycorrhizal species exist entirely underground (they do not form mushrooms), and they completely lack sexual cycles. They may acquire 25% of the photosynthetic product of their hosts in exchange for tremendously expanding the plants' access to soil resources.

Mycorrhizae greatly enhance a plant's uptake of water, as well as minerals such as nitrogen and phosphorus. In addition, the hyphae sequester toxins, and they actually distribute organic substances from one plant to another. Mycorrhizae may join many different plants, even different species, in a vast, nutrient-sharing network.

Thought Question

21.10 Design an experiment to test the hypothesis that the presence of mycorrhizae enhances plant growth in nature.

Wetland Soils

So far, we have considered the interface between ground and water (the benthic sediment beneath oceans and lakes) and the interface between ground and air (aerated soil). An interesting case that combines the two is wetlands (**Fig. 21.44A**). A **wetland** is defined as a region of land that undergoes seasonal fluctuations in water level, so that sometimes the land is dry and oxygenated, and at other times, water saturated and anoxic. Wetlands provide many crucial functions; for example, the Everglades filter much of the water supply for Florida communities.

Wetland soil that undergoes such periods of anoxic water saturation is known as **hydric soil**. Hydric soil is characterized by "mottles," patterns of color and paleness (**Fig. 21.44B**). The reddish brown portions (for example, surrounding a plant root) result from oxidized iron (Fe^{3+}). The gray portions indicate loss of iron in its water-soluble reduced form (Fe^{2+}) generated by anaerobic respiration.

Soil becomes anoxic when the rate of oxygen diffusion is too low to support aerobic metabolism. Anaerobic metabolism allows much lower rates of production than metabolism in the presence of oxygen because anaerobes use oxidants of lower redox potential and limited quantity, such as sulfate and nitrate. Many kinds of anaerobic bacteria inhabit wetlands. For example, denitrifiers (bacteria using nitrate to oxidize organic food) remove nitrate from water before it enters the water table—one of the ways that wetlands protect our water supply.

The alternative oxidants (electron acceptors) are always in limited supply, so further catabolism is carried out by fermentation. Fermentation allows only incomplete breakdown of food molecules, generating a rich and diverse supply of nutrients for a variety of consumers, including aerobic organisms when the water recedes. Thus, despite its lower overall productivity, anoxic soil contributes to the nutritional diversity of wetland ecosystems. The relatively slow rate of decomposition can lead to accumulation of high levels of organic carbon, particularly rich for plant growth.

The anoxic conditions of wetlands also favor methanogenesis. Methanogenesis is performed solely by archaea (discussed in Chapters 14 and 19). Methanogenesis occurs when fermenting bacteria generate H_2, CO_2, and other one- or two-carbon substrates that methanogens convert

A.

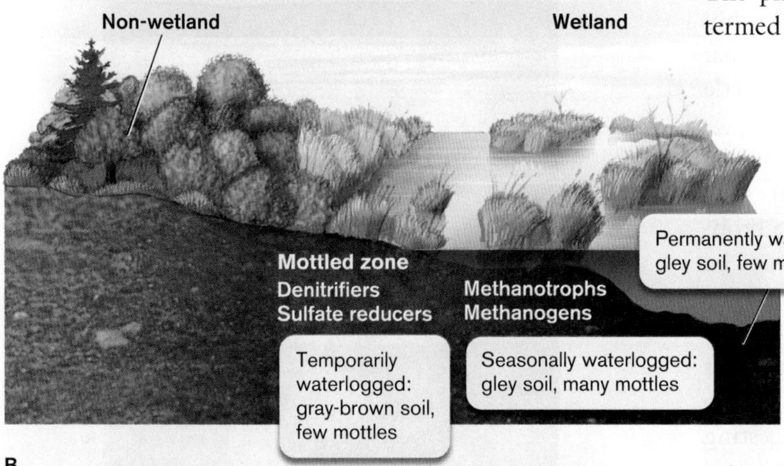

B.

FIGURE 21.44 ■ Wetland soil. A. A true wetland experiences periods of water saturation alternating with dry soil. Soil that is water saturated becomes anaerobic because O_2 diffuses slowly through water. **B.** Alternating periods of water saturation and dryness give the soil a mottled color. The red-orange-colored portions around root holes result from oxidized iron (Fe^{3+}), whereas the gray portions ("gley") indicate that water has washed iron away after reduction (Fe^{2+}).

to methane. Methane is a more potent greenhouse gas than CO_2, and although the current methane concentration in our atmosphere is low, it is rising exponentially.

In water-filled anoxic soils, particularly those of rice fields, substantial quantities of methane escape to the air through air-conducting channels in the roots of the rice plant. The roots of rice plants, like those of other wetland vascular plants, contain channels to carry oxygen. These same channels, however, allow methane to escape from anoxic soil. Fortunately, the rhizosphere of the roots supports methanotrophs that oxidize methane, so research is being done to maximize methanotroph activity and minimize the release of methane. There is some evidence that natural wetlands actually take in more carbon than they put out, whereas disturbed wetlands (wetlands altered by human activity) generate net efflux of CO_2 and CH_4. The role of wetlands in global cycling is discussed further in Chapter 22.

Plant Endophytic Communities

The plant interior provides a special home for microbes termed **endophytes**. The plant's vascular system of phloem tubes conducts photosynthesized sugars down from the leaves, while xylem tubes bring water and minerals up from the roots. Unlike the sterile blood vessels of animals, plant transport vessels are normally colonized by endophytic fungi and bacteria. Thus, when we eat plants, we also consume all their endophytes, or endophytic microbes (**Fig. 21.45**).

Some endophytes are obligate (can live only within the plant), whereas others, such as pseudomonads, have alternative lifestyles in the soil. The microbial partners may grow as mutualists, commensals, or parasites. Plants tolerate endophytes because they confer substantial benefits. For example, prairie grasses called fescue, grazed by cattle in the southeastern United States, host fungal epiphytes. The fungi,

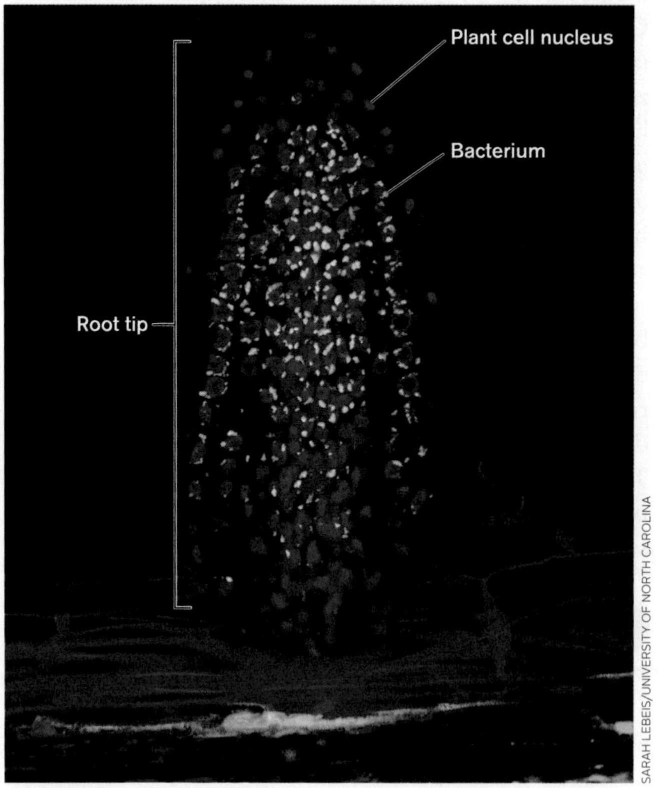

FIGURE 21.45 ■ Endophytic bacteria within plants. Communities of bacteria colonize a new root tip of the plant *Arabidopsis thaliana*. The bacteria (green) and the root cell nuclei (blue) are visualized by FISH with probes hybridized to rRNA, using confocal laser scanning microscopy. Endophytes include species of Actinobacteria and Proteobacteria that may help the plant resist pathogens and survive environmental stresses such as high salt or temperature. Distinct clades of bacteria grow on the root surface or within the plant tissues. *Source:* Derek Lundberg et al. 2012. *Nature* **488**:86.

Neotyphodium coenophialum, produce alkaloids that deter insect predators, pathogens, and root-feeding nematodes. Unfortunately, some of the alkaloids poison cattle, but agricultural scientists have engineered fungal strains that still protect the plant from pathogens while allowing cattle to graze.

Human pathogens such as *E. coli* O157:H7 and *Salmonella enterica* can grow endophytically in crop plants such as spinach and alfalfa. Endophytic pathogens pose a problem for the food industry because the bacteria cannot be "washed off" from raw produce. On the other hand, endophytes such as *Stenotrophomonas* species have potentially valuable uses. *Stenotrophomonas* bacteria produce enzymes that deter many plant pathogens. The bacteria also absorb and concentrate toxic metals such as arsenic, suggesting a possible use for bioremediation of metal-contaminated soil. They produce promising antibiotics and proteases for cleansing agents. Other kinds of endophytes protect plants from heat, salt, and drought—contributions of growing importance as our global climate changes.

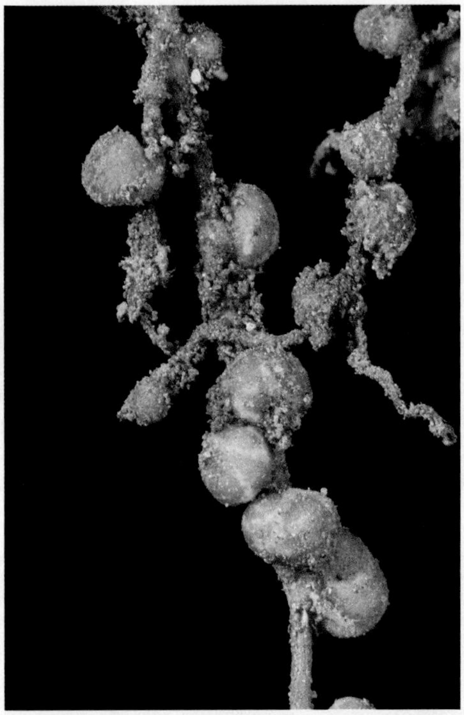

FIGURE 21.46 ■ *Rhizobium* nodules on pea plant roots. Rhizobia induce legume roots to form nitrogen-fixing nodules.

Rhizobia Fix Nitrogen for Legumes

One symbiotic relationship that is critical for agriculture is the mutualism between plants and nitrogen-fixing **rhizobia** (singular, **rhizobium**), a group of soil-dwelling Alphaproteobacteria discussed in Chapter 18. Major rhizobial genera include *Rhizobium*, *Bradyrhizobium*, and *Sinorhizobium*. Rhizobia, associated with legumes such as peas and beans, fix more nitrogen than the plants absorb from soil, actually increasing the soil's nitrogen content. For this reason, farmers often alternate crops such as corn with soybeans to restore nitrogen to the soil. The rhizobial bacteria develop specialized forms within plant cells, called **bacteroids**. Bacteroids lack cell walls and are unable to reproduce; their function is specialized for nitrogen fixation. The rhizobial infection of root hairs induces the formation of nodules within which the nitrogen-fixing bacteroids are sequestered (**Fig. 21.46**).

> **Thought Question**
>
> 21.11 How do you think symbiotic rhizobia reproduce? Why do bacteroids develop if they cannot proliferate?

The initiation, development, and maintenance of the rhizobia-legume symbiosis poses intriguing questions of genetic regulation. How does the association begin? How do host plant and bacterium recognize each other as suitable partners? The legume exudes signaling molecules called flavonoids into its rhizosphere. Flavonoids resemble steroid hormones such as estrogen and have similar effects on animals; they are also called phytoestrogens. The flavonoids are detected by rhizobial bacteria, which respond by chemotaxis, swimming toward the root surface. Flavonoids then induce bacterial expression of Nod factors, molecules composed of chitin with lipid attachments. Nod factors communicate with the host plant and help establish species specificity between bacterium and host.

The entry of the bacteria into the host involves a fascinating interplay between bacterial and plant cells (**Fig. 21.47A**). First, a bacterium is attracted by flavonoids to the surface of a root hair extended by a root epidermal cell (**Fig. 21.47A**, step 1). The bacterial Nod factor induces the root hair to grow in a curl around it and ultimately surround the bacterium with plant cell envelope (steps 2 and 3). The bacterium then induces growth of a tube poking into the plant cell (step 4). The tube growth is directed by the plant nucleus, which migrates toward the plant cortex (step 5). As the tube grows, bacteria proliferate, forming a column of cells that projects down the tube (steps 6 and 7). This column of cells is known as the infection thread. The infection thread can be visualized by light microscopy (**Fig. 21.47B**).

As the infection thread develops, signals from the bacteria induce the cortical cells (below the epidermis) to prepare to receive the bacteria. The bacteria induce further tube formation into the cortical cells and continue penetration as they grow. The penetration of the infection thread into the cortex is shown in fluorescence micrographs (**Fig. 21.47C**) in which the bacteria are engineered to express a fluorescent protein.

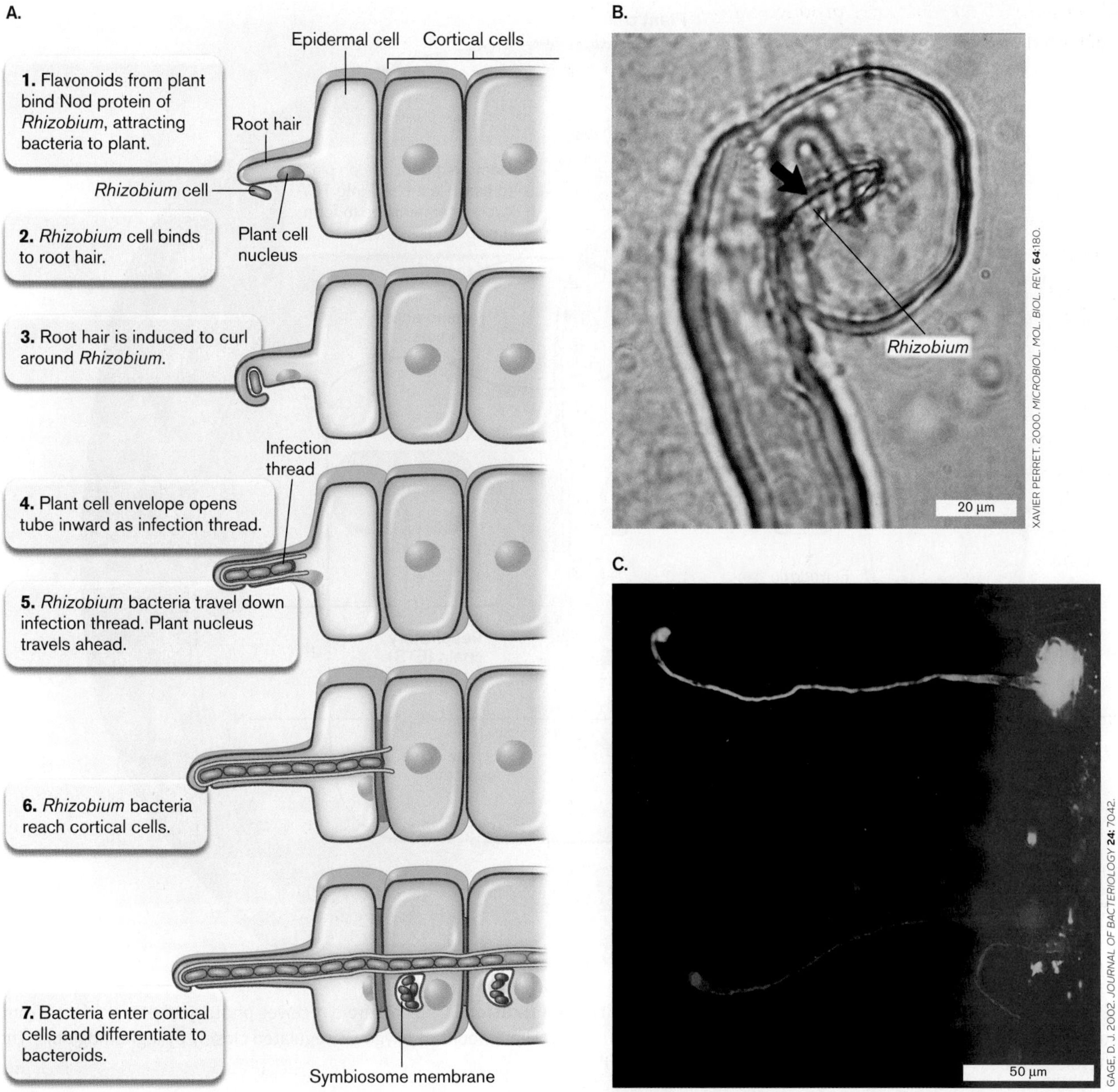

FIGURE 21.47 ▪ The infection thread. A. Rhizobia are attracted to the legume by chemotaxis toward exuded flavonoids. A bacterium induces an epidermal root hair to curl around it and take it up into the infection thread, a tube of plant cell wall material. The thread eventually penetrates cortical cells, where the bacteria lose their cell walls and become nitrogen-fixing bacteroids. **B.** Root hair curling around *Rhizobium*, and the formation of an infection thread (LM). **C.** Infection threads invading the cortex (fluorescence microscopy). Bacteria express either DsRed (pink) or green fluorescent protein (green).

The cortical cells invaded by bacteria are induced to proliferate in an organized manner, forming nodules. Within the nodules, most of the infecting bacteria differentiate into wall-less bacteroids that will fix nitrogen. A few bacteria fail to differentiate; their fate is unclear. The bacteroids remain sequestered within a sac of plant-derived membrane known as the symbiosome. The symbiosome membrane contains special transporters that mediate the exchange of nutrients between the bacteroid and its host cell, sustaining bacteroid metabolism while preventing harm to the host.

From the plant cytoplasm, the bacteroid receives catabolites such as malate, which enter the TCA cycle and donate electrons for respiration. The oxygen for respiration comes from the plant's photosynthesis, regulated by plant-derived leghemoglobin to maintain levels low enough that the bacteroids can fix nitrogen (**Fig. 21.48**). Bacterial nitrogen fixation consumes about a fifth of the plant's photosynthetic products. As discussed in Chapter 15, bacteria fix nitrogen gas (N_2) into ammonium ion (NH_4^+) in a reaction catalyzed by the enzyme nitrogenase:

$$N_2 + 10H^+ + 8e^- + 16\text{ ATP} \longrightarrow 2NH_4^+ + H_2 + 16\text{ ADP} + 16\text{ P}_i$$

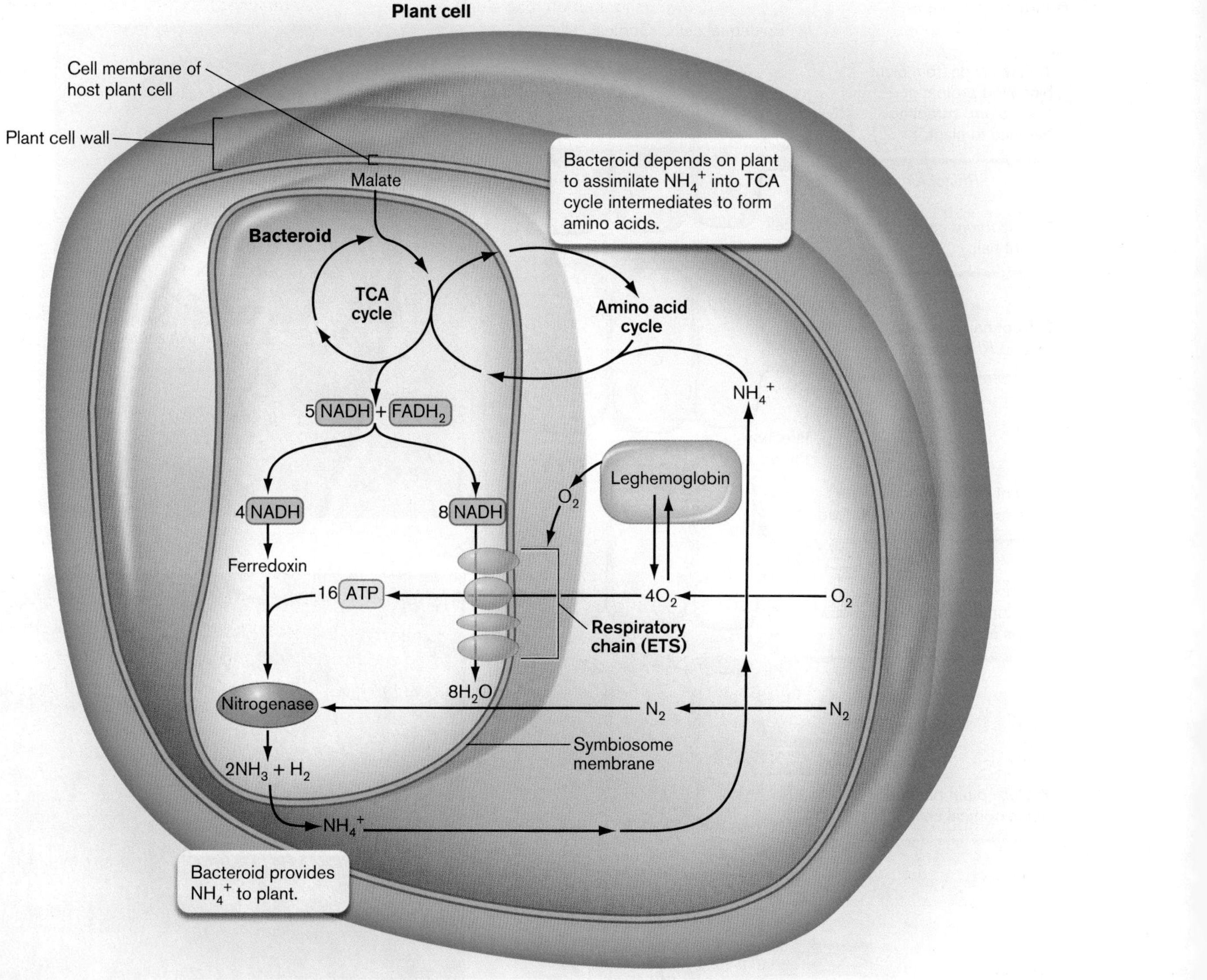

FIGURE 21.48 ■ Energy and oxygen regulation during nitrogen fixation. The bacteroid receives photosynthetic products from the plant, such as malate and oxygen, to generate ATP for nitrogen fixation. The amount of oxygen is regulated closely by leghemoglobin. The bacteroid provides nitrogen (fixed as ammonium ion) to the plant cell.

The reaction requires expenditure of eight NADPH or NADH, plus 16 ATP, which are generated by respiration. But respiration requires oxygen, which poisons nitrogenase. Thus, oxygen needs to be delivered to the bacteroid only as needed, and in an amount just enough to run respiration. The oxygen is sequestered and brought to the bacteroid by leghemoglobin, an iron-bearing plant protein related to blood hemoglobin.

Overall, the nitrogen fixation symbiosis is kept in balance by several regulatory mechanisms. The presence of ammonium or nitrate ions inhibits symbiosis and nitrogen fixation. The bacteroids cannot synthesize their own amino acids; instead, they must provide ammonium to the plant cytoplasm for assimilation into amino acids, some of which cycle back to the bacteroid. But what is known of regulation is dwarfed by the unanswered questions: How is the infection thread formed? How does the plant allow infection while preventing uncontrolled growth of bacteria? What determines how much of the plant's photosynthetic products are harvested by the bacteria? Why is hydrogen gas released, and how can this loss of potential energy be prevented? What limits the host specificity of rhizobia to legumes, and can it be extended to other crop plants, such as corn? Research on these questions is critical for agriculture.

Thought Question

21.12 High levels of nitrate or ammonium ion corepress the expression of Nod factors (see **Fig. 21.47**). What is the biological advantage of Nod regulation?

A. Virus infection of tulips

B. Crown gall tumor on rose stem

C. Anthracnose fungus on dogwood leaf

FIGURE 21.49 ▪ Plant diseases range from innocuous to devastating. **A.** Striped tulips result from a virus. **B.** Crown gall tumor on rose stem, caused by *Agrobacterium tumefaciens*. **C.** Dogwood leaf spotted by anthracnose fungus.

Plant Pathogens

We have seen many ways in which bacteria and fungi interact positively with plants, but some species act as pathogens. In any environment, pathogens are always outnumbered by the vast community of neutral or helpful microbes. Nevertheless, when a pathogen does colonize a plant, its growth can have effects ranging from minimal to devastating (**Fig. 21.49**). A relatively harmless plant virus was associated with a famous historical phenomenon: the sixteenth-century tulip craze in the Netherlands. The virus caused streaking of tulip petals (**Fig. 21.49A**), a pattern much admired by tulip fanciers. Other viruses, however, can cause devastating blights and epidemics. (For more on viruses, see Chapters 6 and 11.)

A pathogenic relative of rhizobia is *Agrobacterium tumefaciens*, a bacterium whose DNA transforms plant cells to form crown gall tumors (**Fig. 21.49B**). *A. tumefaciens* has an unusually broad host range, and its natural genetic transformation system has been applied widely for commercial plant engineering (discussed in Chapter 16). The tumors remain largely confined and have relatively little effect on plant growth. Other bacterial pathogens, particularly species of *Erwinia* and *Xanthomonas*, severely damage plants.

The most common plant pathogens are fungi. Fungal diseases such as anthracnose (**Fig. 21.49C**) cause substantial losses in agriculture, affecting cucumbers, tomatoes, and other vegetables. Dutch elm disease, which has wiped out nearly all the native elms of the United States, is caused by the fungus *Ophiostoma novo-ulmi*. The fungus is carried by bark beetles, which bore into the xylem, damaging the plant's transport vessels and allowing access for fungal spores.

Some fungal pathogens generate specialized structures to acquire nutrients from plants. As a hypha grows across the plant epidermis, its tip can penetrate the plant cell wall, followed by ingrowth of a bulbous extension called a **haustorium** (plural, **haustoria**; **Fig. 21.50**). The haustorium never penetrates the plant cell membrane, thus avoiding leakage and loss of plant cytoplasm. Instead, it causes the membrane to invaginate, while expanding into the volume of the plant cell. The haustorium takes up nutrients such as sucrose, generated by adjacent chloroplasts. Depending on the species of fungus, haustorial parasitism can lead to mild growth retardation, or it can rapidly kill the plant.

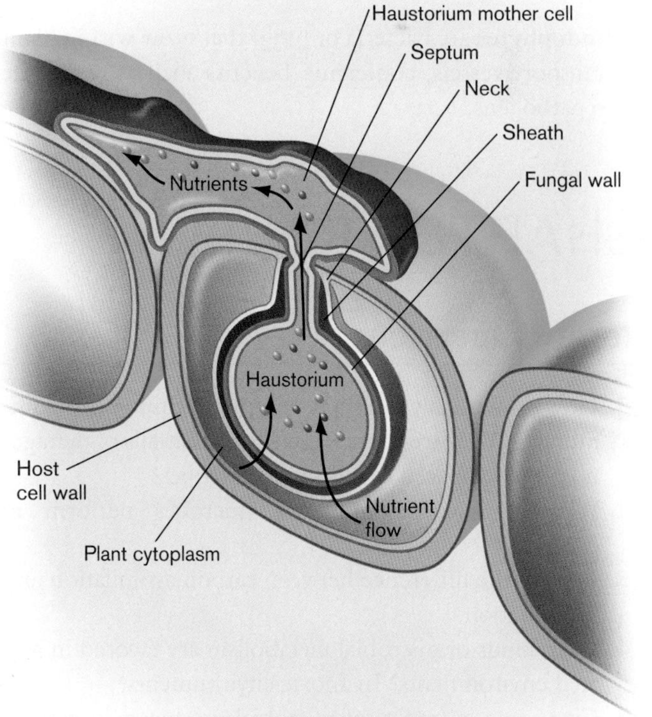

FIGURE 21.50 ▪ A fungal pathogen inserts haustoria into plant cells. The haustorium is surrounded by an invagination of the plant's cell membrane. Plant nutrients such as sucrose flow into the haustorium and are transferred out to the fungal mycelia.

Thought Question

21.13 Compare and contrast the processes of plant infection by rhizobia and by fungal haustoria.

To Summarize

- **The uppermost horizons of soil** consist of detritus and are largely aerated. Below the aerated layers, the eluviated horizon experiences water saturation. Lower layers are anoxic.
- **The soil food web** includes a complex range of microbial producers, consumers, predators, decomposers, and mutualists. Microbial mats and biofilms are complex multispecies communities of bacteria, fungi, and other microbes.
- **Fungi decompose lignin**, a complex aromatic tree component that is challenging to digest. Lignin decomposition forms humus.
- **Certain fungi form symbiotic associations** with plant roots called mycorrhizae. Mycorrhizae transport soil nutrients among many different kinds of plants.
- **Wetland soils** alternate aerated (dry) with anoxic (water-saturated) conditions. Anoxic wetland soil favors methanogenesis. Wetland soils are among the most productive ecosystems.
- **Endophytes** are bacteria or fungi that grow within plant transport vessels, conferring benefits such as resistance to pathogens.
- **Rhizobia induce legume roots** to nodulate for nitrogen fixation. The bacteria enter the root as infection threads. Some of the bacteria enter root cells and develop into nitrogen-fixing bacteroids, which gain energy from plant cell respiration but must remain anaerobic.
- **Plant pathogens** include bacteria, fungi, and viruses. Some pathogens only mildly affect the plant, whereas others cause devastation. Some fungi invade plants using haustoria, which grow into the plant cell by penetrating the cell wall and invaginating the cell membrane to avoid leakage of cytoplasm.

Concluding Thoughts

This chapter has introduced the challenge of characterizing microbial ecosystems, with the promise and pitfalls of the metagenomes. We have described how microbes colonize a vast array of habitats, from the deserts and oceans to the roots of plants and the digestive tracts of animals. Microbes are found from the upper layers of Earth's atmosphere down to the deepest rock strata we can reach. Wherever found, microbes both respond to and modify the environment that surrounds them. Thus, microbes both fill available niches and construct their own. While this chapter has focused on food webs in local habitats, in Chapter 22 we take a global perspective of microbial ecology and its roles in the cycling of Earth's essential elements, such as nitrogen and iron. We will see how microbes interact with global climate change and examine the evidence for existence of microbial life beyond Earth.

CHAPTER REVIEW

Review Questions

1. How do we analyze a metagenome of a microbial community? What are the advantages and pitfalls of metagenomics, compared to culturing microbes?
2. What unique functions do microbes perform in ecosystems?
3. Explain the difference between carbon assimilation and dissimilation.
4. What kinds of microbial metabolism are favored in aerated environments? In anoxic environments?
5. Give some examples of microbial producers in ecosystems. Include phototrophs as well as lithotrophs. Can a microbe be both a producer and a consumer? Explain.
6. Explain the microbial relationships in various forms of symbiosis, including mutualism, commensalism, and parasitism. Outline an example of each, detailing the contributions of each partner.
7. Compare and contrast the digestive communities of the bovine rumen and the human colon with respect to taxa, metabolic pathways, and contributions to the host.
8. Compare and contrast the marine food web with the soil food web. What kinds of organisms are the producers and consumers? How many trophic levels are typically found?
9. Explain how microbes interact with each other in marine and soil habitats. Are these habitats typically uniform or patchy? How does "patchiness" affect microbial growth?

10. Compare and contrast the roles of microbes in photic and aphotic marine communities.
11. Compare and contrast the microbial activities in aerated and waterlogged soils.
12. Explain how anaerobic microbial metabolism can enrich soil for plant cultivation.
13. What are mycorrhizae, and how are they important for plant growth?
14. Explain how bacterial mutualists fix nitrogen for plants.

Thought Questions

1. Explain what you can learn about a marine microbial community from a metagenome, as compared with a metatranscriptome. How and why might the two approaches yield different results?
2. Explain, with specific examples, what mutualism and parasitism have in common, and how they differ. Describe an example of a relationship that combines aspects of both.
3. The photic zone and the benthic zone pose challenges and opportunities for marine microbes. What challenges do they have in common, and how do they differ?

Key Terms

algal bloom (865)
amensalism (850)
amplicon (833)
assembly (831)
assimilation (842)
autofluorescence (837)
bacteroid (876)
barcoding (831)
barophile (862)
benthic organism (858)
bin (833)
binning (833)
biochemical oxygen demand (biological oxygen demand) (BOD) (858)
biomass (842)
cannulated cow (854)
cell sorting (837)
coastal shelf (858)
commensalism (850)
community (828)
computational pipeline (832)
consumer (843)
contig (831)
coral bleaching (849)
dead zone (865)
decomposer (843)
detritus (843)
dissimilation (842)
ecosystem (828)
ectomycorrhizae (873)
endolith (869)
endomycorrhizae (873)
endophyte (875)
enrichment culture (840)
euphotic zone (858)
eutrophic (865)
extremophile (844)
fistulated cow (854)
flow cytometry (837)
fluorescence in situ hybridization (FISH) (839)
food web (842)
grazer (843)
haustorium (879)
holobiont (851)
humic material (humus) (870)
hydric soil (874)
hydrothermal vent (864)
library construction (831)
lichen (848)
lignin (869)
limiting nutrient (866)
marine snow (861)
metabarcoding (831)
metatranscriptomics (835)
microbiome (828)
microplankton (861)
mixotroph (862)
mutualism (848)
mycorrhizae (872)
niche (839)
niche construction (839)
oligotrophic (865)
omics (828)
operational taxonomic unit (OTU) (831)
parasitism (850)
pelagic zone (857)
photic zone (858)
phytoplankton (862)
piezophile (862)
plankton (861)
population (828)
predator (843)
primary producer (842)
psychrophile (862)
rare biosphere (838)
read (831)
rhizobium (876)
rhizosphere (869)
rumen (853)
scaffold (832)
succession (833)
symbiosis (845)
synergism (850)
syntrophy (853)
target community (830)
thermal vent (864)
thermocline (858)
thermophile (864)
trophic level (842)
vesicular-arbuscular mycorrhizae (VAM) (873)
water table (869)
wetland (874)

Recommended Reading

Brune, Andreas. 2014. Symbiotic digestion of lignocellulose in termite guts. *Nature Reviews. Microbiology* **12**:168–180.

Bryant, Jessica A., Tara M. Clemente, Donn A. Viviani, Allison A. Fong, Kimberley A. Thomas, et al. 2016. Diversity and activity of communities inhabiting plastic debris in the North Pacific Gyre. *mSystems* **1**:e00024-16.

Conway, Tyrrell, and Paul S. Cohen. 2014. Commensal and pathogenic *Escherichia coli* metabolism in the gut. *Microbiology Spectrum* **3**:1–15.

Donaldson, Gregory P., S. Melanie Lee, and Sarkis K. Mazmanian. 2016. Gut biogeography of the bacterial microbiota. *Nature Reviews. Microbiology* **14**:20–32.

Dubilier, Nicole, Claudia Bergin, and Christian Lott. 2008. Symbiotic diversity in marine animals: The art of harnessing chemosynthesis. *Nature Reviews. Microbiology* **6**:725–740.

Evans, Paul N., Donovan H. Parks, Grayson L. Chadwick, Steven J. Robbins, Victoria J. Orphan, et al. 2015. Methane metabolism in the archaeal phylum Bathyarchaeota revealed by genome-centric metagenomics. *Science* **350**:434–438.

Fierer, Noah. 2017. Embracing the unknown: Disentangling the complexities of the soil microbiome. *Nature Reviews. Microbiology* **15**:579–590.

Gao, Cheng, Liliam Montoya1, Ling Xu1, Mary Madera1, Joy Hollingsworth, et al. 2019. Strong succession in arbuscular mycorrhizal fungal communities. *ISME Journal* **13**:214–226. https://www.nature.com/articles/s41396-018-0264-0.

Huq, Anwar, Mohammed Yunus, Syed Salahuddin Sohel, Abbas Bhuiya, Michael Emch, et al. 2010. Simple sari filtration is sustainable and continues to protect villagers from cholera in Matlab, Bangladesh. *mBio* **1**:e00034-10.

Hurwitz, Bonnie L., and Matthew B. Sullivan. 2013. The Pacific Ocean Virome (POV): A marine viral metagenomic dataset and associated protein clusters for quantitative viral ecology. *PLoS One* **8**:e57355.

International Glossina Genome Initiative. 2014. Genome sequence of the tsetse fly (*Glossina morsitans*): Vector of African trypanosomiasis. *Science* **344**:380–386.

Meadow, James, Adam E. Altrichter, Ashley C. Bateman, Jason Stenson, G. Z. Brown, et al. 2015. Humans differ in their personal microbial cloud. *PeerJ* **3**:e1258.

Redmond, Molly C., and David L. Valentine. 2012. Natural gas and temperature structured a microbial community response to the *Deepwater Horizon* oil spill. *Proceedings of the National Academy of Sciences USA* **109**:20292–20297.

Rettedal, Elizabeth A., Heidi Gumpert, and Morten O. A. Sommer. 2018. Cultivation-based multiplex phenotyping of human gut microbiota allows targeted recovery of previously uncultured bacteria. *Nature Communications* **5**:4714.

Rodrigues, Jorge L. M., Vivian H. Pellizari, Rebecca Mueller, Kyunghwa Baek, Ederson da C. Jesus, et al. 2013. Conversion of the Amazon rainforest to agriculture results in biotic homogenization of soil bacterial communities. *Proceedings of the National Academy of Sciences USA* **110**:988–993.

Sato, Tomoyuki, Yuichi Hongoh, Satoko Noda, Satoshi Hattori, Sadaharu Ui, et al. 2009. *Candidatus* Desulfovibrio trichonymphae, a novel intracellular symbiont of the flagellate *Trichonympha agilis* in termite gut. *Environmental Microbiology* **11**:1007–1015.

Sharon, Itai, Michael Kertesz, Laura A. Hug, Dmitry Pushkarev, Timothy A. Blauwkamp, et al. 2015. Accurate, multi-kb reads resolve complex populations and detect rare microorganisms. *Genome Research* **25**:534–543.

Slaby, Beate M., Thomas Hack, Hannes Horn, Kristina Bayer, and Ute Hentschel. 2017. Metagenomic binning of a marine sponge microbiome reveals unity in defense but metabolic specialization. *ISME Journal* **11**:2465–2478.

Thaiss, Christoph A., David Zeevi, Maayan Levy, Gili Zilberman-Schapira, Jotham Suez, et al. 2014. Transkingdom control of microbiota diurnal oscillations promotes metabolic homeostasis. *Cell* **159**:514–529.

Vigneron, Adrien, Perrine Cruaud, Patricia Pignet, Jean-Claude Caprais, Marie-Anne Cambon-Bonavita, et al. 2013. Archaeal and anaerobic methane oxidizer communities in the Sonora Margin cold seeps, Guaymas Basin (Gulf of California). *ISME Journal* 7:1595–1608.

Yoshida, Shosuke, Kazumi Hiraga, Toshihiko Takehana, Ikuo Taniguchi, Hironao Yamaji, et al. 2016. A bacterium that degrades and assimilates poly(ethylene terephthalate). *Science* **351**:1196–1199.

CHAPTER 22

Element Cycles and Environmental Microbiology

22.1 The Carbon Cycle and Climate Change

22.2 The Hydrologic Cycle and Wastewater Treatment

22.3 The Nitrogen Cycle

22.4 Sulfur, Phosphorus, and Metals

22.5 Our Built Environment

22.6 Astrobiology

Microbes throughout the biosphere recycle carbon, nitrogen, and other elements essential for all life. Through their biochemical transformations, diverse microbial activities largely determine the quality of soil, air, and water. Today, all of these geochemical cycles are altered profoundly by human activity. Microbes can help us manage environmental change—globally as well as locally, in our "built environment." From local wastewater treatment to the control of greenhouse gases, microbes are our hidden partners on Earth. And Earth's microbial cycles lead us to wonder whether biospheres exist on other worlds.

CURRENT RESEARCH highlight

Red algae melt Arctic snow. As Greenland snow starts to melt, it turns red with algae. The snow algae are cold-adapted species of *Chlamydomonas* whose red color arises from carotenoid pigments. Liane Benning's team at the German Research Center for Geosciences examines how dark-pigmented microbes melt snow and ice by absorbing light energy, which is released as heat. Benning showed that algae on snow and ice decrease the albedo (proportion of light reflected) and accelerate warming. Thus, Arctic microbial communities influence global climate and rising sea levels.

Source: Stefanie Lutz et al. 2016. *Nat. Commun.* **7**:11968.

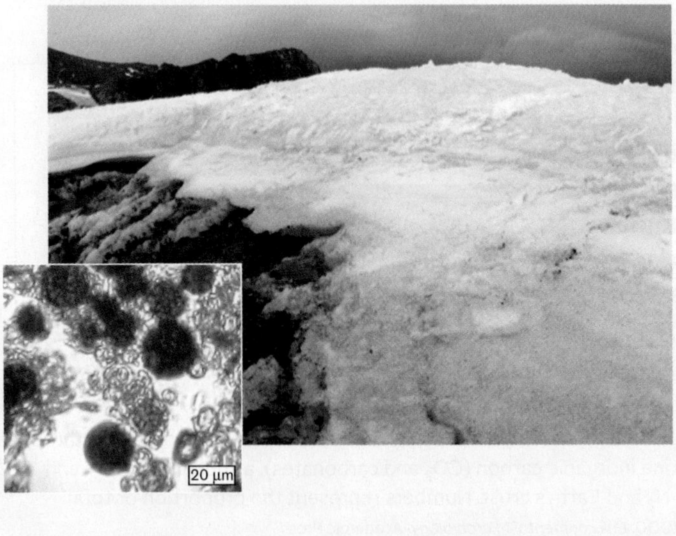

STEFANIE LUTZ, GFZ

COURTESY OF LIANE BENNING

AN INTERVIEW WITH

LIANE BENNING, GEOBIOLOGIST, HELMHOLTZ CENTRE POTSDAM

COURTESY OF PHILIP DERA

Will "red snow" accelerate the rising sea level?

Snow algae develop when melting starts on the surface snow covering a glacier. They belong to the Chlamydomonaceae. They are green, yellow, orange, or fully red depending on the type and developmental stages (mostly linked to carotenoid development and the xanthophyll cycle). They are the primary producers that affect the albedo and help increase the melt rates during snow melt. After the snow melts away, the purplish ice algae become the main carbon cyclers and, through their pigmented blooms, also the main albedo reducers. A lower albedo, in turn, accelerates the snow/ice melt. Thus, indeed, the snow and ice algae have a positive feedback on sea-level rise.

Throughout most of this book, we present microbial biochemistry in the growth of individual organisms. Here in Chapter 22, we show how the collective metabolic activities of microbial populations generate global cycles of elements throughout Earth's biosphere. Gases such as CO_2 trap solar radiation and release it as heat—a phenomenon known as the **greenhouse effect**. Greenhouse gases are generated by bacteria, methanogenic archaea, and other life forms. But human technology accelerates the release of greenhouse gases at a rate that the planet's ecosystem never evolved to tolerate. As snow and ice melt earlier, the associated growth of microbes may, in turn, accelerate the melting and warming (see the Current Research Highlight). Will the human-induced global climate change cause mass extinctions of a majority of Earth's species—as did the rise of ancient cyanobacteria? Or can we use our knowledge of microbial ecology to channel microbial activities into recovering the balance—for example, by increasing microbial CO_2 fixation?

Microbes acquire their elements either from nonliving components of their environment, such as by fixing atmospheric CO_2, or from other organisms, by grazing, predation, or decomposition. Furthermore, all organisms recycle their components back to the biosphere. The partners in this recycling include **abiotic** entities such as air, water, and minerals, as well as **biotic** entities such as predators and decomposers. All elements used for life flow in **biogeochemical cycles** of nutrients throughout the biotic and abiotic components of the biosphere. Collectively, the metabolic interactions of microbial communities with the biotic and abiotic components of their ecosystems are known as **biogeochemistry** or **geomicrobiology**.

The challenge in studying biogeochemical cycles is quantification, adding up all the inputs and outputs of key molecules on a global scale. The parts of the biosphere that contain major amounts of an element needed for life are called **reservoirs** (**Fig. 22.1**). Each reservoir acts both as a **source** of that element for living organisms and as a **sink** to which the element returns. The transfer of elements between sources and sinks, via biotic and abiotic processes, forms the global biogeochemical cycles.

The rate of cycling between sources and sinks determines the importance of a reservoir for the biosphere. For example, Earth's crust is the planet's largest reservoir for carbon (**Fig. 22.1A**), but the carbon cycles extremely slowly. For the biosphere, the major source of carbon (CO_2 in equilibrium with HCO_3^-, bicarbonate ion) is the ocean, which cycles with CO_2 in the atmosphere. The dissolving of atmospheric CO_2 into marine water is the most important abiotic portion of the carbon cycle. Biotic cycling occurs when organotrophs release CO_2 into the atmosphere, and when phototrophs and lithotrophs fix CO_2. Participating

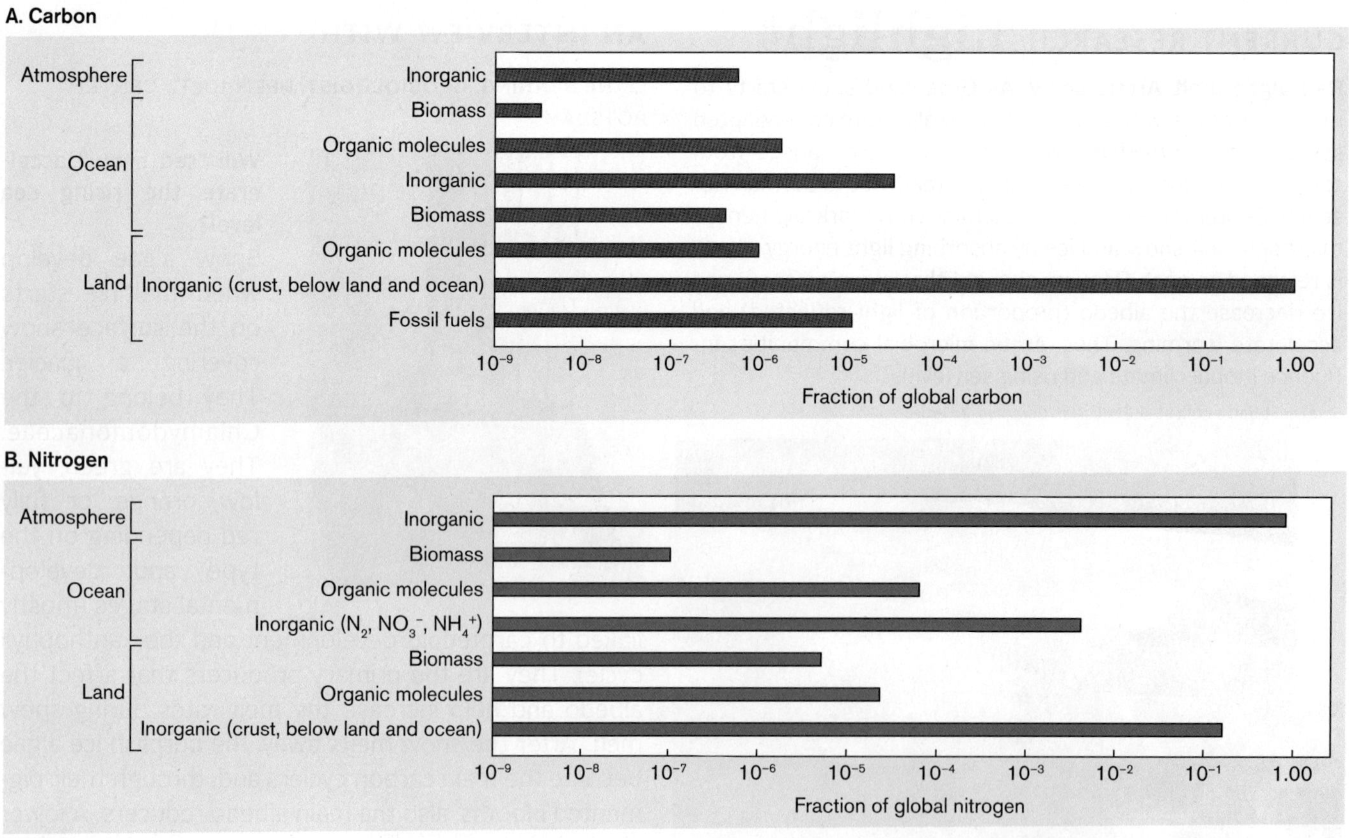

FIGURE 22.1 ■ Global reservoirs of carbon and nitrogen. **A.** The largest reservoir of carbon is Earth's crust, but carbon cycles through the biosphere extremely slowly. The major reservoirs for cycling are marine inorganic carbon (CO_2 and carbonates), atmospheric CO_2, and buried fossil fuels. **B.** The largest reservoirs of nitrogen are atmospheric N_2 and Earth's crust. Numbers represent the proportion of total global carbon and nitrogen, respectively. *Source:* Modified from R. Maier et al. 2000. *Environmental Microbiology.* Academic Press.

organisms include microbes, as well as multicellular plants and animals (discussed in Section 22.1).

The relative importance of global sources and sinks varies with each element. For nitrogen (**Fig. 22.1B**), Earth's dominant source is nitrogen gas (N_2) in the atmosphere. The only biotic processes that fix N_2 into biomass, and that return N_2 to the atmosphere, are performed by bacteria and archaea (Section 22.3). Thus, the nitrogen cycles of all ecosystems require microbes.

Over the past century, however, Earth's cycles of carbon and nitrogen have been altered by human technology. The result has been a rapid rise of the greenhouse gases CO_2, methane, and nitrous oxide (**Fig. 22.2A**). The burning of fossil fuels transfers terrestrial carbon into the atmosphere, with an accelerating greenhouse effect on Earth's temperature. And about half of atmospheric nitrogen is now fixed by human chemical factories into agricultural fertilizer (Section 22.3). The excess reduced nitrogen is then oxidized by lithotrophic microbes (Chapter 14), some of which release nitrous oxide. In 2019, the United Nations Environment Programme found that greenhouse gas emissions need to decrease by half in order to avoid a global temperature increase with potentially catastrophic effects on the planet. Concern about the scientific reports has inspired an international youth movement led by activists such as Swedish teenager Greta Thunberg (**Fig. 22.2B**).

A.

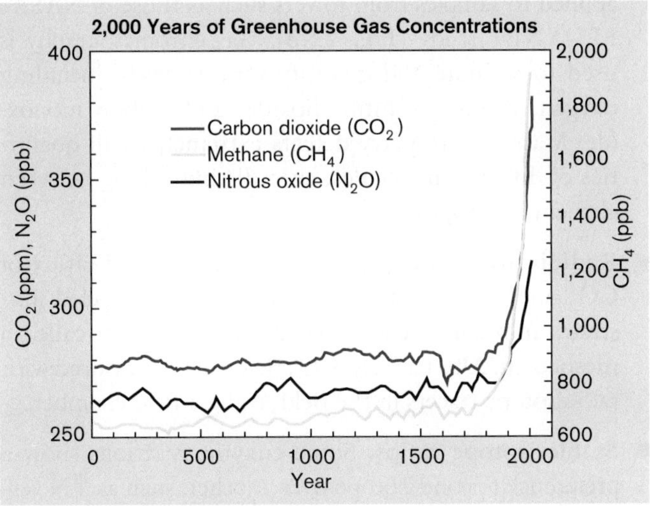

B.

UPI/ALAMY STOCK PHOTO

FIGURE 22.2 ■ Global greenhouse gases. (**A**) Atmospheric levels of CO_2, methane, and nitrous oxide were measured in Antarctic ice cores. The increase of these greenhouse gases since 1750 accompanies the rise in fossil fuel burning and fertilizer-intensive agriculture. (**B**) At the United Nations Climate Action Summit, teen activist Greta Thunberg demanded that world leaders take action to halt carbon emissions. *Source for (A):* Modified from U.S. Global Change Research Program. 2009. *Global Climate Change Impacts in the United States.* Cambridge University Press.

On a more local level, we also consider the role of microbial communities in our **built environment** (Section 22.5). The built environment consists of human-made constructs that we inhabit on a daily basis, from our homes to our workplaces and parks. Our concept of the built environment expands as we recognize the growing impact of human activity throughout the biosphere. The one place left where life may remain untouched by humans is other planets, if life exists there (Section 22.6). The discovery in 2018 of liquid water buried deep on Mars suggests surprising possibilities for life on that planet.

22.1 The Carbon Cycle and Climate Change

The foundation of all food webs involves influx and efflux of carbon. In theory, carbonate rock forms the largest reservoir of carbon (see **Fig. 22.1A**). But Earth's crust is the source least accessible to the biosphere as a whole. Crustal rock provides carbon only to organisms at the surface and to subsurface microbes that grow extremely slowly. Thus, subsurface carbon turnover is very slow. The carbon reservoir that cycles most rapidly is that of the atmosphere, a source of CO_2 for photosynthesis and chemolithoautotrophy. The atmosphere also acts as a sink for CO_2 produced by heterotrophy and by geological outgassing from volcanoes.

The atmospheric reservoir of carbon is much smaller than other sources, such as the oceans, crustal rock, and fossil fuels. For this reason, the industrial burning of fossil fuels has perturbed the balance between atmospheric CO_2 and larger reservoirs, such as the ocean. The ocean actually absorbs a good part of the extra CO_2 in equilibrium with carbonates, and microbial photosynthesis traps carbon in biomass. But despite the ocean's carbon capture, atmospheric CO_2 is rising at an ever-faster rate, unprecedented in Earth's history (**Fig. 22.2**). Today, CO_2 level increases at an annual rate of 2 parts per million (ppm), which is twice the rate that was observed less than a century ago, in the 1960s. In 2013, the Mauna Loa Observatory in Hawaii measured a CO_2 level of 400 ppm—the highest level on Earth for the past 3 million years—and it continues to increase.

TABLE 22.1 Oxidation States of Carbon Compounds

Oxidation state	Carbon molecule	
−4	CH_4	Methane
−2	CH_3OH, $(CH_2)n$	Methanol, hydrocarbon
0	$(CH_2O)n$	Carbohydrate
+2	CHOOH	Formic acid
+4	CO_2, HCO_3^-	Carbon dioxide, bicarbonate ion

Oxidation-State Changes during Carbon Flux

As elements cycle from sources to sinks, microbial metabolism generates a series of redox changes (discussed in Chapter 14). Since the planet was formed, oxidation of carbon has had a profound impact on the biosphere. Two billion years ago, ancient cyanobacteria were the first to photolyze water and produce molecular oxygen (Chapter 17). Oxygen is a powerful oxidant, lethal to most life in that anoxic era, when all living organisms were anaerobic microbes. To survive, microbes evolved defenses such as antioxidant molecules and enzymes. Since then, microbes have shaped our biosphere by releasing oxygen, by fixing nitrogen and returning it to the atmosphere, and by fixing and producing carbon dioxide.

The major oxidation states of carbon are summarized in **Table 22.1.** Biospheric carbon can be found as CH_4 generated by methanogens (completely reduced, −4), as CO_2 produced by respiration and fermentation (completely oxidized, +4), or as one of various intermediate states of oxidation (carbohydrates, alcohols, acids). Near Earth's surface, carbon is found in more oxidized states, as living organisms rapidly combine it with oxygen to yield energy. Reduced forms of carbon, such as methane and hydrocarbons (petroleum) are found buried deep underground, where oxygen gas has been unavailable for millions of years.

The results of carbon cycling differ greatly, depending on the presence of molecular oxygen. Thus, the global cycle of carbon is closely linked to the cycles of oxygen and hydrogen, elements to which most carbon is bonded. Overall, carbon cycles between carbon dioxide (CO_2) and various reduced forms of carbon, including biomass (living material).

Thought Question

22.1 Why is oxidation state important for microbes to use and cycle compounds? Cite examples based on your study of microbial metabolism (see Chapter 14).

Experimental Measurement of Element Cycling

How do we study the cycling of elements on a global scale? How do we figure out whether ecosystems are net sources or sinks of CO_2? Does microbial activity enhance or limit availability of nitrogen? Rates of flux of elements in the biosphere are very difficult to measure, yet the questions have enormous political and economic implications.

To measure environmental carbon, nitrogen, and other elements, various methods are used. These methods fall under the following categories:

- **Chemical and spectroscopic analysis.** Bulk quantities of CO_2, nitrates, and other chemicals can be determined by sophisticated chemical instrumentation. Atmospheric CO_2 is measured by infrared absorption spectroscopy, applied to samples from towers such as those of NASA's FLUXNET study (**Fig. 22.3**). Gas chromatography is used to separate and quantify various gases, including oxygen, nitrogen, sulfur dioxide, and carbon monoxide. Mass spectrometry detects extremely small quantities of different molecules, even distinguishing between elemental isotopes.
- **Radioisotope incorporation.** The influx and efflux of CO_2 can be measured by the uptake of ^{14}C-labeled substrates in a small, controlled model ecosystem called a **mesocosm**. Alternatively, CO_2 flux can be measured with radioisotope tracers in the field, using a field chamber.
- **Stable isotope ratios.** Some enzyme reactions show a preference for one isotope over another, such as ^{14}N versus ^{15}N. Both isotopes are stable (nonradioactive). For example, denitrifiers (bacteria that metabolize nitrate) strongly prefer the ^{14}N isotope, leaving behind nitrate enriched in ^{15}N. The $^{14}N/^{15}N$ ratio is measured using mass spectrometry. Measuring nitrogen isotope ratios can indicate whether denitrifiers could have conducted metabolism in the sample.

Most flux measurements apply to gases, which are the easiest forms to sample, but the data leave unanswered many questions about the deep ocean and subsurface. Subsurface studies require drilling for samples, or more exotic kinds of remote sensing, such as airborne imaging of magnetic resistivity to reveal underground hydrology in Antarctica. On land, terrestrial plants, particularly forest trees, sequester significant amounts of carbon—perhaps 10%–20% of the CO_2 released by burning fossil fuels. Forest carbon sequestration is shown in the international FLUXNET data (**Fig. 22.3B**). This experiment compares the seasonal patterns of net CO_2 uptake (the negative values on the *y*-axis) for forests at four different latitudes. It shows that the higher the latitude of the forest, the later in summer its CO_2 uptake peaks. Such measurements provide the basis

A.

DENNIS BALDOCCHI, BERKELEY

B.

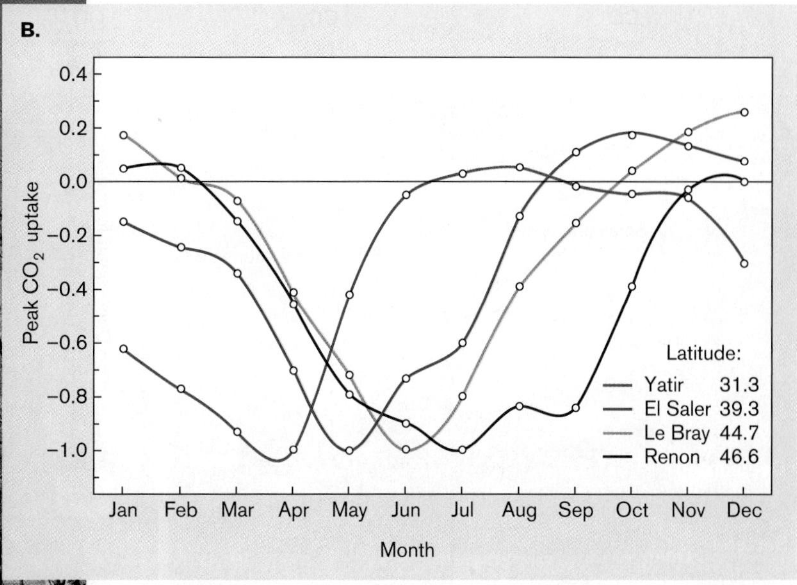

FIGURE 22.3 ■ Measuring flux of elements in the biosphere. **A.** A U.S. FLUXNET tower at Tonzi Ranch, California is used for atmospheric CO_2 sampling as part of a global effort to monitor carbon flux. **B.** Peak CO_2 uptake by forests, recorded across the seasons by FLUXNET towers at Yatir, Israel; El Saler, Spain; Le Bray, France; and Renon, Italy. *Source:* Part B modified from Kadmiel Maseyk. 2013. *FluxLetter* **5**:15.

for modeling global climate change, and for negotiating agreements to "trade" pollution for forest growth.

Note, however, that being a greenhouse gas does not make CO_2 inherently "bad" for the environment. In fact, if heterotrophic production of CO_2 were to cease altogether, phototrophs would run out of CO_2 in roughly 300 years, despite the vast quantities of carbon present in the ocean and crust. Thus, both CO_2 fixers and heterotrophs need each other for a continuous cycle.

Carbon Cycling by Different Ecosystems

Different kinds of ecosystems, such as marine and terrestrial ecosystems, cycle carbon in different ways. Access to oxygen is a major factor in determining the rate of carbon cycling and the form of carbon stored in sinks.

Marine carbon cycling. The largest aerated ecosystem is the photic zone of oceans (**Fig. 22.4A**). In the aerated (or oxic) habitat of the marine photic zone, the ecosystem absorbs enough light for the rate of photosynthesis to exceed the rate of heterotrophy. Photosynthesis drives what is called the **biological carbon pump**. Microbial and plant photosynthesis fixes CO_2 into biomass, designated by the shorthand $[CH_2O]$. Photoautotrophs that fix carbon include bacteria and protists. Marine phototrophs such as diatoms and coccolithophores trap a substantial amount of carbon in biomass (**Fig. 22.4B**). A portion of their biomass sinks to the ocean floor through the weight of their silicate or carbonate exoskeletons.

Photosynthetic CO_2 fixation is accompanied by release of O_2. The O_2 is then used by heterotrophs (such as bacteria, protists, and animals) to convert CH_2O back to CO_2. In the presence of light, a net excess of O_2 is released. Biomass is also produced through lithotrophy or chemolithoautotrophy—the oxidation of hydrogen, hydrogen sulfide, ferrous iron (Fe^{2+}) and other reduced minerals, and even carbon monoxide.

Marine cycling of methane. At the benthos, marine microbial metabolism generates a redox gradient from oxic to anoxic (discussed in Chapter 14). The anoxic region supports methanogenic archaea. The methane released forms methane hydrates (discussed in Chapters 19 and 21). Warming of methane hydrates may lead to release of gaseous methane, which in turn amplifies the greenhouse effect that warms the planet, in a positive feedback loop.

Can methane release be prevented or limited? At the ocean floor, some of the methane hydrates are oxidized by microbial mats of sulfate-reducing bacteria and anaerobic methane-oxidizing archaea (ANME), discussed in Section 19.4. The sulfate reducers plus the methane oxidizers

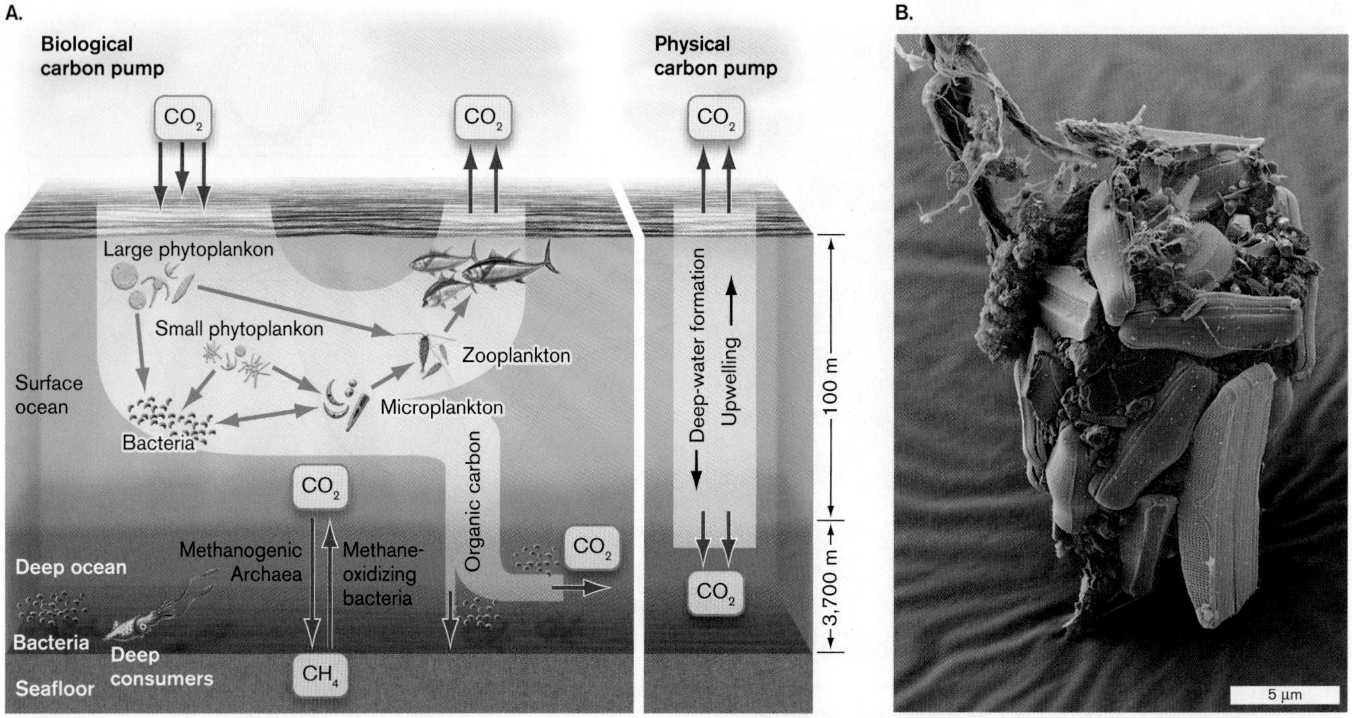

EYE OF SCIENCE/SCIENCE SOURCE

FIGURE 22.4 ▪ The ocean's biological carbon pump. A. CO_2 enters equilibrium with carbonates, releasing H^+ ions; as a result, the ocean acidifies. In the biological carbon pump, CO_2 is fixed by phytoplankton. Aerobic respiration by zooplankton and mixotrophs converts some biomass back to CO_2, while other biomass sinks to the deep ocean. **B.** Diatoms in lake sediment store carbon (colorized SEM).

conduct a syntrophic reaction, for which the overall ΔG value is negative:

$$CH_4 + SO_4^{2-} \longrightarrow HCO_3^- + HS^- + H_2O$$

In this metabolism, methane is the initial electron donor oxidized by the ANME partner, and sulfate is the terminal electron acceptor reduced by the bacteria. The high sulfate concentration of marine water drives this reaction in anoxic sediment, where all of the O_2 has been consumed by microbes that oxidize upwelling reduced minerals (such as sulfide oxidizers at thermal vents and cold seeps, discussed in Chapter 21).

In 2012, researchers reported a surprising source of marine methane from aerated water: the aerobic ammonia-oxidizing archaea, such as *Nitrosopumilus* (discussed in Chapter 19). William W. Metcalf and colleagues at the University of Illinois at Urbana-Champaign showed that *Nitrosopumilus* species conduct reactions that degrade phosphonates, organic compounds containing a direct carbon-phosphorus bond. The reactions release methylphosphonate, a compound that many kinds of bacteria convert to methane in order to acquire the scarce phosphate for phospholipids and nucleic acids. More recent research shows that many bacteria, as well as archaea, release methane from methylphosphonate in the aerated water column.

Terrestrial carbon cycling. Another large aerobic habitat is the oxygenated layer of soil (discussed in Chapter 21; **Fig. 22.5A,** top). While plants perform most of the terrestrial photosynthesis, microbial phototrophs found in soil also fix CO_2 into biomass. Microbial lithotrophy fixes carbon in soil and in weathered areas of crustal rock. Lithotrophy is performed solely by bacteria and archaea, essential microbial partners in these ecosystems. Unlike photosynthesis, lithotrophy usually consumes O_2 instead of producing it. Much carbon is stored as carbon polymers such as cellulose and lignin. But microbial and animal consumers release a lot of CO_2 through respiration.

In 2013, another important microbial carbon sink was discovered by Karina Clemmensen, Björn Lindahl, and colleagues at Uppsala BioCenter, Sweden: the mycorrhizal fungi (presented in Chapter 21). Mycorrhizal fungi (**Fig. 22.5B**) form mutualistic associations with forest plant roots and are especially important for the roots of trees. Clemmensen showed that in some forests, the roots and fungi can sequester as much as 22 kilograms of carbon per square meter of forest soil, which may be 70% of the total carbon sequestered. Thus, mycorrhizal fungi play an important role in minimizing release of the greenhouse gas CO_2.

Beneath the aerated soil, microbial metabolism generates a steep redox gradient down into anoxic layers (discussed in Chapters 14 and 21). Anoxic regions support lower rates of biomass production than do oxygen-rich environments because they depend on oxidants of lower redox potential and limited quantity, such as Fe_3^+. Anaerobic conversion of CO_2 to biomass is done mainly by bacteria and archaea. Vast, permanently anaerobic habitats extend several kilometers below Earth's surface, encompassing a greater volume

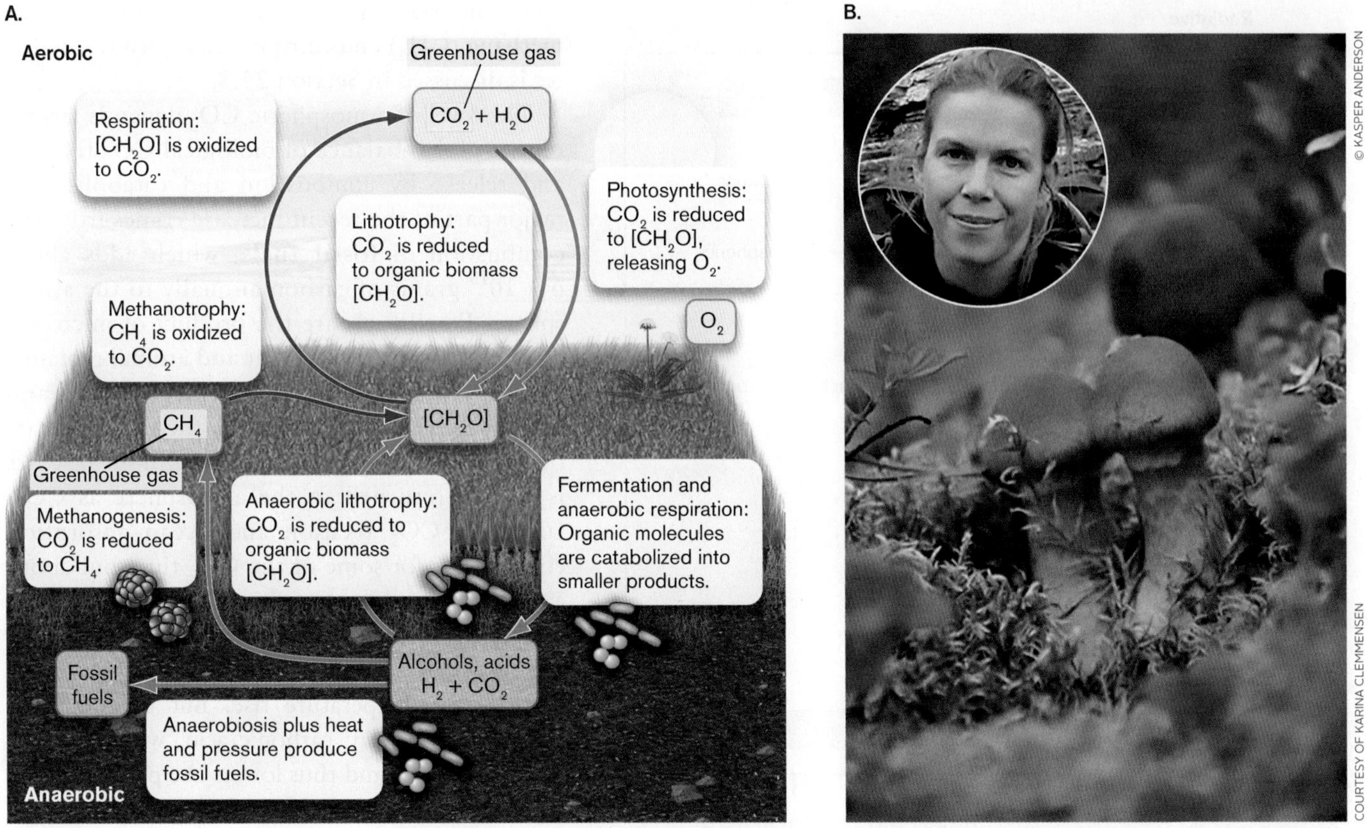

FIGURE 22.5 ▪ The terrestrial carbon cycle: aerobic and anaerobic. A. Aerobic and anaerobic conversions of carbon. Blue = reduction of carbon; red = oxidation of carbon; orange = fermentation; $[CH_2O]$ = organic biomass. In an aerobic environment (top), photosynthesis generates molecular oxygen (O_2), which enables the most efficient metabolism by heterotrophs, methanotrophs, and lithotrophs. B. *Cortinarius armillatus* mushrooms are fruiting bodies of mycorrhizal fungi that share mutualism with birch trees. **Inset:** Karina Clemmensen studies mycorrhizal sequestration of carbon.

than the rest of the biosphere put together (**Fig. 22.5A**, bottom). In these habitats, endolithic bacteria inhabit the interstices of rock crystals (discussed in Chapter 21).

In soil and water, anaerobic metabolism includes fermentation of organic carbon sources, as well as respiration and lithotrophy with alternative electron acceptors such as nitrate, ferric iron (Fe_3^+), and sulfate. Anaerobic decomposition by microbes is one stage in the formation of fossil fuels such as oil and natural gas (primarily methane). In soil, anoxic conditions (extremely low levels of O_2) favor incomplete catabolism. Recall from Chapter 13 that glycans are broken down to sugars, which undergo glycolysis to pyruvate. Without oxygen, pyruvate is rearranged via fermentation to products such as lactate, ethanol, and acetate, as well as the gases CO_2 and H_2. The partly decomposed matter becomes available for further decomposition, when terminal electron acceptors appear.

But anoxic environments also favor production of methane by methanogenic archaea (discussed in Chapters 14 and 19). Methanogenesis involves H_2 reduction of CO_2, acetate, and other bacterial fermentation products. Some methane is oxidized by methanotrophs, but an unknown amount rises to the atmosphere (**Fig. 22.6**). Geological evidence suggests that rapid methane release accompanied the retreat of the glaciers during ice age transitions. A rapid methane release today could accelerate global warming.

The Global Carbon Balance and Temperature Change

Since the beginning of the industrial age, atmospheric CO_2 has risen steeply (see **Fig. 22.2**). The rise of CO_2 is supported by several forms of evidence, most notably the

FIGURE 22.6 ▪ Researchers ignite a bubble of methane on Alaska's Seward Peninsula.

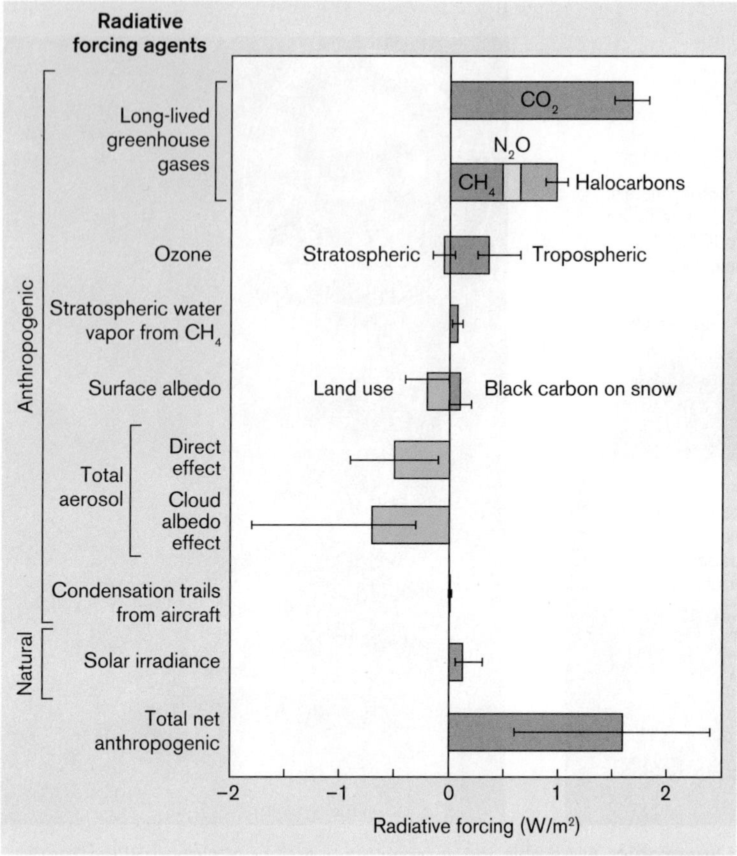

FIGURE 22.7 ■ Sources of radiative forcing that lead to global warming. Radiative forcing is the increase in warming of Earth's atmosphere (in watts per square meter) associated with a particular climate factor. The total radiative forcing ascribed to human activity (anthropogenic) is calculated at 1.6 watts per square meter (W/m^2). *Source:* Modified from Intergovernmental Panel on Climate Change. 2007. *Climate Change 2007.* Cambridge University Press, fig. SPM.2.

measurement of CO_2 concentrations in ice cores drilled from polar glaciers. Results of climate modeling are consistent with the greenhouse gas effect causing our global rise of temperature by 1°C since 1800. Another half-degree increase is likely within the next decades, and there's no sign of stopping. A total increase of 2°C will likely eradicate corals and make many population centers uninhabitable because of extreme heat. What are the roles of microbes in the rise of global temperature?

Radiative forcing. Through their release of CO_2, organotrophic organisms (including microbes, as well as multicellular eukaryotes) are at present the major agent of global temperature rise. The rise in temperature results from radiative forcing agents (**Fig. 22.7**). **Radiative forcing** is defined as the difference between the sunlight energy absorbed by Earth and the energy radiated out to space. Forcing is measured in units of solar radiation energy (watts) absorbed per square meter of the planet's surface. Climate scientists combine many sources of data to generate computed models of how the climate is changing, and which factors have the greatest effect. Besides CO_2, important agents of radiative forcing include methane (CH_4) and nitrous oxide (N_2O); the latter is discussed in Section 22.3.

The level of atmospheric CO_2 depends largely on the global balance of biological CO_2 fixation and release by combustion and catabolism. A major part of the recent increase comes from the combustion of **fossil fuels**, which adds about 6×10^{15} grams of carbon annually to the atmosphere. Fossil fuels are the product of microbial anaerobic digestion of plant and animal remains, reduced to hydrocarbons by the pressure and heat of Earth's crust. When burned as fuel, carbon that had accumulated over millions of years is rapidly returned to the atmosphere as CO_2. Increased CO_2 fixation and ocean absorption compensate for some of the CO_2 flux, but about a tenth of the carbon remains in the atmosphere.

Oceans dissolve more than half of the CO_2 entering Earth's atmosphere, which is important for limiting temperature rise. But the dissolved CO_2 reacts to form carbonic acid, which releases hydronium ions and thus lowers the pH:

$$CO_2 + H_2O \rightleftharpoons H_2CO_3 \rightleftharpoons HCO_3^{2-} + H^+ \rightleftharpoons CO_3^- + 2H^+$$

Ocean acidification is harmful to corals and zooplankton that have shells of calcium carbonate, which dissolves with acid.

Recent data show that land use for beef production is a major source of CO_2. Substantial carbon reductions could occur if consumers switch from beef burgers to plant products engineered for beef taste, such as the Impossible Burger (presented in Chapter 18).

Arctic methanogens accelerate global warming. A major agent of radiative forcing is atmospheric methane (**Fig. 22.7**). Methane is a relatively small component of the atmosphere (0.00017%), but its radiative forcing factor is 25 times that of CO_2. Methane release is increasing at exceptional rates, especially in the Arctic tundra—which is estimated to store more than twice the amount of carbon than exists in the atmosphere. What will happen as all that carbon enters the atmosphere, as CO_2 or as CH_4?

The Arctic tundra stores large quantities of dead plant matter in permanently frozen soil, called permafrost. But today the permafrost is thawing. As permafrost thaws, it accelerates microbial decomposition of material that was accumulated over tens of thousands of years (**Fig. 22.8A).** Bacterial fermentation products include CO_2 and acetate, which methanogens reduce to methane. The acceleration of methane release is amplified by a positive feedback loop (**Fig. 22.8B**).

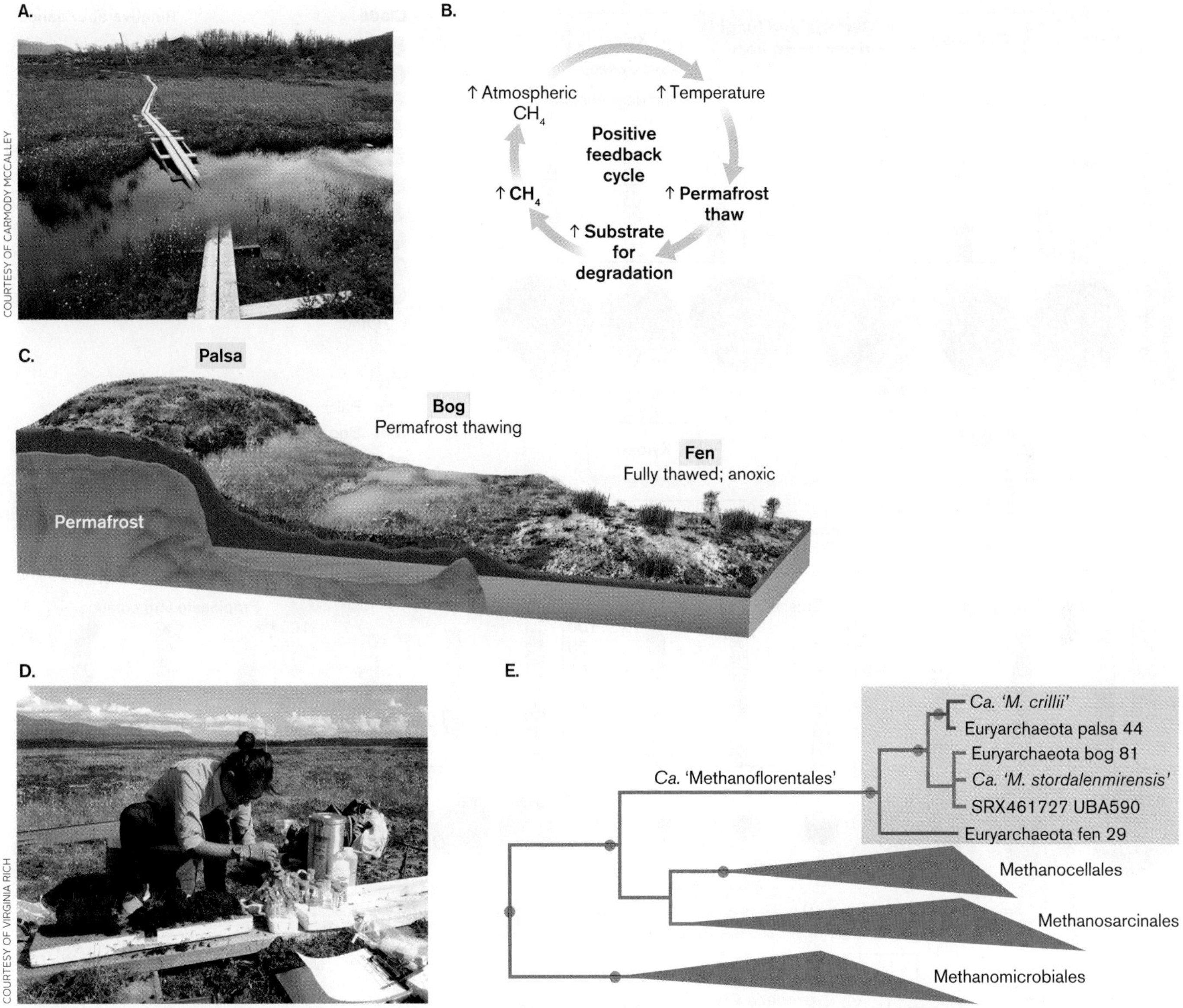

FIGURE 22.8 ▪ Thawing permafrost accelerates methane release by methanogenic archaea. **A.** Stordalen Mire, Sweden, a model site for study of thawing permafrost. **B.** Atmospheric CH_4 raises global temperature, which then accelerates thawing and methane release by methanogens, in a positive feedback cycle. **C.** Palsa, with permafrost; bog, with soil partly melted; fen, fully melted and anoxic. **D.** Eun-Hae Kim, a student of Virginia Rich, collects samples from Stordalen Mire. **E.** Phylogeny of metagenome-assembled genomes (MAGs) of methanogenic archaea identified from Stordalen bogs and fens. The scale is based on the mean number of substitutions per nucleotide. *Sources:* Parts C and E modified from Ben Woodcroft et al. 2018. *Nature* **560**:49–54, fig. 1 (C) and extended data fig. 1 (E).

As CH_4 emerges from the soil and enters the atmosphere, its radiative forcing (greenhouse effect) increases temperature, which further accelerates the thawing of permafrost.

Virginia Rich at the Ohio State University and Scott Saleska at the University of Arizona study the process of thawing permafrost and the resulting succession of microbial communities. Their study at Stordalen Mire addresses questions such as: Which microbes decompose thawed organic matter? Which species form methane? And which species consume methane before it enters the atmosphere?

The Stordalen Mire in Sweden is a model site for thawing permafrost because it includes examples of three stages: "palsa," soil above permafrost; "bog," partially thawed, with some permafrost underneath; and "fen," fully thawed, anoxic wetland (**Fig. 22.8C**). These three tundra types—palsa, bog, and fen—are color-coded brown, green, and blue, respectively, throughout the diagrams of **Figures 22.8** and **22.9.**

At Stordalen Mire, Rich's students collect samples of tundra (**Fig. 22.8D**). From the samples, the students sequence DNA and obtain metagenome-assembled genomes (MAGs) by procedures described in Chapter 21. Thawing permafrost yields diverse communities of bacteria and archaea, with key species predominating in each of the three stages. Methanogenic archaea (**Fig. 22.8E**) are most abundant in the anoxic fens, including candidate species such as *Methanoflorens stordalenmirensis*, which are unique to the Stordalen site.

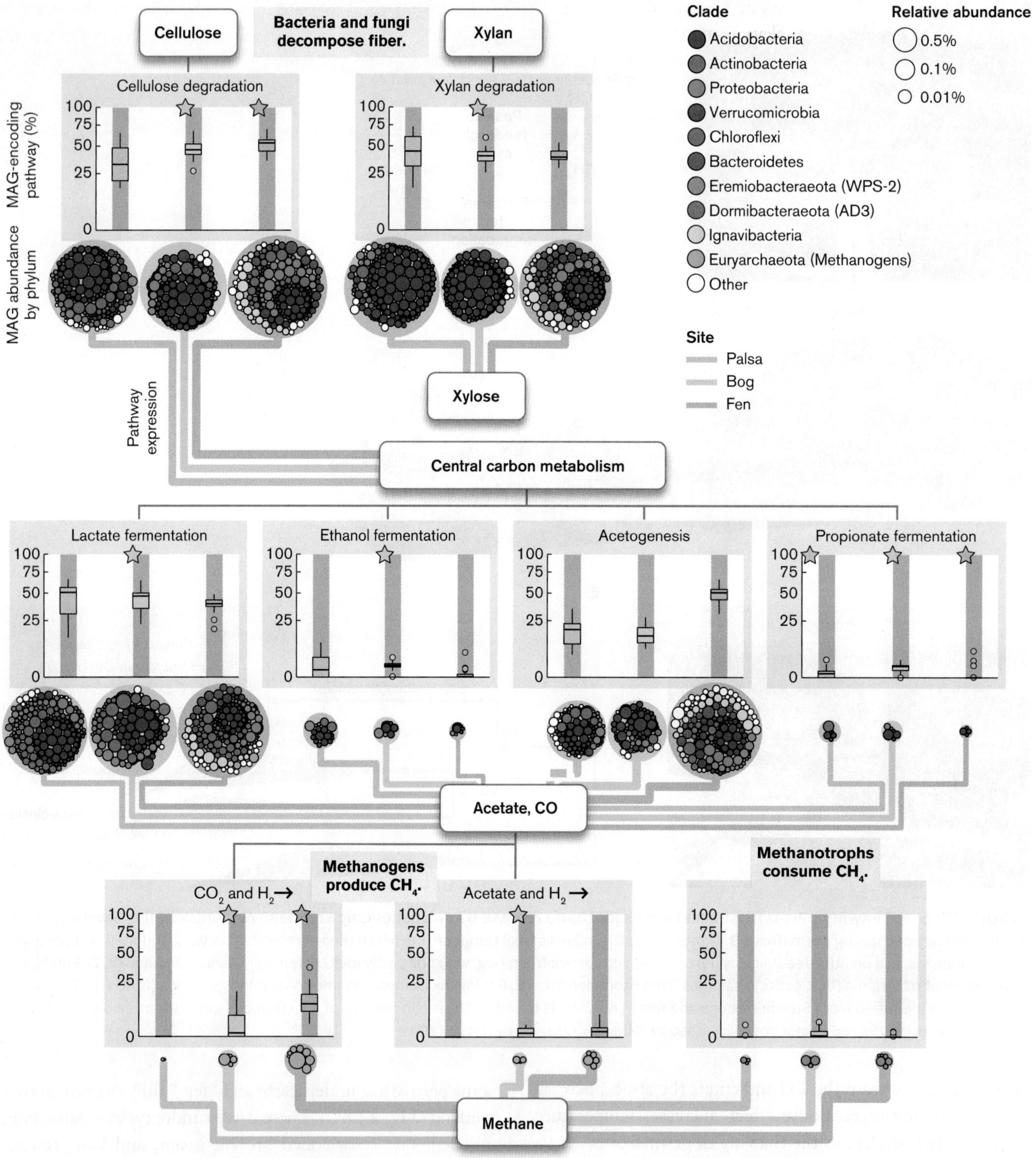

FIGURE 22.9 ■ Decomposition, methanogenesis, and methanotrophy in microbial communities of thawing permafrost (palsa, bog, and fen). Box plots show relative abundance of MAGs that encode the pathway indicated. Bubble plots contain circles colored by phylum; circle size represents the relative abundance of each MAG. *Source:* Modified from Ben Woodcroft et al. 2018. *Nature* **560**:49–54, fig. 2.

By sampling all three thawing stages (**Fig. 22.9**), Rich has mapped a picture of the overall microbial community succession (shift in predominant taxa, a concept discussed in Chapter 21). For each metabolic process, such as cellulose degradation or xylan degradation, box plots show the relative abundance of metagenome-assembled genomes (MAGs) that encode enzymes for that pathway. Each box plot is color-coded for palsa (pale orange), bog (light green) or fen (light purple). The bubble plots indicate the relative abundance of MAG taxa (coded at right).

For plant matter breakdown and fermentation, MAGs appear in all three stages of thawing. But the fens show lack of ethanol production, with elevated acetogenesis (release of acetate). The fens also show elevated methanogenesis, both by CO_2 reduction and by acetate reduction (that is, they are acetoclastic). Most interesting, Rich found MAGs of methanotrophs, methane-oxidizing bacteria. These bacteria oxidize methane to yield energy when molecular oxygen is available. Calculations show that methanotrophs consume a substantial portion of methane before it reaches the air. Thus, understanding the methanotrophic bacteria (Chapter 14) may help us find ways to mitigate the Arctic release of methane.

Human attempts to alter or mitigate global temperature rise. Could human technology provide a fix to climate change? Some human-associated factors can decrease radiative forcing, such as the production of sulfate aerosols (suspended particles or droplets) in the upper atmosphere. Measurements of these factors have led some scientists to propose highly controversial experiments to reverse or prevent global warming. However, the unpredictable potential effects of such global experiments have discouraged such attempts.

From a political standpoint, ecosystems such as forests that act as carbon sinks by fixing carbon into stable biomass are considered desirable because they lessen the rate of CO_2 input into the atmosphere. The question is complex because CO_2-generating ecosystems provide other environmental benefits. For example, wetlands are among Earth's most productive ecosystems, supporting vast amounts of plant and animal life, although they also release significant amounts of CO_2 and methane.

To Summarize

- **Microbes cycle essential elements in the biosphere.** Many key cycling reactions are performed only by bacteria and archaea.
- **Elements cycle between organisms and abiotic sources and sinks.** The most accessible source of carbon and nitrogen is the atmosphere. Earth's crust stores large amounts of key elements, but their availability to organisms is limited.
- **Carbon cycling involves changes in oxidation state.** The more oxidized forms of carbon are found near Earth's surface or in the atmosphere, whereas the more reduced forms accumulate deep underground. Atmospheric carbon cycles are severely perturbed by the burning of fossil fuels.
- **Environmental flux of elements is measured through chemistry.** Methods include infrared spectroscopy and mass spectrometry, gas chromatography, radioisotope incorporation, and the measurement of stable isotope ratios.
- **Cyanobacteria and other phytoplankton consume much of the biospheric CO_2 and release O_2.** Oxygen released by phototrophs is used by aerobic heterotrophs and lithotrophs. Marine plankton, as well as terrestrial trees and mycorrhizal fungi, serve as major carbon sinks.
- **Anaerobic environments cycle carbon through bacteria and archaea.** Bacteria conduct fermentation and anaerobic respiration, while methanogens convert CO_2 and H_2 to methane. Methane is oxidized to CO_2 by methane-oxidizing bacteria and archaea.
- **Radiative forcing is the increase of global warming associated with a particular chemical agent.** At present, the major source of radiative forcing is the net rise of atmospheric CO_2. Microbial decomposition returns CO_2 and CH_4 to the atmosphere and accelerates global warming.
- **Thawing of Arctic permafrost is a growing source of radiative forcing.** Permafrost thawing converts tens of thousands of years' accumulation of organic matter to CO_2 and CH_4.

22.2 The Hydrologic Cycle and Wastewater Treatment

The distribution of complex carbon compounds is largely a function of the **hydrologic cycle,** or **water cycle,** the cyclic exchange of water between atmospheric water vapor and Earth's ecosystems (**Fig. 22.10A**). In the hydrologic cycle, water precipitates as rain, which is drawn by gravity into groundwater, rivers, lakes, and ultimately the ocean. The ocean supplies a vast reservoir of water. All along the cycle, evaporation returns water to the air. Human communities interact with the hydrologic cycle by drawing water for drinking and other purposes, and by returning wastewater. Before wastewater can be safely returned to the hydrologic cycle, organic contaminants must be removed. Key parts of that treatment are performed by microbes.

Biochemical Oxygen Demand

What determines the health of an aquatic ecosystem? As discussed in Chapter 21, a major factor in the health of ecosystems is the balance between the level of oxygen and the levels of reduced organic nutrients. Organic contamination destabilizes marine, freshwater, and terrestrial ecosystems. Water passing through soil and aquatic ecosystems carries organic carbon material from humus, sewage, and fertilizer runoff. A sudden influx of rich carbon substrates accelerates respiration by aquatic microbes. Microbial respiration then competes with respiration by fish, invertebrates, and amphibians for the limited supply of oxygen dissolved in water, raising the **biochemical oxygen demand (BOD),**

A. Hydrologic cycle

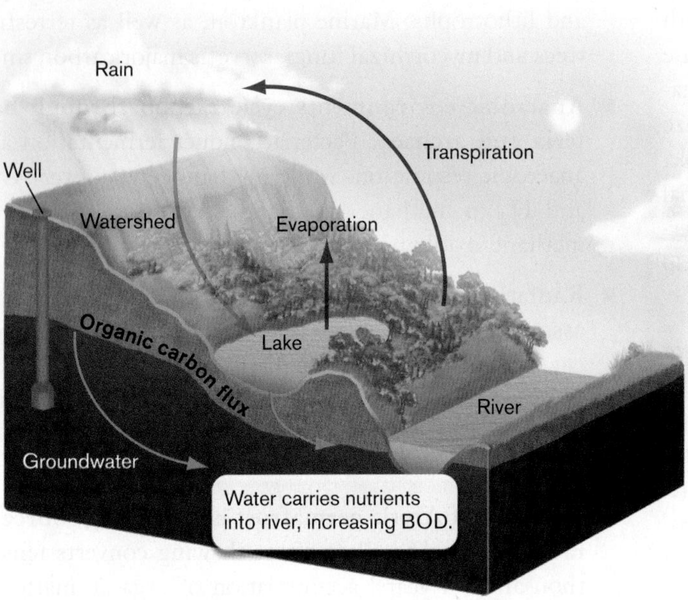

B. BOD measurement

AP PHOTO/THE CASPER STAR-TRIBUNE, ALAN ROGERS

C. BOD analyzer

ENVITECH LTD.

FIGURE 22.10 ■ The hydrologic cycle interacts with the carbon cycle. **A.** The hydrologic cycle carries bacteria and organic carbon into groundwater and aquatic systems. **B.** Bottled water samples are measured for dissolved oxygen over time; the rate of decrease of dissolved oxygen indicates biochemical oxygen demand (BOD). The rate of decrease of dissolved oxygen in water samples is approximately proportional to the concentration of organic matter available for respiration. **C.** A microprocessor-controlled BIOX-1010 BOD analyzer measures rate of respiration. Water samples are mixed with a concentrated microbial biomass, and a dissolved-oxygen (DO) sensor measures small rates of oxygen decrease over time.

also called the biological oxygen demand. The higher the concentration of organic substances, the higher the BOD arising from microbial oxygen consumption.

High BOD can cause massive die-off of fish and other aquatic animals. Thus, a routine part of monitoring the health of lakes and streams is the measurement of BOD from water samples (**Fig. 22.10B**). Oxygen uptake is observed in a BOD analyzer (**Fig. 22.10C**), which detects dissolved oxygen in water. The rate of decrease of dissolved oxygen measured by the BOD analyzer is approximately proportional to the amount of dissolved organic matter available for respiration. Note, however, that the BOD in a natural environment will depend on the microbes actually present, as well as the plants and animals competing for oxygen and other resources.

Until recently, BOD was considered a local issue, affecting the health of lakes and rivers in a community. But today we recognize huge impacts of rising BOD in the oceans (**Fig. 22.11**). Ocean oxygen levels are high near the surface, where phototrophs release oxygen, but organic nutrients are so scarce that respiration is limited. Oxygen is also high in the deep benthos, because most organic nutrients have been consumed, and because the sheer volume of water can hold dissolved oxygen in large amounts. But near the coastal shelf, currents may carry sediment up to a middle region where organic nutrients meet the oxygen. This combination supports rapid bacterial respiration. The result is an **oxygen minimum zone (OMZ)**, a region of low or near-zero oxygen sandwiched between the upper and lower oxygenated layers. Above and below the anoxic water, there is a steep oxycline (gradient of oxygen concentration). A well-known OMZ is located off the coast of Oregon, where crabs and other animals are found dying as they try to escape asphyxiation.

Oxygen minimum zones are worsened by increasing temperatures, which accelerate respiration, and by influx of sewage and agricultural waste. The zone expands upward toward the surface and downward to the sediment, trapping crabs and fish. Today, large regions of ocean have become **dead zones,** or **zones of hypoxia**, devoid of most fish and invertebrates. A major dead zone is a region in the Gulf of Mexico off the coast of Louisiana where the Mississippi

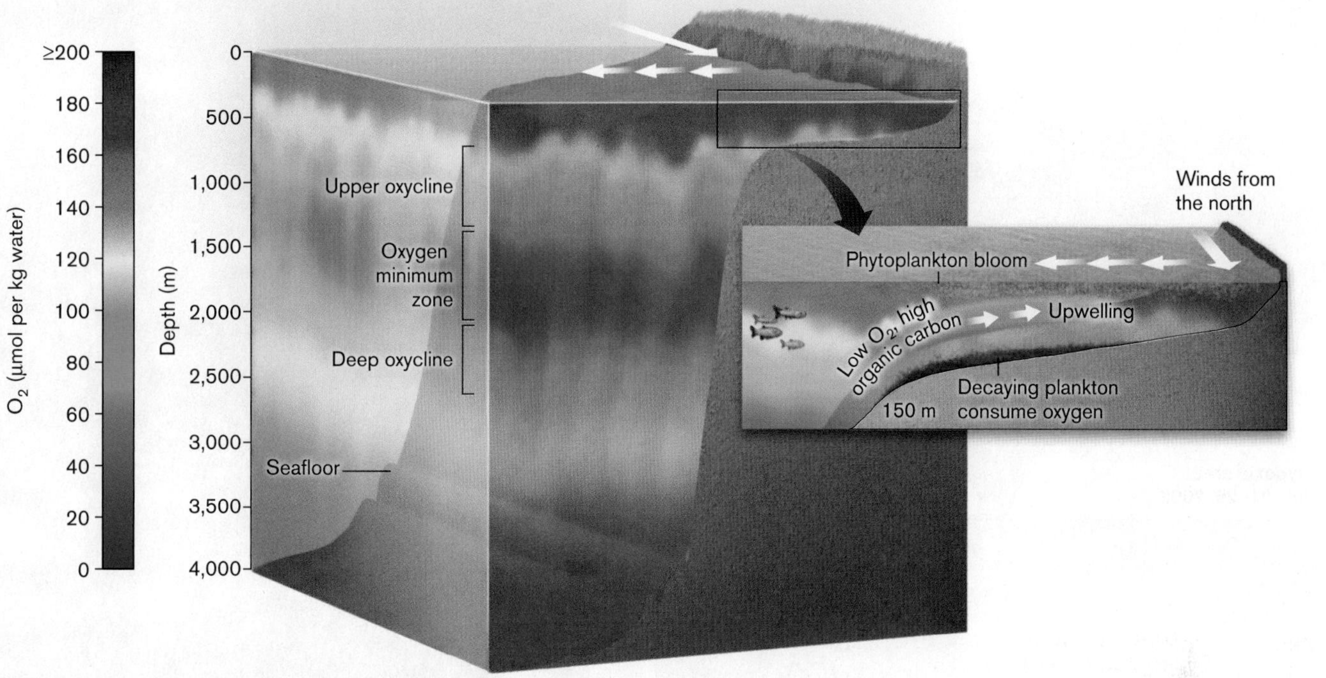

FIGURE 22.11 ■ Oxygen minimum zone in the ocean. The oxygen minimum zone (OMZ) occurs in the region of the coastal shelf where nutrients from below meet the oxygen produced by phototrophs above. In this region, microbial respiration consumes oxygen faster than the organic nutrients. The OMZ can expand upward when phytoplankton blooms decay, consuming the surface oxygen.

River releases about 40% of the U.S. drainage into the sea (**Fig. 22.12A**). Over its long, meandering course, which includes inputs from the Ohio and Missouri rivers, the Mississippi builds up high levels of organic pollutants, as well as nitrates from agricultural fertilizer. When these nitrogen-rich substances flow rapidly out to the Gulf in the spring, they lift the nitrogen limitation on algal growth and feed massive algal blooms. The algal population then crashes, and their sedimenting cells are consumed by heterotrophic bacteria. The heterotrophs use up the available oxygen, causing **hypoxia**. Hypoxia kills off the fish, shellfish, and crustaceans over a region equivalent in size to the state of New Jersey.

In 2010 the Gulf of Mexico dead zone was expanded by the unprecedented spill of oil from the *Deepwater Horizon* oil well blowout (**Fig. 22.12B**; discussed in Chapter 21). Many of the pollutants were metabolized to CO_2 by oil-eating bacteria that feed naturally on oil seeping slowly from the Gulf sediment. But the sudden large input of oil raised microbial respiration and BOD levels throughout the Gulf, thus lowering the amount of oxygen available to wildlife. Dead zones now occur along the coasts of industrial and developing countries throughout the world, including Australia and India. Dead zones deplete habitat for much marine life—for example, forcing sharks to swim out of hypoxic regions and nearer the shore. Dead zones contribute to the crash of fisheries worldwide, removing a critical food resource for human populations. To avoid such dead zones, release of industrial pollutants and fertilizers must be prevented. In addition, all the communities throughout the river drainage areas must treat their wastes to eliminate nitrogenous wastes before disposal.

There are two common approaches to community wastewater treatment, both of which involve microbial partners: wastewater treatment plants and wetland filtration. Both approaches depend on microbes to remove organic carbon and nitrogen from water before it returns to aquatic systems and ultimately the ocean.

Wastewater Treatment in Our Built Environment

In industrialized nations, we supply our "built environment" with clean water, free of toxicants and pathogens. All municipal communities use some form of **wastewater treatment** (see **Fig. 22.13**). The human sewage component of wastewater contains liquid and solid wastes produced by the human digestive and excretory systems. The purpose of wastewater treatment is to decrease the BOD and the level of human pathogens before water is returned to local rivers. The treatment process includes microbial metabolism. The process requires a highly sophisticated facility with complex engineering to maximize the efficiency of microbial catabolism. A wastewater treatment plant can convert sewage into water that exceeds all government standards for humans to drink.

The wastewater treatment plant is the final destination for all household and industrial liquid wastes passing

A. Mississippi River basin with Gulf of Mexico hypoxia

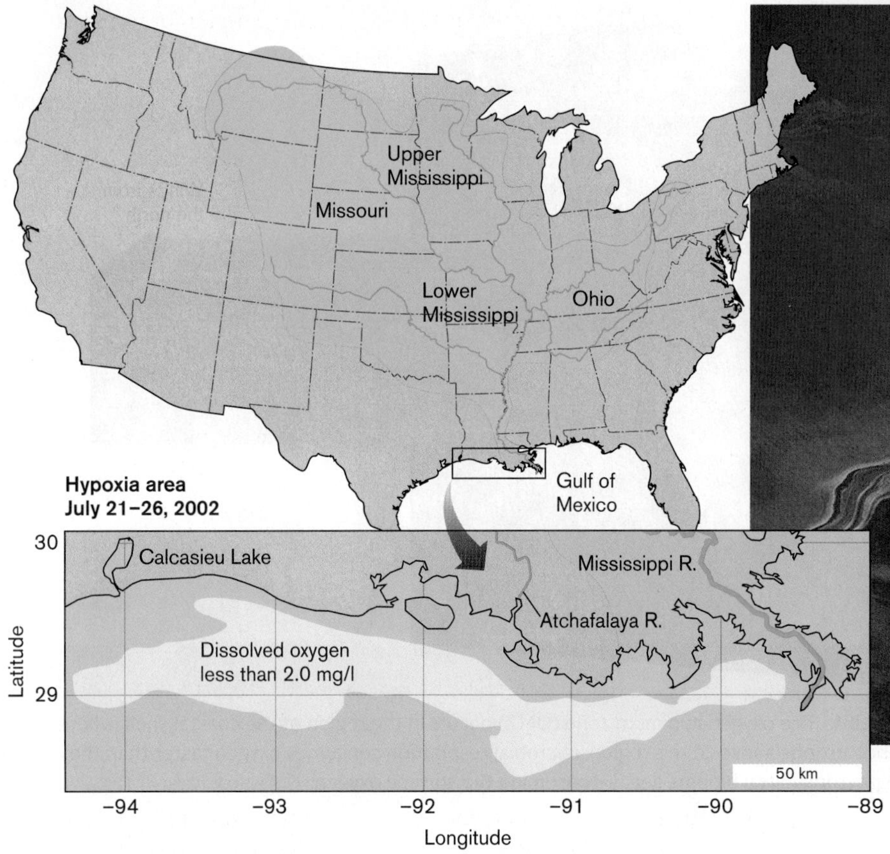

B. *Deepwater Horizon* oil spill

FIGURE 22.12 ■ The dead zone in the Gulf of Mexico. **A.** The Mississippi River drainage (shaded green) empties into the Gulf of Mexico. Every summer, drainage high in organic carbon and nitrogen causes algal blooms, leading to hypoxia and death of fish. **B.** Aerial view of the oil leaked from the *Deepwater Horizon* oil wellhead in the Gulf of Mexico in 2010. *Source:* Part A based on data from the National Oceanographic and Atmospheric Administration.

through the municipal sewage system. A typical plant includes the following stages of treatment, illustrated in **Fig. 22.13A.**

Preliminary and primary treatment. Preliminary treatment consists of screens that remove solid debris, such as sticks, dead animals, and feminine hygiene items. Primary

A.

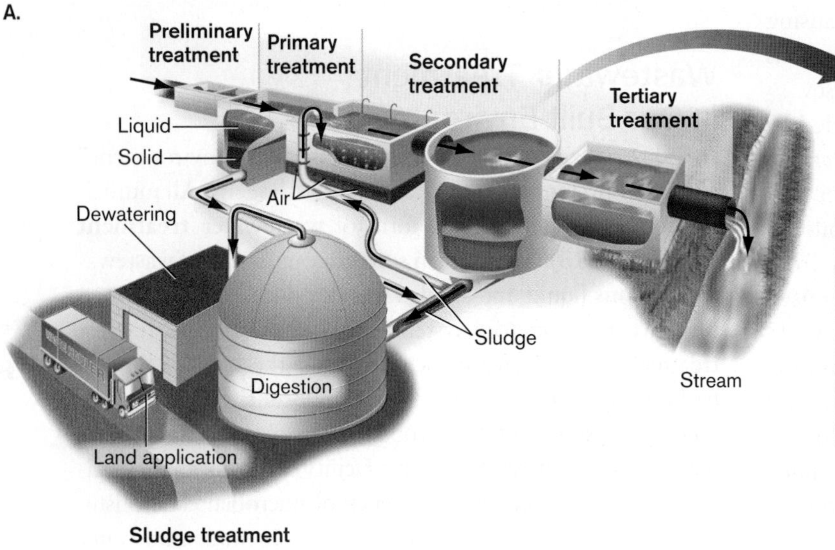

B.

FIGURE 22.13 ■ Wastewater treatment with bioremediation. **A.** In a municipal treatment plant, wastewater undergoes primary treatment (filtering and settling), secondary treatment (bioremediation by microbial decomposition), and tertiary treatment (chemical treatments including chlorination). **B.** Aeration basin for secondary treatment with microbes.

treatment includes fine screens and sedimentation tanks that remove insoluble particles. The particles eventually are recombined with the solid products of wastewater treatment to form what is known as **sludge**. The sludge ultimately is used for fertilizer or landfill.

Secondary treatment. Secondary treatment consists primarily of microbial ecosystems that decompose the soluble organic content of wastewater, by aerobic and anaerobic respiration. The nutrient removal process may include biological removal of nitrogen and phosphorus. If included at this point, nitrogen is removed by nitrifying bacteria that oxidize ammonium, and phosphorus is removed by polyphosphate-accumulating bacteria. The microbes form particulate **flocs** of biofilm. The flocs are sedimented as sludge and are also known as activated sludge, owing to their microbial activity.

The microbial ecosystems of secondary treatment require continual aeration to maximize the breakdown of molecules to carbon dioxide and nitrates. The floc size and composition must be monitored for optimal performance. Floc microbes typically include bacilli such as *Zoogloea*, *Flavobacterium*, and *Pseudomonas*, as well as filamentous species such as *Nocardia* (**Fig. 22.14A**). Optimal treatment depends on the ratio of filamentous to single-celled bacteria: enough filaments to hold together the flocs for sedimentation, but not so many as to trap air and cause flocs to float and foam, preventing sedimentation.

Besides bacteria, the ecosystem of activated sludge includes filamentous methanogens (discussed in Chapter 19), which metabolize short-chain molecules such as acetate within the anaerobic interior of flocs. Thus, wastewater treatment generates methane, often in quantities that can be recovered as fuel.

In addition, the bacteria are preyed on by protists such as stalked ciliates (**Fig. 22.14B**), swimming ciliates, and amebas, as well as invertebrates such as rotifers and nematodes. The predators serve the valuable function of limiting the numbers of planktonic single-celled bacteria, enabling the bulk of the biomass to be removed by sedimentation.

Pathogens in wastewater. The biological and chemical treatment of wastewater can be highly effective at removing bacterial pathogens such as *Salmonella* or *Vibrio cholerae*. It is less effective at removing viral pathogens, which survive higher levels of chlorination. Important viral pathogens include noroviruses, sapoviruses, and enteroviruses. Noroviruses and sapoviruses are major causes of acute viral gastroenteritis (intestinal inflammation; see Chapter 26). Enteroviruses are highly prevalent in wastewater and may cause a variety of illnesses, including poliolike symptoms. The removal of such viruses is a growing concern, as human populations enlarge and concentrate around shared water sources.

Viruses and other human pathogens are most effectively removed by membrane bioreactors (**Fig. 22.15**). The bioreactor consists of hollow-fiber membranes submerged in the biologically active floc suspension during secondary treatment. The influent undergoes microfiltration (pore size 0.1–0.4 μm) or ultrafiltration (0.01–0.04 μm). These pore sizes are not small enough for the complete removal of viruses, such as enteroviruses, which are typically 0.03 μm in size. Nevertheless, the pores exclude floc particles to which the viruses adhere, and they greatly decrease the virus titer in the effluent. Note that disposal of the virus-contaminated sludge requires further disinfection.

Tertiary (advanced) treatment. Tertiary treatment (see **Fig. 22.13A**) includes filtration of particulates from the microbial flocs of secondary treatment, and may include chemical processes to decrease nitrogen and phosphorus. The final step involves disinfection to eliminate pathogens,

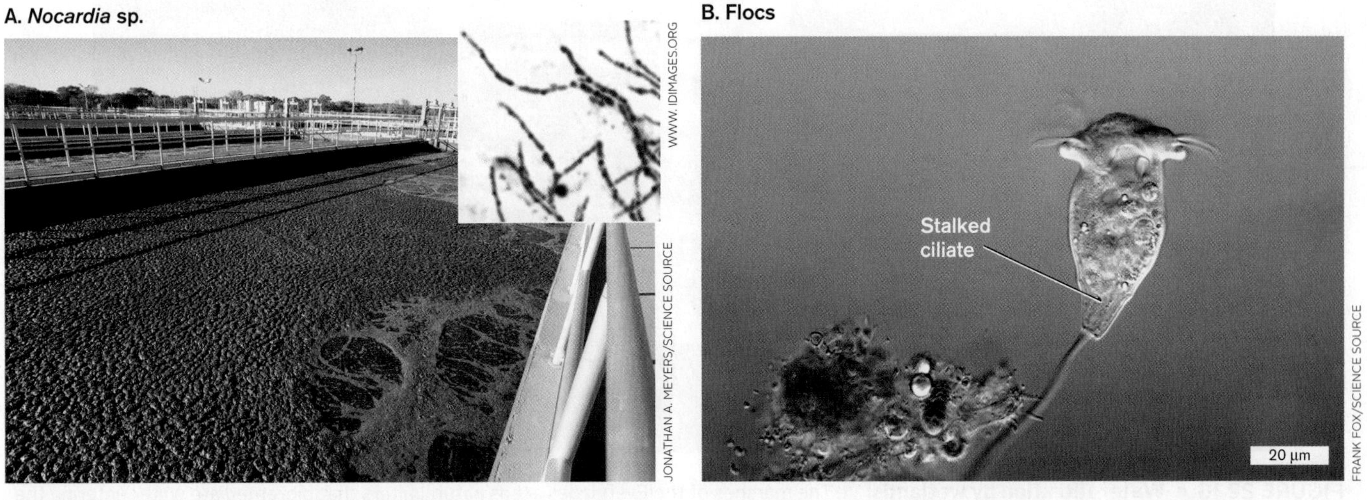

FIGURE 22.14 ■ Microbes in wastewater bioremediation. **A.** Foaming of water in secondary treatment is caused by excess growth of filamentous *Nocardia* sp. bacteria (LM, inset). **B.** Stalked ciliates in flocs prey on bacteria (LM).

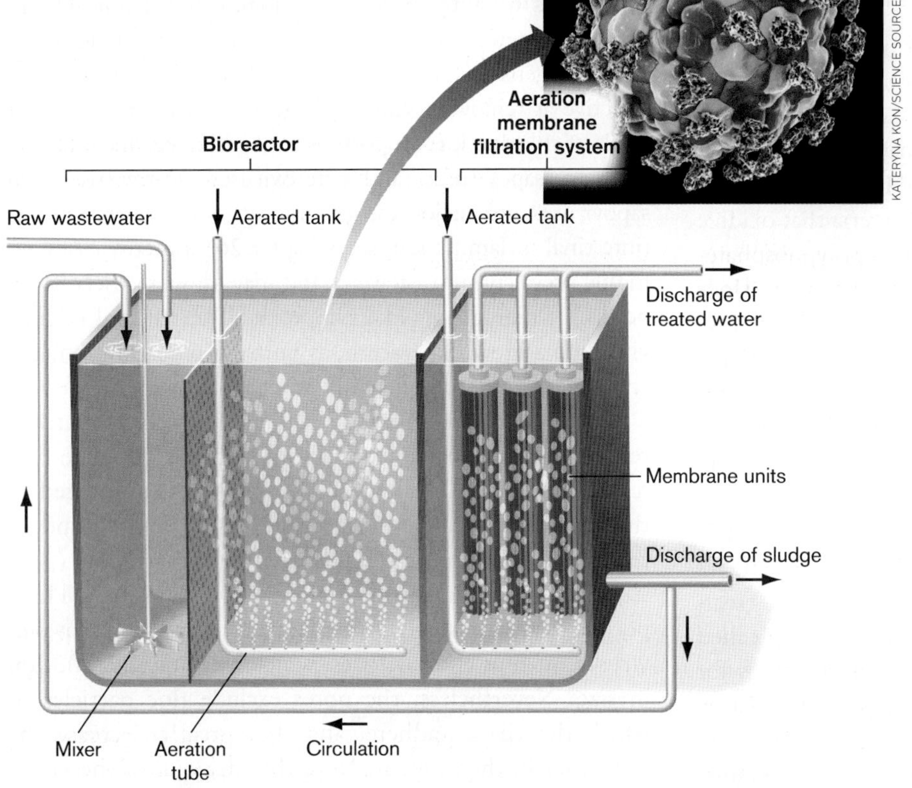

FIGURE 22.15 ▪ Membrane bioreactor filtration removes enteric pathogens. During wastewater aeration (secondary treatment), the effluent is filtered to remove sludge particles and viruses. **Inset:** Enterovirus D68; model based on cryo-electron microscopy.

usually by a chemical process such as chlorination—that is, treatment with hypochlorite ion (ClO^-). The water must then undergo dechlorination, commonly with sulfur dioxide, to convert remaining hypochlorite to chloride ion (Cl^-) and avoid formation of chlorinated organic toxicants.

Following the full wastewater treatment cycle, the treated water is then returned to local streams and lakes. Few people realize that their familiar freshwater sources are receptacles for microbiologically treated wastewater.

Thought Question

22.2 What would happen if wastewater treatment lacked microbial predators? Why would the result be harmful?

Agricultural treatment. Wastewater treatment plants are impractical for purifying the runoff from large agricultural operations. For large-scale alternatives to treatment plants, communities and agricultural operations are looking to wetland restoration. Much of our current water supply is already filtered and purified by natural wetlands, such as the Florida Everglades (**Fig. 22.16A**). Wetlands remove nitrogen through the action of denitrifying bacteria. In wetlands, rainwater and river water trickle slowly through vast stretches of soil, where microbial conversion acts as the foundation of a macroscopic ecosystem that includes trees and vertebrates.

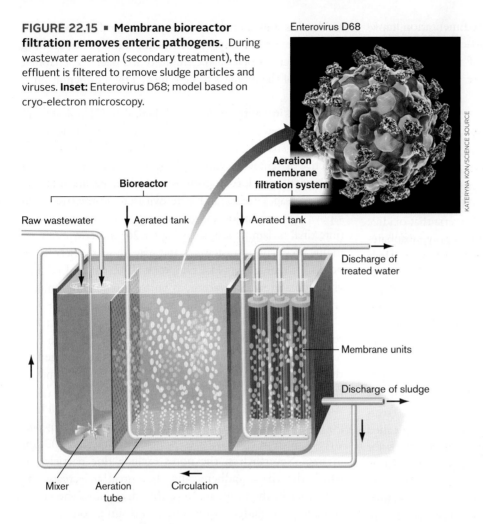

FIGURE 22.16 ▪ Water filtration by wetlands. A. The marshes of the Everglades act as natural filters that bioremediate water entering the aquifers of southern Florida. **B.** The filtering system that Steve Kerns installed on his hog farm in Taylor County, Iowa, consists of a series of hillside terraces that form constructed wetlands. The constructed wetlands contain microbial communities that metabolize hog manure and wastewater.

Thus, much of the carbon and nitrogen is fixed into valuable biomass. In the wetlands of the Everglades, the remaining water filters slowly through limestone into underground aquifers, which ultimately provide water through wells to human communities.

Some agricultural operations are building artificial wetlands, known as **constructed wetlands**, to replace treatment plants. A constructed wetland is a built environment designed to provide environmental services comparable to those of natural wetlands. **Figure 22.16B** shows a hog farm where a series of terraced wetlands was built to drain liquefied manure. The wetlands were found to produce fewer odors and to remove organics more efficiently and at a lower cost than do traditional filtration plants.

A much greater challenge is that of industrial runoff, in which toxic wastes from factory chemistry enter an aquifer at levels endangering human health. These chemicals may poison the treatment plants designed for human effluents. Large regions of land surrounding such a plant may require remediation that is expensive or impractical. And traces of pollutants such as chlorinated aromatics reach every spot on the globe; for example, dioxins are found in the tissues of Antarctic penguins.

How can we remediate industrial pollutants before they poison the local countryside and spread throughout the globe? One way is to harness naturally occurring bacteria for bioremediation. The bacterial community's reaction rate can be enhanced through selection by enrichment culture (described in Chapter 4). In the culture, organisms from the polluted site are cultivated repeatedly in the presence of added pollutant, and then tested for activity using radiolabeled substrates. An example of successful bioremediation is the case of Aberdeen Proving Ground, where a microbial mat was used to catabolize chlorinated hydrocarbons left by weapons manufacture (**eTopic 22.1**).

To Summarize

- **The hydrologic cycle is the cyclic exchange of water between the atmosphere and the biosphere.** Water precipitates as rain, which enters the ground and ultimately flows to the oceans. Along the way, some of the water evaporates, returning to the atmosphere.
- **Water carries organic carbon that generates biochemical oxygen demand (BOD).** High BOD accelerates heterotrophic respiration and depletes oxygen needed by fish.
- **Oxygen minimum zones are hypoxic regions of ocean sandwiched between upper and lower oxygenated layers.** Low oxygen levels are found at middle depth, where the oxygen from phototrophs meets upwelling organic nutrients.
- **Dead zones occur where sewage and agricultural runoff expand oxygen minimum zones.** Dead zones exclude all aerobic life.
- **Wastewater treatment cuts down BOD before returning water to aquatic systems and our built environment.** Secondary treatment involves the formation of flocs, microbial communities that decompose the soluble organic content.
- **Removal of viral pathogens from wastewater is improved by filtration.** Filtration removes flocs with adherent viruses.
- **Wetlands filter water naturally.** Natural and constructed wetlands can purify groundwater entering aquifers.
- **Industrial effluents are highly toxic and can reach all parts of the globe.** Bioremediation with microbes may eliminate such toxins.

22.3 The Nitrogen Cycle

Besides carbon, oxygen, and hydrogen, another major element that cycles largely by microbial conversion is nitrogen (**Fig. 22.17**). The nitrogen cycle is notable for its dependence on prokaryotes; several steps of the cycle, such as N_2 fixation and nitrous oxide production, are performed solely by bacteria and archaea. Nitrogen cycles are highly perturbed by human technology.

Sources of Nitrogen

Nitrogen is found in Earth's crust, in the form of ammonium salts in rock (see **Fig. 22.1B**). Until recently, the nitrogen found in rock was thought to be inaccessible to the biosphere. In 2018, Ben Houlton and Randy Dahlgren at UC Davis showed how up to a quarter of the nitrogen obtained by terrestrial plants could come from weathered rock, in the form of ammonium ion and nitrate. Nonetheless, the largest accessible source of nitrogen is the atmosphere.

The N_2 molecule is highly stable, requiring an enormous input of reducing energy before assimilation is possible (see Chapter 15). Thus, for many natural ecosystems, and most forms of agriculture, nitrogen is the limiting nutrient for primary productivity. Until recently in Earth's history, N_2 was fixed entirely by nitrogen-fixing bacteria and archaea. In the twentieth century, however, the Haber process was invented for artificial nitrogen fixation to generate fertilizers for agriculture. The process was devised by German chemist Fritz Haber (1868–1934), who won the 1918 Nobel Prize in Chemistry. In the Haber process, N_2 is hydrogenated by hydrogen derived from methane in the form of natural gas. Hydrogenation of N_2 requires extreme

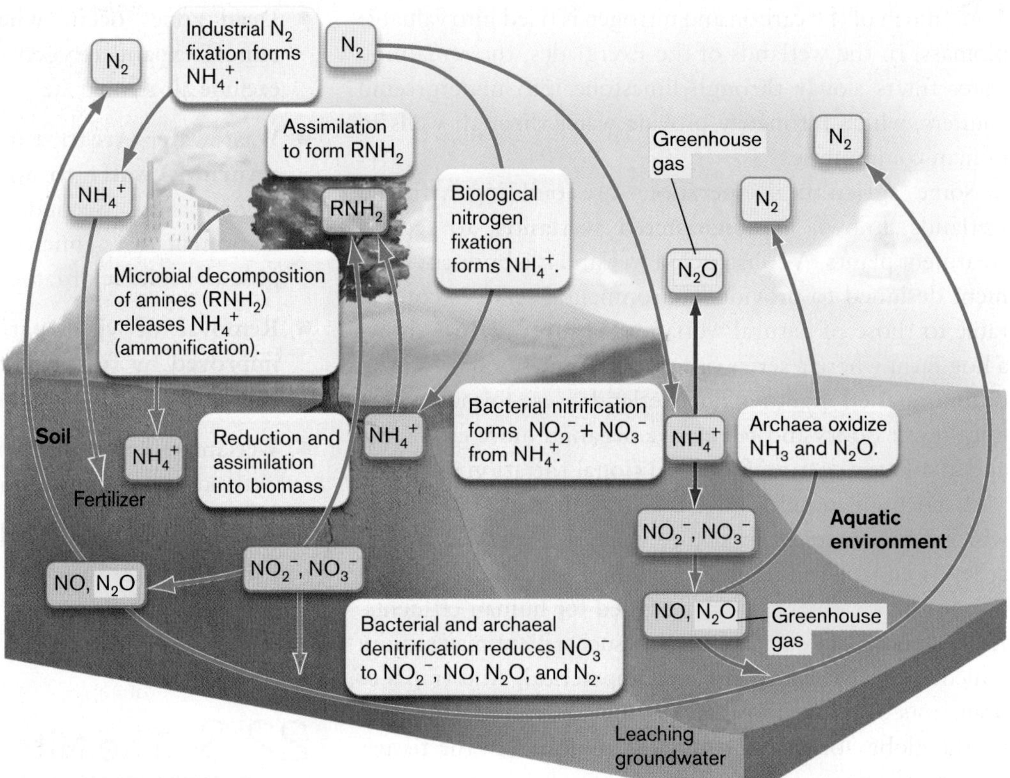

FIGURE 22.17 ■ The global nitrogen cycle. Bacteria and archaea interconvert forms of nitrogen throughout the biosphere. Blue = reduction; red = oxidation; orange = redox-neutral.

heat and pressure. Today, the Haber process for producing fertilizers accounts for approximately 30%–50% of all nitrogen fixed on Earth, and it consumes a substantial portion of fossil fuels. Other human activities, such as fuel burning and use of nitrogenous fertilizers, contribute to oxidized nitrogen pollutants such as nitrous oxide (N_2O), a potent greenhouse gas (**Fig. 22.17**, highlighted yellow).

Because of extensive fertilizer use, the nitrogen cycle today is the most perturbed of the major biogeochemical cycles. Half the nitrogen in the biosphere now comes from anthropogenic (human-generated) sources. The ultimate effects of perturbation are not yet clear. Some models show that increased nitrogen fixation could fertilize CO_2 fixation by marine and terrestrial ecosystems, partly decreasing net CO_2 emissions. The amount, however, will be too small to prevent global warming. Other models show that accelerated nitrogen use could limit nitrogen availability in the global biosphere, with unknown consequences.

Oxidation States: The "Nitrogen Triangle"

Biological nitrogen exists in a larger number of redox states than any other element (**Table 22.2**). The multiple oxidized forms of nitrogen available for microbial metabolism generate a complex cycle of conversions. We can envision a "nitrogen triangle" (**Fig. 22.18A**) whose corners include three key oxidation states:

- **NH_3, fully reduced** (ammonia, which ionizes to ammonium ion, NH_4^+)
- **NO_3^-, fully oxidized** (nitrate)
- **N_2, oxidation state of zero** (nitrogen gas)

At the base of the triangle, both reduced and oxidized forms of nitrogen are assimilated into biomass.

Thought Question

22.3 Which kinds of biomolecules can you recall that contain nitrogen? What are the usual oxidation states for nitrogen in biological molecules?

TABLE 22.2 Oxidation States of Nitrogen Compounds

Oxidation state	Nitrogen molecule	
−3	NH_3, NH_4^+	Ammonia, ammonium ion
−2	$H_2N\text{-}NH_2$	Hydrazine
−1	NH_2OH	Hydroxylamine
0	N_2	Nitrogen
+1	N_2O	Nitrous oxide
+2	NO	Nitric oxide
+3	HNO_2, NO_2^-	Nitrous acid, nitr<u>ite</u> ion
+4	NO_2	Nitrogen dioxide
+5	HNO_3, NO_3^-	Nitric acid, nitr<u>ate</u> ion

A. The nitrogen triangle

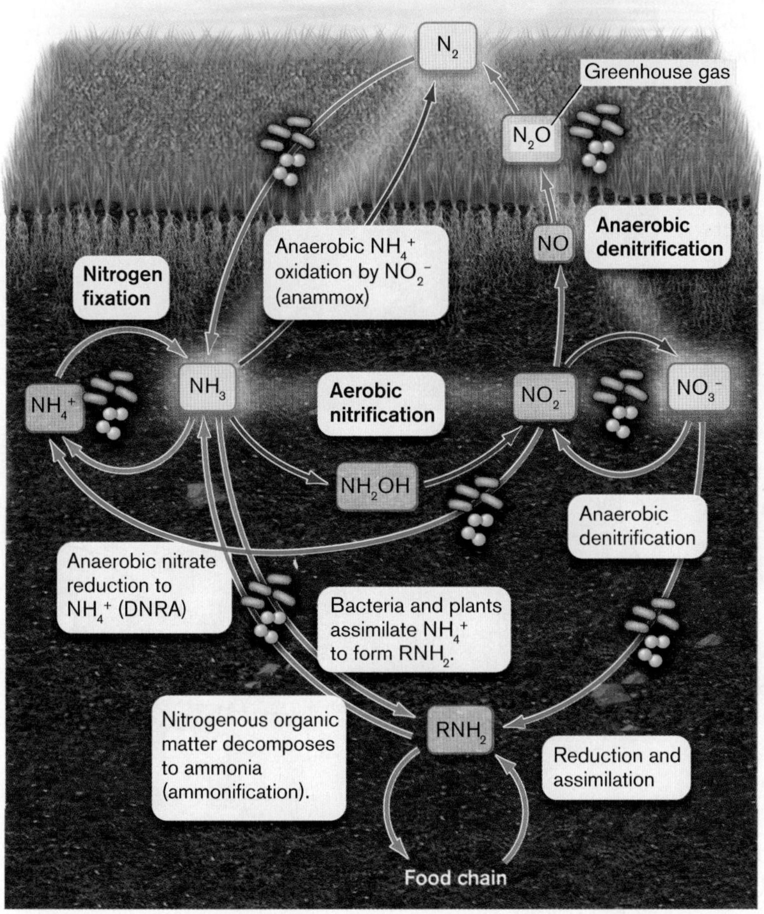

B. *Nitrobacter winogradskyi*

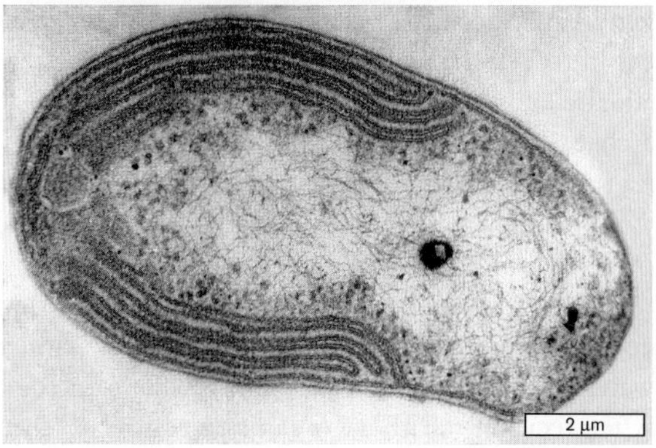

FIGURE 22.18 ■ The nitrogen cycle: fixation, nitrification, denitrification. **A.** The "nitrogen triangle" consists of nitrogen fixation and assimilation, reductive dissimilation of nitrate (denitrification; blue), and oxidation (nitrification; red). Denitrification includes production of the potent greenhouse gas nitrous oxide (N_2O). Assimilation into biomass is often reductive; virtually all nitrogen in biomolecules is highly reduced. Oxidation of ammonia generates nitrites and nitrates, whose runoff can pollute water supplies. DNRA = dissimilatory nitrate reduction to ammonia. **B.** *Nitrobacter winogradskyi* oxidizes nitrite to nitrate. Folded layers of membrane contain the electron transport complexes.

The three sides of the nitrogen triangle are defined by these microbial reactions:

- **N_2 fixation (reduction) to NH_4^+**, a form assimilated into biomass by microbes and plants.
- **Ammonia (NH_3) oxidation to nitrite (NO_2^-) and nitrate (NO_3^-).** Nitrite and nitrate can then be assimilated by microbes and plants. Aerobic oxidation to nitrite or nitrate is called nitrification. In anoxic habitats, anaerobic NH_3 oxidation by nitrite generates N_2—a reaction called anammox.
- **Denitrification of NO_2^- and NO_3^- back to N_2** (or, in carbon-rich habitats, microbial reduction by anaerobic respiration to NH_3).

Nitrogen fixation. The main avenue for entry of nitrogen into the biosphere is bacterial and archaeal **nitrogen fixation**—specifically, fixation of dinitrogen, or nitrogen gas (N_2), into NH_3. (At neutral pH, most of NH_3 is protonated to NH_4^+). Ammonium ion is rapidly assimilated by bacteria and plants, typically by combination with TCA cycle intermediates such as 2-oxoglutarate, to form key amino acids such as glutamate and glutamine (discussed in Chapter 15). Most kinds of living organisms can assimilate NH_4^+ into nitrogenous organic molecules.

Fixation of nitrogen requires enormous energy because the triple bond of N_2 is exceptionally stable. Breaking the triple bond to generate ammonia requires a series of reduction steps involving high input of energy:

Dinitrogen

$$N{\equiv}N \xrightarrow[2H^+ + 2e^-]{} HN{=}NH \xrightarrow[2H^+ + 2e^-]{} H_2N{-}NH_2 \xrightarrow[2H^+ + 2e^-]{} 2NH_3$$

All the intermediate reactions of nitrogen fixation are tightly coupled, so the intermediate compounds rarely become available for other uses. The final product, ammonia, exists in ionic equilibrium with ammonium ion:

$$NH_3 + H_2O \rightleftharpoons NH_4^+ + OH^-$$

Half the ammonia is protonated at about pH 9.3, so at near-neutral pH, most of it exists in the protonated form, NH_4^+.

Nitrogen fixation is catalyzed by the enzyme nitrogenase (discussed in Chapter 15). The highly reductive reactions of nitrogenase require exclusion of oxygen, yet they also require a tremendous input of energy, usually provided by aerobic metabolism. Therefore, nitrogen-fixing bacteria generally isolate anaerobic nitrogen fixation from aerobic metabolism by one of several mechanisms, such as the heterocysts of cyanobacteria (also discussed in Chapter 15).

Given the energy expense and the need to exclude oxygen, many species of bacteria, as well as all eukaryotes, lack the nitrogen fixation pathway. But all ecosystems, both aquatic and terrestrial, include some species of bacteria and archaea that fix N_2 into ammonia. Nitrogen-fixing bacteria in the soil include obligate anaerobes, such as *Clostridium* species, and facultative Gram-negative enteric species of *Klebsiella* and *Salmonella*, as well as obligate respirers such as *Pseudomonas*. The rhizobia form nitrogen-fixing endosymbionts of legume plants. In oceans and in freshwater systems, cyanobacteria are the major nitrogen fixers. After fixation, all these organisms assimilate the reduced nitrogen into essential components of their cells. Within an ecosystem, nitrogen fixers ultimately make the reduced nitrogen available for assimilation by nonfixing microbes and plants, either directly through symbiotic association (such as that of rhizobia and legumes, discussed in Chapter 21) or indirectly through predation (marine cyanobacteria) and decomposition (soil bacteria).

If nitrogen fixation is ubiquitous in soil and water, then why is nitrogen limiting? Nitrogen fixation is extremely energy intensive; thus, the rate of fixation usually fails to meet the potential demand of other members of the ecosystem. An exception is legume symbiosis with rhizobia, which provide ample nitrogen for their hosts.

In agriculture, symbiotic nitrogen fixation by rhizobial bacteria increases the yield of crops such as soybeans (**Fig. 22.19A**). To enhance colonization by nitrogen fixers, farmers apply molecules called isoflavonoids, which mimic the natural plant-derived attractants for rhizobia.

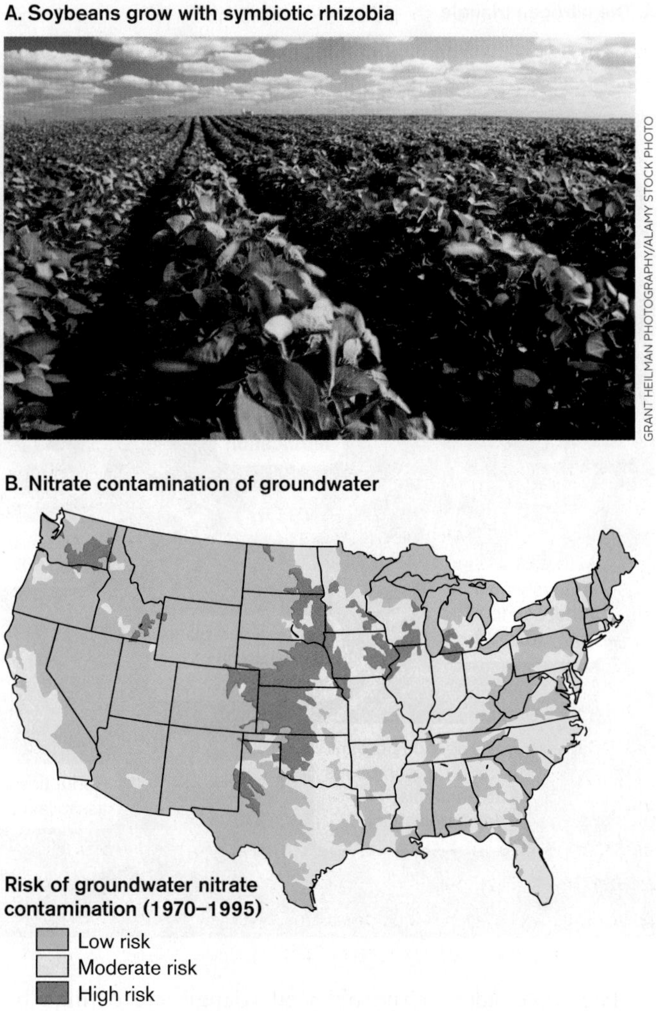

FIGURE 22.19 ■ Agricultural benefits and consequences of microbial nitrogen metabolism. **A.** Soybean field in Maine. Rhizobial nitrogen fixation enhances growth of soybeans and other major crops. **B.** Agricultural fertilization releases ammonium, converted by lithotrophs to nitrate, which enters the water supply. Nitrate in drinking water is especially prevalent in agricultural regions of the United States. Nitrate and nitrite runoff from oxidized nitrogenous fertilizers pollutes streams and groundwater.

Thought Question

22.4 The nitrogen cycle has to be linked with the carbon cycle, since both contribute to biomass. How might the carbon cycle of an ecosystem be affected by increased input of nitrogen?

Nitrification. Free ammonia in soil or water is quickly oxidized for energy by **nitrifiers**, bacterial species that possess enzymes for oxidation of ammonia to nitrite (NO_2^-), or of nitrite to nitrate (NO_3^-). This process is called **nitrification**. Nitrification of ammonia is a form of lithotrophy, an energy generation pathway involving oxidation of minerals. The nitrification pathway generates the red base of the triangle in **Figure 22.18A**. The pathway of nitrification includes:

$$\underset{\text{Ammonia}}{NH_3} \xrightarrow[+\frac{1}{2}O_2]{} NH_2OH \xrightarrow[+O_2]{} \underset{+\ H_2O + H^+}{\overset{\text{Nitrite}}{NO_2^-}} \xrightarrow[+\frac{1}{2}O_2]{} \overset{\text{Nitrate}}{NO_3^-}$$

The pathway contains two separate energy generation mechanisms: (1) ammonia through NH_2OH to nitrite, and (2) nitrite to nitrate. Typically, soil contains both kinds of microbes, collaborating to complete the conversion to nitrate, while some species of *Nitrospira* perform the entire pathway, oxidizing ammonia to nitrate. Nitrifying genera include *Nitrosomonas*, which oxidizes ammonia to nitrite, as well as *Nitrobacter* (**Fig. 22.18B**) and *Nitrospira*, which oxidize nitrite to nitrate. Nitrite can also serve as an electron donor for photosynthesis by *Thiocapsa* species. Note that the production of both nitrite and nitrate generates acid, which can acidify the soil.

Thought Question

22.5 In the laboratory, which bacterial genus would likely grow on artificial medium including NH_2OH as the energy source: *Nitrosomonas* or *Nitrobacter*? Why?

Nitrate produced in the soil is assimilated by plants and bacteria nearly as quickly as ammonium ion, although extra energy is needed to reduce nitrate to NH_4^+ for incorporation into biomass. Nitrate assimilation to biomass is called **assimilatory nitrate reduction**. Assimilatory nitrate reduction differs from the nitrate reduction involved in anaerobic respiration, which releases the reduced nitrogen and yields energy. In agriculture, intensive fertilization generates a large excess of ammonia resulting from **ammonification**, the breakdown of organic nitrogen by catabolism. Catabolism may release ammonia. Ammonia is oxidized rapidly by lithotrophic bacteria (nitrifiers)—a dissimilatory reaction. The excess ammonia leads to a buildup of nitrites and nitrates, which are highly soluble in water, and they readily diffuse into aquatic systems. Aquatic nitrate reacts with organic compounds to form toxic nitrosamines. Nitrate influx also relieves the nitrogen limit on algae, causing algal blooms and raising BOD. Chronic nitrate influx leads to eutrophication and die-off of fish.

Consumption of nitrate in drinking water can lead to methemoglobinemia, a blood disorder in which hemoglobin is inactivated. Methemoglobinemia is a problem for infants because their stomachs are not yet acidic enough to inhibit growth of bacteria that convert nitrate to nitrite. Nitrite oxidizes the iron in hemoglobin, eliminating its capacity to carry oxygen. The failure to carry oxygen leads to a bluish appearance—one cause of "blue baby syndrome." Nitrite-induced blue baby syndrome is a problem in intensively cultivated agricultural regions, such as Kansas and Nebraska (**Fig. 22.19B**).

Agricultural runoff also contributes to oxygen minimum zones and dead zones, such as those off the Oregon coast and in the Gulf of Mexico. The amines in sewage are removed as ammonia (ammonification), and the ammonia is oxidized by lithotrophs. The major marine ammonia oxidizers include the Thaumarchaeota, such as *Nitrosopumilus* (discussed in Chapter 19).

Denitrification. N_2 is regenerated by anaerobic respiration (see **Fig. 22.18A**), in which an oxidized form of nitrogen, such as nitrate or nitrite, receives electrons from organic electron donors (dissimilatory nitrogen reduction). Bacteria and archaea reduce nitrate through a series of decreased oxidation states back to atmospheric nitrogen:

$$\begin{array}{ccccccc} \text{Nitrate} & & \text{Nitrite} & & \text{Nitric oxide} & & \\ 2NO_3^- & \longrightarrow & 2NO_2^- & \longrightarrow & 2NO & \longrightarrow & \\ & 4H^+ + 4e^- & + & 4H^+ + 2e^- & + & 2H^+ + 2e^- & \\ & & 2H_2O & & 2H_2O & & \end{array}$$

$$\begin{array}{ccc} \text{Nitrous oxide} & & \text{Dinitrogen} \\ N_2O & \longrightarrow & N_2 \\ + & 2H^+ + 2e^- & + \\ H_2O & & H_2O \end{array}$$

In terms of elemental flux through ecosystems, nitrate and nitrite reduction are known as **denitrification** or **dissimilatory nitrate reduction**. (In contrast, assimilatory nitrate reduction incorporates nitrogen into biomass.) Many anaerobic respirers in soil or water use an oxidized form of nitrogen as an alternative electron acceptor in the absence of O_2 (discussed in Chapter 14). All types of nitrogen-based anaerobic respiration are repressed in the presence of oxygen, a more favorable electron acceptor; therefore, denitrification is limited to anoxic habitats.

In undisturbed environments, the products of nitrate respiration rarely build up to levels that harm the ecosystem. Heavy fertilization, however, causes buildup of excess nitrate and so increases the environmental rate of denitrification. During this process, some of the nitrogen escapes as nitrous oxide gas (N_2O). A highly potent greenhouse gas, N_2O generates 200 times the warming effect of CO_2; thus, relatively small amounts of N_2O can make a disproportionate contribution to global warming. Furthermore, N_2O in the upper atmosphere reacts catalytically with ozone, depleting the ozone layer. Thus, the atmospheric effects of bacterial denitrification are a serious concern in agricultural and waste treatment processes.

Nitrous oxide also builds up in marine dead zones, where the nitrates from ammonium oxidation are subsequently used as electron acceptors. Syed Wajih Naqvi and colleagues at the National Institute of Oceanography in Goa, India, investigated denitrification in the zone of hypoxia off the coast of India (**Fig. 22.20A**). The study revealed unexpectedly high levels of N_2O production at a series of stations within the zone. In one experiment, introduction of nitrate into water samples from the dead zone led quickly to production of nitrite and N_2O (**Fig. 22.20B**), a process conducted by denitrifying bacteria. Studies in 2012 suggested additional N_2O production by ammonia-oxidizing archaea such as *Nitrosopumilus*. Thus, bacterial denitrification and archaeal ammonia oxidation in polluted ocean waters may contribute significantly to global warming.

Dissimilatory nitrate reduction to ammonia. Most environmental nitrate and nitrite are reduced via N_2O as described previously. Certain conditions, however, favor the alternate route of dissimilatory nitrate reduction to ammonia (DNRA; see **Fig. 22.18A**). Nitrate reduction to ammonia is a form of anaerobic lithotrophy in which nitrate serves as an electron acceptor. An electron donor such as hydrogen gas (H_2) reduces nitrate as follows:

$$NO_3^- + 4H_2 + H^+ \longrightarrow NH_3 + 3H_2O$$

Bacteria reduce nitrate to ammonia mainly in anaerobic environments rich in organic carbon and H_2 generated by fermentation, but low in reduced nitrogen—such as sewage sludge and stagnant water. Another carbon-rich habitat favoring this pathway is the rumen, the digestive tract of cattle, goats, and other ruminant animals (see Chapter 21). Ruminants consume grasses whose cellulose requires digestion by mutualistic microbes. Ruminants also depend on their digestive microbes to assimilate nitrate into ammonia and synthesize amino acids.

Anaerobic ammonium oxidation (anammox). For many years, denitrification was considered the main way that nitrogen compounds return N_2 to the atmosphere. In 2003, two research groups from Europe and Costa Rica, led by Tage Dalsgaard and Marcel Kuypers, showed that in anoxic deep-sea water, the major source of N_2 is bacteria conducting anaerobic ammonium oxidation by nitrite (the anammox reaction, discussed in Chapter 14):

$$NH_4^+ + NO_2^- \longrightarrow N_2 + 2H_2O$$

In the **anammox reaction**, ammonium ion serves as electron donor, and nitrite serves as anaerobic electron acceptor—an unusual combination of two different oxidation states of the same key element, nitrogen. The reaction is a kind of anaerobic lithotrophy.

Anammox is performed by bacteria in all kinds of anaerobic habitats, including terrestrial soil and aquatic sediment. The reaction accounts for a majority of all N_2 returned to the atmosphere. Among bacteria, the main anammox contributors are planctomycete genera such as *Kuenenia* and *Scalindua* (**Fig. 22.21**). Anammox planctomycetes (discussed in Chapter 18) are an unusual kind of cell wall–less bacteria with a special interior membrane that segregates the toxic intermediates of the anammox reaction. Promising anammox microbes have been identified from wastewater sludge, in the hope of using them for more effective removal of excess nitrogen from wastewater.

A.

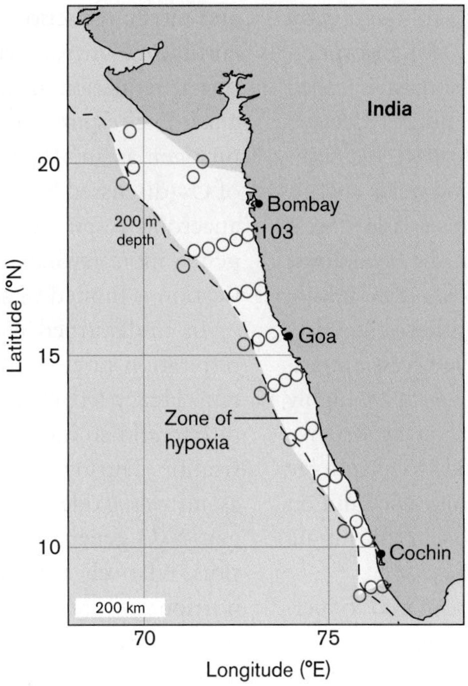

B.

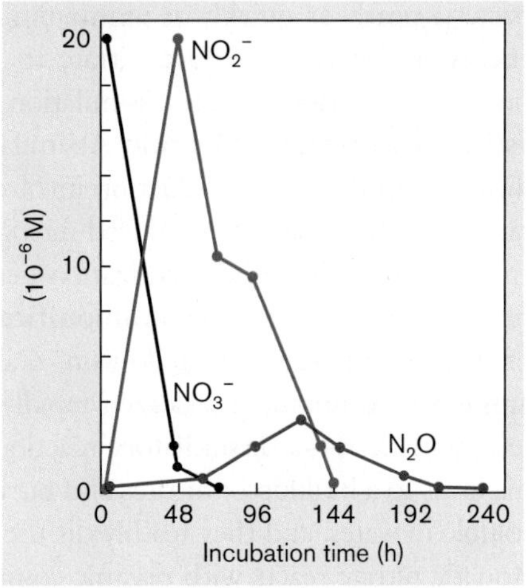

FIGURE 22.20 ■ N_2O production from a coastal dead zone. **A.** Zone of hypoxia off the coast of India. Circles represent sample-collection stations. **B.** Levels of NO_3^-, NO_2^-, and N_2O following addition of NO_3^- to a water sample from the zone of hypoxia. The sequential rise of NO_2^- and N_2O indicates metabolism by denitrifying bacteria.

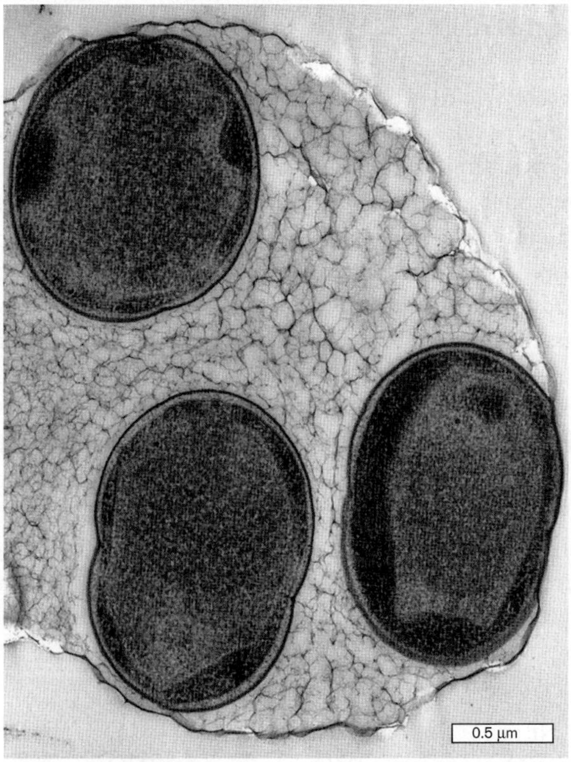

AWATA ET AL. 2013. *APPL. ENVIRON. MICROBIOL.* **79**:4145, FIG. 2A.

FIGURE 22.21 ■ Anammox bacterium: *Scalindua*. The planctomycete *Scalindua* has interior compartments to undergo anammox lithotrophy (discussed in Chapter 14; TEM).

To Summarize

- **Nitrogen in ecosystems is found in many different oxidation states.** Interconversion of most oxidation states requires bacteria or archaea. The nitrogen triangle includes conversions among NH_3 (fully reduced), NO_3^- (fully oxidized), and N_2 (oxidation state is zero).
- **The main source and sink of nitrogen is the atmosphere.** N_2 is fixed into NH_4^+ by some bacteria and archaea. Denitrifying bacteria reduce NO_3^- successively back to N_2 and return it to the atmosphere.

- **Nitrogen fixation is conducted by symbiotic bacteria in association with specific plants.** Legume-associated nitrogen fixation is critical for agriculture.
- **Nitrification ($NH_4^+ \rightarrow NO_2^- \rightarrow NO_3^-$)** is aerobic oxidation of ammonia to nitrite and nitrate. Nitrification yields energy for lithotrophic bacteria in soil and water, and consumes oxygen, thus expanding marine oxygen minimum zones.
- **Bacteria conduct anaerobic respiration**, reducing nitrate to N_2 and nitrous oxide (N_2O), a greenhouse gas. In the deep ocean, NO_3^- may be reduced by hydrogen gas to NH_3.
- **Under anaerobic conditions, bacteria oxidize NH_4^+ with NO_2^-, generating nitrogen gas (the anammox reaction).**

22.4 Sulfur, Phosphorus, and Metals

Besides carbon and nitrogen, many other elements participate in biochemical cycles that have important consequences for the biosphere, as well as for human environments. Sulfur is a major component of biomass, including proteins and cofactors. Like carbon and nitrogen, sulfur can be assimilated in either mineral or organic form. At the same time, assimilation competes with dissimilation by reactions yielding energy such as sulfide oxidation. Phosphorus, unlike other biological elements, is generally assimilated in the oxidized state, and phosphate is often a limiting nutrient for plants. Iron cycles in complex interactions with sulfur and phosphate. Beyond these macronutrients, many toxic metals in the environment, such as mercury and arsenic, are either bioactivated or detoxified by microbes.

The Sulfur Cycle

Sulfur undergoes a "triangle" of redox conversions analogous to that of nitrogen (**Table 22.3**). The major redox states are:

- **H_2S and SH^-, fully reduced** (sulfides)
- **S^0, oxidation state of zero** (elemental sulfur)
- **SO_4^{2-}, fully oxidized** (sulfate ion)

Reduced forms of sulfur offer electron donors for lithotrophy; these energy-yielding reactions are dissimilatory. H_2S and SH^- can also participate in photolysis through reactions analogous to those of H_2O (Chapter 14). Oxidized forms of sulfur such as sulfate, sulfite, and thiosulfate serve as electron acceptors for anaerobic respiration. As we saw for nitrogen, most of these redox reactions are performed solely by bacteria and archaea. The biochemistry of sulfur cycling has important consequences for our "built environment," such as the corrosion of concrete and iron (discussed shortly).

TABLE 22.3 Oxidation States of Sulfur Compounds

Oxidation state	Sulfur molecules	
−2	H_2S, HS^-	Sulfides
0	S^0	Elemental sulfur
+2	$S_2O_3^{2-}$	Thiosulfate
+4	SO_3^{2-}	Sulfite
+6	H_2SO_4, SO_4^{2-}	Sulfuric acid, sulfate

In the ocean, sulfate is the second most common anion after chloride. Marine sulfate turns over slowly and constitutes an essentially limitless supply. Thus, sulfur is rarely a limiting nutrient. Marine algae release various reduced forms of sulfur, such as dimethylsulfoniopropionate (DMSP). Bacteria metabolize DMSP to dimethyl sulfide [$(CH_3)_2S$], which causes part of the "salty sea smell." Dimethyl sulfide enters the atmosphere, where it becomes oxidized to aerosols that may help water condense and form clouds. In the atmosphere, the overall amount of sulfur (mainly sulfur dioxide) is small. Nevertheless, some sulfur compounds generate toxic effects, as well as acidic pollution. Thus, both biochemical and industrial sources of atmospheric sulfur are of concern.

Assimilatory and dissimilatory sulfur reactions by microbes interconvert H_2S, S^0, and SO_4^{2-} (**Fig. 22.22A**). The "triangle" of oxidation and reduction pathways is analogous to that of nitrogen (compare **Fig. 22.18A**). Bacterial sulfur metabolism includes additional options of anaerobic phototrophy, such as H_2S phototrophy. In an aquatic system, H_2S arises from spring waters welling up from the sediment and from decomposition of detritus. During decomposition, anaerobic respirers such as *Desulfovibrio* species convert sulfate to sulfur, then to H_2S. As H_2S rises to the oxygenated surface water, it is readily oxidized by sulfur-oxidizing bacteria such as *Acidithiobacillus* and *Beggiatoa*.

Beggiatoa species are known as "white sulfur bacteria" because they form white mats (**Fig. 22.22B**). Their appearance is due to the sulfur granules generated by sulfide oxidation. Microbial sulfide oxidation is helpful for environments because it removes H_2S, which is highly toxic to most nonsulfur bacteria and plants. Because of the toxicity of H_2S, autotrophs more readily assimilate SO_4^{2-} into biomass, despite the extra energy needed to reduce SO_4^{2-} to the thiol form found in proteins.

If light is available, anaerobic phototrophs such as *Rhodopseudomonas* species will use light energy to oxidize H_2S. Some phototrophic bacteria further oxidize the S^0

A. The sulfur triangle

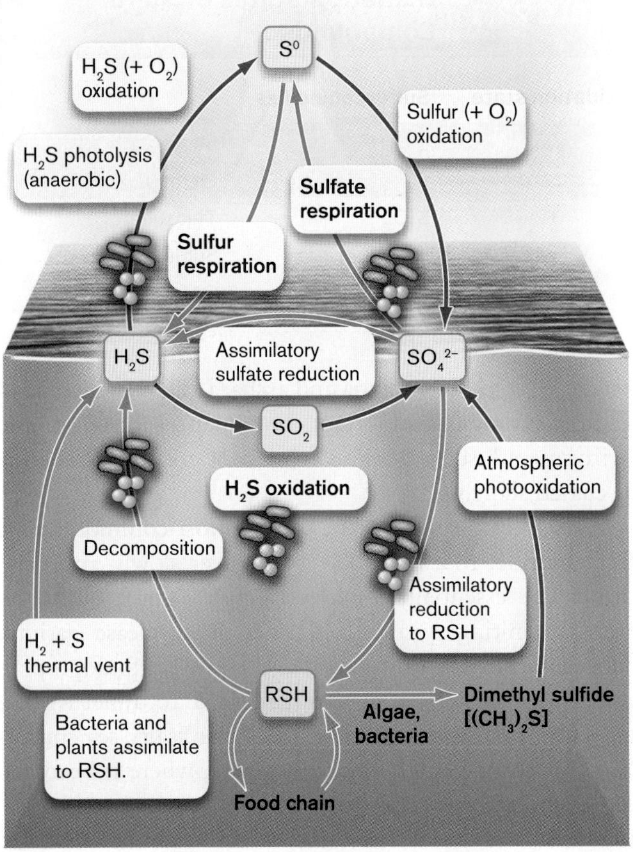

B. White sulfur bacteria

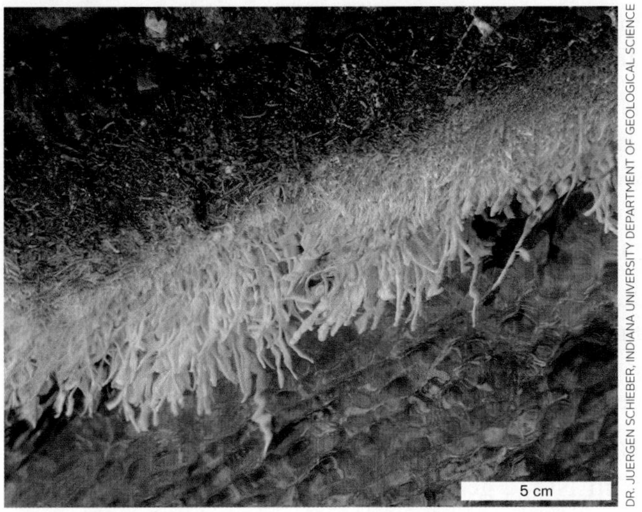

FIGURE 22.22 ▪ The sulfur cycle. **A.** The "sulfur triangle." In the presence of oxygen, bacteria and archaea oxidize H_2S to sulfur dioxide, and then to sulfate. Anaerobically, H_2S may be photolyzed to sulfur. Sulfate serves as an electron acceptor for anaerobic respiration, or it may be assimilated into biomass, with reduced sulfur groups (RSH). Algae form DMSP (dimethylsulfoniopropionate), which bacteria convert to dimethyl sulfide. **B.** Microbial mat of white, sulfur-oxidizing bacteria (probably *Beggiatoa*) growing at a sulfide spring.

generated to sulfate. In some sulfur-rich lakes, such as the Russian Lake Sernoye, underground springs pump in so much H_2S that most of the sulfur is photolytically converted to S^0, which forms up to 5% of the sediment. This elemental sulfur can be mined commercially.

In thermal vent ecosystems, several sulfur-based reactions drive metabolism. For example, thermal vent archaea such as *Pyrodictium* species use sulfur to oxidize hydrogen gas to H_2S:

$$H_2 + S^0 \longrightarrow H_2S$$

This type of sulfur-based hydrogenotrophy is enhanced by the vent conditions of extreme pressure and temperature (100°C)—conditions under which elemental sulfur exists in a molten state more accessible to microbes than is solid sulfur.

Decomposition of biomass generates various organic sulfur compounds, many of which are volatile. Odors of certain microbial sulfur products contribute to the smell of rotting eggs, but others enhance the taste of cheeses (discussed in Chapter 16).

Sulfur is reduced by sulfur-reducing bacteria in many subsurface environments, such as beneath deposits of oil and coal (**Fig. 22.23A**). Oil and coal contain sulfur in the form of SO_4^{2-} and provide a carbon source for sulfate respirers (sulfate-reducing bacteria that respire using sulfate as a terminal electron acceptor). The products of sulfate respiration include S^0 and organic sulfur. When the fuel is burned, these forms of sulfur cause severe pollution. Before burning, however, the S^0 and organic sulfur can be oxidized by sulfur-oxidizing bacteria. Fuel processors are now turning to sulfur-oxidizing microbes for experimental use in "desulfuration," the removal of sulfur from coal.

One habitat that exhibits the entire range of sulfur oxidation states is a sewer pipe. In a concrete sewer pipe, the alternation between anaerobic and oxygenated sulfur biochemistry causes severe corrosion (**Fig. 22.23B**). The microbial decomposition of sewage yields large quantities of toxic H_2S, which then volatilizes to high levels that endanger sewer workers. The H_2S is then oxidized to sulfuric acid by *Acidithiobacillus ferrooxidans*, a bacterium that colonizes the surface of the concrete. The sulfuric acid (H_2SO_4) decreases the pH at the concrete surface to pH 2. In the concrete surface, sulfuric acid converts calcium hydroxide to calcium sulfate, which dissolves in water. Over several years, this corrosion can eat away half the thickness of a sewer pipe.

Sulfur metabolism shows important connections with metabolism of metals. For example, sulfur-oxidizing bacteria such as *A. ferrooxidans* oxidize iron as well (discussed shortly).

Thought Question

22.6 Compare and contrast the cycling of nitrogen and sulfur. How are the cycles similar? How are they different?

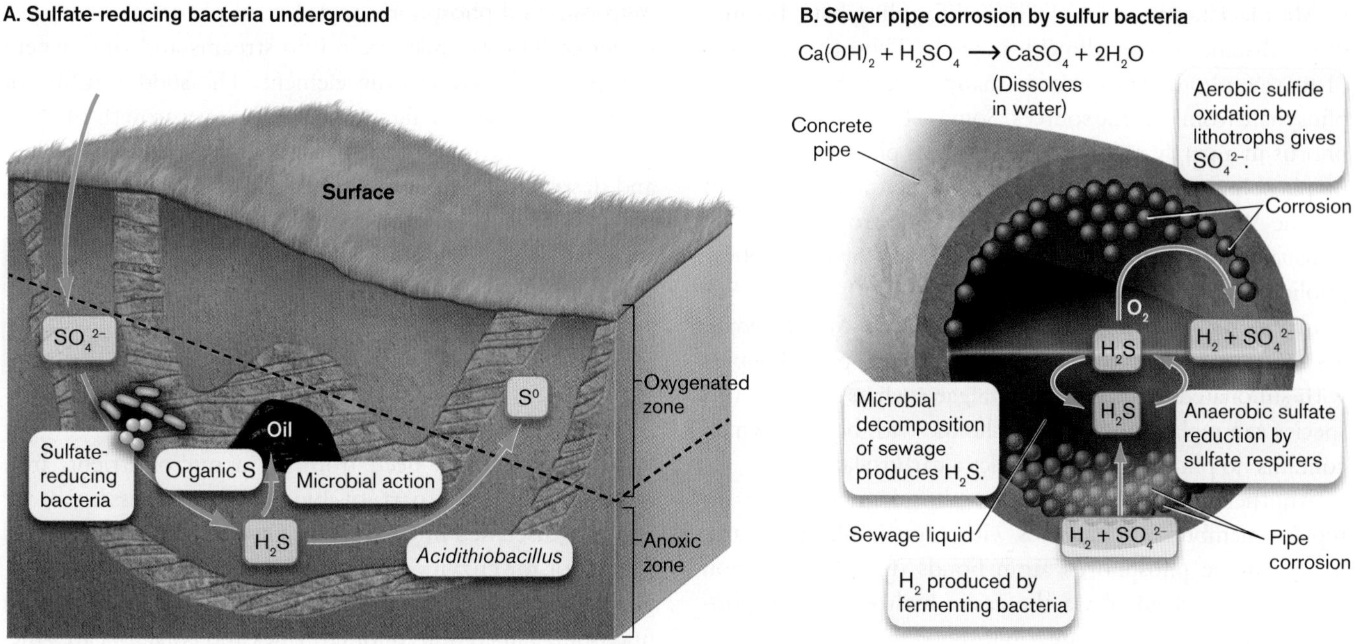

FIGURE 22.23 ■ Consequences of the sulfur cycle. **A.** Oil- or coal-bearing geological strata provide rich electron donors for sulfate-reducing bacteria (sulfate respirers). Eventually, various forms of sulfur contaminate the oil or coal, which, when burned as fuel, generates first SO_2 and ultimately sulfuric acid (acid rain). **B.** In a sewer pipe, sulfate-reducing bacteria (sulfate respirers) generate H_2S. Sulfur-oxidizing bacteria then oxidize H_2S to sulfate in the form of sulfuric acid. The sulfuric acid reacts with calcium hydroxide in the concrete, thus corroding the interior surface of the pipe.

The Phosphate Cycle

Phosphorus is a fundamental element of nucleic acids, phospholipids, and phosphorylated proteins. Unlike sulfur and nitrogen, which are found in several different oxidation states, phosphorus is cycled mainly in the fully oxidized state of phosphate (**Fig. 22.24**). The absence of fully reduced phosphorus in ecosystems may be due to the fact that reduced phosphorus (phosphine, PH_3) undergoes spontaneous combustion in the presence of oxygen. Nevertheless, some anaerobic decomposers use phosphate as a terminal electron acceptor, reducing it to phosphine. In marshes and graveyards, where extensive decomposition occurs, phosphine emanates from the ground, where it ignites with a green glow. Microbial respiration of phosphate might be the cause of such "ghostly" apparitions.

Where is phosphate available? Although phosphate is abundant in Earth's crust, its availability in ecosystems is limited by its tendency to precipitate with calcium, magnesium, and iron ions. Thus, dissolved phosphate in water and soil is often a limiting nutrient for productivity. In natural ecosystems, the available phosphate is taken up rapidly by bacteria and phytoplankton, and then consumed by grazers and predators and dispersed by decomposers.

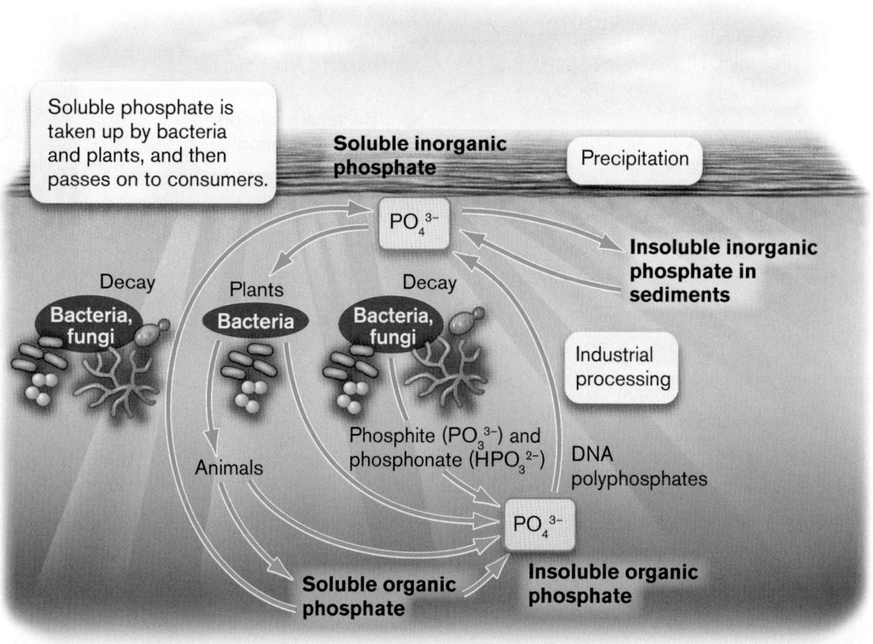

FIGURE 22.24 ■ The phosphate cycle. Phosphorus in the biosphere occurs entirely in the form of inorganic or organic phosphate. Most phosphate precipitates as insoluble salts in sediment. The small amount of soluble phosphate is taken up by plants and bacteria, which may then be taken up by consumers. Decomposers return phosphate to the environment.

Marine water is extremely limited for phosphate, because of the distance from sediment minerals. Thus, the genomes of marine phototrophs encode many systems for acquiring phosphate from organic sources. Some actually acquire phosphorus in a partly reduced form, such as phosphite (PO_3^{3-}) or phosphonate (HPO_3^{2-}), both of which are available in marine water. In addition, marine phototrophs show an unusual ability to substitute nonphosphorus lipids for phospholipids, thus cutting their phosphorus needs by half. For example, the cyanobacteria *Prochlorococcus*, *Synechococcus*, and *Trichodesmium* can replace membrane phospholipids with sulfonated lipids. Similarly, algae such as *Thalassiosira* species can replace phosphatidylcholine with betaine, which contains no phosphate but includes a carboxylate group.

Another microbial response to phosphate scarcity is to replace membrane phosphates with organic phosphonates, in which the phosphorus atom bonds directly to carbon ($R–PO_3^{2-}$), instead of via the easily hydrolyzed phosphoester bond. These organic phosphonates, however, are cleaved by marine archaea, the ammonia oxidizers such as *Nitrosopumilus*. The archaea release methylphosphonate, which many bacteria can cleave to obtain phosphate. Phosphate scavenging from methylphosphonate releases methane, thus interacting with the carbon cycle.

In agriculture, phosphate is often added as a fertilizer. Phosphate fertilizer is obtained by treating calcium phosphate rock with sulfuric acid, producing calcium sulfate (gypsum) and phosphoric acid. Excess phosphate from fertilizer or industry may drain into streams and lakes where phosphorus is the limiting element. The sudden influx of phosphate causes an algal bloom. The overgrowth of algae leads to overgrowth of heterotrophs, depletion of oxygen, and destruction of the food chain.

Thought Question

22.7 Compare and contrast the cycling of nitrogen and phosphorus. How are the cycles similar? How are they different?

The Iron Cycle

Why do organisms need iron? As a micronutrient, iron forms a negligible part of biomass but is essential for growth (discussed in Chapter 4). Organisms require iron as a cofactor for enzymes and an essential component of oxygen carrier molecules such as hemoglobin. Other common micronutrients required for enzymes and cofactors are zinc, copper, and selenium.

Iron is a major component of Earth's crust, and substantial quantities are present in most soil and aquatic sediment. Yet the availability of iron to organisms is limited by its extremely low solubility in the oxidized form. In microbial biochemistry, the oxidized form, ferric iron (Fe^{3+}), interconverts with the reduced form, ferrous iron (Fe^{2+}; **Fig. 22.25A**). In the presence of oxygen, iron metal (Fe^0)

A. The iron cycle

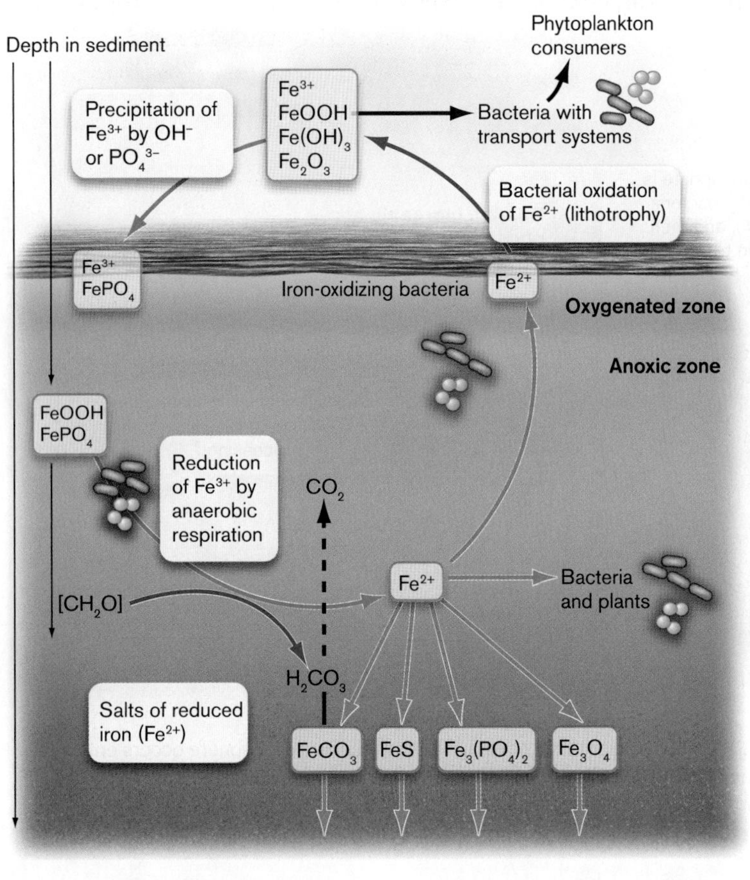

B. Iron precipitation due to lithotrophy

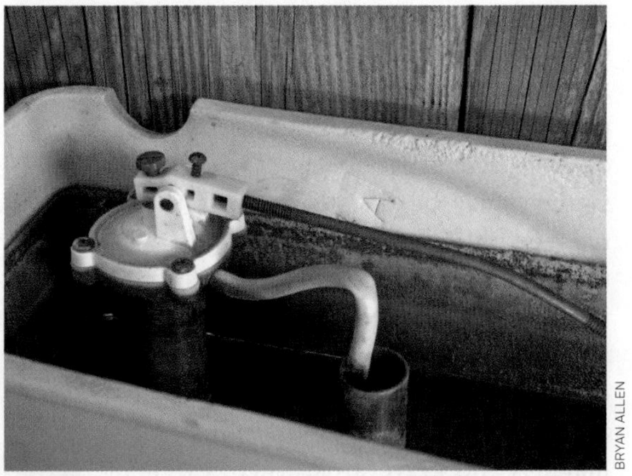

FIGURE 22.25 ■ The iron cycle. **A.** Ferric iron (oxidized iron, Fe^{3+}) precipitates with hydroxide or phosphate. Only bacteria can assimilate Fe^{3+}. In anoxic sediment, bacterial respiration reduces Fe^{3+} to Fe^{2+}, a more soluble form available to plants. **B.** Home plumbing shows signs of lithotrophic oxidation of Fe^{2+} to Fe^{3+} (orange precipitate).

"rusts" and is therefore available to organisms mainly as Fe^{3+}. Ferric iron is especially insoluble at high pH, precipitating with hydroxide ions as ferric hydroxide [$Fe(OH)_3$] or with phosphate ions as ferric phosphate ($FePO_4$). At neutral pH, iron is oxidized by bacteria such as *Gallionella, Leptothrix,* and *Mariprofundus.*

Marine iron. In the upper layers of most oceans, iron is extremely scarce. This is because the benthic sediment containing iron is so distant that its iron is largely inaccessible. The main source of marine iron is eolian (wind-borne) dust from dry land, such as from windstorms in the Sahara desert. Most of the wind-borne iron is particulate, oxidized, and unavailable to eukaryotic phytoplankton (algae). Thus, bacteria that acquire and reduce ferric iron provide the main entry of iron into the ecosystem. This may explain why so many marine algae are mixotrophs: Although their photosynthesis can fix plenty of carbon for biomass, they consume bacteria as a source of iron. Another source of marine iron, discovered in 2013, is hydrothermal vents.

Iron is thus a limiting nutrient for marine phytoplankton. As discussed in Chapter 21, when higher quantities of a limiting nutrient enter an ecosystem, the limited populations rapidly increase. The rise of algal populations following iron addition was demonstrated in an "iron fertilization" experiment conducted in 2007 by Phillip Boyd and colleagues at the University of Otago, New Zealand. Several thousand kilograms of ferrous sulfate ($FeSO_4$) was released in the Southern Ocean, near Antarctica. Within a week after the iron release, there was a bloom of phytoplankton. The dominant phytoplankton in the bloom were diatoms of the species *Fragilariopsis kerguelensis.* The diatom bloom was so large that it was detected by a NASA satellite as a region of increased color reflected by chlorophyll.

The dramatic effects of iron fertilization led some researchers to propose that increasing iron throughout the oceans could cause blooms of diatoms, removing CO_2 from the atmosphere. If a sufficiently large proportion of the diatoms were to escape predation and fall to the ocean floor, they would effectively remove their carbon from circulation. An alternative outcome, however, would likely be eutrophication, causing an ecological collapse comparable to a dead zone.

Iron in soil and sediment. In anoxic habitats, such as benthic sediment, Fe^{3+} is largely reduced to Fe^{2+} by anaerobic respiration. Reduction leads to loss of the reddish color, generating gray-colored sediment, as in wetland soil (discussed in Chapter 21). The reduction of Fe^{3+} to Fe^{2+} is one of the enriching contributions of anaerobic sediment to freshwater wetlands and coastal estuaries, because reduced iron is more available to plants and bacteria. Oxidized iron, in aerobic soils, can be taken up only by bacteria. Bacteria synthesize special iron uptake systems, including molecules called **siderophores** that bind ferric ion outside and then are taken up by the cell (see Section 4.2). Bacterial iron then becomes available to consumers in the ecosystem.

In iron mines, where pyrite (FeS_2) is exposed to air, spontaneous oxidation releases sulfuric acid:

$$2FeS_2 + 7O_2 + 2H_2O + 4H^+ \longrightarrow 2Fe^{2+} + 4H_2SO_4$$

Lithotrophs such as *Acidithiobacillus ferrooxidans* can catalyze iron oxidation to yield energy, thus increasing the rate of production of sulfuric acid. Rapid sulfuric acid production leads to acid mine drainage. Where acid mine drainage enters an aquatic system, the reduced iron is oxidized by lithotrophs to ferric hydroxide, forming an orange precipitate. A similar precipitate from bacteria in iron-rich water occurs in home plumbing (**Fig. 22.25B**). Iron mine drainage causes severe pollution of streams (**Fig. 22.26**).

Iron cycling often connects with the sulfur cycle in ways that can prove unfortunate for our built environment. The most serious problem is anaerobic corrosion of iron, such as the iron structures of bridges. Iron corrodes spontaneously in the presence of oxygen. In anoxic environments, however, little spontaneous corrosion takes place. Instead, iron is corroded by sulfate-reducing bacteria (**Fig. 22.27**). Bacteria such as *Desulfovibrio* and

THOMAS R. FLETCHER/ALAMY

FIGURE 22.26 ▪ Mine drainage enters a stream. Lithotrophic oxidation of reduced iron yields the orange material polluting this stream in Preston County, West Virginia.

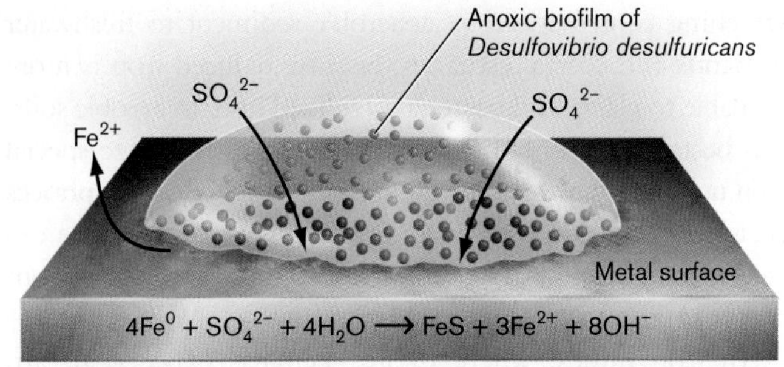

FIGURE 22.27 ▪ Anaerobic corrosion. Iron-sulfur bacteria convert Fe^0 to FeS. FeS flakes off, exposing iron to further oxidation and rapid corrosion. *Source:* Modified from Hang T. Dinh et al. 2004. *Nature* **427**:829–832.

Desulfobacter can grow on the iron surface, forming an anaerobic biofilm. Within the biofilm, iron is oxidized to Fe^{2+}, and sulfate dissolved in the water is reduced to ferrous sulfide:

$$4Fe^0 + SO_4^{2-} + 4H_2O \longrightarrow FeS + 3Fe^{2+} + 8OH^-$$

The FeS flakes away, exposing more iron to react with water, resulting in cyclic corrosion.

Other Metals in the Environment

Another concern for environmental management is the problem of toxic metals and metalloids such as mercury and arsenic. Besides iron, numerous trace metals interact with bacterial species as either electron donors or acceptors (**Table 22.4**). Bacterial conversion of metals can either produce toxic species or remove toxic metals from ecosystems. For example, chromium-6 [Cr(VI)] is an extremely toxic pollutant that can be reduced by soil bacteria to the much less toxic Cr(III). Microbial metal cycling is discussed further in **eTopic 22.2**.

How can bioremediation remove toxic minerals from groundwater? Microbial metabolism can convert a toxic metal to an insoluble form, which then precipitates out of groundwater. For example, the anaerobic respiration by *Geobacter metallireducens* is used to convert toxic uranium ion, U^{6+} (VI) to the insoluble form U^{4+} (IV). Bioremediation by *Geobacter* was described in Chapter 14. A different example of bioremediation, the bioconversion of arsenite to arsenate, is described in **eTopic 22.3**.

Multiple Limiting Factors Modulate Complex Ecosystems

A common assumption of terrestrial and aquatic ecology is that a single nutrient, such as phosphorus or iron, is limiting for a given ecosystem. In marine water, however, more than one factor maybe limiting. The requirement for more than one limiting factor is known as **resource colimitation**. In highly oligotrophic marine water, populations

TABLE 22.4 Microbial Metabolism of Metals

Metal	Major conversions	Microbial genera (examples)	Effects within environment
Arsenic (As)	$AsO_3^{3-} \longrightarrow AsO_4^{3-}$	*Alcaligenes, Pseudomonas*	Oxidation of poisonous AsO_3^{3-} via use as terminal e^- acceptor, converting to insoluble AsO_4^{3-}.
	$AsO_4^{3-} \longrightarrow AsO_2^-$	*Bacillus, Chrysiogenes, Pyrobaculum*	Formation of AsO_4^{3-} (arsenate) used as terminal e^- acceptor.
	$AsO_4^{3-} \longrightarrow (CH_3)_3As$	*Candida* (a fungus), *Scopulariopsis* (a fungus)	Formation of methylarsines. Poisonous; inhalation of moldy wallpaper with arsenic pigment.
Chromium (Cr)	$CrO_4^{2-} \longrightarrow Cr^{3+}$	*Aeromonas, Arthrobacter, Desulfovibrio*	CrO_4^{2-} [Cr(VI)] is mutagenic and carcinogenic; as e^- acceptor, reduced to Cr^{3+} [Cr(III)], less toxic.
Manganese (Mn)	$Mn^{2+} \longrightarrow Mn^{4+}$	*Hyphomicrobium, Arthrobacter*	Mn is a trace element required for enzymes.
	$Mn^{4+} \longrightarrow Mn^{2+}$	*Geobacter, Pseudomonas*	Anaerobic respiration in sediment.
Mercury (Hg)	$Hg^{2+} \longrightarrow Hg^0$	*Acidithiobacillus*	Hg^0 volatilizes; little harm.
	$Hg^{2+} \longrightarrow (CH_3)Hg^+$	*Desulfovibrio*	$(CH_3)Hg^+$ is a severe neurotoxin; accumulates at higher trophic levels, such as fish.
Selenium (Se)	$Se^0 \longrightarrow SeO_3^{2-}$	*Bacillus, Micrococcus*	Se is a trace element; small amounts in food help remove mercury. Larger amounts are toxic.
Uranium (U)	$UO_2^{2+} \longrightarrow UO_2$	*Veillonella, Shewanella, Geobacter*	UO_2^{2+} [U(VI)], soluble, is respired to uranium dioxide, UO_2 [U(IV)], insoluble; used for cleanup of radioactive uranium.
Vanadium (V)	$VO_3^- \longrightarrow VO(OH)$	*Veillonella, Desulfovibrio, Clostridium*	V is a trace element for some nitrogenases and invertebrate blood pigments. VO_3^- (vanadate) is oxidized as an e^- donor.

often depend on multiple limiting factors. For example, in the Baltic Sea, both nitrogen and phosphorus are present in such low concentrations that they must be added together in order to stimulate a phytoplankton bloom. In another example, the North Atlantic is limited for both phosphorus and iron. Addition of both P and Fe stimulates growth of nitrogen-fixing cyanobacteria.

To Summarize

- **Oxidized and reduced forms of sulfur are cycled in ecosystems.** Sulfate and sulfite serve as electron acceptors for respiration. Hydrogen sulfide serves as an electron donor. Sulfur oxidation to sulfuric acid causes acid mine drainage and pipe erosion.
- **Phosphorus cycles primarily in the fully oxidized form (phosphate).** Phosphate limits growth of phototrophic bacteria and algae in some freshwater and marine systems.
- **Iron cycles in oxidized and reduced forms.** Oxidized iron (Fe^{3+}) serves as a terminal electron acceptor in anaerobic soil and water. Reduced iron (Fe^{2+}) from rock is oxidized through weathering or mining. Bacterial lithotrophy accelerates iron oxidation, leading to acidification.
- **Metal toxins can be metabolized by bacteria.** Bacterial metabolism may either increase or decrease toxicity. Metabolic conversion to an insoluble form offers an effective means of bioremediation.
- **Marine habitats show resource colimitation.** Multiple resources may be limiting for the phytoplankton community or for different microbial populations.

22.5 Our Built Environment

Humans increasingly live in a habitat that was built by our technology, rather than existing in a "natural" environment. In developed countries, people are estimated to spend 90% of their time indoors. By shaping environments for our own comfort and convenience, we unwittingly have shaped the microbial communities that cohabit with us. Microbiologists study our built environment with the aim of understanding how the associated microbial communities interact with human inhabitants, in positive as well as negative ways.

The Microbiome of Building Interiors

Built structures include homes, offices, and transportation vehicles such as automobiles and subway cars. Questions we ask about the interiors of such structures include: What kinds of microbes are prevalent? What are the natural or human sources of the microbes? How do the microbes' activity affect human life?

Figure 22.28 presents the results of a study by Marzia Miletto and Steven Lindow at the UC Berkeley, in which air was sampled from 29 different home interiors. From the air samples, 16S rDNA was amplified and sequenced. In this study, the most abundant organism found was *Diaphorobacter* species (**Fig. 22.28** inset), family Comamonadaceae. These are Betaproteobacteria commonly found in air, water, and soil; they are capable of aerobic or anaerobic catabolism on various substrates. Yet *Diaphorobacter* represents only 10% of the community found. Other prominent members include Sphingomonadaceae (Alphaproteobacteria) and Propionibacteriaceae, which are Actinobacteria that colonize human skin. (See Chapter 18 for a review of bacterial diversity).

The abundance values of various taxa of indoor air samples were plotted and compared to those from samples of other interior sources, such as human skin, countertops, and tap water. The highest proportions of corresponding taxa were found in outdoor air and tap water. Pet animals showed many taxa in common with those of indoor and outdoor air, including the top five taxa found in indoor air. Pets showed additional taxa not found in air (not shown in **Fig. 22.28**). Human skin showed one highly represented taxon, the Propionibacteriaceae, which may cause skin infections such as acne. Another airborne taxon shared by human skin was Corynebacteriaceae, a family including commensals, as well as pathogens such as *Corynebacterium diphtheriae.* Other airborne bacteria do not normally reside in skin. Surprisingly low airborne representation was seen for taxa found in human saliva.

Indoor environments vary in their microbial community structure. Environments largely limited to human occupation, such as hospitals and public transport, often show high abundance of bacteria derived from human skin and mucosal sources (**Fig. 22.29**). These include the genera *Staphylococcus* and *Streptococcus*, as well as *Cutibacterium.* Hospitals show additional taxa associated with pathogens such as *Pseudomonas* and *Acinetobacter.* On the other hand, well-ventilated buildings show relatively little airborne content from the human occupants, and they share greater content with external air.

An interesting finding is that farms and home spaces exposed to pet animals show a more diverse microbiome with a broader range of harmless, animal-associated taxa. Studies suggest that the farm microbiome offers a protective effect against allergies, particularly in children. At the same time, the farm microbiome includes exposure to potential pathogens such as *Bartonella* (cause of cat scratch disease) and *Enterococcus.* Thus, increasing study focuses on the mechanisms of microbial transfer from sources and sinks of harmful microbes—as well as those that are beneficial.

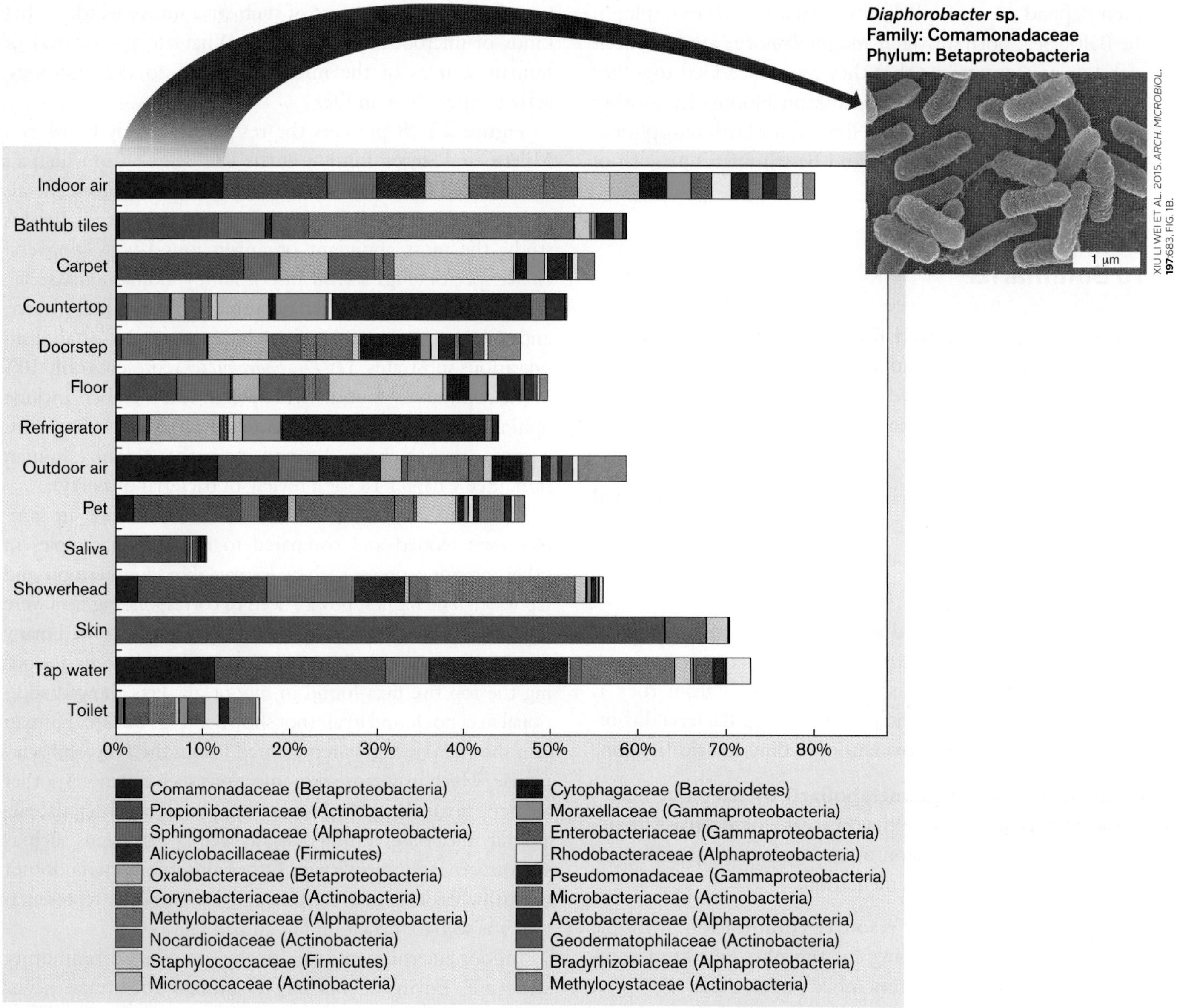

FIGURE 22.28 ■ Indoor air microbiome. Air samples from indoor environments in San Francisco were tested for bacterial taxa at the family level. The relative abundance of the top 20 taxa found was compared with respect to the same taxa from other sources. **Inset:** *Diaphorobacter* species from a soybean root nodule. *Source:* Modified from Miletto and Lindow. 2015. *Microbiome* **3**:61, fig. 1.

Transmission of Microbes between Humans and Our Built Environment

Studies have addressed the means of transmission of microbes from humans to our home and office interiors. One study estimates that a human body disperses millions of bacteria and fungal spores per hour, largely via skin contact and shedding of particles. A patient room in a hospital rapidly acquires the microbiome of its occupant. Such microbes commonly include difficult-to-treat pathogens such as methicillin-resistant *Staphylococcus aureus* (MRSA) and *Candida auris*, an emerging fungus (**Fig. 22.30**). Pathogens may be transmitted via inhalation of droplets and desiccated particles from other patients or from hospital staff. One study showed that an important source of pathogens could be the doctor's necktie.

In the home, various common objects may be surprising sources of microbes. A remarkably fertile source is the kitchen sponge (**Special Topic 22.1**).

What practical lessons do we learn from the built environment's microbiome? The results suggest some recommendations for further study:

- **Increased ventilation** may promote greater airborne content of health-promoting microbes from outdoor air, and less exposure to human pathogens.

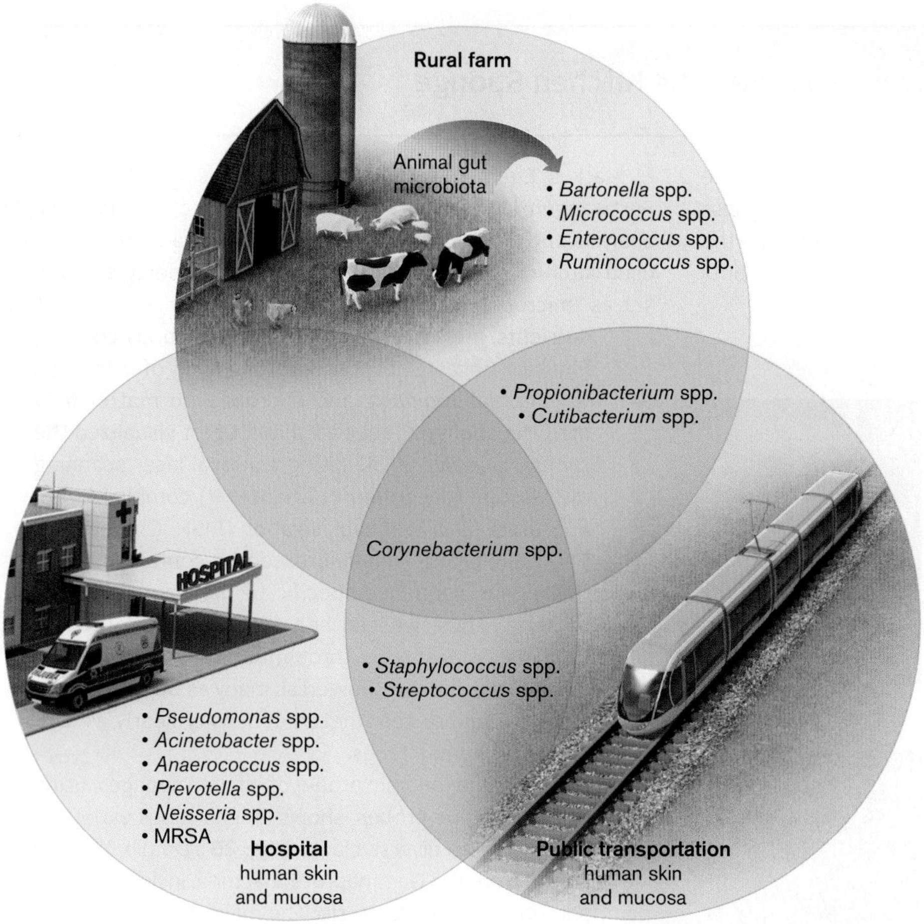

FIGURE 22.29 ■ Bacteria found in the built environment. Human-built habitats such as hospitals and public transport show distinctive microbial communities that differ from those of farm environments. *Source:* Modified from Gilbert and Stephens. 2018. *Nat. Rev. Microbiol.* **16**:661–670, fig. 1.

- **Controlled exposure to farms and pet animals** may increase the diversity of health-enhancing microbes in the indoor microbial community.
- **Decreasing surface moisture and dampness** may help control the growth of mold and release of allergenic fungal spores.
- **Sensors could be developed** to monitor the microbiome and detect potential pathogens.

On a larger scale, how do we define human versus "natural" environments? Even environments far from human habitation are so affected by human processes that pristine, untouched ecosystems no longer exist. Our view of Earth's biosphere increasingly is moving toward environmental management—the concept that wilderness as such no longer exists, but is only a biosphere to be managed for better or worse. Microbes will be our partners in managing the planet's environment. Perhaps the only environments left that are untouched by humans are those of other planets—if life exists there.

To Summarize

- **Building interiors acquire microbes** from exterior air, from human occupants, from pet animals, and from fomites.
- **Built environments host diverse microbial taxa,** including Alphaproteobacteria, Betaproteobacteria, Actinobacteria, Firmicutes, and others.
- **Pet animals and farm environments** can transmit diverse microbes that promote health.
- **Confined hospital spaces promote transmission of pathogens** among patients, staff, and fomites handled by both.

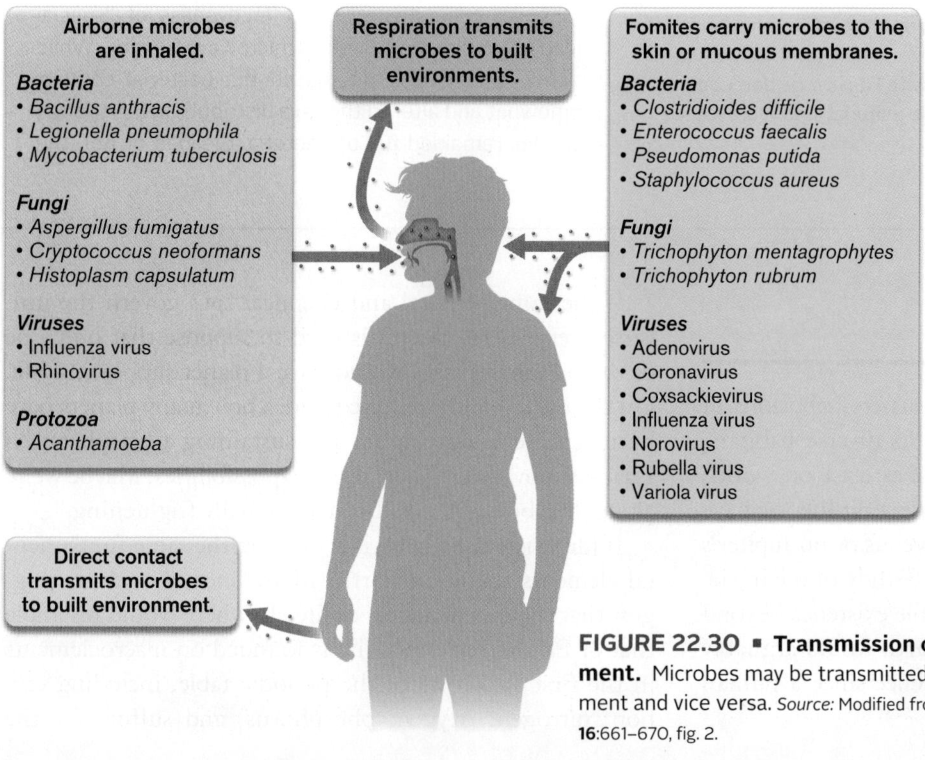

FIGURE 22.30 ■ Transmission of microbes within the built environment. Microbes may be transmitted from human occupant to the built environment and vice versa. *Source:* Modified from Gilbert and Stephens. 2018. *Nat. Rev. Microbiol.* **16**:661–670, fig. 2.

SPECIAL TOPIC 22.1 A Microbial Jungle: The Kitchen Sponge

Which object would you guess hosts the largest, most diverse source of bacteria in your home? Many of us would first guess the bathroom toilet. The answer, however, turns out to be the kitchen sponge—the implement we rely on to scrub our pots clean (**Fig. 1**).

A.

B.

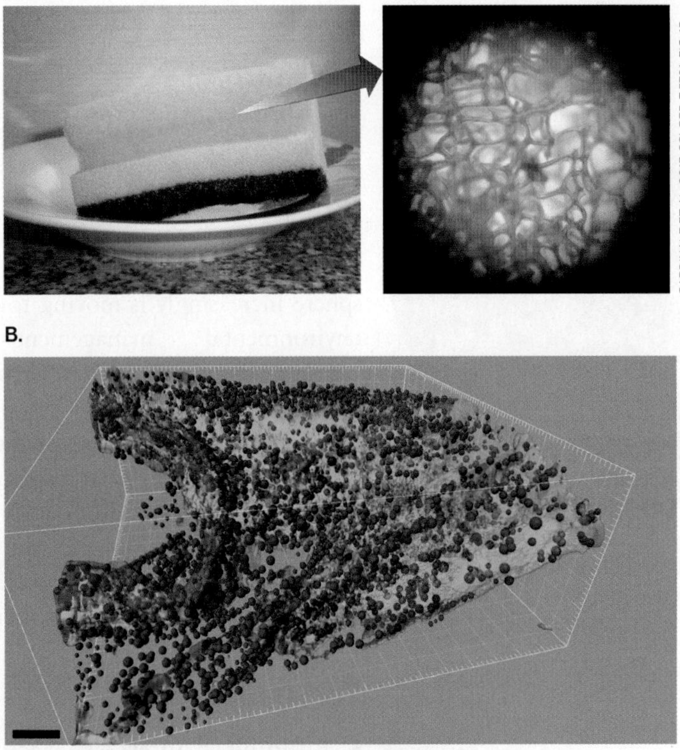

CARDINALE ET AL. 2017. *SCI. REP.* **7**:5791, FIG. 1B.

CARDINALE ET AL. 2017. *SCI. REP.* **7**:5791, FIG. 1B.

CARDINALE ET AL. 2017. *SCI. REP.* **7**:5791, FIG. 4D.

FIGURE 1 ■ Microbiome of a kitchen sponge. A. Kitchen sponge. **Inset:** Microscopic view of pores in a sponge sample (LM). **B.** Kitchen sponge sample, 3D confocal microscopy, with FISH probe detection of Gammaproteobacteria (red) and sponge material autofluorescence (cyan).

Markus Egert (**Fig. 2A**) and colleagues at Hochschule Furtwangen University, Germany, conducted a taxonomic analysis and 3D microscopic visualization of a collection of used kitchen sponges. They reasoned that sponges would act as "microbial incubators," continually exposed to fresh nutrients, moisture, and microbes from body contact. **Figure 1** (inset) shows the porous form of a sponge, which traps moisture and microbes, no matter how hard the sponge is squeezed out. Egert visualized the sponge samples in 3D, using confocal laser scanning microscopy (described in Chapter 2) combined with fluorescence in situ hybridization (FISH; Chapter 21). The FISH probe (red fluorescence) contained a 16S rRNA gene sequence specific to the Gammaproteobacteria. Thus, we can see that Gammaproteobacteria are widespread throughout the used sponge. Overall, the sponge showed as many as 50 billion bacteria per square centimeter—a density nearly as high as that of human feces.

The full taxonomic profile of kitchen sponges, using 16S rRNA gene probes, shows an enormous variety of bacteria, from diverse clades (**Fig. 2B**). Dominating the profile were the Gammaproteobacteria, including *E. coli* and other enteric bacteria. The major family, however, was not Enterobacteriaceae but Moraxellaceae, representing a third of the sponge community. Moraxellaceae includes a respiratory pathogen, *Moraxella catarrhalis*, but also contains numerous commensals of humans and other animals. Another family abundant in the sponge was Pseudomonadaceae, which may include pathogens and soil-dwelling bacteria.

Egert's team tested the effectiveness of "special cleaning" procedures such as microwave heating. While microwave cleaning decreased the bacterial content somewhat and altered the taxa distribution, the sponge samples remained full of bacteria. Cleaned or not, the

22.6 Astrobiology

As we come to appreciate the ubiquitous contributions of microbes to shaping the planet, in all its diverse habitats, increasingly we wonder whether microbes exist on worlds beyond Earth. Is Earth unique in supporting life, or have living cells evolved as well on Mars or Venus or on Jupiter's planet-sized moons? Astrobiology is the study of life in the universe, including its origin and possible existence beyond Earth. The discovery of life beyond Earth would arguably be the most significant advance in science since a human set foot on the moon.

If the same physical and chemical laws govern the universe everywhere, then it is hard to suppose that only one of the billions of stars would have a planet supporting life. On the other hand, we have no idea how many planets have been capable of developing and sustaining a biosphere. As Isaac Asimov said, "There are two possibilities. Maybe we're alone. Maybe we're not. Both are equally frightening."

If life exists elsewhere, is it built on the same fundamental elements as life on Earth? Many lines of evidence suggest that the biochemistry of life elsewhere would resemble that of Earth. Terrestrial life is founded on macroelements in the first two rows of the periodic table, including carbon, nitrogen, oxygen, phosphorus, and sulfur (for the

A.

B.

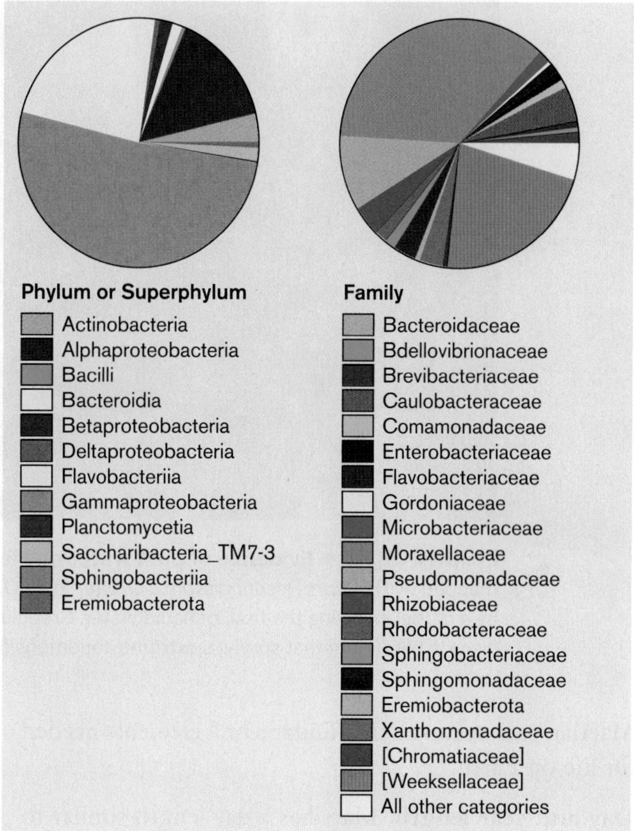

FIGURE 2 ■ Markus Egert studies the kitchen sponge microbiome. A. Markus Egert. B. Bacterial taxa detected by sequencing 16S rRNA gene amplicons. *Source:* Part **B** modified from Cardinale et al. 2017. *Sci. Rep.* **7**:5791, fig. 1C.

kitchen sponge retains its title as the most dense and diverse source of bacteria in the human built environment.

RESEARCH QUESTION

How does the kitchen sponge microbiome interact with that of the home's human inhabitants? Do sponge microbiomes influence the human microbiome, or vice versa?

Cardinale, Massimiliano, Dominik Kaiser, Tillmann Lueders, Sylvia Schnell, and Markus Egert. 2017. Microbiome analysis and confocal microscopy of used kitchen sponges reveal massive colonization by *Acinetobacter, Moraxella* and *Chryseobacterium* species. *Scientific Reports* 7:5791.

periodic table, see Appendix 1). The valence numbers (the number of electrons in the outer electron shell) of these elements from the middle of the periodic table enable them to form complex molecular structures with strong covalent bonds. The same fundamental molecules that appear in early-Earth simulation experiments, such as adenine and glycine, also appear in meteorites. Thus, we suspect that the fundamental building blocks for biochemistry are universal. On the other hand, if life could indeed be founded on some other basis, how would we recognize it?

If life is found on other planets, it will almost certainly include microbes. In fact, a case can be made that the majority of biospheres in our galaxy would consist entirely of microbial life, since microbes inhabit a wider range of conditions than do multicellular plants and animals. Even on Earth itself, the largest bulk of the biosphere—including deep sediments and rock strata—consists of microbial ecosystems.

Could Mars Support a Biosphere?

For several reasons, the most studied candidate for extraterrestrial life has been the planet Mars:

- **Geology.** Of all the solar planets, Mars seems the most similar to Earth in its topography; indeed, some areas of Mars remarkably resemble desertscapes on Earth. And

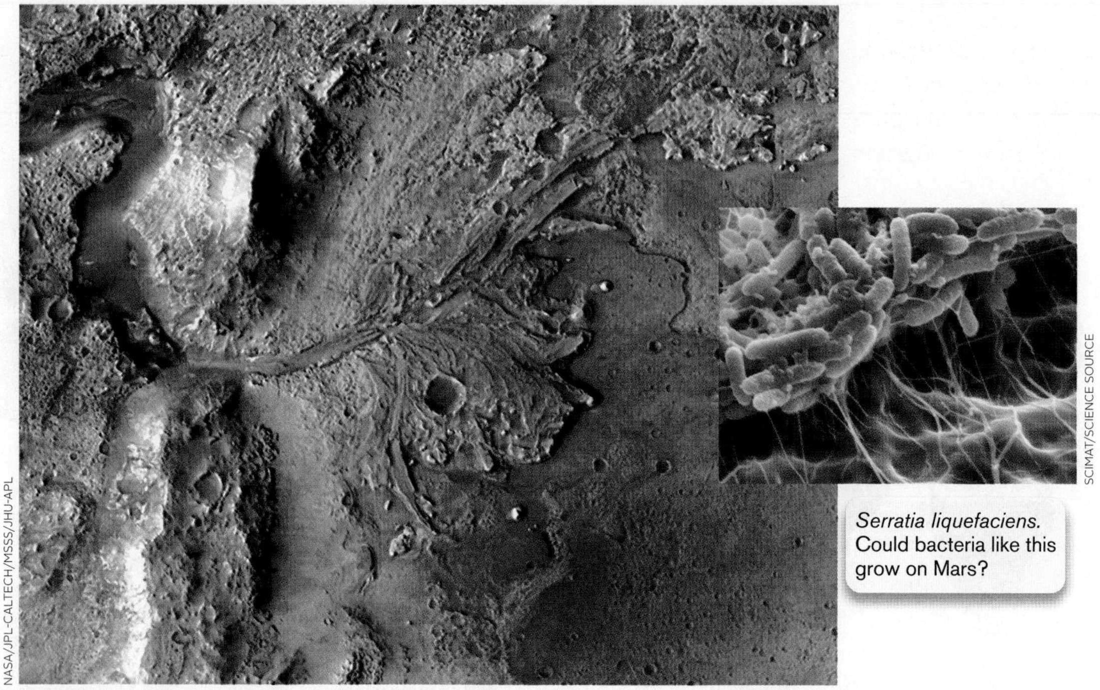

FIGURE 22.31 ■ Evidence of past water on Mars. Sedimentary deposits in the Mars Jezero crater, mapped by the *Mars Reconnaissance Orbiter* in 2007. The flow patterns resemble a river delta, best explained by a model involving the flow of liquid water. Green indicates claylike mineral deposits. **Inset:** *Serratia liquefaciens*, a bacterium that survives extreme conditions found on Mars.

Martian rock contains the fundamental elements needed for life on Earth.

- **Day and year length.** Mars has a day length similar to that of Earth, and a year only twice as long as Earth's.
- **Temperature.** The average temperature on Mars is 220 K (–53°C), too cold for most biochemistry on Earth, but its temperature rises above freezing at the equator. By contrast, the torrid heat of Venus (460°C) would exclude stable macromolecules.
- **Atmosphere.** Overall, Mars has an atmospheric pressure of 6 millibars (mbar), barely a hundredth that of Earth (1,013 mbar), and it lacks molecular oxygen. Thus, aerobes could not grow. But the Martian atmosphere does include carbon dioxide, actually at 20 times the CO_2 content on Earth, so there would be plenty for photoautotrophic production of biomass.
- **Water.** Surface water freezes out of the Martian atmosphere, without existing as a liquid. But mineral formations suggest that liquid water flowed in the past. Liquid water may yet exist deep underground, supporting life forms similar to the endoliths of Earth's crustal rock.

The existence of liquid water is a key question because on Earth, wherever liquid water exists—even brine (concentrated salt) at –20°C—there we find microbial life. In 2015, NASA's *Mars Reconnaissance Orbiter* found evidence of flowing water on Mars (see Chapter 1). Water condenses and freezes out of the Martian atmosphere, generating ice clouds and snow. The Martian soil composition showed high levels of perchlorate, a chemical that attracts water to form liquid solution and can support the metabolism of some Earth microbes.

Other evidence for liquid water in the past on Mars comes from geological formations surveyed by the *Mars Reconnaissance Orbiter* in 2007. The orbiter mapped sedimentary deposits in Mars's Jezero crater (**Fig. 22.31**). The crater formations reveal flow patterns typical of a river delta flowing into a lake. Claylike mineral deposits (false-colored green in **Fig. 22.31**) may have trapped organic compounds needed for life. The layered patterns are best explained by a model involving fluid flow, such as the flow of water. In 2019, geologists reported evidence that an ancient groundwater system once existed on Mars.

If life once existed on Mars, what became of it? Two main possibilities are considered:

- **Life developed and existed until the planet froze.** Under this scenario, life originated much as it did on Earth, during a time of heavy bombardment from space and outgassing of nitrogen and carbon dioxide. Unfortunately, however, life failed to generate sufficient atmosphere for a greenhouse effect to sustain temperate conditions. No oxygenic organisms produced molecular oxygen and an ozone layer.
- **Life developed and still exists underground.** In the absence of an ozone layer, the Martian surface is ster-

ilized by cosmic radiation. Nevertheless, microbes may yet exist deep underground, similar to the endolithic prokaryotes on Earth, or the subsurface brine communities of Antarctic lakes. Subsurface microbes are protected from cosmic radiation.

In 2018, a team of Italian astronomers reported radar evidence for liquid water beneath the ice cap of the Martian south pole. The team analyzed data from the Mars Advanced Radar for Subsurface and Ionosphere Sounding (MARSIS), an instrument on the *Mars Express* orbiter sent by the European Space Agency. The analysis indicates a large subglacial lake. The lake probably consists of perchlorate brine whose antifreeze property keeps the water liquid. On Earth, some microbes have been found to grow in such brines.

Do Biosignatures Indicate Life?

If microbes do exist on Mars, they should be detectable. But the detection of unknown life, even on Earth, remains a challenge. The advent of PCR sequence detection of species based on ribosomal genes has revealed thousands of unknown species, most of which cannot be grown or recognized by other methods. Other species may well be missed if their rRNA sequences fail to amplify with our probes. On Mars, assuming life evolved independently of Earth life forms, we would have no way to define sequence probes for detection.

Instead, researchers try to define **biosignatures**, chemical and physical signs that only life could have formed. Most of the proposed biosignatures are based on types of evidence for life on Earth, either as fossils of ancient life (discussed in Chapter 17) or as signs of current life in extreme habitats. Proposed biosignatures include the following:

- **Microfossils**. The mineralization of microbial cells leads to the formation of structures that can be visualized under a microscope. The cell fossils must be sufficiently distinct to establish that no abiotic process could have formed them.
- **Isotope ratios**. Certain biochemical reactions preferentially use one isotope of an atom over another; for example, Rubisco, the key enzyme of carbon dioxide fixation, uses ^{12}C in preference to ^{13}C (discussed in Chapter 17). Thus, both photosynthetic and chemolithoautotrophic use of ^{12}C can decrease the $^{12}C/^{13}C$ ratio in subsequent carbonate deposits. Isotope ratios of nitrogen, oxygen, and sulfur are also used as biosignatures.
- **Mineral deposits.** Certain mineral formations are observed to be caused only by microbial activity. For example, insoluble manganese oxides such as Mn_2O_3 are almost always the result of microbial oxidation of reduced manganese (discussed in Chapter 14). Reduced manganese ions are very stable, and their abiotic oxidation rate is extremely slow.
- **Metabolic activity.** Samples of soil can be incubated with radioactive tracer substances such as $^{14}CO_2$ and tested for metabolic conversion or incorporation into biomass. It must be established that the conversion could not have occurred abiotically and that no organisms from Earth were present.

Various kinds of evidence for life on Mars have been reported, such as possible microfossils within a Martian meteorite that landed in Antarctica. Metabolic activity was tested in samples obtained by the NASA *Viking* lander in 1975. As of this writing, however, no evidence has proved conclusive for active or past living microorganisms on Mars.

What kinds of microbes might survive on Mars? Terrestrial bacteria in extreme habitats suggest possibilities. For example, the halophilic archaeon *Halorubrum lacusprofundi*, isolated from Deep Lake, Antarctica, grows in high salt at temperatures below 0°C. Other Haloarchaea, such as *Haloarcula argentinensis*, grow in solutions of up to 0.5-M perchlorate, a powerful oxidant found in high concentration on Mars. The Gram-negative bacterium *Serratia liquefaciens* (**Fig. 22.31** inset) was shown to survive subfreezing temperatures (–50°C), ultraviolet exposure, and low atmospheric pressures—all conditions comparable to those found on Mars.

Could Mars Be Terraformed to Support Earth Life?

If no life exists on Mars—or if it exists only in the form of microbes deep underground—should we consider human intervention, or **terraforming,** to make Mars habitable for life from Earth? Scenarios for terraforming, long explored in science fiction, are now receiving serious thought among space scientists. Terraforming Mars would require increasing the temperature and air pressure. The temperature of the atmosphere might be increased by release of greenhouse gases. In addition, sufficient carbon dioxide and nitrogen gases must be made available from the Mars surface rock. In principle, microbes from Earth could be seeded to grow and generate an atmosphere containing nitrogen, oxygen, and CO_2.

The dilemmas and consequences of terraforming are depicted in the novel *Red Mars* (1993), by Kim Stanley Robinson. In favor of terraforming, it is argued that Mars offers enormous natural resources of potential benefit for humanity, especially as terrestrial resources are used up. Human settlements on Mars would be a major step forward for space exploration. On the other hand, it is argued that the planet Mars is a natural monument, a place with its own right to exist as such, and should be allowed to remain in its natural state for future generations to appreciate. As a practical matter, terraforming remains unfeasible for the near future. For example, the amount of chlorofluorocarbons required to raise Martian temperature is calculated to be 100 times greater than our global capacity to produce such substances.

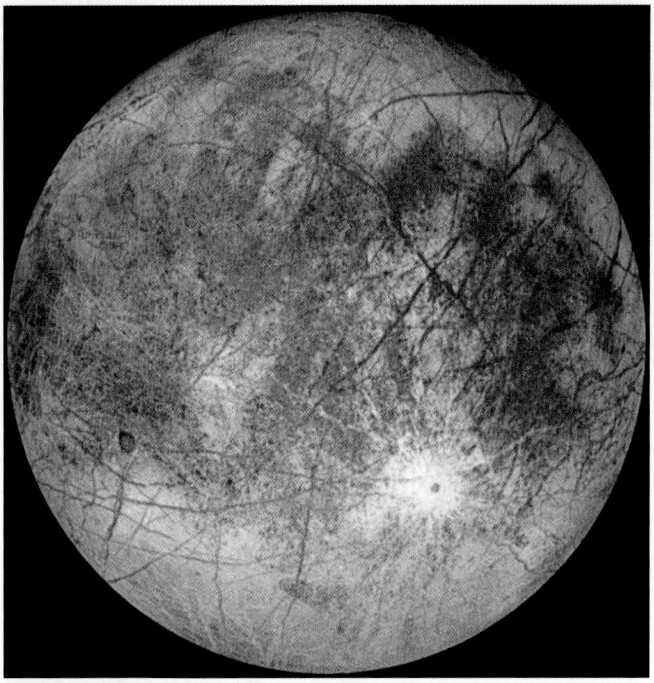

FIGURE 22.32 ■ Jupiter's moon Europa. Does life exist beneath Europa's ice, in a salty sea?

Does Europa Have an Ocean?

Farther out in the solar system, surprising candidates for microbial life are the moons of Jupiter. In 2000, the *Galileo* space probe passed several of Jupiter's moons, including Ganymede, Callisto, and Europa (**Fig. 22.32**). While their distance from the Sun results in extreme cold, these bodies receive extra heat from friction generated by tidal forces from the giant planet Jupiter. In the case of Europa, the tidal forces have been calculated to provide enough heat to liquefy water without boiling it off. Furthermore, measurement of Europa's magnetic field suggests that its composition includes a dense iron core surrounded by 15% water. Most of the water must be locked in ice, but tidal heating could melt enough water for an underlying ocean of brine. On Earth, similar brine lakes beneath the ice of Antarctica harbor halophilic archaea that grow at –20°C.

If such oceans do exist, how could they support life without photosynthesis? Photosynthesis is impossible at Jupiter's distance from the Sun. However, an alternative source of chemical energy might be the influx of charged particles accelerated in Jupiter's magnetic field. Charged particles entering Europa's ice can react with water to form hydrogen peroxide (H_2O_2). The H_2O_2 then breaks down, releasing molecular oxygen. Oxygen reaching the brine layer below could combine with electron donors from crustal vents and power metabolism. Alternatively, molecular hydrogen (H_2) could be generated from water ionized by decay of radioisotopes. A similar source of H_2 for life has been proposed to occur on Earth in rock strata several kilometers below the surface, where it may support lithotrophs. On Europa, the hydrogen could then combine with oxygen gas or other oxidants to power life. These schemes are highly speculative—but tantalizing enough to encourage future NASA missions to take a closer look at Europa and its sister moons.

Does Life Exist on Planets of Distant Stars?

The past decade of astronomy has seen an extraordinary growth in our knowledge of solar systems beyond our own. Now that we know that so many other stars possess planets, we can only wonder whether they also possess biospheres of life forms we would recognize.

In recent years, astronomers have found several hundred "extrasolar" planets orbiting distant stars, including some the size of Earth. Could we ever hope to detect signs of life on an extrasolar planet? A possible means of detection is suggested by William Sparks and colleagues at the Space Telescope Science Institute in Baltimore, and collaborating institutions. They show that light scattered by marine phytoplankton exhibits a property called "circular polarization." Circular polarization arises from substances that are homochiral—that is, present in only one of two mirror forms, such as the L and D forms of an amino acid. Homochirality is a strong biosignature, typical of proteins and metabolites. If someday we have telescopes capable of detecting light from distant planets, circular polarization might provide evidence of life.

Figure 22.33 shows an infrared photograph through the Spitzer Space Telescope of nebula RCW-49, a gaseous cloud full of newborn stars. Recall from Chapter 17 that as stars form, they take up dust of supernovas that includes all the elements needed to form biomolecules. Spectroscopic observation of RCW-49 reveals stars surrounded by disks coalescing into planets. The planetary disks contain icy particles full of organic molecules such as methanol, glycine, and ethylene glycol, a reduced form of sugar. Could there be biospheres in the making?

To Summarize

- **Astrobiology is the study of life in the universe, including possible habitats outside Earth.**
- **Mars is the planet whose geology most closely resembles that of Earth.** Geological features strongly support the past existence of flowing water, a prerequisite for microbial life.
- **The search for extraterrestrial life is based on methods similar to those used to seek early life on Earth.** Evidence includes chemical and physical biosignatures, isotope ratios, microfossils, and metabolic activity.

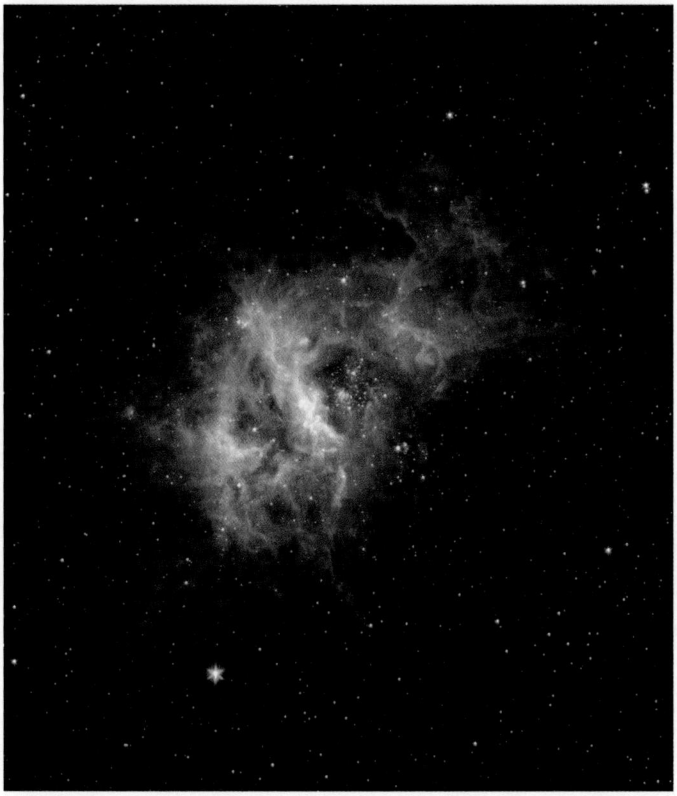

FIGURE 22.33 ▪ A stellar nursery. Nebula RCW-49 contains more than 300 newborn stars, from which NASA's Spitzer Space Telescope detected spectroscopic signals of common organic constituents of life (infrared photograph).

- **Terrestrial extremophile microbes are found that survive some of the conditions found on Mars.** Haloarchaea survive high salt and perchlorate, while certain bacteria survive ultraviolet exposure and low atmospheric pressures.
- **Jupiter's moon Europa is proposed as another possible site for life.** Europa is bathed in a sea of brine similar to terrestrial habitats for halophiles.

Concluding Thoughts

Our observations of distant molecules, as well as our analysis of meteorites (discussed in Chapter 17), suggest that the biomolecules of Earth fit into a universal pattern of interstellar chemistry. Whether life exists elsewhere or we are alone, we must remember that Earth is the only place we know of at this time that can support humans and the forms of life we require for our own survival. The survival of Earth's entire biosphere depends on our microbial partners cycling key elements and acquiring energy to drive the food web. For the first time in history, human technology now rivals the ability of microbes to alter fundamental cycles of biogeochemistry. But to manage and moderate our alterations—for our own survival and that of the biosphere—our fate still depends on the microbes.

CHAPTER REVIEW

Review Questions

1. Identify the major sources and sinks of carbon, nitrogen, and sulfur. Which sources recycle rapidly, and why?
2. Explain how the carbon cycle differs in oxygenated and anoxic environments.
3. Explain two different chemical methods of measuring the environmental levels of carbon and nitrogen.
4. Outline the hydrologic cycle. Explain the role of biochemical oxygen demand (BOD) in water quality and how it may be perturbed by human pollution.
5. Outline the functions of a wastewater treatment plant. Include the phases of primary, secondary, and tertiary treatment. Explain the roles of microbes in these phases of water treatment.
6. Outline the main transformations of the nitrogen cycle. Which reactions are carried out only by microbes?
7. Outline the main transformations of the sulfur cycle. Which features are comparable to the nitrogen cycle, and which are unique to the sulfur cycle?
8. For the iron cycle, explain aerobic and anaerobic processes of microbial transformation. Explain how microbial iron transformation may be linked to sulfur transformation.
9. Explain how bacteria may convert metals to toxic forms. Alternatively, explain how bacteria may detoxify metal-polluted sediment.
10. Explain the sources of microbes in the human built environment. How are humans affected by members of the built-environment microbiome?
11. Offer several arguments for and against the existence of life on planets beyond Earth.

Thought Questions

1. How does influx of nitrogen-rich fertilizers to soil ecosystems increase the rate of CO_2 efflux from wetlands?
2. In the past 50 years, most of the wetlands off the coast of Louisiana have been destroyed. How is wetland destruction related to formation of the dead zone in the Gulf of Mexico? How could the dead zone be revived, and why would the cost of restoration be projected to be billions of dollars?
3. What nutrients are limiting in the marine benthos, and why?

Key Terms

abiotic (884)
ammonification (903)
anammox reaction (904)
assimilatory nitrate reduction (903)
biochemical oxygen demand (BOD) (893)
biogeochemical cycle (884)
biogeochemistry (884)
biological carbon pump (887)
biosignature (917)
biotic (884)
built environment (885)
constructed wetland (899)
dead zone (894)
denitrification (903)
dissimilatory nitrate reduction (903)
floc (897)
fossil fuel (890)
geomicrobiology (884)
greenhouse effect (884)
hydrologic cycle (893)
hypoxia (895)
mesocosm (886)
nitrification (902)
nitrifier (902)
nitrogen fixation (901)
oxygen minimum zone (OMZ) (894)
radiative forcing (890)
reservoir (884)
resource colimitation (910)
siderophore (909)
sink (884)
sludge (897)
source (884)
terraforming (917)
wastewater treatment (895)
water cycle (893)
zone of hypoxia (894)

Recommended Reading

Arrigo, Kevin R. 2009. Marine microorganisms and global nutrient cycles. *Nature* **437**:349–355.

Canfield, Donald E., Alexander N. Glazer, and Paul G. Falkowski. 2010. The evolution and future of Earth's nitrogen cycle. *Science* **330**:192–193.

Clemmensen, Karina E., Adam Bahr, Otso Ovaskainen, Anders Dahlberg, Alf Ekblad, et al. 2013. Roots and associated fungi drive long-term carbon sequestration in boreal forest. *Science* **339**:1615–1618.

Gilbert, Jack A., and Brent Stephens. 2018. Microbiology of the built environment. *Nature Reviews. Microbiology* **16**:661–670.

Heinz, Jacob, Janosch Schirmack, Alessandro Airo, Samuel P. Kounaves, and Dirk Schulze-Makuch. 2018. Enhanced microbial survivability in subzero brines. *Astrobiology* **18**:1171–1180.

King, Gary M. 2015. Carbon monoxide as a metabolic energy source for extremely halophilic microbes: Implications for microbial activity in Mars regolith. *Proceedings of the National Academy of Sciences USA* **112**:4465–4470.

Melton, Emily D., Elizabeth D. Swanner, Sebastian Behrens, Caroline Schmidt, and Andreas Kappler. 2014. The interplay of microbially mediated and abiotic reactions in the biogeochemical Fe cycle. *Nature Reviews. Microbiology* **12**:797–808.

Montzka, S. A., E. J. Dlugokencky, and J. H. Butler. 2011. Non-CO_2 greenhouse gases and climate change. *Nature* **476**:43–50.

Salese, Francesco, Monica Pondrelli, Alice Neeseman, Gene Schmidt, and Gian G. Ori. 2019. Geological evidence of planet-wide groundwater system on Mars. *Journal of Geophysical Research. Planets.* **124**:374–395.

Sohm, Jill A., Eric A. Webb, and Douglas G. Capone. 2011. Emerging patterns of marine nitrogen fixation. *Nature Reviews. Microbiology* **9**:499–508.

Vuitton, Dominique Angèle, and Jean-Charles Dalphin. 2017. From farming to engineering: The microbiota and allergic diseases. *Engineering* **3**:98–109.

Woodcroft, Ben J., Caitlin M. Singleton, Joel A. Boyd, Paul N. Evans, Joanne B. Emerson, et al. 2018. Genome-centric view of carbon processing in thawing permafrost. *Nature* **560**:49–54.

Wright, Jody J., Kishori M. Konwar, and Steven J. Hallam. 2012. Microbial ecology of expanding oxygen minimum zones. *Nature* **10**:381–394.

CHAPTER 23

The Human Microbiome and Innate Immunity

23.1 The Human Microbiome

23.2 Benefits and Risks of Microbiota

23.3 Overview of the Immune System

23.4 Physical and Chemical Defenses against Infection

23.5 Innate Immunity: Surveillance, Cytokines, and Inflammation

23.6 Complement and Fever

The human body teems with microbes (the microbiome or microbiota) vital to our existence, and is under constant attack from pathogens. How do we survive? Obstacles such as skin and stomach acid, which are nonspecific defenses, will repel most microorganisms. But for microbes able to breach these physical barriers, there await powerful innate and adaptive immune defenses. Chapter 23 introduces the human microbiome and its effects on human health, and describes the series of barriers and elaborate innate immune defenses that keep our microbiota (and invading pathogens) at bay.

CURRENT RESEARCH highlight

Tumorigenic bacteria. Can bacteria cause colon cancer? Cynthia Sears and Franck Housseau use mice to study how Bft enterotoxin, produced by gut microbe *Bacteroides fragilis* causes colon microadenomas. Mice colonized with Bft-producing *B. fragilis* develop 10- to 20-fold more tumors than do mice colonized with Bft-deficient *B. fragilis*. Bft stimulates colonic epithelial cell proliferation and triggers an immunological cascade that enhances tumor growth. In response to Bft, colon-resident T lymphocytes release inflammatory cytokine IL-17, which activates colonic epithelial cells to secrete a chemokine that attracts myeloid-derived suppressor cells (MDSCs). MDSCs migrate to the colon and supply growth factors that accelerate tumor growth. In the future, chemically stopping the recruitment of MDSCs could prevent tumor formation.

Source: Liam Chung, et al. 2018. *Cell Host Microbe* **23**:203–214.

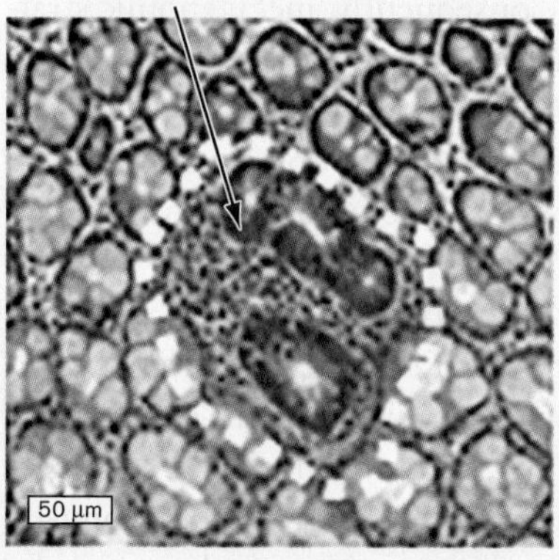

CHUNG, ET AL. 2018. *CELL HOST MICROBE* **23**:203–214, FIG. 4F.

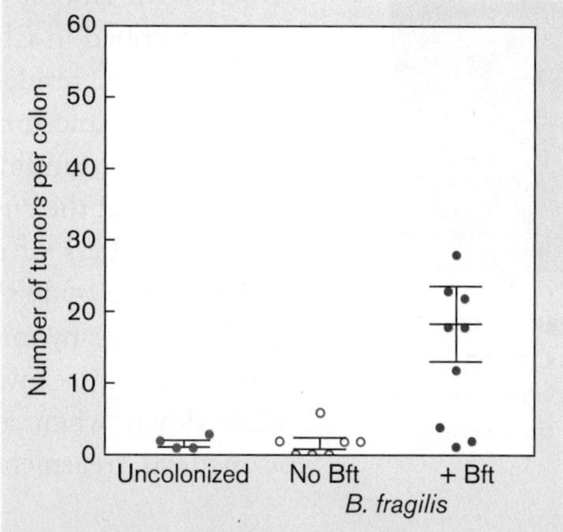

AN INTERVIEW WITH

CYNTHIA SEARS, PHYSICIAN AND MICROBIOLOGIST, JOHNS HOPKINS UNIVERSITY

COURTESY OF CYNTHIA SEARS

How will your work change our thinking about colon cancer?

Two trends are alarming—increasing colon cancer among people 20–50 years old, even in countries traditionally thought to be at low risk of colon cancer. We hope to harness knowledge about specific microbes, likely bacteria, or features of the composite microbiome as new tools for colon cancer prevention. We imagine a cost-effective and accurate stool test that predicts a person's risk for colon cancer. Such a test would facilitate early diagnosis of this preventable cancer.

Most scientists once thought that the microbes in our gut offered us minimal benefit, and no one suspected that some might cause colon cancer. We now know that microbes living in and on our bodies have myriad positive effects on human health, but also some negative ones. As described in the Current Research Highlight, Cynthia Sears and Franck Housseau (**Fig. 23.1**) discovered a series of events triggered by *Bacteroides fragilis* (a member of the gut microbiome) that ends with colon cancer. In the early sections of Chapter 23 we will describe the human microbiome and examine how these organisms affect human health and survival.

One critical function of our microbiome is to shape and sharpen our immune system. Later in this chapter we will explore the immune system: how it works and how it can fail. For example, imagine two people swimming in the Gulf of Mexico—a healthy 12-year-old girl and an alcoholic 66-year-old man. While they swam, Gram-negative marine bacteria called *Vibrio vulnificus* entered their bodies through tiny cuts. Four days later, the girl returned to her Mississippi home, oblivious to the immunological battle recently waged in her bloodstream. The man, however, lay in a morgue, dead of an aggressive *V. vulnificus* blood infection. Why did one live and the other die? A variety of nonspecific, innate immune factors present in the girl's body killed the invading pathogen before it could multiply. The alcoholic man with liver disease was deficient in several of those defense mechanisms—mechanisms that made the difference between life and death.

This chapter introduces our bodies' front-line, innate defense systems. But first it explores the human microbiome, a vital ecosystem that trains our immune system.

COURTESY OF FRANK HOUSSEAU

FIGURE 23.1 ▪ Immunologist Franck Housseau, Johns Hopkins University. Housseau collaborated with Cynthia Sears to elucidate the elegant immunological pathway that enterotoxigenic *Bacteroides fragilis* uses to drive colon cancer (see the Current Research Highlight).

23.1 The Human Microbiome

From the moment of our birth to the time of our death, we are constantly populated by microbes. These include bacteria, archaea, fungi, viruses, and even some protozoa. The consortium of colonizing microbes has been dubbed the human **microbiota**, or **microbiome**. These microbes have a stake in keeping us, their hosts, alive. The fact that the human microbiome has been conserved over millennia is testament to their importance.

Note: Use of the terms "microbiota" and "microbiome" has blurred. "Microbiota" originally referred to actual microbes inhabiting a body site, whereas "microbiome" referred to the gene pool within those microbes. We use the terms interchangeably in this book. "Metagenome" is the term now used to designate the gene pool, or set of complete genomes present within a microbiome.

To illustrate the significance of the microbiome, consider that our bodies carry about as many bacterial cells (3.8×10^{13}) as human cells and about 100 times more nonredundant bacterial genes than human genes.

Microbes colonize wherever our body meets the external environment (for example, skin, mouth, gastrointestinal tract, and parts of the genitourinary tract). Most internal organs, blood, and cerebrospinal fluid are sterile. The presence of any microbes at these sites is considered an infection. The majority of species that make up our microbiota are unknown and until now have never been grown in the laboratory. Consequently, metagenomic strategies (described in Chapters 7 and 21) are being employed to identify our resident microbes. In 2016, the U.S. government announced the National Microbiome Initiative to advance understanding of how microbiomes contribute to our health and the environment.

Can members of our microbiome cause us harm? By and large, the body's barriers of defense work well to prevent incursions by microbiota or, in the event of an incursion, to kill the invader. Unfortunately, these defenses break down when an immune system is compromised by medical treatments (such as anticancer drugs) or by diseases (for instance, a deficiency in complement factors, discussed in Section 23.6). Such a person is described as a **compromised host** and can be repeatedly infected by certain normal biota. Organisms causing disease in this situation are called **opportunistic pathogens**. Later we will discuss how an imbalance in the relative numbers of species in the microbiome (dysbiosis) can negatively impact health.

We now understand that bacteria colonizing our bodies may be as important collectively as a kidney or liver. Microbiota in the gut, for instance, can communicate chemically with the brain and affect the function of the endocrine, nervous, and immune systems, while also affecting metabolism in a variety of ways. The communication pathway (called the microbiome-gut-brain axis) takes place via the vagus nerve that connects the brain to the intestine and a number of other organs. Another more pervasive mode of communication was described by Gary Siuzdak and his colleagues at the Scripps Research Institute. They discovered that our body cells are continually bathed in, and respond to, microbial metabolites that circulate in our blood.

Are all of these host-microbe communications important? A variety of studies suggest that microbiota and their products have roles in allergies, liver disease, obesity, gastrointestinal syndromes (such as inflammatory bowel disease), and possibly psychiatric syndromes such as autism and depression. Its vast impact on human health has led some to call our microbiota a new organ-like system, and the term "holobiont" is used to describe a human paired with his/her microbiota.

Microbiome Acquisition and the Hygiene Hypothesis

When and how does a person develop a microbiome? Recent evidence suggests that the microbiome begins to develop before birth. The first stool (meconium) passed by either a natural-birth newborn or a cesarean newborn <u>before</u> its first meal contains some bacteria. The source of those microbes is the placenta, which has its own microbiome.

Once the baby breaks out of the embryonic membrane, however, it is exposed to a dizzying array of microbes residing in the birth canal and the outside world. Contributions to the baby's skin, mouth, gut, and genitourinary-tract microbiomes come from the food they eat, the air they breathe, and the people, places, and things they touch. The initial makeup of a neonate's microbiome is shaped, in part, by the mode of delivery. Natural-birth babies are initially colonized by microbes acquired by passage through the mother's vagina. Babies born by cesarean section, however, start with microbes donated by the delivery room and by contact with the mother's skin. Young babies actually have microbiomes that are more diverse than those of adults. But by 3 years old, the diversity decreases in complexity to assume an adultlike composition.

Figure 23.2 presents the major human body sites (skin, respiratory, digestive, and genitourinary tracts) that are colonized by microbes and illustrates the relative makeup of bacteria, fungi, and viruses that populate each system. As you view the figure, notice the dramatic difference in levels of Bacteroidetes found in nasal and intestinal microbiomes, or of *Candida* in the microbiomes of the mouth and vagina. Although Archaea are missing from this figure, they are present at various host sites (intestine, skin, nose, and lung). Representatives from five archaeal phyla have been identified, but they comprise only a small portion of the entire microbiome. Their contributions to human health are unknown but under investigation. With all of its complexity, an individual host's microbiome remains relatively constant over time but can significantly fluctuate with diet, age, geography, or drug use.

The hygiene hypothesis. For millennia, human microbiota were shaped by human contact with natural environments composed of animals, caves, dirt, poop, and bugs. This natural outdoor world harbored a vast array of microbial taxa that could compete to populate our skin and mucosal surfaces. Today we are mostly an indoor species, spending almost all of our time inside closed buildings, segregated from nature—an arguably less diverse microbial environment. Add to that our use of soaps, antibiotics, and disinfectants and you can appreciate how severely we have restricted our access to microbes. As a result, our microbiota appear less diverse than those of our long-ago ancestors.

Studies that support this idea have compared the modern-day human microbiome with those of closely related wild African apes and of uncontacted Amerindians (see Chapter 27). Although improved hygiene limits exposure to pathogens, recent studies suggest that narrowing the diversity of our microbiome can contribute to inflammatory diseases such as asthma, inflammatory bowel disease, colorectal cancer, and obesity. Several intriguing studies have found that, even today, children who grow up on dairy farms are less likely to develop allergies and asthma. The reason is that our microbiome helps train our immune system (discussed later and in Chapter 24). Exposure to more microbes and other environmental antigens, especially early in life, may produce a more tolerant, well-controlled immune system less prone to inflammatory and autoimmune diseases (disorders in which the immune system reacts against the self).

The sections that follow describe the microbiota of various body sites and discuss the benefits and risks of a microbiome.

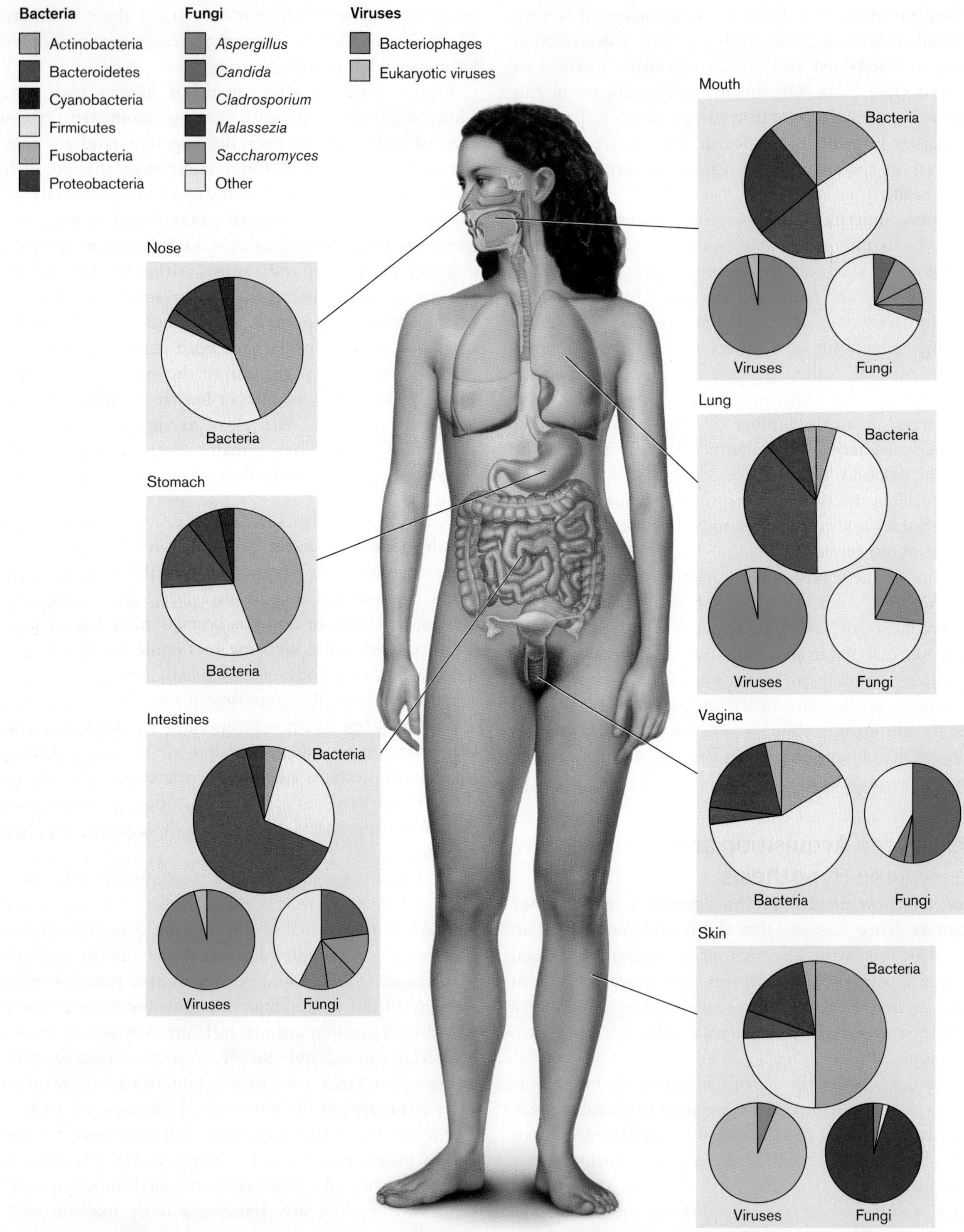

FIGURE 23.2 ■ Relative amounts of microbial phyla and families present at various colonizing sites. Each pie chart presents the relative compositions of bacterial phyla, fungal genera, or viruses present in each microbiome as determined by the sequencing of genes encoding 16S ribosomal RNA (see Sections 7.6 and 15.5). Microbial taxa are identified by color in the legend. *Source:* Data from Marsland and Gollwitzer. 2014. Nat. Rev. Immunol. **14**:827.

Skin

The human adult, on average, is covered with 2 square meters (over 21 square feet) of skin (**epidermis**) populated by 10^{11} microorganisms. Skin microbiota include aerobes, anaerobes, and facultative bacteria. There are, for instance, approximately 10^4–10^5 microbes per sweat gland, at a ratio of 1 aerobe to 10 facultative or anaerobic species. As with all colonized body sites, resident (normal) and transient members of the microbiota inhabit the skin. But even a resident microbe exhibits diversity as different strains colonize at different times.

Several features of epidermis make it difficult to colonize. The skin has an acidic pH (pH 4–6) owing to the secretion of organic acids by oil and sweat glands. As noted in Section 5.3, organic acids inhibit microbial growth by lowering bacterial cytoplasmic pH. Epidermal secretions are also high in salt and low in water activity (see Section 5.2), and they contain enzymes, such as lysozyme, that degrade bacterial peptidoglycan. Despite these hurdles, many species of bacteria manage to colonize the epidermal habitat. Most of these are Gram-positive organisms, because they tend to be more resistant to salt and dryness. Large expanses of dry skin tend to support the growth of Betaproteobacteria. Moister areas, such as scalp, ear, armpit, genital, and anal regions, are colonized primarily by *Corynebacterium* species and *Staphylococcus epidermidis*.

Skin microbiota benefit us by prompting the expression of epithelial cell tight-junction proteins, modulating functions of the immune system (immunomodulation), and secreting antibacterial peptides. For instance, *Staphylococcus hominis* and *Staphylococcus epidermidis*, friendly members of the skin microbiome, can kill *Staphylococcus aureus*, a pathogen that can form abscesses and a variety of other infections.

Some members of the skin microbiome, however, can be problematic. One such member is the Gram-positive, anaerobic rod *Cutibacterium acnes* (formerly *Propionibacterium acnes*), which causes acne, a very visible plague of adolescence. Increased hormonal activity in teenagers stimulates oil production by the sebaceous glands (**Fig. 23.3**). *C. acnes* readily degrades the triglycerides in this oil, turning them into free fatty acids that then promote inflammation of the gland. One consequence of the inflammatory response is the formation of a blackhead, a plug of fluid and keratin that forms in the gland duct. The result is the typical skin eruptions of acne. Because of its microbial basis, treatments for acne include tetracycline (oral) or clindamycin (topically applied) to kill the bacteria.

A.

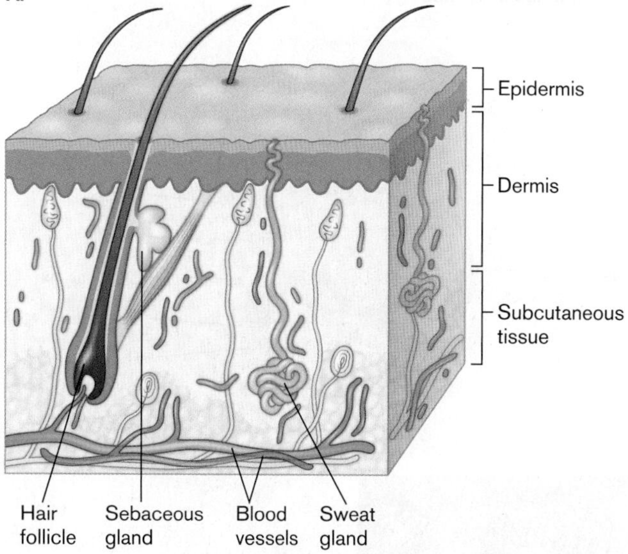

B.

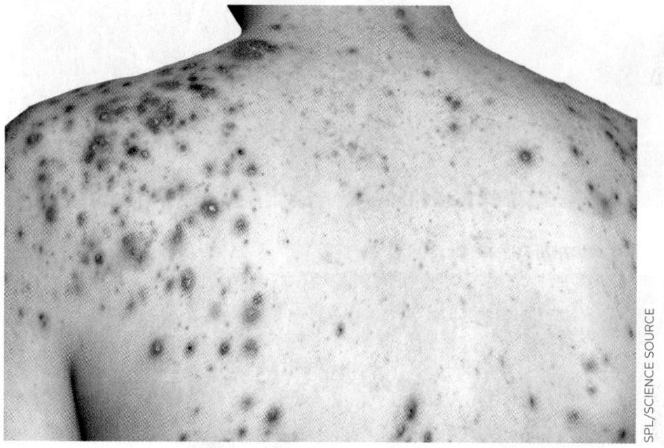

FIGURE 23.3 ▪ Microbiology of skin and the development of acne. A. The location of sebaceous glands in skin. *Cutibacterium acnes* can produce inflammation and blockage of the gland duct, leading to blackheads and acne. **B.** Acne.

Eye

The eye is exposed to the outside environment, so is it heavily colonized? Actually, it's not, because colonization is inhibited by the presence of antimicrobial factors, such as lysozyme, in the tears that continually rinse the eye surface (conjunctiva). Despite this protection, a few transient commensal bacteria can be found on the conjunctiva. Skin microbiota such as *Staphylococcus epidermidis* and diphtheroids (Gram-positive rods that look like clubs), as well as some Gram-negative rods, such as *Escherichia coli*, *Klebsiella*, and *Proteus*, manage, at least temporarily, to make the eye their home without causing damage. (The traits of these genera were discussed in Chapter 18.)

Oral and Nasal Cavities

Within hours after birth, a human infant's mouth becomes colonized with nonpathogenic *Neisseria* species (Gram-negative cocci), *Streptococcus*, *Actinomyces*, *Lactobacillus* (all Gram-positive), and some yeasts. These organisms come from the environment surrounding the newborn, such as the mother's skin and garments. As teeth emerge

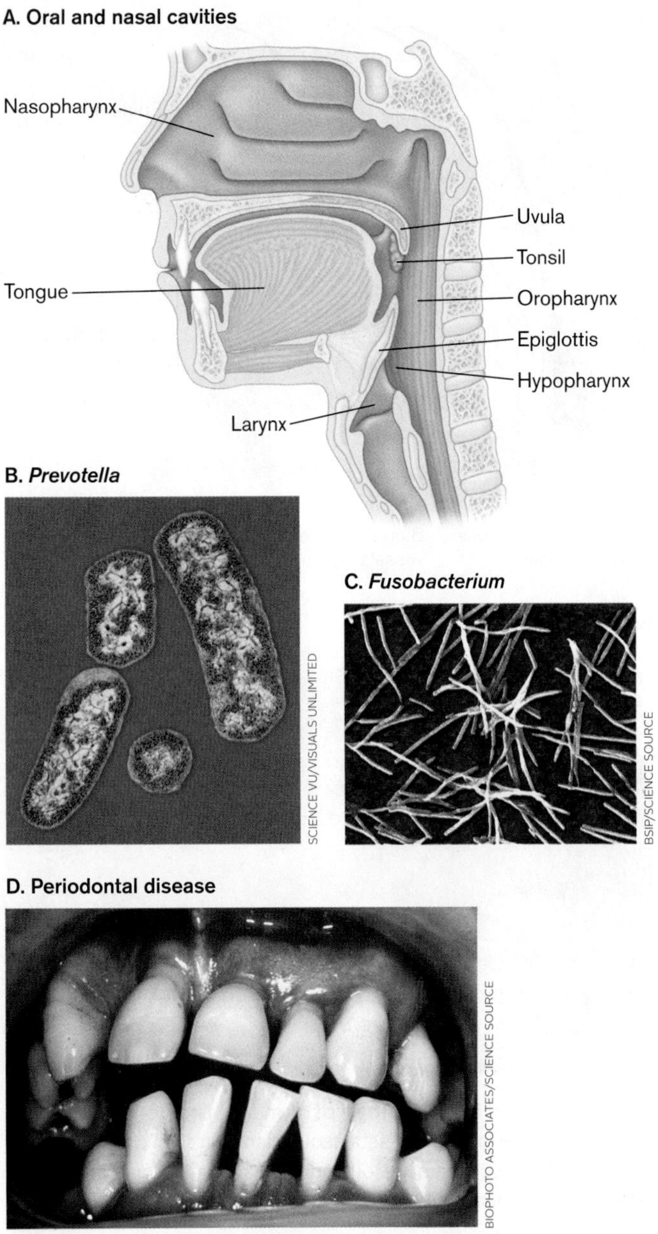

Nose
Corynebacterium
Cutibacterium
Staphylococcus aureus
Moraxella spp.

Nasopharynx
Streptococcus pneumoniae
Haemophilus influenzae
Moraxella catarrhalis
Dolosigranulum
Neisseria

Oropharynx
Streptococcus salivarious
Streptococcus pyogenes
Streptococcus agalactiae
Streptococcus pneumoniae
Veillonella
Moraxella catarrhalis
Fusobacterium
Neisseria
Bacteroidetes
Prevotella

FIGURE 23.4 ▪ Oral microbiota and periodontal disease. **A.** Structures of the oral and nasal cavities with examples of normal microbiota. **B, C.** Two anaerobic bacteria associated with periodontal disease. **B.** *Prevotella* (colorized TEM; each cell approx. 2 μm long). **C.** *Fusobacterium* (SEM; each cell approx. 1–5 μm long). **D.** Symptoms of periodontal disease include red, swollen gums; bleeding gums; gum shrinkage; and teeth drifting apart.

in the newborn, the anaerobic space between teeth and gums supports the growth of anaerobes, such as *Prevotella* and *Fusobacterium* (**Fig. 23.4B** and **C**). Whatever the organism, colonizers of the oral cavity must be able to adhere to surfaces, like teeth and gums, to avoid mechanical removal and flushing into the acidic stomach. The teeth and gingival crevices are colonized by 500–700 species of bacteria.

Organisms such as *Streptococcus mutans* (which attaches to tooth enamel) and *Streptococcus salivarius* (which binds gingival surfaces) form a glycocalyx that enables them to firmly adhere to oral surfaces and to each other. They are but two of the microbes that lead to dental plaque formation. The acidic fermentation products of these organisms demineralize teeth and cause dental caries (tooth decay).

Important microbial habitats of the throat include the **nasopharynx**, which is the area leading from the nose to the oral cavity, and the **oropharynx**, which lies between the soft palate and the upper edge of the epiglottis (**Fig. 23.4A**). Organisms like *Staphylococcus aureus* and *Staphylococcus epidermidis* populate these sites. The nasopharynx and oropharynx can also harbor relatively harmless streptococci, such as *Streptococcus salivarius* and *Streptococcus oralis*, as well as *Streptococcus mutans*. Other oropharyngeal organisms include a large number of diphtheroids and the small, Gram-negative rod *Moraxella catarrhalis*. Within the tonsillar crypts (small pits along the tonsil surface) lie anaerobic species such as *Prevotella*, *Porphyromonas*, and *Fusobacterium*.

The oral microbiota are normally harmless, but they can cause disease. Dental procedures, for instance, will often cause these organisms to enter the bloodstream, producing what is called **bacteremia** (infection of the bloodstream). Normal immune mechanisms typically clear these transient bacteremias quite easily, but in patients who have a mitral valve prolapse (heart murmur), microbes can become trapped in the defective valve and form bacterial vegetations. Vegetations are biofilms that contain a large number of bacterial cells encased within glycocalyx (a polysaccharide or peptide polymer secreted by the organism) and fibrin (produced by clotting blood). Because the onset of disease is often insidious (slow), it is called subacute bacterial endocarditis (**Fig. 23.5**). Once ensconced within a vegetation, the microbes are extremely difficult to kill with antibiotics. Immunocompetent individuals are at very low risk of valve infection. Immunocompromised patients, however, are at much higher risk and receive prophylactic treatment with antibiotics before any dental procedure.

Thought Question

23.1 How can an anaerobic microorganism grow on skin or in the mouth, both of which are exposed to air?

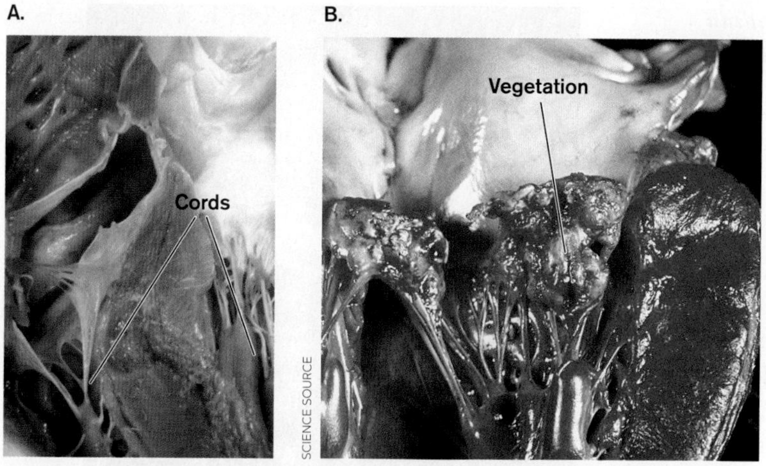

FIGURE 23.5 ▪ Gross pathology of subacute bacterial endocarditis involving the mitral valve. A. An open, normal mitral valve shows cords that tether it to the heart wall. **B.** The left ventricle of the heart has been opened to show mitral valve fibrin vegetations due to infection. These growths are not present in a normal heart.

Respiratory Tract

The respiratory tract has a surface area of approximately 246 square feet (75 square meters) directly exposed to the environment. Originally thought to be sterile, the lungs and trachea are now recognized to harbor normal microbiota. Estimates in the lower respiratory tract are 10–100 bacteria per 1,000 human cells, the most prominent members being *Prevotella* (Bacteroidetes), *Streptococcus* (Firmicute), and *Veillonella* (Firmicute).

Many organisms entering the nasopharynx become trapped in the nose by cilia that beat toward the pharynx. The microbes are propelled toward the acidic stomach and death. Microorganisms that slip into the trachea are trapped by mucus produced by ciliated epithelial cells lining airways. Cilia usher the microbes up and away from the lungs. The ciliated mucous lining of the trachea, bronchi, and bronchioles makes up the **mucociliary escalator** (**Fig. 23.6**), which constantly sweeps foreign particles up and out of the lungs.

The mucociliary escalator is extremely important for preventing respiratory infections. When it fails, as when it is covered with tar from years of heavy smoking, or when it is overwhelmed by the inhalation of too many infectious microbes, infections such as the common cold (for example, rhinovirus) or pneumonia (for example, *Streptococcus pneumoniae*) can result. Patients who smoke and develop chronic obstructive pulmonary disease (COPD) have a significantly altered lung microbiome. Decreased mucociliary clearance leads to enrichment with *Pseudomonas* species. In contrast, patients with asthma have a microbiome enriched with Proteobacteria. The significance of these changes is not clear.

Genitourinary Tract

Much of the genitourinary tract is normally free from microbes. These areas include the kidneys (which remove waste products from the blood) and the ureters (which remove urine from the kidneys). The urinary bladder, which holds urine until it is excreted, was thought to be sterile but is now known to harbor microbes, mainly anaerobes. The distal urethra, however—because of its proximity to the outside world—normally contains *Staphylococcus epidermidis*, *Enterococcus* species, and some members of Enterobacteriaceae. Some of these organisms can cause bladder disease (known as urinary tract infection, or UTI) if they make their way into the bladder (for example, via catheterization).

The large surface area and associated secretions of the female genital tract make it a rich environment for microbes, including bacteria and yeasts. Firmicutes are the largest bacterial contributor to the vaginal microbiome; however, the microbiome's composition changes with the menstrual cycle, owing to changing nutrients and pH. The mildly acidic nature of vaginal secretions (approximately pH 4.5) discourages the growth of many bacteria. As a result, the acid-tolerant *Lactobacillus crispatus* is among the most populous vaginal species.

Healthy women appear to fall into two broad categories: 70% have lactobacillus as the primary member of their vaginal microbiota, while 30% have mixed species with few lactobacilli. Women in the latter group appear to be more susceptible to sexually transmitted infections. A recent study of the endometrium, where fertilized eggs are implanted, found that a *Lactobacillus* dominance of at least 90% correlates well with reproductive success.

The balance between different species that comprise vaginal and endometrial microbiota is crucial to preventing disease. Antibiotic therapy to treat an infection anywhere in the body can also affect the vaginal microbiota.

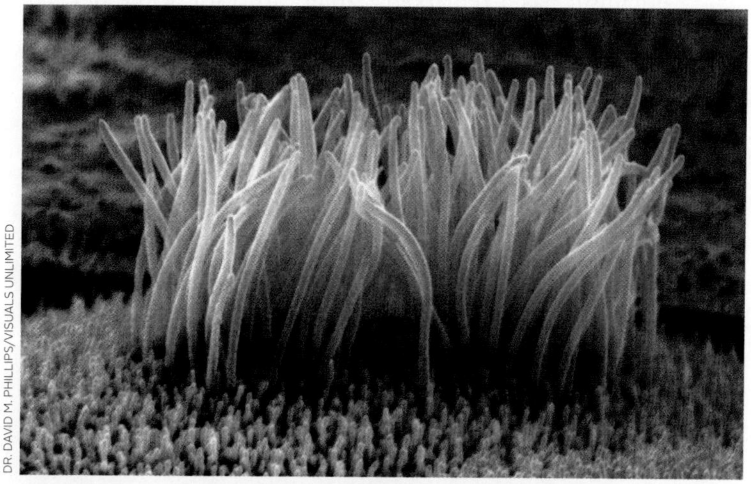

FIGURE 23.6 ▪ Mucociliary escalator. Movement of these hairlike cilia ushers particles up and out of the trachea and lungs (colorized SEM; diameter between 0.5 and 1 μm).

The resulting imbalance can allow overgrowth of *Candida albicans*, otherwise known as a yeast infection. *C. albicans*, as a fungus, is not susceptible to antibiotics designed to kill bacteria.

Stomach

We have known for a hundred years that the stomach contents are acidic and that gastric acidity can kill bacteria. Just how important that acidity is for protection against microbes is illustrated by the infection caused by *Vibrio cholerae*, the causative agent of cholera. Cholera is a severe diarrheal disease endemic to many of the poorer countries of the world. Although cholera actually affects the intestines, not the stomach, the bacteria must survive passage through the stomach to reach the intestines. The organism, however, is extremely acid sensitive. Healthy volunteers must ingest a trillion organisms before contracting disease.

Despite the high infectious dose, cholera epidemics in developing countries kill tens of thousands of people every year. Part of the reason is that the poor, malnourished populations in these countries suffer from hypochlorhydria (decreased stomach acid). The less acidic stomach gives ingested microbes more time to enter the intestine, where they can thrive and cause devastating disease. Compare the astronomical number of acid-sensitive *V. cholerae* needed to cause disease to the mere ten organisms needed to develop disease from *Shigella*, a very acid-resistant pathogen.

Although the stomach contents are very acidic, the mucous lining of the stomach is much less so. It is there that some bacteria can take refuge, primarily Actinobacteria and Firmicutes (see **Fig. 23.2**). In fact, the stomach harbors a diverse microbiota, as detected by cultural and molecular techniques. Estimates are between 10 and 1,000 organisms per gram.

A classic stomach pathogen is *Helicobacter pylori*. This organism has a remarkable ability to resist acidic pH. (It survives at pH 1 using the enzyme urease to generate ammonia, which neutralizes acid.) However, *H. pylori* will not grow in strong acid pH conditions, but it can grow in the mucous lining of the stomach, where the pH is closer to 5 or 6 (**Fig. 23.7**). The U.S. Centers for Disease Control and Prevention (CDC) estimates that *Helicobacter* colonizes the stomachs of half the world's population. Most of the time, *Helicobacter* does not cause any apparent problem, but on occasion, the organism can produce gastric ulcers and even cancer. On the plus side, recent data suggest that *H. pylori* colonization in children is associated with a reduced risk for allergic disease.

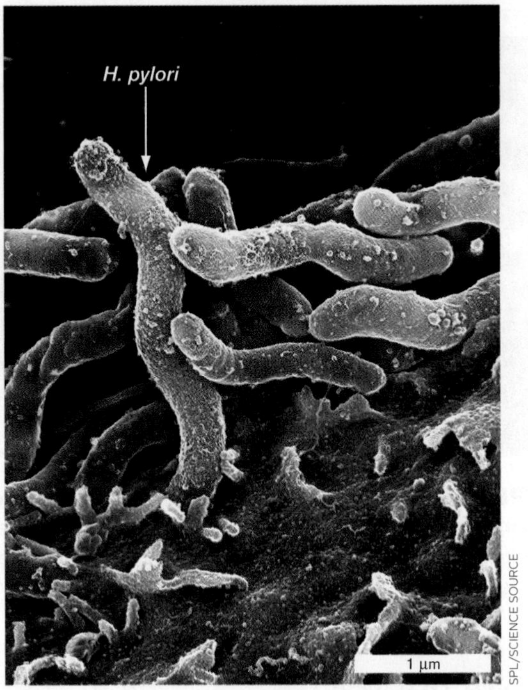

FIGURE 23.7 ■ *Helicobacter pylori* defies stomach acidity. *H. pylori* growing in the mucus on stomach epithelium (colorized SEM). *H. pylori* bacteria attach to gastric epithelial cells and induce specific changes in cell function, such as an increase in the expression of laminin receptor 1, a protein associated with malignancy. Peptic ulcers can result.

Intestine

The intestine is an extremely long tube (approximately 25 feet) consisting of several sections, each of which supports the growth of different combinations of bacterial species. Pancreatic secretions (at pH 10) enter the intestine at a point just past the stomach (pH 1–2) and raise the intestinal pH to about pH 8. The relatively high pH and bile content of the duodenum and jejunum allow colonization by only a few resident, mostly Gram-positive, bacteria (enterococci, lactobacilli, and diphtheroids). These particular Gram-positive organisms (Firmicutes phylum) possess a bile salt hydrolase that helps them grow in the presence of intestinal bile salts.

The distal parts of the human intestine (ileum and colon) have a slightly acidic pH (pH 5–7) and a lower concentration of bile salts—conditions that support a more diverse ecosystem. The intestine actually contains roughly 10^{11}–10^{13} bacteria per gram of feces, mostly anaerobes (1,000 anaerobes to 1 facultative organism). Why is the intestinal lumen anaerobic? The small amount of oxygen that diffuses from the intestinal wall into the lumen is immediately consumed by the facultative bacteria, such as *E. coli*, thereby rendering the environment anaerobic (discussed in Chapter 21).

Most of the data used to inventory members of the gut microbiome, as well as microbiota from other systems, comes from culture-independent molecular techniques. Segments of 16S ribosomal RNA genes are randomly amplified by PCR from DNA extracted from fecal specimens. Sequences from individual amplicons are used to identify the phyla, species, or operational taxonomic unit (OTU;

Chapter 21) of bacteria in the gut. As noted earlier, culture-independent sequencing was necessary because most members of the microbiome were considered unculturable. However, a new culture enrichment strategy devised by Michael Surette (McMaster University) and colleagues may change how we inventory a microbiome.

These scientists collected fecal material from volunteers and smeared dilutions from each sample onto 66 different kinds of highly enriched and pre-reduced agar media. Duplicate plates were incubated under aerobic and anaerobic conditions. The colonies that grew were removed from the agar plates and mixed together, and their DNA was extracted. PCR was used to amplify a common segment of the many bacterial 16S rRNA genes in the mixture, and the individual amplicons were sequenced to determine distinct species or OTUs. Using this culture enrichment strategy, Surette and his group managed to culture 95% of the gut microbiota that were predicted by culture-independent techniques, and actually revealed much greater bacterial diversity than was observed from previous culture-independent techniques. Altogether, there are approximately 1,500 different species of bacteria in the human gut microbiome, over 90% of which are from the Proteobacteria, Firmicutes, Actinobacteria, and Bacteroidetes phyla. This new culture enrichment approach will enable scientists to grow previously unknown (or uncultured) microbiome members and test their importance to human health.

Acquisition of gut microbiota. Infant intestines are initially colonized by large numbers of *E. coli* and streptococci that quickly generate a reducing environment able to support growth of strict anaerobic species—mainly *Bifidobacterium*, *Bacteroides*, *Clostridium*, and *Ruminococcus*. The intestines of breast-fed babies are dominated by bifidobacteria, possibly because of growth factors present in breast milk. In contrast, the microbiota of formula-fed infants is more diverse, with high numbers of Enterobacteriaceae, enterococci, bifidobacteria, *Bacteroides*, and clostridia. As seen in **Figure 23.2**, by adulthood over 90% of the bacteria in the intestine are composed of just two phyla: Bacteroidetes (for example, *Bacteroides* species) and Firmicutes (for example, Clostridiales clusters XIV and IV), followed by Proteobacteria (for example, *Escherichia coli*) and Actinobacteria (for example, *Bifidobacterium* species). Besides bacteria, other inhabitants are the yeast species *Candida albicans*, and protozoa such as *Trichomonas hominis* and *Entamoeba hartmanni*.

Some 1,500 different bacterial species and a few methanogenic archaea comprise the intestinal microbiome. One reason the intestine can support such a large and eclectic mix of species is that different bacteria attach to different host cell receptors. Another reason is that many different food sources are available to support diverse groups of microbes (discussed previously, in Chapter 21). Even human breast milk evolved to encourage the growth of specific subpopulations of microbiota. Breast milk contains carbohydrates (lactose) that nourish babies but also includes complex carbohydrates that only our microbiota can digest.

Makeup of the gut microbiome within a single individual is relatively constant but still varies over time. **Figure 23.8** follows the fecal biota from a single individual over a period of 14 months. The relative proportions of bacterial families fluctuate significantly but tend to return to typical adult composition. Across the spectrum of the human population, however, the composition of intestinal microbiota is a continuum of many different combinations of microbes (enterotypes). At one end of the spectrum are microbiomes dominated by *Bacteroides*, while at the other end are communities dominated by *Prevotella* (another Gram-negative anaerobic species). One study showed, for example, that people with high protein and animal-fat diets have fecal communities enriched with the *Bacteroides*-dominant enterotype, while those with carbohydrate-rich diets have fecal communities dominated by *Prevotella*. The significance of this difference is not yet clear.

Does a host's genetic makeup influence composition of the gut microbiome? The most recent evidence

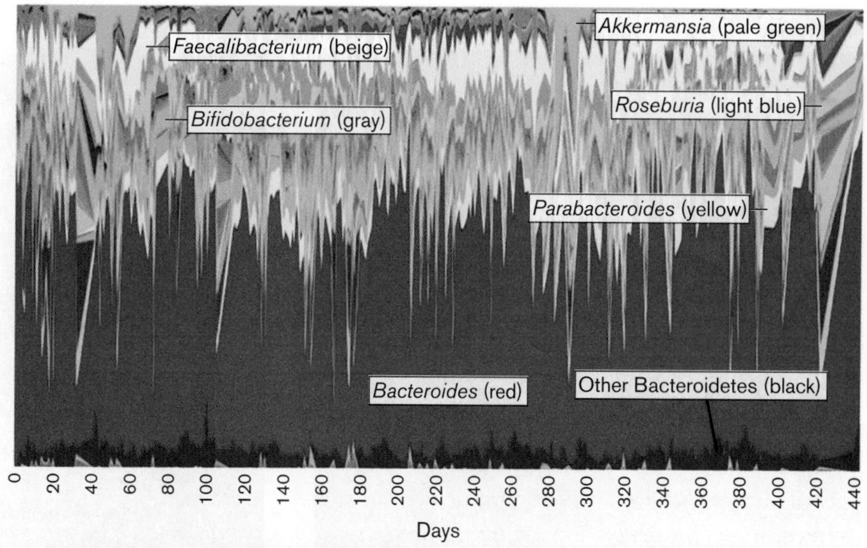

FIGURE 23.8 ■ Variability of gut microbiota over time. Daily fecal samples taken from one individual were analyzed to identify and quantify resident bacterial families by sequencing 16S rRNA genes. Selected genera are identified. Colored segments at each time point represent the proportions of specific genera relative to the entire population. The depth of a colored segment along the *y*-axis reflects that microbe's proportional contribution to the whole microbiome. Thus, the major genus at most time points is *Bacteroides* (red).

indicates that host genetics plays only a small role. Environmental considerations such as diet, cholesterol levels, and blood glucose are more predictive of gut microbiota. However, a small number of bacterial taxa are strongly heritable. For instance, *Christensenella minuta* (a Gram-negative anaerobe) is the most highly heritable bacterial species among identical twins (Chapter 7). "Heritability" here does not mean that the organism was passed from mother to siblings. The term reflects the ability of a host's genome to sustain the presence of this organism. The organism is more likely to be maintained in two identical twins than in two non-twin siblings from the same mother.

Keeping our microbiome at bay. Clearly, our intestines are packed with microbes. Why don't they cause chronic intestinal inflammation? Part of the answer was discovered by Lora Hooper (**Fig. 23.9A**) and colleagues at the University of Texas Southwestern Medical Center. Whenever bacteria contact epithelial cells, the human cells secrete lectins (carbohydrate-binding proteins) into the mucus layer lining the intestine. The lectins kill bacteria that get too close (**Fig. 23.9B**). Hooper's group discovered that one of the antimicrobial lectins, RegIIIα (**Fig. 23.9B** inset; cryo-EM map), forms a membrane-penetrating pore in Gram-positive microbes. The enforced separation of microbiome and host mucosal cells minimizes the likelihood that members of the microbiome will activate an immune response that could severely damage the intestinal lining. Other immune mechanisms that control microbiota in the intestinal mucosa, such as secretory IgA antibodies, are described in Chapter 24.

Thought Questions

23.2 Why do many Gram-positive microbes that grow on the skin, such as *Staphylococcus epidermidis*, grow poorly or not at all in the gut?

23.3 How might you provide evidence that the intestinal microbiome communicates with the brain via the vagus nerve? *Hint:* The vagus nerve includes afferent nerves that send information from organs to the brain and efferent nerves that regulate gastrointestinal secretion and gut endocrine activity.

To Summarize

- **Opportunistic pathogens** infect only compromised hosts.
- The **normal microbiota** present on skin and mucosal surfaces is acquired at birth but changes over a lifetime. **Skin microbiota** consists primarily of Gram-positive microbes, including *Cutibacterium acnes*, which can cause acne. **Oral and nasal surfaces** are colonized by aerobic and anaerobic microbes. **Vaginal microbiota** influences susceptibility to sexually transmitted infections.
- **Microbe-free areas of the body** include blood, cerebrospinal fluid, and internal organs.
- **Normal microbiota can cause disease** if organisms gain access to the circulation or deeper tissues.
- **The intestine** is populated by 10^{11}–10^{13} microbes per gram of feces. Principal phyla are Firmicutes,

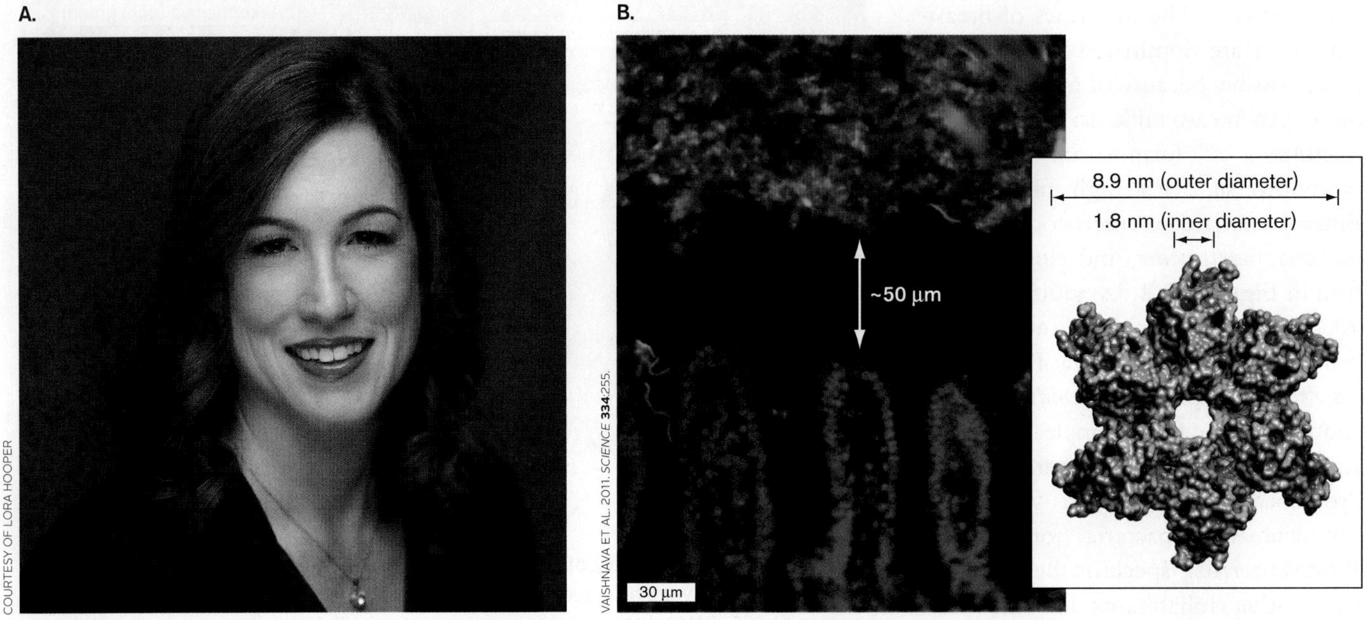

FIGURE 23.9 ■ Keeping the gut microbiome at bay. A. Lora Hooper discovered the mechanism of an antimicrobial lectin. **B.** Separation of the microbiome from the intestinal mucosal surface. Bacteria are green (FISH using a DNA probe that hybridizes to bacterial 16S rRNA genes), and the nuclei of intestinal mucosal cells are blue (DAPI stained). **Inset:** Model of the Reg III α pore that forms in Gram-positive cell membranes.

Bacteroidetes, Proteobacteria, and Actinobacteria. The ratio of anaerobes to facultative bacteria is 1,000:1. The gut microbiome includes bacteria, archaea, fungi, and viruses.

23.2 Benefits and Risks of Microbiota

Our colonizing microbes are not hostile armies camped at our body's gates (also called portals of entry) waiting to invade. These microorganisms are of great benefit to us, and we, as good hosts, reciprocate. Microbes in our gut, for instance, have enzymes to catabolize foods that we cannot digest. Some bacteria synthesize vitamins that we cannot make (*E. coli*, for instance, makes vitamin B_{12}). In addition, chemical signals released by members of the microbiome promote host tissue development.

A recent study from Rockefeller University found that certain lipids made by gut bacteria interact with host membrane receptors and affect host metabolism, immune cell differentiation, immune cell traffic through the body, and tissue repair. Even the short-chain fatty acid products of fermentation affect host gene expression and can influence host immune system function. Acetate, for instance, tends to promote inflammation, whereas butyrate tends to inhibit inflammation.

We, for our part, maintain our gut microbiome by secreting complex carbohydrates that members of the microbiome can use, and chemical factors, such as hormones, that can alter microbiome gene expression.

Examples of Beneficial Gut Microbiota

An important gut microbe that promotes host tissue rejuvenation is *Akkermansia muciniphila*, a Gram-negative, anaerobic bacterium (Verrucomicrobia phylum) that degrades mucin, the glycosylated protein layer that covers gut epithelium. Mucin degradation by *A. muciniphila* fosters continuous rejuvenation of the protective mucin layer. The short-chain fatty acids produced by this organism feed epithelial metabolism and modulate inflammatory responses of immune cells.

Another important microbe, the anaerobe *Bacteroides thetaiotaomicron*, metabolizes many of the complex carbohydrates we eat, breaking them down into products that can be absorbed by the body or used by other members of the microbiome. We humans actually absorb 15%–20% of our daily caloric intake in this way. An example is discussed in Chapter 21.

Our gut microbiota also limits infections by competing with pathogens for attachment receptors on host cells and for food sources. Symbiotic bacteria can also make antimicrobial compounds that limit growth of potential pathogens. For example, nonpathogenic gut *E. coli* produces bacteriocin (a secreted protein toxin), which directly inhibits growth of the related pathogen enterohemorrhagic *E. coli* (EHEC; see Chapter 25).

Gut microbes also influence the development and efficiency of our immune system. The intestinal microbe *Faecalibacterium prausnitzii*, a member of the order Clostridiales (Firmicutes phylum), comprises up to 5% of total fecal microbiota in adults. The organism elicits powerful anti-inflammatory effects on human immune cells through secreted bacterial factors and interactions with Toll-like receptors (see Section 23.5). For example, it inhibits induction of the inflammatory cytokine IL-8 (cytokines are small, secreted host proteins that regulate immune cell function; Section 23.5); and increases the population of an anti-inflammatory T-cell lymphocyte population (regulatory T cells; see Chapter 24). The absence of this organism in human intestines has been linked to painful inflammatory intestinal diseases such as Crohn's disease.

Another important group of intestinal anaerobes consists of the segmented filamentous bacilli (SFBs) in mice and probably humans (see Special Topic 18.1). These spore formers are important for developing a healthy immune system. We will discuss the connections between gut microbiota and the immune system more fully in Chapter 24. The anti-inflammatory effects of these and other microbes protect the intestinal mucosa from an overzealous immune system.

Containment Breaches and Dysbiosis

Everything works well, as long as the composition of our microbiota is balanced and stays where it belongs. However, the accidental penetration of certain organisms beyond a site of colonization or an imbalance in microbiome composition, called **dysbiosis**, can cause infections and inflammatory diseases, respectively. A cancerous lesion in the colon, for example, can provide a passageway for microbes to enter deeper tissues and cause infection. *Bacteroides fragilis*, a harmless anaerobe in the gut, and even *Escherichia coli* can invade tissues through surgical wounds, causing intra-abdominal abscesses and even gangrene after abdominal surgery (**Fig. 23.10**). Other infections caused by specific gut microbes escaping the intestine include urinary tract infections (cystitis), septicemia, and meningitis.

Emotional stress, a change in diet, or antibiotic therapy can alter gut microbiome balance. The resultant dysbiosis can lead to poor digestion; to pseudomembranous enterocolitis, an infectious disease caused by *Clostridioides difficile* (formerly *Clostridium difficile*; **Fig. 23.11**); or to an extremely painful inflammatory condition such

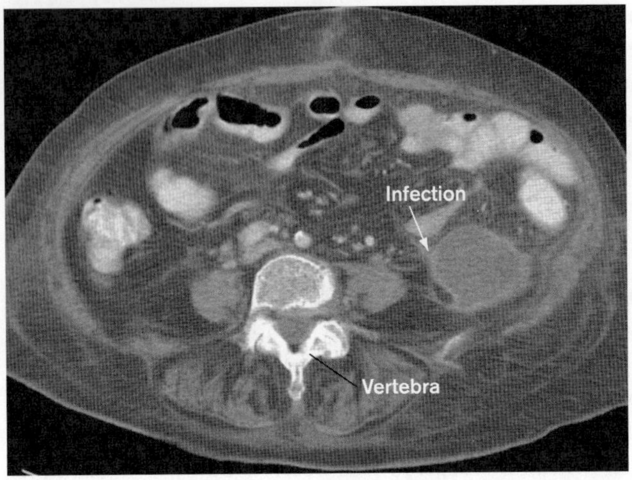

2005 CONSULTANT 45:1531. © HMP.

FIGURE 23.10 ■ Abscess of colon caused by *Escherichia coli*. This infection, shown by computed tomography (CT) scan, was an unfortunate result of bowel surgery. Organisms escaped from the intestine and initiated infection in the abdominal wall (arrow). CT scan shown is cross sectional view of the abdominal region.

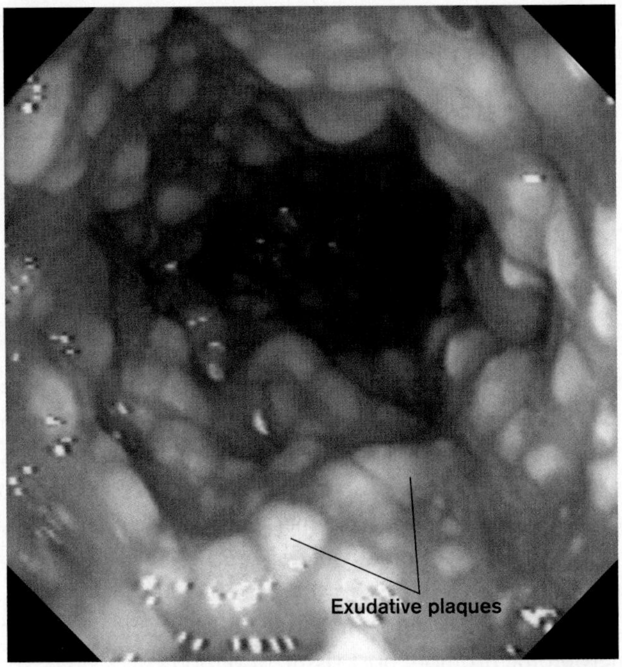

DAVID M. MARTIN M. D./SCIENCE SOURCE

FIGURE 23.11 ■ Pseudomembranous enterocolitis caused by *Clostridioides difficile*. *C. difficile* is often a part of the normal intestinal microbiota, but antibiotic treatment can kill off competing microbes, leaving it to grow unabated. *C. difficile* toxins kill host cells, causing exudative plaques to form on the intestinal wall. The small plaques coalesce to form a large pseudomembrane that can slough off into the intestinal contents.

as inflammatory bowel disease (IBD) or Crohn's disease. Even non-antibiotic drugs such as proton pump inhibitors, nonsteroidal anti-inflammatory drugs (NSAIDs), anti-diabetics (metformin), and atypical antipsychotics can perturb microbiome balance. One study that screened 1,000 drugs found that 24% inhibited growth of one or more gut microbes.

Some people favor restoring natural microbial balance by orally ingesting living microbes such as the lactobacilli present in yogurt. These supplements, called **probiotics** (from the Greek meaning "for life"), are thought to restore balance to the microbial community and return the host to good health. The most commonly used probiotic genera are *Lactobacillus* and *Bifidobacterium* (**Fig. 23.12**). The potential mechanisms by which they improve intestinal health include competitive bacterial interactions (normal biota prevent growth of pathogenic bacteria), production of antimicrobial compounds, and immunomodulation (an ability to change the activity of cells of the immune system). One study even found that the probiotic organism *Lactobacillus rhamnosus* GG does not alter microbiome composition but orchestrates broad transcriptional changes

A. B.

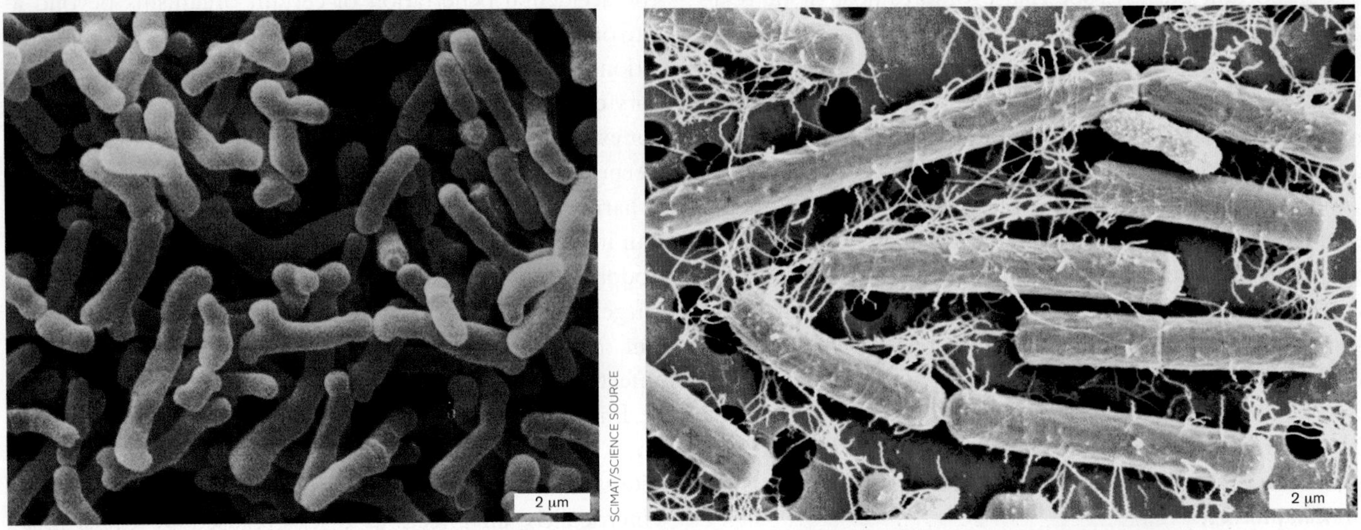

SCIMAT/SCIENCE SOURCE
SCIMAT/SCIENCE SOURCE

FIGURE 23.12 ■ Commonly used probiotic microorganisms. **A.** *Bifidobacterium* (note the Y shape of some cells). **B.** *Lactobacillus acidophilus*. Colorized SEMs.

in other members of the microbiome—functions that could promote anti-inflammatory pathways in the resident microbes.

Probiotics are now being tested for treating several gastrointestinal disorders, including IBD. Realize, though, that using probiotics such as live-culture yogurt to treat complex diseases like IBD is still controversial.

An extreme, but clinically proven, form of delivering probiotic microbes to the intestine is the so-called fecal transplant (fecal bacteriotherapy). In its original forms, fecal transplants transferred the intestinal microbiome of a healthy person to a relative suffering from a severe intestinal disease such as "pseudomembranous enterocolitis." Restoring a "normal" microbiome in this way has successfully cured patients with repeated *C. difficile* infections that did not respond to antibiotics. Today the procedure uses a "superdonor" rather than relatives, and it may someday be a personalized bacterial cocktail delivered in pill form. Probiotics in the form of vaginal suppositories are also used to treat patients prone to vaginal infections.

The Link between Obesity and Dysbiosis

Obesity is epidemic in industrialized and developing countries. Over 500 million people worldwide are obese and predisposed to developing related diseases, such as diabetes, cardiovascular disease, nonalcoholic fatty liver disease, cancer, and some immune disorders. Obesity involves complex, puzzle-like interactions among genes, diet, and a long-term imbalance between energy intake and expenditure, which together yield an excessive increase in body fat. Compelling evidence now points to gut microbiota playing a prominent role in obesity by influencing nutrient acquisition, regulating energy metabolism, and affecting fat storage.

Many studies showing connections between gut microbiota and obesity use gnotobiotic animals. A **gnotobiotic animal** is a germ-free animal or an animal in which all the microbial species present are known because they were added by the experimenter. To develop a gnotobiotic colony of animals, offspring must be delivered by cesarean section under aseptic conditions in an isolator. The newborn is moved to a separate isolator where all entering air, water, and food are sterilized. Once gnotobiotic animals are established, the colony is maintained by normal mating between the members. Fecal organisms and fecal content can be transplanted orally to these animals by syringe and stomach tube (gavage). After the transplant, the effects on animal physiology and disease can be observed. Fecal transplants include mouse-to-mouse transfers or even human-to-mouse transfers (discussed shortly).

Tantalizing evidence suggesting a link between the gut microbiome and obesity goes back decades. For instance, livestock farmers used to add antibiotics in low doses to animal feed as a way to promote animal growth (a practice no longer permitted because of the potential for generating antibiotic-resistant microbes). In another example, treating obese mice with high doses of antibiotics to remove microbiota reduces mouse body weight and improves glucose metabolism. Such studies suggested that altering an animal's microbiome with antibiotics can increase or decrease weight, depending on circumstances.

Newer metagenomic studies find that obese humans and mice carry a less diverse gut microbiome and have an increased capacity to absorb energy compared to their lean counterparts. The ratios of major bacterial phyla in the colon also seem to differ between obese and lean individuals. For instance, the gut microbiomes of obese children have Firmicutes/Bacteroidetes phylum ratios that are higher than those of lean children. Studies looking at this ratio among adults are less clear. Some find that the Firmicutes/Bacteroidetes ratio is higher in obese adults than in lean adults. Other studies indicate the opposite, and some show no difference. The real key to microbiome effects on obesity may be the difference in species makeup and metabolic output rather than broad phylum differences.

Can your microbiome keep you thin? A landmark study from Jeffrey Gordon (**Fig. 23.13**; Washington University School of Medicine in St. Louis) and colleagues demonstrated that obesity in humans is clearly linked to their microbiomes. The scientists collected fecal contents from sets of monozygotic human twins, one of whom was obese and the other lean, and then orally transplanted the fecal

JAMES BYARD/WUSTL PHOTOS

FIGURE 23.13 ■ Jeffrey I. Gordon, MD. Gordon (second from left) is a recipient of the 2015 Keio Medical Science Prize. He stands here with some of the researchers who work in his lab.

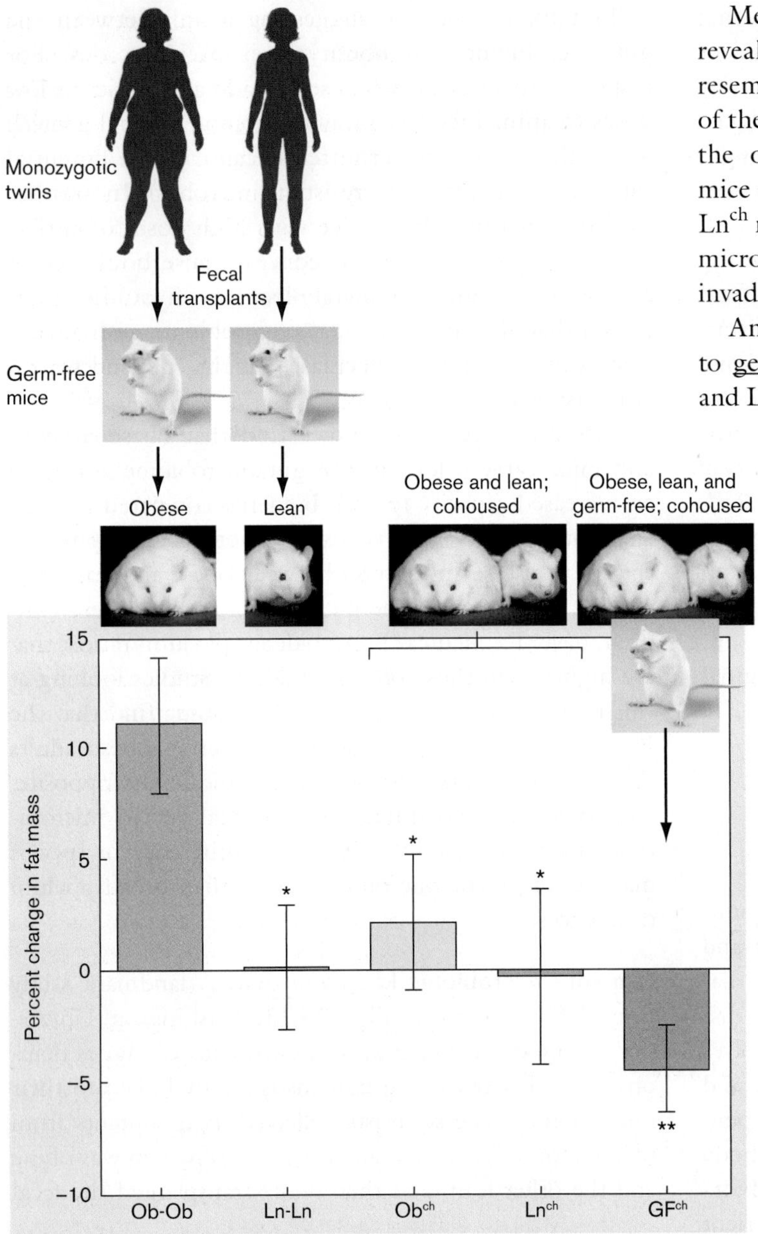

FIGURE 23.14 ■ Adiposity of germ-free mice transplanted with fecal preparations from human twins. Fecal contents from sets of monozygotic human twins—one obese (Ob), the other lean (Ln)—were transplanted orally into germ-free (GF) mice. Mice were cohoused (ch) in different combinations. Changes in fat mass were measured after 10 days. * = *P* value ≤0.05; ** = *P* ≤0.01.
Source: (Germ-free mice) tiripero/Getty Images; (obese and lean mice) Human Genome wall for SC99 at http://www.ornl.gov or Wikimedia.

preps into germ-free mice (**Fig. 23.14**). Mice receiving fecal microbiota from the lean human co-twin remained lean, as measured by change in fat mass ("Ln" in **Fig. 23.14**). However, mice receiving fecal microbiota from the obese co-twin gained adiposity over a mere 10-day period ("Ob" in **Fig. 23.14**). Even more remarkable, when Ob and Ln co-twin fecal-transplanted mice were cohoused ("ch") in a single cage immediately after transplant, the Ob^{ch} mice gained less weight than the control Ob mice.

Metagenomic analysis of fecal contents from the mice revealed that the microbiome of Ob^{ch} mice came to resemble that of the Ln^{ch} mice. Essentially, the microbiota of the Ln^{ch} mouse invaded the Ob^{ch} mouse and displaced the original microbiome. Transfer took place because mice are naturally coprophagous (eat feces). Even though Ln^{ch} mice also consumed Ob^{ch} mouse feces, the Ln^{ch} fecal microbiome did not change. The most successful Ln^{ch} invaders of the Ob^{ch} microbiota were *Bacteroides* species.

Another aspect of this study asked what would happen to germ-free (GF) mice if they were cohoused with Ob and Ln fecal-transplanted mice. Up to 5 days after cohousing, the microbiome initially acquired by GF^{ch} mice resembled that of non-cohoused Ob mice, but surprisingly, after 5 days the profile dramatically shifted to that of their Ln^{ch} cage mates. GF^{ch} mice then lost weight (see **Fig. 23.14**). This result confirmed that the Ln-derived microbial taxa had greater fitness over the Ob-derived taxa. From these studies, Gordon concluded that a higher Firmicutes/Bacteroidetes ratio favors obesity, while the opposite favors leanness. His studies also found that as fat people lose weight, the proportion of Bacteroidetes to Firmicutes in the gut microbiome increases.

How might the microbiome influence obesity? Microbiota can influence obesity by affecting two major processes: the harvesting of energy from ingested foods, and the triggering of intestinal inflammation.

As noted in Chapter 21, the metabolic dexterity of intestinal microbes allows them to digest many foods that we cannot. In the process, these microbes produce short-chain fatty acids (SCFAs) that our cells use in a variety of ways. The major SCFAs are butyrate, propionate, and acetate. Numerous investigations suggest that the amounts and ratios of these SCFAs influence obesity (**Fig. 23.15**).

Generally, acetate is thought to promote obesity. Acetate is absorbed well by intestinal epithelial cells, travels to the liver where it is used for lipid synthesis, and is distributed throughout the body to be used as a substrate for cholesterol synthesis (particularly in adipose tissues). Butyrate, however, seems to prevent obesity. It is used primarily by intestinal epithelial cells for energy and stimulates production of gut hormones that limit hunger. Butyrate also dampens synthesis of the pro-inflammatory cytokine interferon-gamma (IFN-gamma), described in Sections 23.5 and 24.3. Propionate, for its part, has effects that can either stimulate or prevent obesity. This SCFA stimulates IFN-gamma synthesis, which promotes inflammation associated with obesity. But propionate is also involved in gluconeogenesis, an anti-obesity process that diverts fatty acids away

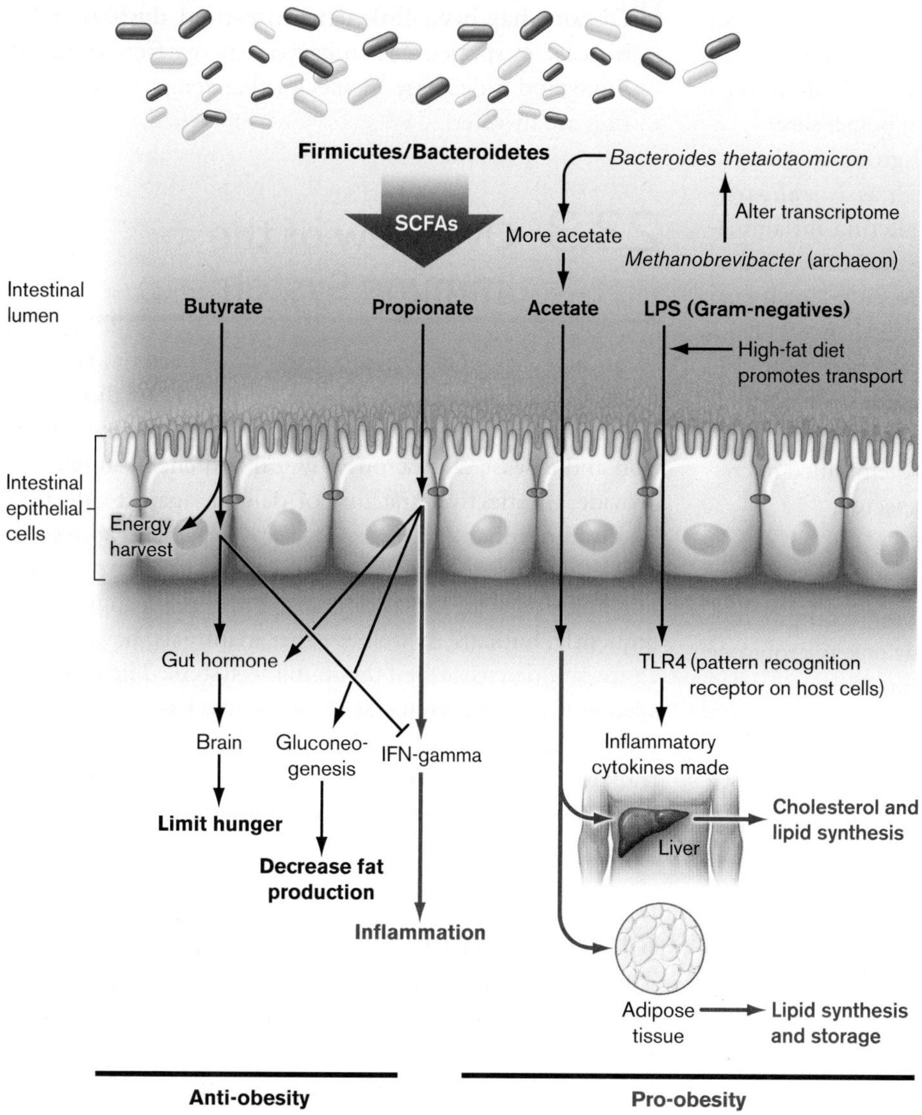

FIGURE 23.15 ■ Effects of microbiome products on weight gain. Members of the gut microbiome ferment ingested complex carbohydrates and produce short-chain fatty acids (SCFAs). SCFAs have varied effects on hunger, fat production, and inflammation. LPS from Gram-negative bacteria can be transported across the mucosal barrier and interact with pattern recognition receptor TLR4. The inflammatory cytokines made can affect liver metabolism. Arrows and terms in red indicate processes that promote weight gain.

from cholesterol synthesis and toward glucose synthesis. In addition, propionate stimulates the production of anti-obesity gut hormones that limit hunger. So, is there a correlation between SCFA levels in the colon and obesity?

A study from Elena Comelli's lab at the University of Toronto found that a cohort of obese or overweight patients generally had higher levels of SCFAs (propionate, butyrate, and acetate) than did healthy, lean patients on similar diets, and that SCFA levels did correlate with Firmicutes/Bacteroidetes ratios. The more Firmicutes there were, the more SCFAs were produced. However, obese and lean subjects actually had similar Firmicutes/Bacteroidetes ratios, suggesting that the most important factor is which species of Firmicutes and Bacteroidetes are present (and what they are doing), rather than the overall phylum ratios.

Archaea in the gut might also tilt metabolic balance toward obesity. About 50% of humans have significant numbers of methane-producing archaea, primarily *Methanobrevibacter smithii* (discussed in Chapter 21). This archaeon can comprise up to 10% of the anaerobes in the colons of healthy adults. Bruce Rittmann and colleagues from Arizona State University found that a small group of obese patients had significantly more archaea than either lean or gastric-bypass patients had. This finding raised the possibility that Archaea may play a role in obesity. But how?

Jeffrey Gordon's laboratory provides a possible explanation. His team discovered that the archaeon *M. smithii* dramatically altered the transcriptome of *Bacteroides thetaiotaomicron* when the two organisms co-colonized germ-free mice. The methanogen shifted the carbohydrate priority of *Bacteroides* from glucans to fructans, which stimulated more acetate production. Higher levels of acetate could enhance adiposity as outlined already. How the archaeon modulated the anaerobe's metabolism is not known. Another hypothesis is that methanogens in the gut can rapidly consume the H_2 end product of bacterial fermentation. Removing H_2 removes feedback control over fermentation. The resultant increase in carbohydrate fermentation yields even more acetate that can contribute to weight gain.

In addition to affecting the harvesting of energy, inflammation induced by gut microbes appears to influence the development of obesity and diabetes. Obese individuals are in a chronic state of low-grade inflammation, characterized by elevated blood and tissue levels of pro-inflammatory cytokines and markers of inflammation such as C-reactive protein (discussed in Section 23.6). The source of inflammation is thought to be outer membrane lipopolysaccharides (LPS), also called endotoxin, produced by Gram-negative members of the gut microbiome. A high-fat diet promotes absorption of endotoxin across the intestinal epithelium and its distribution to tissues via blood. LPS interacts with a cell receptor called TLR4, present in a variety of host cells (see Section 23.5). LPS-TLR4 interaction induces production of pro-inflammatory cytokines, such as IL-6, and reactive oxygen species that interfere with the normal metabolism of liver, adipose, and skeletal muscle, all of which are important to the regulation of glucose and lipid homeostasis.

While the presence of some bacterial groups in the gut correlate with energy intake, others, such as *Faecalibacterium prausnitzii*, influence obesity by altering inflammation. For example, the presence of *F. prausnitzii*, a major producer of butyrate in the gut, reduces the low-grade inflammation associated with obesity and diabetes. Even culture supernatants of *F. prausnitzii* fed to mice were able to lower inflammation in chemically induced colitis models. Butyrate in the supernatant was one anti-inflammatory factor, but others, including a 15-kDa protein called microbial anti-inflammatory molecule, are suspected.

Can individual members of a lean-associated microbiome treat obesity? We have gained tremendous insight into host-microbe relationships from the various microbiome projects under way. The data obtained thus far suggest that different blends of microbiota influence the efficiency of calorie harvesting from the diet, alter how derived energy is stored, and modulate chronic inflammatory processes. But can we pick out individual microbes or blends of microbes to treat obesity or other gastrointestinal diseases?

Research is under way to identify specific members of the microbiome that might individually, or in combination, treat diseases such as IBS (irritable bowel syndrome), IBD, or perhaps obesity. For example, Section 7.6 describes the identification of the intestinal Firmicute *Christensenella minuta* from human twins, which prevents weight gain when transferred to germ-free mice. The idea is that select species can be harvested from a healthy microbiome and administered in pill form to patients—a probiotic-like treatment. There is great hope that we may one day cure or prevent these disorders using this novel method of microbiome manipulation.

To Summarize

- **Benefits of the microbiome** include interfering with pathogen colonization, producing immunomodulatory proteins, metabolizing foods that the host cannot process (energy harvesting), producing vitamins that the host cannot make, and honing our immune system.
- **Infections such as septicemia, abscesses, cystitis, and meningitis** can be caused by some members of the microbiota if they breach the body's containment mechanisms.
- **Dysbiosis of the microbiome** can contribute to infection, obesity, and inflammatory and autoimmune diseases.
- **Probiotics** are foods or solutions containing helpful members of a microbiome. Different probiotics can restore balance to the intestinal or vaginal microbiomes.
- **Obesity has been linked to intestinal dysbiosis.** An increase in species able to harvest energy from ingested foods and a decrease in species that temper inflammation are involved.

23.3 Overview of the Immune System

Surrounded by, and a host to, trillions of bacteria, how are we not constantly infected? Here we begin our discussion of the many layers of protection designed to prevent infection and disease. Numerous physical and chemical barriers provide an effective first line of defense against infection by invading microbes. As such, they play a critical role in managing our resident ecosystems. But these barriers are not unbreachable. Organisms can still slip through. Consequently, humans, as well as other mammals, have a more aggressive defense called the immune system. The **immune system** is an integrated system of organs, tissues, cells, and cell products that differentiates self from nonself and neutralizes potentially pathogenic organisms or substances. This complex collection of cells and soluble proteins is capable of responding to nearly any foreign molecular structure.

Innate and Adaptive Immunity

There are two broad types of immunity: **innate immunity** (often called **nonadaptive immunity**) and **adaptive immunity** (discussed in detail in Chapter 24). Innate immune mechanisms are ancient, having evolved early in invertebrates, and persist in vertebrates, including humans. Active immunity is more recent, evolving only in vertebrates. Innate and adaptive immunity have several key differences. Innate immunity includes physical barriers such as skin, chemical barriers such as stomach acid, and relatively nonspecific (innate) cellular responses to infection that engage if the physical and chemical barriers are breached. The cellular innate responses are triggered by microbial structures such as peptidoglycan and lipopolysaccharides. Innate immunity is essentially "hardwired" into the body. It is present at birth, so that mechanisms of innate immunity exist before the body ever encounters a microbe. The protection afforded by innate immunity is nonspecific, capable of blocking or attacking many different types of foreign substances and organisms.

In contrast to innate immunity, adaptive immunity evolved to react to very specific structures called **antigens**. An antigen is any chemical, compound, or structure foreign to the body that elicits an immune response. The adaptive immune response to a specific antigen is not launched until the body "sees" that antigen. Amazingly, adaptive immune mechanisms can recognize at least 10^{10} different antigenic structures and specifically launch a directed attack against each one. Once such an attack has been activated, the organism keeps a "memory" of the exposure in the form of specific memory cells. An encounter with the microbe years later will reactivate the memory cells specific to that antigen.

The two types of immunity are illustrated by the response of the immune system to infection by the microorganism *Neisseria gonorrhoeae*, which causes the sexually transmitted disease gonorrhea. A component of the innate immune response is **complement**, composed of several soluble protein factors constantly present in the blood. Within moments of an initial infection, especially upon entering the bloodstream, complement proteins form holes in the bacterial membrane, thereby killing the microbe (see Section 23.6). Later, over time, the adaptive immune response will generate specific antibodies against cells of *N. gonorrhoeae* that escaped the innate mechanisms. These antibodies, however, are not made until well after the organism infects a person. Together, innate and adaptive immunity can help ward off disease caused by this organism. Unfortunately, *N. gonorrhoeae* infection does not generate long-lasting "memory," so reinfection with this organism is possible.

It is important to know that the innate and adaptive immune systems are interconnected. Antibodies made by the adaptive immune system will trigger parts of the innate immune system, such as the complement cascade that we discuss in Section 23.6. Likewise, activation of innate resistance mechanisms will cause the release of small immunomodulatory peptides (cytokines or chemokines) that influence the type and strength of adaptive immunity brought to bear. In military terms, the cooperation between innate and adaptive immunities is similar to an army coordinating its actions with those of the air and naval forces.

Infection versus Disease

Contact with an infectious agent does not guarantee that a person will actually contract the disease. If the number of infecting organisms is small and the immune system (innate and adaptive) is effective, the individual may not develop disease. However, a person known to have been exposed to certain microorganisms will be treated with antibiotics as a preventive measure. Chapter 26 more fully discusses the difference between being infected and developing disease.

Any microbe that causes disease must first breach the host's physical and chemical barriers to gain entrance to the body. It must then survive the innate defense mechanisms and begin to multiply. Finally, the microbe must surmount the last line of defense—adaptive immunity—which begins to respond as the microbe struggles to overcome innate immune defenses. The rest of this chapter will discuss the various innate defense mechanisms. Adaptive immunity is revealed in Chapter 24.

Although innate and adaptive immunity are often treated as separate entities, certain kinds of cells and organs play a role in both types of immunity. We thus introduce various cells and organs of the immune system as a whole before focusing on innate immunity.

Cells of the Immune System

Blood is composed of red blood cells, white blood cells (also generally known as **leukocytes**), and platelets (**Fig. 23.16**). The many types of white blood cells are formed by the differentiation of stem cells produced in bone marrow (**Fig. 23.17**). Stem cells differentiate into two main lineages: myeloid cells that mainly produce the leukocytes of innate immunity, and lymphoid cells that primarily develop into the leukocytes (primarily lymphocytes) of adaptive immunity.

The leukocytes of innate immunity include:

- Polymorphonuclear leukocytes (PMNs)
- Monocytes
- Macrophages
- Dendritic cells
- Mast cells

PMNs, also called granulocytes, are known for their conspicuous multilobed nuclei and enzyme-rich lysosome organelles. PMNs differentiate from an intermediate cell called the myeloblast. There are several types of PMNs, named for their different staining characteristics. Each cell type has a different function. **Neutrophils** (**Fig. 23.18A**), making up the vast majority of white cells in the blood, can engulf microbes by phagocytosis. Phagocytosis involves the extrusion of pseudopods that attach to and envelop the pathogen, which ends up in a phagosome vacuole. The

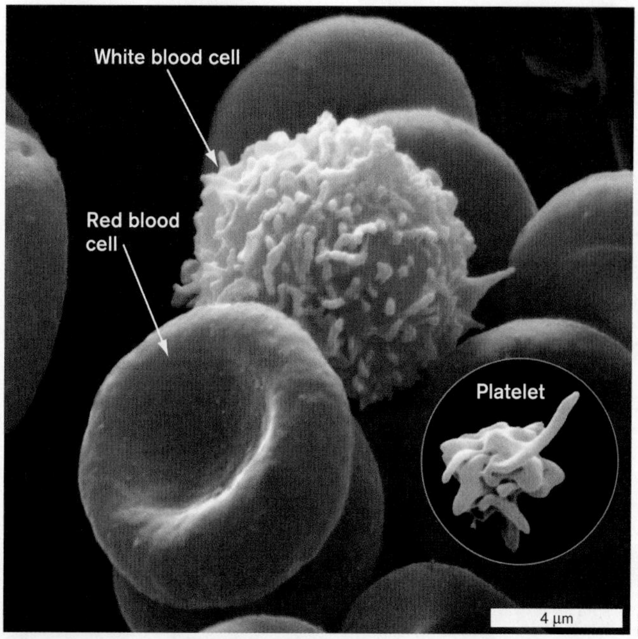

FIGURE 23.16 ■ Red blood cells, white blood cell, and a platelet. This colorized scanning electron micrograph illustrates the relative sizes and 3D morphologies of these components of blood.

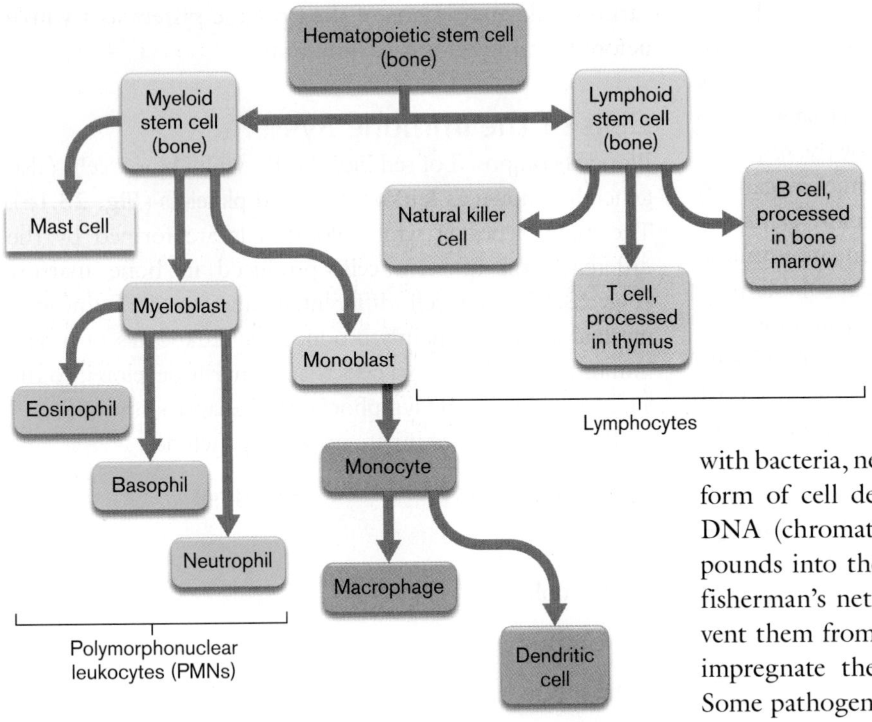

FIGURE 23.17 ■ Development of white blood cell components of the immune system. Pluripotent hematopoietic stem cells in bone marrow divide to form two lineages: myeloid cells that develop into PMNs and monocytes; and lymphoid cells that differentiate into B cells, T cells, and natural killer cells. Final maturation into B cells and T cells (the principal cells involved in adaptive immunity) occurs in the bone marrow and thymus, respectively. Colors indicate a group of differentiated cells that arise from the same progenitor.

phagocyte then kills the organism by fusing enzyme-gorged lysosomes with the **phagosomes** (**Fig. 23.19**). Enzymes (described later) spilling from the lysosome into the phagosome will destroy various components of the microbe and, ultimately, the microbe itself (see Section 23.5).

Neutrophils also "throw" **neutrophil extracellular traps** (**NETs**) around nearby pathogens. After interacting with bacteria, neutrophils can undergo NETosis, an unusual form of cell death, in which cells spew a latticework of DNA (chromatin) impregnated with antimicrobial compounds into the immediate area (**Fig. 23.20**). Much as a fisherman's net traps fish, NETs trap pathogens and prevent them from spreading. Antimicrobial compounds that impregnate the NET then kill the captured microbes. Some pathogens, such as group B *Streptococcus* and *Neisseria gonorrhoeae*, can escape NETs by secreting extracellular nucleases. NETs have also been implicated in certain

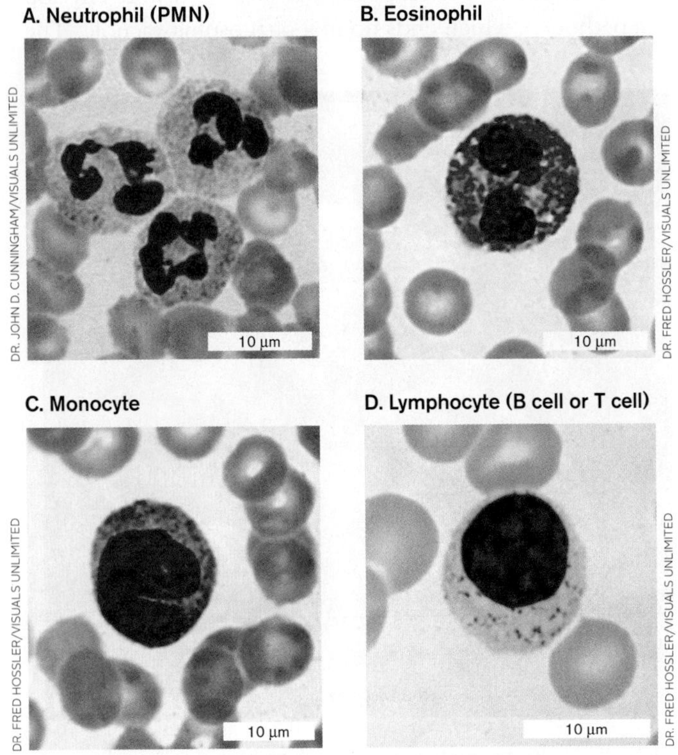

FIGURE 23.18 ■ Types of white blood cells. Basophils (not shown) look similar to eosinophils.

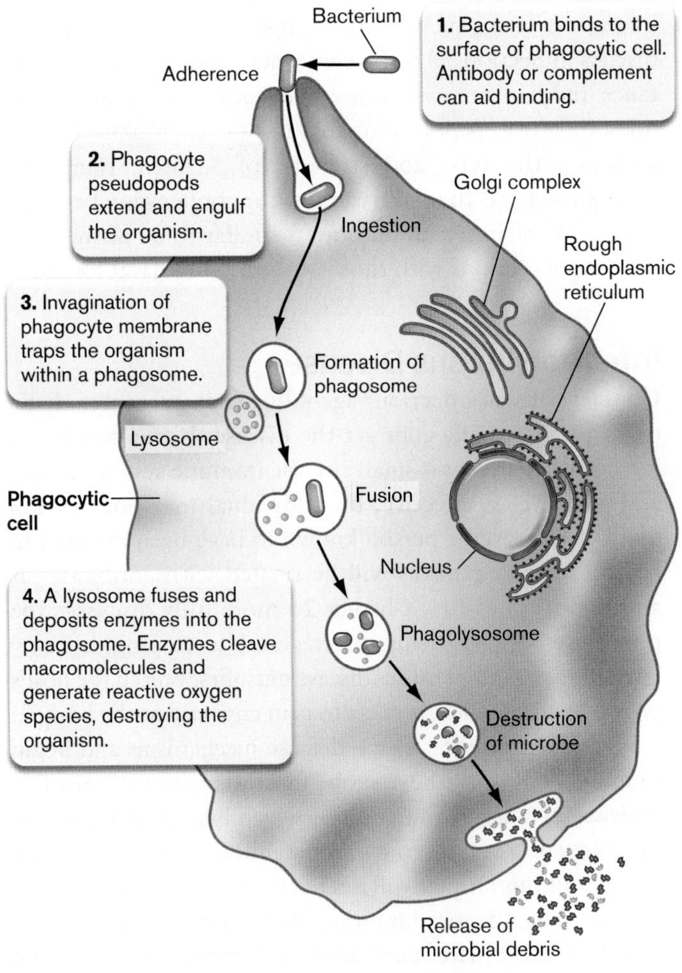

FIGURE 23.19 ■ Phagocytosis and phagosome-lysosome fusion.

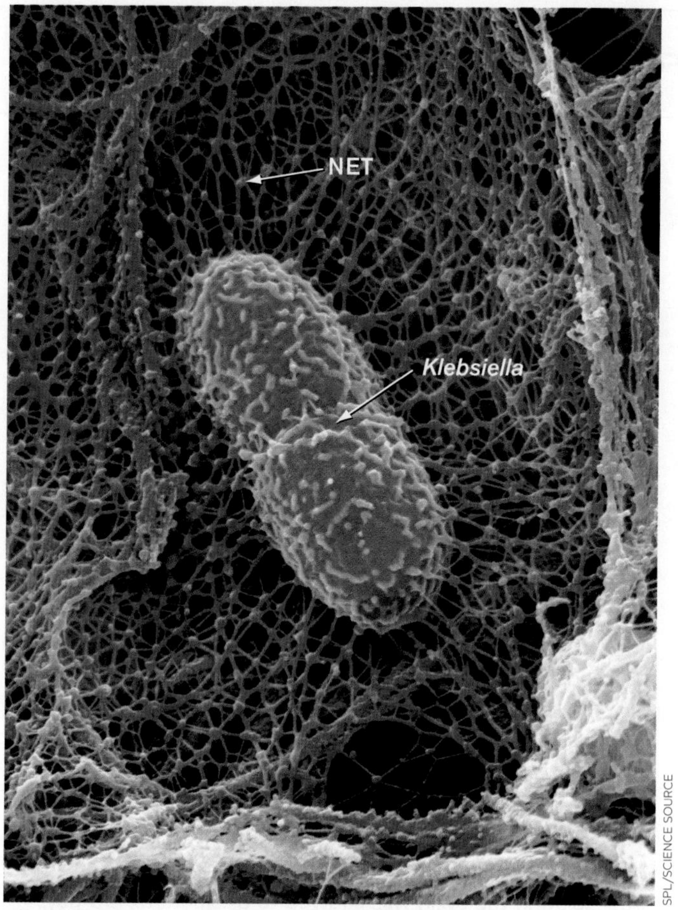

FIGURE 23.20 ■ Neutrophil extracellular trap (NET). An infected mouse lung shows a *Klebsiella pneumoniae* bacterium (colorized; length approx. 1 μm) snared in a NET (green), a web of decondensed chromatin released by neutrophils to catch and kill pathogens.

autoimmune diseases, such as systemic lupus erythematosus (**eTopic 23.1**).

Thought Question

23.4 **Figure 23.20** shows how a neutrophil extracellular trap can ensnare a nearby pathogen. Which bacterial structure might blunt the microbicidal effect of NETs?

Basophils, which stain with basic dyes, and **eosinophils** (**Fig. 23.18B**), which stain with the acidic dye eosin, do not phagocytose microbes or throw NETs, but release products, such as major basic protein, that are toxic to the microbe. These two types of blood cells also release chemical mediators (called vasoactive agents) that affect the diameter and permeability of blood vessels (the significance of which is discussed in Section 23.5). Eosinophils play a major role in killing multicellular parasites such as helminth worms. **Mast cells** are similar to basophils in structure but differentiate in a lineage separate from PMNs (**Fig. 23.17**). Unlike PMNs, mast cells are residents of connective tissues and mucosa and do not circulate in the bloodstream. Basophils and mast cells also contain high-affinity receptors for a class of antibody, called immunoglobulin E (IgE), that is associated with allergic responses (detailed in Section 24.7).

Note: Because neutrophils constitute the vast majority of PMNs, the terms "neutrophil" and "PMN" are often used interchangeably.

Monocytes (**Fig. 23.18C**) are white blood cells with a single nucleus (not multilobed like a PMN); they engulf (phagocytose) foreign material. Monocytes circulating in the blood can migrate out of blood vessels into various tissues and differentiate into **macrophages** and **dendritic cells** (**Fig. 23.21**). Macrophages are phagocytic and form a major part of the

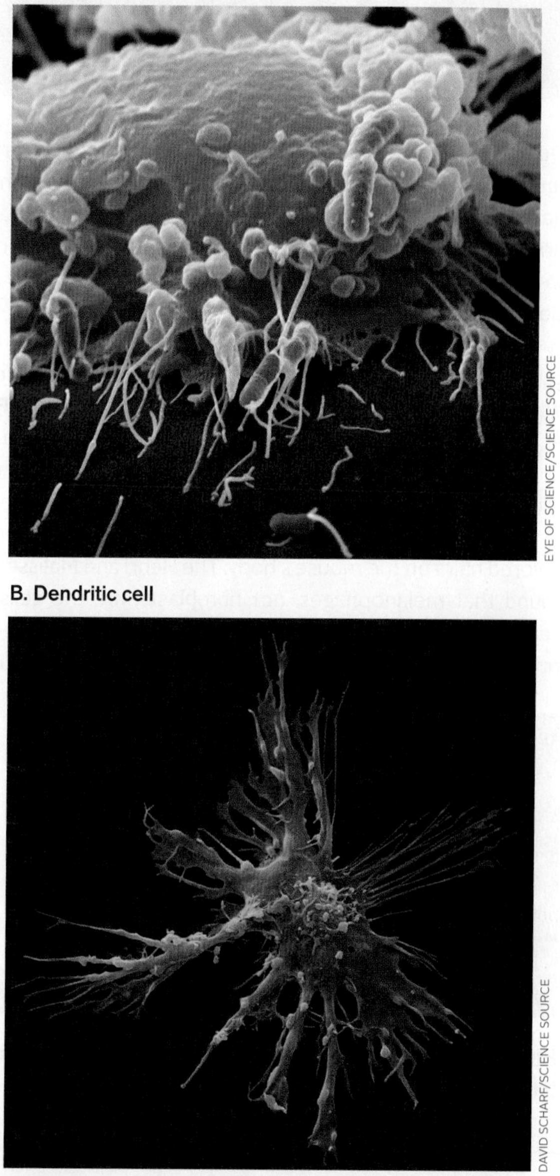

FIGURE 23.21 ■ The innate immune system depends on white blood cells called macrophages and dendritic cells. A. Membrane protrusions from a macrophage (20 μm long) detecting and engulfing bacteria (pink; *E. coli*, 1.5 μm long). **B.** Dendritic cell. Colorized SEMs.

SPECIAL TOPIC 23.1 Why Do Tattoos Last Forever?

CLEM ONOJEGHUO/PEXELS

Do you or your friends have a tattoo? It seems everyone under 40 has one. Many are beautiful, but some are unmitigated disasters, like the misspelled tattoo on one teenager's arm ironically proclaiming "No Regrits." Unfortunately for him, tattoos are forever. The question we address here is: Why don't they fade?

In the past, fibroblasts were considered the primary long-term reservoir of tattoo pigment granules. But now, research groups headed by Sandrine Henri and Bernard Malissen (**Fig. 1**) at Aix-Marseille University in France have uncovered a new culprit: a type of macrophage called a melanophage. In mice, melanophages normally ingest melanin produced by melanocytes and deliver the melanin to keratinocytes that become colored hairs on the mouse's body. The Henri and Malissen labs found that melanophages, not fibroblasts, are the cells that capture and retain tattoo pigment particles. Melanophages do not migrate from the skin to lymph nodes, but they will eventually die in place. If a melanophage is laden with dye particles when it dies, the particles are released and should be drained via the lymphatic vessels. But contrary to expectation, they are not drained. So, what keeps the particles there?

COURTESY OF SANDRINE HENRI

COURTESY OF BERNARD MALISSEN

FIGURE 1 ▪ Sandrine Henri (left) and Bernard Malissen (right).

Figure 2 examines cells taken from the skin of a mouse tail that was tattooed with a green pigment. Pigment particles were engulfed only by melanophage cells (**Fig. 2A**); the particles were not seen in other macrophages or dendritic cells (**Fig. 2B** and **C**). Next the scientists asked what would happen to the particle if they selectively killed the melanophages (**Fig. 3**). To delete, or ablate, melanophages, the mice used had been genetically engineered to express diphtheria toxin receptor (DTR), but only on mouse melanophages. Mouse cells normally lack the DTR gene. **Figure 3A** shows the green pigment particles residing in melanophages after tattooing. Diphtheria toxin was then injected into the same mice. Two days later the melanophages were gone and the green particles were seen free, not within any cells (**Fig. 3B**). Why aren't these particles drained from the site, causing the tattoo to disappear? Do new melanophages enter the tattooed region and reingest the particles?

Figure 3C (control) again shows tail skin melanophages laden with green particles 3 weeks after tattooing. Melanophages were then ablated using diphtheria toxin for 2 days. After ablation the toxin was allowed to dissipate for 90 days so that the mice could restore the melanophage population. Once again the tattoo did not fade. After the 90-day period, skin cells taken from the tattoo site were examined. **Figure 3D** shows that, indeed, new melanophages had entered the area and reingested the particles.

The data indicate that a cycle of pigment capture by melanophages, pigment release, and then recapture by new melanophages can account for long-term tattoo persistence. These results also explain why the "No Regrits" tattoo will taunt our teenager for many years to come.

RESEARCH QUESTION

Laser pulses are used to remove tattoos, but multiple treatments are often required. How might the research presented in this paper be used to explain this phenomenon and facilitate tattoo removal?

Anna Baranska, Alaa Shawket, Mabel Jouve, Myriam Baratin, Camille Malosse, et al. 2018. Unveiling skin macrophage dynamics explains both tattoo persistence and strenuous removal. *Journal of Experimental Medicine* **215**:1115–1133. http://doi.org/10.1084/jem.20171608.

amorphous **reticuloendothelial system**, which is widely distributed throughout the body. The reticuloendothelial system is a group of cells with the ability to take up and sequester particles. In addition to macrophages, the system is composed of specialized endothelial cells lining the sinusoids (special capillaries) of the liver (Kupffer cells), spleen, and bone marrow, as well as reticular cells of lymphatic tissue (macrophages) and of bone marrow (fibroblasts). The function of macrophages and

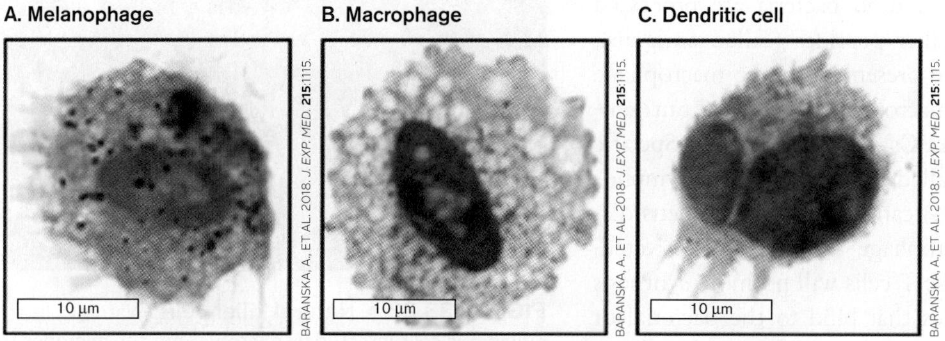

FIGURE 2 ■ Cells sorted from green-tattooed mouse tail.

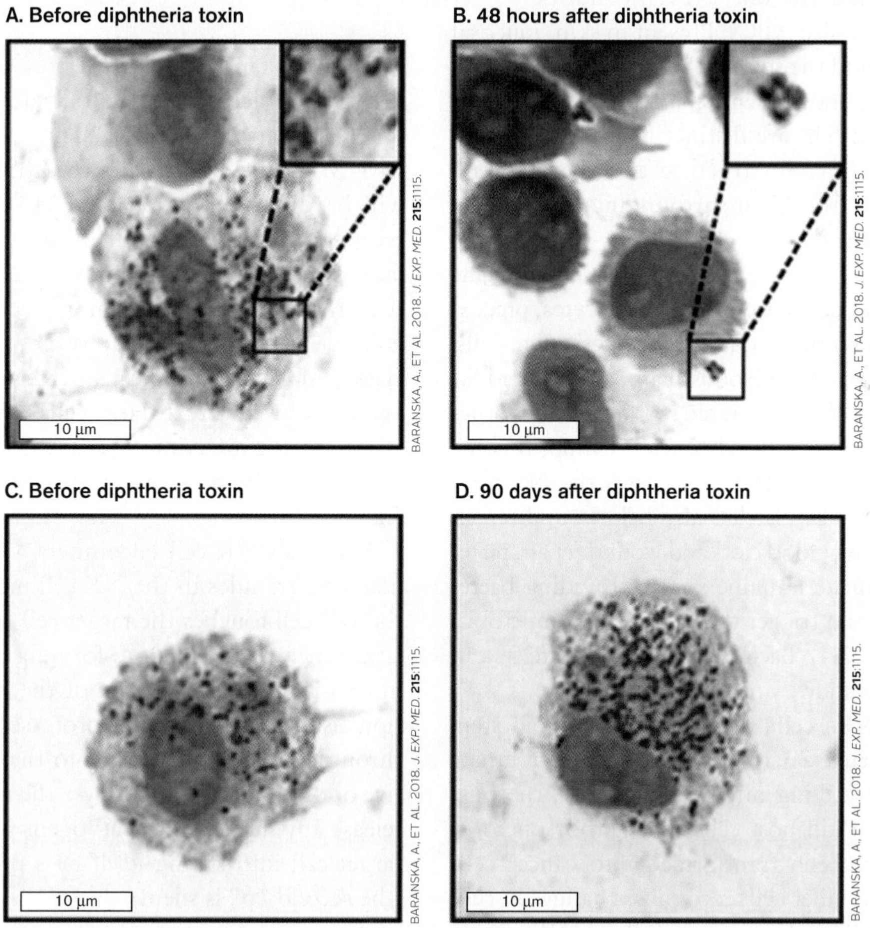

FIGURE 3 ■ Fate of green tattoo particles when melanophages are ablated. A, C. Melanophages contain green particles at the tattoo site before ablation. **B.** Forty-eight hours after ablation with diphtheria toxin, melanophages are absent and green particles are found only extracellularly. **D.** Ninety days after ablation, melanophages have returned to the tattoo site and recaptured green particles.

the reticuloendothelial system is to phagocytose microorganisms and other foreign particles. In fact, recent research indicates that a certain type of macrophage in skin contributes to the longevity of tattoos (**Special Topic 23.1**).

Macrophages are present in most tissues of the body and are the cells most likely to make first contact with invading pathogens. They have two functions. As part of innate immunity, they kill invaders directly. Protrusions from the macrophage surface extend and clasp nearby bacteria, pulling them into the cell (**Fig. 23.21A**). Once ingested by the macrophage, the bacteria are destroyed.

Subsequently, as the first step in adaptive immunity, the remnants of dead bacteria are processed (degraded) into smaller peptides (called antigens), and the antigens are presented on the macrophage cell surface. Thus, macrophages are called **antigen-presenting cells** (**APCs**; see Chapter 24). Specific white blood cells called T cells (a type of lymphocyte; see Chapter 24) can bind to the antigens displayed on the macrophage and become activated. Once activated, these T cells will promote synthesis of specific antibodies that bind to the bacterial or viral antigen, as well as stimulate cytotoxic cells to kill any host cell infected by the pathogen.

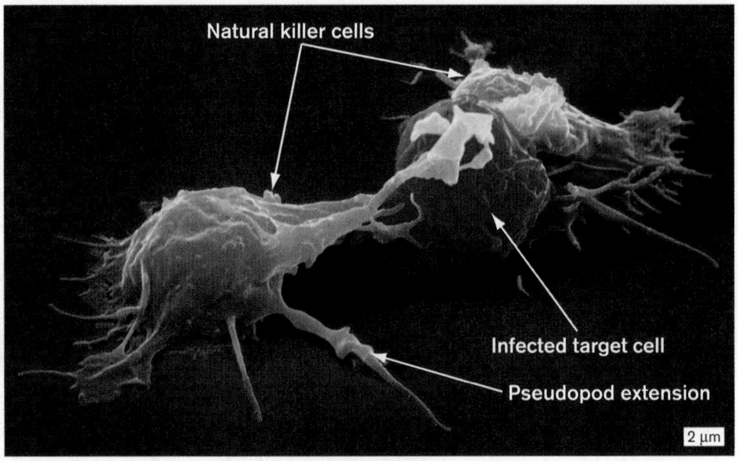

EYE OF SCIENCE/SCIENCE SOURCE

FIGURE 23.22 ■ **Natural killer cells.** Natural killer (NK) cells attack eukaryotic cells infected by microbes, not the microbes themselves. Perforin produced by the NK cell punctures the membranes of target cells, causing them to burst (colorized SEM).

Macrophages are not the only APCs; dendritic cells (**Fig. 23.21B**) are also APCs. Present in skin, mucosal tissue, lymph nodes, and the spleen, they, like macrophages, can take up, process, and present small antigens on their cell surface. Dendritic cells are distinct from macrophages in that they have a different structure and they take up small soluble antigens from their surroundings in addition to phagocytosing whole bacteria.

Platelets are small, puzzle piece–shaped cell fragments that lack a nucleus and derive from megakaryocytes, precursor cells that also differentiate from hematopoietic stem cells. Platelets, which circulate in the bloodstream, are required for efficient blood clotting. When activated by damaged endothelial cells (cells lining blood vessels), platelets clump, become trapped by fibrin, and form plugs that stop the bleeding. Several factors produced by platelets also help wound repair. In addition to mediating blood clots and wound repair, platelets are part of the innate immune system. Invading bacteria can bind platelets and trigger the release of antimicrobial peptides (see Section 23.4). Bacterially activated platelets can also induce NET formation by neutrophils.

Natural killer (**NK**) **cells** derive from lymphoid stem cells, as shown in **Figure 23.17**, and are also part of innate immunity. Instead of killing microbes, however, the mission of NK cells is to kill host cells that harbor microorganisms or that have been transformed into cancer cells (**Fig. 23.22**). Natural killer cells recognize changes in cell-surface proteins of infected or cancer cells (MHC class I molecules; described below). This recognition causes the NK cells to degranulate—that is, to release chemicals that kill the target host cells.

Natural killer cells recognize their targets in two basic ways: by the absence of major histocompatibility complex (MHC) class I molecules on host cells, or by the presence of antibodies on host cells. The **major histocompatibility complex** (**MHC**) consists of proteins found on the surfaces of cells that help the immune system recognize self versus foreign substances. A normal host cell displays two classes of MHC molecules on the outside of the cell membrane. MHC I is an indicator of "self." All nucleated host cells have MHC I on their surface. (MHC I and II molecules will be discussed in Chapter 24.) NK cells have specific receptors that bind to self MHC I molecules on the surfaces of other cells in the body. An NK cell that "touches" a self MHC I molecule–containing cell from the same person will not attack that cell. However, if a host cell lacks MHC class I molecules, NK cells perceive the target as foreign and a potential threat. Host cells can lose their MHC molecules during infection or as a result of malignant transformation. (MHC proteins are discussed in Section 23.5 and Chapter 24.)

When an NK cell encounters a host cell lacking these markers, granules in the NK cell move (polarize) to where the NK cell touches the target cell (**Fig. 23.23**). Degranulation then releases a pore-forming protein (**perforin**) that inserts into the membrane of the target cell. Degranulation also releases cytotoxic proteases (granzymes) that pass through the perforin pore into the target cell. The objective of the NK cell is not to lyse the target cell, which would release any intracellular pathogens present, but to coax the target cell into killing itself by a process called **apoptosis** (the second "p" is silent).

Apoptosis is triggered when granzymes damage mitochondria. Cytochromes spilled from damaged mitochondria activate cytosolic host proteases called caspases (cysteine-dependent aspartate-directed proteases). Activated caspases cleave a variety of host molecules, including signal transduction proteins, cytoskeletal proteins, DNA repair proteins, and inhibitors of endonucleases. The consequence of this complex process is that host cell DNA becomes fragmented, cell membranes bleb, and the cell breaks into small, membrane-encased blobs

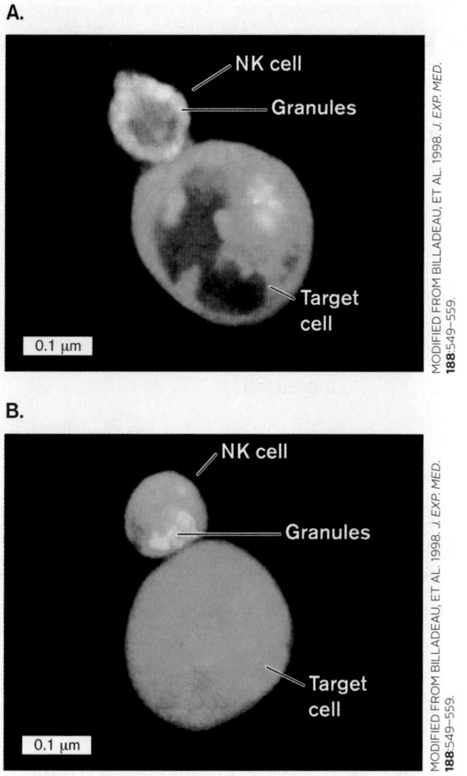

FIGURE 23.23 ▪ Polarization of NK cell granules at the interface between NK cell and target cell. A. First contact between cells. Granules are dispersed throughout the NK cell. **B.** Polarization of granules at the NK cell–target cell interface before they are released from the NK cell. Cytoplasmic granules of NK cells were labeled with acridine orange.

(apoptotic bodies) that can be cleared by neutrophils via phagocytosis without causing inflammation, a major goal of apoptosis. Phagocytosis of the apoptotic bodies also kills the infectious agent.

Leaving nothing to chance, another granzyme (granzyme B), also delivered by NK cells into infected cells, will enter intracellular bacteria and disrupt their macromolecular processes, thereby killing the pathogens directly. Farokh Dotiwala, Sriam Chandrasekaran, and Judy Lieberman recently discovered that granzyme B protease prevents protein synthesis in *E. coli*, *Listeria monocytogenes*, and *Mycobacterium smegmatis* (**Fig. 23.24**). The enzyme does so by digesting key ribosomal proteins and aminoacyl-tRNA synthetases.

The second killing mechanism wielded by NK cells is called **antibody-dependent cell-mediated cytotoxicity (ADCC)**. Besides receptors that detect MHC I molecules, NK cells also contain receptors on their cell surface that bind to a part of an antibody called the Fc region. The Fc region of an antibody does not bind to antigens but can bind to specific membrane Fc receptors on other cells (discussed in Chapter 24, Fig. 24.7). ADCC is activated when the Fc receptor on the NK cell links to an antibody-coated host cell.

Why would host cells be coated with antibodies? During their replication, many viruses place viral proteins in the membrane of the infected cell. Antibodies to those viral proteins will coat the compromised cell, tagging it for ADCC. Once the compromised cell has been targeted, it is killed by the NK cell in the same way as described for cells lacking MHC—namely, by insertion of a perforin molecule and injection of granzymes to initiate apoptosis. This killing mechanism is an example of cooperation between innate immunity (NK cells) and adaptive immunity (antibody-producing lymphocytes).

Diagnostic value of white blood cell ratios. Many diseases, including infections, can alter the ratios of white blood cells (WBCs). A WBC differential is a test that physicians often order to help them diagnose infections and other syndromes. **Table 23.1** presents general guidelines for

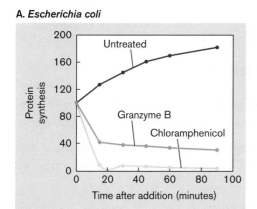

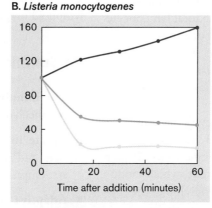

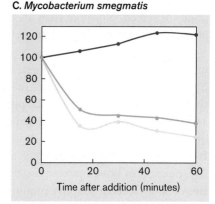

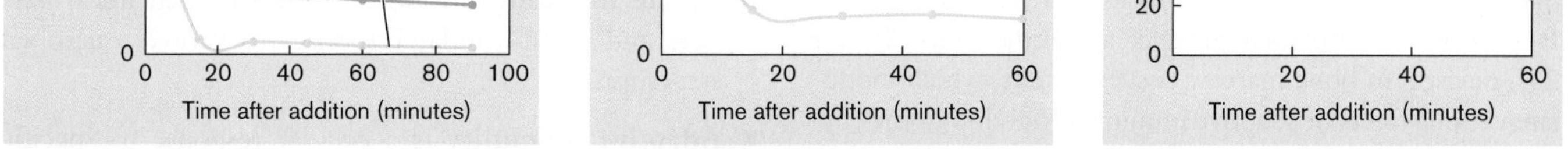

FIGURE 23.24 ▪ Granzyme B disrupts bacterial protein synthesis. Granzyme B and sublytic amounts of the pore-forming protein granulysin that delivers granzyme B into bacteria were added to *Escherichia coli* **(A)**, *Listeria monocytogenes* **(B)**, and *Mycobacterium smegmatis* **(C)**. Protein synthesis was measured as incorporation of ^{35}S-radiolabeled methionine into trichloroacetic acid precipitable material (percentage, compared to untreated controls). The protein synthesis inhibitor chloramphenicol was added as a positive control. *Source:* Modified from Dotiwala et al. 2017. *Cell* **171**:1125–1137, fig. 3A.

TABLE 23.1 Guidelines for Interpreting White Blood Cell (WBC) Counts with Differential*

	Normal	Acute bacterial infection	Viral infection	Parasitic infection	Allergy
Total WBC count	4,500–11,000/mm^3	Elevated: 12,000–30,000	Elevated	Elevated	Elevated
Differential					
Neutrophils	54%–62%	Increased numbers and more immature forms (band cells)	Can be reduced		
Eosinophils	1%–3%			Increased	
Basophils	0%–0.75%				Increased
Lymphocytes	25%–33%		Increased		
Monocytes	3%–7%	Typically increase with chronic infections (for example, tuberculosis, brucellosis, subacute bacterial endocarditis, *Rickettsia*, many protozoan infections)			

*This table provides general guidelines. Individual organisms or certain noninfectious medical conditions or immunological defects can alter the findings. A blank space indicates little change in the parameter.

interpreting a WBC differential, indicating which cell types increase or decrease in response to infections with bacteria, viruses, or parasites (protozoa or worms). Notice that total WBC counts increase in each case, but the type of WBC that increases in number differs with the infectious agent.

Lymphoid Organs

As **Figure 23.17** shows, lymphoid stem cells produce lymphocytes as well as natural killer cells, described earlier. **Lymphocytes** (**Fig. 23.18D**), which are the main participants in adaptive immunity, are present in blood at about 2,500 cells per microliter, accounting for about one-third of all peripheral white blood cells. However, an individual lymphocyte spends most of its life within specialized solid tissues (lymphoid organs) and enters the bloodstream only periodically, where it migrates from one place to another, surveying tissues for possible infection or foreign antigens. Consequently, most lymphocytes are found in the lymph nodes or spleen. No more than 1% of the total lymphocyte population circulates in the blood at any one time.

The tissues of the immune system, where the great majority of lymphocytes are found, are classified as primary or secondary lymphoid organs or tissues, depending on their function (**Fig. 23.25**). The primary lymphoid organs and tissues are where immature lymphocytes made in bone marrow mature into antigen-sensitive B cells and T cells. **B cells**, which ultimately produce antibodies (see Chapter 24), develop in bone marrow tissue; **T cells**, which modulate various facets of adaptive immunity, develop in the thymus, an organ located above the heart.

The secondary lymphoid organs serve as stations where lymphocytes can encounter antigens. These encounters lead to the differentiation of B cells into antibody-secreting plasma cells, and of T cells into antigen-specific helper cells, as discussed in Chapter 24. The spleen is an example of a secondary lymphoid organ. It is designed to filter blood directly and detect microorganisms. Macrophages in the spleen engulf these organisms and destroy them, and then migrate to other secondary lymphoid organs and present pieces of the microbe (called antigens) to the B and T cells, which then become activated.

The **lymph nodes** are another kind of secondary lymphoid organ. Lymph nodes are arranged to trap organisms that drain from local (nearby) tissues. These are sometimes called draining lymph nodes. Organisms travel to a draining lymph node through lymphatic vessels rather than through blood vessels. Lymph nodes are situated at various sites in the body where lymphatic vessels converge (for example, in the armpits). Lymphoid tissues are also present in the mucosal regions of the gut and respiratory tracts (for example, Peyer's patches and gut-associated lymphoid tissue, or GALT, discussed in the next section). Other secondary lymphoid organs are the tonsils, adenoids, and appendix.

To Summarize

- **The immune system** consists of both innate and adaptive mechanisms that recognize and eliminate pathogens.
- **Innate immunity** includes physical and chemical barriers and some cellular responses to various microbial structures.
- **Adaptive immunity** is a cellular response to specific structures (antigens) in which a memory of exposure is produced.
- **Myeloid bone marrow stem cells** differentiate to form cells of the innate immune system—namely, phagocytic

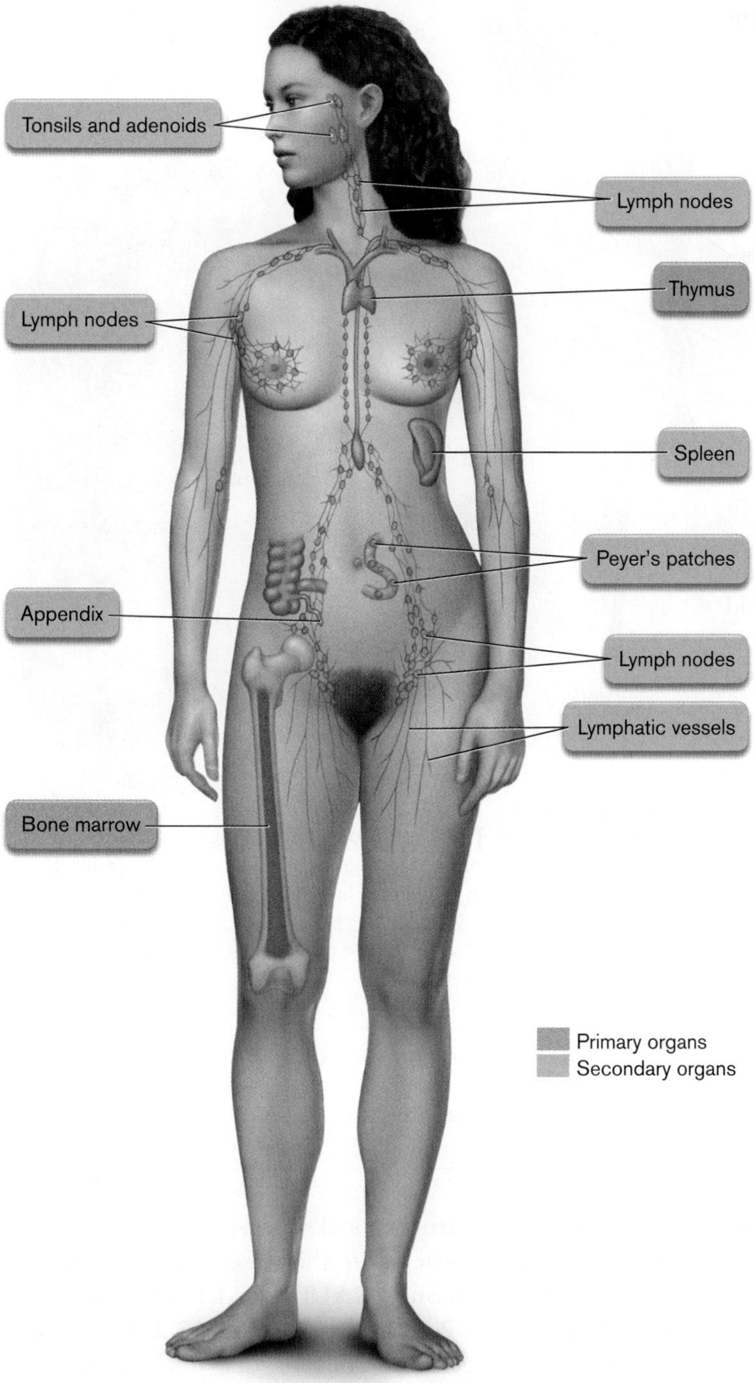

FIGURE 23.25 ■ Lymphoid organs.

PMNs, monocytes, macrophages, antigen-presenting dendritic cells, and mast cells. **Platelets** are derived from a different cell line.

- **Lymphoid stem cells** differentiate into natural killer cells (part of the innate immune system) and B cells and T cells (part of the adaptive immune system). B cells ultimately produce antibodies, whereas T cells regulate adaptive immunity.
- **Natural killer (NK) cells** are a class of white blood cells that continually patrol tissues for cancer cells, cells infected with microbes, or host cells coated with antibody (antibody-dependent cell-mediated cytotoxicity, ADCC). **NK cells kill by inserting pores of perforin and granzymes** into target cell membranes.
- **Primary lymphoid organs** include bone marrow (where B cells develop) and the thymus (where T cells develop). **Secondary lymphoid organs** (spleen, lymph nodes, Peyer's patches, tonsils, appendix) are where lymphocytes encounter antigens.

23.4 Physical and Chemical Defenses against Infection

In defending a castle during medieval times, the first lines of defense included physical barriers (the castle wall), chemical barriers (boiling oil tossed onto invaders trying to scale the wall), and finally, hand-to-hand combat once the wall was breached. Similarly, the body's initial defenses against infectious disease are composed of physical, chemical, and cellular barriers designed to prevent a pathogen's access to host tissues. Although generally described as nonspecific, some innate defense systems are more specific than others.

Physical Barriers to Infection

The first line of defense against any potential microbial invader (either commensal or pathogenic) is found where parts of the body interface with the environment. These interfaces (skin, lung, gastrointestinal tract, genitourinary tract, and oral cavities) have similar defense strategies, although each has unique characteristics. A defense common to all host surfaces involves **tight junctions**—watertight adhesions that link adjacent epithelial cells at mucosal surfaces and endothelial cells lining blood vessels. Tight junctions prevent bacteria, and even host cells, from moving between internal and external host compartments. The "glue" that holds tight junctions together is a series of interconnecting glycoprotein molecules (**Fig. 23.26**).

Skin. Few microorganisms can penetrate skin, because of the thick keratin armor produced by closely packed cells called keratinocytes. Keratin protein is a hard substance (hair and fingernails are made of it) that is not degraded by known microbial enzymes. An oily substance (sebum) produced by the sebaceous glands will cover and protect the skin. Its slightly acidic pH inhibits bacterial growth. Skin secretions also contain antimicrobial peptides, such as cathelicidin LL-37 (**eTopic 23.3**), that disrupt membrane integrity, and antimicrobial RNases that can destabilize bacterial membranes and penetrate to cleave intracellular mRNA. Competition between species also limits colonization by pathogens, and microorganisms that manage to adhere are continually removed by the constant shedding of outer epithelial skin layers.

Other, more specialized cells just under the skin can recognize microbes managing to slip through the physical barrier. They are part of a consortium of cells called **skin-associated lymphoid tissue** (**SALT**). **Langerhans cells** make up a significant portion of SALT. They are specialized dendritic cells that can phagocytose microbes. Once a lymphoid Langerhans cell has ingested a microbe, the cell migrates by ameboid movement to nearby lymph nodes and presents parts of the microbe to the immune system to activate antimicrobial immunity.

Note: Do not confuse phagocytic Langerhans cells with the pancreatic "islets of Langerhans," which secrete insulin.

Mucous membranes. Mucosal surfaces form the largest interface (200–300 square meters) between the human host and the environment (the intestine alone is 7–8 meters, or 23–26 feet, long). Mucous membranes in general present a containment problem. They must be selectively permeable in order to exchange nutrients, as well as to export products and waste components. At the same time, they must constitute a barrier against invading pathogens. Mucosal membranes are covered with special tightly knit epithelial layers that support this barrier function. The mucus secreted from stratified squamous epithelial cells coats mucosal surfaces and traps microbes. Compounds within the mucus can serve as a food source for some microbiome organisms, but other secreted compounds—like the enzymes lysozyme (which cleaves cell wall peptidoglycan) and lactoperoxidase (which produces superoxide radicals)—can kill an organism trapped in the mucus.

Semispecific innate immune mechanisms are also associated with mucosal surfaces. Host cells, even epithelial cells, in mucosa have evolved mechanisms to distinguish harmless compounds from microorganisms. Patterns of conserved structures on microbes, called **microbe-associated molecular patterns** (**MAMPs**), are recognized by host cell-surface receptors such as various Toll-like receptors and CD14 (discussed in Section 23.5). Once a MAMP has been recognized, the host cell sends out chemicals that can activate immune system cells that perform innate and adaptive immune functions.

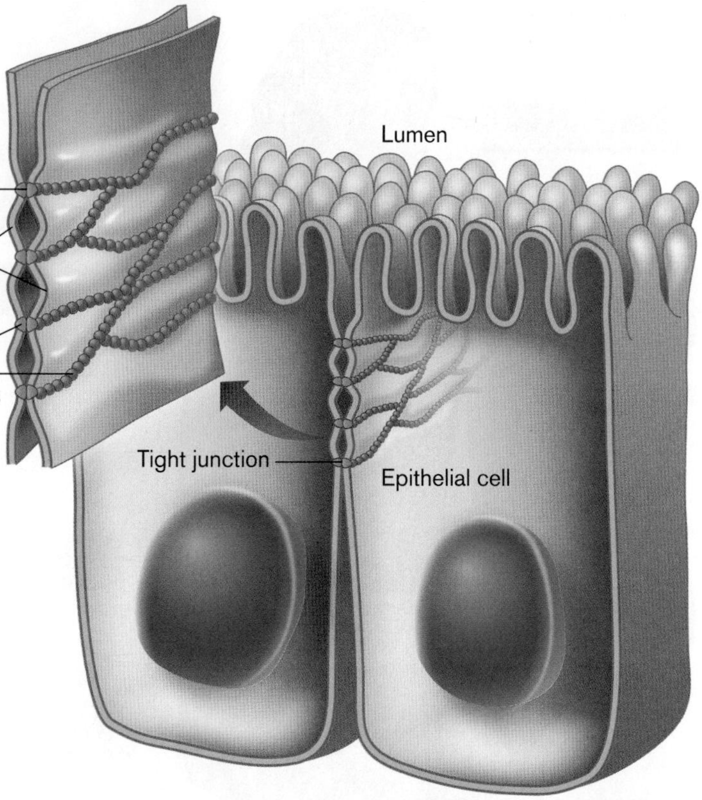

FIGURE 23.26 ▪ Tight junctions hold adjacent cells together. Tight junctions are formed by an interconnected series of glycoproteins (blowup). By tightly linking adjacent cell membranes together, tight junctions produce a barrier through which bacteria, viruses, and other host cells cannot easily pass.

Note: MAMPs were previously called PAMPs, for "pathogen-associated molecular patterns." Because the structures recognized are also present on nonpathogenic bacteria and viruses, the term was changed to MAMP.

Like skin, the gastrointestinal system possesses an innate mucosal immune system, in this case called **gut-associated lymphoid tissue** (**GALT**). GALT includes tonsils, adenoids, and Peyer's patches (**Fig. 23.27A**). These tissues contain specialized **M cells** that dot the intestinal surface and are wedged between epithelial cells. "M" stands for "microfold," which describes their appearance (**Fig. 23.27B**). These are immobilized cells that take up microbes (microbiota or pathogens) from the intestine and release them, or pieces of them, into a pocket formed on the opposite, or basolateral, side of the cell. Other cells of the innate immune system, such as macrophages (which migrate through tissues), gather here and collect the organisms that emerge. Macrophages engulf and try to kill the organism. If successful, the macrophage will place small, degraded components of the microbe onto cell-surface MHC I and MHC II molecules (antigen presentation). Other immune system cells can recognize the presented antigens and initiate adaptive immune functions

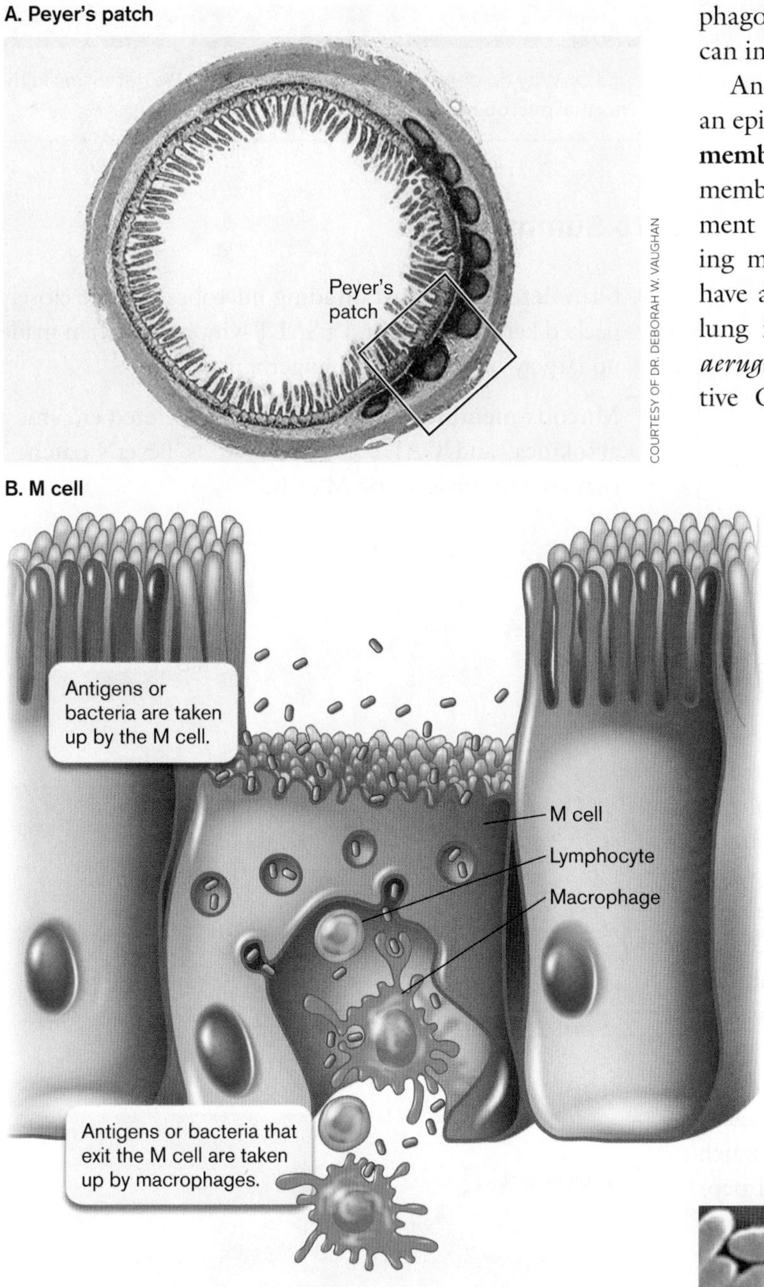

FIGURE 23.27 ■ Gut-associated lymphoid tissue (GALT). **A.** A Peyer's patch located on the small intestine. **B.** Diagram of an M cell (microfold cell).

such as antibody production (described in Chapter 24). As a result, M cells are extremely important for the development of mucosal immunity to pathogens. However, M cells can also serve as a portal for some pathogens to gain entry to the body.

The lungs. The lungs also have a formidable defense. In addition to the mucociliary escalator discussed in Section 23.1, microorganisms larger than 100 μm become trapped by hairs and cilia lining the nasal cavity and trigger a forceful expulsion of air from the lungs (a sneeze). The sneeze is designed to clear the organism from the respiratory tract. Organisms that make it to the alveoli are met by phagocytic cells called **alveolar macrophages**. These cells can ingest and kill most bacteria.

Another important factor that prevents lung infections is an epithelial membrane protein called **cystic fibrosis transmembrane conductance regulator** (**CFTR**). CFTR is a membrane chloride channel that regulates chloride movement across the membrane—an essential part of hydrating mucus in healthy individuals. Cystic fibrosis patients have a defective CFTR and are much more susceptible to lung infections, especially those caused by *Pseudomonas aeruginosa*. Diminished chloride secretion due to a defective CFTR produces highly viscous airway mucus that impedes the mucociliary escalator. The *P. aeruginosa* strains posing the most serious threat produce a thick slimy material (alginate) that further impedes lung clearance mechanisms (**Fig. 23.28**).

Chemical Barriers to Infection

Some examples of chemical barriers were mentioned earlier, such as the acidic pH of the stomach, lysozyme in tears, and generators of superoxide. In addition, a variety of human cells produce small antimicrobial, cationic (positively charged) peptides called defensins (**Table 23.2**). These antimicrobial peptides are important components of innate immunity against microbial infections.

Defensins range in length from 29 to 47 amino acids and are found in mammals, birds, amphibians, and plants. Defensins (and other antimicrobial peptides) destroy an invading microbe's cytoplasmic membrane and are effective against Gram-positive and Gram-negative

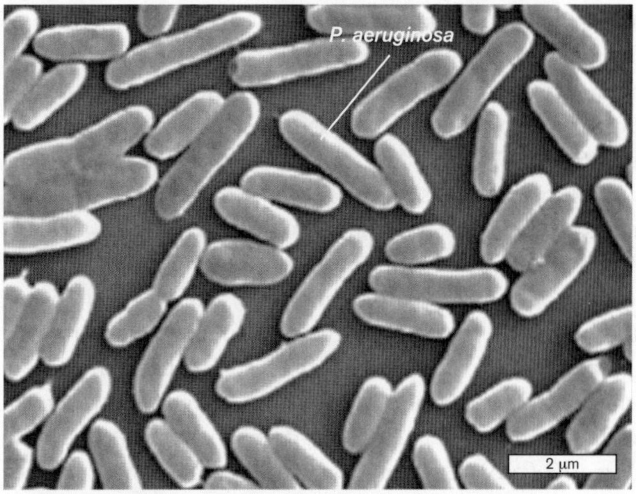

REPUBLISHED WITH PERMISSION OF DOVE PRESS. DOSUNMU, E., ET AL. 2015. *INT J NANOMEDICINE* **10**:5025–5034.

FIGURE 23.28 ■ Mucoid strains of *Pseudomonas aeruginosa* that infect cystic fibrosis patients. The organism can grow as biofilms in tissue and secrete an extracellular polymeric substance (EPS) that gives colonies of these strains a very mucoid appearance. (SEM.)

TABLE 23.2 Categories of Natural Antimicrobial Peptides

Class (Examples)	Major presence in humans (Source)
Alpha-defensins (-1, -2, -3, -4)[a]	Neutrophils (Stored in granules)
Alpha-defensins (-5, -6)	Paneth cells, small intestine (Stored in granules)
Cathelicidins (LL-37, hCAP18)	Neutrophils (Secreted)
Histatins	Saliva (Secreted)
Beta-defensins (HBD-1, -2)	Epithelia (Secreted)
Kinocidins (tPMP,[b] PF-4)	Platelets (Secreted)
Other species	
Maganins	Frogs
Protegrins	Pigs
Indolicidin	Cattle

[a]Alpha-defensins are named alpha-defensin-1, alpha-defensin-2, and so on.

[b]tPMP = thrombin-induced platelet microbicidal protein.

bacteria, fungi, and even some viruses (those with membranes, like HIV). To kill Gram-negative bacteria, the peptides must first bind the negatively charged outer membrane lipopolysaccharides (LPS) of Gram-negative bacteria and move into the periplasm. Defensins are then pulled into the cytoplasmic membrane of either Gram-positive or Gram-negative bacteria by the transmembrane electrical potential, about –150 mV in bacteria (that is, the cell interior is more negative than the exterior). The peptides assemble into channels that destroy the cytoplasmic membrane barrier, killing the bacterial cell. Defensins generally do not affect eukaryotic cells, which have a much lower membrane potential (–15 mV). Antimicrobial peptides are produced by many human cells, including cells of the skin, lungs, genitourinary tract, and gastrointestinal tract (**Fig. 23.29**). Specific defensins produced by different animals may partially explain pathogen-host specificity (see **eTopic 23.2**).

Vertebrate defensins of the alpha variety are stored in membrane-enclosed granules within neutrophils and in Paneth cells in the small intestine (**Fig. 23.29A**). When stimulated, these cells **degranulate** (release their granule contents) by fusing their granule membranes to cytoplasmic or vacuolar membranes, dumping their contents into the surroundings or into phagocytic vacuoles, where the alpha-defensins can destroy engulfed microbes (**Fig. 23.29B**). In contrast, the beta-defensins are not stored in cytoplasmic granules. The synthesis of beta-defensins is activated only after contact with bacteria or their products. Other cationic antimicrobial peptides are listed in **Table 23.2**. Neutrophil cathelicidins are discussed in **eTopic 23.3**.

Thought Question

23.5 Why do defensins have to be so small? Do defensins kill normal microbiota?

To Summarize

- **Skin defenses** against invading microbes include closely packed keratinocytes and a SALT lymphoid system made up largely of phagocytic Langerhans cells.
- **Mucous-membrane defenses** involve secreted enzymes, cytokines, and GALT tissues, such as Peyer's patches, that contain phagocytic M cells.

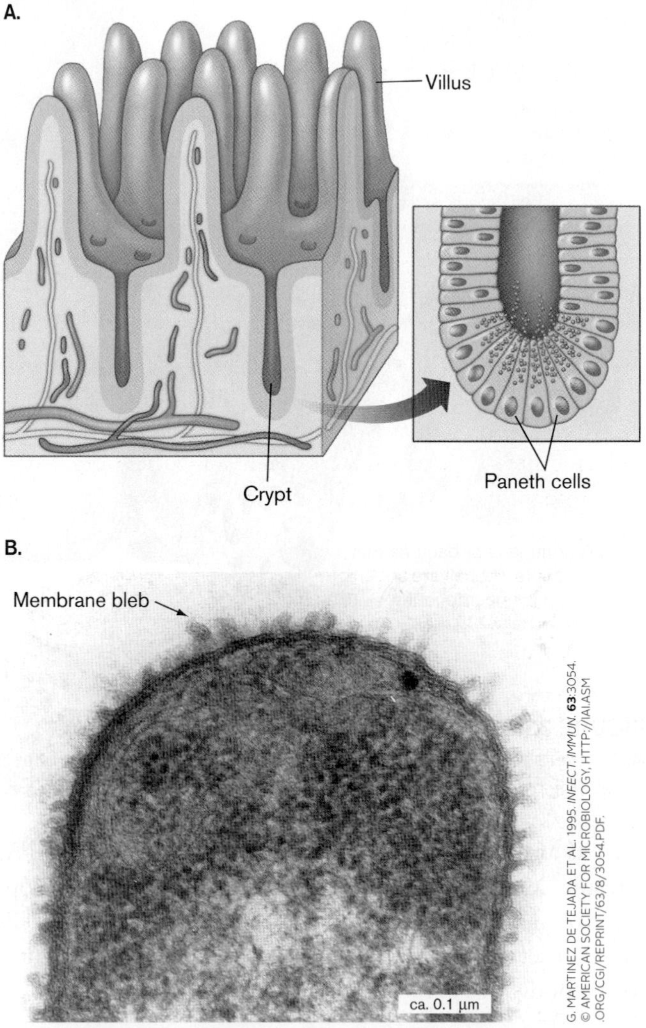

FIGURE 23.29 ■ Defensins. A. Certain defensins are produced in the crypts of the intestine. The crypts contain granule-rich Paneth cells (blowup) that discharge their granules into the crypt lumen in response to the entry of bacteria or as a result of food-related stimulation by acetylcholine. **B.** The effect of cationic peptides on *E. coli* O111. Polymyxin B is a small cationic peptide antibiotic that mimics the action of defensins. In this micrograph, polymyxin B causes blebs of membrane to ooze from the surface of the cell.

- **Microbe-associated molecular patterns (MAMPs)** are recognized by Toll-like receptors found on many host cells, such as macrophages. Binding triggers release of chemical signaling molecules that activate innate and adaptive immune mechanisms.
- **M cells** in gut-associated lymphoid tissues sample bacterial cells at their surface and release pieces of them to awaiting immune system cells.
- **Phagocytic alveolar macrophages** inhabit lung tissues, contributing to nonspecific (innate) defense.
- **Chemical barriers against disease** include cationic defensins, acid pH in the stomach, and superoxide produced by certain cells.

23.5 Innate Immunity: Surveillance, Cytokines, and Inflammation

The boil shown in **Figure 23.30** is an inflammatory response triggered by infection with the organism *Staphylococcus aureus*. Inflammation is a critical innate defense in the war between microbial invaders and their hosts. It provides a way for phagocytic cells (such as neutrophils) normally confined to the bloodstream to gain access to infected sites within tissues. Movement of these cells out of blood vessels is called **extravasation** or diapedesis, which we discuss shortly. Once at the infection site, the neutrophils begin engulfing microbes. The white pus associated with an infection is teeming with these white blood cells. In this section we describe the pathophysiology of inflammation that leads to the five cardinal (major) signs of inflammation, first described more than 2,000 years ago. Those signs are:

- Heat (warmth at the site from increased blood flow)
- Edema (swelling from fluid accumulating outside of blood vessels)
- Redness (due to dilated blood vessels)
- Pain (pharmacological stimulation of nerve endings)
- Altered function at the affected site (due to pain or damage to an inflamed organ)

Although many things can trigger inflammation, we focus here on how microbes cause the response. The process begins with the infection itself. Microorganisms introduced into the body—for example, on a wood splinter—will begin to grow and produce compounds that host cells sense or that damage host cells (**Fig. 23.31** ▶). Resident macrophages that wander into the infected area engulf these organisms and then release inflammatory mediators that orchestrate the inflammatory response. These mediators include **chemokines** (chemoattractants) that "call out" to neutrophils for more help, and **vasoactive factors** such as leukotrienes, platelet-activating factor, and prostaglandins, which act on blood vessels of the microcirculation, increasing blood volume and capillary permeability to help deliver white blood cells to the area. In addition, small protein molecules called **cytokines** are secreted, diffusing to the vasculature and stimulating the expression of specific receptors (selectins) on the endothelial cells of capillaries and venules. But how are cytokines made in response to an infection?

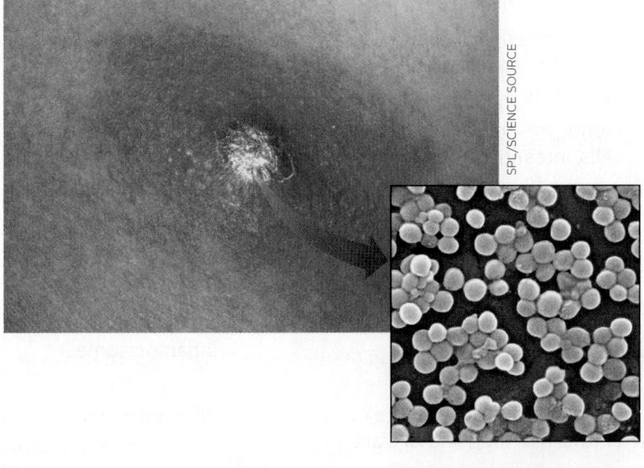

FIGURE 23.30 ■ Inflammation caused by infection. Boil resulting from infection of a hair follicle by *Staphylococcus aureus* (size 0.5–1.0 μm). **Blowup:** Colorized SEM.

Sensing the Invader: Pattern Recognition Receptors and Cytokines

The faster the body can detect a pathogen, the more quickly it can deal with it. The more quickly it deals with the pathogen, the better the outcome of an infection. It takes time, however, for the adaptive immune system to make antibodies specific for a microbe (see Chapter 24). All the while, the pathogen can grow and cause disease. Fortunately, bacteria and viruses possess unique structures that immediately tag them as foreign. Structures such as peptidoglycan, flagellin, lipoteichoic acids, and double-stranded RNA are not present in tissues unless bacteria or viruses are present. These structures have microbe-associated molecular patterns (MAMPs) that can be recognized by Toll-like or NOD-like receptors (TLRs or NLRs) present on or in various host cell types (**Fig. 23.32**; **Table 23.3**). TLRs and NLRs are tantamount to burglar alarm systems that activate

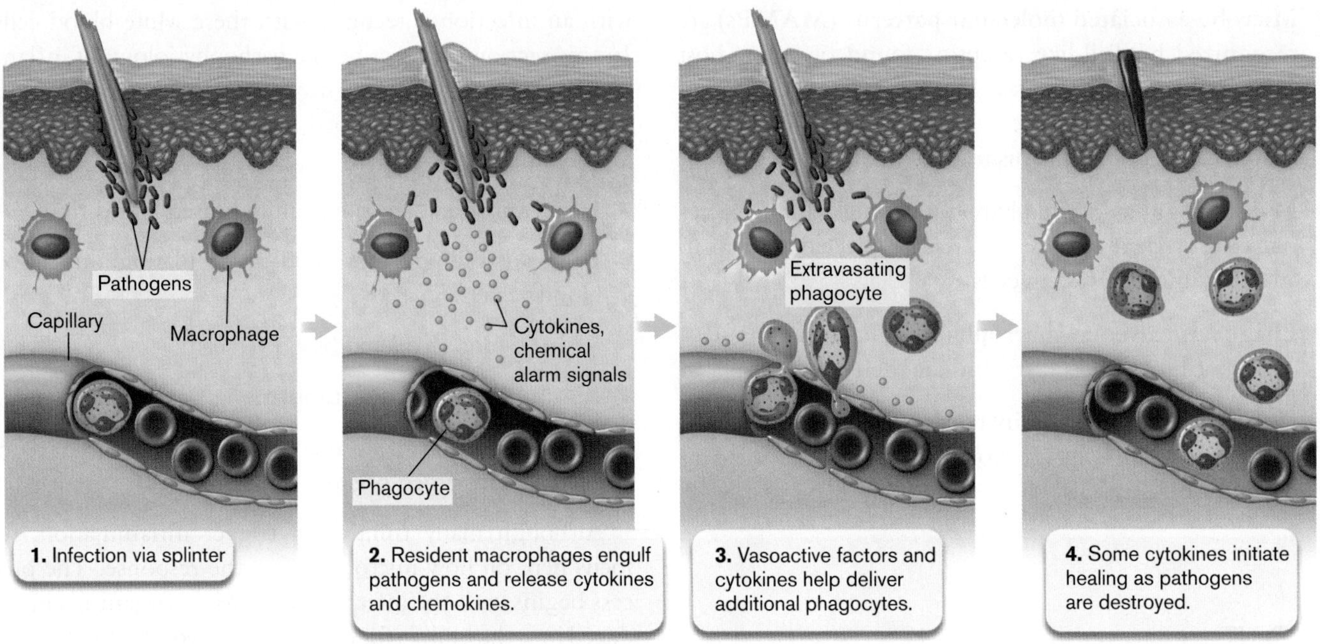

FIGURE 23.31 ■ Basic inflammatory response. Neutrophils (a type of phagocyte) circulate freely through blood vessels and can squeeze between cells in the walls of a capillary (extravasation) to the site of infection. They then engulf and destroy any pathogens they encounter.

upon encountering an intruder. Collectively, TLRs, NLRs, and similar proteins are called **pattern recognition receptors** (**PRRs**) because they recognize MAMPs.

Toll-like receptors. First discovered in insects and named Toll receptors, **Toll-like receptors** (**TLRs**) are evolutionarily conserved cell-surface glycoproteins present on the cells of many eukaryotic genera (**Fig. 23.32**). The term "Toll" came from Christiane Nüsslein-Volhard's 1985 exclamation, "That's crazy!" (in German, "Das ist ja toll!"), when shown the fungus-destroyed posterior of a mutant fruit fly (*Drosophila*). Thus the gene was dubbed "*toll*." TLRs in mammals

TABLE 23.3 Examples of Toll-like Receptors and NOD-like Receptors

Receptor	MAMPs recognized	Source	Host cells	Location
TLR1	Lipopeptides	Bacteria	Monocytes/macrophages, dendritic cells, B cells	Cell surface
TLR2	Glycolipids, lipoteichoic acids, viral capsid	Bacteria, viruses	Monocytes/macrophages, dendritic cells, mast cells	Cell surface
TLR3	Double-stranded RNA	Viruses	Dendritic cells, B cells	Cell compartment
TLR4	Lipopolysaccharide, heat-shock proteins	Bacteria	Monocytes/macrophages, dendritic cells, mast cells, intestinal epithelium	Cell surface
TLR5	Flagellin	Bacteria	Monocytes/macrophages, dendritic cells, intestinal epithelium	Cell surface
TLR6	Diacyl lipopeptides	*Mycoplasma*	Monocytes/macrophages, mast cells, B cells	Cell surface
TLR9	Unmethylated CpG (cytosine-phosphate-guanine) residues in DNA	Bacteria	Monocytes/macrophages	Cell compartment
NOD 1	Component of Gram-negative peptidoglycan	Gram-negative bacteria	Many cell types	Inflammasome
NOD 2	Peptidoglycan component	Bacteria	Macrophages, dendritic cells, epithelia of lung and GI tract	Inflammasome
NLRP-3	Peptidoglycan	Bacteria	Many cell types	Inflammasome
NLRP-4	Flagellin, CpG, ATP, dsRNA	Bacteria	Many cell types	Inflammasome

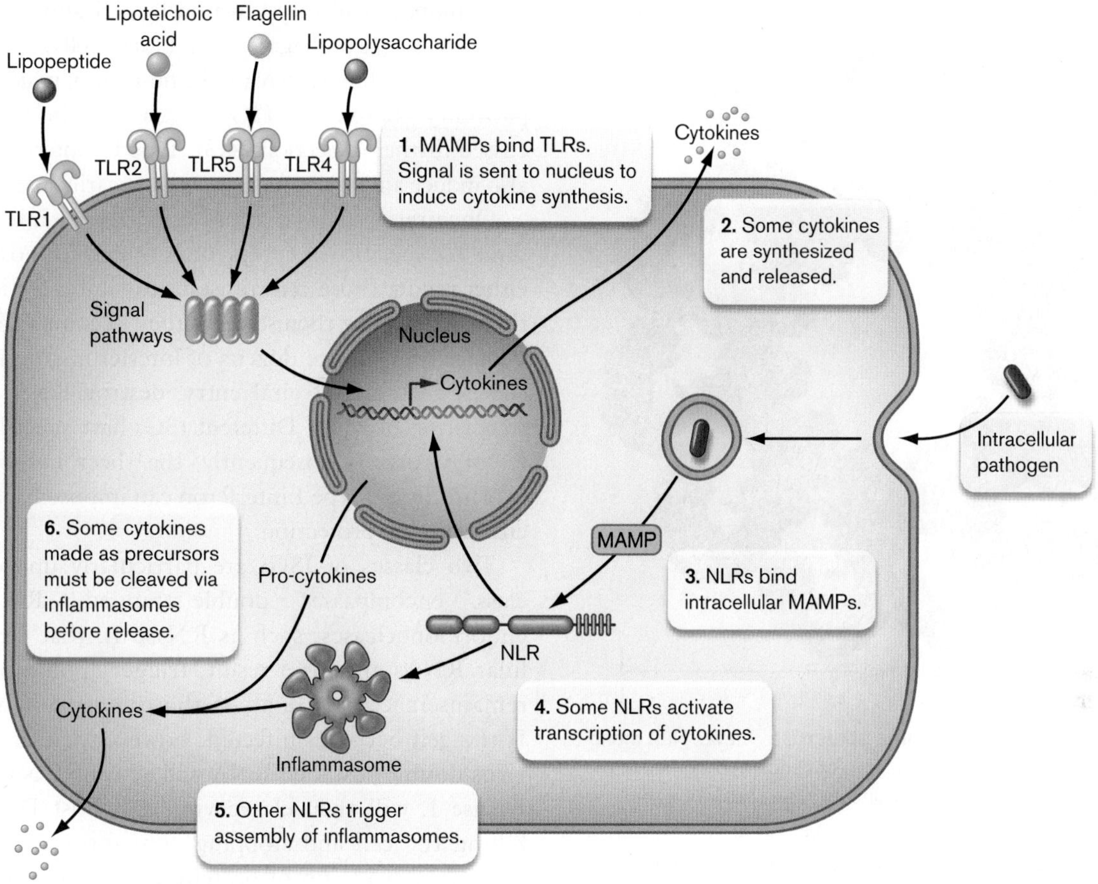

FIGURE 23.32 ■ TLRs, NLRs, and inflammasomes. Stimulation of TLRs and some NLRs activates transcription factors to induce production of cytokines and other factors. Some cytokines are directly released (for instance, IL-1, IL-8). Others are made as inactive precursors (procytokines) that must be cleaved by inflammasomes. Inflammasome assembly is orchestrated by certain NLRs.

display sequence similarities to the *toll* genes involved with insect embryogenesis (Nüsslein-Volhard was co-winner of the 1995 Nobel Prize in Physiology or Medicine for her research on the control of embryonic development).

TLRs are transmembrane proteins found mostly on the cell surface, although some are found in endosomes (**Table 23.3**). TLRs have an extracellular (or intravesicular) MAMP-binding domain and an intracellular Toll/interleukin 1 receptor domain (TIR domain). Humans have numerous Toll-like receptors, each of which recognizes different MAMPs present on pathogenic microorganisms, making them an innate defense mechanism with some degree of specificity. For example, TLR2 binds to lipoarabinomannan from mycobacteria, zymosan from yeasts, lipopolysaccharide (LPS) from spirochetes, and peptidoglycan. TLR4, on the other hand, binds LPS from Gram-negative bacteria, as well as host proteins released at sites of infection (for example, heat-shock protein 60). CD14, another host cell-surface protein, serves as a coreceptor for LPS. Note that these receptors bind fragments of structures after they are released from the microbe. They don't interact with the whole organism.

Once bound to a MAMP, the TLRs trigger an intracellular transcription regulatory cascade via their TIR domain, causing the host cell to make and release cytokines that diffuse away from the site, bind to receptors on various cells of the immune system (see **Table 23.3**), and direct them to engage the invader. Cytokines are discussed further here and in Chapter 24. The cells that respond to cytokines can be part of innate immunity, adaptive immunity, or both. TLR recognition of MAMPs can also trigger autophagy in infected cells. Bruce Beutler and Jules Hoffman shared the 2011 Nobel Prize in Physiology or Medicine for their work on the role of TLRs in immunity.

NOD-like receptors. While TLRs are important sensors of external MAMPs (or of MAMPs inside endosomes), **NOD-like receptor** (**NLR**) proteins are important cytoplasmic sensors of MAMPs (see **Table 23.3**). NLRs are structurally similar to a family of plant proteins called NODs (nucleotide-binding oligomerization domains) that provide resistance to pathogens. In mammals, some NLRs bound to a MAMP send a signal to the nucleus to activate cytokine production. Other NLRs become part of large, intracellular, multimeric, disklike complexes of proteins, called **inflammasomes**, on which a key protease called caspase-1 oligomerizes (**Fig. 23.33**). When activated, caspase-1 processes (cleaves) inactive cytokine precursors (procytokines) to make a smaller but active cytokine

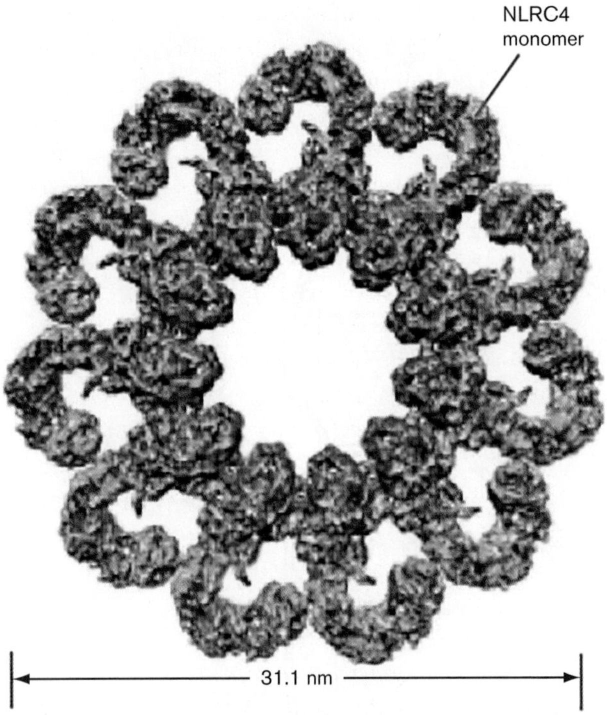

FIGURE 23.33 ■ Inflammasome structure. Eleven NLR monomers form an inflammasome. Caspase-1 (not shown) oligomerizes at the base. *Source:* Modified from Zhang et al. 2015. *Science* **350**:404–409.

(for instance, IL-1β and IL-18). The cytokine array that is ultimately secreted at the site of infection stimulates inflammation and activates adaptive immune mechanisms.

Inflammasome assembly starts with a single NOD-like receptor (NLR) protein binding to its cognate MAMP. The NLR-MAMP initiates the circular assembly of 10–11 monomeric blades of another NLR to form a disk. Caspase-1 protease oligomerizes at the base of the disk.

Type I Interferons Are "Intruder-Alert" Cytokines

When a community is threatened by a thief, a neighbor who has been robbed alerts others to take precautions. Something similar happens during viral infections. In 1957, it was discovered that cells exposed to inactivated viruses produce at least one soluble factor that can "interfere" with viral replication when applied to newly infected cells. The term **interferon** was coined to represent these molecules. Interferons are low-molecular-weight cytokines (14–20 kDa) produced by many eukaryotic cells in response to intracellular infection. The action of interferons is usually species specific but virus nonspecific; that is, an interferon from mice works only on mice but will protect mice from different viruses.

There are three general types of interferons, which differ in the receptors they bind and the responses they generate. Type I interferons are made by almost all cells in the body and have high antiviral potency. Type I interferons include IFN-alpha (IFN-α), IFN-beta (IFN-β), and others (κ, ε, ω). The lone type II interferon, IFN-gamma (IFN-γ), has more of an immunomodulatory function by activating various white blood cells (macrophages, natural killer cells, and T cells) to, among other things, increase the number of **major histocompatibility complex** (**MHC**) antigens on their surfaces (discussed further in Section 24.3). Type III interferons (IFN-λ) also induce antiviral responses, but primarily in epithelial cells.

Once released by infected cells, type I interferons will bind to specific receptors on uninfected host cells and either render those cells resistant to viral infection or cause them to sacrifice themselves if they become infected. Type I interferons induce dozens of interferon-stimulated genes (ISGs) that inhibit viral entry, destroy RNA, or prevent translation of RNA. Different ISGs have specificity for different viruses. Consequently, the sheer number of ISGs explains how a type I interferon can impose broad, nonspecific antiviral protection.

Two classes of ISGs are particularly important. One class encompasses double-stranded, RNA-activated endoribonucleases, such as RNase L, that cleave all cellular RNAs and, as a result, trigger apoptosis. RNase L remains inactive as long as the cell remains uninfected. If the cell becomes infected, however, the appearance of virus double-stranded RNA will activate RNase L. Active RNase L will then cleave viral and host DNA, thereby killing the cell and stopping viral spread. Another class of ISGs is made up of protein kinases that become activated by binding viral double-stranded RNA. One of the activated protein kinases phosphorylates eukaryotic initiation factor 2 (eIF2), rendering ribosomes unable to translate viral or cellular RNA. These mechanisms can affect both RNA and DNA viruses because protein synthesis is required for the propagation of all viruses. However, RNA viruses will induce more interferon I than DNA viruses induce. Type I interferons are also used medically to treat certain viral infections (for example, chronic hepatitis C).

Type II interferon functions by activating various white blood cells—for example, macrophages, natural killer cells, and T cells—to, among other things, increase the number of MHC antigens on their surfaces. MHC proteins are important for recognizing self and for presenting foreign antigens to the adaptive immune system. They are discussed more fully in Chapter 24.

Acute Inflammation

In medicine, the term "acute" means "rapid onset." Acute inflammation, for instance, is inflammation that develops rapidly, usually within a day, after a foreign object such as a splinter or infectious agent is introduced into the body. Chronic inflammation develops over long periods of time. The function of acute inflammation is to wall off, kill, or digest the intruder, thereby quickly resolving an infection and promoting healing of the affected area. Neutrophils play a central role in the inflammatory process.

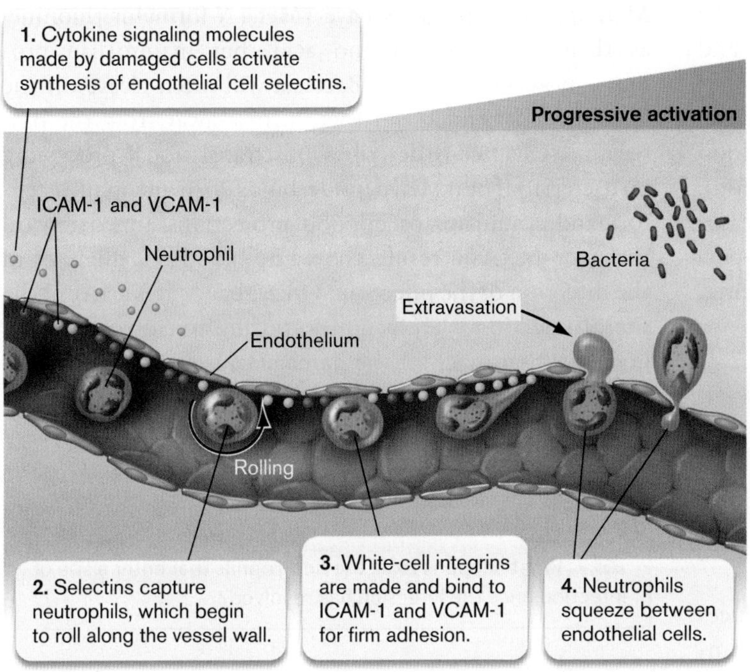

FIGURE 23.34 ■ Mechanism of extravasation. Extravasation is the process by which neutrophils move from the bloodstream into surrounding tissues.

Neutrophils, however, normally reside in the circulation. How do they find the infection site? Certain cytokines synthesized after MAMPs bind to PRRs (pattern recognition receptors) are called **chemokines** (examples are IL-8 and monocyte chemoattractant protein, MCP-1). Chemokines diffuse from a site of infection toward nearby capillaries and form a concentration gradient. Then, like a fox following a prey's scent, neutrophils and macrophages leave the bloodstream and follow the gradient back to the infection. But how do these white blood cells pass through blood vessel walls?

The events leading to extravasation begin when the cytokines **interleukin 1 (IL-1)** and **tumor necrosis factor alpha (TNF-alpha)**, released by macrophages, stimulate the production of adhesion molecules (**selectins**) on the inner lining of the capillaries (**Fig. 23.34**, step 1 ▶). P-selectin is produced first, followed by E-selectin. The selectins snag neutrophils zooming by in the bloodstream, slow them down, and cause them to roll along the endothelium (step 2). Rolling neutrophils that encounter inflammatory mediators are activated to produce and display integrin adhesion molecules on their surface (step 3). Integrins on neutrophils lock onto the endothelial adhesion molecules ICAM-1 (intercellular adhesion molecule 1) and VCAM-1 (vascular cell adhesion molecule 1). Binding to these endothelial adhesion molecules stops the neutrophils from rolling and initiates extravasation in which the white blood cells squeeze through the endothelial wall and into the tissues (step 4).

The passage of neutrophils through vascular walls also requires loosening adhesions between endothelial cells. Vasoactive factors such as **bradykinin**, a nine-amino-acid polypeptide released by damaged tissue cells and macrophages, increase vascular permeability so that neutrophils can pass through vessel walls (**Fig. 23.35**, step 5). Increased vascular permeability also allows blood plasma to escape into tissues, causing swelling (edema). Even though tight junctions are loosened, to actually pass between endothelial cells activated neutrophils need the enzyme sialidase to temporarily break carbohydrate linkages that hold tight junctions together. Bradykinin also triggers degranulation of mast cells.

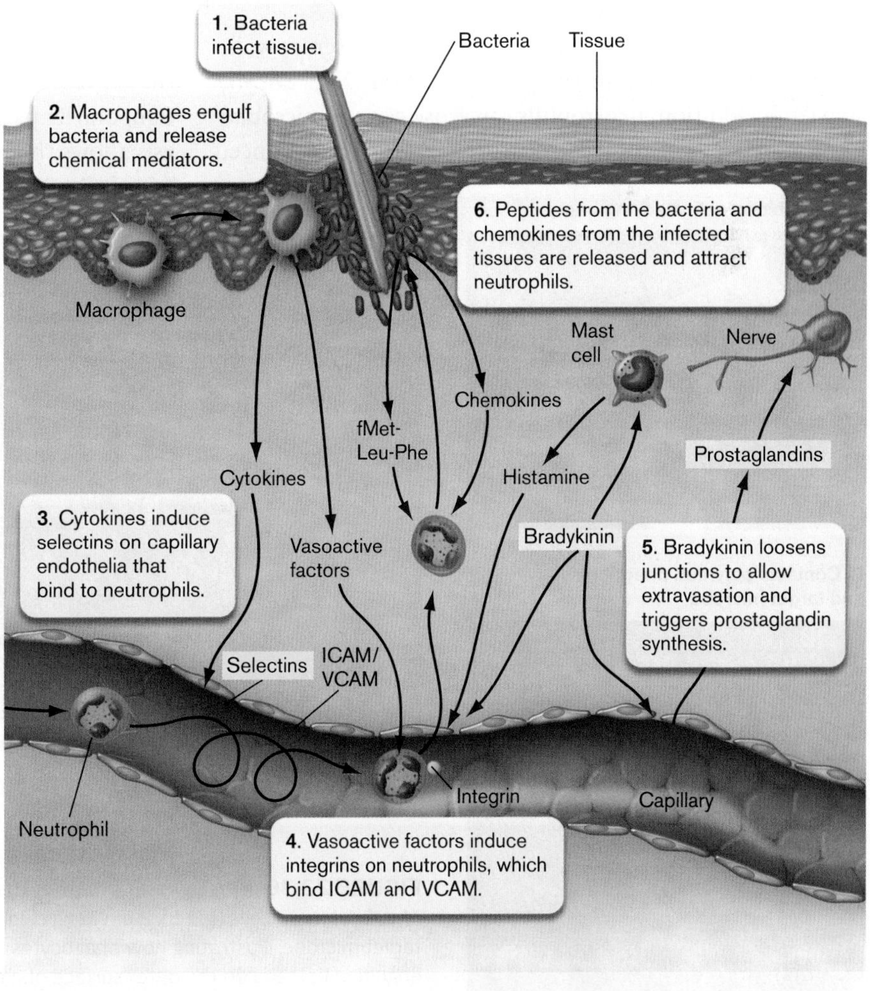

FIGURE 23.35 ■ Summary of the inflammatory response. Steps 1–4 are the same as in Figure 23.31.

The histamine released from mast cells further loosens the endothelial cell junctions. More fluid enters tissues and accumulates.

In addition to increasing vascular permeability, bradykinin and histamine relax smooth muscles within blood vessel walls so that vessel diameter increases (vasodilation). Vasodilation slows blood flow and, as a result, increases blood volume in the affected area. Increased vascular permeability and vasodilation cause the localized swelling, redness, and heat associated with inflammation.

Thought Question

23.6 As illustrated in **Figure 23.34**, integrin is important for neutrophil extravasation. Some individuals, however, produce neutrophils that lack integrin. What is the likely consequence of this genetic disorder?

What causes the pain of inflammation? Bradykinin induces capillary cells to make prostaglandins that cause pain by stimulating nerve endings in the area (**Fig. 23.35**, step 5). A key enzyme involved in prostaglandin synthesis is cyclooxygenase (COX). Aspirin, ibuprofen, and the anti-inflammatory agent naproxen are COX inhibitors that prevent the synthesis of prostaglandins and thus reduce inflammatory pain.

Once neutrophils have passed through the vascular wall, chemokines lure them to the proper location (**Fig. 23.35**, step 6). In addition, neutrophils can sense certain microbial chemoattractants—fMet-Leu-Phe peptide, for instance. Many bacterial proteins have fMet (*N*-formylmethionine) as their N-terminal amino acid, but mammalian proteins, in general, do not. Bacteria will often cleave off the fMet peptide, which can then diffuse away from the bacterium. (Chapter 8 describes posttranslational processing in bacteria.) The fMet peptide binds to neutrophil receptors and stimulates pseudopod projections aimed toward the microbe. As a result, the white blood cell migrates in the direction of the infection. Once phagocytes arrive at the site of infection, they begin devouring microbes. Realize, however, that much of the damage caused by an infection is not due directly to the microbe but is the result of the body's inflammatory reaction to the microbe's presence.

Thought Question

23.7 What happens to all the neutrophils that enter a site of infection once the infection has resolved?

Phagocytes Recognize Alien Cells and Particles

For phagocytosis to proceed safely, macrophages and neutrophils must first recognize the surface of a particle as foreign (**Fig. 23.36**). When a phagocyte surface interacts with the surface of another body cell, the phagocyte becomes temporarily paralyzed (unable to form pseudopods). Paralysis allows the phagocyte to evaluate whether the other cell

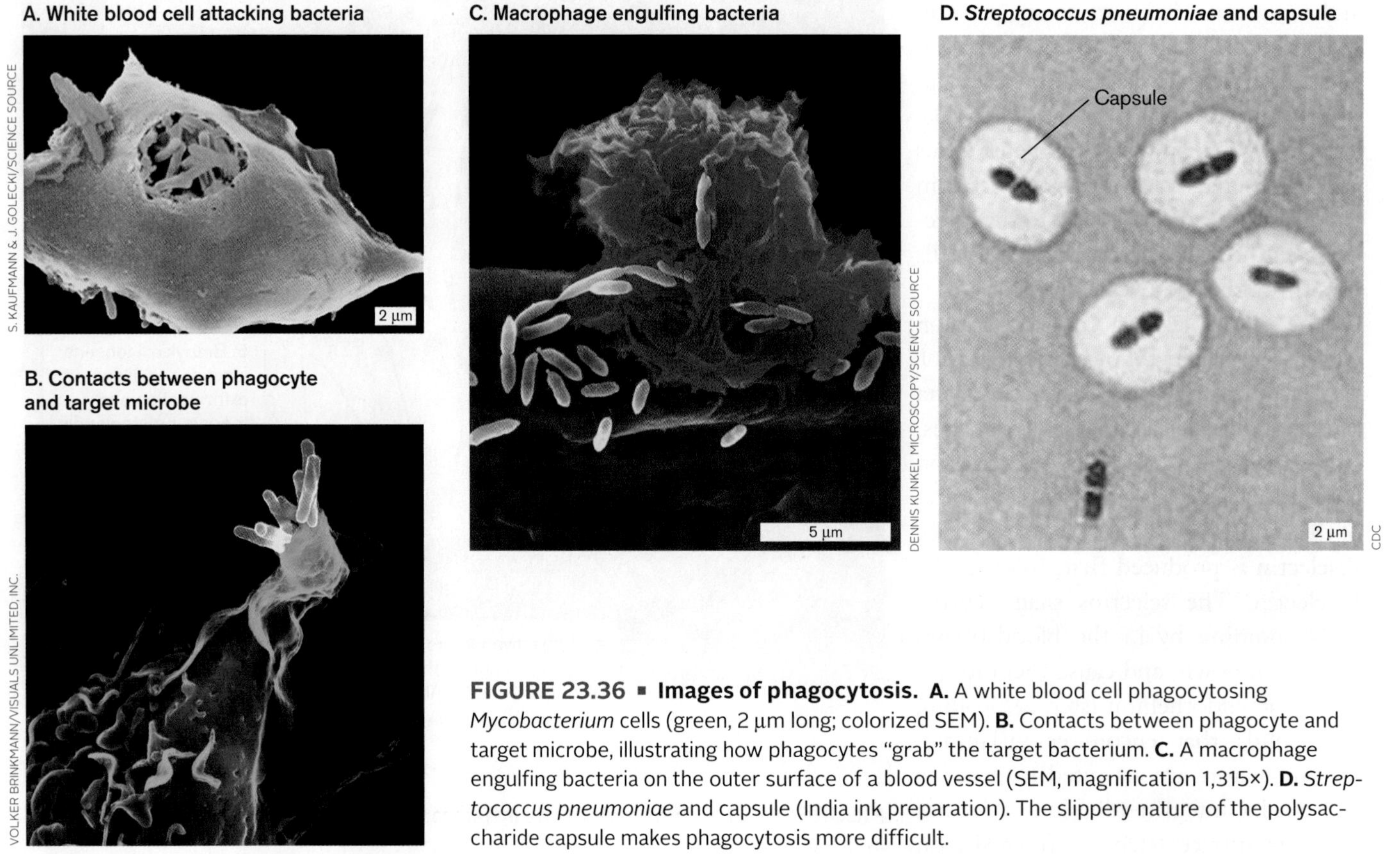

FIGURE 23.36 ■ Images of phagocytosis. A. A white blood cell phagocytosing *Mycobacterium* cells (green, 2 µm long; colorized SEM). **B.** Contacts between phagocyte and target microbe, illustrating how phagocytes "grab" the target bacterium. **C.** A macrophage engulfing bacteria on the outer surface of a blood vessel (SEM, magnification 1,315×). **D.** *Streptococcus pneumoniae* and capsule (India ink preparation). The slippery nature of the polysaccharide capsule makes phagocytosis more difficult.

is friend or foe, self or nonself. To recognize self, glycoproteins located on the white blood cell membrane must bind to inhibitory glycoproteins present on other host cell membranes. The inhibitory glycoprotein on human cells is called CD47. Because invading bacteria lack these inhibitory surface molecules, they can readily be engulfed.

Although many bacteria, such as *Mycobacterium* or *Listeria* species, are easily recognized and engulfed by phagocytosis (**Fig. 23.36A–C**), others, such as *Streptococcus pneumoniae*, possess polysaccharide capsules that are too slippery for pseudopods to grab (**Fig. 23.36D**). This is where innate immunity and adaptive immunity join forces. Adaptive immunity produces antibodies that bind to bacterial capsules. These anticapsular antibodies can aid the innate immune mechanism of phagocytosis through a process known as **opsonization** (**Fig. 23.37**). An **opsonin** is any factor, such as antibodies, that can promote phagocytosis. The anticapsular antibodies coat the surface of the bacterium, leaving the tail end of the antibodies, called the Fc region (see Section 24.2), pointing outward. These bacteria are said to be "opsonized." The Fc regions of these antibodies are recognized and bound by specific receptors on phagocyte cell surfaces. As a result, the antibodies stick the opsonized bacteria to phagocytic Fc receptors, thereby helping the phagocyte to more easily engulf the invader.

Thought Question

23.8 If NK cells can attack infected host cells coated with antibody, can neutrophils do the same?

Oxygen-Independent and -Dependent Killing Pathways

During phagocytosis, the cytoplasmic membrane of the phagocyte flows around the bacterium and then engulfs it, producing an intracellular phagosome, as described earlier (see **Fig. 23.19**). Subsequent phagosome-lysosome fusion (producing a phagolysosome) results in both oxygen-independent and oxygen-dependent killing pathways. Mechanisms independent of oxygen include enzymes like lysozyme to destroy the cell wall; compounds such as lactoferrin to sequester iron away from the microbe; and defensins, small cationic antimicrobial peptides (described in Section 23.4).

Oxygen-dependent mechanisms are activated through Toll-like receptors (discussed earlier) and kill by producing various oxygen radicals that can damage macromolecules. NADPH oxidase, myeloperoxidase, and nitric oxide synthetase in the phagosome membrane are extremely important. NADPH oxidase yields superoxide ion ($^{\bullet}O_2^-$), and superoxide dismutase converts $^{\bullet}O_2^-$ to hydrogen peroxide (H_2O_2). Ferrous ion (Fe^{2+}) reacts with H_2O_2 to produce hydroxyl radicals ($^{\bullet}OH$) and hydroxide ions (OH^-). Myeloperoxidase, present only in neutrophils, converts hydrogen peroxide and chloride ions to hypochlorous acid (HOCl) and OH^-.

$$NADPH + 2O_2 \xrightarrow[\text{oxidase}]{\text{NADPH}} NADP^+ + H^+ + 2^{\bullet}O_2^- \xrightarrow[\text{dismutase}]{\text{superoxide}}$$

$$H_2O_2 + Cl^- \xrightarrow{\text{myeloperoxidase}} OH^- + HOCl$$

Macrophages, mast cells, and neutrophils also generate reactive nitrogen intermediates that serve as potent cytotoxic agents. Nitric oxide (NO) is synthesized from arginine by NO synthetase. Further oxidation of NO by oxygen yields nitrite (NO_2^-) and nitrate (NO_3^-) ions.

All of these reactive oxygen and nitrogen species attack bacterial membranes and proteins. The mechanisms generating these molecules greatly increase oxygen consumption during phagocytosis and produce what is called the **oxidative burst**. The reactive chemical species formed during the oxidative burst do little to harm the phagocyte because

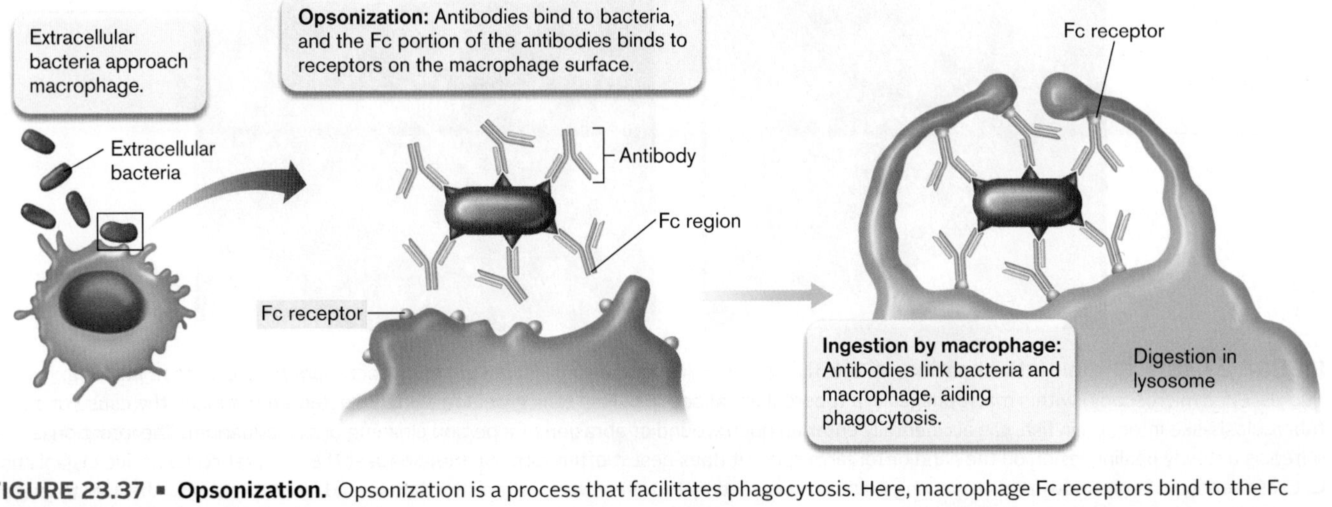

FIGURE 23.37 ■ Opsonization. Opsonization is a process that facilitates phagocytosis. Here, macrophage Fc receptors bind to the Fc region of antibodies attached to bacteria.

the burst is limited to the phagosome (where the bacteria are) and because the various reactive oxygen species, such as superoxide, are very short-lived. Although phagocytes are very good at clearing infectious agents, many bacteria have developed ways to outsmart this aspect of innate immunity.

Autophagy and Intracellular Pathogens

Intracellular pathogens that grow in eukaryotic cytoplasm can be a serious problem for the host. Many pathogens, such as *Mycobacterium tuberculosis*, the cause of tuberculosis, enter the host cell in ways that bypass endosome formation or, if they do enter via an endosome, can escape from that compartment. These pathogens block normal host cell clearance pathways. To circumvent this problem, eukaryotic host cells (not just phagocytes) use a process that normally degrades damaged organelles (called **autophagy**) to clear themselves of intracellular pathogens. During autophagy, the cell constructs a double membrane around the organism (or damaged organelle). This structure, called the autophagosome, sequesters the microbe from the nutrient-rich cytosol. Lysosomes then fuse with the autophagosome, depositing degradative enzymes that digest the organism. Ever-adapting intracellular microbes, however, have found ways to suppress autophagy and survive.

Chronic Inflammation Causes Permanent Damage

Inflammation that persists over months or years is called **chronic inflammation** and is provoked by the long-term presence of a causative stimulus. Chronic inflammation inevitably causes permanent tissue damage, even though the body attempts repair. The causes of chronic inflammation are many. For example, infectious organisms such as *Mycobacterium tuberculosis*, *Actinomyces bovis*, and various protozoan parasites can avoid or resist host defenses (**Fig. 23.38**). As a result, they persist at the site and continually stimulate the basic inflammatory response. The continual stimulation of an inflammatory response leads to chronic inflammation.

Nonliving, irritant material like wood splinters, inhaled asbestos particles, or surgical implants can also cause

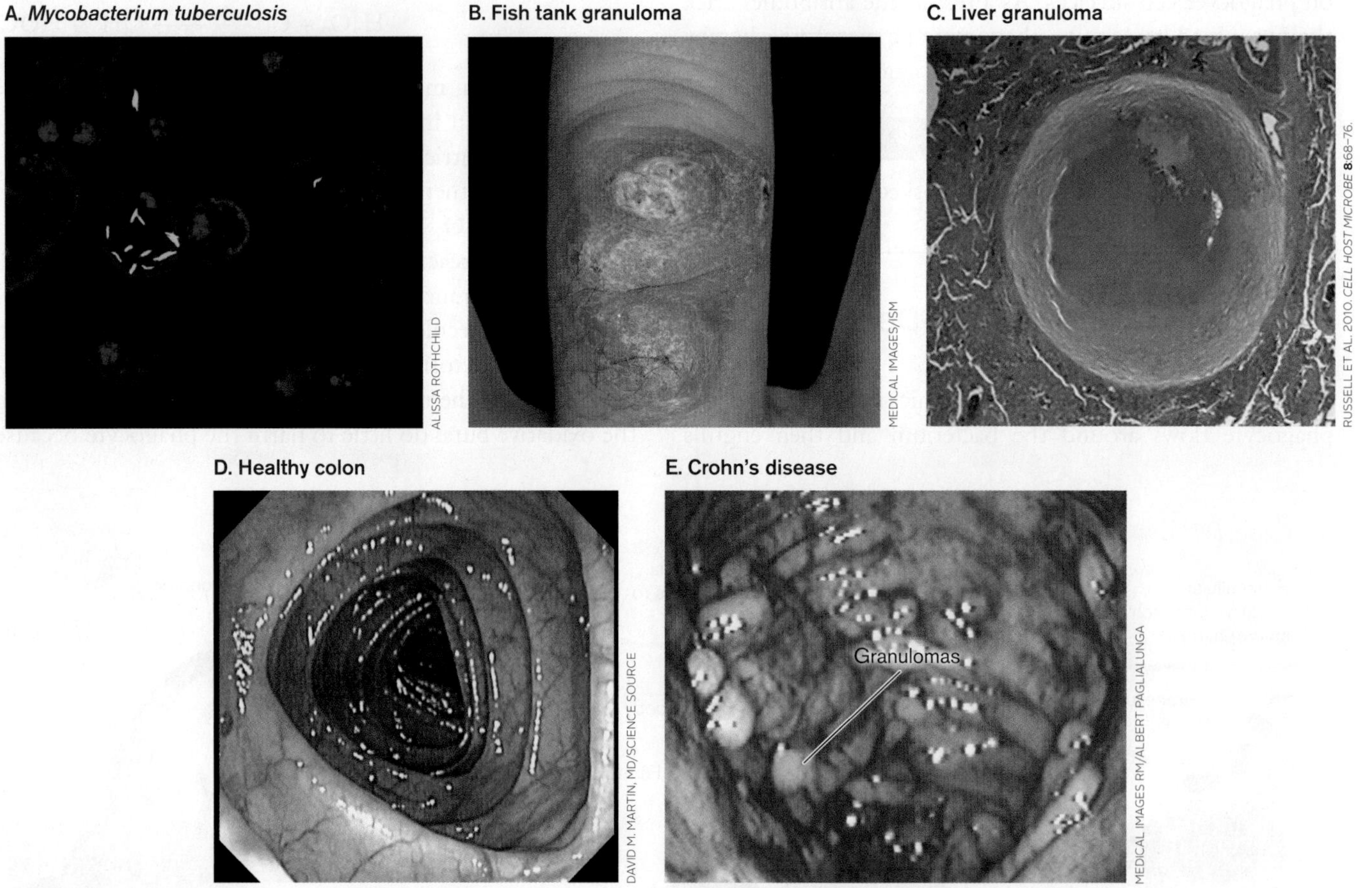

FIGURE 23.38 ▪ Chronic inflammation. A. The fluorescent green organisms shown are *Mycobacterium tuberculosis* (2 μm long; fluorescence microscopy) within macrophages in a tuberculosis abscess. **B.** Fish tank granuloma. *Mycobacterium marinum*, the cause of a tuberculosis-like infection in fish, can accidentally enter an open wound or abrasion of a person cleaning out an aquarium. The infection is first noted as a slowly healing lesion on the hand or forearm. Once it does heal, it often forms a granuloma at the site that contains live organisms. **C.** Cross section of liver showing a necrotizing granuloma caused by *M. tuberculosis* organisms that spread from the lung to the liver through the bloodstream. **D.** Healthy colon. **E.** Intestinal granulomas of Crohn's disease.

chronic inflammation. Autoimmune diseases are another important cause. Autoimmunity (reaction against self) occurs when there is a failure to regulate some aspect of adaptive immunity (see Chapter 24). As a result, the immune system recognizes a part of the body as foreign (not self) and begins to react against it. Rheumatoid arthritis is one example of the body attacking itself.

Whatever causes chronic inflammation, macrophages and lymphocytes are continually recruited from the circulation. The body may attempt to "wall off" the site of inflammation by forming a **granuloma**. A granuloma begins as an aggregation of the mononuclear inflammatory cells surrounded by a rim of lymphocytes. The body then deposits fibroconnective tissue around the lesion, causing tissue hardening known as fibrosis.

Several forms of granulomas are shown in **Figure 23.38**. *M. tuberculosis*, for example, has a thick, waxy cell wall that protects mycobacteria against the mechanisms used by macrophages to destroy microorganisms. As a result, the organisms live for prolonged periods within macrophages (**Fig. 23.38A**). Resistance to host defenses can produce long-term chronic infections and granulomas. For example, skin infections caused by *Mycobacterium marinum* will produce skin granulomas (**Fig. 23.38B**), while *M. tuberculosis* can cause liver granulomas (**Fig. 23.38C**). Another disease thought to involve granuloma formation is Crohn's disease, which commonly manifests as abdominal pain, frequent bowel movements, and rectal bleeding. Crohn's disease has been attributed to an autoimmune reaction possibly activated by intestinal microbiota. In this case, the intestinal bacteria are thought to cause a chronic inflammation resulting in characteristic granulomas (compare **Fig. 23.38D** and E).

To Summarize

- **Acute inflammation begins** when host cells are damaged or infected. The process walls off and digests invading microbes, and initiates the healing of affected areas. Damaged cells or tissue macrophages release vasoactive factors that produce vasodilation and increased vascular permeability, cytokines that stimulate the production of blood vessel selectin receptors, and chemoattractant molecules that cause neutrophil movement (extravasation) from the bloodstream into infected tissues. **Bradykinin** causes the release of prostaglandins, which produce pain in the affected area. The cardinal signs of inflammation include redness, warmth, swelling, pain, and loss of function.
- **Surface Toll-like receptors (TLRs)** and cytoplasmic **NOD-like receptors (NLRs)** in most host cells recognize microbe-associated molecular patterns (MAMPs) and synthesize cytokine proteins that diffuse and activate other cells of the immune system.
- **Interferons are one group of cytokines** that can nonspecifically interfere with viral replication (type I and type III) or modulate the immune system (type II).
- Phagocytic cells have **oxygen-independent and oxygen-dependent mechanisms of killing** that are initiated by the fusion of lysosomes and bacteria-containing phagosomes.
- **The oxidative burst**, a large increase in oxygen consumption during phagocytosis, results in the production of superoxide ions, nitric oxide, and other reactive oxygen species.
- **Autophagy** is a process by which intracellular bacteria can be sequestered from the cytoplasm (via an autophagosome) and killed following fusion with a lysosome.
- **Chronic inflammation** results from the persistent presence of a foreign object.

23.6 Complement and Fever

White blood cells (WBCs) engulf and kill pathogens, but can simple serum proteins also kill microbes? Yes, a series of 20 serum proteins (complement factors) that make up the complement cascade can also attack bacterial invaders. Complement was first discovered as a heat-labile component of blood that enhances (or complements) the killing effect of antibodies on bacteria. Several complement factors are proteases that sequentially form and cleave other complement factors. (The liver is the main source of complement proteins.) Once a complement cascade is triggered, several things happen. Pores are inserted into bacterial membranes, causing cytoplasmic leaks, while pieces of some complement proteins attract WBCs and facilitate phagocytosis (opsonization). Complement has also been implicated in the killing of pathogens trapped by neutrophil NETs, as described earlier (Section 23.3).

Complement Activation Pathways

The three routes to complement activation are officially known as the classical pathway, the alternative pathway, and the lectin pathway. The classical complement pathway is discussed in the next chapter (Section 24.4) because triggering the pathway depends on antibody, so it is part of both adaptive immunity and innate immunity. The lectin pathway requires the synthesis of mannose-binding lectin by the liver in response to certain macrophage cytokines. Lectin coats the surfaces of invading microbes and activates complement without needing antibody. However, the lectin pathway has more in common with the classical complement pathway and is also described in Chapter 24. We focus here in Chapter 23 on the alternative pathway because this pathway does not involve adaptive immunity and can attack invading microbes long before a specific immune response can be launched.

The alternative complement pathway begins with the complement factor C3. In blood, C3 slowly cleaves into C3a and C3b (**Fig. 23.39**, step 1). C3b, under normal circumstances, is rapidly degraded—a process that thwarts inadvertent complement activation. However, if C3b meets LPS on an invading Gram-negative microbe, the bound C3b becomes stable and binds another factor, designated factor B (step 2), and makes factor B susceptible to cleavage by yet another protein, factor D (step 3). The resulting complex, called C3bBb, has two roles: It can quickly cleave more C3 to amplify the cascade and is changed by another serum protein (properdin) into what is called C5 convertase (step 4).

From this point on, all complement pathways are identical. C5 convertase cleaves C5 in serum to C5a and C5b (**Fig. 23.39**, step 5), and C5b then forms a prepore complex by binding to C6 and C7 (step 6). The resulting C5bC6C7 complex binds to target membranes. Finally, C8 and C9 factors join in to form the **membrane attack complex** (**MAC**), becoming a destructive pore (step 7).

In Gram-negative bacteria, MAC pores first form in the outer membrane (**Fig. 23.39**, step 7). Lysozyme (present in serum) enters through the MAC outer membrane pores and cleaves peptidoglycan, making the cytoplasmic membrane more susceptible to the membrane attack complex. The inner membrane MAC pores destroy membrane integrity and, with it, proton motive force. Gram-positive bacteria are resistant to complement because they lack an outer membrane (and therefore have no LPS to efficiently start the cascade) and have a thick peptidoglycan layer that hinders access of complement components. Even in the absence of LPS, however, there are ways to activate complement that involve antigen-antibody complexes (see Section 24.4).

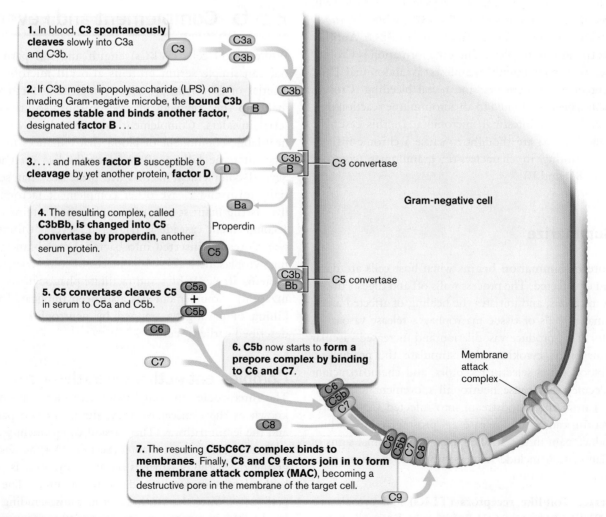

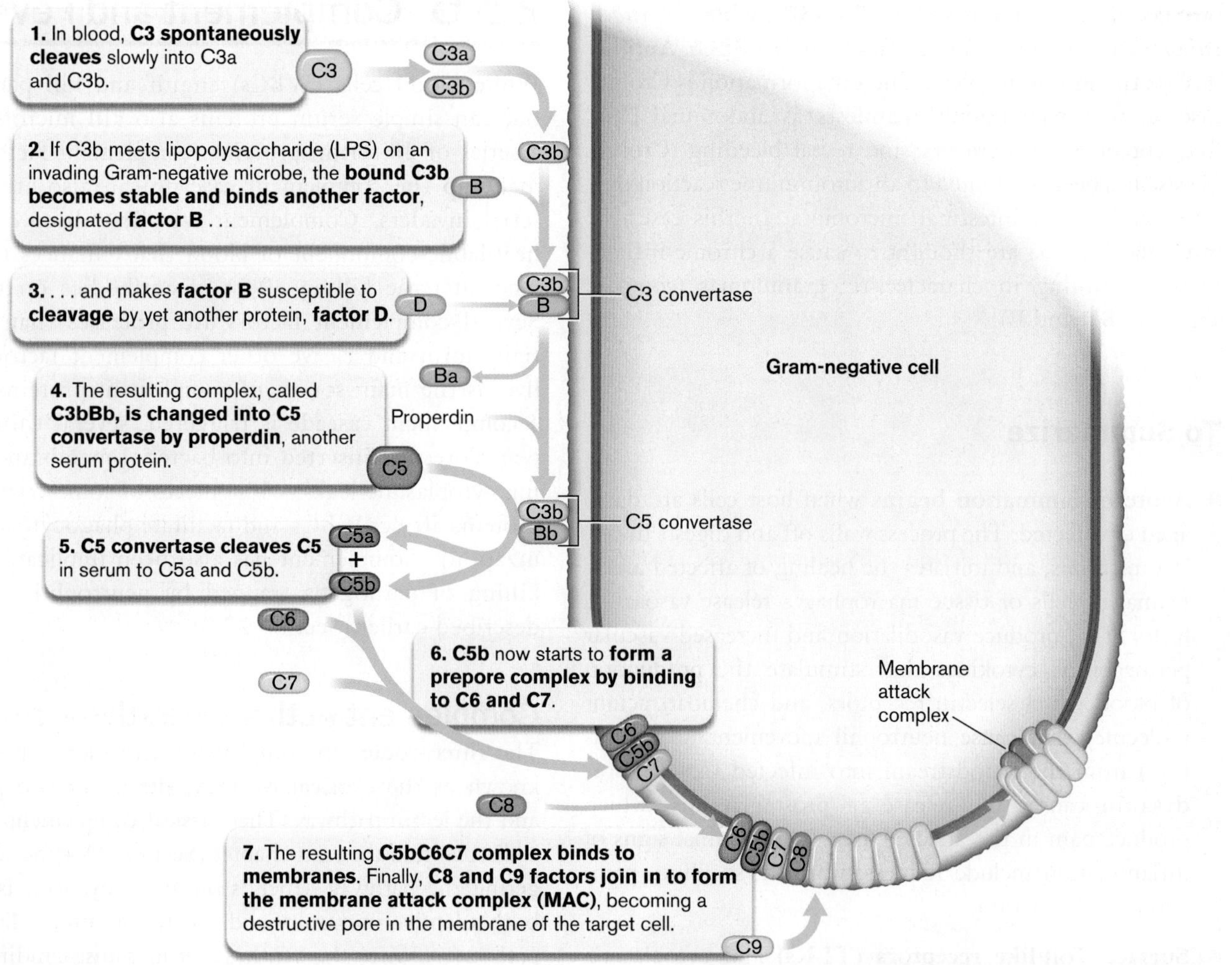

FIGURE 23.39 ▪ The alternative complement pathway. Although called "alternative," this complement cascade is part of the first-line innate defense system.

Thought Question

23.9 **Figure 23.39** shows how the complement cascade can destroy a bacterial cell. Factor H (not shown) is a blood protein that regulates complement activity. Factor H binds host cells and inhibits complement from attacking our cells by accelerating degradation of C3b and C3bBb. How could bacteria take advantage of factor H?

Other Roles for Complement Peptides in Innate Immunity

Factor C3b, in addition to initiating the complement cascade, is a potent opsonin. As described earlier, an opsonin is any factor that can promote phagocytosis. PMNs (neutrophils) have specific C3b receptors on their surface. Thus, when C3b binds to a bacterial cell surface, it tags that cell and makes it easier for PMNs to grab and engulf the organism.

The complement fragments C5a (which forms part of the prepore complex) and C3a have many roles in immune function. They are anaphylatoxins, which trigger degranulation of vasoactive factors such as histamine from endothelial cells, mast cells, or phagocytes. They can also stimulate chemotaxis of immune cells. C5a and C3a bind to separate receptors on WBC plasma membranes and activate separate signal cascade pathways. C5a triggers Ca^{2+} release from intracellular stores and stimulates the actin polymerization needed for cell migration. Conversely, C3a triggers Ca^{2+} influx, which facilitates extravasation. These peptide mediators also stimulate the release of certain cytokines, such as IL-4 from monocytes and mast cells, that prepare capillary endothelium for the rolling and adhesion of neutrophils. C5a will also up-regulate P-selectin and ICAM-1. Thus, some complement factors not only help directly destroy target bacterial cells, but also facilitate phagocytosis and contribute directly to inflammation.

Acute-Phase Reactants and Complement

As noted earlier, inflammation is associated with the production of various cytokines by macrophages. Some of these cytokines (such as IL-1, TNF-alpha, and IL-6) travel to the liver, where they stimulate synthesis of several so-called acute-phase reactant proteins, including **C-reactive protein**. Acute-phase reactants circulate in the bloodstream and can be detected during laboratory testing. Named for its ability to activate complement, C-reactive protein will bind to components of bacterial cell surfaces but not to host cell membranes. Once linked to the bacterial cell surface, C-reactive protein will bind complement factor C1q of the classical complement pathway (see Chapter 24), ultimately converting factor C3 to C3b, propagating the complement cascade. C-reactive protein accelerates C3b production at the bacterial surface, where it can do the most damage. Note that an elevated level of C-reactive protein in serum is considered a general indicator of inflammation, not just inflammation caused by infection.

Fever

What is fever, and why is it a good thing? To understand fever, we first need to know how the body controls its temperature (thermoregulation). Heat in a human body is produced as a consequence of metabolic reactions. The liver and muscles are the major generators of heat and will warm the blood that passes through them. Body heat is generated (thermogenesis) by lipolysis in certain adipose tissues through the combustion of lipid substrates and the ability to uncouple the electron transport chain in mitochondria from the synthesis of ATP. Protons then leak from a high concentration at the mitochondrial intermembrane space into the matrix, releasing heat energy as the protons make new bonds. In a healthy person, body temperature is kept between 36°C and 38°C (97°F and 100°F), despite large differences in surrounding temperature and physical activity. Fever is defined as an oral temperature above 100.4°F (38°C).

How do we normally maintain a body temperature of 37°C? Heat sensors located throughout the skin and large organs and along the spinal cord send information about the body's temperature to the thermoregulatory center in the hypothalamus, a small organ located in the brain near the brain stem. The hypothalamus acts as a thermostat by controlling lipolysis in adipose tissue, as well as blood flow through the skin and subcutaneous areas. Vasoconstriction (tightening of blood vessel diameter) prevents the release of body heat when we are cold, whereas vasodilation secures its quick release when we are hot. If skin temperature is too high, the hypothalamus directs vasodilation to accelerate heat release. If body temperature is too low, blood flow will decrease to conserve heat, and shivering begins as a way to generate more heat.

Fever is a natural reaction to infection and is usually accompanied by general symptoms such as sweating, chills, and the sensation of being cold. Substances that cause fever are known as **pyrogens**. Exogenous pyrogens (for example, certain bacterial toxins) originate outside the body, whereas internal or endogenous pyrogens (such as tumor necrosis factor and the interferon IL-6) are made by the body itself. External pyrogens generally cause fever by inducing the release of internal cytokine pyrogens.

Pyrogenic cytokines cross the blood-brain barrier and bind to neurons in the thermoregulatory center of the anterior hypothalamus. Cytokine-receptor interaction stimulates the production of phospholipase A2, an enzyme required to make prostaglandins. Prostaglandin E2 is made and changes the responsiveness of thermosensitive neurons. In other words, prostaglandin E2 turns up the thermostat. The body "thinks" it is cold, so the hypothalamus sends signals via the autonomic nervous system (which acts below the level of consciousness) to increase lipolysis and constrict peripheral blood vessels (vasoconstriction). Heat is not released and builds up to cause fever. Involuntary muscle contractions (shivering and chills) also generate heat.

What are the advantages of fever? Because the ideal growth temperature for many microbes is 37°C, elevated temperature can place the pathogenic organism outside its "comfort zone" of growth. There is also evidence that fever decreases iron availability to bacteria (cytokine release causes an increase in iron storage protein). Slower growth of the pathogen "buys" the immune system time to subdue the infection before it is too late. Consequently, interventions that reduce a moderate fever caused by infection may actually slow recovery.

Thought Question

23.10 If increased fever limits bacterial growth, why do bacteria make pyrogenic toxins?

To Summarize

- **Complement** is a series of 20 proteins naturally present in serum.
- **Activation of the complement cascade** results in a pore being introduced into target membranes.
- **The three pathways for activation** are the classical, alternative, and lectin pathways.
- **The alternative activation pathway** begins when complement factor C3b is stabilized by interaction with the LPS of an invading microbe.
- **The cascade of protein factors**, C3b ⟶ factor B ⟶ factor D ⟶ properdin ⟶ C5 ⟶ C6 ⟶ C7 ⟶ C8 and C9, results in the formation of a membrane attack complex in target membranes.
- **C-reactive protein** in serum is activated when bound to microbial structures and will convert C3 to C3b, which can start the complement cascade.
- **The hypothalamus** acts as the body's thermostat.
- **Exogenous and endogenous pyrogens** elevate body temperature by stimulating prostaglandin production.
- **Prostaglandins** change the responsiveness of thermosensitive neurons in the hypothalamus.

Concluding Thoughts

Immunology has been defined as the science of discriminating self from nonself. However, the human body plays host to a vast constellation of microbes (our microbiota) that are, immunologically speaking, nonself, and yet indispensable for human health. Our immune system has evolved to tolerate our microbiota (you might call them "frenemies") up to a point, but it will exact revenge if a member penetrates into tissues that must remain sterile. The various innate immune mechanisms described here in Chapter 23 provide an effective first line of defense against "impertinent" microbiota and potential pathogens. The next chapter addresses what happens when microbes breach these innate defenses. Unlike the general protective mechanisms just discussed, the system of adaptive immunity generates a molecular defense specifically tailored to a given pathogen, or member of the microbiome. Keep in mind that innate and adaptive immune systems "talk" to each other and collaborate to present the most effective response to a given organism, whether bacterial, viral, or parasitic.

CHAPTER REVIEW

Review Questions

1. Name some sterile body sites.
2. Which body sites are colonized by normal microbiota?
3. Under what circumstances can microbiota cause disease?
4. Why are commensal organisms beneficial to the host?
5. Name and describe various types of innate immunity.
6. What are probiotics? How do they help maintain health?
7. How do the lungs avoid being colonized?
8. What is a gnotobiotic animal?
9. Describe GALT and SALT.
10. Describe some chemical barriers to infection.
11. Discuss the different types of white blood cells.

12. What is a lymphoid organ?
13. Outline the process of inflammation.
14. Explain why phagocytes do not indiscriminately phagocytose body cells.
15. What is interferon?
16. Describe antibody-dependent cell-mediated cytotoxicity.
17. How does complement kill bacteria?
18. Why might fever be helpful in fighting infection?

Thought Questions

1. Is it common for microbial pathogens to pass the placental barrier (called transplacental transmission) and infect the fetus? Consider papillomavirus, *Listeria monocytogenes*, *Escherichia coli*, HIV, *Treponema pallidum*, *Neisseria gonorrhoeae*, and *Staphylococcus aureus*.
2. The vagina contains competing, commensal microbes that contribute to the health of the organ. So, is "cleaning" the vagina by douching actually unhealthy and likely to lead to an increased chance of infection (vaginosis)?
3. Why have microbes not altered their structures to avoid being recognized by Toll-like receptors?
4. The inflammatory response kills invading pathogens but can also damage bystander host cells and tissues. To prevent excessive damage, how does the host reset its innate immune system once an infection resolves?

Key Terms

adaptive immunity (936)
alveolar macrophage (947)
antibody-dependent cell-mediated cytotoxicity (ADCC) (943)
antigen (936)
antigen-presenting cell (APC) (942)
apoptosis (942)
autophagy (956)
B cell (944)
bacteremia (926)
basophil (939)
bradykinin (953)
C-reactive protein (959)
chemokine (949, 953)
chronic inflammation (956)
complement (937)
compromised host (923)
cystic fibrosis transmembrane conductance regulator (CFTR) (947)
cytokine (949)
defensin (947)
degranulate (948)
dendritic cell (939)
dysbiosis (931)
eosinophil (939)
epidermis (925)
extravasation (949)
gnotobiotic animal (933)
granuloma (957)
gut-associated lymphoid tissue (GALT) (946)
immune system (936)
inflammasome (951)
innate immunity (936)
interferon (952)
interleukin 1 (IL-1) (953)
Langerhans cell (946)
leukocyte (937)
lymph node (944)
lymphocyte (944)
M cell (946)
macrophage (939)
major histocompatibility complex (MHC) (942, 952)
mast cell (939)
membrane attack complex (MAC) (958)
microbe-associated molecular pattern (MAMP) (946)
microbiota (microbiome) (922)
monocyte (939)
mucociliary escalator (927)
nasopharynx (926)
natural killer (NK) cell (942)
neutrophil (937)
neutrophil extracellular trap (NET) (938)
NOD-like receptor (NLR) (951)
nonadaptive immunity (936)
opportunistic pathogen (923)
opsonin (955)
opsonization (955)
oropharynx (926)
oxidative burst (955)
pattern recognition receptor (PRR) (950)
perforin (942)
phagosome (938)
platelet (942)
probiotic (932)
pyrogen (959)
reticuloendothelial system (940)
selectin (953)
skin-associated lymphoid tissue (SALT) (946)
T cell (944)
tight junction (945)
Toll-like receptor (TLR) (950)
tumor necrosis factor alpha (TNF-alpha) (953)
vasoactive factor (949)

Recommended Reading

Azzouz, Louiza, Ahmed Cherry, Magdalena Riedl, Meraj Khan, Fred G. Pluthero, et. al. 2018. Relative antibacterial functions of complement and NETs: NETs trap and complement effectively kills bacteria. *Molecular Immunology* **97**:71–81.

Blaser, Martin. 2018. The past and future biology of the human microbiome in an age of extinctions. *Cell* **172**:1173–1177.

Bloes, Dominik Alexander, Dorothee Kretschmer, and Andreas Peschel. 2015. Enemy attraction: Bacterial agonists for leukocyte chemotaxis receptors. *Nature Reviews. Microbiology* **13**:95–104.

Chow, Jonathan, Kate M. Franz, and Jonathan C. Kagan. 2015. PRRs are watching you: Localization of innate sensing and signaling regulators. *Virology* **479–480**:104–109.

Eldridge, Matthew J., and Avinash R. Shenoy. 2015. Antimicrobial inflammasomes: Unified signalling against diverse bacterial pathogens. *Current Opinions in Microbiology* **23**:32–41.

Gasteiger, Georg, Andrea D'Osualdo, David A. Schubert, Alexander Weber, Emanuela M. Bruscia, et al. 2017. Cellular innate immunity: An old game with new players. *Journal of Innate Immunity* **9**:111–125.

Katsimichas, Themistoklis, Alexios S. Antonopoulos, Alexandros Katsimichas, Tomohito Ohtani, et al. 2019. The intestinal microbiota and cardiovascular disease. *Cardiovascular Research* **cvz135**, https://doi.org/10.1093/cvr/cvz135.

Koskinen, Kaisa, Manuela R. Pausan, Alexandra K. Perras, Michael Beck, Corinna Bangd, Maximilian Mora, et al. 2017. First insights into the diverse human archaeome: Specific detection of Archaea in the gastrointestinal tract, lung, and nose and on skin. *mBio* **8**:e00824-00817.

Lau, Jennifer T., Fiona J. Whelan, Isiri Herath, Christine H. Lee, Stephen M. Collins, et al. 2016. Capturing the diversity of the human gut microbiota through culture-enriched molecular profiling. *Genome Medicine* **8**:72.

Leiva-Juárez, Miguel Martin, Jay K. Kolls, and Scott E. Evans. 2018. Lung epithelial cells: Therapeutically inducible effectors of antimicrobial defense. *Mucosal Immunology* **11**:21–34.

Maier, Lisa, Mihaela Pruteanul, Michael Kuhn, Georg Zeller, Anja Telzerow, et al. 2018. Extensive impact of non-antibiotic drugs on human gut bacteria. *Nature* **555**:623–628.

Maurice, Corinne, F. 2019. Considering the Other Half of the Gut Microbiome: Bacteriophages. *mSystems* **4**:4/3/e00102–19.

Milani, Christian, Sabrina Duranti, Francesca Bottacini, Eoghan Casey, Francesca Turroni, et al. 2017. The first microbial colonizers of the human gut: Composition, activities, and health implications of the infant gut microbiota. *Microbiology and Molecular Biology Reviews* **81**:e00036-17.

Moustafa, Ahmed, Chao Xie, Ewen Kirkness, William Biggs, Emily Wong, et al. 2017. The blood DNA virome in 8,000 humans. *PLoS Pathogens* **13**:e1006292. http://doi.org/10.1371/journal.ppat.1006292.

Orvedahl, Anthony, and Beth Levine. 2009. Eating the enemy within: Autophagy in infectious diseases. *Cell Death and Differentiation* **16**:57–69.

Raniga, Kavita, and Chen Liang. 2018. Interferons: Reprogramming the metabolic network against viral infection. *Viruses* **10**:36. http://doi.org/10.3390/v10010036.

Rathinam, Vijay A. K., and Francis Ka-Ming Chan. 2018. Inflammasome, inflammation, and tissue homeostasis. *Trends in Molecular Medicine* **24**:304–318.

Sender, Ron, Shai Fuchs, and Ron Milo. 2016. Are we really vastly outnumbered? Revisiting the ratio of bacterial to host cells in humans. *Cell* **164**:337–340.

Thursby, Elizabeth, and Nathalie Juge. 2017. Introduction to the human gut microbiota. *Biochemical Journal* **474**:1823–1836.

Wang, Baohong, Mingfei Yao, Longxian Lv, Zongxin Ling, and Lanjuan Li. 2017. The human microbiota in health and disease. *Engineering* **3**:71–82.

Wang, Miao, Zequin Gao, Zhongwang Zhang, Li Pan, and Yonguang Zhang. 2014. Roles of M cells in infection and mucosal vaccines. *Human Vaccine Immunotherapy* **10**:3544–3551.

Wikoff, William R., Andrew T. Anfora, Jun Liu, Peter G. Schultz, Scott A. Lesley, et al. 2009. Metabolomics analysis reveals large effects of gut microflora on mammalian blood metabolites. *Proceedings of the National Academy of Sciences USA* **106**:3698–3703.

CHAPTER 24

The Adaptive Immune Response

24.1 Overview of Adaptive Immunity

24.2 Antibody Structure, Diversity, and Synthesis

24.3 T Cells Link Antibody and Cellular Immune Systems

24.4 Complement as Part of Adaptive Immunity

24.5 Gut Mucosal Immunity and the Microbiome

24.6 Immunization

24.7 Hypersensitivity and Autoimmunity

Once an infecting pathogen eludes innate immune defenses, it faces an even more daunting foe: an adaptive immune system that stirs only when provoked. Adaptive immunity is composed of special lymphocytes called B cells and T cells that multiply during an infection and "remember" the specific invader for years. These specially trained, memory lymphocytes are poised to respond quickly to a subsequent infection by the same pathogen. Left unregulated, however, adaptive immunity can turn on us, producing autoimmune disease.

CURRENT RESEARCH highlight

Making dead vaccines look alive. Live vaccines beat dead vaccines at eliciting antibodies. Julie Blander and colleagues used live and heat-killed *E. coli* to identify viability signals that the immune system uses to recognize live vaccines. One viability signal is bacterial RNA. Bacterial RNA in live vaccines causes monocytes to release cytokines that stimulate *E. coli*–specific B cells to become IgG-secreting plasma cells. Shown here is a lymph node germinal center after vaccination with heat-killed *E. coli* plus RNA. Fluorescent antibody staining reveals B-cell (blue) and T-cell (gray) zones. B cell–T cell interaction (red) forms only with added RNA. B cells expressing anti-*E.coli* IgG antibodies on their surfaces are green.

Source: Barbet, G., et al. 2018. *Immunity* **48**:584–598.

Vaccine: **Heat-killed *E. coli* + RNA**

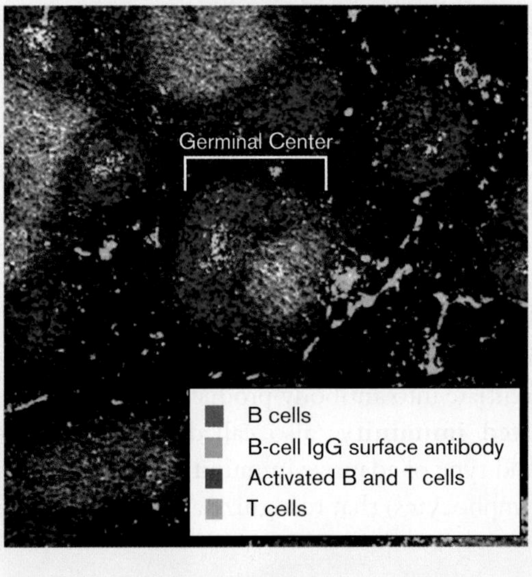

MODIFIED FROM BARBET, G., ET AL. 2018. *IMMUNITY* **48**:584–598.

AN INTERVIEW WITH

JULIE BLANDER, IMMUNOLOGIST, WEILL CORNELL MEDICINE, CORNELL UNIVERSITY

COURTESY OF JULIE MAGARIAN BLANDER

How might understanding pathogen viability signals influence future vaccines?

We envision future vaccines would incorporate a simple change—the addition of a signature-of-pathogen-viability, what we call a vita-PAMP (pathogen-associated molecular pattern). The modified vaccine would gain tremendous efficacy with a single dose. Vita-PAMPs trick the immune system into thinking an infection is under way, so the best protective response gets mobilized. We may also be able to resurrect old vaccines that were shelved because of inefficacy. While we have identified two vita-PAMPs, we need a concerted effort to find more and use them as vaccine adjuvants.

As recently as 200 years ago, infectious diseases were thought to result from inhaling poisonous vapors, called "miasmas," produced by decaying organic matter. According to theory, disease resulted when miasmas invaded the body and disturbed vital functions. Although physicians of the seventeenth and early eighteenth centuries were wrong about this, they did recognize that humans reacted to disease with fevers and "humors" that somehow overcame the miasmas—at least most of the time. Today we know that miasmas are actually infectious pathogens and that the innate and adaptive immune systems collaboratively react to their presence. The Current Research Highlight illustrates how twenty-first-century scientists manipulate these immune systems with finely tuned vaccines that can better prevent infectious diseases.

The immune response is a stunningly complex biological system controlled by many checks and balances that prevent overreaction. Unfortunately, with complexity comes the potential for catastrophic genetic defects. Consider the following case: A visibly ill 9-month-old baby boy was admitted to the hospital with intractable diarrhea, inappropriate weight loss, and oral thrush caused by *Candida albicans* (**Fig. 24.1**). This infection was just the latest of several that the boy had suffered during his short life. Laboratory results found that he was deficient in all classes of his antibodies (a syndrome called hypogammaglobulinemia) and, most disturbingly, his antigen-presenting cells (monocytes, macrophages, and B cells) lacked major histocompatibility complex (MHC) II molecules.

Because MHC II molecules are essential for generating protective levels of different antibody classes, their absence explained the boy's hypogammaglobulinemia. He was diagnosed with bare lymphocyte syndrome, a disease caused by a defect in the gene that activates the transcription of MHC II genes. The only treatment is a bone marrow transplant to replace "bare" lymphocytes with compatible donor lymphocytes covered with MHC II molecules.

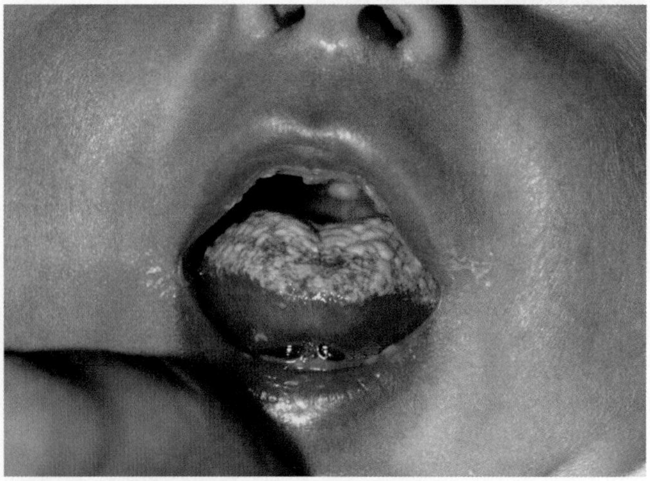

DR. P. MARAZZI/SCIENCE SOURCE

FIGURE 24.1 ■ Baby with thrush. The yeast *Candida albicans* can infect the oral cavities of babies, especially children with genetic immunodeficiencies.

This case dramatically illustrates the importance and fragility of the human immune system. All it takes is a small defect in a single gene to subvert the entire process. Another dramatic example is the case of David Philip Vetter, better known as "the bubble boy," who was born in 1971 without an immune system (see **eTopic 24.1**).

In Chapter 24 we describe the two types of adaptive immunity: antibody-dependent immunity, in which B cells differentiate into plasma cells that make antibodies; and cell-mediated immunity, whereby specific T-cell lymphocytes develop to directly kill infected host cells. You will also learn how certain types of T cells balance antibody and cell-mediated responses to a given infection. Along the way we will explore how gut mucosal immunity mediates coexistence with the microbiome and how vaccines are used medically to train the immune system. Ultimately, you will appreciate that adaptive immunity is a major reason why humans still exist.

24.1 Overview of Adaptive Immunity

Recall from Chapter 23 that the immune system has both nonadaptive and adaptive mechanisms. Nonadaptive (innate) immune mechanisms are present from birth, whereas adaptive immunity develops as the need arises. For instance, adaptive immunity against malaria does not develop until the individual has encountered the plasmodial parasite that causes the disease. The adaptive immune response—**adaptive immunity**—is a complex, interconnected, and cross-regulated defense network.

Note: The terms "adaptive immune response" and "immune response" are often used interchangeably.

Antibody-Dependent versus Cell-Mediated Immunity

Two types of adaptive immunity are recognized: humoral immunity and cell-mediated immunity. In **humoral immunity** (also called **antibody-dependent**), **antibodies** are produced that directly target microbial invaders. The term "humoral" means "related to body fluids." Thus, antibodies are proteins that circulate in the bloodstream and recognize foreign structures called antigens. An **antigen** (also called an **immunogen**) is any molecule that, when introduced into a person, elicits the synthesis of antibodies that specifically bind the antigen. Antigens stimulate B cells (B lymphocytes) to differentiate into antibody-producing **plasma cells**.

Cell-mediated immunity (also called cellular immunity), the second type of adaptive immunity, employs teams of T cells (T lymphocytes) that recognize antigens and then destroy host cells infected by the microbe possessing the antigen. The humoral and cellular immune responses are intertwined, each relying on some facet of the other to work efficiently. T cells serve a central role in adaptive immunity by determining whether humoral or cell-mediated mechanisms predominate in response to a specific antigen.

What triggers an immune response, and how long does it take? Adaptive immunity develops over a period of 3–4 days after exposure to an invading microbe. The immune system does not recognize the whole microbe or even a whole protein, but it will recognize innumerable tiny pieces of them. Each small segment of a protein or other molecule that is capable of eliciting an immune response is called an **antigenic determinant** or **epitope**. Upon phagocytosis, the microbe and its structures are broken into smaller segments that can be recognized as epitopes.

Even distinct tertiary (3D) shapes within a protein may be counted as antigenic determinants if they produce a specific response. Immune responses to 3D shapes are possible when stretches of amino acids far removed from each other in a protein's primary sequence align side by side in 3D space after folding (**Fig. 24.2**). Such a 3D structure may be recognized by the immune system as a single entity or antigen. Besides proteins, other structures in the cell, such as complex polysaccharides, can have linear and 3D epitopes. So, the immune response to a microbe is really a composite of thousands of B-cell responses to different epitopes. The antibody response to each individual epitope is **clonal**; that is, the response to an individual epitope gives rise to a population of identical cells (clones) that originate from a single B cell. Each B cell within that clone targets the same epitope.

The humoral immune response requires several cell types and cell-to-cell interactions. What are those interactions, and where do they take place? As illustrated in **Figure 24.3**, the process begins with an infection somewhere in the body (the area labeled "Periphery" in **Fig. 24.3A**). Dendritic cells and macrophages patrolling the area gather up the foreign antigens and present them on their cell surface (**Fig. 24.3A**, step 1). Any phagocytic cell that degrades large antigens into smaller antigenic determinants and places those determinants on their cell surface is called an **antigen-presenting cell** (**APC**). Many types of cells can be antigen presenters. They include "professional" APCs, such as macrophages (monocytes), mast cells, dendritic cells, and B lymphocytes; and, under the right circumstances, "nonprofessional" APCs (for example, endothelial cells or fibroblasts). Because dendritic cells are so important to the immune response, their discoverer, Ralph Steinman (1943–2011), was awarded the Nobel Prize in Physiology or Medicine in 2011 (**Fig. 24.3B**).

Once it has been decorated with antigen, the professional antigen-presenting cell travels to secondary lymphoid organs (lymph nodes), where B cells and T cells await (**Fig. 24.3A**, step 2). Specific T cells in the node then link to the antigen presented on the APC and become activated T cells. One type of activated T cell then binds to and activates a lymph node B cell that encountered, and bound, a different molecule of the same antigen (step 3). Note that the B cell–antigen and T cell–APC binding events are actually independent of each other.

The T cell–B cell interaction authorizes B cells to secrete antibody and transform into plasma cells (usually short-lived) that will pump out large amounts of specific antibody (**Fig. 24.3A**, step 4). (Remember that plasma cells are not B cells. They are terminally differentiated cells derived from B cells.) The activated B cells also produce long-lived memory B cells that remember antigen exposure

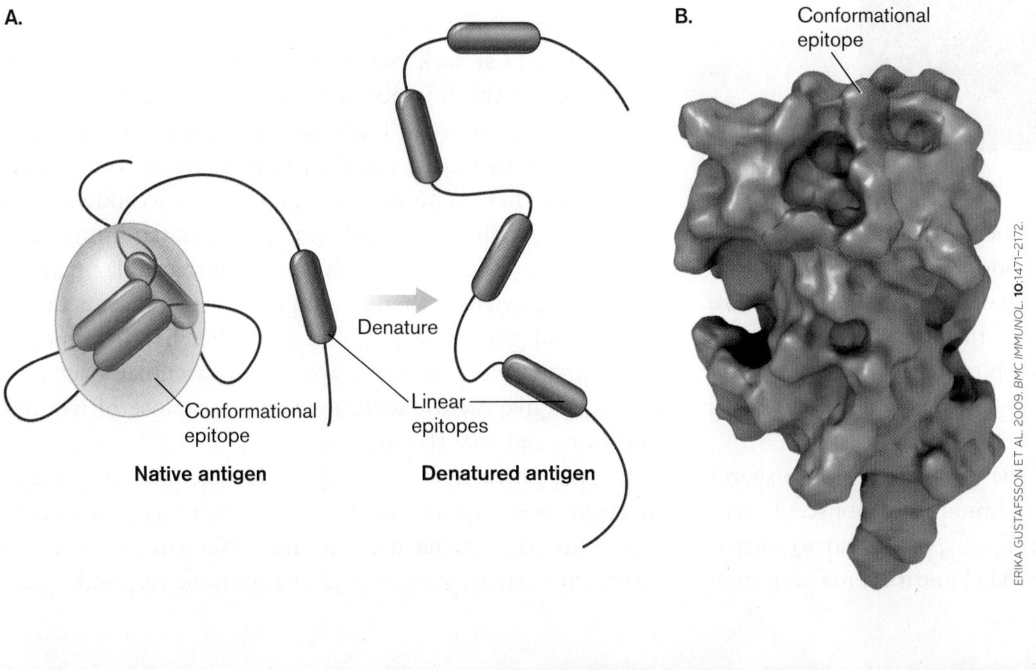

FIGURE 24.2 ■ Antigens and epitopes. A. Native proteins fold into a 3D shape, where several regions separated in the linear sequence can reside next to each other to form a conformational epitope. Denaturing the protein with the detergent sodium dodecyl sulfate (SDS) and reducing agents like dithiothreitol (to remove disulfide bonds) will unfold the protein and separate the various amino acid stretches that formed the conformational epitope. **B.** A 3D protein structure, showing (in red) four amino acids that form a conformational epitope.

A.

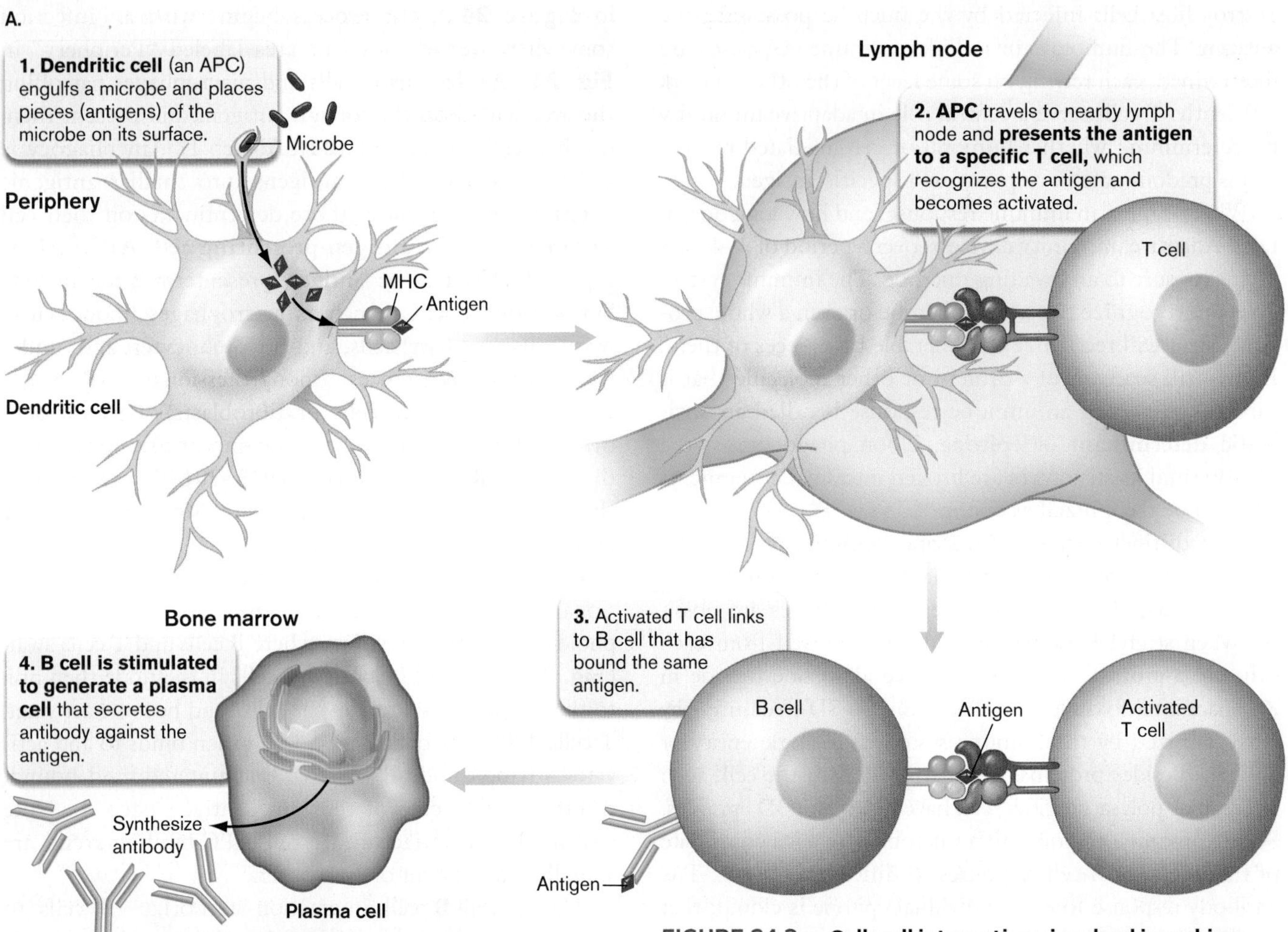

B.

FIGURE 24.3 ▪ Cell-cell interactions involved in making antibody. **A.** Two basic cell-cell interactions are required to make an antibody: In the first, an antigen-presenting cell (APC; dendritic cell) presents antigen to a helper T cell. In the second, the activated T cell links to and activates a B cell bound to the same antigen. The two interactions can be broken down into four steps, as illustrated. **B.** Ralph Steinman discovered dendritic cells.

and stand ready to quickly generate plasma cells, should the antigen be encountered months or years later. Once formed, the memory B cells and most plasma cells leave the lymph node and migrate to the bone marrow. Other plasma cells remain in the lymph node. Memory B cells in the bone marrow patiently wait to be called to future sites of infection.

The cell-mediated arm of the immune response shares some aspects of the humoral immune response. T cells of one type, called cytotoxic T cells, also bind to microbial antigens presented on an APC and become activated. Activated cytotoxic T cells (also called cytotoxic lymphocytes, or CTLs) seek out and directly kill any host cell infected with the microbe. In addition to directly killing infected cells, cytotoxic T cells synthesize and secrete small peptide growth factors called cytokines (see Sections 23.2 and 23.5), that incite nearby macrophages to indiscriminately attack cells in the local area. Thus, cellular immunity, in general, is critical for dealing with intracellular pathogens such as viruses (herpes simplex, for instance, a cause of oral and genital ulcers). On the other hand, humoral immunity is most effective against extracellular bacterial pathogens like *Streptococcus pyogenes*, one cause of wound infections and sore throat.

As we proceed through this chapter, we will reveal in layers how the immune system functions, with each layer building on the previous one. We will periodically return to a particular aspect of the immune response—for example, B-cell differentiation into plasma cells—to integrate seemingly distinct parts of the immune system into a unified concept of immunity.

Thought Question

24.1 Two different stretches of amino acids in a single protein form a 3D antigenic determinant. Will the specific immune response to that 3D antigen also recognize the same two amino acid stretches if they are removed from the whole protein?

Immunogenicity

Immunogenicity measures the relative effectiveness by which an antigen elicits an immune response. For example, proteins are the strongest antigens, but carbohydrates can also elicit immune reactions. Nucleic acids and lipids are usually weaker antigens, in part because these molecules are both made of relatively uniform repeating units that are very flexible. The flexible units present a variable 3D structure that does not easily interact with antibodies. Proteins are more effective antigens for three reasons: Different proteins have different shapes, they maintain their tertiary structure, and they are made of many different amino acids that can be assembled in many different combinations. These features provide stronger interactions with antibodies in the bloodstream and enable better recognition by lymphocytes, the cellular workhorses of the immune system.

Several other factors contribute to the immunogenicity of proteins (see **eTopic 24.2**). For example, the larger the antigen, the more likely it is that phagocytic cells will "see" and engulf it. This is important because, as noted earlier, an immune response cannot occur until phagocytic antigen-presenting cells (such as macrophages and dendritic cells) first engulf and degrade large antigens, presenting the smaller epitopes on their cell surface.

For an APC to present antigen to a T cell, the antigenic determinant must first bind to an APC protein structure called the **major histocompatibility complex**, or **MHC**. MHC proteins are initially found on the inner surface of membranes that form intracellular vesicles. The vesicles carry the antigens bound to MHC to the cell surface (**Fig. 24.3A**). The more tightly an antigen can bind to these MHC surface proteins, the more immunogenic it is. The stronger the binding, the easier it is for T cells to recognize the complex.

Each specific antigen shows a different **threshold dose** needed to generate an optimal response. A dose higher or lower than that threshold will generate a weaker immune response. Lower doses activate only a few B cells, whereas exceedingly high doses of antigen can cause **B-cell tolerance**, a state in which B cells have been overstimulated to the point that they do not respond to subsequent antigen exposures and make antibody. Tolerance is part of the reason your immune system does not react against your own protein antigens.

As you might expect, the body must regulate the immune system carefully so that a response is not leveled against itself. In effect, the immune system must become "blind" to its own antigens; as a result, the host will often be blind to foreign antigens that resemble epitopes of its own cells. Therefore, the more complex the foreign protein is, the more likely it will possess antigenic determinants that a lymphocyte can recognize as nonself. The farther an antigen is from "self," the greater its immunogenicity will be.

Immunological Specificity

The earliest clues about the nature of immunity came from smallpox. Smallpox is a devastating disease, caused by the variola virus, that inflicted enormous suffering and killed millions of people in the seventeenth and eighteenth centuries (see Chapter 1 and **Fig. 24.4**). There was no cure, and the only available preventive treatment was to take dried material from the lesions of a previous smallpox sufferer, place it on a healthy person, and hope the person survived. Survivors were protected from subsequent bouts of smallpox but were still susceptible to other diseases. This early observation gave rise to the idea of **immunological specificity**,

A. Smallpox patient

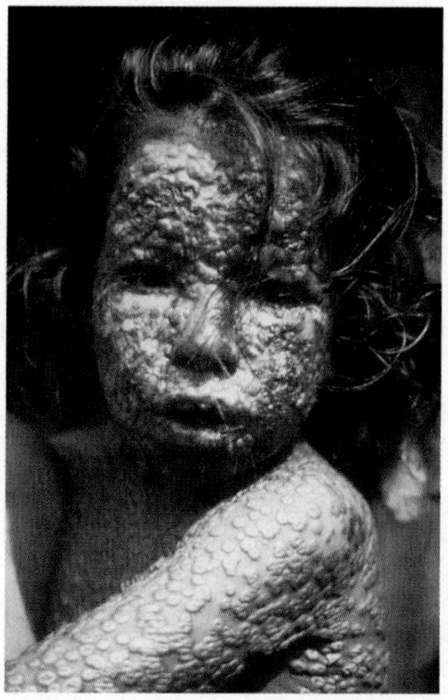

B. Variola major

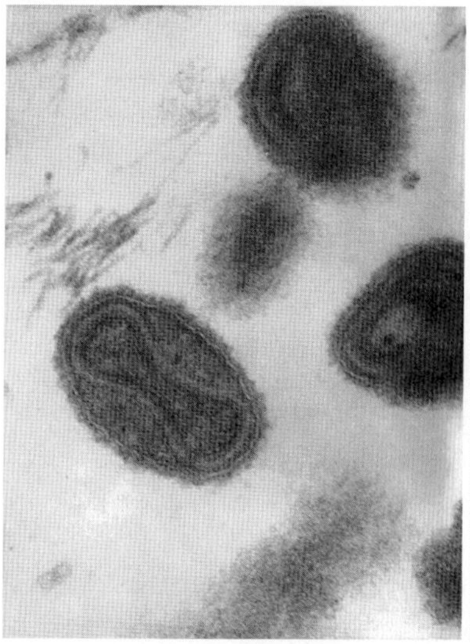

FIGURE 24.4 ■ Immunological specificity is the basis of vaccination. **A.** Smallpox patient covered with pox pustules. **B.** The smallpox virus, variola major (300 nm long; TEM). The photo shows the dumbbell-shaped, membrane-enclosed nucleic acid core. Edward Jenner recognized the similarity between the deadly smallpox and less severe cowpox diseases and used cowpox scrapings to vaccinate humans against smallpox.

in which an immune response to one antigen is not effective against a different antigen. In other words, the immune response to smallpox will not protect someone against the plague bacillus (*Yersinia pestis*), which is antigenically different from the smallpox virus.

While immunological specificity is important, it is not absolute. As described in Chapter 1, an English country physician named Edward Jenner (see Fig. 1.20B) in the late eighteenth century (long before viruses were discovered) learned to protect townsfolk from deadly smallpox disease (**Fig. 24.4B**) by inoculating them with scrapings from lesions produced by cowpox (a tamer disease caused by the genetically related vaccinia virus). This story illustrates that an immune reaction against a less virulent organism or virus may be sufficient to cross-protect against an antigenically related but more virulent pathogen. The technique of protecting individuals from virulent microbes by exposing them to a less virulent version of the pathogen, now generally called **vaccination**, has been used to safeguard humans against many bacterial and viral pathogens (**Table 24.1**). Most vaccinations today involve administering either crippled (live but attenuated) strains of the pathogenic microbe, or inactivated microbial toxins (for example, diphtheria toxin).

Cross-protection, in which immunization against one microbe protects against a second, will work only if two proteins critical to the pathogenesis of the two different microorganisms share key antigenic determinants. Cross-protection will not take place if the structures of these determinants differ too much from one another. A good example is the common cold, which is caused by hundreds of closely related rhinovirus strains (rhinitis, a runny nose, is one of the symptoms of this viral disease). Infection with one strain will not immunize the victim

TABLE 24.1 Examples of Vaccines against Viral and Bacterial Pathogens

Disease	Vaccine	Vaccination recommended for:
Viral diseases		
Chickenpox	Attenuated strain (will still replicate)	Children 12–18 months
Hepatitis A	Inactivated virus (will not replicate)	Children 12 months
Hepatitis B	Viral antigen	Newborns
Influenza	Inactivated virus or antigen	Everyone, after 6 months old, yearly
Measles, mumps, rubella (MMR)	Attenuated viruses; MMR combined vaccine	Children 12 months
Polio	Inactivated (injection, Salk)	Children 2–3 months
Rabies	Inactivated virus	Persons in contact with wild animals
Yellow fever	Attenuated virus	Military personnel
Bacterial diseases		
Anthrax	*Bacillus anthracis*, toxin components; unencapsulated strain	Agricultural and veterinary personnel; key health care workers
Cholera	Killed *Vibrio cholerae*, toxin components	Travelers to endemic areas
Diphtheria	Toxoid (inactivated toxin)	Children 2–3 months
Lyme disease	*Borrelia burgdorferi*, lipoproteins OspA and OspC surface antigens	Canines; human vaccine discontinued
Meningitis caused by *Haemophilus influenzae* type b (Hib)	Bacterial capsular polysaccharide	Children under 5 years
Meningococcal disease	*Neisseria meningitidis*, bacterial capsular polysaccharides	Children >2 years; adults >50 years
Pertussis	Acellular *Bordetella pertussis*	Children 2–3 months
Pneumococcal pneumonia	*Streptococcus pneumoniae*, bacterial capsular polysaccharides	Children; adults >50 years
Tetanus	Toxoid	Children 2–3 months
Tuberculosis (*Mycobacterium tuberculosis*)	Attenuated *Mycobacterium bovis* [BCG (Bacille Calmette-Guérin) vaccine]	Exposed individuals
Typhoid fever	Killed *Salmonella* Typhi	Individuals in endemic areas
Typhus	Killed *Rickettsia prowazekii*	Medical personnel in endemic areas; scientists; discontinued

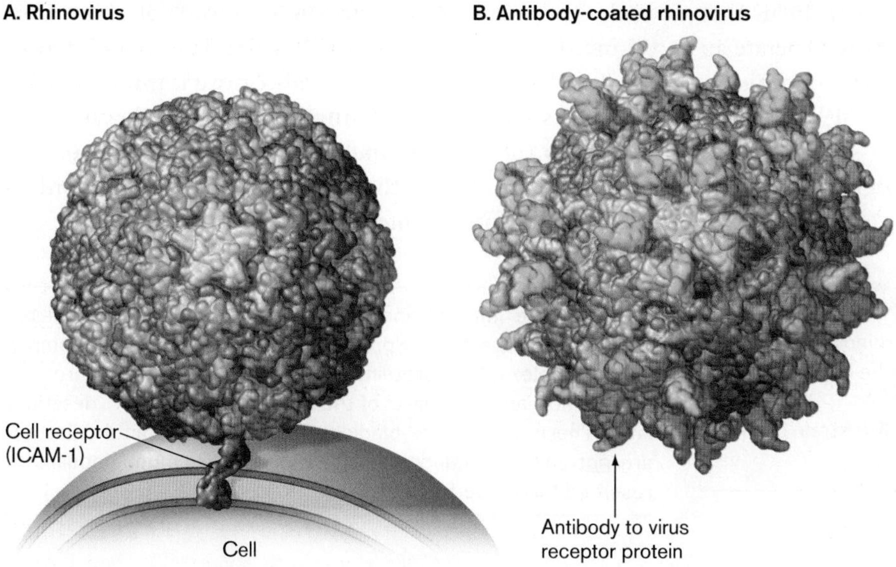

FIGURE 24.5 ▪ Antibodies prevent rhinovirus attachment to cell receptors. **A.** The complex rhinovirus capsid is pictured here attaching to the cell-surface molecule ICAM-1 (intercellular adhesion molecule, shown in reddish brown). (PDB code: 1rhi) **B.** Rhinovirus coated with protective (neutralizing) antibodies (green) block the ICAM-1 receptors on the virus. As a result, the virus fails to attach to and infect the host cell. (PDB code: 1RVF)

against a second strain. The reason is that the structures of rhinovirus proteins that attach to the ICAM-1 surface protein on host cells differ dramatically between different strains of rhinovirus (**Fig. 24.5A**). Antibodies called neutralizing antibodies, which bind to the attachment protein on one strain of rhinovirus, will prevent infection by that strain (**Fig. 24.5B**) but will not bind an antigenically distinct ICAM-1 receptor protein from a different strain. A key to one lock will not work in a different lock.

Thought Question

24.2 How does a neutralizing antibody that recognizes a viral coat protein prevent infection by the associated virus?

Antigens and Immunogens

Antigens that, by themselves, can elicit antibody production are called **immunogens**. Molecules of molecular weight less than 1,000, however, are generally not immunogenic, because they do not bind MHC molecules. Nevertheless, these small molecules, called **haptens** (from the Greek word meaning "to fasten" and the German word for "stuff"), will elicit the production of specific antibodies if they are covalently attached to a larger carrier protein or other molecule (**Fig. 24.6**). Haptens can be thought of as small, incomplete antigens. An example of a hapten is the antibiotic penicillin, a serious cause of immune hypersensitivity reactions in some individuals (see Section 24.7).

The following example illustrates the hapten concept. A protein carrier such as bovine serum albumin (BSA) injected into a mouse elicits antibodies that react against BSA (**Fig. 24.6**). BSA antigen is an immunogen because it elicits an immune response to itself. In contrast, a mouse injected with the hapten benzene sulfate fails to produce antibodies to benzene sulfate. However, when the hapten chemical is attached to BSA protein and the hapten-BSA complex is injected, the mouse produces not only antibodies that react to the carrier (BSA), but also other antibodies that react against the benzene hapten. The reason for this carrier effect will become evident later in the chapter, when we describe how antigens are processed by cells of the immune system. Thus, antigens include immunogens that

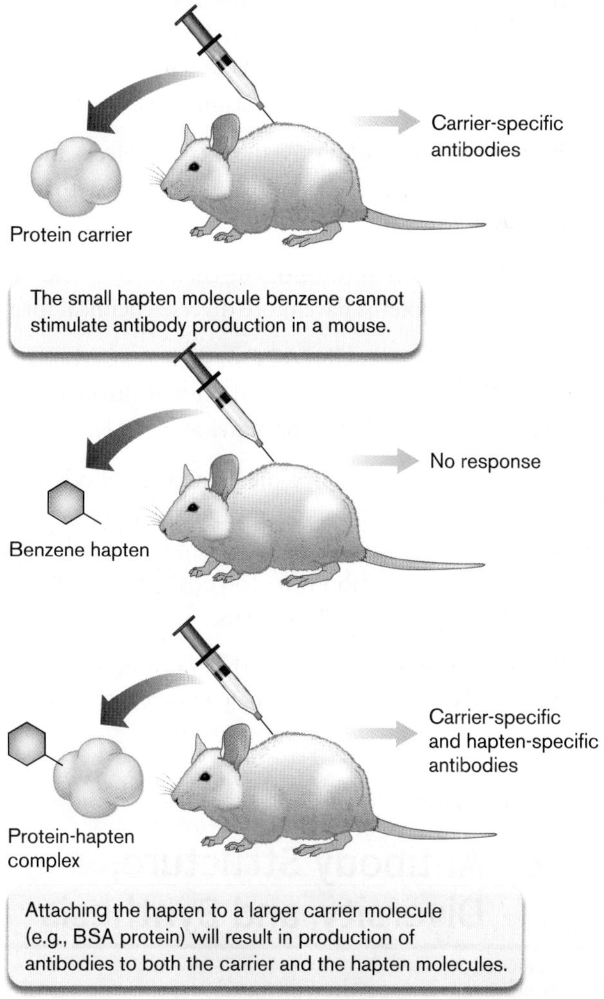

FIGURE 24.6 ▪ Basic schematic showing how haptens can elicit antibody production.

elicit an immune response by themselves, and haptens that must be attached to an immunogen in order to generate an immune response. Karl Landsteiner uncovered the antigen-immunogen-hapten relationship in the early 1900s, along with discovering the ABO blood group system, which first defined the concept of immunological specificity (eTopic 24.3).

> **Thought Question**
>
> **24.3** The attachment proteins of different rhinovirus strains all bind to ICAM-1. How can all these proteins be immunologically different if they find the same target (ICAM-1)? Why won't antibodies directed against one rhinovirus strain block the attachment of other rhinovirus strains?

To Summarize

- **An antigen can elicit an antibody response.** An antigen usually consists of many different epitopes (antigenic determinants), each of which binds to a different, specific antibody.

- **Humoral immunity** against infection is the result of antibody production originated by B cells. Cells belonging to a subgroup of T-cell lymphocytes stimulate B cells to become antibody-secreting plasma cells.

- **Cell-mediated (cellular) immunity** involves another subgroup of T cells, called cytotoxic T cells, that can directly kill host cells.

- **Proteins are better immunogens** than nucleic acids and lipids because proteins have more diverse chemical forms.

- **Antigen-presenting cells (APCs)**, such as macrophages, degrade microbial pathogens and present distinct pieces on their cell-surface MHC proteins.

- **Immunological specificity** means that antibody made to one epitope will not bind to different epitopes. However, antibodies can weakly cross-bind to similar epitopes; for example, antibody to cowpox virus will bind to a similar epitope on smallpox virus.

- **A hapten** is a small compound that must be conjugated to a larger carrier antigen to elicit the production of an antibody.

24.2 Antibody Structure, Diversity, and Synthesis

The broad overview of the immune response focused on how antibodies are made. But what are antibodies, and why are they important? Antibodies are proteins made by the body in response to foreign antigens, such as those from an invading pathogen. Antibodies circulate though blood, ready to bind a foreign antigen to remove it from the body. They are sometimes called **immunoglobulins** because they belong to the larger immunoglobulin superfamily of proteins. All members of this superfamily have a 110-amino-acid domain with an internal disulfide bond.

Note: The immunoglobulin superfamily of proteins includes many cell-surface-binding proteins, such as the major histocompatibility complex (MHC) proteins, various cytokine receptor proteins, and the non-antibody parts of the B-cell receptor (BCR), described later. These cell-surface-binding proteins are not antibodies and are not called immunoglobulins. The term "immunoglobulins" is reserved for antibodies.

Like miniature "smart bombs," antibodies individually circulate through blood, ignoring all antigens except for the one they were selected to bind. When an antibody finds its antigenic match, it binds to the antigen and initiates several events that destroy the target. Antibodies not only are free-floating in blood, but also are strategically situated on the surfaces of B cells as the antigen-binding part of B-cell receptors (described later).

A typical antibody consists of four polypeptide chains. There are two large **heavy chains** and two smaller **light chains** (Fig. 24.7). The four polypeptides combine to form a Y-shaped tetrameric structure held together by disulfide bonds. Two bonds connect the two identical heavy chains to each other. One light chain is then attached near its carboxyl end to the middle of each heavy chain by a single disulfide bond. The antigen-binding sites are formed at the amino-terminal ends of the light and heavy chains. One antibody molecule possesses two identical antigen-binding sites, one on each "arm" of the molecule. The two binding sites allow a single antibody to bind to two identical antigens. When enough antibodies bind to identical antigens on enough different molecules, a large, cross-linked matrix will form and precipitate out of solution in a process called **immunoprecipitation**. Immunoprecipitation is the basis for a number of important molecular and clinical assays (see eAppendix A.3.11).

Antibodies Have Constant and Variable Regions

There are five classes of antibodies—defined by five different types, or isotypes, of heavy chains—called alpha (α), mu (μ), gamma (γ), delta (δ), and epsilon (ϵ). The heavy-chain classes are distinguished one from another by regions of highly conserved amino acid sequences, known as **constant regions** (denoted C_H for the heavy chain; **Fig. 24.8**). Antibodies containing gamma heavy chains are called IgG;

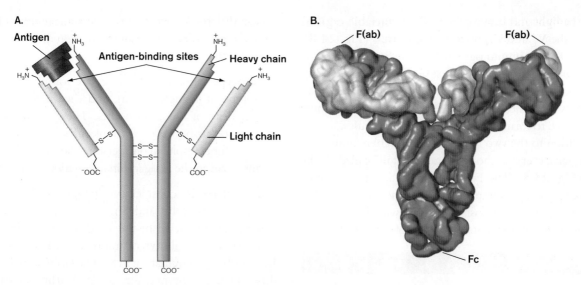

FIGURE 24.7 • Basic antibody structure. A. Each antibody contains two heavy chains and two smaller, light chains held together by disulfide bonds. The Y-shaped structure contains two antigen-binding sites, one at each arm of the molecule. These two sites are formed by the amino-terminal regions of the heavy- and light-chain pairs. **B.** The 3D structure of an antibody. The heavy chains are shown in green, the light chains in yellow. The F(ab) regions represent the antigen-binding sites. The Fc portion points downward and is used to attach the antibody to different cell-surface molecules. (PDB code: 1R70)

those with alpha, mu, delta, and epsilon heavy chains are called IgA, IgM, IgD, and IgE, respectively. Each antibody class serves a specific purpose in the immune system.

In contrast to heavy chains, there are only two classes of light chains—kappa (κ) and lambda (λ)—which are defined by their own constant regions (C_L; **Fig. 24.8**). A single antibody of any heavy-chain class (for example, IgG) may contain two kappa light chains or two lambda light chains, but never one of each. In humans, two-thirds of all antibody molecules carry kappa chains; the rest have lambda chains.

The antigen-binding part of an antibody is formed by highly variable amino acid sequences at the amino-terminal

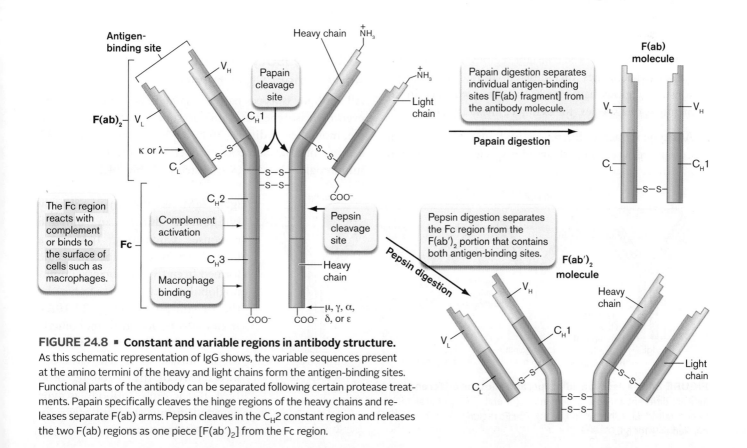

FIGURE 24.8 • Constant and variable regions in antibody structure. As this schematic representation of IgG shows, the variable sequences present at the amino termini of the heavy and light chains form the antigen-binding sites. Functional parts of the antibody can be separated following certain protease treatments. Papain specifically cleaves the hinge regions of the heavy chains and releases separate F(ab) arms. Pepsin cleaves in the C_H2 constant region and releases the two F(ab) regions as one piece [F(ab')$_2$] from the Fc region.

ends of the light and heavy chains. These **variable regions** are called the V_L and V_H regions, respectively (**Fig. 24.8**). The rest of each immunoglobulin chain is composed of highly conserved constant regions; C_H1, C_H2, and C_H3 in each heavy chain; and C_L in each light chain. Shortly, we will discuss the genes that code for these regions and how they assemble to form different antibody molecules.

In addition to the two "arms," called F(ab) regions, that bind antigens, every antibody contains a "tail" called the **Fc region** (**Fig. 24.8**). The Fc region is not involved in antigen recognition but is important for anchoring antibodies to the surface of certain host cells (those with Fc receptors) and for binding components of the complement system.

Thought Question

24.4 Can an $F(ab')_2$ antibody fragment prevent the binding of rhinovirus to the ICAM-1 receptor on host cells? And can an $F(ab')_2$ antibody fragment facilitate phagocytosis of a microbe?

Isotypes, Allotypes, and Idiotypes

For the immune system to work properly, antibodies must contain amino acid differences that distinguish different species, different individuals within a species, and different antibodies within a single individual. The following terms are important for understanding how antibodies carry out their different functions in a host.

- **Isotype** sequence differences define the various heavy chains within, and between, species. These amino acid differences occur in the constant regions of heavy-chain classes (what makes IgG different from IgE). Isotype constant regions have nearly the same sequence in all members of a species (for example, *Homo sapiens*).
- **Allotype** sequence differences are found within a given isotype between individuals of a species. These amino acid differences occur in the constant regions of an antibody that make IgG from one person different from IgG in another person.
- **Idiotype** sequence differences are found within a single isotype of a single person, and they reflect the antigen specificity of those antibodies. These amino acid differences occur between two molecules of the same isotype (IgG, for example) in one person. Idiotype differences usually occur in the antigen-binding region of an antibody.

In real terms, consider two people, John and Sherrie. Blood of both John and Sherrie carries the human IgG isotype. However, all of the IgG molecules circulating in John have small, allotypic amino acid differences from the IgG antibodies circulating in Sherrie (John and Sherrie have different IgG allotypes; **Fig. 24.9**). Within Sherrie herself, the antigen-binding site of an IgG molecule that binds a herpesvirus epitope possesses idiotypic amino acid differences from an IgG molecule in her that binds to a rhinovirus epitope. In the next several sections we will describe the synthesis and functional purpose of these different levels of antibody diversity.

Thought Question

24.5 (refer to **Fig. 24.9**) Because antibodies are proteins, they are also antigens and can stimulate an immune response. What types of antibodies—anti-isotype, anti-allotype, or anti-idiotype—will IgG taken from Sherrie raise when injected into John?

Antibody Isotype Functions and "Super" Structures

All antibody isotypes have the same basic structure. However, each isotype has a unique superstructure (for example, monomer or dimer), and each is designed to carry out a different task. Some key properties of the five different immunoglobulin classes are listed in **Table 24.2**.

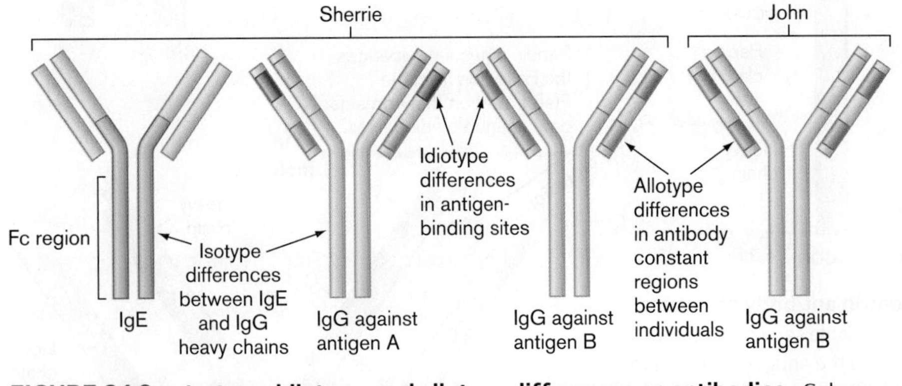

FIGURE 24.9 ■ Isotype, idiotype, and allotype differences on antibodies. Colors indicate different amino acid sequences (or epitopes) within an individual (Sherrie) or between individuals (Sherrie and John). The Fc region is present in all of these antibodies but is marked only for IgE.

IgM. IgM is the first antibody synthesized during an immune response. Circulating **IgM** is a huge, Ferris wheel–shaped molecule formed from five monomeric immunoglobulins tethered together by the J-chain protein (**Fig. 24.10A**). It can also be found in monomeric form attached by its Fc receptor to the surface of B cells, where it forms part of the B-cell receptor. IgM is the first antibody isotype detected during the early stages of an immune response.

TABLE 24.2 Properties of Human Immunoglobulins

Property	Antibody isotype: IgM	IgG: IgG1	IgG: IgG2	IgG: IgG3	IgG: IgG4	IgA	IgD	IgE
Serum half-life (days)	10	21	20	7	21	6	3	2
% Total serum Ig	5%–10%	70%				15%–20%	0.2%	0.002%
Antigen-binding sites	2–10	2				2–4	2	2
Produced by fetus	Yes	Poorly, if at all				Poorly, if at all	?	Poorly, if at all
Transmitted across placenta	No	Yes				No	No	No
Binds complement	Yes	Yes				No	No	No
Opsonizing	No	Yes				No	No	No
Binds mast cells	No	No				No	No	Yes

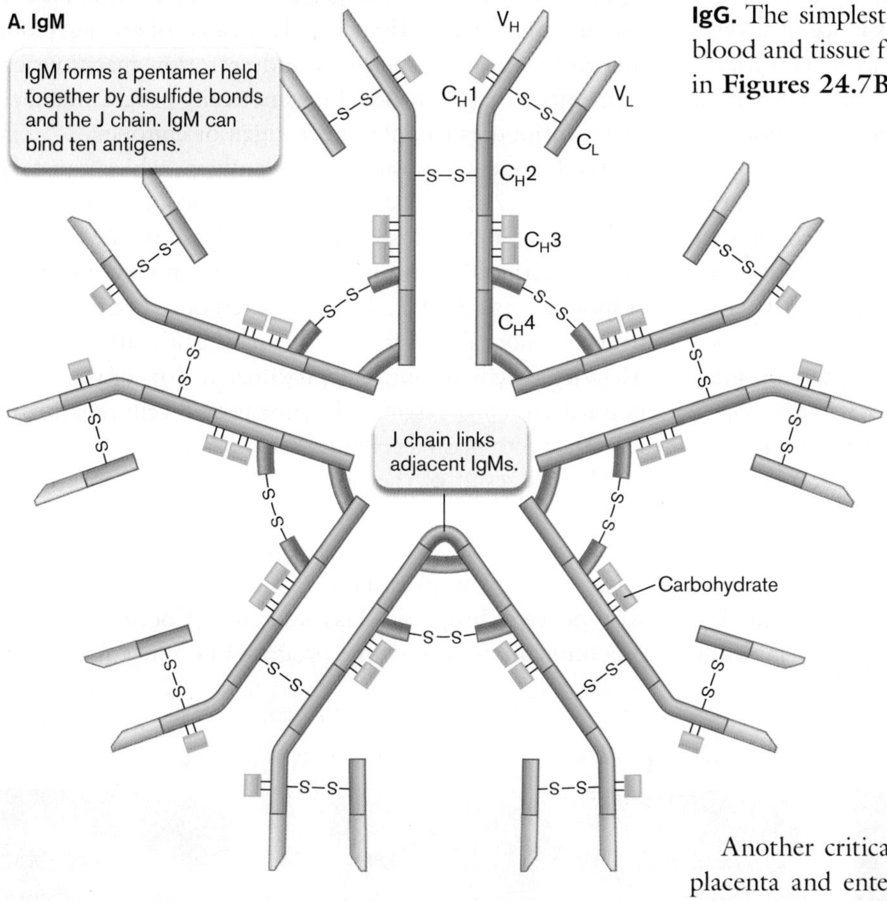

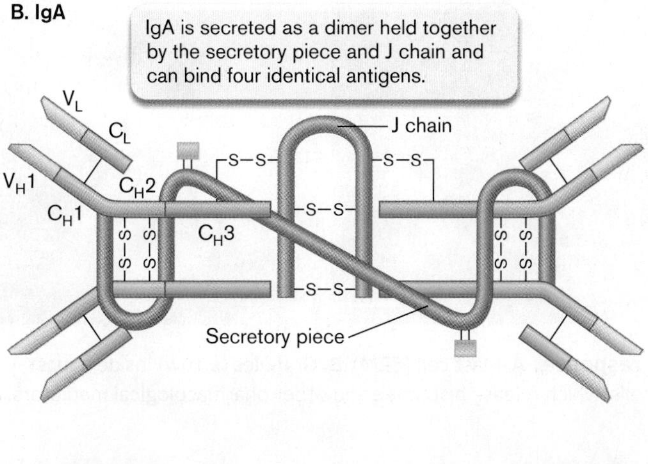

IgG. The simplest and most abundant antibody isotype in blood and tissue fluids is **IgG** (its superstructure was shown in **Figures 24.7B** and **24.8**). It is made as a monomer but has four subclasses (see **Table 24.2**). Each subclass varies in its amino acid composition and number of interchain cross-links. IgG molecules carry out several missions for the immune system. First, they bind and **opsonize** microbes; that is, they make microbes more susceptible to phagocytes. Opsonizing IgG antibodies use their antigen-binding sites to stick to microbes, thereby causing their Fc regions to point away from the microbe. Phagocytes possess surface Fc receptors that can attach to the Fc region of the antibody to gain a firmer "grip" on the microbe, facilitating phagocytosis (discussed in Section 23.5). IgG can also directly neutralize viruses by binding to virus attachment sites, and it is one of only two antibody types that can activate complement by the classical pathway (described in Section 24.4).

Another critical feature of IgG is that it can cross the placenta and enter a fetus. Other antibody isotypes cannot, because the antibody transport system in the placenta, called the neonatal Fc transporter, is specific for the Fc region of IgG.

IgA. IgA is secreted across mucosal surfaces (linings of the respiratory, gastrointestinal, urinary, and reproductive tracts) and is most commonly found as a dimer

FIGURE 24.10 ▪ Structures of IgM and IgA. The antibodies are made as multimers of immunoglobulin molecules: five in the case of IgM **(A)**; two in the case of IgA **(B)**.

(**Fig. 24.10B**). This conformation explains why **IgA** can bind four molecules of antigen (each monomer can bind two). The components of the IgA dimer are linked by disulfide bonds to a protein called the J chain, which joins two IgAs by their Fc regions. In addition to the eight molecules that comprise the two IgA monomers and the J chain, a tenth molecule—the secretory piece, is wrapped around the IgA dimer during the secretion process. The secreted molecule, now called sIgA (secretory IgA), is found in tears, breast milk, and saliva, and on other mucosal surfaces. The molecule sIgA is important for mucosal immunity against pathogens that infect mucosal linings.

IgD. The last two antibody isotypes, IgD and IgE, are present at very low levels in the blood. **IgD** is a monomer that can neither bind complement nor cross the placenta. IgD molecules, however, are abundant on the surface of B cells. IgD, as with monomeric IgM, is attached to B-cell surfaces by its Fc region. IgD and IgM act as receptors that bind antigen and signal B cells to differentiate and make antibody.

IgE. While **IgE** is also present in only trace amounts in the blood, it is more prominently found on the surfaces of mast cells and basophils. Mast cells and basophils contain granules loaded with inflammatory mediators. The primary role of IgE is to amplify the body's response to invaders. Once secreted into serum, IgE attaches to Fc receptors on mast cells (**Fig. 24.11A** and **B**)—again by way of its Fc region—and like a Venus flytrap, waits until its matched antigen binds to its antigen-binding site. When two adjacent surface IgE molecules on a mast cell are cross-linked by antigen, a signal is sent internally that triggers degranulation (see Section 24.7). The release of histamine and other pharmacological mediators helps orchestrate an acute inflammation response early during a microbial infection (that is, while the antibody response is gearing up). The system can also cause severe allergic hypersensitivities (such as anaphylaxis) and milder forms like hay fever (**Fig. 24.11C**).

Primary and Secondary Antibody Responses

Once you have been infected with a microorganism or have been given a vaccine, what happens? After a lag period of several days, antibodies begin to appear in the **serum** (the fluid that remains after the blood clots). During the lag period, called the **primary antibody response** (**Fig. 24.12**), a series of molecular and cellular events causes a distinct subset of B cells located in lymph nodes and the spleen to proliferate and differentiate into antibody-secreting **plasma cells** and **memory B cells**. Plasma cells are much larger than B cells because of an enormous increase in protein synthesis and secretion machineries. Remember that each B cell is genetically programmed to make antibodies to only one antigen or epitope.

A subsequent exposure to the antigen, which can take place months or years after the initial encounter, will trigger a rapid, almost instantaneous increase in the production of antibodies and is called the **secondary antibody response** (**Fig. 24.12**). This quick response occurs thanks to the memory B cells formed during the primary response. Memory B cells comprise approximately 40% of the B-cell population. Once restimulated, memory B cells rapidly differentiate into plasma cells and secrete antibody.

The net result of the primary antibody response is the early synthesis and secretion of pentameric IgM molecules specifically directed against the antigen (or immunogen). Later during the primary response, a process known as **isotype switching** (or **class switching**) occurs, and the predominant antibody type produced becomes IgG rather

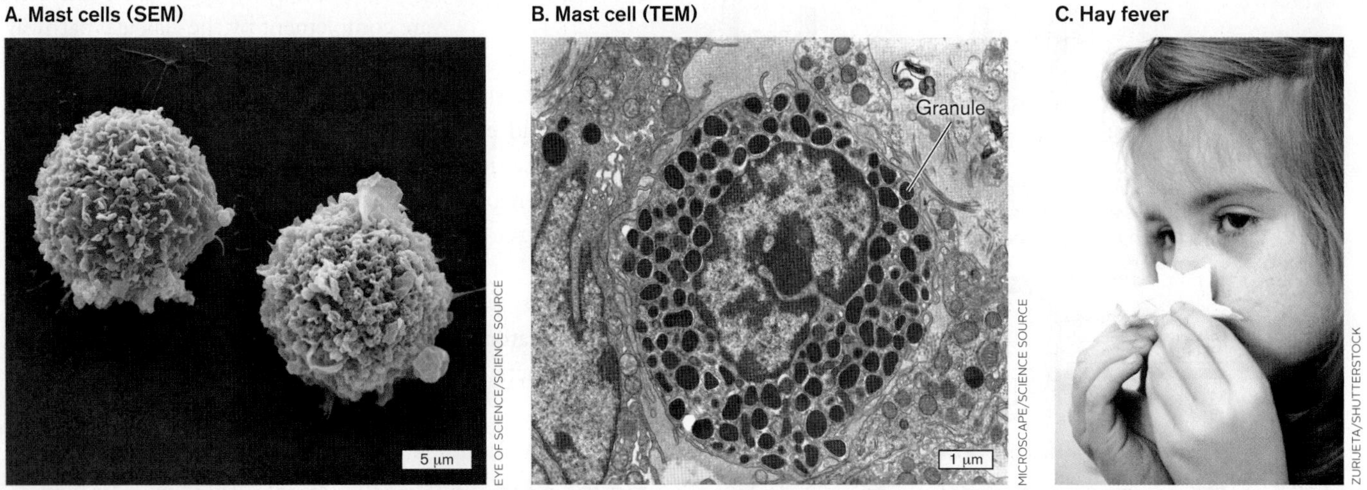

FIGURE 24.11 ■ Mast cells are major players in the inflammation response. **A.** Mast cell (SEM). **B.** Granules (arrow) inside a mast cell (TEM). **C.** Hay fever is the result of degranulation of IgE-coated mast cells, which release histamine and other pharmacological mediators.

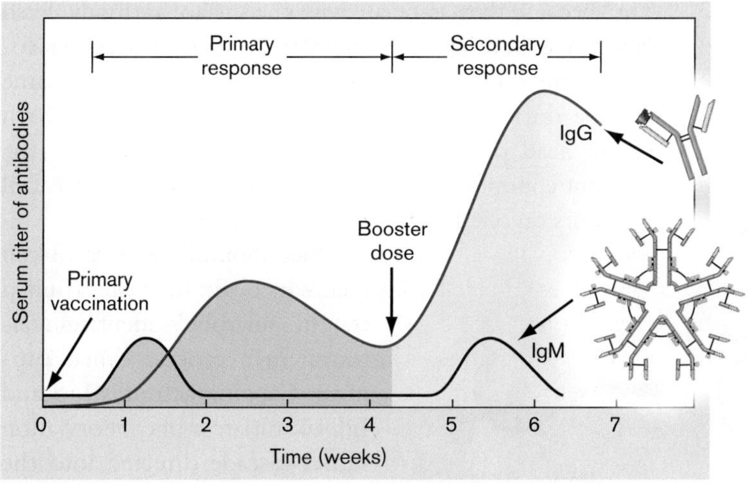

FIGURE 24.12 ▪ Primary versus secondary antibody response. Primary vaccination or infection leads to the early synthesis of IgM, followed by IgG. Either reinfection or a second, booster dose of a vaccine results in a more rapid antibody response, consisting mainly of IgG, because of memory B cells formed during the primary response. Note that the time course and level of antibody made vary with the immunogen and the host.

infection, but pathogens can do considerable harm during the primary-response lag phase. To avoid this harm, an innocuous version of a pathogen, or a harmless piece of it, can be injected into a person to trigger a primary response without producing disease (or, at worst, producing only a mild form of the illness). Immunization thus primes the immune system to respond efficiently and without delay upon encountering the real pathogen (see the Current Research Highlight). **Table 24.1** lists a variety of viral and bacterial diseases for which immunizations are available.

> **Thought Question**
>
> **24.6** The mother of a newborn was found to be infected with rubella, a viral disease. Infection of the fetus could lead to serious consequences for the newborn. How could you determine whether the newborn was infected in utero?

than IgM (discussed shortly). Antibodies made during the primary response, while specific for the immunogen, are actually not of the highest affinity. Later responses by memory B cells increase antibody affinity (see the next section).

As the immunogen is cleared from the body during the primary immune response, the levels of both IgG and IgM decline because the plasma cells that produced them die. Plasma cells have an average life span of only 100 days. The antibodies themselves are slowly removed from serum by pinocytosis or renal clearance. The memory B cells, meanwhile, are maintained in lymph nodes or bone marrow and continue to divide, albeit slowly. If the immune system encounters the antigen again at a later date, the rapid secondary response (or "anamnestic response," from the Greek *anamnesis*, meaning "remembrance") ensues.

During the secondary response, memory B cells that have undergone isotype switching from IgM to IgG become plasma cells that secrete copious amounts of IgG antibody. These antibodies have a higher specificity for the antigen than do the antibodies produced during the primary response (see the paragraph "Making memory B cells" at the end of this section). The higher specificity—a result of hypermutation—causes some plasma cells to produce antibodies that bind their antigens more strongly than primary-response antibodies do. Small amounts of IgM are also produced from the few memory cells that did not undergo isotype switching during the primary response.

The secondary antibody response is why vaccinations work. Before vaccinations, the only way to be protected from an infection by a given pathogen was to have been infected earlier by that pathogen. Memory cells made during the primary response will protect you against a second

How B Cells Differentiate into Plasma Cells

As mentioned earlier, each B cell circulating throughout the body or nestled in a lymphoid organ is programmed to synthesize antibody that reacts with a single epitope (a small portion of a protein). **Clonal selection** is the process whereby a foreign antigen selects which B-cell clone proliferates to large numbers and differentiates into antibody-producing plasma cells or memory B cells. The clonal expansion of the matching B cell enables large amounts of antibody specific for the antigen to be made. As illustrated in **Fig. 24.13**, clonal selection begins when an antigen binds to an antigen-specific B-cell receptor on a matching B cell (see the discussion of B-cell receptors that follows here).

Mature but naive B cells (those that have not previously encountered antigen) can produce only IgM and IgD, which have identical antigen specificities. These two antibody classes are displayed like tiny satellite dishes on the B-cell surface, anchored by their Fc regions through hydrophobic transmembrane segments. These surface antibodies (**B-cell receptors**, or **BCRs**) are the keys to stimulating the proliferation and differentiation of B cells into antibody-secreting plasma cells or memory B cells. Upon binding to its corresponding antigen via these surface antibodies, the B cell is said to become activated, whereby it multiplies and differentiates into a plasma cell that ultimately synthesizes only one antibody isotype (for example, IgG1). Clonal selection has begun. In addition to antigen binding, most B cells require help from T cells to become plasma cells and memory B cells (discussed later).

Each BCR is composed of antibody and two other membrane proteins, called Igα and Igβ (**Fig. 24.14**). Igα and Igβ are not immunoglobulins, but they are designated

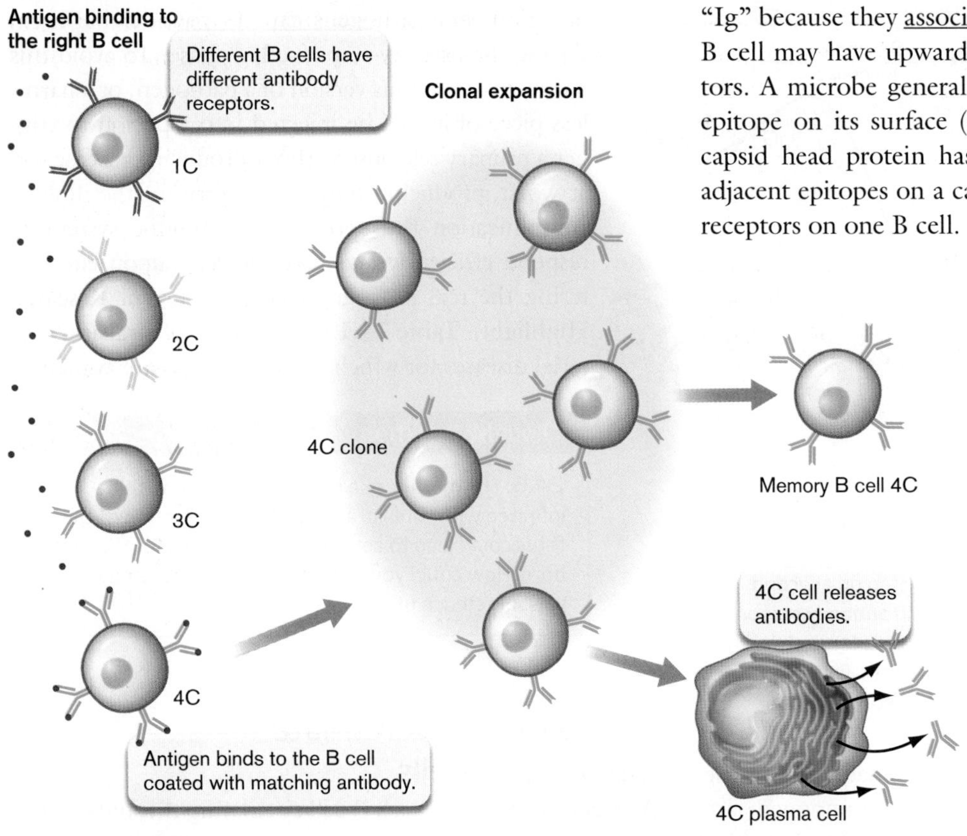

FIGURE 24.13 ■ Clonal selection. The B-cell population is composed of individuals that have antibody specificity for different antigens (represented by differently colored surface antibodies). When a B cell contacts its cognate antigen, an intracellular signal is generated, leading to proliferation and differentiation of that clone (clonal expansion). Plasma cells and memory B cells result.

"Ig" because they associate with the surface antibody. Each B cell may have upwards of 50,000 identical B-cell receptors. A microbe generally has multiple copies of the same epitope on its surface (think about a virus capsid). Each capsid head protein has the same epitope. As such, the adjacent epitopes on a capsid head can bind adjacent B-cell receptors on one B cell.

Once bound, surface B-cell receptors begin to cluster (linked by the microbe's identical epitopes) in a process called **capping**. Capping activates Igα and Igβ to initiate a phosphorylation signal cascade directed into the nucleus (**Fig. 24.15**). In a phosphorylation cascade, a phosphate group donated by ATP is passed from one protein to another, usually ending up on, and activating, a transcriptional regulator. The transcription factors at the end of the BCR cascade stimulate the transcription of genes that contribute to cell proliferation. Some of these now activated B cells can differentiate into plasma cells and secrete IgM antibody as part of the primary immune response.

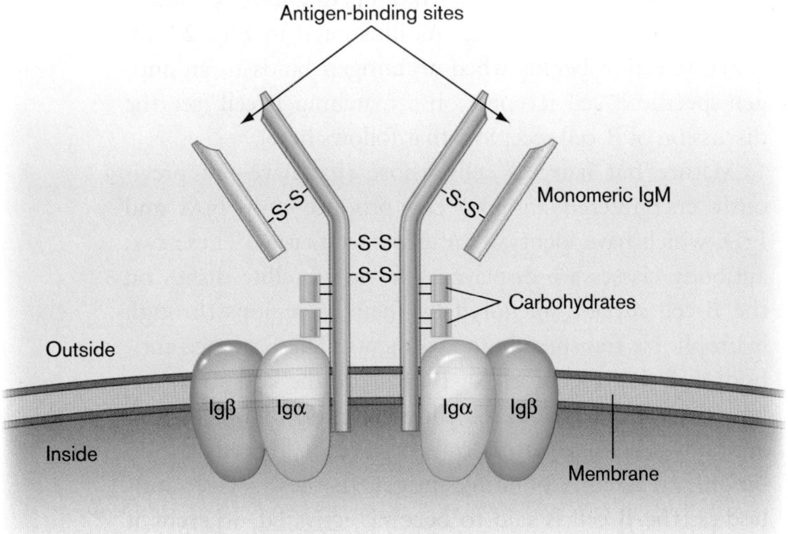

FIGURE 24.14 ■ B-cell receptor. The B-cell receptor is formed as a complex consisting of a monomeric IgM plus Igα and Igβ in the membrane. The Igαβ complex initiates the phosphorylation signal cascade (not shown).

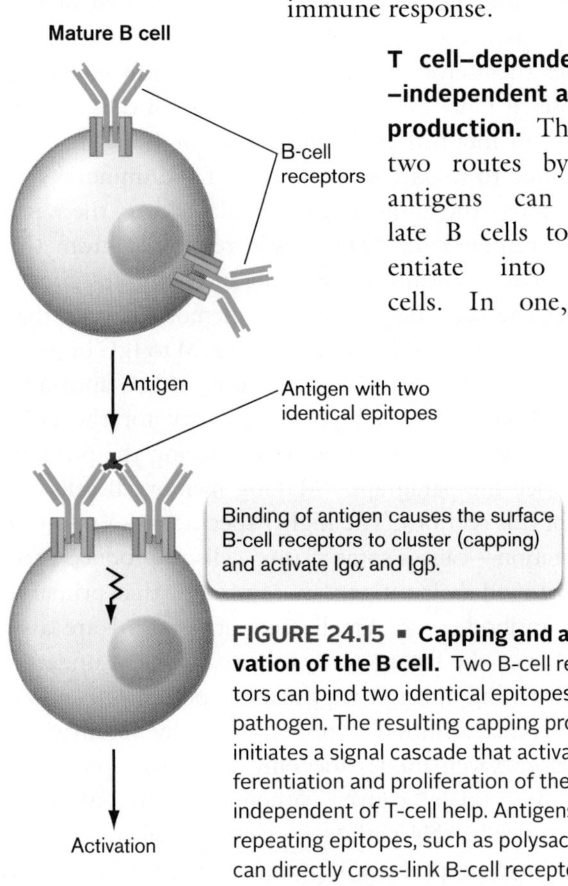

FIGURE 24.15 ■ Capping and activation of the B cell. Two B-cell receptors can bind two identical epitopes on a pathogen. The resulting capping process initiates a signal cascade that activates differentiation and proliferation of the B cell independent of T-cell help. Antigens with repeating epitopes, such as polysaccharides, can directly cross-link B-cell receptors.

T cell–dependent and –independent antibody production. There are two routes by which antigens can stimulate B cells to differentiate into plasma cells. In one, called the T cell–independent route, antigens that possess multiple repeating epitopes (for example, polysaccharide antigens) can directly cross-link B-cell receptors (the capping process)—a step necessary for triggering differentiation (see **Fig. 24.15**). Proteins, however, which are the largest group of antigens, do not contain multiple repeating units. A single protein possesses many small, discrete, single epitopes, making the cross-linking of B-cell receptors difficult. B-cell responses to these types of antigens require help from specific T cells, and this constitutes the second, T cell–dependent, route to B-cell activation. Thus, B cells usually require multiple signals to initiate a primary response. How T cells help foster B-cell activation will be discussed in Section 24.3.

Note: A single protein with multiple, nonidentical epitopes will not cross-link B-cell receptors. However, a microbe with many copies of that protein on its surface (think virus capsid) can cross-link B-cell receptors if the protein copies lie close enough together. To achieve cross-linking, the antigen-binding sites of a single B-cell receptor antibody must reach epitopes on adjacent proteins.

Figure 24.16 summarizes the basic steps of antibody formation leading to the production of plasma cells and

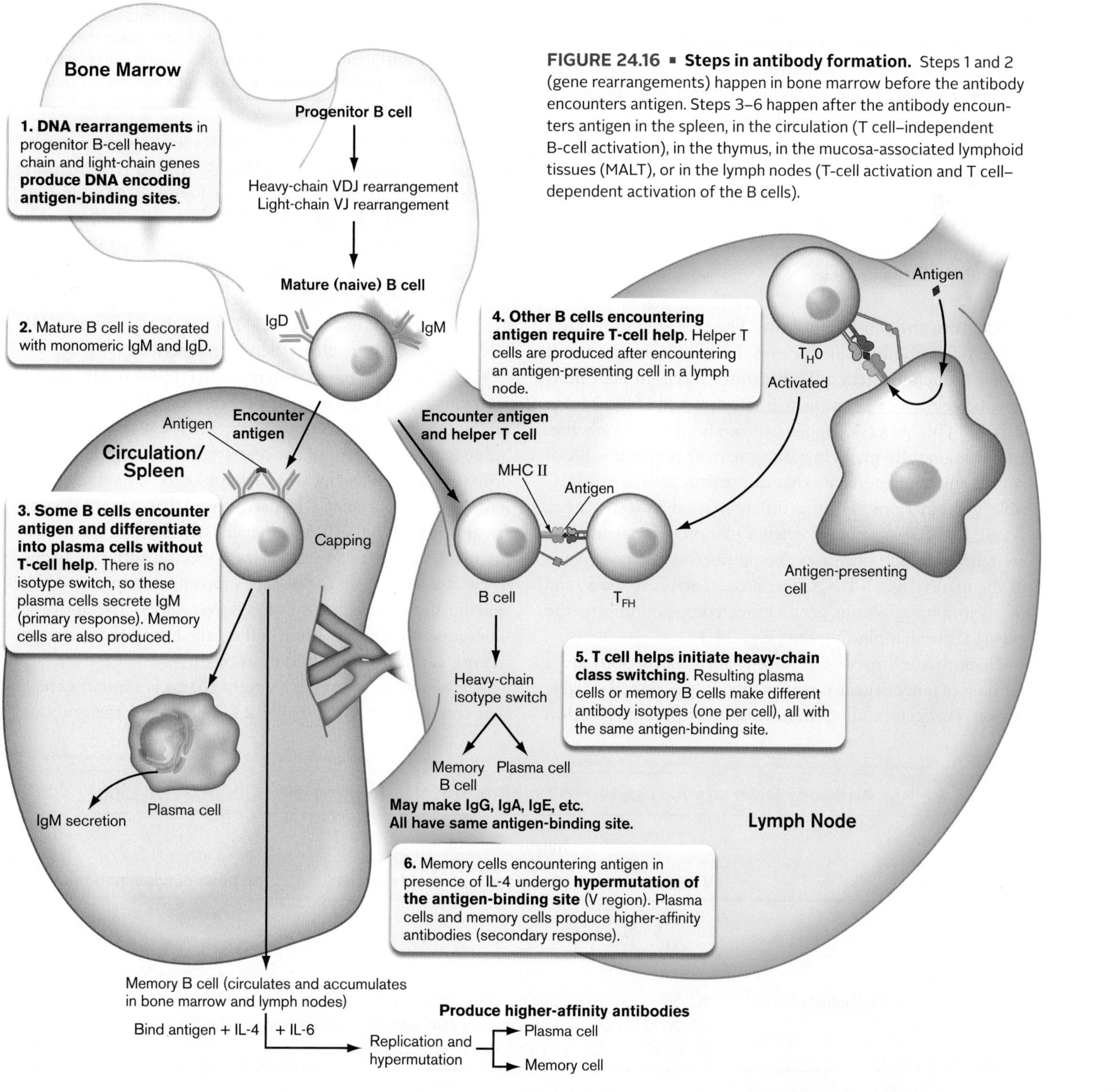

FIGURE 24.16 ■ Steps in antibody formation. Steps 1 and 2 (gene rearrangements) happen in bone marrow before the antibody encounters antigen. Steps 3–6 happen after the antibody encounters antigen in the spleen, in the circulation (T cell–independent B-cell activation), in the thymus, in the mucosa-associated lymphoid tissues (MALT), or in the lymph nodes (T-cell activation and T cell–dependent activation of the B cells).

memory cells. More detailed descriptions of these events follow in the next section.

Genetics of Antibody Production

Before we discuss how T cells influence antibody production, let's jump ahead and look at the process leading to antibody diversity. It is estimated that each human can synthesize 10^{11} different antibodies. Given that each B cell displays antibodies to only one antigenic determinant, it follows that there are 10^{11} different B cells in the body. We have learned, however, that each person possesses only about a thousand genes or gene segments involved in antibody formation. How are 10^{11} different antibodies made from only 10^3 genes?

Susumu Tonegawa was awarded the 1987 Nobel Prize in Physiology or Medicine for discovering that antibody genes can move and rearrange themselves within the genome of a differentiating cell. Three steps are involved: (1) the rearrangement of antibody gene segments (or cassettes), (2) the random introduction of somatic mutations, and (3) the generation of different codons during antibody gene (DNA) splicing. In humans, antibody diversity is generated continually over a lifetime.

Making the antigen-binding site. The first step in making a specific antibody occurs during the formation of a B cell from a progenitor stem cell (progenitor B cell) in bone marrow, before a foreign antigen is encountered (**Fig. 24.16**, step 1). This process happens throughout a person's life. Immunoglobulin genes in a bone marrow progenitor B cell have many gene segments that can rearrange in many possible combinations. During differentiation into a mature B cell, DNA segments are deleted in a process called **gene switching**, which decreases the number of gene segments in the mature B-cell DNA. The process starts at the 5′ end of an immunoglobulin gene cluster corresponding to the eventual antigen-binding site (**Fig. 24.17**).

In both the heavy- and light-chain gene regions are a number of tandem gene cassettes encoding potential variable regions (antigen-binding sites), separated by **recombination signal sequences** (**RSSs**). RSSs allow recombination to bring two widely separated gene segments together. There are approximately 170 "variable"-region (V-region) gene segments for the heavy and light chains. The light-chain V-region gene cluster lies upstream of a cluster of J-region ("J" for "joint" or "joining") genes that will eventually join a variable region to a light-chain constant region (C). Genes for the constant region reside farther downstream in the DNA. The arrangement of the heavy-chain genes is slightly more complex. In this case, the heavy-chain V cluster is followed by a D ("diversity") cluster and then the J region.

Note: Do not confuse the J-region gene segments used to make heavy- and light-chain proteins with the J-chain protein that holds together IgM and IgA multimers. They are completely different and unrelated. The J-region gene segments do not encode the J chain.

A summary of the genetic processes leading to antibody formation is shown in **Figure 24.17**. Antibody formation begins with a recombination event between RSS sites at one heavy-chain D segment and one heavy-chain J segment, which deletes all the intervening D and J segments (**Fig. 24.17**, step 1). Next, the new DJ region joins to one of the heavy-chain V segments, deleting all of the intervening V and D segments (step 2). The result is a joined VDJ DNA sequence, which can then be transcribed (step 3).

Each V segment has its own promoter. However, if extra V segments remain upstream of the rearranged VDJ sequence on the DNA, then the only promoter that will fire is the one immediately upstream of the rearranged VDJ segment. The primary RNA transcript will then undergo RNA splicing to remove any J-segment RNA sequences that remain downstream of the VDJ RNA sequence (step 3). The result is a mature B cell that can synthesize a specific antibody (steps 4–6). Remember, all of the DNA recombination and RNA splicing happen before the B cell ever "sees" the antigen. Consequently, the mature B cell is called "naive" because it has not yet been stimulated by antigen.

The sequence of events for light chains is similar, except that the product is VJ. **Table 24.3** illustrates the amount

TABLE 24.3 Antibody Diversity Attributed to Combinatorial Joining in the Human Germ Line

	Number of:			
Chain type	**V regions**	**D regions**	**J regions**	**Number of combinations**
λ Light chains	30	0	4	30 × 4 = 120
κ Light chains	40	0	5	40 × 5 = 200
Heavy chains	100	27	6	100 × 27 × 6 = 16,200
Number of possible antibodies	16,200 heavy-chain combinations × 120 λ-chain combinations = 1.94×10^6 16,200 heavy-chain combinations × 200 κ-chain combinations = 3.24×10^6 $(1.94 \times 10^6) + (3.24 \times 10^6) = 5.18 \times 10^6$ combinations			

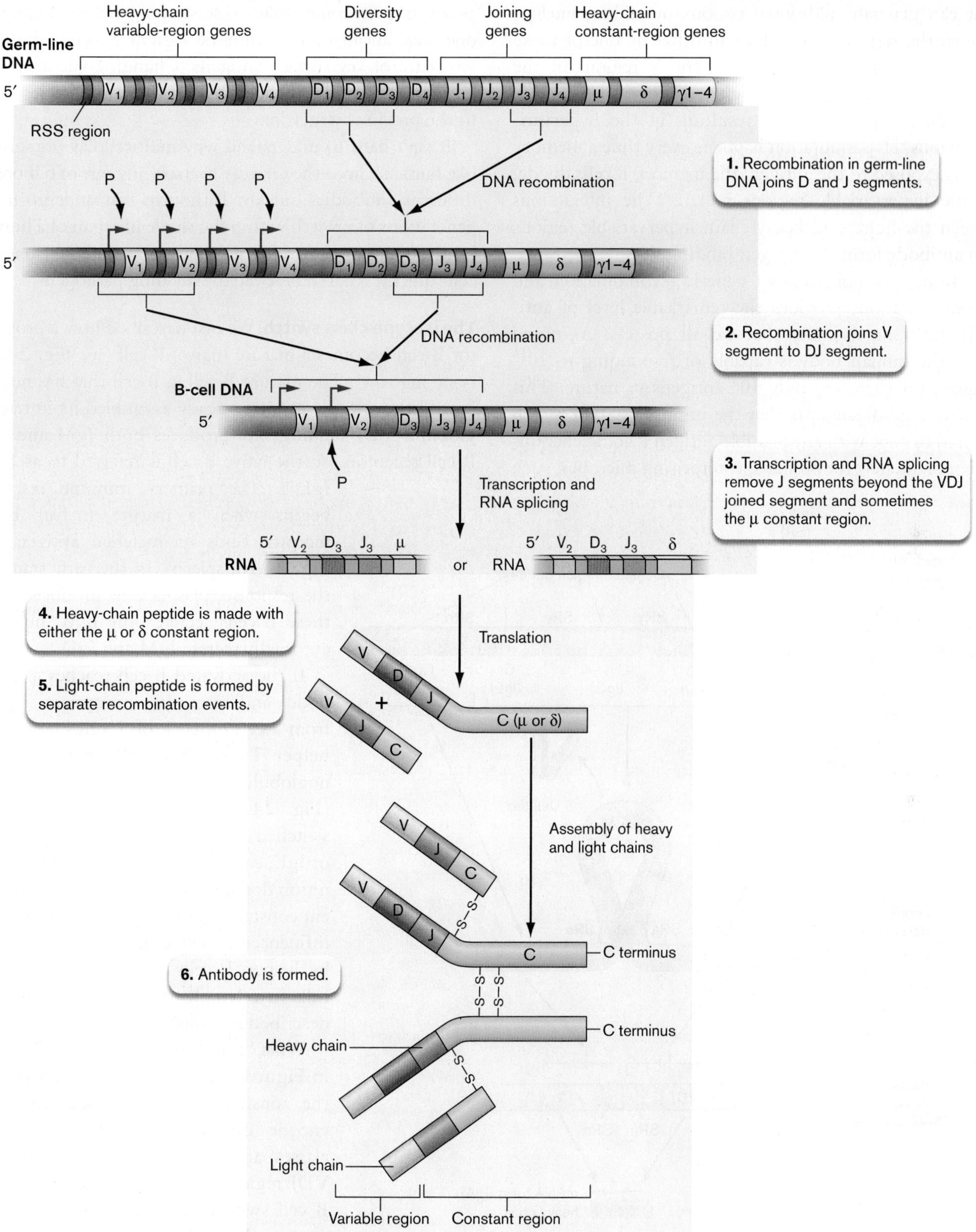

FIGURE 24.17 ■ Formation of the VDJ regions of heavy chains. Note that only a small subset of the V, D, and J genes listed in Table 24.3 is shown in this model. RSS = recombination signal sequence.

of antibody diversity that can be achieved in humans simply by this combinatorial re-joining (a total of about 5×10^6 antigens can be recognized). However, that is only about 0.0001% of the total possible diversity.

Where does the rest of antigen-binding-site diversity arise? Additional diversity comes from the junctions of VJ and VDJ, where recombinational joining can occur between different nucleotides. Each genetic recombination event can generate additional codons by adding nucleotides, so the resulting peptides will differ by one or more amino acids. After recombination, the V regions of the germ lines are susceptible to high levels of somatic mutation (called hypermutation), resulting in the hypervariable regions. Hypermutation happens every time a memory B cell is exposed to the antigen; the memory B cells divide, and the hypervariable regions mutate. The interactions between the light- and heavy-chain hypervariable regions in an antibody form the antigen-binding sites.

In sum, a combination of genetic recombination and random mutations provides the remarkable level of antigen-binding-site diversity that we all possess. As noted earlier, the human body is capable of responding to 10^{11} antigens, yet there are only 10^8 antigens in nature. This apparent overkill suggests that the immune system is well prepared to cope with any possible antigen it could encounter. Unfortunately for humans, enterprising microbes, such as the trypanosomes that cause sleeping sickness, can stay one step ahead of the immune system by changing the structure of key surface antigens. Changing the antigenic structure of a protein renders useless those antibodies made to the previous structure.

It isn't hard to understand why multicellular organisms, like humans, need the capacity to make any one of billions of different antibodies quickly. Pathogens can undergo many generations of growth within the single life span of a human host. So, humans have to generate recombinant clones of cells quickly to overcome rapidly dividing pathogens.

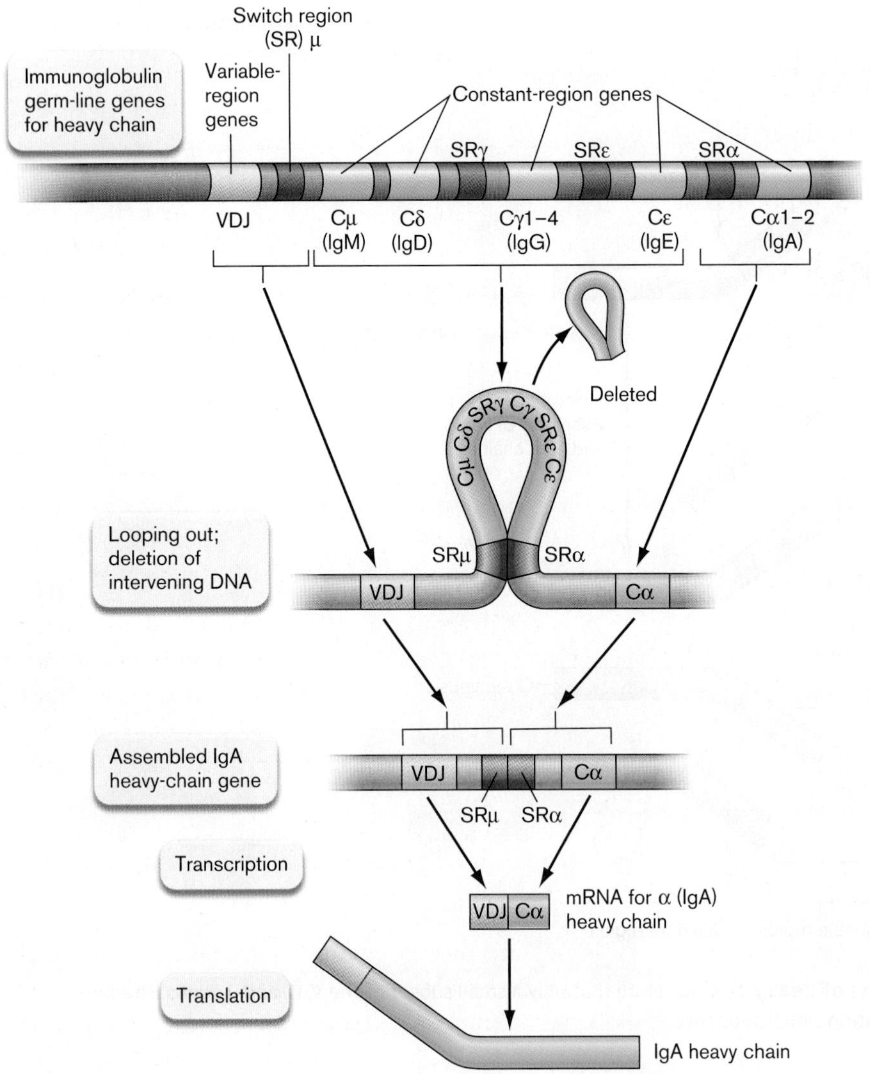

FIGURE 24.18 ■ Heavy-chain class switching. As a B cell becomes activated, a switch in antibody isotype will occur. The switch involves recombination between isotype cassettes that brings one heavy-chain constant region (Cα in the example, for IgA) in tandem with a VDJ sequence.

The isotype class switch. We just described how a progenitor B cell becomes a mature (naive) B cell (see **Fig. 24.16**, bone marrow). The mature B cell (a B cell that has not yet "seen" the antigen but has already assembled its immunoglobulin VDJ binding site) produces both IgM and IgD B-cell receptors (so the naive B cell is referred to as IgM^+ IgD^+). The primary immune response begins when a mature (naive) B-cell receptor finds its matched antigen (see **Fig. 24.16**, spleen). In the early stages of the response, plasma cells produced from these B cells secrete only IgM, but they eventually secrete IgM and IgD.

If the activated B cell reaches a lymph node and receives appropriate signals from a certain type of T cell known as a **helper T cell (T_H cell)**, further immunoglobulin isotype switching will occur (**Fig. 24.16**, lymph node, step 5). The switched B cell may then make IgG, IgA, or IgE, each with the same antigen recognition domain (variable region) but different constant regions. The type of switch is influenced by **cytokines** that are secreted by helper T cells. How T cells meet B cells in a lymph node germinal center is described in **eTopic 24.4**.

What "flips" the isotype switch? Notice in **Figures 24.17** (step 2) and **24.18** that the constant-region gene segments that encode different antibody heavy-chain classes are arranged in tandem after a VDJ region. The mechanism by which a B cell switches to make IgG (C regions gamma 1–4), IgE (C-epsilon), or IgA (C-alpha 1 and 2) is very similar to VDJ formation (**Fig. 24.18**). Each constant segment, except delta, contains a repeating DNA base sequence called a **switch region**. Recombination between these switch regions will delete the intervening DNA between the VDJ region and either the IgG, IgE, or IgA constant regions. Because the VDJ region is the same regardless of which C_H gene is selected, the antibody produced will have the same antigenic specificity as the original IgM.

However, the heavy-chain switch selection process is not random. The type of cytokine present at the time of the switch will influence which C_H gene is selected. Consequently, cytokines control whether IgG is made for circulation or, instead, IgA is made for secretion. Note that any heavy-chain peptide (alpha, gamma, delta, and so on) can combine with any light chain (lambda or kappa). However, a single, mature B cell can make only one type of heavy chain and one type of light chain.

B-cell selection. We just described how the immune system can make antibodies against any antigen. But why aren't antibodies made against our own antigens? Normally, antibodies against self are not made, because immature B cells in bone marrow undergo three forms of negative selection if their BCR binds to a self antigen. The immature B cell can undergo apoptosis, completely removing it from the B-cell repertoire. The cell can perform receptor editing in which the immature B cell can develop a new BCR that does not respond to self. Or it can induce a state of anergy in which the mature B cell is released to the circulation but cannot react to self antigen because intracellular signaling pathways have been blocked. Another reason self-reacting antibodies are not made involves the T-cell selection process that kills T cells, including helper T cells, that recognize self antigens. Without a self-reacting helper T cell, a B cell with matching self specificity cannot be activated (described in Section 24.3).

Thought Questions

24.7 B cells in early stages have both IgM and IgD surface antibodies, but the delta region has no switch region. Why does the delta region have no switch region?

24.8 Why do individuals with type A blood have anti-B and not anti-A antibodies?

Making memory B cells. Having encountered an antigen, the antigen-activated B cell will divide to make memory B cells, as well as plasma cells (see **Figs. 24.13** and **24.16**). Memory B cells are B cells that have already undergone class switching (committed to making IgG, for instance) but, unlike other B cells, are very long-lived. Their long lives provide immunological memory.

During their lifetimes, memory B cells, like other B cells, hypermutate the antigen-binding regions of antibody genes (the VDJ and VJ regions). Mutations that increase the affinity of the antibody are clonally selected during a secondary response. That is, the hypermutation helps sharpen the antigen-binding site, making it even more specific toward a given microbe or antigen. Note, too, that as an immune response clears an infection, the antigen becomes scarce. This means that only B cells with the highest-affinity antibodies as part of their B-cell receptors will remain activated. This process is known as **affinity maturation**.

To Summarize

- **Antibodies, or immunoglobulins, are Y-shaped molecules** that contain two heavy chains and two light chains. **There are five classes (isotypes) of antibodies defined by the structure of the heavy chains.** IgM is composed of five antibody monomers. IgG is a monomer and the main antibody type in blood. IgA is mostly secreted as a dimer at mucosal areas, while IgD is mostly found bound to B cells. IgE is a circulating antibody that attaches to mast cells and mediates some allergic reactions. Antibodies are members of the immunoglobulin superfamily of proteins.
- **Each antibody molecule contains two antigen-binding sites.** Each binding site is formed by the hypervariable ends of a heavy- and light-chain pair. **The Fc portion** ("tail") of an antibody can bind to specific receptors on host cells. This binding is antigen independent.
- **Immunological distinctions between antibodies are characterized by a hierarchy of protein sequence differences. Isotypes** reflect species-specific differences in the heavy chains of major antibody classes (for instance, IgM versus IgG). Antibody **allotypes** are sequence differences in a heavy-chain isotype (for example, IgG) between individuals of a single species. **Idiotypes** represent sequence differences in the antigen-binding sites of antibodies found within a single individual.
- **Antibody diversity occurs within B-cell precursor cells** via a complex series of splicing events between adjacent DNA cassettes, as well as mutational events in DNA sequences encoding the hypervariable regions of heavy and light chains. A **mature (naive) B cell** has randomly completed VDJ rearrangements in its DNA to make the specific antigen-binding site for antibodies. The naive B cell places IgM and IgD antibodies on its surface (part of the B-cell receptor).
- **A B-cell receptor** consists of a membrane-embedded antibody in association with the Igα and Igβ proteins. BCRs on a single B cell are made of a single antibody isotype that specifically binds one epitope. Binding of antigen to the B-cell receptor triggers B-cell proliferation and differentiation called **clonal selection**, which occurs during the primary and secondary antibody responses.

- **The primary antibody response** to an antigen begins when B cells differentiate into antibody-producing plasma cells and memory B cells. IgM antibodies are generally the first class of antibodies made during the primary response.
- **Isotype switching** (or class switching) from IgM production to other antibody isotypes begins once the B-cell receptors (BCRs) on a naive B cell bind their target antigen.
- **The secondary antibody response** occurs during subsequent exposures to an antigen. Memory B cells are activated, hypermutate the VDJ region, and then rapidly proliferate and differentiate into antibody-secreting plasma cells. IgG is the predominant antibody made.

24.3 T Cells Link Antibody and Cellular Immune Systems

Does each arm of the adaptive immune system (antibody-dependent or cell-mediated) know that the other exists? Because different types of infections tilt the immune response one way or the other, the systems must communicate with each other to provide balance. T-cell lymphocytes, with integral roles in both antibody production and cell-mediated immunity, manage the balance.

Although derived from the same progenitor stem cell as B cells (see Fig. 23.17), T cells develop in the thymus (rather than in the bone marrow where B cells develop), and they contain surface protein markers (antigens) different from those of B cells. T cells come in several varieties marked by the presence of different cell differentiation (CD, or "cluster of differentiation") surface proteins and by the types of cytokines they produce. Two critically important groups are helper T cells (T_H cells, already mentioned) and **cytotoxic T cells (T_C cells)**. Helper T cells display the surface antigen CD4, while cytotoxic T cells display CD8. Cytotoxic T cells are the "enforcers" of the cell-mediated immune response. They destroy the membranes of host cells infected with viruses or bacteria.

Our understanding of helper T cells has evolved over the last few years. Helper T cells come in several models, the first of which is the T_H0 cell. T_H0 cells are precursors to the other types, including:

- T_{FH} (follicular helper T cells), a heterogeneous set of CD4 T cells that drive B-cell differentiation into antibody-secreting plasma cells. Different T_{FH} cell subsets secrete different combinations of cytokines that trigger antibody isotype (class) switching.
- T_H1 cells assist in the activation of cytotoxic T cells.
- T_H2 cells recruit eosinophils to combat parasitic infections and can inhibit T_H1 proliferation.
- T_H17 cells stimulate inflammation by secreting interleukin 17 (IL-17), a pro-inflammatory cytokine. IL-17 helps trigger recruitment of neutrophils and macrophages to an infection.
- Treg cells (or Tregs, regulatory T cells) dampen the inflammatory response and secrete the anti-inflammatory cytokine IL-10.

The ratios of these cells and the cytokines they produce tilt the immune system toward either more humoral or more cell-mediated responses. To be of any use, however, T cells first must be activated by antigen.

T-Cell Activation Requires Antigen Presentation

Unlike B cells, T cells never bind free-floating antigen. T cells are activated only by antigen bound to another cell's surface. These other cells are collectively called **antigen-presenting cells (APCs)** because they present antigens to T cells. The APC surface proteins that hold and present the antigen are known as **major histocompatibility complex (MHC)** proteins (**Fig. 24.19A**). MHC proteins differ between species and between individuals within a species. They help determine whether a given antigen is recognized as coming from the host (a self antigen) or from another source (a foreign antigen), in a phenomenon called histocompatibility (hence the name "major histocompatibility complex").

The salient feature of all MHC molecules is that they bind antigen that has entered the host cell, and then present the antigen back on the cell surface. MHC molecules are critical to the immune system because in order to be activated, the T cell must recognize the complex of a foreign antigen attached to an MHC molecule on an APC.

Two classes of MHC molecules are found on cell surfaces. Both classes belong to the immunoglobulin family of proteins, but they are not immunoglobulins (antibodies). **Class I MHC molecules** are found on all nucleated cells, whereas **class II MHC molecules** have a more limited distribution on professional APCs such as dendritic cells, B cells, and macrophages. MHC I is a heterodimer made of an alpha chain and β_2-microglobulin. The antigen-binding cleft is within the alpha chain. MHC II has alpha and beta subunits (different from MHC I), both of which contribute to the antigen-binding cleft.

The surface CD proteins CD8 and CD4, mentioned earlier, help T cells distinguish between MHC class I and class II molecules on antigen-presenting cells. The CD8 molecules on T_C cells selectively bind MHC class I, while CD4 molecules on T_H cells selectively bind MHC class II. Antigens presented on class I MHC molecules generally

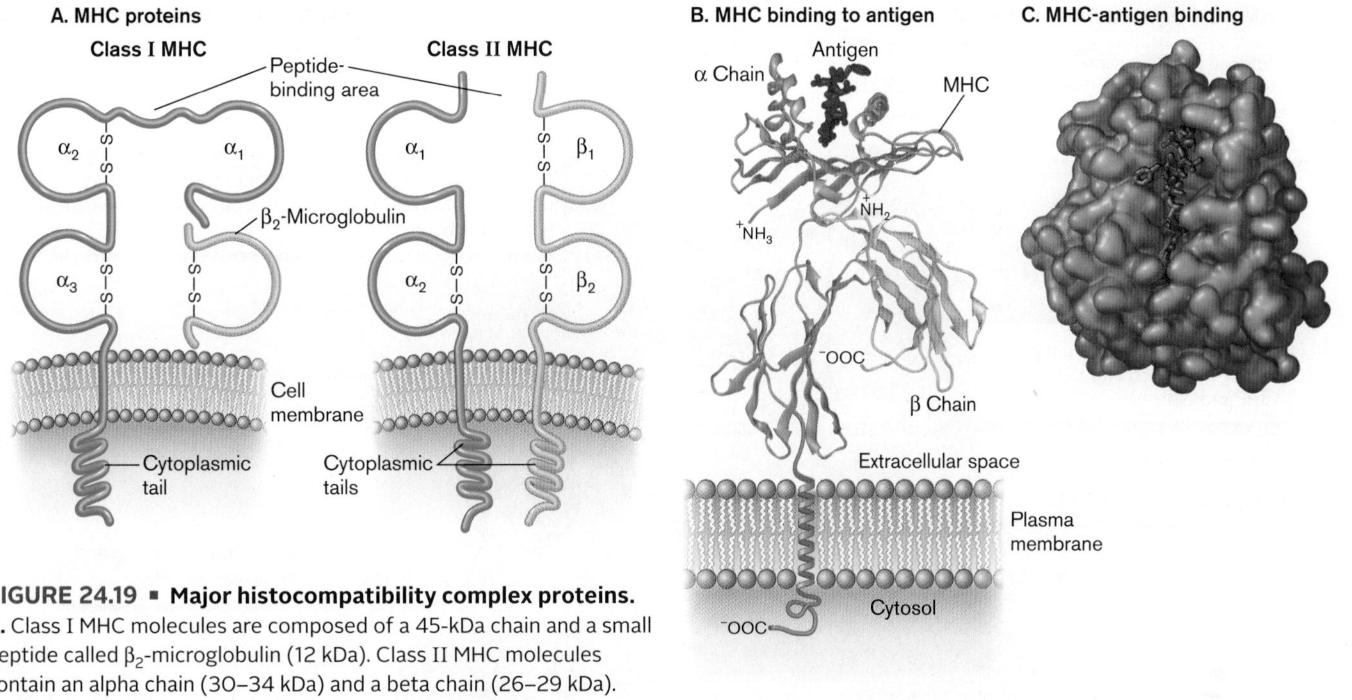

FIGURE 24.19 ■ Major histocompatibility complex proteins. **A.** Class I MHC molecules are composed of a 45-kDa chain and a small peptide called β_2-microglobulin (12 kDa). Class II MHC molecules contain an alpha chain (30–34 kDa) and a beta chain (26–29 kDa). The peptide-binding regions of both classes show variability in amino acid sequence that yields different shapes and grooves. CD8 T cells recognize antigen peptides associated with class I molecules, while CD4 T cells recognize peptides bound to class II molecules. Peptide antigens nestle in grooves formed by these molecules and are held there awaiting interaction with T-cell receptors. **B.** Antigen binding to an MHC I molecule. **C.** Top view of antigen (red) nestled in the MHC peptide-binding site. (PDB code: 1BII)

arise from intracellular pathogens, such as viruses and some bacteria that require cell-mediated immunity for resolution. In contrast, antigen peptides presented on class II MHC molecules originate from extracellular infections, which are resolved by antibody.

APCs Receive, Process, and Present Antigens by Two Paths

How are foreign peptide antigens placed, or presented, on host cell surfaces? In the initial stages of an immune response, antigen-presenting cells internalize the pathogen, such as a virus or a bacterium. Inside the APCs, pathogen proteins are degraded into smaller peptides (epitopes). These epitopes are placed within the MHC-binding clefts of MHC I or II molecules and transported back to the cell surface.

Whether an antigen peptide binds to class I or class II MHC molecules generally depends on how the antigen initially entered the cell (**Fig. 24.20**). Endogenous antigens, which are synthesized by viruses and intracellular bacteria as they grow within the cytoplasm of an APC, will attach to class I MHC molecules on the endoplasmic reticulum and are moved to the cell surface (**Fig. 24.20**, left). In contrast, exogenous antigens, which are produced outside of the APC (as are most bacterial antigens), enter the cell via phagocytosis (**Fig. 24.20**, right). The antigen-containing phagosome then fuses with a vacuole whose interior is lined with MHC II molecules. There, antigens bind to MHC II clefts and the MHC class II–peptide complex is carried to the cell surface. Once the antigen is presented on the surface of the APC, T cells can interact with the antigen-MHC complex via their T-cell receptors (discussed shortly). APCs and naive T cells interact within the lymph nodes, the spleen, or Peyer's patches in the gut. So, APCs must make their way to those locations to generate an immune response.

Note: MHC molecules present only peptide antigens. Carbohydrate antigens cannot attach to the binding clefts of MHCs but will trigger antibody production independent of T cells.

The antigen-binding clefts of MHC molecules are shown in **Fig. 24.19B** and **C**. It is important to realize that the antigen-binding specificity of any single MHC molecule is very broad, unlike the highly antigen-specific targeting of an antibody molecule or T-cell receptor (described next). In contrast to the many billions of different antibody molecules made as a result of multiple gene rearrangements and somatic mutations, a person possesses genes that express only six MHC I molecules and six to eight MHC II molecules, with each parent donating half of the genes necessary (all the MHC genes are on chromosome 6 and are codominantly expressed). MHC genes do not undergo rearrangement or somatic mutation. However, the promiscuous binding capability of each MHC binding site, plus the independent assortment of maternal and paternal alleles, means

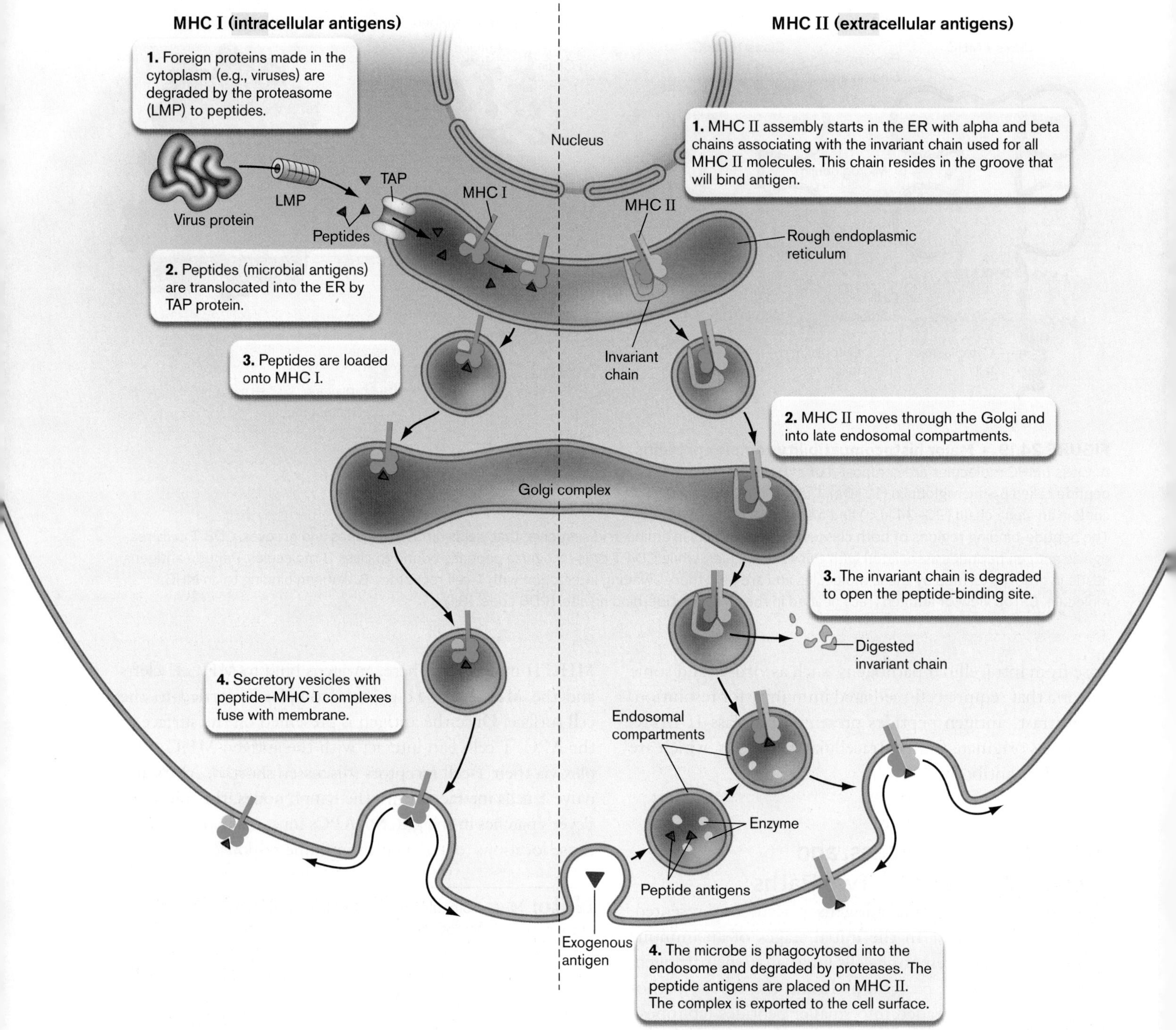

FIGURE 24.20 ■ APC processing and presentation of peptide antigens on class I and class II MHC proteins. Left: Microbial proteins made in the host cytoplasm are degraded, and peptides are placed on class I MHC molecules in the endoplasmic reticulum (ER). **Right:** Microbial proteins made outside the cell are phagocytosed (bottom), degraded in a phagosome or endosome, and placed on class II MHC molecules. LMP = low-molecular-mass polypeptide component of the proteasome; TAP = transporter of antigen peptides.

that this limited number of MHC molecules can bind most peptide antigens. Despite that diversity, not all possible antigens can be recognized by the MHCs in a single person. Among the human population, however, there are thousands of alleles for each MHC locus, each of which encodes a slightly different binding cleft that shifts specificity. This form of MHC diversity ensures that the population as a whole will not succumb to a new or mutated pathogen.

T-Cell Receptors

T-cell receptors (**TCRs**) are the antigen-binding molecules present on the surfaces of T cells (**Fig. 24.21**). Unlike BCRs, TCRs are not antibodies. A TCR on helper T cells will bind only to antigens attached to MHC II surface proteins on antigen-presenting cells, as noted earlier. However, the TCRs of cytotoxic T cells can bind viral antigens attached to MHC I surface proteins present on any

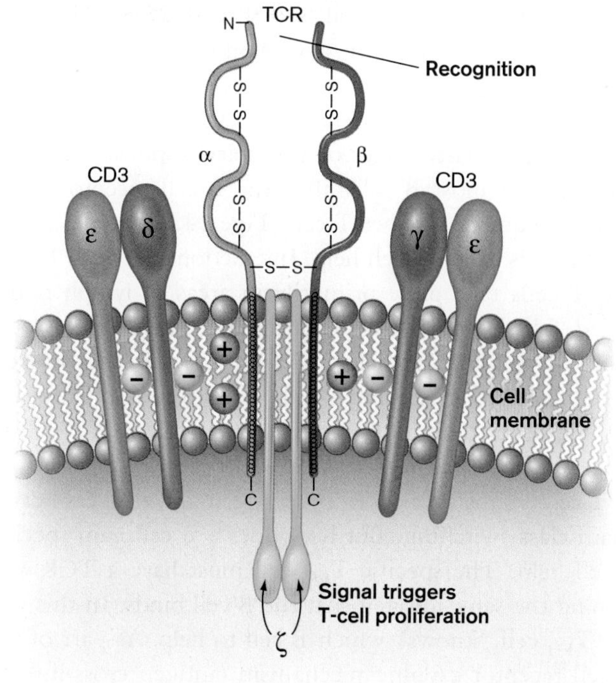

FIGURE 24.21 ■ The T-cell receptor (TCR) and CD3 complex. T-cell receptor proteins are associated with CD3 proteins at the cell surface. Antigen binds to the alpha and beta subunits. The positive and negative charges holding the complex together come from amino acids in the peptide sequences. Once bound to antigen, the complex transduces a signal into the cell. This signal triggers T-cell proliferation.

virus-infected cell. Remember that all nucleated cells have surface MHC I molecules.

The T-cell receptor is composed of several transmembrane proteins. Two molecules—alpha and beta—make up the part that recognizes antigen. The TCR alpha and beta proteins are found in a complex with four other peptides, which together form the CD3 complex (see **Fig. 24.21**). When stimulated by binding to antigen, these ancillary CD3 complex proteins recruit and activate intracellular protein kinases and launch a phosphorylation cascade that triggers proliferation of the T cell and initiates key immunological events that we describe later in the chapter.

How is antigen diversity generated in TCR molecules? Much like antibodies, the alpha and beta proteins of TCRs are formed from gene clusters that undergo gene rearrangements analogous to, but different from, those of the immunoglobulin genes. For instance, the antigen-binding site of an alpha chain is formed by VJ regions, while those of the beta chain are composed of VDJ regions. The alpha and beta V, D, and J genes are distinct from those of antibodies.

Why don't your T cells see your body's antigens? Are they tolerant somehow? Actually, T cells that bind self antigens are not made tolerant; they are killed in a process called T-cell education or T-cell selection.

T-Cell Education and Deletion

Greek mythology tells of Narcissus, a young man punished by the gods for scorning the women who fell in love with him. One day Narcissus saw a beautiful face in a pool of water. Not knowing it was his own reflection, he fell in love, tumbled into the pool, and drowned. His inability to recognize himself is analogous to the danger posed by an immune system. Immune systems must be able to distinguish what is self (meaning antigens present in the body's own tissues) from what is not self. Otherwise, the immune system would constantly attack the body's own cells.

Earlier we examined B-cell selection within bone marrow. Here we describe how T cells are selected in the thymus. The selection process has two steps. First, T cells whose TCR can bind self MHC proteins are positively selected. Second, surviving T cells that can bind self MHC are negatively selected (deleted) if they bind other self antigens. The goal is to educate T cells to distinguish self from nonself.

In the first step (**positive selection**), precursor T cells are passed by special thymus epithelial cells bearing MHC molecules. T cells with TCRs that bind these self MHC proteins are allowed to live, and they continue through the thymus. However, T cells that do not recognize self MHC peptides, or recognize them too strongly, are killed (deleted). Positive selection is important because T cells must recognize self MHC in order to bind antigen-MHC complexes on APCs (MHC restriction is discussed in **eTopic 24.5**).

Why kill T cells that strongly bind MHC molecules? If the immune system's T-cell repertoire included cells that bound self MHC too tightly, then the T cells would constantly react to the body's own MHC molecules, regardless of which antigen peptides were attached (self or nonself). Note, however, that some strongly self-reactive T cells are allowed to survive and are converted into **regulatory T cells** (**Tregs**). Regulatory T cells can block the activation of harmful, self-reactive (autoimmune) lymphocytes that escape deletion and enter the circulation. Note that this is the first way that Tregs can be made. The second way involves APC activation of T_H0 cells, discussed shortly.

The second stage of T cell "education" is called **negative selection**. T cells that were positively selected for binding self MHC molecules move on to another section of the thymus, where their TCRs are further screened for an ability to bind self antigens—an undesirable property. But wait—doesn't the thymus express only thymus antigens? How can this organ screen for T-cell recognition of antigens expressed on other host cells (for example, heart cells)?

The answer is a special gene activator called AIRE in certain thymus epithelial cells. AIRE stimulates the synthesis of all human proteins in small amounts. A constellation of self proteins is then presented on the epithelial cells'

MHC I and MHC II molecules. T cells that originally survived positive selection are screened by these MHC–self antigen complexes to test for TCR recognition. If a TCR on one of the T cells binds to a self antigen, that T cell is instructed to kill itself (via apoptosis). Almost 95% of T cells entering the thymus die during these positive and negative selection processes. What remains are T cells that can recognize MHCs bound to foreign antigens. Having been "educated," these T cells leave the thymus to seed secondary lymphoid organs such as the spleen.

What about thymectomy? You might ask how someone whose thymus has been removed (a treatment for myasthenia gravis) can live if the organ is critical for T-cell maturation and for deleting self-reactive T cells. Actually, the thymus begins losing function shortly after birth, such that very little function remains in adults. Fortunately, a large amount of T-cell education occurs during fetal development. Once a T cell is educated, reserve pools of these T cells are maintained throughout life outside of the thymus. Some T cells, apparently, can also mature in secondary lymphoid tissues. Thus, adults without a thymus can live relatively normal lives. Babies born without a thymus, however, have a severe, life-threatening T-cell deficit.

APCs Activate T_H0 Helper T cells

Figure 24.22 summarizes the steps that lead to T-cell activation in the lymph node and highlights how activated T cells influence the two types of adaptive immunity: humoral (antibody) and cellular. The different classes of T cells—T_H and T_C—require different but interrelated activation programs. Let's start with T_H cell activation. T_H cells require two molecular signals to become active. First, T-cell receptors along with CD4 on precursor T_H0 cells specifically recognize and link to antigen-MHC II complexes on antigen-presenting cells (APCs; **Fig. 24.22**, step 1a). The second signal needed to activate a T_H0 cell is the binding of a CD28 molecule present on the T_H0 cell surface to a B7 protein (aka CD80) on the APC cell surface (also step 1a). Different cytokines produced by mast cells, macrophages, epithelial cells, and so on during infection then convert activated T_H0 cells to T_{FH}, T_H1, T_H2, T_H17, or Treg cells (step 2a). The activation of cytotoxic T cells (T_C cells) will be examined later.

T_{FH} Cells Activate B Cells

During the early phase of a primary response, one type of B cell in the lymph node can become activated after binding antigen if the B cell also binds to complement C3 on a bacterium (see Section 23.6), or if a microbe-associated molecular pattern (MAMP) binds to a Toll-like or NOD-like receptor on or in the B cell (see Section 23.5). This activation does not result in heavy-chain class switching; only IgM is secreted by the resulting plasma cell. This is partly why the primary response starts with IgM (see **Fig. 24.12**).

During the later stages of a primary response, however, activation of most B cells after the B-cell receptor binds antigen requires a helper T cell (**Fig. 24.22**, step 3). This assistance is called T-cell help. Interactions between B cells and T cells take place primarily in areas of lymph nodes called follicles, also called germinal centers (see **eTopic 24.4**). The follicular helper T cell (T_{FH}) is the primary T cell involved in driving B-cell activation. (Actually, any T_H cell other than T_H0 and Treg can activate a B cell to make antibody, but to a lesser extent.)

T-cell help is specific to the antigen and triggers heavy-chain class switching. But how does a B cell gain specific T-cell help? The specific T_{FH} cell must have a TCR able to bind the same antigen that the B cell binds. In this way the T_{FH} cell "knows" which B cell to help. As part of the B-cell receptor capping mechanism (antigen cross-linking of BCR), some antigen bound by B-cell receptors becomes internalized and processed to be presented back on the B cell's surface MHC II receptor (**Fig. 24.22**, step 3). For instance, a T_{FH} cell that was activated by dendritic cells presenting a particular antigen (step 1a) can also use its T-cell receptor to bind that same antigen presented on a B-cell MHC II molecule (step 3). This contact allows CD40 on the B cell to bind CD154 on the T cell—an interaction that completes activation of the B cell. The B cell will now differentiate into a plasma cell (also step 3).

The isotype class switch from IgM to another antibody type is directed by cytokines secreted by the T_{FH} cell. For instance, IL-4 and interferon-gamma (IFN-gamma or IFN-γ) trigger the switch from IgM to IgE and IgG, respectively. However, as revealed in the Current Research Highlight, the innate immune system can tell the adaptive system that it will be facing a living pathogen and then better prepare T_{FH} cells to trigger the B-cell isotype switch.

During the secondary immune response, too, memory B cells need T-cell help to become plasma cells, but direct contact with helper T cells is not required. Memory B cells that have antigen bound to their B-cell receptors can respond to the soluble IL-4 and IL-6 cytokines secreted by activated helper T cells without having direct contact with the T_H cell (**Fig. 24.22**, step 4). IL-4 stimulates B-cell proliferation, while IL-6 directs differentiation into antibody-secreting plasma cells.

Thought Question

24.9 What would happen to someone lacking CD154 on T_{FH} cells because of a gene mutation?

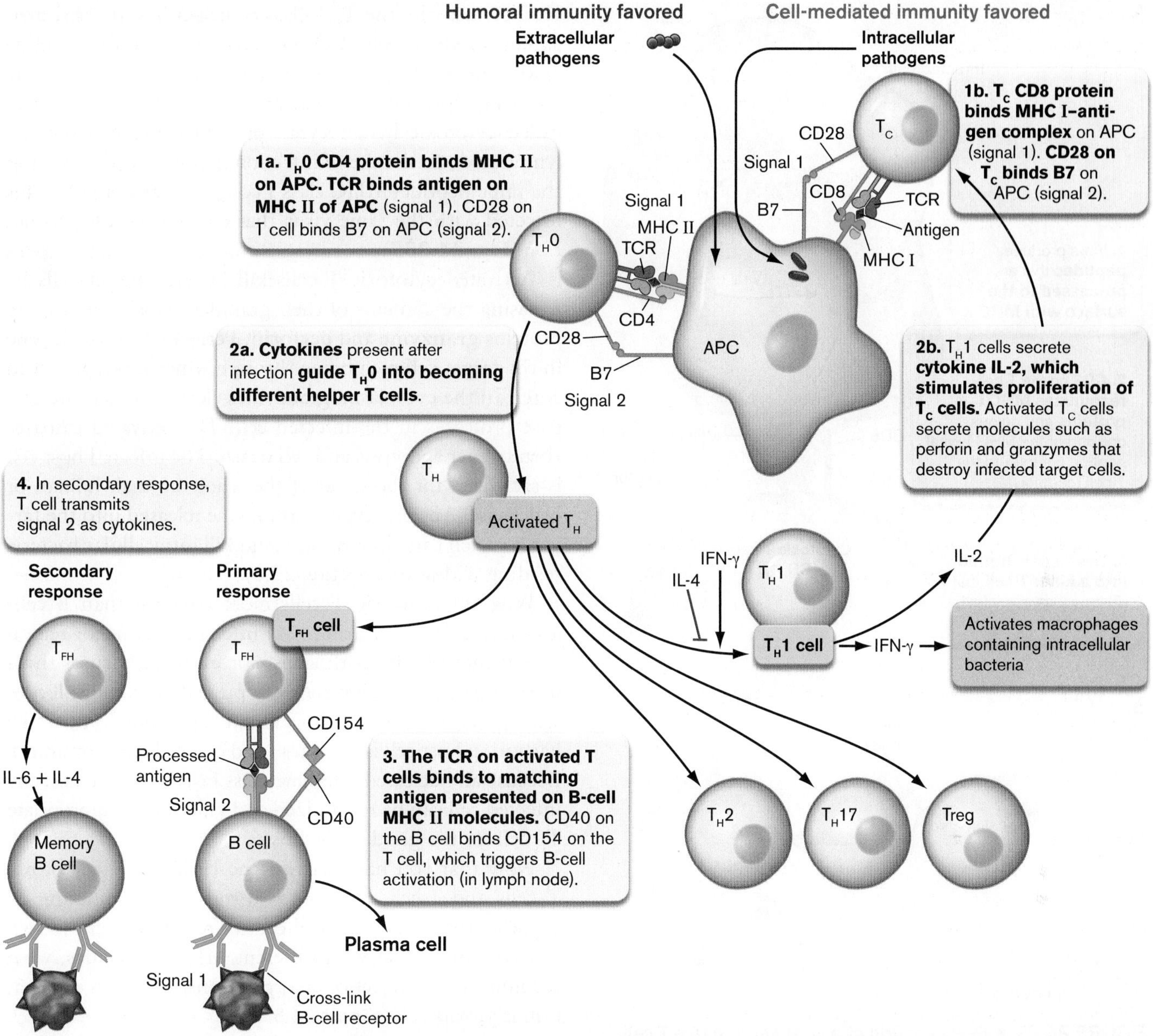

FIGURE 24.22 ▪ **Summary of the activation of humoral and cell-mediated pathways. Left:** Extracellular pathogens tend to activate humoral immunity (B cells). **Right:** Intracellular pathogens generally activate cell-mediated immunity by stimulating cytotoxic T cells (CD8 marker). The balance between cell-mediated and humoral immune responses to a given infection is regulated by the balance between the production of T_H1 (cell-mediated) versus other T_H (antibody) helper T cells. This balance is influenced by whether the foreign antigen was made by intracellular pathogens (T_H1-favored) or extracellular pathogens (refer to Fig. 24.20). T_H1 cells will encourage activation of cytotoxic T cells (cell-mediated immunity; best for killing intracellular pathogens), while T_{FH} cells will promote antibody production (humoral immunity; best for attacking extracellular pathogens). See text for more details.

Thought Question

24.10 How can a stem cell be differentiated from a B cell at the level of DNA?

T_H1 Cells Activate Cytotoxic T Cells

Because they directly attack host cells, CD8 T_C cells are the major "enforcers" of the cellular immune system, along with macrophages and natural killer (NK) cells. T_C cells, however, are not actually cytotoxic until they are activated. As with helper T cells, activation of cytotoxic T cells takes place in the lymph node and requires two signals. The first signal is the binding of T-cell receptors (TCRs) to antigen–MHC class I complexes on an antigen-presenting cell (**Fig. 24.22**, step 1b). CD8 on the T_C cell recognizes an MHC I–antigen complex on the APC. Because class I MHC molecules are found on all nucleated cells,

A.

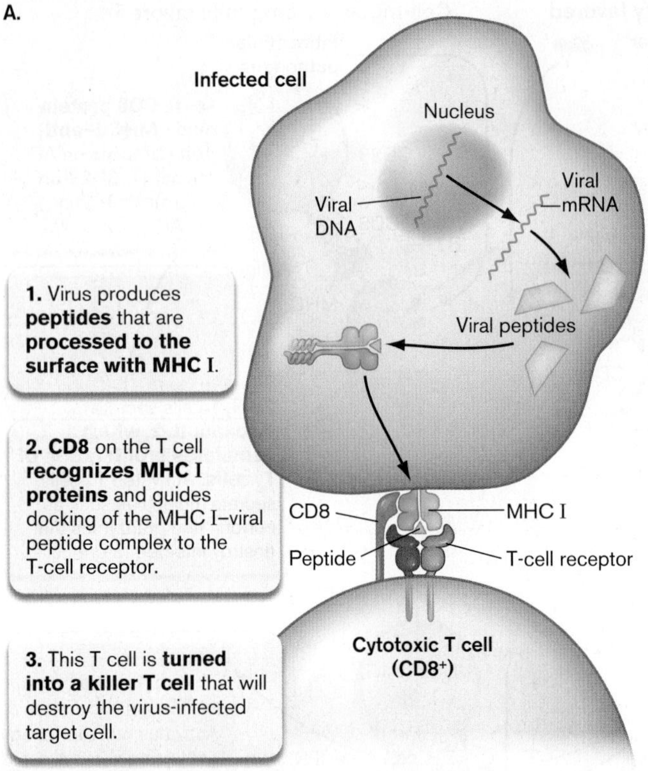

B.

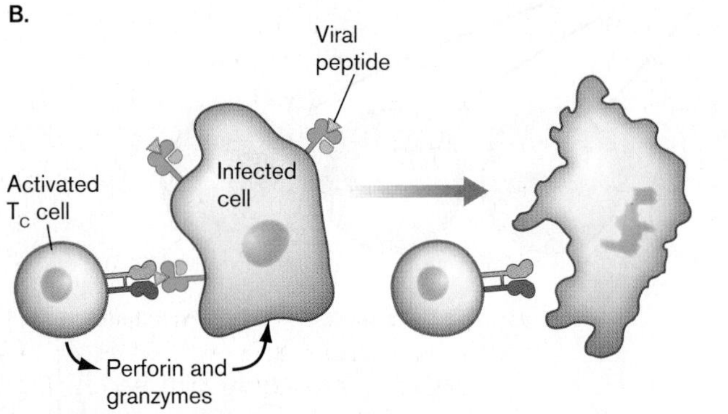

FIGURE 24.23 ▪ Presentation of a viral antigen to a T cell, and cytotoxic T-cell action. A. CD8 protein on a T_C cell directs the interaction between a T-cell receptor and a viral antigen bound to a class I MHC protein on an antigen-presenting cell. Signal 2 for activation of cytotoxic T cells is an interaction between B7 (on an APC) and CD28 on a T_C cell, described in the text (not shown here). **B.** The activated T_C cell (also called cytotoxic lymphocyte, or CTL) then leaves the lymph node and migrates to the site of infection, where it can recognize the viral peptide presented on the class I MHC receptor on an infected cell. This interaction authorizes the cytotoxic T cell to kill the infected cell.

any infected cell can potentially activate T_C cells, converting them to cells that ultimately kill the infected cell. **Figure 24.23A** takes a closer look at signal 1 interaction.

The second signal needed to activate T_C cells is a B7-to-CD28 interaction that further links the APC to the T_C cell. This is the same second signal described earlier for helper T cells. Once activated, the T_C cell gains cytotoxic activity and places receptors for the cytokine IL-2 on its surface. IL-2 secreted by the T_H1 class of helper T cells (activated earlier) binds to the IL-2 receptors and stimulates proliferation of the cytotoxic T cell (**Fig. 24.22**, step 2b). The resulting "platoon" of activated cytotoxic T cells (also called cytotoxic lymphocytes, or CTLs) moves from the lymph node to the site of infection, killing any cell bearing the same peptide–MHC class I complex (for example, cells infected with the same virus that triggered CTL production; **Fig. 24.23B**).

Activated cytotoxic T cells kill infected target cells by releasing the contents of their granules, which contain the proteins **granzyme** and **perforin**. Perforin produces a pore in the target cell membrane through which granzymes can enter. In the cytoplasm, granzymes cleave and activate caspase proteases in the infected cell. The activated caspases then trigger apoptosis and cell death. The infected host cell is sacrificed for the good of the whole animal, human or otherwise. Another type of granzyme released into the target host cell can directly enter and kill intracellular bacteria residing within (see Section 23.3).

Why are cytotoxic T cells more effective than B cells and antibodies at clearing viral infections? A major reason is that intracellular pathogens (for example, viruses) hide inside host cells, where they are protected from antibody. Consequently, these pathogens are best killed when the harboring host cell is also sacrificed (via cellular immunity). Because all nucleated cells have class I MHC molecules, any infected cell can be recognized and killed by an appropriate T_C cell for the good of the host.

Now that you have learned the basics about cytotoxic T cells and T-cell selection, you might be wondering: If T_C cells from one individual cannot recognize cells from a different individual with different MHC molecules, what accounts for the rejection of organ transplants? An explanation is provided in **eTopic 24.5**.

T_H17 Cells Promote, and Treg Cells Limit, Inflammation

The immune system is full of checks and balances. Two sets of T cells critical to regulating the immune response are **T_H17 cells** that amplify inflammation, and regulatory T cells (Tregs) that dampen inflammation. T_H17 cells are derived from T_H0 cells and are characterized by their secretion of pro-inflammatory cytokine IL-17. T_H17 cells are the first subset of T cells generated during an infection. Receptors for IL-17 are expressed on fibroblasts, epithelial cells, and keratinocytes.

Contact with IL-17 causes the production of several cytokines. These include IL-6 to initiate fever, chemokines like CXCL8 (formerly called IL-8) to recruit neutrophils and macrophages, and granulocyte-macrophage colony-stimulating factors (GM-CSFs) to enhance the synthesis of neutrophils and macrophages in bone marrow. IL-22, also produced by T_H17 cells, cooperates with IL-17 to induce the synthesis of antimicrobial peptides, such as beta-defensins, in epidermal keratinocytes. Altogether, T_H17 cells enhance the innate acute inflammatory response to infection. However, the problem with inflammation is that it also causes damage to local tissues. This is where Treg cells become crucial.

Treg cells can form during T-cell selection in the thymus (mentioned earlier), or differentiate from T_H0 cells in a lymph node. Treg cells help to shut down immune responses after an invading organism has successfully been eliminated, and they can prevent autoimmunity. Once activated by an antigen presented to its T-cell receptor, a Treg cell secretes the anti-inflammatory cytokine IL-10. IL-10 inhibits inflammatory cytokine secretion by macrophages [tumor necrosis factor alpha (TNF-alpha or TNF-α), IL-1, and IL-12], reduces MHC II expression in macrophages, blocks macrophage activation by IFN-gamma, and blocks cytokine IL-2 production from T_H1 cells, thereby limiting the production of cytotoxic T cells (see **Fig. 24.22**).

Ratios of T_H17 to Treg cells vary depending on the severity of an infection, either exacerbating or quelling inflammation. The cytokines produced by APCs and other cells during an infection can tilt the balance one way or the other. In addition, Treg cells themselves can inhibit T_H17 cell functions.

CD4 Helper T Cells and Cytokines Balance the Immune Response

Antigens presented on class I MHC molecules elicit only cell-mediated immunity (via CD8 T cells). However, antigens presented on class II MHC molecules can stimulate both arms of the immune system by activating CD4 helper T cells. CD4 T_H cells do not kill host cells directly, but instead release various cytokines that incite other cells to do the killing. IL-8, for instance, attracts additional white blood cells to the area. Activated T_H1 cells secrete cytokines that stimulate cytotoxic T cells (see **Fig. 24.22**, step 2b), as well as macrophages and natural killer cells. However, sets of activated T_{FH} cells secrete different combinations of cytokines to trigger B cells to switch immunoglobulin classes and differentiate into plasma cells that secrete antibody (see **Fig. 24.22**, step 3). **Table 24.4** lists a fraction of the many different cytokines produced by various cell types and describes their influence over the immune system.

Figure 24.24 outlines the myriad effects that one cardinal cytokine, interferon-gamma (IFN-gamma or IFN-γ, the type II interferon) has on innate and adaptive immune cells. The cascade begins when an APC encounters microbes and releases IFN-alpha (Chapter 23). IFN-alpha activates NK

TABLE 24.4 Select Cytokines that Modulate the Immune Response*

Cytokine	Sample sources	General functions
IL-1	Many cell types, including endothelial cells, fibroblasts, neuronal cells, epithelial cells, macrophages	Pro-inflammatory; affects differentiation and activity of cells in inflammatory response; acts as endogenous pyrogen in the central nervous system
IL-2	T_H1 cells	Stimulates T-cell and B-cell proliferation
IL-3	T cells, mast cells, keratinocytes	Stimulates production of macrophages, neutrophils, mast cells, others
IL-4	T_H2 cells, mast cells	Promotes differentiation of CD4 and T cells into T_H2 helper T cells; promotes proliferation of B cells, class switch to IgE
IL-5	T_H2 cells	Chemoattracts eosinophils; activates B cells and eosinophils
IL-6	T_H2 cells, macrophages, fibroblasts, endothelial cells, hepatocytes, neuronal cells	Stimulates T-cell and B-cell growth; stimulates production of acute-phase proteins
IL-8 (CXCL8)	Monocytes, endothelial cells, T cells, keratinocytes, neutrophils	Chemoattracts PMNs; promotes migration of PMNs through endothelium
IL-10	T_H2 cells, B cells, macrophages, keratinocytes	Inhibits production of IFN-γ, IL-1, TNF-α, and IL-6 by macrophages
IL-17	T_H17 cells (helper T cells unique from T_H1 and T_H2)	Recruits macrophages and neutrophils to sites of inflammation
IFN-α/β	T cells, B cells, macrophages, fibroblasts	Promotes antiviral activity
IFN-γ	T_H1 cells, cytotoxic T cells, NK cells	Activates T cells, NK cells, macrophages, B-cell class switch to IgG
TNF-α	T cells, macrophages, NK cells	Exerts wide variety of immunomodulatory effects
TNF-β	T cells, B cells	Exerts wide variety of immunomodulatory effects

*IFN = interferon; IL = interleukin; NK = natural killer; PMN = polymorphonuclear leukocyte; TNF = tumor necrosis factor.

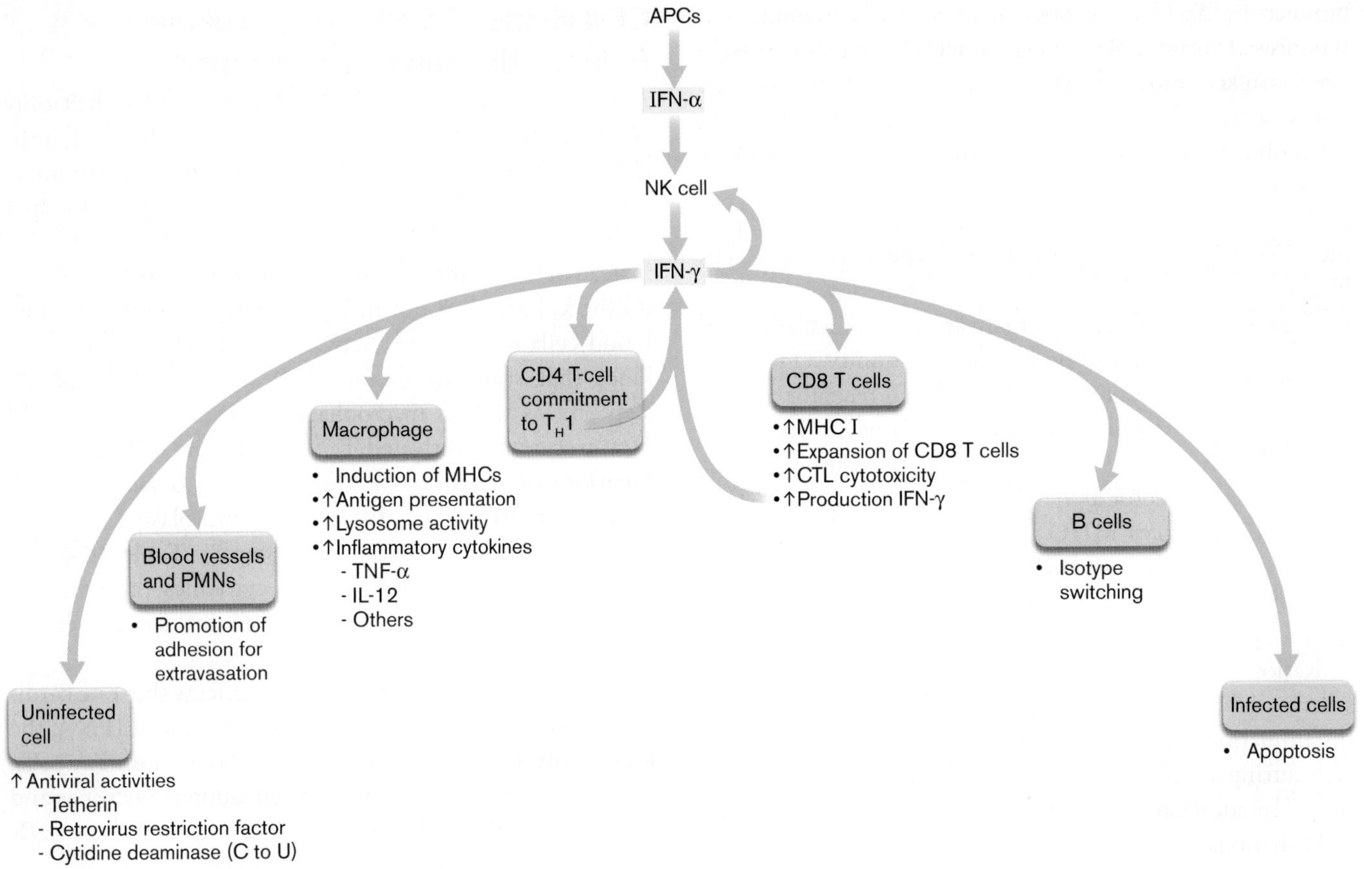

FIGURE 24.24 ■ Effects of interferon-gamma on the immune system. Interferon-alpha secreted by antigen-presenting cells (APCs) stimulates natural killer (NK) cells (the major producers of IFN-gamma) to secrete IFN-gamma. Binding of this cytokine to IFN-gamma receptors on a variety of cells can influence various aspects of innate and adaptive immunity, as well as trigger cell death of infected cells and activate antiviral activities in uninfected cells. CTL = cytotoxic lymphocyte; PMN = polymorphonuclear leukocyte.

cells to secrete IFN-gamma. IFN-gamma interacts with key immune cell types to influence their functions. These functions include increasing antigen presentation and the secretion of inflammatory cytokines by macrophages, promoting CD4 T-cell differentiation into T_H1 cells, enhancing the cytotoxicity of CD8 CTL cells, stimulating isotype switching by plasma B cells, and promoting PMN adhesion to blood vessels before extravasation. IFN-gamma secreted from NK cells also stimulates additional IFN-gamma secretion from T_H1 cells, as well as CD8 T cells, amplifying the response. In addition, this cytokine will trigger apoptosis of infected cells (in collaboration with TLR3 interactions with viral dsRNA) and stimulate antiviral activities of uninfected bystander cells. The immunomodulatory reach of IFN-gamma is remarkable.

But what determines which class of T_H cell predominates during an infection? Part of the answer is found in the blend of cytokines generated by macrophages or mast cells during different types of infection. Infections by viruses and intracellular bacteria, for instance, generate cytokines, such as IFN-gamma (as just discussed), that favor the production of T_H1 cells. T_H1 cells promote cell-mediated immunity. Infections by extracellular bacteria and parasites, however, generate IL-4, which inhibits T_H0 differentiation into T_H1 cells, so antibody-mediated immunity is favored (see **Fig. 24.22**, step 2a).

As a result, once the immune system starts to go down one pathway, the other pathway is held back. But this is not an all-or-none response. For instance, a viral infection will also result in T_{FH} cells that stimulate antibody production. The tilt toward T_H1 or T_{FH} predominance occurs because the body realizes which pathway—cell-mediated (T_H1) or humoral (T_{FH})—will more effectively clear a particular infection.

The importance of CD4 T cells in immunity is tragically illustrated by acquired immunodeficiency syndrome (AIDS, discussed in Chapter 11). The human immunodeficiency virus (HIV) binds to CD4 molecules to infect CD4 T cells. As the disease progresses, the number of CD4 T cells declines below its normal level of about 1,000 per microliter. When the number of CD4 T cells drops below 400 per microliter, the ability of the patient to mount an immune response is jeopardized. The patient not only becomes hypersusceptible to infections by pathogens, but also becomes susceptible to infections by commensal organisms. Most AIDS patients actually die from these opportunistic infections.

Superantigens Do Not Require Processing to Activate T Cells

Superantigens are a family of microbial toxins that induce an abnormally aggressive T-cell response. Normally, antigens require processing by antigen-presenting cells to stimulate T-cell responses. Each peptide epitope produced during antigen processing must be placed on an MHC molecule and presented to a cognate T cell. When an antigen is introduced into a host, only a few T cells have the proper TCR needed to recognize that antigen—in the range of 1–100 cells per million. Proliferation of the T cells through antigen-dependent activation increases their number and increases the immune response to that antigen.

Superantigens, such as staphylococcal toxic shock syndrome toxin (TSST), bypass the normal route of antigen processing. In fact, antigen recognition is not even involved. As illustrated in **Figure 24.25**, superantigens can bind simultaneously to the outside of T-cell receptors on T cells and to the MHC molecules on APCs (for example, macrophages). This promiscuous joining of T cells and macrophages activates many more T cells than a typical immune reaction does, stimulating the release of massive amounts of inflammatory cytokines from both cell types.

The effect of superantigens can be devastating because cytokines, like **tumor necrosis factor** (**TNF**), will overwhelm the host immune system's regulatory network and cause severe damage to tissues and organs. The result is disease and sometimes death. Jim Henson, the creator of Kermit the Frog and other Muppet characters, died in 1990 from complications of pneumonia caused by a potent superantigen produced by *Streptococcus pyogenes*. This example is but one of many ways that overreaction by the immune system causes morbidity (disease) and mortality (death)—effects more pronounced than the direct effects of the pathogen involved.

Microbial Evasion of Adaptive Immunity

As efficient as the immune system is, numerous viral and bacterial pathogens have developed effective means for avoiding the adaptive immune response. Many viruses produce proteins that down-regulate the production of

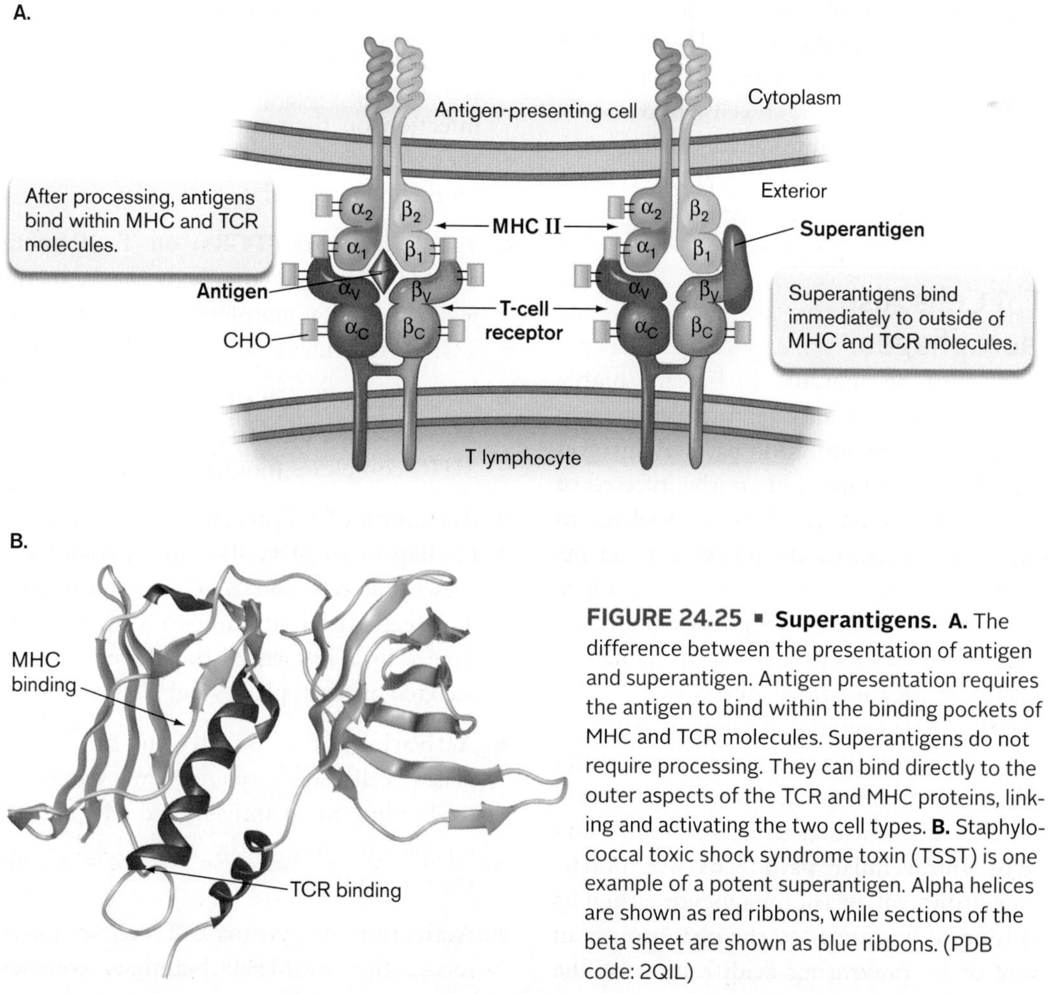

FIGURE 24.25 ■ Superantigens. A. The difference between the presentation of antigen and superantigen. Antigen presentation requires the antigen to bind within the binding pockets of MHC and TCR molecules. Superantigens do not require processing. They can bind directly to the outer aspects of the TCR and MHC proteins, linking and activating the two cell types. **B.** Staphylococcal toxic shock syndrome toxin (TSST) is one example of a potent superantigen. Alpha helices are shown as red ribbons, while sections of the beta sheet are shown as blue ribbons. (PDB code: 2QIL)

class I MHC molecules on infected cell surfaces. Down-regulation of MHC I will limit antigen presentation, since MHC I is needed for that process. On the other hand, losing MHC I exposes the infected host cell to natural killer cells because NK cells attack peers that lack MHC I (discussed in Section 23.5). To surmount this obstacle, human cytomegalovirus, for example, places a decoy MHC I–like molecule on the surface of infected cells. These decoys are thought to bind inhibitory receptors on NK surfaces that block NK cell cytotoxicity.

Like viruses, bacteria are masters of illusion when it comes to the immune system. A major cause of gastric ulcers, *Helicobacter pylori*, expresses proteins from a cluster of pathogenicity genes that trigger apoptosis (programmed cell death) of T cells. Other bacteria have evolved mechanisms that interfere with signal transduction pathways controlling the expression of cytokines. For example, YopP from *Yersinia enterocolitica*, one cause of gastroenteritis, inhibits a specific signal transduction pathway needed to produce TNF, IL-1, and IL-8. Thus, *Yersinia* avoids the detrimental effects of those pro-inflammatory cytokines.

Another method of immune evasion is employed by various mycobacteria, some of which cause tuberculosis and leprosy. These bacteria induce the production of anti-inflammatory cytokines, which dampen the immune response. *Mycobacterium*-infected macrophages produce IL-6, which inhibits T-cell activation, and IL-10, which down-regulates the production of MHC II molecules needed to present antigens to helper T cells. Fortunately for humans, in most cases the immune system catches on to these tricks and, through redundant humoral and cellular mechanisms, manages to resolve these infections.

Activated T_H1 Cells Also Activate Macrophages

Macrophages, too, must be activated to become highly effective killers of microbes. Activation of macrophages also requires two signals. One activation pathway involves interferon-gamma (IFN-γ) produced by nearby infected or damaged cells, followed by binding of the macrophage to lipopolysaccharide (LPS) or other microbial components via Toll-like or NOD-like receptors (TLRs or NLRs). Once activated, the macrophage becomes aggressive in terms of phagocytosis and increases its production of numerous antimicrobial reactive oxygen intermediates.

As noted earlier, some bacterial pathogens, such as mycobacteria, the causative agents of tuberculosis and leprosy, grow primarily in the phagolysosomes of macrophages. There they are shielded from antibodies and cytotoxic T cells. Intracellular pathogens live in the usually hostile environment of the phagocyte either by inhibiting the fusion of lysosomes to the phagosomes in which they grow or by preventing acidification of the vesicles needed to activate lysosomal proteases. However, a macrophage activated by a T_H1 cell can rid itself of such pathogens. Even unactivated macrophages are able to process some of the intracellular bacteria and place antigen from them on their class II MHC molecules. T_H1 cells then activate these macrophages by binding their TCR molecules to the macrophage MHC II–antigen complex and through secretion of IFN-gamma. The activated macrophages can then kill any bacteria that may be growing within them.

Given that activated macrophages are such effective assassins of microbial pathogens, why are macrophages not always kept in an active state? A major reason is that once activated, macrophages also damage nearby host tissue through the release of reactive oxygen radicals and proteases. Thus, effective killing of microbial pathogens comes at the expense of host tissue damage.

To Summarize

- **The major histocompatibility complex (MHC)** consists of membrane proteins with variable regions that can bind antigens. Class I MHC molecules are on all nucleated cells, while class II MHC molecules are found only on professional antigen-presenting cells.
- **Antigen-presenting cells (APCs)** such as dendritic cells present antigens synthesized during an intracellular infection on their surface class I MHC molecules, but they place antigens from engulfed microbes or allergens on their class II MHC molecules.
- **T-cell receptors (TCRs) on T-cell membranes bind to MHC-antigen complexes** on the membranes of professional and nonprofessional antigen-presenting cells. TCR-MHC interactions trigger T-cell differentiation.
- **T-cell education in the thymus** selects for T cells whose TCRs will not bind to self-antigens but will bind to self-MHC complexes (not too tightly).
- **Activation of a T_H0 cell** requires two signals: TCR-CD4 binding to an MHC II–antigen complex on an antigen-presenting cell, and B7-CD28 interaction. **Subsequent T_H0 cell differentiation** to T_{FH}, T_H1, T_H2, T_H17, or Treg cells is influenced by different cytokine "cocktails" secreted by macrophages and NK cells during an infection.
- **Activation of a B cell** into an antibody-producing plasma cell usually requires two signals: a B-cell receptor binding to an antigen, and a B-cell MHC II-antigen complex binding to the TCR of a T_{FH} cell activated by the same antigen.
- **Activation of cytotoxic T cells** requires two signals: recognition of MHC I–antigen complexes on APCs

by TCR CD8 molecules and B7-CD28 interactions. IL-2 secreted from activated T_H1 cells stimulates cytotoxic-T-cell proliferation. The activated T_C cells, in turn, destroy infected host cells.

- **Measures that balance immune responses** include T_H17 cells and Treg cells that, respectively, promote and limit inflammation; and cytokine cocktails secreted by CD4 helper T cells that direct predominantly antibody-mediated or cell-mediated immunity.
- **Superantigens** abnormally stimulate T cells by directly linking TCRs on T cells with MHCs on APCs without undergoing APC processing and surface presentation.

Thought Questions

24.11 Transplant rejection is a major consideration in the transplantation of most tissues, because host T_C cells can recognize allotypic MHC on donor cells. Why, then, are corneas easily transplanted from a donor to just about any other person?

24.12 Why does attaching a hapten to a carrier protein enable antihapten antibodies to be produced?

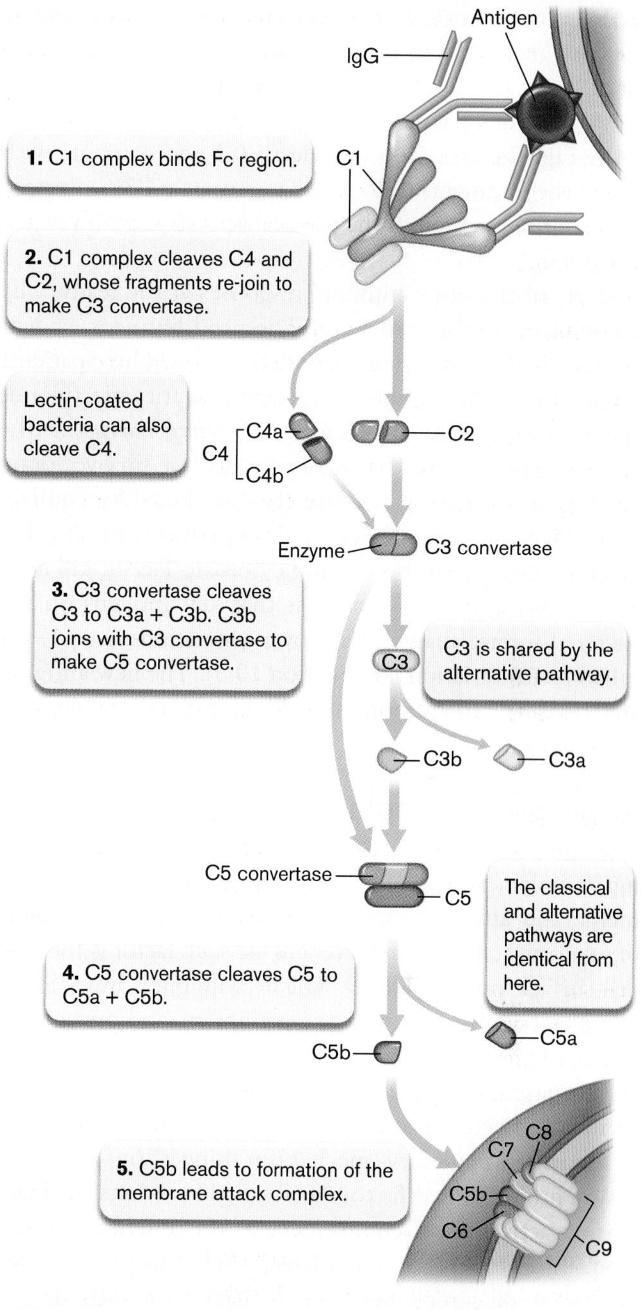

FIGURE 24.26 ■ Classical complement cascade.

24.4 Complement as Part of Adaptive Immunity

Chapter 23 describes how complement, in what is called the alternative pathway, can attack invading microbes before an adaptive immune response is launched (see Section 23.6). Factor C3b binds to LPS and sets off a reaction cascade ending with a membrane attack complex (MAC) pore composed of C5b, C6, C7, C8, and C9 proteins (see Fig. 23.39). However, antibody made during the adaptive response to a pathogen offers another route to activate complement, called the **classical complement pathway**. Dubbed "classical" because it was the first complement pathway to be discovered, it requires a few additional proteins before reaching C3, the linchpin factor connecting the two pathways.

The classical cascade begins when a complement C1 protein complex binds to the Fc region of an antibody bound to a bacterial or viral pathogen (**Fig. 24.26**). The bound C1 complex then cleaves two other complement factors, C2 and C4, not used in the alternative pathway. Two fragments, one each from C2 and C4, combine to form another protease, called C3 convertase, that cleaves C3 into C3a and C3b. C3b in the alternative pathway is stabilized by interacting with LPS. In the classical pathway, however, C3b combines with C3 convertase to make a C5 convertase (note that this C5 convertase is different from the C5 convertase formed in the alternative pathway; see Fig. 23.39). The subsequent steps leading to formation of a membrane attack complex (MAC) are the same as in the alternative pathway. As before, C5b binds to a target membrane and is joined by C6, C7, and C8. Multiple C9 proteins then assemble around the MAC and form a pore that compromises the integrity of the target cell.

In Section 23.6 we noted the existence of two complement activation pathways that do not require antigen-antibody complexes. One of them—the alternative complement pathway—was explained in detail, but discussion of the lectin activation pathway was deferred until now because it more closely resembles the classical pathway. The lectin activation pathway begins when lectins (produced by the liver) recognize and coat the sugar structures that decorate the surfaces of infectious organisms. The lectin-coated polysaccharides trigger cleavage of factor C4 to C4a and C4b (**Fig. 24.26**). C4b can cleave C2 and combine with one of the fragments to form C3 convertase. C3 convertase then continues through the classical activation pathway just described.

With all the other immune responses at the ready, why is complement also necessary? The need becomes evident in individuals with complement deficiencies. These patients are extremely susceptible to recurrent septicemic (blood) infections by organisms, such as *Neisseria gonorrhoeae* (the cause of gonorrhea), that normally do not survive forays into the bloodstream, because they are killed by complement. Why, in a complement-deficient patient, won't other defenses, such as antibodies and cytotoxic T cells, kill reinfecting *Neisseria*? The reason is that key antigens on the bacterial surface change shape over generations in a process called phase variation (see Section 10.5). The new antigens are "invisible" to antibodies made against earlier versions.

Regulating Complement Activation

How do normal body cells prevent self-destruction following complement activation? The sequential assembly of the membrane attack complex provides several places where regulatory factors can intervene. One such factor is the host cell-surface protein CD59. CD59 will bind any C5bC8 complex trying to form in the membrane and prevent C9 from polymerizing. Thus, no pore is formed and the host cell is spared. (Complement will not normally attack uninfected or infected host cells, since both contain CD59.)

Another regulatory mechanism hinges on a normal serum protein called **factor H**. Factor H prevents the inadvertent activation of complement in the absence of infection. If factor H is unavailable, then the uncontrolled activation of complement can damage host cells despite the presence of CD59. To short-circuit the cascade, factor H binds to C3bBb, displaces Bb from the complex, and acts as a cofactor for factor I protease, which then cleaves C3b. Without C3b, the complement cascade stops. Some bacteria, such as certain strains of *Neisseria gonorrhoeae* and *Streptococcus pyogenes*, have learned to protect themselves from complement by binding factor H. Factor H bound to the microbe provides a protective "force field" against local complement activation.

Microbiota can also assist host cells in resisting complement. The intestinal microbe *Bacteroides thetaiotaomicron* protects host cells from complement-mediated cytotoxicity by up-regulating a **decay-accelerating factor** present in host cell membranes. Decay-accelerating factor stimulates decay of complement factors and prevents their deposition at the cell surface.

Other Roles for Complement Fragments

The C3a, C4a, and C5a cleavage fragments generated by the complement cascade do not participate in MAC formation but are important for amplifying the immune reaction. These peptides act as chemoattractants to lure more inflammatory cells into the area. C3b, which does participate in MAC formation, also can act as an opsonin when it is bound to a bacterial cell. An opsonin is a protein factor that can facilitate phagocytosis. Because phagocytes contain C3b receptors on their surfaces, it is easier for them to grab and engulf cells coated with C3b. Additional roles for complement fragments were discussed in Section 23.6.

To Summarize

- **The classical pathway for complement activation** begins with an interaction between the Fc portion of an antibody bound to an antigen and C1 factor in blood. (The alternative pathway does not need antibody but begins when C3b binds to LPS on a bacterium.)
- **The Fc-C1 complex** reacts with C2 and C4, leading to production of C3b and a novel C5 convertase specific to the classical pathway.
- **After C5 convertase cleaves C5**, the classical pathway is the same as the alternative pathway, resulting in the formation of a membrane attack complex (MAC).
- **CD59 and/or factor H** prevents inappropriate MAC formation in host cells.

24.5 Gut Mucosal Immunity and the Microbiome

The mucosal surface of the gastrointestinal tract is exposed to a vast number of food, chemical, and other ingested antigens. The gut also harbors an expansive microbiome whose members must be controlled and nurtured. Monitoring this large antigen load requires localized and highly integrated innate and adaptive immune systems, and a system of gut-associated lymphoid tissues, or GALT (see Section 23.4). The innate immune mechanisms that keep gut microbes at bay were described in Chapter 23, but the intestine also has a dedicated adaptive immune system that monitors and shapes the microbiota, usually without causing excessive inflammation. In this section we describe how the innate and adaptive immune systems in the gut shape the microbiome while protecting the host.

The Gut Immune System

We discussed in Chapter 21 how the gut microbial community helps digest food and outcompetes pathogens (see Fig. 21.22). In Chapter 23 we described how gut microbiota are acquired and how dysbiosis can lead to intestinal inflammatory diseases and obesity. Now in Chapter 24, we will explain how the microbiome trains the immune system and, in turn, how the immune system shapes the human microbiome. But first we must discuss the gut immune system, where the training and shaping take place.

As described in Chapter 23, the intestinal epithelial barrier is one cell thick and composed mainly of columnar epithelial cells: intestinal epithelial cells (IECs) at site 1 in **Figure 24.27**, and goblet cells at site 2. Goblet cells secrete mucin and antimicrobial peptides. Epithelial cells express numerous pattern recognition receptors (PRRs), including TLRs and NLRs that can be activated by MAMPs shed from resident microbes (**Fig. 24.27** blowup). Once activated, epithelial cells will synthesize and secrete a variety of cytokines that stimulate cells of the immune system—for instance, chemokines that attract neutrophils and lymphoid cells. The epithelial barrier is also punctuated by cells called **intraepithelial lymphocytes (IELs)**, mostly T cells, capable of secreting cytokines, including IFN-gamma. IELs exert protective and inflammatory effects in the intestine (**Fig. 24.27**, site 3).

Beneath the epithelial layer is the lamina propria, rich in B cells and T cells. Epithelial cells can process antigens they encounter and place them onto MHC I and MHC II receptors for presentation to T cells residing in the lamina propria. (Although normally epithelial cells are not thought of as having MHC II, they will express the molecule if IFN-gamma is made during an infection.) After being presented with gut antigens, T cells can rapidly stimulate inflammatory responses via differentiation to T_H1, T_H2, and T_H17 cells (**Fig. 24.27**, site 4).

Alternatively, the T cells can inhibit inflammatory responses by converting to Treg cells. In addition, dendritic cells in the lamina propria use their dendrite extensions to "reach" between epithelial cells and capture antigens in the intestinal lumen (site 5). One class of dendritic cells stimulates inflammation by producing cytokines TNF-alpha and IL-6. A separate set of dendritic cells, however, promotes the generation of regulatory T cells whose cytokine repertoire can dampen inappropriate inflammatory responses that might be triggered by microbiota.

Other important cells in the lamina propria include macrophages and **innate lymphoid cells (ILCs)**. Macrophages were discussed earlier. ILCs look like lymphocytes, but because they lack antigen-specific B-cell or T-cell receptors, they are really part of the innate immune system (**Fig. 24.27**, site 6). A diverse range of stimuli can activate ILCs, including neuropeptides, hormones, and cytokines produced by epithelial cells responding to microbiome MAMPs. ILCs include natural killer cells (described in Chapter 23). Other ILCs secrete the pro-inflammatory cytokine

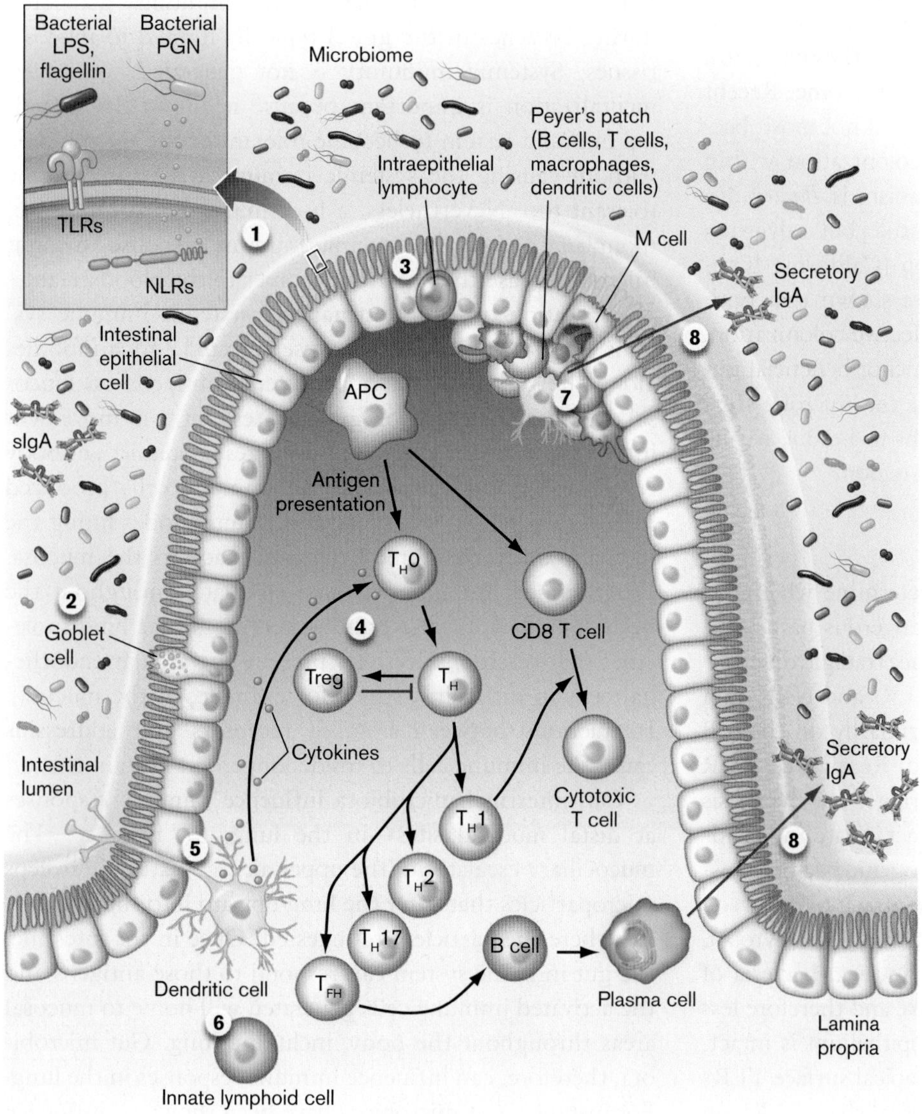

FIGURE 24.27 ■ Mucosal immune system. Depicted is the region around one intestinal villus. Mucosal immunity functions independently of regional lymph nodes to control inflammation induced by microbiota or to stimulate inflammation when pathogens are present. See text for details. PGN = peptidoglycan; sIgA = secretory IgA.

IL-17 to summon neutrophils, and cytokine IL-22 to stimulate epithelial cells to secrete antimicrobial peptides. IL-22 also promotes tissue repair and regeneration in an inflamed intestine. Consequently, ILCs contribute to immunity, inflammation, tissue homeostasis, and microbiome control.

Peyer's patches (**Fig. 24.27**, site 7), first described in Section 23.4, are lymphoid tissues made up of IgA^+ B-cell and T-cell centers. Peyer's patches are ideal sites for adaptive immune responses along the intestinal tract. M cells in the epithelium above a Peyer's patch will sample microbes and other antigens floating in the gut and offer them to underlying macrophages for engulfment and presentation to Peyer's patch T cells.

T_{FH}–B cell interactions taking place in sites 4 and 7 of **Figure 24.27** are influenced by local cytokines to trigger a B-cell class switch to IgA production and B-cell differentiation into IgA-secreting plasma cells. Secretion of dimeric sIgA (secretory IgA; site 8) into the intestinal lumen from Peyer's patches and from plasma cells present in lamina propria inhibits penetration by microbes through the epithelium and modulates the composition of the microbiome. Recent evidence also suggests that IgA coats many, if not most, bacteria in the gut and fosters microbial colonization within specific intestinal niches. One such organism is *Bacteroides fragilis*, which produces a thick, fuzzy capsule, but only when growing in the gut. Gregory Donaldson (California Institute of Technology) and colleagues have shown that both the capsule and IgA are necessary for effective colonization. They propose that secretory IgA helps anchor beneficial gut microbiota to the intestine. Binding of the microbe-sIgA complex to the mucous lining is thought to be due to the secretory component of sIgA, not the Fc region.

Determining Friend or Foe

How might the gut immune system determine whether a microbe is indigenous microbiota or dangerous pathogen? The distinction, in large part, involves the strategic distribution of pattern recognition receptors (TLRs and NLRs) on and within epithelial cells. TLRs are primarily on the cell surface, whereas NLRs are cytoplasmic. Recall that TLR and NLR interactions with MAMPs trigger the synthesis and release of various cytokines that impact innate and adaptive immune systems (Chapter 23). Pathogenic bacteria possess virulence factors that enable them to attach to or invade host cells, thereby introducing MAMPs into the cytosol, where they are recognized by NLRs. However, bacteria of the indigenous microbiota are noninvasive and therefore less potent activators of NLRs. When the epithelium is intact, these "friendlies" interact only with the apical surface TLRs of epithelial cells, which are less responsive than TLRs on the host side (basolateral side) of epithelial cells.

However, an interaction with TLRs on the basolateral side of epithelial cells signals a barrier breach, which could be caused by either a pathogen or an indigenous microbe. The resultant cytokine cocktail will trigger inflammation. Nevertheless, members of the microbiome evolved to lessen inflammation by dampening epithelial cell TLR signaling. The repertoire of cytokines secreted by epithelium also supports microbiome tolerance by conditioning a subset of dendritic cells to become tolerogenic. Tolerogenic dendritic cells will present antigen to T cells but convert those T cells into regulatory T cells that suppress inflammatory responses. Treg cells are essential for maintaining tolerance to the microbiota. Without Treg cells, effector T-cell responses (for instance, those of cytotoxic T cells) go unopposed and produce inflammatory bowel diseases.

Compartmentalization of the Gut Immune System

It is important to note that adaptive immunity generated during breaches in the gut is typically limited to mucosal tissues. Systemic immunity is not generated. Compartmentalization is important because it allows the intestinal immune system to become tolerant of the microbiome without causing the systemic immune system to become tolerant too. Nevertheless, a host must remain capable of mounting a vigorous systemic immune response to a gut microbe in case the organism escapes to the bloodstream.

Separation of the intestinal and systemic immune systems is possible, in part, because secreted IgA antibodies along the intestine will trap bacterial antigens on mucosal surfaces. The trapping decreases antigen movement into lymph nodes, which minimizes systemic antibody responses. In addition, adaptive immune cells produced in Peyer's patches or in mesenteric lymph nodes lining the intestine are programmed to travel back to the mucosa. Programmed immune cells can circulate throughout the body, but they express homing receptors that bind molecules (**addressins**) selectively present on vascular endothelial cells in different tissues (for example, in the mucosa). Interactions between homing receptors and addressins cause the immune cells to reside longer in a certain tissue.

Can intestinal microbiota influence immune responses at distal mucosal sites—in the lung, for instance? The mucociliary escalator of the upper airways naturally sweeps microparticles that enter the lung upward into the oral cavity, where the particles are ingested. Once in the intestine, the gut immune system can respond to those antigens and the activated immune cells generated will home to mucosal areas throughout the body, including lung. Gut microbiota, therefore, can influence immune responses in the lung. For instance, gut microbiota have been shown to influence the generation of influenza virus–specific CD4 and CD8 T cells and affect antibody responses to respiratory influenza virus infections.

The Microbiome Shapes Gut Immunity and Human Health

We have known for decades that germ-free animals devoid of intestinal microbiota exhibit major defects in the organization and activity of immune structures in the gut. Germ-free mice display an "underdeveloped" innate and adaptive immune system that includes reduced expression of antimicrobial peptides, reduced IgA production, fewer T-cell types, and increased susceptibility to microbial infections. Proper gut immunity can be restored, however, by simple microbial stimulation brought about by fecal transplant.

Several gut microbiota with beneficial effects for human health were discussed in Chapters 13, 21, and 23. Organisms such as *Akkermansia muciniphila* rejuvenate the mucin layer. *Bacteroides thetaiotaomicron* contributes greatly to carbohydrate utilization and to the production of short-chain fatty acids (SCFAs) that induce Treg cell production. Also contributing to Treg cell production are organisms such as *Bacteroides fragilis*, via its capsular polysaccharide antigen (PSA). *Faecalibacterium prausnitzii* and a group of Clostridiales called segmented filamentous bacilli (SFBs) induce IgA-secreting cells in Peyer's patches and stimulate the production of T_H17 cells. *Bifidobacterium* and *Lactobacillus* species also contribute to the shaping of the gut immune system.

As described earlier, the gut microbiome influences **mucosal immunity** through encounters with TLRs, NLRs, and other pattern recognition receptors on epithelial cells and a variety of immune cells. Influence is also provided through antigen interactions with dendritic cells and subsequent antigen presentation to T cells. The various blends of cytokines released can tilt immunity toward tolerance or inflammation and can even affect how strongly immune cells react to different antigens. For example, antibiotic-induced alterations in the microbiota will change not only the number of Treg cells, but also the T-cell receptor (TCR) repertoire on the remaining Treg cells, suggesting that gut microbiome composition can influence the response of Treg cells to various antigens. Evidence that microbiota can dampen inflammation is described in **eTopic 24.6**.

It has also become clear that the influence of the gut microbiome extends beyond the intestinal tract. Immune-related disorders such as inflammatory bowel disease (IBD), cancer, diabetes, allergies, and even obesity (see Chapter 23) may result from dysbiosis of the commensal microbial communities. Bacteria can produce neurotoxic metabolites such as D-lactic acid and ammonia. Even beneficial metabolites such as SCFAs may exert neurotoxicity. Gut microbes can also produce hormones and neurotransmitters that are identical to those produced by humans (discussed in Chapters 13 and 21). Consequently, gut bacteria can directly stimulate afferent neurons of the enteric nervous system to send signals to the brain via the vagus nerve. Through these varied mechanisms, gut microbes can shape the architecture of sleep and stress reactivity of the hypothalamic-pituitary-adrenal axis. Believe it or not, your gut microbes can influence your memory, mood, and cognition.

To Summarize

- **The gut immune system** includes intestinal epithelial cells (IECs); goblet cells; innate lymphoid cells (ILCs); lamina propria, which contains B cells, T cells, and dendritic cells; and Peyer's patches, which contain M cells, IgA^+ B cells, and T cells.
- **Secretion of sIgA** helps control the composition and balance of the gut microbiome.
- **Intestinal epithelial cell TLRs and NLRs** can distinguish normal microbiota from pathogens.
- **Microbiota composition** influences the ratios of T_{FH}, T_H1, T_H2, T_H17, and Treg cells to tilt the gut immune response between inflammatory and anti-inflammatory.
- **Mucosal immunity** is an immune system compartment that is mostly separate from the rest of the immune system.
- **Development of the mucosal immune system** is guided by bacterial members of the gut microbiota.

24.6 Immunization

Most of you reading this have been vaccinated with many of the vaccines listed in **Table 24.1**. The adaptive immune response is why vaccines work. In this section we will describe the various types of vaccines, note when and how they are given, and discuss why some children are not vaccinated.

Vaccines

Vaccines come in three basic types. As noted in **Table 24.1**, some vaccines utilize killed organisms (examples are the hepatitis A vaccine and the Salk inactivated polio vaccine), while others contain live, but attenuated, microbes (BCG for tuberculosis, or the Sabin live polio vaccine). The third type of vaccine consists of purified components (subunits) of an infectious agent (such as capsular antigens from *Streptococcus pneumoniae* and *Haemophilus influenzae* type b, or virus capsids such as those used in the Gardasil vaccine for human papillomavirus, a cause of genital warts and cancers of the cervix and penis). Some vaccines are injected in combination as polyvalent vaccines to control multiple diseases (for example, MMR for measles, mumps, and rubella), while others are given individually but may change from year to year (influenza viral envelope proteins).

Multiple vaccines can be given simultaneously and safely. Most vaccines are administered during childhood, when the diseases can be the most devastating. **Table 24.5** provides the current immunization schedule recommended for children and adolescents by the Centers for Disease Control and Prevention. Notice that all of these vaccines are given in multiple doses, called booster doses. The exception to this rule is the influenza vaccine, which is given in a single dose but changes every year. The reason for multiple doses, as noted earlier, is that secondary exposure to an antigen provides a more robust and long-lasting immunity. However, most vaccines are not administered until 2 months of age, because maternal antibody crossing through the placenta to the fetus (or to a newborn through breast milk) will persist for a short time in the newborn, temporarily protecting the baby from disease and possibly dampening the response to a vaccine antigen administered during that time.

TABLE 24.5 Recommended Immunization Schedule for Children and Adolescents[a]

<table>
<tr><th rowspan="2">Vaccine</th><th colspan="11">Age of administration</th></tr>
<tr><th>Birth</th><th>1 month</th><th>2 months</th><th>4 months</th><th>6 months</th><th>9 months</th><th>12 months</th><th>15–18 months</th><th>24 months</th><th>4–6 years</th><th>11–12 years</th></tr>
<tr><td>Hepatitis B[b]</td><td>HepB</td><td colspan="2">HepB</td><td></td><td colspan="4">HepB</td><td colspan="3">HepB series</td></tr>
<tr><td>Rotavirus</td><td></td><td></td><td>Rota</td><td>Rota</td><td>Rota</td><td></td><td></td><td></td><td></td><td></td><td></td></tr>
<tr><td>Diphtheria, tetanus, acellular pertussis[c]</td><td></td><td></td><td>DTaP</td><td>DTaP</td><td>DTaP</td><td></td><td></td><td>DTaP</td><td></td><td>DTaP</td><td>Tdap</td></tr>
<tr><td>Haemophilus influenzae type b[d]</td><td></td><td></td><td>Hib</td><td>Hib</td><td>Hib</td><td></td><td colspan="2">Hib</td><td></td><td></td><td></td></tr>
<tr><td>Inactivated poliovirus</td><td></td><td></td><td>IPV</td><td>IPV</td><td colspan="4">IPV</td><td></td><td>IPV</td><td></td></tr>
<tr><td>Measles, mumps, rubella</td><td></td><td></td><td></td><td></td><td></td><td></td><td colspan="2">MMR</td><td></td><td>MMR</td><td>MMR</td></tr>
<tr><td>Varicella</td><td></td><td></td><td></td><td></td><td></td><td></td><td colspan="2">Varicella</td><td></td><td>Varicella</td><td></td></tr>
<tr><td>Meningococcal (ACWY)[e]</td><td></td><td></td><td></td><td></td><td></td><td></td><td colspan="4">Administer to high-risk children</td><td>MCV4</td></tr>
<tr><td>Meningococcal (B)</td><td></td><td></td><td></td><td></td><td></td><td></td><td></td><td></td><td></td><td></td><td>MenB (10–12 yr)</td></tr>
<tr><td>Pneumococcal[f]</td><td></td><td></td><td>PCV13</td><td>PCV13</td><td>PCV13</td><td></td><td colspan="2">PCV13</td><td colspan="2">PPSV23</td><td></td></tr>
<tr><td>Influenza[g]</td><td></td><td></td><td></td><td></td><td colspan="7">Influenza (yearly)</td></tr>
<tr><td>Hepatitis A[h]</td><td></td><td></td><td></td><td></td><td></td><td></td><td colspan="3">HepA (2-dose series)</td><td colspan="2">HepA</td></tr>
<tr><td>Human papillomavirus[i]</td><td></td><td></td><td></td><td></td><td></td><td></td><td></td><td></td><td></td><td></td><td>HPV</td></tr>
</table>

[a]This schedule indicates the recommended ages for routine administration of currently licensed childhood vaccines, as of January 1, 2019.
■ Range of recommended ages for each vaccine dose. ■ Catch-up immunization. ■ Assessment at age 11–12 years.

[b]Hepatitis B vaccine (HepB). At birth: All newborns should receive monovalent HepB, administered soon after birth and before hospital discharge.

[c]Diphtheria, tetanus, and acellular pertussis (DTaP) vaccine. Tdap is a modified vaccine with lower doses of diphtheria and tetanus toxoids.

[d]*Haemophilus influenzae* type b (Hib) conjugate vaccine.

[e]Meningococcal conjugate vaccine (MCV4). MCV4 should be administered to all children at age 11–12 years, as well as to unvaccinated adolescents at high school entry (age 15 years). The vaccine contains four types of capsules: A, C, W, and Y.

[f]Pneumococcal vaccine. The 13-valent pneumococcal conjugate vaccine (PCV) is recommended for all children aged 2–23 months and for certain children aged 24–59 months. The final dose in the series should be administered at age ≥12 months. Pneumococcal polysaccharide vaccine (PPSV) is a 23-valent vaccine recommended in addition to PCV for certain high-risk groups. PPSV is also recommended for people over 65.

[g]Inactivated influenza vaccine should be administered annually starting at age 6 months. Live, attenuated influenza vaccine should not be given until 2 years, and not to immunocompromised individuals.

[h]Hepatitis A vaccine (HepA). HepA is recommended for all children at age 1 year (12–23 months).

[i]Human papillomavirus (HPV) vaccine: HPV4 (Gardasil) and HPV2 (Cervarix). A 3-dose series of HPV vaccine should be administered on a schedule of 0, 1–2, and 6 months to all adolescents aged 11–12 years. Either HPV4 or HPV2 may be used for females; only HPV4 may be used for males.

Source: Adapted from the Centers for Disease Control and Prevention website (http://www.cdc.gov).

Herd Immunity

You might wonder whether all members of a community must be vaccinated against a given microbe to lower the risk of disease for every individual in that community. In fact, the risk of an infected person spreading disease to an unvaccinated person can be lowered dramatically even when only about three-fourths of the community is vaccinated. Vaccinating a large percentage of a community effectively conveys community (or herd) immunity by interrupting transmission of the disease. If one individual contracts the disease, the chance that he/she will come into contact with another unvaccinated person and transmit the disease is much reduced. Thus, the risk of disease to any single unvaccinated person is lessened as a result of community immunity. Gardasil, the vaccine against human papillomavirus (the cause of genital warts, and cervical and throat cancers), is a good example of a vaccine that can provide herd immunity (see **eTopic 26.1**).

Herd immunity works well for diseases such as diphtheria, whooping cough (pertussis), measles, and mumps, all of which are infections spread by person-to-person contact. However, herd immunity will not lower the risk that an unvaccinated person will contract tetanus, which is not spread by person-to-person contact. *Clostridium tetani*, the agent whose toxin causes tetanus, is a ubiquitous soil organism transmitted through punctured skin. The risk of tetanus for an unvaccinated person doesn't change, even if every other person in the community is vaccinated against tetanus.

Vaccines and Immune System Compartments

What difference does it make which type of vaccine you take? Attenuated, killed, subunit—don't they all generate immunity? Yes, but generally, a live, attenuated vaccine is better because when a crippled but live microbe replicates at its normal body target site, an immune response most appropriate to that site develops and both the innate and adaptive responses are engaged (discussed later). Take the case of the Salk (killed) and Sabin (attenuated) polio vaccines. Poliovirus typically enters the body through ingestion, replicates in the mucosa, moves to the regional lymph nodes, and produces a viremia. Eventually, the virus can attack the central nervous system and cause paralysis.

The killed Salk vaccine (inactivated polio vaccine, or IPV) is injected and generates an antibody response in the bloodstream. The antibody response is capable of preventing the viremia and paralytic consequences of a natural infection, but it does not generate a mucosal immunity capable of preventing mucosal replication of the natural virus, in case the natural virus is ingested. Consequently, mild disease can result if wild poliovirus infects someone vaccinated with the Salk vaccine. These immunized people will shed poliovirus in their feces, enabling the virus to spread to others who might not be vaccinated. In contrast, the live virus in the Sabin vaccine, which is administered orally (oral polio vaccine, or OPV), will replicate in the mucosal lymphatic system of the gut, where secretory IgA antibodies are best generated (discussed earlier). Anti-polio IgA antibodies secreted in these mucosal areas will prevent wild-type poliovirus from replicating in the intestinal mucosa of a vaccinated person, so no virus is shed and there are no symptoms. However, OPV has not been used in the United States since the year 2000 for two reasons: first, polio in this country is rare, so fecal spread is unlikely, and secondly, the live attenuated virus could produce polio in an immunocompromised, vaccinated person.

Building better vaccines. The Current Research Highlight posed the question: Why are live vaccines better than dead, or component, vaccines at producing effective antibody responses? Julie Blander's group discovered that components of the innate immune system in mice can sense a "live" signal (bacterial RNA) during an infection (or vaccination). Once engaged, the innate cells secrete a blend of cytokines (IFN-β and IL-1β) that fosters the production of T_{FH} cells. The T_{FH} cells drive antibody class switching in B cells and differentiation of those B cells into plasma cells.

Leif Sander (Charité–Berlin University of Medicine) made a similar discovery using human tissue cultures. His group found that bacterial RNA from a live vaccine strain of *E. coli* binds to TLR8 in monocytes. The activated monocytes secrete IL-12, which promotes the formation of fully functional human T_{FH} cells. The viability trigger (bacterial RNA) that links innate and adaptive responses for mice and humans is the same, but the differentiation pathways are not. The data suggest that adding viability MAMPs, like bacterial RNA, to inactivated or component vaccines might greatly improve their efficacies in the future.

Figure 24.28 illustrates how flow cytometry was used to determine how efficiently live and heat-killed *E. coli* vaccines drive the differentiation of human T cells into T_{FH} cells. Flow cytometers can identify and count cell types that display different surface antigens (review the flow cytometer depicted in Figure 4.19 and discussed in Section 21.1). As immune cells differentiate, they express unique surface antigens. All helper T cells, for instance, express CD4 on their cell surfaces. However, when a T_H cell differentiates into a T_{FH} cell, the new cell will also express the CXCR5 surface marker. CXCR5 is a receptor for a specific cytokine. Multiple surface markers on a single cell can be detected by the use of antibodies tagged with different fluorescent dyes.

To observe whether live organisms are better than heat-killed organisms at promoting T_{FH} production, Sander's group stimulated human monocytes (APCs) with either control medium, live *E. coli*, or heat-killed *E. coli* (**Fig. 24.28**). The stimulated APCs were then mixed with human T cells

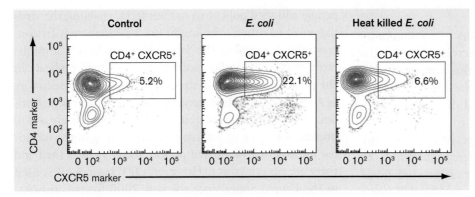

FIGURE 24.28 ■ **Live *E. coli* drives differentiation of CD4^{pos}-CXCR5^{neg} T cells into CD4^{pos}-CXCR5^{pos} T$_{FH}$ cells.** Human APC monocytes were stimulated with control medium, live *E. coli* (EC), or heat-killed *E. coli* (HKEC). Stimulated APCs were then mixed with T cells and incubated for 5 days with staphylococcal enterotoxin B (superantigen). The x- and y-axes measure, respectively, the amount of surface CXCR5 marker and CD4 marker per cell. All T cells are plotted. Red boxes mark the graphical locales of CD4^{pos}-CXCR5^{pos} T$_{FH}$ cells. Numbers in boxes indicate the percentage of total CD4^{pos} cells (T cells) that are also CXCR5^{pos} (T$_{FH}$ cells).
Source: Modified from Ugolini et al. 2018. *Nature Immunol.* **19**:386–396, fig. 1C.

preincubated with staphylococcal enterotoxin B (superantigen). Addition of superantigen circumvented the need for TCR specificity. After incubation, the cells were tagged with fluorescently labeled anti-CD4 and anti-CXCR5 antibodies (wearing different fluorescent tags) and passed through a flow cytometer. Lasers of different wavelengths were used to activate the two different fluors, and detectors measured the levels of fluorescence for each marker.

Figure 24.28 shows that treating monocytes with live *E. coli* greatly increased the monocytes' ability to promote T-cell differentiation into CD4^{pos}-CXCR5^{pos} T$_{FH}$ cells (the cell population present within the rectangles). The live vaccine converted 22.1% of CD4^{pos} cells into T$_{FH}$ cells, as compared to a conversion rate of only 6.6% achieved by the heat-killed vaccine (subtracting the control of 5.2% from each of these percentages makes the difference 1.4 for the heat-killed vaccine versus 16.9 for the live vaccine, a 12-fold increase). This example shows how valuable flow cytometry is for dissecting immune responses.

We just discussed how vaccines can best generate antibody responses, but can vaccines also stimulate cell-mediated immunity? Live, attenuated vaccines can, but killed-organism vaccines generally do not. A discussion of why this is and a possible solution is provided in **eTopic 24.7**.

> **Thought Question**
>
> **24.13** Why do immunizations lose their effectiveness over time?

Are Vaccines Dangerous?

Successful vaccination programs carried out in the United States have come close to eradicating many once-feared diseases, such as measles, polio, rubella, and diphtheria, and they have dramatically lowered the morbidity and mortality of many others. The vast majority of people who receive vaccines suffer no, or only mild, reactions, such as fever or soreness at the injection site. Very rarely do more serious side effects, such as allergic reactions or disease, occur. However, vaccinating individuals who have a defect in their immune system—for example, severe combined immune deficiency (SCID; described in **eTopic 24.1**) or bare lymphocyte syndrome (mentioned at the start of the chapter)—can have severe consequences.

Immunocompromised people should never receive live, attenuated vaccines. Although innocuous to someone with a healthy immune system, a live, attenuated vaccine strain will replicate unchecked in someone lacking critical immune system components and can cause severe disease, even death. Subunit vaccines, in contrast, cannot replicate, and they remain safe for immunocompromised individuals. Unfortunately, certain immunodeficiencies prevent subunit vaccines from generating an immune response. Infections in immunocompromised individuals must be treated with antibiotics or intravenous administration of immunoglobulin (IVIG; pooled antibody collected from a thousand or more random individuals).

Despite the unassailable proof that vaccines are extremely safe for people with healthy immune systems, people known as anti-vaxxers circulate unsettling, erroneous information on the Internet purporting a link between vaccinations and other diseases, such as diabetes or autism. No well-controlled scientific study supports any of these claims. In fact, the fraudulent 1998 *Lancet* article written by Andrew Wakesfield that spawned this movement has been disavowed by all its coauthors (except Wakesfield) and was retracted by the journal itself.

The risk of disease and death from preventable infectious diseases far outweighs the minimal risk associated with being vaccinated against them. As proof of this point, the undervaccination of children in the United States starting in the 1980s has led to an increase in cases of measles. Although measles was declared eradicated from the United States in 2000, cases of the disease have risen dramatically in the last few years. In 2002, there were only 50 cases. The estimated number of cases for 2019 is more than 1,600 (the first 5 months of 2019 produced 971 confirmed cases). That level of measles has not be seen in over 25 years.

Likewise, vaccination efforts decreased the incidence of whooping cough from over 250,000 cases per year in the

1930s and '40s to 6,586 cases in 1993. In part because of decreased vaccination rates, whooping cough cases slowly rose to 28,639 in 2013. Fortunately, renewed vaccination efforts and improvements in vaccine effectiveness have started to reverse this trend. There were only 13,439 cases of whooping cough in 2018.

Sometimes designing vaccines that stimulate adaptive immunity is very difficult, especially if the infectious agent undergoes rapid antigenic shifts or drifts. A good example is the influenza virus, for which yearly vaccines are required. **Special Topic 24.1** discusses the heroic, ongoing search for therapeutic antibodies to treat Ebola and describes an alternative to adaptive-immunity vaccines. The technique is called passive immunity, in which agent-specific antibodies are injected directly into patients. Protection is short-lived but can be essential if a vaccine is unavailable or a patient is immunocompromised.

> **Thought Question**
>
> **24.14** How might you design/construct a more effective vaccine for an antigen (for instance, the *Yersinia pestis* F1 antigen, or hepatitis A) that would harness the power of a Toll-like receptor (TLR)? *Hint:* Look at TLR5 in Table 23.3.

To Summarize

- **Vaccines** can be made from live, attenuated organisms; killed organisms; or purified microbe components.
- **Herd immunity** can help protect unimmunized persons from diseases transmitted person-to-person.
- **Vaccine makeup and delivery** influence whether the immune response is primarily humoral or cellular in nature.
- **Serious side effects** from immunizations are very rare.

24.7 Hypersensitivity and Autoimmunity

The immune system is not perfect. Sometimes the system overreacts to certain foreign antigens and causes more damage than the antigen (microbe) alone might cause. In addition, some foreign antigens can possess epitopes that look like host epitopes and can trick the immune system into reacting against self. These immune miscues are called allergic hypersensitivity reactions, and the antigen causing the reaction is called an **allergen**. There are four types of hypersensitivity reactions (**Table 24.6**). Types I–III are antibody mediated; type IV is cell mediated.

Case History: Type I Hypersensitivity

A bee stings a 9-year-old boy walking with his mother at the zoo. Within minutes, the boy begins sweating and itching. His chest then starts to tighten, and he has tremendous difficulty breathing. Terrified, he looks to his equally frightened mother for help.

This is a classic and severe example of **type I** (**immediate**) **hypersensitivity**, called **anaphylaxis**, in which smooth muscle contracts and capillaries dilate in response to the release of pharmacologically active substances. Type I hypersensitivity occurs within minutes of a <u>second</u> exposure to an allergen when the allergen reacts with IgE-coated mast cells. Recall that the primary role of IgE is to amplify the body's response to invaders by causing mast cells to release inflammatory mediators (see Section 24.2). Allergic individuals with type I hypersensitivity produce an excessive amount of IgE to the allergen. On initial exposure, an allergen elicits the production

TABLE 24.6 Summary of Hypersensitivity Reactions

Type	Description	Time of onset	Mechanism[a]	Manifestations
I	IgE-mediated hypersensitivity	2–30 min	Antigen induces cross-linking of IgE bound to mast cells with release of vasoactive mediators.	Systemic anaphylaxis, local anaphylaxis, hay fever, asthma, eczema
II	Antibody-mediated cytotoxic hypersensitivity	5–8 h	Antibody directed against cell-surface antigens mediates cell destruction via ADCC or complement.	Blood transfusion reactions, hemolytic disease of the newborn, autoimmune hemolytic anemia
III	Immune complex–mediated hypersensitivity	2–8 h	Antigen-antibody complexes deposited at various sites induce mast cell degranulation via Fc receptor; PMN degranulation damages tissue (localized reaction; eTopic 24.8).	Systemic reactions, disseminated rash, arthritis, glomerulonephritis
IV	Cell-mediated hypersensitivity	24–72 h	Memory T_H1 cells release cytokines that recruit and activate macrophages.	Contact dermatitis, tubercular lesions

[a]ADCC = antibody-dependent cell-mediated cytotoxicity; PMN = polymorphonuclear leukocyte.

SPECIAL TOPIC 24.1 — A Monoclonal Magic Bullet for Ebola?

Ebola virus, found primarily in Africa, causes a devastating, deadly hemorrhagic fever whose symptoms include widespread damage to blood vessels and extensive internal and external bleeding (Chapter 26). Without an approved drug or vaccine available, mortality can reach as high as 90% of those infected. Ebola virus comes in three clinically relevant but antigenically distinguishable strains: EBOV (formerly Zaire), Bundibugyo (BDBV), and Sudan (SUDV). Efforts to develop a single vaccine that could protect against all three strains have been stymied by these antigenic differences.

However, another way to cure or prevent Ebola disease would be to isolate broadly neutralizing anti-Ebola antibodies from Ebola survivors. Broadly neutralizing antibodies would enable clinicians to cure patients infected with any Ebola strain, and to passively immunize people who have been in contact with those patients. James Crowe (**Fig. 1A**) from Vanderbilt University Medical Center and Alexander Bukreyev (**Fig. 1B**) from the University of Texas Medical Branch have searched for these antibodies and found two promising candidates.

The researchers and their colleagues started their search by screening plasma from 17 survivors of the 2014 Ebola outbreak in West Africa, looking for patients whose antibodies would cross-react with the surface glycoprotein (GP) of the three Ebola virus strains. GP mediates viral attachment and entry into host cells. Antibodies that bind to GP can neutralize the virus by preventing virus entry. Plasma from two of the survivors contained antibodies able to react against GP from all three Ebola strains (**Fig. 2**). The scientists hoped that blood of these two survivors had circulating memory B cells that produced broadly neutralizing, anti-Ebola antibodies (bNAbs).

Six hundred individual B-cell lines made from the two survivors produced antibodies that reacted against EBOV GP. Six of those antibodies were able to bind to all three Ebola strains in ELISA tests. These six short-lived B-cell lines were then fused to immortal myeloma cancer cells to make six immortal B-cell hybridoma cell lines. Each B-cell hybridoma produced a single antibody isotype (for instance, IgG) with binding specificity to a single epitope. Once made, a hybridoma can be stimulated to make large amounts of that antibody. Because all of the antibodies made from a hybridoma are identical, they are called monoclonal antibodies.

Two of the monoclonal antibodies generated from the study (EBOV-515 and -520) neutralized all three Ebola strains, limiting their abilities to form plaques on tissue culture cells. The antibodies also enhanced the survival of mice infected with mouse-adapted EBOV (**Fig. 3**), wild-type SUDV in mice, and BDBV in ferrets (not shown).

Electron microscopy and other studies then revealed the binding sites (epitopes) for each antibody [F(ab) regions] on the GP ΔTM (**Fig. 4**). Three F(ab) molecules of each antibody bound to the GP trimer, one at each monomer. The antibodies bind to quaternary-structure epitopes formed by spatially aligned residues in the GP1 and GP2 subunits of a GP monomer. The reason these antibodies are broadly neutralizing is that the residues critical for binding EBOV-515 or -520 are identical among the EBOV, BDBV, and SUDV Ebola viruses.

People living on the African continent are continually threatened by random outbreaks of Ebola disease that can quickly spread to become epidemics. The authors propose

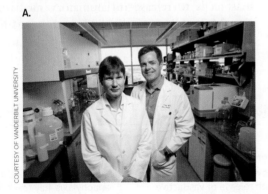
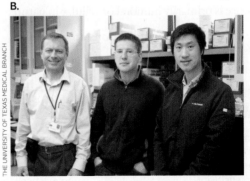

FIGURE 1 ■ **Ebola disease and the scientists searching for a cure. A.** James Crowe (right) and Pavlo Gilchuk (left). **B.** From left to right: Alexander Bukreyev, Philipp Ilinykh, and Kai Huang.

of IgE antibodies specific to the allergen (bee venom in this example). The Fc portions of these antibodies bind to Fc receptors on the surfaces of mast cells, leaving the antigen-binding sites waving away from the cell, "looking" for antigen. IgE-coated mast cells are then said to be sensitized.

During a second exposure to the allergen, identical antigenic sites on the allergen bind to adjacent surface IgE molecules affixed to the mast cell. The result is a bridge between adjoining binding sites (**Fig. 24.29**). This cross-linked complex launches a complex signaling cascade that causes mast cell granules to quickly migrate to the

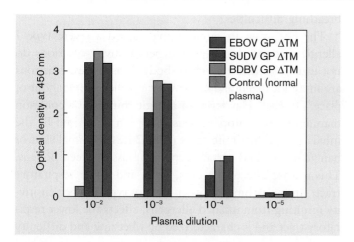

FIGURE 2 ▪ **Binding of Ebola survivor antibodies to modified GPs from the three Ebola viruses.** Enzyme-linked immunosorbent assay (ELISA), as described in Section 28.2, was used to determine binding. Dilutions of survivor plasma were added to microtiter plate wells coated with different GP antigens whose transmembrane domains were deleted (ΔTM) to increase solubility. The level of absorbance at 450 nm is proportional to the amount of anti-Ebola GP present in the survivor's plasma. *Source:* Modified from Gilchuk et al. 2018. *Immunity* **49**:363–374, fig. 1A.

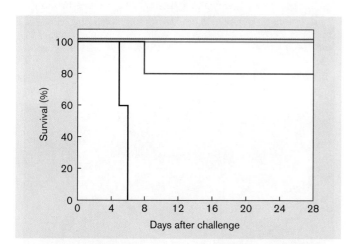

FIGURE 3 ▪ **In vivo protection of bNAbs against EBOV-MA (mouse-adapted EBOV).** Mice (in groups of 5) were infected with EBOV-MA. One day after infection, different groups were treated with the following antibodies: EBOV-515 (blue); EBOV-520 (red); EBOV-442 (green). The control mice (black) were treated with no antibodies. *Source:* Modified from Gilchuk et al. 2018. *Immunity* **49**:363–374, fig. 2D.

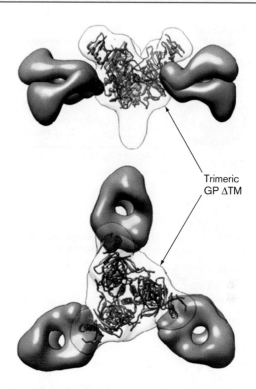

FIGURE 4 ▪ **3D reconstructions of F(ab)–EBOV ΔTM complexes.** Electron density of F(ab) of EBOV-515 (blue) or -520 (orange) antibodies. GP1 and GP2 subunits of each protomer of the GP1 trimer are not highlighted. *Source:* Modified from Gilchuk et al. 2018. *Immunity* **49**:363–374, fig. 6A.

that these broadly neutralizing monoclonal antibodies are promising candidates for development as pan–Ebola virus therapeutic molecules.

RESEARCH QUESTION

Viruses can evolve to evade the immune system by randomly changing one or more residues of an epitope so that a neutralizing antibody can no longer bind. How might therapeutic strategies be designed to minimize this risk?

Gilchuk, Pavlo, Natalia Kuzmina, Philipp A. Ilinykh, Kai Huang, Bronwyn M. Gunn, et al. 2018. Multifunctional pan-ebolavirus antibody recognizes a site of broad vulnerability on the ebolavirus glycoprotein. *Immunity* **49**:363–374.

---

cell surface and release their contents in a process called **degranulation**.

Mast cell degranulation releases chemicals with potent pharmacological activities. The most important of these is histamine, which binds to histamine receptors (H1 receptors) present on most body cells. Antihistamines have a structure similar to histamine and work as an antagonist by preventing histamine from binding the H1 receptors. Histamine bound to H1 receptors on smooth muscle will trigger synthesis of a signaling molecule (inositol trisphosphate) that initiates smooth-muscle contraction to constrict small blood vessels. Histamine also weakens contacts between

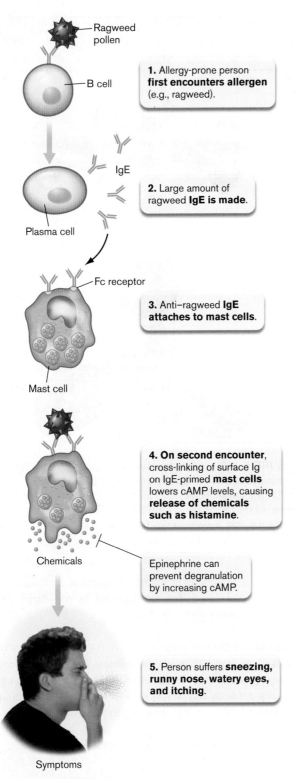

FIGURE 24.29 ▪ **Events leading to type I hypersensitivity reactions.**

swelling (**edema**) in the joints and around the eyes and a rash (similar to hives), with burning and itching of the skin due to nerve involvement. In the case of the boy stung by a bee, the contraction of lung smooth muscles also led to breathing difficulties.

The bee sting in our example caused a severe type I allergic reaction, but type I hypersensitivity reactions do not usually involve the whole body. Most type I reactions are more localized and cause what is called atopic ("out of place") disease. Hay fever, or allergic rhinitis, is a common manifestation of atopic disease that can be caused by the inhalation of dust mite feces (**Fig. 24.30**), animal skin or hair (dander), and certain types of grass or weed pollens. This disease affects the eyes, nose, and upper respiratory tract. Atopic asthma, another form of type I hypersensitivity resulting from inhaled allergens, affects the lower respiratory tract and is characterized by wheezing and difficulty breathing.

The administration of antihistamines is an effective treatment of allergic rhinitis, but the more important chemical mediators of asthma are the leukotrienes produced during what is called **late-phase anaphylaxis**. In late-phase anaphylaxis, mast cells release chemotactic factors that call in

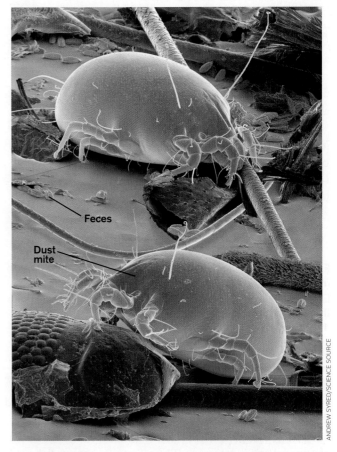

FIGURE 24.30 ▪ **Inhaled allergen.** Dust mite (200–600 μm) and dust mite feces (colorized SEM).

adhesion proteins (VE-cadherin) on vascular endothelial cells, causing gaps between cells through which blood fluids can seep. As a result, histamine-induced constriction of small blood vessels causes fluid to be forced from the circulation into the tissues. The immediate consequence is eosinophils. Eosinophils entering the affected area produce large amounts of leukotrienes that, like histamine, cause vasoconstriction and inflammation. At this point, antihistamines have little effect. Effective treatment includes inhaled steroids to minimize inflammation, and bronchodilators (for example, albuterol) to widen bronchioles and facilitate breathing. One medication, Singulair, works by blocking leukotriene receptors in the lung. During severe asthma attacks, injected epinephrine is critical. Epinephrine will open airways to ease breathing through direct hormonal action.

People sensitized to allergens are not necessarily doomed to suffer with the allergy their entire life. A clinical treatment called **desensitization** can sometimes be used to prevent anaphylaxis. Desensitization involves injecting small doses of allergen over a period of months. This process is thought to produce IgG molecules that circulate, bind, and neutralize allergens before they contact sensitized mast cells. Desensitization has been useful in cases of asthma and bee stings. Desensitization by injection has not proved useful for food allergies; however, studies of oral desensitization are under way.

Another treatment for allergic asthma, omalizumab (Xolair), can actually prevent sensitization. Omalizumab is a monoclonal antibody (an antibody preparation of one antibody type that binds only one epitope). Administered by injection, omalizumab selectively binds IgE. Given to allergic patients once a month, the monoclonal antibody prevents IgE from binding to Fc receptors on mast cells, and thus prevents sensitization. A subsequent encounter with an allergen will not trigger an asthma attack.

Why don't all antigens generate type I hypersensitivity? It turns out that most antigens do not elicit high levels of IgE antibodies. It is unclear what gives certain antigens this capability and what makes different individuals prone to different allergies.

Case History: Type IV Hypersensitivity

The patient is an 8-year-old girl from Argentina. She received the BCG vaccine for tuberculosis about 4 months before coming to the United States. Upon entering school, she is required to take a skin test for tuberculosis. She tries to refuse but is told it is a requirement and allows the nurse to apply the test to her arm. Three days later, the test site has a large, red lesion and the skin is starting to slough (peel off). Does she have tuberculosis?

Type IV hypersensitivity, also known as **delayed-type hypersensitivity (DTH)**, is the only class of hypersensitivity triggered by antigen-specific T cells. Because T cells have to react and proliferate to cause a response, it generally takes 24–48 hours before a reaction is noticed. This delay distinguishes type IV hypersensitivity from the more rapid antibody-mediated allergic reactions (types I–III). The girl in the case study was initially sensitized by being vaccinated with BCG (an attenuated strain of *Mycobacterium bovis*, a close relative of *M. tuberculosis*). Because the organism is intracellular, the vaccination produced a cell-mediated immunity, complete with preactivated memory T cells (similar to memory B cells). When the girl was reinoculated by the skin test, the memory T cells activated and elicited a localized reaction at the site of injection. (Note: About 6 months after vaccination, the hypersensitivity usually diminishes and the tuberculin skin test becomes useful once again.)

Type IV hypersensitivity develops in two stages. In the first stage (sensitization), antigen is processed and presented on cutaneous dendritic cells (called Langerhans cells). These APCs travel to the lymph nodes, where T_H0 cells can react to them as described earlier (see **Fig. 24.22**), generating activated T cells and a subset of memory T cells. On second exposure, two routes leading to DTH are possible. In the first pathway (**Fig. 24.31A**), memory T_H1 cells bind antigen that is complexed to class II MHC receptors (this happens at the site of infection) and release IFN-gamma, TNF-beta, and IL-2. These cytokines recruit macrophages and PMNs to the site and activate macrophages and natural killer cells to release inflammatory mediators that damage innocent, uninfected bystander host cells. In the second pathway (**Fig. 24.31B**), memory T_C cells recognize antigen on class I MHC receptors, become activated, and directly kill the host cell presenting the antigen. A hallmark of type IV hypersensitivity is that white blood cells, not serum, can transfer the sensitivity to a naive animal. This is because T cells, not serum antibody, cause the reaction.

Forms of DTH reactions include contact dermatitis (such as poison ivy rash) and allograft rejection. The antigens involved with contact dermatitis are usually small haptens that have to bind and modify normal host proteins to become antigenic (**Fig. 24.32**). In this case, the hapten-modified protein is processed, and fragments containing the bound hapten are presented on the surfaces of antigen-presenting cells. DTH reactions also play important roles in causing tissue damage during chronic infectious diseases, such as tuberculosis.

There are two other types of hypersensitivity, both of which involve antigen-antibody complexes. **Type II hypersensitivity** occurs when antibody binds to host cell-surface antigens. ABO blood group incompatibility, discussed in **eTopic 24.3**, is an example of type II hypersensitivity. **Type III hypersensitivity** happens when large complexes are formed between antibody and small, soluble

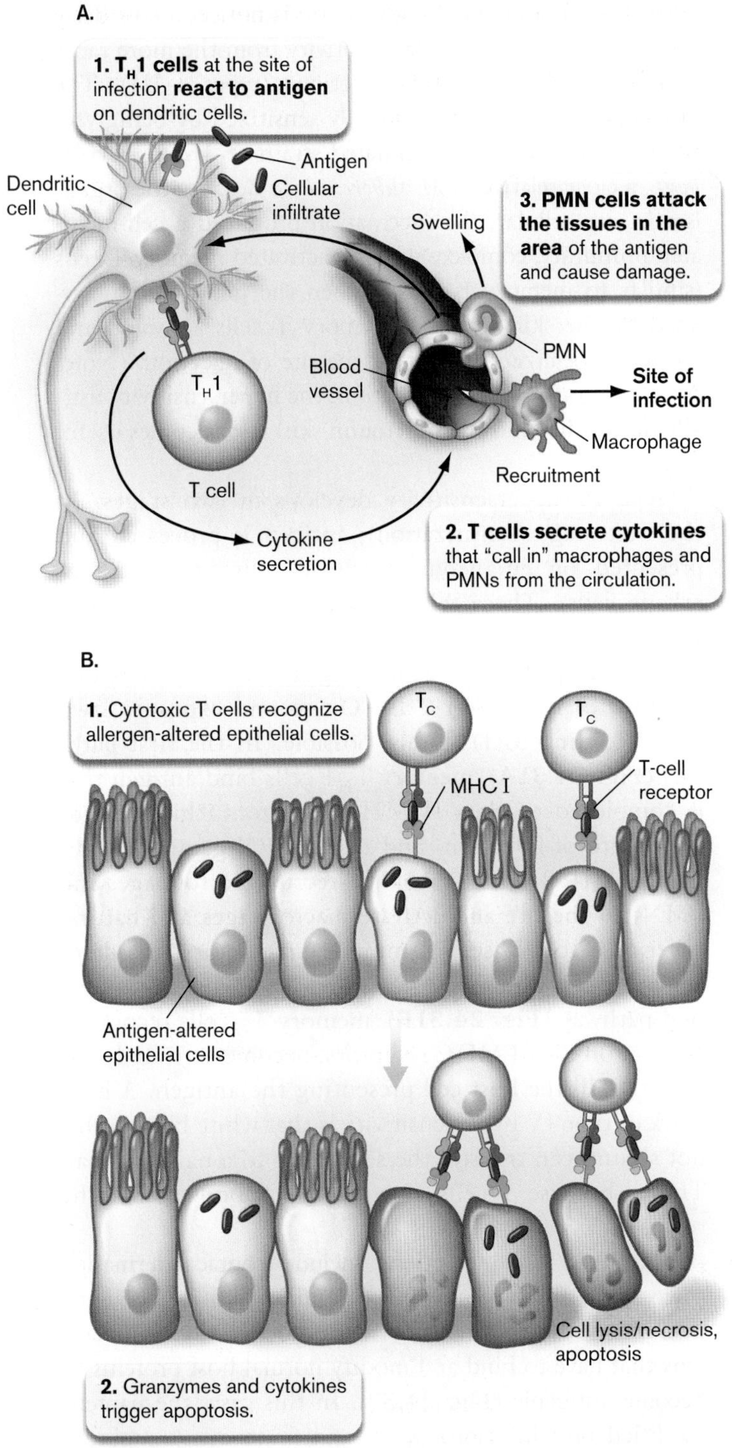

FIGURE 24.31 ▪ Mechanisms of damage in delayed-type hypersensitivity (type IV). A. T_H1 cells react to antigen on dendritic cells. **B.** Antigen-sensitized cytotoxic T cells recognize allergen haptens attached to proteins of other cells. This recognition triggers the release of granzymes and the production of cytokines, which cause apoptosis of the target cell.

antigens. The complexes can trigger complement activation that leads to rash formation. Specific case histories and discussion of these types of hypersensitivity can be found in **eTopic 24.8**.

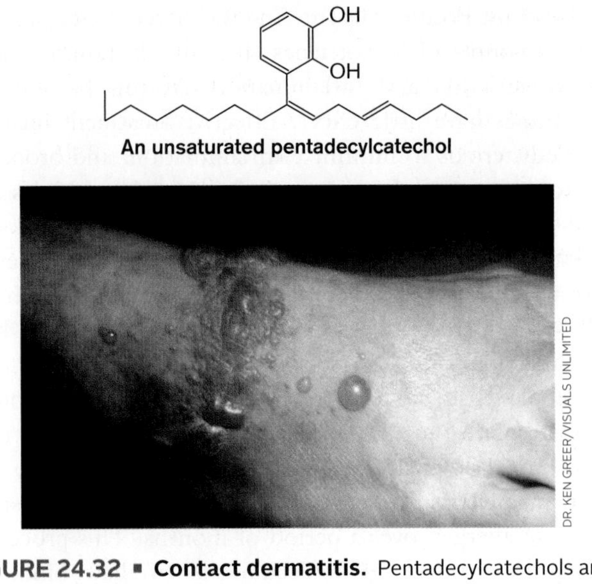

FIGURE 24.32 ▪ Contact dermatitis. Pentadecylcatechols are chemicals present on the surface of poison ivy leaves. They are haptens that bind to proteins in dermal cells, where they can activate T cells. The result of the ensuing delayed-type hypersensitivity is contact dermatitis.

Autoimmunity: An Inability to Recognize Self

The ability to distinguish between self antigens and foreign antigens is crucial to human survival; without this ability, the immune system would constantly attack us from within (see **eTopic 24.5**). Normally, the body develops tolerance to self; occasionally, however, an individual loses immune tolerance against some self antigens, and the body attacks its own tissues. The attack can involve antibodies or T cells and is called an **autoimmune response**. Autoimmune responses may or may not be associated with pathological changes (autoimmune disease). Almost 30% of humans will have an autoimmune antibody by age 65, but many will not exhibit disease. The mechanisms that lead to autoimmune disease are essentially hypersensitivity reactions.

Autoimmune disease results when an autoantibody or autoimmune lymphocyte (cytotoxic T cell) damages tissue components. Tissue damage develops when an autoantibody triggers antibody-dependent cell cytotoxicity (see Section 23.5). In this scenario, NK cell Fc receptors bind to autoantibodies affixed to host cells. The NK cell "thinks" the target cell is infected and kills it. In contrast, autoreactive cytotoxic T cells that escaped negative selection in the thymus can also directly kill host cells that express the cognate self antigen.

How might autoimmune antibodies be formed? One proposed mechanism starts with a self-reacting B cell occasionally escaping the negative selection process. A renegade autoreactive B cell is usually not a problem, because the specific helper T cell needed to activate it was most likely deleted from the T-cell population. However, the B-cell receptor on the autoreactive B cell can, of course, take up and process a self antigen (to no avail) or a foreign antigen (from a pathogen, perhaps) that mimics a self antigen. A foreign antigen that resembles self will also contain one or more nonself epitopes that "piggyback" with the foreign epitope into the self-reacting B cell. As a result, the nonself epitope is presented on a B-cell MHC molecule. A helper T cell specific to the nonself epitope (this T cell was not deleted in the thymus) will then bind to the nonself epitope-MHC complex presented by the self-reacting B cell. The helper T cell is essentially tricked into activating the B cell to become a plasma cell that secretes the autoantibody. Once made, the self-reacting antibodies begin attacking whatever host tissues express the self antigen.

Rheumatic fever is a good example of an autoimmune disease caused by molecular mimicry. After a throat infection with some strains of *Streptococcus pyogenes* ("strep throat"), the patient's heart may be attacked by autoantibodies to heart antigen. The cause appears to be an *S. pyogenes* surface protein called M protein. The M protein contains an epitope that resembles cardiac antigen, but the cardiac-like epitope is flanked by epitopes unrelated to the human host. The cardiac-like epitope can bind to the B-cell receptor of a cardiac self-reacting B cell and be taken up. The nonself epitope linked to the heart-like epitope will "piggyback" its way into the B cell. The nonself epitope is then processed and placed on a B-cell surface MHC II molecule. The TCR of T cells specific to the nonself epitope will bind to the MHC–nonself peptide complex and activate the B cell. But because the B cell is preprogrammed to make anti-cardiac antibody, the resultant plasma cells make autoantibodies to cardiac tissue. Cardiac tissue is damaged, and rheumatic fever results.

Other autoimmune diseases involve the production of autoantibodies that bind and block certain cell receptors. Graves' disease (hyperthyroidism), for example, occurs when an autoantibody is made that binds to the thyroid-stimulating hormone (TSH) receptor. This receptor binding mimics actual TSH binding and continually stimulates the production of two hormones—triiodothyronine (T3) and thyroxine (T4)—that increase metabolic rate. The thyroid is not destroyed but, in fact, enlarges to form a goiter. Examples of other important autoimmune diseases are listed in **Table 24.7**.

T-cell engineering: a way to fight autoimmune disease. Loss of self-tolerance leads to autoimmune disease. So, the ultimate goal of autoimmune therapy must be to restore self-tolerance. Because regulatory T cells are an important part of self-tolerance, scientists have been exploring ways to boost Treg numbers and alter their function. One of the most promising approaches is the adoptive transfer of Tregs with TCRs that have been reengineered to suppress aspects of innate and adaptive immunity. The basic strategy is to convert conventional T cells into Tregs by using retrovirus or lentivirus vectors to cotransfer the Treg-specific regulatory gene *FoxP3* (needed to convert a T cell into a Treg) with a TCR gene that would confer antigen specificity on the Treg-like cell. TCR interaction with a cognate autoantigen would stimulate secretion of anti-inflammatory cytokines that could control T_H1-driven cytotoxic T cells, pro-inflammatory T_H17 cells, or T_{FH} cells that promote

TABLE 24.7 Examples of Autoimmune Diseases

Disease	Autoantigen(s)	Pathology
Type II hypersensitivity—mediated by antibody to cell-surface antigens		
Acute rheumatic fever	Streptococcal M protein, cardiomyocytes	Myocarditis, scarring of heart valves
Autoimmune hemolytic anemia	Rh blood group	Destruction of red blood cells by complement, phagocytosis
Goodpasture's syndrome	Basement membrane collagen	Pulmonary hemorrhage, glomerulonephritis
Graves' disease	Thyroid-stimulating hormone (TSH) receptor	Antibody stimulation of T3, T4 production; hyperthyroidism
Myasthenia gravis	Acetylcholine receptor	Interruption of electrical transmission, progressive muscular weakness
Type III hypersensitivity—mediated by antibody complexes with small soluble antigens (immune complex)		
Systemic lupus erythematosus	DNA, histones, ribosomes	Arthritis, vasculitis, glomerulonephritis
Type IV hypersensitivity—mediated by antigen-specific T cells		
Type 1 diabetes	Pancreatic beta cell antigen	Beta cell destruction
Multiple sclerosis	Myelin protein	Demyelination of axons

autoantibody formation. Although these approaches have been shown to work only in mice, the prospect of producing tailor-made cellular therapies with disease-specific function could revolutionize the treatment of autoimmunity.

To Summarize

- **Allergens** cause the host immune system to overrespond or react against self.
- **Type I hypersensitivity** involves IgE antibodies bound to mast cells by the antibody Fc region. Binding of antigen to the mast cell–attached IgE causes mast cell degranulation. This type of hypersensitivity can occur within minutes of exposure.
- **Type IV hypersensitivity** (delayed-type hypersensitivity) involves antigen-specific T cells. T_H1 cells release cytokines that activate macrophages and NK cells. T_C cells can directly kill cells that present the antigen. Reaction is seen within a few days of exposure.
- **Autoimmune disease** is caused by the presence of lymphocytes that can react to self. These autoreactive lymphocytes escaped negative selection in the bone marrow or thymus.
- **Autoreactive B cells** (B cells that make antibody directed against self epitopes) can be activated if their BCR takes up a self-mimic epitope linked to a nonself epitope and presents the nonself epitope on the surface MHC. T cells that recognize the nonself epitope can then activate the B cell, which secretes antibody against the self epitope. NK cells can target host cells decorated with those autoantibodies.
- **Cytotoxic T cells can produce autoimmune disease** by killing host cells expressing the cognate self antigen.

Concluding Thoughts

The idea that any one person's immune system can recognize and respond to virtually any molecular structure and yet remain selectively "blind" to his/her own antigens is a stunning evolutionary feat—one that is crucial to human survival in our microbe-laden world. What is alarming, however, is that pathogenic microbes have also evolved ways to outmaneuver the many redundancies and safeguards of the immune system. In some cases, the pathogen evolves to become less harmful and coexists with the host; indeed, pathogens in nature are far outnumbered by closely related strains that are harmless or even beneficial to their hosts. In other cases, however, evolution generates a never-ending arms race between pathogen and host. In the next chapter we describe some of the microbial strategies that contribute to the success of a pathogen.

CHAPTER REVIEW

Review Questions

1. Define "antigen," "epitope," "hapten," and "antigenic determinant."
2. What is the basic difference between humoral immunity and cellular immunity?
3. Why are proteins better immunogens than nucleic acids are?
4. What makes IgA antibody different from IgG?
5. Explain isotypic, allotypic, and idiotypic differences in antibodies.
6. How is IgE involved in allergic hypersensitivity?
7. Discuss differences in the primary and secondary antibody responses.
8. Outline the basic steps that turn a B cell into a plasma cell.
9. What is isotype switching, and how is antibody diversity achieved?
10. What signals are needed to activate helper T cells?
11. What signals activate cytotoxic T cells?
12. Discuss the differences between transplant rejections of nucleated cells and the rejection of blood cells.
13. How do superantigens activate T cells?
14. Discuss the differences between the alternative and classical pathways of complement activation.
15. How does the host prevent membrane attack complexes from being formed in host cells?
16. Describe the development of type I hypersensitivity.
17. How does a B cell programmed to make an antibody against self become activated in the absence of specific T-cell help?
18. Explain key features of the gut immune system.
19. Discuss the types of immune responses generated by different types of vaccines.

Thought Questions

1. Cytotoxic T cells lyse the membranes of host cells carrying viruses or bacteria. Why doesn't this lysis facilitate the spread of those organisms rather than help clear the infection?
2. Immunity raised by the poliovirus vaccine will prevent subsequent infection by poliovirus, but someone with a *Staphylococcus aureus* infection (and the attendant immune response) can be reinfected many times by *S. aureus.* Why does the immune system work so well against infection by some pathogens (for example, poliovirus) but not others (for example, *S. aureus*)?
3. (refer to **eTopic 24.3**) Why are blood-type-O people "universal donors" (meaning they can donate their red blood cells to type O, type A, type B, or type AB individuals), but not universal acceptors of red blood cells from type O, A, B, or AB individuals? Which blood-type person might be considered a universal recipient?
4. IgM is the first antibody produced during a primary immune response. Of the five different types of antibodies produced, why would the body want IgM to be the first?
5. Some pathogens, such as rubella virus, can cross the placental barrier from mother to fetus and cause a dangerous infection. Which antibody isotype against rubella virus would you look for in a newborn to diagnose congenital rubella syndrome: IgG or IgM?
6. Sketch a graph of flow cytometry data that shows the percentage of T cells in a population that are CD4 positive versus CD8 positive. Which cell markers would you use?

Key Terms

adaptive immunity (964)
addressin (996)
affinity maturation (981)
allergen (1001)
allotype (972)
anaphylaxis (1001)
antibody (964)
antigen (964)
antigen-presenting cell (APC) (965, 982)
antigenic determinant (965)
autoimmune response (1006)
B-cell receptor (BCR) (975)
B-cell tolerance (967)
capping (976)
cell-mediated immunity (964)
class I MHC molecule (982)
class II MHC molecule (982)
class switching (974)
classical complement pathway (993)
clonal (965)
clonal selection (975)
constant region (970)
cytokine (980)
cytotoxic T cell (T_C cell) (982)
decay-accelerating factor (994)
degranulation (1003)
delayed-type hypersensitivity (DTH) (1005)
desensitization (1005)
edema (1004)
epitope (965)
factor H (994)
Fc region (972)
gene switching (978)
granzyme (988)
hapten (969)
heavy chain (970)
helper T cell (T_H cell) (980)
humoral immunity (antibody-dependent) (964)
idiotype (972)
immediate hypersensitivity (1001)
immunogen (964, 969)
immunogenicity (967)
immunoglobulin (970)
  IgA (974)
  IgD (974)
  IgE (974)
  IgG (973)
  IgM (972)
immunological specificity (967)
immunoprecipitation (970)
innate lymphoid cell (ILC) (995)
intraepithelial lymphocyte (IEL) (995)
isotype (972)
isotype switching (974)
late-phase anaphylaxis (1004)
light chain (970)
major histocompatibility complex (MHC) (967, 982)
memory B cell (974)
mucosal immunity (997)
negative selection (985)
opsonize (973)
perforin (988)
plasma cell (964, 974)
positive selection (985)
primary antibody response (974)
recombination signal sequence (RSS) (978)
regulatory T cell (Treg) (985)
secondary antibody response (974)
serum (974)
superantigen (991)
switch region (980)
T-cell receptor (TCR) (984)
T_H17 cell (988)
threshold dose (967)
tumor necrosis factor (TNF)(991)
type I hypersensitivity (1001)
type II hypersensitivity (1005)
type III hypersensitivity (1005)
type IV hypersensitivity (1005)
vaccination (968)
variable region (972)

Recommended Reading

Artis, David, and Hergen Spits. 2015. The biology of innate lymphoid cells. *Nature* **517**:293–301.

Attaf, Meriem, Mateusz Legut, David K. Cole, and Andrew K. Sewell. 2015. The T cell antigen receptor: The Swiss army knife of the immune system. *Clinical and Experimental Immunology* **181**:1–18.

Chen, Yuezhou, Neha Chaudhary, Nicole Yang, Alessandra Granato, Jacob A. Turner, et al. 2018. Microbial symbionts regulate the primary Ig repertoire. *Journal of Experimental Medicine* **215**:1397. https://doi.org/10.1084/jem.20171761.

Cyster, Jason G. 2010. B cell follicles and antigen encounters of the third kind. *Nature Immunology* **11**:989–996.

Donaldson, Gregory P., Mark S. Ladinsky, Kristie B. Yu, Jon G. Sanders, Bryan B. Yoo, et al. 2018. Gut microbiota utilize immunoglobulin A for mucosal colonization. *Science* **360**:795–800.

Dunston, Christopher R., Rebecca Herbert, and Helen R. Griffiths. 2015. Improving T cell-induced response to subunit vaccines: Opportunities for a proteomic systems approach. *Journal of Pharmacy and Pharmacology* **67**:290–299.

Gieseck, Richard L., III, Mark S. Wilson, and Thomas A. Wynn. 2018. Type 2 immunity in tissue repair and fibrosis. *Nature Reviews. Immunology* **18**:63–76.

Hale, J. Scott, and Rafi Ahmed. 2015. Memory T follicular helper CD4 T cells. *Frontiers in Immunology* **6**:16.

Hevia, Arancha, Susana Delgado, Borja Sanchez, and Abelardo Margolles. 2015. Molecular players involved in the interaction between beneficial bacteria and the immune system. *Frontiers in Microbiology* **6**:1285.

Iwasaki, Aikiko, and Rusian Medzhitov. 2015. Control of adaptive immunity by the innate immune system. *Nature Immunology* **16**:343–353.

Joffre, Olivier P., Elodie Segura, Ariel Savina, and Sebastian Amigorena. 2012. Cross-presentation by dendritic cells. *Nature Reviews. Immunology* **12**:557–569.

Marciano, Beatriz E., and Steven M. Holland. 2017. Primary immunodeficiency diseases: Current and emerging therapeutics. *Frontiers in Immunology* **8**:937. https://doi.org/10.3389/fimmu.2017.00937.

McCoy, Kathy D., Regula Burkhard, Markus B. Geuking. 2019. The microbiome and immune memory formation. *Immunology and Cell Biology* **97**: https://doi.org/10.1111/imcb.12273.

McDermott, Andrew J., and Gary B. Huffnagle. 2014. The microbiome and regulation of mucosal immunity. *Immunology* **142**:24–31.

McGovern, Jenny L., Graham P. Wright, and Hans J. Stauss. 2017. Engineering specificity and function of therapeutic regulatory T cells. *Frontiers in Immunology* **8**:1517. https://doi.org/10.3389/fimmu.2017.01517.

O'Keeffe, Meredith, Wai Hong Mok, and Kristen J. Radford. 2015. Human dendritic cell subsets and function in health and disease. *Cellular and Molecular Life Sciences* **72**:4309–4325.

Spencer, Jo, Linda S. Klavinskis, and Louise D. Fraser. 2012. The human intestinal IgA response; burning questions. *Frontiers in Immunology* **3**:108. https://doi.org/10.3389/fimmu.2012.00108.

Ugolini, Matteo, Jenny Gerhard, Sanne Burkert, Kristoffer Jarlov Jensen, Philipp Georg, et al. 2018. Recognition of microbial viability via TLR8 drives TFH cell differentiation and vaccine responses. *Nature Immunology* **19**:386–396.

Wieczorek, Marek, Esam T. Abualrous, Jana Sticht, Miguel Álvaro-Benito, Sebastian Stolzenberg, et al. 2017. Major histocompatibility complex (MHC) class I and MHC class II proteins: Conformational plasticity in antigen presentation. *Frontiers in Immunology* **8**:292. https://doi.org/10.3389/fimmu.2017.00292.

CHAPTER 25
Pathogenesis

25.1 Host-Pathogen Interactions

25.2 Microbial Attachment: First Contact

25.3 Toxins Subvert Host Functions

25.4 Deploying Toxins and Effectors

25.5 Surviving within the Host

25.6 Tools Used to Probe Pathogenesis

How does a pathogen differ from natural microbiota? The answer lies in the molecular tools that pathogens use to avoid or subvert the immune system, as well as to exploit host cell functions. Some pathogens, for instance, avoid phagocytosis, whereas others interfere with cytokine synthesis or function. Other pathogens develop a mysterious undetectable latent stage in the host, only to emerge later to cause disease. Some infectious agents, such as Ebola virus, efficiently slay their victims. The more patient pathogen, however, can persist for many years, causing only minor chronic symptoms. In Chapter 25 we explore the diverse strategies that pathogenic bacteria and viruses use to infect hosts, subvert immune responses, and cause disease.

CURRENT RESEARCH highlight

Bacterium inflicts pain to survive. *Streptococcus pyogenes* causes necrotizing fasciitis (NF); the aggressive infection of soft tissues sometimes called "flesh-eating disease." Initial tissue damage is minor, but the pain is great. Why? Isaac Chiu and colleagues discovered that the streptolysin S hemolysin of *S. pyogenes* activates host nociceptor neurons to produce pain. The pain benefits the pathogen because activated neurons release neuropeptide (CGRP) that suppresses neutrophil recruitment, which enables *S. pyogenes* to spread more quickly through tissues. Botulism toxin, known for preventing neurotransmitter release, inhibited CGRP release. When injected around mouse hindquarter NF lesions (left-hand images), Botox hastened healing without antibiotics.

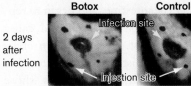

Source: Pinho-Ribiero et al. 2018. *Cell* **173**:1083–1097.

AN INTERVIEW WITH

ISAAC CHIU, NEUROIMMUNOLOGIST/MICROBIOLOGIST, HARVARD MEDICAL SCHOOL

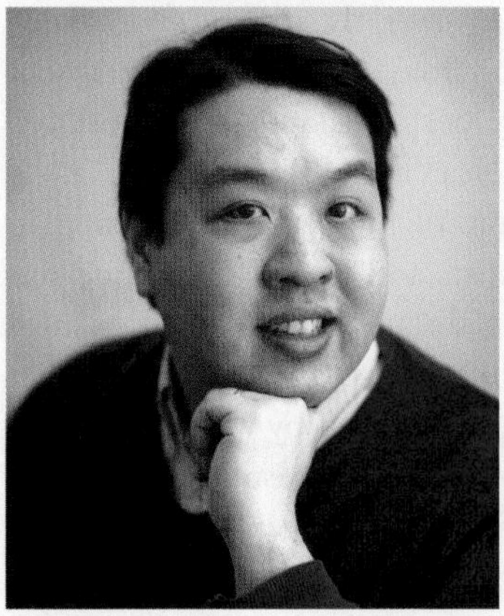

Would combining antibiotics with botulism toxin promote healing even better than antibiotics alone?

That is a great experiment. We have not yet tried combining antibiotics and botulinum toxin vs. antibiotics alone. It is something we will try as we translate our preclinical bench-side findings into bedside clinical trials. One of the big translational questions we would like to explore is whether botulinum toxins can be used to treat multiple types of bacterial pathogens that cause serious skin and soft-tissue infections—for example, Gram-negative pathogens such as *Pseudomonas aeruginosa*, or another important (Gram-positive) agent of necrotizing fasciitis, methicillin-resistant *Staphylococcus aureus* (MRSA).

A 21-year-old construction worker from Idaho accidentally hit his right knee against a cement wall and sustained a seemingly minor abrasion. The next day he started complaining of knee pain and had to stop working. Overnight the pain had intensified, causing him to visit the emergency department, where he received intravenous benzylpenicillin and flucloxacillin antibiotics (see Chapter 27). The ED referred him to the orthopedic unit the next day for a follow-up. On arrival, now 70 hours postinjury, he was in considerable pain, even though the wound appeared superficial. He was afebrile (had no fever), but the wound area was reddened, a little swollen, and warmer than normal. Surprisingly, the patient's white blood cell count was 18,000 per microliter (normal is 4,000–11,000) and C-reactive protein was 63.1 mg/l (normal is <3 mg/l). Sure signs of an infection.

The clinicians suspected septic arthritis, but a Gram stain and culture of synovial fluid were negative. Meanwhile, the man's pain and tenderness continued to expand up his thigh, despite continued antibiotic use. Suspicion then quickly turned to necrotizing fasciitis. Emergency surgery revealed necrotic fasciae and muscles. Fasciae (singular, fascia) are sheets of connective tissue lying below the skin that enclose muscles and organs. Surgeons carefully debrided (removed) the necrotic tissue, and new antibiotics were added to the treatment regimen. Fortunately, the patient slowly recovered without needing his leg amputated. The causative agent was found to be *Streptococcus pyogenes* (also called group A streptococci, or GAS), the same organism that causes strep throat.

S. pyogenes is a Gram-positive coccus that is visually indistinguishable from many other nonpathogenic microbes, such as *S. salivarius*, a normal member of the oral microbiome. So, why does one Gram-positive coccus cause life-threatening necrotizing fasciitis, while the other is a normally harmless host symbiont? In the case of *S. pyogenes*, the difference between friend and foe lies with numerous secreted toxins (absent from *S. salivarius*) that enter host cells and hijack host cell functions (see the Current Research Highlight).

Pathogens use a vast array of molecular tools to gain access to nutrients within hosts and to avoid immune retribution. In Chapter 25 we discuss various relationships between pathogens and their victims, and the factors that contribute to **pathogenesis**, the process by which microbes cause disease. The degree of harm that results depends on the virulence mechanisms that the pathogen wields and the immune response to the pathogen's presence.

25.1 Host-Pathogen Interactions

How long have we humans suffered with infections? Millions of years, it turns out. Paleopathologists found evidence of brucellosis in a skeleton from an *Australopithecus africanus* male, a predecessor of *Homo sapiens* that lived over 2 million years ago. The vertebrae of this "person" exhibited damage that is characteristic of disease caused by a pathogenic species of *Brucella*. Even though we have long been plagued by infectious diseases, the idea that these diseases can be caused by tiny living organisms invading our bodies became apparent only 150 years ago, when Robert Koch discovered the microbial cause of anthrax.

The Language of Pathogenesis

Before discussing the microbial mechanisms of disease, we should establish the vocabulary of infectious disease. In its broadest sense, the term **parasite** (defined as an organism that receives benefits at the expense of a host), includes bacteria, viruses, fungi, protozoa, and worms that colonize and harm their hosts. In practice, however, the term "parasite" is usually reserved for disease-causing protozoa and worms. Bacterial, viral, and fungal agents of disease are referred to as **pathogens**. Parasitic protozoans can also be call pathogens. Pathogens and parasites infect their animal and plant hosts in a variety of ways and enter into a variety of host-pathogen relationships, depending on the site of colonization. For example, organisms that live on the surface of a host are called ectoparasites (here, "parasite" is used in its broadest sense). The fungus *Trichophyton rubrum*, one cause of athlete's foot, is an ectoparasite (**Fig. 25.1**). *Wuchereria bancrofti*, the worm parasite that causes elephantiasis, is an endoparasite because it lives inside the body (**Fig. 25.2**).

An infection occurs when a pathogen or parasite enters or begins to grow on a host. But the term **infection** does not necessarily imply overt disease. Any potential pathogen growing in or on a host is said to cause an infection, but that infection may be only transient, if immune defenses kill the pathogen before noticeable disease results. Indeed, most infections go unnoticed. For example, every time you have your teeth cleaned by a dentist, your gums bleed and your resident oral microbes transiently enter the bloodstream, but you rarely suffer any consequences.

Primary pathogens are disease-causing microbes with the means to breach the defenses of a healthy host. *Shigella flexneri*, the cause of bacillary dysentery, is a primary pathogen. When ingested, it can survive the natural barrier of an acidic (pH 2) stomach, enter the intestine, and begin to replicate. **Opportunistic pathogens**, on the other hand, cause disease only in a compromised host. *Pneumocystis jirovecii* (formerly *P. carinii*) is an opportunistic yeast pathogen that causes life-threatening infections in AIDS patients whose immune systems have been eroded by HIV (**Fig. 25.3A**). Another opportunist is *Pseudomonas aeruginosa*, a bacterial pathogen that commonly infects burn victims who have compromised skin barriers. Some microbes even enter into a **latent state** during infection, in which the organism cannot be found by culture. Herpesvirus, for instance, can

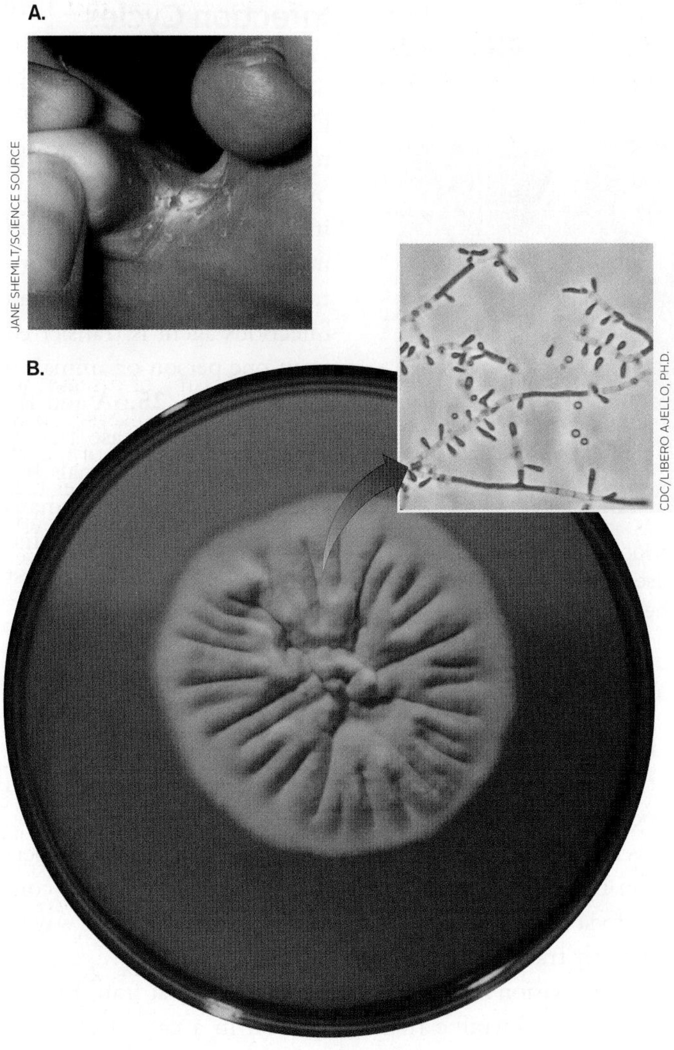

FIGURE 25.1 ■ **An ectoparasite.** A. Athlete's foot can be caused by the fungus *Trichophyton rubrum*. B. Colony morphology and microscopic, branching conidia (blowup) of *T. rubrum*. Conidia are asexual spores that grow on stalks called conidiophores (see Chapter 20).

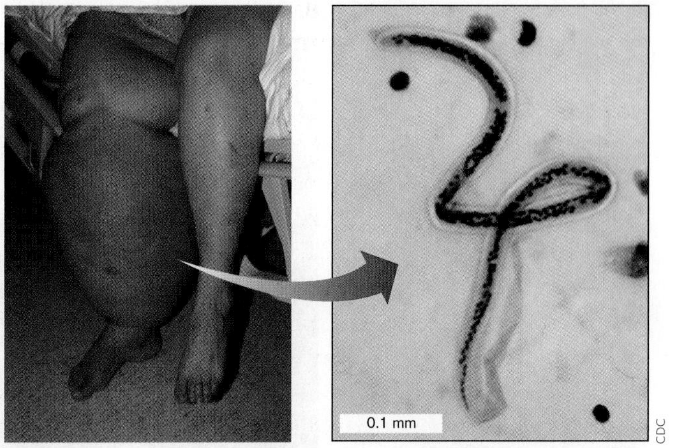

FIGURE 25.2 ■ **An endoparasite.** The disease filariasis, commonly known as "elephantiasis" for obvious reasons, is caused by the worm *Wuchereria bancrofti* (inset), which enters the lymphatics and blocks lymphatic circulation. Adult worms are threadlike and measure 4–10 cm in length. The young microfilariae (inset) are approximately 0.5 mm in length. Though not a problem in the United States, *W. bancrofti* and elephantiasis are found throughout Asia and middle Africa.

enter the peripheral nerves and remain dormant for years, and then suddenly emerge to cause cold sores (**Fig. 25.3B**). The bacterium *Rickettsia prowazekii* causes epidemic typhus, but it can also enter a latent phase and then, months or years later, cause a disease relapse called recrudescent typhus.

Pathogenicity is an organism's ability to cause disease. It is defined in terms of how easily an organism causes disease (infectivity) and how severe that disease is (virulence). Pathogenicity, overall, is shaped by the genetic makeup of the pathogen. In other words, an organism is more—or less—pathogenic, depending on the tools at its disposal (such as toxins) and their effectiveness.

Virulence is a measure of the degree, or severity, of disease. For instance, Ebola virus has a case fatality rate near 50%, so the virus is highly virulent (**Fig. 25.4**). By contrast, rhinovirus, the cause of the common cold, is very effective at causing disease but almost never kills its victims, so it is highly infective but has low virulence. Both organisms are pathogenic, but one lets you live while the other may kill you.

One way to measure virulence is to determine how many bacteria or virions are required to kill 50% of an experimental group of animal hosts. This value is called the **lethal dose 50% (LD_{50})**. A pathogen with a low LD_{50}, in which very few organisms (or viruses) are required to kill 50% of the hosts, is more virulent than one with a high LD_{50} (**Fig. 25.5**). For organisms that colonize but do not kill the host, the infectious dose needed to colonize 50% of the experimental hosts—that is, the **infectious dose 50% (ID_{50})**—can be measured. ID_{50} is the dose required to cause disease symptoms in half of an experimental group of hosts.

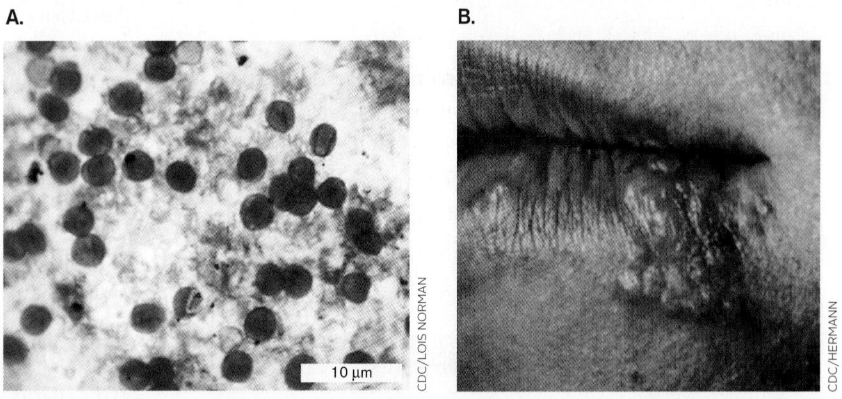

FIGURE 25.3 ■ **Opportunistic and latent infections.** A. *Pneumocystis jirovecii* cysts in bronchoalveolar material. Notice that the fungi look like crushed Ping-Pong balls. B. Cold sore produced by a reactivated herpesvirus hiding latent in nerve cells.

A.

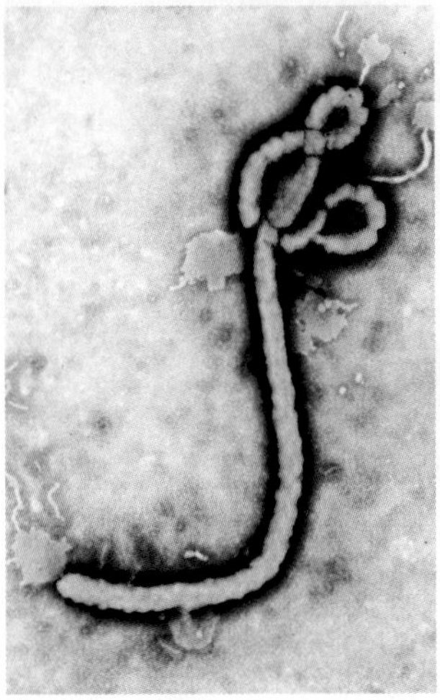

B.

FIGURE 25.4 ■ Highly virulent viruses. A. Ebola virus (approx. 1 μm long; TEM). B. A health care worker tends to an Ebola patient in Kenema, Sierra Leone. Ebola causes hemorrhagic infections in which patients bleed from the mouth, nose, eyes, and other orifices. The mortality rate for this disease is approximately 50%.

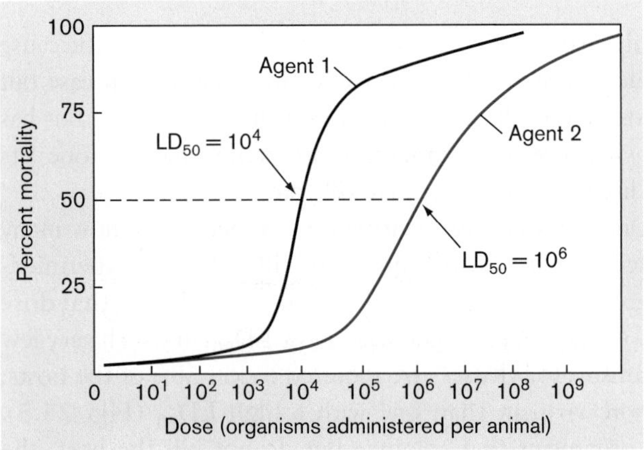

FIGURE 25.5 ■ Measurement of virulence. Each LD_{50} measurement requires infecting small groups of animals with increasing numbers of the infectious agent and observing how many animals die. The number of microbes that kill half the animals is called the LD_{50} dose. In this example, agent 1 is more virulent than agent 2.

Although it might be possible to measure the infectious dose rather than the lethal dose for a lethal pathogen, it is not typically done. LD_{50} gives a clear end point, so it is much easier to use when trying to determine the effectiveness of a given treatment (an antibiotic, for example) or to quantify the role of a given gene in pathogenesis.

Thought Question

25.1 **Figure 25.5** presents the association between LD_{50} and virulence. What does this figure tell you about infectivity?

Infection Cycles

Pathogens must pass from one person or animal to another, if a disease is to spread. The route of transmission an organism takes is called its **infection cycle**. A cycle of infection can involve **horizontal transmission**, in which an infectious agent is transferred from one person or animal to the next (**Fig. 25.6A** and **B**). An infectious disease that is seen primarily in animals but can be transmitted to humans is called a **zoonotic disease**; examples include West Nile disease and Lyme disease. Infection cycles can also involve **vertical transmission**, whereby the agent is transferred from an infected mother to her fetus (transplacental transmission) or during birth (parturition; **Fig. 25.6C**). Organisms that spread directly from person to person, such as rhinovirus, have a simple infection cycle. Rhinovirus can be spread directly from person to person by a sneeze (**airborne transmission**) or by direct contact via handshaking. Handshaking is a very efficient way to directly transfer pathogens through a community.

To envision how effective handshakes are at transmitting pathogens, imagine that one person in a city of 100,000 people has a cold and sneezes on her hands. Then, without washing them, she goes through the day shaking hands with 10 people, and each of those people shakes hands with another 10 people per day, and so on. If there were no repeat handshakes, and if none of the contacts washed their hands, it would take only 4 days to spread the virus throughout the population. In this example, the entire populace of the city would eventually come in contact with the virus, but not everyone would actually contract disease. Additional factors influence whether the virus successfully replicates in a given individual.

Rhinovirus, and other pathogens, can also be transmitted indirectly through the sharing of inanimate objects (**fomites**) such as contaminated utensils (fork or pen), towels, cloth handkerchiefs, and doorknobs (**Fig. 25.6B**). Indirect transmission via fomites, food, or water is also called **vehicle transmission**. Many gastrointestinal infectious agents can be indirectly transmitted between people or animals when food or water becomes contaminated with fecal material containing the pathogen. This is why handwashing after using the bathroom is an important control measure.

More complex infection cycles often involve **vectors**, usually insects or ticks (arthropods), as intermediaries (**Fig. 25.6D**).

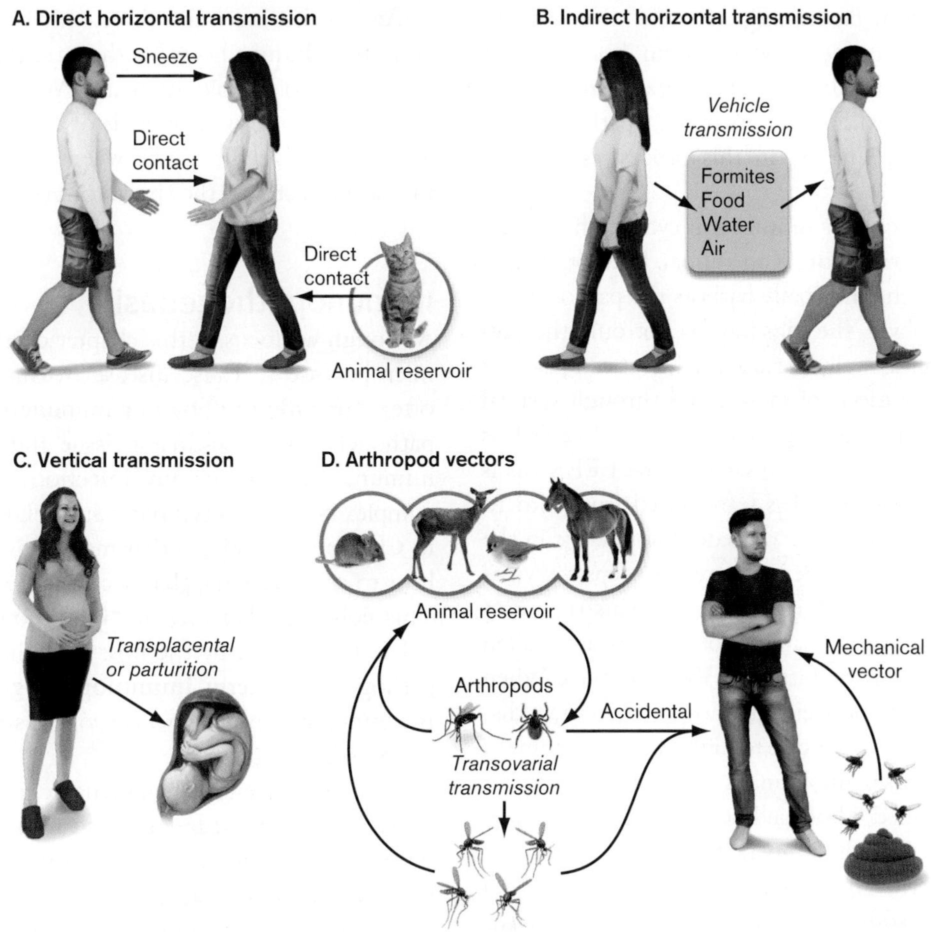

FIGURE 25.6 ▪ Infection cycles. Infectious agents can be transmitted by a variety of means. **A.** Horizontal transmission most commonly involves direct contact between two people (sneezing or touch) or direct contact between a person and an animal reservoir, such as cats harboring *Bartonella hensalae* (cat scratch disease). **B.** Horizontal transmission can also involve indirect transmission by vehicles such as inanimate objects (fomites), air, food, or water. **C.** In vertical transmission, a pathogen passes directly from parent to offspring. In humans, vertical transmission takes place when the organism passes through the placenta to the fetus or from mother to newborn during birth. **D.** Transmission can also occur through arthropod vectors (insect or tick) feeding on animal reservoir hosts. Some insects also practice vertical transmission wherein the egg itself is infected during formation. Accidental transmission happens when a host that is not part of the normal infection cycle unintentionally encounters an animal host-insect vector cycle. In mechanical transmission, a pathogen is transported on the body surface of a vector (for instance, a fly after landing on fecal matter) to a susceptible host.

Vectors carry infectious agents from one animal to another. A mosquito vector (*Aedes*), for example, transfers the virus causing yellow fever horizontally from infected to uninfected individuals when it feeds on a new host (**Fig. 25.7**). The mosquito can also bequeath this particular virus to its offspring via infected eggs in a form of vertical transmission called **transovarial transmission**. Although yellow fever is not a problem in the United States today, a flavivirus closely related to yellow fever virus, called West Nile virus (WNV), claims several victims each year in the United States. West Nile virus is also transmitted to humans by a mosquito vector (*Culex*) that transovarially passes the virus though its eggs.

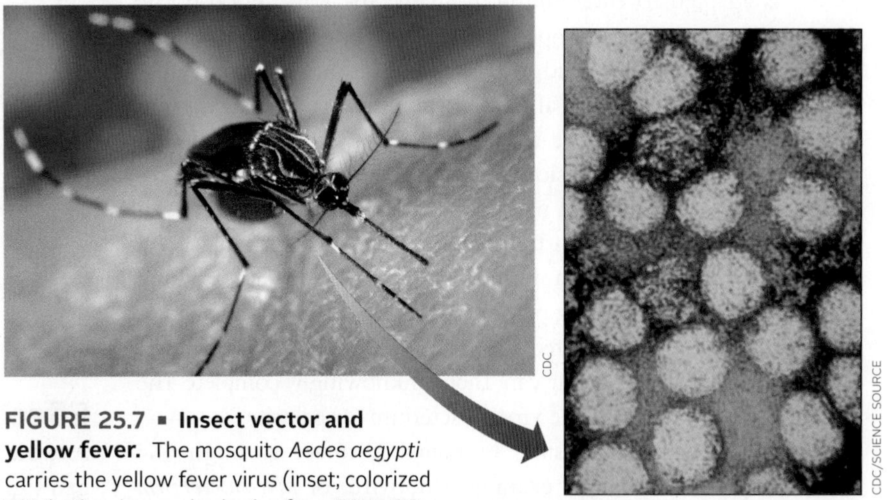

FIGURE 25.7 ▪ Insect vector and yellow fever. The mosquito *Aedes aegypti* carries the yellow fever virus (inset; colorized TEM). The virus varies in size from 50 to 90 nm. Yellow fever remains endemic in the northern part of South America and in central Africa.

Because insects and ticks can transmit many pathogens, strategies to repel or kill arthropods, or prevent them from laying eggs, are effective ways to halt the spread of disease. The common housefly can also transmit disease, by landing on contaminated material (for example, feces) and then carrying pathogens on its surface to a host. The fly in this process is called a **mechanical vector**.

Another critical factor in an infection cycle is the "reservoir" of infection. A **reservoir** is an animal, bird, or arthropod (insect or tick) that normally harbors the pathogen. In the case of yellow fever, the mosquito is not only the vector but the reservoir as well, because the insect can pass the virus to future generations of mosquitoes through vertical transmission. The virus causing eastern equine encephalitis (EEE), however, uses birds as a reservoir. The EEE virus is normally a bird pathogen and is transmitted from bird to bird via a mosquito vector. The virus does not persist in the insect, but transmission by the insect vector keeps the virus alive by passing it to new avian hosts. Humans or horses entering geographic areas that harbor the disease (called endemic areas) can also be bitten by the mosquito. When this happens, they become accidental hosts and contract disease. EEE virus does not replicate to high titers in mammals, which means that horses and humans are poor reservoirs for the virus and are called "dead-end hosts." EEE virus, however, does replicate to high numbers in the avian host.

Reservoirs are critically important for the survival of a pathogen and as a source of infection. If the EEE virus had to rely on humans to survive, it would cease to exist because of limited replication potential and limited access to mosquitoes. Note that the reservoir of a given pathogen might not exhibit disease.

Portals of Entry

How do infectious agents enter the body? Each organism is adapted to enter the body in specific ways. Food-borne pathogens (for example, *Salmonella*, *E. coli*, *Shigella*, and rotavirus) are ingested by mouth and ultimately colonize the intestine. They have an oral portal of entry. These gastrointestinal microbes are also described as having a **fecal-oral route of transmission**, in which pathogens or parasites excreted in the fecal matter of an infected person are indirectly ingested by an uninfected person. For example, fecal organisms are often found on hands or surfaces. Those organisms, including pathogens, can be transferred through touch or contaminated foods to someone else. That newly contaminated person can then unknowingly complete the cycle by ingesting the virus, bacterium, or parasite.

Airborne organisms, in contrast, infect through the respiratory tract (for example, rhinovirus or *Mycobacterium tuberculosis*). Other microbes enter through the conjunctiva of the eye or through the mucosal surfaces of the genital and urinary tracts.

Agents that are transmitted only by mosquitoes or ticks enter their human hosts via the **parenteral route**, meaning injection into the bloodstream. Wounds and needle punctures can also serve as portals of entry for many microbes. Thus, shared needle use between drug addicts has been an important factor in the spread of HIV.

Immunopathogenesis

Although we focus in this chapter on the mechanisms that microbes use to cause disease (toxins, for example), it is often "friendly fire" by our immune system reacting to a pathogen that causes major tissue and organ damage. The immune response to any infection involves activating a complex network of cell types and soluble factors (discussed in Chapters 23 and 24) that may inadvertently damage the host to such a degree that it causes illness and even death. This collateral damage, or "immunopathology," is a calculated risk taken by the host in its haste to eradicate the pathogen. The term **immunopathogenesis** applies when the immune response to a pathogen is a contributing cause of pathology and disease.

The disease dengue hemorrhagic fever is a case in point. Caused by the dengue virus and transmitted by the *Aedes* mosquito, dengue fever manifests as a severe headache, muscle and joint pain, fever, and rash. The symptoms can also include abdominal pain, nausea, and vomiting. However, these symptoms are more a consequence of immunopathogenesis than they are a direct result of viral replication. Replication of the virus in host cells will produce a massive activation of T cells ($CD4^+$ and $CD8^+$). Activation triggers a cytokine cascade (some call it a "storm") that targets vascular endothelial cells and produces an endothelial "sieve" effect leading to fluid and protein leakage. Cytokines such as TNF-alpha, IL-1, and IFN-gamma (discussed in Chapters 23 and 24), along with many others that are released, contribute to inflammation by attracting and activating neutrophils and macrophages. The disease symptoms mentioned earlier can result.

So, to fully understand any infectious disease, you must be aware of the pathogenic mechanisms wielded by the pathogen, but also realize that some disease symptoms may be due to immunopathogenesis. The immunopathogenic features of various diseases will be described in the next chapters.

Effect of Infections on Microbiota

As you learn more about pathogens, realize that a pathogen's growth and the resulting immune response will also affect the host's microbiota. For example, diarrhea can reduce overall numbers of gut microbiota by causing more forceful and frequent evacuations. Intestinal pathogens can change microbiome diversity by occupying limited host binding sites or by altering available nutrients. Microbiota may also be killed more effectively than the pathogen by infection-induced inflammation, depending on the pathogen's virulence mechanisms. Aspects of inflammation can even generate nutrients that feed the pathogen but not normal microbiota (for example, tetrathionate for *Salmonella*, discussed later). All of these mechanisms affect competition between microbial species and, ultimately, species diversity. Finally, as victims recover from a disease, their gut microbiota may not always achieve preinfection balance (for instance, following cholera; see Fig. 26.12). The resulting dysbiosis, as discussed in Chapters 23 and 24, can have negative health effects beyond those of the infection itself.

Virulence Factors and How to Find Them

To cause disease, all pathogens must enter a host; find their unique niche; avoid, circumvent, or subvert normal host defenses; multiply; and eventually be transmitted to a new susceptible host. Pathogens employ **virulence factors**, encoded by virulence genes, to accomplish these goals. Virulence factors include toxins, attachment proteins, capsules, and other devices used to avoid host innate and adaptive immune systems. The next section will describe some of these factors. But first, how are virulence genes identified?

Molecular Koch's postulates. Identifying genes for a metabolic or biosynthetic pathway is relatively easy because it requires merely observing a clear phenotype on an agar plate. If a gene involved with an amino acid biosynthetic pathway is defective, then the amino acid is not made and must be added to the medium, or else the mutant will not grow. Finding a virulence gene is more difficult because the screen involves growth in a host. There is no straightforward way to identify a virulence phenotype using an agar plate. True virulence genes can be recognized only if mutants defective in the gene fail to survive or cause disease in animals.

Numerous clever techniques have been developed over past decades to identify potential virulence genes. Examples are discussed in **eTopic 25.1** (signature-tagged mutagenesis), **eTopic 25.2** (in vivo expression), and Section 25.6 (dual RNAseq). Regardless of how the suspected virulence gene was found, it can be confirmed as having a role in virulence or pathogenicity only if it fulfills a set of "molecular Koch's postulates" originally formulated by Stanley Falkow (1934–2018), a preeminent infectious disease scientist. The molecular postulates are as follows:

1. The phenotype under study should be associated with pathogenic strains of a species.
2. Specific inactivation of the suspected virulence gene(s) should lead to a measurable loss in virulence or pathogenicity. The gene(s) should be isolated by molecular methods.
3. Reversion or replacement of the mutated gene should restore pathogenicity.

A variety of other molecular questions can suggest the role of a virulence protein in pathogenesis. For example:

- Does moving the virulence gene into an avirulent strain impart a pathogenicity trait on the avirulent strain? For instance, does moving a gene for attachment from a pathogen to a nonpathogen allow the nonpathogen to attach to host cells?
- Does the suspected virulence protein bind to important host proteins? Binding to a host protein could indicate the target of the microbial protein.
- Does the microbial protein resemble the sequence or structure of an important host protein? Such resemblance might indicate that the microbial protein mimics the function of the orthologous host protein.
- Does introducing the suspected virulence protein (or its encoding gene) into a host cell alter the host cell's physiology, or does a version of the virulence protein tagged with green fluorescent protein (GFP) localize to a specific host cell compartment or organelle? A positive finding in either case could reveal a host target for the suspected virulence protein.

In addition to the experimental approaches just described, there are bioinformatic ways to identify potential pathogenicity genes.

Pathogenicity Islands

Extensive sequencing efforts have enabled us to compare the genomes of many pathogens and expose some "footprints" of their evolution. For example, in bacterial pathogens, most chromosomes are dotted with clusters of pathogenicity genes that encode virulence functions. These gene clusters, called **pathogenicity islands**, can be considered the toolboxes of pathogens (originally discussed in Sections 9.4 and 9.5). Many virulence genes reside in pathogenicity islands, although many others do not. Some virulence genes reside on plasmids (for example, the genes for the diarrhea-producing labile toxin of certain *E. coli* strains) or in phage genomes (such as the genes encoding the diphtheria toxin of *Corynebacterium diphtheriae*).

Most pathogenicity islands appear to have been horizontally transmitted (via conjugation or transduction; discussed in Section 9.5) from long-extinct organisms into the ancestors of today's pathogens. Horizontal gene transfers move whole blocks of DNA (more than 10 kb) from one organism to another, placing the blocks directly in

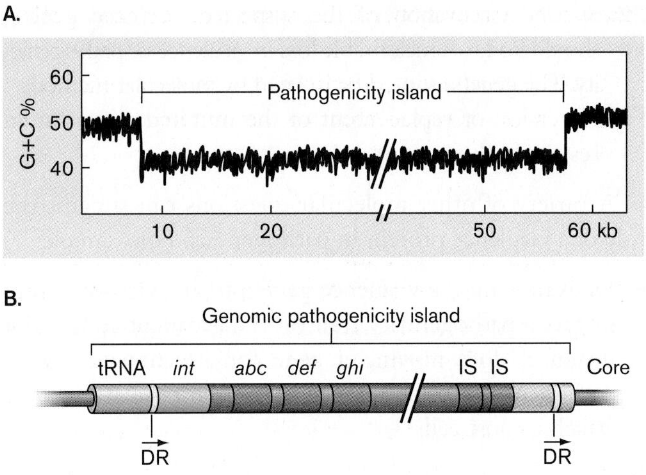

FIGURE 25.8 ■ Model pathogenicity island. **A.** The guanine + cytosine (G+C) content of the island is different from that of the core genome. **B.** Schematic model of a pathogenicity island. The DNA block is linked to a tRNA gene and flanked by direct repeats (DRs) that may be "footprints" of a transposon or viral-mediated transfer. The integrase gene (*int*) and insertion sequences (ISs) may also be remnants of transposition.

the chromosome in what is called a **genomic island** (see Chapter 9). If the island increases the "fitness" (virulence) of a microorganism (pathogen) that interacts with a host, it is called a pathogenicity island. Genomic islands generally reveal themselves by several anomalies that they possess with respect to the rest of the host genome:

- **A very different GC/AT ratio.** For example, a plot of GC content along the length of a chromosome may reveal that most of the genome has a 50% GC content. But somewhere in the middle, a 50-kb region sticks out on the graph, showing a content of 40% (**Fig. 25.8A**). This deviation probably reflects the GC content of the microbe that donated the island.
- **Linkage to a tRNA gene.** The reason for this linkage, however, is not clear. One hypothesis is that the conserved secondary structure of tRNA facilitates integration by an integrase.
- **Association with genes homologous to phage or plasmid genes.** Typically, genomic islands are flanked by genes that show homology to phage or plasmid genes (**Fig. 25.8B**). This arrangement is thought to reflect the transfer vector used to move the island from one organism to another.

Figure 25.9 provides examples of pathogenicity islands from different pathogens.

But what do the pathogenicity genes actually do? Some genes encode molecular "grappling hooks," such as pili that attach to host cells. Once attached, microbes can secrete toxins that injure the host cell. Other bacteria wall themselves off to prevent damage by host inflammatory responses. Some bacterial pathogens are even capable of what could be called "host cell reprogramming." These organisms inject proteins directly into the host cell to disrupt normal signaling pathways. The reprogrammed target cell can be made to do one of several things: engulf the bacterium; "commit suicide" (undergo apoptosis); engineer a tighter, more intimate attachment

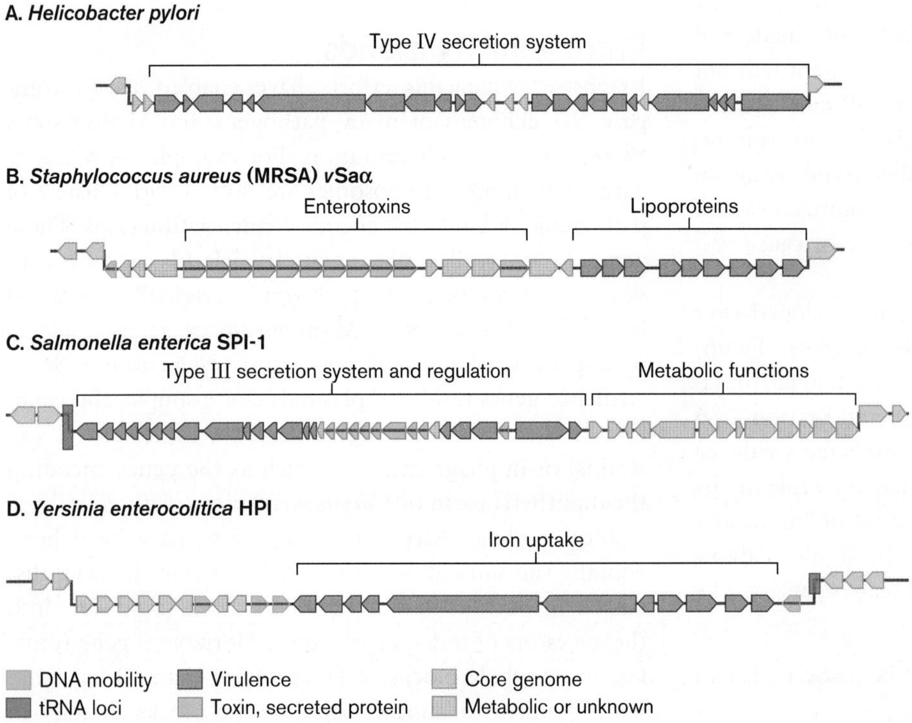

FIGURE 25.9 ■ Examples of bacterial pathogenicity islands. **A.** The *cag* island of *Helicobacter pylori* (stomach ulcers) harbors genes for a type IV secretion system that can translocate the toxin CagA into human gastric cells, causing an inflammatory response. **B.** The *v*Saα island of a particularly virulent strain of *Staphylococcus aureus* (MRSA) encodes a remarkably high number of enterotoxins. **C.** The SPI-1 island of *Salmonella enterica* (enteritis disease) encodes a type III secretion system (gray), secreted effector proteins (dark gray), and regulatory proteins. The island includes genes for metabolic proteins unrelated to virulence. **D.** The high-pathogenicity island (HPI) of *Yersinia enterocolitica* (enteritis) carries genes for a high-affinity iron uptake system (dark gray) needed for extracellular growth during host colonization.

platform at the cell surface; or alter the amounts or types of cytokines that the affected cell produces. Detailed functions of various pathogenicity genes will be described as the chapter proceeds.

Caught in the Act: Examples of Pathogen Evolution by Horizontal Gene Transfer

Pathogens and hosts continually coevolve, with each trying to gain the upper hand. The process is typically slow, but scientists sometimes catch a pathogen in the act. Examples include *Escherichia coli* and *Streptococcus pyogenes*.

Escherichia coli. *E. coli* is a member of the normal gut microbiota but includes many pathovars that evolved through horizontal gene transfers. A pathovar represents one strain of an organism that causes disease in a specific organ system. Different *E. coli* pathovars cause diseases ranging from urinary tract infections to diarrhea, sepsis, and meningitis. A recent example of *E. coli* evolving via horizontal gene transfer involves the enteroaggregative hemorrhagic strain O104:H4, which caused the frightening 2011 outbreak of diarrhea and hemolytic uremic syndrome that began in Germany and spread through much of Europe (see **eTopic 26.5**). This pathovar included genes for a powerful Shiga toxin (described later), resistance genes against many antibiotics, and several new virulence traits that were absent in its closest relative. Horizontal gene transfers were instrumental in its evolution.

Streptococcus pyogenes. Another dangerous pathogen caught in the act of evolving is *Streptococcus pyogenes*, otherwise known as group A streptococci (GAS), a strict human pathogen that can cause sore throat, scarlet fever, and necrotizing fasciitis (the so-called "flesh-eating disease"; see the Current Research Highlight and Section 26.1). For a century, GAS strains have been sorted into serological types by amino acid differences present in a cell-surface molecule called M protein. James Musser from the Houston Methodist Research Institute and his colleagues determined through large-scale genome sequencing that epidemics of streptococcal disease are caused by horizontal gene transfers rather than by the simple reemergence of older strains. For instance, the origin of the most recent GAS M1 global pandemic strain was traced to about 1983, when a horizontal gene transfer event introduced a 36-kb chromosomal region encoding several proteins. Two of those proteins, NAD glycohydrolase (or SPN; the gene is *nga*) and streptolysin O (SLO; gene *slo*), are potent toxins (**Fig. 25.10A**).

Compared to preepidemic strains, all of the new epidemic strains contained the same three single nucleotide polymorphisms (SNPs) within the 36-kb region. Two SNPs were located in the upstream promoter region of the SPN operon. These two promoter mutations increased expression of the SPN and SLO proteins (**Fig. 25.10B**) and dramatically increased virulence. The third change was a missense mutation in SPN that changed a glycine to aspartate and restored NAD glycohydrolase enzymatic activity. The restored activity of NAD glycohydrolase was required to maximize virulence of the epidemic GAS

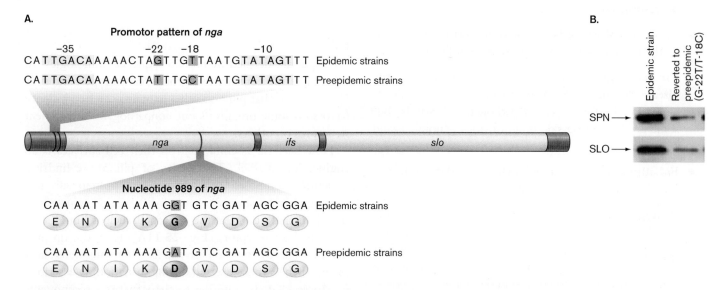

FIGURE 25.10 ■ **Emergence of a pandemic clone of group A streptococci.** A. Sequence comparison in the *nga-slo* operons of preepidemic and epidemic strains of *Streptococcus pyogenes* revealed three consistent single nucleotide polymorphisms (SNPs). The *nga* and *slo* genes encode NAD glycohydrolase (SPN) and streptolysin O (SLO), respectively. The *ifs* gene encodes a regulator of SPN. B. Western blot of secreted proteins, showing that changing the SNPs in the promoter of an epidemic strain back to those found in a preepidemic strain reduced secretion of SPN and SLO. *Source:* Micrographs in part B from Luchang Zhu et al. 2015. *J. Clin. Invest.* **125**(9):3545–3559, fig. 3C. ©2015, American Society for Clinical Investigation.

strain. Up-regulating SPN and SLO in the pandemic strain enhanced tissue destruction, heightened resistance to killing by polymorphonuclear leukocytes, and helped propagate GAS infections around the world.

Evolutionary features of two other pathogens—*Streptococcus agalactiae* and *Staphylococcus aureus*—are described in **eTopic 25.3**.

Sections 25.2 through 25.5 describe some of the specific tools that pathogens have evolved to undermine the integrity of the body. From attachment to toxins to intracellular invasion, the infection process is like a chess match, with each side, human and microbe, trying to outmaneuver the other.

To Summarize

- **Infection does not equal disease.** The immunocompetence of the host and the virulence potential of the pathogen influence whether an infection causes disease.
- **Primary pathogens** have mechanisms that help the organism circumvent host defenses in a healthy host, whereas **opportunistic pathogens** cause disease only in a compromised host.
- **Pathogenicity** refers to the mechanisms a pathogen uses to produce disease and how efficient the organism is at causing disease, whereas **virulence** is a measure of disease severity.
- **Diseases can be spread** by direct or indirect contact between infected and uninfected persons/animals or by insect vectors.
- **Pathogens use portals of entry** best suited to their mechanisms of pathogenesis.
- **Immunopathogenesis** is damage to host tissues caused by the immune system's response to an infection.
- **Virulence genes** encode products that enhance the disease-causing ability of the organism. Many virulence genes can be found within pathogenicity islands, but some are located outside of an obvious genomic island or reside in plasmids.
- **Fulfilling the molecular Koch's postulates** validates the identity of a virulence gene.
- **Pathogenicity islands** are DNA sequences within a species that are acquired by horizontal gene transfer from a different species. They contain distinct features, such as GC content and the remnants of phages or plasmids, that mark them as being different from the rest of the genome.
- **Horizontal gene transfer mechanisms** move virulence factor genes and pathogenicity islands among bacterial strains and species.

25.2 Microbial Attachment: First Contact

Regardless of the disease, pathogens must reach a colonization site either through their own motility or by hitchhiking with a vector. Once at the site, the pathogen needs attachment mechanisms to stay there.

The human body has many ways to exclude pathogens. The lungs use a mucociliary escalator (see Fig. 23.6) to rid themselves of foreign bodies, the intestine uses peristaltic action to ensure that its contents are constantly flowing, and the bladder uses contraction to propel urine through the urethra with tremendous force. How do bacteria ever manage to stick around long enough to cause disease? Like a person grasping a telephone pole during a hurricane, successful pathogens moving through the body manage to grab and tenaciously hold on to host cells. Thus, the first step toward infection is attachment, also called adhesion. Any microbial factor that promotes attachment is called an **adhesin**.

Viruses attach to the host through their capsid or envelope proteins, which bind to the specific host cell receptors discussed in Chapters 6 and 11. Bacteria have a variety of similar strategies. They can use hairlike appendages called **pili** (also called **fimbriae**; composed of protein monomers), whose tips contain receptors for mammalian cell-surface structures. Or they can use a variety of adherence proteins or other molecules that are not part of a pilus. Sometimes they use both. **Table 25.1** provides examples of bacterial attachment strategies. Note that different pili in the same species can impart different tissue specificity for attachment (for example, uropathogenic versus diarrheagenic *E. coli*).

Pili

Pili types are defined by protein sequence homologies. In this chapter we consider two groups, called type I and type IV pili. Note that pili and other adhesins can be virulence factors on some organisms but nonpathogenic attachment structures on others.

How bacteria assemble pili on their cell surfaces is an engineering marvel. The shafts of pili are cylindrical structures composed of identical pilin protein subunits (described in Chapter 3). Several different proteins adorn the tip, including one at the very apex, such as FimH, that binds to host receptors (**Fig. 25.11A**). How the tip protein manages to "hold on" in the face of tremendous shear forces (such as those exerted during urination) is described in **eTopic 25.4**. In addition to the structural components of pili, numerous other proteins collaborate to assemble the structure. Genes encoding a given pilin protein and the associated assembly apparatus are typically arranged on the chromosome as an operon.

TABLE 25.1 Examples of Bacterial Adhesins

Bacterium	Adhesin	Host receptor	Attachment site	Disease
Streptococcus pyogenes	Protein F	Amino terminus of fibronectin	Pharyngeal epithelium	Sore throat
Streptococcus mutans	Glucan	Salivary glycoprotein	Pellicle of tooth	Dental caries
Staphylococcus aureus	Clumping factors A and B	Fibronectin	Mucosal epithelium	Various
Neisseria gonorrhoeae	*N*-methylphenylalanine pili	Glucosamine galactose carbohydrate	Urethral/cervical epithelium	Gonorrhea
Uropathogenic *E. coli*	Type I fimbriae (pili)	Complex carbohydrate	Urethral epithelium	Urethritis
	P pili (pyelonephritis-associated pili)	P blood group	Upper urinary tract	Pyelonephritis
Bordetella pertussis	Pili ("filamentous hemagglutinin")	Galactose on sulfated glycolipids	Respiratory epithelium	Whooping cough
Chlamydia	Lipooligosaccharide, OmcB surface protein	Sulfonated glycosaminoglycans	Conjunctival or urethral epithelium	Conjunctivitis or urethritis

Assembly of type I pili. Pyelonephritis-associated pili (Pap) of the uropathogenic *E. coli* are type I pili that bind to a digalactoside present on host urinary tract surfaces called the P-blood-group antigen. "Pyelonephritis" is the medical term for kidney infection. Pap pili are essential for uropathogenic *E. coli* to cause this disease. **Figure 25.11B** illustrates how the type I pilus from uropathogenic *E. coli* (Pap) is assembled. The mechanism is representative of other type I pili; only the names of the proteins will differ for each system. Protein components synthesized in the cytoplasm are secreted into the periplasm by the SecA-dependent general secretory system (discussed in Section 8.5). Once in the periplasm, the subunits are chaperoned one at a time by PapD to the membrane site of assembly, which is marked

A. FimH adhesin tip

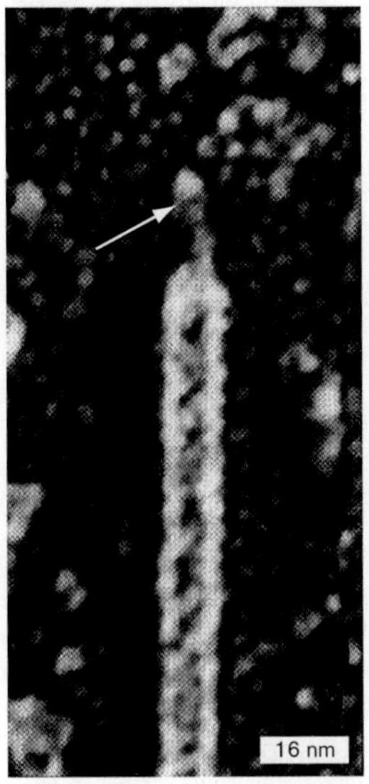

C. H. JONES ET AL. 1995. *PNAS* **92**:2081–2085, FIG. 3

B.

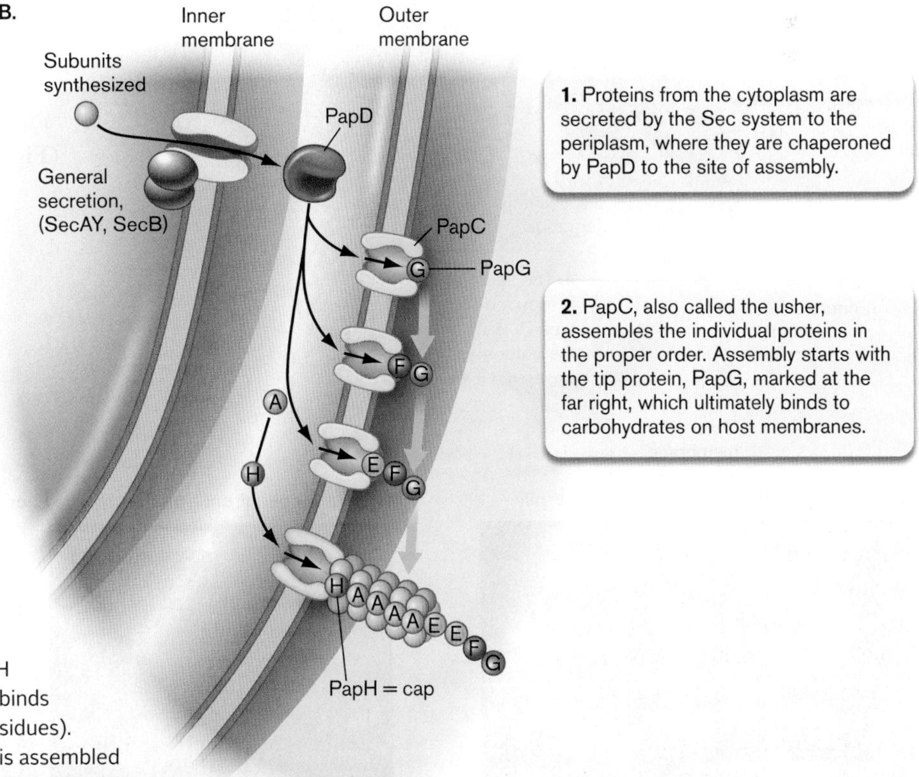

FIGURE 25.11 ▪ Attachment pilus and pilus assembly. **A.** High-resolution micrograph showing a type I pilus (TEM). The FimH adhesin at the tip (arrow) is the protein that binds to the cell receptor (cell-surface mannose residues). **B.** The pyelonephritis-associated pilus (Pap) is assembled in a stepwise fashion. The protein subunits fit together like pieces of a jigsaw puzzle. New protein subunits are added to the bottom of the growing structure.

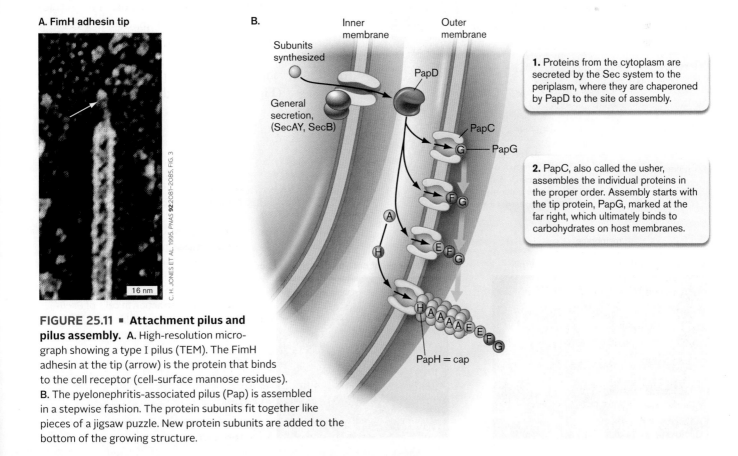

by the presence of the usher protein PapC. Like an usher in a theater, PapC directs the subunits to their proper places.

Assembly of pili at the usher site starts with the tip protein, PapG, which will ultimately bind to carbohydrates on host membranes after the pilus is complete. After PapG, the ushers add PapF and PapE, forcing PapG farther away from the surface. Then, identical PapA pilin subunits are strung together in a series to form the shaft. PapA subunits assemble by sequentially sharing a domain with one another, linking together like pieces of a jigsaw puzzle. Once assembled, type I pili are static. They simply stick to the host receptor. Type IV pili are more dynamic, continually extending and contracting.

Type IV pili. Another group of pili with important roles in pathogenesis is the type IV pili. These pili are dynamic, not static, and are found in a broad spectrum of Gram-negative pathogens (*Pseudomonas* and *Neisseria*, for instance). What makes these pili amazing is their ability to continually assemble and disassemble—a feat that produces a remarkable type of cell movement called twitching motility (described later).

The assembly machinery for type IV pili involves at least a dozen proteins (**Fig. 25.12**). A major difference between type IV pilus assembly and that of type I pili is that type IV pilus proteins are never free in the periplasm; instead, they are inserted and assembled at the cytoplasmic membrane, after which the assembled pilus is "pushed" outside the cell through a channel in the outer membrane (**Fig. 25.12A**). Over the course of evolution, the genes for type IV pili were duplicated and modified into a protein secretion mechanism called type II secretion that exports virulence proteins unrelated to pili (discussed in Section 25.4).

How does the type IV pilus assembly mechanism make bacteria move by twitching motility? The assembly process involves the reiterative elongation and retraction of the pili. The pilus elongates, attaches to a surface, and then depolymerizes from the base, which shortens the pilus and pulls the cell forward (**Fig. 25.12B**). This mechanism is akin to using a grappling hook to scale a building. Similarly, the slime mold *Myxococcus xanthus* uses type IV pili to mediate gliding motility. The type IV pili of *Neisseria meningitidis* are essential for crossing the blood-brain barrier and causing bacterial meningitis.

As noted for *N. meningitidis*, type IV pili are also important to the pathogenesis of some organisms. The retraction of type IV pili in diarrhea-causing strains of *E. coli*, for example, helps disrupt the tight junctions connecting adjacent host cells that line the intestine (**Fig. 25.12C**). Tight junctions normally

A.

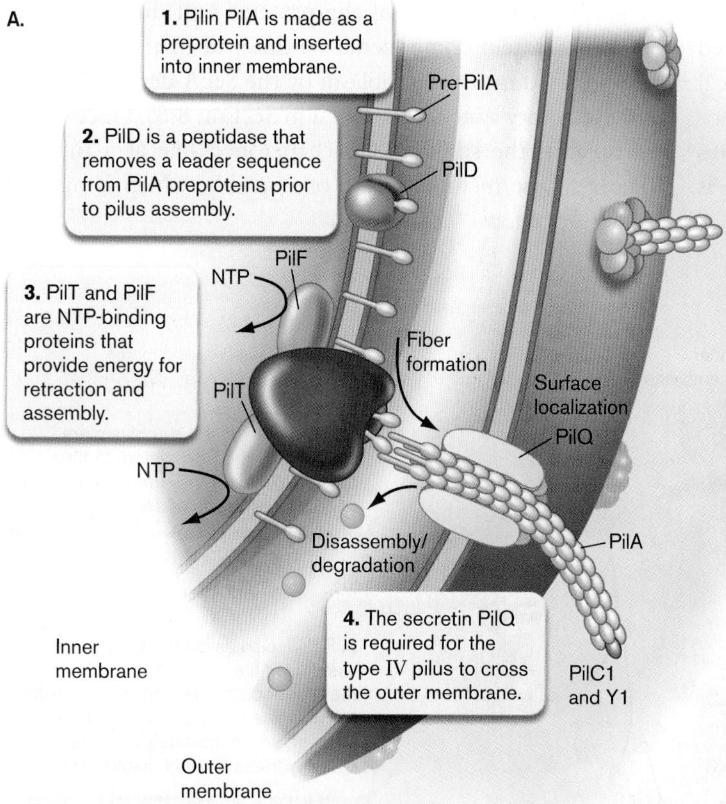
1. Pilin PilA is made as a preprotein and inserted into inner membrane.
Pre-PilA
2. PilD is a peptidase that removes a leader sequence from PilA preproteins prior to pilus assembly.
PilD
NTP
PilF
3. PilT and PilF are NTP-binding proteins that provide energy for retraction and assembly.
PilT
NTP
Fiber formation
Surface localization
PilQ
Disassembly/ degradation
PilA
4. The secretin PilQ is required for the type IV pilus to cross the outer membrane.
PilC1 and Y1
Inner membrane
Outer membrane

B. Extension and contraction of type IV pili

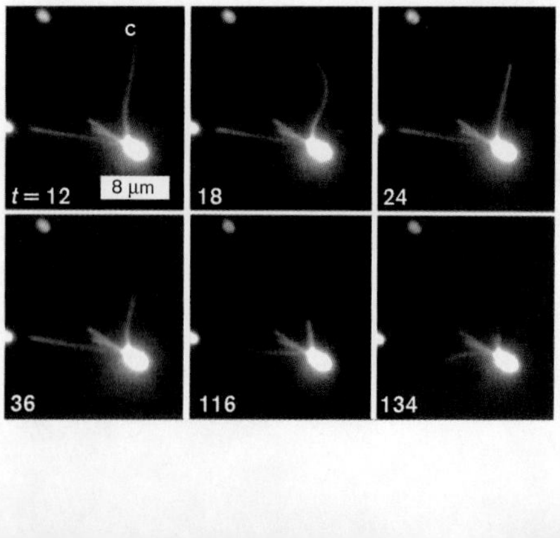
c
t = 12
8 μm
18
24
36
116
134

SKERKER, J., ET AL. 2001. *PNAS* **98**: 6901.

C. Type IV pili of *enterohemorrhagic E. coli*

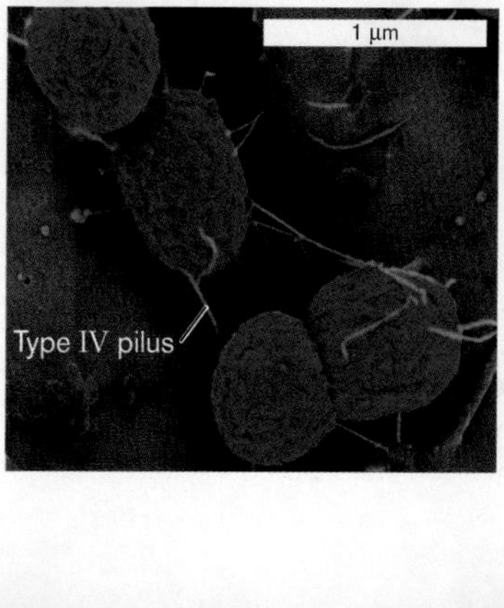
1 μm
Type IV pilus

ROBINSON, G. C. 2007. **117**: 3519. USED WITH PERMISSION OF AMERICAN SOCIETY FOR CLINICAL INVESTIGATION.

FIGURE 25.12 ■ Type IV pili. **A.** Model of pilus assembly and disassembly. In this example, PilA is the pilin protein, and PilC1 and Y1 form the attachment tip. Filament approx. 6 nm in diameter. Assembly and disassembly require the hydrolysis of nucleoside triphosphate (NTP) and take place at the inner membrane, not in the periplasm. **B.** Photographic evidence of type IV pilus retraction in cells of *Pseudomonas aeruginosa*. Filament c attaches briefly at its distal tip (note straightening at 24 seconds) and then begins to retract. Fluorescent microscopy. *t* = time, in seconds. **C.** Type IV pili (green) are essential for *enterohemorrhagic E. coli* to attach to brain endothelial cells. SEM. *Source:* Part A modified from Bardy et al. 2003. *Microbiology* **149:**295–304.

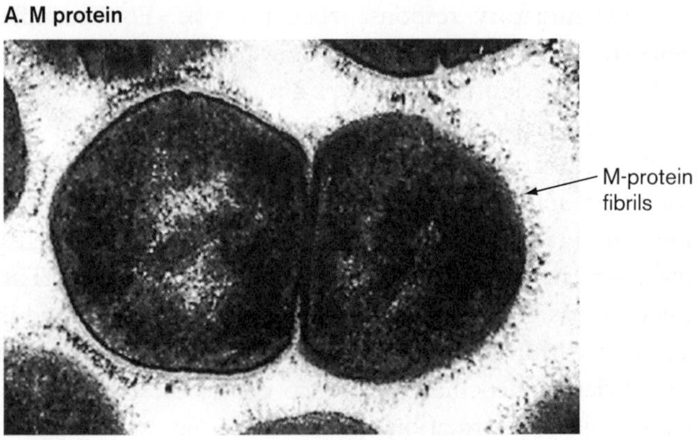

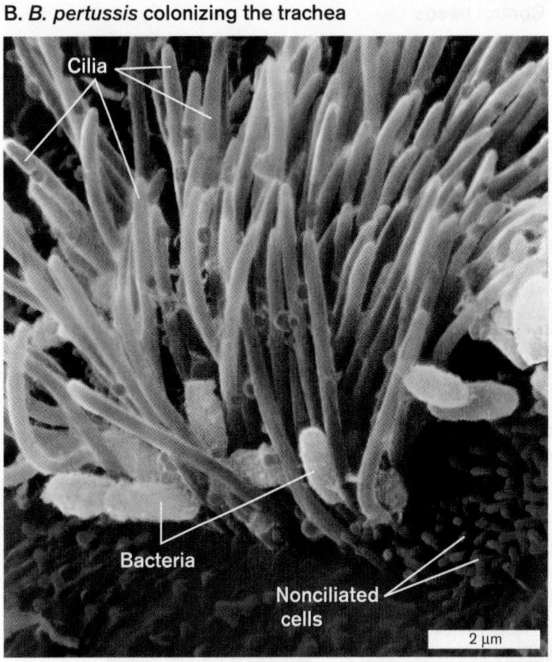

FIGURE 25.13 ▪ **Nonpilus adhesins.** **A.** M-protein surface fibrils on *Streptococcus pyogenes* (TEM). Cells 0.5–1 μm in diameter. **B.** Colonization of tracheal epithelial cells by *Bordetella pertussis* (colorized SEM). This organism uses a surface protein called pertactin, as well as a pilus called filamentous hemagglutinin (FHA), to bind bronchial cells.

form a permeability barrier between the intestinal lumen and the intestine, through which nutrients, ions, and water are absorbed. Disrupting tight junctions prevents the absorption of water and electrolytes, which accumulate in the intestine and contribute to the diarrhea.

Nonpilus Adhesins

Bacteria also sport proteins that bind host tissues but are not pili (**Fig. 25.13**). Some examples include *Bordetella* pertactin that binds to host cell integrin, and *Streptococcus pyogenes* M protein that binds to fibronectin. Many Gram-positive bacteria have other surface-exposed proteins with serine-rich repeats able to bind host sialic acid or keratin.

Initial binding between bacterium and host commonly involves pili, after which a more intimate attachment is formed by a nonpilus attachment protein. In the case of *Neisseria gonorrhoeae*, once the type IV pilus has attached to the surface of the mucosal epithelial cell, the filamentous pilus contracts, pulling the bacterium down onto the host cell membrane. Tight secondary interactions are then mediated by the neisserial Opa membrane proteins—another example of nonpilus adhesins. (Opa gets its name from the opacity it adds to colony appearance.)

An interesting nonpilus adhesin widely distributed among Gram-negative bacterial pathogens is called multivalent adhesion molecule 7 (MAM7). It is an outer membrane protein that binds phospholipids and fibronectin in host cell membranes. Because of its wide distribution among pathogens, it is considered to be an attractive target for antimicrobial development. Kim Orth (University of Texas Southwestern Medical Center) and Anne Marie Krachler (University of Texas McGovern Medical School, Houston) developed microbeads coated with fragments of MAM7 that can bind host tissues and competitively inhibit binding by multidrug-resistant *Pseudomonas aeruginosa*. When used in a burn model of infection, daily topical applications of the inhibitor-coated microbeads prevented the spread of infection into adjacent tissues (**Fig. 25.14**).

Host receptors dictate susceptibility to pathogens. Why are some people susceptible to certain infections, while others are not? Part of the reason is immunocompetence, but another is receptor availability. Pathogens rely on key host surface structures such as gangliosides to recognize and attach to the correct host cell by the mechanisms just described. But a host species can evolve to become resistant to infection when the gene encoding the receptor (or receptor synthesis) mutates. The mutation could lead to complete loss of the protein or a change in its shape to prevent recognition or alter its function. An example is the T-cell surface protein CCR5, which acts as a receptor for HIV (see Section 11.3). Individuals with a genetic defect that eliminates CCR5 are resistant to HIV infection, so even without methods for preventing and curing HIV infection, humans would eventually evolve a level of resistance to HIV. Differences in attachment receptors also explain, at least in part, why some pathogens have broad host specificity while others more narrowly target their host. It can also explain aspects of tissue specificity.

Biofilms and Infections

As first discussed in Section 4.5, bacteria in most environments form organized, high-density communities of cells, called biofilms, that are embedded in self-produced exopolymer

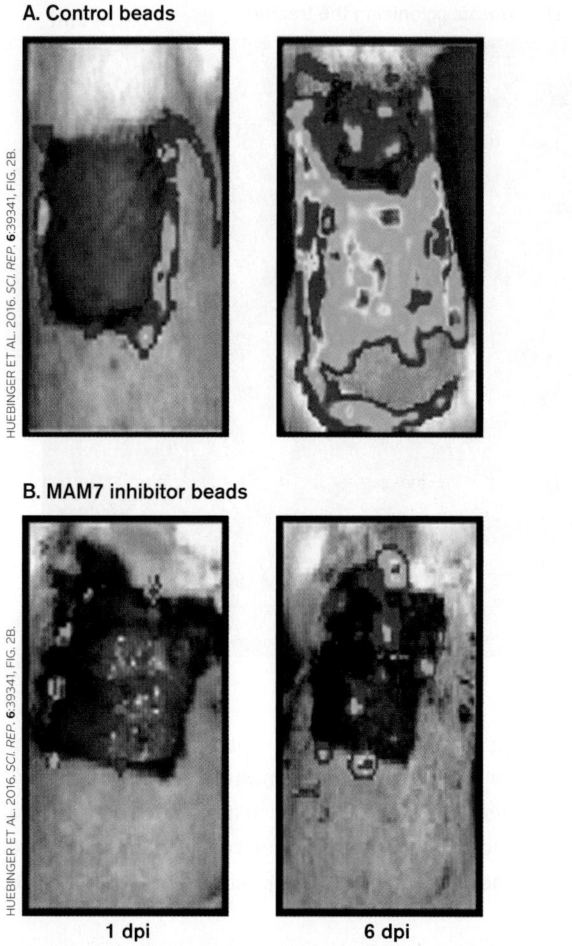

FIGURE 25.14 ■ Treatment of burn wounds with inhibitor MAM7 beads. Shaved areas of heavily anesthetized rats were burned with 100°C water for 12 seconds. Two days later, dead skin over the burn was excised and 5×10^6 colony-forming units (CFUs) of bioluminescent *Pseudomonas aeruginosa* were applied to the exposed area. **A.** Control rats were treated with a daily application of control microbeads over a period of 6 days postinfection (dpi). **B.** Test rats were treated with a daily application of inhibitor microbeads. Infections were monitored by fluorescence using an in vivo imaging system. Green and orange sections over the burn area indicate the presence of *P. aeruginosa*.

matrices (Fig. 4.27). Biofilm development is an ancient prokaryotic adaptation that enables microorganisms to adhere to any surface, living or nonliving, and facilitates survival in hostile environments. Within a single biofilm you can find localized differences in the expression of surface molecules, antibiotic resistance, nutrient utilization, and virulence factors. Bacteria in biofilms also coordinate their behavior through cell-cell communication using secreted chemical signals.

Biofilms, once they form, tend to cause chronic infections that may linger for months, years, or even a lifetime. The lingering presence of the pathogen in a chronic infection continually stimulates innate immune mechanisms through interactions with Toll-like receptors. The result is chronic inflammation. However, the reason chronic infections persist is that biofilms can stunt the effectiveness of the inflammatory response they provoke. For example, **Special Topic 25.1** describes how *Staphylococcus aureus* growing as a biofilm will trick invading neutrophils into killing themselves through NETosis.

Biofilms are important features in chronic infections found on oral, lung, and urogenital (bladder) tissues. *Pseudomonas aeruginosa* causes a life-threatening, chronic lung infection in individuals with cystic fibrosis (CF). This microbe has been found growing as aggregates enclosed in a matrix within mucus from CF patients. It is thought that insufficient mucociliary clearance contributes to *P. aeruginosa* biofilm formation. Biofilms are also important in periodontitis (gum disease), indwelling catheter infections, infections of artificial heart valves, chronic urinary tract infections, recurrent tonsillitis, rhinosinusitis, chronic otitis media (middle ear infection), chronic wound infections, and osteomyelitis (bone infection). **Figure 25.15A** shows a scanning EM of *Campylobacter jejuni* on intestinal mucosa. The confocal fluorescent image in **Figure 25.15B** shows top and side views of a biofilm on adenoid tissue, with live cells stained green and dead cells stained red. The side view illustrates biofilm depth.

Biofilm infections can also form on implanted medical devices such as heart valves, artificial knees, or indwelling venous catheters (discussed in **eTopic 4.3**). The bacteria appear to use their pili, nonpilus adhesins, and exopolysaccharides to attach to host factors such as matrix proteins that coat the device. Once infected, the implanted medical device may have to be replaced. Today, vascular catheter-related bloodstream infections are the most serious and costly health care–associated infections.

Biofilm infections are also important clinically because bacteria in biofilms exhibit tolerance to antimicrobial compounds and persistence in spite of sustained host defenses. Thus, biofilm infections are hard to cure. Tolerance to antibiotics may be caused by poor nutrient penetration through the exopolymer matrix into the deeper regions of the biofilm, leading to a stationary phase–like dormancy (discussed in Section 27.3). Bacterial factors important to biofilm formation include type IV pili, structural genes and regulators controlling cell-cell signaling (quorum sensing), and extracellular matrix synthesis. Interfering with cell-cell signaling is effective in preventing or limiting biofilm formation and may provide a target for new antimicrobial therapies.

To Summarize

- **Bacteria use pili and nonpilus adhesins** to attach to host cells.
- **Type I pili** produce a static attachment to the host cell, whereas **type IV pili** continually assemble and disassemble. Pili assemble starting at the tip.

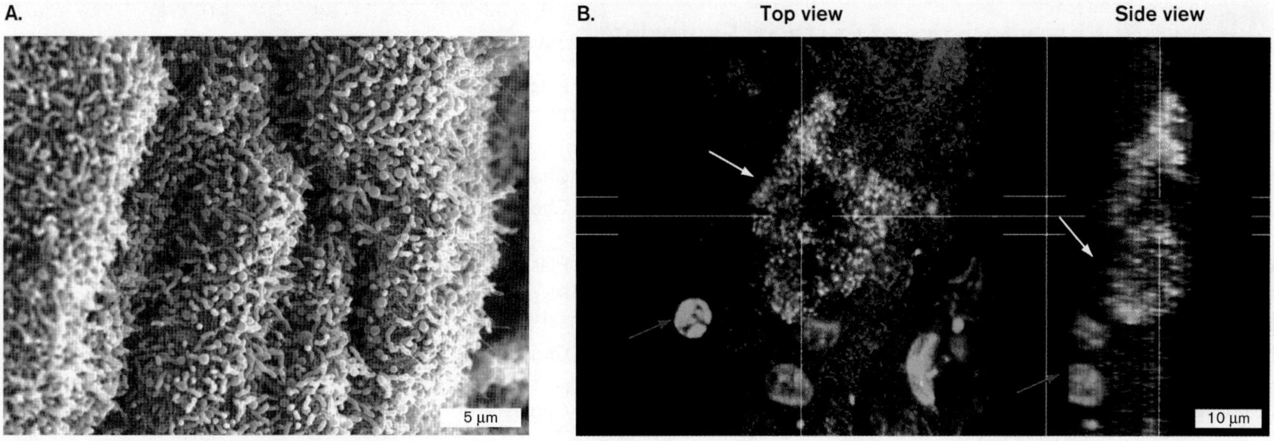

FIGURE 25.15 ▪ A bacterial biofilm infection. A. Scanning EM showing a biofilm (red arrow) of *Campylobacter jejuni* adhering to human intestinal mucosa. B. Confocal micrograph (described in Chapter 2) showing top and side views of biofilm clusters (white arrows) consisting of rods and cocci on the mucosa of a pediatric adenoid. Removal of the adenoid is a routine treatment for recurrent otitis media (middle ear infection). Specimens were treated with nucleic acid stains using the LIVE/DEAD *Bac*Light Bacterial Viability Kit, in which live bacteria stain green and dead bacteria stain red. Host inflammatory cells (red arrows) were also stained green, but their nuclei appear much larger than the bacteria. The mucosal surface (blue) was imaged using reflected light.

- **Nonpilus adhesins** are bacterial surface proteins, or other molecules, that can tighten interactions between bacteria and target cells.
- **Biofilms** play an important role in chronic infections by enabling persistent adherence and resistance to bacterial host defenses and antimicrobial agents.

25.3 Toxins Subvert Host Functions

Following attachment, many microbes secrete protein toxins (called **exotoxins**) that kill host cells and unlock their nutrients (because dead host cells ultimately lyse). Bacterial pathogens have developed an impressive array of toxins that take advantage of different key host proteins or structures. Gram-negative bacteria also possess a toxic compound called **endotoxin**, which is an integral component of lipopolysaccharides (LPS). Endotoxin can hyperactivate host immune systems to harmful levels.

Note: Do not confuse endotoxins from lipopolysaccharides in Gram-negative bacteria with protein exotoxins that can be secreted by Gram-negative and Gram-positive bacteria.

Categories of Microbial Exotoxins

Microbial exotoxins fall into several categories based on their mechanisms of action (**Table 25.2**). These classes are summarized here, and several are illustrated in **Figure 25.16**.

- **Plasma membrane disruption.** Toxins that damage membranes are exemplified by alpha (α) toxin of *Staphylococcus aureus*. These toxins form pores in host cell membranes and cause leakage of cell constituents (**Fig. 25.16A**).
- **Cytoskeleton alterations.** These toxins alter actin polymerization.
- **Protein synthesis disruption.** Diphtheria and Shiga toxins target eukaryotic ribosomes and destroy host protein synthesis (**Fig. 25.16B**).
- **Cell cycle disruption (cyclomodulins).** These toxins either stop (*E. coli* cytolethal distending toxin, or CLDT) or stimulate (*Pasteurella multocida* toxin) host cell division.
- **Signal transduction disruption.** These toxins alter host cell second messenger pathways. *E. coli* ST (stable toxin), for instance, stimulates synthesis of cyclic guanosine monophosphate (cGMP; **Fig. 25.16C**) in target cells, which alters ion transport and fluid movement.
- **Cell-cell adherence.** These toxins cleave adhesion proteins that bind host cells together. Exfoliative toxin of *S. aureus*, for example, breaks adhesion between dermis and epidermis, causing the appearance of scalded skin.
- **Vesicle traffic.** The major toxin in this class (VacA of *Helicobacter pylori*) has several modes of action, depending on the host cell. The most visually striking effect is its ability to cause vacuolization, which is the fusion of numerous intracellular vesicles.

TABLE 25.2 Characteristics of Bacterial Exotoxins[a,b]

Toxin	Organism	Mode of action	Host target	Disease
Damage membranes				
Perfringolysin O	*Clostridium perfringens*	Pore former	Cholesterol	Gas gangrene[c]
Listeriolysin O	*Listeria monocytogenes*	Pore former	Cholesterol	Food-borne systemic illness, meningitis
Alpha toxin	*Staphylococcus aureus*	Pore former	Plasma membrane	Abscesses[c]
Panton-Valentine leukocidin	*Staphylococcus aureus*	Pore former	Plasma membrane	Abscesses, necrotizing pneumonia
Pneumolysin	*Streptococcus pneumoniae*	Pore former	Cholesterol	Pneumonia[c]
Streptolysin S	*Streptococcus pyogenes*	Pore former	Cholesterol	Strep throat, scarlet fever
Disrupt cytoskeletons				
Iota toxin	*Clostridium perfringens*	ADP-ribosyltransferase	Actin	Gas gangrene[c]
Inhibit protein synthesis				
Diphtheria toxin	*Corynebacterium diphtheriae*	ADP-ribosyltransferase	Elongation factor 2	Diphtheria
Shiga toxins	*E. coli/Shigella dysenteriae*	*N*-glycosidase	28S rRNA	HC and HUS
Disrupt cell cycle				
Pasteurella multocida toxin	*Pasteurella multocida*	Mitogen (also activates Rho GTPases)	Nucleus (encourages cell division)	Wound infection
Activate second messenger pathways				
LT	*E. coli*	ADP-ribosyltransferase	G proteins	Diarrhea
ST	*E. coli*[d]	Stimulates guanylate cyclase	Guanylate cyclase receptor	Diarrhea
Edema factor	*Bacillus anthracis*	Adenylyl cyclase	ATP	Anthrax
Pertussis toxin	*Bordetella pertussis*	ADP-ribosyltransferase	G protein(s)	Pertussis (whooping cough)
Toxin A and B	*Clostridioides difficile*	Glucosyltransferase	Rho G protein(s)	Diarrhea/PC
Cholera toxin	*Vibrio cholerae*	ADP-ribosyltransferase	G protein(s)	Cholera
Lethal factor	*Bacillus anthracis*	Metalloprotease	MAPKK1/MAPKK2	Anthrax
Disrupt cell-cell adherence				
Exfoliative toxins	*Staphylococcus aureus*	Serine protease, superantigen	Desmoglein; TCR, and MHC II (superantigen)	Scalded skin syndrome[c]
Alter vesicle traffic				
VacA	*Helicobacter pylori*	Large vacuole formation, apoptosis	Receptor-like protein tyrosine phosphatase, sphingomyelin	Gastric ulcers, gastric cancer
Block exocytosis				
Neurotoxins A–G	*Clostridium botulinum*	Zinc metalloprotease	VAMP/synaptobrevin, SNAP-25 syntaxin	Botulism
Tetanus toxin	*Clostridium tetani*	Zinc metalloprotease	VAMP/synaptobrevin	Tetanus
Superantigens (activate immune response)				
Enterotoxins	*Staphylococcus aureus*	Superantigen	TCR and MHC II, medullary emetic center (vomit center)	Food poisoning[c]
Toxic shock syndrome toxin	*Staphylococcus aureus*	Superantigen	TCR and MHC II	Toxic shock syndrome[c]
Pyrogenic exotoxins	*Streptococcus pyogenes*	Superantigens	TCR and MHC II	Toxic shock syndrome, scarlet fever
Lethal factor	*Bacillus anthracis*	Metalloprotease	MAPKK1/MAPKK2	Anthrax

[a]**Abbreviations:** HC = hemorrhagic colitis; HUS = hemolytic uremic syndrome; LT = heat-labile toxin; MAPKK = mitogen-activated protein kinase kinase; MHC II = major histocompatibility complex class II; PC = antibiotic-associated pseudomembranous colitis; SNAP-25 = synaptosomal-associated protein; ST = heat-stable toxin; TCR = T-cell receptor; VAMP = vesicle-associated membrane protein.

[b]All toxins have a known role in pathogenesis, as shown in an animal model or appropriate cell culture.

[c]Other diseases are also associated with the organism.

[d]Toxin is also produced by other genera of bacteria.

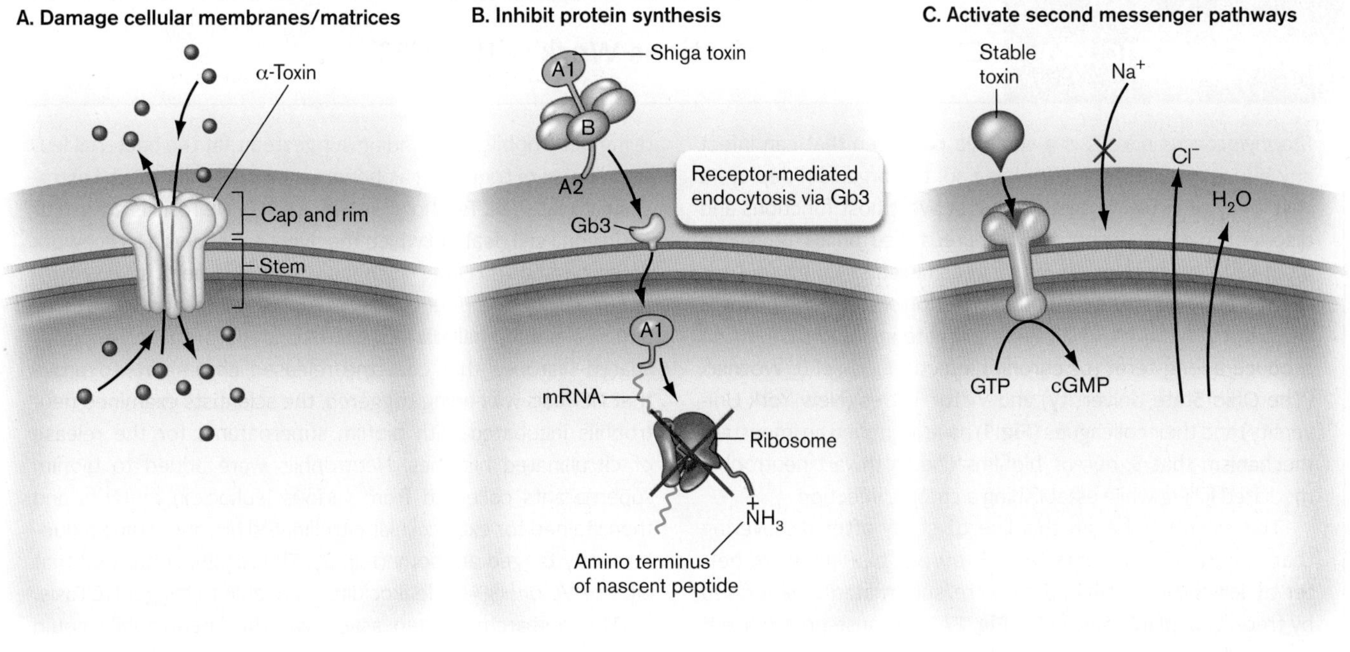

FIGURE 25.16 ▪ Three classes of microbial exotoxins. These classes are defined by mode of action. **A.** Pore-forming toxins assemble in target membranes and cause leakage of compounds into and out of cells. **B.** Shiga toxin attaches to ganglioside Gb3, enters the cell, and cleaves 28S rRNA in eukaryotic ribosomes to stop translation. **C.** Enterotoxigenic *E. coli* heat-stable toxin affects cGMP production. The result is altered electrolyte transport: inhibition of Na^+ uptake and stimulation of Cl^- transport. In response to the resulting electrolyte imbalance, water leaves the cell.

- **Inhibit exocytosis.** Tetanus and botulism toxins are proteases that prevent exocytosis of neurotransmitters to cause spastic and flaccid paralysis, respectively.
- **Superantigens.** These toxins, exemplified by toxic shock syndrome toxin (TSST), activate the immune system without being processed by antigen-presenting cells (discussed in Section 24.3).

Mechanisms of selected exotoxins are described in the following sections. We focus on toxins that disrupt membranes and a set of exotoxins collectively called AB-subunit exotoxins that target either protein synthesis or signal transduction. Superantigens were described in Section 24.3, and exotoxins that affect exocytosis (tetanus and botulism toxins) will be discussed in Section 26.6.

Membrane Disruption

Toxins that disrupt membranes include pore-forming proteins that bind cholesterol and insert themselves into target membranes, and phospholipases that hydrolyze membrane phospholipids into fatty acids. General descriptive terms for these toxins are **hemolysins**, which lyse red blood cells (but other cells as well), and **leukocidins**, which more specifically lyse white blood cells (leukocytes). One example of a hemolysin that is also a leukocidin is streptolysin S, produced by *Streptococcus pyogenes*. This pathogen can cause pharyngitis (sore throat) and necrotizing fasciitis.

A classic example of a pore-forming exotoxin is the hemolytic alpha toxin produced by *Staphylococcus aureus*, an organism that causes boils and blood infections. Alpha toxin forms a transmembrane, oligomeric (seven-member) beta barrel pore in target cell plasma membranes (**Fig. 25.17A** and **B**). It is easy to see how the resulting leakage of cell constituents and influx of fluid cause the target cell to burst. Diagnostic microbiology laboratories visualize hemolysins such as alpha toxin by inoculating bacteria onto agar plates containing sheep red blood cells (**Fig. 25.17C**). The clear, yellow zones around the *S. aureus* colonies growing on blood agar indicate that the microbe secretes a hemolysin.

A particularly potent pore-forming leukocidin produced by most strains of methicillin-resistant *Staphylococcus aureus* (MRSA) is Panton-Valentine toxin. This leukocidin contributes to the formation of chronic staphylococcal infections by triggering neutrophil extracellular traps (NETs; **Special Topic 25.1**).

Phospholipase toxins are different from pore-forming toxins. For instance, phospholipase C of *Clostridium perfringens*, a cause of gas gangrene, cleaves phosphatidylcholine in host plasma membranes. At high concentrations the toxin causes membrane disruption, but at sublethal concentrations the exotoxin will generate signaling molecules from membrane lipids that can activate cytokine production. Some phospholipases, also called lecithinases, can increase the permeability of capillaries to cause edema (fluid accumulation in tissues).

SPECIAL TOPIC 25.1 Chronic Staph Infections Work with a NET

Staphylococcus aureus is a versatile pathogen that can infect any human organ system. In doing so, the organism deploys a vast array of virulence factors that subvert host functions and disorient immune responses. *S. aureus* also builds tenacious biofilms that repel innate immune mechanisms. Neutrophils, for example, can migrate to a biofilm infection site but will fail to kill the pathogen. Such a failure in the innate immune system can produce a long-term (or chronic) infection. Daniel J. Wozniak (The Ohio State University) and Victor Torres (New York University) and their colleagues (**Fig. 1**) have identified an important mechanism that *S. aureus* biofilms use to thwart neutrophil-mediated killing while establishing a chronic infection.

The scientists began this line of study after discovering that culture supernatants from *S. aureus* biofilms were better at killing neutrophils than were supernatants generated by free-living planktonic cells (**Fig. 2A**). Because proteinase K and heat (100°C) treatments of the supernatants destroyed the neutrophil-killing activity, the researchers suspected that a protein secreted by the biofilm was involved.

S. aureus secretes five different leukocidins that can form pores and damage neutrophil cell membranes. Wondering whether one or more of these leukocidins was involved in killing the neutrophils, the scientists prepared biofilm supernatants from *S. aureus* mutants that lacked one or more of the enzymes, and they tested the fluids for neutrophil-killing ability (**Fig. 2B**). The results showed that eliminating two of the leukocidins (PVL and HlgACB) prevented biofilm-specific neutrophil death.

The authors noticed that the leukocidin-containing supernatants also caused DNA (chromatin) to be released from the dying neutrophils. This finding suggested that the bacterial leukocidins were triggering a phenomenon called neutrophil extracellular traps (NETs; shown in Fig. 23.20). NETosis is a form of neutrophil cell death in which the dying cell spews a latticework of chromatin into the surrounding environment. Laced with antimicrobial compounds, the NETs can trap and kill nearby pathogens. One hallmark of NETosis is the presence of citrullinated histones that coat the released chromatin. To prove that NETosis was being triggered, the scientists examined neutrophils incubated with biofilm supernatants for the release of citrullinated histones. Neutrophils were added to biofilm supernatants collected from various leukocidin mutants and then stained for extracellular citrullinated histones using a fluorescently tagged antibody (**Fig. 3**). The results confirmed that either PVL or HlgACB leukocidins were able to trigger NETosis.

The researchers then asked whether neutrophils could even penetrate the leukocidin-producing biofilms. Surprisingly, neutrophils penetrated biofilms erected by wild-type and leukocidin-defective mutants (**Fig. 4**). However, the neutrophils invading the wild-type, leukocidin-producing biofilms underwent NETosis and were no longer able to effectively kill the bacteria. In contrast, the neutrophils that penetrated a leukocidin-deficient mutant retained their nuclear structure and their ability to kill biofilm bacteria.

Why, then, don't the chromatin NETs triggered by leukocidins kill *S. aureus* cells in the biofilm? The authors propose that the combination of a known *S. aureus*–secreted DNA nuclease and an adenosine synthase can degrade NET chromatin and destroy its antimicrobial activity.

A. B.

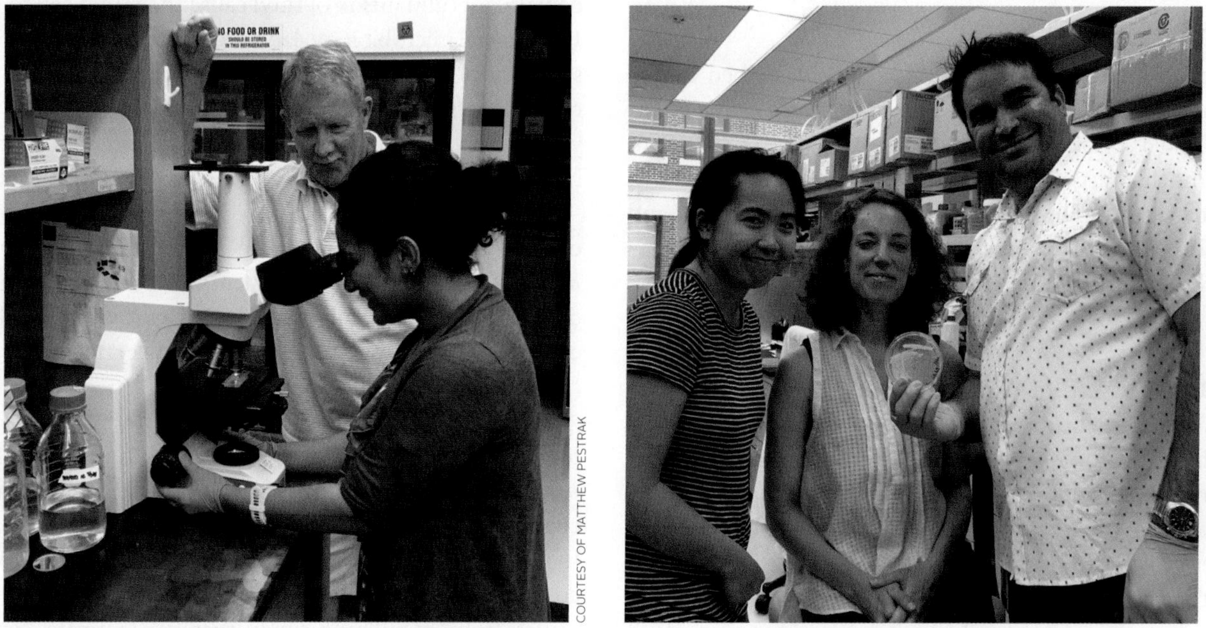

COURTESY OF MATTHEW PESTRAK

COURTESY OF VICTOR J. TORRES

FIGURE 1 ■ Wozniak and Torres laboratories. **A.** Daniel Wozniak (back) and graduate student Mohini Bhattacharya. **B.** Victor Torres (right, holding a Petri dish with MRSA), next to lab tech Evelien Berends (center) and postdoctoral researcher Rita Chan (left).

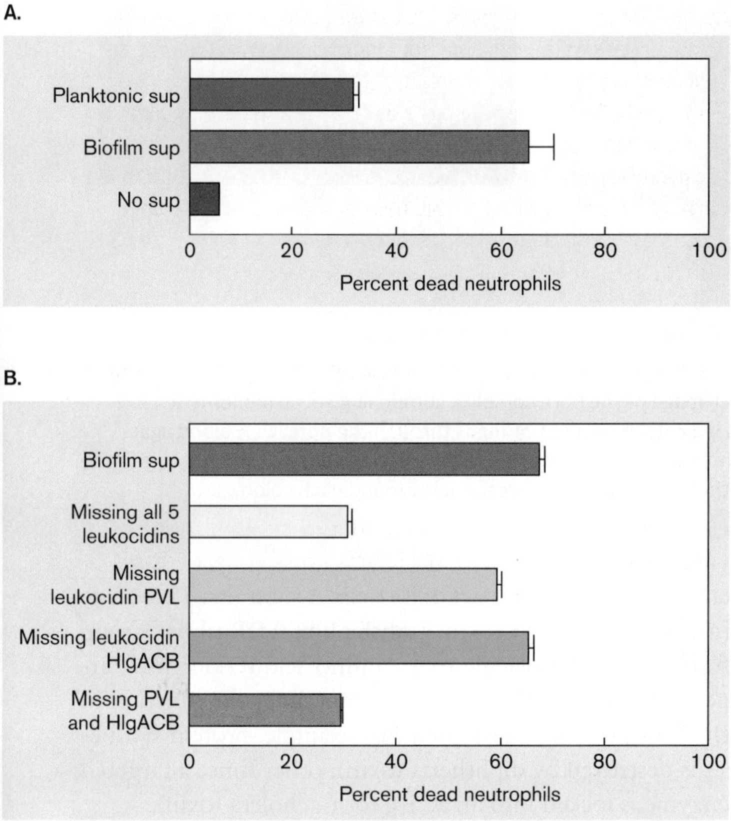

FIGURE 2 ■ Leukocidins PVL and HlgACB mediate biofilm-dependent neutrophil killing. Cell-free supernatants (sup) were incubated with neutrophils for 90 minutes. Neutrophil death was monitored by LIVE/DEAD fluorescent staining. **A.** Comparison of biofilm and planktonic cell supernatants. **B.** Biofilm supernatants used from various leukocidin mutants. *Source:* Modified from Bhattacharya et al. 2018. *PNAS* **115**:7416–7421, figs. 1B (part A) and 2A (part B).

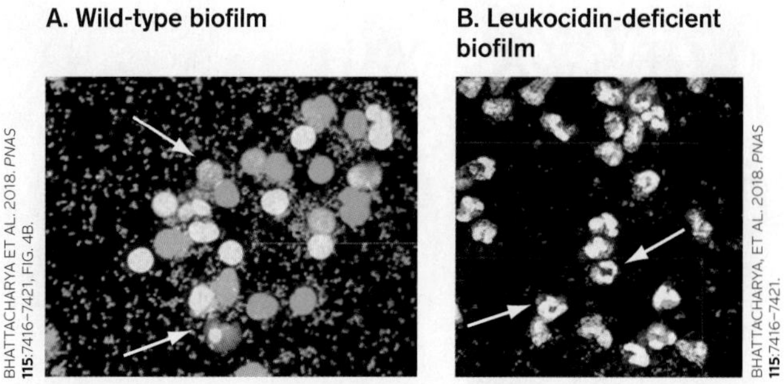

FIGURE 4 ■ Neutrophils penetrate biofilms of wild-type and leukocidin-deficient *S. aureus*. Neutrophils prestained with CellTracker Blue dye were added to 24-hour biofilms. After 1 hour, biofilm cross sections were stained for viable bacteria (Syto-9 dye; small green cells) and for DNA of damaged or dead neutrophils (ethidium homodimer-1 dye; red). **A.** Neutrophils penetrating the wild-type biofilm underwent NETosis (seen as red and yellow cells). **B.** Neutrophils penetrating the leukocidin-deficient biofilm retained intact nuclei. Examples are marked with white arrows.

These remarkable findings suggest a possible new therapeutic strategy for treating chronic *S. aureus* infections. By administering an anti-leukocidin antibody in combination with antibiotic agents, clinicians could enhance a patient's ability to eliminate biofilms—the root cause of chronic infections.

RESEARCH QUESTION

How would you test whether the *S. aureus* nuclease NucA is required to protect leukocidin-producing biofilms from the neutrophil extracellular traps that they trigger?

Mohini Bhattacharya, Evelien T. M. Berends, Rita Chan, Elizabeth Schwab, Sashwati Royd, et al. 2018. *Staphylococcus aureus* biofilms release leukocidins to elicit extracellular trap formation and evade neutrophil-mediated killing. *Proceedings of the National Academy of Sciences USA* **115**: 7416–7421.

	Leukocidin present (+) or absent (–)			
HglACB	+	+	–	–
PVL	+	–	+	–

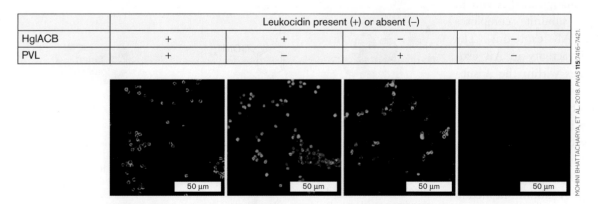

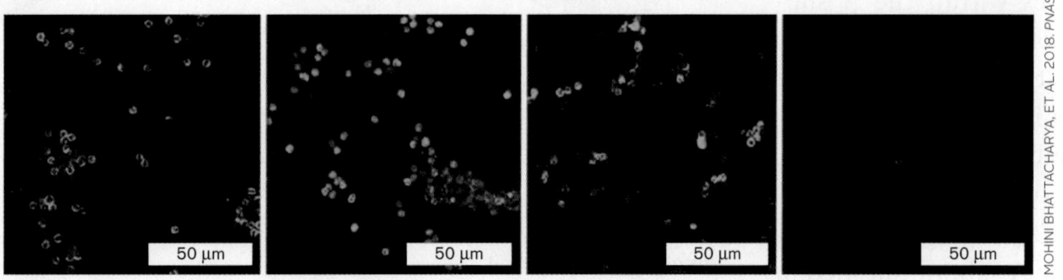

FIGURE 3 ■ Biofilm culture supernatants containing leukocidins PVL or HlgACB trigger NETosis. Chromatin released by neutrophils undergoing NETosis (containing citrullinated histones) appears green. Scale bar = 10 µm.

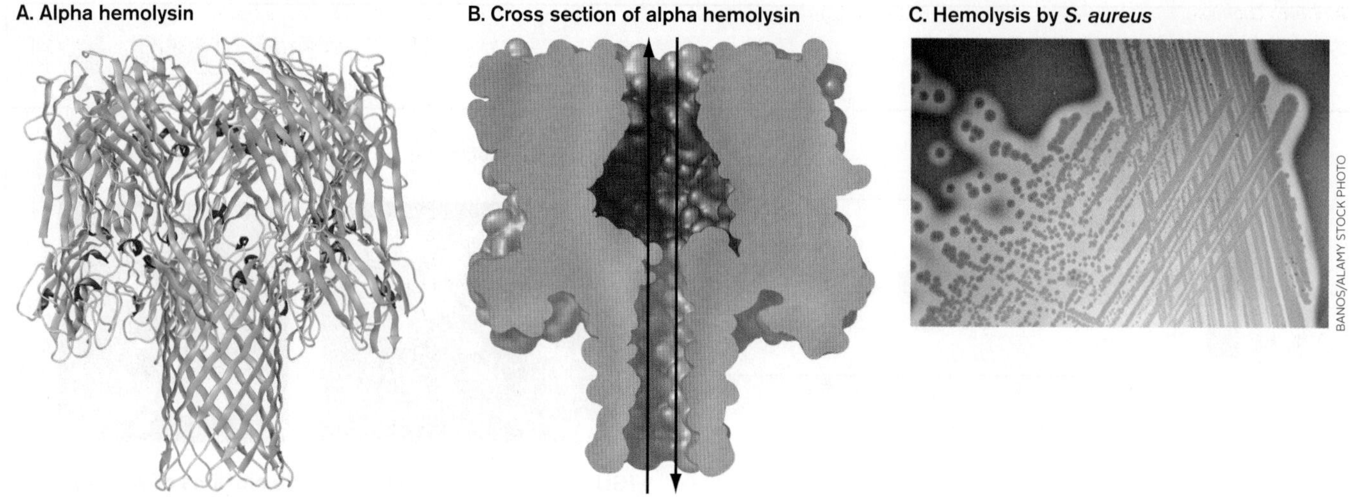

FIGURE 25.17 ▪ Alpha hemolysin of *Staphylococcus aureus*. **A.** 3D image of the pore complex, comprising seven monomeric proteins. (PDB code: 7AHL) **B.** Cross section showing the channel. Arrows indicate movement of fluids through the pore. **C.** A blood agar plate inoculated with *S. aureus*. The alpha toxin is secreted by the organism and diffuses away from the producing colony. It forms pores in the red blood cells embedded in the agar, causing the cells to lyse—a process that causes the clear area visible around each colony.

Other membrane-disrupting exotoxins specifically target vacuolar membranes. Many bacterial pathogens enter eukaryotic host cells by inducing phagocytosis and then end up in a phagosome vacuole. Some of these intracellular pathogens need to break out of the phagosome to grow in the cytoplasm. One example is *Listeria monocytogenes*, a gastrointestinal pathogen that can also cause meningitis (described later). This Gram-positive rod uses listeriolysin (a pore-forming exotoxin) and two phospholipases to escape the phagosome to grow in the cytoplasm.

Two-Subunit AB Exotoxins

As noted earlier, exotoxins target a variety of different host mechanisms. But despite the diversity of their targets, many exotoxins share a common structural theme; they have two subunits, usually called A and B. These two-subunit complexes are called AB exotoxins. The actual toxic activity in AB exotoxins resides within the A subunit. The role of the B subunit is to bind host cell receptors. Thus, the B subunit for each toxin delivers the A subunit to the host cell. Many AB toxins have five identical B subunits arranged as a ring with a single A subunit, nestled in the center, called AB5 exotoxins (**Figs. 25.16B** and **25.18A**).

One major subclass of AB exotoxins has **ADP-ribosyltransferase** enzymatic activity as part of its toxic A subunits. These toxins, such as cholera toxin, transfer the ADP-ribose group from an NAD molecule to an amino acid residue in a target host protein (**Fig. 25.18B**). Sometimes the function of the host protein is destroyed (for example, protein synthesis is destroyed by diphtheria toxin); other times a targeted enzyme is locked into an active form (cholera toxin).

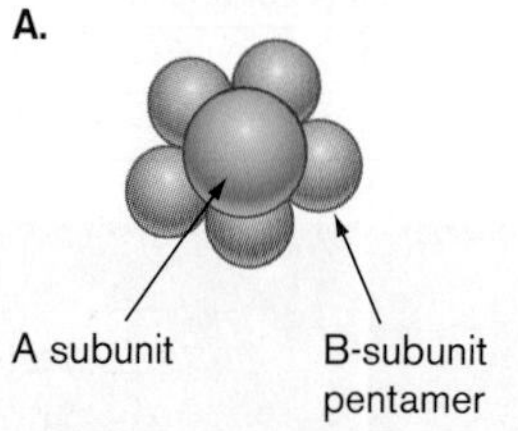

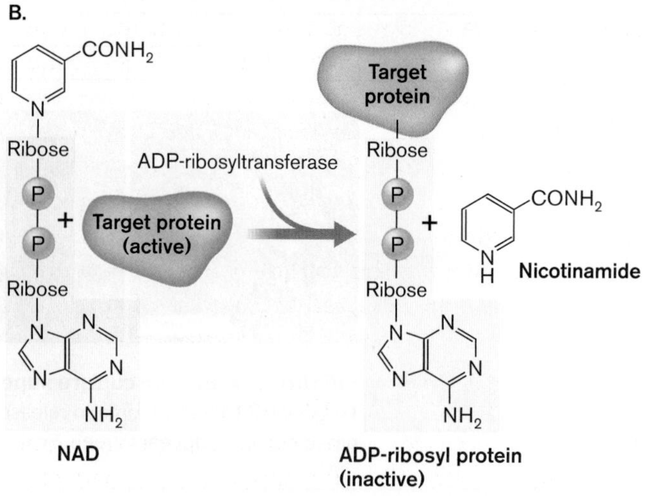

FIGURE 25.18 ▪ AB toxins. **A.** A typical AB toxin consists of an A subunit and a pentameric B subunit joined noncovalently. **B.** Many AB toxins are ADP-ribosyltransferase enzymes that modify protein structure and function.

Cholera toxin. *Vibrio cholerae* (**Fig. 25.19A**), a waterborne pathogen commonly found in marine environments, produces a severe diarrheal disease called cholera that generally afflicts malnourished people in developing countries like Bangladesh, or countries where access to clean water has been disrupted by natural disasters or war, as is happening now (2019) across central and western Africa. After being ingested, *V. cholerae* colonizes the brush border of the victim's small intestine (**Fig. 25.19B** and **C**) and secretes

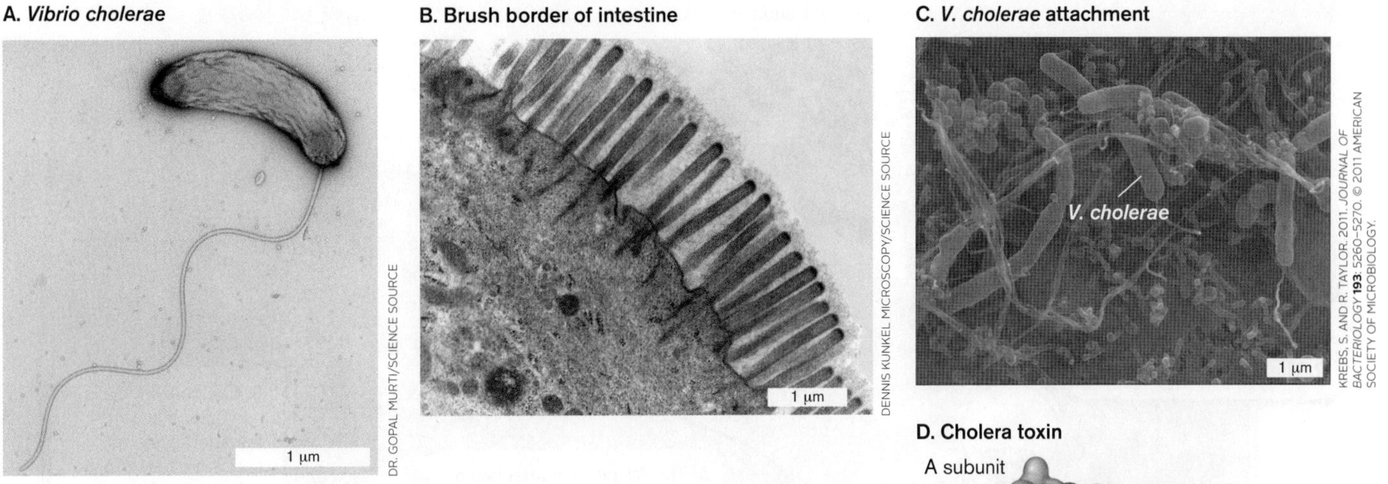

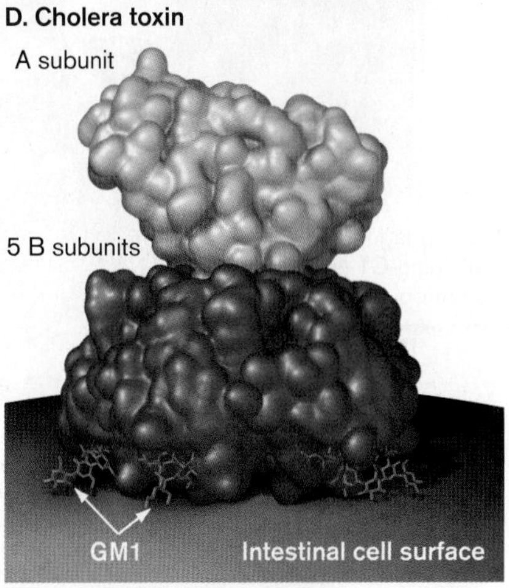

FIGURE 25.19 ■ Pathogenesis of cholera. **A.** *Vibrio cholerae* (SEM). Note the slight curve of the cell and the presence of a single polar flagellum. **B.** Brush border of intestine (TEM). *V. cholerae* binds to the fingerlike villi on the apical surface. **C.** *V. cholerae*, binding to the surface of a host cell (SEM). Note that *V. cholerae* does not invade the host cell. **D.** 3D structure of cholera toxin, binding ganglioside GM1 on the intestinal cell surface. (PDB code: 1S5F).

cholera enterotoxin (**Fig. 25.19D**). Cholera toxin causes diarrhea by reversing an important intestinal process. Normally, the intestine absorbs NaCl and other ions (electrolytes), as well as water from food material moving through the intestine. The result is well-formed feces with very little water and salt content. Cholera toxin, however, reverses this process by causing intestinal cells to secrete water and electrolytes into the intestinal lumen. So, how does cholera toxin accomplish this feat?

Cholera toxin is an AB5 exotoxin. The B subunits attach to ganglioside GM1 on intestinal cell membranes to trigger endocytosis and the formation of a toxin-containing vacuole (**Fig. 25.20**, steps 1 and 2). The vacuole is then transported to the endoplasmic reticulum (ER) and the A1 subunit containing ADP-ribosyltransferase activity is released into the ER. The A1 subunit is then exported from the ER into the cytoplasm (steps 3 and 4).

The mission of the A1 peptide is to modify (that is, ADP-ribosylate) an arginine residue in a membrane-associated GTPase (or G factor) called G_s. When bound to GTP, G_s stimulates host adenylyl cyclase to make cAMP (**Fig. 25.20**, step 5). Normally, the GTPase activity in G_s quickly hydrolyzes GTP, which halts stimulation and limits cAMP production. (A discussion of adenylyl cyclase G-factors is found in **eTopic 25.5**). Disease results when cholera toxin removes this control by ADP-ribosylating G_s (step 5). The modified G_s factor does not hydrolyze GTP, so adenylyl cyclase is constantly stimulated and cAMP levels sharply rise (step 6).

The sharp rise in cAMP stimulates a host protein kinase that activates various ion transport channels, including the cystic fibrosis transmembrane conductance regulator (CFTR), so named because a defect in this protein manifests as the lung disease cystic fibrosis. CFTR controls chloride transport in several cell types, including intestinal epithelia (discussed in Section 23.4). As a result of CFTR activation, chloride, sodium, and other ions leave the cell, and in an attempt to equilibrate osmolarity, water leaves as well. Because the affected cells line the intestine, the escaping water enters the intestinal lumen, leading to watery stools, or diarrhea (**Fig. 25.20**, step 6).

Note that some pathovars of *Escherichia coli*, called enterotoxigenic *E. coli*, also make an enterotoxin close in sequence but identical in function to cholera toxin. The *E. coli* enterotoxin is called **labile toxin (LT)** because it is easily destroyed by heat.

Thought Question

25.2 **Figure 25.20** illustrates how cholera toxin works to cause diarrhea. To develop a vaccine that generates protective antibodies, which subunit of cholera toxin should be used to best protect a person from the toxin's effects?

How does diarrhea benefit the pathogen? For one thing, diarrhea can decrease competition with resident microbiota as the microbiota are swept away. For example, the vast majority of organisms found in the diarrhea of cholera

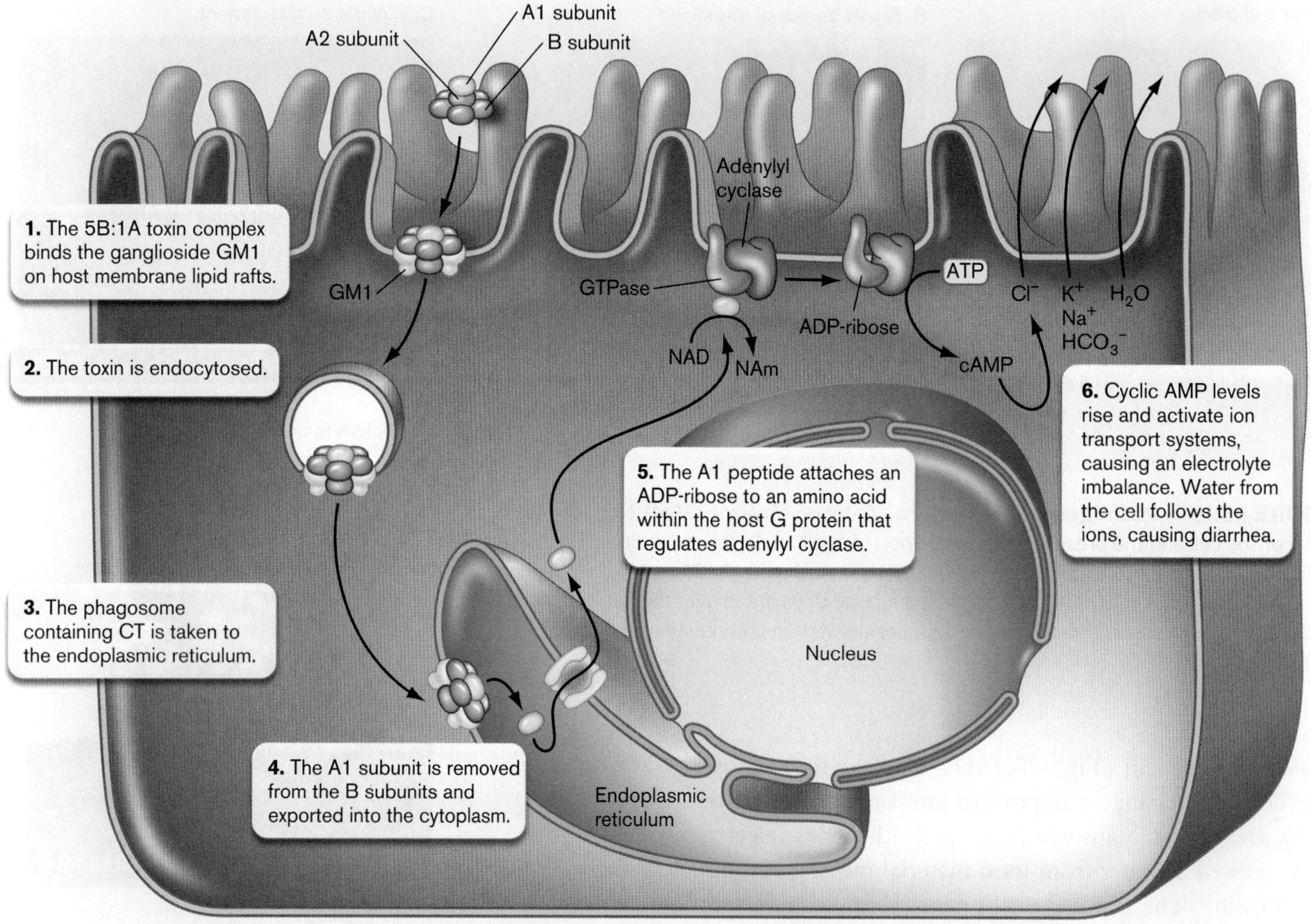

FIGURE 25.20 ■ Cholera toxin mode of action. Delivery of cholera toxin (CT) into target cells and deregulation of adenylyl cyclase activity. NAm = nicotinamide.

patients are *V. cholerae* bacteria. Few normal microbiota are present. Diarrhea also distributes the pathogen throughout the environment, thereby increasing the chance that another host will ingest the organism and perpetuate the species.

Anthrax. A century ago, anthrax (caused by *Bacillus anthracis*; **Fig. 25.21A**) was mainly a disease of cattle and sheep. Humans acquired the disease only accidentally. Today we fear the deliberate shipment of *B. anthracis* through the mail (as happened in 2001) or its dispersion from the air ducts of heavily populated buildings. What makes this Gram-positive, spore-forming microbe so dangerous? In large part, its lethality is due to the secretion of a plasmid-encoded tripartite toxin (a variant of the AB exotoxin theme).

The core subunit of the toxin is called **protective antigen** (**PA**) because immunity to this protein protects hosts from disease. PA is the B subunit of anthrax toxin. Protective antigen binds to a host cell surface (there are multiple receptors), where a human protease cleaves off a fragment (**Fig. 25.21B**). The remaining part of PA autoassembles in the membrane to form seven- and eight-membered pores. The other two components of anthrax toxin—edema factor (EF) and lethal factor (LF)—bind to the PA rings and are carried into the cell (**Fig. 25.21C**). EF and LF represent different A subunits of anthrax toxin. After the complex is endocytosed, EF and LF are passed through the pore into the host cytoplasm. Proton motive force helps unfold the exotoxins and translocates them across the endocytic membrane.

Edema factor and lethal factor are enzymes that attack the signaling functions of the cell. Edema factor is an adenylyl cyclase that remains inactive until entering the cytoplasm, where it binds host calmodulin. Binding to calmodulin activates adenylyl cyclase, resulting in a huge production of cAMP, and inactivates calmodulin from its normal function in the cell.

Lethal factor is a protease that cleaves several host protein kinase kinases, each of which is part of a critical regulatory cascade affecting cell growth and proliferation. A protein kinase kinase is an enzyme that phosphorylates, and thereby activates, another protein kinase that can then

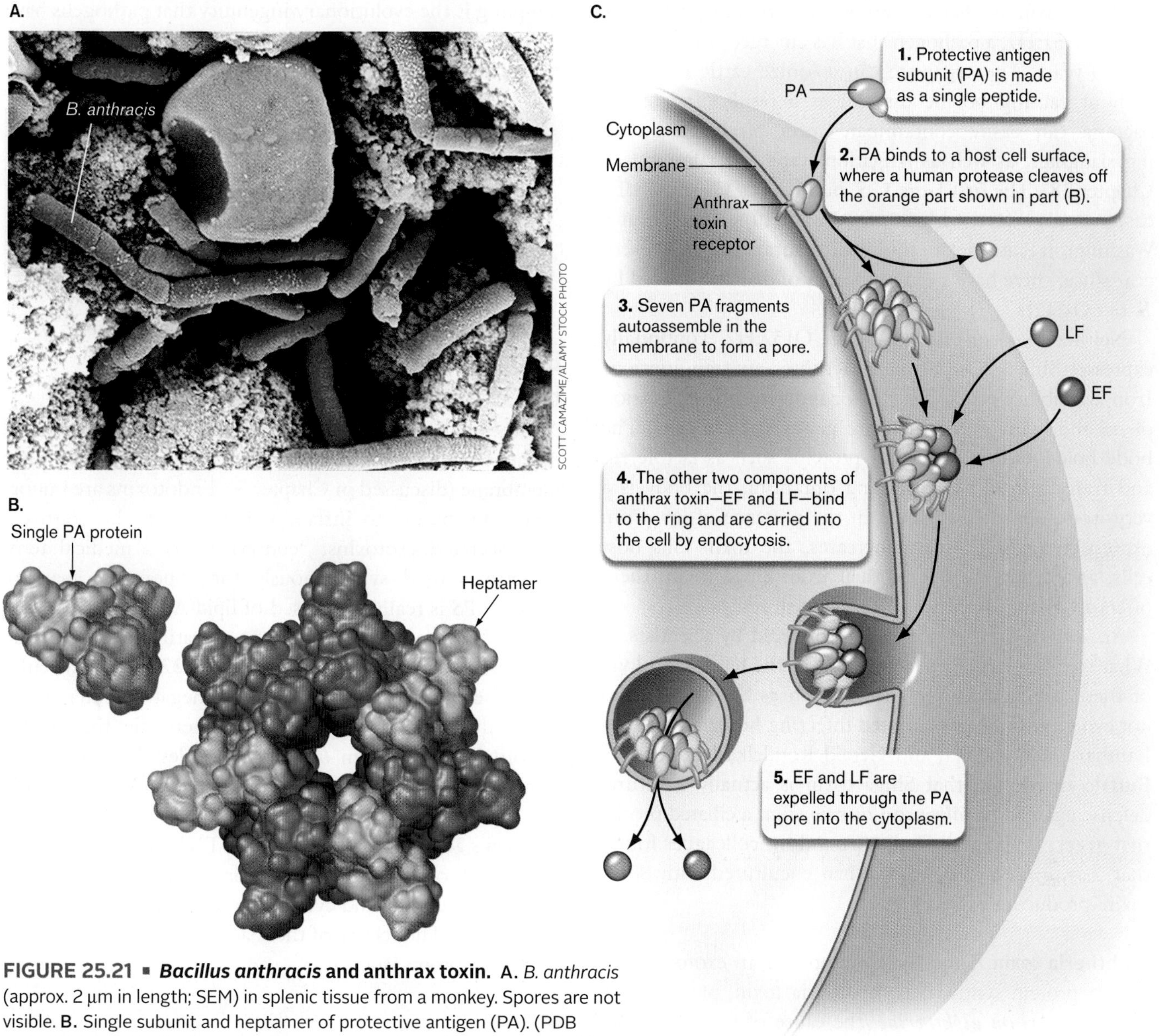

FIGURE 25.21 ■ *Bacillus anthracis* and anthrax toxin. **A.** *B. anthracis* (approx. 2 μm in length; SEM) in splenic tissue from a monkey. Spores are not visible. **B.** Single subunit and heptamer of protective antigen (PA). (PDB code: 1TZO) **C.** Mechanism of toxin entry. EF = edema factor; LF = lethal factor.

phosphorylate one or more subsequent target proteins. One consequence of subverting these phosphorylation cascades is a failure to produce signals that recruit immune cells to fight the infection.

Thought Question

25.3 How might you experimentally determine whether a pathogen secretes an exotoxin? (*Hint:* You can learn more about how new toxins are found in **eTopic 25.7**).

AB Exotoxins That Target Protein Synthesis

Shiga toxin. *Shigella dysenteriae* and *E. coli* O157:H7 (also known as enterohemorrhagic *E. coli*) cause food-borne diseases whose symptoms include bloody diarrhea. These organisms produce an important toxin known as Shiga toxin (or Shiga-like toxin). The gene (*stx*) encoding Shiga toxin is part of a phage genome integrated into the bacterial chromosome. The toxin is an AB toxin with five B subunits for binding and one A subunit imbued with toxic activity. The A subunit, upon entry, destroys protein synthesis by cleaving 28S rRNA in eukaryotic ribosomes. Strains that produce high levels of this toxin are associated with acute kidney failure, known as hemolytic uremic syndrome.

Note: *Shigella dysenteriae* is the only *Shigella* species that produces Shiga toxin. Also recall that the O and H designations for strains within a species of Enterobacteriaceae reflect antigenic differences in LPS and flagella, respectively.

Shiga toxin is also an important virulence factor for *E. coli* O157:H7, a pathogen that has emerged over the past 30–40 years. The organism can colonize cattle intestines without causing bovine disease; as a result, undetected bacteria can easily contaminate meat products following slaughter, as well as irrigation water from farms (see Chapter 28). The first large U.S. outbreak of O157:H7 disease was associated with fast-food hamburgers served at a Washington State Jack in the Box restaurant in 1993. Every year since, there have been outbreaks of diarrhea caused by *E. coli* O157:H7.

Neither *Shigella* nor *E. coli* O157:H7 continually expresses Shiga toxin, so what activates *stx* transcription? Iron availability is a key factor for inducing the expression of *stx* and many other virulence genes in pathogens. The body holds its iron tightly in proteins such as lactoferrin and transferrin. To an invading organism, the body is a very iron-poor environment. In the presence of low iron, expression of Shiga toxin increases, the toxin kills host cells, and dead cells release their iron. Shiga toxin, then, offers a way to rob the host of its iron stores.

An intriguing question often pondered by scientists is: What roles do virulence factors play in the natural ecology of these bacteria? Surely, factors such as Shiga toxin did not evolve after *Shigella* started infecting humans. William Lainhart, Gino Stolfa, and Gerald Koudelka from SUNY Buffalo discovered that Shiga toxin is actually a natural defense against *Tetrahymena thermophila*, a ciliated protist that grazes on bacteria. Lainhart and his colleagues found that *Tetrahymena* was killed when cocultured with Shiga toxin–producing bacteria.

Diphtheria toxin. The classic example of an exotoxin that targets protein synthesis is diphtheria toxin, produced by *Corynebacterium diphtheriae*, the cause of the respiratory disease diphtheria. The two-component diphtheria exotoxin, discussed more fully in **eTopic 25.6**, kills cells by ADP-ribosylating eukaryotic protein synthesis elongation factor 2 (eEF-2). The vaccine used to prevent diphtheria (the "D" in DTaP) is an inactivated form of this exotoxin (see Section 24.6). *C. diphtheriae* is another example of a pathogen whose toxin gene (*dtx*) is regulated by iron availability and is part of a prophage genome integrated into the bacterial chromosome.

Thought Question

25.4 Would patients with iron overload (excess free iron in the blood) be more susceptible to infection?

We have examined only a few of the many protein exotoxins employed by pathogens. Some of the others, including tetanus and botulism toxins, will be described in the next chapter. What should be apparent from our brief sampling is the evolutionary ingenuity that pathogens have used to try to subdue their hosts. A discussion of how new microbial toxins are discovered is found in **eTopic 25.7**.

Thought Question

25.5 Use an Internet search engine to determine which other toxins are related to the cholera toxin A subunit. (*Hint:* Start by searching on "protein" at PubMed to find the protein sequence; then use a BLAST program.)

Endotoxin (LPS) Is Made Only by Gram-Negative Bacteria

Another important virulence factor common to all Gram-negative microorganisms is endotoxin present in the outer membrane (discussed in Chapter 3). Endotoxins are important contributors to inflammation. Not to be confused with secreted exotoxins, "endotoxin" is a medical term sometimes used synonymously for "lipopolysaccharide" (LPS). LPS is really composed of lipid A (the actual endotoxic factor), core glycolipid, and a repeating polysaccharide chain known as the O antigen (**Fig. 25.22**). LPS molecules form the outer leaflet of the Gram-negative outer membrane (discussed in Chapter 3). As bacteria die, they release endotoxin in the form of LPS molecules. Endotoxin is a microbe-associated molecular pattern (MAMP) molecule that can bind to certain Toll-like receptors on macrophages or B cells and trigger the release of TNF-alpha, interferon, IL-1, and other pro-inflammatory cytokines (MAMPs, Toll-like receptors, and cytokines are discussed in Chapters 23 and 24). The release of these active agents causes a variety of symptoms, such as:

- Fever
- Activation of clotting factors, leading to disseminated intravascular coagulation
- Activation of the alternative complement pathway
- Vasodilation, leading to hypotension (low blood pressure)
- Shock due to hypotension
- Death, when other symptoms are severe

The major differences between exotoxins and endotoxins are summarized in **Table 25.3**.

The role of endotoxin in disease can be seen in infections with the Gram-negative diplococcus *Neisseria meningitidis* (**Fig. 25.23A**), a major cause of bacterial meningitis. *N. meningitidis* has, as part of its pathogenesis, a septicemic phase in which the organism can replicate to high numbers in the bloodstream. The large amount of endotoxin present causes a massive depletion of clotting factors, which leads to

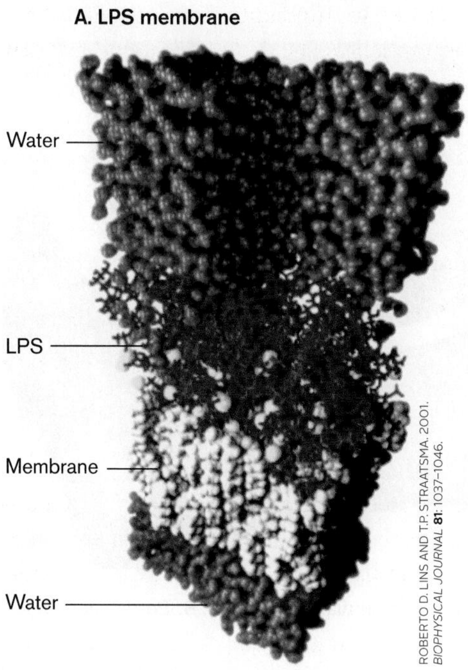

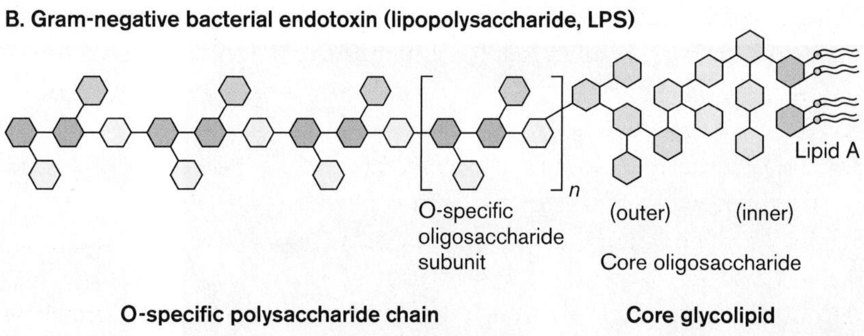

FIGURE 25.22 ■ **Endotoxin.** **A.** Model of a lipopolysaccharide (LPS) membrane of *Pseudomonas aeruginosa*, consisting of 16 lipopolysaccharide molecules (red) and 48 ethylamine phospholipid molecules (white). **B.** Basic structure of endotoxin, showing the repeating O-antigen side chain that faces out from the microbe and the membrane proximal core glycolipid and lipid A (contains endotoxic activity).

internal bleeding, most prominently displayed to a physician as small pinpoint hemorrhages called **petechiae** on the patient's hands and feet (**Fig. 25.23B**). Capillary bleeding near the surface of the skin causes petechiae. One danger of treating massive Gram-negative sepsis with antibiotics is that the enormous release of endotoxin from dead bacteria could well kill the patient. Untreated Gram-negative sepsis is, however, almost always fatal, so we need to apply antibiotics despite the risk.

An approach to prevent endotoxic shock currently under study is based on the knowledge that LPS must bind to Toll-like receptor TLR4 to cause endotoxic shock. What if we could neutralize TLR4 by antibody and prevent it from binding LPS during an infection? Several investigators have shown that injecting antibody raised to TLR4 (anti-TLR4) into infected mice will successfully block TLR4 and protect the mice from *E. coli*–induced septic shock. Human trials are under way.

To Summarize

- **There are nine categories of protein exotoxins** based on mode of action. These include toxins that disrupt membranes, inhibit protein synthesis, or alter the synthesis of host cell signaling-molecules, as well as toxins that are superantigens or target-specific proteases.

TABLE 25.3 Major Distinctions between Bacterial Exotoxins and Endotoxins

Property	Exotoxins	Endotoxins
Producing organism	Gram-positive or Gram-negative	Gram-negative only
Chemical	Protein (size 50–1,000 kDa)	Lipopolysaccharide (lipid A moiety; size 10 kDa)
Denatured by boiling	Yes, if boiled long enough	No
Mode of action	Some exotoxins target specific features of eukaryotic cells (membrane, protein synthesis, signal transduction, etc.); others are superantigens	Bind Toll-like receptor 4; activate cytokine production
Enzyme activity	Often	No
Toxicity	High (1-μg quantities)	Low (>100-μg quantities), but the primary cause of Gram-negative sepsis (serious blood infections)
Immunogenicity	Highly antigenic	Poorly antigenic
Vaccine	Toxoids can be made for some	Toxoids cannot be made
Fever production (pyrogenicity)	Occasionally	Yes

A.

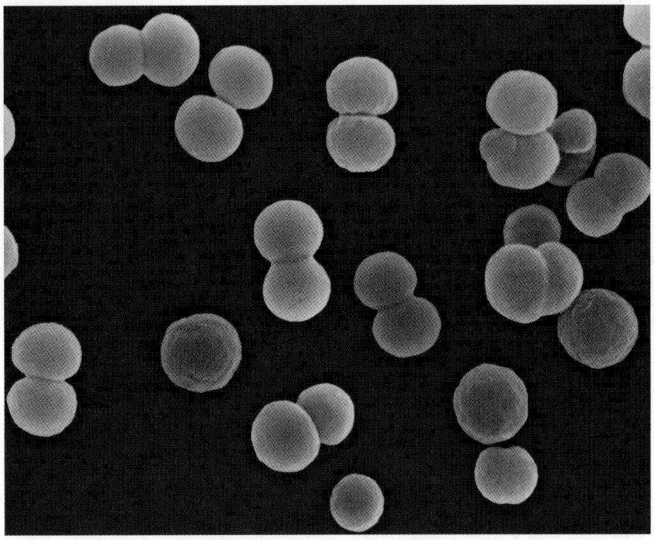

B.

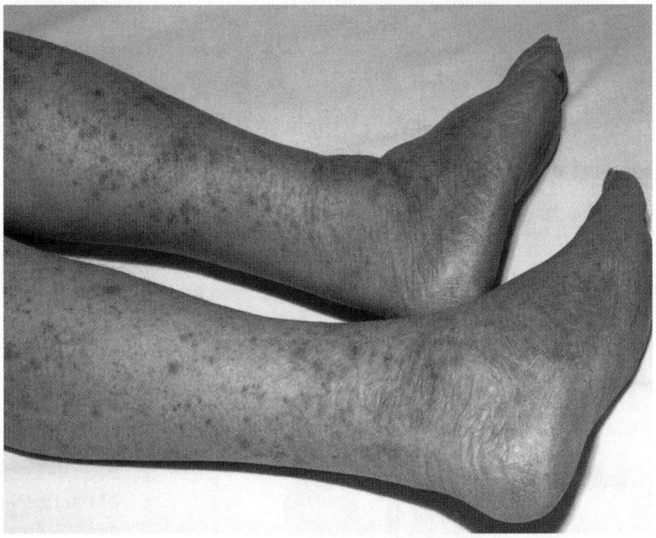

FIGURE 25.23 ■ Effect of *Neisseria meningitidis* endotoxin. **A.** *N. meningitidis* (cell 0.8–1 μm in diameter; SEM). **B.** Petechial rash caused by *N. meningitidis*.

- **Staphylococcus aureus alpha toxin** forms pores in host cell membranes.
- **Many bacterial toxins are two-component, AB-subunit toxins.** The B subunit promotes penetration through host cell membranes, while the A subunit has toxic activity.
- **Cholera toxin,** *E. coli* labile toxin, and pertussis toxin are AB5 toxins that alter host cAMP production by adding ADP-ribose groups to different G-factor proteins.
- **Anthrax toxin** is a three-part AB toxin with one B subunit (protective antigen) and two different A subunits that affect cAMP levels (edema factor) and cleave host protein kinases (lethal factor).
- **Shiga toxin** is an AB toxin that cleaves host cell 28S rRNA in host cell ribosomes.
- **Lipopolysaccharide (LPS),** also known as endotoxin, is an integral component of Gram-negative outer membranes and an important virulence factor that triggers massive release of cytokines from host cells. The indiscriminate release of cytokines can trigger fever, shock, and death.

25.4 Deploying Toxins and Effectors

A recurring theme among bacterial pathogens is the secretion of proteins that destroy, cripple, or subvert host target cells. The bacterial toxins described in the previous section are secreted into the surrounding environment, where they float randomly until chance intervenes and they hit a membrane-binding site. However, many pathogens attach to tissue cells and inject bacterial proteins (called effectors) directly into the host cell cytoplasm. The proteins may not kill the cell, but they redirect host signaling pathways in ways that benefit the microbe.

Protein secretion pathways were introduced in Section 8.5, which focused on ATP-binding cassette (ABC) proteins as a model. Additional secretion models are described here in their critical role of delivering pathogenicity proteins such as toxins. A particularly interesting aspect of these secretory systems is that many of them evolved from, and bear structural resemblance to, other cell structures that serve fundamental cell functions. The secretion systems and the molecular processes that share an evolutionary history include:

- Type II protein secretion (homologous to type IV pilus biogenesis)
- Type III protein secretion (homologous to flagellar synthesis)
- Type IV protein secretion (homologous to DNA transfer by conjugation)
- Type VI protein secretion (homologous to T4 phage tail structures)

Table 25.4 lists features of nine export systems of bacteria and examples of associated virulence effector proteins. We will focus on secretion system types II, III, and IV as model delivery systems.

Type II Secretion Resembles Type IV Pilus Assembly

Cholera toxin, discussed in Section 25.3, is a well-known example of a toxin secreted by a **type II secretion system**.

TABLE 25.4 Secretion Systems for Bacterial Toxins[a]

Secretion type	Features	Examples
I	SecA dependent, one effector per system	*E. coli* alpha hemolysin, *Bordetella pertussis* adenylyl cyclase
II	SecA dependent, similar to type IV pili	*Pseudomonas aeruginosa* exotoxin A, elastase, cholera toxin
III	SecA independent, multiple effectors secreted, syringe mechanism injects effectors into target cells, related to flagella	*Yersinia* Yop proteins, *Salmonella* Sip proteins, enteropathogenic *E. coli* (EPEC) EspA proteins, TirA
IV	Related to conjugational DNA transfers, multiple effectors secreted, some systems inject effectors into target cells	*B. pertussis* toxin, *Helicobacter* CagA
V	Autotransporter, SecA dependent to periplasm, self-transport through outer membrane, one effector per system	Gonococcal and *Haemophilus influenzae* IgA proteases
VI	Related to phage tails, single effector, harpoon mechanism	*Burkholderia* and *Vibrio cholerae* VgrG
VII	Unrelated to other systems	*Mycobacterium tuberculosis* Esx and Esp
VIII	Unrelated to other systems	*E. coli* Curli pili subunits
IX	Unrelated to other systems	*Porphyromonas gingivalis* (gingipain proteases)

[a]Systems I–VI and VIII are found in Gram-negative bacteria; type VII is found in Gram-positive bacteria and *Mycobacterium tuberculosis*; type IX is restricted to Bacteroidetes.

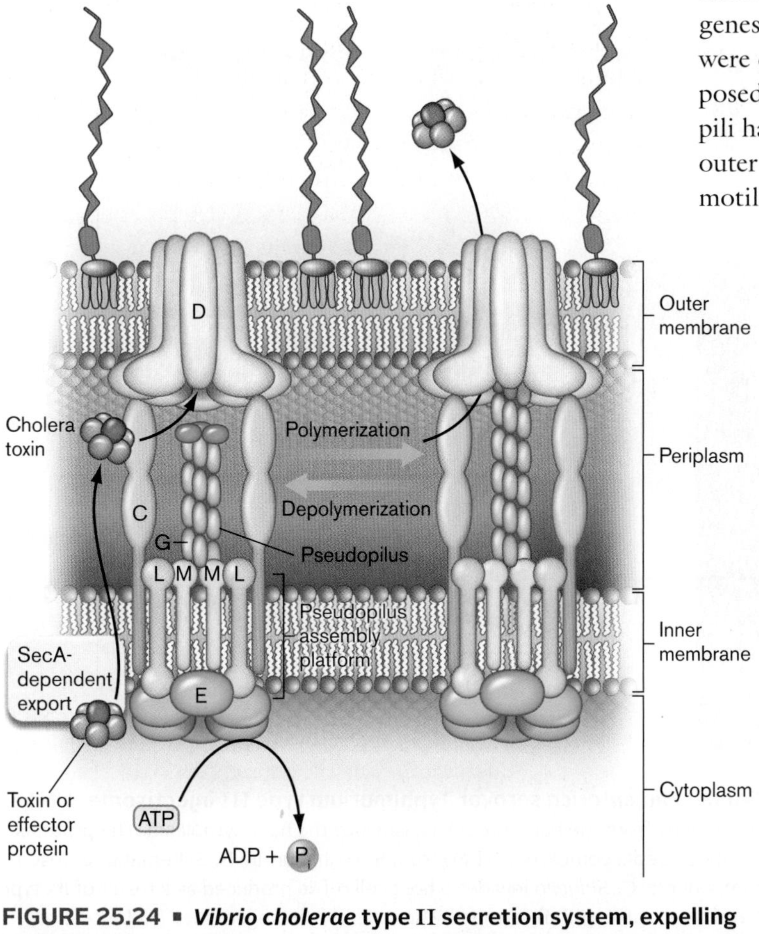

FIGURE 25.24 ■ *Vibrio cholerae* type II secretion system, expelling cholera toxin. C, D, E, G, L, M, and N are protein components of the secretion system.

Type II secretion offers a clear example of how nature has modified the blueprints of one system to do a very different task. DNA sequence analysis has revealed that the genes used for type IV pilus biogenesis (see Section 25.2) were duplicated at some point during evolution and repurposed to serve as a protein secretion mechanism. Type IV pili have the unusual ability to extend and retract from the outer membrane—a property that produces the gliding motility of *Myxococcus* (see Section 4.6) and the twitching motility of *Neisseria* and *Pseudomonas*. As you might guess, assembly/disassembly of these appendages is quite complex.

Type II protein secretion mechanisms mirror this complexity. Proteins to be secreted first make their way, via the SecA-dependent general secretion pathway, to the periplasm, where they are folded and then encounter the appropriate type II secretion system. Type II secretion systems cyclically assemble and disassemble the pilus-like structure (pseudopilus), using it as a piston to ram folded toxins or effector proteins through an outer membrane pore structure and into the surrounding void (**Fig. 25.24**).

Type III Secretion Is an Injection Machine

Yersinia, *Salmonella*, and *Shigella* are the etiological agents of Black Death and various forms of diarrhea. In the 1990s it was discovered that these organisms could somehow take the bacterial virulence proteins (effector proteins) made in their cytoplasm and drive them directly into the eukaryotic cell cytoplasm without the protein ever getting into the extracellular environment. Direct delivery is a good idea because it eliminates the dilution that happens when a toxin is secreted into an extracellular environment. Another advantage of this strategy is that it avoids the need to tailor the toxin to fit a preexisting host receptor.

What kind of molecular machine can directly deliver cytoplasmic bacterial proteins into target cells? Some microbes use tiny molecular syringes (injectisomes) embedded in their membranes to inject proteins directly into the host cytoplasm; this mechanism of delivery is called a **type III secretion system** (**T3SS**). **Figure 25.25** shows an electron micrograph of type III secretion needles and a model of the complex spanning the cytoplasmic and outer bacterial membranes. Genes encoding type III systems are actually related to flagellar genes, whose products export the flagellin proteins through the center of a growing flagellum (discussed in Section 3.6). It appears that a duplicated set of flagellar genes was evolutionarily reengineered to encode proteins that act more like molecular syringes (**Fig. 25.25A** and **B**).

The bacterial virulence proteins (effectors) secreted by type III systems subvert normal host cell signaling pathways, some of which cause dramatic rearrangements of host cytoskeleton at the cell membrane that lead to engulfment of the microbe (**Fig. 25.25C**). The genes encoding type III systems in modern-day pathogens are usually located within pathogenicity islands inherited long ago via horizontal transfer from ancestral microbial sources. Many bacterial pathogens use this type of secretion system, including plant pathogens such as *Pseudomonas syringae* (the cause of blight, a disease of many plants in which leaves or stems develop brown spots). Secretion is normally triggered by cell-cell contact between host and bacterium.

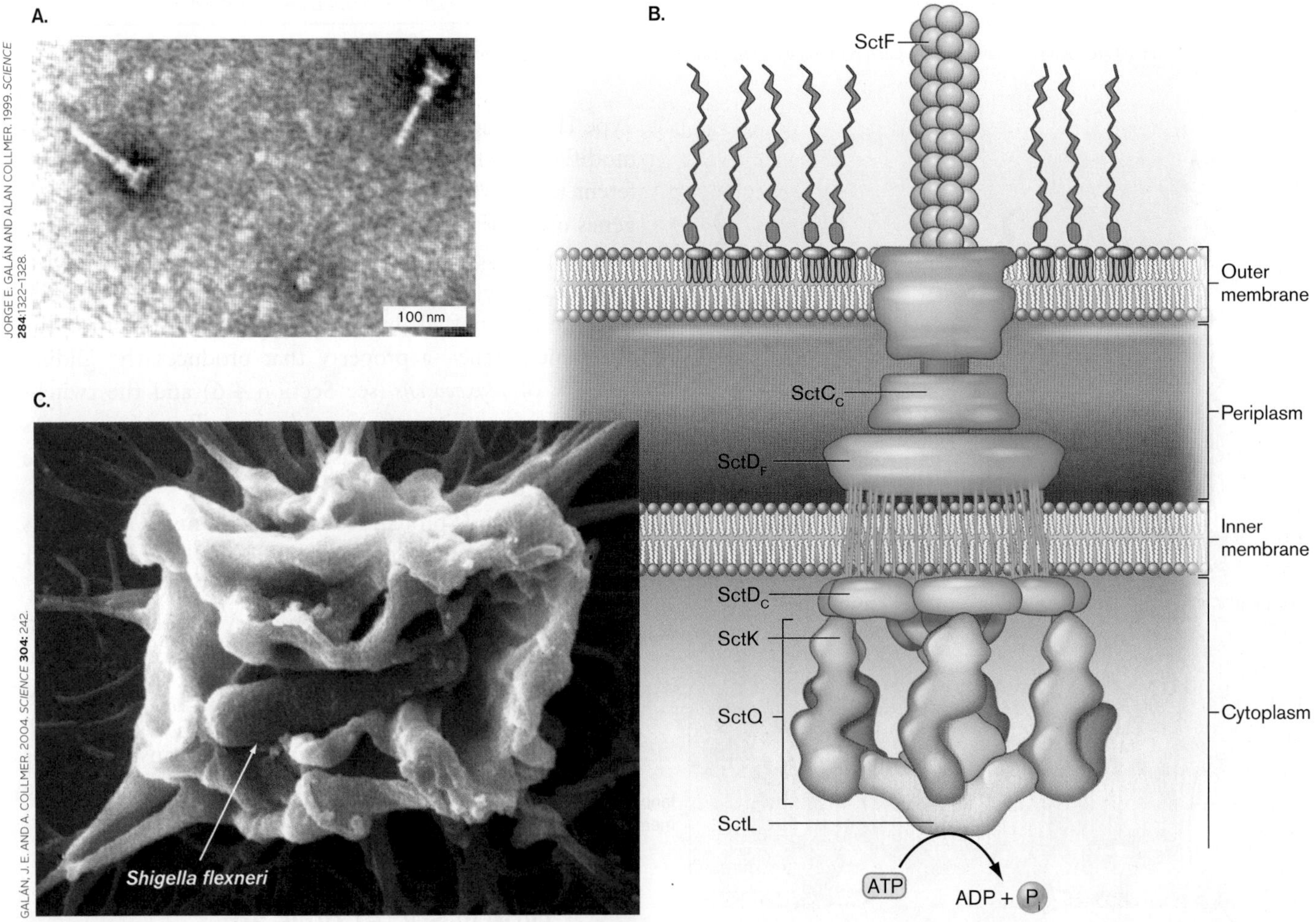

FIGURE 25.25 ■ The type III secretion complex from *Salmonella enterica* serovar Typhimurium type III injectisome. Unlike other secretion systems, the type III mechanism injects proteins directly from the bacterial cytoplasm into the host cytoplasm. The proteins in these systems are related to flagellar assembly proteins. **A.** Purified needle complexes (TEM) from S. Typhimurium **B.** Schematic representation of the S. Typhimurium needle complex and its putative components. **C.** *Shigella* invades a host cell ruffle produced as a result of its type III secretion system. *Shigella flexneri* (approx. 2 μm) entering a HeLa cell ruffle (SEM) formed by host actin rearrangements. (HeLa cells are an immortal cancer cell line.) The ruffle engulfs the bacterium and eventually disassembles, internalizing the bacterium. *Source:* Part B modified from Galán and Waksman. *Cell* **172**:1306–1318.

A.

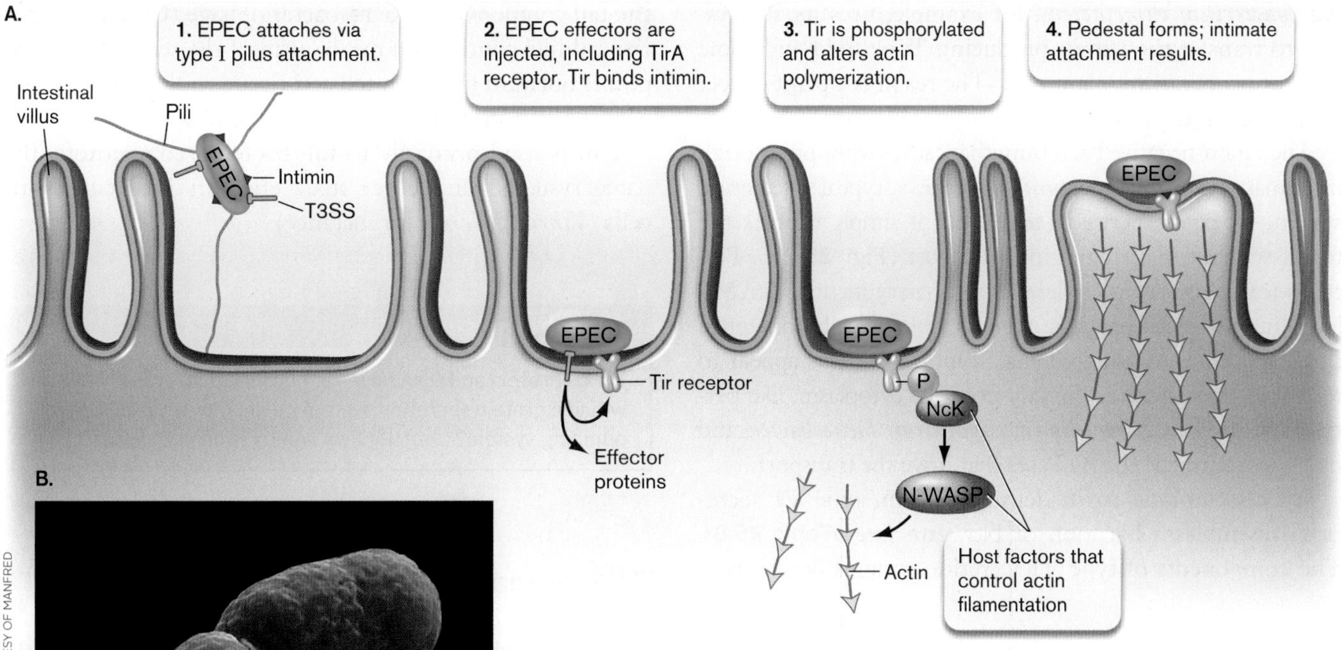

B.

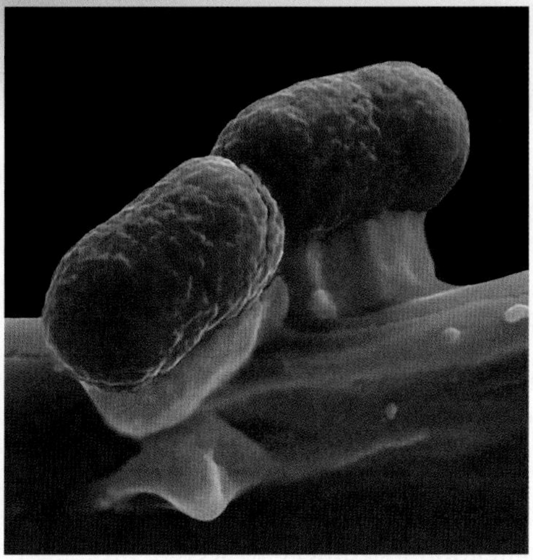

CELL HOST MICROBE. 2009. 5(3), COVER. IMAGE COURTESY OF MANFRED ROSE (HZI; BRAUNSCHWEIG, GERMANY).

FIGURE 25.26 ■ ***E. coli* type III secretion and cell-cell interaction.** **A.** Model of enteropathogenic *E. coli* (EPEC) attachment and pedestal formation on intestinal epithelial cells. (1) EPEC attaches first, using type I pili. (2) Bound EPEC uses a T3SS to inject Tir protein into the host cell. Tir inserts into the host membrane and acts as a receptor for the EPEC surface protein intimin. (3) Tir also communicates through phosphorylation with other host factors that control actin filamentation and cytoskeleton formation. Actin polymerization raises the host membrane to produce a pedestal upon which EPEC sits (4). **B.** Pedestal formation (colorized SEM).

***E. coli* pathogens use a T3SS to "inject" their own receptor into host cells.** Enteropathogenic *E. coli* (EPEC) and enterohemorrhagic *E. coli* (EHEC) are two diarrhea-producing forms of *E. coli*. These pathogens use pili to initially bind to the host's intestinal epithelial cells (**Fig. 25.26A**, step 1); however, the bacterium must establish a more intimate attachment to these cells to cause disease. The bacterial outer membrane protein called **intimin** mediates this intimate attachment. The only problem is that the host lacks a receptor for intimin. To solve this problem, EPEC and EHEC use a type III secretion system to insert their own receptor, Tir (for translocated intimin receptor), into the target cells (**Fig. 25.26A**, step 2). Brett Finlay and his colleagues in Vancouver, British Columbia, discovered this system and found that the genes encoding intimin, Tir, and the secretion apparatus are all part of an EPEC pathogenicity island.

Once injected and placed in the host membrane, Tir binds intimin on the bacterial surface. Think of Tir as a wall anchor that you poke into a board in order to attach something to it. The result is a tighter, more intimate adherence between the bacterium and host cell surface, which is required for infection to proceed. In addition, host protein kinases phosphorylate Tir at tyrosine residue 474 (**Fig. 25.26A**, step 3). Phosphorylated Tir directly triggers a remarkable reorganization of host cell cytoskeletal components (actin, alpha-actinin, ezrin, talin, and myosin light chain) such that a membrane "pedestal" is formed, raising the microbe up (**Fig. 25.26A**, step 4, and **Fig. 25.26B**). The result of this attachment is the characteristic attaching and effacing (A/E) lesion, characterized by destruction of the microvilli and pedestal formation. By placing itself on a "pedestal," EPEC avoids engulfment and the perils of the phagolysosome.

Because of their importance to virulence, type III secretion systems are the subject of intensive research designed to exploit them as potential drug targets.

Type IV Secretion Resembles Conjugation Systems

Many bacteria can transfer DNA from donor to recipient cells via a cell-cell contact system known as conjugation (see Section 9.3). The conjugation systems of some pathogens have been modified, through evolution, into new systems, called **type IV secretion systems** (**T4SSs**), that transport proteins, or proteins plus DNA, directly into target cells.

Agrobacterium tumefaciens, for example, uses its Vir system to transfer the tumor-producing Ti plasmid and some effector proteins into plant cells. The result is a plant cancer called crown gall disease.

The Gram-negative bacterium that causes whooping cough in humans, *Bordetella pertussis*, also uses a type IV secretion system, to export pertussis toxin, but it simply exports the toxin, without injecting it into the host (**Fig. 25.27**). Pertussis toxin, similar to cholera toxin, sharply increases cAMP levels in lung epithelial cells. Water leaves the cells and enters the interstitium (causing edema). Type IV systems appear to recruit effector proteins directly from the cytoplasm, like pertussis toxin from *Bordetella* or CagA from *Helicobacter*, and pass them through the ATPases that drive the transport.

Another unique toxin delivery system, type VI secretion, resembles a harpoon or blowgun (see **eTopic 25.8**). The components of type VI secretion systems derive from the tail components of a T4 bacteriophage (Chapter 6), but the tail-like contraction mechanism of the secretion system points outward from the cell and fires a rodlike core capped with a toxin protein into target cells, like a harpoon. The system is used primarily to kill bacterial competitors, but some systems can deliver toxic effectors into eukaryotic cells (*Vibrio cholerae*, for instance).

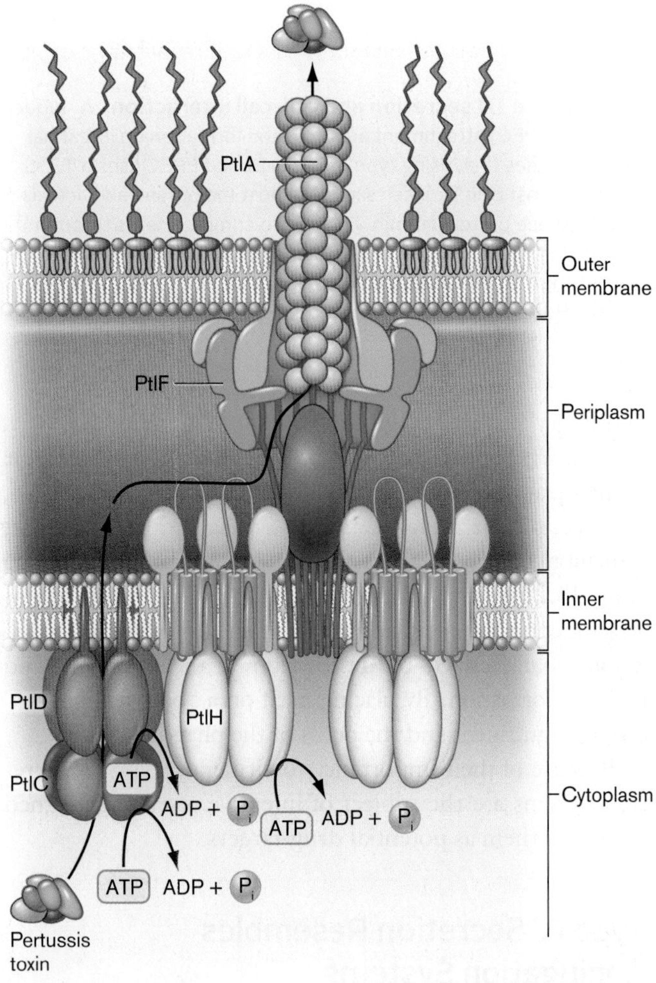

FIGURE 25.27 ■ Type IV secretion of pertussis toxin. Evolutionarily related to conjugation systems, this type IV system in *Bordetella pertussis* takes pertussis toxin from the cytoplasm (moved through the PtlC/PtlH ATPases to the PtlF secretion channel). Effector proteins exit to the extracellular milieu through the pilus-like extension (this pseudopilus does not extend and retract as happens in type II secretion).

Thought Question

25.6 Protein and DNA have very different structures. Why would a protein secretion system be derived from a DNA-pumping system? (*Hint:* Review conjugation in Chapter 9.)

To Summarize

- **Many pathogens** use specific protein secretion pathways to deliver toxins.
- **Type II secretion** systems use a pilus-like extraction/retraction mechanism to push proteins out of the cell.
- **Type III secretion** uses a molecular syringe to inject proteins from the bacterial cytoplasm into the host cytoplasm.
- **Type IV secretion** utilizes a group of proteins homologous to conjugation machinery to secrete proteins from either the cytoplasm or the periplasm.

25.5 Surviving within the Host

Once inside a host, how does a successful pathogen avoid detection and destruction? Many of the virulence factors in the pathogen's arsenal help the microbe escape or resist innate immune mechanisms. Others are dedicated to stealth—that is, hiding from the immune system. But before discussing how these organisms survive in a host, we must ask how the pathogen knows it is in a host.

Where Am I?

A pathogen that can grow either outside or inside a host must adjust its physiology to match its whereabouts. Why make a type III secretion system if there are no host cells around? But how do microbial pathogens know whether they are in a host or in a pond? And which bacterial genes are expressed exclusively while in a host?

The same types of regulatory mechanisms that sense environmental conditions in a pond are used by the microbe to determine its whereabouts in a host. That is, various sensing systems act in concert to recognize a specific environmental niche. Two-component signal transduction systems, discussed in Section 10.1, are used to monitor magnesium concentrations, which are characteristically low in a host cell vacuole. Other regulators measure pH, which will be low (acidic) in the same vacuole. The point is that there is no single in vivo sensing system. The various regulators collaborate to trigger the expression of virulence genes.

Enterohemorrhagic *E. coli* uses positive and negative signals to determine where in a host it should express virulence genes. As mentioned earlier, EHEC uses a T3SS to inject effector proteins into gastrointestinal cells, so synthesis of the T3SS apparatus is best induced only in the intestine. One signal that induces T3SS synthesis is fucose. Fucose ends up in the intestine because fucosidases from the nonpathogenic gut microbe *Bacteroides thetaiotaomicron* cleave fucose from the host glycans present on intestinal epithelial cells. EHEC senses the fucose via a two-component system and responds by synthesizing the T3SS. However, other metabolites in the host can inhibit expression of the T3SS apparatus in unfavorable body sites. Andrew Roe and his team at the University of Glasgow found that the host metabolite d-serine can inhibit EHEC T3SS expression. d-Serine is found at high concentrations at extraintestinal sites such as the urinary tract and brain, two sites that EHEC does not infect. How d-serine inhibits T3SS gene expression is not clear.

Many bacterial pathogens regulate virulence genes by sensing the concentration of free iron, typically very low in the host. For instance, *Corynebacterium diphtheriae* sensing a low iron environment will induce synthesis of diphtheria toxin (see **eTopic 25.6**). The toxin kills host cells to release their iron, which the bacteria can now use. For other pathogens, iron concentration alone is usually not enough to provoke virulence. Regulators sensitive to other in vivo signals must be activated to achieve a successful infection.

Cell-cell communication is also important during infections. *Pseudomonas aeruginosa*, for example, has at least three quorum-sensing systems that detect secreted autoinducers (quorum sensing is discussed in Chapter 10). As the number of bacteria in a given space increases, so, too, does the concentration of the chemical autoinducer. When the autoinducer reaches a critical concentration, it diffuses back into the bacterium or binds to a surface receptor and triggers the expression of bacterial target genes. Genes included in the *P. aeruginosa* quorum-sensing regulons encode *Pseudomonas* exotoxin A and other secreted proteins, such as elastase, phospholipase, and alkaline protease.

Why would a pathogen employ quorum sensing to regulate virulence factors? One reason may be to prevent alerting the host that it is under attack before enough microbes can accumulate through replication. Tripping the host's alarms too early would make eliminating infection easy. Waiting until a large number of bacteria have amassed before releasing toxins and proteases will increase the chance that the host can be overwhelmed.

Extracellular Immune Avoidance

The topic of extracellular immune avoidance was first covered in the discussions of host defense (Chapter 23) and immunology (Chapter 24). Many bacteria, such as *Streptococcus pneumoniae* and *Neisseria meningitidis*, produce a thick polysaccharide **capsule** that envelops the cell. Capsules help organisms resist phagocytosis in several ways. Recall that phagocytes must recognize bacterial cell-surface structures or surface-bound C3b complement factor to begin phagocytosis (see Section 23.6). A capsule will cover bacterial cell wall components and mannose-containing carbohydrates that phagocytes normally use for attachment. The uniformity and slippery nature of capsule composition make it difficult for phagocytes to lock on to the bacterial cell.

But what about complement factor C3b, which can bind to the bacterial cell? Phagocytes have surface C3b receptors that latch on to C3b molecules fixed to an invading pathogen. Capsules, however, can envelop and hide any C3b complement factor that binds to the bacterial surface. Fortunately for us, immune defense mechanisms can eventually circumvent this avoidance strategy by producing opsonizing antibodies (IgG) against the capsule itself (see Section 24.2). The Fc regions of antibodies that bind to the capsule point away from the bacterium, so they are free to bind Fc receptors on phagocyte membranes. Binding of the Fc region to the phagocyte's Fc receptor triggers phagocytosis.

Pathogens can also use proteins on their cell surface to avoid phagocytosis. *Staphylococcus aureus* has a cell wall protein called **protein A** that binds to the Fc region of antibodies, hiding the bacteria from phagocytes in two ways. Protein A on the cell surface can bind to the Fc region of any antibody and prevent phagocytes from binding to the antibody. When the Fc region is bound to protein A, the antigen-binding site is pointing away from the microbe. However, even if an antibody-binding site finds its antigen target on a bacterium, protein A from a second bacterium can bind to that Fc region, once again blocking phagocyte recognition.

Some extracellular microbes can trigger apoptosis in target host cells. Proteins made by the pathogen can slip into a host cell—a macrophage, for instance—and trigger apoptosis. How does this help the pathogen? If the duped macrophage self-destructs, it cannot destroy the microbe.

Another immune avoidance strategy used by microorganisms, both the extracellular and intracellular types, is to change their antigenic structure. Genes encoding flagella, pili, and other surface proteins often use site-specific gene inversions to express alternative proteins (for example,

Salmonella phase variation) or slipped-strand mispairing (see **eTopic 10.4**) to add or remove amino acids from a sequence. You can think of these processes as shape-shifting to avoid recognition.

Intracellular Immune Avoidance

In an effort to escape both innate and humoral immune mechanisms (see Chapters 23 and 24), many bacterial pathogens, called **intracellular pathogens**, seek refuge by invading host cells. (Viruses, by definition, are intracellular pathogens.) Hiding within a host cell temporarily provides the pathogen safe harbor from antibodies and phagocytic cells. Some bacteria dedicate their entire lifestyle to intracellular parasitism and are called **obligate intracellular pathogens**. *Rickettsia*, for example—for reasons unknown—will not grow outside a living eukaryotic cell. Other microbes, such as *Salmonella* and *Shigella*, are considered **facultative intracellular pathogens** because they can live either inside host cells or free. We have already discussed how intracellular pathogens get into cells, but how do they withstand intracellular attempts to kill them?

Once inside the phagosome, intracellular pathogens have three options to avoid being killed by a phagolysosome (**Fig. 25.28**). They can prefer growth inside the phagolysosome, prevent phagosome-lysosome fusion, or simply escape the phagosome.

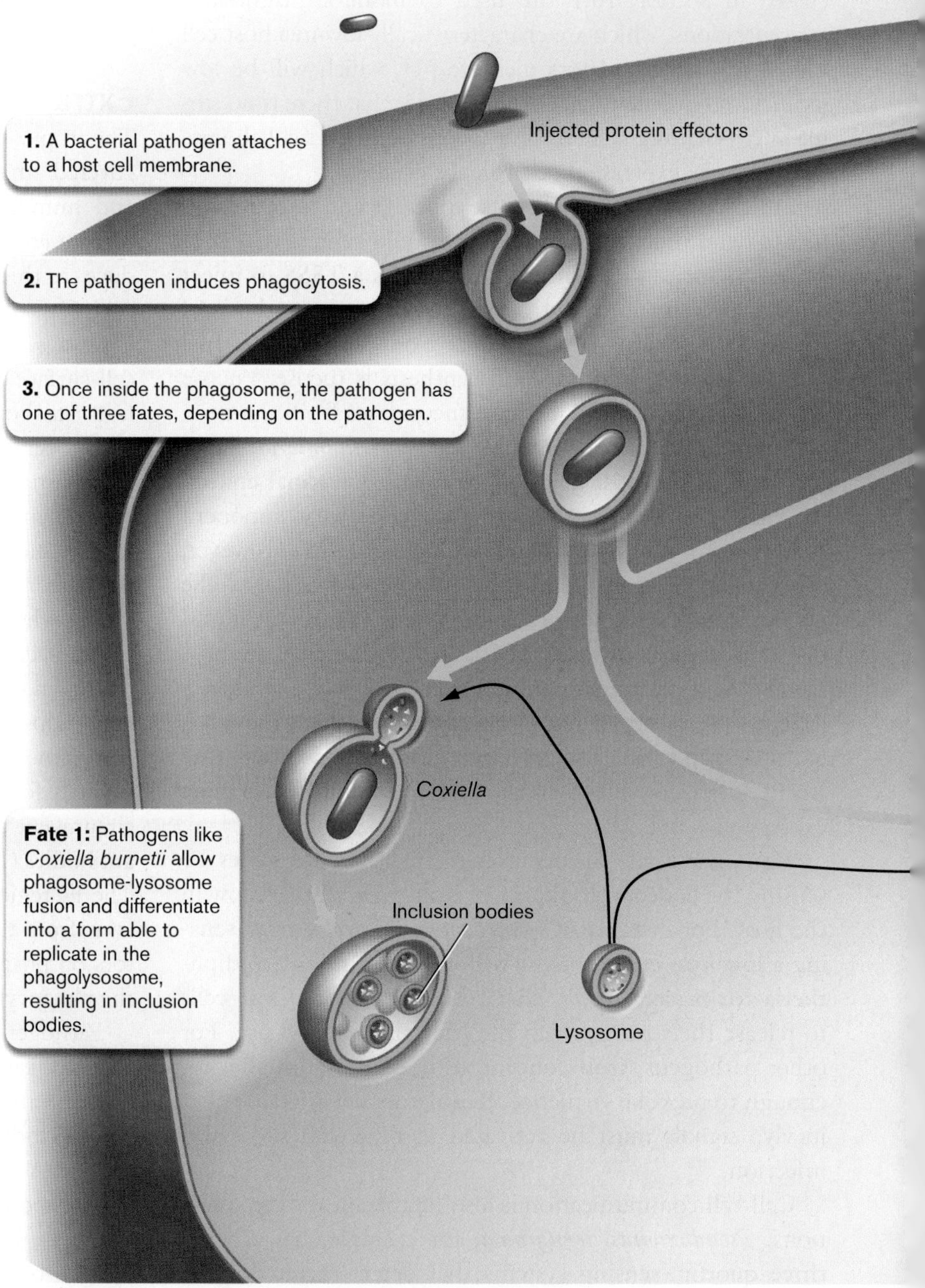

FIGURE 25.28 ■ Alternative fates of intracellular pathogens. Different pathogens have different strategies for surviving in a host cell. Some tolerate phagolysosome fusion (for example, *Coxiella*), others prevent phagolysosome fusion (*Salmonella*), and still others escape the phagosome to replicate in the cytoplasm (*Shigella* and *Listeria*).

Thriving under stress. In what could be called the "bring it on" strategy, some intracellular pathogens prefer the harsh environment of the phagolysosome (**Fig. 25.28**, fate 1). *Coxiella burnetii*, for example, grows well in the very acidic phagolysosome environment (**Fig. 25.29A**). This obligate intracellular organism (an organism that grows only inside another living cell) causes a flu-like illness called Q fever (query fever). The symptoms of Q fever include sore throat, muscle aches, headache, and high fever. The illness has a mortality rate of about 1%, so most people recover to good health. The organism allows phagosome-lysosome fusion because the acidic environment that results is needed for it to survive and grow.

Thought Question

25.7 How can you determine whether a bacterium is an intracellular pathogen?

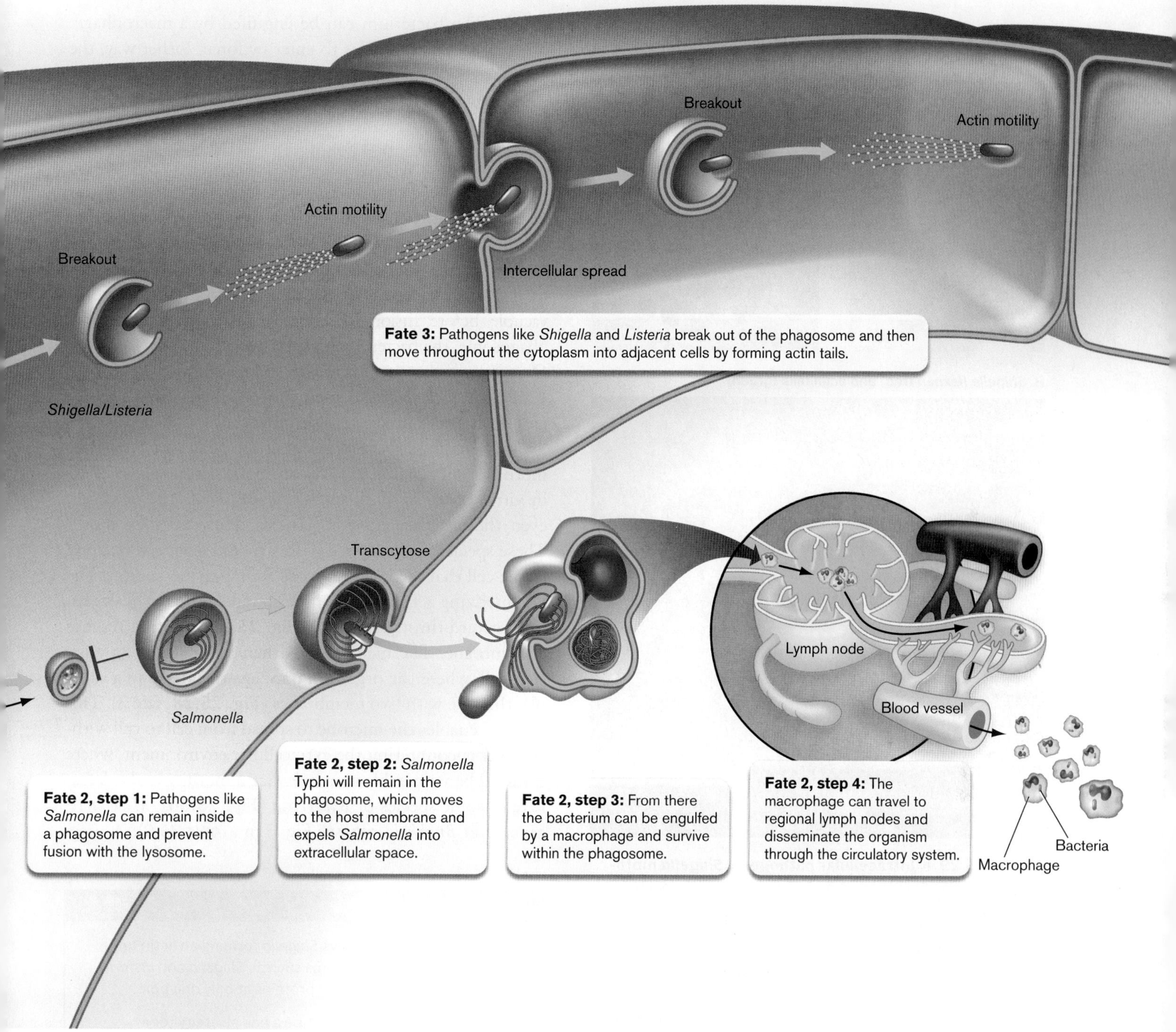

Why some bacteria are obligate intracellular pathogens is unclear. One intracellular bacterium, *Rickettsia prowazekii*, a cause of epidemic typhus, appears to be an "energy parasite" that can transport ATP from the host cytoplasm and exchange it for spent ADP in the bacterium's cytoplasm. But this does not explain its obligate intracellular status, since giving *Rickettsia* ATP outside a host does not allow the bacterium to grow. Other factors remain to be discovered.

Inhibiting phagosome-lysosome fusion. Some intracellular pathogens avoid lysosomal enzymes by preventing lysosomal fusion with the phagosome. *Salmonella*, *Mycobacterium*, *Legionella*, and *Chlamydia* are good examples. For example, *Legionella pneumophila* grows inside alveolar macrophage phagosomes and produces the potentially fatal Legionnaires' disease, so named for the veterans group that suffered the first recognized outbreak, in 1976. The organism, once inside a phagosome (called a *Legionella*-containing vacuole, or LCV), uses a type IV secretion system to secrete proteins through the vesicle membrane and into the cytoplasm.

The bacterial proteins secreted by *Legionella* interfere with host cell signaling pathways that cause phagosome-lysosome

A. *Coxiella burnetii*

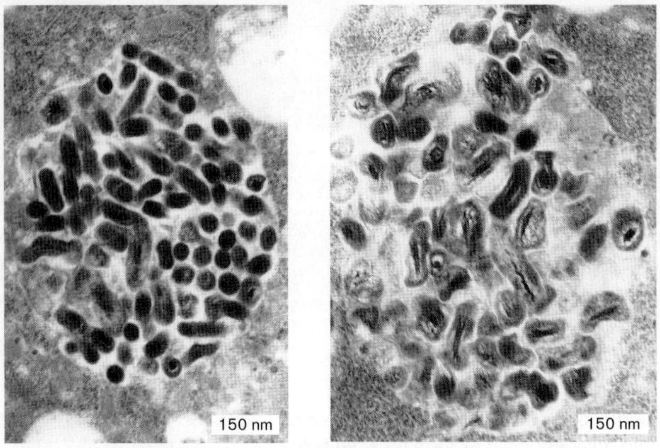

B. *Shigella flexneri* (red) and actin tails (green)

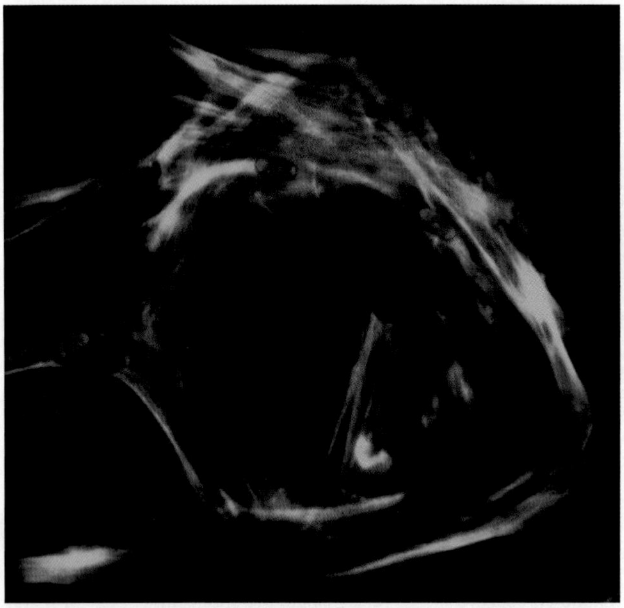

FIGURE 25.29 ■ Intracellular pathogens: *Shigella* motility and *Coxiella* development. **A.** A typical vacuole in J774A.1 mouse macrophage cells infected with *Coxiella burnetii* at 2 hours (left) and 6 hours (right) postinfection (TEM). The organism lives in an acidified vacuole and undergoes a form of differentiation that changes its shape and alters its interactions with the host cell. **B.** Intracellular *Shigella flexneri* (fluorescence microscopy), fluorescently stained red (1 μm in length), moves through the host cytoplasm propelled by actin tails, stained green.

fusion. The bacterial protein LegC7, for instance, inhibits endosome trafficking that would lead to lysosome fusion. The result is that *L. pneumophila* can grow in a friendlier vesicle. Interestingly, *L. pneumophila* is actually a soil and water microbe. Its ability to survive inside macrophages evolved from its ability to survive in amebas, which serve as a natural reservoir for this pathogen.

Salmonella Typhi is another pathogen that prevents phagosome-lysosome fusion, but this organism eventually leaves the host cell (via exocytosis) and enters extracellular tissue spaces where macrophages await (**Fig. 25.28**, fate 2). The bacterium can be engulfed by a macrophage, or use T3SS mechanisms to enter by force. Either way, the pathogen will survive inside the macrophage phagosome. The infected macrophage travels to a regional lymph node, enters the bloodstream, and, like a Trojan horse, disseminates *Salmonella* throughout the body. *Salmonella* as a model of pathogenesis will be discussed in more detail later.

Escaping the phagosome. The Gram-negative bacillus *Shigella dysenteriae* and the Gram-positive bacillus *Listeria monocytogenes*, both of which cause food-borne gastrointestinal disease, use hemolysins to break out of the phagosome vacuole before fusion (described earlier). By escaping the phagosome, the bacteria completely avoid lysosomal enzymes. Once free in the cytoplasm, they enjoy unrestricted growth. Yet even in the cytoplasm, these microbes have found a way to redirect host cell function to their own ends.

A fascinating aspect of escaping the phagosome involves motility. *Shigella* and *Listeria* are both nonmotile at 37°C in vitro; however, both move around inside the host cell, even though they have no flagella. How do they move? These species are equipped with a special device at one end of the cell that mediates host cell actin polymerization. The polymerizing actin, called a "rocket tail," propels the organism forward through the cell (**Fig. 25.29B**) until it reaches a membrane. The membrane is then pushed into an adjacent cell, where the organism once again ends up in a vacuole, this one with two membranes (**Fig. 25.28**, fate 3). This strategy enables the microbe to spread from cell to cell without ever encountering the extracellular environment, where it would be vulnerable to attack. Actin motility is also a feature of some species of *Rickettsia* (**Fig. 25.30**), *Mycobacterium*, and *Burkholderia* (see Fig. 3 in **eTopic 25.8**).

Thought Questions

25.8 **Figure 25.29B** shows *Shigella* forming an actin tail at one pole. Why do organisms such as *Shigella* and *Listeria* assemble actin-polymerizing proteins at only one pole?

25.9 Why might killing a host be a bad strategy for a pathogen?

Sleeping with the Enemy

As just described, many bacterial and, of course, viral pathogens find safe haven by growing inside host cells. However, from our knowledge of innate and adaptive immune responses, it is not intuitively obvious why intracellular growth provides safety. After all, infected cells present microbial antigens on their class I or class II MHC receptors to alert the innate and adaptive immune systems that the infected host cell must be killed to resolve the infection. In addition, pieces of intracellular microbes (flagella, LPS, peptidoglycan)

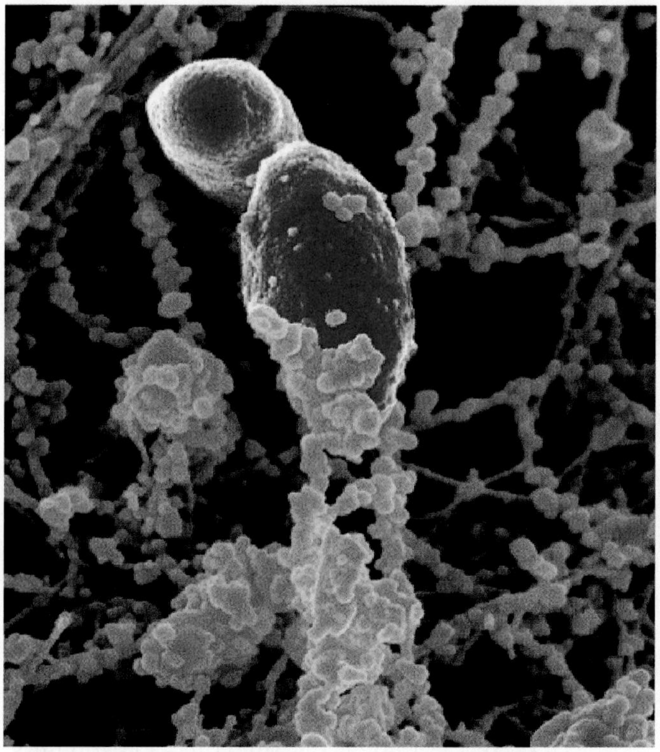

D. ELLISON ET AL. 2008. *INFECT. IMMUN.* **76**(2):542–550. COURTESY OF E. R. FISCHER AND T. HACKSTADT.

FIGURE 25.30 ■ The obligate intracellular pathogen *Rickettsia rickettsii*. This SEM shows *R. rickettsii* (blue, approx. 0.7 μm in length), the cause of Rocky Mountain spotted fever, in association with host actin (gold). Several pathogens propel themselves through host cytoplasm by polymerizing host actin at one pole of the bacterial cell.

will bind pattern recognition receptors (Toll-like and NOD-like receptors—TLRs and NLRs) that activate intracellular inflammasomes. Inflammasomes trigger production of pro-inflammatory cytokines that mediate inflammation.

So, how do intracellular pathogens avoid destruction? It turns out that these invaders employ a variety of molecular tricks that misdirect the immune system much as a magician misdirects an audience. All of these strategies, summarized in **Figure 25.31**, buy the microbe more time to overwhelm the host.

Molecular mimicry and subverting antigen presentation. A variety of bacteria and viruses use mimicry to confuse the immune system (discussed in Section 24.7). In some cases, microbial proteins are made that look like cytokines, or that bind to host cells and hitchhike via normal host trafficking to the nucleus, where the bacterial protein interferes with cytokine gene expression. These factors can manipulate the balance of helper T cells, for example, and send immunity down the wrong path for combating the microbe.

Microbes, especially viruses, can also interfere with antigen presentation on the surfaces of infected cells. Recall from Chapter 24 that proteins made by infectious agents growing in the cytoplasm of a host cell are broken down in the cytosol and transported into the endoplasmic reticulum by the transporter of antigen peptides (TAP). The viral or bacterial antigens are loaded onto MHC I molecules found on the ER membrane, and the complexes are sent to the cell surface. Surface MHC I presents those peptide antigens to roaming CD8 cytotoxic T cells that will then kill the infected cell. Any scheme that interrupts MHC I presentation will spare the infected cell and its infectious cargo from destruction. So, how do pathogens derail MHC presentation?

Collectively, pathogens have four basic ways to subvert antigen presentation: (1) Make proteins that resist digestion by host proteasomes so that antigens are not processed. (2) Make a protein that blocks the TAP protein so that processed antigens cannot be loaded onto MHC I (herpesviruses such as HSV, CMV, and VZV are famous for this approach). (3) Make a protein that induces TAP degradation via ubiquitylation (*Pseudomonas aeruginosa*). (4) Induce degradation of MHC molecules via ubiquitylation. Ubiquitylation strategies that coerce the host to degrade MHC proteins are described later.

Flipping cytokine profiles. Many pathogens defy death by growing inside macrophages that are "armed" with numerous antimicrobial weapons. Pathogens that can grow inside macrophages include bacteria such as *Salmonella*, *Yersinia*, *Listeria*, *Mycobacterium tuberculosis*, *Francisella tularensis*, and *Chlamydia*, as well as protozoa such as *Leishmania* and *Trypanosoma cruzi*. How do they survive? These, and many

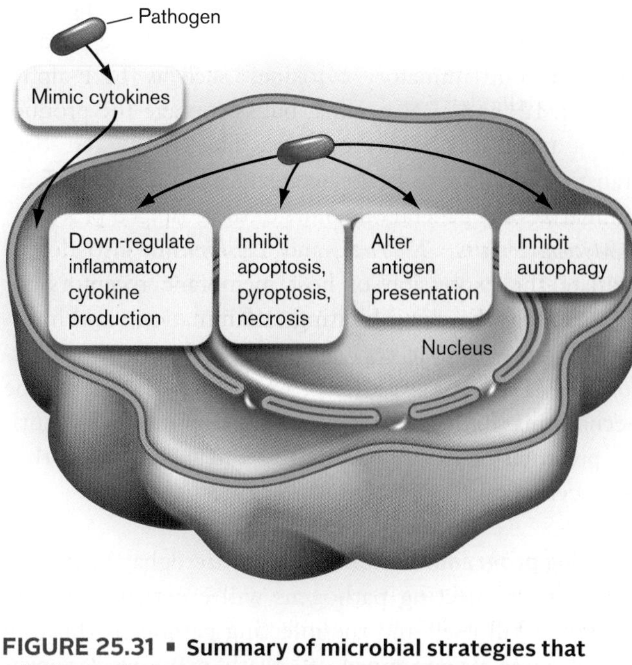

FIGURE 25.31 ■ Summary of microbial strategies that misdirect the immune system. Bacteria and viruses produce different molecules that can mimic cytokines or transcriptional regulators, alter cytokine production, prevent programmed cell death, alter antigen presentation, or inhibit autophagy.

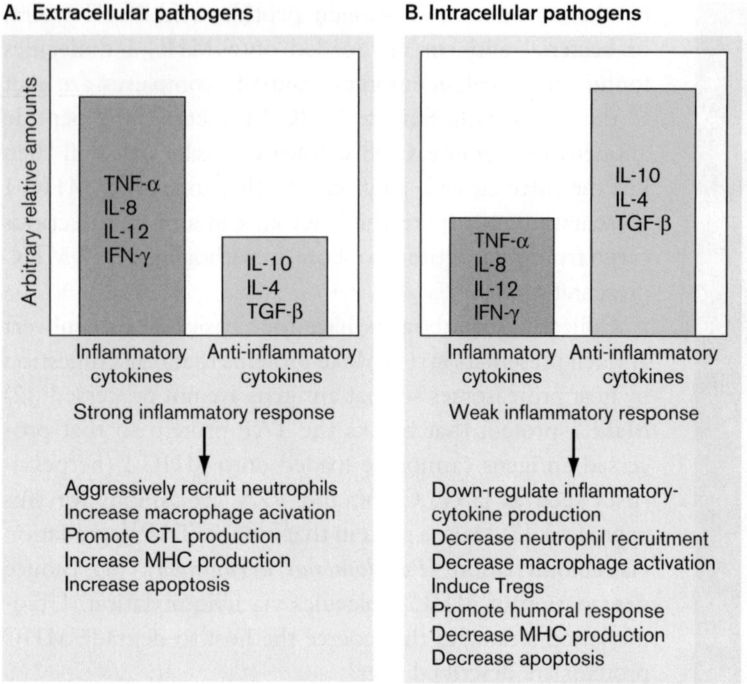

FIGURE 25.32 ■ General cytokine profiles for bacterial pathogens. The cytokine secretion profile during infection generally differs in extracellular pathogens **(A)**, compared to intracellular pathogens **(B)**. Not all cytokines are shown, and not all pathogens follow this scheme exactly. Extracellular pathogens generally permit high-inflammatory cytokine production, whereas intracellular pathogens promote anti-inflammatory cytokine production. Bars are not meant to be quantitative—only relative. CTL = cytotoxic T lymphocyte; MHC = major histocompatibility complex.

other pathogens, interrupt host cell signaling pathways that activate macrophage antimicrobial mechanisms. To perform this trick, most intracellular pathogens inhibit the production of pro-inflammatory cytokines, such as TNF-alpha, IL-8, IL-12, and IFN-gamma, but encourage the production of anti-inflammatory cytokines like IL-10, TGF-beta (transforming growth factor beta), and IL-4. **Figure 25.32** summarizes the general outcomes of these flipped profiles.

Mycobacterium, *T. cruzi*, and *Leishmania* also down-regulate the expression of host membrane receptors for IFN-gamma (thereby inhibiting inflammation), and interfere with downstream regulators that activate the production of MHC class I proteins, which dulls host immune mechanisms. Reduced MHC I production means less antigen presentation, fewer activated T_H1 helper cells, and thus fewer cytotoxic T cells to attack the infected cell.

Stopping programmed cell death. A macrophage that fails to eliminate infecting pathogens will eventually give up and try to kill itself and the infecting pathogens through one of three programmed-cell-death pathways (apoptosis, necrosis, or pyroptosis). This is a last resort to clear the infection. Apoptotic cells maintain membrane integrity during apoptotic death and are engulfed by nearby phagocytes. Inflammation is not provoked, because cytokines are not released. In contrast, necrosis and pyroptosis initiate rapid inflammatory responses by secreting inflammatory cytokines. All of these mechanisms can kill the intracellular pathogen.

To "keep hope alive" and retain their intracellular niche, intracellular microbes can prevent host cell suicide by interfering with the molecular signals that initiate host death programs, or they can activate pro-survival mechanisms. Some pathogens even synthesize microbial mimics of host anti-apoptotic proteins. *Yersinia enterocolitica* and *Mycobacterium tuberculosis*, for instance, prevent pyroptotic cell death by inhibiting inflammasome formation and caspase-1 activation. Caspase-1 is a protease that initiates pyroptosis.

Some intracellular pathogens (*Trypanosoma cruzi*, *Leishmania*, and *Yersinia pseudotuberculosis*) actually tilt macrophage suicide pathways toward apoptosis to promote pathogen dissemination (recall that apoptotic cells are engulfed by other phagocytes). Some intracellular pathogens both inhibit and activate host cell death—just not at the same time. Early in infection, *M. tuberculosis* (the cause of tuberculosis) orchestrates the inhibition of host cell suicide pathways to enable the organism to grow, but then later it promotes suicide as a way to disseminate.

Neisseria gonorrhoeae uses a surprising strategy to activate apoptosis of macrophages. *N. gonorrhoeae* is mainly an extracellular pathogen that infects the mucosa of the genitourinary tract. The mucosa, however, is surveilled by resident tissue macrophages. *N. gonorrhoeae* triggers macrophage apoptosis to kill as many of these phagocytes as possible before the phagocytes engulf and kill the bacteria. The bacterial protein that triggers apoptosis is PorB, the major outer membrane porin of *N. gonorrhoeae*.

Until now, no one knew how PorB gets from the bacterium to the mitochondria. Thomas Naderer (**Fig. 25.33A**) and colleagues at Monash University in Australia recently discovered that *N. gonorrhoeae* pinches off outer membrane vesicles (OMVs), containing PorB (**Fig. 25.33B**), that enter macrophages via endocytic pathways and ultimately deliver PorB into macrophage mitochondria (OMV formation and function are discussed in Chapter 3). After OMVs make contact with mitochondria, PorB appears to insert into the outer mitochondrial membrane (remember that mitochondria evolved from a Gram-negative ancestor) and then into the inner mitochondrial membrane, causing loss of membrane potential and leakage of cytochrome *c* into the cytosol, with the latter triggering apoptosis. **Figure 25.33C** shows the characteristic blebbing of apoptotic cells 15 hours after macrophages were mixed with *N. gonorrhoeae* OMVs. This is the first report of a membrane protein identified to target mitochondria, but others are believed to exist.

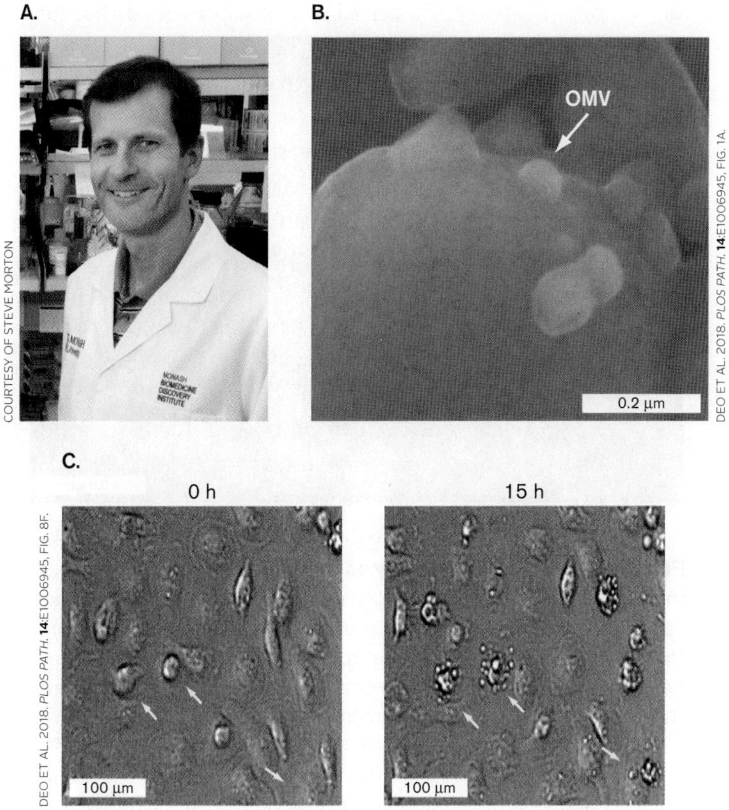

FIGURE 25.33 ▪ *Neisseria gonorrhoeae* OMVs trigger macrophage apoptosis. **A.** Thomas Naderer studies the pathogenesis of *N. gonorrhoeae*. **B.** Scanning EM of OMV formation by *N. gonorrhoeae*. The white arrowhead points to a budding OMV. **C.** Macrophages undergo apoptotic cell death, noted by the blebbing (arrows), after 12–15 hours of contact with OMV (fluorescence microscopy).

Autophagy is a highly regulated surveillance mechanism by which eukaryotic cells form vesicles around damaged organelles to scavenge them for nutrients. One of the discoverers of autophagy, Yoshinori Ohsumi, won the 2016 Nobel Prize in Physiology or Medicine. Autophagy is also used as a universal innate defense mechanism to fight intracellular pathogens (**Fig. 25.34A**). Autophagic vacuoles (autophagosomes) can encase these pathogens and deliver them to degradative lysosomes for destruction. Pathogen components are then sent to endosomes, where microbial structures are recognized by endosomal Toll-like receptors that trigger the innate immune system (see Chapter 23). The microbial components (antigens) are also sent to cell compartments rich in major histocompatibility complex II (MHC II) molecules. MHC II molecules then rise to the cell surface and present the microbial antigens to the adaptive immune system as described in Chapter 24.

Intracellular pathogens, however, have evolved mechanisms that can prevent autophagy and prolong their survival. For example, the Nef protein of human immunodeficiency virus (HIV) and protein M2 of influenza prevent autophagosome formation by targeting beclin 1, a protein central to autophagosome production. RNA viruses, as a group, encode proteins that can interact with 35% of autophagy-associated proteins, suggesting that autophagy is widely targeted by pathogens.

Shigella avoids autophagy by motoring through host cytoplasm using actin tails (described earlier). This "keep moving" tactic may sometimes work, but the host can stop the bacterium from making actin tails by wrapping *Shigella* in septin filaments, as shown in the 3D rendering in **Figure 25.34B**. Septin cages initially require actin to form, but then they inhibit actin polymerization. The septin-caged microbe is now trapped and marked for autophagy.

Redirecting host ubiquitylation signals. One important tool deployed by many pathogens forces the host to destroy or inactivate its own immune system regulators. The result is misdirection of the immune system. Pathogens accomplish this misdirection by taking over a ubiquitous host protein modification system—namely, ubiquitylation (ubiquitination). Ubiquitin is a highly conserved, 76-amino-acid polypeptide in eukaryotes that can be covalently attached to other proteins. Attachment involves an enzymatic cascade of three enzymes: E1, E2, and E3 (reviewed in **eTopic 8.6**). Depending on where ubiquitin is placed on a protein, the protein can be activated or, alternatively, tagged for destruction by the host proteasome. While there are only a few E1 and E2 enzymes, there are hundreds of E3 ubiquitin ligase enzymes that recognize and tag target proteins with polyubiquitin. There are also deubiquitylation enzymes that can reverse the process.

In contrast to their eukaryotic hosts, viral and bacterial pathogens lack ubiquitylation systems but have evolved E3 ligases and deubiquitylases that effectively subvert normal host ubiquitylation pathways (**Fig. 25.35**). Like someone changing signs on a highway, the viral enzymes cause host signaling systems, and the immune system, to veer off course.

There are several ways that ubiquitylation normally directs the immune system. In the innate immune system, TLR pathway regulators become ubiquitylated and activated after a TLR encounters a MAMP. Simultaneously, proteins that inhibit the TLR pathway are marked by ubiquitin for destruction. The result is an activated signal pathway that leads to the formation of inflammatory cytokines (**Fig. 25.35**, step 1).

To subvert this system, some pathogens produce their own E3 ligases that divert the normal signal induction pathways. For example, rotavirus, a major cause of infant diarrhea, produces an E3 ligase that adds ubiquitin to activators of NF-kappaβ, the transcription regulator that induces cytokine production (**Fig. 25.35**, step 2). The ubiquitylated

A.

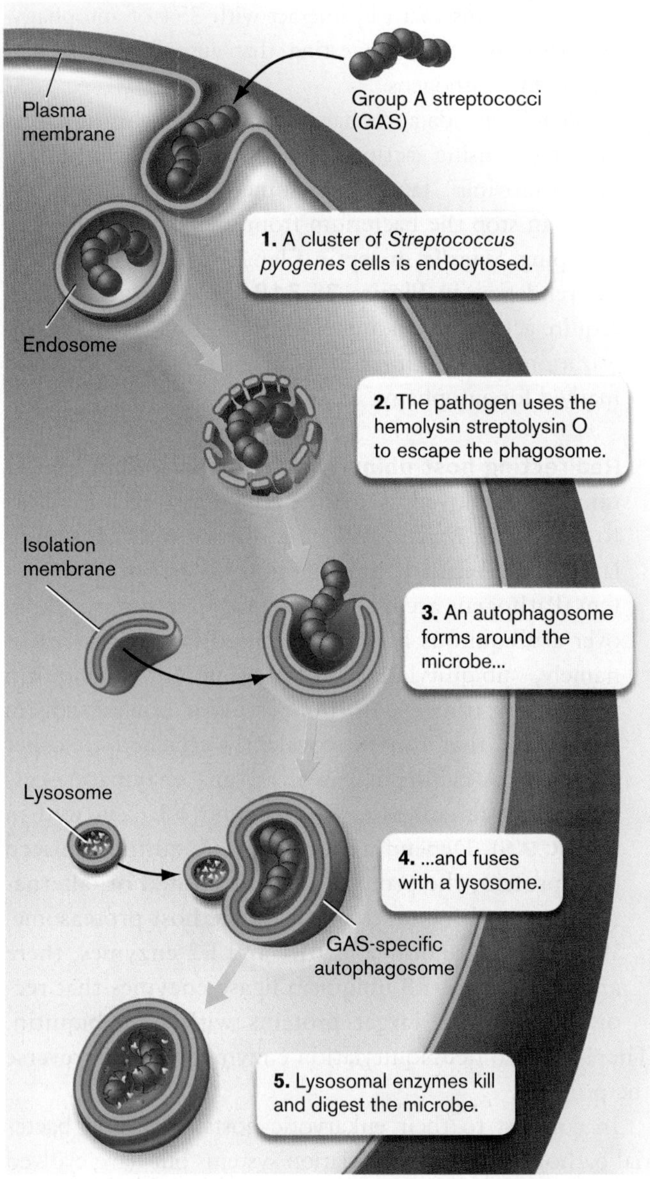

B.

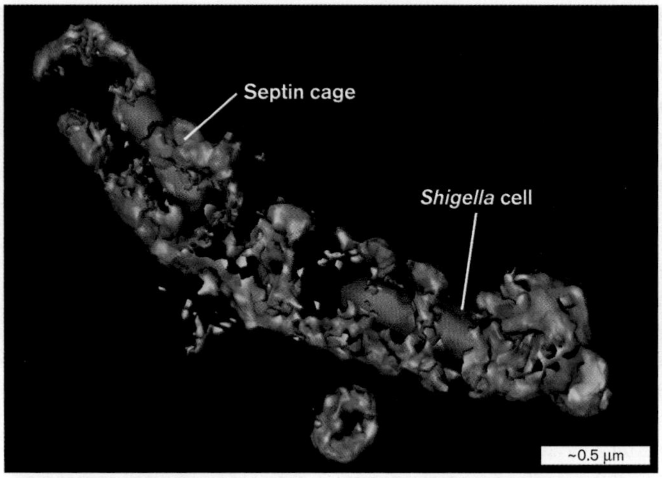

SERGE MOSTOWY ET AL. 2010. *CELL HOST MICROBE* **8**(5):433–444.

FIGURE 25.34 ■ Autophagy as an innate immune mechanism. **A.** When a pathogen escapes the phagosome, the host cell will try to form an intracellular vacuole (autophagosome) around the organism in a second attempt to kill it. **B.** *Shigella* may move in the host cytoplasm by actin tails to escape autophagy, or it can become trapped by a septin cage and succumb to autophagy.

proteins are destroyed, and NF-kappaβ is not activated. Alternatively, the bacterial pathogen *Salmonella enterica* produces a deubiquitylase that removes ubiquitin from a normal inhibitor of NF-kappaβ. Because the inhibitor is not degraded, NF-kappaβ is not activated, and inflammatory cytokines are not made (step 3). In both instances, inflammation is minimized, allowing the pathogen to survive.

Ubiquitylation is an important mechanism for adaptive immunity too. For example, the process determines when cell-surface MHC class I and II molecules are expressed in dendritic cells (**Fig. 25.36**). Before dendritic cells mature, MHC molecules are polyubiquitylated and degraded, limiting their placement on cell surfaces. After maturation, however, the MHC molecules are no longer ubiquitylated, and as a result they are not degraded. The MHC molecules can now accumulate on the cell surface to present antigens to T cells. A number of viruses exploit this process by producing viral E3 ligases that polyubiquitylate MHC proteins, marking them for destruction (**Fig. 25.36**). The result is fewer MHC molecules that are able to present viral proteins to the immune system.

Some pathogens even make E3 ligases that ubiquitylate host proteins that are not normally tagged. One example is found in the interferon signal cascade. Interferons are secreted by virus-infected host cells as a signal to protect nearby uninfected cells from virus infection. Paramyxoviruses such as mumps and measles viruses produce E3 ligases that ubiquitylate key regulatory components (JAK-STAT) of the interferon signal cascade, which marks them for destruction. The host cell then becomes quite vulnerable to virus attack.

These examples demonstrate that ubiquitylation is a crucial part of immune regulation and a popular target of pathogens. Gaining an understanding of the ways that ubiquitylation directs the immune system and how pathogens influence that direction could lead to new ways of enhancing host defense.

Thought Question

25.10 Can antibodies against a viral E3 ligase stop an infection by that virus? Why or why not? What might be an alternative strategy that does not involve antibodies?

Salmonella Is the Very Model of a Major GI Pathogen

When looking for a pathogen that employs the widest array of the virulence mechanisms just discussed, you can't go wrong with the Gram-negative enteric pathogen *Salmonella*

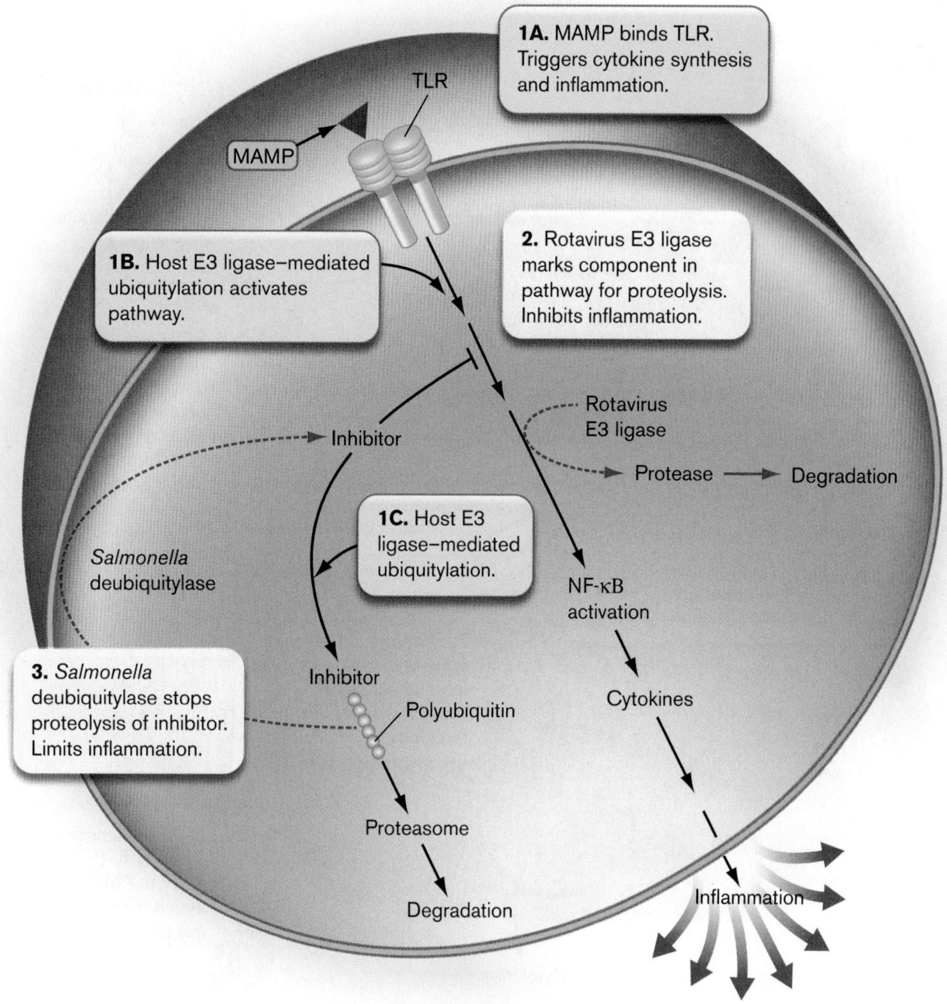

FIGURE 25.35 ■ Microbial E3 ligases and deubiquitylases alter innate immune systems. When a MAMP binds to a Toll-like receptor (TLR; 1A), a series of events, including a host E3 ligase–mediated ubiquitylation (1B and 1C), activates the transcriptional regulator NF-kappaβ, which activates transcription of inflammatory cytokine genes. The cytokines are secreted and initiate inflammatory processes. Microbial E3 ligases (for example, rotavirus E3 ligase) can ubiquitylate other components of the NF-kappaβ activation pathway and mark them for destruction (2). As a result, cytokine synthesis is inhibited and inflammation is limited. *Salmonella* makes a deubiquitylase that removes polyubiquitin from an inhibitor of NF-kappaβ activation (3). Deubiquitylation saves the inhibitor from destruction, enabling continued inhibition of NF-kappaβ activation.

enterica serovar Typhimurium. *Salmonella enterica* (a facultative intracellular, Gram-negative pathogen) is currently the most common food-borne bacterial pathogen in the United States, causing approximately 1.2 million cases of diarrhea per year. Outbreaks in 2018 were linked to precut melons, dried coconut, poultry (eggs or chicken), dried cereals, snack crackers, and even turtles. Here we describe *S.* Typhimurium as a model of bacterial pathogenesis.

Following ingestion of contaminated food or water, *S.* Typhimurium must survive passage through the acidic stomach and establish a presence in the small intestine. Attachment to intestinal epithelial cells involves numerous pili and nonpilus adhesins. As part of its pathogenesis, *Salmonella* uses type III secretion systems to invade the eukaryotic host cell and replicate intracellularly. The bacterium

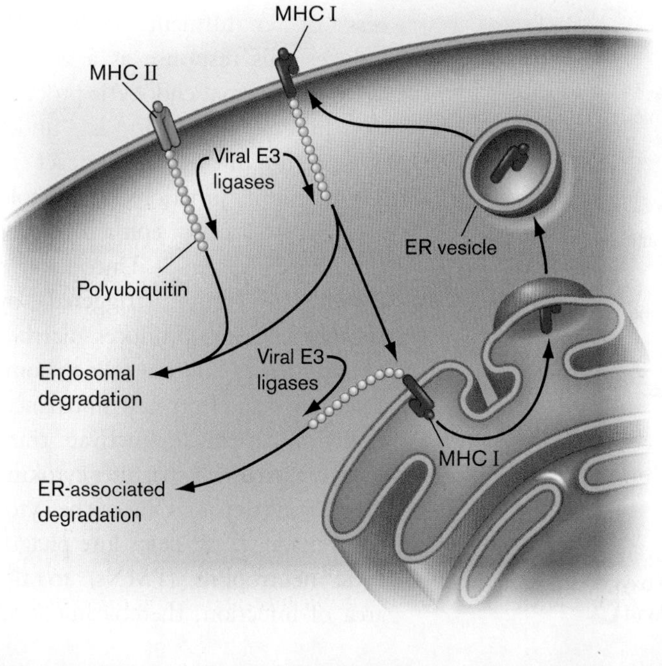

FIGURE 25.36 ■ Microbial E3 ligases and deubiquitylases alter adaptive immune systems. MHC class I and class II molecules present antigens on host cell surfaces to helper T cells. Some viral E3 ligases can ubiquitylate the MHC molecules, marking them for degradation via endosomal pathways. Other E3 ligases can ubiquitylate MHC I molecules while in the endoplasmic reticulum (ER), leading to degradation of the MHC I receptors before they can be placed on the cell surface.

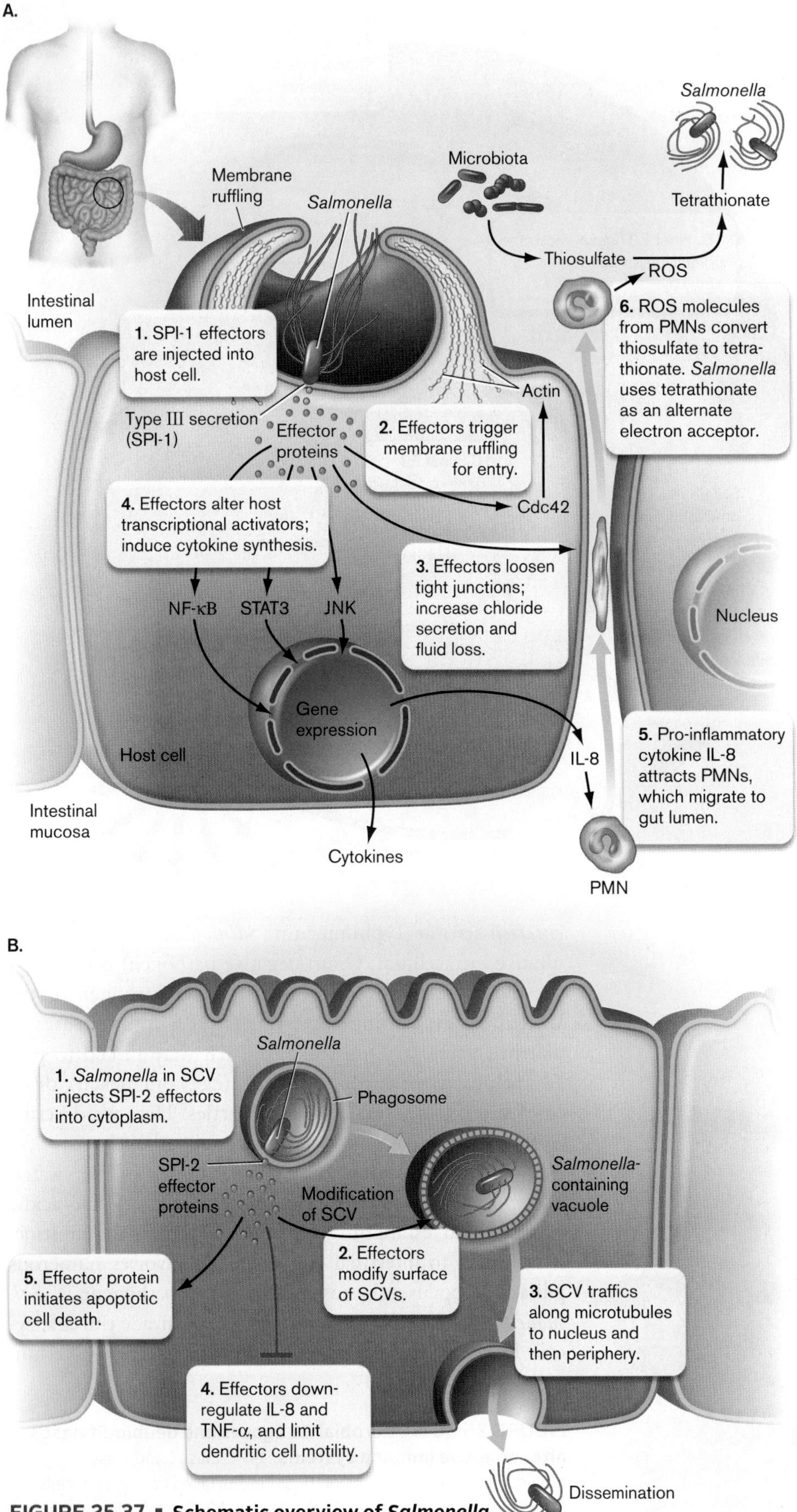

FIGURE 25.37 ■ Schematic overview of *Salmonella* pathogenesis. Effector proteins injected by *Salmonella* into a host M cell affect many aspects of host physiology. **A.** SPI-1 effector functions. **B.** SPI-2 effector functions. PMN = polymorphonuclear leukocyte; ROS = reactive oxygen species; SCV = *Salmonella*-containing vacuole; SPI = *Salmonella* pathogenicity island.

primarily attaches to and invades M cells that are interspersed along the wall of the small intestine. M cells are specialized intestinal epithelial cells (see Fig. 23.27B) that sample normal intestinal microbes and transfer pathogens across the epithelial barrier for recognition by the immune system. *Salmonella* subverts the normal function of M cells and causes an inflammatory response that leads to diarrhea.

During its evolutionary journey toward becoming a pathogen, *Salmonella* has acquired as many as 23 pathogenicity islands. Five of those islands can be found in all *S. enterica* serovars, but only two will be discussed here. *Salmonella* pathogenicity island 1 (SPI-1) encodes a type III protein secretion system that delivers a cocktail of at least 13 different protein toxins (called effector proteins) directly into the cytosol of host epithelial cells in the gut (**Fig. 25.37A**, step 1). Inside epithelial cells, these effector proteins interfere with signal transduction cascades and modulate the host response. One mission of these effectors is to induce cytoskeletal actin rearrangements that cause ruffling of the eukaryotic membrane around the microbe (step 2; also see **Fig. 25.25C**). The membrane ruffling starts the process of engulfment. *Salmonella* induces this response as a way to avoid the normal endocytic process.

While engulfment is under way, some SPI-1 effectors act to loosen tight junctions that hold together adjacent epithelial cells (**Fig. 25.37A**, step 3). Uncontrolled chloride secretion (triggered by another effector) produces diarrhea as water leaves infected cells to compensate for an electrolyte imbalance. Other SPI-1 effectors activate transcription pathways that alter cytokine expression (step 4). One of the cytokines made, IL-8, helps lure phagocytic neutrophils (PMNs) to the area of infection, thereby initiating inflammation (step 5). Another effector is an E3 ligase that ubiquitylates and activates host proteins that induce interferon-beta expression, once again enhancing inflammation.

Initiating inflammation may seem counterintuitive, but *Salmonella*'s strategy is to employ the neutrophils as "mercenaries" to kill competitors among the microbiota. The neutrophils squeeze between epithelial cells to reach the gut lumen, where they begin to engulf native members of the microbiome and produce reactive oxygen species (ROS) as part of their oxidative burst (see Chapter 23). Besides killing competitors, the oxidative burst converts thiosulfate, a compound made by gut microbiota, into tetrathionate, which *Salmonella*, but not its competitors, can use as an alternative electron acceptor (**Fig. 25.37A**, step 6). Tetrathionate provides a competitive advantage to *Salmonella* growing in the lumen.

Once *Salmonella* enters an epithelial cell or a macrophage, it finds itself in a vacuole called the *Salmonella*-containing vacuole, or SCV (**Fig. 25.37B**, step 1). SPI-1 is down-regulated in this environment. In the normal course of events, an enzyme-packed lysosome would fuse with the phagosome and release its contents in an effort to kill the invader. *Salmonella*, however, possesses a second pathogenicity island, called SPI-2, which subverts this host response. SPI-2 is induced in the vacuole and uses another type III secretion system to inject effector proteins that alter vesicle remodeling (step 2) and vesicle trafficking (step 3), thereby reducing phagosome-lysosome fusion so that the intracellular bacteria are spared.

Some SPI-2 effectors, one of which is the bacterial deubiquitylase, down-regulate production of IL-8 and TNF-alpha to limit inflammation; other effectors inhibit migration of dendritic cells, which will subvert antigen presentation (**Fig. 25.37B**, step 4). *Salmonella* eventually escapes from the initial host cell when the SCV traffics to the cell periphery and fuses with the host membrane. In addition, some SPI-2 effectors will trigger host cell apoptosis (step 5), a programmed cell death that kills the host cell in a way that prevents further inflammation. Freed from the initial cell, *Salmonella* can now infect other cells, including macrophages. The combined effect of SPI-1 and SPI-2 effectors is the leakage of fluid and blood through a damaged intestinal lining and into the intestinal lumen to produce diarrhea that contains varying amounts of blood.

This is but one model of pathogenesis. Realize that every pathogen has a different but equally fascinating story to tell.

To Summarize

- **Two-component signal transduction systems** can regulate virulence gene expression in response to the host environment.
- **Quorum sensing** may prevent pathogens from releasing toxic compounds too early during infection.
- **Extracellular pathogens** evade the immune system by hiding in capsules, by changing their surface proteins, or by triggering apoptosis.
- **Intracellular bacterial pathogens** attempt to avoid the immune system by growing inside host cells. They use different mechanisms to avoid intracellular death.
- **Inhibiting phagosome-lysosome fusion** is one way that pathogens can survive in phagosomes.
- **Hemolysins** are used by certain pathogens to escape from the phagosome and grow in the host cytoplasm.
- **Actin tails** are used by some microbes to move within and between host cells.
- **Molecular mechanisms for avoiding the immune system** include molecular mimicry, altering cytokine profiles, stopping programmed host cell death, interfering with autophagy, and redirecting ubiquitylation signals.
- **Specialized physiologies** enable some organisms to survive in the normally hostile environment of fused phagolysosomes.

25.6 Tools Used to Probe Pathogenesis

For any scientist studying pathogenic microbes, it is essential to identify virulence genes and determine what they do. As mentioned in Section 25.1, Stanley Falkow's molecular Koch's postulates argue that once a gene has been singled out as a possible virulence factor, proof requires knocking out the gene and observing a decrease in virulence, followed by restoring virulence by replacing the mutant gene. But how do you identify which genes to test? Also in Section 25.1 (see **eTopics 25.1** and **25.2** as well), we discussed how we can identify virulence genes in microbes by using some clever selection techniques. Today there exist amazing molecular techniques that can broadly, and quickly, reveal a microbe's overall pathogenic strategy and help scientists focus on likely virulence genes. In addition, powerful techniques can now deeply probe how pathogens and hosts respond to each other during an infection.

Genomics

The sequencing of a pathogen's genome, followed by bioinformatic analysis of the sequence (both techniques described in Chapter 12), can yield valuable information about a pathogenic organism's metabolism and can identify potential pathogenicity islands and virulence genes.

We can gain clues about potential virulence genes also by comparing the sequences of virulent and attenuated strains of a pathogen. For example, *Leptospira* species are spirochetes that cause a zoonotic renal disease in humans called leptospirosis. Dereck Fouts (J. Craig Venter Institute), Joseph Vinetz (UC San Diego), and colleagues compared the genome sequences of numerous *Leptospira* species differentially classified as pathogenic, intermediate, and nonpathogenic. The distributions of several genes were restricted to the pathogenic strains, suggesting involvement in *Leptospira* pathogenesis or host adaptation. The pathogen-unique genes included, among others, a catalase, a protease able to degrade complement, a large family of virulence-modifying proteins of unknown function, and a CRISPR-Cas system. The results provided many new directions for research that probes leptospiral pathogenesis.

DNA sequencing can also reveal how to grow so-called obligate intracellular pathogens in the laboratory. The Gram-positive microorganism *Tropheryma whipplei* is an intracellular pathogen that causes the gastrointestinal illness called Whipple's disease. The symptoms of Whipple's disease include diarrhea, intestinal bleeding, abdominal pain, loss of appetite, weight loss, fatigue, and weakness. Though identified, the causative agent had never been grown outside of fibroblast cells, making study of its physiology nearly impossible. Once the complete genome sequence of *T. whipplei* became available, however, Didier Raoult and colleagues at the Marseille School of Medicine discovered that the organism lacks the machinery to make several amino acids. Using this information, the investigators designed a cell-free culture medium that supported the growth of *T. whipplei*. Similar approaches were used to coax cell-free growth of the so-called obligate intracellular pathogen *Coxiella burnetii*. *C. burnetii* causes the respiratory infection Q fever. Genomic strategies like these should lead to other successes in growing previously unculturable intracellular pathogens.

Transcriptomics

Other major goals of infectious disease research include learning how a pathogen causes an infection, how the host responds to the infection, and how the pathogen responds to the host's response. The situation is much like watching two armies wage war, where the strategies of each side continually shift in response to each other. How can we map the similar cascading dynamics of attack and counterattack between pathogen and host?

Dual RNA sequencing. RNAseq analysis (Chapter 12) is now used to separately monitor transcript modulation in pathogens during the course of an infection and view the host's transcriptional response to a pathogen. Jörg Vogel (**Fig. 25.38A**) at the University of Würzburg and colleagues recently used RNAseq to <u>simultaneously</u> profile the changing host and pathogen transcriptomes during *Salmonella* infection of human host cells. The scientists

A.

FIGURE 25.38 ▪ Dual RNAseq captures the full transcript repertoire of *Salmonella*-infected human cells. **A.** Jörg Vogel uses dual RNAseq to study the interactions between pathogens and their hosts. **B.** Transcript profile of *Salmonella* during infection. Compared to extracellular *Salmonella* (0 hours), intracellular bacteria at 4 hours postinfection repress SPI-1 and induce SPI-2 effector genes. **C.** Transcript profile of infected host cells. Invaded (GFP^+) host cells at 24 hours postinfection activate NF-kappaβ-associated immunity genes relative to uninfected cells (GFP^-).
Source: Parts B and C modified from Alexander J. Westermann et al. 2016. *Nature* **529**:496–501, fig. 1A, C, E.

B. 4 hours postinfection

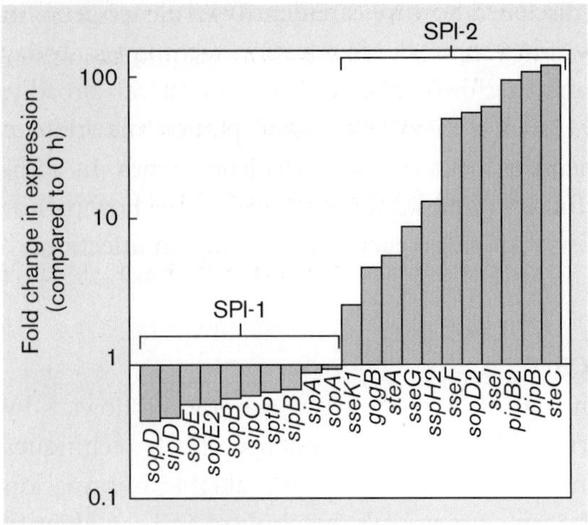

C. 24 hours postinfection

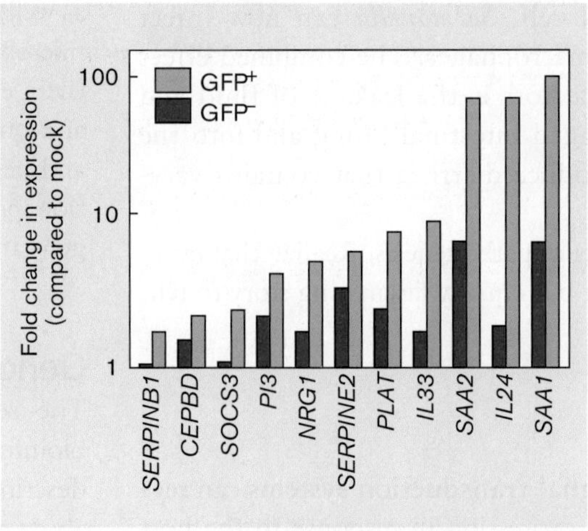

started by infecting human cells with GFP-labeled *Salmonella* and used fluorescence-activated cell sorting (FACS; see Section 4.3) to select host cells at 4 hours (10 bacteria per cell) and 24 hours (75 bacteria per cell) postinfection. Simultaneously, they used next-generation RNAseq techniques (see Section 12.1) to sequence pathogen and host cell transcripts.

Their analysis showed that 4 hours postinfection, bacterial transcripts of SPI-1 genes decreased relative to levels just prior to infection, while the transcript levels of SPI-2 genes increased (**Fig. 25.38B**). They also found that host genes activated by NF-kappaβ were elevated at 24 hours postinfection in infected cells but not in uninfected cells (**Fig. 25.38C**). NF-kappaβ is a transcriptional activator of cytokine genes. The NF-kappaβ result illustrates the host response to *Salmonella* invasion.

The scientists then identified *Salmonella* small RNA transcripts that were induced during infection. The most highly induced was the 80-nt sRNA called PinT (PhoP-induced sRNA in intracellular *Salmonella*). This sRNA was found to control the transition from SPI-1 to SPI-2 expression. By altering the timing of this transition, PinT influenced many aspects of the infected host cell's physiology, including production of cytokine IL-8 and mitochondrial gene expression.

The simultaneous examination of host and pathogen transcriptional profiles by dual-RNAseq techniques will reveal other hidden gene functions in pathogens with profound effects on host cells.

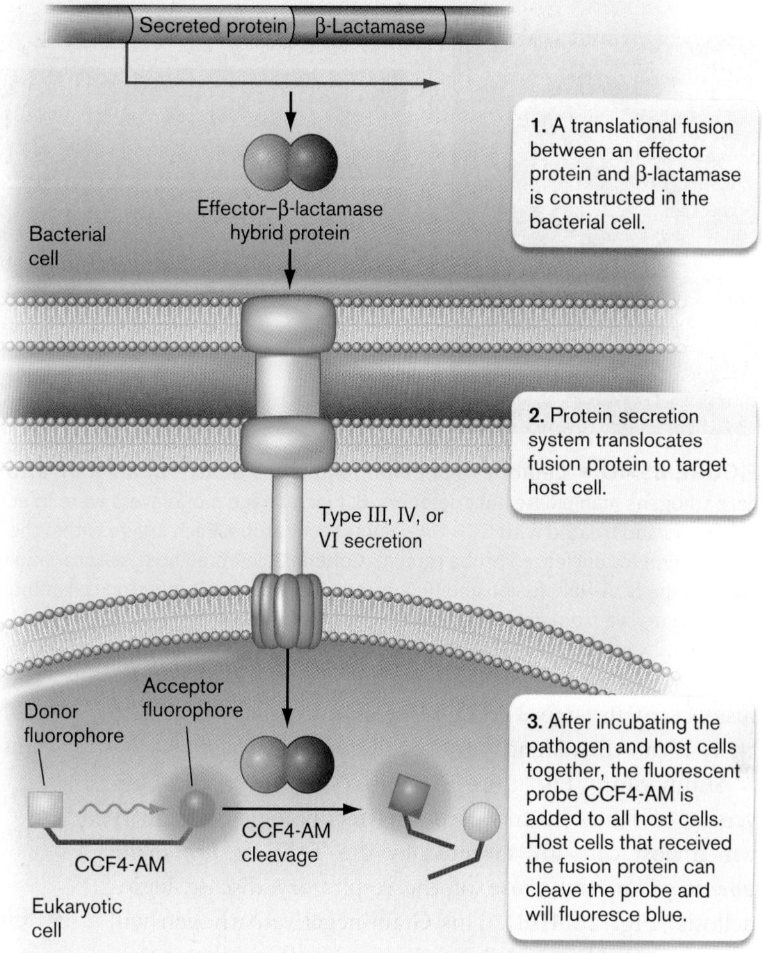
Secreted protein
β-Lactamase
Effector–β-lactamase hybrid protein
Bacterial cell
1. A translational fusion between an effector protein and β-lactamase is constructed in the bacterial cell.
2. Protein secretion system translocates fusion protein to target host cell.
Type III, IV, or VI secretion
Donor fluorophore
Acceptor fluorophore
CCF4-AM
CCF4-AM cleavage
Eukaryotic cell
3. After incubating the pathogen and host cells together, the fluorescent probe CCF4-AM is added to all host cells. Host cells that received the fusion protein can cleave the probe and will fluoresce blue.

FIGURE 25.39 ▪ FRET detection of bacterial effector protein translocation to host cells. The emission wavelength of the released donor fluorophore is detected (blue). The emission wavelength of the acceptor fluorophore (green) is different from that of the donor.

Cell Biology

You have seen numerous examples throughout this and previous chapters in which fluorescent stains, proteins (GFP), and antibodies combined with fluorescent microscopy were used to identify dramatic alterations in host cell structures, as well as the movement/location of bacterial or host cell proteins during the course of an infection. A relatively new fluorescent technology has been developed that can identify which host cells during an infection have been targeted by translocated microbial effector proteins. The affected host cells fluoresce and can be sorted by a FACS machine for closer examination.

The technique starts with a translational fusion between a gene encoding a bacterial effector protein and a beta-lactamase gene (**Fig. 25.39**, step 1; fusion proteins are described in Chapter 12). Beta-lactamase is an enzyme that cleaves the beta-lactam ring of the antibiotic penicillin (see Chapter 27). Once the effector–beta-lactamase fusion protein is expressed, it can be translocated from the pathogen directly into host cells using a type III, IV, or VI secretion system (step 2). Host cells that receive the effector protein can be identified by addition of the fluorescent reporter CCF4-AM.

CCF4-AM is composed of two fluorescent parts (fluorophores) linked by a beta-lactam ring. One fluorophore can fluoresce blue when excited, but the blue emission wavelength is absorbed (or quenched) by the second fluorophore, which will fluoresce green (**Fig. 25.39**, step 3). The process is called **fluorescence resonance energy transfer (FRET)**. However, if the beta-lactam bond linking the two molecules is cleaved by beta-lactamase, blue fluorescence is no longer quenched and can be seen. CCF4-AM can enter all host cells, but fluorophore and quencher can be separated only by host cells that received the beta-lactamase–effector

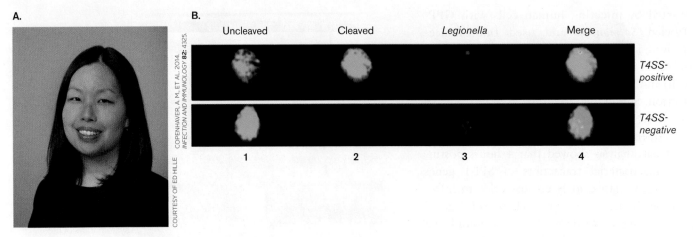

FIGURE 25.40 ■ *Legionella pneumophila* translocates BlaM-RalF into mouse macrophages. **A.** Sunny Shin studies how bacterial pathogens manipulate host defenses. **B.** Macrophage monolayers were infected with dot^+ (T4SS-positive) and Δdot (T4SS-negative) *Legionella* and treated with CCF4-AM fluorescent probe. Each image within the panels represents a single macrophage. **Column 1:** Uninfected cell scanned for uncleaved probe (green). **Column 2:** Infected host cell scanned for cleaved probe (blue). Only the T4SS-positive *Legionella* injected the BLA-RalF fusion and fluoresced blue. **Column 3:** Infected cell immunostained for *Legionella* (red). The immunostain does not differentiate live from dead bacteria. **Column 4:** Merged images from columns 1, 2, and 3.

fusion protein (step 3). Fluorescence microscopy or FACS analysis can then identify those cells.

Sunny Shin (**Fig. 25.40A**) and colleagues at the University of Pennsylvania used this technology to identify which host cells are targeted by the T4SS of *Legionella pneumophila*, the cause of the respiratory disease legionellosis (**Fig. 25.40B**). This Gram-negative pathogen will invade host cells to inhabit a phagosome. From the phagosome it uses a T4SS to send effector proteins into the host cell cytoplasm. Shin's group found that macrophages and neutrophils pulled from the airway space of infected mice were the primary recipients of T4SS-translocated effector RalF (fused to beta-lactamase) and that host cells receiving the fusion were the only cells harboring viable bacteria. **Figure 25.40B** (column 2) shows that the T4SS system (encoded by the *dot* genes) was required to translocate the BlaM-RalF fusion protein into the host cell. Only the T4SS-positive *Legionella* fluoresced blue. FRET–beta-lactamase technology can also be used to recognize mutant bacteria defective in effector secretion or host factors that contribute to effector translocation.

> **Thought Question**
>
> **25.11** How could you use the FRET–beta-lactamase technique to identify chemical inhibitors of type III, IV, or VI translocation?

To Summarize

- **Genome sequence analysis** of pathogenic and nonpathogenic strains of a species can identify potential virulence genes.

- **DNA sequencing** revealed growth requirements for the intestinal pathogen *Tropheryma whipplei*.

- **Dual RNA sequencing** of pathogen and host transcripts from infected and uninfected host cells can expose a complex, yet orchestrated, response to infection.

- **Fluorescence resonance energy transfer (FRET)** approaches can identify host cells targeted by pathogens for type III, type IV, and type VI effector protein delivery.

Concluding Thoughts

This chapter has only scratched the surface of all that is known about microbial pathogenesis. Transmission, attachment, immune avoidance, and subversion of host signaling pathways are common goals of most successful pathogens, whether bacterial, fungal, viral, or parasitic. But how these pathogens achieve those goals varies greatly, and there is much we still do not know. You will discover in reading the next chapters that many pathogens are still hard to detect and difficult, if not impossible, to kill. You will also learn that new pathogens are constantly emerging. Over the last 40–50 years, we have seen the development of Zika virus, HIV, SARS, avian flu virus, hantavirus, West Nile virus, and *E. coli* O157:H7, as well as the reemergence of flesh-eating streptococci, to name but a few examples. Can we ever stop pathogens from emerging? Probably not. For every countermeasure we develop, nature designs a counter-countermeasure. Our hope is that continued research into the molecular basis of pathogenesis and antimicrobial pharmacology will keep us one step ahead.

CHAPTER REVIEW

Review Questions

1. Describe the differences between infection and disease; pathogenicity and virulence; LD_{50} and ID_{50}.
2. What is meant by direct versus indirect routes of infection?
3. What are the characteristics of a good reservoir for an infectious agent?
4. Name the various portals of entry for infectious agents, and a disease associated with each.
5. Describe the basic features of a pathogenicity island.
6. Explain various ways in which bacteria can attach to host cell surfaces.
7. Describe the basic steps by which pili are assembled on the bacterial cell surface. How do type I and type IV pili differ?
8. Explain the nine broad categories of toxin mode of action.
9. What is ADP-ribosylation, and how does it contribute to pathogenesis?
10. Explain the differences between exotoxins and endotoxins.
11. Explain the mechanisms of secretion carried out by type II and type III protein secretion systems. What are the paralogous origins of these systems?
12. Describe the key features of *Salmonella* pathogenesis.
13. How can genomic approaches help identify pathogens in an infection?
14. What different mechanisms do intracellular pathogens use to survive within the infected host cell?
15. Describe different molecular strategies that microbes use to avoid the immune system.
16. How do bacteria determine whether they are in a host environment?
17. Discuss the relationships between ubiquitylation and intracellular pathogens.

Thought Questions

1. Using the Internet as an investigative tool, describe how you might interrupt the infection cycle of Zika virus without using chemical pesticides.
2. How would you determine whether a particular pilus on group A streptococci (GAS) is required for an organism's pathogenesis? Use a tissue culture model.
3. Why have humans not developed resistance to microbial toxins?
4. You want to make a live oral vaccine for cholera. But, because *Vibrio cholerae* is an acid-sensitive organism, the person to be immunized would have to ingest a large number of organisms. *E. coli*, on the other hand, is very acid resistant and able to survive stomach acidity for long periods of time. You think that moving the acid resistance system from *E. coli* to *V. cholerae* will solve this problem. What ethical issue should you consider when trying to move the acid resistance system from *E. coli* into *V. cholerae*?
5. You have discovered a Gram-negative lung pathogen that uses a T3SS to inject an effector protein into a subset of lung cells. Design an in vivo experiment to identify which lung cells are targeted by the pathogen, and then determine the transcriptional responses of pathogen and host, but only among the infected cells.

Key Terms

adhesin (1020)
ADP-ribosyltransferase (1030)
airborne transmission (1014)
autophagy (1047)
capsule (1041)
endotoxin (1025)
exotoxin (1025)
facultative intracellular pathogen (1042)
fecal-oral route of transmission (1016)
fimbria (1020)
fluorescence resonance energy transfer (FRET) (1053)
fomite (1014)
genomic island (1018)
hemolysin (1027)
horizontal transmission (1014)
immunopathogenesis (1016)
infection (1012)
infection cycle (1014)
infectious dose 50% (ID_{50}) (1013)
intimin (1039)
intracellular pathogen (1042)
labile toxin (LT) (1031)
latent state (1012)
lethal dose 50% (LD_{50}) (1013)
leukocidin (1027)
mechanical vector (1016)

obligate intracellular pathogen (1042)
opportunistic pathogen (1012)
parasite (1012)
parenteral route (1016)
pathogen (1012)
pathogenesis (1012)
pathogenicity (1013)
pathogenicity island (1017)
petechia (1035)
pilus (1020)
primary pathogen (1012)
protective antigen (PA) (1032)
protein A (1041)
reservoir (1016)
transovarial transmission (1015)
type II secretion system (1036)
type III secretion system (T3SS) (1038)
type IV secretion system (T4SS) (1039)
vector (1014)
vehicle transmission (1014)
vertical transmission (1014)
virulence (1013)
virulence factor (1017)
zoonotic disease (1014)

Recommended Reading

Anderson, Christopher J., and Mellissa M. Kendall. 2017. *Salmonella enterica* serovar Typhimurium strategies for host adaptation. *Frontiers in Microbiology* 8:1983. https://doi.org/10.3389/fmicb.2017.01983.

Aprianto, Rieza, Jelle Slager, Siger Holsappel, and Jan-Willem Veening. 2017. Time-resolved dual RNA-seq reveals extensive rewiring of lung epithelial and pneumococcal transcriptomes during early infection. *Genome Biology* 17:198. https://doi.org/10.1186/s13059-016-1054-5.

Basler, Marek, Martin Pilhofer, Gregory P. Henderson, Grant J. Jensen, and John J. Mekalanos. 2012. Type VI secretion requires a dynamic contractile phage tail-like structure. *Nature* 483:182–186.

Campbell, Tessa Mollie, Brian Patrick McSharry, Megan Steain, Tiffany Ann Russell, David Carl Tscharke, et al. 2019. Functional paralysis of human natural killer cells by alphaherpes viruses. *PLoS Pathogens* 15(6):e1007784.

Cianciotto, Nicholas P., and Richard C. White. 2017. Expanding role of type II secretion in bacterial pathogenesis and beyond. *Infection and Immunity* 85:e00014-17.

Deo, Pankaj, Seong H. Chow, Iain D. Hay, Oded Kleifeld, Adam Costin, et al. 2018. Outer membrane vesicles from *Neisseria gonorrhoeae* target PorB to mitochondria and induce apoptosis. *PLoS Pathogens* 14:e1006945. https://doi.org/10.1371/journal.ppat.1006945.

Didelot, Xavier, A. Sarah Walker, Tim E. Peto, Derrick W. Crook, and Daniel J. Wilson. 2016. Within-host evolution of bacterial pathogens. *Nature Reviews. Microbiology* 14:150–162.

Fouts, Derrick E., Michael A. Matthias, Haritha Adhikarla, Ben Adler, Luciane Amorim-Santos, et al. 2016. What makes a bacterial species pathogenic?: Comparative genomic analysis of the genus *Leptospira*. *PLoS Neglected Tropical Diseases* 10:e0004403.

Galán, Jorge E., and Gabriel Waksman. 2018. Protein-injection machines in bacteria. *Cell* 172:1306–1318.

Golovkine, Guillaume, Emeline Reboud, and Philippe Huber. 2018. *Pseudomonas aeruginosa* takes a multi-target approach to achieve junction breach. *Frontiers in Cellular and Infection Microbiology* 7:532. https://doi.org/10.3389/fcimb.2017.00532.

Huibregtse, Jon, and John R. Rohde. 2014. Hell's BELs: Bacterial E3 ligases that exploit the eukaryotic ubiquitin machinery. *PLoS Pathogens* 10:e1004255.

Keilberg, Daniella, and Karen M. Ottermann. 2016. How *Helicobacter pylori* senses, targets and interacts with the gastric epithelium. *Environmental Microbiology* 18:791–806. https://doi.org/10.1111/1462-2920.13222.

Pechous, Roger D., and William E. Goldman. 2015. Illuminating targets of bacterial secretion. *PLoS Pathogens* 11:e1004981. https://doi.org/10.1371/journal.ppat.1004981.

Santos, José Carlos, and Jost Enninga. 2016. At the crossroads: Communication of bacteria-containing vacuoles with host organelles. *Cellular Microbiology* 18:330–339. https://doi.org/10.1111/cmi.12567.

van de Weije, Michael L., Rutger D. Luteijn, and Emmanuel J. H. J. Wiertz. 2015. Viral immune evasion: Lessons in MHC class I antigen presentation. *Seminars in Immunology* 27:125–137.

Zhang, Wenchao, Xiaofeng Jiang, Jinghui Bao, Yi Wang, Huixing Liu, et al. 2018. Exosomes in pathogen infections: A bridge to deliver molecules and link functions. *Frontiers in Immunology* 9:90. https://doi.org/10.3389/fimmu.2018.00090.

CHAPTER 26

Microbial Diseases

26.1 Skin, Soft-Tissue, and Bone Infections

26.2 Respiratory Tract Infections

26.3 Gastrointestinal Tract Infections

26.4 Genitourinary Tract Infections

26.5 Cardiovascular and Systemic Infections

26.6 Central Nervous System Infections

The impact of infectious disease on the world's population is staggering. Combining the number of infection-related deaths of immunocompetent individuals with those of people immunocompromised by conditions such as cancer, smoking, or drug abuse, makes infectious disease the leading contributor to death in the world. There are over 1,400 different species of viruses, bacteria, fungi, and protozoa that infect humans. In Chapter 26 we explore the major types and etiologies of infections and introduce the art of diagnosis.

CURRENT RESEARCH highlight

For *Cryptococcus*, size matters. *Cryptococcus neoformans*, an encapsulated fungus, causes meningitis in immunocompromised individuals. Once inhaled, some cells transition from small, haploid yeast cells to huge, polyploid cells called Titans. Titans produce thick capsules, alter PAMPs, and become antibiotic resistant, all of which enables dissemination to the brain. The signal(s) that initiate this striking morphogenesis are unclear. Elizabeth Ballou and colleagues discovered that Titanization begins when the fungus encounters muramyl dipeptide (a bacterial peptidoglycan subunit). They suggest that Titanization is triggered in vivo when *C. neoformans* and the lung microbiome interact—a finding that may impact efforts to control this disease.

Source: Dambuza et al. 2018. *PLoS Pathog.* **14**(5):e1006978

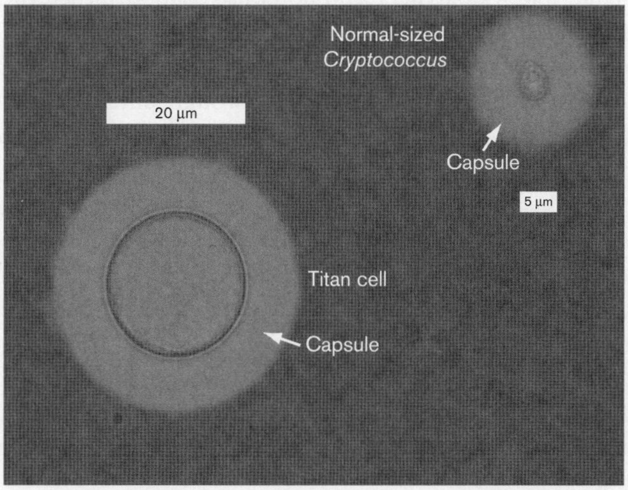

MODIFIED FROM DAMBUZA ET AL. 2018. *PLOS PATHOG.* **14**(5):E1006978, FIG. 1A.

AN INTERVIEW WITH

ELIZABETH R. BALLOU, FUNGAL GENETICIST, UNIVERSITY OF BIRMINGHAM, UNITED KINGDOM

COURTESY OF LLEWELYN MORON

Why are undergraduate and graduate students crucial to your research?

Sometimes discoveries happen by accident. For instance, I asked a new student, Tom Drake, to grow *Cryptococcus* under a new condition and then add serum to trigger capsule formation—a classically stunning response to a host condition. However, when Tom looked at the result, he knew something was wrong. Rather than discard the experiment, he asked me to peek. And there they were . . . Titan cells; something previously seen only in patient lungs. This young novice had discovered a new technique to probe the pathogenesis of *Cryptococcus*.

Infectious diseases have challenged human existence for millions of years. Our tenacity as a species is due in large part to our immune system, but also to our growing ability to understand infectious diseases. The Current Research Highlight, for instance, describes how Elizabeth Ballou and two students in her lab—Tom Drake (**Fig. 26.1**) and Ivy Dambuza (photo unavailable)—discovered an interaction between *Cryptococcus neoformans* and the lung microbiome that enhances the yeast's pathogenicity and exposes a potential target for intervention.

Our knowledge of infectious diseases has also enhanced our capacity to diagnose, treat, and prevent known infections, and to identify newly emerging ones. An example of the latter is Tatjana Avšič Županc and neuropathologist Mara Popović's discovery of a viral genome in the brain of a microcephalic fetus in 2015, which provided a direct link between Zika virus and emerging microcephaly in newborns.

In Chapter 26 we discuss a wide variety of infectious microbes by describing their routes of infection, the pathologies they cause, and the symptoms that result. We also examine prevention strategies such as vaccinations when available. Antimicrobial treatments are surveyed in Chapter 27, while laboratory diagnostics and epidemiological considerations are examined in Chapter 28.

What we present in this chapter is not an exhaustive compendium of microbial illnesses, but a representative sampling of infections that illustrate key aspects of microbial disease. The pathogens selected are presented as a clinician views them, by the organ system affected. The advantage of this strategy is diagnostic. When examining a sick patient, a clinician must first determine which organ system is affected and then mentally sort through the pathogens known to affect that system. Taking this approach will help you better integrate your knowledge of microbiology and immunology within the framework of the practice of medicine.

TOM DRAKE

FIGURE 26.1 ■ **Tom Drake.**

So, how is a diagnosis made? It is not a simple matter of taking a specimen, ordering a test, and prescribing a drug. The clinician first needs to figure out which microbes, out of thousands, are possible causes. When you go to the clinic, what does a clinician first ask? "What brings you here today?" followed by "How long have you had these symptoms?" and perhaps "Where have you traveled recently?" This is not idle conversation; the clinician is taking a patient history.

Because many infectious diseases display similar symptoms, a patient history can provide clues about the likely culprit. For example, *Vibrio cholerae* and enterotoxigenic *E. coli* both produce diarrheal diseases characterized by cramps, lethargy, and liters of watery stool each day. Here is where a good patient history can make all the difference. Cholera is not commonly seen in the United States, but a clinician might suspect cholera if the patient recently traveled to, or emigrated from, an endemic area where cholera is regularly observed. Clinicians learn this information by talking with the patient and asking about their recent travels .

Many questions that the clinician asks while taking a patient's social history can seem irrelevant or intrusive to the patient, who only wants relief from the symptoms. For instance: Do you have any hobbies? What is your occupation? What foods have you eaten recently? Does your ill child attend day care? Has anyone in your family had similar symptoms? All of those questions address possible sources of the infectious agent. Other questions can reveal high-risk behavior that can lead to certain infections. Do you smoke, drink, or take recreational drugs? Have you had multiple sex partners? Do you use contraception? If a woman seeking help for abdominal pain reveals that she has multiple sex partners who don't use condoms, then the **differential**, the list of possible causes of her symptoms, should include gonorrhea, syphilis, and chlamydia. A good patient history combined with a thorough physical exam helps the clinician decide which further tests to order and which procedures to perform.

Because patient histories are so helpful in diagnosing infectious diseases, we use case histories throughout this chapter to segue into discussions of the various microbes that can infect each organ system. Key aspects of infectious diseases will emerge as we proceed.

But why do the symptoms and severity of a disease differ between patients? People infected with the same pathogen can display somewhat different symptoms. For

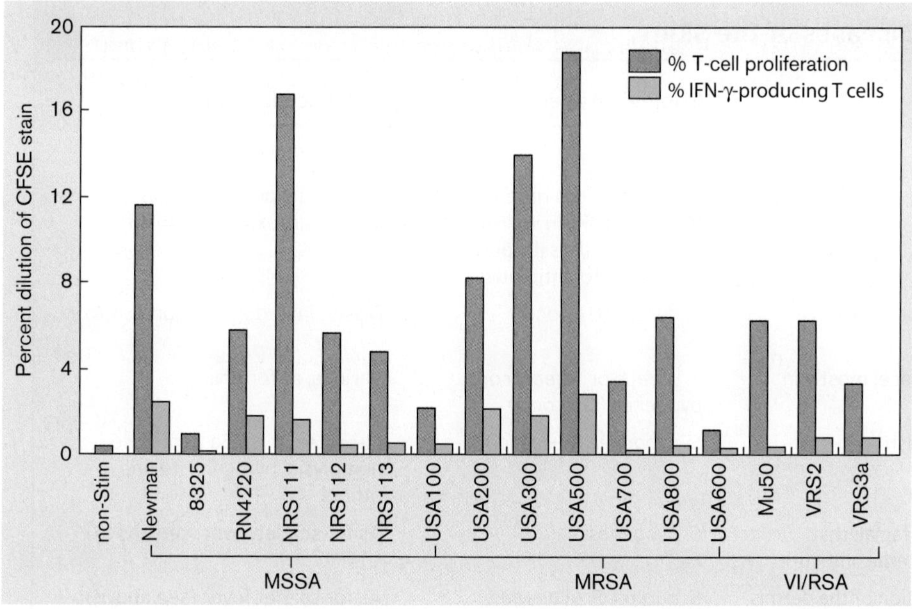

FIGURE 26.2 ▪ The immune response of a host varies with pangenomic differences in *Staphylococcus aureus* strains. Lymphocytes from a single host were tested in vitro for the ability to respond to 16 different heat-killed strains of *S. aureus* (designated by number or name along the *x*-axis). MSSA = methicillin-sensitive *S. aureus*; MRSA = methicillin-resistant *S. aureus*; VRSA = vancomycin-resistant *S. aureus*. T-cell proliferation was determined by staining T cells with CFSE (carboxyfluorescein succinimidyl ester) stain and monitoring dilution of the stain per cell as the cells proliferated. Every time a cell containing stain replicated, the stain was divided (diluted) into the two new cells. T cells that produced IFN-gamma were identified by fluorescent antibody followed by FACS analysis. non-Stim = T cells not stimulated by heat-killed *S. aureus*. *Source:* Modified from Sela et al. 2018. *PLoS Pathog.* **14**:e1006726, fig. 1A.

instance, some patients infected with measles virus develop muscle pain; other patients do not. A small number of patients infected with measles virus develop life-threatening encephalitis, but most do not. Part of the explanation is immunocompetence. Those with more effective immune systems can better defend themselves against the pathogen. As a result, some infected people may not develop any symptoms. However, genetic differences in immunocompetence are only part of the story.

Uri Sela (Rockefeller University) and colleagues recently discovered that pangenomic gene differences between different strains of a pathogen will also alter a patient's immune response (the concept of pangenome versus core genome was introduced in Chapter 17). In one experiment the scientists examined 16 different heat-killed *Staphylococcus aureus* strains (clinical isolates) for their abilities to stimulate T-cell proliferation and interferon-gamma (IFN-gamma) production in vitro using lymphocytes from a single volunteer (**Fig. 26.2**). T-cell proliferation and IFN-gamma production are markers of the immune response to infection (Chapter 24). Even though the bacteria were dead, the different strains stimulated different levels of T-cell proliferation and IFN-gamma production. That is, a single host displayed a heterogeneous response to the different strains of *S. aureus*. This finding and other data indicate that the final immune response of a host is due as much to interstrain (pathogen) variability as it is to interindividual (host) variability. Both of these parameters can complicate diagnosis, treatment, and patient outcome.

Keep these aspects of infectious diseases in mind as we describe the infections that plague each organ system.

26.1 Skin, Soft-Tissue, and Bone Infections

Bones, skin, and soft tissues such as muscle and underlying connective tissues collectively shape and maintain body architecture. Because these structures are interconnected, infections that threaten the integrity of bones, skin, and soft tissues will be discussed together. Infections that affect one of these tissues can also spread directly to an adjacent tissue.

Skin and soft-tissue infections range from simple boils to severe, complicated, so-called flesh-eating diseases that can be caused by a variety of bacteria, fungi, and viruses (see **Table 26.1**). Recall that the integrity of the skin, as well as the presence of normal skin microbiota, prevents most infections. However, even minor insults to the skin (such as a paper cut) can result in infections, most of which are caused by the Gram-positive pathogen *Staphylococcus aureus*. Healthy individuals develop infections of the skin only rarely, whereas people with underlying immunosuppressive diseases, such as diabetes, are at much higher risk.

Boils

Staphylococcus aureus (**Fig. 26.3A**) is a common cause of painful skin infections called boils or furuncles. This Gram-positive organism, often a normal inhabitant of the nares (nostrils), can infect a cut or gain access to the dermis via a hair follicle. It possesses a number of enzymes that contribute to disease, including coagulase, which helps coat the organism with fibrin, thereby walling off the infection from the

TABLE 26.1 Common Infectious Diseases of the Skin

Disease	Symptoms	Etiological agent(s)[a]	Virulence factors
Bacterial			
Folliculitis	Boils	*Staphylococcus aureus* (G+ cocci); fibrin wall around abscess renders it poorly accessible to antibiotics	Coagulase, protein A, TSST, leukocidin, exfoliative toxin
Scalded skin syndrome	Peeling skin on infants, systemic toxin	*S. aureus* (G+ cocci)	Disseminated exfoliation exotoxin
Impetigo	Skin lesions on the face, mostly in children	*S. aureus* or *Streptococcus pyogenes* (G+ cocci)	Various, as for boils
Scarlet fever	Sore throat, fever, rash	*S. pyogenes* (G+ cocci)	M-protein pili, C5a peptidase, hemolysin, pyrogenic toxins, others
Erysipelas	Skin lesions, usually facial, that spread to cause systemic infection	*S. pyogenes*	As for scarlet fever (see above)
Cellulitis	Uncomplicated infection of the dermis	*S. aureus*, *S. pyogenes*	As for scarlet fever (see above)
Necrotizing fasciitis	Rapidly progressive cellulitis	*S. aureus*, *S. pyogenes*, *Clostridium perfringens*	As for scarlet fever (see above); a variety of toxins
Viral			
Rubella[b]	Discolored, pimply rash; mild disease unless congenital	Rubella virus [ssRNA(+)]	Envelope proteins E1 and E2
Measles[b]	Severe disease, fever, conjunctivitis, cough, rash	Rubeola virus [ssRNA(−)]	V protein (interferes with interferon signaling)
Chickenpox[b]	Generalized discolored lesions	Varicella-zoster (dsDNA)	Glycoprotein B (fusion of viral and cell membranes)
Shingles	Pain and skin lesions, usually on trunk in adults	Varicella-zoster (dsDNA)	Glycoprotein E (required for cell-cell fusion)
Smallpox[c]	Raised, crusted skin rash, highly contagious	Variola major (dsDNA)	SPICE (smallpox inhibitor of complement enzymes)
Warts[b]	Rapid growth of skin cells	Papillomaviruses (dsDNA)	E6 and E7 oncoproteins (Chapter 25)
Fungal			
Dermatophytosis	Dry, scaly lesions like athlete's foot (tinea pedis)	Dermatophytes (*Epidermophyton*, *Tricophyton*, *Microsporum*)	Unclear
Sporotrichosis	Granulomatous, pus-filled lesions; can disseminate to lungs or other organs	*Sporothrix schenckii*	Melanin
Blastomycosis	Granulomatous, pus-filled lesions; can disseminate to lungs or other organs	*Blastomyces dermatitidis*	BAD1 adherence
Candidiasis	Patchy inflammation of mouth (thrush) or vagina; can disseminate in immunocompromised patients	*Candida albicans*; *Candida glabrata*	Proteinase, phospholipase, Ssn6/Tup1 regulators, others
Aspergillosis	Infected wounds, burns, cornea, external ear	*Aspergillus* spp.	PacC/FOS1 regulators, gliotoxin
Zygomycosis	Oropharyngeal infections; affects mainly diabetic patients; can rapidly disseminate	*Mucor* and *Rhizopus* spp.	Iron acquisition (rhizoferrin); rhizoxin

[a]G+ = Gram-positive.

[b]Vaccine is available against the causative agent.

[c]Vaccine is no longer in use, because the disease has been eradicated. The exceptions are highly restricted laboratories.

A.

B.

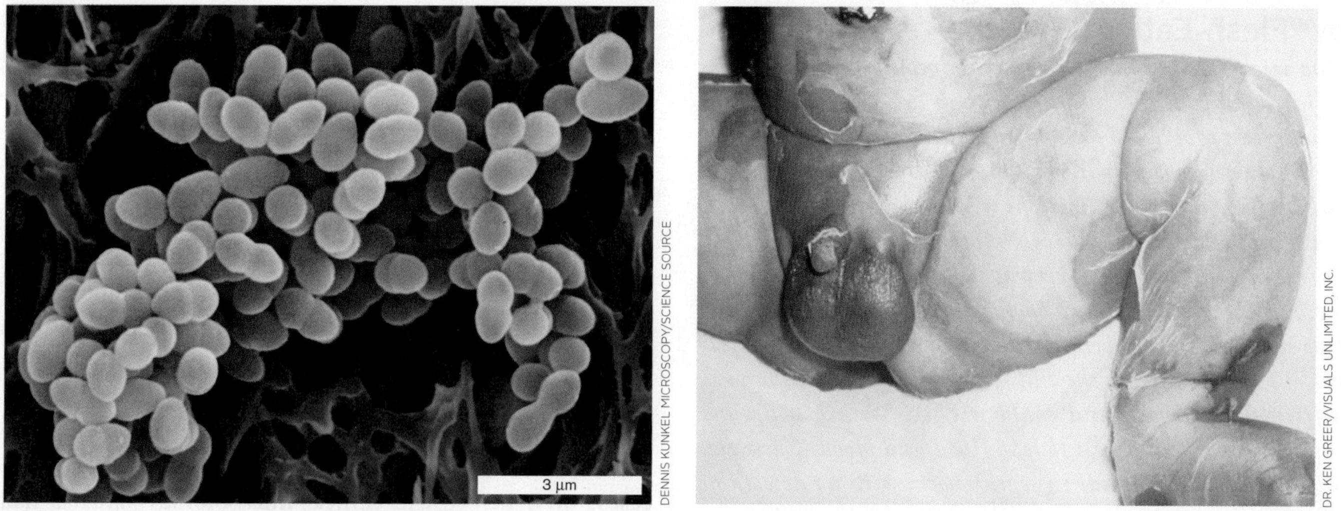

FIGURE 26.3 ■ *Staphylococcus aureus*. **A.** *S. aureus* (colorized SEM). **B.** Exfoliative toxin from some strains of *S. aureus* causes scalded skin syndrome.

immune system and antibiotics. As a result, boils generally require surgical drainage in addition to antibiotic therapy.

As noted in Chapters 24 and 25, some strains of *S. aureus* also produce toxic shock syndrome toxin (TSST), a superantigen that can lead to serious systemic symptoms. Recall that a superantigen links and activates antigen-presenting cells and T cells by binding to the outside of MHC class II receptors and T-cell receptors. Antigen recognition is not required. As a result, many different T cells become activated to release a flood of cytokines.

A particularly dangerous strain of *Staphylococcus aureus* is called methicillin-resistant *S. aureus* (MRSA). Today, infections suspected to be caused by *S. aureus* are treated with penicillin-like drugs, such as oxacillin (similar to methicillin). MRSA strains, however, have developed resistance to methicillin, oxacillin, and many other penicillin-like drugs through an altered penicillin-binding protein (PBP) involved in cell wall synthesis (see the discussion of MecA in Section 27.2). So, using oxacillin or amoxicillin to treat what turns out to be a MRSA infection will result in treatment failure that can become life-threatening. Once the mistake is discovered, immediate use of alternative drugs, such as vancomycin, is required.

MRSA strains, originally seen mainly as causing hospital-acquired (**nosocomial**) infections, are no longer found only in hospitals. MSRA and other infections can develop during a visit to any health care facility, such as hospitals, nursing homes, doctor offices, or rehabilitation facilities. These infections are now referred to as **health care–associated infections (HAIs)**. Another category, called community-acquired infections (CAIs), especially MSRA, can develop in individuals who have not recently encountered health care. HAI-MSRA and CAI-MRSA infections are occurring at an epidemic rate in the United States (an incidence of approximately 20 per 100,000 population).

Seventy percent of staphylococcal skin infections are now caused by MRSA. Thus, a physician can no longer assume that a patient walking into the office with a staphylococcal infection will respond to methicillin-like drugs such as oxacillin. The doctor must assume that the infection may be caused by MRSA. As a result, treatment regimens around the country, and the world, are being forced to change. Today, vancomycin and linezolid are initially used to treat serious infections where MRSA is a suspected cause. Once the clinical laboratory rules out MRSA, the antibiotic treatment is changed. Antibiotics are discussed in Chapter 27.

Other staphylococcal diseases are caused by toxin-producing strains in which the organism remains localized but the toxin disseminates. We have already mentioned TSST, but there are other toxins. For example, some strains of *S. aureus* produce a toxin called exfoliative toxin, which causes a blistering disease in children called staphylococcal scalded skin syndrome (**Fig. 26.3B**). Like TSST, exfoliative toxin is a superantigen (described in Section 24.3). In addition, exfoliative toxin cleaves a skin cell adhesion molecule that, when severed, results in the epithelium separating from the underlying dermis (blisters).

Thought Question

26.1 Does *Staphylococcus aureus* have to disseminate through the circulation to produce the symptoms of scalded skin syndrome (SSS)? Explain why or why not.

Case History: Necrotizing Fasciitis by "Flesh-Eating" Bacteria

One weekend in June, Cassi was camping with her three children. She suffered a minor cut on her finger, which she bandaged properly. She also injured the left side of her body while playing sports with her kids. Not thinking much of either of her minor injuries, she went to bed. Two days later, Cassi was extremely ill. Her symptoms included vomiting, diarrhea, and a fever. She was also in severe pain where she had injured her side, and the area had begun to bruise (the skin was not broken). By the next day she could barely get out of bed, and by the end of that night she was having difficulty breathing and could not see. Within hours her side worsened and began to leak fluid and blood. Cassi was admitted to the hospital in septic shock, with no detectable blood pressure. An infectious disease specialist looking at the wound (similar to ***Fig. 26.4A****) diagnosed the problem as necrotizing fasciitis. Cassi also developed sepsis, which led to dangerously low blood pressure. Because necrotizing fasciitis is often deadly, Cassi was rushed into surgery. In an effort to save her life, the surgeons removed about 7% of her body surface. The large wound infection in her side would need to resolve before a skin graft could be performed to repair it. So, the hole in Cassi's body was left wide open for days. After nearly 3 months and several operations, Cassi recovered.*

What kind of organism can cause this type of devastating disease? The disease **necrotizing fasciitis**, also known incorrectly as "flesh-eating disease," is rare and is often caused by the Gram-positive coccus *Streptococcus pyogenes* (**Fig. 26.4B**), a microbe normally associated with throat infections (pharyngitis). Although sometimes described as a recently emerging infectious disease, necrotizing fasciitis was first discovered in 1783, in France. Its incidence may have risen recently owing to the increased use of nonsteroidal, anti-inflammatory drugs (such as ibuprofen), which increase a person's susceptibility to infection (see Section 25.1).

In this case history, Cassi probably had this organism on her skin when the injury to her side occurred. The injured area probably suffered an invisible microabrasion, providing a good growth environment for the organism, leading to the secretion of potent toxins and death of surrounding tissues. The bacteria will spread through subcutaneous tissue, destroying fat and fascia without initially harming the skin itself. Fascia is the sheath of thin, fibrous tissue that covers muscles and organs.

Rapid, aggressive surgical removal of affected tissue and treatment with antibiotics is required in these extreme cases, even before the clinical microbiology lab has had time to identify the organism. In this approach to antibiotic treatment, called **empiric therapy**, one or more antibiotics are given to "cover" (kill) the most likely causative agents. Therapy can include clindamycin and metronidazole (which act against anaerobes and Gram-positive cocci), and gentamicin or piperacillin (drugs particularly effective against Gram-negative microbes). (Chapter 27 further discusses these and other antibiotics.) Often, however, antibiotic treatment of patients with necrotizing fasciitis is difficult because of insufficient blood supply to affected dead tissues.

Other bacteria can also cause necrotizing fasciitis, such as *Staphylococcus aureus* and the marine organism *Vibrio vulnificus*. In rare cases, most recently in 2018, *Capnocytophaga canimorsus*, a Gram-negative bacterium unique to the mouths of dogs and cats, has caused necrotizing fasciitis in people who have been licked or bitten by a dog. *C. canimorsus* infections are very rare, however, because those most at risk are immunocompromised in some way.

A. B.

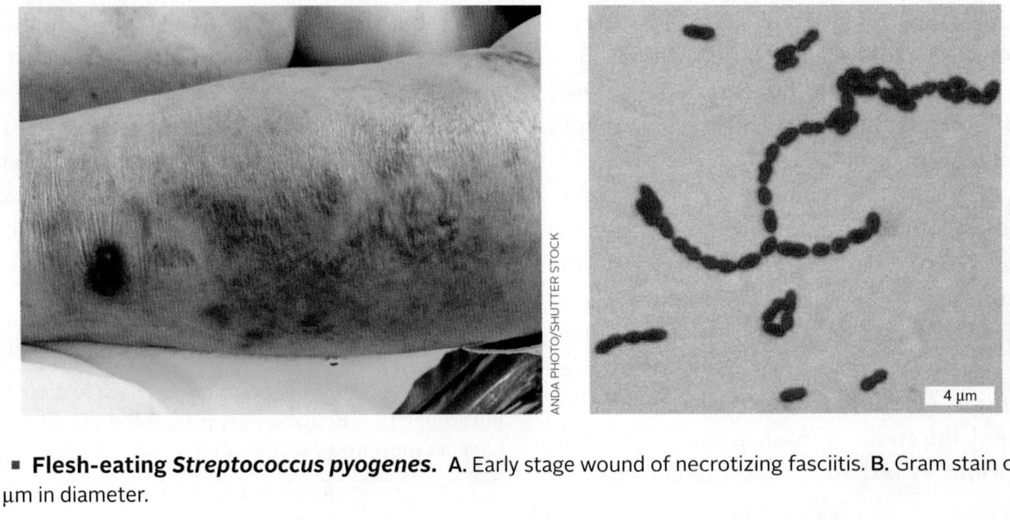

FIGURE 26.4 ▪ Flesh-eating *Streptococcus pyogenes*. **A.** Early stage wound of necrotizing fasciitis. **B.** Gram stain of *S. pyogenes*. Each cell is approx. 1 µm in diameter.

Cellulitis is another form of skin infection, but one that is less aggressive than necrotizing fasciitis. Cellulitis does not involve the fascia or muscles, but is characterized by localized pain, swelling, tenderness, erythema, and warmth. *Streptococcus pyogenes* is the most frequent cause of cellulitis in immunocompetent adults, but a number of other bacteria, including *Staphylococcus aureus*, Gram-negative bacilli, and anaerobes, can also cause this skin infection. Note that cellulitis can progress to necrotizing fasciitis if left untreated.

Streptococcus pyogenes wields many different virulence factors, including the M protein used for nonpilus attachment and inflammasome activation, superantigen exotoxins, and secreted enzymes such as hyaluronidase and DNase (some of these are described in Chapter 25). Many established virulence factor genes in *Streptococcus pyogenes* are located on prophages (phage genomes) integrated into the bacterial genome. Prophages constitute approximately 10% of the organism's genome. One study found that a soluble factor produced by human pharyngeal cells could facilitate activation of at least some of these phages and cause horizontal transfer of the associated virulence factors between strains of this pathogen. Prophage activation and phage production are discussed in Chapter 6.

Thought Question

26.2 Why would treatment of some infections require multiple antibiotics?

Osteomyelitis

Osteomyelitis is a bone infection with accompanying inflammation and bone destruction, caused by bacteria. All bones can be infected, but the lower extremities and vertebrae are most commonly involved. Bone can be infected in several ways. Acute trauma or surgery can directly introduce organisms into affected bone, or organisms from an adjacent soft-tissue infection can spread to nearby bone. Organisms can also spread through the bloodstream from peripheral sites of infection (an abscess, for instance) to bone (called hematogenous seeding). Hematogenous seeding most commonly affects vertebrae because they are well vascularized.

The most common causes of osteomyelitis include staphylococci (*Staphylococcus aureus*), streptococci (*Streptococcus pyogenes*), Gram-negative bacilli (*Pseudomonas*, *Escherichia*), and some anaerobes (*Bacteroides*). Bone biopsy is usually needed to identify the organism. Host factors that contribute to osteomyelitis include diabetes (when a diabetic foot wound goes untreated), children with sickle-cell disease, joint prosthetics, and intravenous drug use. Antibiotic treatment of osteomyelitis is challenging because most bones are not well vascularized, so debridement (removal of damaged tissue) is also necessary.

Viral Diseases Causing Skin Rashes

Several viruses can produce skin rashes, although their route of infection is usually through the respiratory tract. Measles, for example (see Section 6.1), is a highly contagious viral infection caused by a paramyxovirus, whose hallmark symptom is a maculopapular skin rash (see Fig. 6.2D). In a maculopapular rash, flat, red (macular) spots are intermixed with raised bumps (papules). The virus is transmitted by coughing and sneezing, but the first signs of measles, also known as rubeola, are fever, cough, runny nose, and red eyes occurring 9–12 days after exposure. A few days later, telltale spots (Koplik's spots) appear in the mouth, along with a sore throat. Then a skin rash develops that typically starts on the face and spreads down the body. The virus replicates in the lymph nodes and spreads to the bloodstream (viremia), where it can infect endothelial cells of the blood vessels. The rash occurs when T cells begin to interact with these infected cells.

Although skin rash is the main symptom of measles, infection can also cause respiratory symptoms and serious complications, including pneumonia, bronchitis, croup, and even a fatal encephalitis in immunocompromised patients. Approximately one out of 4 cases of measles requires hospitalization. One out of 1,000 die. In the United States, measles had been almost completely eliminated by the measles, mumps, and rubella (MMR) multivalent vaccine. Unfortunately, a disturbing, misinformed anti-vaccine movement has led to an increase in U.S. measles cases over the past two decades. Case numbers in the U.S. for 1999 were 100. In 2018 the number rose to 349. The number for 2019 as of November 7 is 1,261. Worldwide, measles remains a serious problem, including in places where vaccinations are routine. Europe alone had 82,896 cases in 2018, compared to 24,000 for 2017. The startling rise is once again attributed to fewer people being vaccinated.

Rubella virus, a togavirus, causes a maculopapular rash known as German measles or three-day measles. The rash is similar to but less red than that of measles (**Fig. 26.5**). German measles is an infection of primarily the skin and lymph nodes and is usually transmitted from person to person by aerosolization of respiratory secretions. It is not dangerous in adults or children; the virus can, however, cross the placenta in a pregnant woman and infect her fetus. If the virus crosses the placenta within the first trimester, the result is congenital rubella syndrome, which can cause death or serious congenital defects in the developing fetus.

Other viruses affecting the skin, such as chickenpox and the related disease shingles, smallpox, and human papillomavirus, are included in **Table 26.1** and discussed in **eTopic 26.1**. The vaccination schedule for preventing these diseases is provided in Table 24.5.

To Summarize

- ***Staphylococcus aureus*** **and** ***Streptococcus pyogenes*** are common bacterial causes of skin infections. The organisms usually infect through broken skin.
- **Methicillin-resistant** ***Staphylococcus aureus*** **(MRSA)** has become an important cause of community-acquired staphylococcal infections (CAI-MRSA).
- **Necrotizing fasciitis** is usually caused by *Streptococcus pyogenes*, but it can be the result of other infections.
- **Infections of the skin** can disseminate via the bloodstream to other sites in the body.
- **Osteomyelitis**, usually caused by *Staphylococcus aureus*, begins with direct bone trauma, hematogenous seeding, or contact with a nearby soft-tissue infection.
- **Rubeola and rubella viruses** infect through the respiratory tract, but their main manifestation is the production of similar maculopapular skin rashes.

26.2 Respiratory Tract Infections

Lung and upper respiratory tract infections are among the most common diseases of humans. Many different bacteria, viruses, and fungi are well adapted to grow in the lung. Successful lung pathogens come equipped with appropriate attachment mechanisms and countermeasures to avoid various lung defenses (such as alveolar macrophages). One reemerging bacterial pathogen, *Bordetella pertussis*, the cause of whooping cough, inhibits the mucociliary escalator by binding to lung cilia (**eTopic 26.2**). Although many microbes can infect the lung, most respiratory diseases are of viral origin, and most viral respiratory infections (such as the common cold) do not spread beyond the lung. Fortunately, viral diseases by and large are self-limiting and typically resolve within 2 weeks; however, the damage caused by a primary viral infection can lead to secondary infections by bacteria.

Bacterial infections of the lung, whether of primary or secondary etiology, require antibiotic therapy. Before the advent of antibiotics, the only recourse was to insert a tube into the patient's back to drain fluid accumulating in the pleural cavity around the lung (a pathological process known as pleural effusion). Unless released, the pressure on the lung will collapse the alveoli and make breathing difficult.

Viral infections predispose patients to secondary bacterial infections in several ways. Viral lung infections cause the patient to dehydrate, which increases mucus viscosity in the airways. Increased mucus viscosity limits motility of the mucociliary escalator (described in Section 23.1), making it harder to eliminate bacterial pathogens. To keep the escalator moving, cold sufferers are advised to drink plenty of fluids to decrease mucus viscosity. Another factor leading to secondary bacterial infections is that viruses can inhibit key aspects of lung innate immune mechanisms that prevent bacterial growth. Note that many deaths resulting from viral influenza are caused by secondary bacterial infections.

A.

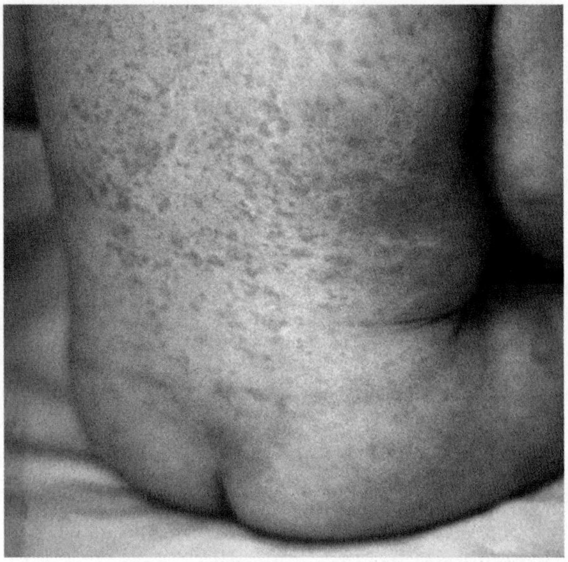

B.

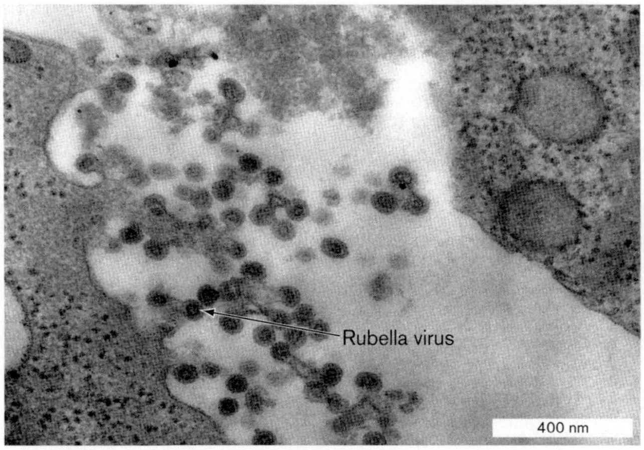

FIGURE 26.5 ■ German measles. **A.** Skin rash caused by rubella virus. **B.** Rubella virus budding from the ace to form an enveloped virus particle (approx. 50–70 nm; TEM).

Case History: Bacterial Pneumonia

In March, James, an 80-year-old resident of a New Jersey nursing home, had a fever accompanied by a productive cough with brown sputum (mucous secretions of the lung that can be coughed up). He reported to the attending physician that he *had pain on the right side of his chest and suffered from night sweats. Blood tests revealed that his white blood cell (WBC) count was 14,000 per microliter (normal is 5,000–10,000), with a makeup of 77% segmented forms (polymorphonuclear leukocytes, PMNs; normal range 40%–60%) and 20% bands (immature PMNs; normal range 0%–5%). The chest radiograph revealed a right-upper-lobe infiltrate (**Fig. 26.6A**). From this information, the clinician diagnosed pneumonia. Microscopic examination of the patient's sputum (Gram stain and capsule stain) revealed Gram-positive cocci in pairs and short chains surrounded by a capsule (**Fig. 26.6B**). Bacteriological culture of the patient's sputum and blood yielded* Streptococcus pneumoniae.

Pneumonia is a disease that can be caused by many different microbes (**Table 26.2**). The pneumococcus *Streptococcus pneumoniae* accounts for about 25% of community-acquired cases of pneumonia, but pneumococcal pneumonia occurs mostly among the elderly and immunocompromised, including smokers, diabetics, and alcoholics. A breakdown of pneumonia cases by causative organism is shown in **Figure 26.6C**.

The noses and throats of 30%–70% of a given population can contain *S. pneumoniae*. The microbe can be spread from person to person by sneezing, coughing, or other close, personal contact. Pneumococcal pneumonia may begin suddenly, with a severe shaking chill usually followed by high fever, cough, shortness of breath, rapid breathing, and chest pains.

Pneumococcal lung infection begins when the pneumococcus is aspirated into the lung. Once in the lung, the microbe grows in the nutrient-rich edema fluid of the alveolar spaces. Neutrophils and alveolar macrophages then arrive to try to stop the infection. They are called into the area from the circulation by chemoattractant chemokines released by damaged alveolar cells. The thick polysaccharide capsule of the pneumococcus, however, makes phagocytosis very difficult (see Section 25.6). In an otherwise healthy adult, pneumococcal pneumonia usually involves one lobe of the lungs; thus, it is sometimes called lobar pneumonia. The infiltration of PMNs and fluid leads to the typical radiological findings of diffuse, cloudy areas. In contrast, infants, young children, and elderly people more commonly develop an infection in other parts of the lungs, such as around the air vessels (bronchi), causing bronchopneumonia.

The white blood cell count in the case history is telling. The patient had an elevated WBC count (normal is 5,000–10,000 per microliter) and an elevated proportion of band cells (normal is 0%–5%). These increases are indicative of a bacterial, not viral, infection. Neutrophils (PMNs), the front-line combatants against infection, rise in response to bacterial infections and are first released from bone marrow as immature band cells, whose presence is a sure sign of bacterial infection.

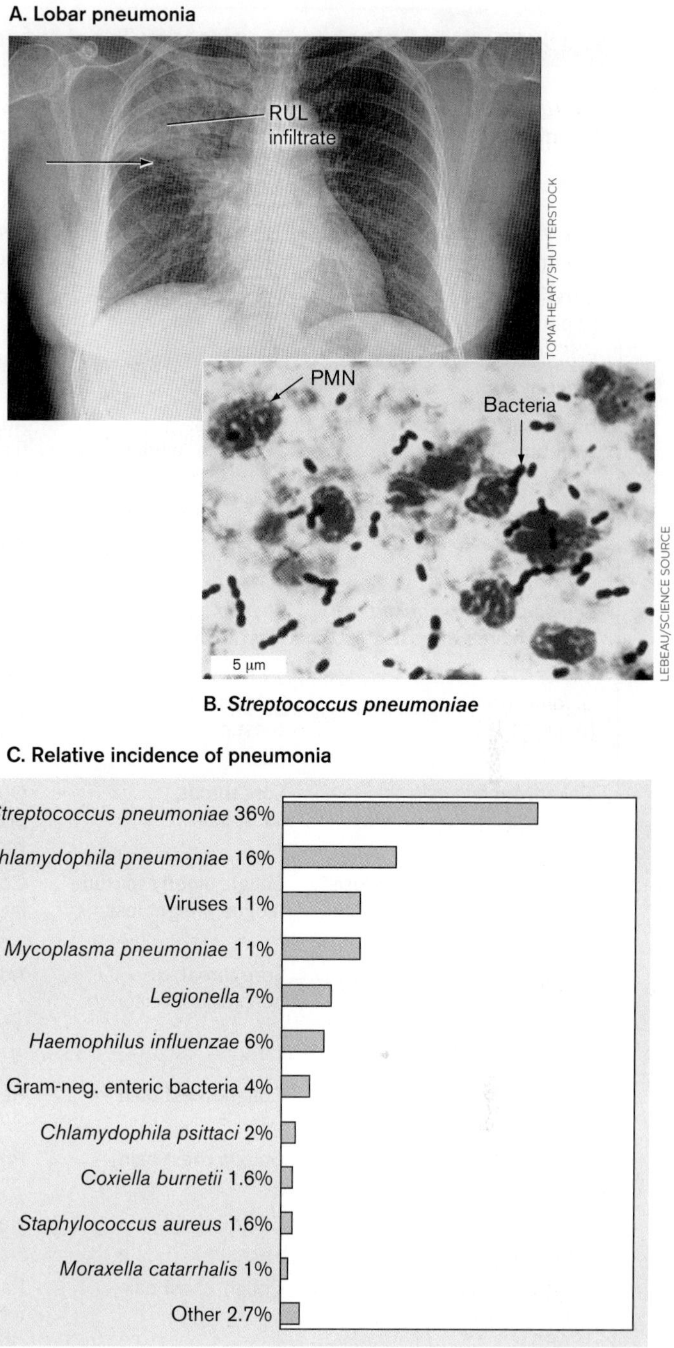

FIGURE 26.6 ▪ Pneumonia caused by *Streptococcus pneumoniae*. **A.** X-ray view of a patient with lobar pneumonia. Infiltrate in the right upper lobe (RUL) is caused by *S. pneumoniae*. The sharp lower border of the infiltrate marks the upper boundary of the right middle lobe (arrow). **B.** Micrograph of *S. pneumoniae*. Gram-stained sputum sample showing numerous PMNs and extracellular diplococci in pairs and short chains. Bacteria range from 0.5 to 1.2 μm in diameter. **C.** Relative incidence of pneumonia caused by various microorganisms.

In recent years several outbreaks of pneumococcal pneumonia have occurred in nursing homes, with numerous residents affected in each case. These incidents underscore the importance of elderly people receiving the pneumococcal polysaccharide vaccine (PPSV) as a hedge against infection.

TABLE 26.2 Selected Respiratory Tract Infectious Diseases

Agent [disease]	Key symptoms	Virulence properties	Source(s)	Treatment options (as of 2017)
Bacterial				
Bacillus anthracis [anthrax]	Hypotension, respiratory failure	Peptide capsule; PA, LF, and EF toxins	Soil/airborne	Ciprofloxacin [vaccine (military)]
Corynebacterium diphtheriae [diphtheria]	Tracheal pseudomembrane	Diphtheria toxin	Humans	Penicillin (vaccine)
Streptococcus pneumoniae [pneumonia]	Fever, chills, cough, chest pain	Capsule, pneumolysin	Humans	Fluoroquinolones, ceftriaxone (vaccine)
Bordetella pertussis [whooping cough]	Violent cough, inhalation "whoop"	Adenylate cyclase toxin, filamentous hemagglutinin (adhesin)	Humans	Azithromycin (vaccine)
Pseudomonas aeruginosa [pneumonia]	Infects cystic fibrosis patients	Exotoxin A, phospholipase C, exopolysaccharide, others	Water, soil	Quinolones, aminoglycosides, carbapenems
Legionella pneumophila [Legionnaire's disease]	Chest pain, cough, muscle pain, vomiting	Intracellular growth, hemolysin, cytotoxin, protease	Water towers/ inhalation	Azithromycin (macrolide), quinolones
Chlamydophila pneumonia [pneumonia]	Sore throat, chest pain	Obligate intracellular growth; prevents phagolysosome fusion	Humans	Tetracycline, macrolides
Chlamydophila psittaci [psittacosis]	Sore throat, chest pain	Obligate intracellular growth; prevents phagolysosome fusion	Bird droppings/ dust	Doxycycline, erythromycin
Mycobacterium tuberculosis [tuberculosis]	Cough, bloody sputum, fatigue, weight loss	Cord factor, wax D, intracellular growth	Humans	Rifampin, isoniazid, ethambutol, pyrazinamide [vaccine (BCG[a])]
Mycoplasma pneumoniae [pneumonia]	Sore throat, nonproductive cough	Adhesin tip	Humans	Azithromycin, doxycycline
Viral				
Cytomegalovirus (CMV) [CMV disease]	Cough, chest pain	Reduces MHC I presentation	Humans	Ganciclovir, valganciclovir
Respiratory syncytial virus (RSV) [RSV disease]	Cough, chest pain	Prevents T-cell activation	Humans	Treat symptoms/ribavirin, palivizumab[b]
Influenza virus [influenza]	Cough, chest pain	Neuraminidase; hemagglutinin	Humans	Oseltamivir, zanamivir (vaccine)
Severe acute respiratory syndrome (SARS) virus [SARS]	Cough, chest pain	Papain-like protease (PLP) inhibits phosphorylation of host IRF-3 to prevent interferon-beta synthesis	Humans	Treat symptoms
Fungal				
Aspergillus spp. [aspergillosis]	Lungs, sinuses; breathing difficulty	Dimorphism; gliotoxin	Environment	Amphotericin B, voriconazole
Histoplasma capsulatum [histoplasmosis]	Flu-like	Dimorphism; calcium-binding protein	Bird, chicken, bat droppings	Amphotericin B, itraconazole
Coccidioides immitis [coccidioidomycosis]	Flu-like	Dimorphism; arginase 1	Environment	Amphotericin B, fluconazole
Blastomyces dermatitidis [blastomycosis]	Flu-like	Dimorphism; BAD1	Environment	Amphotericin B, itraconazole
Pneumocystis jirovecii [pneumocystosis]	Chest pain, cough, skin lesion	Unknown	Environment	Bactrim

[a]BCG = Bacille Calmette-Guérin (a weakened strain of the bovine tuberculosis strain).

[b]Palivizumab = monoclonal antibody to RSV protein.

While there are over 80 antigenic types of pneumococcal capsular polysaccharides, the injected vaccine contains only the 23 types that are most often associated with disease (see Section 24.6). Because it contains antigens from multiple strains, the vaccine is called multivalent. The pneumococcal polysaccharide vaccine is recommended for individuals over 65, as well as for those who are immunocompromised. The patient in this case history failed to receive the vaccine.

In addition to causing serious infections of the lungs, *S. pneumoniae* can invade the bloodstream (bacteremia) and the covering of the brain (meningitis). The death rates for these infections are about one out of every 20 who get pneumococcal pneumonia, about four out of 20 who get bacteremia, and six out of 20 who get meningitis. Individuals who are immunocompromised because of liver disease, AIDS (caused by HIV), or organ transplants, are even more likely to die from the disease.

An emerging infectious disease problem throughout the United States and the world is the increasing resistance of *S. pneumoniae* to antibiotics. At least 40% of the strains isolated are already resistant to macrolides, the former drug of choice for treating pneumonia. Currently, third-generation cephalosporins (ceftriaxone) or fluoroquinolones are typically used to treat these infections. Chapter 27 discusses why antibiotic resistance is on the rise for this and other microbes.

Case History: Fungal Lung Infection

*A 40-year-old salesman named Jaylen presented to an Ohio hospital emergency department with fever, cough, myalgias, and chest pain. One month earlier he had reported similar symptoms and had been treated with azithromycin for suspected bacterial lung infection. His symptoms had not improved, and he now reported increasing weakness, difficulty breathing, abdominal pain, and a weight loss of 8 pounds in the previous month. While taking the patient's history, the physician assistant discovered that Jaylen is a weekend spelunker who frequently explores local caves, and that he is bisexual, having had two male and three female partners in the previous year. Physical exam revealed enlarged cervical (neck) and axillary (armpit) lymph nodes (lymphadenopathy); lung congestion and hepatosplenomegaly (enlarged liver and spleen); and red, bumpy lesions on his legs. A complete blood count showed low numbers of red cells, white cells, and platelets (pancytopenia). A chest X-ray (**Fig. 26.7A**) showed calcified lymph nodes and small clumpy calcifications (opaque white areas) in both lungs. Suspecting tuberculosis, the PA administered a tuberculin skin test. The patient's symptoms and sexual history also dictated that serological tests for syphilis; hepatitis A, B, and C; and HIV be administered. Tests for tuberculosis, syphilis, and hepatitis were all negative, but Jaylen tested positive for HIV. His sputum was cultured for* Mycobacterium tuberculosis, *and his blood was cultured for the presence of bacteria and fungi.*

This is a complicated, but not unusual, case. Health providers often treat a patient for the most likely cause of the chief complaint, which in this case was a suspected bacterial lung infection. Azithromycin was prescribed, and the infection worsened. Antibiotic failure leading to a return visit forced a closer look. The flu-like symptoms and calcifications seen on the lung X-ray then suggested tuberculosis (another lung infection that can disseminate, discussed later). The negative skin test for tuberculosis, however, ruled out that disease, leaving fungus as the probable cause, given the chronic nature of the patient's symptoms. The infection in this patient probably started in the lung (clued by the cough), after which the organism spread throughout the body via the bloodstream to infect the liver and spleen.

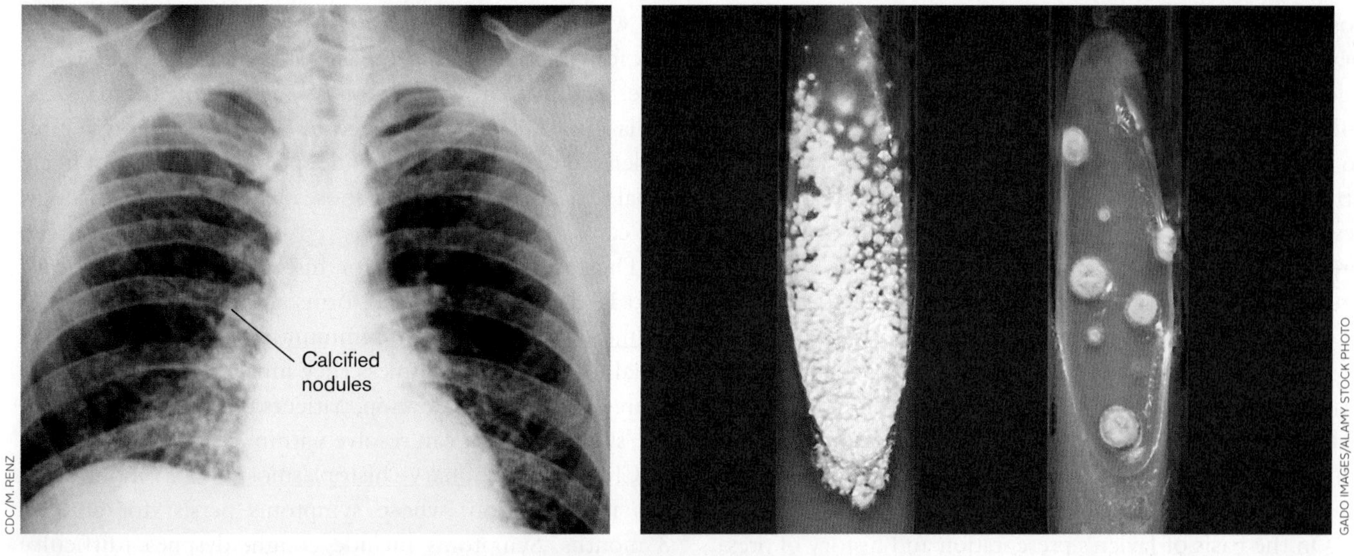

FIGURE 26.7 ■ Lung infection by *Histoplasma capsulatum*. **A.** X-ray of a patient with histoplasmosis. **B.** Colonies of *H. capsulatum*.

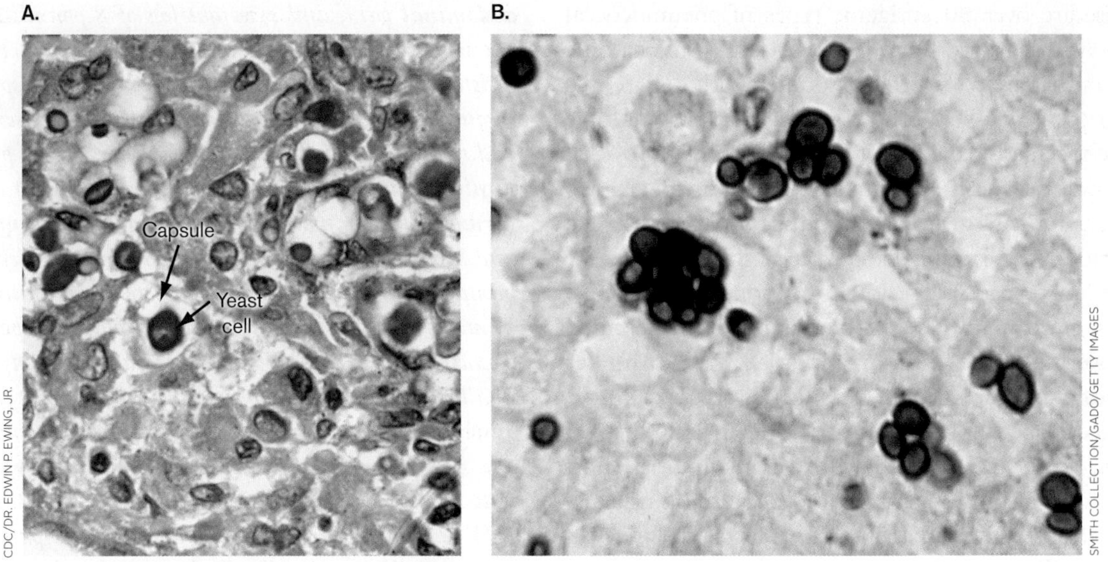

FIGURE 26.8 ■ *Cryptococcus neoformans* and *Histoplasma capsulatum* in lung biopsies. **A.** *C. neoformans* in lung nodule. Periodic acid–Schiff stain. **B.** *H. capsulatum*. Silver stain.

Cryptococcus, an encapsulated yeast, is a potential cause (see the Current Research Highlight). It typically requires an immunocompromised host to cause disease, and Jaylen's positive HIV test indicated he might be immunocompromised (low CD4 T-cell count). *Cryptococcus* usually causes meningoencephalitis, but the organism can also infect the lungs, prostate gland, urinary tract, eyes, myocardium, bones, skin, and joints. The chest X-ray does not rule out *Cryptococcus*, because this yeast can cause similar images when infecting the lung. However, the pancytopenia is not typical of *Cryptococcus*, and a biopsy failed to find yeast cells with a thick capsule (example shown in **Fig. 26.8A**).

The most likely fungal causes of infection in this case history are the endemic mycoses, such as coccidioidomycosis, blastomycosis, and histoplasmosis. This patient had never traveled to the western United States, where coccidioidomycosis is endemic, so exposure to *Coccidioides* was ruled out. *Blastomyces dermatitidis* is a soil fungus endemic to the Ohio and Mississippi river valleys and the southeastern United States. The disease, blastomycosis, begins as a respiratory infection following the inhalation of conidia. From the lung, the organism can disseminate to the skin, bone, and genitourinary tract, but rarely affects the liver or spleen, both of which were affected in Jaylen. (A case of disseminated blastomycosis is presented in **eTopic 26.3**.) Histoplasmosis, on the other hand, most commonly presents as a flu-like pulmonary illness that can progress to more serious pulmonary infection and can disseminate to the liver and spleen in severe cases.

On the basis of Jaylen's presentation and history of present illness, amphotericin B, a powerful antifungal agent (discussed in Section 27.5), was administered intravenously. The patient's fever lowered almost immediately. *Histoplasma* antigen was identified in his urine, and a fungus was found in the cultures of bronchoalveolar lavage fluid (lung washes; **Fig. 26.7B**). The fungus was definitively identified by a DNA probe and tissue biopsy as *Histoplasmosis capsulatum* (tissue biopsy in **Fig. 26.8B**), confirming the diagnosis of histoplasmosis.

Histoplasma is a dimorphic fungus that thrives in damp soil rich in organic material, especially the droppings of birds and bats. Dimorphic fungi take on a mycelial form at 25°C but grow as budding yeast at body temperature (37°C). Although *Histoplasma* is distributed worldwide, endemic areas in the United States include the Ohio and Mississippi river valleys (where Jaylen lives). People like Jaylen who explore caves populated by bats are at particular risk for contracting this disease, sometimes called cave disease. The infectious forms of *H. capsulatum* are microconidia (fungal spores) produced by differentiated mycelium. Jaylen did not wear a mask while spelunking and most likely inhaled a good number of spores while crawling through the caves.

The incubation period for histoplasmosis ranges from 3 to 17 days. Three general forms are recognized. In acute pulmonary histoplasmosis, immunocompetent patients are usually asymptomatic unless they inhale a large number of spores. If symptoms develop, patients typically display cold-like symptoms that can resolve within 3–4 weeks.

Chronic pulmonary histoplasmosis is a more serious manifestation, whose symptoms persist for at least 3 months. Symptoms include cough, dyspnea (difficulty breathing), fever, weight loss, malaise, and red skin nodular lesions called erythema nodosum (a delayed-type hypersensitivity reaction to antigens from various infectious agents, including *Histoplasma*). Jaylen displayed all of these symptoms.

The most serious form of histoplasmosis, progressive disseminated histoplasmosis, includes dissemination to the spleen and liver (enlarging both) and is generally seen in patients who are immunocompromised because of corticosteroid use, organ transplant, or HIV infection. The pancytopenia seen in Jaylen is also common in this form of the disease (70%–90%). Jaylen tested positive for HIV and had a low CD4 T-cell count of 150 per microliter (normal is 500–1,500). Jaylen is immunocompromised. Because of his HIV status, Jaylen was also treated with antiviral agents to control his viral load. Lowering the viral load in his blood would allow his T-cell count to increase and improve his immunocompetence.

Several critical features of this case help differentiate it from the preceding case of pneumococcal pneumonia. First, the initial macrolide antibiotic, azithromycin, should have killed most bacterial sources of infection. Second, the X-ray finding of diffuse pulmonary nodules is more indicative of fungal lung infection than of bacterial infection, which in a patient of this age would likely appear uniformly dense and confined to one lobe (see **Fig. 26.6A**). The blood count was also a clue. Fungal infections do not usually cause an increase in WBCs or an increase in band cells, as happens with pneumococcal pneumonia. The pancytopenia, however, is consistent with disseminated histoplasmosis. Note that many infectious diseases start out as a localized infection but end up disseminating throughout the body to cause new sites of infection.

Thought Question

26.3 Why isn't a dimorphic fungus like histoplasma easily transmitted from person to person via respiratory droplets?

Tuberculosis as a Reemerging Disease

Tuberculosis, caused by the acid-fast bacillus *Mycobacterium tuberculosis* (see Fig. 28.5), is the leading cause of death due to a single pathogen, claiming 1.6 million lives worldwide in 2017, according to the World Health Organization. This fact is shocking, considering that the disease was nearly eradicated in the developed world going into the 1980s.

Then, in 1985, the HIV pandemic arrived. Because HIV kills T cells, patients became immunocompromised and susceptible to many infections, including TB. By 1991, inner-city hospitals were beginning to see highly infectious multidrug-resistant (MDR) strains of *M. tuberculosis* that produced fulminant (rapid-onset) and fatal disease among patients infected with HIV (time from TB exposure to death was 2–7 months). The number of TB cases exploded and the tuberculin skin test conversion rates among exposed U.S. health care workers rose to 50%. (The tuberculin test is described in the next section.)

Tuberculin conversion rate is the percentage of tuberculin-negative health care providers who converted to tuberculin-positive over time, indicating exposure to *M. tuberculosis.* Today the rate is less than 1%. New cases of TB in the United States have decreased over the years (9,105 in 2017, but approximately 1% of the new cases are caused by MDR-TB). Despite progress reducing TB in developed countries, tuberculosis remains a leading cause of global child mortality. Information about mycobacterial structure can be found in Section 18.3, and about mycobacterial pathogenesis, in Section 25.5.

Tuberculin skin test. The tuberculin skin test (also known as the Mantoux test) is the primary screen for tuberculosis. A mixture of *Mycobacterium tuberculosis* proteins (called purified protein derivative, or PPD) is injected under the skin of the lower arm. A person who has been infected will develop a localized delayed-type hypersensitivity reaction (a reddened area of skin with blisters) within 48 hours (Chapter 24). Note, however, that a positive tuberculin skin test does not signify active disease, but only that the person was infected at one time. The bacterium may have been killed by the immune system without having caused disease, or may lie dormant, waiting to reactivate. **Figure 26.9** outlines the disease course, if untreated, of tuberculosis.

Primary tuberculosis. Once inhaled into the lung, the bacilli deposit into alveoli and are subject to three possible outcomes (**Fig. 26.9**, steps 1 and 2): They can die, become latent, or produce progressive primary tuberculosis. In all three cases the bacteria are initially phagocytosed by alveolar macrophages, and a battle ensues. If the bacilli are not killed by the macrophages, the bacteria will survive ensconced within modified phagolysosomes, where they multiply and kill the macrophage via induced apoptosis. The bacilli are then free to infect other macrophages. The primary lesion is called a Ghon focus. During the pitched battle, some infected macrophages from the Ghon focus will travel to a regional lymph node of the lung. The Ghon focus and the infected lymph node are called the Ghon complex. However, the Ghon complex cannot be seen on an X-ray at this early stage. A delayed hypersensitivity develops, and these patients usually become tuberculin-positive.

The body reacts to bacilli in the lung by trying to wall them off in what is called a **granuloma**, a nodule (sometimes

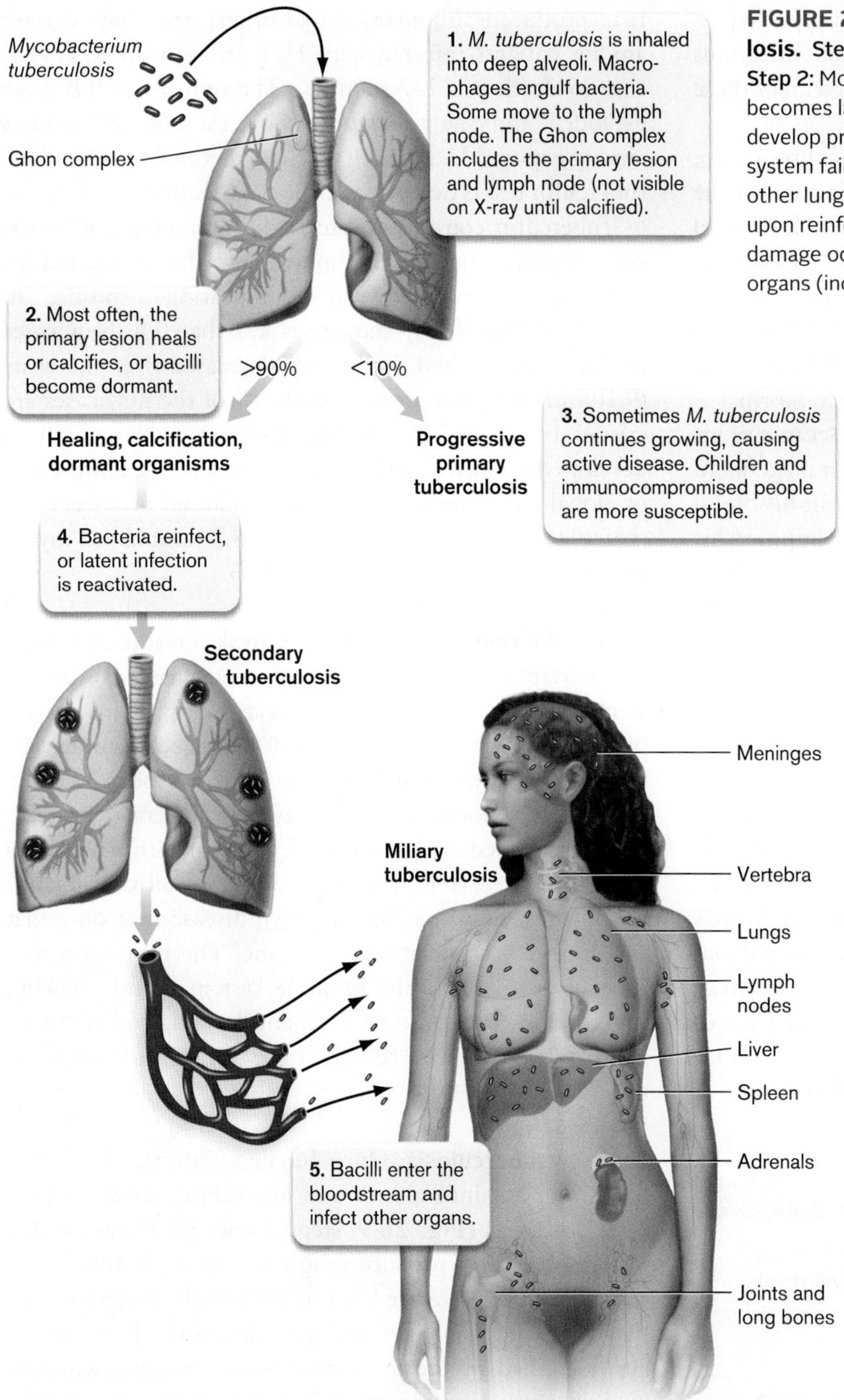

FIGURE 26.9 ■ The disease progression of tuberculosis. **Step 1:** The disease begins as primary tuberculosis. **Step 2:** Most patients never develop disease, or the organism becomes latent within granulomas. **Step 3:** A few patients develop progressive primary tuberculosis because the immune system failed to control the infection, leading to its spread to other lung regions. **Step 4:** Secondary tuberculosis develops upon reinfection or reactivation of latent bacilli. **Step 5:** More damage occurs in the lung, and bacilli disseminate to other organs (including reseeding the lung) to produce miliary TB.

called a tubercle) containing macrophages (live and dead), plus other white blood cells and bacteria (live and dead). The inside of the granuloma can become necrotic with dead host and bacterial cells and develops the consistency of cheese—a process called caseation. Over time, calcium is deposited and the nodules become calcified. At this point the hardened calcified nodules (Ghon complex) become visible on X-rays. Alternatively, the caseous center of the granuloma can liquefy and the bacilli multiply to large numbers. The granuloma can erode (cavitate) to release large numbers of bacilli into the bronchus. Coughing will expel the bacilli into the air for another person to inhale.

Four to six weeks after infection, about 10% of infected people will develop progressive (or active) primary disease in which bacilli in the primary focus grow and caseous material is disseminated to other parts of the lung (**Fig. 26.9**, step 3). Lung X-rays can show consolidation similar to pneumonia in this form of tuberculosis. Symptoms include a productive cough that generates sputum, fever, night sweats, and weight loss.

Secondary tuberculosis. As noted earlier, an alternative to primary disease is an asymptomatic latent tuberculosis infection (LTBI). Here the bacilli remain hidden in granulomas and may even become dormant, remaining so for many years. However, *M. tuberculosis* can sometimes overcome dormancy and the imposed confinement of the immune system (**Fig. 26.9**, step 4). The bacteria begin to multiply and cause **secondary tuberculosis**. Secondary TB due to reactivation commonly occurs in immunocompromised people (such as HIV patients). Secondary tuberculosis can also be caused by the inhalation of new bacilli. Because these patients already developed cell-mediated immunity during primary infection, the symptoms of secondary TB are more serious than those of primary TB and include severe coughing, greenish or bloody sputum, low-grade fever, night sweats, and weight loss. The gradual wasting of the body is what led to the older name for tuberculosis: "consumption."

Extrapulmonary (miliary) TB. *Mycobacterium tuberculosis* bacilli in the lung can sometimes enter the pulmonary vein (**Fig. 26.9**, step 5). Once in the bloodstream, the organism will disseminate to produce abscesses in many different organ systems. Extrapulmonary TB is sometimes called miliary tuberculosis because the size of the infected nodules (1–5 mm), called tubercles, approximates the size of millet seeds. Miliary TB can develop during primary or secondary TB infections.

Treatment. Ten drugs are currently approved by the Food and Drug Administration (FDA) for treatment of tuberculosis. Initial treatment of active disease is aggressive and involves a four-drug regimen of what are called first-line drugs (or drugs of choice)—isoniazid, rifampin, pyrazinamide, and ethambutol—given over a course of several months. MDR strains (about 1% of total TB cases in the United States) are defined as being resistant to two or more first-line drugs. An MDR strain is treated with a regimen of four to five drugs that do not include the first-line drugs to which it is resistant. Extensively drug-resistant tuberculosis (XDR-TB) strains are resistant to three or more second-line drugs (drugs used when drugs of choice fail), as well as two or more first-line drugs. These strains are almost untreatable. About 9% of MDR-TB strains are XDR-TB.

Thought Question

26.4 Explain why patient noncompliance (failure to take drugs as directed) is thought to have led to XDR-TB.

Viral Diseases of the Lung

Numerous viruses can cause lung infections (see **Table 26.2**). Influenza virus and rhinovirus are described in Chapters 6 and 11, and compared in **eTopic 26.4**. The recently emerged infection SARS (severe acute respiratory syndrome) is discussed in Chapter 28. Another important viral lung infection is respiratory syncytial disease, caused by respiratory syncytial virus (RSV). A negative-sense, single-stranded RNA enveloped virus, RSV is the most common cause of bronchiolitis and pneumonia among infants and children under 1 year of age (**Fig. 26.10**).

Illness begins most frequently with fever, runny nose, cough, and sometimes wheezing. RSV is spread from respiratory secretions through close contact with infected persons or by contact with contaminated surfaces or objects. Infection can occur when the virus contacts mucous membranes of the eyes, mouth, or nose and possibly through the inhalation of droplets generated by a sneeze or cough. Unlike rubella or rubeola, which infect the respiratory tract and disseminate through the body, RSV remains localized in the lung.

The majority of children hospitalized for RSV infection are under 6 months of age. RSV can cause repeated infections throughout life, usually associated with moderate to severe cold-like symptoms. Severe lower respiratory tract disease may occur at any age, especially among the elderly or people with compromised cardiac, pulmonary, or immune systems. As yet, a vaccine to control this disease is not available. However, a monoclonal antibody called palivizumab, which binds to an RSV epitope, can prevent RSV infection in high-risk infants who were born either premature or with medical problems such as congenital heart failure.

Table 26.2 presents many other bacterial, fungal, and viral microbes that can cause respiratory tract infection. Be aware that very different diseases can produce similar symptoms. For instance, people constantly confuse influenza (the flu) with the common cold. Symptomatically, they may start out similarly, but there are telling differences. Influenza is characterized by fever, myalgia (muscle aches), pharyngitis (sore throat), and headache (viral infection is discussed in Chapter 11). A runny nose is not one of the symptoms. The common cold, however, manifests as a runny nose, nasal congestion, sneezing, and throat irritation. No myalgia. The observant clinician will note the difference.

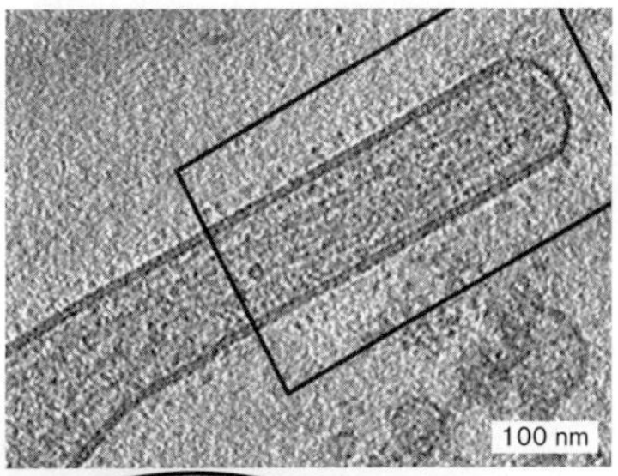

KISS ET AL. 2014. J. VIROL. 88:7602–7617, FIG. 1A.

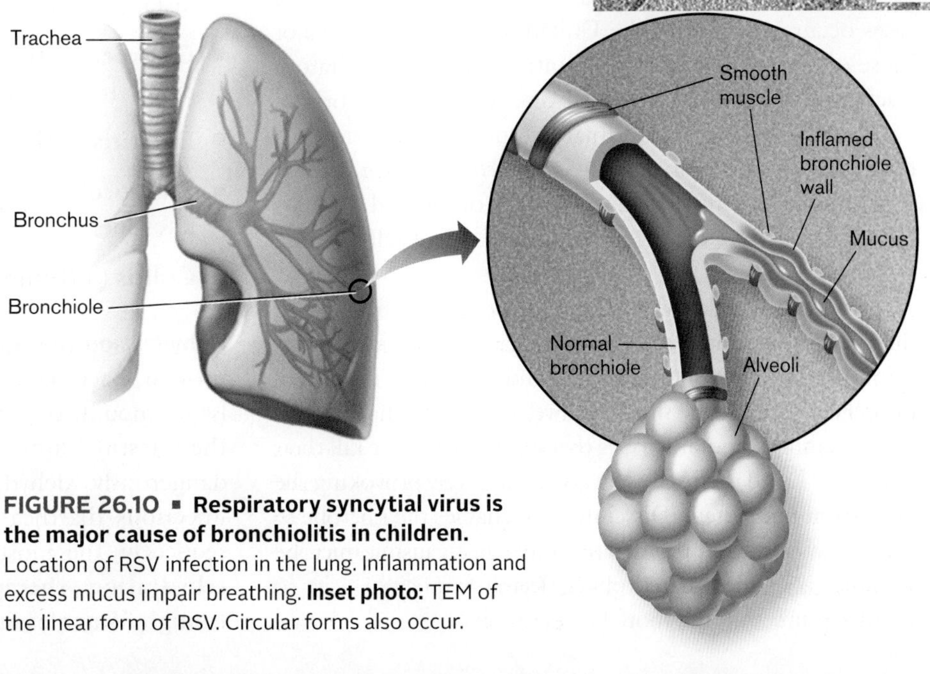

FIGURE 26.10 ▪ Respiratory syncytial virus is the major cause of bronchiolitis in children. Location of RSV infection in the lung. Inflammation and excess mucus impair breathing. **Inset photo:** TEM of the linear form of RSV. Circular forms also occur.

To Summarize

- **Most lung infections are viral**, but most deadly lung infections are caused by bacteria.
- **The mucociliary escalator** is a primary defense mechanism used by the lung to avoid infection.
- **An elevated white cell count** in blood is an indicator of bacterial infection.
- **Pneumococcal vaccine** should be administered to the elderly because they are often immunocompromised.
- **Fungal agents** commonly cause long-term, chronic infections.
- **Localized bacterial infections in the lung can disseminate** via the bloodstream to form lesions at other body sites.
- **Tuberculosis** is an ancient bacterial disease with an increasing mortality rate resulting from multidrug-resistant strains, the susceptibility of HIV patients, and an increasing indigent population.
- **Respiratory syncytial virus** is one of several viruses that can cause lung disease, but it rarely spreads to other organs.

26.3 Gastrointestinal Tract Infections

Nearly everyone has experienced diarrhea, a condition characterized by frequent loose bowel movements accompanied by abdominal cramps. Hundreds of millions of cases occur each year in the United States and are a major cause of death in developing countries. As with respiratory tract infections, most diarrheal disease is viral in origin, with rotavirus being the primary culprit. Among the bacteria, the Gram-negative rod *Salmonella enterica* serovar Typhimurium and the spiral-shaped or curved bacillus *Campylobacter* are the most frequent causes of self-limiting diarrheal disease.

You might wonder why some microbes prefer the chaos of diarrhea to the relative stability of a nice commensal relationship. After all, isn't finding a niche and sticking to it the goal of every microbe? The simple answer is that diarrhea enables the dissemination of microbes that might otherwise kill their host or be killed themselves by too aggressively provoking the host's immune system. Dissemination enables the microbe to reach new hosts and proliferate. A diarrhea-causing microbe is rather like a serial bank robber fleeing from city to city to avoid capture and find more banks to rob.

Types of Diarrhea

What causes diarrhea? Normally, intestinal mucosal cells absorb water (8–10 liters per day) from the intestinal contents, essentially drying out stool. Diarrhea occurs when water is not absorbed or when it actually leaves intestinal cells and enters the intestinal lumen. The excess water loosens stool, and diarrhea results. There are several types of diarrhea.

- **Osmotic diarrhea** happens when nonabsorbable substrates, such as lactulose, a synthetic sugar, are used to treat constipation. Lactulose increases osmolarity in the intestine, causing water to leave mucosal cells. Infectious organisms (such as rotavirus) that prevent nutrient absorption can cause osmotic diarrhea.
- **Secretory diarrhea** develops when microbes cause mucosal cells to increase ion secretion, as seen with cholera toxin. Mucosal cells expel water to try to equilibrate the resulting electrolyte imbalance.
- **Inflammatory diarrhea** forms when an infectious agent triggers the production of inflammatory cytokines that attract PMNs. Subsequent damage to the intestinal wall limits water and nutrient absorption, and causes red and white blood cells to enter the stool (bloody diarrhea or dysentery). *Shigella*, *Salmonella*, and some strains of *E. coli* can cause inflammatory diarrhea.
- **Motility-related diarrhea** can develop when enterotoxins made by pathogens such as rotavirus cause intestinal hypermotility. Food moves through the intestine too fast for water or nutrients to be absorbed.

Terms used to describe the inflammation of different parts of the GI tract include:

- **Gastritis** (inflammation of the stomach lining—for instance, ulcers).
- **Gastroenteritis** (nonspecific term for any inflammation along the gastrointestinal tract).
- **Enteritis** (inflammation mainly of the small intestine).
- **Enterocolitis** (inflammation of the colon and small intestine).
- **Colitis** (inflammation of the large intestine—colon).

Rehydration therapy. Diarrhea and vomiting due to viral growth, bacterial growth, or toxin production can cause large amounts of water to leave the intestinal cells and enter the intestinal lumen. As a result, the patient can become dangerously dehydrated. Most deaths resulting from infectious diarrhea are the result of dehydration. Consequently, the most important treatment for diarrhea is **rehydration therapy**. But giving water alone is not enough. Water leaving tissues causes an osmotic imbalance that compels electrolytes to exit too (as a way to reestablish balance). The resulting electrolyte imbalance severely affects cardiovascular, respiratory, and renal systems, which is the real cause of death. Giving water alone will only make the electrolyte imbalance worse. Therefore, rehydration solutions must also contain glucose, as well as sodium and potassium salts in proper balance (for example, Pedialyte).

Antibiotics are often inappropriate when treating diarrhea. Although antibiotic treatment of infectious gastroenteritis seems intuitive, it is rarely used and actually contraindicated. Most gastrointestinal infections are viral (norovirus or rotavirus), so antibiotics are ineffective. Likewise, gastroenteritis caused by bacteria usually resolves spontaneously, without antibiotic treatment. However, severe systemic disease stemming from gastroenteritis can develop, often in the young or elderly. Diseases like bacillary dysentery (*Shigella dysenteriae*) respond well to antibiotics.

In some cases, antibiotic treatment can actually trigger gastrointestinal disease. For example, many antibiotics used to treat infectious diseases (especially clindamycin) can kill most normal intestinal bacteria, except the naturally resistant Gram-positive anaerobe *Clostridioides difficile* (formerly *Clostridium difficile*), the causative agent of pseudomembranous enterocolitis. Unrestrained by microbial competition, *C. difficile* growing at the epithelial surface of the intestine will produce specific toxins that damage and kill intestinal cells. The organism's growth leads to inflammation and the formation of exudative plaques along the intestinal wall (refer to Fig. 23.11). The plaques eventually coalesce into larger pseudomembrane structures that block the intestinal mucosa. The blockage causes the malabsorption of nutrients and water, resulting in diarrhea. As the pseudomembrane enlarges, it begins to slough off and pass into the stool. Diagnosis of this disease involves PCR identification of the organism or immunological identification of the toxin in fecal samples.

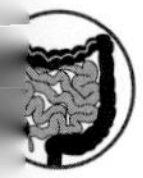

Some patients suffer recurrent *C. difficile* infections. It is not clear why, but one suggestion is that spores lodged in colon folds may escape clearance by peristalsis. Another hypothesis is that patients with recurrent disease have an impaired response to the *C. difficile* toxins. How can recurrences be prevented? In some instances, after vegetative *C. difficile* has been killed by antibiotic treatment, a procedure known as a fecal transplant (see Section 23.2) can be used to restore a healthy gastrointestinal microbiota that prevents the recurrence of *C. difficile* disease.

Staphylococcal Food Poisoning

We have all heard of the local church picnic or restaurant where scores of people become violently ill within hours of eating unrefrigerated potato salad or other food. *Staphylococcus aureus* is the usual cause of these disasters, but it is not an infection. The culprit is an enterotoxin (an exotoxin that affects the gastrointestinal tract) secreted by some strains of *S. aureus* into tainted foods such as pies, turkey dressing, or potato salad. After ingestion, the toxin travels to the intestine, where it enters the bloodstream and stimulates the vagus nerve leading to the vomit center in the brain.

Because the toxin is preformed, symptoms occur quickly after ingestion. Within 2–6 hours, the poisoned patient will begin vomiting and may also experience diarrhea. The disease, though violent, is not life-threatening and usually resolves spontaneously within 24–48 hours. In contrast, diarrhea caused by infectious agents, such as *Salmonella enterica*, that must first grow in the victim, does not occur until 12–24 hours after ingestion, sometimes longer. A clinician noting quick onset of symptoms in a patient will immediately suspect staphylococcal food poisoning. Obviously, antibiotic treatment is not needed for staph food poisoning, but it may be indicated for other gastrointestinal infections. Staphylococcal enterotoxins are also heat resistant, so simply heating a food already containing enterotoxin will not destroy the toxic activity.

Case History: Enterohemorrhagic *E. coli*

Tammy, a 6-year-old girl from Montgomery County, Pennsylvania, arrived at the ER with bloody diarrhea, a temperature of 39°C (102.2°F), abdominal cramping, and vomiting. She was admitted to the hospital 5 days after a kindergarten field trip to the local dairy farm. When questioned about Tammy's activities during the trip, her parents said she had purchased a snack while at the farm. Upon laboratory analysis, a fecal smear was positive for leukocytes, and isolation of organisms confirmed the presence of Gram-negative rods that produced Shiga toxins 1 and 2. In subsequent testing by pulsed-field gel electrophoresis, the isolate was indistinguishable from E. coli *O157:H7. By this time, Tammy had developed additional problems. Her face and hands had become puffy, her urine output had decreased despite being given IV fluids (suggesting kidney damage), and she was beginning to develop some neurological abnormalities. Laboratory analyses of blood samples revealed thrombocytopenia (reduced blood platelet count) and confirmed hemolytic uremic syndrome (HUS; renal failure). Tammy was treated by IV fluid and electrolyte replacement. Antibiotics were not administered. She eventually recovered.*

In this case history, the presence of leukocytes in a fecal smear is a sign that the intestinal pathogen may have invaded the epithelial mucosa of the intestine (or severely damaged it). Breaching this barrier sends out a chemical call (chemokine) to neutrophils, which then enter the area and, in an effort to kill the pathogen, also damage the intestinal cells. *Shigella dysenteriae*, *Salmonella enterica*, and enteroinvasive *E. coli* (EIEC) actually invade enterocytes and are considered intracellular pathogens. Enterohemorrhagic *E. coli* (EHEC), which also produces leukocytes and blood in stools, is not an intracellular parasite (it is not invasive), but causes damaging attachment and effacing lesions, described in Section 25.3, that destroy the mucosal epithelium. The resulting inflammation, in conjunction with damage caused by the Shiga toxin it produces, leads to blood and white cells in the stool. *E. coli* O157:H7, the etiological agent in the case history, is a common serotype of EHEC.

There are at least six different classes (pathovars) of pathogenic *E. coli*, differing in their repertoire of pathogenicity islands, plasmids, and virulence factors. They include the enteroinvasive (EIEC) and enterohemorrhagic (EHEC) pathovars just mentioned, the enterotoxigenic *E. coli* (ETEC), neonatal meningitis-causing *E. coli* (NMEC), and uropathogenic *E. coli* (UPEC) described in Chapter 25, and the enteropathogenic *E. coli* (EPEC) and enteroaggregative *E. coli* (EAEC). All but UPEC and NMEC cause gastrointestinal disease. (A newly emerged diarrheagenic strain is described in **eTopic 26.5**.) To distinguish these strains, each group has telltale O and H antigens that can be identified using serology.

Note: "O antigen" is part of the bacterium's LPS, while "H antigen" is a flagellar protein. Thus, "O157:H7" denotes the specific versions of LPS (O157) and flagellar protein (H7) found on *E. coli* O157:H7. Other pathogenic strains of *E. coli* have different O and H antigens.

Shiga toxin. *Shigella* and EHEC, the agent in the preceding case history, both produce toxins, called Shiga toxins 1 and 2, that are encoded by genes of bacteriophage genomes embedded in the bacterial chromosome (the prophage). The toxins, which are absorbed through the intestine and disseminated via the bloodstream, are AB5 toxins (see Section 25.3). The A subunit, upon entry, destroys protein synthesis by cleaving an adenine from 28S rRNA in eukaryotic ribosomes. These toxins inhibit host protein synthesis and, in the process, damage endothelial cells in the intestine, kidney, and brain. Shiga toxin–induced death of vascular endothelial cells in the intestine causes the breakdown of blood vessel linings, followed by hemorrhage that manifests as bloody diarrhea. Shiga toxin 2 also triggers the release of pro-inflammatory cytokines.

Endothelial damage initiates the formation of platelet-fibrin microthrombi (clots) that occlude blood vessels in the various organs, leading to two major syndromes: hemolytic uremic syndrome (HUS) and thrombotic thrombocytopenic purpura (TTP). HUS develops when the microthrombi are limited to the kidney. The microclots clog the tiny blood vessels in this organ and cause decreased urine output, ultimately leading to kidney failure and death. In TTP, the clots occur throughout the circulation, causing reddish skin hemorrhages called petechiae and purpuras (discussed in Section 26.6). Neurological symptoms (for example, confusion, severe headaches, and possibly coma) then arise from microhemorrhages in the brain. The hemorrhaging occurs because platelets needed for normal clotting have been removed from the circulation as they form the microthrombi. The decreased number of platelets is called thrombocytopenia.

HUS is a common consequence of *E. coli* O157:H7 infection, as in the case history described. Unfortunately, HUS can be treated only with supportive care, such as blood transfusions and dialysis throughout the critical period until kidney function resumes. Antibiotic treatment can increase the release of Shiga toxins from the organisms and actually trigger HUS. Thus, antimicrobial therapy is not recommended.

Enterohemorrhagic *E. coli* (EHEC). *E. coli* O157:H7 is a recently emerged pathogen that can colonize cattle intestines at the recto-anal junction without affecting the animal and, as a result, can contaminate meat products following slaughter. Initially identified in 1982, the organism came to national prominence during a large-scale U.S. outbreak, in 1993, linked to a Washington State Jack in the Box restaurant, in which 732 people were sickened.

E. coli O157:H7 rarely affects the health of the reservoir animal. But when an infected steer is slaughtered, the carcass can become contaminated with EHEC-containing feces despite considerable efforts by slaughterhouses to prevent it. Grinding the tainted meat into hamburger distributes the microbe throughout. Cooking burgers to 160°C is essential to kill any existing EHEC. Cross-contamination between foods is possible too. Using the same cutting board to prepare meat and salad is a great way to contaminate the salad, which will not be cooked.

Despite EHEC's common association with hamburger, vegetarians are not safe from this organism. During heavy rains, waste from a cattle farm can easily wash into nearby vegetable fields unless precautions are taken. If the cattle waste contains *E. coli* O157:H7, the crops become contaminated, and the pathogen can enter the plant through stomates (respiration openings in the leaves; **Fig. 26.11**). One such outbreak occurred in 2018. Romaine lettuce grown in certain areas of Yuma, Arizona, was contaminated with this pathogen. Lettuce shipped to 36 states infected 210 people. Of those, 96 were hospitalized, 25 developed HUS, and 5 died. Whole-genome sequencing genetically linked all of the isolated EHEC strains to various farms in Yuma

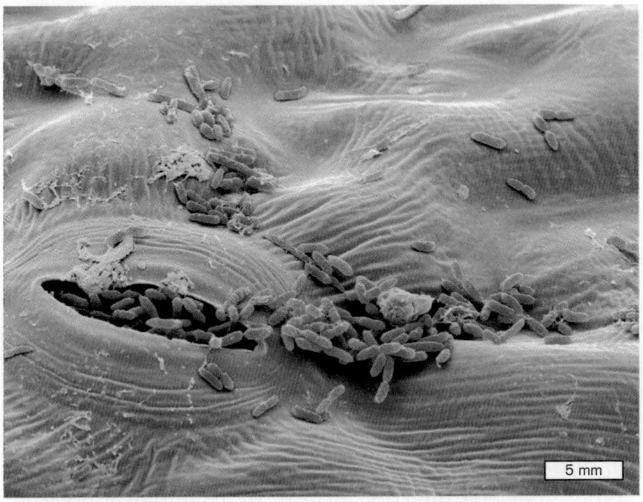

FIGURE 26.11 ■ *E. coli* entering a lettuce leaf. *Escherichia coli* individuals on a lettuce leaf congregate at the stomates and enter the plant veins. Once inside the plant, the organisms resist efforts to wash them out.

and identified the likely source as water from a canal in the Yuma growing area.

Early on, the remarkably low infectious dose of *E. coli* O157:H7 mystified researchers. However, we have since learned that *E. coli* has an impressive level of acid resistance, rivaling that of the gastric pathogen *Helicobacter pylori*. Acid resistance mechanisms permit *E. coli* to survive in the acidic stomach and enable a mere 10–100 individual organisms to cause disease.

In contrast to the case just described, many gastrointestinal infections do not produce fecal leukocytes or blood in the stool. Diarrheal diseases caused by *Vibrio cholerae* (cholera) or enterotoxigenic *E. coli* (ETEC), which produces a cholera-like disease, do not involve invasion of the intestinal lining by the microbe, and they yield copious amounts of watery diarrhea. In these two toxin-driven diseases, the bacteria attach to cells lining the intestine and secrete toxins that are imported into the target cells (see Section 25.3).

Epidemiology of EHEC. Epidemiology is the study of factors and mechanisms involved in the spread of disease. Researchers at the U.S. Centers for Disease Control and Prevention (CDC) used epidemiological principles (discussed in Chapter 28) to identify the risk factors associated with the case study presented earlier. They interviewed 51 infected patients and 92 controls (children who visited the farm but did not become ill). Infected patients were more likely than controls to have had contact with cattle, an important reservoir for *E. coli* O157:H7. All 216 cattle on the farm were sampled by rectal swab, and 13% yielded *E. coli* O157:H7 with a DNA restriction pattern that was identical to those isolated from the patients. This finding indicated that the cattle were the source of infection.

Another contributing factor in this case was that separate areas were not established for eating and interactions with farm animals. Visitors could touch cattle, calves, sheep, goats, llamas, chickens, and a pig while eating and drinking. Handwashing facilities were unsupervised and lacked soap, and disposable hand towels were out of the children's reach. All of these circumstances provided opportunity for infection.

How can we prevent disease caused by enterohemorrhagic *E. coli*? Industry approaches include thoroughly washing carcasses before processing, maintaining cold temperatures, and testing for possible contamination. In addition, the use of gamma irradiation to sterilize beef, spinach, and lettuce has been approved. Recent outbreaks of EHEC disease caused by contaminated hamburger have declined dramatically because of industry practices and USDA inspections. Irradiation could eliminate the problem (described in Chapter 5), but less than 1% of hamburger meat is currently irradiated in the United States.

Rotavirus and Norovirus

Many people wrongly think that most cases of diarrhea are caused by a bacterial agent. Actually, two viruses—**rotavirus** and **norovirus**—cause more intestinal disease than do any bacterial species. Rotavirus is a double-stranded RNA virus (Group III in Table 6.1); norovirus is a positive-sense single-stranded RNA virus (Group IV). Rotavirus is highly infectious, spreading by the fecal-oral route; it is endemic around the globe, and it affects all age groups, although children between 6 and 24 months are the most severely affected. It is estimated that by age 3, all children have had a rotavirus infection.

The incubation period is approximately 2 days, after which the victim commonly suffers frequent watery, dark green, explosive diarrhea. Accompanying symptoms may include nausea, vomiting, and abdominal cramping. Severe dehydration and electrolyte loss due to the diarrhea will cause death unless supportive measures, such as fluid replacement, are undertaken. There is no cure, but most patients recover if rehydrated properly. Few deaths from rotavirus occur in the United States, but each year more than 128,500 children worldwide die from this viral diarrhea (2016 estimate). The mortality and incidence of this disease have decreased because of the introduction in 2006 of a safe and effective vaccine (see Section 24.6). In Mexico alone, the vaccine resulted in a 50% decline in diarrheal deaths.

With the success of the rotavirus vaccine, norovirus is set to become the most common worldwide cause of nonbacterial gastroenteritis. Norovirus infections have already surpassed rotavirus in the United States. The virus is perceived as the scourge of cruise ships and assisted-living facilities, but norovirus can also spread quickly in hotels or anywhere there are many people in a small area. The virus spreads by the fecal-oral route among children or adults, via contaminated food or person-to-person contact. Within 24 hours of infection, the victim experiences sudden vomiting, stomach cramps, and watery diarrhea that mercifully resolves within 12–24 hours. Treatment is similar to that for rotavirus, and there is no vaccine. Death is rare, but when it does occur, it is most common among infants and the elderly. Take notice that newly emerging single-stranded, positive-sense viruses in a group called sapoviruses are now causing increasing cases of norovirus-like disease.

Note: Norovirus infection is sometimes called the "stomach flu," but that is a misnomer. It is not the flu and has nothing to do with influenza virus.

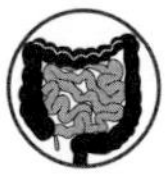

Diarrhea and the gut microbiome. An obvious question to ask about diarrhea is: How does it affect the microbiome? A study headed by Shannon Manning (Michigan State University) found that the composition of intestinal microbiota of patients with diarrhea differs significantly from those of their healthy family members.

For one thing, the gut microbiomes of diarrhea patients who were not given antibiotics were less diverse than those of their uninfected family members. Abundance of Bacteroidetes and Firmicutes was higher in the healthy individuals, whereas Proteobacteria dominated the patient microbiomes. *Escherichia coli*, for instance, predominated in all patients, regardless of the pathogen causing the infection. The composition of diarrheal microbiomes also varied with the bacterial cause of infection. As one example, the microbiome of *Campylobacter*-infected patients differed from that of patients infected with *Salmonella* or *Shigella*.

How does the intestine restore its bacterial population after being decimated by diarrhea or antibiotic treatment? Lawrence David (Duke University; **Fig. 26.12A**) and Peter Turnbaugh (UC San Francisco), along with their colleagues, used metagenomic procedures to characterize the stools of 41 people in Bangladesh (children and adults) who

A.

FIGURE 26.12 ▪ Successive repopulation of the gut microbiome following *Vibrio cholerae* infection. A. Lawrence David (pictured) and Peter Turnbaugh (not shown) unravel the complexities of the human microbiome. B. Fecal samples were taken at 1, 7, and 30 days past presentation (dpp) of diarrhea. Patients, identified by letters, were given a single dose of azithromycin on the day of presentation, which nearly eliminated *V. cholerae* by 1 dpp. Healthy contacts of each patient are identified by numbers and are shown on the left. For example, A-1, A-2, A-3, and A-4 are the healthy contacts of patient A. Colored boxes reflect groups of different genera that are prominent during the infection and afterward, during early, mid, and late stages of repopulation. The *x*-axis values reflect relative abundances of genera. "All swab" microbes were present only in rectal swab samples. *Source:* Part B modified from Harvard Press Office figure.

B.

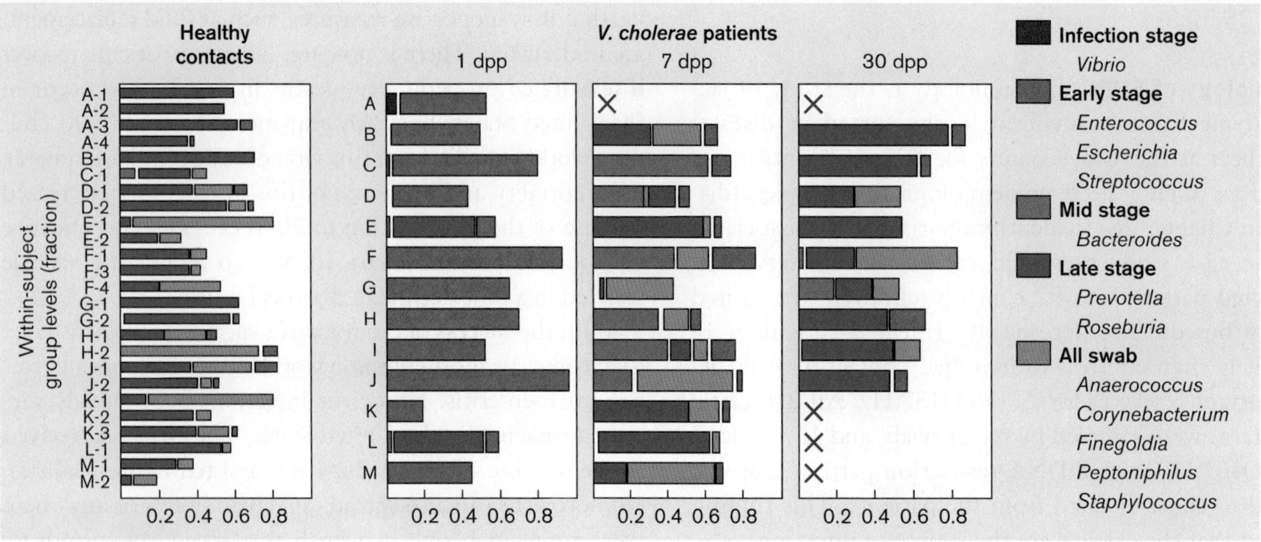

had diarrhea caused by *E. coli* or *Vibrio cholerae*. Stools were monitored before, during, and after diarrhea episodes.

The researchers identified a consistent succession of repopulation events in nearly every case, regardless of the cause of diarrhea (**Fig. 26.12B**). After diarrhea (or antibiotic treatment for the diarrhea) clears out much of the microbiome, carbohydrates and oxygen accumulate in the gut. Carbohydrates and oxygen would normally be metabolized by gut microbiota. During the early stage of repopulation, facultative, oxygen-respiring and carbohydrate-utilizing bacteria (especially those using simple carbohydrates, such as *Escherichia*, *Enterococcus*, and *Streptococcus*) colonize the gut and consume these nutrients. Midstage recovery begins when the lack of simple sugars and oxygen (as well as increased phage predation) leads to a decline in the early-stage species. This decline enables succession to anaerobic, complex carbohydrate-fermenting bacteria (*Bacteroides*). Finally, in late-stage recovery, the gut microbiome once again resembles the complex community that existed prior to infection—the same composition seen in healthy contacts (**Fig. 26.12B**). The entire process takes about 30 days to complete but depends on a variety of factors, such as diet, antibiotic use, and duration of diarrhea.

Repopulation occurs, in part, by microbes being reingested from food. But exciting research suggests that the much maligned and trivialized appendix is, among other things, an important reservoir of gut microbes that can seed the intestine and reestablish the microbiota. Once properly reestablished, gut microbiota are capable of fending off pathogens such as *Clostridioides difficile*. As evidence, researchers found that patients with appendectomies were more than twice as likely to develop repetitive infections with *C. difficile*.

Case History: Ulcers—It's Not What You Eat

Gary was a 34-year-old accountant who had immigrated to Nebraska from Poland 7 years earlier. Since his teenage years, he had been bothered periodically by episodes of epigastric pain (pain around the stomach), nausea, and heartburn. Antacids usually alleviated the symptoms. Over the years, he had received several courses of treatment with Tagamet or Pepcid to reduce acid secretion and provide relief. Recently, an upper-GI endoscopy had been performed, in which a long, thin tube tipped with a camera and light source was inserted into Gary's mouth and threaded down into his stomach. The view through the endoscope showed some reddened areas in the antrum (bottom part) of the stomach. The endoscope was also equipped with a small clawlike structure that obtained a small tissue sample from the lining of Gary's stomach. A urease test performed on the antral biopsy turned positive in 20 minutes. Histological examination of the biopsy confirmed moderate chronic active gastritis (inflammation of the stomach lining) and revealed the presence of numerous spiral-shaped organisms. Cultures of the antral biopsy were positive for Helicobacter pylori.

Painful and sometimes life-threatening gastric ulcers were for many years blamed on spicy foods and stress. In the 1980s, after discovering odd, helical bacteria present in the biopsies of gastric ulcers, Australians J. Robin Warren and Barry Marshall (a medical intern at the Royal Perth Hospital at the time; **Fig. 26.13A**) proposed that bacteria, not pepperoni, cause ulcers (**Fig. 26.13B** and **C**).

Their hypothesis was viewed with skepticism and declared as heresy by the established medical community. Faced with disbelief bordering on ridicule, the young intern drank a vial of the helical organisms and waited. A week later he began vomiting and suffered other painful symptoms of gastritis. Barry Marshall could not have been happier. He had proved his point. We now know that this curly microbe causes the vast majority of stomach ulcers (and has colonized humans for at least 100,000 years).

The discovery of *Helicobacter pylori* and its association with gastric ulcer disease led to a major shift in ulcer treatment, previously limited to suppressing acid production via proton pump inhibitors. Therapy now includes antimicrobial treatment to kill the bacteria, coupled with acid suppression therapy to prevent further inflammation while the ulcer heals. Warren and Marshall, who recovered from his gastritis, received the 2005 Nobel Prize in Physiology or Medicine for their groundbreaking work.

H. pylori can be detected in about half of the world's population, especially in impoverished countries. Why most people colonized with *H. pylori* do not develop gastric ulcers is unclear. The exact mechanism by which *H. pylori* causes gastric ulcers is not known, although a variety of virulence factors have been identified. The basic scheme of *Helicobacter* pathogenesis is shown in **Fig. 26.14**. After the pathogen is ingested, *H. pylori* flagella propel the organism toward the mucosa driven by an ill-defined chemotaxis system that, in part, senses urea produced by the human body (**Fig. 26.14**, step 1). As it approaches the mucosa, the pathogen produces intracellular and extracellular urease, an important virulence factor that converts urea to CO_2 and ammonia. The ammonia neutralizes acid around *Helicobacter*, thereby enabling the organism to survive the extreme acidity of the stomach (step 2).

Two other enzymes—collagenase and mucinase—then soften the mucous lining, helping the bacteria reach the stomach's epithelial lining (**Figs. 26.13C** and **26.14**, step 3). The epithelial lining is much less acidic than the lumen, so the organism can grow and divide. Once at the epithelium, *Helicobacter* produces various adhesins, such as BabA or HpaA, to bind host cells (step 4). After the organism has adhered, tissue damage develops with the release of vacuolating cytotoxin (VacA)

A.
B.
C.

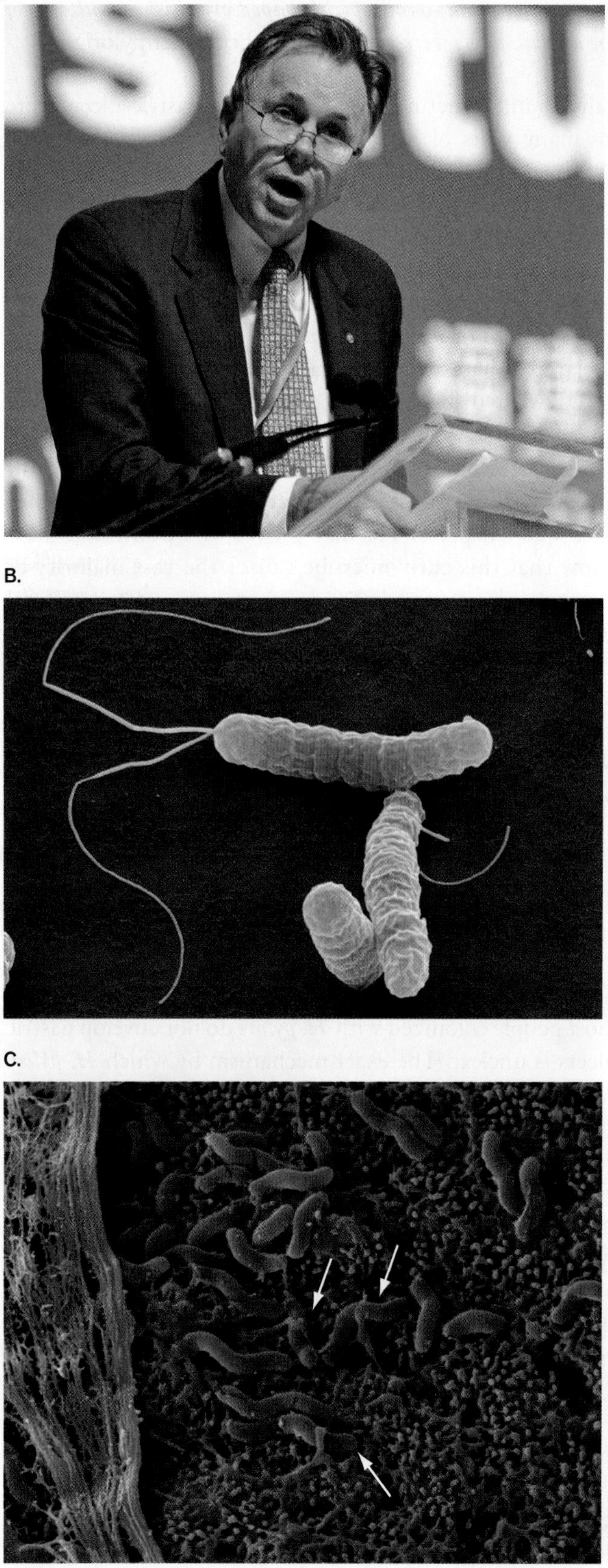

XINHUA/ALAMY STOCK PHOTO

EYE OF SCIENCE/SCIENCE SOURCE

EYE OF SCIENCE/SCIENCE SOURCE

FIGURE 26.13 ■ A bacterial cause of gastric ulcers. **A.** Physician Barry Marshall was so sure he was right about the cause of stomach ulcers that he swallowed bacteria to prove his point. **B.** *Helicobacter pylori* (SEM). Note the tuft of flagella at one pole. Cell length approx. 2 μm. **C.** *Helicobacter* (arrows) attached to gastric mucosa.

and neutrophil-activating protein (NAP). NAP activates neutrophils and mast cells to damage local tissue (step 5). VacA forms a hexameric pore in the host membrane and induces apoptosis (programmed cell death) by damaging mitochondria. Apoptotic cells decrease the immune response, which will stabilize a chronic infection.

Another protein, CagA, is injected (by a type IV secretion system; see Chapter 25) into host epithelial cells, where it becomes phosphorylated. CagA then interacts with host signaling proteins and activates host signal transduction pathways that can stimulate inflammation and growth possibly leading to cancer (**Fig. 26.14**, step 6). Gastric adenocarcinoma associated with *Helicobacter* is the third leading cause of cancer-associated deaths worldwide.

Recent evidence suggests that *H. pylori* might also affect diseases outside the stomach. CagA, for example, has been found in patient serum, packaged within exosome vesicles generated by host exocytosis. *Helicobacter* also sheds outer membrane vesicles (OMVs) that contain CagA. OMVs containing CagA incubated with host cells produced changes in host cell gene expression and function. It is unclear as yet whether the same can happen in an infected person.

Tools useful for diagnosing *H. pylori* include a fecal antigen test, rapid urease testing, and serology [for example, enzyme-linked immunosorbent assay (ELISA) to detect antibody to the CagA antigen]. ELISA is a common immunological tool used to detect the presence, in serum, of antibodies to a specific organism—an indication of infection. The ELISA test is described more fully in Section 28.3.

Protozoan Causes of Diarrheal Disease

As we learned in Chapter 20, some protozoa (also called protists) cause serious human diseases. For instance, *Entamoeba histolytica* (**Fig. 26.15A** and Section 20.6) and *Cryptosporidium* (*C. parvum* and *C. hominis*; **Fig. 26.15B**) cause the diarrheal diseases amebic dysentery and cryptosporidiosis, respectively. In 2016, the CDC tallied 13,453 cases (up from 2,640 in 2005) of cryptosporidiosis, a reportable disease in the United States. *Cryptosporidium* infection begins with oocysts that contain four sporozoites (infectious form of protozoan) being shed in the feces of infected hosts (human and nonhuman animals) to contaminate water (drinking or recreational, such as in water parks). Following ingestion, the sporozoites are released and parasitize intestinal epithelial cells. Asexual and sexual cycles then lead to more oocysts. Treatment of immunocompetent patients usually involves only rehydration therapy, although an antiparasitic drug is available. HIV patients infected with *C. parvum* must also receive antiretroviral therapy to improve CD4 T-cell count.

The flagellated protozoan *Giardia lamblia* is a major cause of diarrhea throughout the world. In the United States alone, *G. lamblia* caused 16,310 reported cases

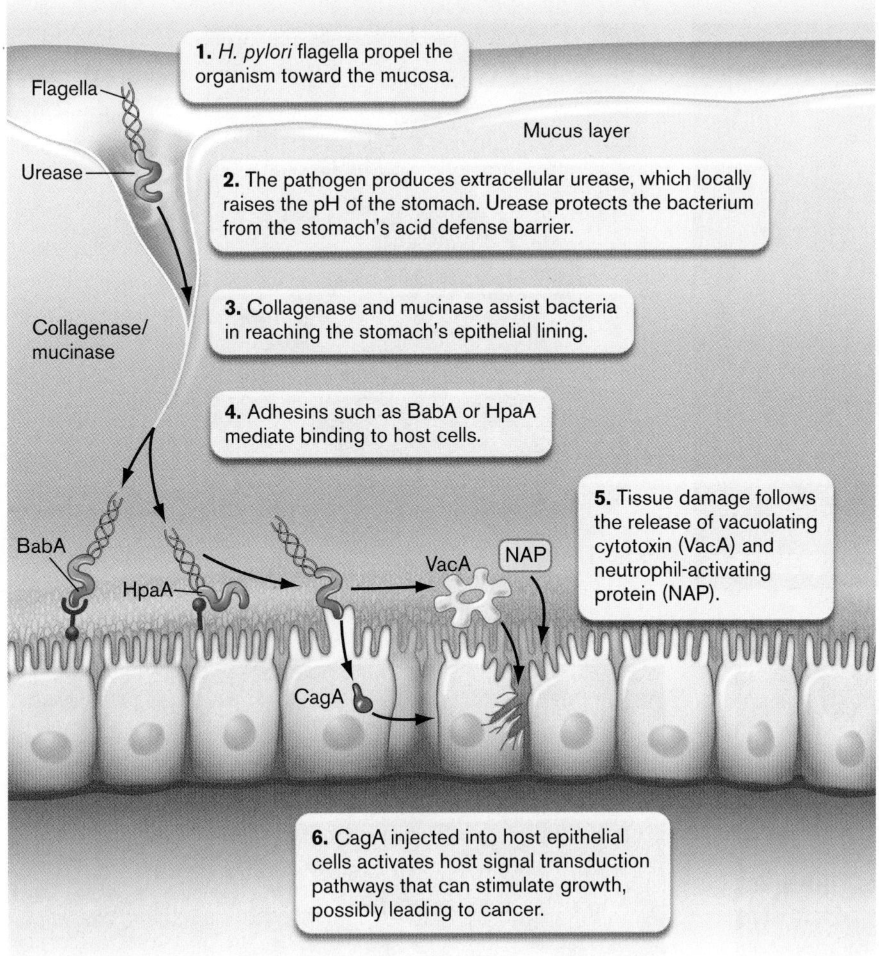

FIGURE 26.14 ■ Steps in *Helicobacter* pathogenesis.

of giardiasis diarrhea in 2016 (over 19,000 in 2010) and likely caused thousands more that were not reported. *G. lamblia* enters a human or other host as a cyst present in drinking water contaminated by feces (**Fig. 26.15C**). A cyst is a dormant form of a protist encased in a protective wall (Chapter 20). Aside from humans, *G. lamblia* can be found in various rodents, deer, cattle, and even household pets. It is very infectious. Ingestion of as few as 25 cysts can lead to disease. Following ingestion, the hard, outer coating of the cyst is dissolved by the action of digestive juices to produce a trophozoite (**Fig. 26.15D**), which attaches itself to the wall of the small intestines and reproduces. Offspring quickly encyst and are excreted out of the host's body.

Asymptomatic carriers of *G. lamblia* are common; it has been estimated that anywhere from 1% to 30% of children in U.S. day-care centers are carriers. Disease usually manifests as greasy stools alternating between a watery diarrhea, loose stools, and constipation. However, some patients will experience explosive diarrhea. Diagnosis usually comes from observing the cysts or trophozoite forms of the protozoan in feces. Metronidazole is a drug often used to cure the disease. To prevent it in the first place, proper treatment of community water supplies is essential.

Although most gastrointestinal disease is intestinal in locale, specialized microbes can also target the stomach, with its harsh acidic environment.

Hepatitis Viruses Target the Liver

The liver is also considered part of the gastrointestinal system, because it provides bile to the intestine and processes the nutrients absorbed from the intestine. **Hepatitis** is a general term meaning "inflammation of the liver." Although many pathogens can affect this organ, we will focus on the major infectious cause, an eclectic group of viruses collectively called the hepatitis viruses. They include a single-stranded, negative-sense RNA virus (hepatitis A, HAV) a double-stranded DNA virus (hepatitis B, HBV), and a single-stranded, positive-sense RNA virus (hepatitis C, HCV). Space does not permit coverage of HDV or HEV.

Table 26.3 compares general features of these viruses. The disease symptoms of infectious hepatitis, regardless of viral cause, include abdominal pain, fever, vomiting, and, often, dark urine, clay-colored stools, and jaundice. Liver damage from hepatitis is marked by the detection of liver transaminases in serum, and an enlarged liver (hepatomegaly).

Hepatitis A virus (HAV) is a single-stranded RNA picornavirus that causes an acute infection spread person-to-person by the fecal-oral route, but hepatitis A can also be contracted by eating undercooked shellfish collected from contaminated waters. The virus replicates in the intestinal endothelium and is disseminated via the bloodstream to the liver. After replicating in hepatocytes, the progeny enter the bile and are released into the small intestine, explaining why stools are so infectious. Though the virus has an early viremic stage after leaving the intestine, it is rarely transmitted by transfusion, because the viremic stage is transient and ends after liver symptoms develop. In contrast, hepatitis B and C viruses produce persistent viremia and are readily transmitted by transfusion.

Many people who are infected with HAV are asymptomatic or exhibit very mild symptoms that include nausea, vomiting, diarrhea, low-grade fever, and fatigue. As the virus attacks the liver, liver transaminases are detected in serum, patients may become jaundiced (from the

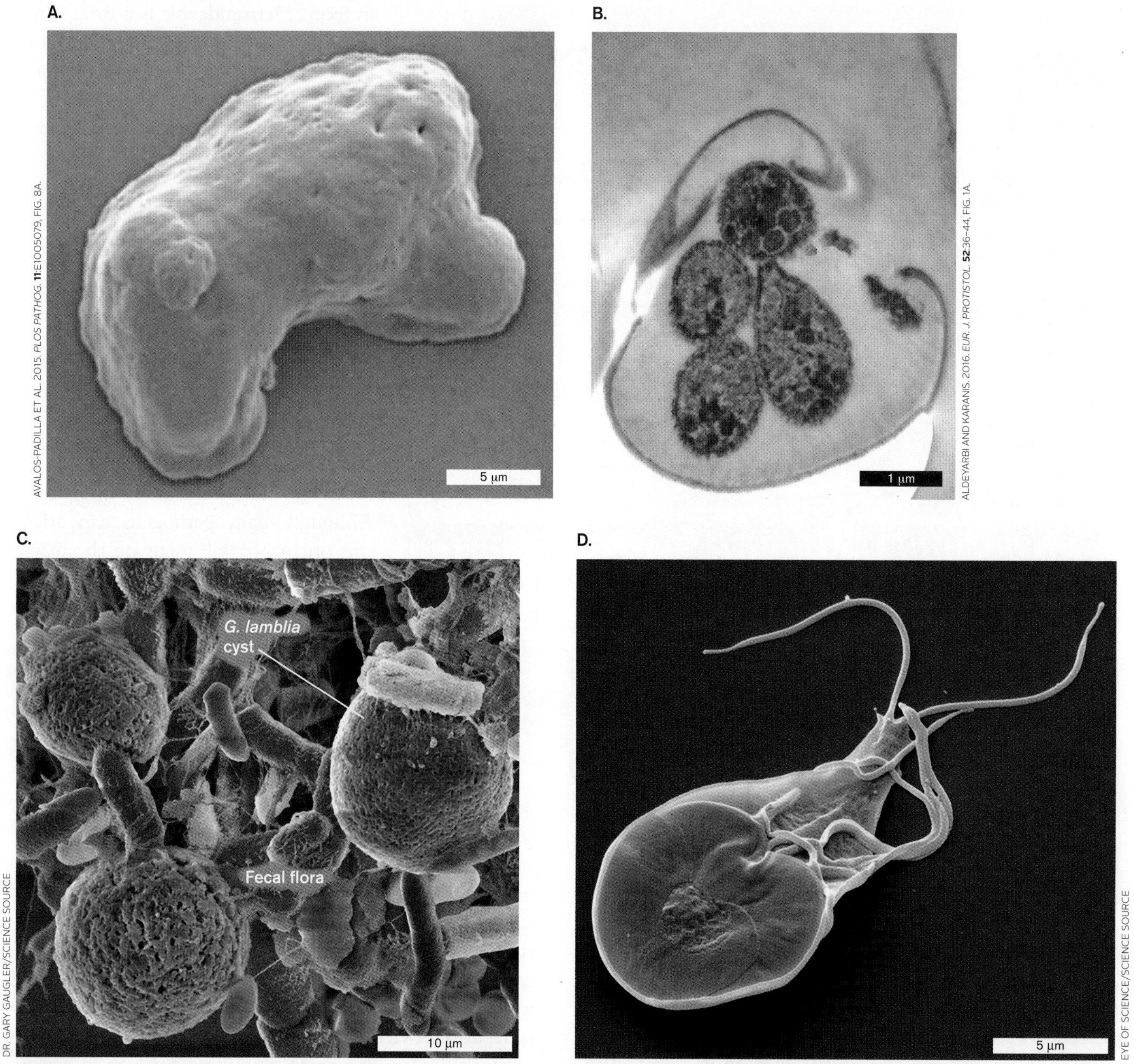

FIGURE 26.15 ■ Protists that cause diarrhea. **A.** *Entamoeba histolytica* trophozoite (SEM, motile form). **B.** *Cryptosporidium parvum* oocyst encasing four infectious sporozoites (TEM). **C.** *Giardia lamblia* cysts (7–14 µm) present in fecal matter (colorized SEM). **D.** Trophozoite form of *G. lamblia* (5–15 µm in length; colorized SEM).

accumulation of bilirubin in the skin), and their urine will turn dark brown. Diagnosis can be confirmed by PCR or the detection of IgM antibody to HAV antigens. There is no specific treatment, but the disease usually lasts for only a few months and then resolves without establishing a carrier state. Disease can be prevented, however, if immunoglobulin is given to someone who has had contact with an infected individual. A hepatitis A vaccine containing inactivated virus (called HepA vaccine) is administered after 1 year of age. For those not vaccinated, frequent handwashing is important for preventing spread of the disease, because it interrupts the fecal-oral cycle.

In contrast to HAV, **hepatitis B virus** (**HBV**) is a partially double-stranded circular DNA virus (family *Hepadnaviridae*) that causes diseases of varying severity. These include acute and chronic hepatitis, cirrhosis, and hepatocarcinoma. The virus is also wrapped in a membrane envelope when progeny viruses are released from infected cells. The virion coat protein, a surface antigen, is called HBsAg. The virus makes an excess amount of HBsAg, so it is sometimes extended as a tubular tail on one side of the virus particle and is often found in the blood of infected individuals in the form of noninfectious filamentous and spherical particles (**Fig. 26.16**). The presence of HBsAg in blood is an

TABLE 26.3 Comparative Features of Hepatitis Viruses and Disease

	Hepatitis A	Hepatitis B	Hepatitis C	Hepatitis D	Hepatitis E
Type	ssRNA, negative sense	DNA, enveloped	ssRNA, positive sense, enveloped	ssRNA, negative sense, circular, enveloped	ssRNA, positive sense, nonenveloped
Incubation period	30 days	90 days	40 days	30 days; requires coinfection with HBV	50 days
Route	Fecal-oral	Parenteral, sexual, perinatal	Parenteral, sexual	Parenteral, sexual	Fecal-oral
Viremia	Transient	Persistent	Persistent	Uncommon	Transient
Severity	Mild	Severe	Mild	Mild to severe	Mild
Chronic	None	90% children 10% adults	50%–60%	80%–90%	None

indicator of HBV infection. There are also nucleic acid and serological tests for HBV RNA and antibodies, respectively.

HBV is transferred primarily via parenteral routes such as blood transfusions, contaminated needles shared by IV drug users, and any human body fluid, including saliva, semen, sweat, breast milk, tears, urine, and feces. It can even be transferred transplacentally to a fetus and can be sexually transmitted. Infection by HBV has two stages: a short-term acute phase and a long-term chronic phase that, if it extends beyond 6 months, may never resolve (chronic infection). Symptoms resemble those of the flu, but with jaundice and brown urine.

Liver damage caused by HBV infection is due in large part to an efficient cell-mediated immune response. Cytotoxic T cells and natural killer cells cause immune lysis of infected liver cells. Over the long term, chronic hepatitis will lead to a scarred and hardened liver (cirrhosis), the only recourse being a liver transplant. Fortunately, about 90% of those infected are able to fight off infection and never proceed to the chronic stage. A HepB vaccine (made from recombinant HBsAg) is available. Its administration is recommended after birth, followed by booster shots administered by 2 months and 18 months of age.

Hepatitis C virus (HCV) causes another form of hepatitis. HCV is a single-stranded positive-sense, linear RNA virus with a lipid coat and is a member of the *Flaviviridae* family. It is transmitted by blood transfusions and causes 90% of transfusion-related cases of hepatitis. It can also be transmitted by needle sticks, razor blades, tattooing, and, less frequently, by sex. Over 100 million people worldwide are infected with HCV. Screening for HCV (serology or PCR) is recommended for anyone who exhibits signs of hepatitis or practices the risky behaviors noted here. In addition, the CDC recommends that anyone born between 1945 and 1965 (baby boomers) be tested because, for unknown reasons, baby boomers are five times more likely to be infected than other adults.

Most HCV-infected individuals (80%) do not exhibit symptoms, and in those who do, symptoms may not appear for 10–20 years. At least 75% of patients who exhibit symptoms ultimately progress to chronic hepatitis requiring a liver transplant, or possibly to liver cancer. Fortunately, infection can be detected using ELISA. Liver biopsies of HCV patients are used to determine the extent of liver damage, which in turn helps establish the stage of disease.

Prevention of HBV or HCV infection for health care personnel includes avoiding inadvertent needle sticks. If such a stick should occur with HBV, anti-HBV immunoglobulin should be administered within 7 days. Currently, no effective post-exposure prophylaxis is recognized for HCV. Chronic hepatitis can be treated with an antiviral protease inhibitor (glecaprevir) that prevents the proteolytic processing of an important HCV polyprotein; and Harvoni, a combination of an RNA chain terminator with an inhibitor of a virus phosphoprotein needed for replication. Though vaccines have been developed for HAV and HBV (see Section 24.6), no vaccine is yet available for HCV.

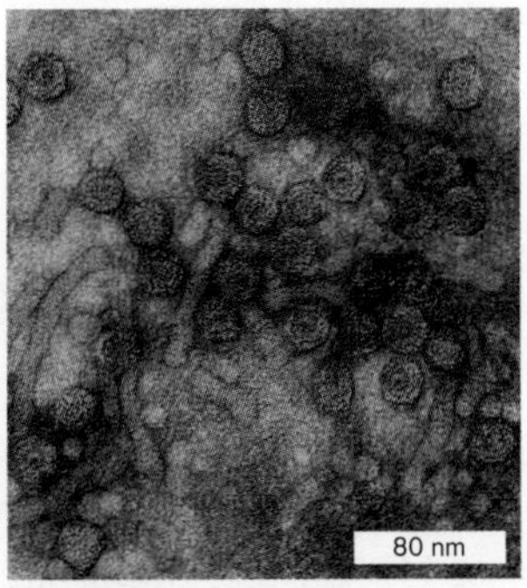

EYE OF SCIENCE/SCIENCE SOURCE

FIGURE 26.16 ■ Structure of hepatitis B virus. Hepatitis B is an enveloped, double-stranded DNA virus in the *Hepadnaviridae* family (TEM). Also shown are the tubular structures made from HepB surface antigen (HBsAg).

Note: Since hepatitis viruses can be spread via contaminated blood products, all blood donations collected by the Red Cross and other agencies are tested for the presence of these viruses, as well as for HIV. Thus, the blood supply is safe.

We have examined in this section only a handful of the bacterial, viral, and protozoan microbes that cause gastrointestinal infection. Others are listed in **Table 26.4**, and some of the protozoan pathogens are described in Chapter 20.

TABLE 26.4 Selected Microbes That Cause Diseases of the Gastrointestinal Tract

Etiological agent[a]	Disease	Symptoms	Virulence factors	Source(s)	Treatment
Bacterial					
Campylobacter jejuni (G–)	Gastroenteritis	Fever, muscle pain, watery diarrhea, blood in stool, headache	Cytotoxin, enterotoxin, adhesin	Poultry, unpasteurized milk	Erythromycin
Clostridioides difficile (G+)	Pseudomembranous enterocolitis	Fever, abdominal pain, diarrhea, pseudomembrane in colon	Cytotoxin, antibiotic resistance	Animals, normal microbiota	Vancomycin
Clostridium botulinum (G+, anaerobe)	Botulism	Fast onset of symptoms; flaccid paralysis	Neurotoxin	Preformed toxin in foods	Antiserum
Clostridium perfringens (G+)	Gastroenteritis	Watery diarrhea, nausea	Alpha toxin	Soil, food	Self-limiting
Enterohemorrhagic *E. coli*	Gastroenteritis	Bloody diarrhea, HUS	Intimin, Tir, type III secretion, Shiga toxin	Contaminated foods (hamburger) and crops	Oral rehydration; antibiotics if severe
Enterotoxigenic *E. coli*	Traveler's diarrhea	Watery diarrhea	Labile and stable toxins	Humans; food, water	Oral rehydration
Helicobacter pylori (G–)	Gastric ulcers	Abdominal pain, bleeding, heartburn	Adhesin, urease, CagA, vacuolating toxin	?	Triple drug protocol (omeprazole, clarithromycin, metronidazole)
Salmonella enterica (G–)	Salmonellosis	Symptoms after 18 h; abdominal pain, diarrhea; invade intestinal M cells	Type III secretion, intracellular growth	Chickens, other animals; fecal-oral route	Oral rehydration; antibiotics if severe
Salmonella enterica serovar Typhi (G–)	Typhoid fever	Headache, fever, chills, abdominal pain, rash (rose spots), hypotension, diarrhea in late stages	Type III secretion, intracellular growth, PhoPQ regulators, Vi antigen capsule	Human carriers (gallbladder reservoir); food, water	Quinolones
Shigella spp. (G–)	Shigellosis	Bloody diarrhea, HUS	Shiga toxin, type III secretion, intracellular growth, actin-based motility, escape phagosome	Humans; fecal-oral route	Oral rehydration; antibiotics if severe
Staphylococcus aureus (G+)	Staphylococcal food poisoning	Symptoms within 4 h of ingestion; nausea, vomiting, diarrhea	Enterotoxin	Preformed toxin in foods	Supportive
Vibrio cholerae (G–)	Cholera	Watery diarrhea	Cholera toxin, toxin-coregulated pili (TCPs), ToxR regulator	Human waste–contaminated water	Oral rehydration, antibiotics
Vibrio parahaemolyticus (G–)	Gastroenteritis	Diarrhea, blood in stool	Enterotoxin	Raw seafood	Self-limiting
Viral					
Norovirus (Norwalk virus)	Stomach "flu"	Nausea, vomiting, diarrhea	VP1	Fecal-oral route	Oral rehydration
Rotavirus (most common cause)	Stomach "flu"	Nausea, vomiting, diarrhea	NSP4	Fecal-oral route	Oral rehydration

[a]G+ = Gram-positive; G– = Gram-negative.

To Summarize

- **Diarrhea leads to dehydration**, for which fluid replacement is a critical treatment. Antibiotic treatment is usually not recommended.
- **Staphylococcal food poisoning** is not an infection. It is a toxigenic disease.
- **Antibiotic treatments can sometimes cause gastrointestinal disease** (for example, pseudomembranous enterocolitis by *Clostridioides difficile*).
- **The presence of red and white blood cells in fecal contents** is an indication of invasive bacterial infection by intracellular pathogens such as *Shigella*, *Salmonella*, and EIEC.
- **Bacteria that do not invade intestinal cells** usually produce watery diarrhea. EHEC is an exception; the attachment and effacing lesions it produces result in bloody stools.
- **Bacterial toxins** produced by bacterial enteric pathogens can cause systemic symptoms.
- **Rotavirus** is still the single greatest cause of diarrhea worldwide. Increasing use of rotavirus vaccine may eventually leave norovirus as the world's leading cause of diarrhea.
- **The bacterium *Helicobacter pylori***, a common cause of gastric ulcers, is highly acid resistant and lives in the stomach of half the world's population. It is also a possible cause of gastric cancer.
- ***Giardia lamblia*** and ***Cryptosporidium parvum*** are major protozoan causes of diarrhea worldwide.
- **Hepatitis** is caused by several unrelated viruses; among them, HAV, HBV, and HCV account for most disease. **HAV** is transmitted by the fecal-oral route and does not establish chronic infection. **HBV** and **HCV** can be transmitted by blood products (such as transfusions) and shared hypodermic needles and can lead to chronic hepatitis. Vaccines for HAV and HBV, but not HCV, are available.

26.4 Genitourinary Tract Infections

The genital and urinary tracts are certainly different organ systems. One is used for procreation, whereas the other filters and excretes waste products from blood. However, their close association in the body makes it useful to discuss their infections in the same section. Despite this link, there are two interesting peculiarities to note about the agents that cause these infections: First, viruses and bacteria can infect the genital tract, but viruses rarely cause urinary tract infections. Second, very few pathogens can infect both organ systems.

Urinary Tract Infections

The urinary tract includes the kidneys, ureters, urinary bladder, and urethra. Infections anywhere along this route are called urinary tract infections (UTIs). UTIs are the second most common type of bacterial infection in humans, ranking in frequency just behind respiratory infections such as bronchitis or pneumonia. In the United States, bladder infections and other UTIs result in over 6 million patient visits annually, mostly by women. Estimates are that at least 25% of women between 20 and 40 years of age have experienced a UTI, and 20%–40% of those infected develop recurrent infections. UTIs result in 100,000 hospital admissions in the United States and $1.6 billion in medical expenses each year.

Urine, as produced in the kidneys, is normally sterile and was thought to be sterile when stored in the bladder. Studies now show that the bladder has a normal microbiome, although its role in human health is unclear. Why didn't we notice this earlier? All known bacterial causes of urinary tract infections are facultative anaerobes that grow under aerobic conditions. The microbiota of our urinary bladder are typically slow-growing and/or anaerobes, and do not grow when urine is plated aerobically on blood agar for 24–48 hours—the normal procedure used to identify urinary tract pathogens. The current thought is that bladder microbiota are not pathogenic, so the fact that they do not grow under conditions used to identify uropathogens is convenient for the clinical laboratory seeking pathogens.

Bacteria that cause UTIs are introduced into the bladder or kidney in one of four ways:

- **Infection from the urethra to the bladder.** This is the most common route for bladder infections (called **cystitis**). Bacteria residing along the urethra can ascend to the bladder. This is a more common occurrence in women than in men. Uropathogenic bacteria colonizing the urethra can also be introduced into the bladder by means of mechanical devices such as catheters or cystoscopes that are passed through the urethra into the bladder.
- **Deposition of bacteria from the bloodstream to the kidney.** Kidney infections (called **pyelonephritis**) can arise when microorganisms from infections elsewhere in the body disseminate via the bloodstream.
- **Descending infection from the kidney to the bladder.** Descending infection occurs when bacteria from an infected kidney are shed into the ureters. The microbes are then carried by urine into the bladder.

- **Ascending infection to the kidney.** In ascending infection, bacteria from an established infection in the bladder ascend along the ureter to infect the kidney. This is the most common route for pyelonephritis.

Thought Question

26.5 Why do you think most urinary tract infections occur in women?

Urine is bacteriostatic to most of the commensal organisms inhabiting the perineum and vagina, such as *Lactobacillus*, *Corynebacterium*, diphtheroids, and *Staphylococcus epidermidis*. In contrast, many Gram-negative organisms thrive in urine. As a result, most urinary tract infections are caused by facultative Gram-negative rods from the GI tract. The most common etiological agents of UTIs are:

- Certain serotypes of *E. coli* that comprise the uropathogenic *E. coli* (75% of all UTIs)
- *Klebsiella*, *Proteus*, *Pseudomonas aeruginosa*, *Enterobacter* (20%)
- *Staphylococcus aureus*, *Enterococcus*, *Chlamydia*, fungi, *Staphylococcus saprophyticus*, other (5%)

Case History: Classic Urinary Tract Infection

Lashandra was 24 years old and had been experiencing back pain, increased frequency of urination, and dysuria (painful or burning urination) over the previous 3 days. This was the first time Lashandra had ever suffered from persisting dysuria. She consulted her general practitioner, who requested a midstream specimen of urine. Upon microscopic examination, the urine was found to contain more than 50 leukocytes per microliter (normal is fewer than 5) and 35 red blood cells per microliter (normal is 3–20). No epithelial squamous cells (skin cells) were seen, indicating a well-collected midstream catch. The urine culture plated on agar medium yielded more than 10^5 colonies per milliliter of urine of a facultative anaerobic Gram-negative bacillus capable of fermenting lactose.

The first question to ask in this case is whether the patient had a significant UTI. The purpose of the midstream urine collection is to provide laboratory data to make this determination. Even though urine in the bladder is normally considered sterile (see the preceding discussion), urine becomes contaminated with normal skin or GI microbiota that may adhere to the urethral wall. In a midstream collection, the patient urinates briefly, stops to position a collection jar, and resumes urinating to collect the sample. This procedure washes away organisms clinging to the urethra before actually collecting the sample. Nevertheless, the collected sample will still contain low numbers of organisms representing normal microbiota of the urethra. (See Section 28.1 for more on urine collection.)

The number of bacteria per milliliter of urine does not have to reach 10^5, as happened in the case history, to diagnose cystitis. A diagnosis of cystitis can be made in a symptomatic patient when the number of bacteria (single colony type) in a sample is at least 1,000 per milliliter. In fact, relying only on the number of bacteria in urine can be misleading. Some people (especially the elderly) can be asymptomatic yet have bacterial counts of 10^5 per milliliter. These patients have asymptomatic bacteriuria and are not usually treated with antibiotics, unless they are pregnant women. The patient in the case history was symptomatic and had more than enough bacteria in her urine to indicate a UTI.

The laboratory found the organism to be a Gram-negative bacillus that ferments lactose, suggesting *E. coli* as the likely culprit. *E. coli* that can colonize the intestine and infect the urinary tract are called uropathogenic *E. coli* (UPEC). Given that this was the first UTI suffered by the patient, the infection was likely the result of an inadvertent introduction of the gut microbe into the urethra. The organism makes its way up the urethra and into the bladder. Another way the bladder can become infected is via a descending route from the kidney. Organisms from a kidney that was infected as a result of sepsis can descend along a ureter to cause a bladder infection.

Gram-negative rods not only thrive in urine, but are adapted to cause urinary tract infections through specialized pili. These pili have terminal receptors for glycolipids and glycoproteins present on urinary tract epithelial cells (**Fig. 26.17A**). UPEC strains of *E. coli*, for example, typically have type I pili whose tips attach to mannose receptors on bladder epithelial cells, and P-type pili whose terminal receptor binds to P antigens present on bladder and kidney cells. P antigen is a so-called blood group marker expressed by approximately 75% of the population. Individuals with P antigen are particularly susceptible to UTIs.

Some patients, predominantly women, suffer with recurrent bladder infections. These infections are thought to be caused in two ways: by a UPEC strain colonizing the intestine that is accidentally reintroduced into the bladder, or by a uropathogenic *E. coli* already in the bladder that invades urinary tract epithelial cells to form compact intracellular biofilms (discussed in **eTopic 26.6**). *E. coli* can emerge from these uropods to "reignite" a UTI. What triggers the emergence? Amanda Lewis's laboratory (Washington University School of Medicine in St. Louis; **Fig. 26.18**), using a mouse model, found that exposing the bladder to *Gardnerella vaginalis*, a Gram-positive bacillus that is part of the vaginal microbiome, can trigger the egress of *E. coli* from

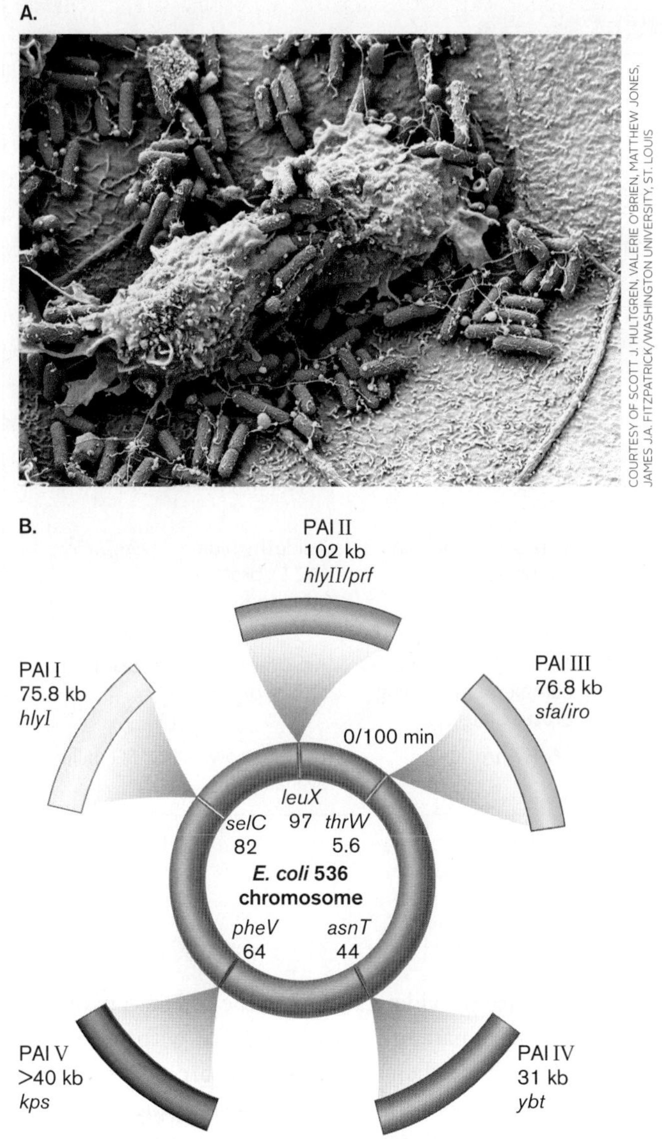

FIGURE 26.17 ▪ Uropathogenic *E. coli*. **A.** Bladder cell with adherent uropathogenic *E. coli* (SEM). White blood cells (yellow) reach out with extracellular traps to immobilize and kill the pathogen **B.** Distribution of pathogenicity islands (PAIs) in uropathogenic *E. coli*. The location of each insert is given in map units within the circle representing the genome. The 0–100 map units are called centisomes. Each centisome is approx. 44 kb of DNA. Zero is arbitrarily placed at the *thr* (threonine) gene. The origin of replication on this map is near 82 centisomes. A chromosomal gene flanking the insert is also provided. The size of each island is shown above the insert. A key virulence gene for each island is listed.

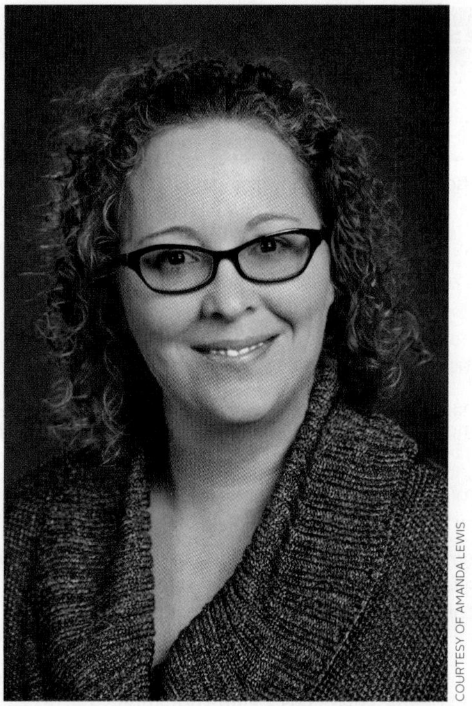

FIGURE 26.18 ▪ Amanda Lewis studies recurrent urinary tract infections.

a uropod. Lewis proposes a new paradigm, called "covert pathogenesis," in which transient exposure of a dormant pathogen to some member of the human microbiome can drive the recurrence of disease.

Urinary tract infections are frequently acquired during a hospital stay (so-called nosocomial, or hospital-acquired, infections). These infections are often precipitated by urinary catheters. In these cases, the causal organism is less likely to be *E. coli* and more likely to be another Gram-negative bacterium or *Staphylococcus*. Many UTIs resolve spontaneously, but others can progress to pyelonephritis (destroying the kidney) or septicemia. As a result, antibiotic therapy is recommended. In older patients, UTIs frequently show atypical symptoms, including delirium, which disappears when the UTI is treated.

How can recurrent UTIs be managed? One way would be to prevent UPEC strains from colonizing the intestine. Scott Hultgren's laboratory (also at Washington University in St. Louis) found that feeding mice a compound (mannoside M4284) that binds UPEC type I pili with high affinity will prevent UPEC strains from colonizing the intestine. The mannoside blocks the pilus tip from binding to host mannose receptors. The idea, not yet tested, is that simultaneously treating a patient with antibiotics to clear the UTI, along with the mannoside to prevent UPEC from colonizing the intestine, will reduce the number of recurrent UTIs.

Thought Question

26.6 Urine samples collected from six hospital patients were placed on a table at the nurses' station awaiting pickup from the microbiology lab. Several hours later, a courier retrieved the samples and transported them to the lab. The next day, the lab reported that four of the six patients had UTIs. Would you consider these results reliable? Would you start treatment based on these results?

What makes uropathogenic *E. coli* different from other strains of *E. coli*? This is a question still under investigation,

TABLE 26.5 Common Sexually Transmitted Infections

Disease	Symptoms	Etiological agent[a]	Virulence factors	Treatment	Reported cases
Gonorrhea	Purulent discharge, burning urination; can lead to sterility	*Neisseria gonorrhoeae* (G–)	Type IV pili, phase variation	Ceftriaxone plus azithromycin	468,514[b]
Syphilis	1°: chancre; 2°: joint pain, rash; 3°: gummata, aneurism, central nervous system damage	*Treponema pallidum* (spirochete)	Motility	Penicillin	88,042[b]
Nongonococcal urethritis	Watery or mucoid urethral discharge, burning urination	*Chlamydia trachomatis*	Intracellular growth; prevents phagolysosome fusion	Azithromycin	1.6 million[b]
Trichomoniasis	Vaginal itching, painful urination, strawberry cervix	*Trichomonas vaginalis* (protozoan)	Cytotoxin	Metronidazole	2.3 million[c]
Chancroid	Painful genital lesion	*Haemophilus ducreyi* (G–)	?	Erythromycin	7[b]
HIV diagnoses ranging from asymptomatic to AIDS	For AIDS, fever, diarrhea, cough, night sweats, fatigue, opportunistic infections	HIV	gp120, Rev, Nef, and Tat proteins	Azidothymidine (AZT), protease inhibitors, zidovudine	34,775[b]
Genital herpes	Painful ulcer on external genitals, painful urination	Herpes simplex 2	Cell fusion protein, complement-binding protein, latency	Acyclovir, iododeoxyuridine	50 million[c]
Genital warts	Warts on external genitals	Human papillomavirus	E6, E7 proteins	Vaccine now available	50 million[c]

[a]G+ = Gram-positive; G– = Gram-negative.

[b]CDC reported cases for 2016.

[c]Total estimated current cases. Nonreportable disease.

but genomic analysis has exposed five pathogenicity islands unique to these strains (**Fig. 26.17B**). The functions of these pathogenicity islands are still under investigation.

Sexually Transmitted Infections

Sexually transmitted infections (STIs) are defined as infections transmitted primarily through sexual contact. The organisms or viruses involved are generally very susceptible to drying and require direct physical contact with mucous membranes for transmission. Because sex can take many forms in addition to intercourse, these microbes can initiate disease in the urogenital tract, rectum, or oral cavities. Condoms can prevent transmission but do so, of course, only when used properly. Examples of common sexually transmitted infections are listed in **Table 26.5**.

Note: Because not all infections progress to diseases, the term "sexually transmitted infection" (STI) is preferred over the older phrase "sexually transmitted disease" (STD).

An important part of limiting the spread of any STI is to identify and treat all sexual partners the patient has had in the previous 60 days. Either the patient (preferably) or the health care provider can notify the affected people. In some instances, patients can be supplied with medication to give directly to their partners. STIs are also reportable to state and federal health agencies, such as the Centers for Disease Control and Prevention.

Case History: Secondary Syphilis

An 18-year-old pregnant woman came to the county urgent-care clinic with a low-grade fever, malaise, and headache. She was sent home with a diagnosis of influenza. She again sought treatment 7 days later, after she discovered a macular rash (flat, red) developing on her trunk, arms, palms of her hands, and soles of her feet. When asked, the patient revealed that 1 year earlier, she had had a painless ulcer on her vagina that healed spontaneously. She was diagnosed with secondary syphilis—a diagnosis confirmed by a serological test. She was given a single intramuscular injection of penicillin and told that her sexual partners had to be treated as well.

The vaginal ulcer, the long latent period, and the secondary development of a rash on the hands and feet described in the case history are classic symptoms of syphilis. Christopher Columbus and/or his crew are thought to have inadvertently delivered the treponeme that causes syphilis

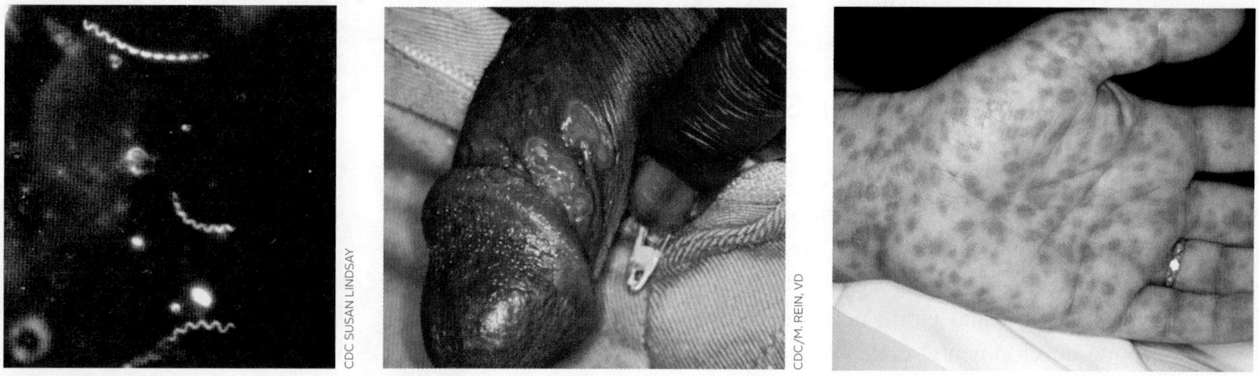

FIGURE 26.19 ▪ Syphilis. **A.** *Treponema pallidum* (dark-field microscopy). Organisms are 10–25 μm long. **B.** Chancre of primary syphilis. **C.** Rash of secondary syphilis.

from the Americas to Europe in the fifteenth century, but syphilis as a disease was not recognized until the sixteenth century. The infectious agent, a spirochete named *Treponema pallidum*, was finally discovered in 1905 (**Fig. 26.19A**). (Chapter 18 describes spirochete structure.) But another 113 years would pass before this anaerobe would be cultured in vitro. Steven Norris (University of Texas Health Science Center at Houston) finally achieved long-term logarithmic growth of *T. pallidum* in 2018, by periodically subculturing (6–7 days) and feeding the treponeme in a microaerobic rabbit epithelial cell coincubation system.

The disease syphilis has several stages. The incubation stage can last from 2 to 6 weeks after transmission, during which time the organism multiplies and spreads throughout the body. **Primary syphilis** is defined by an inflammatory reaction at the site of infection called a **chancre** (**Fig. 26.19B**). About a centimeter in diameter, the chancre is painless and hard, and it contains spirochetes. Patients are often too embarrassed to seek medical attention and, because it is painless, hope it will just go away. It does go away after several weeks, and without scarring. The disease has now entered the primary latent stage. Over the next 5 years, symptoms may be absent, but at any time, as described in the case history, the infected person can develop the rash typical of **secondary syphilis** (**Fig. 26.19C**).

The rash of secondary syphilis can be similar to rashes produced by many different diseases, which contributes to syphilis's nickname as the "great imitator." In this stage the patient remains contagious. The symptoms eventually resolve, and the patient reenters a latent phase of syphilis. Some patients eventually progress over years to **tertiary syphilis** and develop many cardiovascular and neurological symptoms. Neurological symptoms resulting from syphilis at any stage of the disease are referred to as neurosyphilis. The patient can develop dementia and eventually die from the disease.

The presence of *T. pallidum* in tissues can be detected with fluorescent antibody, but the initial screen is usually serological (that is, patient serum is tested for antibodies). Antibiotics are useful for eradicating the organism, but there is no vaccine, and cure does not confer immunity.

The disease is particularly dangerous in pregnant women. The treponeme can cross the placental barrier and infect the fetus to cause **congenital syphilis**. At birth, infected newborns will have notched teeth (visible on X-rays), perforated palates, and other congenital defects. Women should be screened for syphilis as part of their prenatal testing to prevent these congenital infections.

Crossing the placental barrier. Several viral and bacterial pathogens can, like *Treponema pallidum*, cross the placental barrier. Known as the TORCH complex, these pathogens include *Toxoplasma*, rubella virus (German measles), cytomegalovirus (CMV), and herpes simplex 2. The "O" stands for "others," such as *T. pallidum*, *Listeria monocytogenes*, HIV, and varicella-zoster. Bacteria that cross the barrier do so in one of two ways. They move from the mother's blood across the placental villus using intracellular cell-to-cell spread to reach and breach fetal capillaries. Alternatively, the pathogen can initiate an acute inflammatory response that disrupts the integrity of the placental barrier. In contrast, viral pathogens generally use white blood cells (macrophages or lymphocytes) as cellular Uber drivers to carry them across the placenta.

The Tuskegee experiment. Unfortunately, much of what we know about untreated syphilis is the result of the infamous 1930s Tuskegee experiment entitled "Untreated Syphilis in the Negro Male" conducted in Alabama. Through dubious means and deception, a group of African-American males was enlisted in a study that promised treatment but whose real purpose was to observe how the disease progressed without treatment. Today, such experiments are barred, thanks to strict oversight by institutional review boards (IRBs) that require human subjects to sign informed consent forms. A treatise on the Tuskegee experiment can be found on the

Internet at the National Center for Case Study Teaching in Science (search the site for "Bad Blood").

Chlamydial Infections Are Often Silent

Chlamydia is the most frequently reported sexually transmitted infectious disease in the United States, according to the Centers for Disease Control and Prevention, but many people are unaware that they are infected. Three-fourths of infected women, for instance, have no symptoms.

The chlamydias are unusual Gram-negative organisms with a unique developmental cycle. (Chapter 18 describes chlamydial morphology.) They are obligate intracellular pathogens that start as a small, nonreplicating, infectious elementary body that enters target eukaryotic cells. Once inside vacuoles, they begin to enlarge into replicating reticulate bodies (**Fig. 26.20**). As the vacuole fills, the reticulate bodies divide to become new nonreplicating elementary bodies. *Chlamydia trachomatis* causes a sexually transmitted infection called chlamydia but also an eye disease called trachoma, a major worldwide cause of blindness. *Chlamydophila pneumoniae* and *Chlamydia psittaci* are chlamydial species that cause pneumonia but not STIs.

People most at risk of developing genitourinary tract infections with chlamydia are young, sexually active men and women; anybody who has recently changed sexual partners; and anybody who has recently had another sexually transmitted infection. The astute clinician knows that when one STI is discovered, others may also be present.

Left untreated, chlamydia can cause serious health problems. In women, the organism can produce pelvic inflammatory disease, a damaging infection of the uterus and fallopian tubes that can be caused by several different microbial species. The damage produced can lead to infertility, tubal pregnancies, and chronic pelvic pain. Men left untreated can suffer urethral and testicular infections and a serious form of arthritis.

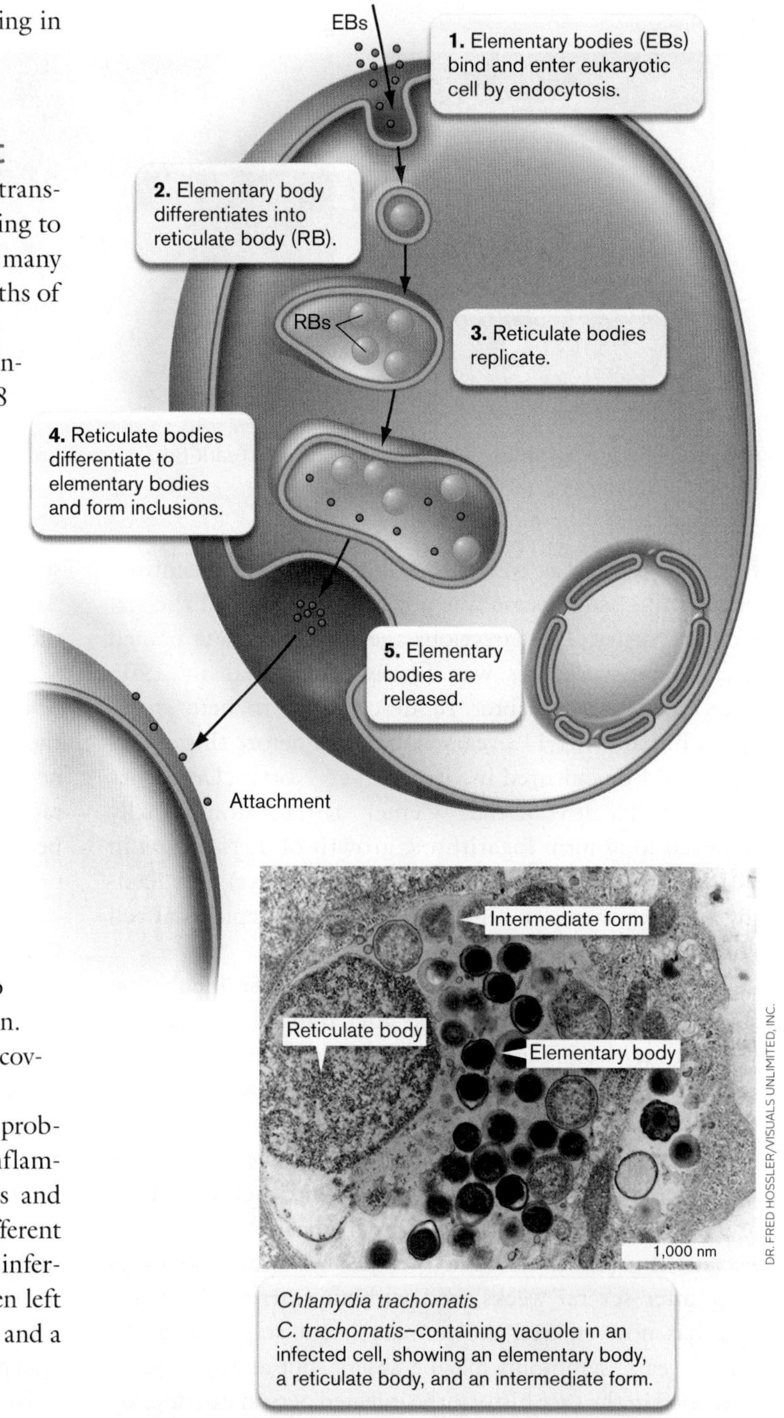

FIGURE 26.20 ■ Replication cycle of *Chlamydia*. Inset: EM of a *C. trachomatis*–containing vacuole in an infected cell, showing a reticulate body, infectious elementary bodies, and intermediate forms.

Case History: Gonorrhea

*A 22-year-old mechanic saw his family doctor for treatment of painful urination and urethral discharge. The patient was sexually active, with three regular and several "one time–good time" partners. Physical examination was unremarkable except for prevalent urethral discharge. The discharge was Gram-stained and sent for culture. The Gram stain revealed many pus cells, some of which contained numerous phagocytosed Gram-negative diplococci (**Fig. 26.21A**). Blood was drawn for syphilis serology, which proved negative. The patient was given a single intramuscular injection of ceftriaxone, and oral doxycycline was prescribed for 7 days. The bacteriology lab was able to recover the bacteria seen in the Gram-stained smear of the urethral discharge. The organism produced characteristic colonies on chocolate agar (agar plates containing heat-lysed red blood cells that turn the medium chocolate brown; **Fig. 26.21B**). The case was subsequently reported to the state public health department. When the patient came back for his return visit, his symptoms had resolved, and a repeat culture was negative.*

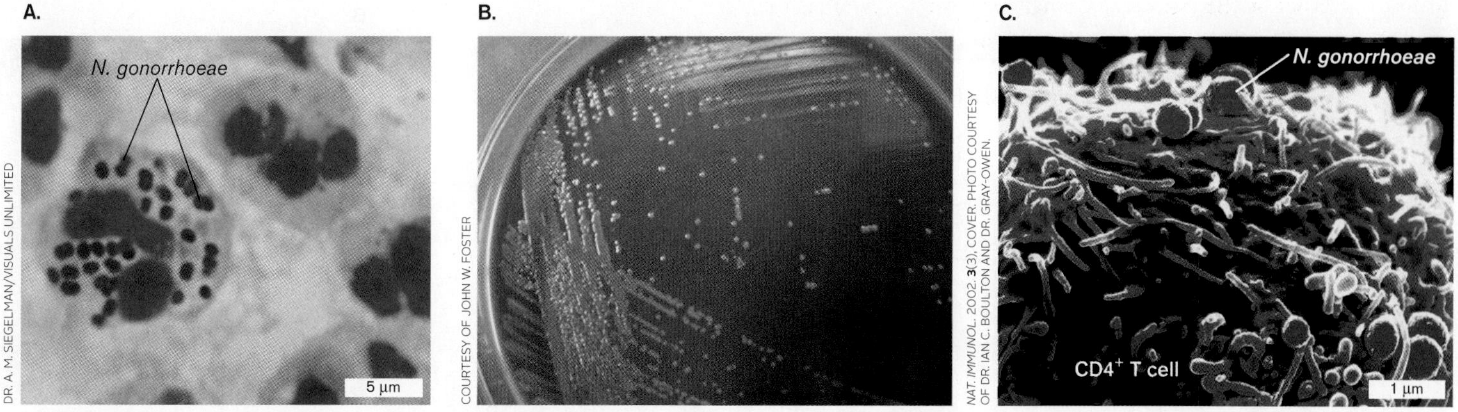

FIGURE 26.21 ■ ***Neisseria gonorrhoeae.*** **A.** Within pus-filled exudates, the Gram-negative diplococci are found intracellularly inside PMNs. The intracellular bacteria in this case are no longer viable, having been killed by the antimicrobial mechanisms of the white cell. **B.** Colonies of *N. gonorrhoeae* growing on chocolate agar. **C.** *N. gonorrhoeae* binding to $CD4^+$ T cells, inhibiting T-cell activation and proliferation, which may explain the ease of reinfection (colorized SEM).

The disease here is classic gonorrhea caused by *Neisseria gonorrhoeae*. A characteristic that distinguishes *Neisseria* infections from *Chlamydia* infections is that bacterial cells are visible in gonorrheal discharges but not in chlamydial discharges, even though they are there. Gonorrhea has been a problem for centuries and remains epidemic in this country today. Symptoms generally occur 2–7 days after infection, but they can take as long as 30 days to develop. Most infected men (85 to 90%) exhibit symptoms that can include painful urination, yellowish white discharge from the penis, and in some cases, swelling of the testicles and penis. The Greek physician Galen (CE 129–ca. 199) originally mistook the discharge for semen. This mistake led to the name *gonorrhea*, which means "flow of seed."

In contrast to men, most infected women (80%) do not exhibit symptoms and constitute the major reservoir of the organism. If they are asymptomatic, they have no reason to seek treatment and can unknowingly spread the disease. When symptoms are present, they are usually mild. A symptomatic woman will experience a painful burning sensation when urinating and will notice vaginal discharge that is yellow or occasionally bloody. She may also complain of cramps or pain in her lower abdomen, sometimes with fever or nausea. As the infection spreads throughout the reproductive organs (uterus and fallopian tubes), pelvic inflammatory disease occurs (see earlier discussion of chlamydia). There is no serological test or vaccine for gonorrhea because the organism frequently changes the structure of its surface antigens (phase variation is discussed in Section 10.5 and **eTopic 10.4**).

Although *N. gonorrhoeae* is generally serum sensitive, owing to its sensitivity to complement (Section 23.6), certain serum-resistant strains can make their way to the bloodstream and carry infection throughout the body. As a result, both sexes can develop purulent arthritis (joint fluid containing pus), endocarditis, or meningitis. An infected mother can also infect her newborn during parturition (birth), leading to a serious eye infection called ophthalmia neonatorum. Because of this risk and because most infected women are asymptomatic, all newborns receive antimicrobial eyedrops at birth.

Because adults engage in a variety of sexual practices, *N. gonorrhoeae* can also infect the anus or the pharynx, where it can develop into a mild sore throat. These infections generally remain unrecognized until a sex partner presents with a more typical form of genitourinary gonorrhea. Because no lasting immunity is built up, reinfection with *N. gonorrhoeae* is possible. Reinfection occurs, in part, because there is phase variation in various surface antigens and because the organism can apparently bind to $CD4^+$ T cells, inhibiting their activation and proliferation to become memory T cells (**Fig. 26.21C**).

Over the decades, *N. gonorrhoeae* has incrementally developed resistance to many antibiotics used in its treatment, but there has always been a new, effective drug ready to take the place of the old drug. Soon, this may no longer be the case. To prevent treatment failures, the CDC recommends dual antibiotic therapy that includes an intramuscular injection of ceftriaxone and oral azithromycin or tetracycline. However, the incidence of ceftriaxone-resistant strains of *N. gonorrhoeae* overseas is increasing. The alarm has been raised that we must develop new antibiotics if we are to prevent an uncontrollable explosion of cases of this already epidemic disease.

Thought Question

26.7 Aside from the CDC guidelines for treating gonorrhea, why else do you suppose the patient in this case was treated with doxycycline (a derivative of tetracycline)?

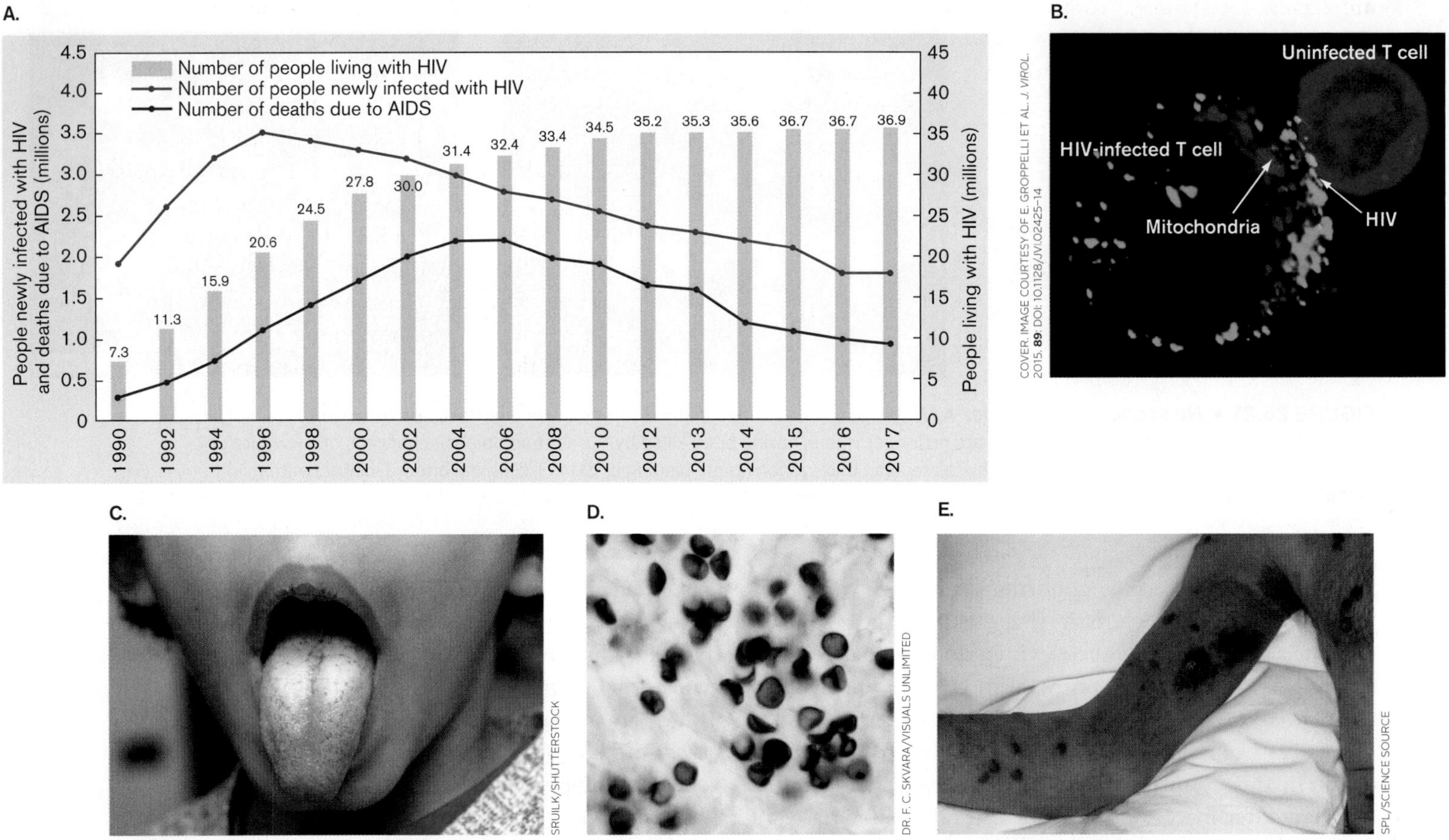

FIGURE 26.22 ■ Acquired immunodeficiency syndrome. **A.** HIV infections and AIDS deaths worldwide. The number of people living with HIV continues to increase, yet the number of people newly infected with HIV and the number of deaths due to AIDS have decreased. **B.** HIV (green) can directly transfer from an infected to an uninfected T cell by virological synapse. **C.** Oral candidiasis (thrush). The white patches are caused by secondary infection by the yeast *Candida albicans*. **D.** *Pneumocystis jirovecii* infection of the lung. Note the cuplike appearance of the fungus, almost like crushed Ping-Pong balls. Organisms range from 2 to 6 μm in diameter. **E.** Kaposi's sarcoma (oval spots) and periorbital cellulitis infection. *Source:* Part A based on data from the World Health Organization.

HIV Causes AIDS, a Sexually Transmitted and Blood-Borne Disease

Though HIV (human immunodeficiency virus) is believed to have originated around the year 1900, it was not discovered until 1981, when the virus caused the greatest pandemic of the late twentieth century. HIV remains a serious problem today, especially in Africa, which is home to nearly two-thirds of the people living with HIV worldwide. HIV has claimed the lives of almost 2 million people per year worldwide (about 14,000 per year in the United States). Women make up about 19% of new AIDS cases per year in the United States, the majority resulting from heterosexual sex. The molecular biology and virulence of HIV are discussed in Chapters 11 and 25, and its pathogenesis is covered in **eTopic 26.7**. This section focuses on the disease that HIV causes—namely, acquired immunodeficiency syndrome (AIDS).

HIV, a lentivirus in the retroviral family, is a prominent example of viruses that can be transmitted either sexually (vaginally, orally, anally—homosexually or heterosexually) or through direct contact with body fluids, such as occurs with blood transfusion or the sharing of hypodermic needles by intravenous drug users. HIV is not transmitted by kissing, tears, or mosquito bites. It can, however, be transferred from mother to fetus through the placenta (transplacental transfer). We discuss HIV infections in this section because sexual contact remains the major route of transmission.

Figure 26.22A shows the worldwide decrease in the number of newly infected HIV patients and deaths due to AIDS. The number of people living with HIV has increased, however, because of the development of more effective antiviral treatments. The proportion of AIDS cases by sex and ethnicity are described in **eTopic 26.7**.

Once HIV enters the bloodstream, it infects $CD4^+$ T cells and macrophages (which also have CD4 on their surface). The virus replicates very rapidly, producing a billion particles per day, and it can spread directly from cell to cell via virological synapses (**Fig. 26.22B**). When infected and uninfected T cells make contact, HIV virions and mitochondria (to supply energy) line up at the contact interface where transmission occurs. Viral replication starts to kill the $CD4^+$ T cells, which progressively decrease in number.

HIV infection has four stages:

1. A primary stage beginning with seroconversion (finding antibodies to the virus)
2. Clinical latency (slow steady loss of T cells without symptoms, or sometimes swollen lymph nodes)
3. Early symptomatic disease (formerly called AIDS-related complex)
4. Acquired immunodeficiency syndrome (AIDS)

Symptoms in the early symptomatic stage can include fever, headache, macular rash, and weight loss. The symptoms can manifest within a few months of infection, resolve within a few weeks, and then recur. As T-cell numbers decline, the debilitated immune system leaves the victim susceptible to secondary infections such as thrush (**Fig. 26.22C**), caused by the yeast *Candida albicans* (candidiasis) and related species.

Monitoring an HIV patient involves assaying blood for viral load using quantitative PCR to detect HIV-specific genes, detecting anti-HIV antibodies, and determining the $CD4^+$ T-cell count. Remember, a person who is HIV-positive does not necessarily have AIDS, which may take years to develop. The CDC defines a patient with AIDS as someone who is HIV positive with a $CD4^+$ cell count of less than 200 cells per microliter or who has contracted an AIDS indicator disease. For example, once the $CD4^+$ T-cell population falls below 500 cells per microliter, opportunistic infections start to arise. Opportunistic infections include pneumonia by *Mycobacterium avium-intracellulare* or *Pneumocystis jirovecii* (**Fig. 26.22D**), cryptococcal meningitis, *Histoplasma capsulatum* infection, and tuberculosis. The presence of these diseases indicates that the patient has AIDS, regardless of the $CD4^+$ count.

Various cancers are also AIDS indicator diseases. AIDS patients are more susceptible to cancers because their depressed immune systems cannot detect and destroy cancer cells generated by secondary agents. Kaposi's sarcoma (**Fig. 26.22E**), for instance, is a common cancer seen in AIDS patients that is caused by human herpesvirus type 8 (HHV8). Kaposi's sarcoma originates in endothelial or lymphatic cells, but the resulting tumors can develop anywhere—gastrointestinal tract, mouth, lungs, skin, or brain.

A vaccine is not yet available to prevent AIDS, in part because the envelope proteins of the virus (see Fig. 11.24) frequently change their antigenic shape, and because the targets of HIV (CD4 T cells), are required for the immune response. However, progression of the disease can be controlled by antiretroviral drugs that inhibit several critical features of HIV biology: entry, integration, reverse transcription, and protein processing via protease. Antiretroviral drugs are discussed further in Chapter 27.

New treatment regimens (antiretroviral therapy, or ART; see Chapter 27) have made HIV an increasingly survivable infection. Consider this: In the past, nearly all HIV-infected individuals eventually became ill and died from AIDS-related diseases, but in the United States at least half of, if not most, HIV-positive people now die from diseases unrelated to HIV (for example, heart attack). Because HIV infection is now considered a treatable disease, the CDC recommends routine screening for HIV.

Can HIV infection be cured? The only person "cured" of HIV is Timothy Ray Brown, a man from Berlin who was diagnosed in 2006 with leukemia and HIV. In an attempt to eradicate both diseases, doctors destroyed Brown's white cells with chemotherapy and transplanted bone marrow from a donor who lacked CCR5, the coreceptor needed for HIV to enter T cells. The transplant successfully repopulated Brown's blood with HIV-resistant T cells. Mr. Brown has not needed antiretroviral therapy for more than 10 years, even though traces of the HIV genome remain. Unfortunately, this approach repeatedly failed with other patients, until recently, when a patient from London with Hodgkin's lymphoma received a stem cell transplant from another CCR5-deficient donor. As of 2019, this patient has remained HIV-free for 18 months without needing antiretroviral therapy.

Because HIV DNA integrates into the genome of an infected cell, it has been impossible to completely cure anyone of the infection. However, a study by Kamel Khalili proved that the CRISPR-Cas9 system described in Chapters 9 and 12 can be engineered to "surgically" remove proviral HIV DNA from infected T cells. Preventive strategies would use CRISPR-Cas9 as a vaccine, introducing it as a stable part of uninfected cells, waiting to attack HIV DNA made by reverse transcription during a future infection. Whether these strategies can be used in a patient remain to be seen.

Thought Question

26.8 Like the cause of plague, HIV is a blood-borne pathogen. Why, then, do you think fleas and mosquitoes fail to transmit HIV?

The Protozoan *Trichomonas vaginalis* Produces a Common Vaginal Infection

Trichomonas vaginalis is a flagellated protozoan (**Fig. 26.23**) that causes an unpleasant, sexually transmitted vaginal disease called trichomoniasis. Approximately 2–3 million infections occur each year in the United States. Both men and women can be infected; however, men are usually asymptomatic. Even among infected women, 25%–50% are considered asymptomatic carriers.

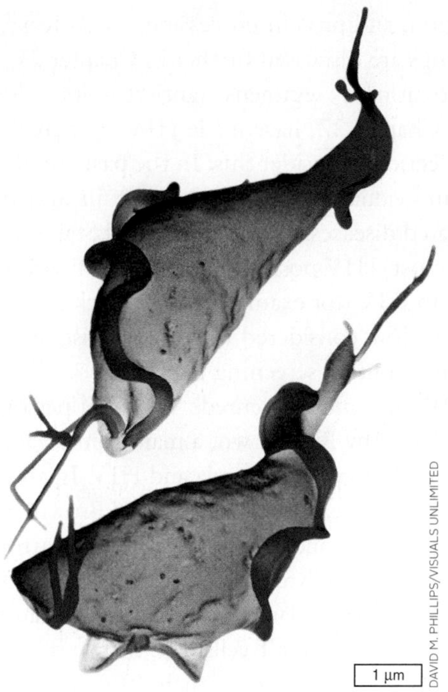

FIGURE 26.23 ■ *Trichomonas vaginalis.* *T. vaginalis* is a protozoan that causes a common sexually transmitted infection (SEM).

There is no cyst in the life cycle of *T. vaginalis*, so transmission is via the trophozoite stage only (the form of a protozoan in the feeding stage). The female patient with trichomoniasis may complain of vaginal itching and/or burning and a musty vaginal odor. An abnormal vaginal discharge also may be present. Males will complain of painful urination (dysuria), urethral or testicular pain, and lower abdominal pain.

Owing to colonization by lactobacilli (which produce large amounts of lactic acid), the normal, healthy vagina has a pH of less than 4.5. However, since *T. vaginalis* feeds on bacteria, the pH of the vagina rises as the numbers of lactobacilli decrease. Definitive diagnosis requires microscopically identifying the flagellated protozoan in vaginal secretions. PMNs, which are the primary host defense against the organism, are also usually present. Like giardiasis, this disease is treated with metronidazole.

To Summarize

- **Urinary tract infections (UTIs)** include cystitis (bladder) and pyelonephritis (kidney). ***E. coli*** is the most common cause of UTIs.
- **Cystitis can develop** from bacteria ascending up the urethra (most common route) or descending down a ureter from an infected kidney.
- **Pyelonephritis can develop** from bacteria ascending along a ureter from an infected bladder (most common route) or from bacteria in the bloodstream disseminating from an infection elsewhere in the body.
- **Syphilis, chlamydia, and gonorrhea** are the most common sexually transmitted infections.
- **A patient with one STI** often has another STI too.
- **Complement prevents bloodstream dissemination** of *Neisseria gonorrhoeae*, which lacks a carbohydrate capsule. Because the organism frequently changes the structure of its surface antigens, no vaccine is available for *N. gonorrhoeae*.
- **HIV depletion of $CD4^+$ T cells** results in lethal secondary infections and cancers.
- ***Trichomonas vaginalis*** is a flagellated protozoan that causes a sexually transmitted vaginal infection. The reservoirs for this organism are the male urethra and female vagina.

26.5 Cardiovascular and Systemic Infections

The cardiovascular system, which includes the heart, arteries, veins, and capillaries, delivers oxygen, nutrients, and immune system components to all tissues of the body. There are pathogens that can directly infect the heart or the endothelial cells lining blood vessels. However, the blood itself can also become infected or serve as a mass transit system to spread pathogens throughout the body. Pathogens disseminated in this way can cause infections in many different organ systems. We will examine both types of infections in this section.

Infections of the cardiovascular system include septicemia, endocarditis (inflammation of the heart's inner lining), pericarditis (inflammation of the heart's outer lining), myocarditis (inflammation of heart muscle), and, possibly, atherosclerosis (deposition of fatty substances along the inner lining of arteries; **eTopic 26.8**). These are all life-threatening diseases.

Septicemia is, by strict definition, the presence of bacteria or viruses in the blood. The presence of viruses is a condition more specifically called **viremia**, while bacteria in the circulation is more specifically called **bacteremia**. In practice, however, the terms "septicemia" and "bacteremia" are often used interchangeably. Septicemia can develop from a local tissue infection situated anywhere in the body. Once in the bloodstream, bacteria can travel to other organs and cause infections there. To prevent septicemia, regional lymph nodes waiting to receive organisms from the initial infection site recruit massive numbers of neutrophils to ambush the bacteria when they arrive (**Special Topic 26.1**). Should some of the organisms escape the lymph node, blood factors such as complement can nonspecifically kill many types of bacteria that enter the blood. Despite these failsafe mechanisms,

Gram-positives, Gram-negatives, aerobes, and anaerobes can all produce septicemia under the right conditions.

Endocarditis can be either viral or bacterial in origin. It can be a consequence of many bacterial diseases, such as brucellosis, gonorrhea, psittacosis, staphylococcal and streptococcal infections, candidiasis, and Q fever (*Coxiella*). Among the many viral causes are coxsackievirus, echovirus, Epstein-Barr virus, and HIV. Bacterial infections of the heart are always serious. Viral infections, although common, are rarely life-threatening in healthy individuals and are usually asymptomatic.

Case History: Bacterial Endocarditis

Elizabeth was 38 years old and had a history of mitral valve prolapse (a common congenital condition in which a heart valve does not close properly). She was also on immunosuppressive therapy following a kidney transplant. Recently, Elizabeth was admitted to the hospital complaining of fatigue, intermittent fevers for 5 weeks, and headaches for 3 weeks—symptoms the physician recognized as possible indications of endocarditis. Elizabeth reported having had a dental procedure a few weeks prior to the onset of symptoms. A sample of her blood was cultured in a liquid bacteriological medium. The culture grew Gram-positive cocci, which turned out to be Streptococcus mutans, *a member of the viridans streptococci. With the finding of bacteria in the bloodstream, the diagnosis of bacterial endocarditis was confirmed. Elizabeth began a 1-month course of intravenous penicillin G and gentamicin therapy and eventually recovered to normal health.*

Endocarditis (inflammation of the inner lining of the heart) is traditionally classified as acute or subacute, depending on the pathogenic organism involved and the speed of clinical presentation. Subacute bacterial endocarditis (SBE) has a slow onset with vague symptoms. It is usually caused by bacterial infection of a heart valve (**Fig. 26.24**). SBE infections are usually (but not always) caused by a viridans streptococcus from the oral microbiota (for example, *Streptococcus mutans*, a common cause of dental caries). "Viridans streptococci" is a general term used for commensal streptococci whose colonies produce green alpha hemolysis on blood agar ("viridans" is from the Greek *viridis*, "green"). Many patients who develop infective endocarditis (50%) have mitral valve prolapse or mitral valve damage from rheumatic fever. Also at high risk are intravenous drug abusers or patients who develop hospital-acquired (nosocomial) infections.

Subacute bacterial endocarditis can begin at the dentist's office, as it did in the case history presented here, although it very rarely does (our patient was also immunosuppressed). Following a dental procedure (such as tooth restoration) or even while brushing your teeth, oral bacteria can transiently enter the bloodstream and circulate.

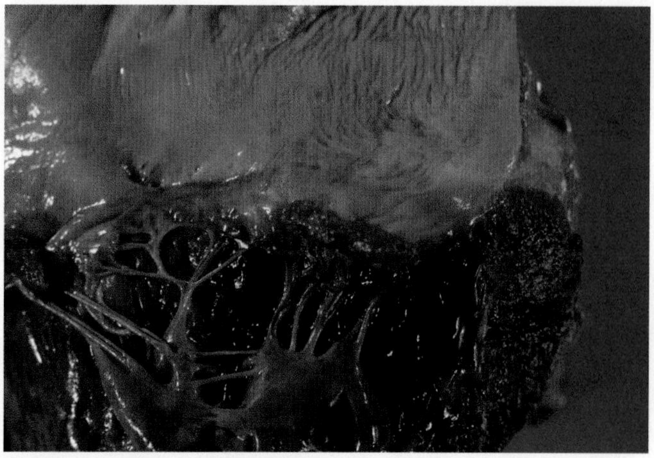

DR. E. WALKER/SCIENCE SOURCE

FIGURE 26.24 ■ View of bacterial endocarditis. Close-up of mitral valve endocarditis, showing vegetation.

Streptococcus mutans, a member of the normal oral microbiome, can become lodged onto damaged heart valves, grow as a biofilm, and secrete a thick glycocalyx coating that encases the microbes and forms a vegetation on the valve, damaging it further. If untreated, the condition can be fatal within 6 weeks to a year.

A rapidly progressive (acute) and highly destructive infection can develop when more virulent organisms, such as *Staphylococcus aureus*, gain access to cardiac tissue. Symptoms of acute endocarditis include fever, pronounced valvular regurgitation (backflow of blood through the valve), and abscess formation.

Most patients with subacute bacterial endocarditis present with a fever that lasts several weeks. They also complain of nonspecific symptoms, such as cough, shortness of breath, joint pain, diarrhea, and abdominal or flank pain. Endocarditis is suspected in any patient who has a heart murmur and an unexplained fever for at least a week. It should also be considered in an intravenous drug abuser with a fever, even in the absence of a murmur. In either case, definitive diagnosis requires blood cultures that grow bacteria. Blood cultures involve taking samples of a patient's blood from two different locations (such as two different arms). Once collected, the blood is added to liquid culture medium (two bottles per site) and incubated at 37°C. One bottle should be incubated aerobically, the other anaerobically. Growth of the same organism in cultures taken from two body sites rules out inadvertent contamination with skin microbes, which would likely yield growth in only one culture.

Curing endocarditis is difficult because the microbes are usually ensconced in a nearly impenetrable glycocalyx. Consequently, eradicating microorganisms from the vegetations almost always requires hospitalization, where high doses of intravenous antibiotic therapy can be administered and monitored. Antibiotic therapy usually continues for at least a month, and in extreme cases, surgery may be necessary to repair or replace the damaged heart valve.

SPECIAL TOPIC 26.1 How Neutrophils Ambush *Staphylococcus aureus* in a Lymph Node

Every day, bacteria breach the human skin barrier through cuts, abrasions, insect bites, and burns. Yet, in healthy people these pathogens rarely disseminate from the initial site of infection to the systemic circulation. This inability to spread seems odd, since bacteria always arrive at the site of infection well before neutrophils do. Yet the bacteria rarely make it into the bloodstream to cause sepsis or infect other sites. What stops them? Paul Kubes's lab (University of Calgary) found that macrophages, obvious gatekeepers that reside at infection sites or in regional lymph nodes, can slow dissemination, but not by much. Could there be an alternative intercept strategy?

One day, Ania Bogoslowski (**Fig. 1**), a novice graduate student in the Kubes laboratory, infected a mouse's footpad with *Staphylococcus aureus* and examined the lymph nodes. She found that the organisms had traveled to the regional lymph nodes behind the knee (popliteal lymph nodes, or popLNs) but no farther (**Fig. 2A**). Why didn't they break into the circulation? Bogoslowski examined the nodes using in vivo microscopy (IVM) and found that, within 5–6 hours of infection, neutrophils were streaming into the nodes, increasing in number from 1,000 per node to over 250,000. IVM enables the observation of processes occurring in a live animal. **Figure 2B** shows the increase in neutrophils after just 4 hours. The strategy seems to be this: get neutrophils to the lymph node quickly and wait for lymph to bring the microbes to them. This way the neutrophils can ambush, rather than chase, any pathogens attempting to disseminate.

Bogoslowski and colleagues showed that the neutrophils migrated into the nodes from the bloodstream, passing through tiny balloon-like structures called high endothelial venules spaced along capillaries situated within the node. But what called the neutrophils to the lymph node? It wasn't the bacteria; neutrophils swarmed to the node even when heat-killed bacteria were injected into the footpad. Macrophages were not involved, since deleting them did not affect neutrophil recruitment. The scientists then examined complement factor C5a, which can also act as a chemoattractant for neutrophils (Section 23.6). When they blocked receptors for C5a receptors (C5aR) on neutrophils with anti-C5aR antibody, neutrophil recruitment to the lymph nodes stopped (**Fig. 3**).

Bogoslowski proposes that a lot of C5a is formed at the site of infection (footpad). The edema that forms in response to the infection causes fluid containing C5a to move from the footpad through the lymphatic vessels to the regional lymph node. The complement factor arrives at the node well before bacteria do and can form a chemotactic gradient that lures neutrophils into the structure to await the arrival of bacteria. The strategy seems to hold also for other pathogens known to cause neutrophil recruitment to regional lymph nodes (*Pseudomonas aeruginosa* and *Mycobacterium bovis*, for instance).

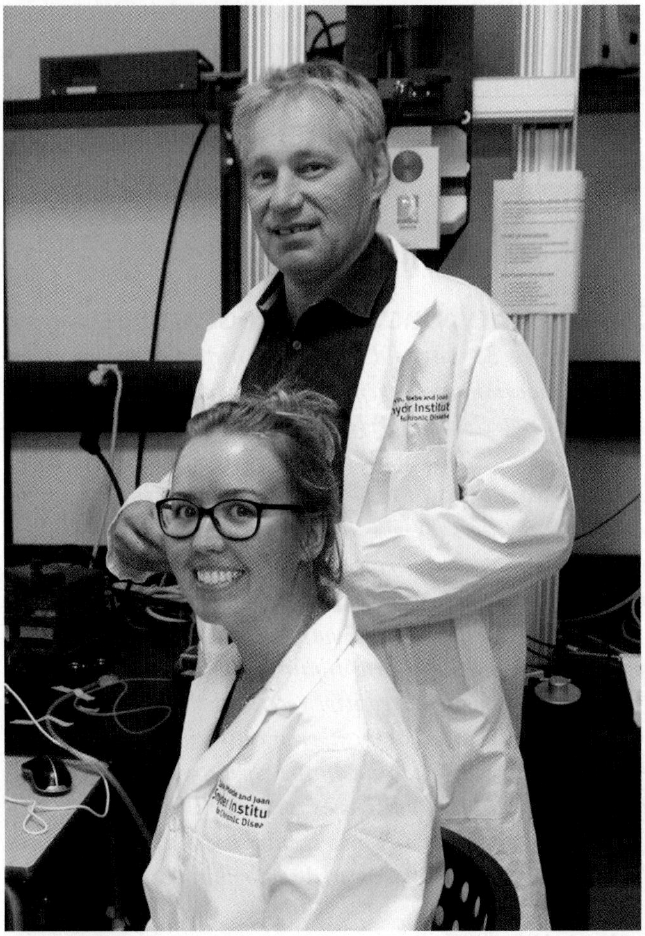

FIGURE 1 ▪ Ania Bogoslowski (seated) and Paul Kubes study the innate immune response to infection.

RESEARCH QUESTION

How would you use flow cytometry to confirm the in vivo microscopy results? Search the Internet for antigens you would use to screen for neutrophils and draw out (graphically) the results of the flow cytometry experiment.

Bogoslowski, Ania, Eugene C. Butcher, and Paul Kubes. 2018. Neutrophils recruited through high endothelial venules of the lymph nodes via PNAd intercept disseminating *Staphylococcus aureus*. *Proceedings of the National Academy of Sciences USA* **115**:2449–2454.

A. Bacterial dissemination

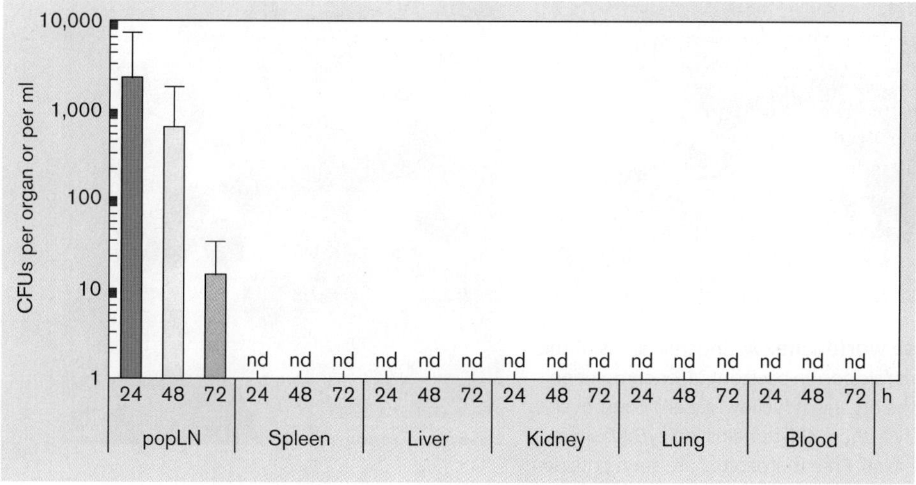

FIGURE 2 ▪ Neutrophil recruitment to popliteal lymph nodes (popLNs) stops *S. aureus* dissemination. **A.** Bacterial dissemination from a footpad infection site. Data represent counts of colony-forming units (CFUs) of *Staphylococcus aureus* in various organ systems 24, 48, and 72 hours after initial infection (mean ± standard deviation; nd = none detected). **B.** In vivo microscopy of neutrophils present in the lymph node before and 4 hours after footpad infection. Outline of node is marked with a dotted line. Steady-state control shows neutrophils present in an uninfected node. *Source:* Modified from Ania Bogoslowski et al. 2018. *PNAS* **115**:2449–2454, fig. 1A.

B. Neutrophil accumulation in lymph nodes

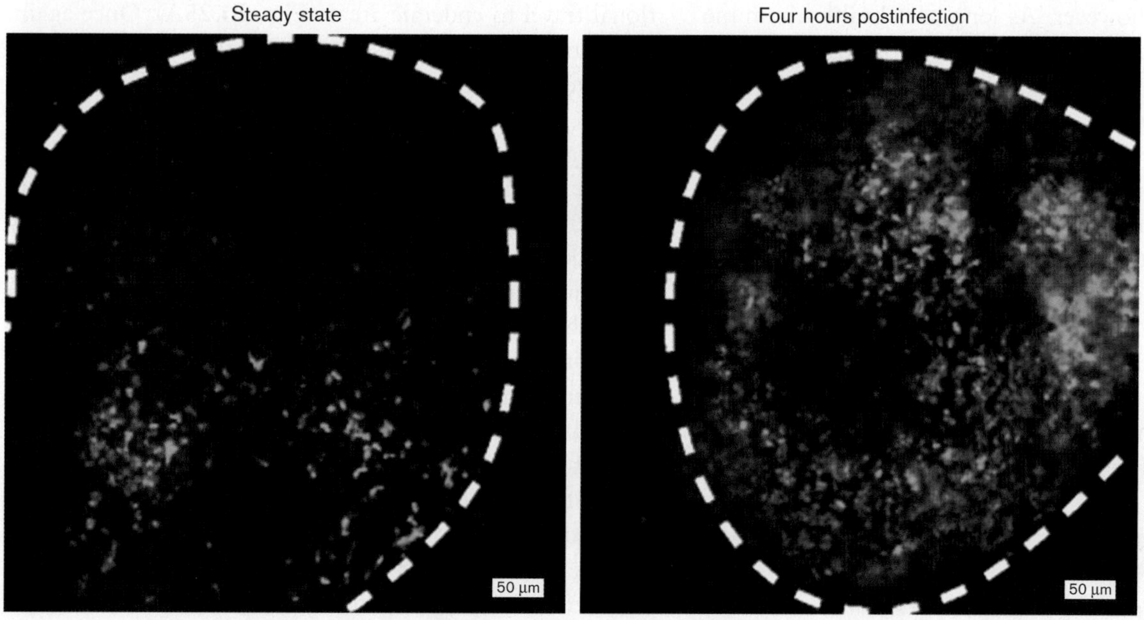

ANIA BOGOSLOWSKI ET AL. 2018. *PNAS* **115**:2449–2454, FIG. 1B.

Condition	Neutrophils/LN
Phosphate-buffered Saline	1.6×10^5
C5aR inhibitor	$<1\times10^1$

FIGURE 3 ▪ Complement factor C5a recruits neutrophils to the regional lymph node after infection. Mice were treated with phosphate-buffered saline (control) or with blocking antibody to C5aR. *Source:* Modified from Ania Bogoslowski et al. 2018. *PNAS* **115**:2449–2454, fig. 3E.

A.

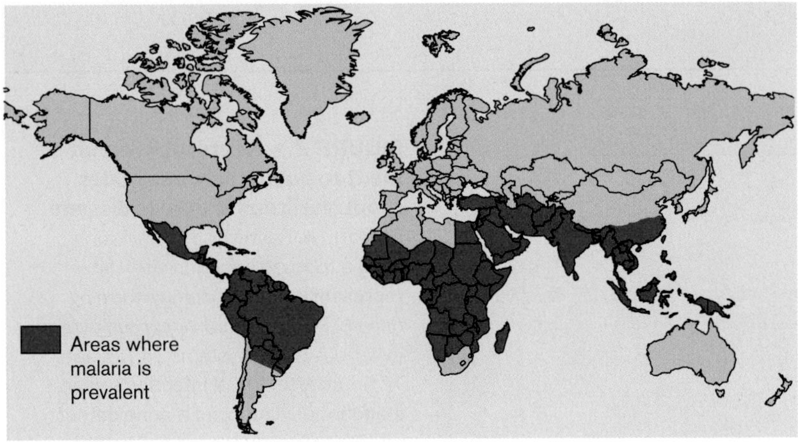

B.

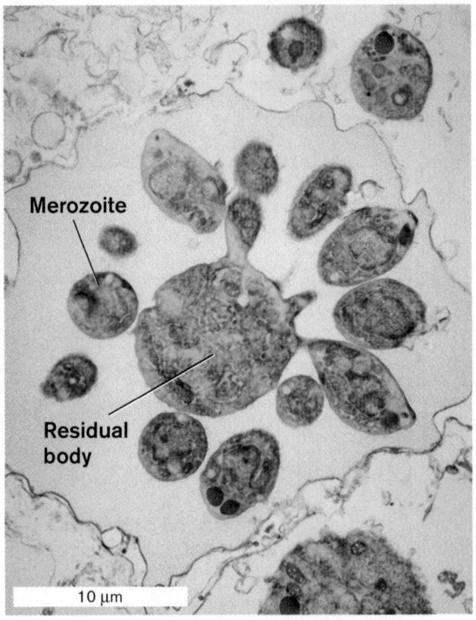

FIGURE 26.25 ■ Malaria is a major disease worldwide. **A.** Endemic areas of the world where malaria is prevalent. **B.** *Plasmodium falciparum* schizont after completion of division (colorized TEM). A residual body of the organism (yellow-green) is left over after division. The erythrocyte has lysed and only a ghost cell remains; no cytoplasm is seen surrounding the merozoites just being released. Free merozoites are seen outside the membrane.

Patients with heart valve prolapse should not shy away from the dentist, however. As long as a healthy immune system is in place, the risk of infection is low. In fact, maintaining good oral hygiene can keep the risk of developing infectious endocarditis very low. Immunocompromised patients, on the other hand, are at increased risk and should be treated prophylactically with oral amoxicillin or azithromycin 1 hour before a procedure to kill oral bacteria that enter the bloodstream. Note, too, that prosthetic heart valves can develop endocarditis (usually caused by *Staphylococcus aureus*). The treatment of prosthetic valve endocarditis often requires valve replacement.

Viruses such as adenovirus and some enteroviruses can also cause endocarditis, as well as a condition known as myocarditis, an inflammation of the heart muscle.

Thought Question

26.9 A patient presenting with high fever and in an extremely weakened state is suspected of having septicemia. Two sets of blood cultures are taken from different arms. One bottle from each set grows *Staphylococcus aureus*, yet the laboratory report states that the results are inconclusive. New blood cultures are ordered. What would make these results inconclusive?

Malarial Parasites Target Red Blood Cells

Malaria is the most devastating infectious disease known. Each year, 300–500 million people develop malaria worldwide, and 1–3 million of these people, mostly children, die. In the United States only about a thousand cases occur annually, and almost all are acquired as a result of international travel to endemic areas (**Fig. 26.25A**). Once again, this correlation illustrates the diagnostic value of knowing a patient's travel history.

The disease is caused by four species of *Plasmodium*: *P. falciparum* (the most deadly), *P. malariae*, *P. vivax*, and *P. ovale*. The life cycle of *Plasmodium*, discussed in detail in Chapter 20, is complex and involves two cycles: an asexual erythrocytic cycle in the human, and a sexual cycle in the mosquito (see Fig. 20.39).

In the erythrocytic cycle, an infected female *Anopheles* mosquito bites the victim and injects a small amount of saliva containing *Plasmodium* into the bloodstream. The haploid sporozoites go directly to the liver and undergo asexual fission to produce merozoites. The merozoites attach to and penetrate red blood cells, where *Plasmodium* consumes hemoglobin and enlarges into a trophozoite. The protist nucleus divides so that the cell, now called a schizont, contains up to about 20 nuclei. The schizont then divides to make the smaller, haploid merozoites (**Fig. 26.25B**). The glutted red blood cell eventually lyses, releasing merozoites that can infect new red blood cells (see Fig. 20.39).

Sudden, synchronized release of the merozoites and red blood cell debris triggers the telltale symptoms of malaria: violent, shaking chills followed by high fever and sweating. The erythrocytic cycle, and thus the symptoms, repeats every 48–72 hours. After several cycles, the patient goes into remission lasting several weeks to months, after which there is a relapse.

Much of today's research focuses on malarial relapse. Why does the immune system fail to eliminate the parasite after the first episode? When *Plasmodium* invades the red blood cells, it lines the blood cells with a protein, PfEMP1, that causes RBCs to stick to blood vessels, removing the parasite from circulation. But the protein cannot protect the parasite from patrolling macrophages, which eventually detect the invader and recruit other immune cells to fight it. However, a fraction of each generation of parasite expresses a different version of PfEMP1 that the immune system never encountered. In its new disguise, *Plasmodium* can invade more red blood cells and cause another wave of fever, headaches, nausea, and chills. The body now has to repeat its recognition and attack responses all over again. The parasite has 60 of these cloaking genes, called *var*, that can be turned on and off individually, changing the organisms' antigenic structure, like a criminal repeatedly changing his disguise to elude police.

The *var* genes are regulated by chromosome packaging, which unwraps one gene for expression while packing away inactive genes. DNA can be encased so securely that transcription proteins cannot access the packed nucleic acid—a process known as epigenetic silencing. Epigenetic protein modifications (acetylation or methylation) alter the affinity of histones toward different regions of the plasmodial chromosomes. Becoming immune to all the types of malaria can take upwards of 5 years and requires constant exposure; otherwise immunity is lost. Many children afflicted with malaria do not live long enough to gain immunity to the disease in all its forms.

Diagnosis of malaria involves microscopic demonstration of the protist within erythrocytes (Wright stain) or serology to identify antimalarial antibodies. Treatment regimens include artemisinin, whose mechanism of action is unknown; chloroquine or mefloquine, which kill the organisms in their erythrocytic asexual stages; and primaquine, effective in the exoerythrocytic stages by disrupting *Plasmodium* mitochondria.

The chloroquine family of drugs prevents the detoxification of heme generated from hemoglobin digestion. Malaria parasites accumulate and digest the hemoglobin released from red blood cells in plasmodial lysosomes and use the amino acids to grow. However, the heme released is toxic. The organism normally detoxifies heme via polymerization, which produces a black pigment. Many antimalarial drugs, such as chloroquine, prevent polymerization by binding to the heme. As a result, the increased iron level (from heme) kills the parasite. Unfortunately, *Plasmodium* has been developing resistance to these drugs, forcing the development of new ones.

An effective vaccine has so far proved elusive because of the antigenic shape-shifting carried out by this parasite. To prevent infection, antimalarial drugs are given prophylactically to persons traveling to endemic areas. Which drug is given depends on the destination. New strategies to control the spread of malaria include using anti-mosquito bacteria to eradicate mosquito vectors, and a genetic approach, called gene drive, that is designed to eliminate female mosquitoes. Female mosquitoes are targeted because only female mosquitoes take blood, which they need to make eggs. In one gene drive approach, a CRISPR-Cas9 system would be introduced into the mosquito Y chromosome. The guide RNAs of this system will direct cleavage of sequences within the X chromosome. As a result, sperm from these males should all contain the Y chromosome. No X chromosomes will be passed to eggs, so no female mosquitoes will be produced. Various gene drive systems are currently being tested.

Bear in mind that we have described only selected organisms that cause cardiovascular infections; there are many more (for example, *Rickettsia typhi*). There is even evidence that *Chlamydophila pneumoniae* may have a role in coronary artery disease (see **eTopic 26.8**).

Note: Babesiosis is an emerging disease caused by a protozoan (*Babesia microti*) that, like *Plasmodium*, infects red blood cells. *B. microti* is transmitted by the deer tick, the same insect that transmits the agent for Lyme disease. Typically, babesiosis is a mild, flu-like disease, but in its severe form it can present with symptoms similar to those of malaria.

Systemic Infections

Many pathogenic bacteria can produce septicemia as a way to disseminate throughout the body and infect other organs. These organisms cause what are considered systemic infections.

Case History: The Plague

*A 25-year-old New Mexico rancher was admitted to an El Paso hospital because of a 2-day history of headache, chills, and fever (40°C; 104°F). The day before admission, he had begun vomiting. The day of admission, an orange-sized, painful swelling in the right groin area was noted (**Fig. 26.26A**). A lymph node aspirate and a smear of peripheral blood were reported to contain Gram-negative rods that exhibited bipolar staining (**Fig. 26.26B**). The patient's white blood cell count was 24,700 per microliter (normal is 5,000–10,000), and his platelet count was 72,000 per microliter (normal is 130,000–400,000). In the 2 weeks prior to becoming ill, the patient had trapped, killed, and skinned two prairie dogs, four coyotes, and one bobcat. The patient also mentioned that he had cut his left hand shortly before skinning a prairie dog. The clinical laboratory isolated a Gram-negative rod from blood cultures. PCR and biochemical testing identified the organism as* Yersinia *pestis, the bacterium that causes plague. The severely ill rancher received an antibiotic cocktail of gentamicin and tetracycline and eventually recovered, after 6 weeks in intensive care.*

Plague is caused by the bacterium *Yersinia pestis*, which can infect both humans and animals. During the Middle Ages, the disease, known as the Black Death, decimated over a third of the population of Europe. Such was the horror it evoked that invading armies would actually catapult dead plague victims into embattled fortresses. This was probably the first reported case of biowarfare.

Y. pestis is present in the United States and is endemic in 17 western states. Typically, the bite of infected fleas transmits the microbe between animals such as rats and even prairie dogs (**Fig. 26.26C**). **Figure 26.27** illustrates the various infection cycles of the plague bacillus. Humans are not part of the natural infection cycle. In the absence of an animal host, however, the flea can take a blood meal from humans and thereby transmit the disease to them. During the Middle Ages, urban rats venturing back and forth to the countryside became infected by the fleas of wild rodents that served as a reservoir. Upon returning to the city, the rat fleas passed the organism on to other rats, which then died in droves. The rat fleas, deprived of their normal meal, were forced to feed on city dwellers, passing the disease on to them.

Individuals bitten by an infected flea or accidentally infected through a cut while skinning an infected animal first exhibit the symptoms of the plague form known as **bubonic plague**. Bubonic plague emerges as the organism moves from the site of infection to the regional lymph nodes, producing characteristically enlarged nodes called buboes (see **Fig. 26.26A**). From the lymph nodes, the pathogen can enter the bloodstream, causing **septicemic plague**. In this phase the patient can go into shock (sudden drop in blood pressure) from the massive amount of endotoxin in the bloodstream. In the case history, the rancher's low platelet count (thrombocytopenia) was an indication that his platelets were being consumed by the considerable clotting taking place. Neither bubonic nor septicemic plague is passed from person to person. As the organism courses through the bloodstream, however, it will invade the lungs and produce **pneumonic plague** (see **Fig. 26.26D**), which can be easily transmitted from person to person through aerosol droplets generated by coughing.

Pneumonic plague is the most dangerous form of the disease because it can kill quickly and spread rapidly through a population. Pneumonic plague is so virulent that an untreated patient can die within 24–48 hours. The organism is usually identified postmortem. The rapid spread of plague during the Middle Ages was most likely the result of person-to-person transmission through respiratory aerosols.

Y. pestis has numerous virulence factors. For instance, YadA is a surface adhesin that binds collagen. Another factor, the F1 protein capsule surface antigen, plays a part in blocking phagocytosis in mammalian hosts. Certain biofilms formed by *Y. pestis* are also important. An extracellular matrix synthesized by *Y. pestis* produces an adherent biofilm in the flea midgut that contributes to flea-to-mammal transmission. The biofilm blocks the flea's digestion, making the flea feel "starved" even after a blood meal. Therefore, the

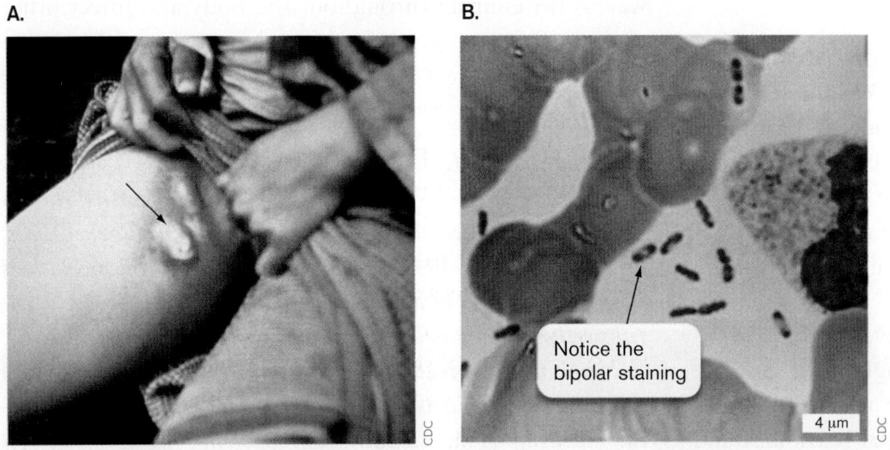

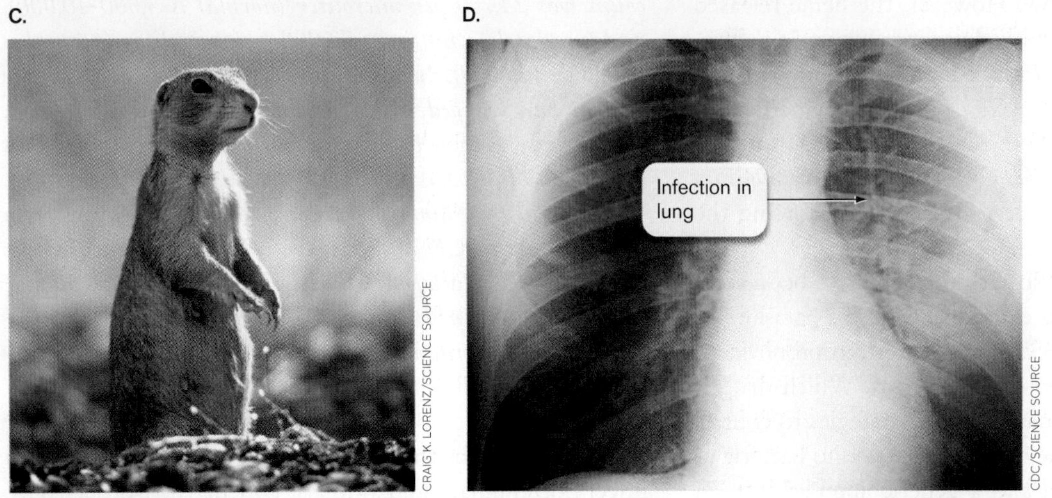

FIGURE 26.26 ▪ The plague. **A.** Classic bubo (swollen lymph node) of bubonic plague. **B.** *Yersinia pestis*, bipolar staining (length 1–3 µm). **C.** Prairie dogs are often hosts to fleas that carry plague bacilli. **D.** X-ray of pneumonic plague, showing pulmonary infection.

FIGURE 26.27 ▪ The cycles of plague. The sylvatic cycle occurs in the wild, where fleas transmit the organism between rodents. An accidental interaction with urban rats can trigger a similar urban cycle. Humans can be infected through contact with infected fleas coming from either cycle. Flea bite transmission initiates bubonic plague symptoms that can progress to pneumonic plague. Because it is highly infectious, pneumonic plague can cause epidemic spread of the disease.

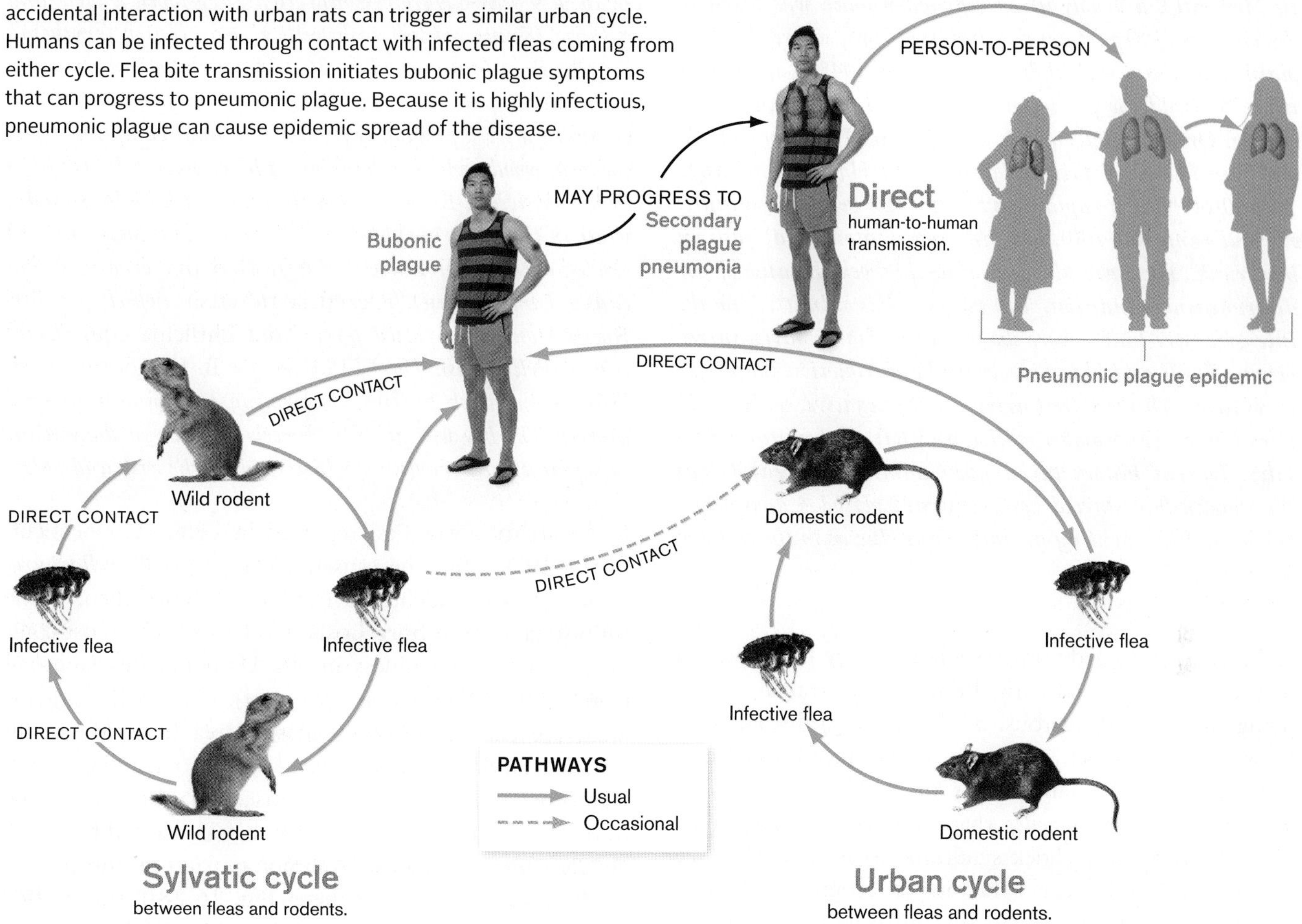

flea jumps from host to host in a futile effort to feel full. As the flea tries to take a blood meal, the blockage causes the insect to regurgitate bacteria into the wound. This curious effect of *Y. pestis* on the insect vector is another unique aspect of how plague spreads so quickly.

Y. pestis also uses type III secretion systems to inject virulence proteins (YopB and YopD) into host cell membranes. Unlike *Salmonella*, which uses type III–secreted proteins to gain entrance into host cells, *Y. pestis* is not primarily an intracellular pathogen, although it can survive in macrophages. Injection of the Yop proteins disrupts the actin cytoskeleton, thereby helping the organism evade phagocytosis. By evading phagocytosis, the organism avoids triggering an inflammatory response and produces massive tissue colonization.

The last major outbreak of plague in Europe occurred in 1772. Why the disease disappeared from Europe is unclear, but part of the reason was probably human intervention. Long before doctors understood how germs could cause disease (1800s), Europeans recognized that plague was contagious and could be carried from one area to another. Beginning in the 1600s, governments established a medical boundary, or *cordon sanitaire*, between Europe and the areas to the east from which epidemics came. Ships traveling west from the Ottoman Empire were forced to wait in quarantine before passengers and cargo could be unloaded. Those who attempted to evade medical quarantine were shot. We don't shoot them anymore, but we do quarantine groups of people exposed to dangerous infectious agents (Ebola virus, for example). Anyone who gets sick is then placed in isolation.

Sepsis and Toxic Shock

Many bacterial pathogens can, under the right circumstances, infect the bloodstream to cause septicemia and ultimately sepsis—a life-threatening condition involving high fever, high white blood cell counts, rapid heart rate and/or rapid breathing, and low platelet counts when severe. The offending bacteria are typically found in the blood (see Chapter 28). Organisms that can cause sepsis include *Neisseria meningitidis*, *Escherichia coli*, *Enterococcus* species, and many others, including *Yersinia pestis* from the preceding case. In severe cases sepsis can progress to septic shock, in which the patient's blood pressure drops to dangerous levels.

Case History: Toxic Shock

In May 2017, a visibly ill 24-year-old woman was taken to the emergency department by her boyfriend, where she complained of diarrhea, high fever (40°C; 104°F), and vomiting. She told the physician that she had become ill 2 days earlier. On examination, the doctor noticed that the woman had very low blood pressure (80/40 mm Hg; normal range 90–130/60–90), a rapid heart rate (122 beats per minute; normal range 60–100), and an erythematous (red) rash on her trunk. These are all signs of septic shock. Because of her deteriorating condition, the patient was admitted to the intensive care unit, where she was immediately given intravenous fluids and broad-spectrum IV antibiotics. Blood cultures taken when she first arrived were negative, no bacteria were found. The woman rallied and left the hospital 4 days later. Patient history taken upon admission revealed that the woman had started her menstrual period 4 days before becoming ill, providing an important clue as to the cause of the disease.

This potential tragedy reflects a larger story that emerged in the late 1970s and early 1980s, when women started dying from this dangerous, new (emerging) disease, now known as toxic shock syndrome. We've since learned that it is caused by certain strains of *Staphylococcus aureus* and *Streptococcus pyogenes* that produce a superantigen type of toxin (toxic shock syndrome toxin, or TSST; see Section 24.3). Why wasn't the disease recognized in previous decades? The answer turned out to be the use of one brand of superabsorbent tampons (since removed from the market). The tampons produced a rich growth environment for *S. aureus.* So, if the patient was colonized by a TSST-producing strain, huge amounts of toxin were released to circulate in the bloodstream.

Even though the superabsorbent tampons are no longer available, toxic shock syndrome is still sometimes associated with menstruation, as in the case history. Today, however, we recognize that toxic shock syndrome can occur in both men and women and is a possible consequence of any *S. aureus* infection involving a TSST-producing *S. aureus.* The toxin is secreted by *S. aureus* at the infection site and spreads rapidly through the circulation to cause generalized, potentially lethal symptoms.

Knowledge that a protein toxin was the cause of the disease led to a recommendation that treatment of toxic shock syndrome include the antibiotic clindamycin. As you will learn in Chapter 27, clindamycin is an inhibitor of bacterial protein synthesis. By blocking protein synthesis, clindamycin decreases the amount of TSST made, while other antibiotics (vancomycin, for instance) kill the pathogen. In our case study, the combination saved the woman's life. In Chapter 25 we discuss other weapons that *S. aureus* uses to cause disease.

Case History: Lyme Disease

*Brad, a 9-year-old from Connecticut, developed a fever and a large (8-cm) reddish rash with a clear center (**erythema migrans**) on his trunk (**Fig. 26.28A**). He also had some left-facial-nerve palsy (partial paralysis of his face). Brad had returned a week previously from a Boy Scout camping trip to the local woods, where he had done a lot of hiking. When asked by his physician, Brad recalled finding a tick on his stomach while in the woods but thinking little of it. The doctor ordered serological tests for* Borrelia burgdorferi *(the organism that causes Lyme disease),* Rickettsia rickettsii *(which produces Rocky Mountain spotted fever), and* Ehrlichia equi *(which causes ehrlichiosis). The ELISA test for* B. burgdorferi *antibodies came back positive, confirming a diagnosis of Lyme disease. The boy was given a 3-week regimen of doxycycline (a tetracycline derivative), which resolved the rash and palsy.*

Lyme arthritis was first reported in Lyme, Connecticut, in the 1970s, but the causative organism, *Borrelia burgdorferi*, was not identified until 1982. Since then, **Lyme disease** (a form of **borreliosis**) has become the most common vector-borne illness in the United States (approximately 20,000 new cases per year). The main endemic areas are the northeastern coastal area from Massachusetts to Maryland, Wisconsin and Minnesota, and northern California and Oregon, but the disease has been spreading into the southern US states. Lyme disease caused by different *Borellia* species is also common in parts of Europe.

B. burgdorferi is a spirochete (**Fig. 26.28B**) transmitted to humans by ixodid ticks (hard ticks; **Fig. 26.28C** and **D**). In the northeastern and central United States, where most cases occur, the deer tick *Ixodes scapularis* transmits the spirochete, usually during the summer months. In the western United States, *I. pacificus* is the tick vector.

During its nymphal stage (the stage after taking its first blood meal), *I. scapularis* is the size of a poppy seed. Its bite is painless, so it is easily overlooked. Infection takes place when the tick feeds, because the spirochete is regurgitated into the host. However, the organism grows in the tick's digestive tract and takes about 2 days to make its way to the tick's salivary gland, so if the tick is removed before that time, the patient will not be infected. Once the pathogen is transferred to the human, the microbe can travel rapidly via the bloodstream to any area in the body, but it prefers to grow in skin, nerve tissue, synovium (joint lining), and the conduction system of the heart.

Lyme disease has three stages. Similar to syphilis caused by the spirochete *Treponema pallidum*, Lyme disease has three general stages. Stage 1 involves the localized spread of *Borrelia burgdorferi* between 3 and 30 days after the initial exposure. Approximately 75% of patients experience an erythema migrans rash, usually at the site of the tick bite. The appearance of the rash varies but is classically erythematous

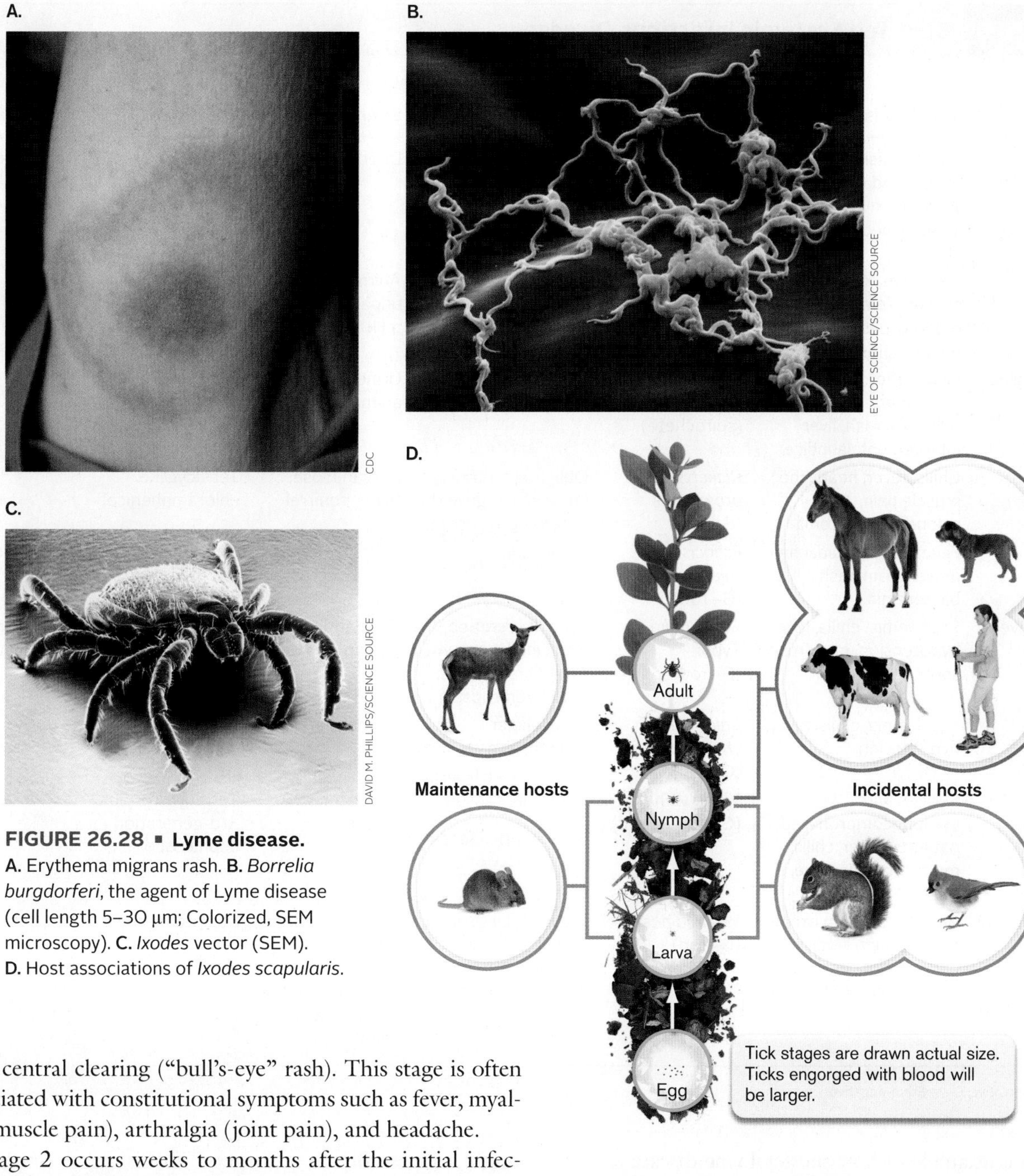

FIGURE 26.28 ■ Lyme disease. **A.** Erythema migrans rash. **B.** *Borrelia burgdorferi*, the agent of Lyme disease (cell length 5–30 μm; Colorized, SEM microscopy). **C.** *Ixodes* vector (SEM). **D.** Host associations of *Ixodes scapularis*.

with central clearing ("bull's-eye" rash). This stage is often associated with constitutional symptoms such as fever, myalgia (muscle pain), arthralgia (joint pain), and headache.

Stage 2 occurs weeks to months after the initial infection as *B. burgdorferi* spreads from blood to other organs. In this stage, the patient can be quite ill with malaise, myalgia, and arthralgia, as well as neurological or cardiac involvement. Common neurological manifestations include Bell's palsy (facial paralysis), inflammation of spinal nerve roots, and chronic meningitis. The most common cardiac manifestation is an irregular heart rhythm.

Stage 3 borreliosis occurs months to years later and can involve the synovium, nervous system, and skin. Arthritis occurs in the majority of previously untreated patients; it is usually intermittent, involves the large joints, particularly the knee, and lasts from weeks to months in any given joint. Joint-fluid analysis typically shows a WBC count of 10,000–30,000 per microliter (normal is fewer than 200). Late neurological involvement may include peripheral neuropathy and encephalopathy, manifested as memory, mood, and sleep disturbances.

Treatment with antibiotics (usually doxycycline) is recommended for all stages of Lyme disease but is most effective in the early stages. Lyme arthritis is typically slow to respond to antibiotic therapy, so treatment lasts from 2 to 4 weeks. However, the disease in some people does not resolve at all; why is not known.

Curiously, 25% of people infected with *B. burgdorferi* never experience erythema migrans, and many infected individuals are also unsure of tick bites. Hence, patients with Lyme disease may present with arthritis as their first complaint. Although this makes diagnosis extremely difficult, knowing that the patient lives in or recently traveled to an endemic area can provide a critical clue.

TABLE 26.6 Selected Systemic Infectious Diseases

Disease	Symptoms	Etiological agent[a]	Virulence properties	Source(s)	Treatment	Vaccine
Lyme disease	Stage 1: rash; stage 2: chills, headache, malaise, systemic involvement; stage 3: neurological changes	*Borrelia burgdorferi* (spirochete)	Antigenic variation, OspE (binds complement)	Deer tick	Penicillin, doxycycline	No longer available
Brucellosis	Fever, weakness, sweats, splenomegaly, osteomyelitis, endocarditis, others	*Brucella abortis* (G– rod)	Intracellular, growth in monocytes	Animal products, unpasteurized milk	Doxycycline	Yes, for animals
Leptospirosis	Fever, photophobia, headache, abdominal pain, skin rash, liver involvement, jaundice	*Leptospira interrogans* (spirochete)	Burrowing motility	Urine of infected animals	Penicillins, erythromycin	Yes, for animals
Epidemic typhus	Chills, fever, headache, muscle pain, splenomegaly, coma	*Rickettsia prowazekii* (G– rod)	Obligate intracellular growth, escapes phagosome	Human louse, flying-squirrel flea	Tetracycline, chloramphenicol	
Tularemia	Fever, chills, headache, muscle pain, rash, bacteremia	*Francisella tularensis* (G– rod)	Intracellular	Rabbits, rodents, insect vectors	Gentamicin, streptomycin	No longer available
Typhoid fever	Septicemia, chills, fever, hypotension, rash (rose spots)	*Salmonella* Typhi (G– rod)	Type III secretion, intracellular growth, PhoPQ regulators, Vi antigen capsule	Gallbladder of human carrier	Ciprofloxacin, ceftriaxone	Vi antigen
	Septicemia, chills, fever, hypotension	*Salmonella choleraesuis* (G– rod)	Intracellular growth, invasin	Animals, poultry	Ceftriaxone	
Vibriosis	Serious with immunocompromised patients; fever, chills, multi-organ damage, death	*Vibrio vulnificus* (G– curved rod)	Cytolysin, capsule	Seawater, raw oysters	Tetracycline plus 3rd-generation cephalosporin	
Bubonic plague	Buboes (swollen lymph glands), high fever, chills, headache, cough, pneumonia, septicemia	*Yersinia pestis* (G– rod)	Intracellular growth, type III secretion of YOPs (*Yersinia* outer proteins), phospholipase D, toxin	Rodents, rodent fleas, human respiratory aerosol, potential bioterrorism agent	Streptomycin or tetracycline	Yes, but not available in the U.S.

[a]G+ = Gram-positive; G– = Gram-negative.

A startling example of how sinister Lyme disease can be is the case of Kris Kristofferson, a prolific singer, songwriter, and actor, now in his eighties. Kristofferson suffered for years with severe memory loss presumed to be a manifestation of Alzheimer's disease—until someone decided to test him for Lyme disease. To his doctor's surprise, and relief, the singer tested positive. Kristofferson immediately began antibiotic treatments and has since recovered much of his memory.

Many other bacteria can cause septicemia and systemic illness (**Table 26.6**). Gram-negative organisms like *E. coli*, *Salmonella* Typhi, and *Francisella*, and Gram-positive microbes like *Staphylococcus aureus*, *Enterococcus*, and *Bacillus anthracis*, can grow in the bloodstream if they can gain entrance. Even anaerobes that are normal inhabitants of the intestine (for example, *Bacteroides fragilis*) can be lethal if they escape the intestine and enter the blood, as might happen following surgery. This is one reason surgical patients are given massive doses of antibiotics immediately before and after surgery.

Ebola—The Perfect Pathogen or Too Deadly for Its Own Good?

How would you define the perfect pathogen? Would it be an organism that can kill its host with terrifying ease and quickness? If so, Ebola virus would fit the description. Ebola virus, a lipid-enveloped, threadlike RNA virus (*Filoviridae*; see Fig. 25.4), was first associated with an outbreak of 318 cases of a hemorrhagic disease in Zaire in 1976. Of the 318 people who contracted the disease, 280 died within days. The disease was characterized by acute (rapid) onset of fever, headache, diarrhea, and severe muscle pains. Some patients exhibited horrible bleeding from multiple orifices (nose, mouth, anus, and vagina), and ultimately died.

The Ebola virus has a frightening reputation. It spreads like wildfire through the body after infection, causing severe hemorrhagic fever, and typically kills 20%–90% of its victims, depending on the strain involved and the level of care received. Internal bleeding results in shock and acute respiratory distress, leading to death. The major 2014–2015 epidemic of Ebola in West Africa involved 28,683 cases, 40% of whom died (see Chapter 28). As of this writing, another outbreak is taking place in the Democratic Republic of the Congo. As of March 2019, the outbreak involved 872 cases and 548 deaths.

The symptoms of Ebola (and of a related disease caused by the Marburg virus) reflect subversion of the innate immune system, coupled with uncontrolled viral replication, particularly in macrophages and dendritic cells. Ebola virus infection of these cells enhances production of pro-inflammatory cytokines, such as TNF-alpha, and inhibits stimulation of T-cell maturation by dendritic cells. Thus, Ebola infections stimulate inflammatory processes leading to tissue damage, but they shut down early immune responses and prevent activation of adaptive immune responses, thereby allowing unfettered viral replication.

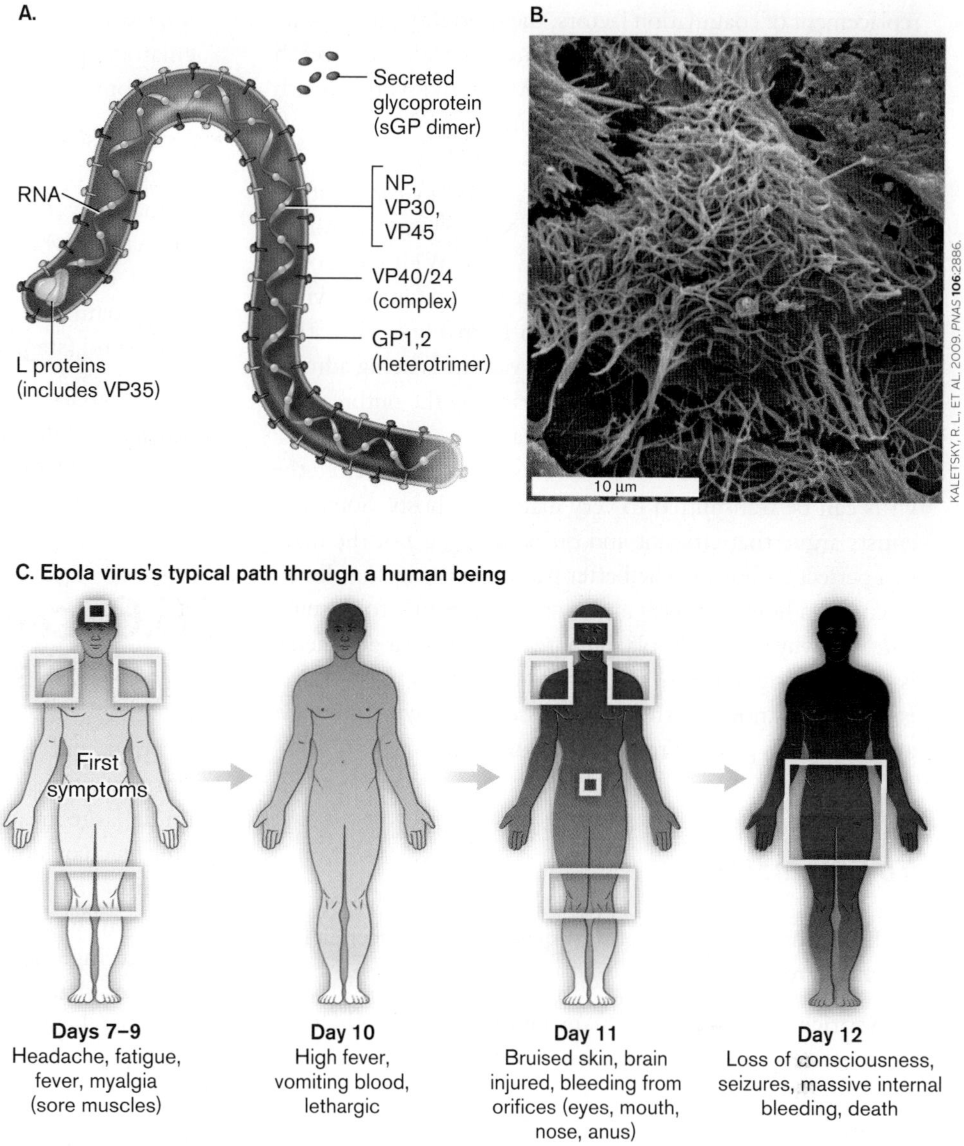

FIGURE 26.29 ■ Ebola virion. **A.** Composition of the virus. The ribonucleoprotein complex consists of the nucleoprotein (NP), the structural proteins VP30 and VP35, and the virion-associated RNA-dependent RNA polymerase (L proteins). The glycoprotein (GP-sGP) is an integral membrane protein that can be secreted. **B.** Threadlike Ebola virions budding from a cell (center). **C.** Progression of the disease. *Source:* Part C modified from http://themindbodyshift.com/index.php/2014/10/16/when-the-threat-of-ebola-hits-home.

Ebola viral proteins and their locations in the virion are shown in **Figure 26.29A**. Ebola VP35 protein is a component of the viral RNA polymerase complex, but it is also a potent inhibitor of host interferon (IFN) production. The cellular response to whichever IFN is made is inhibited by VP24, which blocks the nuclear accumulation of a regulatory protein called STAT1. STAT1 is critical to IFN-stimulated gene expression. These and other strategies enable rapid replication of the virus.

After replicating, Ebola offspring sprout from the cell surface in a mass of tangled threads (**Fig. 26.29B**). These new virions go on to attack new cells, riddling blood vessels and organs with damage as they go. Rapid release of new virions involves subverting another host mechanism, tetherin, designed to slow viral spread. Paul Bates and his colleagues at the University of Pennsylvania discovered that the cell protein tetherin essentially tethers mature virus particles inside a cell so that they are unable to spread. Tetherin is IFN induced and can restrict the spread of structurally diverse enveloped viruses, including HIV (thus, it is part of the innate response). Ebola glycoprotein, however, counteracts tetherin, so that nothing slows down viral spread. The result is rapid release of massive numbers of virus particles that can then quickly spread infection to other organs and tissues.

The outlook for a patient infected with Ebola is typically dire (occasionally a person infected with Ebola virus can develop antibodies but remain asymptomatic). The incubation period is 4–16 days, and death occurs within 7–16 days (**Fig. 26.29C**). With good supportive care, including the replacement of coagulation factors, the mortality rate appears near 20%. In the absence of such care, mortality can reach over 80%. For the latest outbreaks in Africa, the World Health Organization has approved use of three experimental treatments and a vaccine. The treatments include a recombinant monoclonal antibody (ZMapp) that has Ebola virus–neutralizing activity, an RNA-dependent RNA polymerase inhibitor, and an RNA chain terminator. The WHO-approved vaccine (ERVEBO) is an attenuated vesicular stomatitis virus engineered to express a glycoprotein from Ebola virus that generates neutralizing antibody. The vaccine is being administered to uninfected people in areas ringing the outbreak, in an attempt to limit spread of the disease

Death comes very quickly with this disease and before the virus can be transmitted to very many new hosts. Some scientists argue that efficient and quick killing is not the mark of a perfect pathogen. The better pathogen lets its host linger to ensure a home for itself and more opportunity to disseminate. So, why doesn't Ebola die out? Where does it go when it is not infecting humans? The natural ecology of these viruses is largely unknown, although an association with monkeys and/or bats as possible reservoirs is suggested.

To Summarize

- **Blood cultures** are useful in diagnosing septicemia and endocarditis.
- **Septicemia** is caused by many Gram-positive and Gram-negative bacterial pathogens. It can start with the bite of an infected insect, introduction via a wound, escape from an abscess, or penetration of the mucosal epithelium by the pathogen (intestine or vagina), and can lead to disseminated, systemic disease.
- **Endocarditis** can have acute or subacute onsets. **Subacute bacterial endocarditis** is usually an endogenous infection of a heart valve caused by *Streptococcus mutans.*
- **Malaria, caused by *Plasmodium* species**, manifests as repeated episodes of chills, fever, and sweating, owing to the organism's ability to alter the antigenic appearance of its surface proteins and evade the immune response.
- **Plague, caused by *Yersinia pestis***, has sylvatic and urban infection cycles involving transmission between fleas. The bite of an infected flea leads to bubonic plague. Bubonic plague can progress to septicemic and pneumonic stages. Pneumonic plague can be spread directly from person to person (no insect vector) by **aerosolized respiratory secretions**.
- **Toxic shock syndrome is a sepsis-like, systemic disease caused by an exotoxin (toxic shock syndrome toxin, TSST)** secreted by some strains of *Staphylococcus aureus* and *Streptococcus pyogenes.* It is caused by the circulatory spread of TSST made from bacteria that infect or colonize a localized tissue.
- **Lyme disease** is caused by the spirochete *Borrelia burgdorferi*, which is transmitted from animal reservoirs to humans by the bite of *Ixodes* ticks. The three stages of Lyme disease are characterized by a bull's-eye rash, called erythema migrans (stage 1); joint, muscle, and nerve pain (stage 2); and arthritis with WBCs in the joint fluid (stage 3).
- **Ebola virus** spreads from human to human via body fluids and kills its victims quickly. Its viral proteins alter cytokine production and facilitate virus release from infected cells.

26.6 Central Nervous System Infections

The brain and spinal cord are especially well protected against infection. Microbes cannot gain easy access to the brain, in large measure because of the **blood-brain barrier**, a filter mechanism that allows only selected substances into the brain. The blood-brain barrier works to our advantage when harmful substances, such as bacteria, are prohibited from entering. However, it works to our disadvantage when substances that we want to enter the brain, such as antibiotics, are kept out. The barrier is not a single structure but a function of the way blood vessels, especially capillaries, are organized in the brain. Furthermore, the endothelial cells in those vessels have tight junctions that do not allow most compounds or microbes to cross. And yet, brain infections do occur.

Case History: Meningitis

*In April 2001, Laila, a 4-month-old infant from Saudi Arabia, was hospitalized with fever, tender neck, and purplish spots (purpuric spots) on her trunk (**Fig. 26.30A**). Suspecting meningitis, the clinician took a cerebrospinal fluid (CSF) sample and examined it by Gram stain. The smear revealed Gram-negative diplococci inside PMNs. The CSF was turbid with 900 leukocytes per microliter, and* Neisseria meningitidis *was confirmed by culture. The child was treated with cefotaxime (a cephalosporin antibiotic; see Chapter 27) and made a full recovery. Her father, Abdul, the person who brought her in, was clinically well. However, the meningococcus was isolated from his oropharynx, as well as from the throat of the patient's 2-year-old brother. Isolates from the patient, her father, and her brother were positive by agglutination with meningococcal A/C/Y/W135 polyvalent reagent. Records showed that the father had previously received a quadrivalent*

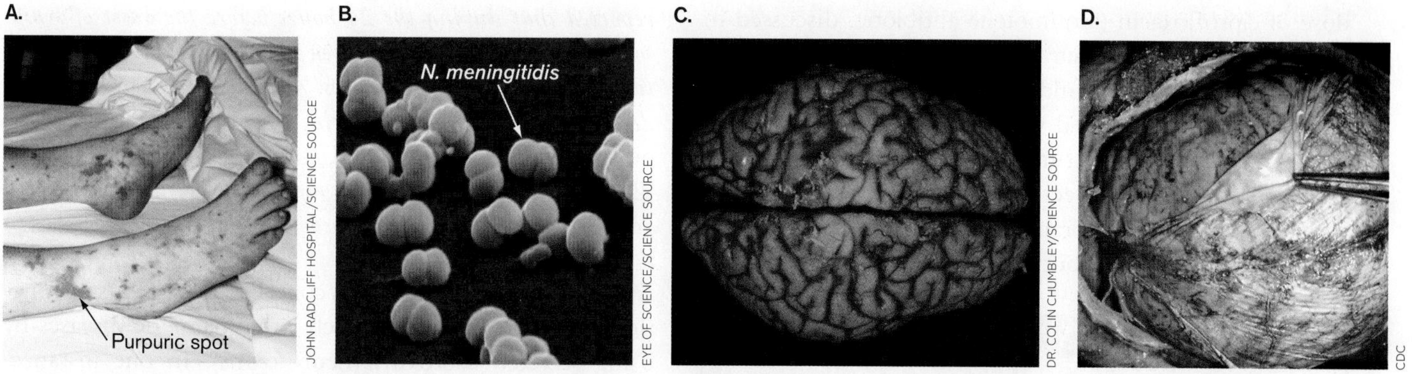

FIGURE 26.30 ■ Bacterial meningitis. **A.** Purpuric spots produced by local intravascular coagulation due to *Neisseria meningitidis* endotoxin. The rash in meningitis typically has petechial (small) and purpuric (large) components. **B.** *N. meningitidis* (diameter approx. 1 μm; SEM). **C.** Normal brain. **D.** Autopsy specimen of meningitis due to *Streptococcus pneumoniae*. Note the greening of the brain, compared with the pink normal brain.

meningococcal vaccine. All three isolates were confirmed to be meningococcus serogroup W135. DNA analysis found the three isolates to be indistinguishable, meaning that the father and his children were infected with the same strain of N. meningitidis. *Why did Laila's brother not have meningitis? And why was Laila's vaccinated father colonized?*

Meningitis is an inflammation of the meninges, the membranes that surround the brain and spinal cord. Meningitis can be either bacterial or viral in origin. Sinus and ear infections can extend directly to the meninges, whereas septicemic spread requires passage through the blood-brain barrier. Viral meningitis is serious but rarely fatal in people with a normal immune system. The symptoms generally persist for 7–10 days and then completely resolve. Bacterial meningitis is usually caused by *Streptococcus pneumoniae*, *Neisseria meningitidis*, or *Haemophilus influenzae*. Symptoms of bacterial meningitis can include sudden onset of fever, headache, neck pain or stiffness, painful sensitivity to strong light (photophobia), vomiting (often without abdominal complaints), and irritability. Prompt medical attention is extremely important because the disease can quickly progress to convulsions and death.

The meningococcus *N. meningitidis* (**Fig. 26.30B**) can colonize the human oropharynx, where it causes mild, if any, disease. At any given time, 10%–20% of the healthy population can be colonized and asymptomatic. The organism spreads directly by person-to-person contact or indirectly via droplet nuclei from sneezing or fomites. The problem arises when this organism enters the bloodstream. Unlike *N. gonorrhoeae*, the cause of gonorrhea, *N. meningitidis* is very resistant to complement, owing to its production of a polysaccharide capsule. Protected by this capsule, the microbe can produce a transient blood infection (bacteremia) and reach the blood-brain barrier.

How does *N. meningitidis* enter the bloodstream from the nasopharynx and then cross the blood-brain barrier? The organism initially uses type IV pili to adhere to and enter nasopharyngeal endothelial cells (type IV pili are discussed in Section 25.2). The bacteria then cross the endothelial cell layer by **transcytosis**, a process by which an internalized pathogen passes through a host cell along microtubules to the opposite side. Ultimately, the pathogen passes into the capillary lumen and is swept away to the brain. Once in the brain, type IV pili again adhere to endothelial cells, but this time they trigger the recruitment of host proteins that destabilize intercellular junctions. The bacteria then slip between the loosened junctions and enter the cerebrospinal fluid. Once in the CSF, microbes can multiply almost at will. **Figure 26.30C** and **D** shows the remarkable damage (greening) that *N. meningitidis* and other microbes, such as *S. pneumoniae*, cause in the brain.

Several antigenic types of capsules, called type-specific capsules, are produced by different strains of pathogenic *N. meningitidis*: types A, B, C, W135, and Y. Types A, C, Y, and W135 are usually associated with epidemic infections seen among people kept in close proximity, such as college students or military personnel. Type B meningococcus is typically involved in sporadic infections. However, large outbreaks of type B in the United States occurred at Princeton University and UC Santa Barbara in 2013.

Antibodies to the capsular antigens are used to classify the capsular types of the organisms causing an outbreak. Knowing the capsular type of organism involved in each case helps determine whether the disease cases are related and where the infection may have started. In Abdul's case, a polyvalent reagent containing antibodies to the four main capsular types was used to confirm *N. meningitidis*, and antisera to individual capsule types specifically identified the strain as W135. Why was Abdul colonized but not ill? W135 likely colonized Abdul's throat but failed to spread before circulating anti-W135 antibodies developed.

Meningococcal meningitis is highly communicable. As a result, close contacts, such as the parents or siblings of any patient with meningococcal disease, should receive antimicrobial prophylaxis within 24 hours of diagnosis; a single dose of ciprofloxacin (a quinolone antibiotic, discussed in Section 27.4) can be given to adults, and 2 days of rifampicin can be given to children. Highly susceptible populations can be immunized with a vaccine containing four of the five capsular structures. Because the type B capsule is not immunogenic (Section 24.6), a separate vaccine containing outer membrane vesicles and proteins from *N. meningitidis* type B (but not its capsule) is used.

Thought Question

26.10 Normal cerebrospinal fluid is usually low in protein and high in glucose. The protein and glucose content does not change much during a case of viral meningitis, but bacterial infection leads to greatly elevated protein and lowered glucose levels. What could account for this difference?

Case History: Botulism—It Is What You Eat

In June, a 47-year-old resident of Oklahoma was admitted to the hospital with rapid onset of progressive dizziness, blurred vision, slurred speech, difficulty swallowing, and nausea. Findings on examination included drooping eyelids, facial paralysis, and impaired gag reflex. He developed breathing difficulties and required mechanical ventilation. The patient reported that during the 24 hours before the onset of symptoms, he had eaten home-canned green beans and a stew containing roast beef and potatoes. Analysis of the patient's stool detected botulinum type A toxin, but no Clostridium botulinum *organisms were found. The patient was hospitalized for 49 days, including 42 days on mechanical ventilation, before recovering and being discharged.*

Imagine a disease that causes complete loss of muscle function. Two microbes cause such lethal paralytic diseases by using secreted exotoxins (neurotoxins). In one instance, botulism, the victim suffers a flaccid paralysis in which the muscles go limp, as in the case history just presented, causing paralysis and respiratory difficulty. **Botulism** is typically a food-borne disease caused by an anaerobic, Gram-positive, spore-forming bacillus, named *Clostridium botulinum*, that produces botulism toxin.

In striking contrast to botulism is tetanus, a very painful disease in which muscles continually and involuntarily contract (called tetany or spastic paralysis). Tetanus is caused by **tetanospasmin**, a potent exotoxin made by another anaerobic, Gram-positive spore-forming bacillus, called *Clostridium tetani* (**Fig. 26.31A**). Tetanospasmin alters neural transmission, but in contrast to botulism toxin (which inhibits neurotransmission), tetanospasmin causes excessive nerve signaling to muscles,

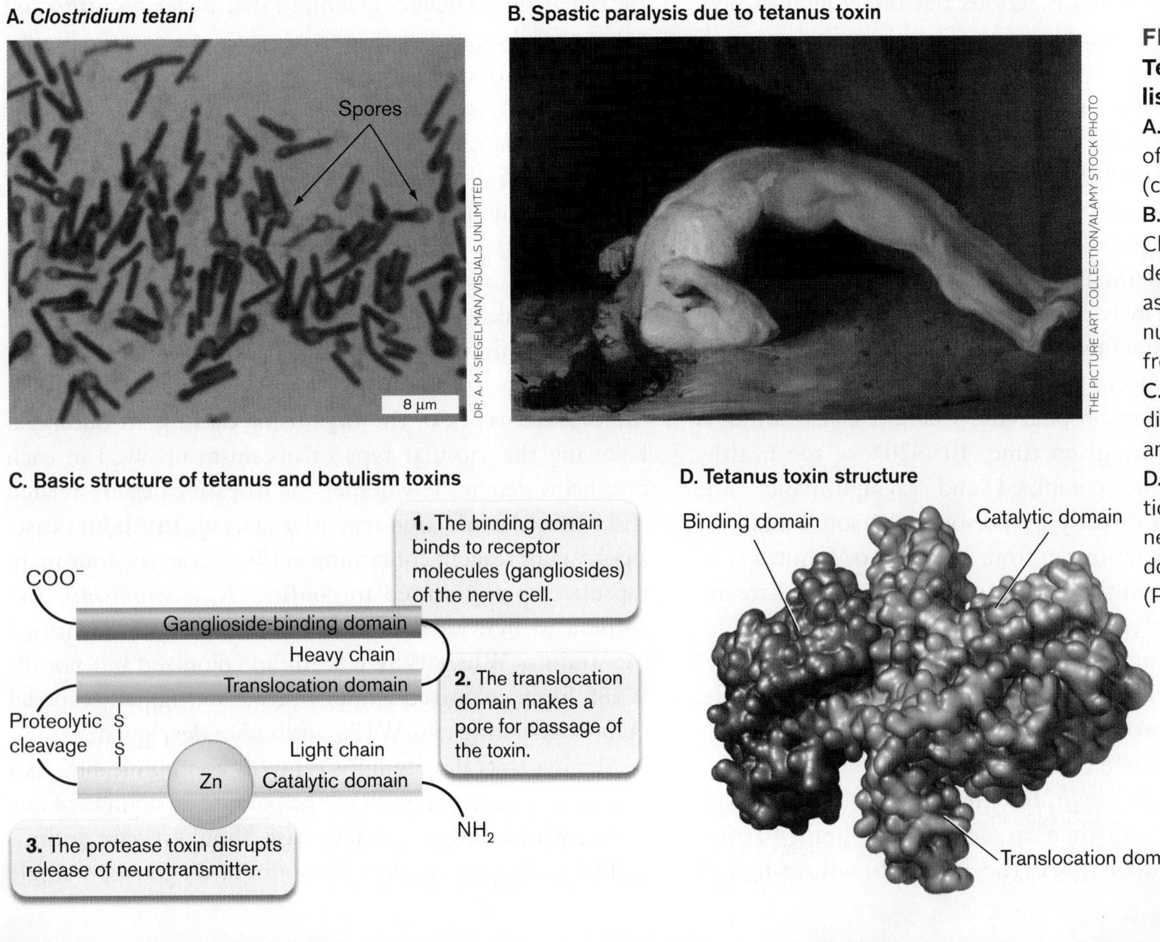

FIGURE 26.31 ■ Tetanus and botulism toxins. **A.** Photomicrograph of *Clostridium tetani* (cell length 4–8 µm). **B.** Painting from Charles Bell that depicts the spasms associated with tetanus in a soldier dying from tetanus. **C.** Schematic diagram of tetanus and botulism toxins. **D.** 3D representation of the tetanus neurotoxin, with the domains marked. (PDB code: 3BTA)

forcing the victim's back to arch grotesquely while the arms flex and legs extend (**Fig. 26.31B**). The patient remains locked this way until death. Spasms can even be strong enough to fracture the patient's vertebrae. In both botulism and tetanus, death can result from asphyxiation.

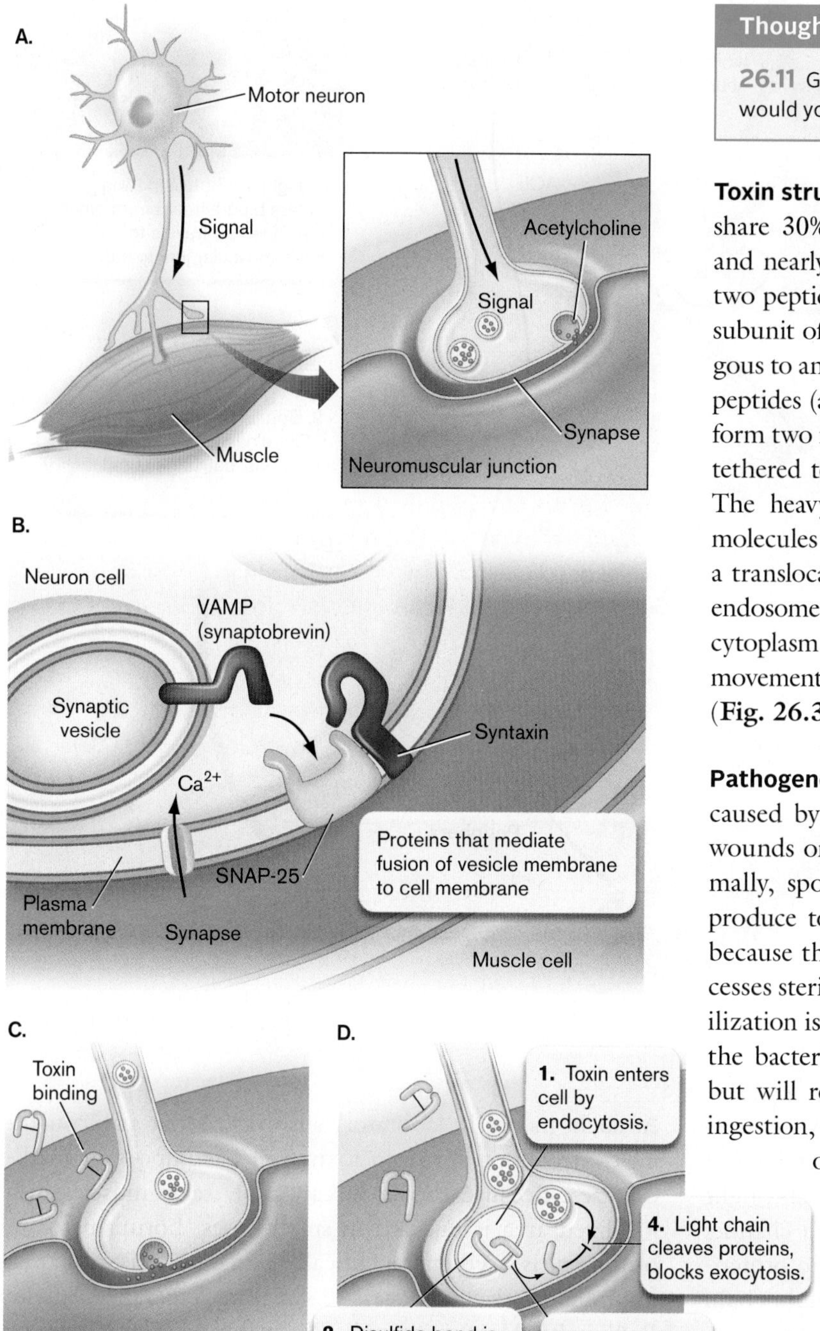

FIGURE 26.32 ■ Mechanism of action of botulism toxin. **A.** The neuromuscular junction. The blowup shows vesicles filled with neurotransmitters. **B.** A series of proteins within the nerve is needed to allow synaptic vesicles to bind to the nerve endings. Fusion of the membranes releases acetylcholine into the neuromuscular junction. Botulism toxin types A and E cleave SNAP-25. Botulism toxins B, D, F, and G cleave VAMP. Botulism toxin C1 cleaves syntaxin and SNAP-25. **C, D.** Toxin binds via the heavy chain and is endocytosed into the nerve terminal. Once the toxin cleaves its target, the nerve terminal is no longer able to release acetylcholine.

Thought Question

26.11 Given the symptoms of tetanus, what kind of therapy would you use to treat the disease?

Toxin structure and function. Botulism and tetanus toxins share 30%–40% identity, and they have similar structures and nearly identical modes of action. Each is composed of two peptides—a large, or heavy, fragment analogous to a B subunit of AB toxins and a small, or light, fragment analogous to an A subunit. Both toxins are initially made as single peptides (about 150 kDa) that are cleaved after secretion to form two fragments (the heavy and light chains) that remain tethered to each other by a disulfide bond (**Fig. 26.31C**). The heavy chains possess binding domains for receptor molecules (gangliosides) on the nerve cell membrane, and a translocation domain that makes a pore in the nerve cell endosome through which the toxic light chains pass into the cytoplasm. The light chains are proteases that disrupt the movement of exocytic vesicles containing neurotransmitters (**Fig. 26.31D**).

Pathogenesis of botulism neurotoxin. Botulism is typically caused by ingestion of preformed toxin. Rarely, infected wounds or ingested spores can also produce disease. Normally, spores germinate in contaminated food and cells produce toxin, but an anaerobic environment is required because the organism is an anaerobe. Home-canning processes sterilize the food but also remove oxygen. If the sterilization is incomplete, surviving spores will germinate and the bacteria will secrete toxin. The toxin is heat-sensitive but will remain active in improperly cooked food. After ingestion, the toxin is absorbed from the intestine. The organism is often absent from stool samples, but we can detect the toxin in stool.

A rare form of botulism, called infant botulism or "floppy head syndrome," can occur when infants (not older children) are fed honey. Honey can harbor *Clostridium botulinum* spores that can germinate in the gastrointestinal tract, after which the growing vegetative cells will secrete toxin.

Figure 26.32A and **B** illustrates normal neurotransmission at a neuromuscular junction. A signal sent from a motor neuron to the neuromuscular junction releases the neurotransmitter molecule acetylcholine (ACh) from vesicles in the axon terminal (**Fig. 26.32A**). **Figure 26.32B** shows the proteins that enable the vesicle and plasma membranes to fuse. Once released, ACh traverses the synapse and causes muscle contraction.

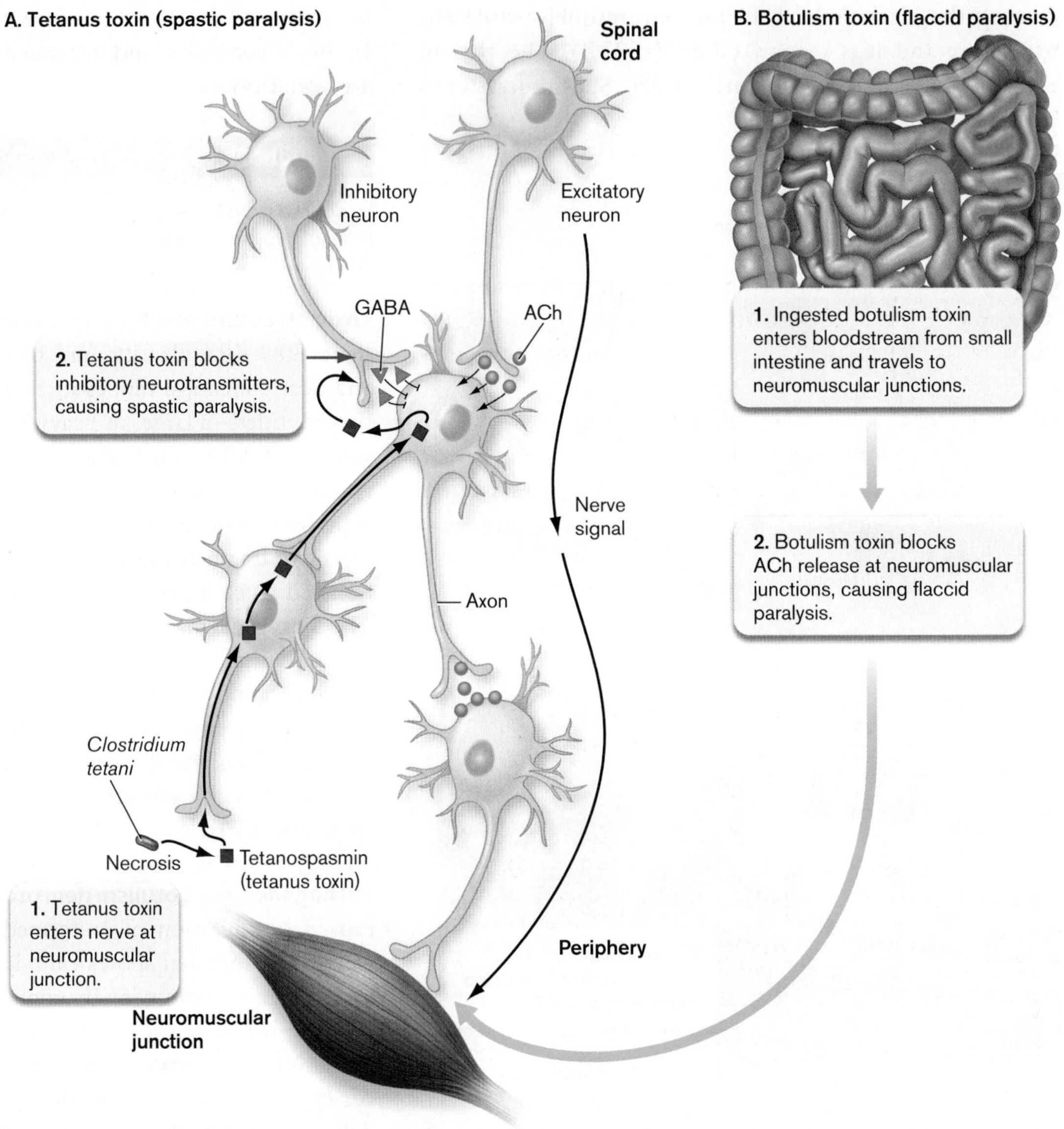

FIGURE 26.33 ■ Action of tetanus and botulism toxins. **A.** Tetanus toxin enters the nervous system at the neuromuscular junction (A1) and travels retrogradely up the axons until reaching an inhibitory neuron located in the central nervous system (A2). There it cleaves VAMP protein (Fig. 26.32B) associated with the exocytosis of vesicles containing inhibitory neurotransmitters (GABA). ACh = acetylcholine; GABA = gamma-aminobutyric (4-aminobutanoic acid). **B.** Botulism toxin is absorbed by the small intestine, enters the bloodstream, and acts at the neuromuscular junction to prevent release of acetylcholine.

Botulism toxin enters a peripheral nerve by endocytosis (**Fig. 26.32C**), and acidification of the endosome reduces the disulfide bonds holding the heavy and light chains together. The heavy chain assembles as a channel in the endosome membrane through which the proteolytic light chain moves into the cytoplasm. The light chain then cleaves key host proteins, such as synaptobrevin (a vesicle-associated membrane protein, or VAMP), syntaxin, or synaptosomal-associated protein (SNAP-25), involved in the exocytosis of vesicles containing acetylcholine (**Fig. 26.32D**). Without acetylcholine to activate nerve transmission, muscles will not contract and the patient is paralyzed. Currently, there is no FDA-approved vaccine for botulism. Treatment involves a multivalent antitoxin that can neutralize all seven toxin serotypes, coupled with supportive care.

Although botulism is now rare, the toxin is considered a select biological agent of potential use to bioterrorists. As such, its use in laboratories is under strict governmental control. But botulism toxin also has important medical uses. Because the toxin can safely relax muscles in a localized area if injected in small doses, botulism toxin, or Botox, is used therapeutically by neurologists to treat migraine headaches, excessive sweating, crossed eyes, and Bell's palsy. Most commonly, Botox is used cosmetically by plastic surgeons to reduce facial wrinkles.

Pathogenesis of tetanus neurotoxin. With such a high degree of structural and mechanistic similarity, why do tetanus and botulism toxins have such drastically different effects? The answer is based on where each toxin acts in the nervous system (see **Fig. 26.33**). Tetanus, in contrast to botulism, is not a food-borne disease. *Clostridium tetani* spores are introduced into the body by trauma (such as by stepping on the wrong end of a nail). Necrotic tissue provides the anaerobic environment required for germination. Growing bacteria release tetanospasmin, which enters the peripheral nerve cells at the site of injury. But rather than cleaving targets here, the toxin travels up axons in the direction opposite to nerve signal transmission until it reaches the spinal column, where it becomes fixed at the presynaptic inhibitory motor neuron (**Fig. 26.33A** ▶).

There the toxin also cleaves proteins like VAMP, but the vesicles in these nerves release inhibitory neurotransmitters (GABA and glycine) that dampen nerve impulses. Tetanus toxin blocks release of these inhibitors into the synaptic cleft, leaving nerve impulses unchecked. As a result, impulses come too frequently and produce the generalized muscle spasms characteristic of tetanus. A tetanus toxoid vaccine is available and is administered to children as part of the DTaP (diphtheria, tetanus, and acellular pertussis) vaccine (see Section 24.6).

Thought Questions

26.12 **Figure 26.33** demonstrates that tetanus toxin has a mode of action (spastic paralysis) opposite to that of botulism toxin (flaccid paralysis). Since they have opposing modes of action, can botulism toxin be used to save a patient with tetanus?

26.13 How do the actions of tetanus toxin and botulism toxin actually help the bacteria colonize or obtain nutrients?

26.14 If *Clostridium botulinum* is an anaerobe, how might botulism toxin get into foods?

Case History: Eastern Equine Encephalitis

In August, Mr. C took his 21-year-old son, Rich, to a New Jersey emergency department. Rich appeared dazed and had trouble responding to simple commands. When questioned about his son's activities over the previous few months, Mr. C told the physician that Rich had spent the month of July relaxing and sunning himself on New Jersey beaches and visiting a pond in a wooded area near a horse farm. On the afternoon prior to admission, Rich had become lethargic and tired. He returned home and went to bed. That evening, his father woke him for supper, but Rich was confused and had no appetite. By 11 p.m., Rich had a fever of 40.7°C (103.5°F) and could not respond to questions. A few hours later, when his father had trouble rousing him, he took Rich to the ED. Over the next week, Rich's condition worsened to the point where his limbs were paralyzed. Two weeks later he died. Serum samples taken when he entered the hospital and a few days before he died showed a sixfold rise in antibody titer to eastern equine encephalitis (EEE) virus. Brain autopsy showed many small foci of necrosis in both the gray and white matter.

The EEE virus, a member of the family *Togaviridae*, is transmitted from bird to bird by mosquitoes. Horses contract the disease in the same way. Rich contracted it inadvertently while visiting the pond. The disease, an encephalitis, is often fatal (35% mortality) but fortunately rare. One reason human disease is rare is that the species of mosquito that usually transmits the virus between marsh birds does not prey on humans. But sometimes a human-specific mosquito bites an infected bird and then transmits EEE virus to humans. Another reason human disease is rare is that the virus generally does not fare well in the body. Many persons infected with EEE virus have no apparent illness, because the immune system thwarts viral replication. However, as already noted, mortality is high in individuals that do develop disease.

An interesting point in the case history is that diagnosis relied on detecting an increase in antibody titer to the virus. Typically, at the point disease symptoms first appear the body has not had time to generate large amounts of specific antibodies. After a week or so, often when the patient is nearing recovery (convalescence), antibody titers have risen manyfold. The rule of thumb is that a greater-than-fourfold rise in IgG antibody titer between acute disease and convalescence (or in this case death) indicates that the patient has had the disease. While this knowledge could not help the patient in the case described here, it was valuable in terms of public health and prevention strategies.

Several bacterial, fungal, and viral causes of encephalitis and meningitis are listed in **Table 26.7**.

Prions Are Infectious Proteins

Imagine slowly losing your mind and knowing there is nothing you can do about it. An unusual infectious agent called the prion has been implicated as the cause of a series of relatively rare but invariably fatal brain diseases (**Table 26.8**). Prions are infectious agents that do not have a nucleic acid genome. The protein alone can mediate an infection. The **prion** is now recognized as an infectious, misfolded protein that resists inactivation by procedures that destroy proteins.

The discovery that proteins alone can transmit an infectious disease came as a surprise to the scientific community. Diseases caused by prions are especially worrying, since prions resist destruction by many chemical agents and remain active after heating at extremely high temperatures. There have even been documented cases in which sterilized neurosurgical instruments, originally used on a prion-infected person, still held infectious agent and transmitted the disease to a subsequent surgical patient. Extremely rigorous decontamination procedures, such as autoclaving surgical instruments immersed in 1-M sodium hydroxide, will destroy prions.

How can a nonliving entity without nucleic acid be an infectious agent? Prions associated with human brain disease are thought to be aberrantly folded forms of a

TABLE 26.7 Selected Microbes That Cause Meningitis or Encephalitis

Type of disease	Etiological agent(s)[a]	Virulence factors	Typical initial infection or source	Treatment	Vaccine
Bacterial (septic) meningitis	*Streptococcus pneumoniae* (G+)	Capsule, pneumolysin	Lung	Ampicillin	Multivalent, capsule
	Neisseria meningitidis (G–)	Capsule, IgA protease, endotoxin	Throat	Cephalosporin (3rd generation), ceftriaxone	Multivalent, capsule
	Haemophilus influenzae type b (G–)	Polyribitol capsule, IgA protease, endotoxin	Ear infection	Cephalosporin (3rd generation), ceftriaxone	Type b polysaccharide
	Other G– bacilli	Endotoxin	Septicemia		
	Group B streptococci (G+)	Sialic acid capsule, streptolysin, inhibition of alternate complement path	Neonate infected during parturition	Ampicillin	Capsular
	Listeria monocytogenes (G+)	Intracellular growth, PrfA regulator, actin-based motility	Mild GI disease of mother	Ampicillin plus gentamicin	
	Mycobacterium tuberculosis	Cord factor, wax D, intracellular growth	Lung	Combination therapy (rifampin, isoniazid, ethambutol, pyrazinamide)	BCG[b]
	Staphylococcus aureus (G+)	Coagulase, protein A, TSST, leukocidin	Septicemia	Oxacillin, vancomycin	
	Staphylococcus epidermidis (G+)	Biofilm, slime	Complication of surgical procedure	Vancomycin, methicillin	
Aseptic meningitis[c]	Fungi (e.g., *Coccidioides*, *Cryptococcus*)		Sinusitis, direct spread to meninges	Ketoconazole, fluconazole	
	Amebas (e.g., *Naegleria*)		Swimming in contaminated waters	Amphotericin B	
	Treponema pallidum		Syphilis		
	Mycoplasmas	Adhesin tip	Respiratory	Erythromycin	
	Leptospira	Burrowing motility	Septicemia, water contaminated with animal urine	Erythromycin	Killed whole cell, animals only
Viral meningitis or encephalitis	Viruses (90% caused by enteroviruses)			Self-limiting	
Eastern equine encephalitis	EEE virus	?	Mosquito bite	None; often fatal	
West Nile disease	West Nile virus	?	Mosquito bite	Supportive therapy[d]	

[a]G+ = Gram-positive; G– = Gram-negative.

[b]BCG = Bacille Calmette-Guérin (a weakened strain of the bovine tuberculosis strain).

[c]For aseptic meningitis, bacterial sources cannot be isolated by ordinary means.

[d]Supportive therapy = hospitalization, intravenous fluids, airway management, respiratory support, and prevention of secondary infections.

normal brain protein (PrP^C). The theory is that when a disease-related prion, designated PrP^{Sc} or PrP^D, is introduced into the body and manages to enter the brain, it will cause normally folded forms of the protein to refold incorrectly (**Fig. 26.34A**). The improperly folded proteins fit together like Lego blocks to produce damaging aggregated structures within brain cells.

Prion diseases are often called **spongiform encephalopathies** because the postmortem appearance of the brain includes large, spongy vacuoles in the cortex and cerebellum. These are visible in a brain sample from a victim of one of these diseases, called Creutzfeldt-Jakob disease (CJD; **Fig. 26.34C**). Most mammalian species appear to be susceptible to these diseases.

TABLE 26.8 Prion Diseases

Disease	Susceptible animal	Incubation period	Disease characteristics
Creutzfeldt-Jakob [sporadic, familial, new variant (vCJD)]	Human	Months (vCJD) to years	Spongiform encephalopathy (degenerative brain disease)
Kuru	Human	Months to years	Spongiform encephalopathy
Gerstmann-Straussler-Scheinker syndrome	Human	Months to years	Genetic neurodegenerative disease
Fatal familial insomnia	Human	Months to years	Genetic neurodegenerative disease with untreatable insomnia
Mad cow disease	Cattle	5 years	Spongiform encephalopathy
Wasting disease	Deer	Months to years	Spongiform encephalopathy

Since 1996, mounting evidence has pointed to a causal relationship between outbreaks in Europe of a disease in cattle called bovine spongiform encephalopathy (BSE, or "mad cow disease"), which is caused by prions, and a disease in humans called variant Creutzfeldt-Jakob disease (vCJD). Both disorders are invariably fatal. For humans to contract disease, prions from contaminated meat are ingested and penetrate the intestinal mucosa via the antigen-sampling M cells in Peyer's patches. The circulation then traffics the agent to the brain.

As of this writing, only four cases of BSE in cattle have been detected in the United States since 1993. Because of aggressive surveillance efforts in the United States and Canada (where 20 cases have been found), it is unlikely, but not impossible, that BSE will be a food-borne hazard to humans in this country. The CDC monitors the trends and current incidence of typical CJD (approximately 300 cases per year) and variant CJD (3 cases total) in the United States (see **eTopic 26.9**). The worldwide incidence of CJD is one per 1 million population.

There is no cure for CJD or vCJD. A test to detect PrP^{Sc} in asymptomatic primates has been developed and is being tested for human screening. The test is called protein-mediated cyclic amplification (PMCA). PMCA has successfully identified the presence of very tiny amounts of PrP^{Sc} in blood or urine. The process involves adding large amounts of normal PrP (PrP^{C}) to samples containing tiny amounts of PrP^{Sc} prions. The prions, if present in the sample, will seed the misfolding of PrP^{C} to PrP^{Sc} molecules that aggregate. Sonication is then used to partially break up the newly formed PrP^{Sc} aggregates into pieces that can further seed the misfolding and aggregation of normal PrP. Repeated cycles of sonication-incubation will exponentially amplify the amount of PrP^{Sc} present in a sample. The PrP^{Sc} is then identified by western blot (see **eTopic 12.1**).

To Summarize

- ***Neisseria meningitidis*** is resistant to serum complement because it produces a type-specific capsule that enables the organism to reach and then cross the blood-brain barrier.
- In contrast to *Neisseria gonorrhoeae*, a **vaccine for *N. meningitidis* is available.**
- **Botulism toxin** causes flaccid paralysis.
- **Tetanospasmin** causes spastic paralysis.

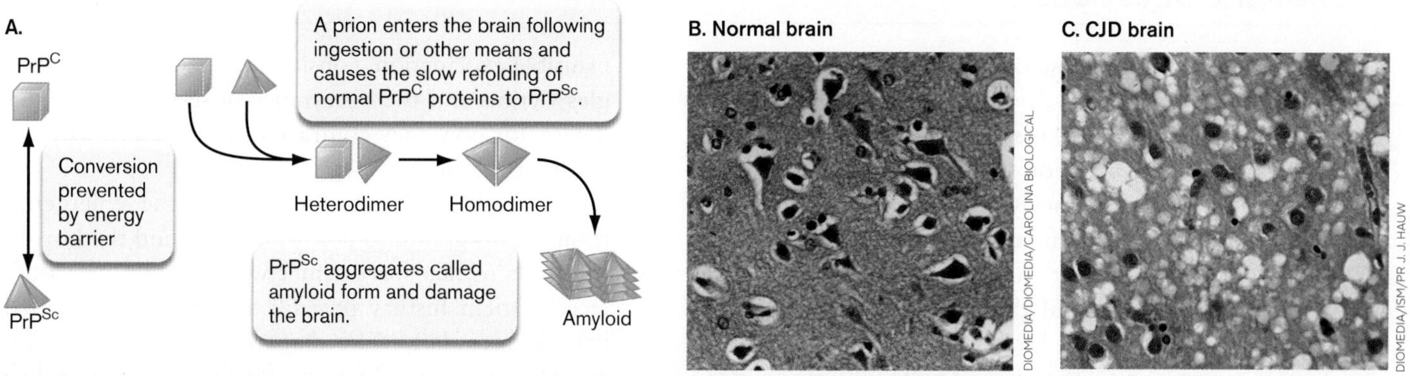

FIGURE 26.34 ■ Spongiform encephalopathies. **A.** Refolding model of prion diseases. PrP is a brain protein that can take two forms: the normal form (PrP^{C}), which is a natural brain protein; and the prion form (PrP^{Sc}). **B.** Normal brain section. **C.** Section of brain taken from a CJD victim. Note the Swiss-cheese appearance, indicating brain damage.

- **Serological diagnosis of many infectious diseases** is possible if the specific pathogen IgG antibody titer rises fourfold between the acute and convalescent stages of disease (eastern equine encephalitis virus, for instance).
- **Spongiform encephalopathies** are believed to be caused by nonliving proteins called prions.

Concluding Thoughts

In this chapter we have described key concepts of infectious disease using just a small sample of disease-causing pathogens. Other diseases are considered elsewhere in the book or on the accompanying website. These include anthrax, cholera, the common cold (caused by rhinovirus), dengue fever, diphtheria, gastric ulcers (*Helicobacter pylori*), herpes, AIDS, warts (human papillomavirus), influenza, salmonellosis, *Streptococcus agalactiae* infections, West Nile disease, and Whipple's disease (*Tropheryma whipplei*). The principal goal in this and the previous chapter was to illustrate how microbial metabolism can undermine the physiology of a human host. In the next chapter we describe how humans fight back by using pharmacology to sabotage the physiology of the infecting microbes.

CHAPTER REVIEW

Review Questions

1. Organize a list of pathogens by their mechanism of transmission (fecal-oral, aerosol, and so on).
2. Discuss some common skin infections.
3. What causes boils?
4. What are the symptoms of necrotizing fasciitis?
5. What is the difference between primary and secondary infections?
6. How do pneumococci avoid engulfment by phagocytes?
7. What causes the clouding seen in X-rays of infected lungs?
8. What are the key features of the pneumococcal vaccine?
9. Name the common fungal causes of lung disease.
10. What is the Ghon complex?
11. Why is diarrhea watery?
12. What is the most common microbial cause of diarrhea?
13. List common bacterial agents that cause diarrhea.
14. Would you suspect *Salmonella* infection in a cluster of nauseated patients rushed to the hospital directly from a church picnic? Why or why not?
15. What is the significance of finding leukocytes in stool?
16. What is one reservoir of *E. coli* O157:H7?
17. How is a UTI diagnosed?
18. What is an important virulence determinant of uropathogenic *E. coli*?
19. What is the most common sexually transmitted infection?
20. How is gonorrhea different in men and women?
21. Why will *Neisseria gonorrhoeae* not usually disseminate in the bloodstream, while *N. meningitidis* will?
22. Name the major causes of bacterial meningitis. What are the two routes of infection?
23. If tetanus and botulism toxins have the same mode of action, why do they cause opposite effects on muscles?
24. What are some virulence factors of *Yersinia pestis*?
25. Members of one class of infectious agent cannot be treated with antimicrobial chemotherapy. Explain why.

Thought Questions

1. Chickenpox is a disease of children and young adults caused by herpesvirus 3 (varicella). It is characterized by a rash of fluid-filled vesicles that eventually become crusty. The rash starts on the trunk and spreads to the extremities. Illness usually resolves in 7–10 days, and the patient becomes immune to the disease. However, some individuals later in life (usually over 60 years old) develop a painful disease, called shingles, that is caused by the same virus—even if they have never been reexposed to the virus. The lesions are tender, persistent vesicles that form on the skin. Propose a plausible explanation, considering the age of the shingles victims and the occurrence of severe pain in this illness, for how they contracted shingles and why the lesions are painful.
2. A 5-year-old male was brought to the emergency room by his grandmother, who had found the boy on the floor of her apartment covered in bloody, loose feces. Patient history revealed that the boy attended day care regularly. The diagnostic laboratory determined the etiological agent to be a Gram-negative rod. This organism is a facultative intracellular

pathogen that escapes the host cell vacuole and moves in and between cells by actin polymerization. From your reading of Chapters 25 and 26, what do you think are the most likely genus and species involved in this case? Why is the fact that the boy attended day care significant?

3. Why are urinary tract infections among the most commonly acquired nosocomial (hospital-acquired) infections?
4. On the Internet, find the MMWR (Morbidity and Mortality Weekly Report) that provides the "notifiable infectious diseases" weekly tables (the URL for data as of this writing is http://wonder.cdc.gov/nndss/nndss_weekly_tables_menu.asp). Study the tables that summarize the weekly incidence of gonorrhea and West Nile disease. For the latter, you may need to view one of the arboviral tables to find "West Nile virus disease" or "WNV disease." Explain the differences in the weekly incidence of these two diseases, and then view the incidence of those diseases in your state.
5. Why do new versions of swine and avian flu often originate in Asia?

Key Terms

bacteremia (1092)
blood-brain barrier (1104)
borreliosis (1100)
botulism (1106)
bubonic plague (1098)
cellulitis (1063)
chancre (1087)
chlamydia (1088)
congenital syphilis (1087)
cystitis (1083)
differential (1058)
empiric therapy (1062)
endocarditis (1093)
erythema migrans (1100)
granuloma (1069)
health care–associated infection (HAI) (1061)
hepatitis (1079)
hepatitis A virus (HAV) (1079)
hepatitis B virus (HBV) (1080)
hepatitis C virus (HCV) (1081)
Lyme disease (1100)
necrotizing fasciitis (1062)
norovirus (1075)
nosocomial (1061)
pneumonic plague (1098)
primary syphilis (1087)
prion (1109)
pyelonephritis (1083)
rehydration therapy (1072)
rotavirus (1075)
secondary syphilis (1087)
secondary tuberculosis (1070)
septicemia (1092)
septicemic plague (1098)
spongiform encephalopathy (1110)
tertiary syphilis (1087)
tetanospasmin (1106)
transcytosis (1105)
viremia (1092)

Recommended Reading

Aragon, Isabel M., Bernardo Herrera-Imbroda, Maria I. Queipo-Ortuno, Elizabeth Castillo, Julia Sequeira-Garcia Del Moral, et al. 2018. The urinary tract microbiome in health and disease. *European Urology Focus* **4**:128–138.

Backert, Steffen, Matthias Neddermann, Gunter Maubach, and Michael Naumann. 2016. Pathogenesis of *Helicobacter pylori* infection. *Helicobacter* **21**(Suppl.1):19–25.

Coureuil, Mathieu, Hervé Lécuyer, Sandrine Bourdoulous, and Xavier Nassif. 2017. A journey into the brain: Insight into how bacterial pathogens cross blood-brain barriers. *Nature Reviews. Microbiology* **15**:149–159.

David, Lawrence A., Ana Weil, Edward T. Ryan, Stephen B. Calderwood, Jason B. Harris, et al. 2015. Gut microbial succession follows acute secretory diarrhea in humans. *mBio* **6**:e00381-15.

Gilbert, Nicole M., Valerie P. O'Brien, and Amanda L. Lewis. 2017. Transient microbiota exposures activate dormant *Escherichia coli* infection in the bladder and drive severe outcomes of recurrent disease. *PLoS Pathogens* **13**:e1006238. https://doi.org/10.1371/journal.ppat.1006238.

Hammond, Andrew M., and Roberto Galizi. 2017. Gene drives to fight malaria: Current state and future directions. *Pathogens and Global Health* **111**:412–423.

Harper, Kristin N., Molly K. Zuckerman, and George J. Armelagos. 2014. Syphilis: Then and now. *Scientist* (February 1).

Kaminski, R., Y. Chen, T. Fischer, E. Tedaldi, A. Napoli, et al. 2016. Elimination of HIV-1 genomes from human T-lymphoid cells by CRISPR/Cas9 gene editing. *Scientific Reports* **6**:22555.

Misasi, John, Morgan S. A. Gilman, Masaru Kanekiyo, Miao Gui, Alberto Cagigi, et al. 2016. Structural and molecular basis for Ebola virus neutralization by protective human antibodies. *Science* **351**:1343–1346.

Noyan, Kajsa, Son Nguyen, Michael R. Betts, Anders Sönnerborg, and Marcus Buggert. 2018. Human immunodeficiency virus type-1 elite controllers maintain low co-expression of inhibitory receptors on CD4+ T cells. *Frontiers in Immunology* **9**:19. https://doi.org/10.3389/fimmu.2018.00019.

Olson, Patrick D., and David A Hunstad. 2016. Subversion of host innate immunity by uropathogenic *Escherichia coli*. *Pathogens* **5**:2. https://doi.org/10.3390/pathogens5010002.

Rhodes, Johanna. 2019. Rapid Worldwide Emergence of Pathogenic Fungi. *Cell Host Microbe* **26**:12–14.

Sela, Uri, Chad W. Euler, Joel Correa da Rosa, and Vincent A. Fischetti. 2018. Strains of bacterial species induce a greatly varied acute adaptive immune response: The contribution of the accessory genome. *PLoS Pathogens* **14**:e1006726. https://doi.org/10.1371/journal.ppat.1006726.

Wang, Gang, Na Zhao, Ben Berkhout, and Atze T. Das. 2017. CRISPR-Cas based antiviral strategies against HIV-1. *Virus Research* **244**:321–332. https://doi.org/10.1016/j.virusres.2017.07.020.

CHAPTER 27

Antimicrobial Therapy and Discovery

27.1 Fundamentals of Antimicrobial Therapy

27.2 Antibiotic Mechanisms of Action

27.3 Challenges of Drug Resistance and Discovery

27.4 Antiviral Agents

27.5 Antifungal Agents

Imagine a world without antibiotics. Suddenly, simple infections would turn deadly, food supplies would be threatened, and human life expectancy would shorten . . . dramatically. This may sound like science fiction, but we face this prospect today. Antibiotic resistance has developed in many pathogens because of indiscriminant antibiotic use in medicine, and in farming to boost animal growth. For the moment we still have effective antibiotics, but their number is dwindling, and the discovery of new antibiotics has been exceedingly slow. In Chapter 27 we discuss the way antimicrobial agents work, the countermeasures microbes use to resist them, and the ingenious techniques by which new drugs are discovered.

CURRENT RESEARCH highlight

Last-resort resistance. Most antibiotic resistance systems stop drugs from finding or binding to their targets. But what if an antibiotic still manages to bind its target? In one newly recognized form of antibiotic resistance—ribosome protection—proteins knock antibiotics off the ribosome. Yong-Gui Gao and Liang Yang used cryo-EM to show that one such protein, MsrE, binds a ribosome's tRNA exit site (E site) and projects a needlelike structure into the peptidyltransferase center and peptide exit tunnel, where macrolide antibiotics bind. Shape changes in MsrE and the ribosome dislodge the macrolide and restore translation. The cell is saved.

Source: Weixin Su et al. 2018. *PNAS* **115**:5157–5162.

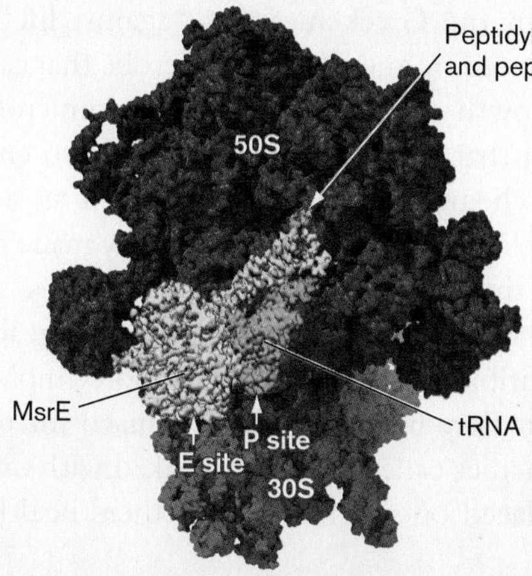

MODIFIED FROM WEIXIN SU ET AL. 2018. *PNAS* **115**:5157–5162, FIG. 1A.

AN INTERVIEW WITH

YONG-GUI GAO, STRUCTURAL BIOLOGIST, NANYANG TECHNOLOGICAL UNIVERSITY, SINGAPORE

LOH KWONG PING

Does MsrE, an ATP-binding cassette protein, requireATPhydrolysis to function? And might "antiresistance" antibiotics be designed to inactivate MsrE?

MsrE does require ATP hydrolysis, but not for entry or exit from ribosomes, at least in vitro. The possibility of designing "antiresistance proteins" to target MsrE is intriguing. An inhibitor that prevents ATP hydrolysis by MsrE could function as an antiresistance drug. Of course, any antiresistance drug must be used with the real antibiotic to be useful. Other scientists are conducting these studies.

Antibiotics undeniably have been a benefit to modern society. Infections that we consider minor today often killed their victims just 50 or 60 years ago. But there has been a downside too: the evolution of genetically adapted, antibiotic-resistant pathogens that threaten to make current antibiotics obsolete. One statistic underscores the seriousness of the situation: The number of patients that contract antibiotic-resistant infections in the United States (currently 2 million per year) has doubled since 2002. Many of these patients die. A variety of antibiotic resistance mechanisms can prevent an antibiotic from getting into the bacterial cell or, if it gets in, prevent it from binding to its target. The Current Research Highlight describes a mechanism, newly discovered by Yong-Gui Gao and Liang Yang (**Fig. 27.1**) via cryo-EM imaging (Section 2.6), that knocks an antibiotic off its target.

How serious is the antibiotic resistance problem? Consider the case of a 56-year-old man who entered the hospital for a heart transplant. The operation went well, but 1 week after the surgery, he developed a severe chest wound infection that exuded pus. Treatment with oxacillin, a penicillin derivative that can be effective against suspected Gram-positive infections, failed, and the patient became comatose. The diagnostic microbiology laboratory ultimately identified the agent as a methicillin (oxacillin)–resistant *Staphylococcus aureus* (MRSA), prompting the surgeon to immediately place the patient on intravenous vancomycin for 6 weeks. Vancomycin is an antibiotic, structurally different from methicillin and oxacillin, that can usually kill methicillin-resistant bacteria. Fortunately, this patient recovered; too often, they do not.

FIGURE 27.1 ▪ Molecular microbiologist Liang Yang. Yang collaborated with Yong-Gui Gao in discovering how MsrE protects ribosomes from macrolide antibiotics.

Where did the MRSA strain come from? Nasal swabs taken of all hospital personnel revealed that several members of the surgical team harbored drug-resistant *S. aureus* as part of their resident microbiota. But which individual was the actual source of the patient's infection? In what could be described as "forensic microbiology," genomic DNA from each strain was first treated with an endonuclease that targets rare DNA sequences and then subjected to pulsed-field gel electrophoresis. The resulting genomic restriction patterns were compared with that of the isolate from the infected patient. A single match was found. The unwitting culprit turned out to be the perfusionist who had manipulated the tubing used for cardiopulmonary bypass.

Health care–associated infections (HAIs), which include hospital-acquired (also known as nosocomial) infections, are relatively common. As many as 5%–10% of all patients admitted to acute-care hospitals will contract hospital-acquired infections, resulting in about 80,000 deaths each year. The deaths are due in part to the poor health of the patient and in part to the antibiotic-resistant nature of bacteria lurking in hospitals. In fact, 50%–60% of staph infections that develop in a hospital setting are caused by methicillin-resistant *S. aureus* (MRSA). One study in the United States found that the noses of nearly one in four nonhospitalized people are also colonized by MRSA. Antibiotic resistance is a problem that will only get worse.

We begin Chapter 27 with a discussion of the golden age of antibiotic discovery (1940–1960) and the basic concepts of antimicrobial use. We then explore various modes of antibiotic action and the microbial countermeasures of resistance. In the process, we will examine novel approaches that scientists use to search for new antibiotics and targets. The urgency of this search cannot be overstated. In 2016, for instance, an *E. coli* uropathogen resistant to a drug of last resort, colistin, had already arrived in the United States.

27.1 Fundamentals of Antimicrobial Therapy

Antibiotics (from the Greek meaning "against life") are compounds produced by one species of microbe that can kill or inhibit the growth of other microbes. "Antimicrobials" is a broader term that includes antibiotics but also encompasses synthetic chemotherapeutic agents, such as sulfonamides, that are clinically useful but chemically made in the laboratory. Over time, the distinction between these words has blurred to where they are often used interchangeably.

We think of antibiotics as being a recent biotechnological development, but they have actually been used for centuries. Ancient remedies called for cloths soaked with organic material to be placed on wounds to help them heal faster.

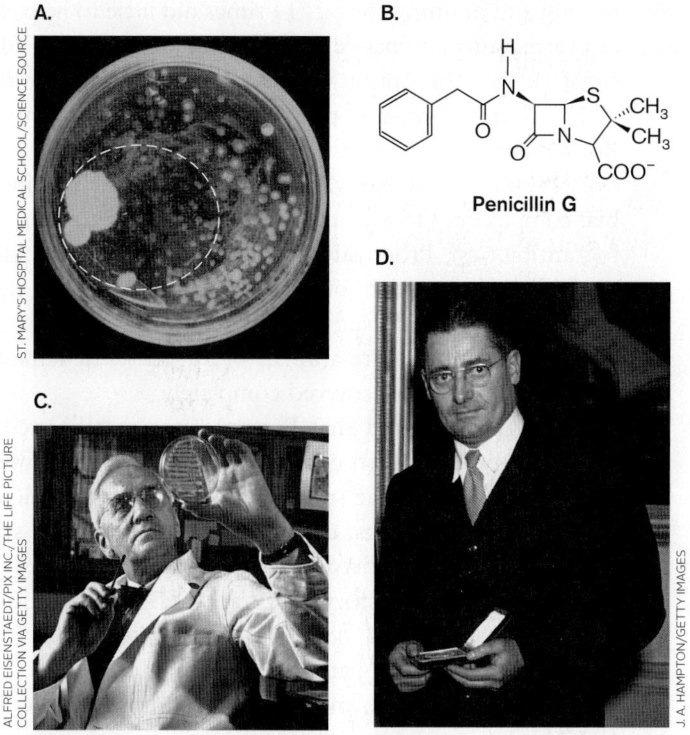

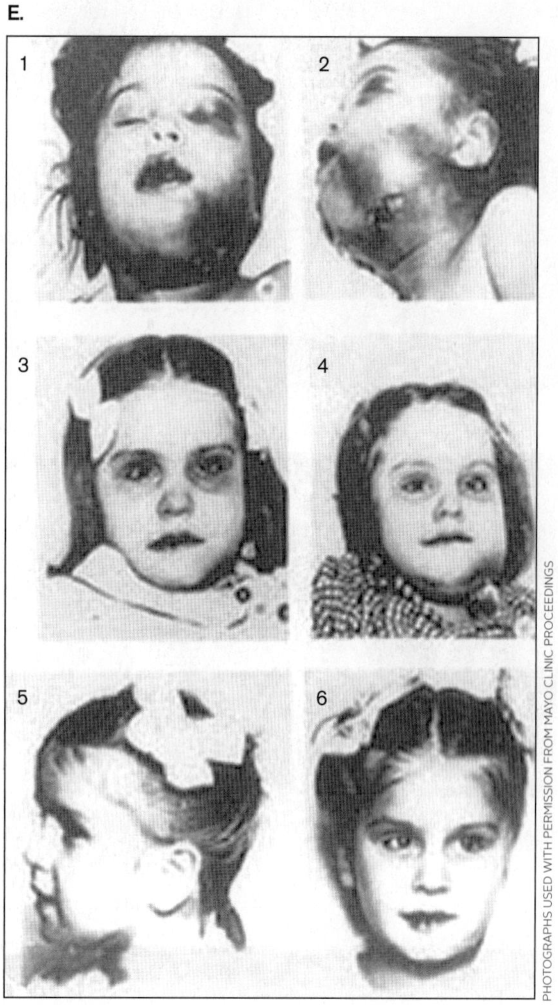

FIGURE 27.2 ■ The dawn of antibiotics. **A.** Alexander Fleming's photo of the dish with bacteria and penicillin mold. The ring highlights the area of decreased growth of *Staphylococcus aureus* colonies. **B.** The chemical structure of penicillin G. **C.** Alexander Fleming at work in his laboratory. **D.** Howard Florey. **E.** These pictures, taken in 1942, shortly after the introduction of penicillin, show the improvement in a child suffering from an infection 4 days (panel 2) and 9 days (panel 4) after treatment. Panels 5 and 6 show her fully recovered.

This organic material likely contained natural antibiotics that killed bacteria and prevented further infection. The medicinal properties of molds were also recognized for centuries. The ancient Chinese successfully treated boils with warm soil and molds scraped from cheeses, and in England a paste of moldy bread was a home remedy for wound infections until the beginning of the twentieth century.

The Golden Age of Antibiotic Discovery

The modern antibiotic revolution began with the discovery of **penicillin** in 1928 by Sir Alexander Fleming (1881–1955). This discovery was actually a rediscovery, and it is arguably one of the greatest examples of serendipity in science. Although Fleming generally receives the credit for discovering penicillin, a French medical student, Ernest Duchesne (1874–1912), originally discovered the antibiotic properties of *Penicillium*, in 1896.

Duchesne observed that Arab stable boys at the nearby army hospital kept their saddles in a dark and damp room to encourage mold to grow on them. When asked why, they told him the mold helped heal saddle sores on the horses. Intrigued, Duchesne prepared a solution from the mold and injected it into guinea pigs infected with typhoid fever bacillus. All recovered.

Penicillium was forgotten in the scientific community until Fleming rediscovered it one day in the late 1920s. Petri dishes were glass in those days and could be rewashed and sterilized. Fleming was preparing to wash a pile of old Petri dishes he had used to grow the pathogen *Staphylococcus aureus*. He opened and examined each dish before tossing it into a cleaning solution. He noticed that one dish had grown contaminating mold, which in itself was not unusual in old plates, but all around the mold the staph bacteria had failed to grow (**Fig. 27.2A**).

Fleming took a sample of the mold and found that it was from the penicillium family; the sample was later identified as *Penicillium notatum*. The mold appeared to have synthesized a chemical, now known as penicillin (**Fig. 27.2B**), which diffused through the agar, killing cells of *S. aureus* before they could form colonies. Fleming (**Fig. 27.2C**) presented his findings in 1929, but they raised little interest, since penicillin appeared to be unstable and was not active in the body long enough to kill pathogens.

As World War II began, Oxford professor Howard Florey (**Fig. 27.2D**) and his colleague Ernst Chain, rediscovered

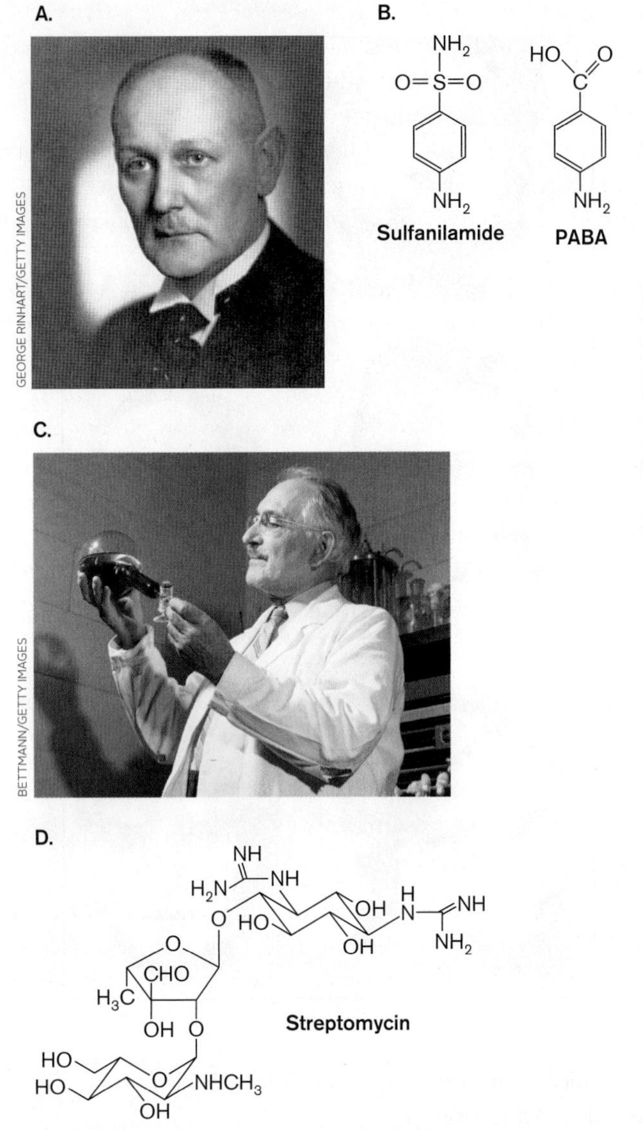

FIGURE 27.3 ■ The discoverers of sulfanilamide and streptomycin. **A.** Gerhard Domagk discovered sulfanilamide. **B.** Chemical structure of sulfanilamide, an analog of *para*-aminobenzoic acid (PABA), a precursor of the vitamin folic acid, which is necessary for growth. Sulfanilamide inhibits one of the enzymes that converts PABA into folic acid. **C.** Selman Waksman discovered streptomycin in 1944. **D.** Chemical structure of streptomycin.

Fleming's work, thought it held promise, and set about purifying penicillin. To their amazement, the purified penicillin cured mice infected with staphylococci or streptococci. Subsequent human trials proved successful (**Fig. 27.2E**), and penicillin gained wide use, saving countless lives during the war. Fleming, Florey, and Chain received the 1945 Nobel Prize in Physiology or Medicine for their work.

The next landmark discovery in antibiotics was made by Gerhard Domagk (1895–1964; **Fig. 27.3A**), a German physician at the Bayer Institute of Experimental Pathology and Bacteriology. In 1935, Domagk's 6-year-old child developed a serious streptococcal infection induced by an innocent pinprick to the finger. The infection spread to her axillary (armpit) lymph nodes and became so severe that lancing and draining the pus 14 times did little to help. The only remaining alternative was to amputate the arm. Unfortunately, even this option would probably not save her life.

Frustrated, Domagk took what would appear to be drastic measures. He gave his daughter a dose of a red dye (Prontosil) that he was investigating as an antimicrobial compound. On agar plates (the usual medium for testing antibiotics), Prontosil had shown absolutely no ability to inhibit the growth of streptococcus. Domagk, however, had tested the drug in animals, not agar plates, and found that it could cure animals of infection. Remarkably, Domagk's daughter recovered completely.

Domagk discovered that Prontosil was metabolized by the body into another compound, sulfanilamide, which was clearly lethal to the streptococcus. This finding led to an entire class of drugs, called the sulfonamides, or **sulfa drugs**, that saved hundreds of thousands of lives. The take-home message of this story is that an antibiotic's activity on a plate, or lack thereof, does not necessarily correlate with the drug's activity in a patient.

Sulfanilamide is an analog of *para*-aminobenzoic acid (PABA), a precursor of folic acid, a vitamin necessary for nucleic acid synthesis (**Fig. 27.3B**). Sulfanilamide and other sulfa drugs bind to and inhibit the enzyme that converts PABA to folic acid. Without folic acid to make nucleic acid precursors, the pathogen stops growing. Sulfa drugs inhibit bacterial growth without affecting human cells because folic acid is not synthesized by humans (it is a dietary supplement instead), and because bacteria do not transport folic acid (they must make it themselves).

In a dark turn, Domagk's possible participation in human experimentation with concentration camp prisoners during World War II (ordered by his German employer) was a source of controversy that haunted him long after the war ended. Yet his contributions to medicine continued; he also developed two effective chemotherapeutic agents for tuberculosis—the thiosemicarbazones and isoniazid—which are still used today.

During the same period of history, Selman Waksman (1888–1973; **Fig. 27.3C**) at Rutgers University began screening 10,000 strains of soil bacteria and fungi for their ability to inhibit growth or kill bacteria. In 1944, this herculean effort paid off with the discovery of streptomycin, an antibiotic produced by the actinomycete *Streptomyces griseus* (**Fig. 27.3D**). Waksman's discovery of streptomycin triggered the antibiotic gold rush and earned him the 1952 Nobel Prize in Physiology or Medicine.

Antibiotics Exhibit Selective Toxicity

As early as 1904, the German physician Paul Ehrlich (1854–1915) realized that a successful antimicrobial compound would be a "magic bullet" that would selectively kill or inhibit the pathogen but not the host. This seemingly obvious premise was innovative at the time. Ehrlich made several discoveries based on this concept, the most celebrated of which was the arsenical compound known as Salvarsan. Salvarsan proved to be quite effective in killing the syphilis agent *Treponema pallidum* (this was long before penicillin was discovered). Syphilis, a sexually transmitted infection, had been untreatable and the source of considerable long-term suffering. Ehrlich's "magic bullet" concept is now known as **selective toxicity**. Salvarsan, however, was not as selectively toxic as Ehrlich thought. This arsenical compound did harm the host, but usually it killed off the treponemes before killing the patient.

Selective toxicity is possible because key aspects of a microbe's physiology are different from those of eukaryotes. For example, suitable bacterial antibiotic targets include peptidoglycan, which eukaryotic cells lack, and ribosomes, which are structurally distinct between Bacteria and Eukarya. Thus, chemicals like penicillin, which prevents peptidoglycan synthesis, and tetracycline, which binds to bacterial 30S ribosomal subunits, inhibit bacterial growth but are essentially invisible to host cells because these drugs will not interact with host structures (at low doses).

Although their intended targets are bacterial cells, some antibiotics, particularly at high doses, can interact with elements of eukaryotic cells and cause side effects that harm the patient. For example, chloramphenicol, a drug that targets bacterial 50S ribosomal subunits, can interfere with the development of blood cells in bone marrow—a phenomenon that may result in aplastic anemia (failure to produce red blood cells). The toxicity of an antibiotic can also depend on the age of the patient. Ciprofloxacin, for instance, can cause defects in human bone growth plates and should not be administered to children. Another problem caused by drugs is that of allergic reactions. For example, many people develop an extreme allergic sensitivity to penicillin, in which case the treatment of an infection may end up being worse than the infection itself. Physicians must be aware of these allergies and use alternative antibiotics to avoid harming their patients.

Note: As a student of microbiology, you must be able to properly distinguish between the terms "drug susceptibility" and "drug sensitivity." A microbe is susceptible to the drug's action, but a human can develop an allergic sensitivity to the drug.

Spectrum of Activity

Because no single antimicrobial drug affects all microbes, antimicrobial drugs are classified by the types of organisms they affect. Thus, we have antifungal, antibacterial, antiprotozoan, and antiviral agents. The term "antibiotic" is usually reserved for compounds that affect bacteria. Even within a group, one agent might have a very narrow **spectrum of activity**, meaning that it affects only a few species, while another antibiotic inhibits many species. For instance, penicillin has a relatively narrow spectrum of activity, killing primarily Gram-positive bacteria. However, ampicillin is penicillin with an added amino group that enables the drug to more easily penetrate the Gram-negative outer membrane. As a result, ampicillin kills Gram-positive and Gram-negative organisms, giving it a broader spectrum of activity than penicillin has. There are antimicrobials as well that exhibit extremely narrow activities. One example is isoniazid, which is clinically useful only against *Mycobacterium tuberculosis*, the agent of tuberculosis. The spectrum of activity of select antibiotics is explored in **eTopic 27.1**.

Patients typically believe that all antibiotics kill their intended targets. This is a misconception. Many drugs simply prevent growth of the organism and let the body's immune system dispatch the intruding microbe. Thus, antimicrobials are also classified on the basis of whether or not they kill the microbe. An antibiotic is **bactericidal** if it kills the target microbe, but **bacteriostatic** if it merely prevents bacterial growth. A bacteriolytic antibiotic, such as penicillin, is bactericidal and also results in cell lysis.

Measuring Drug Susceptibility

One critical decision a clinician must make when treating an infection is which antibiotic to prescribe for the patient. There are several factors to consider, including:

- **The relative effectiveness of different antibiotics on the organism causing the infection.** The less effective a drug is at stopping growth, the less effective it will be at treating the infection.
- **The average attainable tissue levels of each drug.** An antibiotic may work on an agar plate, but the concentration at which it affects bacterial growth may be too high to be safe in the patient.
- **The route of administration.** The easiest way to deliver an antibiotic is orally. However, some antibiotics are not absorbed well from the intestine. For instance, ceftriaxone, a third-generation cephalosporin, is not absorbed well in the gut but is effectively distributed to tissues when given intravenously (IV). Cefixime, another third-generation cephalosporin, can be taken orally and will distribute to tissues.

Minimal inhibitory concentration. The in vitro effectiveness of an antimicrobial agent is determined by measuring how little of it is needed to stop growth. This amount is classically measured in terms of an antibiotic's **minimal inhibitory concentration** (**MIC**), defined as the lowest concentration of the drug that will prevent the growth of an organism. But the MIC for any one drug will differ among different bacterial species. For example, the MIC of ampicillin needed to stop the growth of *Staphylococcus aureus* will be different from that needed to inhibit *Shigella dysenteriae*. The reasons that a drug may be more effective

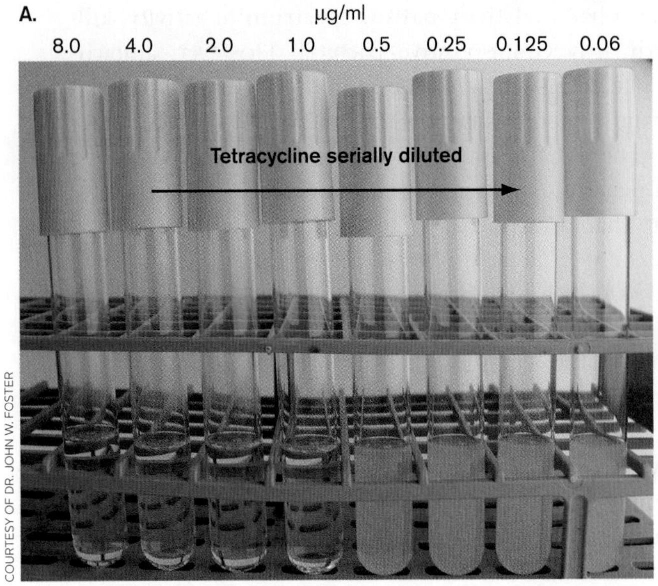

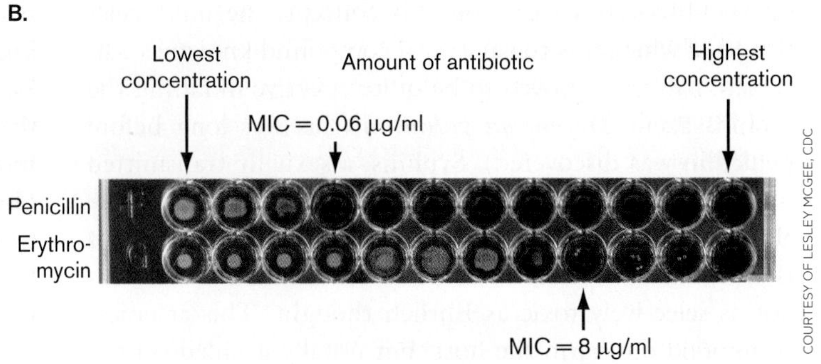

FIGURE 27.4 ■ Determining minimal inhibitory concentration (MIC). **A.** In this series of tubes, tetracycline was diluted serially starting at 8 µg/ml. Each tube was then inoculated with an equal number of bacteria (*Salmonella enterica*). Turbidity indicates that the antibiotic concentration was insufficient to inhibit growth. The MIC in this example is 1.0 µg/ml. **B.** Microdilution, using a microtiter plate, to determine MIC. Pictured are two rows of a 96-well plate in which dilutions of penicillin (upper row) and erythromycin (lower row) were tested against group B streptococci. The principle is the same as in (A), except smaller volumes (200 ml) are used. The MIC in each row is circled.

against one organism than another include the ease with which the drug penetrates the bacterial cell and the affinity of the drug for its molecular target.

So, how do we measure MIC? As shown in **Figure 27.4A**, an antibiotic is serially diluted along a row of test tubes containing nutrient broth. After dilution, the organism to be tested is inoculated at low, constant density into each tube, and the tubes are usually incubated overnight. Growth of the organism is seen as turbidity. In **Figure 27.4**, the tubes with the highest concentration of drug are clear, indicating no growth. The tube containing the MIC is the tube with the lowest concentration of drug that shows no growth. Note, however, that the MIC does not indicate whether a drug is bacteriostatic or bactericidal. Today, clinical laboratories use microtiter plates read by automated systems to determine MICs (**Fig. 27.4B** and Chapter 28).

Thought Questions

27.1 Figure 27.4 illustrates how MICs are determined. Test your understanding of how MICs are measured in the following example. The drug tobramycin is added to a concentration of 1,000 µg/ml in a tube of broth from which serial twofold dilutions are made. Including the initial tube (tube 1), there are a total of ten tubes. Twenty-four hours after all the tubes are inoculated with *Listeria monocytogenes*, turbidity is observed in tubes 6–10. What is the MIC?

27.2 What additional test performed on an MIC series of tubes will tell you whether a drug is bacteriostatic or bactericidal?

MIC determinations are very useful for estimating a single drug's effectiveness against a single bacterial pathogen isolated from a patient, but they are not very practical to a technician trying to screen 20 or more different drugs. Dilutions take time—time that the technician, not to mention the patient, may not have. The time required to evaluate antibiotic effectiveness can be reduced by use of a strip test (like the ETEST shown in **Fig. 27.5**) that avoids the need for dilutions. The strip, containing a gradient of antibiotic, is placed on an agar plate freshly seeded with a dilute lawn of bacteria. While the bacteria are trying to grow, the drug diffuses out of the strip and into the media. The drug will diffuse at equal rates from all points along the strip. However, drug diffusing away from the most concentrated part of the strip will maintain a higher

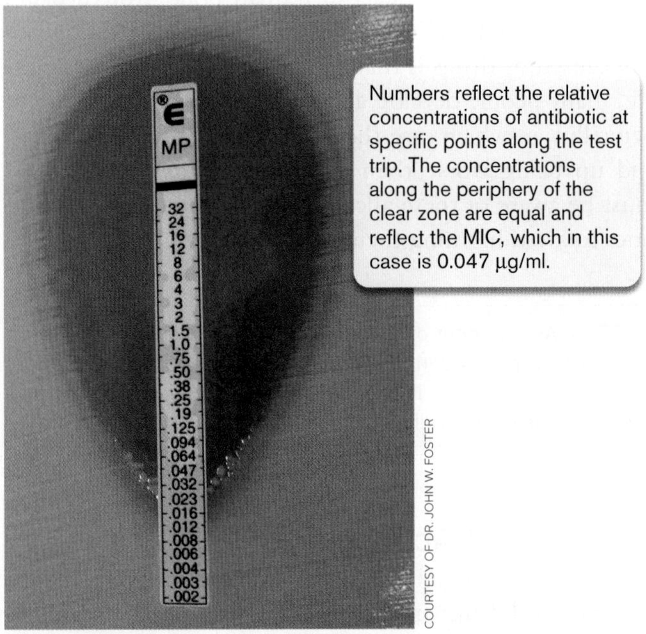

FIGURE 27.5 ■ An MIC strip test. The ETEST (manufactured by bioMérieux) is a commercially prepared strip that produces a gradient of antibiotic concentration (in µg/ml) when placed on an agar plate. The MIC corresponds to the point where bacterial growth crosses the numbered strip.

concentration in the agar per unit of time compared to that diffusing from less concentrated parts of the strip. Thus, the drug's effect (killing or inhibiting the growth of cells) will extend farther away from the strip at locations of high concentration than at locations of lower concentration. The result is a **zone of inhibition** where the antibiotic has stopped bacterial growth. The MIC is the point at which the elliptical zone of inhibition intersects with the strip.

Kirby-Bauer disk susceptibility test. Although the strip test eliminates the time and effort needed to make dilutions, it would take 20 or more plates to test an equal number of antibiotics for just one bacterial isolate. Clinical labs can receive up to 100 or more isolates in one day, so individual MIC determinations are impractical. A simplified agar diffusion test, however, which can test 12 antibiotics on one plate, makes evaluating antibiotic susceptibility a manageable task.

Named for its inventors, the **Kirby-Bauer assay** uses a series of round filter paper disks impregnated with different antibiotics. A dispenser (**Fig. 27.6A**) delivers up to 12 disks simultaneously to the surface of an agar plate covered by a bacterial lawn. Each disk is marked to indicate the drug used. During incubation, the drugs diffuse away from the disks into the surrounding agar and prevent growth of the lawn (**Fig. 27.6B–D**). The zones of inhibition vary in width, depending on the antibiotic used, the concentration of the drug in the disk, and the susceptibility of the organism to the drug. The diameter of the zone correlates to the MIC of the antibiotic against the organism tested. **Figure 27.6B** and **C** show the results for methicillin-sensitive and methicillin-resistant *Staphylococcus aureus* (MSSA and MRSA, respectively). Note the lack of inhibition by the oxacillin disk (compare arrows in the two photos). This strain is resistant to both methicillin and oxacillin because they are structurally similar.

Correlations between MIC values determined in broth and Kirby-Bauer zone sizes on agar are made empirically. Every disk containing a given antibiotic is impregnated with a standard concentration of that drug, but the standard concentration used may be different for each drug. The outermost ring of the no-growth zone in a Kirby-Bauer disk test must, by definition, contain the minimal concentration of drug needed to prevent growth on agar. However, if species A and B have MIC values for penicillin of 4 μg/ml and 40 μg/ml, respectively, then species A will exhibit a proportionally larger zone of inhibition than species B in the disk test. A graph plotting MIC on one axis and zone diameter on the other provides the correlation.

After the agar plates are incubated, the diameters of the zones of inhibition around each disk are measured, and the results are compared with a table listing whether a zone is wide enough (meaning the MIC is low enough)

A.

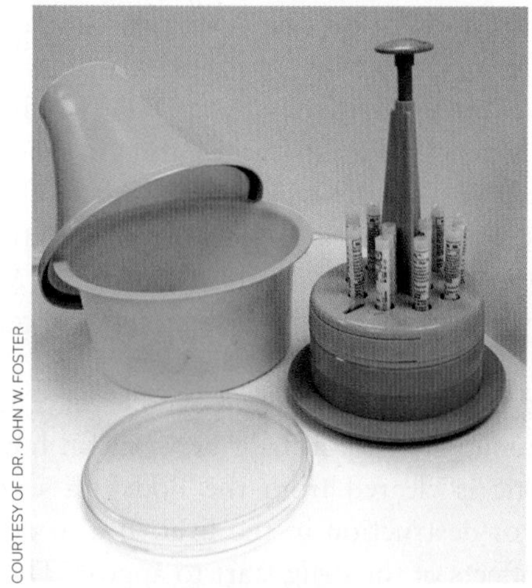

FIGURE 27.6 ■ The Kirby-Bauer disk susceptibility test. **A.** This device delivers up to 12 disks to the surface of a Mueller-Hinton plate. **B–D.** Disks impregnated with different antibiotics are placed on a freshly laid lawn of bacteria and incubated overnight. The clear zones around certain disks indicate growth inhibition. C, chloramphenicol; CC, clindamycin; CZ, cefazolin; E, erythromycin; NOR, norfloxacin; OX, oxacillin; P, penicillin; RA, rifampin; SAM, sulbactam-ampicillin; SXT, sulfa-trimethoprim; TE, tetracycline; VA, vancomycin. Results are shown for methicillin-sensitive *Staphylococcus aureus* (MSSA) **(B)**, methicillin-resistant *S. aureus* (MRSA) **(C)**, and *Streptococcus pneumoniae* **(D)**. The arrow in (C) points out MRSA's resistance to oxacillin, compared to the lack of resistance exhibited by MSSA in (B). The brownish tint of the blood agar plates outside the zones of bacterial inhibition is caused by a hemolysin secreted by the lawn of pneumococci.

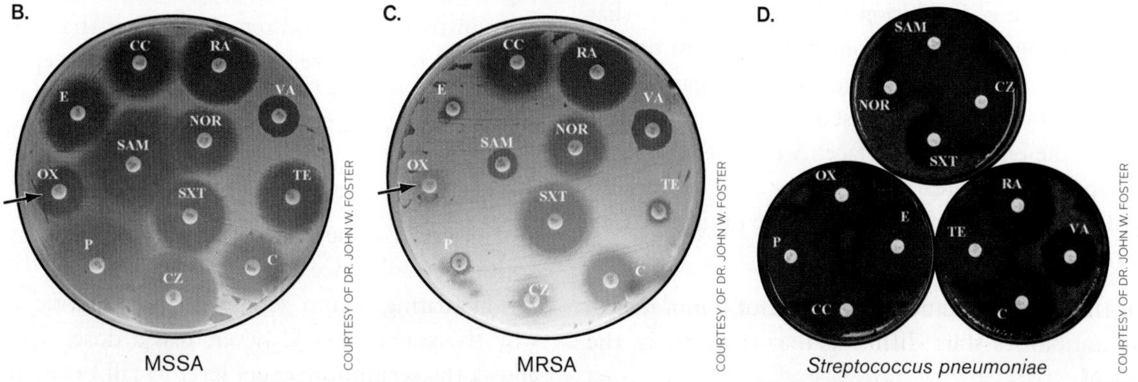

TABLE 27.1 Susceptibility Results for *Staphylococcus aureus*

		Zone of inhibition diameter (mm)		
Antibiotic	**Quantity in disk (μg)**	**Resistant**	**Intermediate**	**Susceptible**
Ampicillin	10	<12	12–13	>13
Chloramphenicol	30	<13	13–17	>17
Erythromycin	15	<14	14–17	>17
Gentamicin	10	≤12.5		>12.5
Streptomycin	10	<12	12–14	>14
Tetracycline	30	<15	15–18	>18

to be clinically useful. **Table 27.1** shows susceptibility data for *S. aureus*. The zone diameter (MIC) that indicates an organism will be clinically susceptible to the antibiotic correlates to the average attainable tissue level for that antibiotic (discussed later). For the antibiotic to remain effective in vivo, the tissue concentration of the drug must remain above the MIC; otherwise, the bacteria can grow, and spontaneous mutations providing drug resistance could develop.

To ensure reproducibility, the Kirby-Bauer test was standardized over a half century ago. Reproducibility means that results from a laboratory in California will match those in Alabama, Ohio, or any other location. The following are standardizations used to make the test reproducible and easier.

- **Size of the agar plate.** 150 mm.
- **Depth of the media.** 4 mm. Because antibiotics diffuse out of disks in three dimensions, the zone of inhibition measured on a thinly poured agar plate will be larger than the zone from a thick agar plate.
- **Media composition.** Media should lack PABA. Sulfonamide antibiotics inhibit PABA synthesis in bacteria, so the presence of PABA in testing media bypasses any block imposed by the sulfonamides. The standardized medium used for the Kirby-Bauer test, called **Mueller-Hinton agar**, contains no PABA.
- **The number of organisms.** The more organisms that are spread on an agar plate, the less time an antibiotic has to diffuse before visible growth develops. Consequently, the zone of inhibition will appear smaller that it should. To prevent this phenomenon, standard optical density solutions (0.06 at 600 nm) of each organism are prepared and a cotton swab is used to deliver organisms evenly over the entire agar surface. A commercially prepared suspension of latex particles (the new McFarland standards) is used to visually estimate the optical density of culture dilutions of the test organism.
- **Size of the disks.** A standard diameter of 6 mm means that all antibiotics start diffusing into the agar at the same point.
- **Concentrations of antibiotics in the disks.** The zone of inhibition for an antibiotic is proportional to the concentration of antibiotic in the disk. The higher the concentration of antibiotic in a disk, the farther the drug can diffuse and maintain a concentration above the MIC. To avoid differences between labs, the concentration of each drug impregnating a disk has been standardized.
- **Incubation temperature.** Incubation temperature will not affect growth and diffusion equally. To avoid differences, a temperature of 37°C is standard (body temperature).

Automated MIC determinations. The methods just described to determine MIC are effective but take at least 24 hours to complete (the amount of time needed to see visible turbidity in broth or growth on agar). Modern clinical laboratories today are equipped with automated machines (described in Chapter 28) that can determine an MIC within 6 hours. This speed is critical to a clinician wanting to quickly treat a serious infection with the most effective drug. The instruments work by monitoring growth of the infectious agent in microtiter plate wells and extrapolating the MIC from growth curves recorded at different antibiotic concentrations.

Correlating antibiotic MIC with tissue level. The average attainable tissue level for a drug depends on how quickly the antibiotic is cleared from the body via secretion by the kidney or destruction in the liver. It also depends on when side effects of the drug start to appear. The graph in **Figure 27.7** shows that as long as the concentration of the drug in tissue or blood remains higher than the MIC, the drug will be effective. The clinician can keep the concentration at sufficient levels either by initially administering a higher dose (running the risk of side effects) or by giving a second dose before the blood levels from the first dose decline below the MIC. This is why patients are told to take doses of some antibiotics four times a day and other antibiotics only once a day.

Fluctuating serum levels of an antibiotic also explain why the patient should never miss a dose. Missing a dose allows the serum (or tissue) level to fall below the MIC for

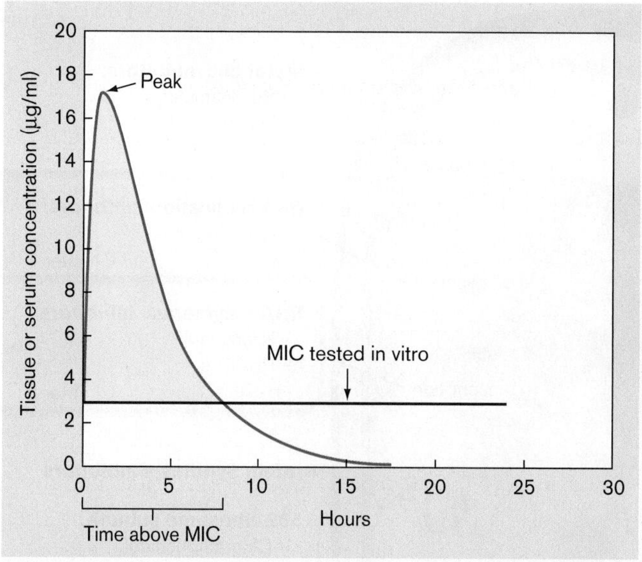

FIGURE 27.7 ■ Correlation between MIC and serum or tissue level of an antibiotic. This graph illustrates the serum level of ampicillin over time. The important consideration here is how long the serum level of the antibiotic remains higher than the MIC. Once the concentration falls below the MIC, owing to destruction of the drug in the liver or clearance through the kidneys and secretion, the infectious agent fails to be controlled by the drug—in this case, 7–8 hours after the initial dose. To maintain a serum level higher than the MIC, a second dose would be taken. The shaded area of the curve represents time above MIC.

the pathogen. When that happens, any bacteria still living can once again grow and possibly develop spontaneous antibiotic resistance mutations. These antibiotic-resistant mutants will continue to grow and cause disease even after the serum level of antibiotic is restored. The result is called treatment failure.

Thought Questions

27.3 A patient with a bacterial lung infection was given the antibiotic represented in **Figure 27.7** and was told to take one pill twice a day. The pathogen is susceptible to this drug. Will the prescribed treatment be effective? Explain your answer.

27.4 You are testing whether a new antibiotic will be a good treatment choice for a patient with a staph infection. The Kirby-Bauer test using the organism from the patient shows a zone of inhibition of 15 mm around the disk containing this drug. Clearly, the organism is being affected by the drug in vitro. But you conclude from other studies that the drug would be ineffective in the patient. What would make you draw this conclusion?

To Summarize

- **The importance of antimicrobials** in treating disease was recognized in the early 1940s. Some antimicrobials are naturally produced by living organisms (antibiotics); others are synthetically made through chemical engineering.
- **Selective toxicity** is the ability of an antibiotic to attack a unique component of microbial physiology that is missing or distinctly different from eukaryotic physiology. However, **antibiotic side effects** (host toxicity) can limit the clinical usefulness of an antimicrobial agent.
- **Antibiotic spectrum of activity** is the range of microbes affected by a given drug.
- **Bactericidal antibiotics, or antimicrobials,** kill microbes; **bacteriostatic antibiotics** inhibit microbial growth; and **some antimicrobial agents are initially inactive** until converted by the body to an active agent.
- **An antibiotic's spectrum of activity and the infectious agent's susceptibility to the antibiotic** are critical points of information required before an antibiotic therapy is prescribed.
- **Minimal inhibitory concentration (MIC)** of a drug, when correlated with average attainable tissue levels of the antibiotic, can predict the effectiveness of an antibiotic in treating disease.
- **MIC is measured** by tube dilution techniques, but it can be approximated by the Kirby-Bauer disk diffusion technique.

27.2 Antibiotic Mechanisms of Action

As noted in the preceding section, selective toxicity of an antibiotic depends on enzymes or structures unique to the bacterial target cell. The following aspects of a microbe's physiology are classic targets:

- Cell wall synthesis
- Cell membrane integrity
- DNA synthesis
- RNA synthesis
- Protein synthesis
- Metabolism

Table 27.2 summarizes the general targets of common antibiotics. Chapters 3, 7, and 8 describe these cell components and provide the basis for understanding how antibiotics work. **Because the mechanisms of action for antibiotics affecting DNA, RNA, and protein synthesis are described in Chapters 7 and 8, they receive only brief mention here. Figure 27.8** summarizes the general mechanisms of action for these drugs.

FIGURE 27.8 ▪ Summary of antimicrobial agents and their targets. Antibiotics that target DNA synthesis, RNA synthesis, and protein synthesis are described in Chapters 7 and 8. *Source:* Modified from Foster et al. 2018. *Microbiology: The Human Experience.* W. W. Norton.

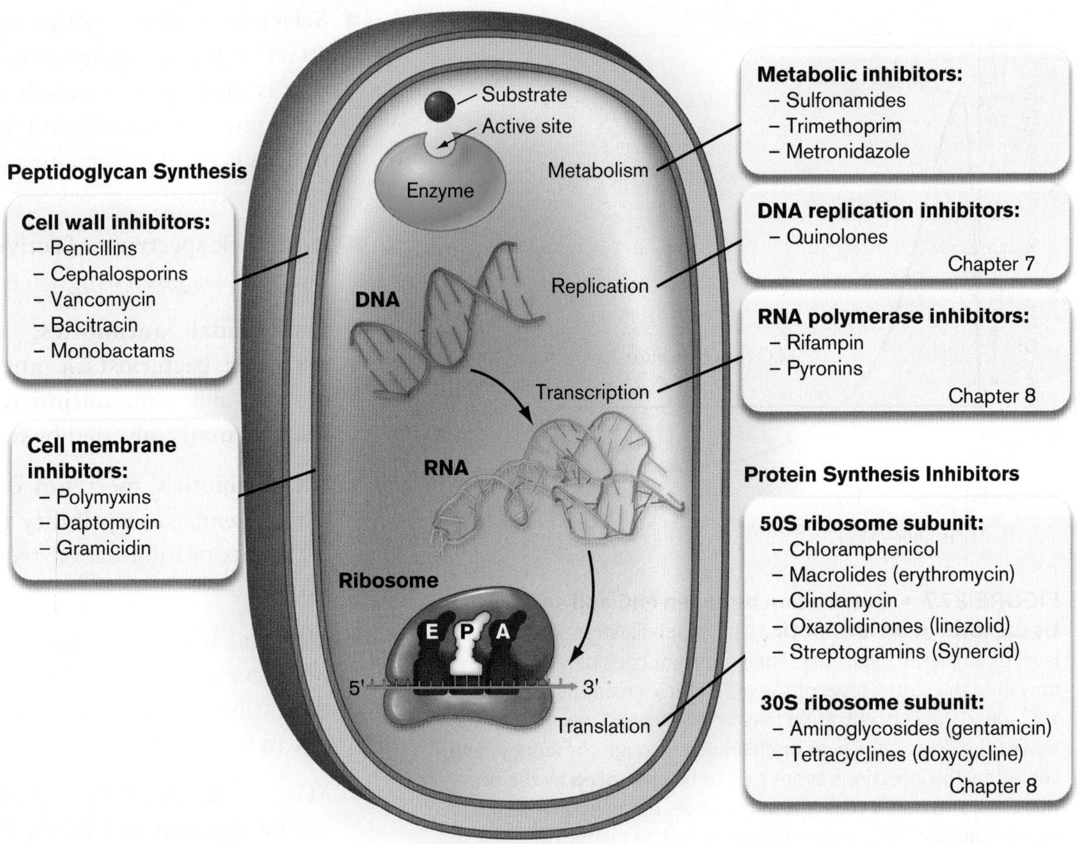

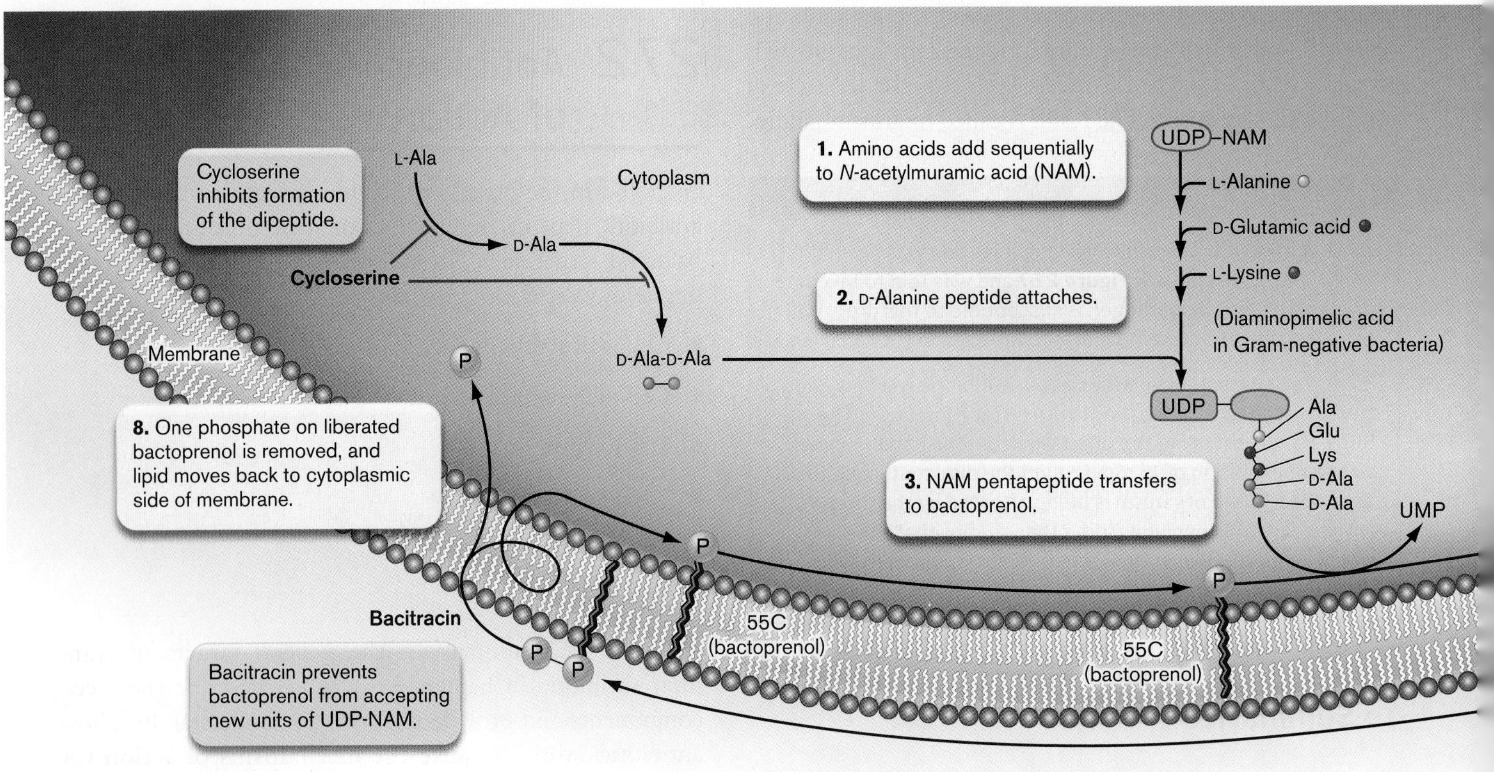

FIGURE 27.9 ▪ Peptidoglycan synthesis in a Gram-positive bacterium, and the targets of antibiotics. Several small-molecular-weight compounds are sequentially joined to form a disaccharide unit that will be added to preexisting extracellular chains of this unit. Red lines indicate inhibition. Cycloserine inhibits ligation of the two D-alanines (step 2); bacitracin inhibits linking of the disaccharide units; vancomycin and the beta-lactams, such as penicillin, inhibit the peptide cross-linking of peptidoglycan side chains.

TABLE 27.2 Targets of Antimicrobial Agents

Target	Antibiotic examples
Cell wall synthesis	Penicillins, cephalosporins, bacitracin, vancomycin
Protein synthesis	Chloramphenicol, tetracyclines, aminoglycosides, macrolides, lincosamides
Cell membrane integrity	Polymyxin, daptomycin, amphotericin, imidazoles (vs. fungi)
Nucleic acid function	Nitroimidazoles, nitrofurans, quinolones, rifampin; some antiviral compounds, especially antimetabolites
Intermediary metabolism	Sulfonamides, trimethoprim

Cell Wall Antibiotics

Bacterial cell walls are the basis of selective toxicity for some antibiotics because peptidoglycan does not exist in mammalian cells; thus, antibiotics that target the synthesis of these structures should selectively kill bacteria. The following case history illustrates the use of two cell wall–targeting antibiotics and also reveals how bacteria can evolve to escape destruction.

Case History: Meningitis

A 3-year-old child was brought to the emergency room crying, with a stiff neck and high fever. Gram stain of cerebrospinal fluid revealed Gram-positive cocci, generally in pairs. The diagnosis was meningitis. The physician immediately prescribed intravenous ampicillin. Unfortunately, the child's condition worsened, so antibiotic treatment was changed to a third-generation cephalosporin (which more easily crosses the blood-brain barrier). The patient began to improve within hours and was released after 2 days. A report from the clinical microbiology laboratory identified the organism as Streptococcus pneumoniae.

Both of the antibiotics used in this case kill bacteria by targeting cell wall synthesis (introduced in Section 3.3). To synthesize peptidoglycan, sugar molecules called *N*-acetylglucosamine (NAG) and *N*-acetylmuramic acid (NAM) are made by the cell and linked together by a **transglycosylase** enzyme into long chains assembled at the cell wall (**Fig. 27.9**). *N*-acetylmuramic acid contains a short side chain of amino acids that is assembled enzymatically, not by a ribosome. The side chains from adjacent strands are cross-linked to make the structure rigid. The enzyme **transpeptidase** (D-alanyl-D-alanine carboxypeptidase/transpeptidase) catalyzes the cross-link. Several antibiotics target various stages of this assembly process.

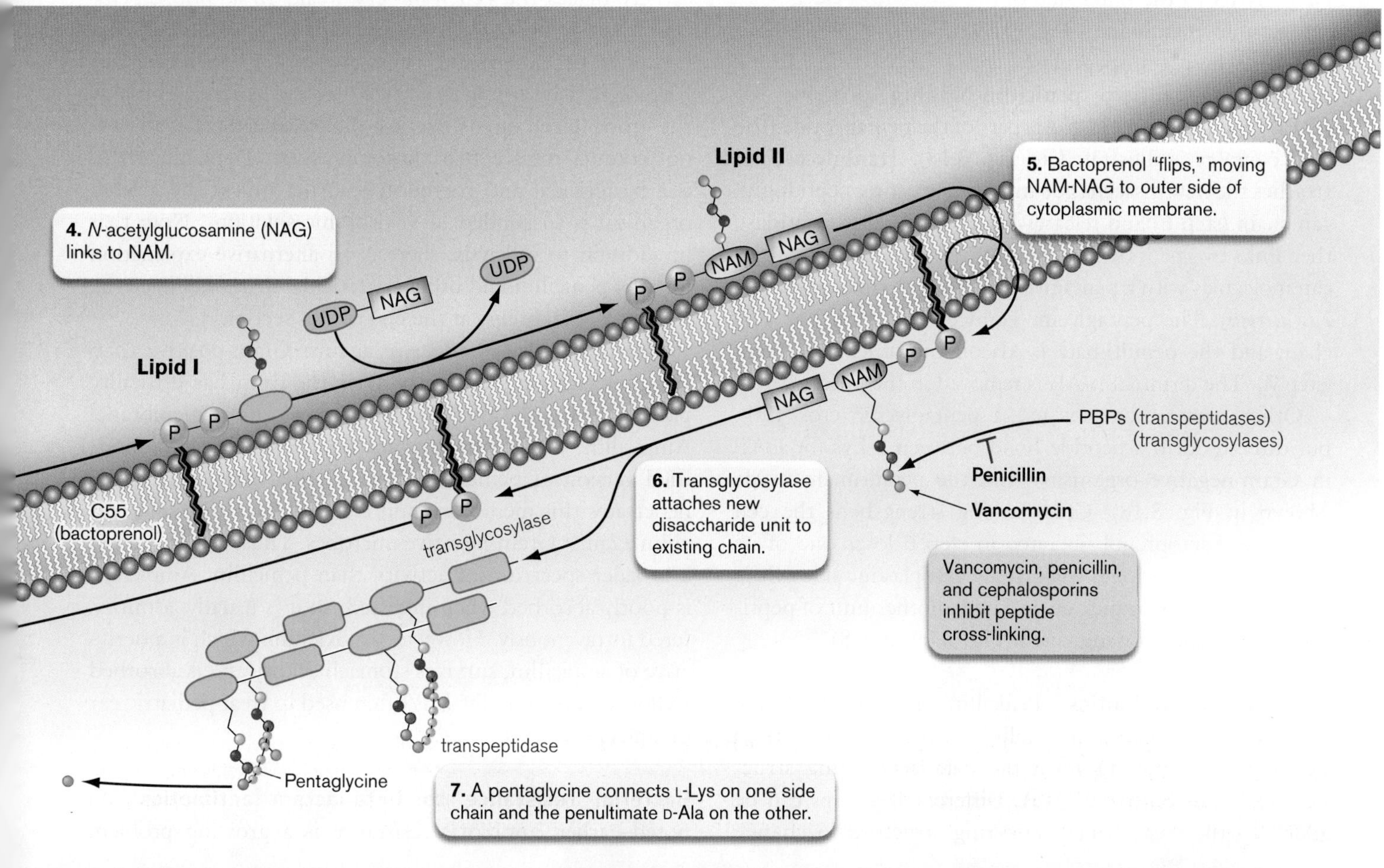

Peptidoglycan synthesis. Before we can explain how cell wall antibiotics work, you need to know how the cell wall is made. Synthesis of peptidoglycan starts in the cytoplasm with a uridine diphosphate (UDP)–NAM molecule. The amino acids L-alanine, D-glutamic acid, and L-lysine [or diaminopimelic acid (DAP) in Gram-negative organisms] are individually and sequentially added to NAM (**Fig. 27.9**, step 1), and then a dipeptide of D-alanine is attached (step 2). Next, the NAM pentapeptide is transferred to a membrane-situated, 55-carbon lipid molecule called bactoprenol (step 3), releasing uridine monophosphate (UMP). This structure is also called lipid I. Another sugar molecule, NAG, is then linked to NAM—once again through a UDP intermediate (step 4)—to make the bactoprenol structure called lipid II. All of this takes place on the cytoplasmic side of the membrane.

Bactoprenol then "flips," moving NAM-NAG to the outer side of the cytoplasmic membrane (**Fig. 27.9**, step 5), where transpeptidases and transglycosylases (two **penicillin-binding proteins**, or **PBPs**) bind to the D-Ala-D-Ala part of the pentapeptide (the PBP complex is illustrated in Fig. 3.13). Transglycosylase attaches the new disaccharide unit to an existing peptidoglycan chain (step 6) and releases bactoprenol. Transpeptidase then links two peptide side chains from adjacent peptidoglycan molecules with a pentaglycine cross-link (in *Staphylococcus aureus*). The pentaglycine connects L-Lys on one side chain and the penultimate D-Ala on the other side chain (step 7). The terminal D-Ala is removed in the process.

Other bacteria do not use a pentaglycine cross-link, but directly form a peptide bond between L-Lys (or DAP in Gram-negative organisms) and the penultimate D-Ala (shown in Fig. 3.18). Cross-linking strengthens the cell wall. The bactoprenol liberated in step 6 loses one of its phosphates and recycles back to the cytoplasmic side of the membrane, ready to pick up and taxi another unit of peptidoglycan to the growing chain (**Fig. 27.9**, step 8).

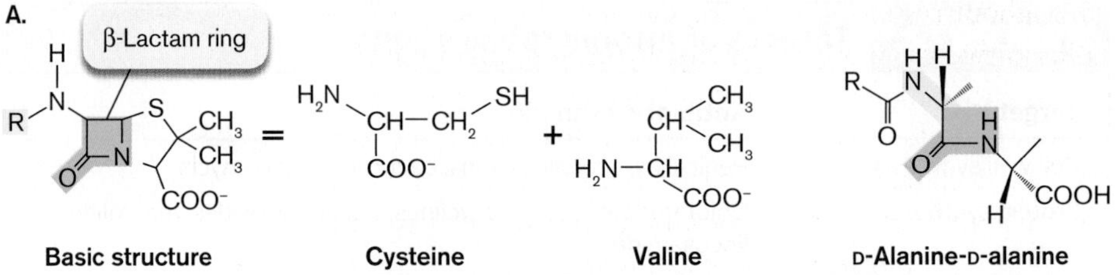

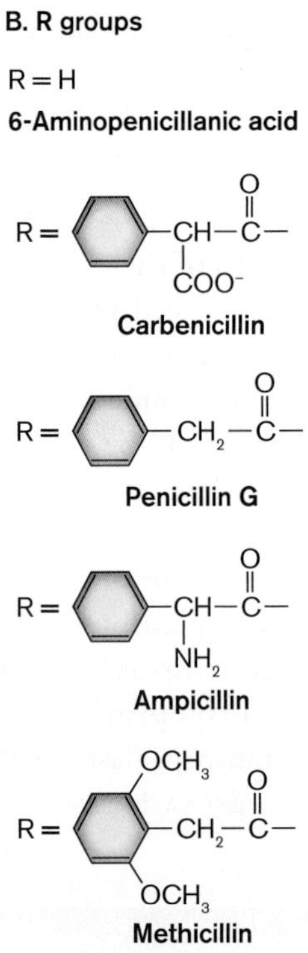

FIGURE 27.10 ■ The structure of penicillins. **A.** Penicillanic acid (R=H) is derived from cysteine and valine. Also shown is the D-alanine-D-alanine structure of peptidoglycan (far right), which is structurally similar to the beta-lactam ring of penicillins (green shading). **B.** The R group highlighted in (A) can be any one of a number of different groups, some of which are shown here. Modifying this group changes the pharmacological properties and antimicrobial spectrum of the drug.

Beta-lactam antibiotics. Penicillin is an antibiotic derived from cysteine and valine, which are condensed by fungal enzymes to form the beta-lactam ring structure shown in **Figure 27.10A**. Different R groups can be added to the basic beta-lactam ring structure to change the antimicrobial spectrum and stability of the derivative penicillin (**Fig. 27.10B**). Note that the beta-lactam ring of penicillin chemically resembles the D-Ala-D-Ala piece of peptidoglycan, as highlighted by green shading in **Figure 27.10A**. This molecular mimicry enables penicillin and other beta-lactam antibiotics to target so-called penicillin-binding proteins, including transpeptidase and transglycosylase, that carry out cell wall synthesis and remodeling.

Penicillins act by inhibiting transpeptidase-mediated cross-linking between adjacent peptidoglycan chains. This activity makes the cell wall very weak. In addition, penicillins can somehow activate proteins in the cell wall that hydrolyze peptidoglycan. The consequence is a disaster for bacteria that are trying to grow larger and larger. Eventually, the growing cell bursts because the weakened cell wall cannot counter intracellular turgor pressure. Penicillin, then, is a bactericidal and bacteriolytic drug (unless the treated organism is suspended in an isotonic solution). Note that in addition to cell lysis, there is an alternative explanation for why penicillin and other bactericidal drugs kill bacteria, which we will discuss at the end of this section.

Penicillin is more effective against Gram-positive than Gram-negative organisms because the drug has difficulty passing through the Gram-negative outer membrane. Ampicillin, which was used in the case history, is a modified version of penicillin (**Fig. 27.10B**) that more easily penetrates this membrane and is more effective than penicillin against Gram-negative microbes. Thus, ampicillin has a broader spectrum of activity than penicillin. Ampicillin is poorly absorbed when ingested, so it is usually administered intravenously. However, amoxicillin, which is a derivative of ampicillin, survives stomach acidity and is absorbed well when taken orally. It is often used to treat pediatric ear or sinus infections.

Bacterial resistance to beta-lactam antibiotics. As noted earlier, antibiotic resistance is a growing problem throughout the world. Bacteria develop resistance to penicillin in two basic ways. The first is through inheritance of a gene encoding one of the beta-lactamase enzymes, which cleave the critical ring structure of this class of antibiotics. Beta-lactamase is transported out of the cell and into the surrounding medium (for Gram-positives) or the periplasm (for Gram-negatives), where it can destroy penicillin before the drug even gets to the cell. Bacteria that produce beta-lactamase are still susceptible to certain modified penicillins and cephalosporins engineered to be poor substrates for the enzyme. Methicillin, for example, works well against beta-lactamase-producing microbes.

Unfortunately, a type of beta-lactamase called New Delhi metallo-beta-lactamase-1 (NDM-1) has emerged that confers resistance to almost all beta-lactam antibiotics. NDM-1-containing plasmids, which originated in India, are promiscuously transferred, being found in various enterobacterial species, such as *Klebsiella pneumoniae* and *E. coli*, as well as the nonenteric pathogens *Pseudomonas aeruginosa* and *Acinetobacter baumannii*.

Aside from beta-lactamases, the second way a microbe can become resistant to beta-lactam antibiotics is by acquiring a gene that encodes an altered penicillin-binding protein that no longer binds penicillin. Methicillin-resistant *Staphylococcus aureus* (MRSA) uses this strategy. Resistance to methicillin in *S. aureus* is mediated by the *mecA* gene, which is part of a mobile genetic element called staphylococcal cassette chromosome *mec* (SCC*mec*). The *mecA* gene encodes an altered penicillin-binding protein (PBP2A or PBP2′) with low affinity for beta-lactam antibiotics. The low affinity provides resistance to all beta-lactam antibiotics, rendering them useless. Hospitals take special interest in MRSA because very few drugs can kill it.

One of the few remaining antibiotics effective against MRSA is vancomycin. Unfortunately, resistance is developing to this drug too. Vancomycin-resistant *S. aureus* strains are called VRSA. The penicillin-resistant *Streptococcus pneumoniae* in the preceding case history actually had an altered penicillin-binding protein. No beta-lactamase-producing *S. pneumoniae* has yet been found.

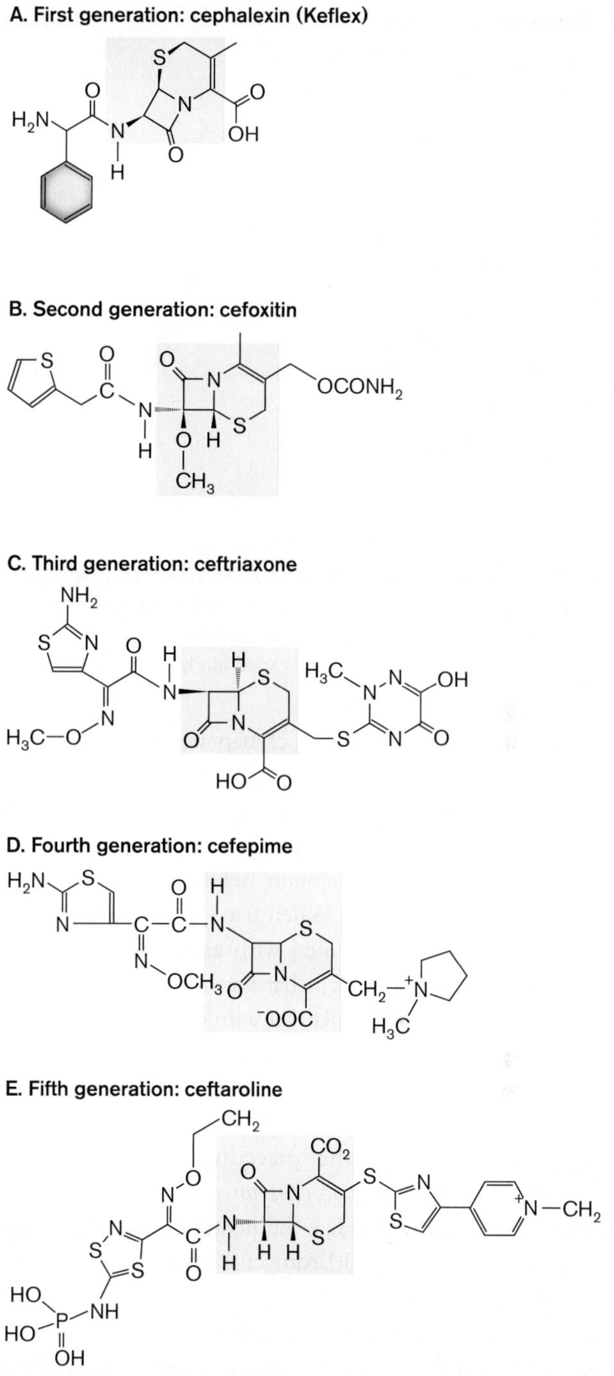

FIGURE 27.11 ■ Cephalosporin generations. Representative examples. With each successive generation, the side groups become more complex. Highlighted areas indicate the core structure of each of the cephalosporins, with beta-lactam rings.

Beta-lactamase-resistant antibiotics. Cephalosporins are another type of beta-lactam antibiotic originally discovered in nature but modified in the laboratory to fight microbes that are naturally resistant to penicillins (especially *Pseudomonas aeruginosa*). Over the years, the structure of cephalosporin has undergone a series of modifications to improve its effectiveness against penicillin-resistant pathogens. Each modification adds complexity and produces what is called a new "generation" of cephalosporins. There are currently five generations of this semisynthetic antibiotic (**Fig. 27.11**).

Unfortunately, the microbial world continually adapts and eventually becomes resistant to new antibiotics. In the case of the cephalosporins, new beta-lactamases have evolved that can attack the sterically buried beta-lactam rings in these molecules. It is also important to note that because the core feature of these drugs is the beta-lactam ring, persons who are allergic to penicillins may also suffer a hypersensitivity reaction to lower-generation cephalosporins.

Carbapenems (such as imipenem) are another subclass of beta-lactam antibiotics, called extended-spectrum penicillins. These antibiotics are resistant to most beta-lactamase enzymes made by pathogenic bacteria. To minimize the

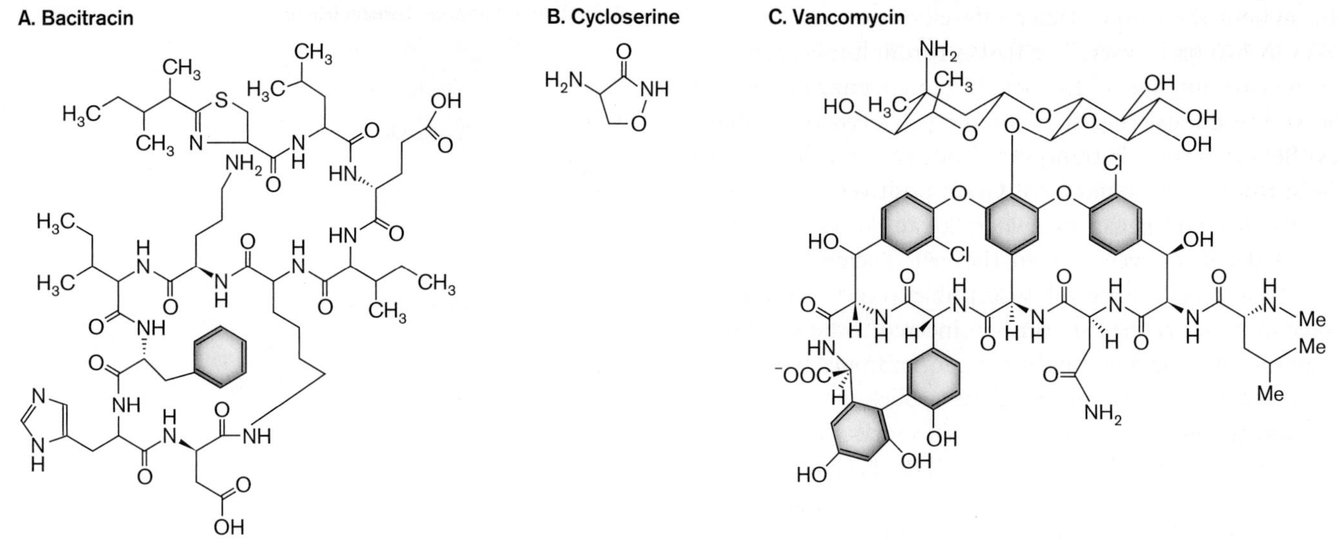

FIGURE 27.12 ■ Other antibiotics that affect peptidoglycan synthesis. **A.** Bacitracin is produced by *Bacillus subtilis*. It is generally used only topically to prevent infection. **B.** Cycloserine, an analog of D-alanine, is one of several drugs used to treat tuberculosis. **C.** Vancomycin is a cyclic polypeptide made by *Amycolatopsis orientalis*, previously classified as a streptomycete. These antibiotics, especially bacitracin and vancomycin, are synthesized by exceedingly complex biochemical pathways in the producing organisms. Me = methyl.

chance of developing resistance, carbapenems are reserved for severe or high-risk infections caused by multidrug-resistant, usually Gram-negative bacteria, including *Pseudomonas*. However, carbapenems such as piperacillin are often used empirically in seriously ill patients before the identity of the infectious agent is known. When used empirically, the carbapenem is typically combined with another drug, such as vancomycin (described later), that has broad activity against Gram-positive bacteria. Once the clinical lab identifies the actual organism, the carbapenem should be replaced with an antibiotic that has a narrower spectrum of activity (see the paragraph "Antibiotic stewardship" in Section 27.3).

Treatment note: In the preceding case history, the infecting strain of *Streptococcus pneumoniae* was initially treated with ampicillin. Had the hospitalized patient been an adult, a macrolide (azithromycin) or a fluoroquinolone (see Section 7.2) might have been the initial treatment choices because their respective targets, the ribosome and a type II topoisomerase, are unrelated to cell wall synthesis. Quinolones are not recommended for children, however, because of potential side effects. Other beta-lactam antibiotics, such as the third-generation cephalosporins, may still work on penicillin-resistant *Streptococcus pneumoniae*, because the modified antibiotic often can still bind the altered PBP. Unfortunately, cephalosporin-resistant strains of *S. pneumoniae* are now common.

Note: Archaeal pseudopeptidoglycan contains talosaminuronic acid instead of muramic acid and lacks the D-amino acids found in bacterial peptidoglycan. Archaea are thus insensitive to penicillins, which interfere with bacterial transpeptidases. This natural resistance is not a problem, because no archaea are known to be pathogens.

Antibiotics that target other steps in peptidoglycan synthesis. Another antibiotic that affects cell wall synthesis is **bacitracin**, a large polypeptide molecule produced by *Bacillus subtilis* and *Bacillus licheniformis* (**Fig. 27.12A**). The antibiotic inhibits cell wall synthesis by binding to the bactoprenol lipid carrier molecule that normally transports monomeric units of peptidoglycan across the cell membrane to the growing chain (see **Fig. 27.9**). Bacitracin binds to and inhibits dephosphorylation of the carrier, which prevents the carrier from accepting a new unit of UDP-NAM. The inability to make peptidoglycan causes growth arrest. Resistance to bacitracin can evolve if the organism can rapidly recycle the phosphorylated lipid carrier molecule through dephosphorylation or if the organism possesses an efficient drug export system (discussed in Section 27.3). Normally, bacitracin is used only topically because of serious side effects, such as kidney damage, that can occur if bacitracin is ingested. For instance, the drug is often applied topically to skin that has been tattooed, to prevent infection.

Lantibiotics are cyclic peptide antibiotics produced by some Gram-positive bacteria (such as *Streptomyces*). Members of the clinically useful subclass of lantibiotics, including Duramycin, interact with a membrane phospholipid (phosphotidylethanolamine) to impede lipid II function (see **Fig. 27.9**), resulting in inhibition of peptidoglycan synthesis.

Cycloserine (made by *Streptomyces garyphalus*) is one of several antimicrobials used to treat tuberculosis (**Fig. 27.12B**). Relative to bacitracin, it acts at an even earlier step in peptidoglycan synthesis. Cycloserine inhibits the two enzymes that make the D-Ala-D-Ala dipeptide. As a result, the complete pentapeptide side chain on *N*-acetylmuramic acid cannot be made (see **Fig. 27.9**). Without these alanines, cross-linking cannot occur and peptidoglycan integrity is compromised.

Vancomycin, a very large and complex glycopeptide produced by *Amycolatopsis orientalis* (formerly *Streptomyces orientalis*; **Fig. 27.12C**), binds to the D-Ala-D-Ala terminal end of the disaccharide unit and prevents the action of transglycosylases and transpeptidases (see **Fig. 27.9**). The mechanism of resistance is very different for vancomycin and penicillin, which makes vancomycin particularly useful against penicillin-resistant bacteria. To prevent the development and spread of vancomycin-resistant bacteria, this antibiotic is typically used only as a drug of last resort.

Resistance to vancomycin can develop when products from a cluster of *van* genes collaborate to make D-lactate and incorporate it into the ester D-Ala-D-lactate, to which vancomycin cannot bind. Another enzyme in the *van* gene cluster prevents the accumulation of D-Ala-D-Ala; as a result, D-Ala-D-lactate replaces D-Ala-D-Ala in peptidoglycan. Peptidoglycan containing D-Ala-D-lactate functions just fine, but the organism is resistant to the antibiotic because vancomycin cannot bind the D-lactate form. However, the antibiotic teixobactin, discussed in Special Topic 4.1, will bind to the D-Ala-D-lactate form and stop the growth of vancomycin-resistant Gram-positive bacteria.

Note that antibiotics targeting cell wall biosynthesis generally kill only growing cells. These drugs do not affect static or stationary-phase cells, because in this state the cell has no need for new peptidoglycan.

Thought Question

27.5 When treating a patient for an infection, why would combining a drug such as erythromycin with a penicillin be counterproductive? (Erythromycin is described in Section 8.3.)

Drugs That Affect Bacterial Membrane Integrity

Poking holes in a bacterial cytoplasmic membrane is an effective way to kill bacteria. A few compounds are useful in this regard, among them a group called the peptide antibiotics, of which **gramicidin** is an example. Produced by *Bacillus brevis*, gramicidin is a cyclic peptide composed of 15 alternating D- and L-amino acids. It inserts into the membrane as a dimer, forming a cation channel that disrupts membrane polarity (**Fig. 27.13**). Polymyxin (from *Bacillus polymyxa*), another polypeptide antibiotic, has a positively charged polypeptide ring that binds to the outer (lipid A) and inner membranes of bacteria, both of which are negatively charged. Its major lethal effect seems to be to destroy the inner membrane, much as a detergent does. Loss of the cell's osmotic barrier causes the cell to lyse.

These peptide antibiotics are used only topically to treat or prevent infection. Because they can also form channels across human cell membranes, they should never be ingested. Polymyxin has been fused to some bandage materials used to treat burn patients, who are particularly susceptible to Gram-negative bacterial infections (for example, *Pseudomonas aeruginosa*). Despite the drug's toxicity, polymyxin or colistin can be injected as a drug of last resort to treat certain multidrug-resistant bacterial infections (for instance, those caused by *Klebsiella pneumoniae*).

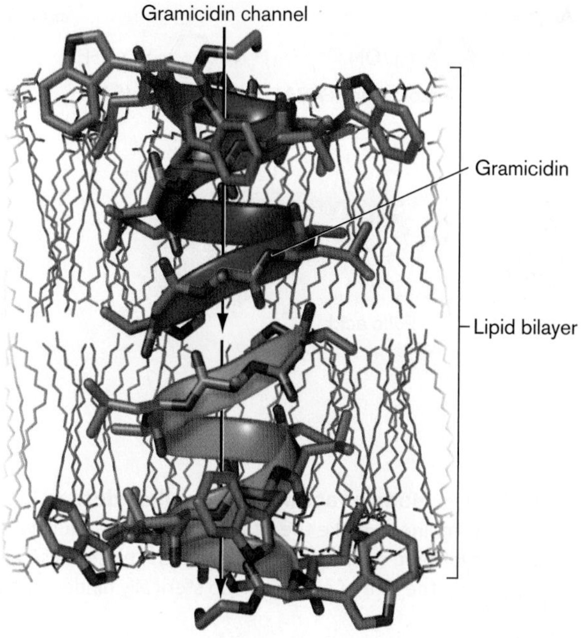

FIGURE 27.13 ▪ Gramicidin is a peptide antibiotic that affects membrane integrity. As a dimer, gramicidin forms a cation channel across cell membranes through which H^+, Na^+, or K^+ can freely pass. (PDB code: 1GRM)

Daptomycin is a lipopeptide made by *Streptomyces roseosporus* using nonribosomal peptide synthetases. The drug aggregates in the membranes of Gram-positive bacteria to form an ion channel that leaks potassium ions. The resulting membrane depolarization leads to cell death. This drug is very effective against MRSA.

Drugs That Affect DNA Synthesis and Integrity

Bacteria generally make and maintain their DNA using enzymes that closely resemble those of mammals. Thus, you might think it impossible to selectively target bacterial DNA synthesis, but it is possible, as you will see.

Case History: Pneumonia Due to a Gram-Negative Anaerobe

A 23-year-old woman arrived at the emergency room by ambulance with fever, chills, and severe muscle aches. She developed a nonproductive (dry) cough, had difficulty breathing, had pleuritic chest pain (stabbing pain when inhaling or

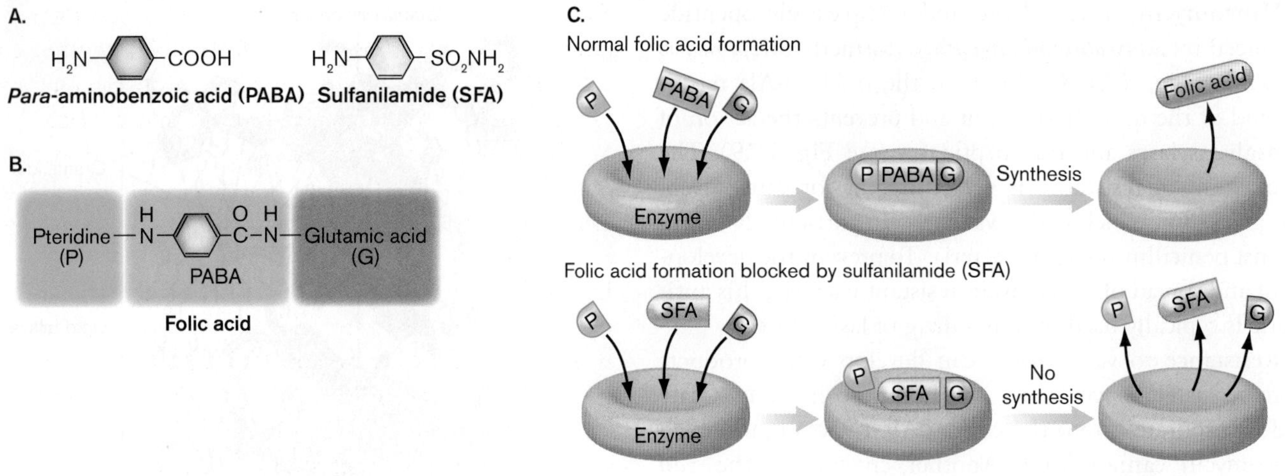

FIGURE 27.14 ■ **Mode of action of sulfanilamides.** **A.** The structures of PABA and sulfanilamide are very similar. **B.** PABA, pteridine, and glutamic acid combine to make the vitamin folic acid. **C.** Normal synthesis of folic acid requires that all three components engage the active site of the biosynthetic enzyme. The sulfa drugs replace PABA at the active site. The sulfur group, however, will not form a peptide bond with glutamic acid, and the size of sulfanilamide sterically hinders the binding of pteridine, so folic acid cannot be made.

exhaling), and became hypotensive (had low blood pressure). An X-ray showed lower-lobe infiltrate in the lungs, and the clinical laboratory reported the presence of the Gram-negative anaerobe Fusobacterium necrophorum *in blood cultures. The patient was diagnosed with pneumonia and treated with metronidazole, a DNA-damaging agent specific for anaerobes. She fully recovered.*

There are several classes of drugs, including sulfa drugs, quinolones, and metronidazole, that selectively affect the synthesis or integrity of DNA in microorganisms.

Sulfa drugs. The sulfa drugs belong to a group of drugs known as antimetabolites because they interfere with the synthesis of metabolic intermediates. Ultimately, sulfa drugs inhibit the synthesis of nucleic acids. Drugs such as sulfamethoxazole or sulfanilamide work at the metabolic level to prevent the synthesis of tetrahydrofolic acid (THF), an important cofactor in the synthesis of nucleic acid precursors (**Fig. 27.14**).

All organisms use THF to synthesize nucleic acids, so why are the sulfa drugs selectively toxic to bacteria? The selectivity occurs because mammals do not synthesize folic acid, a precursor of THF. Higher mammals generally rely on bacteria and green leafy vegetables as sources of folic acid. Bacteria make folic acid from the combination of PABA, glutamic acid, and pteridine. Sulfanilamide (SFA), a structural analog of PABA, competes for one of the enzymes in the bacterial folic acid pathway and inhibits both folic acid and THF production (**Fig. 27.14C**). Because humans lack that pathway, sulfa drugs are selectively toxic toward bacteria. A commonly used drug called Bactrim combines sulfamethoxazole with trimethoprim to inactivate two different enzymes in the folic acid pathway—a strategy that limits the development of antibiotic resistance.

Quinolones. Another group of drugs inhibits DNA synthesis by targeting microbial topoisomerases such as DNA gyrase and topoisomerase IV (the mechanism is discussed in Section 7.2). Because these enzymes are structurally distinct from their mammalian counterparts, drugs can be designed to selectively interact with them while not interfering with mammalian DNA metabolism. One such drug, nalidixic acid, was discovered in 1963. The drug targets bacterial DNA gyrase but has a very narrow antimicrobial spectrum, covering only a few Gram-negative organisms. However, adding various chemical modifications to nalidixic acid, such as fluorine and amine groups, has increased its antimicrobial spectrum and its half-life in the bloodstream. The result is the class of drugs known as the **quinolones**. (The mode of action of quinolones and fluoroquinolones was discussed in Section 7.2.)

Mounting evidence, however, suggests that quinolones can also affect human mitochondria, which evolved from bacterial progenitors. Quinolones have been reported to increase mitochondrial production of reactive oxygen species, whose damaging effects can lead to muscle weakness, tendonitis, and other conditions. Some people are more prone than others.

Metronidazole. Also known as Flagyl, metronidazole is an example of a prodrug—a drug that is harmless until activated. Metronidazole is activated after it receives an electron (is reduced) from the microbial protein cofactors flavodoxin and ferredoxin, found in microaerophilic and anaerobic bacteria such as *Bacteroides* (**Fig. 27.15**). Once activated, the compound begins nicking DNA at random, thus killing the cell. Because the etiological agent in our case history was an anaerobe, metronidazole was an effective therapy. Metronidazole is also effective against protozoa such as *Giardia*, *Trichomonas*, and *Entamoeba*, all of

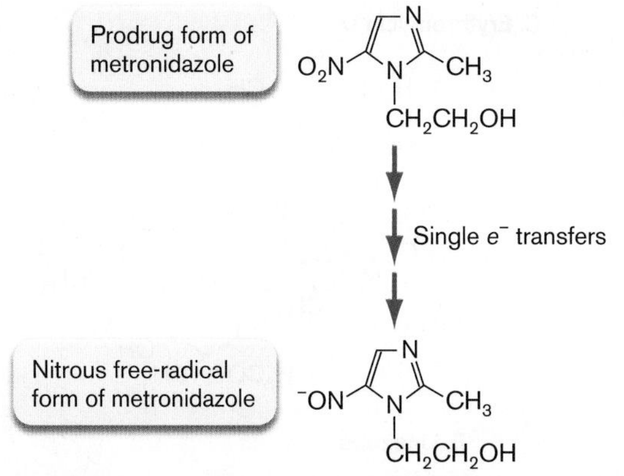

FIGURE 27.15 ▪ Activation of metronidazole. Single-electron transfers are made by ferredoxin and flavodoxin from anaerobes. Ferredoxin and flavodoxin are reducing agents capable of reducing oxidized molecules in cells such as thioredoxin. They can also reduce the prodrug form of metronidazole.

which use ferredoxin and are anaerobes. Aerobic microbes also possess ferredoxin, but they are incapable of reducing metronidazole, presumably because oxygen is reduced in preference to metronidazole.

RNA Synthesis Inhibitors

The mode of action of antibiotics that inhibit transcription, such as rifampin and actinomycin D (Fig. 27.16), was described in Chapter 8. These drugs are bactericidal and are most active against growing bacteria. The tricyclic ring of actinomycin D binds DNA from any source. As a result, it is not selectively toxic and not used to treat infections. Rifampin (also called rifampicin), on the other hand, is selectively toxic for bacterial RNA polymerase and is often prescribed to treat tuberculosis or meningococcal meningitis. Rifampin binds to the exit tunnel of RNA polymerase. Binding stops transcription by blocking RNA from exiting the polymerase (**Fig. 27.16B**; described in Chapter 8). Curiously, rifampin, which is reddish orange, turns bodily secretions, including breast milk, orange. The patient should be warned of this highly visible but harmless side effect to avoid unnecessary anxiety when the patient's urine changes color.

Although rifampin is clinically useful, bacterial resistance to the drug has developed. Fortunately, two new classes of RNA polymerase–targeting antibiotics have been discovered. Pyronins, represented by myxopyronin (produced by *Myxococcus fulvis*), prevent RNA polymerase from ever starting polymerization. The pyronins bind to RNA polymerase at a site called the hinge region, which is needed to separate (melt) DNA strands—a requirement to begin transcription.

The second drug, lipiarmycin, is an unusual macrolide antibiotic made by the actinomycete bacterium *Actinoplanes deccanensis*. Lipiarmycin binds to the same region as myxopyronin but completely stops RNA polymerase–DNA closed complexes from transitioning to the open forms. Rifampin-resistant RNA polymerase molecules are still sensitive to the new antibiotics because the binding site for rifampin is different from the binding site for the other two antibiotics. This is exciting because pathogens such as rifampin-resistant strains of *Mycobacterium tuberculosis* or *Clostridioides difficile* (formerly *Clostridium difficile*) can now be treated with a new drug that targets the same enzyme. Lipiarmycin has been approved by the FDA to treat *C. difficile*–associated diarrhea.

Thought Question

27.6 Given the mechanism by which rifampin stops transcription, what limitation does the drug have? Consider initiation versus elongation.

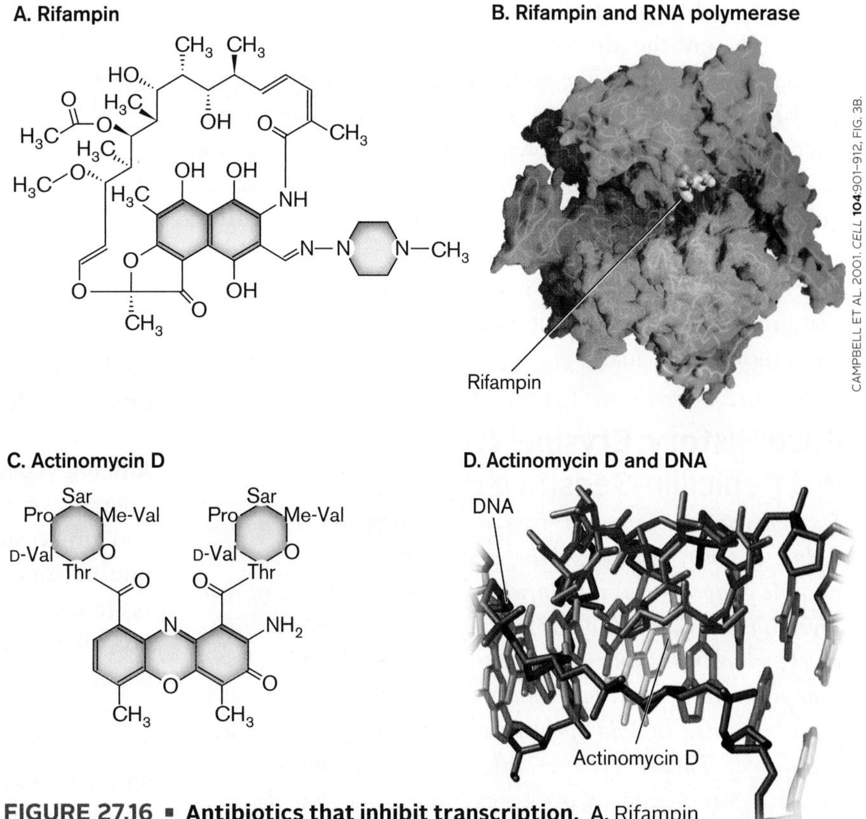

CAMPBELL ET AL. 2001. *CELL* **104**:901–912, FIG. 3B.

FIGURE 27.16 ▪ Antibiotics that inhibit transcription. **A.** Rifampin structure. **B.** Rifampin-binding site on the RNA polymerase beta subunit. cyan = beta subunit; pink = beta-prime subunit; alpha subunits are behind the complex and not shown; the Mg^{2+} ion chelated at the active site is shown as a magenta sphere. **C.** Actinomycin D structure. **D.** Actinomycin D (yellow and red) interacting with DNA. Covalent intercalation of actinomycin (yellow) between bases (gray) interferes with DNA synthesis and transcription. (PDB code: 1DSC)

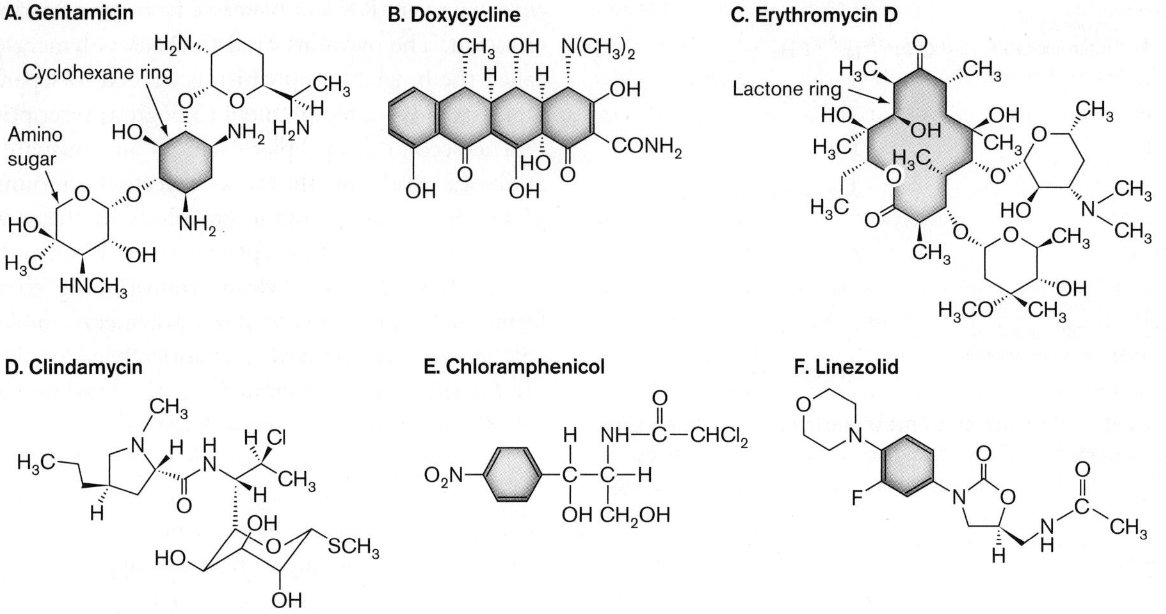

FIGURE 27.17 ▪ Protein synthesis inhibitors. **A.** The aminoglycoside gentamicin. **B.** The tetracycline doxycycline. **C.** The macrolide erythromycin. **D.** The lincosamide clindamycin. **E.** Chloramphenicol. **F.** The oxazolidinone linezolid; F = fluoride. Purple shading highlights chemical groups common to different members within a given antibiotic class.

Protein Synthesis Inhibitors

Antibiotics that specifically inhibit bacterial protein synthesis rely on the differences between prokaryotic and eukaryotic ribosomes for selective toxicity. **How various antibiotics inhibit protein synthesis was discussed in Section 8.3.** Recall that protein synthesis inhibitors can be classified into several groups based on structure and function (**Fig. 27.17**). Most of these antibiotics work by binding and interfering with the function of bacterial rRNA, which differs from the function of eukaryotic rRNA. Recall, too, that protein synthesis inhibitors are, by and large, bacteriostatic (not bactericidal).

Case History: Erysipelas in a Penicillin-Sensitive Patient

Sixteen-year-old Jamal arrived at the emergency room after 2 days of fever, malaise, chills, and neck stiffness. His most notable symptom was a painful, red, rapidly spreading rash covering the right side of his face. The rash covered his entire cheek, which was swollen, and extended into his scalp. About 7 days earlier, Jamal had had a severe sore throat. Because it had subsided in 2 days, however, he had not been clinically evaluated. Throat cultures taken on admission revealed group A Streptococcus pyogenes, *suggesting that the rash was a case of erysipelas caused by this organism. Although penicillin would be the drug of choice, Jamal was known to be allergic to this antibiotic.*

When a patient is known to be allergic to the usual drug of choice, it is best to prescribe a structurally different drug. Often, that drug will be one that inhibits protein synthesis; in this case, the drug chosen was the macrolide azithromycin. Drugs that inhibit protein synthesis can be subdivided into different groups based on their structures and on which part of the translation machine is targeted.

Targeting the 30S Subunit

Recall that bacterial 50S and 30S ribosomal subunits must assemble around mRNA to produce a functional 70S ribosome. Antibiotics affecting protein synthesis are classified by which ribosomal subunit is targeted. Thus, one class of antibiotics interferes with 30S subunit function, and the other impedes 50S subunit activities.

Aminoglycosides. Different aminoglycosides vary considerably in structure, but all contain a cyclohexane ring and amino sugars (**Fig. 27.17A**). The aminoglycosides are unusual among protein synthesis inhibitors in that they are bactericidal rather than bacteriostatic. Most of them bind 16S rRNA and cause translational misreading of mRNA, which is why these drugs are bactericidal (another possible explanation is presented later in the chapter). The resulting synthesis of jumbled or truncated peptides wreaks havoc with physiology and kills the cell.

Streptomycin and gentamicin (**Fig. 27.17A**) are two widely used drugs in this class. Ototoxicity (hearing damage) is a major, but uncommon, side effect of these antibiotics (approximately 0.5%–3% of patients treated with gentamicin suffer from this toxicity). Hearing is generally affected at frequencies above 4,000 Hz. The toxicity of aminoglycosides appears related to their ability to inhibit the function of mitochondrial ribosomes, which are evolutionarily related to bacterial ribosomes. Individuals with specific mutations in mitochondrial rRNA are more susceptible to aminoglycoside toxicity.

Tetracyclines. Tetracycline antibiotics are characterized by a structure with four fused cyclic rings—hence the name. **Figure 27.17B** shows one important, clinically used example, called doxycycline. Tetracyclines are bacteriostatic and act by binding to and distorting the ribosomal A site that accepts incoming charged tRNA molecules. Doxycycline is used to treat early stages of Lyme disease (caused by *Borrelia burgdorferi*), acne (*Cutibacterium acnes*), and other infections. An important adverse side effect of tetracyclines is that they can interfere with bone development in a fetus or young child. Tetracycline use by pregnant mothers will also cause yellow discoloration of the infant's teeth. As a result, this drug is not recommended for pregnant women or nursing mothers.

Targeting the 50S Subunit

Five classes of drugs subvert translation by binding to the 50S ribosomal subunit. **Most of these drugs were discussed in Chapter 8 and are recapped here only briefly.**

- **Macrolides**, all of which contain a 14- to 16-member lactone ring (**Fig. 27.17C**), inhibit translocation of the growing peptide (bacteriostatic action). Commonly prescribed examples are erythromycin and azithromycin. Azithromycin was the antibiotic used to treat the *Streptococcus pyogenes* infection in our case history, although other drugs could have been used. Because it is structurally dissimilar to any of the beta-lactam antibiotics, such as penicillin, azithromycin can be used safely in patients who are penicillin sensitive.
- **Lincosamides** (**Fig. 27.17D**), such as clindamycin, are similar to macrolides in function but have a different structure.
- **Chloramphenicol** (**Fig. 27.17E**) inhibits peptidyltransferase activity (bacteriostatic). Bone marrow depression leading to aplastic anemia is the most common serious side effect and limits its clinical use.
- **Oxazolidinones** (**Fig. 27.17F**) are a class of synthetic antibiotics discovered in the 1990s that are effective against many antibiotic-resistant microbes. In fact, this was the first new class of antibiotics discovered since the "golden age" of antibiotic discovery over 35 years ago. Oxazolidinones such as linezolid bind to the 23S rRNA in the 50S subunit of the prokaryotic ribosome and prevent formation of the protein synthesis 70S initiation complex. This is a novel mode of action; other protein synthesis inhibitors either block polypeptide extension or cause misreading of mRNA. Linezolid binds to the 50S subunit near where chloramphenicol binds, but it does not inhibit peptidyltransferase. Resistance is limited because most bacterial genomes have multiple operons encoding 23S rRNA. Usually more than one of these genes must mutate to confer high-level resistance. The more mutant 23S rRNA genes there are relative to native 23S genes, the more oxazolidinone-resistant ribosomes will be present. Oxazolidinones are useful primarily against Gram-positive bacteria. Gram-negative bacteria are intrinsically resistant because of multidrug efflux pumps (see Section 27.3) and decreased permeability due to the outer membrane.
- **Streptogramins** (**Fig. 27.18**), produced by some *Streptomyces* species, fall into two groups, designated A and B. Streptogramins belonging to group A have a large nonpeptide ring (**Fig. 27.18A**), whereas streptogramin B members are cyclic peptides (**Fig. 27.18B**). The two groups differ in their modes of action, although both inhibit bacterial protein synthesis by binding to the peptidyltransferase site. Group A streptogramins bind to the peptidyltransferase site and prevent binding of tRNA to the ribosome A site. In contrast, group B streptogramins are thought to narrow the peptide exit channel, preventing exit of the peptide and thereby blocking translocation. Natural streptogramins are produced as a mixture of A and B, the combination of which is more potent

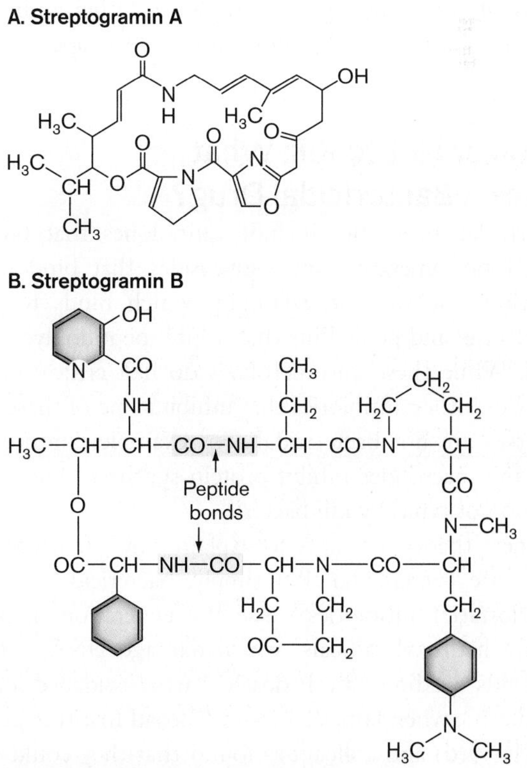

FIGURE 27.18 ■ The streptogramins. **A.** Streptogramin A is a large nonpeptide ring structure. **B.** Streptogramin B is a cyclic peptide.

than either individual compound alone (an example of synergy). In tribute to this synergistic action, the drug combination is marketed under the name Synercid. Synergy between the two drugs occurs because the A-type streptogramin alters the binding site for the B-type drug, increasing its affinity. Bacteria can develop resistance through ribosomal modification (the modification in 23S rRNA is the same one that provides resistance to macrolides), via the production of inactivating enzymes, or by active efflux of the antibiotic.

Targeting Aminoacyl-tRNA Synthetases

Aminoacyl-tRNA synthetases attach amino acids to the 3′ CCA end of tRNA molecules (described in Chapter 8). The catalytic domains of these enzymes include three distinct pockets; one recognizes the cognate amino acid, another recognizes adenylate, and the third binds to the cognate tRNA. An antibiotic called mupirocin (made by *Pseudomonas fluorescens*) is a structural mimic of isoleucyl-AMP, the intermediate used by bacterial isoleucyl-tRNA synthetase to link isoleucine to isoleucyl-tRNA.

Mupirocin binds to the enzyme's isoleucyl-AMP pocket and inhibits synthetase activity. As a result, protein synthesis stops. The drug is selective for bacterial isoleucyl-tRNA synthetase and does not affect the human form of the enzyme. Mupirocin (sold as Bactroban) is used topically in creams to treat skin infections caused by Gram-positive pathogens. Mupirocin cannot be used internally, because it is rapidly degraded in blood. Other aminoacyl-tRNA inhibitors exist, but none are used therapeutically.

To Kill or Not to Kill: What Makes a Bactericidal Drug?

Bactericidal antibiotics include quinolones that bind to DNA topoisomerases; aminoglycosides that bind to the 30S ribosome subunit; rifampin, which binds to RNA polymerase; and penicillins that inhibit peptidoglycan synthesis. While these antimicrobials do halt critical cellular processes, other antibiotics that inhibit some of these same processes are not bactericidal. For example, aminoglycosides and macrolides inhibit protein synthesis, but macrolides do not typically kill bacteria.

A new theory attempts to explain why particular antibiotics are bactericidal. Put simply, bactericidal (but not bacteriostatic) antibiotics cause the generation of highly reactive hydroxyl radicals, which damage DNA, protein, and lipids, leading to cell death. Partial evidence for this model came when James J. Collins (Broad Institute of MIT and Harvard) and colleagues found that they could make cells more sensitive to killing by bactericidal antibiotics by preventing induction of the SOS response that limits and repairs oxidative damage to DNA (see Section 9.2). Note that preventing the generation of reactive oxygen species (ROS) will limit the killing effect of bactericidal antibiotics, but not their bacteriostatic effect by, for instance, halting protein synthesis.

The proposed mechanism leading to ROS accumulation starts with drug-target interactions stimulating NADH oxidation via the electron transport system (ETS). How drug-target interactions stimulate the ETS varies with the antibiotic. Hyperactivation of the ETS induces formation of superoxide and hydrogen peroxide. The superoxide damages iron-sulfur clusters in proteins to release ferrous ions (Fe^{2+}). Fe^{2+} then reacts with hydrogen peroxide via the Fenton reaction (Fig. 5.17) to produce the highly reactive hydroxyl radicals. A practical consequence of this finding is that determining how to inhibit the SOS response by pathogens could make them more susceptible to bactericidal antibiotics. It should be noted that while this model is still controversial, it is gaining acceptance.

Thought Question

27.7 Why might a combination therapy of an aminoglycoside antibiotic and cephalosporin be synergistic?

To Summarize

- **Antibiotic specificity** for bacteria can be achieved by targeting a process present in bacteria but not host cells; by targeting small structural differences between components of a process shared by bacteria and hosts; or by exploiting a physiological growth condition unique to bacteria, such as anaerobiosis.
- **Antibiotic targets** include cell wall synthesis, cell membrane integrity, DNA synthesis, RNA synthesis, protein synthesis, and metabolism.
- **Antibiotics targeting the cell wall** bind to the transglycosylases, transpeptidases, and lipid carrier proteins involved with peptidoglycan synthesis and cross-linking.
- **Antibiotics interfering with DNA** include the antimetabolite sulfa drugs that inhibit nucleotide synthesis; quinolones that inhibit DNA topoisomerases; and a drug, metronidazole, that, when activated, randomly nicks the phosphodiester backbone.
- **Inhibitors of RNA synthesis** target RNA polymerase (rifampin and pyronins) or bind DNA and inhibit polymerase movement (actinomycin D).
- **Different protein synthesis inhibitors target prokaryotic ribosomes.** Some target the 30S subunit to cause misreading of mRNA (aminoglycosides) or prevent

tRNA binding (tetracyclines). Others target the 50S subunit to inhibit translocation (macrolides, lincosamides), peptidyltransferase activity (chloramphenicol), formation of the 70S complex (oxazolidinones), or peptide exit through the ribosome exit channel (streptogramins). Still others target aminoacyl-tRNA synthetases (mupirocin) to prevent charging of tRNAs.

27.3 Challenges of Drug Resistance and Discovery

Why do microbes make antibiotics, and how do they avoid killing themselves in the process? The answers provide insight into the origin of antibiotic-resistant pathogens and our fight to halt their spread throughout the world.

Risky Business: Why Do Microbes Make Antibiotics?

Antibiotics are considered **secondary metabolites** because they often have no apparent primary use in the producing organism. Certainly, antibiotic production today can help one microbe compete favorably with another in nature. Antibiotic production can also forge a mutualistic relationship between a microbe and a colonized host by protecting the host from deadly pathogens (as *Streptomyces* does for leaf-cutter ants; see Section 18.3 and eTopic 17.6). Whatever their use today, growth inhibition by antibiotics may not have been the original purpose of secondary metabolite production. The complexity of the biosynthetic pathways involved in making antibiotics suggests a more immediate purpose—for example, cell-cell signaling—that evolved into cross-species inhibition. Features of antibiotic biosynthetic pathways are discussed in eTopic 27.2 (penicillin synthesis) and Section 15.3 (vancomycin biosynthesis).

Given that microbes continue to make antibiotics for a reason, how does the producing microorganism avoid committing suicide? Fungi that make penicillin face no consequence for having done so, because the organism does not contain peptidoglycan. Actinomycetes that produce compounds such as streptomycin or chloramphenicol, however, could be susceptible to their own secondary metabolite. Ribosomes isolated from *Streptomyces griseus*, for example, are fully sensitive to the streptomycin that the organism itself produces.

S. griseus avoids killing itself in two ways. First, the organism synthesizes an inactive precursor of streptomycin, 6-phosphorylstreptomycin, that is secreted from the cell and, once outside the mycelium, becomes activated by a specific phosphatase. In addition, this streptomycete has an enzyme that inactivates any streptomycin that may leak back into the mycelium. Other organisms protect themselves by methylating key residues on their rRNA to prevent drug binding or by setting up permeability barriers that thwart reentry of the antibiotic.

These and other strategies of self-preservation employed by antibiotic-producing microbes are clever. Unfortunately, these mechanisms have been shared via horizontal gene transfers (plasmids, transposons, transduction), making many pathogens antibiotic resistant. Horizontal gene transfers are discussed in Sections 9.5 and 25.1.

Case History: Multidrug-Resistant Pneumonia

A 14-year-old boy with fever (39°C; 102.2°F), chills, and left-sided pleuritic chest pain was referred to a hospital emergency department by his general practitioner. A chest X-ray showed left-lower-lobe pneumonia. The boy reported that he was allergic to amoxicillin and cephalosporins (as a child he had developed a rash in response to these agents) and had been taking daily doxycycline (tetracycline) for the previous 3 months to treat mild acne. He was admitted to the hospital and treated with intravenous azithromycin (a macrolide antibiotic) because of his reported beta-lactam allergies, but he continued to feel sick. The day after admission, both sputum and blood cultures grew Streptococcus pneumoniae. *After 48 hours, antibiotic susceptibility results indicated that the microbe was resistant to penicillin, azithromycin, and tetracycline. Armed with this information, the clinician immediately changed antibiotic treatment to vancomycin. The boy's fever resolved during the next 12 hours, and he made a slow but full recovery over the next week.*

Unfortunately, the scenario presented in this case is far too common and has become an extremely serious concern. **Figure 27.19** shows the rapid rise of penicillin resistance among *Streptococcus pneumoniae* strains in the world. Another instance of emerging antibiotic resistance is unfolding in Europe and the Far East: The non-Enterobacteriaceae Gram-negative rod *Acinetobacter baumannii* is increasingly seen as a dangerous cause of nosocomial infections. It commonly colonizes hospitalized patients, particularly those in intensive care units. Before 1998, there were almost no cases of multidrug-resistant *A. baumannii*. The rate is now as high as 8%. The organism is resistant to drugs as diverse as ciprofloxacin (a quinolone), amikacin (an aminoglycoside), penicillins, third-generation cephalosporins, tetracycline, and chloramphenicol. Imipenem, one of a relatively new class of beta-lactam drugs (carbapenems), is currently useful, but resistance to it has already developed.

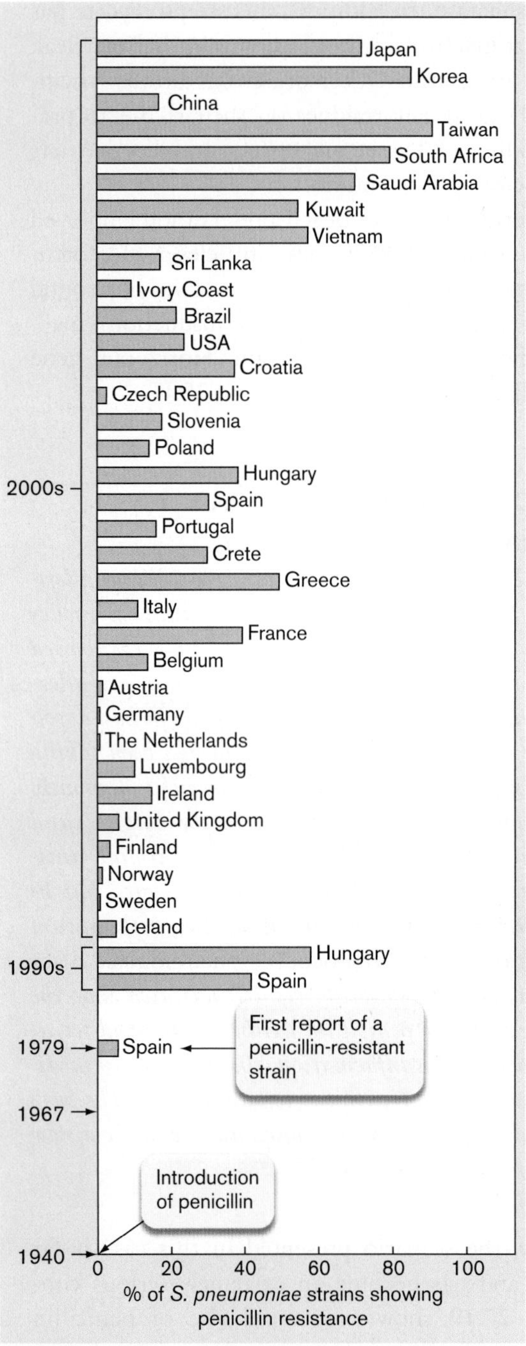

FIGURE 27.19 ■ The rise of penicillin-resistant *Streptococcus pneumoniae* throughout the world. Numbers reflect the number of penicillin-resistant strains among clinical isolates (strains of disease-causing bacteria isolated from patients from different countries). No resistance among clinical isolates was noted until after 1967.

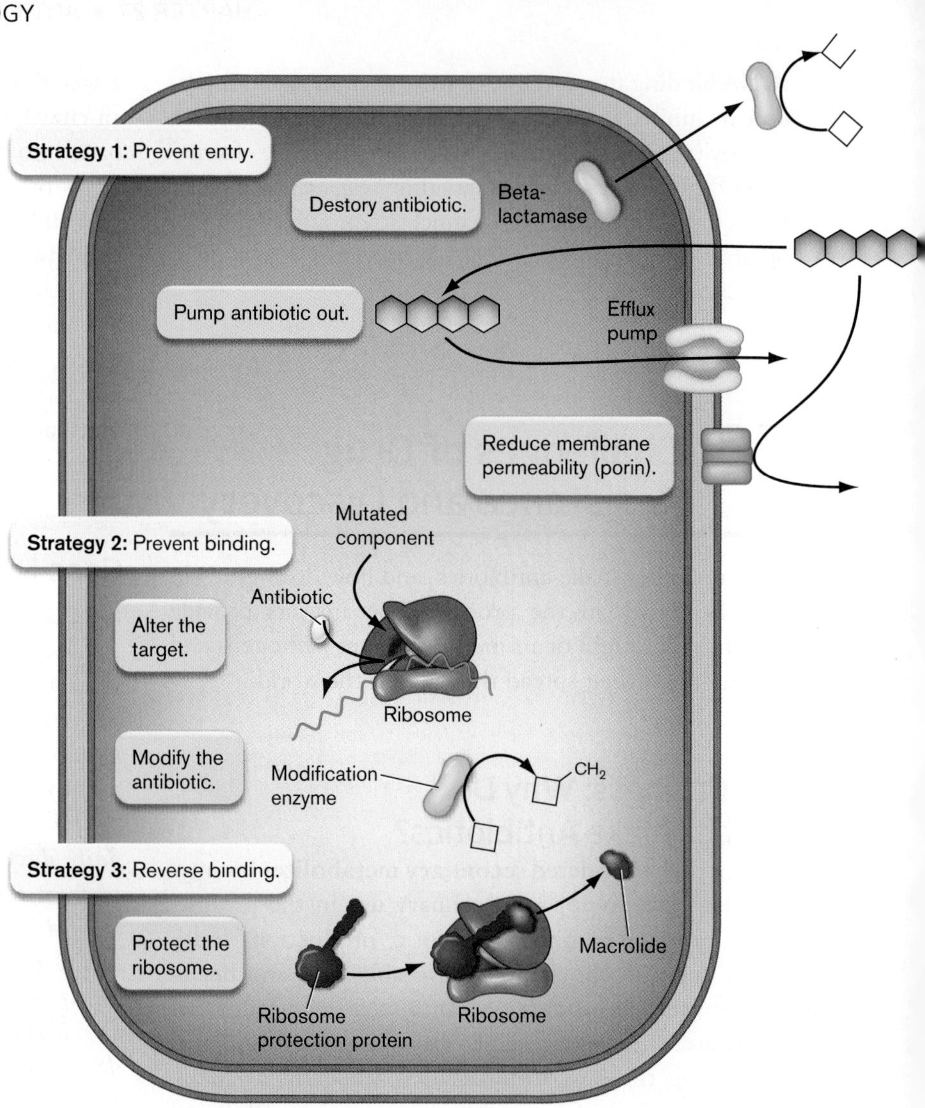

FIGURE 27.20 ■ Alternative mechanisms of antibiotic resistance. Antibiotic resistance genes can be plasmid-borne, or they can be part of the chromosome. Each specific antibiotic resistance gene product will use only one of the six mechanisms shown.

There are three basic strategies for antibiotic resistance (**Fig. 27.20**). The resistant organism can keep the antibiotic out of the cell, prevent the antibiotic from binding to its target, or dislodge bound antibiotic from its target. The three strategies involve six antibiotic resistance mechanisms.

Strategy 1: Keep antibiotics out of the cell.

- **Destroy the antibiotic before it enters the cell.** For example, the enzyme beta-lactamase (or penicillinase) is made exclusively to destroy penicillins (see Section 27.2). The sites of ring cleavage and the structure of the enzyme are illustrated in **Figure 27.21**.
- **Pump the antibiotic out of the cell via specific transporters (for example, tetracycline export).** This strategy works because the pumps bail drugs out of the cell faster than the drugs can enter. Some are single-component pumps present in the cytoplasmic membrane of Gram-negative and Gram-positive bacteria (for example, NorA in *Staphylococcus aureus*, PmrA in *Streptococcus pneumoniae*, and the TetA and B proteins in Gram-negative organisms). Other drug efflux pumps are multicomponent systems present in Gram-negative bacteria only (discussed shortly). Efflux in either case is usually energized by proton motive force.

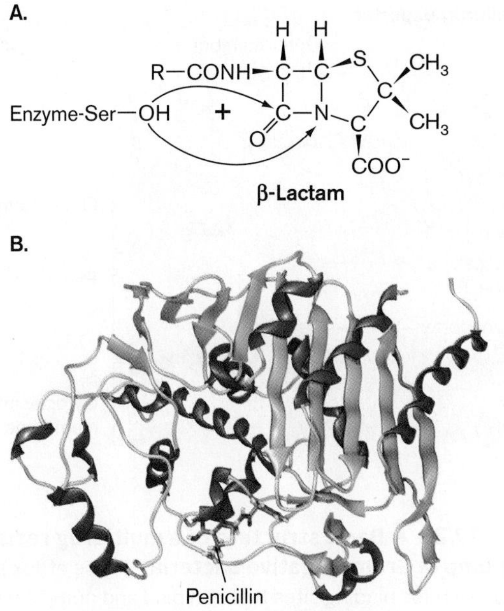

FIGURE 27.21 ■ Destroying penicillin. A. Beta-lactamase (or penicillinase) cleaves the beta-lactam ring of penicillins and cephalosporins. There are two types of penicillinases, based on where the enzyme attacks the ring. In either type, a serine hydroxyl group launches a nucleophilic attack on the ring. B. Structure of a beta-lactamase and location of the penicillin-binding site. (PDB code: 1XX2)

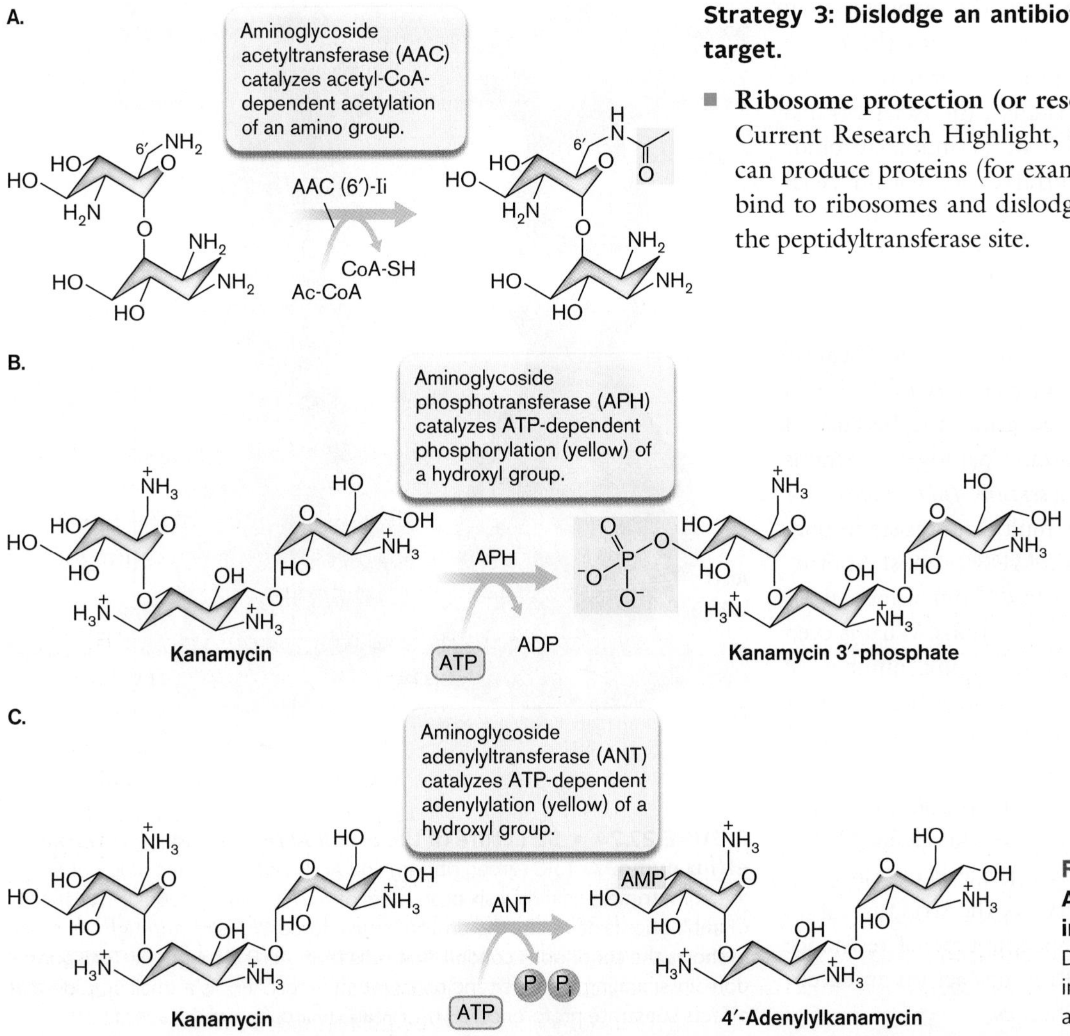

FIGURE 27.22 ■ Aminoglycoside-inactivating enzymes. Different enzymes can inactivate aminoglycoside antibiotics.

- **Decrease membrane permeability across the outer membrane.** Many antibiotics, such as beta-lactams, tetracyclines, and fluoroquinolones, have to pass through outer membrane porins to gain access to the target cell. Gram-negative bacteria, however, can express alternative porins with pores too narrow to allow penetration.

Strategy 2: Prevent antibiotics from binding the target.

- **Modify the target so that it no longer binds the antibiotic.** Mutations in key penicillin-binding proteins and ribosomal proteins, for instance, can confer resistance to methicillin and streptomycin, respectively. These mutations occur spontaneously and are not typically transferred between organisms. An exception to this rule is MecA, the plasmid-borne, beta-lactam-resistant transpeptidase found in strains of *Staphylococcus aureus* described earlier.
- **Add modifying groups that inactivate the antibiotic.** For instance, three classes of enzymes modify and inactivate aminoglycoside antibiotics. The results of these types of enzyme modifications are illustrated for kanamycin in **Figure 27.22**. The modifications decrease the antibiotic's ability to bind to its target (increase the MIC).

Strategy 3: Dislodge an antibiotic already bound to its target.

- **Ribosome protection (or rescue).** As described in the Current Research Highlight, Gram-positive organisms can produce proteins (for example, MsrE or TetO) that bind to ribosomes and dislodge antibiotics bound near the peptidyltransferase site.

Thought Questions

27.8 Fusaric acid is a cation chelator that normally does not penetrate the *E. coli* membrane, which means *E. coli* is typically resistant to this compound. Curiously, cells that develop resistance to tetracycline become sensitive to fusaric acid. Resistance to tetracycline is usually the result of an integral membrane efflux pump that pumps tetracycline out of the cell. What might explain the development of fusaric acid sensitivity?

27.9 Mutations in the ribosomal protein S12 (encoded by *rpsL*) confer resistance to streptomycin. Would a cell containing both $rpsL^{+}$ and $rpsL^{R}$ genes be streptomycin resistant or sensitive? (Recall that genes encoding ribosome proteins for the small subunit are designated *rps*, and "+" indicates the wild-type allele, while "R" indicates a gene whose product is resistant to a certain drug.)

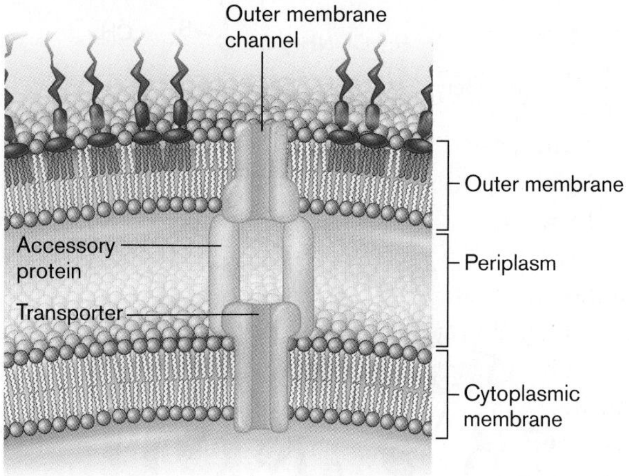

FIGURE 27.23 ■ Basic structure of a multidrug resistance efflux pump in Gram-negative bacteria. These efflux systems have promiscuous binding sites that can bind and pump a wide range of drugs out of the bacterial cell.

A particularly dangerous type of drug resistance is mediated by what are called **multidrug resistance (MDR) efflux pumps** (**Fig. 27.23**). A single pump in this class can export many different kinds of antibiotics with little regard to structure. MDR pumps of Gram-negative microbes are similar to the ABC export systems described in Section 4.2. They include three proteins: an inner membrane pump protein (fueled by proton motive force—a distinction from true ABC exporters), an outer membrane channel connected to the pump protein, and an accessory protein that may link the other two proteins. For instance, the AcrAB transporter (**Fig. 27.24**) almost indiscriminately binds antibiotics in a large central cavity within AcrB (a promiscuous binding site) and uses proton motive force to move those compounds through the AcrB pore and out a funnel (AcrA) that connects to an outer membrane channel, TolC.

Antibiotic efflux pumps contribute significantly to bacterial antibiotic resistance because of the broad variety of substrates they recognize and because of their expression in important pathogens. Strains of the pathogen *Mycobacterium tuberculosis*, for instance, have developed multidrug-resistant phenotypes in part because of MDR pumps. Approximately 2 million people die from tuberculosis annually, mostly in developing nations. What is even more alarming is that an increasing number of *M. tuberculosis* strains isolated from patients exhibit multidrug resistance. Although most of the antibiotic resistance in the majority of *M. tuberculosis* multidrug-resistant strains is due to the accumulation of independent mutations in several genes, MDR pumps are thought to increase the level of resistance. Chemists typically try to tweak the structure of an antibiotic to overcome a specific type of resistance mechanism, but the MDR pumps act on an exceptionally wide range of antibiotics.

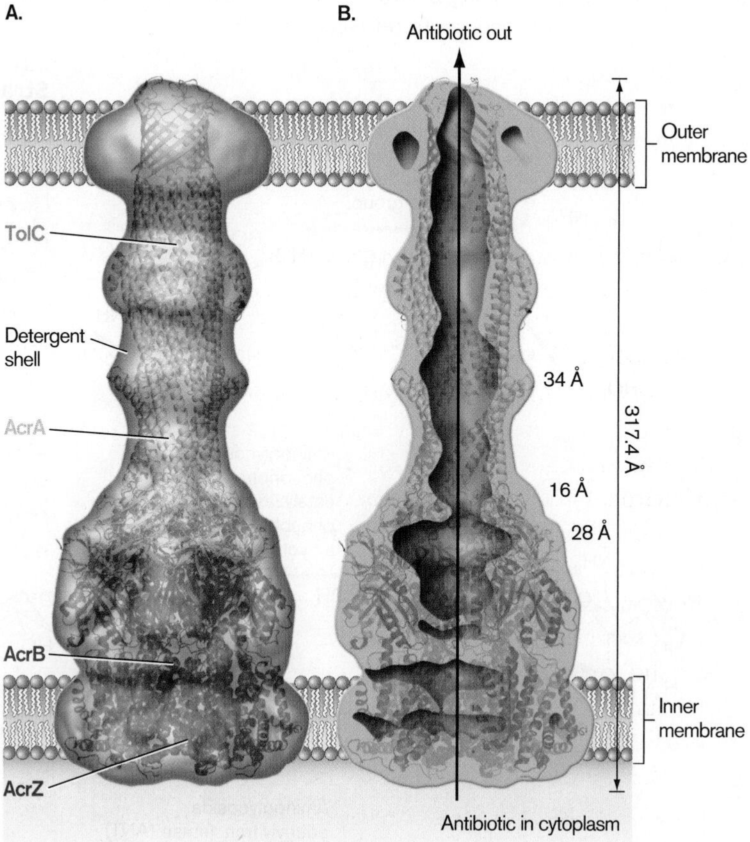

FIGURE 27.24 ■ Structure of the *E. coli* AcrAB multidrug resistance efflux pump. **A.** TolC (green ribbon) and AcrB and Z (red and blue ribbons) are homotrimers linked by six protomers of AcrA (yellow ribbons). Transport of antibiotics is driven by the proton motive force. **B.** A slice through the model shows the continuous conduit that runs from AcrB through the TolC porin domain, spanning the inner and outer membranes. AcrZ is a small peptide that affects substrate preference. *Source:* Modified from Du et al. *Nature* **509**:512–515.

Evolving, and Sharing, Drug Resistance Genes

As discussed in previous chapters, nature has engineered a certain degree of flexibility in the way genomes are replicated and passed from one generation to the next. DNA repair pathways that involve lesion bypass DNA polymerases (for example, UmuDC; see Section 9.2) are thought to contribute significantly to randomized adaptive and evolutionary processes. For instance, at some point during evolution, gene duplication and mutational reshaping produced a beta-lactamase gene, and an organism resistant to penicillin.

However, de novo antibiotic resistance through gene duplication and/or mutation does not occur in all species. Why reinvent the wheel—or, in this case, drug resistance? Gene transfer mechanisms such as conjugation, described in Chapter 9, can move antibiotic resistance genes from one organism to another and from one species to another. In fact, several drug resistance genes found in pathogenic bacteria actually had their start in the chromosomes of antibiotic-producing organisms and were passed on through gene transfer. For instance, the bacterium *Streptomyces clavuligerus* produces penicillin but also makes a protective beta-lactamase (encoded by the *bla* gene) to prevent killing itself. These antibiotic resistance genes can be incorporated into plasmids that transfer into new species by transformation, conjugation, transduction, or nanotubes (see Chapter 5 Current Research Topic).

Antibiotic resistance within uncontacted communities. Insight about the origins of antibiotic resistance in the human microbiome was provided in 2015 by Maria Dominguez-Bello (New York University and the University of Puerto Rico) and her collaborators, who studied the fecal, oral, and skin microbiomes of uncontacted Amerindians in the Amazon. The Amerindians had no previous contact with pharmaceutical antibiotics or Westerners who could be a source of antibiotic-resistant microbes. The researchers predicted that the microbiomes of Amerindians would lack antibiotic resistance genes. They were wrong.

DNA sequencing revealed that the Amerindian microbiome did contain antibiotic resistance–like genes, possibly obtained by exchange with antibiotic-producing soil microbes. The scientists cloned the genes, placed them under the control of a constitutive plasmid promoter to ensure expression, and transferred the plasmids to *E. coli*. Many of these resistance-like genes conferred resistance to natural as well as synthetic antibiotics. The authors, however, suspect that many of these genes are naturally silenced in the Amerindian microbiome, but that exposure to antibiotics could readily select for regulatory mutations that would activate those genes.

It is also worth noting that the microbiomes of the Amerindians were unprecedented in their broad species diversity. The diversity is thought to have developed, at least in part, because of the more intimate contact that Amerindians have with the natural environment (soils, animals, bodies of water) compared to the contact Westerners have.

Multidrug-resistant pathogens. An interesting case study of antibiotic resistance is provided by the Gram-positive bacterium *Enterococcus faecalis*, a natural inhabitant of the mammalian gastrointestinal tract that can cause life-threatening disease if granted access to other body sites (as in subacute bacterial endocarditis; see Section 26.5). *E. faecalis* is naturally resistant to numerous antibiotics, making disease treatment particularly challenging.

Vancomycin is one of the last lines of defense for treating serious *E. faecalis* infections. Unfortunately, increasing numbers of vancomycin-resistant strains of *Enterococcus* (called VREs) have arisen in recent years. The completed genome sequence of one vancomycin-resistant strain illustrates the reason: the incorporation of numerous mobile genetic elements that encode drug resistance. About a quarter of the genome consists of mobile or exogenously acquired DNA, including 7 probable phages, 38 insertion elements, numerous transposons, and integrated plasmid genes. One such mobile element encodes vancomycin resistance.

Multidrug resistance in various microbes (for example, *Salmonella enterica*) has also been attributed to the presence of integrons. Integrons are gene expression elements that account for rapid transmission of drug resistance because of their mobility and ability to collect resistance gene cassettes (see **eTopic 9.6**).

We should note that the development of antibiotic resistance is not without consequence to the bacterium. For example, the altered DNA gyrase that affords resistance to quinolones may not function as well as the "normal" gyrase does. Thus, when resistant and susceptible organisms cohabit the same environment, the wild-type (sensitive) strain may grow faster and eventually overwhelm the mutant strain—unless fluoroquinolone is present.

Methods to Identify Drug-Resistant Pathogens

The proportion of antibiotic-resistant infections has doubled since 2002, rising from 5.2% to 11% of all infections. So, the faster a clinical laboratory can identify a pathogen's antibiotic susceptibility pattern, the more quickly a clinician can prescribe an appropriate, and more narrow-spectrum, antibiotic. The traditional MIC method described in Section 27.1 can take 3 days—one day to grow the organism on an agar plate from a clinical sample, one day to prep the MIC tubes, and another day for the organisms to grow and for the technician to read the results. Automated MIC determinations can cut one day from that timeline. Meanwhile, the seriously ill patient is subjected to empiric therapy in which very broad-spectrum antibiotics—sometimes two or three drugs—are used to "cover" as many pathogens as possible until the organism's identity and antibiotic susceptibility pattern are known. Quickly replacing a broad-spectrum antibiotic with a narrow-spectrum antibiotic will slow the development of resistance to the broad-spectrum antibiotic.

Fortunately, more rapid tests, able to provide answers in less than a day, are being introduced into the clinical laboratory. For instance, multiplex PCR platforms are available that work directly from respiratory tract or stool samples. The technology can detect pathogen-specific or drug resistance gene DNA sequences within an hour. A miniature magnetic resonance machine has also been developed that can detect pathogens at concentrations as low as one organism per milliliter of blood within a few hours. Combining these technologies could mean that a sample brought to the lab at 8 a.m. could be ready by lunch. Such speed could quickly lead to more focused, pathogen-directed therapy. Unfortunately, we have a long way to go before realizing this ideal.

How Did We Get into This Mess?

Consider the following case: A grandmother takes her 4-year-old grandchild to the physician. The child is screaming because he has an extremely painful sore throat. Simply looking at the throat is not diagnostic. The raw tissue could mean the child is suffering from a bacterial infection, in which case antibiotics are needed. Alternatively, a virus could be the cause—a situation in which antibiotics do nothing but pacify the grandparent, or parent. Even today, a clinician will often prescribe an antibiotic without ever knowing the cause of disease. Sometimes when an infection is serious, it is necessary, because time is of the essence. However, blindly administering an antibiotic for a minor infection, such as a sore throat, is inappropriate. The problem is this: The more an antibiotic is used, the more opportunities there are to select for an antibiotic-resistant organism. The extent of the problem was made clear by the Centers for Disease Control and Prevention, which reported in 2016 that nearly one in three antibiotic prescriptions in the United States is inappropriate. For acute respiratory infections, only half of the prescriptions for antibiotics were deemed appropriate.

Of course, the presence of a drug does not cause resistance, but it will kill off or inhibit the growth of competing bacteria that are sensitive, thereby allowing a resistant organism to grow to detectable numbers. A 2018 study, for instance, found that administering ceftriaxone (a third-generation cephalosporin) intravenously to hospitalized patients enabled the emergence of intestinal Enterobacteriaceae that produce a powerful beta-lactamase (AmpC). AmpC can inactivate penicillins, as well as second- and third-generation cephalosporins. The

COURTESY OF OANA CIOFU

FIGURE 27.25 ▪ Oana Ciofu studies bacterial evolution in biofilms.

danger, then, is that the gene imparting resistance might be horizontally transferred to other bacteria—some of them pathogens.

Oana Ciofu (**Fig. 27.25**) and colleagues at the University of Copenhagen have even shown that exposing biofilms to a subinhibitory concentration of ciprofloxacin can promote the development of resistance in *Pseudomonas aeruginosa*.

Antibiotics and the microbiome. Antibiotics are good for treating infectious diseases, but our microbiomes pay the price. As discussed in several prior chapters, we are increasingly aware that our natural microbiota contribute in important ways to human health and development. Many studies are now exploring the impact of antibiotic use on host-microbiota interactions (discussed in Chapters 23 and 24). We have known for decades that antibiotics—especially broad-spectrum antibiotics—can destroy the ecological balance of bacterial species in the gut (as well as at other body sites) and lead to gastrointestinal disease. Pathology can result when one species resistant to the antibiotic gains a growth advantage over various drug-susceptible species that ordinarily keep the pathogen in check (see the discussion of *Clostridioides difficile* in Section 26.3). More recent research suggests that disturbing the microbial balance of power in the intestine with antibiotics can contribute to a vast range of diseases, such as irritable bowel disease, vitamin deficiency, obesity, and even asthma.

Antibiotic stewardship. When should antimicrobials be used? Certainly, in life-threatening situations where time is of the essence, antibiotics (usually broad-spectrum antibiotics) should be administered even before the cause of the infection is known (this is called empiric therapy). On the other hand, the most prudent course to take when a patient has a simple infection is to confirm a bacterial etiology and then prescribe a suitable drug. An exception may be in an elderly or otherwise immunocompromised individual, who may be more susceptible to secondary bacterial infections that can occur subsequent to viral disease.

Clinicians, pharmacists, and laboratory personnel are now being trained in antibiotic stewardship to slow, if not eliminate, the development of new antibiotic-resistant pathogens. Antibiotic stewardship is defined as coordinated interventions that improve and measure antibiotic use. The goal is to cure patients of infections while minimizing drug toxicity and the selection of antibiotic-resistant strains. Proper antibiotic stewardship includes adhering to the following guidelines:

- Do not use antibiotics to treat what are most likely viral infections (for example, typical respiratory tract infections). Consider using antibiotics only if the patient worsens or does not improve.
- Do not use an antibiotic to treat an infection if the patient's microbiome includes strains that are already resistant to the antibiotic. Doing so can promote transfer of the resistance gene to the pathogen. This practice is difficult to achieve because it requires screening the patient's microbiome before antibiotic treatment.
- Know which antibiotic-resistant strains are prevalent in the community or hospital before prescribing an antibiotic regimen.
- Consider how long a patient needs to take the antibiotic. The chance of antibiotic resistance increases the longer the patient is being treated.
- De-escalate antibiotic usage whenever possible. For instance, a patient being treated empirically with broad-spectrum antibiotics (sometimes involving two or more drugs) should be switched to a suitable, more narrow-spectrum antibiotic once an infectious agent and its resistance pattern are identified. In addition, discontinue any unnecessary antibiotics.

Thought Question

27.10 A clinician admits a seriously ill patient with sepsis to the hospital. She treats the patient empirically with piperacillin and vancomycin until the lab identifies the infectious agent. The next day the agent is identified as *Escherichia coli* sensitive to third-generation cephalosporins. What should the clinician do now? Why were piperacillin and vancomycin used initially?

Antibiotics in animal feed. Another proposed source of antibiotic resistance is the widespread practice of adding antibiotics to animal feed. Giving animals subtherapeutic doses of antibiotics in their food makes for larger, and therefore more profitable, animals. The reason antibiotics promote livestock growth is unclear, but the stimulation in growth may be the result of altering the diversity of gut microbiota. Some estimates suggest that 80% of all antibiotics used in the United States (up until 2017) were fed to healthy livestock. The consequence is that the animals may serve as incubators for the development of antibiotic resistance. Even if the resistance develops in nonpathogens, the antibiotic genes produced can be transferred to pathogens. Fortunately, in 2017 the FDA banned the use of medically important antibiotics for improving livestock weight gain. These drugs can still be used to treat animals for infectious diseases.

The continued use of nonmedically important antibiotics in cattle feed to promote growth causes other problems. While this practice does not select for resistance to therapeutic antibiotics, it can stimulate the spread of pathogenicity genes between bacteria. The antibiotics trigger SOS responses (see Chapter 9) that reactivate prophages embedded in bacterial chromosomes. If those phages carry toxin genes, the new phage can transfer toxin production to new strains or species that cohabit the intestine. For example, growth-promoting antibiotics such as carbadox [methyl-3-(2-quinoxalinylmethylene)-carbazate-N^1,N^4-dioxide] or monensin (not used in humans) can reactivate prophages that carry Shiga toxin (*stx*) and foster the distribution of *stx* to other strains. Bradley Bearson of the National Laboratory for Agriculture and the Environment (Ames, Iowa) showed that carbadox will induce prophage replication in pathogenic *Salmonella* and *Shigella* strains. The phage can subsequently transfer virulence or antibiotic resistance genes to other bacteria (**Fig. 27.26**).

Many of the situations that we have noted here have conspired to produce incredibly dangerous bacteria that are resistant to almost every antibiotic known. For example, a multidrug-resistant strain of *Klebsiella pneumoniae* was first isolated in 2008 in Örebro, Sweden, from a Swedish citizen returning from New Delhi, India (**Fig. 27.27**). This organism is resistant to most commonly used antibiotics, such as aminopenicillins, beta-lactam/beta-lactamase inhibitors, aminoglycosides, fluoroquinolones, cephalosporins, tigecycline (structurally similar to tetracycline), and carbapenems. It carries the NDM-1 plasmid noted earlier (Section 27.2), and numerous other antibiotic resistance genes. Despite attempted treatment with linezolid, the Swedish patient died.

The Infectious Diseases Society of America (IDSA) coined the term "ESKAPE pathogens" almost a decade

A.

COURTESY OF BRAD BEARSON

B.

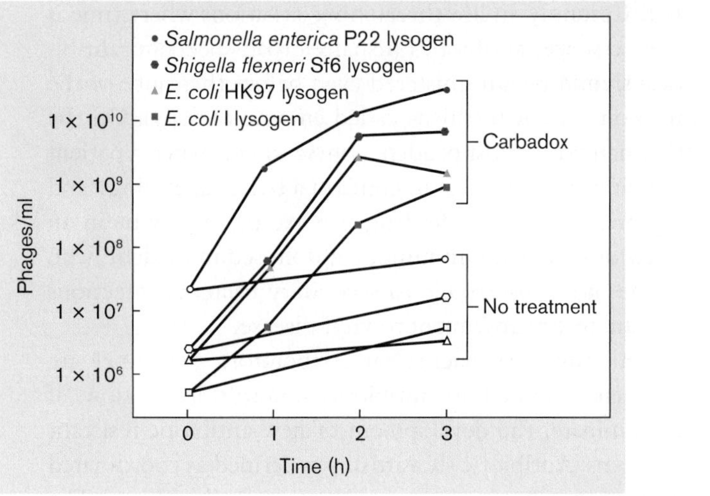

FIGURE 27.26 ■ Carbadox induces prophage replication in enteric pathogens. **A.** Bradley Bearson showed that carbadox, an antibacterial compound formally used to promote growth in swine, activates prophage replication in several enteric bacteria. A phage example is shown on the screen behind him. **B.** Carbadox (0.5 μg/ml) was added to growing cultures of lysogens for 3 hours (closed symbols). At the indicated times, cells were killed with chloroform, and supernatants were titrated for phage on susceptible hosts. The data illustrate that carbadox-treated lysogens generated 100- to 1,000-fold higher titers of phage.

ago, referring to the six bacterial species (including *K. pneumoniae*) that collectively cause about two-thirds of all U.S. nosocomial infections and are highly resistant to many existing drugs. These bacteria are *Enterococcus faecium*, *Staphylococcus aureus*, *Klebsiella pneumoniae*, *Acinetobacter baumannii*, *Pseudomonas aeruginosa*, and *Enterobacter* species. Considerable effort has been made to identify new antibiotics capable of killing or stopping the growth of these pathogens.

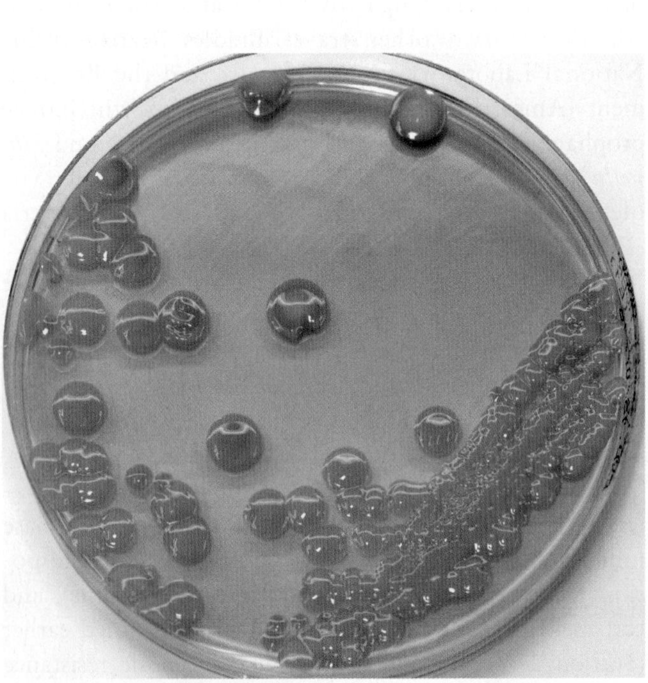

SIRIRAT SHUTTERSTOCK

FIGURE 27.27 ■ MacConkey agar plate containing a sputum culture of *Klebsiella pneumoniae*. *K. pneumoniae* carrying the antibiotic resistance gene $bla_{NDM\text{-}1}$ is emerging as a dangerous, drug-resistant pathogen.

Thought Questions

27.11 Figure 27.27 shows a colony with a distinctive morphology. Why are the colonies on this agar plate red and mucoid?

27.12 Could genomics ever predict the drug resistance phenotype of a microbe? If so, how?

Biofilms, Persisters, and the Mystery of Antibiotic Tolerance

Why do some infections return after bactericidal antibiotic treatment is discontinued? At least part of the reason is a subpopulation of dormant organisms, called **persister cells**, that arise within a population of antibiotic-susceptible bacteria. The stalled metabolism of persisters renders them tolerant to bactericidal antibiotics during treatment. Removing the drug allows persisters to grow and reestablish infection. The strategy is analogous to a hiker's tactic, on accidentally encountering a bear in the woods, of "playing dead" to avoid being attacked.

Persistence is a long-recognized mystery of microbiology. Joseph Bigger in 1944 noticed that penicillin would lyse a growing culture of *Staphylococcus aureus*, but a small number of persister cells always survived. These persisters were not mutants made permanently resistant through mutation; they acted as though dormant. Persister cells that tolerate antibiotic treatment can be found in any biofilm or population of late-exponential-phase cells. In addition to causing antibiotic treatment failures, persistence may be the reason for latent bacterial infections such as recrudescent typhus or latent tuberculosis.

The mechanisms that cause persistence in the face of bactericidal antibiotics appear to be varied. The laboratory of Kim Lewis (Northeastern University), for instance, published evidence that stochastic depletion of ATP in cells of *Staphylococcus aureus* and other species underlies persister formation. In one example, the Lewis lab artificially increased ATP levels in *Pseudomonas aeruginosa* and nearly abolished persister formation. Lowering ATP levels, in contrast, increased the number of persisters.

Other laboratories have focused on a link between antibiotic persister cells and so-called toxin-antitoxin modules encoded by chromosomal genes. More on toxin-antitoxin modules in plasmids can be found in **eTopic 7.3**.

One example of a toxin-antitoxin module thought to be involved in persistence relies on the *hipA* and *hipB* genes in uropathogenic *E. coli*. HipA is a toxin that is neutralized by the antitoxin HipB (**Fig. 27.28**, step 1). A delicate counterbalance between HipA and HipB levels allows cells to grow normally. However, because HipB antitoxin is less stable than HipA, a portion of HipA toxin can become active if antitoxin degrades or synthesis lags (step 2). When freed of antitoxin, HipA will phosphorylate and inactivate glutamyl-tRNA synthetase (step 3). The subsequent lack of charged glutamyl-tRNA stalls translation. The stalled ribosome synthesizes ppGpp (see the subsection "The Stringent Response" in Section 10.4), which is proposed to stimulate Lon protease to degrade ten or more different antitoxin proteins, including HipB (step 4). The toxins that are unleashed encode additional mRNA nucleases thought to further inhibit translation and cell growth (step 5). The result will be a dormant cell. Note that the toxin-antitoxin model of dormancy and antibiotic persistence is hotly debated in today's literature. Nevertheless, most scientists in the field agree that there are multiple routes to persistence, but disagree as to what they are.

How might dormancy explain antibiotic tolerance? If persisters are dormant and have little or no cell wall synthesis, translation, or topoisomerase activity, then even if bactericidal antibiotics can bind to their targets, target function cannot be corrupted. Tolerance, then, provides antibiotic resistance at the price of not growing. A bactericidal antibiotic will kill all susceptible bacteria in an infection, but the remaining persister cells serve as a source of population regrowth (and reinfection) once the antibiotic is removed.

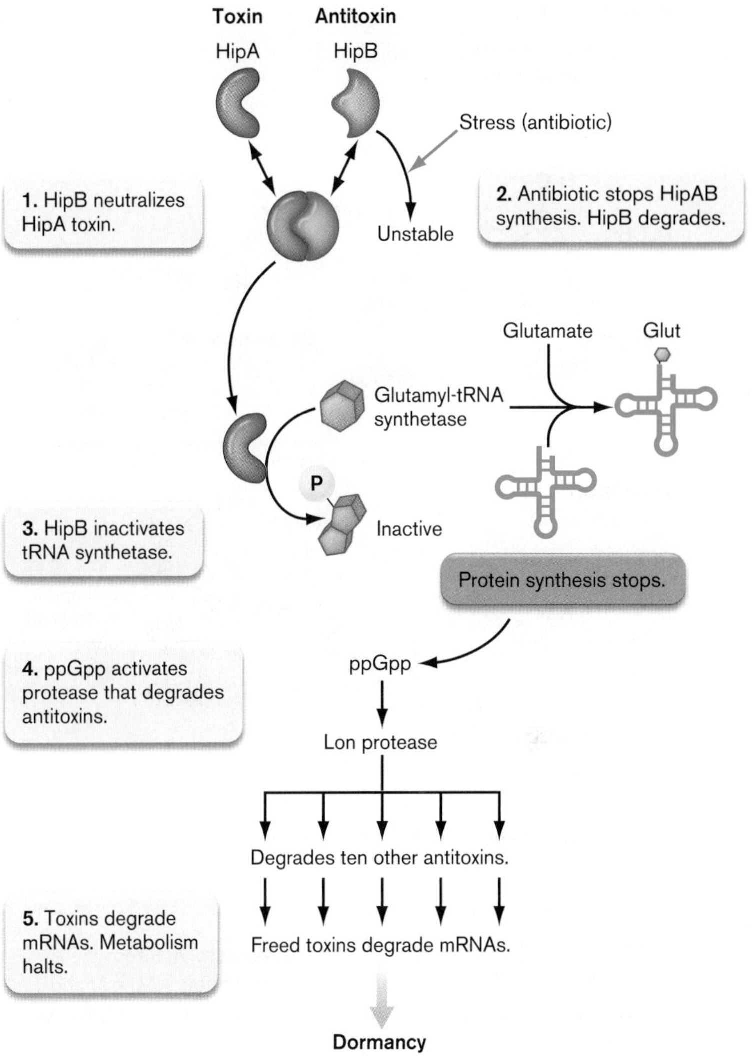

FIGURE 27.28 ▪ HipA (toxin) and HipB (antitoxin) control over dormancy in *Escherichia coli*.

Fighting Resistance and Finding New Drugs

The pervasive nature of bacterial resistance to antibiotics has led many to declare that we are in the "postantibiotic era." But is humankind really doomed to a future in which antibiotics will no longer work? The hope is that the prudent use of current antibiotics and innovative strategies for finding new ones will enable us to continue to control evolving bacterial pathogens.

Directly countering drug resistance. In addition to antibiotic stewardship (discussed earlier), several strategies are being used to counter drug resistance. In some instances, dummy target compounds that bind to and inactivate resistance enzymes have been developed. Clavulanic acid, for example, is a compound used in combination with penicillins such as amoxicillin. Clavulanic acid is a beta-lactam compound with no antimicrobial effect. It is, however, a chemical decoy that competitively binds to beta-lactamases secreted from penicillin-resistant bacteria. Because the enzyme releases bound clavulanic acid very slowly, the amoxicillin remains free to enter and kill the bacterium. Another strategy is to alter the structure of the antibiotic in a way that sterically hinders the access of bacterial modifying enzymes that could inactivate the drug. **Figure 27.29** illustrates how adding a side chain to gentamicin that converts it to amikacin blocks the activity of various aminoglycoside-modifying enzymes. Of course, we are now seeing the development of resistance to amikacin.

Finding new antibiotics. How do you find new antibiotics? Certainly, the classic approach—in which microbes, plants,

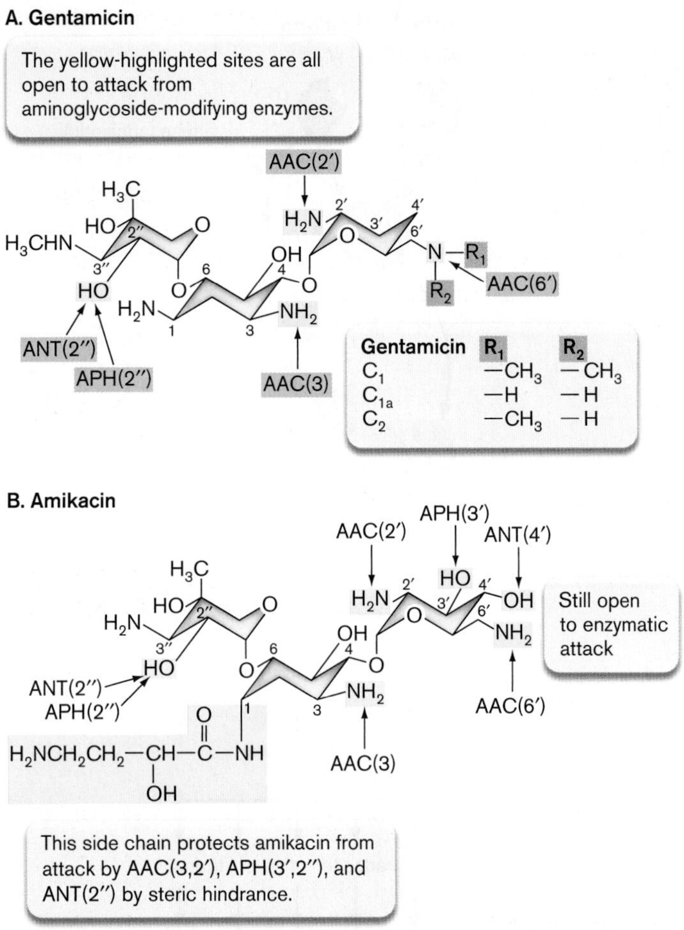

FIGURE 27.29 ▪ Fighting drug resistance. **A.** Sites where gentamicin is vulnerable to enzymatic inactivation. AAC = aminoglycoside acetyltransferase; ANT = aminoglycoside adenylyltransferase; APH = aminoglycoside phosphotransferase. The inset shows the R groups for different gentamicin compounds: C_1, C_{1a}, and C_2. **B.** Gentamicin can be chemically modified at the highlighted sites to prevent loss of activity due to enzyme action. The side groups block access to enzyme active sites by steric hindrance (that is, the added groups prevent the active site from interacting with its target structure) but do not inactivate the antibiotic.

and even animals collected from around the world are screened for their abilities to make new antibiotics—is still valid and remains a fruitful source of new potential drugs. Even previously unculturable soil bacteria have been screened for new antibiotics, as described in Special Topic 4.1.

Recent brute-force screening techniques have revealed new classes of antibiotics. One example is platensimycin, made by *Streptomyces platensis.* The screening method, although laborious, was novel. Merck scientists screened 250,000 natural product extracts for an ability to specifically inhibit bacterial fatty acid biosynthesis. Fatty acid biosynthesis is an attractive target because the bacterial process is different from that of eukaryotes. Scientists engineered a strain of *Staphylococcus aureus* to contain a gene expressing an antisense RNA to *fabF* mRNA. The protein FabF is essential for bacterial fatty acid synthesis. When the antisense RNA was induced, it bound to *fabF* mRNA, spanning and sequestering the ribosome-binding site to prevent efficient translation. As a result, the level of FabF protein in the cell decreased. The strain could still grow but would be exquisitely sensitive to any compound that targeted the remaining FabF protein. This novel screening method led to the discovery of platensimycin.

Platensimycin binds FabF and exhibits bacteriostatic, broad-spectrum activity, acting on Gram-positive and Gram-negative bacteria. It is only the fourth entirely new class of antibiotic developed in the last four decades [lipopeptides (daptomycin), oxazolidinones (linezolid), and bedaquiline (see below) are the other three]. The novel chemical structure of platensimycin and its unique mode of action provide a great opportunity to develop a new class of critically needed antibiotics—a class that selectively targets fatty acid biosynthesis. As of this writing, however, platensimycin has not yet won FDA approval for use. More information on fatty acid biosynthesis and inhibitors can be found in Chapter 15.

Another recent success story paired brute-force screening with combinatorial chemistry to identify a novel antibiotic. *Mycobacterium tuberculosis* is a reemergent, slow-growing pathogen that causes tuberculosis in 9 million people each year (Chapter 26). What makes the situation even more desperate is that, of those 9 million cases, 500,000 are caused by multidrug-resistant (MDR) *M. tuberculosis.* Unfortunately, the antibiotics for this pathogen had not changed for 40 years.

To find new anti-tuberculosis agents, a team led by Belgian scientist Koen Adries screened 70,000 compounds for antimicrobial effects on *Mycobacterium smegmatis*, a fast-growing relative of *M. tuberculosis.* One compound that significantly affected growth was chemically modified to increase its efficacy. After exhaustive clinical trials, the new drug, now called bedaquiline, was approved by the FDA in 2013 to treat MDR-TB. The antibiotic selectively targets the organism's energy-generating ATP synthase—a novel mode of action—and starves the pathogen of energy. The hope is that this new antibiotic can finally stem the rising tide of drug-resistant TB. Tempering that hope, however, is the knowledge that bedaquiline-resistant mutants of *M. tuberculosis* have already been isolated in the laboratory.

Newer strategies of drug discovery center on genome sequence analysis to identify potential bacterial molecular targets. Once a target is identified, clever screening techniques are used to find natural antibiotics, and molecular modeling is used to synthetically design potential inhibitor molecules. High-throughput biochemical screens of large collections of synthetic chemicals have also been attempted. Although many promising drugs have been identified, unfortunately only a rare few have proved therapeutically useful. Examples of metagenomic and peptidogenomic approaches to drug discovery are described in the Chapter 15 Current Research Highlight and Special Topic 15.1, respectively.

Virulence genes and proteins of pathogens also hold great promise as targets for new drugs because these genes are required for in vivo growth. However, these proteins are conditionally essential, so inhibitors fail to inhibit growth in vitro. Virulence proteins are relatively easy to find by measuring loss of virulence when the gene is missing (for instance, the gene encoding the tip of a type III secretion system), but how do you screen for specific inhibitors of these proteins? You can't measure effects on in vitro growth, because inhibiting a virulence factor will not affect growth on agar. Currently, it is easier to make monoclonal antibodies that target virulence proteins than it is to design and find chemical inhibitors. Synthetically engineered, bispecific antibodies are now being tested as pathogen-specific antimicrobials (see **Special Topic 27.1**).

Another intriguing idea involves using photosensitive chemicals that can penetrate the microorganism and generate toxic reactive oxygen species (such as superoxide) when exposed to specific wavelengths of visible light (obviously, good only for topical use). Interfering with the quorum-sensing mechanisms of pathogens is yet another clever approach (see **eTopic 27.3**). Finally, the promise and flexibility of synthetic-biology approaches discussed in Chapter 12 could revolutionize the process of antibiotic discovery and production. For instance, CRISPR-based strategies are being considered as pathogen-specific antimicrobials, and as a tool to reverse antibiotic resistance in a pathogen by excising the resistance gene in vivo (CRISPR technology was examined in Chapter 12).

Antipersister and antibiofilm approaches. Because persister cells and biofilms play prominent roles in recurrent infections and treatment failures, antimicrobials that target these features would be a major breakthrough in treating infections. Several approaches targeting persisters are possible. Some compounds can directly kill persisters—for example, compound HT61, a quinolone derivative that depolarizes the cell membrane and destroys the cell wall; or clofazimine, a compound that increases production of destructive reactive oxygen species. Other compounds can prevent persister formation by interfering with ppGpp synthesis, or reverse the persister state by stimulating the cell's metabolism, once again making them susceptible to existing antibiotics.

What about biofilms? Biofilms not only harbor metabolically dormant persister cells, but their sheer density and complex architecture provide their own level of antibiotic tolerance. There are two main antibiofilm approaches: interfere with the synthesis and secretion of extracellular polymeric substances that coat biofilms and hold them together, or induce biofilm dispersal. Small molecule inhibitors that limit synthesis of cyclic-di-GMP, cyclic-di-AMP, and ppGpp signaling molecules are being sought and tested (discussed in Chapters 4 and 10). These signaling molecules control several aspects of extracellular polymeric substance metabolism and biofilm dispersal.

Recently, Lindsey Shaw at the University of South Florida codiscovered an antimicrobial diterpene chemical, called darwinolide, that was extracted from the Antarctic sponge *Dendrilla membranosa* by his colleague Bill Baker. Shaw's students showed that darwinolide has the novel ability to kill biofilms of methicillin-resistant *Staphylococcus aureus* (MRSA) fourfold better than it can kill planktonically growing MRSA. The scientists predict that the darwinolide structure could provide a unique scaffold upon which more potent anti-MRSA therapies can be built.

Although many of the strategies just described have yielded promising antimicrobial candidates, none of the compounds have been approved yet for clinical use.

Some scientists are even thinking "outside the box" to develop pathogen-specific antimicrobials. Section 5.6 described how bacteriophages that target specific pathogens are regaining favor in the treatment of some infections. A more recent strategy is chronicled in **Special Topic 27.1**. It's an exciting tactic of engineering bispecific antibodies to contain binding sites for two different virulence factors deployed by a single pathogen.

Why has progress been so slow in finding new antimicrobial compounds? Part of the answer is that the road to FDA approval is long (8–10 years) and expensive, involving numerous animal and human trials. The process is necessary for safety reasons, but it discourages the pharmaceutical industry from investing in antibiotic discovery. However, because of the public health implications of a dwindling pipeline of new antibiotics, policy makers are trying to incentivize investment.

To Summarize

- **Certain microbes make antibiotics** to eliminate competitors in the environment and prevent self-destruction by means of various antibiotic resistance mechanisms. Genes encoding some of these drug resistance mechanisms have been transferred to pathogens.
- **Antibiotic resistance can arise spontaneously** through mutation, can be inherited by gene exchange mechanisms, or can arise de novo through gene duplication and mutational reengineering.
- **Antibiotic resistance involves three basic strategies.** (1) Keep the antibiotic out of the cell by destroying the drug, reducing permeability, or pumping the drug out. (2) Prevent the antibiotic from binding to its target by altering the target or the drug. (3) Knock the drug off its target. **Multidrug resistance efflux pumps** use promiscuous binding sites to bind antibiotics of diverse structure.
- **Indiscriminate use of antibiotics** has significantly contributed to the rise in antibiotic resistance.

SPECIAL TOPIC 27.1 Are Designer Antibodies the Next Antibiotics?

The use of broad-spectrum antibiotics to treat an infectious disease could be called the "nuclear option" of antimicrobial therapy. Not only do these antibiotics kill the intended pathogen, but they also wipe out much of our own microbiota. Nearly indiscriminate clinical use of broad-spectrum antibiotics has spawned widespread resistance to all classes of antimicrobials, sparking fear that we are fast approaching a postantibiotic world in which traditional antibiotics no longer work. The situation, though dire, has fostered interest in developing new, more focused, pathogen-specific antimicrobial strategies to treat infections—something that would minimize the use of broad-spectrum antibiotics. One such option is monoclonal antibody technology.

A monoclonal antibody (mAb) originates from a single antibody-producing B-cell clone and is specific for a single epitope. Once identified in the laboratory, a B cell that produces a desired antibody is fused to an immortal line of cancer cells. The result, called a hybridoma, can live indefinitely and secrete antigen-specific monoclonal antibodies on demand. Monoclonal antibodies directed against virulence factors can disrupt a target microbe's pathogenesis and facilitate clearance of the pathogen by the immune system (for example, via opsonization).

A recent advance in this technology is the ability to construct <u>bispecific</u> antibodies that can bind two <u>different</u> epitopes. Bispecific antibodies are usually made by engineering single-chain variable fragment (scFv) domains that are specific for different epitopes and fusing them to the free termini of mAb heavy-chain or light-chain sequences. Because two different epitopes are targeted on a pathogen, bispecific antibodies should be more effective antimicrobials than are monoclonal antibodies.

C. Kendall Stover and colleagues (**Fig. 1**), at MedImmune, LLC, in Gaithersburg, Maryland (now part of AstraZeneca) have constructed novel bispecific antibodies that target the dangerous, multiantibiotic-resistant pathogen *Pseudomonas aeruginosa*, a major cause of lung and burn infections. The chimeric antibody, called BiS4αPa (or MEDI3902), is shown in **Figure 2**.

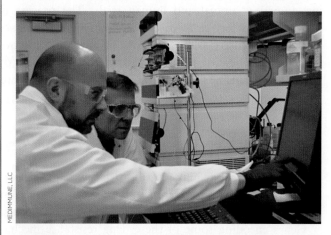

FIGURE 1 ▪ Developing bispecific antibodies. MedImmune's senior director Kendall Stover (right) and senior scientist Antonio DiGiandomenico developed the bispecific antibody directed against *Pseudomonas aeruginosa*.

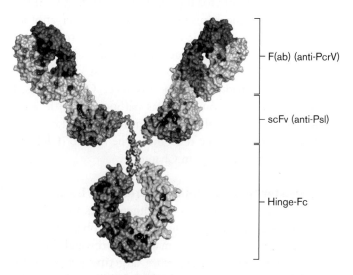

FIGURE 2 ▪ Anti-*Pseudomonas* bispecific antibody BiS4αPa (MEDI3902). This bispecific antibody has binding sites for Psl (an extracellular sugar polymer implicated in biofilm formation) and PcrV (a component of the *Pseudomonas* type III secretion system). It also includes an Fc region that can promote opsonization and complement binding. F(ab) light chain is light blue; F(ab) heavy chain is dark blue; PcrV binding site is red; scFv variable light chain (V_L) is light orange, and variable heavy chain (V_H) is dark orange; Psl-binding site is salmon; hinge and Fc are green. Linkers between V_H and V_L in scFv and between scFv and IgG sequences are gray.

- **Measures to counter antibiotic resistance** include synthetically altering the antibiotic, using combination antibiotic therapy, and adding a chemical decoy.
- **Persister cells** in a population have stopped actively growing, making them tolerant to bactericidal antibiotics.
- **The quest to discover novel antibiotics** includes **designing candidate antimicrobial compounds** to interact with and inhibit the active site of a known microbial enzyme, and **screening previously uncultured microbes** for new antibiotics.
- **Potential targets for new antimicrobials** include persister cells, proteins expressed only in vivo (virulence proteins), and pathways contributing to biofilm formation and dispersal.

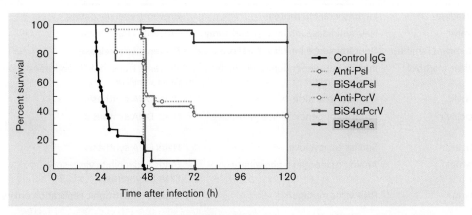

FIGURE 3 ▪ **Effect of bispecific antibody BiS4αPa on survival in an acute-pneumonia model.** Mice were treated with molar-equivalent doses of the indicated antibodies, followed by intranasal infection with wild-type *Pseudomonas* strain 6206 (1.1×10^6 colony-forming units, CFUs). Mouse survival results were compiled from six independent experiments. BiS4αPa is the bispecific antibody.

BiS4αPa includes antibody-binding sites for PcrV (a component of the *Pseudomonas* type III secretion system) and Psl (an extracellular sugar polymer that coats *Pseudomonas* and is implicated in biofilm formation). The bispecific antibody was constructed by taking the monoclonal antibody against PcrV and genetically inserting a Psl-binding scFv domain within the hinge region that connects the anti-PcrV F(ab) segment to the Fc region.

BiS4αPa inhibits *Pseudomonas* biofilm formation by binding to the Psl surface polymer, and interferes with effector protein secretion by blocking the type III secretion system (via PcrV). Furthermore, because it has an Fc region, this bispecific antibody promotes opsonization of *Pseudomonas* by phagocytes. All of these properties should translate into effective antipseudomonad activity for the bispecific antibody. **Figure 3** shows that this antibody readily protected mice from *Pseudomonas* in a mouse model of acute pneumonia. Notice, too, how much better BiS4αPa was at protecting mice than was either anti-PcrV or anti-Psl alone. BiS4αPa was also superior to an engineered BiS4 antibody containing only the PcrV- or Psl-binding sites.

Why was the bispecific antibody better than the monospecific anti-PcrV at protecting mice? The idea proposed was that bispecific antibody initially binds to the plentiful Psl polysaccharide coating the organism and produces a "cloud" of antibody. The antibody cloud then increases the likelihood that the anti-PcrV part of the bispecific antibody will find and inactivate the less abundant surface PcrV molecules. This hypothesis was tested by comparison of how well BiS4αPa and anti-PcrV prevented the *Pseudomonas* type III secretion–dependent killing (cytotoxicity) of host cells. *P. aeruginosa* injects a phospholipase (ExoU) through its type III secretion system into host cells. ExoU then kills host cells by lysing host cell membranes. Consistent with the "cloud" hypothesis, the bispecific antibody prevented the cytotoxic effect at far lower concentrations than did monospecific anti-PcrV antibody.

These studies suggest that pathogen-targeted, bispecific antibodies could be used in at least two ways: prophylactically in severely ill patients at high risk for contracting lethal infections, or as a direct treatment for acute drug-resistant *Pseudomonas* infections for which few antibiotic options remain. Who knows? Designer antimicrobial antibodies may, indeed, be our next-generation antibiotics.

RESEARCH QUESTION

How would you test whether *Pseudomonas aeruginosa* can develop resistance to BiS4αPa? Discuss whether resistance could develop, and why or why not.

DiGiandomenico, Antonio, Ashley E. Keller, Cuihua Gao, Godfrey J. Rainey, Paul Warrener, et al. 2014. A multifunctional bispecific antibody protects against *Pseudomonas aeruginosa*. *Science Translational Medicine* **6**:262ra155.

27.4 Antiviral Agents

A father pleading with a physician to give his child antibiotics when the infant is suffering with a cold is an all-too-common dilemma faced by the general practitioner, but there is nothing of substance the physician can do. The common cold is caused by the rhinovirus, and no antibiotic designed for bacteria can touch it. Why are there so few antiviral agents in the clinician's arsenal? The reason is that it is much harder to apply the principle of selective toxicity for viruses than it is for bacteria. Viruses routinely commandeer host cell functions to make copies of themselves. Thus, a drug that hurts the virus is likely to also harm the patient. Nevertheless, there are several useful antiviral agents for which selective viral targets have been found and

TABLE 27.3 Examples of Antiviral Agents

Virus	Agent	Mechanism of action	Result
Influenza virus	Amantadine	Inhibits viral M2 protein	Prevents viral uncoating
	Zanamivir	Neuraminidase inhibitor (nasal spray)	Prevents viral release
	Oseltamivir (Tamiflu)	Neuraminidase inhibitor (oral prodrug)	Prevents viral release
	Baloxavir marboxil	Inhibits cap-dependent endonuclease (oral prodrug)	Prevents transcription and translation of viral mRNA
Herpes simplex virus and varicella-zoster virus (shingles)	Acyclovir	Guanosine analog	Halts DNA synthesis
	Famciclovir	Prodrug of penciclovir, a guanosine analog	Halts DNA synthesis
Cytomegalovirus	Ganciclovir	Similar to acyclovir	Halts DNA synthesis
	Foscarnet	Analog of inorganic phosphate	Binds and inhibits virus-specific DNA polymerase
Respiratory syncytial virus and chronic hepatitis C virus	Ribavirin	RNA virus mutagen	Causes catastrophic replication errors
Hepatitis C virus	Sofosbuvir (Sovaldi)	Analog inhibitor of HCV polymerase	Halts RNA synthesis
	Simeprevir (Olysio)	Protease inhibitor	Prevents viral maturation
HIV	Zidovudine (AZT)	Nucleoside analog; resembles thymine	Inhibits reverse transcriptase
	Nevirapine	Binds to allosteric site	Inhibits reverse transcriptase
	Tenofovir	Nucleotide analog, resembles AMP	Inhibits reverse transcriptases of HIV and hepatitis B virus
	Nelfinavir	Protease inhibitor	Prevents viral maturation
	Raltegravir	Integrase inhibitor	Prevents integration into host genome
	Maraviroc	CCR5 entry inhibitor	Prevents virus entry into host cells

exploited. Some of these agents are listed in **Table 27.3**. Select examples are discussed in this section. All of the molecular mechanisms of viruses presented in Chapter 11 are studied as potential drug targets.

Target Virus Uncoating or Release

Membrane-coated viruses are vulnerable at two stages. The first is when the virus is invading the host cell. The second is after viral propagation, when the progeny viruses release from the host cell. The flu virus presents a good example of both.

Case History: Antiviral Treatment of Infant Influenza

A 9-month-old infant arrived at the Johns Hopkins Hospital with an acute onset of fever, cough, regurgitation from his gastrostomy feeding tube, and dehydration. This illness followed a series of chronic problems, including bronchiolitis (infection and inflammation of the bronchioles) caused by respiratory syncytial virus, and neonatal group B streptococcal sepsis. (Neonatal sepsis is often caused by Lancefield group B Streptococcus agalactiae; Lancefield group classification is described in Chapter 28.) Physical exam revealed fever, a severe cough resulting in respiratory distress, a rapid heart rate, and moderate dehydration. Nasopharyngeal aspirate was positive for influenza A antigen. The patient was treated with oseltamivir when influenza was diagnosed. He gradually improved and was discharged home 4 days after admission.

In 2017–2018, an unusually severe form of influenza (H3N2) spread across the United States. Flu-related hospitalizations in the United States were anticipated to exceed 700,000, and most states reported a higher-than-normal number of influenza-related deaths of children and young adults. Several factors, however, helped keep the outbreak from becoming an epidemic of larger proportions. The administration of flu vaccine afforded the population what is called herd immunity (discussed in Section 24.6). Herd immunity takes place when a majority of the population is immunized to a pathogen. The vaccinated individuals decrease the risk that an unvaccinated person will have direct contact with an infected person. At the very least, herd immunity slows the spread of infection throughout the population.

Note: It is impossible to immunize all humans against any given disease. However, it is estimated that immunizing about 80% of a population for influenza can halt an epidemic by cutting off transmission. (See the discussion of herd immunity in Section 24.6.) Unfortunately, this level of immunization is rarely achieved.

Another factor that limited the scope and severity of recent influenza epidemics was the availability of antiviral agents that can limit the disease course. There are now three viral targets. Hemagglutinin is required for virus entry; neuraminidase is needed for viral exit; and a viral endonuclease, the newest target, is critical for viral replication.

Recall that influenza virus (200 nm) is encased in a membrane envelope acquired when the virion buds from an infected cell. As described in Section 11.2, the envelope contains the viral proteins neuraminidase (NA) and hemagglutinin (HA). Spikes of hemagglutinin bind to sialic acid receptors on the host cell and trigger receptor-mediated endocytosis. Once the virus is inside an endosome, proton pumps in the membrane acidify endosomal contents. The structure of hemagglutinin on the viral membrane changes and binds to receptors on the endocytic membrane. The viral and endocytic membranes fuse and the virion is released into the cytoplasm.

For hemagglutinin structure to change, the interior of the enveloped virion must also be acidified. Acidification is mediated by a membrane channel formed by the virus-encoded M2 protein. The drug amantadine (**Fig. 27.30A**) is a specific inhibitor of the influenza M2 protein that prevents M2 channel formation, which, in turn, prevents viral uncoating. Unfortunately, amantadine-resistant strains of influenza have developed, in part because of the widespread use of amantadine by Chinese poultry farmers. As a result, the drug is no longer recommended as a treatment for influenza.

The second target of the influenza virus is the envelope protein neuraminidase. The newer antiflu drugs, such as zanamivir (Relenza) and oseltamivir (Tamiflu), are **neuraminidase inhibitors** that act against types A and B influenza strains (**Fig. 27.30B** and **C**). Sialic acid residues on the host cell surface can bind the hemagglutinin on flu viruses trying to exit the cell. Neuraminidase on the same viral envelope will cleave the cell-surface sialic acid and free the virus. Unencumbered, the virus can leave the cell surface and infect another cell. Neuraminidase inhibitors, however, prevent sialic acid cleavage, causing the virus particles to aggregate at the cell surface and reduce the number of virus particles released. The contributions of different NA and HA genes to the severity of influenza are discussed in **eTopic 27.4**.

The neuraminidase inhibitors, when used within 48 hours of disease onset, decrease shedding and reduce the duration of influenza symptoms by approximately one day. However, flu symptoms generally last only 3–10 days. While this does not sound like a substantial benefit, shortening the course of the flu in the elderly can minimize damage to the lungs, which in turn reduces the chance of developing life-threatening secondary bacterial infections such as pneumonia and bronchitis.

Note: RNA viruses, such as influenza, use RNA-dependent RNA polymerases that lack proofreading capability. The consequence of an error-prone polymerase is a high mutation rate and the generation of antiviral-resistant virions. Widespread use of Tamiflu during the 2008 flu epidemic quickly led to Tamiflu resistance. Fortunately, the mutations that resulted in resistance to Tamiflu also slowed virus replication rate. Instituting more judicious use of the drug caused the resistant mutants to lose competitive advantage over the faster-replicating sensitive mutants, and the resistant mutants disappeared.

Target Virus Cap-Snatching

In 2018, the FDA approved a new antiviral agent that is effective against influenza and has a unique mode of action. Host mRNA molecules contain a 5′ cap composed of the modified base 7-methylguanosine, which protects host message from nucleases and is required for translation. As described in Chapter 11, influenza virus steals host 5′ caps (cap-snatching), along with 10–20 downstream nucleotides, using a cap-dependent endonuclease.

Once the cap is stolen, it is used by RNA polymerase to initiate transcription of virus mRNA from the negative-sense RNA strand of the infecting flu virus. The stolen cap is also required for translation of the viral mRNA. The new drug, called baloxavir marboxil (sold as Xofluza) inhibits the activity of the influenza cap–dependent endonuclease, which will prevent cap-snatching and stop viral replication. Xofluza is effective against the growing number of flu viruses that

A. Amantadine

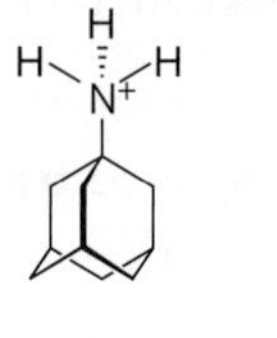

B. Zanamivir

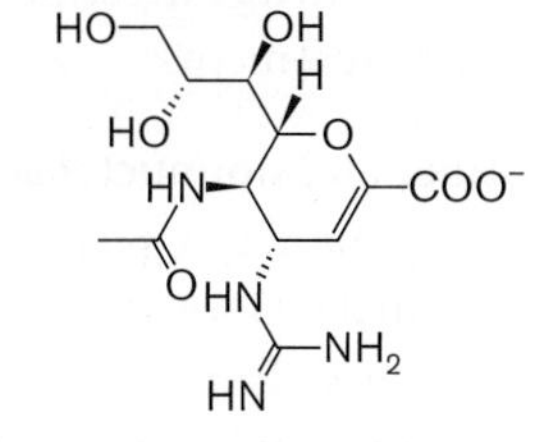

C.

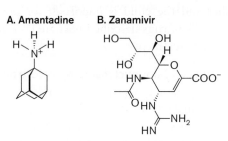

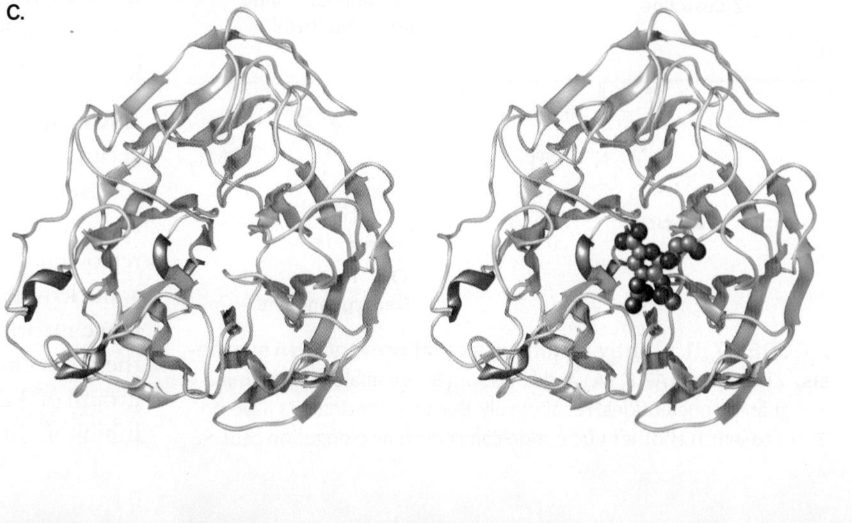

FIGURE 27.30 ▪ Inhibitors of influenza proteins. **A.** Amantadine inhibits the M2 protein. **B.** Zanamivir inhibits neuraminidase. **C.** Neuraminidase without (left) and with (right) bound inhibitor. (PDB: 2HTQ)

are resistant to M2 and neuraminidase inhibitors, and it has been approved to treat influenza in adolescents and adults.

Targeting Viral DNA Synthesis

Most antiviral agents work by inhibiting viral DNA synthesis. These drugs chemically resemble normal DNA nucleosides, molecules containing deoxyribose and analogs of adenine, guanine, cytosine, or thymine. Viral enzymes then add phosphate groups to these deoxynucleoside analogs to form deoxynucleotide analogs. The deoxynucleotide analogs are then inserted into the growing viral DNA strand in place of a normal nucleotide. Once inserted, however, new nucleotides cannot attach to the nucleotide analogs, and DNA synthesis stops.

These DNA chain–terminating analogs (**Fig. 27.31**) are selectively toxic because viral polymerases are more prone to incorporate nucleotide analogs into their nucleic acid than are the more selective host cell polymerases. Antiviral DNA synthesis inhibitors work on DNA viruses or retroviruses, but not on viruses such as influenza, with its RNA genome.

Antiretroviral Therapy

Retroviruses are RNA viruses that use viral reverse transcriptase to make DNA and then use viral integrase to insert that DNA into the eukaryotic host cell genome (see Chapter 11). The integrated provirus can then be activated to make retroviral RNA. The retroviral RNA travels to the cytoplasm and directs synthesis of more virus particles. One of the most devastating retroviruses is human immunodeficiency virus (HIV), the cause of acquired immunodeficiency syndrome (AIDS; see Section 11.3).

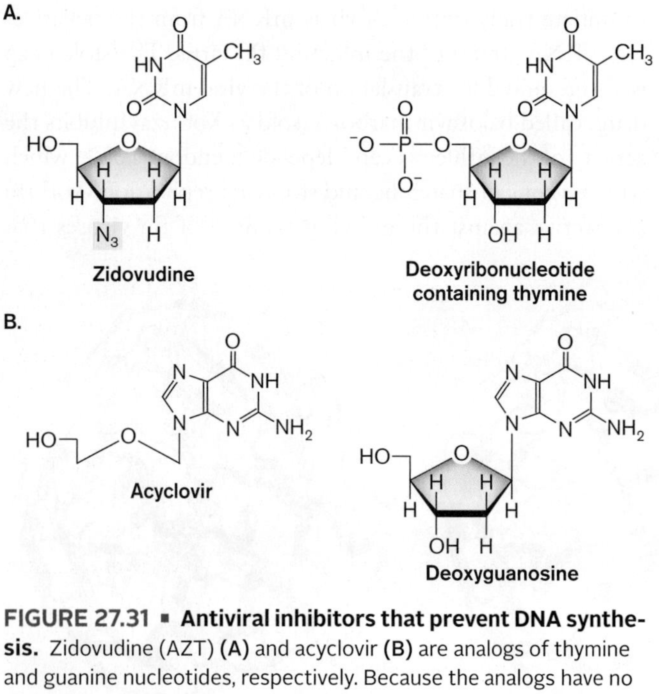

FIGURE 27.31 ▪ Antiviral inhibitors that prevent DNA synthesis. Zidovudine (AZT) **(A)** and acyclovir **(B)** are analogs of thymine and guanine nucleotides, respectively. Because the analogs have no 3′ OH to which another nucleotide can add, chain elongation ceases.

Case History: Treatment of HIV

A married couple came to the community clinic for prenatal care. He was 20 years old. She was 19 and reportedly 2 months pregnant with her first child. She denied intravenous (IV) drug use or a history of other sexual partners, and she had no history of sexually transmitted infection; however, a routine prenatal HIV antibody screen was reported as positive for HIV-1. Careful questioning of the patient and her husband elicited from him a history of IV drug use 5 years earlier. An HIV antibody screen for him was also positive. The laboratory results indicated that the wife had a low viral load—that is, less than 1,000 copies per milliliter of blood—and a high CD4 T-cell count. The husband had a higher HIV viral load (10,000 copies per milliliter) and a lower CD4 count (150 cells per microliter). Antiretroviral therapy was initiated for both husband and wife (even though her CD4 count was high), but because she was pregnant, the wife's regimen avoided drugs that could potentially harm the fetus. Their regimens included two nucleotide reverse transcriptase inhibitors plus a protease inhibitor.

Being diagnosed with HIV is no longer a death sentence. Advances in antiretroviral therapy (ART) over the past 10 years have transformed HIV into a manageable chronic condition. At least half, if not most, of the HIV-positive people in the United States now live long enough to die from diseases of aging, such as heart attacks or strokes. Because it is such a treatable infection, the CDC recommends that everyone should be tested for HIV, and further recommends that everyone with HIV, regardless of their CD4 T-cell count, receive antiretroviral therapy.

The reasons for these recommendations are that many people do not know they are infected, because they are asymptomatic, but treating an asymptomatic, HIV-positive person will reduce the risk of sexually transmitting the virus. Antiretroviral therapy can also prevent transplacental transmission from an asymptomatic, HIV-positive pregnant woman to her fetus, as in our case history. Because HIV transmission from the mother to the neonate can also occur at delivery or by breast-feeding, postdelivery treatment of the mother and the child is important for preventing transmission. The drugs described next are crucial components of effective treatment.

Nucleoside, nucleotide, and nonnucleoside reverse transcriptase inhibitors. As an RNA retrovirus, HIV uses a reverse transcriptase to make DNA that then integrates into host nuclear DNA (discussed in Section 11.3). The antiretroviral drug zidovudine (abbreviated ZDV or AZT) is a nucleoside analog recognized by reverse transcriptase. Once incorporated into a replicating HIV DNA molecule, the DNA chain–terminating property (lack of a 3′ hydroxyl group) of AZT prevents further DNA synthesis. Nucleoside inhibitors must undergo three successive phosphorylation steps inside the cell to become an active trinucleotide form of the drug. Nucleotide inhibitors are essentially monophosphorylated analogs that require only two phosphorylation steps for activation. Tenofovir is another nucleotide analog (similar to AMP) that is recognized by HIV reverse transcriptase and also by the reverse transcriptase of hepatitis B virus. Tenofovir is often given following needle stick accidents to prevent HIV or HBV transmission.

In addition to nucleoside inhibitors, there are nonnucleoside reverse transcriptase inhibitors. For example, the drug delavirdine binds directly to reverse transcriptase and allosterically inactivates the enzyme.

Protease inhibitors. To make optimal use of its limited provirus DNA sequence, HIV generates long, nonfunctional polypeptide chains that are proteolytically cleaved into the actual proteins and enzymes used to replicate and produce new virions. For example, the *gag* and *pol* genes reside next to each other in the HIV genome and are transcribed as a single mRNA molecule (see Section 11.3). The Gag and Pol open reading frames overlap but are offset by one base. This mRNA produces two polyproteins, called Gag and Gag-Pol, the latter being the result of a shift in reading frames that takes place during translation. Once made, both polyproteins are cleaved by HIV protease. The Gag protein is proteolytically cleaved to make different capsid components (p17, p24, and p15, the latter of which is further cleaved to make nucleocapsid protein p7; **Fig. 27.32A**). Gag-Pol is cleaved to make reverse transcriptase and integrase.

Protease inhibitors such as Viracept and Lopinavir belong to a powerful class of drugs that block the HIV protease (**Fig. 27.32B**). When the protease is inactivated, the polyproteins remain uncleaved and the virus cannot mature, even though new virus particles are made. Because immature HIV particles cannot infect other cells, progress of the disease stalls. Note that protease inhibitors do not cure AIDS; they can only decrease the number of infectious copies of HIV.

Entry inhibitors. Another way to stop HIV is to prevent the virus from infecting cells in the first place. Drugs called entry inhibitors do just that: They stop entry. There are two types of entry inhibitors. CCR5 inhibitors block virus envelope protein gp120 (also known as SU) from binding to host surface protein CCR5, a coreceptor that, together with CD4, is needed for virus binding (see Chapter 11 and Fig. 11.24 for details). As a result of CCR5 inhibition, the virus never attaches. Fusion inhibitors, in contrast, do not prevent initial binding; they prevent HIV membranes from fusing with T-cell membranes and thereby halt viral entry. Imagine trying to enter a room through a door. A CCR5 inhibitor is like removing the doorknob from the door so that there is nothing to grab. Fusion inhibitors, however, are like gluing the door shut. You can grab the knob but still cannot open the door.

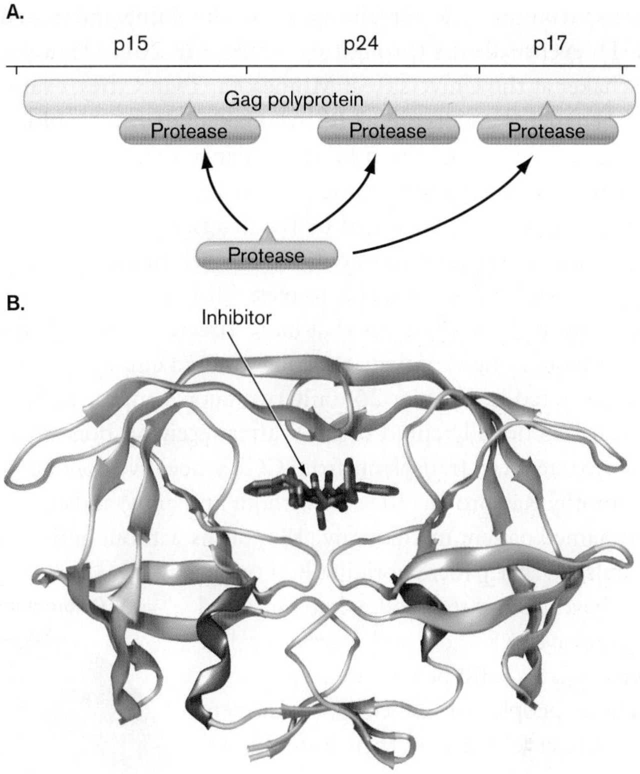

FIGURE 27.32 ■ HIV protease inhibitor. **A.** Representation of HIV protease cleavage of a single Gag polyprotein into multiple, smaller proteins. **B.** The protease enzyme is shown here as a ribbon structure, while the protease inhibitor BEA 369 is shown as a stick model. (PDB code: 1EBY)

A new member of the entry inhibitor class of HIV drugs is actually a postattachment inhibitor. The monoclonal antibody ibalizumab binds to, and blocks, the host cell coreceptors for HIV called CCR5 and CXCR4. To enter a cell, HIV must bind to CD4 and one of these other two coreceptors. The FDA approved this drug in 2018 to treat patients with multidrug-resistant HIV.

Treatment regimens and HIV controllers. Because HIV can mutate rapidly and become resistant to single-drug therapies (see Section 11.3), HIV treatments today involve combinations of three or more antiretroviral drugs. This therapeutic strategy was originally called highly active antiretroviral therapy (HAART), but the name has been changed to simply **antiretroviral therapy (ART)**. Current ART regimens include three drugs—usually two nucleoside reverse transcriptase inhibitors, plus a protease inhibitor, a nonnucleoside reverse transcriptase inhibitor, or an integrase inhibitor. Integrase inhibitors block the enzyme needed to insert viral DNA into the host genome.

You might wonder why HIV cannot be eliminated from an infected person if the available antiretroviral drugs are so effective. One reason is that HIV is a retrovirus whose cDNA genome has integrated into host genomes. But if ART prevents HIV from replicating and spreading to new cells, wouldn't the remaining cells containing integrated HIV eventually die through senescence? In 2011, Timothy Schaker at the University of Minnesota, Twin Cities, examined patients undergoing ART who had undetectable blood levels of HIV. His group found evidence that the virus in these individuals still remained trapped in lymphatic tissues that were poorly penetrated by the drugs. So, even though ART can lower HIV to undetectable levels in blood, tissue pockets of HIV remain, able to reestablish infection if ART is stopped. New strategies that more effectively force drugs into tissues might provide the long-awaited cure.

As noted in Chapter 26, only two people appear to have been functionally cured of HIV after receiving bone marrow or stem cell transplants from CCR5-negative donors: Timothy Ray Brown (10 years without antivirals) and an unnamed patient in Germany (18 months without antivirals). Other patients originally thought to have been cured following early and aggressive ART do still have very low viral loads (<400 per milliliter of blood). These people are called posttreatment controllers. Somehow their immune systems are able to keep the virus in check. Another group of individuals, called elite controllers, manage to prevent the progression of HIV in the absence of any treatment—possibly through the presence of newly discovered inhibitory receptors on their CD4 T cells.

HIV treatment as prevention. AIDS can be a devastating disease. Fortunately, we have effective antivirals that can prevent HIV replication and AIDS. Could treating at-risk populations with antivirals before exposure be effective at preventing infection? This strategy is known as preexposure prophylaxis (PrEP). In fact, the FDA has approved the daily use of an HIV medicine, Truvada (tenofovir/emtricitabine), by healthy but high-risk people hoping to lower their risk of infection by a sexual partner. Both drugs are nucleoside analogs of adenosine and cytosine, respectively, and they are reverse transcriptase inhibitors. Although this is an approved strategy, it is controversial even among physicians. Some fear that the drug will encourage risky behavior.

Future Antivirals May Target Host Functions

As you can see from the preceding discussion, most antiviral drugs approved for clinical use target viral proteins (proteases, polymerases, entry proteins) because they afford some measure of selective toxicity. However, a perceived limitation of these direct-acting antivirals is their narrow spectrum of virus coverage. Narrow-spectrum antivirals cannot provide adequate protection against newer, rapidly emerging viral threats. Examples include the flavivirus dengue, coronaviruses SARS-CoV and MERS-CoV, and the filovirus Ebola. Finding new broad-spectrum antivirals could address this need. But what do we target?

A novel approach is to ignore the virus and, instead, target host cell pathways that are required by multiple viruses for replication (**Fig. 27.33**). For example, cyclophilin A is involved in host and viral protein folding. Cyclophilin A inhibitors such as

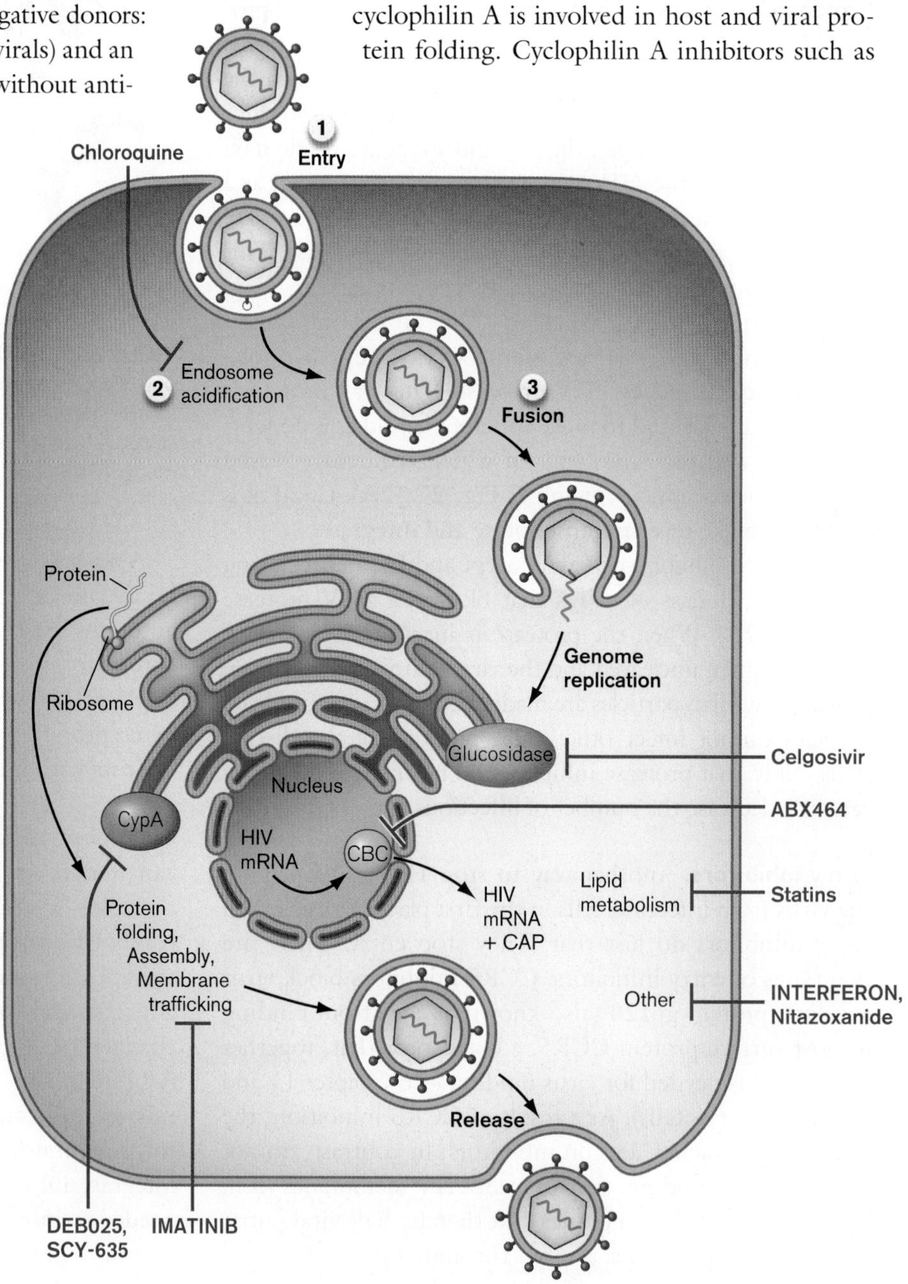

FIGURE 27.33 ■ Potential host targets for broad-spectrum antivirals. Different stages of viral development are shown (entry, fusion, genome replication, assembly, release). Examples of broad-spectrum compounds are connected to the corresponding targeted proteins or pathways by blunt arrows. CypA = cyclophilin A.

alisporivir (DEB025) impair the folding of viral proteins and augment innate immune responses. These drugs affect a variety of DNA and RNA viruses (dengue, hepatitis C, HIV, SARS-CoV). Another potential target is the enzyme alpha-glucosidase in the endoplasmic reticulum. Many virus glycoproteins depend on host glucosidases for proper folding. The glucosidase inhibitor celgosivir has, indeed, proved effective against many unrelated viruses in vitro and in rodent models, but not yet in humans. Nevertheless, the enzyme warrants further investigation. Host kinases that regulate intracellular virus trafficking are also potential drug targets.

A more pathogen-specific antiviral agent (ABX464) affects HIV replication by binding to the host cap-binding complex (CBC). ("Cap" refers to the 5′ 7-methylguanosine cap added to eukaryotic and viral mRNAs.) The CBC controls the export of host mRNA from the nucleus to the cytoplasm, where mRNA is translated to protein. The HIV Rev protein brings HIV mRNA to the CBC. ABX464 binds to the CBC and specifically blocks Rev-mediated export of viral RNA, while not interfering with host transcripts. Because the drug binds to a host protein rather than a viral protein, HIV is less likely to develop resistance. This antiviral agent is currently in clinical trials.

Figure 27.33 also shows some host-targeted drugs already approved for other purposes, such as chloroquine (antimalarial agent) and statins (anticholesterol metabolism). These drugs are now being evaluated for their usefulness against emerging viruses. Chloroquine inhibits endosome acidification, which is needed by some enveloped viruses to escape the vacuole and enter the host cytoplasm (dengue virus and Zika virus, for instance). Statins interfere with lipid metabolism needed for the life cycles of viruses such as hepatitis C virus. How therapeutically effective these drugs will be as antiviral agents is unclear.

A particularly exciting strategy is predicted to provide broad antiviral protection. The idea is to stimulate a specific group of pattern recognition receptors (PRRs) called Rig-1-like receptors (RLRs). When stimulated, RLRs activate interferon production. Interferon, as described in Chapter 23, signals neighboring cells to express a series of host proteins that degrade viral nucleic acids and inhibit viral protein synthesis. A compound capable of modulating RLRs would be a broad-spectrum antiviral able to thwart replication of many different viruses. For example, short-nucleotide RNA molecules that retain the 5′ triphosphate end will activate RLRs and protect cells against vesicular stomatitis virus, vaccinia virus, dengue virus, and influenza virus. However, delivering these molecules to a host is challenging. Naked RNA molecules cannot be taken up by host cells, so they may have to be attached to nanoparticles or wrapped within lipid vesicles to be introduced into cells.

There are many challenges to the host-as-target approach to antiviral therapy. Host proteins function in a complex network of interactions, so elucidating drug mechanisms of action is difficult. Toxicity is another worry, although it may be possible to identify a therapeutic concentration window in which the drug inhibits virus replication but has minimal toxic effects on the patient.

Thought Question

27.13 The text states that cells and viruses would have difficulty developing resistance to antiviral drugs that target host proteins rather than viral proteins. Why would targeting a host protein decrease the likelihood of developing resistance? Propose a mechanism by which the drug could still become ineffective against a virus.

To Summarize

- **Fewer antiviral agents** than antibacterial agents are available because it is harder to identify viral targets that provide selective toxicity.
- **Preventing viral attachment to, or release from, host cells** is a mechanism of action for antiviral agents, such as amantadine and zanamivir, used to treat influenza virus.
- **Inhibiting DNA synthesis** is the mode of action for most antiviral agents, although it works only for DNA viruses and retroviruses.
- **HIV treatments** include reverse transcriptase inhibitors that prevent the synthesis of DNA, protease inhibitors that prevent the maturation of viral polyproteins into active forms, and entry inhibitors that either prevent HIV from binding to host membranes or inhibit fusion of the HIV envelope to the host cell membrane.
- **Future antiviral agents** may target host proteins needed by the virus to replicate or stimulate interferon-dependent innate immune mechanisms that can repel multiple viral pathogens.

27.5 Antifungal Agents

Fungal infections are much more difficult to treat than bacterial infections, in part because fungal physiology is more similar to that of humans than bacterial physiology is. The other reason is that fungi have an efficient drug detoxification system that modifies and inactivates many antibiotics. Thus, to have a fungistatic effect, repeated applications of antifungal agents are necessary to keep the level of unmodified drug above MIC levels.

Case History: Blastomycosis

A 37-year-old male presented to the emergency department of a Florida hospital with persistent fever, malaise, and a painful right-arm mass. He denied trauma to the arm. White blood cell count was elevated at 27,000 per microliter (leukocytosis), and chest X-ray revealed a left-lung infiltrate. Bronchoscopy revealed granulomatous inflammation containing a single yeastlike mass. Incision and drainage were performed on the arm mass, and cultures were obtained. Serum cryptococcal antigen tests were negative, as were tests for Bartonella henselae *and* Toxoplasma. *Cultures from the right arm grew out a fungal form similar to that identified from the bronchoscopy specimens. A tentative diagnosis of* Blastomyces dermatitidis *was confirmed using PCR. The patient was placed on amphotericin B, and his fevers and leukocytosis subsequently subsided. His medication was changed to fluconazole for a recommended duration of 6 months.*

Superficial mycoses (fungal infections) such as athlete's foot, and systemic mycoses such as blastomycosis, require very different treatments. Imidazole-containing drugs (clotrimazole, miconazole) are often used topically in creams for superficial mycoses (**Fig. 27.34A**). Others, such as itraconazole, are administered orally. Superficial mycoses include infections of the skin, hair, and nails, as well as *Candida* infections of moist skin and mucous membranes (for example, vaginal yeast infections). The imidazole-containing drugs appear to disrupt the fungal membrane by inhibiting sterol synthesis. Lamisil (a terbinafine compound) is a different class of agent that selectively inhibits ergosterol synthesis by fungi. Lamisil targets fungi because humans do not make ergosterol or use it in their cell membranes. This antifungal agent is also used to treat superficial mycoses.

More chronic dermatophytic infections typically require another antifungal agent, called **griseofulvin**, produced by a *Penicillium* species (**Fig. 27.34B**). Griseofulvin disrupts the mitotic spindle and derails cell division (called metaphase arrest). This action does not kill the fungus, but as the hair, skin, or nails grow and are replaced, the fungus is shed.

Vaginal yeast infections caused by *Candida* are often treated with nystatin, a polyene antifungal agent synthesized by *Streptomyces* that forms membrane pores in yeast cells (**Fig. 27.34C**). The name "nystatin" came about because two of the people who discovered it worked for the laboratory of the New York State Public Health Department (now the Wadsworth Center).

The serious, sometimes fatal, consequences of systemic mycoses require more aggressive therapy. The drugs used in these instances include **amphotericin B** (produced by *Streptomyces*; **Fig. 27.34D**) and fluconazole. Amphotericin B binds to the sterols in fungal membranes and destroys membrane integrity. It has a high affinity for ergosterol,

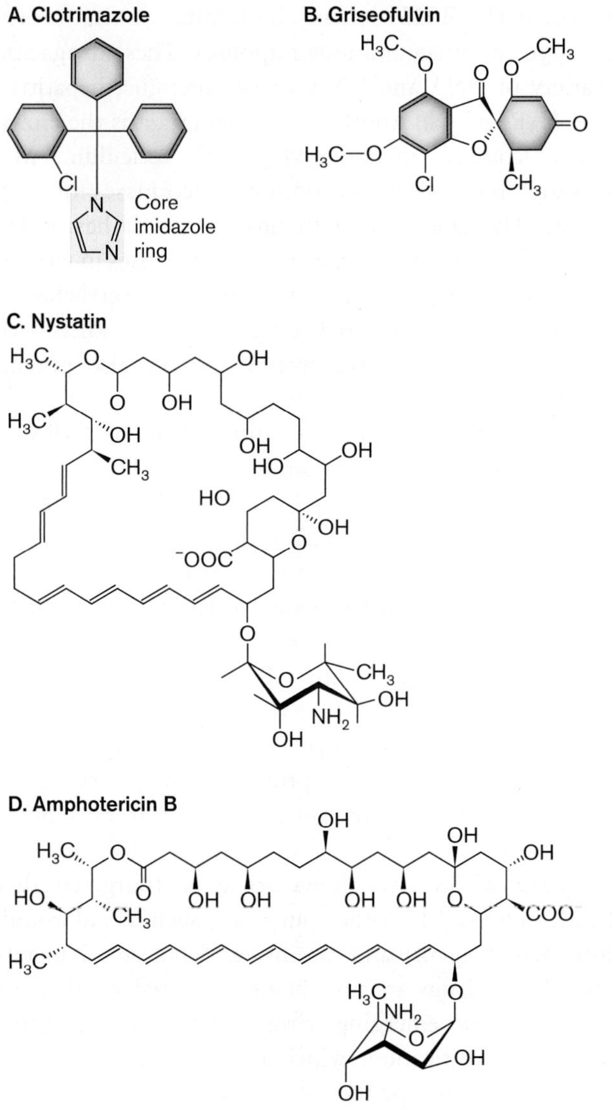

FIGURE 27.34 ■ Examples of antifungal agents. **A.** Clotrimazole belongs to the group of imidazole antifungals, so named because they all contain an imidazole ring. **B.** Griseofulvin is produced by *Penicillium griseofulvum*. **C.** Nystatin is a polyene macrolide produced by *Streptomyces noursei*. **D.** Amphotericin B is a polyene produced by *Streptomyces nodosus*.

which is prevalent in fungal but not mammalian membranes. Fluconazole, on the other hand, inhibits the synthesis of ergosterol. Thus, fungal cells grown in the presence of fluconazole make defective membranes. Typically, curing systemic fungal infections requires long-term treatment to prevent disease relapse. **Table 27.4** lists a number of other commonly used antifungal agents.

A new class of antifungal drugs, called orotomides, was discovered in 2015. These drugs are unique among antifungals because they stop pyrimidine biosynthesis in fungi. The drug specifically inhibits fungal dihydroorotate dehydrogenase, an enzyme that converts dihydroorotate to orotate, a precursor to pyrimidine. These drugs are more effective than other antifungals, especially against

TABLE 27.4 Major Antifungal Agents and Their Common Uses

	Clinical applications[a]								
	Systemic mycoses					Opportunistic mycoses			
Drug/Mode of administration	**Coccidioidomycosis**	**Histoplasmosis**	**Blastomycosis**	**Paracoccidioidomycosis**	**Aspergillosis**	**Candidiasis[b]**	**Cryptococcosis**	**Dermatophytosis**	**Other**
Polyenes (fungicidal)									
Amphotericin B/Intravenous	+	+	+	+	+	+	+	–	
Nystatin/Oral; topical	–	–	–	–	–	+ (mc)	–	–	
Natamycin/Topical; eyedrops	–	–	–	–	+ (superficial)	+ (superficial)	–	–	Mycotic keratins
Azoles (fungistatic): imidazoles									
Clotrimazole/Topical	–	–	–	–	–	+ (mc)	–	+	
Miconazole (Monistat)/Topical	–	–	–	–	–	+ (mc)	–	+	
Ketoconazole/Topical	–	–	+	–	–	+	–	+	Dandruff shampoo
Azoles (fungistatic): triazoles									
Itraconazole/Oral	+	+	+	+	+	+ (mc)	+	+	Sporotrichosis
Fuconazole/Oral	+	+	+	?[c]	–	+	+	+	Sporotrichosis
Voriconazole/Oral	–	–	–	–	+	+	–	–	*Fusarium*
Allylamines (fungicidal)									
Terbinafine (Lamisil)/Topical or oral	–	–	–	–	–	–	–	+	
Griseofulvin (fungistatic)									
Griseofulvin/Oral	–	–	–	–	–	–	–	+	
Echinocandins									
Caspofungin/Intravenous	–	–	–	–	+ (fungistatic)	+ (fungicidal)	–	–	
Antimetabolites (fungistatic or fungicidal)									
5-Fluorocytosine[d] (flucytosine)/Oral	–	–	–	–	+	+	+	–	Phaeohyphomycosis

[a]+ indicates that the drug inhibits growth of the disease-causing agent; – indicates that the drug is not useful for the disease.

[b]mc = mucocutaneous but not systemic candidiasis.

[c]Insufficient data.

[d]Used only in combination with amphotericin B.

Aspergillus, which can cause life-threatening lung infections in immunocompromised patients. As of this writing, one member of this class, F2G, is under development but not yet approved for clinical use.

To Summarize

- **Fungal infections** are difficult to treat because of similarities in human and fungal physiologies.
- **Imidazole-containing** antifungal agents inhibit sterol synthesis.
- **Griseofulvin** inhibits mitotic spindle formation.
- **Nystatin** produces membrane pores.
- **Amphotericin B** binds to membranes and destroys membrane integrity.

Concluding Thoughts

Antibiotics have greatly improved the health and well-being of humans and animals around the globe. Millions of people are alive today because antimicrobial drugs cured them of deadly infectious diseases. Unfortunately, a sense of complacency about infections and antibiotics developed among patients and health care workers. This complacency led to the irresponsible use of antibiotics and to our current antibiotic resistance crisis. Despite considerable effort over the past four decades, very few new, clinically useful, antibiotics with novel mechanisms of action have been discovered, and when one is found, pathogens eventually develop resistance to it. Nevertheless, new approaches to antimicrobial discovery are continually being conceived and tested. With any luck, and with effective antibiotic stewardship, the human race will remain one step ahead of the microbes.

CHAPTER REVIEW

Review Questions

1. What is selective toxicity? Provide examples.
2. Explain the difference between antibiotic susceptibility and antibiotic sensitivity.
3. What does the term "spectrum of antibiotic activity" mean?
4. Provide examples of bacteriostatic and bactericidal antibiotics.
5. What is the Kirby-Bauer test? Does it indicate whether a drug is bacteriostatic or bactericidal?
6. Give examples of drugs that target each of the following: cell wall synthesis, RNA synthesis, protein synthesis, and DNA replication. What are their modes of action?
7. What mechanism do producing organisms use to synthesize peptide antibiotics?
8. How do antibiotic-producing microorganisms prevent "suicide"?
9. Why is antibiotic resistance a growing problem?
10. What are the four basic mechanisms of antibiotic resistance?
11. Explain the basic concept of an MDR efflux pump.
12. Discuss the current concepts of the origin of antibiotic resistance.
13. What are some mechanisms used to combat the development of drug resistance?
14. Why are there few antiviral agents available to treat disease?
15. What is herd immunity?
16. How does oseltamivir inhibit influenza?
17. Discuss the general modes of action of antifungal agents.

Thought Questions

1. A cephalosporin and clindamycin are often used to treat a patient with toxic shock syndrome caused by a TSST-producing *Staphylococcus aureus*. One agent is bactericidal; the other is bacteriostatic. Usually this combination is discouraged because the bacteriostatic agent dampens the effectiveness of the bactericidal agent, which kills only growing cells. Why would an exception to this rule be considered in the treatment of toxic shock?
2. A patient presented to the emergency room complaining of a nonproductive cough (no sputum) that had persisted for 6 weeks. The clinician prescribed a 7-day course of a cephalosporin. After day 7 the patient returned, no better than before he started the

treatment. Laboratory tests later showed that the infection was caused by *Mycoplasma pneumoniae.* Explain why the antibiotic did not work.

3. How would you determine the MIC of an obligate intracellular pathogen such as *Rickettsia prowazekii*, the cause of typhus?

4. You already know that some antimicrobial compounds must be converted by human metabolism into an active drug. Imagine how else an antibiotic might be more effective in vivo than in vitro, even if you do not know of any specific examples of your hypothesis.

Key Terms

amphotericin B (1154)
antibiotic (1116)
antiretroviral therapy (ART) (1151)
bacitracin (1128)
bactericidal (1119)
bacteriostatic (1119)
chloramphenicol (1133)
cycloserine (1128)
daptomycin (1129)
gramicidin (1129)
griseofulvin (1154)
Kirby-Bauer assay (1121)
lincosamide (1133)
macrolide (1133)
minimal inhibitory concentration (MIC) (1119)
Mueller-Hinton agar (1122)
multidrug resistance (MDR) efflux pump (1138)
neuraminidase inhibitor (1149)
oxazolidinone (1133)
penicillin (1117)
penicillin-binding protein (PBP) (1126)
persister cell (1142)
quinolone (1130)
secondary metabolite (1135)
selective toxicity (1119)
spectrum of activity (1119)
streptogramin (1133)
sulfa drug (1118)
transglycosylase (1125)
transpeptidase (1125)
vancomycin (1129)
zone of inhibition (1121)

Recommended Reading

Cameron, David R., Yue Shan, Eliza A. Zalis, Vincent Isabella, and Kim Lewis. 2018. A Genetic determinant of persister cell formation in bacterial pathogens. *Journal of Bacteriology* **200**:e00303-18

Clemente, Jose C., Erica C. Pehrsson, Martin J. Blaser, Kuldip Sandhu, Zhan Gao, et al. 2015. The microbiome of uncontacted Amerindians. *Science Advances* **1**:e500183.

Conlon, Brian P., Sarah E. Rowe, Autumn Brown Gandt, Austin S. Nuxoll, Niles P. Donegan, et al. 2016. Persister formation in *Staphylococcus aureus* is associated with ATP depletion. *Nature Microbiology* **1**:art. 16051.

Crofts, Terence S., Andrew J. Gasparrini, and Gautam Dantas. 2017. Next-generation approaches to understand and combat the antibiotic resistome. *Nature Reviews. Microbiology* **15**:422–434.

Defraine, Valerie, Maarten Fauvart, and Jan Michiels. 2018. Fighting bacterial persistence: Current and emerging anti-persister strategies and therapeutics. *Drug Resistance Updates* **38**:12–26.

DiGiandomenico, Antonio, and Brett R. Sellman. 2015. Antibacterial monoclonal antibodies: The next generation? *Current Opinion in Microbiology* 27:78–85.

Fleming-Dutra, Katherine E., Adam L. Hersh, Daniel J. Shapiro, Monina Bartoces, Eva A. Enns, et al. 2016. Prevalence of inappropriate antibiotic prescriptions among US ambulatory care visits, 2010–2011. *Journal of the American Medical Association* **315**:1864–1873.

Flentie, Kelly, Gregory A. Harrison, Hasan Tükenmez, Jonathan Livny, James A. D. Good, et al. 2019. Chemical disarming of isoniazid resistance in *Mycobacterium tuberculosis. Proceedings of the National Academy of Sciences USA* **116**:10510–10517.

Hauser, Alan R., Joan Mecsas, and Donald T. Moir. 2016. Beyond antibiotics: New therapeutic approaches for bacterial infections. *Clinical Infectious Diseases* **63**:89–95. https://doi.org/10.1093/cid/ciw200.

Hayden, Frederick G., Norio Sugaya, Nobuo Hirotsu, Nelson Lee, Menno D. de Jong, et al. 2018. Baloxavir marboxil for uncomplicated influenza in adults and adolescents. *New England Journal of Medicine* **379**:913–923.

Jung, Eric H., David J. Meyers, Jürgen Bosch, and Arturo Casadevall. 2018. Novel antifungal compounds discovered in medicines for malaria venture's malaria box. *mSphere* **3**:e00537-17.

Kim, Wooseong, Wenpeng Zhu, Gabriel Lambert Hendricks, Daria Van Tyne, Andrew D. Steele, et al. 2018. A new class of synthetic retinoid antibiotics effective against bacterial persisters. *Nature* **556**:103–107.

Koo, Hyun, Raymond N. Allan, Robert P. Howlin, Paul Stoodley, and Luanne Hall-Stoodley. 2017. Targeting microbial biofilms: Current and prospective therapeutic strategies. *Nature Reviews. Microbiology* **15**:740–755.

Munita, Jose M., and Cesar A. Arias. 2016. Mechanisms of antibiotic resistance, p. 481–511. *In* Indira T. Kudva, Nancy A. Cornick, Paul J. Plummer, Qijing Zhang, Tracy L. Nicholson, et al. (eds.), *Virulence Mechanisms of Bacterial Pathogens*, 5th ed. ASM Press, Washington, DC. https://doi.org/10.1128/microbiolspec.VMBF-0016-2015.

Pursey, Elizabeth, David Sünderhauf, William H. Gaze, Edze R. Westra, and Stineke van Houte. 2018. CRISPR-Cas antimicrobials: Challenges and future prospects. *PLoS Pathogens* **14**:e1006990.

Sharkey, Liam K. R., Thomas A. Edwards, and Alex J. O'Neill. 2016. ABC-F proteins mediate antibiotic resistance through ribosomal protection. *mBio* **7**:e01975-15.

Van Acker, Heleen, and Tom Coenye. 2017. The role of reactive oxygen species in antibiotic-mediated killing of bacteria. *Trends in Microbiology* **25**:456–466.

Yong, Hui Yee, and Dahai Luo. 2018. RIG-I-like receptors as novel targets for pan-antivirals and vaccine adjuvants against emerging and re-emerging viral infections. *Frontiers in Immunology* **9**:1379. https://doi.org/10.3389/fimmu.2018.01379.

Yount, Nannette Y., and Michael R. Yeaman. 2012. Emerging themes and therapeutic prospects for anti-infective peptides. *Annual Review of Pharmacology and Toxicology* **52**:337–360.

Zipperer, Alexander, Martin C. Konnerth, Claudia Laux, Anne Berscheid, Daniela J. Anek, et al. 2016. Human commensals producing a novel antibiotic impair pathogen colonization. *Nature* **535**:511–516.

CHAPTER 28
Clinical Microbiology and Epidemiology

28.1 Clinical Specimen Collection and Handling

28.2 Pathogen Identification by Culture and Phenotype

28.3 Molecular and Serological Identification of Pathogens

28.4 Epidemiology

28.5 Detecting Emerging Microbial Diseases

Over the course of human history more people have died from infectious diseases than by all wars combined. Today, global health agencies can quickly detect disease outbreaks and swiftly implement containment measures that limit death. To identify outbreaks, the global agencies monitor data from regional clinical microbiology laboratories that help clinicians diagnose infectious diseases. Here in Chapter 28 we discuss the principles of clinical microbiology and epidemiology that are used to identify, treat, and contain outbreaks, and to predict emerging diseases.

CURRENT RESEARCH highlight

Emerging pathogen? Botulism neurotoxin (BoNT), made by *Clostridium botulinum*, can produce flaccid paralysis and death in humans, if ingested. Fortunately, BoNT genes were found only in *Clostridium* . . . until now. Min Dong and Andrew Doxey discovered genes for a BoNT-like toxin (called BoNT/En) in a strain of *Enterococcus faecium* isolated from cow feces. *E. faecium*, however, is also part of the human gut microbiome. Luckily, BoNT/En toxin does not affect humans (or mice, as shown here), but a chimeric toxin combining the toxic component of enterococcal BoNT/En with the transport subunit of human BoNT can paralyze a mouse leg (lower right). Could a version of BoNT/En toxin affecting humans evolve in *E. faecium* and enter the human microbiome?

Source: Modified from Zhang et al. 2018. *Cell Host Microbe* **23**:1–8, fig. 1N

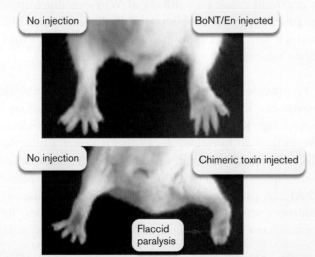

AN INTERVIEW WITH

MIN DONG, MD, HARVARD MEDICAL SCHOOL

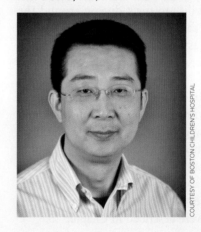

Have you found BoNT/En-containing strains in the human microbiome yet? Should we be concerned?

We have not yet detected BoNT/En-containing strains in human microbiomes. However, we have identified genes resembling the BoNT/En genes in insect metagenomes. This discovery, together with the observed low toxicity of BoNT/En toward human neurons, suggests that BoNT/En may have a nonhuman host specificity. Although there is little evidence suggesting that BoNT/En is a human health risk, the potential for neurotoxin gene transfer between *Clostridium* and other strains/species is still a legitimate concern and outlines the need for continual pathogen biomonitoring by DNA sequencing.

On January 12, 2010, the small country of Haiti on the Caribbean island of Hispaniola was violently shaken by a magnitude 7.0 earthquake. Over 100,000 people died, thousands of buildings collapsed, and the already fragile sanitation system was severely crippled. Almost immediately, volunteers from around the world poured in to help rescue the living, recover the dead, and rebuild what little infrastructure remained. Then, in October, a cholera epidemic struck. Over the next 5 years, 700,000 cases of severe diarrhea (see Chapter 25) ravaged the populace, and more than 9,000 died. Haiti had never experienced cholera before. Where did it come from?

Extensive epidemiological studies relying on clinical laboratory diagnostics and whole-genome sequencing strategies eventually traced the Haitian strain of *Vibrio cholerae* halfway around the world to a September 2010 cholera outbreak in Nepal. How did this faraway strain make it to Haiti so fast? Unwittingly, the United Nations sent troops infected with the organism from Nepal to Haiti to assist in the recovery. The Nepalese troops set up camp and used a nearby river for sanitation. The river became contaminated with *V. cholerae* and carried the pathogen through the town of Mirebalais, the site of the first reported cholera case. From there, the infection spread.

Such compelling stories underscore the integral roles that clinical microbiology and epidemiology play in health care. The Haitian cholera epidemic demonstrates how modern methods of laboratory diagnostics and DNA analysis have improved our ability to track disease around the world. These techniques can even uncover new emerging pathogens as they evolve, as Min Dong and Andrew Doxey (**Fig. 28.1**) have demonstrated (see the Current Research Highlight).

We start Chapter 28 by explaining the basic strategies and methodologies used to diagnose infectious diseases, and we end by discussing how epidemiologists track epidemics and identify emerging pathogens. Our goal is not to catalog every infectious disease, but to demonstrate general principles and problem-solving approaches. Recall that basic concepts of infectious disease—such as transmission, vectors, vehicles, and reservoirs—were described in Chapter 25.

28.1 Clinical Specimen Collection and Handling

As in any good detective mystery, the first step in investigating an infectious disease is to identify the most likely suspects. Presented with a patient, a clinician can develop a list of possible suspects (called the differential) by observing the patient's symptoms, and then recall which organisms can produce those symptoms. Awareness of a similar disease outbreak under way in the community is also helpful.

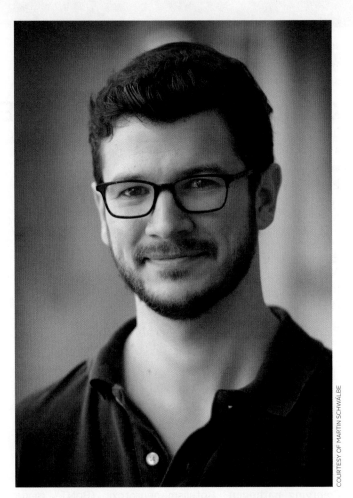

FIGURE 28.1 ▪ **Andrew Doxey, University of Waterloo.** Doxey collaborated with Min Dong to find the botulism toxin–like gene cluster in *Enterococcus faecium*.

Beyond these clues, the clinician must rely on biochemical, molecular, serological, or antigen detection strategies to identify the etiological agent.

Why Take the Time to Identify an Infectious Agent?

Do we really need to identify the genus and species of an organism causing an infection? Why not simply treat the patient with an antibiotic and be done with it? This approach may sound appealing, but there are several compelling reasons to identify an infectious agent.

1. To provide effective treatment and limit antibiotic resistance. "Know your enemy" is a good rule of war, and of medicine. Knowing the pathogen informs the clinician how to treat an infection. Recall that different antibiotics affect different microorganisms; that is, each antibiotic has its own spectrum of activity. Simply being aware of whether a bacterial pathogen is Gram-positive or Gram-negative influences which antibiotic will be used. Macrolides, for instance, are effective against most Gram-positive bacteria, but only a few Gram-negatives. And viruses are not at all susceptible to antibiotics. Using antibiotics to treat a patient with a viral disease will not effect a cure but can select for antibiotic resistance in the microbiome.

Identifying any bacterial disease agent usually includes characterizing its antibiotic resistance profile, which informs the choice of therapy. Antibiotic resistance profiles also help global health organizations track the spread of antibiotic-resistant strains around the world. For example, before 1970 most strains of *Neisseria gonorrhoeae* were susceptible to penicillin. Today, most strains are resistant to penicillin and to many other antibiotics.

2. To prevent pathogen-specific disease complications. Many diseases have serious complications that are common to a given organism or strain of organism. For example, children whose sore throats are caused by certain strains of *Streptococcus pyogenes* can develop serious complications long after the infection has resolved. These complications, called **sequelae** (singular, **sequela**) because they occur after the infection itself is over, are the immunological consequence of bacterial and host antigen cross-reactivity. Life-threatening sequelae, such as rheumatic fever and acute glomerulonephritis caused by certain strains of *S. pyogenes*, produce severe damage to the heart and kidney, respectively. Knowing early on that *S. pyogenes* has caused a child's sore throat enables the physician to prescribe penicillin or penicillin-like antibiotics that will quickly eradicate the infection and prevent the development of sequelae.

Note: Contrary to what you might expect, *Streptococcus pyogenes* has not evolved to become penicillin resistant. It is not understood why.

3. To track disease spread through a population. Consider a situation in which ten infants scattered throughout a city develop bloody diarrhea. The clinical laboratory identifies *Shigella sonnei*, a Gram-negative bacillus, as the cause in each case. Are the cases linked in some way?

Clinical microbiology labs use immunological or nucleic acid amplification tests to subtype each diarrheal isolate (these methods are discussed later). Finding the same strain (or subtype) of *S. sonnei* in all of the cases would suggest that they come from the same source. Carriers of *Shigella* shed this organism in their feces, but inadequate hand-washing after defecation can leave bacteria on hands. Contaminated hands can then transfer the pathogen to foods, utensils, or directly to another person. The challenge is to find the source.

Informed that all of the infants have the same strain of *Shigella*, public health officials question the parents and learn that all of the children attend the same day-care center. By testing the other children and the workers in that center, officials can confirm the source and stop the infection from spreading. This investigative process is called **epidemiology** (covered in Section 28.4).

Specimen Collection and Processing

To diagnose an infectious disease, health care workers must collect, and the laboratory must process, a wide variety of clinical specimens. The types of specimens range from simple cotton swabs of sore throats to urine and fecal samples. Here we describe how these samples should be collected. Note that, upon receiving and processing these specimens, the clinical microbiologist must wear protective gloves and use a laminar flow biosafety hood (see Fig. 5.28) as protection against self-inoculation with potential infectious agents.

Some body sites should not contain any microorganisms (they are sterile) when collected from a healthy individual. These include blood, cerebrospinal fluid (CSF), pleural fluid from the space that lines the outside of the lungs, synovial fluid from joints, and peritoneal fluid from the abdominal cavity. Because these sites are normally sterile, specimens can be plated onto nonselective agar media as well as selective media. Nonselective media, such as blood or chocolate agars, can be used because any organism found in these specimens is considered significant.

Here is how samples are collected from sterile body sites:

Cerebrospinal fluid (CSF). Lumbar puncture (spinal tap; Fig. 28.2A) is used to collect CSF. A long, thin needle is inserted between lumbar vertebrae L3 and L5, which are safely located below the end of the spinal cord. Spinal fluid that slowly drips out of the needle is collected in a sterile tube.

Blood samples. Blood is generally taken by syringe from two body sites and placed in liquid media for aerobic and anaerobic culture. An organism isolated from both sites is considered the likely etiological agent.

Pleural, synovial, and peritoneal fluid. Samples are aspirated by syringe. The fluid can be viewed microscopically for inflammatory cells and inoculated to agar plates for aerobic and anaerobic incubation.

Identifying pathogens present at body sites that contain normal microbiota is more challenging. A stool or fecal sample, for instance, is normally teeming with microbiota. These specimens are typically plated onto selective media—for example, MacConkey, Hektoen, or colistin–naladixic acid (CNA) agar (described in Section 28.2)—to eliminate or decrease the number of normal microbiota that might contaminate the specimen. The following techniques are used to collect specimens from nonsterile body sites:

Swabs. Throat swabs (Fig. 28.2B), for example, should be placed in specialized liquid nutrient transport medium.
Sputum. Deep lung secretions are expectorated for oral collection (Fig. 28.2C).

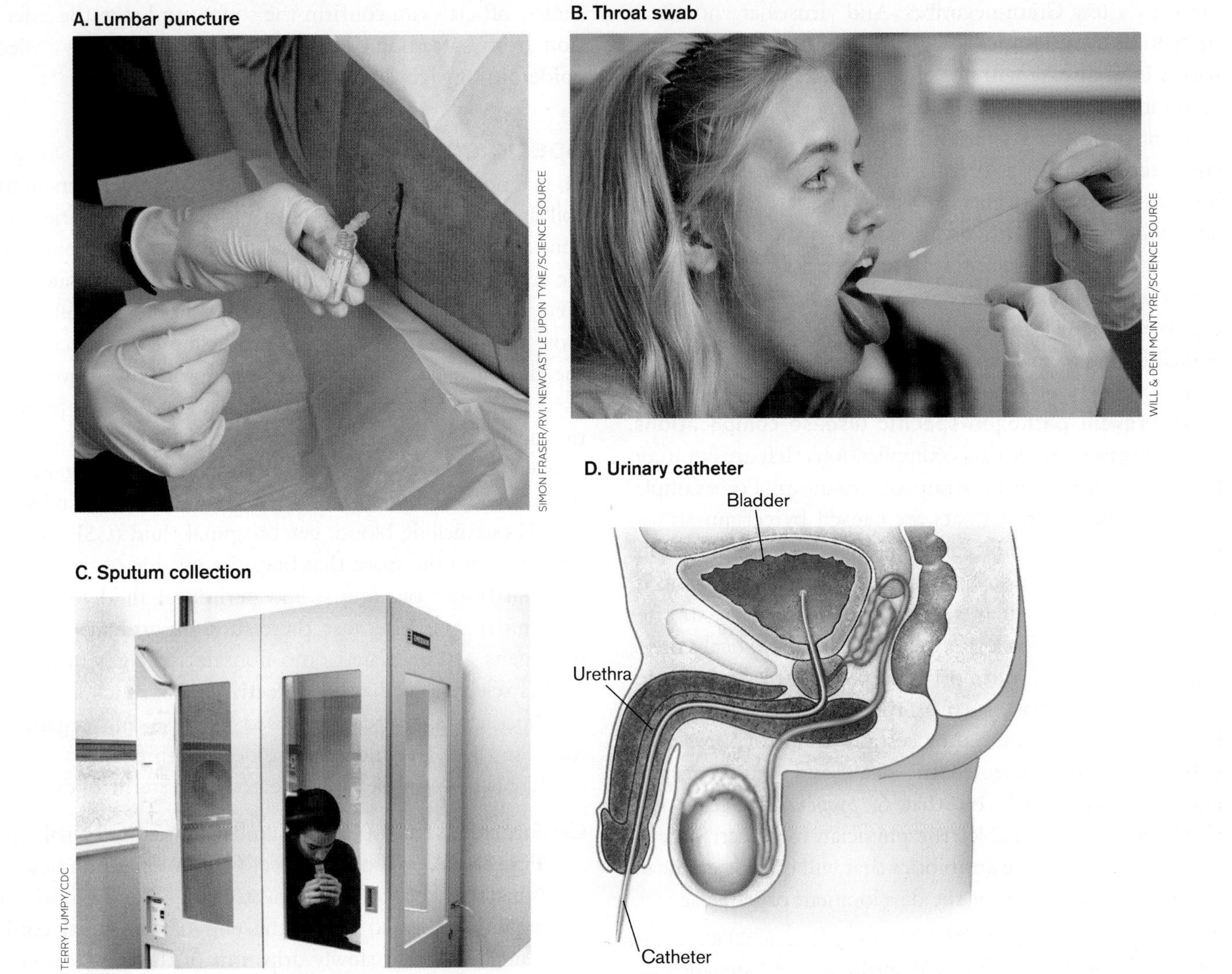

FIGURE 28.2 ■ Specimen collection. **A.** Lumbar puncture to obtain cerebrospinal fluid. **B.** Throat swab. **C.** Sputum collection. A TB patient has coughed up sputum and is spitting it into a sterile container. The patient is sitting in a special sputum collection booth that prevents the spread of tubercle bacilli. The booth is decontaminated between uses. **D.** Urinary catheter, showing placement in the urethra.

Stool samples. Stool is collected by cup or rectal swab, for identifying diarrhea-causing microbes.

Abscesses. Samples from abscesses are collected by needle aspirations.

Urine samples. Samples for diagnosing urinary tract infections (UTIs) are obtained via midstream clean catch (which may contain some microbiota from the urethra) or collected from catheters placed in the bladder (should be sterile).

Urine is a special case. Urine in the bladder of a healthy individual was once considered sterile. Studies now show that the bladder does, indeed, have normal microbiota (at low numbers), although their role in human health is unclear (discussed in Section 26.4). Even though we now know that the bladder contains normal microbiota, for practical purposes clinical microbiologists still consider urine to be "sterile," or nearly so. The symptoms of urinary tract infections (which include increased frequency, painful urination, and suprapubic or flank pain) occur only when significant numbers of easily grown, aerobic or facultative microbes are present.

Urine is sampled in several ways. When collected from a catheterized patient, urine should contain few, if any, aerobically culturable organisms. Catheterization involves passing thin, sterile tubing through the urethra and directly into the bladder (**Fig. 28.2D**). (Catheterization is used primarily to assist urination by immobilized patients, but it also provides a convenient way to collect urine for bacteriological examination.) Unfortunately, the simple process of inserting the catheter through the nonsterile urethra can sometimes introduce organisms into the bladder and precipitate an infection. In addition, urine that will be cultured from a catheterized patient should be collected from a port in the catheter, <u>never</u> from the collection bag. Urine may sit for hours in the collection bag, so organisms initially present at insignificant numbers have time to replicate to high numbers even if the patient does not have a UTI.

When a catheter is not in place, urine is most commonly collected by what is called the midstream clean-catch technique, which is performed by the patient. In this procedure, the external genitals are first cleaned with a sterile wipe containing an antiseptic. The patient then partially urinates to wash as many organisms as possible out of the urethra and, on resuming urination, collects 5–15 milliliters of the midstream urine in a sterile cup. This urine sample will usually not be sterile, because of urethral contamination, but the number of bacteria will be low. The clinical laboratory determines how many organisms per milliliter are present in the midstream catch and tells the physician whether an infection is present. In a symptomatic patient, finding more than 1,000 organisms of a single species per milliliter of midstream clean-catch urine is now considered indicative of an infection in a symptomatic patient and is treated. However, finding 100,000 organisms or more per milliliter in an asymptomatic patient is considered asymptomatic bacteriuria and is usually not treated unless found in a pregnant woman.

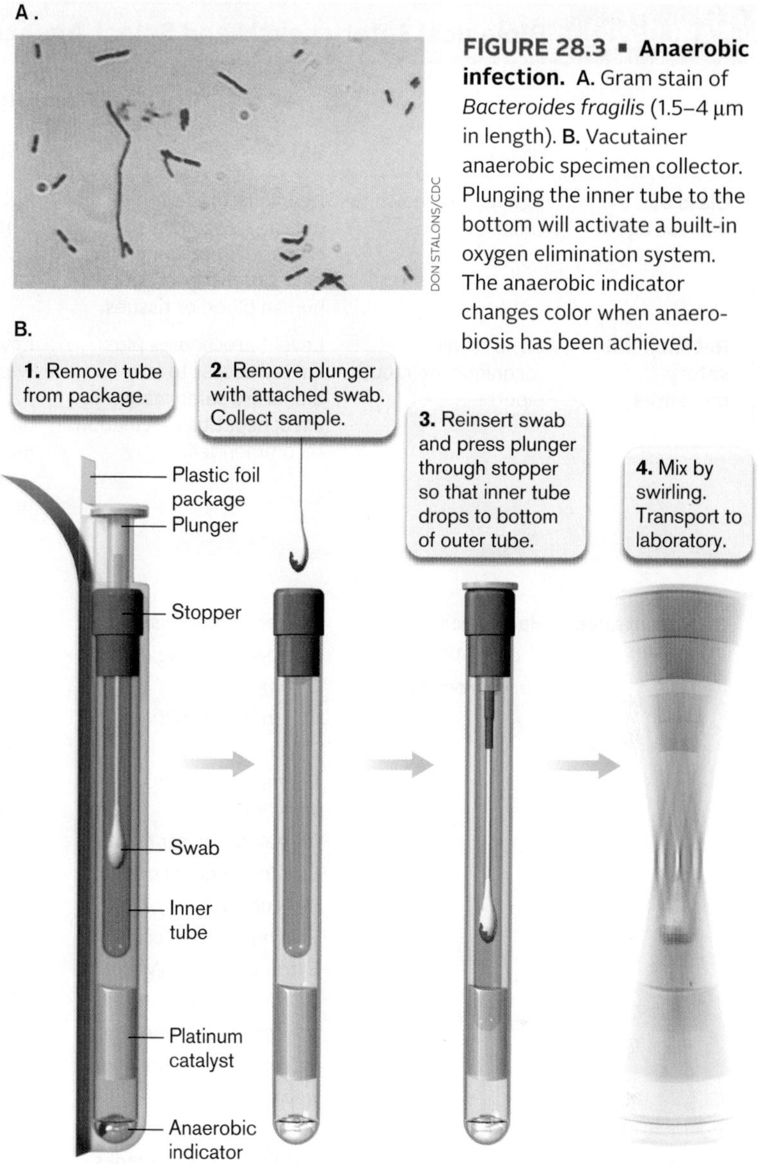

FIGURE 28.3 ▪ Anaerobic infection. **A.** Gram stain of *Bacteroides fragilis* (1.5–4 μm in length). **B.** Vacutainer anaerobic specimen collector. Plunging the inner tube to the bottom will activate a built-in oxygen elimination system. The anaerobic indicator changes color when anaerobiosis has been achieved.

Note: The minimum standard for diagnosing a UTI used to be a finding of greater than 100,000 organisms per milliliter of urine. As noted in the text, this is no longer the case.

Once a specimen from any body site has been properly collected, it must be transported to the clinical laboratory under conditions that will not undermine the viability of the pathogen. Then, after arriving in the lab, the sample must be processed quickly using protocols that ensure the growth of likely pathogens. More on UTIs can be found in Section 26.4.

Case History: Abdominal Abscess

A 4-year-old boy was admitted to the hospital for evaluation and treatment of persistent pain in the rectal area. His problem had begun about a week earlier with ill-defined pain in that area. He had a white blood cell count of 24,900 per microliter (normal is 4,000–11,000) with 87% neutrophils (normal is 60%). An abdominal computed tomography (CT) scan revealed an abscess adjacent to his rectum. A needle aspiration drained 20 milliliters of yellowish, foul-smelling fluid from the abscess. Aerobic cultures of this specimen plated on blood and MacConkey agars were negative. Why didn't the infectious agent grow?

The problem in this instance is related to specimen collection and processing. Internal abscesses located near the gastrointestinal tract are usually anaerobic infections. In this case the infection was caused by the Gram-negative rod *Bacteroides fragilis*, a strict anaerobe (**Fig. 28.3A**). Section 5.4 discusses anaerobes. Intestinal microbes, the majority of which are anaerobic, can sometimes escape the intestine if the organ is damaged in some way. The specimen in this instance should have been collected under anaerobic conditions by aspiration into a nitrogen-filled tube before transport to the clinical laboratory. Alternatively, a swab of the abscess material can be inserted into a special transport tube that has a built-in oxygen elimination system (**Fig. 28.3B**). Because the specimen in the case presented here was collected aerobically, many anaerobic microbes, including *B. fragilis*, were probably killed by the oxygen.

Nevertheless, *B. fragilis* has a stress response system that permits survival of this anaerobe for 1 or 2 days in oxygen,

TABLE 28.1 Biological Safety Levels and Select Agents[a]

	Biosafety level (BSL)			
	BSL-1	**BSL-2**	**BSL-3**	**BSL-4**
Class of disease agent	Agents not known to cause disease.	Agents of moderate potential hazard; also required if personnel may have potential contact with human blood or tissues.	Agents may cause disease by inhalation route.	Dangerous and exotic pathogens with high risk of aerosol transmission; only 13 labs in the United States handle these.
Recommended safety measures	Basic sterile technique; no mouth pipetting.	Level 1 procedures plus limited access to lab; biohazard safety cabinets used; hepatitis vaccination recommended.	Level 2 procedures plus full body wrap around gowns, respiratory protection, ventilation providing directional airflow (ventilation air into room, exhaust air outdoors); restricted access to lab (no unauthorized persons).	Level 3 procedures plus complete clothing change, one-piece positive-pressure suits; lab personnel must shower before leaving; lab is completely isolated from other areas present in the same building or is in a separate building.
Representative organisms in class	*Bacillus subtilis* *E. coli* K-12 *Saccharomyces* spp.	*Bordetella pertussis* *Campylobacter jejuni* *Chlamydia* spp. *Clostridioides difficile* *Clostridium* spp. *Corynebacterium diphtheriae* *Cryptococcus neoformans* *Cryptosporidium parvum* Dengue virus Diarrheagenic *E. coli* *Entamoeba histolytica* *Giardia lamblia* *Haemophilus influenzae* *Helicobacter pylori* Hepatitis virus *Legionella pneumophila* *Listeria monocytogenes* *Mycoplasma pneumoniae* *Neisseria* spp. *Salmonella* spp. *Shigella* spp. *Staphylococcus aureus* *Toxoplasma* Pathogenic *Vibrio* spp. *Yersinia enterocolitica*	*Bacillus anthracis* (anthrax) *Brucella* spp. (brucellosis) *Burkholderia mallei* (glanders) California encephalitis virus *Coxiella burnetii* (Q fever) EEE (eastern equine encephalitis) virus *Francisella tularensis* (tularemia) Japanese encephalitis virus La Crosse encephalitis virus LCM (lymphocytic choriomeningitis) virus *Mycobacterium tuberculosis* Rabies virus *Rickettsia prowazekii* (typhus fever) Rift Valley fever virus SARS (severe acute respiratory syndrome) virus Variola major (smallpox) and other poxviruses VEE (Venezuelan equine encephalitis) virus West Nile virus Yellow fever virus *Yersinia pestis*	Ebola virus Guanarito virus Hantavirus Junin virus Kyasanur Forest disease virus Lassa fever virus Machupo virus Marburg virus Tick-borne encephalitis viruses

[a]Organisms in blue are on the list of CDC select agents that are considered possible agents of bioterrorism. Modified from *Biosafety in Microbiological and Biomedical Laboratories*, 5th ed. Centers for Disease Control and Prevention.

so some of the bacteria may have survived transport. The laboratory still had a chance to find the organism, which raises the second problem in the case. After receiving the specimen, the lab cultured it only under aerobic conditions. The laboratory should have also incubated a series of plates anaerobically (see Section 5.4, Fig. 5.19). This case, therefore, illustrates the importance of both proper specimen collection and proper processing.

Biosafety: Proper Handling of Clinical Specimens

Medical and laboratory personnel are exposed to extremely dangerous pathogens on a daily basis. When working with dangerous pathogens, clinical microbiologists must protect themselves from accidental infection and at the same time be certain the pathogen does not escape from the lab.

Case History: Fatal Meningitis

On July 15, an Alabama microbiologist was taken to the emergency room with acute onset of generalized malaise, fever, and diffuse myalgias. She was given a prescription for oral antibiotics and released. On July 16, she became tachycardic and hypotensive and returned to the hospital. She died 3 hours later. Blood cultures were positive for Neisseria meningitidis *serogroup C. Three days before the onset of symptoms, the microbiologist had prepared a Gram stain from the blood culture of a patient subsequently shown to have meningococcal disease; she had also handled agar plates containing cerebrospinal fluid (CSF) cultures from the same patient. Co-workers reported that fluids were aspirated from blood culture bottles at the open laboratory bench. No biosafety cabinets, eye protection, or masks were used for this procedure. Testing at the CDC indicated that the isolates from both patients were indistinguishable. The laboratory at the hospital infrequently processed isolates of* N. meningitidis *and had not processed another meningococcal isolate during the previous 4 years.*

The microbiologist in this case did not take appropriate measures to protect herself and ended up with a laboratory-acquired infection leading to meningitis. The CDC has published a series of regulations designed to protect workers at risk of infection by human pathogens. Infectious agents are ranked by the severity of disease and ease of transmission. The more severe the disease or the more easily it is transmitted, the higher the risk category. On the basis of this ranking, four levels of biological containment are employed (**Table 28.1**).

- **Biosafety level 1.** BSL-1 organisms have little to no pathogenic potential and require the lowest level of containment. Standard sterile techniques and laboratory practices are sufficient.
- **Biosafety level 2.** BSL-2 agents have greater pathogenic potential, but vaccines and/or therapeutic treatments (for example, antibiotics) are readily available. The pathogen in the case described here, *Neisseria meningitidis*, is in this risk group. These agents require more rigorous containment procedures, such as limiting laboratory access when experiments are in progress and using biological laminar flow cabinets if aerosolization is possible.
- **Biosafety level 3.** BSL-3 pathogens produce a serious or lethal human disease. Vaccines or therapeutic agents may be available. To safely handle these organisms, level 2 procedures are supplemented with wraparound gowns, respiratory protection, a lab design ensuring that ventilation air flows only <u>into</u> the room and that exhaust air vents directly to the outside, thus producing negative pressure. Negative pressure will keep any organism that may aerosolize from escaping into hallways. In addition, access to the lab is strictly regulated and includes double-door air locks at the entrance.

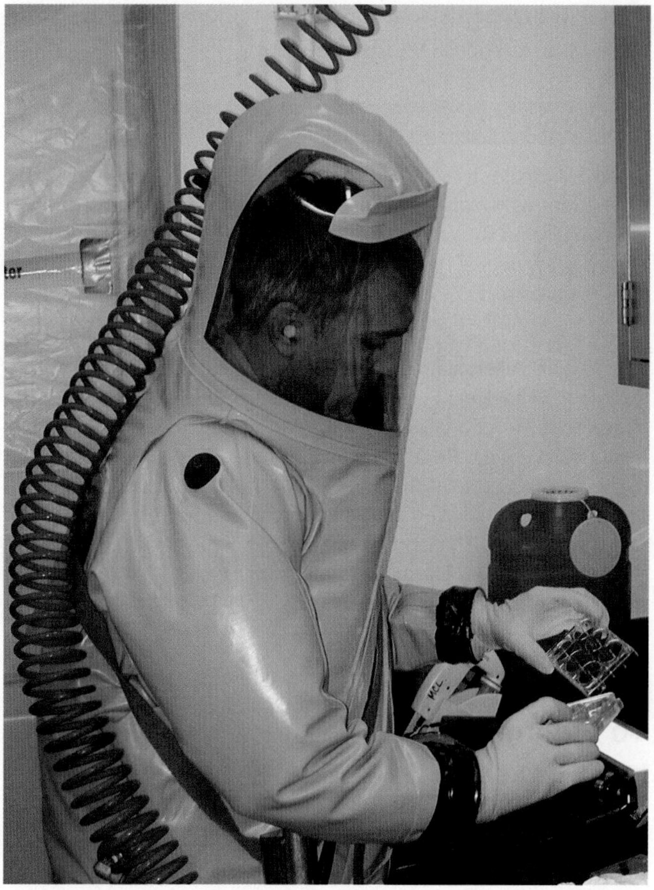

FIGURE 28.4 ■ Biosafety level 4 containment. Dr. Kevin Karem at the CDC performs viral plaque assays to determine the neutralization potential of serum from smallpox vaccination trials. He is protected by a positive-pressure suit while working in a BSL-4 laboratory. The airflow into his suit is so loud that he must wear earplugs to protect his hearing. Note that this virus is used under extremely tight security at the CDC, one of only two places in the world allowed to work with the virus.

- **Biosafety level 4.** BSL-4 is required by law to study extremely dangerous pathogens for which there is no treatment or vaccine (for instance, the Ebola virus). Practices here dictate that lab personnel change clothes upon entering and exiting the lab, shower before exiting, and wear positive-pressure lab suits connected to a separate air supply (**Fig. 28.4**). The positive pressure ensures that if the suit is penetrated, organisms will be blown away from the breach and not sucked into the suit.

As reasonable as these regulations may seem, they were not always in effect. Before 1970, liquid cultures containing live organisms were routinely transferred from one vessel to another by mouth pipetting (essentially using a glass or plastic pipette as a straw). This practice is now forbidden, for obvious reasons. As of this writing, there are 13 BSL-4 laboratories planned or operating in the United States.

It is important to note that clinical microbiology laboratories in the United States are equipped to handle BSL-2 organisms. Patient samples suspected to contain an agent at level 3 or higher are sent directly to regional reference laboratories or to the CDC for analysis.

Thought Questions

28.1 Two blood cultures, one from each arm, were taken from a patient with high fever. One culture grew *Staphylococcus epidermidis*, but the other blood culture was negative (no organisms grew out). Is the patient suffering from septicemia caused by *S. epidermidis*?

28.2 A 30-year-old woman with abdominal pain went to her physician. After examining the patient, the doctor asked her to collect a midstream urine sample that would be sent to the lab across town for analysis. The woman complied and handed the standard urine collection cup to the nurse. The nurse placed the cup on a table at the nurses' station. Three hours later, a courier service picked up the specimen and transported it to the laboratory. The next day the report came back: "Greater than 200,000 CFUs/ml; multiple colony types; sample unsuitable for analysis." Why was this determination made?

To Summarize

- **Identifying a pathogen** enables clinicians to prescribe appropriate **antibiotics**, anticipate possible **sequelae**, and **track the spread** of the disease.
- **Specimen collection** is a critical first step in the process of identifying a pathogen. **Common specimens** include blood, pus, urine, sputum, throat swabs, stool, and cerebrospinal fluid.
- **Specimens** from sites containing normal microbiota must be collected, handled, and processed differently from specimens taken from normally sterile body sites. Suspecting that anaerobes might be present also affects how specimens are handled.
- **A specimen from an abscess or other infection that might contain anaerobes** must be collected and processed under anaerobic conditions.
- **Various levels of protective measures** are used in handling potentially infectious biological materials. **Biosafety level 1** agents are generally not pathogenic and require the lowest level of containment.
- **Biosafety level 2** agents are pathogenic but not typically transmitted via the respiratory tract. Laminar flow hoods are required.
- **Biosafety level 3** agents are virulent and transmitted by the respiratory route. They require laboratories with special ventilation and air-lock doors.
- **Biosafety level 4** agents are highly virulent and require the use of positive-pressure suits.

28.2 Pathogen Identification by Culture and Phenotype

Once a specimen is collected, how are pathogens identified? A variety of techniques can be used, such as staining of clinical specimens to reveal the presence of organisms, nucleic acid assays, serology (testing a patient's serum for the presence of antibodies reactive against a specific microbe), and detection of biochemical clues left by the pathogen in vivo or in vitro. In this section we focus on the classical methods of identification: staining and metabolic profiles. In Section 28.3 we consider molecular and serological methods of identification.

Staining

Some specimens, such as CSF and sputum, can be directly stained with the Gram stain procedure (described in Section 2.3) or the acid-fast stain. Knowing that an organism is Gram-positive, Gram-negative, or acid-fast guides which additional tests the clinical microbiologist must run. The case that follows illustrates how the acid-fast stain is critical for presumptively identifying mycobacteria.

Case History: Tuberculosis

A 31-year-old male presented to an emergency department (ED) in New York City after experiencing gross hemoptysis (blood in sputum). He had a 2-month history of productive cough, a 25-pound weight loss, night sweats, and fatigue. A chest X-ray revealed bilateral cavitary infiltrates. The initial sputum specimen was negative by Gram stain but positive for acid-fast bacilli. The specimen was submitted for a nucleic acid amplification test (NAAT) to detect 16S rRNA, as well as for culture and sensitivity. The patient had a history of heavy alcohol and drug use.

The likely suspect in this case is *Mycobacterium tuberculosis*, although other mycobacterial species are possible causes. *M. tuberculosis* is presumptively diagnosed in the clinical laboratory by the acid-fast stain, a technique first described in 1882 (called the Ziehl-Neelsen stain) that is used to find the tubercle bacillus in a patient's sputum. The acid-fast stain enables a technician to visualize bacteria such as *Mycobacterium* species that are not stained by the Gram stain. Mycobacteria have a waxy outer coat composed of mycolic acid that resists penetration by most dyes—an obstacle the acid-fast stain was designed to overcome.

The original Ziehl-Neelsen acid-fast stain used phenol and heat to drive carbolfuchsin (a red dye) into mycobacterial cells on glass slides. Destaining with an acidic alcohol solution removes the stain from all cell types except mycobacteria. The slide is subsequently counterstained with methylene blue, after which the mycobacteria will be seen as curved, red rods (the acid-fast bacilli), while everything else will appear blue (**Fig. 28.5A**). A more modern version of the acid-fast stain uses the fluorochrome auramine O to stain the mycolic acid (**Fig. 28.5B**). This dye also resists removal by an acidic alcohol wash, so the mycobacteria will fluoresce bright yellow-green when observed under a fluorescence microscope.

Although the acid-fast stain is very useful, it detects organisms in the sputum of tuberculosis patients only about 60% of the time; and even if they are found, confirmatory tests are needed for a definitive diagnosis. Growth-dependent identification of *M. tuberculosis* typically starts with inoculating the sputum sample to a blue-green Löwenstein-Jensen medium (selective for mycobacteria) and waiting several weeks for the organism to grow before additional tests can be done (**Fig. 28.5C**). While the organism grows, however, a rapid, nucleic acid amplification test can be used to quickly detect even small amounts (Section 28.3).

A. Acid-fast stain

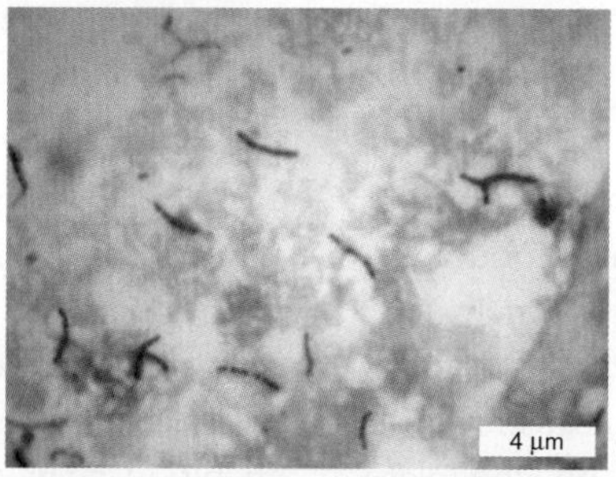

B.

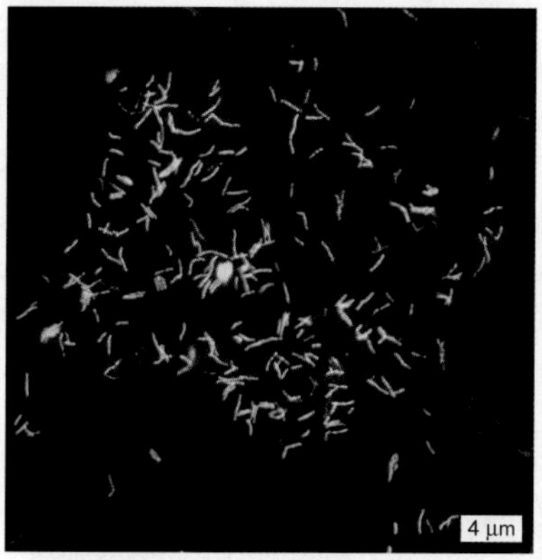

C. ***M. tuberculosis*** **colonies on Löwenstein-Jensen agar**

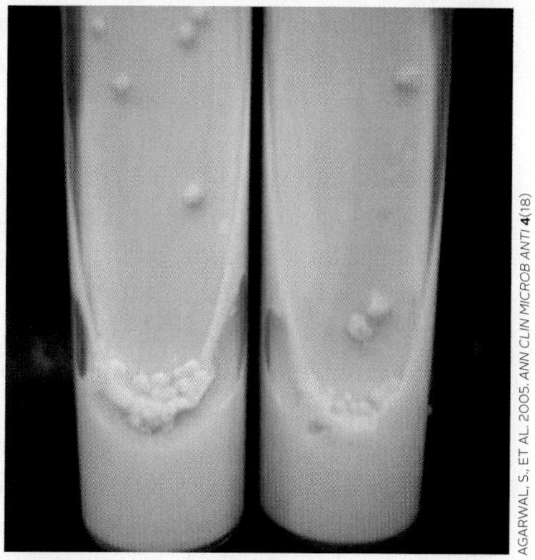

FIGURE 28.5 ■ Acid-fast stain and growth of *Mycobacterium tuberculosis*. **A.** Acid-fast Ziehl-Neelsen stain of *M. tuberculosis*. **B.** Auramine O fluorescent acid-fast bacilli (AFB) stain. **C.** Löwenstein-Jensen medium enables growth of mycobacterial species, some of which grow extremely slowly. The colonies have a "bread crumb–like" appearance.

Growth and Biochemical Testing

Once stained samples have been viewed and the Gram reaction of the organisms is known, the conventional laboratory approach used to identify bacterial pathogens requires understanding basic microbial physiology and its many variations. Thousands of bacterial species are capable of causing disease, but no two species have the same biochemical "signature." The clinical microbiologist can look for reactions, or combinations of reactions, that are unique to a given species, as in the following case history.

Case History: Meningitis

A 4-week old girl with no significant previous medical history came to the emergency room crying uncontrollably, particularly when the physician tried to move her head. The day before coming to the ER she was vomiting. She was not on any medications, and lived with her parents and two older brothers, all of whom were well. Cerebrospinal fluid (CSF) was collected from a spinal tap. The CSF appeared cloudy (it should be clear) and contained 871 white blood cells per microliter (normal is 0–10), the glucose level was 1 milligram per deciliter (normal is 50–80), and the total protein level was 417 milligrams per deciliter (normal is less than 45). Gram stain of a CSF smear revealed Gram-negative rods. The CSF sample was sent to the diagnostic laboratory for microbial identification.

As discussed in Thought Question 26.10, low glucose and elevated protein levels in CSF are indicators of bacterial (not viral) infection. The increase in white blood cells revealed that the baby's immune system was trying to fight the disease. The presence of Gram-negative rods in the CSF smear confirmed a diagnosis of bacterial meningitis, since CSF should be sterile. Now it was up to the clinical laboratory to determine the etiological agent.

Algorithms to identify bacteria. Over the years, clinical microbiologists have developed algorithms (step-by-step problem-solving procedures) that expose the most likely cause of a given infectious disease. For instance, only a limited number of microbes are known to cause meningitis. The microbiologist poses a series of binary yes/no questions about the clinical specimen in the form of biochemical or serological tests. This type of tool is called a dichotomous key. Typical questions in this case might include: Is an organism seen in the CSF of a patient with symptoms of meningitis? Is the organism Gram-positive or Gram-negative? Does it stain acid-fast? Answers to a first round of questions will then dictate the next series of tests to be used.

Because speed is of the essence in deciding how to treat the patient, a slew of tests are carried out simultaneously, but the results are interpreted sequentially using the algorithm. In our case history, for instance, consider the most common causes of bacterial meningitis: *Neisseria meningitidis*, *Streptococcus pneumoniae*, *Haemophilus influenzae*, and *Escherichia coli*.

The CSF sample was Gram-stained and simultaneously plated onto three media: chocolate agar, blood agar, and Hektoen agar. Chocolate agar is an extremely rich medium that looks brown, owing to the presence of heat-lysed red blood cells (**Fig. 28.6A** and **B**). Because it is so nutrient-rich, all four organisms will grow on chocolate agar. However, nutritionally fastidious organisms such as *N. meningitidis* and *H. influenzae* will not grow well, if at all, on ordinary blood agar, because these bacteria cannot lyse red blood cells and release required nutrients. Less fastidious organisms, such as *S. pneumoniae* and *E. coli*, will grow on blood agar, but of these two, only *E. coli* can grow on Hektoen agar (**Fig. 28.6C–E**), which is a selective and differential medium for enteric Gram-negative rods. Hektoen is selective because bile salts and dyes inhibit the growth of Gram-positives. It is differential because the medium reveals organisms that ferment lactose or sucrose and produce hydrogen sulfide. Differential and selective media are described in Section 4.3.

In our case history, the Gram stain of the CSF revealed Gram-negative rods, which ruled out *N. meningitidis* (a Gram-negative diplococcus) and *S. pneumoniae* (a Gram-positive diplococcus). The organism in CSF did grow on blood agar, which eliminated *H. influenzae* (a Gram-negative, nonenteric rod) as a candidate. It also grew on Hektoen, where it produced orange, lactose-fermenting colonies. Thus, the organism was a Gram-negative, enteric rod, and likely *E. coli*, probably a strain of neonatal meningitis *E. coli* (NMEC). Additional biochemical tests confirming the identity of the organism had to be carried out, but this simple example shows how simultaneous tests can be interpreted.

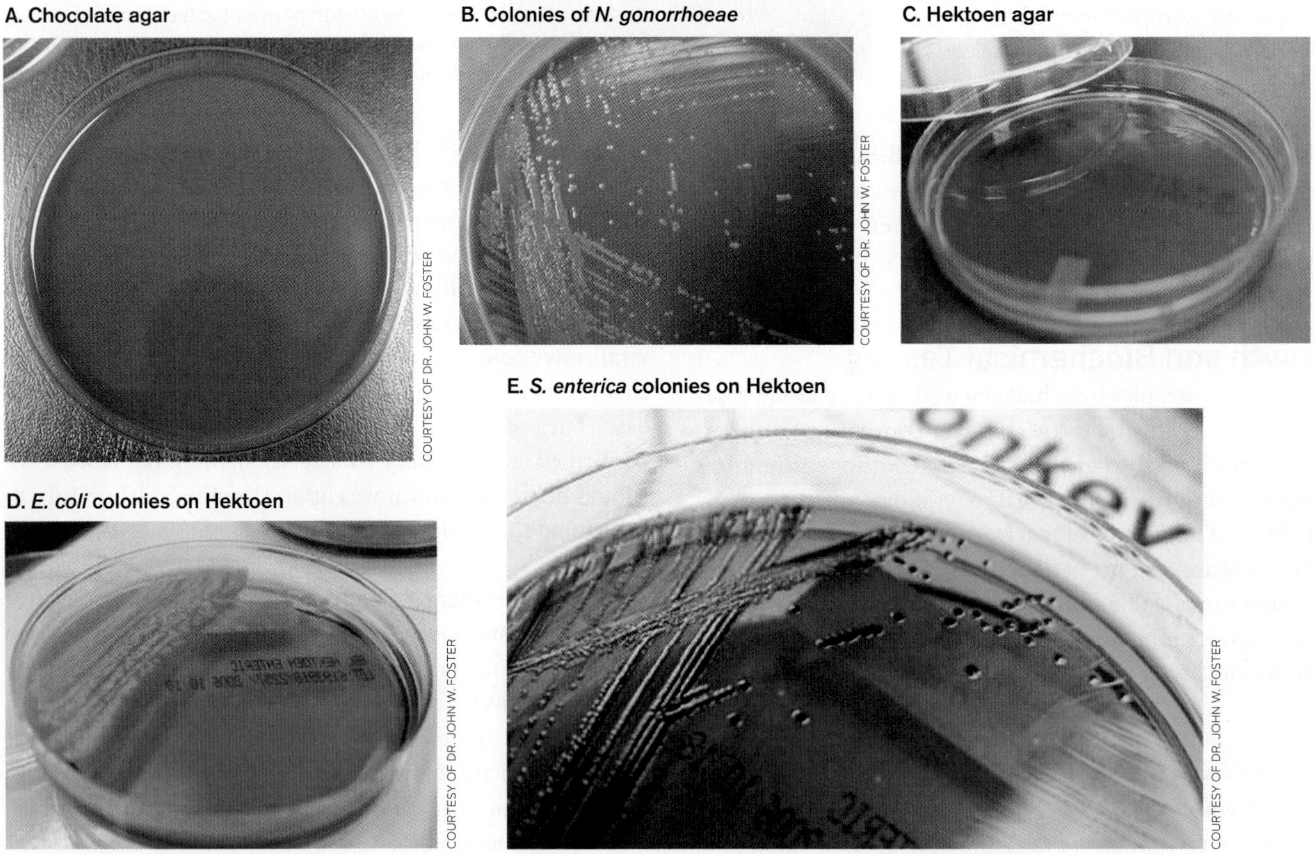

FIGURE 28.6 ■ Chocolate agar (A and B) and Hektoen agar (C–E), two widely used clinical media. **A.** Uninoculated chocolate agar. Its color is due to gently lysed red blood cells that provide a rich source of nutrients for fastidious bacteria. **B.** Chocolate agar inoculated with *Neisseria gonorrhoeae*. This organism will not grow well on typical blood agar, because important nutrients remain locked within intact red blood cells. **C.** Uninoculated Hektoen agar, which contains lactose, peptone, bile salts, thiosulfate, an iron salt, and the pH indicators bromothymol blue and acid fuchsin; the bile salts prevent growth of Gram-positive microbes. **D.** Hektoen agar inoculated with *Escherichia coli*. This organism ferments lactose to produce acid-fermentation products that give the medium an orange color, owing to the pH indicators acid fuchsin and bromophenol blue. **E.** Hektoen agar inoculated with *Salmonella enterica*. This organism does not ferment lactose but grows instead on the peptone amino acids. The resulting amines are alkaline and produce a more intense blue color with bromothymol blue. *Salmonella* species also produce hydrogen sulfide gas from the thiosulfate. Hydrogen sulfide reacts with the medium's iron salt to produce an insoluble, black iron sulfide precipitate visible in the center of the colonies.

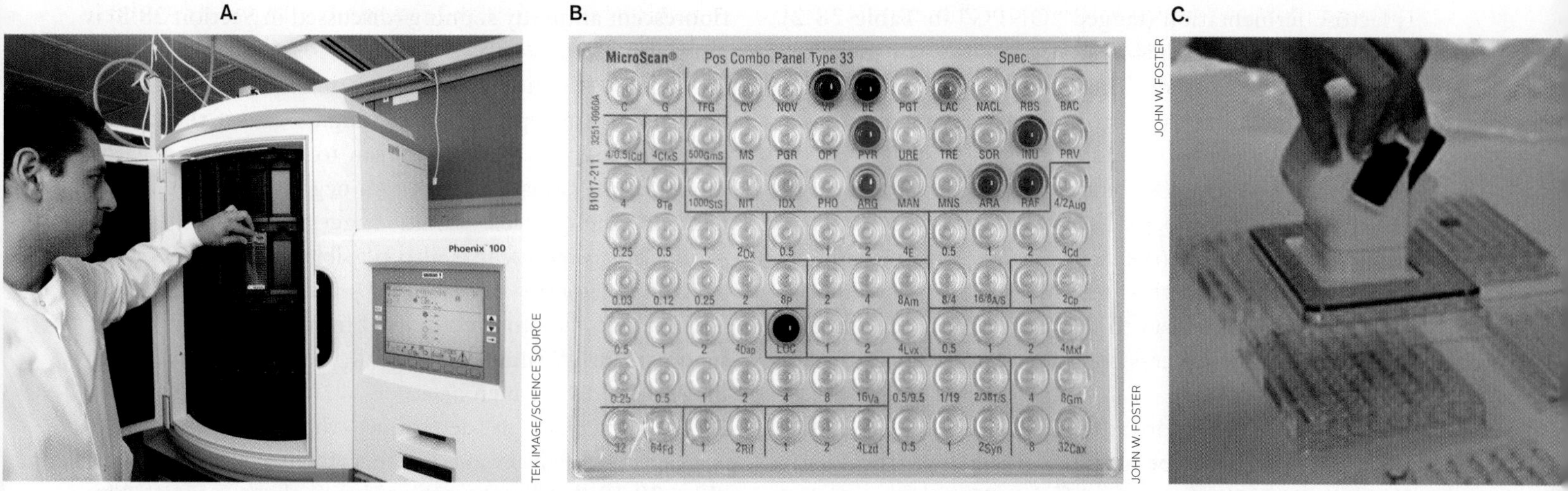

FIGURE 28.7 ■ Automated microbiology system. A. The BD Phoenix 100 system uses plates with numerous reaction wells and a computerized plate reader to automatically identify pathogenic bacteria. This type of instrument automatically generates and evaluates the numbers. B. Microbial identification plate. C. Loading a multiwall ID plate with a multichannel pipettor. All wells are simultaneously loaded with the same volume and number of bacteria.

Identifying Gram-negative bacteria. The Gram-negative bacterium in this case was subjected to a battery of 36 biochemical tests in an automated microbial identification instrument (**Fig. 28.7A**). Most clinical laboratories in the United States and Europe now use these automated identification systems. The organism is inoculated into the wells of a prepared microtiter dish in which each well contains materials that test an organism's ability to ferment different carbon sources or make different metabolic end products (**Fig. 28.7B** and **C**). By monitoring the growth or change in color produced in different wells, the instrument can quickly identify the pathogen's metabolic profile, sometimes within 5–6 hours. For each bacterial species, the system's software "knows" the probability that a given reaction will be positive or negative for a given species. Known as a **probabilistic indicator**, the computer program will integrate all the metabolic reaction results for an unknown bacterial isolate and determine whether the overall probabilities for these reactions match those of a specific pathogen. Identification of bacteria using a probabilistic indicator was described in Section 17.5.

A simplified example of the pathogen identification process is shown in **Figure 28.8**. The analytical profile index (API) 20E strip can test 20 metabolic processes (listed in **Table 28.2**). An uninoculated control and strips for two organisms are pictured in **Figure 28.8**. Overnight incubation is needed before the chamber reactions can be read, so the API strip does not deliver results as quickly as the automated systems do. Different-colored reactions in each chamber are scored as positive or negative, depending on the color (**eTopic 28.1**). For example, in the indole chamber (well 9), a red reaction at the top of the tube is positive and indicates that the organism can produce indole from tryptophan (**Fig. 28.8B**). A colorless chamber (**Fig. 28.8C**) would be a negative result. Go to **eTopic 28.1** to see how API strips can also be used to generate a seven-digit number that identifies the bacterium.

We can use the results of the API chambers as a dichotomous key, by making a stepwise interpretation that begins with a key reaction. A simplified dichotomous key using a limited number of enteric Gram-negative species is shown in **Figure 28.9**. Often, the first reaction examined

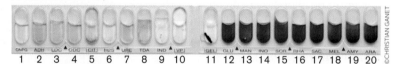

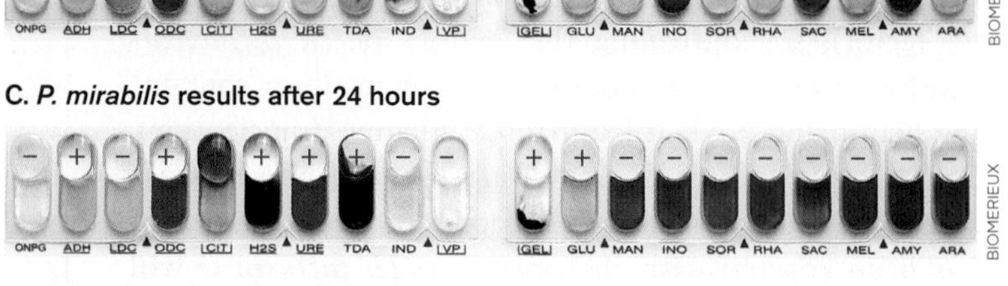

FIGURE 28.8 ■ API 20E strip technology for the biochemical identification of Enterobacteriaceae. A. Uninoculated API strip. Each well contains a different medium that tests for a specific biochemical capability. The well numbers correspond to Table 28.2. The color of each medium after 24-hour incubation indicates a positive or negative reaction (see Table 28.2). B, C. API results for *E. coli* **(B)** and *Proteus mirabilis* **(C)**. Plus (+) and minus (−) indicate positive and negative reactions, respectively.

is lactose fermentation (tagged "ONPG" in **Table 28.2**). The lactose reaction is read as positive or negative, depending on the color. Then, following a printed flowchart, the technician goes to the next key reaction—say, indole production—and reads it as positive or negative. If the organism is a lactose fermenter and the indole test is positive, then the choices have been narrowed to *E. coli* or *Klebsiella* species (the other lactose-positive, indole-positive organisms in the figure do not cause meningitis). Another reaction is read to distinguish between the next two choices. The process continues until a single species is identified.

In reality, many more reactions than the 11 shown in **Figure 28.9** have to be used to make a definitive identification, because of species differences with respect to a single reaction. For example, the figure shows that *Klebsiella pneumoniae* and *Klebsiella oxytoca* (highlighted in yellow) exhibit opposite indole reactions, even though they are of the same genus. Even within a given species, only a certain percentage of strains might be positive for a given reaction. The inherent danger in using a dichotomous key, as opposed to a probabilistic system, is that one anomalous result can lead to an incorrect identification. Be aware, too, that the phenotypic dichotomous key does not mirror a phylogeny tree, in which the two *Klebsiella* species would branch from a common ancestor.

Identifying nonenteric Gram-negative bacteria. The procedures we have outlined will accurately identify members of Enterobacteriaceae, but there are pathogenic Gram-negative bacilli found in other, nonenteric, families. For instance, one possibility in the preceding case history is that the Gram-negative bacillus seen in CSF smears would not grow on blood agar or on the other selective media, but would grow as small, glistening colonies on chocolate agar. This result would implicate the Gram-negative rod *Haemophilus influenzae*.

Meningitis caused by *H. influenzae* was a major problem prior to 1988, before the introduction of a vaccine containing *H. influenzae* type b capsular material. Growth of *H. influenzae* requires hemin (X factor) and NAD (V factor), so confirming the identity of *H. influenzae* involves growing the organism on agar medium containing hemin and NAD (nicotinamide adenine dinucleotide). Small filter-paper disks containing these compounds are placed onto a nutrient agar surface (not blood agar) that has been covered with the organism. *H. influenzae* will grow only around a strip containing both X and V factors. Alternatively, X and V factors can be incorporated into Mueller-Hinton agar, as shown in **Figure 28.10A**. The organism grows on chocolate medium because the lysed red blood cells release these factors. Although the XV growth phenotype is still used today for identification, fluorescent antibody staining (discussed in Section 28.3) is more specific.

A completely different identification scheme would have been used in the preceding meningitis case if the laboratory had discovered the organism to be a Gram-negative diplococcus (rather than a Gram-negative rod). A Gram-negative diplococcus would suggest *Neisseria meningitidis*. *N. gonorrhoeae* is also possible, but it is less likely because the gonococcus lacks the protective capsule that *N. meningitidis* uses to protect itself from complement in the bloodstream. Without this capsule, *N. gonorrhoeae* cannot disseminate to the meninges.

The first test to determine whether the organism is a species of *Neisseria* is the cytochrome oxidase test (**Fig. 28.10B**). Cytochrome oxidase is the terminal oxidoreductase for O_2 (discussed in Chapter 14). In this test, a few drops of the colorless reagent *N,N,N′,N′*-tetramethyl-*p*-phenylenediamine dihydrochloride are applied to the suspect colonies. The reaction, which takes place only if the organism possesses both cytochrome oxidase and cytochrome *c*, turns the *p*-phenylenediamine reagent (and the colony) a deep purple/black. Many bacteria possess cytochrome oxidase, but only a few genera, such as *Neisseria*, also contain cytochrome *c* in their membranes.

TABLE 28.2 Reading the API 20E

Well number	Test	Reaction tested
1	ONPG[a]	Beta-galactosidase
2	ADH	Arginine dihydrolase
3	LDC	Lysine decarboxylase
4	ODC	Ornithine decarboxylase
5	CIT	Citrate utilization
6	H_2S	H_2S production
7	URE	Urea hydrolysis
8	TDA	Tryptophan deaminase
9	IND	Indole production
10	VP	Acetoin production
11	GEL	Gelatinase
12	GLU	Glucose fermentation/oxidation
13	MAN	Mannitol fermentation/oxidation
14	INO	Inositol fermentation/oxidation
15	SOR	Sorbitol fermentation/oxidation
16	RHA	Rhamnose fermentation/oxidation
17	SAC	Sucrose fermentation/oxidation
18	MEL	Melibiose fermentation/oxidation
19	AMY	Amygdalin fermentation/oxidation
20	ARA	Arabinose fermentation/oxidation

[a]ONPG = *ortho*-nitrophenyl-beta-D-galactoside.

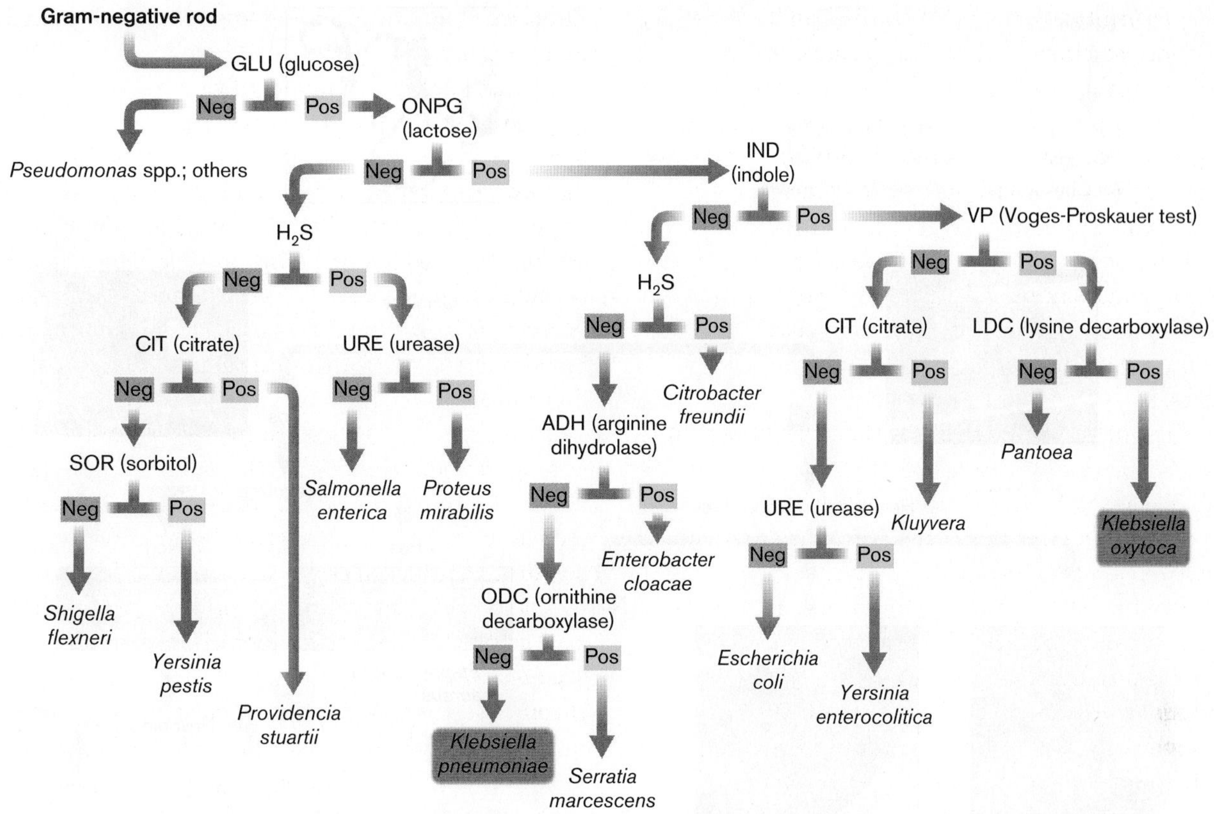

FIGURE 28.9 ▪ **Simplified biochemical algorithm to identify Gram-negative rods.** The diagram presents a dichotomous key using a limited number of biochemical reactions and selected organisms to illustrate how species identifications can be made using biochemistry. Abbreviations and reactions: **ADH:** stepwise degradation of arginine to citrulline and ornithine; **CIT:** citrate utilization as a carbon source; **GLU:** glucose fermentation to produce acid; **H_2S:** production of hydrogen sulfide gas; **IND:** indole production from tryptophan; **LDC:** cleavage of lysine to produce CO_2 and cadaverine; **ODC:** cleavage of ornithine to make CO_2 and putrescine; **ONPG:** lactose fermentation to produce acid (*ortho*-nitrophenyl-beta-D-galactoside, cleaved by beta-galactosidase); **SOR:** sorbitol fermentation to produce acid; **URE:** production of CO_2 and ammonia from urea; **VP:** production of acetoin or 2,3-butanediol. (*Note:* "Neg" for *Pseudomonas* in terms of glucose indicates an inability to ferment glucose. *Pseudomonas* can still use glucose as a carbon source.)

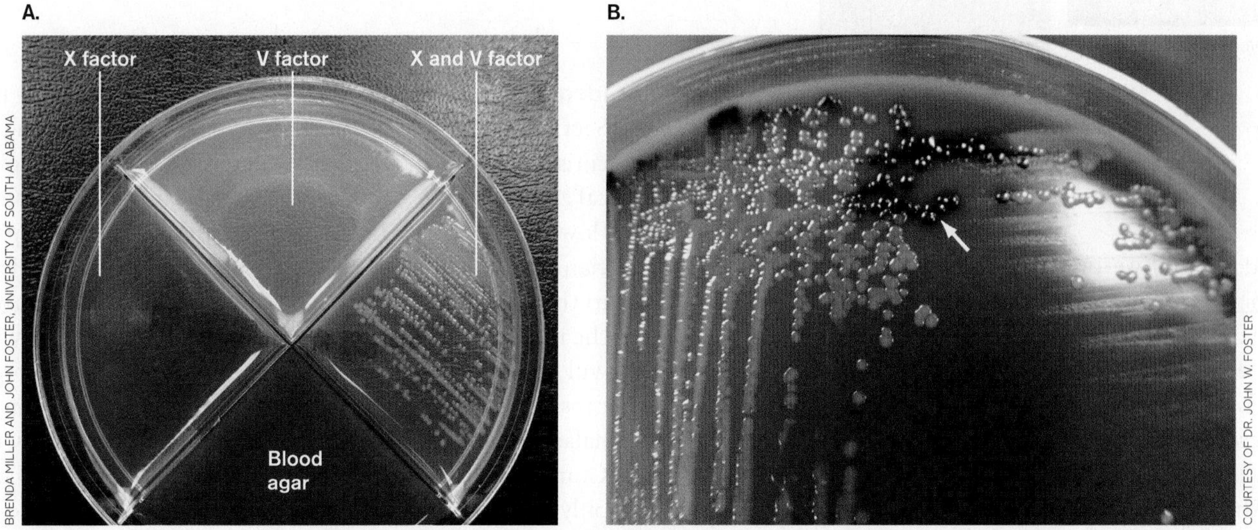

FIGURE 28.10 ▪ ***Haemophilus influenzae*** **growth factors and** ***Neisseria meningitidis*** **oxidase reaction.** **A.** *H. influenzae* will grow on an agar plate (here, Mueller-Hinton agar) only when the medium has been fortified with both X factor (hemin) and V factor (NAD), but not either one alone. Nor will the organism grow on blood agar. **B.** Oxidase-positive reaction for *N. meningitidis*. Oxidase reagent (which is colorless) was dropped onto colonies of *N. meningitidis* grown on chocolate agar (arrow indicates oxidase-positive colonies). The test is called the cytochrome oxidase test, but it really tests for cytochrome *c*.

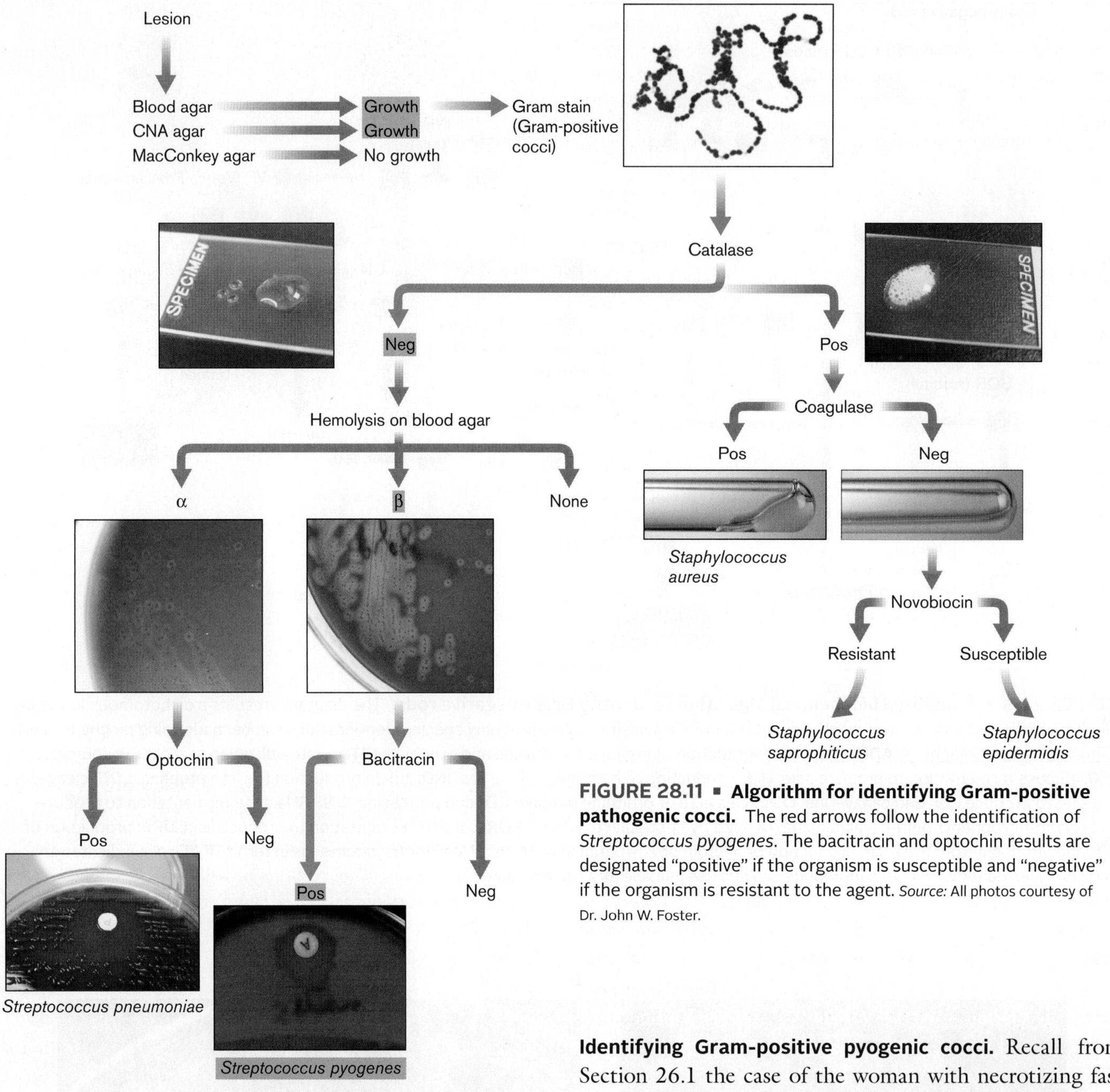

FIGURE 28.11 ■ Algorithm for identifying Gram-positive pathogenic cocci. The red arrows follow the identification of *Streptococcus pyogenes*. The bacitracin and optochin results are designated "positive" if the organism is susceptible and "negative" if the organism is resistant to the agent. *Source:* All photos courtesy of Dr. John W. Foster.

Oxidase-positive organisms use cytochrome oxidase to oxidize cytochrome *c*, which then oxidizes *p*-phenylenediamine. Other oxidase-positive bacteria include *Pseudomonas*, *Haemophilus*, *Bordetella*, *Brucella*, and *Campylobacter* species—all of them Gram-negative rods. None of the Enterobacteriaceae, however, are oxidase-positive, because they lack cytochrome *c*.

An oxidase-positive, Gram-negative diplococcus is very likely a member of *Neisseria*. Differentiation between species of *Neisseria* is based on their ability to grow on certain carbohydrates, or it can be determined using immunofluorescent antibody staining tests that test for the presence of different capsule antigens in *N. meningitidis*.

Identifying Gram-positive pyogenic cocci. Recall from Section 26.1 the case of the woman with necrotizing fasciitis. How did the laboratory determine that the etiological agent was *Streptococcus pyogenes*? A sample algorithm, or flowchart (**Fig. 28.11**), shows how this is done. The physician sends a cotton swab containing a sample from a lesion to the clinical laboratory. The laboratory technician streaks the material onto several media: (1) plain blood agar (which will grow both Gram-positive and Gram-negative organisms); (2) blood agar containing the inhibitors colistin and naladixic acid (called a CNA plate, this agar will grow only Gram-positives); and (3) MacConkey agar (which will grow only Gram-negative organisms; see Section 4.3). The suspect organism in this case grows on the CNA and blood plates. Because it grows in the presence of the Gram-negative inhibitory compounds in CNA, one would immediately suspect the organism to be Gram-positive—an assumption borne out by the Gram stain.

Note: Though the skin is normally populated by many different microorganisms, samples from an infected lesion are overwhelmingly populated by the etiological agent. The pathogen predominates because it outgrows normal microbiota. Using selective media (CNA) to isolate the infectious agent will further simplify diagnosis by decreasing the growth of any normal microbiota that may still be present.

The algorithm tells the laboratory technician that since the organism is a Gram-positive coccus, the next step is to test for catalase production. Remember, Gram stain morphology alone is not enough to definitively differentiate the two genera. Catalase, which converts hydrogen peroxide (H_2O_2) to O_2 and H_2O, clearly distinguishes staphylococci (catalase-positive) from streptococci (catalase-negative). In the catalase test, a colony is mixed with a drop of H_2O_2 on a glass slide. Effusive bubbling due to the release of oxygen indicates catalase activity (see **Fig. 28.11**). Note that many other organisms possess catalase activity, including the Gram-negative rod *E. coli*. According to the algorithm, however, *E. coli* would not be considered, because it does not grow on CNA agar and is not Gram-positive.

Note: When performing a catalase test from colonies grown on blood agar, be sure not to transfer any of the agar, since red blood cells also contain catalase.

Having established that the organism is catalase-negative, the technician examines the blood plate for evidence of hemolysis. Three types of colonies are possible: nonhemolytic, alpha-hemolytic, and beta-hemolytic. Nonhemolytic streptococci do not produce any lytic zone. Alpha-hemolytic strains produce large amounts of hydrogen peroxide that oxidize the heme iron within intact red blood cells to generate a green product. As a result, alpha-hemolytic streptococci produce a green zone around their colonies called alpha "hemolysis"—even though the red blood cells remain intact. (For example, *Streptococcus mutans*, a cause of dental caries and subacute bacterial endocarditis, is alpha-hemolytic.) Still other streptococci produce a completely clear zone of true hemolysis surrounding their colony. This is called beta hemolysis. Red blood cells are hemolyzed completely by exported enzymes, called hemolysins, that lyse red-cell membranes. The flowchart in **Figure 28.11** indicates that the organism from the case history was beta-hemolytic.

The final relevant test in this flowchart is susceptibility to the antibiotic bacitracin, which identifies the most pathogenic group of beta-hemolytic streptococci, called group A streptococci, or GAS. The beta-hemolytic streptococci are subdivided into immunologically distinct Lancefield groups (A–U), based on the carbohydrate antigen anchored to peptidoglycan. Rebecca Lancefield (**Fig. 28.12**), for whom the classification scheme is named,

FIGURE 28.12 ■ Rebecca Lancefield. In 1918, Dr. Lancefield joined the Rockefeller Institute for Medical Research in New York City, where she studied the hemolytic streptococci, known then as *Streptococcus haemolyticus*. She was the first to use serum precipitation methods to classify *S. haemolyticus* into groups according to differences in cell wall carbohydrate antigens. The basic technique is still used today and, in her honor, is known as the Lancefield classification scheme.

was the first to use immunoprecipitation to group the streptococci (immunoprecipitation is described in eAppendix 3). The vast majority of streptococcal diseases are caused by group A beta-hemolytic streptococci (also called GAS), defined as the species *Streptococcus pyogenes*.

Unfortunately, the lengthy Lancefield classification procedure is unsuitable as a rapid identification method. However, GAS are uniformly susceptible to the antibiotic bacitracin, whereas other groups of *Streptococcus* are resistant. Thus, a simple antibiotic disk susceptibility test can be used to indicate GAS (that is, *S. pyogenes*). But beware—many bacteria are bacitracin sensitive, so, like the catalase test, to be useful for identification the bacitracin test must be used in conjunction with an algorithm. The technician must follow the appropriate algorithm before assigning importance to this or any other test result. It is irrelevant, for instance, if an alpha-hemolytic organism is bacitracin susceptible. Some of these may exist, but they are not associated with disease. The organism in our case of necrotizing fasciitis, however, was beta-hemolytic, so bacitracin susceptibility indicated that the organism was *S. pyogenes*.

The other tests named in **Figure 28.11** are equally important for identifying Gram-positive infectious agents. For example, *Streptococcus pneumoniae* is an important cause of pneumonia. Like *S. pyogenes*, *S. pneumoniae* is a catalase-negative, Gram-positive coccus; but unlike *S. pyogenes*, it is alpha-hemolytic. Optochin susceptibility is a property closely associated with *S. pneumoniae*, while other alpha-hemolytic strains of streptococci are resistant to this compound. Thus, an optochin susceptibility disk test is a useful tool for identifying *S. pneumoniae*.

Coagulase catalyzes a key reaction used to distinguish the pathogen *Staphylococcus aureus*, a cause of boils and bone infections, from other staphylococci, such as the normal skin species *Staphylococcus epidermidis*. To test for coagulase, a tube of plasma is inoculated with the suspect organism. If the organism is *S. aureus*, it will secrete the enzyme coagulase to convert fibrinogen to fibrin, which produces a clotted, or coagulated, tube of plasma (a coagulase-positive reaction; **Fig. 28.11**). Coagulase-negative staphylococci can still be medically important, however. *Staphylococcus saprophyticus*, for instance, is an important cause of urinary tract infections. Resistance to novobiocin distinguishes *S. saprophyticus* from *S. epidermidis*.

Thought Question

28.3 Use **Figure 28.11** to identify the organism from the following case: A sample was taken from a boil located on the arm of a 62-year-old man. Bacteriological examination revealed the presence of Gram-positive cocci that were also catalase-positive, coagulase-positive, and novobiocin resistant.

To Summarize

- **Direct Gram or acid-fast stains** are appropriate procedures to perform on some specimens. The results guide the direction of subsequent testing.
- **Specimens** can be cultured on **selective media** to prevent growth of some bacteria while permitting growth of others (such as Gram-positive bacteria versus Gram-negative bacteria). Plating on **differential media** can expose unique biochemical properties that distinguish a pathogen from similar-looking nonpathogens.
- **Growth-dependent pathogen identification** uses numerous, simultaneously run biochemical tests. **An algorithm or a dichotomous key** is applied to the results to identify the species. Different algorithms (or decision trees) are applied to Gram-positive and Gram-negative bacteria.

28.3 Molecular and Serological Identification of Pathogens

Conventional, Petri plate–dependent methods used to identify pathogens take a minimum of 3 days to complete and may take several weeks, depending on the pathogen. *Brucella* species, for instance, can take 14–21 days to grow in blood culture bottles. (Blood culture bottles are not called "negative growth" until after 21 days of incubation.) The delay, however long, is annoying for the physician, but agonizing for the patient awaiting a cure. Today, technologies that offer more rapid identification, sometimes within minutes, have been introduced into the clinical laboratory.

Identification by Mass Spectrometry

Section 8.4 describes how mass spectrometry works and its use in identifying proteins extracted from cells. This technology has now been applied to the rapid identification of bacterial pathogens in a clinical laboratory. The specific technique, shown in **Figure 28.13**, is called matrix-assisted laser desorption/ionization time-of-flight (MALDI-TOF) mass spectrometry and is based on the fact that each pathogen has its own unique protein signature. Bacterial cells from a colony are fixed in a matrix (**Fig. 28.13**, step 1) and irradiated by a laser beam (step 2) that releases proteins from the fixed cells and places a charge on them (ionizes them). The intact, but ionized, proteins enter a vacuum where they are accelerated by an electrostatic field (step 3). The ionized proteins then race toward a detector at different speeds that depend on each ion's charge and mass. Large (heavy) proteins with a high mass-to-charge ratio take longer to reach the detector than do smaller (lighter) proteins having a low mass-to-charge ratio. The resulting pattern of peaks, called the protein signature, is compared against a growing database derived from different species of microorganisms previously subjected to the procedure (step 4).

Many hospital laboratories have invested in MALDI-TOF equipment because it offers significant advantages over classic methods of identification. As noted earlier, it typically takes 2–3 days to identify a pathogen using classic biochemical methods. With mass spectrometry, it takes only 1 day to culture a specimen and find a suspicious colony, then 5 minutes more for MALDI-TOF to identify the culprit. MALDI-TOF can also determine whether an organism has antibiotic resistance determinants. The procedure can be used on positive blood cultures (sepsis) or directly on urine samples (cystitis) and cerebrospinal fluid (meningitis).

Nucleic Acid–Based Detection

Most diagnostic laboratories today use rapid DNA-based methods in addition to conventional Petri dish microbiology. The DNA-based methods take mere hours to detect and identify bacteria and viruses. DNA/RNA detection methods are especially useful for viruses, which otherwise require elaborate electron microscopy to view morphology, or serology to detect an increased presence of antiviral antibodies. The problem with serology is that by the time these antibodies become detectable in blood, the patient is either acutely ill (IgM) or already recovering from the disease (IgG).

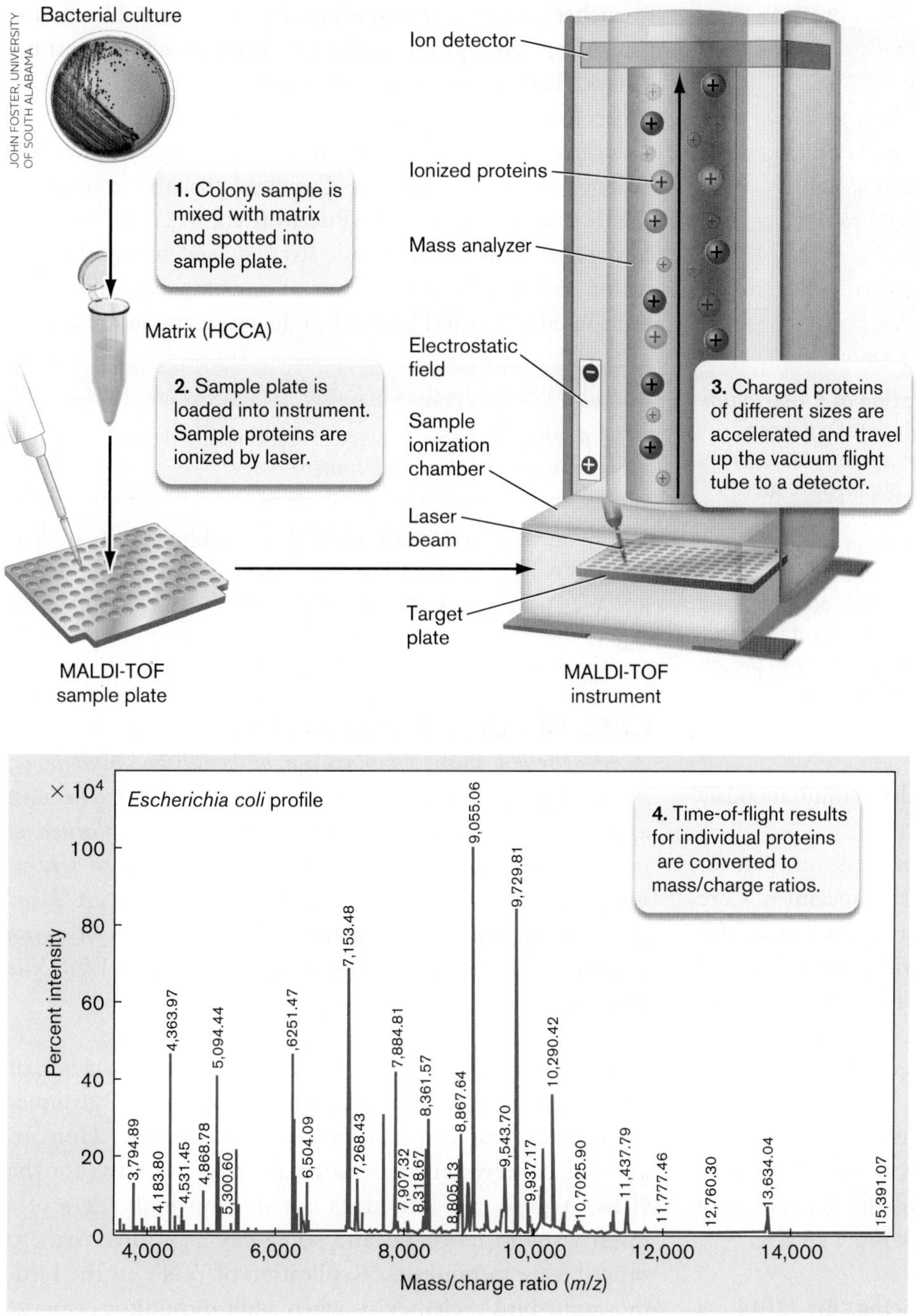

FIGURE 28.13 ■ MALDI-TOF MS identification of bacterial pathogens. **Step 1:** Bacterial colony is spotted into a well of the sample plate. **Step 2:** Laser ionizes proteins. **Step 3:** Charged proteins of different sizes travel at different speeds to the detector. **Step 4:** Mass/charge peaks form a signature used to identify the organism. *Source: E. coli* profile modified from "Identification of Microorganisms" (Shimadzu), fig. 1, https://www.shimadzu.com/an/industry/pharmaceuticallifescience/proteome0207005.htm.

is described later. In general, NAAT assays use oligonucleotide primers to bind and amplify species-specific genes in a pathogen's DNA or RNA genome. Successful amplification is visualized as an appropriately sized DNA fragment (**amplicon**) in agarose gels following electrophoresis, or revealed by accumulation of a fluorescence signal, as in real-time PCR or FRET (fluorescence resonance energy transfer) analysis (Section 12.1).

Why is NAAT needed to detect the presence of these nucleic acids? Without the amplification steps, clinical samples usually provide too little nucleic acid from infecting microorganisms to be detected. For example, sputum samples containing *Mycobacterium tuberculosis*, the cause of tuberculosis, yield minuscule amounts of bacterial DNA. Amplifying *M. tuberculosis* nucleic acid by PCR, however, will turn one copy of a species-specific gene into billions of copies.

NAAT assays are also very quick compared to classic cultural methods (but they take more effort to run than does pathogen identification by mass spectrometry). Extracting the DNA or RNA from a clinical specimen for PCR usually takes an hour. Amplification itself is completed in 1–2 hours. Detection of the PCR-amplified DNA by DNA gel electrophoresis takes an additional 1–2 hours. But fluorescence detection is immediate. So, what might take 2–3 days (or sometimes weeks) using biochemical algorithms may take only an hour or two using NAAT, if the required primer set is on hand.

Nucleic acid amplification tests (NAATs). NAAT assays are the most widely used molecular tools in the diagnostic toolbox. These tests include the polymerase chain reaction (PCR), which uses a heat-resistant DNA polymerase and thermocycling to amplify DNA (see Section 12.1, Fig. 12.2); and several isothermal amplification techniques that require neither. One isothermal amplification technique, called recombinase polymerase amplification (RPA),

The table in eTopic 28.2 lists several instances in which DNA detection tests are useful. The specific example presented in **Figure 28.14** illustrates the use of PCR to type different strains of the anaerobic pathogen *Clostridium botulinum*, the cause of food-borne botulism. Unlike *Streptococcus pneumoniae*, these organisms are not typed serologically. *C. botulinum* species are divided into different types based on the neurotoxin genes they possess. In

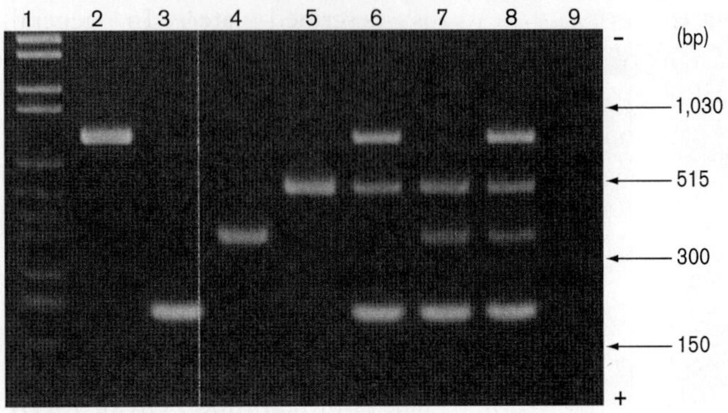

LINDSTROM, M. ET AL. 2001. *APPL AND ENVIRON MICROBIO* **67:** 5694.

FIGURE 28.14 ■ Multiplex PCR identification of *Clostridium botulinum*. *C. botulinum* cells are typed by which toxin genes a strain possesses, not on the basis of surface antigens. Isolates can be typed in a single PCR reaction that includes primer pairs specific for each of the four major toxin genes: types A, B, E, and F. Because multiple PCR products are sought in a single reaction, this technique is called multiplex PCR. Each lane in the agarose gel was loaded with a multiplex reaction made from different isolates of *C. botulinum* and subjected to electrophoresis. The slower-moving fragments (toward the top of the gel) are larger than those moving farther down the gel toward the positive pole. Lane 1 shows DNA size markers.

this example, **multiplex PCR** was used to simultaneously search for these toxin genes.

Multiplex PCR uses multiple sets of primers, one pair for each gene, combined in a single tube with a specimen. Care must be taken to be sure that the primers chosen make different-sized products, do not interfere with each other, and do not produce artifactual (biologically false) products that can confuse interpretation. Multiplex PCR can help identify sets of specific genes present in a single species or can screen for the presence of multiple pathogens in a clinical sample. In the latter, primer sets are designed to amplify genes unique to each pathogen. A novel bioelectronic DNA detection device that couples hybridization to the generation of an electrochemical signal is described in **eTopic 28.3**.

Rapid, next-generation DNA sequencing. In 2014, a 14-year-old boy with severe combined immunodeficiency (SCID) was in a medically induced coma at a Wisconsin hospital. His brain had swollen with fluid as a result of encephalitis, but its cause was unknown. The usual diagnostic tests and antibiotic treatments did not provide answers or relief. In a last attempt to find the infection's cause, the boy's physicians called Joseph DeRisi and Charles Chiu at UC San Francisco, who are renowned for their use of next-generation DNA sequencing (see Chapter 12) to identify pathogens. The scientists used this technology to perform an unbiased metagenomic analysis of all the DNA contained in the boy's cerebrospinal fluid.

After 3 million DNA fragments were sequenced, the answer was apparent. There, staring out from among the sequences of the boy's genome, were DNA sequences of the spirochete *Leptospira santarosai*, an easily treatable organism. After being given intravenous high-dose penicillin for 7 days, the boy eventually recovered.

Leptospirosis is a zoonotic disease. The organism is excreted in the urine of infected animals, usually into nearby bodies of freshwater. The child probably contracted the disease during a trip to Puerto Rico, where he swam in freshwater. This metagenomic strategy to detect pathogens is not widely used yet because of its complexity and cost, but someday it could be used to diagnose any infection.

Thought Question

28.4 Why didn't immunological tests initially identify antibodies to *Leptospira* in the comatose boy?

As you've seen, NAAT assays can quickly identify microbial DNA in clinical samples, but what about RNA? Many RNA viruses cause disease. Can molecular tools identify them?

Case History: West Nile Virus

A 55-year-old man complaining of headache, high fever, and neck stiffness was admitted to a local hospital. The man appeared confused and disoriented. He also complained of muscle weakness. History indicated he had received several mosquito bites approximately 2 weeks previously. A blood specimen was sent to the laboratory. The report the following day indicated that the patient was suffering from West Nile disease.

West Nile virus (WNV) is an RNA-containing virus that infects primarily birds and culicine mosquitoes (a group of mosquitoes that can transmit human diseases). Humans and horses serve only as incidental, dead-end hosts for the virus. A dead-end host does not develop high titers of a given virus in its blood and so cannot pass that virus to other biting mosquitoes. Replication of WNV in the bird-mosquito-bird cycle begins when adult mosquitoes emerge in early spring and continues until fall. Among humans, the incidence of disease peaks in late summer and early fall. Birds provide an efficient means of geographic spread of the virus. As a result, over the past two decades the virus has spread throughout much of the United States.

Isolating a disease-causing virus is extremely challenging. Most laboratories are not equipped for the special tissue culture techniques required to grow viruses. Consequently, the means of diagnosing many viral infections, including human WNV infections, has been to measure the antibody response of the patient (described shortly). For instance, the presence of West Nile virus–specific IgM in cerebrospinal fluid is a good indicator of current WNV infection, but it is indirect and inconclusive. Real-time PCR is a molecular test that can quickly reveal the presence of the viral RNA.

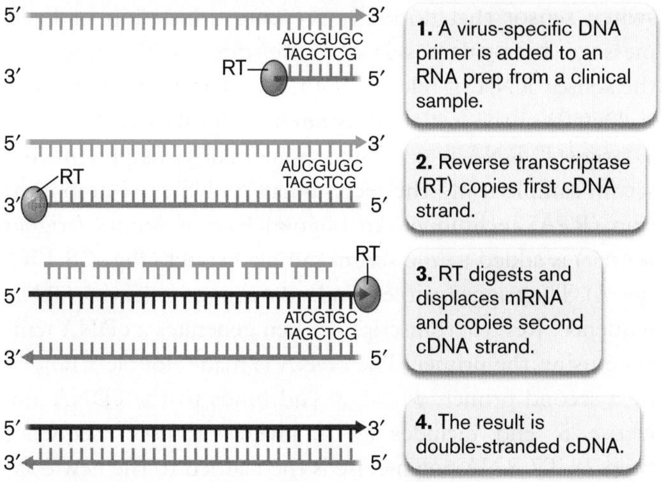

FIGURE 28.15 ▪ Identifying viruses by reverse-transcription PCR. Reverse transcriptase uses viral RNA as a template to make DNA. The DNA can then be amplified by standard PCR methods.

Thought Question

28.5 Why does finding IgM to West Nile virus indicate a current infection? Why wouldn't finding IgG indicate the same?

Quantitative reverse-transcription PCR (qRT-PCR). The qRT-PCR technique is used routinely for the high-throughput diagnosis of many viral pathogens, including West Nile virus. Because West Nile virus (*Flaviviridae* family) contains single-stranded RNA, its RNA must be converted to DNA using reverse transcriptase before PCR can be attempted. The quantitative advantage of qRT-PCR is that the number of viral RNA molecules present in a sample is an indicator of how many virus particles are there.

The basic technique is this: RNA is first extracted from the sample, and a DNA primer specific to a viral RNA sequence is added. Once annealed, the primer enables reverse transcriptase to synthesize the first strand of a complementary DNA (cDNA; **Fig. 28.15**). Reverse transcriptase then digests the initial RNA while synthesizing the opposite DNA strand. The more virus particles there are in the sample (that is, the greater the viral load), the more viral RNA will be present and the more cDNA product will be made. The cDNA is then amplified by PCR using two specific primers and a heat-stable DNA polymerase such as Taq polymerase (see Section 12.1, Fig. 12.3).

The trick is in the method used to quantify the result. In one method, a third, fluorescent oligonucleotide (called the probe) is added to the PCR reaction (see Fig. 12.3A). The probe contains a fluorescent dye at the 3′ end and a chemical dye at the 5′ end that quenches (absorbs) energy emitted from the fluorescent dye. As long as the two chemicals are kept in close proximity by the intact probe, no light is emitted. The probe is designed to anneal to a sequence between the binding sites of the two other oligonucleotide primers (modifications on the ends of the probe prevent it from being used as a primer). So, in a successful amplification, Taq polymerase will synthesize DNA from the two outside primers and degrade the probe oligonucleotide as it passes through that area. This cleavage separates the dye from the quencher, and the dye begins to fluoresce. The more viral RNA there is in a sample, the fewer cycles it takes for cDNA to accumulate and register a fluorescence increase over background (**Fig. 28.16**). The amount of fluorescence

A.

Cycle number	Amount of DNA
0	1
1	2
2	4
3	8
4	16
5	32
6	64
7	128
8	256
9	512
10	1,024
11	2,048
12	4,096
13	8,192
14	16,384
15	32,768
16	65,536
17	131,072
18	262,144
19	524,288
20	1,048,576
21	2,097,152
22	4,194,304
23	8,388,608
24	16,777,216
25	33,554,432
26	67,108,864
27	134,217,728
28	268,435,456
29	536,870,912
30	1,073,741,824
31	1,400,000,000
32	1,500,000,000
33	1,550,000,000
34	1,580,000,000

B.

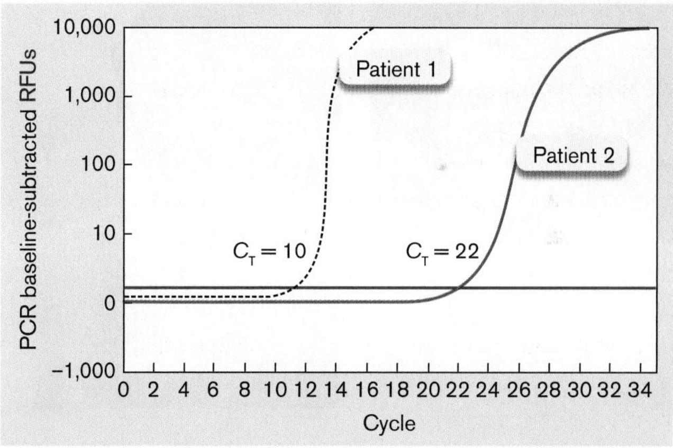

FIGURE 28.16 ▪ Results of quantitative (real-time) PCR. **A.** The exponential increase in PCR products after each cycle of hybridization and polymerization. The switch to yellow indicates the point where the increase in product plateaus because the primers have been exhausted. **B.** There is an increase in relative fluorescence units (RFUs) as the fluorescent dye is released from the dual-labeled probe during quantitative PCR. The red line indicates background fluorescence threshold. The cycle at which fluorescence rises above background is the cycle threshold (C_T). In patient 1, the dashed curve shows that the fluorescent PCR product remained flat for the first ten cycles because the amount of DNA made, and therefore the level of fluorescent dye released remained below background. After cycle 10, fluorescence increased logarithmically: $C_T = 10$. For patient 2, the solid blue curve representing PCR product remained below detection for 22 cycles and then increased: $C_T = 22$. The more starting DNA there is, the sooner RFU values will increase over background (that is, the fewer cycles will be needed to see the increase over background). The slope eventually decreases because the fluorescent probe has become limiting.

emitted is proportional to the amount of cDNA, and thus viral RNA, present in the sample.

> **Thought Question**
>
> **28.6** If the results in **Figure 28.16** came from testing for HIV RNA, then which patient would have the higher viral load in their blood?

Programmable RNA sensors. Zika virus is a fast-emerging, insect-borne pathogen that causes serious developmental defects, such as microcephaly in infected fetal brains (see Section 28.5). The virus has been declared a public health emergency that requires a quick and easy diagnostic test. Keith Pardee (University of Toronto) and Alexander Green (Arizona State University), shown in **Figure 28.17A and B**, with a team of scientists led by James Collins, have designed just such a test using synthetic-biology approaches (see Section 12.5). The test takes only 3 hours to complete. In short, an RNA molecule (trigger) is synthetically made from any Zika virus that may be present in a host sample. The trigger RNA is added to a toehold riboswitch sensor that normally prevents translation of LacZ message (beta-galactosidase). If the trigger RNA anneals to the sensor RNA, it releases the toehold and LacZ is made.

Details of the test are illustrated in **Figure 28.17C**. First a "trigger" RNA is generated from Zika virus present in a serum sample using the recombinase polymerase amplification (RPA) technique. An oligonucleotide primer (trigger primer) is added to the serum sample extract (**Fig. 28.17C**, step 1). The trigger primer binds to a unique Zika virus RNA sequence. Reverse transcriptase then generates a cDNA template using the primer. The cDNA is made double-stranded by a second primer, whose 3′ end binds to the cDNA and whose 5′ end includes a T7 phage promoter sequence (step 2). T7 RNA polymerase is then added to the new double-stranded cDNA to generate multiple copies of Zika trigger RNA (step 3). Recombinase polymerase amplification is a type of isothermal DNA amplification, because thermocycling, as used by PCR, is not needed. The trigger RNA (evidence that the sample contained Zika virus) will then trip the toehold riboswitch (step 4) in the next phase of the test.

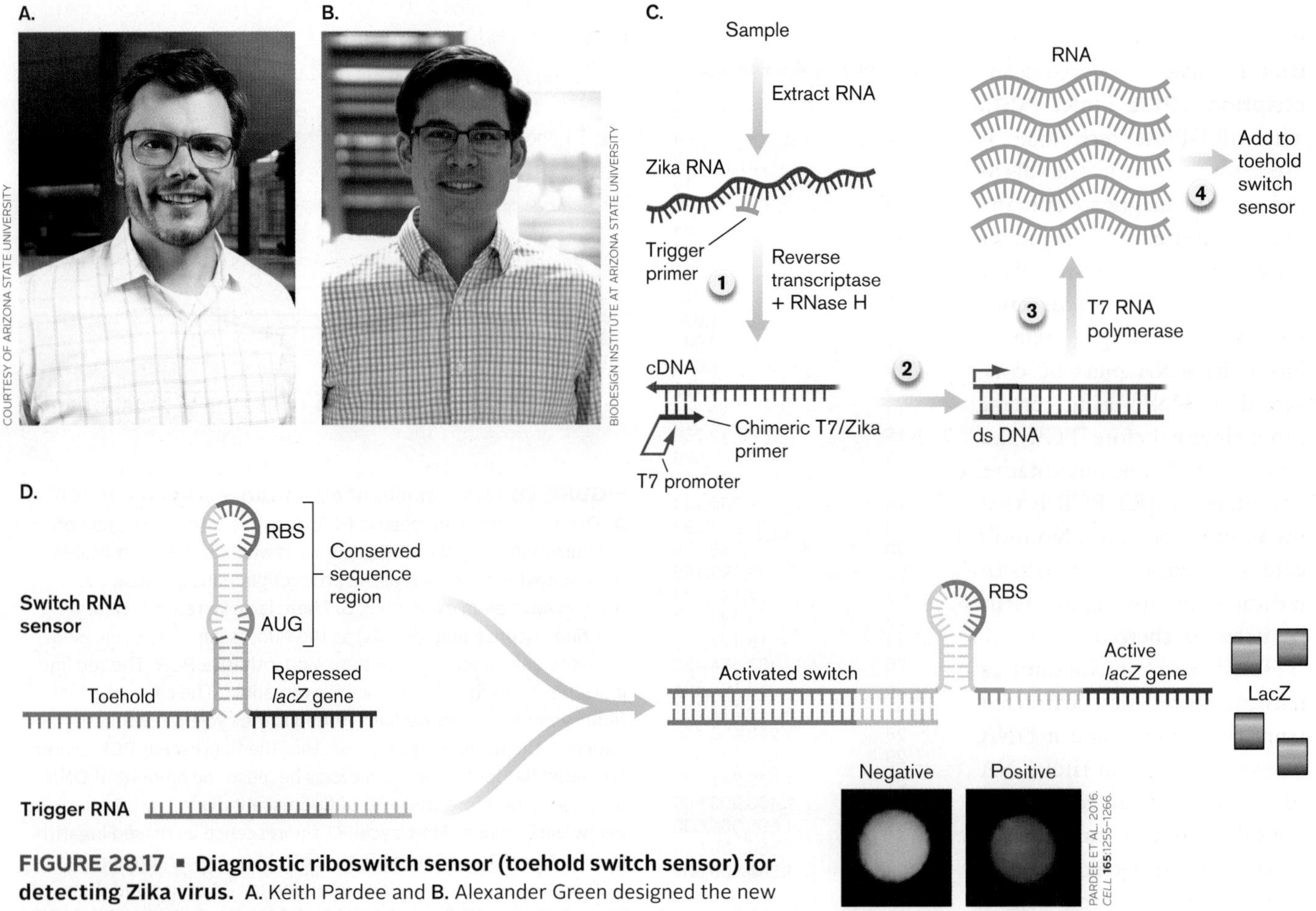

FIGURE 28.17 ■ Diagnostic riboswitch sensor (toehold switch sensor) for detecting Zika virus. **A.** Keith Pardee and **B.** Alexander Green designed the new rapid test for Zika virus. **C.** Method to generate trigger RNA from Zika virus particles present in a clinical sample. **D.** Toehold switch sensor. Binding of trigger RNA to the toehold sensor releases the sequestered ribosome-binding site (RBS) and start codon and enables translation of the LacZ message. **Insets:** Negative (left) and positive (right) tests for Zika virus. *Source:* Parts B and C modified from Pardee et al. 2016. *Cell* **165**:1255–1266.

The synthetically constructed toehold riboswitch sensor is shown in **Figure 28.17D**. The RNA switch contains a hairpin structure that blocks translation of downstream LacZ RNA by sequestering the LacZ ribosome-binding site (RBS) and start codon. This is called a "toehold" switch. When Zika trigger RNA (made previously) is added, the trigger RNA binds to the toehold switch and releases the LacZ RBS and start codon. Translation of the mRNA produces LacZ protein that will convert a yellow substrate to a purple product (a positive test). Once the trigger RNA is added to the switch, the mixture is placed on paper disks containing protein synthesis components, which, after an hour, are analyzed by an electronic reader. This rapid-test product is currently undergoing patient and field trials in Brazil, Ecuador, and Colombia. Its development provides a glimpse of what synthetic biology can bring to the field of rapid diagnostics. **Special Topic 28.1** explores another exciting and novel approach in diagnostics that uses CRISPR technology to detect a pathogen's nucleic acid signature and metaphorically sends out a fluorescent "signal flare."

Profiling gut microbiota by PCR. As noted in Chapter 23, the proper balance between Firmicutes, Proteobacteria, and Bacteroidetes within the gut microbiome is important to human health. Dysbiosis of the gut microbiome has been linked to several serious diseases, including Crohn's, ulcerative colitis, irritable bowel syndrome (IBS), inflammatory bowel disease (IBD), diabetes, and obesity. Consequently, finding ways to identify and correct microbiome imbalances before a patient develops chronic disease has been an important goal of gastroenterologists.

Several companies have been formed to achieve this goal. For example, Christina Casén and colleagues at Genetic Analysis AS devised a diagnostic test for gut dysbiosis. The test uses 54 DNA probes that specifically target taxon-specific variable regions within 16S ribosomal DNA (labeled V3–V7 in **Fig. 28.18A**). Probes that bind to a region are fluorescently labeled by PCR and mixed with bar-coded magnetic pull-down beads containing complementary probes. The level of fluorescence associated with each bar-coded probe reflects the prevalence of a particular taxonomic group. **Figure 28.18B** shows a profile of a healthy microbiome compared to IBS and IBD patient profiles. The data indicate, for instance, that the IBS patient harbors more Firmicutes than a healthy subject does; in contrast, the IBD patient has fewer Firmicutes.

How might profiling gut microbiota contribute to patient care? IBS, for instance, is currently diagnosed in broad subgroups based on clinical symptoms (diarrhea, chronic constipation, or a mix). Patients within subgroups, however, harbor very different microbiota. So, profiling patient microbiota can lead to more specific diagnoses. Profiling microbiota can also indicate how a patient might respond to treatments such as anti–tumor necrosis factor.

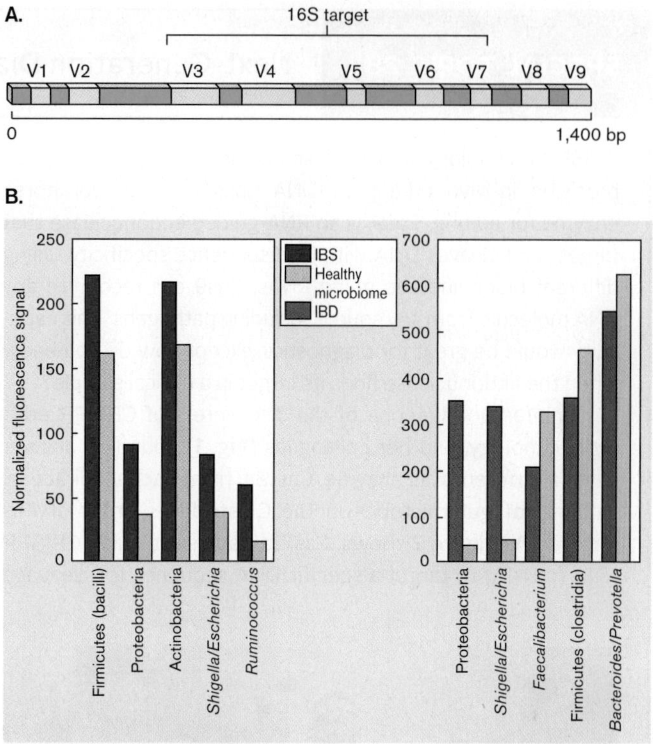

FIGURE 28.18 ▪ Profiling the gut microbiome in healthy and dysbiotic individuals. **A.** The DNA region encoding bacterial 16S ribosomal RNA. Areas labeled V1 to V7 mark the variable regions that differ between species and phyla. **B.** The region encompassing V3–V7 was PCR-amplified from the bacterial DNA extracted from fecal samples of healthy individuals, people with irritable bowel syndrome (IBS), and individuals with inflammatory bowel disease (IBD). Different oligonucleotide probes hybridize to V regions specific to different species. Probes that hybridize were fluorescently labeled by a DNA polymerase and fished out of the mix by bar-coded complementary oligonucleotides attached to magnetic beads. In the two bar graphs, the amount of fluorescence for a specific probe reflects the abundance of the corresponding bacterial species or phylum.

Fecal bacteriotherapy might then convert a nonresponder profile into a responder. Finally, monitoring the profile of a recovered patient could signal a coming relapse. Preventive medication could shorten the period to remission.

Immunology-Based Pathogen Detection

In addition to directly identifying a pathogen by culture isolation or by NAAT detection, clinical laboratories can use immunological techniques to find evidence of a pathogen in patient tissues or in serum, as seen in the following case history.

Case History: Ebola Outbreak

On December 26, 2013, an 18-month-old West African boy in the town of Meliandou, Guinea, fell ill with a mysterious disease of high fever, black stools, and vomiting. He died 2 days later. By January 2014, several members of the boy's family, as well as staff at the nearby Guéckédou hospital, had developed similar

SPECIAL TOPIC 28.1 Next-Generation Diagnostics: CRISPR Launches a "Flare"

CRISPR technology, initially described in Chapter 9, has made precision in vivo editing of DNA possible. The workhorse enzyme for editing, Cas9, is an RNA-guided endonuclease that targets and cleaves DNA with high sequence specificity. Using different bioengineered guide RNAs, Cas9 can recognize any DNA molecule from any source, including pathogens. This capability would be great for diagnostics, except how do you know when the endonuclease finds its target in a clinical sample?

Jennifer Doudna, one of the discoverers of CRISPR editing technology, and her colleagues (**Fig. 1**), found an answer using a different Cas enzyme, Cas12a, from Lachnospiraceae, a family of gut microbes in the Clostridiales order of Firmicutes. As **Figure 2** shows, Cas12a also uses a guide CRISPR RNA (crRNA) to target a specific DNA sequence for cleavage by the enzyme's RuvC endonuclease region. However, when a CRISPR-Cas12a protein cleaves double-stranded DNA in a sequence-specific manner, a robust nonspecific ssDNA nuclease activity is also activated. The researchers predicted that if Cas12a was guided to cleave a specific pathogen's DNA—say, HPV (human papillomavirus)—it would activate the nonspecific endonuclease activity, which could then cleave a separate, reporter ssRNA called a DETECTR (DNA endonuclease–targeted CRISPR trans reporter).

COURTESY OF LUCAS HARRINGTON/MAMMOTH BIOSCIENCES

FIGURE 1 ▪ Members and cofounders of Mammoth Biosciences, a company that develops commercial pathogen detection kits based on DETECTR technology. Janice Chen and Lucas Harrington were principal cofounders of Mammoth Biosciences. Pictured from left to right are Lucas Harrington, James Broughton, Ashley Tehranchi, Andy Lane, Janice Chen, Pedro Galarza, unknown.

The DETECTR ssRNA has a fluorescent tag (F) at one end and a quenching tag (Q) at the other end. The quenching tag absorbs energy released from the fluorescent tag. Consequently, there is no fluorescence when the ssRNA reporter is intact. If Cas12a becomes activated by a specific dsDNA target (pathogen DNA), the nonspecific endonuclease of Cas12a will degrade the ssRNA FQ reporter and release the fluorescent tag. Fluorescence means the pathogen was present in the patient sample. Thus, having detected the pathogen, CRISPR launches a kind of fluorescent "flare." **Figure 3** shows the specificity by which Cas12a loaded with crRNA targeting HPV16 (panel A) or HPV18 (panel B) differentiates between the two viruses. A single DETECTR probe can be used for any pathogen, but different guide RNAs must be used depending on the pathogen. Unleashing the ssDNase activity of Cas12a provides a new strategy to improve the speed, sensitivity, and specificity of molecular diagnostics.

RESEARCH QUESTION

How might you turn a system like this into a multiplex assay—that is, design a single reaction mix that can detect several different target pathogens at the same time. *Hint:* Cas13 enzymes are similar to Cas12a, except that different Cas13 enzymes taken from different species will collaterally cleave different FQ dinucleotide reporters.

Janice S. Chen, Enbo Ma, Lucas B. Harrington, Maria Da Costa, Xinran Tian, et al. 2018. CRISPR-Cas12a target binding unleashes indiscriminate single-stranded DNase activity. *Science* **360**:436–439.

symptoms that included severe bloody diarrhea. All died. Then, in February, a member of the boy's family traveled to the capital city, Conakry, where he, too, fell ill and died. The cause remained unknown, and no precautions were taken. Consequently, over the next month the illness spread to four other prefectures in Guinea, killing at least half of those infected. In March, a World Health Organization (WHO) laboratory at the Pasteur Institute in France finally identified the agent as the Zaire strain of Ebola virus, the most deadly form of this filovirus. Laboratory confirmation tests included qRT-PCR, viral antigen detection, and antibody ELISA tests. Unfortunately, identification of the virus came too late to stop the horrific Ebola epidemic that swept through western Africa in 2014–2015, killing thousands.

Ebola is another RNA virus, but one far more deadly than the West Nile virus identified in the earlier case history. Pathogenesis of Ebola was discussed in Section 26.5 (see Fig. 26.29). This case history includes many elements of epidemiology (discussed later), all of which rely on an ability to identify the presence of the virus. It should be noted that a new series of Ebola outbreaks developed during 2018 in the war-torn Democratic Republic of the Congo. These infections appear to be unrelated to the 2014 epidemic. A new recombinant Ebola vaccine (rVSV-ZEBOV) is being tested during these outbreaks.

Diagnosis of Ebola involves three techniques. We already discussed qRT-PCR, but what serological tests were used?

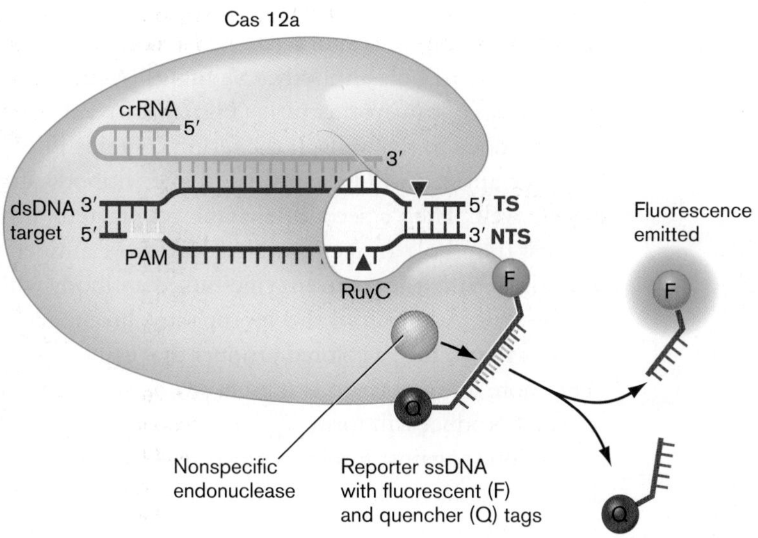

FIGURE 2 ▪ Functions of sequence-specific endonuclease and nonspecific nuclease of Cas12a. crRNA is the guide RNA that directs Cas12a to the DNA target. Pathogen target DNA is prepared from patient samples using RPA isothermal amplification. The RuvC region of Cas12a contains the endonuclease activity that generates a 5′ overhang staggered cut in the target strand (TS) and the nontarget strand (NTS). Activation of the RuvC site also activates a nonspecific DNase that cleaves the ssDNA FQ reporter, which releases a fluorescent signal. PAM = protospacer adjacent motif. *Source:* Modified from Chen et al. 2018. *Science* **360**:436–439, figs. 1A and 4C.

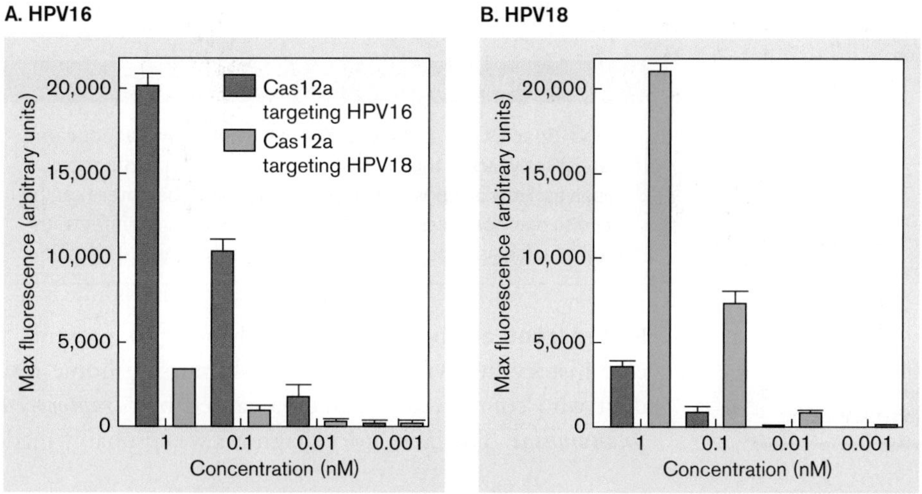

FIGURE 3 ▪ DETECTR distinguishes between two different HPV sequences. Cas12a was assembled with guide RNA targeting either HPV16 **(A)** or HPV18 **(B)**. *Source:* Modified from Chen et al. 2018. *Science* **360**:436–439, fig. S10D and E.

The ELISA test. Enzyme-linked immunosorbent assay (ELISA) is an immunological technique that can detect antigens or antibodies present in picogram quantities. (A picogram is one one-thousandth of a nanogram.) One form of ELISA detects serum antibodies. It is carried out in a 96-well microtiter plate, which allows multiple patient serum samples to be tested simultaneously. An antigen from the virus (Ebola in this case) is attached (adsorbed) to the plastic of the wells (**Fig. 28.19**). Albumin or powdered milk is used to block the remaining sites on the plastic that could result in false positives. Patient serum is then added.

Ebola-specific antibodies present in the serum will react with the antigen attached to the microtiter plate. The antigen-antibody complex is then reacted with rabbit antihuman IgG to which an enzyme has been attached, or conjugated (for example, horseradish peroxidase). The result is a chain of viral antigen connected to patient antibody connected to rabbit antibody-enzyme conjugate. The chain links the enzyme to the well. The chromogenic substrate for the enzyme is added next (for example, tetramethylbenzidine). If enzyme-conjugated antibody has bound to human IgG captured by the antigen in the well, the enzyme will convert the substrate to a colored product (blue for tetramethylbenzidine). Enzyme activity can be measured with an ELISA plate reader. The amount of colored product formed, detected as absorbance with a spectrophotometer, will be an indication of the amount of anti-Ebola antibody present in the patient sample.

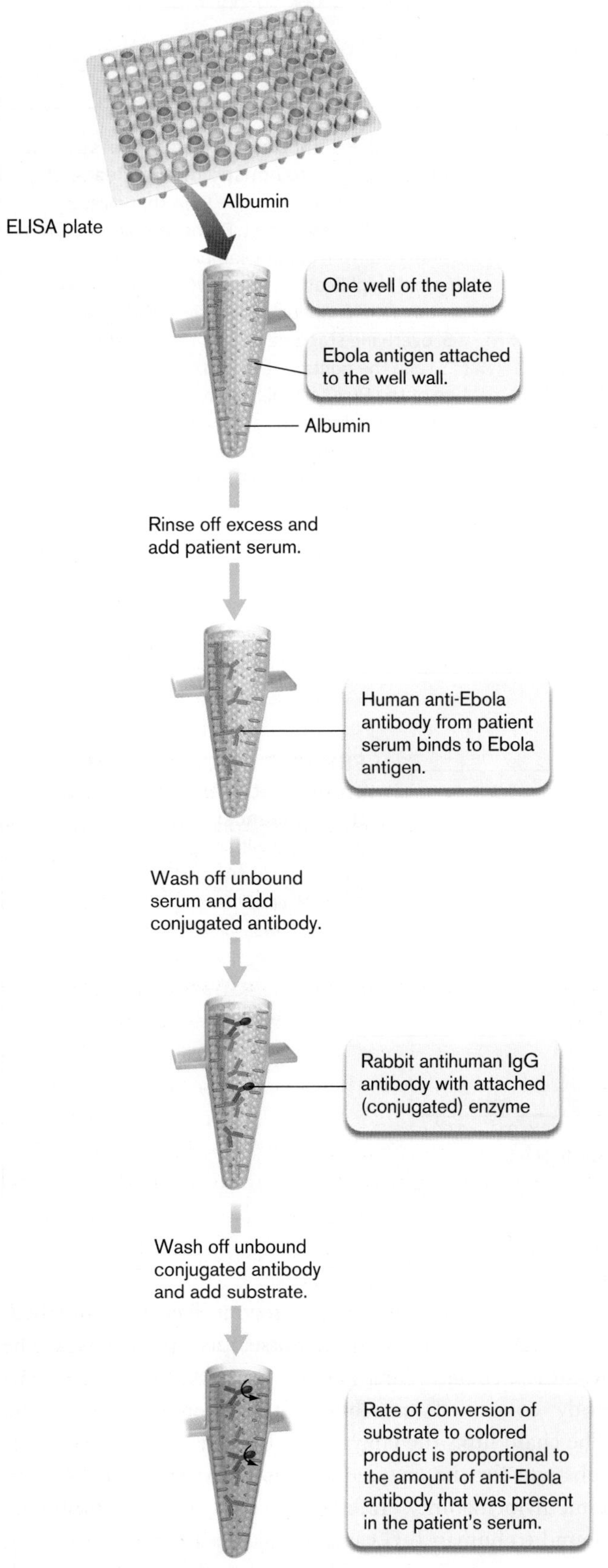

FIGURE 28.19 ■ Enzyme-linked immunosorbent assay (ELISA). ELISA to detect anti-Ebola antibodies circulating in patient serum. The 96-well plate can be used to make dilutions of a single patient's serum to more precisely determine the amount of anti-Ebola antibody, or it can be used to test samples from multiple patients.

Thought Question

28.7 Why does adding albumin or powdered milk prevent false positives in ELISA?

Antigen capture is another ELISA technique, but in this instance, anti-Ebola antibody, not viral antigen, is adsorbed to the wells of a microtiter plate (**Fig. 28.20**). Patient serum is then added to the wells. If the serum contains Ebola antigen, the antigen will be captured by the antibody attached to the well. Then a second, enzyme-conjugated, antibody against the Ebola antigen is added. The more antigen there is in the serum, the more enzyme-linked antibody will affix to the well. Addition of the appropriate chromogenic substrate will produce a colored product that can be measured. The more antigen there is present in the serum, the more colored product will form.

Antibody against Ebola, or any other viral pathogen, may be easier to detect than viral antigen because antibodies will be present at higher levels than is the virus itself. But since there is a delay between the time when the virus is first present in serum and when the body manages to make antibody, directly detecting viral antigen enables earlier diagnosis.

Thought Question

28.8 Specific IgG antibodies against an infectious agent can persist for years in the bloodstream, long after the infection resolves. How is it possible, then, that IgG antibody titers can be used to diagnose recently acquired diseases such as infectious mononucleosis? Couldn't the antibody be from an old infection?

Fluorescent antibody staining. Chapter 26 presents a case history involving an 80-year-old nursing-home resident who contracted pneumonia caused by *Streptococcus pneumoniae*. The laboratory diagnosis was probably made

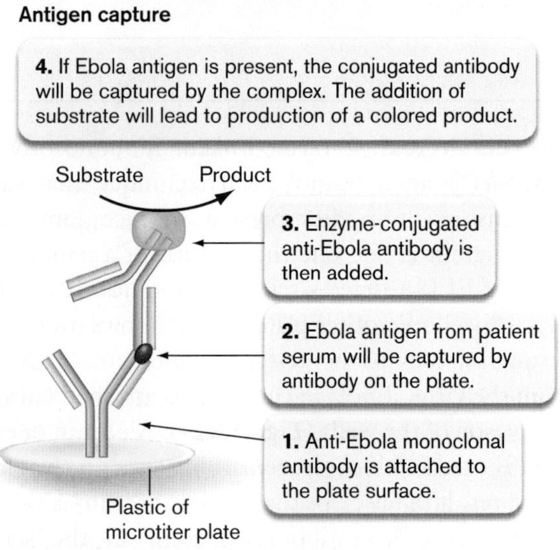

FIGURE 28.20 ■ Antigen-capture ELISA. This ELISA technique captures Ebola antigens circulating in patient serum.

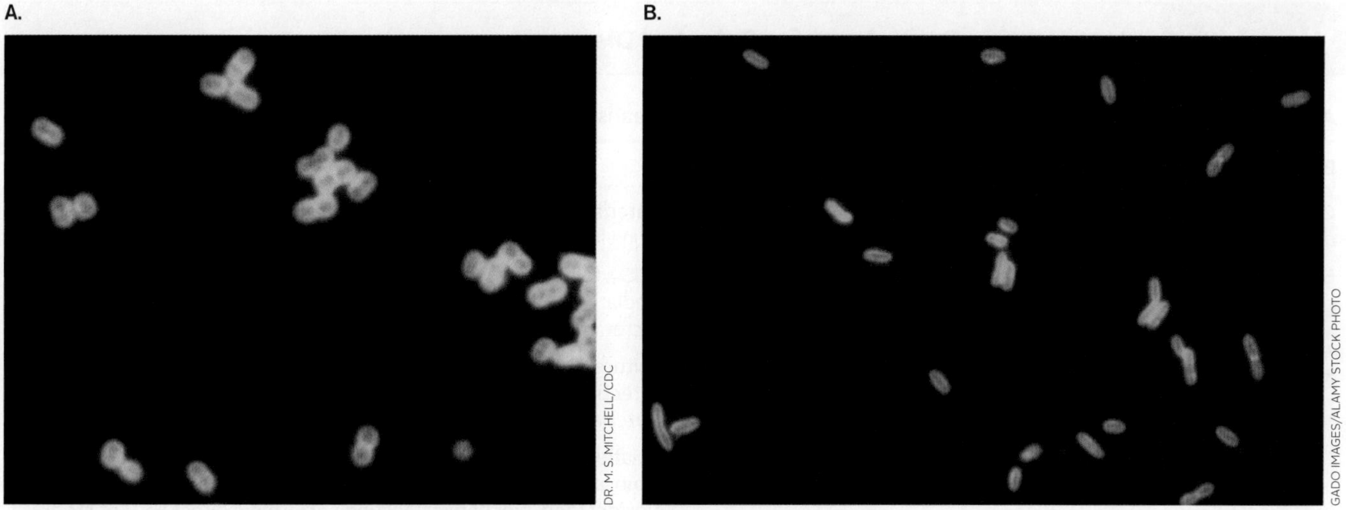

FIGURE 28.21 ■ Fluorescent antibody stain. A. *Streptococcus pneumoniae* capsules (cells approx. 0.8 μm; fluorescence microscopy). The capsule is the green halo. The center of the halo is the cell. B. *Legionella pneumophila* (cells approx. 1 μm in length) from a respiratory tract specimen.

using the biochemical algorithm previously described. However, there are over 80 serological types of *S. pneumoniae*, each one containing a different capsular antigen. How can the lab identify which antigenic type has caused the infection? One way is to stain the organism with antibodies.

Figure 28.21A illustrates the result of staining a smear of the isolated streptococcus with fluorescently tagged antibodies directed against a specific antigenic type of capsule. Viewed under a fluorescence microscope, the organism is "painted" green when the right antibody binds to the capsule. In the case of pneumonia, this knowledge probably will not help in treating the patient, but its broader value is in determining whether a single type of organism is the cause of an outbreak of pneumonia—information that is of epidemiological value for identifying the source of the bacterium.

On the other hand, fluorescent antibody staining techniques are critically important for rapidly identifying organisms that are difficult to grow. Infected tissues can be subjected to direct fluorescent antibody staining. **Figure 28.21B**, for example, shows a direct fluorescent antibody stain of pleural fluid from a patient with Legionnaires' disease.

Other Pathogens Identified by Conventional or Rapid Diagnostics

Table 28.3 presents some additional examples of procedures used to identify pathogens. In general, bacteria that are easily cultured are grown in the laboratory, after which biochemical tests are performed or MALDI-TOF mass spectrometry analysis is conducted. Bacterial, viral, and fungal species that are difficult to grow are typically identified by immunological techniques that either identify a microbial antigen present in infected tissues or measure a rise in antibody titer. Nucleic acid–based methodologies are also used to identify organisms that are difficult to grow. Eukaryotic microbial parasites such as *Plasmodium* species (the cause of malaria), *Giardia lamblia* (which causes giardiasis, a diarrheal disease), and *Entamoeba histolytica* (the cause of amebic dysentery) can be identified via their telltale morphologies under the microscope, making biochemical tests unnecessary.

How do we find pathogens that cannot be cultured? Microbiologists have successfully developed many strategies to identify the causes of infectious diseases. As noted earlier, Robert Koch in the late nineteenth century devised a set of postulates that, when followed, can identify the agent of a new disease (discussed in Sections 1.3 and 25.1). One important tenet of Koch's postulates is that the suspected organism must be grown in pure culture. For some bacterial diseases, however, the agent cannot be cultured. One example is Whipple's disease.

The name sounds made up, but Whipple's disease is very real. The disease is characterized by malabsorption, weight loss, arthralgia (joint pain), fevers, and abdominal pain. Any organ system can be affected, including the heart, lungs, skin, joints, and central nervous system. Although the disease was discovered in 1907 by George Whipple, it took 85 years and modern molecular technology to find the cause. As detailed in **eTopic 28.4**, random amplification of bacterial 16S RNA genes from patient stool samples yielded a unique 16S RNA sequence similar to that of a Gram-positive actinomycete. The new species was named *Tropheryma whipplei*. Other unculturable pathogens exist, and the techniques described throughout this section could help find them.

TABLE 28.3 Identification Procedures for Selected Diseases

Agent	Disease	Means of identification[a]
Bacteria		
Corynebacterium diphtheriae	Diphtheria	Material from nose and throat cultured on a special medium; in vivo or in vitro tests for toxin; NAAT screening for toxin gene
Bordetella pertussis	Pertussis (whooping cough)	ELISA for toxin in respiratory secretions; PCR; culture on special media: MALDI (fluorescent antibody staining of nasopharyngeal secretions no longer recommended)
Legionella pneumophila	Legionellosis (Legionnaires' disease), Pontiac fever	Culture on special medium is preferred method; urine antigen detection by ELISA; direct fluorescent antibody staining of clinical material; NAAT screening of clinical material; MALDI
Campylobacter spp.	Campylobacteriosis	Isolation of bacteria on selective media incubated at 42°C in atmosphere of nitrogen containing 5% oxygen and 10% carbon dioxide; then, biochemical testing performed; direct PCR of stool; MALDI
Leptospira interrogans	Leptospirosis	IgM detected by ELISA; NAAT screening of urine; serology early in the illness and after 2–3 weeks to detect 4× rise in antibody titer
Listeria monocytogenes	Listeriosis	Culture of blood and spinal fluid; selective and enrichment cultures performed on food samples to grow potential pathogens; DNA probe for rapid identification of colonies; MALDI
Chlamydia trachomatis	Chlamydial genital infections	NAAT tests; identification of *C. trachomatis* antigen in urine or pus using monoclonal antibody
Treponema pallidum	Syphilis	Direct fluorescent antibody staining; serological tests
Francisella tularensis	Tularemia	Cultures using cysteine-containing media; fluorescent antibody stain of pus; detection of rise in antibody titer
Yersinia pestis	Plague	Identification of capsular antigen using fluorescent antibody or ELISA
Viruses		
Rhinovirus	Common cold	Strain identification requires use of specific antibodies (not usually done)
Influenza virus	Flu	NAAT assay; influenza antibody levels in acute and convalescent stages of illness
Hantavirus	Hantavirus pulmonary syndrome	NAAT assay; antigen detection in tissues using electron microscopy or monoclonal antibody; ELISA and western blot tests for IgG and IgM antibodies in victim's blood
Herpes simplex virus	Different strains cause cold sores, ocular lesions, and genital lesions	NAAT assay; identifying the viral antigen in clinical material using fluorescent antibody
Mumps virus	Mumps	NAAT; rise in antibody titer, or presence of IgM antibody to mumps virus in patient's blood
Rotavirus	Diarrhea	Electron microscopy or ELISA of diarrheal stool for virus
HIV	AIDS	NAAT assay; detection of antibody to HIV-1 in patient's blood
Fungi		
Coccidioides immitis	Coccidioidomycosis, infection of lung; can disseminate to almost any tissue	NAAT tests; observation of large, thick-walled, round spherules from clinical specimens; MALDI
Histoplasma capsulatum	Histoplasmosis, intracellular infection of lung; sometimes disseminates	Stained material from pus, sputum, tissue, etc., examined for intracellular *H. capsulatum* yeast phase; blood tests for antibody to the organism; MALDI

[a]MALDI = organism is part of the VITEK MS v3 database developed by bioMérieux.

Point-of-Care Rapid Diagnostics

Conventional diagnosis of an infection often requires sending a clinical specimen to a faraway laboratory, followed by a considerable delay in obtaining results. Inevitably, some patients lose patience and fail to attend follow-up appointments. Point-of-care (POC) laboratory tests, by contrast, are used directly at the site of patient care, such as physicians' offices, outpatient clinics, intensive care units, emergency departments, hospital laboratories, and even patients' homes. Most patients are happy to wait 40–50 minutes for a rapid POC test result in order to receive immediate treatment or reassurance.

The typical POC test approved for use involves an immunochromatographic assay. The test for *Streptococcus pneumoniae* capsular antigen shown in **Figure 28.22A** is one example. In this particular immunochromatographic test, called a "red colloidal gold" test, the relevant antigen (for example, capsule antigen) is extracted from a clinical specimen, and a few drops of the extract are placed on a test strip containing rabbit antibodies to the antigen. The antibodies have red colloidal gold particles attached to them. The antigen-antibody complexes, if present, move by capillary action to the upper level of the strip, where they are captured by a line of more anti-antigen antibodies embedded in the strip, forming an antibody-antigen-antibody sandwich. The colloidal gold particles accumulate and eventually produce a red test line, indicating the presence of the antigen. In contrast, rabbit antibodies that are not bound to the antigen pass through the test line but are captured by goat antirabbit IgG antibodies on a control line, once again forming a red control line that indicates the strip components are working. **Figure 28.22B** shows results of the Thermo Scientific™ Xpect™ Legionella Test for *Legionella* antibodies.

Two properties that are critical to any POC test are sensitivity and specificity. **Sensitivity** measures how often a test will be positive if a patient has the disease. Sensitivity reflects how small a concentration of antigen the test can detect. A highly sensitive test will detect lower concentrations of an antigen than a less sensitive test will detect. In contrast to sensitivity, **specificity** measures how often a test will be negative if a patient does not have the disease. Specificity reflects how well a test can distinguish between two closely related antigens (for example, *Streptococcus* strains carrying group A versus group B capsular antigens). Assays with high specificity and sensitivity are valuable diagnostic tools.

Commercial POC tests are widely available for the diagnosis of bacterial and viral infections and of parasitic diseases, including malaria (**Table 28.4**). However, as convenient as these tests are, sensitivity may be compromised in the quest for a speedy result. Some tests exhibit insufficient sensitivity and should therefore be coupled with confirmatory tests when the results are negative (one test that can produce false negatives is the *Streptococcus pyogenes* rapid antigen detection test). Other POC tests need to be confirmed when positive; for instance, a rapid malaria POC test can produce false positives.

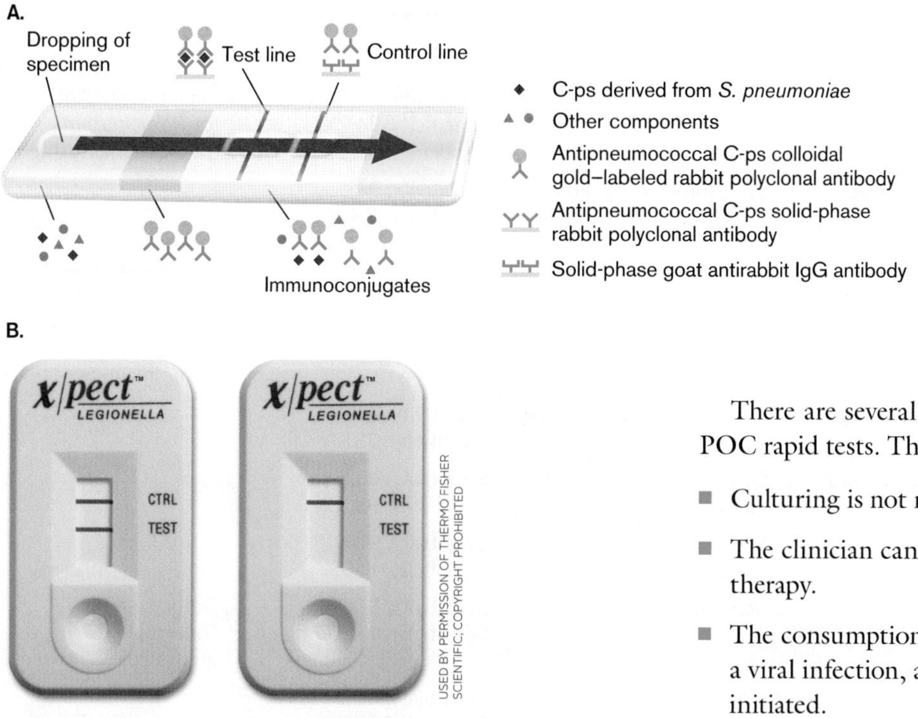

FIGURE 28.22 ▪ Principle of immunochromatographic rapid diagnostic tests. **A.** The example is a test for the presence of *Streptococcus pneumoniae* in sputum. An extract of sputum is placed onto one end of the strip where *S. pneumoniae* capsular antigen (C-ps) will bind antipneumococcal C-ps polyclonal antibodies. The resulting immunoconjugates move by capillary action to the upper membrane and are captured by the antipneumococcal C-ps solid-phase polyclonal antibodies, thereby forming sandwich conjugates in the sample. A positive test result is indicated by the presence of both a test and a control line, whereas a negative result is indicated by the appearance of only a single control line. **B.** Immunochromatography test for anti-*Legionella* antibodies. Serum placed on strip 1 contained antibodies to *Legionella*. (Photo used by permission of Thermo Fisher Scientific.)

There are several advantages and some disadvantages to POC rapid tests. The advantages are:

- Culturing is not required.
- The clinician can immediately initiate specific antibiotic therapy.
- The consumption of antibiotics is avoided in the case of a viral infection, and antiviral therapy, if possible, can be initiated.
- Infection chains among patients with similar symptoms are revealed.
- Compliance by patients who are difficult to reach is improved.

The disadvantages include:

- The tests provide no data on a pathogen's antibiotic sensitivity.
- There is a higher risk of the technician becoming infected.

TABLE 28.4 Examples of Point-of-Care Rapid-Test Kits for Infectious Diseases[a]

Disease (pathogen)	Type of test	Sample	Indication	Performance[b]	Notes
Bacterial pathogens					
Chlamydia (*Chlamydia trachomatis*)	PCR	Vaginal swab, urine	Screening, suspicion of PID	Sensitivity: 98.7% Specificity: 99.4%	
Gonorrhea (*Neisseria gonorrhoeae*)	PCR	Urine	Screening, suspicion of PID	Sensitivity: 98% Specificity: 99%	
Syphilis (*Treponema pallidum*)	ICT	Blood	Screening	Sensitivity: 90%–95% Specificity: 90%–95%	
Strep throat (*Streptococcus pyogenes*)	EIA	Pharyngeal swab	Sore throat	Sensitivity: 92% Specificity: 100%	Confirm negative swabs
Legionellosis (*Legionella* spp.)	ICT	Urine	Severe pneumonia; risk factors for legionellosis	Sensitivity: 97.7% Specificity: 100%	Only serotype 1 reliably detected
Pneumococcal pneumonia (*Streptococcus pneumoniae*)	ICT	Urine; also pleural fluid or CSF	Severe pneumonia; also empyema, meningitis	Sensitivity: 86% Specificity: 94%	Detects capsule antigens
Pseudomembranous enterocolitis (*Clostridioides difficile*)	ICT	Stool	Antibiotic-associated diarrhea	Sensitivity: 92.8% Specificity: 92.6%	Notably less sensitive than cultures or PCR
Neonatal septicemia (*Streptococcus agalactiae*)	PCR	Vaginal swab	Peripartum detection of colonization	Sensitivity: 92% Specificity: 96%	
Protozoan pathogens					
Malaria (*Plasmodium falciparum*)	ICT	Blood	Fever in returning traveler	Sensitivity: 99.7% Specificity: 94.2%	Sensitivity better for *P. falciparum* (pan-malarial tests)
Trichomoniasis (*Trichomonas* spp.)	ICT	Vaginal swab	Symptoms of vaginitis	Sensitivity: 92% Specificity: 100%	
Viral pathogens					
Influenza (influenza virus)	ICT	Nasopharyngeal swab	Flu-like symptoms	Sensitivity: 20%–55% Specificity: 99%	Low sensitivity; probably not helpful during outbreaks; lower in adults
RSV disease (RSV)	ICT	Nasopharyngeal swab	Viral symptoms, especially during winter	Sensitivity: 93% Specificity: 93%	
AIDS (HIV)	ICT	Blood; also oral fluid	Screening, prevention of vertical transmission	Sensitivity: 99.6% Specificity: 99.3%	
Dengue fever (dengue virus)	ICT	Blood	Screening in endemic regions	Sensitivity: 90% Specificity: 100%	
Mononucleosis (Epstein-Barr virus)	ICT	Blood	Screening	Sensitivity: 90% Specificity: 100%	Detects IgM "heterophile antibodies"[c]
Diarrheal disease (rotavirus)	ICT	Stool	Diarrhea	Sensitivity: 88% Specificity: 99%	
Hepatitis B (HBV)	ICT	Blood	Prenatal or transfusion screening; or suspicion of acute or chronic carriage of HBV	Sensitivity: 95% Specificity: 100%	Detects HBs antigen[d]
Rubella (rubella virus)	ICT	Blood	Pregnancy	Sensitivity: 99% Specificity: 99%	

[a]Abbreviations: AIDS = acquired immunodeficiency syndrome; CSF = cerebrospinal fluid; EIA = enzyme immunoassay; HBV = hepatitis B virus; HIV = human immunodeficiency virus; ICT = immunochromatographic test; PCR = polymerase chain reaction; PID = pelvic inflammatory disease; RSV = respiratory syncytial virus.

[b]Sensitivity = proportion of actual positives correctly identified. Specificity = proportion of actual negatives correctly identified.

[c]Epstein-Barr virus randomly infects B cells and causes them to secrete antibodies. Because many thousands of different antibodies are made, they are referred to collectively as "heterophile antibodies."

[d]HBs is a type of hepatitis B antigen that is soluble.

- Double or multiple infections are more likely to be overlooked than in culture.
- The levels of false positive or false negative results can vary for different POC tests.

The newer nucleic acid–based tests described earlier exhibit better sensitivity and specificity than do most immunochromatographic assays, but they require more expensive instrumentation and training. In the coming years, further evolution of POC tests may lead to new diagnostic approaches, such as panel testing that targets all possible pathogens suspected in a specific clinical setting. The development of next-generation serology-based and/or molecular-based multiplex tests will certainly facilitate quicker diagnosis and improved patient care. We also saw earlier how synthetic biology may revolutionize rapid diagnostics, as exemplified by the newly described Zika virus riboswitch test.

To Summarize

- **Pathogens can be identified by molecular techniques** such as MALDI-TOF mass spectrometry (useful for bacteria), nucleic acid amplification tests (useful for bacteria or viruses), and/or **immunological methods** (for example, ELISA; useful for bacteria or viruses).
- **Fluorescent antibody staining** performed directly on a tissue can rapidly identify organisms or antigens present.
- **Point-of-care (POC) diagnostic tests** can rapidly identify or rule out the cause of an infectious disease. Therapy can be initiated rapidly following a positive result. **Sensitivity** (the ability to detect small amounts of the pathogen) and **specificity** (the ability to distinguish one pathogen from another) for any point-of-care test are factors used to evaluate how accurate the test is at eliminating false negative and false positive reactions, respectively.
- **Immunochromatography** is the primary platform for POC testing.
- **A drawback to POC tests** is that simultaneous multiple infections may be missed.

28.4 Epidemiology

In Section 25.1 we discussed basic epidemiological concepts—vectors, transmission cycles, vehicles, and reservoirs, and others—as well as how a new infectious disease can be identified using a version of Koch's postulates. But how do scientists track the spread of a new disease, or find and identify new variants of influenza virus that develop thousands of miles away, and then predict, many months in advance, when that virus will arrive on our "doorstep"? These questions are addressed using the tools of epidemiology. We start our discussion of epidemiology with a case in which scientists had to trace a heinous criminal act back to its source.

Case History: Inhalation Anthrax

On October 16, 2001, a 56-year-old African-American U.S. Postal Service worker became ill with a low-grade fever, chills, sore throat, headache, and malaise. These symptoms were followed by minimal dry cough, chest heaviness, shortness of breath, night sweats, nausea, and vomiting. On October 19, the man arrived at a local hospital, where he presented with a normal body temperature and normal blood pressure. He was not in acute distress, but he had decreased breath sounds and rhonchi (dry sounds in lungs due to congestion). No skin lesions were observed, and he was not a smoker. Total white blood cell count was normal, but there was a left shift in the differential—that is, more polymorphonuclear leukocytes (PMNs; see Section 26.2). A chest X-ray showed bilateral pleural effusions (accumulation of fluids in the lung) and a small, right-lower-lobe air space opacity. Within 11 hours, blood cultures taken upon admission grew Bacillus anthracis. *Ciprofloxacin, rifampin, and clindamycin antibiotic treatments were initiated, and the patient recovered. His job at the post office was simply to sort mail.*

From October 4 to November 2, 2001, the Centers for Disease Control and Prevention (CDC), and various state and local public health authorities, reported 10 confirmed cases of inhalational anthrax and 12 confirmed or suspected cases of cutaneous anthrax in persons who worked in the District of Columbia, Florida, New Jersey, and New York. Many of them were postal workers. It was clear that a biological attack was in progress.

Painstaking detective work by federal agents and epidemiology scientists proved that the strain of *Bacillus anthracis* used in the 2001 anthrax attack had the same genetic signature as a strain used by a scientist working at the army's Fort Detrick biodefense laboratory. Unfortunately, the scientist took his own life before his connection could be proved.

The word "epidemiology" is derived from the Greek meaning "that which befalls man." In scientific parlance, **epidemiology** examines the distribution and determinants of disease frequency in human populations. Put more simply, epidemiologists determine the source of a disease outbreak and the factors that influence how many individuals will succumb to the disease. Epidemiological principles are also used to determine the effectiveness of therapeutic measures and to identify new diseases or syndromes, such as those caused by Zika virus, SFTS virus (severe fever and thrombocytopenia syndrome), *Candida auris* (invasive fungal infections), and emerging influenza animal viruses. Some of the basic concepts of epidemiology were already covered in Chapter 25 when we discussed infection cycles. Now we will explore how those principles are used to track disease.

Epidemiological early-warning systems require an extensive organization that coordinates information from many sources. In the United States, that duty falls to the Centers for Disease Control and Prevention (CDC). On the world stage, the World Health Organization (WHO) bears this responsibility. Any disease considered highly dangerous or infectious is first reported to local public health centers, usually within 48 hours of diagnosis. The local centers forward that information to their state agencies, which then report to the CDC in Atlanta. This is how authorities in 2001 quickly recognized that an outbreak of anthrax was under way.

John Snow, Father of Modern Epidemiology

The science of epidemiology can be traced back to Hippocrates (ca. 460–375 BCE), who noted, for example, that malaria and yellow fever commonly occurred in swampy areas (it took another 2,000 years to make the connection to mosquitoes). Many others after Hippocrates made epidemiological observations about infectious diseases. However, the first case in which the source of a disease outbreak was methodically investigated arose in the mid-nineteenth century.

A serious outbreak of cholera had developed in the Soho district of London in 1854. The source of the infection was unknown. A London physician named John Snow (**Fig. 28.23A**) thought that if the cases of cholera clustered geographically, he might gain a clue as to its source. He visited the addresses of all the diarrheal cases he learned about and drew a map marking each case (**Fig. 28.23B**). The answer jumped out at him. Water in that part of London was pumped from separate wells located in the various neighborhoods. Snow realized there was a close association between the density of cholera cases and a single well located on Broad Street. Simply removing the pump handle of the Broad Street well put an end to the epidemic, proving that the well water was the source of infection (known as the "point source," since all infections originated from that point). Snow stopped the cholera outbreak 50 years before the agent of the disease, *Vibrio cholerae*, was discovered (1905). In the process, he laid the groundwork for descriptive and analytical epidemiological approaches.

Note: Outbreaks of disease that originate from a common source, as in the London cholera outbreak, are called common-source outbreaks. In contrast, a propagated outbreak involves person-to-person transmission (with SARS, for instance) or transmission by insect vectors (such as in West Nile disease).

Endemic, Epidemic, or Pandemic?

The terms "endemic" and "epidemic" are often used when referring to disease outbreaks. A disease is **endemic** if it is always present at a low frequency in a population. For example, Lyme disease, caused by the spirochete *Borrelia burgdorferi*, is endemic to the northeastern United States because the organism has found a reservoir in deer and

A.

THE JOHN SNOW ARCHIVE AND RESEARCH COMPANION

B.

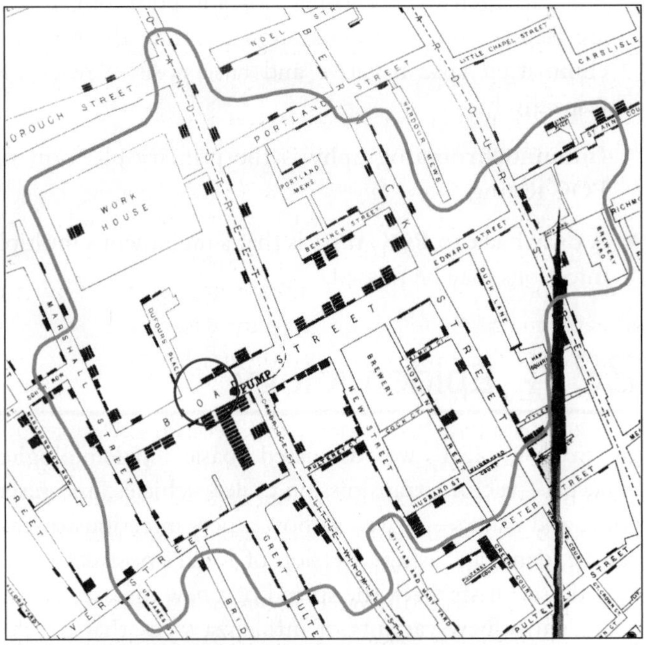

THE JOHN SNOW ARCHIVE AND RESEARCH COMPANION

FIGURE 28.23 ■ Early epidemiology. A. John Snow (1813–1858). B. A map of London commissioned by Snow in 1855 shows the location of cholera victims of the 1854 outbreak. The map illustrates that each victim within the marked area lived closer to the Broad Street pump than to other nearby pumps. The Broad Street well is found within the red circle. Each black bar represents a death from cholera.

ticks. Recall that a **reservoir** is any bird, insect, mammal, or other animal that harbors the infectious agent and is indigenous to a geographic area. Humans become infected only when they come in contact with the reservoir. Thus, the disease incidence is low but relatively constant. A disease is **epidemic**, on the other hand, when larger-than-normal numbers of individuals in a population become infected over a short time. Epidemics arise, in part, because of rapid and direct human-to-human transmission. Food-source epidemics, for instance, are discussed in Chapter 16.

Figure 28.24A illustrates the difference in the frequency of cases observed between endemic and epidemic disease. An endemic disease can become epidemic if the population of the reservoir increases, allowing for more frequent human contact; or if the infectious agent evolves to spread directly from person to person, bypassing the need for a reservoir. This is the concern with the H5N1 avian flu virus, which is endemic in some mammals and birds in Asia (see Section 11.2 and **eTopic 26.4**). A **pandemic** is an epidemic that occurs over a wide geographic area, usually the world. Pandemics may be long-lived, such as the bubonic plague pandemic in the fourteenth century and the AIDS pandemic in the late twentieth and early twenty-first centuries; or they may be short-lived, as with the 1918 flu pandemic.

When discussing a disease, epidemiologists distinguish between the prevalence and incidence of active cases. **Prevalence** describes the total number of active cases of a disease in a given location at a given time, regardless of when a case first developed (old plus new cases). Thus, duration of the disease (acute or chronic) will impact disease prevalence. Prevalence could increase or decrease over time, depending on whether more people are being cured or perhaps die. **Incidence**, however, refers to the number of new cases of a disease in that location over a specified time, and reflects the risk a person has of acquiring the disease. Incidence rates can provide insight into whether efforts to limit a disease are working. Take a city of 100,000 people. If the incidence of new cases of a disease rises from 10 cases per 100,000 population in one year to 30 cases per 100,000 population the next year, then efforts to prevent the disease are failing. Endemic diseases maintain relatively constant prevalence and incidence rates. Outbreaks are typically marked by an increase in both parameters.

A good analogy to describe the relationship between incidence and prevalence is a bathtub. Think of the tub as representing an area's total population, and the volume of water in the tub as the prevalence of a disease. (Filling one-tenth of the tub with water would represent a prevalence of one out of ten people infected.) Water that enters the tub through a faucet signifies new cases of the disease (incidence), whereas

A. Endemic versus epidemic

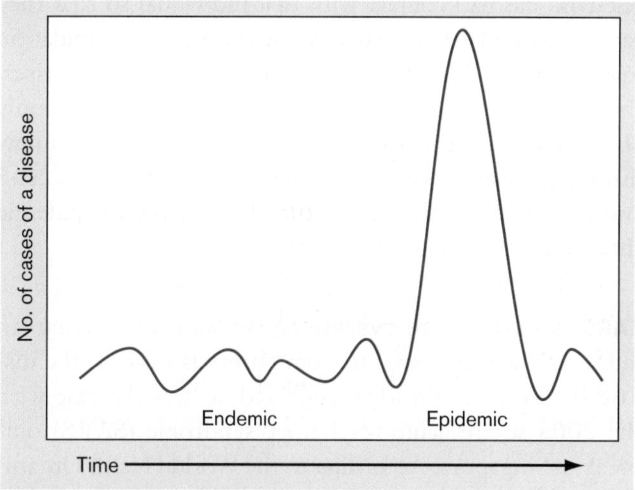

B.

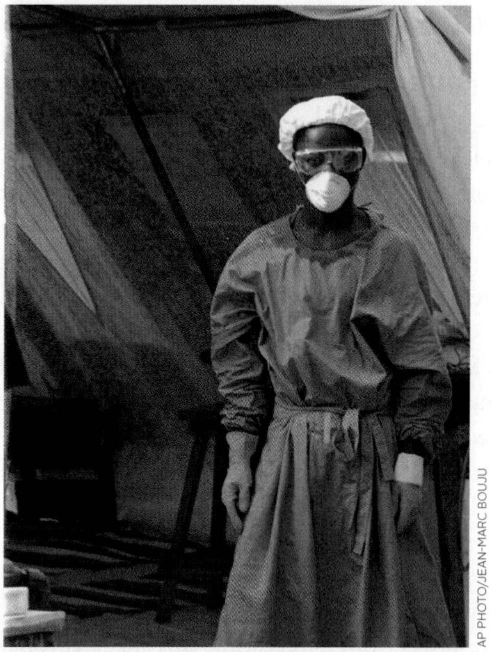

FIGURE 28.24 ■ The difference between endemic and epidemic disease. A. An endemic disease is continually present at a low frequency in a population. A sudden rise in disease frequency constitutes an epidemic. **B.** A health care worker stands outside a quarantined area housing Ebola patients in the Ivory Coast. Epidemics can be minimized if infected persons are kept segregated from the general population (quarantined) to avoid spread of the infectious agent.

water leaving through a drain represents people who are being cured or die from the disease. Thus, the volume of water in the tub (prevalence) changes, depending on how much water enters (incidence) or leaves the tub.

Finding Patient Zero

When trying to contain the spread of an epidemic, it is vital to track down the first case of the disease (known as the **index case** or **patient zero**) and then identify everyone who has had contact with that individual so that they can be treated or separated from the general population (**quarantined**; **Fig. 28.24B**). When a new disease arises, the epidemiological search for the index case starts only after a number of patients have been diagnosed and a new disease syndrome declared. This is what happened in 2015 and 2018 with Ebola, and in 2003 when a SARS epidemic threatened the world.

SARS: a case where everything worked. Identifying an index case within a specific community is easier if the disease syndrome is already recognized, as was the case with the 2003 severe acute respiratory syndrome (SARS) outbreak in Singapore. According to the World Health Organization, a suspected case of SARS is defined as an individual who has a fever greater than 38°C (100.4°F), who exhibits lower respiratory tract symptoms, and who has traveled to an area of documented disease or has had contact with a person afflicted with SARS. The index case in Singapore was a 23-year-old woman who had stayed on the ninth floor of a hotel in Hong Kong while on vacation. A physician from southern China who stayed on the same floor of the hotel during this period is believed to have been the source of her infection, as well as that of the index patients who precipitated subsequent outbreaks in Vietnam and Canada.

During the last week of February, the woman, who had returned to Singapore, developed fever, headache, and a dry cough. She was admitted to Tan Tock Seng Hospital, Singapore, on March 1 with a low white blood cell count and patchy consolidation in the lobes of the right lung. Tests for the usual microbial suspects (*Legionella*, *Chlamydia*, *Mycoplasma*) were negative. Electron microscopy of nasopharyngeal aspirations showed virus particles with widely spaced club-like projections, suggesting this was a coronavirus (**Fig. 28.25A**). At the time of the woman's admission to the hospital, the clinical features and highly infectious nature of SARS were not known. Thus, for the first 6 days of hospitalization, the patient was in a general ward, without barrier infection-control measures. During this period, the index patient infected at least 20 other individuals, including hospital staff, nearby patients, and visitors.

Within weeks, WHO named the disease in China "SARS" and issued travel alerts (**Fig. 28.25B**; discussed further in Section 28.5). These alerts enabled Singapore health officials to rapidly identify the index patient and her contacts. As a result, they were able to limit spread of the illness. When all was said and done, 8,098 people worldwide became ill with SARS, and 770 victims died. Today, SARS remains a threat, but one of lesser concern. Methicillin-resistant *Staphylococcus aureus* (MRSA), H5N1 avian flu, and H1N1 swine flu are considered more pressing dangers.

A.

B.

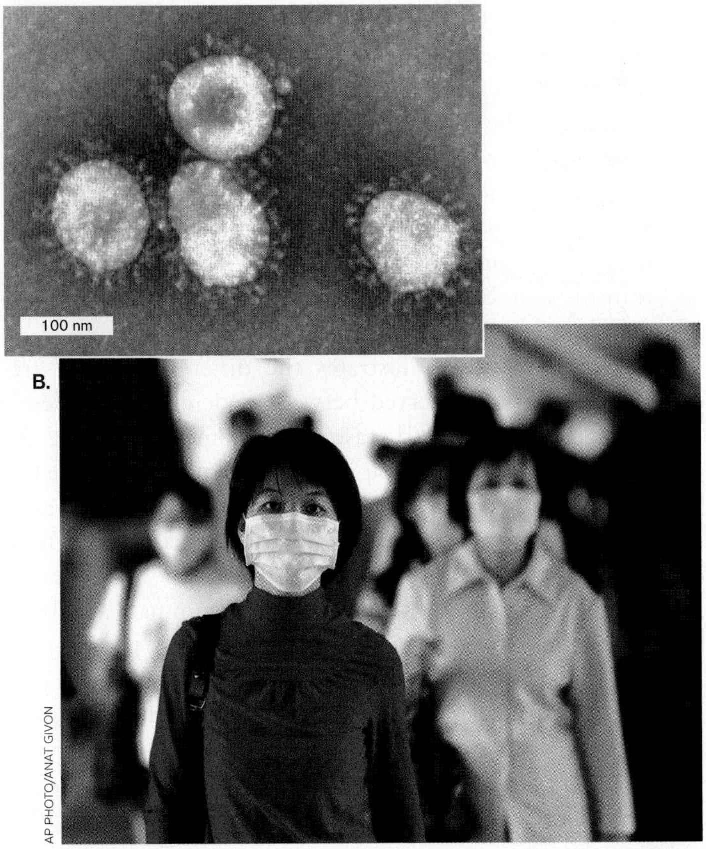

DR. FRED MURPHY/CDC

AP PHOTO/ANAT GIVON

FIGURE 28.25 ■ Severe acute respiratory syndrome (SARS). **A.** The coronavirus that causes SARS (TEM). **B.** Citizens of China, including the military, donned surgical masks in 2003 to slow the spread of SARS.

Identifying Disease Trends

How do epidemiologists first recognize that an epidemic is under way, and then identify the agent and its source? Certain diseases, because of their severity and transmissibility, are called reportable, or notifiable, diseases (**Table 28.5**). In a process known as **systematic surveillance** or **syndromic surveillance**, physicians are required to report instances of these diseases to a central health organization, such as the CDC in the United States and WHO. This reporting enables the incidences of certain diseases within a population to be tracked and upsurges noted.

An emerging disease not on the list of notifiable diseases can be detected as a cluster of patients with unusual symptoms or combinations of symptoms. Such detection is possible because diseases of unknown etiology are also reported to health authorities. A new disease could manifest with common symptoms (for example, the cough and fever of SARS) that cannot be linked to a known disease agent by clinical tests. An upsurge in cases, of either a reportable disease or an emerging disease, will set off institutional "alarms" that initiate epidemiological efforts to determine the source and cause of the outbreak.

TABLE 28.5 Notifiable Infectious Diseases[a]

Bacterial		
Anthrax	Hansen's disease (leprosy)	Salmonellosis (non–typhoid fever types)
Botulism	Legionellosis	Shigellosis
Brucellosis	Leptospirosis	Staphylococcal enterotoxin
Campylobacter infection	Listeriosis	Streptococcal invasive disease
Chlamydia infection	Lyme disease	*Streptococcus pneumoniae*, invasive disease
Cholera	Meningitis, infectious	Syphilis
Diphtheria	Pertussis	Tuberculosis
Ehrlichiosis	Plague	Tularemia
E. coli O157:H7 infection	Psittacosis	Typhoid fever
Gonorrhea	Q fever	Vancomycin-resistant *Staphylococcus aureus* (VRSA)
Haemophilus influenzae, invasive disease	Rocky Mountain spotted fever	
Viral		
Dengue fever	Mumps	Smallpox
Hantavirus infection	Poliomyelitis	Varicella (chickenpox), fatal cases only (all types)
Hepatitis, viral	Rabies	Yellow fever
HIV infection	Rubella and congenital rubella syndrome	
Measles	SARS coronavirus	
Fungal		
Coccidioidomycosis		
Parasitic		
Amebiasis	Giardiasis	Trichinosis
Cryptosporidiosis	Malaria	
Cyclosporiasis	Microsporidiosis	

[a]For the latest list of notifiable agents, search on "CDC notifiable agents 2018" online.

Thought Questions

28.9 Methicillin, a beta-lactam antibiotic, is very useful in treating staphylococcal infections. The emergence of methicillin-resistant strains of *Staphylococcus aureus* (MRSA) is a very serious development because few antibiotics can kill these strains. Imagine a large metropolitan hospital in which there have been eight serious nosocomial infections with MRSA and you are responsible for determining the source of infection so that it can be eliminated. How would you accomplish this task using common bacteriological and molecular techniques?

28.10 What are some reasons why some diseases spread quickly through a population while others take a long time?

Molecular Approaches for Disease Surveillance

A worldwide pandemic of tuberculosis (active and latent forms) is currently estimated to affect over 2 billion people. Many of the *Mycobacterium tuberculosis* infections are caused by multidrug-resistant strains that are difficult, if not impossible, to kill with existing antibiotics (see Section 26.2). This problem is especially serious among refugee populations attempting to flee war-torn countries. As a result, it is important to screen these refugees as they enter neighboring countries that have low incidences of tuberculosis. Although chest X-rays are mandatory in many cases, a positive image will be obtained only if the disease is at a relatively advanced stage. Actively infected individuals who have not developed the characteristic lung tubercles seen on X-ray will not be identified.

Unfortunately, the acid-fast staining of sputum samples (discussed in Section 28.2) also fails to detect individuals at an early stage of active infection. Studies have shown, however, that PCR techniques are much more sensitive for detecting individuals with active tuberculosis. As time progresses, PCR surveillance strategies will be used more often to track the worldwide ebb and flow of microbial diseases.

PCR and restriction fragment length polymorphism (RFLP) strategies are already used for epidemiological purposes to type (that is, determine the relatedness of) different microbial isolates by generating a complex DNA profile that is specific for a particular strain (see Section 28.2). For example, DNA profiles were used to link an outbreak of over 2,000 cases of salmonellosis in 2010 to a single strain of *Salmonella enterica* serovar Enteritidis. The results, conducted by a national network of public health agencies (PulseNet), led to the recall of over a half billion eggs. More recently, whole-genome sequencing was used to track an outbreak of salmonellosis caused by a *Salmonella* strain called *Salmonella* Reading. The organism was found in turkey products such as raw turkey pet foods and live turkeys during the 2018 Thanksgiving season. Contamination was widespread in the industry, and as of this writing (November 2019), a common supplier had not been identified. Reports of people infected came from 41 states: 279 people were infected, 107 were hospitalized, and 1 died.

Bioterrorism

"A wide-scale bioterrorism attack would create mass panic and overwhelm most existing state and local systems within a few days," said Michael T. Osterholm, director of the Center for Infectious Disease Research & Policy at the University of Minnesota, in October 2001. "We know this from simulation exercises."

Less than a month after the September 11 attack on the World Trade Center, a biodefense scientist working at the U.S. Army Medical Research Institute of Infectious Diseases (Fort Detrick, Maryland) allegedly sent weapons-grade anthrax spores through the U.S. mail (see the inhalation anthrax case history described earlier). The anthrax attack was unrelated to 9/11. Five persons died, and a mere 25 became ill. But even though the efficiency of the attack was poor, the impact was enormous. Over 10,000 people took a 2-month course of antibiotics after possible exposure, and mail delivery throughout the country was affected. The simple act of opening an envelope suddenly became a risky business.

As a result of this attack and other events, the CDC and the National Institutes of Health (NIH) assembled a list of select agents (marked in blue in **Table 28.1**) that could potentially be used as bioweapons. A bioweapon is considered to be any infectious agent or toxin that has high virulence and/or mortality rate. Microorganisms considered bioweapons could be used to conduct biowarfare, with the intent of inflicting massive casualties; or bioterrorism, which may result in only a few casualties but cause widespread psychological trauma.

Although the branding of select agents is recent, biowarfare is not new. In the Middle Ages, victims of the Black Death (plague caused by *Yersinia pestis*) were flung over castle walls via catapults. During the French and Indian War in the eighteenth century, British field marshal Jeffrey Amherst distributed smallpox-infected blankets to Native Americans. The Imperial Japanese Army during World War II experimented with infectious disease weapons, using Chinese prisoners as guinea pigs. Even the United States participated by developing weapons-grade anthrax spores after World War II. That project was discontinued in the 1970s.

The 2001 anthrax attack described earlier was not the first act of bioterrorism in the United States. The first documented act occurred in 1984, when followers of the cult leader Bhagwan Shree Rajneesh tried to control a local election in The Dalles, Oregon, by infecting salad bars with *Salmonella*. Over 700 people became ill. Rajneesh was given a 10-year suspended sentence, fined $400,000, and deported.

How effective are bioweapons? The method by which a biological agent is dispersed plays a large role in its effectiveness as a weapon. Only a few people became ill during the 2001 anthrax attack—not only because of the epidemiological surveillance, but because anthrax is inherently difficult to disperse. A person has to inhale thousands of spores to contract the disease, which means that effective dispersal of the spores is critical for the use of anthrax as a weapon. Once spores hit the ground, the threat of infection is limited. Weapons-grade spores are very finely ground so that they stay airborne longer. But, as we saw earlier, the letter-borne dispersal system did not effectively generate large numbers of victims. Nevertheless, the potential threat of weapons-grade anthrax on the battlefield led the U.S. military in 2006 to resume vaccinating all soldiers serving in Iraq, Afghanistan, and South Korea.

An effective bioweapon would capitalize on person-to-person transmission. In an easily transmitted disease, one infected person could disseminate disease to scores of others within 1 or 2 days. So, in terms of generating massive numbers of deaths, anthrax was a poor choice. The goal of most terrorists, however, is not to kill large numbers of people, but to terrorize them. In that regard, the anthrax attack succeeded (**Fig. 28.26A**).

The most effective bioweapon in terms of inflicting death (biowarfare) would have a low infectious dose, would be easily transmitted between people, and would be a potential threat to a large percentage of the population. Smallpox fits these criteria and would be the bioweapon of choice (**Fig. 28.26B**). Fortunately, however, smallpox has been eradicated (almost) from the planet. Two laboratories, however, still harbor the virus—one in the United States and one in Russia. It is believed that the virus has been destroyed in all other laboratories. Since smallpox is the perfect biowarfare agent, it is imperative that the last two smallpox repositories remain secure.

FIGURE 28.26 ■ **Dealing with bioterrorism.** **A.** Members of a hazardous-materials team near Capitol Hill during the anthrax attacks in 2001. **B.** An Illinois man suffering from smallpox in 1912.

While the good news is that smallpox disease has been eradicated, the bad news is that no individual born in the United States after 1970 has been vaccinated, with the exception of some military and laboratory personnel. As a result, anyone under 50 years of age is susceptible to smallpox. Even those of us who received the smallpox vaccination over 45 years ago are at risk, since our protective antibody titers have diminished. A terrorist attack with smallpox would cause terrible numbers of deaths. A new, safer vaccine does exist, however, and has been stockpiled. Were a smallpox attack to be launched, the vaccine would be rapidly administered to limit the spread of disease. Nevertheless, the economic and psychological impact of a smallpox epidemic would be devastating. Interestingly, the smallpox vaccine can also protect people from monkeypox, which is now spreading through central and western Africa, and could evolve into a more deadly pathogen.

In the United States, research with organisms considered to be select agents is tightly regulated. Because *Yersinia pestis*, for instance, is a select agent, laboratory personnel working with it must now possess security clearance with the Department of Justice (the organism must be handled under biosafety level 3 conditions; see **Table 28.1**). The laboratory must also register with the CDC to legally possess this pathogen, and access to the lab and the organism must be tightly controlled.

Much has improved since Michael Osterholm offered his dire assessment of a wide-scale bioterrorism attack. Education and surveillance procedures have been bolstered, and new detection technologies are being developed (**eTopics 28.3 and 28.5**). We will probably never be fully protected from attack, biological or otherwise, but recent efforts have improved the situation.

To Summarize

- **John Snow** founded the discipline of epidemiology.
- **Epidemiology** examines factors that determine the distribution and source of disease.
- **Endemic, epidemic, and pandemic** are terms for different frequencies of disease in different geographic areas.
- **Finding patient zero (the index case)** is important for containing the spread of disease.
- **Molecular approaches using PCR and nucleic acid hybridization** are used to identify nonculturable pathogens and to track disease movements.
- **Bioweapons**, when they have been used, typically kill few people but incite great fear.
- **The CDC** has assembled a list of select agents with bioweapon potential.

28.5 Detecting Emerging Microbial Diseases

News, whether obtained from traditional sources or social media, almost always carries stories of new infectious diseases cropping up in the world, so-called emerging infectious diseases. An **emerging disease** is defined as an infectious disease that has recently appeared in a population. A reemerging disease, in contrast is a known disease that was controlled but whose incidence or geographic

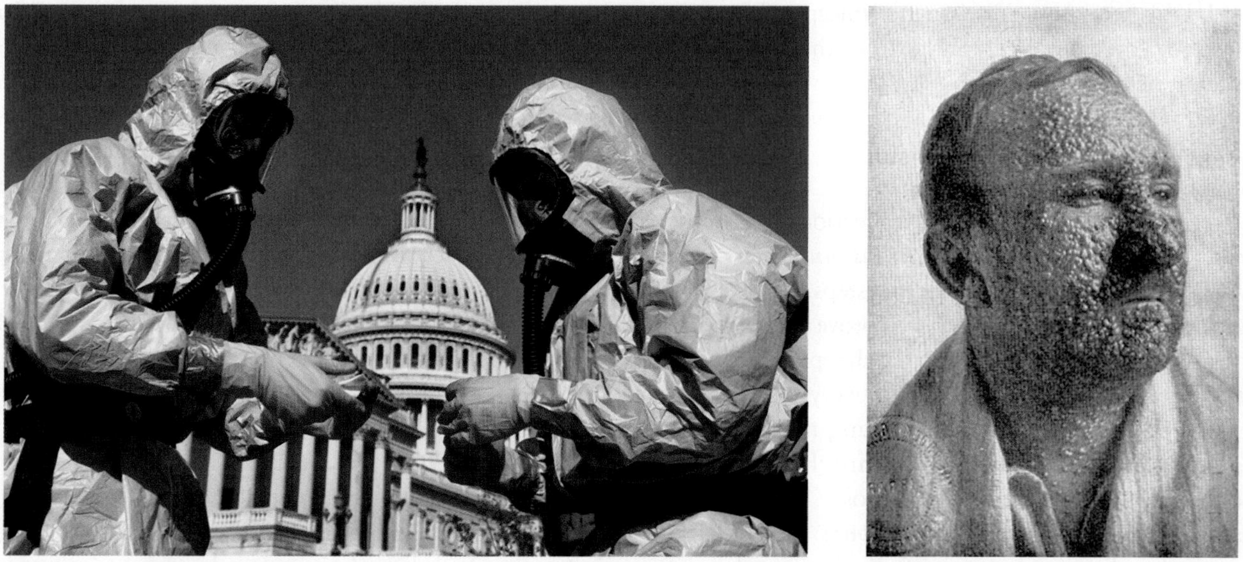

range is increasing or is threatening to increase in the near future. In this section we discuss the concept of emerging diseases, how they develop, and how health organizations detect and track them.

To Catch a Pathogen

How are the microscopic agents of infectious disease discovered? The revelation that *Bacillus anthracis* causes anthrax led Koch to propose a set of steps, or postulates (discussed in Section 1.3), needed to prove that a specific microbe causes a specific disease. Koch's postulates state that the organism must be present in every case of a disease, must be propagated in pure culture, must cause the same disease when inoculated into a naive host, and must be recovered from the newly diseased host.

Viruses, however, cannot grow in pure culture. Thus, viruses causing disease cannot satisfy Koch's postulates. To accommodate viral diseases, Koch's criteria were modified by Thomas Rivers in 1937 to include cultivating the agent in host cells (rather than in pure culture), proving that the agent passes through a 0.2-μm filter (known bacteria did not), and demonstrating an immune response to the virus in patients. These steps were used in 2003 to rapidly discover the virus causing severe acute respiratory syndrome (SARS) and in 2013 to discover a related virus causing Middle East respiratory syndrome (MERS). Both viruses were grown in a macaque monkey model, fulfilling Rivers's postulates.

Fulfilling Koch's and Rivers's postulates remains the most persuasive evidence of causation, but there are some problems with this standard. Many agents cannot be cultured, and there may be no suitable animal model in which disease can be reproduced. In these situations we must resort to a statistical association between organism and disease based on the presence of the agent or its footprints (nucleic acid, antigen, and preferably, an immune response). Statistical association was used to link Zika virus and newborn microcephaly, first observed in South America in 2015. This link was strengthened by the discovery of Zika virus in the tissues of stillborn infants.

Emerging and Reemerging Pathogens

A world map showing the general locations of emerging and reemerging diseases is shown in **Figure 28.27**. As described in Chapter 26, tuberculosis is considered a reemerging infection. Once a worldwide scourge, tuberculosis was thought to be conquered by effective antimicrobial treatments. As evidence, the incidence rate of tuberculosis in the United States dropped sharply starting in the 1950s, falling from 52 per 100,000 population in 1953 to 10 per 100,000 in 1983. During the 30 years since 1983, the rate of decline has slowed to 3 per 100,000 (2013). Meanwhile, the worldwide incidence is still about 150 per 100,000 population. So, while the incidence of tuberculosis is nowhere near the level it was in the 1950s, we are far from eliminating TB completely.

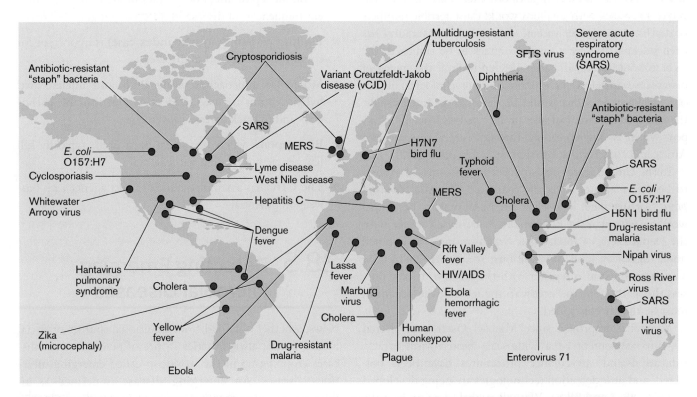

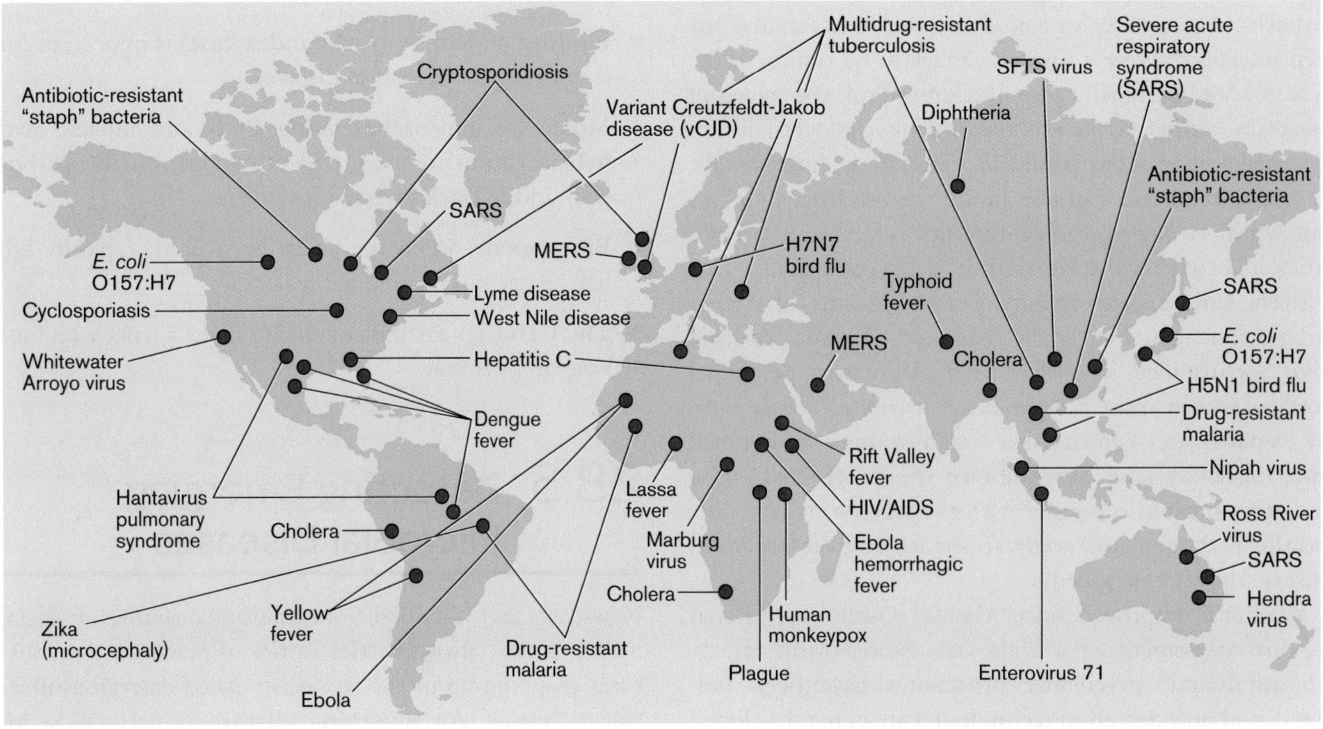

FIGURE 28.27 ■ Locations of some emerging and reemerging infectious diseases and pathogens. The examples given represent extreme increases in the reported cases over the last 20 years. Many of these diseases, such as HIV/AIDS and cholera, are widespread but show alarming increases in the areas indicated.

What happened? The reemergence of tuberculosis is linked to two events: the AIDS pandemic and the development of drug-resistant strains. Because of their highly immunocompromised state, many AIDS patients during the 1980s and '90s developed tuberculosis, which too often became a death sentence.

The second event, drug resistance, developed in *Mycobacterium tuberculosis* largely because of noncompliance on the part of many patients in completing their full courses of antibiotic treatment. Treatment usually involves three or more antibiotics to reduce the risk that resistance will develop to any one drug. Many patients, however, failed to take all three drugs simultaneously, thereby enabling the organism to develop resistance to one drug at a time, until it became resistant to all of them. These multidrug-resistant (MDR) and extensively drug-resistant (XDR) strains are almost impossible to kill, and they are the reason *M. tuberculosis* has also reemerged among non-AIDS patients. The link between noncompliance and the development of drug resistance is the primary reason for the current requirement that tuberculosis patients be in the presence of a medical staff member when taking the multiple antibiotics prescribed.

As described in Section 28.4, highly infectious diseases, like tuberculosis, are identified using aggressive epidemiological surveillance. Communication among local, national, and world health organizations is critical and helped expose the rise in tuberculosis. To see for yourself how emerging diseases are monitored, search the Internet for a website called ProMED-mail. There you will find daily reports, posted from around the world, that describe new outbreaks of infectious diseases.

A posting on June 13, 2019, for example, reported 3 new cases of Lassa fever in Nigeria, making a total of 581 confirmed cases since January 1, 2019—the largest outbreak of this deadly disease in that country. Lassa fever is an acute hemorrhagic disease caused by the Lassa virus (an arenavirus; single-stranded RNA genome). Humans contract disease following contact with urine or feces of *Mastomys* rats (the reservoir) or with body fluids of infected humans. Symptoms begin slowly (fever, weakness) and progress to chest pain, cough, vomiting, diarrhea, abdominal pain, and, in severe cases, bleeding from the mouth, nose, vagina, or GI tract. Death can occur within 14 days. The mortality rate is about 25% for those who develop disease. Lassa fever is endemic in West Africa because infected rats are unaffected and do not die. Unfortunately, there is no vaccine.

ProMED also posted a report, on November 19, 2018, about a 46-year-old biologist who was bitten by a bat while he and his family explored a cave in Costa Rica. Two months later he developed numbness, paralysis, and difficulty swallowing. At that time he entered a San Juan hospital, but he died 10 days later . . . from rabies. Rabies virus, with its single-stranded, negative-sense RNA genome and telltale bullet-shaped capsid, is transmitted by the bite of rabies-infected animals, including bats. The virus is neurotrophic, traveling along peripheral nerves to the central nervous system. As the virus spreads through the brain, nerve cell dysfunction and inflammation causes confusion, coma, seizures, widespread paralysis, and respiratory arrest. However, a vaccine is available. This post, then, serves as a cautionary tale. Had the biologist sought immediate treatment, which is vaccination and the administration of rabies immunoglobulin, he would most likely have lived to go spelunking again. Every month, ProMED posts 100–200 of these alerts from around the world.

Thought Question

28.11 On the ProMED-mail web page (http://www.promedmail.org), click on the interactive world map to view outbreaks recorded by WHO. What outbreaks happened throughout the world during the current year?

Tracking an Emerging Disease

Sometimes world and national health organizations quickly and efficiently identify and contain outbreaks of emerging infectious diseases, as was the case in 2003 with severe acute respiratory syndrome (SARS; **eTopic 28.6**). In other instances, outbreaks take more time to contain, such as the 2014 Ebola epidemic in West Africa, described earlier. Agencies can also face difficulties in clearly identifying the causative agent of an outbreak, as in the 2015–2016 Zika virus outbreak in Brazil. The Zika virus story provides an instructive example of how new diseases are identified.

Case History: Zika Virus

Brazil, 2015. In about the fourth month of her pregnancy, a young woman from Recife, Brazil, awoke one morning with fever, a fiery red rash, and pain in her joints. Her symptoms went away within a few days, and she went on about her life, oblivious to what was to come. Five months later, in September 2015, she delivered her baby and immediately knew from the look on the obstetrician's face that something was terribly wrong. Her son had an abnormally small head and incomplete brain development—a condition called microcephaly. The young mother was devastated.

In August 2015, Dr. Vanessa van der Linden, a neurologist in Recife, started noticing an increasing number of newborns with microcephaly (**Fig. 28.28**). Normally, months would pass without her seeing a single case. Now there were three or four in a single day. Using serological tests, she ruled out the usual suspects, such as rubella (German measles) and toxoplasmosis. Imaging studies, however, revealed

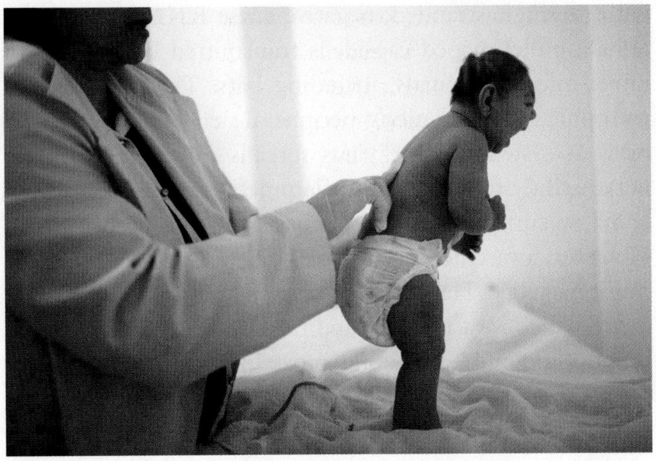

FIGURE 28.28 ▪ A child born with microcephaly in Brazil.

unusual patterns of calcification in the affected brains (a possible sign of virus-induced necrosis). Vanessa knew something odd was going on. Then her mother called.

Vanessa's mother, Dr. Ana van der Linden, a neuropediatrician in another Recife hospital, told her daughter that she had seen seven babies with microcephaly in just one day, and that some of the mothers remembered having a rash early during pregnancy. Could there be an infectious cause to these birth defects? In October, Vanessa alerted the state health secretary about the spike in microcephaly cases. This was the first step in the epidemiological journey linking Zika virus, the cause of a normally benign, mosquito-transmitted disease in adults, to the development of microcephaly in Zika virus–infected fetal brains.

As the number of microcephaly cases grew, suspicion by public health officials initially fell on dengue virus, in part because the symptoms of Zika fever resemble those of dengue fever and because prior to 2014, Zika virus infections were rare in Brazil. Dengue virus was far more common and considered the most likely cause of the women's symptoms. Consequently, individuals who actually had Zika fever were not screened for Zika virus; they were screened by ELISA for anti–dengue virus antibodies. Unfortunately, patient antibodies to Zika virus can cross-react with dengue virus antigen, so a person infected with Zika can appear to test positive for dengue (see the discussion of test specificity in Section 28.3). This technical problem slowed efforts to identify Zika virus as the cause of the women's symptoms and obscured the virus's connection to the alarming rise in newborn microcephaly. Today, there are more specific qRT-PCR tests and virus isolation procedures to confirm Zika virus diagnoses.

Both Zika and dengue viruses have positive-sense RNA genomes and are members of the *Flaviviridae* family of viruses. West Nile virus, mentioned earlier, is also a member of this family. Zika, dengue, and West Nile viruses are transmitted to humans by *Aedes* mosquito vectors in person-to-person or animal-to-human transmission cycles. West Nile virus uses birds as natural reservoirs, whereas Zika and dengue viruses infect primarily human and nonhuman primates. There are also some reports of sexual transmission of Zika virus in areas devoid of the *Aedes* mosquito vectors. As of July 2018, Zika virus was not considered endemic in the United States, even though there are people living in the United States infected with Zika, including over 2,000 pregnant women initially infected elsewhere in the world.

How does a viral pathogen become endemic in a country? In order for Zika, and many other mosquito-borne viruses, to become endemic in a new location, four things must happen:

1. The mosquito vector must already be present in the area.
2. A person infected in an endemic area such as Brazil must travel to the nonendemic area.
3. A local mosquito must bite the infected traveler and transmit the virus to another victim.
4. Finally, the virus must also be transmitted to an animal reservoir that can maintain the virus within a geographic area. For example, West Nile virus, introduced into the United States in 1999, quickly infected the native bird population and is now transmitted from birds to humans by mosquito.

As of January 2019, Zika virus had fulfilled three of the four criteria in the United States. The *Aedes* mosquito vector inhabits areas of at least 30 states, stretching from California through the southern and northeastern states. There are hundreds of people infected with Zika virus now living in the United States. The virus has also been introduced into the native mosquito population around Miami, and the infected mosquitoes have transmitted the Zika virus to several humans in the area. Fortunately, the ability to meet the last requirement—finding a native animal host reservoir—is thought to be unlikely. Zika virus is not known to infect birds or animals other than nonhuman primates native to much of South America and the Caribbean but not to the United States.

How did Zika virus get to Brazil? Zika virus was previously recognized in Africa and Asia as causing mild disease but had never been reported in Brazil. Epidemiologists have used next-generation DNA sequencing of Brazilian Zika virus isolates to explore its origin. The data indicate that the Brazilian strain originated from an Asian strain that circulated in French Polynesia in 2013, but the event that introduced the Asian virus into Brazil is unclear. Regardless of how it got there, once the virus was introduced, native mosquitoes fed on the carrier(s), became infected and began transmitting the virus throughout the country. The disease is now endemic to Brazil. There is no treatment for Zika fever, although furious efforts are under way to develop a vaccine. At the moment, the only protection against Zika is to prevent infection by avoiding or eradicating the *Aedes* mosquitoes.

So, have Koch's postulates been fulfilled? Does Zika virus infection of the mother lead to transplacental transmission of the virus to the fetus? And does virus replication interfere with fetal brain development? Evidence of the virus has been found in the blood of mothers who have delivered babies with microcephaly, and also in umbilical cord blood. The virus itself has been isolated from the brains of fetuses who died from microcephaly. Other studies have recapitulated the neurotropic nature of the virus and indicate that Zika virus can interfere with neurogenesis during human brain development. There are also reports of monkey fetuses infected with Zika virus showing signs of slow brain growth. In sum, the evidence of a link is convincing.

Has the danger passed? Between 2015 and 2018, at least 4,000 babies in Brazil (116 in the United States) were born with microcephaly and other birth defects linked to Zika virus. Estimates are that about one in seven babies exposed to Zika virus in the womb will develop at least one developmental defect within a year. So the risk still exists. Fortunately, the number of Zika infections overall has dropped significantly because of mosquito control measures and personal protection strategies such as using mosquito repellent. The World Health Organization (WHO) and the U.S. Centers for Disease Control and Prevention (CDC) continue to carefully monitor the spread of Zika virus throughout the Americas by intensively screening mosquitoes.

Epidemiological studies were extremely helpful in solving the 2015 epidemic of microcephaly caused by Zika virus. But that's not always the case with new syndromes.

Case History: Acute Flaccid Myelitis Remains an Epidemiological Mystery

Chase, a 1-year-old boy, developed a runny nose in October 2016 that doctors suspected was a simple respiratory infection. However, Chase awoke after a nap the next day with his right arm paralyzed. Chase's parents rushed him to an urgent-care clinic, where he was diagnosed with acute flaccid myelitis (AFM), a rare disease (so far) that resembles polio but whose cause is unknown. Surgeons knew the only way to restore function to Chase's arm was to move unneeded nerves from Chase's rib cage to his shoulder. When told that doctors were going to make his arm work again, Chase gladly explained to his mother that "they are going to put batteries in there." Two years after the operation he finally regained function in his right arm, but he still favors his left.

Here is a situation in which there were few epidemiological insights about cause. The standard case definition for acute flaccid myelitis is simply the presence of illness or fever prior to an acute onset of limb paralysis. Laboratory criteria include an MRI image showing a spinal cord lesion restricted to gray matter that spans one or more vertebral segments, plus CSF containing more than five white blood cells per microliter (indicating infection). Cases of AFM all of a sudden spiked in the United States during the fall of 2014 (51), again in 2016 (44), and once more in 2018 (233). A total of more than 500 cases were reported in 46 states.

Why the case incidence spikes every 2 years is unknown, as is the etiologic cause. Most patients (greater than 90%) were children who contracted a mild respiratory illness or fever before the onset of AFM. Case numbers increased during August–October, a time of the year when many viruses, such as enteroviruses, circulate. Unfortunately, a definitive cause of the disease remains a mystery. Spinal fluid in only 4 out of 430 AFM patients contained coxsackieviruses (a type of enterovirus), the likely cause of AFM in those four patients. No agent was found in the other cases examined. One enterovirus suspect, the infamous poliovirus, was ruled out as the cause. Health organizations continue to collect fecal samples (enteroviruses shed in feces), CSF samples, and respiratory secretions in hopes of finding the causative agent.

Technology Helps New Infectious Agents Emerge and Spread

Despite everything we know of microbes and despite the many ways we have to combat microbial diseases, our species, for all its cleverness, still lives at the mercy of the microbe. Lyme disease, MRSA, SARS, MERS-CoV, Ebola, *E. coli* O157:H7, HIV, "flesh-eating" streptococci, hantavirus, swine flu—all of these and many other new diseases and pathogens have emerged over the last 30 years. Worse yet, forgotten scourges, such as tuberculosis, have reappeared. Yet in the 1970s, medical science was claiming victory over infectious disease. What happened?

Part of the equation has been progress itself. Travel by jet, the use of blood banks, and suburban sprawl have all opened new avenues of infection. The spread of Zika virus from Africa to Asia and then the Americas is but one example. People unwittingly infected by a new disease in Asia or Africa can, traveling by jet, take the pathogen to any other country in the world within hours. A person may not even show symptoms until days or weeks after the trip. This means that diseases can spread faster and farther than ever before. In addition, newly emerged blood-borne pathogens can spread by transfusion. This was a major problem with HIV before an accurate blood test was developed to screen all donated blood.

Although human encroachments into the tropical rain forests have often been blamed for the emergence of new pathogens, one need go no farther than the Connecticut woodlands to find such developments. *Borrelia burgdorferi*, the spirochete that causes Lyme disease, lives on deer and white-footed mice and is passed between these hosts by the deer tick (see Fig. 26.28)—in an infection cycle that has been going on for years. Humans crossed paths with these animals long before the disease erupted in our communities. Why have we suddenly become susceptible? The answer appears to be suburban development. In the wild, foxes and bobcats hunt the mice that carry the Lyme agent. These predators disappear when developers clear land and build roads and houses, leaving the infected mice and ticks to proliferate. Humans in these developed areas are more likely to be bitten by an infected tick and contract the disease than in prior decades. Luckily, many diseases that successfully leap from animals to humans find the new host to be a dead end, unable to spread the disease to others.

There are numerous examples in which technology and progress have had the unintended consequence of breeding disease. Here are just a few:

- **Mad cow disease.** Modern farming practices (in North America and Europe) of feeding livestock the remains of other animals help spread transmissible spongiform encephalopathies, similar to Creutzfeldt-Jakob disease, that are associated with prions. Because the prion is infectious, the brain matter from one case of mad cow disease could end up infecting hundreds of other cattle, thereby increasing the chance that the disease could spread to humans.
- **Lyme disease.** Suburban development in the northeastern United States destroys predators of the mice that carry *Borrelia burgdorferi.*
- **Hepatitis C.** Transfusions and transplants spread this blood-borne disease.
- **Influenza.** Live poultry markets in Asia serve as breeding grounds for avian flu viruses that can jump to humans.
- **Enterohemorrhagic *E. coli* (for example, *E. coli* O157:H7).** Modern meat-processing plants can accidentally grind trace amounts of these acid-resistant, fecal organisms into beef while making hamburger.

Natural environmental events can also trigger upsurges in the incidence of unusual diseases. For example, an unprecedented outbreak of hantavirus pulmonary disease occurred in 1993 in the Four Corners area of the Southwest, where Arizona, New Mexico, Colorado, and Utah meet, when rain led to greater-than-normal increases in plant and animal numbers. The resulting tenfold increase in deer mice, which carries the virus, made it more likely that infected mice and humans would come in contact. A more recent outbreak of hantavirus occurred among campers at Yosemite National Park in 2012.

Climate Change Influences Emerging Diseases

The world's climate has been changing slowly but significantly over the last 50 years—largely as a result of humans pumping enormous amounts of greenhouse gases, such as carbon dioxide and methane, into the atmosphere. Greenhouse gases trapping heat in the atmosphere have caused the world's average yearly temperature to rise by about 1.5°F since 1970. Effects of global warming on climate, depending on where you live, are evident as disappearing glaciers, rising sea levels, mounting ferocity of hurricanes and tornadoes, extended droughts, increased rainfall (floods), dust storms, and heat waves. These climate changes are bad enough, but they can also affect the epidemiology of infectious disease.

Rising air temperatures, for example, can extend the habitat of mosquito and tick vectors to higher mountain elevations and wider latitudes. The result is that Zika virus, dengue virus, and *Plasmodium* (the protozoan cause of malaria), among other infectious agents, have increased their geographic range. Mathematical simulations predict that the area populated by *Anopheles* mosquito vectors that carry malaria will increase between 16% and 49% by 2030, depending on the vector. Extending a pathogen's range can bring it into contact with novel groups of hosts, possibly establishing new vector-pathogen transmission cycles or a situation that promotes pathogen evolution (such as influenza). Warmer temperatures can also accelerate host-parasite cycles that will further increase the incidence and prevalence of infections.

Increased rain and flooding promote the breeding of mosquito vectors that transmit disease—a situation that will worsen with climate change. Floods can also accelerate the spread of diarrheal diseases such as cholera when sanitation systems fail. As these events increase, so, too, will infectious disease.

The Arctic is particularly sensitive to climate change. It is home to diverse populations of plants, animals, and people. Approximately 10 million people live within the Arctic Circle, a geographic area that encompasses parts of nine countries, including Canada and the United States. The Arctic has warmed twice as much as the global average of other parts of the world and is becoming increasingly fragile. The associated health risks for humans and animals include potential changes in pathogen and vector demographics that can affect disease patterns, degradation in the quality and availability of both drinking water and food, and changes in animal and plant species health.

Climate change will continue to provide opportunities for some pathogens to expand their geographic "footprint" and may even limit the transmission of others. To stay ahead of these evolving infections, surveillance programs capable of detecting pathogen or disease emergence are essential. Make no mistake, the effect of climate change on infectious diseases and the rise in antibiotic-resistant microbes pose serious threats to the living world.

The One Health Initiative

We have provided numerous examples of how the natural environment can harbor pathogens and foster the emergence of new ones. Knowledge that pathogens have reservoirs in animals, arthropods, and plants has given rise to a new collaborative effort among clinicians, scientists, veterinarians, and ecologists that is called the One Health Initiative. The goal of the One Health Initiative is to control human health through animal health, and vice versa. For example, vaccinating wild rodents could decrease Lyme disease in humans. The initiative includes plant pathology because some bacteria are pathogens of both plants and humans (for example, *Pantoea agglomerans*, formerly *Enterobacter agglomerans*). In addition, some enteric bacteria can live within a plant vascular system (for example, *E. coli*, *Klebsiella*, and *Salmonella*).

Solving a mystery. An excellent example of how multidisciplinary collaboration resulted in better understanding of an infectious outbreak occurred in 2006 when approximately 200 people in 26 states were diagnosed with a particularly virulent case of *E. coli* O157:H7. Nearly half of the cases were hospitalized, and many suffered from hemolytic uremic syndrome (HUS, kidney failure described in Section 26.4). The source of the infection was contaminated spinach traced to the Salinas Valley of California. It turned out the organisms were contained within the vascular system of the spinach, so washing the spinach would not remove the pathogen. Had this outbreak been viewed through only the narrow lens of human health, efforts would have focused on morbidity, mortality, outbreak investigation, laboratory diagnosis, and clinical treatment. The origin of the disease would have remained a mystery.

Working together, epidemiologists and veterinarians found a genetically identical organism in cattle close to where the spinach was produced and in wild hogs that ran through the same fields. Ecologists and hydrologists understood that the groundwater and surface water in this region were being mixed because of a drought followed by heavy rains, and that irrigation systems were strained in the effort to keep up with intensified agricultural production. Eventually, the same *E. coli* strain was found in one of the water ditches close to the spinach fields in the area. Scientists pondering these facts within the One Health framework deduced that cattle harboring the *E. coli* had defecated in a field, thereby contaminating wild hogs. The wild hogs, by running through the spinach fields, had contaminated those fields with their feces, and the irrigation water had then swept the pathogen into the plant vasculature.

Only by integrating our knowledge of the environment and ecology could this investigation be completely understood and appropriate intervention and prevention strategies implemented. This outbreak exemplifies the fact that human health and animal health are inextricably linked and that a holistic approach is needed to understand, protect, and promote the health of all species.

Stopping zoonotic diseases. A more recent example of the One Health approach is being carried out in the Ruaha region of Tanzania, Africa—a sprawling, wild area with scattered small villages and a rich wildlife. Thousands of children and adults in underdeveloped countries such as Tanzania die every day from diseases arising from the human-animal-environment interface. People in these areas often live in close contact with wild and domestic animals and are brought even closer as water becomes scarcer (**Fig. 28.29**). Sharing the same water sources for drinking, washing, and swimming facilitates zoonotic disease transmission to humans.

The One Health Initiative simultaneously addresses multiple and interacting causes of poor human health to try to stop or limit these zoonotic diseases. One underdiagnosed zoonotic disease that the collaborative is trying to address is bovine tuberculosis (BTB), a disease that normally affects animals but also infects humans. Tanzania suffers 40,000 new cases of tuberculosis per year caused by human, bovine, or atypical strains of mycobacteria. A majority of these TB patients are also infected with HIV.

PHOTO BY MISTY RICHMOND. COURTESY OF HALI PROJECT

FIGURE 28.29 ■ The One Health Initiative. This photograph illustrates the close relationship that some residents of Tanzania have with their livestock, and the opportunities by which zoonotic infections can be transmitted to humans.

BTB in humans often progresses to extrapulmonary TB, making BTB a major focus of the initiative.

The approach in Tanzania, called the Health for Animals and Livelihood Improvement (HALI) Project, includes the testing of wildlife, livestock, and their water sources for zoonotic pathogens and disease. Water quality, availability, and use are monitored, as is the impact of livestock and human disease on farming households. New diagnostic techniques for disease detection are being introduced, and Tanzanians of all educational levels are trained about zoonotic diseases. Finally, new health and environmental policies have been developed to mitigate the impacts of zoonotic diseases.

The HALI Project has identified bovine tuberculosis and brucellosis in livestock and wildlife in the Ruaha ecosystem and has identified geographic areas where water availability increases the risk of transmission among wildlife, livestock, and people. In addition, *Salmonella*, *Escherichia coli*, *Cryptosporidium*, and *Giardia* species that cause disease in humans and animals have been isolated from multiple water sources used by people and frequented by livestock and wildlife. A major lesson of the HALI Project is that the causes and consequences of zoonotic diseases, as well as the interventions to mitigate them, are all cross-sectoral. Effective surveillance, assessments, and interventions are possible only if communication and cooperation improve among institutions that study and manage wildlife, livestock, water, and public health.

To Summarize

- **Koch's and Rivers's postulates** are important for identifying the microbial cause of a new disease, but they must be supplemented with molecular or disease-tracking tools if one of the postulates cannot be satisfied.
- **Emerging diseases** can spread quickly around the world as a result of air travel.
- **Modern technology** and urban growth have provided opportunities for new diseases to emerge.
- **Climate change** has significantly affected the epidemiology of infectious disease and the emergence of new pathogens.
- **Multidisciplinary collaboration** among ecologists, veterinarians, clinicians, and other scientists is necessary for devising appropriate strategies aimed at disease intervention and epidemic prevention.

Concluding Thoughts

In this chapter we have examined the basic principles used to collect, detect, and track pathogenic microorganisms. The task of controlling the spread of disease is daunting and ever changing because microorganisms continue to evolve and the environments in which they live continue to change. Known pathogens, for instance, continually adapt to evade immune systems and antibiotics. Meanwhile, new pathogens keep emerging. The world will depend on the next generation of microbiologists and epidemiologists to face the growing threat of climate change and evolving pathogens.

CHAPTER REVIEW

Review Questions

1. Why is it important to identify the genus and species of a pathogen?
2. What is an API strip, and how is it used in clinical microbiology?
3. Describe three examples of selective media.
4. If a colony on a nutrient agar plate is catalase-positive, does this mean it is made up of Gram-positive microorganisms? Why or why not?
5. Describe the types of hemolysis visualized on blood agar.
6. What is the clinical significance of a group A, beta-hemolytic streptococcus?
7. How does one distinguish *Staphylococcus aureus* from *Staphylococcus epidermidis*?
8. Why are PCR identification tests preferable to biochemical approaches?
9. How is qRT-PCR performed?
10. Describe an ELISA.
11. Name some sterile and nonsterile body sites.
12. List seven common types of clinical specimens collected for bacteriological examination.
13. Describe key features of the four levels of biological containment.
14. How is a pandemic different from an epidemic?
15. How can genomics help identify nonculturable pathogens?
16. List and briefly describe four emerging diseases.
17. Name four select agents and the diseases they cause.

Thought Questions

1. Consider the following hypothetical case: Five small outbreaks of Ebola occurred within days of one another in five different cities in the United States and Canada. Patient histories revealed that one person from each city had been in the Atlanta airport at the same time 2 days prior to becoming seriously ill. None had left the country recently; none flew on the same jet or even crossed paths while in the airport. Imagining yourself to be an epidemiologist who is aware that another outbreak occurred recently in the Congo, conjure up a scenario, excluding bioterrorism, that could account for this scattered outbreak.

2. A patient is brought to the hospital with an intra-abdominal abscess. Aspirates are sent to the laboratory for aerobic and anaerobic culture. What is wrong with this order?

Thought questions 3–6 below are based on the following case history:

An infectious disease physician in Florida telephoned the CDC to report two possible cases of botulism. The two male patients presented with drooping eyelids, double vision, difficulty swallowing, and respiratory problems. The physician had drawn sera and collected stool specimens from the men to test for botulinum toxin, but no results were available.

3. What are the major concerns raised by these two possible cases of botulism?
4. How might you go about swiftly determining whether there is a link between the two cases and whether there are other cases of botulism?
5. The two patients ate only one food item in common: a cured and fermented fish called "moloha." How would you determine whether the men are indeed suffering from botulism and whether the fermented fish is the source of disease?
6. How could this anaerobic pathogen grow and make toxin in this fish product? Propose rational theories for what has hindered development of a vaccine against botulism.

Key Terms

amplicon (1175)
emerging disease (1193)
endemic (1188)
epidemic (1189)
epidemiology (1161, 1187)
incidence (1189)
index case (1190)
multiplex PCR (1176)
pandemic (1189)
patient zero (1190)
prevalence (1189)
probabilistic indicator (1169)
quarantine (1190)
reservoir (1189)
sensitivity (1185)
sequela (1161)
specificity (1185)
syndromic surveillance (1190)
systemic surveillance (1190)

Recommended Reading

Baum, Sarah E., Catherine Machalab, Peter Daszak, Robert H. Salerno, and William B. Karesh. Evaluating One Health: Are we demonstrating effectiveness? *One Health* 3:5–10. https://doi.org/10.1016/j.onehlt.2016.10.004.

Casen, Christina, Heidi C. Vebø, Monika Sekelja, Finn T. Hegge, Magdalena K. Karlsson, et al. 2015. Deviations in human gut microbiota: A novel diagnostic test for determining dysbiosis in patients with IBS or IBD. *Alimentary Pharmacology and Therapeutics* 42:71–83.

Eisenstein, Michael. 2018. Cloudy with a chance of flu. *Nature* 555:S2–S4.

Fairfax, Marilynn Ransom, Martin H. Bluth, and Hossein Salimnia. 2018. Diagnostic molecular microbiology: A 2018 snapshot. *Clinical Laboratory Medicine* 38:253–276.

Faria, Nuno Rodrigues, Raimunda do Socorro da Silva Azevedo, Moritz U. G. Kraemer, Renato Souza, Mariana Sequetin Cunha, et al. 2016. Zika virus in the Americas: Early epidemiological and genetic findings. *Science* 352:345–349.

Green, Manfred S., James LeDuc, Daniel Cohen, and David R. Franz. 2018. Confronting the threat of bioterrorism: Realities, challenges, and defensive strategies. *Lancet Infectious Diseases* 19:E2–E13. https://doi.org/10.1016/S1473-3099(18)30298-6.

Khamesipour, Faham, Kamran Bagheri Lankarani, Behnam Honarvar, and Tebit Emmanuel Kwenti. 2018. A systematic review of human pathogens carried by the housefly (*Musca domestica* L.) *BMC Public Health* 18:1049. https://doi.org/10.1186/s12889-018-5934-3.

Koser, Claudio U., Matthew J. Ellington, Edward J. Cartwright, Stephen H. Gillespie, Nicholas M. Brown, et al. 2012. Routine use of microbial whole genome sequencing in diagnostic and public health microbiology. *PLoS Pathogens* **8**:e1002824.

Morse, Stephen S. 2012. Public health surveillance and infectious disease detection. *Biosecurity and Bioterrorism* **10**:6–16.

Nuismer, Scott L., Ryan May, Andrew Basinski, and Christopher H. Remien. 2018. Controlling epidemics with transmissible vaccines. *PLoS One* **13**:e0196978.

Pardee, Keith, Alexander A. Green, Mellisa K. Takahashi, Dana Braff, Guillaume Lambert, et al. 2016. Rapid, low cost detection of Zika virus using programmable biomolecular components. *Cell* **165**:1255–1266.

Sauget, Marlène, Benoît Valot, Xavier Bertrand, and Didier Hocquet. 2017. Can MALDI-TOF mass spectrometry reasonably type bacteria? *Trends in Microbiology* **25**:447–455.

Schürch, Anita C., S. Arredondo-Alonso, R. J. L. Willems, and R. V. Goering. 2018. Whole genome sequencing options for bacterial strain typing and epidemiologic analysis based on single nucleotide polymorphism versus gene-by-gene-based approaches. *Clinical Microbiology and Infection* **24**:350–354.

Semenza, Jan C., Elisabet Lindgren, Laszlo Balkanyi, Laura Espinosa, My S. Almqvist, et al. 2016. Determinants and drivers of infectious disease threat events in Europe. *Emerging Infectious Diseases* **22**:581–589.

Varo, Rosauro, Xavier Rodó, and Quique Bassat. 2019. Climate change, cyclones and cholera—Implications for travel medicine and infectious diseases. *Travel Medicine and Infectious Disease* **29**:6–7.

Waits, Audrey, Anastasia Emelyanova, Antti Oksanen, Khaled Abassa, and Arja Rautio. 2018. Human infectious diseases and the changing climate in the Arctic. *Environment International* **121**:703–713.

Wilson, Michael R., Samia N. Naccache, Erik Samayoa, Mark Biagtan, Hiba Bashir, et al. 2014. Actionable diagnosis of neuroleptospirosis by next-generation sequencing. *New England Journal of Medicine* **370**:2408–2417.

APPENDIX
Reference and Review

A.1 A Periodic Table of the Elements
A.2 Chemical Functional Groups
A.3 Amino Acids
A.4 The Genetic Code
A.5 Calculating the Standard Free Energy Change, $\Delta G°$, of Chemical Reactions
A.6 Generalized Cells
A.7 Semipermeable Membranes
A.8 The Eukaryotic Cell Cycle and Cell Division

The cell is the basic unit of life. Cells are composed primarily of water and organic molecules: proteins, carbohydrates, nucleic acids, and lipids. The chemistry of life is organic chemistry, based on carbon (C) and several other elements: hydrogen (H), nitrogen (N), oxygen (O), phosphorus (P), and sulfur (S). This Appendix provides reference information on some essential aspects of chemistry, biochemistry, and cell biology that are likely covered in an introductory biology course. For further review of basic chemistry, biochemistry, and cell biology, see eAppendices 1 and 2.

A.1 A Periodic Table of the Elements

A periodic table organizes all of the known elements, grouping them by columns and rows according to their electronic and chemical properties (**Fig. A.1**). Note that the six major elements of life—H, C, N, O, P, and S—are all nonmetals found in the top three periods (rows).

A.2 Chemical Functional Groups

A functional group is a group of atoms that defines the structure and reactions of an organic compound (**Table A.1**). Since organic chemistry underlies so much of biology, it is useful to be able recognize functional groups and their chemical properties.

Main-group elements (groups 1–2 and 13–18); Transitional elements (groups 3–12)

Key: Atomic number / Chemical symbol / Atomic mass (average of all isotopes)

Legend: Metals; Nonmetals; Metalloids; Noble gases

Period	1 1A	2 2A	3 3B	4 4B	5 5B	6 6B	7 7B	8 8B	9 8B	10 8B	11 1B	12 2B	13 3A	14 4A	15 5A	16 6A	17 7A	18 8A
1	1 **H** 1.00794																	2 **He** 4.00260
2	3 **Li** 6.941	4 **Be** 9.01218											5 **B** 10.811	6 **C** 12.011	7 **N** 14.0067	8 **O** 15.9994	9 **F** 18.9984	10 **Ne** 20.1797
3	11 **Na** 22.9898	12 **Mg** 24.3050											13 **Al** 26.9815	14 **Si** 28.0855	15 **P** 30.9738	16 **S** 32.066	17 **Cl** 35.4527	18 **Ar** 39.948
4	19 **K** 39.0983	20 **Ca** 40.078	21 **Sc** 44.9559	22 **Ti** 47.88	23 **V** 50.9415	24 **Cr** 51.9961	25 **Mn** 54.9381	26 **Fe** 55.847	27 **Co** 58.9332	28 **Ni** 58.693	29 **Cu** 63.546	30 **Zn** 65.39	31 **Ga** 69.723	32 **Ge** 72.61	33 **As** 74.9216	34 **Se** 78.96	35 **Br** 79.904	36 **Kr** 83.80
5	37 **Rb** 85.4678	38 **Sr** 87.62	39 **Y** 88.9059	40 **Zr** 91.224	41 **Nb** 92.9064	42 **Mo** 95.94	43 **Tc** (98)	44 **Ru** 101.07	45 **Rh** 102.906	46 **Pd** 106.42	47 **Ag** 107.868	48 **Cd** 112.411	49 **In** 114.818	50 **Sn** 118.710	51 **Sb** 121.76	52 **Te** 127.60	53 **I** 126.904	54 **Xe** 131.29
6	55 **Cs** 132.905	56 **Ba** 137.327	57 ***La** 138.906	72 **Hf** 178.49	73 **Ta** 180.948	74 **W** 183.84	75 **Re** 186.207	76 **Os** 190.23	77 **Ir** 192.22	78 **Pt** 195.08	79 **Au** 196.967	80 **Hg** 200.59	81 **Tl** 204.383	82 **Pb** 207.2	83 **Bi** 208.980	84 **Po** (209)	85 **At** (210)	86 **Rn** (222)
7	87 **Fr** (223)	88 **Ra** 226.025	89 **†Ac** 227.028	104 **Rf** (261)	105 **Db** (262)	106 **Sg** (263)	107 **Bh** (262)	108 **Hs** (265)	109 **Mt** (266)	110 **Ds** (271)	111 **Rg** (272)	112 **Cn** (277)		114 **Fl** (289)		116 **Lv** (293)		

Series														
*Lanthanide series	58 **Ce** 140.115	59 **Pr** 140.908	60 **Nd** 144.24	61 **Pm** (145)	62 **Sm** 150.36	63 **Eu** 151.965	64 **Gd** 157.25	65 **Tb** 158.925	66 **Dy** 162.50	67 **Ho** 164.930	68 **Er** 167.26	69 **Tm** 168.934	70 **Yb** 173.04	71 **Lu** 174.967
†Actinide series	90 **Th** 232.038	91 **Pa** 231.036	92 **U** 238.029	93 **Np** 237.048	94 **Pu** (244)	95 **Am** (243)	96 **Cm** (247)	97 **Bk** (247)	98 **Cf** (251)	99 **Es** (252)	100 **Fm** (257)	101 **Md** (258)	102 **No** (259)	103 **Lr** (260)

Symbol	Name	Symbol	Name	Symbol	Name	Symbol	Name	Symbol	Name
Ac	Actinium	Cn	Copernicium	I	Iodine	Os	Osmium	Si	Silicon
Al	Aluminum	Cu	Copper	Ir	Iridium	O	Oxygen	Ag	Silver
Am	Americium	Cm	Curium	Fe	Iron	Pd	Palladium	Na	Sodium
Sb	Antimony	Ds	Darmstadtium	Kr	Krypton	P	Phosphorus	Sr	Strontium
Ar	Argon	Db	Dubnium	La	Lanthanum	Pt	Platinum	S	Sulfur
As	Arsenic	Dy	Dysprosium	Lr	Lawrencium	Pu	Plutonium	Ta	Tantalum
At	Astatine	Es	Einsteinium	Pb	Lead	Po	Polonium	Tc	Technetium
Ba	Barium	Er	Erbium	Li	Lithium	K	Potassium	Te	Tellurium
Bk	Berkelium	Eu	Europium	Lv	Livermorium	Pr	Praseodymium	Tb	Terbium
Be	Beryllium	Fm	Fermium	Lu	Lutetium	Pm	Promethium	Tl	Thallium
Bi	Bismuth	Fl	Flerovium	Mg	Magnesium	Pa	Protactinium	Th	Thorium
Bh	Bohrium	F	Fluorine	Mn	Manganese	Ra	Radium	Tm	Thulium
B	Boron	Fr	Francium	Mt	Meitnerium	Rn	Radon	Sn	Tin
Br	Bromine	Gd	Gadolinium	Md	Mendelevium	Re	Rhenium	Ti	Titanium
Cd	Cadmium	Ga	Gallium	Hg	Mercury	Rh	Rhodium	W	Tungsten
Ca	Calcium	Ge	Germanium	Mo	Molybdenum	Rg	Roentgenium	U	Uranium
Cf	Californium	Au	Gold	Nd	Neodymium	Rb	Rubidium	V	Vanadium
C	Carbon	Hf	Hafnium	Ne	Neon	Ru	Ruthenium	Xe	Xenon
Ce	Cerium	Hs	Hassium	Np	Neptunium	Rf	Rutherfordium	Yb	Ytterbium
Cs	Cesium	He	Helium	Ni	Nickel	Sm	Samarium	Y	Yttrium
Cl	Chlorine	Ho	Holmium	Nb	Niobium	Sc	Scandium	Zn	Zinc
Cr	Chromium	H	Hydrogen	N	Nitrogen	Sg	Seaborgium	Zr	Zirconium
Co	Cobalt	In	Indium	No	Nobelium	Se	Selenium		

FIGURE A.1 ■ Periodic table of the elements. The atomic number (number of protons) and atomic mass are shown for each element.

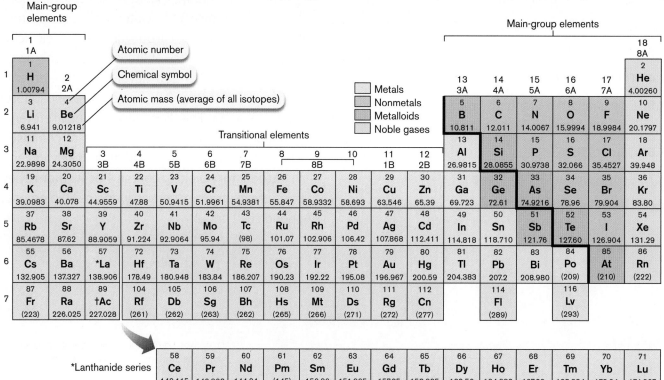

TABLE A.1 Common Functional Groups

Functional group	General structure	Example	Comments
Aldehyde	R–C(=O)–H	Acetaldehyde (H–C(H)(H)–C(=O)–H)	Can react with alcohols $R-C(=O)H + HO-R' \rightarrow R-C(OH)(H)-OR'$
Alkane	R–C(H)(H)–H	Ethane (H–C(H)(H)–C(H)(H)–H)	Nonpolar, tends to make molecules containing it hydrophobic
Amino	R–N(H)(H)	Methylamine (H–C(H)(H)–N(H)(H))	Acts as a base by binding a proton; found in amino acids $R-NH_2 + H^+ \rightarrow R-NH_3^+$
Carboxyl	R–C(=O)–O–H	Acetic acid (H–C(H)(H)–C(=O)–OH)	Acts as an acid by releasing a proton; the ionized-form name ends in "-ate" (e.g., "acetate") $CH_3-C(=O)O^- + H^+$ Acetate Proton
Ester	R–C(=O)–O–R	Phosphodiester (R_1–O–P(=O)(O^-)–O–R_2)	Derived by condensation of an acid and an alcohol releasing H_2O. A diester is formed by condensation of two acids.
Hydroxyl	R–O–H	Ethanol (H–C(H)(H)–C(H)(H)–OH)	Polar, makes compounds more soluble through hydrogen bonding; found in alcohols and sugars
Ketone	R–C(=O)–R	Dihydroxyacetone (CH_2OH–C(=O)–C(H)(OH)–H)	Found in many intermediates of metabolism
Phosphoryl	R–O–P(=O)(O^-)(O^-)	Glyceraldehyde 3-phosphate	When two or more phosphoryl groups are linked, the negative oxygens repel each other

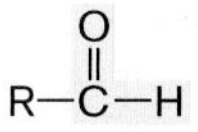

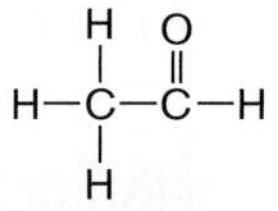

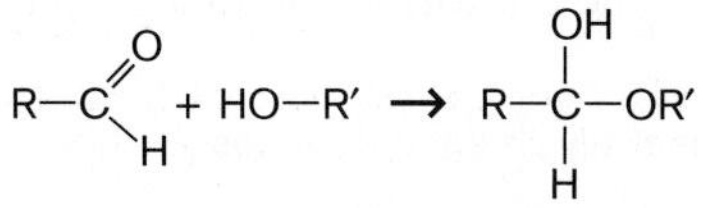

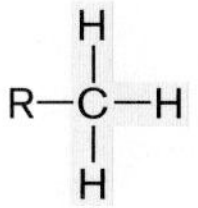

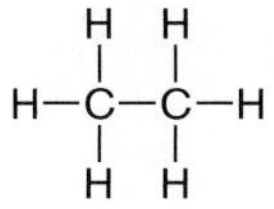

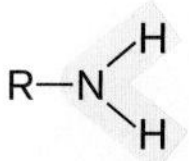

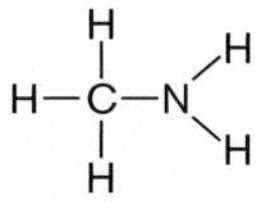

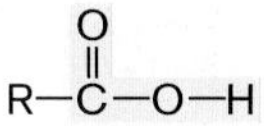

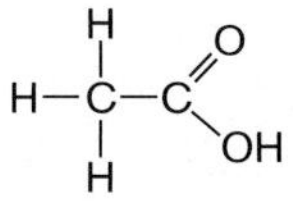

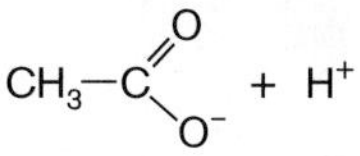

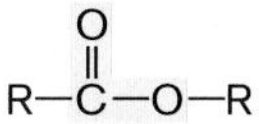

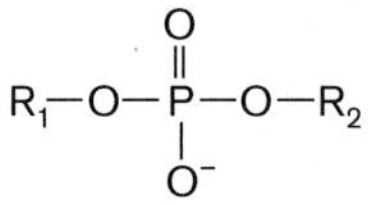

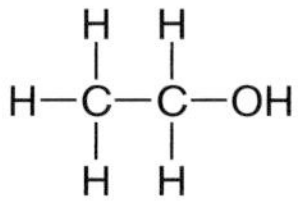

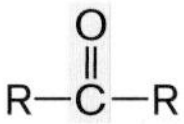

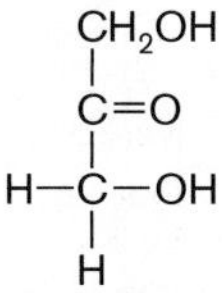

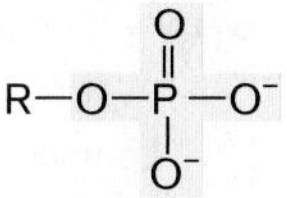

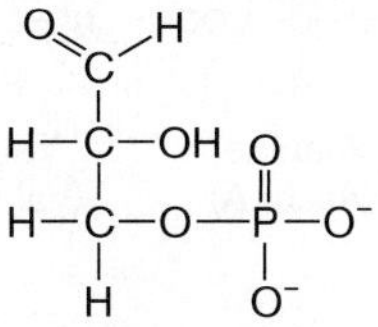

A.3 Amino Acids

Proteins are made of amino acid monomers condensed to form polypeptide chains, which then fold into three-dimensional shapes. The conformation of a protein is determined largely by the side chains of the amino acids in that polypeptide. So, understanding protein structure and function requires knowing the structures and chemical properties of the twenty amino acids. **Figure A.2** shows one way to organize the amino acids.

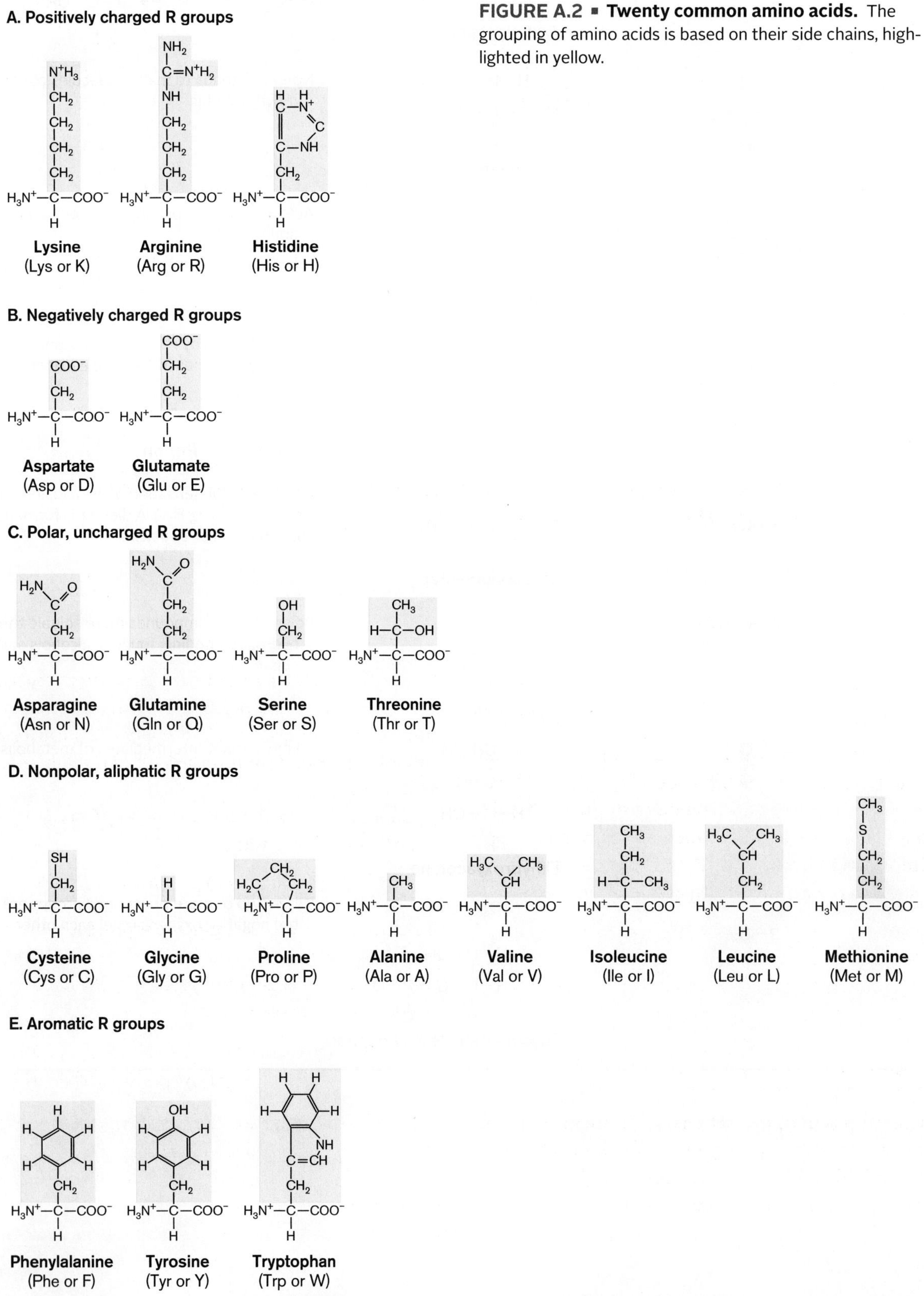

FIGURE A.2 ■ Twenty common amino acids. The grouping of amino acids is based on their side chains, highlighted in yellow.

1st Base	2nd Base U: RNA codon	U: Amino acid	2nd Base C: RNA codon	C: Amino acid	2nd Base A: RNA codon	A: Amino acid	2nd Base G: RNA codon	G: Amino acid	3rd Base
U	U U U	Phe	U C U	Ser	U A U	Tyr	U G U	Cys	U
U	U U C	Phe	U C C	Ser	U A C	Tyr	U G C	Cys	C
U	U U A	Leu	U C A	Ser	U A A	Stop	U G A	Stop	A
U	U U G	Leu	U C G	Ser	U A G	Stop	U G G	Trp	G
C	C U U	Leu	C C U	Pro	C A U	His	C G U	Arg	U
C	C U C	Leu	C C C	Pro	C A C	His	C G C	Arg	C
C	C U A	Leu	C C A	Pro	C A A	Gln	C G A	Arg	A
C	C U G	Leu	C C G	Pro	C A G	Gln	C G G	Arg	G
A	A U U	Ile	A C U	Thr	A A U	Asn	A G U	Ser	U
A	A U C	Ile	A C C	Thr	A A C	Asn	A G C	Ser	C
A	A U A	Ile	A C A	Thr	A A A	Lys	A G A	Arg	A
A	A U G	Met	A C G	Thr	A A G	Lys	A G G	Arg	G
G	G U U	Val	G C U	Ala	G A U	Asp	G G U	Gly	U
G	G U C	Val	G C C	Ala	G A C	Asp	G G C	Gly	C
G	G U A	Val	G C A	Ala	G A A	Glu	G G A	Gly	A
G	G U G	Val	G C G	Ala	G A G	Glu	G G G	Gly	G

Single-letter abbreviations for amino acids

Ala (A)
Arg (R)
Asn (N)
Asp (D)
Cys (C)
Gln (Q)
Glu (E)
Gly (G)
His (H)
Ile (I)
Leu (L)
Lys (K)
Met (M)
Phe (F)
Pro (P)
Ser (S)
Thr (T)
Trp (W)
Tyr (Y)
Val (V)

FIGURE A.3 ■ The genetic code. The major start codon is highlighted white. Stop codons are highlighted red.

A.4 The Genetic Code

The link between protein synthesis and the information encoded in DNA is the genetic code (**Fig. A.3**). Three-letter codons in mRNA have counterparts in the tRNA anticodons. Transfer RNA molecules are the adapters converting information in the mRNA transcript to the amino acid sequence in the growing polypeptide chain.

A.5 Calculating the Standard Free Energy Change, $\Delta G°$, of Chemical Reactions

For a chemical reaction, the net change in free energy ΔG determines whether the reaction will go forward. The value of ΔG at standard conditions of temperature (25°C), pressure (1 atmosphere, P_a), and concentration (each reactant at 1 mole/liter), is designated $\Delta G°$. For a given reaction, how do we find the value of $\Delta G°$? The value is the sum of the individual values for standard energy of formation ($\Delta G_f°$) for all the products, minus the sum of the values for all the reactants.

Consider the oxidation of glucose ($C_6H_{12}O_6$) to form carbon dioxide plus water:

$$C_6H_{12}O_6 + 6O_2 \rightarrow 6CO_2 + 6H_2O$$

We must assume that the glucose is present at 1-molar concentration, that the O_2 and CO_2 are gases at 1 atmosphere, and that the water is liquid at 25°C (298 K). The value of $\Delta G°$ for this reaction is given by:

$$\Delta G° = [(6 \times \Delta G_f° \, CO_2) + (6 \times \Delta G_f° \, H_2O)] \\ - [(\Delta G_f° \, C_6H_{12}O_6) + 6 \times (\Delta G_f° \, O_2)]$$

From **Table A.2**, we can insert the values of $\Delta G_f°$ for each reactant and product. Note that one reactant, O_2, has a value of zero, because oxygen gas is the defined standard state for oxygen:

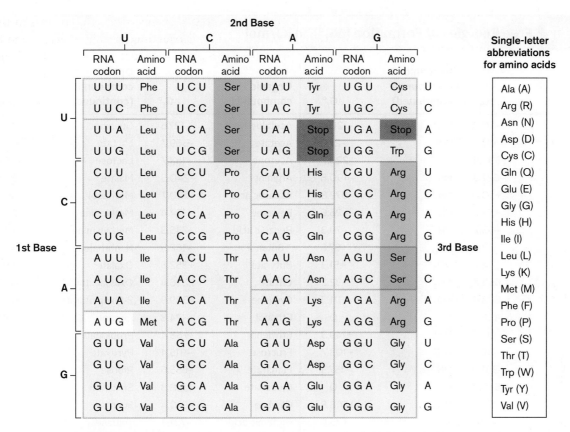

TABLE A.2 Free Energies of Formation ($\Delta G_f°$), in kJ/mol

Inorganic compounds	$\Delta G_f°$	Inorganic compounds (continued)	$\Delta G_f°$	Organic compounds	$\Delta G_f°$	Organic compounds (continued)	$\Delta G_f°$
CH_4	–50.8	H_2O_2	–120.4	Acetaldehyde	–127.6	Guanine	+47.4
CO	–137.2	H_2S	–27.9	Acetate	–369.7	Lactate	–517.8
CO_2	–394.4	Mn^{2+}	–228.1	Acetone	–152.7	Lactose	–1,567.0
CO_3^{2-}	–527.9	$MnCl_2$	–490.8	Arginine	–240.5	Malate	–845.1
Cu^+	+50.0	$MnSO_4$	–972.8	Aspartic acid	–730.7	Methanol	–166.6
Cu^{2+}	+65.5	NH_3	–26.6	Benzene	+124.4	Methionine	–505.8
CuS	–53.7	NH_4^+	–79.3	Benzoic acid	–245.3	Methylamine	+35.7
Fe^{2+}	–78.9	NO	+87.6	1-Butanol	–162.5	Naphthalene	+201.6
Fe^{3+}	–4.7	NO_2	+51.3	Butyrate	+352.6	Oxalate	–697.9
$FeCO_3$	–666.7	NO_2^-	–32.2	Citrate	–1,236.4	Oxaloacetate	–797.2
$FeSO_4$	–823.4	NO_3^-	–111.3	Cysteine	+339.8	2-Oxoglutarate	+797.1
FeS_2	–156.1	N_2	0	Ethanol	–174.8	Phenol	–50.4
H^+ (pH 0)	0	N_2H_4	+149.3	Ethylene	+68.4	Propionate	–361.1
H^+ (pH 7)	–39.7	N_2O	+103.7	Fructose	–915.4	Pyruvate	+474.6
$HCOO^-$	–351.0	OH^-	–157.3	Fumarate	–655.6	Ribose	–757.3
HCO_3^-	–586.9	O_2	0	Gluconate	–1,128.3	Succinate	–690.2
HCl	–131.3	PO_4^{3-}	–1,018.8	Glucose	–910.4	Sucrose	–1,544.7
HS^-	+12.1	S^{2-}	+85.8	Glutamic acid	–731.3	Toluene	+113.8
H_2	0	S^0	0	Glutamine	–529.7	Trimethylamine	+93.0
H_2CO_3	–623.2	SO_3^{2-}	–486.5	Glyceraldehyde	+437.7	Tryptophan	–119.4
H_2O	–237.1	$S_2O_3^{2-}$	–522.5	Glycerol	–477.0	Tyrosine	–385.7

Sources: James G. Speight. 2005. *Lange's Handbook of Chemistry*, 16th ed. McGraw-Hill, New York; WolframAlpha (http://www.wolframalpha.com); Rudolf K. Thauer. 1977. *Bacteriol. Rev.* **41**:100.

$$\begin{aligned}\Delta G° &= [(6 \times -394.4 \text{ kJ/mol}) + (6 \times -237.1 \text{ kJ/mol})] \\ &\quad - [(-910.4 \text{ kJ/mol}) + 6 \times (0 \text{ kJ/mol})] \\ &= -2{,}879 \text{ kJ/mol}\end{aligned}$$

So, the oxidation of glucose has a negative value of ΔG at standard conditions, and the reaction will go forward, releasing energy. The magnitude of the ΔG value suggests that enough energy could be released to form a number of energy carriers, such as ATP; but many additional factors need to be included before we know fully what a cell gains from this reaction.

A.6 Generalized Cells

Prokaryotic cells and eukaryotic cells have several features in common. They each have a cell membrane that defines the boundary of the cell, ribosomes for protein synthesis, and chromosomes made of DNA. Nevertheless, cells from different domains differ in a number of ways. **Figure A.4** shows examples of generalized cells that highlight notable subcellular components.

A.7 Semipermeable Membranes

The major functions of membranes (such as containing cytoplasmic components, regulating which substances enter and leave cells and organelles, and producing energy) depend on the semipermeable nature of membranes. Semipermeable (also called selectively permeable) membranes are permeable to some substances but not to others. In general, the cell membrane is permeable to hydrophobic molecules and impermeable to charged molecules (**Fig. A.5**).

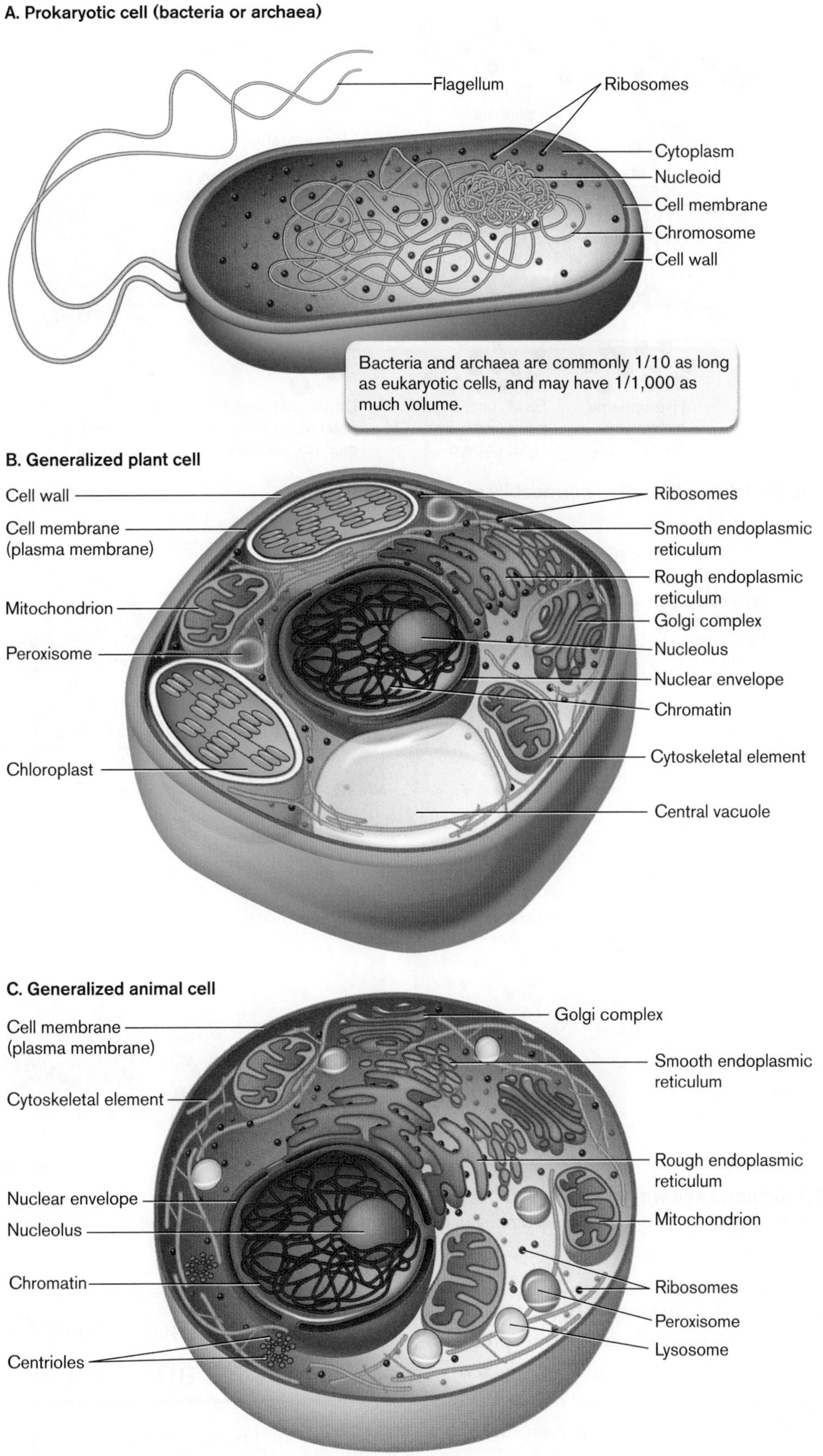

FIGURE A.4 ▪ The prokaryotic cell and the eukaryotic cell. **A.** The prokaryotic cell typically contains a single compartment, and its DNA is organized in the nucleoid region. **B, C.** Eukaryotic cells are typically much larger than prokaryotic cells and contain subcellular compartments formed by membranes.

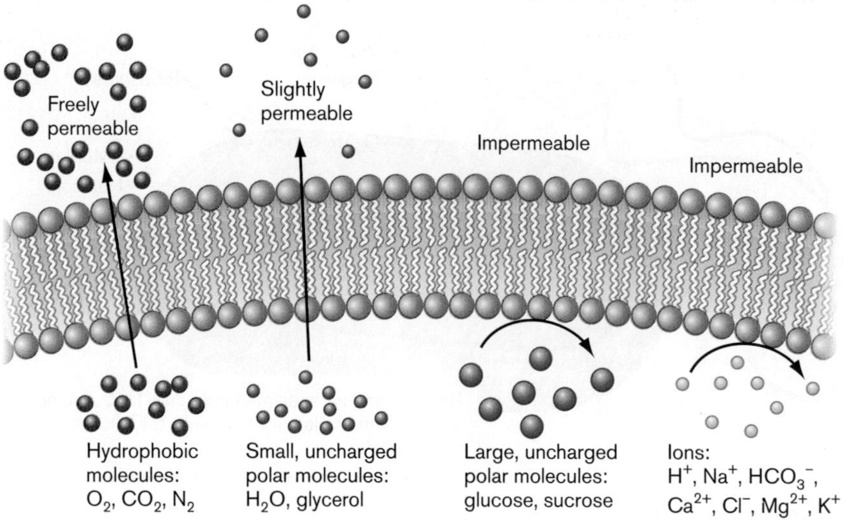

FIGURE A.5 ■ Selective permeability of cell membranes.

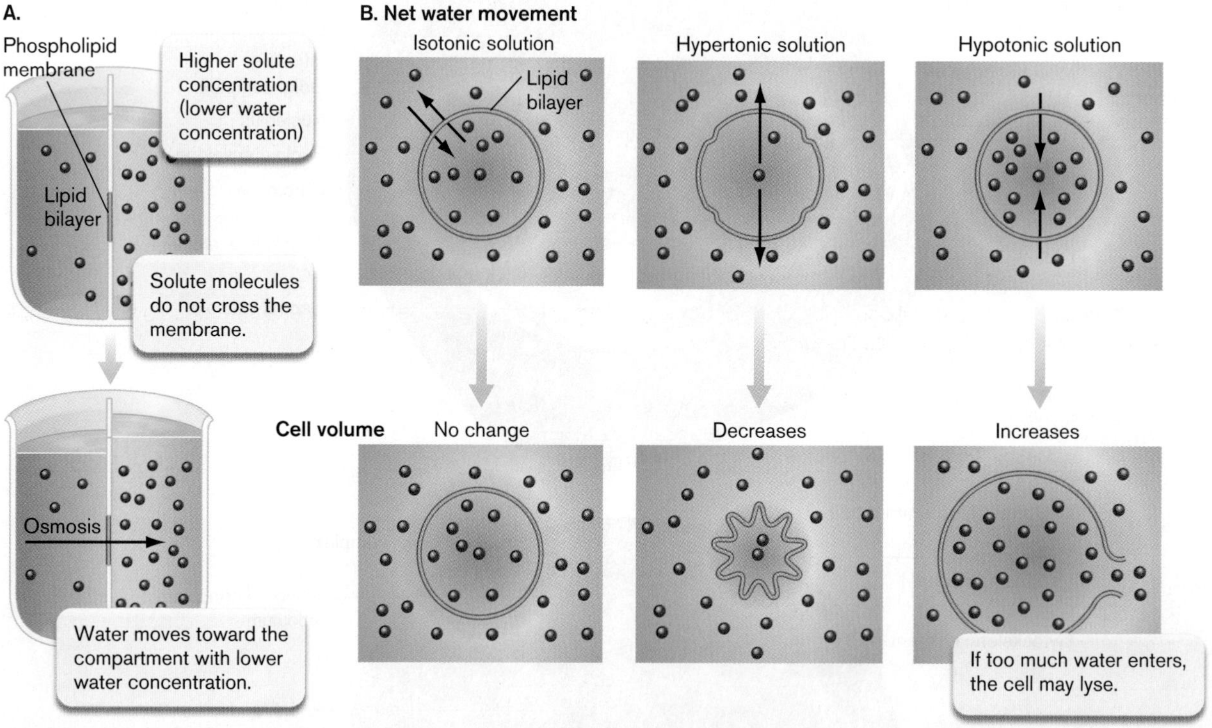

FIGURE A.6 ■ Osmosis and water balance. **A.** Osmosis. **B.** Movement of water across the cell membrane, and shrinkage or expansion of the membrane in isotonic, hypertonic, and hypotonic environments. Black arrows indicate net water movement.

Cells are filled with and surrounded by water. Having water on both sides of the cell membrane presents a challenge for living cells because water moves across a semipermeable membrane from regions of low solute concentration to regions of higher solute concentration (**Fig. A.6**). Cells must maintain osmotic balance with their environment.

A.8 The Eukaryotic Cell Cycle and Cell Division

Mitosis is the process that eukaryotic cells use to apportion their replicated DNA chromosomes evenly to their daughter cells. It is one phase in the cell cycle. The key steps of mitosis are summarized in **Figure A.7**.

Meiosis is a special form of mitosis that generates haploid gametes. The key steps of meiosis are summarized in **Figure A.8**.

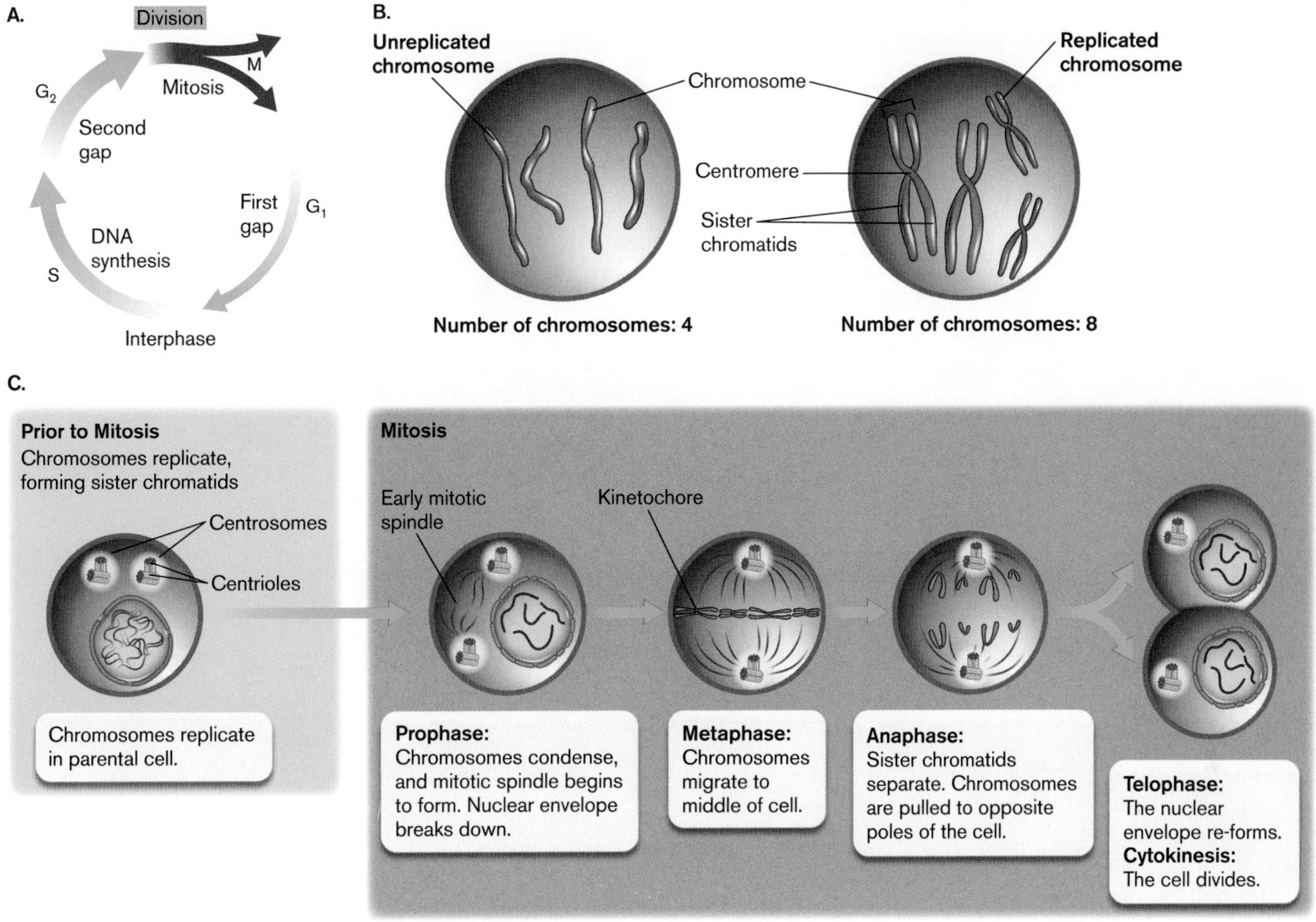

FIGURE A.7 ■ The cell cycle and mitosis. **A.** Stages of the eukaryotic cell cycle. G_1, S, and G_2 make up interphase (in blue). **B.** Duplication of the chromosomes during S phase. **C.** The phases of mitosis.

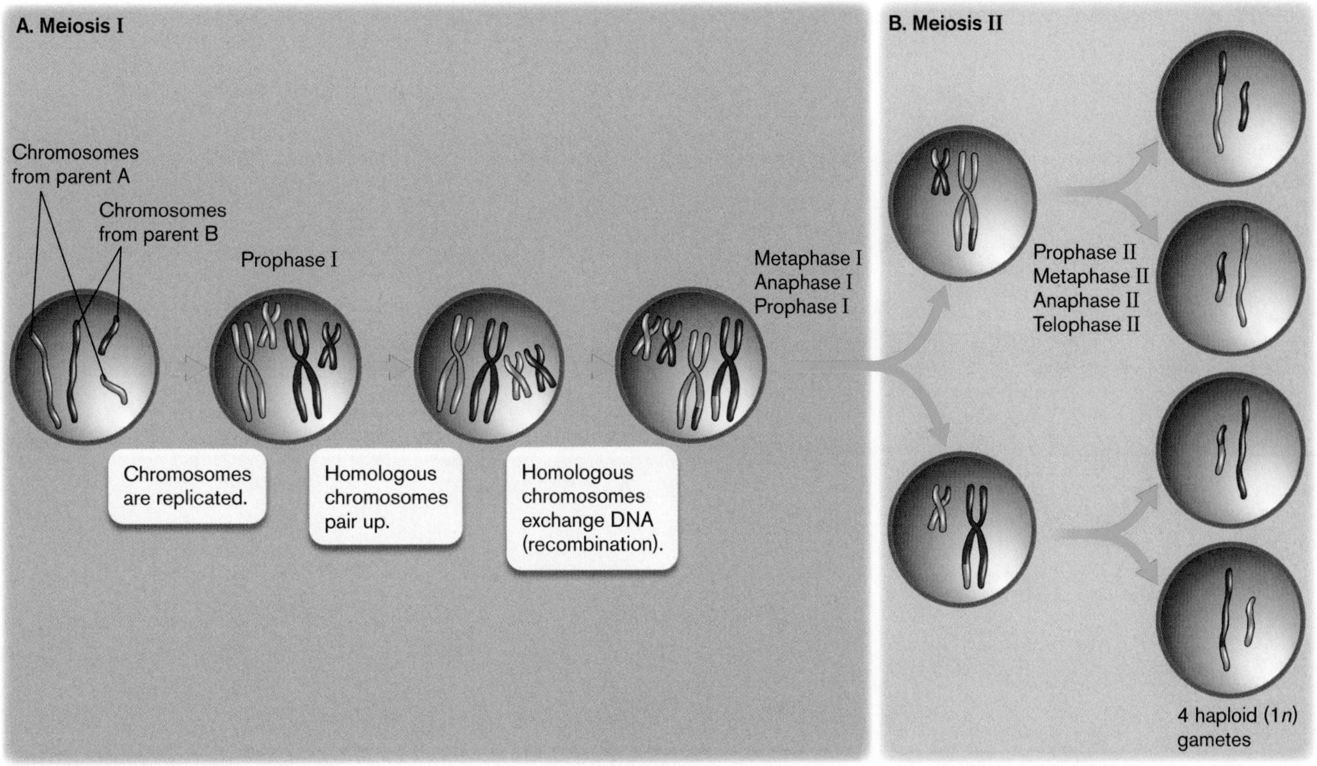

FIGURE A.8 ■ Meiosis. The chromosomes of a diploid (2*n*) cell undergo replication. **A.** In meiosis I, homologous chromosomes exchange DNA, and all the pairs are separated between two daughter cells. **B.** In meiosis II, homologous pairs are separated to produce haploid (1*n*) gametes.

ANSWERS TO THOUGHT QUESTIONS

CHAPTER 1

1.1 The minimum size of known microbial cells is about 0.2 μm. Could even smaller cells be discovered? What factors may determine the minimum size of a cell?

ANSWER: The smallest cells known, about 0.2 μm in length, are cell wall–less bacteria called mycoplasmas—for example, *Mycoplasma pneumoniae*, a causative agent of pneumonia. Bacteria smaller than 0.2 μm have been discovered by passing stream water through a filter of that pore size. It is hard to see how their cell components, such as ribosomes (about a tenth this size), could fit inside such a small cell. The volume required for DNA and the apparatus of transcription and translation probably sets the lower limit on cell size.

1.2 If viruses are not functional cells, are they "alive"?

ANSWER: A traditional definition of a life form includes the capability for metabolism and homeostasis (maintaining internal conditions of its cytoplasm), as well as reproduction and response to its environment. Viruses direct their own replication and respond to the environment of the host cell, but they lack both metabolism and homeostasis outside their host cell. Nevertheless, viruses such as herpesviruses contain numerous metabolic enzymes that participate in the metabolism of the host. Certain large viruses, such as the mimivirus, appear to have evolved from cells. Some microbiologists argue that viruses should be considered alive if reproduction is the main criterion and if the viral "environment" is considered the inside of the host cell.

1.3 Why do you think it took so long for humans to connect microbes with infectious disease? What innovations helped make the connection?

ANSWER: For most of human history, we were unaware that microbes existed. Even after microscopy had revealed their existence, the incredible diversity of the microbial world and the difficulties in isolating and characterizing microbial organisms made it difficult to discern the specific effects of microbes. All healthy people contain microbes, and most disease-causing microbes are indistinguishable from normal microbiota by light microscopy. Not all microbial diseases can be transmitted directly from human to human; they may require complex cycles with intermediate hosts, such as the fleas and rats that carry bubonic plague.

1.4 How could you use Koch's postulates (Fig. 1.19) to demonstrate the causative agent of influenza? What problems not encountered with anthrax would you need to overcome?

ANSWER: Using Koch's postulates to demonstrate the causative agent of influenza would require an animal model host. Secretions from diseased patients could be applied to different animal species, such as monkeys and mice, in order to find an animal showing signs of the disease. To determine the causative agent of disease, the patient's secretions could be filtered in order to separate bacteria and viruses. Only the filtrate would cause disease, because it contains viruses (relevant to Koch's postulates 1 and 3). Viruses, however, are more difficult to isolate in pure culture than are bacteria (postulate 2)—a problem Koch did not address. Furthermore, some viruses, such as HIV (human immunodeficiency virus) have no animal model; they grow only in human cells. Today, viruses are usually isolated in a tissue culture. Once isolated, the virus could be used to inoculate a new host animal (if an animal model exists) or a tissue culture and determine whether infection results (postulates 3 and 4). Another problem Koch did not address was the detection of infectious agents too small to be observed under a microscope. Today, antibody reactions are used to determine whether an individual has been exposed to a putative pathogen. An antibody test could be used to determine whether healthy and diseased individuals have been exposed to the isolated virus.

1.5 Why do you think some pathogens generate immunity readily, whereas others evade the immune system?

ANSWER: Some pathogens (microbes that cause disease) have external coat proteins that strongly stimulate the immune system and induce the production of antibodies. Other pathogens have evolved to avoid the immune system by changing the identity of their external proteins. Immunity also varies greatly with the host's status. The very young and very old generally have weaker immune systems than do people in the prime of life. Some pathogens, such as HIV, will directly attack the host's immune system, limiting the immune response to the pathogen.

1.6 How do you think microbes protect themselves from the antibiotics they produce?

ANSWER: Microbes protect themselves from their antibiotics by producing their own resistance factors. As discussed in later chapters, microbes may synthesize pumps to pump the antibiotics out; or they may make altered versions of the target macromolecule, such as the ribosome subunit; or they may produce enzymes to cleave the antimicrobial substance.

1.7 Why don't all living organisms fix their own nitrogen? Consider the structure of a dinitrogen molecule, N≡N.

ANSWER: Nitrogen fixation requires a tremendous amount of energy, about 30 molecules of ATP per dinitrogen molecule converted to ammonia (discussed in Chapter 15). In a community containing adequate nitrogen sources, organisms that lose the nitrogen fixation pathway make more efficient use of their energy reserves than do those that spend energy to fix nitrogen from the atmosphere. Another consideration is that nitrogenase is an oxygen-sensitive enzyme, whereas plants, animals, and fungi are aerobes. In order to fix nitrogen, aerobic organisms need to develop complex mechanisms to keep oxygen away from nitrogenase.

1.8 Could endosymbiosis occur today; that is, could a small microbe be engulfed by a larger one and evolve into an endosymbiont, and then into an organelle? Explain.

ANSWER: There are many examples today of endosymbiotic associations that look like evolution of an interdependent relationship. For example, a paramecium can acquire internalized chlorella algae that conduct photosynthesis and provide nutrients for the protozoan host. However, in the dark, where light is unavailable, the paramecium may instead digest the chlorella for food. Other bacteria, such as *Wolbachia* species, have evolved as endosymbionts of insect cells. In some cases, the insect absolutely requires bacterial endosymbionts to provide amino acids, but digests the endosymbiont when the nutrients are no longer needed. Still other insects host permanent endosymbiotic bacteria such as *Buchnera*, which are transmitted vertically (from parent to offspring). These permanent endosymbionts may be on their way to evolving into organelles.

1.9 What arguments support the classification of Archaea as a third domain of life? What arguments support the classification of archaea and bacteria together, as prokaryotes, distinct from eukaryotes?

ANSWER: The sequence of 16S rRNA (small-subunit rRNA) and other fundamental genes differs as much between archaea and bacteria as it does between archaea and eukaryotes. The composition of archaeal cell walls and phospholipids is completely distinct from that of bacteria and eukaryotes. Some aspects of gene expression, such as the RNA polymerase complex, are more similar between archaea and eukaryotes than between archaea and bacteria. On the other hand, archaeal and bacterial cells are prokaryotic; they both lack nuclei and complex membranous organelles. Archaeal metabolism and lifestyles are more similar to those of bacteria than to those of eukaryotes. Some archaea and bacteria sharing the same environment, such as high-temperature springs, have undergone horizontal transfer of genes that encode traits such as heat-stable membrane lipids.

1.10 State an argument in favor of patenting a microbial isolate or a gene sequence. What argument can be made against patenting microbes or genes?

ANSWER: A microbe can be patented if it is genetically modified from the "natural" state, and if it has a commercial use or application. Mere discovery of a pollutant-eating organism from nature is not patentable, but modifying an organism for commercial use constitutes technology. If a microorganism consists of a complex of molecules, then modifying the organism is legally equivalent to modifying a drug molecule or other industrial chemical. An argument against patenting organisms might be that life has a special status, in that organisms have their own agency to proliferate. Living organisms have a special spiritual status in the worldview of most major religions. Therefore, it would be inappropriate to patent a microbe, or a mouse—or a human being treated by gene modification therapy.

CHAPTER 2

2.1 As shown in Figure 2.2, the image passing through your cornea and lens is inverted on your retina. Why, then, does the world appear right side up?

ANSWER: The human brain interprets the image from the retina. Based on this interpretation, the brain knows that the image is upside down and inverts it to appear right side up. Researchers have tested what happens if an experimental subject wears special glasses that invert the image before the retina. After several days, the brain inverts the image perceived through the glasses, so it appears right side up. When the glasses are removed, the brain again takes time to restore its perception to right side up.

2.2 (refer to Fig. 2.7) You have discovered a new kind of microbe, never observed before. What kinds of questions about this microbe might be answered by light microscopy? What questions would be better addressed by electron microscopy?

ANSWER: Light microscopy could answer questions such as: What is the overall shape of this cell? Does it form individual cells or chains? Is the organism motile? Only light microscopy can visualize an organism alive. Electron microscopy can answer questions about internal and external subcellular structures. For example, does a bacterial cell possess external filamentous structures, such as flagella or pili? If the dimensions of the unknown microbe are smaller than the lower limits of a light microscope's resolution, EM may be the only way to observe the organism. Viruses are often characterized by shape, and this shape is observed by electron microscopy.

2.3 Explain what happens to the refracted light wave as it emerges from a piece of glass of even thickness. How do its new speed and direction compare with its original (incident) speed and direction?

ANSWER: The part of the wavefront that emerges first travels faster than the portion still in the glass, causing the wavefront to bend toward the surface of the glass. Ultimately, the wave travels in the same direction and with the same speed as it did before entering the glass. The path of the emerging light ray is parallel to the path of the light ray entering the glass and is shifted over by an amount dependent on the thickness of the glass. This refraction will alter the path of the beam of light and decrease the amount of light reaching the lens of the microscope. Immersion oil has the same refractive index as glass and will limit the amount of light lost in this way.

2.4 (refer to Fig. 2.13) For a single lens, what angle θ might offer magnification even greater than 100×? What practical problem would you have in designing a lens to generate this light cone?

ANSWER: In theory, an angle θ (theta) of 90° would produce the highest resolution—even greater than 100°. However, a 90° angle θ generates a cone of 180°, which would require the object to sit in the same position as the objective lens—in other words, to have a focal distance of zero. In practice, the cone of light needs to be somewhat less than 180°, to allow room for the object and to avoid substantial aberrations (light-distorting properties) in the lens material.

2.5 Under starvation, a bacterium such as *Bacillus subtilis* packages its cytoplasm into a spore, leaving behind an empty cell wall. Suppose, under a microscope, you observe what appears to be a hollow cell. How can you tell whether the cell is indeed hollow or is simply out of focus?

ANSWER: You can tell whether the cell is out of focus or actually hollow by rotating the fine-focus knob to move the objective up and down while observing the specimen carefully. If the hollow shape appears to be the sharpest image possible, it is probably a hollow cell. If the hollow shape turns momentarily into a sharp, dark cell, it was probably out of focus before. Alternatively, you could use a confocal microscope to visualize the center of the hollow cell.

2.6 What experiment could you devise to determine the order of events in *Bacillus subtilis* DNA replication?

ANSWER: One way to track the movement of DNA during DNA replication would be to stain the DNA with a dye such as DAPI at various stages of cell division. Alternatively, green fluorescent protein (GFP) fused to a DNA-binding protein could be used to label a specific sequence of DNA and track its position. Another way to determine the order of events in sporulation could be to observe mutant strains of bacteria that contain defects in different proteins of the replication process. (DNA replication is discussed further in Chapter 7.)

2.7 Some early observers claimed that the rotary motions observed in bacterial flagella could not be distinguished from whiplike patterns, comparable to the motion of eukaryotic flagella. Can you design an experiment to distinguish the two and prove that the flagella rotate? *Hint:* Bacterial flagella can get "stuck" to the microscope slide or coverslip.

ANSWER: To prove that flagella rotate, you can "tether" a bacterium to the microscope slide by getting one of its flagella stuck to the slide. A simple way to tether bacteria is by using a slide coated with anti-flagellin antibody. When the flagellum is stuck to the slide, its motor continues to rotate; thus, the entire cell now rotates. The rotation of the cell body can easily be seen by video microscopy. If the flagella moved in a whiplike fashion, the tethered cell would move back and forth, not rotate.

2.8 Compare and contrast fluorescence microscopy with dark-field microscopy. What similar advantage do they provide, and how do they differ?

ANSWER: Both dark-field and fluorescence microscopy enable detection (but not resolution) of objects whose dimensions are smaller than the wavelength of light. Dark-field technique is based on light scattering, which detects all small objects without discrimination. Fluorescence, however, provides a means to label specific parts of cells, such as the cell membrane or DNA, or particular species of microbes, using fluorescent antibody tags.

2.9 Like a light microscope, an electron microscope can be focused at successive powers of magnification. At each level, the image rotates at an angle of several degrees. Given the geometry of the electron beam (see Fig. 2.36), why do you think the image rotates?

ANSWER: The image rotates because the electron beam is not straight, as for photons, but travels in a spiral through the magnetic field lines. As magnification increases, the spiral expands, and it reaches the image plane at a slightly different angle than before.

2.10 What kinds of research questions could you investigate using SEM? What questions could you answer using TEM?

ANSWER: SEM could be used to examine the surface of cells: Do the cells possess a smooth surface, or does their surface contain protein complexes or bulges that serve special functions? How do pathogens attach to the surface of cells? TEM can be used to determine the intracellular structure of attachment sites, as well as of internal organelles. TEM can also visualize the shapes of macromolecular complexes such as flagellar motors or ribosomes.

2.11 How could you use atomic force microscopy to study the effect of an antibiotic on *Pseudomonas aeruginosa* contamination of medical catheters?

ANSWER: Bacteria colonize catheters by forming a biofilm, a structure of cells that grow attached to each other as well as the substrate. The height and volume of a biofilm can be measured by AFM. As shown in Figure 2.46, deflection of the tip of the cantilever indicates very precise measurements of thickness of a biofilm, allowing three-dimensional mapping. An antibiotic can be added at various concentrations, and the height and volume of the biofilm can be measured over time. This procedure will indicate how an antibiotic affects biofilm formation on a catheter.

CHAPTER 3

3.1 Which chemicals do we find in the greatest number in a bacterial cell? The smallest number? Why does a cell contain 100 times as many lipid molecules as strands of RNA?

ANSWER: The chemicals that occur in the greatest number in a prokaryotic cell are inorganic ions (250 million/cell). They are also the smallest in size. DNA molecules are found in the lowest number (one large molecule, branched during replication). A prokaryotic cell contains 100 times as many lipid molecules as strands of RNA because lipids are small structural molecules, highly packed. They are a major component of the cellular membrane. RNA molecules are long macromolecules that either are packed into complexes (such as ribosomal RNA) or are temporary information carriers (messenger RNA), present only as needed to make proteins.

3.2 Suppose we wish to isolate flagellar motors, which are protein complexes that span the envelope from inner membrane to outer membrane (Fig. 3.1). How might we modify the cell fractionation procedure to achieve such isolation?

ANSWER: There is not one answer, but several alternatives can be tried. Flagellar motors associate with both inner and outer membranes, complicating the cell fractionation. It is possible that the motors might associate more strongly with one membrane or the other, so the motor could be found primarily in one of the membrane fractions. A fluorescent antibody could identify fractions that contain the motors. Mild-detergent treatment of the fractions could strip away the membrane. Alternatively, the whole-membrane prep could be treated with mild detergent. The protein fraction could then be centrifuged through a sucrose gradient; because the motor complexes have a very specific size and density, they would be concentrated in one fraction. Electron microscopy could confirm the fraction containing the motors.

3.3 Amino acids have acidic and basic groups that can dissociate. Why are they not membrane-permeant weak acids or weak bases? Why do they fail to cross the phospholipid bilayer?

ANSWER: At neutral pH, an amino acid has both a positively charged amine and a negatively charged carboxylate; that is, it can act as either a weak acid or a weak base. Charged ions, no matter what their size, will not freely pass through a plasma membrane. In an amino acid, if either charged group becomes neutralized by acid or base, the other group remains charged, so the molecule as a whole will never cross the membrane.

3.4 What genetic experiments could you propose to test the model of envelope expansion shown in Figure 3.13?

ANSWER: Use mutagenesis (treatment with a mutagen) to generate *E. coli* strains containing point mutations in *rodA* or *rodZ*. Observe growth of the mutant cells. Prediction: Some of the *rodA* or *rodZ* mutants will grow slowly, as bulging, blob-shaped cells, not as rods. Now test the question, How does RodA or RodZ interact with other components of the envelope extension complex? Introduce new mutations, in genes encoding proteins of the peptidoglycan extension complex (**Fig. 3.13**), such as *mreB*, *mrdA*, and *pbp2*. With the newly introduced mutations, some of the *rodZ* mutants revert to normal rod-shaped growth. This result suggests that the RodZ protein could interact with MreB, MrdA, and Pbp2 in the extension complex. For research published on these questions, see Daisuke Shiomi et al., 2013, *Mol. Microbiol.* 87:1029.

3.5 What other ways can you imagine that bacteria might mutate to become resistant to vancomycin?

ANSWER: A common means of resistance to antibiotics is to pump them out of the cell. A protein pump that exports other molecules might mutate to capture vancomycin and export it from the cell. Another possibility is that an enzyme could modify the vancomycin by adding phosphoryl groups or acetyl groups, which would prevent the antibiotic from binding the alanine dipeptide. Still another possibility is that the bacteria might evolve a thicker cell wall that would exclude the vancomycin from the inner layers of peptidoglycan.

3.6 Figure 3.15 highlights the similarities and differences between the cell envelopes of Gram-negative and Gram-positive bacteria. What do you think are the advantages and limitations of a cell having one layer of peptidoglycan (Gram-negative) versus several layers (Gram-positive)?

ANSWER: Having multiple layers of peptidoglycan increases the cell's resistance to osmotic shock, to desiccation stress, and to enzymes that cleave the cell wall. On the other hand, to build the layers of peptidoglycan requires more energy and biomass. In addition, a thick cell wall could slow the uptake of nutrients. The mycobacteria, which have exceptionally thick cell walls, grow relatively slowly.

3.7 Why would laboratory culture conditions select for evolution of cells lacking an S-layer?

ANSWER: Degeneration of protective traits is a common problem when conducting research on microbes that can produce 30 generations overnight. Their rapid reproductive rate gives ample opportunity for spontaneous mutations to accumulate over an experimental timescale. In the case of the S-layer, in a laboratory test tube free of predators or viruses, mutant bacteria that fail to produce the thick protein layer would save energy compared to S-layer synthesizers, and would therefore grow faster. Such mutants would quickly take over a rapidly growing population.

3.8 Why would proteins be confined to specific cell locations? Why would a protein not be able to function everywhere in the cell?

ANSWER: Proteins have evolved one or more specific functions often optimized for a specific part of the cell. For example, water-conducting porins are found solely in the inner membrane (cell membrane), which is otherwise impermeable to water. The outer membrane, which is water permeable, is the sole location for specific porins that transport small peptides and sugars. The sugars then need to be taken across the inner membrane by transport proteins that have evolved to function best in this location. Similarly, different chaperones (proteins that aid peptide folding) have evolved to function best in the environment of the cytoplasm or periplasm, membrane-enclosed regions that differ substantially in pH and ion concentrations. In a different chemical environment of the cell, a protein may denature and lose its functional structure. A protein may be active only as part of a complex of proteins. If the protein is placed in a different location within the cell, its protein partners may be absent, rendering the protein nonfunctional.

3.9 Suppose a cell has a defect in its *ftsZ* gene. What might happen to the cell during growth? How could such a mutant strain be maintained in the laboratory?

ANSWER: A cell with a defective *ftsZ* gene will fail to septate. As the cell grows, it expands and replicates its DNA, but no septum forms and the daughter cells do not separate. Eventually, the cell's nucleoids will entangle and the long, filamented cell chain will die. There are several ways to maintain an *ftsZ* mutant in a viable state. One is to use a temperature-sensitive mutant, in which the FtsZ protein is functional at the permissive temperature but nonfunctional at the nonpermissive temperature. Another way is to maintain a copy of the *ftsZ* gene fused to a promoter that can be turned on or off by the presence of an inducer molecule such as a sugar (discussed in Chapter 10).

3.10 Figure 3.31 presents data from an experiment that allows the function of the TipN protein of *Caulobacter* to be visualized by microscopy. Can you propose an experiment with mutant strains of *Caulobacter* to test the hypothesis that one of the proteins shown in Figure 3.32 is required for one of the cell changes shown?

ANSWER: The diagram of **Figure 3.32** proposes that PodJ protein is required for a pole to develop a flagellum. Suppose we construct a mutant strain with a deletion of the gene *podJ*. This *podJ* mutant fails to express PodJ protein. When the *podJ* mutant is supplied with nutrients, the stalked cells should grow and fission, but the progeny from the plain pole should fail to grow a flagellum. The stalked progeny will continue to divide, producing a stalked cell and a cell with plain poles, lacking flagellum or stalk. Other results are possible, but the result described would be consistent with a requirement of PodJ for flagellar development.

3.11 Could two bacteria share protein complexes via nanotubes? What about hydrogen molecules (H_2) as electron donors?

ANSWER: Cells have been shown to share proteins via nanotubes, such as enzymes for carbohydrate catabolism. In principle, a nanotube could be wide enough to allow transmission of ribosomes. However, nanotubes could not share dihydrogen molecules, because H_2 is a gas that penetrates membranes and would escape through the nanotube walls.

3.12 Most laboratory strains of *E. coli* and *Salmonella* commonly used for genetic research lack flagella. Why and how do bacterial strains evolve to lose flagella? How can a researcher maintain a motile strain?

ANSWER: The motility apparatus requires 50 different genes generating different protein components. Cells that acquire mutations eliminating expression of the motility apparatus gain an energy advantage over cells that continue to invest energy in motors. In a natural environment, the nonmotile cells lose out in competition for nutrients, despite their energetic advantage; but in the laboratory, cells are cultured in isotropic environments such as a shaking test tube, where motility confers no advantage. These culture conditions lead to evolutionary degeneration of motility, as they do for the S-layer (see Thought Question 3.7). In order to maintain a motile strain, bacteria are cultured on a soft agar medium

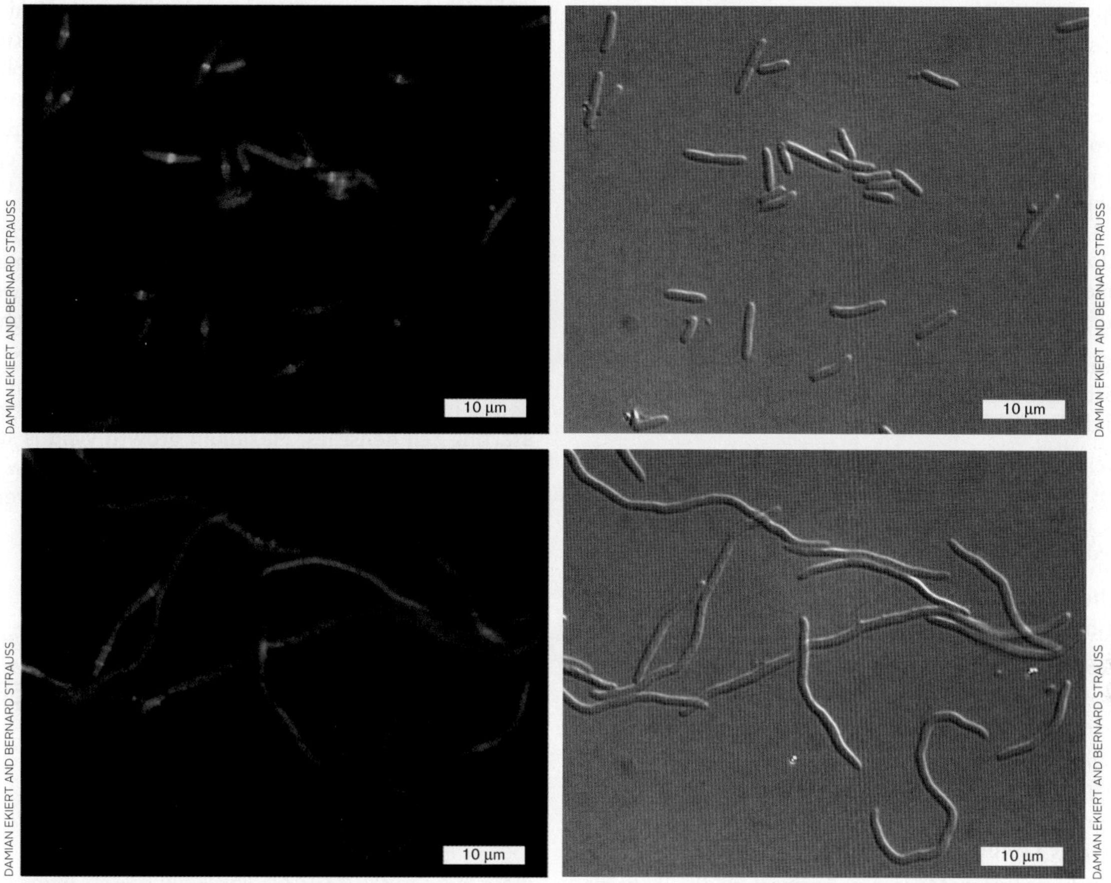

FIG. TQ 3.9.1 ■ **Top panels:** *E. coli* cells expressing an FtsZ-GFP fusion protein. **Bottom panels:** *E. coli* cells when DNA synthesis is disrupted. (Left photos imaged with fluorescence, right photos imaged with brightfield/DIC.)

containing an attractant nutrient. As cells consume the attractant, they generate a gradient, and chemotaxis leads them to swim outward. By subculturing only bacteria from the leading edge of swimming cells, one can maintain a motile strain.

3.13 How would a magnetotactic species have to behave if it were in the Southern Hemisphere instead of the Northern Hemisphere?

ANSWER: In the Northern Hemisphere, the field lines for magnetic north point downward; in the Southern Hemisphere, the opposite is true. Thus, if downward direction is the aim of magnetotaxis, bacteria existing in the two hemispheres would have to respond oppositely to the magnetic field; in the Southern Hemisphere, anaerobic magnetobacteria swim toward magnetic south. Near the equator, the proportions of north-seeking and south-seeking bacteria are roughly equal.

CHAPTER 4

4.1 In a mixed ecosystem of autotrophs and organotrophs, what happens if the autotroph begins to outgrow the organotroph, producing more and more organic food?

ANSWER: As the organotroph begins to grow on the organic material, the growth of the organotroph might overtake and outpace the growth of the autotroph, using the carbon sources faster than the autotroph can make them. As the organic carbon sources diminish through consumption, growth of the organotrophs decreases, but the CO_2 formed by the organotrophs will allow the autotroph to grow and make more organic carbon. Ultimately, the ecosystem comes into balance.

4.2 How could a symport transporter produce electroneutral coupled transport?

ANSWER: Coupled transport can be electroneutral if molecules of opposite charge are being transported—for exammple, Na^+ flux together with Cl^-.

4.3 How might mutations in transporter gene sequences influence bacterial survival under different conditions—for example, normal versus very low glucose concentrations?

ANSWER: A mutation that eliminates the glucose transporter or destroys its function will produce a mutant cell that cannot use glucose. It will die if an alternative carbon source is not available or if a transport system for that carbon source is not present. A mutation in a glucose transporter protein that increases the transporter's affinity for glucose will yield a cell that can grow in very low glucose concentrations. This mutant will outcompete a cell with a normal transporter when glucose concentrations are very low.

4.4 Describe the phenotype (growth characteristic) of a cell that lacks the *trp* genes (genes required for the synthesis of

tryptophan). What would be the phenotype of a cell missing the *lac* genes (genes whose products catabolize the carbohydrate lactose)?

ANSWER: The difference lies in the function of the two pathways. The *trp* operon is a biosynthetic operon. Errors in the biosynthetic pathway will lead to a failure to produce tryptophan. Therefore, a *trp* auxotrophic mutant will grow on defined medium only if tryptophan is added. The lactose operon involves the catabolism of a carbon source, lactose. If any of these genes are damaged, the cells are no longer able to use lactose as a carbon source. A *lac* mutant will not grow on defined medium with lactose as the sole carbon source.

4.5 If lactose were left out of MacConkey medium (Fig. 4.14), would lactose-fermenting *E. coli* bacteria grow, and if so, what color would their colonies be?

ANSWER: Even without lactose in the medium, *E. coli* would grow nonfermentatively on the peptides present. The colonies would appear white because without lactose, the cells do not make acidic products needed to bring neutral red into the colony.

4.6 Use the information in Figure 4.17 to determine the concentration (in cells per milliliter) of bacteria shown.

ANSWER: 1.25×10^6 bacteria per ml.

SOLUTION:

- Each small square is $0.0025\ mm^2$ in size, and the depth from coverslip to surface is 0.2 mm.
- $0.0025\ mm^2 \times 0.2\ mm = 0.0005\ mm^3$ ($1\ mm^3 = 1\ \mu l$). Each square defines a volume of $0.0005\ \mu l$.
- 10 cells observed over 16 squares averages to 0.625 bacteria/$0.0005\ \mu l$ or 0.625 bacteria/square.
- 0.625 bacteria/$0.0005\ \mu l = 1.25 \times 10^3$ cells/μl $= 1.25 \times 10^6$ cells/ml.

4.7 A virus such as influenza virus might produce 800 progeny virus particles from one host cell infected by one virus. How would you mathematically represent the exponential growth of the virus? What practical factors might limit such growth?

ANSWER: In theory, the growth rate of the virus would be proportional to 800^n. In practice, however, it is unlikely that the 800 virus particles released from one host cell will find 800 different host cells to infect. Furthermore, it turns out that only a small proportion of the influenza virus progeny are viable (see Chapter 11).

4.8 Suppose one cell of the nitrogen fixer *Sinorhizobium meliloti* colonizes a plant root. After 5 days (120 hours), there are 10,000 bacteria fixing N_2 within the plant cells. What is the bacterial doubling time?

ANSWER: 9 hours. Note that this generation time is much longer than it would be if these same organisms were grown in a test tube containing suitable medium. In a suitable laboratory medium, the generation time is about 1.5 hours.

4.9 It takes 40 minutes for a typical *E. coli* cell to completely replicate its chromosome and about 20 minutes to prepare for another round of replication. Yet the organism enjoys a 20-minute generation time growing at 37°C in complex medium. How is this possible? *Hint:* How might the cell overlap the two processes?

ANSWER: After the DNA is replicated about halfway around the chromosome, each daughter half-chromosome initiates a second round of replication, so the time needed to divide from one cell to two is effectively halved. Most cells in a log-phase culture in rich medium actually have four copies of the DNA origin of replication, each with a separate attachment site on the cell envelope, the future midpoint of a cell two generations ahead (see Chapter 3).

4.10 The bacterium *Acidithiobacillus thiooxidans* is an extremophile that grows using sulfur as an energy source. (a) Draw the approximate growth curves that you would expect to see, extending from log phase to stationary phase, if four cultures with different starting numbers of bacteria were grown in the same concentration of sulfur. Use Figure 4.21B as the model, and 4×10^5, 4×10^6, 4×10^7, and 4×10^8 as the starting cell densities. Maximum growth yield is 10^9 cells per milliliter. (b) Draw a second graph showing how the curves would change if the initial population density were constant but the concentration of sulfur varied.

ANSWER:

(a)

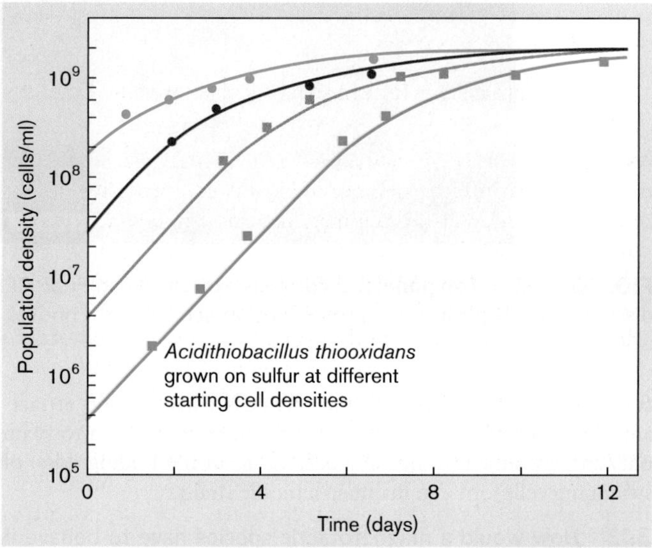

(b)

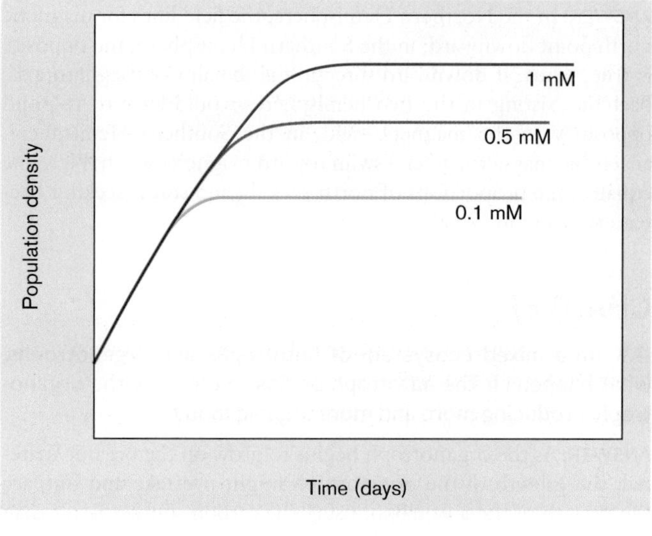

4.11 What can happen to the growth curve when a culture medium contains two carbon sources, if one is a preferred carbon source of growth-limiting concentration and the second is a nonpreferred source?

ANSWER: There are two possibilities. If the enzyme systems needed to utilize both carbon sources are always made, the growth curve will look normal because both will be used simultaneously. Usually, the enzyme system for the nonpreferred carbon source is not produced until the preferred source is used up. In this case, a second lag phase will interrupt the exponential phase. This is called a diauxic growth curve and is commonly seen when cells are grown on both glucose and lactose. Lactose is the nonpreferred carbon source and is used second. The second lag phase marks the exhaustion of one nutrient and the gearing up by the cell to use the other (see Fig. 10.12).

4.12 How would you modify the equations describing microbial growth rate to describe the rate of death?

ANSWER: The death rate applies to a period of declining cell numbers. Therefore, the logarithm of the cell number ratio of N_1 to N_0 will be a negative number, and this factor will need to be preceded by a negative sign to convert it to a positive "halving time," or half-life of the culture.

4.13 Why are cells in log phase larger than cells in stationary phase?

ANSWER: Cells strive to maintain a certain DNA/mass ratio. In so doing, they balance the number of biochemical processes needed to sustain viability. If the cell mass becomes large relative to the number of copies of a given critical gene, the amount of enzyme produced may not be sufficient to keep the cell alive and growing. In addition, the DNA/mass ratio serves as a signal to trigger cell division. Thus, when a cell divides faster than it replicates its chromosome, it must start a second round of replication before it finishes the first. This type of replication ensures that at least one chromosome duplication will be complete at the time of division. Because fast-growing cells contain more than one chromosome, they will increase in size to maintain the desired DNA/mass ratio. If the ratio were not maintained, cell division would not occur when needed.

4.14 How might members of the Actinomycetales such as *Streptomyces* species avoid "committing suicide" when they make their antibiotics?

ANSWER: Bacteria that produce antibiotics need to make defenses against the antibiotic within their own cytoplasm. For example, their genes can express an altered form of the target molecule, such as a ribosomal subunit; or they can make pumps to pump the antibiotic out of the cell.

CHAPTER 5

5.1 Why haven't cells evolved so that all their enzymes have the same temperature optimum? If they did, wouldn't they grow even more rapidly?

ANSWER: An enzyme's function is not determined by temperature alone. There are other physical and chemical constraints, based on the variety and complexity of functions that different enzymes must carry out. The thousands of different enzyme molecules must work in a coordinated fashion to support the basic functions of life. Having some enzymes work below or above their optimal temperatures will alter the rates of the reactions they catalyze. A population's evolution is based on an entire organism's ability to reproduce, not the speed at which each individual chemical reaction is carried out. The primary goal of a microbe is not just to grow fast but also to survive. Growing too fast could deplete food sources and produce toxic by-products too quickly.

5.2 If microbes lack a nervous system, how can they sense a temperature change?

ANSWER: Most bacteria respond to outside stimuli, such as heat, by altering their gene expression. They sense heat by monitoring the concentration of misfolded proteins, a consequence of excessively high temperature. The mechanism does not perceive heat per se, but recognizes the deleterious effects of moving outside the optimal growth temperature range so that the cell can launch an emergency response. The same mechanisms can sense other environmental stresses that misfold proteins, such as acid stress. See Chapter 10.

5.3 Predict how hyperthermophilic microorganisms colonize a newly formed hydrothermal vent (black smoker), which is sterile at birth. How do the microbes get there through ice-cold water (0°C–3°C)?

ANSWER: Experimental data for colonization do not exist, but a hypothetical scenario can be proposed. Hyperthermophiles can persist, if not grow, in low-temperature seawater for long periods, probably because of a cold-adaptation program analogous to heat shock. Hyperthermophilic microbes floating in the ocean could, by chance, encounter a thermal vent, attach to its surface, and form a biofilm.

5.4 How might the concept of water availability be used by the food industry to control spoilage?

ANSWER: Food preservation traditionally includes water exclusion by salt, as seen in hams, back bacon, and salted fish; or by high concentration of sugar, as in canned fruit or jellies. The lower a_w prevents microbial growth. Dehydrating foods will also prevent microbial growth.

5.5 Recall from Section 4.2 that an antiporter couples movement of one ion down its concentration gradient with movement of another molecule uphill, against its gradient. For Na^+/H^+ symporters, that means more sodium inside than out and more protons out than in. If this is true, how could a Na^+/H^+ antiporter work to bring protons into a haloalkaliphile growing in high salt at pH 10? In this situation (high salt and high pH media) there will be more sodium outside than there is in, and more H^+ inside than out. The opposite of what you'd think the cell would need. Both ions would have to move AGAINST their concentration gradients. Sodium moves out, protons move in.

ANSWER: In this situation the cell has to expend some energy to make the antiporter work. The energy involved is rooted in the charge difference between the inside of the cell (negative charge) and the outside of the cell (positive charge), called delta psi, $\Delta\psi$ (see Chapter 14). Delta psi is expended to drive Na^+ out and, thus, H^+ in. The antiporter in this case must also exchange a different number of Na^+ and H^+ ions, maintaining an electrical charge difference across the membrane; for example, export 2 Na^+ and import 1 H^+ to keep delta Psi negative inside.

5.6 If anaerobes cannot live in oxygen, how do they incorporate oxygen into their cell components?

ANSWER: Obligate anaerobes incorporate oxygen from their carbon sources (for example, CO_2 and carbohydrates such as glucose), all of which contain oxygen. This form of oxygen will not damage the cells.

5.7 How can anaerobes grow in the human mouth, where there is so much oxygen?

ANSWER: A synergistic relationship exists between facultatives and anaerobes within a tooth biofilm. The facultative anaerobes consume oxygen within the biofilm microenvironment, thereby allowing the anaerobes to grow underneath them.

5.8 What evidence led people to think about looking for anaerobes? *Hint:* Look up "Spallanzani," "Pasteur," and "spontaneous generation" on the Internet.

ANSWER: The Italian priest Lazzaro Spallanzani (1729–1799), during his quest to disprove spontaneous generation, said: "Every beast on Earth needs air to live, and I am going to show just how animal these little animals are by putting them in a vacuum and watching them die." He dipped a glass tube into a culture, sealed one end, and attached the other to a vacuum. He was astonished to find that the microbes lived for weeks. He then wrote: "How wonderful this is. For we have always believed there is no living being that can live without the advantages air offers it." Fifty years later, Louis Pasteur observed that air could kill some organisms. After looking at a drop of liquid from a fermentation culture, he wrote: "There's something new here—in the middle of the drop they are lively, going every which way. . . . But here at the edge they're not moving, they're lying round stiff as pokers."

5.9 Given a mixture of two microbes, A and B—where organism A can utilize limiting phosphate more efficiently than organism B, but organism B can utilize limiting nitrogen better than organism A—what would happen to the relative growth of the two organisms placed in a limiting nitrogen and phosphate medium if excess nitrogen were added to the mixed culture? What about both excess phosphate and excess nitrogen?

ANSWER: Excess nitrogen would give microbe A a nutrient advantage because it is already better than microbe B at using the limiting phosphate in the medium. Organism A might outgrow microbe B. However, adding excess P and N would favor neither strain. In that case, other factors, such as relative growth rates, relative abilities to use alternative carbon sources, and so on, would play more dominant roles.

5.10 Bacteriostatic antibiotics do not kill bacteria; they only inhibit their growth. Why are they nevertheless effective at treating bacterial infections? *Hint:* Is the human body a quiet bystander during an infection?

ANSWER: The bacteriostatic agent stops growth of the bacteria and allows the host immune response to kill them.

5.11 If a disinfectant is added to a culture containing 1×10^6 CFUs per milliliter and the D-value of the disinfectant is 2 minutes, how many viable cells will be left after 4 minutes of exposure?

ANSWER: 1×10^4 CFUs/ml (90% after 2 min = 1×10^5; 90% after another 2 min = 1×10^4).

5.12 How would you test the killing efficacy of an autoclave?

ANSWER: Construct a death curve by measuring survival of a known quantity of spores (for example, for *Bacillus stearothermophilus*) after autoclaving for various lengths of time. Spores should be used because they are more resistant to heat than is any vegetative cell. Typically, autoclaves are regularly checked with spore strips that change color once the endospores are no longer viable.

CHAPTER 6

6.1 Suppose a certain virus depletes the population of an algal bloom. What will happen if some of the algae are genetically resistant?

ANSWER: If some algae are resistant to the virus, they will reproduce and avoid infection. But their population growth will still face competition from other algal species that were never hosts of the virus. Furthermore, if the resistant algae form another bloom, eventually some other viral species will infect them and cut their population again. The result is an evolutionary arms-race.

6.2 Search for some specific viruses on the Internet. Which viruses do you think have a narrow host range, and which have a broad host range?

ANSWER: Examples of viruses with a narrow host range include poliovirus (poliomyelitis), which infects only humans and chimpanzees; smallpox virus, which infects only humans; and feline leukemia virus, which infects only cats. Examples of viruses with a broad host range include rabies virus, which infects numerous species of mammals; and influenza strains, which show preference for particular species but can jump between various mammals and birds.

6.3 What will happen if a virus particle remains intact within a host cell and fails to release its genome?

ANSWER: In most cases, a virus particle that fails to release its genome will be unable to reproduce, because DNA polymerases cannot reach its genome for reproduction and RNA polymerase cannot transcribe its genes to make gene products. An exception is double-stranded RNA viruses, which keep their genome partly enclosed in order to protect it from recognition by the host cell immune system.

6.4 For a viral capsid, what is the advantage of an icosahedron (20-sided solid), as shown in Figure 6.7, instead of some other polyhedron. such as a cube or a tetrahedron?

ANSWER: The icosahedron is a polyhedron of 20 triangular faces, the largest number possible. Thus, the icosahedron turns out to be the largest and most economical form to enclose space with a small repeating unit. Natural selection probably favors viruses that can build the largest capsid from the smallest amount of genetic information.

6.5 Klosneuvirus shows evidence of integrating genes from cellular hosts. What kind of fitness advantage might favor acquisition of host genes?

ANSWER: Integrating genes from a host leads to a genome of larger size. A larger genome may lead to formation of larger virus particles. The larger virus particles are more effectively phagocytosed by host amebas. Another possible advantage of acquiring cellular genes is that the viral homologs may evolve into variant components of biosynthetic machinery that more effectively produce virus components.

6.6 How could viruses with different kinds of genomes (RNA versus DNA) combine and share genetic content in their progeny?

ANSWER: DNA viruses require messenger RNA intermediates to express their proteins. In a rare event, a DNA virus could mutate and evolve the ability to package its RNA transcript, rather than its DNA, in a capsid. Alternatively, RNA retroviruses form DNA intermediates within their host cell; these DNA intermediates might recombine with the DNA genome of another virus.

6.7 What are the relative advantages and disadvantages (to a virus) of the slow-release strategy, compared with the strategy of a temperate phage, which alternates between lysis and lysogeny?

ANSWER: A disadvantage of slow release is that the phages can never reproduce progeny phages as rapidly as in a lytic burst. The drain on resources of the host cell infected by a slow-release virus causes it to grow more slowly compared with uninfected cells; in contrast, a lysogenized cell suffers little or no reproductive deficit compared with uninfected cells. An advantage of reproduction by slow release is the continuous release of phage, while avoiding the possibility of releasing all particles into an environment where no other host cell exists.

6.8 How else, besides Acr proteins, might a phage evolve resistance to the CRISPR host defense (outlined in Fig. 6.25)?

ANSWER: The phage could have a mutation in its DNA sequence that was cleaved to form the spacer. Now when the Cas-crRNA complex forms, the crRNA will no longer base-pair correctly with the phage DNA and will not cleave the DNA of the infecting phage.

6.9 How might humans undergo natural selection for resistance to rhinovirus infection? Is such evolution likely? Why or why not?

ANSWER: Resistance to all rhinovirus infections might evolve through a mutation in the host gene encoding ICAM-1. The mutation would have to prevent rhinovirus binding without impairing the protein's ability to bind integrin. Such evolution is unlikely because of the importance of integrin binding and because rhinovirus infection is rarely fatal; thus, there is little selection pressure to evolve inherited resistance. Note, however, that the immune system rapidly generates immunity to particular strains of rhinovirus. Over a lifetime, most individuals acquire immunity to many rhinovirus strains but remain susceptible to others.

6.10 From the standpoint of a virus, what are the advantages and disadvantages of replication by the host polymerase, compared with using a polymerase encoded by its own genome?

ANSWER: An advantage of using the host polymerase is energetic: The virus avoids the energetic cost of manufacturing a polymerase to package with each virion. This is an advantage to the virus because its reproductive potential is limited by the energy resources of its host cell. Furthermore, because DNA and RNA polymerases are so central to cell function, the host species is unlikely to evolve a mutant form of the polymerase that resists the virus. On the other hand, the advantage of the virus making its own polymerase is that the viral polymerase can evolve traits that better meet the needs of its own replication, such as high speed and low accuracy to generate frequent variants. One disadvantage of a DNA virus using the host cell DNA polymerase is that the virus must gain access to the host cell nucleus where the polymerase is. In addition, if the host cell is fully differentiated, it has exited the cell cycle and is not replicating. If it is not replicating, a DNA polymerase may not be available unless the virus can force the host cell to start going through the cell cycle. These are two problems that the virus will not have to overcome if it brings in its own DNA polymerase.

6.11 Why does bacteriophage reproduction give a step curve, whereas cellular reproduction generates an exponential growth curve? (Compare Fig. 6.36 with Fig. 4.21.) Could you design an experiment in which viruses generate an exponential curve? Under what conditions does the growth of cellular microbes give rise to a step curve?

ANSWER: Lytic viruses appear to make a step curve because the number of progeny per infected cell is 100 or more, released simultaneously. After two or three generations the cell cycles would fall out of synchrony, and the curve would smooth out, but the later cell cycles are rarely observed in practice because by then, the supply of host cells is exhausted. If, however, an extremely low ratio of viruses to host cells is provided, the growth of virus particles will eventually generate an exponential curve. By contrast, the growth of cellular microbes is rarely observed during the first few doublings. By the time we measure the population, the cells are all undergoing different stages of division, and the population growth overall generates a smooth exponential curve. But if we observe the growth of a synchronized population of cells, we see a step curve of cell division too.

6.12 What kinds of questions about viruses can be addressed in tissue culture, and what questions require infection of an animal model?

ANSWER: Questions that can be answered using tissue culture would be how a virus binds to cell-surface receptors, and how it replicates progeny virions within a cell. Questions of viral transmission, however, would require the organ system of an animal, where the virions must undergo transport within and escape the immune system. Often, a virus passaged through tissue culture, such as influenza virus, will accumulate mutations that allow faster growth in the laboratory but poor infection of an animal host.

CHAPTER 7

7.1 Before the studies by Avery, Hershey, and others, some scientists believed that, unlike plants and animals, bacteria lacked genes. Considering what little was known about the modes of reproduction and the recombination of alleles in bacteria, why was this a reasonable, albeit incorrect, assumption?

ANSWER: No mode of mating was apparent for microbes, and there was no segregation of traits (alleles) in the offspring. Some leading scientists at the time believed that the microbial cell was simply a "dynamic reaction network" that did not require genes to account for cellular activity.

7.2 What do you think happens to two single-stranded DNA molecules isolated from different genes when they are mixed together at very high concentrations of salt? *Hint:* High salt concentrations favor bonding between hydrophobic groups.

ANSWER: In high salt conditions, the stacking of hydrophobic bases is so strongly favored that two single strands of DNA will form a duplex no matter what the sequence of base pairs is.

7.3 How do the kinetics of denaturation and renaturation depend on DNA concentration?

ANSWER: The speed of denaturation does not depend on DNA concentration, but the speed of renaturation does. The higher the concentration of ssDNA, the more likely it is that complementary sequences will find each other and the faster the duplex can re-form.

7.4 DNA gyrase is essential to cell viability. Why, then, are nalidixic acid-resistant cells that contain mutations in *gyrA* still viable?

ANSWER: The *gyrA* mutations alter only the nalidixic acid-binding site on GyrA, not its gyrase activity. In other words, active DNA gyrase is still made, but the drug cannot bind to it.

7.5 Bacterial cells contain many enzymes that can degrade linear DNA. How, then, do linear chromosomes in organisms like *Borrelia burgdorferi* (the causative agent in Lyme disease) avoid degradation?

ANSWER: DNA-digesting exonucleases act on free 5′ or 3′ ends. The *Borrelia* linear chromosomes possess covalently closed hairpin ends called telomeres and do not possess free 5′ or 3′ groups.

7.6 Reexamine Figure 7.15. If you GFP-tagged a protein bound to the *ter* region, where would you expect fluorescence to appear in the cell with two origins?

ANSWER: The *ter* region is the last region of the chromosome to be replicated. It will appear at midcell, between the two origins.

7.7 Would you expect to find genes encoding Topo IV, XerC, and XerD in prokaryotes with exclusively linear chromosomes? Why or why not?

ANSWER: Genes for Topo IV are present in *Streptomyces* with linear genomes. They are not essential for survival, but mutants lacking this enzyme have severe growth defects, suggesting that, like circular chromosomes, linear chromosomes form knots during replication that are optimally resolved by the Topo IV enzyme. However, genes encoding XerC and XerD are not found in the *Streptomyces* genome. This may not be surprising: while an odd number of recombination events produce chromosome dimers in circular chromosomes, recombination does not result in dimers in linear chromosomes.

7.8 Individual cells in a population of *E. coli* typically initiate replication at different times (asynchronous replication). However, depriving the population of a required amino acid can synchronize reproduction of the population. Ongoing rounds of DNA synthesis finish, but new rounds do not begin. Replication stops until the amino acid is once again added to the medium—an action that triggers simultaneous initiation in all cells. Why is this replication synchronized?

ANSWER: Initiation requires synthesis of the initiator protein DnaA. Depriving the population of an amino acid prevents protein synthesis, which precludes synthesis of DnaA. Because DnaA is not required to complete already-initiated rounds of replication, all rounds already started are completed, but reinitiation cannot occur. Adding the amino acid once again will allow all cells to simultaneously make DnaA, so initiation is triggered in every cell at the same time.

7.9 The antibiotic rifampin inhibits transcription by RNA polymerase, but not by primase (DnaG). What happens to DNA synthesis if rifampin is added to a synchronous culture?

ANSWER: Initiation of DNA synthesis requires primer transcription at the origin by RNA polymerase, an enzyme sensitive to rifampin. Primase (DnaG), which synthesizes RNA primers in the lagging strand throughout DNA synthesis, is resistant to rifampin. So adding rifampin to a synchronized culture will prevent new rounds of DNA replication but will not affect already-initiated rounds.

7.10 How would you demonstrate that a gene is essential?

ANSWER: One method to establish that a gene is essential is to construct a complete loss-of-function (null) allele of that gene and show that the organism cannot survive without it. As one might imagine, this is a challenging prospect because, by definition, the organism cannot survive once the mutation is in place. One trick is to conditionally express the gene, such that it can be turned on to let the cells grow in culture, but can be turned off by the researcher's manipulation of the growth conditions to see whether its loss kills the cells. Mechanisms of gene regulation and genetic engineering are described in Chapters 10 and 12. Ideally, multiple environmental conditions are tested such that the gene is found to be essential in as many conditions as possible. Some genes are obviously essential, such as genes that encode RNA polymerase and the ribosomal proteins, because these are single-copy genes in the genome, and their products function in processes essential for cell growth. For other genes, their absolute requirement may be less obvious and require more rigorous testing for confirmation. For instance, the cell has genes that encode several different DNA polymerases, but it is unclear how many of these polymerases are essential for cell growth.

7.11 How might you interpret the discovery of genes for photosynthesis in the metagenome of the human gut? Could it indicate a possible error in the analysis, or could something else be going on?

ANSWER: Photosynthesis does not occur in the gut, because of the absence of a light source, but genes could be present in photosynthetic microbes (bacteria and/or eukaryotes) that came to the gut as a food source or a food contaminant. Though unable to colonize the gut, these microbes might be present long enough to contribute their DNA to the metagenome. Alternatively, it is possible that the photosynthesis genes could be used in a later, sun-exposed stage in the life cycle of the organism, once it leaves the gut and before it colonizes a new human host. Finally, it is also possible that the genes only resemble photosynthesis genes and, through evolution, have changed to functions that are independent of light.

CHAPTER 8

8.1 If each sigma factor recognizes a different promoter, how does the cell manage to transcribe genes that respond to multiple stresses, each involving a different sigma factor?

ANSWER: In these situations, a given gene has multiple promoters. Each promoter is recognized by a different sigma factor and begins transcription at different distances from the start codon of the gene.

8.2 Imagine two different sigma factors with different promoter recognition sequences. What would happen to the overall gene expression profile in the cell if one sigma factor were artificially overexpressed? Could there be a detrimental effect on growth?

ANSWER: Since sigma factors compete for the same site on core polymerase, overexpressing one sigma factor could displace the other sigma factors from the RNA polymerase population and compromise expression of those target genes. If those genes were important to survival, the cell could die.

8.3 Why might some genes contain multiple promoters, each one specific for a different sigma factor?

ANSWER: The gene might need to be expressed under multiple conditions at different levels. If a given condition increases expression of an alternate sigma factor, the target gene will need a promoter that the new sigma factor can recognize. As the need disappears and the sigma factor diminishes in concentration, a promoter that uses the housekeeping sigma factor will be needed. For example, the gene for DnaK heat-shock protein has promoters for RpoH (sigma-32) and RpoD (sigma-70), the housekeeping sigma factor. The level of protein needed during normal growth is supplied by sigma-70. Upon encountering heat stress, the RpoH sigma factor level increases and mediates an increase in DnaK production.

8.4 Figure 8.5 illustrates an operon and its relationship to transcripts and protein products. Imagine that a mutation that stops translation (TAA, for example) was substituted for a normal amino acid about midway through the DNA sequence that encodes gene *A*. What would happen to the expression of the gene *A* and gene *B* proteins?

ANSWER: The part of the gene *A* protein only up until the stop codon would be made. The complete mRNA transcript, however, would be made and would include gene *B* (there are some rare exceptions). Gene *B* has its own translation start codon, so the complete gene *B* protein could still be produced.

8.5 If rifamycin targets bacterial RNA polymerase, why doesn't it also kill its producer, the bacterium *Amycolatopsis mediterranei*?

ANSWER: The RNA polymerase of *A. mediterranei* has evolved to be naturally resistant to this antibiotic. Unfortunately, pathogens such as *Mycobacterium tuberculosis*, the causative agent of tuberculosis, can also evolve rifampicin-resistant RNA polymerase under selection with the antibiotic.

8.6 How might the redundancy of the genetic code be used to establish evolutionary relationships between different species? *Hints:* 1. Genomes of different species have different overall GC content. 2. Within a given genome one can find segments of DNA sequence with a GC content distinctly different from that found in the rest of the genome.

ANSWER: The codon preferences of different microorganisms are based in part on their GC content. Thus, an organism with an AT-rich genome will preferentially use codons for a given amino acid that have As and Ts over those with Gs and Cs. Evolutionarily, finding a long AT-rich region that encodes mRNA with AT codon bias within a chromosome that is otherwise GC-rich suggests that the AT-rich region was inherited by horizontal DNA transfer from another species. (See Chapter 9.)

8.7 How might one gene code for two proteins with different amino acid sequences?

ANSWER: One gene can code for two proteins with different amino acid sequences by having two different translation start sites in different reading frames. While this is not a common occurrence, it happens. Hepatitis B virus is one example.

8.8 Why involve RNA in protein synthesis? Why not translate directly from DNA?

ANSWER: Because transcription enables the cell to amplify the gene sequence information into multiple copies of RNA. Amplification means that more ribosomes can be engaged in translating the same protein, causing the concentration of the protein to rise more quickly than if only a single gene were used. The transcriptional process also provides an additional location to regulate the production of a protein.

8.9 Codon 45 of a 90-codon gene was changed into a translation stop codon, producing a shortened (truncated) protein. What kind of mutant cell could produce a full-length protein from the gene without removing the stop codon? *Hint:* What molecule recognizes a codon?

ANSWER: If a tRNA gene sequence corresponding to an anticodon is altered by mutation so that the anticodon of the tRNA "sees" the stop codon as an amino acid codon, then the mutant cell can produce a full-length protein from the gene. The mutated tRNA molecule will transfer its amino acid to the peptide chain. The stop codon is still there, but now it can direct the addition of an amino acid. The attached amino acid can be used to bridge the gap caused by the stop codon, and a full-length protein is made. These modified tRNAs are called suppressor tRNAs because they suppress the mutant phenotype.

8.10 While working as a member of a pharmaceutical company's drug discovery team, you find that a soil microbe snatched from the jungles of South America produces an antibiotic that will kill even the most deadly, drug-resistant form of *Enterococcus faecalis*, which causes bacterial endocarditis. Your experiments indicate that the compound stops protein synthesis. How could you more precisely determine the antibiotic's mode of action? *Hint:* Can you use mutants resistant to the antibiotic?

ANSWER: One way is to take a culture of bacteria susceptible to the antibiotic and isolate resistant mutants (bacteria that are not killed by the antibiotic), purify their ribosomes, and separate the 30S and 50S ribosomal subunits. Cross-mix subunits from sensitive and resistant cells (for example, mix 30S subunits from sensitive cells with 50S subunits from resistant cells). Then measure the ability of the hybrid ribosome to carry out protein synthesis with and without the drug. If resistance is due to an altered ribosomal protein or RNA, the subunit mix containing the altered component will make protein regardless of whether the drug is present. Once identified, the responsible ribosomal subunits from resistant and sensitive cells can be broken down further into their component parts, reconstituted in hybrid form, and again tested for an ability to make protein in the presence of the drug. This reductive approach will likely, but not always, uncover the target ribosomal protein or rRNA.

8.11 Why do you think evolution by natural selection favors changes in codons, rather than in anticodons?

ANSWER: Changes to codons change the sequence of a single protein, whereas changes to anticodons (in tRNA) change the code itself and can change the sequence of many proteins simultaneously. It is likely that not all of the changes will be adaptive, and some could be lethal.

8.12 A major way that bacteria acquire new functions is through the acquisition and expression of genes from other microbes via horizontal gene transfer (Chapters 9 and 17). How would this mechanism of innovation be affected if the recipient bacterium changed its genetic code?

ANSWER: A change in the recipient's genetic code should not affect transcription of the imported genes, but it will change the amino acid sequence of the genes' proteins, compared to their sequences in the donor organism. As a result, nonfunctional proteins of no value to the recipient could be produced, thus limiting the benefit of horizontal gene transfer.

8.13 The incorporation of pyrrolysine and selenocysteine into the genetic code involved stop-to-sense changes, rather than sense-to-sense. Why do you think this was the case?

ANSWER: This change is unlikely to be due to chance. Of the 64 possible codons in the standard code, only 3 are nonsense, and nonsense codons were reassigned in both cases where the code was expanded. Perhaps, as the initial evolutionary event progresses toward code expansion, changes from stop to sense are less harmful to the cell than are changes from one (old) sense to (a new) sense. In fact, for many genes, additional stop codons using one of the other two stop codon sequences are located downstream of the first; these can provide "backup" signals for translation termination in the event that an amino acid is misincorporated at the first stop codon.

CHAPTER 9

9.1 How could frameshift mutations be used to confirm that codons consist of three bases, versus two or four? *Hint:* Think of how a series of "like" frameshifts (for example, single base-pair additions) along a gene would impact the reading frame.

ANSWER: Francis Crick and colleagues performed experiments in which one, two, or three base pairs were added within the reading frame of a gene. They discovered that three "like" frameshifts (for example, three single base-pair additions) along a protein maintained some protein activity, whereas one or two frameshifts resulted in no protein activity. (Protein activity was measured as an indicator of whether the newly made protein was fully synthesized and folded properly.) Their results provided evidence that each codon is made up of a triplet of bases. The addition of three bases results in one new codon and hence one more amino acid in the protein. Only the addition (or subtraction) of three bases (or a multiple of three bases) maintains the reading frame. Adding one or two bases shifts the reading frame and shuffles the amino acid sequence.

9.2 Does deamination of cytosine to uracil lead to a transition mutation or a transversion mutation? Work this out in a drawing, using Figure 9.3 as a guide.

ANSWER: **Figure 9.4** shows that deamination of cytosine generates a uracil. Uracil cannot base-pair with guanine, as cytosine does. Thus, in the next round of replication the uracil will bind adenine, and eventually an AT base pair will form where a GC base pair once was, as **Figure 9.3** shows. This is a transition mutation.

9.3 Would mutants that lack Dam have a mutator phenotype? Explain why or why not. What about mutants that overexpress Dam?

ANSWER: Both types of mutants have mutator phenotypes. Mutants lacking Dam are unable to discriminate between new and old strands during mismatch repair. Mutants that overexpress Dam have an increased mutation rate because the newly synthesized DNA becomes methylated faster, giving the mismatch repair system less time to find and repair the mutations.

9.4 It has been reported that hypermutable bacterial strains are overrepresented in clinical isolates. Out of 500 isolates of *Haemophilus influenzae*, for example, 2%–3% were mutator strains having mutation rates 100–1,000 times higher than

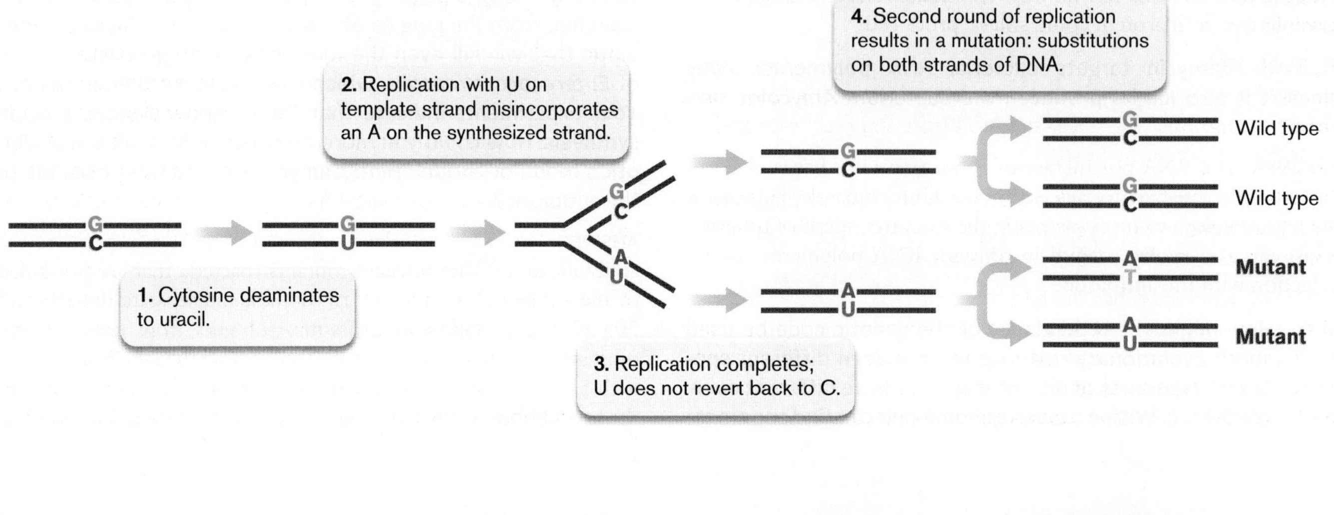

those of lab reference strains. Why might the mutator phenotype be beneficial to pathogens?

ANSWER: The mutator strains may speed microbial evolution, which could help the microbe outwit the immune system or escape the effects of administered antibiotics.

9.5 Transfer of an F factor from an F^+ cell to an F^- cell converts the recipient to F^+. Why doesn't transfer of an Hfr do the same?

ANSWER: The last piece of an Hfr to transfer is the F factor and *oriT*. Only rarely will an entire chromosome transfer from one cell to another, so most Hfr transfers do not result in transfer of *oriT* and thus cannot initiate conjugation.

9.6 In a transductional cross between an $A^+B^+C^+$ genotype donor and an $A^-B^-C^-$ genotype recipient, 100 A^+ recombinants were selected. Of those 100, 15 were also B^+, while 75 were C^+. Is gene *B* or gene *C* closer to gene *A*?

ANSWER: Gene *C* is closer to gene *A*, because gene *C* was cotransduced with *A* at the higher frequency.

9.7 How do you think phage DNA containing restriction sites evades the restriction-modification screening systems of its host?

ANSWER: Phage DNA will survive because sometimes the modification enzyme reaches the foreign DNA before the restriction endonuclease gets there. Once methylated, the phage DNA will be shielded from restriction, and the methylated molecule will replicate unchallenged. If, on the other hand, a foreign DNA fragment (not necessarily a phage) has been modified and the conditions are right, it might recombine into the host chromosome and convey a new character to the strain.

9.8 *E. coli* has several DNA methyltransferases, such as Dam (see Chapter 7), that methylate many bases in the genome. If a researcher wishes to transfer a plasmid from *E. coli* into a new species, the presence of which type(s) of restriction-modification systems in the new species might prompt a researcher to use a Dam-minus mutant of *E. coli*?

ANSWER: Since type IV restriction systems target methylated DNA, the researcher might get higher transfer efficiency if the plasmid were replicated in a Dam-minus mutant of *E. coli* compared to the wild-type *E. coli*.

9.9 Some bacteriophage genomes contain functional CRISPR-Cas arrays. Can you think of a reason why this might be advantageous to the bacteriophage?

ANSWER: Some bacteriophages use this system to turn the tables on the host. During the lytic infection of *Vibrio cholerae*, one phage uses its CRISPR-Cas system to target and inactivate antiviral defense genes of the host.

9.10 Evolutionarily speaking, why might it be advantageous for a transposon to utilize replicative transposition? Why might nonreplicative transposition be advantageous?

ANSWER: Replication has the advantage of increasing the number of transposon copies, thereby enabling the transposon to proliferate within a genome and increase its chances of invading new genomes via horizontal gene transfer if its host genome serves as donor. Nonreplicative transposition may have an advantage because the transposon can move to a location that is more beneficial, or at least less costly, to the host than its current location. By improving fitness of the host, the transposon can therefore increase its own abundance when the host replicates its chromosome.

9.11 Gene homologs of *dnaK* encoding the heat-shock chaperone HSP70 exist in all three domains of life. All bacteria contain HSP70, but only some species of archaea encode a *dnaK* homolog. The archaeal homologs are closely related to those of bacteria. Knowing this information, how do you suppose *dnaK* genes arose in archaea?

ANSWER: The *dnaK* gene may have moved by some type of horizontal gene transfer mechanism from the domain Bacteria to some members of the domain Archaea.

9.12 Every strain of *Prochlorococcus* examined to date has a unique cluster of genes that modifies the chemical properties of the cell surface. What sort of selective pressure(s) might account for the high degree of variation in cell-surface properties for this marine bacterium?

ANSWER: Probably there is strong selective pressure to avoid being eaten by protozoan grazers, or being attacked by bacteriophages. By changing its surface chemistry, *Prochlorococcus* might be able to "taste" different to grazers, or either remove or mask an attachment site of the bacteriophages.

CHAPTER 10

10.1 Why is the "scanning mode" that permits regulatory proteins to search for promoters along DNA more efficient than a random search within the cell's cytoplasm?

ANSWER: Scanning along the DNA strand reduces search space to one dimension, compared to a three-dimensional search within the contents of the cytoplasm.

10.2 Knowing that the affinity of regulators for DNA depends on the DNA sequence, propose how evolution could result in the ability of a preexisting regulator to now regulate a newly acquired gene. Why might it be critical that the sequence encoding the regulator itself does not change during this evolutionary step?

ANSWER: The DNA near the promoter of the newly acquired gene could change such that the regulator would now bind with high affinity. If instead the new gene was unchanged but the regulator was mutated to recognize the gene's promoter, the regulator's ability to regulate other genes might be affected, with possibly dire consequences for the cell.

10.3 Operators and repressors were discovered before promoters (the DNA sequences recognized by the sigma factor of RNA polymerase). Why do you think it took longer to discover the promoters? *Hint:* Finding elements of transcription initiation using genetics involves analysis of mutants; think about the phenotype of a repressor mutant compared to that of a promoter mutant.

ANSWER: Loss-of-function promoter mutations would have the same phenotype as loss-of-function mutations in the open reading frame, and would therefore be difficult to distinguish. They would also be a rare class of mutation, because the promoter region is a much smaller target than the open reading frame. In contrast, loss-of-function mutations in the operator or repressor derepress the operon, such that the phenotype is the gain in expression of the gene, which would be much easier to identify.

10.4 The transmembrane sensor kinase could have a direct role in gene expression, if its cytoplasmic domain has the ability to bind DNA. Why, then, might it be advantageous for the cell to use the response regulator as an intermediate in this signaling process? *Hint:* Consider the spatial organization of the cell.

ANSWER: If the transmembrane receptor were also the transcription factor, it would have to be able to contact the promoter of the gene it regulates. Thus, the chromosomal region for the promoter needs to be able to move to the membrane. In addition, if there are more copies of the response regulator protein than of the sensor kinase, the signal can be amplified because each sensor kinase could activate multiple response regulators. Finally, a single transmembrane receptor can control many genes (in a regulon) if it activates multiple copies of the response regulators that each bind to a different operon.

10.5 Null mutations completely eliminate the function of a given mutated gene. Predict the effects of the following null mutations on the induction of beta-galactosidase by lactose, and predict whether the *lacZ* gene is expressed at high or low levels in each case. The inactivated, mutant genes to consider are *lacI*, *lacO*, *lacP*, *crp*, and *cya* (the gene encoding adenylyl cyclase). What effect will those mutations have on catabolite repression?

ANSWER: Loss of LacI repressor will lead to constitutive expression of *lacZ* and will partially affect catabolite repression. [Explanation: Because LacI is missing, allolactose inducer is not required, so the glucose effect on the LacY permease is irrelevant. What remains relevant is that the mechanism by which glucose transport reduces cAMP synthesis remains (see **eTopic 10.1**). Thus, decreased cAMP levels resulting from growth on glucose will cause a decrease in cAMP-CRP-dependent activation of *lac* operon expression.] A *lacO* mutant will not bind LacI repressor, so the phenotype will mimic that of a *lacI* mutation. A *lacP* mutation will prevent expression of *lacZYA* because RNA polymerase will not bind. Mutations in *crp* or *cya* will partially prevent catabolite repression (see preceding explanation), but *lacZYA* induction by lactose will be normal. Without the cAMP-CRP complex, however, expression can never achieve maximal levels.

10.6 Predict what will happen to the expression of *lacZ* when a second copy of the *lac* operon region containing various mutations is present on a plasmid. The genotypes of these partial diploid strains are presented as chromosomal genes/plasmid gene. (a) *lacI*$^-$ *lacO*$^+$*P*$^+$*Z*$^+$*Y*$^+$*A*$^+$/plasmid *lacI*$^+$; (b) *lacO*$^-$ *lacI*$^+$*P*$^+$*Z*$^+$*Y*$^+$*A*$^+$/plasmid *lacO*$^+$; (c) *crp*$^-$ *lacI*$^+$*O*$^+$*P*$^+$*Z*$^+$*Y*$^+$*A*$^+$/plasmid *crp*$^+$.

ANSWER: (a) The *lacI*$^+$ gene on the plasmid will produce LacI repressor protein that can diffuse through the cytoplasm, bind chromosomal *lacO*, and repress the *lacZYA* operon. Because the complementing gene and the mutant gene are on different DNA molecules, the gene is said to work in *trans*. (b) Because the *lacO* gene does not produce a diffusible product (for example, protein or RNA), the plasmid *lacO*$^+$ cannot complement a *lacO* mutation in *trans*, and the strain will not make beta-galactosidase. Thus, the *lacO* gene functions only in *cis*—that is, when it resides next to the gene it regulates. (c) The *crp* gene produces a diffusible protein product, so it can function in *trans* and complement a *crp* mutation. The strain will make beta-galactosidase to the highest level in the presence of inducer lactose.

10.7 Researchers often use isopropyl-β-D-thiogalactopyranoside (IPTG) rather than lactose to induce the *lacZYA* operon. IPTG resembles lactose, which is why it can interact with the LacI repressor, but it is not degraded by beta-galactosidase. Why do you think the use of IPTG is preferred in these studies?

ANSWER: There are at least two reasons IPTG is used. First, the level of IPTG inducer will not change, but the level of lactose inducer will continually decrease as it is consumed, affecting the kinetics of induction. Second, the act of degrading lactose produces glucose and galactose. Glucose, as the preferred carbon source, will catabolite-repress the *lacZYA* operon, once again affecting the kinetics of induction.

10.8 When the *lacI* gene of *E. coli* is missing because of mutation, the *lacZYA* operon is highly expressed regardless of whether lactose is present in the medium. Judging by the illustration of arabinose operon expression in Figure 10.14, what do you think would happen to *araBAD* expression if AraC were missing? Why?

ANSWER: The operon would be poorly expressed because contact between AraC and RNA polymerase is needed to activate transcription.

10.9 In a newly discovered bacterium, an operon suspected to synthesize an amino acid has what appears to be the following leader sequence: 5′-ATGCCCCAGCAGAGTTGA-3′. Assuming the microbe uses the standard genetic code (Fig. 8.12), predict which amino acid the products of the operon synthesize.

ANSWER: The leader sequence is translated to fMet-Pro-Gln-Gln-Ser-Stop. Because the most frequent codon codes for glutamine, it is likely that the operon's products are involved in synthesizing glutamine.

10.10 The relationship between the small RNA RyhB, the iron regulatory protein Fur, and succinate dehydrogenase is shown in Figure 10.24. Given this regulatory circuit, will a *fur* mutant grow on succinate?

ANSWER: No. Without Fur, RyhB is made whether or not iron is present, and RyhB sRNA causes the continuous degradation of the *sucCDAB* mRNA. Succinate dehydrogenase cannot be made, so the cell cannot grow on succinate.

10.11 What would be the outcome if purified autoinducer were experimentally provided to a low-density culture?

ANSWER: These cells would make light, because autoinducer concentration, and not cell abundance per se, dictates expression of the *lux* operon.

10.12 Genes encoding luciferase can be used as "reporters" of gene expression when placed under the regulatory control of other genes. Luminometers are machines that can quantify light production (luminescence) from luciferase. From the discussion in Section 9.2, propose an experiment to confirm that RecA is induced during the SOS response.

ANSWER: You could fuse the luciferase gene to the *recA* promoter and use a luminometer to demonstrate a real-time increase in luminescence after treatment with a mutagen, such as ultraviolet irradiation, that triggers the SOS response.

10.13 What would happen if a culture were coinoculated with *Vibrio (Aliivibrio) fischeri luxI* and *luxA* mutants, neither of which produces light?

ANSWER: The *luxI* mutant would cause the bacterium to glow, and the *luxA* mutant would still make autoinducer as the bacterium grew. This autoinducer would accumulate in the culture medium and then diffuse and enter the *luxI* mutant cells, where it would trigger induction of the *lux* operon and the production of luciferase.

10.14 Figure 10.36 illustrates the process of transformation in *Streptococcus pneumoniae*. Would a mutant of *Streptococcus* lacking ComD be able to transform DNA?

ANSWER: No. The membrane sensor ComD detects the competence stimulation peptide (CSP) and initiates a cascade of events leading to transformasome construction. A lack of ComD would mean no transformasome and no DNA transformation.

10.15 While viewing Figure 10.41, imagine the phenotype of a cell in which *fljB* has been deleted but *fljA* is still expressed. Would cells be motile? What type of flagella would be produced? Would the cells undergo phase variation? What would happen if *fliC* alone were deleted?

ANSWER: A *fljA* mutant lacks the repressor needed to turn off FliC. Thus, a cell will switch back and forth from making H2 flagellin to making both H1 and H2 flagellins. A *fliC* mutant, however, will switch from being motile to nonmotile. In one orientation, the invertible element will allow H2 flagellin to be made, but in the opposite orientation no flagellin will be made, at which point the cell will not be motile.

10.16 Using antibody to flagella, a cell of *E. coli* can be tethered to a glass slide via a single flagellum. Looking through a microscope, you will then see the bacterium rotate in opposite directions as the flagellar rotor switches from clockwise to counterclockwise rotation and back again (see **Fig. 10.42**). Which way will the bacillus rotate when an attractant is added? What would the interval between clockwise and counterclockwise switching be if you tethered the mutants *cheY*, *cheA*, *cheZ*, and *cheR* to the slide and then added attractant?

ANSWER: When an attractant is added to the slide, the rotor will turn counterclockwise for smooth swimming (CheA becomes less active, so there is less CheY-P and there are less frequent tumbles; thus, motor rotation is biased to counterclockwise). But because the flagellum is fixed to the slide, the bacillus will rotate in the opposite direction (clockwise). You would expect the following rotating phenotypes from mutants fixed to slides: For mutant *cheY*, the rotor will turn mostly counterclockwise because there is no CheY-P, so there will be longer runs, but fixed cells will turn mostly clockwise. Mutant *cheA* will have the same phenotype as *cheY* (no CheY-P, longer runs). For mutant *cheZ*, the rotor will turn mostly clockwise, because there is more CheY-P (so there will be frequent tumbles and shorter runs), but the fixed cells will turn counterclockwise. For mutant *cheR*, no methylation of MCPs will reactivate CheA kinase after attractant is added, so there will be less CheY-P; therefore, the rotor will more frequently switch to clockwise, but fixed cells will turn counterclockwise.

CHAPTER 11

11.1 Plaques from phage lambda quickly fill with resistant lysogens. Could there be a different way for the host cells to become resistant, without forming lysogens?

ANSWER: A gene encoding a host product essential for phage infection could mutate within the host bacteria. A common source of resistance is loss of the phage receptor, a host cell-surface protein. In the case of phage lambda, the receptor protein would be maltose porin. Other ways to become resistant could include specific cleavage of the phage DNA as it enters the cell, failure to interact with phage replication components, and the CRISPR-Cas memory defense system.

11.2 What advantages does rolling-circle replication offer a phage?

ANSWER: An advantage of rolling-circle replication is that many genome copies are made quickly from a single template. No proofreading occurs, and no methylation step distinguishes "old" from "new" DNA. Because viral genomes are relatively small, viruses tolerate a higher error rate per base pair than for cellular genomes.

11.3 A researcher adds phage lambda to an *E. coli* population whose cells fail to express maltose porin. After several days, the *E. coli* are now lysed by phage. What could be the explanation?

ANSWER: Various answers are possible. In practice, the most common explanation is that phage populations contain genetic variants, some of which can occasionally infect *E. coli* by binding a different porin for a different sugar. Such rare events lead to evolution of lambda mutants now adapted to bind a different receptor.

11.4 Suppose a mutant lambda phage lacks holin and endolysin. What will happen when this mutant infects *E. coli*?

ANSWER: The phage can undergo a lytic cycle, filling the *E. coli* cell with progeny phage particles. But the particles are trapped inside the cell wall. They cannot emerge to infect new host cells.

11.5 Could the influenza genome change by recombination of segments, rather than reassortment? What about the lambda phage genome?

ANSWER: The RNA segments of the influenza genome are unlikely to recombine with segments of another genome, because the single-stranded RNA has no mechanism for maintaining homology with another strand. The phage lambda genome can recombine with the genome of a coinfecting phage, by using the host cell's protein complex for homologous recombination (discussed in Chapter 9).

11.6 How could swine play a role in generating a pandemic strain of influenza? How could avian influenza strain H7N9 become a pandemic strain endangering many people?

ANSWER: Strain H7N9 is highly virulent in humans, but the transmission rate among humans is low. Transmission between birds is high because birds have the appropriate receptors in their upper respiratory tract. Poultry can spread H7N9 rapidly among many birds. If birds are reared adjacent to swine, then a mutant form of H7N9 might infect the swine. Suppose H7N9 were to infect a swine that was simultaneously infected with strain H1N1. The H1N1 strain is transmitted between humans with high efficiency. During coinfection of swine, the H7N9 strain might acquire segments from H1N1 that enable efficient transmission to humans. The reassortant strain might then escape into humans and spread, causing deadly disease.

11.7 How do attachment and entry of HIV resemble attachment and entry of influenza virus? How do attachment and entry differ between these two viruses?

ANSWER: Attachment of HIV requires the envelope spike proteins to bind receptors in the host plasma membrane, just as the influenza envelope protein hemagglutinin binds the sialic acid protein in the host membrane. However, the entry processes differ between the two viruses. Influenza virus induces formation of an endocytic vesicle, whose acidification triggers membrane fusion and release of the core contents into the host cytoplasm. HIV virions, however, do not induce endocytosis and do not require acidification to induce fusion of the membrane and release of the core into the cytoplasm. In both cases, receptor binding signals the major rearrangement of a viral envelope protein so that a fusion peptide is inserted into the host membrane. For HIV, it is the plasma membrane; for influenza virus, it is the endocytic membrane.

11.8 What do you think are the arguments for or against lentivectors editing the germ line?

ANSWER: Parents who carry two copies of a deleterious allele such as that for sickle-cell disease might wish to have their sperm or egg cells edited to replace the bad allele, thus enabling production of healthy children. It could be argued that the human genome is full of retroviral modifications that accumulated gradually during the history of our species, so why not intentionally add another one? As a practical matter, it is hard to know what would happen to the lentivector sequences in a germ line. Perhaps the process of sperm or egg maturation and embryogenesis would activate the viral sequences in unexpected ways. The future child would have no choice about this procedure. In addition, parents have the alternative of producing children by selecting a sperm or egg donor.

11.9 Compare and contrast the fate of the HSV genome with that of the HIV genome.

ANSWER: HSV-1 contains a DNA chromosome, which is transported to the nuclear membrane within an intact capsid. HIV contains two RNA chromosomes, which are released in the cytoplasm upon dissolution of the capsid. The RNA chromosomes of HIV are copied to double-stranded DNA for transport into the nucleus. In HSV-1, the DNA chromosome circularizes and generates concatemeric duplicates by rolling-circle replication. In HIV, the replicated DNA circularizes but immediately integrates into the host chromosome. In both cases, the viral DNA can persist for decades as a latent infection.

CHAPTER 12

12.1 PCR is a powerful technique, but a sample can be easily contaminated and the wrong DNA amplified—perhaps sending an innocent person to jail. What can be done to minimize this possibility?

ANSWER: Use sterile, filtered micropipette tips. The filters prevent contamination of the micropipette and subsequent reactions. In addition, perform a control PCR reaction without adding a DNA sample. If the water, buffers, or tubes are contaminated with a given DNA, a PCR product will appear even when DNA template has not been added deliberately to the reaction.

12.2 Figure 12.15B illustrates how agarose gels separate DNA molecules. If you ran a highly supercoiled plasmid in one lane of the gel and the same plasmid with one phosphodiester bond cut in another lane, where would the two molecules end up relative to each other in this type of gel?

ANSWER: Removing one phosphodiester bond from a supercoiled plasmid would unravel (relax) all supercoils. Even though both plasmids would have the same DNA sequence, the highly supercoiled plasmid would be more compact than the relaxed plasmid and would move more quickly through the pores of the agarose gel. As a result, the supercoiled plasmid would be closer to the bottom of the gel than would the relaxed form.

12.3 Given that *E. coli* possesses restriction endonucleases that cleave DNA from other species, how is it possible to clone a gene from one organism to another without the cloned gene sequence being degraded?

ANSWER: The best way to overcome the restriction barrier is to use a mutant strain of *E. coli* in which the restriction system has been inactivated. DNA entering the cell will not be degraded and, if the modification system is still in place, will be modified. Once modified, that DNA can be safely transferred to other *E. coli* strains that still have their restriction endonucleases.

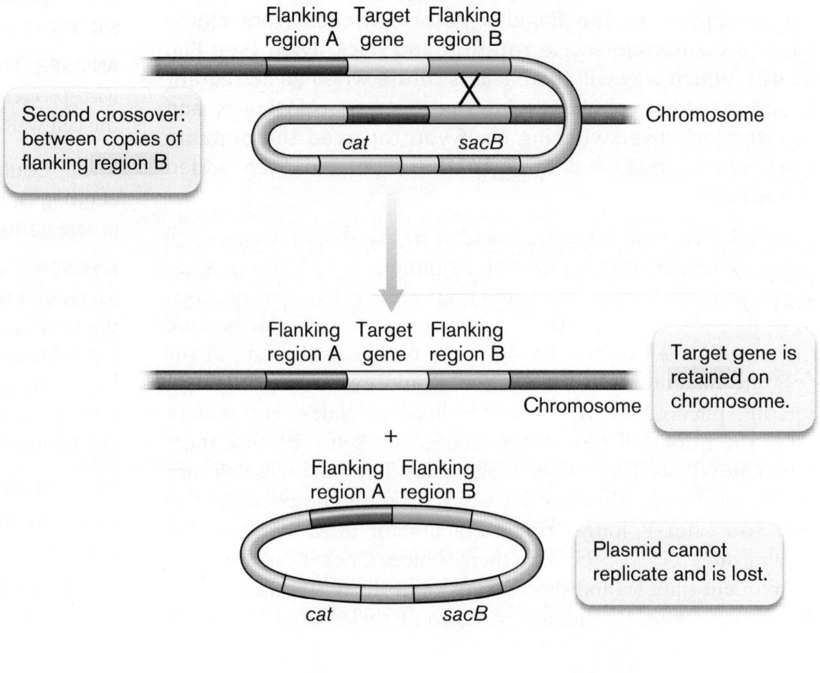

12.4 Refer to **Figure 12.19**. What would be the outcome for the chromosome if the second crossover event occurred between the copies of flanking region B, instead of between the copies of flanking region A?

ANSWER: As the figure below indicates, the excised plasmid would carry the deletion, and the target gene would be retained on the chromosome. The cell would remain a wild-type cell. In fact, when using this type of system to generate deletions, crossover events usually occur at either flanking region. Sucrose-resistant colonies must therefore be screened to confirm the presence of the deletion, since some colonies will not have the deletion but will, in fact, still have the target gene on the chromosome.

12.5 How could you quickly separate cells in the population that express sigma B–YFP from those that do not express this fusion? And what could you learn by separating these subpopulations? *Hint:* A technique presented in Chapter 4 will help.

ANSWER: The population of cells can be passed single file through a fluorescence-activated cell sorter (FACS; see Chapter 4). Cells that fluoresce can be collected in one tube, and nonfluorescent cells can be collected in another. The cells can then be analyzed by deep sequencing to determine the transcriptomes, and by mass spectrometry to determine which proteins are produced in the different cell populations.

12.6 You have monitored expression of a gene fusion in which *gfp* is fused to *gadA*, the glutamate decarboxylase gene of *E. coli*. You find a 50-fold increase in expression of *gadA* (based on fluorescence level) when the cells carrying this fusion are grown in media at pH 5.5, compared to pH 8. What additional experiments must you do to determine whether the control is transcriptional or translational?

ANSWER: To measure regulation of the *gadA* promoter only, you can fuse *gfp* containing its native ribosome-binding site (RBS) to the *gadA* promoter. This fusion will expose only transcriptional control of the *gadA* promoter. Translational control of the *gadA* message is absent because the *gadA* RBS (and all other *gadA* sequences) was replaced by the *gfp* RBS. To measure translational control, you must make a new fusion that starts with a constitutive promoter (a constantly expressed promoter), add a section of *gadA* containing the *gadA* RBS, and follow that with *gfp* (minus its RBS) fused in-frame to the *gadA* sequence. Any condition that alters the expression of the first fusion affects transcription of the *gadA* promoter. Any condition that alters the expression of the second fusion must affect translation of the *gadA* message.

12.7 You suspect that two proteins, A and C, can simultaneously bind to a third protein, B, and that this interaction then allows A to interact with C in the complex. How could you use two-hybrid analysis to determine whether these <u>three</u> proteins can interact?

ANSWER: Place three plasmids in the cell. The first plasmid makes protein A fused to the GAL4 DNA-binding domain, the second plasmid makes protein C fused to the GAL4 activation domain, and the third plasmid makes simply protein B. If the yeast cell contains only the first two plasmids, then no interaction will occur and *lacZ* will not be activated. If all three plasmids are in the same yeast cell, then protein B will be the center of a sandwich, linking the A and C fusion proteins. The fusion proteins can then interact and activate *lacZ*.

12.8 Would insect resistance to an insecticidal protein be a concern if you were developing a transgenic plant? Why or why not? How would you design a transgenic plant to limit the possibility of insects developing resistance?

ANSWER: Insects have, in fact, developed resistance to single insecticidal proteins. The most common resistance mechanism involves a change in the membrane receptors in the midgut to which activated *Bt* toxins bind. Resistance can be due to a reduced number of *Bt* toxin receptors or to a reduced affinity of the receptor for the toxin. Some insects, such as the spruce budworm, can inactivate specific toxins by precipitating them with a protein complex present in the midgut.

While developing a transgenic plant, you could take several steps to limit the development of resistance in an insect population. You could fuse the insecticidal gene to a promoter that is expressed only when the plant is most susceptible to attack. Alternatively, you could fuse the gene to a promoter that is expressed only in a tissue of the plant that is most vulnerable to attack. Either of these approaches would limit the time during which the insects could develop resistance. For instance, cotton plants attacked by bollworms could produce toxin only in young boll tissues, the most important part of the plant. In addition to specifically protecting the critical plant tissue, this strategy would affect only one generation of bollworms, avoiding the constant selection pressure that hastens evolution of resistance. Another technique would be to engineer two different insecticidal proteins into the plant genome that would not exhibit cross-resistance. In other words, even if an insect developed resistance to one toxin, it would still remain susceptible to the second.

12.9 What would happen with the toggle switch shown in Figure 12.29 if the genetic engineer could set repressor 1 and repressor 2 protein levels to be <u>exactly</u> equal? Imagine that this is done without either inducer present.

ANSWER: In a perfect system, both proteins would remain at the same level, but any small deviation in condition that tilts the balance between the two proteins by even a small amount will eventually lock the cell into either GFP on or GFP off. Because of intrinsic noise in the system, the population of cells could be half on and half off.

CHAPTER 13

13.1 Consider glucose catabolism in your blood, where the sugar is completely oxidized by O_2 and converted to CO_2:

$$C_6H_{12}O_6 + 6O_2 \rightarrow 6CO_2 + 6H_2O$$

Do you think this reaction releases greater energy as heat, or by change in entropy? Explain.

ANSWER: At first glance, the breakdown of glucose to six molecules of carbon dioxide seems to incur a large increase in entropy. But the reaction also consumes six molecules of oxygen, so the entropy gain is small. Furthermore, the oxidation reaction is associated with a large enthalpy change, approximately $\Delta H^{\circ\prime} = -2{,}540$ kJ/mol. (The degree symbol followed by prime connotes biochemical standard conditions.) At 37°C, the temperature-entropy term $-T\Delta S^{\circ\prime} = -(310 \text{ K})(0.973 \text{ kJ/mol/K}) = -302$ kJ/mol. The overall value of free energy change $\Delta G^{\circ\prime} = -2{,}540$ kJ/mol $- 302$ kJ/mol $= -2{,}842$ kJ/mol. Thus, the entropy term yields some energy, but less than an eighth as much as the enthalpy ($\Delta H^{o\prime}$) term.

13.2 The bacterium *Lactococcus lactis* was voted the official state microbe of Wisconsin because of its importance for cheese production. During cheese production, *L. lactis* ferments milk sugars to lactic acid:

$$C_6H_{12}O_6 \rightarrow 2C_3H_6O_3 \rightleftharpoons 2C_3H_5O_3^- + 2H^+$$

Large quantities of lactic acid are formed, with relatively small increase in bacterial biomass. Why do you think biomass is limited? Cheese making usually runs more efficiently at high temperature; why?

ANSWER: The lactic acid fermentation reaction does not involve a strong oxidant, but only breakdown of sugar to smaller molecules, so a larger proportion of the free energy yield is in the entropy term $-T\Delta S$. The overall energy yield is low, so a large amount of sugar must be cycled to lactic acid (lactate) for a relatively small amount of bacterial growth, as compared to bacterial growth with oxygen. Because fermentation depends on entropy change ($-T\Delta S$), the free energy yield can be increased by increasing the temperature.

13.3 When ATP phosphorylates glucose to glucose 6-phosphate, what is the net value of $\Delta G^{o\prime}$? What if ATP phosphorylates pyruvate? Can this latter reaction go forward without additional input of energy? (See Table 13.3.)

ANSWER: **Table 13.3** shows that the phosphorylation of glucose by ATP is composed of these two reactions:

Reaction	$\Delta G^{o\prime}$ (kJ/mol)
$ATP + H_2O \rightarrow ADP + P_i + H^+$	–31
$Glucose + P_i \rightarrow Glucose\ 6\text{-}P + H_2O$	+14
ATP + Glucose → ADP + Glucose 6-P	–17

The net energy lost is –17 kJ/mol, so the phosphorylation can go forward. To phosphorylate pyruvate, however, the $\Delta G^{o\prime}$ of ATP hydrolysis (–31 kJ/mol) must be subtracted from the value of phosphoenolpyruvate formation (+62 kJ/mol), giving a net value of +31 kJ/mol. Since $\Delta G^{o\prime}$ is positive, this reaction cannot go forward without additional energy.

13.4 Linking an amino acid to its cognate tRNA is driven by ATP hydrolysis to AMP (adenosine monophosphate) plus pyrophosphate. Why release PP_i instead of P_i?

ANSWER: The formation of aminoacyl-tRNA must be irreversible until the ribosome is ready to release the tRNA. The pyrophosphate from ATP is immediately cleaved into 2 P_i, preventing the reversal of aminoacyl-tRNA formation.

13.5 In the microbial community of the bovine rumen, the actual ΔG value has been calculated for glucose fermentation to acetate:

$$C_6H_{12}O_6 + 2H_2O \rightarrow 2C_2H_3O_2^- + 2H^+ + 4H_2 + 2CO_2 \qquad \Delta G = -318\ \text{kJ/mol}$$

If the actual ΔG for ATP formation is +44 kJ/mol and each glucose fermentation yields 4 ATP, what is the thermodynamic efficiency of energy gain? Where does the lost energy go?

ANSWER: The energy efficiency is $(4 \times 44\ \text{kJ/mol})/(318\ \text{kJ/mol}) \times 100 = 55\%$. The remaining energy is dissipated as heat.

13.6 What would happen to the cell if pyruvate kinase catalyzed PEP conversion to pyruvate but failed to couple this reaction to ATP production?

ANSWER: If the bacterial cell were to convert PEP to pyruvate without coupling to ATP production, it would lose much of the energy available from glucose and other food substrates converted to glucose. The energy would be lost as heat.

13.7 Some bacteria make an enzyme, dihydroxyacetone kinase, that phosphorylates dihydroxyacetone to dihydroxyacetone phosphate (DHAP). How could this enzyme help the cell yield energy?

ANSWER: Bacteria can obtain dihydroxyacetone from their environment using a transporter protein. The bacterial kinase can then phosphorylate the substrate to DHAP and direct it into glycolysis. Some bacteria can grow on dihydroxyacetone as a sole carbon source.

13.8 Explain why the ED pathway generates only 1 ATP, whereas the EMP pathway generates 2 ATP. What is the consequence for cell metabolism?

ANSWER: The EMP pathway primes the six-carbon sugar with two phosphoryl groups. The sugar then splits into two three-carbon units (glyceraldehyde 3-phosphate, G3P), each of which generates two ATPs for one of the original ATPs. By contrast, the ED pathway phosphorylates the sugar only once before it splits in two. The phosphorylated end yields G3P, which enters the EMP pathway to generate ATP, ending up as pyruvate. The unphosphorylated three-carbon unit yields pyruvate directly, with no ATP. The consequence for cell metabolism is that the ED pathway needs to cycle more substrate in order for a cell to grow the same amount of biomass as it would with the EMP pathway. Bacteria growing with the ED pathway may produce greater amounts of a valuable product such as ethanol.

13.9 If a cell respiring on glucose runs out of oxygen and other electron acceptors, what happens to the electrons transferred from the catabolic substrates?

ANSWER: The electrons from the catabolic substrates are transferred to NADH and $FADH_2$ during glycolysis and the TCA cycle. Without a terminal electron acceptor, the cytoplasmic electron carriers cannot use the electron transport system. Instead, they must transfer their electrons back onto pyruvate, acetate, and other products of catabolism in order to complete the reactions of fermentation.

13.10 Compare the reactions catalyzed by pyruvate dehydrogenase and pyruvate formate lyase. What conditions favor each reaction, and why?

ANSWER: The pyruvate dehydrogenase complex (PDC) is favored in the presence of oxygen because the electrons transferred to NADH can enter the electron transport chain, eventually combining with oxygen to release energy. In the absence of oxygen, pyruvate formate lyase is favored to yield fermentation products that can be excreted from the cell without reducing more energy carriers. At high pH, formate and acetate production is especially favorable because the extra acid counteracts alkalinity.

13.11 Suppose a cell is radiolabeled briefly with ^{14}C-acetate (pulse-labeled, then chased with unlabeled acetate). Can you predict what will happen to the level of radioactivity observed

in isolated TCA intermediates? Plot a curve showing your predicted level of radioactivity as a function of the number of rounds of the cycle.

ANSWER: The amount of radioactivity measured in TCA intermediates will rise steeply as labeled acetate is incorporated, and then will decrease by half with each succeeding cycle, as the order of the carbons is randomized by succinate. Succinate is a symmetrical molecule in which the two ends (labeled and unlabeled) are equivalent.

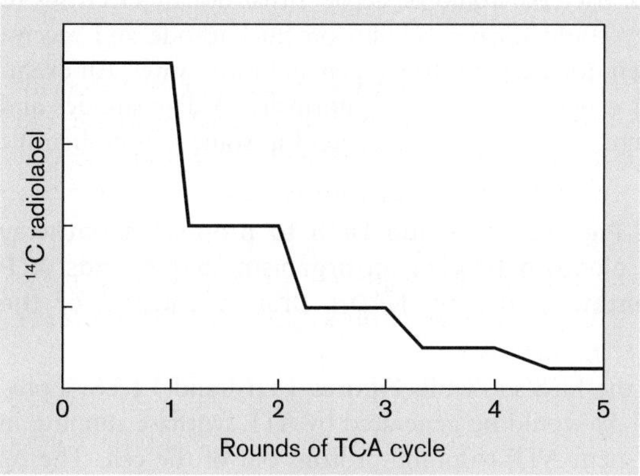

13.12 The thermophilic bacteria *Thermus* species grow in deep-sea hydrothermal vents at 80°C. It was proposed that they metabolize formate to bicarbonate ion and hydrogen gas:

$$HCOO^- + H_2O \rightarrow HCO_3^- + H_2$$

But the standard $\Delta G°'$ is near zero (−2.6 kJ/mol). Under actual conditions, do you think the reaction yields energy? Assume concentrations of 150 mM for formate, 20 mM for bicarbonate ion, and 10 mM for hydrogen gas.

ANSWER: The ability to gain energy from formate will depend on the temperature and the concentrations of reactants and products. Remember that $[H_2O]$ equals 1. Consider the following equation:

$$\Delta G = \Delta G°' + 2.303\ RT \log([\text{products}]/[\text{reactants}])$$

$$= -2.6\ \text{kJ/mol} + 2.303\ [8.315 \times 10^{-3}\ \text{kJ/(mol} \cdot \text{K)}]$$

$$(273\ \text{K} + 80\ \text{K}) \times \log([HCO_3^-][H_2]/[HCO_2^-][H_2O])$$

$$= -2.6\ \text{kJ/mol} + 6.760\ \text{kJ} \times \log[(0.02\ \text{M} \times 0.01\ \text{M})/0.15\ \text{M}]$$

$$= -22\ \text{kJ/mol}$$

The ΔG value is small, but it is enough to drive growth of some species of *Thermus.* Note that this simplified treatment omits the role of gas formation. (Data are based on Yun Jae Kim et al. 2010. *Nature* **467**:352.)

CHAPTER 14

14.1 *Pseudomonas aeruginosa*, a cause of pneumonia in cystic fibrosis patients, oxidizes NADH with nitrate (NO_3^-) to nitrite (NO_2^-) at neutral pH. What is the value of $E°'$?

ANSWER: To calculate the reduction potential $E°'$:

$$NADH + H^+ + NO_3^- \rightarrow NAD^+ + NO_2^- + H_2O$$
$$E°' = 320\ \text{mV} + 420\ \text{mV} = 740\ \text{mV}$$

14.2 Could a bacterium obtain energy from succinate as an electron donor with nitrate (NO_3^-) as an electron acceptor? Explain.

ANSWER: For nitrate reduction:

$$\text{Succinate} + NO_3^- \rightarrow \text{fumarate} + NO_2^- + H_2O$$
$$E°' = -33\ \text{mV} + 420\ \text{mV} = 387\ \text{mV}$$

Although succinate is a relatively poor electron donor, nitrate is a strong electron acceptor. This reaction should provide energy for bacterial metabolism.

14.3 Use Table 14.1 to write the chemical reaction used by cable bacteria. What is its standard reduction potential? Explain why electrons must travel such a long distance within the bacterial cable.

ANSWER: The equation for sulfide oxidation performed by cable bacteria combines two half reactions from **Table 14.1**, those of sulfide oxidation and of water formation:

$$HS^- \rightarrow S^0 + H^+ + 2e^- \qquad E°' = +280\ \text{mV}$$
$$\tfrac{1}{2}O_2 + 2H^+ + 2e^- \rightarrow H_2O \qquad E°' = +820\ \text{mV}$$
$$E°' = \frac{-\Delta G}{nF} = +280\ \text{mV} + 820\ \text{mV} = 1{,}100\ \text{mV}$$

The reason that electrons must flow over such a distance is that the reduced substrate, sulfide, is buried underneath millimeters or centimeters of sediment in which the oxygen has been depleted by respiring organisms. Thus, in order to conserve energy from the reaction, the bacteria must retain the electrons, avoiding their dissipation until they reach the upper portion of the bacterial cable, where enzymes can transfer them to oxygen. Enzymes couple the energy-yielding reactions to energy-spending reactions of the cell.

14.4 What do you think happens to $\Delta\psi$ as the cell's external pH increases or decreases? What could happen to the Δp of bacteria that are swallowed and enter the extremely acidic stomach?

ANSWER: As external pH changes, ΔpH increases or decreases, affecting the magnitude of Δp. As enteric bacteria enter the stomach, they encounter pH values (pH 1.5–3.0) below their growth range (pH 5.0–9.0). At first, the cell's transmembrane ΔpH may be very large, as the cell tries to maintain its cytoplasmic pH above 5.0, the limit for viability. But since the cell no longer grows, it loses its energy supply and can no longer spend energy to maintain ΔpH. One way to maintain cytoplasmic pH homeostasis is to reverse the electrical potential $\Delta\psi$ (inside positive) so as to drive out some H^+ and maintain a small ΔpH. Oppositely directed ΔpH and $\Delta\psi$ enable the cell to keep cytoplasmic pH high enough to survive when external pH is extremely low. On the other hand, when the external pH is raised by pancreatic secretions (alkaline), the cell needs to compensate by inverting its ΔpH and maintaining a relatively large $\Delta\psi$.

14.5 How could a simple experiment provide evidence that a proton pump in the bacterial cell membrane drives efflux of antibiotics such as tetracycline?

ANSWER: Test the ability of the bacteria to form colonies on media buffered at a range of pH values. At lower pH, the ΔpH is increased, and thus a larger Δp is available to pump antibiotics out of the cell. At a higher pH, however, the ΔpH is inverted and actually subtracts from Δp. Thus, we expect the bacteria to grow with greater drug concentrations at low pH than at high pH.

14.6 Suppose that de-energized cells of *E. coli* ($\Delta p = 0$) with an internal pH of 7.6 are placed in a solution at pH 6. What do you predict will happen to the cell's flagella? What does this effect demonstrate about the function of Δp?

ANSWER: The flagella will rotate, driven solely by the ΔpH component of Δp. This result is consistent with the hypothesis that the transmembrane proton potential Δp drives flagellar rotation.

14.7 In Figure 1414, what is the advantage of the oxidoreductase transferring electrons to a pool of mobile quinones, which then reduce the terminal reductase (cytochrome complex)? Why does each oxidoreductase not interact directly with a cytochrome complex?

ANSWER: The mobile quinone pool connects diverse electron donors with diverse electron acceptors. If each oxidoreductase had to interact specifically with a different terminal oxidase, the pathways of electron transport would be limited; for example, NADH might donate electrons only to O_2, whereas succinate might donate electrons only to nitrate. Instead, all potential electron donors can be coupled with all potential acceptors.

14.8 In Figure 14.14, why are most electron transport proteins fixed within the cell membrane? What would happen if they "got loose" in aqueous solution?

ANSWER: If the electron transport proteins came away from the membrane into aqueous solution, they could carry their energized electrons back into the cytoplasm or lose them outside the cell. In either case, they could no longer convert the flow of electrons into a proton gradient.

14.9 Would *E. coli* be able to grow in the presence of an uncoupler that eliminates the proton potential supporting ATP synthesis?

ANSWER: Yes. *E. coli* can grow with the proton gradient eliminated, but only with a rich supply of nutrients for substrate phosphorylation to generate ATP (for example, from glycolysis). In addition, the external pH and salt levels must be maintained close to those of the cytoplasm, to minimize the need for ion transport.

14.10 The proposed scheme for uranium removal requires injection of acetate under highly anoxic conditions, with less than 1 part per million (ppm) dissolved oxygen. Why must the acetate be anoxic?

ANSWER: Oxygen is the strongest terminal electron acceptor. If O_2 is present, bacteria will use it preferentially (instead of U^{6+}) to oxidize the acetate to CO_2.

14.11 What conditions might favor electron transfer by pili, versus by electron shuttles?

ANSWER: Pili would be favored when the bacteria have a medium free of obstacles between the donor bacterium and the electron acceptor. The bacterium could then extend pili to reach the target. Electron shuttles would be favored where a matrix obstructs extension of pili but allows diffusion of shuttle molecules. The disadvantage of the shuttle molecule is that it may be lost in the medium, whereas the pilus may be retracted by the donor bacterium.

14.12 An alternative mechanism for a fuel cell involves the use of a bacterium that receives electrons from an electrode, instead of donating electrons. How might this work?

ANSWER: The bacteria would be aerobes that donate electrons to oxygen. They could form a biofilm on the cathode and receive electrons, then donate them to oxygen and form water. An example is that of cable bacteria, which normally oxidize sulfides and reduce oxygen. Cable bacteria are found in some marine benthic fuel cells.

14.13 Use Figures 14.14 and 14.15 to propose a pathway for reverse electron flow in an organism that spends ATP from fermentation to form NADH. Draw a diagram of the pathway.

ANSWER: In the ETS shown in **Figures 14.14** and **14.15**, a proton potential Δp would be generated by ATP synthase running in reverse as it spent ATP to pump protons out of the cell. The Δp would drive a cytochrome oxidase complex in reverse to transfer electrons from an electron donor onto quinones, forming quinols (**Fig. 14.11D**). The quinols would transfer electrons to an NDH complex, using energy from proton influx to reduce NAD^+ to NADH.

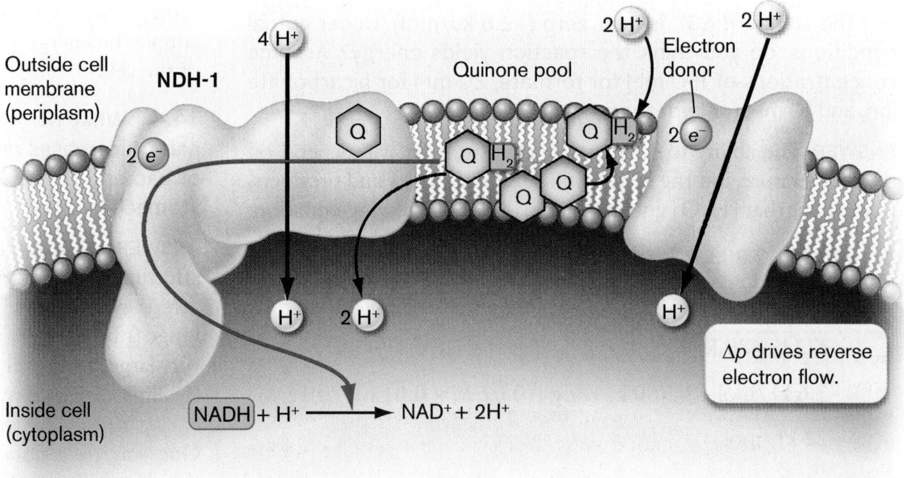

14.14 Hydrogen gas is so light that it rapidly escapes from Earth. Where does all the hydrogen come from to be used for hydrogenotrophy and methanogenesis?

ANSWER: Hydrogen is produced in substantial quantities as a byproduct of fermentation. It may seem surprising that organisms would readily excrete quantities of energy-rich H_2, but in the absence of a good electron acceptor (or the enzymes to utilize electron acceptors), hydrogen may be just another waste product. Hydrogen gas trapped underground supports large communities of methanogens and hydrogenotrophs. The human colonic bacteria generate so much hydrogen that all parts of the body show traces of hydrogen gas.

14.15 Suppose you discover bacteria that require a high concentration of Fe^{2+} for photosynthesis. Can you hypothesize what the role of Fe^{2+} might be? How would you test your hypothesis?

ANSWER: The organism uses reduced iron as an electron donor for its photosystem ($Fe^{2+} \rightarrow Fe^{3+}$). To test this hypothesis, grow the organism on a defined concentration of Fe^{2+}. Measure the amount of iron oxidized and the amount of carbon fixed into biomass; if the Fe^{2+} is an electron donor for the photosystem, the two numbers should show a linear correlation.

CHAPTER 15

15.1 To run the TCA cycle and glycolysis in reverse, does a cell use the same enzymes or different ones? Explain why some enzymes might be used in both directions, whereas other steps require different enzymes for catabolic and anabolic directions.

ANSWER: Most enzymes are capable of catalyzing a reaction in either direction, depending on the relative amounts of substrates and products, and available energy. In glycolysis and the TCA cycle, many individual steps of catalysis involve very small energy transitions, such as the interconversion of glucose 6-phosphate with fructose 6-phosphate. The direction of such reactions may be determined by the relative concentrations of substrates and products. (The contribution of reactant concentrations to free energy change ΔG is discussed in Chapter 13.) However, certain key steps of catabolism require a different enzyme for reversal, regulated by conditions that require catabolism or biosynthesis, respectively. For example, the catabolic enzyme phosphofructokinase phosphorylates fructose 6-phosphate to fructose 1,6-phosphate, spending ATP. The reversal of this key step requires fructose 1,6-bisphosphatase. These two enzymes are regulated by metabolites that signal whether the cell has a greater need for energy or for biosynthesis.

15.2 Speculate on why Rubisco catalyzes a competing reaction with oxygen. Why might researchers be unsuccessful in attempting to engineer Rubisco without this reaction?

ANSWER: The oxygenation reaction might have an essential function in regulation of metabolism. For example, it might help prevent excessive reduction of cell components or fixation of too much carbon to be used in biosynthesis. Given the universal existence of the oxygenation reaction in bacterial and chloroplast Rubiscos, it seems unlikely that oxygenation serves no purpose. For this reason, attempts to engineer Rubisco without oxygenation may not succeed.

15.3 Why does ribulose 1,5-bisphosphate have to contain two phosphoryl groups, whereas the other intermediates of the Calvin cycle contain only one?

ANSWER: Only ribulose 1,5-bisphosphate needs to split into two molecules (3-phosphoglycerate). Each of the two products needs to have its own phosphate as a tag for the enzymes to recognize it within the cycle.

15.4 Which catabolic pathway (see Chapter 13) includes some of the same sugar-phosphate intermediates that the Calvin cycle has? What might these intermediates in common suggest about the evolution of the two pathways?

ANSWER: The pentose phosphate pathway includes ribulose 5-phosphate, erythrose 4-phosphate, and sedoheptulose 7-phosphate in a similar series of carbon exchanges. Perhaps the pentose phosphate pathway and the Calvin cycle pathway evolved from a common amphibolic pathway of sugar consumption and biosynthesis. Alternatively, the one pathway evolved earlier, and then the sugar intermediates were available for evolution of the second pathway.

15.5 For a given species, uniform thickness of a cell membrane requires uniform chain length of its fatty acids. How do you think chain length may be regulated?

ANSWER: One way to regulate chain length might be that the enzyme fits only a limited length of chain. In *E. coli*, the chain length of a growing fatty acid appears to be limited by beta-ketoacyl-ACP synthase, which binds only precursor acyl-ACPs shorter than 18 carbons. Thus, only carbon chains of up to 18 carbons are synthesized. An alternative way to limit chain length might be for another enzyme to cleave carbons that extend out too far for the fatty acyl group to fit in a membrane.

15.6 Suggest two reasons why transamination is advantageous to cells.

ANSWER: Ammonia is toxic to cells. Transamination enables cells to store amine groups in nontoxic form, readily available for biosynthesis. The availability of multiple enzymes of transamination from different amino acids enables cells to quickly recycle existing resources into the amino acids most needed by the cell in a given environment. For example, if a sudden supply of glutamine appears, cells can immediately distribute its amines into all 20 amino acids.

15.7 Which energy carriers (and how many) are needed to make arginine from 2-oxoglutarate?

ANSWER: Arginine biosynthesis requires three ATP molecules and three NADPH molecules (including two for converting two molecules of 2-oxoglutarate to glutamate). An additional ATP is spent converting acetate to acetyl-CoA.

15.8 Why are purines synthesized onto a sugar-base-phosphate?

ANSWER: Purines are highly hydrophobic, insoluble in the cytoplasm. The ribose phosphate component solubilizes the molecule, enabling synthesis to occur in the cytoplasm, where the purines are needed to make RNA and DNA.

15.9 Why are the ribosyl nucleotides synthesized first and then converted to deoxyribonucleotides as necessary? What does this order suggest about the evolution of nucleic acids?

ANSWER: Ribonucleic acid is believed to be the original chromosomal material of cells. Cells evolved to synthesize RNA first; then later, as DNA was used, pathways evolved to synthesize it by modification of RNA, which the cell already had the ability to make.

CHAPTER 16

16.1 Why do the lipid components of food experience relatively little breakdown during anaerobic fermentation?

ANSWER: Lipids are highly reduced molecules, largely hydrocarbon with relatively low oxidizing potential. Thus, lipids cannot undergo as many intramolecular redox reactions as do sugars, which readily generate energy through anaerobic fermentation.

16.2 Why does oxygen allow excessive breakdown of food, compared with anaerobic processes?

ANSWER: Oxygen functions as the terminal electron acceptor for the complete breakdown of all kinds of organic molecules yielding water and CO_2. Anaerobic processes, such as yogurt fermentation, lack oxygen or alternative electron acceptors that would support the further breakdown of metabolic products. Thus, fermented foods retain relatively high-energy organic molecules for the human consumer.

16.3 In an outbreak of listeriosis from unpasteurized cheese, only the refrigerated cheeses were found to cause disease. Why would this be the case?

ANSWER: In the cheeses kept at room temperature, other naturally occurring bacteria outgrew the pathogenic *Listeria*, whereas in the refrigerator only the *Listeria* could grow. (Note, however, that many other potential pathogens, such as *Salmonella*, are inhibited by refrigeration.)

16.4 Cow's milk contains 4% lipid (butterfat). What happens to the lipid during cheese production?

ANSWER: Lipids undergo little catabolism, because the fermentation conditions are anaerobic. During coagulation, lipid droplets become trapped in the network of denatured protein and are largely retained in the bulk of the cheese. "Low-fat" cheeses are made from skim milk, which eliminates the lipids before fermentation.

16.5 In traditional fermented foods, without a pure starter culture, how could someone control the kind of fermentation that occurs?

ANSWER: The fermentation can be controlled by introduction of a crude starter culture obtained from a previous batch of the food product or from a natural source of a particular microbe; for example, rice straw is a source of *Bacillus natto* for natto production. The fermentation type can be manipulated by the addition of factors, such as brine, that retard growth of all but a few strains. In pidan, for example, the high concentration of sodium hydroxide limits bacterial growth to alkali-tolerant strains of *Bacillus*.

16.6 Compare and contrast the role of fermenting organisms in the production of cheese and bread.

ANSWER: In cheese production, fermentation causes major biochemical changes in the food, such as the buildup of acids and the breakdown of proteins to smaller peptides and amino acids. Minor by-products, such as methanethiols and esters, accumulate to levels that confer flavors. In yeast bread, by contrast, the only significant product of fermentation is the carbon dioxide that leavens the dough. The small amount of ethanol produced evaporates during cooking. A form of bread in which extended fermentation does generate flavor is injera, the dough of which ferments for 3 days.

16.7 Compare and contrast the role of low-concentration by-products in the production of cheese and beer.

ANSWER: In both cheese and beer, minor by-products such as esters contribute flavor. Oxidation of esters can lead to off-flavors. In cheese, however, the exclusion of oxygen usually prevents off-flavors. In beer, the yeast requires a low level of oxygen; thus, significant amounts of acetaldehyde and diacetyl are produced and must be eliminated by a secondary fermentation.

16.8 Why would bacteria convert trimethylamine oxide (TMAO) to trimethylamine? Would this kind of spoilage be prevented by exclusion of oxygen?

ANSWER: TMAO acts as a terminal electron acceptor—that is, an alternative to oxygen for anaerobic respiration, as discussed in Chapter 14. Exclusion of oxygen inhibits only aerobic bacteria; TMAO respirers continue to grow and can spoil the fish.

16.9 Is it possible for physical or chemical preservation methods to completely eliminate microbes from food? Explain.

ANSWER: Preservation methods either slow microbial growth or induce microbial death. Microbial death follows a negative exponential curve, as discussed in Chapter 5. In theory, the exponential curve never reaches zero, so total exclusion of microbes is impossible. In practice, there is a high probability of totally eliminating microbes if the treatment time extends several "half-lives" beyond the time at which microbial concentration declines to less than one per total volume.

16.10 Why would different industrial strains or species be used to express different kinds of cloned products?

ANSWER: Different industrial strains have biochemical systems that favor different products. Some fungi naturally possess the highly complex pathways to generate antibiotics, as well as regulatory timing to turn on these pathways after the culture has grown to high population density. On the other hand, bacteria such as *Bacillus subtilis* are the most genetically tractable and predictable in their growth cycles, and the easiest to manipulate to express recombinant products such as human genes.

16.11 Why would an herbicide resistance gene be desirable in an agricultural plant? What long-term problems might be caused by microbial transfer of herbicide resistance genes into plant genomes?

ANSWER: Introduction of an herbicide resistance gene allows application of higher amounts of herbicide to crops in order to control growth of weeds. But the higher concentrations of herbicide may also have greater side effects on animals and on human consumers of the crop. In the long run, the herbicide resistance gene is likely to escape into weed plants through natural gene transfer mechanisms. Thus, eventually the weeds may require still higher concentrations of herbicide. While the costs versus benefits of new gene modifications remain poorly understood, it must be recognized that all modern crops today are the product of many generations of genetic manipulation.

16.12 Why do you think the SIN mutation is important in the lentivector, even though the integrated viral sequence lacks all the genes for virus replication (provided on the original plasmids)?

ANSWER: The human cells containing the integrated lentivector are exposed to other retroviruses, including endogenous retroviruses encoded and expressed by the native human genome. Although highly unlikely, it is possible that a reverse transcriptase from another virus could transcribe the lentivector and recombine with it to constitute an aberrant virus. This unlikely event is made impossible by deletion of the LTR promoter.

CHAPTER 17

17.1 Evolution by natural selection is based on competition, yet the earliest fossil life shows organized structures such as a stromatolite built by cooperating cells. How could this be explained?

ANSWER: Individual microbes compete for resources in a given environment. Part of one's environment consists of other microbes—which may provide resources or favorable conditions. In the stromatolite, the adherence of cells to each other and to the substrate could help maintain their position in a favorable part of the sea, at an elevation with access to light. In addition, adherent microbes might resist predation. We cannot know the biochemistry of cells from ancient fossils, but some cells could have evolved specialized metabolism that led to cross-feeding. In modern stromatolites, phototrophic bacteria photolyze H_2S to sulfate, which is then reduced by lower layers of sulfate-reducing bacteria. Overall, the cells of the stromatolite might outcompete cells that grow individually.

17.2 What would have happened to life on Earth if the Sun were of a different stellar class, substantially hotter or colder than it is?

ANSWER: If the Sun were hotter, too much ultraviolet and gamma radiation would reach Earth, breaking chemical bonds of living organisms so rapidly that life could not be sustained. If the Sun were colder, too little radiation with sufficient energy would be available to drive photosynthesis. In either case, life as we know it could not have evolved on Earth.

17.3 In Figure 17.7A, why does the shallow-ocean CO_2 show a positive value of $\delta^{13}C$?

ANSWER: The shallow ocean has extremely high productivity of photosynthesis, which rapidly fixes dissolved CO_2 into biomass. Thus, the CO_2 that remains in the water may actually show enrichment for ^{13}C resulting from the preferential fixation of ^{12}C.

17.4 Outline the strengths and limitations of the prebiotic soup model and the RNA world model of the origin of living cells. Which aspects of living cells does each model explain?

ANSWER: The two models are complementary. Each explains aspects of modern cells not addressed by other models. The prebiotic soup model accounts for the major classes of compounds used by cells, such as nucleosides, TCA cycle intermediates, amino acids, and fatty acids. It also suggests the origin of membranes as soap bubble–like micelles. It does not, however, account for the evolution of metabolic pathways and replication of genetic information. The RNA world accounts for the central role of RNA in living cells; of all molecular classes, RNA and ribonucleotides probably serve the widest range of functions as information carriers, agents of catalysis, and genetic regulators. Most RNA-world models do not address the origin of membranes.

17.5 Suppose a NASA rover discovered living organisms on Mars. How might such a find shed light on the origin and evolution of life on Earth?

ANSWER: If life on Mars showed a completely different basis than that of Earth—for example, it was based on silicon polymers instead of carbon—such a find would support the view that life originated independently on each planet, rather than traveling from one planet to the other, or that both planets were seeded from somewhere else. If life on Mars were based on similar macromolecules, perhaps even showing the same genetic code, this finding would support the view that life arose on Mars first, or that both planets were seeded from the same source.

17.6 What kinds of DNA sequence changes have no effect on gene function? (*Hint:* Refer to the table of the genetic code, Figure 8.12.)

ANSWER: Base substitutions that do not change the amino acid specified by the codon (silent mutations) have no immediate effect on gene function. For example, CUA → CUG still encodes leucine. In addition, a majority of the amino acids in any given protein can be replaced by an amino acid of similar form (for example, leucine → valine) without significantly affecting function of the gene. Nevertheless, even silent mutations change the DNA sequence in ways that may enable later substitution of amino acids that alter function of the product.

17.7 What are the major sources of error and uncertainty in constructing phylogenetic trees?

ANSWER: Phylogenetic trees are affected by variability in the number of substitutions, or rate of mutation in different strains. The tree is distorted by errors in sequence alignment, and by systematic errors due to failure of the fundamental assumptions of the molecular clock. These assumptions include the constant rate of mutation for all branches, constant generation time, and true orthology of the gene chosen (that is, the encoded product has the same function and hence the same degree of selection pressure in all taxa under consideration).

17.8 What are the limits of evidence for horizontal gene transfer in ancestral genomes? What alternative interpretation might be offered?

ANSWER: Horizontal gene transfer is inferred from the appearance of genes in clade A that are absent from other members of the clade but present in clade B. The degree of similarity between genes in the two clades, however, must be high enough to exclude the possibility that the genes in question were retained from a common ancestor of the two clades but lost from other members of clade A. This possibility is difficult to exclude in the case of deep-branching clades, where all genes have had a long time to diverge. For example, the large number of archaeal genes present in deep-branching thermophilic bacteria such as *Thermotoga* may include some inherited from the last common ancestor, or they may represent archaeal genes that were horizontally transferred to bacteria sharing the high-temperature habitat.

17.9 In the fitness competition between cells labeled by YFP and CFP, how can we rule out fitness differences associated with the bacterial expression of the two different fluorescent proteins?

ANSWER: For each experiment, conduct two classes of replicates, for which the YFP and CFP labels are reversed; that is, in one case the resistant strain has the YFP label, and in the other case the resistant strain has the CFP label. Fitness differences between the two fluorescent proteins should then cancel each other. This model assumes no interaction between fluorophore effects and the antibiotic resistance phenotype.

17.10 The evolution of clones requiring nutrients from other clones may be more common than expected. How might the mechanisms of evolution favor the emergence of dependent clones?

ANSWER: The early mutations that arise under selection pressure confer imperfect phenotypes with deleterious side effects, such as excretion of valuable nutrients. Thus, an evolving population is likely to increase the excreted nutrients available in the medium, such as acetate and citrate. The increase of nutrients provides opportunities for new clones to achieve relative fitness within a population.

17.11 Besides mitochondria and chloroplasts, what other kinds of entities within cells might have evolved from endosymbionts?

ANSWER: Some of the large "megaplasmids" found in bacteria and protists are as large as genomic chromosomes and contain numerous housekeeping genes. These megaplasmids may have originated as endosymbiotic cells that lost all their membranes through reductive evolution. Similarly, some of the giant viruses, such as mimivirus and smallpox virus, as well as phages such as T4, possess a wide spectrum of housekeeping genes. These viruses may have originated as cellular parasites that underwent reductive evolution.

CHAPTER 18

18.1 What taxonomic questions are raised by the apparent high rate of gene transfer between archaea and thermophilic bacteria?

ANSWER: If gene transfer results in a species containing a quarter of its genes from organisms outside its domain, such a mosaic genome raises questions of how to define the species and the domain. How can a species be defined if its genome contains large portions from distantly related sources? Other interesting questions relate to the means of gene transfer. How do such distantly related organisms as bacteria and archaea maintain a compatible mechanism of gene transfer?

18.2 Which taxonomic groups in Table 18.1 stain Gram-positive, and which stain Gram-negative? Which group contains both Gram-positive and Gram-negative species? For which groups is the Gram stain undefined, and why?

ANSWER: Most Firmicutes stain Gram-positive. These bacteria have a relatively thick cell wall that retains the stain. Actinobacteria also have a thick cell wall, but some have a waxy coat that excludes the stain. The Proteobacteria, Nitrospirae, Bacteroidetes, Verrucomicrobia, and Chlorobi groups stain Gram-negative. The Cyanobacteria have an outer membrane and are considered Gram-negative, although their cell wall is thick. The Deinococcus-Thermus phylum includes both Gram-positive and Gram-negative members. For Chlamydiae, the Gram stain is irrelevant because they lack the cell wall that retains the stain. For Spirochetes, many species are too narrow to observe the stain under light microscopy.

18.3 Which groups of bacterial species share common structure and physiology within the group? Which groups show extreme structural and physiological diversity?

ANSWER: Cyanobacteria all carry out oxygenic photosynthesis within thylakoid membranes. Their overall cell structure and organization, however, take diverse forms. All spirochetes are a sheathed flexible spiral with internal flagella; most share anaerobic or facultative heterotrophy. Species of the Chlamydiae and Planctomycetes groups each share general structural features. Other groups, particularly Firmicutes and Actinobacteria, show considerable diversity of form and physiology. The Proteobacteria display more extreme diversity of metabolism than any other division.

18.4 What are the relative advantages and disadvantages of propagation by hormogonia, as compared with akinetes?

ANSWER: Hormogonia are motile, and thus capable of active chemotaxis toward a more favorable environment. On the other hand, hormogonia have active metabolism that requires nutrition; if the environment lacks nutrients, the hormogonia will die. Akinete cells can persist until environmental conditions improve, but they cannot actively seek out a new location.

18.5 What are the relative advantages and disadvantages of the different strategies for maintaining separation of nitrogen fixation and photosynthesis?

ANSWER: Temporal separation has the advantage that all cells possess the ability to perform both nitrogen fixation and photosynthesis. On the other hand, it eliminates the ability of a chain of cells to conduct both processes simultaneously—the benefit of heterocysts. Heterocysts face the problem of operating in close proximity to photosynthetic cells generating toxic oxygen. This problem may be solved by symbiosis with respiring bacteria. Globular clusters of cells can bury their nitrogen fixers within the cluster; this arrangement effectively excludes oxygen, but it may lack flexibility during environmental change. Endosymbiotic nitrogen fixation within a respiring eukaryote is probably the most effective strategy of all, because the host provides oxygen-removing proteins such as leghemoglobin. Endosymbiosis, however, requires the presence of an appropriate host organism.

18.6 Why would *Streptomyces* produce antibiotics targeting other bacteria?

ANSWER: *Streptomyces* species may produce antibiotics to curb the growth of bacterial competitors with smaller genomes and faster rates of reproduction. The lysed cells release nutrients that feed growing mycelia of *Streptomyces*.

18.7 Why might genes for the proteorhodopsin light-powered proton pump be more likely to transfer horizontally than the genes for bacteriochlorophyll-based photosystems PS I and PS II?

ANSWER: Proteorhodopsin requires only the one gene encoding the pump, plus one or two genes to produce retinal. A relatively small amount of sequence has to be transferred, and the encoded products generate proton potential on their own, without requiring interaction with recipient enzymes. By contrast, PS I and PS II each involves multiple electron carriers that must function together and interact with the recipient electron transport chain.

18.8 Can you hypothesize a mechanism for migration of the daughter nucleoid of *Hyphomicrobium* through the stalk to the daughter cell? For possibilities, see Chapter 3 and consider the various molecular mechanisms of cell division and shape formation.

ANSWER: One mechanism might involve polar localization similar to that seen in *Caulobacter*. A DNA-binding protein might pull the nucleoid through the stalk until it binds to a polar localization protein at the end of the daughter cell. Another mechanism might involve formation of a scaffold of cytoskeletal proteins similar to FtsZ or MreB. The cytoskeleton could act as a track for DNA movement through the stalk, perhaps powered by ATP hydrolysis.

18.9 Why do you think it took many years of study to realize that *Escherichia coli* and other Proteobacteria can grow as a biofilm?

ANSWER: *E. coli* and its relatives grow exceptionally well in liquid culture. Liquid culture is attractive because it enables quantitative measurement of defined aliquots of a microbial population. However, repeated subculturing in liquid medium selects for planktonic (nonbiofilm) cells. Eventually, the biofilm-forming property may be lost if nonbiofilm mutants evolve to grow faster than the original genotype in liquid medium.

18.10 Compare and contrast the formation of firmicute endospores, actinomycete arthrospores, and myxococcal myxospores.

ANSWER: An endospore forms as the daughter product (forespore) of a single cell. Within the same cell, endospore development is supported by the mother cell, which disintegrates after release of the endospore. Endospores have tough coatings of calcium dipicolinate; they are heat resistant. By contrast, arthrospores and myxospores are less durable and are not heat resistant, although they can persist in the environment for an extended period. Arthrospores form through binary fission of actinomycete filaments. Myxospores are formed by a multicellular fruiting body. In all three cases, spore formation can be induced by depletion of nutrients, and the spore-producing entity is left behind to die.

CHAPTER 19

19.1 Suppose that two deeply diverging clades each show a wide range of growth temperature. What does this suggest about the evolution of thermophily or psychrophily?

ANSWER: The two clades diverged before temperature adaptation occurred. Adaptations to high or low temperature must have evolved independently in the two clades.

19.2 What might be the advantages of archaellar motility for a hyperthermophile living in a thermal spring or in a black smoker vent? What would be the advantages of growth in a biofilm?

ANSWER: Archaellar motility enables isolated cells to detect a new nutrient source, or an appropriate temperature range, and approach it through chemotaxis. Growth in a biofilm attached to a substrate prevents the microbes from floating away from the nutrient source, or from being carried away in the flow from the vent.

19.3 What problem with cell biochemistry is faced by acidophiles that conduct heterotrophic metabolism?

ANSWER: Heterotrophic metabolism generates fermentation products such as acetate and lactate, which act as permeant acids. Permeant acids become protonated outside the cell, at low pH; the protonated forms then permeate the membrane, returning into the cell. Given the high transmembrane pH difference maintained by *Sulfolobus*, one would expect even small traces of fermentation acids to cross the membrane in the protonated form, and then dissociate and accumulate to toxic levels of organic acids. It is unknown how *Sulfolobus* solves this problem.

19.4 What hypotheses might be proposed about archaeal evolution if viruses of mesophilic archaea are found to have RNA genomes? What if, instead, all archaeal viruses have DNA genomes only?

ANSWER: If mesophilic viruses show RNA genomes but thermophiles do not, then it is likely that only double-stranded DNA is sufficiently stable for viruses to persist in the environment of hyperthermophiles. Finding only double-stranded DNA viruses throughout the archaea would suggest that all archaeal viruses evolved from viruses infecting a common ancestral cell that was a thermophile. The latter hypothesis would require supporting evidence from archaeal cell physiology and phylogeny.

19.5 Anoxic soil contains bacteria and methanogens. What will happen to the microbial populations when the soil is tilled and aerated?

ANSWER: When soil is broken up by tilling, the oxic and anoxic layers become mixed. Overall diversity of the microbial community decreases. The presence of oxygen will limit growth of methanogens and anaerobic bacteria. There will likely be an increase of aerobic or facultative species such as Actinobacteria and *Bacillus* species.

19.6 What do the multiple metal requirements suggest about how and where the early methanogens evolved?

ANSWER: The requirement for so many different metals may suggest that methanogens evolved in habitats such as geothermal vents, where superheated water carries up high concentrations of dissolved metal ions.

19.7 Compare and contrast the metabolic options available for *Pyrococcus* and for the TACK organism *Sulfolobus*.

ANSWER: *Sulfolobus* catabolizes sugars and amino acids aerobically, using O_2 as terminal electron acceptor. *Pyrococcus abyssi* catabolizes sugars and amino acids anaerobically, using S^0 as the terminal electron acceptor. *P. abyssi* can also reduce sulfur lithotrophically with H_2 to form H_2S. By contrast, *Sulfolobus* uses molecular oxygen (O_2) to oxidize sulfur lithotrophically, from S^{2-} to S^0, SO_3^{2-}, and ultimately SO_4^{2-}.

19.8 Compare and contrast sulfur metabolism in *Pyrococcus* and in *Ferroplasma*.

ANSWER: *Pyrococcus* species reduce S^0 with hydrogens from organic substrates, forming HS^- and H_2S. *Ferroplasma* species oxidize sulfur in the form of FeS_2 (using oxidant Fe^{3+}), forming sulfuric acid. The result is extreme acidification of their environment.

CHAPTER 20

20.1 How could you demonstrate that eukaryotic flagella move with a whiplike motion, instead of rotary motion? What experiment might you conduct?

ANSWER: In order to track the motion of eukaryotic flagella, you could raise an antibody against a protein subunit of the flagellum. Attach the antibody to a fluorophore. Combine the antibody-fluorophore with a suspension of flagellated protists, at a concentration ratio that allows approximately one fluorescent antibody to attach to a flagellum. Now it should be possible to observe and video-record protists with individual labeled flagella, and trace their whiplike motion.

20.2 Why would yeasts remain unicellular? What are the relative advantages and limitations of hyphae?

ANSWER: Yeasts grow in environments with sufficient dissolved nutrients to absorb from the medium. The advantage of forming hyphae is that they can penetrate solid substrates, such as soil or a host organism, and hence provide access to nutrients. On the other hand, hyphae formation limits the rate of dispersal of progeny cells. Since yeasts grow in environments where dissolved nutrients can be absorbed from the medium, they do not need to produce hyphae and can proliferate more rapidly than mycelial fungi.

20.3 Why would some fungi avoid sexual reproduction? What are the advantages and limitations of sexual reproduction?

ANSWER: The sexual life cycle involves significant genetic and metabolic costs to the organism. Asexual reproduction enables fungi to eliminate an energy drain and produce more offspring, using fewer resources. Reductive evolution leading to loss of sexual reproduction might eliminate an energy drain and perhaps enable greater proliferation with fewer resources. On the other hand, the sexual life cycle provides a valuable means of generating diversity through genetic recombination, so that the population can respond to environmental change. Fungi reproducing asexually must rely on mutation and gene transfer by viruses and mobile sequence elements to generate genetic diversity.

20.4 What are the advantages and limitations of motile gametes, as compared to nonmotile spores?

ANSWER: Motile gametes have the advantage of rapid dispersal on their own and the potential for chemotaxis toward a food source or toward a gamete of the opposite mating type. On the other hand, motility uses up energy that could alternatively be invested in production of a greater number of nonmotile gametes. Motile gametes are especially useful in a watery habitat but are of little use in a terrestrial habitat, where air currents or animal hosts must be used for dispersal.

20.5 Compare the life cycle of an ascomycete (Fig. 20.13C) with that of a chytridiomycete (Fig. 20.11C). How are they similar, and how do they differ?

ANSWER: Both chytridiomycetes and ascomycetes undergo alternation of generations. Each has an alternative route of an asexual cycle of mitotic cell proliferation. In the chytridiomycete asexual cycle, the diploid form develops a mycelium, motile zoospores, and cysts. In the ascomycete, however, the haploid form undergoes mitotic divisions. The sexual cycles of the two fungal groups differ structurally. In the chytridiomycete, the zoosporangium forms motile zoospores that develop and release motile gametes, which fertilize each other to form motile zygotes. In the ascomycete, there are no motile forms. Instead of motile gametes, haploid hyphal antheridia undergo cytoplasmic fusion and form fruiting bodies, or asci, within which ascospores develop. The ascospores are not motile; they are carried by wind or water.

20.6 What cellular interactions can happen when an ameba phagocytoses algae?

ANSWER: If light is available, the algae may be retained as endosymbionts providing energy through photosynthesis. For example, *Chlorarachnion* possesses obligate chloroplast-bearing endosymbionts descended from green algae. Alternatively, the ameba may digest all but the algal chloroplast, which persists for some time, providing photosynthetic products.

20.7 What kind of habitat would favor a flagellated ameba?

ANSWER: A dilute watery habitat would favor flagella, which allow more rapid propulsion than do pseudopods. Pseudopod motility requires a solid substrate, such as debris in the sediment of a pond.

20.8 What are the relative advantages of being unicellular or multicellular?

ANSWER: Single-celled organisms require minimal nutrients to reproduce and can disperse rapidly, avoiding competition. These important advantages serve the vast majority of organisms on Earth, which are unicellular. On the other hand, a multicellular colony can work together to obtain a larger food source (by predation) and protect its cells from rapid changes in the habitat, such as change in salt concentration or pH. A multicellular organism can protect its offspring from predation, as in the case of *Volvox*. Cell differentiation (for example, nitrogen-fixing heterocysts) can increase the efficiency of an organism exploiting its environment.

20.9 Compare and contrast the process of conjugation in ciliates and bacteria (see Chapter 9).

ANSWER: Conjugation in ciliates is a completely different process from conjugation in bacteria, although the function (gene transfer) is similar. In ciliates, two cells form a bridge allowing cytoplasm to flow directly between them, along with micronuclei containing chromosomes. In bacteria, a donor cell attaches to another (by pili in some cases), and then a protein complex transfers DNA across both cell envelopes, without direct cytoplasmic contact. In bacterial conjugation, DNA is transferred unidirectionally from the donor cell to the recipient, whereas in ciliates there is reciprocal exchange of DNA. A donor bacterium generally transfers only part of its genome, whereas ciliates exchange entire copies of their respective genomes.

20.10 For ciliates, what are the advantages and limitations of conjugation, as compared with gamete production?

ANSWER: The process of conjugation avoids the necessity of dissolving the intricate cell structure of the ciliate in order to form gametes that fuse or fertilize each other. On the other hand, conjugation requires two diploid organisms to find each other and make contact for several hours, during which time feeding is suspended and the pair is vulnerable to predation.

CHAPTER 21

21.1 Suppose you are conducting a metagenomic analysis of soil sampled from different parts of a wetland. Would you use one DNA extraction method or multiple methods?

ANSWER: The advantage of using one DNA extraction method is that it will maximize information about the differences between communities from different parts of the wetland. The advantage of multiple extraction methods applied to a given sample is that they will maximize the detection of diversity within a sample. Your choice of method (or a compromise design) may depend on how different you expect the microhabitats to be within the overall wetland.

21.2 Suppose you plan to sequence a marine metagenome for the purpose of understanding carbon dioxide fixation and release, to improve our model for global climate change. Do you focus your resources on assembling as many complete genomes as possible, or do you focus on identifying all the community's enzymes of carbon metabolism?

ANSWER: To maximize your yield of information about carbon flux, you are likely to learn more about the community as a whole by focusing on those enzymes involved in carbon metabolism. However, you will miss novel genes that were not previously known to participate in carbon flux. If your equipment and computational resources enable genome assembly, it may be informative to learn more about the most abundant species that participate in carbon flux.

21.3 Could you design a metagenome experiment analogous to the oil plume experiment to test which kinds of human gut bacteria digest a certain food, such as hamburger meat? What follow-up experiments would be needed?

ANSWER: Experimental human subjects could be fed a vegetarian diet for several weeks, then switched to a diet full of hamburgers. The gut metagenome could be sampled before and after the influx of hamburger. We might then observe a shift in the taxon profile, to bacteria with increased capability for catabolism of fat and protein, rather than plant fiber. As in the petroleum plume investigation, however, the result would provide only a correlation between diet and taxa prevalence. Additional kinds of experiments, such as isotope labeling, are needed to demonstrate functional connections between microbes and a food source.

21.4 How could you determine whether an organism actually performs the functions predicted by your analysis of its genome and transcriptome, such as metabolizing petroleum components?

ANSWER: To demonstrate microbial metabolism of a petroleum component, one approach would be to add a heavy isotope–enriched amendment to the medium of your microbial sample (as in the Current Research Highlight). This experiment requires that the microbe be capable of metabolism in the laboratory; ideally, a cultured organism. For example, ^{13}C-enriched cyclohexane could be provided to a cultured species of Oceanospirillales, such as *Profundimonas piezophila* (see Yi Cao et al. 2014. *Appl. Environ. Microbiol.* **80**:54). The culture medium could then be tested for the presence of ^{13}C-enriched degradation products, a sign of energy-yielding metabolism. Bacterial proteins could be tested for ^{13}C enrichment, a sign of biosynthetic incorporation of the cyclohexane carbons.

21.5 From Chapters 13–15, give examples of microbial metabolism that fit patterns of assimilation and dissimilation.

ANSWER: Microbes can assimilate carbon either by reducing carbon dioxide or by oxidizing methane, and they can dissimilate carbon by fermentation and respiration. Nitrogen is assimilated by N_2 fixation, and by incorporation of NH_4^+ into glutamine and glutamate. Nitrogen is dissimilated by deamination of amino acids, and by lithotrophic oxidation.

21.6 How does ruminant microbial fermentation provide food molecules that the animal host can use? How is the animal able to obtain nourishment from waste products that the microbes could not use?

ANSWER: The rumen interior is anaerobic. In the absence of oxygen as a terminal electron acceptor, microbes are forced to generate waste products in which the electrons are put back onto the electron donors (fermentation; see Chapter 13). When the short-chain fatty acid wastes enter the animal's bloodstream, the blood is full of oxygen, which enables complete digestion to CO_2 and water.

21.7 How do you think cattle feed might be altered or supplemented to decrease methane production?

ANSWER: Several methods have been proposed to limit methanogenesis. One is to feed cattle an inhibitor of a process that methanogens (but not bacteria) require. An example would be inhibitors of sodium transport, which methanogens need to maintain a sodium potential. Another approach is to feed cattle an organic electron acceptor for H_2, such as fumarate, which bacteria use to generate short-chain fatty acids instead of methane. These approaches have been used with only partial success—not surprisingly, given the complexity of the system.

21.8 How could you design an experiment to test the hypothesis that an animal makes use of amino acids synthesized by its gut bacteria?

ANSWER: Select a type of amino acid, such as lysine, that the animal cannot synthesize and that is therefore essential to the diet of the test animal. The diet can then be altered to exclude the essential amino acid but include a nitrogen source labeled with a heavy isotope. (For a short period, the animal can grow without the added amino acid.) The animal's gut bacteria can be tested for incorporation of the heavy isotope into their amino acids, and the host animal can be tested for protein that incorporates the heavy-isotope-labeled amino acid.

21.9 How do viruses select for increased diversity of microbial plankton?

ANSWER: Since viruses tend to infect only a narrow host range, their existence favors the evolution of a large number of different host species with highly dispersed populations. Highly dispersed populations minimize the chance of viral transmission from one host to another. Over generations, mutations that allow host microbes to "escape" viral infection will be selected for, while viral mutations that allow a virus to infect previously unsusceptible hosts will also be selected.

21.10 Design an experiment to test the hypothesis that the presence of mycorrhizae enhances plant growth in nature.

ANSWER: Such an experiment requires a control based on the natural environment, where various unknown factors may be very different from those in the laboratory. One possibility is to compare the growth of seedlings in natural soil versus sterilized natural soil. However, this experiment would not prove that fungi are the cause of enhanced growth in unsterile soil. The sterilization procedure (usually involving heat and pressure) could break down key nutrients in the soil. A follow-up experiment might be to grow the plants in the presence of a fungus inhibitor in sterilized and unsterilized soil.

21.11 How do you think symbiotic rhizobia reproduce? Why do bacteroids develop if they cannot proliferate?

ANSWER: Various answers have been proposed. Not all of the invading bacteria become bacteroids; some continue to undergo cell division, particularly within senescing tissues of the plant. These bacteria benefit from plant growth, which is sustained by the bacteroids whose genes they share. Alternatively, the entire plant-bacteroid system may benefit rhizobia that grow just outside the plant, in the rhizosphere.

21.12 High levels of nitrate or ammonium ion corepress the expression of Nod factors (see Fig. 21.47). What is the biological advantage of Nod regulation?

ANSWER: Nitrate and ammonium ion are the main forms of nitrogen assimilated by plants. If they are abundant in the soil, the plant does not need rhizobial symbionts to fix N_2—a process that consumes much energy. The energy required to maintain the symbiosis comes from the plant, in the form of sugars and other nutrients; thus, it is more efficient not to have the symbiosis when fixed nitrogen is already available. Therefore, the presence of alternative nitrogen sources inhibits development of the rhizobia-legume symbiosis.

21.13 Compare and contrast the processes of plant infection by rhizobia and by fungal haustoria.

ANSWER: Both rhizobial bacteria and fungal haustoria penetrate the volume of a plant cell, but they keep the plant cell membrane intact, its invagination always surrounding the invading cell. Rhizobia establish a complex, highly regulated exchange of nutrients with the host, receiving catabolites and oxygen in exchange for ammonium, and cycling the components of amino acids. By contrast, haustoria establish one-way removal of nutrients such as sucrose, while providing no nutrients in return. Fungal pathogens weaken the structure of the host plant and decrease or halt its growth.

CHAPTER 22

22.1 Why is oxidation state important for microbes to use and cycle compounds? Cite examples based on your study of microbial metabolism (see Chapter 14).

ANSWER: Elements may be present in the environment at oxidation levels different from those needed for biomass. In the case of carbon, autotrophic microbes may fix CO_2 with substantial reduction by NADPH or other hydrogen donors, through photosynthesis or through lithotrophy. The reduced form CH_4, however, cannot be fixed directly into biomass. Instead, methane must be oxidized by methanotrophic bacteria, usually with oxygen as electron acceptor, or with sulfate (SO_4^{2-}) under anoxic conditions. Nitrogen gas can be assimilated only by nitrogen-fixing bacteria and archaea, with extensive reduction by NADPH. The fully reduced form NH_4^+ can be assimilated into biomass by many plants and microbes. Alternatively, with oxygen present, NH_4^+ can be oxidized for energy by lithotrophs to various forms, including the fully oxidized form nitrate (NO_3^-).

22.2 What would happen if wastewater treatment lacked microbial predators? Why would the result be harmful?

ANSWER: Without predators, too many planktonic bacteria would remain in the wastewater after sedimentation of the sludge. The bacteria could be killed by chlorination, but the treated water would have significant BOD (biochemical oxygen demand) because the bacterial remains provide an organic carbon source for respirers.

22.3 Which kinds of biomolecules can you recall that contain nitrogen? What are the usual oxidation states for nitrogen in biological molecules?

ANSWER: Amino acids, nucleotide bases, polyamines for DNA stabilization, peptidoglycan (both amino sugar and peptide chains), and the heme derivatives of cytochromes, chlorophyll, and vitamin B_{12} all include nitrogen (as do many other biochemicals). The oxidation states of nitrogen in living organisms are nearly always reduced: $R{-}NH_2$, $R{=}NH$, or $R{-}N{=}R$. An exception is the neurotransmitter NO (nitric oxide).

22.4 The nitrogen cycle has to be linked with the carbon cycle, since both contribute to biomass. How might the carbon cycle of an ecosystem be affected by increased input of nitrogen?

ANSWER: One hypothesis is that the injection of nitrogen into an ecosystem accelerates growth of producers (phytoplankton in the ocean, or trees in a forest) and therefore facilitates net removal of CO_2 from the atmosphere. Overall, however, the additional fixed carbon ends up dissipated by consumers and decomposers.

22.5 In the laboratory, which bacterial genus would likely grow on artificial medium including NH_2OH as the energy source: *Nitrosomonas* or *Nitrobacter*? Why?

ANSWER: *Nitrosomonas* is more likely to utilize NH_2OH, since it performs the intermediate oxidation of NH_2OH during nitrification of ammonia.

22.6 Compare and contrast the cycling of nitrogen and sulfur. How are the cycles similar? How are they different?

ANSWER: Cycling of both nitrogen and sulfur involves interconversion between different oxidation states. Most of these interconversion reactions are performed solely by microbes, many of them solely by bacteria. Examples include nitrification of ammonia and denitrification to N_2, as well as sulfide oxidation and photolysis. In both cases, oxidation produces strong acids (HNO_3, H_2SO_4). The major sources and sinks differ; nitrogen is obtained primarily from the atmosphere as N_2, whereas sulfur (in the form of sulfate) is at high levels in the ocean and soil. Sulfur is rarely limiting, whereas nitrogen frequently is. Sulfur participates extensively in phototrophy; nitrogen shows little involvement in phototrophy, although phototrophy based on nitrate reduction has been observed.

22.7 Compare and contrast the cycling of nitrogen and phosphorus. How are the cycles similar? How are they different?

ANSWER: Nitrogen and phosphorus are both limiting nutrients in many ecosystems—marine, freshwater, and terrestrial. Addition of either element into an aquatic system may cause algal bloom and eutrophication. On the other hand, the two elements differ in their major sources: the atmosphere for nitrogen, and crustal rock for phosphate. Within biomass, nitrogen exists almost entirely in reduced form, whereas phosphorus is entirely oxidized. Phosphorus cycles through the biosphere mainly as inorganic or organic phosphates, whereas nitrogen cycles through a broad range of oxidation states, from NH_3 to NO_3^-.

CHAPTER 23

23.1 How can an anaerobic microorganism grow on skin or in the mouth, both of which are exposed to air?

ANSWER: Facultative organisms living in proximity to the anaerobes will deplete oxygen in the environment, especially around nooks and crannies (for example, between teeth and gums, in gingival pockets) that would ordinarily prevent anaerobes from growing. These small spaces have limited access to oxygen.

23.2 Why do many Gram-positive microbes that grow on the skin, such as *Staphylococcus epidermidis*, grow poorly or not at all in the gut?

ANSWER: Bile salts present in the intestine (not on the skin) easily gain access to and destroy cytoplasmic membranes of Gram-positive organisms (unless the organism possesses bile salt hydrolases). Gram-negative microbes have extra protection in the form of an outer membrane and so can survive better in the intestine.

23.3 How might you provide evidence that the intestinal microbiome communicates with the brain via the vagus nerve? *Hint:* The vagus nerve includes afferent nerves that send information from organs to the brain and efferent nerves that regulate gastrointestinal secretion and gut endocrine activity.

ANSWER: One way to determine whether gut microbiota communicate with the brain via the vagus nerve is to selectively cut the afferent parts of the vagus nerve that enervate the intestine, in both gnotobiotic mice and colonized mice. A look at gene expression in the mouse brains would reveal differences between colonized mice that were and were not vagotomized, but not between similarly treated gnotobiotic mice.

23.4 Figure 23.20 shows how a neutrophil extracellular trap can ensnare a nearby pathogen. Which bacterial structure might blunt the microbicidal effect of NETs?

ANSWER: Bacterial capsules can prevent direct contact between the NET and the cell.

23.5 Why do defensins have to be so small? Do defensins kill normal microbiota?

ANSWER: Defensins need to be small so that they can penetrate the outer membrane of Gram-negative organisms and the thick maze of peptidoglycan comprising the Gram-positive cell wall. Defensins do kill normal microbiota and, in fact, are part of what keeps levels of the normal intestinal microbiota in check. Research suggests that a decrease in intestinal defensin production can lead to an imbalance of gastrointestinal microbes in both number and species. This imbalance appears to contribute to conditions such as irritable bowel syndrome (IBS) and inflammatory bowel disease (IBD). Pathogens or normal microbiota that penetrate the intestinal mucosa probably encounter higher concentrations of defensins as they do so.

23.6 As illustrated in Figure 23.34, integrin is important for neutrophil extravasation. Some individuals, however, produce neutrophils that lack integrin. What is the likely consequence of this genetic disorder?

ANSWER: Individuals with neutrophils that lack integrin have leukocyte adhesion deficiency. Their neutrophils are defective in extravasation. These patients are more susceptible to infections because the neutrophils cannot easily get out of the bloodstream.

23.7 What happens to all the neutrophils that enter a site of infection once the infection has resolved?

ANSWER: Once bacteria at the site of infection have been killed, the tissue cells in the area stop making the cytokines and chemokines that attracted neutrophils in the first place, so neutrophils stop coming in, and some wander out. But the majority of neutrophils undergo a self-programmed cell death (called apoptosis) and are cleared by monocytes in the area through phagocytosis. The average life span of a neutrophil is only 5 days.

23.8 If NK cells can attack infected host cells coated with antibody, can neutrophils do the same?

ANSWER: Neutrophils (PMNs) can attack infected cells coated with antibody, but the killing mechanism is different from that of antibody-dependent cell-mediated cytotoxicity (ADCC). Human neutrophils do not make perforin or the other ADCC-related compounds, called granzymes, used by NK cells to kill target cells. In addition to that difference, NK cells possess a type of Fc receptor not found on neutrophils, which means the intracellular signaling pathways are different between NK cells and neutrophils. Neutrophils can be activated, however, when their Fc receptors bind antibody. Activated neutrophils make reactive oxygen products and can release a variety of peptides, including defensins, cathelicidins, and myeloperoxidase, which can all damage target cells.

23.9 Figure 23.39 shows how the complement cascade can destroy a bacterial cell. Factor H (not shown) is a blood protein that regulates complement activity. Factor H binds host cells and inhibits complement from attacking our cells by accelerating degradation of C3b and C3bBb. How could bacteria take advantage of factor H?

ANSWER: Some pathogens have structures on their cell surfaces that can bind factor H. This binding inhibits the alternative pathway from attacking the microbe.

23.10 If increased fever limits bacterial growth, why do bacteria make pyrogenic toxins?

ANSWER: The pyrogenic toxins have other effects that compromise and damage the host. The toxins can induce cytokines that damage local host cells (helping to provide the pathogen with nutrients) or confuse the immune system (allowing the pathogen to delay detection). Pyrogenic toxins include lipopolysaccharide and protein toxins such as toxic shock syndrome toxin (see Chapter 25).

CHAPTER 24

24.1 Two different stretches of amino acids in a single protein form a 3D antigenic determinant. Will the specific immune response to that 3D antigen also recognize the same two amino acid stretches if they are removed from the whole protein?

ANSWER: Most likely no. It is the 3D shape formed by the two stretches that is recognized as an antigen. Separating the two amino acid stretches from the whole protein will alter their 3D shape, making them "invisible" to antibodies that recognize their 3D shape within the protein. However, other specific immune responses involving different subsets of lymphocytes can recognize the separate amino acid stretches that together form the 3D antigenic determinant. As an analogy, take a computer image of a friend's face and shuffle the facial features. Turn the nose upside down, exchange the eyes with the mouth, and lower the ears. Since you are programmed to respond to the original facial configuration, you likely would not recognize the rearranged face as a whole. But you might find that the nose looks familiar.

24.2 How does a neutralizing antibody that recognizes a viral coat protein prevent infection by the associated virus?

ANSWER: Neutralizing antibodies usually bind attachment proteins on the virus and sterically prevent them from binding to host cell receptors (see **Fig. 24.5**). Unable to attach, the virus cannot enter a cell. Some antibodies of enveloped viruses might trigger the complement cascade (see Section 24.4), thus destroying the virus membrane. The damaged virus membrane stops the virus from entering the host via membrane fusion.

24.3 The attachment proteins of different rhinovirus strains bind to ICAM-1. How can all these proteins be immunologically different if they find the same target (ICAM-1)? Why won't antibodies directed against one rhinovirus strain block the attachment of other rhinovirus strains?

ANSWER: The ICAM-1-binding sites on different rhinovirus strains have small, but immunologically significant, differences. ICAM-1 is "promiscuous" in its binding specificity toward the rhinovirus attachment proteins, whereas antibodies are more specific. Thus, ICAM-1 protein is like a master key that can fit dozens of different locks (different rhinovirus-binding sites). Each lock (virus-binding site) also has a very specific key (antibody) that won't unlock any of the other locks. But the master key (ICAM-1) can turn all locks.

24.4 Can an F(ab')$_2$ antibody fragment prevent the binding of rhinovirus to the ICAM-1 receptor on host cells? And can an F(ab')$_2$ antibody fragment facilitate phagocytosis of a microbe?

ANSWER: An F(ab')$_2$ antibody fragment can prevent binding of rhinovirus to the ICAM-1 receptor on host cell surfaces. The antigen-binding sites will block virus receptor access to ICAM-1. However, an F(ab')$_2$ antibody fragment cannot facilitate phagocytosis of a microbe, because an opsonizing antibody needs its Fc region to bind Fc receptors on phagocytes. Thus, an F(ab')$_2$ antibody can bind to the microbe antigen but cannot link the microbe to a phagocyte's cell surface.

24.5 (refer to **Fig. 24.9**) Because antibodies are proteins, they are also antigens and can stimulate an immune response. What types of antibodies—anti-isotype, anti-allotype, or anti-idiotype—will IgG taken from Sherrie raise when injected into John?

ANSWER: Since they are both human, Sherrie's IgG will not elicit anti-isotype antibodies in John. Sherrie's IgG will elicit anti-allotype antibodies and anti-idiotype antibodies in John. Thus, John will develop antibodies that react only to the amino acid differences between John's and Sherrie's IgG antibodies.

24.6 The mother of a newborn was found to be infected with rubella, a viral disease. Infection of the fetus could lead to serious consequences for the newborn. How could you determine whether the newborn was infected in utero?

ANSWER: Since maternal IgM antibodies cannot cross the placenta into the fetus, finding IgM antibodies to rubella antigens in the newborn's circulation indicates that the fetus was infected and initiated its own immune response. If the newborn has only IgG antibodies to rubella (no IgM antibodies), then the child was not infected and maternal IgG crossed the placenta.

24.7 B cells in early stages have both IgM and IgD surface antibodies, but the delta region has no switch region. Why does the delta region have no switch region?

ANSWER: B cells in early stages have both IgM and IgD surface antibodies. If the delta region had a switch region, then the B cell could make IgD only after DNA rearrangement. Then, no single B cell could have IgM and IgD at the same time. Recombination at the DNA level is not involved. Alternative RNA-splicing events after transcription determine whether an IgM or IgD molecule is made.

24.8 Why do individuals with type A blood have anti-B and not anti-A antibodies?

ANSWER: A type A individual does not make anti-A antibodies because self A antigens present during B cell development will trigger the deletion of anti-A antibody-producing B cells. The type A person does not make B antigen, anti-B antibody-producing B cells are not deleted. Those cells can be stimulated to make anti-B antibodies.

24.9 What would happen to someone lacking CD154 on T$_{FH}$ cells because of a gene mutation?

ANSWER: The person's T-cell and B-cell numbers would remain normal, but the B cells would not undergo heavy-chain class switching. Thus, any plasma cells produced would make only IgM, causing serum levels of IgM to rise. No other antibody type would be secreted; the result is called hyper IgM syndrome.

24.10 How can a stem cell be differentiated from a B cell at the level of DNA?

ANSWER: DNA recombination at switch regions will have taken place in the B cell but not the stem cell. Thus, the B cell will have fewer segments for each of the V, D, and J regions, while the stem cell will have all of them. PCR techniques can be used to view these differences.

24.11 Transplant rejection is a major consideration in the transplantation of most tissues, because host T$_C$ cells can recognize allotypic MHC on donor cells. Why, then, are corneas easily transplanted from a donor to just about any other person?

ANSWER: The cornea is not normally vascularized. So, even though corneal cells express MHC proteins, circulating host T cells do not have an opportunity to interact with them. The cornea will not be rejected. The cornea is thus referred to as an "immune-privileged site."

24.12 Why does attaching a hapten to a carrier protein enable antihapten antibodies to be produced?

ANSWER: B cells with antihapten surface antibody (as part of the B-cell receptor) can take up hapten but cannot present the hapten to a helper T cell. The same B cell can also take up the hapten bound to a carrier molecule, and because the carrier molecule is larger than the hapten, the B cell will present a part of the carrier epitope carrying the hapten to the helper T cell. The helper T cell stimulates the B cell, which was already programmed to make antihapten antibody, to differentiate into plasma cells and memory B cells.

24.13 Why do immunizations lose their effectiveness over time?

ANSWER: Immunizations gradually become less effective because memory B cells eventually die. Without some exposure to antigen, those memory cells will not be replaced and the antibody already made will turn over within weeks.

24.14 How might you design/construct a more effective vaccine for an antigen (for instance, the *Yersinia pestis* F1 antigen, or hepatitis A) that would harness the power of a Toll-like receptor (TLR)? *Hint:* Look at TLR5 in Table 23.3.**

ANSWER: By genetically fusing DNA encoding a hepatitis antigen to DNA encoding flagellin, a MAMP (microbe-associated molecular pattern) that is recognized by TLR5 on dendritic cells, you would end up with a chimeric protein that would activate innate immune mechanisms, via TLR5, and improve antigen presentation. The result is a better link between innate and adaptive immune mechanisms.

CHAPTER 25

25.1 Figure 25.5 presents the association between LD_{50} and virulence. What does this figure tell you about infectivity?

ANSWER: It doesn't tell you much. Strain 1 is clearly more lethal (virulent) than strain 2, but strain 2 might still cause non-lethal symptoms of disease at a much lower dose.

25.2 Figure 25.20 illustrates how cholera toxin works to cause diarrhea. To develop a vaccine that generates protective antibodies, which subunit of cholera toxin should be used to best protect a person from the toxin's effects?

ANSWER: Antibodies to the B subunit will be more protective. Inactivating the B subunit will prevent binding of the toxin to cell membranes. The A-subunit active site is typically sequestered in these toxins and inaccessible to antibody. Furthermore, once the A subunit has entered a host cell, antibodies cannot enter and neutralize it.

25.3 How might you experimentally determine whether a pathogen secretes an exotoxin? (*Hint:* You can learn more about how new toxins are found in eTopic 25.7.)

ANSWER: The microbe can be grown in liquid culture and the cells removed by either centrifugation or filtration. If the organism makes an exotoxin, the toxin may well be present in the cell-free supernatant. We can determine the presence of a toxin by injecting the supernatant into an animal model (for example, mice) and examining the result (death or altered function). Alternatively, we can administer the supernatant to a layer of tissue culture cells and record the health of the monolayer.

25.4 Would patients with iron overload (excess free iron in the blood) be more susceptible to infection?

ANSWER: With a few exceptions, withholding iron from potential pathogens is a host defense strategy, because when iron is plentiful, the microbe does not have to expend energy to get it and so can readily grow. On the other hand, low iron can also be a signal to express various virulence genes, so for some organisms, high iron might hinder infection.

25.5 Use an Internet search engine to determine which other toxins are related to the cholera toxin A subunit. (*Hint:* Start by searching on "protein" at PubMed to find the protein sequence; then use a BLAST program.)

ANSWER: Go to PubMed (http://www.ncbi.nlm.nih.gov/pubmed). At the top of the page, select **Protein** from the Search dropdown list. Then type "cholera enterotoxin A subunit" in the text box and click **Search**. Click on the appropriate result (any designated "258 aa protein"). Highlight and copy the entire amino acid sequence at the bottom of the page (the entire string of numbers and letters under "ORIGIN"). Then, at the top of the same page, on the right-hand side under "Analyze this sequence", click on **Run BLAST**. Paste the A-subunit sequence into the box, and click on **BLAST** at the bottom of the page. After some seconds, all the homology results will be displayed. The toxin of *Escherichia coli* possesses an A subunit that is about 80% identical to that of cholera.

25.6 Protein and DNA have very different structures. Why would a protein secretion system be derived from a DNA-pumping system? (*Hint:* Review conjugation in Chapter 9.)

ANSWER: Conjugation systems actually move DNA that is attached to a pilot protein at the 5′ end. A pilot protein is made by the conjugation system and then binds to the 5′ end of the DNA to be transferred and "pilots" the DNA through the conjugation structure. So, a modified conjugation system that moves only protein is not as much of a leap as might initially be thought.

25.7 How can you determine whether a bacterium is an intracellular pathogen?

ANSWER: Microscopic examination to see whether bacteria are found within cultured mammalian cells is usually not satisfactory. The difficulty lies in determining whether the organism is inside the host cell or just bound to its surface—or, if it is inside, whether it is a live or dead bacterium. One commonly used approach is to take an infected cell monolayer and add an antibiotic that can kill the microbe but will not penetrate the mammalian cells. The protein synthesis inhibitor gentamicin is typically used. A bacterium that invades a host cell will gain sanctuary from gentamicin and grow intracellularly. Extracellular bacteria, as well as bacteria attached to the outside of the host cell, are killed. Counting viable colony-forming units of bacteria released from the mammalian cells by gentle detergent treatments at various times will reveal whether the organism grew intracellularly. However, this method works only if the microorganism is not an obligate intracellular pathogen. If viable cells do not plate, you could use reverse-transcription PCR to identify RNA in bacteria that are obligate intracellular pathogens.

25.8 Figure 25.29B shows *Shigella* forming an actin tail at one pole. Why do organisms such as *Shigella* and *Listeria* assemble actin-polymerizing proteins at only one pole?

ANSWER: If actin tails formed at both poles, the organism would spin aimlessly throughout the cell and would likely not be able to protrude into and infect an adjacent host cell. The organisms have mechanisms that localize the key bacterial actin polymerization protein at only one pole (the older pole) of the cell. The molecular basis for this unipolar localization remains unclear but is probably related in some way to polar aging (discussed in Chapter 3).

25.9 Why might killing a host be a bad strategy for a pathogen?

ANSWER: The goal of any microbe is to maintain its species. If a microbe did not have an opportunity to easily spread to a new host, killing its host would be tantamount to suicide.

25.10 Can antibodies against a viral E3 ligase stop an infection by that virus? Why or why not? What might be an alternative strategy that does not involve antibodies?

ANSWER: Because antibodies cannot penetrate infected (or uninfected) host cells, circulating antibodies cannot bind to an intracellularly made viral E3 ligase. In addition, viruses circulating in the blood will not carry the enzyme. An alternative strategy might be to design and use a chemical inhibitor that can penetrate the host membrane and bind to the active site of the enzyme. You would have to be careful that the inhibitor doesn't interfere with normal host cell E3 ligases.

25.11 How could you use the FRET–beta-lactamase technique to identify chemical inhibitors of type III, IV, or VI translocation?

ANSWER: Microtiter plate wells containing human cell cultures can be used to screen a battery of potential inhibitor compounds collected from the environment or made by rational design in the laboratory. A pathogen capable of translocating a beta-lactamase–effector fusion protein into host cells would be added to the wells in the presence or absence of the potential inhibitor. After incubation you would add the fluorescent probe and scan for fluorescence. Wells in which host cells did not fluoresce would have failed to receive the beta-lactamase–effector fusion protein, suggesting that the compound added to that well may have inhibited the bacterial secretion system. You should add controls to test whether the test compound affects pathogen or host cell viability, or beta-lactamase activity.

CHAPTER 26

26.1 Does *Staphylococcus aureus* have to disseminate through the circulation to produce the symptoms of scalded skin syndrome (SSS)? Explain why or why not.

ANSWER: Because SSS symptoms are caused by an exotoxin, the infection can be in a single location [a focus of infection, such as the nares (nostrils)], but the toxin can disseminate through the circulatory system.

26.2 Why would treatment of some infections require multiple antibiotics?

ANSWER: Although not typically recommended for simple infections, with some organisms multiple-antibiotic therapy is required to effect a cure. The organisms that require multidrug therapy are very hardy in vivo and able to grow in the presence of single antibiotics. Active infection with *Mycobacterium tuberculosis*, for example, is typically treated with isoniazid (thought to inhibit the synthesis of mycolic acid) and rifampin (an RNA polymerase inhibitor), and sometimes a third drug, pyrazinamide (which inhibits fatty acid synthesis). Using multiple antibiotics minimizes the chance that the organism will develop resistance to any one antibiotic. Multiple-antibiotic therapy is also recommended for *Helicobacter pylori* (ulcers, treated with lansoprazole, amoxicillin, and clarithromycin) and methicillin-resistant *Staphylococcus epidermidis* (MRSE; treated with vancomycin and gentamicin). MRSE can cause serious infections, such as meningitis.

26.3 Why isn't a dimorphic fungus like histoplasa easily transmitted from person to person via respiratory droplets?

ANSWER: The infectious forms of *Histoplasma capsulatum*, and other dimorphic fungi, are the microconidia produced by mycelial forms of the organism that grow around 25°C, but not at body temperature. As a result, microconidia are not produced in the lung. Although yeast forms of the organism could be present in respiratory aerosols, they will not be numerous enough or travel deep enough into the lung to cause disease in a potential victim.

26.4 Explain why patient noncompliance (failure to take drugs as directed) is thought to have led to XDR-TB.

ANSWER: In most cases, an organism causing an infection starts out being susceptible to a given antimicrobial agent. However, the more times an organism divides, the more likely it is that a spontaneous mutation producing drug resistance will arise. The amount of the antibiotic and the duration of its use are calibrated so that all organisms are killed—either by the drug itself or by the immune system. Often the drug is given to stop growth of the bacterium so that the immune system has enough time to actually do the killing. If a patient stops taking an antibiotic before the prescribed end of the treatment, the organism can start to replicate again, renewing the chance that a spontaneous drug-resistant variant will be produced. Even if drug therapy is resumed, the resistant organism will continue to grow and cause disease—and can be transmitted to other individuals.

26.5 Why do you think most urinary tract infections occur in women?

ANSWER: The major reason is anatomy. Most bladder UTIs come from access through the urethra. Since the urethra in men is longer than in women, it usually takes a catheter to introduce bacteria into a male bladder. The trip for the infecting organism in women is much shorter. In people older than 50, however, UTIs become more common in both men and women, with less difference between the sexes. The reason is as yet unclear.

26.6 Urine samples collected from six hospital patients were placed on a table at the nurses' station awaiting pickup from the microbiology lab. Several hours later, a courier retrieved the samples and transported them to the lab. The next day, the lab reported that four of the six patients had UTIs. Would you consider these results reliable? Would you start treatment based on these results?

ANSWER: Because urine is a good growth medium for many bacteria, the delay of several hours in picking up the samples gave the organisms time to replicate and increase their numbers. Consequently, the lab results should be viewed with suspicion.

26.7 Aside from the CDC guidelines for treating gonorrhea, why else do you suppose the patient in this case was treated with doxycycline (a derivative of tetracycline)?

ANSWER: Because STIs often travel in pairs and because the initial symptomology is similar, the clinician will want to "cover" the patient for possible chlamydia infection. Tetracycline (or azithromycin) will treat chlamydia infections. Chlamydias are not susceptible to ceftriaxone. The large, single dose of ceftriaxone (as opposed to smaller multiple injections given over days) was given because patients with gonorrhea are typically poorly compliant and fail to return for subsequent injections.

26.8 Like the cause of plague, HIV is a blood-borne pathogen. Why, then, do you think fleas and mosquitoes fail to transmit HIV?

ANSWER: When insect vectors take a blood meal, they typically defecate or regurgitate simultaneously. So, theoretically, they could serve as a vector for HIV. Pathogens, such as the West Nile virus, that are transmitted by insect vectors actually grow in their insect hosts; however, HIV does not. Because it is unusually fragile, HIV dies too quickly. There have been no cases of HIV transmitted by insect vectors.

26.9 A patient presenting with high fever and in an extremely weakened state is suspected of having septicemia. Two sets of blood cultures are taken from different arms. One bottle from each set grows *Staphylococcus aureus*, yet the laboratory report states that the results are inconclusive. New blood cultures are ordered. What would make these results inconclusive?

ANSWER: There must be something different about the two strains of staphylococci grown in the separate bottles. For example, they may have different antibiotic susceptibility patterns when tested against a battery of antibiotics. One strain may be susceptible to penicillin, while the other strain is resistant. Since the expectation is that a single strain initiated the infection, both isolates should exhibit the same susceptibility pattern. The laboratory suspects contamination of the bottles from separate sources. These organisms are not the source of infection.

26.10 Normal cerebrospinal fluid is usually low in protein and high in glucose. The protein and glucose content does not change much during a case of viral meningitis, but bacterial infection leads to greatly elevated protein and lowered glucose levels. What could account for this difference?

ANSWER: There are several explanations. Bacteria and infiltrating PMNs will consume glucose, and alterations in the blood-brain barrier can lead to decreased transport of glucose into the spinal fluid. As a result, glucose levels plummet. Protein levels increase during bacterial meningitis because inflammatory processes weaken the blood-brain barrier and allow protein to leak from blood into the subarachnoid space. Growth of the bacteria and infiltration of PMNs may also contribute to the increase in CSF protein levels. Viruses do not produce as much inflammation, so the glucose and protein permeability barriers are maintained. In addition, viral meningitis does not cause a great infiltration of PMNs into the CSF—another reason glucose levels remain high and protein levels remain low.

26.11 Given the symptoms of tetanus, what kind of therapy would you use to treat the disease?

ANSWER: At the first sign of muscle spasm (tetany), antitoxin should be given. If tetany is severe, then muscle relaxants can relieve spasms.

26.12 Figure 26.33 demonstrates that tetanus toxin has a mode of action (spastic paralysis) opposite to that of botulism toxin (flaccid paralysis). Since they have opposing modes of action, can botulism toxin be used to save a patient with tetanus?

ANSWER: When you first think about it, this sounds feasible. Tetanus toxin causes spastic paralysis, while botulinum toxin causes flaccid paralysis. However, the two toxins act at different places in the nervous system. Tetanus toxin produced by *Clostridium tetani* growing in a wound travels to the central nervous system to exert its effect, whereas botulism toxin works at the periphery. If you administer botulinum toxin intravenously to counteract the whole-body effect of tetanus toxin, then it will affect essentially all neuronal junctions, autonomic and voluntary alike. Thus, the dose of botulinum toxin required to counteract the effect of tetanus toxin on voluntary muscle contraction would likely kill the patient by stopping his/her breathing.

26.13 How do the actions of tetanus toxin and botulism toxin actually help the bacteria colonize or obtain nutrients?

ANSWER: This is a difficult question to answer. Few scientists have speculated about it. Recall that the toxins are encoded by genes in resident bacteriophages that became part of the clostridial genome through horizontal transfer from some other source. Since the organisms (vegetative, as well as spores) normally reside in soil, the actual function of these toxins may have something to do with survival in that habitat. The toxin's effect on humans may simply be an unfortunate accident. With tetanus toxin, however, there may be benefit in that muscle spasms could limit oxygen delivery to infected tissues, enabling a more anaerobic environment for growth. Cell death may also release iron or other nutrients useful to *Clostridium tetani*.

26.14 If *Clostridium botulinum* is an anaerobe, how might botulism toxin get into foods?

ANSWER: The most common way botulism toxin gets into food today is via home canning. The canning process involves heating food in jars to very high temperatures. The heat destroys microorganisms and drives out oxygen; both processes help to preserve foods. If the jars are not heated to sterilization conditions, spores of *C. botulinum* will survive. When the jars are cooled for storage at room temperature, the spores germinate; the organism then grows in the anaerobic medium and releases the toxin. When the food is eaten, the toxin is eaten too.

CHAPTER 27

27.1 Figure 27.4 illustrates how MICs are determined. Test your understanding of how MICs are measured in the following example. The drug tobramycin is added to a concentration of 1,000 μg/ml in a tube of broth from which serial twofold dilutions are made. Including the initial tube (tube 1), there are a total of ten tubes. Twenty-four hours after all the tubes are inoculated with *Listeria monocytogenes*, turbidity is observed in tubes 6–10. What is the MIC?

ANSWER: The MIC is 62.5 μg/ml, the concentration in tube 5, the last tube with no growth. (Relative to tube 1, tube 5 has been diluted 2^4 times—a 16-fold dilution: 1,000 → 500 → 250 → 125 → 62.5.)

27.2 What additional test performed on an MIC series of tubes will tell you whether a drug is bacteriostatic or bactericidal?

ANSWER: The appropriate test is to streak a portion of the broth from the dilution tubes that show no growth onto an agar plate. If the drug is bacteriostatic, colonies will form on the agar plate because during streaking, the bacteria are removed from the presence of the drug. If the antibiotic is bactericidal, no colonies will form, because the organisms are dead before plating. This method determines the minimum bactericidal concentration (MBC) of an antibiotic. The MBC is the lowest dilution that does not yield viable cells.

27.3 A patient with a bacterial lung infection was given the antibiotic represented in Figure 27.7 and was told to take one pill twice a day. The pathogen is susceptible to this drug. Will the prescribed treatment be effective? Explain your answer.

ANSWER: No. **Figure 27.7** shows that the antibiotic remains at effective serum levels for about 8 hours. If the patient takes two pills 12 hours apart, there will be two 4-hour periods (a total of 8 hours) during each 24-hour interval when serum levels fall below an effective MIC concentration. Each 4-hour window of low concentration will allow regrowth and a greater opportunity for antibiotic-resistant strains to develop.

27.4 You are testing whether a new antibiotic will be a good treatment choice for a patient with a staph infection. The Kirby-Bauer test using the organism from the patient shows a zone of inhibition of 15 mm around the disk containing this drug. Clearly, the organism is being affected by the drug in vitro. But you conclude from other studies that the drug would be ineffective in the patient. What would make you draw this conclusion?

ANSWER: If the average attainable tissue level of the drug is below the MIC, then the drug will be ineffective.

27.5 When treating a patient for an infection, why would combining a drug such as erythromycin with a penicillin be counterproductive? (Erythromycin is described in Section 8.3.)

ANSWER: Erythromycin, a bacteriostatic drug, will stop growth, which indirectly stops cell wall synthesis and renders the microbe insensitive to penicillin.

27.6 Given the mechanism by which rifampin stops transcription, what limitation does the drug have? Consider initiation versus elongation.

ANSWER: Rifampin binds to the exit channel of RNA polymerase, a length equal to about 14 nucleotides of RNA from the polymerase active center. If the nascent RNA transcript is longer than that, the nascent RNA blocks the rifampin-binding site in the exit channel. So, rifampin will prevent only the initiation of transcription. An RNA polymerase that has transcribed more than 14 bases is no longer susceptible to rifampin.

27.7 Why might a combination therapy of an aminoglycoside antibiotic and cephalosporin be synergistic?

ANSWER: The two drugs given together could act synergistically because the cephalosporin can weaken the cell wall and allow the aminoglycoside easier access to the cell interior, where it can attack ribosomes. This synergism is especially useful in organisms that have some resistance to both drugs.

27.8 Fusaric acid is a cation chelator that normally does not penetrate the *E. coli* membrane, which means *E. coli* is typically resistant to this compound. Curiously, cells that develop resistance to tetracycline become sensitive to fusaric acid. Resistance to tetracycline is usually the result of an integral membrane efflux pump that pumps tetracycline out of the cell. What might explain the development of fusaric acid sensitivity?

ANSWER: Fusaric acid is imported by the tetracycline efflux pump. This phenomenon can be used to isolate mutants with deletions of transposons encoding tetracycline resistance. Transposons, such as Tn*10*, which carries tetracycline resistance, can spontaneously delete at a frequency of about 10^{-6}, but finding one out of a million tetracycline-susceptible cells is impossible without a positive selection. Fusaric acid provides that positive selection, since the cell with the Tn*10* deletion will be resistant to fusaric acid.

27.9 Mutations in the ribosomal protein S12 (encoded by *rpsL*) confer resistance to streptomycin. Would a cell containing both *rpsL*$^+$ and *rpsL*^{R} genes be streptomycin resistant or sensitive? (Recall that genes encoding ribosome proteins for the small subunit are designated *rps*. "+" indicates the wild-type allele, while "R" indicates a gene whose product is resistant to a certain drug.)

ANSWER: This merodiploid cell would contain two sets of ribosomes: One set, containing normal S12, would be sensitive to streptomycin; another set, containing the resistant S12, would be resistant. Because streptomycin causes mistranslation of mRNA on sensitive ribosomes, inappropriate proteins that can kill the cell would still be synthesized. Thus, the cell would remain sensitive to streptomycin. Note, however, that the recessive nature of antibiotic resistance seen in this case is not the norm. Resistance is a dominant trait in the majority of cases.

27.10 A clinician admits a seriously ill patient with sepsis to the hospital. She treats the patient empirically with piperacillin and vancomycin until the lab identifies the infectious agent. The next day the agent is identified as *Escherichia coli* sensitive to third-generation cephalosporins. What should the clinician do now? Why were piperacillin and vancomycin used initially?

ANSWER: Proper antibiotic stewardship would have the clinician immediately discontinue vancomycin treatment (which was used in case of Gram-positive pathogens, including MRSA) and switch the piperacillin (a broad-spectrum beta-lactam antibiotic) to ceftriaxone, a somewhat more narrow-spectrum third-generation cephalosporin. Initially, the clinician did not know the cause of the infection, so empiric therapy had to cover all possibilities.

27.11 Figure 27.27 shows a colony with a distinctive morphology. Why are the colonies on this agar plate red and mucoid?

ANSWER: MacConkey medium contains lactose. As described in Chapter 4, an organism that ferments lactose acidifies the medium and takes up neutral red, turning the colonies red. Organisms that produce a large capsule form slimy, mucoid colonies. *Klebsiella pneumoniae* is one such bacterium that also ferments lactose. A well-trained laboratory technician will immediately suspect *K. pneumoniae* upon seeing these colonies.

27.12 Could genomics ever predict the drug resistance phenotype of a microbe? If so, how?

ANSWER: Yes. If the organism's genome possesses genes whose deduced protein sequences harbor significant similarity to antibiotic resistance proteins from other organisms, then one can predict a similar drug resistance. Definitive proof of drug resistance requires actual in vitro testing.

27.13 The text states that cells and viruses would have difficulty developing resistance to antiviral drugs that target host proteins rather than viral proteins. Why would targeting a host protein decrease the likelihood of developing resistance? Propose a mechanism by which the drug could still become ineffective against a virus.

ANSWER: The drug does not bind to a viral protein, so typical drug resistance mutations used by viruses to prevent drug binding are impossible. Likewise, incorporation of a gene in the virus genome that could inactivate the drug is unlikely because the virus would first have to replicate using the targeted host protein (which would be inactivated by the presence of the drug) in order to produce the drug-inactivating enzyme. A mutation in the host that could reduce host protein affinity for the drug is also unlikely because such a mutation would provide a selective advantage only to the virus, and not to the host. So, what kind of viral mutation could provide the virus with drug resistance against a drug that targets a host protein? The virus could conceivably mutate so that it no longer needed the original host target protein, perhaps using a different host protein. You might come up with other plausible scenarios.

CHAPTER 28

28.1 Two blood cultures, one from each arm, were taken from a patient with high fever. One culture grew *Staphylococcus epidermidis*, but the other blood culture was negative (no organisms grew out). Is the patient suffering from septicemia caused by *S. epidermidis*?

ANSWER: Probably not. *S. epidermidis* is a common inhabitant of the skin and could easily have contaminated the needle when blood was taken from the patient. The fact that only one of the two cultures grew this organism supports this conclusion. If the patient had really been infected with *S. epidermidis*, both blood cultures would have grown this organism.

28.2 A 30-year-old woman with abdominal pain went to her physician. After examining the patient, the doctor asked her to collect a midstream urine sample that would be sent to the lab across town for analysis. The woman complied and handed the standard urine collection cup to the nurse. The nurse placed the cup on a table at the nurses' station. Three hours later, a courier service picked up the specimen and transported it to the laboratory. The next day the report came back: "Greater than 200,000 CFUs/ml; multiple colony types; sample unsuitable for analysis." Why was this determination made?

ANSWER: Although the CFU number is high enough to consider relevant, UTIs are typically caused by a single organism. The fact that the lab found many different colony types suggests a problem with specimen collection. In this case, lack of refrigeration allowed the small number of urethral contaminants in the sample to overgrow the specimen. When examined quickly after collection, urine obtained using standard urine collection cups, as in this example, is suitable for bacterial culture and sensitivity testing and for analytical dipsticks that test for a variety of metabolites in urine. Today, urine for culture and sensitivity is typically collected in cups that include a bacteriostatic preservative such as boric acid. Viability of bacteria is maintained for about 24 hours in these cups, but the preservative will interfere with dipstick results.

28.3 Use Figure 28.11 to identify the organism from the following case: A sample was taken from a boil located on the arm of a 62-year-old man. Bacteriological examination revealed the presence of Gram-positive cocci that were also catalase-positive, coagulase-positive, and novobiocin resistant.

ANSWER: The organism is *Staphylococcus aureus*. The novobiocin test is irrelevant in this situation.

28.4 Why didn't immunological tests initially identify antibodies to *Leptospira* in the comatose boy?

ANSWER: The boy had SCID, so his immunocompromised state likely prevented him from making detectible amounts of anti-*Leptospira* antibodies. DNA, then, was the only way to identify the pathogen. MALDI-TOF MS probably would work if he had a high enough number of spirochetes in his CSF and if the organism's profile were in the database.

28.5 Why does finding IgM to West Nile virus indicate a current infection? Why wouldn't finding IgG indicate the same?

ANSWER: Upon infection with any organism, IgM antibodies are the first to rise. After a short time, the levels of IgM decline as IgG levels rise. IgG, however, can remain in serum for years, making it a poor indicator of current infection. A serum sample that is IgG-positive but IgM-negative for a specific agent probably reflects a past infection.

28.6 If the results in Figure 28.16 came from testing for HIV RNA, then which patient would have the higher viral load in their blood?

ANSWER: Patient 1. The detection threshold was crossed earlier for patient 1 (cycle 10) than for patient 2 (cycle 22), so patient 1's serum must have a higher number of HIV virus particles in it.

28.7 Why does adding albumin or powdered milk prevent false positives in ELISA?

ANSWER: Antibodies are proteins. They can stick to plastic just as easily as to the antigen being tested. If all the possible binding sites on plastic were not blocked with albumin (a major protein ingredient in milk), any antibody from the patient's serum could stick to the plastic instead of to the antigen and react with the secondary enzyme-conjugated antihuman antibody.

28.8 Specific IgG antibodies against an infectious agent can persist for years in the bloodstream, long after the infection resolves. How is it possible, then, that IgG antibody titers can be used to diagnose recently acquired diseases such as

infectious mononucleosis? Couldn't the antibody be from an old infection?

ANSWER: During the course of a disease, the body's immune system increases the amount of antibody made specifically against the infectious agent. Thus, one can compare the IgG antibody titer in a blood sample taken from a patient in the active, or acute, phase of disease with the IgG antibody titer several weeks later, when the patient is in the recovery, or convalescent, stage. Seeing a greater-than-fourfold rise in a specific antibody titer (for example, in mononucleosis) indicates that the patient's immune system was responding to the specific agent. Remember, simply finding IgG against an organism or virus in serum indicates only that the patient was exposed to that microbe at some time in the past.

28.9 Methicillin, a beta-lactam antibiotic, is very useful in treating staphylococcal infections. The emergence of methicillin-resistant strains of *Staphylococcus aureus* (MRSA) is a very serious development because few antibiotics can kill these strains. Imagine a large metropolitan hospital in which there have been eight serious nosocomial infections with MRSA and you are responsible for determining the source of infection so that it can be eliminated. How would you accomplish this task using common bacteriological and molecular techniques?

ANSWER: Samples from all the affected patients and from the hospital staff would be screened for the presence of *S. aureus* resistant to methicillin. Each strain would then be rapidly sequenced in part (multilocus sequencing) or in whole (whole-genome sequencing), using the techniques described in Chapter 12. Strains from all the patients would likely have identical restriction patterns or DNA sequences if they came from the same source. The source would then be identified by determining which staff member possesses MRSA with the same pattern. The source might also be inanimate, such as surgical equipment or ventilation apparatus. A connection between patients and specific staff members or instruments would also have to be demonstrated.

28.10 What are some reasons why some diseases spread quickly through a population while others take a long time?

ANSWER: There are several factors. One is mode of transmission; airborne diseases can spread more quickly than food-borne diseases, for instance. Sexually transmitted infections spread more slowly still. Herd immunity is another factor. Herd immunity is based on the number of individuals within a population that are resistant to a disease. Someone immune to the disease cannot pass it on. The more immune people there are in a population (or herd), the slower the epidemic spreads to susceptible people. Herd immunity can be achieved if at least 75% of people in a population are vaccinated against a disease.

28.11 On the ProMED-mail web page (www.promedmail.org), click on the interactive world map to view outbreaks recorded by WHO. What outbreaks happened throughout the world during the current year?

ANSWER: As of this writing (March 2019), ProMED had reported outbreaks of Lassa fever in West Africa (involving over 1,800 cases that included 16 health care workers); of rabies in South Carolina (32 cases); and of measles (a large outbreak, 426 cases, that had spread over 15 states).

GLOSSARY

A

A site See **acceptor site**.
ABC transporter An ATP-powered transport system that contains an ATP-binding cassette.
aberration An imperfection in a lens.
abiotic Produced without living organisms; occurring in the absence of life.
absorption In optics, the capacity of a material to absorb light.
acceptor site (A site) The region of a ribosome that binds an incoming charged tRNA.
accessory protein A protein found in the viral capsid or tegument that is needed early in the viral life cycle.
acid-fast stain A diagnostic stain for mycobacteria, which retain the dye fuchsin because of mycolic acids in the cell wall.
Acidobacteria A phylum of Gram-negative bacteria with an outer membrane, related to the Proteobacteria; often found in soil habitats.
acidophile An organism that grows fastest in acid (generally defined as below pH 5).
ACP See **acyl carrier protein**.
acquired immunodeficiency syndrome (AIDS) A disease caused by HIV that leads to the destruction of T cells and the inability to fight off opportunistic infections.
actin filament See **microfilament**.
Actinobacteria A phylum of Gram-positive bacteria with high GC content.
actinomycete A member of the group Actinomycetales, an order of Actinobacteria that includes branched spore formers such as *Streptomyces*, as well as irregularly shaped corynebacteria.
activated sludge See **floc**.
activation energy (E_a) The energy needed for reactants to reach the transition state between reactants and products.
activator A regulatory protein that can bind to a specific DNA sequence and increase transcription of genes.
active transport An energy-requiring process that moves molecules across a membrane against their electrochemical gradient.
acyl carrier protein (ACP) A protein that can carry an acetyl group for anabolic pathways such as fatty acid synthesis.
adaptive immunity Immune responses activated by a specific antigen and mediated by B cells and T cells.
ADCC See **antibody-dependent cell-mediated cytotoxicity**.
addressin A tissue-specific protein that is selectively present on vascular endothelial cells in different tissues.
adenosine triphosphate (ATP) A ribonucleotide with three phosphates and the base adenine. It has many functions in the cell, including precursor for RNA synthesis and energy carrier.
adenylate cyclase An enzyme that converts ATP into cyclic adenosine monophosphate (cAMP).
adhesin Any cell-surface factor that promotes attachment of an organism to a substrate.
ADP-ribosyltransferase A bacterial toxin that enzymatically transfers the ADP-ribose group from NAD^+ to target proteins, altering the target protein's structure and function.
aerial mycelium A mass of hyphae (branched filaments) that extend above the surface and produces spores at the tips.
aerobe An organism that grows only in the presence of oxygen.
aerobic respiration The use of oxygen as the terminal electron acceptor in an electron transport chain. A proton gradient is generated and used to drive ATP synthesis.
aerotolerant anaerobe An organism that does not use oxygen for metabolism but can grow in the presence of oxygen.
affinity chromatography A chromatographic technique that uses the ability of biological molecules to bind to certain ligands specifically and reversibly.
affinity maturation The process by which the antigen-binding site of an antibody gains increased affinity for its target antigen or epitope.
AFM See **atomic force microscopy**.
agar A polymer of galactose that is used as a gelling agent.
AIDS See **acquired immunodeficiency syndrome**.
Aigarchaeota A phylum of archaea in the TACK superphylum that contains filamentous aerobic hyperthermophiles in hot springs.
airborne transmission In disease, the transfer of a pathogen via dust particles or on respiratory droplets produced when an infected person sneezes or coughs.
alcoholic fermentation See **ethanolic fermentation**.
alga *pl.* algae A microbial eukaryote that contains chloroplasts.
algal bloom An overgrowth of algae on the water surface, caused by an increase in a limiting nutrient.
alkaline fermentation Bacterial fermentation in conjunction with proteolysis and amino acid catabolism that generates ammonia in amounts that raise pH.
alkaliphile An organism that grows fastest in alkali (generally defined as above pH 9).
allergen An antigen that causes an allergic hypersensitivity reaction.
allosteric site A regulatory site on a biological molecule distinct from the ligand/substrate-binding site.
allotype An amino acid difference in the antibody constant region that distinguishes different individuals within a species.
Alphaproteobacteria A phylum of Proteobacteria that includes, for example, rhizobia and rickettsias.
alternation of generations A life cycle that alternates between populations of haploid cells and diploid cells, which undergo meiosis and fertilization.
alternative sigma factor A sigma factor, distinguished from the housekeeping sigma factor, sigma-70, that has a distinct promoter consensus sequence to which it binds. Genes under control of an alternative sigma factor typically protect the cell from environmental stresses, such as heat shock or starvation.
alveolar macrophage A type of macrophage, located in the lung alveoli, that phagocytoses foreign material.
alveolate A member of the eukaryotic group Alveolata—ciliated or flagellated protists with complex cortical structure.
ameba *or* amoeba A protist that moves via pseudopods.
amensalism An interaction between species that harms one partner but not the other.
amino acid The monomer unit of proteins. Each amino acid contains a central carbon covalently bonded by a hydrogen, an amino group, a carboxyl group, and a side chain. An exception is proline, in which the side chain is cyclized with the central carbon.

aminoacyl-tRNA synthetase An enzyme that condenses a specific amino acid with the 3′ OH group of the correct tRNA, thereby charging the tRNA.

ammonification The generation of ammonia from organic nitrogen.

amoeba See **ameba**.

amphibolic Describing a metabolic pathway that is reversible and can be used for both catabolism and anabolism.

amphipathic Having both hydrophilic and hydrophobic portions.

amphotericin B An antifungal drug that binds the fungus-specific sterol ergosterol and destroys membrane integrity.

amplicon A specific PCR product in which a small DNA sequence is amplified (many copies are synthesized).

anabolism See **biosynthesis**.

anaerobe An organism that grows in an environment lacking oxygen.

anaerobic Lacking oxygen.

Anaerobic Methane-Oxidizing Euryarchaeota (ANME) *or* anaerobic methane oxidizers Also called *Syntrophoarchaeota*. A group of archaea in the superphylum Euryarchaeota that oxidize methane syntrophically with sulfate-reducing bacteria.

anaerobic respiration The use of a molecule other than oxygen as the final electron acceptor of an electron transport chain.

anammox reaction The anaerobic oxidation of ammonium to nitrogen gas, using nitrate as electron acceptor; yields energy.

anaphylaxis A severe type I hypersensitivity reaction caused by chemically induced contraction of smooth muscles and dilation of capillaries.

anaplerotic reaction A type of metabolic reaction, occurring in all organisms, that fixes small amounts of CO_2 to regenerate TCA cycle intermediates.

angle of aperture The width of a light cone (theta, θ) that projects from the midline of a lens. Greater angles of aperture increase resolution.

anion A negatively charged ion.

ANME See **Anaerobic Methane-Oxidizing Euryarchaeota**.

annotation The deciphering of genome sequences, including identification of genes and prediction of gene function.

antenna complex A complex of chlorophylls and accessory pigments in the photosynthetic membrane that collects photons and funnels them to a reaction center.

anti-anti-sigma factor A protein that inhibits an anti-sigma factor, allowing the target sigma factor to participate in initiating transcription.

anti-attenuator stem loop An mRNA secondary structure whose formation prevents assembly of a downstream transcriptional termination (attenuator) stem loop. The anti-attenuator stem loop structure permits transcription of the downstream structural genes.

antibiotic A molecule that can kill or inhibit the growth of selected microorganisms.

antibody A host defense protein produced by B cells in response to a specific antigenic determinant. Antibodies, a type of immunoglobulin, bind to their corresponding antigenic determinants.

antibody-dependent cell-mediated cytotoxicity (ADCC) The process by which natural killer cells destroy viral protein expressing antibody-coated host cells.

antibody tag The attachment of a stain to an antibody to visualize cell components recognized by the antibody with high specificity.

anticodon A set of three nucleotides in the middle loop of a tRNA that base-pairs with a codon in mRNA.

antigen A compound, recognized as foreign by the cell, that elicits an adaptive immune response. See also **immunogen**.

antigen-presenting cell (APC) An immune cell that can process antigens into antigenic determinants and display those determinants on the cell surface for recognition by other immune cells.

antigenic determinant Also called *epitope*. A small segment of an antigen that is capable of eliciting an immune response. An antigen can have many different antigenic determinants.

antigenic shift A major genetic alteration in a virus that results in a new pandemic strain.

antimicrobial agent A chemical substance that can kill microbes or slow their growth.

antiparallel Oriented such that the two strands are in opposite directions. Commonly refers to a nucleic acid double helix with one strand in the 5′-to-3′ orientation and the other strand in the 3′-to-5′ orientation.

antiport Coupled transport in which the molecules being transported move in opposite directions across the membrane.

antiretroviral therapy (ART) A three-drug antiretroviral cocktail that is highly effective at inhibiting the replication of HIV in patients.

antisepsis The removal of pathogens from living tissues.

antiseptic Describing a chemical that kills microbes. Also, the chemical itself.

anti-sigma factor A protein that inhibits a specific sigma factor, preventing transcription initiation.

AP site A position in DNA where no base is attached to the sugar of the backbone.

APC See **antigen-presenting cell**.

apicomplexan A member of the eukaryotic group Apicomplexa—parasitic alveolates that possess an apical complex used for entry into a host cell.

apoptosis A cell death program triggered during tissue differentiation or in certain damaged or infected cells.

apurinic site A DNA site missing a purine base because the bond linking the base to the sugar has been hydrolyzed.

Aquificae A phylum of hyperthermophilic bacteria.

arbuscular mycorrhizae Also called *vesicular-arbuscular mycorrhizae (VAM)* or *endomycorrhizae*. Mutualistic associations between plant roots and certain fungi, involving hyphal penetration of plant root cells.

Archaea One of the three domains of life, consisting of organisms with a last common ancestor not shared with members of Bacteria or Eukarya. Organisms are prokaryotic (lacking nuclei, unlike eukaryotes) and possess ether-linked phospholipid membranes (unlike bacteria).

Archaean eon The second eon (major time period) of Earth's existence, from 4.0 or 3.8 gigayears (Gyr, 10^9 years) to 2.5 Gyr before the present. The earliest geological evidence for life dates to this eon.

archaellum *pl.* archaella A rotary complex for motility in archaea, analogous to the bacterial flagellum.

archaeon *pl.* archaea A prokaryotic organism that is a member of the domain Archaea, distinct from bacteria (domain Bacteria) and eukaryotes (domain Eukarya).

Archaeplastida See **Plantae**.

aromatic Describing a planar, unsaturated, ring-shaped organic molecule whose bonding electrons are delocalized equally around the ring.

ART See **antiretroviral therapy**.

artifact A structure viewed through a microscope that is incorrectly interpreted.

ascomycete A member of the eukaryotic group Ascomycota—fungi whose mycelia form paired nuclei. Haploid ascospores are produced in pods called asci.

ascospore The spore produced by an ascomycete fungus.

ascus *pl.* **asci** A spore-containing pod produced by ascomycete fungi.
aseptic Free of microbes.
asexual reproduction Reproduction of a cell by fission or by mitosis to form identical daughter cells.
asRNA See ***cis*-antisense RNA**.
assembly 1. In a virus, the packaging of a viral genome into the capsid to form a complete virion. 2. In metagenomics, the piecing together of DNA sequence reads into contigs, and of contigs into scaffold.
assimilation An organism's acquisition of an element, such as carbon from CO_2, to build into body parts.
assimilatory nitrate reduction The uptake of nitrate and reduction to NH_4^+ by plants, fungi, archaea, and bacteria for use in biosynthetic pathways.
atomic force microscopy (AFM) A technique that maps the 3D topography of a object using van der Waals forces between the object and a probe.
atomic mass The mass (in grams) of one mole of an element.
atomic number The number of protons in an atom; it is unique for each element.
ATP See **adenosine triphosphate**.
ATP synthase A protein complex that synthesizes ATP from ADP and inorganic phosphate using energy derived from the transmembrane proton potential. It is located in the prokaryotic cell membrane and in the mitochondrial inner membrane.
attenuator stem loop Also called *terminator stem loop*. An intramolecular mRNA structure consisting of a base-paired stem connected by a single-stranded loop. The stem loop structure causes transcription to terminate. Its formation requires efficient translation of a leader peptide sequence.
autoclave A device that uses pressurized steam to sterilize materials by raising the temperature above the boiling point of water at standard pressure.
autofluorescence Fluorescence of living cells with no added fluorophore.
autoimmune response A pathology caused by lymphocytes that can react to self antigens.
autoinducer A secreted molecule that induces quorum-sensing behavior in bacteria.
autoinduction A mode of gene regulation independent of outside intervention, involving the conditioning of the medium by the cells. Usually refers to quorum-sensing control mediated by secreted autoinducer molecules.
autophagy Eukaryotic cell function normally used to degrade damaged organelles. Also used to kill intracellular pathogens.
autoradiography The visualization of a radioactive probe by exposing the probed material to X-ray film and then photographically developing the film.
autotrophy The metabolic reduction of carbon dioxide to produce organic carbon for biosynthesis.
auxotrophic Describing a mutant state in which the cell has lost the ability to synthesize a substance required for growth. An auxotroph has a nutritional requirement not shared by the parent.
azidothymidine (AZT) A nucleotide analog that inhibits reverse transcriptase and was the first drug clinically used to fight HIV infections.
AZT See **azidothymidine**.

B

B cell An adaptive immune cell, developed in bone marrow tissue, that can give rise to antibody-producing cells.
B-cell receptor (BCR) A B-cell membrane protein complex containing an antibody in association with the Igα and Igβ immunoglobulins.
B-cell tolerance The exposure of B cells to a high antigen dose, preventing future antibody production against that antigen.
bacillus *pl.* **bacilli** A rod shaped bacterial or archael cell.
bacitracin A topical antibiotic that affects cell wall synthesis.
bacteremia A bacterial infection of the blood.
Bacteria One of the three domains of life, consisting of organisms with a last common ancestor not shared with members of Archaea or Eukarya. Organisms are prokaryotic (lacking nuclei, unlike eukaryotes) and possess ester-linked phospholipid membranes (unlike archaea).
bactericidal Having the ability to kill bacterial cells.
bacteriochlorophyll The chlorophyll of anaerobic phototrophic bacteria; it absorbs photons most strongly in the far-red end of the light spectrum.
bacteriophage Also called *phage*. A virus that infects bacteria.
bacteriorhodopsin A haloarchaeal membrane-embedded protein that contains retinal and acts as a light-driven proton pump; it is homologous to the bacterial proteorhodopsin.
bacteriostatic Having the ability to inhibit the growth of bacterial cells.
bacterium *pl.* **bacteria** A prokaryotic organism that is a member of the domain Bacteria, distinct from archaea (domain Archaea) and eukaryotes (domain Eukarya).
bacteroid A cell wall–less, undividing, differentiated rhizobial cell within a plant cell. The bacteroid provides fixed nitrogen for the plant.
Bacteroidetes A phylum of Gram-negative bacteria; nearly all members are obligate anaerobes.
banded iron formation (BIF) A geological formation containing layers of oxidized iron (Fe^{3+}), which indicates formation under oxygen-rich conditions.
barcoding The use of a short DNA sequence, such as the gene for SSU rRNA, to identify single taxa.
barophile Also called *piezophile*. An organism that requires high pressure to grow.
base excision repair (BER) A DNA repair mechanism that cleaves damaged bases off the sugar-phosphate backbone. After endonuclease activity at the AP site, a new, correct DNA strand is synthesized complementary to the undamaged strand.
basidiomycete A member of the eukaryotic group Basidiomycota—fungi that form mushrooms.
basidiospore A haploid spore formed by a basidiomycete through meiosis of a basidium, a reproductive cell of a mushroom.
basophil A white blood cell, stained by basic dyes, that secretes compounds that aid innate immunity.
batch culture The growth of bacteria in a closed system without additional input of nutrients.
Bathyarchaeota A phylum of thermophilic archaea in the TACK superphylum, widely distributed in anoxic soil, water, and marine sediments.
BCR See **B-cell receptor**.
benthic organism An organism that lives on the ocean floor or within the sediment.
BER See **base excision repair**.
Betaproteobacteria A phylum of Proteobacteria that includes, for example, species of *Neisseria* and *Burkholderia*.
BIF See **banded iron formation**.
bin A set of sequences composed from metagenomic DNA reads showing a given level of similarity; each bin defines an operational taxonomic unit.
binary fission The process of replication in which one cell divides to form two genetically equivalent daughter cells of equal size.
binning The sorting of metagenomic sequences into taxonomic bins.

biochemical oxygen demand (BOD) Also called *biological oxygen demand*. The amount of oxygen removed from an environment by aerobic respiration.
biocontrol Use of a beneficial organism to control a harmful organism.
biofilm A community of microbes growing on a solid surface.
biogenic Formed by living organisms.
biogeochemical cycle The recycling of elements needed for life (such as carbon or nitrogen) through the biotic and abiotic components of the biosphere.
biogeochemistry Also called *geomicrobiology*. The metabolic interactions of microbial communities with the abiotic (mineral) components of their ecosystems.
bioinformatics A discipline at the intersection of biology and computing that analyzes gene and protein sequence data.
biological carbon pump The fixation of carbon dioxide by phototrophs and gravitational settling of biomass particles within the oceans.
biological oxygen demand See **biochemical oxygen demand**.
biological signature See **biosignature**.
biomass The mass found in the bodies of living organisms.
bioprospecting The search for organisms with potential commercial applications.
bioremediation The use of microbes to detoxify environmental contaminants.
biosignature Also called *biological signature*. A chemical indicator of life.
biosphere The region containing the sum total of all life on Earth.
biosynthesis Also called *anabolism*. The building of complex biomolecules from smaller precursors.
biotic Caused by living organisms.
black smoker An oceanic thermal vent containing high concentrations of dark minerals such as iron sulfide.
blood-brain barrier A selectively permeable membrane made up of tightly packed capillaries that supply blood to the brain and spinal cord. Large molecules and most pathogens cannot permeate the narrow spaces. Fat-soluble (lipophilic) molecules and oxygen can dissolve through the capillary cell membranes and are absorbed into the brain.
BOD See **biochemical oxygen demand**.
borreliosis Any of a variety of diseases caused by *Borrelia* species and transmitted by ticks or lice. Lyme disease is a form of borreliosis.
botulism A food-borne disease caused by a *Clostridium botulinum* toxin, involving muscle paralysis.
bradykinin A cell signaling molecule that promotes extravasation, activates mast cells, and stimulates pain perception.
Braun lipoprotein See **murein lipoprotein**.
bright-field microscopy A type of light microscopy in which the specimen absorbs light and appears dark against a light background.
bubonic plague A disease caused by the bacterium *Yersinia pestis*; it is characterized by swollen lymph nodes that often turn black.
budding A form of reproduction in which mitosis of the mother cell generates daughter cells of unequal size.
built environment Human-made constructs that we inhabit on a daily basis, including homes, workplaces, and parks.
burst See **lysis**.
burst size The number of virus particles released from a lysed host cell.

C

C-reactive protein A peptide that stimulates the complement cascade, induced by cytokines in the liver. Elevated levels in the blood may be associated with heart disease or other inflammatory processes.
calorimeter A device used to measure the amount of heat released or absorbed during a reaction.
Calvin cycle *or* Calvin-Benson cycle *or* Calvin-Benson-Bassham (CBB) cycle Also called *pentose phosphate cycle* or *reductive pentose phosphate cycle*. The metabolic pathway of carbon fixation in which the CO_2-condensing step is catalyzed by Rubisco. Found in chloroplasts and in many bacteria.
Candidate Phyla Radiation (CPR) A recently described group of bacterial phyla whose members represent broad genetic diversity but small genome size and limited metabolic capabilities.
candidate species A newly described microbial isolate that may become accepted as an official species.
cannula *pl.* cannulae A narrow tubule. For *Pyrodictium* species, glycoprotein cannulae interconnect cells at their periplasm.
cannulated cow See **fistulated cow**.
capping The clustering of B-cell receptor molecules on the surface of B cells after binding antigens or epitopes.
capsid The protein shell that surrounds a virion's nucleic acid. Within an enveloped virus, such as HIV, the capsid may be called a *core particle*.
capsule A slippery outer layer composed of polysaccharides that surrounds the cell envelope of some bacteria.
carbon-concentrating mechanism (CCM) The inducible expression of transporters of CO_2 and bicarbonate (HCO_3^-) into carboxysomes to enhance CO_2 levels near Rubisco.
carbon dioxide fixation The enzymatic reduction and covalent incorporation of inorganic carbon dioxide (CO_2) into an organic compound.
carbon sequestration In environmental science, the fixation of carbon into biomass, resulting in removal of a greenhouse gas from the atmosphere.
carboxysome A protein-enclosed compartment containing Rubisco to fix CO_2.
cardiolipin Diphosphatidylglycerol, a double phospholipid linked by glycerol.
carotenoid An accessory photosynthetic pigment that absorbs photons in the green wavelengths of the spectrum.
catabolism The cellular breakdown of large molecules into smaller molecules, releasing energy.
catabolite repression The inhibition of transcription of an operon encoding catabolic proteins in the presence of a more favorable catabolite, such as glucose.
catalytic domain 1. Also called *catalytic site*. The portion of an enzyme that performs catalysis. 2. The A subunit of a toxin, which carries the ADP-ribosyltransferase activity.
catalytic RNA Also called *ribozyme*. An RNA molecule that is capable of catalyzing reactions.
catalytic site See **catalytic domain** (definition 1).
catenane A pair of linked rings of DNA that occurs as a by-product of replication of circular chromosomes.
cation A positively charged ion.
CBB cycle See **Calvin cycle**.
CCM See **carbon-concentrating mechanism**.
CCR See **chemokine receptor**.
cell fractionation A procedure to separate cell components that often includes ultracentrifugation.
cell-mediated immunity A type of adaptive immunity employing mainly T-cell lymphocytes.
cell membrane Also called *cytoplasmic membrane* or *plasma membrane*. The phospholipid bilayer that encloses the cytoplasm.
cell sorting The separation of classes of cells using fluorescent markers, as detected by flow cytometry.
cell-surface receptor A transmembrane protein that senses a specific extracellular signal and may be the docking site for a specific virus.

cell wall A rigid structure external to the cell membrane. The molecular composition depends on the organism; in bacteria, it is composed of peptidoglycan.
cellular slime mold A slime mold in which the individual bacterial cells retain their own cell membranes; not a fungus.
cellulitis A spreading infection (with inflammation) of connective tissue just below the skin.
CFTR See **cystic fibrosis transmembrane conductance regulator**.
chain of infection The serial passage of a pathogenic organism from an infected individual to an uninfected individual, thus transmitting disease.
chancre A painless, hard lesion due to an inflammatory reaction at the site of infection with *Treponema pallidum*, the causative agent of syphilis.
chaperone *or* chaperonin A protein that helps other proteins fold into their correct tertiary structure.
cheddared curd Curd that has been cut and piled in order to remove the liquid whey.
cheese A solid or semisolid food product prepared by coagulating milk proteins, forming curd. Its production commonly involves microbial fermentation.
chemical imaging microscopy A method of microscopy that maps the distribution of specific elements or chemicals within a sample.
chemiosmotic theory A theory stating that the products of oxidative metabolism store their energy in an electrochemical gradient that can drive cellular processes such as ATP synthesis.
chemoheterotrophy See **organotrophy**.
chemokine An attractant for white blood cells that is produced by damaged tissues.
chemokine receptor (CCR) A human T-cell membrane protein that binds chemokine hormones but is also used by HIV for attachment and infection.
chemolithoautotrophy Metabolism in which single-carbon compounds are fixed into organic biomass, using energy from chemical reactions without light absorption.
chemolithotrophy See **lithotrophy**.
chemoorganotrophy See **organotrophy**.
chemoreceptor See **methyl-accepting chemotaxis protein**.
chemostat A continuous culture system in which the introduced medium contains a limiting nutrient.
chemosynthesis The fixation of single-carbon molecules (usually carbon dioxide) into organic biomass, using energy from oxidation of inorganic electron donors.
chemotaxis The ability of organisms to move toward or away from specific chemicals.
chemotrophy Metabolism that yields energy from oxidation-reduction reactions without using light energy.
ChIP See **chromatin immunoprecipitation**.
ChIP sequencing (ChIP-seq) A procedure that scans a genome for all DNA sequences capable of binding to a specific DNA-binding protein.
chiral carbon A carbon bonded to four different types of functional groups; it can thus take two different forms that exhibit mirror symmetry.
chlamydia 1. A pathogenic bacterium lacking a cell wall; chlamydias (or chlamydiae) grow within human or animal cells, transmitting as elementary bodies to other cells. 2. A disease caused by chlamydia cells. The most frequently reported sexually transmitted disease in the United States. Symptoms range from none, to a burning sensation upon urination, to sterility.
Chlamydiae A phylum of intracellular parasitic bacteria that grow only within a host cell and generate multiple spore-like structures that escape to infect the next host.
chloramphenicol A bacteriostatic antibiotic that acts by inhibiting peptidyltransferase activity of the bacterial ribosome.
Chlorobi A phylum of Gram-negative bacteria. They are obligate anaerobes, "green sulfur" phototrophs that photolyze sulfides or H_2.
Chloroflexi A phylum of bacteria that are filamentous phototrophs with chlorosomes.
chlorophyll A magnesium-containing porphyrin pigment that captures light energy at the start of photosynthesis.
chlorophyte See **green alga**.
chloroplast An organelle of endosymbiotic origin (sharing descent with cyanobacteria) that conducts oxygenic photosynthesis; found in algae and plant cells.
chlorosome A membranous photosynthetic organelle found in some "green" bacteria of the phyla Chloroflexi and Chlorobi.
cholesterol A sterol lipid found in eukaryotic cell membranes.
chromatin Chromosomal DNA complexed with proteins. Usually refers to a eukaryotic chromosome.
chromatin immunoprecipitation (ChIP) An experimental method used to determine the DNA-binding sites on a chromosome to which a DNA protein binds.
chromophore A light-absorbing redox cofactor.
chronic inflammation Inflammation that has persisted over long periods of time, usually months or years.
chrysophyte A member of the eukaryotic group Chrysophyceae (also known as "golden algae")—flagellated heterokont protists possessing chloroplasts as secondary endosymbionts.
ciliate An alveolate that has paired cilia.
cilium *pl.* cilia A short, hairlike structure of eukaryotes that is structurally similar to the prokaryotic flagellum. Cilia beat in waves to propel the cell.
circadian clock From the Latin *circa* ("about") and *dies* ("day"). A regulatory machine akin to a clock that controls cellular activity as a function of time of day.
cis-antisense RNA (asRNA) Noncoding RNA from transcription of a DNA sequence complementary to the template strand of a gene that encodes a protein. A *cis*-antisense RNA binds and regulates the coding transcript made from the template strand of the gene. These regulatory RNAs can stop transcription, promote transcript degradation, or prevent translation.
clade Also called *monophyletic group*. A group of organisms that includes an ancestral species and all of its descendants.
class I MHC molecule A membrane surface protein on all nucleated cells of the human body (absent from red blood cells and platelets) that present intracellular foreign antigen epitopes to cytotoxic T cells.
class II MHC molecule A membrane surface protein on antigen-presenting cells (dendritic cells, macrophages) and lymphocytes that present phagocytosed extracellular foreign antigen epitopes to helper T cells.
class switching See **isotype switching**.
classical complement pathway An antibody-mediated pathway for complement activation.
classification The recognition of different forms of life and their placement into different categories.
clonal Giving rise to a population of genetically identical cells, all descendants of a single cell.
clonal selection The rapid proliferation of a subset of B cells during the primary or secondary antibody response.
cloning The insertion of DNA into a plasmid where it can be replicated.
CoA See **coenzyme A**.
coastal shelf Shallow regions of the ocean, less than 200 meters deep, that are adjacent to land.

coccus *pl.* **cocci** A spherically shaped bacterial or archaeal cell.

codon A set of three nucleotides that encodes a particular amino acid.

coenzyme A (CoA) A nonprotein cellular organic molecule that can carry acetyl groups and participates in metabolism.

coevolution The evolution of two species in response to one another.

cofactor A metallic ion or a coenzyme required by an enzyme to perform normal catalysis.

colony A visible cluster of microbes on a plate, all derived from a single founding microbe.

commensalism An interaction between two different species that benefits only one partner.

community The sum of all populations of organisms interacting within an ecosystem.

compatible solute A small molecule that does not disrupt normal cell metabolism even at high intracellular concentrations.

competent Able to take up DNA from the environment.

complement Innate immunity proteins in the blood that form holes in bacterial membranes, killing the bacteria.

complex medium Also called *rich medium.* A nutrient-rich growth solution including undefined chemical components such as beef broth.

compound microscope A microscope with multiple lenses to compensate for lens aberration and increase magnification.

compromised host An animal with a weakened immune system.

computational pipeline A linear series of programs that combine mathematical tools with biological assumptions to propose assembled genomes.

concatemer A long line of tandemly repeated genomes; commonly formed during rolling-circle replication.

condensation In biochemistry, the joining of two molecules to release a water molecule.

condenser In a microscope, a lens that focuses parallel light rays from the light source onto a small area of the specimen to improve the resolution of the objective lens.

confluent Describing a mode of growth that results in a lawn of organisms completely covering a surface.

confocal laser scanning microscopy *or* **confocal microscopy** A type of fluorescence microscopy in which the excitation light from a laser and the emitted light from the specimen are focused together, producing high-resolution images.

congenital syphilis Syphilis contracted in utero.

conjugation Horizontal gene transfer involving cell-to-cell contact. In bacteria, pili draw together the donor and recipient cell envelopes, and a protein complex transmits DNA across. In ciliated eukaryotes, a conjugation bridge forms between two cells connecting their cytoplasm, through which micronuclei are exchanged.

conjugative transposon A transposon that can be transferred from one cell to another via conjugation.

consensus sequence A sequence of nucleotides or amino acids with a common function at many nucleic acid or protein positions. Consists of the base pair or amino acid most frequently found at each position in the sequence.

constant region The region of an antibody that defines the class of a heavy chain or a light chain.

constructed wetland A built environment designed to provide environmental services comparable to those of natural wetlands.

consumer An organism that acquires nutrients from producers, either directly or indirectly.

contig A sequence of overlapping fragments of cloned DNA that are contiguous along a chromosome.

continuous culture A culture system in which new medium is continuously added to replace old medium.

contractile vacuole An organelle in eukaryotic microbes that pumps water out of the cell.

contrast Differential absorption or reflection of electromagnetic radiation between an object and a background that allows the object to be distinguished from the background.

coral bleaching The death or expulsion of coral algal symbionts. One cause is an increase in temperature.

core genome A set of genes shared by a group of related bacterial strains, showing stable inheritance.

core particle A viral capsid that encloses its nucleic acid genome and is surrounded by an envelope.

coreceptor A cell-surface receptor needed for viral entry along with a primary receptor.

corepressor A small molecule that must bind to a repressor to allow the repressor to bind operator DNA.

cortical alveolus One of the vesicles that forms a network in the outer covering of an alveolate protist.

counterstain A secondary stain used to visualize cells that do not retain the first stain.

coupled transport The movement of a substance against its electrochemical gradient (from lower to higher concentration, or from opposite charge to like charge) using the energy provided by the simultaneous movement of a different chemical down its electrochemical gradient.

covalent bond A chemical bond in which two atoms share a pair of electrons.

CPR See **Candidate Phyla Radiation**.

CRISPR Clustered regularly interspaced short palindromic repeats. CRISPR consists of short repeated DNA sequences in a bacterial or archaeal genome, derived from previous bacteriophage or viral infection and conferring protection from future infection; considered a prokaryotic "immune system."

cross-bridge An attachment that links parallel molecules, such as the peptide link between glycan chains in peptidoglycan.

cryo-electron microscopy (cryo-EM) Also called *electron cryomicroscopy.* Electron microscopy in which the sample is cooled rapidly in a cryoprotectant medium that prevents freezing. The sample does not need to be stained.

cryo-electron tomography Also called *electron cryotomography.* A method of cryo-electron microscopy in which the electron beam generates multiple views in parallel planes through the specimen.

cryptogamic crust A low-growing desert ground cover composed of cyanobacteria, lichens, and nonlichenous algae, fungi, and mosses.

curd Coagulated milk proteins produced by the combined action of lactic acid–producing bacteria and stomach enzymes of certain mammals, such as cattle.

Cyanobacteria *or* **Cyanophyta** A phylum of oxygen-producing photoautotrophic bacteria containing chlorophylls *a* and *b.* They share an ancient ancestor of chloroplasts.

cyclic photophosphorylation A photosynthetic process in which chlorophyll serves as both the initial electron donor and the final electron acceptor. ATP is produced via the proton potential from an electron transport system, but no NADPH is generated.

cycloserine A polypeptide antibiotic that inhibits peptidoglycan synthesis.

cystic fibrosis transmembrane conductance regulator (CFTR) A chloride channel found in respiratory epithelia. Mutations in the CFTR gene lead to cystic fibrosis.

cystitis Bladder infection.

cytochrome A membrane protein that donates and receives electrons.

cytochrome system See **electron transport system**.

cytokine A small, secreted host protein that binds to receptors on various endothelial and immune system cells, regulating the cells' responses.

cytoplasm Also called *cytosol*. The aqueous solution contained by the cell membrane in all cells and outside the nucleus (in eukaryotes).

cytoplasmic membrane See **cell membrane**.

cytoskeleton A collection of filamentous proteins that impart structure to and aid movement of cells; in a eukaryote, these include microfilaments, intermediate filaments, and microtubules.

cytosol See **cytoplasm**.

cytotoxic T cell (T_C cell) A T cell that expresses CD8 on its cell surface and can secrete toxic proteins such as perforin and granzymes.

D

D-value See **decimal reduction time**.

daptomycin A lipopeptide antibiotic that forms ion channels in Gram-positive bacteria.

dark-field microscopy A type of light microscopy in which structures that are too small to be resolved by light rays are detected by observing the light they scatter.

dead zone Also called *zone of hypoxia*. An anoxic region of an ocean or freshwater system, devoid of most fish and invertebrates.

death phase A period of cell culture, following stationary phase, during which bacteria die faster than they replicate.

death rate The rate at which cells die; it is exponential during the death phase.

decay-accelerating factor A host cell membrane protein that stimulates the decay of complement factors and prevents their deposition at the cell surface.

decimal reduction time (D-value) The length of time it takes for a treatment to kill 90% of a microbial population, and hence a measure of the efficacy of the treatment.

decomposer An organism that consumes dead biomass.

deep-branching taxon A lineage whose genome sequences diverged early, at or before the well-known phyla.

defensin A type of small, positively charged peptide, produced by animal tissues, that destroys the cell membranes of invading microbes.

degenerative evolution See **reductive evolution**.

degranulate To release antimicrobial granule contents by fusing granule membranes to cytoplasmic or vacuolar membranes.

Deinococcus-Thermus A phylum of bacteria that are resistant to ionizing radiation and high temperature.

delayed-type hypersensitivity (DTH) See **type IV hypersensitivity**.

deletion The loss of nucleotides from a DNA sequence.

Deltaproteobacteria A phylum of Proteobacteria that include, for example, the Desulfobacterales (sulfate reducers) and the myxobacteria.

denature To lose secondary and tertiary structure in a protein or nucleic acid because of high temperature or chemical treatment.

dendritic cell An antigen-presenting white blood cell that primarily takes up small soluble antigens from its surroundings.

denitrification Also called *dissimilatory nitrate reduction*. Energy-yielding metabolism in which nitrate (NO_3^-) is reduced to nitrite (NO_2^-), diatomic nitrogen (N_2), and in some cases ammonia (NH_3).

depth of field In a microscope, a region of the optical column over which a specimen appears in focus.

derepression An increase in gene expression caused by the decrease in concentration of a corepressor.

desensitization A clinical treatment to decrease allergic reactions by exposing patients to small doses of an allergen.

Desulfurococcales A phylum of thermophilic archaea in the TACK superphylum that metabolize sulfur and organic compounds.

detection The ability to determine the presence of an object.

detritus Discarded biomass that can be consumed by decomposers.

diapedesis See **extravasation**.

diatom A member of the eukaryotic group Bacillariophyceae—protists that possess intricate bipartite shells that contain silica.

diauxic growth A biphasic cell growth curve caused by depletion of the favored carbon source and a metabolic switch to the second carbon source.

DIC See **differential interference contrast microscopy**.

dichotomous key A tool for identifying organisms, in which a series of yes/no decisions successively narrows down the possible categories of species.

diderm Enclosed by two phospholipid membranes—an inner membrane and an outer membrane.

differential In disease, the list of possible causes of an infection.

differential interference contrast microscopy (DIC) Also called *Nomarski microscopy*. A form of microscopy based on the interference pattern between two light beams split and polarized, one of which passes through a sample. The interference between the two beams generates contrast in a transparent sample.

differential medium A growth medium that can distinguish between various bacteria on the basis of metabolic differences.

differential RNA sequencing (dRNAseq) An RNA sequencing technique that differentiates the true start of an RNA transcript, which begins with a 5′ triphosphate, from an internal cleavage fragment that begins with a 5′ monophosphate.

differential stain A stain that differentiates among objects by staining only particular types of cells or specific subcellular structures.

diffusion The energy-independent net movement of a substance from a region of high concentration to a region of lower concentration.

dilution streaking A method of spreading bacteria on a plate in order to obtain colonies arising from an individual bacterium.

dinoflagellate A member of the eukaryotic group Dinoflagellata—secondary or tertiary endosymbiont algae, alveolates with two flagella, one of which is wrapped distinctively around the cell equator.

diplococcus The paired cocci of *Neisseria* species.

direct repeat A DNA sequence that is found close to another sequence with the identical form and that is aligned in the same direction (e.g., 5′-ATCGATCGnnnnnnATCGATCG-3′).

disinfection The removal of pathogenic organisms from inanimate surfaces.

dissimilation An organism's catabolism or oxidation of nutrients to inorganic minerals that are released into the environment.

dissimilatory denitrification Metabolic reduction of nitrate or nitrite to yield energy; anaerobic respiration of nitrate or nitrite.

dissimilatory metal reduction A type of anaerobic respiration that uses metal cations as terminal electron acceptors.

dissimilatory nitrate reduction See **denitrification**.

divisome A protein complex that manages the overall process of septation.

DNA-binding protein A protein that binds to DNA and modulates its function.

DNA-dependent RNA polymerase See **RNA polymerase**.

DNA ligase An enzyme that cells use to form a covalent bond at a nick in the phosphodiester backbone. It is also used in molecular biology laboratories to join pieces of DNA.
DNA microarray A technique, used for measuring the amount of specific mRNA molecules transcribed in cells, in which DNA fragments from every open reading frame in a genome are affixed to separate locations on a solid support surface (a DNA microchip), producing a grid, or array.
DNA replication The biological process of making an identical copy of double-stranded DNA using existing DNA as a template.
DNA reverse-transcribing virus See **pararetrovirus**.
DNA sequencing A technique to determine the order of bases in a DNA sample.
DNA shuffling A method of artificially evolving new genes in which fragments of different examples (or orthologs) of a given gene are randomly mixed and PCR-amplified to make a new functional gene. It is an example of directed evolution.
DNase I footprinting A technique that identifies the specific DNA sequence to which regulatory proteins bind. Stretches of DNA bound by the regulatory protein are protected from digestion by DNase I, and this zone of protection shows up as a "footprint" in gel electrophoresis of the DNA.
domain 1. In taxonomy, one of three major subdivisions of life: Archaea, Bacteria, and Eukarya. 2. In protein structure, a portion of a protein that possesses a defined function, such as binding DNA. 3. In membranes, a region of membrane consisting of certain types of phospholipids that are distinct from surrounding lipids.
doubling time The generation time of bacteria in culture. The amount of time it takes for the population to double.
downstream processing The recovery and purification of a commercial product produced by industrial microbes.
dRNAseq See **differential RNA sequencing**.
DTH See **type IV hypersensitivity**.
duplication The production of a second copy of a sequence fragment on a DNA molecule, usually adjacent to the original copy.
dysbiosis An imbalance in microbiome composition that can lead to disease.

E

E site See **exit site**.
early gene A viral gene expressed early in the infection cycle.
eclipse period The time in the viral life cycle after viral genome injection into a host cell but before complete virions are formed.
ecosystem A community of species plus their environment (habitat).
ectomycorrhizae Mycorrhizae that colonize the surface of plant roots. Their mycelia do not penetrate the root cells.
ED pathway See **Entner-Doudoroff pathway**.
edema Tissue swelling due to fluid accumulation.
electrochemical potential A type of potential energy formed by the combined concentration gradient of a molecule and the electrical potential across a membrane.
electrogenic biofilm A biofilm that generates electricity in a human-made device.
electromagnetic radiation Energy radiating in the form of alternating electrical and magnetic waves, quantized in photons.
electron acceptor An oxidized molecule (e.g., NAD^+) that can accept electrons.
electron bifurcation A biochemical reaction in which an electron transfer that yields energy is coupled to an electron transfer that consumes energy.
electron cryomicroscopy See **cryo-electron microscopy**.
electron cryotomography See **cryo-electron tomography**.
electron donor Also called *reducing agent*. A reduced molecule (e.g., NADH) that can donate electrons.
electron microscope A microscope that obtains high resolution and magnification by using magnetic lenses to focus electron beams on samples.
electron microscopy (EM) A form of microscopy in which a beam of electrons accelerated through a voltage potential is focused by magnetic lenses onto a specimen.
electron shuttle An aromatic molecule that allows low-energy transitions for gaining or losing an electron.
electron transport system (ETS) *or* **electron transport chain (ETC)** Also called *cytochrome system*. A series of membrane-embedded proteins that converts the energy of redox reactions into a proton potential.
electronegativity The affinity of an atom for electrons. The greater the electronegativity, the stronger the attraction for electrons.
electrophoresis A technique to separate charged proteins and nucleic acids based on how rapidly they migrate in an electrical field through a gel.
electrophoretic mobility shift assay (EMSA) A technique to observe DNA-protein interactions that is based on the ability of a bound protein to slow the voltage-driven migration of DNA through a gel.
elementary body The endospore-like form of chlamydias transmitted outside host cells.
EM See **electron microscopy**.
Embden-Meyerhof-Parnas (EMP) pathway See **glycolysis**.
emerging Describing an organism or other entity that is newly isolated, defined, or recognized, as in "emerging clade," "emerging pathogen," or "emerging disease."
emission wavelength The wavelength of light emitted by a fluorescent molecule. It is longer than the excitation wavelength and has lower energy.
EMP pathway See **glycolysis**.
empiric therapy An approach to treating infection before the infective organism is known, in which multiple antibiotics are administered in an effort to kill the most likely causative agents.
empty magnification Magnification without an increase in resolution.
EMSA See **electrophoretic mobility shift assay**.
enantiomer See **optical isomer**.
endemic Describing a disease that is always present in a population, although the frequency of infection may be low.
endocarditis An inflammation of the heart's inner lining.
endocytosis The invagination of the cell membrane to form a vesicle that contains extracellular material.
endogenous retrovirus A retroelement (genome sequence descended from a retrovirus) that contains *gag*, *env*, and *pol* genes.
endogenous virus A virus whose genome is encoded within the germ line of a host animal; may be considered a functional part of the host.
endolith A bacterium that grows within the crystals of solid rock.
endomembrane system A series of membranous organelles that organize uptake, transport, digestion, and expulsion of particles through a eukaryotic cell. It includes endosomes, lysosomes, endoplasmic reticulum, and the Golgi complex.
endomycorrhizae Also called *arbuscular mycorrhizae* or *vesicular-arbuscular mycorrhizae (VAM)*. Mycorrhizae whose fungal hyphae penetrate plant root cells.
endophyte An endosymbiont of vascular plants.
endosome A vesicle formed from the pinching in of the cell membrane.

endospore A durable, inert, heat-resistant spore that can remain viable for thousands of years.
endosymbiont An organism that lives as a symbiont inside another organism.
endosymbiosis An intimate association between different species in which one partner population grows within the body of another organism.
endothermic Absorbing heat.
endotoxin A lipopolysaccharide in the outer membrane of Gram-negative bacteria that becomes toxic to the host after the bacterial cell has lysed.
energy The ability to do work.
energy carrier A molecule in the cell, such as ATP or NADH, that serves as energy currency. Energy carriers are produced during catabolic reactions and can be used to drive energy-requiring reactions.
enhancer A noncoding DNA regulatory region in eukaryotes that can lead to activation of transcription when bound by an appropriate transcription factor. Its location on the chromosome can be far removed from the regulated gene.
enriched medium A growth solution for fastidious bacteria, consisting of complex medium plus additional components.
enrichment culture The use of selective growth media to allow only certain microbes to grow.
enthalpy A measure of the heat energy in a system.
Entner-Doudoroff (ED) pathway A glycolytic pathway in which glucose 6-phosphate is initially oxidized to 6-phosphogluconate, and ultimately yields 1 pyruvate, 1 ATP, 1 NADH, and 1 NADPH.
entropy A measure of the disorder in a system.
envelope A structure external to the cell membrane, such as the cell wall or outer membrane of a bacterium. For a virus, the envelope is a membrane enclosing the capsid or core particle.
enzyme A biological catalyst; a protein or RNA that can speed up the progress of a reaction without itself being changed.
eosinophil A white blood cell that stains with the acidic dye eosin and secretes compounds that facilitate innate immunity.
epidemic A disease outbreak in which large numbers of individuals in a population become infected over a short time.
epidemiology The study of factors affecting the health and illness of populations.
epidermis The outer protective cell layer in most multicellular animals.
epifluorescence See **fluorescence**.
epitope See **antigenic determinant**.
EPS See **exopolysaccharides**.
Epsilonproteobacteria A phylum of Proteobacteria that includes, for example, species of *Campylobacter* and *Helicobacter*.
equilibrium *pl.* **equilibria** A dynamic state in which there is no net change in a reaction.
error-prone repair Low-accuracy DNA repair mechanisms that allow mutations.
error-proof repair DNA repair mechanisms that minimize the occurrence of mutations.
erythema migrans A bull's-eye rash characteristic of borreliosis (Lyme disease).
essential gene A gene that is required for cell viability under all environmental conditions.
essential nutrient A compound that an organism cannot synthesize and must acquire from the environment in order to survive.
ETC See **electron transport system**.
ethanolic fermentation Also called *alcoholic fermentation.* A fermentation reaction yielding 2 ethanol and $2CO_2$ as products.
ETS See **electron transport system**.
Eukarya One of the three domains of life, consisting of organisms with a last common ancestor not shared with members of Archaea or Bacteria. Cells possess nuclei, unlike cells of bacteria and archaea.
eukaryote An organism whose cells contain a nucleus. All eukaryotes are members of the domain Eukarya.
Eumycota True fungi, a taxonomic group of opisthokont eukaryotes with chitinous cell walls; the group most closely related to animals.
euphotic zone Also called *photic zone.* The region of the ocean that receives sunlight capable of supporting photosynthesis.
Euryarchaeota A major division of Archaea, containing methanogens, halophiles, and acidophiles.
eutrophic lake A lake in which overgrowth of heterotrophic microbes has eliminated oxygen, leading to a decrease in animal life.
eutrophication A sudden increase of a formerly limiting nutrient in an aquatic environment, leading to overgrowth of algae and grazing bacteria and subsequent oxygen depletion.
excitation wavelength The wavelength of light that must be absorbed by a molecule in order for the molecule to fluoresce. It is shorter than the emission wavelength and has higher energy.
exergonic reaction A spontaneous chemical reaction that releases free energy.
exit site (E site) The region of a ribosome that holds the uncharged, exiting tRNA.
exocytosis The fusion of vesicles with the cell membrane to release vesicle contents extracellularly.
exon An expressed (protein-coding) portion of a eukaryotic gene.
exonuclease An enzyme that cleaves DNA from the end.
exopolysaccharides (EPS) Also called *extracellular polymeric substance.* Polysaccharides and entrapped materials that form a thick extracellular matrix around the microbes in a biofilm.
exothermic Releasing heat.
exotoxin A protein toxin, secreted by bacteria, that kills or damages host cells.
experimental evolution Also called *laboratory evolution.* The repeated culturing of a population in a laboratory under defined environmental conditions, leading to evolution of adapted genotypes.
exponential phase Also called *logarithmic (log) phase.* A period of cell culture during which bacteria grow exponentially at their maximal possible rate, given the conditions.
extracellular polymeric substance (EPS) See **exopolysaccharides**.
extravasation Also called *diapedesis.* The movement of cells of the immune system out of blood vessels and into surrounding infected tissue.
extreme thermophile See **hyperthermophile**.
extremophile An organism that grows only in an extreme environment—that is, an environment including one or more conditions that are "extreme" relative to the conditions for human life.

F

F^- cell The DNA recipient cell in bacterial conjugation.
F^+ cell The DNA donor cell that transmits the fertility factor F^+ to an F^- cell during bacterial conjugation.
F factor See **fertility factor**.
F-prime (F′) factor *or* F-prime (F′) plasmid A fertility factor plasmid that contains some host chromosomal DNA.
facilitated diffusion A process of passive transport across a membrane that is facilitated by transport proteins.
FACS See **fluorescence-activated cell sorting**.
factor H A normal serum protein that prevents the inadvertent activation of complement in the absence of infection.

facultative anaerobe An organism that can grow in either the presence or absence of oxygen.

facultative intracellular pathogen A pathogen that can replicate either inside host cells or outside host cells.

FAD See **flavin adenine dinucleotide**.

Fc region The region of an antibody that binds to specific receptors on host cells in an antigen-independent manner. It is found in the carboxy-terminal "tail" region of the antibody.

fecal-oral route of transmission A method by which pathogens or parasites excreted in the fecal matter of an infected person are then indirectly ingested by an uninfected person.

fermentation Also called *fermentative metabolism*. 1. The production of ATP via substrate-level phosphorylation, using organic compounds as both electron donors and electron acceptors. 2. Industrial fermentation is the production of microbial products that are made by microbes grown in fermentation vessels; it may include respiratory metabolism to maximize microbial growth.

fermentation industry Commercial production of microbial products that are made by microbes grown in fermentation vessels; it may include respiratory metabolism to maximize microbial growth.

fermentative metabolism See **fermentation**.

fermented food Food products that are biochemically modified by microbial growth.

ferredoxin An iron- and sulfur-containing protein that transfers electrons in electron transport systems.

fertility factor (F factor) A specific plasmid (transferred by an F^+ donor cell) that contains the genes needed for pilus formation and DNA export.

filamentous virus A viral structure type consisting of a helical capsid surrounding a single-stranded nucleic acid.

fimbria *pl.* fimbriae See **pilus**.

Firmicutes A phylum of Gram-positive bacteria with relatively low GC content.

FISH See **fluorescence in situ hybridization**.

fistulated cow Also called *cannulated cow*. A cow in which a hole in the skin has been connected surgically to a hole in the rumen and fitted with a cannula, allowing access for experimental analysis of the rumen.

fitness trade-off The situation where a trait improves fitness in one condition, but lowers fitness in a different condition.

fix To incorporate an inorganic compound into biomass, the organic molecules that form the cell.

fixation The adherence of cells to a slide by a chemical or heat treatment.

flagellate A protist that has one or more flagella.

flagellum *pl.* flagella A filamentous structure for motility. In prokaryotes, a helical protein filament attached to a rotary motor; in eukaryotes, an undulating membrane-enclosed complex of microtubules and ATP-driven motor proteins.

flavin adenine dinucleotide (FAD) An energy carrier in the cell that can donate ($FADH_2$) or accept (FAD) electrons.

flesh-eating disease See **necrotizing fasciitis**.

floc Also called *activated sludge*. Particulate matter formed by clumps of microbes during wastewater treatment.

flow cytometry A tool for analyzing cell populations, in which cells of multiple types are detected individually and distinguished by light scatter and fluorescence emission.

fluid mosaic model A model of the cell membrane in which proteins are free to diffuse laterally within the membrane.

fluorescence Also called *epifluorescence*. The emission of light from a molecule that absorbed light of a shorter, higher-energy wavelength.

fluorescence-activated cell sorting (FACS) A technique in which cells are counted (via flow cytometry), and then different classes of cells are collected into fractions according to differences in fluorescence (via cell sorting).

fluorescence in situ hybridization (FISH) A technique to detect individual microbes in an ecological or clinical sample, using a fluorophore-labeled oligonucleotide probe (usually a short DNA sequence) that hybridizes to microbial DNA or rRNA.

fluorescence resonance energy transfer (FRET) The detectable transfer of fluorescent energy from one molecule to another. Since the participating molecules must be near each other, FRET can be used to monitor protein-protein interactions in cells and is also used in real-time PCR.

fluorophore A fluorescent molecule used to stain specimens for fluorescence microscopy.

focal point The position at which light rays that pass through a lens intersect.

focus *pl.* foci The point at which rays of energy converge; in light microscopy, the convergence of light rays maximizes the clarity of the optical image.

folate Folic acid, a heteroaromatic cofactor that is required by some enzymes.

fomite An inanimate object on which pathogens can be transmitted from one host to another.

food poisoning Food contamination, the presence of human disease-causing microbial pathogens or toxins in food.

food spoilage Microbial changes that render a food unfit or unpalatable for consumption.

food web A network of interactions in which organisms obtain or provide nutrients for each other—for example, by predation or by mutualism.

foraminiferan *or* foram An ameba with a calcium carbonate shell and a helical arrangement of chambers.

forespore In sporulation of Gram-positive bacteria, the smaller cell compartment formed through asymmetrical cell division; it develops into the endospore.

fossil fuel Ancient organismal remains that have been converted to hydrocarbons (petroleum and natural gas) or sedimentary rock (coal) through microbial digestion followed by reduction under high pressure underground. Fossil fuels are extracted and burned by humans for energy.

frameshift mutation A gene mutation involving the insertion or deletion of nucleotides that cause a shift in the codon reading frame.

free energy change See **Gibbs free energy change**.

freeze-drying Also called *lyophilization*. The removal of water from food, by freezing under vacuum, to limit microbial growth.

FRET See **fluorescence resonance energy transfer**.

fruiting body A multicellular fungal or bacterial reproductive structure.

frustule The silica bipartite shell produced by a diatom.

functional group A cluster of covalently bonded atoms that behaves with specific properties and functions as a unit.

fungus *pl.* fungi A heterotrophic opisthokont eukaryote with chitinous cell walls. Includes Eumycota, but traditionally may refer to fungus-like protists such as the oomycetes.

fusion peptide A portion of a viral envelope protein that changes shape to facilitate envelope fusion with the host cell membrane.

Fusobacteria A phylum of anaerobic Gram-negative bacteria with an outer membrane, related to the Proteobacteria; includes human pathogens.

G

gain-of-function mutation A mutation that enhances the activity or allows new activity of a gene product.

GALT See **gut-associated lymphoid tissue**.

Gammaproteobacteria A phylum of Proteobacteria that includes, for example, the Enterobacteriales (enteric facultative anaerobes) and the pseudomonads.
gamogony The differentiation of parasitic haploid cells into male and female gametes.
gas vesicle An organelle that traps gases to increase the buoyancy of aquatic microbes.
GC content The proportion of an organism's genome consisting of guanine-cytosine base pairs.
GDH See **glutamate dehydrogenase**.
gene A sequence of nucleotides that has a distinct function (regulatory) or whose encoded product (protein or RNA) has a distinct function. The functional unit of heredity.
gene duplication The formation of an extra copy of a gene within a genome.
gene fusion In biotechnology, the construction of a recombinant gene composed of portions from two different genes, one of which may be a reporter for gene expression.
gene switching Switching between two (out of five) different classes of immunoglobulin genes (e.g., from IgM to IgG) during B-cell development.
gene transfer vector A mobile DNA engineered from a virus or plasmid, designed to insert a genetic sequence into the genome of an organism for experimental study or for medical therapy.
generalized transduction A phage-mediated gene transfer process in which any donor gene can be transferred to a recipient cell.
generation time The species-specific time period for doubling of a population (e.g., by bacterial cell division) in a given environment, assuming no depletion of resources.
genome The complete genetic content of an organism. The sequence of all the nucleotides in a haploid set of chromosomes.
genome reduction The large-scale loss of genes through evolution.
genomic island A region of DNA sequence whose properties indicate that it has been transferred from another genome. Genomic islands usually comprise a set of genes with shared function, such as pathogenicity or symbiosis support.
genotype The genome sequence of an organism.
genus ***pl.*** **genera** A group of closely related species.
Geoarchaeota A clade of Thermoprotei, thermophilic archaea including mats at thermal vents, in the TACK superphylum.
geochemical cycling The global interconversion of various inorganic and organic forms of elements.
geomicrobiology See **biogeochemistry**.
germ theory of disease The theory that many diseases are caused by microbes.
germicidal Able to kill cells but not spores.
germination The activation of a dormant spore to generate a vegetative cell.
Gibbs free energy change ***or*** **Gibbs energy change (ΔG)** Also called *free energy change*. In a chemical reaction, a measure of how much energy available to do work is released or required as the reaction proceeds.
gliding motility The movement of cells individually or as a collective over surfaces using specialized pili.
gluconeogenesis The biosynthesis of glucose from single-carbon compounds.
glutamate dehydrogenase (GDH) An enzyme that condenses NH_4^+ with 2-oxoglutarate to form glutamate. The condensation requires reduction by NADPH.
glutamate synthase Glutamine:2-oxoglutarate aminotransferase (GOGAT), an enzyme that converts 2-oxoglutarate plus glutamine into two molecules of glutamate.
glutamine synthetase (GS) An enzyme that condenses NH_4^+ with glutamate to form glutamine.
glycan A polysaccharide chain composed of oxygen-linked (O-linked) monosaccharides.
glycolysis Also called *Embden-Meyerhof-Parnas (EMP) pathway*. The catabolic pathway of glucose oxidation to pyruvate, in which glucose 6-phosphate isomerizes to fructose 6-phosphate, ultimately yielding 2 pyruvate, 2 ATP, and 2 NADH.
glyoxylate bypass An alternative to the tricarboxylic acid cycle in which isocitrate is converted to glyoxylate and then malate; induced under low glucose conditions.
gnotobiotic animal An animal that is germ-free or colonized by a known set of microbes.
GOGAT See **glutamate synthase**.
Golgi complex ***or*** **Golgi apparatus** A series of membrane stacks that modifies proteins and helps sort them to the correct eukaryotic cell compartment.
Gram-negative Describing cells that do not retain the Gram stain.
Gram-positive Describing cells that retain the Gram stain and appear dark purple after staining.
Gram stain A differential stain that distinguishes cells that possess a thick cell wall and retain a positively charged stain (Gram-positive) from cells that have a thin cell wall and outer membrane and fail to retain the stain (Gram-negative).
gramicidin A peptide antibiotic that acts as a channel for monovalent cations to cross the cell membrane, thus collapsing the transmembrane ion gradients.
granuloma A thick lesion formed around a site of infection.
granzyme An enzyme, secreted by cytotoxic T cells, that damages target cells.
grazer A first-level consumer, feeding directly on producers.
green alga Also called *chlorophyte*. A member of the eukaryotic group Chlorophyta—microbes that have chloroplasts; closely related to plants (Plantae).
greenhouse effect The trapping of solar radiation heat in the atmosphere by CO_2; a cause of global warming.
griseofulvin An antifungal antibiotic that inhibits cell division.
group translocation A form of active transport in which the transported molecule is modified after it enters the cell, thus keeping a favorable inward concentration gradient for the unmodified extracellular molecule.
growth factor A compound needed for the growth of only certain cells.
growth rate The rate of increase in population number or biomass.
growth rate constant The number of organismal generations per unit time; k.
GS See **glutamine synthetase**.
gut-associated lymphoid tissue (GALT) Lymphatic tissues such as tonsils and adenoids that are found in conjunction with the gastrointestinal tract and contain immune cells.
gut microbiome The microbial community normally present in the intestinal lumen of a healthy host.

H

Haber process Industrial nitrogen fixation, in which dinitrogen is hydrogenated by methane (natural gas) under extreme heat and pressure to form ammonia.
Hadean eon The first eon (major time period) of Earth's existence, from 4.5 to approximately 4.0–3.8 gigayears (Gyr, 10^9 years) before the present.
HAI See **health care–associated infection**.
half-life The amount of time it takes for one-half of a radioactive sample to decay.
Haloarchaea A monophyletic clade of Euryarchaeota that contains extremely halophilic archaea, inhabiting high-salt environments. Formerly known as Halobacteria.

halophile An organism that requires a high extracellular sodium chloride concentration for optimal growth.

halorhodopsin A haloarchaeal membrane-embedded protein that contains retinal and acts as a light-driven chloride pump; it is homologous to bacteriorhodopsin.

hamus *pl.* hami An archaeal cell appendage with grappling hooks that enable cells to connect with each other, adhere to a surface, and form biofilms.

hapten A small compound that must be conjugated to a larger carrier antigen in order to elicit production of an antibody that binds to it.

haustorium *pl.* haustoria A bulbous hyphal extension of a fungal plant pathogen into the host cell.

HAV See **hepatitis A virus**.

HBV See **hepatitis B virus**.

HCV See **hepatitis C virus**.

health care–associated infection (HAI) Any infection contracted by a patient while receiving treatment for a medical condition at a health care facility, such as hospitals, nursing homes, doctor offices, or rehabilitation centers.

heat-shock protein (HSP) A chaperone protein whose synthesis is induced by high-temperature stress.

heat-shock response A coordinated response of cells to higher-than-normal temperatures. It includes changes in the membrane and expression of heat-shock genes.

heavy chain The larger of the two protein types that make up an antibody. Each antibody contains two heavy chains and two light chains.

helper T cell (T_H cell) A T cell that expresses CD4 on its cell surface and secretes cytokines that modulate B-cell isotype, or class, switching.

heme An organic molecule containing a ring of conjugated double bonds surrounding an iron atom. It is involved in redox reactions and oxygen binding.

hemolysin A toxin that lyses red blood cells.

hepatitis An inflammation of the liver, caused by infection or by exposure to a toxic substance.

hepatitis A virus (HAV) A single-stranded RNA picornavirus that causes an acute infection of the liver spread person-to-person by the fecal-oral route.

hepatitis B virus (HBV) A partially double-stranded, circular-DNA hepadnavirus that causes diseases of the liver of varying severity, including acute and chronic hepatitis, cirrhosis, and hepatocarcinoma.

hepatitis C virus (HCV) A single-stranded, positive-sense, linear-RNA flavivirus that is transmitted by blood transfusions and causes 90 percent of transfusion-related cases of hepatitis.

heterocyst In filamentous cyanobacteria, a specialized nitrogen-fixing cell that maintains a reducing environment and excludes O_2.

heteroduplex A double-stranded nucleic acid in which the two strands come from different sources.

Heterokonta See **Stramenopiles**.

heterolactic fermentation A fermentation reaction in which the products are lactic acid, ethanol, and CO_2.

heterologous host An organism engineered for expression of a recombinant product from genes obtained from a species that expresses the product in a natural environment.

heterotrophy The use of external sources of organic carbon compounds for biosynthesis.

Hfr strain A high-frequency recombination bacterial strain, caused by the presence of a chromosomally integrated F factor.

histone A protein that binds eukaryotic DNA and compacts chromosomes in nucleosomes.

HIV See **human immunodeficiency virus**.

HMO See **human milk oligosaccharide**.

Holliday junction A cross-like configuration of recombining DNA molecules that forms during generalized recombination.

holobiont An entity composed of multiple types of organisms, including microbes.

homolog *or* homologous gene A gene derived from a common ancestral gene. Homologs may be orthologs or paralogs.

homologous recombination The process by which two DNA molecules exchange arms by cutting and splicing their helix backbones. Exchange occurs between sequences that are identical or nearly identical, as the machinery requires complementary base pairing to exchange the DNA molecules.

hopanoid *or* hopane A five-ringed hydrocarbon lipid found in bacterial cell membranes.

horizontal gene transfer Also called *lateral gene transfer*. The natural movement of genes from one genome into another, nonprogeny genome.

horizontal transmission In disease, the transfer of a pathogen from one organism into another, nonprogeny organism.

hormogonium *pl.* hormogonia A short, motile chain of three to five cells produced by filamentous cyanobacteria to disseminate their cells.

host factor A trait of an individual host that affects susceptibility to disease, in comparison with other individuals.

host range The species that can be infected by a given pathogen.

HSP See **heat-shock protein**.

human immunodeficiency virus (HIV) A human-specific retrovirus that causes AIDS.

human milk oligosaccharide (HMO) A glycan secreted in milk by a lactating mammal.

humic material Also called *humus*. Phenolic molecules, derived from lignin, that are resistant to degradation and hence very stable in soil.

humoral immunity A type of adaptive immunity mediated by antibodies.

humus See **humic material**.

hybridization The annealing of a nucleic acid strand with another nucleic acid strand containing a complementary sequence of bases. The binding of one nucleic acid strand with a complementary strand.

hydric soil Soil that undergoes periods of anoxic water saturation.

hydrogen bond An electrostatic attraction between a hydrogen bonded to an oxygen or nitrogen and a second, nearby oxygen or nitrogen.

hydrogenotrophy The use of molecular hydrogen (H_2) as an electron donor for a variety of electron acceptors.

hydrologic cycle Also called *water cycle*. The cyclic exchange of water between atmospheric water vapor and Earth's bodies of liquid water.

hydrolysis The cleaving of a bond by the addition of a water molecule.

hydrophilic Soluble in water; either ionic or polar.

hydrophobic Insoluble in water; nonpolar.

hydrothermal vent Also called *thermal vent*. An opening in the seafloor through which superheated water arises, carrying high concentrations of reduced minerals such as sulfides.

3-hydroxypropionate cycle A carbon fixation process in which hydrated CO_2 condenses with acetyl-CoA to form 3-hydroxypropionate.

hyperthermophile Also called *extreme thermophile*. An organism adapted for optimal growth at extremely high temperatures, generally above 80°C, and as high as 121°C.

hypertonic Having more solutes than another environment separated by a semipermeable membrane. Water will tend to flow toward the hypertonic solution.

hypha *pl.* **hyphae** The threadlike filament that forms the mycelium of a fungus.
hypotonic Having fewer solutes than another environment separated by a semipermeable membrane. Water will tend to flow away from the hypotonic solution.
hypoxia A state of lower-than-normal oxygen concentration.

I

icosahedral capsid For a virus, a crystalline protein shell with 20 identical faces, enclosing the nucleic acid.
ID_{50} See **infectious dose 50%**.
identification The recognition of the species (or higher taxonomic category) of a microbe isolated in pure culture.
idiotype An amino acid difference in the antigen-binding site (N terminus of heavy or light chains) that distinguishes different antibodies within an individual.
IEF See **isoelectric focusing**.
IEL See **intraepithelial lymphocyte**.
IgA An antibody isotype that contains the alpha heavy chain. It can be secreted and is found in tears, saliva, breast milk, and so on.
IgD An antibody isotype that contains the delta heavy chain. It is found on B-cell membranes.
IgE An antibody isotype that contains the epsilon heavy chain. It is involved in degranulation of mast cells.
IgG An antibody isotype that contains the gamma heavy chain. It is found in serum.
IgM The first antibody isotype detected during the early stages of an immune response. It contains the mu heavy chain and is found as a pentamer in serum.
IL-1 See **interleukin 1**.
ILC See **innate lymphoid cell**.
immediate hypersensitivity See **type I hypersensitivity**.
immersion oil An oil with a refractive index similar to glass that minimizes light-ray loss at wide angles, thereby minimizing wavefront interference and maximizing resolution.
immune complex disease See **type III hypersensitivity**.
immune system An organism's cellular defense system against pathogens.
immunity A body's resistance to a specific disease.
immunization The stimulation of an immune response by deliberate inoculation with a weakened pathogen, in hopes of providing immunity to disease caused by the pathogen.
immunogen An antigen that, by itself, can elicit antibody production.
immunogenicity A measure of the effectiveness of an antigen in eliciting an immune response.
immunoglobulin A member of a family of proteins that contain a 110-amino-acid domain with an internal disulfide bond. Members include antibodies and major histocompatibility proteins.
immunological specificity The ability of antibodies produced in response to a particular epitope to bind that epitope almost exclusively. Antibodies made to one epitope bind only weakly, if at all, to other epitopes.
immunopathogenesis The process by which an immune response or the products of an immune response cause disease.
immunoprecipitation The antibody-mediated cross-linking of antigens to form large, insoluble complexes. It is used in research labs and is normally seen only in vitro.
imperfect fungus See **mitosporic fungus**.
in vivo expression technology (IVET) Techniques that identify bacterial genes that are expressed only when the organism is growing inside a host.
incidence The number of new cases of a disease in a given location over a specified time.
index case Also called *patient zero*. The first case of an infectious disease, and an important piece of data for helping to contain the spread of disease.
indigenous microbiota Microbes found naturally in a particular location, often in association with a food substrate.
inducer A molecule that binds to a repressor and prevents it from binding to the operator sequence.
inducer exclusion The ability of glucose to cause metabolic changes that prevent the cellular uptake of less favorable carbon sources that could cause unnecessary induction.
induction Increased transcription of target genes because an inducer binds to a repressor and prevents repressor-operator binding.
industrial fermentor The equipment used to grow microbes on an industrial scale.
industrial microbiology The commercial exploitation of microbes.
industrial strain A microbial strain whose characteristics are optimized for industrial use.
infection The growth of a pathogen or parasite in or on a host.
infection cycle The route a pathogen takes as it moves from one host into another.
infectious dose 50% (ID_{50}) The number of bacteria or virions required to cause disease symptoms in 50 percent of an experimental group of hosts.
inflammasome A cytoplasmic multiprotein complex that promotes the maturation of inflammatory cytokines IL-1β and IL-18. Inflammasome assembly is triggered by NLR interactions with microbe-associated molecular patterns (MAMPs).
injera A highly fermented Ethiopian flat bread made from the grain teff.
innate immunity Also called *nonadaptive immunity*. Nonspecific mechanisms for protecting against pathogens.
innate lymphoid cell (ILC) A lymphocyte-like cell in the intestinal lamina propria that lacks B- or T-cell receptors but secretes pro-inflammatory cytokines.
inner leaflet The layer of the cell membrane phospholipid bilayer that faces the cytoplasm.
inner membrane *or* **inner cell membrane** In Gram-negative bacteria, the membrane in contact with the cytoplasm, equivalent to the cell membrane.
insertion The addition of nucleotides to the middle of a DNA sequence.
insertion sequence (IS) A simple transposable element consisting of a transposase gene flanked by short, inverted-repeat sequences that are the target of transposase.
integral protein See **transmembrane protein**.
integron A large and complex mobile element that can accept, or capture, different antibiotic resistance gene cassettes to use later, when needed.
interactome All of the proteins that interact with other proteins in a cell.
interference The interaction of two wavefronts. Interference can be additive (amplitudes in phase, constructive) or subtractive (amplitudes out of phase, destructive).
interferon A host-secreted immunomodulatory protein that inhibits viral replication.
interleukin 1 (IL-1) A cytokine release by macrophages.
intermediate filament A eukaryotic cytoskeletal protein that is composed of various proteins depending on the cell type.
internal ribosome entry site (IRES) A site within an mRNA sequence where a ribosome can bind and initiate translation.
intimin A pathogenic *E. coli* adhesion protein that binds tightly to an *E. coli*–produced receptor injected into host cells.
intracellular pathogen A pathogen that lives within a host cell.

intraepithelial lymphocyte (IEL) A lymphocyte embedded among epithelial cells that line the intestine.
intron In eukaryotic genes, an intervening sequence that does not code for protein and is spliced out of the mRNA prior to translation.
inversion A mutation in which a DNA fragment is flipped within a chromosome. It may allow or repress the transcription of a particular gene.
inverted repeat A DNA sequence that is found in an identical but inverted form at two sites on the same double helix (e.g., 5′-ATCGATCGnnnnnnCGATCGAT-3′).
ion An atom or molecule containing negative or positive charge—that is, a number of electrons, respectively, greater than or less than the number of protons.
ion gradient A difference in concentration of an ion across a membrane.
ionic bond A chemical bond between ions of positive and negative charge.
IRES See **internal ribosome entry site**.
IS See **insertion sequence**.
isoelectric focusing (IEF) A technique that separates proteins according to their charge, via the migration of proteins to their isoelectric point in a pH gradient.
isoelectric point The pH at which there is no net charge on an amino acid or a protein.
isolate A microbe that has been obtained from a specific location and grown in pure culture.
isoprenoid A condensed isoprene chain, found in archaeal membrane lipids.
isotonic Being in osmotic balance, having equal concentrations of solutes on both sides of a semipermeable membrane. A cell in an isotonic environment will neither gain nor lose water.
isotope An atom of an element with a specific number of neutrons. For example, carbon-12 (^{12}C), carbon-13 (^{13}C), and carbon-14 (^{14}C) are all isotopes of carbon.
isotope ratio The ratio of amounts of two different isotopes of an element. It may serve as a biosignature if the ratio between certain isotopes of a given element is altered by biological activity.
isotype An antibody class within a species that is defined by the structure of the antibody's heavy chain. IgG, IgA, IgD, and IgE are examples of isotypes. An isotype from one species contains species-specific amino acid sequences that are present in the heavy chain of all members of that species.
isotype switching Also called *class switching*. A change in the predominant antibody type produced by a cell.
iTAG analysis See **metabarcoding**.
IVET See **in vivo expression technology**.

J

joule (J) The standard SI unit for energy.
jumping gene See **transposable element**.

K

kimchi A popular Korean food based on brine-fermented cabbage.
Kirby-Bauer assay A method for determining antibiotic susceptibility. Antibiotic-impregnated disks are placed on an agar plate whose surface has been confluently inoculated with a test organism. The antibiotic diffuses away from the disk and inhibits growth of susceptible bacteria. The width of the inhibitory zone is proportional to the susceptibility of the organism.
knockout mutation A mutation that completely eliminates the activity of a gene product.
Koch's postulates Four criteria, developed by Robert Koch, that should be met for a microbe to be designated the causative agent of an infectious disease.
Korarchaeota A phylum of archaea in the TACK superphylum that contains anaerobic hyperthermophiles in hot springs and deep-sea thermal vents.
Krebs cycle See **tricarboxylic acid cycle**.

L

labile toxin (LT) An *E. coli* enterotoxin, destroyed by heat, that increases cellular cAMP concentrations.
laboratory evolution See **experimental evolution**.
lactic acid fermentation A fermentation reaction that generates lactic acid from reduction of pyruvic acid.
lag phase A period of cell culture, occurring right after bacteria are inoculated into new media, during which there is slow growth or no growth.
laminar flow biological safety cabinet An air filtration appliance that removes pathogenic microbes from within the cabinet.
Langerhans cell A specialized, phagocytic dendritic cell that is the predominant cell type in skin-associated lymphatic tissue.
late gene A viral gene expressed late in the replication cycle.
late-phase anaphylaxis Anaphylaxis caused by leukotrienes that are released by eosinophils recruited by mast cells.
latent period The time in the viral life cycle when progeny virions have formed but are still within the host cell.
latent state A period of the infection process during which a pathogenic agent is dormant in the host and cannot be cultured.
lateral gene transfer See **horizontal gene transfer**.
law of mass action The tendency of a reaction at equilibrium to return to equilibrium after the concentrations of products or reactants are altered.
LD_{50} See **lethal dose 50%**.
leaflet One of the two lipid layers in a phospholipid bilayer. The inner leaflet of the cell membrane faces the cytoplasm.
leaven For bread dough, to cause to rise by generating air spaces, usually through carbon dioxide production by microbial fermentation.
leghemoglobin An iron-bearing plant protein that sequesters oxygen to maintain an anoxic environment for nitrogenase within cells containing bacteroids.
lens An object composed of transparent, refractive material that bends light rays to converge at a focal point (or to diverge from an imaginary point). For electron microscopy, a series of magnets arranged as a magnetic lens bends electron beams to converge or diverge.
lentivector or lentiviral vector A gene transfer vector derived from a lentivirus such as HIV; designed to integrate genes into a host chromosome.
lentivirus A member of a family of retroviruses with a long incubation period. An example is HIV.
lethal dose 50% (LD_{50}) A measure of virulence; the number of bacteria or virions required to kill 50 percent of an experimental group of hosts.
leukocidin A toxin that lyses white blood cells.
leukocyte White blood cell.
library construction The amplification and processing of DNA for sequencing reactions, such as those of next-generation sequencing (NGS).
lichen A simple multicellular organism formed by a mutualistic relationship between a fungus and an alga or cyanobacterium.
light chain The smaller of the two protein types that make up an antibody. Each antibody contains two heavy chains and two light chains.

light microscopy (LM) Observation of a microscopic object based on light absorption and transmission.

lignin A complex aromatic organic compound that forms the key structural support for trees and woody stems.

limiting nutrient A nutrient whose depletion generally restricts growth in a given ecosystem.

lincosamide Any of a class of bacteriostatic antibiotics that include a pyrrolidine ring linked to a pyranose (e.g., clindamycin).

lineage In a phylogenetic tree, a line of individuals, past and present, that descends from one common ancestor.

lipid A The anchor lipid of lipopolysaccharide (LPS), composed of glucosamine plus six lipid chains.

lipopolysaccharides (LPS) Structurally unique phospholipids found in the outer leaflet of the outer membrane in Gram-negative bacteria. Many are endotoxins.

lithotrophy Also called *chemolithotrophy.* The metabolic oxidation of inorganic compounds to yield energy and fix single-carbon compounds into biomass.

LM See **light microscopy**.

logarithmic (log) phase See **exponential phase**.

Lokiarchaeota A phylum in the Asgard superphylum of Archaea that contains deep-sea thermophiles isolated from a thermal vent named Loki's Castle. Formerly known as the Deep-Sea Archaeal Group/Marine Benthic Group B (DSAG/MBGB).

long terminal repeat (LTR) A repeated nucleic acid sequence at the 5′ and 3′ ends of a provirus.

loss-of-function mutation A mutation that eliminates or decreases the function of the gene product.

LPS See **lipopolysaccharides**.

LT See **labile toxin**.

LTR See **long terminal repeat**.

Lyme disease One form of borreliosis. A tick-borne disease caused by *Borrelia burgdorferi*, which may involve skin lesions and arthritis.

lymph node A secondary lymphatic organ, formed by the convergence of lymphatic vessels, that traps foreign particles from local tissue and presents them to resident immune cells.

lymphocyte A mononuclear leukocyte (white blood cell) that is a product of lymphoid tissue and participates in immunity (e.g., B cell and T cell).

lyophilization See **freeze-drying**.

lysate The contents of broken cells; may include virus particles.

lysis Also called *burst.* The rupture of the cell by a break in the cell wall and membrane.

lysogeny A viral life cycle in which the viral genome integrates into and replicates with the host genome, but retains the ability to initiate host cell lysis.

lysosome An acidic eukaryotic organelle that aids digestion of molecules. Not found in plant cells.

M

M cell A phagocytic innate immune cell (microfold cell) found between intestinal epithelial cells.

MAC See **membrane attack complex**.

macrolide Any of a group of antibiotics containing a large lactone ring (e.g., erythromycin).

macronucleus A form of nucleus found in ciliates that is derived from gene amplification and rearrangement of micronuclear DNA; contains actively transcribed genes.

macronutrient A nutrient that an organism needs in large quantity.

macrophage A mononuclear, phagocytic, antigen-presenting cell of the immune system.

magnetosome An organelle that contains the mineral magnetite and thus enables microbes to sense a magnetic field.

magnetotaxis The ability to direct motility along magnetic field lines.

magnification An increase in the apparent size of a viewed object as an optical image.

major histocompatibility complex (MHC) Transmembrane cell proteins important for recognizing self and for presenting foreign antigens to the adaptive immune system.

malaria A disease caused by the apicomplexan *Plasmodium falciparum*, transmitted by mosquitoes.

malolactic fermentation Fermentation of l-malate (a side product of glucose fermentation) by *Oenococcus oeni* bacteria; an important process in winemaking.

MAMP See **microbe-associated molecular pattern**.

marine snow Microbial biofilms on particles suspended in marine water.

mass spectrometry An analytical technique that measures the mass of molecules. Molecules are ionized and sorted according to their mass-to-charge (m/z) ratio.

mast cell A white blood cell that secretes proteins that aid innate immunity. Mast cells reside in connective tissues and mucosa and do not circulate in the bloodstream.

matrix protein A protein, found in some viruses, that is located between the capsid and the membrane envelope.

MCP See **methyl-accepting chemotaxis protein**.

MDR efflux pump See **multidrug resistance efflux pump**.

mechanical transmission A nonspecific means of transfer of a pathogen to a host, such as the transmission of a virus through a wound, or of a bacterial pathogen from the body surface of a vector (such as a fly).

meiosis A form of cell division by which a diploid eukaryotic cell generates haploid sex cells that contain recombinant chromosomes.

membrane attack complex (MAC) A cell-destroying pore produced in the membrane of invading bacteria by the host cell complement cascade.

membrane-permeant weak acid An acid that exists in equilibrium between negatively charged and uncharged forms, such as acetic acid. The uncharged form can penetrate the membrane.

membrane-permeant weak base A base that exists in equilibrium between positively charged and uncharged forms, such as methylamine. The uncharged form can penetrate the membrane.

membrane potential Energy stored as an electrical voltage difference across a membrane.

memory B cell A long-lived type of lymphocyte preprogrammed to produce a specific antibody. After encountering their activating antigen, memory B cells differentiate into antibody-producing plasma cells.

merozoite The form of *Plasmodium falciparum*, the causative agent of malaria, that invades red blood cells.

mesocosm A small, controlled model ecosystem.

mesophile An organism with optimal growth between 20°C and 40°C.

messenger RNA (mRNA) An RNA molecule that encodes a protein.

metabarcoding Also called *iTAG analysis.* The use of a short DNA sequence, such as the gene for SSU rRNA, to screen taxa from a metagenome.

metabolic island A genomic island that contributes genes involved in metabolism.

metagenome The sum of genomes of all members of a community of organisms.

metagenomics The study of community genomes, or metagenomes.

metatranscriptomics The study of all the RNA transcripts expressed by members of a community, known as a metatranscriptome.

methane gas hydrate A crystalline material in which methane molecules are surrounded by a cage of water molecules. This molecular configuration is found in the deep ocean.

methanogen An archaeon that uses hydrogen to reduce CO_2 and other single-carbon compounds to methane, yielding energy. Clades of methanogens branch within the superphylum Euryarchaeota.

methanogenesis An energy-yielding metabolic process that releases methane, commonly from hydrogen gas and oxidized one- or two-carbon compounds. It is unique to archaea.

methanotrophy The metabolic oxidation of methane to yield energy.

methyl-accepting chemotaxis protein (MCP) Also called *chemoreceptor.* A cell-membrane signal transduction protein that becomes methylated during adaptation to a chemotactic signal.

methyl mismatch repair A DNA repair system that fixes misincorporation of a nucleotide after DNA synthesis. The unmethylated daughter strand is corrected to complement the methylated parental strand.

methylome All of the bases in the genome that are chemically modified by methylation.

methylotrophy The metabolic oxidation of single-carbon compounds such as methanol, methylamine, or methane to yield energy.

MGI See **mobilizable genomic island**.

MHC See **major histocompatibility complex**.

MHC restriction The ability of T cells to recognize only those antigens complexed to self MHC molecules.

MIC See **minimal inhibitory concentration**.

microaerophilic Requiring oxygen at a concentration lower than that of the atmosphere, but unable to grow in high-oxygen environments.

microbe An organism or virus too small to be seen with the unaided human eye.

microbe-associated molecular pattern (MAMP) Formerly called *pathogen-associated molecular pattern (PAMP).* Molecules associated with groups of microbes, both pathogenic and nonpathogenic, that are recognized by cells of the innate immune system.

microbial mat A complex biofilm of microbes, usually containing multiple layers.

microbiota *or* microbiome The total community of microbes associated with an organism (such as the human body) or with a defined habitat (such as soil or plants).

microfilament Also called *actin filament.* A eukaryotic cytoskeletal protein composed of polymerized actin.

microfossil A microscopic fossil in which calcium carbonate deposits have filled in the form of ancient microbial cells.

micronucleus A form of nucleus found in ciliates; contains a diploid set of chromosomes and undergoes meiosis for sexual exchange by conjugation.

micronutrient A nutrient that an organism needs in small quantity, typically a vitamin or a mineral.

microplankton Plankton consisting of microbes approximately 20–1,000 mm in diameter.

microscope A tool that increases the magnification of specimens to enable viewing at higher resolution.

microtubule A eukaryotic cytoskeletal protein composed of polymerized tubulin.

minimal defined medium A solution of chemically defined compounds for organismal growth that contains only the minimal components required for growth.

minimal inhibitory concentration (MIC) The lowest concentration of a drug that will prevent the growth of an organism.

miso A Japanese condiment, made from ground soy and rice, salted and fermented by the mold *Aspergillus oryzae*.

missense mutation A point mutation that alters the sequence of a single codon, leading to a single amino acid substitution in a protein.

mitochondrion *pl.* mitochondria An organelle of endosymbiotic origin that produces ATP through the use of an electron transport system to generate a proton potential. O_2 is the final electron acceptor to produce H_2O.

mitosis The orderly replication and segregation of eukaryotic chromosomes, usually prior to cell division.

mitosporic fungus Also called *imperfect fungus.* A species of fungus that generates spores by mitosis and lacks a known sexual cycle.

mixed-acid fermentation A bacterial fermentation process in which pyruvate is converted to several different organic acids, as well as ethanol, CO_2, and H_2O.

mixotrophy Metabolism that includes CO_2 fixation through photosynthesis and catabolism of organic compounds.

mobilizable genomic island (MGI) A genomic island that can be transferred to another cell via the machinery of a conjugative transposon.

modular enzyme A multifunctional enzyme in which several domains or subunits conduct sequential steps to generate a product.

modular synthesis The construction of a polymeric molecule based on incorporation of repeating units.

MOI See **multiplicity of infection**.

molarity A unit of concentration measured as the number of moles of solute per liter of solution.

molecular clock The use of DNA or RNA sequence information to measure the time of divergence among different species.

molecular formula A notation indicating the number and type of atoms in a molecule. For example, H_2O is the molecular formula for water.

Mollicutes See **Tenericutes**.

monocyte A white blood cell with a single nucleus that can differentiate into a macrophage or a dendritic cell.

monosaccharide The monomer unit of sugars. Monosaccharides have a molecular formula of $(CH_2O)_n$.

monoderm Enclosed by a single phospholipid membrane.

monophyletic Having a single evolutionary origin—that is, diverging from a common ancestor.

monophyletic group See **clade**.

mordant A chemical binding agent that causes specimens to retain stains better.

mother cell In sporulation of Gram-positive bacteria, the larger cell that forms during the asymmetrical cell division leading to spore formation. The mother cell will engulf the forespore, but then disintegrates as the forespore matures.

motility The ability of a microbe to direct its own movement.

mRNA See **messenger RNA**.

mucociliary escalator The ciliated mucous lining of the trachea, bronchi, and bronchioles that sweeps foreign particles up and away from the lungs.

mucosal immunity The portion of the innate and adaptive immune systems that protects the mucosa from microbial invasion.

Mueller-Hinton agar A specialized, standardized, *para*-aminobenzoic acid–free medium used for the Kirby-Bauer assay.

multidrug resistance (MDR) efflux pump A transmembrane protein pump that can export many different kinds of antibiotics with diverse structure.

multiplex PCR A polymerase chain reaction that uses multiple pairs of oligonucleotide primers to amplify several different DNA sequences simultaneously.

multiplicity of infection (MOI) The ratio of infecting virions to host cells.

murein See **peptidoglycan**.
murein lipoprotein Also called *Braun lipoprotein*. The major lipoprotein that connects the outer membrane of Gram-negative bacteria to the peptidoglycan cell wall.
mutagen A chemical that damages DNA and increases the rate of mutations.
mutation A heritable change in a DNA sequence.
mutator strain A strain of cells with a high mutation rate, usually due to a mutation in a DNA repair enzyme.
mutualism A symbiotic relationship in which both partners benefit.
mycelium *pl.* **mycelia** A single mass of fungal hyphae that projects into the air (aerial mycelium) or into the growth substrate (surface mycelium).
mycolic acid One of a diverse class of sugar-linked fatty acids found in the cell envelopes of mycobacteria such as *Mycobacterium tuberculosis*.
mycology The study of fungi.
mycorrhizae *sing.* **mycorrhiza** Fungi involved in an intimate mutualism with plant roots, in which nutrients are exchanged.
myxospore A durable spherical cell produced by the fruiting body of myxobacteria.

N

N-terminal rule The tendency of the N-terminal amino acid of a protein to influence protein stability.
NAD See **nicotinamide adenine dinucleotide**.
Nanoarchaeota A phylum in the DPANN superphylum of Archaea that includes very small cells, which are obligate symbionts of vent hyperthermophiles.
nanopore DNA strand sequencing A type of nucleic acid sequence determination that relies on the changes in electric current that occur when a single strand of DNA passes through a nanopore structure.
nanoscale secondary ion mass spectrometry (NanoSIMS) A technique of chemical imaging in which an ionizing beam breaks off organic ions from a sample, which fly off the sample and are captured for analysis by a mass spectrometer.
nanotube A tube of plasma membrane that connects the cytoplasm of two cells, forming a conduit through which intracellular materials or pathogens may pass.
nanowire In microbiology, a bacterial appendage that conducts electric current.
nasopharynx The passage leading from the nose to the oral cavity.
native conformation The fully folded, functional form of a protein.
natto A soybean product, similar to tempeh, produced by alkaline fermentation.
natural killer (NK) cell A lymphocyte that does not need antigen stimulation to kill tumor or infected host cells by inserting granules containing perforin.
natural selection The change in frequency of genes in a population under environmental conditions that favor some genes over others.
necrotizing fasciitis Also known as *flesh-eating disease*. A severe skin infection usually caused by the Gram-positive coccus *Streptococcus pyogenes*.
negative selection In immunology, the destruction of T cells bearing T-cell receptors that bind strongly to self MHC proteins displayed on thymus epithelial cells.
negative stain A stain that colors the background and leaves the specimen unstained.
NER See **nucleotide excision repair**.
NET See **neutrophil extracellular trap**.
neuraminidase inhibitor Any of a class of anti-influenza drugs that target neuraminidase on the viral envelope and decrease the number of virus particles produced.
neutralophile An organism with an optimal growth range in environments between pH 5 and pH 8.
neutrophil A white blood cell of the innate immune system that can phagocytose and kill microbes.
neutrophil extracellular trap (NET) A net of chromatin (including DNA) and antimicrobial peptides expelled by dying neutrophils to trap and injure nearby pathogenic bacteria.
NHEJ See **nonhomologous end joining**.
niche An organism's environmental requirements for existence and its relations with other members of the ecosystem.
niche construction The actions of an organism that alter its environmental niche and change its chance of survival in that niche.
nicotinamide adenine dinucleotide (NAD) An energy carrier in the cell that can donate (NADH) or accept (NAD^+) electrons.
nitrification The oxidation of reduced nitrogen compounds to nitrite or nitrate.
nitrifier An organism that converts reduced nitrogen compounds to nitrite or nitrate.
nitrogen fixation The ability of some prokaryotes to reduce inorganic diatomic nitrogen gas (N_2) to two ammonium ions ($2NH_4^+$).
nitrogen-fixing bacterium A bacterium that can reduce diatomic nitrogen gas (N_2) to two molecules of ammonium ion (NH_4^+).
nitrogenase The enzyme that catalyzes nitrogen fixation.
nitrogenous base See **nucleobase**.
Nitrospirae A phylum of Gram-negative bacteria, many of which are lithotrophs, oxidizing nitrite to nitrate, or ammonia to nitrate.
NK cell See **natural killer cell**.
NLR See **NOD-like receptor**.
NOD-like receptor (NLR) A eukaryotic cytoplasmic protein that recognizes particular microbe-associated molecular patterns (MAMPs) present on microorganisms.
node In a phylogenetic tree, the most recent common ancestor of branching descendents.
noise Variance from a mean in an assay. A system in which individual results vary greatly from a mean that was derived from many results is a noisy system.
Nomarski microscopy See **differential interference contrast microscopy**.
nomenclature The naming of different taxonomic groups of organisms.
nonadaptive immunity See **innate immunity**.
nonhomologous end joining (NHEJ) A pathway that repairs double-strand DNA breaks by direct ligation without the need for large regions of homology.
nonpolar covalent bond A covalent bond in which the electrons in the bond are shared equally by the two atoms.
nonribosomal peptide A peptide with antimicrobial activity synthesized by modular enzymes and not by ribosomes.
nonribosomal peptide synthetase A modular enzyme that synthesizes a peptide without using a ribosome.
nonsense mutation A mutation that changes an amino acid codon into a premature stop codon.
nori A Japanese food obtained from the red algae *Porphyra* species.
norovirus Also known as *Norwalk virus*. A nonenveloped ssRNA virus that causes severe diarrhea in children and adults.
northern blot A technique to detect specific RNA sequences. Sample RNA is subjected to gel electrophoresis, transferred to a blot, and probed with a labeled cDNA that will hybridize to target RNA sequences.

Norwalk virus See **norovirus**.
nosocomial Hospital-acquired; commonly refers to an infectious agent.
NP See **nucleocapsid protein**.
nucleobase Also called *nitrogenous base*. A planar, heteroaromatic nitrogen-containing base that forms a nucleotide of nucleic acids; nucleobases determine the information content of DNA and RNA. There are five nucleobases: adenine, cytosine, guanine, thymine, and uracil.
nucleocapsid protein (NP) A protein that coats a viral genome.
nucleoid The looped coils of a bacterial chromosome.
nucleolus *pl.* nucleoli A region inside the nucleus where ribosome assembly begins.
nucleomorph A vestigial nucleus within a eukaryotic cell, evolved by genetic reduction from the nucleus of an endosymbiont.
nucleotide The monomer unit of nucleic acids, consisting of a five-carbon sugar, a phosphate, and a nitrogenous base.
nucleotide excision repair (NER) A DNA repair mechanism that cuts out damaged DNA. New, correctly base-paired DNA is synthesized by DNA polymerase I.
nucleus *pl.* nuclei A eukaryotic organelle that contains DNA.
numerical aperture The product of the refractive index of the medium and sin θ (where θ is the angle of aperture). As numerical aperture increases, the magnification increases.

O

O antigen *or* O polysaccharide A sugar chain that connects to the core polysaccharide of lipopolysaccharides.
objective lens In a compound microscope, the lens that is closest to the specimen and generates the initial magnification.
obligate intracellular pathogen A pathogen that can replicate only inside host cells.
ocular lens In a compound microscope, the lens situated closest to the observer's eye; part of the eyepiece.
Okazaki fragments Short fragments of DNA that are synthesized on the lagging strand during DNA synthesis.
oligotrophic lake A lake having a low concentration of organic nutrients; the opposite of a eutrophic lake.
omics Analytical tools that probe community samples at the molecular level (metagenomes, metatranscriptomes, and single-cell genomes), as well as whole-cell protein populations (proteomes). Examples include genomics, metagenomics, metatranscriptomics, and proteomics.
OMZ See **oxygen minimum zone**.
oncogenic virus A virus that causes cancer.
oomycete Also called *water mold*. A member of the eukaryotic group Oomycetes—heterokont protists whose life cycle resembles that of fungi; formerly classified as fungi (Oomycota).
open reading frame (ORF) A DNA sequence predicted to encode a protein.
operational taxonomic unit (OTU) A taxonomic group defined by a designated degree of similarity among members based on DNA sequence.
operator A region of DNA to which the repressor protein binds. Operators are usually located near promoters.
operon A collection of genes that are in tandem on a chromosome and are transcribed into a single RNA.
opportunistic pathogen A microbe that normally is not pathogenic but can cause infection or disease in an immunocompromised host organism.
opsonin An antibody that renders its target (e.g., bacteria) susceptible to phagocytosis.
opsonization The coating of pathogens with antibodies that aid pathogen phagocytosis by innate immune cells.
opsonize To bind IgG antibodies to microbes in order to enhance microbial phagocytosis by host immune cells.
optical density A measure of how many particles are suspended in a solution, based on light scattering by the suspended particles.
optical isomer Also called *enantiomer*. Either of two forms of a molecule that are mirror images of each other. Molecules that contain a chiral carbon can have optical isomers.
oral groove A mouthlike structure of a ciliate cell, for food uptake.
ORF See **open reading frame**.
organelle A membrane-enclosed compartment within eukaryotic cells that serves a specific function.
organic molecule A molecule that contains a carbon-carbon bond.
organotrophy Also called *chemoorganotrophy* or *chemoheterotrophy*. The metabolic oxidation of organic compounds to yield energy without absorption of light.
origin (oriC) The region of a bacterial or archaeal chromosome where DNA replication initiates.
oropharynx The area between the soft palate and the upper edge of the epiglottis.
ortholog *or* orthologous gene A gene present in more than one species that derived from a common ancestral gene and encodes the same function.
osmolarity A measure of the concentration of solute molecules in solution.
osmosis The diffusion of water from regions of high water concentration (low solute) to regions of low water concentration (high solute) across a semipermeable membrane.
osmotic pressure Also called *turgor pressure*. Pressure exerted by the osmotic flow of water through a semipermeable membrane.
OTU See **operational taxonomic unit**.
outer leaflet The layer of the cell membrane phospholipid bilayer that faces away from the cytoplasm.
outer membrane In Gram-negative bacteria, a membrane external to the cell wall.
outer membrane–associated protein A protein linked to or embedded in the outer membrane.
outgroup In a phylogenetic tree, a taxon that diverged from a lineage before the ancestor shared by all other taxa in the tree.
oxazolidinone One of a class of synthetic antibiotics, containing an oxazole ring, that inhibit protein synthesis.
oxidative burst A large increase in the oxygen consumption of immune cells during phagocytosis of pathogens as the immune cells produce oxygen radicals to kill the pathogen.
oxidative phosphorylation A process of an electron transport chain that uses diatomic oxygen as a final electron acceptor and generates a proton gradient across a membrane for the production of ATP via ATP synthase.
oxidoreductase An electron transport system protein that accepts electrons from one molecule (oxidizing that molecule), and donates electrons to a second molecule, thereby reducing the second molecule.
oxygen minimum zone (OMZ) The region of the marine water column in which oxygen is depleted by respiration; usually at a mid level, between the aerated upper water and the deeper oxygenated water where organic food is scarce.
oxygenic Z pathway An ATP-producing photosynthetic pathway consisting of photosystems I and II. Water serves as the initial electron donor (generating O_2), and $NADP^+$ is the final electron acceptor (generating NADPH).

P

P site See **peptidyl-tRNA site**.

PA See **protective antigen**.

palindrome A DNA sequence in which the top and bottom strands have the same sequence in the 5′-to-3′ direction.

PAMP See **microbe-associated molecular pattern**.

pandemic An epidemic that occurs over a wide geographic area.

pangenome All the genes possessed by all individual members of a species.

panspermia The hypothesis that life forms originated elsewhere in the universe and "seeded" life on Earth.

paralog *or* **paralogous gene** A gene that arises by gene duplication within a species and evolves to carry out a different function from that of the original gene.

paraphyletic group A group of organisms that contains members of larger clades that do not share a unique common ancestor excluding members of other groups.

pararetrovirus Also called *DNA reverse-transcribing virus.* A virus with a double-stranded DNA genome that generates an RNA intermediate and thus requires reverse transcriptase to generate progeny DNA genomes.

parasite Any bacterium, virus, fungus, or protozoan (protist) that colonizes and harms its host; the term commonly refers to protozoa and to invertebrates.

parasitism A symbiotic relationship in which one member benefits and the other is harmed.

parenteral route of transmission A method by which an infectious agent enters the body via injection into the bloodstream, often by a mosquito or other insect.

parfocal In a microscope with multiple objective lenses, having the objective lenses set at different heights that maintain focus when switching among lenses.

passive transport Net movement of molecules across a membrane without energy expenditure by the cell.

pasteurization The heating of food at a temperature and time combination that will kill spore-like structures of *Coxiella burnetii.*

pathogen A bacterial, viral, or fungal agent of disease.

pathogen-associated molecular pattern (PAMP) See **microbe-associated molecular pattern**.

pathogenesis The processes through which microbes cause disease in a host.

pathogenicity The ability of a microorganism to cause disease.

pathogenicity island A type of genomic island in which the stretch of DNA contains virulence factors and may have been transferred from another genome.

patient zero See **index case**.

pattern recognition receptor (PRR) A protein receptor that recognizes microbe-associated molecular patterns (MAMPs) and signals production of cytokines.

PBP See **penicillin-binding protein**.

PCM See **phase-contrast microscopy**.

PCR See **polymerase chain reaction**.

PDC See **pyruvate dehydrogenase complex**.

pelagic zone The water column of the open ocean, away from the shore and the ocean floor.

penicillin An antibiotic, produced by the *Penicillium* mold, containing a beta-lactam ring; it blocks cross-bridge formation during peptidoglycan synthesis.

penicillin-binding protein (PBP) A bacterial protein, involved in cell wall synthesis, that is the target of the penicillin antibiotic.

pentose phosphate cycle See **Calvin cycle**.

pentose phosphate pathway (PPP) *or* **pentose phosphate shunt (PPS)** An alternate glycolytic pathway in which glucose 6-phosphate is first oxidized and then decarboxylated to ribulose 5-phosphate, ultimately generating 1 ATP and 2 NADPH.

peptide bond The covalent bond that links two amino acid monomers.

peptidoglycan Also called *murein.* A polymer of peptide-linked chains of amino sugars; a major component of the bacterial cell wall.

peptidyltransferase A ribozyme that catalyzes the formation of peptide bonds.

peptidyl-tRNA site (P site) The region of a ribosome that contains the growing protein attached to a tRNA.

perforin A cytotoxic protein, secreted by T cells, that forms pores in target cell membranes.

peripheral membrane protein A protein that is associated with a membrane but does not span the phospholipid bilayer.

periplasm In Gram-negative bacteria, the gel-like solution between the outer and inner membrane; it contains the cell wall.

permease A substrate-specific carrier protein in the membrane.

persister cell Any of a subpopulation of dormant organisms that arise within a population of antibiotic-susceptible bacteria and are tolerant to bactericidal antibiotics during treatment. On removal of the drug, persisters can grow and reestablish infection.

peroxisome A eukaryotic organelle that converts hydrogen peroxide to water.

petechia *pl.* **petechiae** A pinpoint capillary hemorrhage due to the absence of clotting factors. Petechiae may indicate the presence of endotoxin.

Petri dish *or* **Petri plate** A round dish with vertical walls covered by an inverted dish of slightly larger diameter. The smaller dish can be filled with a substrate for growing microbes.

PFU See **plaque-forming unit**.

phage See **bacteriophage**.

phage display A technique in which a phage particle contains recombinant coat proteins expressed by genes encoding a coat protein fused to the protein of interest, such as a vaccine antigen.

phagocytosis A form of endocytosis in which a large extracellular particle is brought into the cell.

phagosome A large intracellular vesicle that forms as a result of phagocytosis.

phase-contrast microscopy (PCM) Observation of a microscopic object based on the differences in the refractive index between cell components and the surrounding medium. Contrast is generated as the difference between refracted light and transmitted light shifts out of phase.

phase variation A gene regulatory mechanism that reversibly changes the DNA sequence within or near the gene. One mechanism involves site-specific recombination that flips a DNA sequence in a chromosome.

phenol coefficient test A test of the ability of a disinfectant to kill bacteria; the higher the coefficient, the more effective the disinfectant.

phenotype The observable characteristics of an organism.

phosphodiester bond The phosphoryl bond that includes ester links with two adjacent nucleotides in a nucleic acid.

phospholipid The major component of membranes. A typical phospholipid is composed of a core of glycerol to which two fatty acids and a modified phosphate group are condensed.

phospholipid bilayer Two layers of phospholipids; the hydrocarbon fatty acid tails face the interior of the bilayer, and the charged phosphate groups face the cytoplasm and extracellular environment. The cell membrane is a phospholipid bilayer.

phosphorimaging The use of phosphors for detecting radiolabeled nucleic acids, typically in Southern or northern blots.
phosphorylation The enzyme-catalyzed addition of a phosphoryl group onto a molecule.
phosphotransferase system (PTS) A group translocation system that uses phosphoenolpyruvate to transfer phosphoryl groups onto the incoming molecule.
photic zone See **euphotic zone**.
photoautotrophy The fixation of single-carbon compounds into organic biomass, using light as an energy source.
photoexcitation Light absorption that raises an electron to a higher energy state, as in bacteriorhodopsin or in chlorophyll.
photoferrotrophy Photosynthesis in which light absorption provides energy to separate an electron from reduced iron (Fe^{2+}).
photoheterotrophy Metabolism that includes gain of energy from light absorption with biosynthesis from preformed organic compounds. Usually also includes organotrophy, gain of energy from reactions of organic compounds.
photoionization Light absorption that causes electron separation.
photolysis The first energy-yielding phase of photosynthesis; the light-driven separation of an electron from a molecule coupled to an electron transport system.
photoreactivation A light-induced, photolyase-catalyzed repair of pyrimidine dimers.
photosynthesis The metabolic ability to absorb and convert solar energy into chemical energy for biosynthesis. Autotrophic photosynthesis, or photoautotrophy, includes CO_2 fixation.
photosystem I (PS I) A protein complex that harvests light from a chlorophyll or bacteriochlorophyll, donates an electron to an electron transport system, receives an electron from a small molecule such as H_2S or H_2O, and stores energy in the form of NADPH.
photosystem II (PS II) A protein complex that harvests light from a chlorophyll or bacteriochlorophyll, donates an electron to an electron transport system, and stores energy in the form of a proton potential.
phototrophy The use of chemical reactions powered by the absorption of light to yield energy.
phylogenetic tree A diagram depicting estimates of the relative amounts of evolutionary divergence among different species.
phylogeny A measurement of genetic relatedness. The classification of organisms on the basis of their genetic relatedness.
phylum *pl.* phyla The taxonomic rank one level below domain; a group of organisms sharing a common ancestor that diverged early from other groups.
phytoplankton Phototrophic marine bacteria, algae, and protists, the primary producers in pelagic food webs.
picornavirus A member of a medically important group of RNA viruses. An example is poliovirus.
piezophile See **barophile**.
pilus *pl.* pili Also called *fimbria*. A straight protein filament composed of a tube of protein monomers that extend from the bacterial cell envelope.
pinocytosis A form of endocytosis in which only extracellular fluid and small molecules are brought into the cell.
Planctomycetes A phylum of free-living bacteria that have stalked cells and reproduce by budding. Their nucleoid is surrounded by a membrane.
plankton Organisms that float in water.
planktonic cell An isolated cell, growing individually in a liquid without connections to other cells.
Plantae Also called *Archaeplastida*. A eukaryotic superphylum that includes plants, as well as green and red primary algae.
plaque A cell-free zone on a lawn of bacterial cells caused by viral lysis.
plaque-forming unit (PFU) A measure of the concentration of phage particles in liquid culture.
plasma cell A short-lived antibody-producing cell.
plasma membrane See **cell membrane**.
plasmid An extrachromosomal genetic element that may be present in some cells. Plasmids carry no essential genes.
plasmodesma *pl.* plasmodesmata A membrane channel in plants that connects adjacent plant cells.
plasmodial slime mold A slime mold in which a fertilized zygote undergoes multiple nuclear divisions, generating a multinucleate single cell (plasmodium).
plasmodium *pl.* plasmodia The giant, multinucleate cell formed by a plasmodial slime mold.
platelet A small, cell fragment without a nucleus found in blood that is involved in clotting.
PMF See **proton potential**.
pneumonic plague A highly virulent and contagious *Yersinia pestis* lung infection.
point mutation A change in a single nucleotide within a nucleic acid sequence.
polar covalent bond A covalent bond in which the electrons in the bond are distributed unequally between two atoms.
poliomyelitis *or* polio A paralytic disease caused by the poliovirus.
poliovirus A human-specific RNA virus that is the causative agent of poliomyelitis.
poliovirus receptor (PVR) A cell membrane glycoprotein, also known as CD155, found on cells of the intestinal epithelium and on the surface of motor neurons.
polyketide A polymer that consists of alternating carbonyl and CHR groups (–CO–CHR–) in which many diverse R groups are possible; often has antimicrobial properties.
polymerase chain reaction (PCR) A method to amplify DNA in vitro using many cycles of DNA denaturation, primer annealing, and DNA polymerization with a heat-stable polymerase.
polyphyletic Having multiple evolutionary origins.
polyprotein A long peptide translated from one open reading frame but later cleaved into separate proteins with different functions.
polysaccharide A polymer of sugars. See also **glycan**.
polysome A cell structure consisting of multiple ribosomes performing translation on the same mRNA molecule.
population A group of individuals of one species living in a common location.
porin A transmembrane protein complex that allows movement of specific molecules across the cell membrane or the outer membrane.
positive selection In immunology, the survival of T cells bearing T-cell receptors that don't recognize self MHC proteins displayed on thymus epithelial cells.
pour plate technique A procedure in which organisms are added to melted agar cooled to 45°C–50°C and the mixture is poured into an empty Petri plate. Colonies grow in the agar after the agar solidifies.
PPP See **pentose phosphate pathway**.
PPS See **pentose phosphate pathway**.
prebiotic soup A model for the origin of life based on the abiotic formation of fundamental biomolecules and cell structures such as membranes out of a "soup" of nutrients present on early Earth.
pre-catenane A knot of intertwined DNA that is generated during chromosome replication.
predator A consumer that feeds on grazers.
preliminary mRNA transcript (pre-mRNA) A eukaryotic messenger RNA prior to intron removal.

pre-mRNA See **preliminary mRNA transcript**.
prevalence The total number of active cases of a disease in a given location regardless of when the case first developed.
primary algae Algae that are derived from a single endosymbiotic event; closely related to green plants (Plantae).
primary antibody The first antibody added in an immunological assay, such as enzyme-linked immunosorbent assay (ELISA) or western blot. This antibody binds to the antigen of interest.
primary antibody response The production of antibodies upon first exposure to a particular antigen. B cells become activated and differentiate into plasma cells and memory B cells.
primary pathogen A disease-causing microbe that can breach the defenses of a healthy host.
primary producer An organism that produces biomass (reduced carbon) from inorganic carbon sources such as CO_2.
primary recovery The initial isolation of commercial product from industrial microbes.
primary structure The first level of organization of polymers, consisting of the linear sequence of monomers—for example, the sequence of amino acids in a protein or of nucleotides in a nucleic acid.
primary syphilis The initial inflammatory reaction (chancre) at the site of infection with *Treponema pallidum.*
primase An RNA polymerase that synthesizes short RNA primers complementary to a DNA template to launch DNA replication.
primer An oligonucleotide that anneals to a complementary DNA sequence and serves as the substrate to initiate DNA synthesis by DNA polymerase. The primer is RNA for chromosomal replication, and DNA for PCR applications. Also refers to a primer for RNA virus replication, which can be RNA or protein.
primer extension A technique to determine the 5′ end of an RNA transcript. A primer is used in a reverse-transcription reaction (reverse transcriptase makes DNA from RNA) that extends the primer to the 5′ end (the start) of the message.
prion An infectious agent that causes propagation of misfolded host proteins; usually consists of a defective version of the host protein.
probabilistic indicator A method used to identify an unknown strain of bacteria. The results of a battery of biochemical tests performed on the unknown strain are compared to the probabilities that a known species will have the same results.
probiotic A food or nutritional supplement that contains live microorganisms and aims to improve health by promoting beneficial bacteria.
processivity The ability of a DNA polymerase to elongate a new DNA strand before falling off the template strand.
programmed cell death Cell death mediated by a regulated intracellular process.
prokaryote An organism whose cell or cells lack a nucleus. Both bacteria and archaea are prokaryotes.
promoter A noncoding DNA regulatory region immediately upstream of a structural gene that is needed for transcription initiation.
proofreading An enzymatic activity of some nucleic acid polymerases that attempts to correct mispaired bases.
prophage A phage genome integrated into a host genome.
propionic acid fermentation The fermentation of lactic acid to propionic acid by *Propionibacterium* species; used in the production of Swiss cheese.
protease inhibitor A molecule that inhibits a protease enzyme; some are used as anti-HIV drugs to block the virally encoded protease needed to complete HIV assembly.
protective antigen (PA) The core subunit of anthrax toxin, so called because immunity to this protein protects against disease.
protein A A *Staphylococcus aureus* cell wall protein that binds to the Fc region of antibodies, hiding the *S. aureus* cells from phagocytes.
Proteobacteria A large, metabolically and morphologically diverse group of Gram-negative bacteria.
proteome All the proteins expressed in a cell at a given time. The "complete proteome" includes all the proteins the cell can express under any condition. The "expressed proteome" represents the set of proteins made under a given condition.
proteorhodopsin A bacterial membrane-embedded protein that contains retinal and acts as a light-driven proton pump; it is homologous to the archaeal protein bacteriorhodopsin.
protist A single-celled eukaryotic microbe, usually motile; not a fungus.
proton potential *or* proton motive force (PMF) The potential energy of the concentration gradient of protons (hydrogen ions, H^+) plus the charge difference across a membrane.
protozoan *pl.* protozoa A heterotrophic eukaryotic microbe, usually motile, that is not a fungus.
provirus A viral genome that is integrated into the host cell genome.
PRR See **pattern recognition receptor**.
PS I See **photosystem I**.
PS II See **photosystem II**.
pseudogene A non-functional gene-like sequence that evolved by degenerative evolution.
pseudopeptidoglycan *or* pseudomurein A peptidoglycan-like molecule composed of sugars and peptides that is found in some archaeal cell walls.
pseudopod A locomotory extension of cytoplasm enclosed by the cell membrane.
psychrophile An organism with optimal growth at temperatures below 20°C.
psychrotroph A cold-resistant organism that can grow at temperatures between 0°C and 7°C but shows optimal growth between 20°C and 35°C.
PTS See **phosphotransferase system**.
pure culture A culture containing only a single strain or species of microorganism. A large number of microorganisms that all descended from a single individual cell.
purine A nitrogenous base with fused rings (that is, a bicyclic nucleobase) found in nucleotides; examples are adenine and guanine.
putrefaction Food spoilage due to the decomposition of proteins and amino acids.
PVR See **poliovirus receptor**.
pyelonephritis Kidney infection.
pyrimidine A single-ring nitrogenous base (that is, a monocyclic nucleobase) found in nucleotides; examples are cytosine, thymine, and uracil.
pyrogen Any substance that induces fever.
pyruvate dehydrogenase complex (PDC) The multisubunit enzyme that couples the oxidative decarboxylation of pyruvate to acetyl-CoA and NADH production.

Q

qPCR See **quantitative PCR**.
quantitative PCR (qPCR) Also called *real-time PCR*. A technique using fluorescence to detect the products of PCR amplification as the reaction progresses, in order to quantify the amount of DNA in a sample.
quarantine The separation of infectious individuals from the general population to limit the spread of infection.

quasispecies A collection of isolates (usually viruses) from a common source of infection that have evolved into many different types within one host.
quaternary structure The fourth (and highest) level of organization of proteins, in which multiple polypeptide chains interact and function together.
quinol A reduced electron carrier that can diffuse laterally within membranes.
quinolone A type of antibiotic drug that inhibits DNA synthesis by targeting bacterial topoisomerases such as DNA gyrase.
quinone An oxidized electron carrier that can diffuse laterally within membranes.
quorum sensing The ability of bacteria to sense the abundance of other bacteria via secreted chemical signals called autoinducers.

R

radial immunodiffusion A technique in which a ring of precipitation is visualized in an agarose gel impregnated with antibody. Antigen placed within a well diffuses outward until reaching a zone of equivalence where antigen-antibody complexes precipitate and form a ring.
radiative forcing The increase in warming of Earth's atmosphere (in watts per square meter) associated with a particular climate factor, which is calculated as the difference between the sunlight energy absorbed by Earth and the energy radiated out to space.
radiolarian A member of the eukaryotic group Radiolaria—amebas with a silicate shell penetrated by filamentous pseudopods.
rancidity Food spoilage due to the oxidation of fats; it may or may not involve microbial activity.
rare biosphere Species that are present on Earth in such low abundance that metagenomic screens may not pick them up.
rarefaction curve A graph representing the number of different operational taxonomic units (species) found in a metagenome, as a function of increasing sample size.
RC See **reaction center**.
reaction center (RC) The complex containing a chlorophyll molecule that donates its excited electron to an electron transport system.
read A short DNA sequence that is generated by shotgun or next-generation sequencing methods.
real-time PCR See **quantitative PCR**.
reassortment The packaging of viral chromosome segments from two different viruses into one progeny virion. Refers to separate segments from a segmented genome, without helix recombination.
receptor-binding domain The part of a secreted protein (e.g., exotoxin) that binds to a target cell membrane receptor.
recombination See also *homologous recombination*. The process by which two DNA molecules exchange arms by cutting and splicing their helix backbones.
recombination signal sequence (RSS) A DNA region downstream of antibody heavy- and light-chain genes that allows recombination between widely separated gene segments.
recombinational repair A DNA repair mechanism that relies on recombination between an undamaged chromosome and a gap that occurred during replication of damaged DNA.
red alga Also called *rhodophyte*. An alga of the eukaryotic group Rhodophyta, which contain chloroplasts as primary algae, with red accessory photopigments.
redox couple The oxidized and reduced states of a compound. For example, NAD^+ and NADH form a redox couple.
redox reaction A chemical reaction in which one molecule or functional group becomes reduced and another becomes oxidized.
reducing agent See **electron donor**.
reductive acetyl-CoA pathway A carbon assimilation pathway in which two CO_2 molecules are condensed and reduced by two H_2 molecules to form an acetyl group.
reductive evolution Also called *degenerative evolution*. The loss or mutation of DNA encoding unselected traits.
reductive pentose phosphate cycle See **Calvin cycle**.
reductive TCA cycle Also called *reverse TCA cycle*. A CO_2 fixation pathway that generates acetyl-CoA through reversal of TCA cycle reactions. It requires ATP and NADPH.
reflection The deflecting of an incident light ray by an object, at an angle equal to the incident angle.
refraction The bending and slowing of light as it passes through a substance.
refractive index The degree to which a substance causes the refraction of light; a ratio of the speed of light in a vacuum to its speed in another medium.
regulatory protein A protein that can bind DNA and modulate transcription in response to a metabolite.
regulatory T cell (Treg) A T cell that regulates the activity of another T cell, usually by suppressing its activity.
regulon A group of genes and operons located at different positions in a genome that are coordinately regulated and share a common function.
rehydration therapy A medical treatment for dehydration, in which a liquid solution of salts and glucose is delivered orally.
relative fitness The ability of one strain of an organism to yield more progeny than a competitor does.
release factor A molecule that enters a ribosome A site containing an mRNA stop codon and initiates protein cleavage from the tRNA.
replication complex See **virus factory**.
replication fork During DNA synthesis, the region of the chromosome that is being unwound.
replisome A complex of DNA polymerase and other accessory molecules that performs DNA replication.
reporter gene A gene whose protein product can be easily quantified; commonly used in gene fusions. Two examples are *lacZ* (beta-galactosidase) and *gfp* (green fluorescent protein).
repression The down-regulation of gene transcription.
repressor A regulatory protein that can bind to a specific DNA sequence and inhibit transcription of genes.
reservoir 1. The major part of the biosphere that contains a significant amount of an element needed for life. 2. An organism that maintains a virus or bacterial pathogen in an area by serving as a high-titer host.
resistance island A genomic island that contributes genes involved in antibiotic resistance.
resolution The smallest distance that two objects can be separated and still be distinguished as separate objects.
resource colimitation A situation in which a population size is limited by the lack of two different resources, such as nitrogen and phosphorus.
respiration The oxidation of reduced organic electron donors through a series of membrane-embedded electron carriers to a final electron acceptor. The energy derived from the redox reactions is stored as an electrochemical gradient across the membrane, which may be harnessed to produce ATP.
response regulator A cytoplasmic protein that is phosphorylated by a sensor kinase and modulates gene transcription depending on its phosphorylation state.
restricted transduction See **specialized transduction**.
restriction endonuclease *or* **restriction enzyme** A bacterial enzyme that cleaves double-stranded DNA within a specific short sequence, usually a palindrome.

restriction site A DNA sequence recognized and cleaved by a restriction endonuclease.

reticulate body The metabolically and reproductively active form of chlamydias.

reticuloendothelial system A collection of cells that can phagocytose and sequester extracellular material.

retinal A vitamin A–related cofactor in opsin proteins; it undergoes a conformational change upon absorbing a photon.

retrotransposon A retroelement that contains only partial retroviral sequences but may encode reverse transcriptase to allow further movement into the host genome.

retrovirus Also called *RNA reverse-transcribing virus.* A single-stranded RNA virus that uses reverse transcriptase to generate a double-stranded DNA.

reverse electron flow An enzyme-catalyzed redox reaction that couples an electron transfer with a positive ΔG (such as NAD^+ reduction to NADH) to a source of energy with a larger negative ΔG (such as a proton potential generated by an electron transport system).

reverse TCA cycle See **reductive TCA cycle**.

reverse transcriptase (RT) An enzyme that produces a double-stranded DNA molecule from a single-stranded RNA template.

reversion A mutation that changes a previous mutation back to its original state.

Rhizaria A clade of eukaryotic microbes that have filamentous pseudopods.

rhizobium *pl.* rhizobia A bacterial species of the order Rhizobiales that forms highly specific mutualistic associations with plants in which the bacteria develop into intracellular bacteroids that fix nitrogen for the plant.

rhizosphere The soil environment surrounding plant roots.

Rho factor A bacterial protein involved in terminating transcription.

rhodophyte See **red alga**.

ribosomal RNA (rRNA) An RNA molecule that includes the scaffolding and catalytic components of ribosomes.

ribosome A large enzyme, composed of RNA and protein subunits, that translates mRNA into protein.

ribosome-binding site Also called *Shine-Dalgarno sequence.* In bacteria, a stretch of nucleotides upstream of the start codon in an mRNA that hybridizes to the 16S rRNA of the ribosome, correctly positioning the mRNA for translation.

riboswitch A secondary structure (hairpin) within some mRNA transcripts that can interact with metabolites or antisense RNA molecules, change structure, and affect the production or translation of the mRNA.

ribozyme See **catalytic RNA**.

ribulose 5-phosphate The five-carbon sugar ribulose, phosphorylated at carbon 5.

rich medium See **complex medium**.

ripening The aging of cheese.

rise period The period of time in the viral life cycle when cells lyse and viral progeny are liberated.

RNA-dependent RNA polymerase An enzyme that produces an RNA complementary to a template RNA strand.

RNA polymerase Also called *DNA-dependent RNA polymerase.* An enzyme that produces an RNA complementary to a template DNA strand.

RNA reverse-transcribing virus See **retrovirus**.

RNA world A model of early life in which RNA performed all the informational and catalytic roles of today's DNA and proteins.

rolling-circle replication A form of DNA replication that proceeds in one direction around a circular template, making tandem copies in a linear array (concatemer). The copies are later cleaved and circularized.

root In a phylogenetic tree, the earliest common ancestor of all members shown.

rotavirus One of a group of nonenveloped dsRNA viruses that cause severe diarrhea in children.

rRNA See **ribosomal RNA**.

RSS See **recombination signal sequence**.

RT See **reverse transcriptase**.

Rubisco Ribulose 1,5-bisphosphate carboxylase oxygenase, the enzyme that catalyzes the carbon fixation step in the Calvin cycle.

rumen The first chamber of the digestive tract of ruminant animals such as cattle; the main site for microbial digestion of feed.

S

S-layer A crystalline protein surface layer replacing or external to the cell wall in many species of archaea and bacteria.

sacculus *pl.* sacculi The bacterial cell wall, consisting of a single covalent molecule.

SALT See **skin-associated lymphoid tissue**.

Sanger sequencing A method of sequencing DNA based on the incorporation of chain-terminating dideoxynucleotides by a DNA polymerase.

sanitation The safe disposal of wastes hazardous to humans.

sarcina *pl.* sarcinae A cubical octad cluster of cells formed by septation at right angles to the previous cell division.

sargassum weed An unrooted kelp that floats in marine water and forms kelp forests.

scaffold The assembly of contigs (regions of contiguous sequence) into a large segment of a draft genome.

scanning electron microscopy (SEM) Electron microscopy in which the electron beams scan across the specimen's surface to reveal the 3D topology of the specimen.

scanning probe microscopy (SPM) A type of microscopy in which a physical probe scans the surface of a specimen and maps the topography by detecting a property such as electron tunneling current (scanning tunneling microscopy) or atomic force (atomic force microscopy).

scattering Interaction of light with an object that results in propagation of spherical light waves at relatively low intensity.

schizogony Mitotic reproduction of parasitic cells to achieve a large population within a host tissue.

screen assay A genetic assay, used to identify mutants that have lost a particular function, in which all mutants can grow but only the mutants involved in the function of interest show a phenotype different from the wild type.

second messenger A regulatory molecule such as cAMP that is produced in response to a primary signal. Second messengers typically affect the expression of numerous genes.

secondary algae Algae that evolved by engulfing primary algae in a second endosymbiotic event.

secondary antibody The second antibody added in an immunological assay, such as enzyme-linked immunosorbent assay (ELISA) or western blot. A secondary antibody carries a fluorescent or enzymatic tag and binds to the primary antibody in the assay.

secondary antibody response A memory B cell–mediated rapid increase in the production of antibodies in response to a repeat exposure to a particular antigen.

secondary chromosome A plasmid-like small chromosome that carries at least one essential gene.

secondary metabolite *or* secondary product An organic product of biosynthesis that does not have essential functions but enhances nutrient uptake under certain conditions, or inhibits competing species (e.g., an antibiotic). Often produced during stationary phase.

secondary structure The second level of organization of polymers, consisting of regular patterns that repeat, such as the double helix in DNA or the beta sheet in proteins.

secondary syphilis A rash that may appear at some point after the primary latent stage of syphilis.

secondary tuberculosis A new round of serious disease that is caused by *Mycobacterium tuberculosis* in patients with latent tuberculosis who have become immunocompromised. Symptoms include severe cough, blood sputum, night sweats, and weight loss.

sedimentation rate The rate at which particles of a given size and shape travel to the bottom of a tube under centrifugal force. The rate depends on the particle's mass and cross-sectional area.

segmented genome A viral genome that consists of more than one nucleic acid molecule.

selectin One of a family of cell adhesion molecules.

selection assay A genetic assay, used to identify mutants that have lost a particular function, in which only the mutants involved in the function of interest can grow, while all other mutants and the wild-type parent die.

selective medium A medium that allows the growth of certain species or strains of organisms but not others.

selective toxicity The ability of a drug, at a given dose, to harm the pathogen and not the host.

selectively permeable membrane See **semipermeable membrane**.

SEM See **scanning electron microscopy**.

semiconservative Describing the mode of DNA replication whereby each new double helix contains one old, parental strand and one newly synthesized daughter strand.

semipermeable membrane Also called *selectively permeable membrane*. A membrane that is permeable to some substances but impermeable to other substances.

sensitivity In diagnostic testing, a measure of how often a test will be positive if a patient has a particular disease, reflecting how small a concentration of antigen the test can detect.

sensor kinase A transmembrane protein that phosphorylates itself in response to an extracellular signal, and transfers the phosphoryl group to a receiver protein.

septation The formation of a septum, a new section of cell wall and envelope to separate two daughter cells.

septicemia An infection of the bloodstream.

septicemic plague Infection of the bloodstream by *Yersinia pestis*.

septum ***pl.*** **septa** A plate of cell wall and envelope that forms to separate two daughter cells.

sequela ***pl.*** **sequelae** A serious, harmful immunological consequence of bacterial and host antigen cross-reactivity that occurs after the infection itself is over. An example is rheumatic fever.

sequencing by synthesis A method of DNA sequence determination in which a template DNA strand is used for polymerization of the second strand. The incorporated nucleotides in the synthesized strand are identified by methods that depend on the instrument used. Illumina sequencing is one example.

serial endosymbiosis theory The theory that mitochondria and chloroplasts were originally free-living prokaryotes that formed an internal symbiosis with early eukaryotes.

serum The noncellular, liquid component of the blood.

sexual reproduction Reproduction involving the joining of gametes generated by meiosis.

Shine-Dalgarno sequence See **ribosome-binding site**.

siderophore A high-affinity iron-binding protein used to scavenge iron from the environment and deliver it to a siderophore-producing organism.

sigma factor A protein needed to bind RNA polymerase for the initiation of transcription in bacteria.

signal recognition particle (SRP) A receptor that recognizes the signal sequence of peptides undergoing translation. The complex attaches to the cell membrane of prokaryotes (or the rough endoplasmic reticulum of eukaryotes), where it docks the protein-ribosome complex to the membrane for protein membrane insertion or secretion.

signal sequence A specific amino acid sequence on the amino terminus of proteins that directs them to the endoplasmic reticulum (of a eukaryote) or the cell membrane (of a prokaryote).

silent mutation A mutation that does not change the amino acid sequence encoded by an open reading frame. The changed codon encodes the same amino acid as the original codon.

simple stain A stain that makes an object more opaque, increasing its contrast with the external medium or surrounding tissue.

single-cell genomics The study of the genome of a single cell.

single-celled protein An edible microbe of high food value, such as *Spirulina* or some yeasts.

single-molecule sequencing A method of nucleic acid sequence determination that can be applied to single nucleic acid molecules. The method does not require a prior PCR amplification step.

sink A part of the biosphere that can receive or assimilate significant quantities of an element; may be biotic (as in plants fixing carbon) or abiotic (as in the ocean absorbing carbon dioxide).

site-directed mutagenesis A molecular biology technique that intentionally alters a specific sequence in a DNA molecule.

site-specific recombination Recombination between DNA molecules that do not share long regions of homology but do contain short regions of homology specifically recognized by the recombination enzyme.

skin-associated lymphoid tissue (SALT) Immune cells, such as dendritic cells, located under the skin that help eliminate bacteria that have breached the skin surface.

sliding clamp A protein that keeps DNA polymerase affixed to DNA during replication.

sludge The solid products of wastewater treatment.

small RNA (sRNA) A non-protein-coding regulatory RNA molecule that modulates translation or mRNA stability.

small-subunit rRNA (SSU rRNA) In bacteria and archaea, 16S rRNA; in eukaryotes, 18S rRNA. A ribosomal RNA found in the small subunit of the ribosome. Its gene is often sequenced for phylogenetic comparisons.

SOS response A coordinated cellular response to extensive DNA damage. It includes error-prone repair.

source A part of the biosphere that stores a significant quantity of a given element; may be biotic (as in tree biomass, a source of carbon) or abiotic (as in carbonate rock).

sourdough An undefined yeast population, derived from a previous batch of dough, that is used in bread production.

Southern blot A technique (named for its inventor, Edward M. Southern) to detect specific DNA sequences. Sample DNA segments are separated by gel electrophoresis, transferred to a blot, and probed with a labeled DNA that will hybridize to complementary DNA sequences.

space-filling model A molecular model that represents the volume of the electron orbitals of the atoms, usually to the limit of the van der Waals radii.

specialized transduction Also called *restricted transduction*. Transduction in which the phage can transfer only a specific, limited number of donor genes to the recipient cell.

species A single, specific type of organism, designated by a genus and species name.

specificity In diagnostic testing, a measure of how often a test will be negative if a patient does not have a particular disease, reflecting how well a test can distinguish between two closely related antigens.

spectrum of activity The range of pathogens for which an antimicrobial agent is effective.

spheroplast A cell whose peptidoglycan is degraded by lysozyme; thus the cell loses its shape, forming a sphere.

spike protein A viral glycoprotein that connects the membrane to the capsid or the matrix and may be involved in viral binding to host cell receptors.

spirochete A bacterium with a tight, flexible spiral shape; a species of the phylum Spirochetes (Spirochaeta).

Spirochetes *or* Spirochaeta A phylum of bacteria with a unique morphology: a flexible, extended spiral that twists via intracellular flagella.

SPM See **scanning probe microscopy**.

spongiform encephalopathy A brain-wasting disease caused by a prion.

spontaneous generation The theory, much debated in the eighteenth century, that under current Earth conditions life can arise spontaneously from nonliving matter.

sporangiospore A haploid spore that can germinate to form a haploid mycelium.

sporangium *pl.* sporangia A fungal organ that releases nonmotile spores.

spore stain A type of differential stain that is specific for the endospore coat of various bacteria, typically a firmicute species.

spread plate A method to grow separate bacterial colonies by plating serial dilutions of a liquid culture.

sRNA See **small RNA**.

SRP See **signal recognition particle**.

SSU rRNA See **small-subunit rRNA**.

staining The process of treating microscopic specimens with a stain to enhance their detection or to visualize specific cell components.

stalk An extension of the cytoplasm and envelope that attaches a microbe to a substrate.

stalked ciliate A ciliate that adheres to a substrate and uses its cilia to obtain prey.

standard reduction potential (E°) The reduction potential (tendency of a chemical to gain electrons and thereby become reduced) under standard conditions of 1-M concentration, 25°C temperature, and 1-atm pressure.

staphylococcus A hexagonal arrangement of cells formed by septation in random orientations.

starch A glucose polymer in which an acetal (O–COH) of each glucose has condensed with a hydroxyl group of the next glucose, releasing H_2O.

start codon A codon (usually AUG) that signals the first amino acid of a protein.

starter culture A mixture of fermenting microbes added to a food substrate to generate a fermented product.

stationary phase A period of cell culture, following exponential phase, during which there is no net increase in replication.

sterilization The destruction of all cells, spores, and viruses on an object.

stick model A molecular model in which stick lengths represent the distances between bonded pairs of atomic nuclei.

Stickland reaction An energy-yielding reaction between two amino acids in which one oxidizes the other. The reaction typically produces short organic acids plus $2NH_4^+$; it may also produce CO_2 and H_2.

stop codon One of three codons (UAA, UAG, UGA) that do not encode an amino acid, and thus trigger the end of translation.

Stramenopiles Also called *Heterokonta*. A superphylum of eukaryotic microbes that usually possess a pair of differently shaped flagella.

streptogramin An antibiotic that binds 23S rRNA and blocks elongation of protein synthesis in bacteria.

strict aerobe An organism that performs aerobic respiration and can grow only in the presence of oxygen.

strict anaerobe An organism that cannot grow in the presence of oxygen.

stringent response A cellular response to idle ribosomes (often indicating low carbon and energy stores) that includes a decrease in rRNA and tRNA production.

stromatolite A mass of sedimentary layers of limestone produced by a marine microbial community over many years.

structural formula A representation of molecular structure in which each covalent bond is shown as a line between atoms.

structural isomer A molecule with the same molecular formula as a different molecule but a different arrangement of atoms.

substrate-level phosphorylation The formation of ATP by the enzymatic transfer of phosphate from a substrate molecule onto ADP.

succession Change of species composition over time in a community or microbiome.

sulfa drug An antibiotic that inhibits folic acid synthesis and, thus, nucleotide synthesis.

Sulfolobales A phylum of thermophilic archaea in the TACK superphylum that includes sulfur oxidizers.

superantigen A molecule that directly stimulates T cells without undergoing antigen-presenting-cell processing and surface presentation.

super-resolution imaging Techniques of microscopy that pinpoint the location of an object with a precision greater than the resolution of ordinary optical or fluorescence microscopy.

Svedberg coefficient A measure of particle size based on the particle's sedimentation rate in a tube subjected to a high g force.

swarming A behavior in which some microbial cells differentiate into large swarmer cells and swim together as a unit.

switch region A repeating DNA sequence interspersed between antibody constant-region genes that serves as a recombination site during isotype, or class, switching.

symbiont An organism that lives in a close association with another organism.

symbiosis *pl.* symbioses The intimate association of two different species.

symbiosis island A type of genomic island in which the stretch of DNA expresses proteins that enable a symbiotic relationship with another organism.

symport Coupled transport in which the molecules being transported move in the same direction across the membrane.

syndromic surveillance See **systemic surveillance**.

synergism Cooperation between species in which both species benefit but can grow independently. The cooperation is less intimate than symbiosis.

synthetic biology The genetic construction of novel organisms with useful functions.

Syntrophoarchaeota See **Anaerobic Methane-Oxidizing Euryarchaeota**.

syntrophy Metabolic cooperation between two different species; usually one member releases a product whose removal by the second species enables the pair to metabolize with a negative value of ΔG.

systemic surveillance Also called *syndromic surveillance*. The policy, established by central health organizations such as the U.S. Centers for Disease Control and Prevention or the World Health Organization, whereby physicians are required to report instances of diseases that have particularly high levels of severity and transmissibility, and thus are known as notifiable diseases.

T

T cell An adaptive immune cell, developed in the thymus, that can give rise to antigen-specific helper cells and cytotoxic T cells.

T-cell receptor (TCR) A surface receptor on T cells that binds MHC-bound antigen on antigen-presenting cells.

T3SS See **type III secretion system**.

T4SS See **type IV secretion system**.

tailed phage A phage such as T4 that contains a genome delivery device called the tail.

tandem repeat A stretch of directly repeating DNA sequence (direct repeats) without any intervening DNA.

TaqMan A real-time PCR technique in which Taq polymerase, in the process of synthesizing DNA along a template, degrades a downstream fluorescent oligonucleotide probe. The increase in fluorescence indicates the production of an amplified DNA product.

target community A community whose genomes are sequenced for metagenomic analysis.

taxon *pl.* taxa A category of organisms with a shared genetic ancestor.

taxonomy The description of distinct life forms and their organization into different categories.

T_C cell See **cytotoxic T cell**.

TCA cycle See **tricarboxylic acid cycle**.

TCR See **T-cell receptor**.

tegument The contents of a virion between the capsid and the envelope.

teichoic acid A chain of phosphodiester-linked glycerol or ribitol that threads through and reinforces the cell wall in Gram-positive bacteria.

telomerase A reverse transcriptase enzyme complex that reads RNA as a template to synthesize DNA.

telomere The DNA segment at either end of a eukaryotic chromosome.

TEM See **transmission electron microscopy**.

tempeh A mold-fermented soy product, popular as a food in parts of Asia.

temperate phage A phage capable of lysogeny.

template strand A DNA strand (or an RNA strand in some viruses) that is used as a template for the synthesis of mRNA.

Tenericutes Also called *Mollicutes*. A phylum of bacteria lacking a cell wall; closely related to Firmicutes.

terminal electron acceptor The final electron acceptor at the end of an electron transport system.

termination (ter) site A sequence of DNA that halts replication of DNA by DNA polymerase.

terminator stem loop See **attenuator stem loop**.

terpenoid A branched lipid derived from isoprene that is found in hydrocarbon chains of archaeal membranes.

terraforming The idea of transforming the environment of another planet to make it suitable for life from Earth.

tertiary structure The third level of organization of polymers; the unique 3D shape of a polymer.

tertiary syphilis A final stage of syphilis, manifested by cardiovascular and nervous system symptoms.

tetanospasmin The tetanus-causing potent exotoxin produced by *Clostridium tetani*.

tetraether A molecule containing four ether links. An example is found in archaeal membranes, when two lipid side chains form ether linkages with a pair of side chains from the other side of the bilayer.

T_H cell See **helper T cell**.

T_H17 cell A class of helper T cell that secretes the inflammatory cytokine IL-17.

Thaumarchaeota A phylum of archaea in the TACK superphylum that includesammonia oxidizers and marine invertebrate symbionts. Formerly known as Marine Group I.

thermal vent See **hydrothermal vent**.

thermocline A region of the ocean where temperature decreases steeply with depth, and water density increases.

thermophile An organism adapted for optimal growth at high temperatures, usually 55°C or higher.

Thermoplasmatales An order of extreme thermoacidophiles in the Euryarchaeota superphylum.

Thermoproteales A clade of Thermoprotei, thermophilic archaea including hyperthermoacidophiles, in the TACK superphylum.

Thermotogae A phylum of thermophilic bacteria; in some species, the cell is enclosed by a toga-shaped outer covering.

threshold dose The concentration of antigen needed to elicit adequate antibody production.

thylakoid An intracellular chlorophyll-containing membrane folded within a phototrophic bacterium or a chloroplast.

Ti plasmid A plasmid found in tumorigenic strains of *Agrobacterium tumefaciens* that can be used as a vector to introduce DNA into plant cells.

tight junction A type of junction between the membranes of two adjacent vertebrate cells that form an impermeable barrier.

TLR See **Toll-like receptor**.

tmRNA A molecule resembling both tRNA and mRNA that rescues ribosomes stalled on damaged mRNAs lacking a stop codon.

TNF See **tumor necrosis factor**.

TNF-alpha See **tumor necrosis factor alpha**.

Toll-like receptor (TLR) A member of a eukaryotic transmembrane glycoprotein family that recognizes a particular microbe-associated molecular pattern (MAMP) present on pathogenic microorganisms.

tomography The acquisition of projected images of a transparent specimen from different angles that are digitally combined to visualize the entire specimen.

topoisomerase An enzyme that can change the supercoiling of DNA.

total magnification The magnification of the ocular lens multiplied by the magnification of the objective lens.

transamination The transfer of an ammonium ion between two metabolites.

transcript An RNA copy of a DNA template.

transcription The synthesis of RNA complementary to a DNA template.

transcriptional attenuation A regulatory mechanism that terminates transcription in progress, before it even reaches the first gene.

transcriptional fusion A technique to monitor transcriptional regulation by fusing a reporter gene containing its own ribosome-binding site to the 3′ end of an operon. Unlike the translational fusion technique, only the promoter of the target gene directs expression of the reporter.

transcriptome The set of transcribed genes in a cell at a given time. The "complete transcriptome" includes all the possible RNA transcription products from a given genome. The "expressed transcriptome" is the set of RNAs present during a given condition.

transcytosis The movement of a cell or substance from one side of a polarized cell to the other side, using an intracellular route.

transduction The transfer of host genes between bacterial cells via a phage head coat.

transfection In biotechnology, the transfer of DNA (usually viral) into cells.

transfer RNA (tRNA) An RNA that carries an amino acid to the ribosome. The anticodon on the tRNA base-pairs with the codon on the mRNA.
transform To cause bacteria to take up exogenous DNA. In eukaryotes, to convert cultured cells into cancer cells.
transformation The internalization of free DNA from the environment into bacterial cells.
transgene A gene that has been transferred by genetic engineering techniques from one organism to another.
transglycosylase An enzyme that condenses *N*-acetylglucosamine and *N*-acetylmuramic acid into chains during bacterial cell wall synthesis.
transition A point mutation in which a purine is replaced by a different purine or a pyrimidine is replaced by a different pyrimidine.
translation The ribosomal synthesis of proteins based on triplet codons present in mRNA.
translational control A regulatory mechanism that modulates protein production by influencing the translation of mRNA.
translational fusion A technique to measure control of gene transcription or translation by inserting a reporter gene into a target gene. Expression of the reporter relies on both the promoter and the ribosome-binding site of the target gene.
translocation The energy-dependent movement of a ribosome to the next triplet codon along an mRNA.
transmembrane domain The membrane-spanning amino acid sequence within a membrane protein that connects intracellular and extracellular parts of the membrane protein.
transmembrane protein Also called *integral protein.* A protein with a membrane-spanning region.
transmission electron microscopy (TEM) Electron microscopy in which electron beams are transmitted through a thin specimen to reveal internal structure.
transovarial transmission The transfer of a pathogen from parent to offspring by infection of the egg cell. Typically seen in insects.
transpeptidase An enzyme that cross-links the side chains from adjacent peptidoglycan strands during bacterial cell wall synthesis.
transport protein *or* transporter A membrane protein that moves specific molecules across a membrane.
transposable element Also called *jumping gene.* A segment of DNA that can move from one DNA region to another.
transposase A transposable element–encoded enzyme that catalyzes the transfer of the transposable element from one DNA region to another.
transposition The process of moving a transposable element from one DNA region to another.
transposon A transposable DNA element that contains genes in addition to those required for transposition. Examples of additional genes include those that encode resistance to antibiotics.
transversion A point mutation in which a purine is replaced by a pyrimidine or vice versa.
Treg See **regulatory T cell**.
tricarboxylic acid (TCA) cycle Also called *Krebs cycle.* A metabolic cycle that catabolizes the acetyl group from acetyl-CoA to $2CO_2$ with the concomitant production of NADH, $FADH_2$, and ATP.
tRNA See **transfer RNA**.
trophic level A level of a food web representing the consumption of biomass of organisms from another level, usually closer to producers.
tropism The ability of a virus to infect a particular tissue type.
true fungi See **Eumycota**.
trypanosome A parasitic excavate protist that has a cortical skeleton of microtubules culminating in a long flagellum.
tumor necrosis factor (TNF) A cytokine released by several cell types (e.g., macrophages) in response to cell damage.
tumor necrosis factor alpha (TNF-alpha) A cytokine involved in systemic inflammation.
turgor pressure See **osmotic pressure**.
twitching motility A type of bacterial movement on solid surfaces in which a specific pilus extends and retracts.
two-component signal transduction system A message relay system composed of a sensor kinase protein and a response regulator protein that regulates gene expression in response to a signal (usually an extracellular signal).
two-hybrid analysis An in vivo technique to determine protein-protein interactions in which DNA sequences encoding proteins of interest are fused separately to the DNA-binding and activation domains of a transcription factor. The recombinant organism is then tested for expression of a reporter gene.
type I hypersensitivity Also called *immediate hypersensitivity.* An IgE-mediated allergic reaction that causes degranulation of mast cells within minutes of exposure to the antigen. The severe reaction known as anaphylaxis is triggered by type I hypersensitivity.
type II hypersensitivity An immune response in which antibodies bind to the patient's own cell-surface antigens or to foreign antigens adsorbed onto the patient's cells. Antibody binding triggers cell-mediated cytotoxicity or activation of the complement cascade.
type II secretion system A bacterial protein secretion system that uses a type IV pilus–like extraction/retraction mechanism to push proteins out of the cell.
type III hypersensitivity Also called *immune complex disease.* An immune reaction triggered when IgG antibody binds to an excess of soluble foreign antigen in the blood. The immune complexes deposit in small blood vessels, where they interact with complement to initiate an inflammatory response.
type III secretion system (T3SS) A bacterial protein secretion system that uses a molecular syringe to inject bacterial proteins into the host cytoplasm.
type IV hypersensitivity Also called *delayed-type hypersensitivity (DTH).* An immune response that develops 24–72 hours after exposure to an antigen that the immune system recognizes as foreign. The response is triggered by antigen-specific T cells. It is delayed because the T cells need time to proliferate after being activated by the allergen.
type IV secretion system (T4SS) Protein secretion system of Gram-negative bacteria whose components exhibit sequence homology with the components of conjugation systems.

U

ubiquitin (Ub) In eukaryotes, a 76-amino-acid peptide that tags proteins, marking them for degradation by the proteasome.
ultracentrifuge A machine that subjects samples to high centrifugal forces and can be used to separate subcellular components.
uncoating The release of a viral genome from its capsid, following entry of the virion into a host cell.
uncoupler A molecule that makes a membrane permeable to protons, dissipating the proton motive force and uncoupling electron transport from ATP synthesis.
uncultured *or* unculturable Describing an organism whose requirements for culture remain unknown.
upstream processing The culturing of industrial microbes to produce large quantities of a desired product.

V

vaccination Exposure of an individual to a weakened version of a microbe or a microbial antigen to provoke immunity and prevent development of disease upon reexposure.

VAM See **arbuscular mycorrhizae**.

van der Waals force A weak, temporary electrostatic attraction between molecules caused by shifting electron clouds.

vancomycin A glycopeptide antibiotic that inhibits bacterial cell wall synthesis in a mechanism distinct from penicillin inhibition.

variable region The amino-terminal portions of antibody light and heavy chains that confer specificity to antigen binding and define the antibody idiotype.

vasoactive factor A cell signaling molecule that increases capillary permeability.

vector 1. An organism (e.g., insect) that can carry infectious agents from one animal to another. 2. In molecular biology, a molecule of DNA into which exogenous DNA can be inserted to be cloned.

vegetative cell A metabolically active, replicating bacterial cell.

vegetative mycelium A mass of hyphae (branched filaments) produced by vegetative cells that expand into the substrate.

vehicle transmission In disease, the transfer of a pathogen when an infected person deposits it on a surface or in food or drink that another person touches or consumes.

Verrucomicrobia A phylum of free-living aquatic bacteria with wart-like, protruding structures containing actin.

vertical gene transfer The generational movement of genes from parent to offspring through reproduction.

vertical transmission In disease, the transfer of a pathogen from parent to offspring. See also **transovarial transmission**.

vesicle A small, membrane-enclosed sphere found within a cell.

vesicular-arbuscular mycorrhizae (VAM) See **arbuscular mycorrhizae**.

viable Capable of replicating—for instance, by forming a colony on an agar plate.

viral shunt The release by viral lysis of cell contents as organic material available for microbial consumers in the upper region of the ocean.

viremia The presence of large numbers of virions in the bloodstream.

virion A virus particle.

viroid An infectious naked nucleic acid.

virome The genomes of all the viruses that inhabit a particular organism or environment.

virulence A measure of the severity of a disease caused by a pathogenic agent.

virulence factor A trait of a pathogen that enhances the pathogen's disease-producing capability.

virus A noncellular particle containing a genome that can replicate only inside a cell.

virus factory Also called *replication complex*. An intracellular complex of membranes and proteins that forms progeny virions.

W

wastewater treatment A series of wastewater transformations designed to lower biological oxygen demand and eliminate human pathogens before water is returned to local rivers.

water activity A measure of the water that is not bound to solutes and is available for use by organisms.

water cycle See **hydrologic cycle**.

water mold See **oomycete**.

water table The layer of soil that is permanently saturated with water.

western blot A technique to detect specific proteins. Proteins are subjected to gel electrophoresis, transferred to a blot, and probed with enzyme-linked or fluorescently tagged antibodies that specifically bind the protein of interest.

wet mount A technique to view living microbes with a microscope by placing the microbes in water on a slide under a coverslip.

wetland A region of land that undergoes seasonal fluctuations in water level and aeration.

whey The liquid portion of milk after proteins have precipitated out of solution, usually during cheese production.

Winogradsky column A glass tube containing a stratified environment that causes specific microbes to grow at particular levels; a type of enrichment culture for the growth of microbes from wetland environments.

X

X-ray crystallography *or* X-ray diffraction analysis A technique to determine the positions of atoms (atomic coordinates) within an array of identical molecules or molecular complexes, based on the diffraction of X-rays by the molecule.

Y

yeast A unicellular fungus.

yogurt A semisolid food produced through acidification of milk by lactic acid–producing bacteria.

Z

Z pathway See **oxygenic Z pathway**.

zone of hypoxia See **dead zone**.

zone of inhibition A region of no bacterial growth on an agar plate that is due to the diffusion of a test antibiotic. Correlates to the minimal inhibitory concentration.

zoonotic disease An infection that normally affects animals but can be transmitted to humans.

zoospore A flagellated reproductive cell produced by chytridiomycete fungi.

zygomycete A member of the eukaryotic group Zygomycota—fungi forming nonmotile haploid gametes that grow toward each other, fusing to form the zygospore.

zygospore In zygomycetes, the diploid structure formed by the fusion of two gamete-bearing hyphae.

CREDITS

TEXT PERMISSIONS

Figure 3.35B: Republished with permission of Elsevier, from Gyanendra P. Dubey and Sigal Ben-Yehuda, "Intercellular nanotubes mediate bacterial communication," *Cell 144*(4):P590–600, 2011; permission conveyed through Copyright Clearance Center.

Figure 4.23B: From Kiernan B. Pechter, et al. "Molecular basis of bacterial longevity," *mBio 8*:e01726-17 (2017). Reprinted by permission of the authors.

Figure 6.18: From Didier Raoult et al., "The 1.2-megabase genome sequence of mimivirus," *Science 306*(5700): 1344–1350, 2004. Reprinted with permission from AAAS.

Figure ST 6.1.1A: From S. Nguyen et al., "Bacteriophage transcytosis provides a mechanism to cross epithelial cell layers," *mBio 8*:e01874-17 (2017). Reprinted by permission of the authors.

Figure 7.27B: Republished with permission of Elsevier, from Julia K. Goodrich, et al., "Human genetics shape the gut microbiome," *Cell 159*(4):788–799, 2014; permission conveyed through Copyright Clearance Center, Inc.

Figures 7.28 and 7.29: From Shinichi Sunagawa et al., "Structure and function of the global ocean microbiome," *Science 348*(6237), 2015. Reprinted with permission from AAAS.

Figure 8.10: From Claudia Steglich et al., "Short RNA half-lives in the slow-growing marine cyanobacterium Prochlorococcus," *Genome Biology 11*:R54, 2010. Reprinted by permission of the authors.

Figure 9.41: From J. Grote et al., "Streamlining and core genome conservation among highly divergent members of the SAR11 clade," *mBio 3*(5):e0025-12, 2012. Reprinted by permission of the authors.

Figure 9.43: Material from Gabrielle Rocap et al., "Genome divergence in two Prochlorococcus ecotypes reflects oceanic niche differentiation," *Nature 424*:1042–1047, published 2003 by Nature Publishing Group. Reproduced with permission of SNCSC.

Figure 9.44: From Alastair Crisp et al., "Expression of multiple horizontally acquired genes is a hallmark of both vertebrate and invertebrate genomes," *Genome Biology 16*:50, 2015. DOI: 10.1186/ s13059-015-0607-3. This article is licensed under a Creative Commons Attribution 4.0 International License.

Figure 10.21: From Jenny L. Baker et al., "Widespread genetic switches and toxicity resistance proteins for fluoride," *Science* 335(6065): 233–235, 2012. Reprinted with permission from AAAS.

Figure 10.39: Republished with permission of Elsevier, from Mark A. Woelfle et al., "The adaptive value of circadian clocks: An experimental assessment in cyanobacteria," *Current Biology 14*(16): 1481–1486, 2004; permission conveyed through Copyright Clearance Center, Inc.

Figure 11.13A: Material from Amie Eisfeld et al., "At the centre: Influenza A virus ribonucleoproteins," *Nature Reviews Microbiology 13*:28-41, published 2015 by Macmillan Publishers Limited. Adapted with permission of SNCSC.

ST 11.1.1C: From Sander Herfst et al., "Airborne transmission of influenza a/h5n1 virus between ferrets," Science 336 (6088):1534–1541, 2012. Reprinted with permission from AAAS.

Figure 12.11: Material from Benjamin A. Flusberg et al., "Direct detection of DNA methylation during single-molecule, real-time sequencing," *Nature Methods 7*(6): 461–467, published 2010 by Macmillan Publishers Limited. Adapted with permission of SNCSC.

Figure 12.12BC: From Lacey L. Westphal et al., "Genome dam methylation in escherichia coli during long-term stationary phase," *mSystems 1*(6): e00130-16, 2016. Reprinted by permission of the authors.

Figures 12.32: Jarred M. Callura, et al. Figure 1 from "Genetic switchboard for synthetic biology applications," *Proceedings of the National Academy of Sciences* 109(15), 2012. Reprinted by permission of the National Academy of Sciences.

Figure 12.33: Adapted from Figure 3 from Jarred M. Callura, et al., "Genetic switchboard for synthetic biology applications," *Proceedings of the National Academy of Sciences,* 109(15): 5850–5855, 2012. Reprinted by permission of the National Academy of Sciences.

Figure 12.34: Adapted from Figure 3A from Jarred M. Callura, et al., "Tracking, tuning, and terminating microbial physiology using synthetic riboregulators," *Proceedings of the National Academy of Sciences* 107(36): 15898–15903, 2010. Reprinted by permission of the National Academy of Sciences.

Figure ST 12.1.2: From Clyde A. Hutchison III et al., "Design and synthesis of a minimal bacterial genome" Research Article Summary, Science *351*(6280): 1414, 2016. Reprinted with permission from AAAS.

Figure 13.28: Material from Jan-Hendrik Hehemann et al., "Transfer of carbohydrate-active enzymes from marine bacteria to Japanese gut microbiota" *Nature 464*, 908–912, published 2010 by Macmillan Publishers Limited. Adapted with permission of SNCSC.

Figure 13.34B: From Nidal Laban et al., "Anaerobic benzene degradation by Gram-positive sulfate-reducing bacteria," *FEMS Microbiological Ecology 68*(3):300–311, 2009. Reprinted with permission of Oxford University Press on behalf of the Federation of European Microbiological Societies.

Figure ST 13.1.2B: From Berkley Luk et al., "Postnatal colonization with human 'infant-type' Bifidobacterium species alters behavior of adult gnotobiotic mice," *PLoSOne* 13(5):e0196510, 2018. Reprinted by permission of the authors.

Figure 14.21A: From Gemma Reguera, "Microbial nanowires and electroactive biofilms," *FEMS Microbiology Ecology 94*(7), 2018. Reprinted with permission of Oxford University Press on behalf of the Federation of European Microbiological Societies.

Figure ST 14.1.2: Adapted from Clare E. Reimers and Michael Wolf, "Power from benthic microbial fuel cells drives autonomous sensors and acoustics modems," *Oceanography 31*(1): 98–103, 2018. DOI: 10.5670/ oceanog.2018.115. This article is licensed under a Creative Commons Attribution 4.0 International License.

Figure 16.38: Upper portion from Michael C. Milone and Una O'Doherty, "Clinical use of lentiviral vectors," *Leukemia 32*: 1529–1541, 2018. This material is licensed under a Creative Commons Attribution 4.0 International license.

Figure 17.20: From W. Ford Doolittle, "Phylogenetic Classification and the Universal Tree," *Science 284*(5432): 2124–2128, 1999. Reprinted with permission from AAAS.

Figure 17.23BC: From Erik Gullberg et al., "Selection of Resistant Bacteria at Very Low Antibiotic Concentrations," *PLoS Pathogens* 7(7):e1002158 (2011). Reprinted by permission of the authors.

Figure 17.25B: Figure 6 from Richard E. Lenski and Michael Travisano, "Dynamics of adaption and diversification: A 10,000-generation experiment with bacterial populations," *Proceedings of the National Academy of Sciences,* 91(15): 6808–6814, 1994. Copyright (1994) National Academy of Sciences, U.S.A. Reprinted by permission of the National Academy of Sciences.

Figure 17.26A: Adapted from Figure 1 from Zachary Blount et al., "Historical contingency and the evolution of a key innovation in an experimental population of Escherichia coli," *Proceedings of the National Academy of Sciences,* 105(23): 7899–7906, 2008. Copyright (2008)

National Academy of Sciences, U.S.A. Reprinted by permission of the National Academy of Sciences.

Figure 18.8: Material from Dennis Claessen et al., "Bacterial solutions to multicellularity: a tale of biofilms, filaments and fruiting bodies," *Nature Reviews Microbiology 12*:115–124, published 2014 by Macmillan Publishers Limited. Adapted with permission of SNCSC.

Figure 19.13B: From SIB Swiss Institute of Bioinformatics, ViralZone. Reprinted with permission from Swiss Institute for Bioinformatics.

Figure 19.19C: Material from Markus B. Karner, Edward F. DeLong, and David M. Karl, "Archaeal dominance in the mesopelagic zone of the Pacific Ocean," *Nature 409*: 507–510, published 2001 by Macmillan Magazines Ltd. Adapted with permission of SNCSC.

Figure 19.21: Matsutani, Naoki, Tatsunori Nakagawa, Kyoko Nakamura, Reiji Takahashi, Kiyoshi Yoshihara, et al. 2011. Figure 2 from "Enrichment of a novel marine ammonia-oxidizing archaeon obtained from sand of an eelgrass zone." *Microbes and Environments 26*:23–29. © The Japanese Society of Microbial Ecology/ The Japanese Society of Soil Microbiology.

Figure ST 19.1.2: Figures 1A and B from Elizabeth Hansen, et al., "Pan-genome of the dominant human gut-associated archaeon, *Methanobrevibacter smithii*, studied in twins," *Proceedings of the National Academy of Sciences*, 108(Suppl 1): 4599–4606, 2011. Reprinted by permission of the National Academy of Sciences.

Figure 21.6B: Figure 1 from Molly C. Redmond and David L. Valentine, "Natural gas and temperature structured a microbial community response to the Deepwater Horizon oil spill," *Proceedings of the National Academy of Sciences*, 109(50): 20292–20297, 2012. Reprinted by permission of the National Academy of Sciences.

Figures 21.7B and 21.8: Material from Olivia U. Mason et al., "Metagenome, metatranscriptome and single-cell sequencing reveal microbial response to Deepwater Horizon oil spill," *The ISME Journal 6*, 1715–1727, published 2012 by International Society for Microbial Ecology. Adapted with permission of SNCSC.

Figure 21.9A: From Flow Cytometry Guide, creative-diagnostics.com. Reprinted by permission of Creative Diagnostics.

Figure 21.9B: Material from Floriane Delpy et al., "Pico- and nanophytoplankton dynamics in two coupled but contrasting coastal bays in the NW Mediterranean Sea (France)," *Estuaries and Coasts 41*(7): 2039–2055, published 2018 by Coastal and Estuarine Research Federation. Adapted with permission of SNCSC.

Figure 21.12C: Shosuke Yoshida et al., "A bacterium that degrades and assimilates poly(ethylene terephthalate)," *Science 351*(6278): 1196–1199, 2018. Reprinted with permission from AAAS.

Figure 21.17: From International *Glossina* Genome Initiative, "Genome sequence of the Tsetse Fly (*Glossina morsitans*): Vector of African Trypanosomiasis," *Science 344*(6182):380–386, 2014. Reprinted with permission from AAAS.

Figure 21.18A: Figure from *Symbiosis in Cell Evolution* by Lynn Margulis. Copyright © 1981, 1993 by W.H. Freeman and Company. Reprinted with permission from the Lynn Margulis Estate.

Figure 21.23: Material from Elizabeth A. Rettedal, Heidi Gumpert, and Morten O. A. Sommer, "Cultivation-based multiplex phenotyping of human gut microbiota allows targeted recovery of previously uncultured bacteria," *Nature Communications 5*(4714), published 2014 by Macmillan Publishers Limited. Adapted with permission of SNCSC.

Figure 21.26: Figure 2 from Pedro Flombaum, et al., "Present and future global distributions of the marine Cyanobacteria Prochlorococcus and Synechococcus," *Proceedings of the National Academy of Sciences*, 110(24): 9824–9829, 2013. Reprinted by permission of the National Academy of Sciences.

Figure 21.31: Raina Maier, et al.: Figure from *Environmental Microbiology*, Copyright © 2000 by Academic Press. Reprinted by permission of Elsevier, Ltd.

Figure 21.38: Material from Noah Fierer, "Embracing the unknown: disentangling the complexities of the soil microbiome," *Nature Reviews Microbiology 15*, 579–590, published 2017 by Macmillan Publishers Limited. Adapted with permission of SNCSC.

Figures 22.8E and 22.9: Material from Ben J. Woodcroft et al., "Genome-centric view of carbon processing in thawing permafrost," *Nature 560*, 49–54, published 2018 by Springer Nature Limited. Adapted with permission of SNCSC.

Figure 22.28: From Marzia Miletto and Steven E. Lindow, "Relative and contextual contribution of different sources to the composition and abundance of indoor air bacteria in residences," *Microbiome 3*:61, 2015. Reproduced by permission of the authors.

Figure ST 22.1.2B: From Massimiliano Cardinale et al., "Microbiome analysis and confocal microscopy of used kitchen sponges reveal massive colonization by Acinetobacter, Moraxella, and *Chryseobacterium* species," *Scientific Reports 7*:5791, 2017. Reproduced by permission of the authors.

Figure 24.28: Material from Matteo Ugolini et al., "Recognition of microbial viability via TLR8 drives TFH cell differentiation and vaccine responses," *Nature Immunology 19*: 386–396, published 2018 by Macmillan Publishers Limited. Adapted with permission of SNCSC.

Figures ST 24.1.3 and ST 24.1.4: Republished with permission from Elsevier, from Pavlo Gilchuk et al., "Multifunctional Pan-ebola virus Antibody Recognizes a Site of Broad Vulnerability on the Ebloavirus Glycoprotein," *Immunity 49*: 363–374, 2018; permission conveyed through Copyright Clearance Center, Inc.

Figure 25.9: Republished with permission from the American Society for Microbiology, from Herbert Schmidt and Michael Hensel, "Pathogenicity Islands in Bacterial Pathogenesis," *Clinical Microbiology Reviews 17*(1): 14–56, 2004; permission conveyed through Copyright Clearance Center, Inc.

Figure 25.10A: Republished with permission from the American Society for Clinical Investigation, from Luchang Zhu et al., "A molecular trigger for intercontinental epidemics of group A Streptococcus," *Journal of Clinical Investigation 125*(9), 2015; permission conveyed through Copyright Clearance Center, Inc.

Figure 25.38BC: Material from Alexander J. Westermann et al., "Dual RNA-seq unveils noncoding RNA functions in host–pathogen interactions." *Nature 529*:49, published 2016 by Macmillan Publishers Limited. Adapted with permission of SNCSC.

Figure 26.29C: Melina Yingling/TNS. © 2014. All rights reserved. Reprinted with permission from Tribune News Service.

Figure ST 26.1.2: Figure 1A from Ania Bogoslowski, Eugene C. Buthcher, and Paul Kubes, "Neutrophils recruited through high endothelial venules of the lymph nodes via PNAd intercept disseminating Staphylococcus aureus," *Proceedings of the National Academy of Sciences,* 115(10): 2449–2454, 2018. Reprinted by permission of the National Academy of Sciences.

Figures ST 27.1.2 and ST 27.1.3: From Antonio DiGiandomenico et al., "A multifunctional bispecific antibody protects against *Pseudomonas aeruginosa*," *Science Translation Medicine 6*: 262ra155, 2015. Reprinted with permission from AAAS.

Figure 28.13: MS spectrum of Escherichia coli NBRC 3972 reprinted courtesy of Shimadzu Corporation.

Figure ST 28.1.3: From Janice S. Chen et al., "CRISPR-Cas12a target binding unleashes indiscriminate single-stranded DNase activity," *Science 360*(6387): 436–439, 2018. Reprinted with permission from AAAS.

PHOTO CREDITS

Front Matter

Figures: PO.1: Veronika Burmeister/Visuals Unlimited, Inc.; PO.2: Eye of Science/Science Source; PO.3: David Scharf/Science Source; PO.4: Joan Slonczewski and Rachel Morgan-Kiss, NSF; PO.5: Scimat/Science Source; Title Page: Eye of Science/Science Source

Chapter 1

Figures: CO1.A: Professor Jason Biggs; CO1.A (inset): Jessica Blanton; CO1.B: Dr. Michelle Schorn; 1.1: Erik Zinser, University of Tennessee, Knoxville; 1.2A: Scimat/Science Source; 1.2B: Blickwinkel/Alamy Stock Photo; 1.2C: Eye of Science/Science Source; 1.2D: Louise Hughes/

Science Source; 1.3: AP Photo/Laurent Cipriani; 1.4A: From "Dense Populations of a Giant Sulfur Bacterium in Namibian Shelf Sediments" by H.N. Schulz et al.,*Science*, 1999, Volume 284, Issue 5413, Cover, DOI: 10.1126/science.284.5413.493. Reprinted with permission from AAAS; 1.4B: Leibniz-Institut für Ostseeforschung Warnemünde/D. Gohlke; 1.5: Bettmann/Getty Images; 1.6: CNRI/Science Source; 1.7: Institute of Genome Sciences, University of Maryland School of Medicine; 1.8B: Hans L. Bonnevier, Johner/Johner Images/Getty Images; 1.8A: Bettmann/Getty Images; 1.9A: Chronicle of Aegidius Li Muisis/Private Collection/Bridgeman Images; 1.9B: Vanessa Vick/Getty Images; 1.10A: Everett Historical/Shutterstock; 1.10B: Smith Collection/Gado/Getty Images; 1.11: Library of Congress; 1.12A: Bettmann/Getty Images; 1.12C: Brian J. Ford; 1.13A: The Print Collector/Alamy Stock Photo; 1.14: J. William Schopf; 1.15A: Jim Sugar/Getty Images; 1.15B: Album/Alamy Stock Photo; 1.17: Sueddeutsche Zeitung Photo/Alamy Stock Photo; 1.18A: Courtesy of the U.S. National Library of Medicine; 1.18B: Dr. Michael Gabridge/Visuals Unlimited, Inc; 1.20A: Hulton Archive/Getty Images; 1.20B: Popperfoto/Getty Images; 1.20C: Library of Congress; 1.21: Institut Pasteur; 1.22A: Davies/Getty Images; 1.22B: Mediscan/Visuals Unlimited, Inc.; 1.23A: Dennis Kunkel Microscopy/Science Source; 1.24: Timothy A. Wilkerson; 1.26A: Courtesy of Rita Colwell; 1.26B: From "Extending the Upper Temperature Limit for Life" by K. Kashefi and D.R. Lovley, *Science*, 2003, Volume 301, Issue 5635, p. 934. Reprinted with permission from AAAS; 1.26C: Kazem Kashefi/Michigan State University; 1.27: "Colonization of Mucin by Human Intestinal Bacteria and Establishment of Biofilm Communities in a Two-Stage Continuous Culture System" by S. Macfarlane et al., *Applied and Environmental Microbiology*, 2005, Volume 71, Issue 11, pp. 7483–7492, DOI: 10.1128/aem.71.11.7483-7492.2005. Reproduced with permission from American Society for Microbiology; 1.28: NASA/JPL-CALTECH/MSSS; 1.29B: Nancy R. Schiff/Getty Images; 1.30A: OAR/NURP/NOAA; 1.30B: *J. Bacteriol.* September 2008 Vol. 190 NO. 18 6039–6047. Fig. 4; 1.31A: AP Photo; 1.32: Hagley Museum and Archive/Science Source; 1.33: "*Chlorobium tepidum* Mutant Lacking Bacteriochlorophyll *c* Made by Inactivation of the *bchK* Gene, Encoding Bacteriochlorophyll *c* Synthase" by N. Frigaard et al. 2002, *Journal of Bacteriology*, Volume 184, Issue 12, pp. 3368, DOI: 10.1128/jb.184.12.3368-3376.2002. Reproduced with permission from American Society for Microbiology; 1.34A: Science Source; 1.34B: Omikron/Science Source; 1.34C: A. Barrington Brown/Science Source; 1.36A: Jeremy Papasso/Digital First Media/Boulder Daily Camera via Getty Images; 1.36C: Alexander Heinl/picture-alliance/dpa/AP Images; 1.37: Joan Slonczewski, Kenyon College; ST1.1.1: European Society for Medical Oncology via AP Images; ST1.1.2: From "Gut microbiome influences efficacy of PD-1–based immunotherapy against epithelial tumors" by B. Routy et al., *Science*, 2018, Volume 359, Issue 6371, pp. 91–97, DOI: 10.1126/science.aan3706. Reprinted with permission from AAAS.

Chapter 2

Figures: CO2.A: Mark S. Ladinsky, California Institute of Technology; CO2.B: Deborah Williams-Hedges; Caltech; 2.1: Reprinted by permission from Springer Nature: *Nature*, "Bacterial colonization factors control specificity and stability of the gut microbiota", by M. Lee et al., Volume 501, Issue 7467, pp. 426–429, 2013, DOI: 10.1038/nature12447; 2.3A: Joan Slonczewski, Kenyon College; 2.3B: C.M. Lucy Joseph, UC Davis Department of Viticulture and Enology Wine Yeast and Bacteria Collection; 2.5A: Andrew Syred/Science Source; 2.5B: CDC/DPDx; 2.6A: Dr. Arthur Siegelman/Visuals Unlimited, Inc.; 2.6B: Dennis Kunkel Microscopy/Science Source; 2.6C: Michael Abbey/Visuals Unlimited, Inc.; 2.6D: Dennis Kunkel Microscopy/Science Source; 2.6E: Dr. Edward Chan/Visuals Unlimited, Inc.; 2.6F: Dennis Kunkel Microscopy/Science Source; 2.7A: Ed Reschke/Getty Images; 2.7B: Biodisc/Visuals Unlimited, Inc.; 2.7C: Dr. Terry Beveridge/Visuals Unlimited, Inc.; 2.7D: Dr. Donald Fawcett, Kiseleva/Visuals Unlimited, Inc.; 2.16: Biotechs Flow Cell (Photo by J. Slonczewski; 2.17: C.M. Lucy Joseph, UC Davis Department of Viticulture and Enology Wine Yeast and Bacteria Collection); 2.20A: CDC/Dr. Mike Miller; 2.20B: CDC/Science Source; 2.22A: CDC/Dr. George P. Kubica; 2.22B: Gado Images/Alamy Stock Photo; 2.23A: Alicia M. Muro-Pastor; 2.23B: "Spatial organization of the gut microbiota" by Jessica L. Mark Welch, Yuko Hasegawa, Nathan P. McNulty, Jeffrey I. Gordon, Gary G. Borisy. *Proceedings of the National Academy of Sciences* Oct 2017, 114 (43) E9105 E9114; DOI: 10.1073/pnas.1711596114; 2.27A: Courtesy of Melanie Berkmen, Suffolk University; 2.27B: Courtesy of Melanie Berkmen, Suffolk University; 2.28A: Reprinted by permission from Springer Nature: Gahlmann, A., Moerner, W. Exploring bacterial cell biology with single-molecule tracking and super-resolution imaging. *Nat Rev Microbiol* 12, 9–22 (2014) doi:10.1038/nrmicro3154; 2.28C: Linda A. Cicero/Stanford News Service; 2.29A: Lucy Shapiro; 2.29B: Reprinted by permission from Springer Nature: Ptacin, J., Lee, S., Garner, E. et al. A spindle-like apparatus guides bacterial chromosome segregation. *Nat Cell Biol* 12, 791–798 (2010) doi:10.1038/ncb2083; 2.30: Republished with permission of American Society for Clinical Investigation, from "Pharmacological inhibition of quorum sensing for the treatment of chronic bacterial infections" by M. Hentzer and M. Givskov, *Journal of Clinical Investigation*, Issue 112(9), pp. 1300–1307, 2003. Permission conveyed through Copyright Clearance Center, Inc.; 2.31B: "Host-compound foraging by intestinal microbiota revealed by single-cell stable isotope probing" by David Berry et al., *Proceedings of the National Academy of Sciences* Mar 2013, 110 (12) 4720–4725; DOI: 10.1073/pnas.1219247110; 2.32A-B: "Examination of bacterial flagellation by dark-field microscopy" by R.M. Macnab, *Journal of Clinical Microbiology*, 1976, Volume 4, Issue 3, pp. 258–265. Reproduced with permission from American Society for Microbiology; 2.33: Science Photo Library/Alamy Stock Photo; 2.38B: Colin Cuthbert/Science Source; 2.39A: "The Capsule and S-Layer: Two Independant and Yet Compatible Macromolecular Structures in *Bacillus anthracis*" by S. Mesnage et al., *Journal of Bacteriology*, 1998, Volume 180, Issue 52, pp. 52–58. Reproduced with permission from American Society for Microbiology; 2.39B: Thomas C. Marlovits and Oliver Schraidt; 2.40A: Alexandra K. Perras et al. 2014. *Front. Microbiol.* 5: 397; 2.40B: Veronika Burmeister/Visuals Unlimited, Inc.; 2.41A-B: "Three-Dimensional Structure of the Ultraoligotrophic Marine Bacterium "*Candidatus* Pelagibacter ubique"" by X. Zhao et al., *Applied and Environmental Microbiology*, 2017, Volume 83, Issue 3, pp. 1–14, DOI: 10.1128/aem.02807-16. Reproduced with permission from American Society for Microbiology; 2.43A: Lingpeng Cheng, Tsinghua University and Hongrong Liu, Hunan Normal University; 2.43B: From "Cryo-EM shows the polymerase structures and a nonspooled genome within a dsRNA virus" by H. Liu et al., *Science*, 2015, Volume 349, Issue 6254, pp. 1347–1350, DOI: 10.1126/science.aaa4938. Reprinted with permission from AAAS; 2.44: Chaban et al. *Scientific Reports* volume 8, Article number: 97 (2018); 2.44 inset: Bonnie Chaban Quigley; 2.45A: From "Magnetosomes Are Cell Membrane Invaginations Organized by the Actin-Like Protein MamK" by A. Komeili et al., *Science*, 2006, Volume 311, Issue 5758, pp. 242–245, DOI: 10.1126/science.1123231. Reprinted with permission from AAAS; 2.45B: NIH, The Jensen Laboratory; 2.45C: NIH, The Jensen Laboratory; 2.46B: Reprinted by permission from Springer Nature: Malfatti, F., Samo, T. & Azam, F. High-resolution imaging of pelagic bacteria by Atomic Force Microscopy and implications for carbon cycling. *ISME J* 4, 427–439 (2010) doi:10.1038/ismej.2009.116; 2.47C: Alfred Pasieka/Science Source; 2.48A: J L Finney Private Collection; 2.48B: age fotostock/Alamy Stock Photo; ST2.1.1 William DePas; ST2.1.2 William DePas, Ana Gallego, Dianne Newman, Fitnat H. Yildiz.

Chapter 3

Figures: CO 3.A: Republished with permission of Annual Reviews, Inc. from "Cellular Electron Cryotomography: Toward Structural Biology In Situ" by C.M. Oikonomu et al., *Annual Review of Biochemistry*, Volume 86, pp. 873–896, 2017, DOI:10.1146/annurev-biochem-061516-044741; permission conveyed through Copyright Clearance Center, Inc.; CO 3.B: Institute of Biology Leiden; 3.2: "Cardiolipin microdomains localize to negatively curved regions of Escherichia coli membranes" by Lars D. Renner, Douglas B. Weibel, *Proceedings of the National Academy of Sciences* Apr 2011, 108 (15) 6264–6269; DOI: 10.1073/pnas.1015757108. Copyright (2011) National Academy of Sciences, U.S.A.; 3.3: Reprinted with permission from "Efficient Subfractionation of Gram-Negative Bacteria for Proteomics Studies" by M. Thein et al.,

Journal of Proteome Research, 2010, Volume 9, Issue 12, pp. 6135–6147, DOI: 10.1021/pr1002438. Copyright 2010 American Chemical Society; 3.6: Stetter and R. Rachel, University of Regensburg, Germany; 3.8C: Republished with permission of John Wiley and Sons Inc., from "Cardiolipin promotes polar localization of osmosensory transporter ProP in *Escherichia coli*" by T. Romantsov et al., Molecular Microbiology, Volume 64, Issue 6, pp. 1455–1465, DOI: 10.1111/j.1365-2958.2007.05727.x; permission conveyed through Copyright Clearance Center, Inc.; 3.12A: W. Vollmer et al., "Peptidoglycan structure and architecture", *FEMS Microbiology Reviews*, 2007, Volume 32, Issue 2, pp. 149–167, DOI:10.1111/j.1574-6976.2007.00094.x, by permission of Oxford University Press; 3.14A: Republished with permission of John Wiley and Sons Inc., from "In Situ Probing of Newly Synthesized Peptidoglycan in Live Bacteria with Fluorescent D- Amino Acids" by E. Kuru et al., *Angewandte Chemie International Edition*, Volume 51, Issue 6, pp. 12519–12523, 2012; permission conveyed through Copyright Clearance Center, Inc.; 3.14B: © Jean-Francois Gout; 3.15A: Benoît Zuber 3.15B: Benoît Zuber; 3.17A-B: Republished with permission of Royal Society of Chemistry, from "Immobilization of microorganisms for AFM studies in liquids", by T. J. Günther et al., *RSC Advances*, Issue 93, 2014, DOI: 10.1039/C4RA03874F; permission conveyed through Copyright Clearance Center, Inc.; 3.17C: "Self-catalyzed growth of S layers via an amorphous-to-crystalline transition limited by folding kinetics" by S. Chung et al., *Proceedings of the National Academy of Sciences* Sep 2010, 107 (38) 16536–16541; DOI: 10.1073/pnas.1008280107. Copyright (2010) National Academy of Sciences, U.S.A.; 3.18B: "Role of Murein Lipoprotein in Morphogenesis of the Bacterial Division Septum: Phenotypic Similarity of lkyD and lpo Mutants" by J. Fung et al., *Journal of Bacteriology*, 1978, Volume 133, Issue 3, pp. 1467–1471. Reproduced with permission from American Society for Microbiology; 3.21B: "Disclosure of the mycobacterial outer membrane: Cryo-electron tomography and vitreous sections reveal the lipid bilayer structure" by Christian Hoffmann, Andrew Leis, Michael Niederweis, Jürgen M. Plitzko, Harald Engelhardt. *Proceedings of the National Academy of Sciences* Mar 2008, 105 (10) 3963–3967; DOI: 10.1073/pnas.0709530105. Copyright (2008) National Academy of Sciences, U.S.A.; 3.22A: "The actin homologue MreB organizes the bacterial cell membrane," by Strahl, H., Bürmann, F., & Hamoen, L. *Nat Commun* 5, 3442 (2014). doi: 10.1038/ncomms 4442. https://www.nature.com/articles/ncomms 4442; https://creative commons.org/licenses/by/3.0/us/legalcode; 3.22B: "Mutations in the Lipopolysaccharide Biosynthesis Pathway Interfere with Crescentin-Mediated Cell Curvature in Caulobacter crescentus" by M.T. Cabeen et al., *Journal of Bacteriology*, 2010, Volume 192, Issue 13, pp. 3368–3378, DOI: 10.1128/jb.01371-09. Reproduced with permission from American Society for Microbiology; 3.23A: "FtsZ Dynamics during the Division Cycle of Live *Escherichia coli* Cells" by Q. Sun and W. Margolin, Journal of Bacteriology, 1998, Volume180, Issue 8, pp. 2050–2056. Reproduced with permission from American Society for Microbiology; S.23B: Duplication and segregation of the actin (MreB) cytoskeleton during the prokaryotic cell cycle" by P. Vats, L. Rothfield. *Proceedings of the National Academy of Sciences* Nov 2007, 104 (45) 17795–17800; DOI: 10.1073/pnas.0708739104. Copyright (2010) National Academy of Sciences, U.S.A.; 3.23C: Reprinted from *Cell*, Volume 115/Issue 6, N. Ausmees et al., "The Bacterial Cytoskeleton: An Intermediate Filament-Like Function in Cell Shape" pp. 705–713, 2003, DOI: 10.1016/S0092-8674(03)00935-8, with permission from Elsevier; 3.24A: From "Treadmilling by FtsZ filaments drives peptidoglycan synthesis and bacterial cell division" by A.W. Bisson-Filho et al., *Science*, 2017, Volume 355, Issue 6326, pp. 739–743, DOI: 10.1126/science.aak9973. Reprinted with permission from AAAS; 3.24B: Sandee Milhouse; 3.25A: Republished with permission of John Wiley and Sons Inc., from "Roles for both FtsA and the FtsBLQ subcomplex in FtsN-stimulated cell constriction in *Escherichia coli*" by B. Liu et al., *Molecular Microbiology*, Volume 95, Issue 6, pp. 945–970, 2015, DOI: 10.1111/mmi.12906; permission conveyed through Copyright Clearance Center, Inc.; 3.25B: Republished with permission of John Wiley and Sons Inc., from "Roles for both FtsA and the FtsBLQ subcomplex in FtsN-stimulated cell constriction in *Escherichia coli*" by B. Liu et al., *Molecular Microbiology*, Volume 95, Issue 6, pp. 945–970, 2015, DOI: 10.1111/mmi.12906; permission conveyed through Copyright Clearance Center, Inc.; 3.26: "Attaching and effacing enteropathogenic *Escherichia coli* O18ab invades epithelial cells and causes persistent diarrhea" by IC Scaletsky et al., *Infection and Immunity*, 1996, Volume 64, Issue 11, pp. 4876–4881, DOI: 10.1006/mpat.1996.0051. Reproduced with permission from American Society for Microbiology; 3.28 (insets): Republished with permission of John Wiley and Sons Inc., from "Spatial and temporal organization of replicating *Escherichia coli* chromosomes" by I. Lau et al., *Molecular Microbiology*, Volume 49, Issue 3, pp. 731–743, 2003, DOI: 10.1046/j.1365-2958.2003.03640.x; permission conveyed through Copyright Clearance Center, Inc.; 3.28 (inset 3): J. Buss et al. (2015) "A Multi-layered Protein Network Stabilizes the Escherichia coli FtsZ-ring and Modulates Constriction Dynamics" *PLOS Genetics* 11(4): e1005128. https://doi.org/10.1371/journal.pgen.1005128; 3.29A-C: "Atomic Force Microscopy of Cell Growth and Division in *Staphylococcus aureus*" by A. Touhami et al., *Journal of Bacteriology*, 2004, Volume 186, Issue 11, pp. 3286–3295, DOI: 10.1128/jb.186.11.3286-3295.2004. Reproduced with permission from American Society for Microbiology; 3.29D: Kwangshin Kim/Science Source; 3.30: Yves Brun; 3.31A-B: Reprinted from *Cell*, Volume 124/Issue 5, H. Lam et al., "A Landmark Protein Essential for Establishing and Perpetuating the Polarity of a Bacterial Cell", pp. 891–893, 2006, DOI: 10.1016/j.cell.2005.12.040, with permission from Elsevier; 3.31C: © Jason Varney/Varneyphoto.com; 3.34A-B: From "Bacterial Vesicles in Marine Ecosystems" by S.J Biller et al., *Science*, 2014, Vol. 343, Issue 6167, pp. 183–186, DOI: 10.1126/science.1243457. Reprinted with permission from AAAS; 3.34C: Courtesy of Sallie Chisholm. Photo credit: Richard Howard; 3.35A: Reprinted from *Cell*, Volume 144/Issue 4, G.P. Dubey et al., "Intercellular Nanotubes Mediate Bacterial Communication", pp. 590–600, 2011, DOI: 10.1016j.cell.2011.0, with permission from Elsevier; 3.36A: Reprinted by permission from Springer Nature: *Nature Reviews Microbiology*, "The importance of culturing bacterioplankton in the 'omics' age" by S. Giovannoni and U. Stingl, Volume 5, pp. 820–826, 2007, DOI: 10.1038/nrmicro1752; 3.36B: "The Gas Vesicle Gene Cluster from Microcystis aeruginosa and DNA Rearrangements That Lead to Loss of Cell Buoyancy" by A. Mlouka et al., *Journal of Bacteriology*, 2004, Volume 186, Issue 8, pp. 2355–2365, DOI: 10.1128/jb.186.8.2355-2365.2004. Reproduced with permission from American Society for Microbiology; 3.37: Republished with permission of Microbiology Society, from *International Journal of Systematic and Evolutionary Microbiology* 57 (Pt 7): 1429–34, August 2007; permission conveyed through Copyright Clearance Center, Inc.; 3.38: Reprinted from *Molecular Cell*, Volume 23/Issue 5, L. Craig et al., "Type IV Pilus Structure by Cryo-Electron Microscopy and Crystallography: Implications for Pilus Assembly and Functions", pp. 651–662, 2006, DOI: 10.1016/j.molcel.2006.07.004, with permission from Elsevier; 3.39A: Kwangshin Kim/Science Source; 3.39B: "Examination of bacterial flagellation by dark-field microscopy" by R.M. Macnab, *Journal of Clinical Microbiology*, 1976, Volume 4, Issue 3, pp. 258–265. Reproduced with permission from American Society for Microbiology; 3.40A: Reprinted from *Journal of Molecular Microbiology*, Volume 235/Issue 4, N.R. Francis et al., "Isolation, Characterization and Structure of Bacterial Flagellar Motors Containing the Switch Complex", pp. 1261–1270, 1994, DOI: 10.1006/jmbi.1994.1079, with permission from Elsevier; 3.41A: Republished with permission of John Wiley and Sons Inc., from "Charged residues in the cytoplasmic loop of MotA are required for stator assembly into the bacterial flagellar motor" by Y. V. Morimoto et al., *Molecular Microbiology*, Volume 78, Issue 5, pp. 1117–1129, 2010, DOI: 10.1111/j.1365-2958.2010.07391.x; permission conveyed through Copyright Clearance Center, Inc.; 3.41B: Courtesy of Protonic Nanomachine Group, Osaka University; ST 3.1.1A-B: "Uncharacterized Bacterial Structures Revealed by Electron Cryotomography" by M.J. Dobro et al., *Journal of Bacteriology*, 2017, Volume 199, Issue 17, pp.1–14, DOI: 10.1128/jb.00100-17. Reproduced with permission from American Society for Microbiology; ST3.1.1C: Martin Pilhofer; ST3.1.2A-B: "Uncharacterized Bacterial Structures Revealed by Electron Cryotomography" by M.J. Dobro et al., *Journal of Bacteriology*, 2017, Volume 199, Issue 17, pp.1–14, DOI: 10.1128/jb.00100-17. Reproduced with permission from American Society for Microbiology; TQ3.9.1: Damian Ekiert and Bernard Strauss.

Chapter 4

Figures: CO4.A: From "Coupling between distant biofilms and emergence of nutrient time-sharing" by J. Liu et al., *Science*, 2017, Volume 356, Issue 6338, pp. 638,Äì342, DOI: 10.1126/science.aah4204. Reprinted with permission from AAAS; CO4.B: Gürol Süel; 4.1C: Courtesy of Satoshi Hanada; 4.3A: Dr Jeremy Burgess/Science Source; 4.3B: Inga Spence/Science Source; 4.6C: Courtesy of Ronald Kaback; 4.8B: Tim S. Bugni, Ph.D.; 4.10: Courtesy Wikimedia Commons. https://commons.wikimedia.org/wiki/File:VCU_agar_plate_colonies.jpg; https://creativecommons.org/licenses/by/3.0/us/legalcode; 4.11B: John Foster, University of South Alabama; 4.12: "Biofilm Formation on Reverse Osmosis Membranes Is Initiated and Dominated by *Sphingomonas* spp." by L.A. Bereschenko et al., *Applied and Environmental Microbiology*, 2010, Volume 76, Issue 8, pp. 2623–2632, DOI: 10.1128/aem.01998-09. Reproduced with permission from American Society for Microbiology; 4.13B: John Foster, University of South Alabama; 4.14: John Foster, University of South Alabama; 4.15: "Short Peptide Induces an 'Uncultivable' Microorganism To Grow In Vitro" by D. Nichols et al., *Applied and Environmental Microbiology*, 2008, Volume 74, Issue 15, pp. 4889–4897, DOI: 10.1128/aem.00393-08. Reproduced with permission from American Society for Microbiology; 4.16A: "In Vitro Studies on Rickettsia-Host Cell Interactions: Intracellular Growth Cycle of Virulent and Attenuated *Rickettsia prowazeki* in Chicken Embryo Cells in Slide Chamber Cultures" by C.L. Wisseman, Jr. and A.D. Waddell, *Infection and Immunity*, 1975, Volume 11, Issue 6, pp. 1391–1401. Reproduced with permission from American Society for Microbiology; 4.16B: Jonathon Audia, University of South Alabama; 4.18: "Determination of the spatiotemporal dependence of *Pseudomonas aeruginosa* biofilm viability after treatment with NLC-colistin" by Sans-Serramitjana et al., *International Journal of Nanomedicine* 2017 12 4409–4413. Used with permission by Dove Medical Press.; 4.20A: Dennis Kunkel Microscopy/Science Source; 4.20B: Ellen Quardokus, Indiana University; 4.22A: "Constant activity of stationary-phase bacteria" by Orit Gefen, Ofer Fridman, Irine Ronin, Nathalie Q. Balaban. *Proceedings of the National Academy of Sciences* Jan 2014, 111 (1) 556–561; DOI: 10.1073/pnas.1314114111 ; 4.22B: Courtesy of Hebrew University; 4.23A: Troy Morris; 4.24B: Sebastian Kopf, University of Colorado, Boulder; 4.26A: USGS; 4.26B: Dr. Armen Taranyan/Science Source; 4.27: "SagS Contributes to the Motile-Sessile Switch and Acts in Concert with BfiSR To Enable *Pseudomonas aeruginosa* Biofilm Formation" by O.E. Petrova and K. Sauer, Journal of Bacteriology, 2011, Volume 193, Issue 23, pp. 6614–6628, DOI: 10.1128/jb.00305-11. Reproduced with permission from American Society for Microbiology; 4.28A: "Amyloid fibers provide structural integrity to *Bacillus subtilis* biofilms" by D. Romero et al. *Proceedings of the National Academy of Sciences* Feb 2010, 107 (5) 2230–2234; DOI: 10.1073/pnas.0910560107; 4.29A-B: Republished with permission of John Wiley and Sons Inc., from "Stress responses go three dimensional – the spatial order of physiological differentiation in bacterial macrocolony biofilms" by D. O. Serra and R. Hengge, *Environmental Microbiology*, Volume 16, Issue 6, pp. 1455–1471, 2014, DOI: 10.1111/1462-2920.12483; permission conveyed through Copyright Clearance Center, Inc.; 4.30A-B: Cesar de la Fuente-Nunez, et al. 2014. PLoS Pathog. 10(5):E1004152; 4.30C: Tamea Burd Photography; 4.31A: CDC HTTP://PHIL.CDC.GOV/PHIL/HOME.ASP; 4.31B: Peter Setlow, Ph.D.; 4.32A: John Walsh/Science Source; 4.32B: William J. Buikema; 4.33: "Fruiting Body Morphogenesis in Submerged Cultures of *Myxococcus xanthus*" by J.M. Kuner and D. Kaiser, *Journal of Bacteriology*, 1982, Volume 151, Issue 1, pp. 458–461. Reproduced with permission from American Society for Microbiology; 4.34A: David Scharf/Science Sourc; 4.34B: Andrew Davis and Mervyn Bibb, John Innes Centre; 4.34C: Andrew Davis, John Innes Centre; ST4.1.1B: Photo by Slava Epstein; ST4.1.2A: Kim Lewis, Northeastern University; ST4.1.2C: Kim Lewis, PhD.

Chapter 5

Figures: CO5.A-B: Courtesy of Sigal Ben-Yehuda; 5.2A, C: Courtesy of Asim K. Bej; 5.2B inset: Dr. Alfonso Davila; 5.3A: Robert Harding/Alamy Stock Photo; 5.3B: Courtesy of Phillip Brumm/C5-6 Technologies LLC; 5.3C: "Rupture of the Cell Envelope by Decompression of the Deep-Sea methanogen *Methanococcus jannaschii*" by C.B. Park and D.S. Clark, *Applied and Environmental Microbiology*, 2002, Volume 68, Issue 3, pp. 1458–1463, DOI: 10.1128/aem.68.3.1458,Äì1463.2002. Reproduced with permission from American Society for Microbiology; 5.4: Photo provided by Robert Kelly; 5.5B: "Extremely Barophilic Bacteria Isolated from the Mariana Trench, Challenger Deep, at a Depth of 11,000 Meters" by C. Kato et al., *Applied and Environmental Microbiology*, 1998, Volume 64, Issue 4, pp. 1510–1513. Reproduced with permission from American Society for Microbiology; 5.8A: Wayne P. Armstrong, Palomar College; 5.8B-C: Courtesy of Shiladitya DasSarma. University of Maryland School of Medicine; 5.12A: Jennifer Coulter/Alamy Stock Photo ; 5.12B: Republished with permission of John Wiley and Sons Inc., from "Archaea Morphology, Physiology, Biochemistry and Applications" by E. Jayashantha, *Australian Journal of Public Administration*, 2015, DOI: 10.13140/RG.2.1.4382.9281; permission conveyed through Copyright Clearance Center, Inc.; 5.13A: Birute Vijeikiene/Shutterstock; 5.13B: Republished with permission from John Wiley and Sons Inc., from "Natronobacterium" by A. Oren, Wiley Books, 2018, DOI: 10.1002/9781118960608.gbm00491; permission conveyed through Copyright Clearance Center, Inc.; 5.13C: Wild Horizons/Universal Images Group via Getty Images; 5.19A: Dr. Jack Bostrack/Visuals Unlimited, Inc.; 5.19B: Joan Slonczewski; 5.20A: John Foster, University of South Alabama; 5.20B: Last Refuge/ardea.com/age footstock; 5.21B: John Emerson/Rutgers University; 5.22A: Courtesy of Ada Yonath. Nature Communications. http://creativecommons.org/licenses/by/4.0; 5.22B: Courtesy of Mee-Ngan Yap; 5.23: NASA; 5.28A: Will & Deni McIntyre/Science Source; 5.28C: Republished with permission of Royal Society of Chemistry from "Prompt and synergistic antibacterial activity of silver nanoparticle-decorated silica hybrid particles on air filtration" Y. Ko et al., *Journal of Materials Chemistry B*, 2014, Issue 39, DOI: 10.1039/C4TB01068J; permission conveyed through Copyright Clearance Center, Inc.; 5.29A-B: John R. Battista; 5.31: Romain Briandet; 5.32A-D: Reprinted from *Molecular Cell*, Volume 48, Issue 5, Z. Yao et al., "Distinct Single-Cell Morphological Dynamics under Beta-Lactam Antibiotics", pp. 705–712, 2012, DOI: 10.1016/j.molcel.2012.09.016, with permission from Elsevier; ST5.1.1: Ronen Hazan; ST5.1.2-ST5.1.4: "Targeting Enterococcus faecalis Biofilms with Phage Therapy" by L. Khalifa et al., *Applied and Environmental Microbiology*, 2015, Volume 81, Issue, 8, pp. 2696–2705; DOI: 10.1128/AEM.00096-15. Reproduced with permission from American Society for Microbiology.

Chapter 6

Figures: CO6.B: Photo Courtesy Ohio State University; 6.2A: Lee D. Simon/Stammers/Science Source; 6.2B: Dr. Edward Chan/Visuals Unlimited, Inc.; 6.2C: NIBSC/Science Source; 6.2D: Lowell Georgia/Science Source; 6.2E: Republished with permission of Rockefeller University Press, from "Assembly and Aggregation of Tobacco Mosaic Virus in Tomato Leaves" by T.A. Shalla, *The Journal of Cell Biology*, 1964, Volume 21, Issue 2, pp. 253–264, DOI: 10.1083/jcb.21.2.253; permission conveyed through Copyright Clearance Center, Inc.; 6.2F: Norm Thomas/Science Source; 6.4B: National Science Foundation; 6.4C: Reprinted from *Virology*, Volume 466–467, A. Highfield et al., "How many *Coccolithovirus* genotypes does it take to terminate an *Emiliania huxleyi* bloom?", pp. 138–145, 2014, DOI: 10.1016/j.virol.2014.07.017, with permission from Elsevier; 6.5B: Photo Credit: Oregon State University; 6.8A: National Institute of Arthritis and Musculoskeletal and Skin Diseases/CDC; 6.9A: Ami Images/Science Source; 6.9B-C: Tim Booth/PHAC/ASPC; 6.10A: Laguna Design/Science Source; 6.10A inset: Dr. John Finch/Science Source; 6.11A: Republished with permission of Rockefeller University Press, from "Deep-etch EM reveals that the early poxvirus envelope is a single membrane bilayer stabilized by a geodetic 'honeycomb' surface coat" by J. Heuser, Volume 169, Issue 2, 2005; permission conveyed through Copyright Clearance Center, Inc.; 6.11C: Peter Morenus/UConn Photo; 6.12B: Dr. George Chapman/Getty Images; 6.14B: Ralph C. Eagle Jr./Science Source; 6.16A: Dr. Raoult/Science Source; 6.16B: © Didier RAOUL T/Leon ESPINOSA/Bernard CAMPANAIIRD 198/URMITE/CNRS Photothèque; 6.17A: Jenny E. Ross/Getty Images; 6.17B: © Julia BARTOLI/Chantal ABERGEL/AMU/IGS/CNRS Photothèque;

6.19B: Paul Slade/Paris Match via Getty Images; 6.23A: From "The Bacteriophage T7 Virion Undergoes Extensive Structural Remodeling During Infection" by B. Hu et al., *Science*, 2013, Volume 339, Issue 6119, pp. 576–579, DOI: 10.1126/science.1231887. Reprinted with permission from AAAS; 6.28A: Reprinted from *Cell Host & Microbe*, Volume 8, Issue 1, J.A. den Boon et al., "Cytoplasmic Viral Replication Complexes", pp. 77–85., 2010, DOI: 10.1016/j.chom.2010.06.010, with permission from Elsevier; 6.29A: Dr. Ken Greer/Visuals Unlimited, Inc.; 6.31 inset: J-Y SGRO/PHANIE/age footstock; 6.32 inset: EMBL/Mattei/Briggs. Structure of the HIV-1 capsid core determined inside the virus by electron-cryo tomography (Mattei et al., Science, 2016); 6.33A: Centre for Bioimaging, Rothamsted Research/Science Source; 6.33B: Scott Bauer, Agricultural Research Service, USDA; 6.33C: C. Avilez/Alamy Stock Photo; 6.33D: Nature and Science/Alamy Stock Photo; 6.38: "Principles of Viriology, Fourth Edition, Bundle" by J. Flint et al., ASM Book, 2015, DOI: 10.1128/9781555819521. Reproduced with permission from American Society for Microbiology; 6.40A: Lynn Thomason; 6.40B: Courtesy of Kema Malki; 6.41B: Effect of plaque purification Fig. 5 from Plaque assay for human coronavirus NL63 using human colon carcinoma cells. Petra Herzog,Christian Drosten and Marcel A Müller. Virology Journal 20085:138 doi: 10.1186/1743-422X-5-138; ST6.1.1B: Courtesy of Jeremy Barr; ST6.1.2A-B: Jeremy J. Barr and Benjamin S. Padman; Table 6.1.a: Biophoto Associates/Science Source; Table 6.1.b: Cindy L. McKenzie/USDA; Table 6.1.c: CDC/James Gathany; Table 6.1.d: Scott Camazine/Visuals Unlimited, Inc.; Table 6.1.e: Eye of Science/Science Source; Table 6.1.f: James Cavallini/Science Source; Table 6.1.g: Science Photo Library/Science Source.

Chapter 7

Figures: CO7.A: Courtesy of Elizabeth Hutchison; CO7.B: Keith Walters; 7.1: Courtesy of Carol von Dohlen; 7.7: Dr. Gopal Murti/Visuals Unlimited, Inc.; 7.15: Arthur Kornberg/U.S. National Library of Medicine; 7.16: Bettmann/Getty Images; 7.21A: Science VU/Drs. H. Potter-D. Dressler/Visuals Unlimited, Inc.; 7.24 inset: Republished with permission of John Wiley and Sons Inc., from "Prokaryotic DNA segregation by an actin-like filament" by J. Møller-Jensen et al., *The EMBO Journal*, Volume 12, Issue 12, pp. 3119–3127, 2002; permission conveyed through Copyright Clearance Center, Inc.; 7.27A: Cornell Brand Communications; 7.31A: Courtesy of Karen Lloyd; ST7.1.1: Wyss Institute at Harvard University; ST7.1.2B-C: Reprinted by permission from Macmillan Publishers Ltd: *Nature*, "CRISPR-Cas encoding of a digital movie into the genomes of a population of living bacteria" by S.L. Shipman et al., Volume 547, pp. 345–349, 2017, DOI: 10.1038/nature23017.

Chapter 8

Figures: CO8.A-B: Mikhail Shapiro; 8.1A: Courtesy of National Library of Medicine/National Institute of Health; 8.1B: National Cancer Institute; 8.8B: David Scharf/Science Source; 8.10: Courtesy of Claudia Steglich; 8.24: Republished with permisssion of John Wiley and Sons, Inc., from "Electron microscopic visualization of a discrete class of giant translation units in salivary gland cells of chironomus tentans," by C. Francke, J.E. Edstrom, A.W. Mcdowall, O.L. Miller, Jr. *The Embo Journal* (1982) 1:59–62; 8.26: Republished with permission of John Wiley and Sons Inc., from "Superresolution Imaging of Ribosomes and RNA polymerase in live *Escherichia coli* cell" by S. Bakshi et al., *Molecular Microbiology*, Volume 85, Issue 1, pp. 21–38, 2012, DOI: 10.1111/j.1365-29.2012.08081.x; permission conveyed through Copyright Clearance Center, Inc.; 8.30A: Courtesy of Thermo Fisher Scientific; 8.30B: Simko/Visuals Unlimited, Inc.; 8.35: Republished with permission of John Wiley and Sons Inc., from "The ExPortal: an organelle dedicated to the biogenesis of secreted proteins in *Streptococcus pyogenes*" by M.G. Caparon and J.W. Rosch, *Molecular Microbiology*, Volume 58, Issue 4, pp. 959–968, 2005; permission conveyed through Copyright Clearance Center, Inc.; 8.38: Reprinted by permission from Springer Nature: Koronakis, V., Sharff, A., Koronakis, E. et al. Crystal structure of the bacterial membrane protein TolC central to multidrug efflux and protein export. *Nature* 405, 914–919 (2000) doi:10.1038/35016007; ST 8.1.1: Thomas Steitz.

Chapter 9

Figures: CO9.A: From "Mutation dynamics and fitness effects followed in single cells" by L. Robert et al., *Science*, 2018, Volume 359, Issue 6381, pp. 1283–1286, 10.1126/science.aan0797. Reprinted with permission from AAAS; CO9.B: Lydia Robert; 9.14A: Courtesy of the University of Wisconsin-Madison Archives (ID S04387); 9.15: Dennis Kunkel Microscopy/Science Source; 9.16A: Dr. Robert Calentine/Visuals Unlimited, Inc.; 9.16B: Martha Hawes, University of Arizona; 9.22: Reprinted with permission from Springer Nature: *Nature Microbiology*, "Retraction of DNA-bound type IV competence pili initiates DNA uptake during natural transformation in Vibrio cholerae" by C. Ellison et al., 2018, DOI: 10.1038/s41564-018-0174-y; 9.28A: Peter Steffen/picture-alliance/dpa/AP Images; 9.28B: Keegan Houser, UC Berkeley; 9.30: "Defining Gene-Phenotype Relationships in *Acinetobacter baumannii* through One-step Chromosomal Gene Inactivation" by A.T. Tucker et al., *Journal of Microbiology*, 2014 Volume 5, Issue 4, pp. 1–9, DOI: 10.1128/mBio.01313.14.Reproduced with permission from American Society for Microbiology; 9.36A: Alison Buchan, University of Tennessee, Knoxville, College of Arts & Sciences, Department of Microbiology; 9.36B: Photo Courtesy of Alison Buchan. "Production of the Antimicrobial Secondary Metabolite Indigoidine Contributes to Competitive Surface Colonization by the Marine *Roseobacter Phaeobacter* sp. Strain Y4I" by W.N. Cude et al., *Applied and Environmental Microbiology*, 2012, Volume 78, Issue 14, pp. 4774, DOI: 10.1128/AEM.00297-12. Reproduced with permission from American Society for Microbiology; 9.36 B inset: Concept: W. Nathan Cude, Design: Ashley M. Frank, Photography: Mohammad Moniruzzaman; ST9.1.1: Courtesy of Jacquline Barton/California Institute of Technology; ST9.1.3: Reprinted with permission from American Chemical Society. "DNA Charge Transport with the Cell" by J.K. Barton et al., *Biochemistry*, Volume 54, Issue 4, pp. 962–973, DOI: 10.1021/bi501520w. Copyright 2015 American Chemical Society.

Chapter 10

Figures: CO10.A-B: Vincent Cassone Paulose, et al. 2016. *Plos One* 11: e0146643; 10.5A: Keystone Press/Alamy Stock Photo; 10.19A: AP Photo/J. Scott Applewhite; 10.21A: Dr. Ronald R. Breaker, Yale University; 10.22A: Dale Lewis, NIH/NCI; 10.22B: NIH Medical Arts; 10.28C: "Genetic reductionist approach for dissecting individual roles of GGDEF proteins within the c-di-GMP signaling network in *Salmonella*" by C. Solano et al. *Proceedings of the National Academy of Sciences* May 2009, 106 (19) 7997–8002; DOI:; 10.1073/pnas.0812573106 ; 10.29A: Edward G. Ruby and Margaret McFall-Ngai, University of Hawaii; 10.29B: Edward G. Ruby, University of Hawaii; 10.29C-D: Courtesy J.W. Hastings, Harvard University, through E.G. Ruby, University of Hawaii; 10.30A: K. Nealson, Univ. S. Cal.; 10.32: E. Peter Greenberg, University of Washington; 10.33: Zach Donnell; 10.35: Ian Joint; 10.38: From "Reconstitution of Circadian Oscillation of Cyanobacterial KaiC Phosphorylation in Vitro" by M. Nakajima et al., *Science*, 2005, Volume 308, Issue 5720, pp. 414–415, DOI: 10.1126/science.1108451. Reprinted with permission of AAAS; 10.39A: Carl Johnson, Vanderbilt University; ST10.1.1: Chirstopher Lennon and Marlene Belfort; ST10.1.2B: Lennon et al. *Genes Dev*. 30:2663–2668.

Chapter 11

Figures: CO11.A: Reprinted by permission from Springer Nature: Ke, Z., Strauss, J.D., Hampton, C.M. et al. Promotion of virus assembly and organization by the measles virus matrix protein. *Nat Commun* 9, 1736 (2018) doi:10.1038/s41467-018-04058-2; CO11.B Courtesy of Robert O. Davies; 11.1: Science Source/Science Source; 11.2: The Esther M. Zimmer Lederberg Trust - www.estherlederberg.com; 11.3A-C: Courtesy of Gökhan Tolun; 11.4A: Courtesy of Robert Duda; 11.9A: NIH; 11.12 inset: Dr. Fred Murphy/Visuals Unlimited, Inc.; 11.13B: Reprinted by permission from Springer Nature: Noda, T., Sugita, Y., Aoyama, K. et al. Three-dimensional analysis of ribonucleoprotein complexes in influenza A virus. *Nat Commun* 3, 639 (2012) doi:10.1038/ncomms1647; 11.15A: CDC/James Gathany; 11.16A: Zero Creatives/Cultura/Corbis; 11.16B: Christopher Morris/Corbis via Getty Images; 11.16C: Dimas Ardian/Getty Images; 11.19: Courtesy of Isabel Fernández de Castro and Cristina Riasco, Centro Nacional de Biotecnología, CSIC; 11.20A: John Mottern/AFP/Getty Images; 11.20B: Bob Strong/Reuters/ Newscom; 11.20C: Reprinted from *Structure*, Volume 14, Issue1, J.A.G Briggs et al., "The Mechanism of HIV-1 Core Assembly: Insights from Three-Dimensional Reconstructions of Authentic Virions", pp. 15–21, 2006, DOI: 10.1016/j.str.2005.09.010, with permission from Elsevier; 11.20D: Reprinted by permission from Springer Nature: Campbell, E., Hope, T. HIV-1 capsid: the multifaceted key player in HIV-1 infection. *Nat Rev Microbiol* 13, 471–483 (2015) doi:10.1038/nrmicro3503; 11.21C: AP Photo/Koji Sasahara; 11.27A: Reprinted by permission from Springer Nature: Neil, S., Zang, T. & Bieniasz, P. Tetherin inhibits retrovirus release and is antagonized by HIV-1 Vpu. *Nature* 451, 425–430 (2008) doi:10.1038/nature06553; 11.27B: Reprinted by permission from Springer Nature: Sowinski, S., Jolly, C., Berninghausen, O. et al. Membrane nanotubes physically connect T cells over long distances presenting a novel route for HIV-1 transmission. *Nat Cell Biol* 10, 211–219 (2008) doi:10.1038/ncb1682; 11.29: Courtesy of Joanna Wysocka; 11.29 inset: Reprinted by permission from Springer Nature: Grow, E., Flynn, R., Chavez, S. et al. Intrinsic retroviral reactivation in human preimplantation embryos and pluripotent cells. *Nature* 522, 221–225 (2015) doi:10.1038/nature14308; 11.30 inset: Emily Whitehead Foundation; 11.32: Dr P. Marazzi/Science Source; ST11.1.1A: From "Airborne Transmission of Influenza A/H5N1 Virus Between Ferrets" by S. Herfst et al., *Science*, 2012, Volume 336, Issue 6088, pp. 1534–1541, DOI: 10..1126/science.1213362. Reprinted with permission from AAAS; ST11.1.1B: Dirk-Jan Visser/The New York Times/Redux; ST11.1.1.2: Science History Images/Alamy Stock Photo.

Chapter 12

Figures: CO12.A-B: Courtesy of Roberto Di Leonardo; 12.1A: "Thermus aquaticus gen. n. and sp. n., a Nonsporulating Extreme Thermophile" by T.D. Brock and H. Freeze, *Journal of Bacteriology*, Volume 98, Issue 1, pp. 289–297. Reproduced with permission from American Society for Microbiology; 12.1B: Eye of Science/Science Source; 12.4A: Erik Zinser; 12.6: From "Accurate Multiplex Polony Sequencing of an Evolved Bacterial Genome" by J. Shendure et al., *Science*, 2005, Vol.309, Issue 5741, pp. 1728–1732, DOI: 10.1126/ science.1117389. Reprinted with permission from AAAS; 12.9: Sarah Johnson; 12.12A: Courtesy of Steven E. Finkel; 12.15: Wellcome Images/Science Source; 12.16A-B: Steve Northup/The LIFE Images Collection via Getty Images/Getty Images; 12.22: From "Stochastic Pulse Regulation in Bacterial Stress Response" by J C.W. Locke et al., *Science*, 2011, Vol. 334, Issue 6054, pp. 366–369. Reprinted with permission from AAAS.; 12.23A-C: Reprinted from *Virology*, Volume 309, Issue 1, K. Krishnamurthy et al., "The Potato virus X TGBp3 protein associates with the ER network for virus cell-to-cell movement", pp. 135–151, DOI: 10.1016/s0042-6822(02)00102-2, with permission from Elsevier; 12.24: Republished with permission of American Association for Clinical Chemistry Inc., from "Next-Generation Sequencing: From Basic Research to Diagnostics", Voelkerding et al, Volume 55, Issue 4, pp. 641–658, 2009, DOI: 10.1373/clinchem.2008.112789; permission conveyed through Copyright Clearance Center, Inc.; 12.26A-B: Dr. George Chapman/ Visuals Unlimited, Inc.; 12.35: The iGEM Foundation and Justin Knight; ST12.1.1: J. Craig Venter Institute; ST12.1.4A: Tom Deerinck and Mark Ellisman of the National Center for Imaging and Microscopy Research at the University of California at San Diego.

Chapter 13

Figures: CO13.A: Reprinted from *Biophysical Journal*, Volume 115, Issue 2, H.H. Tuson et al., "The Starch Utilization System Assembles around Stationary Starch-Binding Proteins", pp. 242–250, 2018, DOI: 10.1016/j.bpj.2017.12.015, with permission from Elsevier; CO13.B: University of Michigan; 13.1A: Fredrik Bäckhed; 13.4A: Urs von Stockar; 13.4B: Dennis Kunkel Microscopy/Science Source; 13.5: Photo by Keith Weller, USDA HTTP://WWW.ARS.USDA.GOV/IS/ GRAPHICS/PHOTOS/JUL98/K6010-15.HTM; 13.12D.1: B.A.E. Inc./Alamy Stock Photo; 13.12D.2: Cornishman/iStockphoto; 13.18A: "An *Escherichia coli* MG1655 Lipopolysaccharide Deep-Rough Core Mutant Grows and Survives in Mouse Cecal Mucus but Fails To Colonize the Mouse Large Intestine" by A.K. Møller et al., Infection and Immunity, 2003, p. 2142–2152, DOI: 10.1128/IAI.71.4.2142-2152.2003. Reproduced with permission from American Society for Microbiology; 13.18B: Courtesy of Tyrrell Conway; 13.21A: Food Collection/ Superstock; 13.21B: age fotostock/Alamy Stock Photo; 13.21C: Lee Hacker/Alamy Stock Photo; 13.21D: D. Hurst/Alamy Stock Photo; 13.23A: W.W. Norton & Company; 13.23B: Courtesy and © Becton, Dickinson and Company; 13.28B: © Tristan Barbeyron; 13.31B: Uchechukwu Okere; 13.32B: Win McNamee/Getty Images; 13.34A: AP Photo/Ed Andrieski; 13.35A: Neil Q. Wofford, University of Oklahoma; 13.35B: Courtesy of Jessica Sieber; 13.36B: Courtesy of H.J.M. Harmsen, Univ. of Groningen, The Netherlands; ST13.1.1A: Dr. James Versalovic; ST13.1.1B; ST13.1.2B: Luk, al. 2018. Postnatal colonization with human "infant-type" PLoS ONE 13:e0196510.

Chapter 14

Figures: CO14.A: Rob Cornelissen, University of Hasselt; CO14.B: Filip Meysman; 14.1: Bjerg et al. 2018 PNAS 115:5786; 14.2A: Reprinted by permission from Springer Nature: Liu, Y., Hidaka, E., Kaneko, Y. et al. J Gastroenterol (2006) 41: 569. https://doi.org/10.1007/s00535-006-1813-2; 14.2B: Republished with permission of The Society of Neuroscience, from "The Micro-Architecture of Nitochondria at Active Zones: Electron Tomography Reveals Novel Anchoring Scaffolds and Cristae Structured for High-Rate Metabolism" by G.A. Perkins et al., *The Journal of Neuroscience*, Volume 30, Issue 3, 1981 DOI: 10.1523/JNEUROSCI.1517-09.2010; permission conveyed through Copyright Clearance Center, Inc.; 14.6A: From the photograph collection of the former Glynn Research Foundation Ltd. Republished with permission of Portland Press, Ltd., from *Bioscience Reports*, Volume 11, Issue 6, p. 293, 1991, DOI: 10.1007/BF01130210. Permission conveyed through Copyright Clearance Center, Inc.; 14.20A: Derek Lovley; 14.20C: RGB Ventures/ SuperStock/Alamy Stock Photo; 14.21B: Michigan State University; 14.22B: S. Pirbadian, et al. Shewanella oneidensis MR–1 nanowires are outer membrane and periplasmic extensions of the extracellular electron trasport components. *Proceedings of the National Academy of Sciences of the United States of America* 111: 12883–12888. 2014. Courtesy of Moh El-Naggar; 14.25A: Reprinted from *Journal of Structural Biology*, Volume 161, Issue 3, L. van Niftrik et al., "Combined structural and chemical analysis of the anammoxosome: A membrane-bounded intracytoplasmic compartment in anammox bacteria", pp.401-410, 2008, DOI: 10.1016/j.jsb.2007.05.005, with permission from Elsevier; 14.26A: Paramount/Courtesy: Everett Collection; 14.26B: Tuul & Bruno Morandi/ Getty Images; 14.26C: "Subunit cell wall of Sulfolobus acidocaldarius" by R.L. Weiss, *Journal of Bacteriology*, Volume 118, Issue 1, 1974, pp. 275–84. Reproduced with permission from American Society for Microbiology; 14.27A: Environmental Microbe-Metal Interactions by D. Lovley, 2000, ASM Press; 14.28A: Robert Harding/Alamy Stock Photo; 14.28B: Reprinted by permission from Springer Nature: Murr, L.E. JOM (2006) 58: 23. https://doi.org/10.1007/s11837-006-0136-3; 14.29B: Lorenz Adrian; 14.31: Maria Stenzel/National Geographic; 14.34A: Republished with permission of The Royal Society (UK), from "Roger Yate Stanier, 22 October - 29 January 1982" by P.H. Clarke, *Biographical Memoirs of fellows of the Royal Society*, Volume 32, pp. 542–568, 1982; permission conveyed through Copyright Clearance Center, Inc. ;14.34B: Reprinted by permission from Springer Nature: Govindjee, Allen, J.F. & Beatty, J.T. *Photosynthesis Research* (2004) 80: 1. https://doi.org/10.1023/ B:PRES.0000030564.59043.ca; 14.34C: Gerd Guenther/Science Source; 14.37B: Biophoto Associates/Science Source; ST14.1.1: James Swift; ST14.1.1 inset: Clare Reimers.

Chapter 15

Figures: CO15.A-B: Courtesy of Paul Jensen; 15.1: Scripps Institute of Oceanography at UC San Diego; 15.4A: Andrew Allen/J. Craig Venter Institute and Scripps institution of Oceanography (UCSD); 15.4B: Greg Wanger and the Allen Lab at JCVI; 15.5: Reprinted by permission from Springer Nature: Pande, S., Shitut, S., Freund, L. et al. Metabolic cross-feeding via intercellular nanotubes among bacteria. *Nat Commun* 6, 6238

(2015) doi:10.1038/ncomms7238; 15.6A: Michael Clayton (emeritus), University of Wisconsin-Madison; 15.6B: Courtesy of Jill Zeilstra-Ryalls; 15.6C: Reprinted from *Hydrometallurgy*, Volume 119, S.N. Tan and M. Chen, "Early stage adsorption behaviour of *Acidithiobacillus. ferrooxidans* on minerals I: An experimental approach", pp. 87–94, 2012, DOI: 10.1016/j.hydromet.2012.02.001, with permission from Elsevier; 15.10A-B: Courtesy of Sabine Heinhorst and Dr. Gordon C. Cannon; 15.11B: "*Chlorobium tepidum* Mutant Lacking Bacteriochlorophyll *c* Made by Inactivation of the *bchK* Gene, Encoding Bacteriochlorophyll *c* Synthase" by N. Frigaard et al., *Journal of Bacteriology*, Volume 184, Issue 12, 2002, pp. 3368–3376, DOI: 10.1128/JB.184.12.3368-3376.2002. Reproduced with permission from American Society for Microbiology; 15.12B: "Rupture of the Cell Envelope by Decompression of the Deep-Sea Methanogen *Methanococcus jannaschii*" by C.B. Park et al., *Applied and Environmental Microbiology*, 2002, Volume 68, Issue 3, pp. 1458–1463, DOI: 10.1128/AEM.68.3.1458-1463.2002. Reproduced with permission from American Society for Microbiology; 15.15A: The Society for Actinomycetes (by Y. Gyobu). http://atlas.actino.jp; 15.15B inset: Dr. Barry Slaven/Visuals Unlimited, Inc.; 15.19: Elif Bayraktar/Shutterstock; 15.20 inset: Dartmouth Electron Microscope Facility; 15.24: The Natural History Museum/Alamy Stock Photo; ST15.1.1: Courtesy of Pieter Dorrestein/University of California, San Diego; ST15.1.3 inset: Courtesy of © Leibniz-Institut DSMZ-Deutsche Sammlung.

Chapter 16

Figures: CO16.A: Freedomnaruk/Shutterstock; CO16.B: Courtesy of Roane Freitas Schwan; 16.1A: Album/Art Resource, NY; 16.1B: Courtesy of © Novozymes, 2019; 16.2: Kelly Ivors; 16.3A: Xvision/Getty Images; 16.3B: Stockfood, Ltd./Alamy Stock Photo; 16.5A: Picture Partners/Alamy Stock Photo; 16.5A inset: Scimat/Science Source; 16.6A: Svetlana Foote/Alamy Stock Photo; 16.6B: Dorling Kindersley ltd/Alamy Stock Photo; 16.6C: Maximilian Stock Ltd./Jupiter Images; 16.6D: Barmalini/Shutterstock; 16.8A: RossHelen Editorial/Alamy Stock Photo; 16.8B: Juice Images/Alamy Stock Photo; 16.8C: Oliver Ring/imageBROKER/Shutterstock; 16.8D: ITAR-TASS News Agency/Alamy Stock Photo; 16.10A: Christy Liem/Shutterstock; 16.10B: Gregory G. Dimijian/Science Source; 16.11A: The Yarvin Pantry/Alamy Stock Photo; 16.11B: Goolia Photography/Shutterstock; 16.12A: Pulsar Images/Alamy Stock Photo; 16.13: Mariusz Szczawinski/Alamy Stock Photo; 16.14A: Aflo Co., Ltd/Alamy Stock Photo; 16.14B: age fotostock/Alamy Stock Photo; 16.15A: David Scharf/Science Source; 16.15B NEW CREDIT: From "Yeast Cells Provide Insight into Alpha-Synuclein Biology and Pathobiology" by T.F. Outeiro and S. Lindquist, *Science*, 2003, Volume 302, Issue 5651, pp. 1772–1775, DOI: 10.1126/science.1090439. Reprinted with permission from AAAS; 16.16: John Foxx/Getty Images; 16.17: Susan C. Bourgoin/FoodPix/Jupiterimages; 16.18A: Jim Sugar/Getty Images; 16.18B: Bert de Ruiter/Alamy stock photo; 16.19A: Borromeo/Art Resource, NY; 16.19B: Lior + Lone/Stocksy; 16.22: U.S. Department of Agriculture; 16.23B: CDC; 16.24A-B: Republished with permission of John Wiley and Sons Inc., from "The persistence of Salmonella following desiccation under feed processing environmental conditions: a subject of relevance" by O. Habimana et al., *Letters in Applied Microbiology*, Volume 59, Issue 5, pp. 464–470, 2014, DOI: 10.1111/lam.12308; permission conveyed through Copyright Clearance Center, Inc.; 16.25A: Courtesy of Julia DeNiro; 16.27: Republished with permission of Microbiology Society, from "Production of attacking-effacing lesions in ligated large intestine loops of 6-month-old sheep by *Escherichia Coli* 0157:H7," by A. D. Wales, et al. *The Journal of Medical Microbiology* 2002. Volume S1, Issue 9; permission conveyed throgh Copyright Clearnce Centre, inc.; 16.29A: Dennis Kunkel Microscopy/Science Source; 16.32A: Joanne Lawton/Washington Business Journal; 16.33B: Maximilian Stock Ltd/Science Source; 16.35A: From "Production of type II ribotoxins by Aspergillus species and related fungi in Taiwan" by S.S. Tzean et al., *Toxicon*, 1995, Volume 33, Issue 1, pp.105-110, DOI: 10.1016/0041-0101(94)00140-4. Reprinted with permission from AAAS; 16.35B: Dennis Kunkel Microscopy/Science Source; 16.37A: Colin Cuthbert/Science Source; ST16.1.1A: Stephen Barnes/Medical/Alamy Stock Photo ; Table 16.4a: Jana Milin/Shutterstock; Table 16.4b: Duane Braley/Star Tribune via Getty Images; Table 16.4c: Monticello/Shutterstock; Table 16.4d: TaurusPhotography/Alamy Stock Photo; Table 16.4e: Ed Endicott/Alamy Stock Photo; Table 16.4f: Vitalii Borovyk/Alamy Stock Photo; Table 16.4g: Henadzi Pechan/Alamy Stock Photo; Table 16.4h: Lydia Vero/Shutterstock; Table 16.4i: Francesco Cantone/Shutterstock; Table 16.4j: Helen Sessions/Alamy Stock Photo; Table 16.4k: The Photo Works/Alamy Stock Photo.

Chapter 17

Figures: CO17.A-B: Roberto Nespolo; 17.1A: Francois Gohier/Science Source; 17.1B: Jane Gould/Alamy Stock Photo; 17.3 insets: Reprinted by permission from Springer Nature: *Nature*, "Geobiology of a microbial endolithic community in the Yellowstone geothermal environment" by J. Walker et al., Volume 434, pp. 1011–1014, 2005, DOI:10.1038/nature03447; 17.4 inset a: Reprinted by permission from Springer Nature: *Nature*, "Large colonial organisms with coordinated growth in oxygenated environments 2.1 Gyr ago", by A. Albani et al., Volume 466, pp. 100–104, 2010, DOI:10.1038/nature09166; 17.4 inset b: Andrew Knoll; 17.4 inset c: Shuhai Xiao at Virginia Tech; 17.4 inset d: Sinclair Stammers/SPL/Science Source; 17.5A-B: Nicholas J. Butterfield, University of Cambridge; 17.6: Reprinted by permission from Springer Nature: *Nature*, "Questioning the evidence for Earth's oldest fossils" by M.D. Brasier et al., Volume 416, pp. 76–81, 2002, DOI: 10.1038/416076a; 17.7B: Christian Knudsen; 17.8A-C: Reprinted from *Precambrian Research*, Volume 304, Z. Guo et al., "Cellular taphonomy of well-preserved Gaoyuzhuang microfossils: A window into the preservation of ancient cyanobacteria", pp. 88–98, 2018, with permission from Elsevier; 17.9A-B: Dr. Kurt Konhauser/USGS; 17.11B: National News/ZUMA Press/Newscom; 17.12A: Dianne Newman, Caltech; 17.17 inset: Ryuichi Koga; 17.18 inset: Institute for Genomic Biology, University of Illinois at Urbana-Champaign; 17.19: "Morphological variation of new *Thermoplasma acidophilum* isolates from Japanese hot springs" by m. Yasuda et al., *Applied and Environmental Microbiology*, 1995, Volume 61, Issue 9, pp. 3482–3485. Reproduced with permission from American Society for Microbiology;; 17.20 inset: Reprinted from *Systematic and Applied Microbiology*, Volume 15, Issue 3, R. Huber et al., "*Aquifex pyrophilus* gen. nov. sp. nov., Represents a Novel Group of Marine Hyperthermophilic Hydrogen-Oxidizing Bacteria", pp. 340–351, 1992, with permission from Elsevier; 17.22A: Gao et al. *PLOS Pathogens* June 10, 2010; 17.22B-C: Y. Lin et al., "Emergence of a small colony variant of vancomycin-intermediate *Staphylococcus aureus* in a patient with septic arthritis during long-term treatment with daptomycin", *Journal of Antimicrobial Chemotherapy*, Volume 71, Issue 7, July 2016, Pages 1807–1814, DOI: 10.1093/jac/dkw060, by permission of Oxford University Press; 17.22D: Gao et al. *PLOS Pathogens* June 10, 2010; 17.24: Zachary Blount; 17.24 inset: Michigan State University Communications and Brand Strategy; 17.25C: Sarin Images/GRANGER; 17.28A: Panel A: Figure 1B from Hastings et al. 2004. *PLoS Biology* 2:e399. https://doi.org/10.1371/journal.pbio.0020399; 17.28B: Susan Rosenberg; 17.30: Gerd Guenther/Science Source; 17.31: CDC; 17.32: Pfarr K., Hoerauf A. (2005) *PLoS Med* 2(4): E110; 17.33 inset: Tommaso Bonaventura/Contrasto/Redux; ST17.1.1-ST17.1.2: Courtesy of Harvard Medical School and Technion–Israel Institute of Technology.

Chapter 18

Figures: CO18.A: Nika Pende et al.2014. *Nature Communications* 5:4803.; CO18.B: Silvia Bulgheresi; 18.1 inset: Image Courtesy of The Franklin Institute; 18.2: Christine M. Dejea et al. 2014. PNAS 111:18321; 18.3A: Durk Talsma/Alamy Stock Photo; 18.3B: L.L. Jahnke/NASA; 18.3C: Science VU/Visuals Unlimited, Inc.; 18.4: Reprinted by permission from Springer Nature: Luef, B., Frischkorn, K., Wrighton, K. et al. Diverse uncultivated ultra-small bacterial cells in groundwater. *Nat Commun* 6, 6372 (2015) doi:10.1038/ncomms7372; 18.5A: Dennis Kunkel Microscopy/Science Source; 18.5B: Republished with permission of John Wiley and Sons Inc., from "Chroococcoid cyanobacteria in the sea: A ubiquitous and diverse phototrophic biomass" by P.W. Johnson and J.McN. Sieburth, Limnology and Oceanography, Volume24, Issue 5, pp. 928–935, 1979, DOI:

10.4319/lo.1979.24.5.0928; permission conveyed through Copyright Clearance Center, Inc.; 18.6A: Slonczewski, Kenyon College; 18.6B; Dr. Robert Calentine/Visuals Unlimited, Inc.; 18.6C: M. I. Walker/ Science Source; 18.7A: Michael Abbey/Science Source; 18.7B: Dr. Peter Siver/Visuals Unlimited, Inc.; 18.7C: Burkhard Büdel; 18.9: Reprinted by permission from Springer Nature: Overmann J., Garcia-Pichel F. (2013) The Phototrophic Way of Life. In: Rosenberg E., DeLong E.F., Lory S., Stackebrandt E., Thompson F. (eds) *The Prokaryotes.* Springer, Berlin, Heidelberg; 18.10A: CDC; 18.10B: "An in vivo membrane fusion assay implicates SpoIIIE in the final stages of engulfment during Bacillus subtilis sporulation" by Marc D. Sharp, Kit Pogliano. *Proceedings of the National Academy of Sciences* Dec 1999, 96 (25) 14553–14558; DOI: 10.1073/pnas.96.25.14553. Copyright (1999) National Academy of Sciences, U.S.A.; 18.11A: Dr. George Chapman/ Visuals Unlimited, Inc.; 18.11B: SCIMAT/Science Source; 18.12A: Dennis Kunkel Microscopy/Science Source; 18.12B: John M. Hilinski, M.D.; 18.13A-B: Esther Angert, Cornell University; 18.14A: Esther Angert, Cornell University; 18.15A: Julie A. Theriot and Timothy J. Mitchison; 18.16A (a): Eye of Science/Science Source; 18.16A (b): David Scharf/Science Source; 18.16B (a): Eye of Science/Science Source; 18.16B (b): David Scharf/Science Source; 18.17A: Reprinted by permission from Springer Nature: C. Holliger et al. 1998 Arch. Holliger, C., Hahn, D., Harmsen, H. et al. *Arch Microbiol* (1998) 169: 313. https://doi.org/10.1007/s002030050577; 18.18A-B: Mitchell F. Balish, Miami University (Ohio); 18.18C: Dr. Michael Gabridge/ Visuals Unlimited, Inc.; 18.19A: Juan Pablo Gomez-Escribano, John Innes Centre, Norwich, UK; 18.19B: NCRS/USDA; 18.19C: K. Furihata, A. Shimazu & T. Shomura, http://atlas.actino.jp/; 18.20A: Roberto Nistri/Alamy Stock Photo; 18.20B: Alexis Rosenfeld/ Getty Images; 18.20C: © In-Depth Images Kwajalein; 18.20D: Karen Gowlett-Holmes/Getty Images; 18.21A: CDC/Dr. George P. Kubica; 18.21B: CDC; 18.22A: DIOMEDIA/DIOMEDIA/James Webb; 18.23A: Dr. Gary Gaugler/Science Source; 18.23B: "Myceloid growth of Arthrobacter globiformis and other Arthrobacter species" by J.J. Germida and L.E. Casida, Jr, *Journal of Bacteriology*, 1980, Volume 144, Issue 3, pp. 1152–1158. Reproduced with permission from American Society for Microbiology; 18.25A: Republished with permission of John Wiley and Sons Inc., from "Connectivity of centermost chromatophores in *Rhodobacter sphaeroides* bacteria" by J. Noble et al., *Molecular Microbiology*, Vol. 109, Issue 6, pp. 812–825, 2018, DOI: 10.1111/ mmi.14077; permission conveyed through Copyright Clearance Center, Inc.; 18.25B: "*Rhodopseudomonas acidophila*, sp. n., a New Species of the Budding Purple Nonsulfur Bacteria" by N. Pfennig, *Journal of Bacteriology*, 1969, Volume 99, Issue 2, pp. 597–602. Reproduced with permission from American Society for Microbiology; 18.25C: "Aerobic Anoxygenic Phototrophic Bacteria" by V.V. Yurkov and J.T.Beatty, *Microbiology and Molecular Biology Reviews*, 1998, Volume 62, Issue 3, pp. 695–724. Reproduced with permission from American Society for Microbiology; 18.26A: Reprinted from *Systematic and Applied Microbiology*, Volume 28, Issue 6, S.A. Moosvi et al., "Molecular detection and isolation from Antarctica of methylotrophic bacteria able to grow with methylated sulfur compounds", pp. 541–554, 2005, DOI: 10.1016/j. syapm.2005.03.002, with permission from Elsevier; 18.27A: Dr Jeremy Burgess/Science Source; 18.27B: John D. Ivanko/Alamy Stock Photo; 18.27C: Republished with permission of American Society of Plant Biologists, from "Plant physiology" by J. Fournier et al., pp. 1985–1995, 2008, DOI: 10.1104/pp.108.125674; permission conveyed through Copyright Clearance Center, Inc.; 18.28: Nigel Cattlin/Science Source; 18.29A: Dr. James L. Castner/Visuals Unlimited, Inc.; 18.29B: Science VU/Visuals Unlimited, Inc.; 18.29C: Vsevolod Popov, University of Texas Medical Branch and the Microbe Library; 18.30B: Science VU/S. Watson/Visuals Unlimited, Inc.; 18.31: Jean E. Roche/naturepl.com; 18.32A: Romano, P., Blázquez, M. L., et al (2001). Selective copper–iron dissolution from a molybdenite concentrate using bacterial leaching. *Journal of Chemical Technology and Biotechnology*, 76(7), 723–728. https://doi.org/10.1002/jctb.443 ; 18.32B: Royce Bair/Getty Images; 18.33A: Reprinted by permission from Springer Nature: Schlegel, H.G. & Pfennig, N. *Archiv. Mikrobiol.* (1960) 38: 1. https://doi. org/10.1007/BF00408405; 18.33B: "Anaerobic oxidation of ferrous iron by purple bacteria, a new type of phototrophic metabolism" by A. Ehrenreich et al. *Applied and environmental microbiology* 60 12 (1994): 4517–26. Reproduced with permission from American Society for Microbiology; 18.34A: Reprinted from *Trends in Microbiology*, Vol. 22/ Issue 9, R. Belas, "Biofilms, flagella, and mechanosensing of surfaces by bacteria", pp. 517–527, 2014, with permission from Elsevier; 18.35: Npj Biofilms and Microbiomes volume 2, Article number: 2 (2016); 18.36A: CDC/ Dr. Barry S. Fields; 18.36B: CDC; 18.37: Nigel Cattlin/Science Source; 18.38A: "The Junctional Pore Complex and the Propulsion of Bacterial Cells" by C.W. Wolgemuth and G. Oster, *Journal of Molecular Microbiology and Biotechnology*, Volume 7, pp. 72–77, DOI: 10.1159/000077871. Copyright © 2004 Karger Publishers, Basel, Switzerland; 18.38B: "Fruiting body morphogenesis in submerged cultures of *Myxococcus xanthus*" by Kuner et al., *Journal of Bacteriology*, 1982, Issue 151, pp 458–459. Reproduced with permission from American Society for Microbiology; 18.40A: Reprinted from *Biophysical Journal*, Volume 84, Issue 5, M.E. Núñez et al., "Investigations into the Life Cycle of the Bacterial Predator Bdellovibrio bacteriovorus 109J at an Interface by Atomic Force Microscopy" pp. 3379–3388, 2003, DOI: 10.1016/S0006-3495(03)70061-7, with permission from Elsevier; 18.41: CAMR/A. Barry Dowsett/Science Source; 18.42A-B: "Description of *Treponema azotonutricium* sp. nov. and *Treponema primitia* sp. nov., the First Spirochetes Isolated from Termite Guts" by J.R. Graber et al., *Applied and Environmental Microbiology*, 2004, Volume 70, Issue 3, pp. 1315–1320, DOI: 10.1128/ AEM.70.3.1315-1320.2004. Reproduced with permission from American Society for Microbiology; 18.42C: Michael Abbey/Visuals Unlimited, Inc.; 18.42D: CDC/James Gathany; 18.43B: "Electron Microscopy of Axial Fibrils, Outer Envelope, and Cell Division of Certain Oral Spirochetes" by M.A. Listgarten and S.S. Socransky, *Journal of Bacteriology*, 1964, Volume 88, Issue 4, pp. 1087–1103. Reproduced with permission from American Society for Microbiology; 18.44A-B: Republished with permission of John Wiley and Sons Inc., from "Isolation of novel bacteria, including a candidate division, from geothermal soils in New Zealand" by P.F. Dunfield et al., *Environmental Microbiology*, Volume 10, Issue 8, pp. 2030–2041, 2008, DOI: 10.1111/j.1462-2920.2008.01621.x; permission conveyed through Copyright Clearance Center, Inc.; 18.45: DIOMEDIA/Medical Images RM/Claude Bucau; 18.46A: Reprinted by permission from Springer Nature: Overmann J. (2010) The Phototrophic Consortium "*Chlorochromatium aggregatum*" – A Model for Bacterial Heterologous Multicellularity. In: Hallenbeck P. (eds) Recent Advances in Phototrophic Prokaryotes. *Advances in Experimental Medicine and Biology*, vol 675. Springer, New York, NY; 18.46B: "*Chlorobium tepidum* Mutant Lacking Bacteriochlorophyll c Made by Inactivation of the bchK Gene, Encoding Bacteriochlorophyll c Synthase" by N. Frigaard et al., *Journal of Bacteriology*, 2002, Volume 184, Issue 12, pp. 3368–3376; DOI: 10.1128/JB.184.12.3368-3376.2002. Reproduced with permission from American Society for Microbiology; 18.47A-B: John Fuerst, University of Queensland; 18.48: Dennis Kunkel Microscopy/Science Source; 18.49A: CNRI/Science Source; ST18.1.1A: Reprinted from *Cell*, Volume 139, Issue 3, I.I. Ivanov et al., "Induction of Intestinal Th17 Cellsby Segmented Filamentous Bacteria", pp. 485–498, 2009, DOI: 10.1016/j.cell.2009.09.033, with permission from Elsevier; ST18.1.1B: Patrick Allard/REA/Redux; ST18.1.2B: Reprinted by permission from Springer Nature: Schnupf, P., Gaboriau-Routhiau, V., Gros, M. et al. Growth and host interaction of mouse segmented filamentous bacteria in vitro. *Nature* 520, 99–103 (2015) doi:10.1038/ nature14027.

Chapter 19

Figures: CO19.A: Courtesy of Thomas Heimerl; CO19.B: Courtesy of Reinhard Rachel; 19.2B: Courtesy of John Reeve; 19.3 inset: Courtesy of Thijs Ettema; 19.5: Mark Thiessen/National Geographic; 19.6: Hsueh-Yi Chen/Alamy Stock Photo; 19.7: "*Desulfurococcus fermentans* sp. nov., a novel hyperthermophilic archaeon from a Kamchatka hot spring, and emended description of the genus *Desulfurococcus*, Fig. 1a. A. PEREVALOVA1, ET AL. *INTERNATIONAL JOURNAL OF*

SYSTEMATIC AND EVOLUTIONARY MICROBIOLOGY 55, 995–999; © 2005, https://www.microbiologyresearch.org/contenUjournal /ijsem/10.1099/ijs.0.63378-0; 19.8: Republished with permission of Microbiology Society, from "Ignicoccus gen. nov., a novel genus of hyperthermophilic, chemolithoautotrophic Archaea, represented by two new species, Ignicoccus islandicus sp nov and Ignicoccus pacificus sp nov. and Ignicoccus pacificus sp. Nov", H.Huber et al., Volume 50, Issue 6, pp. 2093–2100, 2000; permission conveyed through Copyright Clearance Center, Inc.; 19.9A: Courtesy of Ken Stedman; 19.9B: Eye of Science/Science Source; 19.10B: Eye of Science/Science Source; 19.11A: Courtesy of Li Huang; 19.12A: "The structure of a thermophilic archaeal virus shows a double-stranded DNA viral capsid type that spans all domains of life", Fig. 1; George Rice, Liang Tang, Kenneth Stedman, Francisco Roberto, Josh Spuhler, Eric Gillitzer, John E. Johnson, Trevor Douglas, and Mark Young . *PNAS* May 18, 2004 101 (20) 7716–7720; https://doi.org/10.1073/pnas.0401773101. Copyright (2004) National Academy of Sciences, U.S.A; 19.12B: "Particle Assembly and Ultrastructural Features Associated with Replication of the Lytic Archaeal Virus Sulfolobus Turreted Icosahedral Virus" by S.K. Brumfield et al., *Journal of Virology*, 2009, Volume 83, Issue 12, pp. 5964–5970, DOI: 10.1128/JVI.02668-08. Reproduced with permission from American Society for Microbiology; 19.12C: "Particle Assembly and Ultrastructural Features Associated with Replication of the Lytic Archaeal Virus Sulfolobus Turreted Icosahedral Virus" by S.K. Brumfield et al., *Journal of Virology*, 2009, Volume 83, Issue 12, pp. 5964–5970, DOI: 10.1128/JVI.02668-08. Reproduced with permission from American Society for Microbiology; 19.13A: Reprinted by permission from Springer Nature: Prangishvili, D., Bamford, D., Forterre, P. et al. The enigmatic archaeal virosphere. *Nat Rev Microbiol* 15, 724‚Äì739 (2017) doi:10.1038/nrmicro.2017.125; 19.14A: Dr. Michael Perfit, University of Florida, and NOAA VENTS program; 19.15A: © 2006 MBARI; 19.15B: Todd Walsh © 2006 MBARI; 19.16: "Cell Surface Structures of Archaea, Äù by S.Y.M. Ng et al., Journal of Bacteriology, 2008, Volume 190, Issue 18, pp. 6039–6047, DOI: 10.1128/JB.00546-08. Reproduced with permission from American Society for Microbiology; 19.17: "In Vivo Observation of Cell Division of Anaerobic Hyperthermophiles by Using a High-Intensity Dark-Field Microscope" by C. Horn et al., *Journal of Bacteriology*, 1999, Volume 181, Issue 16, pp. 5114–5118. Reproduced with permission from American Society for Microbiology;; 19.18A-B: "A korarchael genome reveals insighs into the evolution of the Archaea" by J.G. Elkins et al., Proceedings of the National Academy of Sciences Jun 2008, 105 (23) 8102–8107; DOI: 10.1073/pnas.0801980105; 19.19A: Courtesy of Jed Fuhrman; 19.19B: Courtesy of Edward F. DeLong; 19.20: Reprinted by permission from Springer Nature: Könneke, M., Bernhard, A., de la Torre, J. et al. Isolation of an autotrophic ammonia-oxidizing marine archaeon. *Nature* 437, 543–546 (2005) doi:10.1038/nature03911; 19.22A-B: "Thaumarchaeotes abundant in refinery nitrifying sludges express amoA but are not obligate autotrophic ammonia oxidizers" by Marc Mußmann et al. *Proceedings of the National Academy of Sciences* Oct 2011, 108 (40) 16771-16776; DOI: 10.1073/pnas.1106427108; 19.23A: Courtesy of Edward F. DeLong; 19.23B: Christina Preston; 19.24A Ralph Robinson/Visuals Unlimited, Inc.; 19.24B-C, 19.26B: Reprinted from Bioresource Technology, Volume 89, Issue 3, P.J. Sallis and S. Uyanik, "Granule development in a split-feed anaerobic baffled reactor", pp. 255–265, 2003, DOI: 10.1016/S0960-8524(03)00071-3, with permission from Elsevier; 19.26C: Courtesy of Natuschka Lee; 19.27A: Reprinted by permission from Springer Nature: Cavicchioli, R. Microbial ecology of Antarctic aquatic systems. *Nat Rev Microbiol* 13, 691–706 (2015) doi:10.1038/nrmicro3549; 19.27B: Courtesy Rick Cavicchioli; 19.31A: Courtesy of Don Green Photography; 19.32: Permission from Baltimore Sun Media. All Rights Reserved.; 19.33A: Eye of Science/Science Source; 19.33B: "Ultrastructure of Square Bacteria from a Brine Pool in Southern Sinai" by M. Kessel and Y. Cohen, *Journal of Bacteriology*, 1982, Volume 151, Issue 2, pp. 851–860. Reproduced with permission from American Society for Microbiology; 19.33C: "Walsby's square bacterium: fine structure of an orthogonal procaryote" by W. Stoeckenius, *Journal of Bacteriology*, 1981, Volume 148, Issue 1, pp. 352–360. Reproduced with permission from American Society for Microbiology; 19.35: Reprinted by permission from Springer Nature: González, J., Masuchi, Y., Robb, F. et al. Extremophiles (1998) 2: 123. https://doi.org/10.1007/s007920050051; 19.36A: Reprinted by permission from Springer Nature: Pivovarova, T.A., Kondrat'eva, T.F., Batrakov, S.G. et al. *Microbiology* (2002) 71: 698. https://doi.org/10.1023/A:1021436107979; 19.36B: From "An Archaeal Iron-Oxidizing Extreme Acidophile Important in Acid Mine Drainage" by K.J. Edwards et al., *Science*, 2000, Volume 287, Issue 5459, pp. 1796–1799, DOI: 10.1126/science.287.5459.1796. Reprinted with permission from AAAS; 19.37A-B: "*Nanoarchaeum equitans* and ignicoccus hospitalis: New Insights into a Unique, Intimate Association of Two Archaea,Äù by U. Jahn et al., *Journal of Bacteriology*, 2008, Volume 190, Issue 5, pp. 1743–1750, DOI: 10.1128/JB.01731-07. Reproduced with permission from American Society for Microbiology; 19.38: Alexandra K. Perras et al. 2014. *Front. Microbiol.* 5: 397; 19.38 inset: Martin Wiesner; 19.39A Source: Alexandra K. Perras et al. 2014. Front. Microbiol. 5: 397; 19.39B: Republished with permission of John Wiley and Sons Inc., from "The unique structure of archaeal 'hami', highly complex cell appendages with nano-grappling hooks" by R. Huber et al., *Molecular Microbiology*, Volume 56, Issue 2, pp. 361–370, 2005, DOI: 10.1111/j.1365-2958.2005.04294.x; permission conveyed through Copyright Clearance Center, Inc.; ST19.1.1A: "Genomic and metabolic adaptations of *Methanobrevibacter smithii* to the human gut" by B.S. Samuel et al. *Proceedings of the National Academy of Sciences* Jun 2007, 104 (25) 10643–10648; DOI: 10.1073/pnas.0704189104. Copyright (2007) National Academy of Sciences, U.S.A.

Chapter 20

Figures: CO20.A: Courtesy of Todd LaJeunesse; CO20.B: Courtesy of Monica Medina; 20.1A: The Natural History Museum/Alamy Stock Photo; 20.1B: Dennis Kunkel Microscopy/Science Source; 20.1C: CDC/Janice Haney Carr; 20.2A: Courtesy of Grant Jensen; 20.3A: Diarmuid/Alamy Stock Photo; 20.3B: Nastasic/Getty Images; 20.5A: Thibaut Brunet (Howard Hughes Medical Ins\tute/University of California Berkeley); 20.6C: Dennis Kunkel Microscopy/Science Source; 20.7A: M.I. Walker/Science Source; 20.7B: Richard Allen (University of Hawaii) (2011) CIL:39127, Paramecium caudatum, cell by organism, eukaryotic cell, Eukaryotic Protist, Ciliated Protist. CIL. Dataset. https://doi.org/doi:10.7295/W9CIL39127; 20.9A: Republished with permission of John Wiley and Sons Inc., from "Kinesin is essential for cell morphogenesis and polarized secretion in *Neurospora crassa*" by S. Seiler et al., *The EMBO Journal*, Volume 16, Issue 11, pp. 3025–3034, 1997, DOI: 10.1093/emboj/16.11.3025; permission conveyed through Copyright Clearance Center, Inc.; 20.10A: Eye of Science/Science Source; 20.11A: Dr. Elizabeth Davidson/Visuals Unlimited, Inc.; 20.11B: Dr Alex Hyatt, CSIRO Australian Animal Health Laboratory; 20.12A: Gregory G. Dimijian/Science Source; 20.12B: Dr. James Richardson/Visuals Unlimited, Inc.; 20.13A-B: Ed Reschke/Peter Arnold/Getty Images; 20.14A: Dr. David M. Phillips/Visuals Unlimited, Inc.; 20.14B: Science Photo Library/Science Source; 20.14C: Joan Slonczewski; 20.15A: El_cigarrito/Shutterstock; 20.15B: Arco Images GmbH/Alamy Stock Photo; 20.17A: USDA; 20.18A: M. I. Walker/Science Source; 20.20A: David Scharf/Science Source; 20.21A: Eye of Science/Science Source; 20.21B: Greg Antipa/Science Source; 20.22A: Dr. Peter Siver/Visuals Unlimited, Inc.; 20.23B: AMI IMAGES/Science Source; 20.24A: Frank Fox/Science Source; 20.24B: Courtesy of William Radcliff; 20.25A: Biophoto Associates/Science Source; 20.25B: M. I. Walker/Science Source; 20.25C: M. I. Walker/Science Source; 20.26A: Biophoto Associates/Science Source; 20.26A inset: Republished with permission of John Wiley and Sons Inc., from "Factors limiting the biodegradation of *Ulva* sp cell-wall polysaccharides" by C. Bobin-Dubigeon et al., Journal of the Science of Food and Agriculture, Vol. 75, Issue 3, pp. 341–351, 1997; permission conveyed through Copyright Clearance Center, Inc.; 20.27A: Premaphotos/Alamy Stock Photo; 20.27B: FoodCollection/SuperStock; 20.28A: Biophoto Associates/Getty Images; 20.28B: Jan Hinsch/Science Source; 20.28C: Republished with permission of John Wiley and Sons Inc., from "Phytoplankton defence mechanisms: traits and trade-offs" by T. Kiørboe and M. Pancic, *Biological Reviews*, 2018, Volume 93, Issue 2, pp. 1269–1303, DOI: 10.1111/brv.12395. Reprinted with permission from AAAS; 20.29: Wim Van Egmond/Science Source; 20.30: Trustees of the Natural History Museum, London; 20.31: Masa Ushioda/Visuals Unlimited, Inc.; 20.32A: Biophoto Associates/Science Source; 20.32C: M. I. Walker/Science Source; 20.33A: Michael Abbey/Visuals Unlimited, Inc.; 20.34A: Blickwinkel/Alamy Stock Photo; 20.34B: Greg Antipa/Science Source; 20.35A: Dr. David M. Phillips/Visuals Unlimited, Inc.; 20.35B: Dennis Kunkel Microscopy/Science Source; 20.35D: Bill Bachman/Science Source; 20.36A-B: Courtesy of Todd LaJeunesse; 20.37: Ed Reschke/Getty Images; 20.40A: Courtesy of CDC/Dr. D.S. Martin; 20.40B: Dennis Kunkel Microscopy/Science Source; 20.40C: Eye of Science/Science Source; 20.41A: Dr. Tony Brain/Science Source; 20.42A: Dr. Gary Gaugler/Visuals Unlimited, Inc.; 20.42B: Biophoto Associates/Science Source; 20.42C: Charles Stratton/Visuals Unlimited, Inc.; 20.42D: "Laboratory Identification of the Microsporidia" by L.S. Garcia, *Journal of Clinical Microbiology*, 2002, Volume 40, Issue 6, pp. 1892–1901; DOI: 10.1128/JCM.40.6.1892-1901.2002. Reproduced with permission from American Society for Microbiology; ST20.1.1: Shutterstock; ST20.1.2A-C: From "Functional Links Between Aβ Toxicity, Endocytic Trafficking, and Alzheimer's Disease Risk Factors in Yeast" by S. Treusch et al., *Science*, 2011, Volume 334, Issue 6060, pp. 1241–1245, DOI: 10.1126/science.1213210. Reproduced with permission from AAAS; ST20.1.3A: Photo by Jan-Olof Yxell, Chalmers University of Technology; ST20.1.3B: Captured by Dr. Nisha Agarwal. Fig. 2C from Chen X, Bisschops MMM, Agarwal NR, Ji B, Shanmugavel KP and Petranovic D (2017) "Interplay of Energetics and ER Stress Exacerbates Alzheimer's Amyloid-β (Aβ) Toxicity in Yeast" *Front. Mol. Neurosci* .10:232. doi:10.3389/fnmol.2017.00232.

Chapter 21

Figures: CO21.A: Zakee Sabree; CO21.A (inset): Marco Uliana/Alamy Stock Photo; CO21.B: Zakee Sabree; 21.1A: Mitch Jones/Jillian Banfield; 21.1B: "Evaluation of *Leptospirillum ferrooxidans* for Leaching" by W. Sand et al., *Applied and Environmental Microbiology* Jan 1992, 58 (1) 85–92; Reproduced with permission from American Society for Microbiology; 21.1C: Michael Hertel/picture-alliance/dpa/AP Images; 21.1D: "Cell Surface Enzyme Attachment Is Mediated by Family 37 Carbohydrate-Binding Modules, Unique to *Ruminococcus albus*" by A. Ezer et al., *Journal of Bacteriology* 2008; 190 (24): 8220–8222. Reproduced with permission from American Society for Microbiology; 21.2 (top left): FLPA/Shutterstock; 21.2 (top right): Photobret2014/Shutterstock; 21.2 (bottom): ImageBROKER/Alamy Stock Photo; 21.6A: Saul Loeb/AFP/Getty Images; 21.6C: "Novel *Psychropiezophilic Oceanospirillales* Species *Profundimonas piezophila* gen. nov., sp. nov., Isolated from the Deep-Sea Environment of the Puerto Rico Trench" by Y. Cao et al., *Applied and Environmental Microbiology*, 2013, Volume 80, Issue 1, pp. 54–60, DOI: 10.1128/AEM.02288-13. Reproduced with permission from American Society for Microbiology; 21.6D: Republished with permission of John Wiley and Sons Inc., from "Deep-sea bacteria enriched by oil and dispersant from the Deepwater Horizon spill" by J.K. Jansson et al., *Environmental Microbiology*, Volume 14, Issue 9, pp. 2405–2416, 2012, DOI: 10.1111/j.1462-2920.2012.02780.x; permission conveyed through Copyright Clearance Center, Inc.; 21.7A: Olivia Mason; 21.10B: Cruaud P. & Vigneron A., IFREMER; 21.11A: Duncan Phillips/Alamy Stock Photo; 21.11B: Shosuke Yoshida; 21.12A-B: From "A bacterium that degrades and assimilates poly(ethylene terephthalate)" by S. Yoshida et al., *Science*, 2016, Vol. 351, Issue 6278, p. 1196–1199, DOI: 10.1126/science.aad6359. Reprinted with permission from AAAS; 21.14: Reprinted by permission from Springer Nature: Gwynne, P. Microbiology: There's gold in them there bugs. *Nature* 495, S12-S13 (2013) doi:10.1038/495S12a; 21.15A: Joan Slonczewski; 21.15B: Joan Slonczewski; 21.15C: Biodisc/Visuals Unlimited, Inc.; 21.16A: Oxford Scientific/Getty Images; 21.16B: Renn Tumlison, Henderson State University; 21.16B inset: Eunsoo Kim; 21.18B: Reprinted from *European Journal of Protistology*, Volume 39, Issue 1, M. Wenzel et al., "Identification of the ectosymbiotic bacteria of *Mixotricha paradoxa* involved in movement symbiosis", pp. 11–23, 2003, DOI: 10.1078/0932-4739-00893, with permission from Elsevier; 21.18C: Reprinted by permission from Springer Nature: Brune A. (2013) Symbiotic Associations Between Termites and Prokaryotes. In: Rosenberg E., DeLong E.F., Lory S., Stackebrandt E., Thompson F. (eds) *The Prokaryotes*. Springer, Berlin, Heidelberg; 21.20B: Dennis Kunkel Microscopy/Science Source; 21.25B: NOAA/Science Source; 21.27: Courtesy of Rita Colwell; 21.28A: Ami Images/Science Source; 21.28A inset: James M. Bell/Science Source; 21.28B: Nitin Kanotra/Hindustan Times via Getty Images; 21.30 inset: Scripps Institution of Oceanography at UC San Diego; 21.31A-B: OAR/National Undersea Research Program (NURP); 21.32A: Biophoto Associates/Science Source; 21.32B: Ty Wright for The Washington Post via Getty Images; 21.39A: Eye of Science/Science Source; 21.39B: "Cell Invasion and Matricide during *Photorhabdus luminescens* Transmission by *Heterorhabditis bacteriophora* Nematodes" by T. Ciche et al., *Applied and Environmental Microbiology*, 2008, Volume 74, Issue 8, pp. 2275–2287; DOI: 10.1128/AEM.02646-07. Reproduced with permission from American Society for Microbiology; 21.40B: Gail Jankus/Science Source; 21.42B: Paula Flynn; 21.43B: Mark Brundrett; 21.44B: Courtesy of Chien-Lu Ping, University of Alaska, Anchorage; 21.45: Sarah Lebeis; 21.46: Custom Life Sciene Images/Alamy Stock Photo; 21.47B: "Molecular Basis of Symbiotic Promiscuity" by X. Perret et at., *Microbiology and Molecular Biology Reviews*, Volume 64, Issue 1, 2000, pp. 180–201, DOI: 10.1128/mmbr.64.1.180-201.2000. Reproduced with permission from American Society for Microbiology; 21.47C: "Analysis of Infection Thread Development Using Gfp- and DsRed- Expressing *Sinorhizobium meliloti*", by D.J.Gage, *Journal of Bacteriology*, Volume 24, Issue 5, 2002, pp, 7042–7046, DOI: 10.1128/JB.184.24.7042-7046.2002. Reproduced with permission from American Society for Microbiology; 21.49A: Volodymyr Kaluskyi/Alamy Stock Photo; 21.49B: Nigel Cattlin/Alamy Stock Photo; 21.49C: FLPA/Alamy Stock Photo; ST21.1.1A: Joan Slonczewski and Rachel Morgan-Kiss, NSF; ST21.1.1B: Steve Gschmeissner/Science Source; ST21.1.2 inset: Joan Slonczewski.

Chapter 22

Figures: CO22.A: Liane Benning; CO22.A inset: Stefanie Lutz; CO22.B: Photo: Phil Dera; 22.2B UPI/Alamy Stock Photo; 22.3A Dennis Baldocchi; 22.4B Eye of Science/Science Source; 22.5B Karina Clemmensen; 22.5B inset: Kasper Andersen; 22.6: Luis Sinco/Los Angeles Times via Getty Images; 22.8A: Carmody McCalley; 22.8D: Virginia Rich, Ohio State University; 22.10B: AP Photo/The Casper Star-Tribune, Alan Rogers; 22.10C: Envitech Ltd.; 22.12B: © Daniel Beltra/Greenpeace; 22.13B: Desintegrator/Alamy Stock Photo; 22.14A: Jonathan A. Meyers/Science Source; 22.14A inset: www.idimages.org; 22.14B: Frank Fox/Science Source; 22.15: Kateryna Jon/Science Source; 22.16A: Joan Slonczewski; 22.16B: Tim McCabe/USDA/NRCS; 22.18B: Science VU/S. Watson/Visuals Unlimited, Inc.; 22.19A: Grant Heilman Photography/Alamy Stock Photo; 22.21: "Physiological Characterization of an Anaerobic Ammonium-Oxidizing Bacterium Belonging to the 'Candidatus Scalindua' Group" by T. Awata et al., *Applied and Environmental Microbiology*, Volume 79, Issue 13, 2013, pp. 4145–4148, DOI: 10.1128/AEM.00056-13. Reproduced with permission from American Society for Microbiology; 22.22B: Dr. Juergen Schieber, Indiana University Department of Geological Sciences; 22.26: Thomas R. Fletcher/Alamy Stock Photo; 22.28 inset: Reprinted by permission from Springer Nature: Wei, X.L., Han, M.S., Xia, C.C. et al. *Arch Microbiol* (2015) 197: 683. https://doi.org/10.1007/s00203-015-1102-7; 22.31: NASA/JPL-Caltech/MSSS/JHU-APL; 22.31 inset: Scimat/Science Source; 22.32: NASA/DRL; 22.33: NASA; ST22.1.1A: Cardinale et al. 2017. *Scientific Reports* 7:5791 ST22.1.2A: Britt Schilling for Markus Egert; ST22.1.2B: Cardinale et al. 2017. *Scientific Reports* 7:5791.

Chapter 23

Figures: CO23.A: Reprinted from *Cell Host & Microbe*, Volume 23, Issue 2, L. Chung et al., from "Bacteroides fragilis Toxin Coordinates a Pro-carcinogenic Inflammatory Cascade via Targeting of Colonic Epithelial Cells", pp. 203–214, 2018, DOI: 10.1016/j.chom.2018.01.007, with permission from Elsevier; CO 23.B: Courtesy of Cynthia L. Sears; 23.1: Courtesy of Franck Housseau; 23.3B: SPL/Science Source; 23.4B: Science VU/Visuals Unlimited, Inc.; 23.4C: BSIP/Science Source; 23.4D: Biophoto Associates/Science Source; 23.5A: VideoSurgery/Getty Images; 23.5B: CDC/Science Source; 23.6: Dr. David M. Phillips/Visuals Unlimited, Inc.; 23.7: SPL/Science Source; 23.9A: Courtesy of Lora Hooper; 23.9B: From "The Antibacterial Lectin RegIIIgamma Promotes the Spatial Segregation of Microbiota and Host in the Intestine" by S. Vaishnava et al., *Science*, 2011, Volume 334, Issue 6053, pp. 255–258, DOI: 10.1126/science.1209791. Reprinted with permission from AAAS; 23.10: Postoperative intra-abdominal abscess. Consultant. 2005;45(13):1531. https://www.consultant360.com/articles/photoclinic-postoperative-intra-abdominal-abscess. ©HMP.; 23.11: David M. Martin, M. D./Science Source; 23.12A-B: SCHMAT/Science Source; 23.13: James Byard/WUSTL Photos; 23.14A: Tiripero/Getty Images; 23.14B-C: Courtesy of Oak Ridge National Laboratory, U.S. Department of Energy; 23.16: Eye of Science/Science Source; 23.16 inset: Scimat Scimat/Getty Images; 23.18A Dr. John D. Cunningham/Visuals Unlimited, Inc.; 23.18B-D: Dr. Fred Hossler/Visuals Unlimited, Inc.; 23.20: SPL/Science Source; 23.21A: Eye of Science/Science Source; 23.21B: David Scharf/Science Source; 23.22: Eye of Science/Science Source; 23.23A-B: Reprinted with permission of Rockefeller University Press, from "The Vav-Rac1 Pathway in Cytotoxic Lymphocytes Regulates the Generation of Cell-mediated Killing" by Billadeau et al., *The Journal of Experimental Medicine*, Volume 188, pp. 549–559, 1998, DOI: 10.1084/jem.188.3.549; permission conveyed through Copyright Clearance Center, Inc.; 23.27A: Courtesy of Dr. Deborah W. Vaughan; 23.28: Republished with permission of Dove Press, from "Silver-coated carbon nanotubes downregulate the expression of *Pseudomonas aeruginosa* virulence genes: a potential mechanism for their antimicrobial effect" by E. Dosunmu et al., *International Journal of Nanomedicine*, 2015, Volume 10, Issue 1, pp. 5025–5034, DOI: 10.2147/IJN.S85219; permission conveyed through Copyright Clearance Center, Inc.; 23.29B: "The outer membranes of Brucella spp, areresistant to bactericidal cationic peptides" by G. Martínez de Tejada et al., *Infection and Immunity*, Volume 63, Issue 8, 1995, pp. 3054–3061. Reproduced with permission from American Society for Microbiology; 23.30: SPL/Science Source; 23.30 inset: Image Source/Alamy Stock Photo; 23.36A: S. Kaufmann & J. Golecki/Science Source; 23.36B: Volker Brinkmann/Visuals Unlimited, Inc.; 23.36C: Dennis Kunkel Microscopy/Science Source; 23.36D: CDC; 23.38A: Courtesy of Alissa Rothchild; 23.38B: DIOMEDIA/ISM; 23.38C: Reprinted from *Cell Host & Microbe*, Volume 8, Issue 1, D.G. Russell et al., "Mycobacterium tuberculosis Wears What It Eats", pp. 68–76, 2010, DOI: 10.1084/j.chom.2010.06.002, with permission from Elsevier; 23.38D: David M. Martin, M.D/Science Source; 23.38E: DIOMEDIA/Medical Images RM/Albert Paglialunga; ST23.UN: Clem Onojeghuo/Pexels; ST23.1.1A: Courtesy of Sandrine Henri; ST23.1.1 B: Courtesy of Bernard Malissen; ST23.1.2-ST23.1.3: Republished with permission of *Journal of Experimental Medicine*, "Unveiling skin macrophage dynamics explains both tattoo persistence and strenuous removal" by A. Baranska et al. Volume 215, Issue 4, pp. 1115–1133, 2018, DOI: 10.1084/jem.20171608; permission conveyed through Copyright Clearance Center, Inc.

Chapter 24

Figures: CO24.A: Reprinted from *Immunity*, Volume 48, Issue 3, G. Barbet et al., "Sensing Microbial Viability through Bacterial RNA Augments T Follicular Helper Cell and Antibody Responses", pp. 584–598, 2018, DOI: 10.1016/j.immuni.2018.02.015, with permission from Elsevier; CO24.B: Courtesy of Julie Margarian Blander; 24.1: Dr. P. Marazzi/Science Source; 24.3B: The Rockefeller University; 24.4A: CDC/ James Hicks; 24.4B: CDC/ Dr. Fred Murphy; Sylvia Whitfield; 24.11A: Eye of Science/Science Source; 24.11B: MicroScape/Science Source; 24.11C: Zurijeta/Shutterstock; 24.30: Andrew Syred/Science Source; 24.32: Dr. Ken Greer/Visuals Unlimited, Inc.; ST24.1.1B: Courtesy of Vanderbilt University; ST24.1.1C: The University of Texas Medical Branch; ST24.1.4: Courtesy of Alexander Bukreyev. Adapted from Fig. 6A in *Immunity*, Vol. 49/Issue 2, P. Gilchuk et al., "Multifunctional Pan-ebolavirus Antibody Recognizes a Site of Broad Vulnerability on the Ebolavirus Glycoprotein", pp. 363–374, 2018.

Chapter 25

Figures: CO25.A: Reprinted from *Cell*, Volume 173/Issue 5, Pinho-Ribiero, et al., "Blocking Neuronal Signaling to Immune Cells Treats Streptococcal Invasive Infection", pp. 1083–1097, 2018, DOI: 10.1016/j.cell.2018.04.006, with permission from Elsevier; CO25.B:Gretchen Ertl; 25.1A: Jane Shemilt/Science Source; 25.1B: CDC; 25.2A: John Greim/Science Source; 25.2B: CDC; 25.3A: CDC/Lois Norman; 25.3B: CDC/Hermann; 25.4A: CDC/ Frederick A. Murphy; 25.4B: Tommy Trenchard/Alamy Stock Photo; 25.7A: CDC; 25.7B: Science Source/Science Source; 25.10B: Republished with permission of American Society for Clinical Investigation, from "A molecular trigger for intercontinental epidemics of group A *Streptococcus*" by L. Zhu et al., Volume 125, Issue 9, pp. 3545–3559, 2015, DOI: 10.1172/JCI82478; permission conveyed through Copyright Clearance Center, Inc.; 25.11A: "FimH adhesin of type 1 pili is assembled into a fibrillar tip structure in the Enterobacteriaceae" by C H Jones, J S Pinkner, R Roth, J Heuser, A V Nicholes, S N Abraham, S J Hultgren, Proceedings of the National Academy of Sciences Mar 1995, 92 (6) 2081–2085; DOI: 10.1073/pnas.92.6.2081 ; 25.12B: "Direct observation of extension and retraction of type IV pili" by Jeffrey M. Skerker, Howard C. Berg, Proceedings of the National Academy of Sciences Jun 2001, 98 (12) 6901–6904; DOI: 10.1073/pnas.121171698; 25.12C: Used with permission of American Society for Cliniical Investigation from "Intestinal adherence associated with type IV pili of enterohemorrhagic *Escherichia coli* O157:H7" by G.C. Robinson, 2007, Volume 117, Issue 11, pp. 3519–3529; permission conveyed through Copyright Clearance Center, Inc.; 25.13A: Maria Fazio and Vincent A. Fischetti, Ph.D. with permission. The laboratory of Bacterial Pathogenesis and Immunology, Rockefeller University; 25.13B: NIBSC/Science Source; 25.14: Huebinger, R. M. et al. Targeting bacterial adherence inhibits multidrug-resistant *Pseudomonas aeruginosa* infection following burn injury. *Sci. Rep.* 6, 39341; doi: 10.1038/srep39341 (2016); 25.15A: Photo courtesy Paul Everest. G. Haddock et al. *Microbiology* (2010), 156, 3079–3084; 25.15B: Republished with permission of John Wiley and Sons Inc., from "Evolving concepts in biofilm infections" by P. Stoodley and L. Hall-Stoodley, *Cellular Microbiology*, Volume 11, Issue 7, pp. 1034–1043, 2009, DOI: 10.1111/j.1462-5822.2009.01323.x; permission conveyed through Copyright Clearance Center, Inc.; 25.17C: Banos/Alamy Stock Photo ; 25.19A: Dr. Gopal Murti/Science Source; 25.19B: Dennis Kunkel Microscopy/Science Source; 25.19C: "Protection and Attachment of Vibrio cholerae Mediated by the Toxin-Coregulated Pilus in the Infant Mouse Model" by S. Krebs and R. Taylor, *Journal of Bacteriology*, 2011, Volume 193, Issue 19, pp. 5260–5270, DOI: 10.1128/JB.00378-11. Reproduced with permission from American Society for Microbiology; 25.21A: Scott Camazine/Alamy Stock Photo; 25.22A: Reprinted from *Biophysical Journal*, Vol. 81/Issue 2, R. Lins and T.P. Straatsma, "Computer Simulation of the Rough Lipopolysaccharide Membrane of *Pseudomonas aeruginosa*", pp. 1037–1046, 2001, DOI: 10.1016/S0006-3495(01)75761-X, with permission from Elsevier; 25.23A: Dennis Kunkel Microscopy/Science Source; 25.23B: Mediscan/Visuals Unlimited, Inc.; 25.25A: From "Type III Secretion Machines: Bacterial Devices for Protein Delivery into Host Cells" by J.E. Galán and A. Collmer, *Science*, 1999, Volume 284, Issue 5418, pp. 1322–1328, DOI: 10.1126/science.284.5418.1322. Reprinted with permission from AAAS; 25.25C: From "Bacterial Invasion: The Paradigms of Enteroinvasive Pathogens" by P. Cossart and P. Sansonetti, *Science*, 2004, Volume 304, Issue 5668, pp. 242–248, DOI: 10.1126/science.1090124. Reprinted with permission from AAAS; 25.29A: "*Coxiella burnetii* Exhibits Morphological Change and Delays Phagolysosomal Fusion after Internalization by J774A.1 Cells" by D. Howe and L.P. Mallavia, *Infection and Immunity*, Volume 68, Issue 7, 2000, pp. 3815–3821, DOI: 10.1128/IAI.68.7.3815-3821.2000. Reproduced with permission from American Society for Microbiology; 25.29B: "Actin-Based Motility of Intracellular Microbial Pathogens" by M.B. Goldberg, *Microbiology and Molecular Biology Reviews*, 2001, Volume 65, Issue 4, pp. 595–626, DOI: 10.1128/MMBR.65.4.595-626.2001. Reproduced with permission from American Society for Microbiology; 25.30: "Genomic Comparison of Virulent *Rickettsia rickettsii* Sheila Smith and Avirulent *Rickettsia rickettsii* Iowa" by D.W. Ellison et al., *Infection and Immunity*, 2008, Volume 76, Issue 2, pp. 542–550, DOI: 10.1128/IAI.00952-07. Reproduced with permission from American Society for Microbiology; 25.33A: Courtesy of Steven Morten; 25.33B-C: Deo P, Chow SH, Hay ID, Kleifeld O, Costin A, Elgass KD, et al. (2018) "Outer membrane vesicles from *Neisseria gonorrhoeae* target PorB to mitochondria and induce apoptosis". *PLoS Pathog* 14(3): e1006945. https://doi.org/10.1371/journal.ppat.1006945; 25.34B: Reprinted from *Cell Host & Microbe*, Volume 8, Issue 5, S. Mostowy et al., "Entrapment of Intracytosolic Bacteria by Septin Cage-like Structures", pp. 433–444, 2010, DOI: 10.1016/j.chom.2010.10.009, with permission from Elsevier; 25.38A: Courtesy of Jörg Vogel; 25.40A: Courtesy of Ed Hille; 25.40B: "Alveolar Macrophages and Neutrophils Are the Primary Reservoirs for Legionella pneumophila and Mediate Cytosolic Surveillance of Type IV Secretion" by A.M. Copenhaver et al., *Infection and Immunity*, Volume 82, Issue 10, 2014, pp. 4325–4336, DOI: 10.1128/IAI.01891-14. Reproduced with permission from American Society for Microbiology; ST25.1.1A: Courtesy of Matthew Pestrak; ST25.1.1B: Courtesy of Victor J. Torres; ST25.1.3-ST25.1.4: From "Staphylococcus aureus biofilms release leukocidins to elicit extracellular trap formation and evade neutrophil-mediated killing" by Mohini Bhattacharya et al., July 10, 2018. *Proceedings of the National Academy of Science.*

Chapter 26

Figures: CO26.A: Adapted from Fig.1A: Dambuza IM, Drake T, Chapuis A, Zhou X, Correia J, Taylor-Smith L, et al. (2018) The Cryptococcus neoformans Titan cell is an inducible and regulated morphotype underlying pathogenesis. PLoS Pathog 14(5): e1006978. https://doi.org/10.1371/journal.ppat.1006978; CO26.B: Courtesy of Llewelyn Moron; 26.1: Tom Drake; 26.2: Fig 1A from Sela U, Euler CW, Correa da Rosa J, Fischetti VA (2018) Strains of bacterial species induce a greatly varied acute adaptive immune response: The contribution of the accessory genome. *PLoS Pathog* 14(1): e1006726. https://doi.org/10.1371/journal.ppat.1006726; 26.3A: Dennis Kunkel Microscopy/Science Source; 26.3B: Dr. Ken Greer/Visuals Unlimited, Inc.; 26.4A: Anda-Photo/Shutterstock; 26.4B Courtesy of Dr. William Schwan 26.4C: Courtesy of Dr. William Schwan; 26.5A: Centers for Disease Control and Prevention/Science Source; 26.5B: CDC/ Dr. Fred Murphy; Sylvia Whitfield ; 26.6A: Tomatheart/Shutterstock; 26.6B: Lebeau/Science Source; 26.7A: CDC/ M. Renz; 26.7B: Gado Images/Alamy Stock Photo; 26.8A: CDC/ Dr. Edwin P. Ewing, Jr.; 26.8 B: Smith Collection/Gado/Getty Images; 26.10 inset: "Structural Analysis of Respiratory Syncytial Virus Reveals the Position of M2-1 between the Matrix Protein and the Ribonucleoprotein Complex" by G. Kiss et al., *Journal of Virology*, 2014, Volume 88, Issue 13, pp. 7602–7617, DOI: 10.1128/JVI.00256-14. Reproduced with permission from American Society for Microbiology; 26.11: Scimat/Science Source; 26.12A: Courtesy of Lawrence David ; 26.13A: Xinhua/Alamy Stock Photo; 26.13B: Eye of Science/Science Source; 26.13C: Eye of Science/Science Source; 26.15A: Fig 8A from: Y. Avalos-Padilla et al. (2015) "EhVps32 Is a Vacuole-Associated Protein Involved in Pinocytosis and Phagocytosis of *Entamoeaba histolytica". PLOS Pathogens.* https://doi.org/10.1371/journal.ppat.1005079; 26.15B: Reprinted from *European Journal of Protistology*, Volume 52, M. Hebatalla et al., "Electron microscopic observation of the early stages of Cryptosporidium parvum asexual multiplication and development in in vitro axenic culture", pp.36-44, 2016, DOI: 10.1016/j.ejop.2015.07.002, with permission from Elsevier; 26.15C: Dr. Gary Gaugler/Science Source; 26.15D: Eye of Science/Science Source; 26.16 : Eye of Science/Science Source; 26.17A: Courtesy of Scott J. Hultgren, Valerie O'Brien, Matthew Joens, James A.J. Fitzpatrick/Washington University, St. Louis; 26.18: Courtesy of Amanda Lewis; 26.19A: CDC/Susan Lindsay; 26.19B: CDC/M. Rein, VD 26.19C: CDC/M. Rein, VD; 26.20 inset: Dr. Fred Hossler/Visuals Unlimited, Inc.; 26.21A: Dr. A. M. Siegelman/Visuals Unlimited, Inc.; 26.21B: Courtesy of Dr. John W. Foster; 26.21C: Reprinted by permission from Springer Nature: Boulton, I., Gray-Owen, S. Neisserial binding to CEACAM1 arrests the activation and proliferation of CD4+ T lymphocytes. *Nat Immunol* 3, 229–236 (2002) doi:10.1038/ni769; 26.22B: Groppelli E, Starling S, Jolly, C. 2015. Contact-induced mitochondrial polarization supports HIV-1 virological synapse formation. *J of Virol*, 89 (1):14-24, cover. DOI: 10.1128/JVI.02425-14; 26.22C: Sruilk/Shutterstock; 26.22D: Dr. F. C. Skvara/Visuals Unlimited, Inc.; 26.22E: SPL/Science Source; 26.23: David M. Phillips/Visuals Unlimited, Inc.; 26.24: Dr. E. Walker/Science Source; 26.25B: Dennis Kunkel Microscopy/Science Source; 26.26A: CDC; 26.26B: CDC; 26.26C: Craig K. Lorenz/Science Source; 26.26D: CDC/Science Source; 26.28A: CDC; 26.28B: Eye of Science/Science Source; 26.28C: David M. Phillips/Science Source; 26.29B: "Tetherin-mediated restriction of filovirus budding is antagonized by the Ebola glycoprotein" by R.L. Kaletsky et al., *Proceedings of the National Academy of Sciences* Feb 2009, 106 (8) 2886–2891; DOI: 10.1073/pnas.0811014106; 26.30A: John Radcliffe Hospital/Science Source; 26.30B: Eye of Science/Science Source; 26.30C: Dr. Colin Chumbley/Science Source; 26.30D: CDC; 26.31A: Dr. A. M. Siegelman/Visuals Unlimited, Inc.; 26.31B: The Picture Art Collection/Alamy Stock Photo; 26.34B: DIOMEDIA/DIOMEDIA/Carolina Biological; 26.34C: DIOMEDIA/ISM/Pr J.J. Hauw; ST26.1.1: Courtesy of Caitlyn MacQueen; ST26.1.2: "Neutrophils in lymph node intercept *S. aureus*" by Ania Bogoslowski, Eugene C. Butcher, Paul Kubes. *Proceedings of the National Academy of Sciences* Jan 2018, 201715756; DOI: 10.1073/pnas.1715756115.

Chapter 27

Figures: CO27.B: Kwong Ping Loh; 27.1: Courtesy of Liang Yang; 27.2A: St. Mary's Hospital Medical School/Science Source; 27.2C: Alfred Eisenstaedt/Pix Inc./The LIFE Picture Collection via Getty Images; 27.2D: J.A. Hampton/Getty Images; 27.2E: Photographs used with permission from Mayo Clin Proc. 1,2; 1. Mayo Clinic Proceedings legacy – May 2019: The research and practical use of antibiotics [published online only May 1, 2014]. https://els-jbs-prod-cdn.literatumonline.com/pb/assets/raw/Health%20Advance/journals/jmcp/jmcp_la_89_5_1_2-1398374782743.pdf. Accessed November 8, 2019.; 2. Herrell WE, Smith HL. Further observations on the clinical use of penicillin. Proc Staff Meet Mayo Clin. 1943;18 (5)65-76.; 27.3A: George Rinhart/Corbis via Getty Images; 27.3C: Bettmann/Getty Images; 27.4A: Courtesy of Dr. John W. Foster; 27.4B: Courtesy of Lesley McGee, CDC; 27.5: Courtesy of Dr. John W. Foster; 27.6A-D: Courtesy of Dr. John W. Foster; 27.25: Courtesy of Oana Ciofu; 27.26A: Courtesy of Brad Bearson ; 27.27: Sirirat/Shutterstock; ST27.1.1: MedImmune LLC.

Chapter 28

Figures: CO28.A: Reprinted from *Cell Host & Microbe*, Volume 23/Issue 2, S. Zhang et al., "Identification of a Botulinum Neurotoxin-like Toxin in a Commensal Strain of *Enterococcus faecium*" pp. 169–176, 2018, DOI: 10.1016/j.chom.2017.12.018, with permission from Elsevier; CO 28.B: Boston Children's Hospital; 28.1: Martin Schwalbe; 28.2A: Simon Fraser/RVI, Newcastle upon Tyne/Science Source; 28.2B: Will & Deni McIntyre/Science Source; 28.2C: CDC/Terry Tumpy; 28.3A: CDC/Don Stalons; 28.4: CDC/Dr. Kevin Karem; 28.5A: CDC; 28.5B: Sittipong Sreechat/© 123RF.com; 28.5C: Reprinted by permission from Springer Nature: Agarwal, S., Caplivski, D. & Bottone, E.J. Disseminated tuberculosis presenting with finger swelling in a patient with tuberculous osteomyelitis: a case report. *Ann Clin Microbiol Antimicrob* 4, 18 (2005) doi:10.1186/1476-0711-4-18; 28.6A-E: Courtesy of Dr. John W. Foster; 28.7A: Tek Image/Science Source; 28.7B-C: John W. Foster; 28.8A-C: BIOMERIEUX and the BIOMERIEUX logo and API are used pending and/or registered trademarks belonging to bioMérieux, or one of its subsidiaries, or one of its companies; 28.10A: Brenda Miller and John Foster, University of South Alabama; 28.10B: Courtesy of Dr. John W. Foster; 28.11: Courtesy of Dr. John W. Foster; 28.12: PF-(bygone1)/Alamy Stock Photo; 28.13A: John Foster, University of South Alabama; 28.14: "Multiplex PCR Assay for Detection and Identification of Clostridium botulinum Types A, B, E, and F in Food and Fecal Material" by M. Lindström et al., *Applied and Environmental Microbiology*, 2001, Volume 67, Issue 12, pp. 5694–5699; DOI: 10.1128/AEM.67.12.5694-5699.2001. Reproduced with permission from American Society for Microbiology; 28.17A.1: Courtesy Steve Southon/University of Toronto; 28.17A.2: Courtesy of Arizona State University; 28.17 inset: Reprinted from *Cell*, Volume 165, Issue 5, K. Pardee et al., "Rapid,

Low-Cost Detection of Zika Virus Using Programmable Biomolecular Components", pp. 1255–1266, 2016, DOI: 10.1016/j.cell.2016.04.059, with permission from Elsevier; 28.21A: CDC/Dr. M.S. Mitchell; 28.21B: Gado Images/Alamy Stock Photo ; 28.22B: Used by permission of Thermo Fisher Scientific, the copyright owner; 28.23A-B: The John Snow Archive and Research Companion; 28.24B: AP Photo/Jean-Marc Bouju; 28.25A: CDC/Dr. Fred Murphy; 28.25B: AP Photo/Anat Givon; 28.26A: AP Photo/Kenneth Lambert; 28.26B: Illinois Department of Public Health; 28.28: AP Photo/Felipe Dana, File; 28.29: Photo by Misty Richmond. Courtesy of Hail Project; ST28.1.1: Courtesy of Lucas Harrington/Mammoth Biosciences.

End Papers

Figures: EP.front [[A]]: Courtesy of Cynthia L. Sears; EP.front [[D]]: Cornell Brand Communications; EP.front [[E]]: Courtesy of Lora Hooper; EP.front [[F]]: Courtesy of Jörg Vogel; EP.front [[G]]: Dr. James Versalovic; EP.front [[background]]: Kateryna Kon/Shutterstock; EP.back [[B]]: Dr. Michelle Schorn; EP.back [[C]]: Courtesy of Sallie Chisholm. Photo credit: Richard Howard; EP.back [[D]]: Courtesy of Rita Colwell; EP.back [[E]]: Photo Courtesy The Ohio State University; EP.back [[F]]: Olivia Mason; EP.back [[G]]: K. Nealson, Univ. S. Cal.; EP.back [[background]]: NASA.

INDEX

Page numbers followed by f or t denote figures or tables, respectively. Page numbers set in bold type refer to key terms.

A

ABC transporters, **127–28,** 128f
ABC (ATP-binding cassette) transporters, **127–28,** 128f, 311–12, 311f
Abdominal abscess, 1163–64, 1163f
Aberrations, lens, **46, 47**
AB exotoxins, 1030–33, 1030f–1033f
Abiotic cycling, 884
Abiotic entities, **883**
Abiotic factors, 844–45, 844t, 845f
Abiotic reactions, **669**
Abiotrophia, 134t
ABO blood group system, 970, 1005, eTopic 24.3
Abscesses, 1162, 1163–64, 1163f
Absorption, **43,** 43f
Absorption zone, 794f, 795
Abu Laban, Nidal, 527
ABX464, 1153
Acanthamoeba, 805
Acanthocorbis unguiculata, 790f
Acceptor site (A site), **294–95**
Accessory proteins, **424,** 424t
Accidental transmission, 1015f
Acetate, 506
  methanogenesis, 773, 774f
  obesity, 934–35, 935f
Acetic acid, 506
Acetic acid bacteria, 627–28, 628f
Acetohalobium arabaticum, 301
Acetone, 515
Acetylation, 302, 303f
Acetyl coenzyme A (acetyl-CoA)
  activation, 595, 595f
  biosynthesis, 582, 584f
  entry into TCA cycle, 517–20, 517f, 519f
  pyruvate conversion to, 516, 516f
  reductive pathway, 586t, **593–94,** 593f
Acid-fast stain, **52,** 53f, **725,** 1166–67, 1167f
Acid fermentation
  dairy products, 621t, 623–25, 624f–626f
  vegetables, 621t, 625–27, 626f, 627f
Acidianus, 753t
Acidianus brierleyi, 594
Acidification, food preservation, 643
Acidithiobacillales, 704t
Acidithiobacillus, 732–33, 733f
  photolysis, 569
  sulfur cycle, 905
Acidithiobacillus ferrooxidans
  acidification of mine water, 733, 733f
  biomining, 562–63, 564f
  Calvin cycle, 586, 587f
  classification, 704t
  iron cycle, 909
  lithotrophy, 558–59
  metabolism, 728t
  pyrite oxidation, 180
  sulfur cycle, 906
Acidithiobacillus thiooxidans, 728t
Acidity, ecosystem, 845
Acid mine drainage, 180
Acidobacteria, 704, **705,** 739, 740f
Acidophiles, 161t, 169f, **170,** 170f, 844t
Acid resistance, pH homeostasis and, 171–73, 172f, eTopic 5.4
Acid resistance chaperone, 77f
Acid tolerance, 172–73
Acineta, 789t, 818f
Acinetobacter
  cocoa bean fermentation, 627–28, 628f
  indoor environments, 911
Acinetobacter baumannii, 189
  antibiotic resistance, 1127, 1135, 1142
Acinetobacter baylyi, 107
Acinetobacter calcoaceticus, 187
Acne, 925, 925f
ACP (acyl carrier protein), **595**
Acquired immunodeficiency syndrome (AIDS), 1090–91, 1090f. *See also* Human immunodeficiency virus (HIV)
  CD4 T cells, 990
  defined, **421**
  history, 7, 7f, 421–22
  Koch's postulates, 16–17
  point-of-care rapid diagnostics, 1186t
Acridine derivative, 318t
Acr proteins, 219
Actinoalloteichus, 723
Actinobacteria
  associated with animals and plants, 723–24, 723f
  classification, 704t, 705
  cyanobacterial mats, 847
  defined, **705**
  gut microbiome, 929
  nonmycelial, 724–25, 724f, 725f
  plant endophytic community, 875f
  structure, 714
Actinomyces
  anaerobe, 174t
  oral and nasal cavity microbiome, 925

Actinomyces bovis, 956
Actinomycetaceae, 704t
Actinomycetales
  cell wall growth, 90f
  filaments, 714
Actinomycetes, 721–26
  actinobacteria associated with animals and plants, 723–24, 723f
  amensalism, 850
  antibiotics, 297, 299f
  branching filaments, 712, 713f
  cell differentiation, 153–55, 154f
  cell envelope, 90, 95
  defined, **705**
  filaments, 714
  growth asymmetry, 105
  irregularly shaped, 725, 726f
  Micrococcaceae, 726
  nonmycelial actinobacteria, 724–25, 724f, 725f
  spores, 711t
  Streptomyces, 721, 722–23, 722f
Actinomycin D, 285–86, 285f, 1131, 1131f
Actinoplanes deccanensis, 1131
Actinovate, 650–52, 652f
Activation energy (*E*a), **499,** 499f
Activators, **358–60,** 358f–360f
Active transport, **85, 126–27,** 126f
Acute flaccid myelitis (AFM), 227, 1197
Acute inflammation, 952–54, 953f
Acute-phase reactants, 959
Acyclovir, 1148t, 1150f
Acyl carrier protein (ACP), **595**
Adaptation, 681–90
  bacterial resistance to antibiotics, 682–83, 684–85, 684f, 685f
  evolution and growth cycles, 689–90, 689f, 690f
  experimental environment, 686–87
  experimental evolution in laboratory, 685–89, 686f, 688f
  genomic analysis, 682–84
Adaptive immunity, 963–1008
  antibody-dependent *vs.* cell-mediated, 964–67, 964f, 966f
  antibody properties, 972–74, 973t
  antibody structure, 970–74, 971f–974f
  antigens and immunogens, 969–70, 969f
  autoimmunity, 1006–8, 1007t
  bacteria and archaea, 341–43, 341f–353f
  complement, 993–94, 993f
  defense against viruses, 232
  defined, **936–37, 964**
  development, 965–66, 965f, 966f
  genetics of antibody production, 978–82, 978t, 979f, 980f
  gut mucosal immunity and microbiome, 994–97, 994f
  how B cells differentiate into plasma cells, 975–78, 976f, 977f
  hypersensitivity, 1001–6, 1001t, 1004f, 1006f
  immunogenicity, 967
  immunological specificity, 967–69, 967f, 968t, 969f
  microbial evasion, 991–92
  overview, 964–70
  primary and secondary antibody responses, 974–75, 975f
  T cells link antibody and cellular immune systems, 982–93
  vaccines, 963, 963f, 968, 968t, 997–1001, 1002–3
ADCC (antibody-dependent cell-mediated cytotoxicity), **943**
Addressins, **996**
Adenine, 243f, 244
Adenoids, 945f
Adenosine diphosphate (ADP)-ribosyltransferase, **1030,** 1030f
Adenosine triphosphate (ATP)
  ADP phosphorylation to, 494–95, 494f, 495t
  defined, **494**
  energy carried by, 494–96, 494f, 495t
  produced by glucose catabolism, 495–96
Adenosine triphosphate (ATP)-binding cassette (ABC) transporters, **127–28,** 128f, 311–12, 311f
Adenosine triphosphate (ATP) synthase
  bacterial membrane, 77f, 123, 123f
  defined, **123**
  proton potential drives, 549–50, 550f
  structure, 549, 549f
Adenosine triphosphate (ATP) synthesis, 77f
  chemiosmotic theory, 540–41, 541f
Adenoviruses, 224f, 225, 436t
Adenylate cyclase, **1002**
Adenylation, 302, 303f
Adenylyl cyclase G-factors, 1031, eTopic 25.5
Adhesins, **1020,** 1021t, 1023, 1023f, 1024f
  nonpilus, 1023, 1024f
Adhesion, microbial, 1020–25, 1021f–1025f, 1021t
ADP (adenosine diphosphate), 494–95, 494f
ADP (adenosine diphosphate)-ribosyltransferase, **1030,** 1030f
Adrenaline, 1003
Aedes aegypti
  dengue hemorrhagic fever, 1016
  yellow fever, 1015, 1015f
Aedes mosquito, Zika virus, 1196–97
Aenigmarchaeota, 753t
Aerated (A) horizon, 868, 868f, 870f
Aeration basin, 896f
Aerial mycelia, **722–23,** 722f
Aerobes
  vs. anaerobes, 174–75, 174f, 175f
  defined, **173**
  obligate, 551
  strict, 161t, **173**
Aerobic benzene catabolism, 525–26, 526f
Aerobic metabolism, 843–44, 843t
Aerobic respiration, **173,** 173f
  electronic transport system and ATP synthase, 544–51
Aerobic rods, gammaproteobacteria, 734f, 735, 735f
Aeromonadales, 704t
Aeromonas hydrophila, 704t
Aeropyrum, 679
Aeropyrum pernix, 753t, 757, 757t
Aerotolerant anaerobes, **175,** 175f
Affinity maturation, **981**
AFM (acute flaccid myelitis), 227, 1197
AFM (atomic force microscopy), 41f, **42, 69–70,** 70f
Agar, **15–16,** 16f
Agaricus bisporus, 619, 619f
Agarose gels, 458
Agar plate culture
  animal virus, 236, 236f
  bacterial, 131, 131f
  bacteriophage, 235–36, 235f

Agricultural inoculants, 647, eTopic 16.2
Agricultural runoff, 902f, 903
Agricultural wastewater treatment, 898–99, 898f
Agriculture
food-borne pathogens, 638–41, 638f, 639f, 640t, 641f
symbiotic nitrogen fixation, 902, 902f
Agrobacterium
divergence, 678
endosymbiosis, 730
parasitism, 737
Agrobacterium tumefaciens
conjugation, 330, 331f
crown gall tumor, 330, 331f, 701f, 730, 879, 879f
genome, 241t
plant genetic modification, 652–54, 653f
quorum sensing, 384
toxins, 1040
Aguilar, Abelardo, 597
AGXX, 187
A (aerated) horizon, 868, 868f, 870f
AIDS. *See* Acquired immunodeficiency syndrome (AIDS)
Aigarchaeota, 752–54, 753t, **762**
Air, sterilization, 183, 184f
Airborne transmission, **1014,** 1016
AIRE, 985–86
Airplane travel, emerging infectious agents, 1197
Akkermansia, gut microbiome, 856, 929f
Akkermansia muciniphila, 25, 931, 997
D-Alanine fluorophores, septation, 98, 98f
Alba proteins, 267
Albedo reducers, 883
Alcaligenes, nitrogen cycle, 124
Alcaligenes faecalis, fermentation, 515, 515f
Alcohol(s), 186f
Alcohol dehydrogenase, 631
Alcoholic beverages, 631–33, 631f–633f
Alcoholic fermentation, 629–33
alcoholic beverages (beer and wine), 631–33, 631f–633f
bread dough, 621t, 630–31, 630f, 631f
defined, **512,** 622
overview, 620, 621t
Aldehydes, 186, 186f
Aleuria, 802f
Alga(e), 808–15
brown, 792, 812, 814, 814f
classification, 789t, 791–92, 791f
defined, **787, 791**
edible, 619–20, 619f
golden, 789t, 812
green, 789t, **791,** 791f, 808–11, 809f–811f
historical overview, 787
primary, **788, 791,** 808–11
red, 789t, **791,** 791f, 811, 812f
secondary, **788, 791–92,** 808, 811–14
siphonous, 789t
snow, 883, 883f
terminology, 788
yellow-green, 812
Algal blooms
cyanobacteria, 710, 712
dead zones, 894–95, 896f
defined, **865**
eutrophication, 179, 179f
freshwater microbial community, 865–66, 865f
viruses, 199–200, 199f
Algorithms, pathogen identification, 1167–69, 1168f
Aliivibrio fischeri, 382–84, 382f.383f, 848
Alisporivir, 1153
Alkaline fermentation, 621t, **622,** 628–29, 629f
Alkaline soda lakes, 776
Alkaliphiles, 161t, 169f, **171,** 171f, 172f, 844t
Alkalithermophile, 715
Alkane monoxygenases, *Deepwater Horizon* spill, 835
Alkylating agent, mutagenic effects, 318t
Allelic exchange, 333–34, 334f, eTopic 9.2
Allen, Andrew, 585
Allergen, **1001**
inhaled, 1004, 1004f
Allergic rhinitis, 974f, 1004
Allergy, 1004–5
Allograft rejection, 1005
Allolactose, 363, 363f
Allomyces, 789t
Allosteric site, **500,** 500f
Allotypes, **972,** 972f
Allylamines, 1155t
Alpha-amantin, 801, 802f
Alphaproteobacteria, 728–30
aquatic and soil oligotrophs, 729
classification, 704t, 705
defined, **705**
endosymbionts, 730, 731f, 732f
great Pacific garbage patch, 860
metabolism, 728t
methyltrophy and methanotrophy, 729, 730f
photoheterotrophs, 728–29, 729f
photosystem II, 574–75, 574f
Alphareovirus, 421t
Alpha toxin, 1026t, 1027, 1030f
Alternation of generations, 795f, **798**
Alternative complement pathway, 957, 958–59, 958f
Alternative sigma factors, **370–71,** 371f
Altiarchaeales, 753t, 781, 781f
Altiarchaeum, 753t, 781, 781f
Altiarchaeum hamiconexum, 755, 781
Altman, Sidney, 32, 671
Altruism, 105
Alu sequence, 432
ALV (avian leukosis virus), 208, 208f, 212t
Alveolar macrophages, **947**
Alveolata, 815–21
classification, 789t, 792, 792f
Alveolate(s), 815–21
apicomplexans, 817–20, 819f, 820f
ciliates, 815–16, 816f–818f, eTopic 20.2
classification, 789t, 792, 792f
defined, **792**
dinoflagellates, 816–17, 818f, 819f
Alveoli, cortical, **792,** 792f
Alzheimer's disease, beta-amyloid toxicity, 796–97, 797f
Amanita, 789t, 801–2, 802f

Amanita phalloides, 802f
Amantadine, 1148t, 1149, 1149f
Ambystoma maculatum, 849, 849f
Amebas, 805–8
  classification, 789t, 791
  defined, **791**
  filamentous and shelled, 807, 808f
  overview, 805
  pseudopods, 805–6, 805f, 806f
  slime molds, 806–7, 807f
Amensalism, 845t, **850**
Ames, Bruce, 321
Ames test, 321, 321f
Amherst, Jeffrey, 1192
Amino acid(s)
  aromatic, 611–13, 613f
  biosynthesis, 520, eTopic 13.1
  decarboxylation, 501, 522–23, 522f, 524–25
  degradation, 501, 502f
  D-form and L-form, 89
  prebiotic soup, 669–70
Amino acid biosynthesis, 608–13
  assimilation of $NH4^+$, 609–10, 610f
  complex, 611–13, 612f, 613f
  evolution, 608–9, 608f, 609f
  glutamate and glutamine signal nitrogen availability, 610–11, 610f
  major pathways, 608, 608f
Amino acid catabolism, flavor generation, 625, 626f
Aminoacyl-tRNA synthetases, **291,** 291f, 292f
  antibiotics targeting, 1134
4-Aminobutanoic acid, 522–23, 522f, 524–25
Aminoglycosides, 1132–33, 1132f
  antibiotic resistance, 1137, 1137f
Ammonia ($NH4^+$)
  amino acid biosynthesis, 609–11, 610f
  nitrification, 902
  nitrogen oxidation, 561
  nitrogen triangle, 900
Ammonia-oxidizing archaea (AOA), 763–65, 764f, 765f
Ammonia-oxidizing thaumarchaeotes, 763–65, 764f, 765f
Ammonification, **903**
Ammonium compounds, quaternary, 186, 186f
Ammonium oxidation, anaerobic, 904, 904f
Ammonium oxidizers, betaproteobacteria, 731
Amoeba proteus, 789t, 805f
  light microscopy, 39–40, 40f
Amoebozoa, 789t, 791, 805
Amoxicillin, 1126
Amphibolic pathway, **508–9**
Amphipathic cells, 670
Amphotericin B, **1154,** 1154f, 1155t
Ampicillin, 94, 1126, 1126f
Amplicons, **833, 1175**
Amplification(s)
  bridge, 449f, 450
  polymerase chain reaction, 445–46, 445f, 446f
Amycolatopsis orientalis, 598–99, 598f
  vancomycin, 1129
Amycolatopsis rifamycinia, 283, 284f
Anabaena
  akinetes, 711–12, 712t
  fluorescence microscopy, 54, 54f
  genome, 241t
  multicellularity, 712
  niche, 839
  nitrogen fixation, 601–2
  phototropic, 709t
Anabaena flos-aquae, 277
Anabaena spiroides, 604f
Anabolic pathway repression, 368, 369f
Anabolism. *See* Biosynthesis
Anaerobe(s), **173**
  vs. aerobes, 174–75, 174f, 175f
  aerotolerant, **175,** 175f
  culture, 175–76, 176f
  facultative, **173,** 174t, **175,** 175f
  strict, 161t, **175**
Anaerobe jar, 175–76, 176f
Anaerobic ammonium oxidation, 904, 904f
Anaerobic benzene catabolism, 526–27, 527f
Anaerobic corrosion, 562, 563f, 909–10, 910f
Anaerobic dechlorinators, 718–19, 719f
Anaerobic environments, **173**
Anaerobic glove box, 176, 176f
Anaerobic haloarchaea, 778
Anaerobic metabolism, 843t, 844
  wetland soil, 874
Anaerobic methane-oxidizing euryarchaea (ANME), **769**
  classification, 753t
  marine cycling of methane, 887–88
  methane hydrate metabolism, 552
  spatial organization in habitat, 839, 840f
Anaerobic respiration, 551–53
  anoxic or low-oxygen environments, 552–53, 553f
  classes of anaerobes, **175**
  defined, **551**
  dissimilatory metal reduction, 552–53, 553f
  electron donors and acceptors, 551, 551f
  energy storage, 537
  Escherichia coli, 551, 551f
  Gibbs free energy exchange, **488–89**
  oxidized forms of nitrogen, 551–52
  oxidized forms of sulfur, 552
Anaerobiosis and nitrogen fixation, 606
Analytical profile index (API), 1169–70, 1170f, eTopic 28.1
Anammox reaction, **561,** 562f, **904,** 904f
Anamnestic response, 975
Anapaena, 151–52, 153f
Anaphylaxis, **1001–5,** 1001t, 1004f
  late-phase, **1005**
Anaplerotic reactions, **592**
Anchor bacteria, 852–53, 852f
Andersson, Dan, 685
Andries, Koen, 1144
Anelloviruses, 212t, 229–30
Anemia, autoimmune hemolytic, 1007t
Anfinsen, Christian, 303
Angert, Esther, 716, 717f
Angle of aperture, **46,** 46f
ANI (average nucleotide identity), 691

Aniline, catabolism, 526f
Animal digestive microbiomes, 851–57
  bovine rumen fermenter, 853–55, 853f, 854f
  human colon, 855–57, 855f, 856f
  termite wood-digesting microbiome, 852–53, 852f, 853f
Animal feed, antibiotics, 1141–42, 1142f
Animal vectors, 230, 232
Animal viruses, 222–30, 232
  chronic, 229–30
  DNA viruses, 226–27, 226f, 227f
  host defenses, 232
  oncogenic, 229, eTopic 11.4
  plaque isolation and assay, 236, 236f
  replication cycles, 225–30, 225f–230f, eTopic 11.3
  RNA retroviruses, 228–29, 229f
  RNA viruses, 227–28, 228f, eTopic 11.3
  tissue culture, 234, 234f
  tissue tropism, 222–25, 224f
Anionic detergents, 186
ANME. *See* Anaerobic methane-oxidizing euryarchaea (ANME)
Annotation, DNA sequences, **455–57,** 455f, 456f
Annular ring, phase-contrast microscopy, 62–63, 63f
Anopheles mosquito
  climate change, 1198
  malaria, 1096
Anoxic environment, 844
  carbon cycling, 888–89, 889f
Anoxic water, 866, 866f
Antarctic lake mats, 846–47, 846f, 847f
Antenna complex, chlorophyll, **570–71,** 571f, 572
Anthony, Katey, 755f
Anthrax
  AB exotoxins, 1032–33, 1033f
  bioterrorism, 1192, 1193f
  causative agent, 1194
  culture, 14–15
  inhalational, 1066t, 1187
Anthrax lethal factor, X-ray crystallography, 72, 72f
Anthrax vaccine, 968t
Anti-anti-sigma factors, **372**
Anti-attenuation stem loop, 373f, **374**
Antibiotic(s), 1115–35
  animal feed, 1141–42, 1142f
  bactericidal, **1119,** 1134
  bacteriostatic, **1119**
  beta-lactam, 1126–28, 1126f, 1127f
  biosynthesis, 596–600, 597f–600f
  broad-spectrum *vs.* narrow-spectrum, 1140, 1141, 1146
  cell membrane inhibitors, 1124f, 1125t, 1129, 1129f
  cell wall inhibitors, 1124f–1129f, 1125–29, 1125t
  defined, **188, 1116**
  designer, 1146–47, 1146f, 1147f
  diarrhea, 1073
  DNA replication inhibitors, 1124f, 1125t, 1129–31, 1130f
  finding new, 1143–45, 1144f, 1146–47, eTopic 27.3
  fundamentals, 1116–23
  golden age of discovery, 1117–18, 1117f, 1118f
  history, **19,** 19f
  Kirby-Bauer disk susceptibility test, 1121–22, 1121f, 1122t
  measuring drug susceptibility, 1119–23, 1120f, 1121f, 1122t, 1123f
  mechanism of action, 1123–35, 1124f, 1125t
  and microbiome, 1140
  minimal inhibitory concentration, 1119–21, 1120f, 1122–23, 1123f
  mining bacterial genomes for, 599, 602–3
  nonribosomal peptide, **598–99,** 598f–600f, 602–3
  overview, 1116
  peptidoglycan synthesis inhibitors, 89–90, 89f, 90f, 1124f–1128f, 1125–29, 1125t
  polyketides, **594,** 597–98, 597f
  protein synthesis inhibitors, 1124f, 1125t, 1132–34, 1132f, 1133f
  revealing RNA synthesis machines, 283–86, 284f, 285f
  RNA synthesis inhibitors, 1124f, 1125t1131, 1131f
  selective control of bacterial growth, 188–89, 188f
  selective toxicity, 1118–19
  spectrum of activity, 1119, eTopic 27.1
  stewardship, 1141
  that affect translation, 297, 299f
  unculturable microbes, 136–37, 136f, 137f
  why microbes make, 1135, eTopic 17.6, eTopic 27.2
Antibiotic efflux pumps, 1138, 1138f
Antibiotic reporter bacterium, 409, 410f
Antibiotic resistance, 1135–47
  beta-lactam antibiotics, 1126–27
  biofilms, persisters, and antibiotic tolerance, 1142–43, 1143f, 1145
  broad-spectrum antibiotics, 1140, 1141, 1146
  epidemiology, 1116
  evolution, 682–83, 684–85, 684f, 685f
  evolving and sharing genes for, 1139, eTopic 9.6
  fighting, 1143
  and finding new drugs, 1143–45, 1144f, 1146–47, eTopic 27.3
  history, 19, 1140–42, 1140f, 1142f
  horizontal gene transfer, 339, 339f
  methods to identify, 1139–40
  multidrug-resistant pneumonia, 1135, 1136f
  overprescribing and, 1140
  plasmids, 261, 263, 331
  polar aging, 105, eTopic 3.3
  ribosome protection, 1115, 1115f, 1116, 1116f
  strategies, 1136–37, 1136f, 1137f
  in uncontacted communities, 1139
Antibiotic selection, 856, 856f
Antibiotic stewardship, 1141
Antibiotic tolerance, 1142–43
Antibody(ies)
  adaptive immune response, 937
  auto-, 1007
  conjugated, 56
  constant and variable regions, 970–72, 971f
  defense against viruses, 232
  defined, **964**
  diversity, 978–80, 978t
  isotype functions and "super" structures, 972–74, 973f, 973t, 974f
  isotypes, allotypes, and idiotypes, 972, 972f
  monoclonal, 1005
  structure, 970–71, 971f
  T cells link cellular immune systems and, 982–93

Antibody-dependent cell-mediated cytotoxicity (ADCC), **943**
Antibody-dependent immunity, 964, 965–66, 965f, 966f
Antibody-mediated cytotoxic hypersensitivity, 1001t, 1005, eTopic 24.3
Antibody production
antigen-binding site, 978–80, 978f, 979f
genetics, 978–82
haptens, 969–70, 969f
isotype class switching, 980–81, 980f
memory B cells, 981
steps, 977–78, 977f
T cell–dependent and –independent, 976–77
Antibody response
how B cells differentiate into plasma cells, 975–78, 976f, 977f
primary and secondary, 974–75, 975f
Antibody tags, **53**
Anticodons, **289,** 290, 290f
Antifungal agents, 1153–56, 1154f, 1155t
Antigen(s), **936–37, 964**
immune response, 965, 965f
immunogens, 969–70, 969f
super-, **991,** 991f
Antigen-binding site, 978, 979f
Antigen-capture ELISA, 1182, 1182f
Antigen cross-linking, 986
Antigenic determinant, **965,** 965f
Antigenic drift, **413**
Antigenic shift, **413**
Antigenic structure change, extracellular immune avoidance, 1041–42
Antigen presentation, 982–84, 983f, 984f
subversion, 1045
Antigen-presenting cells (APCs)
adaptive immune response, **965,** 966f
defined, **942**
T cell activation, **982–83,** 983f, 986, 987f
Antihistamines, 1004–5
Antimetabolites, 1155t
Antimicrobial agents, 1115–58
food preservation, **633**
fundamentals, 1116–23
overview, 1116
Antimicrobial coating, 187
Antimicrobial control
cell death at logarithmic rate, 181–82, 181f
classification, 180–81
terminology, 180–81
Antimicrobial peptides, 947–48, 948f, 948t, eTopic 23.2
Antimicrobial touch surfaces, 187
Antiparallel strands, **244,** 252
Antiport, **126–27,** 126f
Antiretroviral therapy (ART), 201, 1091, 1150–52
case history, 1150
defined, **1151**
entry inhibitors, 1151
history, 421–22
HIV attachment to host cell, 426, 426f
nucleoside, nucleotide, and nonnucleoside reverse transcriptase inhibitors, 1150–51
as prevention, 1152
protease inhibitors, 1151, 1151f
replication cycle, 429
reverse transcriptase, 427
treatment regimens and HIV controllers, 1151–52
Antisense RNA (asRNA), **378–79,** 379f
Antisepsis, **181**
Antiseptics, **18–19,** 186f
Anti-sigma factors, **371–72,** 372f, eTopic 10.2
antiSMASH, 596
Anti-vaxxers, 1000, 1063
Antiviral agents, 201, 1147–53
antiretroviral, 1150–52, 1151f
examples, 1148, 1148t
potential targets of future, 1152–53, 1152f
that target viral DNA synthesis, 1150, 1150f
that target virus cap-snatching, 1149–50
that target virus uncoating or release, 1148–49, 1149f, eTopic 27.4
AOA (ammonia-oxidizing archaea), 763–65, 764f, 765f
APCs (antigen-presenting cells)
adaptive immune response, **965,** 966f
defined, **942**
T cell activation, **982–83,** 983f, 986, 987f
Aphotic zone, 857f, 858
API (analytical profile index), 1169–70, 1170f, eTopic 28.1
Apicomplexa, 789t
Apicomplexans, **792,** 817–20, 819f, 820f
APOBEC3G protein, 427
Apoptosis, **942–43,** 943f
triggering, 1041
Aporepressors, inactive, 368, 369f
Appendix, 945f
Applied biotechnology, 444, 472–75
bacterial genes save crops, 472–73, 472f
biocontainment, 474–75
phage display technology, 473–74, 474f
vaccine antigens produced in plants, 473, 473t
AP site, **326**
Apurinic site, **320,** 320f
Aquaporins, 125, 167, 167f
Aquifex, 704t, 737
Aquifex aeolicus, 83–84, 84f, 691
Aquifex pyrophilus, 680, 680f, 706
Aquificae, 704, **706,** eTopic 18.1
Arabidopsis thaliana, 875f
Arabinans, 95, 96f
Arabinogalactans, 95, 96f
Arabinose-binding protein, 77f
Arabinose catabolism, 366–68, 366t, 367f
AraC/XylS family of transcriptional regulators, 366–68, 366t, 367f
Arbuscular mycorrhizae, **803,** 803f
Arbuscules, 803, 803f, 873, 874f
Archaea, **5,** 747–84
Asgard, 753t, 755, 781–82, 782f
vs. bacteria and eukaryotes, **27,** 28f
biofilms, 150
characteristic traits, 748–50, 749t
chromosomes and genomes, 267–68
common traits, 76
divergence, 677–78, 677f, 679t
diversity, 747–56, 753t

DPANN symbionts, 753t, 755, 780–82, 780f–782f
ether-linked isoprenoid membranes, 748–50, 748f, 749t
Euryarchaeota, 753t, 754–55, 755f, 766–80
evolution, 329
gene structure and regulation, 749t, 750–51, 750f
genomes, 241t
glucose catabolism, 754, 754f
Haloarchaea, 753t, 774–78
human microbiome, 923
membrane lipids, 87, 87f
metabolism, 749t
obesity, 935
phylogeny, 751, 752f, 753t
protein secretion, 309
scanning electron microscopy, 65, 66f
size and shape, 40, 41f
small RNA, 378
TACK superphylum, 752–54, 753t, 754f, 756–63
Thaumarchaeota, 753t, 754, 763–65, 764f, 765t
Thermococcales, 753t, 778–79, 778f
Thermoplasmatales, 753t, 779–80, 779f
transcription, 287–88, 288f
transcription regulators, 368–69
Archaean eon, **663–64,** 664f, 665t
Archaellum(a), **750**
Archaeoglobales, 753t, 779
Archaeoglobus fulgidus, 753t, 779, eTopic 19.2
Archaeon(a), 2, 3f, 4, **5**
Archaeplastida, 789t, 791–92, 791f, 808–15
Arctic, climate change, 1198
Arctic methanogens, global warming, 890–92, 891f, 892f
Arginine biosynthesis, 611, 612f
Arnold, Frances H., 473
Aromatic amino acids, 611–13, 613f
Aromatic catabolism, 523–27, 523f, 526f–528f
Aromatic degrading enzymes, *Deepwater Horizon* spill, 835
Aromatic molecules, catabolism, 501–3, 502f, 503f
Aromatic ring, **497**
Arrhenius equation, 161–62
Arsenic, microbial metabolism, 910t
Arsenite, bioremediation, 910, eTopic 22.3
ART. *See* Antiretroviral therapy (ART)
Arthomitus, 720
Arthritis, rheumatoid, 957
Arthrobacter, 725, 726f
Arthropod vectors, 1014–16, 1015f
Arthrospore, 711t
Artifact, **66**
Ascomycete(s), 789t, **799–801,** 800f, 801f
Ascomycota, 789t, 799–801, 800f, 801f
Ascospores, **799,** 800f
Ascus(i), 795f, **798**
Aseptic, **19**
Aseptic meningitis, 1110t
Asgard, 753t, 755, 781–82, 782f
A site (acceptor site), **294–95**
Aspergillosis, 1060t, 1066t
antifungal agents, 1155t
Aspergillus
antifungal agents, 1154–56
classification, 789t
dairy food spoilage, 635
emerging pathogen, 803
reproductive cycle, 798, 801, 801f
respiratory tract infection, 1066t
skin infections, 1060t
Aspergillus fumigatus, 804t
Aspergillus oryzae, 626–27
Aspirin, 85f
asRNA (antisense RNA), **378–79,** 379f
Assembly, **831–32,** 831f, 832f
Assimilation, **842**
Assimilatory nitrate reduction, **903**
Astrobiology, 914–18
biosignatures, 917
Europa, 918, 918f
life on planets of distant stars, 918, 919f
Mars biosphere, 915–17, 916f
terraforming, 917
Asymmetrical virions, 205, 206f
Asymptomatic infection, 411
Athalassic lakes, 776
Atherosclerosis, 8, 1092, eTopic 26
Athlete's foot, 1012, 1013f
Atmosphere, Earth, 662
Atomic force microscopy (AFM), 41f, **42, 69–70,** 70f
Atopic disease, 1004
ATP. *See* Adenosine triphosphate (ATP)
Attachment
HIV, 425–27, 426f
influenza virus, 413–14, 415f
microbial, 1020–25, 1021f–1025f, 1021t
phage, 406–8, 406f, 407f
Attenuation, 17, 18
transcriptional, **372–74,** 373f
Attenuator stem loop, 373f, **374**
Auto-, 122
Autoantibodies, 1007
Autoclave, **13,** 182, 182f
Autofluorescence, **837,** 837f
Autoimmune hemolytic anemia, 1007t
Autoimmune response, 957, **1006–8,** 1007t, eTopic 24.5
Autoimmunity, 957, **1006–8,** 1007t, eTopic 24.5
Autoinducer, **382–84**
Autoinduction, **382**
Automated microbial identification instrument, 1168, 1168f, 1169t
Autooxidation, food spoilage, 634, 634f, 635
Autophagy, **956, 1047,** 1048f
Autotrophs, **121,** 121f, **122**
Autotrophy, **120–21, 122,** 488, 582, 583f
Auxotrophic mutant, **321,** 330, 330f
Auxotrophy, synthetic, 475
Average nucleotide identity (ANI), 691
Avery, Oswald, 30, 240, 337
Avian influenza strain H5N1, 223–25, 411, 414, 418–19
Avian leukosis virus (ALV), 208, 208f, 212t
Axinella mexicana, 764, 764f
Azam, Farooq, 861, 863f
Azidothymidine (AZT), **427**

Azithromycin, 1128, 1133
Azoarcus tolulyticus, 174t
Azoles, 1155t
Azotobacter, 174t
nitrogen fixation, 601, 605, 606
Azotobacter vinelandii, 551
AZT (zidovudine), 1148t, 1150–51, 1150f

B
Babesia microti, 1097
Babesiosis, 1097
Bacillales, 714–15, 714f, 715f
Bacillariophyceae, 789t, 812, 812f, 813–14, 813f
Bacillary dysentery, 1073
Bacillus, 714–15, 714f
alkaline fermentation, 622, 628–29, 629f
endospore formation, 105, 714
Gram stain, 52
multicellularity, 712, 713f
oxidized forms of nitrogen, 552
species definition, 691
spore stain, 53
transformation of naked DNA, 337
Bacillus(i), **40,** 41f
Bacillus alkalophilus, 715
Bacillus anthracis
causative agent, 1194
epidemiology, 1187
facultative microbe, 174t
respiratory tract infection, 1066t
S-layer, 92
species definition, 691, 692
spore stain, 53
sporulation, 151
systemic illness, 1102
toxins, 1026t, 1032–33, 1033f
transmission electron microscopy, 65, 66f
Bacillus brevis, 1129
Bacillus cereus, 691–92, 691f
Bacillus halodurans, 715
Bacillus licheniformis, 1128
Bacillus megaterium, 277
nanotubes, 159
replication, 101
size, 40f
Bacillus natto, 628–29, 629f
Bacillus pestis, 26
Bacillus polymyxa, 1129
Bacillus subtilis, 714f, 715
active transport, 127
anti-sigma factor, 371
bacitracin, 1128
biofilm communication, 149
biofilm dissolution, 150
biological safety cabinet, 184f
cell wall growth, 90, 90f
classification, 704t
culture medium, 132
cytoskeleton, 96, 97f
disinfectant resistance, 187, 188f
exopolysaccharides, 148, 149f
fluorophore labeling, 56, 57f
intercellular nanotubes, 107, 107f
mesophile, 162
nanotubes, 159
segregation of sister chromosomes, 259–60
septation, 98, 98f
sporulation, 151
targeted mutagenesis, 464
transcriptional fusion, 468, 468f
Bacillus thermophilus, 715
Bacillus thuringiensis, 715, 715f
cell envelope, 90
insecticidal, 472–73, 472f, 650
species definition, 691
spore stain, 53
Bacitracin, 1124f, **1128,** 1128f
Bacteremia, **926,** 1067, **1092**
Bacterial blooms. *See* Algal blooms
Bacterial cell(s)
counting, 132f, 135–40, 138f, 139f
microbial nutrition, 120–24
nutrient uptake, 124–30
Bacterial cell differentiation, 103–5, 103f, 104f, 151–55
actinomycetes, 153–55, 154f
cyanobacteria, 151–52, 153f
endospores, 151, 152f, eTopic 10.2
myxococcus, 153, 153f
Bacterial cell division, 97–103
DNA organization in nucleoid, 98–100, 99f, 100f
regulation by DNA replication, 100–102, 101f, 102f
septation, 98, 98f, 99f
Bacterial cell membrane, 82–87
defined, 77f, **78**
lipid diversity, 85–87, 86f, 87f
structure, 82–84, 82f–84f
and transport, 82–85, 82f–85f
Bacterial cell structure and function, 75–117
biochemical composition, 78–80, 79t
cell division, 97–103
cell fractionation, 80–82, 80f, 81f, eTopic 3.1
cell membrane and transport, 82–87
cell polarity, membrane vesicles, and nanotubes, 103–8
envelope and cytoskeleton, 87–97
model, 78
overview, 76f–77f, 77–82
specialized structures, 108–15
Bacterial culture, 130–35
anaerobes, 175–76, 176f
complex *vs.* synthetic media, 132, 133t
culture media, 130
dilution streaking and spread plates, 131–32, 131f, 132f
growth factors and uncultured microbes, 134–35, 134t, 135f, 136–37
pure, 130
selective and differential media, 133–34, 133f
Bacterial endocarditis, 1093–96, 1093f
subacute, 926, 927f, 1093
Bacterial growth curves, 142f
Bacterial growth cycle, 140–47
binary fission, 141, 141f
continuous culture, 146–47, 146f

exponential growth, 141
generation time, 141–43, 142f
stages of growth in batch culture, 143–45, 144f, 145f
symmetrical and asymmetrical cell division, 141, 141f
Bacterial leaching, 7
Bacterial meningitis, 1104–6, 1105f, 1110t
Bacterial microcolonies, 131, 131f
Bacterial pneumonia, 1064–67, 1065f
Bacterial recorder, 409, 410f
Bactericidal antibiotics, **181, 1119,** 1134
Bacteriochlorophylls, 470f, **569–70**
Bacteriophage(s), 215–22
attachment and infection, 406–8, 406f, 407f
bacterial defenses, 219, 220f
batch culture, 233–34, 234f
cell-surface receptors, 215
defense against, 32
defined, **196**
enteric, 402–10
Escherichia coli, 196, 197–98, 197f
gut, 215, 219–20, 221f, 222–23
infection of host cell, 196, 196f, 215–18, 215f, 216f
lysogeny, 216f, 217, 403, 408–9, 408f
lytic cycle, 216f, 217
membrane vesicles, 106
overview, 215
phage therapy, 189
plaque isolation and assay, 235–36, 235f
Prochlorococcus, 195, 195f, 196
replication, 215–18, 216f, 218f, 404f, 405–6, 405f
slow-release cycle, 218, 218f
structure, 404–5, 404f, eTopic 11.1
synthetic biology, 409, 410f
temperate *vs.* virulent, 403
transcytosis, 220, 221, 222–23
transduction, 217, 218f, 334–37, 335f, 336f
Bacteriophage lambda, 197, 212t
Bacteriophage M13, 212t
Bacteriorhodopsin, **567–68,** 567f, 568f, **570**
Haloarchaea, **755, 777–78,** 777f
horizontal gene transfer, 680
Bacteriostatic antibiotics, **181, 1119**
Bacterium(ia), 4, **5,** 701–46
biochemical composition, 78–80, 79t
Candidate Phyla Radiation (CPR), 704t, 706–8, 707f
cell envelope, **78,** 90–96, 91f–96f
cell size, 102
cell wall, 702
common traits, 76, 702–3
Cyanobacteria, 703, 704t, 708–13
cytoskeleton, 96, 97f
deep-branching Gram-negative, 704t, 705, 706f, 738–41
deep-branching thermophiles, 704t, 706, 706f
divergence, 677–78, 677f, 679t, 702–3
diversity, 76–77, 701–3, 703f, 704t
electron transfer, 107
Firmicutes, Tenericutes, and Actinobacteria (Gram-positive), 704t, 705, 714–26
gene expression, 702
giant, 678
long-term storage, 183
membrane proteins, 83–84, 84f
phyla, 703–6
phylogeny, 702–3, 703f
Planctomycetes-Verrucomicrobia-Chlamydiae (PVC), 704t, 705–6, 741–44
Protobacteria, 704t, 705, 726–38
size and shape, 40, 41f
ultrasmall, 707, 707f
Bacterium pestis, 26
Bacteroid(s), 604, **730, 876–78,** 877f, 878f
Bacteroides, 739–40
anaerobe, 174t
catabolism, 501, 502f, 504, 505f
cell structure, 78
classification, 705
commensalism, 850
Enterobacteriaceae and, 734
Gram stain, 51
gut microbiome, 929, 929f, 1077
horizontal gene transfer, 339, 339f
human colon, 855, 856
membrane vesicles, 107
metronidazole, 1130
microbial endosymbiosis, 23
obesity, 934, 935
osteomyelitis, 1063
transposon mutants, 347–48
Bacteroides acidifaciens, 60, 60f
Bacteroides cellulosilyticus, 54, 54f
Bacteroides fragilis, 739–40, 740f
aerotolerant, 175
benefits, 997
colon cancer, 921, 921f, 922, 922f
containment breaches, 931, 932f
cryo-electron tomography, 37, 37f
fluorescence microscopy, 38, 38f
glycosylation, 303
gut immune system, 996
specimen collection, 1163–64, 1163f
systemic illness, 1102
Bacteroides ovatus, 521
Bacteroides plebeius, 521, 521f
Bacteroides thetaiotaomicron
benefits, 931, 997
catabolism, 504
classification, 704t, 739–40
complement, 994
digestion, 770–71
disease category, 693
fluorescence microscopy, 54, 54f
gut bacteria, 485, 485f, 486, 486f
manipulation of gene expression in intestine, 466
obesity, 935
survival within host, 1041
Bacteroides vulgatus, 54f
Bacteroidetes, 739–40, 740f
cell structure, 78
classification, 704t
defined, **705**
gut microbiome, 929, 929f
human microbiome, 923

Bactoprenol, 1124f, 1126
Baeocytes, 712
Bait protein, 471, 471f, eTopic 12.7
Baker, Bill, 1145
Baker's yeast. *See Saccharomyces cerevisiae*
Balaban, Nathalie, 144, 144f
Balantidium coli, 823, 823f
Ballou, Elizabeth, 1057, 1057f, 1058
Baloxavir marboxil (Xofluza), 1148t, 1149–50
Baltimore, David, 210, 211f
Baltimore virus classification, 210–13, 211f, 212t
BAM (beta barrel assembly machine), 310, 311f
Banded iron formations (BIFs), 665t, **667–68,** 668f, 669f
Banfield, Jillian, 102, 702, 703f, 707, 707f, 751, 828, 828f
Barcoding, **831**
Bare lymphocyte syndrome, 964, 964f
Barns, Susan, 762
Barophiles, 161t, **165–66,** 165f, **756,** 844t, **862**
Barophilic vent hyperthermophiles, 760–62, 761f, 762f
Barotolerant, 161t
Barr, Jeremy, 222–23, 222f
Barré-Sinoussi, Françoise, 421, 422f
Barry, Clifton, 646
Barton, Jacqueline, 324, 324f
Bartonella, 911
Bartonella henselae, 729, 1015f
Base analog, mutagenic effects, 318t
Base excision repair (BER), 322t, **323–26,** 326f
Basidiomycete(s), 787f, 789t, **801–3,** 802f, 803f
Basidiomycota, 787f, 789t, **801–3,** 802f, 803f
Basidiospores, **802,** 802f
Basophils, 938f, **939**
Bassham, James, 586
Bassler, Bonnie, 385, 385f
Batch culture, **233–34,** 234f
  growth stages, 142f, **143–45,** 144f, 145f
Bates, Paul, 1103
Bathyarchaeota, **752,** 753t
Batrachochytrium dendrobatidis, 789t, 798, 799t, 804t
Battista, John, 185f
Baym, Michael, 682, 683f
B cell(s), 938f, **944**
  capping and activation, 976, 976f
  differentiated into plasma cells, 975–78, 976f, 977f
  humoral immune response, 965–66, 966f
  lymph node germinal center, 980, eTopic 24.4
  memory, 965–66, 966f, **974,** 975, 975f, 981
B-cell activation, 986, 987f
B-cell receptors (BCRs), **975–76,** 976f
B-cell selection, 981
B-cell tolerance, **967**
Bdellovibrio, 736–37, 737f
Bdellovibrio(s), 736–37, 737f
Bdellovibrio bacteriovorus, 737f
  chemoreceptors, 75, 75f
  cryo-electron tomography, 68–69, 68f
Bdellovibrionales, 704t
Beadle, George, 799
Bearson, Bradley, 1141, 1142f
Bedaquiline, 549, 1144
Bedrock, 868f, 869
Beer, 621t, 631–33, 631f–633f
Beggiatoa, 21, 732
  commensalism, 845t, 850
  nitrogen fixation, 601
  sulfur cycle, 905, 906f
Beggiatoa alba, 728t
Beijerinck, Martinus, 19, 22, 197, 601
Bej, Asim, 163f
Béjà, Oded, 567
Belfort, Marlene, 390–91, 390f
Benning, Liane, 883, 883f
Benson, Andrew, 586
Benthic microbial fuel cell, 558–59, 559f
Benthic organisms, **858**
Benthic sediment, 866, 867f
Benthos, 857f, 858, 862–65, 864f
  cold seep endosymbiosis, eTopic 21.2
  marine cycling of methane, 887–88
Ben-Yehuda, Sigal, 107, 159, 159f
Benzalconium chloride, 186f
Benzene catabolism
  aerobic, 525–26, 526f
  anaerobic, 526–27, 527f
Benzene derivatives, catabolism, 523–27, 523f, 526f–528f
BER (base excision repair), 322t, **323–26,** 326f
Berends, Evelien, 1028f
Berg, Howard, 61, 111
Berg, Paul, 5, 33
Berkmen, Melanie, 57f
Bernal, John, 71, 71f
Beta-amyloid toxicity, 796–97, 797f
Beta barrel assembly machine (BAM), 310, 311f
Beta-defensins, 948
Betadine, 186
Beta-galactosidase, 362
Beta-lactam antibiotics, 1126, 1126f
  bacterial resistance, 1126–27
Beta-lactamase, 89, 1127, 1136, 1137f
Beta-lactamase-resistant antibiotics, 1127–28, 1127f
Betapropiolactone, 186–87, 186f
Betaproteobacteria, 731–32
  classification, 704t, 705
  defined, **705**
  lithotrophs, 731, 732f
  metabolism, 728t
  pathogens, 731–32
Betaretrovirus, 421t
Betzig, Eric, 57
Beutler, Bruce, 951
Beyth, Nurit, 190f
Bft enterotoxin, 921, 921f
Bhagwat, Arvind, 634f
Bhattacharya, Mohini, 1028f
B horizon, 868f
"Biased random walk," 113f, 114
BIF(s) (banded iron formations), 665t, **667–68,** 668f, 669f
Bifidobacterium
  benefits, 997
  biological control of microbes, 189
  classification, 704t
  gut microbiome, 856, 929, 929f

human milk oligosaccharides, 521
probiotic, 932, 932f
Bifidobacterium dentium, 522–23, 524–25
Bigger, Joseph, 1142
Bin, **833,** 833f
Binary fission, **141,** 141f
Binding agent, 50, 52f
Binning, **833,** 833f
BioBricks, 479–80, 481f
Biochemical assays, 140
Biochemical composition, 78–80, 79t
Biochemical oxygen demand (BOD), **858, 893–95,** 894f–896f
Biochemical pathogen identification, 1167–74
Biochemistry, 8t
Biocomplexity Initiative, 22
Biocontainment, 474–75
Biocontrol, **189–91,** 190f, 191f
Biocrystallization, 100
Biofilms, **23,** 23f, 147–51
antibiotic resistance, 187, 188f, 1142–43, 1145
bacterial evolution, 1140, 1140f
bacteriophages, 221
breakup (dissolution), 150, 150f
confocal laser scanning microscopy, 58, 58f
CRISPR, 343, 343f
cyclic di-GMP and, 381, 381f
defined, **23, 147**
differentiation and communication, 148–50, 149f
E. coli, 1084, eTopic 26.6
electrogenic, **553**
floating, 148, 149f
and infections, 1023–24, 1025f, eTopic 4.3
life cycle, 147–48, 147f–149f
medical devices, 1023, eTopic 4.3
microbial attachment, 1023–24, 1025f
mixed, 187, 188f
multispecies, 705, 706f
nutrient time-sharing, 119, 119f, 120
pathogenesis and environmental degradation, 147
phage therapy, 190–91, 190f, 191f
prevention, 147, eTopic 4.3
scanning electron microscopy, 65–66, 66f
shower curtain, 675, eTopic 17.2
Biogenic, **664**
Biogeochemical cycles, **884**
Biogeochemistry, **884**
Biogeography, 59, 59f
Bioinformatic analysis, 161, 271, **455–56,** 455f
Biological carbon pump, **887,** 888f
Biological oxygen demand (BOD), **858, 893–95,** 894f–896f
Biological safety, 1164–66, 1164t, 1165f
Biological safety cabinets, **183,** 184f
Biological safety levels (BSLs), 1164t, 1165
Biological signatures (biosignatures)
Archaean eon, 663–64, 664f, 665t
defined, **663**
early life, 663–67
Hadean eon, 663, 664f
isotope ratios, 665–67, 665t, 666f, 667f
life on other planets, **917**
microfossils, 664–65, 665t, 666f, 667f
organic, 665t, 667
Biological thermodynamics, 490–91, 490f
Bioluminescence, 382–84, 382f, 383f
Biomass, 488, 488f, **842**
free energy change, 488–89, 489f
planktonic communities, 862
Biomining, 562–63
Biomolecular motor, 123, 123f
Biopanning, 473–74, 474f
Bioprospecting, **647,** 650
Bioremediation, **185**
groundwater, 910, eTopic 22.3
wastewater treatment, 896f–898f, 897–99
Biosignatures. *See* Biological signatures (biosignatures)
Biosphere, **662**
rare, **838**
Biosynthesis, 581–616
amino acids, 608–13, 608f–610f, 612f, 613f
antibiotics, 596–600, 597f–600f, 602–3
Calvin cycle, 586–91, 587f–590f
carbon dioxide fixation, 586–94, 586t
carbon dioxide uptake and concentration, 591–92, 591f
competition and predation, 585
defined, **582**
energy expenditure, 583–86, 585f
fatty acids, 594–96, 595f
genome loss and cooperation, 585
3-hydroxypropionate cycle, 594, eTopic 15.4
nitrogen fixation and regulation, 584, 600–607, 601f, 604f–607f
nitrogenous bases, 613–14, 614f
overview, 582–86, 583f
reductive acetyl-CoA pathway, 593–94, 593f
reductive (reverse) TCA cycle, 592–93, 592f
requirements, 582–83, 584f
secondary products, 582, 582f
substrates, 582–83, 584f
Biosynthetic pathways
reductive acetyl-CoA, 593–94, 593f
repression, 368, 369f
Biotechnology, 33, 443–75
applied, 444, 472–75
DNA amplification and sequence analysis, 444–57
gene expression analysis, 466–72
genetic manipulation of microbes, 457–66
history, 444
progression through techniques, 444, eTopic 12.1
Bioterrorism, 5, 1164t, 1192–93, 1193f, eTopic 28, eTopic 28.3
Biotic cycling, 884–85
Biotic entities, **883**
Birnaviruses, 212t
Bispecific antibody, 1005, 1145, 1146–47, 1146f, 1147f
Biteen, Julie, 485, 485f
Black Death, 7f
Black forest mushrooms, 619
Black smoker, **760–62,** 761f, 762f
Blander, Julie, 963, 963f, 999
Blanton, Laura, 933
Blastochloris, 709t, 866f

Blastochloris viridis, 570, 571f
photoheterotrophy, 727
photosystem II, 574
Blastomyces, 821
Blastomyces dermatitidis, 797
antifungal agent, 1154
respiratory tract infection, 1066t, 1068, eTopic 26.3
skin infections, 1060t
Blastomycosis, 1060t, 1066t, 1068, eTopic 26.3
antifungal agents, 1154, 1155t
Blattabacterium, 850t
Bleach, liquid, 186
Blood, 937, 937f
Blood agar, 1168, 1168f
Blood-brain barrier, **1104**
Blood samples, 1161
Blount, Zackary, 685, 686f, 687
Blue baby syndrome, 903
Blue-green algae. *See* Cyanobacteria
BOD (biochemical/biological oxygen demand), **858, 893–95,** 894f–896f
Bog, 890–91, 891f
Bogoslowski, Ania, 1094, 1094f
Boil(s), 949, 949f, 1059–61, 1061f
Boiling to kill microbes, 182
Bone infections, 1063
Bone marrow, 945f
BoNT (botulism neurotoxin). *See* Botulinum toxin (Botox)
Booth, Timothy, 204
Bordetella
growth factors, 134t
microbial attachment, 1023
pathogen identification, 1172
Bordetella pertussis
microbial attachment, 1021t
pathogen identification, 1184t
toxins, 1026t, 1037t, 1040, 1040f
whooping cough, 1064, 1066t, eTopic 26.2
bor gene, 409
Borrelia, 704t, 739
Borrelia burgdorferi, 738f, 739
classification, 704t
dark-field microscopy, 61
doxycycline, 1133
epidemiology, 1188–89
genome, 241t, 242
genomic islands, 353, 354f
Koch's postulates, 16
light microscopy, 41f
Lyme disease, 1100–1102, 1101f, 1102t
nutrient supplies, 120
technology-aided emergence, 1197–98
Borrelia recurrentis, 739
Borreliosis. *See* Lyme disease
Botulinum toxin (Botox), 715, 716f
antibiotics combined with, 1011, 1011f
food poisoning, 634
mechanism of action, 1107–8, 1107f
medical use, 1108
structure and function, 1106f, 1107
vs. tetanus toxin, 1108–9, 1108f
Botulism, 1081t, 1106–9, 1106f–1108f
canned goods, 182
case history, 1106
defined, **1106**
infant, 1107
Botulism neurotoxin (BoNT). *See* Botulinum toxin (Botox)
Botulism toxin. *See* Botulinum toxin (Botox)
Bovine rumen fermenter, 853–55, 853f, 854f
Bovine rumen microbiome, 832–33, 833f
Bovine spongiform encephalopathy (BSE), 1111, 1111t
Bovine tuberculosis (BTB), 1199–1200, 1199f
Boyd, Phillip, 909
Boyer, Herbert, 33, 459, 459f
Boyle, Robert, 10
Bradykinin, **953,** 953f, 954
Bradyrhizobium, 123, 730, 876
Bradyrhizobium japonicum, 601, 604, 605f
Brasier, Martin, 664
Braun lipoprotein, 77f, 93
Bread(s)
industrial microbiology, 648t
yeast, 621t, 630–31, 630f, 631f
Bread dough, alcoholic fermentation (leavening), 621t, 630–31, 630f, 631f
Breaker, Ron, 375, 375f
Breast milk, glycans, 521
Bregoff, Herta, 601
Brewer's yeast, 629
Briandet, Romain, 187
Bridge amplifications, 449f, 450
Briegel, Ariane, 75, 75f
Bright-field microscopy, 46–53
compound microscope, 47–48, 48f
defined, **40, 46**
fixation and staining, 50–53, 50f–53f
focusing, 49–50, 49f
magnification, 46–47, 46f, 47f
prokaryotes, 40, 41f
range of resolution, 40, 41f
specimen preparation, 49, 49f
Brined cheeses, 624, 625f
Brine pools, 776
Broad-spectrum antibiotics, 1140, 1141, 1146
Brocadia, 704t
Brocadia anammoxidans, 728t
Brock, Thomas, 445, 761
Bronchiolitis, 1071, 1071f, 1148
Brown, Timothy Ray, 1091, 1152
Brown algae, 792, 812, 814, 814f
Brucella, 729
paleopathology, 1012
pathogen identification, 1172, 1174
Brucella abortus, 1102t
Brucellosis, 1102t
Brun, Yves, 90, 90f, 98, 98f
Bryant, Jessica, 859–60
BSE (bovine spongiform encephalopathy), 1111, 1111t
BSLs (biological safety levels), 1164t, 1165
BTB (bovine tuberculosis), 1199–1200, 1199f
Bt toxin, 472–73, 472f
Bubble boy, 964, eTopic 24.1

Bubo, 1098f
Bubonic plague, 7–10, 7f, 26, **1098,** 1102t
Buchan, Alison, 347, 347f
Buchnera, 676–77, 707, 850t
Budding, 795f, **796**
Buffer gate, 476, 476f
Bugni, Tim, 129, 129f
Built environment, **885,** 911–13
  microbiome of building interiors, 911, 912f, 913f
  transmission of microbes between humans and, 912–13, 913f, 914–15
  wastewater treatment, 895–99, 896f–898f
Bukreyev, Alexander, 1002, 1002f
Bulgheresi, Silvia, 701, 701f
Burkholderia, 596, 732, 1037t, 1044
Burkholderia cepacia, 732
Burkholderiales, 704t
Burkholderia pseudomallei, 450, 704t
Burn wounds, 1924f
Burst, bacteriophage, 217
Burst size
  bacteria, **217**
  bacteriophage, 234, **234**
Bush, George W., 373f, 421
Butanol, 515
Buttermilk, 621t
Button mushrooms, 619
Butyrate, obesity, 934–35, 935f

C

^{14}C (carbon isotope), 587, eTopic 15.2
  depletion, 665–67, 666f
C1 complex, 993, 993f
C3a fragment, 959, 994
C3b fragment, 994
C3 convertase, 993, 993f
C4a fragment, 994
C5a fragment, 959, 994, 1094, 1095f
C5 convertase, 993, 993f
Cabbage, fermentation, 627, 627f
Cable bacteria, 533, 533f, 534, 534f
CagA protein, 1079
CAI(s) (community-acquired infections), 1061
CAI-1 (cholera autoinducer 1), 385, 385f
Calcitonin gene–related peptide (CGRP), 1011, 1012
Caldisphaera, 753t, 760
Caldisphaerales, 753t, 760
Calorimeter, **491,** 491f
Calvin, Melvin, 586, 587, 808
Calvin-Benson-Bassham (CBB) cycle. *See* Calvin cycle
Calvin-Benson cycle. *See* Calvin cycle
Calvin cycle, 586–91, 586t, 587f
  defined, **586**
  overview, 587–88, 588f–590f
  ribulose 1,5-bisphosphate, 588–91, 588f–590f
cAMP (cyclic adenosine monophosphate), stimulation of transcription, 363–64, 364f, 365f, eTopic 10.1
Campi, Carlo, 766
cAMP (cyclic adenosine monophosphate) receptor protein (CRP), stimulation of transcription, 363–64, 364f, 365f, eTopic 10.1
Campylobacter, 737
  food contamination, 640t
  gastrointestinal tract infections, 1072
  gut microbiome, 1076
  microaerophilic microbe, 174t
  pathogen identification, 1172, 1184t
  S-layer, 95
Campylobacterales, 704t
  cryo-electron tomography, 68–69, 68f
Campylobacter jejuni, 704t
  culturing, 176
  gastrointestinal tract disease, 1081t
CaMV (cauliflower mosaic virus), 208, 208f, 230–32, 231f
Cancer, gut bacteria and, 24–25, 24f, 25f
Candida
  antifungal agents, 1154
  cocoa bean fermentation, 627
  extended fermentation, 631
  human microbiome, 923
Candida albicans, 796–97
  AIDS, 1090f, 1091
  bare lymphocyte syndrome, 964, 964f
  genitourinary tract microbiome, 928
  gut microbiome, 929
  skin infections, 1060t
Candida auris, 1187–88
Candida glabrata, 1060t
Candidate Phyla Radiation (CPR), 704t, **705,** 706–8, 707f
Candidate species, **693**
Candidatus Korarchaeum cryptofilum, 762
Candidatus Savagella, 716, 720–21, 720f, 721f
Candidiasis, 1060t
  AIDS, 1090f, 1091
  antifungal agents, 1155t
Canning, 642
Cannula, **762**
Cannulated cows, **854**
Cap-binding complex (CBC), 1153
Capnocytophaga canimorsus, 1062
Capping, **976,** 976f, 986
Capsid, **196,** 196f, **198**
  classification, 210
  HIV, **422–23,** 422f, 423f
  icosahedral, **202–3,** 203f
Cap snatching, 416
Capsule
  extracellular immune avoidance, **1041**
  firmicutes, **92**
  proteobacterial cell envelope, 95
Carbadox, 1141, 1142f
Carbapenems, 1127–28
Carbenicillin, 1126f
Carbohydrate catabolism, 501, 502f, 503f
Carbolfuchsin, acid-fast stain, 52, 53f
Carbolic acid, 18
Carbon
  global reservoirs, 884–85, 884f
  greenhouse gas, 885, 885f
  metabolism, 843t
  methanogenesis from, 767, 771–73, 772f
Carbon assimilation, **842**

Carbon balance, global, 889–93, 890f–892f
Carbon concentrating mechanism, **591–92,** 591f
Carbon cycles and cycling, 885–93
different ecosystems, 887–89, 888f, 889f
experimental measurement, 886, 887f
global carbon balance and temperature change, 889–93, 890f–892f
and hydrologic cycle, 893, 894f
marine, 887, 888f
microbial nutrition, 120–21, 121f
oxidation-state changes during carbon flux, 885–86, 886t
radiative forcing, 890, 890f
terrestrial, 888–89, 889f
Carbon dioxide (CO_2) concentration, 591–92, 591f
Carbon dioxide (CO_2) fixation, 586–94, 586t
Calvin cycle, 586–91, 586t, 587f–590f
carbon dioxide uptake and concentration, 591–92, 591f
defined, **586**
3-hydroxypropionate cycle, 586t, 594, eTopic 15.4
reductive acetyl-CoA pathway, 586t, 593–94, 593f
reductive (reverse) TCA cycle, 586t, 592–93, 592f
Carbon dioxide (CO_2) uptake, 591–92, 591f
Carbon (CO_2) dissimilation, **842**
Carbon (CO_2) flux
experimental measurement, 886–87, 887f
oxidation-state changes, 885–86, 886t
Carbon isotope (^{14}C), 587, eTopic 15.2
depletion, 665–67, 666f
Carbon-nitrogen-oxygen (CNO) cycle, 661
Carbon pump, biological, **887,** 888f
Carbon reservoirs, 885
Carbon sequestration, **586**
Carbon sinks, global warming, 893
Carbon starvation, 178
Carboxydothermus, 714
Carboxylate, 526
Carboxylic acid, 501, 506
Carboxysomes, **109,** 110f, **591,** 591f
Cardiolipin, **85–86,** 86f
Cardiovascular infections, 1092, eTopic 26.8
endocarditis, 1093–96, 1093f
malaria, 1096–97, 1096f
septicemia, 1092–93, 1094–95
Caries, 926
Carotenoids, **569, 570,** 570f
Carroll, Lewis, 686f, 687
CAR-T therapy, 654–56, 654f–656f
Cas12a, 1180–81, 1181f
Caseation, 1070
Casein(s), 623
Casein catabolism, flavor generation, 625, 626f
Casèn, Christina, 1179
Caspase(s), 942, 951–52, 952f
Caspase-1, 951–52, 952f
Caspofungin, 1155t
Cassette, 311–12
Cassone, Vincent, 357, 357f
Catabolism, **486,** 486f, 501–5
aromatic, 523–27, 523f, 526f–528f
benzene derivatives, 523–27, 523f, 526f–528f
glucose, 506–10
molecular and cell biology, 504, 505f
niche adaptation, 520–21, 521f, 522f
products, 504
substrates, 501–3, 502f, 503f
Catabolite repression, **365,** 365f
via inducer exclusion, 365–66, 366f, eTopic 4.1, eTopic 10.1
Catalysis, metabolic reactions, 499–500, 500f
Catalytic RNA, 32, 32f, **286,** 286t, 292
Catenane, **259,** 259f
Cathelicidin LL-37, 946, eTopic 23.3
Cathelicidins, neutrophil, 948, eTopic 23.3
Cationic detergents, 186
Cat scratch disease, 1015f
Caulerpa taxifolia, 4
Cauliflower mosaic virus (CaMV), 208, 208f, 230–32, 231f
Caulimovirus, 212t
Caulobacter, 729
flagella, 111
S-layer, 95
stalk, 103f, 110–11
symmetrical cell division, 141
turret-shaped structures, 109
Caulobacterales, 704t
Caulobacter crescentus, 729
asymmetrical cell division, 103, 103f, 104, 104f, eTopic 3.2
cell differentiation, 151
classification, 704t
cytoskeleton, 96, 97f
super-resolution imaging, 58, 58f
Cavicchioli, Ricardo, 769, 770f
CAZyme gene, 521, 521f
CBB (Calvin-Benson-Bassham) cycle. *See* Calvin cycle
CBC (cap-binding complex), 1153
CCF4-AM, 1053–54
CCR(s) (chemokine receptors), **426**
CCR5 inhibitors, 1151
CD3 complex, 985, 985f
CD4 surface antigen, 425–26, 426f, 982–83, 989, 990, 1091
CD8 surface antigen, 982–83, 987, 988f
CD47, 955
c-di-GMP (cyclic dimeric guanosine monophosphate), 150, 380–81, 381f
cDNA (complementary DNA), 860–61, 861f
Cech, Thomas, 32, 32f, 671
Cefepime, 1127f
Cefoxitin, 1127f
Ceftaroline, 1127f
Ceftriaxone, 1127f
Celgosivir, 1152f, 1153
Cell(s), formation of first, 669–72, 669f–671f
Cell biology
and DNA revolution, 29–34
history, 8t
to probe pathogenesis, 1053–54, 1053f
Cell-cell adherence, 1025, 1026t
Cell-cell communication, quorum sensing, 381–84, 382f, 383f, 384t
Cell-cell interactions, humoral immune response, 965–66, 966f
Cell counting, 132f, 135–40, 138f, 139f

Cell cycle
  disruption, 1025, 1026t
  natural selection, 140–47
Cell cycle control, oncogenes, 229
Cell density, 144
Cell differentiation, 103–5, 103f, 104f, 151–55
  actinomycetes, 153–55, 154f
  biofilms, 148–49, 149f
  cyanobacteria, 151–52, 153f
  endospores, 151, 152f, eTopic 10.2
  myxococcus, 153, 153f
Cell division, 97–103
  DNA organization in nucleoid, 98–100, 99f, 100f
  regulation by DNA replication, 100–102, 101f, 102f
  septation, 98, 98f, 99f
  symmetrical *vs.* asymmetrical, 103–5, 103f–105f, 141, 141f, eTopic 3.2
Cell envelope, 87–96
  bacterial, 90–96, 91f–96f
  defined, **78**
  firmicute (Gram-positive), 90, 91–92, 91f, 92f
  mycobacterial, 90, 95, 96f
  proteobacterial (Gram-negative), 90, 91f, 92–96, 93f–95f
  structure, 88–90, 88f–90f
Cell fission, 12
Cell fractionation, 80–82
  cell wall lysis and spheroplast formation, 80–81
  defined, **80,** 80f
  spheroplast lysis, 81, 82f
  ultracentrifugation, 81
Cell-mediated hypersensitivity, 1001t, 1005, 1006f
Cell-mediated immunity, **964–65,** 966
  activation, 986–88, 987f
Cell membrane(s), 29–30, 29f, 82–87
  defined, 77f, **78**
  lipid diversity, 85–87, 86f, 87f
  protein export to, 307, 307f
  structure, 82–84, 82f–84f
  and transport, 82–85, 82f–85f
Cell membrane inhibitors, 1124f, 1125t, 1129, 1129f
Cell polarity, 103–5, 103f–105f, eTopic 3.2
Cell sorting, **837–38,** 837f
Cell structure and function, 75–117
  biochemical composition, 78–80, 79t
  cell division, 97–103
  cell fractionation, 80–82, 80f, 81f, eTopic 3.1
  cell membrane and transport, 82–87
  cell polarity, membrane vesicles, and nanotubes, 103–8
  envelope and cytoskeleton, 87–97
  model, 78
  overview, 76f–77f, 77–82
  specialized structures, 108–15
Cell-surface receptors, **215**
Cellular immunity, **964–65,** 966
  activation, 986–88, 987f
Cellular slime mold, **806**
Cellulitis, 1060t, **1063**
Cellulose, catabolism, 502f
Cell wall, 77f, **78**
  bacteria, 702
  lysis, 80–81
Cell wall inhibitors, 1124f–1129f, 1125–29, 1125t
Cenarchaeales, 753t
Cenarchaeum symbiosum, 753t, 764, 764f
Central nervous system infections, 1104–12
  blood-brain barrier, 1104
  botulism, 1106–9, 1106f–1108f
  encephalitis, 1109, 1110t
  meningitis, 1104–6, 1105f, 1110t
  prions, 1109–11, 1111f, 1111t
Centrifugation, 30
Cephalexin (Keflex), 188f, 1127f
Cephalosporins, 1127, 1127f
Ceratium, 818f
Cerebrospinal fluid (CSF), specimen collection, 1161, 1162f
Cetylpyridinium chloride, 186f
CFP (cyan fluorescent protein), 56, 57f
CFTR (cystic fibrosis transmembrane conductance regulator), **947,** 1031
CGRP (calcitonin gene–related peptide), 1011, 1012
Chaban, Beverly, 68, 68f
Chagas' disease, 822
Chain, Ernst, 19, 1117–18
Chain of infection, **15**
Chalfie, Martin, 53
Chan, Rita, 1028f
Chancre, **1087,** 1087f
Chancroid, 1086t
Chandler, Jeremy, 447, 448f
Chandrasekaran, Sriam, 943
Chang, Annie, 459
Chang Li, 475
Changsheng Li, 768
Chaperone(s), 78
  protein folding, **304–5,** 304f, eTopic 8.5
  thermophiles, 164
Chaperone GroEL, 77f
Chaperonins, 304–5, 304f, eTopic 8.5
Charbonneau, Mark, 933
Charge difference, proton motive force, 541–42, 542f
Charge transport (CT), DNA repair, 324–25, 324f, 325f
Charophyta, 789t
Charpentier, Emmanuelle, 341, 341f
Chase, Martha, 215, 240
CheA protein, 114, 114f
Cheddaring, **624–25,** 625f
Cheese, 621t
  brined, 624, 625f
  curd formation, 623
  defined, **623**
  flash pasteurization, 624
  flavor generation, 625, 626f
  hard, 623
  mold-ripened, 624
  production, 624–25, 624f, 625f
  propionic acid fermentation in Swiss, 622, 622f
  ripening, 624–25, 625f
  semihard, ripened, 623
  soft, unripened, 623
  spoilage, 634
  varieties, 623–24, 624f
Chemical affinity, fluorophores, 56

Chemical agents
  mutagenic, 318t
  that kill microbes, 185–88, 186f, 188f
Chemical analysis, element cycling, 886
Chemical gradients, 126
Chemical imaging microscopy, **42,** 58–60, 60f
Chemiosmotic theory, **30,** 540, 540f
Chemo-, 122
Chemoautotrophy, **582**
Chemokine(s), **949,** 950f, **953**
Chemokine receptors (CCRs), **426**
Chemolithoautotrophs, **122,** 535
Chemolithoautotrophy, **121, 122**
Chemolithotrophs, 7, 7f, 21, 121
Chemolithotrophy. *See* Lithotrophy
Chemoorganoheterotrophy, **122**
Chemoorganotrophs, 121–22
Chemoorganotrophy. *See* Organotrophy
Chemoreceptors, 75, 78, 113, 396, 396f
Chemostat, **146–47,** 146f
Chemosynthesis, **582**
Chemotaxis, **113–15,** 113f, 114f
  defined, **395**
  diverse mechanisms, 397
  memory, 396–97
  posttranslational regulation of cell behavior, 395–97
  purposeful movement, 395–96, 396f
Chemotrophy, **121–22, 487,** 488
Chen, Janice, 1180f
CheR protein, 114, 114f
CheW protein, 114, 114f
CheY protein, 114, 114f
Chickenpox, 1060t, 1063, eTopic 26.1
Chickenpox vaccine, 968t, 998t
Chimeric antibody, 1146–47, 1146f, 1147f
Chimeric sites, 336
ChIP (chromatin immunoprecipitation), 469t, **470**
ChIP-seq (chromatin immunoprecipitation sequencing), 469t, **470**
Chirality, 13
Chisholm, Sally, 106, 106f, 195, 860
Chitin, 793f, 794
Chiu, Charles, 1176
Chiu, Isaac, 1011, 1011f
Chlamydia, 742–43, 743f
  flipping cytokine profile, 1045
  intracellular immune avoidance, 1043
  microbial attachment, 1021t
Chlamydia(e), 742–43, 743f
  classification, 704t, 706
  defined, **706**
  vs. gonorrhea, 1089
  point-of-care rapid testing, 1186t
  sexually transmitted infection, 1086t, **1088,** 1088f
  spores, 711t
Chlamydia pneumoniae, 1088
Chlamydia psittaci, 1088
Chlamydia trachomatis, 743
  classification, 704t, 706
  culture, 16
  nongonococcal urethritis, 1086t
  pathogen identification, 1184t
  point-of-care rapid testing, 1186t
  sexually transmitted infection, 1088, 1088f
Chlamydomonas, 789t, 808–9, 809f
  snow algae, 883, 883f
Chlamydomonas reinhardtii, 808–9, 809f
Chlamydophila pneumoniae, 743
  cardiovascular infections, 1097, eTopic 26.8
  respiratory tract infection, 1066t
Chlamydophila psittaci, 1066t
Chloracidobacterium thermophilum, 739
Chloramphenicol, 297, 299f, 1132f, **1133**
Chlorella, 808
  endosymbiosis, 694–95, 695f, 696
Chlorination, 897, 898
Chlorine, 186
Chlorobenzenes, 564, 565f
Chlorobenzoate catabolism, 525–26, 526f
Chlorobi, 740–41, 741f
  classification, 704t, 705
  defined, **705**
  photosystem I, 573–74, 573f
  phototropic, 709t
Chlorobium, 740, 741f
  electron microscopy, 29–30, 29f
  lake phototrophy, 866
  photosystem I, 573–74, 573f
  phototropic, 709t
Chlorobium tepidum, 704t
  reductive TCA cycle, 592, 592f
Chloroflexi, 704t, **706,** 709t
Chloroflexus, 706, 707f, 709t
Chloroflexus aggregans, 121f
Chloroflexus aurantiacus, 594, 704t
Chlorophyll(s), 708
  defined, **569, 570**
  light absorption, 569–70, 570f
  photoexcitation, 568–72, 569f–572f
  structure, 569, 570f
Chlorophyta, 789t, 791, 791f, 808–11, 809f–811f
Chlorophyte(s), 789t, **791,** 791f
Chloroplast(s), **791,** 791f
  Calvin cycle, 586
  endosymbiosis, 696, 697–98, 697f
  genome, 697–98, 697f
  oxygenic phototrophs, 708
  proton motive force, 540
  stealing, 792
Chloroquine(s), 1097, 1152f, 1153
Chlorosomes, 573, 573f, **706**
Chloroviruses, 212t
Choanocytes, 790, 790f
Choanoflagellata, 789t, 790, 790f
Choanoflagellates, 789t, 790, 790f
Chocolate, 617, 621t, 627–28, 628f
Chocolate agar, 1168, 1168f
Cholera, 1081t
  diagnosis, 1058
  epidemiology, 1160, 1188, 1188f
  exotoxin, 1026t, 1030–32, 1031f–1033f, 1037t
  stomach microbiome, 928

Cholera autoinducer 1 (CAI-1), 385, 385f
Cholera toxin, 1026t, 1030–32, 1031f–1033f, 1037f
Cholera vaccine, 968t
Cholesterol, membrane, **86**
Chorismate, 612
Chrichton, Michael, 33
Christensenella minuta
  gut microbiome, 930
  heritable, 268, 269f
  obesity, 936
Chromatin immunoprecipitation (ChIP), 469t, **470**
Chromatin immunoprecipitation sequencing (ChIP-seq), 469t, **470**
Chromatium, 733, 733f
  metabolism, 728t
  phototropic, 709t
Chromium, 910t
Chromophore, **569**
Chromosome(s)
  archaeal, 267–68
  compacted into nucleoid, 245–46, 245f
  eukaryotic, 266–67
  giant bacteria, 239, 239f
  secondary, 263–65, 265f
Chronic inflammation, 952, **956–57,** 956f
Chronic viral infections, 229–30
Chroococcales, 704t
Chroococcus, 712
Chrysophyceae, 789t, 812
Chrysophytes, **792**
Church, George, 246, 246f
Chytrid(s), 789t, 791, 798, 799f
Chytridiomycetes, 789t, 791, 798, 799f, 854
Chytridiomycota, 789t, 791, 798, 799f
Ciliates, **792,** 815–16, 816f–818f, eTopic 20.2
Ciliophora, 789t, 815–16, 816f–818f, eTopic 20.2
Cilium(ia), **792**
Ciofu, Oana, 1140, 1140f
Ciprofloxacin, 249
  toxicity, 1119
Circadian clocks, **388–89,** 388f, 389f, 392f
Circadian rhythms, melatonin, 357, 357f
Circular polarization, 918
CI repressor, 407, 407f, 408f, 409
cis-antisense RNA (asRNA), **378–79,** 379f
Citrate, 518, 519f, 520
  evolution of aerobic catabolism, 687, 688f
Citromicrobium, 728, 729f
CJD (Creutzfeldt-Jakob disease), 207, 207f, 1110, 1111f, 1111t
Clades, **673**
  emerging, **692–93,** 704t, **706–8,** 707f
CLARITY technique, 58, 59, 59f
Classical complement pathway, 957, **993,** 993f
Classification, **690,** 692, 692t
Class II MHC molecules, **982–83,** 983f, 984f
Class I MHC molecules, **982–83,** 983f, 984f
Class switching, **974–75,** 980–81, 980f, 986
Claverie, Jean-Marie, 209
Clavulanic acid, 1143
Clemmensen, Karina, 888, 889f
Climate change, 884
  Arctic methanogens, 890–92, 891f, 892f
  emergence of viral pathogens, 232, eTopic 6.3
  emerging diseases, 1198–99
  global carbon balance, 889–93, 890f–892f
  human attempts to alter or mitigate, 892
  microbial ecosystems, 180, 180f
  radiative forcing, 890, 890f
Clindamycin, 1132f, 1133
Clinical microbiology, 1159–1202
  detecting emerging pathogens, 1193–1200
  epidemiology, 1187–93
  molecular and serological identification of pathogens, 1174–87
  pathogen identification by culture and phenotype, 1166–74
  specimen collection and handling, 1160–66
Clock(s), molecular, 673–75, 674f
Clofazimine, 1145
Clonality, **965**
Clonal selection, **975,** 976f
Cloned human gene products, 647
Cloning, **459–60,** 459f
Clostridia, 715–16, 716f, 717f
Clostridiales, variant sporulation and "live birth," 716, 717f, 720–21
Clostridioides difficile, 716
  antibiotic resistance, 1140
  biological control, 189
  containment breaches, 931–32, 932f
  diarrhea, 1073
  endospore formation, 152f, 714
  fecal transplant, 933
  glycosylation, 303
  Gram stain, 52
  gut microbiome, 1077
  lipiarmycin, 1131
  nutrient sources, 120
  point-of-care rapid diagnostics, 1186t
  probiotics, 933
  pseudomembranous enterocolitis, 931, 932f, 1081t
  recurrent, 1073
  toxins, 1026t
  vancomycin, 89, 598
Clostridium
  anaerobe, 174t
  bovine rumen, 854
  endospore formation, 105
  Gram stain, 52
  gut microbiome, 929
  human colon, 855, 856
  nitrogen fixation, 902
Clostridium acetobutylicum, 715
  cis-antisense RNA, 379, 379f
Clostridium botulinum, 715, 716f
  botulism, 1106–9, 1106f–1108f
  canning, 182
  classification, 704t
  emerging pathogens, 1159, 1159f, 1160
  food contamination, 634, 639, 640t, 641, 641f, 1081t
  integrated virus, 218
  multiplex PCR identification, 1175–76, 1176f
  sporulation, 151
  toxins, 1026t

Clostridium difficile. See Clostridioides difficile
Clostridium pasteurianum, 604
Clostridium perfringens
  gastroenteritis, 1081t
  skin infections, 1060t
  toxins, 1026t, 1027
Clostridium tetani, 715
  classification, 704t
  herd immunity, 999
  spore stain, 52–53, 53f
  sporulation, 151
  tetanus, 1106–9, 1106f, 1108f
  toxins, 1026t
  vaccine antigen, 473t
Clostridium thermoaceticum, 593
Clotrimazole, 1154, 1154f, 1155t
Clp proteases, 305, 305f
ClpS protein, 305
Clustered regularly interspaced short palindromic repeats. *See* CRISPR (clustered regularly interspaced short palindromic repeats)
CMV (cytomegalovirus), 213
  antiviral agents, 1148t
  respiratory tract infection, 1066t
CNO (carbon-nitrogen-oxygen) cycle, 661
CO_2. *See* Carbon dioxide (CO_2)
CoA (coenzyme A), **514,** 514f
Coastal shelf region, 857f, **858**
Coccidioides immitis
  meningitis, 1110t
  pathogen identification, 1184t
  respiratory tract infection, 1066t
Coccidioidomycosis, 1066t
  antifungal agents, 1155t
Coccolith(s), 199
Coccolithophores, 199, 199f, 814, 814f
Coccolithovirus, 199
Coccolithus pelagicus, 786f
Coccus(i), **40,** 41f
Cocoa bean fermentation, 617, 627–28, 628f
Coding strand, 281, 282f
Codon(s)
  codon-anticodon pairing, 290, 290f
  defined, **288**
  genetic code, 289–90, 289f
  start, 282f, 283f, **293**
  stop, 282f, 283f, **289**
  translation, 288, 289–90, 290f, 293
Coenzyme A (CoA), **514,** 514f
Coevolution, **694,** eTopic 17.6
Cofactors, **120**
  electron transport, **544–45,** 544f, 545f
Cohen, Stanley, 459, 459f
Cohen-Bazire, Germaine, 568, 569f
Cold seep endosymbiosis, eTopic 21.2
Cold sores, 436, 1012–13, 1013f
Cold temperature, to slow and preserve microbes, 183
Colitis, 1072
Collins, James, 478, 1178
Collinsella aerofaciens, 54f
Colon, 855–57, 855f, 856f
Colon cancer, 921, 921f, 922, 922f
Colonic crypts, 38, 38f
Colony(ies), **15, 131**
Colony counts, 139–40
Columbus, Christopher, 1086
Colwell, Rita, 14, 22, 23f, 834, 859, 860f, eTopic 1.1
Colwellia, Deepwater Horizon spill, 834, 834f
Comelli, Elena, 935
Commensalism, 845t, **850**
Communication, biofilms, 149–50
Community, **828**
Community-acquired infections (CAIs), 1061
Compatible solutes, **167**
Competence, **386**
Competence stimulation peptide (CSP), 386, 387f
Competition, biosynthesis, 585
Complement, **937,** 957–60, 958f
  adaptive immunity, 993–94, 993f
Complement activation
  pathways, 959–61, 960f
  regulation, 994
Complementary DNA (cDNA), 860–61, 861f
Complement cascade, 959–61, 960f
Complement fragments, 994
Complex medium, **132,** 133t
Compound HT61, 1145
Compound microscope, **47–48,** 48f
Compromised host, **923**
Computational pipeline, **832**
Concatemer, **405**
Concentration gradients, 126, 498–99, 498f
Concentration ratio, 492–93, 492t
Condenser, **47**
Condylostoma magnum, 301
Cone cells, 38, 39f
Confluent growth, **131,** 132f
Confocal laser scanning microscopy, **58,** 58f, eTopic 2.1
Congenital syphilis, **1087**
Conidiophores, 801, 801f, 1013f
Conjugated antibody, 56
Conjugated double bonds, 544
Conjugation, 261, 815, 817f
  defined, **330, 815**
  gene transfer, 330–34, 330f–334f
Conjugative transposons, **345,** 345f, 346t
Conlon, Brian, 137f
Consensus sequence, **279–80**
Constant region, **970–72,** 971f
Constitutive enzymes, 361
Constructed wetlands, 898f, **899**
Consumers, **843,** 843f
Contact dermatitis, 1005, 1006f
Containment breaches and dysbiosis, 931–33, 932f
Contigs, **271,** 271f, **831,** 831f
Continuous culture, **146–47,** 146f
Contractile vacuole, **815**
Contrast, **43, 47**
  bright-field microscopy, 46
  compound microscope, 47
Contreras, Lydia, 185
Controlled atmosphere, food preservation, 642

Control proteins, phage lambda, 406–7
Conus pulicarius, 723f
Convergent evolution, 788
Conway, Tyrrell, 371, 509, 509f
Coomassie blue staining, 468, 469t
Cooperation, biosynthesis, 585, 585f
Copeland, Herbert, 26, 787
Copepods, mutualism, 859, 860f
Coral, mutualism with dinoflagellates, 849, 849f
Coral bleaching, 200, **849**
Coral endosymbionts, 785, 785f, 786, 817, 819f
Coral reef ecosystems, viruses, 200, 200f
Core, Earth, 662, 663f
Coreceptor, **426**
Core genes, 6
Core genome, **692**
Core particle, 203
  HIV, **422–23,** 422f, 423f
Corepressor, **359,** 360f
Corethron criophilum, 813f, 814
Coronaviruses, 212t, 232
Corrosion, anaerobic, 562, 563f, 909–10, 910f
Cortical alveoli, **792,** 792f
Cortinarius armillatus, 889f
Corynebacteriaceae, 704t
Corynebacterium, 721, 724, 725, 725f
  skin microbiome, 925
Corynebacterium diphtheriae, 725
  classification, 704t
  DNA-binding protein, 244, 244f
  indoor air, 911
  integrated virus, 218
  pathogenicity islands, 1017
  pathogen identification, 1184t
  respiratory tract infection, 1066t
  survival within host, 1041, eTopic 25.6
  toxins, 1026t, 1034, eTopic 25.6
Corynebacterium glutamicum, 105
Cotranslational export, 307, 307f
Cottage cheese, 624f
Counterillumination, 381–82, 382f
Counterselection gene, 461–64, 464f
Counterstain, **50,** 52f
Coupled transcription-translation, 299, 300f
Coupled transport, **126–27,** 126f
Covert pathogenesis, 1085
Cowpox, 17–18, 17f, 967f, 968
COX (cyclooxygenase), 954
Coxiella, 1093
Coxiella burnetii, 735
  genomics to probe pathogenesis, 1052
  intracellular immune avoidance, 1042
  pasteurization, 182–83
  thriving under stress, 1042
CPR (Candidate Phyla Radiation), 704t, **705,** 706–8, 707f
C-reactive protein, **959**
Crenarchaeota, 753t, 763, 764f
Crescentin (CreS), bacterial cytoskeleton, 96, 97f
Cresols, 186
Creutzfeldt-Jakob disease (CJD), 207, 207f, 1110, 1111f, 1111t
Crick, Francis, 30, 31f
Crinipellis perniciosa, 636
CRISPR (clustered regularly interspaced short palindromic repeats), 32, 341–43
  anatomy, 341–42, 341f, 342f
  bacterial defense, **219,** 220f
  defined, **341**
  DNA as digital storage, 246
  function, 342–43, 343f
  novel functions, 343, 343f
  rapid diagnostics, 1179, 1180–81
  targeted gene editing, 464–66, 465f
Cristae, 538
Crohn disease
  Bacteroides forming membrane vesicles, 37
  dysbiosis, 932
  granulomas, 957
Cro protein, 407
Crops, bacterial genes save, 472–73, 472f
Cross-bridges, 88f, **89**
Crossover events, allelic exchange, 333–34, 334f, eTopic 9.2
Cross-protection, 968–69, 969f
Cross-species communication, 385–86, 385f, 386f
Crowe, James, 1002, 1002f
Crown gall disease, 330, 331f, 879, 879f
CRP (cAMP receptor protein), stimulation of transcription, 363–64, 364f, 365f, eTopic 10.1
Crust, Earth, 662, 663f
Cryocrystallography, 72
Cryo-electron microscopy (cryo-EM), **66–67,** 67f
  modeling cell parts, 68–69, 68f
  model of cell, 69, 69f
Cryo-electron tomography, 67–69, 68f, 69f
  defined, **67**
  gut bacteria, 37, 37f
Cryomicroscopy, electron. *See* Cryo-electron microscopy (cryo-EM)
Cryophile, 844t
Cryotomography, electron. *See* Cryo-electron tomography
Cryptococcosis, 1155t
Cryptococcus, 1110t
Cryptococcus neoformans, 789t, 796, 804t
  lung infection, 1068, 1068f
  Titan cells, 1057, 1057f, 1058
Cryptogamic crust, **811**
Cryptomonads, 791, 791f
Cryptophytes, 791, 791f
Cryptosporidium, 817, 1077
  transmission from animals, 1200
Cryptosporidium hominis, 1077
Cryptosporidium parvum, 789t, 823, 823f, 1077, 1078f
Crystal violet, 50, 52f
CSF (cerebrospinal fluid), specimen collection, 1161, 1162f
CSP (competence stimulation peptide), 386, 387f
CT (charge transport), DNA repair, 324–25, 324f, 325f
CTLs (cytotoxic lymphocytes), 966
CTXphi, 204
Culex, 1015
Culture-independent sequencing of gut microbiome, 928–29

Culture medium(a), 130
complex *vs.* synthetic, 132, 133t
enriched, **132**
liquid (broth), 130
minimal defined, **132,** 133t
nonselective, 1161
selective and differential, 133–34, 133f, 1161
solid, 130
Culturing bacteria, 130–35
anaerobes, 175–76, 176f
complex *vs.* synthetic media, 132, 133t
culture media, 130
dilution streaking and spread plates, 131–32, 131f, 132f
growth factors and uncultured microbes, 134–35, 134t, 135f, 136–37
pure, **130**
selective and differential media, 133–34, 133f
uncultured, 838
Culturing viruses, 233–36, 1194
batch culture, 233–34, 234f
plaque isolation and assay of animal viruses, 236, 236f
plaque isolation and assay of bacteriophages, 235–36, 235f
tissue culture of animal viruses, 234, 234f
Cupriavidus metallidurans, 844, 844f
Curd, **623**
Cutibacterium, indoor environments, 911
Cutibacterium acnes
doxycycline, 1133
skin microbiome, 925, 925f
CXCR surface marker, 999, 1000f
Cyan fluorescent protein (CFP), 56, 57f
Cyanobacteria, 708–13
akinetes, 711–12, 711t
algal blooms, 865, 865f
biofilms, 150
Calvin cycle, 586
cell envelope, 90–91
cell structure, 710, 710f
circadian clocks, 388–89, 388f, 389f, 392f
classification, 703, 704t
colonial, 712, 712f
communities, 712, 713f
defined, **703**
edible, 619–20, 619f
energy production, 122
environmental effect, 710–11, 711f
eutrophication, 179, 179f
filamentous, 711–12, 711f
fluorescence microscopy, 54, 54f
great Pacific garbage patch, 860
Hormoscilla spongeliae, 1
multicellularity, 712, 713f
nitrogen fixation, 151–52, 153f
oxygenic Z pathway, 575, 576f
photoexcitation and photolysis, 568–72, 569f–572f
phototrophic, 708–10, 709t
phytoplankton, 862
single-celled, 710
Cyanobacterial mats, 846–47, 846f, 847f
Cyanophyta, 708
Cyclic adenosine monophosphate (cAMP), stimulation of transcription, 363–64, 364f, 365f, eTopic 10.1
Cyclic adenosine monophosphate (cAMP) receptor protein (CRP), stimulation of transcription, 363–64, 364f, 365f, eTopic 10.1
Cyclic dimeric guanosine monophosphate (c-di-GMP), 150, 380–81, 381f
Cyclic photophosphorylation, **575**
Cyclization, membrane lipids, 86
Cyclomodulins, 1025, 1026t
Cyclooxygenase (COX), 954
Cyclopentane rings, 87f
Cyclophilin A, 423, 423f
Cyclophilin A inhibitors, 1152–53
Cyclopropane, membrane lipids, 86, 86f
Cycloserine, 1124f, **1128,** 1128f
Cymopolia, 789t
Cypovirus, 68, 68f
Cysteine, 301, 301f
Cystic fibrosis, 947, 947f
Cystic fibrosis transmembrane conductance regulator (CFTR), **947,** 1031
Cystitis, **1083**
Cystoviruses, 212t
Cytochrome(s)
apoptosis, 942
defined, **538**
electron transport system, 538–39, 539f, 546–47, 546f
extracellular, 554–55, 556f
Cytochrome oxidase test, 1170–72, 1171f
Cytochrome system, 173, 173f
Cytokine(s), **949,** 950f, 952
acute inflammation, 953
immune response, 966
isotype class switching, **980**
modulation of immune response, 989–90, 989t
Cytokine profiles, flipping, 1045–46, 1046f
Cytomegalovirus (CMV), 213
antiviral agents, 1148t
respiratory tract infection, 1066t
Cytophaga, 704t
Cytoplasmic extensions, 556, 556f
Cytoplasmic proteins, 77f
export, 307
Cytosine, 243f, 244
Cytoskeleton
alterations, 1025, 1026t
bacterial, 96, 97f
Cytotoxic lymphocytes (CTLs), 966
Cytotoxic T cells (TC cells), 966, **982**
activation, 986–87, 987f
Czjzek, Mirjam, 521f

D

dae gene, 352–53, 354f
Dahlgren, Randy, 899
Dairy products
acid fermentation, 621t, 623–25, 624f–626f
spoilage, 634–35, 635t
Dalgarno, Lynn, 293
Dalia, Ankur, 337

Dalsgaard, Tage, 904
Daly, Michael, 185
Dambuza, Ivy, 1058
DAPI (4',6-diamidino-2-phenylindole), fluorescence microscopy, 54, 56, 56f
Daptomycin, **1129**
Dark-field microscopy, **61–62,** 61f, 62f
Dark reactions. *See* Calvin cycle
Darwin, Charles, 660
Darwin, Erasmus, 660
Darwinolide, 1145
DasSarma, Shiladitya, 168f, 775, 776f
Daughter cells, 141, 141f
David, Lawrence, 1076, 1076f
da Vinci, Leonardo, 443, 443f
Dawadawa, 621t
Dead zone, **866, 894–95,** 896f
  denitrification, 903, 904f
Deaminating agent, mutagenic effects, 318t
Deamination, 501
  mutations arising from, 319, 320f
Death curve, 181, 181f
Death phase, 142f, **144–45**
Death profile, 181, 181f
Death rate, **145**
DEB025, 1152f
de Bruijn graph, 832, 832f
Decarboxylation of amino acids, 501, 522–23, 522f, 524–25
Decay-accelerating factor, **994**
Decimal reduction time (D-value), **181,** 181f
  food preservation, 642
Decolorizer, Gram stain, 50, 52f
Decomposers, **843,** 843f
Decomposition, 869–70, 872f
Deep-branching Gram-negative bacteria, 704t, **705**
Deep-branching taxa, **702**
Deep-branching thermophiles, 704t, **706,** 707f
Deep-Sea Archaeal Group/Marine Benthic Group B (DSAG/MBGB), 755
Deepwater Horizon spill, 833–36, 834f–836f, 840, 858, 895
Defensins, **947–48,** 948f, 948t, 955, eTopic 23.2
Degenerative evolution, 198–99, **673, 683,** 788
Degranulation, **948,** 1003
Degrons, 305
Dehalobacter restrictus, 718–19, 719f
Dehalococcoides, 564, 565f
Dehalorespiration, 564, 565f
Dehydration, food preservation, 641
Deinococcus, 706
Deinococcus radiodurans
  classification, 704t, 706
  extremophile, 160
  resistance to ionizing radiation, 185, 185f, eTopic 9.2
Deinococcus-Thermus, 704t, **706**
Delavirdine, 1151
Delayed-type hypersensitivity (DTH), 1001t, **1005,** 1006f
Deletions, **316,** 317, 317f
DeLong, Edward, 763, 764f, 859–60
Delta endotoxin, 715
Deltaproteobacteria, 735–37
  bdellovibrios, 736–37, 737f
  classification, 704t, 705
  defined, **705**
  metabolism, 728t
  myxobacteria, 736, 736f
Deltaretrovirus, 421t
DeMontigny, Bree, 770
Denaturation, **244–45**
Dendritic cells, 938f, **939,** 939f, 942, 955
Dengue hemorrhagic fever, 1016
  climate change, 1198
  point-of-care rapid diagnostics, 1186t
DeNiro, Julia, 638, 638f
Denitrification, **124**
  dissimilatory, **552**
  nitrogen cycle, 901f, **903,** 904f
De novo assembly, 832
Density-dependent light production, 382–84, 382f, 383f
Dental caries, 926
Dental procedures, bacteremia, 926
Deoxyribonucleic acid. *See* DNA
DePas, William, 59, 59f
Depolarization, 127
Depth of field, **48**
Derepression, **359,** 360f
DeRisi, Joseph, 1176
Dermatitis, 1005, 1006f
Dermatophytes, 1060t, 1154
Dermatophytosis, 1060t
  antifungal agents, 1155t
Desensitization, **1005**
Desulfobacter, 910
Desulfobacterales, 704t
Desulfobacterium autotrophicum, 593
Desulfobulbus, 557
Desulfovibrio, 174t
  iron cycle, 909
  metabolism, 728t
  sulfur cycle, 905
  syntrophy, 528, 529f, 529t
Desulfuration, 906
Desulfurococcales, **752,** 753t, **756–57,** 757f
Desulfurococcus, 757, 769
Desulfurococcus fermentans, 753t, 757, 757f, 757t
Detection *vs.* resolution, **39,** 39f
DETECTR (DNA endonuclease-targeted CRISPR trans reporter), 1180–81, 1181f
Detergent, industrial microbiology, 648t
Detritus, **843**
Deubiquitylases, 1048, 1049f
DGCs (diguanylate cyclases), 380–81
DHAP (dihydroxyacetone phosphate), 508f
d'Herelle, Félix, 189, 197
Diabetes, type I, 1007t
Diagnosis, 1058–59
4',6-Diamidino-2-phenylindole (DAPI), fluorescence microscopy, 54, 56, 56f
Diapedesis, 949
Diapers, industrial microbiology, 648t
Diapherotrites, 753t
Diaphorobacter, indoor air, 911

Diarrhea, 1072–77
antibiotics, 1073
benefit to pathogen, 1031–32
enterohemorrhagic *E. coli,* 1073–75, 1075f, eTopic 26.5
and gut microbiome, 1076–77, 1076f
inflammatory, 1072
motility-related, 1072
osmotic, 1072
point-of-care rapid diagnostics, 1186t
protozoan causes, 1077
rehydration therapy, 1072–73
rotavirus and norovirus, 1075–76
secretory, 1072
traveler's, 1081t
types, 1072
Diatom(s), **792,** 812, 812f, 813–14, 813f
Diauxic growth, **365,** 365f
DIC (differential interference contrast microscopy), 63, eTopic 2.2
Dichotomous key, **694,** 1169–70, 1171f, eTopic 17.5
Dictyostelium, 789t
Dictyostelium discoideum, 151, 806–7, 807f
Dideoxy chain termination method, 448, 448f, 450
Dideoxynucleotide, 448, 448f
2',3'-Dideoxyribonucleotide, 448, 448f
Diderm, 703, **714**
Didinium, 815, 816f
Dietrich, Lars, 555
Differential, **1058**
Differential interference contrast microscopy (DIC), 63, eTopic 2.2
Differential media, **133–34,** 133f
Differential stains, **50–54,** 51f–53f
Differentiation, bacterial cell, 103–5, 103f, 104f
Diffusion
facilitated, **125–26,** 125f
passive, 84
and transport, 498–99, 498f
Digestibility of fermented foods, 620
Digestive enzymes, 125
Digestive methanogenic symbionts, 770–71, 770f, 772–73
DiGiandomenico, Antonio, 1146f
Diglycerol tetraether, 87f
Diguanylate cyclases (DGCs), 380–81
Dihydrouridine, 292f
Dihydroxyacetone phosphate (DHAP), 508f
Di Leonardo, Roberto, 443, 443f
di Lodi, Agostino Bassi, 12, 14
Dilution streaking, **131,** 131f
Dimethylsulfoniopropionate (DMSP), 905
DinG protein, 325, 325f
Dinitrogen, 901
Dinoflagellata, 789t, 816–17, 818f, 819f
Dinoflagellates, 816–17, 818f, 819f
defined, **792**
mutualism with coral, 849, 849f
secondary algae, 808
Diotomaceous earth, 813
Dioxin (TCCD) catabolism, 502f, 503
Diphosphatidylglycerol, 85–86, 86f
Diphtheria, 1066t
tetanus, acellular pertussis vaccine, 998t
Diphtheria toxin, 1026t, 1034, eTopic 25.6
Diphtheria vaccine, 968t, 998t
Diplococci, **731**
Direct count, 135–39, 138f, 139f
Direct repeats, 348, 349f, 350f, **393**
Disaccharides, 501
Discoba, 789t, 792
Discosphaera tubifera, 814, 814f
Disease
categories, 693
vs. infection, 937, 1012
Disease surveillance, molecular approaches, 1191–92
Disease transmission, 15
causative agent, 16–17, 16f
Disease trends, 1190, 1191t
Disinfectants, 185–88
commercial, 186–87
resistance, 187–88, 188f
Disinfection, **181**
chemical agents, 185–88, 186f, 188f
Dissimilation, **842**
Dissimilatory denitrification, **552**
Dissimilatory metal reduction, **552–53,** 553f
Dissimilatory nitrate reduction, **903–4**
Dissimilatory nitrate reduction to ammonia (DNRA), 901f, 903–4
Disulfide bond protein (DsbA), 77f
Divergence
genome evolution, 349, 350f
of three domains of life, 677–78, 677f, 679t
through mutation and natural selection, 673
Diversity, preliminary screen, 830–31, 833–34
Divisome, septation, **98,** 99f
DivJ protein, 103, 104f
DIY (do-it-yourself) synthetic biology, 479–80, 481f
DMSP (dimethylsulfoniopropionate), 905
DNA, 30, 240
cell structure, 77f, 78
complementary, 860–61, 861f
denaturation, 244–45
as digital storage, 246–47, 246f, 247f
exchange, 333
imported, 329
isolation and purification, 444–45, 830, eTopic 12.2
naked, 337–38, 337f, 338f
noncoding, 242–43, 243f
recombinant, 9t, 32–33, 458, 459–61, 459f–461f
renaturation, 245
vs. RNA, 245
structure, 30–31, 31f
structure and function, 243–45, 243f, 244f
supercoils and supercoiling, 247–50, 248f–250f, 257–58
DnaA, 253f, 254
DNA-binding proteins, 77f, **100,** 100f, 244, 244f, 254
DNA-bridging protein H-NS, 77f
DNA-bridging protein HU, 77f
DNA charge transport, 324–25, 324f, 325f
DNA-dependent RNA polymerase, **278**
DNA detection tests, 1175–76, 1176f, eTopic 28.2, eTopic 28.3

DNA endonuclease-targeted CRISPR trans reporter (DETECTR), 1180–81, 1181f
DNA gyrases
supercoiling, 249, 250f
thermophiles, 164
DNA helicase, 253f, 254, 256f
DNA hybridization, fluorophores, 56
DnaJ chaperone, 304
DnaK chaperone, 304, 304f
DNA ligase, **257,** 258f, 459, 459f
DNA methylation, 252–54, 253f
DNA sequencing to detect, 452–55, 453f, 454f
DNA microarrays, 466t, 467
DNA modification, 340–41, 340f, 340t
DNA organization, bacterial cell division, 98–100, 99f, 100f
DNA photolyases, 323
DNA polymerase(s)
DNA organization, 100
DNA replication, 252, 253f, 254–57, 256f–257f
"sloppy," 327, 328f
DNA primase, 253f, **254,** 256f, 257
DNA rearrangements, altered gene expression, 392–95, 394f, eTopic 10.4
DNA regulatory sequences, 359, 359f
DNA repair, 322–29
base excision, 322t, 323–26, 326f
error-prone, 322, 327–28, 327f, 328f
error-proof, 322–27, 323f, 326f
homologous recombination, 322t, 326–27, 327f
methyl mismatch, 322–23, 322t, 323f
nonhomologous end joining, 322t, 328, 329f
nucleotide excision, 322t, 323, eTopic 9.1
photoreactivation, 322t, 323, eTopic 9.1
SOS ("Save Our Ship"), 327–28, 328f
translesion bypass synthesis, 322t
types, 322, 322t
using electrons, 322, 324–25, 324f, 325f
DNA-replicating enzyme, 3
DNA replication, 251–60
elongation, 251, 255–58, 256f–258f
fluorophore labeling, 56, 57f
initiation, 251, 252–55, 253f–255f
overview, 251–52, 251f, 252f
plasmids, 261–63
regulation of cell division, 100–102, 101f, 102f
rolling-circle, 262, 262f
secondary chromosomes, 263–65, 264f, 265f
semiconservative, 251, 251f
speed, 258
termination, 251, 258–60, 259f, 260f
DNA replication errors, repair, 322–23, 323f
DNA replication inhibitors, 1124f, 1125t, 1129–31, 1130f
DNA restriction, 340–41, 340f, 340t
DNA reverse-transcribing viruses, 211f, 212t, **213,** 230–32, 231f
DNA revolution, microbial genetics and, 30–33, 31f, 32f
DNase I footprinting, 469t
DNA sequencing, 448–49, 448f
annotation, 455–57, 455f, 456f
to detect DNA methylation, 452–55, 453f, 454f
dideoxy chain termination method, 448, 448f
history, 5–6, 5f, 6f, **30–31**
methods, 454–55, 454t
next-generation, 449–50, 449f, 830, 831f, 1176
at remote locations (nanopore), 451–52, 452f, 453f
by synthesis, 448–49, 449f
third-generation (single-molecule), 450–52, 451f–453f
DNA shuffling, 474, eTopic 12.11
DNA synthesis, antiviral agents that target, 1150, 1150f
DNA transcription, 99–100, 100f
DNA transfer, membrane vesicles, 106
DNA vaccines, 472, eTopic 12.8
DNA viruses, 20
double-stranded, 211, 211f, 212t
examples, 435, 436t
molecular biology, 435–40
replication, 226–27, 226f, 227f, eTopic 26.1
single-stranded, 211, 211f, 212t
DNRA (dissimilatory nitrate reduction to ammonia), 901f, 903–4
Do-it-yourself (DIY) synthetic biology, 479–80, 481f
Domagk, Gerhard, 1118, 1118f
Domains
cell, 85
DNA, 99–100, 100f
DNA-binding, **359**
of life, **677–78,** 677f, 679t
lipid, 100f
protein, **99–100,** 100f
taxonomic, 100f
Dominguez-Bello, Maria, 1139
Dong, Minh, 1159, 1159f, 1160
Doohan, Douglas, 638
Doolittle, Ford, 680
Dormancy, 145, 145f
Dorrenstein, Pieter, 602–3, 602f
Dotiwala, Farokh, 943
Double helix, 30–32, 31f, 32f
Double-strand breaks, repair, 326–27, 327f
Double-stranded DNA viruses, 211, 211f, 212t
Double-stranded RNA viruses, 211, 211f, 212t
Doubling time, **141–43,** 142f
Doudna, Jennifer, 32, 32f, 341, 341f, 1180
Downstream processing, **649–50,** 651f
Doxey, Andrew, 1159, 1160, 1160f
Doxycycline, 1132f, 1133
DPANN, 753t, 755, 780–82, 780f–782f
Drake, Tom, 1057, 1058, 1058f
Drosophila, endosymbiosis, 695
Drug resistance. *See* Antibiotic resistance
Drug susceptibility
automated determinations, 1122
correlation of MIC with tissue level, 1122–23, 1123f
Kirby-Bauer disk susceptibility test, 1121–22, 1121f, 1122t
measurement, 1119–23
minimal inhibitory concentration, 1119–21, 1120f, 1122–23, 1123f
DSAG/MBGB (Deep-Sea Archaeal Group/Marine Benthic Group B), 755
DsbA (disulfide bond protein), 77f
DTH (delayed-type hypersensitivity), 1001t, **1005,** 1006f
Dual RNA sequencing, 1017, 1052–53, 1052f
Dubey, Gyanendra, 159f

Duchesne, Ernest, 1117
Dulbecco, Renato, 210, 236
Duplication, **316,** 317f, 348–49, 349f, 350f
Dust mites, 1004, 1004f
D-value (decimal reduction time), **181,** 181f
  food preservation, 642
Dysbiosis
  containment breaches and, 931–33, 932f
  defined, **931**
  obesity and, 933–36, 933f–935f
Dysentery, bacillary, 1073

E

E3 ligase, 1047, 1049f
*E*a (activation energy), **499,** 499f
Early genes, bacteriophage, 217
Early log phase, 142f, **143–44**
Earth
  atmosphere, 662
  elemental composition, 662, 663f
  temperature, 663
Eastern equine encephalitis (EEE) virus, 1016, 1109, 1110t
Ebola virus, 204, 204f
  ELISA, 1181–82, 1182f
  epidemiology, 1189f
  immunology-based pathogen detection, 1179–80
  perfect pathogen, 1102–4, 1103f
  tracking, 1195
  tropism, 223
  virulence, 1013, 1014f
Ebola virus vaccine, 1001, 1002–3
Echinocandins, 1155t
Eclipse period, batch culture, **233,** 233f, 234
Ecological categories, 693
Ecological niche, 691
Ecology. *See* Microbial ecology
Ecosystems, **828**
  carbon cycling by different, 887–89, 888f, 889f
  global warming, 893
  limiting factors, 910–11
  marine, 858–59, 858f–860f
  support by environmental microbes, 21–22, 21f–23f
Ecotype, 691
Ectomycorrhizae, **873,** 873f
Ectoparasite, 1012, 1013f
Ectosymbionts, 701, 701f
Edema, **1004**
Edema factor (EF), 1026t, 1032, 1033f
Edible algae, 619–20, 619f
Edible cyanobacteria, 619–20, 619f
Edible fungi, 618–19, 619f
ED (Entner-Doudoroff) pathway, **506,** 506f, **509,** 509f, 510f
EDTA (ethylenediaminetetraacetic acid), 80
Edwards, Katrina, 562
Edwards, Rob, 214
EEE (eastern equine encephalitis) virus, 1016, 1109, 1110t
EF (edema factor), 1026t, 1032, 1033f
EF (elongation factor), 296–97, 296f, 298, 298f
EFDG bacteriophage, 190–91, 190f, 191f
Effectors, toxins, 1036–40, 1037f–1040f, 1037t
Egert, Marcus, 914–15, 915f
EHEC (enterohemorrhagic *E. coli*). *See Escherichia coli* O157:H7
E (eluviated) horizon, 868f, 869
Ehrlich, Garth, 316
Ehrlich, Paul, 1118–19
Ehrlichia equi, 1100
EIEC (enteroinvasive *E. coli*), 1074
Electrical engineering, synthetic biology, 476
Electrically conductive pili, 553–54, 554f, 555f
Electrical potential, 541–42, 542f
Electrochemical potential, **123**
Electrogenic biofilms, **553**
Electromagnetic radiation, **42,** 42f
  mutagenic, 318t, 320, 321f
Electromagnetic spectrum, 42, 42f
Electron(s), to find DNA damage, 322, 324–25, 324f, 325f
Electron acceptor(s), **494, 535**
  anaerobic respiration, 551, 551f
  anoxic or low-oxygen environments, 552–53, 553f
  concentration, 537–38
  ecosystem, 843–44, 843t, 844f
  terminal, **497**
Electron beams, food irradiation, 184
Electron bifurcation, **772**
Electron cryomicroscopy. *See* Cryo-electron microscopy (cryo-EM)
Electron cryotomography. *See* Cryo-electron tomography
Electron donor(s), **494, 535**
  anaerobic respiration, 551, 551f
  concentration, 537–38
Electroneutral coupled transport, 127
Electron hopping, 554
Electron microscopes and microscopy (EM), 63–66
  cryo-, 66–69, 67f
  defined, **29, 40, 63**
  history, 29–30, 29f
  physics, 63–64, 64f
  range of resolution, 40
  sample preparation, 65–66, 66f
  scanning, 40–42, 41f, **64,** 64f, 65f
  transmission, 41f, **42, 64,** 64f, 65f
Electron shuttles, **555,** 556f
Electron tower, 497, 497t, 536–37, 536t
Electron transfer, 493–500
Electron transport chain (ETC). *See* Electron transport system (ETS)
Electron transport system (ETS), **497,** 535–57
  anaerobic respiration, 551–53
  concentrations of donors and acceptors, 537–38
  cytochromes, 538–39, 539f
  defined, **534**
  donors and acceptors, 535
  energy storage, 536–38, 536t
  environmental modulation, 538, eTopic 14.1
  functioning within membrane, 538–39, 538f, 539f
  glucose fermentation and respiration, 518, 519f
  mitochondrial respiration, 548–49, 589f
  nanowires, shuttles and fuel cells, 553–57
  oxygen, **173,** 173f
  photolysis, 571–72
  proton motive force, 539–43, 540f–543f
  reduction potential, 536–37, 536t
  respiratory, 544–51

Electrophoresis, **79**
Electrophoretic mobility shift assay (EMSA), 469t
Eleftheria terrae, 136, 137f
Elemental composition, Earth, 662, 663f
Elemental cycles, 883–911
  carbon, 885–93
  experimental measurement, 886, 887f
  hydrologic, 893–99
  iron, 908–10
  nitrogen, 899–905
  other metals, 910
  phosphate, 907–8
  sulfur, 905–6
Elementary body, 711t, **743,** 743f
Elements of life, 661–63, 662f, 663f
Elephantiasis, 695–96, 695f, 1012, 1013f
ELISA (enzyme-linked immunosorbent assay), 1181–82, 1182f
Elite controllers, 1152
Elliot, Marie, 154
El-Naggar, Moh, 553, 556
Elongation
  replicating DNA, 251, 252f, 255–58, 256f–258f
  transcription, 282, 282f–283f, 283
  translation, 295–96, 295f
Elongation factor (EF), 296–97, 296f, 298, 298f
Elowitz, Michael, 468
Eluviated (E) horizon, 868f, 869
EM. *See* Electron microscopes and microscopy (EM)
Embden-Meyerhof-Parnas (EMP) pathway, **506–9,** 506f–508f
Emerging clades, **692–93,** 704t, **706–8,** 707f
Emerging pathogens and diseases, 1193–1200
  acute flaccid myelitis, 1197
  climate change, 1198–99
  Clostridium botulinum, 1159, 1159f, 1160
  defined, **1193**
  discovery, 1194
  map, 1194–95, 1194f
  One Health Initiative, 1199–1200, 1199f
  technology helps, 1197–98
  tracking, 1195, eTopic 28.6
  Zika virus, 1195–97, 1196f
Emiliana huxleyi, 199, 199f, 814
Emission wavelength, fluorescence microscopy, **54–55,** 55f
Emmentaler cheese, 622, 622f, 624f
Empiric therapy, **1062,** 1140, 1141
EMP (Embden-Meyerhof-Parnas) pathway, **506–9,** 506f–508f
Empty magnification, **45**
EMSA (electrophoretic mobility shift assay), 469t
Enagrostis tef, 631
Encephalitis, 1109, 1110t
Encephalitozoon, 789t
Encephalitozoon intestinalis, 804, 804t, 823
Encephalopathies, spongiform, 1110, 1111f
Endemic, **1188–89,** 1189f, 1190f, 1196
Endemic typhus, 1102t
Enders, John F., 234
Endocarditis, **1093–96,** 1093f
  subacute bacterial, 926, 927f, 1093
Endocytic vesicle, influenza virus, **415,** 416f
Endocytosis, 130, eTopic 4.2
  genome uncoating, 224f, **225**
  influenza virus, **415,** 416f
Endogenous retrovirus, 218t, 219, **421, 431–35,** 432f
Endogenous virus, **197**
Endoliths, **662,** 663f, 844t, **869**
Endomycorrhizae, **873–74,** 874f
Endonucleases, restriction, **340–41,** 340f, 340t
Endoparasite, 1012, 1013
Endopeptidases, 305
Endophytes, **639**
Endoproteases, 305
Endopytes, plant, **875–76,** 875f
Endoribonucleases, 952
Endosomes, 224f
Endospores, 711t
  bacillales, 714–15
  clostridia, 715–16
  early research, 13
  Firmicutes, **705,** 712, 714–16
  formation, 151, 152f
  growth asymmetry, 105–6
  variant sporulation, 716, 717f
Endosymbionts, **22**
  alphaproteobacteria, 730, 731f, 732f
  coral, 785, 785f, 786, 817, 819f
  evolution, 694
  photosynthetic, 848–49, 849f
Endosymbiosis, **22–24,** 23f
  classification of eukaryotes, **788**
  cold seep, eTopic 21.2
  defined, **694**
  evolution of, 694–96, 695f
  evolution of eukaryotes through, 26–27, 27f
  insects with intracellular bacteria, 849–50, 850t, 851f
  mitochondria and chloroplasts, 696–98, 696f–698f
Endothermic reaction, **490**
Endotoxins, 1034–35, 1035f, 1035t, 1036f
  defined, **1025**
  vs. exotoxins, 1034, 1035t
  proteobacterial cell envelope, **93**
Energy, 487–93
  activation, **499,** 499f
  biosynthesis, 583
  defined, **487**
  enthalpy and entropy in metabolism, 490–91, 490f, 491f
  Gibbs free energy change, 488–90, 489f
  reaction conditions, 491–93, 492f, 492t
  solar, 488, 488f
  sources, 487–88, 487t
  used by microbes to build order, 488, 488f
Energy carriers, 493–500
  ATP, 494–95, 494f, 495t
  defined, **494**
  FAD, 496f, 498
  NADH, 496–98, 496f, 497t
  use of different, 498
Energy production, 121–22

Energy storage, 122–23, 123f, 536–38
  concentration gradients, 498–99, 498f
  concentrations of donors and acceptors, 537–38
  reduction potential, 536–37, 536t
Engineering, synthetic biology, 476
Enhancers, **242,** 368–69, 369f
Enoki mushrooms, 619
Enriched medium, **132**
Enrichment culture, **21, 840**
Entamoeba, 1130
Entamoeba coli, 821
Entamoeba hartmanni, 929
Entamoeba histolytica, 821, 823, 823f
  classification, 789t
  diarrhea, 1077, 1078f
  genome sequencing, 806
Enteric bacteriophage, 402–10
Enteritis, 1072
Enterobacter, 734
  antibiotic resistance, 1142
Enterobacter aerogenes, circadian rhythm, 357, 357f, 389
Enterobacter agglomerans, 1199
Enterobacteriaceae, 734–35, 734f
  pathogen identification, 1169f, 1170
Enterobacteriales, 704t
Enterochelin, 128, 129, 129f
Enterococcus, 718
  genitourinary tract microbiome, 927
  Gram stain, 52
  gut microbiome, 1077
  indoor environments, 911
  sepsis, 1099
  systemic illness, 1102
Enterococcus faecalis, 718
  emerging pathogens, 1159, 1159f, 1160
  Gram stain, 52
  multidrug resistance, 1139
  phage therapy, 190
Enterococcus faecium, antibiotic resistance, 1142
Enterocolitis, 1072
  pseudomembranous, 931, 932f, 1081t, 1186t
Enterohemorrhagic *E. coli* (EHEC). *See Escherichia coli* O157:H7
Enteroinvasive *E. coli* (EIEC), 1074
Enteromorpha, interspecies communication, 386, 386f
Enteropathogenic *E. coli* (EPEC), 1074
  DNA organization, 99, 99f
  toxins, 1039, 1039f
Enterotoxigenic *E. coli* (ETEC), 1074, 1081t
  diagnosis, 1058
  toxins, 1026t, 1027f, 1031
  vaccine antigen, 473t
Enterotoxins, 1026t, 1027f, 1031
Enteroviruses, wastewater, 897
Enthalpy, **489,** 490–91
Entner-Doudoroff (ED) pathway, **506,** 506f, **509,** 509f, 510f
Entropy, **488, 489,** 490–91
Entry inhibitors, 1151
Envelope, 87–96
  bacterial, 90–96, 91f–96f
  defined, **78**
  firmicute (Gram-positive), 90, 91–92, 91f, 92f
  mycobacterial, 90, 95, 96f
  proteobacterial (Gram-negative), 90, 91f, 92–96, 93f–95f
  structure, 88–90, 88f–90f
  synthesis, 416–18
  viral capsid, **203,** 203f, 210
Environment(s), 20–26, 21f–25f
  built, **885,** 911–13, 912f, 913f, 914–15
  food-borne pathogens, 638–41, 638f, 639f, 640t, 641f
  strongly selective, 684–85, 684f, 685f
Environmental classification, 161, 161t
Environmental influences on microbial growth, 159–94
  antimicrobials, 180–82, 188–89
  biological control, 189–91
  chemical agents, 185–87
  extremophiles, 160–61
  nutrient deprivation and starvation, 176–80
  osmolarity, 161t, 167–68
  oxygen, 161t, 173–76
  pH and hydroxide ion concentrations, 161t, 168–73
  physical agents, 182–85
  pressure, 161t, 165–66
  temperature, 160–65, 161t
Environmental microbes, support of ecosystems, 21–22, 21f–23f
Environmental sample, 693
Environmental Sample Processor, 761, 761f
Environmental stress, 178–79
  membrane lipid diversity, 85–87, 86f, 87f
Enzyme(s)
  catalysis of metabolic reactions, **499–500,** 499f, 500f
  digestive, 125
  industrial microbiology, 649t
  modular, **597**
Enzyme cofactors, 79
Enzyme-linked immunosorbent assay (ELISA), 1181–82, 1182f
Eosinophils, 938f, **939**
EPEC (enteropathogenic *E. coli*), 1074
  DNA organization, 99, 99f
  toxins, 1039, 1039f
Epidemic, **1189,** 1189f, 1190f
Epidemiology, 1187–93
  bioterrorism, 1192–93, 1193f
  case history, 1187
  defined, **1161, 1187**
  disease trends, 1190, 1191t
  early-warning systems, 1188
  endemic, epidemic, or pandemic, 1188–89, 1189f, eTopic 26.4
  history, 1188, 1188f
  molecular approaches for disease surveillance, 1191–92
  overview, 1187–88
  patient zero, 1189–90, 1190f
  SARS, 1190, 1190f
Epidermis, **925**
Epidermophyton, 1060t
Epifluorescence microscopy. *See* Fluorescence microscopy
Epinephrine, 1003
Epitope, **965,** 965f, 983
  nonself, 1007
EPS(s) (exopolysaccharides), **148**
EPS(s) (extracellular polymeric substances), **148**
Epsilonproteobacteria, 704t, **705,** 737, 737f
Epsilonretrovirus, 421t

Epstein-Barr virus, 436t
  infectious mononucleosis, 201
  lymphoma, 229
  molecular evolution, 213
  point-of-care rapid diagnostics, 1186t
Epulopiscium fishelsoni, 678, 716, 717f
Equilibrium density gradient centrifugation, 445, eTopic 12.2
Ergosterol, 794
Error-prone repair, 322t, **327–28**
  nonhomologous end joining, 322t, 328, 329f
  SOS ("Save Our Ship"), 327–28, 328f
Error-proof repair, 322–27
  defined, **322**
  DNA replication errors (methyl mismatch), 322–23, 322t, 323f
  double-strand breaks (homologous recombination), 322t, 326–27, 327f
  replacement of damaged bases (base excision), 322t, 323–26, 326f
  types, 322t
  UV damage (photoreactivation), 322t, 323, eTopic 9.1
Erwinia, 734–35, 879
Erwinia carotovora, 676, 734–35
Erysipelas, 1060t, 1132
Erythema migrans, **1100,** 1101f
Erythromicrobium ramosum, 729
Erythromycin, 297, 299f, 597–98, 597f, 1133
Erythromycin D, 1132f
Erythrose 4-phosphate, 583
Escherichia coli, 2, 3f, 732, 734
  ABC transporter, 127
  alternative sigma factor, 370–71, 371f
  amino acid sharing, 585, 585f
  ampicillin, 94
  anaerobic respiration, 551, 551f
  antibiotic resistance, 682–83, 682f, 1127, 1139, 1143, 1143f
  asymmetrical cell division, 103, 105
  ATP synthase, 123
  bacteriophage, 196, 197–98, 197f, 206f, 235, 235f
  barophile, 166
  benefits, 931
  biochemical composition, 78–80, 79t
  biofilm differentiation, 148–49, 149f
  biofilms, 1084, eTopic 26.6
  building better vaccine, 999–1000, 1000f
  cell differentiation, 151
  cell envelope, 90
  cell structure, 77, 77f, 78
  cell wall growth, 90, 90f
  cephalexin, 188f
  chemotaxis, 395–96, 396f
  classification, 705
  cloning for gene expression of insulin, 460
  cotranslational export, 307, 307f
  culture medium, 132, 133
  defensins, 948, 948f
  diagnosis, 1058
  diarrheagenic strain, 1074, eTopic 26.5
  disease category, 693
  DNA supercoiling, 247, 249f
  doubling time, 141
  electron transport system, 537, 538, 546–47, 546f, 547f, 548
  enterochelin, 128, 129, 129f
  enterohemorrhagic. *See Escherichia coli* O157:H7
  enteroinvasive, 1074
  enteropathogenic, 99, 99f, 1039, 1039f, 1074
  enterotoxigenic, 473t, 1026t, 1027f, 1031, 1074, 1081t
  evolution of aerobic citrate catabolism, 687, 688f
  exopolysaccharides, 148
  eye microbiome, 925
  facilitative diffusion, 125, 125f
  facultative microbe, 174t
  fermentation, 515, 515f
  flagella, 111
  fusaric acid, 1138
  genome, 241t, 242
  genome evolution, 329, 348
  Gram stain, 51
  granzyme B, 943
  group translocation, 130f
  gut microbiome, 856, 928, 929, 1076, 1077
  heat-shock sigma factor, 389–90, 393f
  horizontal gene transfer, 339, 678, 679, 1019
  inflammatory diarrhea, 1072
  intercellular nanotubes, 107
  mesophile, 162
  microbial attachment, 1020, 1021, 1021t
  microbial endosymbiosis, 23, 23f
  nutrient supplies, 120
  osteomyelitis, 1063
  painting with, 443, 443f
  parasitism, 737, 737f
  pathogenicity islands, 1017
  pathogen identification, 1168, 1168f, 1169, 1169f, 1173, 1175f
  pH, 169, 170f, 171–72, 172f
  phage lambda, 402–10, 402f
  pH lowering, 642, 642f
  plant vascular system, 1199
  plasmid, 331–34, 332f–334f
  portals of entry, 1016
  promoters, 279, 280f
  replication rate, 120
  sacculus, 88, 88f
  sepsis, 1099
  septation, 98
  sequencing to detect DNA methylation, 452–55, 454f
  sexual recombination, 330, 330f
  size, 4t, 40f
  small RNA, 378, 378f
  starvation, 177, 177f, 178f
  stationary phase, 144, 144f
  strain O157:H7, 52
  synthetic biology, 475, 476, 477f
  systemic illness, 1102
  toxins, 1026t, 1027f, 1031, 1037t, 1039, 1039f
  transcriptional attenuation, 373–74, 373f
  transmission from animals, 1200
  ultrasound gut bacteria detection, 277, 277f
  uropathogenic, 1074, 1084–86, 1085f
  viability signals, 963, 963f

Escherichia coli O157:H7, 734, 1073–75
case study, 1073–74
cell envelope, 93
epidemiology, 1075
fermentation, 515, 515f
food contamination, 639, 640t, 641, 1074–75, 1075f
food irradiation, 184
gastroenteritis, 1081t
One Health Initiative, 1199
phage therapy, 189
plant endophytic community, 876
survival within host, 1041
technology-aided reemergence, 1198
toxins, 1033–34, 1039, 1074
virulence factor, 220
E site (exit site), **295**
ESKAPE pathogens, 1141–42
Essential gene, **263–64,** 265f
Essential nutrients, **120**
Esters, food preservation, 643
ETEC. *See* Enterotoxigenic *E. coli* (ETEC)
Ethambutol, 646–47, 646f
Ethanol, 186f
alcoholic fermentation, 631–32, 631f, 632f
Gram stain, 50, 52f
Ethanolic fermentation, **512,** 620, 621t, 629–33
alcoholic beverages (beer and wine), 631–33, 631f–633f
bread dough, 621t, 630–31, 630f, 631f
defined, **622**
Ether-linked isoprenoid membranes, 748–50, 748f
Ethylenediaminetetraacetic acid (EDTA), 80
Ethylene oxide gas (EtO), 186, 186f
ETS. *See* Electron transport system (ETS)
Ettema, Thijs, 751, 752f
Eudistoma, 723f
Eudorina, 810
Eugenol, food preservation, 643
Euglena, 789t
Eukarya, 4, **5**
divergence, 677–78, 677f, 679t
Eukaryotes, 2, 3f, **4, 5,** 785–825
Amoebozoa (amebas and slime molds), 789t, 791, 805–8
chromosomes and genomes, 266–67
classification, 26, 788–89
emerging, 793
evolution through endosymbiosis, 26–27, 27f, 788
Excavata, 789t, 821–23
genome, 788
historical overview, 786–88, 787f
light microscopy, 39–40, 40f
Opisthokonta (fungi and metozoan animals), 789t, 790–91, 790f, 793–805
other protist clades, 792–93, 792f
overview, 786, 786f
phylogeny, 786–93, 787f, 788f, 789t
Plantae or Archaeplastida (primary endosymbiotic algae and plants), 789t, 791–92, 791f, 808–15
protein secretion, 309, 309f
SAR (Stramenopiles, Alveolata, Rhizaria), 789t, 815–21
transcription, 287–88, 288f
transcription regulators, 368–69, 369f
translation, 299–300
Eukaryotic cells, common traits, 76–77
Eumycota, 798–803
Ascomycota, 789t, 799–801, 800f, 801f
Basidiomycota, 789t, 801–3, 802f, 803f
Chytridiomycota, 789t, 791, 798, 799f
classification, 789t, 791
defined, **790**
Glomeromycota, 789t, 803, 803f
Zygomycota, 789t, 798–99, 800f
Euphotic zone, 857f, **858**
Euprymna scolopes, 381–82, 382f, 848
Europa, 918, 918f
Euryarchaeota, 766–80, 766f
classification, 753t, 754–55, 766, 766f
Haloarchaea, 774–78
methanogens, 754–55, 755f, 766–74, 768t
Thermococcales and Archaeoglobales, 778–79, 779f
Thermoplasmatales, 779–80, 779f
Eutrophication, **179,** 179f, **865**
Evolution, 673–81
antibiotic resistance, 682–83, 684–85, 684f, 685f
convergent, 788
divergence of three domains of life, 677–78, 677f, 679t
divergence through mutation and natural selection, 673
experimental (laboratory), **685–89,** 686f, 688f
from first cells, 672
gene transfer, 329
genome, 348–54
growth cycles, 689–90, 689f, 690f
horizontal gene transfer, 678–80, 680f
industrial, 689, eTopic 17.3
microbial, 316
molecular clocks based on mutation rate, 673–75, 674f
phylogenetic trees, 675–77, 675f, 676f, eTopic 17.2
reconciling vertical and horizontal gene transfer, 680–81, 681f
reductive (degenerative), 198–99, **673, 683,** 788
Excavata, 789t, 821–23
Excitation wavelength, fluorescence microscopy, **54–55,** 55f
Exfoliative toxins, 1026t, 1061
Exit site (E site), **295**
Exocytosis, inhibition, 1026t, 1027
Exon(s), **457**
Exonuclease, **255,** 257
Exopolysaccharides (EPSs), **148**
Exothermic reaction, **490**
Exotoxins, 1025–34
categories, 1025–27, 1027f
characteristics, 1025, 1026t
defined, **1025**
deploying, 1036–40, 1037f–1040f, 1037t
diphtheria, 1034, eTopic 25.6
discovery of new, 1034, eTopic 25.7
vs. endotoxins, 1034, 1035t
membrane disruption, 1027–30, 1030f
neutrophil extracellular traps, 1028–29, 1028f, 1029f
two-subunit AB, 1030–33, 1030f–1033f
Experimental environment, adaptation, 686–87
Experimental evolution, **685–89,** 686f, 688f
Exponential death profile, 181, 181f

Exponential growth, 141–43
Exponential phase, 142f, **143–44**
ExPortal, 308–9, 309f
Exserohilum rostratum, 803, 804t
Exteins, 390–91, 391f
Extender modules, 598
Extensively drug-resistant tuberculosis (XDR-TB), 1071
Extracellular cytochromes, 554–55, 556f
Extracellular environment sensing, 360–61, 361f
Extracellular immune avoidance, 1041–42
Extracellular polymeric substances (EPSs), **148**
Extrapulmonary tuberculosis, 1070–71, 1070f
Extravasation, **949,** 950f, 953, 953f
Extreme thermophiles, 162f, 164–65, 164f, 165f
Extremophiles, **22,** 23f, 28f, **160,** 715, **844,** 844t
Eye, 38, 39f
  human microbiome, 925

F

F' (F-primes), 334, eTopic 9.4
F_1F_0 ATP synthase, 123
F2G, 1156
Facilitated diffusion, **125–26,** 125f
FACS (fluorescence-activated cell sorting), **138–39,** 139f, 836, 837–38, 837f
Factor C3b, 959, 1041
Factor H, **994**
Facultative anaerobes, **173,** 174t, **175,** 175f
Facultative intracellular pathogens, **1042**
Facultative lithotrophs, 535
Facultative microbes, 161t, 174t, 175f
FAD (flavin adenine dinucleotide), 174, 174f, 496f, **498**
Faecalibacterium, gut microbiome, 929f
Faecalibacterium prausnitzii
  benefits, 931, 997
  obesity, 936
Fairy ring, 787f
Falkow, Stanley, 459, 1017, 1051
Famciclovir, 1148t
Fasciitis, necrotizing, 1011, 1012, 1060t, **1062,** 1062f
Fatal familial insomnia, 1111t
Fatty acid(s)
  biosynthesis, 594–96, 595f
  catabolism, 501, 502f
  phospholipids, 86, 86f
  unsaturation, 596
Fatty acid synthase complex, 594–95
Fbp (fibronectin-binding protein), 95, 96f
F- cell, **331–34,** 332f–334f
F+ cell, **331–34,** 332f–334f
Fc region, 971f, **972**
$Fe(OH)_3$ (iron hydroxide), 169
Fecal bacteriotherapy, 24–25, 25f, 933
Fecal microbial therapy, 24–25, 25f, 933
Fecal-oral route of transmission, **1016**
Fecal transplant, 24–25, 25f, 933
Feline leukemia virus (FeLV), 212t, 228
Femtoplankton, 861
Fen, 890–91, 891f
Feng, Lihui, 933
Fen Hu, 316
Fermentation, 510–15
  acid-, 621t, 623–28, 624f–628f
  alkaline, 621t, **622,** 628–29, 629f
  anaerobes, **175**
  cocoa beans, 617, 627–28, 628f
  colonic, 855–57, 855f, 856f
  defined, **504**
  ethanolic (alcoholic), **13, 512,** 620, 621t, **622,** 629–33, 630f–633f
  glucose, 620
  heterolactic, **622**
  industrial and clinical applications, 515, 515f, 649–50, 650f.651f
  lactic acid, **512, 622,** 623
  malolactic, 632–33, 633f
  mixed-acid, **514–15,** 514f
  pathways, 510–14, 513f
  propionic acid, **622,** 622f
  reactions in bacteria, 510, 512t
  rumenal, 853–55, 853f, 854f
  soy, 625–27, 625f
  waste products, 512, 513–14
  wetland soil, 874
Fermentation industry, **618,** 618f
Fermentation reactions, 620–22, 620f
Fermentative metabolism, **175**
Fermented fish, 621t
Fermented foods
  acid-, 621t, 623–28, 624f–628f
  alkaline, 621t, 628–29, 629f
  chemical conversions, 620, 620f
  defined, **620**
  ethanolic, 621t, 629–33, 630f–633f
  examples, 620, 621t
  indigenous microbiota, 620
  overview, 620–22, 620f, 621t, 622f
  purpose, 620
  "traditional," 620
Ferredoxin, **573,** 574f
Ferric iron, 908, 908f
Ferric uptake regulator (Fur) repressor, 378, 378f
Ferroplasma, 88, 748, 755, 779, 779f
Ferroplasma acidarmanus, 562, 751, 753t
Ferroplasma acidiphilum, 753t, 779
Ferrous iron, 908, 908f
Fertility (F) factor, **331–34,** 332f–334f
Fertilizer use, 900, 902f, 903
Fescue, 875–76
Feta cheese, 624, 624f
Fever, 959–60
F (fertility) factor, **331–34,** 332f–334f
Fibrobacter flavefaciens, 854
Fibronectin-binding protein (Fbp), 95, 96f
50S ribosomal subunit, antibiotics targeting, 1133–34, 1133f
Filamentous bacteriophages, 473
Filamentous rods, 40, 41f
Filamentous viruses, **204,** 204f, 205f
Filariasis, 695–96, 695f, 1012, 1013f
Filopodia, 805
Filoviridae, 1102
Filoviruses, 212t

Filtration to remove microbes, 183, 183f, 184f
Fimbria(e), **110,** 111
  microbial attachment, **1020–23,** 1021f
Finkel, Steve, 453, 454f
Finlay, Brett, 1039
Firmicutes, 704t, 705, 714–19
  anaerobic dechlorinators, 718–19, 719f
  Bacillales, 714–15, 714f, 715f
  cell envelope, 90, 91–92, 92f, 714
  Clostridia, 715–16, 716f, 717f
  defined, **705**
  endospore formation, 105, 711t, 714–16
  Gram stain, 51, 52
  lactic acid bacteria, 717–18
  Listeria, 716–17, 717f
  non-spore-forming, 716–19
  phototrophic, 708t
  Staphylococcus and *Streptococcus,* 718, 719f
Fish, fermented, 621t
FISH (fluorescence in situ hybridization), **54,** 54f, **839,** 840f
Fish tank granuloma, 956f, 957
Fission. *See* Cell division
Fistulated cows, **854**
Fitness competition, 685, 685f, 686–87, 689f
Fitness trade-off, **684**
Fix, **582**
Fixation, **50**
Flagellates, **792**
Flagellin, 111
Flagellum(a), 77f, **78, 111**
  chemotaxis, 113–15, 113f, 114f
  cryo-electron tomography, 68–69, 68f
  dark-field microscopy, **61,** 61f
  eukaryotic, 113, **792**
  motility, 111–13, 111f, 112f
  rotary, 111–15, 111f–114f
Flagyl (metronidazole), 1130–31, 1131f
Flammulina velutipes, 619
Flash pasteurization, cheese, 624
Flavin(s), prefolded proteins, 309
Flavin adenine dinucleotide (FAD), 174, 174f, 496f, **498**
Flavin mononucleotide (FMN), electron transport, 544, 544f, 545, 545f
Flaviviridae, 1082, 1177, 1196
Flaviviruses, 212t
Flavobacterium, 739
  wastewater treatment, 897
Flavonoids, 876, 877f
Flavor generation, cheese, 625, 626f
Fleming, Alexander, 19, 19f, 188, 444, 1117–18, 1117f
FlgM, 372, 372f
FliC, 111
FliG, 111
Flocs, **897**
Floppy head syndrome, 1107
Florey, Howard, 19, 1117–18, 1117f
Flow cell, 49, 49f
Flow cytometry, **138–39,** 139f, **837–38,** 837f
  fitness competition, 685, 685f
Flu. *See* Influenza virus
Fluconazole, 1154, 1155t
Flucytosine (5-fluorocytosine), 1155t
Fluorescence, **43,** 53–55, 55f
  immuno-, 56
Fluorescence-activated cell sorting (FACS), **138–39,** 139f, 836, 837–38, 837f
Fluorescence in situ hybridization (FISH), **54,** 54f, **839,** 840f
Fluorescence microscopy, 53–61
  Bacteroides fragilis, 38, 38f
  chemical imaging, 58–60, 60f
  defined, **53**
  excitation and emission, 54–55, 55f
  labeling, 56–57, 56f, 57f
  mechanism, 53–54, 54f
  planktonic communities, 861
  super-resolution imaging, 57–58, 57f–59f, 59
Fluorescence resonance energy transfer (FRET), **447, 1053–54,** 1054f
Fluorescent antibody staining, 1182–83, 1183f
5-Fluorocytosine (Flucytosine), 1155t
Fluorophores, **54,** 56, 56f, 468
Fluorophore-tagged antibody labeling, 469t
Fluoroquinolone, 1128
Fluoxetine (Prozac), 85f
Flux measurement, 886–87, 887f
FLUXNET, 886, 887f
fMet (*N*-formylmethionine), 302
FMN (flavin mononucleotide), electron transport, 544, 544f, 545, 545f
Focal point, **44,** 44f, 45f
Focus, **38**
  bright-field microscopy, 49–50, 49f
  compound microscope, 48
Folate, **594**
Folliculitis, 1060t
Fomites, **1014,** 1015f
Food(s), 617–43
  acid-fermented, 623–28, 624f–628f
  alkaline-fermented, 628–29, 629f
  edible algae and cyanobacteria, 619–20, 619f
  edible fungi, 618–19, 619f
  ethanolic fermentation (bread and wine), **512,** 620, 621t, **622,** 629–33, 630f–633f
  fermentation industry, **618,** 618f
  irradiation, 184–85
  microbial, 618–23, 619f, 620f, 621t, 622f
  overview of fermented, 620–22, 620f, 621t, 622f
  processed, 636
Food-borne pathogens, 638–41, 638f, 639f, 640t, 641f
Food contamination. *See* Food poisoning
Food poisoning, **634**
  E. coli O157:H7, 639, 640t, 641, 1074–75, 1075f
  pathogens, 636–41, 637f–639f, 640t, 641f
  staphylococcal, 640t, 1073, 1081t
Food preservation, 620, 641–43, 641f, 642f
Food Safety Modernization Act (2011), 636
Food spoilage, 633–41
  defined, **634**
  examples and mechanisms, 634–36, 635t
  pathogens, 636–41, 637f–639f, 640t, 641f
Food supplements, industrial microbiology, 649t

Food webs, **842–43,** 843f
planktonic, 862, 863f
soil, 869–70, 870f–872f
Foot-and-mouth disease virus, vaccine antigen, 473t
Foram(s), **807,** 808f
Foraminifera, 805
Foraminiferans, **807,** 808f
Forespore, **151,** 152f, **715**
Formaldehyde, 186, 186f
N-Formylmethionine (fMet), 302
Foscarnet, 1148t
Fossil(s), micro-, **664–65,** 665t, 666f, 667f
Fossil fuels, 885, 885f, **890**
Fossilization, 660
Fossil record, 687
Fouchier, Ron, 418, 418f
Fouts, Dereck, 1052
Fovea, 38, 39f
F-primes (F'), 334, eTopic 9.4
Fracastoro, Girolamo, 14
Fractionation, 80–82
cell wall lysis and spheroplast formation, 80–81
defined, **80,** 80f
spheroplast lysis, 81, 82f
ultracentrifugation, 81
Fragilariopsis kerguelensis, 909
Frameshift mutation, 317f, **318**
Francisella
growth factors, 134t
systemic illness, 1102
Francisella tularensis
flipping cytokine profile, 1045
pathogen identification, 1184t
tularemia, 1102t
Frank, Albert, 872
Frankia, 721, 724
Franklin, Rosalind, 20, 30, 31f, 71
Fraser, Claire, 5, 6f
Free energy change, 488–90, 489f
Freeze, Hudson, 445
Freeze-drying, **641**
Freezing, food preservation, 641–42, 641f
Freshwater microbial communities, 865–67, 865f–867f
FRET (fluorescence resonance energy transfer), **447, 1053–54,** 1054f
Fruit
alcoholic fermentation, 632–33, 632f, 633f
spoilage, 636
Fruiting body, **736,** 736f, **791,** 801, 802, 802f
Frustule, 812f, **813**
FtsK, 259, 260f
FtsZ protein
bacterial cytoskeleton, 96, 97f
replication, 101
septation, 98
Fuel cells, 556–57, 557f, 558–59
Fuhrman, Jed, 763, 764f
Fumarate, 518
Functional ecology, 838–45
Antarctic lake mats, 846–47, 846f, 847f
carbon assimilation and dissimilation, 842–43, 843f
niche concept, 839
oxygen and other electron acceptors, 843–44, 843t, 844f
requirement for microbes, 839–42, 840f, 841f
spatial organization of microbes in habitat, 839, 840f
temperature, salinity, and pH, 844–45, 844t, 846–47
Fungal lung infection, 1066t, 1067–69, 1067f, 1068f, eTopic 26.3
Fungicide, 650–52, 652f, 1155t
Fungistatic agents, 1155t
Fungus(i), 793–805
classification, 789t, 790–91, 790f
decomposition, 869–70, 872f
defined, **786–87**
edible, 618–19, 619f
emerging, 803–4, 804t, eTopic 20.1
historical overview, 786–87, 787f
hyphae, 793–95, 793f, 794f
mitosporic, **798**
mycelia, mushrooms, and Mycorrhizae, 798–803, 799f–803
shared traits, 793–94, 793f
true, **790,** 801
unicellular, 795–98, 795f
Fur (ferric uptake regulator) repressor, 378, 378f
Furuncles, 949, 949f, 1059–61, 1061f
Fusaric acid, 1138
Fusarium solani, 804t
Fusellovirus, 759–60, 760f
Fusidic acid, 297
Fusion inhibitors, 1151
Fusion peptides, **413, 415,** 416f
Fusobacteria, 704t, **705,** 741
Fusobacterium, oral and nasal cavity microbiome, 926, 926f
Fusobacterium necrophorum, 1129–30
Fusobacterium nucleatum, 704t, 741

G

G3P (glyceraldehyde 3-phosphate), 506, 507, 508f, 510f
Calvin cycle, 587, 588, 590f
GABA (gamma-aminobutyric acid), 522–23, 522f, 524–25
Gagnon, Matthieu, 298f
Gain-of-function mutation, **317**
Gain-of-function research, 419
GAL4 transcription factor, 471, 471f
Galactans, 95, 96f
Galen, 1089
Gallionella, 909
Gallo, Robert, 421, 422f
GALT (gut-associated lymphoid tissue), **946–47,** 947f
Galvani, Luigi, 534
Gametocytes, 820, 820f
Gamma-aminobutyric acid (GABA), 522–23, 522f, 524–25
Gammaproteobacteria, 732–35
aerobic rods, 734f, 735, 735f
classification, 704t, 705
defined, **705**
Enterobacteriaceae (intestinal fermenters and respirers), 734–35, 734f
metabolism, 728t
sulfur and iron phototrophs, 733, 733f
sulfur lithotrophs, 732–33, 733f

Gamma rays
  food irradiation, 184
  mutagenic effects, 318t
Gammaretrovirus, 421t
Gamogony, **820,** 820f
Ganciclovir, 1148t
Gao, Yong-Gui, 1115, 1115f, 1116, 1116f
Gardasil vaccine, 997, 998t, 999
Gardnerella vaginalis, 1084–85
Garner, Ethan, 98
GAS (group A streptococci)
  horizontal gene transfer, 1019–20, 1019f
  pathogen identification, 1173–74, 1173f
Gas chromatography–mass spectrometry (GC-MS), 527
Gas discharge plasma sterilization, 187
Gas gangrene, 1027
Gas sterilants, 186–87, 186f
Gastric ulcer disease, 1077–79, 1079f, 1080f, 1081
Gastritis, 1072
Gastroenteritis, 1072, 1081t
Gastrointestinal tract infections, 1072–83, 1081t
  diarrhea and gut microbiome, 1076–77, 1076f
  enterohemorrhagic *E. coli,* 1073–75, 1075f, eTopic 26.5
  hepatitis viruses, 1080–82, 1081f, 1081t
  physical barriers, 946–47, 947f
  protozoan causes of diarrheal disease, 1077
  rotavirus and norovirus, 1075–76
  staphylococcal food poisoning, 1073
  types of diarrhea, 1072–73
  ulcers, 1077–80, 1078f, 1079f
Gas vesicles, **109,** 110f, **775**
Gattaca (film), 446
GC content, **705**
GC-MS (gas chromatography–mass spectrometry), 527
GDH (glutamate dehydrogenase), **609–10,** 610f
Geminivirus, 212t
Gemmata obscuriglobus, 742f
Gene(s)
  core, 6
  defined, **240,** 281
  essential, **263–64,** 265f
  housekeeping, 691
  nonessential, 462–63
  operational, 681, 681f
  paralogous, **349,** 350f, **683**
  pseudo-, **267, 350**
  structural, 281, 282f
Gene cloning, 32, 459–60, 459f
  for expression, 271–72
Gene duplication, **683–84**
  homologous recombination, 348–49, 349f
  mutation, **316,** 317f
  paralogs, 349, 350f
Gene editing, targeted, 464–66, 465f
Gene expression
  bacteria, 702
  DNA rearrangements that alter, 392–95, 394f, eTopic 10.4
  single cells, 468, 468f
Gene expression analysis, 466–72
  protein abundance and function, 468–71, 469f–471f, 469t
  RNA and transcription, 466t, 467–68, 467f, 468f
Gene fusion, 467f, **469–70,** 469f, 469t
Gene fusion reporter, 56
Gene inversion, 392–94, 394f
Generalized transduction, **335,** 335f
Generation time, **141–43,** 142f
  molecular clock, 674, 674f
Gene reduction, 683–84
Gene regulation, 32
Gene switching, **978,** 979f
Genetech, 33
Gene therapy, lentiviral vectors, 654–56, 654f–656f
Genetically engineered virus, 472, eTopic 12.9
Genetically modified organisms (GMOs), 474–75
Genetic analysis, 82
  transposable elements, 346–48, 347f, 348f
Genetic code, 289–90, 289f, 290f
  modified, 300–301, 301f
Genetic defect, genetically engineered virus to repair, 472, eTopic 12.9
Genetic discrimination, 446
Genetic Information Nondiscrimination Act (2008), 446
Genetic manipulation of microbes, 457–66
  cloning by PCR (Gibson assembly), 460–61, 460f, 461f, 462–63
  gene cloning, 459–60, 459f
  random mutagenesis, 461, 462–63, eTopic 12.1
  restriction endonucleases and recombinant DNA, 458, 458f
  targeted gene editing with CRISPR-Cas9 technology, 464–66, 465f
  targeted (site-directed) mutagenesis, 461–64, 464f, eTopic 12.4
Genetic resistance, 232
Genetics, history, 8t
Gene transfer, 329–43
  additional modes, 339–40, 339f
  advantages, 329
  Agrobacterium tumefaciens, 652–54, 653f
  conjugation, 330–34, 330f–334f
  CRISPR interference, 341–43, 341f–343f
  DNA restriction and modification, 340–41, 340f, 340t
  examples, 316
  F factor, 333–34, 334f
  horizontal, 339, 339f, 351–53, 351f–354f, **673, 678–80,** 680f
  phage-mediated, 336, 336f
  phage transduction, 334–37, 335f, 336f
  transformation of naked DNA, 337–38, 337f, 338f
  vertical, **351, 678**
Gene transfer agents, 338
Gene transfer vectors, **432–35,** 433f, 433t, 434f
Gene vectors, microbial, 652–57
Genital herpes, 1086t
Genital warts, 1086t
Genitourinary tract, 927–28
Genitourinary tract infections, 1083–92
  sexually transmitted, 1086–92, 1086t
  urinary tract, 1083–86, 1085f
Genome(s)
  archaeal, 267–68
  bacterial, 76
  core, **692**

defined, **5, 240**
eukaryotic, 266–67
mosaic nature, 329
organization, 241–50
reference, 832
segmented, **411–13,** 411f, 412f
variations in size, 241–43, 241t, 242f, 243f
Genome assembly, 831–32, 831f, 832f
Genome entry, animal virus, 224f, 225
Genome evolution, 348–54
genome reduction, 349–50, 350f, 351f
genomic islands, 351–53, 352f–354f
homology, duplications, and divergence, 348–49, 349f, 350f
horizontal gene transfer, 351–53, 352f–354f
Genome integration, oncogenes, 229
Genome loss, biosynthesis, 585
Genome mining, 647–49
Genome reduction, **349–50,** 350f, 351f
Genome replication. *See* Replication
Genome sequencing, 5–6, 5f, 6f
Genome uncoating, animal virus, 224f, 225
Genomic analysis, 682–84
Genomic islands, **351–53,** 352f, **1018**
Genomics
history, 9t
meta-, 160, **269–71,** 271f, 272f
single-cell, 271–72, 272f
used to probe pathogenesis, 1051–52
Genotype, **318**
Gentamicin, 1132–33, 1132f, 1143, 1144f
Genus(era), **692**
Geoarchaeota, **752,** 753t
Geobacillus stearothermophilus, 100
Geobacter
electron shuttles, 555
electron transport system, 539
extracellular cytochromes, 554–55
nanowires, 553–54, 554f, 555f
uranium reduction, 552, 910
Geobacter metallireducens, 735–36
anaerobic benzene catabolism, 526–27
uranium reduction, 552, 910
Geobacter sulfurreducens
electron shuttles, 555
nanowires, 553–54, 554f
Geochemical cycling, **21**
Geogemma, 22, 23f, 27
Geological biosignatures, 663–67, 664f, 665t
Geomicrobiology, **884**
Geomyces destructans, 804t
German measles, 1063, 1064f
Germicidal agent, **181**
Germination, **151,** 152f
Germ theory of disease, **14,** eTopic 1.1
Gerstmann-Straussler-Scheinker syndrome, 1111t
G factors, 1031, eTopic 25.5
GFP (green fluorescent protein)
energy stress, 468
labeling, 56, 56f, 57
protein tracking, 469–70, 469f, eTopic 12.6
Ghon complex, 1070
Giant bacterium, 678
Giardia, 786, 788
metronidazole, 1130
transmission from animals, 1200
Giardia intestinalis, 786f, 789t, 792–93, 822, 822f
diarrhea, 1077, 1078f
pathogen identification, 1183
Giardia lamblia. See Giardia intestinalis
Gibbs energy change, 488–90, 489f
Gibbs free energy change, **488–90,** 489f
Gibson, Daniel, 460
Gibson Assembly, 460–61, 460f
Gilbert, Walter, 5
Gilchuk, Pavlo, 1002f
Glaucocystophyta, 696
Glaucophyta, 789t
Gliding motility, **153,** 153f
Global carbon balance, temperature change, 889–93, 890f–892f
Global warming. *See* Climate change
Globigerinella aequilateralis, 808f
Gloecapsa, 712, 712f
Glomeromycota, 789t, 803, 803f
Glomus, 803
endomycorrhizae, 873–74
Glossina morsitans, 849–50, 851f
GlpF, 125–26, 125f
Glucans, 501
Gluconeogenesis, **583, 593**
Glucose
oxidation, 518, 519f
phosphorylation and splitting, 507, 508f
repression of *lac* operon, 365, 365f
Glucose 6-phosphate, 509, 510f
Glucose catabolism, 506–10
archaea, 754, 754f
ATP produced by, 495–96
Entner-Doudoroff pathway, 506, 506f, 509, 509f, 510f
glycolysis (Embden-Meyerhof-Parnas pathway), 506–9, 506f–508f
pentose phosphate pathway, 506, 506f, 509–10, 511f
Glucose fermentation, 505–20, 620
Glucosidase inhibitor, 1153
Glutamate, amino acid biosynthesis, 609–11, 610f
Glutamate dehydrogenase (GDH), **609–10,** 610f
Glutamine, amino acid biosynthesis, 609–11, 610f
Glutamine synthase (GOGAT, glutamine:2-oxoglutarate aminotransferase), **609–10,** 610f
Glutamine synthetase (GS), **609–10,** 610f
Glutaraldehyde, 186f
Gluten, 630
Glycan(s), **486, 501**
catabolism, 485, 485f, 486, 486f, 501, 502f, 504
cell wall, 88f, **89**
mammalian breast milk, 521
Glyceraldehyde 3-phosphate (G3P), 506, 507, 508f, 510f
Calvin cycle, 587, 588, 590f
Glycerol, catabolism, 501, 502f
Glycerol 3-phosphate, 583
Glycerol channel, 125–26, 125f
Glycerol diether, 87f
Glycolysis, **506–9,** 506f–508f

Glycoproteins, 303
Glycosylases, base excision repair, 326
Glycosylation, 302–3, 303f
Glyoxylate bypass, **518–20,** 519f
GMOs (genetically modified organisms), 474–75
Gnotobiotic animal, **933**
Goblet cells, 995, 995f
GOGAT (glutamine synthase), **609–10,** 610f
Golden algae, 789t, 812
Golden Gate cloning, 460
Gold particle–tagged antibody labeling, 468, 469t
Gonium, 810
Gonorrhea, 1086t, 1088–89, 1089f
  phase variation, 1089, eTopic 10.4
  point-of-care rapid testing, 1186t
Goodpasture's syndrome, 1007t
Gordon, Jeffrey, 347, 933–34, 933f, 935
Gottesman, Susan, 376, 376f
Grains, spoilage, 636
Gram, Hans Christian, 50
Gramicidin, **1129,** 1129f
Gram-negative bacilli, pathogen identification, 1170, 1171f
Gram-negative bacteria, 703
  cell envelope, 90, 91f, 92–96, 93f–95f
  identifying, 1169–72, 1169f, 1170t, 1171f
  identifying nonenteric, 1170–72, 1171f
  staining, **50–52,** 52f
Gram-negative diplococcus, pathogen identification, 1170–72, 1171f
Gram-positive bacteria, 703, 704t
  cell envelope, 90, 91–92, 91f, 92f
  quorum sensing, 386–87, 387f
  staining, **50–52,** 52f
Gram-positive pyogenic cocci, identifying, 1172–74, 1172f, 1173f
Gram stain, **50–52,** 51f, 52f
Gram-variable bacteria, 703
Granulibacter bethesdensis, 729
Granulocytes, 937–39, 938f
Granuloma, 956f, **957**
  tuberculosis, **1070**
Granzymes, 942, 943, 943f, **989**
Grappling hooks, 781, 781f
Graves' disease, 1007, 1007t
Gray (Gy), 184
Grazers, **843,** 843f
Great Pacific garbage patch, 860
Green, Alexander, 1178, 1178f
Green algae, 789t, **791,** 791f, 808–11, 809f–811f
Greenberg, Peter, 384, 384f
Green fluorescent protein (GFP)
  energy stress, 468
  labeling, 56, 56f, 57
  protein tracking, 469–70, 469f, eTopic 12.6
Greenhouse effect, **663, 884**
Greenhouse gases, 884, 885, 885f
  nitrous oxide, 885, 885f, 903
Griffith, Frederick, 30, 240, 337
Griseofulvin, **1154,** 1154f, 1155t
GroEL chaperone, 304, 304f
GroES chaperone, 304, 304f
Groundwater, bioremediation, 910, eTopic 22.3
Group A streptococci (GAS)
  horizontal gene transfer, 1019–20, 1019f
  pathogen identification, 1173–74, 1173f
Group translocation, **129–30,** 130f
Growth
  exponential, 141–43
  nutrient supplies limit, 120
Growth asymmetry, 105–6, 105f
Growth curves, 142f
Growth cycle(s), 140–47
  binary fission, 141, 141f
  continuous culture, 146–47, 146f
  evolution and, 689–90, 689f, 690f
  exponential growth, 141
  generation time, 141–43, 142f
  stages of growth in batch culture, 143–45, 144f, 145f
  symmetrical and asymmetrical cell division, 141, 141f
Growth factors, **134,** 134t
Growth rate, **141,** 159
  temperature, 161–62, 162f, eTopic 5.1
Growth rate constant, **142**
Growth stages, batch culture, 142f, 143–45, 144f, 145f
Growth testing, pathogen identification, 1167–74
GS (glutamine synthetase), **609–10,** 610f
GTP (guanosine triphosphate), 496
Guanine, 243f, 244
Guanosine tetra- and pentaphosphate [(p)ppGpp], 150
Guanosine tetraphosphate (ppGpp), 380, 380f
Guanosine triphosphate (GTP), 496
Guillardia theta, 697–98, 697f
Gulf of Mexico dead zone, 894–95, 896f
Gullberg, Erik, 685
Gut-associated lymphoid tissue (GALT), **946–47,** 947f
Gut bacteria
  biological control, 189
  and cancer, 24
  Candidatus Savagella, 716, 720–21, 720f, 721f
  catabolism of glycans, 485, 485f, 486, 486f, 504, 505f
  circadian rhythm, 357, 357f
  cryo-electron tomography, 37, 37f
  Entner-Doudoroff pathway, 509, 509f
  fluorescence microscopy, 38, 38f, 54f
  Gram stain, 52
  manipulating gene expression, 466
  membrane vesicles, 107
  phylogenetic tree, 676, 676f
  ultrasound detection, 277, 277f
Gut bacteriophage community, 215, 219–20, 221f, 222–23
Gut immune system, 994–97, 995f
Gutman, Antoinette, 215
Gut microbiome, 520–23, 855–57, 928–30
  acquisition, 929–30, 929f
  amino acid decarboxylation, 522–23, 522f, 524–25
  bacteria-host interactions, 855f, 856–57
  beneficial, 931
  biological control, 189
  and cancer, 24–25, 24f, 25f
  catabolism and niche adaptation, 520–21, 521f, 522f
  colon cancer, 921, 921f, 922, 922f
  control, 930, 930f

culture, 856, 856f
culture-independent sequencing, 928–29
defined, **486**
diarrhea and, 1076–77, 1076f
endosymbiosis, 23, 23f
horizontal gene transfer, 339, 339f
kept at bay, 930, 930f
and obesity, 933–36, 933f–935f
ocean *vs.,* 268, 269f, 271f
overview, 928
polymerase chain reaction, 1179, 1179f
shaping of gut immunity and human health, 997, eTopic 24.6
taxa, 855–56
variability over time, 929, 929f
Gut pathogen, biogeography, 59, 59f
Gy (gray), 184
Gymnodinium, 818f

H

H^+ (hydrogen ions), 168–69
H1N1 influenza virus, 411
H1 (histamine) receptors, 1003–4
H_2O_2 (hydrogen peroxide), 174, 174f
H3O+ (hydronium ions), 168–69
H5N1 influenza virus
endemic *vs.* epidemic, 1189, eTopic 26.4
transmission, 414, 418–19
tropism, 223–25
H7N9 influenza virus, 232
HA. *See* Hemagglutinin (HA)
Haber, Fritz, 899
Haber process, **604,** 899–900
Habitat(s)
marine, 857–58, 857f, eTopic 21.1
spatial organization of microbes, 839, 840f
Hadean eon, **663,** 664f
Hadesarchaea, 753t
Haeckel, Ernst, 26
Haemophilus
growth factors, 134t
pathogen identification, 1172
transformation of naked DNA, 337
Haemophilus ducreyi, 1086t
Haemophilus influenzae
capsule, 95
genome sequencing of, 5, 6f
meningitis, 1105, 1110t
pathogen identification, 1168, 1170, 1171f
toxins, 1037t
Haemophilus influenzae type b (Hib) vaccine, 968t, 997, 998t
HAIs (health care–associated infections), **1061,** 1116
Halanaeroarchaeum, 778
HALI (Health for Animals and Livelihood Improvement) Project, 1200
Haloarchaea, 774–78
anaerobic, 778
classification, 753t, 755, 776t
defined, **774**
form and physiology, 775–76, 776f, 777f, eTopic 19.1
hypersaline habitats, 776
retinal-based photoheterotrophy, 777–78, 777f
in salt, 774–75, 775f
Haloarcheal vaccine, 775, 776f
Haloarcula, 753t, 776
Haloarcula argentinensis, 917
Haloarcula marismortui, 241t
Halobacteria, 775
Halobacterium, 753t, 754, 755, 775, 776, 777f
Halobacterium halobium, 755
Halobacterium NRC-1, 775, 776f, eTopic 19.1
Halobacterium salinarum, 567, 568
Haloferax, 753t
Haloferax mediterranei, 775, 777f
Haloferax volcanii, 378, 539, 539f
Halogenated aromatics, 503
Halophiles, 3f, 161t, **168,** 168f, 844t
Haloquadra, 753t
Haloquadratum, 753t
Haloquadratum walsbyi, 692, 776, 777f
Halorhodopsin, **777–78,** 777f
Halorubrum lacusprofundi, 753t, 776, 917
Halothiobacillus neapolitanus, 591f
Halotolerant, 161t
Hamus(i), **781,** 781f
scanning electron microscopy, 65, 66f
Hancock, Robert, 150, 150f
"Handcuffing," 262, 263f
Handelsman, Jo, 269, 829
Hansen, Elizabeth, 772–73
Hantavirus, 1184t, 1198
Haptens, **969–70,** 969f
Harrington, Lucas, 1180f
Harwood, Caroline, 575, 726
Hasty, Jeff, 477
Hatfull, Graham, 402
Haustorium(ia), **879,** 879f
HAV (hepatitis A virus), **1080,** 1082t
Hawkins, Aaron, 165f
Hay fever, 974f, 1004
Hazan, Ronen, 190–91, 190f
HBsAg, 1080–82
HBV (hepatitis B virus), **1080–81,** 1082f, 1082t
point-of-care rapid diagnostics, 1186t
vaccine antigen, 473t
HCR (hybridization chain reaction), 59, 59f
HCV (hepatitis C virus), 224f, 225, 229, **1082,** eTopic 11.4
antiviral agents, 1148t
technology-aided reemergence, 1198
Head, bacteriophage, 404, 404f
Health care–associated infections (HAIs), **1061,** 1116
Health for Animals and Livelihood Improvement (HALI) Project, 1200
Heart valve prolapse, 1093, 1096
Heat, to kill microbes, 182, 182f
Heat-shock proteins (HSPs), **304–5,** 304f
Heat-shock response, **165**
Heat-shock sigma factor, 389–90, 393f
Heat-stable bacterial DNA polymerase, 3
Heavy chains, **970–71,** 971f
Hehemann, Jan-Hendrik, 521, 521f
Heimdallarchaeota, 753t, 781

Hektoen agar, 1168, 1168f
Helicobacter, 737, 737f
  toxins, 1040
Helicobacter pylori, 17, 737, 737f
  classification, 704t
  culturing, 176
  electron transport system, 538, 538f
  gastric ulcer disease, 1077–79, 1079f, 1080f, 1081t
  immune evasion, 992
  lithotrophy, 557
  microaerophilic microbe, 174t
  mixed-acid fermentation, 514
  pathogenicity islands, 1018f
  pH, 172
  scanning electron microscopy, 65–66, 66f
  species definition, 691
  stomach microbiome, 928, 928f
  toxins, 1026t
Heliobacteriaceae, 716
Heliobacterium, 709t
Hell, Stefan, 57
Helling, Robert, 459
Helper T cell (TH cell), **980,** 982, 986–90
  activation by antigen-presenting cells, 986, 987f
  activation of B cells, 986, 987f
  activation of cytotoxic T cells, 987–88, 987f, 988f
  and cytokines, 989–90, 989t, 990f
  promotion and limitation of inflammation, 988–89
Hemagglutinin (HA)
  antiviral agents that target, 1149, 1149f, eTopic 27.4
  envelope synthesis, 416
  viral structure, 411
Hematopoietic stem cell, 938f
Heme, **539,** 539f
  electron transport, 544, 544f, 546–47, 546f
Hemin, 1170
Hemolysin (HlyA), 311f, 312, **1027**
Hemolytic anemia, autoimmune, 1007t
Hemolytic uremic syndrome (HUS), 1074
Henri, Sandrine, 940, 940f
Henson, Jim, 991
Hepadnaviridae, 1080
Hepadnaviruses, 212t
HEPA (high-efficiency particulate air) filters, **183,** 184f
Hepatitis, **1080**
Hepatitis A (HepA) vaccine, 968t, 998t, 1080
Hepatitis A virus (HAV), **1080,** 1082t
Hepatitis B (HepB) vaccine, 968t, 998t, 1082
Hepatitis B virus (HBV), **1080–81,** 1082f, 1082t
  point-of-care rapid diagnostics, 1186t
  vaccine antigen, 473t
Hepatitis C virus (HCV), 224f, 225, 229, **1082,** eTopic 11.4
  antiviral agents, 1148t
  technology-aided reemergence, 1198
Hepatitis D virus, 1082t
Hepatitis E virus, 1082t
Hepatitis viruses, 1080–82, 1081f, 1081t
HepA (hepatitis A) vaccine, 968t, 998t, 1080
HepB (hepatitis B) vaccine, 968t, 998t, 1082
Herd immunity, 999, 1148, eTopic 26.1
Herpes, genital, 1086t
Herpes simplex virus (HSV), 2, 3f, 402, 436t
  antiviral agents, 1148t
  attachment and host cell entry, 438, 439f
  chronic infection, 201
  epidemiology, 436
  genital herpes, 1086t
  infection of oral or genital mucosa, 436, 437f
  molecular biology, 435–40
  molecular evolution, 213
  pathogen identification, 1184t
  persistent viral infections, 438–40, 439f
  replication, 438, 439f
  structure, 202, 203f, 436–38, 437f
Herpesviruses, 201, 436, 436t
  classification, 212t
  engineered to kill tumors, 440
  latent infection, 1012–13, 1013f
  molecular evolution, 213–14, 213f
  persistent viral infections, 438–40, 439f
  structure, 202, 203f
Hershey, Alfred, 215, 240
HERVs (human endogenous retroviruses), 218t, 219, 431–35
Hess, Matthias, 832–33
Hesse, Angelina, 15–16, 16f
Hesse, Walther, 15–16, 16f
Hetero-, 122
Heteroaromatic rings, 544
Heterocysts
  cyanobacteria, 152, 153f, **710,** 712
  nitrogen fixation, **604,** 604f, 606
Heterokont(s), 789t, 792, 812–13
Heterokonta, 789t, 792
Heterolactic fermentation, **622**
Heterologous host, **643**
Heterorhabditis bacteriophora, 869, 872f
Heterotroph(s), **121,** 121f, **122**
  membrane vesicles, 106
Heterotrophy
  biomass, **120–21**
  biosynthesis, 582, 583f
  energy source, **122, 487,** 488
Hexachlorophene, 186f
HflX, 178, 179f
HGT. *See* Horizontal gene transfer (HGT)
HHV8 (human herpesvirus type 8), 1091
Hibernation-promoting factor (HPF), 177–78, 179f
Hib (*Haemophilus influenzae* type b) vaccine, 968t, 997, 998t
High-efficiency particulate air (HEPA) filters, **183,** 184f
High-frequency recombination (Hfr) strain, **333–34,** 334f, eTopic 9.3
High temperature, short time (HTST) method, 183
Hippocrates, 1188
Histamine, 522, 522f, 523, 954, 1003–4
Histamine (H1) receptors, 1003–4
Histones, **267, 751**
Histoplasma capsulatum, 803, 804t
  AIDS, 1091
  pathogen identification, 1184t
  respiratory tract infection, 1066t, 1067–69, 1067f, 1068f
Histoplasmosis, 1066t, 1067–69, 1067f, 1068f
  antifungal agents, 1155t

HIV. *See* Human immunodeficiency virus (HIV)
HlyA (hemolysin), 311f, 312, **1027**
HMOs (human milk oligosaccharides), **521**
Hodgkin, Dorothy Crowfoot, 30, 71–72, 71f
Hoffman, Jules, 951
Hollandina, 704t
Holliday junction, 334, eTopic 9.2
Holobiont, 1, **851–52**
Homolog(s), **349**
Homologous recombination
  circular chromosomes, **259,** 260f
  DNA repair, 322t, **326–27,** 327f
  genome evolution, 348–49, 349f
Homology, 348–49
Homoserine lactones, 384
Honey, floppy head syndrome, 1107
Hong Lian, 165f
Hongrong Liu, 68, 68f
Hooke, Robert, 10–11, 11f
Hooper, Lora, 930, 930f
Hopanes, 82f, **86–87,** 87f
Hopanoids, 82f, **86–87,** 87f
Horizons, soil, 868–69, 868f
Horizontal gene transfer (HGT), 351–53
  between bacteria and eukarya, 352–53, 353f, 354f, eTopic 9.7, eTopic 25.8
  defined, **351, 673**
  evolution, 678–80, 680f
  genomic islands, 351–53, 352f
  gut microbiome, 339, 339f
  hypervariable region, 350, 351f
  pathogen evolution, 1019–20, 1019f
  reconciled with vertical gene transfer, 680–81, 681f
Horizontal transmission, **1014,** 1015f
Hormogonium(ia), **711**
Hormoscilla spongeliae, 1
Horseshoe-shaped vesicles, 109, 109f
Hospital-acquired infections, **1061,** 1116
Host(s)
  pathogen survival within, 1040–51
  viruses, 108f, 196–97
Host cell entry
  HIV, 425–27, 426f
  influenza virus, 414–15, 416f
Host cell reprogramming, 1018–19
Host defenses against viruses, 232
Host factor, **413**
Host-pathogen interactions, 1012–20
  effect of infections on microbiota, 1016–17
  immunopathogenesis, 1016
  infection cycles, 1014–16, 1014f, 1015f
  pathogen evolution by horizontal gene transfer, 1019–20, 1019f
  pathogenicity islands, 1017–19, 1018f
  portals of entry, 1016
  terminology, 1012–14, 1013f, 1014f
  virulence factors, 1017, eTopic 25.1, eTopic 25.2
Host range, **201,** 210
Hot springs, 756, 756f
Houlton, Ben, 899
Housekeeping, 327
Housekeeping genes, 691
Housseau, Franck, 921, 922f
HPF (hibernation-promoting factor), 177–78, 179f
HPV. *See* Human papillomavirus (HPV)
HSPs (heat-shock proteins), **304–5,** 304f
HSV. *See* Herpes simplex virus (HSV)
HTST (high temperature, short time) method, 183
Huang, Kai, 1002f
Huber, Rudolph, 781
Hultgren, Scott, 1085
Human colon, 855–57, 855f, 856f
Human endogenous retroviruses (HERVs), 218t, 219, 431–35
Human gene therapy, lentiviral vectors, 654–56, 654f–656f
Human herpesvirus type 8 (HHV8), 1091
Human immunodeficiency virus (HIV), 402, 420–31, 1090–91, 1090f
  accessory proteins, **424,** 424t
  antiretroviral therapy, 201, 421–22, 426, 426f, 427, 429, 1091, 1148t, 1150–52, 1151f
  attachment and host cell entry, 425–27, 426f
  classification, 1086t
  cure, 1091
  defined, **421**
  epidemiology, 1090, eTopic 26.7
  genome, 423–24, 423f, 424t
  history, 421–22, 422t, 1090
  host, 201
  Koch's postulates, 16
  monitoring, 1091
  opportunistic infections, 1090f, 1091
  origin and evolution, 424–25, 425f
  pathogenesis, 1090–91, eTopic 26.7
  pathogen identification, 1184t
  point-of-care rapid diagnostics, 1186t
  preexposure prophylaxis, 1152
  replication cycle, 228–29, 229f, 429–31, 430f, 431f
  resistance, 232
  reverse transcriptase, 420, 427–29, 428f
  stages, 1091
  structure, 212t, 422–23, 423f
  symptoms, 1091
  testing, 421–22
  transmission, 421, 1090
  tuberculosis, 1069, 1195
  types, 421
  vaccine, 422, 1091
Human influences, microbial ecosystems, 179–80, 179f, 180f, eTopic 5.6
Human microbiome, 922–31
  acquisition, 923, 924f
  benefits and risks, 931–36, 932f–935f
  containment breaches and dysbiosis, 931–33, 932f
  eye, 925
  genitourinary tract, 927–28
  hygiene hypothesis, 923
  intestine, 928–30, 929f, 930f
  link between obesity and dysbiosis, 933–36, 933f–935f
  oral and nasal cavities, 925–26, 926f, 927f
  overview, 922–23
  respiratory tract, 927, 927f
  skin, 925, 925f
  stomach, 928, 928f

Human Microbiome Project, 268, 856
Human milk oligosaccharides (HMOs), **521**
Human papillomavirus (HPV), 436t, 1063, eTopic 26.1
  classification, 212t
  disease, 201
  genital warts, 1086t
  oncogenes, 227
  replication, 226–27, 226f, 227f
  skin infections, 1060t
  tropism, 223
Human papillomavirus (HPV) vaccine, 997, 998t, 999
Humic material, **870**
Humics, 555
Humoral immunity, 065f, **964,** 965–66, 966f
  activation, 986–88, 987f
Humulin, 460
Humus, **870**
Hungate, Robert, 854
Huq, Anwar, 859
HUS (hemolytic uremic syndrome), 1074
Hutchison, Clyde, III, 462, 462f
Hutchison, Elizabeth, 239, 239f
Hybridization, **245**
Hybridization chain reaction (HCR), 59, 59f
Hybridoma, 1146
Hydric soil, **874,** 875f
Hydrocarbon catabolism enzymes, *Deepwater Horizon* spill, 835, 835f, 836
Hydrogenimonas, 737
Hydrogen ions (H^+), 168–69
Hydrogenotrophy, **564,** 564t, 565f, 726
Hydrogen peroxide (H_2O_2), 174, 174f
Hydrologic cycle, 893–99
  biochemical oxygen demand, 893–95, 894f–896f
  and carbon cycle, 893, 894f
  defined, **893**
  wastewater treatment, 895–99, 896f–898f
Hydrolysis, ATP, **495,** 495t
Hydronium ions (H_3O^+), 168–69
Hydrothermal vents, 760–62, 761f, 762f, **864–65,** 864f
Hydroxide ions (OH^-), 168–69
Hydroxyl radical (*OH), 174, 174f
3-Hydroxypropionate cycle, 586t, **594,** eTopic 15.4
Hygiene hypothesis, 923
Hymeniacidon perleve, 723
Hypersaline habitats, 776
Hypersensitivity, 1001–6
  type I (immediate), **1001–5,** 1001t, 1004f
  type II, 1001t, **1005,** eTopic 24.3
  type III, 1001t, **1005–6,** eTopic 24.8
  type IV (delayed-type), 1001t, **1005,** 1006f
Hyperthermophiles, **164–65**
  barophilic vent, 760–62, 761f, 762f
  classification, 161t
  endomembranes, 747, 747f
  environmental conditions, 164, 164f, 844t
  methanogens, 767
  TACK, 756–63
  temperature and growth rate, 162f
Hyperthyroidism, 1007, 1007t
Hypervariable region, 350, 351f
Hypha(e), **153–54,** 154f, **791,** 793–95, 793f
  nutrient absorption, 794–95, 794f
Hyphomicrobium, 729
  metabolism, 728t
  symmetrical cell division, 141, 141f
Hyphomicrobium sulfonivorans, 729, 730f
Hypoxanthine, base excision repair, 326
Hypoxia, **895**
Hypoxic zones, 866, **894–95,** 896f
  denitrification, 903, 904f

I

Ibalizumab, 1151
IBD (inflammatory bowel disease)
  Bacteroides forming membrane vesicles, 37
  dysbiosis, 932, 933
ICAM-1 (intracellular adhesion molecule 1)
  acute inflammation, 953, 953f
  rhinovirus, 969, 969f
  tropism, 222, 224f
iChip (isolation chip), multichannel, 136, 136f
Icosahedral viruses, **202–3,** 203f
ICSP (International Committee on Systematics of Prokaryotes), 693
ICTV (International Committee on Taxonomy of Viruses), 210
ID50 (infectious dose 50%), **1013–14**
Identification, **693–94**
Ideonella sakaiensis, 841–42, 841f
Idiotypes, **972,** 972f
IECs (intestinal epithelial cells), 995, 995f
IELs (intraepithelial lymphocytes), **995,** 995f
IF(s) (initiation factors), 295f, 296
IFNs (interferons), **952**
  antiviral agents, 1152f, 1153
  modulation of immune response, 989, 989t, 990f
Ig(s). *See* Immunoglobulin(s) (Igs)
iGEM (International Genetically Engineered Machine) competition, 480, 481f
Ignicoccus, 748, 757
  classification, 753t
  endomembranes, 747, 747f
  symbionts, 780–81, 780f
Ignicoccus hospitalis
  classification, 755
  endomembranes, 747, 747f
  symbionts, 780f, 781
Ignicoccus islandicus, 753t, 757, 757t
IL(s) (interleukins), modulation of immune response, 989, 989t
IL-1 (interleukin 1), **953**
ILCs (innate lymphoid cells), **995–96,** 995f
Ilinykh, Philipp, 1002f
Illumina process, 449–50, 449f, 454t, 831
Imatinib, 1152f
Imidazoles, 1154, 1154f, 1155t
Imipenem, 1127–28
Immediate hypersensitivity, **1001–5,** 1001t, 1004f
Immersion oil, **47,** 47f
Immune avoidance
  extracellular, 1041–42
  intracellular, 1042–44, 1042f–1045f
Immune complex–mediated hypersensitivity, 1001t, **1005–6,** eTopic 24.8

Immune modulation, 989–90, 989t, 990f
Immune system, **18,** 936–60
  adaptive immunity, 936–37, 963–1008
  cells, 937–44
  chemical barriers to infection, 947–48, 948f, 948t
  defense against viruses, 232
  defined, **936**
  importance, 964, 964f
  infection *vs.* disease, 937
  innate immunity, 936–37, 949–57
  lymphoid organs, 944, 945f
  overview, 936–45
  physical barriers to infection, 945–47, 946f, 947f
Immunity, **18**
  adaptive, 936–37
  herd, 999, 1148, eTopic 26.1
  innate, 936–37, 949–57
Immunization, **17–18,** 17f, 997–1001. *See also* Vaccination; Vaccine(s)
Immunochromatographic assay, 1185, 1185f
Immunocompromised people, vaccines, 1000, eTopic 24.1
Immunofluorescence, 56
Immunogen(s), **936–37, 964, 969**
Immunogenicity, **967,** eTopic 24.2
Immunoglobulin(s), intravenous, 1000
Immunoglobulin(s) (Igs), **970,** 972–74, 973f, 973t
Immunoglobulin A (IgA), 971, **973–74,** 973f, 973t
Immunoglobulin D (IgD), 971, 972t, **973**
Immunoglobulin E (IgE), 971, 972t, **973**
Immunoglobulin E (IgE)-mediated hypersensitivity, 1001–5, 1001t, 1004f
Immunoglobulin G (IgG), 970, **973,** 973t
  immune response, 974–75, 975f
Immunoglobulin M (IgM), 971, **972,** 973f, 973t
  B-cell activation, 986, 987f
  immune response, 974–75, 975f
Immunological specificity, **967–69,** 967f, 968t, 969f, eTopic 24.3
Immunology-based pathogen detection, 1179–83
  case study, 1179–80
  ELISA test, 1181–82, 1182f
  fluorescent antibody staining, 1182–83, 1183f
Immunopathogenesis, **1016**
Immunoprecipitation, **970**
Impetigo, 1060t
Inactivated polio vaccine, 968t, 998t, 999
Inactive aporepressors, 368, 369f
Incidence, **1189**
Index case, **1189–90,** 1190f
Indigenous microbiota, fermented foods, **620**
Indoor air microbiome, 911, 912f
Indoor environments, 911, 912f, 913f
Inducer, **359,** 360, 360f
Inducer exclusion, **365–66,** 366f, eTopic 4.1, eTopic 10.1
Induction, **359,** 360f, 362
Industrial evolution, 689, eTopic 17.3
Industrial fermentation, 618, 618f
Industrial fermentor, **649,** 650f
Industrial microbiology, 643–52
  case example, 646–47, 646f
  commercial success, 643–46
  defined, **643**
  fermentation systems, 649–50, 650f, 651f
  microbial products, 647–49, 648t–649t, 650–52, 651f, Topic 16.2
Industrial runoff, 899, eTopic 22.1
Industrial strain, **647–49**
Infection(s)
  asymptomatic, 411
  biofilms and, 1023–24, 1025f, eTopic 4.3
  cardiovascular and systemic, 1092–1104
  central nervous system, 1104–12
  chain of, **15**
  chemical barriers, 947–48, 948f, 948t
  community-acquired, 1061
  defined, **1012**
  vs. disease, 937, 1012
  effect on microbiota, 1016–17
  gastrointestinal tract, 1072–83
  genitourinary tract, 1083–92
  health care–associated, **1061,** 1116
  latent, 1012–13, 1013f
  multiplicity of, **233**
  nosocomial, **1061,** 1116
  opportunistic, **1012,** 1013f
  phage, 406–8, 406f, 407f
  physical barriers, 945–47, 946f, 947f, eTopic 23.3
  reservoir, **1016**
  respiratory tract, 1064–72
  sexually transmitted, 1086–92, 1086t
  skin, soft-tissue, and bone, 1059–64
Infection cycles, **1014–16,** 1014f, 1015f
Infection thread, 876, 877f
Infectious dose 50% (ID50), **1013–14**
Inflammasomes, **951–52,** 951f, 952f
Inflammation, 949, 949f, 950f
  acute, 952–54, 953f
  chronic, 952, **956–57,** 956f
  gut microbiota, 997, eTopic 24.6
  promotion and limitation, 988–89
Inflammatory bowel disease (IBD)
  Bacteroides forming membrane vesicles, 37
  dysbiosis, 932, 933
Inflammatory diarrhea, 1072
Influenza, 1066t
  point-of-care rapid diagnostics, 1186t
  Tamiflu resistance, 1149
  technology-aided reemergence, 1198
Influenza receptor, 413–14, 415f
Influenza vaccine, 968t, 998t, 1148
Influenza virus, 201, 402, 410–20
  antiviral agents, 1148–49, 1148t, 1149f, eTopic 27.4
  attachment to host cell, 413–14, 415f
  design of deadly, 418–19, 418f, 419f
  emergence of new, 232
  H1N1, 411
  H5N1 (avian), 223–25, 411, 414, 418–19
  H7N9, 232
  host entry, 414–15, 416f
  mutation and reassortment, 413, 413f
  overview, 410–11
  pathogen identification, 1184t

Influenza virus (*continued*)
  point-of-care rapid diagnostics, 1186t
  replication cycle, 415–19, 417f, 420f
  respiratory tract infection, 1066t
  vs. rhinovirus, 1071, eTopic 26.4
  segmented genome, 411–13, 411f, 412f
  structure, 205, 411, 411f
  swine, 411
  transmission, 413–14, 415f, 418–19
Inhaled allergen, 1004, 1004f
Initiation
  replication, 251, 252–55, 252f–255f
  transcription, 282–83, 282f–283f
  translation, 293–94, 294f, 295
Initiation factors (IFs), 295f, 296
Injectisomes, 65, 66f
Injera, 621t, **630–31,** 631f
Innate immunity, 949–57
  acute inflammation, 952–54, 953f
  autophagy and intracellular pathogens, 956
  cells, 937–44
  chronic inflammation, 956–57, 956f
  complement and fever, 957–60, 958f
  defense against viruses, 232
  defined, **936**
  overview, 936–37, 949, 949f, 950f
  oxygen-independent and -dependent killing pathways, 955–56
  pattern recognition receptors and cytokines, 949–52, 950t, 951f, 952f
  phagocytes, 954–55, 954f, 955f
  type 1 interferons, 952
Innate lymphoid cells (ILCs), **995–96,** 995f
Inner membrane, 77f, **78, 93**
Inoculation, 17, 17f, 18
Inosine, 613, 614f
Inositol hexaphosphate, 625
Insect(s), intracellular bacteria, 849–50, 850t, 851f
Insecticides, 472–73, 472f, 650
InSeq technique, 347, 348f
Insertions, **316,** 317, 317f
Insertion sequence (IS), **344,** 344f
Insulin, gene cloning, 460
Integrase inhibitor, 1151
Integrated viral genomes, 3Topic 6.2, 197–98, 198f, 200–201, 218–19
Integrin, 222, 224f
Integrons, 345, eTopic 8.6
Inteins, 390–91, 391f
Interactome, **470–71,** 471f, eTopic 12.7
Intercellular nanotubes, **107,** 107f
Interference, **45,** 45f
  phase-contrast microscopy, 62, 62f
Interferons (IFNs), **952**
  antiviral agents, 1152f, 1153
  modulation of immune response, 989, 989t, 990f
Interferon-stimulated genes (ISGs), 952
Interleukin(s) (ILs), modulation of immune response, 989, 989t
Interleukin 1 (IL-1), **953**
International Committee on Systematics of Prokaryotes (ICSP), 693
International Committee on Taxonomy of Viruses (ICTV), 210
International Genetically Engineered Machine (iGEM) competition, 480, 481f
Interspecies communication, 385–86, 385f, 386f
Intestinal epithelial cells (IECs), 995, 995f
Intestinal microbiota. *See* Gut microbiome
Intestinal parasites, 823, 823f
Intimin, **1039**
Intracellular adhesion molecule 1 (ICAM-1)
  acute inflammation, 953, 953f
  rhinovirus, 969, 969f
  tropism, 222, 224f
Intracellular bacteria, insects, 849–50, 850t, 851f
Intracellular immune avoidance, 1042–44, 1042f–1045f
Intracellular pathogens, 956, **1042,** 1042f–1043f
  facultative, **1042**
  obligate, **1042,** 1045f
Intraepithelial lymphocytes (IELs), **995,** 995f
Intravenous immunoglobulin (IVIG), 1000
Intrinsic termination, **283,** 284f
Introns, 243f, **267,** 299, **457**
Inversion, **316,** 317f, 318
Inverted repeats (IRs), 213, **393**
In vivo expression, 1017, eTopic 25.2
Iodine solution, Gram stain, 50, 52f
Iodophor, 186
Ion(s)
  bacterial composition, 79
  secondary, 60
Ion gradient, **85**
Ionizing probe, 60
Ionizing radiation
  food preservation, 642
  resistance, 185, 185f, eTopic 9.2
Iota toxin, 1026t
IPTG (isopropyl thiogalactoside), 466, 477
IR(s) (inverted repeats), 213, **393**
Iron
  anaerobic respiration, 552
  metabolism, 843t
Iron cycle, 908–10, 908f–910f
Iron fertilization, 909
Iron hydroxide [Fe(OH)3], 169
Iron mine drainage, 909, 909f
Iron oxidation, 557–61, 560f
Iron phototrophs, gammaproteobacteria, 733, 733f
Iron phototrophy, 670, 670f
Iron-sulfur cluster, electron transport, 544, 544f
Iron transport, 128–29, 129f
Irradiation to kill microbes, 183–85, 185f
IS (insertion sequence), **344,** 344f
Isaacs, Farren, 475
ISGs (interferon-stimulated genes), 952
Ishiwata, Shigetane, 715
Isoacceptors, 291
Isocitrate dehydrogenase, 517f
Isolate, **693**
Isolation chip (iChip), multichannel, 136, 136f
Isoniazid, 1119
N-isopentenyladenine, 292f
Isoprenoid, 748f, **749**
Isoprenoid membranes, archaea, 748–50, 748f

Isopropanol, 186f
Isopropyl thiogalactoside (IPTG), 466, 477
Isothermal amplification techniques, 1175
Isotope ratios, **665–67,** 665t, 666f, 667f
  biosignatures, 917
Isotype(s), **972–74,** 972f, 973f, 973t
Isotype switching, **974–75,** 980–81, 980f, 986
Iterons, 262, 263f
Itraconazole, 1154, 1155t
Ivanovsky, Dmitri, 19, 197
IVIG (intravenous immunoglobulin), 1000
Ixodes pacificus, 1100
Ixodes scapularis
  genomic islands, 353, 354f
  Lyme disease, 1100, 1101f

J

J (joule), **490**
Jacob, François, 361–62, 361f, 403
Jacobs-Wagner, Christine, 103, 104f
Jagendorf, André, 540
Jannasch, Holger, 858, 858f, 864
Jansson, Janet, 834, 835–36, 835f
JCVI-syn1.0, 461, 462–63
Jenner, Edward, 17–18, 17f, 967f, 968
Jensen, Paul, 581, 581f
Jinzhong Lin, 298f
Jizhong Zhous, 180
Johnson, Carl, 389, 392f
Johnson, Sarah, 451
Joint, Ian, 386
Joule (J), **490**
Joyce, Gerald, 32, 671
JumpStart, 650
Jupiter's moons, 918, 918f

K

Kaback, H. Ronald, 126, 126f
Kamen, Martin, 587, 601, 808
Kaposi's sarcoma, 1090f, 1091
Kaposi's sarcoma-associated herpesvirus, 213
Kappa chains, 971
Karem, Kevin, 1165f
Kashefi, Kazem, 22, 23f
Kasugamycin, 297
Keasling, Jay, 475
Kefir, 621t
Keflex (cephalexin), 188f, 1127f
Keilin, David, 538–39, 539f, 540
Kelly, Robert, 165, 165f
Kelps, 792, 812, 814, 814f
Kenkey, 621t
Keratinocytes
  human papillomavirus, 227
  physical barriers to infection, 945–46
Kerns, Steve, 898f
Ketoconazole, 1155t
Khalifa, Leron, 190f
Khalili, Kamel, 1091
Khorana, Har Gobind, 289
Kill switches, 479, 481f
Kim, Eun-Hae, 891f
Kimchi, 621t, **627.**627f
"Kinetoplast," 821
Kirby-Bauer disk susceptibility test, **1121–22,** 1121f, 1122t
Kishony, Roy, 682
Kitchen sponge, 912, 914–15
Klebsiella, 734
  eye microbiome, 925
  nitrogen fixation, 902
  phylogenetic tree, 676
  plant vascular system, 1199
Klebsiella pneumoniae
  antibiotic resistance, 1127, 1142–43, 1142f
  antibiotics in animal feed, 1141–42
  horizontal gene transfer, 339
  nitrogen fixation, 605, 607
  pathogen identification, 1170, 1171f
  polymyxin, 1129
Kleptoplasty, 792
Kloeckera, 627
Klosneuvirus, 209
k-mers, 832, 832f
Knockout mutation, **317**
Koala retrovirus (KoRV), 200
Koch, Robert, 14–17, 15f, 16f, 50, 828, 858, 1012, 1183, 1187, 1194, 1197
Koch's postulate, **16–17,** 16f
Kombu, 619
Kondo, Akihiko, 466
Konstantinidis, Konstantinos, 180
Koplik's spots, 1063
KOPS sequences, 259, 260f
Korarchaeota, 752–54, 753t, **762,** 763f
Korarchaeum cryptofilum, 763f
Kornberg, Arthur, 254, 255, 255f
Kornberg, Sylvy, 255f
KoRV (koala retrovirus), 200
Kost, Christian, 107
Koudelka, Gerald, 1034
Krachler, Anne Marie, 1023
Krebs, Hans, 30, 515–16
Krebs cycle, 30, 121f, **515–16,** 516f
Kristofferson, Kris, 1102
Kube, Paul, 1094, 1094f
Kuenenia, 904
Kuru, 207, 1111t
Kuru, Erkin, 90, 90f
Kustu, Sydney, 607
Kuypers, Marcel, 904

L

Labeled antibodies, fluorophores, 56
Labile toxin (LT), 1026t, **1031**
Laboratory evolution, **685–89,** 686f, 688f
lac. See Lactose operon *(lac)*
Lachnospiraceae, 705
lacO operator, 466
Lactate, 506
Lactic acid
  fermentation, **512, 622,** 623
  and lactate, 506
  organic acid stress, 169

Lactic acid bacteria, 717–18
  cocoa bean fermentation, 627, 628f
Lactobacillus, 714, 717
  benefits, 997
  binary fission, 141f
  biological control of microbes, 189
  cheese, 622, 622f, 623, 625
  cocoa bean fermentation, 627, 628f
  genitourinary tract microbiome, 927
  histamine, 523
  microaerophilic microbe, 174t
  oral and nasal cavity microbiome, 925
  probiotic, 932, 932f
Lactobacillus acidophilus, 718
  light microscopy, 41f
  probiotics, 189, 932f
Lactobacillus crispatus, 927
Lactobacillus helveticus, 622, 622f
Lactobacillus lactis, 41f, 704t
Lactobacillus plantarum, 627, 628f
Lactobacillus rhamnosus, 933
Lactococcus, 201, 717, 856
Lactoferrin, 955
Lactose, 623
  catabolism, 362, 362f
  induction of lactose operon, 362–63, 363f, 364f
Lactose operon *(lac),* 361–66
  cAMP and cAMP receptor protein stimulates transcription, 363–64, 364f, 365f, eTopic 10.1
  catabolite repression via inducer exclusion, 365–66, 366f
  discovery, 361–62, 361f
  glucose repression, 365, 365f
  lactose catabolism, 362, 362f
  lactose induces, 362–63, 363f, 364f
LacZ enzyme, 362, 363f, 364f
Lagging strand, 101, 253f, 256f–257f, 257
Lag phase, 142f, **143**
Lainhart, William, 1034
LaJeunesse, Todd, 817, 819f
Lake(s), 865–67, 865f–867f
Lake, James, 680
Lambda chains, 971
Lambda phage, 197, 212t, 336–37, 336f
Lambda receptor, 406
Lambda switch, 402–4, 409
Lamellae, electron transport system, 538
Lamellodysidea, 1, 22
Laminar flow biological safety cabinets, **183,** 184f
Lamisil, 1154
Lamisil (terbinafine), 1155t
Lancefield, Rebecca, 1173, 1173f
Landfills, methanogenesis, 768
Landsteiner, Karl, 970
Langerhans cells, **946,** 1005
Lantibiotics, 1128
Lassa fever, 1195
Late gene, bacteriophage, 217
Late log phase, 142f, 144
Latent period, batch culture, **233,** 233f, 234
Latent state, **1012–13,** 1013
Latent tuberculosis infection (LTBI), 1069, 1070, 1070f
Late-phase anaphylaxis, **1005**
Lau, Susanna, 450
LD50 (lethal dose 50%), **1013–14,** 1014f
Leaching, 563
Leading strand, 101, 253f, 255, 256f–257f
Leaf-cutter ants, coevolution, 694, 724, 851, 1135, eTopic 17.6
Leaflets, **83**
Leavening, 621t, **630–31,** 630f, 631f
Lecithinases, 1027
Lectin(s), 625
Lectin complement pathway, 957–58, 993–94
Lederberg, Esther, 403, 403f
Lederberg, Joshua, 330, 330f, 333
Leeuwenhoek, Antonie van, 11, 11f, 38
Leghemoglobin, **730,** 877, 878f
Legionella
  growth factors, 134t
  intracellular immune avoidance, 1043–44
  point-of-care rapid testing, 1185, 1185f, 1186t
Legionellales, 704t
Legionella pneumophila
  classification, 704t
  fluorescence resonance energy transfer, 1054, 1054f
  fluorescent antibody staining, 1183f
  intracellular immune avoidance, 1043–44
  life cycle, 735, 735f, 805
  parasitism, 845t
  pathogen identification, 1184t
  polyester granules, 596
  respiratory tract infection, 1066t
Legionellosis, point-of-care rapid testing, 1185, 1185f, 1186t
Legionnaire's disease, 1066t
Legumes, nitrogen fixation, 876–78, 876f–878f
Leishmania, 821, 821f
  flipping cytokine profile, 1045, 1046
  stopping programmed cell death, 1046
Leishmania major, 821, 821f
Lennon, Christopher, 390–91, 390f
Lens(es), **44**
  aberrations, **46, 47**
  magnetic, 64, 64f
  magnification, 44–45, 44f
  objective, **46–47,** 46f–48f
  ocular, **47**
  parfocal, **48**
  quality, 46
Lenski, Richard, 685–89, 686f, 842, eTopic 17.4
Lentinula edodes, 619
Lentiviral vectors (lentivectors), **433–35,** 433f, 434f
  human gene therapy, **654–56,** 654f–656f
Lentiviruses, 212t, **420–21,** 421f
Leptospira, 738f, 739
  classification, 704t
  genomics to probe pathogenesis, 1052
  meningitis, 1110t
Leptospira interrogans
  leptospirosis, 1102t
  light microscopy, 41f
  pathogen identification, 1184t
Leptospira santarosai, 1176

Leptospirillum, 741
ecology, 828
Leptospirosis, 1102t, 1176
Leptothrix, 909
Lethal dose 50% (LD50), **1013–14,** 1014f
Lethal factor (LF), 1026t, 1032–33, 1033f
Lettuce, contamination, 638, 638f
Leucine transporter (LeuT), 83–84, 84f
Leuckhart, Rudolf, 787f
Leuconostoc, 717
Leuconostoc mesenteroides, 627
Leukocidins, **1027,** 1028–29, 1029f
Leukocytes, **937,** 937f
development, 937, 938f
types, 937–42, 938f
LeuT (leucine transporter), 83–84, 84f
Levansucrase, 464
Levine, Bruce, 654
Lewis, Amanda, 1084–85, 1085f
Lewis, Kim, 136, 137f, 838, 1143
Ley, Ruth E., 268, 269f
LF (lethal factor), 1026t, 1032–33, 1033f
LH (light-harvesting complex), 570, 571f
Liang Yin, 145, 145f
Library construction, next-generation sequencing, 830f, **831**
Lichens, **811, 848,** 848f
Lieberman, Judy, 943
Life, origins, 13–14, 13f, 14f
Ligands, 359, 360, 360f
Light
absorption, 43, 43f
bright-field microscopy, 46
compound microscope, 48
contrast, 43
electromagnetic radiation, 42, 42f
information, 43
interaction with object, 43–44, 43f
magnification, 43, 44–45, 44f
optics and properties, 42–46
reflection, 43, 43f
refraction, 43, 43f, 44, 44f
resolution of detail, 45, 45f
scattering, 43–44, 43f
speed, 43
wavelength, 43
Light chains, **970,** 971, 971f
Light-dark periodicity, 388–89, 388f, 389f, 392f
Light-harvesting complex (LH), 570, 571f
Light-independent reactions. *See* Calvin cycle
Light microscopy (LM), 38–63
bright-field, 46–53
chemical imaging, 58–60, 60f
dark-field, 61–62, 61f, 62f
defined, **40**
fluorescence, 53–61
observing microbes, 38–42
optics and properties of light, 42–46
phase-contrast, 62–63, 62f, 63f
range of resolution, 40, 41f
super-resolution imaging, 57–58, 57f–59f, 59
Light production, density-dependent, 382–84, 382f, 383f
Lignin
aromatic carbon, 523
catabolism, **501–3,** 502f, 503f
degradation products, 555
microbial degradation, **869,** 872f
soil profile, 868
termite wood-digesting microbiome, 852–53, 852f, 853f
Limiting factors, 910–11
Limiting nutrient, **866**
Lincoln, Tracey, 32, 671
Lincosamides, 1132f, **1133**
Lindahl, Björn, 888
Lindow, Steven, 911
Lindquist, Susan, 796, 796f, 797
LINE (long interspersed nuclear element), 432, 432f
Lineage, **676**
Linezolid, 1132f, 1133
Lingpeng Cheng, 68, 68f
Linnaeus, Carolus, 26, 828
Linné, Carl von, 26, 828
Lipiarmycin, 1131
Lipid(s), catabolism, 501, 502f, 503f
Lipid 1, 1126
Lipid A, **93,** 94f
Lipidation, 302, 303f
Lipid autooxidation, 634, 634f
Lipid bilayers, 164
Lipid domains, 100f
Lipid monolayers, thermophiles, 164
Lipmann, Fritz, 515
Lipopeptides, 603
Lipopolysaccharides (LPS)
cell envelope, **78**
endotoxins, 1034–35, 1035f, 1036f
obesity, 935–36, 935f
proteobacterial cell envelope, **93–94,** 94f
Lipoprotein, proteobacterial cell envelope, 93–94, 94f
Liquid bleach, 186
Lister, Joseph, 18–19, 186, 716
Listeria, 716–17, 718f
cheese, 624
endosymbiosis, 730
flipping cytokine profile, 1045
food contamination, 641
phagocytosis, 955
Listeria monocytogenes, 716–17, 718f
cold temperature, 183
crossing placental barrier, 1087
food contamination, 638, 639f, 640t
food irradiation, 184
granzyme B, 943
intracellular immune avoidance, 1044
meningitis, 1110t
pathogen identification, 1184t
phage therapy, 189
psychrophile, 163
RNA thermometer, 392
toxins, 1026t, 1030
Listeria pneumophila, 187
Listerolysin O, 1026t

List of Prokaryotic Names with Standing in Nomenclature (LPSN), 26
Litho-, 122
Lithoautotrophic bacteria, 587
Lithotrophs
  betaproteobacteria, 731, 732f
  Calvin cycle, 587f
  energy, **121–22**
  facultative *vs.* obligate, 535, 557
  gammaproteobacteria, 732–33, 733f
  history, 7, 7f, 21
Lithotrophy, 557–66
  defined, **534, 535, 557**
  electron donors and acceptors, 557, 560t
  energy acquisition, 487, 487t, 488
  example, 534
  hydrogenotrophy using H2 as electron donor, 564, 564t, 565f
  iron oxidation, 557–61, 560f
  methanogenesis, 564–65, 566f
  methylotrophy, 565
  nitrogen oxidation, 561, 562f
  Proteobacteria, 727, 728t, eTopic 18.2
  sulfur and metal oxidation, 561–63, 563f, 564f
Live/dead stain, 138, 138f
Lloyd, Karen, 271, 272f
LM. *See* Light microscopy (LM)
"Loading module," 598
Lobaria pulmonaria, 848, 848f
Loder, Andrew, 165f
Logarithmic (log) phase, 142f, **143–44**
Logic gates, 476, 476f
Lokiarchaeota, 751, 753t, **755,** 781, 782
Long branch attraction, 782, 782f
Longevity genes, 145, 145f
Long interspersed nuclear element (LINE), 432, 432f
Long-Term Evolution Experiment (LTEE), 685–89, 686f, 688f
Long terminal repeats (LTRs), **429**
  lentiviral vectors, 655, 655f
López-García, Purificación, 751
Lopinavir, 1151
Loss-of-function mutation, **317**
Lovley, Derek, 553, 554f
Low temperature, long time (LTLT) method, 183
LPS. *See* Lipopolysaccharides (LPS)
LPSN (List of Prokaryotic Names with Standing in Nomenclature), 26
LT (labile toxin), 1026t, **1031**
LTBI (latent tuberculosis infection), 1069, 1070, 1070f
LTEE (Long-Term Evolution Experiment), 685–89, 686f, 688f
LTLT (low temperature, long time) method, 183
LTRs (long terminal repeats), **429**
  lentiviral vectors, 655, 655f
Luciferase, 24, 382–84, 466
Lumbar puncture, 1161, 1162f
Lungs, physical barriers to infection, 947
Luria Bertani medium, 133t
LuxI, 382–84
Lwoff, André, 361f, 362, 403
Lycoperdon, 789t
Lyme borreliosis. *See* Lyme disease
Lyme disease, **1100–1102,** 1101f, 1102t
  epidemiology, 1188–89
  infection cycle, 1014
  Koch's postulate, 16
  technology-aided emergence, 1197–98
Lyme disease vaccine, 968t
Lymphatic vessels, 945f
Lymph node(s), **944,** 945f
Lymph node germinal center, 980, eTopic 24.4
Lymphocytes, 938f, 942, **944**
  cytotoxic, 966
  intraepithelial, **995,** 995f
Lymphoid organs, 944, 945f
Lymphoid stem cell, 938f
Lyngbya, 712
Lyophilization, **183, 641**
Lysate, bacteriophage, **234**
Lysine, 301, 301f
Lysinibacillus sphaericus, 92, 92f
Lysis
  cell wall, **80–81**
  spheroplast, 81, 82f
  virus, 216f, **217**
Lysogen, 407
Lysogeny, 216f, **217**
  phage lambda, **403–4,** 408–9, 408f
Lysozyme, 955, 958
Lytic cycle, 216f, 217, 408–9, 408f

M

M1 (matrix protein 1), **411**
M9 medium, 133t
mAb (monoclonal antibody), 1005, 1145, 1146–47, 1146f, 1147f
MAC (membrane attack complex), **958–59,** 958f, 993
MacConkey medium, 133–34, 133f
MacElroy, Robert, 160
MacLeod, Colin, 240, 337
Macnab, Robert, 61, 111
Macrolide antibiotics, 1128, 1131, 1132f, **1133**
Macromolecules, 29–30, 29f, **79**
Macronuclei, **815,** 817f
Macronutrients, **120**
Macrophages, 938f, **939–42,** 939f
  activation, 992
  alveolar, **947**
Maculopapular rash, 1063
Mad cow disease, 207, 1111, 1111t, 1198
MAG(s) (metagenome-assembled genomes), 832, 891–92, 892f
Magnaporthe oryzae, 801, 804t
Magnetic lens, 64, 64f
Magnetosomes, **115**
Magnetospirillum magneticum, 69, 69f
Magnetotaxis, **115**
Magnification, **39,** 39f, **43**
  bright-field microscopy, 46–47, 46f, 47f
  empty, **45**
  by lens, **43,** 44–45, 44f
  total, **47**
Major histocompatibility complex (MHC)
  immunogenicity, **967**
  natural killer cells, **942**

T-cell activation, **982–84,** 983f, 984f
type I interferons, **952**
Major histocompatibility complex (MHC) restriction, 985, eTopic 24.5
Malachite green spore stain, 52–53, 53f
Malaria, **817–19,** 818f, 819f, 1096–97, 1096f
climate change, 1198
point-of-care rapid diagnostics, 1186t
Malate, 518, 519f, 520
MALDI-TOF (matrix-associated laser desorption/ionization–time-of-flight), 602–3, 602f
pathogen identification, 1174, 1175f
Malissen, Bernard, 940, 940f
Malolactic fermentation, **632–33,** 633f
Malonyl-ACP, 595
Maltose-binding protein, 307
MAM7 (multivalent adhesion molecule 7), 1023, 1024f
Mamavirus, 209f
MAMPs. *See* Microbe-associated molecular patterns (MAMPs)
Manganese
anaerobic respiration, 552
metabolism, 843t
microbial metabolism, 910t
Manning, Shannon, 1076
Mantle, Earth, 662, 663f
Mapping to reference, 832
Maraviroc, 426, 426f, 1148t
Marburg virus, 1103
March of Dimes, 33, 201
Margulis, Lynn, 26, 27f, 696, 696f, 787
Marine carbon cycling, 887, 888f
Marine ecosystems, 858–59, 858f–860f
Marine habitats, 857–58, 857f, eTopic 21.1
Marine iron, 909
Marine metagenomes, 859–61, 861f
Marine methane cycling, 887–88
Marine microbiologists, 2, 2f
Marine snow, 150, **861**
Marine sponges, 1, 1f
Mariprofundus, 909
Mars
biosignatures, 917
biosphere, 915–17, 916f
terraforming, 917
Mars Advanced Radar for Subsurface and Ionosphere Sounding (MARSIS), 917
Mars *Curiosity* rover, 25, 26f
Marshall, Barry, 17, 66, 737, 1078, 1079f
Massion, Gene, 761f
Mass spectrometry, 58–60, 60f
pathogen identification, 1174, 1175f
protein and translation, **303,** 304f, 468, 469t
Mast cells, 938f, **939,** 974, 974f
Mastomys rats, 1195
Mather, Cotton, 17
Matic, Ivan, 178
Matrix-associated laser desorption/ionization–time-of-flight (MALDI-TOF), 602–3, 602f
pathogen identification, 1174, 1175f
Matrix protein 1 (M1), **411**
Matthaei, Heinrich, 278, 278f
MazE-MazF antitoxin-toxin module, 177, 178f
Mazmanian, Sarkis, 37, 37f
McCarty, Maclyn, 240, 337
McClintock, Barbara, 343
McCloskey, Robert, 618–19, 619f
McDougall, Andrew, 769
M cells, **946,** 947f
MCG (Miscellaneous Crenarchaeote Group), 752, 753t
McInerny, Michael, 527, 528f
MCPs (methyl-accepting chemotaxis proteins), **396,** 396f
MDA (multiple displacement amplification), 836
MDR. *See* Multidrug resistance (MDR)
MDSCs (myeloid-derived suppressor cells), 921, 921f
Measles, 1059, 1060t, 1063
mumps, rubella (MMR) vaccine, 968t, 998, 1063
Measles virus, 196–97, 197f, 224f, 225
assembly, 401, 401f, 402
Meat spoilage, 635–36, 635t
Mechanical transmission, **230,** 1015f, **1016**
Mechanical vector, **230,** 1015f, **1016**
Mechanosensitive channels, 167
Mechnikov, Ilya, 189
Mediator connector, 369, 369f
Medical microbiology, 14–20, 15f–20f
Medical research, 33, 33t, 34f, eTopic 1.2
Medical statistics, 10, 10f
Medina, Mónica, 785, 785f
Medium(a). *See* Culture medium(a)
MedxCD-Opr efflux system, 187
Megasphaera, 854
Melanophages, tattoos, 940–41, 940f, 941f
Melatonin, circadian rhythms, 357, 357f
Membrane, 82–87
defined, 77f, **78**
lipid diversity, 85–87, 86f, 87f
structure, 82–84, 82f–84f
and transport, 82–85, 82f–85f
Membrane attack complex (MAC), **958–59,** 958f, 993
Membrane bioreactors, wastewater treatment, 897, 898f
Membrane disruption, toxins, 1027–30, 1030f
Membrane filtration devices, 183, 183f
Membrane fusion, influenza virus, **415,** 416f
Membrane lipids
diversity, 85–87, 86f, 87f
structure, 82–83, 83f
Membrane-permeant weak acids and bases, **84–85,** 85f
Membrane potential, **122–23**
Membrane proteins, 83–84, 84f
Membrane-spanning, multidrug efflux pumps, resistance, 187
Membrane vesicle(s), 106–7, 106f, 338
protein analysis, 81, 81f
Memory B cells
cell-cell interaction, 965–66, 966f
making, 981
primary and secondary antibody responses, **974,** 975, 975f
Meningitis
cell wall antibiotics, 1125
meningococcal, 1104–6, 1105f, 1110t, 1165
pathogen identification, 1167
specimen handling, 1165–66, 1165f
Streptococcus pneumoniae, 1067

Meningitis vaccine, 968t, 998t
Meningococcal meningitis, 1104–6, 1105f, 1110t, 1165
Meningococcal vaccine, 968t, 998t
Meningococcus, cell fractionation, 80
Mercury, 910t
Merismopedia, 712, 712f
Merozoite, **819,** 820, 820f
MERS (Middle East respiratory syndrome), 1194
Mesocosm, **886**
Mesophiles, 161t, **162–63,** 162f
Mesophyllum, 789t
Mesorhizobium loti, 316
Messenger RNA (mRNA), **286,** 286t
  cell structure, 78
  preliminary transcript, 457
Metabacterium polyspora, 716, 717f
Metabarcoding, **831**
Metabolic activity, biosignatures, 917
Metabolic inhibitors, 1124f, 1125t
Metabolic islands, **351,** 352f
Metabolic reactions, enzyme catalysis, 499–500, 500f
Metabolites, secondary, **1135**
Metagenome(s), **6, 24, 269–71,** 269f–271f
  marine, 859–61, 861f
Metagenome analysis, 829–33
  bovine rumen microbiome, 832–33, 833f
  Deepwater Horizon spill, 833–36, 834f, 835f
  examples, 829–30, 829t
  isolating DNA, 830, 830f
  library construction for next-generation sequencing, 830f, 831
  metagenome sequencing and assembly, 830f, 831–32, 832f
  preliminary screen for diversity, 830–31
  sampling target community, 830, 830f
Metagenome-assembled genomes (MAGs), 832, 891–92, 892f
Metagenome sequencing, 830f, 831–32, 832f
Metagenomics, 160, **269–71,** 271f, 272f, **693**
Metal(s), microbial metabolism, 910, 910t, eTopic 22.2, eTopic 22.3
Metalloproteins, prefolded, 309
Metal oxidation, 561–63, 563f, 564f
Metal reduction, dissimilatory, **552–53,** 553f
Metamonad(s), 789t, 792–93, 822, 822f
Metamonada, 789t, 792–93, 822, 822f
Metatranscriptomes, **835–36,** 835f, 860
Metatranscriptomics, 467, **835**
Metazoa, 789t
Metcalf, William W., 888
Meteorite, amino acids, 608, 608f
Methane, biogenic paths, 766–67, 767f
Methane atmosphere, 672
Methane cycling, marine, 887–88
Methane hydrates, 552, 565, **769**
Methanobacteriales, 753t, 766, 766f
Methanobacterium, 767, 771, 774f
Methanobacterium thermoautotrophicum, 766f, 767
Methanobrevibacter, 773
  bovine rumen, 854
Methanobrevibacter ruminantium, 767
Methanobrevibacter smithii
  classification, 753t, 766
  gut microbiome, 770, 772–73, 772f, 856
  obesity, 935
Methanocaldococcus jannaschii
  cell form, 767
  classification, 753t
  gene transfer, 329
  genome, 241t
  synthetic auxotrophy, 475
  thermophile, 164f
Methanococcales, 753t, 766
Methanococcus, 768, 771
Methanococcus maripaludis, 528
Methanoflorens stordalenmirensis, 891
Methanogen(s), 766–74, 766f
  biochemistry of methanogenesis, 771–73, 774f
  biogenic paths to methane, 766–67, 767f
  classification, 753t
  culture, 767
  defined, **766**
  digestive methanogenic symbionts, 770–71, 770f, 772–73
  diverse cell forms, 766f, 767–68, 767f, 769f
  Euryarchaeota, 754–55, 755f
  filamentous, 768, 769f
  global warming, 890–92, 891f, 892f
  growth temperatures, 767
  methane hydrates and Syntrophoarchaeota, 769
  psychrophilic, 769, 770f
  reductive acetyl-CoA pathway, 594
  representative species, 768t
  soil and landfills, 768–69, 770f
Methanogenesis, **564–65,** 566f
  from acetate, 773
  biochemistry, 771–73, 774f
  from CO_2, 767, 771–73, 772f
  cofactors, 771, 771f
  early life, 670–71
  Euryarchaeota, **754–55,** 755f
  soil and landfills, 768–69, 770f
  wetland soil, 874–75
Methanogenic symbionts, digestive, 770–71, 770f, 772–73
Methanomicrobiales, 753t, 766
Methanomicrobium, 768
Methanopterin (MPT), 594
Methanopyrales, 753t, 765
Methanopyrus, 751, 767
Methanosaeta, 768, 769f
Methanosarcina, 767, 771, 774f
  bovine rumen, 854
Methanosarcina barkeri, 490, 490f
Methanosarcinales, 753t, 766, 766f
Methanosarcina mazei, 766f, 767
Methanosarcina thermophila, 305f
Methanosphaera stadtmanae, 753t, 766, 770
Methanospirillum, 528, 529t
Methanospirillum hungatei, 767
Methanothermus fervidus, 750f, 766f, 767, 781
Methanotrophy, **565**
  alphaproteobacteria, **729,** 730f
Methemoglobinemia, 903
Methicillin, 1126f

Methicillin-resistant *Staphylococcus aureus* (MRSA), 1061
antibiotics combined with botulinum toxin, 1011
antimicrobial coating, 187
darwinolide, 1145
evolution of antibiotic resistance, 19, 684, 684f
example, 1116
mechanism of resistance, 1127
pathogenicity islands, 1018f
susceptibility test, 1121, 1121f
toxins, 1027
vancomycin, 598, 1127
Methicillin-sensitive *Staphylococcus aureus* (MSSA), susceptibility test, 1121, 1121f
Methionine aminopeptidase, 302
Methionine deformylase, 302
35S-Methionine labeling, 468, 469t
Methyl-accepting chemotaxis proteins (MCPs), **396,** 396f
Methylation, 302, 303f
DNA sequencing to detect, 452–55, 453f, 454f
restriction endonucleases, 341
Methylation agents, mutations due to, 320
Methylbenzene, catabolism, 525–26, 526f
3-Methylcytosine, 292f
Methylene blue, 50, 51f
Methyl mismatch repair, **322–23,** 322t, 323f
Methylobacterium, 729
Methylococcus capsulatus, 728t
Methylome, **452**
Methylotrophy, **565,** 728t
alphaproteobacteria, **729,** 730f
Methylphosphonate, 908
Metronidazole (Flagyl), 1130–31, 1131f
Meysman, Filip, 533, 533f
MGIs (mobilizable genomic islands), **345,** 346f
MHC. *See* Major histocompatibility complex (MHC)
MHET (monomer monoethylene terephthalate), 841f, 842
Miasmas, 964
MIC. *See* Minimum inhibitory concentration (MIC)
Micelles, 623, 670
Michel, Gurvan, 521f
Micklem, Gos, 352
Miconazole (Monistat), 1154, 1155t
Microaerophile, 161t
Microaerophilic microbes, 174t, **175,** 175f
Microbe(s)
classification, 26
defined, **4–5**
functions, 2
growth in pure culture, 14–16, 16f
and human history, 7–14, 8f, 9t–10t, 11f–14f
observing, 38–42
overview, 1–3
pathogens, 2–3, 3f
production and destruction by, 7, 7f
representative kinds, 2, 3f
shape, 40, 41f
size, 4, 4t, 5f, 39–40, 40f
unculturable, 134–35, 135f, 136–37, 692–93, 1183
Microbe-associated molecular patterns (MAMPs)
B-cell activation, 986, 987f
endotoxin, 1034
mucous membranes, **946**
pattern recognition, 949–52, 950t, 951f, 952f
Microbial acidification, 7
Microbial attachment, 1020–25
adhesins, **1020,** 1021t, 1023, 1023f, 1024f
biofilms, 1023–24, 1025f
pili, 1020–23, 1021f
Microbial cell(s)
observing, 37–74
super-size, 4, 5f
Microbial communities, 4, 827–82
animal digestive microbiomes, 851–57
functional ecology, 838–45
marine and freshwater microbes, 857–67
metagenome and single-cell sequencing, 829–38
soil and plant, 867–80
symbiosis, 845–51
Microbial disease(s), 1057–1114
cardiovascular and systemic infections, 1092–1104
central nervous system infections, 1104–12
diagnosis, 1058
difference in symptoms, 1058–59, 1059f
gastrointestinal tract infections, 1072–83
genitourinary tract infections, 1083–92
in history and culture, 7–10, 7f, 8t–9t, 10f
overview, 1058–59, 1059f
respiratory tract infections, 1064–72
skin, soft-tissue, and bone infections, 1059–64
Microbial ecology, 20–26, 21f–25f, 827–82
animal digestive microbiomes, 851–57
functional, 838–45
marine and freshwater microbes, 857–67
metagenome and single-cell sequencing, 829–38
overview, 828, 828f
soil and plant microbial communities, 867–80
symbiosis, 845–51
Microbial ecosystems, human influences, 179–80, 179f, 180f, eTopic 5.6
Microbial endosymbiosis, 22–24, 23f
Microbial family tree, 26–28, 27f, 28f
Microbial foods, 618–23
acid-fermented, 623–28, 624f–628f
alkaline-fermented, 628–29, 629f
chocolate from cocoa bean fermentation, 617, 627–28, 628f
edible algae and cyanobacteria, 619–20, 619f
edible fungi, 618–19, 619f
ethanolic fermented, **512,** 620, 621t, **622,** 629–33, 630f–633f
overview of fermented, 620–22, 620f, 621t, 622f
Microbial genetics and DNA revolution, 30–33, 31f, 32f
Microbial gene vectors, 652–57
Agrobacterium for plant genetic modification, 652–54, 653f
baculovirus, 652, eTopic 16.3
lentiviral, 654–56, 654f–656f
Microbial genome sequencing, 5–6, 5f, 6f
Microbial identification after passive CLARITY technique (MiPACT), 58, 59, 59f
Microbial life, 1–35
cell biology and DNA revolution, 29–34
definition of microbe, 4–5
environment and ecology, 20–26

Microbial life (*continued*)
medical microbiology, 14–20
microbes and human history, 7–14
microbial family tree, 26–28
microbial genome sequencing, 5–6
on other planets, 25, 26f
Microbial mats, **660,** 712, 713f
Microbial nutrition, 120–24, 121f, 123f, 124f
Microbial products, 647–49, 648t–649t, 650–52, 651f, Topic 16.2
Microbial species, 690–94
Microbiology, medical, 14–20, 15f–20f
Microbiome(s)
antibiotics and, 1140
defined, **23,** 23f, 24, **268, 922**
effect of infections on, 1016–17
heritability, 268, 269f
human. *See* Human microbiome
indigenous, **620**
microbial ecology, **828**
Microbiota. *See* Microbiome(s)
Microcephaly, Zika virus, 1195–97, 1196f
Micrococcaceae, 704t, 726
Micrococcus, 726
Micrococcus luteus, 704t
Micrococcus tetragenus, 102f
Microcoleus, 847
Microcolonies, 131, 131f
Microcystis
blooms, 710
classification, 710
gas vesicles, 109, 110f
Microfluidic "mother machines," 315, 315f
Microfossils, **664–65,** 665t, 666f, 667f
biosignatures, 917
origins of life, 13, 13f
Micrographia (Hooke), 11, 11f
Micromonas, 789t
Micromonospora, 723f
Micronuclei, **815,** 817f
Micronutrients, **120**
Microplankton, **861**
Micropore filters, 183, 183f
Microscopes and microscopy
atomic force, 41f, **42, 69–70,** 70f
bright-field, 46–53
chemical imaging, **42,** 58–60, 60f
compound, **47–48,** 48f
confocal laser scanning, **58,** 58f, eTopic 2.1
dark-field, **61–62,** 61f, 62f
defined, **37**
differential interference contrast, 63, eTopic 2.2
at different size scales, 40–42, 41f
early, 8t
electron, **29–30,** 29f, 63–66
fluorescence, 53–61
"golden age," 8t
history, 8t, 10–12, 11f
light, 38–63
phase-contrast, **62–63,** 62f, 63f
scanning probe, **42, 69–70,** 70f
Microsporidia, 789t, 804
Microsporum, 801
skin infections, 1060t
Middle East respiratory syndrome (MERS), 1194
Mid-stream clean-catch technique, 1163
Miletto, Marzia, 911
Miliary tuberculosis, 1070–71, 1070f
Milk
pasteurization, 182–83
spoilage, 634
Miller, Stanley, 13–14, 14f, 25, 669
Mimivirus, 208, 209f, 210f
Minamino, Tohru, 112, 112f
Mineral deposits, biosignatures, 917
Minimal defined medium, **132,** 133t
Minimum inhibitory concentration (MIC), 685, 685f
automated, 1122
correlation with tissue level, 1122–23, 1123f
defined, **1119**
determining, 1119–21, 1120f
variation among bacterial species, 1119–20
Minitransposons, 346, 347f
Minus (-) strand RNA virus, 410–20
MiPACT (microbial identification after passive CLARITY technique), 58, 59, 59f
Miscellaneous Crenarchaeote Group (MCG), 752, 753t
Mismatch repair machinery, 315, 315f
Miso, 621t, **626–27**
Missense mutation, **317,** 317f
Mitchell, Peter, 30, 540, 540f
Mitochondrial genome, 696–97, 697f
Mitochondrial respiration, 548–49, 548f
Mitochondrion(ia)
endosymbiosis, 696–97, 696f, 697f
respiration, 540
Mitosis, 141, **820**
Mitosporic fungi, **798**
Mitral valve prolapse, 1093, 1096
Mitral valve vegetations, 926, 927f
Mixed-acid fermentation, **514–15,** 514f
Mixotricha paradoxa, 852–53, 852f, 853f
Mixotrophs, **862,** 863
Mobile electron carriers, 544f, 545
Mobilizable genomic islands (MGIs), **345,** 346f
Modified atmosphere, food preservation, 642
Modular enzyme, **597,** 599
Modular synthesis, **594**
Moerner, William, 57–58, 57f
MOI (multiplicity of infection), **233**
Moissl-Eichinger, Christine, 764–65, 781, 781f
Mojica, Francisco, 219
Mold(s), 799–801, 800f, 801f
slime, 789t, 791, 806–7, 807f
"water," 789t, 792, 804
Mold-ripened cheeses, 624
Molecular approaches to disease surveillance, 1191–92
Molecular biology, history, 9t
Molecular clocks, 27, **673–75,** 674f
Molecular ecology, 9t
Molecular Koch's postulate, 1017, 1051
Molecular mimicry, 297, 1007, 1045, 1046

Molecular pathogen identification, 1174–87
ELISA test, 1181–82, 1182f
fluorescent antibody staining, 1182–83, 1183f
immunology-based, 1179–83, 1182f, 1183f
mass spectrometry, 1174, 1175f
nucleic acid amplification tests (NAATs), 1175–76, 1176f, eTopic 28.2
nucleic acid–based, 1174–79, 1176f–1179f, 1180–81
point-of-care rapid diagnostics, 1184–87, 1185f, 1186t
profiling gut microbiota by PCR, 1179, 1179f
programmable RNA sensors, 1178–79, 1178f, 1180–81
quantitative reverse-transcription PCR (qRT-PCR), 1177–78, 1177f
rapid, next-generation DNA sequencing, 1176
selected diseases, 1183, 1184t, eTopic 28.4
Molecular regulation, 357–99
alternative sigma factors and anti-sigma factors, 370–72
chemotaxis, 395–97
clocks, thermometers, and switches, 388–95
overview, 358
by RNA, 372–79
second messengers, 380–87
transcription repressors and activators, 358–70
Mollicutes, 719–21, 720f
classification, 704t, 705
defined, **705**
Mollivirus sibericum, 209
Monera, 26
Monoblast, 938f
Monocistronic RNA, 281
Monoclonal antibody (mAb), 1005, 1145, 1146–47, 1146f, 1147f
Monocytes, 938f, **939**
Monod, Jacques, 361–62, 361f, 568
Monoderm, **714**
Monomer monoethylene terephthalate (MHET), 841f, 842
Mononucleosis, point-of-care rapid diagnostics, 1186t
Monophyletic ancestry, **27, 673**
Monosaccharides, 501
Montagnier, Luc, 421, 422f
Montagu, Mary Wortley, 17, 17f
Moraxella catarrhalis, 926
Morchella hortensis, 799
Mordant, **50,** 52f
Morels, 799, 800f
Morgan-Kiss, Rachael, 846–47
Morrill, Kathleen, 403f, 404
Morris, Jeff, 859
Mother cell, **151,** 152f, **715**
"Mother machines," microfluidic, 315, 315f
Motile bacteria, painting with, 443, 443f, 444
Motility
flagellar, **111–13,** 111f, 112f
gliding, **153,** 153f
twitching, **148**
Motility-related diarrhea, 1072
Mougous, Joseph, 352
Moyle, Jennifer, 30, 540, 540f
M protein, 1007, 1019
MPT (methanopterin), 594
MreB proteins, bacterial cytoskeleton, 96, 97f
mRNA (messenger RNA), **286,** 286t
cell structure, 78
preliminary transcript, 457
MRSA. *See* Methicillin-resistant *Staphylococcus aureus* (MRSA)
MsrE protein, 1115, 1115f
MSSA (methicillin-sensitive *Staphylococcus aureus*), susceptibility test, 1121, 1121f
Mucociliary escalator, **927,** 927f, 1064, eTopic 26.2
Mucor, 798
skin infections, 1060t
Mucosal immune system, 994–97, 995f, **997**
Mucous membranes, physical barriers to infection, 946–47, 947f
Mueller-Hinton agar, **1222**
Multicellularity, origins, 659, 659f
Multichannel isolation chip (iChip), 136, 136f
Multicloning site, 459
Multidrug resistance (MDR) efflux pumps, **1138,** 1138f
Multidrug-resistant (MDR) *Mycobacterium tuberculosis,* 19, 1069, 1071, 1138, 1195
Multidrug-resistant (MDR) pneumonia, 1135, 1136f
Multiple displacement amplification (MDA), 836
Multiple sclerosis, 1007t
Multiplex polymerase chain reaction, 446, **1176–77,** 1177f
Multiplicity of infection (MOI), **233**
Multivalent adhesion molecule 7 (MAM7), 1023, 1024f
Multivalent vaccine, 1067
Mumps virus, 1184t
Mupirocin, 1134
Murchison meteorite, 608f
Murein, **88–89,** 88f
Murein lipoprotein, 77f, **93**
Mushrooms, 801–3, 802f, 803f
classification, 789t
edible, 618–19, 619f
fairy ring, 787f
Musser, James, 1019
Mutagen(s)
defined, **318**
effects, 316–20, 318t, 319f–321f, eTopic 4.1
identified by using bacterial "guinea pigs," 321, 321f
Mutagenesis
random, 461, 462–63, eTopic 12.1
signature-tagged, 1017, eTopic 25.1
targeted (site-directed), 461–64, 464f, eTopic 12.4
Mutation(s), 316–22
auxotrophic, **321,** 330, 330f
causes and effects, 316–20, 318t, 319f–321f, eTopic 4.1
classes, 316–18, 317f
defined, **316**
divergence through, 673
frameshift, 317f, **318**
gain-of-function, **317**
influenza virus, 413
knockout, **317**
loss-of-function, **317**
missense, **317,** 317f
nonsense, **317–18,** 317f
point, **316,** 317f
potentiating, 687, 688f
silent, **317**

Mutation(s) (*continued*)
spontaneous, 319–20, 319f, 320f
starvation stress, 689–90, 690f
Mutation rate, molecular clock, 673–75, 674f
Mutator strain, **323**
Mutualism, 845t, 848, 848f
alphaproteobacteria, 730, 731f
cell-cell communication, 382
defined, **694**
insects with intracellular bacteria, 849–50, 850t, 851f
marine phototrophs, 585, 585f
vs. parasitism, 850–51
photosynthetic endosymbionts of animals, 848–49, 849f
Myasthenia gravis, 1007t
Mycelium(ia), **153,** 154f, 793f, **794**
aerial, **722–23,** 722f
nutrient absorption, 795
vegetative, **722**
Mycena, 869
Mycetozoa, 789t
Mycobacteria
cell envelope, 90, 95, 96f
growth asymmetry, 105
Mycobacteriaceae, 704t
Mycobacterium, 721, 724–25
flipping cytokine profile, 1046
growth factors, 134t
immune evasion, 992
intracellular immune avoidance, 1043, 1044
phagocytosis, 955
Mycobacterium avium-intracellulare, 1091
Mycobacterium bovis, 1005, 1094
Mycobacterium leprae, 725, 725f
acid-fast stain, 52
cell envelope, 90, 95
classification, 705
genome reduction, 350
Mycobacterium marinum, 956f, 957
Mycobacterium smegmatis, 725, 943
Mycobacterium tuberculosis, 724–25, 724f, 1069–71
acid-fast stain, 52, 53f
ATP synthase, 549
bedaquiline, 1144
cell envelope, 90, 95
chronic inflammation, 956–57, 956f
classification, 704t, 705
disease surveillance, 1191
drug development, 646–47, 646f
flipping cytokine profile, 1045
genome, 241t
growth asymmetry, 105
history, 10
intracellular, 956
isoniazid, 1119
lipiarmycin, 1131
meningitis, 1110t
modular synthesis, 594
multidrug-resistant, 19, 1138, 1195
nonhomologous end joint, 328, 329f
nucleic acid amplification tests, 1175
pathogen identification by staining, 1166–67, 1167f
programmed cell death, 177, 178f
protein degradation, 305
respiratory tract infection, 1066t
stopping programmed cell death, 1046
structure, 714
toxins, 1037t
transmission, 15, 16, 1016
vaccine, 968t
Mycolic acids, 95, **724**
Mycology, **787**
Mycomembrane, 95, 96f
Mycoplasma, 719
Mycoplasma genitalium, 241t
Mycoplasma mobile, 719–21, 720f
Mycoplasma penetrans, 719, 720f
Mycoplasma pneumoniae, 693, 704t, 1066t
Mycorrhiza(e), 802f, **803, 872–74,** 873f, 874f
arbuscular, **803,** 803f
carbon cycling, 888, 889f
Mycoses
opportunistic, 1155t
superficial, 1154, 1154f
systemic, 1154, 1155t
Mycrocystis, 865, 865f
Mycrocystis aeruginosa, 865, 865f
Myelitis, acute flaccid, 227, 1197
Myeloblast, 937, 938f
Myeloid-derived suppressor cells (MDSCs), 921, 921f
Myeloid stem cell, 938f
Myeloperoxidase, 955
Myrothecium, 189
Myxobacteria, 711t, 713f, 736, 736f
Myxococcales, 704t
Myxococcus, 153, 153f
Myxococcus, 736
toxins, 1037
Myxococcus fulvus, 1131
Myxococcus xanthus, 736, 736f
cell differentiation, 153, 153f
genome, 242
microbial attachment, 1022
Myxopyronin, 1131
Myxosarcina, 712
Myxospores, 711t, **736,** 736t

N

N_2 fixation. *See* Nitrogen (N2) fixation
N_2O (nitrous oxide) gas, 885, 885f, 903, 904f
NA (neuraminidase), 411, 416–18, 1149, eTopic 27.4
NA (numerical aperture) and resolution, **46–47,** 46f
NAATs (nucleic acid amplification tests), 1175–76, 1176f, eTopic 28.2
N-acetyl-glucosamine (NAG), 1125, 1126
N-acetylmuramic acid (NAM), 1124f, 1125, 1126
Nacy, Carol, 646, 646f
NAD. *See* Nicotinamide adenine dinucleotide (NAD)
Naderer, Thomas, 1046, 1047f
Naegleria, 789t
meningitis, 1110t
Naegleria fowleri, 805
Naffakh, Nadia, 418

NAG (*N*-acetyl-glucosamine), 1125, 1126
Na^+/H^+ antiporter, 127
NA (neuraminidase) inhibitors, **1149**
Nakagawa, Tatsunori, 763
Naked DNA, 337–38, 337f, 338f
Nalidixic acid, 249, 1130
NAM (*N*-acetylmuramic acid), 1124f, 1125, 1126
Nanoarchaeota, 753t, **755,** 780–81, 780f
Nanoarchaeum equitans, 753t, 755, 780, 780f
Nanohaloarchaeota, 753t
Nanoplankton, 861
Nanopore DNA strand sequencing, **451–52,** 452f, 453f, 454t
Nanoscale secondary ion mass spectrometry (NanoSIMS), **60,** 60f, 667, 667f
Nanotubes, **107,** 107f, 159, 338
  amino acid sharing, 585, 585f
Nanowires, **553–54,** 554f, 555f
NAP (neutrophil-activating protein), 1079
Na^+ (sodium) pumps, 550
Naqvi, Syed Wajih, 903
Narrow-spectrum antibiotics, 1140
Nasal cavity, 925–26, 926f, 927f
Nasopharynx, **926,** 927
Natamycin, 1155t
National Center for Biotechnology Information (NCBI), 693
Native microbial products, 647
Natronobacterium gregoryi, 171, 171f
Natronococcus, 753t, 755
Natto, 621t, **628–29,** 629f
Natural killer (NK) cells, 938f, **942,** 942f, 943f
Natural selection, 681–90
  bacterial resistance to antibiotics, 682–83, 684–85, 684f, 685f
  cell cycle, 140–47
  defined, **682**
  divergence through, 673
  evolution and growth cycles, 689–90, 689f, 690f
  experimental evolution in laboratory, 685–89, 686f, 688f
  genomic analysis, 682–84
Natural transformation, activation, 386–87, 387f
Nautilia, 737
NCBI (National Center for Biotechnology Information), 693
NDH-1 (NADH:quinone oxidoreductase complex), 545–46, 545f
NDM-1 (New Delhi metallo-beta-lactamase-1), 1127
Nealson, Ken, 382, 383f, 553, 556
Nebula RCW-49, 918, 919f
Necrotizing fasciitis (NF), 1011, 1012, 1060t, **1062,** 1062f
Nef protein, 424t
Negative selection, **985**
Negative stain, 52, **53,** 53f, **65**
Neidhardt, Fred, 78
Neisseria, 731
  aerobe, 174t
  oral and nasal cavity microbiome, 925
  pathogen identification, 1172
  toxins, 1037
  transformation of naked DNA, 337
Neisseria gonorrhoeae, 731
  antibiotic resistance, 1089, 1161
  classification, 704t
  complement, 994
  gonorrhea, 1086t, 1088–89, 1089f
  immune response, 937
  iron scavenging, 128
  microbial attachment, 1021t, 1033
  neutrophil extracellular traps, 938
  oxidized forms, 551–52
  pathogen identification, 1170
  pili, 110, 111f
  point-of-care rapid testing, 1186t
  slipped-strand mispairing, 394–95, eTopic 10.4
  stopping programmed cell death, 1046, 1047f
Neisseriales, 704t
Neisseria meningitidis, 731
  biological safety level, 1164t, 1165–66, 1165f
  cell fractionation, 80
  classification, 704t
  endotoxin, 1034–35, 1036f
  extracellular immune avoidance, 1041
  meningitis, 1104–6, 1105f, 1110t
  microbial attachment, 1022, 1022f, 1023
  pathogen identification, 1168, 1168f, 1170, 1171f, 1172
  sepsis, 1099
  specimen handling, 1165–66, 1165f
Neisseria sicca, 732
Nelfinavir mesylate (Viracept), 1148t, 1151
Nelmes, Sarah, 17, 17f
Nematode ectosymbionts, 701, 701f
Neocallimastix, 789t
  bovine rumen, 854
Neonatal septicemia, 1186t
Neotyphodium coenophialum, 876
Nespolo, Roberto, 659, 659f
NET(s) (neutrophil extracellular traps), **938–39,** 939f, 1027, 1028–29, eTopic 23.1
Neuraminidase (NA), 411, 416–18, 1149, eTopic 27.4
Neuraminidase (NA) inhibitors, **1149**
Neurospora, 798, 799
Neurotoxins A–G, 1026t
Neurotransmitters, amino acid decarboxylation, 522–23, 522f, 524–25
Neuston, 857f, 858
Neutralophiles, 161t, **169–70,** 169f, 170f
Neutrophil(s), **937–39,** 938f, 939f
  inflammatory response, 949, 950f
  Staphylococcus aureus, 1094–95
Neutrophil-activating protein (NAP), 1079
Neutrophil cathelicidins, 948, eTopic 23.3
Neutrophil extracellular traps (NETs), **938–39,** 939f, 1027, 1028–29, eTopic 23.1
Nevirapine, 1148t
New Delhi metallo-beta-lactamase-1 (NDM-1), 1127
Newman, Dianne, 59, 555, 670, 670f
Newton, Isaac, 10
Next-generation sequencing (NGS), 449–50, 449f
  library construction, 830f, 831
  pathogen identification, 1176
NF (necrotizing fasciitis), 1011, 1012, 1060t, **1062,** 1062f
$NH4^+$. *See* Ammonia ($NH4^+$)
NHEJ (nonhomologous end joining), 322t, **328,** 329f
Niccol, Andrew, 446

Niche, **160,** 161, 691, **839**
Niche adaptation, 520–21, 521f, 522f
Niche classification, 161, 161t
Niche construction, **839**
Nicotinamide adenine dinucleotide
  oxidized (NAD^+), 496–97, 496f
  reduced (NADH), 496–97, 496f
   aerobic oxidation, 537, 546, 547, 547f, 548
   electron transport system, 497
   structure and function, 497, 497t
  reduced (NADH):quinone oxidoreductase complex (NDH-1), 545–46, 545f
Nicotinamide adenine dinucleotide (NAD)
  elongation of replicating DNA, 257, 258f
  energy carrier, **496**
  Hemophilus influenzae, 1170
Nicotinamide adenine dinucleotide phosphate
  reduced (NADPH), 509, 510f
  reduced (NADPH) oxidase, 955
Nightingale, Florence, 10, 10f
Nirenberg, Marshall, 278, 278f, 289
Nitazoxanide, 1152f
Nitrate, 900
Nitrate reduction
  assimilatory, **903**
  dissimilatory, **903–4**
Nitric oxide (NO), 955
Nitrification, **124,** 901f, **902–3**
Nitrifiers, **902**
  betaproteobacteria, 731, 732f
  enrichment culture, 21
  nitrogen cycle, 124
  nitrogen oxygenation, **561**
Nitrites
  food preservation, 643
  nitrogen oxidation, 561
Nitrobacter, 124, 902
Nitrobacter winogradskyi, 728t, 901f
Nitrobenzene, 525–26, 526f
Nitrogen
  assimilation, 601, 601f
  eutrophication, 866
  global reservoirs, 884–85, 884f
  metabolism, 843t
  oxidized forms, 551–52
  sources, 899–900, 900f
Nitrogenase, **604–5,** 605f, 606f, 901
Nitrogen cycle, 899–905
  anaerobic ammonium oxidation, 904, 904f
  denitrification, 901f, 903, 904f
  dissimilatory nitrate reduction to ammonia, 903–4
  global, 21, 22f, 900, 900f
  microbial nutrition, 123–24, 124f
  nitrification, 901f, 902–3
  nitrogen fixation, 901–2, 901f, 902f
  oxidation states in nitrogen triangle, 900–904, 900t, 901f
  sources of nitrogen, 899–900, 900f
Nitrogen (N2) fixation, **21,** 22, 22f, 600–607, 601f
  agriculture, 902, 902f
  anaerobiosis and, 606
  cyanobacteria, 151–52, 153f
  early discoveries, 601–4, 604f
  mechanism, 604–6, 605f, 606f
  molecular regulation, 606–7, 607f
  mutualism, 694
  nitrogen cycle, **901–2,** 901f, 902f
  overview, 600–601
  rhizobium, 123, 601, 876–78, 876f–878f
Nitrogen-fixing bacteria, **123,** **551**
Nitrogen gas, 900
Nitrogenous bases, **243**
  biosynthesis, 613–14, 614f
Nitrogen oxidation, 561, 562f
Nitrogen triangle, 900–904
  denitrification, 901f, 903, 904f
  nitrification, 901f, 902–3
  nitrogen fixation, 901–2, 901f, 902f
Nitrosococcus, 731
Nitrosolobus, 731
Nitrosomonas, 731
  carbon assimilation and dissimilation, 842
  nitrification, 902
  nitrogen cycle, 124
Nitrosomonas europaea, 728t, 732f
Nitrosopumilales, 753t
Nitrosopumilus, 763–64, 764f, 908
  denitrification, 903
Nitrosopumilus maritimus, 753t, 763, 888
Nitrososphaera gargensis, 753t
Nitrosovibrio, 731
Nitrospira, 741
  electron transport system, 538
  metabolism, 728t
  nitrification, 902
Nitrospirae, 741
  classification, 704t, 705
  defined, **705**
  metabolism, 728t
Nitrospira marina, 704t
Nitrous oxide (N_2O) gas, 885, 885f, 903, 904f
NK (natural killer) cells, 938f, **942,** 942f, 943f
NO (nitric oxide), 955
Nocardia, 725
  wastewater treatment, 897, 897f
Node, **675,** 675f
Nod factors, 876
NOD-like receptors (NLRs), 950, 950t, **951–52,** 951f, 952f
  gut immune system, 995, 995f, 996
Noise, system, **478,** 478f
Nomenclature, 26, **690,** 692, 692t
Nonadaptive immunity. *See* Innate immunity
Nonbiodegradable molecules, 840–41, 840f
Nonessential genes, 462–63
Nongenetic categories, 693
Nongonococcal urethritis, 1086t, 1088, 1088f
Nonhomologous end joining (NHEJ), 322t, **328,** 329f
Nonnucleoside reverse transcriptase inhibitors, 1151
Nonreplicative transposition, 344, 344f
Nonribosomal peptide(s), **594**
Nonribosomal peptide antibiotics, **598–99,** 598f–600f, 602–3
Nonribosomal peptide synthetase (NRPS), **599**
Nonselective media, 1161

Nonself epitope, 1007
Nonsense mutation, **317–18,** 317f
Nonsterile sites, specimen collection, 1161, 1162f
Nonsteroidal anti-inflammatory drugs (NSAIDs)
  dysbiosis, 932
  inflammation, 954
Nori, **619,** 619f, 811, 812f
Noroviruses, **1075–76,** 1081t
  food contamination, 639, 640t
  wastewater, 897
Norris, Steve, 1087
Northern blot, 466t, 467, eTopic 12.1
Norwalk-like virus, 639, 640t
Norwalk virus
  food contamination, 639, 640t
  vaccine antigen, 473t
Nosema, 804t
Nosocomial infections, **1061,** 1116
Nostoc, 198, 789t
  classification, 704t
  filaments, 711, 711f
  multicellularity, 712
Nostocales, 704t
"NOT gate," 476, 476f
NPs (nucleocapsid proteins), **411**
  HIV, 423, 423f
NRPS (nonribosomal peptide synthetase), **599**
NSAIDs (nonsteroidal anti-inflammatory drugs)
  dysbiosis, 932
  inflammation, 954
N-terminal amino acid, 302
N-terminal rule, **305**
NtrB-NtrC system, 607, 607f
Nucleic acid amplification tests (NAATs), 1175–76, 1176f, eTopic 28.2
Nucleic acid–based detection, 1174–79
  nucleic acid amplification tests (NAATs), 1175–76, 1176f, eTopic 28.2
  profiling gut microbiota by PCR, 1179, 1179f
  programmable RNA sensors, 1178–79, 1178f, 1180–81
  quantitative reverse-transcription PCR (qRT-PCR), 1177–78, 1177f
  rapid, next-generation DNA sequencing, 1176
Nucleobase, **243**
Nucleocapsid proteins (NPs), **411**
  HIV, 423, 423f
Nucleoid, 77f, **78**
  bacterial chromosomes compacted into, 245–46, 245f
  defined, **246**
  DNA organization, 98–100, 99f, 100f
  DNA supercoiling in, 247–50, 248f–250f
Nucleomorph, **791**
Nucleoside inhibitors, 1150–51
Nucleotide(s), 243–44, 243f
Nucleotide excision, 322t
Nucleotide excision repair, 322t, 323, eTopic 9.1
Nucleotide inhibitors, 1151
Numerical aperture (NA) and resolution, **46–47,** 46f
NusG protein, 283
Nüsslein-Volhard, Christiane, 950–51
Nutrient(s)
  essential, **120**
  limiting, **866**
  macro-, **120**
  micro-, **120**
Nutrient deprivation, 176–80
  climate change, 180, 180f
  eutrophication, 179–80, 179f
  stress responses, 178–79
  survival genes, 177–78, 177f, 178f
Nutrient supplies, microbial growth, 120
Nutrient time-sharing, biofilms, 119, 119f, 120
Nutrient uptake, 124–30
  ABC transporters, 127–28, 128f
  active transport, 126–27, 126f
  facilitated diffusion, 125–26, 125f
  group translocation, 129–30, 130f
  selective permeability, 125
  siderophores, 128–29, 129f
Nutrition
  microbial, 120–24, 121f, 123f, 124f
  phage lambda replication, 406–7
Nystatin, 1154, 1154f, 1155t

O

O antigen, **93–94,** 94f, 1034, 1035f
Obama, Barack, 421, 636
Obesity and dysbiosis, 933–36, 933f–935f
Objective lens, **46–47,** 46f
  compound microscope, 47–48, 48f
Obligate aerobes, 551
Obligate intracellular bacteria, 135, 135f
Obligate intracellular pathogens, **1042,** 1045f
Obligate lithotrophs, 535, 557
Ocean acidification, 890
Ocean floor, 857f, 858, 862–65, 864f, eTopic 21.2
Ocean microbiome, 269, 270f, 271f
Oceanospirillales, *Deepwater Horizon* spill, 834f, 836, 836f
Ocular lens, **47**
Oenococcus oeni
  fermentation, 633f
  magnification, 39, 39f
Ogiri, 621t
OH- (hydroxide ions), 168–69
*OH (hydroxyl radical), 174, 174f
O (organic) horizon, 868, 868f
Ohsumi, Yoshinori, 1047
Oil spills
  Deepwater Horizon, 833–36, 834f–836f, 840, 858, 895
  microbial syntrophy, 527–29, 528f, 529f, 529t
Okazai, Reiji, 252
Okazaki, Tsuneko, 252
Okazaki fragments, **252,** 252f, 257, 257f
Okere, Uchechukwu, 523f, 524
Oleic acid, 86f
Oligosaccharides, 501
Oligotroph(s), 844t, **865**
  alphaproteobacteria, 729
Olysio (simeprevir), 1148t
Omalizumab (Xolair), 1005
omc (outer membrane–associated proteins), **555**

"omic" approaches, **828**
  protein abundance and function, 468
  RNA and transcription analysis, 467
OMP(s) (outer membrane proteins), 310, 311f
OmpF porin, 94, 95f
OMPs (outer membrane proteins), proteobacterial cell envelope, 94, 95f
OMZ (oxygen minimum zone), **894–95,** 895f
Oncogenes, 227, 229
Oncogenic viruses, 227, **229–30,** eTopic 11.4, eTopic 26.1
Oncolytic virus, 440
One Health Initiative, 1199–1200, 1199f
Onesimus, 17
Oomycetes, 789t, 792, **804,** eTopic 20.1
Oophila amblystomatis, 849, 849f
Oparin, Aleksandr, 669
Open reading frame (ORF), **281,** 282f, 456–57, 456f
Operational genes, 681, 681f
Operational taxonomic units (OTUs), 269, **831**
Operators, **359**
Operon, **281,** 282f
Operon fusion, 466t, **467,** 467f
Ophiostoma novo-ulmi, 879
Opines, 330, 653, 653f
Opisthokont(s), 789t, 790–91, 790f
Opisthokonta, 789t, 790–91, 790f
Opportunistic infections, 1090f, 1091
Opportunistic mycoses, 1155t
Opportunistic pathogens, **923, 1012,** 1013f
Opsonin, **955,** 955f, 959
Opsonization, **955,** 955f, **973**
Optical density, **140**
Oral cavity, 925–26, 926f, 927f
Oral groove, **815,** 816f
Oral polio vaccine (OPV), 999
ORF (open reading frame), **281,** 282f, 456–57, 456f
Organic biosignatures, 665t, 667
Organic (O) horizon, 868, 868f
Organo-, 122
Organotrophs, **121–22**
Organotrophy, **121,** 487–88, 487t, **534, 535**
Organ transplant rejection, 988, eTopic 24.5
OR gate, 476, 476f
Origin *(oriC),* replication, **252,** 252f, 254, 254f
Origins of life, 659–700
  Archaean eon, 663–64, 664f, 665t
  banded ion formations reveal oxidation, 667–68, 668f, 669f
  conditions, 661
  divergence of three domains of life, 677–78, 677f, 679t
  divergence through mutation and natural selection, 673
  early life, 660–61, 661f
  early oxidation-reduction reactions, 670–71, 670f
  elements of life, 661–63, 662f, 663f
  endosymbiosis, 694–96, 695f, eTopic 17.6
  evolution, 673–81
  evolution and growth cycles, 689–90, 689f, 690f
  experimental evolution in laboratory, 685–89, 686f, 688f, eTopic 17.3, eTopic 17.4
  first cell formation, 669–72
  genomic analysis, 682–84
  geological biosignatures, 663–67, 664f, 665t
  Hadean eon, 663, 664f
  horizontal gene transfer, 678–80, 680f
  isotope ratios, 665–67, 665t, 666f, 667f
  microbial species and taxonomy, 690–94, 691f, 692t, eTopic 17.5
  microfossils, 664–65, 665t, 666f, 667f
  mitochondria and chloroplasts, 696–98, 696f, 697f
  molecular clocks based on mutation rate, 673–75, 674f
  multicellularity, 659, 659f
  natural selection and adaptation, 681–90
  organic biosignatures, 665t, 667
  overview, 660
  phylogenetic trees, 675–77, 675f, 676f, eTopic 17.2
  prebiotic soup, 669–70, 669f
  proposed timeline, 668, 669f
  reconciling vertical and horizontal gene transfer, 680–81, 681f
  RNA world, 671, 671f, eTopic 17.1
  strongly selective environments: antibiotic selection, 684–85, 684f, 685f
  unresolved questions, 671–72
Ornithine, 611
Oró, Juan, 14, 14f, 25
Oropharynx, **926**
Orotomides, 1154–56
Orth, Kim, 1023
Ortholog(s), **213, 349,** 350f
Orthologous genes, **213**
Orthomyxoviruses, 212t
Orthophenylphenol, 186
Osborn, Mary Jane, 80
Oscillatoria, 587f, 704t, 711, 711f
Oscillatoriales, 704t
Oscillator switch, 477, 477f
Oscillobacter ruminantium, 856, 856f
Oseltamivir (Tamiflu), 1148t, 1149
Osmolarity, 161t, **167–68,** 167f, 168f
Osmosis, 84
Osmotic diarrhea, 1072
Osmotic pressure, **84**
Osmotic stress, 167–68, 167f
Osteomyelitis, 1063
Osterholm, Michael T., 1192, 1193
Ostreococcus, 789t
Ostreococcus tauri, 786, 787f
OTUs (operational taxonomic units), 269, **831**
Outer envelope, 76
Outer membrane, 77f, **78**
  protein journeys through, 310–11
  protein journeys to, 310, 311f
Outer membrane–associated proteins (Omc), **555**
Outer membrane proteins (OMPs), 310, 311f
  proteobacterial cell envelope, 94, 95f
Outgroup, 477n, **676**
Oxaloacetate, 519
Oxaloacetic acid, 518
Oxazolidinones, 1132f, **1133**
Oxidation
  food spoilage, 634, 634f
  origins of life, 665t, 667–68, 668f, 669f
Oxidation-reduction reactions, 670–71, 670f

Oxidation-state changes, carbon flux, 885–86, 886t
Oxidative burst, **956**
Oxidative phosphorylation, **549**
TCA cycle, **518,** 519f
Oxidoreductase(s), **539, 545**
initial substrate, 545–46, 546f
Oxidoreductase protein complexes, 545–49, 546f–548f
Oxoglutarate, 518
2-Oxoglutarate, 609–10, 610f
Oxolinic acid, 249
Oxygen, 161t, 173–76
aerobes *vs.* anaerobes, 174–75, 174t, 175f
benefits *vs.* risks, 173–74, 173f, 174f
culturing anaerobes, 175–76, 176f
ecosystem, 843–44, 843t, 844f
Oxygen demand, biochemical (biological), **858, 893–95,** 894f–896f
Oxygen-dependent killing pathway, 955–56
Oxygenic photolysis, 575, 576f
Oxygenic Z pathway, **572,** 573, 575, 576f
Oxygen-independent killing pathway, 955–56
Oxygen minimum zone (OMZ), **894–95,** 895f
Oxyrase, 175
Oyster mushrooms, 619

P

PA (Protective antigen), **1032,** 1033f
PABA (para-aminobenzoic acid), 1130, 1130f
Pace, Norman, 268, 762, 829
Paenibacillus dendritiformis, 177f
PAHs (polycyclic aromatic hydrocarbons), 503, 523
Painting with motile bacteria, 443, 443f, 444
Palindrome, **458**
Palmitic acid, 86f
Palsa, 890–91, 891f
PAM (protospacer adjacent motif), 465
PAMP (pathogen-associated molecular pattern), 963
Pande, Samay, 585
Pandemic, **1189**
Pandoraviruses, 4
Pandorina, 810
Paneth cells, 948, 948f
Pangenome, **691–92,** 691f
Pangenomic gene differences, 1059, 1059f
Panspermia, **672**
Pantoea agglomerans, 1199
Pantoea carbekii, 827, 827f
Panton-Valentine leukocidin, 1026t, 1027
PaP (pyelonephritis-associated pili), 1021, 1021f
Paper chromatography, 587, eTopic 15.3
Paper products, industrial microbiology, 648t
Papillomaviruses, 436t. *See also* Human papillomavirus (HPV)
Papovavirus, 436t
Para-aminobenzoic acid (PABA), 1130, 1130f
Parabacteroides, 929f
Paracoccidioidomycosis, 1155t
Paracoccus, 124
Paracoccus denitrificans, 549
Paracoccus pantatrophus, 728t
Paralogous genes (paralogs), **349,** 350f, **683**
Paramecium, 815, 816f
classification, 789t, 792, 792f
phase-contrast microscopy, 62, 62f
Paramecium bursaria, 694–95, 695f, 696
Paramecium tetraurelia, 792f
Paramyxoviruses, 212t, 224f
Paraphyletic group, **692**
Pararetroviruses, 211f, 212t, **213,** 230–32, 231f
Parasites and parasitism, 845t
alphaproteobacteria, 730, 731f, 732f
bdellovibrios, 736–37, 737f
defined, **1012**
intestinal, 823, 823f
vs. mutualism, 850–51
protozoa, 821–23, 821f–823f
symbiosis, **694, 850**
Pardee, Keith, 1178, 1178f
Parenteral route of transmission, **1016**
Parfocal lenses, **48**
parsABS system, 260
Parvarchaeota, 753t
Parvoviruses, 212t
Passive diffusion, 84
Passive transport, **85**
Pasteur, Louis, 12–13, 12f, 18, 18f, 182, 642
Pasteurella multocida, 1026t
Pasteurella pestis, 26
Pasteurization, **182–83**
flash, 624
food preservation, 642
Pathogen(s), 2–3, 3f
intracellular, 956, **1042,** 1042f–1043f
opportunistic, **1012,** 1013f
primary, **1012**
survival within host, 1040–51
susceptibility to, 1023
Pathogen-associated molecular pattern (PAMP), 963
Pathogenesis, 1011–57
adhesins, 1021t, 1023, 1023f, 1024f
biofilms, 1023–24, 1025f
covert, 1085
defined, **1012**
deploying toxins and effectors, 1035–40, 1036f–1039f
effect of infections on microbiota, 1016–17
endotoxins, 1025, 1034–35, 1035f, 1035t, 1036f
exotoxins, 1025–34, 1026t, 1027f–1033f
host-pathogen interactions, 1012–20
immuno-, 1016
infection cycles, 1014–16, 1014f, 1015f
microbial attachment, 1020–25, 1021f–1025f, 1021t
overview, 1012
pathogen evolution by horizontal gene transfer, 1019–20, 1019f
pathogenicity islands, 1017–19, 1018f
pathogen survival within host, 1040–51, 1042f–1050f
pili, 1020–23, 1021f
portals of entry, 1016
quorum sensing, 384, 384f
terminology, 1012–14, 1013f, 1014f
tools used to probe, 1051–54, 1052f–1054f, eTopic 25.1, eTopic 25.2
virulence factors, 1017, eTopic 25.1, eTopic 25.2

Pathogen-host specificity, 948, eTopic 23.2
Pathogenicity, **1013**
Pathogenicity islands, **1017–19**
food, **639,** 639f
horizontal gene transfer, **351–52,** 352f
model, 1018, 1018f
Pathogen identification
vs. normal biota, 1173
algorithms, 1167–69, 1168f
automated instrument, 1168, 1168f, 1169t
culture and phenotype, 1166–74
ELISA test, 1181–82, 1182f
fluorescent antibody staining, 1182–83, 1183f
Gram-negative bacteria, 1169–70, 1169f, 1170t, 1171f
Gram-negative diplococcus, 1170–72, 1171f
Gram-positive pyogenic cocci, 1172–74, 1172f, 1173f
growth and biochemical testing, 1167–74
immunology-based, 1179–83, 1182f, 1183f
mass spectrometry, 1174, 1175f
molecular and serological, 1174–87
nonenteric Gram-negative bacteria, 1170–72, 1171f
nucleic acid amplification tests (NAATs), 1175–76, 1176f, eTopic 28.2
nucleic acid–based, 1174–79, 1176f–1179f, 1180–81
point-of-care rapid diagnostics, 1184–87, 1185f, 1186t
profiling gut microbiota by PCR, 1179, 1179f
programmable RNA sensors, 1178–79, 1178f, 1180–81
quantitative reverse-transcription PCR (qRT-PCR), 1177–78, 1177f
rapid, next-generation DNA sequencing, 1176
selected diseases, 1183, 1184t, eTopic 28.4
staining, 1166–67, 1167f
Patient history, 1058
Patient zero, **1189–90,** 1190f
Pattern recognition receptors (PRRs), 949–52
antiviral agents, 1153
defined, **950**
examples, 950t
gut immune system, 995, 995f, 996
and inflammasomes, 951f
Patterson, Tom, 189
PBPs (penicillin-binding proteins), **89,** 89f, **1126**
PCM (phase-contrast microscopy), **62–63,** 62f, 63f
PDC (pyruvate dehydrogenase complex), **516,** 516f, eTopic 13.2
Pechter, Kieran, 145, 145f
Pectin, 501, 502f
Pediococcus, 717–18
Pelagibacter
cryo-electron microscopy, 67, 67f
genome reduction, 349–50, 351f
photoheterotrophy, 727
phototropic, 709t
Pelagic marine food web, 862, 863f
Pelagic zone, **857–58,** 857f
Pellicle, 148, 149f
Pelomyxa, 4t
Pelotomaculum, 527
Pelvic inflammatory disease, 1088
Penicillin(s), 1126
allergy, 1132
control of bacterial growth, 188, 188f
discovery, 19, 19f, **1117–18,** 1117f
efficacy, 1126
mechanism of action, 1126
resistance, 1126–27
structure, 85f, 1126, 1126f
Penicillinase, 1136, 1137f
Penicillin-binding proteins (PBPs), **89,** 89f, **1126**
Penicillin G, 1126f
Penicillin-resistant *Streptococcus pneumoniae,* 1135, 1136f
Penicillium, 786
cheese, 625, 635
classification, 789t
discovery of penicillin, 1117–18, 1117f
reproductive cycle, 798, 801, 801f
Penicillium bilaii, 650
Penicillium griseofulvum, 1154, 1154f
Penicillium notatum, 19, 19f, 188, 786f
Pentose phosphate cycle, reductive. *See* Calvin cycle
Pentose phosphate pathway (PPP), **506,** 506f, **509–10,** 511f
Pentose phosphate shunt (PPS), 506, 506f, 509–10, 511f
PEP (phosphoenolpyruvate), 129, 499, 500f
PEPFAT (President's Emergency Plan for AIDS Relief), 421
Peptidases, 305
Peptide(s), 77f
catabolism, 501, 502f, 503f
Peptidogenomics, 602–3, 602f, 603f
Peptidoglycan(s)
cell membrane, 88–90, 88f–90f, 702
defined, **79**
pseudo-, 702
Peptidoglycan synthesis, 1124f, 1125, 1126
Peptidoglycan synthesis inhibitors, 89–90, 89f, 90f, 1124f–1128f, 1125–29, 1125t
Peptidyltransferase, **292**
Peptidyl-tRNA site (P site), 294f, **295**
Peptostreptococcus, 854
Perchloroethene, 564, 565f
Perforin, **942, 989**
Perfringolysin O, 1026t
Peridinium, 789t
Periodontal disease, 926, 926f
Periplasm, 77f, **78**
cell wall lysis, 80
export of prefolded proteins to, 309–10, 310f
protein export to, 307–9, 308f, 309f
proteobacterial cell envelope, 93f, 94
Periplasmic proteins, 307, 308
Peritoneal fluid, 1161
Permafrost, thawing, 890–92, 891f, 892f
Permeases, **125**
Persistence, 145, 145f
Persister cells, **1142–43,** 1145
Personal care products, 648t
Pertussis toxin, 1026t, 1040, 1040f
Pertussis vaccine, 968t, 998t
Pesticides, 472–73, 472f
PET (polyethylene terephthalate), 841–42, 841f
Petechiae, **1035,** 1036f
Petranovic, Dina, 796, 797f
Petri, Julius Richard, 15
Petri dish, **15**

Petroff-Hausser counting chamber, 135–38, 138f
Petromyzon marinus, 466
Peyer's patches, 945f, 995f.996
PfEMP1 protein, 1097
PFUs (plaque-forming units), **235**
PGA (3-phosphoglyerate), 587, 588, 588f
pH, 161t, 168–73
  classification, 169, 169f
  ecosystem, 845
  food preservation, 642, 642f
  homeostasis and acid resistance, 171–73, 172f, eTopic 5.4
  neutralophiles, acidophiles, and alkaliphiles, 169–71, 170f–172f
  nutrient availability, 169
  optima, minima, and maxima, 169, 169f, eTopic 5.3
PHA (poly-3-hydroxyalkanoate), 110
Phaeophyceae, 792, 812, 814, 814f
Phage(s). *See* Bacteriophage(s)
Phage display, **473–74,** 474f, eTopic 12.10
Phage lambda, 402–10
  attachment and infection, 406–8, 406f, 407f
  classification, 212t
  discovery, 403–4, 403f
  genome replication, 404f, 405–6, 405f
  integrated viral genome, 197
  lysogeny, 403, 408–9, 408f
  overview, 402, 402f
  specialized transduction, 336–37, 336f
  structure, 404–5, 404f, eTopic 11.1
  synthetic biology, 409, 410f
Phage M13, 406, eTopic 11.2
Phage therapy, 189, 190–91, 190f, 191f
Phage transduction, 334–37, 335f, 336f
Phagocytes, 954–55, 954f, 955f
Phagocytosis, 938–39, 939f, 954–55, 954f
  defined, **805**
  opsonization, 955, 955f
  resistance, 1041
Phagolysosome, 938f, 955
Phagosome(s), **938,** 938f
  escape from, 1044, 1045f
Phagosome-lysosome fusion, 938f, 955
  inhibiting, 1043–44
Phanizomenon, 153f
Phase-contrast microscopy (PCM), **62–63,** 62f, 63f
Phase interference, 62, 62f
Phase variation, 392–95
  defined, **392**
  gene inversion, 392–94, 394f
  slipped-strand mispairing, 394–95, eTopic 10.4
PHB (polyhydroxybutyrate), 110
pH difference, proton motive force, 541–42, 542f
Phenanthrene bioremediation, 523–25, 523f
Phenazines, 555
Phenol, 186, 186f
Phenol coefficient, **186,** eTopic 5.7
Phenolic(s), 186f
Phenolic glycolipids, 95, 96f
Phenotype, **318**
  actualization of novel, 687–88, 688f
  radically new, 687, 688f
  refinement of novel, 688–89, 688f
Phenotypic categories, 693
Phenylalanine, 612–13, 613f
Phipps, James, 17, 17f
Phosphate(s)
  eutrophication, 866
  food preservation, 643
  hydrolysis releasing, 495
Phosphate cycle, 907–8, 907f
Phosphatidate, 82
Phosphatidylethanolamine, 82–83, 83f
Phosphatidylgylcerol, 82, 83f
Phosphite, 908
Phosphodiester bond, 243f, **244**
Phosphoenolpyruvate (PEP), 129, 499, 500f
Phosphofructokinase, 77f, 508
6-Phosphogluconate, 509, 510f
3-Phosphoglyerate (PGA), 587, 588, 588f
Phospholipase C, 1027
Phospholipase toxins, 1027
Phospholipids
  bacterial composition, 79
  catabolism, 502f
  cell membrane, 82–83, 82f
  defined, **82**
  fatty acid component, 86, 86f
  side chains, 86, 86f
  structure and types, 82–83, 83f
Phosphonate, 908
5-Phosphoribosyl-1-pyrophosphate (PRPP), 613, 614f
Phosphorylation, 302, 303f
  ADP to ATP, **494–95,** 494f
  glucose, 507, 508f
  organic molecule, 495
  oxidative, **518,** 519f, **549**
  response regulators, 361
  substrate-level, **512**
Phosphoryl groups, hydrolysis, 495, 495t
Phosphotransferase system (PTS), **129–30,** 130f, eTopic 4.1
  inducer exclusion, 365–66, 366f, eTopic 4.1, eTopic 10.1
Photic zone, **858**
  carbon cycling, 887, 888f
Photo-, 122
Photoautotroph(s), **121,** 535
  Calvin cycle, 586
Photoautotrophy, **122, 582**
Photobacterium profundum, 676
Photoexcitation, 567–68
  chlorophyll, 568–72, 569f–572f
  defined, **568**
Photoferrotrophy, **670,** 670f
Photoheterolithotrophs, 726
Photoheterotroph(s), **122,** 535, 575
  alphaproteobacteria, 728–29, 729f
  Calvin cycle, 586
Photoheterotrophy, **568,** 727
  Haloarchaea, 777–78, 777f
Photoionization, **568**
Photolithoautotroph(s), **122**
Photolithoautotrophy, 122
Photolithoheterotroph, 575, eTopic 14.2

Photolysis, 568–72
  antenna complex and reaction center, 570–71, 571f, 572f
  chlorophylls absorb light, 569–70, 570f
  defined, **568**
  electron transport system, 571–72
  overview, 569
  oxygenic, 575, 576f
Photons, 43
Photoorganotroph(s), **122**
Photoorganotrophy, **122**
Photophosphorylation, cyclic, **575**
Photoreactivation, 322t, **323**
Photoreceptors, 38, 39f
Photorhabdus luminescens, 676, 869, 872f
Photosynthesis, **21, 568,** 569, **582**
Photosynthetic endosymbionts of animals, 848–49, 849f
Photosynthetic membranes, 571, 572f
Photosystem I (PS I), 572–75
  Chlorobia, 573–74, 573f
  defined, **572**
  overview, 573
  oxygenic photolysis, 575, 576f
Photosystem II (PS II), 572–75
  alphaproteobacteria, 574–75, 574f
  defined, **572**
  overview, 573
  oxygenic photolysis, 575, 576f
Phototrophs, 708, 709t
  Calvin cycle, 587f
  endosymbiotic, 22
  gammaproteobacteria, 733, 733f
  lake, 866, 866f
  organelles, 108–9, 110f
Phototrophy, 566–77
  chlorophyll photoexcitation and photolysis, 568–72, 569f–572f
  defined, **534, 535,** 566
  energy acquisition, **121–22, 487,** 487t, 488
  overview, 566, 567f
  photosystems I and II, 572–75, 573f, 574f, 576f
  retinal-based proton pumps, 567–68, 567f, 568f
Phycobilisome, 571, 575, 576f
Phylloquinone/phylloquinol (PQ), 573, 573f
Phylogenetic relatedness, 691
Phylogenetic trees, **675–77,** 675f, 676f, 703f, eTopic 17.2
Phylogeny, **673**
Phylum(a), **702**
Physarum polycephalum, 789t, 807
Physical agents that kill microbes, 182–85, 182f–185f
Phytate, 625
Phytoestrogens, 876
Phytophthora cinnamomi, 189
Phytophthora infestans, 444, 804
Phytophthora ramosum, 804
Phytoplankton, **808,** 861, **862**
Pickled foods, 621t
Picoeukaryotes, 789t
Picoplankton, 861
Picornaviruses, 227–28, 228f
Pidan, 621t, **629,** 629f
Piezophiles, 161t, **165–66,** 165f, 844t, **862**
Pilhofer, Martin, 108–9, 108f
Pilot protein, 308, 308f
Pilus(i), **110,** 111f
  electrically conductive, 553–54, 554f, 555f
  microbial attachment, **1020–23,** 1021f
  pyelonephritis-associated, 1021, 1021f
  sex, 330, 331f
  type I, 1021–22, 1021f
  type IV, 1022–23, 1022f
Piptoporus, 802
Pirellula, 704t
Pithovirus, 209, 209f
Placental barrier, pathogens that cross, 1087
Plague, 7–10, 7f, 26, 1097–99, 1098f, 1099f
Planctomyces, 742, 742f
Planctomyces bekefii, 742
Planctomycetes, 742, 742f
  classification, 704t, 705–6
  defined, **705**
  metabolism, 728t
Planctomycetes-Verrucomicrobia-Chlamydia (PVC), 741–44, 742f, 743f
  classification, 704t, 705–6
  defined, **705**
Plankton, **861**
Planktonic cells, **140**
Planktonic communities, 861–62
Planktonic food webs, 862, 863f
Plantae, 808–15
  classification, 789t, 791–92, 791f
  defined, **791**
Plant bacteria, human pathogens, 1199
Plant endophytic communities, 875–76, 875f
Plant foods, spoilage, 635t, 636
Plant gall, 730, 731f
Plant microbial communities, 867–80
  microbes associated with roots, 872, 873f
  mycorrhizae, 872–74, 873f, 874f
  nitrogen fixation by rhizobia, 876–78, 876f–878f
  plant endophytic communities, 875–76, 875f
  plant pathogens, 879, 879f
  soil food web, 869–70, 870f–872f
  soil microbiology, 868–69, 868f
  wetland soils, 874–75, 875f
Plant pathogens, 879, 879f
Plant viruses
  DNA pararetroviruses, 230–32, 231f
  entry to host cells, 230, 230f
  host defenses, 232
  replication cycles, 230–32, 230f, 231f
  transmission through plasmodesmata, 230, 231f
Plaque, **196, 235**
Plaque-forming units (PFUs), **235**
Plaque isolation and assay
  animal virus, 236, 236f
  bacteriophage, 235–36, 235f
Plasma cells, **964,** 965–66, **974**
  B cells differentiated into, 975–78, 976f, 977f
Plasma membrane, **78**
  disruption, 1025, 1026t

Plasmids, 261–63
antibiotic resistance, 261, 263, 331
autonomous, 334
defined, **261**
lentiviral vectors, 655, 655f
mobilizable, 333
regulation of replication, 262, 262f
secondary chromosome evolution from, 264–65, 266f
segregation, 263, 264f
size, copy number, and cargo, 261, 261f
transferable, 330–34, 332f–334f
transmission between cells, 261–62
tricks to ensure inheritance, 263, 263f
Plasmodesma(ta), **230,** 231f
Plasmodial slime mold, **807**
Plasmodium
climate change, 1198
pathogen identification, 1183
Plasmodium(ia), **807**
Plasmodium falciparum, 818–20, 819f, 820f
classification, 789t, 792
malaria, 1096–97, 1096f
point-of-care rapid diagnostics, 1186t
Plasmodium malariae, 1096
Plasmodium ovale, 1096
Plasmodium vivax, 1096
Plate culture
animal virus, 236, 236f
bacteria, 131–32, 131f, 132f
bacteriophage, 235–36, 235f
Platelets, 937, 937f, **942**
Platensimycin, 1144
Pleodorina, 810
Pleural fluid, 1161
Pleurocapsa, 712, 712f
Pleurocapsales, 704t
Pleurotus, 619
Plum pox virus, 230, 230f
Pluripotent stem cells, 938f
Plus (+) strand mRNA, 415–16, 417f
PMCA (protein-mediated cyclic amplification), 1111, eTopic 12.1
PMF. *See* Proton motive force (PMF)
PMNs (polymorphonuclear leukocytes), 937–39, 938f
Pneumococcal lung infection, 1064–67, 1065f
Pneumococcal pneumonia, point-of-care rapid diagnostics, 1185, 1185f, 1186t
Pneumococcal polysaccharide vaccine (PPSV), 1064–66
Pneumococcal vaccine, 998t
Pneumocystis carinii. See Pneumocystis jirovecii
Pneumocystis jirovecii, 796, 803–4, 804t
AIDS, 1090f, 1091
opportunistic infection, 1012, 1013f
respiratory tract infection, 1066t
Pneumocystosis, 1066t
Pneumolysin, 1026t
Pneumonia
bacterial, 1064–67, 1065f, 1066t
multidrug-resistant, 1135, 1136f
Pneumonic plague, **1098**
PodJ protein, 103–5, 104f
Point mutation, **316,** 317f
Point-of-care (POC) rapid diagnostics, 1184–87, 1185f, 1186t
Polar aging, 105–6, 105f, eTopic 3.3
Polar head group, 82
Polarity, cell, 103–5, 103f–105f, eTopic 3.2
Poliovirus
classification, 212t
culture, 234, 234f
disease, 201
replication, 227–28
tropism, 223
Poliovirus vaccine, 968t, 998t, 999
Poly-3-hydroxyalkanoate (PHA), 110
Polyamines, 79
Polybrominated toxins, 1
Polychlorinated aromatics, 502f
Polychlorinated biphenyls, 503
Polycistronic RNA, 281
Polycyclic aromatic hydrocarbons (PAHs), 503, 523
Polydnaviruses, 200–201, 201f
Polyenes, 1155t
Polyesters, 596
industrial microbiology, 648t
Polyethylene terephthalate (PET), 841–42, 841f
Polyhydroxybutyrate (PHB), 110
Polyketide antibiotics, **594,** 597–98, 597f
Polymerase, 251, 253f
Polymerase chain reaction (PCR), **3, 32**
amplification of specific genes, 445–46, 445f, 446f
defined, **449**
disease surveillance, 1192
gut microbiota, 1179, 1179f
Koch's postulates, 16
multiplex, 446, 1176–77, 1177f
pathogen identification, 1175
quantitative (real-time), **446–48,** 447f, 448f, 1176–78, 1177f
technique, 445–46, 445f, 446f
Polymorphonuclear leukocytes (PMNs), 937–39, 938f
Polymyxin, 1129
Polypeptide synthesis, 294–97, 294f–296f
Polyphyletic ancestry, **27**
Polysaccharides, **501,** 502f
Polysaccharide utilization locus (PUL), 504, 505f, 521, 522f
Polysomes, **299,** 299f
Polyvinyl chloride (PVC), 564
Popovi, Mara, 1058
Population, **828**
Pore-forming exotoxin, 1027, 1030f
Porins, **94,** 95f
Porphyra, 619, 789t, 811, 812f
Porphyrans, 501, 502f, 521, 522f
Porphyromonas, 926
Porphyromonas gingivalis, 1037t
Portals of entry, 931, 1016
Portobello mushrooms, 619
Positive selection, **985**
Postattachment inhibitor, 1151
Posttreatment controllers, 1152
Potassium ions, 541, 541f
Potato spindle tuber viroid, 206, 207f
Potyvirus, 230, 230f

Poultry, spoilage, 635–36, 635t
Pour plate technique, **139**
Poxviruses, 205, 206f, 436t
PPD (purified protein derivative), 1069
ppGpp (guanosine tetraphosphate), 380, 380f
(p)ppGpp (guanosine tetra- and pentaphosphate), 150
PPP (pentose phosphate pathway), **506,** 506f, **509–10,** 511f
PPS (pentose phosphate shunt), 506, 506f, 509–10, 511f
PPSV (pneumococcal polysaccharide vaccine), 1064–66
PQ (phylloquinone/phylloquinol), 573, 573f
Prebiotic soup, **669–70,** 669f
Pre-catenanes, **258**
Predation, 585
Predators, **843,** 843f
Preexposure prophylaxis (PrEP), 1152
Preliminary mRNA transcript (pre-mRNA), **457**
President's Emergency Plan for AIDS Relief (PEPFAT), 421
Pressure
  growth rate, 161t, 165–66, 165f, 166f
  to kill microbes, 182, 182f
Pressure-sensitive channels, 167
Prevalence, **1189**
Prevotella
  cell structure, 78
  gut microbiome, 855, 929
  oral and nasal cavity microbiome, 926, 926f
  respiratory tract biome, 927
"Prey" proteins, 471, 471f, eTopic 12.7
Primary algae, **788, 791,** 808–11
Primary antibody response, **974–75,** 975f
Primary pathogens, **1012**
Primary producers, **842,** 843f
Primary recovery, **649–50,** 651f
Primary syphilis, **1087,** 1087f
Primary tuberculosis, 1069–70, 1070f
Primase, 253f, **254**
Primer(s), **446,** 446f
Primer extension, 466t, 467, eTopic 12.1
Primer fragment, 253f, 254, 255
Priming oligonucleotide, 447
Prions, **207,** 207f
  diseases, **1109–11,** 1111f, 1111t
  food irradiation, 184
Probabilistic indicator, 694, **1169**
Probiotics, **189, 932–33.**932f
Processed food, 636
Processivity, **450**
Prochlorales, 704t
Prochlorococcus
  bacteriophages, 195, 195f, 196
  classification, 703, 704t, 710
  culturing, 838
  evolution and range expansion, 323
  food web, 843
  gene reduction, 684
  genomic islands, 352, 353f
  marine ecosystem, 859, 859f
  membrane vesicles, 106, 106f
  phosphate cycle, 908
  phytoplankton, 862
  RNA stability, 287, 287f
  size, 4t
Prochlorococcus marinus
  organelles, 110f
  pangenome, 692
  photoexcitation and photolysis, 569
  size, 40f
Programmable RNA sensors, 1178–79, 1178f, 1180–81
Programmed cell death, **177–78,** 178f
  stopping, 1046–47, 1047f–1049f
Programmed immune cells, 996
Prokaryotes, **4**
  classification, 26, 27
  size and shape, 40, 41f
ProMED, 1195
Promoters, **242, 279–81,** 279f, 280f, 282, 282f
Prontosil, 1118
Proofreading, **255**
Prophages, **197,** 198, 217, 218t, **403**
Propionate, obesity, 934–35, 935f
Propionibacteriaceae, 704t
Propionibacterium, 625
Propionibacterium acnes, 925, 925f
Propionibacterium freudenreichii, 204, 622
Propionibacterium shermanii, 644, 704T
Propionic acid fermentation, **622,** 622f
Prostaglandins, 953f, 954
Prostheca, 109
Prosthecobacter, 704t
Prosthecobacter debontii, 108–9, 108f
Prosthecobacter dejongeii, 742
Protease inhibitors, **429,** 1151, 1151f
Proteasomes, 77f, 305, eTopic 8.6
Protective antigen (PA), **1032,** 1033f
Protein(s)
  abundance and function, 468–71, 469f–471f, 469t
  membrane, 83–84, 84f
  transport, 77f, 82f, 83–84, 84f, **85**
Protein A, **1041**
Protein analysis, 468–71, 469f–471f, 469t
  membrane vesicle proteins, 81, 81f
Protein-coding region, **281,** 282f
Protein Data Bank, 72
Protein degradation, 305–6, 305f, 306f
Protein domains, 100f
Protein folding, 303–5, 304f
Protein H, 404–5
Protein hypothesis, 240
Protein kinases, 952
Protein-mediated cyclic amplification (PMCA), 1111, eTopic 12.1
Protein modification, 302–6
  degradation, 305–6, 305f, 306f
  folding, 303–5, 304f
  processing after translation, 302–3, 303f, 304f
Protein processing after translation, 302–3, 303f, 304f
Protein-protein interactions, 470–71, 471f, eTopic 12.7
Protein secretion, 306–12
  defined, 306–7
  export of prefolded proteins to periplasm, 309–10, 310f
  export out of cytoplasm, 307

export to cell membrane, 307, 307f
export to periplasm, 307–9, 308f, 309f
general pathway, 307–9, 308f, 309f
journeys through outer membrane, 310–11
journeys to outer membrane, 310, 311f
type I, 311–12, 311f
Protein synthesis
AB exotoxins that target, 1033–34
disruption, 1025, 1026t
elongation, 295–96, 295f
initiation, 293–94, 294f, 295
stages, 294–97, 294f–296f
termination, 296–97, 296f
Protein synthesis inhibitors, 1124f, 1125t, 1132–34, 1132f
case study, 1132
targeting 30S subunit, 1132–33
targeting 50S subunit, 1133–34, 1133f
Proteobacteria, 704t, 705, 726–38
Alpha-, 728–30, 729f–732f
Beta-, 731–32, 732f
cell envelope, 90, 91f, 92–96, 93f–95f
cell structure, 78
decomposition, 869
defined, **705**
Delta-, 735–37, 736f, 737f
Epsilon-, 737, 737f
Gamma-, 732–35, 733f–735f
Gram stain, 51
metabolism, 726–27, 727f, 728t, eTopic 14.2
phototropic, 709t
plant endophytic community, 875f
Proteolysis, sigma factors, 390–91
Proteome, **214,** 214f
Proteomics, 214, 214f, 303
Proteorhodopsin, **567, 727**
Proteus, 925
Proteus mirabilis, 734, 734f
Gram stain, 50, 51f
pathogen identification, 1169f
Proteus vulgaris, 734
Protist(s)
classification, 792–93, 792f
defined, **787**
historical overview, 787, 787f
light microscopy, 40, 41f
terminology, 787–88
Protist-algae, 697–98, 697f
Proton circulation, and pH homeostasis, 171–72, 172f
Proton-driven ATP synthase, 82f
Proton-driven transport, 126
Proton motive force (PMF), 539–43
active transport, 127
cell functions, 543
charge difference across membrane, 541–42, 542f
defined, **535, 539**
discovery, 540–41, 540f, 541f
dissipation, 542–43, 542f
electron potential and pH difference, 541–42, 542f
electron transport system, 173, 173f
energy storage, **123**
overview, 539–40
water, 540
Proton potential. *See* Proton motive force (PMF)
Proton pumps, retinal-based, 567–68, 567f, 568f
Protospacer adjacent motif (PAM), 465
Protozoa(n)
defined, **787, 788**
diarrheal disease, 1077
historical overview, 787
parasitic, 821–23, 821f–823f
terminology, 787–88
Protrophic cells, 330
Provirus, **197, 201, 429**
Prozac (fluoxetine), 85f
PRPP (5-phosphoribosyl-1-pyrophosphate), 613, 614f
PRRs. *See* Pattern recognition receptors (PRRs)
Prymnesiophyceae, 789t, 814, 814f
Pseudogenes, **267, 350**
Pseudomembranous enterocolitis, 931, 932f, 1081t, 1186t
Pseudomonad(s), 735
Pseudomonadaceae, 735
Pseudomonadales, 704t
Pseudomonas, 734f, 735
antibiotic resistance, 1128
biofilm differentiation, 149
biofilm dissolution, 150, 150f
extracellular cytochromes, 555
indoor environments, 911
microbial attachment, 1022
nitrogen fixation, 902
osteomyelitis, 1063
parasitism, 737
pathogen identification, 1172
respiratory tract biome, 927
toxins, 1037
wastewater treatment, 897
Pseudomonas aeruginosa, 735
ABC transporter, 127
antibiotic resistance, 1127, 1140, 1142, 1143
antibiotics combined with botulinum toxin, 1011
bacteriophages, 221
biofilms, 147, 343, 343f, 1024
bispecific antibody, 1146–47, 1146f, 1147f
burn wound, 1924f
classification, 704t
confocal laser scanning microscopy, 58, 58f
cystic fibrosis, 947, 947f
electron shuttles, 555
electron transport system, 539, 539f
exopolysaccharides, 148, 148f
microbial attachment, 1022f, 1023, 1024f
neutrophil recruitment, 1094
pili, 110
polymyxin, 1129
quorum sensing, 384
resistance, 187
respiratory tract infection, 1066t
siderophore, 129
subverting antigen presentation, 1045
survival within host, 1041
toxins, 1035f, 1037t

Pseudomonas dentrificans, 644
Pseudomonas fluorescens, 174t, 734f, 735
mupirocin, 1134
Pseudomonas syringae, 1038
Pseudopeptidoglycan, 702
Pseudopods, **791,** 805–6, 805f, 806f
Pseudotyping, 655
Pseudouridine, 292f
PS I. *See* Photosystem I (PS I)
PS II. *See* Photosystem II (PS II)
P site (peptidyl-tRNA site), 294f, **295**
Psittacosis, 1066t
Psychrophiles, 161t, 162f, **163–64,** 164f
early life, 672
environmental conditions, 844t
ice formation, 163, eTopic 5.2
ocean floor, **862**
Psychrophilic marine thaumarchaeotes, 753t
Psychrophilic methanogens, 769, 770f
Psychrotolerant bacteria, 163–64, 164f, eTopic 5.2
Psychrotrophs, **163**
Ptashne, Mark, 404, 407, 407f
PTS (phosphotransferase system), **129–30,** 130f, eTopic 4.1
inducer exclusion, 365–66, 366f, eTopic 4.1, eTopic 10.1
Puccinia graminis, 804t
Puerperal fever, 18
PUL (polysaccharide utilization locus), 504, 505f, 521, 522f
Pure culture, **15, 130**
microbe growth, 14–16, 16f
Purified protein derivative (PPD), 1069
Purines, **244**
biosynthesis, 613, 614f
Puromycin, 297
Purple bacteria
Calvin cycle, 587
chlorophylls, 569, 570f
photoheterotrophy, 575, 727
thylakoids, 571
Purple membrane, 568, 568f
Purpura, thrombotic thrombocytopenic, 1074
Putrefaction, **634**
PVC (Planctomycetes-Verrucomicrobia-Chlamydia), 741–44, 742f, 743f
classification, 704t, 705–6
defined, **705**
PVC (polyvinyl chloride), 564
Pyelonephritis, 1021, **1083**
Pyelonephritis-associated pili (PaP), 1021, 1021f
Pyrimidine(s), **244**
biosynthesis, 613, 614f
Pyrimidine dimer, 320, 321f
Pyrobaculum, 592, 753t
Pyrobaculum calidifontis, 267
Pyrococcus, 679, 778–79
Pyrococcus abyssi, 753t, 779
Pyrococcus furiosus
classification, 753t
polymerase chain reaction, 445, 445f
sulfur reduction and oxidation, 778–79
unique metabolic pathways, 754
Pyrococcus horikoshii, 390–91, 391f, 778f
Pyrococcus woesei, 778
Pyrococcus yayanosii, 166, 378
Pyrodictium, 761–62, 762f
sulfur cycle, 906
Pyrodictium abyssi
barophilic vents, 28f, 752, 761–62, 762f, 779
classification, 753t
TACK hyperthermophile, 757t
Pyrodictium brockii, 761
Pyrodictium occultum
classification, 753t
energy source, 487
phylogenetic tree, 675
thermal vents, 761, 864–65
Pyrogen(s), **959–60**
Pyrogenic exotoxins, 1026t
Pyrolobus fumarii, 753t
Pyronins, 1131
Pyrophosphate
DNA replication, 251
hydrolysis releasing, 495
Pyrrolysine, 301, 301f
Pyrsonympha, 789t
Pyruvate
ATP generation, 507, 508f
biosynthesis, 583
conversion to acetyl-CoA, 516, 516f
Entner-Doudoroff pathway, 509, 510f
glucose catabolism, 506
Pyruvate dehydrogenase complex (PDC), **516,** 516f, eTopic 13.2
Pyruvate kinase, 77f, 499, 500f
Pyruvic acid, 506

Q

Q (quinones), **544,** 545, 545f, 546, 547f
Q (query) fever, 183, 735, 1042, 1052, 1093
Quantitative polymerase chain reaction (qPCR), **446–48,** 447f, 448f, 1176–78, 1177f
Quantitative reverse-transcription polymerase chain reaction (qRT-PCR), 466t, 467, 1177–78, 1177f
Quarantine, **1190**
Quasispecies, **425,** 425f
Quaternary ammonium compounds, 186, 186f
Query (Q) fever, 183, 735, 1042, 1052, 1093
Quinol(s), **544,** 545, 546–47, 546f, 547f
Quinolones, **249, 1130**
Quinone(s) (Q), **544,** 545, 545f, 546, 547f
Quinone pool, **546,** 546f
Quorum sensing, 144, **148,** eTopic 4.4
cell-cell communication, 381–84, 382f, 383f, 384t
defined, **382**
Gram-positive organisms, 386–87, 387f
pathogenesis, 384, 384f

R

Rabies immunoglobulin, 1195
Rabies vaccine, 18, 18f, 968t, 1195
Rabies virus, 212t, 232, 1195
vaccine antigen, 473t
RadA protein, 390–91, 391f
Radial phylogenetic tree, 675f, 676

Radiation and desiccation response regulon, 185
Radiative forcing, **890,** 890f
Radioisotope incorporation, element cycling, 886
Radiolabeled substrates, planktonic communities, 862, eTopic 21.1
Radiolaria, 805
Radiolarians, **807,** 808f
Ragsdale, Stephen, 770
Rajneesh, Bhagwan Shree, 1192
Ralstonia eutropha
  horseshoe-shaped vesicles, 109, 109f
  metabolism, 728t
Raltegravir, 1148t
Raman spectroscopy, origins of life, 665, 665t, 667, 667f
Rancidity, **634,** 635
Random mutagenesis, 461, 462–63, eTopic 12.1
Raoult, Didier, 208, 1052
Rapid diagnostics, 1179, 1180–81
  point-of-care, 1184–87, 1185f, 1186t
Rare biosphere, **838**
Rashes, viral, 1063, 1064f
Ratcliff, William, 809, 810f
RBS (ribosome-binding site), 292f, 293–94, eTopic 8.2
  roboswitches, 374–75, 375f
Reactant concentrations, 492–93, 492t
Reaction, intrinsic properties, 491–92, 492f
Reaction center (RC), chlorophyll, **570–71,** 571f, 572
Reaction conditions, 491–93
  concentrations of reactants and products, 492–93, 492t
  standard, 491–92, 492f
Reactive oxygen species (ROS), 174, 174f
  mutations due to, 320, 320f
Read(s), **831,** 831f, 832f
Reading frames, 293
Real-time polymerase chain reaction, **446–48,** 447f, 448f, 1176–78, 1177f
Reassortment, **411, 412,** 413, 414f
RecA protein, 327, 328f
Receptor binding, animal virus, 222, 224f
Recognition sites, 340
Recombinant DNA, 9t, 32–33, 458, 459–61, 459f–461f
Recombinase polymerase amplification (RPA), 1175, 1178
Recombination
  homologous, **259,** 260f, **326–27,** 327f, 348–49, 349f
  prokaryotes, 330, 330f
  reassortment *vs.,* **412**
  site-specific bacteriophage, **217**
Recombinational repair, 323, eTopic 9.1
Recombination signal sequences (RSSs), **978,** 979f
Rectangular phylogenetic tree, 675f, 676
Red algae, 789t, **791,** 791f, 811, 812f
Red blood cells, 937
Red colloidal gold test, 1185, 1185f
Red giant, 662, 662f
Redi, Francesco, 12
Redmond, Molly, 833, 836
Redox center, DNA repair, 324, 324f
Redox changes, carbon flux, 885–86, 886t
Redox couple(s), **536,** 537
Redox gradients, benthic sediment, 866, 867f
Redox levels, 498
Redox reaction, 537
Red Queen's Race, 686f, 687
Red tide, 816, 818f
Reducing agent, 494
Reductases, anaerobic respiration, 551
Reduction
  biosynthesis, 582
  nitrogen fixation, 605–6, 606f
Reduction/oxidation. *See* Redox
Reduction potential, 497, 497t, 536–37, 536t
Reductive acetyl-CoA pathway, 586t, **593–94,** 593f
Reductive evolution, 198–99, **673, 683,** 788
Reductive pentose phosphate cycle. *See* Calvin cycle
Reductive TCA cycle, 586t, **592–93,** 592f
Reemerging pathogens and diseases, 1193–1200
  defined, 1193–94
  discovery, 1194
  map, 1194–95, 1194f
Reeve, John, 750f, 751
Reference genomes, 832
Reflection, **43,** 43f
Refraction, **43,** 43f, 44, 44f
Refractive index, **43**
  phase-contrast microscopy, 62, 62f
  and resolution, 46–47, 46f
Refrigeration, food preservation, 641–42, 641f
Regiella, 676
Reguera, Gemma, 553, 554, 555f
Regulatory proteins, **358–60,** 358f–360f
  small RNA molecules to expand reach, 377–78, 378f
Regulatory RNA, 286, 286t
  untranslated, 375–79, 376f–379f, 376t
Regulatory T cells (Tregs), **985**
  autoimmune disease, 1007–8
  limitation of inflammation, 988–89
Regulon, **370**
  sigma S, 370–71, 371f
Rehydration therapy, **1072–73**
Reimers, Clare, 558, 558f
Reith, Frank, 844, 844f
Relative fitness, **686–87**
Release factors, **296–97,** 296f
Relenza (zanamivir), 1148t, 1149, 1149f
Remote location sequencing, 451–52, 452f, 453f
Renaturation, 245
Rennet, 623
Reoviruses, 212t
Repeats
  direct, 348, 349f, 350f, **393**
  inverted, 213, **393**
  tandem, **393**
Replication, 251–60
  bacteriophage, 215–18, 216f, 218f, 404–5, 404f, eTopic 11.1
  elongation, 251, 252f, 255–58, 256f–258f
  fluorophore labeling, 56, 57f
  initiation, 251, 252–55, 252f–255f
  overview, 251–52, 251f, 252f
  plasmids, 261–63
  regulation of cell division, 100–102, 101f, 102f
  rolling-circle, 262, 262f, **405,** 405f
  secondary chromosomes, 263–65, 264f, 265f

Replication (*continued*)
semiconservative, 251, 251f
speed, 258
termination, 251, 252f, 258–60, 259f, 260f
Replication bubble, 251f, 252, 252f
Replication complexes, 225, 225f
Replication cycles
animal viruses, 225–30, 225f–230f, eTopic 11.3
HIV, 228–29, 229f, 429–31, 430f, 431f
influenza virus, 415–19, 417f, 420f
plant viruses, 230–32, 230f, 231f
Replication error repair, 322–23, 323f
Replication forks, 101, **251,** 251f, 252, 252f
Replication initiator protein, 253f, 254, 254f
Replicative transposition, 344, 344f
Replisomes, 101, **101, 255–57**
fluorophore labeling, 56, 57f
Reporter gene, **467,** 467f
Repression, **359,** 360f
Repressors, transcriptional control, **358–60,** 358f–360f
Reprogramming, host cell, 1018–19
Research, 33, 33t, 34f, eTopic 1.2
Resequencing, 832
Reservoir(s)
biosphere, **884–85,** 884f
infection, **1016, 1189**
Resistance
antibiotic. *See* Antibiotic resistance
genetic, 232
Resistance islands, **351,** 352f
Resolution, **38,** 39f
bright-field microscopy, 46
detail, 45, 45f
detection *vs.,* 39, 39f
numerical aperture and, 46–47, 46f
refractive index and, 46–47, 46f
Resource colimitation, **910–11**
Respiration, **504,** 505–20
aerobic, **173,** 173f, 544–51
anaerobic, **175, 488–89,** 537, 551–53
mitochondrial, 540, 548–49, 548f
use of term, **535**
Respiratory electron transport system, 544–51
cofactors, 544–45, 544f
Na^+ pumps, 550
oxidoreductase and protein complexes, 545–49, 545f–548f
proton potential drives ATP synthase, 549–50, 549f, 550f
Respiratory membranes, 538–39, 538f, 539f
Respiratory syncytial virus (RSV), 1066t, 1071, 1071f
antiviral agents, 1148t
point-of-care rapid diagnostics, 1186t
Respiratory tract, 927, 927f
Respiratory tract infections, 1064–72, 1066t
bacterial pneumonia, 1064–67, 1065f
fungal lung infection, 1067–69, 1067f–1068f
tuberculosis, 1069–71, 1070f
viral, 1064
Response regulator, **361,** 361f
Restricted transduction, **335,** 336, 336f
Restriction and modification, 340–41, 340f, 340t
Restriction endonucleases
bacterial defense, 219
DNA restriction and modification, **340–41,** 340f, 340t
genetic manipulation, **458,** 458f
Restriction fragment length polymorphism (RFLP), 1192
Restriction sites, **458**
Reticulate body, **743,** 743f
Reticulitermes santonensis, 852–53, 852f, 853f
Reticuloendothelial system, **940–41**
Retina, 38, 39f
Retinal, **567**
Retinal-based photoheterotrophy, 777–78, 777f
Retinal-based proton pumps, 567–68, 567f, 568f
Retortamonas intestinalis, 821
Retraction, 337, 337f
Retroelements, 431, 432f
Retrotransposons, **431–32,** 433f
Retroviruses, 420–31
classification, 211f, 212t
defined, **213, 420**
endogenous, 218t, 219, **421, 431–35,** 432f
examples, 420–21, 421f
molecular biology, 420–31
replication, 228–29, 229f
simple, 421t
Reverse electron flow, **559–61**
Reverse TCA cycle, 586t, **592–93,** 592f
Reverse transcriptase (RT), **213, 420, 427–29,** 428f
Reversion, **316**
Rev protein, 424t
RFLP (restriction fragment length polymorphism), 1192
Rhabdovirus, 212t
Rheumatic fever, 1007, 1007t
Rheumatoid arthritis, 957
Rhinitis, allergic, 974f, 1004
Rhinovirus, 201
classification, 212t
cross-protection, 968–69, 969f
evolution, 232
vs. influenza virus, 1071, eTopic 26.4
pathogen identification, 1184t
transmission, 1014, 1016
tropism, 222, 224f
Rhizaria, 789t, **791,** 805, 807, 808f
Rhizobiales, 90f, 704t
Rhizobial nitrogen fixation, 902, 902f
Rhizobium, 730, 737
aerobe, 174t
nitrogen fixation, 123, 601, 876–78, 876f–878f
plasmids, 261
Rhizobium(a), 22, 731f, **876–78,** 876f–878f
Rhizobium leguminosarum, 692
Rhizoplane, 872, 873f
Rhizopus, 798, 800f
skin infections, 1060t
Rhizopus oligosporus, 625–26, 626f
Rhizosphere, **869,** 872, 873f
Rho-dependent termination, **283,** 284f
Rhodobacter, 569, 570f, 572
Rhodobacteriales, 704t

Rhodobacter sphaeroides, 728, 729f
Calvin cycle, 586
chemotaxis, 397
CO_2 uptake and concentration, 591–92
Rhodococcus, 725
quantitative polymerase chain reaction, 447–48, 448f
Rhodoferax, 847
Rhodomicrobium, 728
Rhodomicrobium vannielii, 729f
Rhodomonas salina, 791, 791f
Rhodophyta, 789t, 791, 791f, 811, 812f
Rhodophyte(s), 789t, **791,** 791f
Rhodopseudomonas
lake phototrophy, 866
sulfur cycle, 905
Rhodopseudomonas palustris
iron phototrophy, 670
lake phototrophy, 866
longevity genes, 145, 145f
metabolism, 726–27, eTopic 14.2
photolithoheterotroph, 575, eTopic 14.2
Rhodospiralles, 704t
Rhodospirillum
nitrogen fixation, 601
photoexcitation, 569, 570f, 572
phototropic, 709t
Rhodospirillum rubrum, 728
flagella, 111
lake phototrophy, 866
photoexcitation, 569, 572
photoheterotrophy, 122
photosystem II, 574
Rubisco, 588
Rho factor, **283,** 284f
Rho-independent termination, **283,** 284f
Ribavirin, 1148t
Riboflavin, 648t
Ribonucleic acid. *See* RNA
Ribonucleoprotein (RNP) complexes, 412, 412f
Ribonucleoside monophosphate (rNMP), 283
Ribonucleoside triphosphate (rNTP), 282–83, 282f
Ribose-5-phosphate, 613
Ribosomally synthesized and posttranslationally modified peptides (RiPPs), 598
Ribosomal RNA (rRNA), **286,** 286t
cell structure, 78
modified bases, 291, 292f
small-subunit, 674, 674f, 691
translation, 292–93
Ribosome(s), 77f
analysis, 81, eTopic 3.1
defined, **291–92**
structure, 292–93, 292f, 293f, eTopic 8.1
transcription, 282f
translation, 288, 291–94, 293f
Ribosome-binding site (RBS), 292f, 293–94, eTopic 8.2
roboswitches, 374–75, 375f
Ribosome protection, 1115, 1115f, 1137
Riboswitchboards, engineered, 478, 480f
Riboswitches, **374–75,** 375f
engineered, 478, 479f
Riboswitch sensor, diagnostic, 1178–79, 1178f
Ribozymes
defined, **286, 292, 671**
discovery, 32
structure, 292, 292f
viral, 207
Ribulose 1,5-bisphosphate, 588–91, 588f–590f
Ribulose 1,5-bisphosphate carboxylate/oxygenase. *See* Rubisco
Ribulose 5-phosphate, **509**
Rice blast, 801
Rich, Virginia, 890–91, 891f
Richardson, Jane S., 72
Richardson diagram, 72, 72f
Rickettsia, 732f
Rickettsia, 730, 732f
classification, 705
intracellular immune avoidance, 1042
Rickettsiales, 704t
Rickettsia prowazekii
culture medium, 135, 135f
endemic typhus, 1102t
intracellular immune avoidance, 1043
latent state, 1013
Rickettsia rickettsii, 730, 732f
classification, 1704t
intracellular immune avoidance, 1044, 1045f
Rocky Mountain spotted fever, 1100
Rickettsia typhi, 1097
Rifampin, 285, 1131, 1131f
Rifamycin B, 283–86, 284f
Riftia, 864f, 865
Ringworm, 801
Ripening, cheese, **624–25,** 625f
RiPPs (ribosomally synthesized and posttranslationally modified peptides), 598
Rise period, batch culture, 233f, 234, **234**
Rittmann, Bruce, 935
Rivers, Thomas, 1194
R loops, DNA repair, 325, 325f
RNA
catalytic, 32, 32f, **286,** 286t, 292, 671
cell structure, 77f, 78
classes, 286, 286t
DNA *vs.,* 245
messenger, 78, **286,** 286t, 457
monocistronic *vs.* polycistronic, 281
origins of life, 671, 671f, eTopic 17.1
regulatory, **286,** 286t, 372–79, 376f0379f, 376t
ribosomal, 78, **286,** 286t, 291–93, 292f, 674, 674f, 691
single guide, 466
small, **286,** 286t, 376–78, 376t, 378t, eTopic 10.3
stability (half-life), 286–87, 287f
structure and function, 243–44, 243f, 245
tm, **286,** 286t
transfer, 78, **286,** 286t, 289–92, 290f–292f, 294, 294f
RNA analytical techniques, 466t, 467–68, 467f, 468f
RNA-dependent RNA polymerase, **211**
RNAi (RNA interference), 232, 678
RNA interference (RNAi), 232, 678

RNA polymerase
archaea and eukaryotes, 287, 288f
cell structure, 77f
defined, **278**
DNA-dependent, **278**
initiation, 253f, 254
RNA-dependent, **211**
subunit structure, 279, 279f
transcription, 278–83, 280f, 282f–284f
RNA polymerase inhibitors, 1124f, 1125t1131, 1131f
RNA primer, 253f, 254, 255, 256f, 257, 258f
RNA retroviruses. *See* Retroviruses
RNA reverse-transcribing viruses, 211f, 212t, **213**
RNA sensors, programmable, 1178–79, 1178f, 1180–81
RNA sequencing (RNAseq), 371, 371f, 466t, 467, eTopic 12.5
dual, 1017, 1052–53, 1052f
RNA splicing, 299
RNA synthesis inhibitors, 1124f, 1125t1131, 1131f
RNA thermometers, 389–92, 393f
RNA translation, 100, 100f
RNA viruses
discovery, 20
double-stranded, 211, 211f, 212t
minus (-) strand, 410–20
replication, 227–28, 228f, eTopic 11.3
single-stranded, 211–13, 211f, 212t
RNA world, **31–32,** 287, **671,** 671f, eTopic 17.1
rNMP (ribonucleoside monophosphate), 283
RNP (ribonucleoprotein) complexes, 412, 412f
rNTP (ribonucleoside triphosphate), 282–83, 282f
Robbins, Frederick, 234
Robert, Lydia, 315, 315f
Robinson, Kim Stanley, 917
Rocky Mountain spotted fever, 730, 732f
Rod(s), 2, 3f, 40, 41f
Roe, Andrew, 1041
Rohwer, Forest, 214
Rolling-circle replication, 262, 262f
phage lambda, **405,** 405f
single-molecule sequencing, 450, eTopic 12.3
Ronin, Irene, 144f
Ronsom, Clive, 316
Roosevelt, Franklin, 201
Root(s)
microbes associated with, 872–74, 873f, 874f
phylogenetic tree, 675f, **676**
Roquefort cheeses, 7f, 624f
ROS (reactive oxygen species), 174, 174f
mutations due to, 320, 320f
Roseburia, 929f
ROSE element, 390
Rosenberg, Susan, 689, 690f
Roseobacter litoralis, 728t
Roseobacters, 347, 347f
Rotary flagella, 111–15, 111f–114f
Rotavirus, **1075,** 1081t
classification, 212t
pathogen identification, 1184t
point-of-care rapid diagnostics, 1186t
portals of entry, 1016
Rotavirus vaccine, 998t
Rotem, Eitam, 144f
Rous, Peyton, 197–98
Rous sarcoma virus (RSV), 212t
RPA (recombinase polymerase amplification), 1175, 1178
rRNA. *See* Ribosomal RNA (rRNA)
RSSs (recombination signal sequences), **978,** 979f
RSV (respiratory syncytial virus), 1066t, 1071, 1071f
antiviral agents, 1148t
point-of-care rapid diagnostics, 1186t
RSV (Rous sarcoma virus), 212t
RT (reverse transcriptase), **213, 420, 427–29,** 428f
Rubbing alcohol, 186f
Rubella virus, 1060t, 1063
point-of-care rapid diagnostics, 1186t
Rubeola virus, 1059, 1060t, 1063
Rubisco
Calvin cycle, 587–91, 588f, 590f
defined, **587**
iron phototrophy, 670
isotope ratios, 665–66
marine microbial community, 860
mechanism, 587–89, 589f
structure, 588–89, 589f
Rudy, Edward, 382f
Rumen, **853–55,** 853f, 854f
Ruminococcaceae, 60, 60f
Ruminococcus
bovine rumen, 854
catabolism, 504
gut microbiome, 929
human colon, 855
Ruminococcus albus, 854
Ruminococcus torques, 54f
Ruska, Ernst, 29, 197
Rust, 562, 563f, 908–9

S

S15 protein, 272, 272f
Sabin polio vaccine, 999
Sabree, Zakee, 827, 827f, 828
Saccharomyces, 619
cocoa bean fermentation, 627
Saccharomyces cerevisiae, 174t, 795–96, 795f
classification, 789t, 790
ethanolic fermentation, 629, 630f
model organism for research, 795f, 796–97
origins of multicellularity, 659, 659f
protein secretion, 309
reproductive cycles, 795f, 798
Saccharopolyspora erythaea, 597, 597f
Sacculus, **88,** 88f
Safety cabinets, **183,** 184f
Safranin, 50, 52f
Saint-Ruf, Claude, 178
Sake, 621t
Saleska, Scott, 890
Salih, Tarek, 933
Salinipostin G, 582, 582f
Salinispora, 581, 581f, 721–22
Salinispora arenicola, 581
Salinispora pacifica, 582, 582f

Salinity, ecosystem, 844–45
Salk polio vaccine, 999
Salmonella
acid resistance, 173, eTopic 5.4
antibiotics in animal feed, 1141
anti-sigma factor, 372, 372f
bioterrorism, 1192
cell envelope, 93
culture medium, 134
dual RNA sequencing, 1052–54, 1052f
extracellular immune avoidance, 1042, 1042f–1043f
flagella, 111, 111f
flipping cytokine profile, 1045
food contamination, 636–38, 637f–639f, 639, 640t
food irradiation, 184, 185
gut microbiome, 1076
horizontal gene transfer, 679
inflammatory diarrhea, 1072
intracellular immune avoidance, 1042, 1042f–1043f, 1043, 1044
model GI pathogen, 1048–51, 1050f
nitrogen fixation, 902
nutrients, 1017
phage transduction, 335
phylogenetic tree, 676
plant vascular system, 1199
portals of entry, 1016
probiotics, 189
toxins, 1037–38, 1042
transmission from animals, 1200
ubiquitylation, 1048, 1049f
wastewater, 897
Salmonella choleraesuis, 1102t
Salmonella enterica, 734
bacteriophages, 215, 215f
classification, 705
culture medium, 133
dark-field microscopy, 61, 61f
dilution streaking, 131f
disease surveillance, 1192
flagellar motility, 112–13, 112f
food contamination, 636–37, 637f, 1073
gastrointestinal tract infections, 1072, 1081t
gene inversion, 392–94, 394f
minimal inhibitory concentration, 1120f
model GI pathogen, 1048–51, 1050f
multidrug resistance, 1139, eTopic 9.6
mutant, 321
neutralophile, 169
pathogenicity islands, 1018f
pathogen identification, 1168f
periplasmic protein, 307
phage therapy, 189
plant endophytic community, 876
toxins, 1038f, 1074
transmission electron microscopy, 65, 66f
ubiquitylation, 1048, 1049f
Salmonella pathogenicity island 1 (SPI-1), 1050–51, 1050f
Salmonella pathogenicity island 2 (SPI-2), 1050f, 1051
Salmonella typhi
antibiotics, 1073
intracellular immune avoidance, 1044
systemic illness, 1102
typhoid fever, 1081t, 1102t
Salmonella typhimurium, 1038f
Salmonellosis, 1081t
Salt, food preservation, 643
SALT (skin-associated lymphoid tissue), **946**
Salterns, 775, 775f
solar, 776
Salvarsan, 1119
Sander, Leif, 999
Sanger, Fred, 5, 5f, 448
Sanger method, 448, 448f, 450, 454t
Sanitation, **181**
Sansonetti, Philippe, 720, 720f
Sapoviruses, wastewater, 897
Saprophytes, 869–70, 872f
SAR11, 67, 349–50, 351f
Sarcina(e), 102, **726**
SAR (Stramenopiles, Alveolata, Rhizaria) clade, 789t, 792, 815–21
Sargasso Sea, 812, 812f
Sargassum natans, 812f
Sargassum weeds, **812**
Sari cloth filtration, 859, 860f
SARS. *See* Severe acute respiratory syndrome (SARS)
Sauerkraut, 621t
Sausage, 621t
Savage, Dwayne, 720
Savagella, 716, 720–21, 720f, 721f
"Save Our Ship" (SOS) repair, **327–28,** 328f
SBE (subacute bacterial endocarditis), 926, 927f, 1093
Scaffolds, 831f, **832**
Scalded skin syndrome, 1060t
Scalindua, 904, 904f
Scanning electron microscopy (SEM), **40, 64,** 65f
prokaryotes, 40, 41f
range of resolution, 40–42, 41f
Scanning probe microscopy (SPM), **42, 69–70,** 70f
Scarlet fever, 1060t
Scattering, **43–44,** 43f
SCC*mec* (staphylococcal cassette chromosome *mec*), 1127
SCFAs (short-chain fatty acids), obesity, 934–35, 935f
SCG (single-cell genomics), **271–72,** 272f
Schaal, Alex, 710–11, 711f
Schaker, Timothy, 1152
Schizogony, **820,** 820f
Schizont, 820, 820f
Schnupf, Pamela, 720, 720f
Schopf, William, 664
Schorn, Michelle, 1, 1f
Schulz-Vogt, Heide, 5f
Schwan, Rosane Freitas, 617, 617f
SCID (severe combined immune deficiency), 1000, eTopic 24.1
Scrapie, 207
Screen assay, **346**
SCY-635, 1152f
Scytonema hofmanii, 219
Seafood spoilage, 635t, 637
Sears, Cynthia, 921, 921f, 922, 922f
Seaweed, 619

Sebum, 946
Secondary algae, **788, 791–92,** 808, 811–14
Secondary antibody response, **974–75,** 975f
Secondary chromosomes, **263–65,** 265f
Secondary ions, 60
Secondary metabolites, **1135**
Secondary products, 581, **582,** 582f, 583f, eTopic 15.1
Secondary symbiont algae, 697–98, 697f
Secondary syphilis, **1087,** 1087f
Secondary tuberculosis, **1070,** 1070f
Second messenger(s), 380–87
  cyclic diGMP and biofilm formation, 380–81, 381f
  defined, **380**
  interspecies communication, 385–86, 385f, 386f
  quorum sensing and cell-cell communication, 381–84, 382f, 383f, 384t
  quorum sensing and pathogenesis, 384, 384f
  quorum sensing in Gram-positive organisms and activation of natural transformation, 386–87, 387f
  stringent response, 380, 380f
Second messenger pathways, activation, 1025, 1026t
Secretion systems, bacterial toxins, 1036–40, 1037f–1040f, 1037t
Secretory complex, 77f
Secretory diarrhea, 1072
Sediment, iron in, 909–10, 909f, 910f
Segmented filamentous bacilli (SFBs), 931
Segmented genome, **210, 411–13,** 411f, 412f
Segregation
  plasmids, 263, 264f
  sister chromosomes, 258–60, 259f, 260f, eTopic 7.2
Sela, Uri, 1059
Selectins, **953,** 953f
Selection assay, **346**
Selective media, **133–34,** 133f, 1161
Selective permeability, 125
Selective pressure, 685, 685f
Selective toxicity, antibiotics, **1118–19**
Selenium, 910t
Selenocysteine, 301, 301f
Selenoproteins, 301
Seliberia stellata, 729
SEM (scanning electron microscopy), **40, 64,** 65f
  prokaryotes, 40, 41f
  range of resolution, 40–42, 41f
Semiconservative replication, **251,** 251f
Semmelweis, Ignaz, 18
Senescence zone, hyphae, 795
Sense strand, 281, 282f
Sensitivity, **1185**
Sensitization, 1005
Sensor kinase, **360–61,** 361f
Sepsis, 1099–1100
Septation, **98,** 98f, 99f
  spherical cells, 101–2, 102f
Septicemia, **1092–93,** 1094–95
  neonatal, 1186t
Septicemic plague, **1098**
Septic meningitis, 1104–6, 1105f, 1110t
Septic shock, 1099–1100
Septum(a), **96**
Sequela(e), **1161**
Sequencing
  DNA. *See* DNA sequencing
  RNA. *See* RNA sequencing (RNAseq)
  by synthesis, **448–49,** 449f
Serratia liquefaciens, 916f, 917
Serratia marcescens, 636
Serum, **974**
Setlow, Peter, 151, 152f
Severe acute respiratory syndrome (SARS), 1066t, 1071
  cultivation, 1194
  patient zero, 1190, 1190f
  tracking, 1195, eTopic 28.6
Severe combined immune deficiency (SCID), 1000, eTopic 24.1
Sex pilus(i), 110, 330, 331f
Sexually transmitted infections (STIs), 1086–92, 1086t
  chancroid, 1086t
  chlamydia, 1086t, 1088, 1088f
  genital herpes, 1086t
  genital warts, 1086t
  gonorrhea, 1086t, 1088–89, 1089f
  HIV, 1086t, 1090–91, 1090f, eTopic 26.7
  syphilis, 1086–88, 1086t, 1087f
  trichomoniasis, 1086t, 1091–92, 1092f
SFBs (segmented filamentous bacilli), 931
sgRNA (single guide RNA), 466
Shapiro, Lucy, 58, 58f
Shapiro, Mikhail, 277
Shaw, Lindsey, 1145
Shelley, Mary, 534
Sheperd, Elizabeth, 521
Shewanella
  donor-acceptor system, 537
  phylogenetic tree, 676
Shewanella oneidensis, 735
  cytoplasmic extensions, 556, 556f
Shewanella violacea, 165f
Shiga-like toxin, 1033
Shiga toxins, 1033–34
  antibiotics in animal feed, 1141
  case history, 1073–74
  food contamination, 641
  gut bacteriophage community, 220
  mechanism of action, 1026t, 1027f
Shigella
  antibiotics in animal feed, 1141
  avoidance of autophagy, 1047, 1048f
  culture medium, 134
  food contamination, 640t
  gastrointestinal tract disease, 1081t
  growth factors, 134t
  gut microbiome, 1076
  inflammatory diarrhea, 1072
  intracellular immune avoidance, 1042
  phylogenetic tree, 676
  portals of entry, 1016
  stomach microbiome, 928
  toxins, 1037–38, 1038f, 1074
  virulence factor, 220
Shigella dysenteriae
  antibiotics, 1073

intracellular immune avoidance, 1044
minimal inhibitory concentration, 1119–20
toxins, 1026t, 1033–34, 1074
Shigella flexneri, 734
primary pathogen, 1012
toxins, 1038f
Shigella sonnei, 1161
Shigellosis, 1081t
Shiitake, 619
Shilts, Randy, 421
Shimomura, Osamu, 53
Shin, Sunny, 1054, 1054f
Shine, John, 293
Shine-Dalgarno sequence, 292f, **293–94,** eTopic 8.2
Shingles, 1060t, 1063, eTopic 26.1
antiviral agents, 1148t
Shipman, Seth, 246, 246f
"Shmoos," 795f, 798
Shock
septic, 1099–1100
toxic, 1099–1100
Short-chain fatty acids (SCFAs), obesity, 934–35, 935f
Short interspersed nuclear element (SINE), 432, 432f
Shotgun sequencing, 269–70
Shower curtain biofilm, 675, eTopic 17.2
Siderophores, **128–29,** 129f, **909**
Sieber, Jessica, 527, 528f
Sigma factor(s), 279–80, 279f, 280f, 282f
alternative, **370–71,** 371f
controlled by RNA thermometers and proteolysis, 389–92, 393f
defined, **278**
Sigma H, 389–90, 393f
Sigma S, 370–71, 371f
Signal recognition particle (SRP), **307,** 307f
Signal sequences, **307**
Signal transduction disruption, 1025, 1026t
Signature-tagged mutagenesis, 1017, eTopic 25.1
Silent mutations, **317**
Silver, Pamela, 409
Simeprevir (Olysio), 1148t
Simian immunodeficiency virus (SIV), 201, 424, 425f
Simian virus 40 (SV40), 436t
Simple stains, **50,** 51f
SINE (short interspersed nuclear element), 432, 432f
Singer, Maxine, 33, 278, 278f
Single-celled protein, **620**
Single-cell genome sequencing, 836, 836f
Single-cell genomics (SCG), **271–72,** 272f
Single guide RNA (sgRNA), 466
Single Molecule, Real-Time (SMRT) system, 450–51, 451f, 452f, 454t
Single-molecule sequencing, **450–52,** 451f–453f
Single nucleotide polymorphisms (SNPs), 1019, 1019f
Single-stranded DNA (ssDNA), 390–91, 391f
Single-stranded DNA (ssDNA) viruses, 211, 211f, 212t
Single-stranded RNA viruses, 211–13, 211f, 212t
Sink, **884–85**
Sinorhizobium, 730
envelope, 78
nitrogen fixation, 123, 605, 876
Sinorhizobium meliloti, cell envelope, 90
Sinorhizobium meliloti, 551, 704t, 731f
Siphonous algae, 789t
Sister chromosome segregation, 258–60, 259f, 260f, eTopic 7.2
Site-directed mutagenesis, 461–64, 464f, eTopic 12.4
Site-specific recombination, bacteriophage, **217**
Siuzdak, Gary, 923
SIV (simian immunodeficiency virus), 201, 424, 425f
Skin
human microbiome, 925, 925f
physical barriers to infection, 945–46, eTopic 23.3
Skin-associated lymphoid tissue (SALT), **946**
Skin infections, 1059–64, 1060t
Skin rashes, viral, 1063, 1064f
S-layer
Firmicutes, **92,** 92f
proteobacterial cell envelope, 95
Sulfolobus, 758, 758f
Sliding clamp, 253f, **255,** 257, 257f, eTopic 7.1
Slime molds, 789t, 791, 806–7, 807f
Slipped-strand mispairing, 394–95, eTopic 10.4
Slonczewski, Joan, 847
Slow-release cycle, bacteriophage, 218, 218f
Sludge, **897**
Slug, 806
Small molecules, bacterial composition, 79
Small nucleolar RNAs (snoRNAs), 378
Small-peptide affinity tag, 471, eTopic 12.7
Smallpox, 1060t, 1063, eTopic 26.3
bioterrorism, 1192–93, 1193f
history, 10
immunological specificity, 967–68, 967f
Smallpox vaccination, 17, 17f, 18
Smallpox virus, 17
Small RNA (sRNA), **286,** 286t
archaea, 378
classes, 376, 376t
defined, **376**
expanding reach of regulatory proteins, 377–78, 378f
mechanism of function, 376–77, 376t, 377f, eTopic 10.3
Small-subunit rRNA (SSU rRNA), **674,** 674f, 691
Smith, George P., 473
Smith, Hamilton, 5, 462, 462f
SMRT (single Molecule, Real-Time) system, 450–51, 451f, 452f, 454t
snoRNAs (small nucleolar RNAs), 378
Snow, John, 1188, 1188f
Snow algae, 883, 883f
SNPs (single nucleotide polymorphisms), 1019, 1019f
SOD (superoxide dismutase), 307
Soda lakes, 171, 171f
Sodalis, 850, 850t, 851f
Sodium hypochlorite, 186
Sodium (Na+) pumps, 550
Sofosbuvir (Sovaldi), 1148t
Soft tissue infections, 1059–64
Soil
iron, 909–10, 909f, 910f
methanogenesis, 768
Soil food web, 869–70, 870f–872f

Soil microbial communities, 867–80
microbes associated with roots, 872, 873f
mycorrhizae, 872–74, 873f, 874f
nitrogen fixation by rhizobia, 876–78, 876f–878f
plant endophytic communities, 875–76, 875f
plant pathogens, 879, 879f
soil food web, 869–70, 870f–872f
soil microbiology, 868–69, 868f
wetland soils, 874–75, 875f
Soil microbiology, 868–69, 868f
Soil profile, 868, 868f
Solar energy, 488, 488f
Solar salterns, 776
Solutes, compatible, **167**
Sommer, Morten, 856
Sonication, 80–81
Sonnenberg, Justin, 521
Sorangium cellulosum, 736
Sorokin, Dmitri, 778
Sortase, 90–91
SOS ("Save Our Ship") repair, **327–28,** 328f
SOS response, **327–28,** 328f
Source, **884–85**
Sour cream, 621t
Sourdough, 621t, **631**
Soy fermentation, 625–27, 625f
Soy sauce, 620, 621t
Spallanzani, Lazzaro, 12, 13
Sparks, William, 918
Spatial organization of microbes in habitat, 839, 840f
Specialized structures, 108–15
thylakoids, carboxysomes, and storage granules, 108–10, 110f
turrets and horseshoes, 108–9, 108f–109f
Specialized transduction, **335,** 336, 336f
Species, **692**
candidate, **693**
defining, 690, 691–92, 691f
identification, 693–94
naming, 693
Specificity, **1185**
Specimen collection, 1160–64, 1162f, 1163f
Specimen handling, 1164–66, 1164t, 1165f
Specimen positioning, compound microscope, 48
Specimen preparation, bright-field microscopy, 49, 49f
Specimen processing, 1161–64
Spectroscopic analysis, element cycling, 886, 887f
Spectrum of activity, antibiotics, **1119,** eTopic 27.1
Spherical cells, septation, 101–2, 102f
Spheroplast
defined, **80**
formation, 80–81
lysis, 81, 82f
Sphingomonadales, 704t
Sphingomonas, 704t
SPI-1 (*Salmonella* pathogenicity island 1), 1050–51, 1050f
SPI-2 (*Salmonella* pathogenicity island 2), 1050f, 1051
Spike proteins, **203**
HIV, **422,** 423, 423f, 426, 426f
Spirillum(a), 40
Spirochaeta, 704t, 738–39, 738f
Spirochaeta, 704t, 738
Spirochetes, 738–39, 738f
cell envelope, 90–91
cell structure, 739, 739f
classification, 704t, 705
dark-field microscopy, 61, 61f
defined, **705**
diversity, 738–39, eTopic 13.1
size and shape, **40,** 41f
Spirogyra, 4t, 789t, 811–12, 811f
Spiroplasma, 850t
Spirulina, 171, 171f, 619–20
Spleen, 945f
"Spliced-up" regulation, 390–91, 391f
SPM (scanning probe microscopy), **42, 69–70,** 70f
Sponges, 1, 1f
Spongiform encephalopathies, **1110,** 1111f
Spontaneous generation, **12–13,** 12f, 660
Sporangiospores, **799,** 800f
Sporangium(ia), 151, 152f, 153, **798–99,** 800f
Spores, 13, 711–12, 711t
Spore stain, **52,** 53f
Sporothrix schenckii, 1060t
Sporotrichosis, 1060t
Sporozoa, 789t
Sporulation, 151, 152f
Spread plate, **131–32,** 132f
Sputnik virophage, 208, 209f
Sputum specimen collection, 1161, 1162f
SQ109, 646–47, 646f
SRB (sulfate-reducing bacteria)
marine cycling of methane, 887
spatial organization in habitat, 839, 840f
SRB (sulfur-reducing bacteria), 552, 778–79, 906, 907f
sRNA. *See* Small RNA (sRNA)
SRP (signal recognition particle), **307,** 307f
ssDNA (single-stranded DNA), 390–91, 391f
ssDNA (single-stranded DNA) viruses, 211, 211f, 212t
SSU rRNA (small-subunit rRNA), **674,** 674f, 691
Stable isotope ratios, element cycling, 886
Stachybotrys, 789t, 801, 801f
Stachybotrys chartarum, 804t
Stahl, David, 763
Stains and staining, 50–53
acid-fast (Ziehl-Neelsen), 52, 53f
antibody tags, 53
chemical structure, 50, 50f
defined, **50**
differential, **50–54,** 51f–53f
Gram, **50–53,** 51f, 52f
negative, 52, **53,** 53f, **65**
pathogen identification, 1166–67, 1167f
simple, **50,** 51f
spore, 52–53, 53f
Stalk(s), 103f, **110–11**
Stalked ciliates, **816,** 818f
Standard reduction potential, 497, 497t, **536–37,** 536t
Stanier, Roger, 568, 569f
Stanley, Wendell, 19–20, 197
Staphylococcal cassette chromosome *mec* (SCC*mec*), 1127
Staphylococcal food poisoning, 640t, 1073, 1081t
Staphylococcal scalded skin syndrome, 1061, 1061f

Staphylococcal skin infections, 1059–61, 1060t, 1061f
Staphylococcus, 718, 719f
biological control, 189
cell wall growth, 90f
colonies, 132
discovery of penicillin, 19
facultative microbe, 174t
growth factors, 134t
indoor environments, 911
quorum sensing, 384
urinary tract infection, 1085
Staphylococcus(i), **102**
Staphylococcus aureus, 718
antibiotic resistance, 1136, 1137, 1142–43
bacterial endocarditis, 1093
biofilms, 1024
boils, 949, 949f, 1059–61, 1061f
cellulitis, 1063
cell wall, 702
disinfectant resistance, 187, 188f
evolutionary features, 1020, eTopic 23.1
extracellular immune avoidance, 1041
food contamination, 640t, 1073, 1081t
hibernation-promoting factor, 179f
meningitis, 1110t
methicillin-resistant. *See* Methicillin-resistant *Staphylococcus aureus* (MRSA)
microbial attachment, 1021t
necrotizing fasciitis, 1062
neutrophil recruitment, 1094–95
oral and nasal cavity microbiome, 926
pangenomic gene differences, 1059, 1059f
pathogen identification, 1174
penicillin, 1117–18, 1117f
peptidoglycan structure, 89
peptidoglycan synthesis, 1126
phage therapy, 189
platensimycin, 1144
prophages, 198
septation, 101–2, 102f
size, 40f
skin infections, 1059, 1060t
skin microbiome, 925
small RNA molecules, 377
SOS induction, 328
susceptibility test, 1119, 1122, 1122t
systemic illness, 1102
toxic shock, 1100
toxins, 1026t, 1027, 1028–29, 1030f
vancomycin-resistant, 1127
Staphylococcus epidermidis, 718
eye microbiome, 925
genitourinary tract microbiome, 927
meningitis, 1110t
oral and nasal cavity microbiome, 926
pathogen identification, 1174
skin microbiome, 925
Staphylococcus hominis, 925
Staphylococcus saprophyticus, 1174
Star(s), origins of life, 661–62, 662f
Starch catabolism, **501,** 502f
Starch utilization systems (Sus complexes), 485, 485f, 486, 486f, 504, 505f
Start codon, 282f, 283f, **293**
Starter culture, cheese, **624**
Starvation, 176–80
climate change, 180, 180f
eutrophication, 179–80, 179f
membrane lipids, 85–86
stress responses, 178–79
survival genes, 177–78, 177f, 178f
Starvation stress, mutation, 689–90, 690f
Statins, antiviral agents, 1152f, 1153
Stationary phase, 142f, **144,** 144f
Steam autoclave, 182, 182f
Stedman, Ken, 758
Steglich, Claudia, 287, 287f
Steinman, Ralph, 965, 966f
Steitz, Thomas, 298, 298f
Stem cells, 937, 938f
Stendomycin, 602–3, 603f
Stenotrophomonas, 876
Stentor, 2, 3f, 816, 818f
Stephanodiscus astraea, 812f
Sterile sites, specimen collection, 1161, 1162f
Sterilization, **180**
filtration, 183, 183f
high temperature and pressure, 182, 182f
irradiation, 184–85
Stetter, Karl, 706, 747
STI(s). *See* Sexually transmitted infections (STIs)
Stigmatella, 736, 736f
Stink bug, 827, 827f
STIV (*Sulfolobus* turreted icosahedral virus), 759, 759f
Stolfa, Gino, 1034
Stomach "flu," 1081t
Stomach microbiome, 928, 928f
Stonewashed jeans, 649t, 801
Stool samples, 1162
Stop codon, 282f, 283f, **289**
Storage granules, 109–10, 110f
Storage zone, hyphae, 794f, 795
Stordalen Mire, 890–91, 891f, 892f
Storz, Gisela, 376, 376f
Stover, C. Kendall, 1146, 1146f
Stramenopiles, 789t, **792,** 811, 815–21
Streak plating, 131, 131f
Strecker, Jonathan, 219
Strep throat, 1007
point-of-care rapid testing, 1186t
Streptococcus, 714, 718, 719f
cell count, 140
cell wall growth, 90f
colonies, 132
curd formation, 622, 622f, 623
gut microbiome, 1077
indoor environments, 911
neutrophil extracellular traps, 938
oral and nasal cavity microbiome, 925
quorum-sending regulation of transformation, 386, 387f
respiratory tract biome, 927

Streptococcus (*continued*)
septation, 102
transformation of naked DNA, 337
Streptococcus(i), ExPortal, 308–9, 309f
Streptococcus agalactiae
evolutionary features, 1020, eTopic 23.1
point-of-care rapid diagnostics, 1186t
Streptococcus haemolyticus, 1174f
Streptococcus mutans
bacterial endocarditis, 1093
microbial attachment, 1021t
oral and nasal cavity microbiome, 926
pathogen identification, 1173
Streptococcus oralis, 926
Streptococcus pneumoniae, 718
antibiotic resistance, 1067, 1127, 1128, 1136
extracellular immune avoidance, 1041
fluorescent antibody staining, 1182–83, 1183f
gene transfer, 316
Gram stain, 50, 51f
light microscopy, 41f
meningitis, 1105, 1105f, 1110t
multidrug-resistant, 1135, 1136f
pangenome, 692
pathogen identification, 1168, 1173–74
phagocytosis, 954f, 955
pneumonia, 1064–67, 1065f, 1066t
point-of-care rapid diagnostics, 1185, 1185f, 1186t
respiratory tract biome, 927
SOS induction, 328
toxins, 1026t
transformation of naked DNA, 337
vaccine, 997
Streptococcus pyogenes, 718
antibiotics combined with botulinum toxin, 1011
azithromycin, 1133
cell envelope, 90
cellulitis, 1063
complement, 994
growth factors, 134, 134t
horizontal gene transfer, 1019–20, 1019f
immune response, 966
microbial attachment, 1021t, 1023, 1024f
necrotizing fasciitis, 1011, 1012, 1062, 1062f
osteomyelitis, 1063
pathogen identification, 1172, 1172f, 1173
point-of-care rapid testing, 1185, 1186t
rheumatic fever, 1007
sequelae, 1161
skin infections, 1060t
superantigen, 991
targeted gene editing, 465, 465f
toxic shock, 1100
toxins, 1026t, 1027
Streptococcus salivarius
light microscopy, 41f
oral and nasal cavity microbiome, 926
S. pyogenes vs., 1012
Streptococcus thermophilus, 622f
Streptogramins, **1133–34,** 1133f
Streptolysin S, 1026t, 1027
Streptomyces, 721, 722–23, 722f
amensalism, 845t
antibiotics, 297, 596
cell differentiation, 153–55, 154f
coevolution, 694, 724, 851, 1135, eTopic 17.6
decomposition, 869, 872f
lantibiotics, 1128
streptogramins, 1133
Streptomyces clavuligerus, 1139
Streptomyces coelicolor, 722, 722f, 723
classification, 704t
mycelia, 154f
taxonomy, 692
Streptomyces erythraeus, 297
Streptomyces garyphalus, 1128
Streptomyces griseus, 722f
decomposition, 872f
streptomycin, 188–89, 1118, 1135
Streptomyces hygroscopicus, 602–3, 603f
Streptomyces lydicus, 650–52, 652f
Streptomyces nodosus, 1154, 1154f
Streptomyces noursei, 1154, 1154f
Streptomyces orientalis, vancomycin, 1129
Streptomyces platensis, 1144
Streptomyces roseosporus, daptomycin, 1129
Streptomycin, 297, 299f, 1132–33
antibiotic resistance, 1138
discovery, 1118, 1118f
Stress, thriving under, 1042–43, 1044f
Stress responses, bacterial, 178–79, eTopic 5.5
Strickland reaction, **515**
Strict aerobes, 161t, **173**
Strict anaerobes, 161t, **175**
Stringent response, **380,** 380f
Stromatolites, **660,** 661f, 665t
Strongly selective environments, 684–85, 684f, 685f
Structural biology, 9t
Structural gene, 281, 282f
ST toxin, 1026t
Subacute bacterial endocarditis (SBE), 926, 927f, 1093
Substrate-level phosphorylation, **512**
Succession, **833,** 834f
Succinate dehydrogenase, 548
Succinyl-CoA, 518
Succinyl-CoA synthetase, 518
Suctorians, 816, 818f
Süel, Gürol, 119, 119f
Sugar mycolates, 95
Sugar porin, 77f
Suicide module, 479, 481f
Suicide vector, 461
Sulfa drugs, **1118,** 1130, 1130f
Sulfanilamides, 1118, 1118f, 1130, 1130f
Sulfate ion, 905
Sulfate-reducing bacteria (SRB)
marine cycling of methane, 887
spatial organization in habitat, 839, 840f
Sulfides, 905
Sulfites, food preservation, 643
Sulfolobales, **758–60,** 758f–760f
cell membrane and S-layer, 758, 758f, 759f

classification, **752,** 753t
sulfide oxidation, 758–59
Sulfolobus, 758, 759f
cell membrane and S-layer, 758, 758f, 759f
classification, 753t, 758
conjugation, 330
gene structure, 750, 750f, 751
horizontal transfer, 779
hot springs, 756
isolating, 758, 758t
reaction conditions, 493
small nucleolar RNA, 378
sulfide oxidation, 561–62, 563f, 758–59
unique metabolic pathways, 754
viruses, 759–60, 759f, 760f
Sulfolobus acidocaldarius, 170, 170f
Sulfolobus metallicus, 594
Sulfolobus solfataricus, 757t, 758, 758f, 759, 759f
Sulfolobus turreted icosahedral virus (STIV), 759, 759f
Sulfonamides, 1118
Sulfur
metabolism, 843t
oxidized forms, 552
Sulfur and iron phototrophs, 733, 733f
Sulfur cycle, 905–6, 905t, 906f, 907f
and iron cycle, 909–10, 910f
Sulfurisphaera, 753t
Sulfur lithotrophs, 732–33, 733f
Sulfur oxidation
betaproteobacteria, 731
Sulfolobus, 561–62, 563f, 758–59
Thermococcales, 778–79
Sulfur oxidizers, 133t
Sulfur particles, 110, 110f
Sulfur-reducing bacteria (SRB), 552, 778–79, 906, 907f
Sulfur triangle, 905–6, 906f
Sullivan, John, 316
Sullivan, Matthew, 195, 195f, 196
Sunagawa, Shinichi, 269, 270f, 859
Sun bacteria, 709t, 716
Superantigens, **991,** 991f, 1026t, 1027
Supercoils and supercoiling, DNA, 247–50, 248f–250f, 257–58
Superficial mycoses, 1154, 1154f
Supergiant, 662, 662f
Supernova, 662, 662f
Superoxide, 174, 174f
Superoxide dismutase (SOD), 307
Super-resolution imaging, **45, 57–58,** 57f–59f, 59
Super-size microbial cells, 4, 5f
Surette, Michael, 929
Surveillance, systematic (syndromic), **1190**
Survival genes, 177–78, 177f, 178f
Susceptibility to pathogens, 1023
Sus complexes (starch utilization systems), 485, 485f, 486, 486f, 504, 505f
SV40 (simian virus 40), 436t
Svedberg, Theodor, 30
Swabs, specimen collection, 1161, 1162f
Swanson, Robert, 33
Swarmer cell, 103, 104f
Swarming, **734,** 734f
Swine influenza (swine flu), 411
Swiss cheese, 622, 622f, 624f
Switch region, **980–81,** 980f
SXT transposon, 345, 345f, 346f
SYBR Green, 447
Sylvatic cycle, 1099f
Symbiodiniaceae, 785
Symbiodinium, 817, 819f
classification, 789t
climate change, 785, 785f, 786
mutualism, 849, 849f
Symbiogenesis, 697
Symbionts, 701, 701f
defined, **123**
digestive methanogenic, 770–71, 770f, 772–73
photosynthetic coral, 785, 785f, 786
Symbiosis(es), 845–51
amensalism, 845t, 850
commensalism, 845t, 850
defined, **845**
evolution, **694–96,** 695f
insects with intracellular bacteria, 849–50, 850t, 851f
mutualism, 845t, 848, 848f
photosynthetic endosymbionts of animals, 848–49, 849f
stink bug, 827, 827f
synergism, 845t, 850
types, 845, 845t
varying degrees of cooperation and parasitism, 845t, 850–51, eTopic 17.6
Symbiosis islands, **351–52,** 352f
Symbiosome membrane, 877
Symbiotic ammonia-oxidizing archaea, 763–64, 764f
Symmetrical virions, 202–4, 203f–205f
Symport, **126–27,** 126f
Symptoms, difference in, 1058–59
Syncytin, 219
Syndromic surveillance, **1190**
Synechococcus
circadian clock, 389, 392f
classification, 703, 710
marine ecosystem, 859, 859f
phosphate cycle, 908
phytoplankton, 862
S-layer, 92
Synergism, 845t, **850**
Synovial fluid, 1161
Synteny, 504, 505f
Synthetic auxotrophy, 475
Synthetic biology, 33, 475–81
BioBricks and do-it-yourself, 479–80, 481f
defined, **475**
engineered riboswitches and switchboards, 478, 479f, 480f
examples, 475
first synthetic organism, 461, 462–63
kill switches, 479, 481f
lambda switch, 409, 410f
oscillator switch, 477, 477f
principles, 476, 476f, eTopic 12.12
system noise, 478, 478f
toggle switches, 476, 477f
Synthetic medium, 132

Syntrophoarchaeota, 769
Syntrophomonas wolfei, 528, 529t
Syntrophus aciditrophicus, 528, 529f, 529t
Syntrophy, 527–29, 528f, 529f, 529t
  defined, **493, 527**
  methanogens, **767**
  termite wood-digesting microbiome, **853,** 853f
Syphilis, 739, 1086–88, 1086t, 1087f
  congenital, **1087**
  point-of-care rapid testing, 1186t
  primary, **1087,** 1087f
  secondary, **1087,** 1087f
  selective toxicity, 1119
  tertiary, **1087**
  Tuskegee experiment, 1087–88
Systematic surveillance, **1190**
Systemic infections, 1097–1104, 1102t
  Ebola, 1102–4, 1103f
  Lyme disease, 1100–1102, 1101f
  plague, 1097–99, 1098f, 1099f
  sepsis and septic shock, 1099–1100
Systemic lupus erythematosus, 1007t
Systemic mycoses, 1154, 1155t
System noise, 478, 478f
Szostak, Jack, 32, 670

T

TACK superphylum, 754f, 756–63, 757t
  barophilic vent hyperthermophiles, 760–62, 761f, 762f
  classification, 752–54, 753t
  Desulfurococcales, 756–57, 757f
  hot springs, 756, 756f
  Korarchaeota and Aigarchaeota, 762, 763f
  Sulfolobales, 758–60, 758f–760f
  unique metabolic pathways, 754
Tae protein, 353, eTopic 25.8
Tailed viruses, 205, 206f, eTopic 11.1
Tail tube, 404–5, 404f, eTopic 11.1
Tamiflu (oseltamivir), 1148t, 1149
Tandem repeat, **393**
Tannerella forsythia, 92
Tape measure protein, 404–5
Tappeiner, Hermann von, 766
Taq DNA polymerase, 447, 447f
TaqMan, **447,** 447f
Taq polymerase, 32
Target-AID system, 466
Target community, **830,** 830f
Targeted approaches
  protein abundance and function, 468
  RNA and transcription analysis, 467
Targeted gene editing, 464–66, 465f
Targeted mutagenesis, 461–64, 464f, eTopic 12.4
TA (toxin-antitoxin) systems, 177, 178f, 1143, 1143f
TAT (twin arginine translocase), 309–10, 310f
TATA-binding protein (TBP), 287, 288f, 368
TATA box, 287, 288f, 368
Tat protein, 424t
Tattoos, 940–41, 940f, 941f
Tatum, Edward, 330, 799
Tautomeric shifts, 319, 319f
Taxon(a), **692,** 692t
Taxonomic domains, 100f
Taxonomy, **690, 692,** 692t
Taylor, Pat, 150f
TB. *See* Tuberculosis (TB)
TBP (TATA-binding protein), 287, 288f, 368
TCA cycle. *See* Tricarboxylic acid cycle (TCA cycle)
TCCD (dioxin) catabolism, 502f, 503
T_C cells (cytotoxic T cells), 966, **982**
  activation, 986–87, 987f
T cell(s), 938f, 942, **944,** 964–65
  cytotoxic, 966, **982,** 986–87, 987f
  education and deletion, 985–86
  helper, **980,** 982, 986–90
  humoral immune response, 964–65, 966f
  linking antibody and cellular immune systems, 982–93
  lymph node germinal center, 980, eTopic 24.4
  recognition, 985–86
  regulatory, **985,** 988–89, 1007–8
T-cell activation, 982–83, 983f
T cell–dependent and –independent antibody production, 976–77
T-cell engineering, autoimmune disease, 1007–8
T-cell receptors (TCRs), **984–85,** 985f
TCRs (T-cell receptors), **984–85,** 985f
Technology, emerging infectious agents, 1197–98
Teff, 631
Tegument, 203, 203f, **436–38,** 437f
Teichoic acids, **90,** 91f
Teixobactin, 136–37, 137f
Telomerase, **266–67,** eTopic 6.1
Telomeres, **266,** 815
TEM (transmission electron microscopy), 41f, **42, 64,** 64f
Temin, Howard, 210
Tempeh, 621t, **625–26,** 626f
Temperate phage, **217,** 403
Temperature, 160–65, 161t
  classification, 161t, 162–65, 163f–165f
  Earth, 663
  ecosystem, 844, 844t
  effects on physiology, 161
  growth rate, 161–62, 162f, eTopic 5.1
  heat-shock response, 165
  to kill microbes, 182, 182f
Temperature change, global carbon balance, 889–93, 890f–892f
Template strand, **278,** 281, 282f
Tenericutes, 719–21, 720f
  classification, 704t, 705
  defined, **705**
  structure, 714
Tenofovir, 1148t, 1151
Tequila, 621t
Terbinafine (Lamisil), 1155t
Terminal electron acceptor, **497**
Terminal oxidase, 546–48, 547f
Terminase, 408
Termination
  replication, 251, 252f, 258–60, 259f, 260f
  transcription, 282, 283, 284f
  translation, 296–97, 296f, eTopic 8.4
Termination sites, replication, **252,** 252f

Terminator stem loop, 373f, 374
Termite wood-digesting microbiome, 852–53, 852f, 853f
Terpenoids, **87,** 87f
Terraforming, **917**
Terrestrial carbon cycling, 888–89, 889f
Tertiary syphilis, **1087**
Tests, 807
Tetanospasmin (tetanus toxin), 1026t, **1106–7,** 1106f, 1108–9, 1108f
Tetanus vaccine, 998t
Tetrachloroethene, 564, 565f
Tetracyclines, 85f, 297, 299f, 1132f, 1133
Tetraether, 748f, **749**
Tetrahydrofolate (THF), 594, 1130
Tetrahymena, 671
Tetrahymena thermophila, 1034
Tetrathionate, 1017
Textiles, 648t
T_{FH} cell activation, 986, 987f
T_H0 helper T cell activation, 986, 987f
T_H1 cells
  activation of cytotoxic T cells, 986–87, 987f
  activation of macrophages, 992
T_H17 cells, promotion of inflammation, **988–89**
Thalassic lakes, 776
Thalassospira, 908
Thalassospira profundimaris, 129
Thalassospira rotula, 812f
Thaumarchaeota, 753t, **754,** 763–65, 764f, 765t
Thaumarchaeotes
  ammonia-oxidizing, 763–65, 764f, 765f
  psychrophilic marine, 753t
Thawing permafrost, methane release, 890–92, 891f, 892f
T_H cell. *See* Helper T cell (TH cell)
Theobroma cacao, 627
Thermal vents, 760–62, 761f, 762f, **864–65,** 864f
Thermincola, 555
Thermocline, **858,** eTopic 21.1
Thermococcales, 753t, 778–79, 778f
Thermococcus, 778
Thermococcus litoralis, 778–79
Thermococcus piezophilus, 166, 753t
Thermodynamics, biological, 490–91, 490f
Thermometers, RNA, 389–90, 393f
Thermophiles, **164–65,** 164f, 844t, **864–65**
  bioreactor to grow, 165, 165f
  classification, 161t
  methanogens, 767
  microbial ecology, 22, 23f
  origins of life, 672
  temperature and growth rate, 162f
Thermoplasma, 779
  divergence, 677, 677f
  unique metabolic pathways, 754
Thermoplasma acidophilum, 753t, 779
Thermoplasmata, 753t
Thermoplasmatales, 753t, **779–80,** 779f
Thermoproteales, **752,** 753t, 760
Thermoprotei, 753t, 760
Thermoproteus, 753t, 760
  reductive TCA cycle, 592
Thermosphaera aggregans, 757, 757f, 757t
Thermotoga, 704t, 751
Thermotogae, 704t, **706**
Thermotoga maritima
  ABC transporter, 127
  classification, 706
  horizontal gene transfer, 679
  paralogous genes, 683
Thermozymes, 164
Thermus aquaticus, 164f
  polymerase chain reaction, 32, 164, 445, 445f
  sulfur reduction and oxidation, 778
  transcription elongation, 284f
Thermus thermophilus, 294f, 298
Theta form, 405
Theta replication, 407, 407f
THF (tetrahydrofolate), 594, 1130
Thiobacillus dentrificans, 728t
Thiocapsa, 733, 902
Thioectosymbionts, 701
Thioglycolate, 175
Thiomargarita namibiensis, 4, 5f, 102
Thiomicrospira, 865
Thioploca, 732, 733f
Thiotrichales, 704t
Thiovolum, 737
Third-generation sequencing, 450–52, 451f–453f
30S ribosomal subunit, antibiotics targeting, 1132–33
Thomas, Torsten, 770f
Thomason, Lynn, 403f, 404
Three-day measles, 1063, 1064f
Three-domain model, 27, 28f
Threshold dose, **967**
Throat swabs, 1161, 1162f
Thrombotic thrombocytopenic purpura (TTP), 1074
Thrush, 964, 964f
  AIDS, 1090f, 1091
Thunberg, Greta, 885f
Thurber, Rebecca Vega, 200, 200f
Thylakoids, **108–9,** 110f
  photoexcitation and photolysis, **571,** 572f, 708, 710
Thymectomy, 986
Thymine, 243f, 244
Thymus, 945f, 985–86
Tight junctions, **945,** 946f
Tilex, 186
Ti plasmid, **653–54,** 653f
TipN, 103–5, 104f, eTopic 3.2
Tir receptor, 1039, 1039f
Tissue culture, animal virus, 234, 234f
Tissue tropism, animal viruses, 222–25, 224f
Titan cells, 1057, 1057f
Titong, Allison, 206f
TLRs (Toll-like receptors), **950–51,** 950t, 951f
  gut immune system, 995, 995f, 996
T lymphocytes. *See* T cells
tmRNA (transfer-messenger RNA), **286,** 286t, eTopic 8.4
TMV (tobacco mosaic virus)
  classification, 212t
  host infection, 197, 197f
  structure, 19, 20f, 204, 205f

TNF. *See* Tumor necrosis factor (TNF)
TnSeq technique, 347, 348f
TNT (trinitrotoluene), 502f
Tobacco mosaic virus (TMV)
  classification, 212t
  host infection, 197, 197f
  structure, 19, 20f, 204, 205f
Toehold switch sensor, 1178–79, 1178f
Togaviridae, 1109
Toggle switches, 476, 477f
Toll-like receptors (TLRs), **950–51,** 950t, 951f
  gut immune system, 995, 995f, 996
Toluene, 525–26, 526f
Tomatoes, contamination, 638
Tomography, 37, 37f, **67**
  cryo-electron, **67–69,** 68f, 69f
Tonegawa, Susumu, 978
Tonsils, 945f
Tooth decay, 926
Topoisomerases, **249–50,** 249f, 250f, 258, 259f
TOPO TA cloning, 460
TORCH complex, 1087
Torres, Victor, 1028, 1028f
Total magnification, **47**
Tower of power, 497, 497t, 536–37, 536t
Toxic shock, 1099–1100
Toxic shock syndrome toxin (TSST), 991, 991f, 1026t, 1100
Toxin(s). *See* Endotoxins; Exotoxins
Toxin A, 1026t
Toxin-antitoxin (TA) systems, 177, 178f, 1143, 1143f
Toxin B, 1026t
Toxoplasma, 1087
Toxoplasma gondii, 789t, 817–18
  food contamination, 640t
ToxR, 83
TPEF (two-photon excitation fluorescence), 796, 797f
Trachoma, 1088
Transamination, **610,** 611
Transcript(s), **278**
Transcription, 99–100, 100f, 281–88
  archaea and eukaryotes, 287–88, 288f
  consensus sequence, 279–80
  coupled translation and, 299, 300f
  defined, **278**
  elongation, 282, 282f–283f, 283
  initiation, 282–83, 282f–283f
  promoters, 279–81, 279f, 280f
  RNA polymerase, 278–83, 280f, 282f–284f
  sigma factor, **278,** 279–80, 279f, 280f, 282f
  stages, 282–83, 282f–284f
  termination, 282, 283, 284f
Transcriptional attenuation, **372–74,** 373f
Transcriptional fusion, 466t, **467–68,** 467f
Transcription analytical techniques, 466t, 467–68, 467f, 468f
Transcription regulators, 358–70
  AraC/XylS family, 366–68, 366t, 367f
  of eukaryotes and archaea, 368–69, 369f
  lactose operon, 361–66, 361f–366f
  repression of anabolic (biosynthetic) pathways, 368, 369f
  repressors and activators, 358–60, 358f–360f
  sensing extracellular environment, 360–61, 361f
Transcriptomics, 467
  to probe pathogenesis, 1052–53, 1052f
Transcriptosome, **370–71,** 371f
Transcytosis, **220,** 221, 222–23, **1105**
Transduction
  bacteriophage, **217,** 218f, 334–37, 335f, 336f
  generalized, **335,** 335f
  specialized (restricted), **335,** 336, 336f
Transfection, **435**
Transfer-messenger RNA (tmRNA), **286,** 286t, eTopic 8.4
Transfer RNA (tRNA), **286,** 286t
  binding, 294, 294f
  cell structure, 78
  isoacceptors, 291
  modified bases, 291, 292f
  structure, 290, 290f
  translation, 289–92, 290f–292f
Transformation
  activation of natural, 386–87, 387f
  defined, **30, 337**
  discovery, **240**
  naked DNA, 337–38, 337f, 338f
  plasmids, 261
  viral, **229**
Transformed-focus assay, 236
Transgene, **434–35**
Transgenic plants, 473, 473t
Transglycosylase, **1125,** 1126
Transition, **316**
Translation, 100, 100f, 288–302
  aminoacyl-tRNA synthetases attach amino acids to tRNA, 291, 291f, 292f
  antibiotics that affect, 297, 299f
  codons, 288, 289–90, 289f, 290f, 293
  coupled transcription and, 299, 300f
  defined, **288,** 289
  elongation, 295–96, 295f
  eukaryotes, 299–300
  genetic code and tRNA molecules, 289–90, 289f, 290f
  initiation, 293–94, 294f, 295
  modified genetic codes, 300–301, 301f
  polysomes, 299, 299f
  protein processing after, 302–3, 303f, 304f
  ribosomes, 288, 291–94, 292f, 293f
  stages, 294–97, 294f–296f
  termination, 296–97, 296f, eTopic 8.4
  translocation, 296, 298, 298f
Translational control, **380**
Translational fusion, 467f, **469–70,** 469f, 469t
Translation analytical techniques, 468–71, 469f–471f, 469t
Translesion bypass synthesis, 322t
Translocation, **296,** 298, 298f
  group, **129–30,** 130f
  protein export to periplasm, 308
Transmembrane oxygen gradients, 85
Transmission
  accidental, 1015f
  airborne, **1014,** 1016
  fecal-oral route, **1016**
  horizontal, **1014,** 1015f
  influenza virus, 413–14, 415f, 418–19

mechanical, **230,** 1015f, **1016**
parentereal route, **1016**
transovarial, **1016**
vehicle, **1014**
vertical, **1014,** 1015f
Transmission electron microscopy (TEM), 41f, **42, 64,** 64f
Transovarial transmission, **1016**
Transpeptidase, **1125,** 1126
Transport
active, **85, 126–27,** 126f
coupled, **126–27,** 126f
passive, **85**
Transporter proteins, 77f, 82f, 83–84, 84f, **85**
Transposable elements, 343–47
conjugative transposons, 345, 345f, 346f
defined, **343–44**
genetic analysis, 346–48, 347f, 348f
random mutagenesis, 461
transposons and transposition, 344, 344f.eTopic 9.5
Transposase, **344**
Transposition, **316, 344**
nonreplicative *vs.* replicative, 344, 344f
Transposons, **344,** 344f, 345f
conjugative, 345, 345f, 346t
mini-, 346, 347f
mutagenesis, 346–48, 347f, 348f
random mutagenesis, 461
Transversion, **316**
Traveler's diarrhea, 1081t
Travisano, Michael, 809
Trebouxia, 789t
Tregs (regulatory T cells), **985**
autoimmune disease, 1007–8
limitation of inflammation, 988–89
Tremblaya princeps, 241–42, 241t, 242f
Treponema, 704t, 739
Treponema azotonutricium, 738, 738f
Treponema pallidum, 739, eTopic 13.1
classification, 704t
crossing placental barrier, 1087
dark-field microscopy, 61
gene loss, 683
meningitis, 1110t
microaerophilic microbe, 174t
pathogen identification, 1184t
point-of-care rapid testing, 1186t
selective toxicity, 1119
syphilis, 1086t, 1087
Triazoles, 1155t
Tricarboxylic acid cycle (TCA cycle), **515–16,** 516f
acetyl-CoA entry into, 517–20, 517f, 519f
biosynthesis, 583, 584f
carbon cycle, 121f
discovery, 30
oxidative phosphorylation, 518, 519f
reductive (reverse), 586t, **592–93,** 592f
Trichoderma, 801
Trichodesmium
algal bloom, 712
phosphate cycle, 908
phytoplankton, 862
Trichomonas
metronidazole, 1130
point-of-care rapid diagnostics, 1186t
Trichomonas hominis, 929
Trichomonas vaginalis, 1091–92, 1092f
Trichomoniasis, 1086t, 1091–92, 1092f
point-of-care rapid diagnostics, 1186t
Trichophyton, 801
skin infections, 1060t
Trichophyton rubrum, 1012, 1013f
Triclosan, 187
Trigger element, 409, 410f
Trigger factor, 304, 307, 308, 308f
Triglycerides, 502f
Trimethoprim, 682–83, 682f
Trinitrotoluene (TNT), 502f
Trioses, 507, 508f
Tripeptidases, 305
tRNA. *See* Transfer RNA (tRNA)
Tropheryma whipplei, 1052, 1183, eTopic 28.4
Trophic levels, **842–43**
ocean ecosystems, 862
-trophy, 122
Tropism, animal viruses, **222–25,** 224f
trp operon, 373–74, 373f
True fungi, **790,** 801
Truffles, 619, 799, 872–73
Trypanosoma, 789t
Trypanosoma brucei, 821–22, 821f
light microscopy, 40, 40f
Trypanosoma cruzi, 822, 1045, 1046
Trypanosomes, **821–22,** 821f, eTopic 20.3
Tryptophan
biosynthesis, 612–13, 613f
transcriptional attenuation, 373–74, 373f
Tryptophan biosynthetic pathway, 368, 369f
Tsetse fly, 849–50, 851f
Tsien, Roger, 53
TSST (toxic shock syndrome toxin), 991, 991f, 1026t, 1100
TTP (thrombotic thrombocytopenic purpura), 1074
Tuber aestivum, 799
Tuberculin conversion rate, 1069
Tuberculin skin test, 1069
Tuberculosis (TB), 1066t, 1069–71
bovine, 1199–1200, 1199f
diagnosis, 1069
disease progression, 1069, 1070f
drug development, 646–47, 646f
extensively drug-resistant, 1071
extrapulmonary (miliary), 1070–71, 1070f
latent, 1069, 1070, 1070f
multidrug-resistant, 1069, 1071
pathogen identification by staining, 1166–67, 1167f
primary, 1069–70, 1070f
reemergence, 1069, 1194–95
secondary, **1070,** 1070f
transmission, 15
treatment, 1071
Tuberculosis (TB) vaccine, 968t
Tube worm, 864f, 865
Tubulinosema, 823f

Tularemia, 1102t
Tumor-causing viruses, 197–98
Tumor necrosis factor (TNF), superantigens, **991**
Tumor necrosis factor alpha (TNF-alpha), **953**
  modulation of immune response, 989t
Tumor necrosis factor beta (TNF-beta), modulation of immune response, 989t
Tumpey, Terrence, 414f
Tundra, thawing permafrost and methane release, 890–92, 891f, 892f
Turgor pressure, **84**
Turnbaugh, Peter, 1076, 1076f
Turrets, 108–9, 108f
Tuskegee experiment, 1087–88
T-VEC, 440
12D value, food preservation, 642
Twin arginine motif, 309–10, 310f
Twin arginine translocase (TAT), 309–10, 310f
Twitching motility, **148**
Two-component signal transduction systems, **360–61,** 361f
Two-hybrid analysis, 469t, **471,** 471f
Two-photon excitation fluorescence (TPEF), 796, 797f
Twort, Frederick William, 197
Tyndall, John, 13
Type I hypersensitivity, **1001–5,** 1004f
Type II hypersensitivity, 1001t, **1005,** eTopic 24.3
Type II secretion system, **1036–37,** 1037f, 1037t
Type III hypersensitivity, 1001t, **1005–6,** eTopic 24.8
Type III secretion system (T3SS), **1037–39,** 1037t, 1038f, 1039f
Type IV hypersensitivity, 1001t, **1005,** 1006f
Type IV secretion system, 1037t, **1039–40,** 1040f, eTopic 25.8
Typhoid fever, 1073, 1081t, 1102t
Typhoid fever vaccine, 968t
Typhus, endemic, 1102t
Typhus vaccine, 968t
Tyrosine, biosynthesis, 612–13, 613f

U

Ubiquinol, 544f, 545
Ubiquinone, electron transport, 544f
Ubiquitylation, 1045, 1047–48, 1048f, 1049f, eTopic 8.6
UHT (ultra-high temperature) method, 183
Ulcers, gastric, 1077–79, 1079f, 1080f, 1081t
Ultracentrifuge and ultracentrifugation, **29,** 30, 79f, **80**
Ultra-high temperature (UHT) method, 183
Ultrasmall bacteria, 707, 707f
Ultrasound, gut bacteria detection, 277, 277f
Ultraviolet (UV) damage repair, 322t, 323
Ultraviolet (UV) rays, mutagenic effects, 318t, 320, 321f
Ulva, 789t, 812, 812f
UMP (uridine monophosphate), 613, 614f
Unclassified bacteria, 692–93
Uncoating, animal virus, 224f, **225**
Uncouplers, **542–43,** 542f
Unculturable microbes, 134–35, 135f, 136–37, 838, 1183
  emerging clades, 692–93
Underground salt deposits, 776
Unsaturation, fatty acids, 596
UPEC (uropathogenic *E. coli*), 1074, 1084–86, 1085f
Upstream processing, **649,** 651f
Uracil, 243f, 245
  base excision repair, 326
  biosynthesis, 613, 614f
Uranium
  anaerobic respiration, 552
  bioremediation, 910
  microbial metabolism, 552, 910, 910t
Uranium-238 decay, 869
Urban cycle, 1099f
Ureaplasma urealyticum, 719
Urethritis, nongonococcal, 1086t, 1088, 1088f
Urey, Harold C., 13–14, 669
Uridine monophosphate (UMP), 613, 614f
Urinary catheter, 1162, 1162f
Urinary tract infections (UTIs), 927, 1083–86, 1085f
Urine samples, specimen collection, 1162–63, 1162f
Uropathogenic *E. coli* (UPEC), 1074, 1084–86, 1085f
Ustilago maydis, 789t
UTIs (urinary tract infections), 927, 1083–86, 1085f
UV (ultraviolet) damage repair, 322t, 323
UV (ultraviolet) rays, mutagenic effects, 318t, 320, 321f

V

VacA (vacuolating cytotoxin), 1079
VacA toxin, 1026t
Vaccination, **17–18,** 17f, **968**
  memory B cells, 975
Vaccine(s), 997–1001
  anti-vaxxers, 1000, 1063
  building better, 999–1000, 1000f
  danger, 1000–1001, eTopic 24.1
  Ebola virus, 1001, 1002–3
  examples, 968, 968t
  herd immunity, 999, eTopic 26.1
  HIV, 422, 1091
  immune system compartments, 999–1000, 1000f
  live, attenuated, 999
  malaria, 1097
  multivalent, 1067
  polyvalent, 997–98
  recommended schedule, 998, 998t
  types, 997
  viability signals, 963, 963f
Vaccine antigens produced in plants, 473, 473t
Vaccinia virus, 17–18, 17f, 205, 206f, 436t
Vacuolating cytotoxin (VacA), 1079
Valentine, David, 833, 836
Valley, John, 667
VAM (vesicular arbuscular mycorrhizae), **873**
Vampirella, 869
Vanadium, 910t
Vancomycin, 89, 598–99, 598f, 600f, 1128, **1129**
Vancomycin-resistant *Enterococcus* (VRE), 1139
Vancomycin-resistant *S. aureus* (VRSA), 1127
van der Linden, Ana, 1196
van der Linden, Vanessa, 1195–96
van Niel, Cornelis B., 839–40
var genes, 1097
Variable regions, 971f, **972**
Variant Creutzfeldt-Jakob disease (vCJD), 1111, eTopic 26.9

Varicella vaccine, 968t, 998t
Varicella-zoster virus (VZV), 436, 436t, 1060t
  antiviral agents, 1148t
  molecular evolution, 213, 213f
  size, 4t
Variola major virus, 436t, 1060t
  history, 17
  immunological specificity, 967–68, 967f
Variolation, 17, 17f
Vascular cell adhesion molecule 1 (VCAM), 953, 953f
Vasoactive agents, 939
Vasoactive factors, **949,** 950f
  acute inflammation, 953, 953f
Vasodilation, 954
VCAM (vascular cell adhesion molecule 1), 953, 953f
vCJD (variant Creutzfeldt-Jakob disease), 1111, eTopic 26.9
Vector(s), 230, 232
  arthropod, **1014–16,** 1015f
  mechanical, **230,** 1015f, **1016**
  microbial gene, 652–57
Vegetables
  acid fermentation, 621t, 625–27, 626f, 627f
  spoilage, 636
Vegetations, mitral valve, 926, 927f
Vegetative cells, **715**
Vegetative mycelia, **722**
Vehicle transmission, **1014**
Veillonella, 927
Venkatesh, Sid, 933
Venter, Craig, 5
Verardi, Paolo, 205, 206f
Vernet, Maria, 813–14, 813f
Verrucomicrobia, 704t, **705,** 742, 743f
Verrucomicrobium spinosum, 743f
Versalovic, James, 522–23, 524, 524f
Vertical gene transfer, **351, 678,** 680–81, 680f, 681f
Vertical transmission, **1014,** 1015f
Vesicle traffic, 1025, 1026t
Vesicular arbuscular mycorrhizae (VAM), **873**
Vesicular stomatitis virus (VSV-G), 655
Vetter, David Philip, 964, eTopic 24.1
V factor, 1170
Viable bacteria, **131,** 132f
Viable counts, 132f, 139–40
Vibrio anguillarum, 386, 386f
Vibrio cholerae
  ATP synthesis, 542
  classification, 704t
  conjugative transposons, 345, 345f, 346f
  diagnosis, 1058
  electron transport system, 550
  epidemiology, 1160, 1188, 1188f
  facultative microbe, 174t
  filamentous phage, 204
  gastrointestinal tract infection, 1081t
  genome, 241t, 242, 242f
  gut microbiome, 1076f, 1077
  marine habitat, 857
  membrane protein, 83
  MiPACT imaging, 59, 59f
  mutualism, 859, 860f
  phage-encoded genes, 198
  quorum sensing, 384
  secondary chromosomes, 263–64, 265, 265f
  stomach microbiome, 928
  toxins, 1026t, 1030–32, 1031f–1033f, 1037f
  transformation of naked DNA, 337–38, 337f
  vaccine antigen, 473t
  wastewater, 897
Vibrio fischeri, 382–84, 382f, 383f, 848
Vibrio harveyi, 385, 385f
Vibrionales, 704t
Vibrio parahaemolyticus, 1081t
Vibriosis, 1102t
Vibrio vulnificus
  bioinformatics, 455–56, 455f
  food contamination, 640t
  human microbiome, 922
  vibriosis, 1102t
Vif protein, 424t, 427
Vinetz, Joseph, 1052
Viracept (nelfinavir mesylate), 1148t, 1151
Viral disease, 201
Viral endonuclease, 1149–50
Viral gene therapy, 432–34, 433f, 433t
Viral genomes
  classification, 210–13, 211f, 212t
  integrated, 197–98, 198f, 200–201, 218–19, eTopic 6.2
  small *vs.* large, 208–9, 208f–210f
Viral infections, persistent or latent, 438–40, 439f
Viral lung diseases, 1066t, 1071
Viral meningitis, 1110t
Viral molecular biology, 401–42
  endogenous retroviruses and gene therapy, 431–35
  herpes simple virus (DNA virus), 435–40
  human immunodeficiency virus (retrovirus), 420–31
  influenza virus [(-) strand RNA virus], 410–20
  phage lambda (enteric bacteriophage), 402–10
Viral pathogens, emergence of new, 232, eTopic 6.3
Viral respiratory tract infections, 1066t
Viral ribozymes, 207
Viral shunt, **200,** 862
Viremia, **1092**
Virgin, Herbert, 438
Viridans streptococci, 1093
Virion(s), **196, 198,** 198f
  asymmetrical, 205, 206f
  structure, 202–8
  symmetrical, 202–4, 203f–205f
Viroids, **205–7,** 207f
Virome, **199, 219**
Virulence, **234, 1013,** 1034f
Virulence factors, 220, **1017,** 1034, eTopic 25.1, eTopic 25.2
Virulence genes, 1144–45
Virulence proteins, 1144–45
Virulent phage, 403
Virus(es), 195–238
  acute, 199
  animal, 222–30, 232
  asymmetrical virions, 205, 206f
  bacterial defenses, 219, 220f
  bacteriophages, 196, 196f, 215–22

Baltimore classification, 210–13, 211f, 212t
chronic, 33, 201
culturing, 233–36, 233f–236f, 1194
defined, **4, 196**
discovery, 19–20, 20f, 197
dynamic nature, 198–99, 198f
ecological roles, 199–201, 199f–201f
ecosystems, 196–202
emergence of new, 232, eTopic 6.3
endogenous, **197**
filamentous, 204, 204f, 205f
food irradiation, 184
giant, 198–99, 208–9, 209f, 210f
icosahedral, 202–3, 203f
infection of specific hosts, 196–97, 197f
integrated, 197–98, 198f, 200–201, 218–19, eTopic 6.2
International Committee on Taxonomy, 210
marine, 195, 195f, 196, 199–200, 199f
molecular evolution, 213–14, 213f, 214f
oncogenic, 227, **229–30,** eTopic 11.4, eTopic 26.1
oncolytic, 440
origins, 198–99, eTopic 6.1
overview, 196
persistent, 200
plant, 230–32
prions, 207, 207f
representative, 2, 3f
RNA *vs.* DNA, 20
sexually transmitted, 201
size, 19, 210
skin rashes, 1063, 1064f
SOS induction, 328
structure, 202–8
symmetrical virions, 202–4, 203f–205f
tailed, 205, 206f, eTopic 11.1
tumor-causing, 197–98
viral disease, 201
viral genomes, 197–98, 208–9, 208f–210f
viroids, 205–7, 206f
Virus cap-snatching, antiviral agents that target, 1149–50
Virus factories, **225,** 225f
Virus particle, **198,** 198f
Virus release, antiviral agents that target, 1148–49, 1149f
Virus uncoating, antiviral agents that target, 1148–49, 1149f
Vitamin B_2, 648t
Vitamin B_{12}
industrial microbiology, 644–45, 644f, 645f
mutualism, 585, 585f
Vita-PAMP, 963
Vitis vinifera, 632, 633f
Vogel, Jörg, 1052
Volta, Alessandro, 766
Volvox, 789t, 810, 810f
von Stockar, Urs, 490, 490f
Voriconazole, 1155t
Vorticella, 789t, 816
Vpr protein, 424t
Vpu protein, 424t
VRE (vancomycin-resistant *Enterococcus*), 1139
VRSA (vancomycin-resistant *S. aureus*), 1127
VSV-G (vesicular stomatitis virus), 655
Vulcanisaeta, 753t
VZV. *See* Varicella-zoster virus (VZV)

W

Wagner, Michael, 60
Wah Chiu, 195
Wakame, 619
Wakesfield, Andrew, 1000
Waksman, Selman, 1118, 1118f
Wang, Zhong, 847
Warfare, disease in, 10, 10f
Warren, J. Robin, 17, 737, 1078
Warts, 1060t
genital, 1086t
Wastewater treatment, 895–99
agricultural, 898–99, 898f
defined, **895**
industrial runoff, 899, eTopic 22.1
pathogens, 897, 898f
preliminary and primary, 896–97, 896f
secondary, 896f, 897, 897f
tertiary (advanced), 896f, 897–98
wastewater treatment plant, 895, 896f
Wasting disease, 1111t
Water, bacterial composition, 79
Water activity, **167**
Water cycle, 893–99
biochemical oxygen demand, 893–95, 894f–896f
and carbon cycle, 893, 894f
defined, **893**
wastewater treatment, 895–99, 896f–898f
Water molds, 789t, 792, 804
Water-saturated horizon, 868f, 869
Water table, 868f, **869**
Watson, James, 30, 31f
Wavelength, **43,** 46
WBCs. *See* White blood cell(s) (WBCs)
Weak acids, 169
membrane-permeant, **84–85,** 85f
Weak bases, membrane-permeant, **84–85,** 85f
Weizmann, Chaim, 515
Weller, Thomas J., 234
Wescodyne, 186
Western blotting, 468, 469t
West Nile virus (WNV), 1014, 1015
emergence, 232
host range, 201
meningitis, 1110t
rapid, next-generation DNA sequencing, 1176–78, 1177f
Westphal, Lacey, 453, 454f
Wetland(s), **874**
constructed, 898f, **899**
Wetland restoration, 898–99, 898f
Wetland soils, 874–75, 875f
Wet mount, **49,** 49f
Whey, **623**
Whipple, George, 1183
Whipple's disease, 1052, 1183, eTopic 28.4
Whiskey, 621t
White blood cell(s) (WBCs), 937, 937f

development, 937, 938f
types, 937–42, 938f
White blood cell (WBC) differential, 943–44, 944t
White blood cell (WBC) ratios, 943–44, 944t
Whitehead, Emily, 433f, 434
Whittaker, Robert, 26, 787
Whole-genome DNA-binding analysis, 470, 470f
Whooping cough, 1064, 1066t
Wigglesworthia, 850, 850t, 851f
Wilkins, Maurice, 30
Wine, 621t, 632–33, 632f, 633f
Winogradsky, Sergei, 21, 732, 828
Winogradsky column, **21,** 21f
Winter, Gregory P., 473
WNV. *See* West Nile virus (WNV)
Woese, Carl, 27, 28f, 677, 677f, 748
Wolbachia, 695–96, 695f, 696, 850, 850t, 851f
Wolbachia pipientis, 695
Wolinella succinogenes, 68–69, 68f
Wong-Staal, Flossie, 423, 423f
Woo, Patrick, 450
Woychik, Nancy, 177, 178f
Woyke, Tanja, 751
Wozniak, Daniel, 1028, 1028f
Wright, Elizabeth, 401, 401f
Wuchereria bancrofti, 1012, 1013f
Wybutosine, 291, 292f
Wysocka, Joanna, 432f

X

Xanthomonas, 735, 735f, 879
Xanthophyceae, 812
Xavier, Karina, 385
XDR-TB (extensively drug-resistant tuberculosis), 1071
Xenobiotic molecules, 840–41, 840f
XerC, 259, 260f
XerD, 259, 260f
Xerophile, 844t
X factor, 1170
Xofluza (baloxavir marboxil), 1148t, 1149–50
Xolair (omalizumab), 1005
X-ray(s)
food irradiation, 184
mutagenic effects, 318t
X-ray crystallography, 70–72
application, 71–72, 71f, 72f
defined, **42, 70**
history, 19–20, 20f, 30–31, 31f
limitation, 72
physics, 70–71, 71f
range of resolution, 41f, 42
X-ray diffraction analysis. *See* X-ray crystallography
Xylobolus frustulatus, 872f
Xyloglucans catabolism, 501, 502f, 504, 505f
Xylose catabolism, 366–68, 366t, 367f
XylS activator, 366–68, 366t, 367f

Y

Yamamori, Tetsuo, 165
Yang, Liang, 1115, 1116, 1116f
Yanofsky, Charles, 373, 373f, 613
Yap, Mee-Ngan F., 178, 179f
Yeast(s), 795–98, 795f
cocoa bean fermentation, 627, 628f
defined, **795**
discovery, 13
edible, 619
model organism for research, 795f, 796–97
reproductive cycles, 795f, 798
Yeast breads, 621t, 630–31, 630f, 631f
Yejun Han, 165f
Yellow fever, 1015, 1015f
Yellow fever vaccine, 968t
Yellow fluorescent protein (YFP), 56, 57f, 468, 468f
Yellow-green algae, 812
Yersinia
flipping cytokine profile, 1045
toxins, 1037–38, 1037t
Yersinia enterocolitica
pathogenicity islands, 1018f
stopping programmed cell death, 1046
Yersinia pestis
bioterrorism, 1192, 1193
classification, 705
electron transport system, 550
history, 10
immune evasion, 992
immunological specificity, 968
nomenclature, 26
pathogen identification, 1184t
plague, 1097–99, 1098f, 1099f, 1102t
quorum sensing, 384
sepsis, 1099
transcription regulation, 366
Yersinia pseudotuberculosis
stopping programmed cell death, 1046
synthetic biology, 475
YFP (yellow fluorescent protein), 56, 57f, 468, 468f
Yildiz, Fitnat, 59
Yoghurt, 189, 621t, **623**
Yosef, Ido, 409
Yoshida, Shosuke, 841–42, 841f
Yura, Takashi, 165

Z

Zanamivir (Relenza), 1148t, 1149, 1149f
ZDV (zidovudine), 1148t, 1150–51, 1150f
Zernike, Frits, 62
Zero-mode waveguide (ZMW) chambers, 450–51, 451f
Zhang, Dawei, 644
Zidovudine (AZT, ZDV), 1148t, 1150–51, 1150f
Ziehl-Neelsen stain, 52, 53f, 1166–67, 1167f
Zika virus, 1058
climate change, 1198
programmable RNA sensors, 1178–79, 1178f
tracking, 1195–97, 1196f
Zinder, Norton, 334–35
Zinser, Erik, 684
Zitvogel, Laurence, 24, 24f
ZMW (zero-mode waveguide) chambers, 450–51, 451f

Zobellia galactanivorans, 521
Zone of hypoxia, 866, **894–95,** 896f
  denitrification, 903, 904f
Zone of inhibition, **1121,** 1122t
Zoogloea, 897
Zoonotic disease, **1014**
  One Health Initiative, 1199–1200, 1199f
Zoosporangium(ia), 798, 799f
Zoospores, **790,** 798, 799f
  interspecies communication, 386, 386f
Zooxanthellae, 817, 819f
Z pathway, oxygenic, **572,** 573, 575, 576f
Z-rings
  bacterial cytoskeleton, 96, 97f
  formation, 101
Županc, Tatjana Avšič, 1058
z-values, food preservation, 642
Zygomycetes, 789t, **798–99,** 800f
Zygomycosis, 1060t
Zygomycota, 789t, 798–99, 800f
Zygospores, **799,** 800f
Zymomonas, 509

Marine Microbes
The Ocean's Majority

Michelle Schorn
Symbiotic bacteria make toxins in sponges
Page 1, Ch 1

A coccolithophore bloom
Page 199, Ch 6

Sallie Chisholm
Cyanobacteria release membrane vesicles
Page 106, Ch 3